W9-BDT-503 2013

WITHDRAWN

CHILTON®

ASIAN SERVICE MANUAL
2012 EDITION
VOLUME V
SCION
TOYOTA

ELISHA D. SMITH PUBLIC LIBRARY
MENASHA, WISCONSIN

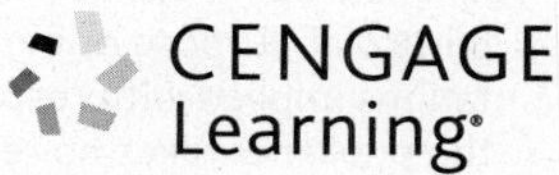

Australia • Brazil • Japan • Korea • Mexico • Singapore • Spain • United Kingdom • United States

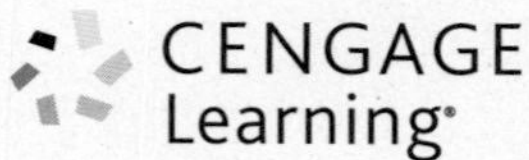

CHILTON®
Asian Service Manual
2012 Edition
Volume V
Scion, Toyota

Vice President, Technology & Trades Professional Business Unit:
Gregory L. Clayton

Publisher:
David Koontz

Director of Marketing:
Beth A. Lutz

Senior Production Director:
Wendy Troeger

Production Manager:
Sherondra Thedford

Senior Marketing Manager:
Jennifer Barbic

Associate Marketing Manager:
Rachael Torres

Chilton Content Specialist:
Paula Baillie

Graphical Designer:
Melinda Fantozzi

Art Director:
Benjamin Gleeksman

Sr. Content Project Manager:
William Tubbert

Senior Editors:
Eugene F. Hannon, Jr., A.S.E.
Ryan Lee Price
Christine L. Sheeky

Editors:
Jim Bailey
Nicholas D'Andrea
Steven D. Junker, A.S.E.
Maureen Lazarz
Kyla White

© 2013 Chilton, a part of Cengage Learning

ALL RIGHTS RESERVED. No part of this work covered by the copyright herein may be reproduced, transmitted, stored, or used in any form or by any means graphic, electronic, or mechanical, including but not limited to photocopying, recording, scanning, digitizing, taping, Web distribution, information networks, or information storage and retrieval systems, except as permitted under Section 107 or 108 of the 1976 United States Copyright Act, without the prior written permission of the publisher.

For product information and technology assistance, contact us at
Professional & Career Group customer Support, 1-800-648-7450.
For permission to use material from this text or product, submit all requests online at
www.cengage.com/permissions.
Further permissions questions can be e-mailed to
permissionrequest@cengage.com

ISBN-13: 978-1-2854-7109-9
ISBN-10: 1-2854-7109-1
ISSN: 2161-8755

Chilton
5 Maxwell Drive
Clifton Park, NY 12065-2919
USA

Chilton products are represented in Canada by Nelson Education, Ltd.

NOTICE TO THE READER

Publisher does not warrant or guarantee any of the products described herein or perform any independent analysis in connection with any of the product information contained herein. Publisher does not assume, and expressly disclaims, any obligation to obtain and include information other than that provided to it by the manufacturer.

The reader is expressly warned to consider and adopt all safety precautions that might be indicated by the activities described herein and to avoid all potential hazards. By following the instructions contained herein, the reader willingly assumes all risks in connection with such instructions.

The publisher makes no representations or warranties of any kind, including but not limited to, the warranties of fitness for particular purpose or merchantability, nor are any such representations implied with respect to the material set forth herein, and the publisher takes no responsibility with respect to such material. The publisher shall not be liable for any special, consequential, or exemplary damages resulting, in whole or part, from the readers' use of, or reliance upon, this material.

Printed in the United States of America
1 2 3 4 5 6 7 XX 17 16 15 14 13

Contents

Sections

Model Index

USING THIS INFORMATION

Organization

To find where a particular model section or procedure is located, look in the Table of Contents. Main topics are listed with the page number on which they may be found. Following the main topics is an alphabetical listing of all of the procedures within the section and their page numbers.

Manufacturer and Model Coverage

This product covers 2011-2012 Asian models that are produced in sufficient quantities to warrant coverage, and which have technical content available from the vehicle manufacturers before our publication date. Although this information is as complete as possible at the time of publication, some manufacturers may make changes which cannot be included here. While striving for total accuracy, the publisher cannot assume responsibility for any errors, changes, or omissions that may occur in the compilation of this data.

Part Numbers and Special Tools

Part numbers and special tools are recommended by the publisher and vehicle manufacturer to perform specific jobs. Before substituting any part or tool for the one recommended, you must be completely satisfied that neither your personal safety, nor the performance of the vehicle will be endangered.

ACKNOWLEDGEMENT

Portions of materials contained herein have been reprinted under license from Toyota Motor Sales, U.S.A., Inc., License Agreement TMS1005

All information contained herein about Toyota, Lexus, and Scion vehicles is based on the latest product information available at the time of publication, is provided "as is" without warranty of any kind, and is intended for service providers and other interested parties in Canada, Mexico, and the United States of America, including Guam, Puerto Rico, and the U.S. Virgin Islands.

Specifications and procedures are subject to change without notice. This information is provided expressly for the purpose of use by professional automobile technicians who have special techniques and certifications. Repair or service by non-specialized or uncertified technicians using only this information, or without proper equipment or tools, may cause severe injury to the individual or other individuals and could possibly cause damage to the vehicle. Certain procedures or content elements may make reference to Toyota Warranty policy or practice - these policies or practices are only applicable to Toyota Lexus, or Scion dealers.

PRECAUTIONS

Before servicing any vehicle, please be sure to read all of the following precautions, which deal with personal safety, prevention of component damage, and important points to take into consideration when servicing a motor vehicle:

- Always wear safety glasses or goggles when drilling, cutting, grinding or prying.
- Steel-toed work shoes should be worn when working with heavy parts. Pockets should not be used for carrying tools. A slip or fall can drive a screwdriver into your body.
- Work surfaces, including tools and the floor should be kept clean of grease, oil or other slippery material.
- When working around moving parts, don't wear loose clothing. Long hair should be tied back under a hat or cap, or in a hair net.
- Always use tools only for the purpose for which they were designed. Never pry with a screwdriver.
- Keep a fire extinguisher and first aid kit handy.
- Always properly support the vehicle with approved stands or lift.
- Always have adequate ventilation when working with chemicals or hazardous material.
- Carbon monoxide is colorless, odorless and dangerous. If it is necessary to operate the engine with vehicle in a closed area such as a garage, always use an exhaust collector to vent the exhaust gases outside the closed area.
- When draining coolant, keep in mind that small children and some pets are attracted by ethylene glycol antifreeze, and are quite likely to drink any left in an open container, or in puddles on the ground. This will prove fatal in sufficient quantity. Always drain the coolant into a sealable container.
- To avoid personal injury, do not remove the coolant pressure relief cap while the engine is operating or hot. The cooling system is under pressure; steam and hot liquid can come out forcefully when the cap is loosened slightly. Failure to follow these instructions may result in personal injury. The coolant must be recovered in a suitable, clean container for reuse. If the coolant is contaminated it must be recycled or disposed of correctly.
- When carrying out maintenance on the starting system be aware that heavy gauge leads are connected directly to the battery. Make sure the protective caps are in place when maintenance is completed. Failure to follow these instructions may result in personal injury.
- Do not remove any part of the engine emission control system. Operating the engine without the engine emission control system will reduce fuel economy and engine ventilation. This will weaken engine performance and shorten engine life. It is also a violation of Federal law.
- Due to environmental concerns, when the air conditioning system is drained, the refrigerant must be collected using refrigerant recovery/recycling equipment. Federal law requires that refrigerant be recovered into appropriate recovery equipment and the process be conducted by qualified technicians who have been certified by an approved organization, such as MACS, ASI, etc. Use of a recovery machine dedicated to the appropriate refrigerant is necessary to reduce the possibility of oil and refrigerant incompatibility concerns. Refer to the instructions provided by the equipment manufacturer when removing refrigerant from or charging the air conditioning system.
- Always disconnect the battery ground when working on or around the electrical system.

• Batteries contain sulfuric acid. Avoid contact with skin, eyes, or clothing. Also, shield your eyes when working near batteries to protect against possible splashing of the acid solution. In case of acid contact with skin or eyes, flush immediately with water for a minimum of 15 minutes and get prompt medical attention. If acid is swallowed, call a physician immediately. Failure to follow these instructions may result in personal injury.

• Batteries normally produce explosive gases. Therefore, do not allow flames, sparks or lighted substances to come near the battery. When charging or working near a battery, always shield your face and protect your eyes. Always provide ventilation. Failure to follow these instructions may result in personal injury.

• When lifting a battery, excessive pressure on the end walls could cause acid to spew through the vent caps, resulting in personal injury, damage to the vehicle or battery. Lift with a battery carrier or with your hands on opposite corners. Failure to follow these instructions may result in personal injury.

• Observe all applicable safety precautions when working around fuel. Whenever servicing the fuel system, always work in a well-ventilated area. Do not allow fuel spray or vapors to come in contact with a spark, open flame, or excessive heat (a hot drop light, for example). Keep a dry chemical fire extinguisher near the work area. Always keep fuel in a container specifically designed for fuel storage; also, always properly seal fuel containers to avoid the possibility of fire or explosion. Do not smoke or carry lighted tobacco or open flame of any type when working on or near any fuel-related components.

• Fuel injection systems often remain pressurized, even after the engine has been turned OFF. The fuel system pressure must be relieved before disconnecting any fuel lines. Failure to do so may result in fire and/or personal injury.

• The evaporative emissions system contains fuel vapor and condensed fuel vapor. Although not present in large quantities, it still presents the danger of explosion or fire. Disconnect the battery ground cable from the battery to minimize the possibility of an electrical spark occurring, possibly causing a fire or explosion if fuel vapor or liquid fuel is present in the area. Failure to follow these instructions can result in personal injury.

• The EPA warns that prolonged contact with used engine oil may cause a number of skin disorders, including cancer! You should make every effort to minimize your exposure to used engine oil. Protective gloves should be worn when changing oil. Wash your hands and any other exposed skin areas as soon as possible after exposure to used engine oil. Soap and water, or waterless hand cleaner should be used.

• Some vehicles are equipped with an air bag system, often referred to as a Supplemental Restraint System (SRS) or Supplemental Inflatable Restraint (SIR) system. The system must be disabled before performing service on or around system components, steering column, instrument panel components, wiring and sensors. Failure to follow safety and disabling procedures could result in accidental air bag deployment, possible personal injury and unnecessary system repairs.

• Always wear safety goggles when working with, or around, the air bag system. When carrying a non-deployed air bag, be sure the bag and trim cover are pointed away from your body. When placing a non-deployed air bag on a work surface, always face the bag and trim cover upward, away from the surface. This will reduce the motion of the module if it is accidentally deployed.

• Electronic modules are sensitive to electrical charges. The ABS module can be damaged if exposed to these charges.

• Brake pads and shoes may contain asbestos, which has been determined to be a cancer-causing agent. Never clean brake surfaces with compressed air. Avoid inhaling brake dust. Clean all brake surfaces with a commercially available brake cleaning fluid.

• When replacing brake pads, shoes, discs or drums, replace them as complete axle sets.

• When servicing drum brakes, disassemble and assemble one side at a time, leaving the remaining side intact for reference.

• Brake fluid often contains polyglycol ethers and polyglycols. Avoid contact with the eyes and wash your hands thoroughly after handling brake fluid. If you do get brake fluid in your eyes, flush your eyes with clean, running water for 15 minutes. If eye irritation persists, or if you have taken brake fluid internally, immediately seek medical assistance.

• Clean, high quality brake fluid from a sealed container is essential to the safe and proper operation of the brake system. You should always buy the correct type of brake fluid for your vehicle. If the brake fluid becomes contaminated, completely flush the system with new fluid. Never reuse any brake fluid. Any brake fluid that is removed from the system should be discarded. Also, do not allow any brake fluid to come in contact with a painted or plastic surface; it will damage the paint.

• Never operate the engine without the proper amount and type of engine oil; doing so will result in severe engine damage.

• Timing belt maintenance is extremely important! Many models utilize an interference- type, non freewheeling engine. If the timing belt breaks, the valves in the cylinder head may strike the pistons, causing potentially serious (also time-consuming and expensive) engine damage.

• Disconnecting the negative battery cable on some vehicles may interfere with the functions of the on-board computer system(s) and may require the computer to undergo a relearning process once the negative battery cable is reconnected.

• Steering and suspension fasteners are critical parts because they affect performance of vital components and systems and their failure can result in major service expense. They must be replaced with the same grade or part number or an equivalent part if replacement is necessary. Do not use a replacement part of lesser quality or substitute design. Torque values must be used as specified during reassembly.

SCION

IQ

1

SPECIFICATIONS AND MAINTENANCE CHARTS

ENGINE AND VEHICLE IDENTIFICATION

Engine							Model Year	
Code ①	Liters (cc)	Cu. In.	Cyl.	Fuel Sys.	Engine Type	Eng. Mfg.	Code ②	Year
1NR-FE	1.3 (1329)	81	4	EFI	DOHC	Toyota	C	2012

EFI Electronic Fuel Injection

① Engine code designation

② 10th position of VIN

71099_SCIQ_C0001

GENERAL ENGINE SPECIFICATIONS

All measurements are given in inches.

Year	Model	Engine Displacement Liters (cc)	Engine ID/VIN	Fuel System Type	Net Horsepower @ rpm	Net Torque @ rpm (ft. lbs.)	Bore x Stroke (in.)	Com-pression Ratio	Oil Pressure @ rpm
2012	Scion iQ	1.3 (1329)	1NR-FE	EFI	94@6000	89@4400	2.85x3.17	11.5:1	22-80@3000

71099_SCIQ_C0002

ENGINE TUNE-UP SPECIFICATIONS

Year	Engine Displacement Liters	Engine ID/VIN	Spark Plug Gap (in.)	Ignition Timing (deg) AT	Fuel Pump (psi)	Idle Speed (rpm) AT	Valve Clearance Intake	Valve Clearance Exhaust
2012	1.3	1NR-FE	0.039-0.043	8-12 BTDC	44-50	600-700	HYD	HYD

HYD Engines equipped with hydraulic lash adjusters

71099_SCIQ_C0003

CAPACITIES

Year	Model	Engine Displacement Liters	Engine ID/VIN	Engine Oil with Filter	Transaxle (pts.) Auto.	Fuel Tank (gal.)	Cooling System (qts.)
2012	iQ	1.3	1NR-FE	3.7	NS	8.5	5.2

NOTE: All capacities are approximate. Add fluid gradually and ensure a proper fluid level is obtained.

NS Not Specified

71099_SCIQ_C0004

FLUID SPECIFICATIONS

Year	Model	Engine Disp. Liters	Engine Oil	Auto. Trans.	Brake Master Cylinder	Cooling System
2012	Scion iQ	1.3	0W-20	①	DOT 3	②

DOT: Department Of Transpotation

① Toyota Genuine CVT Fluid TC

② Toyota Super Long Life Coolant (SLLC)

71099_SCIQ_C0005

VALVE SPECIFICATIONS

Year	Engine Displacement Liters	Engine ID/VIN	Spring Test Pressure (lbs. @ in.)	Spring Free-Length (in.)	Spring Installed Height (in.)	Stem-to-Guide Clearance (in.)		Stem Diameter (in.)	
						Intake	Exhaust	Intake	Exhaust
2012	1.3	1NR-FE	NS	2.015	①	0.00098-0.00236	0.00118-0.00256	0.21535-0.21594	0.21516-0.21575

NS Not Specified

① Front: 1.634 inches minimum
Rear: 1.516 inches minimum

71099_SCIQ_C0006

CAMSHAFT SPECIFICATIONS

All measurements in inches unless noted

Year	Engine Displacement Liters	Engine Code/VIN	Journal Diameter	Brg. Oil Clearance	Shaft End-play	Runout	Journal Bore	Lobe Height Intake	Lobe Height Exhaust
2012	1.3	1NR-FE	①	0.00118-0.00264	NS	0.00118	NS	1.6417-1.6456	1.6417-1.6456

NS: Not Specified

① Journal No. 1: 1.3565-1.3571 inches
Other journals: 0.9037-0.9043 inches
(Same for both intake and exhaust camshafts)

71099_SCIQ_C0007

CRANKSHAFT AND CONNECTING ROD SPECIFICATIONS

All measurements are given in inches.

	Engine		Crankshaft				Connecting Rod		
Year	Displacement Liters	Engine ID/VIN	Main Brg. Journal Dia.	Main Brg. Oil Clearance	Shaft End-play	Thrust on No.	Journal Diameter	Oil Clearance	Side Clearance
2012	1.3	1NR-FE	2.04724-2.04787	0.000748-0.00169	NS	NS	1.77165-1.77259	0.000787-0.002047	NS

NS: Not Specified

71099_SCIQ_C0008

PISTON AND RING SPECIFICATIONS

All measurements are given in inches.

	Engine			Ring Gap			Ring Side Clearance		
Year	Disp. Liters	Engine ID/VIN	Piston Clearance	Top Compression	Bottom Compression	Oil Control	Top Compression	Bottom Compression	Oil Control
2012	1.3	1NR-FE	0.00217-0.00307	0.00709-0.00984	0.0118-0.0177	0.00394-0.01570	0.0008-0.0028	0.0008-0.0024	0.0008-0.0024

71099_SCIQ_C0009

TORQUE SPECIFICATIONS

All readings in ft. lbs.

Year	Engine Disp. Liters	Engine ID/VIN	Cylinder Head Bolts	Main Bearing Bolts	Rod Bearing Bolts	Crankshaft Damper Bolts	Drive Plate Bolts	Manifold Intake	Manifold Exhaust	Spark Plugs	Oil Pan Drain Plug
2012	1.3	1NR-FE	①	②	③	94	65	24	24	15	22

① Step 1: 24 ft. lbs.
Step 2: Additional 90 degrees
Step 3: Additional 90 degrees

② Step 1: 22 ft. lbs.
Step 2: Additional 90 degrees

③ Step 1: 11 ft. lbs.
Step 2: Additional 90 degrees

71099_SCIQ_C0010

WHEEL ALIGNMENT

Year	Model		Caster Range (+/-Deg.)	Caster Preferred Setting (Deg.)	Camber Range (+/-Deg.)	Camber Preferred Setting (Deg.)	Toe-in (in.)
2012	Scion iQ	Front	0.75	7.45	0.75	-0.43	0.067+/-0.118
		Rear	—	—	0.5	-0.87	0.122+/-0.118

71099_SCIQ_C0011

TIRE, WHEEL AND BALL JOINT SPECIFICATIONS

Year	Model	OEM Tires		Tire Pressures (psi)		Wheel	Ball Joint	Lug Nut
		Standard	Optional	Front	Rear	Size	Inspection	(ft. lbs.)
2012	Scion iQ	175/60R16	NA	33	32	16	NA	76

OEM: Original Equipment Manufacturer

PSI: Pounds Per Square Inch

NA: Information not available

71099_SCIQ_C0012

BRAKE SPECIFICATIONS

All measurements in inches unless noted

Year	Model		Brake Disc			Brake Drum Diameter			Minimum Pad/Lining Thickness		Brake Caliper	
			Original Thickness	Minimum Thickness	Max. Runout	Original Inside Diameter	Max. Wear Limit	Maximum Machine Diamter	Front	Rear	Bracket Bolts (ft. lbs.)	Mounting Bolts (ft. lbs.)
2012	Scion	F	0.787	0.709	0.00197	—	—	—	0.0394	—	79	25
	iQ	R	—	—	—	7.09	7.13	NA	—	0.0394	—	—

F: Front

R: Rear

NA: Information not available

71099_SCIQ_C0013

SCHEDULED MAINTENANCE INTERVALS

SCION - iQ

TO BE SERVICED	TYPE OF SERVICE	VEHICLE MILEAGE INTERVAL (x1000)												
		5	10	15	20	25	30	35	40	45	50	55	60	65
Air cleaner filter	R						✓						✓	
Transmission fluid	S/I						✓						✓	
Ball joints & dust covers	S/I			✓			✓			✓			✓	
Bolts & nuts on chassis & body	S/I													
Brake line pipes & hoses	S/I			✓			✓			✓			✓	
Brake pads & discs/linings & drums (front & rear)	S/I	✓	✓	✓	✓	✓	✓	✓	✓	✓	✓	✓	✓	✓
Drive belts	S/I												✓	
Driveshaft boots	S/I			✓			✓			✓			✓	
Engine coolant	S/I			✓			✓			✓			✓	
Engine coolant	R	Replace at 100,000 miles												
Engine oil & filter	R	✓	✓	✓	✓	✓	✓	✓	✓	✓	✓	✓	✓	✓
Exhaust pipes & mountings	S/I			✓			✓			✓			✓	
Fuel lines & connections	S/I						✓						✓	
Propeller shaft bolt	S/I			✓			✓			✓			✓	
Radiator core & condenser	S/I			✓			✓			✓			✓	
Front differential fluid	S/I						✓						✓	
Rotate Tires	S/I	✓	✓	✓	✓	✓	✓	✓	✓	✓	✓	✓	✓	✓
Spark plugs	R	Replace at 120,000 miles												
Steering linkage & gear box	S/I			✓			✓			✓			✓	

R: Replace S/I: Service or Inspect

Drivebelts: After initial inspection at 60,000 miles, inspect every 15,000 miles thereafter.

FREQUENT OPERATION MAINTENANCE (SEVERE SERVICE)

If a vehicle is operated under any of the following conditions it is considered severe service:

- Desert/Extremely dusty areas.

- Trailer towing usage.

Air cleaner filter: service or inspect every 5000 miles

Ball joints & dust covers: service or inspect every 5000 miles.

Bolts & nuts on chassis & body: service or inspect every 5000 miles.

Driveshaft boots: service or inspect every 5000 miles.

Steering linkage: service or inspect every 5000 miles.

Transmission and Front differential fluid: replace every 30,000 miles.

71099_SCIQ_C0014

PRECAUTIONS

Before servicing any vehicle, please be sure to read all of the following precautions, which deal with personal safety, prevention of component damage, and important points to take into consideration when servicing a motor vehicle:

- Never open, service or drain the radiator or cooling system when the engine is hot; serious burns can occur from the steam and hot coolant.
- Observe all applicable safety precautions when working around fuel. Whenever servicing the fuel system, always work in a well-ventilated area. Do not allow fuel spray or vapors to come in contact with a spark, open flame, or excessive heat (a hot drop light, for example). Keep a dry chemical fire extinguisher near the work area. Always keep fuel in a container specifically designed for fuel storage; also, always properly seal fuel containers to avoid the possibility of fire or explosion. Refer to the additional fuel system precautions later in this section.
- Fuel injection systems often remain pressurized, even after the engine has been turned **OFF**. The fuel system pressure must be relieved before disconnecting any fuel lines. Failure to do so may result in fire and/or personal injury.
- Brake fluid often contains polyglycol ethers and polyglycols. Avoid contact with the eyes and wash your hands thoroughly after handling brake fluid. If you do get brake fluid in your eyes, flush your eyes with clean, running water for 15 minutes. If eye irritation persists, or if you have taken brake fluid internally, IMMEDIATELY seek medical assistance.
- The EPA warns that prolonged contact with used engine oil may cause a number of skin disorders, including cancer. You should make every effort to minimize your exposure to used engine oil. Protective gloves should be worn when changing oil. Wash your hands and any other exposed skin areas as soon as possible after exposure to used engine oil. Soap and water, or waterless hand cleaner should be used.
- All new vehicles are now equipped with an air bag system, often referred to as a Supplemental Restraint System (SRS) or Supplemental Inflatable Restraint (SIR) system. The system must be disabled before performing service on or around system components, steering column, instrument panel components, wiring and sensors. Failure to follow safety and disabling procedures could result in accidental air bag deployment, possible personal injury and unnecessary system repairs.
- Always wear safety goggles when working with, or around, the air bag system. When carrying a non-deployed air bag, be sure the bag and trim cover are pointed away from your body. When placing a non-deployed air bag on a work surface, always face the bag and trim cover upward, away from the surface. This will reduce the motion of the module if it is accidentally deployed. Refer to the additional air bag system precautions later in this section.
- Clean, high quality brake fluid from a sealed container is essential to the safe and proper operation of the brake system. You should always buy the correct type of brake fluid for your vehicle. If the brake fluid becomes contaminated, completely flush the system with new fluid. Never reuse any brake fluid. Any brake fluid that is removed from the system should be discarded. Also, do not allow any brake fluid to come in contact with a painted surface; it will damage the paint.
- Never operate the engine without the proper amount and type of engine oil; doing so WILL result in severe engine damage.
- Timing belt maintenance is extremely important. Many models utilize an interference-type, non-freewheeling engine. If the timing belt breaks, the valves in the cylinder head may strike the pistons, causing potentially serious (also time-consuming and expensive) engine damage. Refer to the maintenance interval charts for the recommended replacement interval for the timing belt, and to the timing belt section for belt replacement and inspection.
- Disconnecting the negative battery cable on some vehicles may interfere with the functions of the on-board computer system(s) and may require the computer to undergo a relearning process once the negative battery cable is reconnected.
- When servicing drum brakes, only disassemble and assemble one side at a time, leaving the remaining side intact for reference.
- Only an MVAC-trained, EPA-certified automotive technician should service the air conditioning system or its components.

BRAKES

ANTI-LOCK BRAKE SYSTEM (ABS)

GENERAL INFORMATION

PRECAUTIONS

- Certain components within the ABS system are not intended to be serviced or repaired individually.
- Do not use rubber hoses or other parts not specifically specified for and ABS system. When using repair kits, replace all parts included in the kit. Partial or incorrect repair may lead to functional problems and require the replacement of components.
- Lubricate rubber parts with clean, fresh brake fluid to ease assembly. Do not use shop air to clean parts; damage to rubber components may result.
- Use only DOT 3 brake fluid from an unopened container.
- If any hydraulic component or line is removed or replaced, it may be necessary to bleed the entire system.
- A clean repair area is essential. Always clean the reservoir and cap thoroughly before removing the cap. The slightest amount of dirt in the fluid may plug an orifice and impair the system function. Perform repairs after components have been thoroughly cleaned; use only denatured alcohol to clean components. Do not allow ABS components to come into contact with any substance containing mineral oil; this includes used shop rags.
- The Anti-Lock control unit is a microprocessor similar to other computer units in the vehicle. Ensure that the ignition switch is **OFF** before removing or installing controller harnesses. Avoid static electricity discharge at or near the controller.
- If any arc welding is to be done on the vehicle, the control unit should be unplugged before welding operations begin.

SPEED SENSORS

REMOVAL & INSTALLATION

Front

See Figure 1.

1. Remove front wheel.
2. Remove front wheel opening extension pad.
3. Remove fender rear plate mudguard.
4. Remove front fender splash shield sub-assembly.
5. Disengage the clamp from the No. 3 sensor clamp.
6. Remove front speed sensor

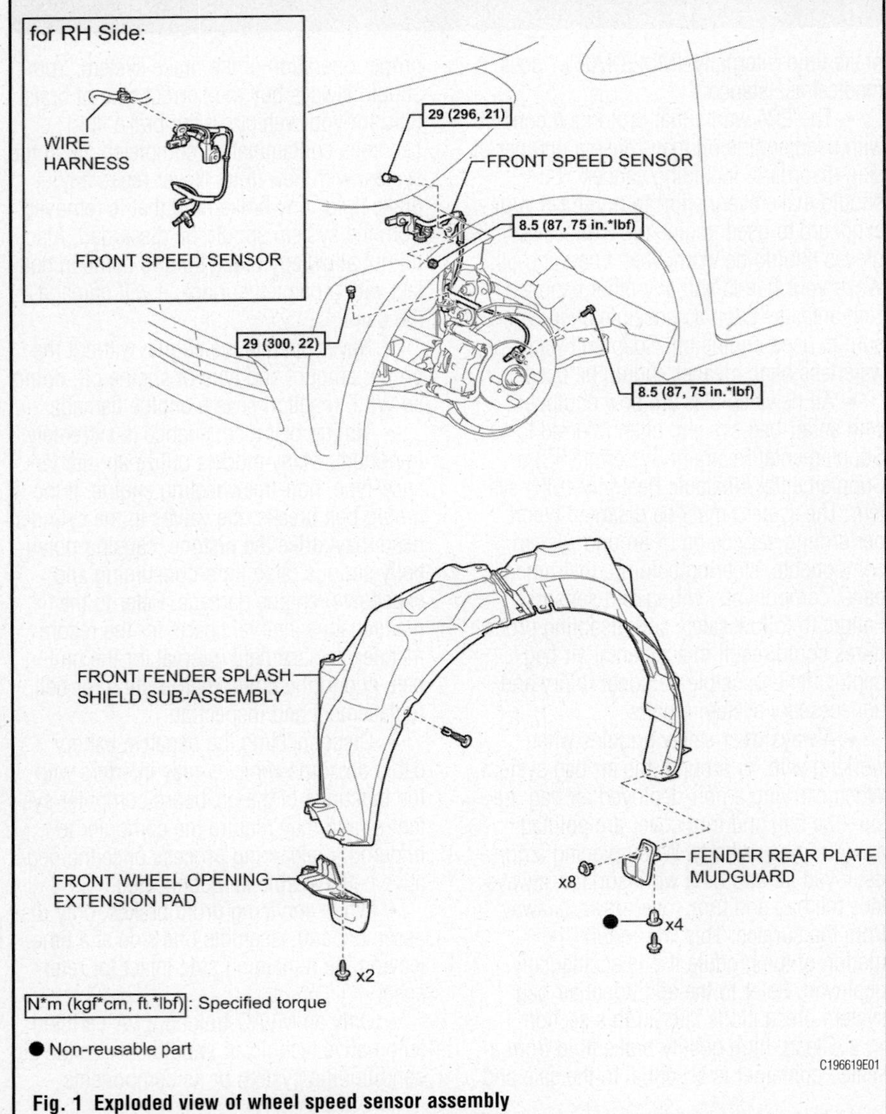

Fig. 1 Exploded view of wheel speed sensor assembly

a. Disconnect the front speed sensor connector.

b. Remove the nut and the No. 3 sensor clamp from the body.

c. Remove the bolt C and separate the No. 2 sensor clamp from the shock absorber.

d. Remove the bolt B and separate the brake flexible hose and No. 1 sensor clamp from the shock absorber.

e. Remove the bolt A and the front speed sensor from the steering knuckle.

➡Prevent foreign matter from attaching to the sensor tip. Clean the installation hole and the surface for the front speed sensor every time the front speed sensor is removed.

To install:

7. Install front speed sensor.

a. Install the front speed sensor with the bolt A.

b. Install the No. 1 sensor clamp and flexible hose with the bolt B.

➡Do not twist the front speed sensor wire harness during installation. Tighten the brake flexible hose and front speed sensor together with the bolt. Make sure that the flexible hose is positioned over the front speed sensor.

c. Install the No. 2 sensor clamp to the shock absorber with the bolt C.

d. Connect the speed sensor connector.

e. Install the No. 3 sensor clamp to the body with the nut.

8. Engage the clamp to the No. 3 sensor clamp.
9. Install front fender splash shield sub-assembly.
10. Install fender rear plate mudguard.
11. Install front wheel opening extension pad.
12. Install front wheel.
13. Check for speed sensor signal.

Rear

See Figure 2.

➡Use the same procedure for the LH side and RH side. The following procedure listed is for the LH side.

1. Remove upper console panel sub-assembly
2. Loosen adjusting nut
3. Remove rear wheel
4. Remove rear brake drum
5. Disconnect skid control sensor wire

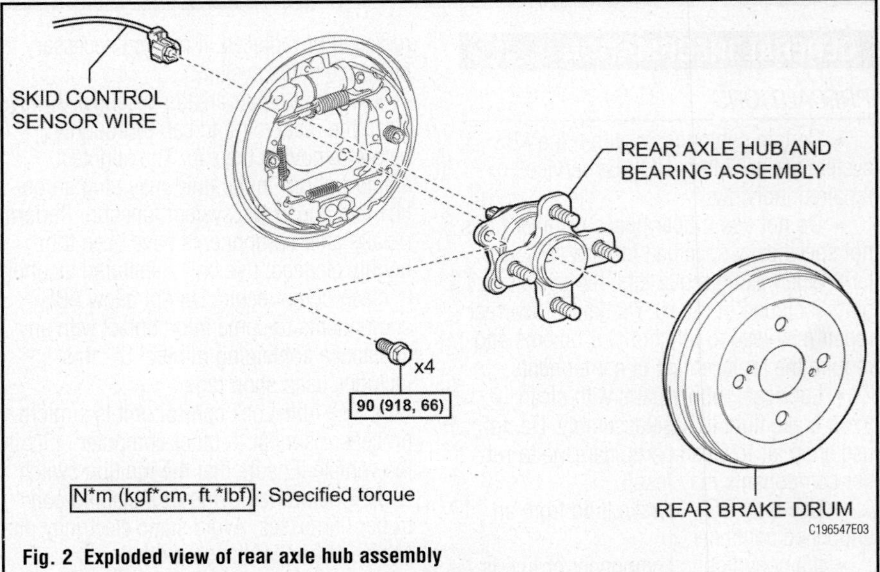

Fig. 2 Exploded view of rear axle hub assembly

6. Remove the 4 bolts and the rear axle hub and bearing assembly.

➡**The rear speed sensor is a component of the rear axle hub and bearing assembly. If the sensor malfunctions, replace the rear axle hub and bearing assembly.**

To install:

7. Install the rear axle hub and bearing assembly with the 4 bolts. Tighten to 66 ft. lbs. 90 Nm).
8. Connect skid control sensor wire.
9. Inspect rear axle hub bearing looseness.
10. Inspect rear axle hub runout.
11. Install rear brake drum.
12. Temporarily tighten adjusting nut.
13. Adjust parking brake lever travel.
14. Install rear wheel.
15. Install upper console panel sub-assembly.
16. Inspect rear wheel alignment.
17. Check for speed sensor signal.

BRAKES

BLEEDING THE BRAKE SYSTEM

BLEEDING PROCEDURE

BLEEDING PROCEDURE

➡**If any work is performed on the brake system or if air in the brake lines is suspected, bleed air from the brake system.**

➡**Move the shift lever to P and apply the parking brake before bleeding the brakes. Add brake fluid to keep the level between the MIN and MAX lines of the reservoir while bleeding the brakes. If brake fluid leaks onto any painted surface, immediately wash it off.**

1. Remove center cowl top ventilator louver.
   a. Disengage the 2 clips and separate the hood to cowl top seal.
   b. Disengage the 2 claws and 6 guides and remove the center cowl top ventilator louver.
2. Fill the reservoir with brake fluid.

➡**Add brake fluid to keep the level between the MIN and MAX lines of the reservoir while bleeding the brakes.**

3. Bleed brake master cylinder.

➡**If the master cylinder is reinstalled or if the reservoir becomes empty, bleed the master cylinder. To prevent brake fluid from damaging painted surfaces, cover any surrounding parts with a piece of cloth.**

   a. Using a union nut wrench, disconnect the 2 brake tubes from the master cylinder.
   b. Slowly depress the brake pedal and hold it.
   c. Cover the 2 outer holes with fingers, and release the brake pedal.
   d. Repeat steps 1 and 2, 3 or 4 times.
   e. Using a union nut wrench, connect the 2 brake tubes to the master cylinder.
4. Bleed brake line.

➡**Bleed the brake line of the wheel farthest from the master cylinder first. Add brake fluid to keep the level between the MIN and MAX lines of the reservoir while bleeding the brakes.**

   a. Connect a vinyl tube to the bleeder plug.
   b. Depress the brake pedal several times, and then loosen the bleeder plug with the pedal depressed.
   c. When fluid stops coming out, tighten the bleeder plug, and then release the brake pedal.
   d. Repeat steps 1 and 2 until all the air in the fluid is completely bled out.
   e. Tighten the bleeder plug completely.
   f. Repeat the above procedure at each wheel to bleed the brake line.
5. Bleed brake actuator.

➡**After bleeding the air from the regular brake system, if the correct height or feel of the brake pedal cannot be obtained, perform air bleeding again by following the procedure below. Bleeding air without the Techstream may lead to an injury or accident. When performing this procedure, always bleed air using the Techstream.**

   a. With the ignition switch off, depress the brake pedal several times to eliminate the vacuum pressure inside the booster.
   b. Connect the Techstream to the DLC3 with the ignition switch off.
   c. Turn the ignition switch to ON.

➡**Do not start the engine.**

   d. Turn the Techstream on.
   e. Enter the following menus: Chassis / ABS/VSC/TRC / Utility / Air Bleeding.
   f. Perform the bleeding at the right front wheel and left rear wheel.
   g. Bleed the Decrease Line.
   h. Select "Decrease Line" on the Techstream menu display.
   i. Depress the brake pedal.
   j. Using the Techstream, operate the solenoid and at the same time, completely depress the brake pedal within 5 seconds.

➡**After operating the solenoid, the motor in the actuator operates for 5 seconds and during this time the brake pedal is pushed out.**

   k. Loosen the bleeder plug at the right front wheel.

➡**Loosen the bleeder plug at only one place. Completely depress the brake pedal.**

   l. Tighten the bleeder plug, and then release the brake pedal.
   m. Bleed the Increase Line.
   n. Perform brake line air bleeding at the right front wheel by depressing the brake pedal.
   o. Repeat step 2 until all air is completely bled out.
   p. Perform the same procedure used at the right front wheel at the left rear wheel.
   q. Repeat step 1 until the air is completely bled out.
   r. Perform the same procedures for the left front wheel and right rear wheel line.
   s. Repeat the brake line air bleeding procedure by depressing the brake pedal.

➡**Repeat the brake line air bleeding procedure at each wheel until the air is bled out.**

6. Inspect for brake fluid leak.
7. Inspect brake fluid level.
8. Install center cowl top ventilator louver.

FLUID FILL PROCEDURE

1. Remove center cowl top ventilator louver.
2. Fill reservoir with brake fluid.
3. Replace brake fluid.

➡**Bleed the brake line of the wheel farthest from the master cylinder first. Add brake fluid to keep the level between the MIN and MAX lines of the reservoir while bleeding the brakes.**

   a. Connect a vinyl tube to the bleeder plug.
   b. Depress the brake pedal several

times, and then loosen the bleeder plug with the pedal depressed.

c. When fluid stops coming out, tighten the bleeder plug, and then release the brake pedal.

d. Repeat steps 1 and 2 until the new brake fluid comes out.

e. Tighten the bleeder plug completely.

f. Repeat the above procedure for each wheel to bleed the brake line.

4. Inspect for brake fluid leak.
5. Inspect brake fluid level.
6. Install center cowl top ventilator louver.

BRAKES

FRONT DISC BRAKES

CAUTION

Dust and dirt accumulating on brake parts during normal use may contain asbestos fibers from production or aftermarket brake linings. Breathing excessive concentrations of asbestos fibers can cause serious bodily harm. Exercise care when servicing brake parts. Do not sand or grind brake lining unless equipment used is designed to contain the dust residue. Do not clean brake parts with compressed air or by dry brushing. Cleaning should be done by dampening the brake components with a fine mist of water, then wiping the brake components clean with a dampened cloth. Dispose of cloth and all residue containing asbestos fibers in an impermeable container with the appropriate label. Follow practices prescribed by the Occupational Safety and Health Administration (OSHA) and the Environmental Protection Agency (EPA) for the handling, processing, and disposing of dust or debris that may contain asbestos fibers.

BRAKE CALIPER

REMOVAL & INSTALLATION

See Figures 3 and 4.

➡Use the same procedure for the LH side and RH side. The following procedure listed is for the LH side.

1. Remove front wheel.
2. Drain brake fluid.
3. Remove the union bolt and gasket, and disconnect the front flexible hose from the disc brake cylinder assembly.
4. Hold the front disc brake cylinder slide pin and remove the 2 bolts and disc brake cylinder assembly.
5. Remove the 2 disc brake pads from the disc brake cylinder mounting.
6. Remove the 2 No. 1 front anti-squeal shims and No. 2 front anti-squeal shim from the brake pads.
7. Remove the No. 1 front disc brake pad support plate and No. 2 front disc brake pad support plate from the front disc brake cylinder mounting.

➡Each front disc brake pad support plate has a different shape. Be sure to put an identification mark on each front disc brake pad support plate so that it can be reinstalled to its original position.

8. Remove the 2 front disc brake cylinder slide pins from the disc brake cylinder mounting.
9. Remove the 2 front disc brake bushing dust boots from the disc brake cylinder mounting.
10. Remove the 2 bolts and front disc brake cylinder mounting from the steering knuckle.
11. Put matchmarks on the disc and the axle hub, and remove the front disc.

To install:

12. Align the matchmarks of the disc and axle hub, and install the front disc.

➡When replacing the disc with a new one, select the installation position where the front disc has the minimal runout.

13. Install the disc brake cylinder mounting to the steering knuckle with the 2 bolts. Tighten to 79 ft. lbs. (107 Nm).

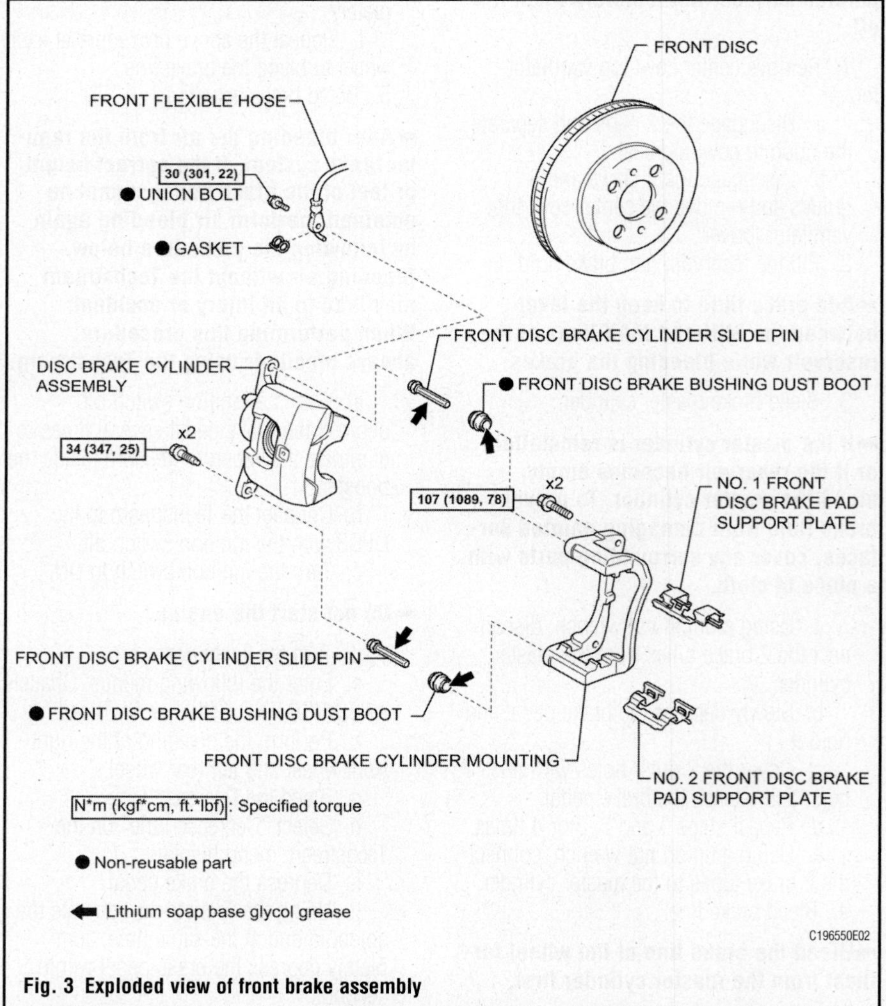

Fig. 3 Exploded view of front brake assembly

14. Install front disc brake bushing dust boot.

a. Apply a light layer of lithium soap base glycol grease to the entire circumference of 2 new front disc brake bushing dust boots.

b. Install the 2 front disc brake bushing dust boots to the disc brake cylinder mounting.

15. Install front disc brake cylinder slide pin.

a. Apply a light layer of lithium soap base glycol grease to the entire circumference of the 2 front disc brake cylinder slide pins.

b. Install the 2 front disc brake cylinder slide pins to the disc brake cylinder mounting.

16. Install the No. 1 front disc brake pad support plate and No. 2 front disc brake pad support plate to the disc brake cylinder mounting.

➡Be sure to install each front disc brake pad support plate in the correct position and direction.

17. Install the 2 No. 1 front anti-squeal shims and No. 2 front anti-squeal shim to each brake pad.

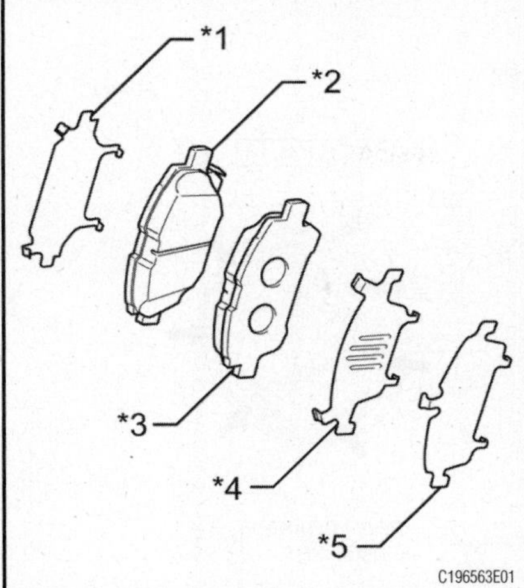

Fig. 4 Install the 2 No. 1 front anti-squeal shims and No. 2 front anti-squeal shim to each brake pad

➡When replacing worn pads, the anti-squeal shims must be replaced together with the pads. Install the shims in the correct positions and directions.

18. Install the 2 disc brake pads to the disc brake cylinder mounting.

➡There should be no oil or grease on the friction surfaces of the disc brake pads or the front disc.

19. Hold the front disc brake cylinder slide pin and install the disc brake cylinder assembly to the disc brake cylinder mounting with the 2 bolts. Tighten to 25 ft. lbs. (34 Nm).

20. Connect the front flexible hose to the disc brake cylinder assembly with a new union bolt and a new gasket. Tighten to 22 ft. lbs. (30 Nm).

➡Install the flexible hose lock securely into the lock hole in the disc brake cylinder.

21. Bleed brake line.

22. Install front wheel.

DISC BRAKE PADS

REMOVAL & INSTALLATION

Refer to Brake Caliper Removal and Installation procedure to service brake pads.

BRAKES — REAR DRUM BRAKES

⁂ CAUTION

Dust and dirt accumulating on brake parts during normal use may contain asbestos fibers from production or aftermarket brake linings. Breathing excessive concentrations of asbestos fibers can cause serious bodily harm. Exercise care when servicing brake parts. Do not sand or grind brake lining unless equipment used is designed to contain the dust residue. Do not clean brake parts with compressed air or by dry brushing. Cleaning should be done by dampening the brake components with a fine mist of water, then wiping the brake components clean with a dampened cloth. Dispose of cloth and all residue containing asbestos fibers in an impermeable container with the appropriate label. Follow practices prescribed by the Occupational Safety and Health Administration (OSHA) and the Environmental Protection Agency (EPA) for the handling, processing, and disposing of dust or debris that may contain asbestos fibers.

BRAKE DRUM

REMOVAL & INSTALLATION

See Figure 5.

➡Use the same procedure for the RH side and LH side. The procedure listed below is for the LH side.

➡The rear wheel brake cylinder assemblies can be disassembled and inspected with the rear wheel brake cylinder assemblies installed to the backing plate. It is not necessary to remove the rear wheel brake cylinder assemblies from the backing plates except when the rear wheel brake cylinder assemblies should be replaced.

1. Remove rear wheel.
2. Drain brake fluid.
3. Remove upper console panel sub-assembly.
4. Loosen the lock nut and adjusting nut to completely release the No. 1 parking brake cable assembly.
5. Remove rear brake drum.

a. Release the parking brake and remove the rear brake drum.

b. If the rear brake drum cannot be removed easily, perform the following procedure.

- Remove the shoe adjusting hole plug and insert a screwdriver through the hole into the backing plate, and hold the automatic adjust lever away from the adjuster.
- Using another screwdriver, contract the brake shoe by turning the adjusting bolt.

To install:

6. Install rear brake drum.
7. Adjust rear drum brake shoe clearance.

a. Temporarily install the 2 wheel nuts.

b. Remove the shoe adjusting hole plug, and turn the adjuster to expand the shoe until the drum locks.

c. Hold the automatic adjust lever away from the adjuster, and contract the brake shoe by turning the adjust bolt using another screwdriver until the drum can rotate smoothly.

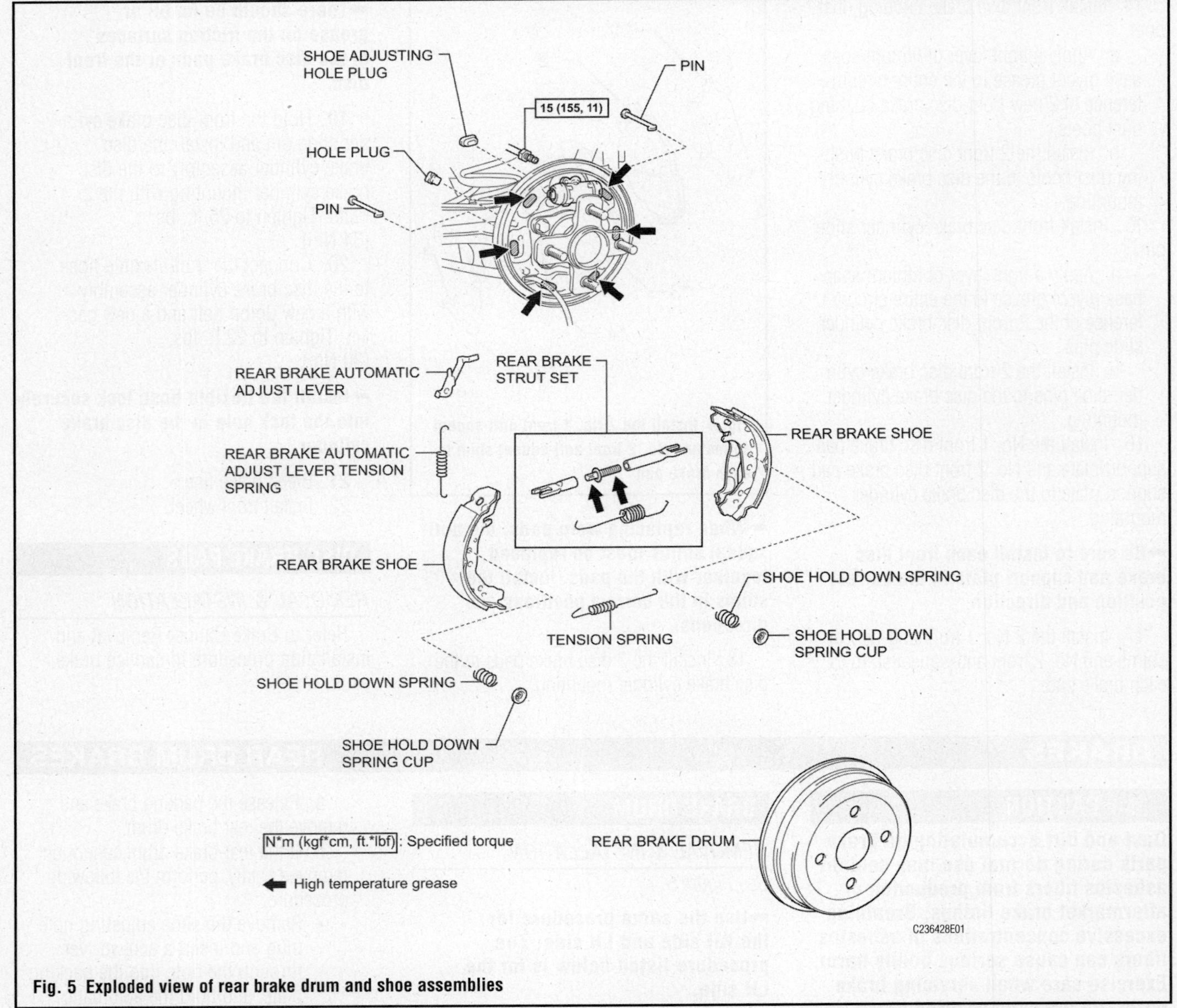

Fig. 5 Exploded view of rear brake drum and shoe assemblies

d. Install the hole plug.
8. Bleed brake line.
9. Install rear wheel.
10. Temporarily install the adjusting nut and the lock nut to the No. 1 parking brake cable assembly.
11. Adjust parking brake lever travel.
12. Install upper console panel sub-assembly.

BRAKE SHOES

REMOVAL & INSTALLATION

Refer to Brake Drum Removal and Installation procedure to service rear brake shoes.

BRAKES

PARKING BRAKE

PARKING BRAKE CABLES

ADJUSTMENT

See Figure 6.

1. Inspect parking brake lever travel.
   a. Pull the parking brake lever firmly.
   b. Release the parking brake lock, and return the parking brake lever to its off position.
   c. Slowly pull the parking brake lever all the way up, and count the number of clicks.
2. Remove console upper panel sub-assembly.
3. Adjust parking brake lever travel.
   a. Completely release the parking brake lever.
   b. Loosen the lock nut and the adjusting nut to completely release the parking brake cable.
   c. Fully depress the brake pedal 3 to 5 times with the engine stopped.
   d. Turn the adjusting nut until the parking brake lever travel is corrected to within the specified range of 6 to 9 clicks at 45 lbs. (200 N) of force.
   e. Using a wrench or an equivalent tool, hold the adjusting nut and tighten the lock nut.
   f. Operate the parking brake lever 3 to 4 times, and check the parking brake lever travel.
   g. Check whether the parking brake drags or not.
4. Inspect brake warning light.

➡When operating the parking brake lever, check that the brake warning light illuminates. If OK, the brake warning light always illuminates at the first click.

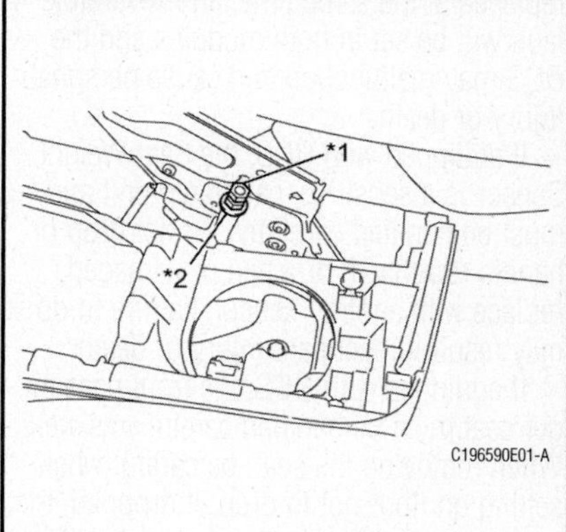

Fig. 6 Loosen the lock nut (1) and the adjusting nut (2) to completely release the parking brake cable

5. Install console upper panel sub-assembly.

CHASSIS ELECTRICAL

AIR BAG (SUPPLEMENTAL RESTRAINT SYSTEM)

GENERAL INFORMATION

⁂ CAUTION

These vehicles are equipped with an air bag system. The system must be disarmed before performing service on, or around, system components, the steering column, instrument panel components, wiring and sensors. Failure to follow the safety precautions and the disarming procedure could result in accidental air bag deployment, possible injury and unnecessary system repairs.

SERVICE PRECAUTIONS

Disconnect and isolate the battery negative cable before beginning any airbag system component diagnosis, testing, removal, or installation procedures. Allow system capacitor to discharge for two minutes before beginning any component service. This will disable the airbag system. Failure to disable the airbag system may result in accidental airbag deployment, personal injury, or death.

Do not place an intact undeployed airbag face down on a solid surface. The airbag will propel into the air if accidentally deployed and may result in personal injury or death.

When carrying or handling an undeployed airbag, the trim side (face) of the airbag should be pointing towards the body to minimize possibility of injury if accidental deployment occurs. Failure to do this may result in personal injury or death.

Replace airbag system components with OEM replacement parts. Substitute parts may appear interchangeable, but internal differences may result in inferior occupant protection. Failure to do so may result in occupant personal injury or death.

Wear safety glasses, rubber gloves, and long sleeved clothing when cleaning powder residue from vehicle after an airbag deployment. Powder residue emitted from a deployed airbag can cause skin irritation. Flush affected area with cool water if irritation is experienced. If nasal or throat irritation is experienced, exit the vehicle for fresh air until the irritation ceases. If irritation continues, see a physician.

Do not use a replacement airbag that is not in the original packaging. This may result in improper deployment, personal injury, or death.

The factory installed fasteners, screws and bolts used to fasten airbag components have a special coating and are specifically designed for the airbag system. Do not use substitute fasteners. Use only original equipment fasteners listed in the parts catalog when fastener replacement is required.

During, and following, any child restraint anchor service, due to impact event or vehicle repair, carefully inspect all mounting hardware, tether straps, and anchors for proper installation, operation, or damage. If a child restraint anchor is found damaged in any way, the anchor must be replaced. Failure to do this may result in personal injury or death.

Deployed and non-deployed airbags may or may not have live pyrotechnic material within the airbag inflator.

Do not dispose of driver/passenger/curtain airbags or seat belt tensioners unless you are sure of complete deployment. Refer to the Hazardous Substance Control System for proper disposal.

Dispose of deployed airbags and tensioners consistent with state, provincial, local, and federal regulations.

After any airbag component testing or service, do not connect the battery negative cable. Personal injury or death may result if the system test is not performed first.

If the vehicle is equipped with the Occupant Classification System (OCS), do not connect the battery negative cable before performing the OCS Verification Test using the scan tool and the appropriate diagnostic information. Personal injury or death may result if the system test is not performed properly.

Never replace both the Occupant Restraint Controller (ORC) and the Occupant Classification Module (OCM) at the same time. If both require replacement, replace one, then perform the Airbag System test before replacing the other.

Both the ORC and the OCM store Occupant Classification System (OCS) calibration data, which they transfer to one another when one of them is replaced. If both are replaced at the same time, an irreversible fault will be set in both modules and the OCS may malfunction and cause personal injury or death.

If equipped with OCS, the Seat Weight Sensor is a sensitive, calibrated unit and must be handled carefully. Do not drop or handle roughly. If dropped or damaged, replace with another sensor. Failure to do so may result in occupant injury or death.

If equipped with OCS, the front passenger seat must be handled carefully as well. When removing the seat, be careful when setting on floor not to drop. If dropped, the sensor may be inoperative, could result in occupant injury, or possibly death.

If equipped with OCS, when the passenger front seat is on the floor, no one should sit in the front passenger seat. This uneven force may damage the sensing ability of the seat weight sensors. If sat on and damaged, the sensor may be inoperative, could result in occupant injury, or possibly death.

DISARMING THE SYSTEM

Wait at least 90 seconds after disconnecting the cable from the negative (-) battery terminal to disable the SRS system.

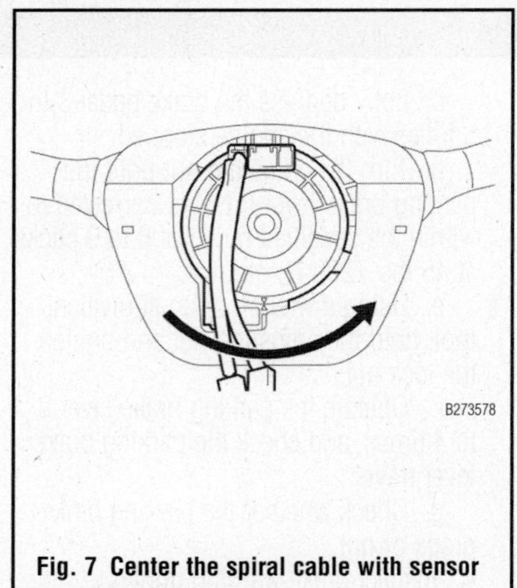

Fig. 7 Center the spiral cable with sensor sub-assembly

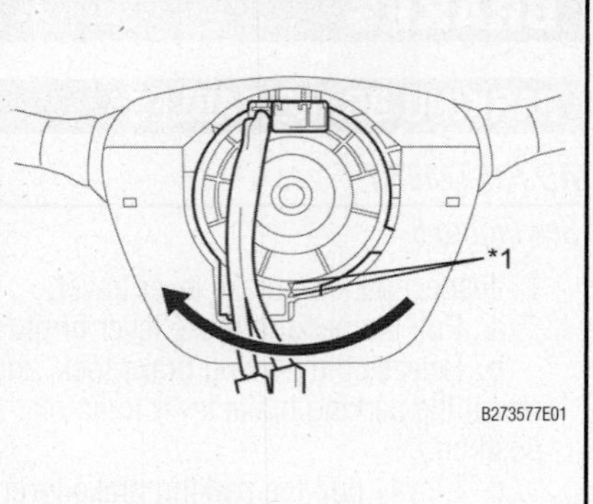

Fig. 8 Turn the spiral cable with sensor sub-assembly 2.5 times clockwise from its locked position and align the alignment marks (1)

ARMING THE SYSTEM

Connect the negative battery cable.

CLOCKSPRING CENTERING

See Figures 7 and 8.

1. Adjust spiral cable with sensor sub-assembly.
2. Center the spiral cable with sensor sub-assembly.
   a. Turn the spiral cable with sensor sub-assembly counterclockwise until it locks.

➡The spiral cable with sensor sub-assembly turns a maximum of approximately 5 times.

   b. Turn the spiral cable with sensor sub-assembly 2.5 times clockwise from its locked position and align the alignment mark.

DRIVE TRAIN

AUTOMATIC TRANSAXLE ASSEMBLY

REMOVAL & INSTALLATION

See Figures 9 and 10.

➡This vehicle is equipped with a Continuously Variable Transaxle (CVT).

1. Remove engine assembly with transaxle.
2. Remove the 2 bolts and remove the drive shaft heat insulator bracket from the engine.
3. Disconnect water by-pass hose assembly.
   a. Disengage the clamp and separate the breather hose from the breather hose clamp.
   b. Remove the bolt and separate the water by-pass hose from the CVT.
   c. Slide the 2 clamps and separate the 2 water by-pass hoses from the oil cooler.
4. Disconnect engine wire.
   a. Disconnect the heated oxygen sensor connector, oil pressure sensor connector and the transmission revolution sensor connector (NOUT).
   b. Disengage the 5 clamps and separate the engine wire from the CVT.
   c. Remove the bolt and separate the engine wire from the CVT.
   d. Disconnect the park/neutral position switch connector, the transmission wire connector, and the 2 transmission revolution sensor connectors.
5. Disengage the 3 clamps and separate the engine wire from the CVT.
6. Remove the 4 bolts and then remove the engine mounting control bracket from the CVT.
7. Remove flywheel housing side cover.
8. Remove starter assembly.
9. Remove drive plate and torque converter setting bolt.
   a. Use SST to hold the crankshaft pulley in place.
   b. Remove the 6 drive plate and torque converter setting bolts.
10. Remove the 9 bolts and the CVT.

To install:

11. Install continuously variable transaxle.
   a. Confirm that 2 knock pins are on the transaxle contact surface of the engine cylinder block before transaxle installation.
   b. Apply clutch spline grease to the round of the crankshaft contact surface with the torque converter centerpiece.
   c. Maintain the engine and CVT in a horizontal position, align the knock pins and holes, and tighten the 9 bolts.
   - Bolt 1: 47 ft. lbs. (64 Nm)
   - Bolt 2: 27 ft. lbs. (37 Nm)
   - Bolt 3: 29 ft. lbs. (39 Nm)

➡Bolts 3 are the shorter bolts.

➡Confirm that there are 2 knock pins on the fitting surface of the engine block before installing the CVT. Do not twist or apply excessive force to the CVT. Check that the torque converter rotates smoothly after installation of the CVT.

12. Install drive plate and torque converter setting bolt.
   a. Clean and degrease the 6 drive plate and torque converter setting bolts.
   b. Apply adhesive to 2 or 3 threads on the ends of the 6 torque converter set bolts.

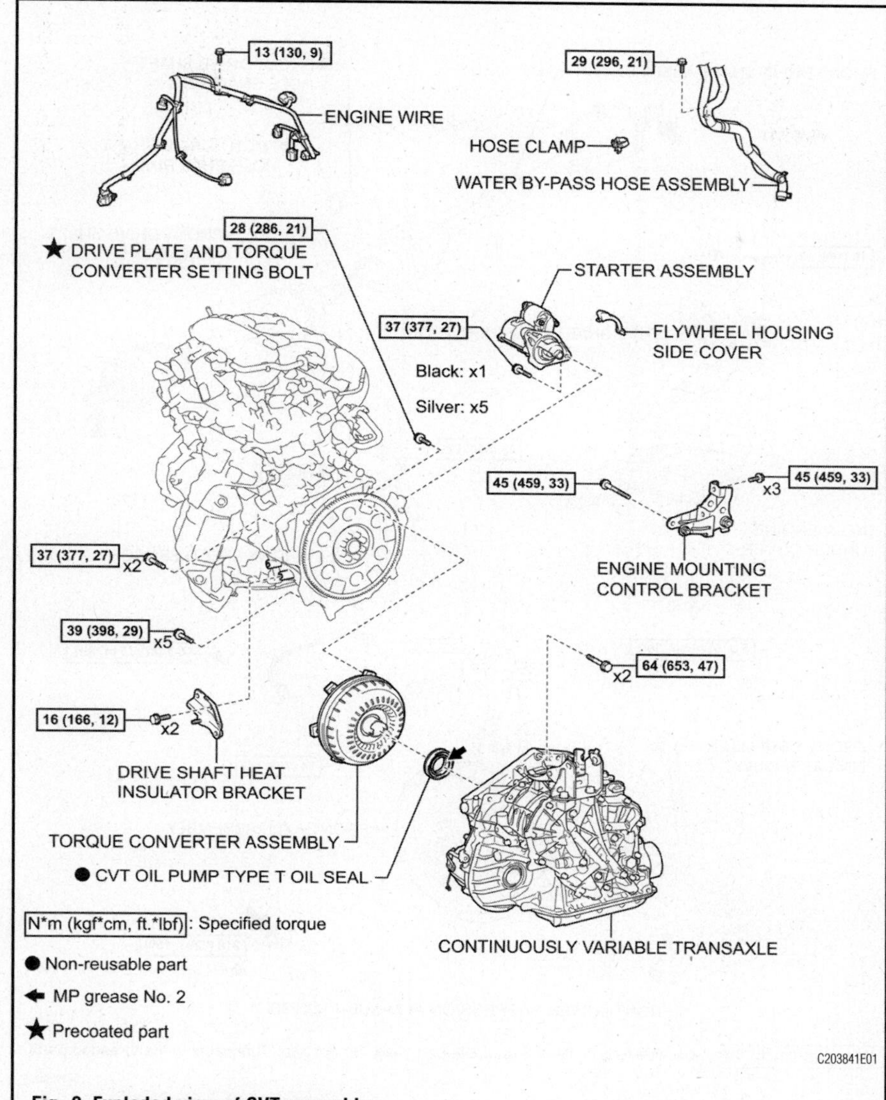

Fig. 9 Exploded view of CVT assembly

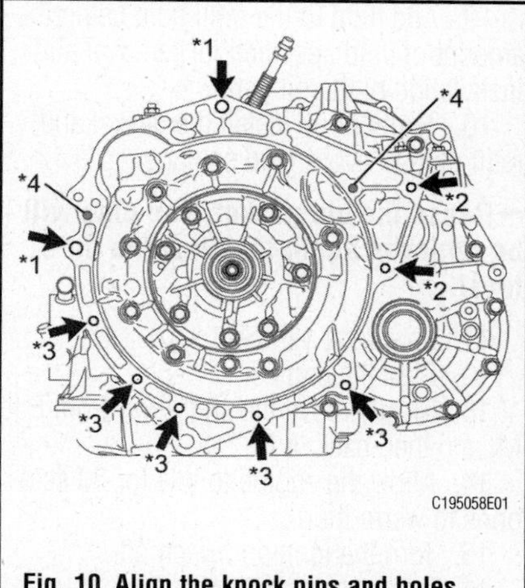

Fig. 10 Align the knock pins and holes, and tighten the 9 bolts

c. Use SST to hold the crankshaft pulley in place.

d. Install the 6 torque converter set bolts. Tighten to 21 ft. lbs. (28 Nm).

➡Tighten the black-colored bolt first, and then tighten the 5 silver-colored bolts.

13. Install starter assembly.

14. Install the engine mounting control bracket onto the CVT with the 4 bolts. Tighten to 33 ft. lbs. (45 Nm).

15. Install engine wire.

a. Engage the 3 clamps and install the engine wire to the CVT.

b. Connect the park/neutral position switch connector, the transmission wire connector, and the 2 transmission revolution sensor connectors.

c. Install the engine wire onto the CVT with the bolt.

d. Engage the 5 clamps and install the engine wire to the CVT.

e. Connect the heated oxygen sensor connector, oil pressure sensor connector and the transmission revolution sensor (NOUT) connector.

16. Install water by-pass hose assembly.

a. Connect the 2 water by-pass hoses to the oil cooler with the 2 clamps.

➡Securely push the hoses over each pipe fitting on the transmission oil cooler until the hoses contact the rib on each pipe fitting.

b. Install the water by-pass hose onto the CVT with the bolt. Tighten to 21 ft. lbs. (29 Nm).

c. Engage the clamp and install the breather hose to the breather hose clamp.

➡Install the breather hose clamp so that the top end of it overlaps the painted area. Install the breather hose clamp so that its protuberance is within the painted area.

17. Install the drive shaft heat insulator bracket onto the engine with the 2 bolts. Tighten to 12 ft. lbs. (16 Nm).

18. Install engine assembly with transaxle.

19. Reset memory (CVT initialization).

➡Perform Reset Memory (CVT initialization) when replacing the CVT.

AUTOMATIC TRANSAXLE FLUID

DRAIN AND REFILL

1. Lift the vehicle.

➡Set the vehicle on a lift so that the vehicle is kept level when it is lifted up (make sure that the tilt angle from the front to rear of the vehicle is within +/-1°).

2. Remove the refill plug and gasket from the CVT.

3. Using a socket hexagon wrench, remove the overflow plug and gasket from the CVT.

4. Using a socket hexagon wrench, remove the No. 1 transmission oil filler tube from the oil pan and drain the CVT fluid.

5. Using a socket hexagon wrench, install the No. 1 transmission oil filler tube to the oil pan.

6. Add fluid to the refill hole until it flows out of the overflow plug hole.

7. Wait until the fluid flow slows and only drops come out.

8. Temporarily install the gasket and overflow plug.

➡Reuse the old gasket. The plug will be removed again to adjust the fluid level.

9. Add fluid to the refill hole using the amount of fluid specified for removal and installation of the oil pan.

10. Temporarily install the gasket and refill plug to avoid fluid spillage.

➡Reuse the old gasket. The plug will be removed again to adjust the fluid level.

11. Lower the vehicle.
12. Start the engine.
13. Slowly move the shift lever from P to M, and then back to P.
14. Allow the engine to idle for 30 seconds to warm it up.
15. Turn the ignition switch off.
16. Repeat steps 1 to 15.
17. Repeat steps 1 to 11.
18. Adjust the fluid level.
19. Lower the vehicle.
20. Operation complete.

FRONT HALFSHAFT

REMOVAL & INSTALLATION

See Figure 11.

1. Remove No. 2 engine under cover.
2. Drain continuously variable transaxle fluid.
3. Remove front wheel.
4. Remove front axle shaft nut.
5. Separate front speed sensor.
6. Separate front stabilizer link assembly.
7. Separate front lower suspension arm sub-assembly.
8. Separate front axle assembly.
9. Remove the 2 bolts and the 2 nuts, and the drive shaft heat insulator.
10. Using a screwdriver and hammer, remove the drive shaft.
11. Remove front drive shaft assembly RH.

➡Use the same procedure for the RH side as for the LH side.

To install:

12. Install front drive shaft assembly LH.
    a. Coat the spline of the inboard joint with gear oil.
    b. Align the inboard joint splines and install the drive shaft with a screwdriver and hammer.

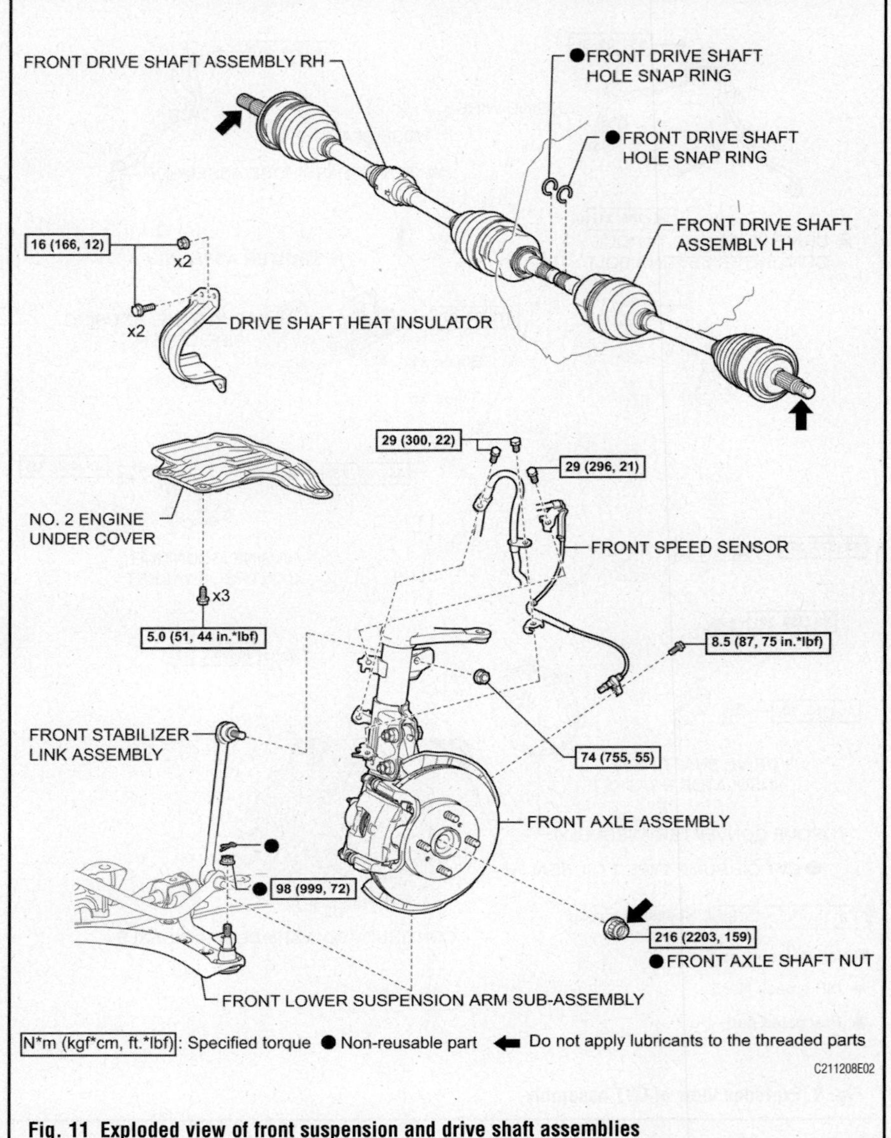

Fig. 11 Exploded view of front suspension and drive shaft assemblies

NOTE

➡Face the cut area of the front drive shaft hole snap ring downward.

13. Install front drive shaft assembly RH.

➡Use the same procedure as for the LH side.

14. Install the drive shaft heat insulator with the 2 bolts and the 2 nuts. Tighten to 12 ft. lbs. (16 Nm).
15. Install front axle assembly.
16. Install front lower suspension arm sub-assembly.
17. Install front stabilizer link assembly.
18. Connect front speed sensor.
19. Install front axle shaft nut. Tighten to 159 ft. lbs. (216 Nm).
20. Adjust continuously variable transaxle fluid.
21. Install No. 2 engine under cover.
22. Install front wheel.
23. Inspect and adjust front wheel alignment.

ENGINE COOLING

ENGINE COOLANT

DRAIN & REFILL PROCEDURE

See Figure 12.

⁂ CAUTION

Be sure that the ignition switch is off if you work near the electric cooling fans or radiator grille. With the ignition switch ON, the electric cooling fans may automatically start if the engine coolant temperature is high or the air conditioning is on.

1. Drain engine coolant.

⁂ CAUTION

To avoid the danger of being burned, do not remove the water filler cap sub-assembly while the engine and radiator assembly are still hot. Thermal expansion will cause hot engine coolant and steam to blow out from the radiator assembly.

a. Loosen the radiator drain cock plug.
b. Remove the water filler cap sub-assembly.

2. Add engine coolant.

a. Tighten the radiator drain cock plug.
b. Loosen the air bleed valve.
c. Remove the water filler cap sub-assembly.
d. Slowly fill the radiator with TOYOTA Super Long Life Coolant (SLLC) from the water filler (Coolant level should be within 0.75 inches (20 mm) from the filler tube end).

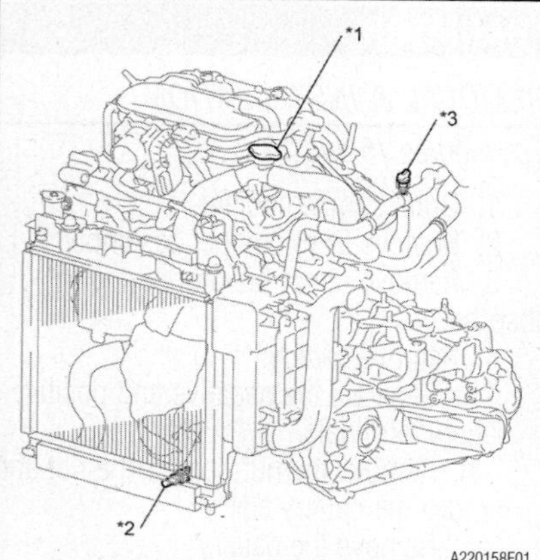

Fig. 12 Coolant filler cap (1), radiator drain cock plug (2), and air bleed valve (3)

⁂ WARNING

Never use water as a substitute for engine coolant.

➡TOYOTA vehicles are filled with TOYOTA SLLC at the factory. In order to avoid damage to the engine cooling system and other technical problems, only use TOYOTA SLLC or similar high quality ethylene glycol based non-silicate, non-amine, non-nitrite, non-borate coolant with long-life hybrid organic acid technology (coolant with long-life hybrid organic acid technology is a combination of low phosphates and organic acids).

e. Remove the reserve tank cap. If the coolant level is low, add coolant.
f. Fill the radiator reserve tank with coolant to the FULL line.
g. Tighten the air bleed valve.
h. Install the water filler cap sub-assembly and reserve tank cap.
i. Start the engine and warm it up.
j. Bleed air from the cooling system.

➡Before starting the engine, turn the A/C switch off. Adjust the air conditioning temperature setting to MAX (HOT). Adjust the air conditioning blower setting to LO.

k. Run the engine intermittently for 7 minutes or more (5 seconds running at 3000 rpm and 45 seconds idling, repeat several times).
l. Stop the engine, and wait until the engine coolant cools down.
m. After the engine is cooled down, check that the coolant level is between the FULL and LOW lines.
n. If the coolant level is low, add coolant to the reservoir tank FULL line.

3. Inspect for coolant leak.

BLEEDING

1. Start the engine and warm it up.
2. Bleed air from the cooling system.

➡Before starting the engine, turn the A/C switch off. Adjust the air conditioning temperature setting to MAX (HOT). Adjust the air conditioning blower setting to LO.

3. Run the engine intermittently for 7 minutes or more (5 seconds running at 3000 rpm and 45 seconds idling, repeat several times).
4. Stop the engine, and wait until the engine coolant cools down.
5. After the engine is cooled down, check that the coolant level is between the FULL and LOW lines.
6. If the coolant level is low, add coolant to the reservoir tank FULL line.
7. Inspect for coolant leak.

ENGINE FAN

REMOVAL & INSTALLATION

See Figures 13 and 14.

1. Drain engine coolant.
2. Remove front bumper assembly.
3. Remove upper radiator support absorber.
4. Remove battery.
5. Remove air cleaner and hose.
6. Remove air cleaner filter element sub-assembly.
7. Disconnect radiator reservoir tank hose.
8. Disconnect No. 3 radiator hose.
9. Disconnect No. 2 radiator hose.
10. Separate hood lock control cable assembly.
11. Remove upper radiator support sub-assembly.
12. Remove fan shroud.
13. Remove the bolt and the cooling fan motor insulator.
14. Remove the nut and the fan.
15. Remove the 3 screws and the cooling fan motor.

To install:

16. Install the cooling fan motor with the 3 screws.
17. Install the fan with the nut.
18. Insert the cooling fan motor insulator into the fan shroud and install it with the bolt.
19. Install fan shroud.
20. Install upper radiator support sub-assembly.
21. Install hood lock control cable assembly.
22. Connect No. 2 radiator hose.
23. Connect No. 3 radiator hose.
24. Connect radiator reservoir tank hose.
25. Install air cleaner filter element sub-assembly.
26. Install air cleaner and hose.
27. Install battery.
28. Install upper radiator support absorber.
29. Install front bumper assembly.
30. Add engine coolant.
31. Warm up engine.
32. Inspect for coolant leak.
33. Inspect hood.
34. Adjust hood.

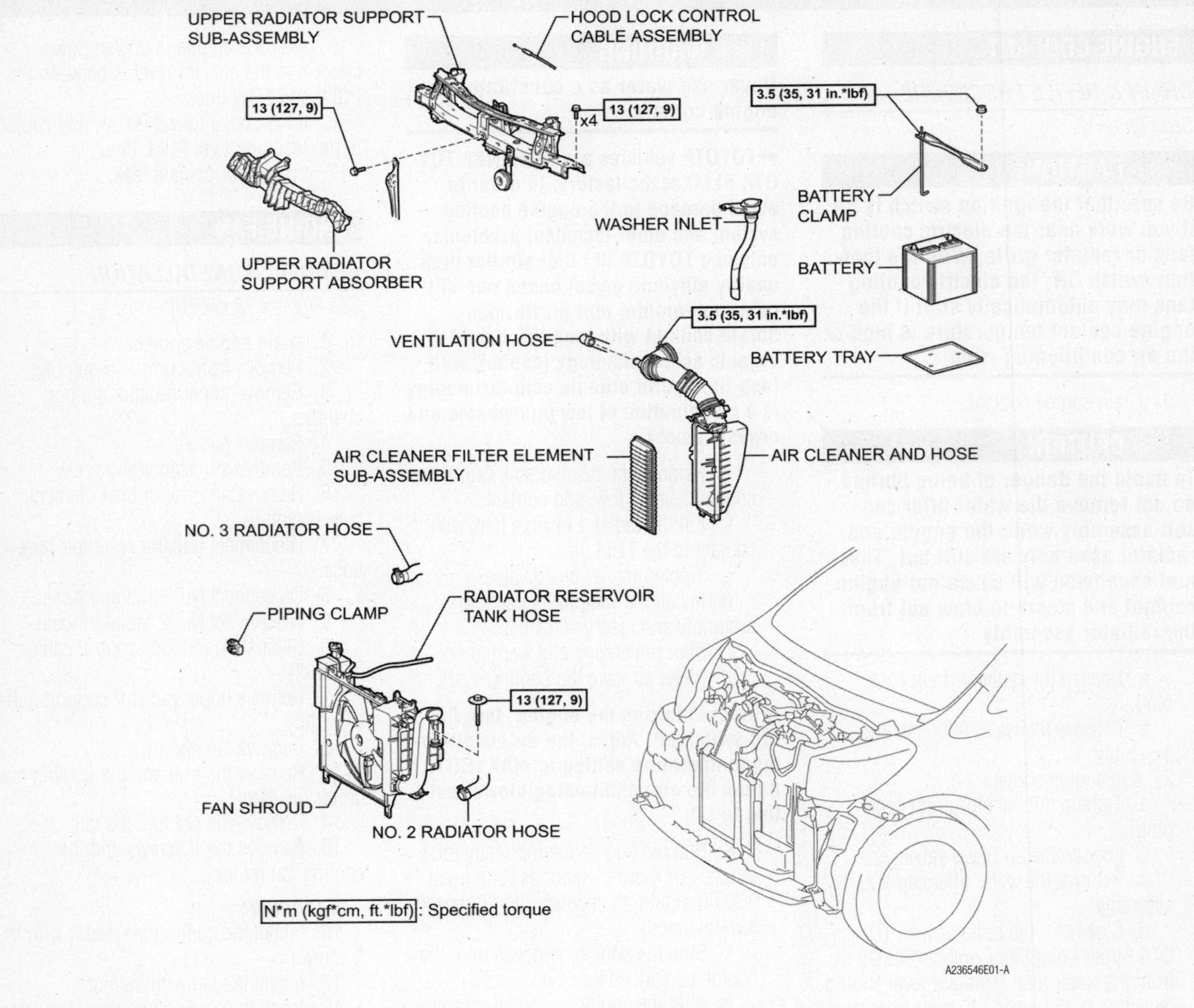

Fig. 13 Exploded view showing assemblies to be removed prior to removing cooling fan assembly

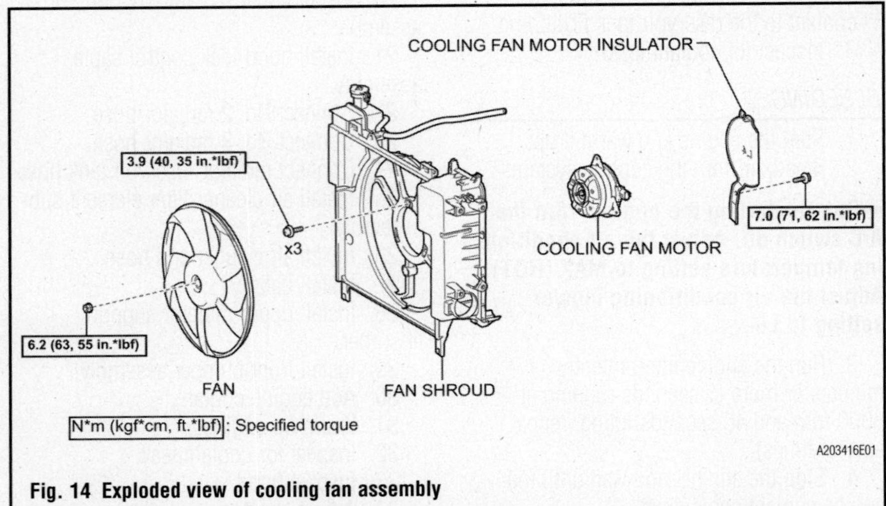

Fig. 14 Exploded view of cooling fan assembly

RADIATOR

REMOVAL & INSTALLATION

See Figures 15 and 16.

1. Drain engine coolant.
2. Remove front bumper assembly.
3. Remove upper radiator support absorber.
4. Remove battery.
   a. Separate the negative and positive battery terminals.
   b. Remove the nut, loosen the nut and remove the battery clamp.
   c. Remove the battery.
   d. Remove the battery tray.
5. Remove air cleaner and hose.
   a. Disengage the wire harness clamp

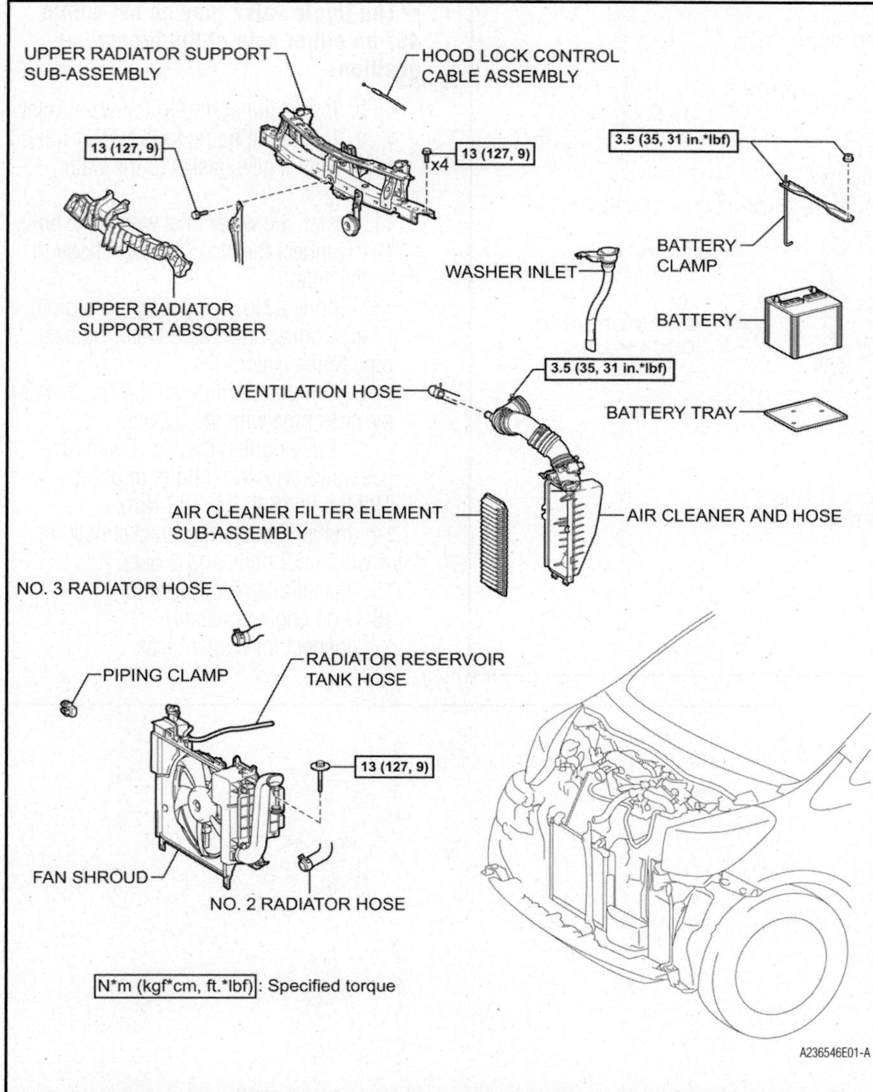

Fig. 15 Exploded view showing assemblies to be removed prior to removing radiator assembly

and disconnect the intake Mass Air Flow (MAF) meter connector.

b. Disconnect the ventilation hose.

c. Loosen the hose clamp and disconnect the air cleaner hose.

d. Disengage the 2 clamps and 2 guides and remove the air cleaner and hose.

6. Remove the air cleaner filter element.

7. Disconnect radiator reservoir tank hose.

a. Disengage the hose clamp.

b. Loosen the clip and disconnect the reserve tank hose from the water filler.

8. Loosen the clip and disconnect the No. 3 radiator hose from the radiator.

9. Loosen the clip and disconnect the No. 2 radiator hose from the radiator.

10. Separate hood lock control cable assembly.

a. Separate the hood lock control cable from the hood lock.

b. Disengage the 2 clamps and separate the hood lock control cable.

11. Remove upper radiator support sub-assembly.

a. Disconnect the hood courtesy switch connector and disengage the wire harness clamp.

b. Disengage the washer inlet clamp.

c. Disconnect the horn connector.

d. Remove the 5 bolts and the upper radiator support.

12. Remove fan shroud.

a. Remove the washer inlet.

b. Disengage the claw and remove the piping clamp.

c. Disengage the 4 wire harness clamps and disconnect the 3 connectors.

d. Remove the bolt from the ECM.

e. Disengage the 2 claws and remove the fan shroud from the vehicle.

13. Disengage the 2 claws on the condenser and remove the radiator from the vehicle.

To install:

14. Install radiator assembly.

a. Install the 2 cushions and 2 grommets.

b. Engage the 2 claws on the condenser and install the radiator assembly onto the vehicle.

15. Install fan shroud.

a. Engage the 2 claws and install the fan shroud onto the vehicle.

b. Install the bolt to the ECM.

c. Engage the 4 wire harness clamps and connect the 3 connectors.

d. Engage the claw and install the piping clamp.

e. Align the locating lip and install the washer inlet.

16. Install upper radiator support sub-assembly.

a. Install the upper radiator support with the 5 bolts.

b. Connect the horn connector.

c. Engage the washer inlet clamp.

d. Connect the hood courtesy switch connector and engage the wire harness clamp.

17. Install the hood lock control cable and engage the 2 clamps.

18. Connect the No. 2 radiator hose to the radiator with the hose clip.

19. Connect the No. 3 radiator hose to the radiator with the hose clip.

20. Connect radiator reservoir tank hose.

a. Connect the reservoir tank hose to the water filler with the hose clip.

b. Engage the hose clamp.

21. Install the air cleaner filter element.

22. Install air cleaner and hose.

a. Align the 2 hinges with the guides on the case and engage them in the grooves.

b. Install the air cleaner and hose with the 2 clamps.

c. Connect the air cleaner hose with the hose clamp.

d. Connect the ventilation hose.

e. Connect the Mass Air Flow (MAF) meter connector and engage the wire harness clamp.

23. Install battery.

a. Install the battery tray.

b. Install the battery.

c. Install the battery clamp with the 2 nuts.

d. Connect the positive and negative battery terminals.

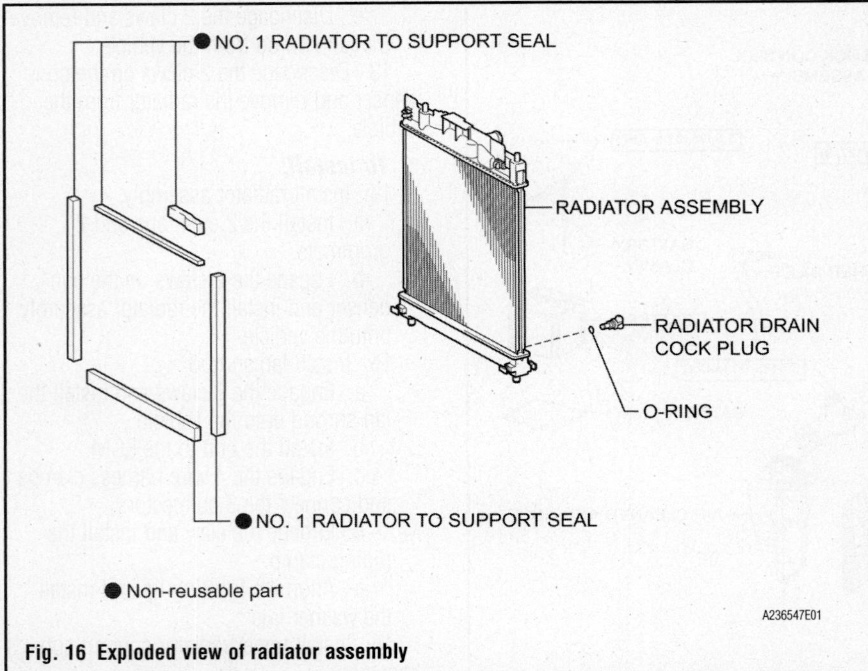

Fig. 16 Exploded view of radiator assembly

24. Install upper radiator support absorber.
25. Install front bumper assembly.
26. Add engine coolant.
27. Warm up engine.
28. Inspect for coolant leak.
29. Inspect hood sub-assembly.
30. Adjust hood sub-assembly.

THERMOSTAT

REMOVAL & INSTALLATION

See Figures 17 and 18.

1. Drain engine coolant.
2. Remove alternator assembly.
3. Remove the 2 bolts, 2 nuts and bearing bracket heat insulator.
4. Remove the 3 bolts and then disconnect the No. 1 water by-pass pipe.
5. Disconnect the No. 2 radiator hose from the water inlet.
6. Remove the 2 bolts and the water inlet.
7. Remove the No. 2 water inlet housing gasket from the water inlet.
8. Remove thermostat.

➡It is not necessary to remove the thermostat unless it is being replaced.

a. Remove the frame from the water inlet.
b. Remove the spring and element.

To install:

9. Install thermostat.
   a. Install the element to the water inlet.

➡The jiggle valve may be set within 45° on either side of the prescribed position.

   b. Install the spring to the water inlet.
   c. Install the frame to the water inlet.
10. Install a new gasket to the water inlet.
11. Install the water inlet with the 2 bolts.
12. Connect the No. 2 radiator hose to the water inlet.
13. Connect No. 1 water by-pass pipe.
   a. Connect the No. 1 water by-pass pipe to the water inlet.
   b. Temporarily tighten the No. 1 water by-pass pipe with the 3 bolts.
   c. Fully tighten the No. 1 water by-pass pipe with the 3 bolts in order. Tighten to 16 ft. lbs. (22 Nm).
14. Install the bearing bracket heat insulator with the 2 bolts and 2 nuts.
15. Install alternator assembly.
16. Add engine coolant.
17. Inspect for coolant leak.

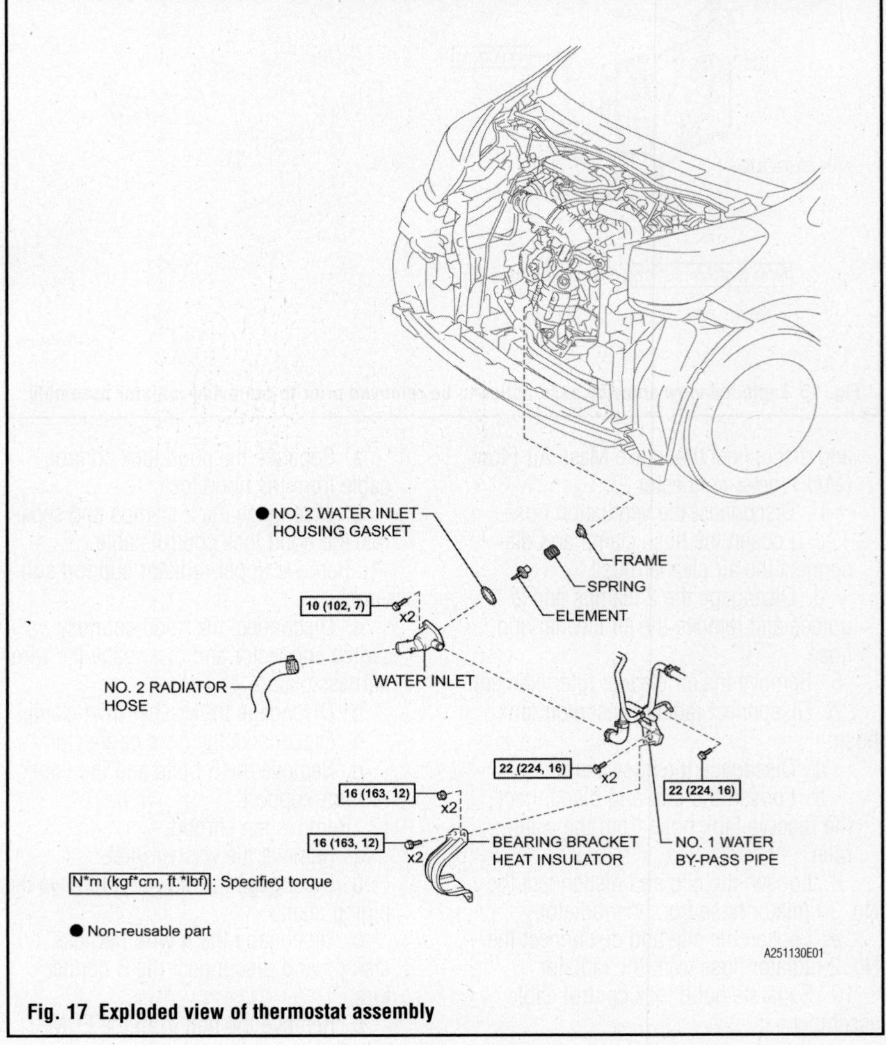

Fig. 17 Exploded view of thermostat assembly

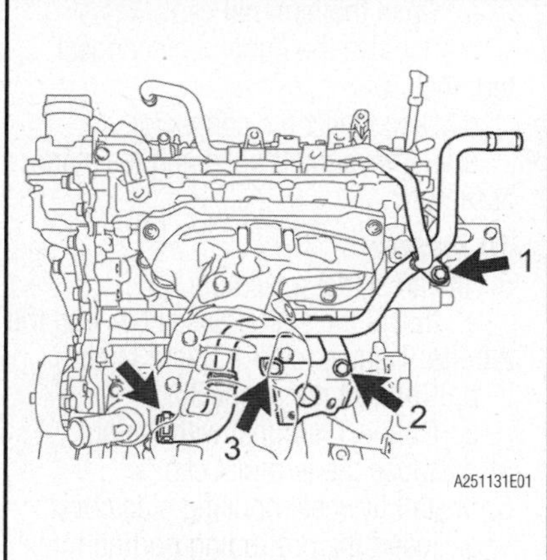

Fig. 18 Fully tighten the No. 1 water by-pass pipe with the 3 bolts in order

Fig. 19 Remove the 5 bolts, engine water pump assembly and water pump gasket from the timing chain cover sub-assembly

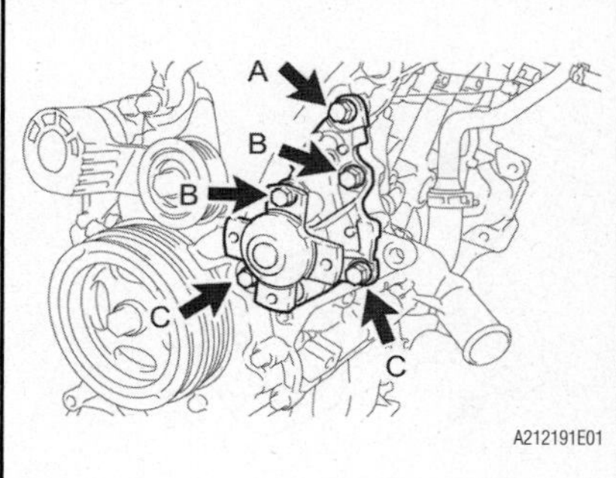

Fig. 20 Install the engine water pump assembly and a new water pump gasket to the timing chain cover sub-assembly with the 5 bolts

WATER PUMP

REMOVAL & INSTALLATION

See Figures 19 and 20.

1. Drain engine coolant.
2. Remove the fan and alternator V belt.
3. Remove alternator assembly.
4. Remove water pump pulley.
   a. Using SST, hold the water pump pulley.
   b. Remove the 4 bolts and the water pump pulley.
5. Remove the 5 bolts, engine water pump assembly and water pump gasket from the timing chain cover sub-assembly.

To install:

6. Install the engine water pump assembly and a new water pump gasket to the timing chain cover sub-assembly with the 5 bolts.
   a. Tighten Bolt A to 52 ft. lbs. (71 Nm).
   b. Tighten Bolts B to 18 ft. lbs. (24 nm).
   c. Tighten Bolts C to 15 ft. lbs. (21 Nm).

➡Bolts C are the shorter bolts.

7. Install water pump pulley.
   a. Using SST, hold the water pump pulley.
   b. Install the water pump pulley with the 4 bolts.
8. Install alternator assembly.
9. Install fan and alternator V belt.
10. Add engine coolant.
11. Inspect for coolant leak.

ENGINE ELECTRICAL — BATTERY SYSTEM

BATTERY

REMOVAL & INSTALLATION

1. Disconnect the negative and positive battery terminals.
2. Remove the nut, loosen the nut and remove the battery clamp.
3. Remove the battery.
4. Remove the battery tray.
5. Installation is the reverse order of removal.

ENGINE ELECTRICAL — STARTING SYSTEM

STARTER

REMOVAL & INSTALLATION

Except For Cold Area

See Figure 21.

1. Disconnect cable from negative battery terminal.

➡After turning the ignition switch off, waiting time may be required before disconnecting the cable from battery terminal. Therefore, make sure to read the disconnecting the cable from the battery terminal notice before proceeding with work.

2. Remove front exhaust pipe assembly.
3. Remove flywheel housing side cover. Disengage the claw by pulling it outward and remove the flywheel housing side cover.
4. Remove starter assembly.
   a. Open the terminal cap.
   b. Remove the nut and disconnect terminal.
   c. Disconnect the connector.
   d. Remove the 2 bolts and the starter assembly.

To install:

5. Install starter assembly.
   a. Install the starter assembly with the 2 bolts. Tighten to 27 ft. lbs. (37 Nm).
   b. Connect the connector.
   c. Connect terminal with the nut.
   d. Close the terminal cap.
6. Install flywheel housing side cover.
   a. Insert the protruding portion into the end of the cylinder block and while pushing it along the cylinder block, fit the claw into the cylinder block.

➡Make sure that the claw makes a click sound, indicating that it fits tightly. Replace the flywheel housing side cover with a new one if it does not fit tightly or is deformed.

7. Install front exhaust pipe assembly.
8. Connect cable to negative battery terminal.
9. Inspect for exhaust gas leak.

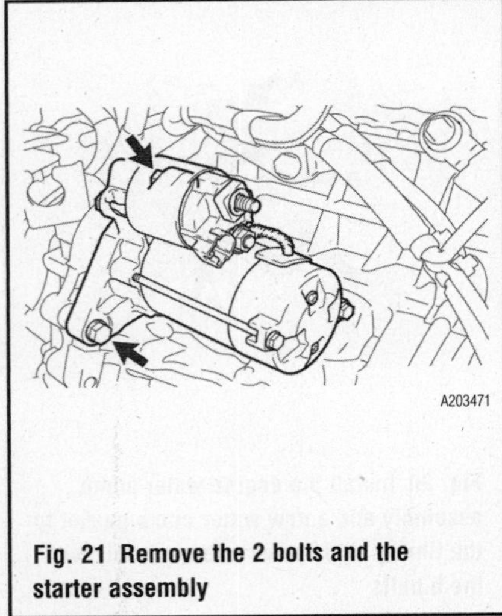

Fig. 21 Remove the 2 bolts and the starter assembly

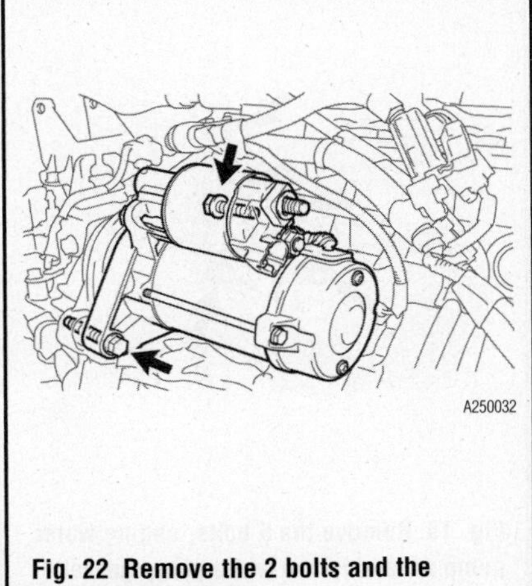

Fig. 22 Remove the 2 bolts and the starter assembly

a. Open the terminal cap.

b. Remove the nut and disconnect terminal 30.

c. Disconnect the connector.

d. Remove the 2 bolts and the starter assembly.

To install:

5. Install starter assembly.

a. Install the starter assembly with the 2 bolts. Tighten to 27 ft. lbs. (37 Nm).

b. Connect the connector.

c. Connect terminal with the nut.

d. Close the terminal cap.

6. Install flywheel housing side cover.

a. Insert the protruding portion into the end of the cylinder block and while pushing it along the cylinder block, fit the claw into the cylinder block.

➡Make sure that the claw makes a click sound, indicating that it fits tightly. Replace the flywheel housing side cover with a new one if it does not fit tightly or is deformed.

7. Install front exhaust pipe assembly.

8. Connect cable to negative battery terminal.

9. Inspect for exhaust gas leak.

For Cold Area

See Figure 22.

1. Disconnect cable from negative battery terminal.

➡After turning the ignition switch off, waiting time may be required before disconnecting the cable from battery terminal. Therefore, make sure to read the disconnecting the cable from the battery terminal notice before proceeding with work.

2. Remove front exhaust pipe assembly.

3. Remove flywheel housing side cover.

a. Disengage the claw by pulling it outward and remove the flywheel housing side cover.

4. Remove starter assembly.

ENGINE MECHANICAL

➡Disconnecting the negative battery cable may interfere with the functions of the on board computer systems and may require the computer to undergo a relearning process, once the negative battery cable is reconnected.

ACCESSORY DRIVE BELTS

ACCESSORY BELT ROUTING

See Figure 23.

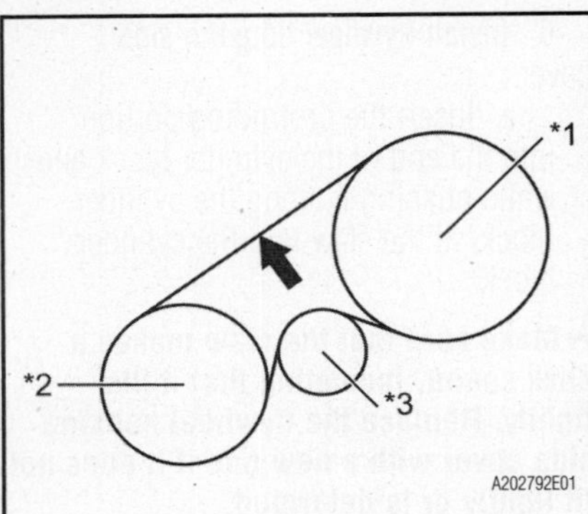

Fig. 23 No. 1 V-ribbed A/C compressor (2) to crankshaft pulley (1) belt routing; tensioner (3)

INSPECTION

1. Remove No. 3 engine under cover.

2. Remove No. 1 cooler cover.

3. Remove No. 1 engine cover.

4. Inspect fan and alternator V-ribbed belt.

a. Check the belt for wear, cracks or other signs of damage. If any of the following defects is found, replace the V-ribbed belt.

- The belt is cracked.
- The belt is worn out to the extent that the cords are exposed.
- The belt has chunks missing from the ribs.

b. Check that the belt fits properly in the ribbed grooves.

➡Check with your hand to confirm that the belt has not slipped out of the grooves on the bottom of the pulley. If it has slipped out, replace the V-ribbed belt. Install a new V-ribbed belt correctly.

5. Inspect No. 1 V-ribbed (cooler compressor to crankshaft pulley) belt.

a. Check the belt for wear, cracks or other signs of damage. If any of the following defects is found, replace the No. 1 V-ribbed belt.

- The belt is cracked.
- The belt is worn out to the extent that the cords are exposed.
- The belt has chunks missing from the ribs.

b. Check that the belt fits properly in the ribbed grooves.

➡Check with your hand to confirm that the belt has not slipped out of the grooves on the bottom to the pulley. If it has slipped out, replace the No. 1 V-ribbed belt. Install a new No. 1 V-ribbed belt correctly.

c. Inspect the No. 1 V-ribbed belt deflection.

6. Install No. 1 engine cover.

7. Install No. 1 cooler cover.

8. Install No. 3 engine under cover.

ADJUSTMENT

A/C compressor to crankshaft pulley belt is equipped with automatic tensioner. No adjustment is needed. For the alternator belt, refer to the installation procedure for tensioning procedure.

REMOVAL & INSTALLATION

See Figures 24 through 26.

1. Remove No. 3 engine under cover.
2. Remove the 2 nuts and the No. 1 cooler cover.
3. Remove the 2 bolts, nut and the No. 1 engine cover.
4. Remove No. 1 V-ribbed (cooler compressor to crankshaft pulley) belt.
    a. Loosen the bolt A.
    b. Turn adjusting bolt B to release the tension and remove the No. 1 V-ribbed belt from the pulleys.
5. Remove fan and alternator V-ribbed belt.
    a. Release the V-ribbed belt tension by turning the V-ribbed belt tensioner counterclockwise, and remove the V-ribbed belt from the V-ribbed belt tensioner.
    b. While turning the V-ribbed belt tensioner counterclockwise, align the holes, and then insert a wrench into the holes to fix the V-ribbed belt tensioner in place.

To install:

6. Install fan and alternator v-ribbed belt.
    a. Install the V-ribbed belt.

➡There are 2 types of V-ribbed belt. The type to use depends on the type of alternator, so be sure to confirm the belt type before installing. Paint mark for 80A alternator is blue. Paint mark for 100A alternator is red.

b. Check that the ribs of the V-ribbed belt are correctly fitted into the grooves of the pulleys.

c. Turn the V-ribbed belt tensioner clockwise and remove the hexagon wrench.

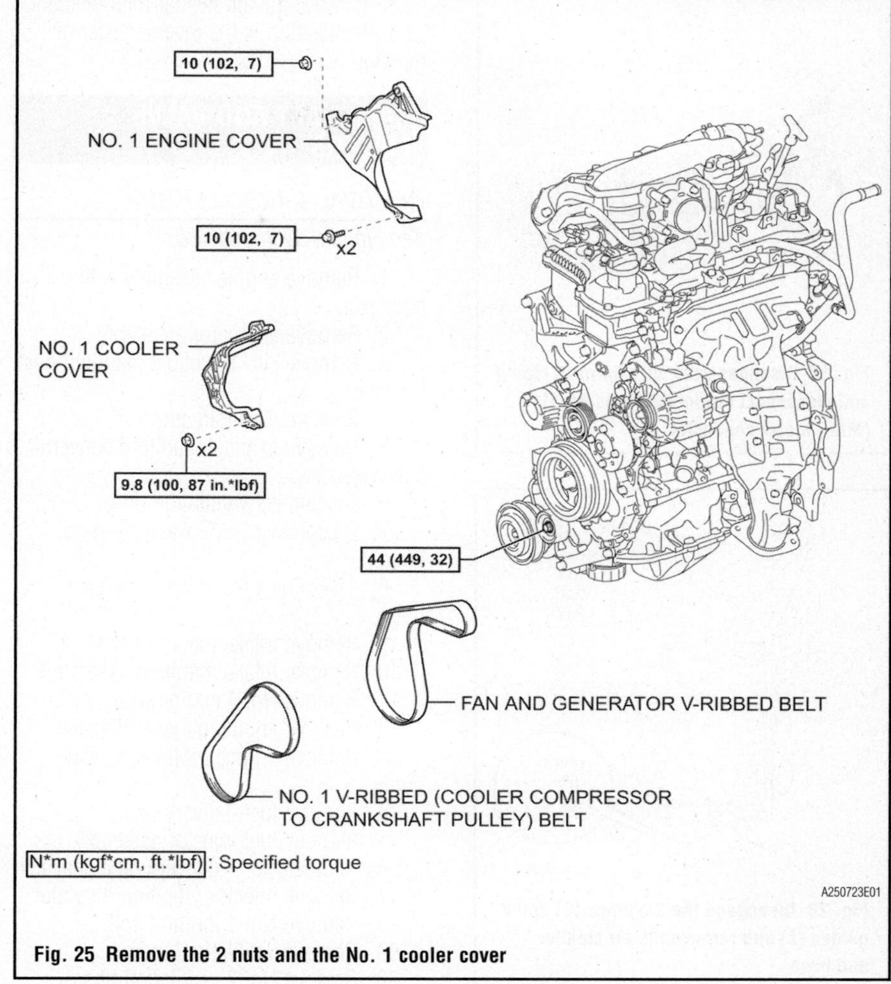

Fig. 25 Remove the 2 nuts and the No. 1 cooler cover

7. Install No. 1 V-ribbed (cooler compressor to crankshaft pulley) belt.
    a. Temporarily install the No. 1 V-ribbed belt onto each pulley.
    b. Turn the adjust bolt to adjust the tension of No. 1 V-ribbed belt.
    c. Tighten bolt to 32 ft. lbs. (44 Nm).
8. Inspect alternator V-ribbed belt.
9. Inspect No. 1 V-ribbed (cooler compressor to crankshaft pulley) belt.
10. Install No. 1 engine cover.
    a. Temporarily install the No. 1 engine cover with the 2 bolts and nut.
    b. Fully tighten the 2 bolts and nut in order (Top bolt first, the bottom bolt second, and middle bolt last).
11. Install No. 1 cooler cover.
    a. Temporarily install the No. 1 cooler cover with the 2 nuts.
    b. Fully tighten the 2 nuts in order (Top bolt first, then the bottom bolt).
12. Install No. 3 engine under cover.

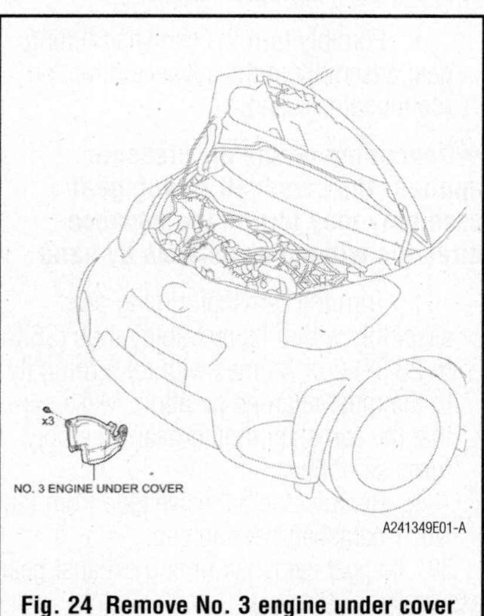

Fig. 24 Remove No. 3 engine under cover

Fig. 26 Turn the adjust bolt (2) to adjust the tension of No. 1 V-ribbed belt, then tighten bolt (1)

AIR CLEANER

REMOVAL & INSTALLATION

See Figures 27 through 29.

1. Remove air cleaner and hose.
    a. Disengage the wire harness clamp and disconnect the intake Mass Air Flow (MAF) meter connector.

Fig. 27 Disengage the wire harness clamp and disconnect the intake Mass Air Flow (MAF) meter connector

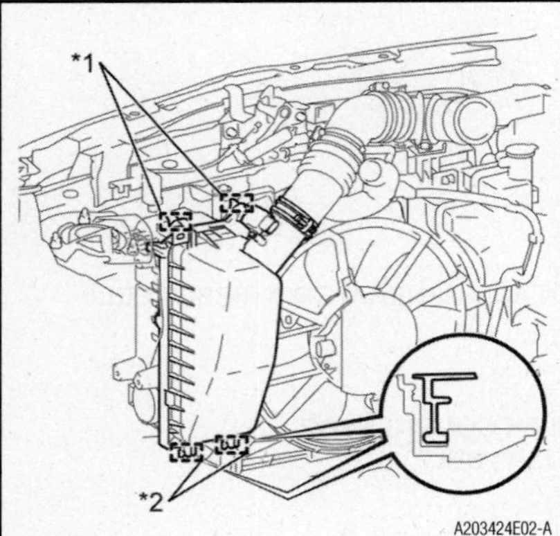

Fig. 28 Disengage the 2 clamps (1) and 2 guides (2) and remove the air cleaner and hose

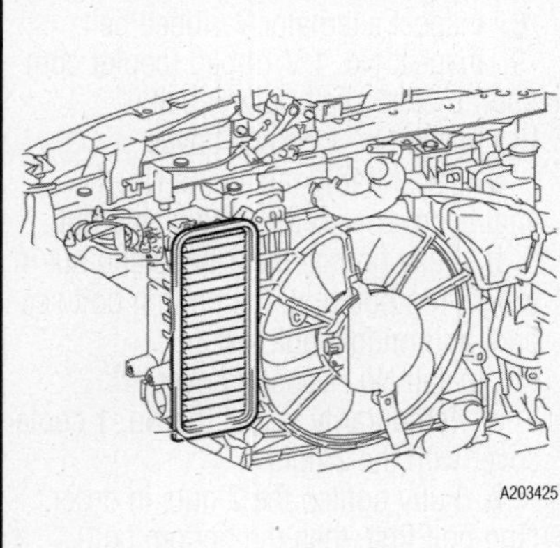

Fig. 29 Remove the air cleaner filter element

b. Disconnect the ventilation hose.

c. Loosen the hose clamp and disconnect the air cleaner hose.

d. Disengage the 2 clamps and 2 guides and remove the air cleaner and hose.

2. Remove the air cleaner filter element.

3. Installation is the reverse order of removal.

CAMSHAFT AND VALVE LIFTERS

REMOVAL & INSTALLATION

See Figures 30 through 36.

1. Remove engine assembly with transaxle.
2. Remove alternator assembly.
3. Remove No. 1 exhaust manifold heat insulator.
4. Remove manifold stay.
5. Remove exhaust manifold converter sub-assembly.
6. Disconnect ventilation hose.
7. Disconnect No. 2 water by-pass hose.
8. Disconnect No. 4 water by-pass hose.
9. Remove intake manifold stay.
10. Remove intake manifold assembly.
11. Remove No. 4 engine wire.
12. Remove engine oil level dipstick.
13. Remove engine oil level dipstick guide.
14. Remove fuel delivery pipe.
15. Remove fuel injector assembly.
16. Remove No. 1 delivery pipe spacer.
17. Remove injector vibration insulator.
18. Remove No. 1 ignition coil.
19. Remove ventilation hose.
20. Remove No. 2 ventilation hose.
21. Remove No. 1 water by-pass pipe.
22. Remove No. 2 idle pulley assembly with bracket.
23. Remove V-ribbed belt tensioner assembly.
24. Remove No. 1 Crankshaft Position (CKP) sensor.
25. Remove cylinder head cover sub-assembly.
26. Remove camshaft bearing cap oil hole gasket.
27. Remove Crankshaft Position (CKP) sensor.
28. Remove water pump pulley.
29. Remove crankshaft pulley.
30. Remove timing chain cover sub-assembly.
31. Remove timing chain cover oil seal.
32. Remove No. 1 chain tensioner assembly.
33. Remove No. 2 chain vibration damper.
34. Remove timing chain tension arm.
35. Remove chain sub-assembly.
36. Remove timing chain guide.
37. Inspect camshaft timing gear assembly.

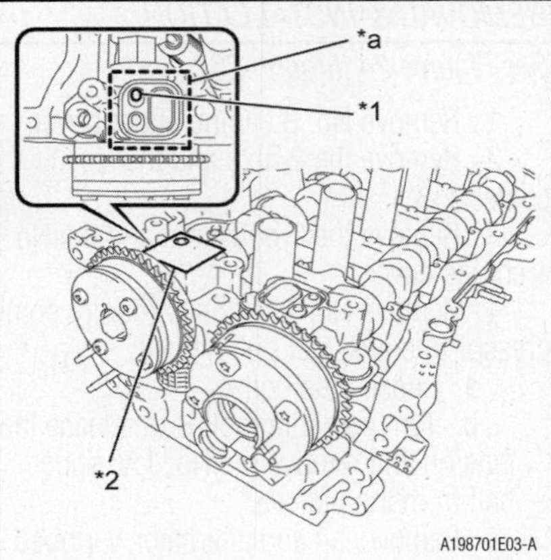

Fig. 30 After cleaning and degreasing the VVT oil hole on the intake side of the No. 1 camshaft bearing cap, completely seal (a) the oil hole (1) with adhesive tape (2) or equivalent to prevent air from leaking

a. Inspect the lock of the camshaft timing gear.

b. After cleaning and degreasing the VVT oil hole on the intake side of the No. 1 camshaft bearing cap, completely seal the oil hole with adhesive tape or equivalent to prevent air from leaking.

➡Be sure to seal the oil hole completely because air leaks due to insufficient sealing will prevent the lock pin from being released.

c. Prick a hole in the tape covering the oil hole (Procedure A).

d. Apply approximately 22 psi (150 kPa) of air pressure to the hole pricked in procedure A to release the lock pin.

➡If air leaks out, reattach adhesive tape. Cover the oil hole with a piece of cloth when applying air pressure to prevent oil from spraying.

e. Forcibly turn the camshaft timing gear assembly in the advance direction (counterclockwise).

➡Depending on the air pressure applied, the camshaft timing gear assembly may turn in the advance direction without assistance by hand.

f. Turn the camshaft timing gear assembly within its movable range (26.5 to 28.5°) 2 or 3 times without turning it to the most retarded position. Make sure that the camshaft timing gear assembly turns smoothly.

g. Remove the adhesive tape from the No. 1 camshaft bearing cap.

38. Inspect camshaft timing exhaust gear assembly.

a. Check the lock of the camshaft timing exhaust gear.

b. After cleaning and degreasing the VVT oil hole on the exhaust side of the No. 1 camshaft bearing cap, completely seal the oil hole with adhesive tape or equivalent to prevent air from leaking.

➡**Be sure to seal the oil hole completely because air leaks due to insufficient sealing will prevent the lock pin from being released.**

c. Prick a hole in the tape covering the oil hole (Procedure B).

d. Apply approximately 28 psi (200 kPa) of air pressure to the hole pricked in procedure B to release the lock pin.

➡**If air leaks out, reattach adhesive tape. Cover the oil hole with a piece of cloth when applying air pressure to prevent oil from spraying.**

e. Using a screwdriver with its tip wrapped with tape, forcibly turn the camshaft timing exhaust gear in the retard direction (clockwise).

➡**Be sure to keep the camshaft timing exhaust gear in the retard direction using a screwdriver. If the gear is released, it will return to the most advanced position automatically due to force from the spring. Do not damage the camshaft timing exhaust gear.**

f. Using a screwdriver with its tip wrapped with tape, turn the camshaft timing exhaust gear within its movable range (19 to 21°) 2 or 3 times without turning it to the most advanced position. Make sure that the camshaft timing exhaust gear turns smoothly.

g. Remove the adhesive tape from the No. 1 camshaft bearing cap.

39. Remove camshaft timing gear assembly.

a. Remove the flange bolt while holding the hexagonal portion of the camshaft, and then remove the camshaft timing gear assembly.

➡**Before removing the camshaft timing gear, make sure that the lock pin has been released. Be sure not to remove the other 4 bolts. Keep the camshaft timing gear assembly horizontal while removing it from the camshaft.**

40. Remove camshaft timing exhaust gear assembly.

a. Remove the flange bolt while holding the hexagonal portion of the camshaft, and then remove the camshaft timing exhaust gear assembly.

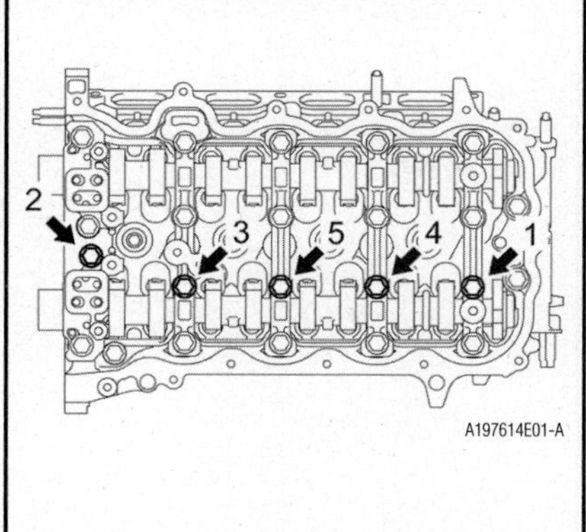

Fig. 31 Uniformly loosen and remove the 5 bearing cap bolts in sequence

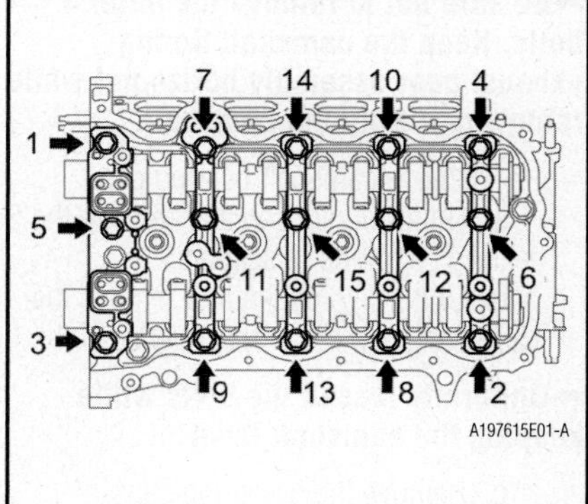

Fig. 32 Uniformly loosen and remove the 15 bearing cap bolts in sequence

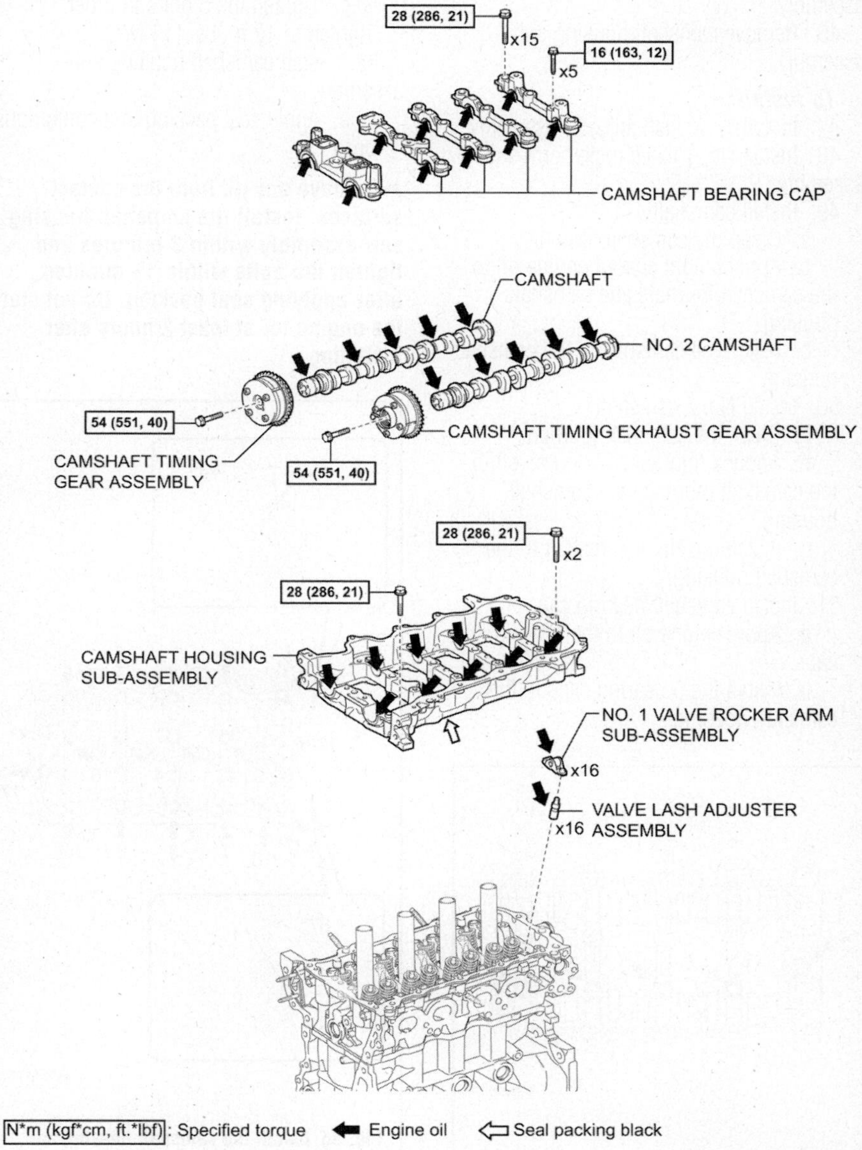

Fig. 33 Exploded view of camshaft assembly

➡Be sure not to remove the other 4 bolts. Keep the camshaft timing exhaust gear assembly horizontal while removing it from the camshaft.

41. Remove camshaft bearing cap.
 a. Uniformly loosen and remove the 5 bearing cap bolts in sequence.
 b. Uniformly loosen and remove the 15 bearing cap bolts in sequence.

➡Uniformly loosen the bolts while keeping the camshaft level.

 c. Remove the 5 bearing caps.
42. Remove No. 2 camshaft.
43. Remove camshaft.
44. Remove No. 1 valve rocker arm sub-assembly.
45. Remove valve lash adjuster assembly.
46. Remove camshaft housing sub-assembly.

To install:

47. Install valve lash adjuster assembly.
48. Install No. 1 valve rocker arm sub-assembly.
49. Install camshaft.
 a. Clean the camshaft journals.
 b. Apply a light coat of engine oil to the camshaft journals and camshaft housing.
 c. Install the camshaft to the camshaft housing.
50. Install No. 2 camshaft.
 a. Clean the camshaft journals.
 b. Apply a light coat of engine oil to the camshaft journals and camshaft housing.
 c. Install the No. 2 camshaft to the camshaft housing.
51. Install camshaft bearing cap.
 a. Apply engine oil to the bearing caps.
 b. Install the 5 bearing caps to the camshaft housing.

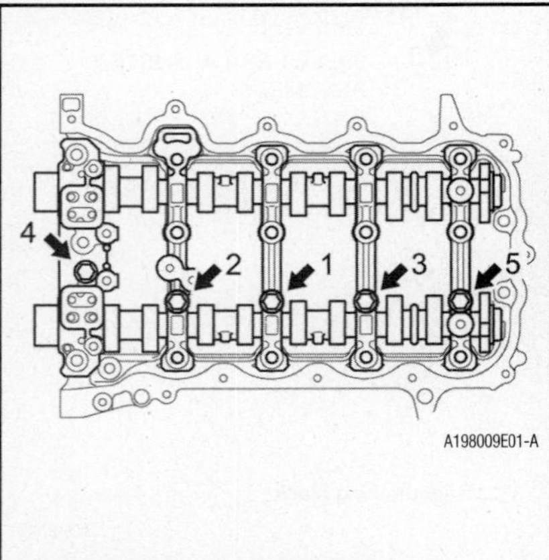

Fig. 34 Tighten the 5 bolts in order

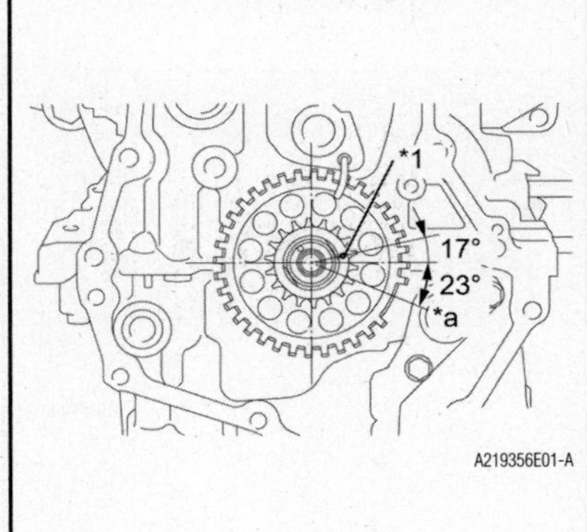

Fig. 35 Set the crankshaft in the position of 40° BTDC

 c. Tighten the 5 bolts in order. Tighten to 12 ft. lbs. (16 Nm).
52. Install camshaft housing sub-assembly.
 a. Apply seal packing in a continuous bead.

➡Remove any oil from the contact surfaces. Install the camshaft housing sub-assembly within 3 minutes and tighten the bolts within 15 minutes after applying seal packing. Do not start the engine for at least 2 hours after installing.

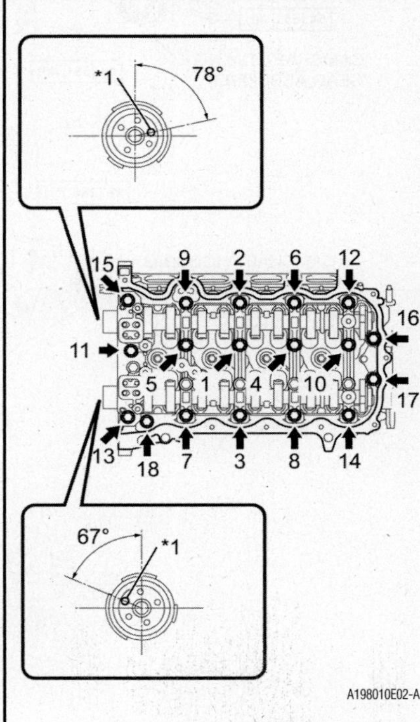

Fig. 36 Install the camshaft housing and tighten the 18 bolts in order; straight pin (1)

 b. Set the crankshaft in the position of 40° BTDC.

➡Turn the crankshaft clockwise when positioning the No. 1 cylinder after the camshaft housing sub-assembly is installed.

 c. Set the camshaft and No. 2 camshaft.
 d. Install the camshaft housing and tighten the 18 bolts in order to 21 ft. lbs. (28 Nm).

➡Note the following:

- After installing the camshaft housing, make sure that the cam lobes are properly positioned.
- If any of the bolts are loosened during installation, remove the camshaft housing, clean the installation surfaces, and reapply seal packing.
- If the camshaft housing is removed because any of the bolts are loosened during installation, make sure that the previously applied seal packing does not enter any oil passages.
- After installing the camshaft housing, wipe off any seal packing that seeped out from between the housing and the cylinder head.

53. Install camshaft timing gear assembly.
 a. Check that the straight pin is installed on the camshaft.
 b. Put the camshaft timing gear and camshaft together by aligning the straight pin hole and straight pin.

➡Do not forcefully push in the camshaft timing gear assembly. This may cause the camshaft straight pin tip to damage the installation surface of the camshaft timing gear assembly. Do not turn the camshaft timing gear in the retard direction (clockwise).

 c. Tighten the flange bolt with the camshaft timing gear secured in place. Tighten to 40 ft. lbs. (54 Nm).
 d. Check that the camshaft timing gear can move in the retard direction (clockwise) and locks in the most retarded position.
54. Install camshaft timing exhaust gear assembly.
 a. Check that the straight pin is installed on the No. 2 camshaft.
 b. Put the camshaft timing exhaust gear and No. 2 camshaft together by aligning the straight pin hole and straight pin.

➡Do not forcefully push in the camshaft timing gear assembly. This may cause the camshaft straight pin tip to damage the installation surface of the camshaft timing gear assembly.

c. Tighten the flange bolt with the camshaft timing exhaust gear secured in place. Tighten to 40 ft. lbs. (54 Nm).

d. Make sure that the camshaft timing exhaust gear is locked.

55. Install timing chain guide.
56. Install chain sub-assembly.
57. Install timing chain tension arm.
58. Install No. 1 chain tensioner assembly.
59. Install timing chain cover oil seal.
60. Install timing chain cover sub-assembly.
61. Install crankshaft pulley.
62. Install water pump pulley.
63. Install Crankshaft Position (CKP) sensor.
64. Install camshaft bearing cap oil hole gasket.
65. Install cylinder head cover sub-assembly.
66. Install V-ribbed belt tensioner assembly.
67. Install No. 2 idle pulley assembly with bracket.
68. Install No. 1 water by-pass pipe.
69. Install No. 2 ventilation hose.
70. Install ventilation hose.
71. Install No. 1 ignition coil.
72. Install injector vibration insulator.
73. Install No. 1 delivery pipe spacer.
74. Install fuel injector assembly.
75. Install fuel delivery pipe.
76. Install engine oil level dipstick guide.
77. Install engine oil level dipstick.
78. Install No. 4 engine wire.
79. Install intake manifold assembly.
80. Install intake manifold stay.
81. Connect No. 4 water by-pass hose.
82. Connect No. 2 water by-pass hose.
83. Connect ventilation hose.
84. Install exhaust manifold converter sub-assembly.
85. Install manifold stay.
86. Install No. 1 exhaust manifold heat insulator.
87. Install alternator assembly.
88. Install engine assembly with transaxle.

CATALYTIC CONVERTER

REMOVAL & INSTALLATION

See Figure 37.

➡The manufacturer does not provide a specific Removal and Installation procedure for this component. Refer to the graphic(s) when servicing this component.

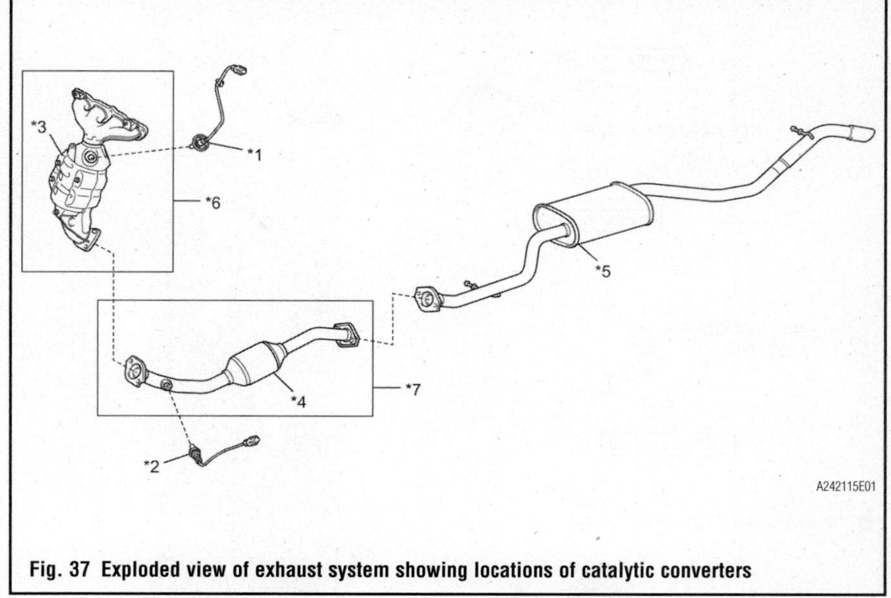

Fig. 37 Exploded view of exhaust system showing locations of catalytic converters

CRANKSHAFT FRONT SEAL

REMOVAL & INSTALLATION

See Figures 38 and 39.

1. Remove front wheel RH.
2. Remove No. 3 engine under cover.
3. Remove No. 1 cooler cover.
4. Remove No. 1 engine cover.
5. Remove No. 1 V-ribbed (cooler compressor to crankshaft pulley) belt.
6. Remove fan and alternator V-ribbed belt.
7. Remove crankshaft pulley assembly.

a. Using SST (09960-10010), remove the crankshaft pulley bolt.

b. Remove the crankshaft pulley from the crankshaft.

8. Remove timing chain cover oil seal.

a. Using a knife, cut off the oil seal lip.

b. Using a screwdriver with its tip wrapped with protective tape, pry out the timing chain cover oil seal.

➡Do not damage the surface of the oil seal press fit hole. After removing, check the crankshaft for damage. If damaged, smooth the surface with 400-grit sandpaper.

To install:

9. Install timing chain cover oil seal.

a. Apply MP grease to the lip of a new timing chain cover oil seal.

b. Using SST and a hammer, tap in the timing chain cover oil seal until its surface is flush with the timing chain cover edge.

➡Do not tap the oil seal at an angle. Wipe off extra grease from the crankshaft.

10. Install crankshaft pulley assembly.

a. Align the pulley set key with the key groove of the crankshaft pulley, and then slide on the crankshaft pulley.

b. Using SST, install the crankshaft pulley bolt. Tighten to 94 ft. lbs. (128 Nm).

11. Install fan and alternator v-ribbed belt.
12. Install No. 1 V-ribbed (cooler compressor to crankshaft pulley) belt.
13. Inspect fan and alternator V-ribbed belt.
14. Inspect No. 1 V-ribbed (cooler compressor to crankshaft pulley) belt.
15. Inspect for engine oil leak.
16. Install No. 1 engine cover.

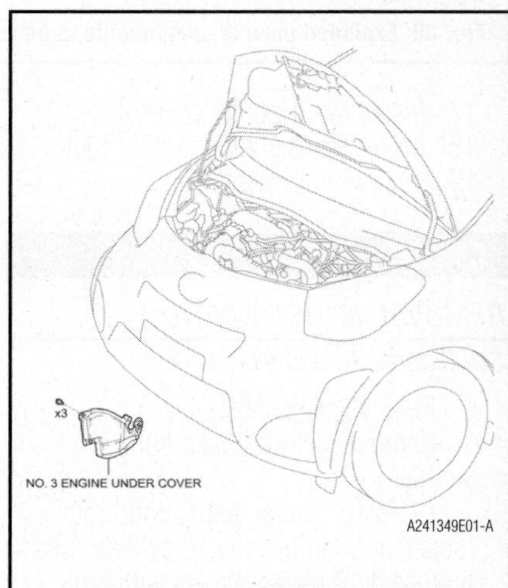

Fig. 38 Remove No. 3 engine under cover

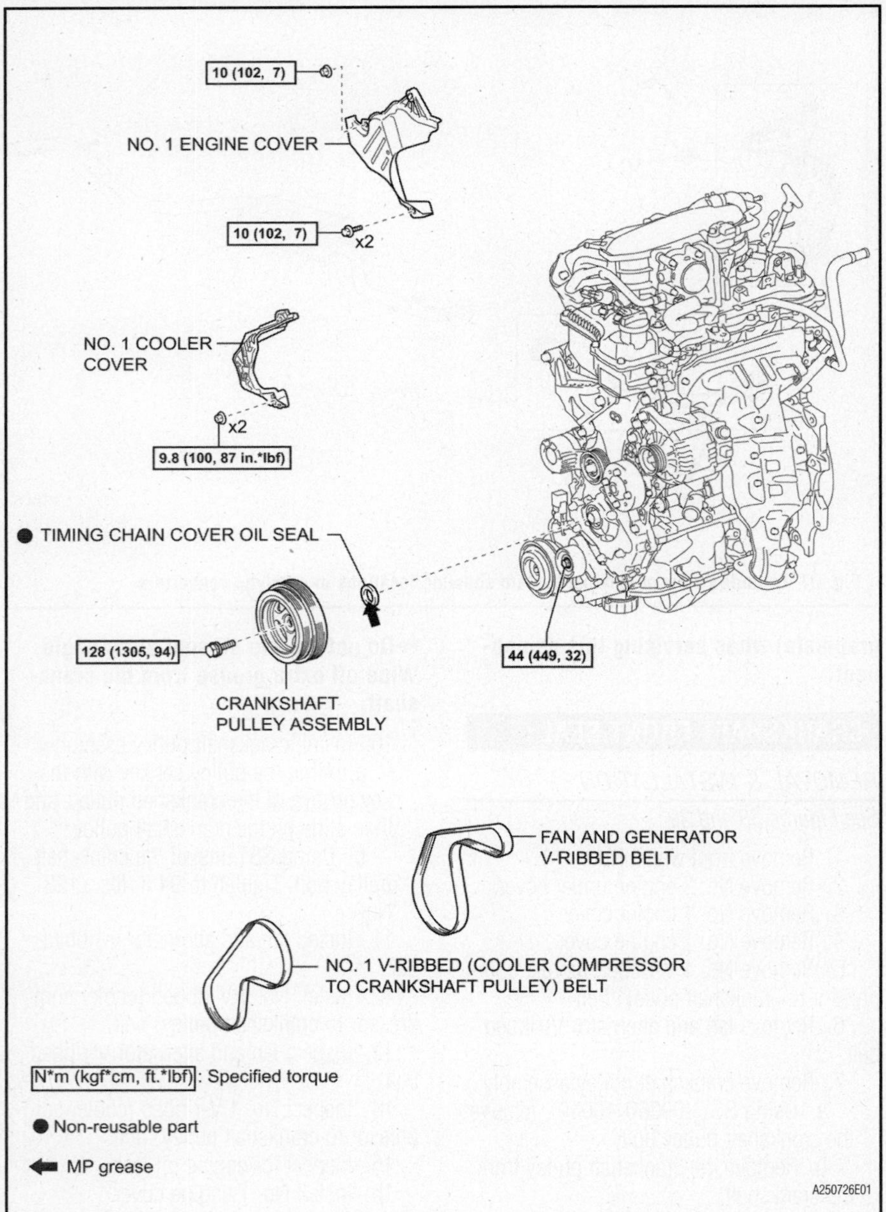

Fig. 39 Exploded view of components to be removed to access oil seal

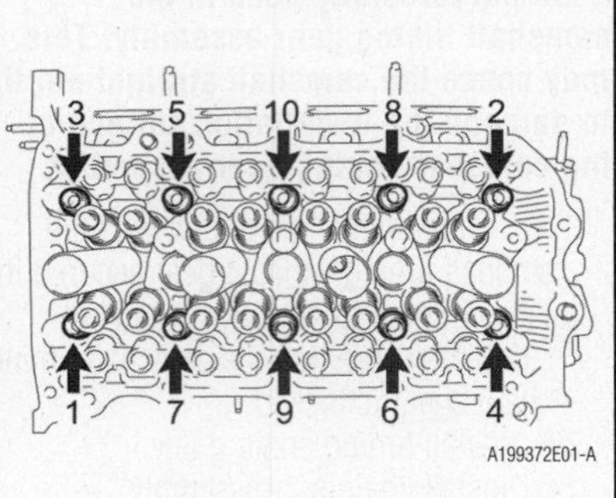

Fig. 40 Using several steps, uniformly loosen and remove the 10 cylinder head bolts and 10 plate washers with a bi-hexagon wrench in sequence

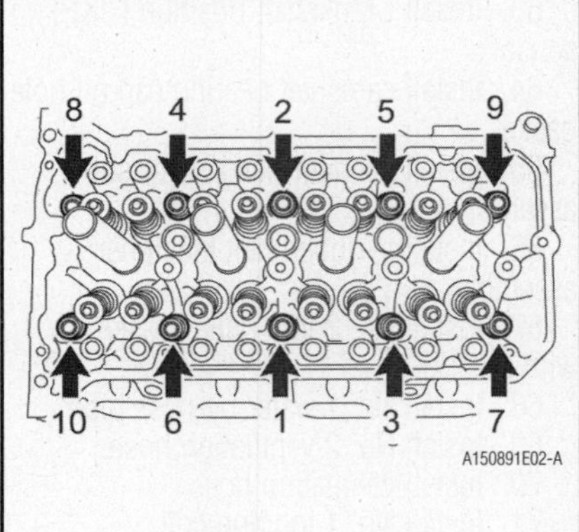

Fig. 41 Using several steps, uniformly install and tighten the 10 cylinder head bolts in order

17. Install No. 1 cooler cover.
18. Install No. 3 engine under cover.
19. Install front wheel RH.

CYLINDER HEAD

REMOVAL & INSTALLATION

See Figures 40 and 41.

1. Remove camshafts.
2. Remove cylinder head sub-assembly.
   a. Using several steps, uniformly loosen and remove the 10 cylinder head bolts and 10 plate washers with a bi-hexagon wrench in sequence.

WARNING

Head warpage or cracking could result from removing the bolts in the wrong order.

   b. Using a screwdriver with its tip wrapped with protective tape, pry between the cylinder head and cylinder block, and remove the cylinder head.
3. Remove the cylinder head gasket.
4. Inspect cylinder head for flatness.
5. Inspect cylinder head for cracks.
6. Inspect cylinder head set bolt.

To install:

7. Install cylinder head gasket.
   a. Apply seal packing to a new cylinder head gasket.

➡Clean the sealing surfaces of the cylinder head gasket, cylinder head and cylinder block. Apply seal packing to the bead on the cylinder head gasket. Install the cylinder head within 3 minutes and tighten the bolts within 15 minutes after applying seal packing.

   b. Place the gasket on the cylinder block surface with the Lot No. stamp facing upward.

➡Remove any oil from the contact surface. Make sure that the gasket is installed in the correct direction.

8. Install cylinder head sub-assembly.

➡The cylinder head bolts are tightened in 2 progressive steps.

   a. Apply a light coat of engine oil to the bolt threads and the area beneath the bolt heads that come in contact with the washers.

b. Install the bolts and plate washers to the cylinder head.

c. Using several steps, uniformly install and tighten the 10 cylinder head bolts and 10 plate washers with a bi-hexagon wrench in order. Tighten to 24 ft. lbs. (32 Nm).

d. Mark the front side of the cylinder head bolts with paint.

e. Retighten the cylinder head bolts an additional 90°, and then once more 90°.

➡Do not tighten the same bolt twice consecutively.

f. Check that the paint mark is now at a 180° angle to the front.

g. After tightening the cylinder head bolts, wipe off the seal packing material that seeped out from the contact surface between the cylinder head and cylinder block.

➡Be sure to wipe off the seal packing from inside to outside, parallel to the joint line. Be sure to avoid clogging the bolt holes when wiping off the seal packing.

9. Install camshafts.

ENGINE ASSEMBLY

REMOVAL & INSTALLATION

See Figures 42 through 44.

1. Recover refrigerant from refrigeration system.
2. Discharge fuel system pressure.
3. Drain engine coolant.
4. Drain continuously variable transaxle fluid.
5. Remove front wheels.
6. Remove front bumper cover.
7. Remove front bumper energy absorber sub-assembly.
8. Remove windshield wiper motor and link.
9. Remove front floor cover LH.
10. Remove front floor cover RH.
11. Remove No. 2 engine under cover.
12. Remove the 3 clips and the No. 3 engine under cover.
13. Remove No. 1 cooler cover.
14. Remove No. 1 engine cover.
15. Remove front exhaust pipe assembly.
16. Remove exhaust pipe gasket (for front side).
17. Remove exhaust pipe gasket (for rear side).
18. Remove No. 1 V-ribbed (cooler compressor to crankshaft pulley) belt.

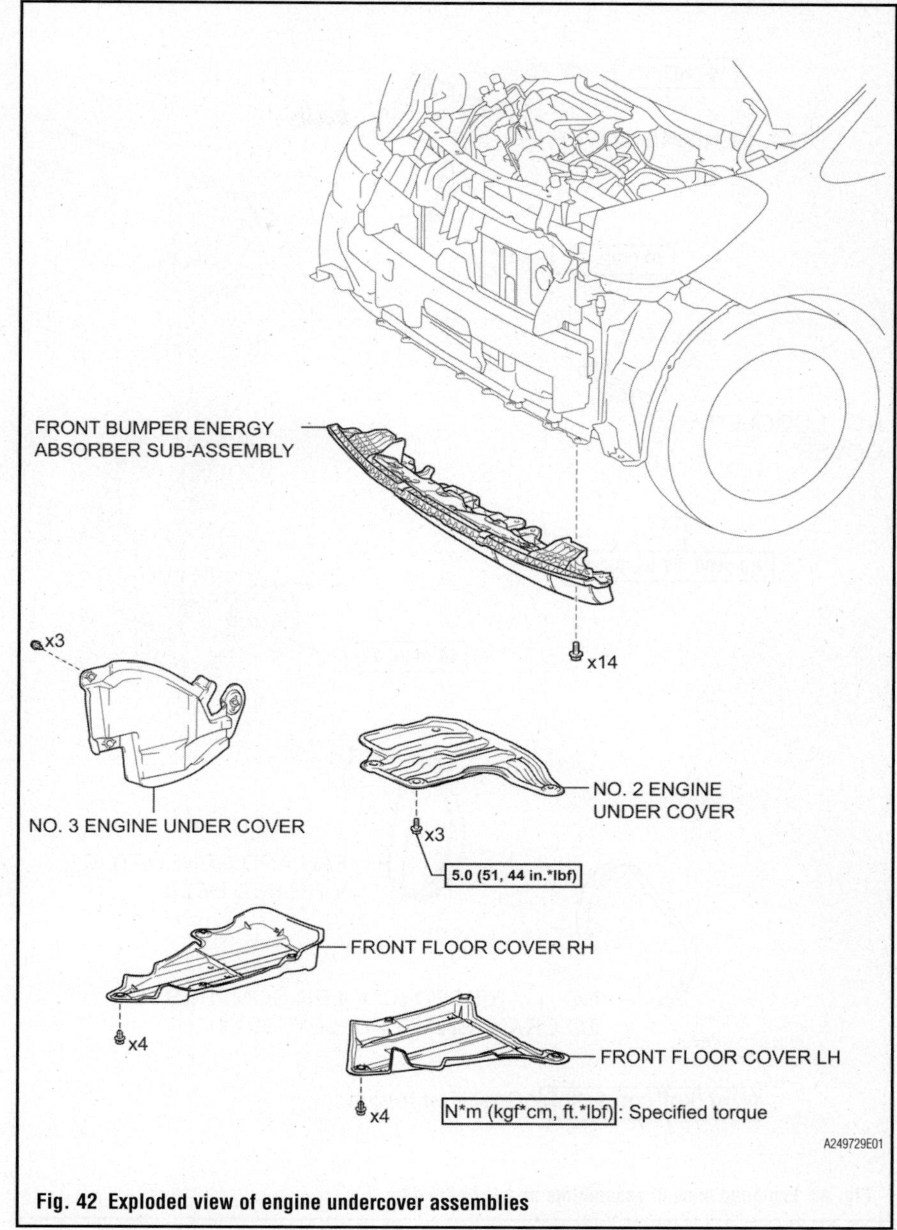

Fig. 42 Exploded view of engine undercover assemblies

19. Remove fan and alternator V-ribbed belt.
20. Remove battery.

a. Disconnect the cables from the battery terminals.

b. Remove the 2 nuts and the battery clamp.

c. Remove the battery.

21. Remove the battery tray.
22. Remove No. 2 battery carrier.

a. Disengage the harness clamp from the No. 2 battery carrier.

b. Remove the 3 bolts and the No. 2 battery carrier.

23. Remove air cleaner and hose.
24. Remove air cleaner filter element sub-assembly
25. Disconnect engine wire

a. Turn the lever and disconnect the ECM connector.

➡Depress the lock button B while depressing A, and turn the lever. Do not turn the lever more than 90°. Do not turn the lever before depressing the lock button B.

b. Disconnect the 2 connectors from the battery terminal.

c. Remove the engine room junction block cover.

d. Remove the 4 connectors and 2 clamps from the engine room junction block

e. Remove the ground bolt and

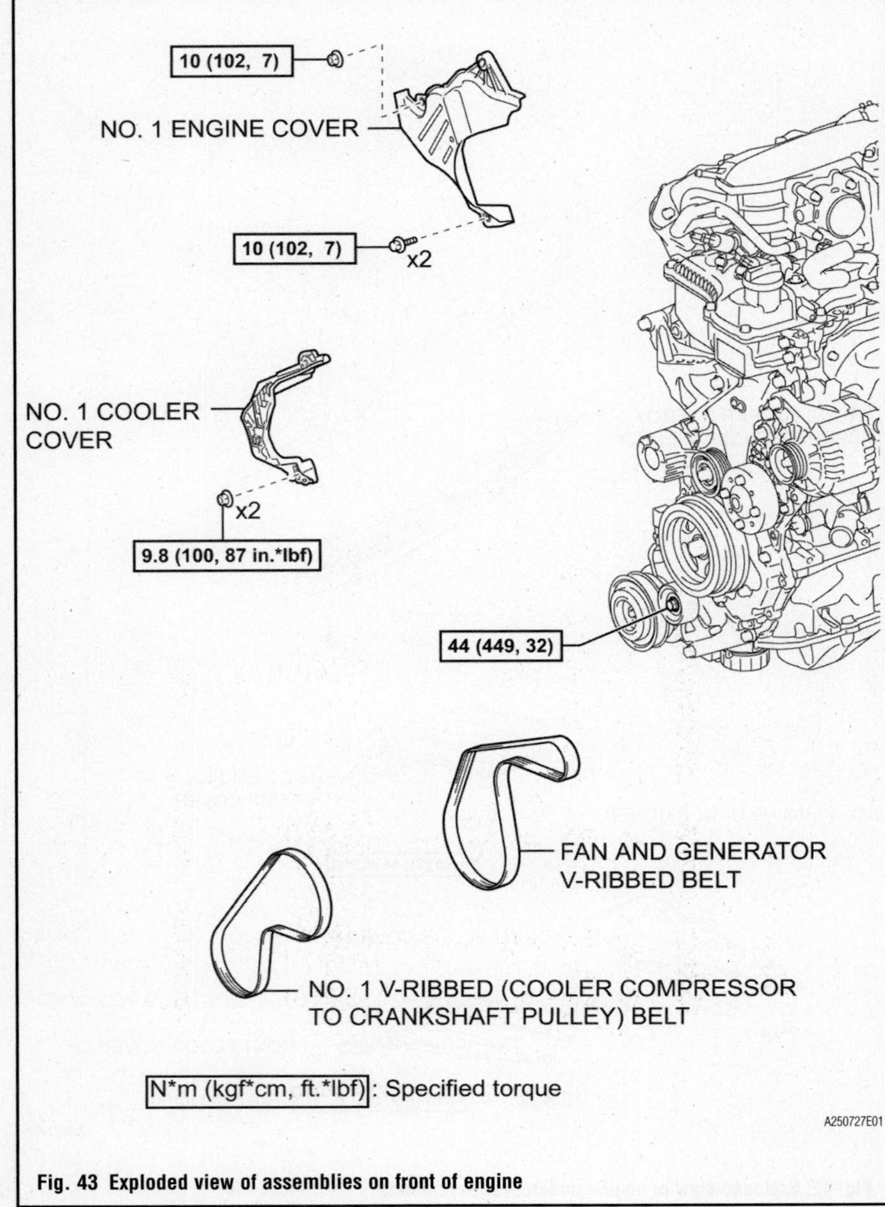

Fig. 43 Exploded view of assemblies on front of engine

harness clamp and disconnect the engine wire.

26. Disconnect radiator reservoir tank hose.
    a. Disengage the radiator reservoir tank hose from the hose clamp.
    b. Loosen the clip and disconnect the radiator reservoir tank hose.
27. Loosen the 2 clips and remove the No. 3 radiator hose.
28. Disconnect No. 2 radiator hose
29. Loosen the clip and disconnect the No. 1 fuel vapor feed hose.
30. Remove EFI fuel pipe clamp.
31. Disconnect fuel tube sub-assembly.
32. Disconnect inlet heater water hose.
33. Disconnect outlet heater water hose.
34. Separate transmission control cable assembly.
    a. Remove the nut and disconnect the transmission control cable from the transmission control shaft lever.
    b. Remove the clip and separate the transmission control cable from the No. 1 transmission control cable bracket.
35. Loosen the clip and disconnect union to check valve hose.
36. Remove front axle shaft LH nut.
37. Remove front axle shaft RH nut.
38. Separate front speed sensor LH.
39. Separate front speed sensor RH.
40. Separate front stabilizer link assembly LH.
41. Separate front stabilizer link assembly RH.
42. Separate front lower suspension arm sub-assembly LH.
43. Separate front lower suspension arm sub-assembly RH.
44. Separate front axle assembly LH.
45. Separate front axle assembly RH.
46. Remove drive shaft heat insulator.
47. Remove front drive shaft assembly LH.
48. Remove front drive shaft assembly RH.
49. Remove front frame assembly.
50. Disconnect discharge hose sub-assembly.
    a. Remove the bolt and disconnect the discharge hose from the compressor.
    b. Remove the O-ring from the discharge hose.

➡Seal the openings of the disconnected parts using vinyl tape to prevent the entry of moisture and foreign matter.

    c. Remove the bolt and the separate the discharge hose.

51. Disconnect suction hose sub-assembly.
52. Remove cooler compressor assembly.
53. Remove engine moving control rod cover (for cold area).
54. Remove engine moving control rod.
55. Remove engine assembly with transaxle.
    a. Set the engine lifter.

➡Install a height adjustment attachment and plate lift attachment onto the engine assembly with transaxle. Securely suspend the engine assembly to prevent it from turning upside down, until it is secured to an engine stand. Do not perform any procedure while the engine assembly is suspended because doing so may cause the engine assembly to drop, resulting in injury. However, the engine assembly needs to be suspended when it is installed to or removed from an engine stand.

    b. Remove the rope that secures the engine assembly and the body.
    c. Remove the bolt, 2 nuts and disconnect the engine mounting insulator RH.
    d. Remove the through bolt and nut and disconnect the engine mounting insulator LH.
    e. Install the engine hangers with the bolts.
    f. Using an engine sling device and a chain block, suspend the engine assembly with transaxle.

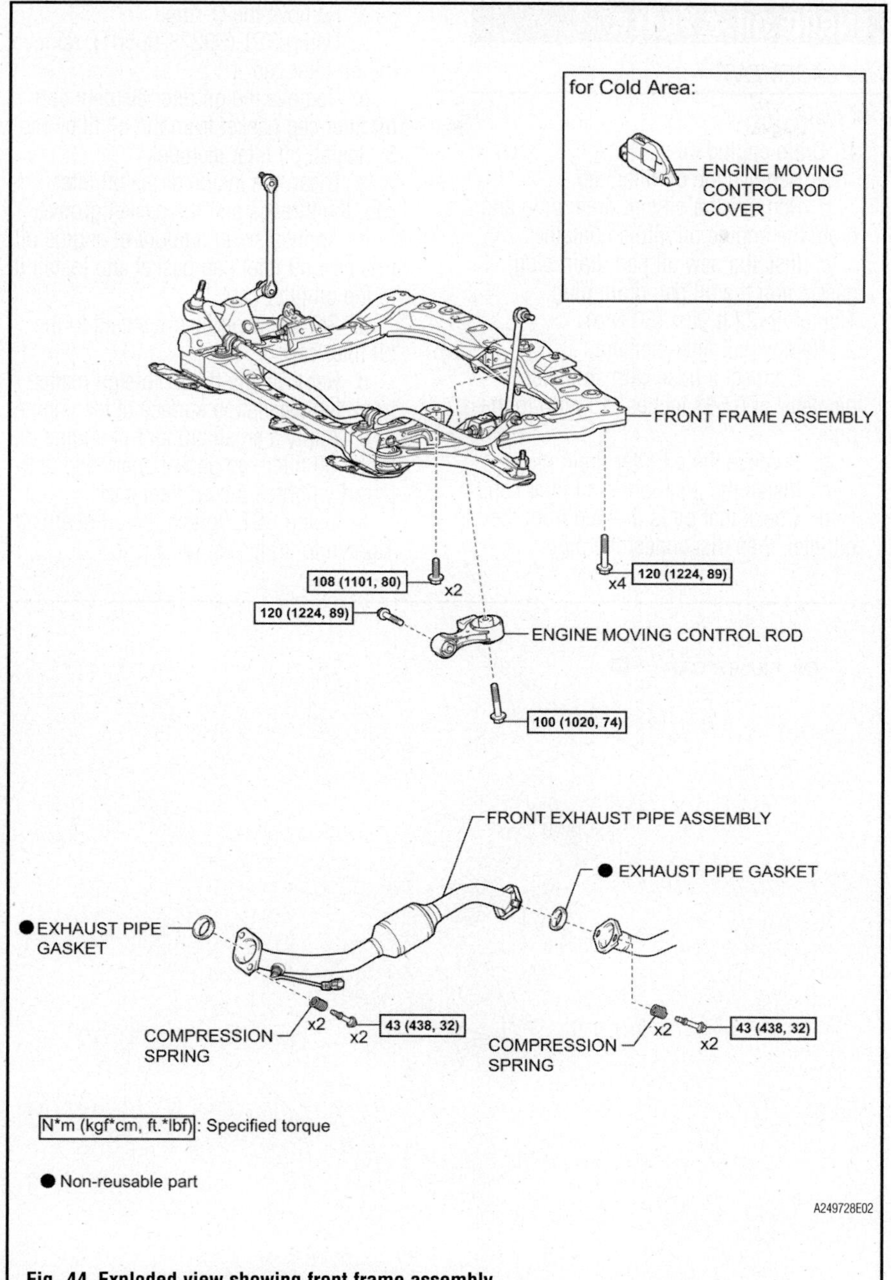

Fig. 44 Exploded view showing front frame assembly

56. Remove water by-pass hose assembly.
   a. Disengage the clamp and separate the breather hose from the water by-pass hose.
   b. Remove the bolt and separate the water by-pass hose from the CVT.
   c. Loosen the 2 clips and separate the 2 water by-pass hoses from the oil cooler.
   d. Disengage the hose clamp.
   e. Loosen the 2 clips and remove the water by-pass hose.
57. Loosen the 2 clips and remove the No. 1 radiator hose.
58. Remove the bolt and the water filler.
59. Loosen the clip and remove the No. 2 radiator hose.
60. Remove engine wire.
   a. Remove the 2 ground bolts and separate the ground wire from the cylinder block.
   b. Remove the nut from the starter.
   c. Remove the bolt and separate the engine wire from the wiring harness bracket.
   d. Remove the terminal cap from the alternator.
   e. Remove the nut from the alternator.
   f. Remove the ground bolt and separate the ground wire from the transaxle.
   g. Disconnect all of the clamps and connectors and remove the engine wire from the engine assembly with transaxle.
61. Remove drive shaft heat insulator bracket.
62. Remove flywheel housing side cover.
63. Remove starter assembly.
64. Remove drive plate and torque converter setting bolt.
65. Remove continuously variable transaxle assembly.
66. Remove drive plate and ring gear sub-assembly.

To install:

67. Install drive plate and ring gear sub-assembly.
68. Install continuously variable transaxle assembly.
69. Install drive plate and torque converter setting bolt.
70. Install starter assembly.
71. Install flywheel housing side cover.
72. Install drive shaft heat insulator bracket.
73. Install engine wire.
74. Install No. 2 radiator hose.
75. Install the water filler with the bolt.
76. Install No. 1 radiator hose.
77. Install water by-pass hose assembly.
78. Install engine assembly with transaxle.
79. Install engine moving control rod.
80. Install engine moving control rod cover (for cold area).
81. Install cooler compressor assembly.
82. Install suction hose sub-assembly.
83. Install discharge hose sub-assembly.
84. Install front frame assembly.
85. Install front drive shaft assembly LH.
86. Install front drive shaft assembly RH.
87. Install drive shaft heat insulator.
88. Install front axle assembly LH.
89. Install front axle assembly RH.
90. Install front lower suspension arm sub-assembly LH.
91. Install front lower suspension arm sub-assembly RH.
92. Install front stabilizer link assembly LH.
93. Install front stabilizer link assembly RH.
94. Install front speed sensor LH.
95. Install front speed sensor RH.
96. Install front axle shaft LH nut.
97. Install front axle shaft RH nut.
98. Install the union to check valve hose with the clip.
99. Install transmission control cable assembly.

100. Connect outlet heater water hose.
101. Connect inlet heater water hose.
102. Connect fuel tube sub-assembly.
103. Install EFI fuel pipe clamp.
104. Connect the No. 1 fuel vapor feed hose with the clip.
105. Connect No. 2 radiator hose
106. Install the No. 3 radiator hose with the 2 clamps.
107. Connect radiator reservoir tank hose.
108. Install engine wire.
109. Install air cleaner filter element sub-assembly.
110. Install air cleaner and hose.
111. Install No. 2 battery carrier.
112. Install the battery tray.
113. Install battery.
114. Install fan and alternator V-ribbed belt.
115. Install No. 1 V-ribbed (cooler compressor to crankshaft pulley) belt.
116. Install exhaust pipe gasket (for front side).
117. Install exhaust pipe gasket (for rear side).
118. Install front exhaust pipe assembly.
119. Install front bumper energy absorber sub-assembly.
120. Install front bumper cover.
121. Install front wheels.
122. Inspect fan and alternator V-ribbed belt.
123. Inspect No. 1 V-ribbed (cooler compressor to crankshaft pulley) belt.
124. Add engine coolant.
125. Add engine oil.
126. Charge refrigerant.
127. Add continuously variable transaxle fluid.
128. Fluid temperature check.
129. Fluid level check.
130. Inspect for engine coolant leak.
131. Inspect reservoir engine coolant level.
132. Inspect for engine oil leak.
133. Inspect continuously variable transaxle fluid leak.
134. Inspect for exhaust gas leak.
135. Inspect for fuel leak.
136. Inspect for refrigerant leak.
137. Inspect ignition timing.
138. Inspect engine idling speed.
139. Inspect CO/HC.
140. Inspect and adjust front wheel alignment.
141. Install No. 1 engine cover.
142. Install No. 1 cooler cover.
143. Install No. 3 engine under cover.
144. Install No. 2 engine under cover.
145. Install front floor cover RH.
146. Install front floor cover LH.
147. Install windshield wiper motor and link.

ENGINE OIL & FILTER

REPLACEMENT

See Figure 45.

1. Drain engine oil.
   a. Remove the oil filler cap.
   b. Remove the oil pan drain plug and drain the engine oil into a container.
   c. Install a new oil pan drain plug gasket and the oil pan drain plug. Tighten to 22 ft. lbs. (30 Nm).
2. Remove oil filter element.
   a. Connect a hose with an inside diameter of 0.591 inches (15 mm) to the pipe.
   b. Remove the oil filter drain plug.
   c. Install the pipe to the oil filter cap.
   d. Check that oil is drained from the oil filter, then disconnect the pipe.
   e. Remove the O-ring.
   f. Using SST (09228-06501), remove the oil filter cap.
   g. Remove the oil filter element and oil filter cap gasket from the oil filter cap.
3. Install oil filter element.
   a. Clean the inside of the oil filter cap, the threads and the gasket groove.
   b. Apply a small amount of engine oil to a new oil filter cap gasket and install it to the oil filter cap.
   c. Set a new oil filter element to the oil filter cap.
   d. Remove any dirt or foreign matter from the installation surface of the engine.
   e. Apply a small amount of engine oil to the oil filter cap gasket again and temporarily tighten the oil filter cap.
   f. Using SST, tighten the oil filter cap. Tighten to 30 ft. lbs. (40 Nm).

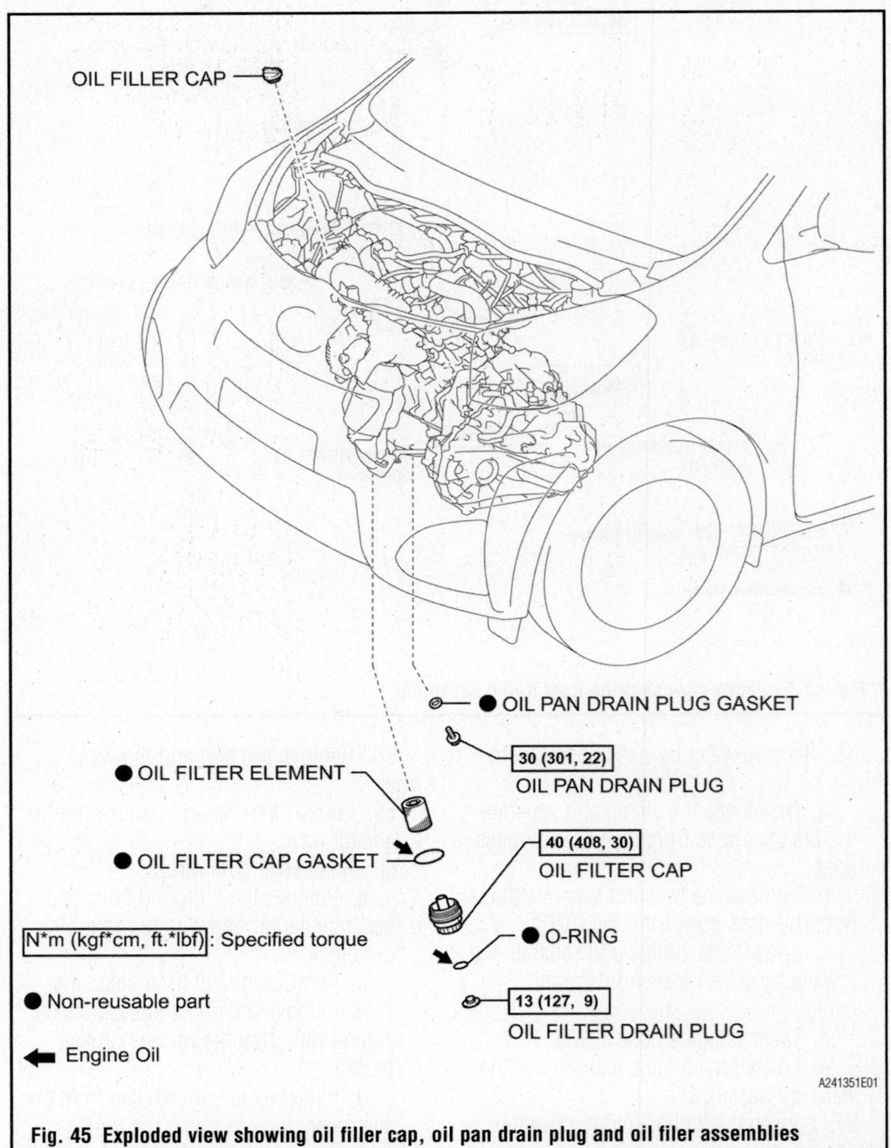

Fig. 45 Exploded view showing oil filler cap, oil pan drain plug and oil filer assemblies

g. Apply a small amount of engine oil to a new O-ring, and install it to the oil filter cap.
 h. Install the oil filter drain plug.
4. Add engine oil.
 a. Add new engine oil.
 b. Install the oil filler cap.
5. Inspect for engine oil leak.
6. Check engine oil level.

EXHAUST MANIFOLD

REMOVAL & INSTALLATION

See Figures 46 and 47.

1. Remove radiator assembly.
2. Remove air fuel ratio sensor.
3. Remove front exhaust pipe assembly.
 a. Disconnect the heated oxygen sensor connector.
 b. Remove the 2 bolts and the 2 compression springs, and disconnect the front exhaust pipe assembly from the tail exhaust pipe assembly.
 c. Remove the 2 bolts and the 2 compression springs, and remove the front exhaust pipe assembly from the exhaust manifold.
4. Remove exhaust pipe gasket (for Front Side).
5. Remove exhaust pipe gasket (for Rear Side).
6. Remove the bolt, the nut and the manifold stay.
7. Remove the 2 nuts, the bolt and the No. 1 exhaust manifold heat insulator.
8. Remove the 3 bolts, the 2 nuts and the exhaust manifold converter.
9. Remove exhaust manifold to head gasket.
10. Remove the 4 bolts, the nut and the No. 2 exhaust manifold heat insulator.
11. Remove the 4 bolts and the No. 3 exhaust manifold heat insulator.

To install:

12. Install the No. 3 exhaust manifold heat insulator with the 4 bolts.

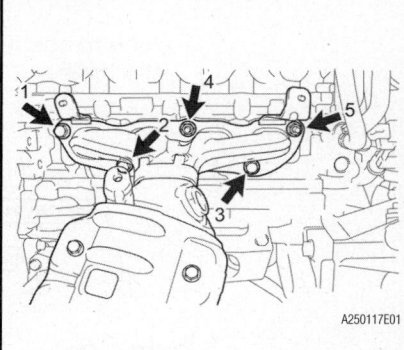

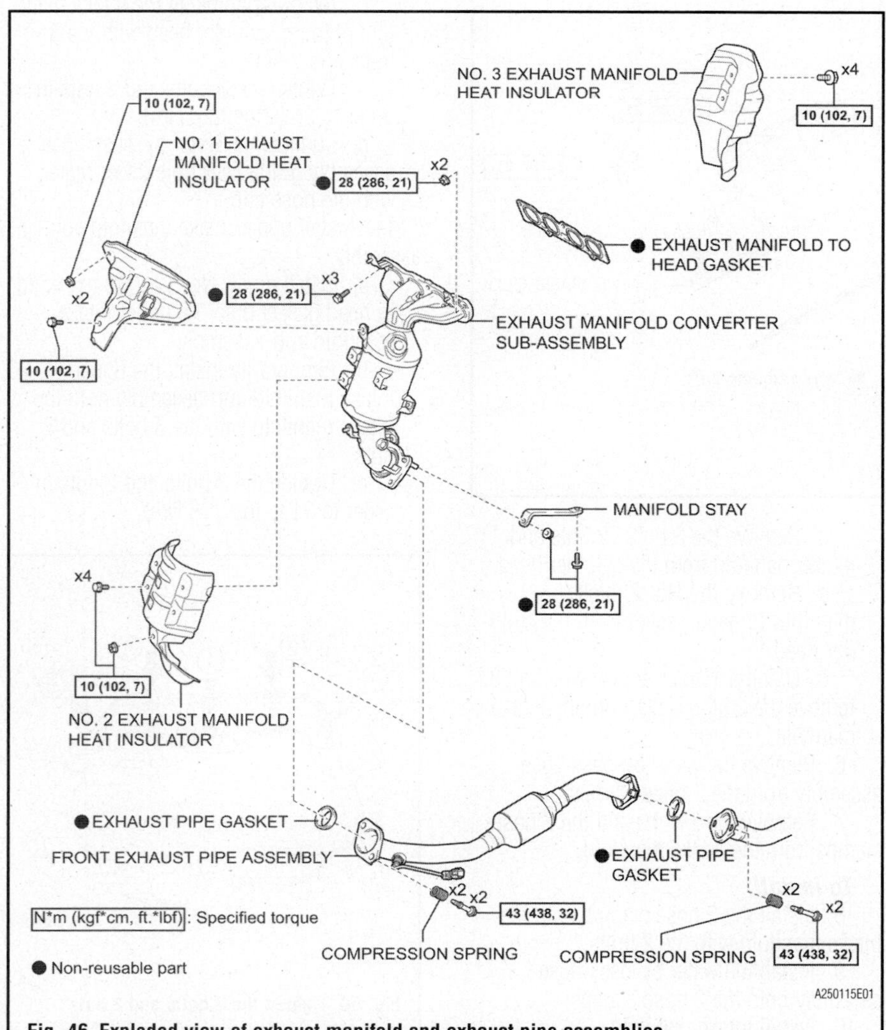

Fig. 46 Exploded view of exhaust manifold and exhaust pipe assemblies

Fig. 47 Install the exhaust manifold converter, tightening the 2 new nuts and 3 new bolts in order

13. Install the No. 2 exhaust manifold heat insulator with the 4 bolts and the nut.
14. Place the new exhaust manifold to head gasket
15. Install the exhaust manifold converter, tightening the 2 new nuts and 3 new bolts in order to 21 ft. lbs. (28 Nm).
16. Install manifold stay.
17. Install the No. 1 exhaust manifold heat insulator with the 2 nuts and the bolt.
18. Install exhaust pipe gasket (for Front Side).
19. Install exhaust pipe gasket (for Rear Side).
20. Install front exhaust pipe assembly.
21. Install air fuel ratio sensor.
22. Install radiator assembly.
23. Inspect for exhaust gas leak.

INTAKE MANIFOLD

REMOVAL & INSTALLATION

See Figures 48 through 50.

1. Remove windshield wiper motor and link.
2. Remove throttle with motor body assembly.
3. Remove the 2 bolts and the intake manifold stay.
4. Remove upper intake manifold sub-assembly.
 a. Loosen the hose clip and disconnect the union to connector tube hose from the upper intake manifold sub-assembly.
 b. Disengage the hose clamp and disconnect the water by-pass hose assembly from the upper intake manifold sub-assembly.
 c. Disconnect the fuel vapor feed hose from the upper intake manifold sub-assembly.

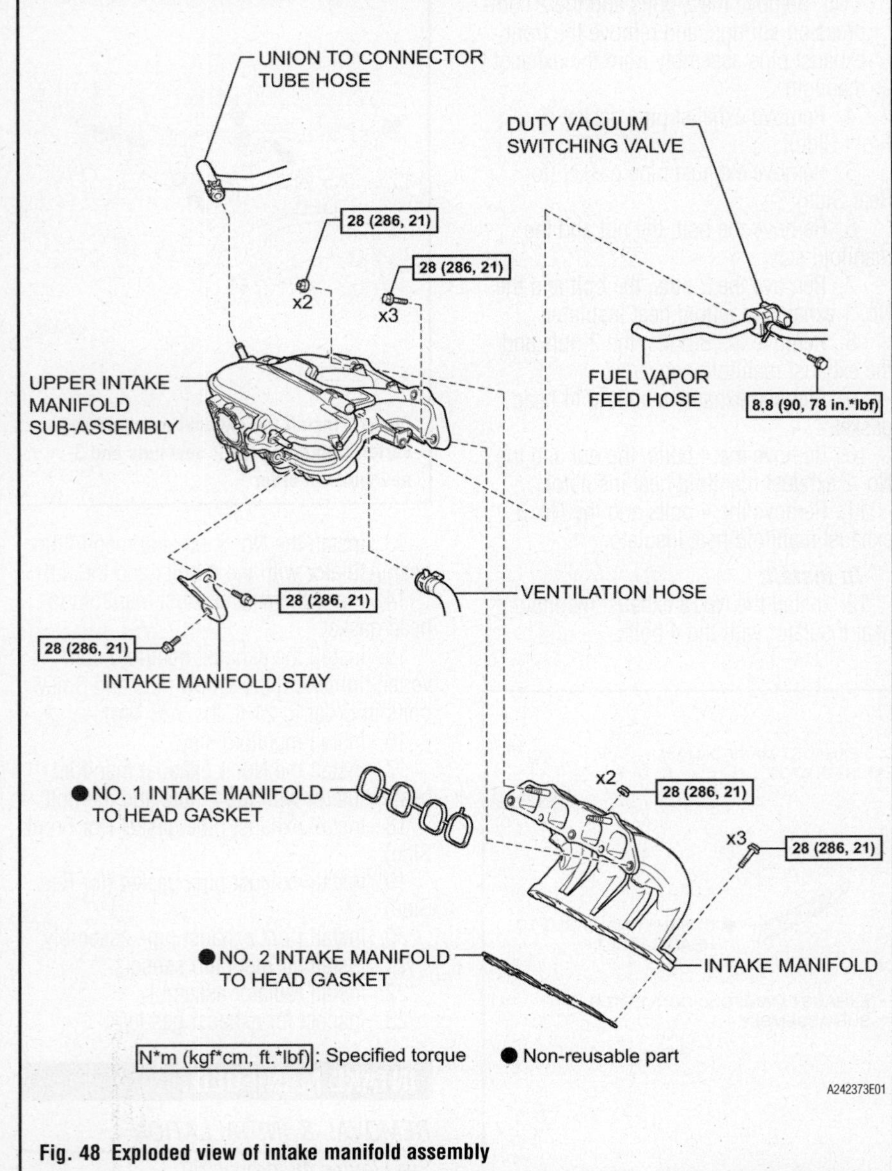

Fig. 48 Exploded view of intake manifold assembly

d. Disengage the hose clamp and disconnect the fuel vapor feed hose from the upper intake manifold sub-assembly.

e. Remove the bolt and separate the duty vacuum switching valve from the upper intake manifold sub-assembly.

f. Loosen the hose clip and disconnect the ventilation hose from the upper intake manifold sub-assembly.

g. Remove the 3 bolts, 2 nuts and upper intake manifold sub-assembly from the intake manifold.

h. Remove the No. 1 intake manifold to head gasket from the upper intake manifold sub-assembly.

5. Remove intake manifold.

a. Loosen the hose clip and disconnect the water by-pass hose assembly from the EGR hole cover plate.

b. Remove the 3 bolts, 2 nuts and intake manifold from the cylinder head.

c. Remove the No. 2 intake manifold to head gasket from the cylinder head.

d. Using a TORX® socket wrench E8, remove the 2 stud bolts from the intake manifold.

6. Remove the water by-pass hose assembly from the 2 hose clamps.

7. Remove the 2 bolts and the 2 hose clamps from the intake manifold.

To install:

8. Install the 2 hose clamps onto the intake manifold with the 2 bolts.

9. Install the water by-pass hose assembly onto the 2 hose clamps.

10. Install intake manifold.

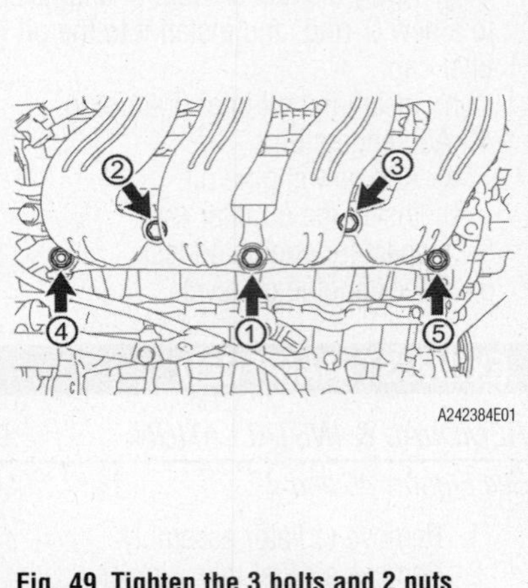

Fig. 49 Tighten the 3 bolts and 2 nuts in order

a. Using a TORX® socket wrench, install the 2 stud bolts onto the intake manifold.

b. Install a new No. 2 intake manifold to head gasket onto the cylinder head.

c. Temporarily install the intake manifold onto the cylinder head with the 3 bolts and 2 nuts.

d. Tighten the 3 bolts and 2 nuts in order to 21 ft. lbs. (28 Nm).

e. Connect the water by-pass hose assembly to the EGR hole cover plate with the hose clip.

11. Install upper intake manifold sub-assembly.

a. Install a new No. 1 intake manifold to head gasket onto the upper intake manifold sub-assembly.

b. Temporarily install the upper intake manifold sub-assembly onto the intake manifold with the 3 bolts and 2 nuts.

c. Tighten the 3 bolts and 2 nuts in order to 21 ft. lbs. (28 Nm).

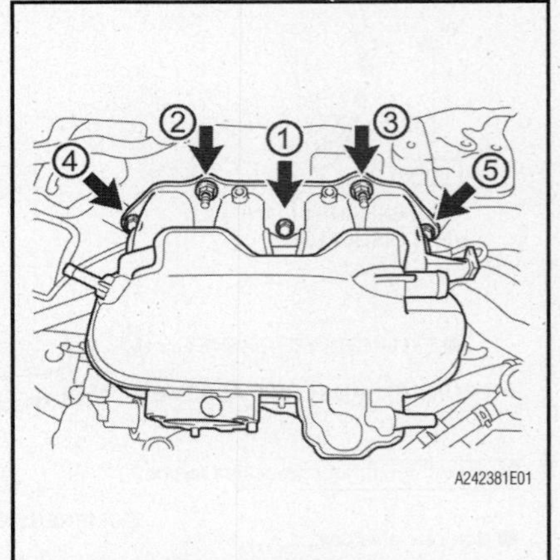

Fig. 50 Tighten the 3 bolts and 2 nuts in order

d. Connect the ventilation hose to the upper intake manifold sub-assembly with the hose clip.

e. Install the duty vacuum switching valve to the upper intake manifold sub-assembly with the bolt.

f. Engage the hose clamp and connect the fuel vapor feed hose to the upper intake manifold sub-assembly.

g. Connect the fuel vapor feed hose to the upper intake manifold sub-assembly.

h. Engage the hose clamp and connect the water by-pass hose assembly to the upper intake manifold sub-assembly.

i. Connect the union to connector tube hose to the upper intake manifold sub-assembly with the hose clip.

12. 5. Install intake manifold stay.

a. Temporarily install the intake manifold stay with the 2 bolts.

b. Tighten the 2 bolts to 21 ft. lbs. (28 Nm).

13. Install throttle with motor body assembly.

14. Install windshield wiper motor and link.

OIL PAN

REMOVAL & INSTALLATION

See Figures 51 through 55.

1. Drain engine oil.
2. Remove No. 2 oil pan sub-assembly.

a. Remove the oil pan drain plug and gasket from the No. 2 oil pan.

b. Remove the 9 bolts.

c. Insert the blade of the oil pan seal cutter between the oil pans. Cut through the applied sealer and remove the No. 2 oil pan.

3. Remove oil pan sub-assembly.

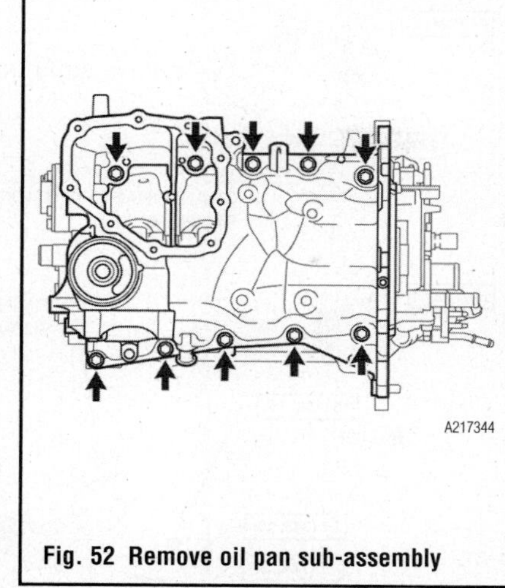

Fig. 52 Remove oil pan sub-assembly

a. Uniformly loosen and remove the 10 bolts.

b. Using a screwdriver with its tip wrapped with protective tape, remove the oil pan by prying between the oil pan and cylinder block.

c. Remove the oil pan O-ring from the cylinder block.

4. Remove the 6 bolts and the No. 1 oil pan baffle plate from the oil pan.

To install:

5. Install the No. 1 oil pan baffle plate to the oil pan with the 6 bolts. Tighten to 7 ft. lbs. (10 Nm).
6. Install oil pan sub-assembly.

a. Install a new oil pan O-ring to the cylinder block.

b. Apply seal packing to the engine block contact areas.

➡Remove any oil from the contact surfaces. Install the oil pan within 3 minutes after applying seal packing. Do not start the engine for at least 2 hours after installing the oil pan.

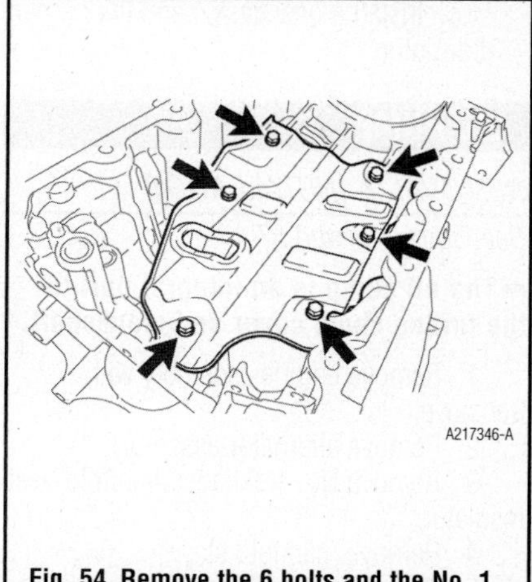

Fig. 54 Remove the 6 bolts and the No. 1 oil pan baffle plate from the oil pan

c. Install the oil pan with the 10 bolts. Tighten to 15 ft. lbs. (21 Nm).

➡Apply adhesive to the threads of bolts C more than 7.0 mm from the bolt tip.

d. Wipe off any excess seal packing with a clean piece of cloth.

7. Install No. 2 oil pan sub-assembly.

➡Do not reuse the No. 2 oil pan.

a. Apply seal packing in a continuous bead to a new No. 2 oil pan.

➡Remove any oil from the contact surfaces. Install the No. 2 oil pan within 3 minutes and tighten the bolts within 10 minutes after applying seal packing.

b. Install the No. 2 oil pan with the 9 bolts. Tighten to 7 ft. lbs. (10 Nm).

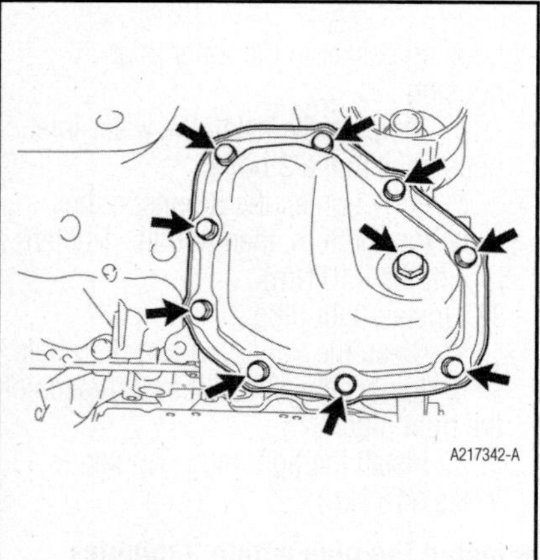

Fig. 51 Remove No. 2 oil pan sub-assembly

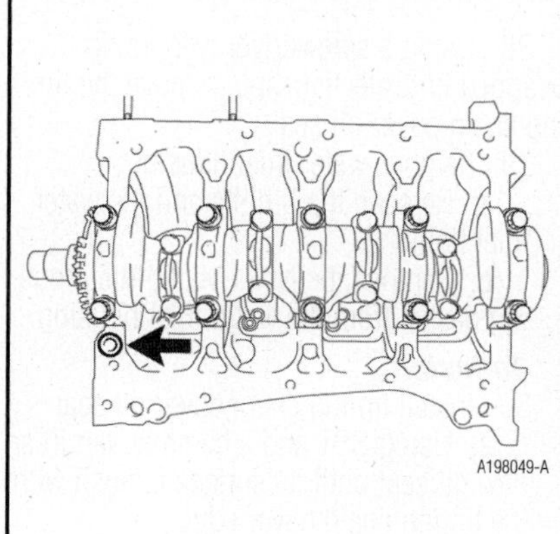

Fig. 53 Remove the oil pan O-ring from the cylinder block

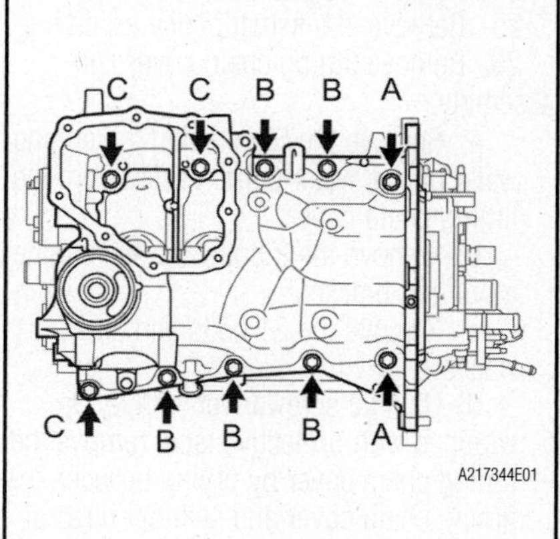

Fig. 55 Install the oil pan with the 10 bolts

c. Install a new gasket and the oil pan drain plug.

OIL PUMP

REMOVAL & INSTALLATION

See Figures 56 and 57.

➡The oil pump is an integral part to the timing chain cover and crankshaft.

1. Remove engine assembly with transaxle.
2. Remove alternator assembly.
3. Remove No. 1 exhaust manifold heat insulator.
4. Remove manifold stay.
5. Remove exhaust manifold converter sub-assembly.
6. Disconnect ventilation hose.
7. Disconnect No. 2 water by-pass hose.
8. Disconnect No. 4 water by-pass hose.
9. Remove intake manifold stay.
10. Remove intake manifold assembly.
11. Remove No. 4 engine wire.
12. Remove engine oil level dipstick.
13. Remove engine oil level dipstick guide.
14. Remove No. 1 ignition coil.
15. Remove ventilation hose.
16. Remove No. 2 ventilation hose.
17. Remove No. 1 water by-pass pipe.
18. Remove No. 2 idle pulley assembly with bracket.
19. Remove V-ribbed belt tensioner assembly.
20. Remove No. 1 Crankshaft Position (CKP) sensor.
21. Remove cylinder head cover sub-assembly.
22. Remove camshaft bearing cap oil hole gasket.
23. Remove crank position sensor.
24. Remove water pump pulley.
25. Remove crankshaft pulley assembly.
26. Remove timing chain cover sub-assembly.
    a. Remove the 5 bolts and the engine water pump, water pump gasket from the timing chain cover.
    b. Remove the 3 bolts and the engine mounting bracket.
    c. Remove the 21 bolts and seal washer.
    d. Using a screwdriver with its tip wrapped with protective tape, remove the timing chain cover by prying between the timing chain cover and cylinder head or cylinder block.
    e. Remove the 2 O-rings.
27. Using an 8 mm socket hexagon wrench, remove the tight plug.

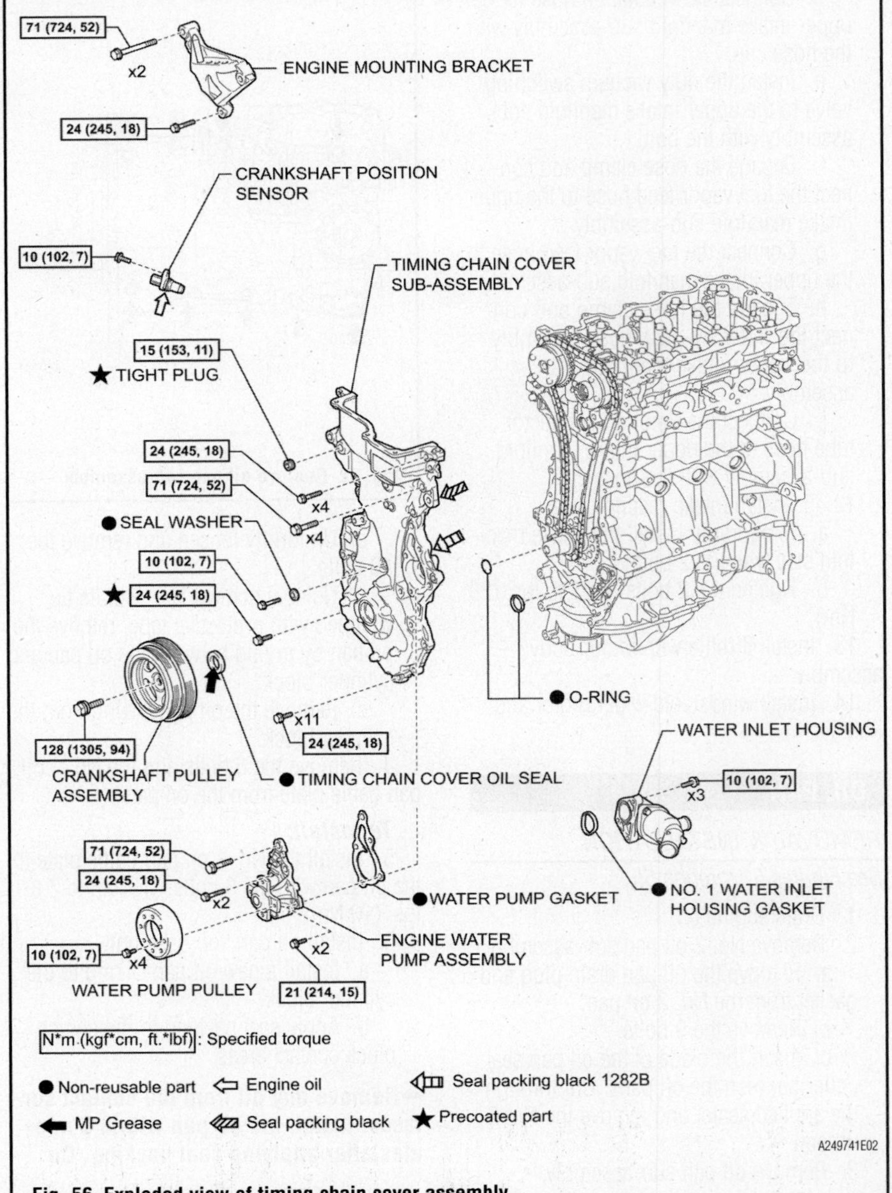

Fig. 56 Exploded view of timing chain cover assembly

28. Using a screwdriver with its tip wrapped in protective tape, remove the timing chain cover oil seal.
29. Remove water inlet housing.
    a. Remove the 3 bolts and the water inlet housing.
    b. Remove the No. 1 water inlet housing gasket from the water inlet housing.

To install:

30. Install timing chain cover oil seal.
    a. Using SST and a hammer, tap in a new oil seal until its surface is flush with the timing chain cover edge.
    b. Apply a light coat of MP grease to the timing chain cover oil seal lip.
31. Install water inlet housing.
    a. Install a new No. 1 water inlet housing gasket to the water inlet housing.
    b. Temporarily install the water inlet housing with the 3 bolts.
    c. Fully tighten the 3 bolts in the order top, bottom, then middle. Tighten to 7 ft. lbs. (10 Nm).
32. Install tight plug.
    a. Clean the tight plug and plug hole.
    b. Apply adhesive to 2 or 3 threads of the tight plug.
    c. Install the tight plug. Tighten to 11 ft. lbs. (15 Nm).

➡Install the plug within 3 minutes after applying adhesive. Do not start the engine within 1 hour after installation.

33. Install timing chain cover sub-assembly.
a. Remove any old packing material remaining on the sealing surfaces before applying seal packing.
b. Clean and degrease the contact surfaces of the timing chain cover, camshaft housing, cylinder head, cylinder block, and oil pan and confirm that no oil, moisture, or other foreign matter remains on the surfaces.
c. Install 2 new O-rings.
d. Apply seal packing to the timing chain cover.

➡When there is oil on contact surfaces, wipe them with oil-free cloth before applying seal packing. Install the chain cover within 3 minutes and tighten the bolts within 10 minutes after applying seal packing. Do not start the engine for at least 2 hours after installing.

e. Align the oil pump drive rotor spline and the crankshaft. Install the chain cover to the crankshaft.

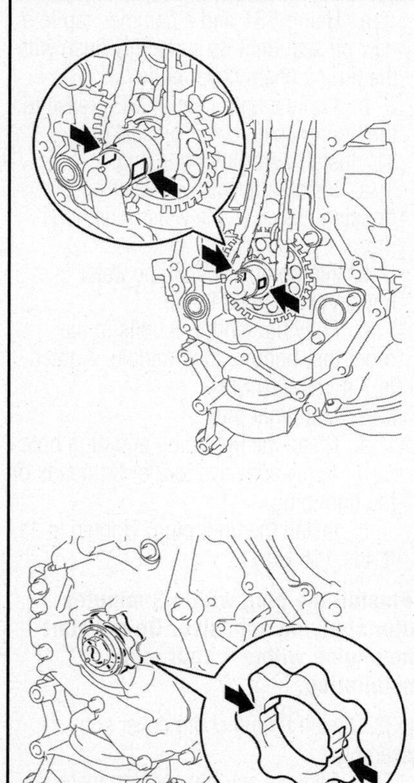

Fig. 57 Align the oil pump drive rotor spline and the crankshaft

f. Temporarily install the engine mounting bracket RH with the 3 bolts.
g. Temporarily install the engine water pump and a new water pump gasket to the timing chain cover with the 5 bolts.
h. Fully tighten the 2 F bolts.
i. Apply adhesive to bolt E more than 0.275 inches (7.0 mm) from the bolt tip.
j. Temporarily tighten the timing chain cover with the 21 bolts and a new seal washer.
k. Fully tighten the timing chain cover with the 27 bolts.
34. Install crankshaft pulley assembly.
35. Install water pump pulley.
36. Install Crankshaft Position (CKP) sensor.
37. Install camshaft bearing cap oil hole gasket.
38. Install cylinder head cover sub-assembly.
39. Install V-ribbed belt tensioner assembly.
40. Install No. 2 idle pulley assembly with bracket.
41. Install No. 1 water by-pass pipe.
42. Install No. 2 ventilation hose.
43. Install ventilation hose.
44. Install No. 1 ignition coil.
45. Install engine oil level dipstick guide.
46. Install engine oil level dipstick.
47. Install No. 4 engine wire.
48. Install intake manifold assembly.
49. Install intake manifold stay.
50. Connect No. 4 water by-pass hose.
51. Connect No. 2 water by-pass hose.
52. Connect ventilation hose.
53. Install exhaust manifold converter sub-assembly.
54. Install manifold stay.
55. Install No. 1 exhaust manifold heat insulator.
56. Install alternator assembly.
57. Install engine assembly with transaxle.

PISTON AND RING

POSITIONING

See Figure 58.

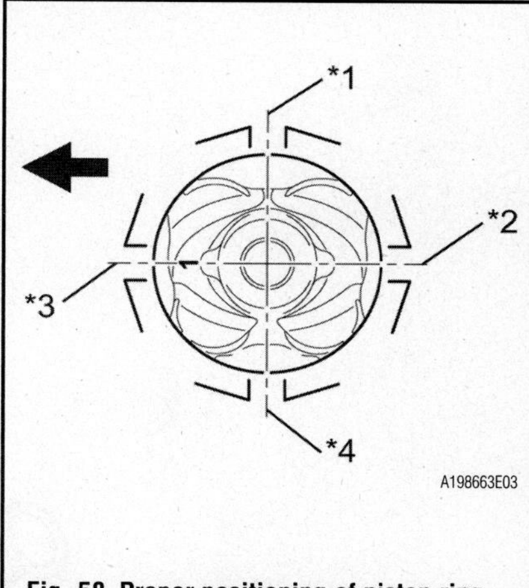

Fig. 58 Proper positioning of piston ring end gaps

REAR MAIN SEAL

REMOVAL & INSTALLATION

See Figure 59.

1. Remove continuously variable transaxle assembly.
2. Remove drive plate and ring gear sub-assembly.
a. Using SST (09960-10010), hold the crankshaft.
b. Remove the 8 bolts, the rear drive plate spacer, the drive plate and the front drive plate spacer.
3. Remove rear engine oil seal.
a. Using a knife, cut off the oil seal lip.
b. Using a screwdriver with its tip wrapped with protective tape, pry out the rear engine oil seal.

To install:

4. Install rear engine oil seal.
a. Remove any old packing material and be careful not to drop any oil on the contact surfaces of the oil pan and cylinder block.
b. Apply seal packing in a continuous bead.

➡Remove any oil from the contact surfaces. Install a new rear engine oil seal within 3 minutes after applying seal packing. Do not start the engine for at least 2 hours after installing.

c. Apply MP grease to the lip of a new rear engine oil seal.
d. Using SST and a hammer, tap in the rear engine oil seal until its surface is flush with the cylinder block and oil pan.
5. Install drive plate and ring gear sub-assembly.
a. Using SST, hold the crankshaft.
b. Install the front drive plate spacer, the drive plate and ring gear and the rear drive plate spacer to the crankshaft.

➡The front spacer is reversible. As the rear drive plate spacer and the drive plate are not reversible, be sure to install them in the proper direction.

c. Clean the 8 bolts and their holes.

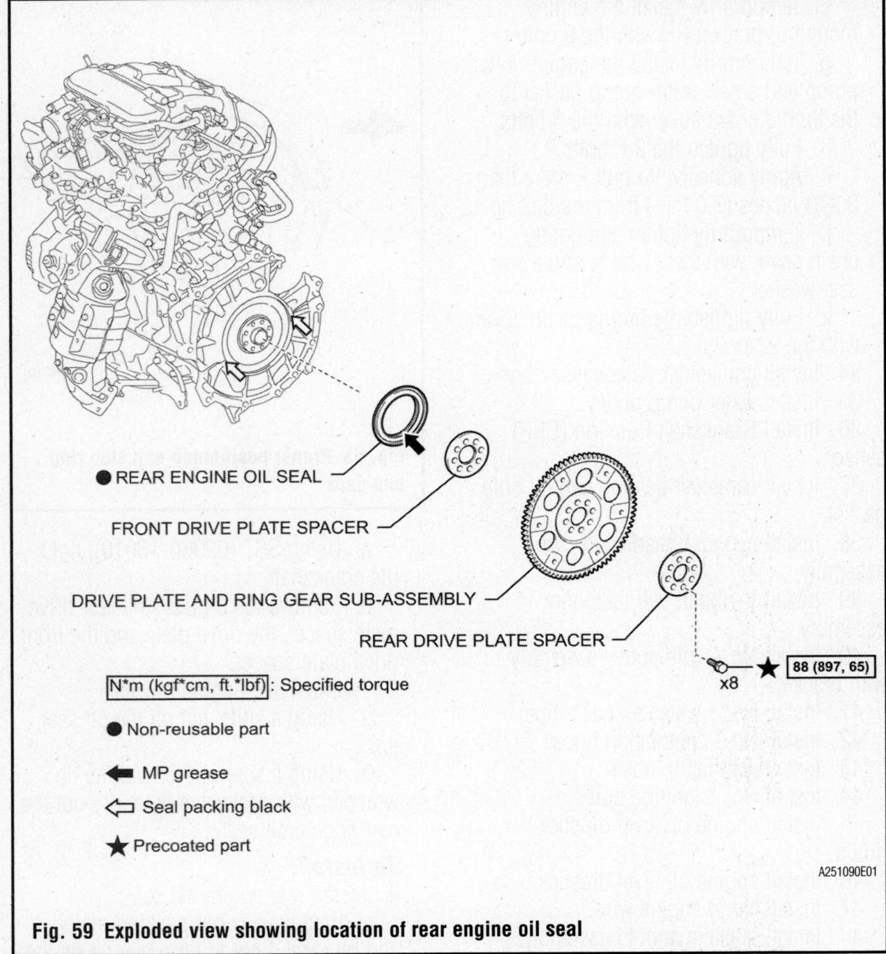

Fig. 59 Exploded view showing location of rear engine oil seal

d. Apply adhesive to 2 or 3 end threads of each of the 8 bolts.

e. Install and uniformly tighten the 8 bolts in several steps in sequence to 65 ft. lbs. (88 Nm) using a crisscross pattern.

➡**Do not start the engine for at least an hour after installing the drive plate.**

6. Install continuously variable transaxle assembly.

TIMING CHAIN FRONT COVER

REMOVAL & INSTALLATION

See Figures 60 and 61.

➡**The oil pump is an integral part to the timing chain cover and crankshaft.**

1. Remove engine assembly with transaxle.
2. Remove alternator assembly.
3. Remove No. 1 exhaust manifold heat insulator.
4. Remove manifold stay.
5. Remove exhaust manifold converter sub-assembly.
6. Disconnect ventilation hose.
7. Disconnect No. 2 water by-pass hose.
8. Disconnect No. 4 water by-pass hose.
9. Remove intake manifold stay.
10. Remove intake manifold assembly.
11. Remove No. 4 engine wire.
12. Remove engine oil level dipstick.
13. Remove engine oil level dipstick guide.
14. Remove No. 1 ignition coil.
15. Remove ventilation hose.
16. Remove No. 2 ventilation hose.
17. Remove No. 1 water by-pass pipe.
18. Remove No. 2 idle pulley assembly with bracket.
19. Remove V-ribbed belt tensioner assembly.
20. Remove No. 1 Crankshaft Position (CKP) sensor.
21. Remove cylinder head cover sub-assembly.
22. Remove camshaft bearing cap oil hole gasket.
23. Remove crank position sensor.
24. Remove water pump pulley.
25. Remove crankshaft pulley assembly.
26. Remove timing chain cover sub-assembly.

a. Remove the 5 bolts and the engine water pump, water pump gasket from the timing chain cover.

b. Remove the 3 bolts and the engine mounting bracket.

c. Remove the 21 bolts and seal washer.

d. Using a screwdriver with its tip wrapped with protective tape, remove the timing chain cover by prying between the timing chain cover and cylinder head or cylinder block.

e. Remove the 2 O-rings.

27. Using an 8 mm socket hexagon wrench, remove the tight plug.
28. Using a screwdriver with its tip wrapped in protective tape, remove the timing chain cover oil seal.
29. Remove water inlet housing.

a. Remove the 3 bolts and the water inlet housing.

b. Remove the No. 1 water inlet housing gasket from the water inlet housing.

To install:

30. Install timing chain cover oil seal.

a. Using SST and a hammer, tap in a new oil seal until its surface is flush with the timing chain cover edge.

b. Apply a light coat of MP grease to the timing chain cover oil seal lip.

31. Install water inlet housing.

a. Install a new No. 1 water inlet housing gasket to the water inlet housing.

b. Temporarily install the water inlet housing with the 3 bolts.

c. Fully tighten the 3 bolts in the order top, bottom, then middle. Tighten to 7 ft. lbs. (10 Nm).

32. Install tight plug.

a. Clean the tight plug and plug hole.

b. Apply adhesive to 2 or 3 threads of the tight plug.

c. Install the tight plug. Tighten to 11 ft. lbs. (15 Nm).

➡**Install the plug within 3 minutes after applying adhesive. Do not start the engine within 1 hour after installation.**

33. Install timing chain cover sub-assembly.

a. Remove any old packing material remaining on the sealing surfaces before applying seal packing.

b. Clean and degrease the contact surfaces of the timing chain cover, camshaft housing, cylinder head, cylinder block, and oil pan and confirm

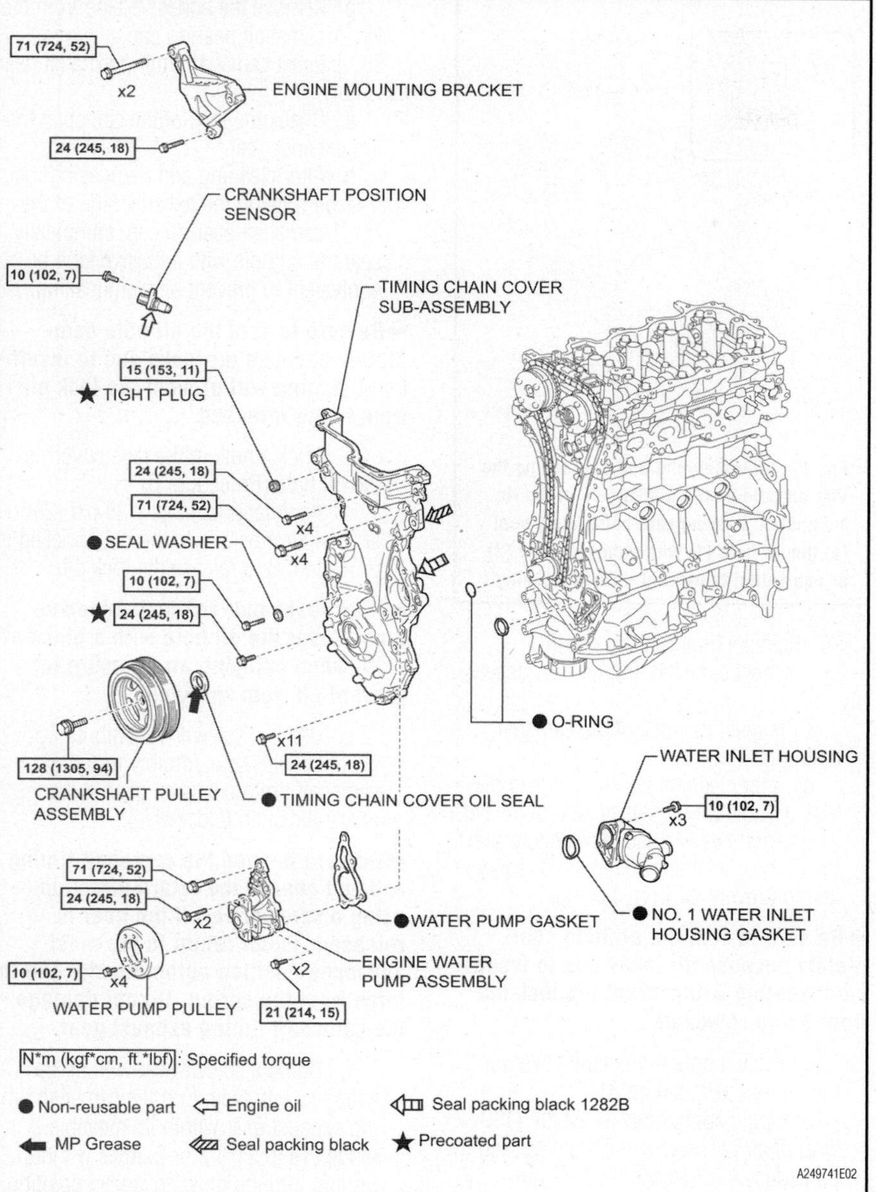

Fig. 60 Exploded view of timing chain cover assembly

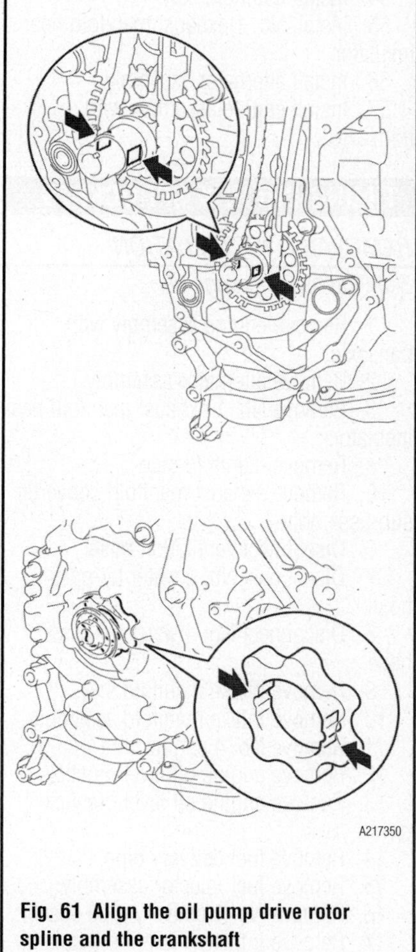

Fig. 61 Align the oil pump drive rotor spline and the crankshaft

that no oil, moisture, or other foreign matter remains on the surfaces.

c. Install 2 new O-rings.

d. Apply seal packing to the timing chain cover.

➡When there is oil on contact surfaces, wipe them with oil-free cloth before applying seal packing. Install the chain cover within 3 minutes and tighten the bolts within 10 minutes after applying seal packing. Do not start the engine for at least 2 hours after installing.

e. Align the oil pump drive rotor spline and the crankshaft. Install the chain cover to the crankshaft.

f. Temporarily install the engine mounting bracket RH with the 3 bolts.

g. Temporarily install the engine water pump and a new water pump gasket to the timing chain cover with the 5 bolts.

h. Fully tighten the 2 F bolts.

i. Apply adhesive to bolt E more than 0.275 inches (7.0 mm) from the bolt tip.

j. Temporarily tighten the timing chain cover with the 21 bolts and a new seal washer.

k. Fully tighten the timing chain cover with the 27 bolts.

34. Install crankshaft pulley assembly.
35. Install water pump pulley.
36. Install Crankshaft Position (CKP) sensor.
37. Install camshaft bearing cap oil hole gasket.
38. Install cylinder head cover sub-assembly.
39. Install V-ribbed belt tensioner assembly.
40. Install No. 2 idle pulley assembly with bracket.
41. Install No. 1 water by-pass pipe.
42. Install No. 2 ventilation hose.
43. Install ventilation hose.
44. Install No. 1 ignition coil.
45. Install engine oil level dipstick guide.
46. Install engine oil level dipstick.
47. Install No. 4 engine wire.
48. Install intake manifold assembly.
49. Install intake manifold stay.
50. Connect No. 4 water by-pass hose.
51. Connect No. 2 water by-pass hose.
52. Connect ventilation hose.
53. Install exhaust manifold converter sub-assembly.

54. Install manifold stay.
55. Install No. 1 exhaust manifold heat insulator.
56. Install alternator assembly.
57. Install engine assembly with transaxle.

TIMING CHAIN & SPROCKETS

REMOVAL & INSTALLATION

See Figures 62 through 68.

1. Remove engine assembly with transaxle.
2. Remove alternator assembly.
3. Remove No. 1 exhaust manifold heat insulator.
4. Remove manifold stay.
5. Remove exhaust manifold converter sub-assembly.
6. Disconnect ventilation hose.
7. Disconnect No. 2 water by-pass hose.
8. Disconnect No. 4 water by-pass hose.
9. Remove intake manifold stay.
10. Remove intake manifold assembly.
11. Remove No. 4 engine wire.
12. Remove engine oil level dipstick.
13. Remove engine oil level dipstick guide.
14. Remove fuel delivery pipe.
15. Remove fuel injector assembly.
16. Remove No. 1 delivery pipe spacer.
17. Remove injector vibration insulator.
18. Remove No. 1 ignition coil.
19. Remove ventilation hose.
20. Remove No. 2 ventilation hose.
21. Remove No. 1 water by-pass pipe.
22. Remove No. 2 idle pulley assembly with bracket.
23. Remove V-ribbed belt tensioner assembly.
24. Remove No. 1 Crankshaft Position (CKP) sensor.
25. Remove cylinder head cover sub-assembly.
26. Remove camshaft bearing cap oil hole gasket.
27. Remove Crankshaft Position (CKP) sensor.
28. Remove water pump pulley.
29. Remove crankshaft pulley.
30. Remove timing chain cover sub-assembly.
31. Remove timing chain cover oil seal.
32. Remove No. 1 chain tensioner assembly.
33. Remove No. 2 chain vibration damper.
34. Remove timing chain tension arm.
35. Remove chain sub-assembly.

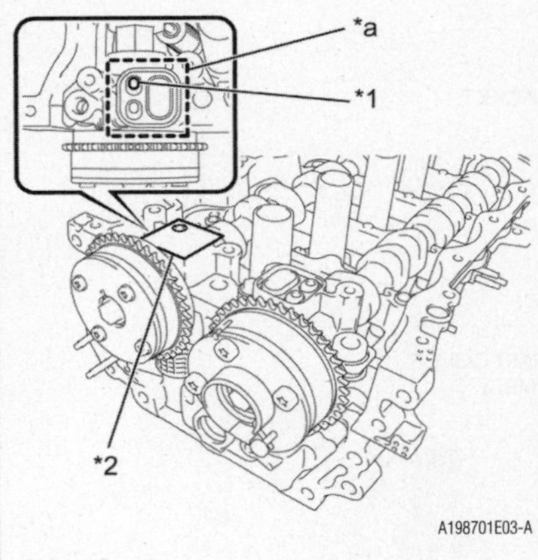

Fig. 62 After cleaning and degreasing the VVT oil hole on the intake side of the No. 1 camshaft bearing cap, completely seal (a) the oil hole (1) with adhesive tape (2) or equivalent to prevent air from leaking

36. Remove timing chain guide.
37. Inspect camshaft timing gear assembly.
   a. Inspect the lock of the camshaft timing gear.
   b. After cleaning and degreasing the VVT oil hole on the intake side of the No. 1 camshaft bearing cap, completely seal the oil hole with adhesive tape or equivalent to prevent air from leaking.

➡Be sure to seal the oil hole completely because air leaks due to insufficient sealing will prevent the lock pin from being released.

   c. Prick a hole in the tape covering the oil hole (Procedure A).
   d. Apply approximately 22 psi (150 kPa) of air pressure to the hole pricked in procedure A to release the lock pin.

➡If air leaks out, reattach adhesive tape. Cover the oil hole with a piece of cloth when applying air pressure to prevent oil from spraying.

   e. Forcibly turn the camshaft timing gear assembly in the advance direction (counterclockwise).

➡Depending on the air pressure applied, the camshaft timing gear assembly may turn in the advance direction without assistance by hand.

   f. Turn the camshaft timing gear assembly within its movable range (26.5 to 28.5°) 2 or 3 times without turning it to the most retarded position. Make sure that the camshaft timing gear assembly turns smoothly.
   g. Remove the adhesive tape from the No. 1 camshaft bearing cap.
38. Inspect camshaft timing exhaust gear assembly.
   a. Check the lock of the camshaft timing exhaust gear.
   b. After cleaning and degreasing the VVT oil hole on the exhaust side of the No. 1 camshaft bearing cap, completely seal the oil hole with adhesive tape or equivalent to prevent air from leaking.

➡Be sure to seal the oil hole completely because air leaks due to insufficient sealing will prevent the lock pin from being released.

   c. Prick a hole in the tape covering the oil hole (Procedure B).
   d. Apply approximately 28 psi (200 kPa) of air pressure to the hole pricked in procedure B to release the lock pin.

➡If air leaks out, reattach adhesive tape. Cover the oil hole with a piece of cloth when applying air pressure to prevent oil from spraying.

   e. Using a screwdriver with its tip wrapped with tape, forcibly turn the camshaft timing exhaust gear in the retard direction (clockwise).

➡Be sure to keep the camshaft timing exhaust gear in the retard direction using a screwdriver. If the gear is released, it will return to the most advanced position automatically due to force from the spring. Do not damage the camshaft timing exhaust gear.

   f. Using a screwdriver with its tip wrapped with tape, turn the camshaft timing exhaust gear within its movable range (19 to 21°) 2 or 3 times without turning it to the most advanced position. Make sure that the camshaft timing exhaust gear turns smoothly.
   g. Remove the adhesive tape from the No. 1 camshaft bearing cap.
39. Remove camshaft timing gear assembly.
   a. Remove the flange bolt while holding the hexagonal portion of the camshaft, and then remove the camshaft timing gear assembly.

➡Before removing the camshaft timing gear, make sure that the lock pin has been released. Be sure not to remove the other 4 bolts. Keep the camshaft timing gear assembly horizontal while removing it from the camshaft.

40. Remove camshaft timing exhaust gear assembly.

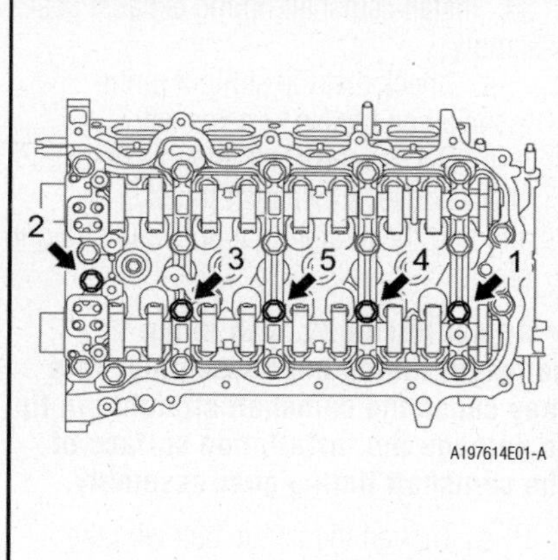

Fig. 63 Uniformly loosen and remove the 5 bearing cap bolts in sequence

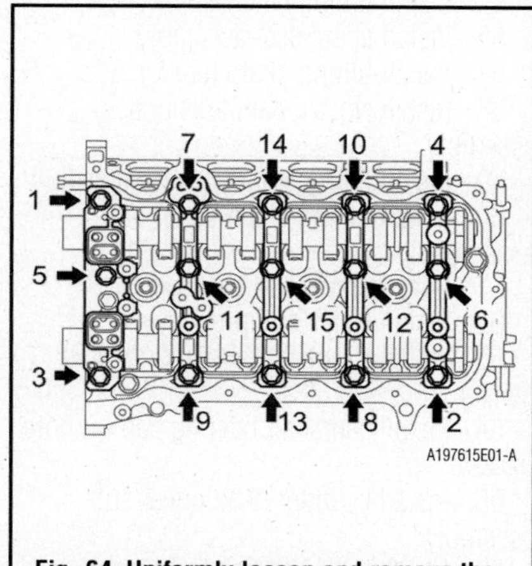

Fig. 64 Uniformly loosen and remove the 15 bearing cap bolts in sequence

a. Remove the flange bolt while holding the hexagonal portion of the camshaft, and then remove the camshaft timing exhaust gear assembly.

➡Be sure not to remove the other 4 bolts. Keep the camshaft timing exhaust gear assembly horizontal while removing it from the camshaft.

41. Remove camshaft bearing cap.
    a. Uniformly loosen and remove the 5 bearing cap bolts in sequence.
    b. Uniformly loosen and remove the 15 bearing cap bolts in sequence.

➡Uniformly loosen the bolts while keeping the camshaft level.

    c. Remove the 5 bearing caps.
42. Remove No. 2 camshaft.
43. Remove camshaft.
44. Remove No. 1 valve rocker arm sub-assembly.
45. Remove valve lash adjuster assembly.

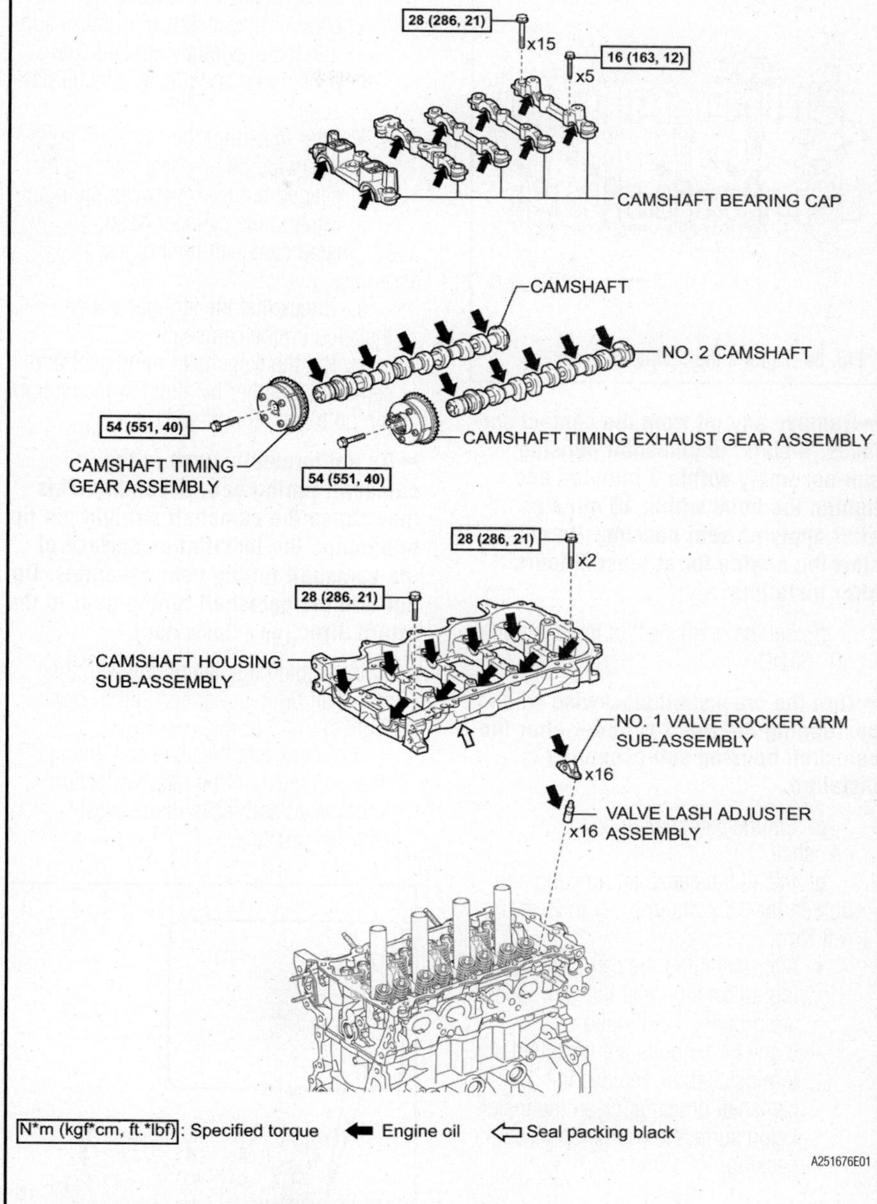

Fig. 65 Exploded view of camshaft assembly

46. Remove camshaft housing sub-assembly.

To install:

47. Install valve lash adjuster assembly.
48. Install No. 1 valve rocker arm sub-assembly.
49. Install camshaft.
    a. Clean the camshaft journals.
    b. Apply a light coat of engine oil to the camshaft journals and camshaft housing.
    c. Install the camshaft to the camshaft housing.
50. Install No. 2 camshaft.
    a. Clean the camshaft journals.
    b. Apply a light coat of engine oil to the camshaft journals and camshaft housing.
    c. Install the No. 2 camshaft to the camshaft housing.
51. Install camshaft bearing cap.
    a. Apply engine oil to the bearing caps.
    b. Install the 5 bearing caps to the camshaft housing.
    c. Tighten the 5 bolts in order. Tighten to 12 ft. lbs. (16 Nm).
52. Install camshaft housing sub-assembly.
    a. Apply seal packing in a continuous bead.

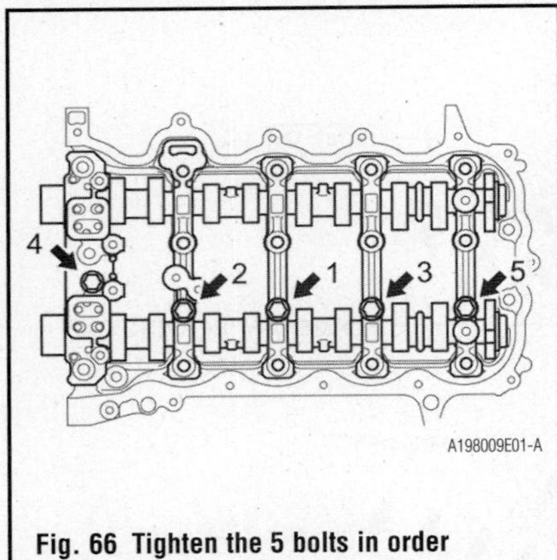

Fig. 66 Tighten the 5 bolts in order

➡Remove any oil from the contact surfaces. Install the camshaft housing sub-assembly within 3 minutes and tighten the bolts within 15 minutes after applying seal packing. Do not start the engine for at least 2 hours after installing.

b. Set the crankshaft in the position of 40° BTDC.

➡Turn the crankshaft clockwise when positioning the No. 1 cylinder after the camshaft housing sub-assembly is installed.

c. Set the camshaft and No. 2 camshaft.

d. Install the camshaft housing and tighten the 18 bolts in order to 21 ft. lbs. (28 Nm).

- After installing the camshaft housing, make sure that the cam lobes are properly positioned.
- If any of the bolts are loosened during installation, remove the camshaft housing, clean the installation surfaces, and reapply seal packing.

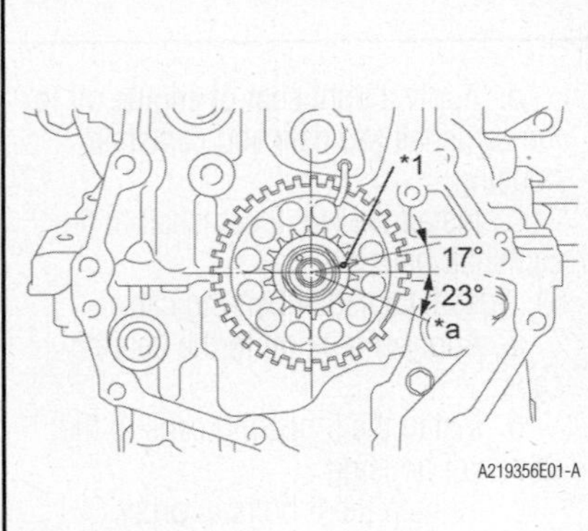

Fig. 67 Set the crankshaft in the position of 40° BTDC

- If the camshaft housing is removed because any of the bolts are loosened during installation, make sure that the previously applied seal packing does not enter any oil passages.
- After installing the camshaft housing, wipe off any seal packing that seeped out from between the housing and the cylinder head.

53. Install camshaft timing gear assembly.

a. Check that the straight pin is installed on the camshaft.

b. Put the camshaft timing gear and camshaft together by aligning the straight pin hole and straight pin.

➡Do not forcefully push in the camshaft timing gear assembly. This may cause the camshaft straight pin tip to damage the installation surface of the camshaft timing gear assembly. Do not turn the camshaft timing gear in the retard direction (clockwise).

c. Tighten the flange bolt with the camshaft timing gear secured in place. Tighten to 40 ft. lbs. (54 Nm).

d. Check that the camshaft timing gear can move in the retard direction (clockwise) and locks in the most retarded position.

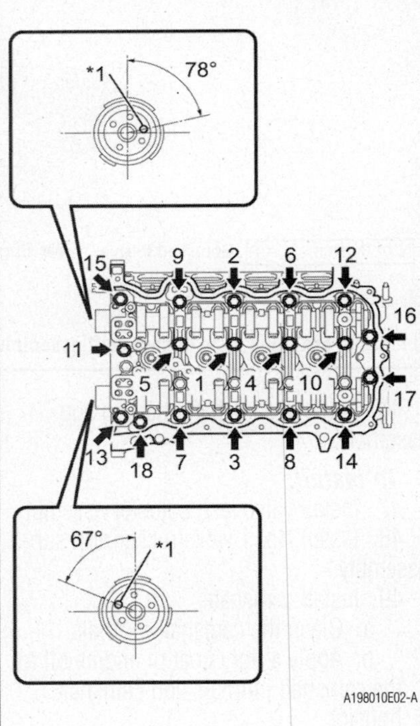

Fig. 68 Install the camshaft housing and tighten the 18 bolts in order; straight pin (1)

54. Install camshaft timing exhaust gear assembly.

a. Check that the straight pin is installed on the No. 2 camshaft.

b. Put the camshaft timing exhaust gear and No. 2 camshaft together by aligning the straight pin hole and straight pin.

➡Do not forcefully push in the camshaft timing gear assembly. This may cause the camshaft straight pin tip to damage the installation surface of the camshaft timing gear assembly.

c. Tighten the flange bolt with the camshaft timing exhaust gear secured in place. Tighten to 40 ft. lbs. (54 Nm).

d. Make sure that the camshaft timing exhaust gear is locked.

55. Install timing chain guide.
56. Install chain sub-assembly.
57. Install timing chain tension arm.
58. Install No. 1 chain tensioner assembly.
59. Install timing chain cover oil seal.
60. Install timing chain cover sub-assembly.
61. Install crankshaft pulley.
62. Install water pump pulley.
63. Install Crankshaft Position (CKP) sensor.
64. Install camshaft bearing cap oil hole gasket.
65. Install cylinder head cover sub-assembly.
66. Install V-ribbed belt tensioner assembly.
67. Install No. 2 idle pulley assembly with bracket.
68. Install No. 1 water by-pass pipe.
69. Install No. 2 ventilation hose.
70. Install ventilation hose.
71. Install No. 1 ignition coil.
72. Install injector vibration insulator.
73. Install No. 1 delivery pipe spacer.
74. Install fuel injector assembly.
75. Install fuel delivery pipe.
76. Install engine oil level dipstick guide.
77. Install engine oil level dipstick.
78. Install No. 4 engine wire.
79. Install intake manifold assembly.
80. Install intake manifold stay.
81. Connect No. 4 water by-pass hose.
82. Connect No. 2 water by-pass hose.
83. Connect ventilation hose.
84. Install exhaust manifold converter sub-assembly.
85. Install manifold stay.
86. Install No. 1 exhaust manifold heat insulator.
87. Install alternator assembly.

88. Install engine assembly with transaxle.

VALVE COVERS

REMOVAL & INSTALLATION

See Figures 69 and 70.

1. Remove the oil filler cap.
2. Remove the gasket from the oil filler cap.
3. Using a 14 mm spark plug wrench, remove the 4 spark plugs.
4. Remove the 2 bolts, 2 O-rings and the 2 camshaft timing oil control valves.
5. Remove the 2 bolts and the 2 No. 1 Crankshaft Position (CKP) sensors.
6. Using a 27 mm deep socket wrench, remove the ventilation valve.
7. Remove cylinder head cover sub-assembly.
    a. Remove the 17 bolts, 2 seal washers and the cylinder head cover.
    b. Remove the cylinder head cover gasket.

To install:

8. Install cylinder head cover sub-assembly.
    a. Install a new gasket to the cylinder head cover.
    b. Apply seal packing to joint between cylinder block and timing chain cover.

➡Remove any oil from the contact surfaces. Install the cylinder head cover within 3 minutes and tighten the bolts within 15 minutes after applying seal packing. Do not start the engine for at least 2 hours after the installation.

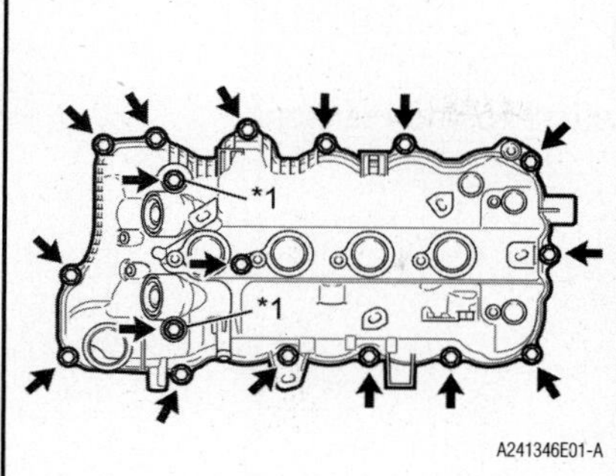

Fig. 69 Remove the 17 bolts, 2 seal washers (1) and the cylinder head cover

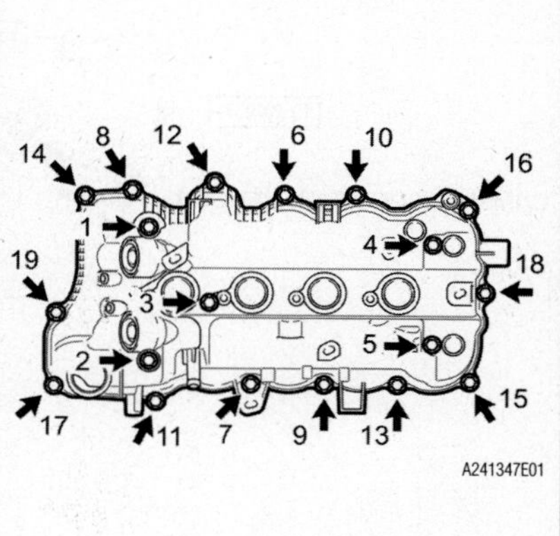

Fig. 70 Fully tighten the 19 bolts in order

    c. Temporarily install the cylinder head cover with 2 new seal washers and the 17 bolts.
    d. Apply a light coat of engine oil to the O-rings of the No. 1 Crankshaft Position (CKP) sensors.
    e. Temporarily install the 2 No. 1 Crankshaft Position (CKP) sensors with the 2 bolts.
    f. Fully tighten the 19 bolts in order. Tighten to 7 ft. lbs. (10 Nm).
9. Using a 27 mm deep socket wrench, install the ventilation valve.
10. Install camshaft timing oil control valve assembly.
    a. Apply a light coat of engine oil to 2 new O-rings, and then install them onto the camshaft timing oil control valves.
    b. Install the 2 camshaft timing oil control valves with the 2 bolts.
11. Using a 14 mm spark plug wrench, install the 4 spark plugs. Tighten to 15 ft. lbs. (20 Nm).
12. Install the gasket to the cap.
13. Install the oil filler cap.

VALVE LASH

ADJUSTMENT

Engine is equipped with hydraulic lash adjusters. No adjustment is necessary.

ENGINE PERFORMANCE & EMISSION CONTROLS

AIR FUEL RATIO SENSOR

LOCATION

See Figure 71.

REMOVAL & INSTALLATION

1. Remove air cleaner and hose.
2. Remove air fuel ratio sensor.
    a. Disengage the 2 clamps and disconnect the air fuel ratio sensor wire from the wiring harness clamp bracket.
    b. Disconnect the connector of the air fuel ratio sensor connector.
    c. Using SST (09224-00010), remove the air fuel ratio sensor from the exhaust manifold.

➡If a component has been dropped or subjected to a strong impact, replace it.

To install:

3. Install air fuel ratio sensor.
    a. Using SST, install the air fuel ratio sensor onto the exhaust manifold.
    b. Connect the connector of the air fuel ratio sensor.

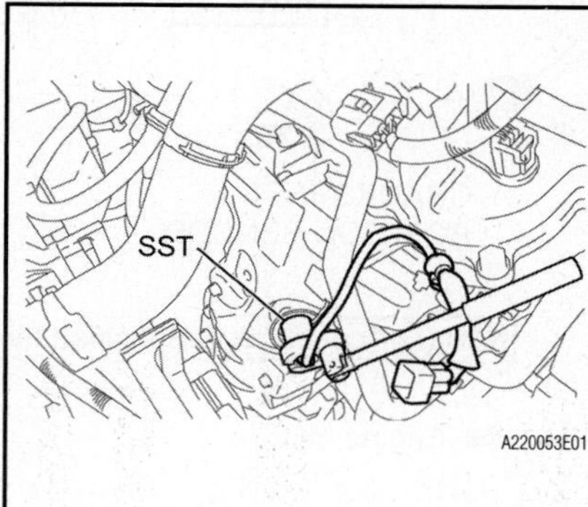

Fig. 71 Air Fuel Ratio sensor location

    c. Engage the 2 clamps and connect the air fuel ratio sensor wire to the wiring harness clamp bracket.
4. Install air cleaner and hose.
5. Inspect for exhaust gas leak.

CAMSHAFT POSITION (CMP) SENSOR

LOCATION

See Figure 72.

REMOVAL & INSTALLATION

1. Remove Camshaft Position (CMP) sensors.
2. For Exhaust Side:
    a. Disconnect the connector of the Camshaft Position (CMP) sensor.
    b. Remove the bolt and Camshaft Position (CMP) sensor from the cylinder head cover sub-assembly.

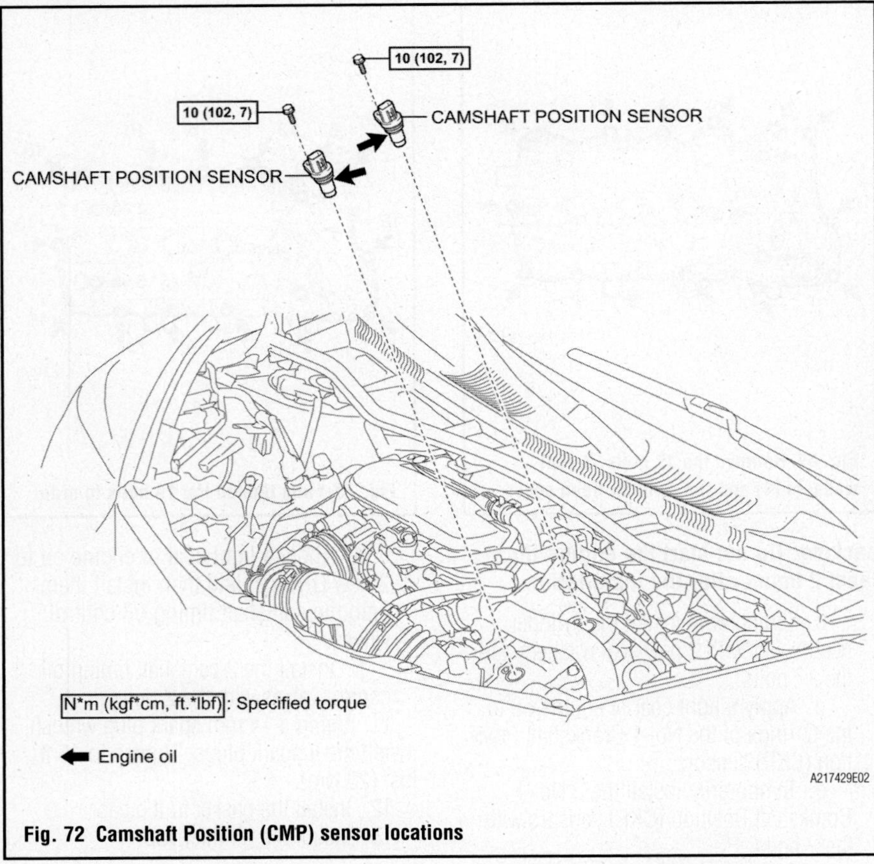

Fig. 72 Camshaft Position (CMP) sensor locations

➡**If a component has been dropped or subjected to a strong impact, replace it.**

3. For Intake Side:
   a. Disconnect the connector of the Camshaft Position (CMP) sensor.
   b. Remove the bolt and Camshaft Position (CMP) sensor from the cylinder head cover sub-assembly.

➡**If a component has been dropped or subjected to a strong impact, replace it.**

To install:

4. Install Camshaft Position (CMP) sensor.
5. For Exhaust Side:
   a. Apply a light coat of engine oil to the O-ring.

➡**If reusing the Camshaft Position (CMP) sensor, be sure to inspect the O-ring.**

   b. Install the Camshaft Position (CMP) sensor onto the cylinder head cover sub-assembly with the bolt. Tighten to 7 ft. lbs. (10 Nm).
   c. Connect the connector of the Camshaft Position (CMP) sensor.
6. For Intake Side:
   a. Apply a light coat of engine oil to the O-ring.

➡**If reusing the Camshaft Position (CMP) sensor, be sure to inspect the O-ring.**

   b. Install the Camshaft Position (CMP) sensor onto the cylinder head cover sub-assembly with the bolt. Tighten to 7 ft. lbs. (10 Nm).
   c. Connect the connector of the Camshaft Position (CMP) sensor.
7. Start the engine and check that there are no oil leaks.

CRANKSHAFT POSITION (CKP) SENSOR

LOCATION

See Figure 73.

REMOVAL & INSTALLATION

1. Remove crankshaft pulley.
2. Remove Crankshaft Position (CKP) sensor.
   a. Disconnect the connector of the Crankshaft Position (CKP) sensor.
   b. Remove the bolt and the Crankshaft Position (CKP) sensor from the timing chain cover sub-assembly.

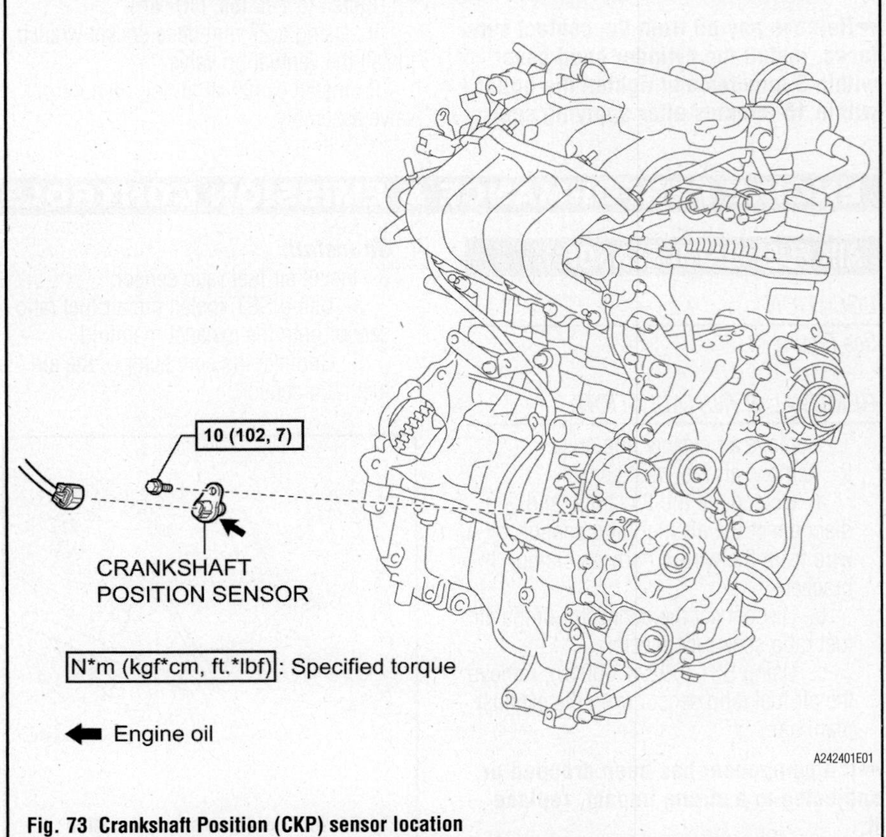

Fig. 73 Crankshaft Position (CKP) sensor location

➡**If a component has been dropped or subjected to a strong impact, replace it.**

To install:

3. Install Crankshaft Position (CKP) sensor.
   a. Apply a light coat of engine oil to the O-ring.
   b. Install the Crankshaft Position (CKP) sensor onto the timing chain cover sub-assembly with the bolt. Tighten to 7 ft. lbs. (10 Nm).
   c. Connect the connector of the Crankshaft Position (CKP) sensor.
4. Install crankshaft pulley.
5. Start the engine and check that there are no oil leaks.

ELECTRONIC CONTROL MODULE (ECM)

LOCATION

See Figure 74.

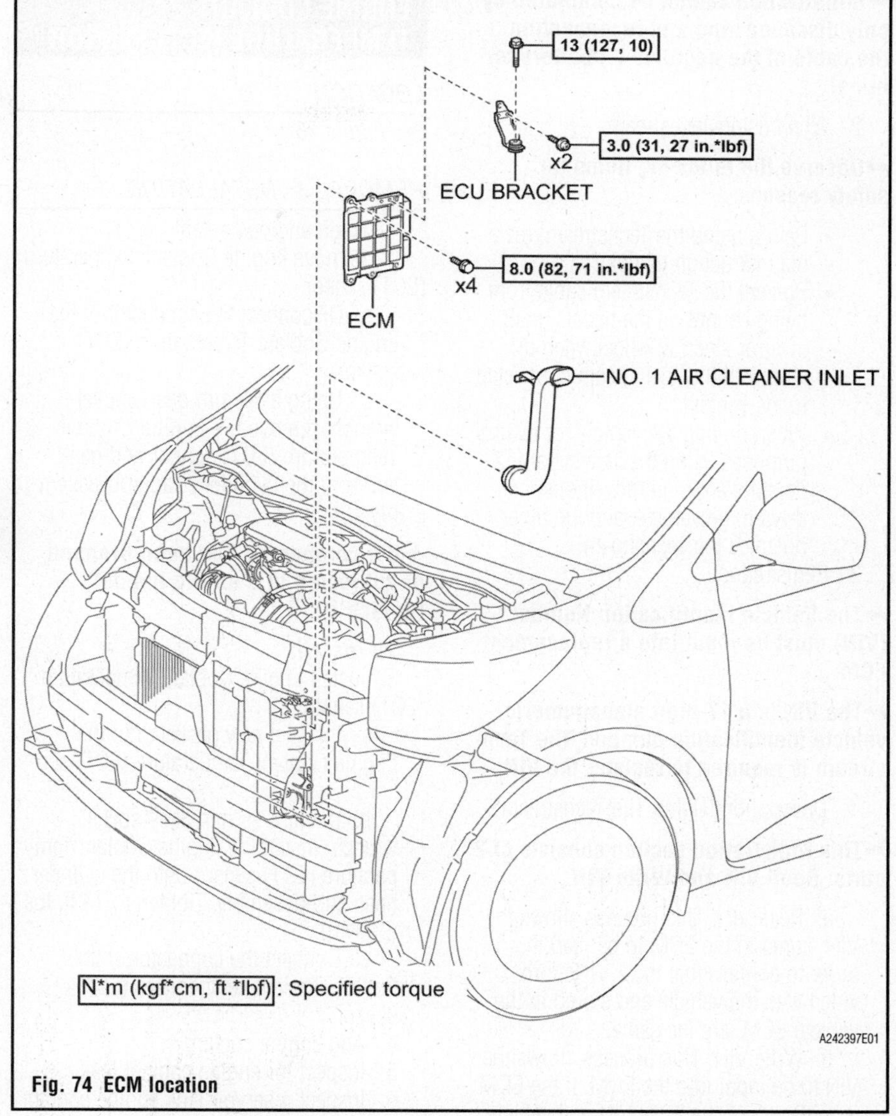

Fig. 74 ECM location

REMOVAL & INSTALLATION

See Figure 75.

1. Remove windshield washer jar assembly.
2. Disengage the 3 claws and remove the No. 1 air cleaner inlet.
3. Remove ECM.
   a. Pull up the lever to loosen the lock and disconnect the 2 connectors from the ECM.
   b. Disengage the clamp and disconnect the wire harness from the ECU bracket.
   c. Remove the 5 bolts and the ECM from the fan shroud.

➡**If a component has been dropped or subjected to a strong impact, replace it.**

4. Remove the 2 screws and the ECU bracket from the ECM.

To install:

5. Install the ECU bracket onto the ECM with the 2 screws.
6. Install ECM.
   a. Temporarily install the ECM with the 5 bolts.
   b. Tighten the 4 bolts A to 71 inch lbs. (8 Nm). Then tighten bolt B to 10 ft. lbs. (13 Nm).
   c. Engage the clamp and connect the wire harness to the ECU bracket.
   d. Pull down the lever to engage the lock and connect the 2 connectors to the ECM.
7. Engage the 2 claws and install the No. 1 air cleaner inlet.
8. Install windshield washer jar assembly.
9. Perform initialization.

INITIALIZATION

1. Initialization:
   a. If the ECM is replaced, register the ECU communication ID for Immobilizer System.
   b. Make sure to perform Reset Memory (AT/CVT initialization), yaw rate sensor 0 point calibration and CVT oil pressure calibration when replacing the ECM.
   c. Perform Registration (VIN registration) when replacing the ECM.
   d. Perform initialization (throttle position) after replacing the throttle with motor body assembly or cleaning any throttle body components.
   e. Perform initialization (throttle position) after reconnecting the battery cable or replacing the ECM.

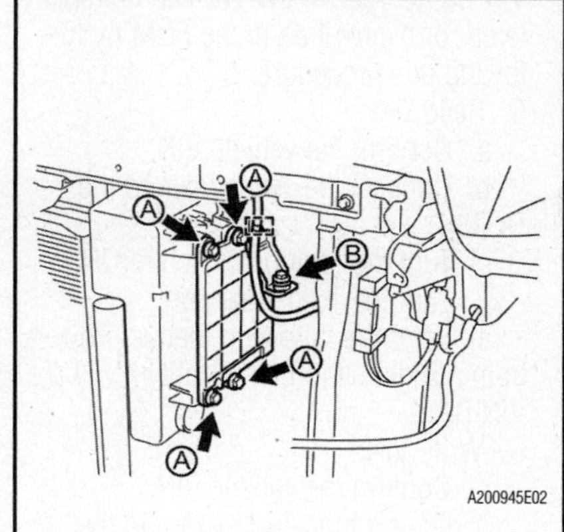

Fig. 75 Tighten the 4 bolts A to 71 inch lbs. (8 Nm), then tighten bolt B to 10 ft. lbs. (13 Nm)

➡Initialization cannot be completed by only disconnecting and reconnecting the cable of the negative (-) battery terminal.

2. When using Techstream:

➡Observe the following items for safety reasons:

- Before using the Techstream, read the instruction manual.
- Prevent the Techstream cable from being caught on the pedals, shift lever or steering wheel when driving with the Techstream connected to the vehicle.
- When driving the vehicle for testing purposes using the Techstream, 2 persons are required. One is for driving the vehicle, and the other operates the Techstream.

3. Registration:

➡The Vehicle Identification Number (VIN) must be input into a replacement ECM.

➡The VIN is a 17-digit alphanumeric vehicle identification number. The Techstream is required to register the VIN.

4. Description (Using The Techstream):

➡This registration section consists of 2 parts: Read VIN and Write VIN.

a. Read VIN: This process allows the VIN stored in the ECM to be read in order to confirm that the 2 VINs, provided with the vehicle and stored in the vehicle ECM, are the same.

b. Write VIN: This process allows the VIN to be input into the ECM. If the ECM is replaced, or the ECM VIN and vehicle VIN do not match, the VIN can be registered, or overwritten in the ECM by following this procedure.

5. Read VIN:

a. Confirm the vehicle VIN.

b. Connect the Techstream to the DLC3.

c. Turn the ignition switch to ON.

d. Turn the Techstream on.

e. Enter the following menus: Powertrain / Engine and ECT / Utility / VIN / VIN Read.

6. Write VIN:

a. Confirm the vehicle VIN.

b. Connect the Techstream to the DLC3.

c. Turn the ignition switch to ON.

d. Turn the Techstream on.

e. Enter the following menus: Powertrain / Engine and ECT / Utility / VIN / VIN Write.

ENGINE COOLANT TEMPERATURE (ECT) SENSOR

LOCATION

See Figure 76.

REMOVAL & INSTALLATION

1. Drain engine coolant.
2. Remove Engine Coolant Temperature (ECT) sensor.

a. Disconnect the connector of the Engine Coolant Temperature (ECT) sensor.

b. Using a 19 mm deep socket wrench, remove the Engine Coolant Temperature (ECT) sensor and gasket from the cylinder head sub-assembly.

➡If a component has been dropped or subjected to a strong impact, replace it.

To install:

3. Install Engine Coolant Temperature (ECT) sensor.

a. Install a new gasket onto the Engine Coolant Temperature (ECT) sensor.

b. Using a 19 mm deep socket wrench, install the Engine Coolant Temperature (ECT) sensor onto the cylinder head sub-assembly. Tighten to 14 ft. lbs. (20 Nm).

c. Connect the connector of the Engine Coolant Temperature (ECT) sensor.

4. Add engine coolant.
5. Inspect for engine coolant leak.
6. Inspect reservoir tank engine coolant level.

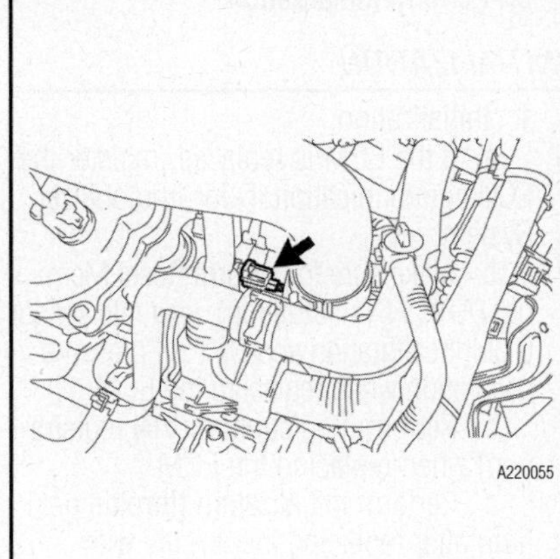

Fig. 76 Engine Coolant Temperature (ECT) sensor location

HEATED OXYGEN SENSOR (HO2S)

LOCATION

See Figure 77.

REMOVAL & INSTALLATION

1. Remove No. 2 engine under cover.
2. Remove heated oxygen sensor.

a. Disengage the clamp and disconnect the heated oxygen sensor wire from the wiring harness clamp bracket.

b. Disconnect the connector of the heated oxygen sensor.

c. Using SST (09224-00010), remove the heated oxygen sensor from the front exhaust pipe assembly.

➡If a component has been dropped or subjected to a strong impact, replace it.

To install:

3. Install heated oxygen sensor.

a. Using SST, install the heated oxygen sensor onto the front exhaust pipe assembly. Tighten to 32 ft. lbs. (44 Nm).

b. Connect the connector of the heated oxygen sensor.

c. Engage the clamp and connect the heated oxygen sensor wire to the wiring harness clamp bracket.

4. Inspect for exhaust gas leak.
5. Install No. 2 engine under cover.

KNOCK SENSOR (KS)

LOCATION

See Figure 78.

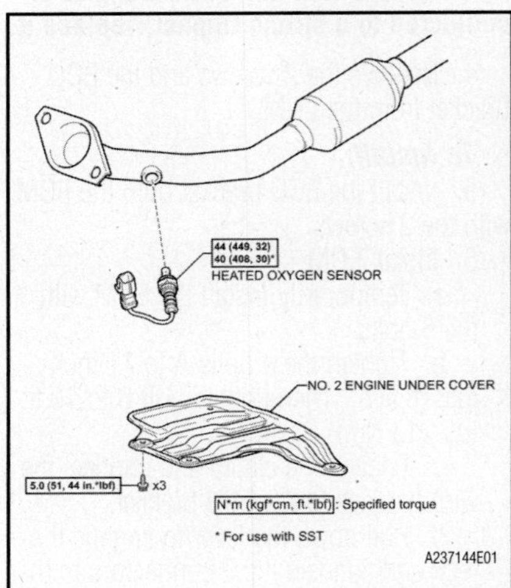

Fig. 77 Heated Oxygen Sensor (HO2S) location

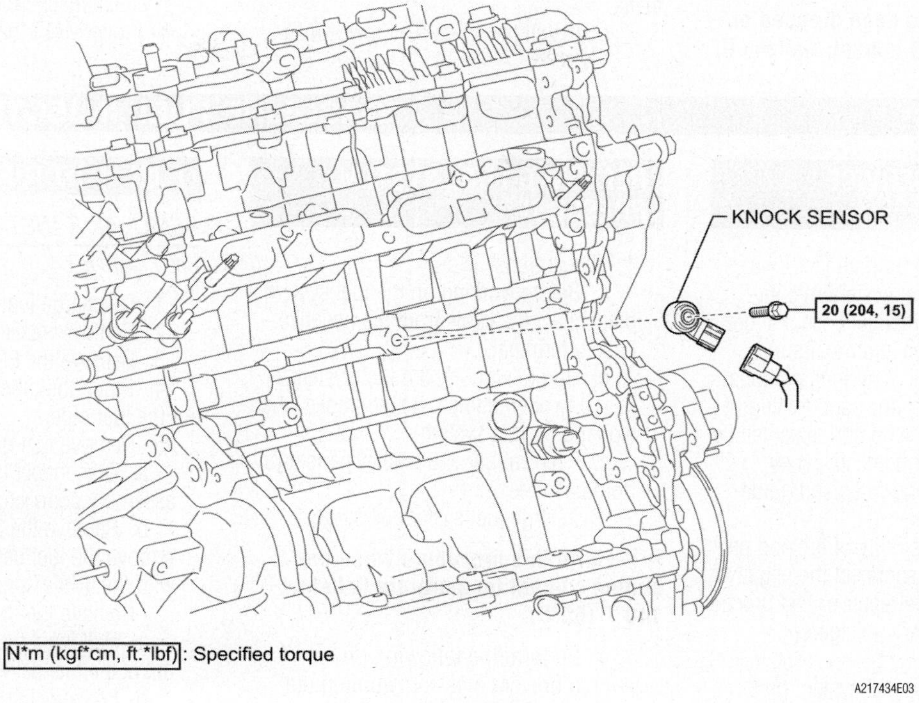

Fig. 78 Knock sensor location

REMOVAL & INSTALLATION

1. Remove intake manifold.
2. Remove Knock Sensor (KS).
   a. Disconnect the connector of the knock sensor.
   b. Remove the bolt and KS from the cylinder block.

➡If a component has been dropped or subjected to a strong impact, replace it.

To install:

3. Install the KS:
   a. Install the KS onto the cylinder block with the bolt. Tighten to 15 ft. lbs. (20 Nm).
   b. Attach the connector to the KS.
4. Install intake manifold.

MASS AIR FLOW (MAF) METER

LOCATION

See Figure 79.

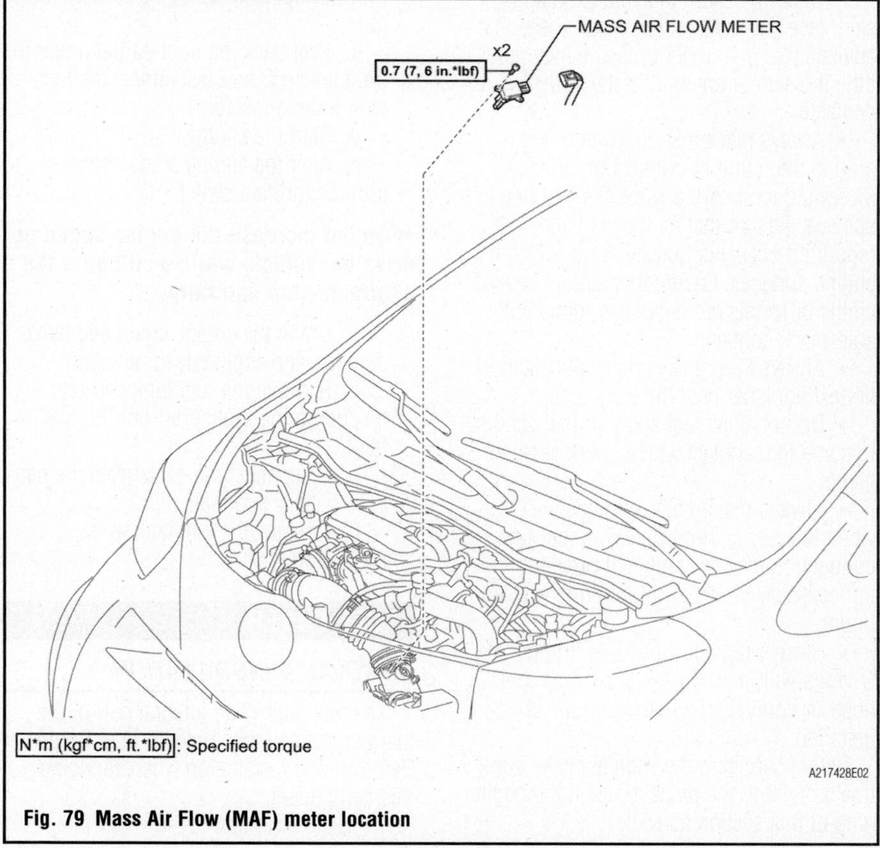

Fig. 79 Mass Air Flow (MAF) meter location

REMOVAL & INSTALLATION

1. Remove Mass Air Flow (MAF) meter.
   a. Disconnect the connector of the Mass Air Flow (MAF) meter.
   b. Remove the 2 screws and the Mass

Air Flow (MAF) meter from the air cleaner cap sub-assembly.

➡**If a component has been dropped or subjected to a strong impact, replace it.**

To install:

2. Install Mass Air Flow (MAF) meter.
 a. Install the Mass Air Flow (MAF) meter onto the air cleaner cap sub-assembly with the 2 screws.
 b. Connect the connector of the Mass Air Flow (MAF) meter.

FUEL

GASOLINE FUEL INJECTION SYSTEM

FUEL SYSTEM SERVICE PRECAUTIONS

Safety is the most important factor when performing not only fuel system maintenance but any type of maintenance. Failure to conduct maintenance and repairs in a safe manner may result in serious personal injury or death. Maintenance and testing of the vehicle's fuel system components can be accomplished safely and effectively by adhering to the following rules and guidelines.

- To avoid the possibility of fire and personal injury, always disconnect the negative battery cable unless the repair or test procedure requires that battery voltage be applied.
- Always relieve the fuel system pressure prior to disconnecting any fuel system component (injector, fuel rail, pressure regulator, etc.), fitting or fuel line connection. Exercise extreme caution whenever relieving fuel system pressure to avoid exposing skin, face and eyes to fuel spray. Please be advised that fuel under pressure may penetrate the skin or any part of the body that it contacts.
- Always place a shop towel or cloth around the fitting or connection prior to loosening to absorb any excess fuel due to spillage. Ensure that all fuel spillage (should it occur) is quickly removed from engine surfaces. Ensure that all fuel soaked cloths or towels are deposited into a suitable waste container.
- Always keep a dry chemical (Class B) fire extinguisher near the work area.
- Do not allow fuel spray or fuel vapors to come into contact with a spark or open flame.
- Always use a back-up wrench when loosening and tightening fuel line connection fittings. This will prevent unnecessary stress and torsion to fuel line piping.
- Always replace worn fuel fitting O-rings with new Do not substitute fuel hose or equivalent where fuel pipe is installed.

Before servicing the vehicle, make sure to also refer to the precautions in the beginning of this section as well.

RELIEVING FUEL SYSTEM PRESSURE

1. Precautions:
 a. Before working on the fuel system, disconnect the cable from the negative battery terminal.
 b. Do not work on the fuel system near any open flames. Never smoke when working on this system.
 c. Keep rubber and leather parts away from gasoline.
2. Discharge fuel system pressure.

➡**Perform this procedure when the engine coolant temperature is below 140°F (60°C).**

 a. Perform the following procedure to prevent fuel from spilling out before removing any fuel system parts.
 b. Even after discharging the fuel pressure, when disconnecting a fuel line, cover it with a piece of cloth to prevent fuel from spraying or coming out.
 c. Peel back the floor carpet under the front left seat, and disconnect the fuel tank wire connector.
 d. Start the engine.
 e. After the engine stops naturally, turn the ignition switch off.

➡**Do not increase the engine speed or drive the vehicle while waiting for the engine to stop naturally.**

 f. Crank the engine again and make sure that the engine does not start.
 g. Remove the fuel tank cap and discharge the pressure from the fuel tank.
 h. Disconnect the cable from the negative battery terminal.
 i. Connect the fuel tank wire connectors.

FUEL FILTER

REMOVAL & INSTALLATION

The fuel filter is an integral part of the fuel pump assembly. Refer to Fuel Pump Removal and Installation when servicing this component.

FUEL INJECTORS

REMOVAL & INSTALLATION

See Figure 80.

1. Discharge fuel system pressure.
2. Remove intake manifold.
3. Remove the EFI fuel pipe clamp.
4. Pinch the tube connector and pull out the fuel pipe.
5. Remove fuel delivery pipe.
 a. Disconnect the 4 fuel injector assembly connectors.
 b. Remove the 2 bolts, and then remove the fuel delivery pipe together with the 4 injectors.
6. Remove fuel injector assembly.
 a. Pull the 4 fuel injector assemblies out of the fuel delivery pipe sub-assembly.
 b. For reinstallation, attach a tag or label to each injector shaft.
 c. Remove the 4 O-rings from the 4 fuel injector assemblies.
7. Remove the 2 No. 1 delivery pipe spacers from the cylinder head.
8. Remove the 4 injector vibration insulators.

To install:

9. Install 4 new injector vibration insulators onto the cylinder head.
10. Install the 2 No. 1 delivery pipe spacers onto the cylinder head.
11. Install fuel injector assembly.
 a. Apply a light coat of gasoline or spindle oil to new O-rings, then install one onto each fuel injector assembly.
 b. Apply a light coat of gasoline or spindle oil to the contact surfaces of the fuel delivery pipe and the O-ring of the fuel injector assembly.
 c. While turning the fuel injector assembly left and right, install it onto the fuel delivery pipe sub-assembly.
12. Install fuel delivery pipe.
 a. Install the fuel delivery pipe with the 4 fuel injectors, and then temporarily install the 2 bolts.
 b. Tighten the 2 bolts to 21 ft. lbs. (28 Nm).
13. Connect fuel tube sub-assembly. Line up the pipe and connector and push them together until a "click" sound is heard.
14. Install the EFI fuel pipe clamp.

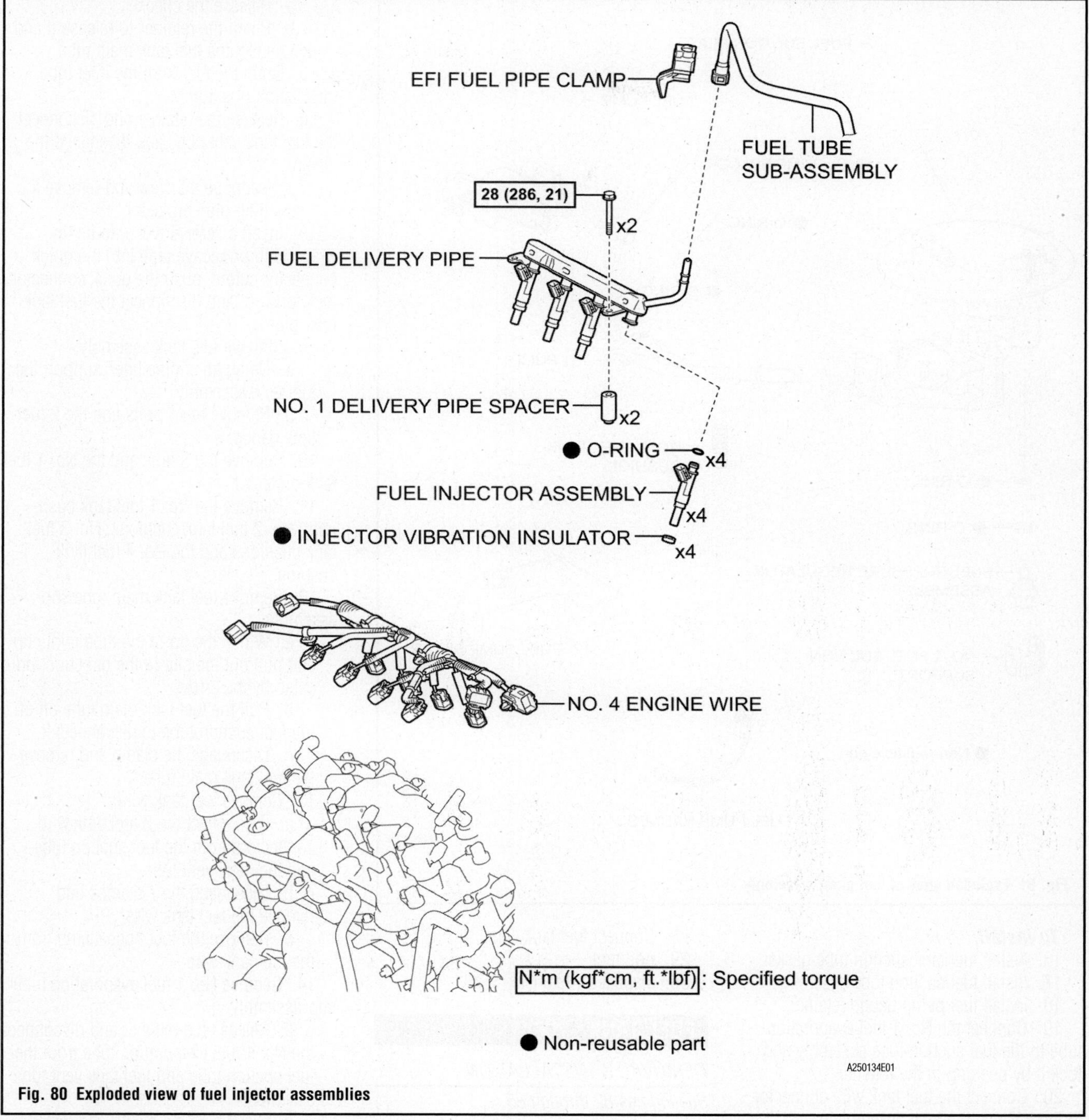

Fig. 80 Exploded view of fuel injector assemblies

15. Install intake manifold.
16. Inspect for fuel leak.

FUEL PUMP

REMOVAL & INSTALLATION

See Figure 81.

1. Discharge fuel system pressure.
2. Separate fuel tank wire.
3. Remove No. 1 fuel tank protector sub-assembly.
4. Disconnect fuel tank main tube sub-assembly.
5. Drain fuel.
6. Remove fuel tank vent hose.
7. Remove fuel tank filler pipe protector.
8. Disconnect fuel tank filler pipe sub-assembly.
9. Remove fuel tank assembly.
10. Disconnect fuel tank main tube sub-assembly.
    a. Widen the tip of the tube joint clip and pull out the clip.
    b. Pull the fuel tank main tube out of the fuel suction tube assembly to disconnect it.
11. Disconnect the fuel tank wire connector from the fuel suction tube assembly.
12. Release the retainer and disconnect the No. 1 fuel evaporation tube from the fuel suction tube assembly.
13. Remove fuel pump gauge retainer.
14. Remove fuel suction tube assembly.
15. Remove fuel tank suction tube gasket.

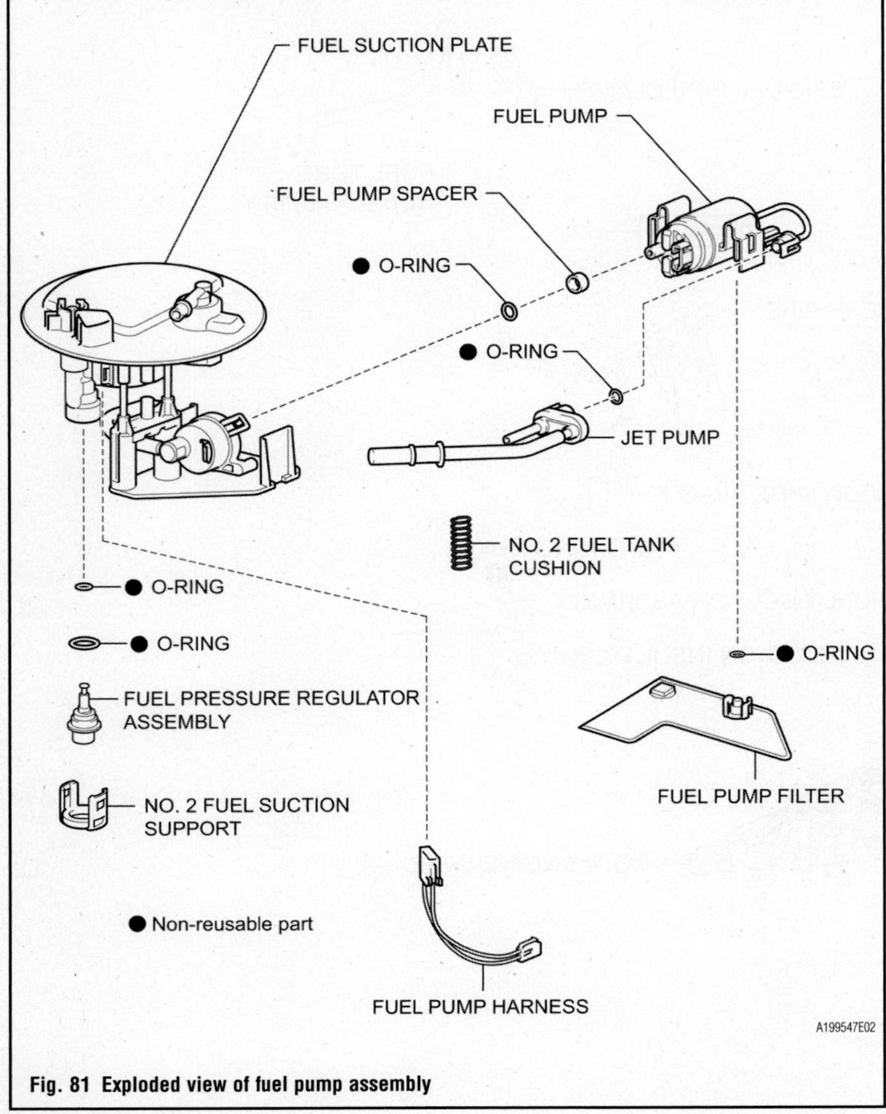

Fig. 81 Exploded view of fuel pump assembly

To install:

16. Install fuel tank suction tube gasket.
17. Install fuel suction tube assembly.
18. Install fuel pump gauge retainer.
19. Connect the No. 1 fuel evaporation tube to the fuel suction tube assembly and lock it by pushing in the retainer.
20. Connect the fuel tank wire connector to the fuel suction tube assembly.
21. Insert the fuel tank main tube into the fuel suction tube assembly plug and lock it with the tube joint clip.
22. Install fuel tank assembly.
23. Install fuel tank filler pipe protector.
24. Connect fuel tank filler pipe sub-assembly.
25. Connect fuel tank vent hose.
26. Install fuel tank main tube sub-assembly.
27. Install No. 1 fuel tank protector sub-assembly.
28. Connect fuel tank wire.
29. Add fuel.
30. Inspect for fuel leak.

FUEL TANK

REMOVAL & INSTALLATION

See Figures 82 through 85.

1. Discharge fuel system pressure.
2. Peel back the floor carpet under the front left seat, disconnect the fuel tank wire connector, and pass the grommet through the floor.
3. Remove No. 1 fuel tank protector sub-assembly.
   a. Remove the 2 clips and 4 nuts.
   b. Disengage the claw and remove the No. 1 fuel tank protector.
4. Disconnect fuel tank main tube sub-assembly.
   a. Release the checker.
   b. Pinch the retainer to release it and disconnect the fuel tank main tube.
5. Drain the fuel from the inlet tube installation position.
6. Release the retainer and disconnect the fuel tank vent hose and disengage the clamp.
7. Disengage the claw and remove the fuel tank filler pipe protector.
8. Insert a screwdriver with its tip wrapped in protective tape into the quick connector cutout, push the quick connector lock forward, and disconnect the fuel tank filler pipe.
9. Remove fuel tank assembly.
   a. Using an engine lifter, support the fuel tank assembly.
   b. Remove the 4 bolts and the 2 fuel tank bands.
10. Remove the 3 nuts and the No. 1 fuel tank protector.
11. Remove the No. 1 fuel tank cushions, No. 2 fuel tank cushions, No. 3 fuel tank cushions and the No. 4 fuel tank cushion.
12. Remove fuel tank main tube sub-assembly.
   a. Widen the tip of the tube joint clip and pull out the clip in the direction indicated by the arrow.
   b. Pull the fuel tank main tube off of the fuel suction tube to disconnect it.
   c. Disengage the clamp and remove the fuel tank main tube.
13. Remove fuel tank wire.
   a. Disconnect the 2 fuel tank wire connectors from the fuel suction tube and fuel tank vent tube.
   b. Disengage the 7 clamps and remove the fuel tank wire.
   c. Remove the fuel hose clamp from the fuel tank wire.
14. Remove No. 1 fuel evaporation tube sub-assembly.
   a. Release the retainer and disconnect the No. 1 fuel evaporation tube from the fuel suction tube and fuel tank vent tube assembly.
   b. Disengage the clamp and remove the No. 1 fuel evaporation tube.
15. Remove No. 2 fuel evaporation tube sub-assembly.
   a. Release the retainer and disconnect the No. 2 fuel evaporation tube from the fuel tank vent tube assembly.
   b. Disengage the clamp and remove the No. 2 fuel evaporation tube.
16. Remove fuel tank vent hose.
   a. Release the retainer and disconnect the fuel tank vent hose from the fuel tank vent tube assembly.

b. Disengage the clamp and remove the fuel tank vent hose.

17. Remove fuel pump gauge retainer.

a. Release the retainer and disconnect the fuel tank breather tube.

b. Set the plate and claws of SST onto the fuel pump gauge retainer and temporarily tighten the bolts.

c. While firmly pressing the claws of SST against the ribs of the retainer, fully tighten each bolt.

d. Set SST handle in place.

e. While pressing down tightly on SST to prevent it from rising up, slowly rotate the handle to loosen the retainer.

➡Note the following:

- Pressing the retainer against the pump with excessive force will make it difficult to turn and may damage the parts, so do not press down too strongly on SST.
- Applying excessive force at an angle may cause SST to slip off, so be sure to keep the handle level while rotating.
- To avoid damaging the parts, do not turn the handle of SST too forcefully, and do not use an impact wrench.
- If SST slips off the retainer, stop turning the handle, loosen the bolts of SST, and perform the SST set procedure again.

f. Remove the fuel pump gauge retainer while holding the fuel tank vent tube assembly by hand.

18. Remove fuel tank vent tube assembly.

a. Rotate the fuel tank vent tube assembly toward the inside of the tank, and remove it in the opposite direction to the arrows for insertion of the fuel tank sender gauge.

b. Disengage the 2 claws of the fuel tank vent tube assembly, disconnect the fuel tank vent tube sub-assembly, and remove the fuel tank vent tube assembly.

19. Remove the gasket from the fuel tank.

20. Remove fuel pump gauge retainer.

a. Set the plate and claws of SST onto the fuel pump gauge retainer and temporarily tighten the bolts.

b. While firmly pressing the claws of SST against the ribs of the retainer, fully tighten each bolt.

c. Set SST handle in place.

d. While pressing down tightly on SST to prevent it from rising up, slowly rotate the handle to loosen the retainer.

➡Note the following:

- Pressing the retainer against the pump with excessive force will make it difficult to turn and may damage the parts, so do not press down too strongly on SST.
- Applying excessive force at an angle may cause SST to slip off, so be sure to keep the handle level while rotating.
- To avoid damaging the parts, do not turn the handle of SST too forcefully, and do not use an impact wrench.
- If SST slips off the retainer, stop turning the handle, loosen the bolts of SST, and perform the SST set procedure again.

e. Remove the fuel pump gauge retainer while holding the fuel suction tube by hand.

21. Pinch the fuel tank vent tube retainer to release it, disconnect the fuel tank vent tube sub-assembly, and then remove the fuel suction tube.

22. Remove the gasket from the fuel tank.

23. Pull the fuel tank vent tube sub-assembly out of the fuel tank.

24. Disengage the claw and remove the No. 1 evaporation vent tube clamp.

To install:

25. Engage the claw and install the No. 1 evaporation tube clamp.

26. Insert the fuel tank vent tube from the sender gauge side, in the direction indicated by the arrows on the fuel tank, run the tube along the bottom edge of the fuel tank, and then pull it out on the fuel pump side.

27. Install a new gasket to the fuel tank.

28. Install fuel suction tube assembly.

a. Line up the pipe and connector and push them together until a "click" sound is heard.

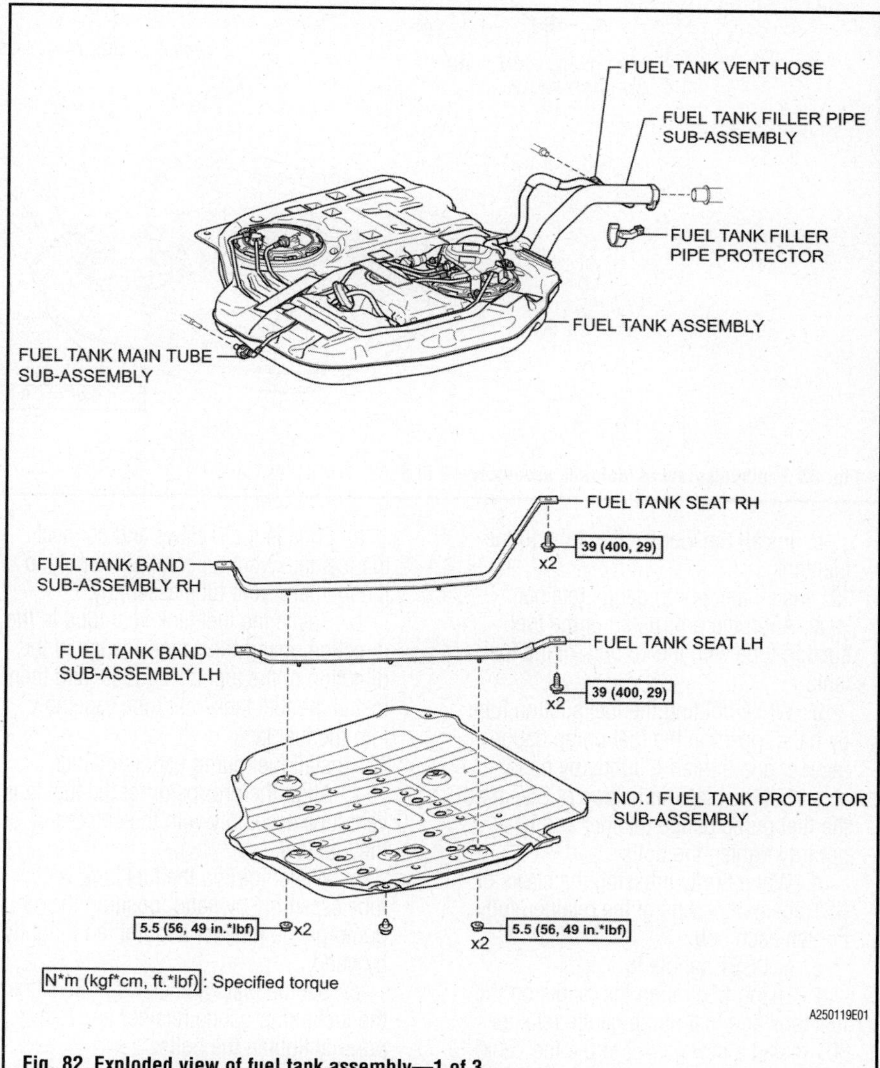

Fig. 82 Exploded view of fuel tank assembly—1 of 3

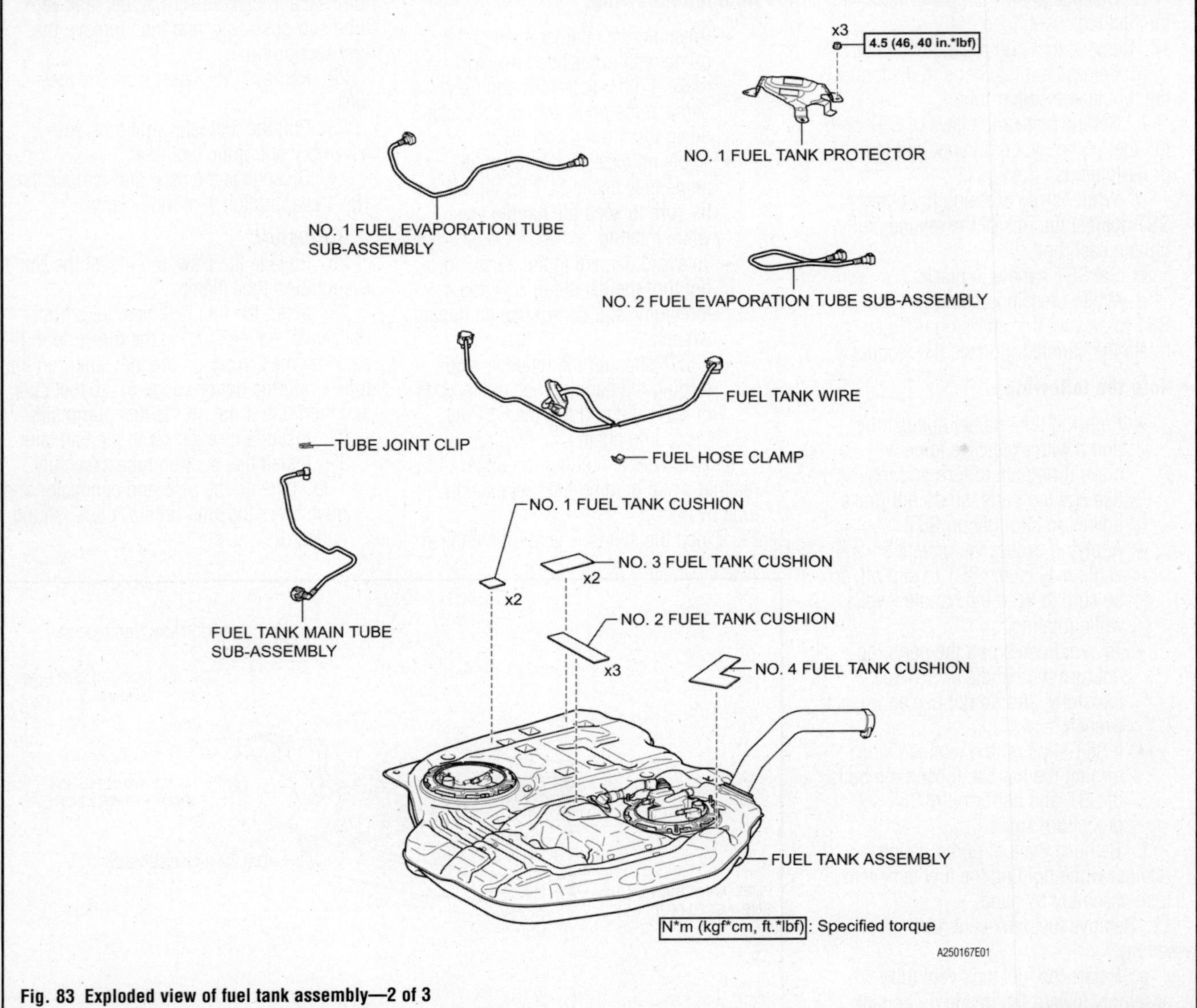

Fig. 83 Exploded view of fuel tank assembly—2 of 3

b. Install the fuel suction tube to the fuel tank.

29. Install fuel pump gauge retainer.

a. Align the protrusion of the fuel suction tube with the cutouts of the fuel tank.

b. While holding the fuel suction tube by hand, position the fuel pump gauge retainer and tighten it lightly by hand.

c. Set the plate and claws of SST onto the fuel pump gauge retainer and temporarily tighten the bolts.

d. While firmly pressing the claws of SST against the ribs of the retainer, fully tighten each bolt.

e. Set SST handle in place.

f. Using SST, align the marks on the fuel tank and fuel pump gauge retainer.

30. Install a new gasket to the fuel tank.

31. Install fuel tank vent tube assembly.

a. Engage the 2 claws and connect the fuel tank vent tube sub-assembly to the fuel tank vent tube assembly.

b. Rotate the fuel tank vent tube in the direction, insert the sender gauge in the direction of the fuel tank gauge, and then install the fuel tank vent tube assembly into the fuel tank.

32. Install fuel pump gauge retainer.

a. Align the protrusion of the fuel tank vent tube assembly with the cutouts of the fuel tank.

b. While holding the fuel tank vent tube assembly by hand, position the fuel pump gauge retainer and tighten it lightly by hand.

c. Set the plate and claws of SST onto the fuel pump gauge retainer and temporarily tighten the bolts.

d. While firmly pressing the claws of SST against the ribs of the retainer, fully tighten each bolt.

e. Set SST handle in place.

f. Using SST, align the marks on the fuel tank and fuel pump gauge retainer.

g. Connect the fuel tank breather tube to the fuel tank vent tube assembly and lock it by pushing in the retainer.

33. Install fuel tank vent hose.

a. Connect the fuel tank vent hose to the fuel tank vent tube assembly and lock it by pushing in the retainer.

b. Engage the fuel tank vent hose.

34. Install No. 2 fuel evaporation tube sub-assembly.

a. Connect the No. 2 fuel evaporation tube to the fuel tank vent tube assembly and lock it by pushing in the retainer.

b. Engage the clamp and install the No. 2 fuel evaporation tube.

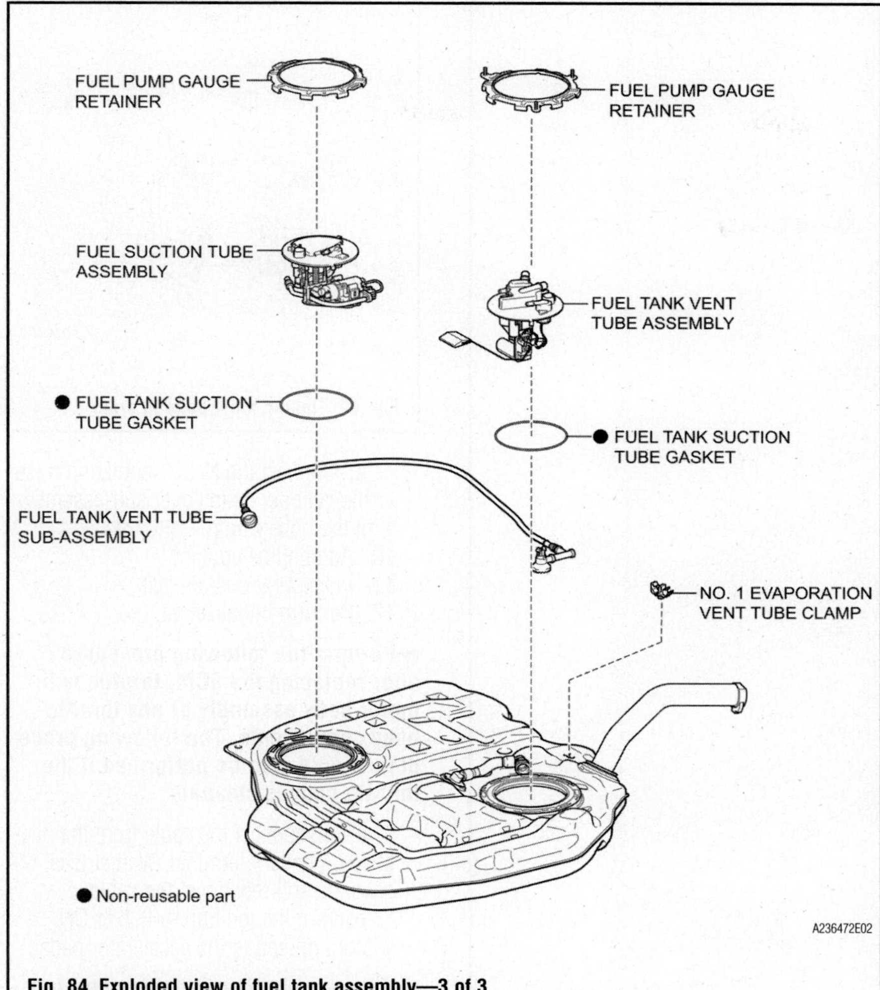

Fig. 84 Exploded view of fuel tank assembly—3 of 3

35. Install No. 1 fuel evaporation tube sub-assembly.
    a. Connect the No. 1 fuel evaporation tube to the fuel suction tube assembly and fuel tank vent tube assembly and lock them by pushing in the retainers.

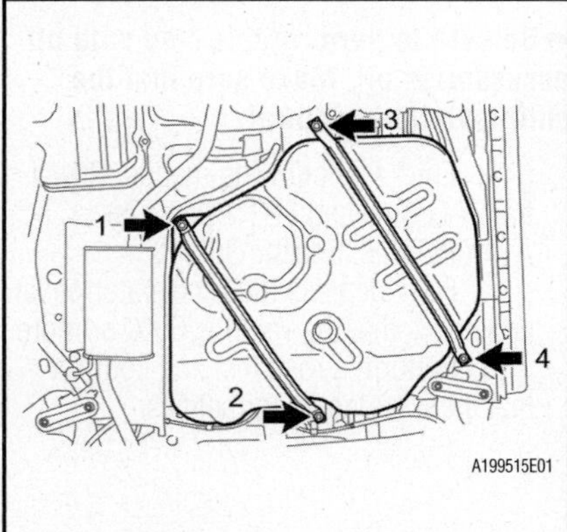

Fig. 85 Tighten the 4 bolts in sequence to 29 ft. lbs. (39 Nm), and secure the 2 fuel tank bands and the fuel tank

    b. Engage the clamp and install the No. 1 fuel evaporation tube.
36. Install fuel tank wire.
    a. Install the fuel hose clamp to the fuel tank wire.
    b. Connect the fuel tank wire connectors to the fuel suction tube and fuel tank vent tube assembly.
    c. Engage the 7 clamps and install the fuel tank wire.
37. Install fuel tank main tube sub-assembly.
    a. Insert the fuel tank main tube into the fuel suction tube plug and lock it with the tube joint clip.
    b. Engage the clamp and install the fuel tank main tube.
38. Install the No. 1 fuel tank cushions, No. 2 fuel tank cushions, No. 3 fuel tank cushions and the No. 4 fuel tank cushion.
39. Install the fuel tank protector with the 3 nuts.
40. Install fuel tank assembly.
    a. Clean and degrease the bolt holes.
    b. Set the fuel tank and 2 fuel tank bands on the engine lifter.
    c. Slowly raise the engine lifter and install the fuel tank onto the vehicle.
    d. Tighten the 4 bolts in sequence to 29 ft. lbs. (39 Nm), and secure the 2 fuel tank bands and the fuel tank.
41. Engage the claw and install the fuel tank filler pipe protector.
42. Connect the fuel tank breather tube and lock it by pushing in the retainer.
43. Connect the fuel tank vent hose and lock it by pushing in the retainer.
44. Install fuel tank main tube sub-assembly.
    a. Line up the pipe and connector and push them together until a "click" sound is heard.
    b. Push the checker in to lock it.
45. Engage the claw and install the No. 1 fuel tank protector with the 2 clips and 4 nuts.
46. Pull the fuel tank wire through the floor, install the grommet in the floor, and connect the fuel tank wire connector.
47. Add fuel
48. Inspect for fuel leak

IDLE SPEED

ADJUSTMENT

Idle speed is controlled by the Engine Control Module (ECM). No adjustment is necessary or possible.

THROTTLE BODY

REMOVAL & INSTALLATION

See Figures 86 and 87.

1. Drain engine coolant.
2. Disconnect No. 1 air cleaner hose.
    a. Loosen the hose clamp and disconnect the No. 2 ventilation hose from the cylinder head cover sub-assembly.
    b. Loosen the hose clamp and disconnect the No. 1 air cleaner hose from the throttle with motor body assembly.
3. Loosen the hose clamp and disconnect the No. 2 water by-pass hose from the throttle with motor body assembly.
4. Loosen the hose clamp and disconnect the water by-pass hose assembly from the throttle with motor body assembly.
5. Remove throttle with motor body assembly.
    a. Disconnect the connector of the throttle with motor body assembly.
    b. Remove the 4 bolts and the throttle with motor body assembly from the upper intake manifold sub-assembly.

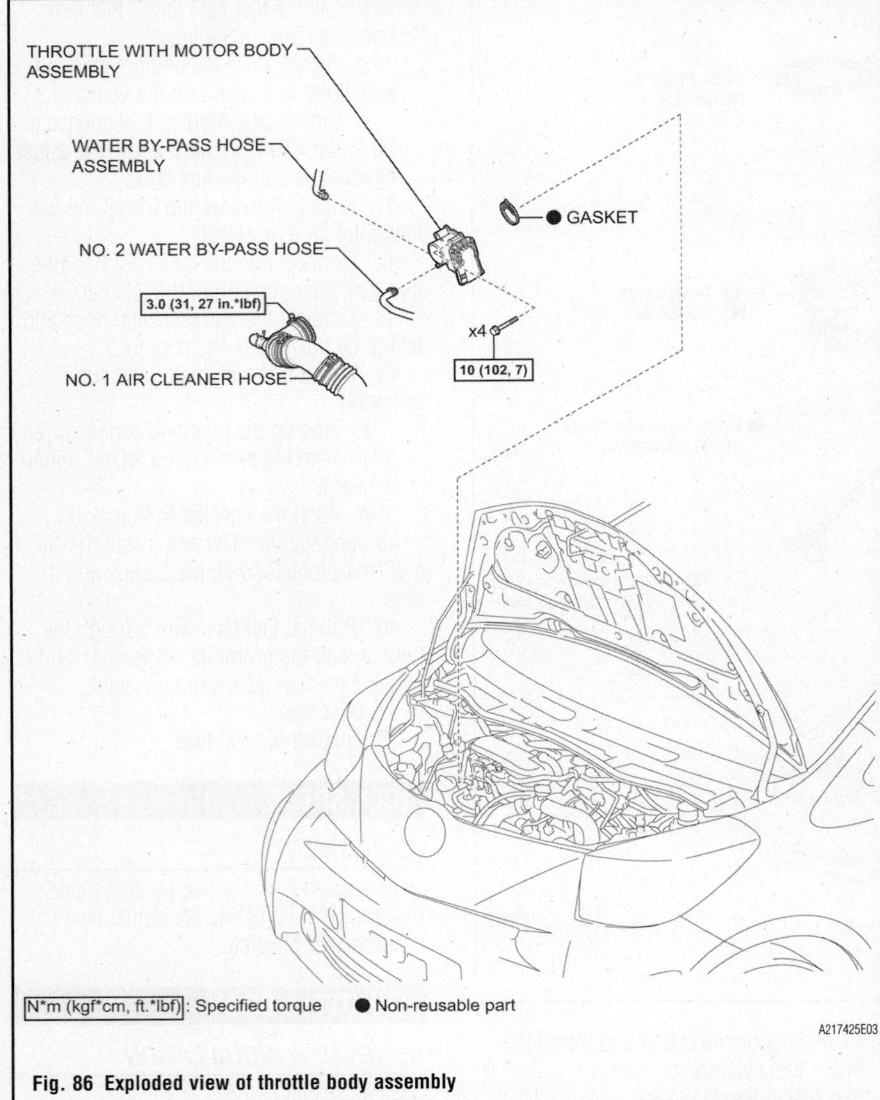

Fig. 86 Exploded view of throttle body assembly

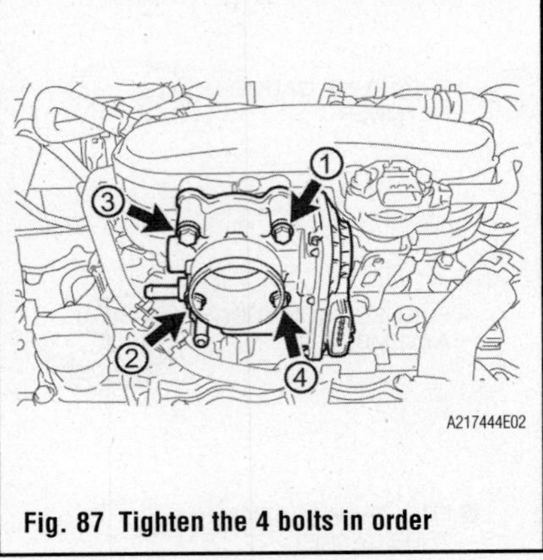

Fig. 87 Tighten the 4 bolts in order

➡If a component has been dropped or subjected to a strong impact, replace it.

c. Remove the gasket from the upper intake manifold sub-assembly.

To install:

6. Install throttle with motor body assembly.

a. Install a new gasket onto the upper intake manifold sub-assembly.

b. Temporarily install the throttle with motor body assembly onto the upper intake manifold sub-assembly with the 4 bolts.

c. Tighten the 4 bolts in order. Tighten to 7 ft. lbs. (10 Nm).

d. Connect the connector of the throttle with motor body assembly.

7. Connect the water by-pass hose assembly to the throttle with motor body assembly with the hose clamp.

8. Connect the No. 2 water by-pass hose to the throttle with motor body assembly with the hose clamp.

9. Connect No. 1 air cleaner hose.

a. Connect the No. 1 air cleaner hose to the throttle with motor body assembly with the hose clamp.

b. Connect the No. 2 ventilation hose to the cylinder head cover sub-assembly with the hose clamp.

10. Add engine coolant.

11. Inspect for coolant leak.

12. Perform initialization.

➡Perform the following procedure after replacing the ECM, throttle with motor body assembly or any throttle body components. The following procedure should also be performed if the throttle body is cleaned.

a. Disconnect the cable from the negative (-) battery terminal. Wait at least 60 seconds and reconnect the cable.

b. Turn the ignition switch to ON without operating the accelerator pedal.

➡If the accelerator pedal is operated, perform the above steps again.

c. Connect the Techstream to the DLC3 and clear the DTCs.

d. Start the engine and check that the MIL is not illuminated and that the idle speed is within the specified range when the A/C is switched off after the engine is warmed up.

➡Be sure to perform this step with all accessories off. Make sure that the shift lever is in neutral.

e. Enter the following menus: Powertrain / Engine and ECT / Data List / Throttle Pos. Sensor Output.

f. Fully depress the accelerator pedal and check that the value is 60% or more.

g. Perform a road test and confirm that there are no abnormalities.

HEATING & AIR CONDITIONING SYSTEM

BLOWER MOTOR

REMOVAL & INSTALLATION

See Figure 88.

1. Remove front door scuff plates LH and RH.
2. Remove cowl side trim board LH and RH.
3. Remove front door opening trim weatherstrip LH and RH.
4. Remove lower instrument panel assembly.
5. Remove No. 3 air duct sub-assembly
   a. Remove the clip.
   b. Disengage the 3 claws and remove the air duct.
6. Remove No. 2 air duct sub-assembly.
7. Remove blower motor with fan sub-assembly.
   a. Disconnect the connector.
   b. Remove the 3 screws and remove the blower motor with fan sub-assembly.

To install:

8. Install blower motor with fan sub-assembly.
   a. Install the blower motor with fan sub-assembly with the 3 screws.
   b. Connect the connector.
9. Install No. 2 air duct sub-assembly.
10. Install No. 3 air duct sub-assembly.
    a. Engage the 3 claws and install the air duct.
    b. Install the clip.
11. Install lower instrument panel assembly.
12. Install front door opening trim weatherstrips.
13. Install cowl side trim boards.
14. Install front door scuff plates.

HEATER CORE

REMOVAL & INSTALLATION

See Figure 89.

➡The manufacturer does not provide a specific Removal and Installation procedure for this component. Refer to the graphic(s) when servicing this component.

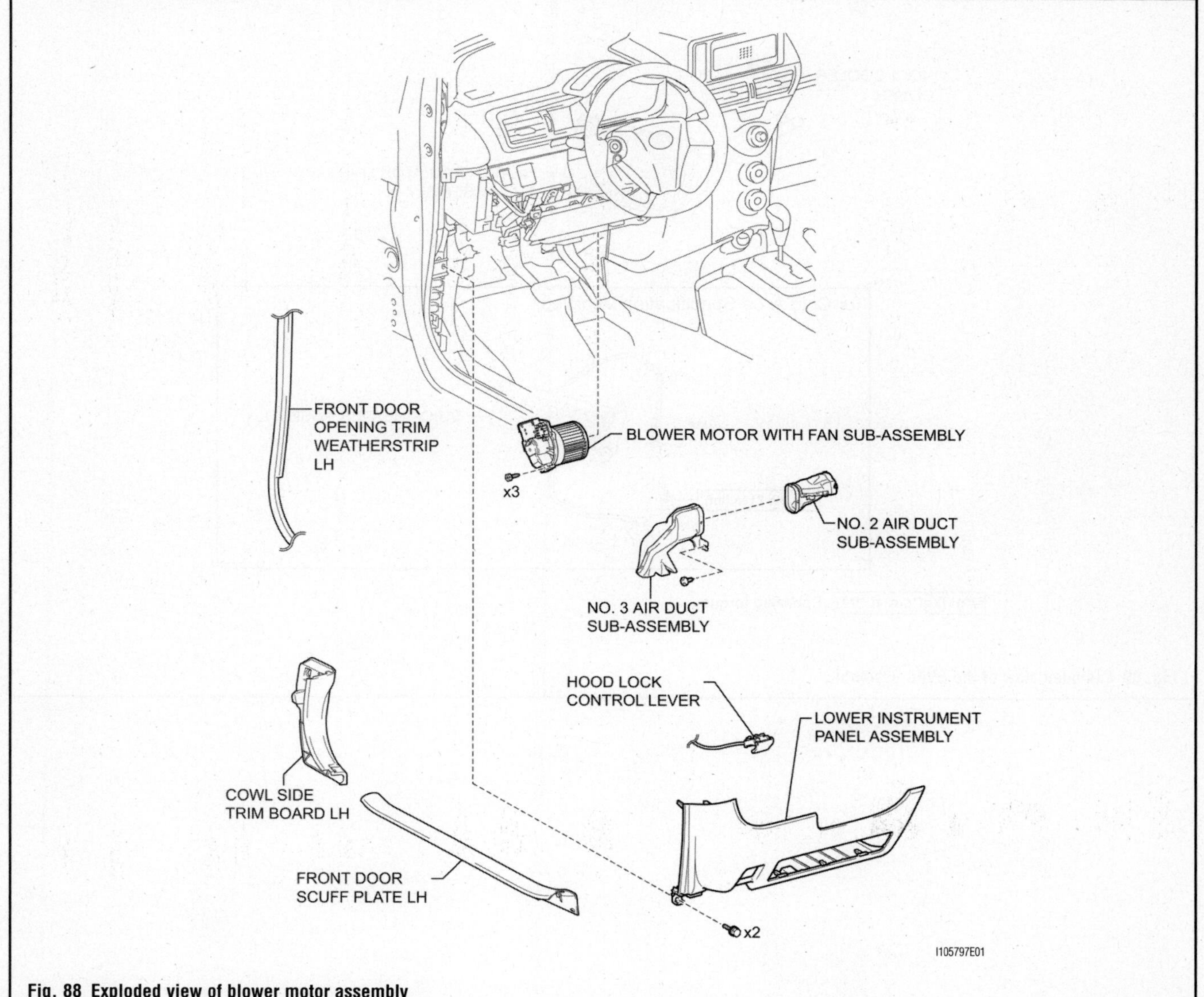

Fig. 88 Exploded view of blower motor assembly

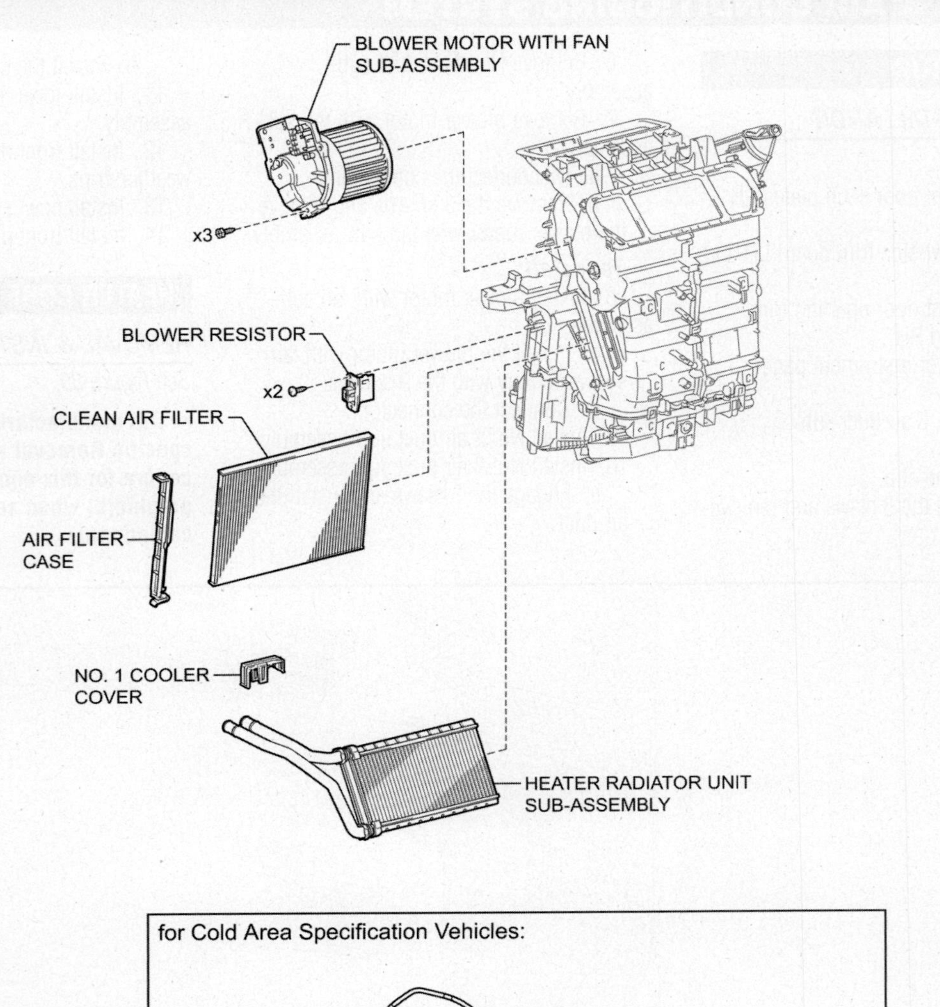

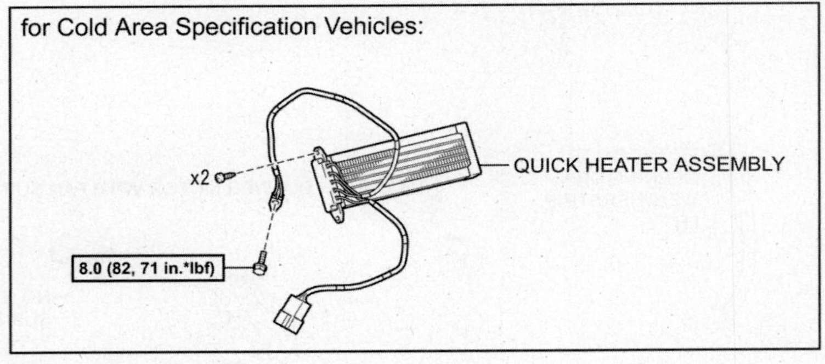

N*m (kgf*cm, ft.*lbf): Specified torque

I105787E01

Fig. 89 Exploded view of the HVAC assembly

STEERING

POWER STEERING GEAR

REMOVAL & INSTALLATION

See Figures 90 and 91.

The power steering system generates torque through the operation of the motor and the reduction gear installed on the column shaft in order to assist power steering effort.

The power steering ECU determines the direction and the amount of assisting power in accordance with the vehicle speed signals and signals from the power steering torque sensor. As a result, steering effort is controlled to be light during low speed driving and moderately high during high speed driving.

➡After turning the ignition switch off, waiting time may be required before disconnecting the cable from the battery terminal. Therefore, make sure to read the disconnecting the cable from the battery terminal notice before proceeding with work.

1. Disconnect cable from negative battery terminal.

CAUTION

Wait at least 90 seconds after disconnecting the cable from the negative (-) battery terminal to disable the SRS system.

2. Place front wheels facing straight ahead.
3. Remove front door scuff plate LH and RH.
4. Remove cowl side trim board LH and RH.
5. Remove front door opening trim weatherstrip LH and RH.
6. Remove lower instrument panel assembly.
7. Remove knee airbag assembly.
8. Remove No. 3 air duct sub-assembly.
9. Separate No. 2 steering intermediate shaft assembly.
   a. Secure the steering wheel with the seat belt in order to prevent it from rotating.

➡This operation is necessary to prevent damage to the spiral cable.

   b. Put matchmarks on the No. 2 steering intermediate shaft assembly and the steering link assembly.
   c. Loosen the bolt A, remove the bolt B and separate the No. 2 steering intermediate shaft assembly.

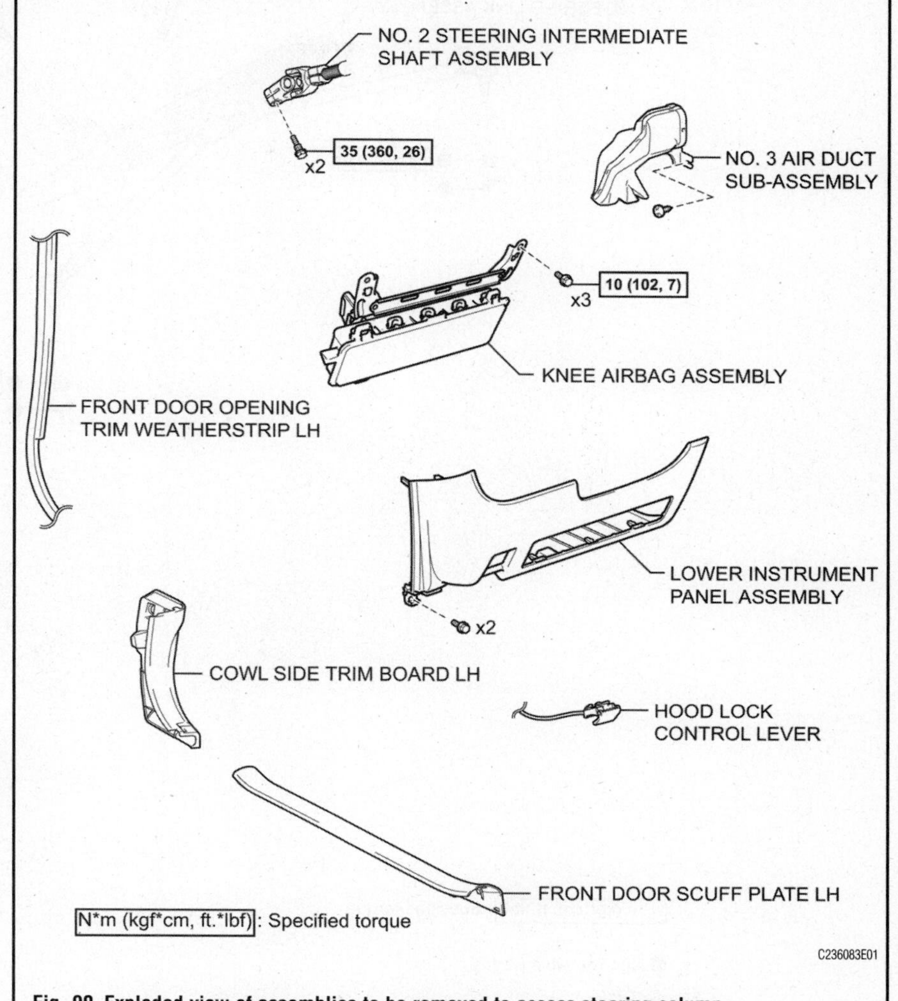

Fig. 90 Exploded view of assemblies to be removed to access steering column

10. Disengage the clip A, detach the clip B from the body and separate the steering column hole cover sub-assembly.
11. Remove front wheels.
12. Remove intake manifold.
13. Separate tie rod end sub-assembly LH.
14. Separate tie rod end sub-assembly RH.
15. Remove steering link assembly.
   a. Disengage the 2 claws and separate the steering column hole cover sub-assembly from the steering link assembly.
   b. Remove the 2 bolts and the steering link assembly from the body.
   c. Remove the steering link assembly through the right wheel housing.

To install:

16. Install steering column hole cover sub-assembly.
17. Install steering link assembly.
   a. Insert the steering link assembly into the engine room through the right wheel housing.
   b. Install the steering link assembly to the body with the 2 bolts. Tighten to 64 ft. lbs. (87 Nm).
   c. Engage the 2 claws and connect the steering column hole cover sub-assembly to the steering link assembly.
18. Connect tie rod end sub-assembly LH.
19. Connect tie rod end sub-assembly RH.
20. Install intake manifold.
21. Install front wheels.
22. Place front wheels facing straight ahead.
23. Align the matchmarks on the No. 2 steering intermediate shaft assembly and the steering link assembly to connect the

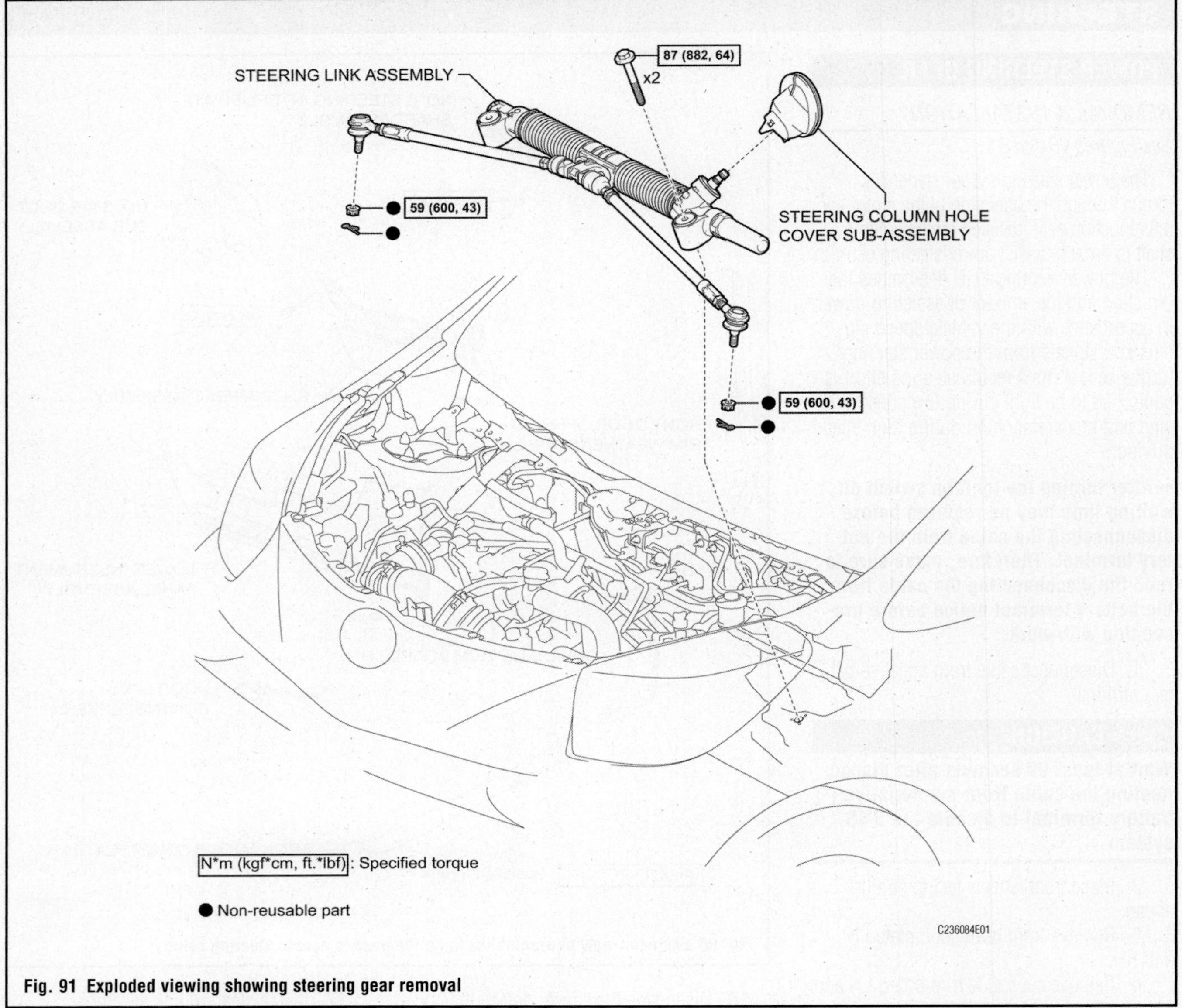

Fig. 91 Exploded viewing showing steering gear removal

No. 2 steering intermediate shaft assembly to the steering link assembly. Tighten the bolts to 26 ft. lbs. (35 Nm).

24. Install No. 3 air duct sub-assembly.
25. Install knee airbag assembly.
26. Install lower instrument panel assembly.
27. Install front door opening trim weatherstrips.
28. Install cowl side trim boards.
29. Install front door scuff plates.
30. Connect cable to negative battery terminal.
31. Inspect SRS warning light.
32. Inspect for exhaust gas leak.
33. Inspect and adjust front wheel alignment.

LOWER CONTROL ARM

REMOVAL AND & INSTALLATION

See Figure 92.

1. Remove front wheel.
2. Remove the 3 clips and the No. 3 engine under cover RH.
3. Remove front lower suspension arm sub-assembly LH.
   a. Remove the clip and castle nut.
   b. Install SST (spacer B) to the threaded section of the lower ball joint.

➡Make sure the upper ends of the threaded section of the lower ball joint and SST (spacer B) are aligned.

   c. Using SST (09960-20010), separate the front lower suspension arm sub-assembly.
   d. Remove the 2 bolts, the nut, and the front lower suspension arm sub-assembly.

To install:

4. Temporarily tighten front lower suspension arm sub-assembly LH.
   a. Provisionally tighten the front lower suspension arm sub-assembly with 2 new bolts and the nut.
   b. Install the front lower suspension arm sub-assembly onto the steering knuckle.
   c. Using a ball joint lock nut wrench, fully tighten a new castle nut to 72 ft. lbs. (98 Nm).
   d. Install a new clip.
5. Install front wheel.
6. Position wheels facing straight ahead.
7. Stabilize suspension.
   a. Lower the vehicle from the jack.
   b. Bounce the vehicle up and down several times to stabilize the suspension.
8. Fully tighten front lower suspension arm sub-assembly LH. Fully tighten the 2 bolts and the nut to 107 ft. lbs. (145 Nm).
9. Install the 3 clips and the No.3 engine under cover.
10. Inspect and adjust front wheel alignment.

STABILIZER BAR

REMOVAL & INSTALLATION

See Figure 93.

1. Remove front frame assembly.
2. Remove the nut and the front stabilizer link assembly LH.
3. Remove front stabilizer link assembly RH.
4. Remove the 2 bolts and the stabilizer bracket LH.
5. Remove front stabilizer bracket RH.
6. Remove front stabilizer bar.

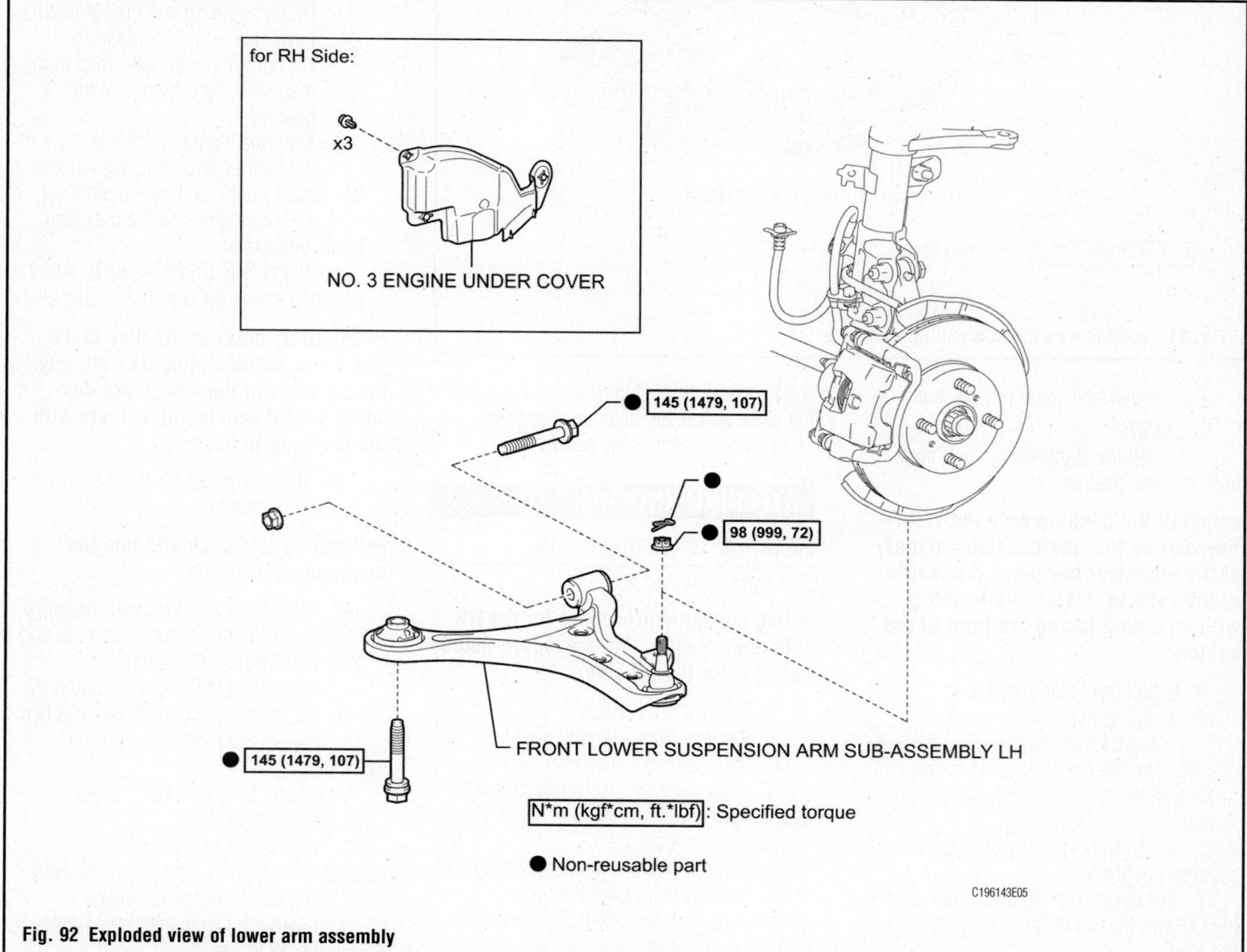

Fig. 92 Exploded view of lower arm assembly

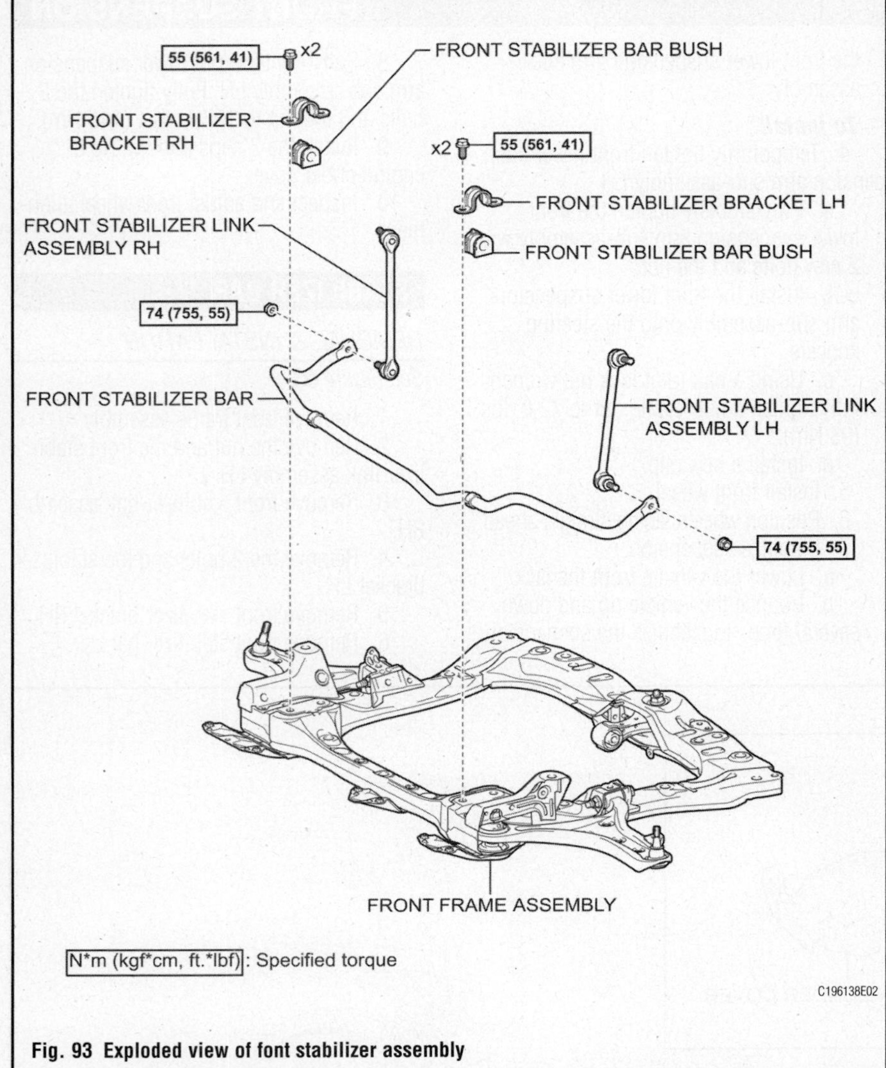

Fig. 93 Exploded view of font stabilizer assembly

7. Remove front stabilizer bar bushing.

To install:

8. Install the stabilizer bar bushings onto the stabilizer bar.

➡Install the bushing onto the stabilizer bar so that the bushing stopper of the stabilizer bar faces the inside of the vehicle. Install the bushing with its cutout facing the front of the vehicle.

9. Install front stabilizer bar.
10. Install front stabilizer bracket LH.
   a. Provisionally tighten the rear bolt.
   b. Tighten the bolts to 41 ft. lbs. (55 Nm), in the order of front bolt, then rear bolt.
11. Install front stabilizer bracket RH same as LH bracket.
12. Install the front stabilizer link assembly LH with the nut. Tighten to 55 ft. lbs. (74 Nm).
13. Install front stabilizer link assembly RH same as LH stabilizer link assembly.
14. Install front frame assembly.

STEERING KNUCKLE

REMOVAL & INSTALLATION

See Figures 94 and 95.

➡Use the same procedure for the RH side and LH side. The procedure listed below is for the LH side.

1. Remove front wheel.
2. Remove front axle hub nut.
3. Separate front speed sensor.
4. Separate front disc brake caliper assembly.
5. Remove front disc.
6. Separate front lower suspension arm.
7. Remove front axle assembly.
8. Remove front axle hub hole snap ring.
9. Remove front axle hub sub-assembly.
10. Remove the dust cover by tapping the back side with a brass bar and hammer.
11. Remove steering knuckle.
   a. Install the removed hub bearing inner race onto the outer side of the hub bearing.
   b. Using SST (09513-36040) and a press, remove the hub bearing from the steering knuckle.

To install:

12. Install steering knuckle.
   a. Using SST (09950-60020 and 09950-70010) and a press, insert a new hub bearing, with its magnetic rotor side facing the inside of the vehicle, until it reaches the end of the steering knuckle.

➡Note the following:

- Do not remove the inner race because the hub bearing is built into the oil seal.
- Do not use bearings that have been removed.
- Do not wipe off any grease that has been applied to new bearings.
- Do not bring magnets close to the magnetic rotor surface of the bearing.
- Keep the magnetic rotor surface of the bearing free of foreign matter.

13. Install front disc brake dust cover.
   a. Provisionally install a new disc brake dust cover.
   b. Using SST (09223-56010) and a hammer, install the disc brake dust cover.

➡Uniformly press in the disc brake dust cover while sliding SST slightly. Surely press in the disc brake dust cover until it comes into contact with the pressing-in base.

   c. Using a chisel, fix the 3 points on the circumference.

➡Securely fold each end into the engagement grooves.

14. Install front axle hub sub-assembly.
15. Install front axle hub hole snap ring.
16. Install front axle assembly.
17. Install front lower suspension arm. Tighten the castle nut to 72 ft. lbs. (98 Nm).
18. Inspect front axle hub bearing looseness.
19. Inspect front axle hub runout.
20. Install front disc.
21. Install front disc brake caliper assembly. Tighten to 79 ft. lbs. (107 Nm).
22. Connect front speed sensor.
23. Install front axle shaft nut. Tighten to 159 ft. lbs. (216 Nm).

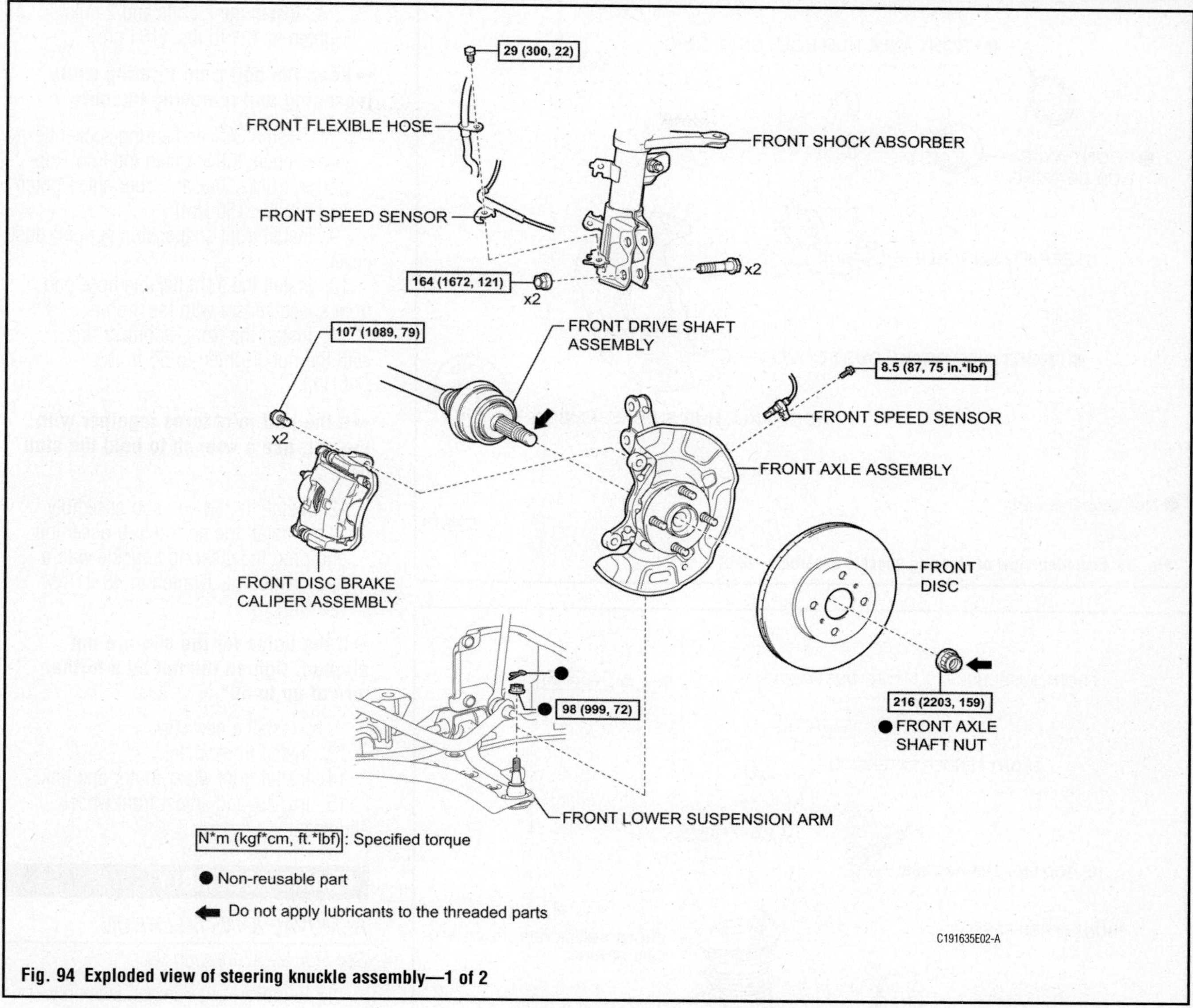

Fig. 94 Exploded view of steering knuckle assembly—1 of 2

24. Install front wheel.
25. Inspect and adjust front wheel alignment.
26. Check for speed sensor signal.

STRUT & SPRING ASSEMBLY

REMOVAL & INSTALLATION

See Figure 96.

1. Remove front wiper motor and link.
2. Remove front wheel.
3. Remove the 3 bolts and separate the front speed sensor and front flexible hose.
4. Remove the nut and separate the stabilizer link assembly from the front shock absorber.

➡If the ball joint turns together with the nut, use a wrench to hold the stud bolt.

5. Separate tie rod end sub-assembly.
   a. Remove the cotter pin and castle nut.
   b. Install SST (spacer B) to the threaded section of the tie rod end sub-assembly.
   c. Using SST, separate the tie rod end from the front axle assembly.
6. Remove front suspension support dust cover.
7. Remove front shock absorber with coil spring.
   a. Using SST and a long socket hexagon wrench, loosen the front support to front shock absorber nut of the front shock absorber.

➡Do not remove the front support to front shock absorber nut. Loosen the nut only when the front shock absorber with coil spring needs to be disassembled.

   b. Remove the 2 nuts and 2 bolts and separate the front shock absorber with coil spring from the steering knuckle.

➡Keep the bolt from rotating while loosening and removing the nuts.

   c. Remove the 3 nuts and the front shock absorber with coil spring.

To install:

8. Install front shock absorber with coil spring.
   a. Loosen the 2 bolts from the front fender extension.
   b. Install the 3 new nuts and the front shock absorber with coil spring. Tighten to 29 ft. lbs. (39 Nm).
   c. Tighten the 2 bolts to 13 ft. lbs. (17 nm).
   d. Install the front shock absorber with coil spring onto the steering knuckle.

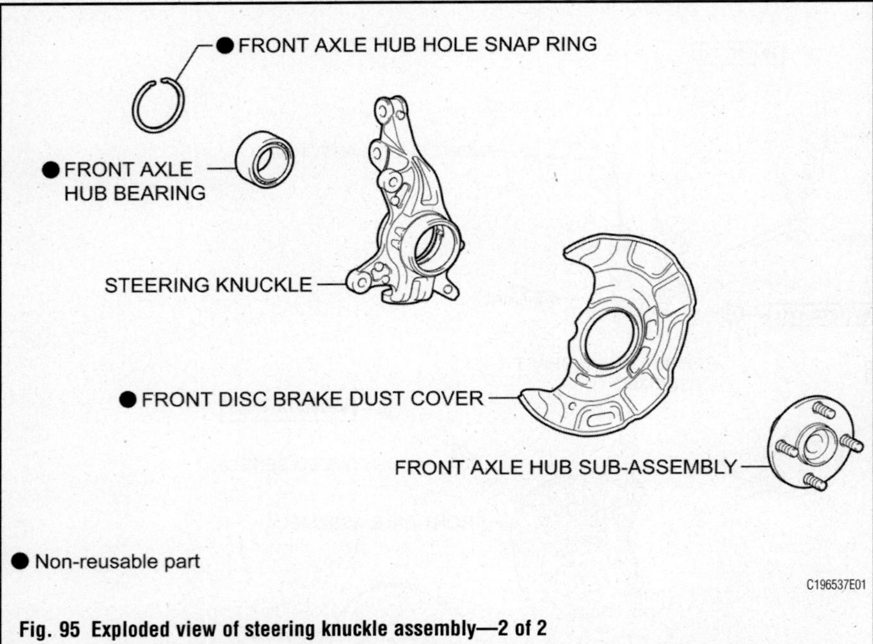

Fig. 95 Exploded view of steering knuckle assembly—2 of 2

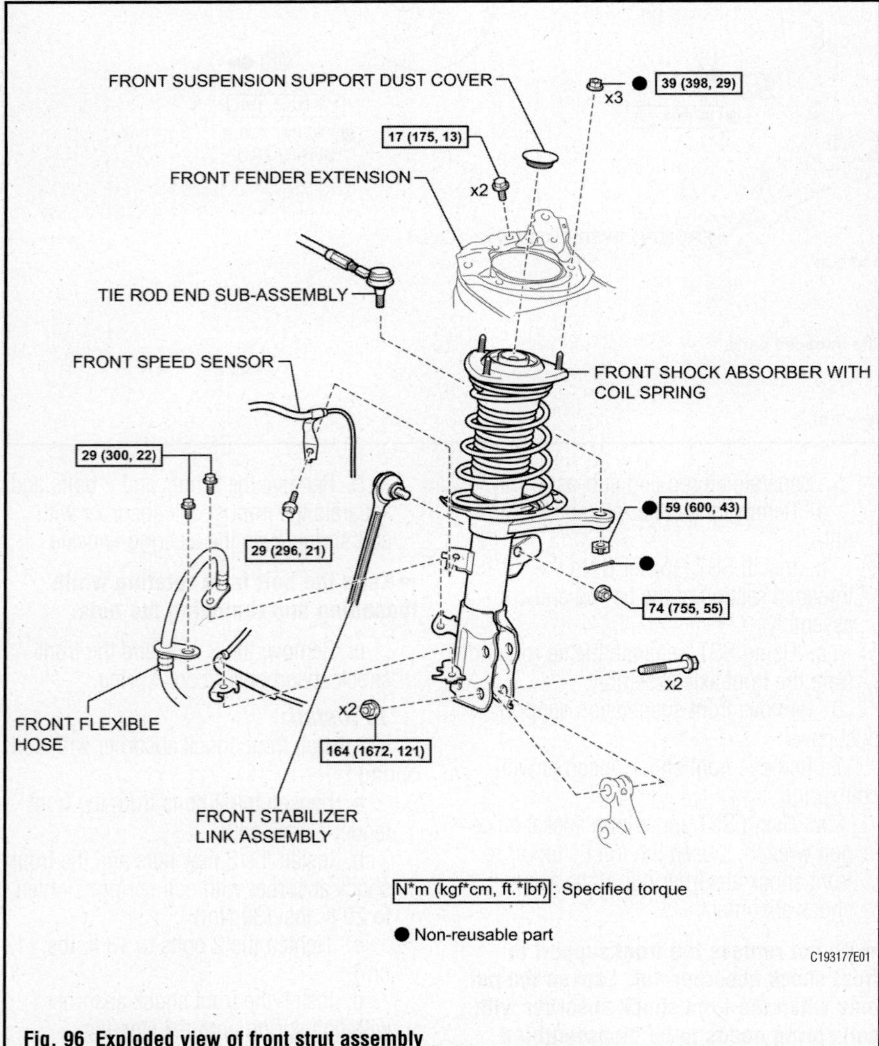

Fig. 96 Exploded view of front strut assembly

e. Install the 2 bolts and 2 nuts. Tighten to 121 ft. lbs. (164 nm).

➡Keep the bolt from rotating while loosening and removing the nuts.

f. Using SST and a long socket hexagon wrench, fully tighten the front support to front shock absorber nut. Tighten to 44 ft. lbs. (59 Nm).

9. Install front suspension support dust cover.
10. Install the front flexible hose and front speed sensor with the 3 bolts.
11. Install the front stabilizer link with the nut. Tighten to 55 ft. lbs. (74 Nm).

➡If the ball joint turns together with the nut, use a wrench to hold the stud bolt.

12. Install tie rod end sub-assembly.
    a. Install the tie rod sub-assembly end onto the steering knuckle with a new castle nut. Tighten to 43 ft. lbs. (59 Nm).

➡If the holes for the clip are not aligned, tighten the nut by a further turn of up to 60°.

    b. Install a new clip.
13. Install front wheel.
14. Install front wiper motor and link.
15. Inspect and adjust front wheel alignment.

FRAME ASSEMBLY

REMOVAL & INSTALLATION

See Figures 97 through 99.

1. Remove front bumper assembly.
2. Remove the 4 clips and the front floor cover LH.
3. Remove the 4 clips and the front floor cover RH.
4. Remove the 3 clips and the No.3 engine under cover.
5. Remove the 3 bolts and the No. 2 engine under cover.
6. Remove front wheel.
7. Separate front stabilizer link assembly LH.
8. Separate front stabilizer link assembly RH.
9. Separate front lower suspension arm sub-assembly LH.
10. Separate front lower suspension arm sub-assembly RH.
11. Remove front frame assembly.
    a. Pass a strong rope through hole in the engine mounting control bracket and tie it securely to the body.

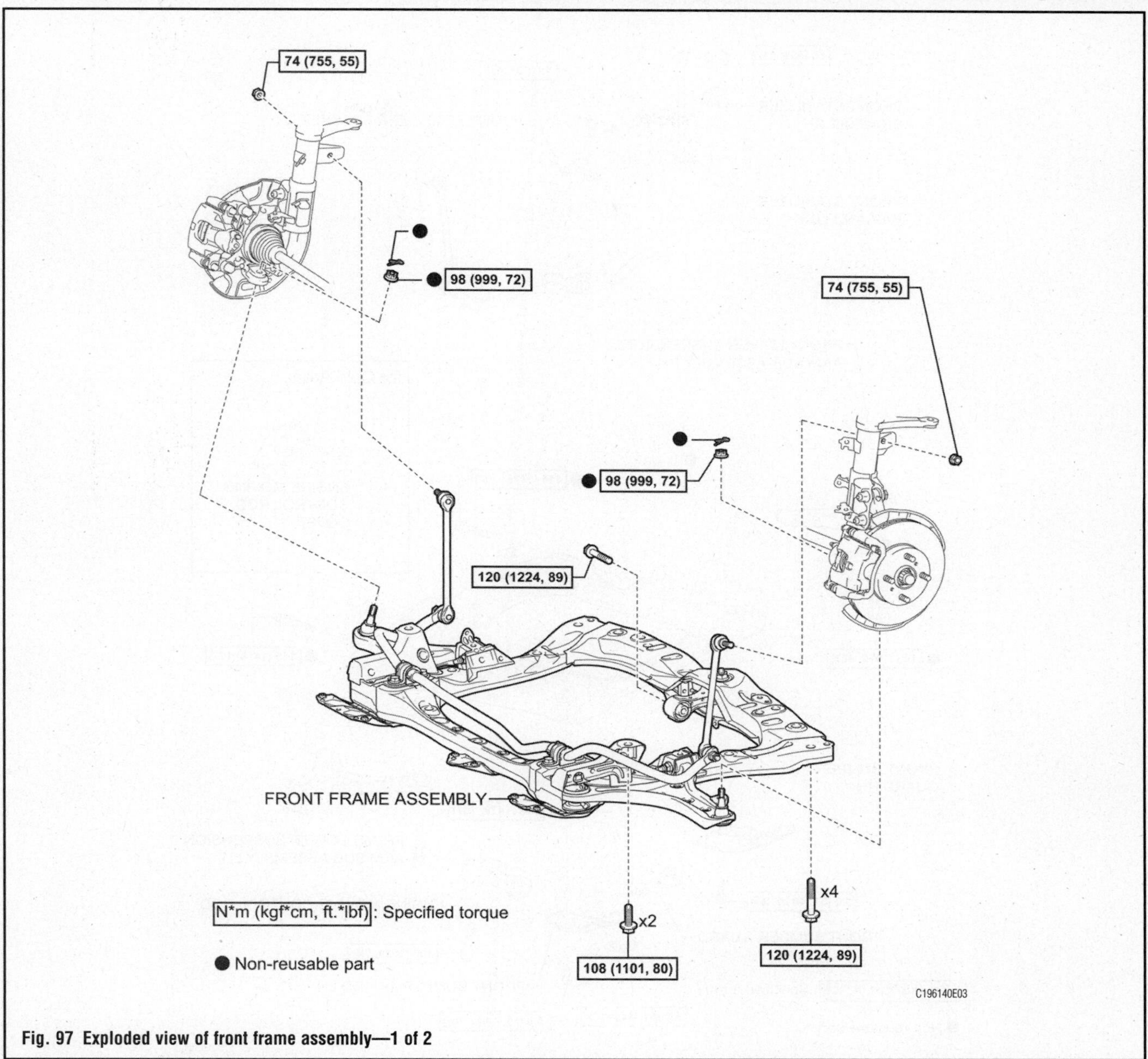

Fig. 97 Exploded view of front frame assembly—1 of 2

WARNING

When the engine moving control rod is separated, the engine assembly moves towards the front of the vehicle and will contact components near the engine if not fastened tightly.

b. Remove the bolt and separate the engine moving control rod from engine mounting control bracket.

c. Remove the 2 bolts.

d. Support the front frame assembly with 4 wooden blocks and the jack.

e. Remove the 4 bolts and remove the front frame assembly.

To install:

12. Install front frame assembly.

a. Support the front frame assembly with 4 wooden blocks and the jack.

b. Provisionally install the front frame assembly onto the body with the 6 bolts.

c. By inserting SST into the datum holes in the front frame assembly RH and LH alternately, tighten bolts A to 80 ft. lbs. (108 Nm) and bolts B to 89 ft. lbs. (120 Nm) on both sides, in several steps.

d. Install the engine moving control rod to engine mounting control bracket with the bolt. Tighten to 89 ft. lbs. (120 Nm).

e. Remove the strong rope from engine mounting control bracket and body.

13. Install front lower suspension arm sub-assembly LH.

14. Install front lower suspension arm sub-assembly RH.

15. Install front stabilizer link assembly LH.

16. Install front stabilizer link assembly RH.

17. Install front wheel.

18. Position wheels facing straight ahead.

19. Stabilize suspension.

20. Fully tighten front lower suspension arm sub-assembly LH.

21. Fully tighten front lower suspension arm sub-assembly RH.

22. Install the No. 2 engine under cover and the 3 bolts.

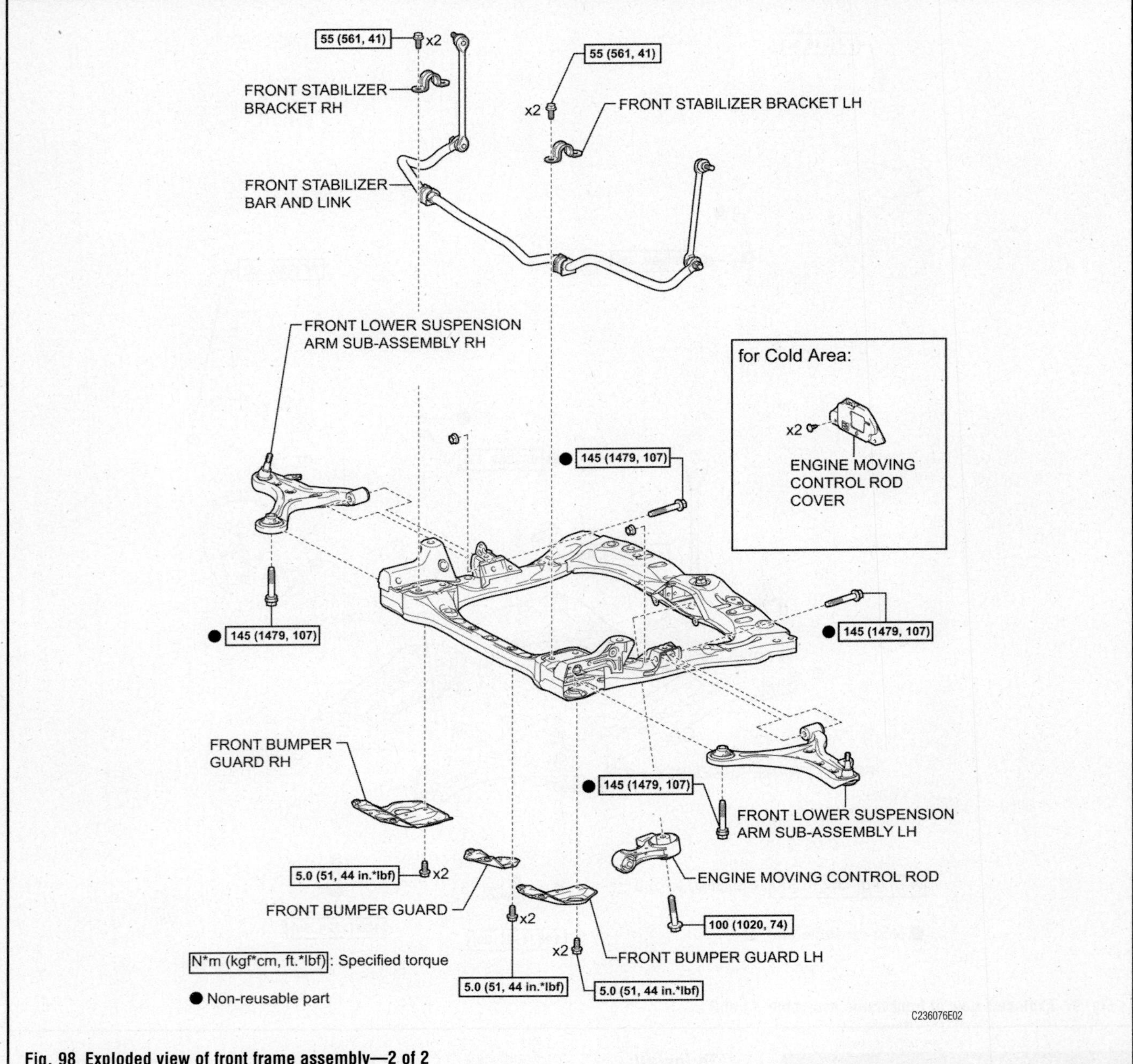

Fig. 98 Exploded view of front frame assembly—2 of 2

23. Install the 3 clips and the No.3 engine under cover.
24. Install the 4 clips and the front floor cover LH.
25. Install the 4 clips and the front floor cover RH.
26. Install front bumper assembly
27. Inspect and adjust front wheel alignment
28. Check for speed sensor signal

WHEEL BEARINGS

REMOVAL & INSTALLATION

See Figures 100 and 101.

➡Use the same procedure for the RH side and LH side. The procedure listed below is for the LH side.

1. Remove front wheel.
2. Remove front axle hub nut.
3. Separate front speed sensor.
4. Separate front disc brake caliper assembly.
5. Remove front disc.
6. Separate front lower suspension arm.
7. Remove front axle assembly.
8. Remove front axle hub hole snap ring.
9. Remove front axle hub sub-assembly.
10. Remove the dust cover by tapping the back side with a brass bar and hammer.
11. Remove steering knuckle.
    a. Install the removed hub bearing inner race onto the outer side of the hub bearing.
    b. Using SST (09513-36040) and a press, remove the hub bearing from the steering knuckle.

To install:

12. Install steering knuckle.
    a. Using SST (09950-60020 and 09950-70010) and a press, insert a new hub bearing, with its magnetic rotor side

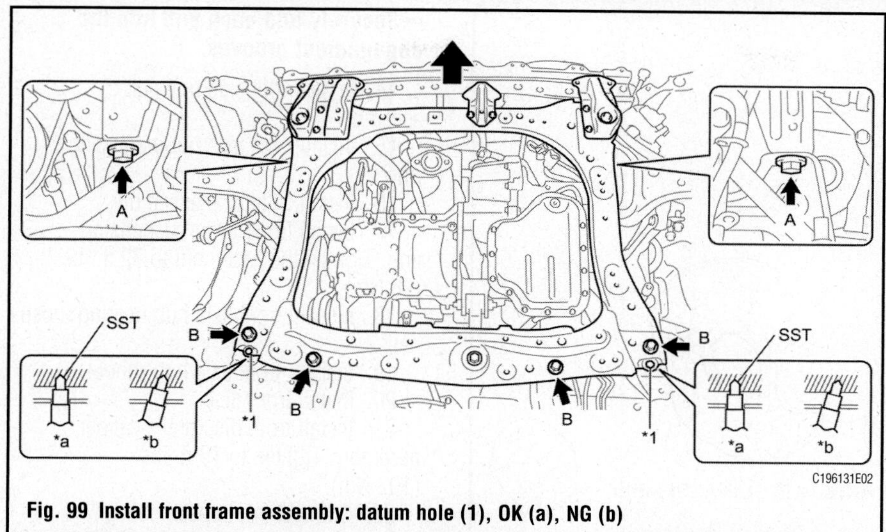

Fig. 99 Install front frame assembly: datum hole (1), OK (a), NG (b)

facing the inside of the vehicle, until it reaches the end of the steering knuckle.

➡

- Do not remove the inner race because the hub bearing is built into the oil seal.
- Do not use bearings that have been removed.
- Do not wipe off any grease that has been applied to new bearings.
- Do not bring magnets close to the magnetic rotor surface of the bearing.
- Keep the magnetic rotor surface of the bearing free of foreign matter.

13. Install front disc brake dust cover.
    a. Provisionally install a new disc brake dust cover.

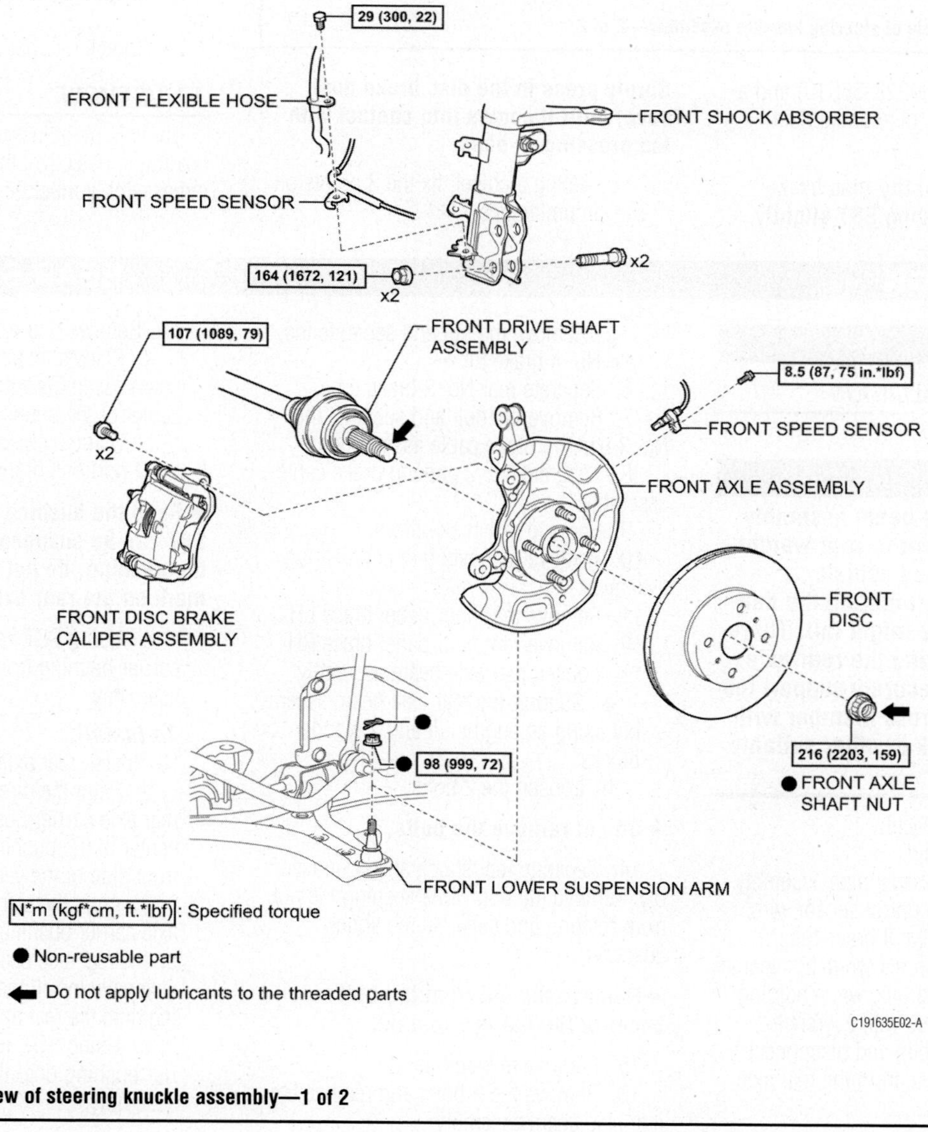

29 (300, 22)
FRONT FLEXIBLE HOSE
FRONT SHOCK ABSORBER
FRONT SPEED SENSOR
164 (1672, 121)
x2
x2
107 (1089, 79)
x2
FRONT DRIVE SHAFT ASSEMBLY
8.5 (87, 75 in.*lbf)
FRONT SPEED SENSOR
FRONT AXLE ASSEMBLY
FRONT DISC BRAKE CALIPER ASSEMBLY
FRONT DISC
98 (999, 72)
216 (2203, 159)
● FRONT AXLE SHAFT NUT
FRONT LOWER SUSPENSION ARM

N*m (kgf*cm, ft.*lbf): Specified torque

● Non-reusable part

⬅ Do not apply lubricants to the threaded parts

C191635E02-A

Fig. 100 Exploded view of steering knuckle assembly—1 of 2

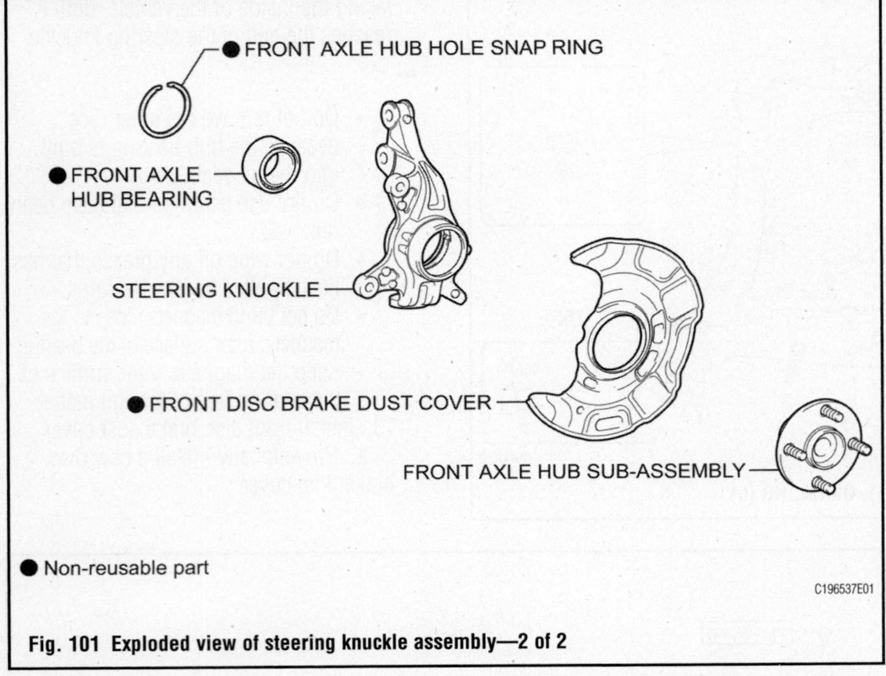

Fig. 101 Exploded view of steering knuckle assembly—2 of 2

b. Using SST (09223-56010) and a hammer, install the disc brake dust cover.

➡**Uniformly press in the disc brake dust cover while sliding SST slightly.**

Surely press in the disc brake dust cover until it comes into contact with the pressing-in base.

c. Using a chisel, fix the 3 points on the circumference.

➡**Securely fold each end into the engagement grooves.**

14. Install front axle hub sub-assembly.
15. Install front axle hub hole snap ring.
16. Install front axle assembly.
17. Install front lower suspension arm. Tighten the castle nut to 72 ft. lbs. (98 Nm).
18. Inspect front axle hub bearing looseness.
19. Inspect front axle hub runout.
20. Install front disc.
21. Install front disc brake caliper assembly. Tighten to 79 ft. lbs. (107 Nm).
22. Connect front speed sensor.
23. Install front axle shaft nut. Tighten to 159 ft. lbs. (216 Nm).
24. Install front wheel.
25. Inspect and adjust front wheel alignment.
26. Check for speed sensor signal.

ADJUSTMENT

There is no adjustment possible. If the bearing is defective, the front axle hub bearing assembly must be replaced.

SUSPENSION

REAR SUSPENSION

AXLE BEAM

REMOVAL & INSTALLATION

See Figures 102 through 104.

⁂ CAUTION

When the rear axle beam assembly is removed, the front-to-rear weight balance of the raised vehicle becomes unstable, creating the danger that the vehicle might fall. Therefore, before removing the rear axle beam assembly, securely support the front suspension cross member with a transmission jack or other suitable lift.

1. Remove rear wheel.
2. Drain brake fluid.
3. Remove tail exhaust pipe assembly.
4. Separate skid control sensor wire.
5. Separate rear No. 4 brake tube.
   a. Using a union nut wrench, separate the rear No. 4 brake tube while holding the rear flexible hose with a wrench.
   b. Remove the clip and disconnect the rear flexible hose from the rear axle beam assembly.
   c. Remove the nut and separate the rear No. 4 brake tube.
6. Separate rear No. 3 brake tube.
7. Remove the bolt and separate the No. 3 parking brake cable assembly.
8. Separate No. 2 parking brake cable assembly.
9. Remove rear brake drum.
10. Remove rear axle hub and bearing assembly.
11. Remove rear floor panel brace LH.
12. Remove rear floor panel brace RH.
13. Loosen rear axle beam assembly
   a. Support the rear axle beam assembly using an engine lift and 2 wooden blocks.
   b. Loosen the 2 bolts.

➡**Do not remove the bolts.**

14. Separate rear shock absorber assembly. Remove the bolt while keeping the nut from rotating and separate the shock absorber.

➡**Remove the nut from the bolt side because the nut is a jam nut.**

15. Remove rear coil spring.
16. Remove the 2 bolts and remove the rear axle beam assembly.
17. Remove rear axle carrier bushing.
   a. Place a matchmark on the rear axle beam assembly with it aligned with the center of the bushing groove.
   b. Using a chisel and hammer, bend the 2 portions of the bushing rib.

➡**Bend the bushing rib until the claw of SST can be suspended. When removing the bushing, do not rub off the matchmark on the rear axle beam assembly.**

   c. Using SST, remove the rear axle carrier bushing from the rear axle beam assembly.

To install:

18. Install rear axle carrier bushing.
   a. Paint the directional mark on a new rear axle carrier bushing, 135° from the center of the bushing groove toward the front side of the vehicle.
   b. Align the painting mark on the rear axle carrier bushing and the hole center on the rear axle beam assembly. And provisionally install the rear axle carrier bushing onto the rear axle beam assembly.
   c. Using SST, install the rear axle carrier bushing onto the rear axle beam assembly.

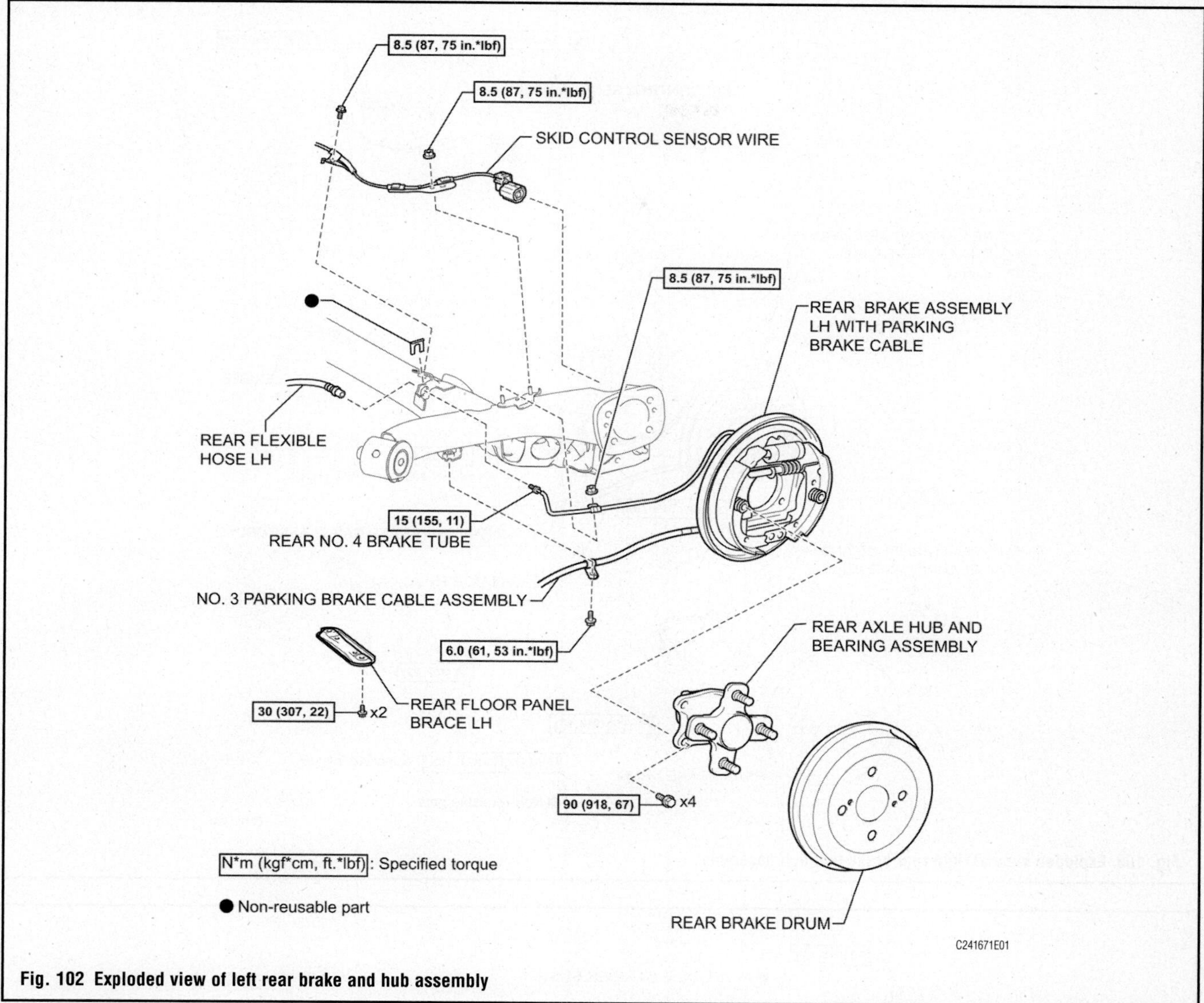

Fig. 102 Exploded view of left rear brake and hub assembly

19. Temporarily tighten rear axle beam assembly.
    a. Support the rear axle beam assembly with an engine lift.
    b. Install the rear axle beam assembly to the vehicle and provisionally tighten the 2 bolts.
20. Install rear coil spring.
21. Provisionally install the shock absorber (lower side) to the rear axle beam assembly with the bolt and nut.
22. Install rear floor panel brace LH.
23. Install rear floor panel brace RH.
24. Install rear axle hub and bearing assembly.
25. Inspect rear axle hub bearing looseness.
26. Inspect rear axle hub runout.
27. Install rear brake drum.
28. Adjust rear drum brake shoe clearance.
29. Install the No. 3 parking brake cable assembly with the bolt.
30. Install No. 2 parking brake cable assembly.
31. Install rear No. 4 brake tube.
    a. Install the rear No. 4 brake tube onto the rear axle beam assembly with the nut.
    b. Connect the rear flexible hose to the rear axle beam assembly with a new clip.
    c. Using a union nut wrench, connect the rear No. 4 brake tube to the rear flexible hose while holding the rear flexible hose with a wrench.
32. Install rear No. 3 brake tube.
33. Install skid control sensor wire.
34. Install tail exhaust pipe assembly.
35. Install rear wheel.
36. Stabilize suspension.
37. Fully tighten rear axle beam assembly. Fully tighten the 2 bolts to 67 ft. lbs. (90 Nm).
38. Fully tighten rear shock absorber assembly.
39. Bleed brake system.
40. Inspect for exhaust gas leak.
41. Inspect rear wheel alignment.
42. Check for speed sensor signal.

COIL SPRING

REMOVAL & INSTALLATION

See Figures 105 and 106.

1. Remove rear wheel.
2. Drain brake fluid.
3. Remove tail exhaust pipe assembly.
4. Separate skid control sensor wire.
    a. Using a screwdriver, disengage the claw of the connector lock and disconnect the skid control sensor wire connector.

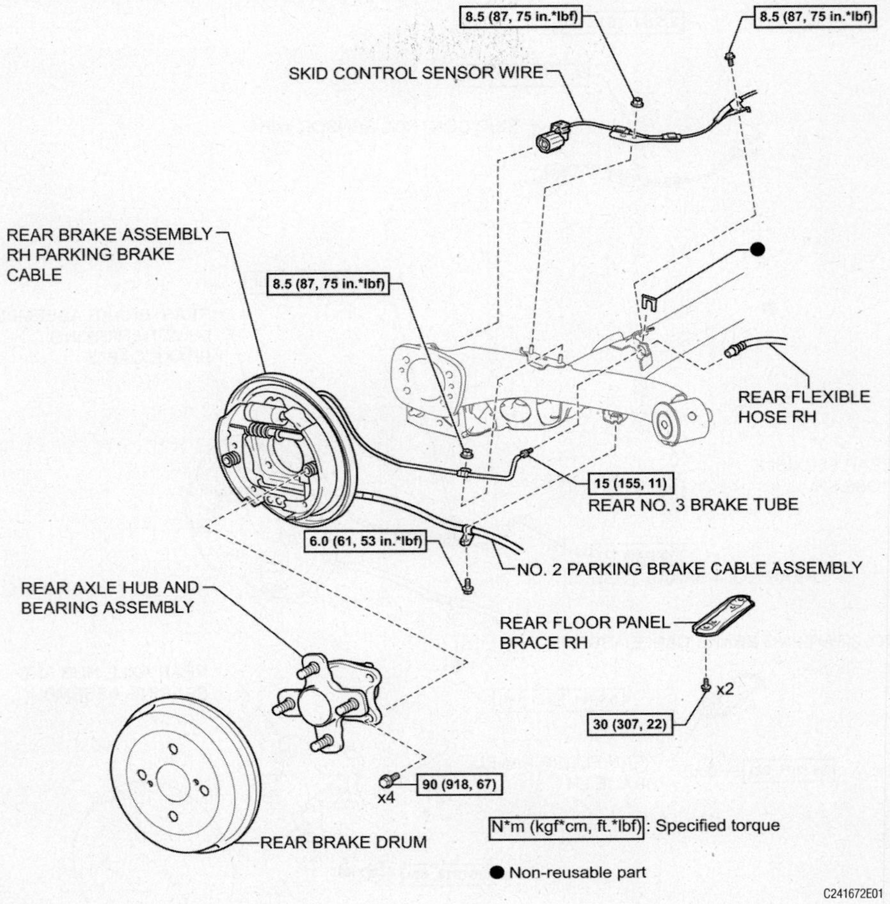

8.5 (87, 75 in.*lbf)
8.5 (87, 75 in.*lbf)
SKID CONTROL SENSOR WIRE
REAR BRAKE ASSEMBLY
RH PARKING BRAKE CABLE
8.5 (87, 75 in.*lbf)
REAR FLEXIBLE HOSE RH
15 (155, 11)
REAR NO. 3 BRAKE TUBE
6.0 (61, 53 in.*lbf)
NO. 2 PARKING BRAKE CABLE ASSEMBLY
REAR AXLE HUB AND BEARING ASSEMBLY
REAR FLOOR PANEL BRACE RH
x2
30 (307, 22)
x4
90 (918, 67)
REAR BRAKE DRUM
N*m (kgf*cm, ft.*lbf): Specified torque
● Non-reusable part
C241672E01

Fig. 103 Exploded view of right rear brake and hub assembly

90 (918, 67)
● REAR AXLE CARRIER BUSH
REAR AXLE BEAM ASSEMBLY
90 (918, 67)
● REAR AXLE CARRIER BUSH
N*m (kgf*cm, ft.*lbf): Specified torque
● Non-reusable part
C236105E01

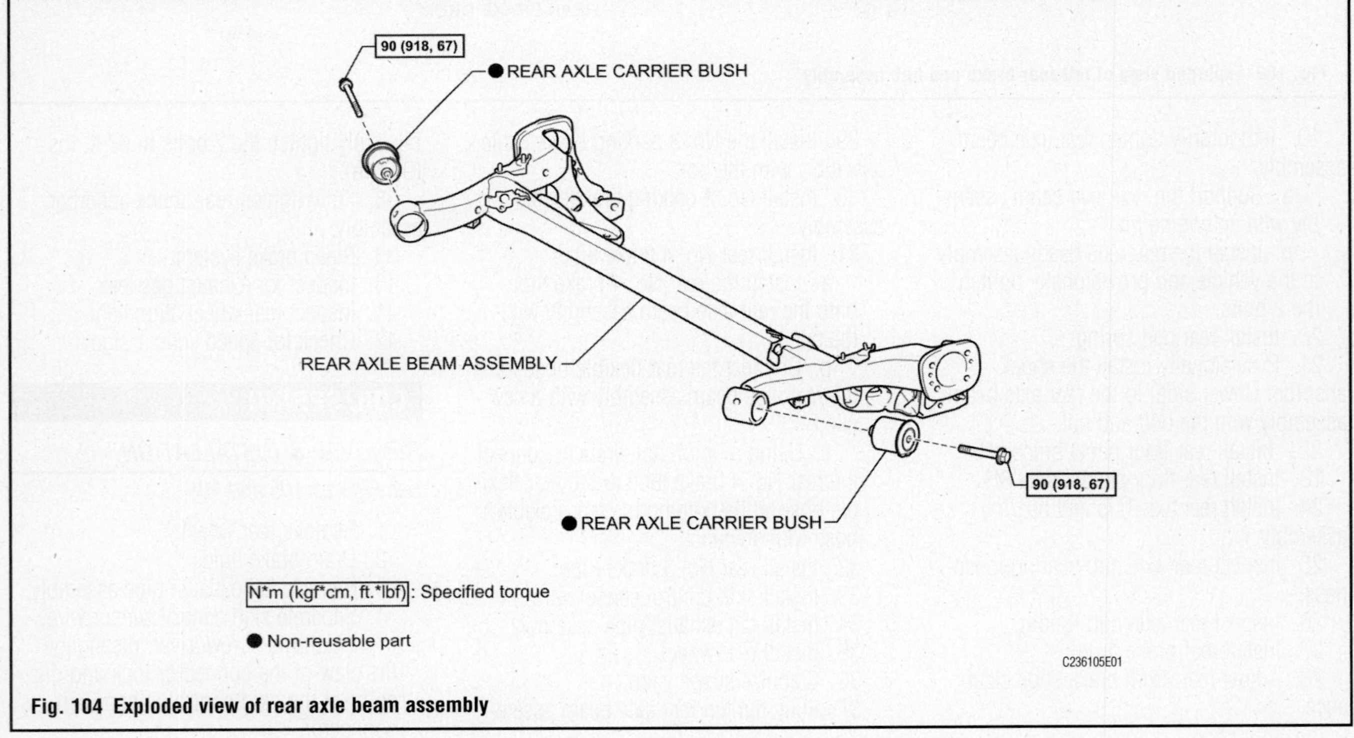

Fig. 104 Exploded view of rear axle beam assembly

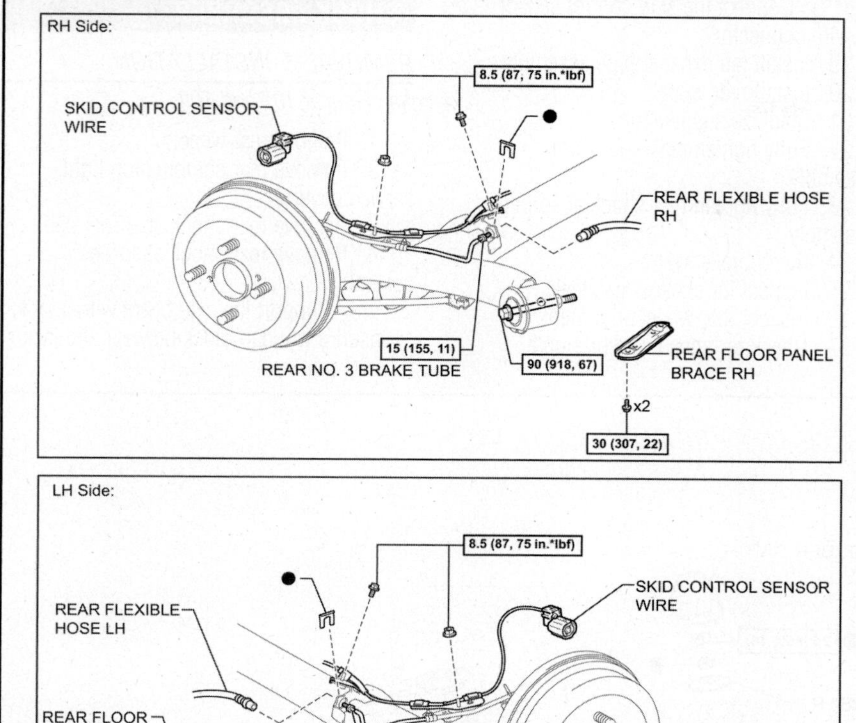

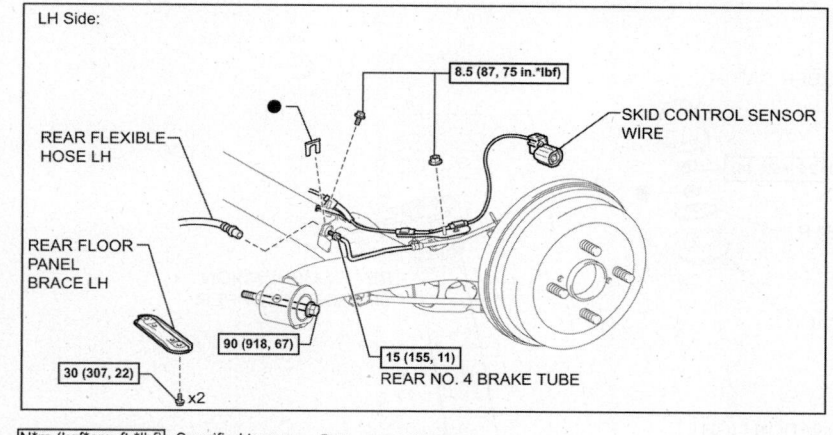

Fig. 105 Exploded view of rear axle and hub assemblies

➡Do not remove the connector cover from the connector because the skid control sensor wire may be damaged.

b. Remove the bolt and the nut, and separate the skid control sensor wire.

5. Separate rear flexible hose LH.

a. Using a union nut wrench, separate the rear No. 4 brake tube while holding the rear flexible hose LH with a wrench.

b. Remove the clip and disconnect the rear flexible hose LH from the axle beam.

6. Separate rear flexible hose RH.

7. Remove the 2 bolts and the rear floor panel brace LH.

8. Remove rear floor panel brace RH.

9. Loosen rear axle beam assembly.

a. Support the rear axle beam assembly using an engine lift and 2 wooden blocks.

⁂ WARNING

Securely support the rear axle beam assembly to prevent it from falling.

b. Loosen the 2 bolts.

➡Do not remove the bolts.

10. Remove the bolt while keeping the nut from rotating and separate the shock absorber.

➡Remove the nut on the bolt side because the one on the lower side is a jam nut.

11. Remove rear coil spring.

a. Lower the engine lift slowly.

b. Remove the rear coil spring, upper coil spring insulator and lower coil spring insulator.

To install:

12. Install rear coil spring.

a. Install the lower coil spring insulator onto the rear axle beam.

b. Install the upper coil spring insulator so that its gap fits onto the end of rear coil spring.

c. Install the rear coil spring onto the rear axle beam.

➡The paint mark of the rear coil spring should be towards the underside and rear side of the vehicle.

13. Provisionally install the shock absorber (lower side) with the bolt and nut to the axle beam.

14. Install the rear floor panel brace LH with the 2 bolts. Tighten to 22 ft. lbs. (30 Nm).

15. Install rear floor panel brace RH.

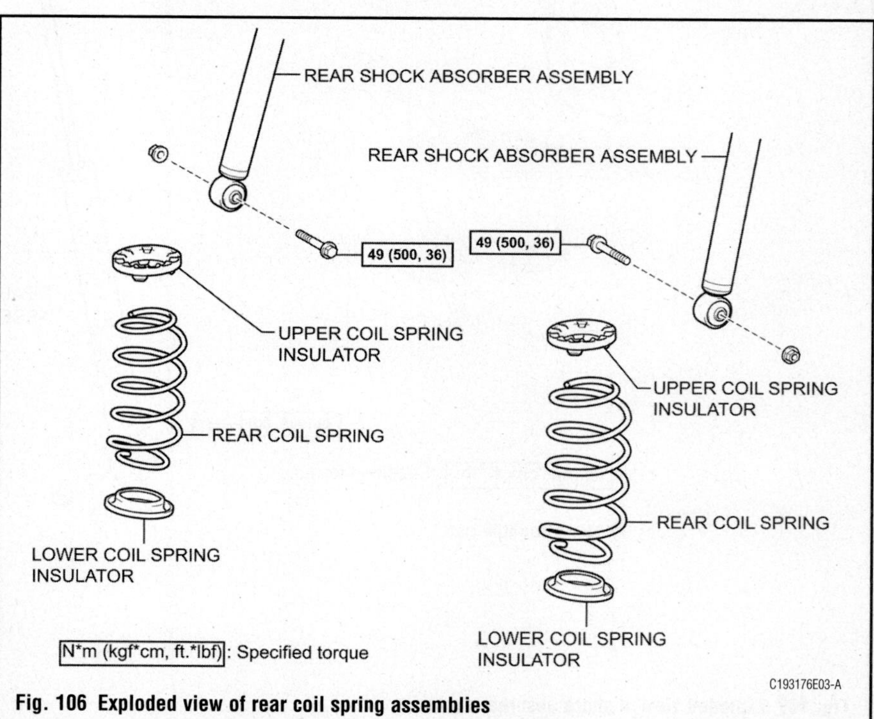

Fig. 106 Exploded view of rear coil spring assemblies

16. Install rear flexible hose LH.
   a. Connect the rear flexible hose LH onto the axle beam with a new clip.
   b. Using a union nut wrench, connect the rear No. 4 brake tube to rear flexible hose LH while holding the rear flexible hose LH with a wrench. Tighten to 11 ft. lbs. (15 Nm).
17. Install rear flexible hose RH.
18. Install skid control sensor wire.
   a. Install the skid control sensor wire onto the axle beam with the bolt and the nut.
   b. Connect the skid control sensor wire connector.
19. Install tail exhaust pipe assembly.
20. Install rear wheel.
21. Stabilize suspension.
22. Fully tighten rear axle beam assembly.
23. Fully tighten rear shock absorber assembly.
24. Bleed brake system.
25. Inspect for exhaust gas leak.
26. Inspect rear wheel alignment.
27. Check for speed sensor signal.

SHOCK ABSORBER

REMOVAL & INSTALLATION

See Figures 107 and 108.

1. Remove rear wheel.
2. Remove rear combination light service cover.
3. Remove rear shock absorber cap.
4. Remove rear shock absorber assembly.
   a. Support the axle beam with a jack. Insert a wooden block between the jack

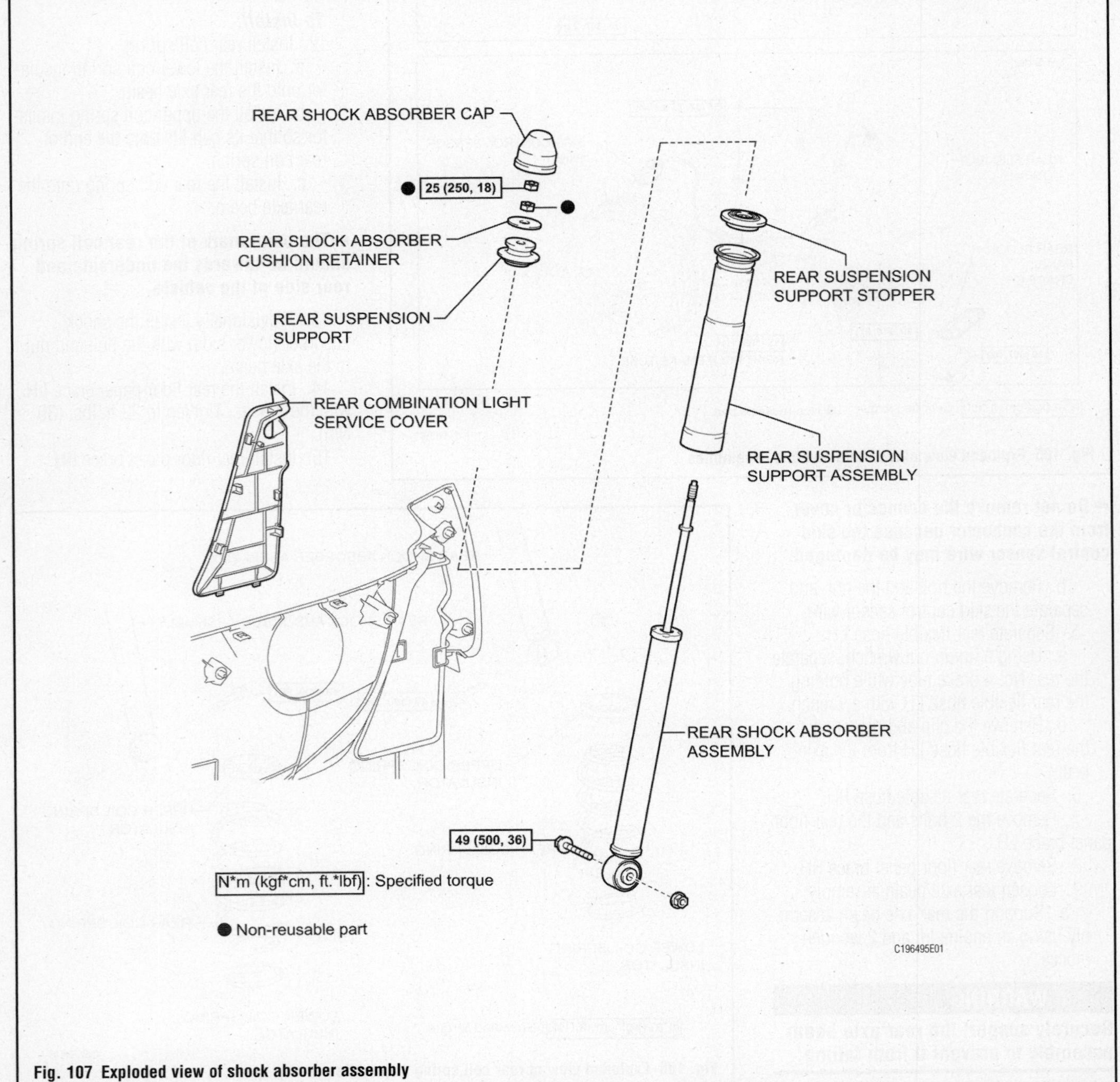

Fig. 107 Exploded view of shock absorber assembly

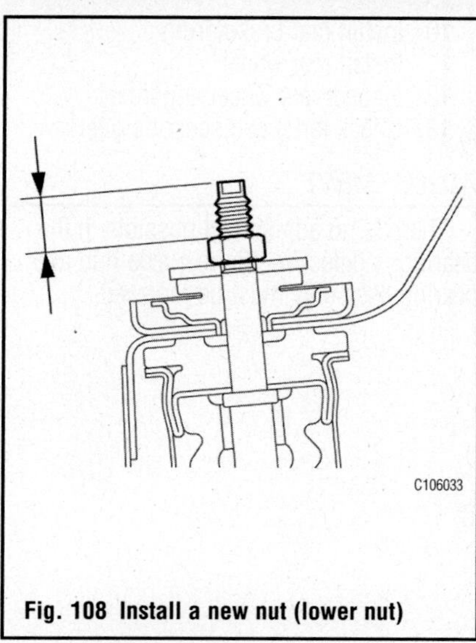

Fig. 108 Install a new nut (lower nut)

and the rear axle spring seat to prevent damage.

b. Hold the lower nut and loosen the upper nut.

c. Remove the 2 nuts.

d. Remove the rear shock absorber cushion retainer and rear suspension support.

e. Remove the bolt while keeping the nut from rotating and remove the rear shock absorber.

➡**Remove the nut from the bolt side because the nut is a jam nut.**

5. Remove rear suspension support stopper.

6. Remove rear suspension support assembly.

To install:

7. Install rear suspension support assembly.

8. Install rear suspension support stopper.

9. Temporarily tighten rear shock absorber assembly.

a. Support the axle beam with a jack. Insert a wooden block between the jack and the rear axle spring seat to prevent damage.

b. Jack up the axle beam slowly, and provisionally install the shock absorber (lower side) with the bolt and nut onto the axle beam.

c. Install the rear suspension support and rear shock absorber cushion retainer.

d. While holding the piston rod, install a new nut (lower nut) to the specified standard of 0.591–0.709 inches 12 to 18 mm).

e. Install a new nut (upper nut) onto the lower nut.

f. Hold the lower nut and tighten the upper nut against it. Tighten to 18 ft. lbs. (25 Nm).

10. Install rear shock absorber cap.

11. Install rear combination light service cover.

12. Install rear wheel.

13. Stabilize suspension.

a. Lower the vehicle from the jack.

b. Bounce the vehicle up and down several times to stabilize the suspension.

14. Fully tighten the rear shock absorber (lower side) with the bolt. Tighten to 36 ft. lbs. (49 Nm).

15. Inspect rear wheel alignment.

WHEEL BEARINGS

REMOVAL & INSTALLATION

See Figure 109.

➡**Use the same procedure for the RH side and LH side. The procedure listed below is for the LH side.**

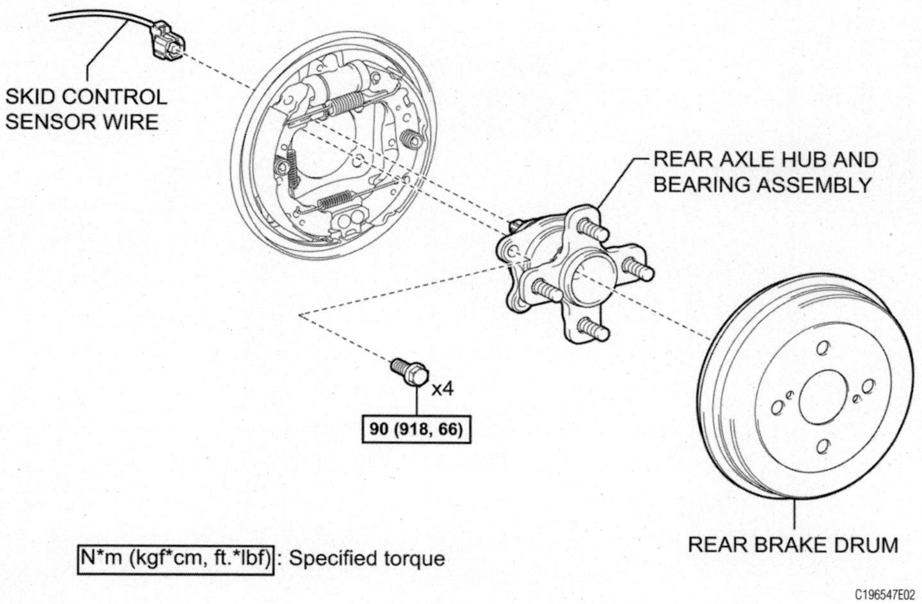

Fig. 109 Exploded view of rear axle hub assembly

1. Loosen adjusting nut.
2. Remove rear wheel.
3. Remove rear brake drum.
4. Using a screwdriver, remove the claw of the connector lock portion and disconnect the skid control sensor wire connector.
5. Remove the 4 bolts and the rear axle hub and bearing assembly.

To install:

6. Install the rear axle hub and bearing assembly with the 4 bolts. Tighten to 66 ft. lbs. (90 Nm).
7. Connect the rear speed sensor wire connector.
8. Inspect rear axle hub bearing looseness
9. Inspect rear axle hub runout
10. Install rear brake drum
11. Install rear wheel
12. Inspect rear wheel alignment
13. Check for speed sensor signal

ADJUSTMENT

There is no adjustment possible. If the bearing is defective, the rear axle hub and bearing assembly must be replaced.

SCION

tC • xB • xD

2

SPECIFICATIONS AND MAINTENANCE CHARTS

ENGINE AND VEHICLE IDENTIFICATION

Engine						
Code	Liters (cc)	Cu. In.	Cyl.	Fuel Sys.	Engine Type	Eng. Mfg.
2AR-FE	2.5 (2494)	153	4	SEFI	DOHC	Toyota
2AZ-FE	2.4 (2362)	147	4	SEFI	DOHC	Toyota
2ZR-FE	1.8 (1798)	110	4	SEFI	DOHC	Toyota

Model Year	
Code ①	Year
B	2011
C	2012

① 10th position of VIN
SEFI Sequential Electronic Fuel Injection

71099_TCXB_C0001

GENERAL ENGINE SPECIFICATIONS

All measurements are given in inches.

Year	Model	Engine Displacement Liters (cc)	Engine ID	Fuel System Type	Net Horsepower @ rpm	Net Torque @ rpm (ft. lbs.)	Bore x Stroke (in.)	Com-pression Ratio	Oil Pressure @ rpm
2011	tC	2.5 (2494)	2AR-FE	SEFI	180@6000	173@4100	3.54x3.86	10.4:1	38@4000
	xB	2.4 (2362)	2AZ-FE	SEFI	158@6000	162@4000	3.48x3.78	9.8:1	24-44@3000
	xD	1.8 (1798)	2ZR-FE	SEFI	128@6000	125@4400	3.17x3.45	10.0:1	22-58@3000
2012	tC	2.5 (2494)	2AR-FE	SEFI	180@6000	173@4100	3.54x3.86	10.4:1	38@4000
	xB	2.4 (2362)	2AZ-FE	SEFI	158@6000	162@4000	3.48x3.78	9.8:1	24-44@3000
	xD	1.8 (1798)	2ZR-FE	SEFI	128@6000	125@4400	3.17x3.48	10.0:1	22-58@3000

71099_TCXB_C0002

ENGINE TUNE-UP SPECIFICATIONS

Year	Engine Displacement Liters	Engine ID	Spark Plug Gap (in.)	Ignition Timing (deg.) MT	Ignition Timing (deg.) AT	Fuel Pump (psi)	Idle Speed (rpm) MT	Idle Speed (rpm) AT	Valve Clearance Intake	Valve Clearance Exhaust
2011	2.5	2AR-FE	0.039-0.043	5-15	5-15	44-50	600-700	600-700	Hyd	Hyd
	2.4	2AZ-FE	0.039-0.043	5-12	5-15	44-50	600-700	600-700	0.0075-0.0114	0.0150-0.0189
	1.8	2ZR-FE	0.039-0.043	8-12	8-12	44-50	650-750	650-750	Hyd	Hyd
2012	2.5	2AR-FE	0.039-0.043	5-15	5-15	44-50	600-700	600-700	Hyd	Hyd
	2.4	2AZ-FE	0.039-0.043	5-12	5-15	44-50	600-700	600-700	0.0075-0.0114	0.0150-0.0189
	1.8	2ZR-FE	0.039-0.043	8-12	8-12	44-50	650-750	650-750	Hyd	Hyd

Hyd Vehicles use hydraulic lifters

71099_TCXB_C0003

CAPACITIES

Year	Model	Engine Displacement Liters	Engine ID	Engine Oil with Filter	Transaxle (pts.) Auto.	Transaxle (pts.) Manual	Fuel Tank (gal.)	Cooling System (qts.)
2011	tC	2.5 (2494)	2AR-FE	4.6	13.8	4.8	14.5	6.1
	xB	2.4 (2362)	2AZ-FE	4.5	①	5.2	14.0	6.3
	xD	1.8 (1798)	2ZR-FE	4.4	②	5.2	11.1	5.3
2012	tC	2.5 (2494)	2AR-FE	4.6	13.8	4.8	14.5	6.1
	xB	2.4 (2362)	2AZ-FE	4.5	①	5.2	14.0	6.3
	xD	1.8 (1798)	2ZR-FE	4.4	②	4.0	11.1	5.3

NOTE: All capacities are approximate. Add fluid gradually and ensure a proper fluid level is obtained.

① Dry Fill: 16.4 pts.
Drain & Refill: 7.4 pts.

② Dry Fill: 13.6 pts.
Drain & Refill: 5.2 pts.

71099_TCXB_C0004

FLUID SPECIFICATIONS

Year	Model	Engine Disp. Liters	Engine Oil	Manual Trans.	Auto. Trans.	Brake Master Cylinder	Cooling System
2011	tC	2.5	0W-20	①	②	DOT 3	③
	xB	2.4	④	⑤	②	DOT 3	③
	xD	1.8	0W-20	①	②	DOT 3	③
2012	tC	2.5	0W-20	①	②	DOT 3	③
	xB	2.4	④	⑤	②	DOT 3	③
	xD	1.8	0W-20	①	②	DOT 3	③

DOT: Department Of Transpotation

① API GL-4 SAE 75W

② Toyota Genuine ATF WS

③ Toyota Super Long Life Coolant

④ 0W-20 or 5W-20

⑤ API GL-3 SAE 75W-90

71099_TCXB_C0005

VALVE SPECIFICATIONS

Year	Engine Displacement Liters	Engine ID	Seat Angle (deg.)	Face Angle (deg.)	Spring Test Pressure (lbs. @ in.)	Spring Free-Length (in.)	Spring Installed Height (in.)	Stem-to-Guide Clearance (in.) Intake	Stem-to-Guide Clearance (in.) Exhaust	Stem Diameter (in.) Intake	Stem Diameter (in.) Exhaust
2011	2.5	2AR-FE	NS	NS	NS	1.970	NS	0.00098-0.00236	0.00118-0.00256	0.2151-0.2157	0.2153-0.2159
	2.4	2AZ-FE	NS	NS	NS	1.867	NS	0.0010-0.0024	0.0012-0.0026	0.2154-0.2159	0.2152-0.2157
	1.8	2ZR-FE	NS	NS	NS	2.1008	NS	0.0010-0.0024	0.0012-0.0026	0.2154-0.2159	0.2152-0.2157
2012	2.5	2AR-FE	NS	NS	NS	1.970	NS	0.00098-0.00236	0.00118-0.00256	0.2151-0.2157	0.2153-0.2159
	2.4	2AZ-FE	NS	NS	NS	1.867	NS	0.0010-0.0024	0.0012-0.0026	0.2154-0.2159	0.2152-0.2157
	1.8	2ZR-FE	NS	NS	NS	2.1008	NS	0.0010-0.0024	0.0012-0.0026	0.2154-0.2159	0.2152-0.2157

NS Not Specified

71099_TCXB_C0006

CAMSHAFT SPECIFICATIONS

All measurements in inches unless noted

Year	Engine Displacement Liters	Engine ID	Journal Diameter	Brg. Oil Clearance	Shaft End-play	Runout	Journal Bore	Lobe Height Intake	Lobe Height Exhaust
2011	2.5	2AR-FE	①	②	0.00236-0.00610	0.00118	NS	1.739-1.744	1.738-1.744
	2.4	2AZ-FE	③	④	⑤	0.0012	NS	1.8654-1.8664	1.8104-1.8143
	1.8	2ZR-FE	⑥	NS	NS	0.0016	NS	1.6857-1.6896	1.755-1.7494
2012	2.5	2AR-FE	①	②	0.00236-0.00610	0.00118	NS	1.739-1.744	1.738-1.744
	2.4	2AZ-FE	③	④	⑤	0.0012	NS	1.8654-1.8664	1.8104-1.8143
	1.8	2ZR-FE	⑥	NS	NS	0.0016	NS	1.6857-1.6896	1.755-1.7494

NS: Not Specified

① No. 1 journal: 1.356-1.357 inches
Other journals: 0.904-0.905 inches

② Intake No. 1 journal: 0.00137-0.00283 inches
Exhaust No. 1 journal: 0.00193-0.00339 inches
All other journals: 0.00098-0.00244 inches

③ No. 1 journal: 1.4162-1.4167 inches
Other journals: 0.9039-0.9045 inches

④ No. 1 journal: 0.0003-0.0015 inches
Other Journals: 0.0010-0.0024 inches

⑤ Intake: 0.0016-0.0037 inches
Exhaust: 0.0150-0.0189 inches

⑥ Journal No. 1: 1.3563-1.3569 inches
Other journals: 0.9035-0.9041 inches

71099_TCXB_C0007

CRANKSHAFT AND CONNECTING ROD SPECIFICATIONS

All measurements are given in inches.

Year	Engine Disp. Liters	Engine ID	Crankshaft Main Brg. Journal Dia.	Crankshaft Main Brg. Oil Clearance	Crankshaft Shaft End-play	Crankshaft Thrust on No.	Connecting Rod Journal Diameter	Connecting Rod Oil Clearance	Connecting Rod Side Clearance
2011	2.5	2AR-FE	2.1649-2.1654	0.00063-0.00154	0.00157-0.0095	NS	2.1649-2.1654	0.00118-0.00248	NS
	2.4	2AZ-FE	2.1649-2.1654	0.00070-0.00160	0.0016-0.0095	NS	NS	0.0009-0.0019	0.0063-0.0142
	1.8	2ZR-FE	1.8893-1.8898	0.00060-0.00150	0.0016-0.0055	NS	①	0.0012-0.0024	NS
2012	2.5	2AR-FE	2.1649-2.1654	0.00063-0.00154	0.00157-0.0095	NS	2.1649-2.1654	0.00118-0.00248	NS
	2.4	2AZ-FE	2.1649-2.1654	0.00070-0.00160	0.0016-0.0095	NS	NS	0.0009-0.0019	0.0063-0.0142
	1.8	2ZR-FE	1.8893-1.8898	0.00060-0.00150	0.0016-0.0055	NS	①	0.0012-0.0024	NS

NS Not Specified

① Mark 1: 1.8504-1.8507 inches
Mark 2: 1.8507-1.8510 inches
Mark 3: 1.8511-1.8513 inches

71099_TCXB_C0008

PISTON AND RING SPECIFICATIONS

All measurements are given in inches.

Year	Engine Disp. Liters	Engine ID	Piston Clearance	Ring Gap Top Compression	Ring Gap Bottom Compression	Ring Gap Oil Control	Ring Side Clearance Top Compression	Ring Side Clearance Bottom Compression	Ring Side Clearance Oil Control
2011	2.5	2AR-FE	0.00016-0.00169	0.00866-0.0106	0.0146-0.0165	0.00394-0.00787	0.000787-0.00276	0.000787-0.00236	0.000787-0.00276
	2.4	2AZ-FE	0.0008-0.0017	0.0094-0.0122	0.0130-0.0169	0.0040-0.0079	0.0008-0.0028	0.0008-0.0024	0.0008-0.0028
	1.8	2ZR-FE	0.0004-0.0017	0.0079-0.0118	0.0138-0.0197	0.0039-0.0157	0.0008-0.0028	0.0008-0.0024	0.0008-0.0026
2012	2.5	2AR-FE	0.00016-0.00169	0.00866-0.0106	0.0146-0.0165	0.00394-0.00787	0.000787-0.00276	0.000787-0.00236	0.000787-0.00276
	2.4	2AZ-FE	0.0008-0.0017	0.0094-0.0122	0.0130-0.0169	0.0040-0.0079	0.0008-0.0028	0.0008-0.0024	0.0008-0.0028
	1.8	2ZR-FE	0.0004-0.0017	0.0079-0.0118	0.0138-0.0197	0.0039-0.0157	0.0008-0.0028	0.0008-0.0024	0.0008-0.0026

71099_TCXB_C0009

TORQUE SPECIFICATIONS

All readings in ft. lbs.

Year	Engine Disp. Liters	Engine ID	Cylinder Head Bolts	Main Bearing Bolts	Rod Bearing Bolts	Crankshaft Damper Bolts	Flywheel Bolts	Manifold Intake	Manifold Exhaust	Spark Plugs	Oil Pan Drain Plug
2011	2.5	2AR-FE	①	②	③	192	72	15	26	18	30
	2.4	2AZ-FE	④	②	⑤	133	⑥	22	27	14	30
	1.8	2ZR-FE	⑦	③	④	140	⑧	21	27	15	27
2012	2.5	2AR-FE	①	②	③	192	72	15	26	18	30
	2.4	2AZ-FE	④	②	⑤	133	⑥	22	27	14	30
	1.8	2ZR-FE	⑦	③	④	140	⑧	21	27	15	27

① Step 1: 27 ft. lbs.
Step 2: 27 ft. lbs.
Step 3: Additional 90 degrees
Step 4:Additional 90 degrees

② Step 1: 15 ft. lbs.
Step 2:30 ft. lbs.
Step 3: Additional 90 degrees

③ Step 1: 30 ft. lbs.
Step 2: Additional 90 degrees

④ Step 1:52 ft. lbs.
Step 2 Additional 90 degrees

⑤ Step 1: 18 ft. lbs.
Step 2: Additional 90 degrees

⑥ Automatic transaxle drive plate: 72 ft. lbs.
Manual transaxle flywheel: 96 ft. lbs.

⑦ Step 1: 36 ft. lbs.
Step 2: Additional 90 degrees
Step 3: Additional 45 degrees

⑧ Automatic transaxle: 65 ft. lbs.
Manual transaxle flywheel: 38 ft. lbs. plus additional 90 degrees

71099_TCXB_C0010

WHEEL ALIGNMENT

Year	Model		Caster Range (+/-Deg.)	Caster Preferred Setting (Deg.)	Camber Range (+/-Deg.)	Camber Preferred Setting (Deg.)	Toe-in (in.)
2011	tC	F	0.75	6.00	0.45	0.75	0.04+/-0.08
		R			0.75	-1.00	0.12+/-0.08
	xB	F	0.75	5.75	0.75	-0.16	0.08+/-0.08
		R			0.5	-1.42	0.08+/-0.12
	xD	F	0.75	5.07	0.75	-0.18	0.05+/-0.08
		R			-0.5	-0.93	0.12+/-0.11
2012	tC	F	0.75	6.00	0.45	0.75	0.04+/-0.08
		R			0.75	-1.00	0.12+/-0.08
	xB	F	0.75	5.75	0.75	-0.16	0.08+/-0.08
		R			0.5	-1.42	0.08+/-0.12
	xD	F	0.75	5.07	0.75	-0.18	0.05+/-0.08
		R			-0.5	-0.93	0.12+/-0.11

F Front
R Rear

71099_TCXB_C0011

TIRE, WHEEL AND BALL JOINT SPECIFICATIONS

Year	Model	OEM Tires Standard	OEM Tires Optional	Tire Pressures (psi) Front	Tire Pressures (psi) Rear	Wheel Size	Ball Joint Inspection	Lug Nut (ft. lbs.)
2011	tC	215/50R17	225/45R18	33	30	17/18	NA	76
	xB	205/55R16	—	35	32	16	NA	76
	xD	195/60R16	—	33	33	16	NA	76
2012	tC	215/50R17	225/45R18	33	30	17/18	NA	76
	xB	205/55R16	—	35	32	16	NA	76
	xD	195/60R16	—	33	33	16	NA	76

OEM: Original Equipment Manufacturer

PSI: Pounds Per Square Inch

NA: Information not available

71099_TCXB_C0012

BRAKE SPECIFICATIONS

All measurements in inches unless noted

Year	Model		Brake Disc Original Thickness	Brake Disc Minimum Thickness	Brake Disc Max. Runout	Brake Drum Diameter Original Inside Diameter	Brake Drum Diameter Max. Wear Limit	Brake Drum Diameter Maximum Machine Diamter	Minimum Pad/Lining Thickness Front	Minimum Pad/Lining Thickness Rear	Brake Caliper Bracket Bolts (ft. lbs.)	Brake Caliper Mounting Bolts (ft. lbs.)
2011	tC	F	1.102	0.984	0.00197	—	—	—	0.0394	—	79	25
		R	0.394	0.335	0.00591	—	—	—	—	0.0394	42	25
	xB	F	0.984	0.866	0.00197	—	—	—	0.0394	—	79	25
		R	0.394	0.335	0.00591	—	—	—	—	0.0394	42	25
	xD	F	0.866	0.748	0.00200	—	—	—	0.0390	—	79	25
		R	—	—	—	9.000	—	9.039	—	0.0390	—	—
2012	tC	F	1.102	0.984	0.00197	—	—	—	0.0394	—	79	25
		R	0.394	0.335	0.00591	—	—	—	—	0.0394	42	25
	xB	F	0.984	0.866	0.00197	—	—	—	0.0394	—	79	25
		R	0.394	0.335	0.00591	—	—	—	—	0.0394	42	25
	xD	F	0.866	0.748	0.00200	—	—	—	0.0390	—	79	25
		R	—	—	—	9.000	—	9.039	—	0.0390	—	—

F: Front

R: Rear

NA: Information not available

71099_TCXB_C0013

SCHEDULED MAINTENANCE INTERVALS

SCION - tC, xB & xD

TO BE SERVICED	TYPE OF SERVICE	VEHICLE MILEAGE INTERVAL (x1000)												
		5	10	15	20	25	30	35	40	45	50	55	60	65
Air cleaner filter	R						✓						✓	
Transmission fluid	S/I						✓						✓	
Ball joints & dust covers	S/I			✓			✓			✓			✓	
Bolts & nuts on chassis & body	S/I													
Brake line pipes & hoses	S/I			✓			✓			✓			✓	
Brake pads & discs/linings & drums (front & rear)	S/I	✓	✓	✓	✓	✓	✓	✓	✓	✓	✓	✓	✓	✓
Drive belts	S/I												✓	
Driveshaft boots	S/I			✓			✓			✓			✓	
Engine coolant	S/I			✓			✓			✓			✓	
Engine coolant	R	Replace at 100,000 miles												
Engine oil & filter	R	✓	✓	✓	✓	✓	✓	✓	✓	✓	✓	✓	✓	✓
Exhaust pipes & mountings	S/I			✓			✓			✓			✓	
Fuel lines & connections	S/I						✓						✓	
Propeller shaft bolt	S/I			✓			✓			✓			✓	
Radiator core & condenser	S/I			✓			✓			✓			✓	
Front differential fluid	S/I						✓						✓	
Rotate Tires	S/I	✓	✓	✓	✓	✓	✓	✓	✓	✓	✓	✓	✓	✓
Spark plugs (tC)	R	Replace at 120,000 miles												
Spark plugs (xB & xD)	R						✓						✓	
Steering linkage & gear box	S/I			✓			✓			✓			✓	

R: Replace S/I: Service or Inspect

Drivebelts: After initial inspection at 60,000 miles, inspect every 15,000 miles thereafter.

FREQUENT OPERATION MAINTENANCE (SEVERE SERVICE)

If a vehicle is operated under any of the following conditions it is considered severe service:

- Desert/Extremely dusty areas.

- Trailer towing usage.

Air cleaner filter: service or inspect every 5000 miles

Ball joints & dust covers: service or inspect every 5000 miles.

Bolts & nuts on chassis & body: service or inspect every 5000 miles.

Driveshaft boots: service or inspect every 5000 miles.

Steering linkage: service or inspect every 5000 miles.

Transmission and Front differential fluid: replace every 30,000 miles.

71099_TCXB_C0014

PRECAUTIONS

Before servicing any vehicle, please be sure to read all of the following precautions, which deal with personal safety, prevention of component damage, and important points to take into consideration when servicing a motor vehicle:

- Never open, service or drain the radiator or cooling system when the engine is hot; serious burns can occur from the steam and hot coolant.
- Observe all applicable safety precautions when working around fuel. Whenever servicing the fuel system, always work in a well-ventilated area. Do not allow fuel spray or vapors to come in contact with a spark, open flame, or excessive heat (a hot drop light, for example). Keep a dry chemical fire extinguisher near the work area. Always keep fuel in a container specifically designed for fuel storage; also, always properly seal fuel containers to avoid the possibility of fire or explosion. Refer to the additional fuel system precautions later in this section.
- Fuel injection systems often remain pressurized, even after the engine has been turned **OFF**. The fuel system pressure must be relieved before disconnecting any fuel lines. Failure to do so may result in fire and/or personal injury.
- Brake fluid often contains polyglycol ethers and polyglycols. Avoid contact with the eyes and wash your hands thoroughly after handling brake fluid. If you do get brake fluid in your eyes, flush your eyes with clean, running water for 15 minutes. If eye irritation persists, or if you have taken brake fluid internally, IMMEDIATELY seek medical assistance.
- The EPA warns that prolonged contact with used engine oil may cause a number of skin disorders, including cancer. You should make every effort to minimize your exposure to used engine oil. Protective gloves should be worn when changing oil. Wash your hands and any other exposed skin areas as soon as possible after exposure to used engine oil. Soap and water, or waterless hand cleaner should be used.
- All new vehicles are now equipped with an air bag system, often referred to as a Supplemental Restraint System (SRS) or Supplemental Inflatable Restraint (SIR) system. The system must be disabled before performing service on or around system components, steering column, instrument panel components, wiring and sensors. Failure to follow safety and disabling procedures could result in accidental air bag deployment, possible personal injury and unnecessary system repairs.
- Always wear safety goggles when working with, or around, the air bag system. When carrying a non-deployed air bag, be sure the bag and trim cover are pointed away from your body. When placing a non-deployed air bag on a work surface, always face the bag and trim cover upward, away from the surface. This will reduce the motion of the module if it is accidentally deployed. Refer to the additional air bag system precautions later in this section.
- Clean, high quality brake fluid from a sealed container is essential to the safe and proper operation of the brake system. You should always buy the correct type of brake fluid for your vehicle. If the brake fluid becomes contaminated, completely flush the system with new fluid. Never reuse any brake fluid. Any brake fluid that is removed from the system should be discarded. Also, do not allow any brake fluid to come in contact with a painted surface; it will damage the paint.
- Never operate the engine without the proper amount and type of engine oil; doing so WILL result in severe engine damage.
- Timing belt maintenance is extremely important. Many models utilize an interference-type, non-freewheeling engine. If the timing belt breaks, the valves in the cylinder head may strike the pistons, causing potentially serious (also time-consuming and expensive) engine damage. Refer to the maintenance interval charts for the recommended replacement interval for the timing belt, and to the timing belt section for belt replacement and inspection.
- Disconnecting the negative battery cable on some vehicles may interfere with the functions of the on-board computer system(s) and may require the computer to undergo a relearning process once the negative battery cable is reconnected.
- When servicing drum brakes, only disassemble and assemble one side at a time, leaving the remaining side intact for reference.
- Only an MVAC-trained, EPA-certified automotive technician should service the air conditioning system or its components.

BRAKES

ANTI-LOCK BRAKE SYSTEM (ABS)

GENERAL INFORMATION

PRECAUTIONS

- Certain components within the ABS system are not intended to be serviced or repaired individually.
- Do not use rubber hoses or other parts not specifically specified for and ABS system. When using repair kits, replace all parts included in the kit. Partial or incorrect repair may lead to functional problems and require the replacement of components.
- Lubricate rubber parts with clean, fresh brake fluid to ease assembly. Do not use shop air to clean parts; damage to rubber components may result.
- Use only DOT 3 brake fluid from an unopened container.
- If any hydraulic component or line is removed or replaced, it may be necessary to bleed the entire system.
- A clean repair area is essential. Always clean the reservoir and cap thoroughly before removing the cap. The slightest amount of dirt in the fluid may plug an orifice and impair the system function. Perform repairs after components have been thoroughly cleaned; use only denatured alcohol to clean components. Do not allow ABS components to come into contact with any substance containing mineral oil; this includes used shop rags.
- The Anti-Lock control unit is a microprocessor similar to other computer units in the vehicle. Ensure that the ignition switch is **OFF** before removing or installing controller harnesses. Avoid static electricity discharge at or near the controller.
- If any arc welding is to be done on the vehicle, the control unit should be unplugged before welding operations begin.

SPEED SENSORS

REMOVAL & INSTALLATION

tC

Front

See Figure 1.

➡Use the same procedure for the RH and LH sides. The following procedure is for the LH side.

1. Remove front wheel.
2. Remove front fender liner LH.

➡It is not necessary to fully remove the fender liner. Partially remove it so that the speed sensor connector can be disconnected in a later step.

3. Remove front speed sensor LH.
   a. Remove the 2 front speed sensor wire harness clamps and disconnect the front speed sensor connector.
   b. Remove the bolt and No. 2 sensor clamp from the body.
   c. Remove the bolt, No. 1 sensor clamp and front brake flexible hose together from the front shock absorber assembly.
   d. Remove the clamp from the front shock absorber assembly.
   e. Remove the bolt and front speed sensor from the steering knuckle.

➡Prevent foreign matter from adhering to the front speed sensor tip. Clean the front speed sensor installation hole and the contact surfaces every time the front speed sensor is removed.

To install:

4. Install front speed sensor LH.
   a. Install the front speed sensor to the steering knuckle with the bolt.

➡Note the following:

- Prevent foreign matter from adhering to the front speed sensor tip.
- Firmly insert the front speed sensor body into the knuckle before tightening the bolt.
- After installing the front speed sensor to the knuckle, make sure that there is no clearance between the front speed sensor stay and knuckle. Also make sure that no foreign matter is stuck between the parts.

   b. Install the clamp to the front shock absorber assembly.
   c. Temporarily install the front brake flexible hose.
   d. Install the front brake flexible hose and No. 1 sensor clamp together to the front shock absorber with the bolt.

➡Do not twist the wire harness for the front speed sensor when installing the front speed sensor. A bolt holds the brake flexible hose and front speed sensor clamp together. Make sure that the front speed sensor clamp is positioned over the front brake flexible hose.

   e. Install the No. 2 sensor clamp to the body with the bolt.
   f. Connect the front speed sensor connector.
   g. Install the 2 front speed sensor wire harness clamps to the body.
5. Install front fender liner LH.
6. Install front wheel.
7. Check for speed sensor signal.

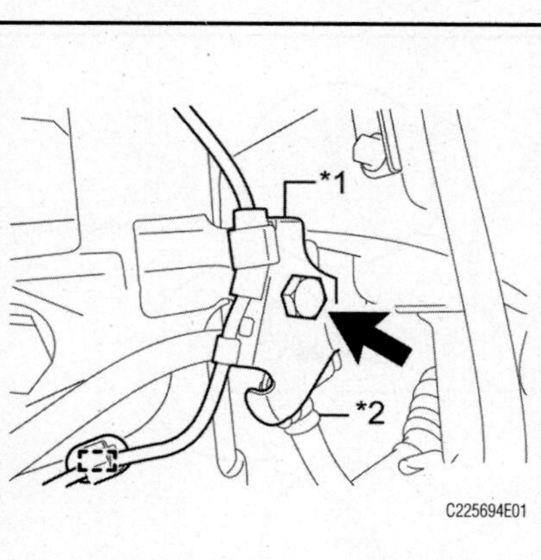

Fig. 1 Remove the bolt, No. 1 sensor clamp (1) and front brake flexible hose (2) together from the front shock absorber assembly

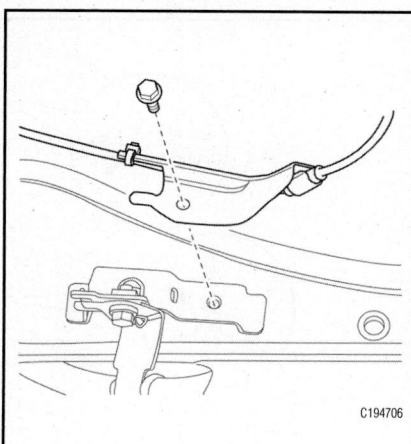

REAR

See Figure 2.

➡Use the same procedure for the RH and LH sides. The procedure listed below is for the LH side.

1. Remove deck trim side panel assembly LH.
2. Remove rear wheel.
3. Disconnect rear disc brake cylinder assembly LH.
4. Remove rear disc.
5. Remove skid control sensor wire LH.
   a. Disconnect the skid control sensor wire connector from the vehicle side connector on the rear floor.
   b. Disconnect the grommet of the skid control sensor wire from the hole of the wheel house.
   c. Remove the bolt and sensor clamp from the wheel house gusset.
   d. Remove the bolt and sensor clamp from the trailing arm bracket.
   e. Detach the sensor clamp.
   f. Disconnect the connector.

Fig. 2 Remove the bolt and sensor clamp from the trailing arm bracket

➡The rear speed sensor is a component of the rear axle hub and bearing assembly. If the sensor malfunctions, replace the rear axle hub and bearing assembly. If the sensor rotor needs to be replaced, replace it together with the rear axle hub and bearing assembly.

To install:

6. Install skid control sensor wire LH.
   a. Connect the skid control sensor wire connector to the skid control sensor.
   b. Attach the sensor clamp.
   c. Install the sensor clamp to the trailing arm bracket with the bolt.
   d. Install the sensor clamp to the wheel house gusset with the bolt.

➡Do not twist the sensor wire when installing the clamp.

   e. Insert the connector and grommet into the inside of the vehicle through the hole in the wheel house.

➡Make sure the band clamp remains on the outside of the vehicle.

7. Hold the grommet and pull it toward the outside of the vehicle. Then fix the grommet in place so that it is not tilted.
   a. Connect the skid control sensor wire connector to the vehicle side connector on the rear floor.
8. Install deck trim side panel assembly LH.
9. Install rear disc.
10. Connect rear disc brake cylinder assembly LH.
11. Install rear wheel.
12. Check speed sensor signal.

xB

FRONT

See Figure 3.

1. Remove front wheel.
2. Remove front side mudguard.
3. Remove front fender liner.
4. Remove front speed sensor.
   a. Disengage clamp A from the body.

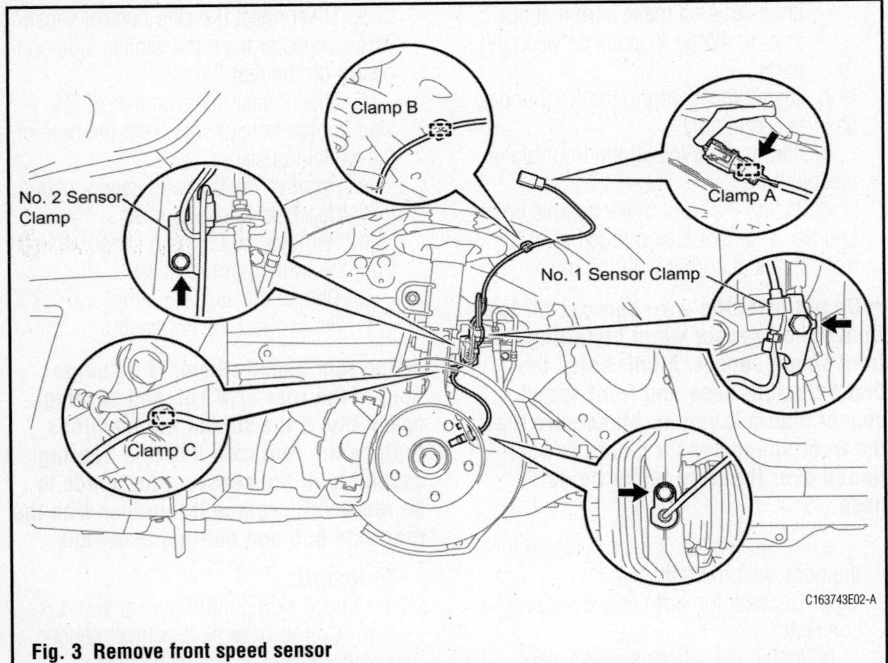

Fig. 3 Remove front speed sensor

b. Disconnect the speed sensor connector.

c. Disengage clamp B from the body.

d. Remove the bolt and No. 2 sensor clamp from the body.

5. Remove the bolt and No. 1 sensor clamp from the shock absorber.

a. Disengage clamp C from the shock absorber.

6. Remove the bolt and speed sensor from the steering knuckle.

➡Keep the speed sensor tip and installation portion free of foreign matter. Remove the speed sensor without turning it from its original installation angle.

To install:

7. Install front speed sensor.

a. Install the speed sensor onto the steering knuckle with the bolt.

➡Check that the speed sensor tip and installation portion are free of foreign matter. Install the speed sensor without turning it from its original installation angle.

b. Engage clamp C onto the shock absorber.

c. Install the No. 1 sensor clamp onto the shock absorber with the bolt.

➡The bolt tightens the flexible hose and front speed sensor together. Make sure that the flexible hose is positioned over the front speed sensor.

d. Install the No. 2 sensor clamp onto the body with the bolt.

e. Engage clamp B onto the body.

f. Connect the speed sensor connector.

g. Engage clamp A onto the body.

8. Install front fender liner.
9. Install front side mudguard.
10. Install front wheel.
11. Inspect speed sensor signal.

xD

FRONT

See Figure 4.

1. Remove front wheel.
2. Remove rocker panel molding cover.
3. Remove side mudguard.
4. Remove front fender liner.
5. Remove front speed sensor.

a. Disengage the clamp A from the body.

b. Disconnect the speed sensor connector.

c. Disengage the clamps B, C and D from the body.

d. Remove the bolt and separate the No. 2 sensor clamp from the body.

e. Remove the bolt and separate the No. 1 sensor clamp from the shock absorber.

f. Remove the bolt and remove the speed sensor from the steering knuckle.

➡Keep the speed sensor tip and installation portion free of foreign matter. Remove the speed sensor without turning it from its original installation angle.

To install:

6. Install front speed sensor.

a. Install the speed sensor onto the steering knuckle with the bolt.

➡Check that the speed sensor tip and installation portion are free of foreign matter. Install the speed sensor without turning it from its original installation angle.

b. Install the No. 1 sensor clamp onto the shock absorber with the bolt.

c. Install the No. 2 sensor clamp onto the body with the bolt.

d. Engage the clamps D, C and B onto the body.

e. Connect the speed sensor connector.

f. Engage the clamp A onto the body.

7. Install front fender liner.
8. Install side mudguard.
9. Install rocker panel molding cover.

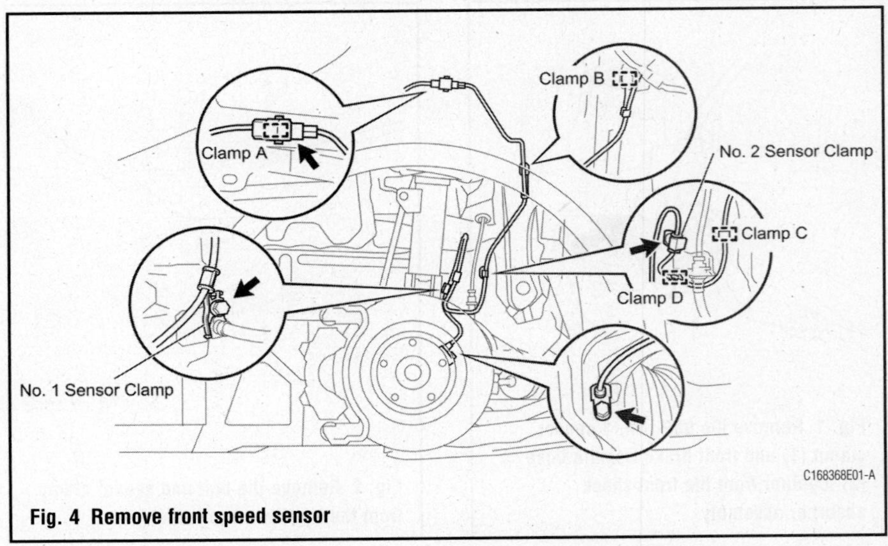

Fig. 4 Remove front speed sensor

10. Install front wheel.
11. Check VSC sensor signal.

REAR

1. Remove rear wheel.
2. Remove rear brake drum sub-assembly.
3. Disconnect speed sensor wire.
4. Remove rear axle hub and bearing assembly (rear speed sensor).
5. Remove the 4 bolts and remove the rear axle hub and bearing from the axle beam.

➡Suspend the backing plate with a piece of rope.

To install:

6. Install rear axle hub and bearing assembly (rear speed sensor)
7. Inspect rear axle hub bearing.
8. Connect speed sensor wire.
9. Install rear brake drum sub-assembly.
10. Adjust rear drum brake shoe clearance.
11. Install rear wheel.
12. Inspect rear wheel alignment.
13. Check VSC sensor signal.

BRAKES

BLEEDING THE BRAKE SYSTEM

BLEEDING PROCEDURE

MASTER CYLINDER BLEEDING

➡If the master cylinder has been disassembled or if the reservoir becomes empty, bleed air from the master cylinder.

1. Fill the brake master cylinder reservoir with brake fluid. Use SAE J1703 or FMVSS No. 116 DOT3 brake fluid.
2. Using SST 09023-00101, or equivalent, disconnect the brake lines from the master cylinder.
3. Slowly depress and hold the brake pedal.
4. Cover the outer holes with your fingers, and release the pedal.
5. Repeat steps (3) and (4) several times.

➡Use a torque wrench with a fulcrum length of 30 cm (11.81 in.) to tighten the brake lines.

6. Connect the brake lines to the master cylinder and tighten to 11 ft. lbs. (15 Nm) without the special tool, or to 10 ft. lbs. (14 Nm) with the special tool.

BRAKE LINE BLEEDING

1. Fill the brake master cylinder reservoir with brake fluid. Use SAE J1703 or FMVSS No. 116 DOT3 brake fluid.
2. Remove the bleeder plug cap.
3. Connect a vinyl tube to either one of the bleeder plugs.
4. Depress the pedal several times, and then loosen the bleeder plug with the pedal depressed.
5. When fluid stops coming out, immediately tighten the bleeder plug. Then release the pedal.
6. Repeat steps (3) and (4) until all the air in the fluid is gone.
7. Tighten the bleeder plug to 73 inch lbs. (8.3 Nm).
8. Install the cap.
9. Repeat the procedure to bleed air from the brake line for each wheel.

FLUID FILL PROCEDURE

➡If any work is performed on the brake system or if air in the brake lines is suspected, bleed the brake system.

1. Apply the parking brake before bleeding the brakes.
2. Add brake fluid to keep the level between the MIN and MAX lines of the reservoir while bleeding the brakes.

➡If brake fluid leaks onto any painted surface, immediately wash it off.

3. Bleed the brake line of the wheel farthest from the master cylinder first.
4. Remove center cowl top ventilator louver.
   a. Slide the hood to cowl top seal and detach the 2 clips and claw.
   b. Detach the 2 claws and 3 guides and remove the center cowl top ventilator louver.
5. Replacement brake fluid
   a. Remove the brake master cylinder reservoir filler cap assembly.
   b. Add brake fluid to keep the level between the MIN and MAX lines of the reservoir while bleeding the brakes.
   c. Remove the bleeder plug cap.
   d. Connect a vinyl tube to the bleeder plug.
   e. Depress the brake pedal several times, and then loosen the bleeder plug with the pedal depressed.
   f. When fluid stops coming out, tighten the bleeder plug, and then release the brake pedal.
   g. Repeat e. and f. until all the air in the fluid is completely bled out.
   h. Tighten the bleeder plug completely.
   i. Install the bleeder plug cap.
   j. Repeat the above procedure for each wheel to bleed the brake line.
   k. Check for brake fluid leaks.
   l. Inspect the fluid level.
6. Install center cowl top ventilator louver
   a. Attach the 2 claws and 3 guides to install the center cowl top ventilator louver.
7. Push the hood to cowl top seal to attach the 2 clips, and slide the hood to cowl top seal to attach the claw.

CAUTION

Dust and dirt accumulating on brake parts during normal use may contain asbestos fibers from production or aftermarket brake linings. Breathing excessive concentrations of asbestos fibers can cause serious bodily harm. Exercise care when servicing brake parts. Do not sand or grind brake lining unless equipment used is designed to contain the dust residue. Do not clean brake parts with compressed air or by dry brushing. Cleaning should be done by dampening the brake components with a fine mist of water, then wiping the brake components clean with a dampened cloth. Dispose of cloth and all residue containing asbestos fibers in an impermeable container with the appropriate label. Follow practices prescribed by the Occupational Safety and Health Administration (OSHA) and the Environmental Protection Agency (EPA) for the handling, processing, and disposing of dust or debris that may contain asbestos fibers.

BRAKE CALIPER

REMOVAL & INSTALLATION

tC

See Figure 5.

➡Use the same procedure for the RH and LH sides. The procedure listed below is for the LH side.

1. Remove front wheel LH.
2. Remove center cowl top ventilator louver.
3. Drain brake fluid.

➡Wash off brake fluid immediately if it comes in contact with any painted surface.

4. Remove the union bolt and gasket and disconnect the front flexible hose from the front disc brake cylinder assembly.
5. Remove the 2 bolts and disc brake cylinder assembly.
6. Remove the 2 front disc brake pads from the front disc brake cylinder mounting.
7. Remove front anti-squeal shim.
   a. Remove the anti-squeal shims from the front disc brake pad.
   b. Remove the 2 pad wear indicator plates from the front disc brake pad (inner and outer).
8. Remove the 2 front No. 1 disc brake pad support plates and 2 front No. 2 disc brake pad support plates from the front disc brake cylinder mounting.

➡Each front disc brake pad support plate has a different shape. Be sure to put an identification mark on each front disc brake pad support plate so that it can be reinstalled to its original position.

9. Remove the 2 front disc brake cylinder slide pins from the front disc brake cylinder mounting.
10. Using a screwdriver, remove the front disc brake cylinder slide bushing from the front disc brake cylinder slide pin.

➡Tape the screwdriver tip before use. Do not damage the front disc brake cylinder slide pin.

11. Remove the 2 front disc brake bushing dust boots from the front disc brake cylinder mounting.
12. Remove the 2 bolts and front disc brake cylinder mounting from the steering knuckle.
13. Remove front disc.
   a. Put matchmarks on the disc and axle hub.
   b. Remove the front disc.

To install:

14. Align the matchmarks of the front disc and axle hub and install the front disc.

➡When replacing the front disc with a new one, select the installation position where the front disc has the smallest runout.

15. Install the front disc brake cylinder mounting to the steering knuckle with the 2 bolts. tighten to 79 ft. lbs. (107 Nm).

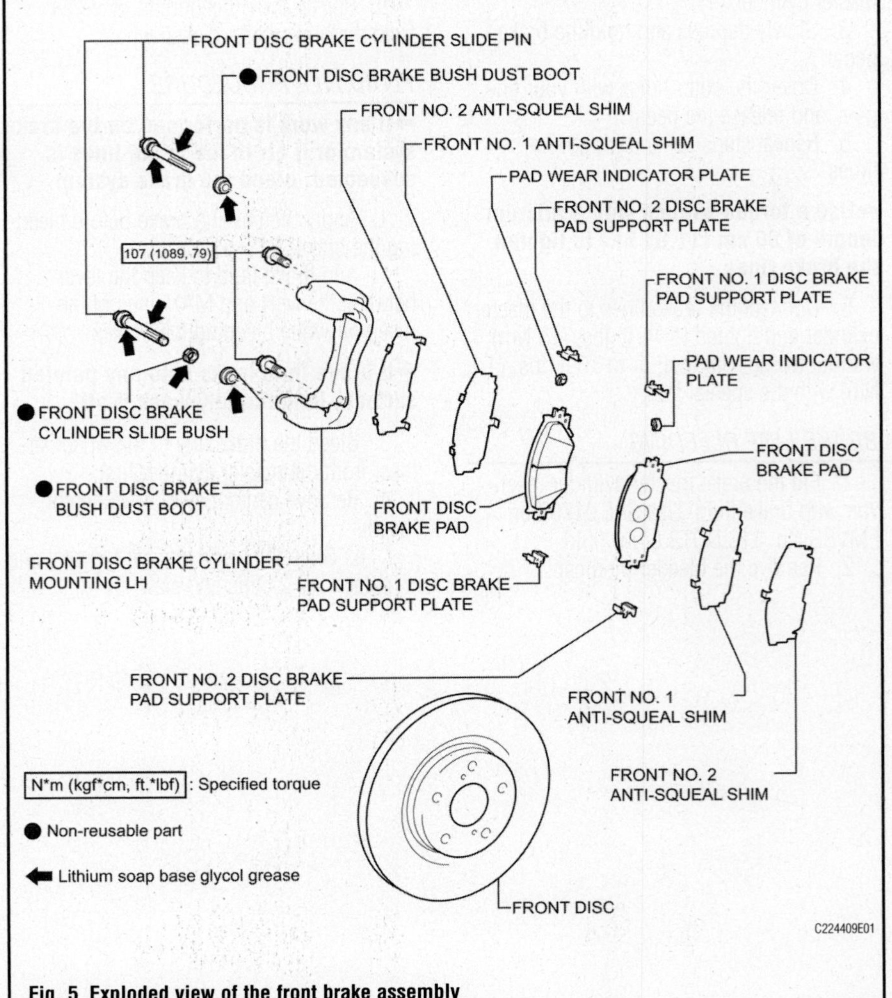

Fig. 5 Exploded view of the front brake assembly

16. Install front disc brake bushing dust boot.

a. Apply a light coat of lithium soap base glycol grease to the entire circumference of 2 new front brake bushing dust boots where they contact the front disc brake cylinder mounting, and the entire inner circumference of both ends of the boots.

➡Apply lithium soap base glycol grease to the front disc brake bushing dust boot.

➡Apply a sufficient amount of lithium soap base glycol grease to the entire circumference of the front disc brake bushing dust boot and front disc brake cylinder mounting contact surfaces.

b. Install the 2 front disc brake bushing dust boots to the disc brake cylinder mounting.

17. Install front disc brake cylinder slide bushing.

a. Apply a light coat of lithium soap base glycol grease to a new front disc brake cylinder slide bushing and front disc brake cylinder slide pin.

b. Install the front disc brake cylinder slide bushing to the front disc brake cylinder slide pin.

18. Install front disc brake cylinder slide pin.

a. Apply a light coat of lithium soap base glycol grease to the sliding part and the seal surface of the 2 front disc brake cylinder slide pins.

b. Install the 2 front disc brake cylinder slide pins to the disc brake cylinder mounting.

19. Install the 2 front No. 1 disc brake pad support plates and 2 front No. 2 disc brake pad support plates to the disc brake cylinder mounting.

➡Be sure to install each front disc brake pad support plate in the correct position and direction.

20. Install front anti-squeal shim.

a. Install the front anti-squeal shims (No. 1 and No. 2) to the front disc brake pad.

➡When replacing worn pads, the shims must be replaced together with the pads. Install the shims in the correct positions and directions.

b. Install the 2 pad wear indicator plates to the pad (inner and outer).

➡Install the pad wear indicator plate in the correct position and direction.

21. Install the 2 disc brake pads to the front disc brake cylinder mounting.

➡Make sure there is no oil or grease on the friction surfaces of the disc brake pads or the front disc.

22. Hold the front disc brake cylinder slide pin and install the front disc brake cylinder assembly to the front disc brake cylinder mounting with the 2 bolts. Tighten to 25 ft. lbs. (34 Nm).

23. Install a new gasket and connect the flexible hose with a new union bolt.

➡Insert the flexible hose lock securely into the lock hole in the brake cylinder.

24. Bleed brake line.

25. Bleed clutch line (for manual transaxle).

26. Install center cowl top ventilator louver.

27. Inspect for fluid leak.

28. Install front wheel LH.

xB and xD

See Figures 6 and 7.

1. Remove front wheel.
2. Drain brake fluid.

➡Immediately wash off any brake fluid that comes into contact with any painted surfaces.

3. Remove the union bolt and gasket and separate the flexible hose from the disc brake cylinder.
4. Hold the slide pin with a spanner, remove the 2 bolts and remove the disc brake cylinder.
5. Remove the 2 brake pads from the disc brake cylinder mounting.
6. Remove the 4 anti-squeal shims from each brake pad.
7. Remove the 4 pad support plates from the disc brake cylinder mounting.

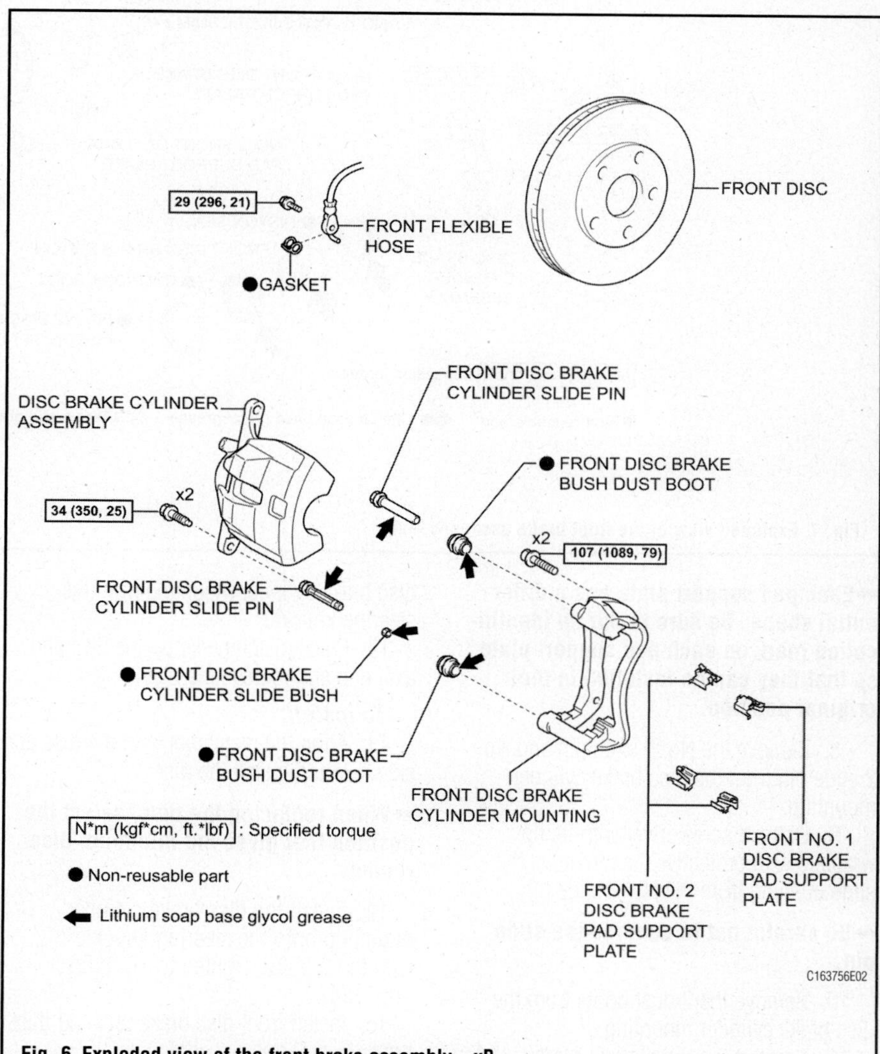

Fig. 6 Exploded view of the front brake assembly—xB

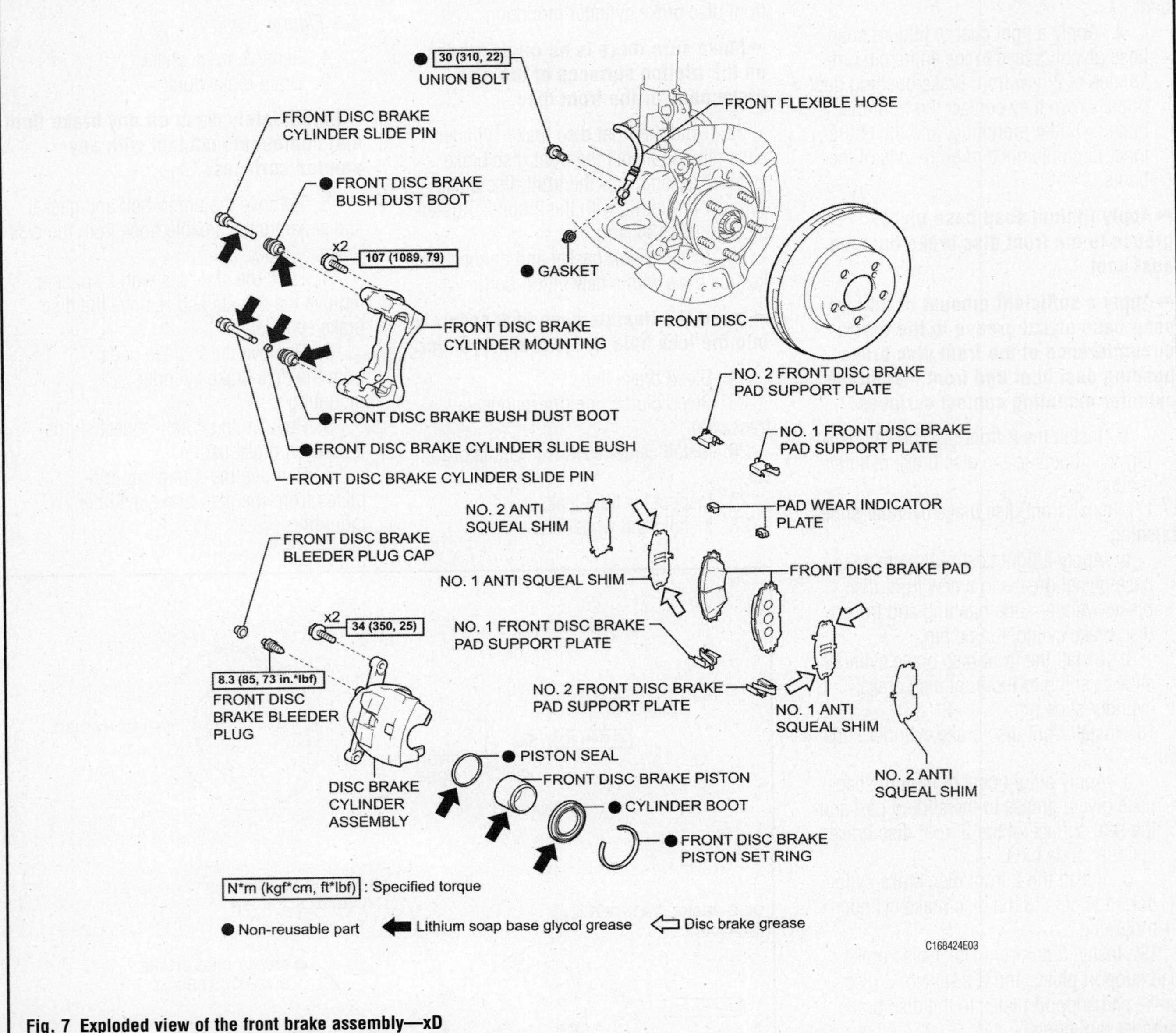

Fig. 7 Exploded view of the front brake assembly—xD

➡**Each pad support plate has a differential shape. Be sure to put an identification mark on each pad support plate so that they can be installed in their original position.**

8. Remove the No. 1 slide pin and No. 2 slide pin from the disc brake cylinder mounting.

9. Using a screwdriver with its tip wrapped with protective tape, remove the slide bushing from the No. 2 slide pin.

➡**Be careful not to damage the slide pin.**

10. Remove the 2 dust boots from the disc brake cylinder mounting.

11. Remove the 2 bolts and remove the disc brake cylinder mounting from the steering knuckle.

12. Place matchmarks on the disc and axle hub and remove the disc.

To install:

13. Align the matchmarks of the disc and axle hub and install the disc.

➡**When replacing the disc, select the position that gives the minimum disc runout.**

14. Install the disc brake cylinder mounting onto the steering knuckle with the 2 bolts. Tighten to 79 ft. lbs. (107 Nm).

15. Install front disc brake bushing dust boot.

a. Apply a light layer of lithium soap base glycol grease to the entire circumference of 2 new dust boots.

b. Install the 2 dust boots onto the disc brake cylinder mounting.

16. Install front disc brake cylinder slide bushing.

a. Apply a light layer of lithium soap base glycol grease to the slide pins and a new slide bushing.

b. Install the slide bushing onto the No. 2 slide pin.

17. Install the No. 1 slide pin and No. 2 slide pin onto the cylinder mounting.

➡**The No. 1 slide pin is for the upper side and the No. 2 slide pin is for the lower side.**

18. Install the 4 pad support plates onto the disc brake cylinder mounting.

➡**Be sure to install each pad support plate to the correct position and direction.**

19. Install front anti-squeal shim.
 a. Apply disc brake grease to both sides of each No. 1 anti-squeal shim.

➡**When replacing worn pads, the anti-squeal shims must be replaced together with the pads. Install the shims in the correct positions and directions. Apply disc brake grease to the area that contacts the anti-squeal shims. Disc brake grease can come out slightly from the area where the anti-squeal shim is installed. Make sure that disc brake grease is not applied onto the lining surface.**

 b. Install the 4 anti-squeal shims to each brake pad.

20. Install the 2 disc brake pads onto the disc brake cylinder mounting.

➡**There should be no oil or grease on the friction surfaces of the disc brake pads or the front disc.**

21. Hold the slide pin with a spanner and install the disc brake cylinder onto the disc brake cylinder mounting with the 2 bolts. Tighten to 25 ft. lbs. (34 Nm).

22. Install the flexible hose with the union bolt and a new gasket. Tighten to 21 ft. lbs. (29 Nm).

➡**Install the flexible hose lock securely into the lock hole in the disc brake cylinder.**

23. Fill reservoir with brake fluid.
24. Bleed master cylinder.
25. Bleed brake line.
26. Bleed brake actuator.
27. Inspect fluid level in reservoir.
28. Inspect for brake fluid leak.
29.. Install front wheel.

DISC BRAKE PADS

REMOVAL & INSTALLATION

Refer to the Front Disc Brakes, Brake Caliper Removal and Installation for procedure to remove and replace the brake pads.

REAR DISC BRAKES

⁂ CAUTION

Dust and dirt accumulating on brake parts during normal use may contain asbestos fibers from production or aftermarket brake linings. Breathing excessive concentrations of asbestos fibers can cause serious bodily harm. Exercise care when servicing brake parts. Do not sand or grind brake lining unless equipment used is designed to contain the dust residue. Do not clean brake parts with compressed air or by dry brushing. Cleaning should be done by dampening the brake components with a fine mist of water, then wiping the brake components clean with a dampened cloth. Dispose of cloth and all residue containing asbestos fibers in an impermeable container with the appropriate label. Follow practices prescribed by the Occupational Safety and Health Administration (OSHA) and the Environmental Protection Agency (EPA) for the handling, processing, and disposing of dust or debris that may contain asbestos fibers.

BRAKE CALIPER

REMOVAL & INSTALLATION

tC

See Figure 8.

➡**Use the same procedure for the RH and LH sides. The procedure listed below is for the LH side.**

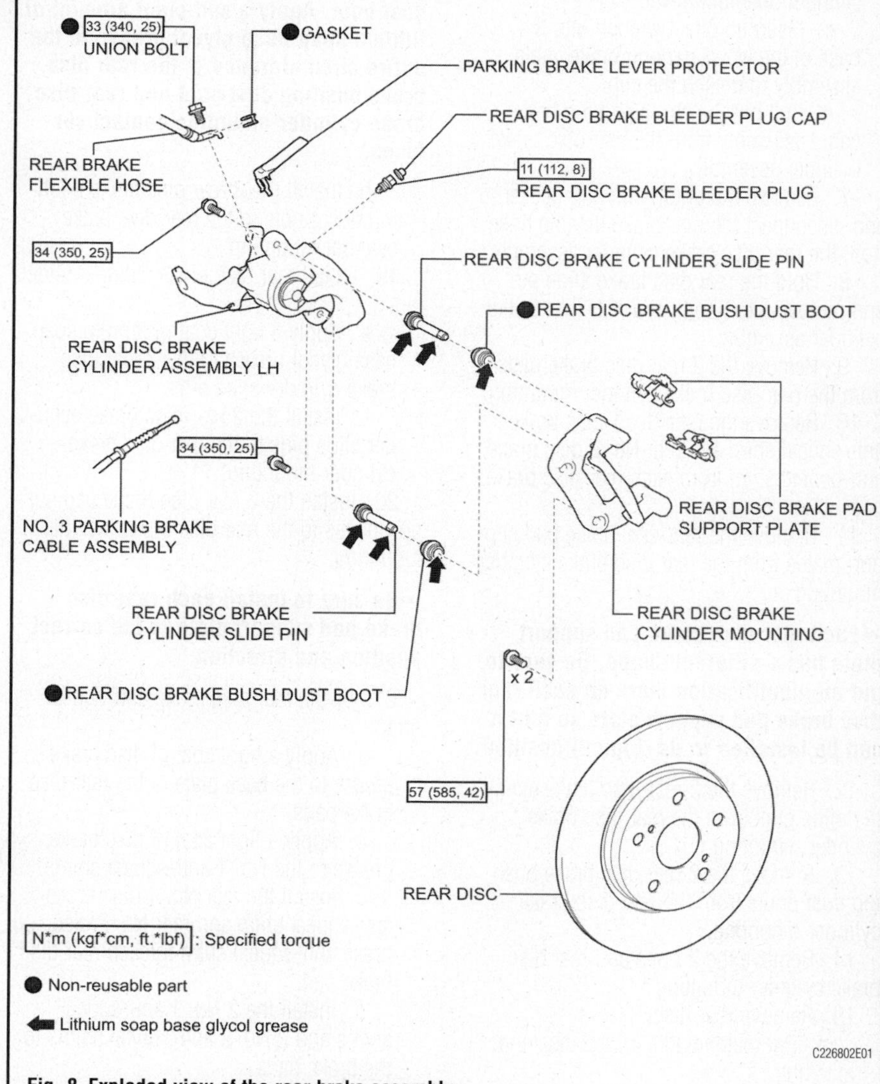

Fig. 8 Exploded view of the rear brake assembly

1. Remove rear wheel.
2. Remove center cowl top ventilator louver.
3. Drain brake fluid.

➡**Immediately wash off any brake fluid that comes into contact with any painted surfaces.**

4. Remove rear console box sub-assembly.
5. Loosen parking brake cable.
   a. Completely release the parking brake.
   b. Loosen the lock nut and adjusting nut to completely release the parking brake cable.
6. Disconnect No. 3 parking brake cable assembly
   a. Remove the parking brake lever protector.
   b. Disconnect the No. 3 parking brake cable assembly from the rear disc brake cylinder operation lever.
   c. Insert an offset wrench into the base of the No. 3 parking brake cable assembly to detach the clip.
   d. Pull out the No. 3 parking brake cable assembly from the rear disc brake cylinder assembly.
7. Remove the union bolt and gasket and disconnect the rear brake flexible hose from the rear disc brake cylinder assembly.
8. Hold the rear disc brake slide pin and remove the 2 bolts and rear disc brake cylinder assembly.
9. Remove the 2 rear disc brake pads from the rear disc brake cylinder mounting.
10. Remove the rear No. 1 disc brake anti-squeal shim and rear No. 2 disc brake anti-squeal shim from each rear disc brake pad.
11. Remove the rear disc brake pad support plates from the rear disc brake cylinder mounting.

➡**Each rear disc brake pad support plate has a different shape. Be sure to put an identification mark on each rear disc brake pad support plate so that it can be installed to its original position.**

12. Remove the 2 rear disc brake cylinder slide pins from the rear disc brake cylinder mounting.
13. Remove the 2 rear disc brake bushing dust boots from the rear disc brake cylinder mounting.
14. Remove the 2 bolts and rear disc brake cylinder mounting.
15. Remove rear disc.
   a. Put matchmarks on the disc and axle hub.
   b. Remove the rear disc.

To install:

16. Align the matchmarks of the rear disc and axle hub and install the rear disc.

➡**When replacing the rear disc with a new one, select the installation position where the rear disc has the smallest runout.**

17. Install the rear disc brake cylinder mounting to the axle beam with the 2 bolts. Tighten to 42 ft. lbs. (57 Nm).
18. 3. Install rear disc brake bushing dust boot.
   a. Apply a light coat of lithium soap base glycol grease to the entire circumference of the 2 new rear disc brake bushing dust boots where they contact the rear disc brake cylinder mounting, and the entire inner circumference of both ends of the boots.

➡**Apply lithium soap base glycol grease to the rear disc brake bushing dust boot. Apply a sufficient amount of lithium soap base glycol grease to the entire circumference of the rear disc brake bushing dust boot and rear disc brake cylinder mounting contact surfaces.**

   b. Install the 2 rear disc brake bushing dust boots to the rear disc brake cylinder mounting.
19. Install rear disc brake cylinder slide pin.
   a. Apply a light coat of lithium soap base glycol grease to the 2 rear disc brake cylinder slide pins.
   b. Install the 2 rear disc brake cylinder slide pins to the rear disc brake cylinder mounting.
20. Install the 2 rear disc brake pad support plates to the rear disc brake cylinder mounting.

➡**Be sure to install each rear disc brake pad support plate in the correct position and direction.**

21. Install rear disc brake anti-squeal shim.
   a. Apply a light coat of disc brake grease to the back plate of the rear disc brake pads.
   b. Apply a light coat of disc brake grease to the No. 1 anti-squeal shims.
   c. Install the rear No. 1 disc brake anti-squeal shim and rear No. 2 disc brake anti-squeal shim to each rear disc brake pad.
   d. Install the 2 No. 1 anti-squeal shims and 2 No. 2 anti-squeal shims to the brake pads.

➡**When replacing worn pads, the anti-squeal shims must be replaced together with the pads. Apply a light coat of disc brake grease to the area that contacts the anti-squeal shim. Disc brake grease may seep out slightly from the areas where the anti-squeal shims are installed. Make sure that disc brake grease is not applied to the lining surface.**

22. Install the 2 disc brake pads to the rear disc brake cylinder mounting.

➡**Make sure there is no oil or grease on the friction surfaces of the disc brake pads or the rear disc.**

23. Install rear disc brake cylinder assembly LH.
   a. To compensate for pad lining thickness, use SST to adjust the protrusion of the rear disc brake piston by turning it.

➡**Place the rear disc between the 2 rear disc brake pads and determine the piston return value. Turn the rear disc brake piston to the position where the protrusion on the rear disc brake pad lines up properly with the piston groove.**

   b. Hold the rear disc brake cylinder slide pin and install the rear disc brake cylinder assembly to the rear disc brake cylinder mounting with the 2 bolts. Tighten to 25 ft. lbs. (33 Nm).

➡**Install the lower bolt first, and then the upper bolt.**

24. Install a new gasket and connect the flexible hose to the rear disc brake cylinder assembly with a new union bolt. Tighten to 25 ft. lbs. (33 Nm).

➡**Insert the flexible hose lock securely into the lock hole in the disc brake cylinder.**

25. Connect No. 3 parking brake cable assembly.
   a. Insert the No. 3 parking brake cable assembly into the rear disc brake cylinder assembly and attach the clip of the No. 3 parking brake cable to the rear disc brake cylinder guide.
   b. Connect the No. 3 parking brake cable end to the rear disc brake cylinder operation lever.
   c. Install the parking brake lever protector.
26. Bleed brake line.
27. Bleed clutch line (for manual transaxle).
28. Install center cowl top ventilator louver.

29. Inspect for fluid leak.

30. Check and adjust parking brake lever travel.

31. Inspect rear disc brake cylinder operation lever and stopper clearance.

32. Install rear console box sub-assembly.

33. Install rear wheel.

xB

See Figure 9.

1. Remove rear wheel.
2. Drain brake fluid.

➡Immediately wash off any brake fluid that comes into contact with any painted surfaces.

3. Remove upper console panel sub-assembly.
4. Loosen wire adjusting nut.
   a. Completely release the parking brake lever.
   b. Loosen the lock nut and adjusting nut to completely release the parking brake cable.
5. Remove rear floor side member cover.
6. Separate parking brake cable assembly.
   a. Remove the bolt and separate the cable bracket.
   b. Disconnect the cable end from the rear disc brake cylinder operation lever.
   c. Insert an offset wrench at the base of the parking brake cable to disengage the clip. Pull out the parking brake cable from the rear disc brake cylinder.
7. Remove the union bolt and gasket and separate the flexible hose from the disc brake cylinder.
8. Hold the guide pin with a spanner, remove the 2 bolts and remove the disc brake cylinder.
9. Remove the 2 brake pads from the disc brake cylinder mounting.
10. Remove the 4 anti-squeal shims from each brake pad.
11. Remove the 2 pad support plates from the disc brake cylinder mounting.

➡Each pad support plate has a differential shape. Be sure to put an identification mark on each pad support plate so that they can be installed to their original position.

12. Remove the 2 guide pins from the disc brake cylinder mounting.
13. Remove the 2 dust boots from the disc brake cylinder mounting.
14. Remove the 2 bolts and remove the disc brake cylinder mounting from the axle beam.
15. Place matchmarks on the disc and axle hub and remove the disc.

To install:

16. Align the matchmarks of the disc and axle hub and install the disc.

➡When replacing the disc, select the position that gives the minimum disc runout.

17. Install the disc brake cylinder mounting onto the axle beam with the 2 bolts. Tighten to 42 ft. lbs. (57 Nm).
18. Install rear disc brake bushing dust boot.
    a. Apply a light layer of lithium soap base glycol grease to the entire circumference of 2 new dust boots.

➡Apply lithium soap base glycol grease to each dust boot.

    b. Install the 2 dust boots onto the disc brake cylinder mounting.

19. Install rear disc brake pad guide pin.
    a. Apply a light layer of lithium soap base glycol grease to 2 guide pins.
    b. Install the 2 guide pins onto the disc brake cylinder mounting.
20. Install the 2 pad support plates onto the disc brake cylinder mounting.

➡Be sure to install each pad support plate to the correct position and direction.

21. Install rear disc brake anti-squeal shim.
    a. Apply disc brake grease to the inside of the 2 No. 1 anti-squeal shims.

➡When replacing worn pads, the anti-squeal shims must be replaced together with the pads. Apply disc brake grease to the area that contacts the anti-squeal shim. Disc brake grease can come out slightly from the area where the anti-

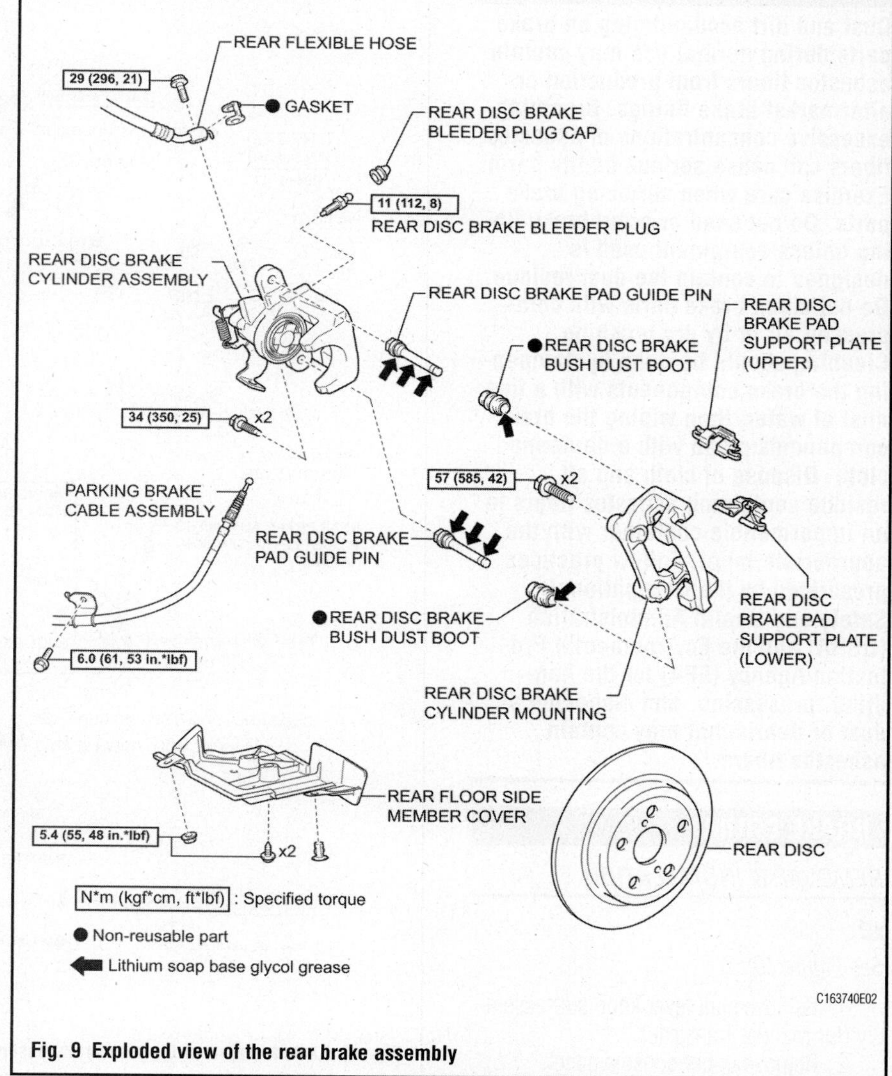

Fig. 9 Exploded view of the rear brake assembly

squeal shim is installed. Make sure that disc brake grease is not applied onto the lining surface.

b. Install the 4 anti-squeal shims onto each brake pad.

22. Install the 2 disc brake pads onto the disc brake cylinder mounting.

➡There should be no oil or grease on the friction surfaces of the disc brake pads or the rear disc.

23. Install rear disc brake cylinder assembly.

a. When reusing the pads, use SST to turn the piston counterclockwise to the position where the protrusion portion of the pad does not protrude beyond the piston groove.

➡Place the disc between the 2 brake pads and determine the piston return value.

b. Hold the guide pin with a spanner and install the disc brake cylinder onto the disc brake cylinder mounting with the 2 bolts. Tighten to 25 ft. lbs. (34 Nm).

24. Install the flexible hose with the union bolt and a new gasket. Tighten to 21 ft. lbs. (29 Nm).

➡Install the flexible hose lock securely into the lock hole in the disc brake cylinder.

25. Install parking brake cable assembly.

a. Insert the parking brake cable into the disc brake cylinder and engage the clip claws of the parking brake cable to the rear disc brake cylinder guide.

b. Connect the cable end to the rear disc brake cylinder operation lever.

c. Install the cable bracket with the bolt.

26. Install rear floor side member cover.
27. Fill reservoir with brake fluid.
28. Bleed master cylinder.
29. Bleed brake line.
30. Bleed brake actuator.
31. Inspect fluid level in reservoir.
32. Inspect for brake fluid leak.
33. Install rear wheel.
34. Adjust parking brake lever travel.
35. Install upper console panel sub-assembly.

BRAKES

REAR DRUM BRAKES

⁂ CAUTION

Dust and dirt accumulating on brake parts during normal use may contain asbestos fibers from production or aftermarket brake linings. Breathing excessive concentrations of asbestos fibers can cause serious bodily harm. Exercise care when servicing brake parts. Do not sand or grind brake lining unless equipment used is designed to contain the dust residue. Do not clean brake parts with compressed air or by dry brushing. Cleaning should be done by dampening the brake components with a fine mist of water, then wiping the brake components clean with a dampened cloth. Dispose of cloth and all residue containing asbestos fibers in an impermeable container with the appropriate label. Follow practices prescribed by the Occupational Safety and Health Administration (OSHA) and the Environmental Protection Agency (EPA) for the handling, processing, and disposing of dust or debris that may contain asbestos fibers.

BRAKE DRUM

REMOVAL & INSTALLATION

xD

See Figure 10.

1. Remove shift lever knob sub-assembly (for Manual Transaxle).
2. Remove upper console panel.

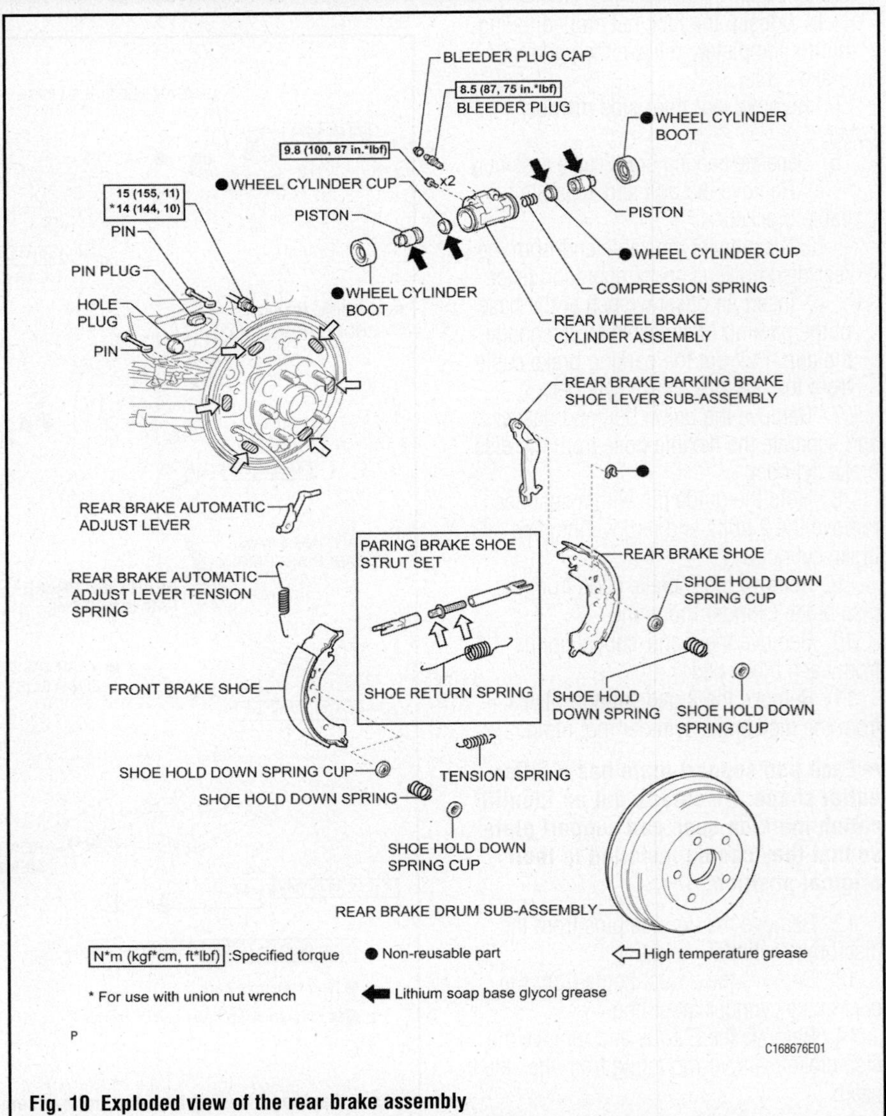

Fig. 10 Exploded view of the rear brake assembly

3. Remove rear console box cover.
4. Remove console box carpet.
5. Remove rear console box sub-assembly.
6. Remove rear wheel.
7. Drain brake fluid.

➡Immediately wash off any brake fluid that comes into contact with any painted surfaces.

8. Remove rear brake drum sub-assembly.
   a. Release the parking brake and remove the rear brake drum.
   If the rear brake drum cannot be removed easily, perform the following procedure.
   b. Remove the hole plug and insert a screwdriver through the hole into the backing plate, and hold the automatic adjust lever away from the adjuster.
   c. Using another screwdriver, contract the brake shoe by turning the adjusting bolt.

To install:

9. Install rear brake drum sub-assembly.
10. Adjust rear drum brake shoe clearance.
    a. Provisionally install the hub nuts.
    b. Remove the hole plug, and turn the adjuster to expand the shoe until the drum locks.
    c. Using a screwdriver, release the adjuster 11 notches.
    d. Install the hole plug.
11. Fill reservoir with brake fluid.
12. Bleed master cylinder.
13. Bleed brake line.
14. Bleed brake actuator.
15. Check fluid level in reservoir.
16. Inspect for brake fluid leak.
17. Install rear wheel.
18. Inspect parking brake lever travel.
19. Adjust parking brake lever travel.
20. Install rear console box sub-assembly.
21. Install console box carpet.
22. Install rear console box cover.
23. Install upper console panel.
24. Install shift lever knob sub-assembly (for Manual Transaxle).

BRAKE SHOES

REMOVAL & INSTALLATION

xD

See Figure 10.

1. Remove rear brake shoe kit.
   a. Using SST, separate the shoe return spring from the front brake shoe.

➡Do not damage the wheel cylinder boot.

   b. Using SST, remove the shoe hold down spring cup, shoe hold down spring, pin and front brake shoe.
   c. Remove the tension spring.
   d. Remove the shoe return spring from the rear brake shoe and remove the parking brake shoe strut set.
   e. Using SST, remove the shoe hold down spring cup, shoe hold down spring, pin and rear brake shoe.
   f. Using needle-nose pliers, separate the parking brake cable.
2. Remove the automatic adjust lever tension spring and remove the automatic adjust lever.
3. Using a screwdriver, remove the C-washer and remove the parking brake shoe lever.

To install:

4. Using needle-nose pliers, install the parking brake shoe lever with a new C-washer.
5. Install the automatic adjust lever and automatic adjust lever tension spring onto the front brake shoe.
6. Install rear brake shoe kit.
   a. Apply high temperature grease to the surface of the backing plate which is in contact with the shoe.
   b. Using needle-nose pliers, install the parking brake cable onto the parking brake shoe lever.
   c. Using SST, install the rear brake shoe, pin, shoe hold down spring and shoe hold down spring cup.
   d. Apply high temperature grease to the adjusting bolt.
   e. Install the parking brake shoe strut set.
   f. Using SST, install the front brake shoe, pin, shoe hold down spring and shoe hold down spring cup.
   g. Using needle-nose pliers, install the tension spring onto the front brake shoe and rear brake shoe.
   h. Using SST, install the shoe return spring onto the front brake shoe.

➡Do not damage the wheel cylinder boot.

7. Check rear drum brake installation.
   a. Check that each part is installed properly.
   b. Measure the brake drum inner diameter and the diameter of the brake shoes. Check that the difference between the diameters is equal to the specified shoe clearance.

BRAKES — PARKING BRAKE

PARKING BRAKE CABLES

ADJUSTMENT

tC and xB

See Figure 11.

1. Check parking brake lever travel.
   a. Pull the parking brake lever to fully engage the parking brake.
   b. Release the lever to disengage the parking brake.
   c. Slowly pull the parking brake lever all the way and count the number of clicks (5 to 8).
2. Remove rear wheel.
3. Adjust parking brake lever travel.

➡Make sure that the brake lines have been bled and no air is present before performing parking brake adjustment.

   a. Remove the rear console box assembly.
   b. Completely release the parking brake lever.
   c. Loosen the lock nut and adjusting nut to completely release the parking brake cable.
   d. Strongly depress the brake pedal 3 to 5 times with the engine stopped.
   e. Turn the adjusting nut until the parking brake lever travel is corrected to within the specified range of 5 to 8 notches at 45 lbs. (200 N).
   f. Tighten the new lock nut.
   g. Operate the parking brake lever 3 to 5 times and check the parking brake lever travel.
   h. Install the rear console box assembly.
4. Inspect rear disc brake cylinder operation lever and stopper clearance.
   a. Release the parking brake lever and check that the clearance measurement between the rear disc brake cylinder operation lever and the stopper is within the specified range of 0.02 inches (0.5 mm).
   b. If the clearance is not within the specified range, replace the rear disc brake cylinder assembly.

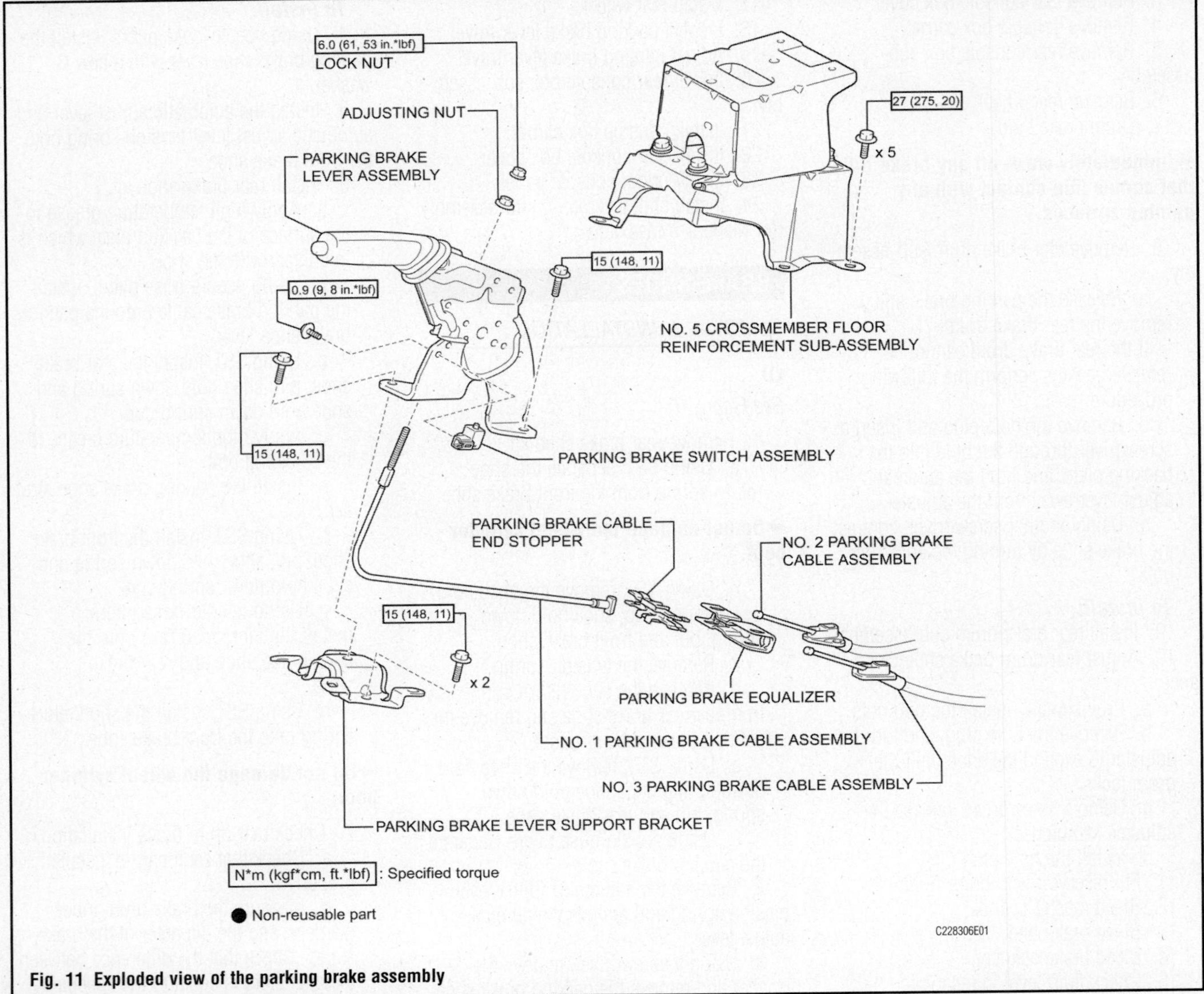

Fig. 11 Exploded view of the parking brake assembly

5. When operating the parking brake lever, check that the brake warning light illuminates.
6. Install rear wheel.

xD

See Figure 12.

1. Adjust rear drum brake shoe clearance.
2. Slowly pull the parking brake lever to the fully applied position, counting the number of clicks (6 to 9 clicks).
3. Remove shift lever knob sub-assembly (for Manual Transaxle).
4. Remove upper console panel.
5. Remove rear console box cover.
6. Remove console box carpet.
7. Remove rear console box sub-assembly.
8. Adjust parking brake lever travel.
   a. Loosen the lock nut and turn the adjusting nut until the parking brake lever travel is corrected to within the specified range of 6 to 9 clicks at 45 lbs. (200 N).
   b. Tighten the lock nut.
   c. Operate the parking brake lever 3 to 4 times, and check the parking brake lever travel.
   d. Check whether the parking brake drags or not.
   e. When operating the parking brake lever, check that the brake warning light illuminates.

➡Brake warning light always illuminates at the first click.

9. Install rear console box sub-assembly.
10. Install console box carpet.
11. Install rear console box cover.
12. Install upper console panel.
13. Install shift lever knob sub-assembly (for Manual Transaxle).

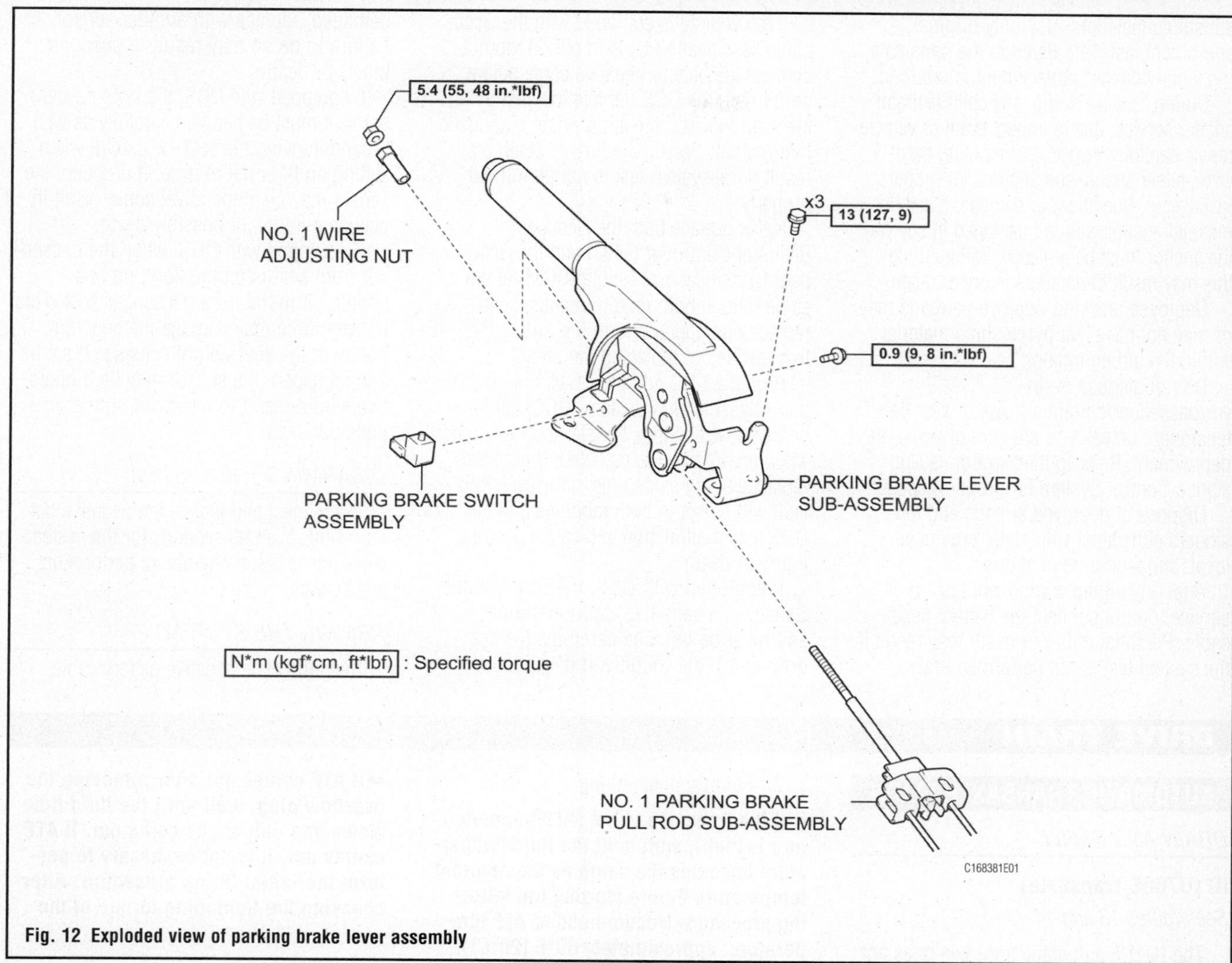

Fig. 12 Exploded view of parking brake lever assembly

CHASSIS ELECTRICAL

AIR BAG (SUPPLEMENTAL RESTRAINT SYSTEM)

GENERAL INFORMATION

CAUTION

These vehicles are equipped with an air bag system. The system must be disarmed before performing service on, or around, system components, the steering column, instrument panel components, wiring and sensors. Failure to follow the safety precautions and the disarming procedure could result in accidental air bag deployment, possible injury and unnecessary system repairs.

SERVICE PRECAUTIONS

Disconnect and isolate the battery negative cable before beginning any airbag system component diagnosis, testing, removal, or installation procedures. Allow system capacitor to discharge for two minutes before beginning any component service. This will disable the airbag system. Failure to disable the airbag system may result in accidental airbag deployment, personal injury, or death.

Do not place an intact undeployed airbag face down on a solid surface. The airbag will propel into the air if accidentally deployed and may result in personal injury or death.

When carrying or handling an undeployed airbag, the trim side (face) of the airbag should be pointing towards the body to minimize possibility of injury if accidental deployment occurs. Failure to do this may result in personal injury or death.

Replace airbag system components with OEM replacement parts. Substitute parts may appear interchangeable, but internal differences may result in inferior occupant protection. Failure to do so may result in occupant personal injury or death.

Wear safety glasses, rubber gloves, and long sleeved clothing when cleaning powder residue from vehicle after an airbag deployment. Powder residue emitted from a deployed airbag can cause skin irritation. Flush affected area with cool water if irritation is experienced. If nasal or throat irritation is experienced, exit the vehicle for fresh air until the irritation ceases. If irritation continues, see a physician.

Do not use a replacement airbag that is not in the original packaging. This may result in improper deployment, personal injury, or death.

The factory installed fasteners, screws and bolts used to fasten airbag components have a special coating and are specifically designed for the airbag system. Do not use substitute fasteners. Use only original equipment fasteners listed in the parts catalog when fastener replacement is required.

During, and following, any child restraint anchor service, due to impact event or vehicle repair, carefully inspect all mounting hardware, tether straps, and anchors for proper installation, operation, or damage. If a child restraint anchor is found damaged in any way, the anchor must be replaced. Failure to do this may result in personal injury or death.

Deployed and non-deployed airbags may or may not have live pyrotechnic material within the airbag inflator.

Do not dispose of driver/passenger/curtain airbags or seat belt tensioners unless you are sure of complete deployment. Refer to the Hazardous Substance Control System for proper disposal.

Dispose of deployed airbags and tensioners consistent with state, provincial, local, and federal regulations.

After any airbag component testing or service, do not connect the battery negative cable. Personal injury or death may result if the system test is not performed first.

If the vehicle is equipped with the Occupant Classification System (OCS), do not connect the battery negative cable before performing the OCS Verification Test using the scan tool and the appropriate diagnostic information. Personal injury or death may result if the system test is not performed properly.

Never replace both the Occupant Restraint Controller (ORC) and the Occupant Classification Module (OCM) at the same time. If both require replacement, replace one, then perform the Airbag System test before replacing the other.

Both the ORC and the OCM store Occupant Classification System (OCS) calibration data, which they transfer to one another when one of them is replaced. If both are replaced at the same time, an irreversible fault will be set in both modules and the OCS may malfunction and cause personal injury or death.

If equipped with OCS, the Seat Weight Sensor is a sensitive, calibrated unit and must be handled carefully. Do not drop or handle roughly. If dropped or damaged, replace with another sensor. Failure to do so may result in occupant injury or death.

If equipped with OCS, the front passenger seat must be handled carefully as well. When removing the seat, be careful when setting on floor not to drop. If dropped, the sensor may be inoperative, could result in occupant injury, or possibly death.

If equipped with OCS, when the passenger front seat is on the floor, no one should sit in the front passenger seat. This uneven force may damage the sensing ability of the seat weight sensors. If sat on and damaged, the sensor may be inoperative, could result in occupant injury, or possibly death.

DISARMING THE SYSTEM

Disconnect and isolate the negative battery cable. Wait 90 seconds for the system capacitor to discharge before performing any service.

ARMING THE SYSTEM

Reconnect the negative battery cable.

DRIVE TRAIN

AUTOMATIC TRANSAXLE FLUID

DRAIN AND REFILL

tC (U760E Transaxle)

See Figures 13 and 14.

The U760E automatic transaxle does not have an oil filler tube or oil level gauge. When adding fluid, add fluid through the refill hole in the transaxle case. The fluid level can be adjusted by draining excess fluid (allowing excess fluid to overflow) through the overflow tube of the oil pan.

➡"Overflow" indicates the condition under which fluid comes out of the overflow plug hole.

Before adjusting the fluid level, add the specified amount of fluid when the engine is cold and warm up the engine to circulate the fluid in the transaxle. Make sure that the fluid temperature is as specified and that the engine is idling.

The U760E automatic transaxle requires Toyota Genuine ATF WS.

1. Before filling transaxle with fluid.
   a. Lift the vehicle.

➡Set the vehicle on a lift so that the vehicle is kept level when it is lifted up.

   b. Remove the rear engine under cover LH.

2. Perform initial filling.

➡If the transaxle is hot (ATF temperature is high), wait until the fluid temperature becomes the same as the ambient temperature before starting the following procedure (recommended ATF temperature: approximately 68°F [20°C]).

   a. Remove the refill plug and gasket from the automatic transaxle.
   b. Using a 6 mm hexagon socket wrench, remove the overflow plug and gasket from the automatic transaxle.

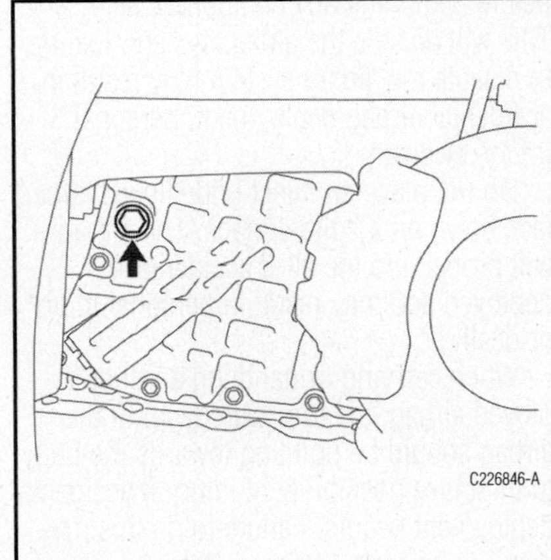

Fig. 13 Remove the refill plug and gasket from the automatic transaxle

➡If ATF comes out after removing the overflow plug, wait until the fluid flow slows and only drops come out. If ATF comes out, it is not necessary to perform the initial filling procedure. After checking the tightening torque of the

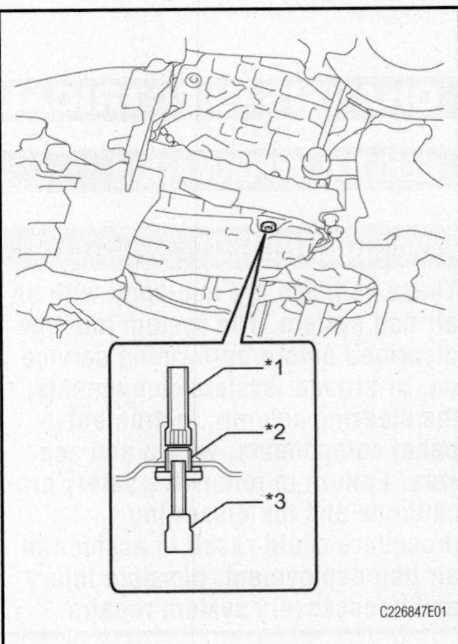

Fig. 14 Using a 6 mm hexagon socket wrench (3), remove the overflow plug (2) and gasket from the automatic transaxle; No. 1 transmission oil filler tube (1)

No. 1 transmission oil filler tube, temporarily install the overflow plug.

c. Using a 6 mm hexagon socket wrench, check that the No. 1 transmission oil filler tube is tightened to 7 inch lbs. (0.8 Nm).

➡If the transmission No. 1 transmission oil filler tube is not tightened to the specified torque, the amount of fluid cannot be precisely adjusted.

d. Perform initial filling. Fill the transaxle through the refill hole until fluid begins to trickle out of the overflow plug hole.
e. Wait until the fluid flow slows and only drops come out.
f. Temporarily install the overflow plug.

➡Reuse the old gasket. The plug will be removed again to adjust the fluid level.

3. Adjust the fluid level.
a. Using a 6 mm hexagon socket wrench, remove the overflow plug and gasket.
b. Check the amount of fluid that comes out of the overflow plug hole.

➡If only a small amount of fluid (approximately 1 cc) comes out of the overflow hole, then only fluid remaining in the No. 1 transmission oil filler tube has come out. This condition is not considered to be overflow, so it is necessary to add fluid.

c. If the amount of fluid that comes out of the overflow plug hole is large, wait until the fluid flow slows and only drops come out.
d. If no fluid comes out of the overflow plug hole, remove the refill plug and gasket. Then add transaxle fluid through the refill hole until fluid comes out of the overflow plug hole. Wait until the fluid flow slows and only drops come out.
e. Check that the fluid flow has slowed and only drops come out.

➡The fluid flow will not completely stop because the fluid expands as its temperature increases.

4. Install a new gasket and the overflow plug. Tighten to 30 ft. lbs. (40 Nm).
5. Install a new gasket and the refill plug. Tighten to 36 ft. lbs. (49 Nm).
6. Lower the vehicle.
7. Turn the ignition switch off.

➡Turning the ignition switch off exits fluid temperature detection mode.

xB (U241E Transaxle) & xD (U341E Transaxle)

1. Inspect automatic transaxle fluid.

➡Drive the vehicle until the engine and transaxle are at normal operating temperature, 158–176°F (70–80°C)

2. Park the vehicle on a level surface and engage the parking brake.
3. With the engine idling and the brake pedal depressed, shift the shift lever into all positions from P to L, and then return it to the P position.
4. Pull out the oil level gauge and wipe it clean.
5. Push it fully back into the pipe.
6. Pull it out and check that the fluid level is within the HOT range.
7. If there is any leakage, repair or replace O-rings, FIPG, oil seals, plugs or other parts.

MANUAL TRANSAXLE FLUID

DRAIN AND REFILL

tC

See Figure 15.

1. Drain manual transaxle oil.
a. Remove the filler plug and gasket.
b. Remove the drain plug and gasket to drain the manual transaxle oil.
c. Install a new gasket and the drain plug. tighten to 29 ft. lbs. (39 Nm).
2. Add manual transaxle oil.
a. Add manual transaxle oil until the oil level is within 0.20 inches (5 mm) from the bottom of the filler plug opening.
b. Install a new gasket and the filler plug. Tighten to 29 ft. lbs. (39 Nm).

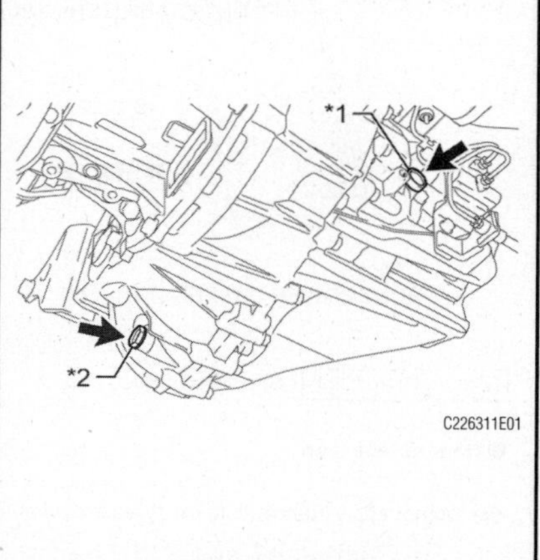

Fig. 15 Remove the filler plug (1) and drain plug (2)

➡When adding transaxle oil, make sure that the vehicle is level. An excessively large or small amount of oil may cause problems. After adding oil, drive the vehicle and recheck the oil level.

3. Inspect for oil leak.

xB

See Figure 16.

1. Stop the vehicle in a level place.
2. Remove the filler plug and the gasket.
3. Check that the oil surface is within 0.20 inches (5 mm) of the bottom of the filler plug opening.

➡Excessively large or small amounts of oil may cause problems. After replacing the oil, drive the vehicle and check the oil level again.

4. Check for oil leakage when the oil level is low.
5. Install the filler plug and a new gasket. Tighten to 36 ft. lbs. (49 Nm).

xD

See Figure 16

1. Stop the vehicle in a level place.
2. Remove the filler plug and the gasket.
3. Check that the oil surface is within 0.20 inches (5 mm) of the bottom of the filler plug opening.

➡Excessively large or small amounts of oil may cause problems. After replacing the oil, drive the vehicle and check the oil level again.

4. Check for oil leakage when the oil level is low.
5. Install the filler plug and a new gasket. Tighten to 29 ft. lbs. (39 Nm).

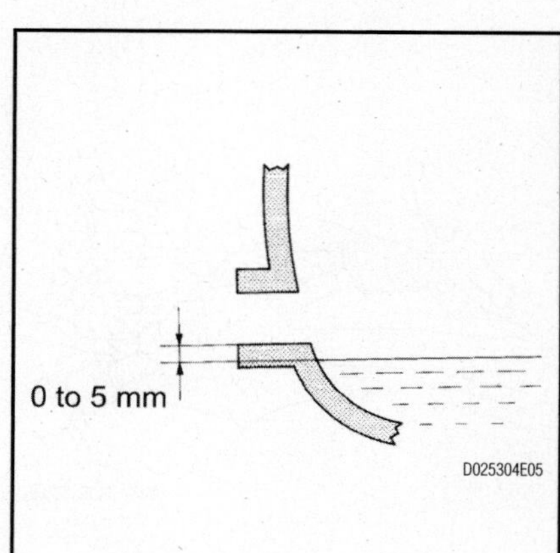

Fig. 16 Check that the oil surface is within 0.20 inches (5 mm) of the bottom of the filler plug opening

FRONT HALFSHAFT

REMOVAL & INSTALLATION

tC and xB

See Figures 17 and 18.

1. Before servicing the vehicle, refer to the Precautions Section.
2. Drain the transaxle fluid.
3. Remove or disconnect the following:
   - Front wheel
   - Engine undercover
   - Hub nut
   - Wheel speed sensor
   - Front stabilizer bar
   - Lower control arm
   - Tie rod end
4. Using Special Tool 09520-01010, tap out the left halfshaft.
5. Remove the right halfshaft as follows:
   a. xB: Using a brass bar and hammer, tap out the right halfshaft.
   b. tC: Remove the two mounting bolts and remove the halfshaft from the transaxle.

To install:

6. Coat the splines of the inboard joint shaft with gear oil (M/T) or ATF (A/T).
7. Align the shaft splines and tap in the left halfshaft with a brass bar and hammer.
8. Install the right halfshaft as follows:
   a. xB: Align the shaft splines and tap in the halfshaft with a brass bar and hammer.
   b. tC: Align the shaft splines and install the halfshaft to the transaxle. Tighten bolts to 47 ft. lbs. (64 Nm).
9. Install or connect the following:
   - Tie rod end. Tighten nut to 36 ft. lbs. (49 Nm).
   - Lower control arm
   - Front stabilizer arm. xB: Tighten nut to 13 ft. lbs. (18 Nm). tC: Tighten nut to 55 ft. lbs. (74 Nm).
   - Wheel speed sensor
   - New hub nut. Tighten to 159 ft. lbs. (216 Nm).
   - Engine undercover
   - Front wheel
10. Refill the transaxle with fluid to the correct level.
11. Check and adjust the alignment if necessary.

xD

See Figure 19.

1. Remove the engine under cover.
2. Drain the automatic transaxle fluid (A/T).

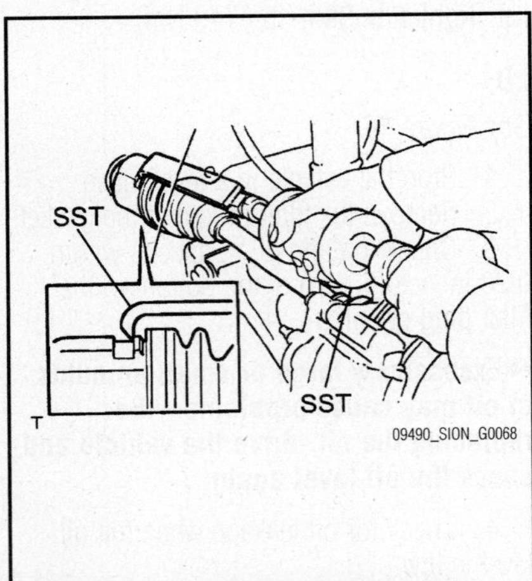

Fig. 17 Use Special Tool 09520-01010 to remove the left halfshaft—xB

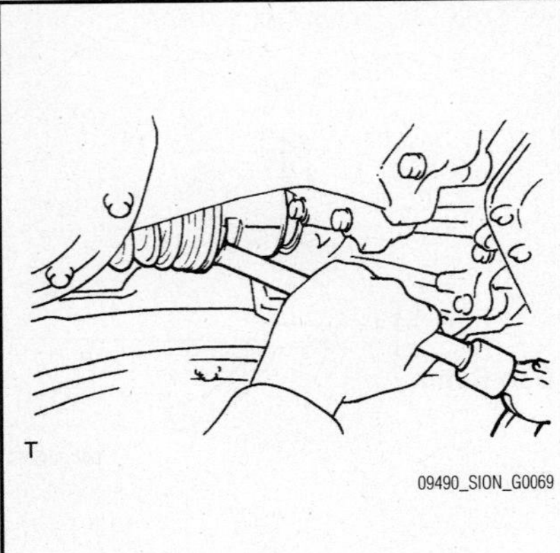

Fig. 18 Tap out the right halfshaft with a brass bar and hammer

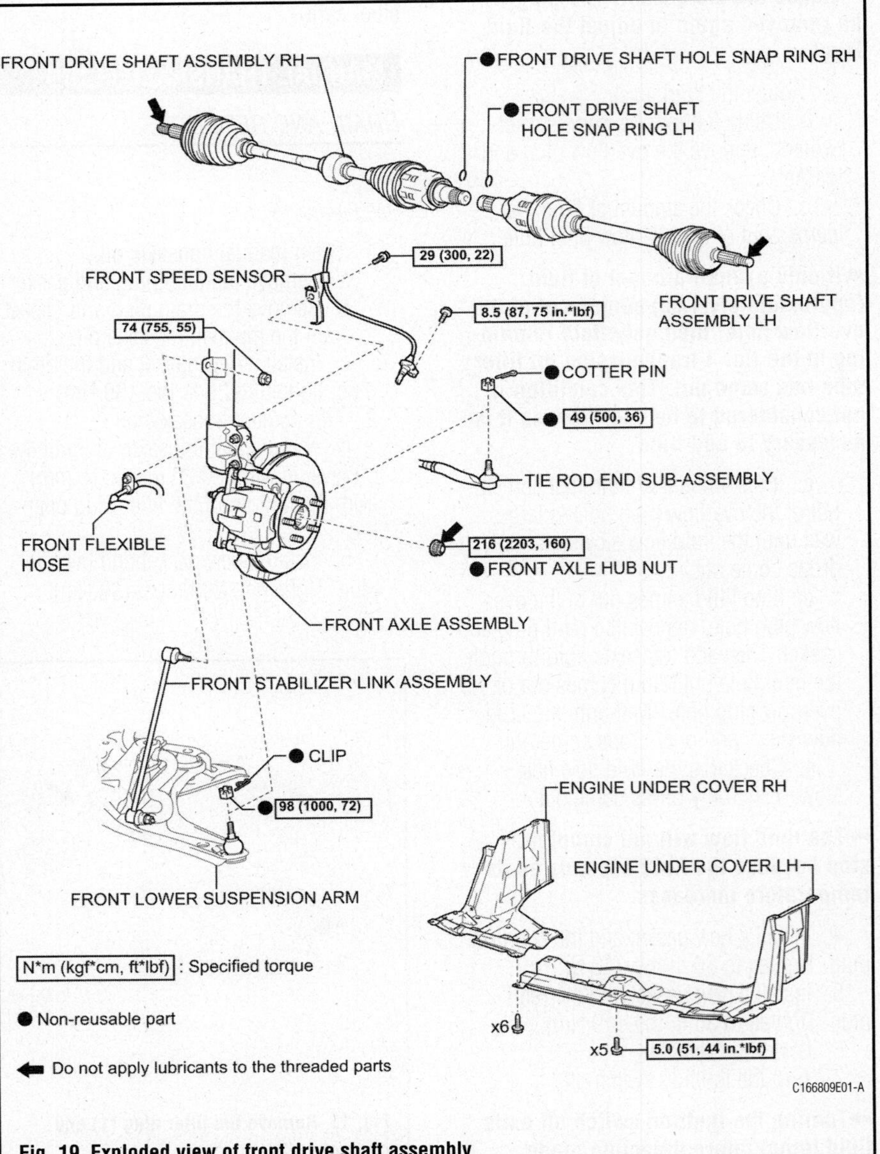

Fig. 19 Exploded view of front drive shaft assembly

3. Drain the manual transaxle oil (M/T).
4. Remove the front wheels.
5. Remove the front axle hub nut.
   a. Using SST and a hammer, release the staked part of the axle hub nut.

➡Insert SST into the groove with the flat surface facing up. Do not damage the tip of SST using grinders. Completely unstake the staked part before removing the axle hub nut. Do not damage the threads of the drive shaft.

   b. Using a 30 mm socket wrench, remove the axle hub nut.
6. Disconnect the front speed sensor.
   a. Remove the bolt and separate the speed sensor and flexible hose.
   b. Remove the bolt and separate the speed sensor from the steering knuckle.

➡Keep the speed sensor tip and installation portion free of foreign matter. Remove the speed sensor without turning it from its original installation angle.

7. Separate the front stabilizer link assembly.
   a. Remove the nut and separate the stabilizer link from the shock absorber.
8. Separate the tie rod end sub-assembly.
   a. Remove the cotter pin and castle nut.
   b. Install SST to the threaded section of the tie rod end.

➡Make sure the upper ends of the threaded section of the tie rod end and SST (spacer B) are aligned.

   c. Using SST, separate the tie rod end from the front axle assembly.

➡Make sure to tie the string of SST to the vehicle to prevent SST from dropping. Install SST so that A and B are parallel. Be sure to place the wrench on the part indicated. Do not damage the ball joint dust cover. Do not damage the front disc brake dust cover.

9. Separate the front lower suspension arm.
   a. Remove the clip and castle nut.
   b. Install SST (spacer B) to the threaded section of the lower ball joint.

➡Make sure the upper ends of the threaded section of the lower ball joint and SST are aligned.

   c. Using SST, separate the front lower suspension arm from the front axle assembly.

➡Make sure to tie the string of SST to the vehicle to prevent SST from dropping. Install SST so that A and B are parallel. Be sure to place the wrench on the part indicated. Do not damage the lower ball joint dust cover. Do not damage the drive shaft outboard joint boots. Do not damage the front disc brake dust cover.

10. Separate the front axle assembly.
   a. Using a plastic hammer, tap the end of the drive shaft and disengage the fitting between the drive shaft and front axle.

➡If it is difficult to disengage the fitting, tap the end of the drive shaft with a brass bar and hammer.

   b. Push the front axle out of the vehicle to remove the drive shaft from the front axle.

➡Do not push the front axle further out of the vehicle than is necessary. Do not damage the outboard joint boot. Do not damage the speed sensor rotor. Suspend the drive shaft with a piece of rope or the equivalent.

11. Remove the front drive shaft assembly LH.
   a. Using SST, remove the drive shaft.

➡Do not damage the oil seal. Do not damage the inboard joint boot. Do not drop the drive shaft.

12. Remove front drive shaft assembly RH.
   a. Using a screwdriver and hammer, remove the drive shaft.

To install:

13. Install the front drive shaft assembly LH.
   a. Coat the spline of the inboard joint with gear oil.
   b. Align the inboard joint splines and install the drive shaft with a screwdriver and hammer.

⁂ CAUTION

Face the cut area of the front drive shaft hole snap ring downward. Do not damage the oil seal. Do not damage the inboard joint boot. Confirm whether the drive shaft is securely driven in by checking the reaction force and sound.

14. Install the front drive shaft assembly RH.
   a. Use the same procedure as for the LH side.
15. Install the front axle assembly.
   a. Push the front axle out of the vehicle to align the spline of the drive shaft with the front axle and insert the front axle.

⁂ CAUTION

Do not push the front axle further out of the vehicle than is necessary. Do not damage the outboard joint boot. Check for any foreign matter on the speed sensor rotor and insertion part. Do not damage the speed sensor rotor.

16. Install the front lower suspension arm.
   a. Install the lower arm onto the steering knuckle with a new castle nut. Torque: 72 ft. lbs. (98 Nm).

➡If the holes for the clip are not aligned, tighten the nut by a further turn of up to 60°.

   b. Install a new clip.
17. Install the tie rod end sub-assembly.
   a. Install the tie rod end onto the steering knuckle with a new castle nut. Torque: 36 ft. lbs. (49 Nm).

➡If the holes for the clip are not aligned, tighten the nut by a further turn of up to 60°.

   b. Install a new cotter pin.
18. Install the front stabilizer link assembly.
   a. Install the stabilizer link with the nut. Torque: 55 ft. lbs. (74 Nm).
19. Install the front speed sensor.
   a. Install the speed sensor onto the steering knuckle with the bolt.

➡Check that the speed sensor tip and installation portion are free of foreign matter. Install the speed sensor without turning it from its original installation angle.

   b. Install the flexible hose and speed sensor with the bolt. Torque: 22 ft. lbs. (29 Nm).
   c. Install the flexible hose and speed sensor without twisting them.
20. Install the front axle hub nut.
   a. Clean the threaded parts on the drive shaft and axle hub nut using a non-residue solvent.

➡Be sure to perform this work for a new drive shaft. Keep the threaded parts free of oil and foreign objects.

   b. Using a 30 mm socket wrench,

install a new axle hub nut. Torque: 160 ft. lbs. (216 Nm).

c. Using a chisel and hammer, stake the axle hub nut.

21. Install the front wheels.
22. Add automatic transaxle fluid (A/T).
23. Inspect A/T fluid leak (A/T).
24. Add manual transaxle oil (M/T).
25. Inspect M/T oil leak (M/T).
26. Inspect and adjust front wheel alignment.
27. Install the engine under cover.

ENGINE COOLING

ENGINE COOLANT

DRAIN & REFILL PROCEDURE

tC

See Figure 20.

1. Remove center engine under cover.
2. Remove No. 1 engine under cover.
3. Drain coolant.

a. Install a vinyl hose to the drain cock plug on the radiator side.

b. Loosen the radiator drain cock plug and drain the coolant.

➡Collect the coolant in a container and dispose of it according to the regulations in your area.

c. Remove the radiator reservoir cap.

⁂ CAUTION

Do not remove the radiator reservoir cap while the engine and radiator are still hot. Pressurized, hot engine coolant and steam may be released and cause serious burns.

4. Add engine coolant.

a. Tighten the radiator drain cock plug by hand.

b. Add TOYOTA Super Long Life Coolant (SLLC) to the radiator reservoir filler opening until it is filled to the B line at the base of the reservoir filler neck.

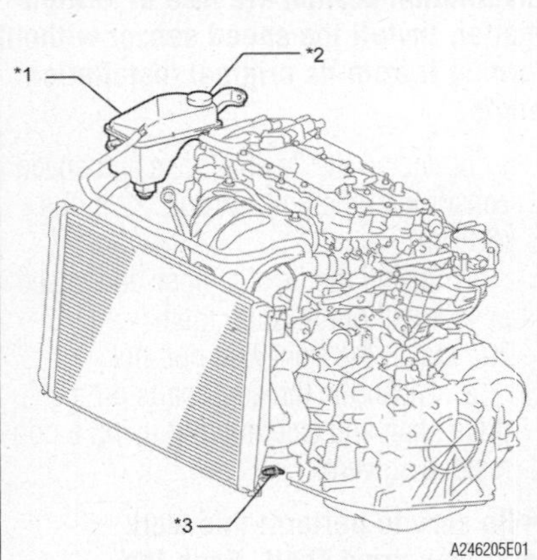

Fig. 20 Radiator reservoir (1), radiator reservoir cap (2), and radiator drain cock plug (3)

➡The B line is the lower edge of the inner wall of the filler neck.

➡TOYOTA vehicles are filled with TOYOTA SLLC at the factory. In order to avoid damage to the engine cooling system and other technical problems, only use TOYOTA SLLC or similar high quality ethylene glycol based non-silicate, non-amine, non-nitrite, non-borate coolant with long-life hybrid organic acid technology (coolant with long-life hybrid organic acid technology consists of a combination of low phosphates and organic acids).

➡Never use water as a substitute for engine coolant.

c. Press the No. 1 and No. 2 radiator hoses several times by hand, and then check the level of the coolant. If the coolant level drops below the B line, add TOYOTA SLLC to the B line.

d. Install the radiator reservoir cap.

e. Start the engine and warm it up until the cooling fan operates. Maintain the engine speed at 2000 to 2500 rpm and warm up the engine until the cooling fan operates.

➡Note the following:

- Make sure that the radiator reservoir still has some coolant in it.
- Pay attention to the needle of the coolant temperature meter. Make sure that the needle does not show an abnormally high temperature.
- If there is not enough coolant, the engine may burn out or overheat.
- After starting the engine, if the radiator reservoir does not have any coolant, perform the following: 1) stop the engine, 2) wait until the coolant has cooled down, and 3) add coolant until the coolant is filled to the B line.
- Run the engine at 2000 rpm until the coolant level has stabilized.

f. Press the No. 1 and No. 2 radiator hoses several times by hand to bleed air.

⁂ CAUTION

When pressing the radiator hoses:

- Wear protective gloves.
- Be careful as the radiator hoses are hot.
- Keep your hands away from the radiator fan.

g. Stop the engine and wait until the coolant cools down to ambient temperature.

h. Check that the coolant level is between the FULL and LOW line. If the coolant level is below the LOW line, repeat all of the procedures above.

i. If the coolant level is above the FULL line, drain coolant so that the coolant level is between the FULL and LOW line.

5. Inspect for coolant leak.
6. Install No. 1 engine under cover.
7. Install center engine under cover.

xB

See Figures 21 and 22.

1. Drain engine coolant.

a. Loosen the radiator drain cock plug.

➡Collect the coolant in a container and dispose of it according to the regulations in your area.

b. Remove the radiator reserve tank cap.

⁂ CAUTION

Do not remove the radiator reserve tank cap while the engine and radiator are still hot. Pressurized, hot engine coolant and steam may be released and cause serious burns.

c. Loosen the cylinder block drain cock plug.

2. Add engine coolant.

➡Be sure to fill the coolant according to the following process in order to bleed air completely, if you drain the coolant while repairing. Improper work may cause engine damage, because air may remain in the cooling system. Do not open the radiator cap when the engine and radiator are hot.

a. Tighten the radiator drain cock plug by hand.

b. Tighten the cylinder block drain cock plug.

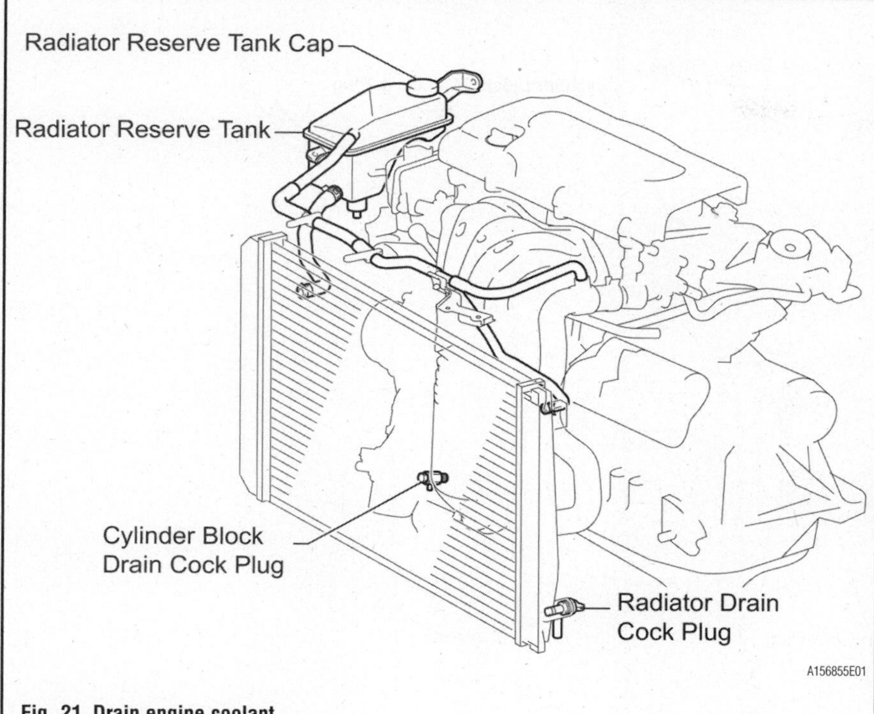

Fig. 21 Drain engine coolant

c. Remove the radiator reserve tank cap, then fill TOYOTA Super Long Life Coolant up to "B" line of reserve tank.

➡TOYOTA vehicles are filled with TOYOTA Super Long Life Coolant at the factory. In order to avoid damage to the engine cooling system and other technical problems, only use TOYOTA Super Long Life Coolant or similar high quality ethylene glycol based non-silicate, non-amine, non-nitrite, non-borate coolant with long-life hybrid organic acid technology (coolant with long-life hybrid organic acid technology consists of a combination of low phosphates and organic acids).

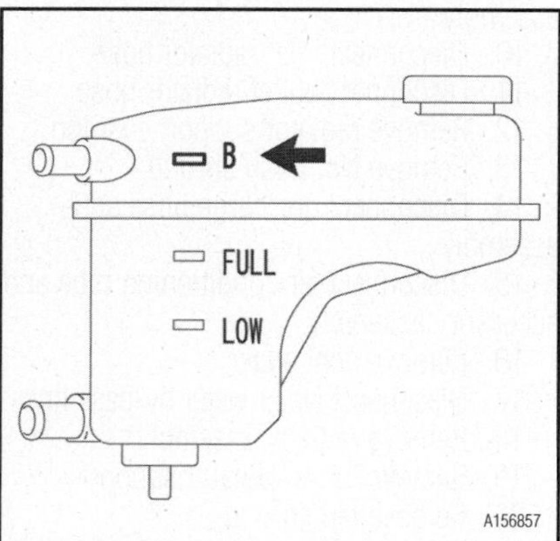

Fig. 22 Remove the radiator reserve tank cap, then fill TOYOTA Super Long Life Coolant up to "B" line of reserve tank

➡Never use water as a substitute for engine coolant.

d. Squeeze the inlet and outlet radiator hoses several times by hand, and then check the level of the coolant. If the coolant level drops below the "B" line, add TOYOTA Super Long Life Coolant to the B line.

e. Bleed air from the cooling system. Warm up the engine until the thermostat opens. While the thermostat is open, allow the coolant to circulate for several minutes.

➡The thermostat opening timing can be confirmed by squeezing the inlet radiator hose by hand, and sensing vibrations when the engine coolant starts to flow inside the hose.

⁂ CAUTION

When squeezing the radiator hose:

- Wear protective gloves.
- Be careful as the radiator hose is hot.
- Keep your hands away from the radiator fan.

f. Run the engine intermittently (5 seconds run at 3000 rpm and 45 seconds idle) for 7 minutes or more. (repeat 8 cycles or more)

g. Squeeze the inlet and outlet radiator hoses several times by hand to bleed air.

h. After the engine cools down, check that the coolant level is between "LOW" and "FULL" line.

i. If the coolant level is below the LOW line, repeat all of the procedures above.

j. If the coolant level is above the FULL line, drain coolant so that the coolant level is between the FULL and LOW line.

3. Inspect for coolant leak.

xD

See Figure 23.

⁂ CAUTION

Be sure that the ignition is off if you work near the electric cooling fans or radiator grille. With the ignition on, the electric cooling fans may automatically start if the engine coolant temperature is high or the air conditioning is on.

1. Drain engine coolant.

⁂ CAUTION

To avoid the danger of being burned, do not remove the radiator cap sub-assembly while the engine and radiator assembly are still hot. Thermal expansion will cause hot engine coolant and steam to blow out from the radiator assembly.

a. Loosen the radiator drain cock plug.

b. Remove the radiator cap sub-assembly.

c. Loosen the cylinder block drain cock plug, then drain the coolant.

2. Add engine coolant.

a. Tighten all the plugs.

b. Pour engine coolant into the radiator assembly until it is completely full.

➡Do not substitute water for engine coolant.

➡Use of improper engine coolant may damage the engine coolant system. Use only Toyota Super Long Life Coolant or similar high quality ethylene glycol based non-silicate, non-amine, non-nitrite, and non-borate engine coolant with long-life hybrid organic acid technology (coolant with long-life hybrid organic acid technology consists of a combination of low phosphates and organic acids).

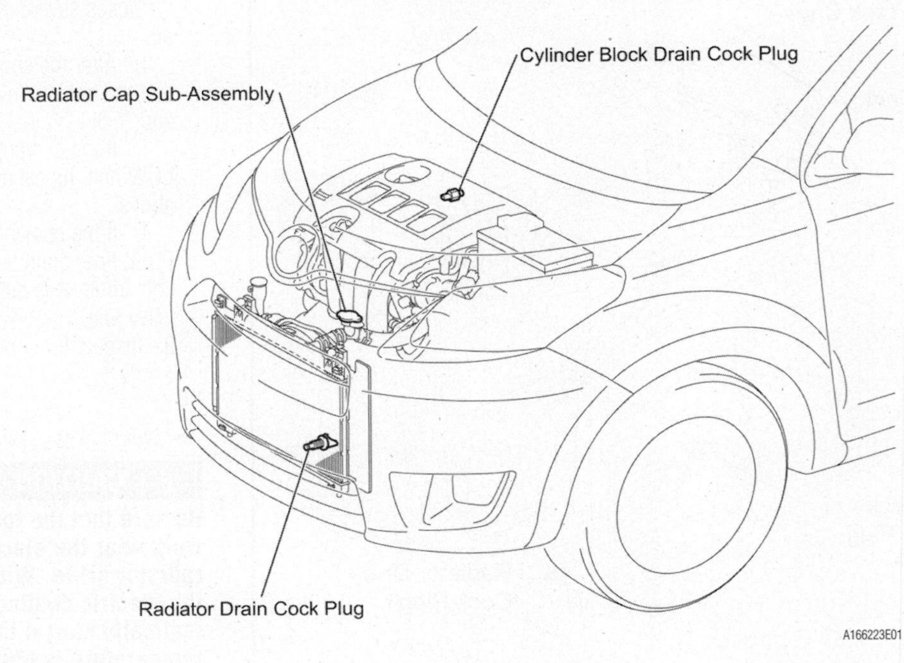

Fig. 23 Drain engine coolant

c. Check the engine coolant level inside the radiator assembly by squeezing the inlet and outlet radiator hoses several times by hand. If the engine coolant level goes down, add engine coolant.

d. Install the radiator cap sub-assembly securely.

e. Slowly pour engine coolant into the radiator reservoir until it reaches the FULL line.

f. Bleed air from the cooling system.

g. Warm up the engine until the thermostat opens. While the thermostat is open, circulate the coolant for several minutes.

➡The thermostat open timing can be confirmed by pressing the inlet radiator hose by hand, and checking when the engine coolant starts to flow inside the hose.

h. Maintain the engine speed at 2500 to 3000 rpm.

i. Press the inlet and outlet radiator hoses several times by hand to bleed air.

CAUTION

When pressing the radiator hoses:

- Wear protective gloves.
- Be careful as the radiator hoses are hot.
- Keep your hands away from the radiator fan.

j. Stop the engine and wait until the coolant cools down.

k. If the engine coolant level is below the full level, perform steps (b) through (g) again and repeat the operation until the engine coolant level stays at the full level.

l. Recheck the engine coolant level inside the radiator reservoir tank assembly. If it is below the full level, add engine coolant.

3. Inspect for engine coolant leak.

BLEEDING

Refer to coolant replacement procedures for bleeding information.

ENGINE FAN

REMOVAL & INSTALLATION

tC

See Figure 24.

1. Remove fan shroud with cooling fan.
2. Remove the 2 nuts, fan and No. 2 fan.
3. Remove No. 1 cooling fan motor.
   a. Detach the clip and 2 clamps.
   b. Remove the 3 screws and fan motor.
4. Remove No. 2 cooling fan motor.
   a. Detach the clip and 3 clamps.
   b. Remove the 3 screws and fan motor.
5. Installation is the reverse order of removal.

xB

See Figure 25.

1. Discharge refrigerant from refrigeration system.
2. Remove front bumper cover.
3. Remove head light assembly LH and RH.
4. Drain engine coolant.
5. Remove battery.
6. Remove No. 2 battery carrier.
7. Separate No. 1 water hose clamp bracket.
8. Separate hood lock control cable assembly.
9. Remove hood lock support sub-assembly.
10. Disconnect inlet radiator hose.
11. Disconnect outlet radiator hose.
12. Remove radiator support cushion.
13. Remove No. 2 fan shroud.
14. Disconnect discharge hose sub-assembly.
15. Disconnect air conditioning tube and accessory assembly.
16. Remove condenser.
17. Disconnect No. 1 water by-pass hose.
18. Remove radiator assembly.
19. Remove lower radiator support.
20. Remove fan shroud.
21. Remove the nut and the No. 2 fan.
22. Remove the nut and the fan.
23. Remove No. 2 cooling fan motor.
    a. Disengage the wire harness clamp.

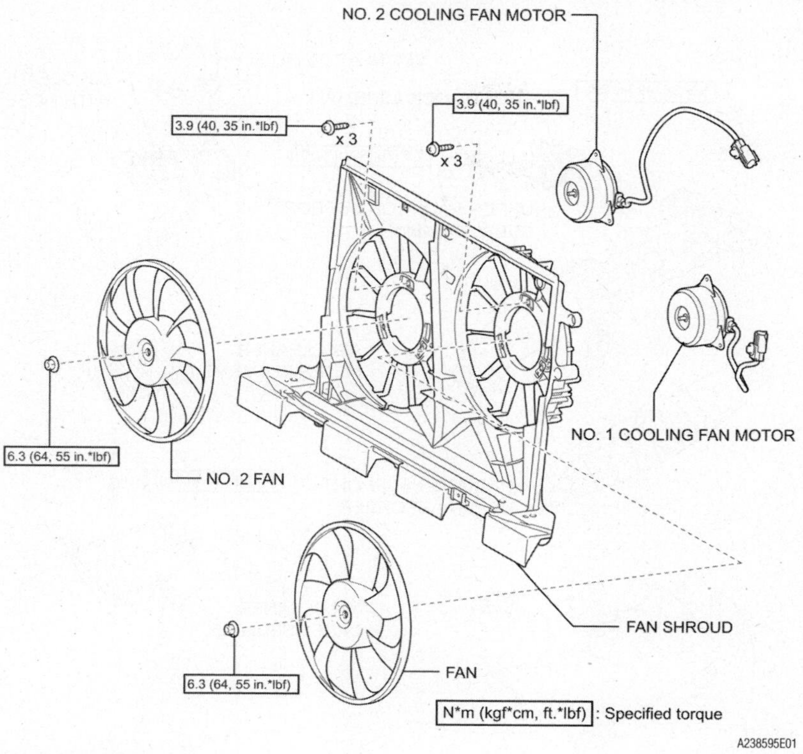

Fig. 24 Exploded view of cooling fan assembly

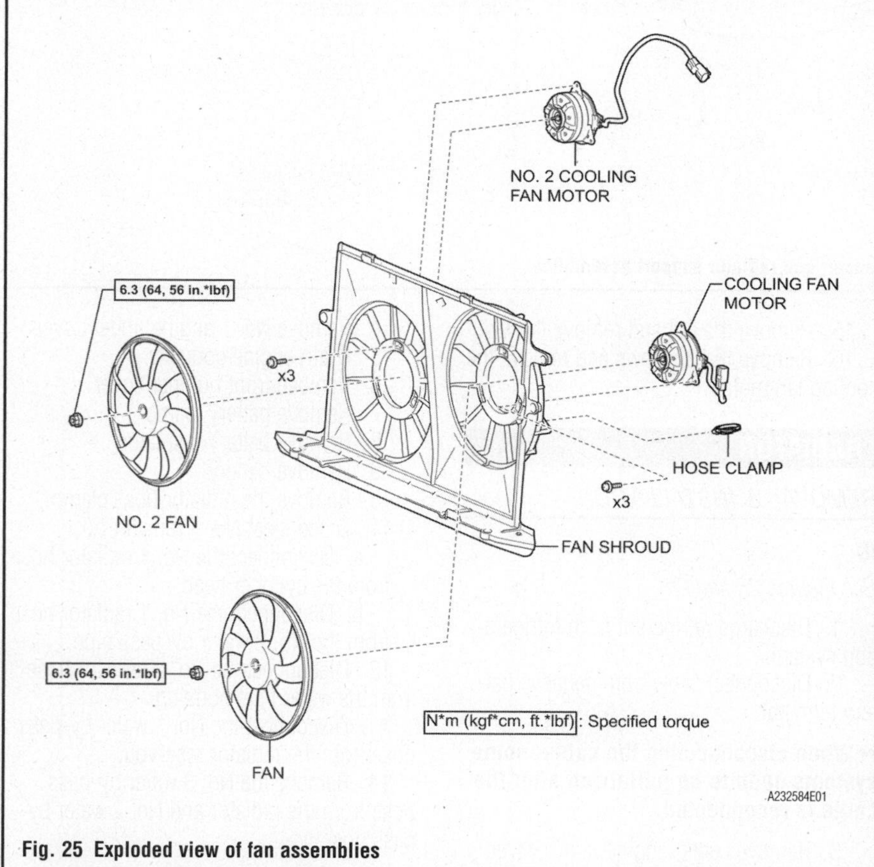

Fig. 25 Exploded view of fan assemblies

b. Remove the 3 screws and the No. 2 cooling fan motor.

24. Remove cooling fan motor.

a. Remove the hose clamp from the fan shroud.

b. Disengage the wire harness clamp.

c. Remove the 3 screws and the cooling fan motor.

To install:

25. Installation is the reverse order of removal.
26. Add engine coolant.
27. Charge refrigerant.
28. Warm up engine.
29. Inspect for coolant leak.
30. Inspect for refrigerant leak.
31. Adjust headlight aiming.
32. Inspect hood.
33. Adjust hood.

xD

See Figures 26 and 27.

CAUTION

Be sure that the ignition is off if you work near the electric cooling fans or radiator grille. With the ignition on, the electric cooling fans may automatically start to run if the engine coolant temperature is high and/or the air conditioning is on.

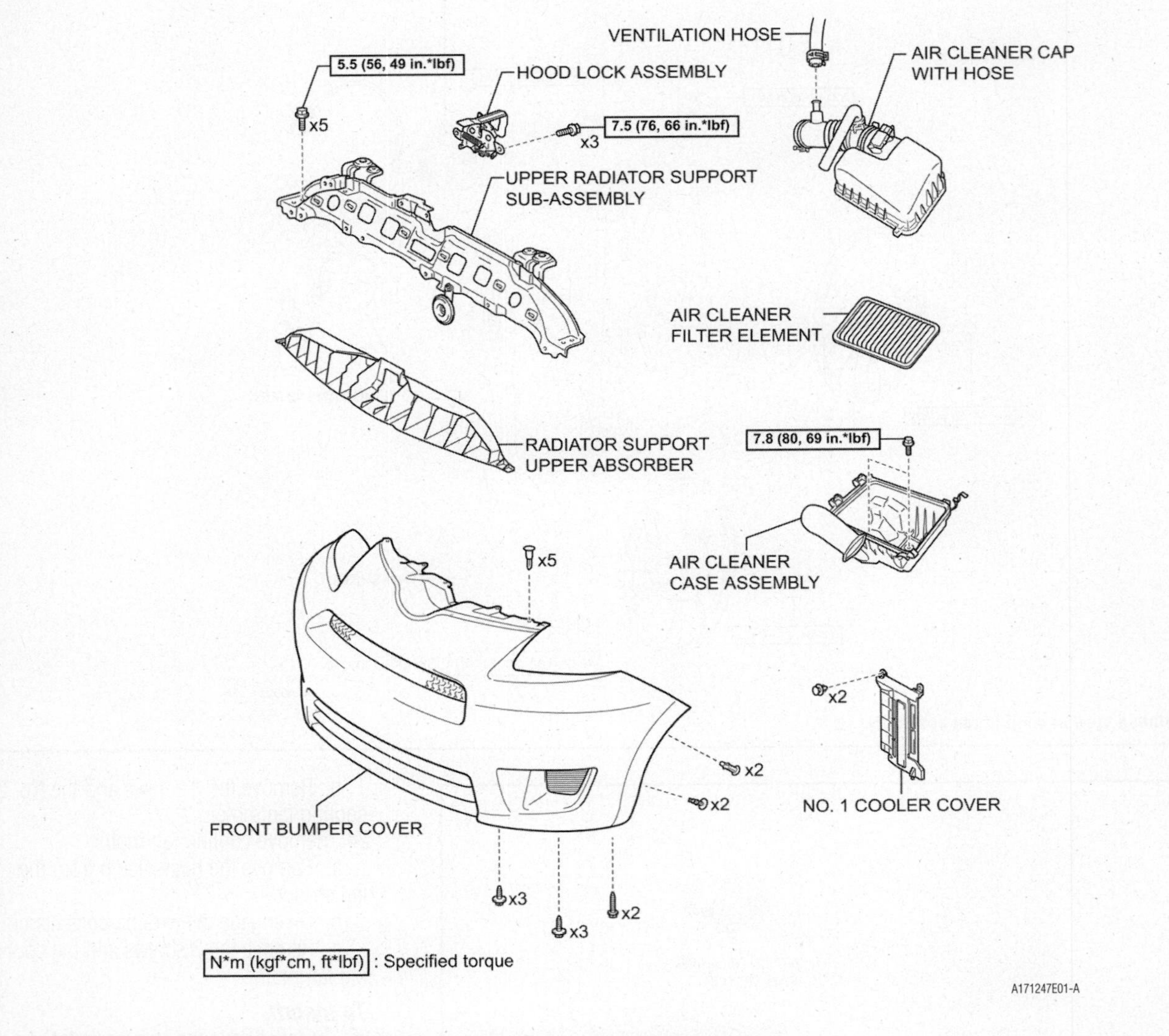

Fig. 26 Exploded view of front bumper cover, air cleaner, and radiator support assemblies

1. Drain coolant.
2. Remove engine under cover LH and RH.
3. Remove front bumper cover.
4. Remove radiator support upper absorber.
5. Remove air cleaner assembly.
6. Remove No. 1 radiator hose.
7. Disconnect inlet No. 2 oil cooler hose (for automatic transaxle).
8. Disconnect outlet No. 1 oil cooler hose (for automatic transaxle).
9. Disconnect No. 2 radiator hose.
10. Remove hood lock assembly.
11. Remove No. 1 cooler cover.
12. Remove upper radiator support sub-assembly.
13. Remove radiator assembly.
14. Remove fan shroud.
15. Remove the nut and remove the fan.
16. Remove the 3 screws and remove the cooling fan motor.

RADIATOR

REMOVAL & INSTALLATION

tC

See Figures 28 and 29.

1. Discharge refrigerant from refrigeration system.
2. Disconnect cable from negative battery terminal.

➡When disconnecting the cable, some systems need to be initialized after the cable is reconnected.

3. Remove center engine under cover.
4. Remove No. 1 engine under cover.
5. Drain engine coolant.
6. Remove front bumper cover.
7. Remove battery clamp.
8. Remove battery insulator.
9. Remove battery.
10. Remove the radiator hose clamp.
11. Disconnect No. 1 radiator hose.
    a. Disconnect the No. 1 radiator hose from the cylinder head.
    b. Disconnect the No. 1 radiator hose from the No. 2 water by-pass pipe.
12. Disconnect the No. 2 radiator hose from the water inlet housing.
13. Disconnect the No. 1 water by-pass hose from the radiator reservoir.
14. Remove the No. 3 water by-pass hose from the radiator and No. 2 water by-pass hose.

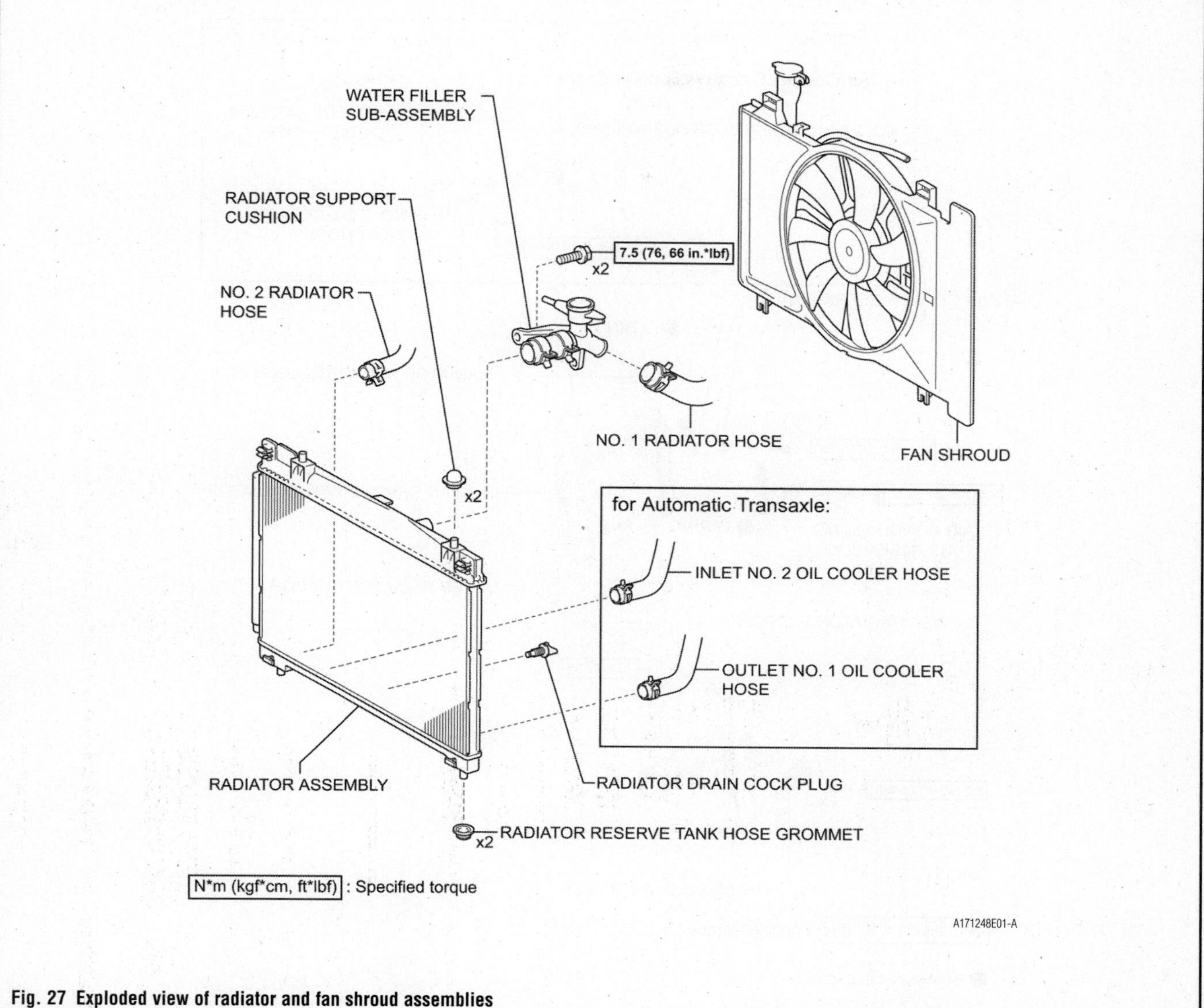

Fig. 27 Exploded view of radiator and fan shroud assemblies

15. Remove the No. 2 water by-pass hose from the radiator reservoir and then remove the No. 2 water by-pass hose from the 2 hose clamps.

16. Remove No. 2 oil cooler tube sub-assembly (for Automatic Transaxle).

a. Disconnect the No. 2 oil cooler inlet hose and No. 2 oil cooler outlet hose from the No. 2 oil cooler tube sub-assembly.

b. Disconnect the No. 3 oil cooler inlet hose and No. 3 oil cooler outlet hose from the radiator.

c. Remove the 2 bolts and No. 2 oil cooler tube sub-assembly.

17. Remove hood lock assembly.

18. Remove upper radiator support.

a. Disconnect the 2 horn connectors.

b. Remove the 4 bolts and upper radiator support.

19. Disconnect No. 1 cooler refrigerant discharge hose.

20. Disconnect air conditioning tube assembly.

21. Remove radiator assembly.

a. Disconnect the 2 fan motor connectors.

b. Remove the radiator together with the 2 lower radiator supports.

➡After removing the radiator, make sure that both lower radiator supports are attached to the fan shroud.

c. Remove the 2 bolts and No. 2 fan shroud.

d. Remove the cooler condenser from the fan shroud with cooling fan.

➡After removing the condenser, make sure that the 4 cooler condenser cushions are attached to the condenser.

e. Remove the No. 1 water by-pass hose and No. 2 radiator hose from the radiator.

f. Disconnect the clamp and remove the No. 1 radiator hose from the radiator.

g. Remove the 2 bolts and radiator from the fan shroud with cooling fan.

To install:

22. Install radiator assembly.

a. Install the radiator to the fan shroud with cooling fan with the 2 bolts.

b. Connect the clamp and install the No. 1 radiator hose to the radiator.

c. Install the No. 1 water by-pass hose and No. 2 radiator hose to the radiator.

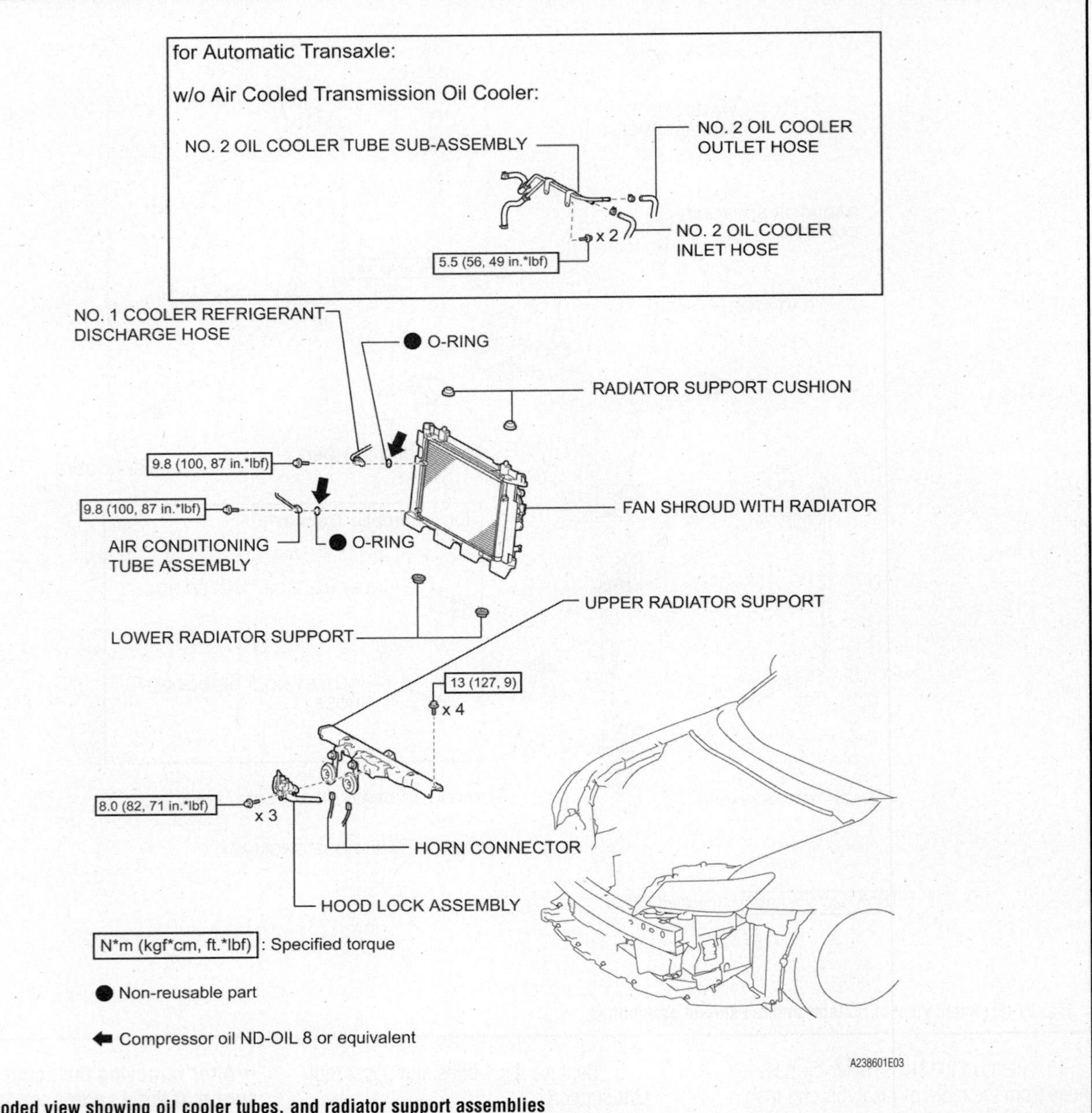

Fig. 28 Exploded view showing oil cooler tubes, and radiator support assemblies

d. Install the cooler condenser to the fan shroud with cooling fan.

➡Before installing the condenser, make sure that all 4 cooler condenser cushions are attached to the condenser.

e. Install the No. 2 fan shroud, and then install the 2 bolts.

f. Install the radiator together with the 2 lower radiator supports.

➡Before installing the radiator, make sure that both lower radiator supports are attached to the fan shroud.

g. Connect the 2 fan motor connectors.

23. Connect No. 1 cooler refrigerant discharge hose.

24. Connect air conditioning tube assembly.

25. Install upper radiator support.

a. Install the upper radiator support with the 4 bolts.

b. Connect the 2 horn connectors.

26. Install hood lock assembly.

27. Install No. 2 oil cooler tube sub-assembly (for automatic transaxle).

a. Connect the No. 3 oil cooler inlet hose and No. 3 oil cooler outlet hose to the radiator.

b. Install the No. 2 oil cooler tube sub-assembly with the 2 bolts.

c. Connect the No. 2 oil cooler inlet hose and No. 2 oil cooler outlet hose to the No. 2 oil cooler tube.

28. Connect the No. 2 water by-pass hose to the radiator reservoir and then install the No. 2 water by-pass hose to the 2 hose clamps.

29. Install the No. 3 water by-pass hose to the radiator and No. 2 water by-pass hose.

30. Connect the No. 1 water by-pass pipe to the radiator reservoir.

31. Connect the No. 2 radiator hose to the water inlet housing.

32. Connect No. 1 radiator hose.

a. Connect the No. 1 radiator hose to the cylinder head.

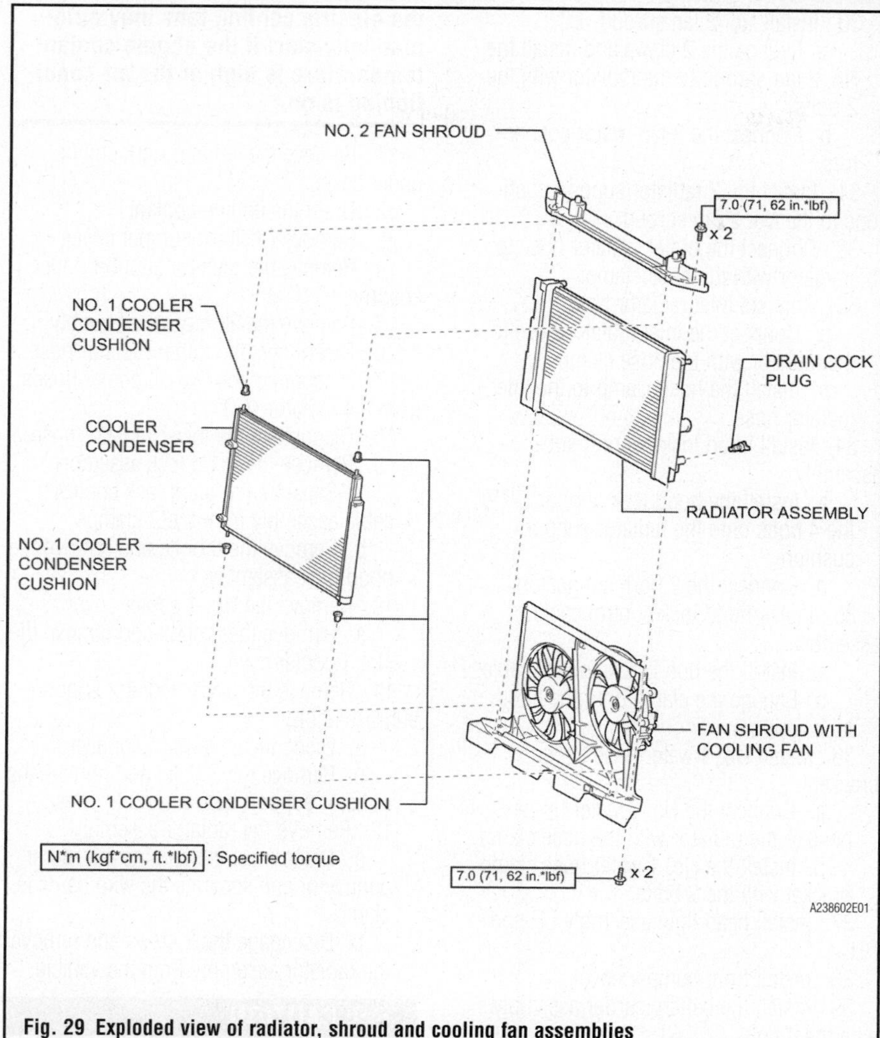

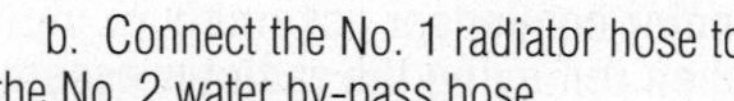

Fig. 29 Exploded view of radiator, shroud and cooling fan assemblies

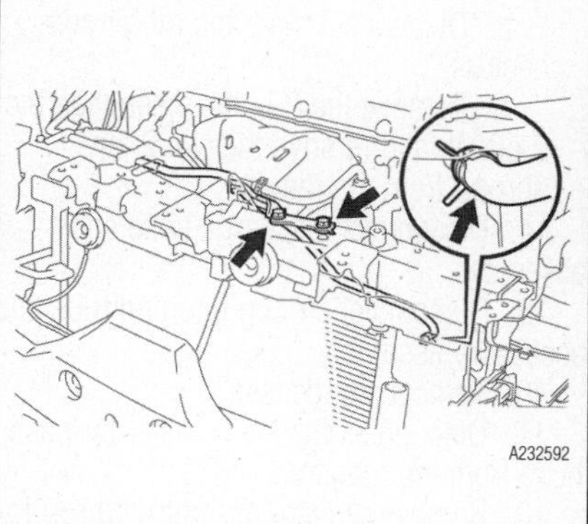

Fig. 30 Loosen the hose clamp and disconnect the No. 3 water by-pass hose from the radiator

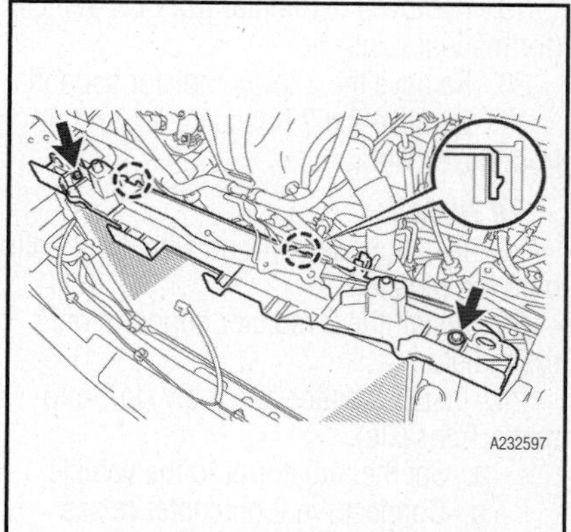

Fig. 31 Remove the 2 bolts, disengage the 2 claws, and remove the No. 2 fan shroud from the radiator

b. Connect the No. 1 radiator hose to the No. 2 water by-pass hose.

33. Install the radiator hose clamp.

➡When installing the clamp, make sure that the air fuel ratio sensor wire harness is not pinched excessively.

34. Install battery.
35. Install battery insulator.
36. Install battery clamp.
37. Install front bumper cover.
38. Connect cable to negative battery terminal.

➡When disconnecting the cable, some systems need to be initialized after the cable is reconnected.

39. Add engine coolant.
40. Charge refrigerant.
41. Warm up engine.
42. Inspect for refrigerant leak.
43. Inspect for coolant leak.
44. Install No. 1 engine under cover.
45. Install center engine under cover.

xB

See Figures 30 through 32.

1. Discharge refrigerant from refrigeration system.
2. Remove front bumper cover.
3. Remove head light assembly LH and RH.
4. Drain engine coolant.
5. Remove battery.
   a. Separate the negative and the positive battery terminals.
   b. Remove the bolt and loosen the nut, and remove the battery carrier.
   c. Remove the battery insulator and the battery.
   d. Remove the battery tray.
6. Remove the 4 bolts and the No. 2 battery carrier.
7. Separate No. 1 water hose clamp bracket.
   a. Loosen the hose clamp and disconnect the No. 3 water by-pass hose from the radiator.
   b. Remove the 2 bolts and separate the No. 1 water hose clamp bracket.
8. Separate hood lock control cable assembly.
   a. Separate the hood lock control cable from the hood lock.
   b. Disengage the clamp and separate the hood lock control cable.
9. Remove hood lock support sub-assembly.
   a. Disconnect the 2 horn connectors.
   b. Remove the 4 bolts and the hood lock support.
10. Disconnect inlet radiator hose.
    a. Disengage the hose clamp from the inlet radiator hose.
    b. Loosen the hose clamp and disconnect the inlet radiator hose from the radiator.
11. Loosen the hose clamp and disconnect the outlet radiator hose from the radiator.
12. Remove the 2 radiator support cushions from the No. 2 fan shroud.

13. Remove No. 2 fan shroud.
   a. Disconnect the 2 fan motor connectors.
   b. Remove the 2 bolts, disengage the 2 claws, and remove the No. 2 fan shroud from the radiator.
14. Disconnect discharge hose sub-assembly.
15. Disconnect air conditioning tube and accessory assembly.
16. Remove condenser.
17. Disconnect the No. 1 water by-pass hose from the radiator.
18. Remove radiator assembly (for Automatic Transaxle).
   a. Disconnect the 2 oil cooler hoses from the radiator.
   b. Remove the radiator from the vehicle.
19. Remove the radiator from the vehicle (for manual transaxle).
20. Remove the 2 lower radiator supports.
21. Remove the 2 bolts and fan shroud from the radiator.

To install:

22. Install the fan shroud with the 2 bolts onto the radiator.
23. Install the 2 radiator supports onto the radiator.
24. Install radiator assembly (for Automatic Transaxle).
   a. Set the radiator onto the vehicle.
   b. Connect the 2 oil cooler hoses onto the radiator.
25. Set the radiator onto the vehicle.
26. Install condenser.
27. Connect air conditioning tube and accessory assembly.
28. Connect discharge hose sub-assembly.
29. Connect the No. 1 water by-pass hose to the radiator with the hose clamp.

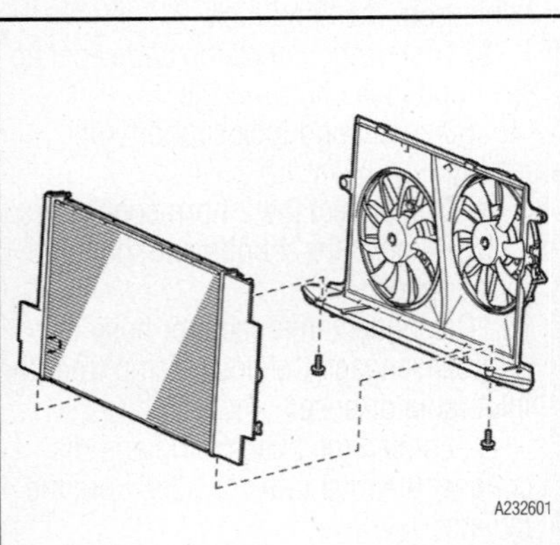

Fig. 32 Remove the 2 bolts and fan shroud from the radiator

30. Install No. 2 fan shroud.
   a. Engage the 2 claws and install the No. 2 fan shroud to the radiator with the 2 bolts.
   b. Connect the 2 fan motor connectors.
31. Install the 2 radiator support cushions to the No. 2 fan shroud.
32. Connect the outlet radiator hose to the radiator with the hose clamp.
33. Connect inlet radiator hose.
   a. Connect the inlet radiator hose to the radiator with the hose clamp.
   b. Install the hose clamp to the inlet radiator hose.
34. Install hood lock support sub-assembly.
   a. Install the hood lock support with the 4 bolts onto the radiator support cushion.
   b. Connect the 2 horn connectors.
35. Install hood lock control cable assembly.
   a. Install the hood lock control cable.
   b. Engage the clamp onto the hood lock support.
36. Install No. 1 water hose clamp bracket.
   a. Connect the No. 3 water by-pass hose to the radiator with the hose clamp.
   b. Install the No. 1 water hose clamp bracket with the 2 bolts.
37. Install head light assembly LH and RH.
38. Install front bumper cover.
39. Install the battery carrier assembly with the 4 bolts.
40. Install battery.
   a. Install the battery tray.
   b. Install the battery and the battery insulator.
   c. Install the battery carrier with the bolt and nut.
   d. Connect the positive and the negative terminals.
41. Add engine coolant.
42. Charge refrigerant.
43. Warm up engine.
44. Inspect for coolant leak.
45. Inspect for refrigerant leak.
46. Adjust headlight aiming.
47. Inspect hood.
48. Adjust hood.

xD

See Figures 26 and 27.

CAUTION

Be sure that the ignition is off if you work near the electric cooling fans or radiator grille. With the ignition on, the electric cooling fans may automatically start if the engine coolant temperature is high or the air conditioning is on.

1. Remove the left and right engine under cover.
2. Drain the engine coolant.
3. Remove the front bumper cover.
4. Remove the radiator support upper absorber.
5. Remove the air cleaner assembly.
6. Disconnect the upper radiator hose.
7. Disconnect the two oil cooler hoses from the radiator (A/T).
8. Disconnect the lower radiator hose.
9. Remove the hood lock assembly.
   a. Separate the hood lock control cable assembly from the 2 clamps.
   b. Remove the 3 bolts and remove the hood lock assembly.
10. Remove the No. 1 cooler cover.
   a. Remove the 2 clips and remove the No. 1 cooler cover.
11. Remove the upper radiator support sub-assembly.
   a. Disconnect the horn connector.
   b. Remove the 5 bolts and remove the radiator support sub-assembly upper.
12. Remove the radiator assembly.
   a. Disconnect the cooling fan motor connector and separate the wire harness clamp.
   b. Disengage the 2 claws and remove the radiator assembly from the vehicle.

CAUTION

Do not apply excessive force to the cooler condenser assembly or piping when removing the radiator assembly.

To install:

13. Install radiator assembly.
   a. Engage the 2 claws and install the radiator assembly onto the vehicle.

CAUTION

Do not apply excessive force to the cooler condenser assembly or piping when installing the radiator assembly.

   b. Connect the cooling fan motor connector and install the wire harness clamp.
14. Install upper radiator support sub-assembly.
   a. Install the radiator support sub-assembly upper with the 5 bolts.
   b. Connect the horn assembly connector.

15. Install No. 1 cooler cover.
a. Install the No. 1 cooler cover with the 2 clips.
16. Install hood lock assembly.
a. Install the hood lock assembly with the 3 bolts.
b. Install the hood lock control cable assembly with the 2 clamps.
17. Connect lower radiator hose.
18. Connect the two oil cooling hoses to the radiator (A/T).
19. Connect upper radiator hose.
20. Install air cleaner assembly.
21. Install radiator support upper absorber.
22. Install front bumper cover.
23. Add engine coolant.
24. Adjust hood lock assembly.
a. Loosen the 3 bolts.
b. Adjust the hood lock position so that the striker can enter it smoothly.
c. Tighten the 3 bolts after the adjustment.
25. Inspect for engine coolant leak.

THERMOSTAT

REMOVAL & INSTALLATION

tC

See Figures 33 and 34.

1. Remove center engine under cover
2. Remove No. 1 engine under cover
3. Drain engine coolant
4. Remove alternator assembly
5. Disconnect No. 2 radiator hose
6. Remove the 2 nuts and water inlet.
7. Remove thermostat
a. Remove the thermostat.
b. Remove the gasket from the thermostat.

Fig. 33 Remove the 2 nuts and water inlet

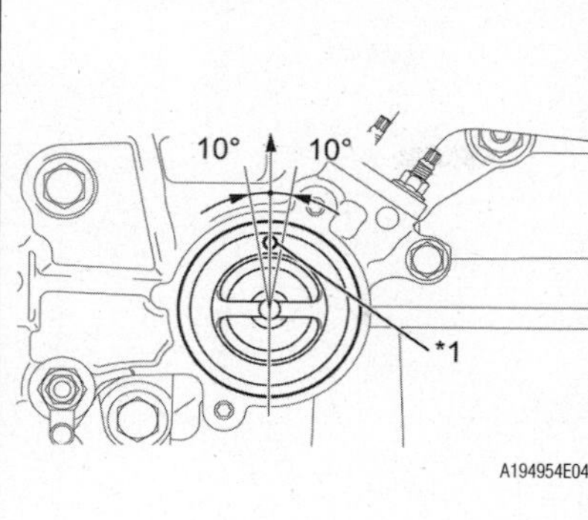

Fig. 34 The jiggle valve (1) may be set to within 10° on either side of the prescribed position

To install:

8. Install thermostat.
a. Install a new gasket to the thermostat.
b. Install the thermostat with the jiggle valve facing upward.

➡The jiggle valve may be set to within 10° on either side of the prescribed position.

9. Install the water inlet with the 2 nuts.
10. Connect No. 2 radiator hose.
11. Install alternator assembly.
12. Add engine coolant.
13. Inspect for coolant leak.
14. Install No. 1 engine under cover.
15. Install center engine under cover.

xB

See Figure 35.

1. Remove No. 1 engine under cover.
2. Drain engine coolant.
3. Loosen the hose clamp and disconnect the outlet radiator hose from the water inlet.
4. Remove the 2 nuts and disconnect the inlet water from the cylinder block.
5. Remove thermostat. Remove the gasket from the thermostat.

To install:

6. Install thermostat.
a. Install a new gasket onto the thermostat.
b. Install the thermostat with the jiggle valve upward.

➡The jiggle valve may be set to within 10° on either side of the prescribed position.

7. Install the inlet water with the 2 nuts.
8. Connect the outlet radiator hose to the water inlet with the hose clamp.

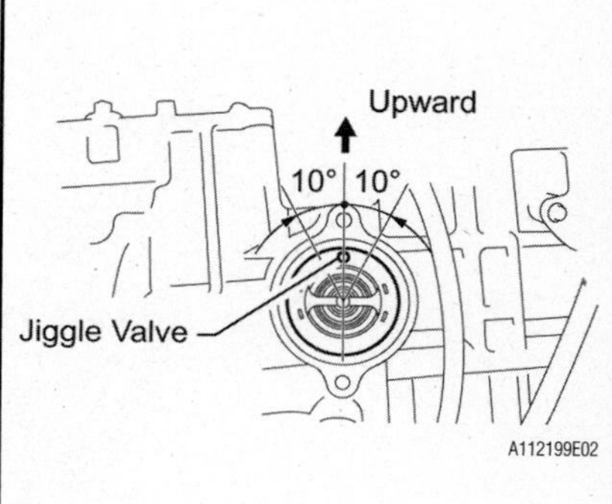

Fig. 35 The jiggle valve (1) may be set to within 10° on either side of the prescribed position

9. Add engine coolant.
10. Inspect for coolant leak.
11. Install No. 1 engine under cover.

xD

See Figures 36 and 37.

1. Drain engine coolant.
2. Remove the 2 nuts and separate the water inlet with radiator hose from the cylinder block.
3. Remove thermostat.
a. Remove the thermostat from the cylinder block.
b. Remove the gasket from the thermostat.

To install:

4. Install thermostat.
a. Install a new gasket onto the thermostat.
b. Install the thermostat with the jiggle valve facing upward.
5. Install the water inlet with radiator hose with the 2 nuts.
6. Add engine coolant.
7. Inspect for engine coolant leak.

WATER PUMP

REMOVAL & INSTALLATION

tC

See Figure 38.

1. Remove the 6 clips and radiator support opening cover.
2. Remove center engine under cover.
3. Remove No. 1 engine under cover.
4. Remove rear engine under cover RH.
5. Drain engine coolant.
6. Remove radiator reservoir.
a. Disconnect the No. 1 and No. 2 water by-pass hoses.

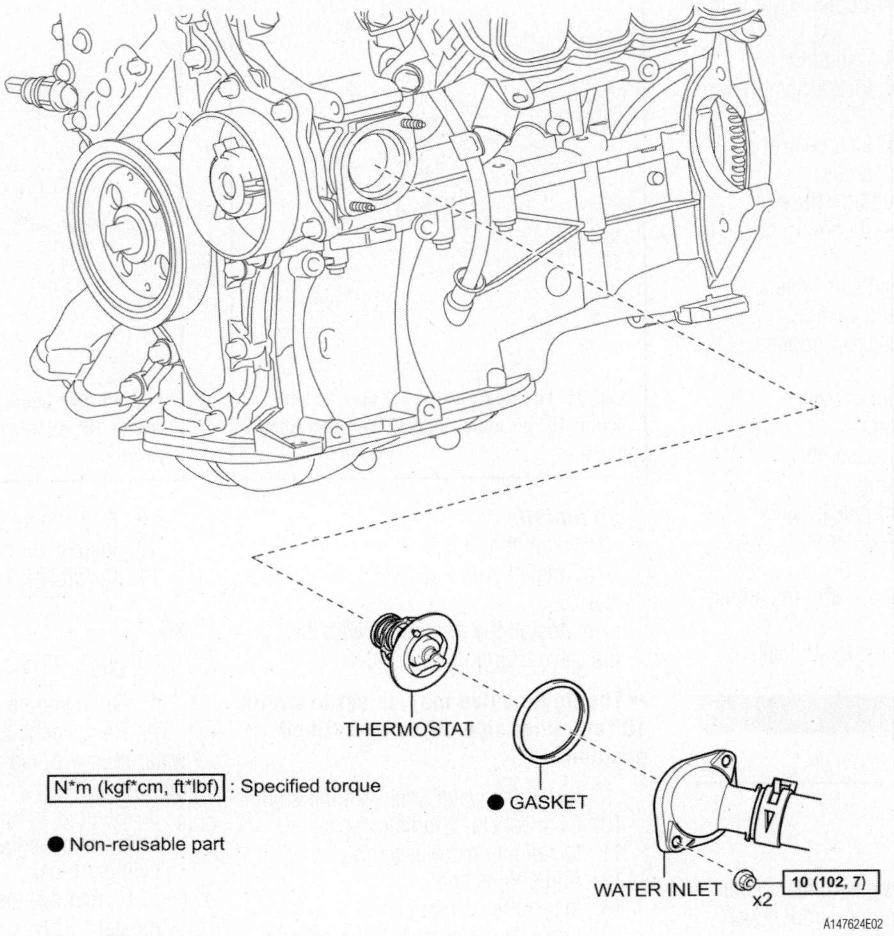

Fig. 36 Exploded view of thermostat assembly

b. Remove the 2 bolts and radiator reservoir.

7. Remove alternator assembly.

8. Remove the bolt and V-ribbed belt tensioner.

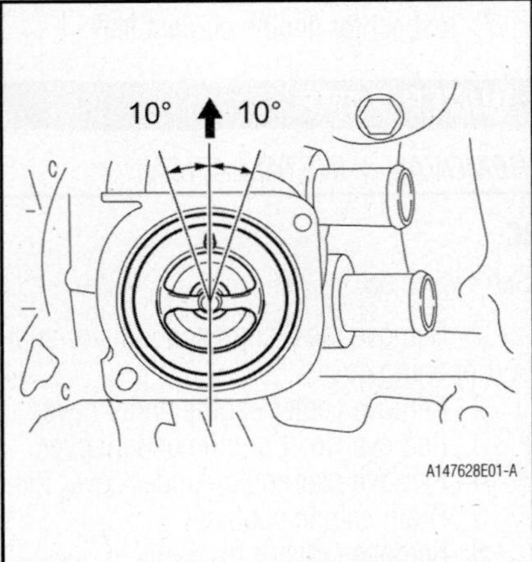

Fig. 37 Install the thermostat with the jiggle valve facing upward

9. Remove water pump assembly.
   a. Detach the clamp of the wire harness from the wire harness bracket.
   b. Remove the bolt and wire harness bracket.
   c. Remove the 7 bolts, water pump and water pump gasket.

To install:

10. Install water pump assembly.
    a. Install a new gasket and the water pump with the 7 bolts. Tighten to 15 ft. lbs. (21 Nm).
    b. Install the wire harness bracket with the bolt.
    c. Attach the clamp of the wire harness to the wire harness bracket.
11. Install the V-ribbed belt tensioner with the bolt. Tighten to 15 ft. lbs. (21 Nm).
12. Install alternator assembly.
13. Install radiator reservoir.
    a. Install the radiator reservoir with the 2 bolts.
    b. Connect the No. 1 and No. 2 water by-pass hoses.
14. Add engine coolant.
15. Inspect for coolant leak.
16. Install rear engine under cover RH.
17. Install No. 1 engine under cover.
18. Install center engine under cover.
19. Install radiator support opening cover. Install the support opening cover with the 6 clips.

xB

See Figure 39.

1. Remove No. 1 engine under cover.
2. Remove rear engine under cover RH.
3. Drain engine coolant.
4. Remove fan and alternator V belt.
5. Remove alternator assembly.
6. Remove water pump pulley. Using SST, remove the 4 bolts and water pump pulley.
7. Remove engine water pump assembly (for TMC Made).
   a. Disengage the Crankshaft Position (CKP) sensor wire from the engine water pump.

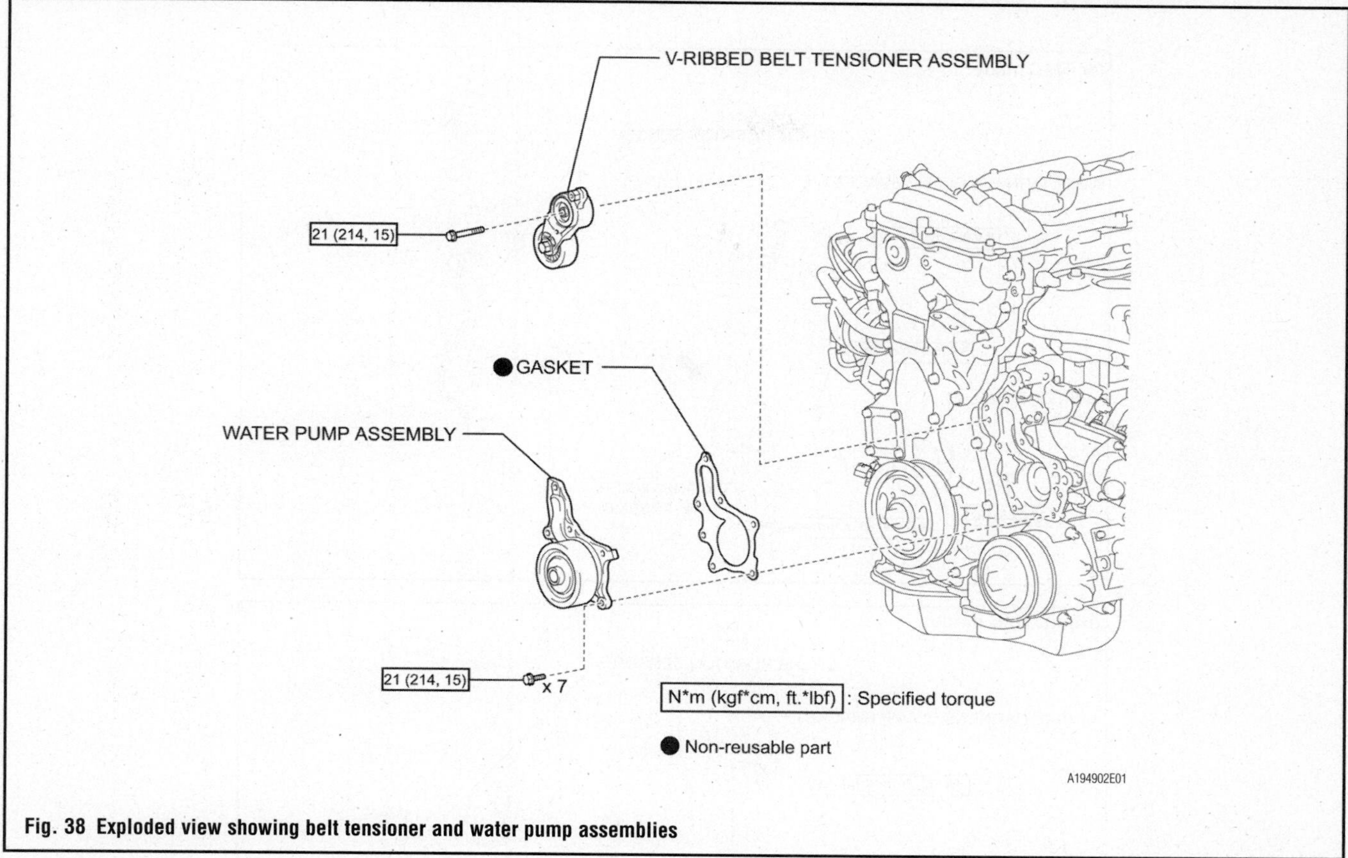

Fig. 38 Exploded view showing belt tensioner and water pump assemblies

b. Disconnect the Crankshaft Position (CKP) sensor wire from the clamp bracket.

c. Remove the 4 bolts, 2 nuts and clamp bracket.

d. Using a screwdriver, pry between the water pump and cylinder block, and then remove the water pump.

8. Remove engine water pump assembly (except TMC Made).

a. Disconnect the connector of the Crankshaft Position (CKP) sensor.

b. Disengage the clamp and disconnect the Crankshaft Position (CKP) sensor connector from the connector clamp bracket.

c. Disconnect the Crankshaft Position (CKP) sensor wire from the clamp bracket.

d. Remove the 4 bolts, 2 nuts and 2 clamp brackets.

e. Using a screwdriver, pry between the water pump and cylinder block, and then remove the water pump.

➡Be careful not to damage the contact surfaces of the water pump and cylinder block.

To install:

9. Install engine water pump assembly (for TMC Made).

a. Remove any old seal packing material from the contact surface.

b. Apply a continuous line of seal packing.

➡Remove any oil from the contact surface. The parts must be set within 3 minutes after applying seal packing. Otherwise, the material must be removed and reapplied.

c. Install the engine water pump and clamp bracket with the 4 bolts and 2 nuts.

d. Connect the Crankshaft Position (CKP) sensor wire to the wire harness clamp bracket.

e. Engage the clamp and connect the Crankshaft Position (CKP) sensor wire to the water pump.

10. Install engine water pump assembly (except TMC Made).

a. Remove any old seal packing material from the contact surfaces.

b. Apply a continuous line of seal packing.

11. Install the water pump assembly and 2 clamp brackets with the 4 bolts and 2 nuts.

12. Engage the clamp and connect the Crankshaft Position (CKP) sensor connector to the connector clamp bracket.

13. Connect the Crankshaft Position (CKP) sensor wire to the wire harness clamp bracket.

14. Connect the connector of the Crankshaft Position (CKP) sensor.

15. Using SST, install the water pump pulley with the 4 bolts. Tighten to 19 ft. lbs. (26 Nm).

16. Install alternator assembly.

17. Install fan and alternator V belt.

18. Add engine coolant.

19. Inspect for coolant leak.

20. Install rear engine under cover RH.

21. Install No. 1 engine under cover.

xD

See Figures 40 and 41.

1. Discharge fuel system pressure
2. Remove battery
3. Remove battery tray
4. Remove front wheels
5. Remove engine under cover LH and RH
6. Drain engine coolant
7. Remove front wiper arm head cap
8. Remove front wiper arm and blade assembly LH
9. Remove front wiper arm and blade assembly RH
10. Remove hood to cowl top seal

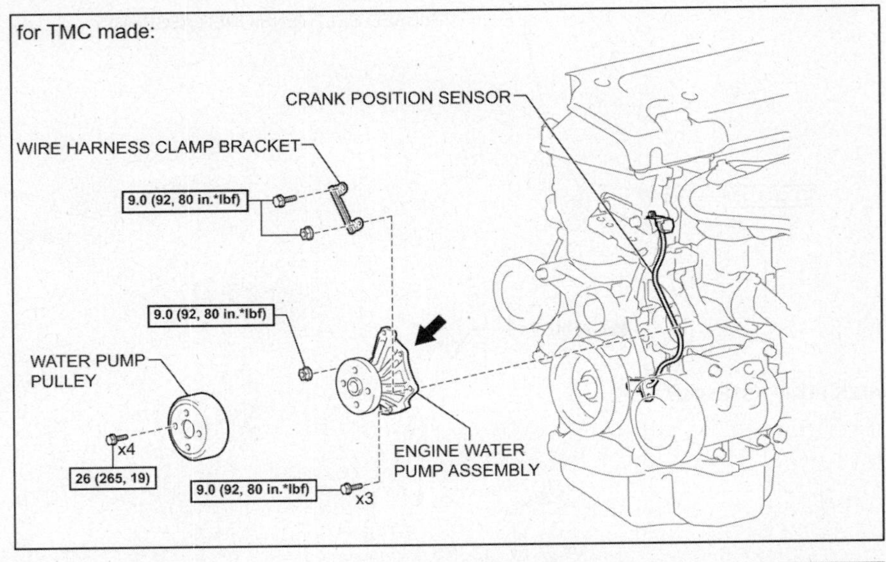

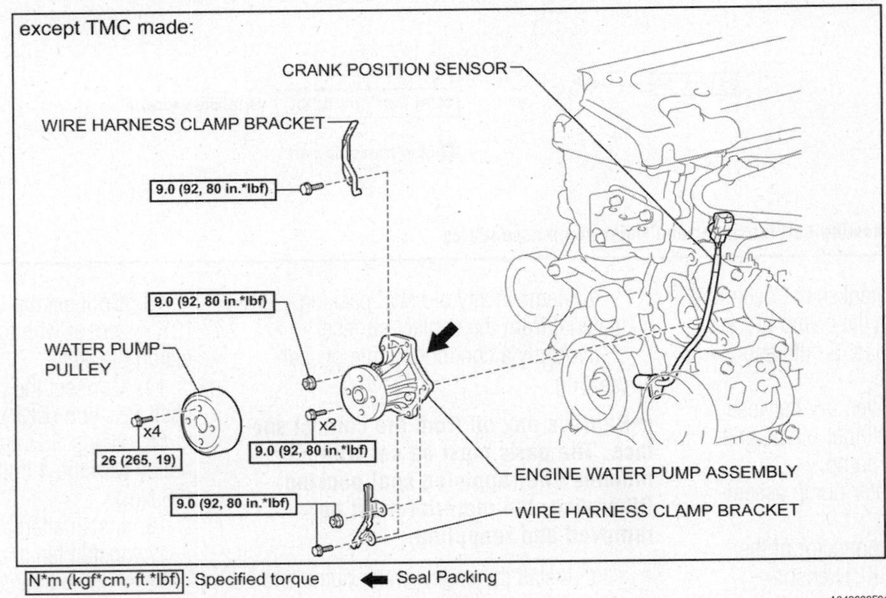

Fig. 39 Exploded view of water pump assemblies

11. Remove cowl top ventilator louver sub-assembly
12. Remove cowl top ventilator louver LH
13. Remove front wiper motor and link
14. Remove front air shutter seal RH
15. Remove outer cowl top panel
16. Remove air cleaner assembly
17. Remove air cleaner bracket
18. Remove battery carrier
19. Remove No. 2 cylinder head cover
20. Remove fan and alternator V belt
21. Remove No. 2 radiator hose
22. Disconnect No. 1 radiator hose
23. Remove front bumper cover
24. Disconnect No. 1 oil cooler outlet hose (for automatic transaxle)
25. Disconnect No. 2 oil cooler inlet hose (for automatic transaxle)
26. Remove hood lock assembly
27. Remove No. 1 cooler cover
28. Remove upper radiator support absorber
29. Remove upper radiator support sub-assembly
30. Remove radiator assembly
31. Separate with pulley compressor assembly
32. Separate transmission control cable assembly (for automatic transaxle)
33. Separate transmission control cable assembly (for manual transaxle)
34. Separate union to check valve hose
35. Separate No. 1 fuel vapor feed hose
36. Disconnect engine wire
37. Disconnect heater water hose outlet A
38. Disconnect heater water hose inlet A
39. Disconnect fuel tube sub-assembly
40. Separate clutch release cylinder assembly (for manual transaxle)
41. Remove column hole cover silencer sheet
42. Separate steering sliding yoke sub-assembly

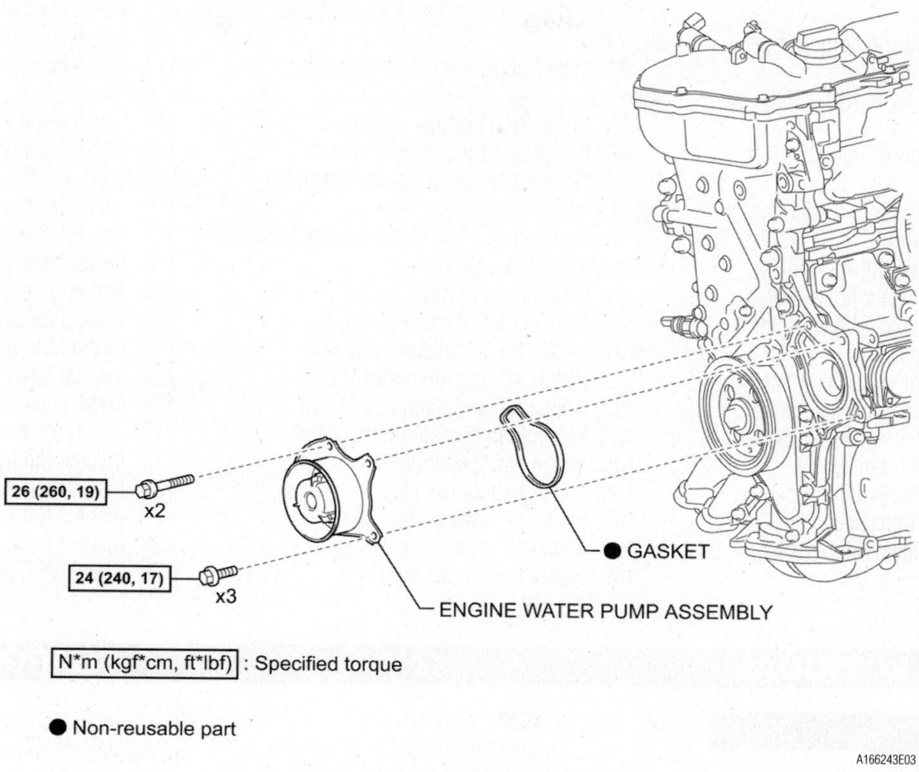

Fig. 40 Exploded view of water pump assembly

43. Separate No. 1 steering column hole cover sub-assembly
44. Remove shift lever knob sub-assembly (for manual transaxle)
45. Remove console panel upper
46. Remove console box cover rear
47. Remove console box carpet
48. Remove rear console box sub-assembly
49. Remove front console box
50. Remove front floor brace center
51. Remove front exhaust pipe assembly
52. Remove front axle shaft LH and RH nut
53. Separate speed sensor front LH and RH
54. Separate tie rod end sub-assembly LH and RH
55. Separate front stabilizer link assembly LH and RH
56. Separate front lower suspension arm sub-assembly LH and RH
57. Separate front axle assembly LH and RH
58. Remove engine assembly with transaxle
59. Remove front suspension cross-member sub-assembly
60. Remove engine water pump assembly
    a. Remove the 5 bolts and remove the engine water pump assembly.
    b. Remove the water pump gasket from the timing chain cover.

To install:

61. Install engine water pump assembly.
    a. Install a new water pump gasket to the timing chain cover.

➡Make sure that the installation surfaces are clean.

    b. Install the engine water pump assembly with the 5 bolts.
    - Bolt A [1.38 inches (35 mm)]: Tighten to 19 ft. lbs. (26 Nm).
    - Bolt B [0.71 inches (18 mm)]: Tighten to 17 ft. lbs. (24 Nm).
62. Install front suspension crossmember sub-assembly.
63. Install engine assembly with transaxle.
64. Install front axle assembly LH and RH.
65. Install front lower suspension arm sub-assembly LH and RH.
66. Install front stabilizer link assembly LH and RH.
67. Install tie rod end sub-assembly LH and RH.
68. Install tie rod end sub-assembly RH.

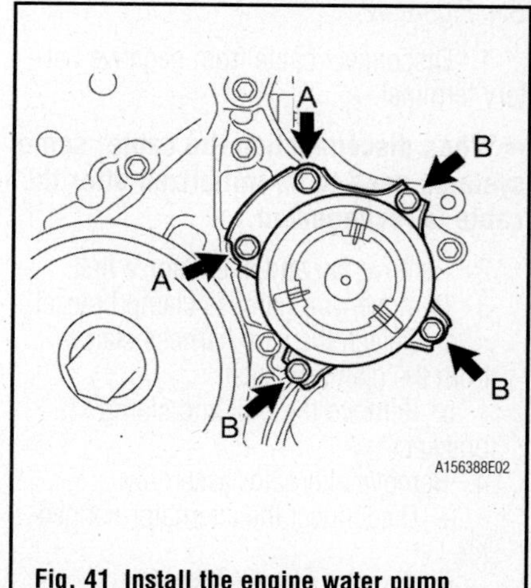

Fig. 41 Install the engine water pump assembly with the 5 bolts

69. Install speed sensor front LH and RH.
70. Install front axle shaft LH and RH nut.
71. Install front exhaust pipe assembly.
72. Install front floor brace center.
73. Install front console box.
74. Install rear console box assembly.
75. Install console box carpet.
76. Install console box cover rear.
77. Install console panel upper.

78. Install shift lever knob sub-assembly (for manual transaxle).
79. Install No. 1 steering column hole cover sub-assembly.
80. Install steering sliding yoke sub-assembly.
81. Install column hole cover silencer sheet
82. Install clutch release cylinder assembly (for manual transaxle).
83. Connect fuel tube sub-assembly.
84. Connect heater water hose inlet A.
85. Connect heater water hose outlet A.
86. Connect engine wire.
87. Install No. 1 fuel vapor feed hose.
88. Install union to check valve hose.
89. Install transmission control cable assembly (for automatic transaxle).
90. Install transmission control cable assembly (for manual transaxle).
91. Install with pulley compressor assembly.
92. Install radiator assembly.
93. Install upper radiator support sub-assembly.
94. Install upper radiator support absorber.
95. Install No. 1 cooler cover.
96. Install hood lock assembly.
97. Connect No. 2 oil cooler inlet hose (for automatic transaxle).
98. Connect No. 1 oil cooler outlet hose (for automatic transaxle).
99. Install front bumper cover.
100. Connect No. 1 radiator hose.
101. Install No. 2 radiator hose.
102. Install fan and alternator V belt.
103. Adjust fan and alternator V belt.
104. Inspect fan and alternator V belt.
105. Install No. 2 cylinder head cover.
106. Install battery carrier.
107. Install air cleaner bracket.
108. Install air cleaner assembly.
109. Install outer cowl top panel.
110. Install front air shutter seal RH.
111. Install front wiper motor and link.
112. Install cowl top ventilator louver LH.
113. Install cowl top ventilator louver sub-assembly.
114. Install hood to cowl top seal.
115. Install front wiper arm and blade assembly LH and RH.
116. Install front wiper arm head cap.
117. Install battery tray.
118. Install battery.
119. Add engine coolant.
120. Inspect for fuel leak.
121. Inspect for exhaust gas leak.
122. Inspect for engine coolant leak.
123. Install engine under cover LH and RH.
124. Install front wheels.
125. Inspect ignition timing.
126. Inspect engine idling speed.
127. Inspect CO/HC.
128. Inspect front wheel alignment.
129. Check VSC sensor signal.

ENGINE ELECTRICAL

CHARGING SYSTEM

ALTERNATOR

REMOVAL & INSTALLATION

tC

See Figure 42.

1. Disconnect cable from negative battery terminal.

➡When disconnecting the cable, some systems need to be initialized after the cable is reconnected.

2. Remove fan and alternator V belt.
3. Remove wire harness clamp bracket.
   a. Detach the wire harness clamp from the clamp bracket.
   b. Remove the bolt and clamp bracket.
4. Remove alternator assembly.
   a. Disconnect the alternator connector.
   b. Remove the terminal cap.
   c. Remove the nut and disconnect the alternator wire.
   d. Remove the bolt and wire harness clamp bracket.
   e. Remove the 2 bolts and alternator.

To install:

5. Install alternator assembly.
   a. Install the alternator with the 2 bolts. Tighten to 38 ft. lbs. (52 Nm).
   b. Install the wire harness clamp bracket with the bolt.
   c. Connect the alternator wire with the nut.
   d. Install the terminal cap.
   e. Connect the alternator connector.
6. Install wire harness clamp bracket.
   a. Install the clamp bracket with the bolt.
   b. Attach the wire harness clamp to the clamp bracket.
7. Install fan and alternator V belt.
8. Connect cable to negative battery terminal.

xB

See Figures 43 and 44.

1. Disconnect cable from negative battery terminal.
2. Remove rear engine under cover RH.
3. Remove fan and alternator V belt.
4. Remove alternator assembly.
   a. Disconnect the alternator connector.
   b. Remove the nut and disconnect the wire harness from terminal B.

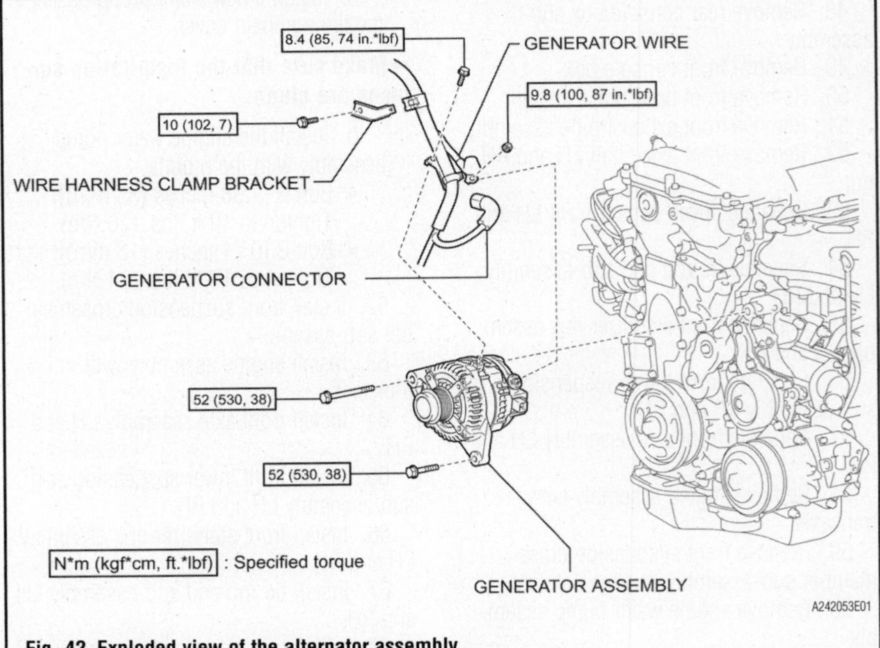

Fig. 42 Exploded view of the alternator assembly

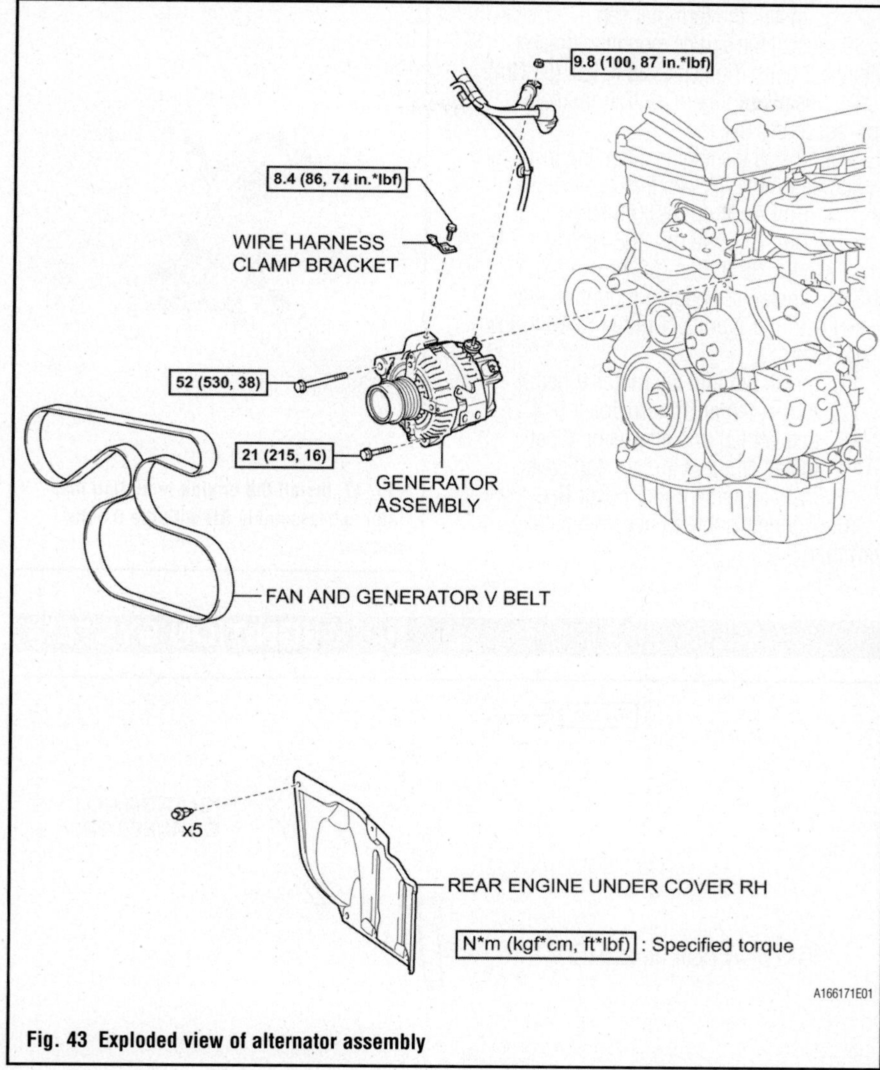

Fig. 43 Exploded view of alternator assembly

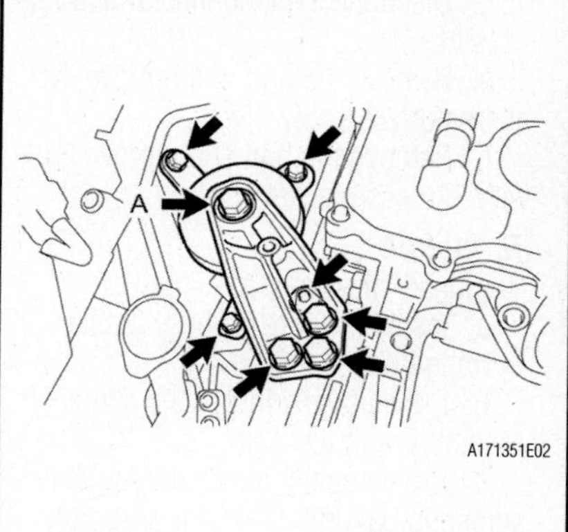

Fig. 45 Remove the 6 bolts and nut and remove the engine mounting insulator sub-assembly RH

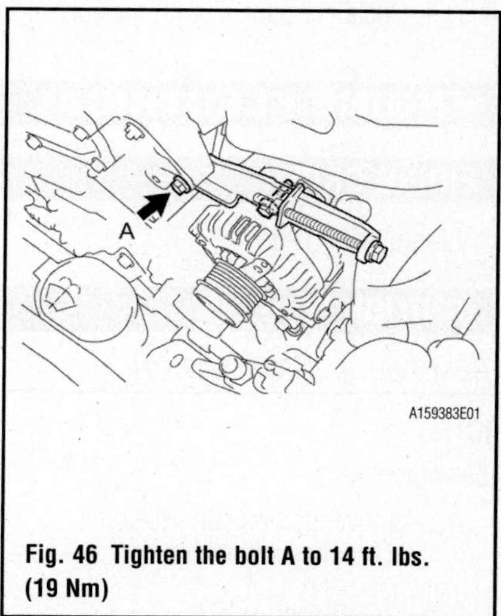

Fig. 46 Tighten the bolt A to 14 ft. lbs. (19 Nm)

c. Separate the 2 wire harness clamps.

d. Remove the 2 bolts and alternator assembly.

5. Remove the bolt and remove the wire harness clamp bracket.

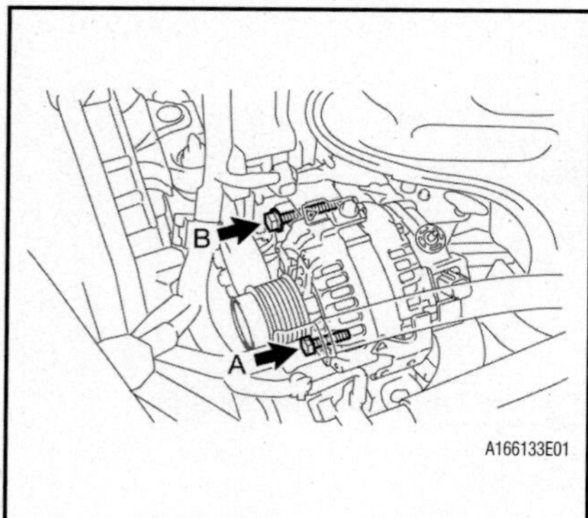

Fig. 44 Install the alternator assembly with the 2 bolts

To install:

6. Install alternator assembly.

a. Install the wire harness clamp bracket with the bolt.

b. Install the alternator assembly with the 2 bolts.

- Bolt A: Tighten to 16 ft. lbs. (21 Nm).
- Bolt B: Tighten to 38 ft. lbs. (52 Nm).

c. Install the alternator wire to terminal B with the nut.

d. Attach the clamp and connect the alternator connector to the alternator.

e. Install the 2 wire harness clamps.

7. Install fan and alternator V belt.
8. Install rear engine under cover RH.
9. Connect cable to negative battery terminal.

xD

See Figures 45 through 47.

1. Disconnect cable from negative battery terminal.
2. Remove engine under cover RH.
3. Remove No. 2 cylinder head cover.
4. Remove fan and alternator V belt.
5. Remove engine mounting insulator sub-assembly RH.

a. Remove 2 bolts and remove the engine mounting stay RH.

b. Place a wooden block on a jack underneath the engine.

c. Remove the 6 bolts and nut and remove the engine mounting insulator sub-assembly RH.

➡Do not remove bolt A.

6. Remove the 3 bolts and remove the engine mounting bracket.
7. Remove alternator assembly.

a. Remove the 2 bolts and remove the fan belt adjusting bar.

b. Remove the terminal cap.

c. Remove the nut and remove terminal B.

d. Disconnect the connector and harness clamp.

e. Remove the bolt and remove the alternator assembly.

f. Remove the bolt and remove the wire harness bracket.

To install:

8. Install alternator assembly.

a. Install the wire harness bracket with the bolt.

b. Provisionally install the alternator assembly with the bolt.

c. Provisionally install the fan belt adjusting bar and alternator assembly with the 2 bolts.

d. Tighten the bolt A to 14 ft. lbs. (19 Nm).

e. Connect the connector and wire harness clamp.

f. Install terminal B with the nut.

g. Install the terminal cap.

9. Install the engine mounting bracket with the 3 bolts. Tighten to 38 ft. lbs. (51 Nm).

10. Install engine mounting insulator sub-assembly RH.

a. Install the engine mounting insulator sub-assembly RH with the 6 bolts and nut.

- Bolt A: 33 ft. lbs. (46 Nm)
- Bolt B: 65 ft. lbs. (88 Nm)
- Nut: 38 ft. lbs. (52 Nm)

b. Install the engine mounting stay RH with the 2 bolts. Tighten to 19 ft. lbs. (26 Nm).

11. Install fan and alternator V belt..

12. Adjust fan and alternator V belt.

13. Inspect fan and alternator V belt.

14. Install No. 2 cylinder head cover.

15. Install engine under cover RH.

16. Connect cable to negative battery terminal.

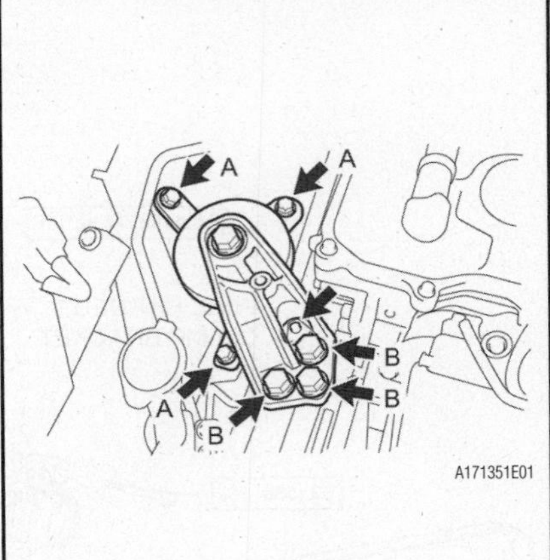

Fig. 47 Install the engine mounting insulator sub-assembly RH with the 6 bolts and nut

ENGINE ELECTRICAL

IGNITION SYSTEM

FIRING ORDER

Firing order: 1–3–4–2

IGNITION COIL

REMOVAL & INSTALLATION

tC

See Figure 48.

1. Remove ignition coil assembly.

a. Disconnect the 4 ignition coil connectors.

b. Remove the 4 bolts and 4 ignition coils.

2. Using a 16 mm spark plug wrench, remove the 4 spark plugs.

To install:

3. Install spark plug using a 16 mm spark plug wrench, install the 4 spark plugs. Tighten to 18 ft. lbs. (25 Nm).

4. Install ignition coil assembly.

5. Connect the 4 ignition coil connectors.

xB

See Figure 49.

1. Remove No. 1 engine cover sub-assembly. Remove the 2 nuts and engine cover.

2. Remove ignition coil assembly.

a. Disconnect the 4 ignition coil connectors.

b. Remove the 4 bolts and 4 ignition coils.

3. Remove the 4 spark plugs.

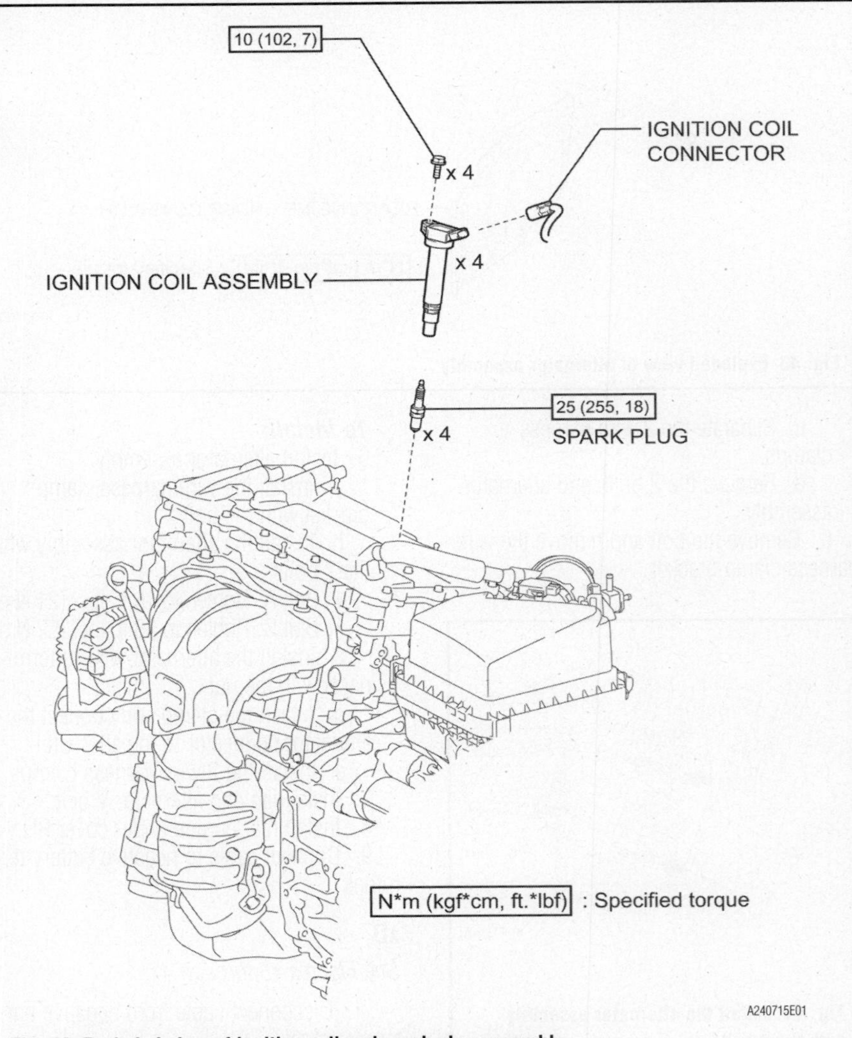

Fig. 48 Exploded view of ignition coil and spark plug assembly

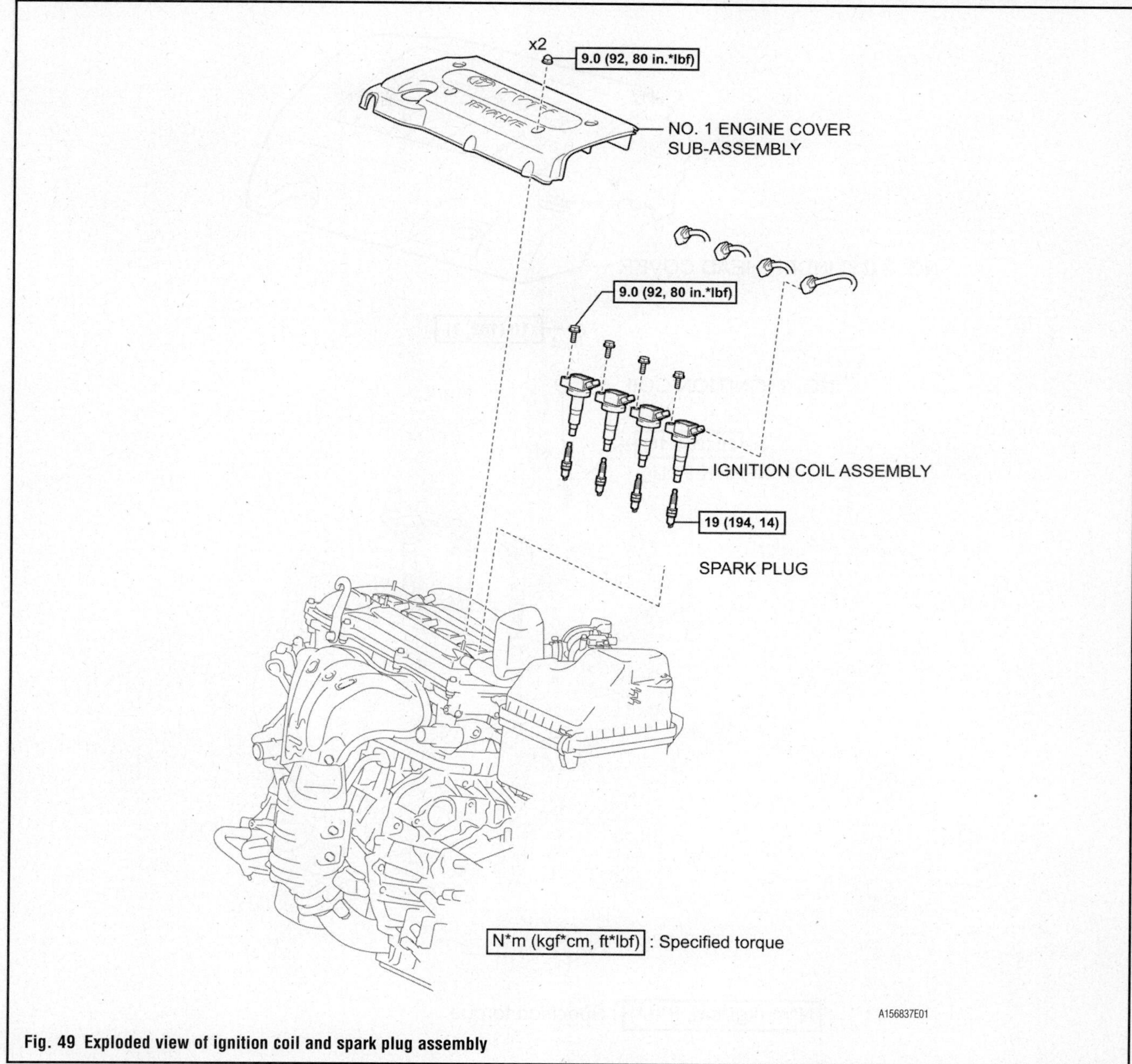

Fig. 49 Exploded view of ignition coil and spark plug assembly

To install:

4. Install the 4 spark plugs. Tighten to 14 ft. lbs. (19 Nm).
5. Install ignition coil assembly.
    a. Install the 4 ignition coils with the 4 bolts.
    b. Connect the 4 ignition coil connectors.
6. Install the engine cover with the 2 nuts.

xD

See Figure 50.

1. Remove No. 2 cylinder head cover.
    a. First, lift up the rear of the No. 2 cylinder head cover to disengage the 2 fittings.
    b. Next, lift the front of the No. 2 cylinder head cover to disengage the 2 fittings. Then remove the No. 2 cylinder head cover.

➡Ensure that the rubber grommets remain attached to the No. 2 cylinder head cover. If any grommets are attached to the bolts, move them to the No. 2 cylinder head cover.

2. Remove ignition coil assembly.
    a. Disconnect the camshaft timing oil control valve connector.
    b. Disconnect the 4 ignition coil connectors.
    c. Remove the 4 bolts and 4 ignition coils.
3. Remove the 4 spark plugs.

To install:

4. Install the 4 spark plugs. Tighten to 15 ft. lbs. (20 Nm).
5. Install ignition coil assembly.
    a. Install the 4 ignition coils with the 4 bolts.

➡Do not damage the rubber ignition coil tips when installing the ignition coils.

    b. Connect the 4 ignition coil connectors.

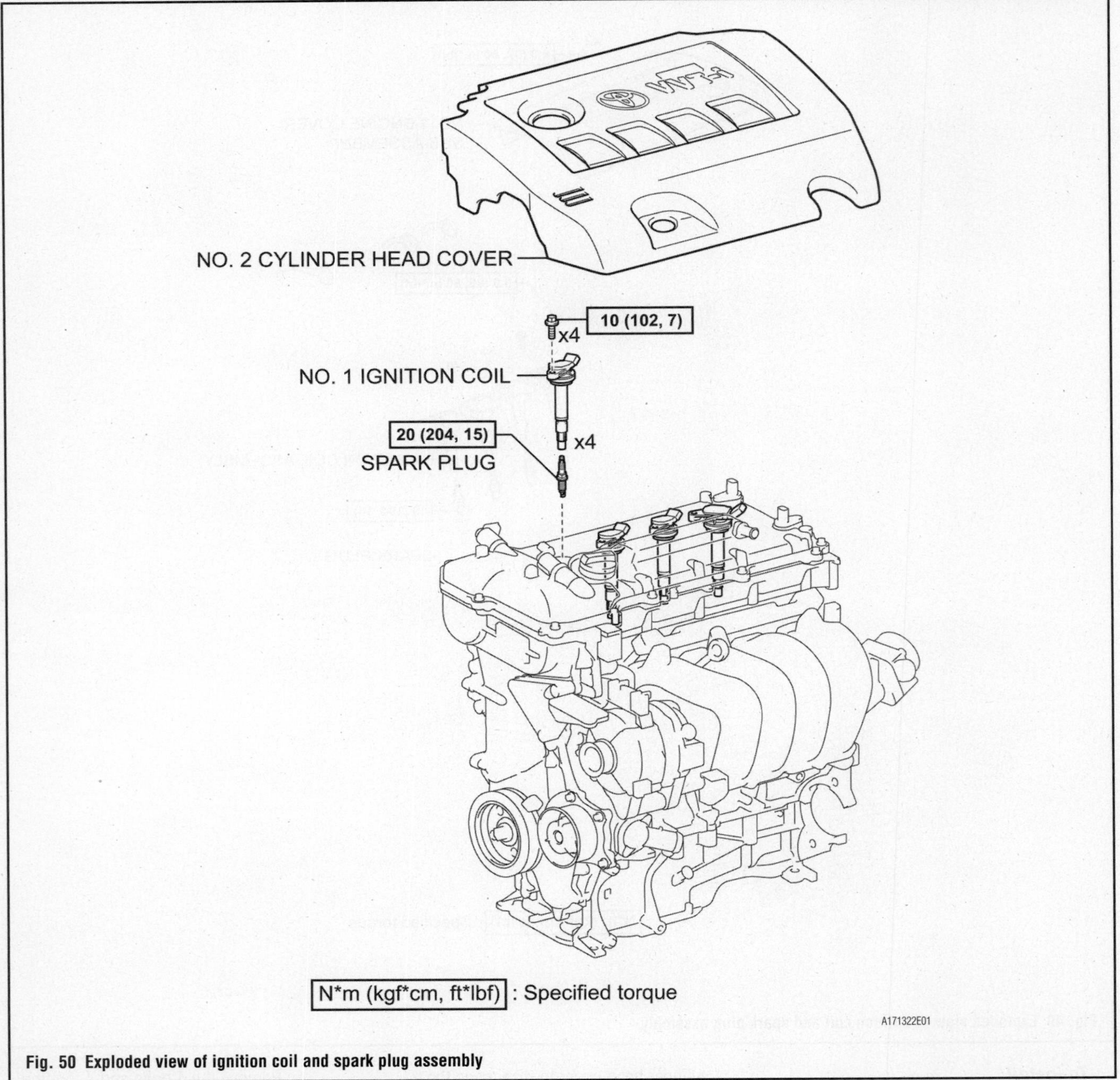

Fig. 50 Exploded view of ignition coil and spark plug assembly

6. Connect the camshaft timing oil control valve connector.
7. Engage the 4 fittings and install the No. 2 cylinder head cover by applying pressure to it.

➡**Make sure that the oil filler cap and the engine oil level dipstick are correctly installed before installing the No. 2 cylinder head cover.**

IGNITION TIMING

ADJUSTMENT

Ignition timing is controlled by the Engine Control Module (ECM). No adjustment is necessary or possible.

SPARK PLUGS

REMOVAL & INSTALLATION

Refer to Ignition Coil Removal and Installation when servicing spark plugs.

BATTERY

REMOVAL & INSTALLATION

See Figure 51.

1. Disconnect cable from negative battery terminal.

➡**When disconnecting the cable, some systems need to be initialized after the cable is reconnected.**

2. Disconnect cable from positive battery terminal.
3. Remove battery clamp.
   a. Remove the bolt and loosen the nut.
   b. Detach the 2 wire harness clamps.
   c. Detach the hook of the battery clamp from the front battery bracket, and then remove the battery clamp.
4. Remove battery insulator.
5. Remove battery.
6. Remove battery tray.
7. Remove the 4 bolts and front battery carrier.
8. Installation is the reverse order of removal.

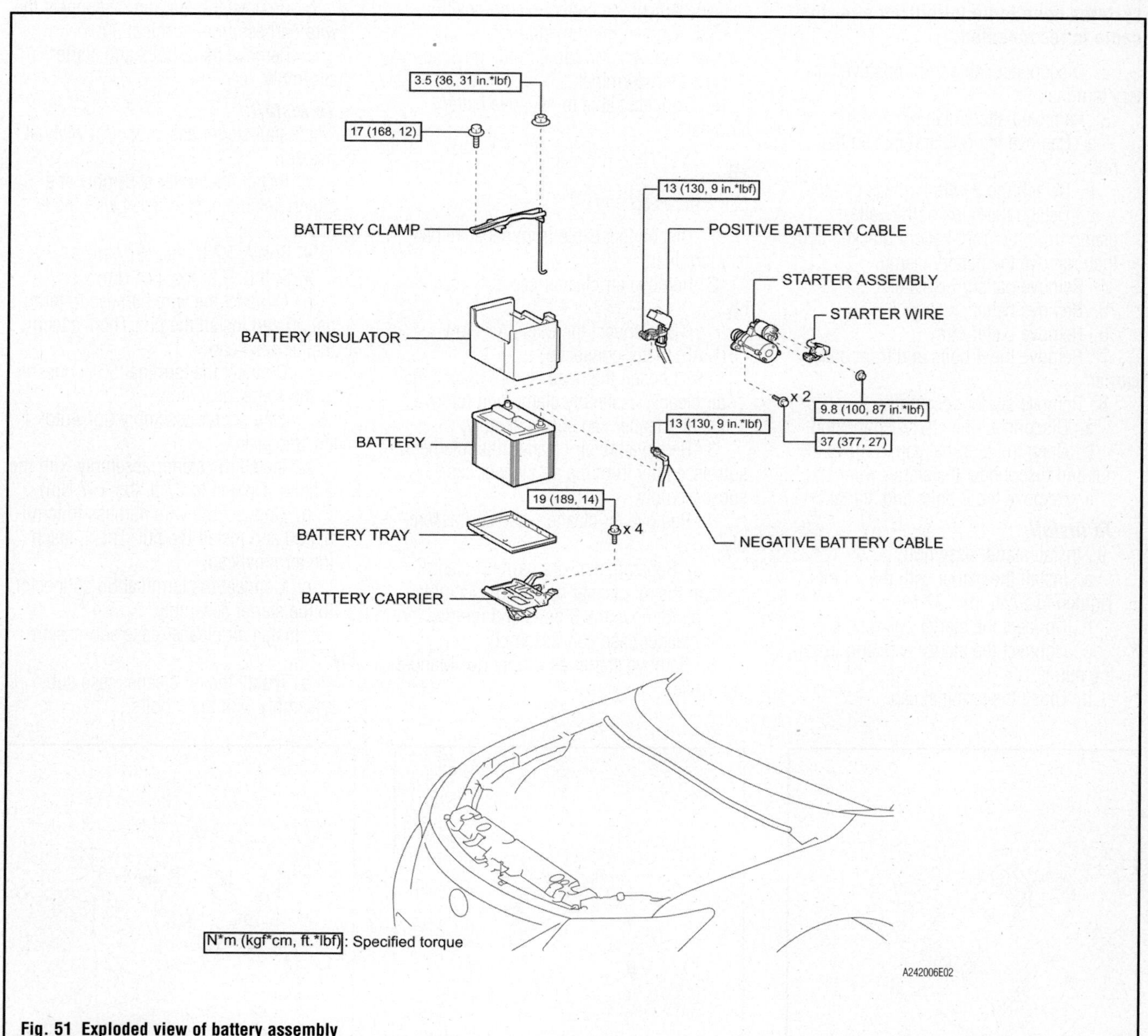

Fig. 51 Exploded view of battery assembly

STARTER

REMOVAL & INSTALLATION

tC

See Figure 52.

1. Disconnect cable from negative battery terminal.

➡**When disconnecting the cable, some systems need to be initialized after the cable is reconnected.**

2. Disconnect cable from positive battery terminal.
3. Remove battery clamp.
    a. Remove the bolt and loosen the nut.
    b. Detach the 2 wire harness clamps.
    c. Detach the hook of the battery clamp from the front battery bracket, and then remove the battery clamp.
4. Remove battery insulator.
5. Remove battery.
6. Remove battery tray.
7. Remove the 4 bolts and front battery carrier.
8. Remove starter assembly.
    a. Disconnect the starter connector.
    b. Open the terminal cap, remove the nut and disconnect the starter wire.
    c. Remove the 2 bolts and starter.

To install:

9. Install starter assembly.
    a. Install the starter with the 2 bolts. Tighten to 27 ft. lbs. (37 Nm).
    b. Connect the starter connector.
    c. Connect the starter wire and install the nut.
    d. Close the terminal cap.
10. Install the front battery carrier with the 4 bolts.
11. Install battery tray.
12. Install battery.
13. Install battery insulator.
14. Install battery clamp.
    a. Attach the hook of the battery clamp to the front battery bracket.
    b. Partially tighten the nut and temporarily install the bolt.
    c. Adjust the battery clamp position.
    d. Tighten the nut and bolt.
    e. Connect the cable to the positive (+) battery terminal.
15. Connect cable to negative battery terminal.

xB

See Figures 53 and 54.

1. Disconnect cable from negative battery terminal.
2. Remove air cleaner cap sub-assembly.
    a. Disconnect the Mass Air Flow (MAF) meter connector.
    b. Loosen the hose clamp, unlock the air cleaner assembly clamp and remove the air cleaner cap sub-assembly.
3. Remove the air cleaner filter element sub-assembly from the air cleaner case sub-assembly.
4. Remove air cleaner case sub-assembly.
    a. Separate the wire harness clamp from the air cleaner case sub-assembly.
    b. Remove the 3 bolts and remove the air cleaner case sub-assembly.
5. Remove starter assembly (for Manual Transaxle).
    a. Disconnect the terminal 50 connector from the starter assembly.
    b. Remove the nut and disconnect the wire harness from terminal 30.
    c. Remove the 3 bolts, clutch flexible hose bracket and starter assembly.
6. Remove starter assembly (for Automatic Transaxle).
    a. Disconnect the terminal 50 connector from the starter assembly.
    b. Remove the nut and disconnect the wire harness from terminal 30.
    c. Remove the 2 bolts and starter assembly.

To install:

7. Install starter assembly (for Manual Transaxle).
    a. Install the starter assembly and clutch flexible hose bracket with the 3 bolts.
        - Bolt A: 27 ft. lbs. (37 Nm)
        - Bolt B: 9 ft. lbs. (12 Nm)
    b. Connect the wire harness to terminal 30 and install the nut. Then, attach the terminal cap.
    c. Connect the terminal 50 connector to the starter assembly.
8. Install starter assembly (for Automatic Transaxle).
    a. Install the starter assembly with the 2 bolts. Tighten to 27 ft. lbs. (37 Nm).
    b. Connect the wire harness to terminal 30 and install the nut. Then, attach the terminal cap.
    c. Connect the terminal 50 connector to the starter assembly.
9. Install air cleaner case sub-assembly.
    a. Install the air cleaner case sub-assembly with the 3 bolts.

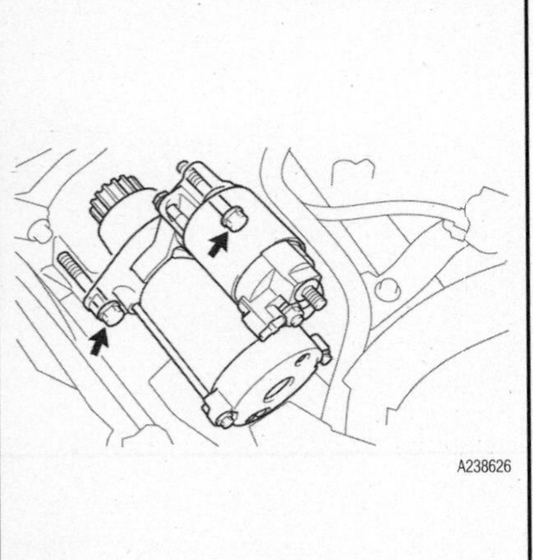

Fig. 52 Remove the 2 bolts and starter

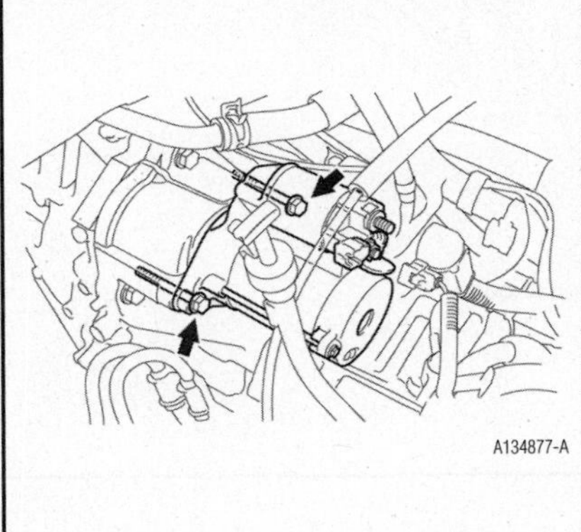

Fig. 53 Remove the 2 bolts and starter assembly

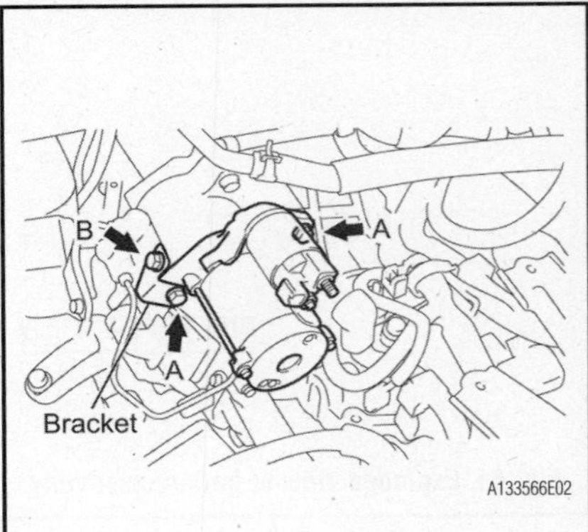

Fig. 54 Install the starter assembly and clutch flexible hose bracket with the 3 bolts

b. Install the wire harness clamp to the air cleaner case sub-assembly.

10. Install the air cleaner filter element sub-assembly onto the air cleaner case sub-assembly.

11. Install air cleaner cap sub-assembly.

a. Install and lock the air cleaner cap sub-assembly and then tighten the hose clamp.

b. Connect the Mass Air Flow (MAF) meter connector.

12. Connect cable to negative battery terminal.

xD

See Figure 55.

1. Disconnect cable from negative battery terminal

2. Remove engine under cover LH and RH.

3. Disengage the claw by pulling it outward and remove the flywheel housing side cover.

4. Remove starter assembly.

a. Remove the terminal cap.

b. Remove the nut and disconnect terminal 30.

c. Disconnect the connector.

d. Remove the 2 bolts and remove the starter assembly.

To install:

5. Install starter assembly.

a. Install the starter assembly with the 2 bolts. Tighten to 27 ft. lbs. (37 Nm).

b. Connect the connector.

c. Connect terminal 30 with the nut.

d. Close the terminal cap.

6. Install flywheel housing side cover. Insert the protruding portion into the end of the cylinder block and while pushing it along the cylinder block, fit the claw into the cylinder block.

➡Make sure that the claw makes a click sound, indicating that it fits tightly. Replace the claw with a new one if it does not fit tightly or is deformed.

7. Install engine under cover RH and LH.

8. Connect cable to negative battery terminal.

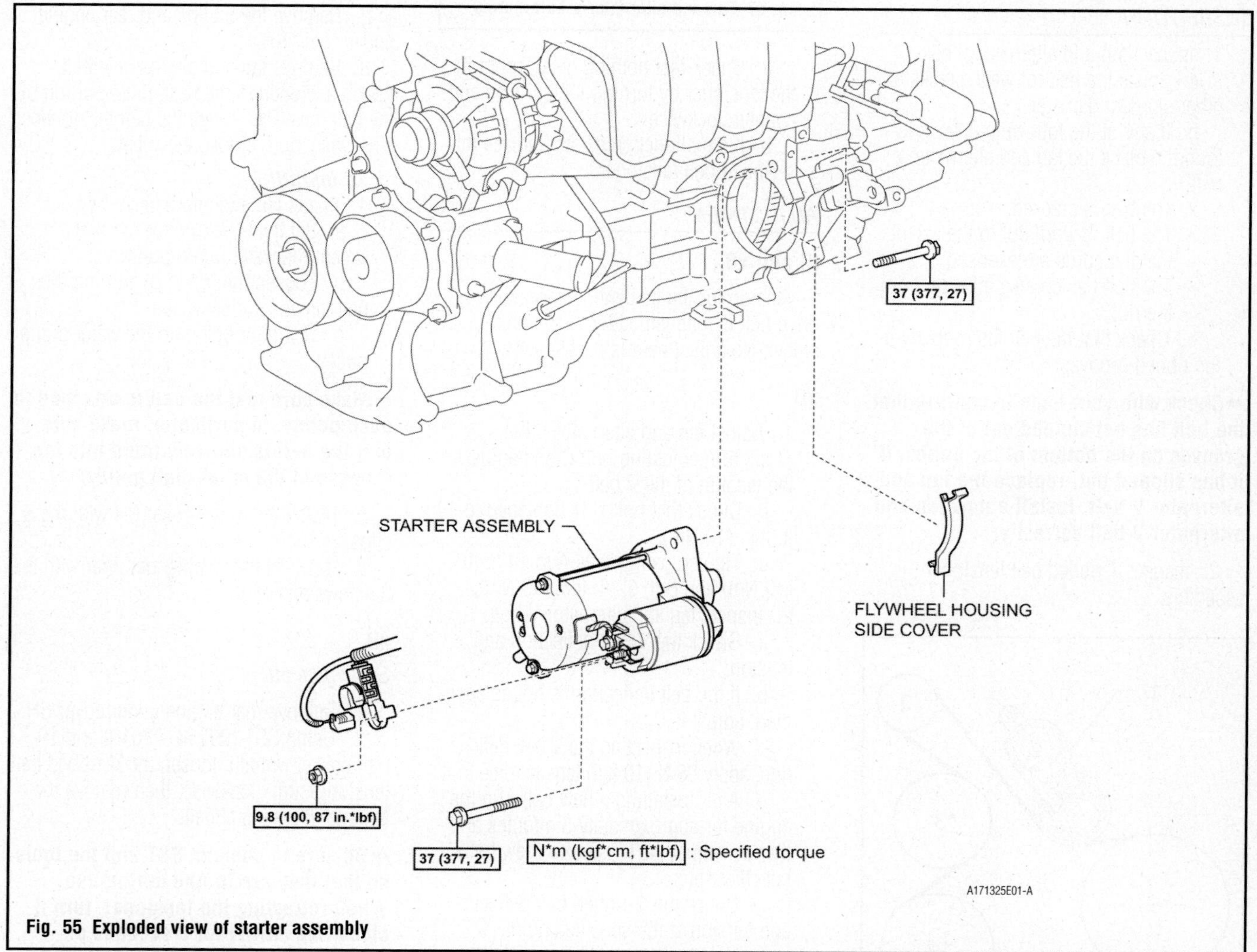

Fig. 55 Exploded view of starter assembly

ENGINE MECHANICAL

➡Disconnecting the negative battery cable may interfere with the functions of the on board computer systems and may require the computer to undergo a relearning process, once the negative battery cable is reconnected.

ACCESSORY DRIVE BELTS

ACCESSORY BELT ROUTING

tC

See Figure 56.

xB

See Figure 57.

INSPECTION

1. Inspect fan and alternator V belt.
   a. Check the belt for wear, cracks or other signs of damage.
   b. If any of the following defects is found, replace the fan and alternator V belt.
      - The belt is cracked.
      - The belt is worn out to the extent that the cords are exposed.
      - The belt has chunks missing from the ribs.

   c. Check that the belt fits properly in the ribbed grooves.

➡Check with your hand to confirm that the belt has not slipped out of the grooves on the bottom of the pulley. If it has slipped out, replace the fan and alternator V belt. Install a new fan and alternator V belt correctly.

2. Inspect V-ribbed belt tensioner assembly.

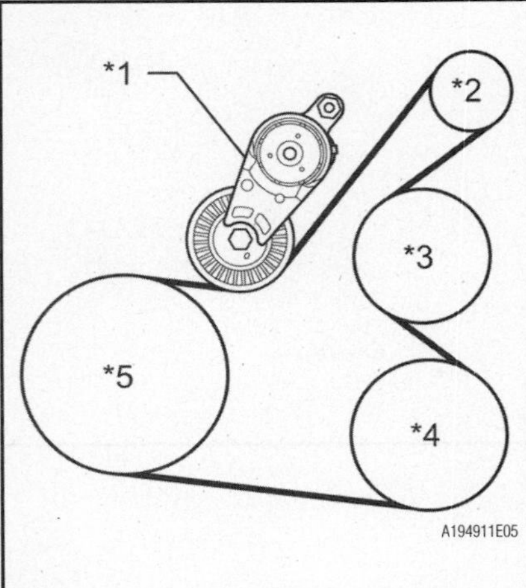

Fig. 56 Fan and alternator V-belt routing

Fig. 57 Fan and alternator V-belt routing

   a. Check that nothing gets caught in the tensioner by turning it clockwise and counterclockwise.
   b. If a malfunction exists, replace the V-ribbed belt tensioner.

ADJUSTMENT

tC and xB

Tension for the serpentine accessory drive belt is maintained by the belt tensioner. No adjustment is necessary.

xD

1. Adjust fan and alternator V belt.
   a. Turn adjusting bolt C and adjust the tension of the V belt.
   b. Check that bolt D is tightened to 14 ft. lbs. (19 Nm).
   c. Tighten fixing bolts A to 14 ft. lbs. (19 Nm) and B to 32 ft. lbs. (43 Nm).
2. Inspect fan and alternator V belt.
   a. Check the V belt deflection and tension.
   b. If the belt deflection is not as specified, adjust it.
   c. When inspecting the V belt deflection, apply 98 N (10 kgf) tensile force to it.
   d. After installing a new belt, run the engine for approximately 5 minutes and then re-adjust the tension to (new belt) specifications.
   e. Check the V-ribbed belt deflection and tension at the specified point.
   f. V-ribbed belt tension and deflection should be checked after 2 revolutions of the engine.
   g. V-ribbed belt tension and deflection should be checked at TDC crank angle and cold condition.
   h. When adjusting a belt, adjust its deflection and tension to the intermediate values of the specification.
   i. When reinstalling a belt which has been used for over 5 minutes, adjust its deflection and tension to the used belt specification.
   j. When using a belt tension gauge, confirm its accuracy by using a master gauge first.

REMOVAL & INSTALLATION

tC

See Figure 58.

1. Remove the 2 screws and clip and disconnect the front fender liner.
2. Remove the 5 clips and rear engine under cover RH.
3. Remove fan and alternator V belt. Attach a wrench to the hexagonal portion of the belt tensioner, rotate the belt tensioner clockwise, and remove the V belt.

To install:

4. Install fan and alternator V belt.
   a. Set the V belt onto each part, except the water pump pulley.
   b. Loosen the V belt by turning the belt tensioner clockwise.
   c. Set the V belt onto the water pump pulley.

➡Make sure that the belt is attached to each pulley. In particular, make sure that the belt is securely fitted into the grooves of the crankshaft pulley.

5. Install the under cover RH with the 5 clips.
6. Connect the front fender liner with the 2 screws and clip.

xB

See Figure 59.

1. Remove rear engine under cover RH.
2. Using SST (09216–42010) and 19 mm socket wrench, loosen the V-ribbed belt tensioner arm clockwise, then remove the fan and alternator V belt.

➡Be sure to connect SST and the tools so that they are in line during use. When retracting the tensioner, turn it clockwise slowly for 3 seconds or more. Do not apply force rapidly. After the tensioner is fully retracted, do not apply force any more than necessary.

To install:

3. Using SST and 19 mm socket wrench, loosen the V-ribbed belt tensioner

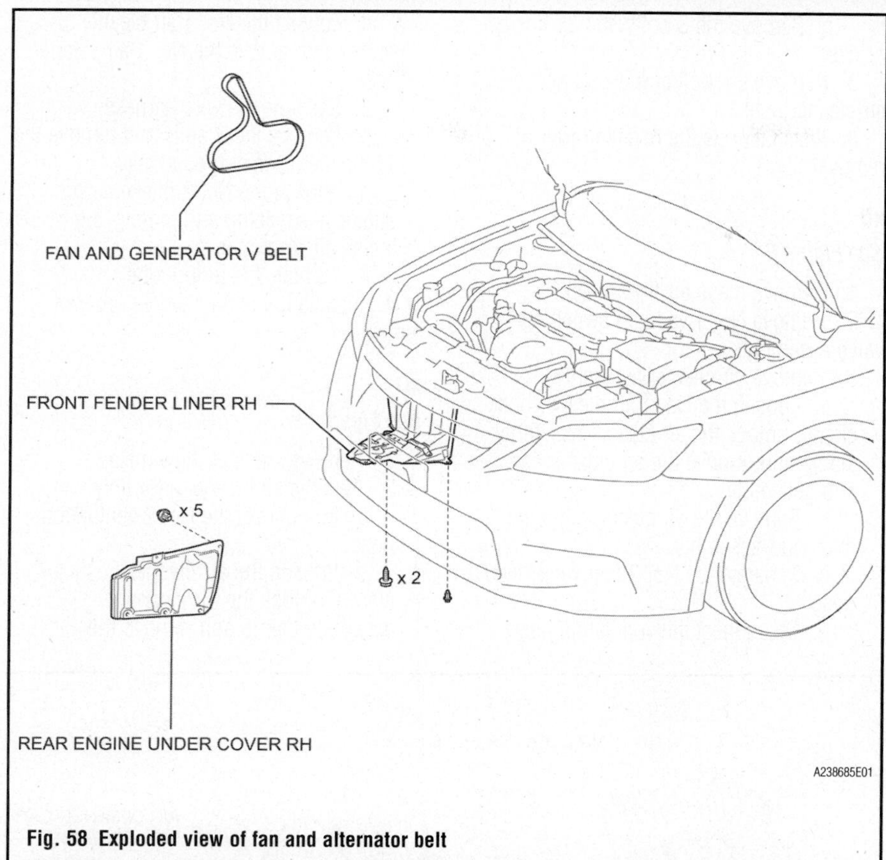

Fig. 58 Exploded view of fan and alternator belt

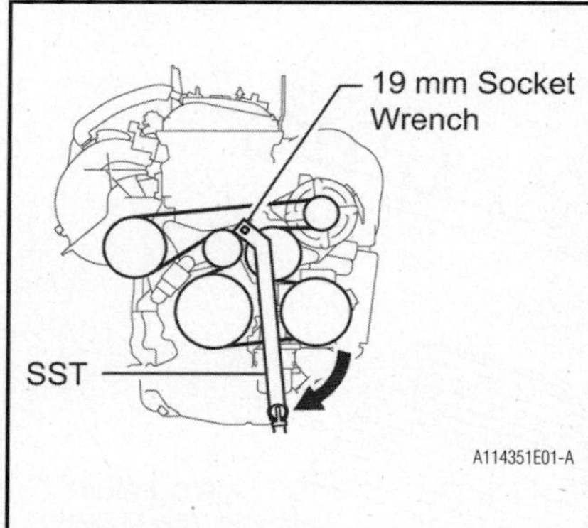

Fig. 59 Using SST (09216–42010) and 19 mm socket wrench, loosen the V-ribbed belt tensioner arm clockwise, then remove the fan and alternator V belt

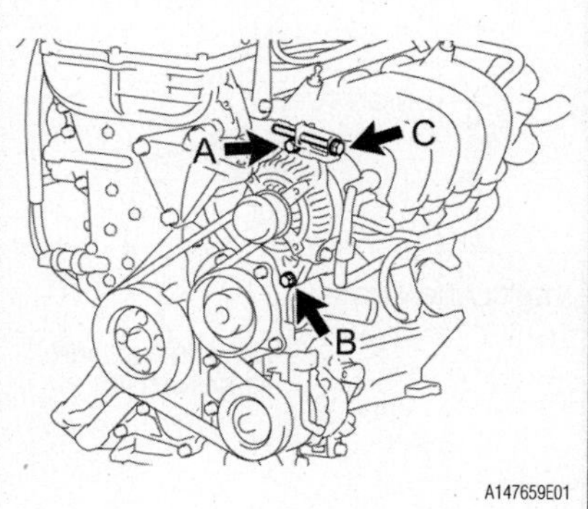

Fig. 60 Loosen bolts A and B, turn adjusting bolt C

arm clockwise, then install the fan and alternator V belt.

4. Install rear engine under cover RH.

xD

See Figure 60.

1. Remove engine under cover RH.
2. Remove No. 2 cylinder head cover.
3. Remove fan and alternator V belt.
   a. Loosen bolts A and B.
   b. Turn adjusting bolt C to release the tension and remove the V belt from the pulleys.
   c. Remove the fan and alternator V belt.

To install:

4. Provisionally install the fan and alternator V belt onto each pulley.

➡Make sure that the V belt is securely fitted into the rib groove of the pulley.

5. Adjust fan and alternator V belt.
   a. Turn adjusting bolt C and adjust the tension of the V belt.
   b. Check that bolt D is tightened to 14 ft. lbs. (19 Nm).
   c. Tighten fixing bolts A to 14 ft. lbs. (19 Nm) and B to 32 ft. lbs. (43 Nm).
6. Inspect fan and alternator V belt.
   a. Check the V belt deflection and tension.
   b. If the belt deflection is not as specified, adjust it.
   c. When inspecting the V belt deflection, apply 98 N (10 kgf) tensile force to it.
   d. After installing a new belt, run the engine for approximately 5 minutes and then re-adjust the tension to (new belt) specifications.
   e. Check the V-ribbed belt deflection and tension at the specified point.
   f. V-ribbed belt tension and deflection should be checked after 2 revolutions of the engine.
   g. V-ribbed belt tension and deflection should be checked at TDC crank angle and cold condition.
   h. When adjusting a belt, adjust its deflection and tension to the intermediate values of the specification.
   i. When reinstalling a belt which has been used for over 5 minutes, adjust its deflection and tension to the used belt specification.
   j. When using a belt tension gauge, confirm its accuracy by using a master gauge first.
7. Install engine under cover RH.
8. Install No. 2 cylinder head cover.

AIR CLEANER

REMOVAL & INSTALLATION

tC

See Figures 61 through 63.

1. Remove air cleaner cap sub-assembly.
   a. Disconnect the Mass Air Flow (MAF) meter connector.
   b. Disconnect the purge VSV connector.
   c. Detach the 2 wire harness clamps.
   d. Disconnect the 2 purge line hoses.
   e. Disconnect the No. 2 ventilation hose from the cylinder head cover.
   f. Disconnect the purge line hose from the clamp.
   g. Lock the No. 1 air cleaner hose clamp, and then disconnect the No. 1 air cleaner hose from the throttle body.

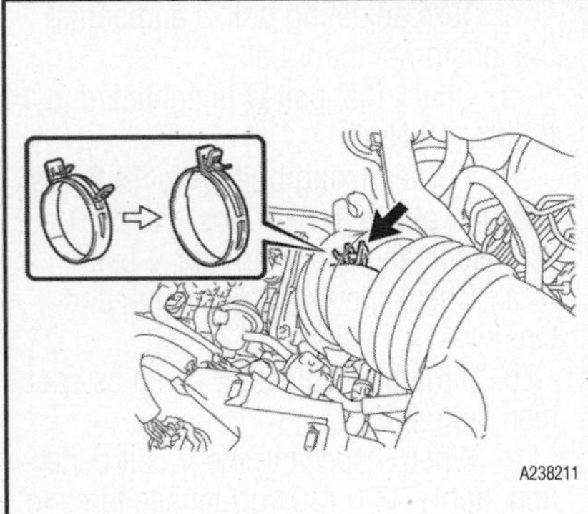

Fig. 61 Lock the No. 1 air cleaner hose clamp, and then disconnect the No. 1 air cleaner hose from the throttle body

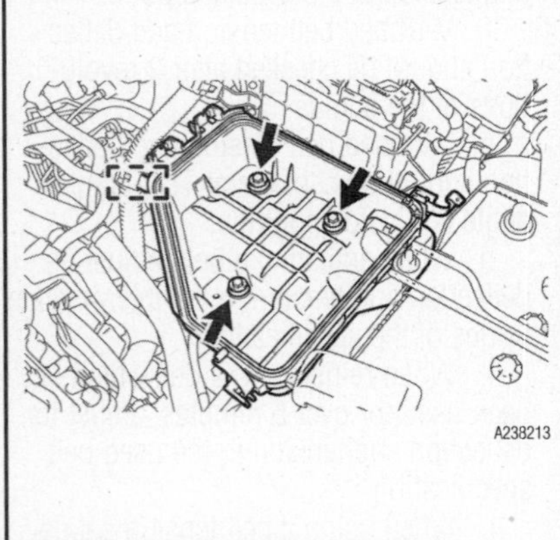

Fig. 62 Remove the 3 bolts and air cleaner case

h. Detach the 2 hook clamps, and then remove the air cleaner cap.

i. Remove the air cleaner filter element from the air cleaner case.

2. Remove air cleaner case.

a. Detach the harness clamp.

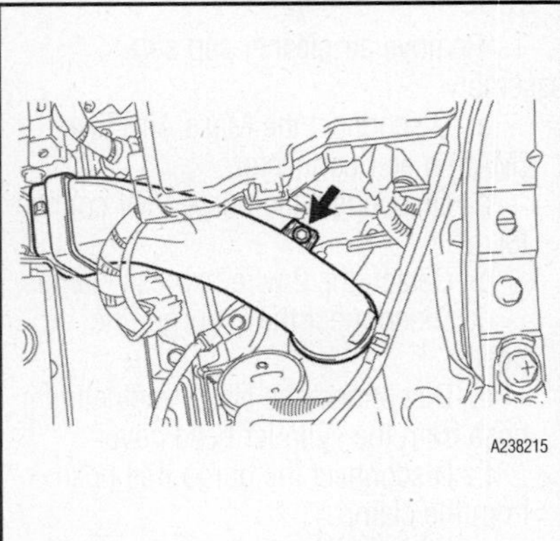

Fig. 63 Remove the bolt and disconnect the air cleaner inlet

b. Remove the 3 bolts and air cleaner case.

3. Remove the bolt and disconnect the air cleaner inlet.

4. Installation is the reverse order of removal.

xB

See Figure 64.

1. Remove mass air flow meter.
2. Remove No. 1 vacuum switching valve assembly.
3. Remove air cleaner assembly.

a. Loosen the No. 1 air cleaner hose clamp, unlock the air cleaner assembly clamp and remove the air cleaner cap sub-assembly.

b. Remove the air cleaner filter element sub-assembly.

c. Separate the No. 2 fuel vapor feed hose.

d. Disconnect the ventilation hose.

e. Loosen the No. 1 air cleaner hose clamp and remove the No. 1 air cleaner hose.

f. Separate the wire harness.

g. Remove the 3 bolts and remove the air cleaner case sub-assembly.

h. Loosen the No. 2 intake air resonator hose clamp and remove the No. 2 intake air resonator.

i. Remove the union tube.

4. Installation is the reverse order of removal.

xD

See Figure 65.

1. Remove mass air flow meter.
2. Remove air cleaner assembly.

a. Disconnect the No. 2 ventilation hose.

b. Loosen the air cleaner hose clamp, unlock the air cleaner assembly clamp and remove the air

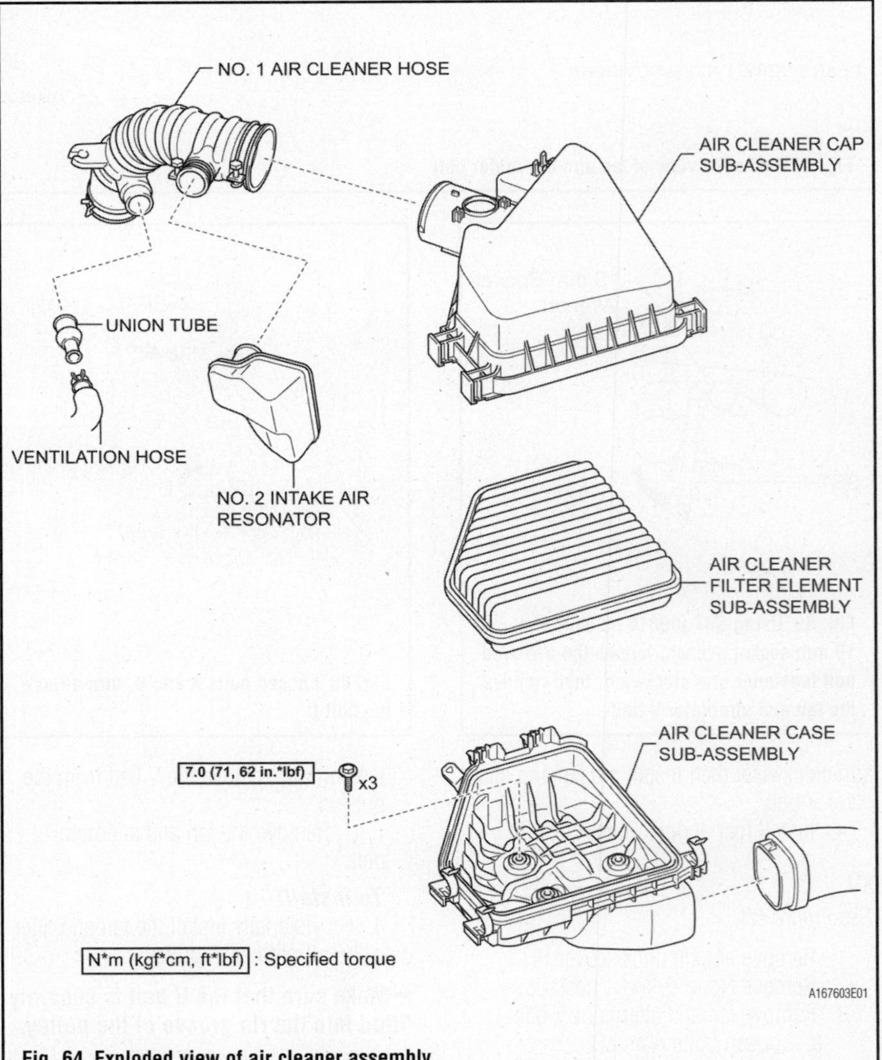

Fig. 64 Exploded view of air cleaner assembly

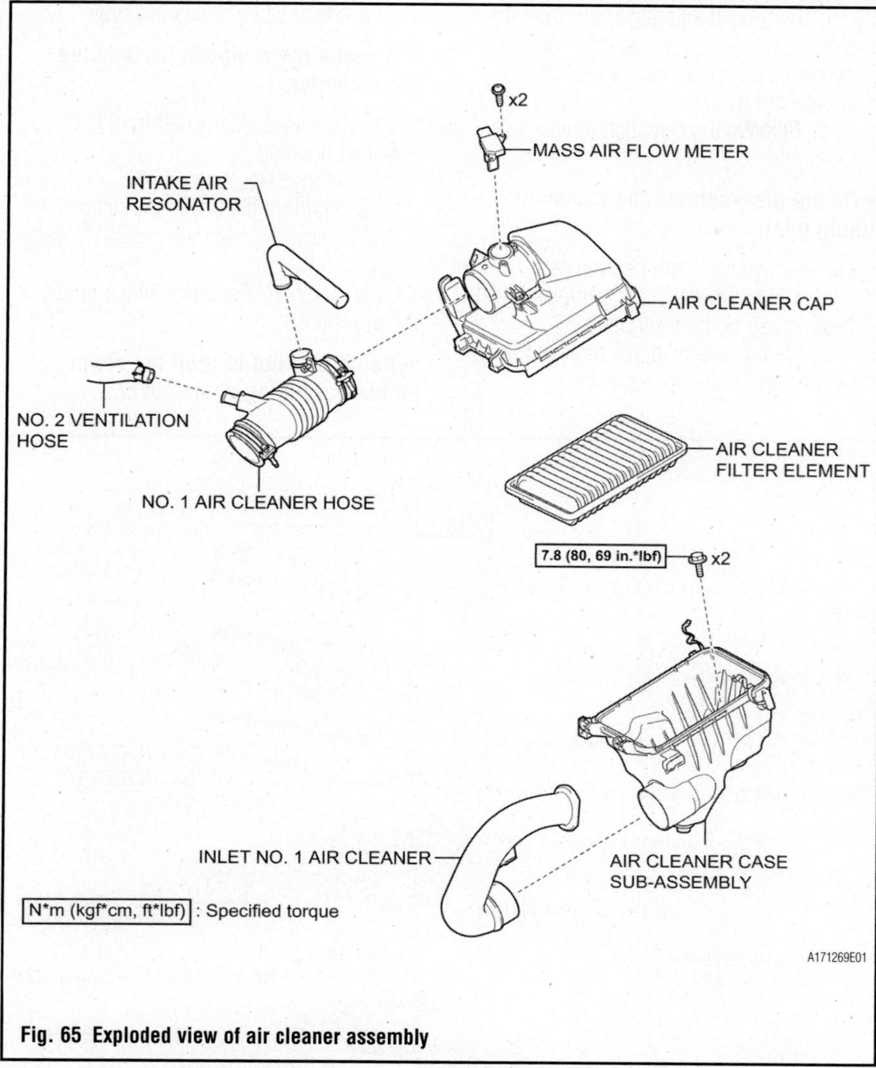

Fig. 65 Exploded view of air cleaner assembly

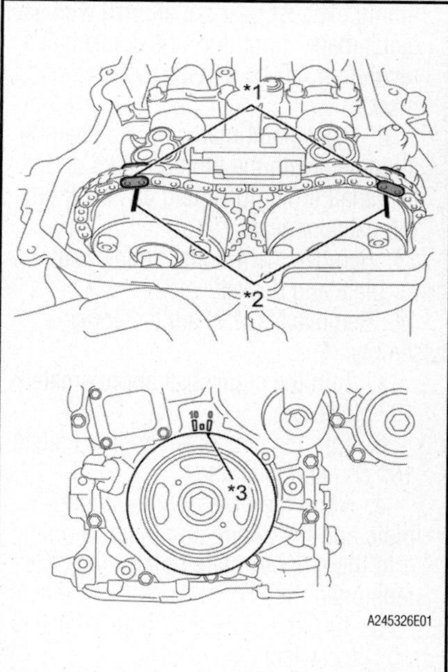

Fig. 66 Turn the crankshaft pulley until its timing notch (3) (groove) and the timing mark "0" of the timing chain cover are aligned; paint mark (1), matchmark (2)

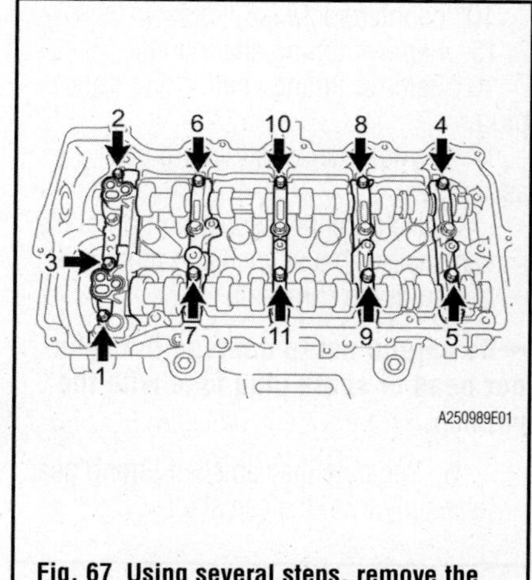

Fig. 67 Using several steps, remove the 11 bearing cap bolts in sequence

cleaner cap sub-assembly with air cleaner hose.

c. Remove the air cleaner filter element.

d. Separate the wire harness.

e. Remove the 2 bolts and remove the air cleaner case with inlet No. 1 air cleaner.

f. Loosen the air cleaner hose clamp and remove the intake air resonator.

g. Loosen the air cleaner hose clamp and remove the air cleaner hose from the No. 1 air cleaner cap sub-assembly.

h. Remove the inlet No. 1 air cleaner from the air cleaner case.

3. Installation is the reverse order of removal.

FILTER/ELEMENT REPLACEMENT

Refer to Air Cleaner Removal and Installation.

CAMSHAFT AND VALVE LIFTERS

REMOVAL & INSTALLATION

tC

See Figures 66 through 74.

1. Disconnect cable from negative battery terminal.

➡When disconnecting the cable, some systems need to be initialized after the cable is reconnected.

2. Remove front wheel RH.
3. Disconnect front fender liner RH.
4. Remove rear engine under cover RH.
5. Remove air cleaner cap sub-assembly.
6. Remove air cleaner case sub-assembly.
7. Disconnect the connectors and clamps, remove the bolts and nuts and disconnect the engine wire from the engine.
8. Remove ignition coil assembly.
9. Remove cylinder head cover sub-assembly.
10. Remove Crankshaft Position (CKP) sensor.
11. Remove the 2 bolts and separate the radiator reserve tank assembly.
12. Set No. 1 cylinder to TDC/compression.

a. Turn the crankshaft pulley until its timing notch (groove) and the timing mark "0" of the timing chain cover are aligned.

b. Check that each matchmark of the camshaft timing gear and camshaft timing exhaust gear are aligned with each matchmark. If not, turn the crankshaft 1 revolution (360°) to align the timing marks.

c. Place paint marks on the chain in alignment with the timing marks on the camshaft timing gear and camshaft timing exhaust gear.

13. Remove the 4 bolts, timing chain cover plate and gasket.

14. Remove No. 1 chain tensioner assembly.

a. Turn the crankshaft approximately 10° clockwise.

b. Turn the crankshaft approximately 10° counterclockwise.

c. Align the holes of the stopper plate and tensioner, and insert a pin into the stopper plate hole to lock the tensioner.

d. Turn the crankshaft approximately 10° clockwise.

e. Remove the 2 bolts, chain tensioner and gasket.

➡Make sure not to drop the gasket inside the timing chain cover.

f. Turn the crankshaft approximately 10° counterclockwise.

15. Remove timing chain guide.

16. Remove timing chain cover tight plug.

17. Remove camshaft timing gear assembly.

a. Hold the hexagonal portion of the camshaft with a wrench and remove the bolt from the camshaft.

➡Be careful not to damage the cylinder head or spark plug tube with the wrench.

b. Separate the camshaft timing gear assembly from the camshaft.

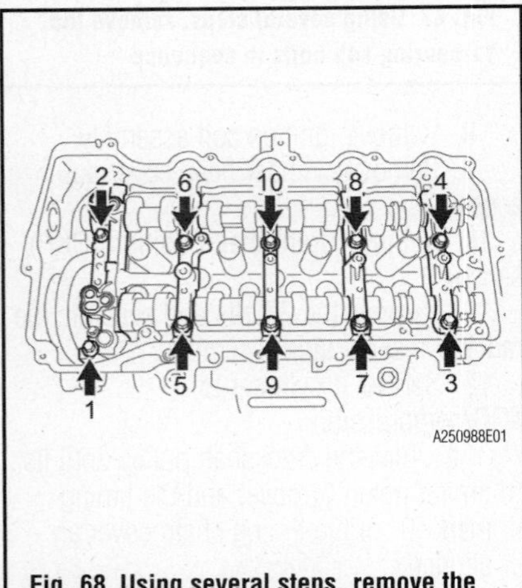

Fig. 68 Using several steps, remove the 10 bearing cap bolts in sequence

c. Remove the timing chain from the camshaft timing gear assembly, and turn the camshaft timing gear assembly approximately 180°.

d. Remove the camshaft timing gear assembly.

➡Do not disassemble the camshaft timing gear.

18. Remove camshaft bearing cap.

a. Using several steps, remove the 11 bearing cap bolts in sequence.

b. Using several steps, remove the 10 bearing cap bolts in sequence.

c. Remove the 5 bearing caps.

➡Arrange the removed parts in the correct order.

19. Remove the camshaft from the camshaft housing.

20. Remove No. 2 camshaft.

a. Hold up the chain and remove the No. 2 camshaft from the camshaft housing.

b. Suspend the chain with a string or equivalent.

➡Be careful not to drop the chain inside the timing chain cover.

12 (122, 9)
x 3
x 3
x 13
CYLINDER HEAD COVER SUB-ASSEMBLY
● CYLINDER HEAD COVER GASKET
● GASKET
16 (163, 12)
27 (275, 20)
● CAMSHAFT BEARING CAP OIL HOLE GASKET
x 11
x 10
x 2
NO. 1 CAMSHAFT BEARING CAP
NO. 3 CAMSHAFT BEARING CAP
OIL CONTROL VALVE FILTER
CAMSHAFT TIMING GEAR ASSEMBLY
NO. 2 CAMSHAFT BEARING CAP
CAMSHAFT
NO. 1 CAMSHAFT BEARING
85 (867, 63)
CAMSHAFT TIMING EXHAUST GEAR ASSEMBLY
85 (867, 63)
NO. 2 CAMSHAFT
21 (214, 15)
NO. 2 CAMSHAFT BEARING
TIMING CHAIN GUIDE
● GASKET
30 (306, 22)
TIMING CHAIN COVER TIGHT PLUG
● GASKET
10 (102, 7)
x 4
TIMING CHAIN COVER PLATE
x 2
● GASKET
10 (102, 7)
NO. 1 CHAIN TENSIONER ASSEMBLY
N*m (kgf*cm, ft.*lbf): Specified torque
● Non-reusable part
A251008E04

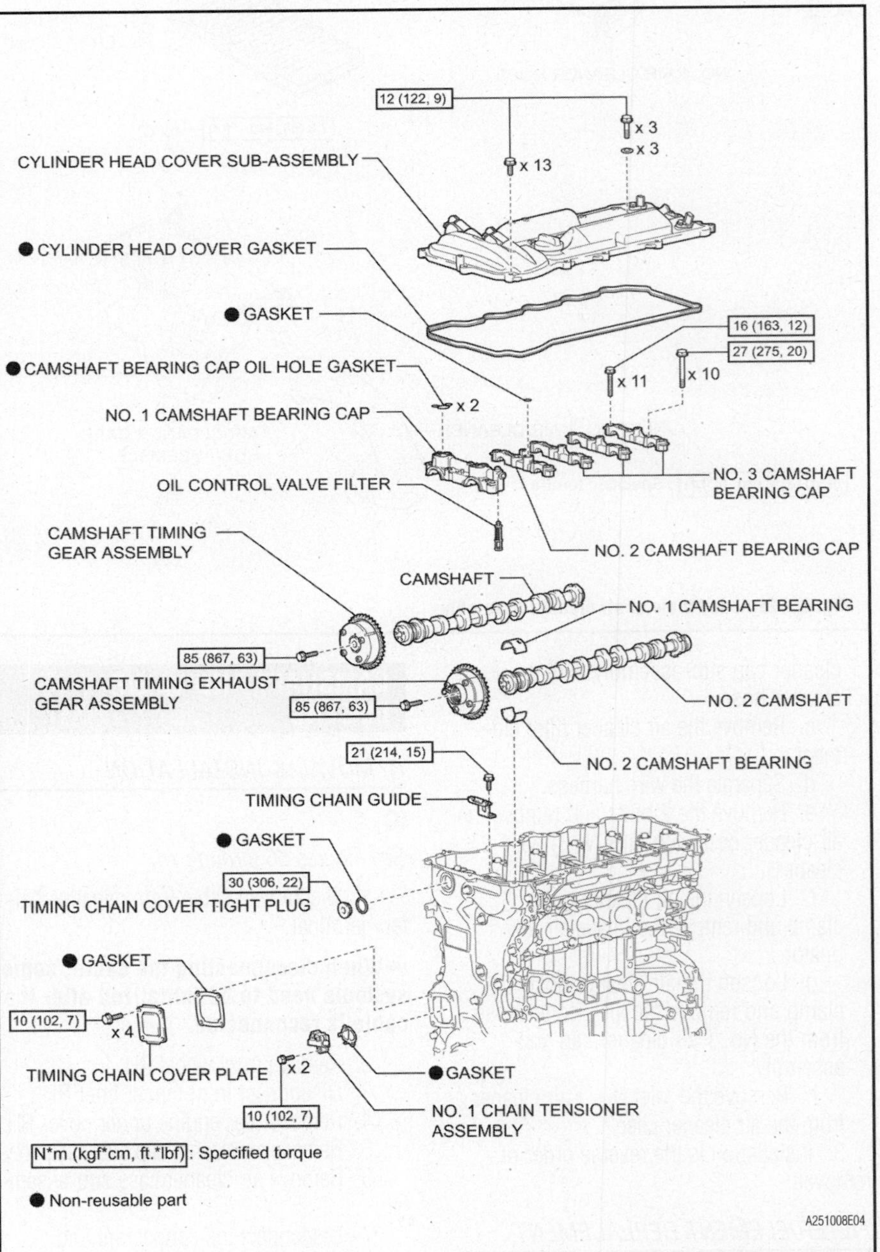

Fig. 69 Exploded view of camshaft assemblies

21. Remove the flange bolt and camshaft timing exhaust gear assembly.

➡Do not disassemble the camshaft timing exhaust gear.

22. Remove oil control valve filter.
23. Remove No. 1 camshaft bearing.
24. Remove No. 2 camshaft bearing.

To install:

25. Install No. 2 camshaft bearing.
26. Install No. 1 camshaft bearing.
27. Install oil control valve filter.
28. Install camshaft timing exhaust gear assembly.
   a. Align and attach the knock pin of the No. 2 camshaft with the pin hole of the camshaft timing exhaust gear assembly.
   b. Check that there is no clearance between the camshaft timing exhaust gear assembly and camshaft flange.
   c. Fix the camshaft timing exhaust gear assembly with the bolt. Tighten to 63 ft. lbs. (85 Nm).

➡Do not disassemble the camshaft timing exhaust gear.

29. Set No. 1 cylinder to TDC/compression. Turn the crankshaft pulley until its timing notch (groove) and the timing mark "0" of the timing chain cover are aligned.
30. Install No. 2 camshaft.
   a. Make sure that the valve rocker arms are properly installed.
   b. Clean the camshaft journals.
   c. Apply a light coat of engine oil to the camshaft journals, camshaft housings and bearing caps.
   d. Hold up the chain and align the matchmark and the paint mark and install the camshaft.

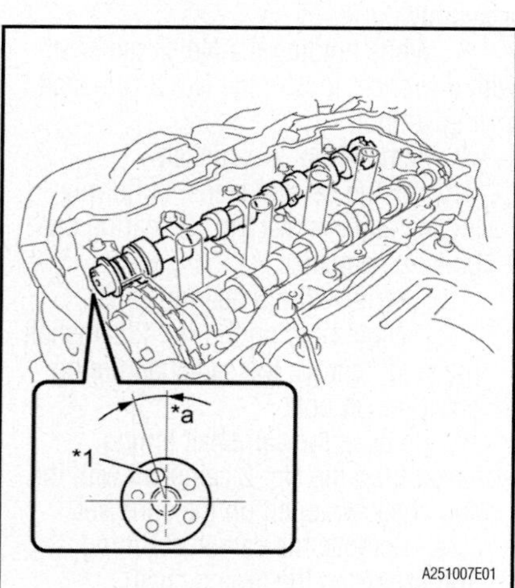

Fig. 70 Install the camshaft to the camshaft housing; knock pin (1), approximately 17° (a)

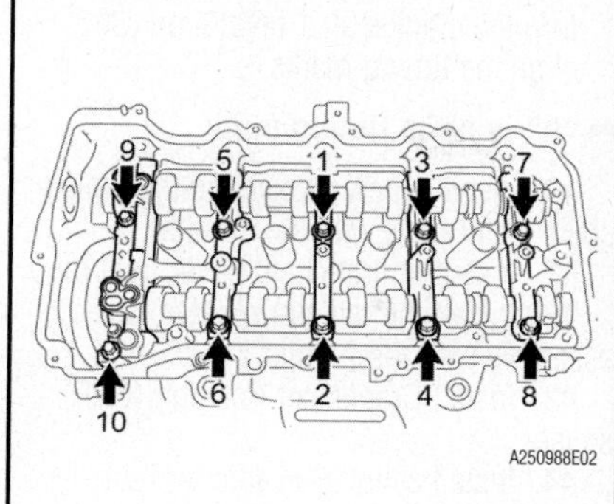

Fig. 71 Using several steps, uniformly tighten the 10 bolts in sequence to 20 ft. lbs. (27 Nm)

31. Install camshaft.
   a. Make sure that the valve rocker arms are properly installed.
   b. Clean the camshaft journals.
   c. Apply a light coat of engine oil to the camshaft journals, camshaft housings and bearing caps.
32. Install the camshaft to the camshaft housing.
33. Install camshaft bearing cap.
   a. Confirm the marks and numbers on the camshaft bearing caps and place them in their proper positions and directions.
   b. Using several steps, uniformly tighten the 10 bolts in sequence to 20 ft. lbs. (27 Nm).
   c. Using several steps, uniformly tighten the 11 bolts in sequence to 12 ft. lbs. (16 Nm).
   d. Check the torque of each bolt again.

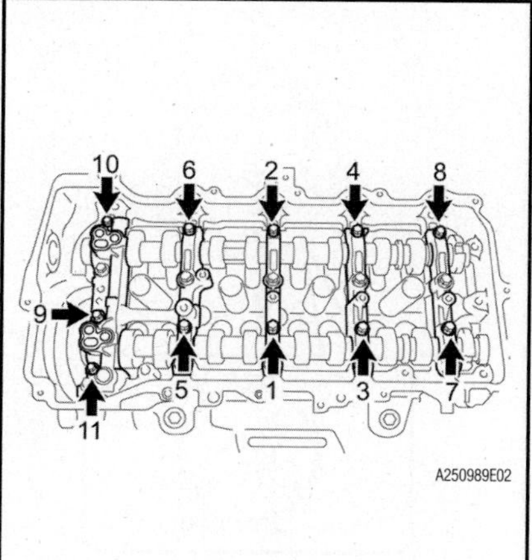

Fig. 72 Using several steps, uniformly tighten the 11 bolts in sequence to 12 ft. lbs. (16 Nm)

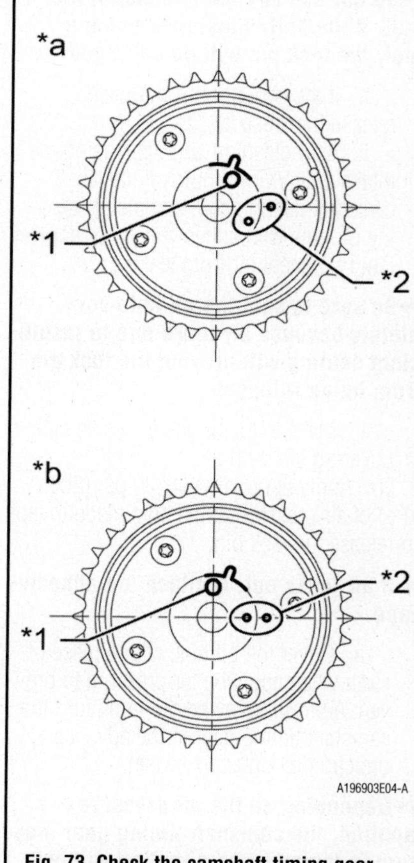

Fig. 73 Check the camshaft timing gear position

34. Install camshaft timing gear assembly.
   a. Check the camshaft timing gear position.

➡If the camshaft timing gear is set to the advanced position, do not let the camshaft timing gear rotate clockwise during installation. If the camshaft timing gear has rotated to the most retarded position, make sure to release the lock pin and set the camshaft timing gear to the most advanced position before tightening the camshaft timing gear.

   b. Install the camshaft timing gear.
   c. Turn the camshaft timing gear approximately 180° counterclockwise.
   d. Align the paint mark with the matchmark to install the chain.
   e. Align and attach the knock pin of the No. 1 camshaft with the pin hole of the camshaft timing gear.
   f. Check that there is no clearance between the camshaft timing gear and camshaft flange.
   g. Secure the camshaft in place by hand, and then install the installation bolt of the camshaft timing gear by hand.

➡Do not use any tools to install the bolt. If the bolt is installed using a tool, the lock pin will be damaged.

h. If the lock pin has not been released, release it.

i. After cleaning and degreasing the intake side VVT oil hole on the No. 1 camshaft bearing cap, completely seal the oil hole with adhesive tape or equivalent to prevent air from leaking.

➡Be sure to seal the oil hole completely because air leaks due to insufficient sealing will prevent the lock pin from being released.

j. Make a hole in the adhesive tape covering the oil hole.

35. Apply approximately 29 psi (200 kPa) of air pressure to the hole made in tape to release the lock pin.

➡If air leaks out, reattach the adhesive tape.

a. Cover the oil hole with a piece of cloth when applying air pressure to prevent oil from spraying. Forcibly turn the camshaft timing gear in the advance direction (counterclockwise).

➡Depending on the air pressure applied, the camshaft timing gear may turn in the advance direction without assistance by hand.

b. Remove the adhesive tape from the No. 1 camshaft bearing cap.

c. Using a wrench to hold the hexagonal portion of the No. 1 camshaft, install the bolt. Tighten to 63 ft. lbs. (85 Nm).

d. Check that each matchmark of the camshaft timing gear and camshaft timing exhaust gear are aligned with each matchmark.

36. Add engine oil.

37. Install timing chain guide.

38. Install No. 1 chain tensioner assembly.

a. Turn the crankshaft approximately 10° clockwise.

b. Install a new gasket and the chain tensioner with the 2 bolts.

➡Make sure not to drop the gasket inside the timing chain cover.

c. Remove the pin from the stopper plate.

39. Check No. 1 cylinder to TDC/compression.

a. Turn the crankshaft pulley until its timing notch (groove) and the timing mark "0" of the timing chain cover are aligned.

b. Check that the timing marks of the camshaft timing gears are correct. If not, turn the crankshaft 1 revolution (360°) to align the timing marks.

➡"A" is not a timing mark.

40. Install a new gasket and the timing chain cover plate with the 4 bolts.

41. Install timing chain cover tight plug.

42. Install the radiator reserve tank assembly with the 2 bolts.

43. Install Crankshaft Position (CKP) sensor.

44. Install cylinder head cover sub-assembly.

45. Install ignition coil assembly.

46. Connect the connectors and clamps, and install the engine wire to the engine with the bolts and nuts.

47. Install air cleaner case sub-assembly.

48. Install air cleaner cap sub-assembly.

49. Connect cable to negative battery terminal.

50. Inspect for oil leak.

51. Install rear engine under cover RH.

52. Connect front fender liner RH.

53. Install front wheel RH.

xB

See Figures 75 through 83.

1. Remove rear engine under cover RH.
2. Remove No. 1 engine cover sub-assembly.
3. Remove ignition coil assembly.
4. Remove spark plug.
5. Remove cylinder head cover sub-assembly.

a. Remove the 2 ventilation hoses from the cylinder head.

Fig. 74 Install the camshaft timing gear; narrow (a), wide (b)

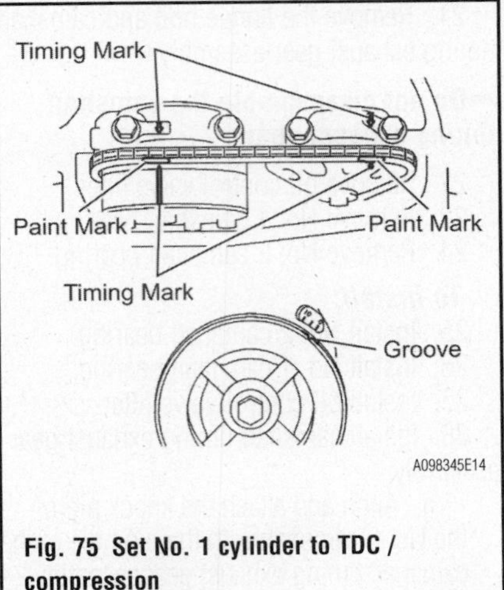

Fig. 75 Set No. 1 cylinder to TDC / compression

b. Remove the 2 bolts and separate the 2 wire harness brackets.

c. Remove the 8 bolts and 2nuts, and then remove the cylinder head cover and gasket.

6. Set No. 1 cylinder to TDC / compression.

a. Turn the crankshaft pulley until the groove and the timing mark "0" on the timing chain cover are aligned.

b. Check that each timing mark on the camshaft timing gear and sprocket is aligned with each timing mark located on the No. 1 and No. 2 bearing caps.

c. If not, turn the crankshaft pulley by 1 revolution (360°) to align the timing marks as above.

d. Place paint marks on the chain in alignment with the timing marks on the camshaft timing gear and camshaft timing sprocket.

7. Remove No. 1 chain tensioner assembly.

8. While holding the No. 2 camshaft with a wrench, loosen the No. 2 camshaft timing set bolt.

9. Remove No. 2 camshaft.

a. Using several steps, uniformly loosen and remove the 10 bearing cap bolts in sequence.

b. Remove the 5 bearing caps.

c. While holding the No. 2 camshaft by hand, remove the camshaft timing sprocket set bolt.

10. Remove the camshaft timing sprocket from the No. 2 camshaft with the timing chain wrapped on the sprocket.

a. Remove the camshaft timing sprocket from the timing chain.

11. Remove camshaft.

a. In several steps, uniformly loosen and remove the 10 bearing caps bolts in sequence.

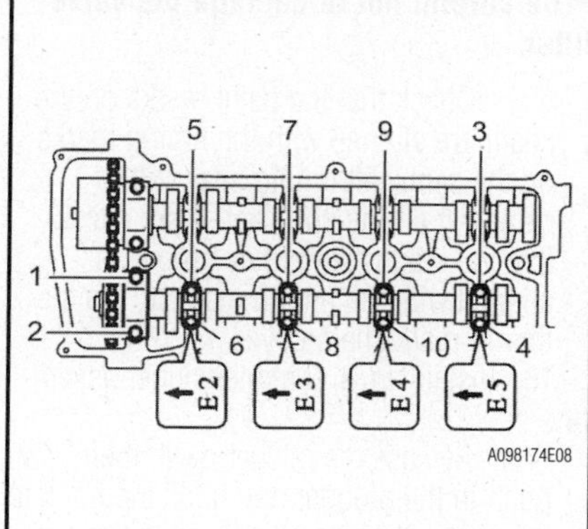

Fig. 76 Using several steps, uniformly loosen and remove the 10 bearing cap bolts in sequence

b. Remove the 5 bearing caps.

c. Remove the camshaft and camshaft timing gear while holding the timing chain by hand.

d. Tie the timing chain with a string.

➡Be careful not to drop anything inside the timing chain cover.

12. Remove camshaft timing gear assembly.

a. Clamp the camshaft in a vise, and make sure that the camshaft timing gear does not rotate.

b. Cover all the oil ports with vinyl tape except the advance side port.

c. Apply air pressure of 22 psi (150 kPa) to the oil path, then turn the camshaft timing gear to the advance direction (counterclockwise) by hand.

➡Depending on the air pressure, the camshaft timing gear may turn to the advance side without applying force by hand. Also, if the pressure is difficult to apply because of air leakage from the port, the lock pin may be difficult to release.

d. Remove the flange bolt of the camshaft timing gear.

➡Be sure not to remove the other 4 bolts. When reusing the camshaft timing gear, release the straight pin lock first, then install the gear.

To install:

13. Install camshaft timing gear assembly.

a. Put the camshaft timing gear and camshaft together with the straight pin and key groove misaligned.

b. Turn the camshaft timing gear pushing it gently against the camshaft.

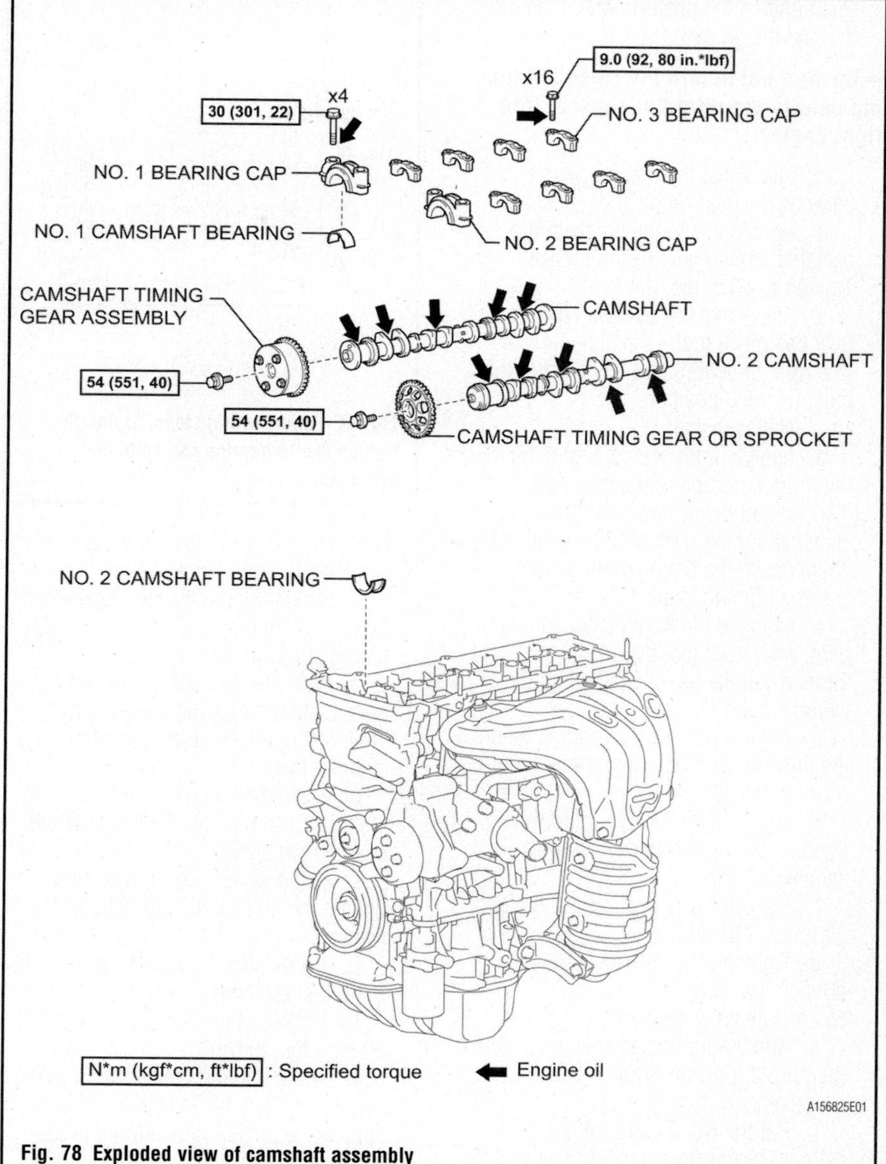

Fig. 78 Exploded view of camshaft assembly

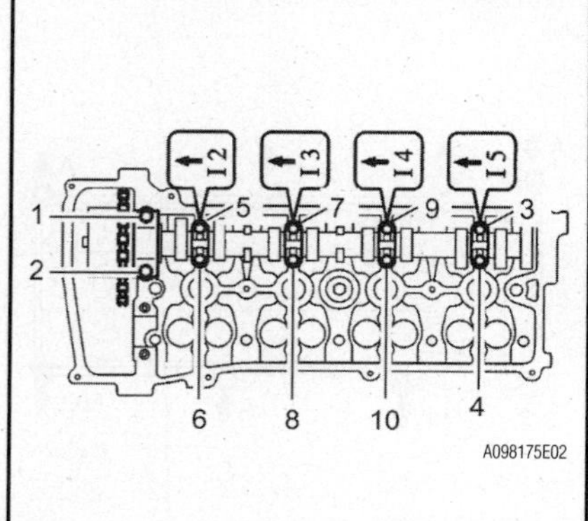

Fig. 77 In several steps, uniformly loosen and remove the 10 bearing caps bolts in sequence

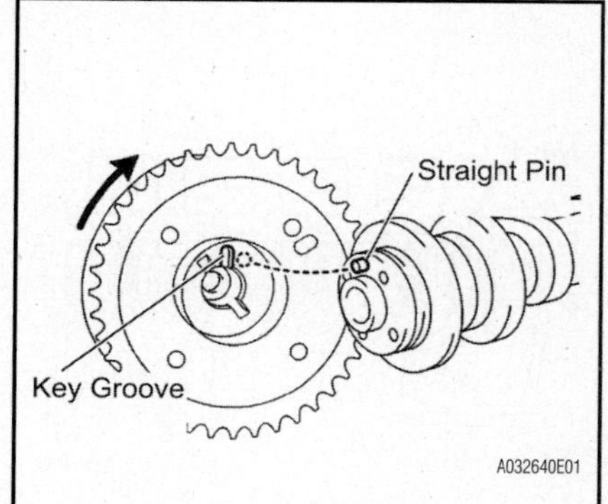

Fig. 79 Put the camshaft timing gear and camshaft together with the straight pin and key groove misaligned

Push until at the position where the pin fits into the groove.

➡**Be sure not to turn the camshaft timing gear to the retard angle side (the right angle).**

c. Check that there is no clearance between the gear fringe and camshaft.
d. Tighten the flange bolt with the camshaft timing gear fixed in place. Tighten to 40 ft. lbs. (54 Nm).
e. Check that the camshaft timing gear can move to the retard angle side (the right direction) and is locked in the most retarded position.
14. Install camshaft.
a. Apply a light coat of engine oil to the journal portion of the camshaft.
b. Install the timing chain onto the camshaft timing gear with the paint mark aligned with the timing mark on the camshaft timing gear.
c. Examine the front marks and numbers, and check that the order is correct. Then install the bearing caps into the cylinder head.
d. Apply a light coat of engine oil on the threads and under the heads of the bearing cap bolts.
e. Using several steps, uniformly tighten the 10 bearing cap bolts in sequence.
f. Tighten No. 1 bearing cap bolts to 22 ft. lbs. (30 Nm).
g. Tighten other bearing cap bolts to 80 inch lbs. (9 Nm).
15. Install No. 2 camshaft.
a. Apply a light coat of engine oil to the journal portion of the No. 2 camshaft.
b. Put the No. 2 camshaft on the cylinder head with the paint mark on the chain aligned with the timing mark on the camshaft timing sprocket.
c. While holding the No. 2 camshaft by hand, temporarily tighten the camshaft timing sprocket set bolt.
d. Examine the front marks and numbers, and check for the correct order. Then install the bearing caps onto the cylinder head.
e. Apply a light coat of engine oil to the threads and under the heads of the bearing cap bolts.
f. Using several steps, uniformly tighten the 10 bearing cap bolts in sequence.
g. Tighten No. 2 bearing cap bolts to 22 ft. lbs. (30 Nm).
h. Tighten other bearing cap bolts to 80 inch lbs. (9 Nm)
i. While holding the camshaft with a wrench, tighten the camshaft timing sprocket set bolt. Tighten to 40 ft. lbs. (54 Nm).

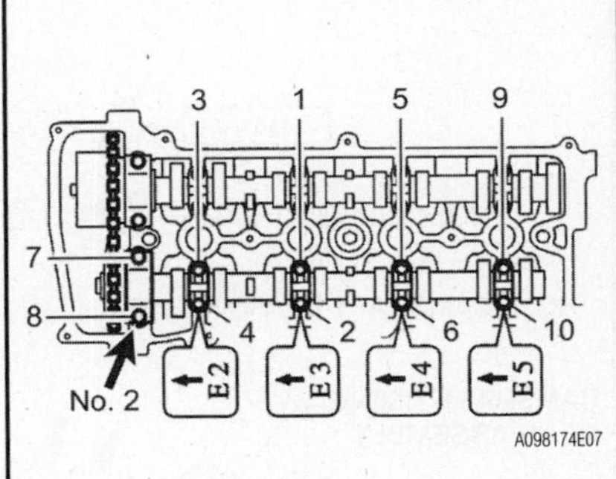

Fig. 81 Using several steps, uniformly tighten the 10 bearing cap bolts in sequence

➡**Be careful not to damage the valve lifter.**

j. Check that the paint marks on the chain are aligned with the timing marks on the camshaft timing gear and camshaft timing sprocket. Also, check that the crankshaft pulley groove is aligned with the timing mark "0" on the timing mark chain cover.
16. Install No. 1 chain tensioner assembly.
a. Release the ratchet pawl, then fully push in the plunger and hook the hook to the pin so that the plunger is in the correct position.
b. Install a new gasket and the chain tensioner with the 2 nuts.

➡**When installing the chain tensioner, set the hook again if the hook releases the plunger.**

c. Turn the crankshaft counterclockwise, then disconnect the plunger knock pin from the hook.
d. Turn the crankshaft clockwise, then check that the plunger is extended.
17. Set No. 1 cylinder to TDC / compression.
18. Check valve clearance.
19. Adjust valve clearance.
20. Install cylinder head cover sub-assembly.
a. Install a new cylinder head cover gasket onto the cylinder head cover.
b. Remove any old packing material from the contact surface.
c. Apply seal packing to the 2 locations.

➡**Remove any oil from the contact surface. Install the oil pan within 3 minutes of applying seal packing. Do not**

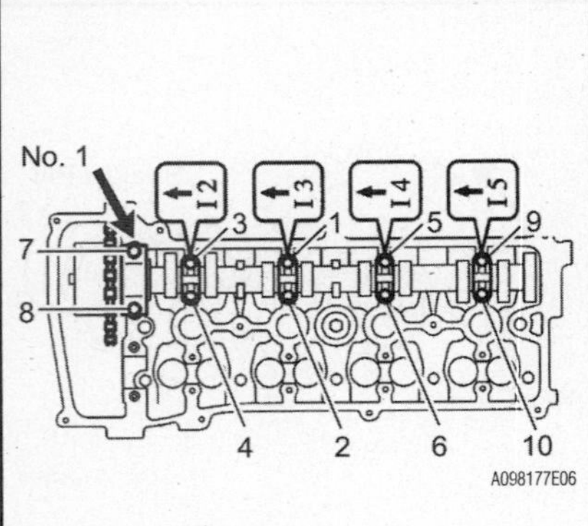

Fig. 80 Using several steps, uniformly tighten the 10 bearing cap bolts in sequence

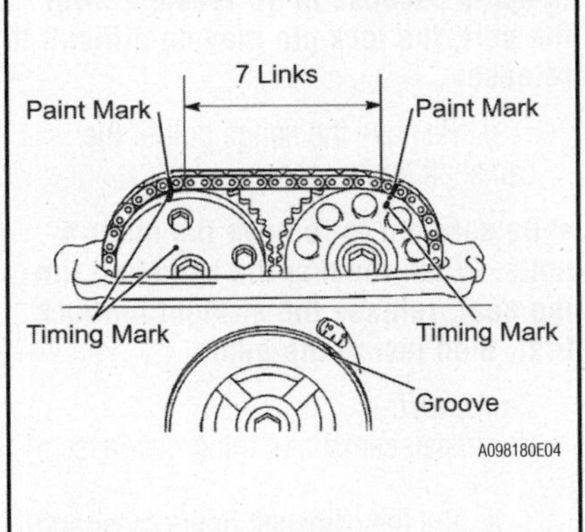

Fig. 82 Check that the paint marks on the chain are aligned with the timing marks on the camshaft timing gear and camshaft timing sprocket

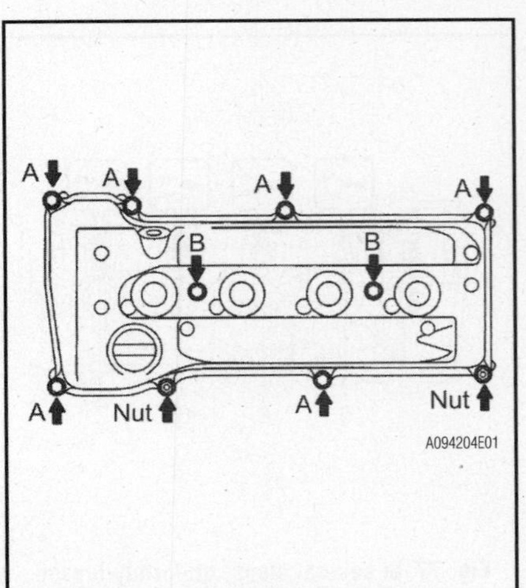

Fig. 83 Install the cylinder head cover with the 8 bolts and 2 nuts

add engine oil for at least 2 hours after installing the oil pan.

d. Install the cylinder head cover with the 8 bolts and 2 nuts.

e. Tighten Bolts A to 8 ft. lbs. (11 Nm).

f. Tighten Bolts B to 10 ft. lbs. (14 Nm).

g. Tighten the nuts to 8 ft. lbs. (11 Nm).

h. Install the 2 engine wires with the 2 bolts.

i. Connect the 2 ventilation hoses to the cylinder head cover.

21. Install spark plug.
22. Install ignition coil assembly.
23. Inspect for engine oil leak.
24. Inspect ignition timing.
25. Remove No. 1 engine cover sub-assembly.
26. Install rear engine under cover RH.

xD

See Figures 84 through 91.

1. Remove front wiper arm head cap.
2. Remove front wiper arm and blade assembly LH and RH.
3. Remove hood to cowl top seal.
4. Remove cowl top ventilator louver sub-assembly.
5. Remove cowl top ventilator louver LH.

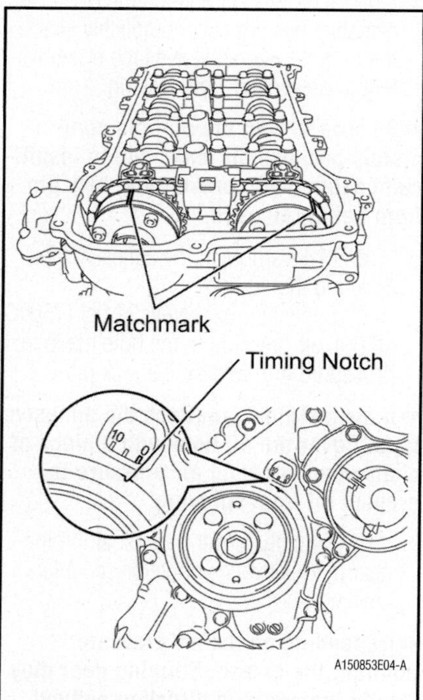

Fig. 84 Turn the crankshaft pulley until its timing notch (groove) and the timing mark "0" of the timing chain cover are aligned

6. Remove front wiper motor and link.
7. Remove front air shutter seal RH.
8. Remove outer cowl top panel.
9. Remove No. 2 cylinder head cover.
10. Remove ignition coil assembly.
11. Remove radio setting condenser.
12. Disconnect the No. 2 ventilation hose.
13. Remove the 2 bolts, 5 connectors, 5 clamps and disconnect the engine wire.
14. Remove the 2 bolts, 4 hoses and air tube assembly.
15. Remove cylinder head cover sub-assembly.
16. Remove cylinder head cover gasket.
17. Set No. 1 cylinder to TDC / compression.

a. Turn the crankshaft pulley until its timing notch (groove) and the timing mark "0" of the timing chain cover are aligned.

b. Check that each matchmark of the camshaft timing gear and camshaft timing exhaust gear are aligned with each matchmark. If not, turn the crankshaft 1 revolution (360°) to align the timing marks.

c. Place paint marks on the chain in alignment with the timing marks on the camshaft timing gear and camshaft timing exhaust gear.

18. Remove service cover.

a. Remove the 5 bolts and service cover.

b. Using a screwdriver with its tip taped, pry and remove the service cover.

➡Do not damage the service cover or the contact surfaces of the timing chain cover.

19. Remove the 2 bolts and No. 2 chain vibration damper from the camshaft bearing cap.

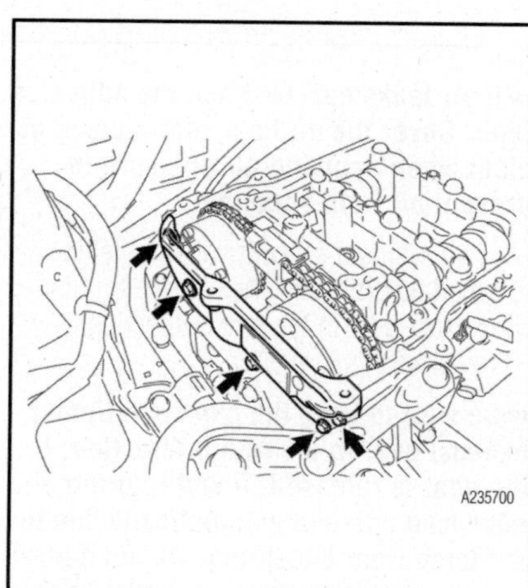

Fig. 85 Remove the 5 bolts and service cover

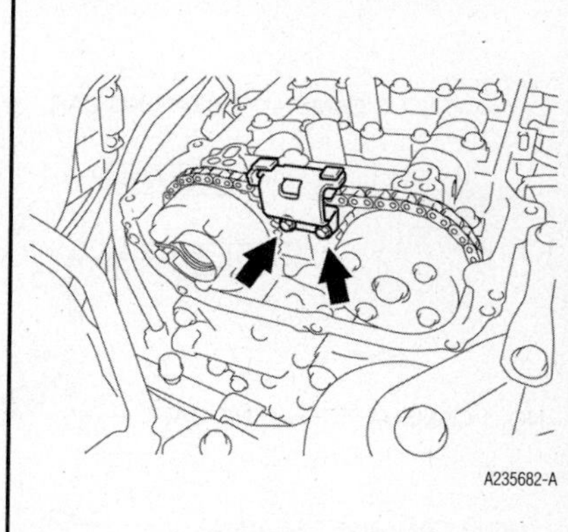

Fig. 86 Remove the 2 bolts and No. 2 chain vibration damper from the camshaft bearing cap

20. Remove No. 1 chain tensioner assembly.
21. While holding the hexagonal portion of the camshaft with a wrench, remove the camshaft timing exhaust gear bolt.

➡Hold the hexagonal portion of the intake camshaft with a wrench and turn it slightly counterclockwise to release the chain.

➡Do not turn the intake camshaft more than necessary.

➡Be sure to loosen the chain because the camshaft timing exhaust gear assembly cannot be removed with the chain tensioned.

22. Remove the camshaft timing exhaust gear assembly.

➡Keep the camshaft timing exhaust gear assembly horizontal while removing it from the camshaft.

23. Inspect camshaft timing exhaust gear assembly.

a. Temporarily install the camshaft timing exhaust gear assembly.

b. Align the knock pin on the No. 2 camshaft with the pin hole in the camshaft timing exhaust gear assembly and temporarily install the camshaft timing exhaust gear assembly to the No. 2 camshaft with the bolt.

➡Do not install the chain onto the gear at this step. Do not allow the chain to interfere with the gear when installing the gear assembly.

c. Inspect the camshaft timing exhaust gear lock.

d. Check that the camshaft timing exhaust gear is locked.

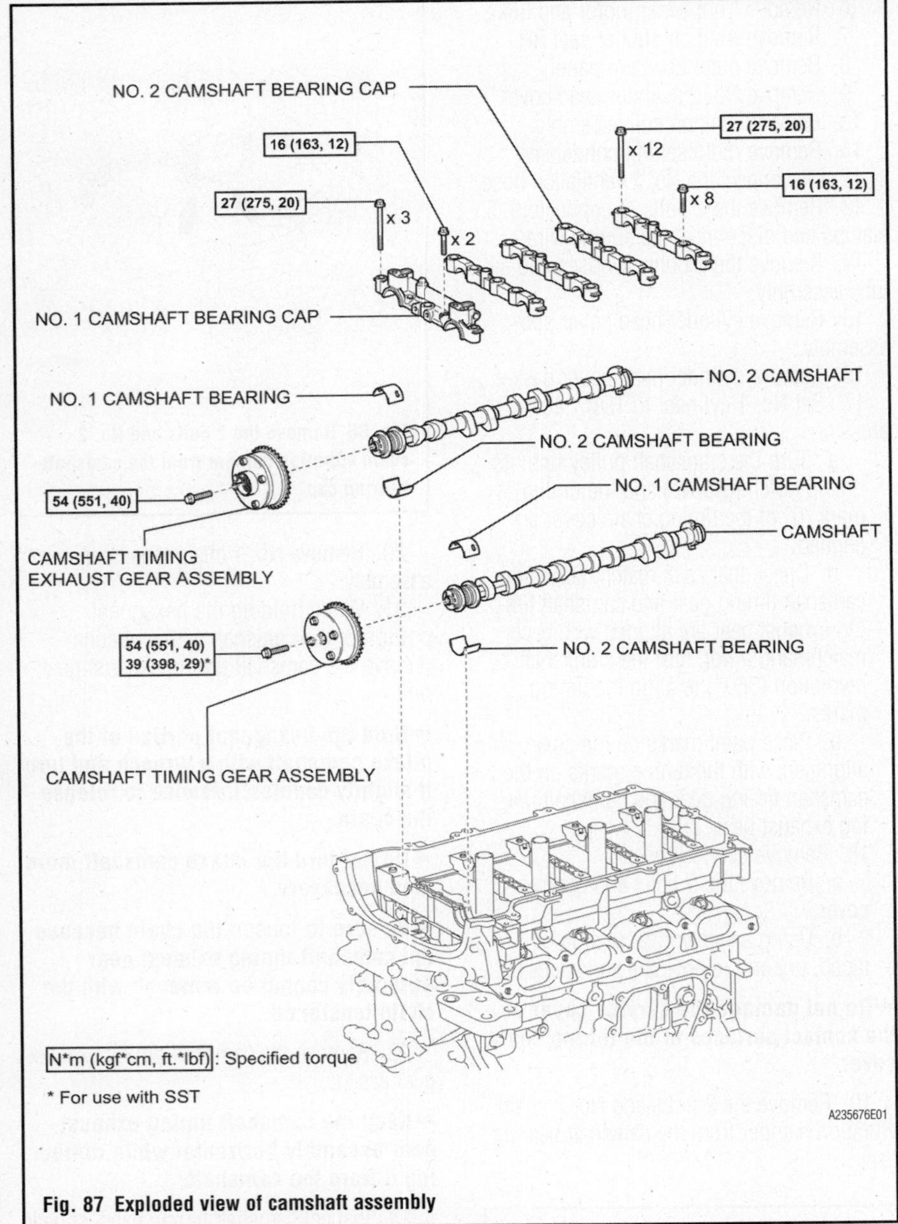

Fig. 87 Exploded view of camshaft assembly

e. Inspect camshaft timing exhaust gear operation.

f. After cleaning and degreasing the exhaust side VVT oil hole on the No. 1 camshaft bearing cap, completely seal the oil hole with adhesive tape or equivalent to prevent air from leaking.

➡Be sure to seal the oil hole completely because air leaks due to insufficient sealing will prevent the lock pin from being released.

g. Make a hole in the adhesive tape covering the oil hole.

h. Apply approximately 28 psi (200 kPa) of air pressure to the hole to release the lock pin.

➡If air leaks out, reattach the adhesive tape. Cover the oil hole with a piece of cloth when applying air pressure to prevent oil from spraying.

i. Using a screwdriver with its tip taped, forcibly turn the camshaft timing exhaust gear in the retard direction (clockwise).

➡Be sure to keep the camshaft timing exhaust gear in the retard direction. If the gear is released, it will return to the advanced position automatically due to the force from the spring. Do not damage the camshaft timing exhaust gear.

➡Depending on the air pressure applied, the camshaft timing exhaust gear may turn in the retard direction without assistance by hand.

j. Using a screwdriver with its tip taped, turn the camshaft timing exhaust gear within its movable range (20°) 2 or 3 times without turning it to the most advanced position. Check that the camshaft timing gear turns smoothly.

k. Lock the camshaft timing exhaust gear.

➡Check that the camshaft timing exhaust gear assembly locks at the most advanced position (the most advanced position of its movable range) and cannot be rotated any further.

l. Remove the adhesive tape from the No. 1 camshaft bearing cap.

m. Remove the camshaft timing exhaust gear assembly.

n. Remove the temporarily installed camshaft timing exhaust gear assembly.

24. Inspect camshaft timing gear assembly.

a. Inspect the camshaft timing gear lock.

b. Check that the camshaft timing gear is locked.

c. Inspect camshaft timing gear operation.

d. After cleaning and degreasing the intake side VVT oil hole on the No. 1 camshaft bearing cap, completely seal the oil hole with adhesive tape or equivalent to prevent air from leaking.

➡Be sure to seal the oil hole completely because air leaks due to insufficient sealing will prevent the lock pin from being released.

e. Make a hole in the adhesive tape covering the oil hole.

f. Apply approximately 22 psi (150 kPa) of air pressure to the hole made in procedure A to release the lock pin.

➡If air leaks out, reattach the adhesive tape. Cover the oil hole with a piece of cloth when applying air pressure to prevent oil from spraying.

g. Forcibly turn the camshaft timing gear in the advance direction (counterclockwise).

➡Depending on the air pressure applied, the camshaft timing gear may turn in the advance direction without assistance by hand.

h. Turn the camshaft timing gear within its movable range (27.5°) 2 or 3 times without turning it to the most

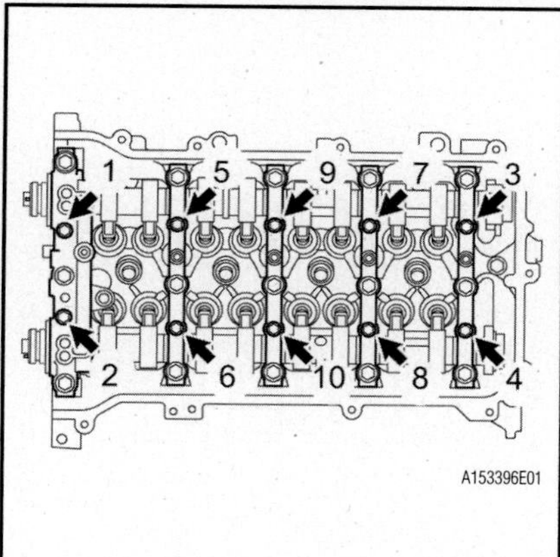

Fig. 88 Uniformly loosen and remove the 10 bearing cap bolts in sequence

retarded position. Check that the camshaft timing gear turns smoothly.

➡Do not lock the camshaft timing gear assembly. If camshaft timing gear assembly is locked, release the lock pin again.

i. Remove the adhesive tape from the No. 1 camshaft bearing cap.

25. Remove camshaft timing gear assembly.

a. While holding the hexagonal portion of the camshaft with a wrench, remove the camshaft timing gear assembly bolt with SST.

➡Before removing the camshaft timing gear, make sure that the lock pin has been released. Be sure not to remove the other 4 bolts. If the camshaft timing gear assembly is to be reused, be sure to use it with the lock pin released.

b. Release the chain and remove the camshaft timing gear assembly

➡Keep the camshaft timing gear assembly horizontal while removing it from the camshaft.

c. Suspend the chain with a string or equivalent.

26. Remove camshaft bearing cap.

a. Uniformly loosen and remove the 10 bearing cap bolts in sequence.

➡Be sure not to loosen the other 15 bearing cap bolts in this step.

b. Remove the 15 bolts and bearing caps in order. Immediately after removing bearing caps, install service bolts and spacers in order.

➡If the bolts are loosened all at once, FIPG on the camshaft housing and cylinder head may peel off, resulting in oil oozing. Therefore, be sure to install the service bolts and spacers to one bearing cap at a time. Do not install the bearing caps when installing the service bolts and spacers.

27. Remove the camshaft from the camshaft housing.

28. Remove the No. 2 camshaft from the camshaft housing.

29. Remove the 2 No. 1 camshaft bearings from the No. 1 camshaft bearing cap.

30. Remove the 2 No. 2 camshaft bearings.

➡Arrange the removed parts in the correct order.

To install:

31. Install No. 1 camshaft bearing.

a. Clean the both surfaces of the 2 No. 1 camshaft bearings.

➡Do not apply engine oil to the bearings or the contact surfaces.

b. Install the 2 No. 1 camshaft bearings to the No. 1 camshaft bearing cap.

c. Using vernier calipers, measure the distance between the bearing cap's edge and the camshaft bearing's edge. Standard: 0.0276 inches (0.7 mm) or less

32. Install No. 2 camshaft bearing.

a. Clean both surfaces of the 2 No. 2 camshaft bearings.

➡Do not apply engine oil to the bearings or the contact surfaces.

b. Install the 2 No. 2 camshaft bearings.

c. Using vernier calipers, measure the distance between the bearing cap's edge and the camshaft bearing's edge. Dimension should be 0.042–0.068 inches (1.05–1.75 mm).

33. Install No. 2 camshaft.

a. Make sure that the valve rocker arm is properly installed.

b. Clean the camshaft journals.

c. Apply a light coat of engine oil to the camshaft journals, camshaft housings and bearing caps.

d. Install the No. 2 camshaft to the camshaft housing.

34. Install camshaft.

a. Make sure that the valve rocker arm is properly installed.

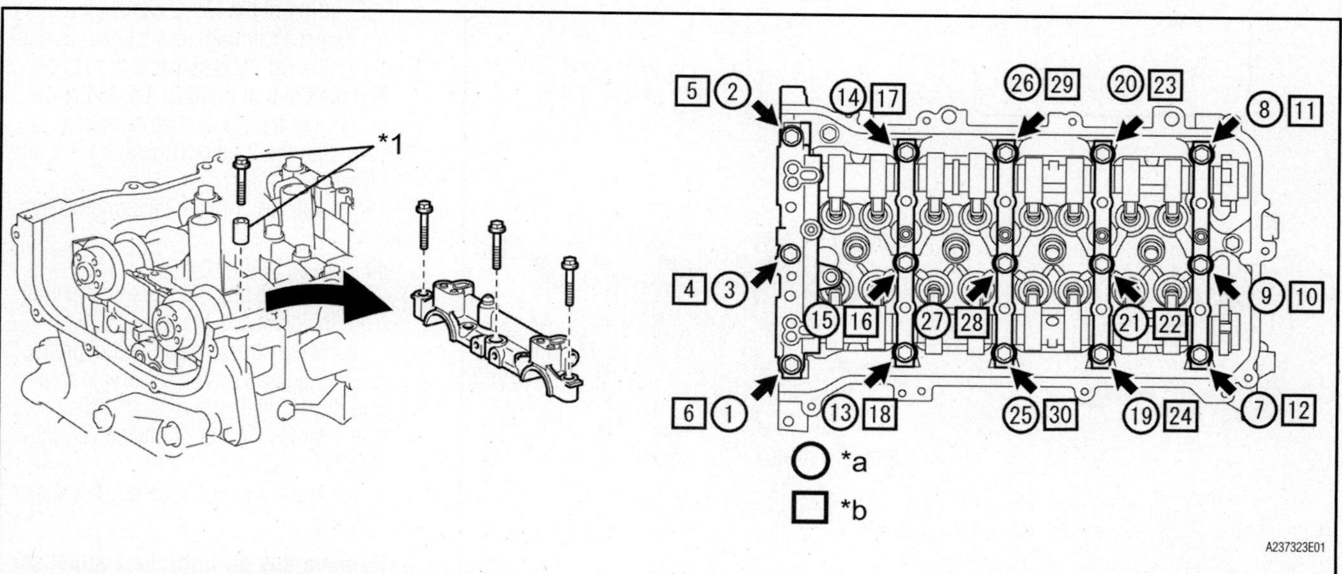

Fig. 89 Remove the 15 bolts and bearing caps in order

b. Clean the camshaft journals.

c. Apply a light coat of engine oil to the camshaft journals, camshaft housings and bearing caps.

d. Install the camshaft to the camshaft housing.

35. Install camshaft bearing cap.

a. Check the marks and numbers on the camshaft bearing caps, and then remove the service bolts and spacers in order. Immediately after removing the service bolts and spacers in the location for bearing caps, install the bearing caps with the bolts in order. Tighten to 20 ft. lbs. (27 Nm).

➡If the bolts are loosened all at once, FIPG on the camshaft housing and cylinder head may peel off, resulting in oil oozing. Therefore, be sure to remove the service bolt and spacer from one bearing cap at a time.

➡Make sure that the knock pin of the camshaft is positioned properly.

b. Tighten the 10 bolts in order. Tighten to 12 ft. lbs. (16 Nm).

c. Check the torque of each bolt again.

36. Install camshaft timing gear assembly.

a. Check that the knock pin is installed on the camshaft.

b. Hold up the chain and align the matchmark on the camshaft timing gear assembly and the paint mark on the chain.

c. Put the camshaft timing gear and camshaft together with the knock pin and key groove misaligned.

➡Do not forcefully push in the camshaft timing gear assembly. This may cause the camshaft knock pin tip to damage the installation surface of the camshaft timing gear assembly.

d. Turn the camshaft timing gear while pushing it gently against the camshaft. Push further at the position where the pin fits into the groove.

➡Do not turn the camshaft timing gear in the retard direction (the right angle).

e. Check that there is no clearance between the camshaft timing gear and camshaft flange.

f. Using SST and a wrench, hold the hexagonal portion of the camshaft and fix the camshaft timing gear assembly to the camshaft with the bolt.

- Without SST: 40 ft. lbs. (54 Nm)
- With SST: 29 ft. lbs. (39 Nm)

➡When tightening the bolts, do not allow the camshaft timing gear assembly to rotate.

g. Check that the camshaft timing gear can move to the retard angle side (the right direction) and is locked in the most retarded position.

37. Install camshaft timing exhaust gear assembly.

a. Check that the knock pin is installed on the camshaft.

b. Hold up the chain and align the matchmark on the camshaft timing gear assembly and the paint mark on the chain.

c. Put the camshaft timing exhaust gear and camshaft together by aligning the key groove and knock pin.

➡Do not forcefully push in the camshaft timing exhaust gear assembly. This may cause the camshaft knock pin tip to damage the installation surface of the camshaft timing exhaust gear assembly.

d. Using a wrench, hold the hexagonal portion of the No. 2 camshaft and fix the camshaft timing exhaust gear assembly to the No. 2 camshaft with the bolt. Tighten the bolt to 40 ft. lbs. (54 Nm).

38. Install the No. 2 chain vibration damper with the 2 bolts. Tighten to 7 ft. lbs. (10 Nm).

39. Install No. 1 chain tensioner assembly.

40. Install service cover.

a. Remove any old packing material from the contact surface.

b. Make sure to clean and degrease the fitting surfaces of the service cover and timing chain cover so that they are free of grease, water or any contamination.

c. Apply a continuous bead of seal packing.

➡Remove any oil from the contact surface. Install the service cover within 3

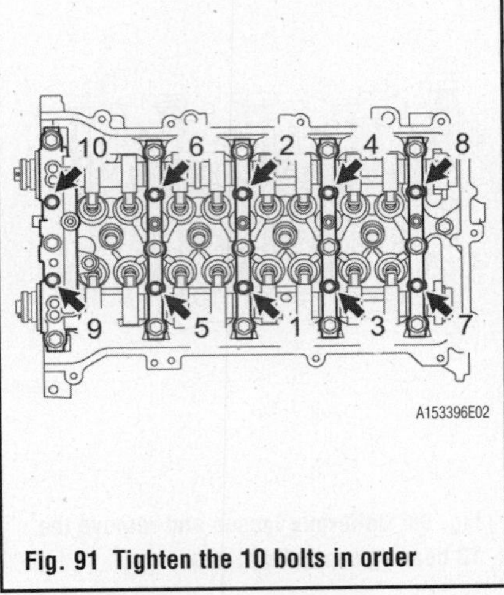

Fig. 91 Tighten the 10 bolts in order

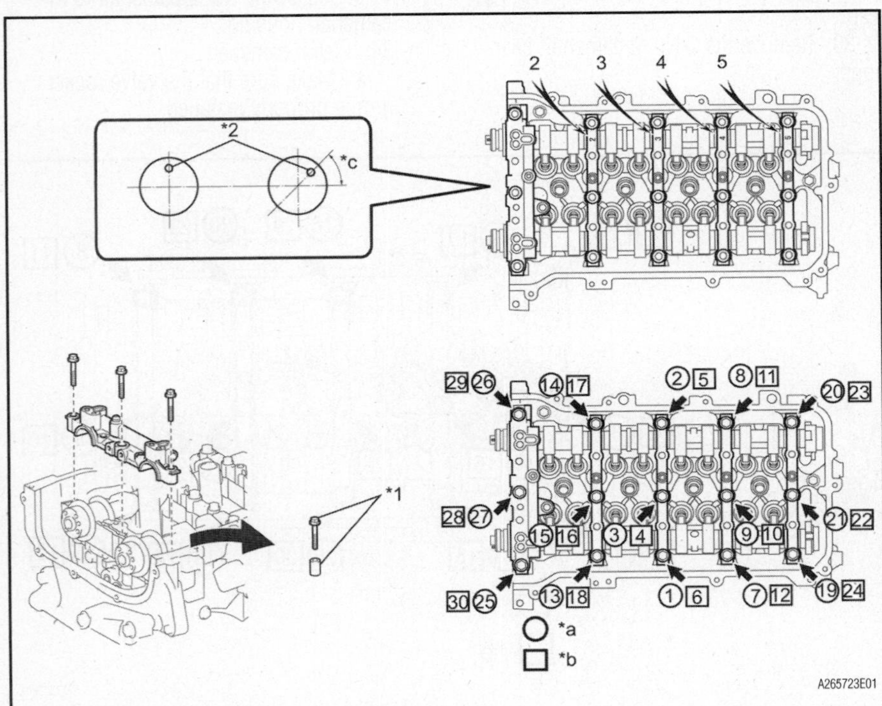

Fig. 90 Install the bearing caps with the bolts in order

minutes after applying seal packing. Do not add engine oil within 2 hours after installing the service cover.

d. Install the service cover to the timing chain cover with the 5 bolts.

41. Set No. 1 cylinder to TDC / compression.

a. Turn the crankshaft pulley until its timing notch (groove) and the timing mark "0" of the timing chain cover are aligned.

b. Check that each matchmark of the camshaft timing gear and camshaft timing exhaust gear are aligned with each matchmark located. If not, turn the crankshaft 1 revolution (360°) to align the timing marks.

42. Remove cylinder head cover gasket.
43. Install cylinder head cover sub-assembly.
44. Install air tube assembly.

a. Install the air tube assembly with the 2 bolts.

b. Connect the 4 hoses.

45. Connect the 5 connectors and install the wire harness with the 2 bolts and 5 clamps.
46. Connect the No. 2 ventilation hose.
47. Install radio setting condenser.
48. Install ignition coil assembly.
49. Inspect for engine oil leak.
50. Install No. 2 cylinder head cover.
51. Install outer cowl top panel.
52. Install front air shutter seal RH.
53. Install front wiper motor and link.
54. Install cowl top ventilator louver LH.
55. Install cowl top ventilator louver sub-assembly.
56. Install hood to cowl top seal
57. Install front wiper arm and blade assembly LH and RH.
58. Install front wiper arm head cap.
59. Inspect ignition timing.
60. Inspect engine idling speed.

CRANKSHAFT FRONT SEAL

REMOVAL & INSTALLATION

tC

See Figure 92.

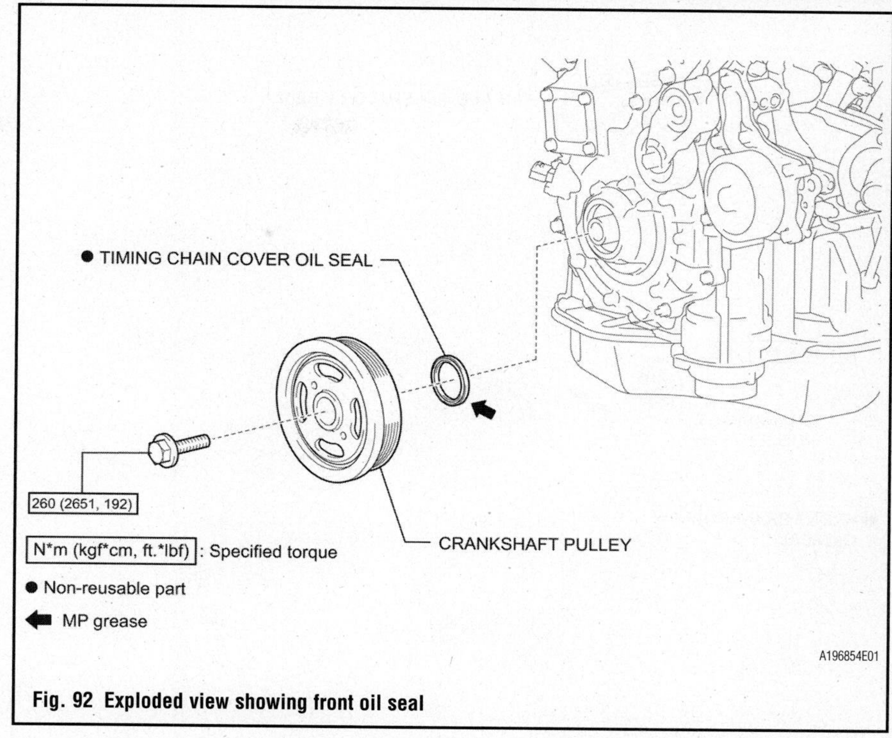

Fig. 92 Exploded view showing front oil seal

1. Remove fan and alternator V belt.
2. Remove crankshaft pulley.
3. Remove timing chain cover oil seal.
4. Using a screwdriver, pry out the oil seal.

➡Do not damage the surface of the oil seal press fit hole or the crankshaft.

To install:

5. Install timing chain cover oil seal.

a. Apply MP grease to the lip of a new oil seal.

➡Do not allow foreign matter to contact the lip of the oil seal. Do not allow MP grease to contact the dust seal.

b. Using SST (09223-22010) and a hammer, tap in the oil seal until its surface is flush with the timing chain cover edge.

➡Keep the lip of the oil seal free from foreign matter. Do not tap in the oil seal at an angle.

6. Install crankshaft pulley.
7. Install fan and alternator V belt.

xB

See Figure 93.

1. Remove front wheel RH.
2. Remove rear engine under cover RH.
3. Remove fan and alternator V belt.
4. Remove crankshaft pulley.

a. Using SST, hold the pulley in place and loosen the pulley bolt.

b. Using SST, remove the crankshaft pulley.

➡If necessary, remove the pulley and pulley bolt using SST.

5. Remove front crankshaft oil seal.

a. Using a knife, cut off the lip of the oil seal.

b. Using a screwdriver, pry out the oil seal.

➡Do not damage the surface of the oil seal press fit hole and crankshaft.

To install:

6. Install front crankshaft oil seal.

a. Apply MP grease to the lip of a new oil seal.

b. Using SST and a hammer, tap in the oil seal until its surface is flush with the rear oil seal retainer edge.

➡Keep the lip free from foreign matter. Do not tap on the oil seal at an angle.

7. Using SST, install the crankshaft pulley in place and tighten the bolt. Tighten to 133 ft. lbs. (180 Nm).
8. Install fan and alternator V belt.
9. Inspect for engine oil leak
10. Inspect engine oil level
11. Install rear engine under cover RH.
12. Install front wheel RH.

xD

See Figure 94.

1. Remove engine under cover RH.
2. Remove No. 2 cylinder head cover.
3. Remove fan and alternator V belt.
4. Remove crankshaft pulley.

a. Using SST, hold the pulley in place and loosen the pulley bolt.

b. Using SST, remove the pulley bolt and pulley.

5. Remove timing chain cover oil seal.

a. Using a knife, cut off the oil seal lip.

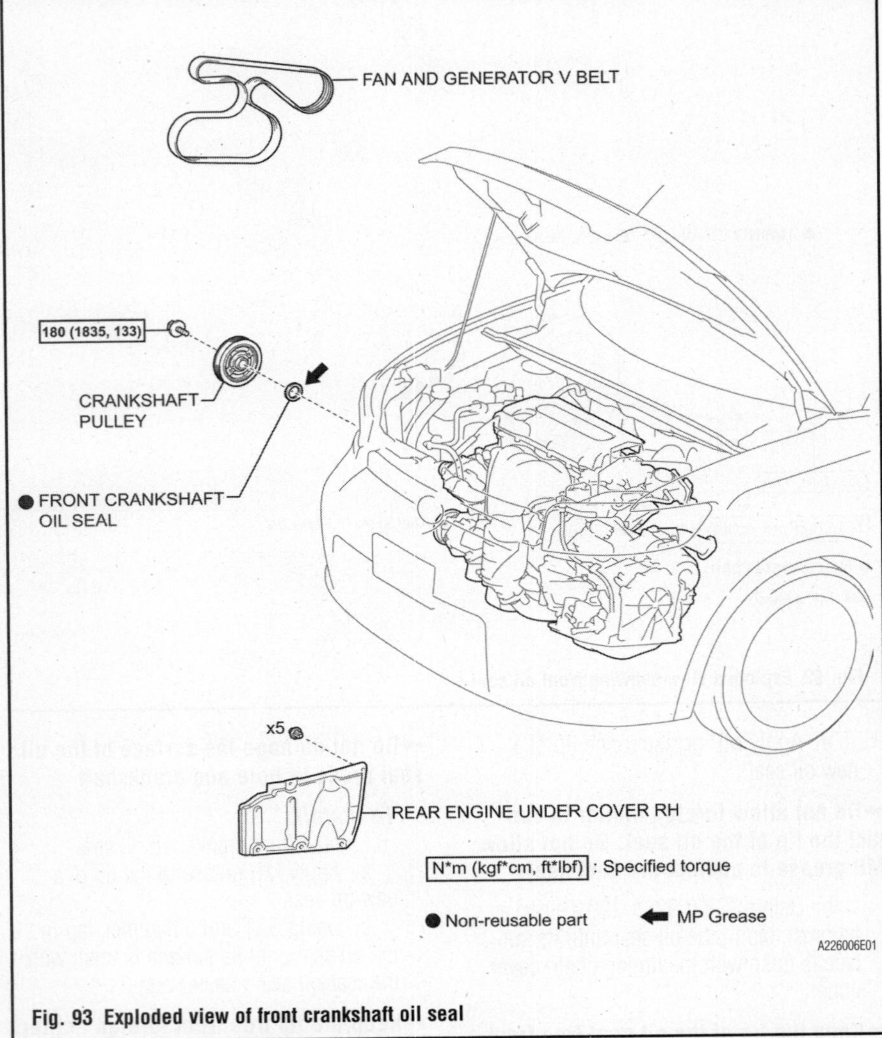

Fig. 93 Exploded view of front crankshaft oil seal

b. Using a screwdriver with its tip taped, pry out the oil seal.

➡After the removal, check the crankshaft for damage. If it is damaged, smooth the surface with 400-grit sandpaper.

To install:

6. Install timing chain cover oil seal.
   a. Apply MP grease to a new oil seal lip.
   b. Using SST and a hammer, tap the oil seal until its surface is flush with the timing chain cover edge.

➡Do not tap the oil seal at an angle. Keep the lip free of foreign matter.

➡The pressing-in depth should be within the range of minus1 mm and plus 1 mm from the chain cover edge.

7. Install crankshaft pulley.
   a. Align the pulley set key with the key groove of the pulley.
   b. Using SST, hold the pulley in place and tighten the bolt.
8. Install fan and alternator V belt.
9. Adjust fan and alternator V belt.
10. Inspect fan and alternator V belt.
11. Install No. 2 cylinder head cover.
12. Inspect for engine oil leak.
13. Install engine under cover RH.

CYLINDER HEAD

REMOVAL & INSTALLATION

xB

See Figures 95 through 105.

➡Determine if the engine is TMC made or not.

1. Discharge fuel system pressure.
2. Disconnect cable from negative battery terminal.

⁂ CAUTION

Wait at least 90 seconds after disconnecting the cable from the negative (-) battery terminal to prevent airbag and seat belt pretensioner activation.

3. Remove hood.
4. Remove front wheel RH.
5. Remove No. 1 engine under cover.
6. Remove rear engine under cover RH.
7. Remove No. 1 engine cover sub-assembly.
8. Drain engine coolant.
9. Drain engine oil.
10. Separate center exhaust pipe assembly.
11. Remove front exhaust pipe assembly.
12. Remove windshield wiper arm cover.
13. Remove front wiper arm and blade assembly LH and RH.
14. Remove hood to cowl top seal.
15. Remove cowl top ventilator louver RH.
16. Remove cowl top ventilator louver LH.
17. Remove front wiper motor and link.
18. Remove outer cowl top panel.
19. Remove fan and alternator V belt.
20. Remove alternator assembly.
21. Separate radiator reserve tank assembly.
22. Remove air cleaner cap sub-assembly with No. 1 air cleaner hose.
23. Remove throttle with motor body assembly.
24. Remove ignition coil assembly.
25. Remove spark plug.
26. Remove cylinder head cover sub-assembly.
27. Remove oil dipstick.
28. Remove engine oil level dipstick guide (for TMC made engine).
29. Remove engine oil level dipstick guide (except TMC made engine).
30. Disconnect fuel main tube.
31. Remove fuel delivery pipe sub-assembly.
32. Remove camshaft timing oil control valve assembly.
33. Remove intake manifold.
34. Remove intake manifold insulator.
35. Remove manifold stay.
36. Remove No. 2 manifold stay.
37. Remove No. 1 exhaust manifold heat insulator.
38. Remove exhaust manifold converter sub-assembly.
    a. Disconnect the air-fuel ratio sensor connector.
    b. Remove the 5 nuts, manifold converter and gasket.

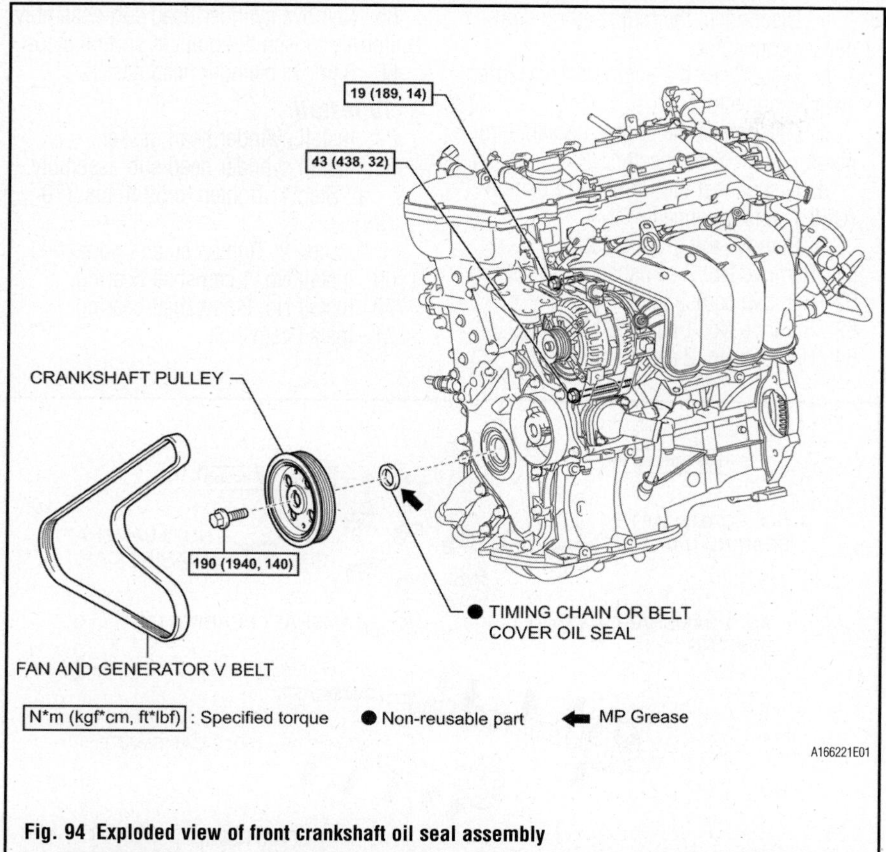

Fig. 94 Exploded view of front crankshaft oil seal assembly

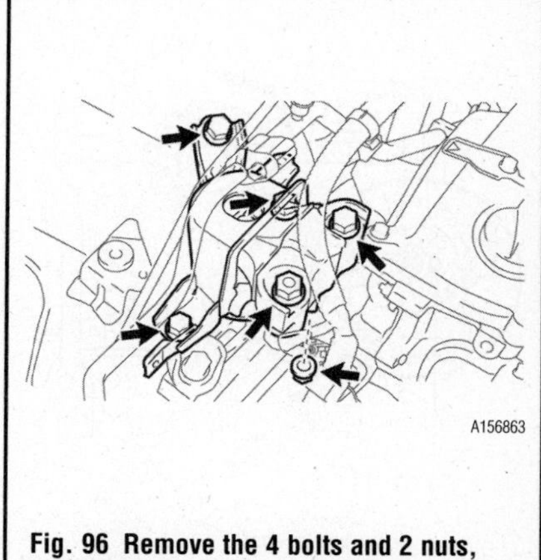

Fig. 96 Remove the 4 bolts and 2 nuts, then remove the engine mounting insulator RH

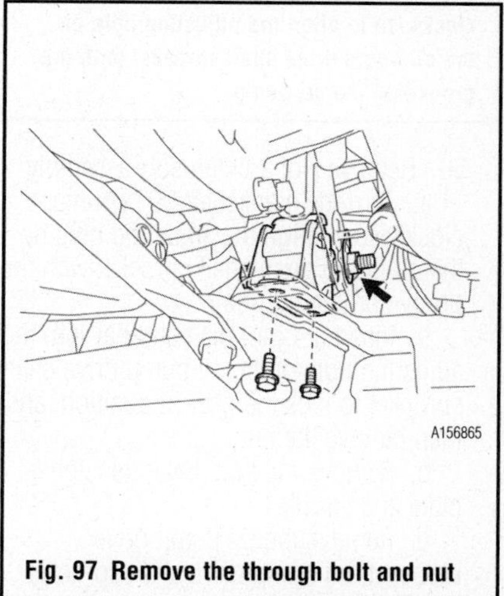

Fig. 97 Remove the through bolt and nut

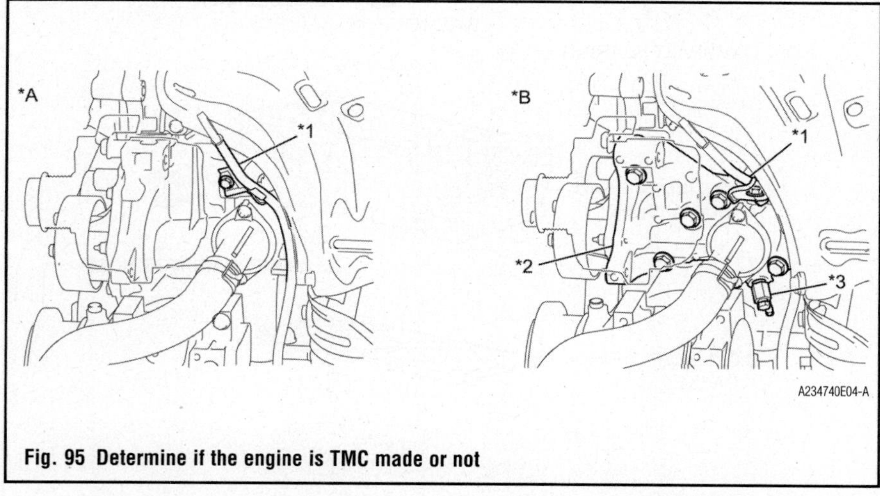

Fig. 95 Determine if the engine is TMC made or not

39. Remove engine mounting insulator sub-assembly RH.
   a. Remove the bolt and cooler bracket.
   b. Remove the bolt and nut, then remove the No. 2 air conditioning pipe bracket.
   c. Remove the 4 bolts and 2 nuts, then remove the engine mounting insulator RH.
40. Remove front engine mounting insulator sub-assembly.
   a. Remove the through bolt and nut.
   b. Remove the 2 bolts and front engine mounting insulator.
41. Remove idler pulley sub-assembly.
42. Remove oil pan sub-assembly.
43. Set No. 1 cylinder to TDC / compression.
   a. Turn the crankshaft pulley until the groove and the timing mark "0" on the timing chain cover are aligned.
   b. Check that each timing mark on the camshaft timing gear and sprocket is aligned with the timing marks located on the No. 1 and No. 2 bearing caps. If not, turn the crankshaft by 1 revolution (360°) to align the timing marks as above.
44. Remove crankshaft pulley.
45. Remove No. 1 chain tensioner assembly.
46. Remove transverse engine mounting bracket.
47. Remove V-ribbed belt tensioner assembly.
48. Remove Crankshaft Position (CKP) sensor.
49. Remove timing chain cover sub-assembly.
50. Remove No. 1 Crankshaft Position (CKP) sensor plate.
51. Remove timing chain guide.
52. Remove chain tensioner slipper.
53. Remove No. 1 chain vibration damper.
54. Remove chain sub-assembly.
55. Remove crankshaft timing gear or sprocket.

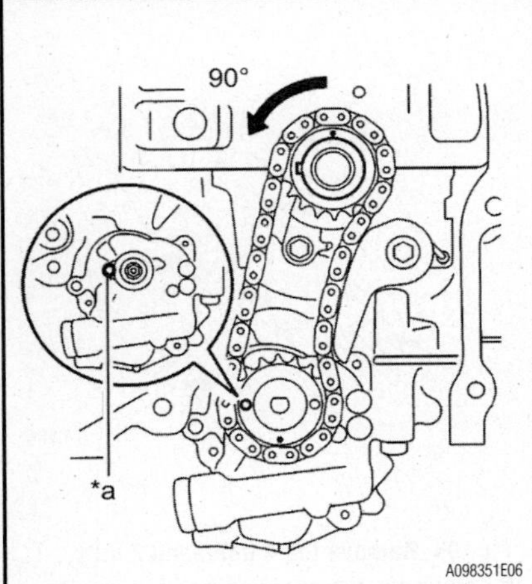

Fig. 98 Turn the crankshaft 90° counter-clockwise to align the adjusting hole on the oil pump drive shaft sprocket with the groove on the oil pump

56. Remove No. 2 chain sub-assembly.
   a. Turn the crankshaft 90° counter-clockwise to align the adjusting hole on the oil pump drive shaft sprocket with the groove on the oil pump.
   b. Insert a 4 mm diameter bar into the adjusting hole of the oil pump drive shaft sprocket to lock the gear in position, and then remove the nut.
   c. Remove the bolt, chain tensioner plate and spring.
   d. Remove the oil pump drive sprocket, oil pump drive shaft sprocket and No. 2 chain.
57. Disconnect inlet radiator hose.
58. Disconnect outlet heater water hose.
59. Disconnect inlet heater water hose.
60. Disconnect engine wire.
   a. Disconnect the radio setting condenser connector.
   b. Disconnect the engine oil pressure switch connector.
   c. Disconnect the engine coolant temperature sensor connector.
   d. Disconnect the Camshaft Position (CMP) sensor connector.
   e. Remove the ground cable.
61. Remove No. 2 camshaft.
62. Remove camshaft.
63. Remove No. 1 camshaft bearing.
64. Remove No. 2 camshaft bearing.
65. Remove cylinder head sub-assembly. Uniformly loosen the bolts in several steps
66. Remove cylinder head gasket.

To install:

67. Install cylinder head gasket.
68. Install cylinder head sub-assembly.
   a. Step 1: Tighten to 52 ft. lbs. (70 Nm).
   b. Step 2: Tighten an additional 90°.
69. Install No. 1 camshaft bearing.
70. Install No. 2 camshaft bearing.
71. Install camshaft.

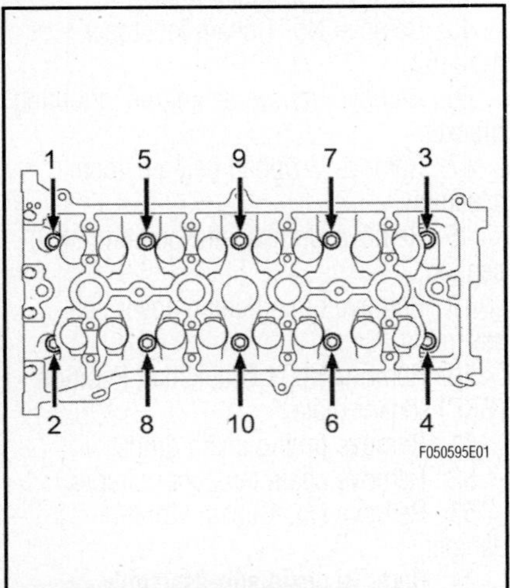

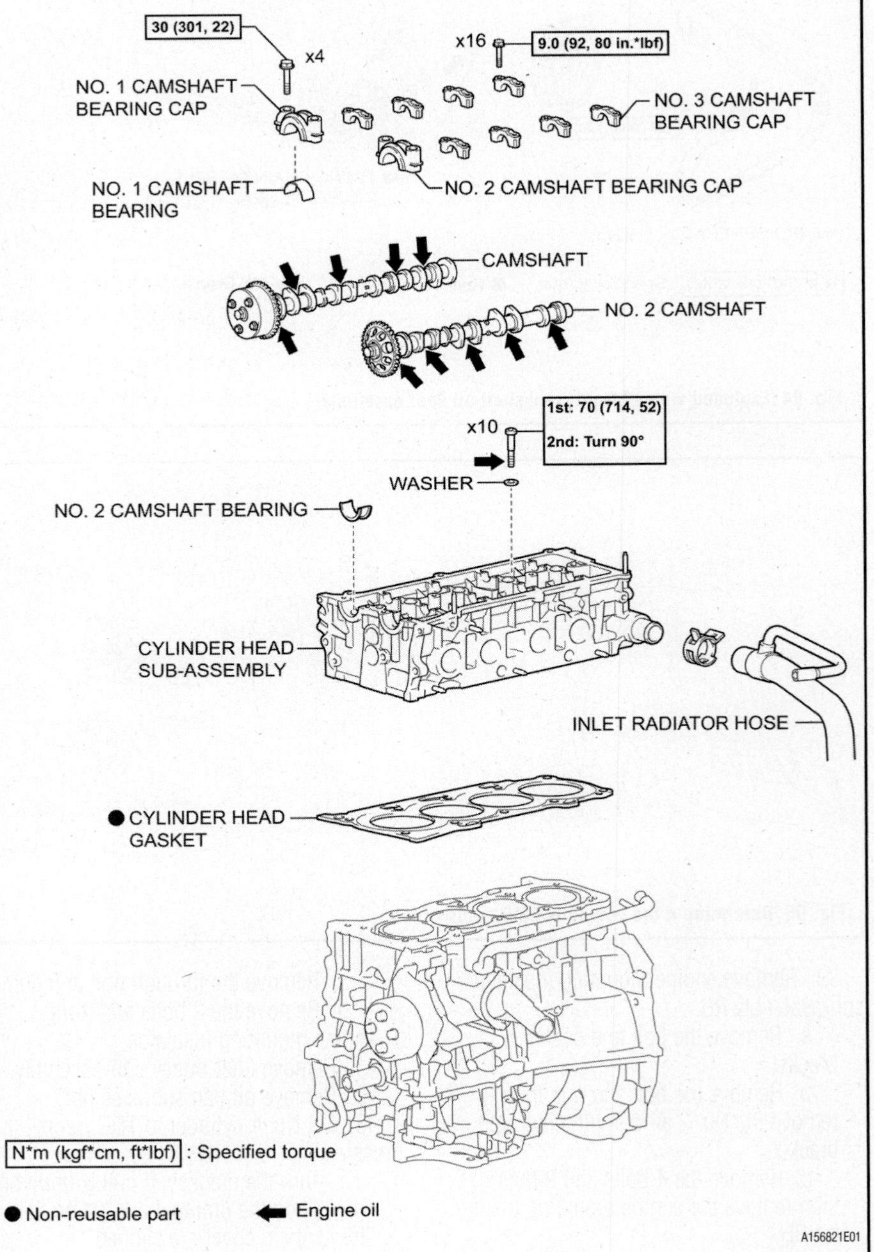

Fig. 99 Cylinder head bolt removal sequence

Fig. 100 Exploded view of the cylinder head assembly

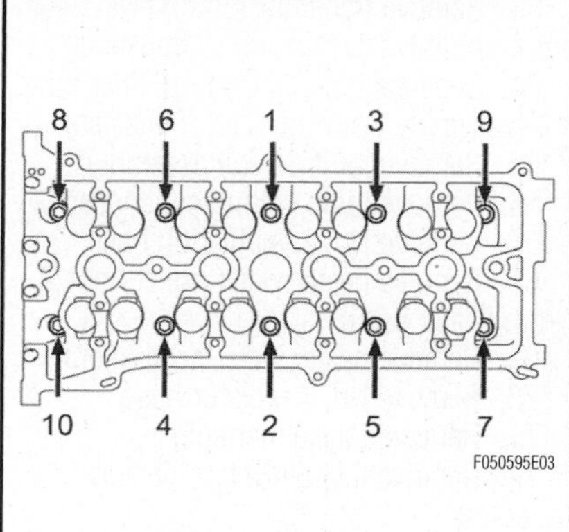

Fig. 101 Cylinder head bolt tightening sequence

72. Install No. 2 camshaft.
73. Connect inlet heater water hose.
74. Connect outlet heater water hose.
75. Connect inlet radiator hose.
76. Connect engine wire.
77. Install No. 2 chain sub-assembly.
78. Install crankshaft timing gear or sprocket.
79. Install No. 1 chain vibration damper.
80. Install chain sub-assembly.
81. Install chain tensioner slipper.
82. Install timing chain guide.
83. Install No. 1 Crankshaft Position (CKP) sensor plate.
84. Install timing chain cover sub-assembly.

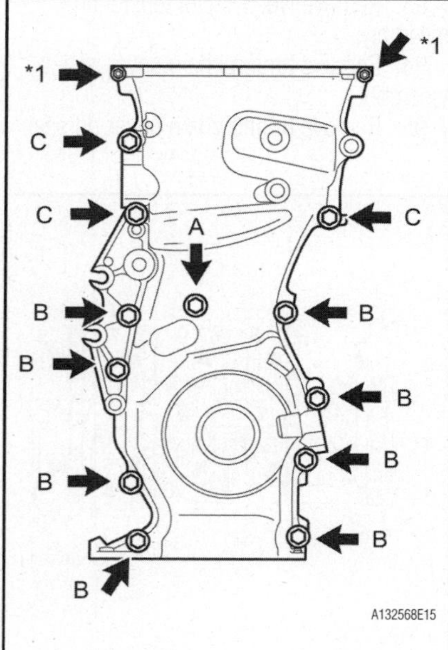

Fig. 102 Install timing chain cover sub-assembly; nuts (1)

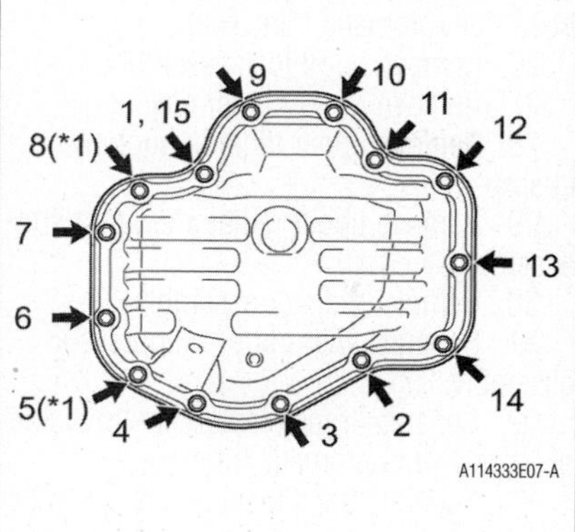

Fig. 103 Oil pan tightening sequence; nuts (*1)

a. Tighten Bolt A to 80 inch lbs. (9 Nm).
b. Tighten Bolts B to 18 ft. lbs. (25 Nm).
c. Tighten Bolts C to 41 ft. lbs. (55 Nm).
d. Tighten the nuts to 8 ft. lbs. (11 Nm).

85. Install No. 1 chain tensioner assembly.
86. Install V-ribbed belt tensioner assembly.
87. Install transverse engine mounting bracket.
88. Install crankshaft pulley.
89. Install oil pan sub-assembly. Tighten to 80 inch lbs. (9 Nm).
90. Install Crankshaft Position (CKP) sensor (for TMC made engine).
91. Install Crankshaft Position (CKP) sensor (except TMC made engine).
92. Install idler pulley sub-assembly.
93. Install engine mounting insulator sub-assembly RH.

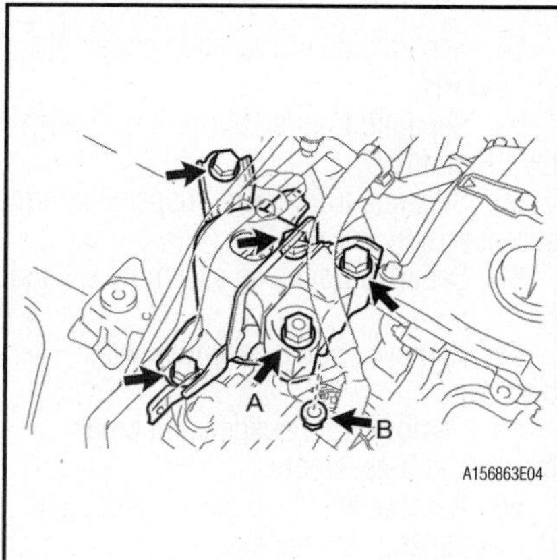

Fig. 104 Install engine mounting insulator sub-assembly RH

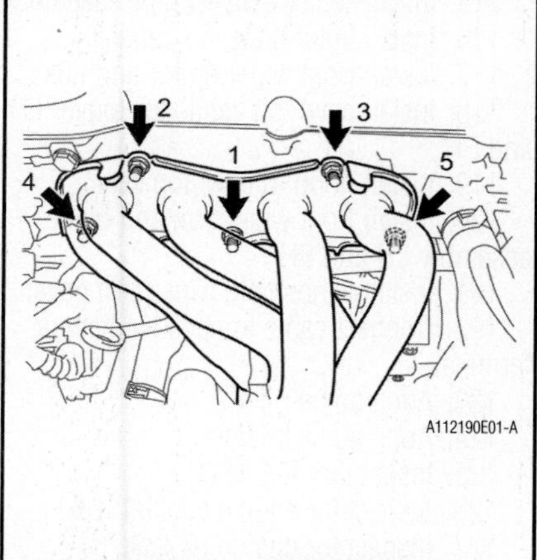

Fig. 105 Exhaust manifold tightening sequence

a. Tighten the bolts to 70 ft. lbs. (95 Nm).
b. Tighten Nut A to 70 ft. lbs. (95 Nm).
c. Tighten Nut B to 38 ft. lbs. (52 Nm).

94. Install front engine mounting insulator sub-assembly.
a. Tighten the two bottom bolts to 70 ft. lbs. (95 Nm).
b. Tighten the through bolt and nut to 107 ft. lbs. (145 Nm).

95. Install exhaust manifold converter sub-assembly. Tighten to 27 ft. lbs. (37 Nm) in sequence.
96. Install No. 1 exhaust manifold heat insulator.
97. Install No. 2 manifold stay.
98. Install manifold stay.
99. Install intake manifold insulator.
100. Install intake manifold.
101. Install camshaft timing oil control valve assembly.
102. Install fuel delivery pipe sub-assembly.
103. Connect fuel main tube.
104. Install engine oil level dipstick guide.
105. Install oil dipstick.
106. Install cylinder head cover sub-assembly.
107. Install spark plug.
108. Install ignition coil assembly.
109. Install throttle with motor body assembly.
110. Install air cleaner cap sub-assembly with No. 1 air cleaner hose.
111. Install radiator reserve tank assembly.
112. Install alternator assembly.
113. Install fan and alternator V belt.
114. Install front exhaust pipe assembly.

115. Install center exhaust pipe assembly.
116. Install outer cowl top panel.
117. Install front wiper motor and link.
118. Install cowl top ventilator louver LH and RH.
119. Install hood to cowl top seal.
120. Install front wiper arm and blade assembly LH and RH.
121. Install windshield wiper arm cover.
122. Connect cable to negative battery terminal.
123. Add engine oil.
124. Add engine coolant.
125. Inspect for fuel leak.
126. Inspect for engine coolant leak.
127. Inspect for engine oil leak.
128. Inspect for exhaust gas leak.
129. Inspect ignition timing.
130. Inspect engine idling speed.
131. Inspect compression.
132. Inspect CO/HC.
133. Install No. 1 engine cover sub-assembly.
134. Install No. 1 engine under cover.
135. Install rear engine under cover RH.
136. Install front wheel RH.
137. Install hood.
138. Adjust hood.

xD

See Figures 106 through 112.

1. Discharge fuel system pressure.
2. Remove battery.
3. Remove battery tray.
4. Remove front wheels.
5. Remove engine under cover LH and RH.
6. Drain engine coolant.
7. Drain engine oil.
8. Remove front wiper arm head cap.
9. Remove front wiper arm and blade assembly LH and RH.
10. Remove hood to cowl top seal.
11. Remove cowl top ventilator louver sub-assembly.
12. Remove cowl top ventilator louver LH.
13. Remove front wiper motor and link.
14. Remove front air shutter seal RH.
15. Remove outer cowl top panel.
16. Remove air cleaner assembly.
17. Remove air cleaner bracket.
18. Remove battery carrier.
19. Remove No. 2 cylinder head cover.
20. Remove fan and alternator V belt.
21. Remove No. 2 radiator hose.
22. Disconnect No. 1 radiator hose.
23. Remove front bumper cover.
24. Disconnect No. 1 oil cooler outlet hose (for automatic transaxle).
25. Disconnect No. 2 oil cooler inlet hose (for automatic transaxle).
26. Remove hood lock assembly.
27. Remove No. 1 cooler cover.
28. Remove upper radiator support absorber.
29. Remove upper radiator support sub-assembly.
30. Remove radiator assembly.
31. Separate with pulley compressor assembly.
32. Separate transmission control cable assembly (for automatic transaxle).
33. Separate transmission control cable assembly (for manual transaxle).
34. Separate union to check valve hose.
35. Separate No. 1 fuel vapor feed hose.
36. Disconnect engine wire.
37. Disconnect heater water outlet hose A (from heater unit).
38. Disconnect heater water hose inlet A.
39. Disconnect fuel tube sub-assembly (for manual transaxle).
40. Separate clutch release cylinder assembly (for manual transaxle).
41. Remove column hole cover silencer sheet.
42. Separate steering sliding yoke sub-assembly.
43. Separate No. 1 steering column hole cover sub-assembly.
44. Remove shift lever knob sub-assembly (for manual transaxle).
45. Remove console panel upper.
46. Remove console box cover rear.
47. Remove console box carpet.
48. Remove rear console box assembly.
49. Remove front console box.
50. Remove front floor brace center.
51. Remove front exhaust pipe assembly.
52. Remove front axle shaft LH and RH nut.
53. Separate speed sensor front LH and RH.
54. Separate tie rod end sub-assembly LH and RH.
55. Separate front stabilizer link assembly LH and RH.
56. Separate front lower suspension arm sub-assembly LH and RH.
57. Separate front axle assembly LH and RH.
58. Remove engine assembly with transaxle.
59. Remove front suspension cross-member sub-assembly.
60. Remove No. 1 oil cooler inlet tube (for Automatic Transaxle).
61. Remove No. 1 oil cooler outlet tube (for Automatic Transaxle).
62. Remove No. 2 oil cooler tube clamp (for Automatic Transaxle).
63. Remove transmission oil level gage sub-assembly (for Automatic Transaxle).
64. Separate transmission oil filler tube sub-assembly (for Automatic Transaxle).
65. Remove ignition coil assembly.
66. Remove radio setting condenser.
67. Remove No. 2 ventilation hose.
68. Remove heater water outlet hose A (from heater unit).
69. Remove heater water hose inlet A.
70. Remove No. 1 radiator hose.
71. Remove intake manifold.
72. Remove fuel delivery pipe sub-assembly.
73. Remove No. 1 delivery pipe space.
74. Remove injector vibration insulator.
75. Remove fuel injector assembly.
76. Remove alternator assembly.
77. Remove fan belt adjusting bar.
78. Remove oil level dipstick.
79. Remove oil level gage guide.
80. Remove water by-pass hose.
81. Remove water inlet hose.
82. Remove No. 1 water by-pass pipe.
83. Remove air tube assembly.
84. Remove No. 1 exhaust manifold heat insulator.
85. Remove manifold stay.
86. Remove exhaust manifold.
87. Remove wire harness clamp bracket.
88. Remove water inlet.
89. Remove thermostat.
90. Remove cylinder head cover sub-assembly.
91. Set No. 1 cylinder to TDC / compression.
92. Remove crankshaft pulley.
93. Remove No. 1 chain tensioner assembly.
94. Remove timing chain cover sub-assembly.
95. Remove timing chain cover oil seal.

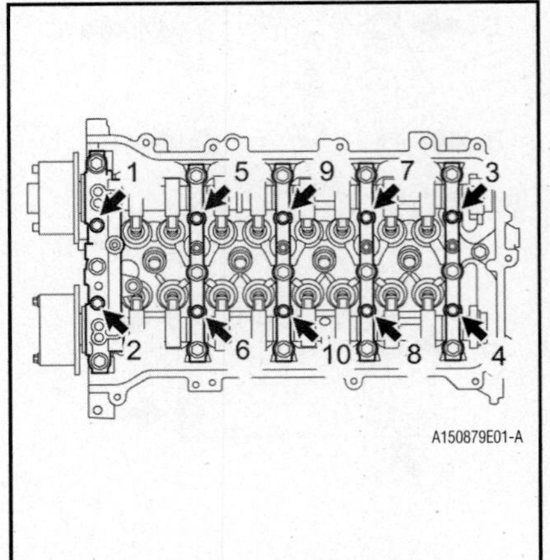

Fig. 106 Uniformly loosen and remove the 10 bearing cap bolts in sequence

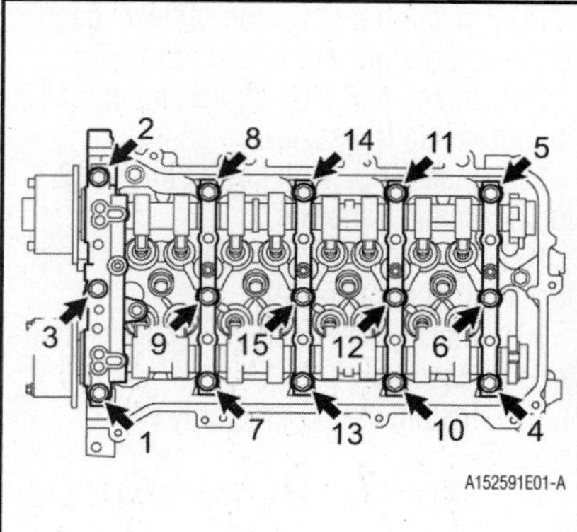

Fig. 107 Uniformly loosen and remove the 15 bearing cap bolts in sequence

96. Remove No. 2 chain vibration damper.
97. Remove chain tensioner slipper.
98. Remove No. 1 chain vibration damper.
99. Remove chain sub-assembly.
100. Remove crankshaft timing gear or sprocket.
101. Remove camshaft bearing cap.
   a. Uniformly loosen and remove the 10 bearing cap bolts in sequence.
   b. Uniformly loosen and remove the 15 bearing cap bolts in sequence.
   c. Remove the 5 bearing caps.
102. Remove camshaft.
103. Remove No. 2 camshaft.
104. Remove No. 1 valve rocker arm sub-assembly.
105. Remove valve lash adjuster assembly.
106. Remove No. 1 camshaft bearing.
107. Remove No. 2 camshaft bearing.
108. Remove camshaft housing sub-assembly.

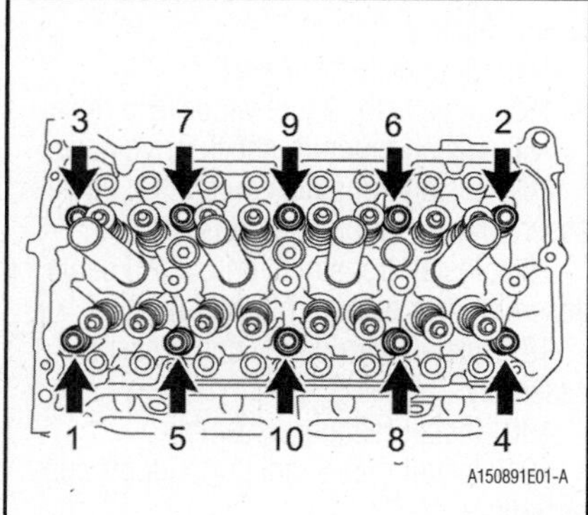

Fig. 108 Using several steps, loosen and remove the 10 cylinder head bolts uniformly in sequence

109. Remove cylinder head sub-assembly.
   a. Using several steps, loosen and remove the 10 cylinder head bolts uniformly in sequence.
   b. Remove the 10 cylinder head bolts and the plate washers.
   c. Remove the cylinder head sub-assembly.
110. Remove cylinder head gasket.

To install:

111. Install cylinder head gasket.
112. Install cylinder head sub-assembly.
   a. Apply a light coat of engine oil to the threads of the cylinder head bolts.
   b. Using several steps, install and tighten the 10 cylinder head bolts and plate washers uniformly in sequence.
   c. Tighten to 36 ft. lbs. (49 Nm).
   d. Retighten the cylinder head bolts by additional 90° and one more additional 45° in sequence.
113. Install valve lash adjuster assembly.
114. Install No. 1 valve rocker arm sub-assembly.
115. Install No. 1 camshaft bearing.

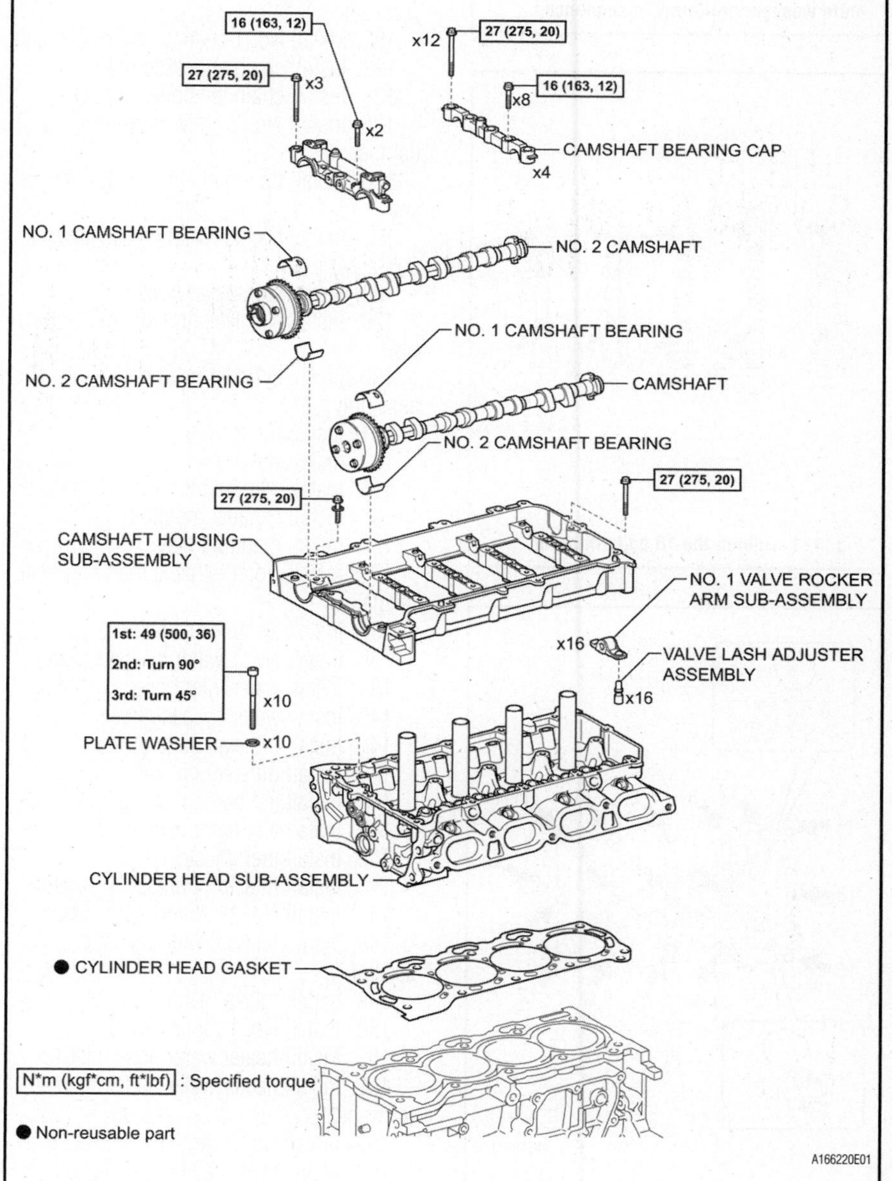

Fig. 109 Exploded view of cylinder head assembly

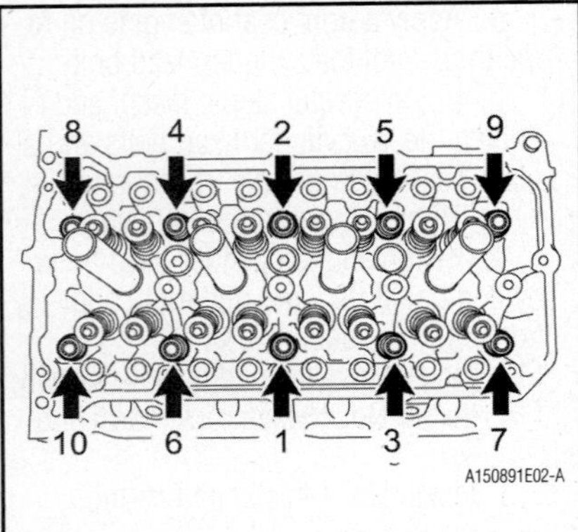

Fig. 110 Using several steps, install and tighten the 10 cylinder head bolts and plate washers uniformly in sequence

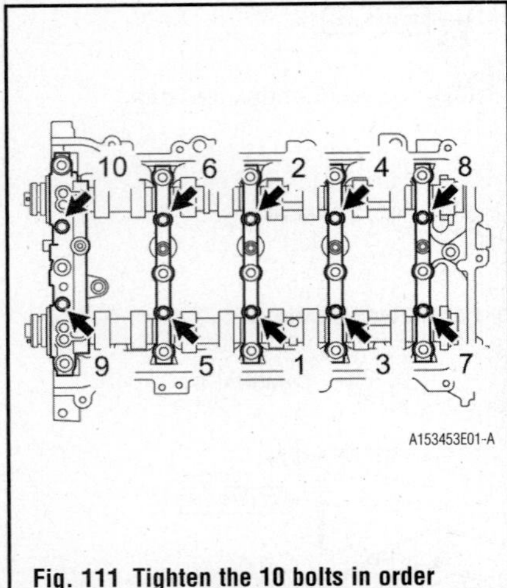

Fig. 111 Tighten the 10 bolts in order

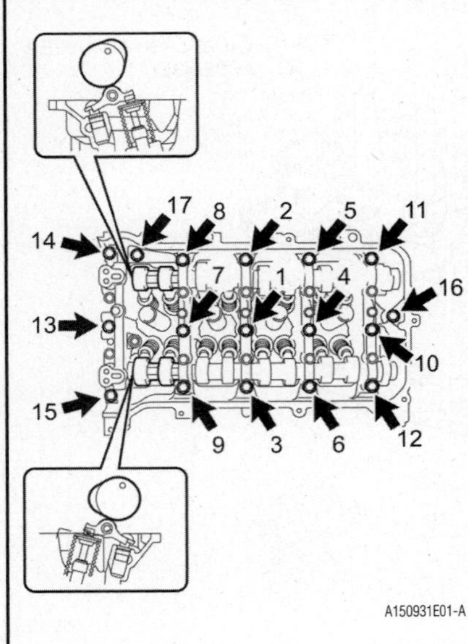

Fig. 112 Install the camshaft housing and tighten the 17 bolts in order

116. Install No. 2 camshaft bearing.
117. Install No. 2 camshaft.
118. Install camshaft.
119. Install camshaft bearing cap.
    a. Apply engine oil to the camshaft journals, camshaft housing and bearing caps.
    b. Tighten the 10 bolts in order to 12 ft. lbs. (16 Nm).
120. Install camshaft housing sub-assembly.
    a. Set the camshaft and No. 2 camshaft.
    b. Install the camshaft housing and tighten the 17 bolts in order to 20 ft. lbs. (27 Nm).
121. Install crankshaft timing gear or sprocket.
122. Install No. 1 chain vibration damper.
123. Install chain sub-assembly.
124. Install chain tensioner slipper.
125. Install No. 2 chain vibration damper.
126. Install timing chain or belt cover oil seal.
127. Install timing chain or belt cover sub-assembly.
128. Install crankshaft pulley.
129. Install No. 1 chain tensioner assembly.
130. Install cylinder head cover sub-assembly.
131. Install thermostat.
132. Install water inlet.
133. Install wire harness clamp bracket.
134. Install exhaust manifold
135. Install manifold stay.
136. Install No. 1 exhaust manifold heat insulator.
137. Install air tube assembly.
138. Install No. 1 water by-pass pipe.
139. Install water inlet hose.
140. Install water by-pass hose.
141. Install oil level gage guide
142. Install oil level dipstick.
143. Install fan belt adjusting bar
144. Install alternator assembly.
145. Install fuel injector.
146. Install injector vibration insulator.
147. Install No. 1 delivery pipe spacer.
148. Install fuel delivery pipe sub-assembly.
149. Install intake manifold.
150. Install No. 1 radiator hose.
151. Install heater water hose inlet A.
152. Install heater water outlet hose A (from heater unit)
153. Install No. 2 ventilation hose.
154. Install radio setting condenser.
155. Install ignition coil assembly.
156. Install transmission oil filler tube sub-assembly (for automatic transaxle).
157. Install transmission oil level gage sub-assembly (for automatic transaxle).
158. Install No. 1 oil cooler outlet tube (for automatic transaxle).
159. Install No. 1 oil cooler inlet tube (for automatic transaxle).
160. Install No. 2 oil cooler tube clamp (for automatic transaxle).
161. Install front suspension crossmember sub-assembly.
162. Install engine assembly with transaxle.
163. Install front axle assembly LH and RH.
164. Install front lower suspension arm sub-assembly LH and RH.
165. Install front stabilizer link assembly LH and RH.
166. Install tie rod end sub-assembly LH and RH.
167. Install speed sensor front LH and RH.
168. Install front axle shaft LH and RH nut.
169. Install front exhaust pipe assembly.
170. Install front floor brace center.
171. Install front console box.
172. Install rear console box assembly.
173. Install console box carpet.
174. Install console box cover rear.
175. Install console panel upper.
176. Install shift lever knob sub-assembly (for manual transaxle).
177. Install No. 1 steering column hole cover sub-assembly.
178. Install steering sliding yoke sub-assembly.
179. Install column hole cover silencer sheet.
180. Install clutch release cylinder assembly (for manual transaxle).
181. Connect fuel tube sub-assembly.
182. Connect heater water hose inlet A.
183. Connect heater water outlet hose A (from heater unit).
184. Connect engine wire.
185. Install No. 1 fuel vapor feed hose.
186. Install union to check valve hose.
187. Install transmission control cable assembly (for automatic transaxle).
188. Install transmission control cable assembly (for manual transaxle).
189. Install with pulley compressor assembly.
190. Install radiator assembly.
191. Install upper radiator support sub-assembly.
192. Install upper radiator support absorber.
193. Install No. 1 cooler cover.
194. Install hood lock assembly.
195. Connect No. 1 radiator hose.

196. Connect No. 2 oil cooler inlet hose (for automatic transaxle).
197. Connect No. 1 oil cooler outlet hose (for automatic transaxle).
198. Install No. 2 radiator hose.
199. Install front bumper cover.
200. Install fan and alternator V belt.
201. Adjust fan and alternator V belt.
202. Inspect fan and alternator V belt.
203. Install No. 2 cylinder head cover.
204. Install battery carrier.
205. Install air cleaner bracket.
206. Install air cleaner assembly.
207. Install outer cowl top panel.
208. Install front air shutter seal RH.
209. Install front wiper motor and link.
210. Install cowl top ventilator louver LH.
211. Install cowl top ventilator louver sub-assembly.
212. Install hood to cowl top seal.
213. Install front wiper arm and blade assembly LH and RH.
214. Install front wiper arm head cap.
215. Install battery tray.
216. Install battery.
217. Add engine coolant.
218. Add engine oil.
219. Inspect for fuel leak.
220. Inspect for engine oil leak.
221. Inspect for exhaust gas leak.
222. Inspect for engine coolant leak.
223. Install engine under cover RH.
224. Install engine under cover LH.
225. Install front wheels.
226. Inspect ignition timing.
227. Inspect engine idling speed.
228. Inspect CO/HC.
229. Inspect front wheel alignment.
230. Check VSC sensor signal.

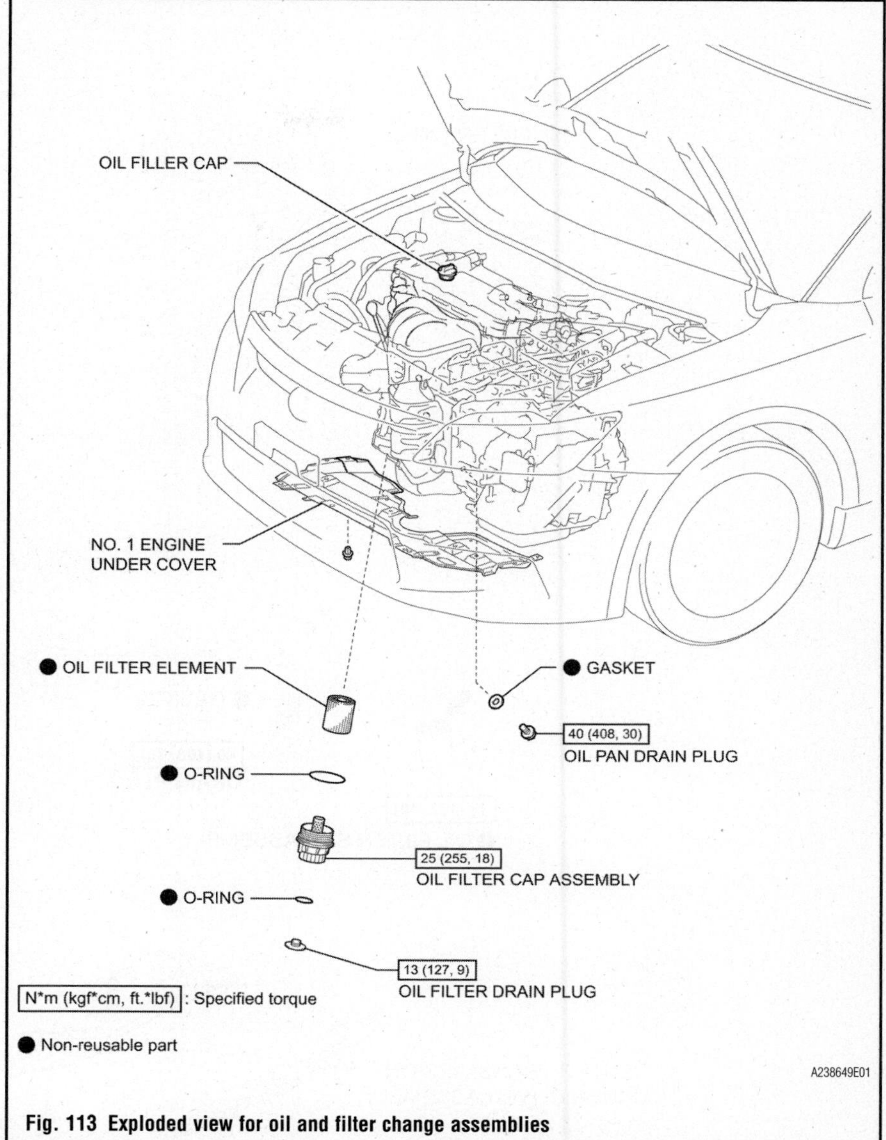

Fig. 113 Exploded view for oil and filter change assemblies

ENGINE OIL & FILTER

REPLACEMENT

tC

See Figure 113.

1. Drain engine oil.
   a. Remove the oil filler cap.
   b. Remove the oil pan drain plug and gasket, and then drain the engine oil into a container.
2. Remove No. 1 engine under cover.
3. Remove oil filter cap assembly.
   a. Connect a hose to the pipe.
   b. Remove the oil filter drain plug.
   c. Install the pipe to the oil filter cap.
   d. Check that the oil is drained from the oil filter. Then disconnect the pipe and remove the O-ring.
   e. Using SST, remove the oil filter cap.
   f. Remove the oil filter element and O-ring from the oil filter cap.
4. Install oil filter cap assembly.
   a. Clean the inside of the oil filter cap, the threads and O-ring groove.
   b. Apply a small amount of engine oil to a new O-ring and install the O-ring to the oil filter cap.
   c. Insert a new oil filter element into the oil filter cap.
   d. Remove any dirt or foreign matter from the installation surface of the engine.
   e. Apply a small amount of engine oil to the O-ring again and temporarily install the oil filter cap.
   f. Using SST, tighten the oil filter cap.
   g. Apply a small amount of engine oil to a new drain plug O-ring and install the O-ring to the oil filter cap.
   h. Install the oil filter drain plug.
5. Add engine oil.
   a. Clean and install a new gasket and the oil pan drain plug. Tighten to 30 ft. lbs. (40 Nm).
   b. Add new oil.
   c. Install the oil filler cap.
6. Inspect for oil leak.
7. Install No. 1 engine under cover.
8. Inspect engine oil level.

xB

See Figure 114.

1. Remove No. 1 engine under cover.
2. Drain engine oil.
   a. Remove the oil filler cap.
   b. Remove the oil drain plug and drain the oil into a container.
   c. Clean and install the oil drain plug

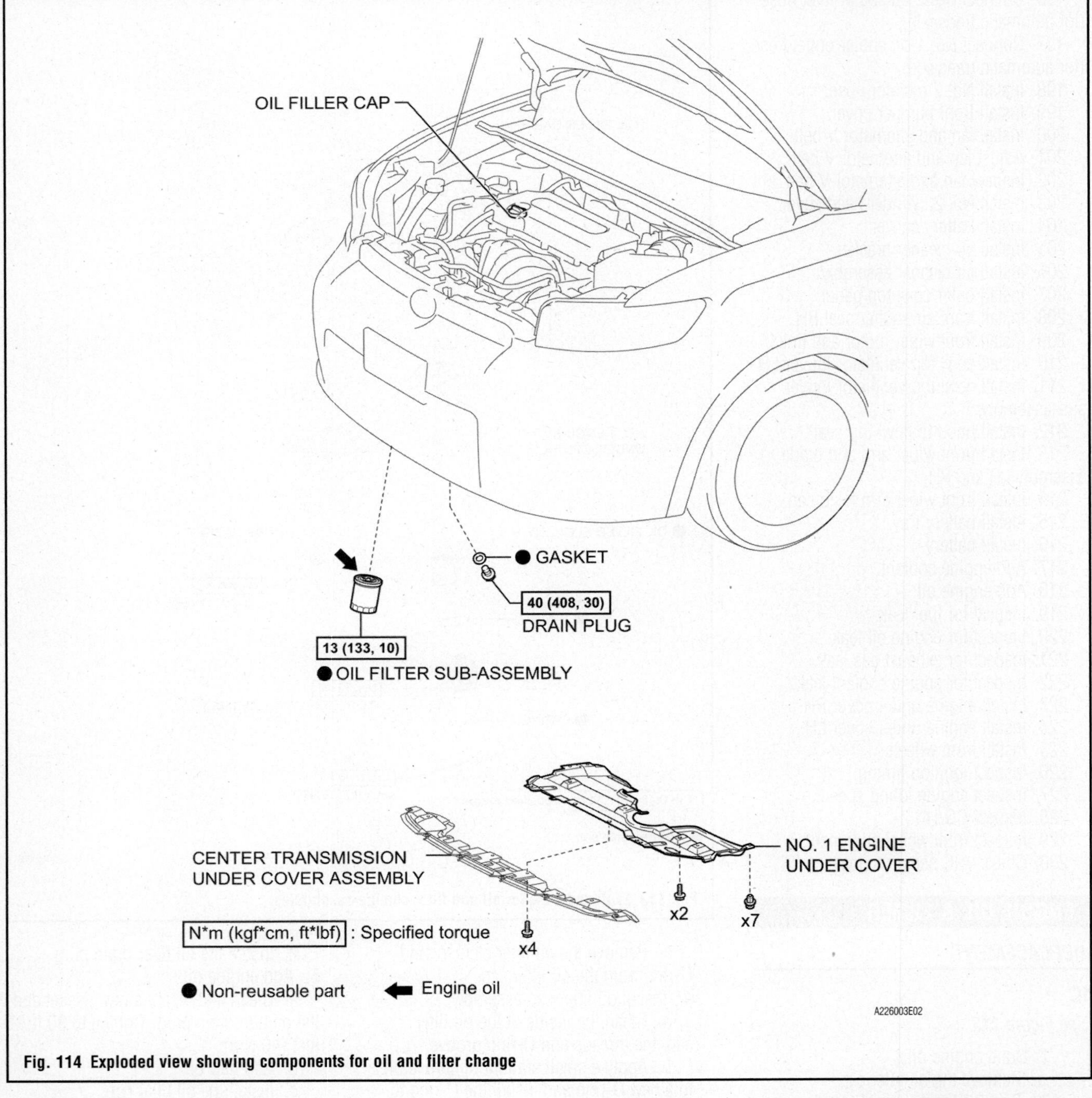

Fig. 114 Exploded view showing components for oil and filter change

with a new gasket. Tighten to 30 ft. lbs. (40 Nm).

3. Remove oil filter sub-assembly.
4. Install oil filter sub-assembly.
   a. If enough space is available, use a torque wrench to tighten the oil filter.
   b. If enough space is not available to use a torque wrench, tighten the oil filter a 3/4 turn by hand or use a common wrench.
5. Add engine oil.
6. Inspect for engine oil leak.
7. Install No. 1 engine under cover.

xD

See Figure 115.

1. Drain engine oil.
   a. Remove the oil pan drain plug and drain the engine oil.
   b. Clean the oil pan drain plug and install it with a new gasket. Tighten to 27 ft. lbs. (37 Nm).
2. Remove oil filter element.
   a. Using SST, remove the oil filter cap.
   b. Remove the oil filter element and O-ring from the oil filter cap.
3. Install oil filter element.
   a. Clean the inside of the oil filter cap, threads and O-ring groove.
   b. Apply a small amount of engine oil to a new O-ring and install it onto the oil filter cap.
   c. Place a new oil filter element into the oil filter cap.
   d. Remove any dirt or foreign matter from the installation surface of the engine.
   e. Apply a small amount of engine oil

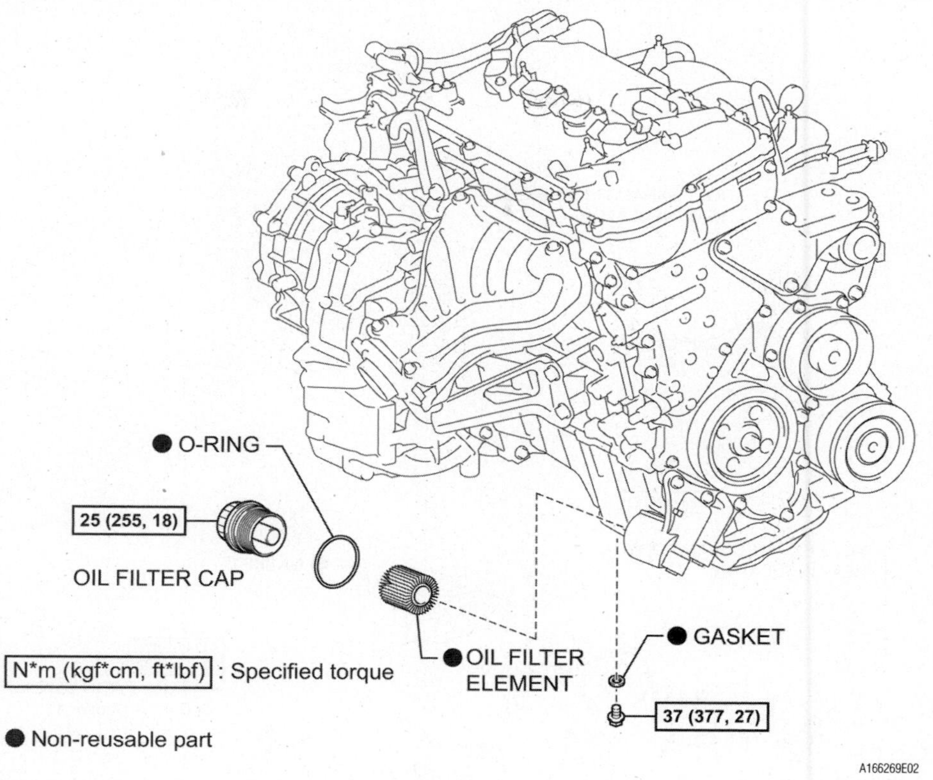

Fig. 115 Exploded view of oil filter assembly

to the O-ring again and install the oil filter cap.

f. Using SST, tighten the oil filter cap.

4. Add engine oil.
5. Inspect for engine oil leak.
6. Inspect for oil level.

EXHAUST MANIFOLD

REMOVAL & INSTALLATION

tC

See Figures 116 and 117.

1. Remove alternator assembly.
2. Remove front exhaust pipe assembly.

a. Disconnect the heated oxygen sensor connector.

b. Remove the 4 bolts, 4 compression springs and front exhaust pipe.

c. Remove the gasket from the front exhaust pipe.

d. Remove the gasket from the exhaust manifold converter.

3. Remove engine oil level dipstick.
4. Remove engine oil level dipstick guide.
5. Remove radiator hose clamp.
6. Remove air fuel ratio sensor (for bank 1 sensor 1).
7. Remove manifold stay.
8. Remove No. 2 manifold stay.
9. Remove No. 1 exhaust manifold heat insulator.
10. Remove No. 2 exhaust manifold heat insulator.
11. Remove exhaust manifold converter sub-assembly.
12. Remove No. 1 manifold converter insulator.

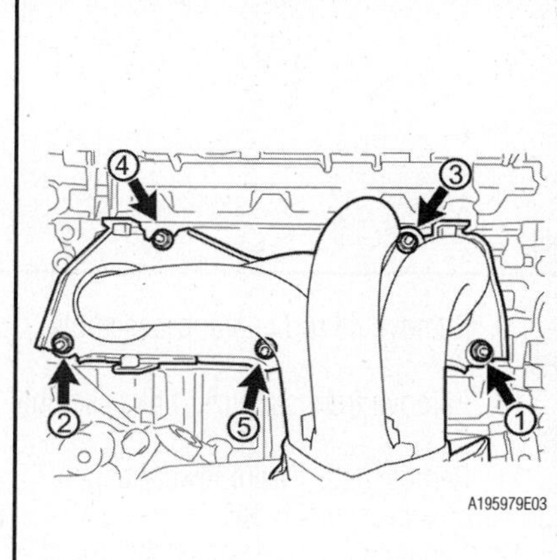

Fig. 116 Remove exhaust manifold converter sub-assembly; remove nuts in sequence

To install:

13. Install No. 1 manifold converter insulator.
14. Install exhaust manifold converter sub-assembly. Tighten nuts in sequence to 26 ft. lbs. (35 Nm).
15. Install No. 2 exhaust manifold heat insulator.
16. Install No. 1 exhaust manifold heat insulator.
17. Install No. 2 manifold stay.
18. Install manifold stay.
19. Install air fuel ratio sensor (for bank 1 sensor 1).
20. Install radiator hose clamp.
21. Install engine oil level dipstick guide.
22. Install engine oil level dipstick.
23. Install front exhaust pipe assembly.
24. Install alternator assembly.
25. Inspect for exhaust gas leak.

xB

See Figure 118.

➡The manufacturer does not provide a specific Removal and Installation procedure for this component. Refer to the graphic(s) when servicing this component.

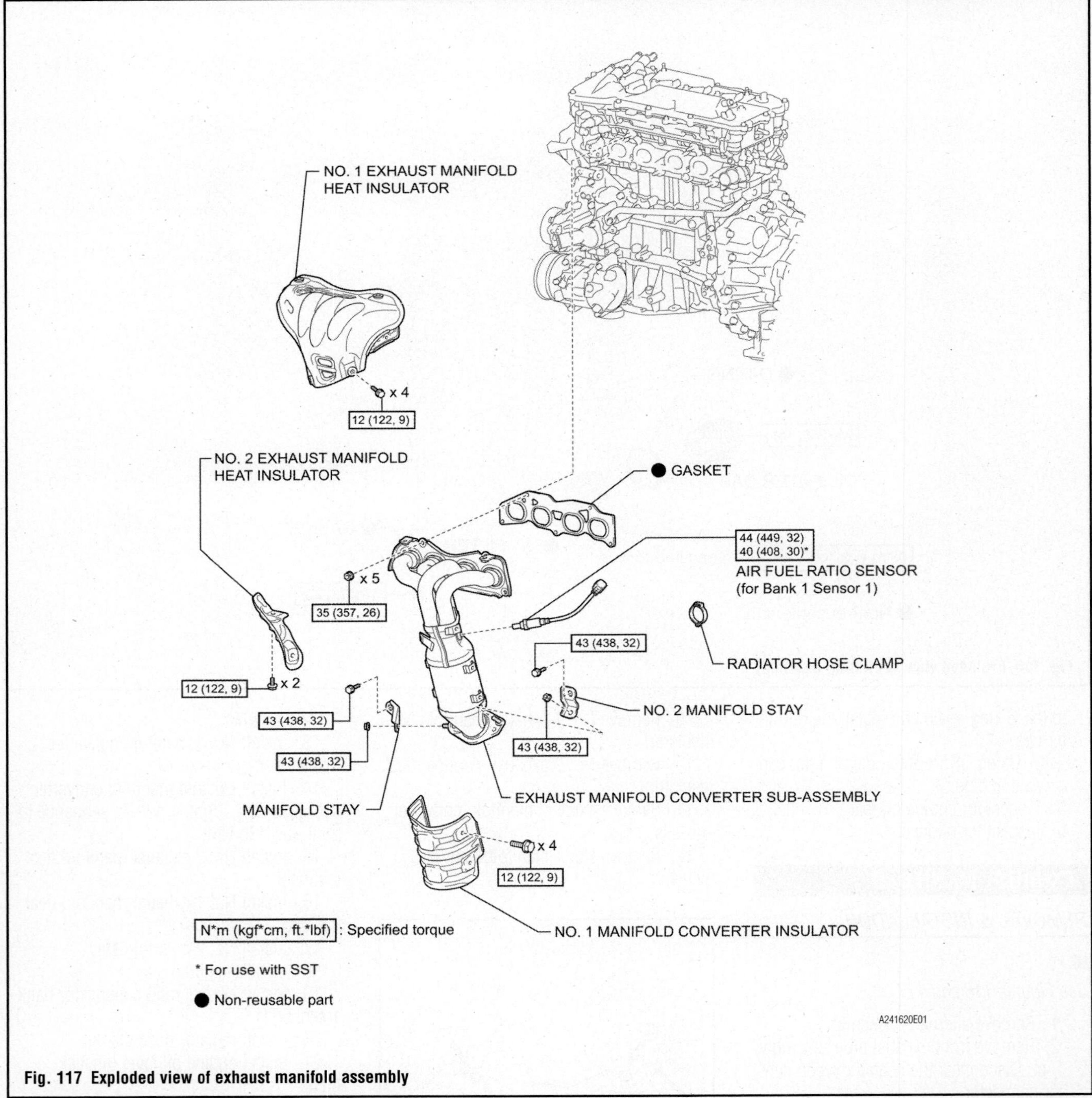

Fig. 117 Exploded view of exhaust manifold assembly

xD

See Figure 119.

1. Remove air fuel ratio sensor.
2. Remove engine under cover RH.
3. Remove front exhaust pipe assembly.
4. Drain automatic transaxle fluid (for automatic transaxle).
5. Drain manual transaxle oil (for manual transaxle).
6. Remove front wheel.
7. Remove front axle shaft RH nut.
8. Remove front speed sensor RH.
9. Remove tie rod end sub-assembly RH.
10. Remove front stabilizer link assembly RH.
11. Remove No. 1 front lower suspension arm sub-assembly RH.
12. Separate front axle assembly RH.
13. Remove drive shaft assembly RH.
14. Remove No. 1 exhaust manifold heat insulator.
15. Remove manifold stay.
16. Remove exhaust manifold.
17. Remove No. 2 exhaust manifold heat insulator.

To install:

18. Install No. 2 exhaust manifold insulator.
19. Install exhaust manifold. Tighten to 16 ft. lbs. (21 Nm).
20. Install manifold support bracket. Tighten to 32 ft. lbs. (43 Nm).
21. Install No. 1 exhaust manifold heat insulator.
22. Install drive shaft assembly RH.

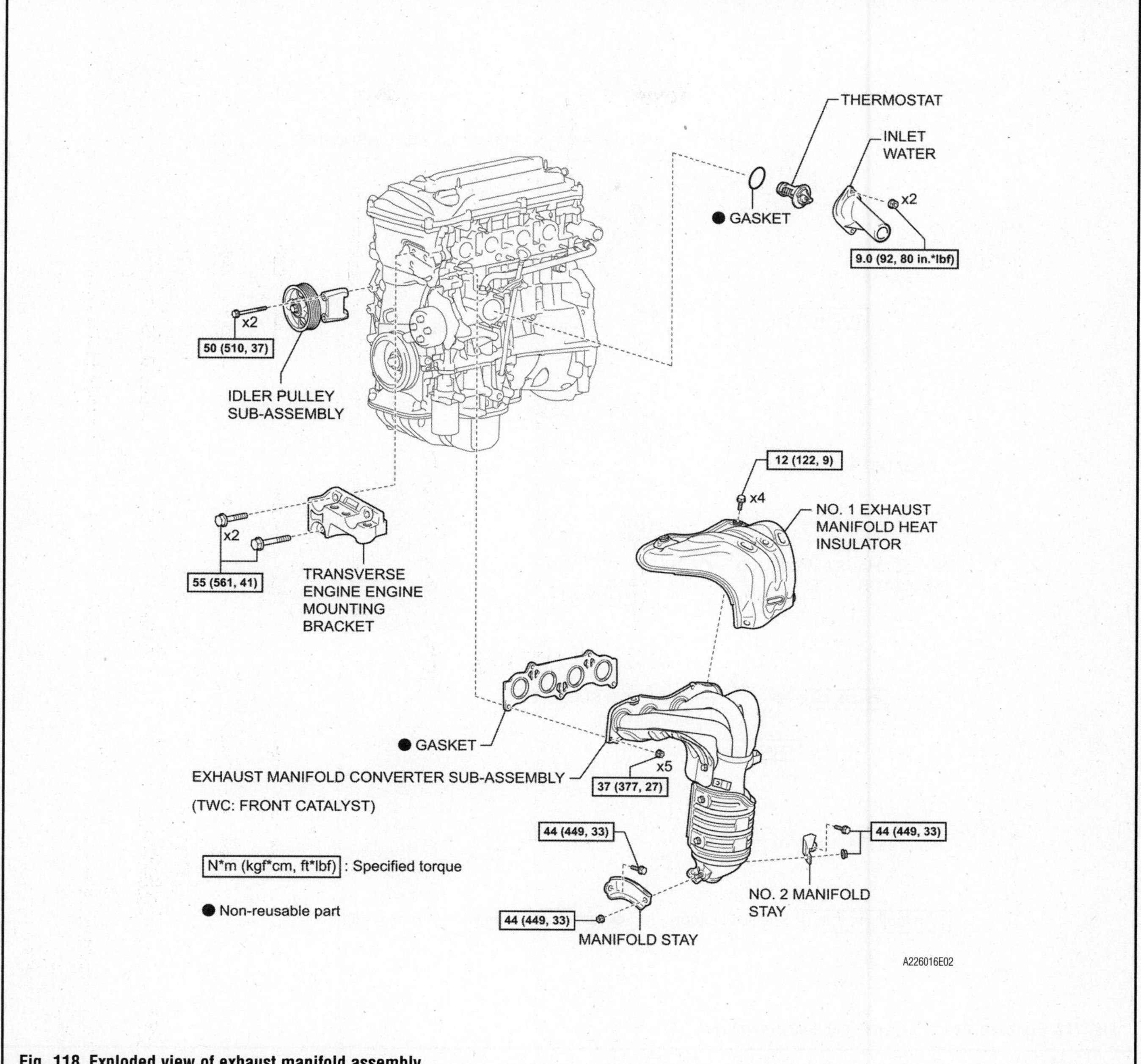

Fig. 118 Exploded view of exhaust manifold assembly

23. Install front axle assembly RH.
24. Install No. 1 front suspension arm sub-assembly RH.
25. Install front stabilizer link assembly RH.
26. Install tie rod end sub-assembly RH.
27. Install front speed sensor RH.
28. Install front axle shaft RH nut.
29. Install front wheel.
30. Install front exhaust pipe assembly.
31. Install air fuel ratio sensor.
32. Add automatic transaxle fluid (for automatic transaxle).
33. Inspect automatic transaxle fluid (for automatic transaxle).
34. Add manual transaxle oil (for manual transaxle).
35. Inspect manual transaxle oil (for manual transaxle).
36. Inspect for automatic transaxle fluid leak (for automatic transaxle).
37. Inspect for manual transaxle oil leak.
38. Inspect for exhaust gas leak.
39. Inspect and adjust front wheel alignment.
40. Install engine under cover RH.

INTAKE MANIFOLD

REMOVAL & INSTALLATION

tC

See Figures 120 and 121.

1. Remove fuel delivery pipe sub-assembly.
2. Remove engine under cover.
3. Drain engine coolant.
4. Remove throttle with motor body assembly.
5. Remove hose to hose tube.
   a. Disconnect the 2 hoses.

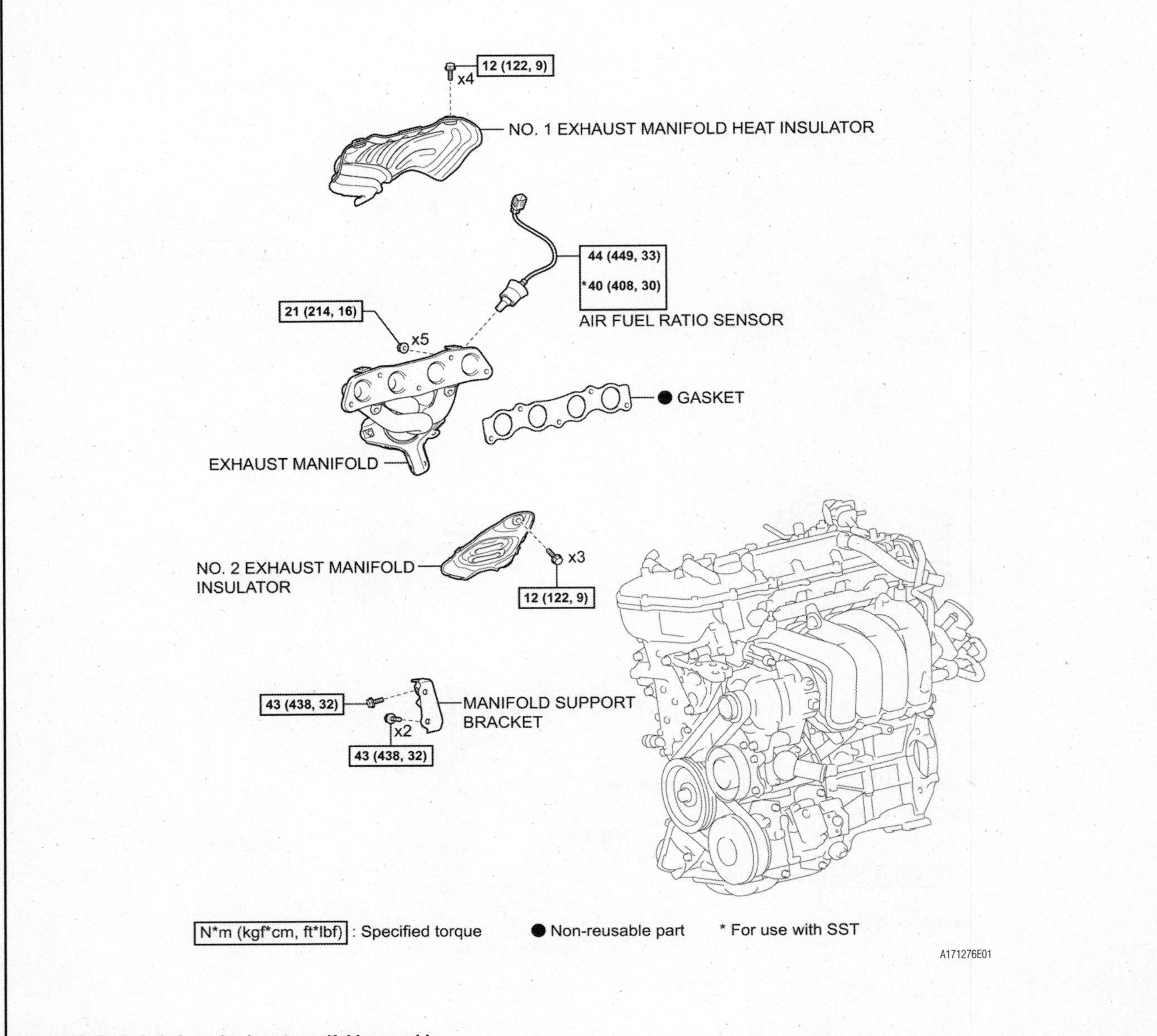

Fig. 119 Exploded view of exhaust manifold assembly

b. Remove the 2 nuts and tube.

6. Disconnect heater hose.

a. Disconnect the inlet heater water hose from the heater radiator unit.

b. Disconnect the outlet heater water hose from the heater radiator unit.

7. Disconnect wire harness.

a. Disconnect the connector, detach the clamp and disconnect the 2 hoses, and then remove the vacuum switching valve.

b. Detach the 2 clamps.

c. Remove the 2 bolts and wiring harness clamp bracket.

d. Disconnect the intake air control valve actuator connector.

e. Remove the bolt and disconnect the bracket.

f. Apply battery voltage to the terminals of the connector to close the tumble control valves.

g. Detach the clamp and disconnect the wire harness.

8. Disconnect No. 2 PCV hose.

9. Remove intake manifold.

10. Remove check valve.

a. Disconnect the 2 vacuum hoses from the intake manifold and remove the check valve.

b. Remove the 2 vacuum hoses from the check valve.

11. Remove purge line hose.

a. Disconnect the purge line hose from the clamp.

b. Remove the purge line hose from the intake manifold.

12. Remove intake air control valve actuator (for TCV).

To install:

13. Install intake air control valve actuator (for TCV).

14. Install purge line hose.

15. Install check valve.

16. Install the intake manifold with the 6 bolts and tighten the bolts in sequence. Tighten to 15 ft. lbs. (21 Nm).

17. Connect No. 2 PCV hose.

18. Connect wire harness.

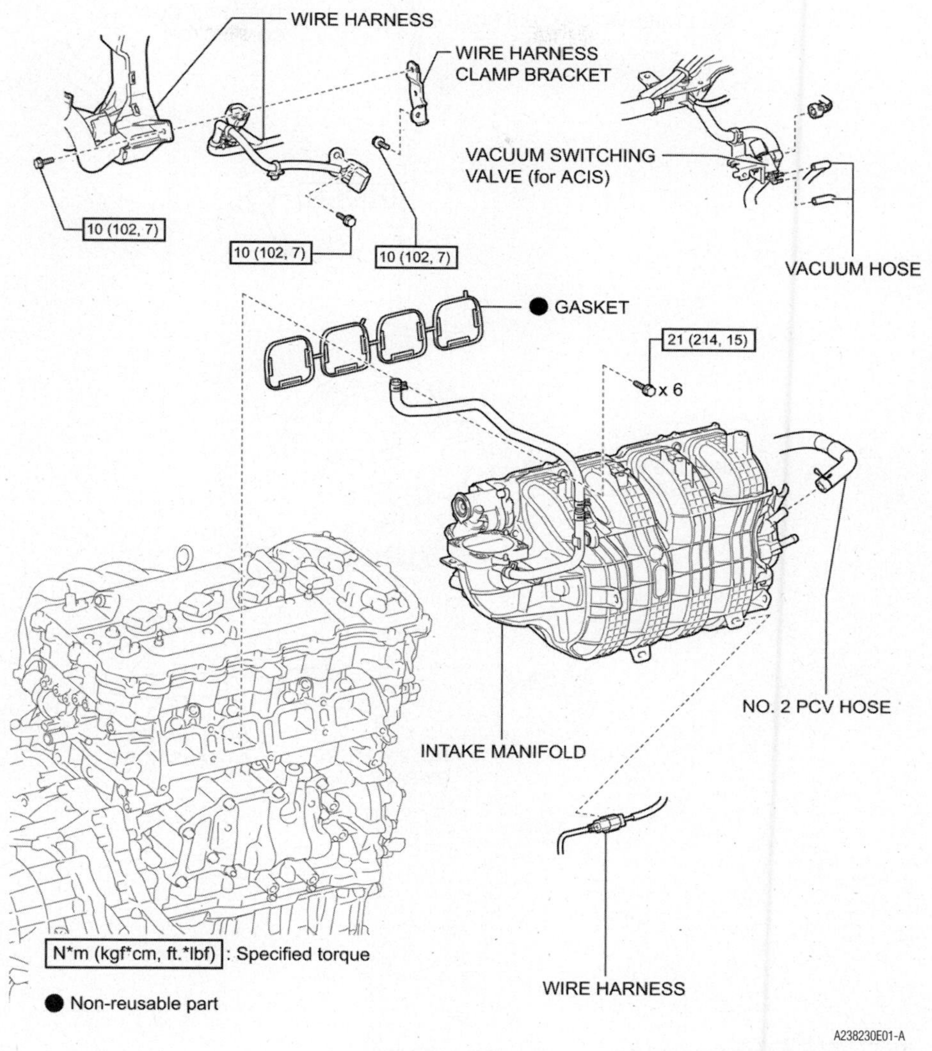

WIRE HARNESS
WIRE HARNESS CLAMP BRACKET
VACUUM SWITCHING VALVE (for ACIS)
10 (102, 7)
10 (102, 7)
10 (102, 7)
VACUUM HOSE
● GASKET
21 (214, 15)
x 6
NO. 2 PCV HOSE
INTAKE MANIFOLD
WIRE HARNESS
N*m (kgf*cm, ft.*lbf) : Specified torque
● Non-reusable part
A238230E01-A

Fig. 120 Exploded view of intake manifold assembly

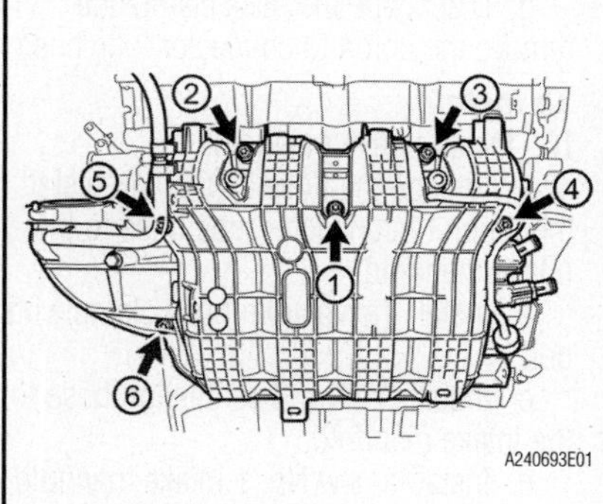

Fig. 121 Install the intake manifold with the 6 bolts and tighten the bolts in sequence

19. Connect heater hose.
20. Install hose to hose tube.
21. Install throttle with motor body assembly.
22. Install engine under cover.
23. Install fuel delivery pipe sub-assembly.
24. Add engine coolant.
25. Inspect for coolant leak.

xB

See Figures 122 through 125.

1. Discharge fuel system pressure.
2. Remove windshield wiper arm cover.
3. Remove front wiper arm and blade assembly LH.
4. Remove front wiper arm and blade assembly RH.
5. Remove hood to cowl top seal.
6. Remove cowl top ventilator louver RH.
7. Remove cowl top ventilator louver LH.
8. Remove front wiper motor and link.
9. Remove outer cowl top panel.
10. Remove No. 1 engine cover sub-assembly.
11. Remove air cleaner cap sub-assembly with No. 1 air cleaner hose.
12. Remove throttle with motor body assembly.
13. Disconnect fuel tube sub-assembly.

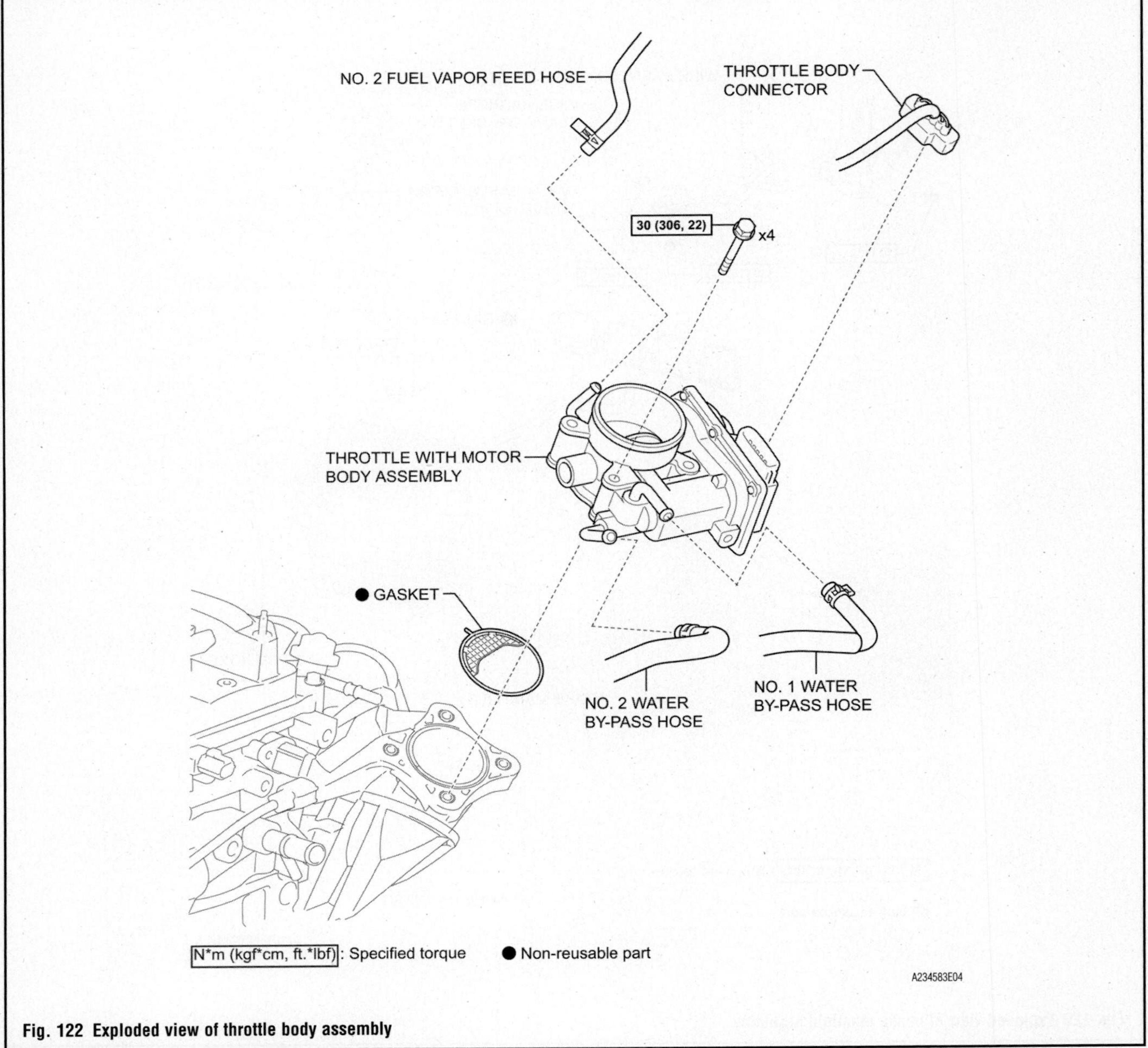

Fig. 122 Exploded view of throttle body assembly

a. Release the claw and remove the EFI fuel pipe clamp.

b. Pinch the retainer, then pull the fuel tube connector out of the pipe.

14. Remove fuel delivery pipe sub-assembly with fuel tube sub-assembly.

a. Disconnect the 4 fuel injector connectors and the oil control valve connector.

b. Disconnect the heated oxygen sensor connector.

c. Disengage the 4 wire harness clamps and separate the wire harness.

d. Disengage the clamp and separate the fuel tube sub-assembly.

e. Remove the 2 bolts and the fuel delivery pipe sub-assembly.

f. Remove the 2 No. 1 delivery pipe spacers.

g. Remove the 4 injector vibration insulators.

15. Remove intake manifold.

a. Disconnect the No. 2 ventilation hose.

b. Disconnect the union to connector tube hose.

c. Remove the 5 bolts, 2 nuts and the intake manifold.

d. Remove the No. 1 intake manifold to head gasket.

e. Remove the No. 2 ventilation hose.

f. Remove the bolt and the vacuum hose clamp.

g. Disengage the hose clamp and remove the union to connector tube hose.

To install:

16. Install intake manifold.

a. Engage the hose clamp and install the union to connector tube hose to the intake manifold.

b. Install the vacuum hose clamp with bolt.

c. Install the No. 2 ventilation hose to the intake manifold.

d. Install a new No. 1 intake manifold to head gasket onto the intake manifold.

e. Install the intake manifold with the 5 bolts and the 2 nuts in sequence. Tighten to 22 ft. lbs. (30 Nm).

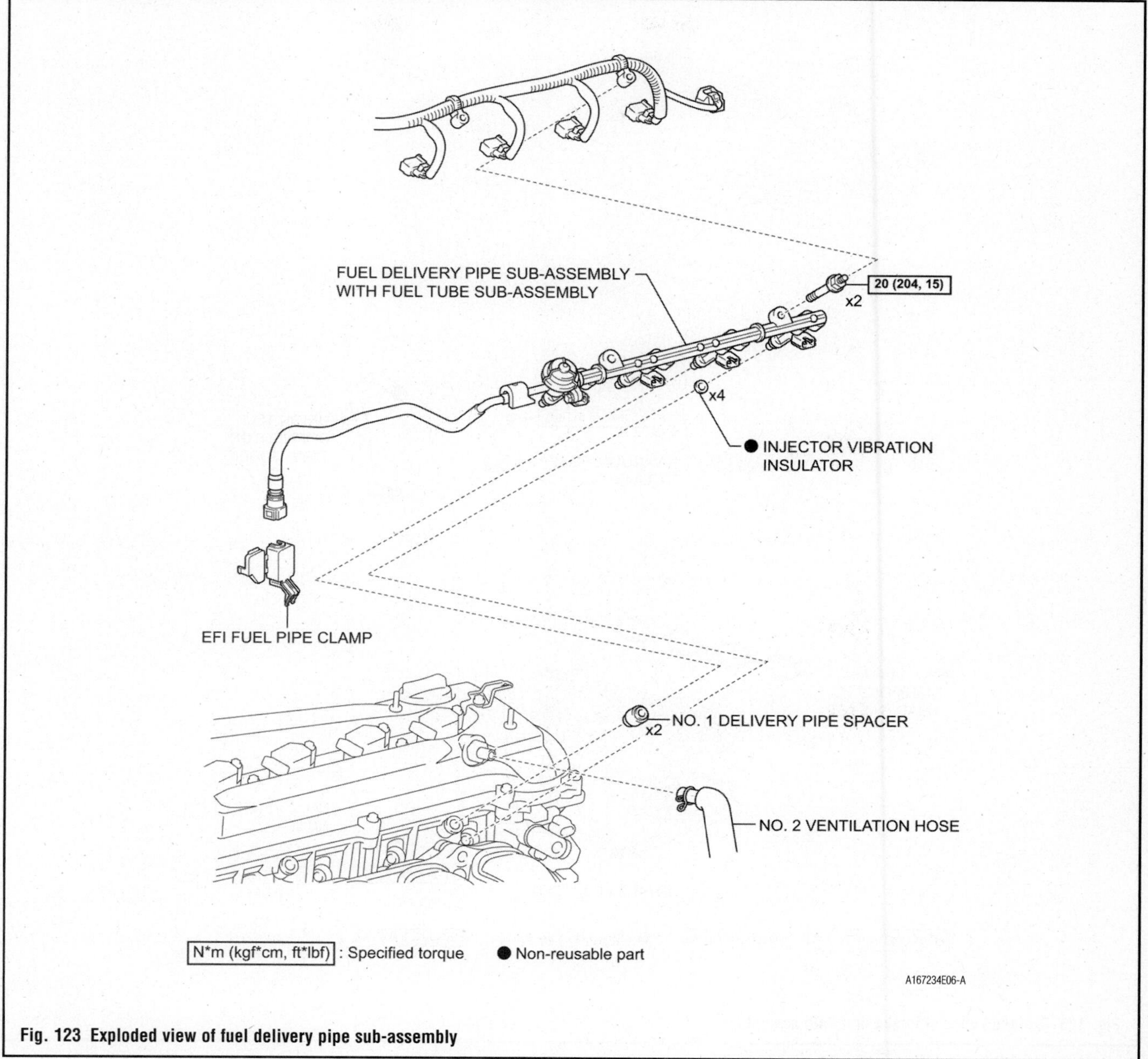

Fig. 123 Exploded view of fuel delivery pipe sub-assembly

f. Connect the union to connector tube hose.

g. Connect the No. 2 ventilation hose to the cylinder head.

17. Install fuel delivery pipe sub-assembly with fuel tube sub-assembly.

a. Install the 4 new injector vibration insulators onto the cylinder head.

b. Install the 2 No. 1 delivery pipe spacers onto the cylinder head.

c. Install the fuel delivery pipe sub-assembly with the 2 bolts. Tighten to 15 ft. lbs. (20 Nm).

d. Engage the clamp and connect the fuel tube sub-assembly.

e. Engage the 4 wire harness clamps and install the wire harness.

f. Connect the heated oxygen sensor connector.

g. Connect the 4 fuel injector connectors and the oil control valve connector.

18. Connect fuel tube sub-assembly.

a. Connect the fuel tube connector and fuel pipe.

b. Engage the claw and install the EFI fuel pipe clamp.

19. Install throttle with motor body assembly.

20. Install air cleaner cap sub-assembly with No. 1 air cleaner hose.

21. Install outer cowl top panel.

22. Install front wiper motor and link.

23. Install cowl top ventilator louver LH.

24. Install cowl top ventilator louver RH.

25. Install hood to cowl top seal.

26. Install front wiper arm and blade assembly LH.

27. Install front wiper arm and blade assembly RH.

28. Install windshield wiper arm cover.

29. Add engine coolant.

30. Inspect for coolant leak.

31. Inspect for fuel leak.

32. Install No. 1 engine cover sub-assembly.

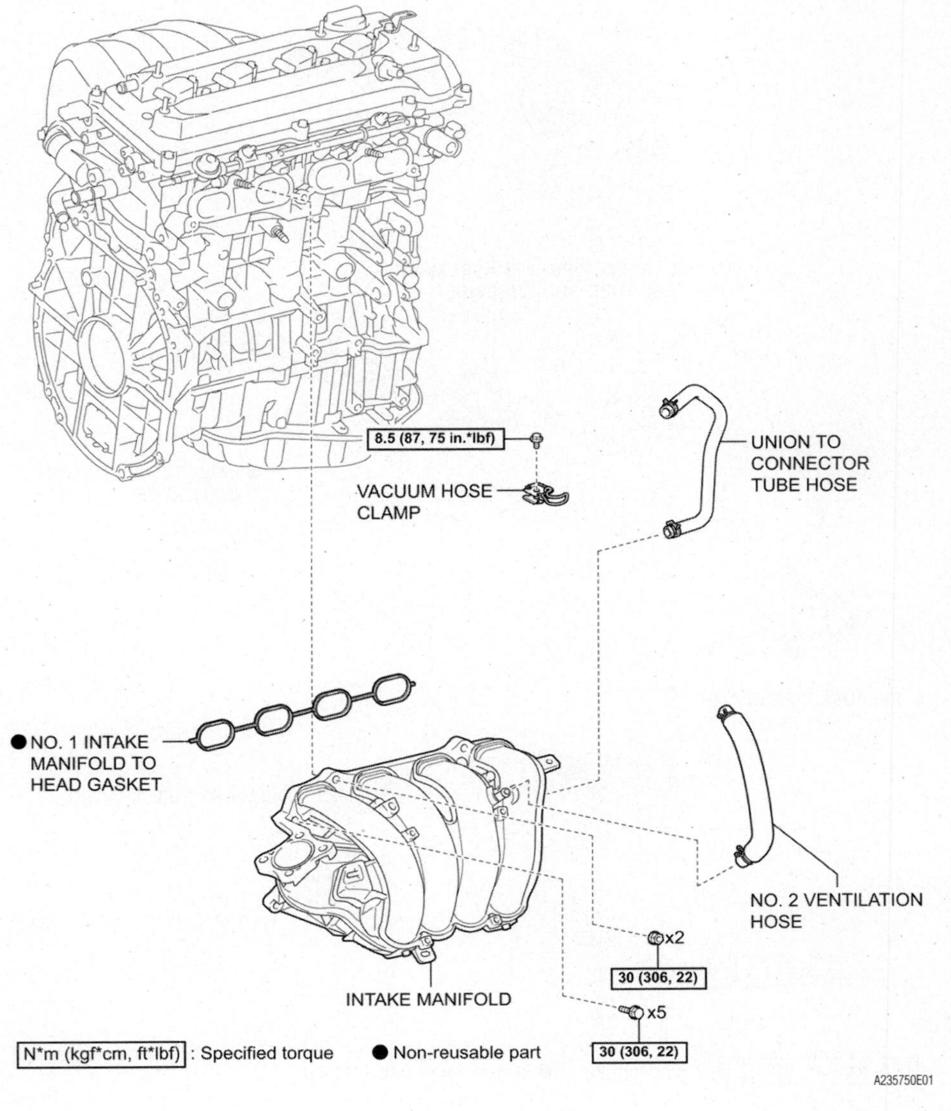

Fig. 124 Exploded view of intake manifold assembly

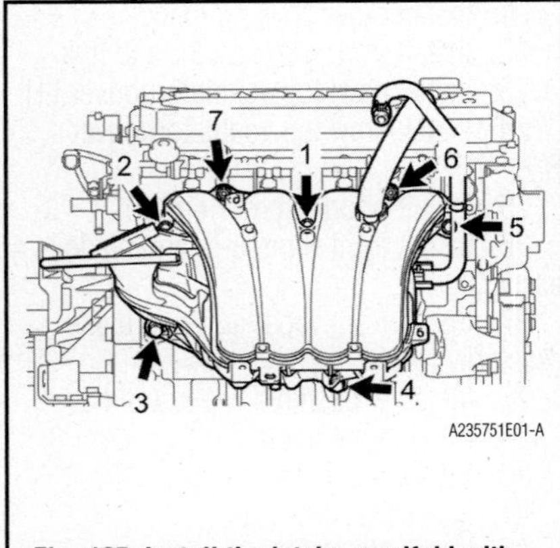

Fig. 125 Install the intake manifold with the 5 bolts and the 2 nuts in sequence

xD

See Figures 126 through 128.

1. Discharge fuel system pressure.
2. Disconnect cable from negative battery terminal.
3. Remove engine under cover LH.
4. Remove engine under cover RH.
5. Drain engine coolant.
6. Remove No. 2 cylinder head cover.
7. Remove air cleaner cap sub-assembly with hose.
8. Separate engine wire harness.
9. Remove harness bracket.
10. Disconnect No. 1 fuel vapor feed hose.
11. Disconnect vacuum hose.
12. Remove fuel pipe clamp.
13. Disconnect fuel tube sub-assembly.
14. Remove fuel delivery pipe sub-assembly.
15. Remove No. 1 delivery pipe spacer.
16. Remove injector vibration insulator.
17. Remove flywheel housing side cover.
18. Remove starter assembly.
19. Remove oil cooler tube.
    a. Using a union nut wrench, separate the inlet No. 1 oil cooler tube while holding the oil cooler tube union with a wrench.
    b. Using a union nut wrench, separate the outlet No. 1 oil cooler tube while holding the oil cooler tube union with a wrench.
    c. Remove the bolt and remove the oil cooler tube clamp.

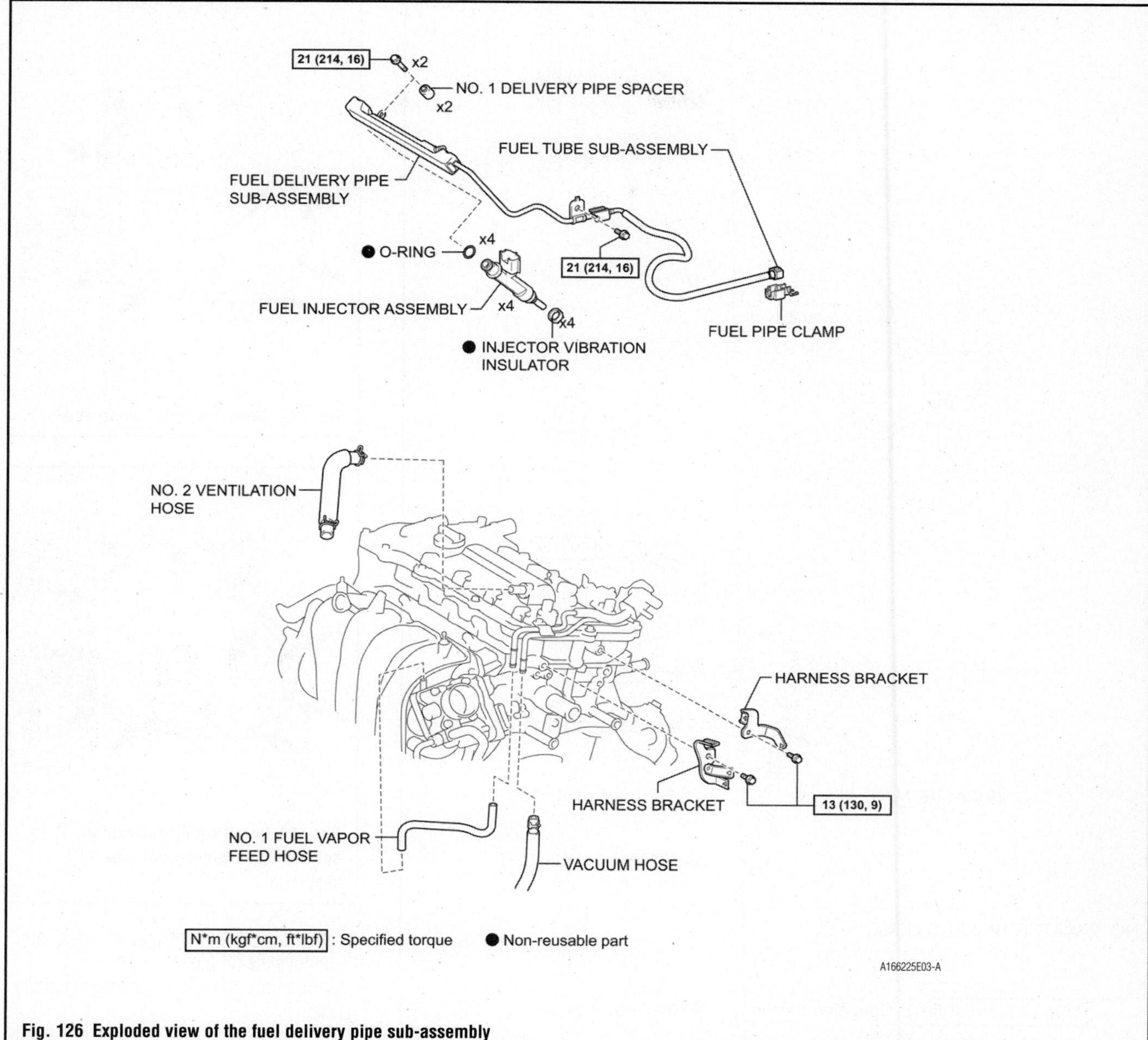

Fig. 126 Exploded view of the fuel delivery pipe sub-assembly

20. Remove transmission oil level gauge sub-assembly.
   a. Remove the transmission oil level dipstick.
   b. Disconnect the breather hose.
   c. Remove the bolt and remove the transmission oil level gauge sub-assembly.
21. Remove engine oil level dipstick.
22. Remove throttle body assembly.
23. Remove intake manifold.
   a. Disconnect the 4 wire harness clamps.
   b. Disconnect the ventilation hose.
   c. Remove the 4 bolts and 2 nuts and remove the intake manifold and intake manifold stay.

To install:

24. Install intake manifold.
   a. Install a new gasket onto the intake manifold.
   b. Install the intake manifold and intake manifold stay with the 4bolts and 2 nuts. Tighten to 21 ft. lbs. (28 Nm).
   c. Connect the ventilation hose.
   d. Connect the 4 wire harness clamps.
25. Install throttle body assembly.
26. Install transmission oil level gauge sub-assembly.
27. Install oil cooler tube.
28. Install engine oil level dipstick.
29. Install starter assembly.
30. Install flywheel housing side cover.
31. Install injector vibration insulator.
32. Install No. 1 delivery pipe spacer.
33. Install fuel delivery pipe sub-assembly.
34. Connect fuel tube sub-assembly.
35. Install fuel pipe clamp.
36. Connect vacuum hose.
37. Connect No. 1 fuel vapor feed hose.
38. Install harness bracket.
39. Install engine wire harness.
40. Install air cleaner cap sub-assembly with hose.
41. Connect cable to negative battery terminal.
42. Add engine coolant.
43. Inspect for coolant leak.
44. Inspect for fuel leak.

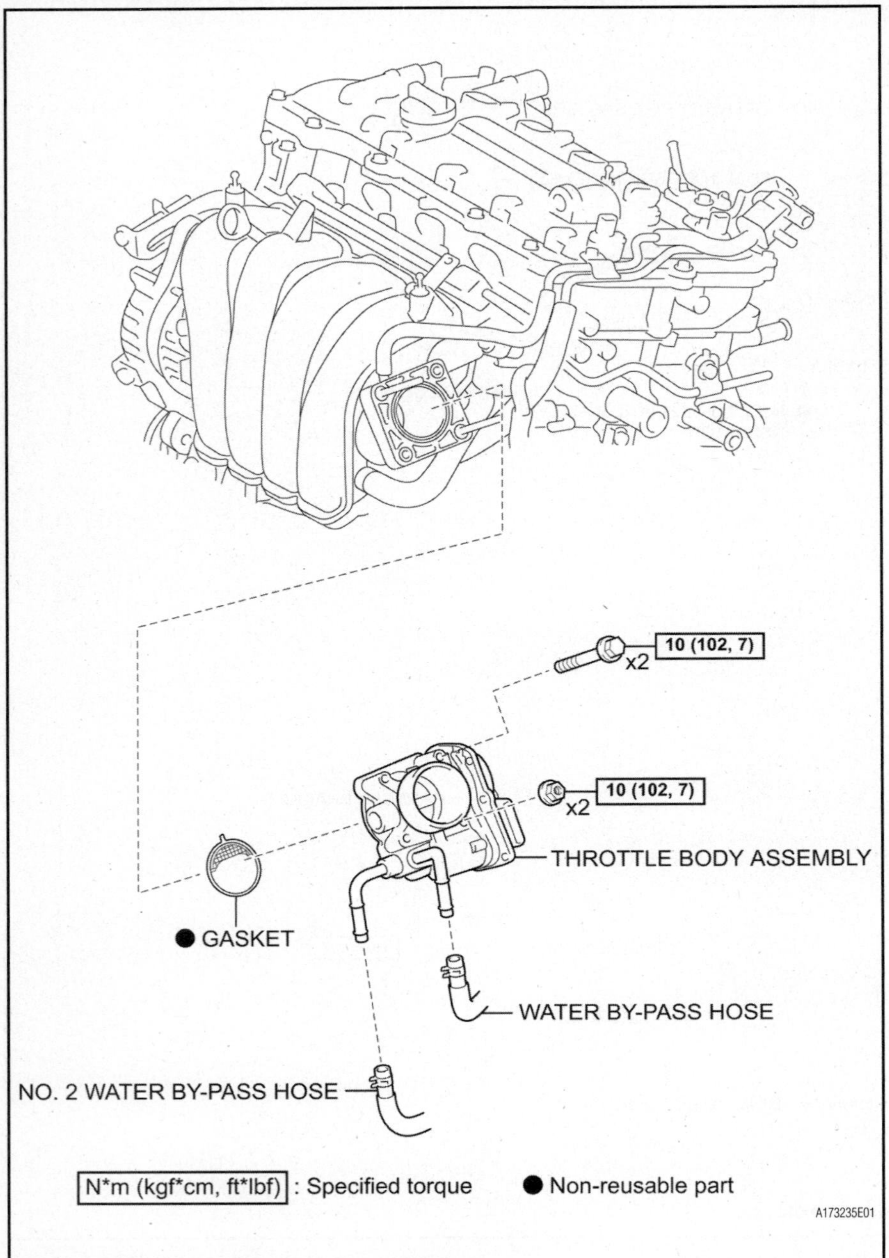

Fig. 127 Exploded view of throttle body assembly

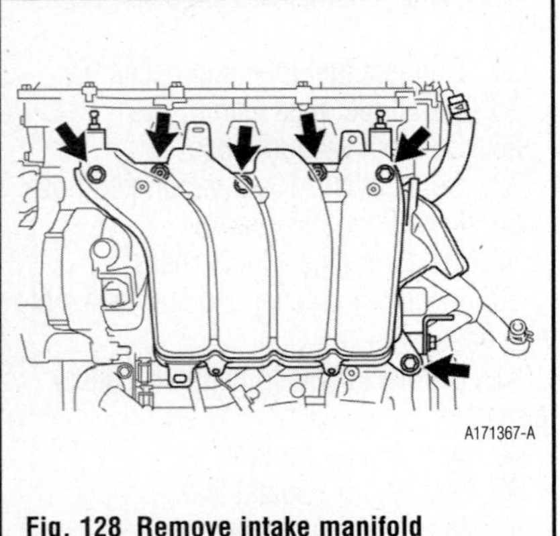

Fig. 128 Remove intake manifold

45. Install No. 2 cylinder head cover.
46. Install engine under cover RH.
47. Install engine under cover LH.

OIL PAN

REMOVAL & INSTALLATION

tC

See Figures 129 and 130.

1. Remove oil pan sub-assembly.
   a. Remove the 11 bolts and 2 nuts.
   b. Insert the blade of an oil pan seal cutter between the oil pan and stiffening

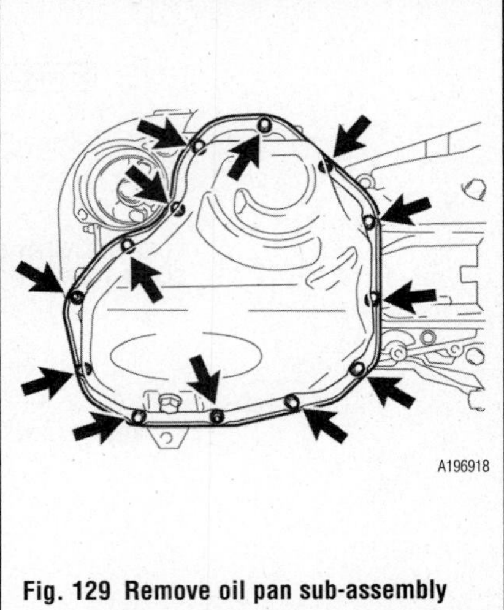

Fig. 129 Remove oil pan sub-assembly

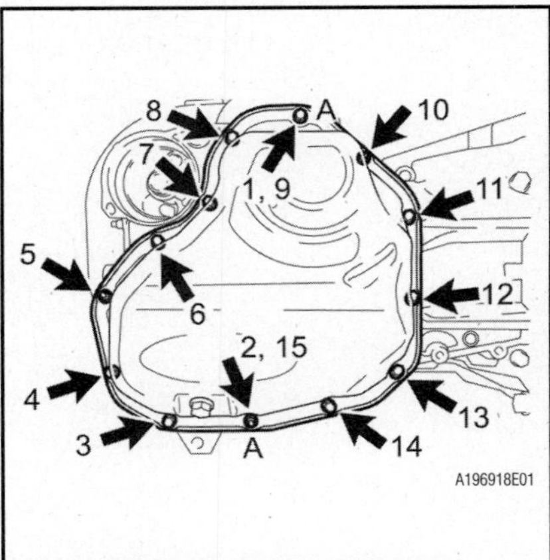

Fig. 130 Install the oil pan with the 11 bolts and 2 nuts in several steps in sequence

crankcase, cut off the applied sealer and remove the oil pan.

2. Remove the 3 bolts, oil strainer and gasket.
3. Remove the 5 bolts and oil pan baffle plate.

To install:

4. Installation is the reverse order of removal.
5. Install the oil pan with the 11 bolts and 2 nuts in several steps in sequence.
6. Tighten the bolts and nuts to 7 ft. lbs. (10 Nm).

xB

See Figure 131.

1. Remove oil pan drain plug.
   a. Remove the oil pan drain plug from the oil pan.
   b. Drain the engine oil.
2. Remove oil pan sub-assembly.
   a. Remove the 12 bolts and 2 nuts.

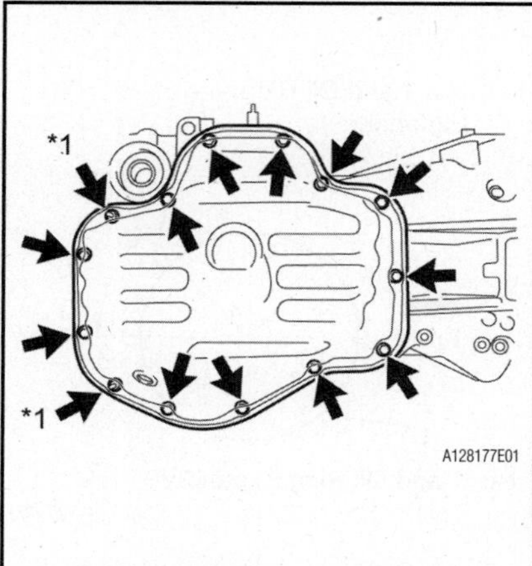

Fig. 131 Remove the 12 bolts and 2 nuts (1)

b. Insert the blade of oil pan seal cutter between the crankcase, chain cover and oil pan, then cut through the applied sealer and remove the oil pan.

➡Be careful not to damage the contact surface of the crankcase, chain cover and oil pan.

To install:

3. Installation is the reverse order of removal.
4. Install oil pan sub-assembly.
   a. Remove any old packing material from the contact surface.
   b. Apply a continuous bead of seal packing to the oil pan mating surface.
   c. Uniformly tighten the 12 bolts and 2 nuts in sequence. Tighten to 80 inch lbs. (9 Nm).
5. Install the oil drain plug with a new gasket. Tighten to 30 ft. lbs. (40 Nm).

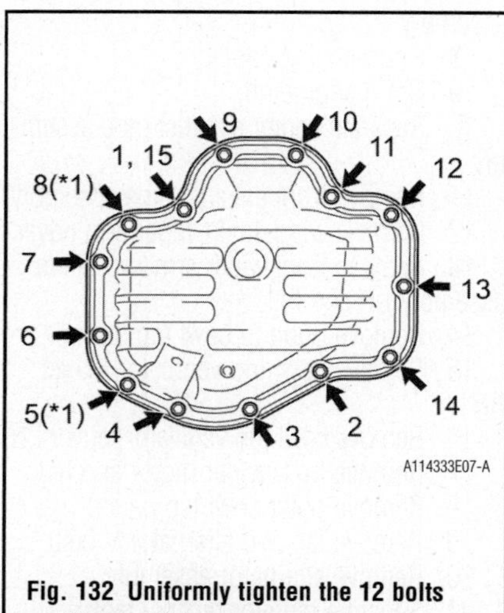

Fig. 132 Uniformly tighten the 12 bolts and 2 nuts (1) in sequence

xD

See Figure 134.

1. Remove the oil pan drain plug and gasket.
2. Remove No. 2 oil pan sub-assembly.
   a. Remove the 10 bolts and 2 nuts.
   b. Install the blade of oil pan seal cutter between the No. 2 oil pan sub-assembly and stiffening crankcase assembly, and cut off the applied sealer and remove the No. 2 oil pan sub-assembly.

➡Do not damage the No. 2 oil pan sub-assembly or stiffening crankcase assembly.

To install:

3. Installation is the reverse order of removal.
4. Install No. 2 oil pan sub-assembly.
   a. Remove any old packing material from the contact surface.
   b. Apply seal packing in a continuous bead.
   c. Install the No. 2 oil pan sub-assembly with the 10 bolts and 2 nuts. Tighten to 7 ft. lbs. (10 Nm).
5. Install a new gasket and the oil pan drain plug. Tighten to 27 ft. lbs. (37 Nm).

OIL PUMP

REMOVAL & INSTALLATION

xB

See Figure 134.

1. Disconnect cable from negative battery terminal.
2. Remove hood.
3. Remove front wheel RH.
4. Remove No. 1 engine under cover.
5. Remove rear engine under cover RH.

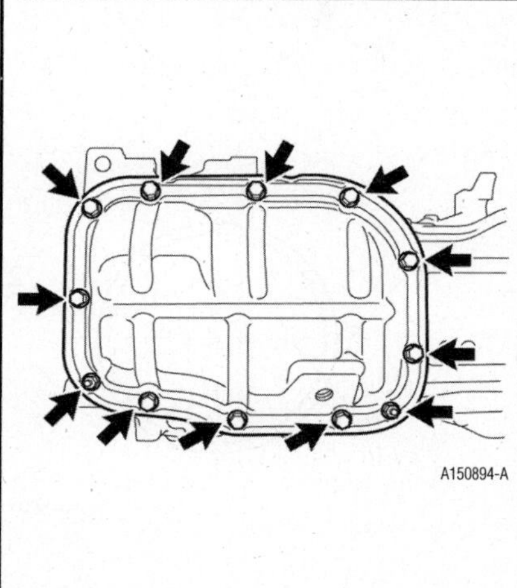

Fig. 133 Remove the 10 bolts and 2 nuts

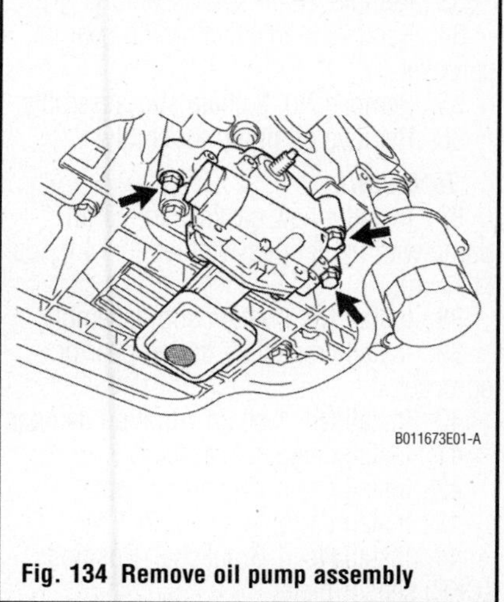

Fig. 134 Remove oil pump assembly

6. Drain engine oil.
7. Separate center exhaust pipe assembly.
8. Remove front exhaust pipe assembly.
9. Remove fan and alternator v belt.
10. Remove radiator reserve tank assembly.
11. Remove No. 1 engine cover sub-assembly.
12. Remove ignition coil assembly.
13. Remove spark plug.
14. Remove cylinder head cover sub-assembly.
15. Remove alternator assembly.
16. Remove engine mounting insulator sub-assembly RH.
17. Remove front engine mounting insulator sub-assembly.
18. Remove idler pulley sub-assembly.
19. Remove oil pan sub-assembly.
20. Set No. 1 cylinder to TDC / compression.
21. Remove crankshaft pulley.
22. Remove No. 1 chain tensioner assembly.
23. Remove transverse engine mounting bracket.
24. Remove V-ribbed belt tensioner assembly.
25. Remove camshaft timing oil control valve assembly.
26. Remove Crankshaft Position (CKP) sensor (for TMC made engine).
27. Remove Crankshaft Position (CKP) sensor (except TMC made engine).
28. Remove timing chain or belt cover sub-assembly.
29. Remove No. 1 Crankshaft Position (CKP) sensor plate.
30. Remove timing chain guide.
31. Remove chain tensioner slipper.
32. Remove No. 1 chain vibration damper.

33. Remove chain sub-assembly.
34. Remove crankshaft timing gear or sprocket.
35. Remove No. 2 chain sub-assembly.
36. Remove oil pump assembly.

To install:

37. Install a new gasket and the oil pump with the 3 bolts. Tighten to 14 ft. lbs. (19 Nm).
38. Install No. 2 chain sub-assembly.
39. Install crankshaft timing gear or sprocket.
40. Install No. 1 chain vibration damper.
41. Install chain sub-assembly.
42. Install chain tensioner slipper.
43. Install timing chain guide.
44. Install No. 1 Crankshaft Position (CKP) sensor plate.
45. Install timing chain or belt cover sub-assembly.
46. Install No. 1 chain tensioner assembly.
47. Install V-ribbed belt tensioner assembly.
48. Install transverse engine mounting bracket.
49. Install crankshaft pulley.
50. Install oil pan sub-assembly.
51. Install Crankshaft Position (CKP) sensor (for TMC made engine).
52. Install Crankshaft Position (CKP) sensor (except TMC made engine).
53. Install camshaft timing oil control valve assembly.
54. Install idler pulley sub-assembly.
55. Install engine mounting insulator sub-assembly RH.
56. Install front engine mounting insulator sub-assembly.
57. Install cylinder head cover sub-assembly.
58. Install spark plug.
59. Install ignition coil assembly.
60. Install radiator reserve tank assembly.
61. Install alternator assembly.
62. Install fan and alternator v belt.
63. Install front exhaust pipe assembly.
64. Install center exhaust pipe assembly.
65. Connect cable to negative battery terminal.
66. Add engine oil.
67. Inspect for engine oil leak.
68. Inspect for exhaust gas leak.
69. Install No. 1 engine cover sub-assembly.
70. Install front wheel RH.
71. Install rear engine under cover RH.
72. Install No. 1 engine under cover.
73. Install hood.
74. Inspect hood.
75. Adjust hood.

PISTON AND RING

POSITIONING

tC

See Figure 135.

xB

See Figure 136.

xD

See Figure 137.

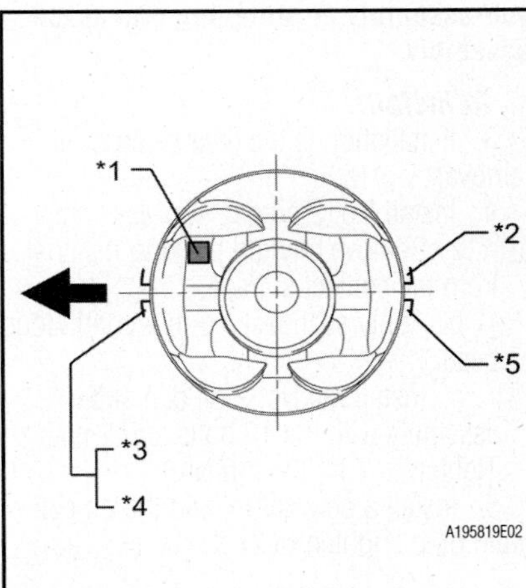

Fig. 135 Proper position for piston ring ends

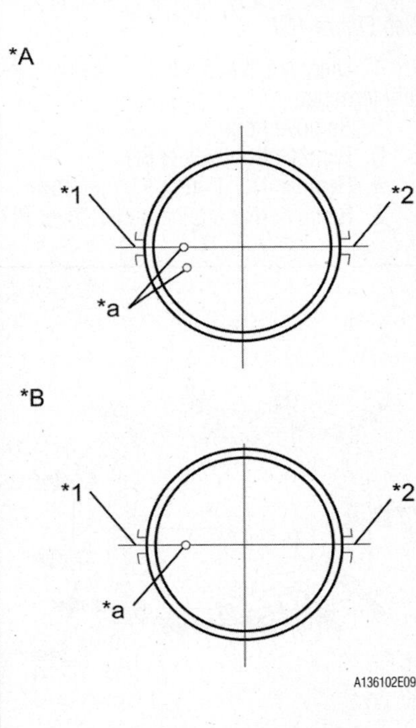

Fig. 136 Proper position for piston ring ends

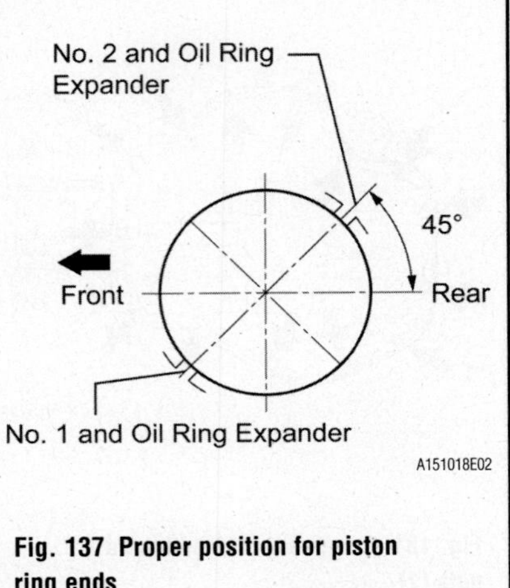

Fig. 137 Proper position for piston ring ends

TIMING CHAIN FRONT COVER

REMOVAL & INSTALLATION

xB

See Figures 138 through 144.

➡Determine if the engine is TMC made or not.

1. Discharge fuel system pressure.
2. Disconnect cable from negative battery terminal.

CAUTION

Wait at least 90 seconds after disconnecting the cable from the negative (-) battery terminal to prevent airbag and seat belt pretensioner activation.

3. Remove hood.
4. Remove front wheel RH.
5. Remove No. 1 engine under cover.
6. Remove rear engine under cover RH.
7. Remove No. 1 engine cover sub-assembly.
8. Drain engine coolant.
9. Drain engine oil.
10. Separate center exhaust pipe assembly.
11. Remove front exhaust pipe assembly.
12. Remove windshield wiper arm cover.
13. Remove front wiper arm and blade assembly LH and RH.
14. Remove hood to cowl top seal.
15. Remove cowl top ventilator louver RH.
16. Remove cowl top ventilator louver LH.
17. Remove front wiper motor and link.
18. Remove outer cowl top panel.
19. Remove fan and alternator V belt.
20. Remove alternator assembly.
21. Separate radiator reserve tank assembly.

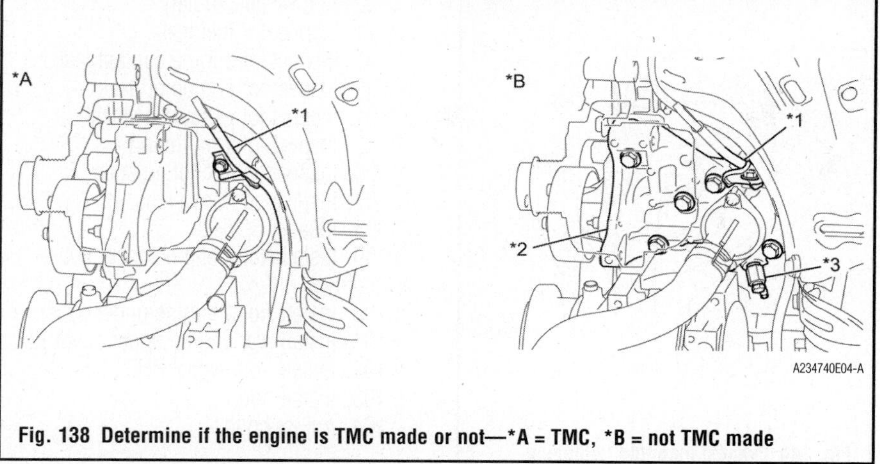

Fig. 138 Determine if the engine is TMC made or not—*A = TMC, *B = not TMC made

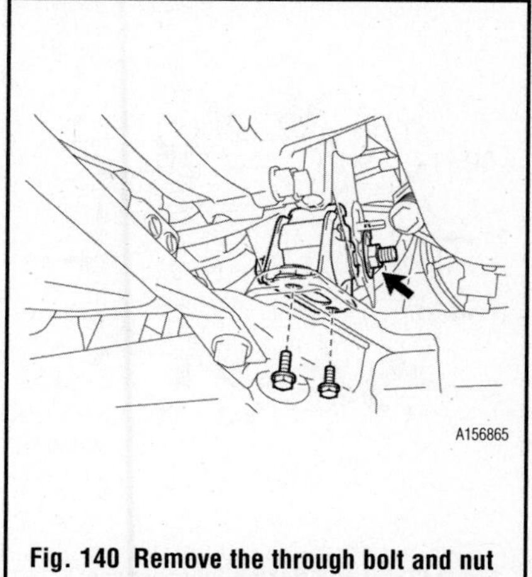

Fig. 140 Remove the through bolt and nut

22. Remove air cleaner cap sub-assembly with No. 1 air cleaner hose.
23. Remove throttle with motor body assembly.
24. Remove ignition coil assembly.
25. Remove spark plug.
26. Remove cylinder head cover sub-assembly.
27. Remove oil dipstick.
28. Remove engine oil level dipstick guide (for TMC made engine).
29. Remove engine oil level dipstick guide (except TMC made engine).
30. Disconnect fuel main tube.
31. Remove fuel delivery pipe sub-assembly.
32. Remove camshaft timing oil control valve assembly.
33. Remove intake manifold.
34. Remove intake manifold insulator.
35. Remove manifold stay.
36. Remove No. 2 manifold stay.
37. Remove No. 1 exhaust manifold heat insulator.
38. Remove exhaust manifold converter sub-assembly.
   a. Disconnect the air-fuel ratio sensor connector.
   b. Remove the 5 nuts, manifold converter and gasket.
39. Remove engine mounting insulator sub-assembly RH.
   a. Remove the bolt and cooler bracket.
   b. Remove the bolt and nut, then remove the No. 2 air conditioning pipe bracket.
   c. Remove the 4 bolts and 2 nuts, then remove the engine mounting insulator RH.
40. Remove front engine mounting insulator sub-assembly.
   a. Remove the through bolt and nut.
   b. Remove the 2 bolts and front engine mounting insulator.
41. Remove idler pulley sub-assembly.
42. Remove oil pan sub-assembly.
43. Set No. 1 cylinder to TDC / compression.
   a. Turn the crankshaft pulley until the groove and the timing mark "0" on the timing chain cover are aligned.
   b. Check that each timing mark on the camshaft timing gear and sprocket is aligned with the timing marks located on the No. 1 and No. 2 bearing caps. If not, turn the crankshaft by 1 revolution (360°) to align the timing marks as above.
44. Remove crankshaft pulley.
45. Remove No. 1 chain tensioner assembly.
46. Remove transverse engine mounting bracket.
47. Remove V-ribbed belt tensioner assembly.
48. Remove Crankshaft Position (CKP) sensor.

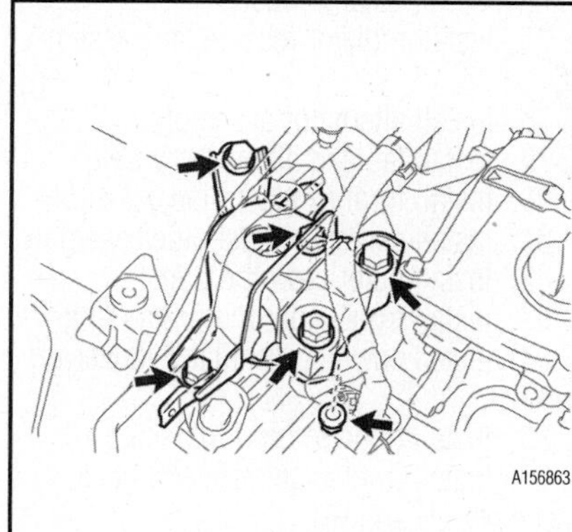

Fig. 139 Remove the 4 bolts and 2 nuts, then remove the engine mounting insulator RH

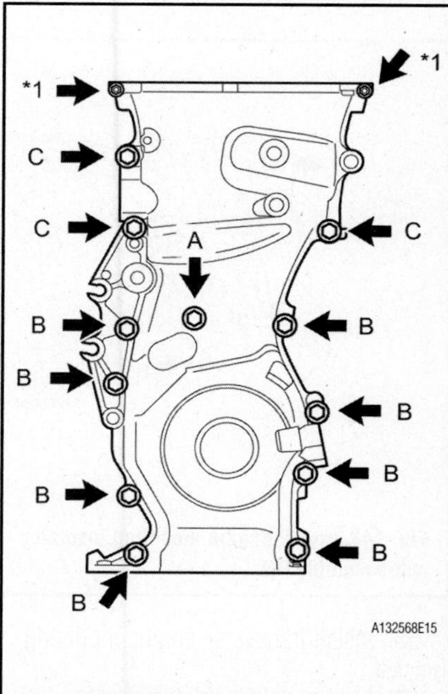

Fig. 141 Install timing chain cover sub-assembly; nuts (1)

49. Remove timing chain cover sub-assembly.

To install:

50. Install timing chain cover sub-assembly.
   a. Tighten Bolt A to 80 inch lbs. (9 Nm).
   b. Tighten Bolts B to 18 ft. lbs. (25 Nm).
   c. Tighten Bolts C to 41 ft. lbs. (55 Nm).
   d. Tighten the nuts to 8 ft. lbs. (11 Nm).
51. Install No. 1 chain tensioner assembly.
52. Install V-ribbed belt tensioner assembly.

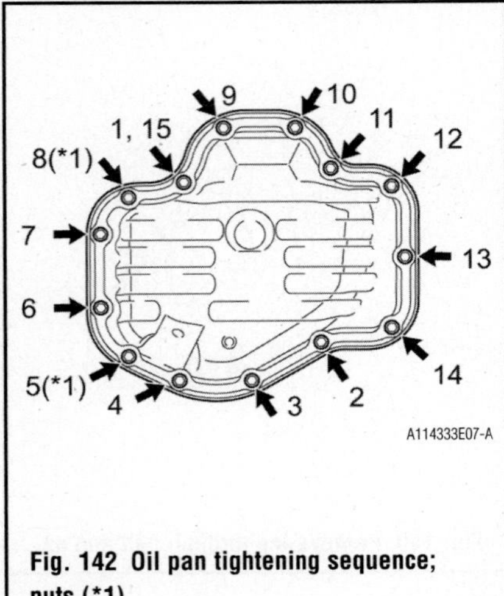

Fig. 142 Oil pan tightening sequence; nuts (*1)

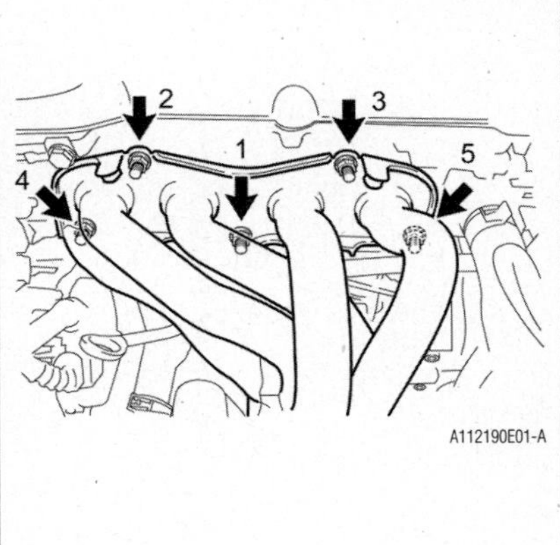

Fig. 144 Exhaust manifold tightening sequence

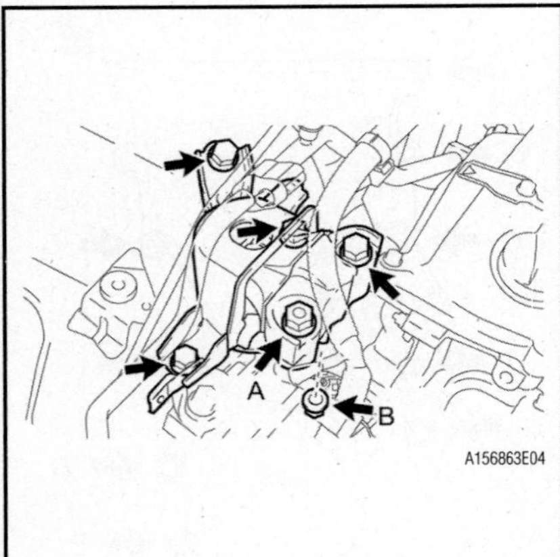

Fig. 143 Install engine mounting insulator sub-assembly RH

53. Install transverse engine mounting bracket.
54. Install crankshaft pulley.
55. Install oil pan sub-assembly. Tighten to 80 inch lbs. (9 Nm).
56. Install Crankshaft Position (CKP) sensor (for TMC made engine).
57. Install Crankshaft Position (CKP) sensor (except TMC made engine).
58. Install idler pulley sub-assembly.
59. Install engine mounting insulator sub-assembly RH.
    a. Tighten the bolts to 70 ft. lbs. (95 Nm).
    b. Tighten Nut A to 70 ft. lbs. (95 Nm).
    c. Tighten Nut B to 38 ft. lbs. (52 Nm).
60. Install front engine mounting insulator sub-assembly.
    a. Tighten the two bottom bolts to 70 ft. lbs. (95 Nm).
    b. Tighten the through bolt and nut to 107 ft. lbs. (145 Nm).
61. Install exhaust manifold converter sub-assembly. Tighten to 27 ft. lbs. (37 Nm) in sequence.
62. Install No. 1 exhaust manifold heat insulator.
63. Install No. 2 manifold stay.
64. Install manifold stay.
65. Install intake manifold insulator.
66. Install intake manifold.
67. Install camshaft timing oil control valve assembly.
68. Install fuel delivery pipe sub-assembly.
69. Connect fuel main tube.
70. Install engine oil level dipstick guide.
71. Install oil dipstick.
72. Install cylinder head cover sub-assembly.
73. Install spark plug.
74. Install ignition coil assembly.
75. Install throttle with motor body assembly.
76. Install air cleaner cap sub-assembly with No. 1 air cleaner hose.
77. Install radiator reserve tank assembly.
78. Install alternator assembly.
79. Install fan and alternator V belt.
80. Install front exhaust pipe assembly.
81. Install center exhaust pipe assembly.
82. Install outer cowl top panel.
83. Install front wiper motor and link.
84. Install cowl top ventilator louver LH and RH.
85. Install hood to cowl top seal.
86. Install front wiper arm and blade assembly LH and RH.
87. Install windshield wiper arm cover.
88. Connect cable to negative battery terminal.
89. Add engine oil.
90. Add engine coolant.
91. Inspect for fuel leak.
92. Inspect for engine coolant leak.
93. Inspect for engine oil leak.
94. Inspect for exhaust gas leak.
95. Inspect ignition timing.
96. Inspect engine idling speed.
97. Inspect compression.
98. Inspect CO/HC.
99. Install No. 1 engine cover sub-assembly.
100. Install No. 1 engine under cover.
101. Install rear engine under cover RH.
102. Install front wheel RH.
103. Install hood.
104. Adjust hood.

xD

See Figures 145 through 149.

➡The manufacturer does not provide a specific Removal and Installation procedure for this component. Refer to the following when servicing this component. This procedure is done with the engine assembly removed from the vehicle.

1. Remove timing chain or belt cover sub-assembly.
    a. Remove the 3 bolts and remove the engine mounting bracket RH.
    b. Remove the 4 bolts and remove the oil filter bracket.
    c. Remove the 2 O-rings from the timing chain cover sub-assembly.
    d. Remove the 19 bolts.
    e. Remove the timing chain or belt cover sub-assembly by prying between the timing chain or belt cover sub-assembly and cylinder head or cylinder block with a screwdriver wrapped in protective tape.

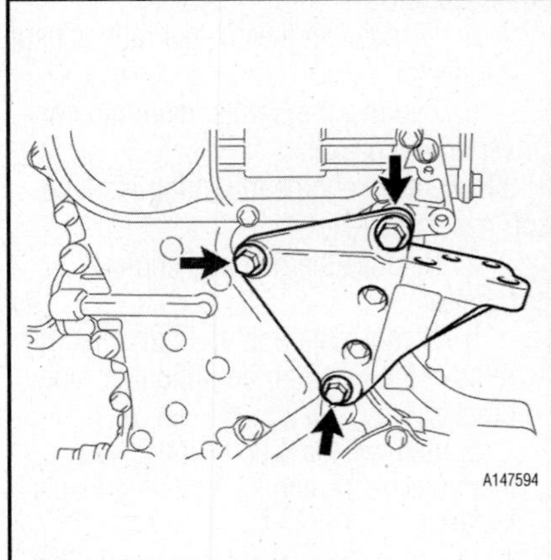

Fig. 145 Remove the 3 bolts and remove the engine mounting bracket RH

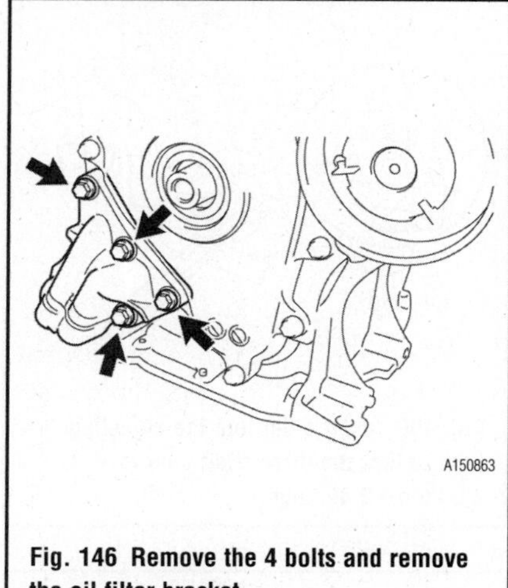

Fig. 146 Remove the 4 bolts and remove the oil filter bracket

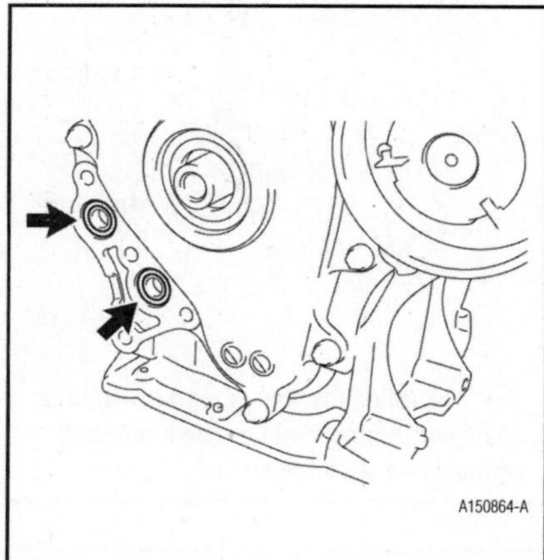

Fig. 147 Remove the 2 O-rings from the timing chain cover sub-assembly

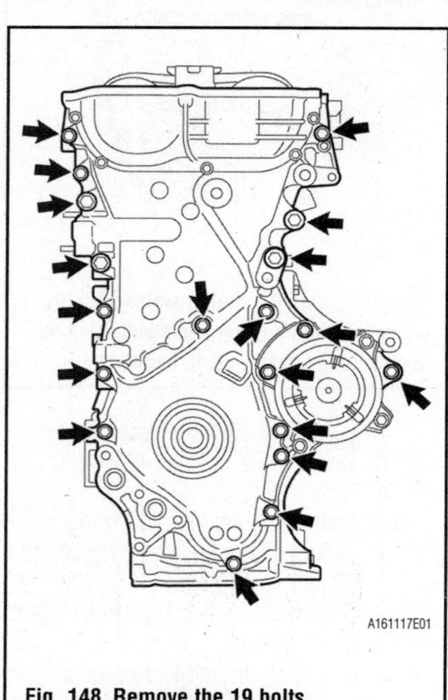

Fig. 148 Remove the 19 bolts

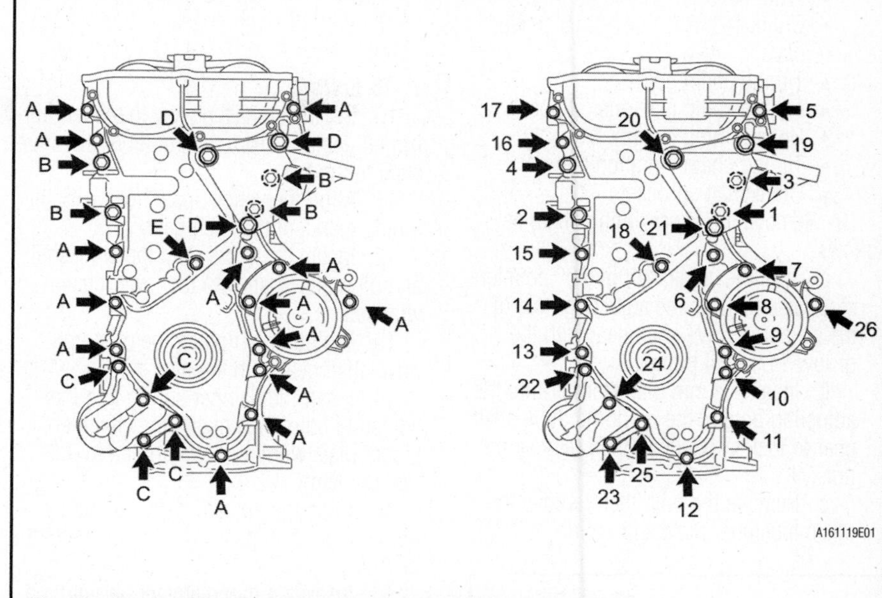

Fig. 149 Install the timing chain cover with the 19 bolts in sequence

To install:

2. Installation is the reverse order of removal.
3. Install the timing chain cover with the 19 bolts in sequence.
   a. Tighten Bolts A and C to 19 ft. lbs. (26 Nm).
   b. Tighten Bolts B and D to 38 ft. lbs. (51 nm).
   c. Tighten Bolt E to 7 ft. lbs. (10 Nm)

TIMING CHAIN & SPROCKETS

REMOVAL & INSTALLATION

tC and xB Models

See Figures 150 through 157.

1. Drain the engine oil.
2. Drain the cooling system.
3. Support the engine with a suitable jack.
4. Remove or disconnect the following:
   - Negative battery cable
   - Hood
   - Right-hand front wheel
   - Engine undercover
   - Front fender apron
   - Engine appearance cover
   - Front exhaust pipe
   - Accessory drive belt
   - Alternator
   - Power steering pump and reservoir
   - Right-hand engine mount
   - Ignition coil
   - Cylinder head cover
   - Accessory drive belt tensioner

➡Lift the engine with a suitable jack to gain access to the tensioner mounting bolts.

   - Crankshaft Position (CKP) sensor
   - Oil pan
5. Set the No. 1 cylinder to TDC.
6. Using Special Tool 09213-54015, remove the crankshaft pulley bolt.
7. Remove the crankshaft pulley. Use Special Tool 09950-50013 and the pulley bolt if necessary.
8. Remove or disconnect the following:
   - Timing chain tensioner

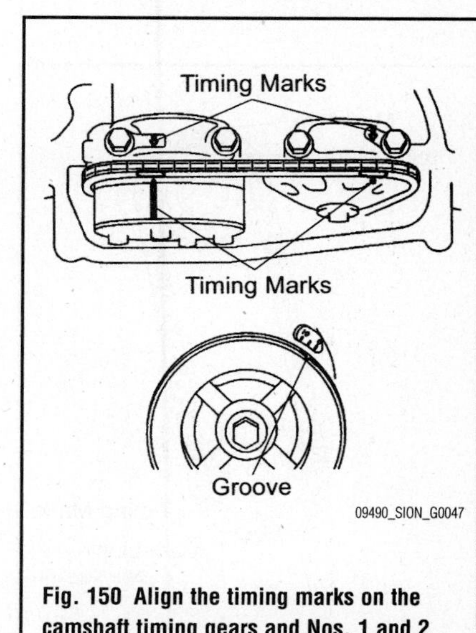

Fig. 150 Align the timing marks on the camshaft timing gears and Nos. 1 and 2 camshaft bearing caps to ensure the No. 1 cylinder is at TDC compression.

- Front cover
- Crankshaft Position (CKP) sensor plate
- Timing chain guide
- Timing chain tensioner slipper
- Upper timing chain
- Timing chain vibration damper
- Crankshaft sprocket

9. Remove the lower timing chain as follows:
   a. Turn the crankshaft by 90° counter-clockwise to align the adjusting hole of the oil pump drive shaft gear with the groove of the oil pump.
   b. Insert a 4 mm diameter bar into the adjusting hole of the oil pump drive shaft gear to lock in position, then remove the nut.
   c. Remove the bolt, then remove the chain tensioner plate and spring.

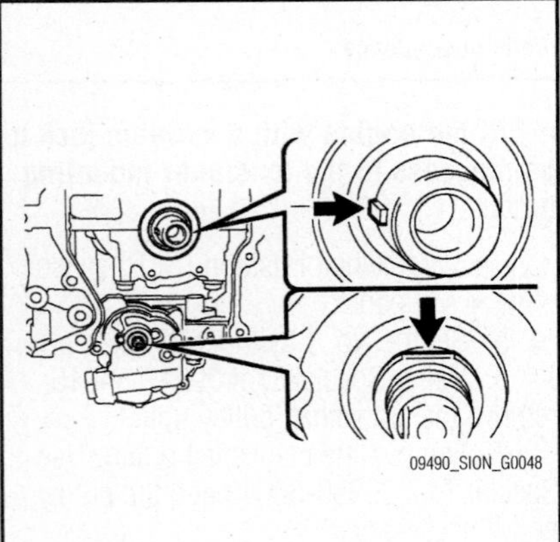

Fig. 151 Set the crankshaft key and drive shaft cutout—2.4L Engine

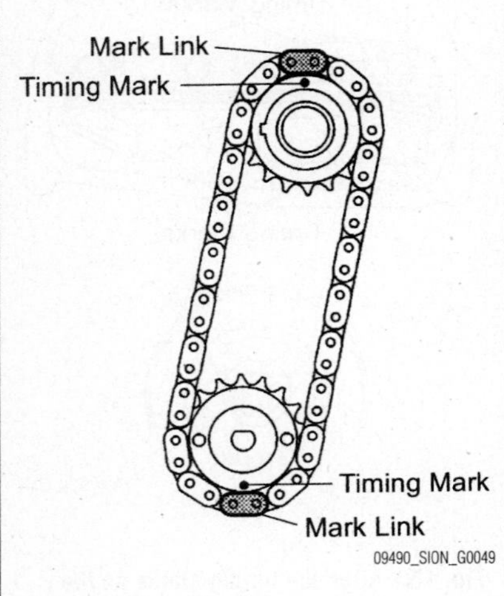

Fig. 152 Align the marked links with the timing gear marks—2.4L Engine

   d. Remove the oil pump drive gear, oil pump drive shaft gear and lower timing chain.

To install:

10. Set the crankshaft key in the left horizontal position. Turn the cutout of the drive shaft to the top.
11. Align the yellow mark links with the timing marks of the each gear.
12. Install the gears onto the crankshaft and oil pump shaft with the lower chain wrapped.
13. Temporarily tighten the oil pump drive shaft gear with the nut.
14. Insert the damper spring into the adjusting hole, then install the chain tensioner plate with the bolt. Tighten to 9 ft. lbs. (12 Nm).
15. Align the adjusting hole of the oil pump drive shaft gear with the groove of the oil pump.
16. Insert a 4 mm diameter bar into the adjusting hole of the oil pump drive shaft gear to lock in position, then tighten the nut to 22 ft. lbs. (30 Nm).
17. Turn the crankshaft clockwise by 90° to position the crankshaft key upward.
18. Install the crankshaft sprocket.
19. Turn the camshafts with a wrench on the hexagonal lobe to align the timing marks of the camshaft timing gear with each timing mark located on the No. 1 and No. 2 bearing caps.
20. Using the crankshaft pulley bolt, turn the crankshaft to position the key on the crankshaft upward.
21. Install the upper timing chain onto the crankshaft timing gear with the gold or orange mark link aligned with the timing mark on the crankshaft sprocket.
22. Using Special Tool 09309-37010, tap in the crankshaft timing gear.
23. Align the gold or yellow mark links with each timing mark located on the camshaft timing gears, then install the upper timing chain.
24. Install or connect the following:
    - Timing chain tensioner slipper. Tighten to 14 ft. lbs. (19 Nm).
    - Timing chain guide. Tighten to 80 inch lbs. (9 Nm).
    - Crankshaft Position (CKP) sensor plate
25. Remove any old sealant from the front cover.
26. Apply a continuous bead of sealant to the front cover contact surfaces.
27. Install the front cover. Tighten the bolts as follows:
    a. Bolt A to 80 inch lbs. (9 Nm).
    b. Bolts B to 18 ft. lbs. (25 Nm).

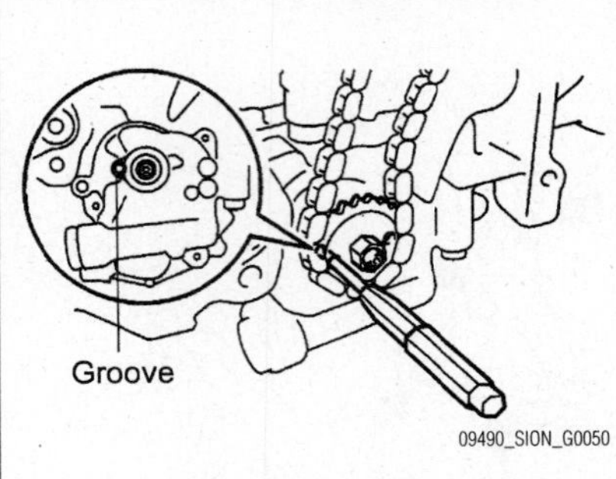

Fig. 153 Insert a bar into the adjusting hole to lock the drive shaft gear in position—2.4L Engine

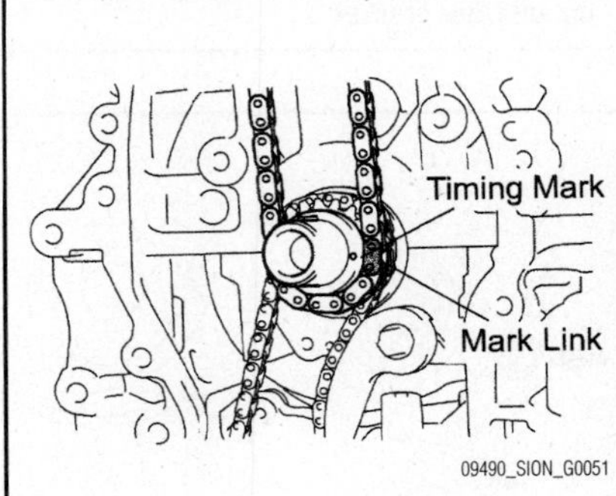

Fig. 154 Align the timing marks installing the upper timing chain on the crankshaft sprocket—2.4L Engine

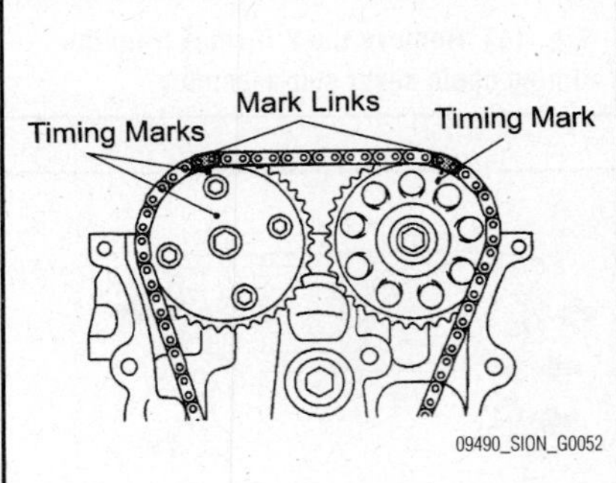

Fig. 155 Align the timing marks with the camshaft timing gears when installing the upper timing chain—2.4L Engine

    c. Bolts C to 41 ft. lbs. (55 Nm).
    d. Bolt D to 32 ft. lbs. (43 Nm).
    e. Nuts to 8 ft. lbs. (11 Nm).
28. Install or connect the following:
    - Timing chain tensioner. Tighten to 80 inch lbs. (9 Nm).
    - Crankshaft pulley. Tighten to 133 ft. lbs. (180 Nm).
    - Accessory drive belt tensioner. Tighten to 44 ft. lbs. (60 Nm).

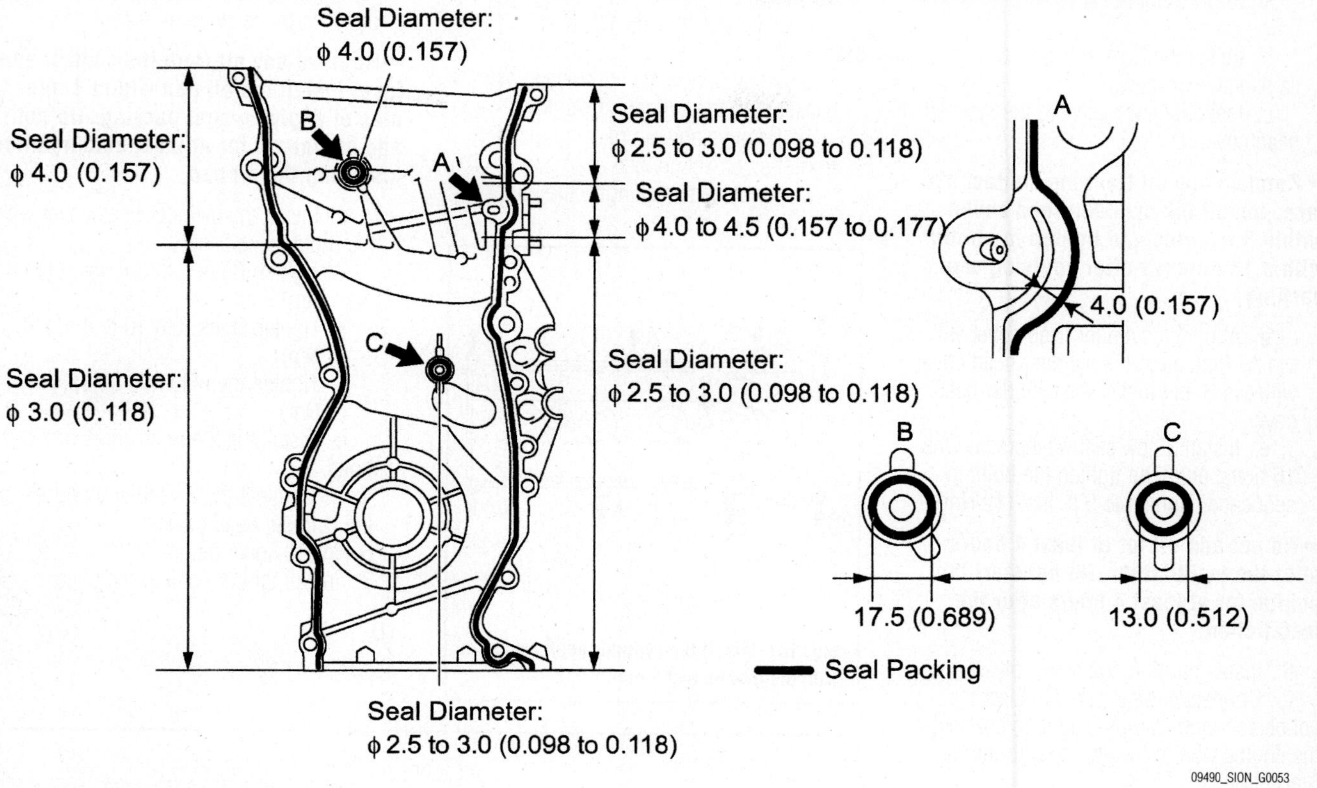

Seal Diameter:
ϕ 4.0 (0.157)
Seal Diameter:
ϕ 4.0 (0.157)
B
A
Seal Diameter:
ϕ 2.5 to 3.0 (0.098 to 0.118)
Seal Diameter:
ϕ 4.0 to 4.5 (0.157 to 0.177)
A
4.0 (0.157)
C
Seal Diameter:
ϕ 3.0 (0.118)
Seal Diameter:
ϕ 2.5 to 3.0 (0.098 to 0.118)
B
C
17.5 (0.689)
13.0 (0.512)
Seal Packing
Seal Diameter:
ϕ 2.5 to 3.0 (0.098 to 0.118)
09490_SION_G0053

Fig. 156 Apply sealant to the front cover contact surfaces—2.4L Engine

- Cylinder head cover
- Ignition coil
- Right-hand engine mount
- Power steering pump and reservoir
- Alternator

Nut
D
Nut
C
C
C
C
A
B
B
B
B
B
B
B
B
B
09490_SION_G0054

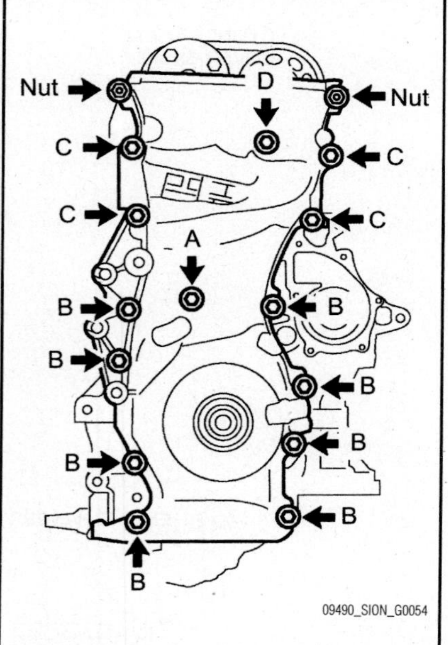

Fig. 157 Front cover bolt identification—2.4L Engine

- Accessory drive belts
- Front exhaust pipe
- Engine undercover
- Right-hand front fender apron
- Front wheel
- Negative battery cable

29. Refill the engine with oil to the correct level.
30. Refill the cooling system to the correct level.
31. Start the engine and check for leaks.

VALVE COVERS

REMOVAL & INSTALLATION

tC

See Figures 158 and 159.

1. Disconnect the connectors and clamps securing the engine wire to the engine, remove the bracket bolts and disconnect the engine wire from the engine.
2. Remove ignition coil assembly.
3. Remove cylinder head cover sub-assembly.
4. Remove the 16 bolts, 3 seal washers, cylinder head cover and gasket.

To install:

5. Install cylinder head cover sub-assembly.

A195876

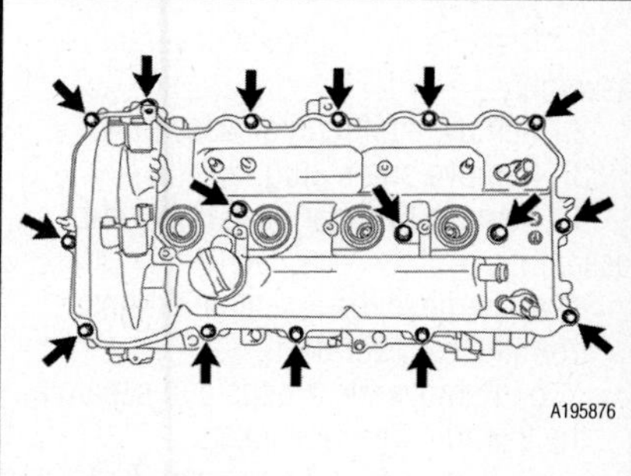

Fig. 158 Remove cylinder head cover sub-assembly

A
10
11
4, 12
13
9
14
2, 20
1, 19
3, 21
15
8
B
7
16
6
5, 18
17
A195876E01

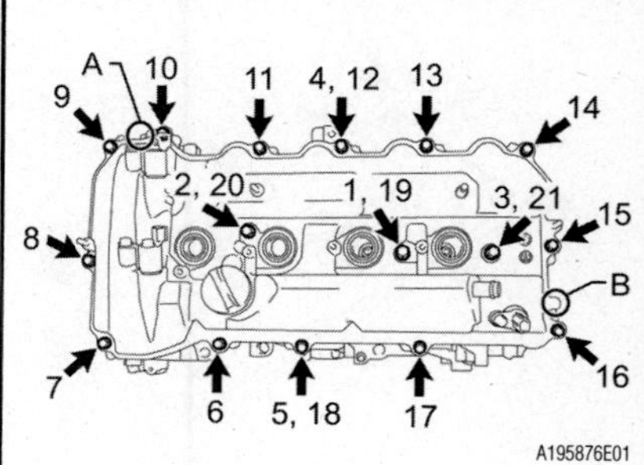

Fig. 159 Install 3 new seal washers and the 16 bolts, and then tighten the bolts in sequence

a. Apply a light coat of engine oil to 3 new gaskets.

b. Install the 3 gaskets to the camshaft bearing caps.

c. Install a new gasket to the cylinder head cover.

➡Remove any oil from the contact surface. Install the cylinder head cover within 3 minutes and tighten the bolts within 15 minutes after applying seal packing.

d. Align the cylinder head cover with pin A. Then align the cylinder head cover with pin B and install the cylinder head cover.

e. Install 3 new seal washers and the 16 bolts, and then tighten the bolts in sequence. Tighten to 9 ft. lbs. (12 Nm).

➡Do not add oil for at least 4 hours after the installation. Do not start the engine for at least 4 hours after the installation.

6. Install ignition coil assembly.
7. Connect engine wire. Connect the connectors and clamps, and then connect the engine wire to the engine with the bracket bolts.

xB

See Figures 160 and 161.

1. Remove No. 1 engine cover sub-assembly.
2. Remove ignition coil assembly.
3. Remove spark plug.
4. Remove cylinder head cover sub-assembly.

a. Remove the 2 ventilation hoses from the cylinder head.

b. Remove the 2 bolts and separate the 2 wire harness brackets.

c. Remove the 8 bolts and 2nuts, and then remove the cylinder head cover and gasket.

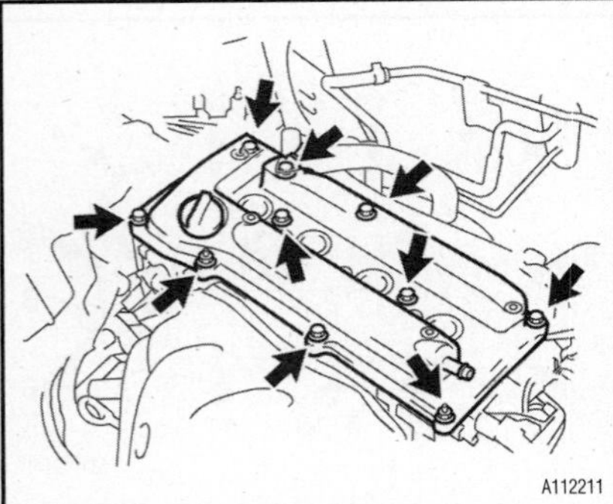

Fig. 160 Remove the 8 bolts and 2nuts, and then remove the cylinder head cover and gasket

To install:

5. Install cylinder head cover sub-assembly.

a. Install a new cylinder head cover gasket onto the cylinder head cover.

b. Remove any old packing material from the contact surface.

c. Apply seal packing to the 2 locations where the timing chain cover attaches to the cylinder head.

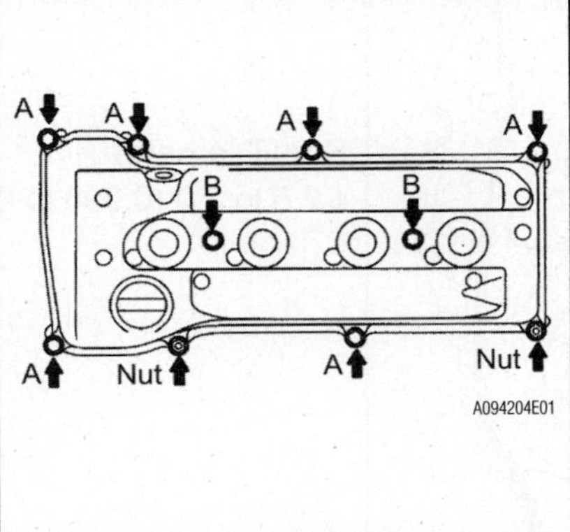

Fig. 161 Install the cylinder head cover with the 8 bolts and 2 nuts

➡Remove any oil from the contact surface. Install the oil pan within 3 minutes of applying seal packing. Do not add engine oil for at least 2 hours after installing the oil pan.

d. Install the cylinder head cover with the 8 bolts and 2 nuts.

- Tighten Bolts A to 8 ft. lbs. (11 Nm).
- Tighten Bolts B to 10 ft. lbs. (14 Nm)
- Tighten the nuts to 8 ft. lbs. (11 Nm)

e. Install the 2 engine wires with the 2 bolts.

f. Connect the 2 ventilation hoses to the cylinder head cover.

6. Install spark plugs.
7. Install ignition coil assemblies.

xD

See Figure 162.

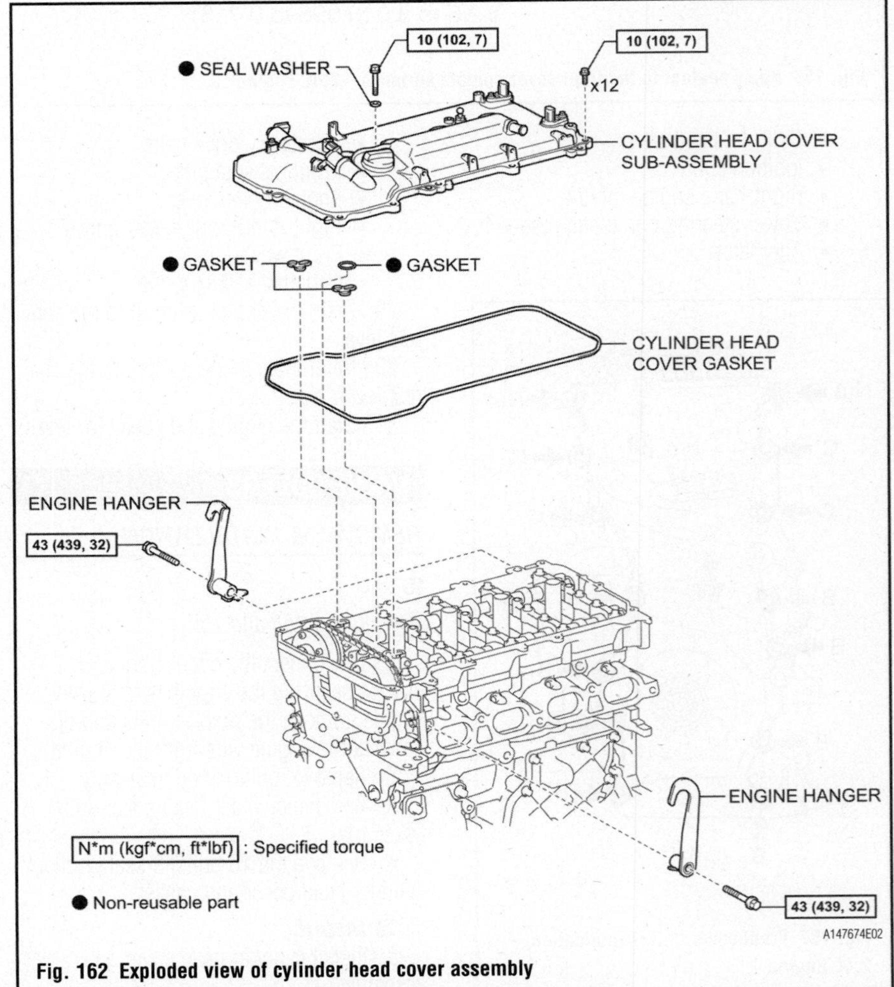

Fig. 162 Exploded view of cylinder head cover assembly

➡**The manufacturer does not provide a specific Removal and Installation procedure for this component. Refer to the graphic(s) when servicing this component.**

VALVE LASH

ADJUSTMENT

All three engines use hydraulic valve lash adjusters. No adjustment is necessary.

ENGINE PERFORMANCE & EMISSION CONTROLS

CAMSHAFT POSITION (CMP) SENSOR

LOCATION

tC

See Figure 163.

xB

See Figure 164.

xD

See Figure 165.

REMOVAL & INSTALLATION

tC

1. Remove Camshaft Position (CMP) sensor (for Exhaust Side).
   a. Disconnect the sensor connector.
   b. Remove the bolt and sensor.

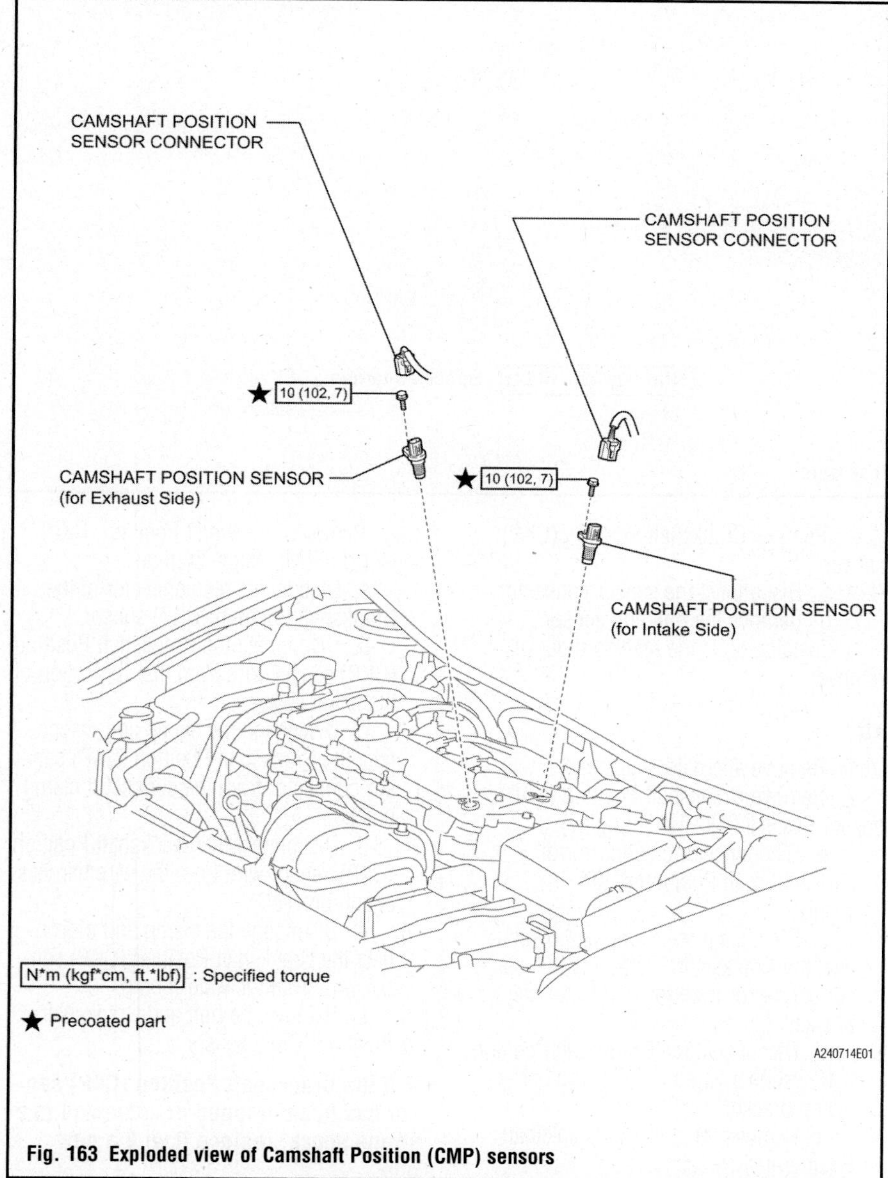

Fig. 163 Exploded view of Camshaft Position (CMP) sensors

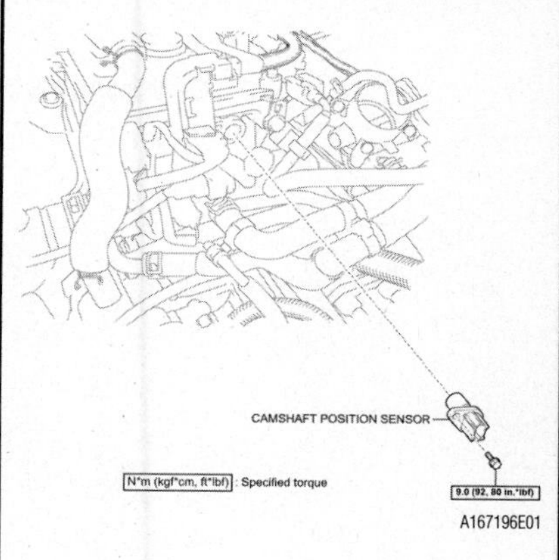

Fig. 164 Exploded view showing Camshaft Position (CMP) sensor location

2. Remove Camshaft Position (CMP) sensor (for Intake Side).
   a. Disconnect the sensor connector.
   b. Remove the bolt and sensor.
3. Installation is the reverse order of removal.

xB

1. Remove No. 1 engine cover sub-assembly
2. Remove air cleaner cap sub-assembly with No. 1 air cleaner hose
3. Remove Camshaft Position (CMP) sensor
   a. Disconnect the sensor connector.
   b. Remove the bolt and sensor.
4. Installation is the reverse order of removal.

xD

1. Remove No. 2 cylinder head cover.
2. Remove Camshaft Position (CMP) sensor.
   a. Disconnect the duty vacuum switching valve connector and 3 engine wire harness clamps (for exhaust camshaft side).
   b. Remove the bolt and remove the Camshaft Position (CMP) sensor (for exhaust camshaft side).
   c. Remove the bolt and remove the Camshaft Position (CMP) sensor (for intake camshaft side).

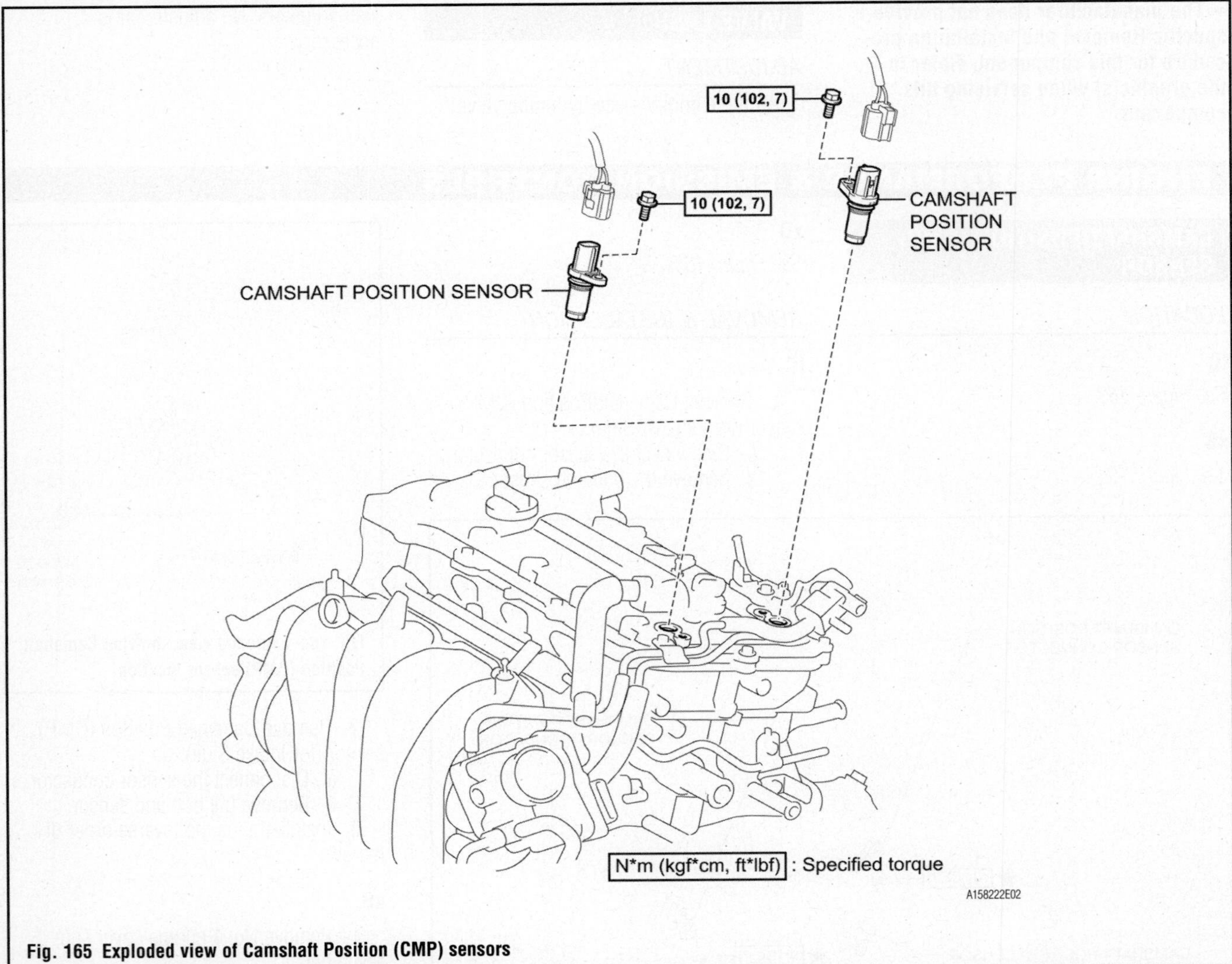

Fig. 165 Exploded view of Camshaft Position (CMP) sensors

3. Installation is the reverse order of removal.

CRANKSHAFT POSITION (CKP) SENSOR

LOCATION

tC

See Figure 166.

xB

See Figure 167.

xD

See Figure 168.

REMOVAL & INSTALLATION

tC

See Figure 169.

1. Remove rear engine under cover RH.
2. Remove Crankshaft Position (CKP) sensor.
   a. Disconnect the sensor connector.
   b. Remove the bolt and sensor.
3. Installation is the reverse order of removal.

xB

1. Remove alternator assembly
2. Remove Crankshaft Position (CKP) sensor (except TMC Made Engine)
   a. Disconnect the connector of the Crankshaft Position (CKP) sensor.
   b. Disengage the clamp and disconnect the Crankshaft Position (CKP) sensor connector from the connector clamp bracket.
   c. Disconnect the Crankshaft Position (CKP) sensor wire from the wire harness clamp bracket.
   d. Remove the bolt and Crankshaft Position (CKP) sensor.
3. Remove Crankshaft Position (CKP) sensor (for TMC Made Engine)
   a. Disconnect the connector of the Crankshaft Position (CKP) sensor.
   b. Disconnect the Crankshaft Position (CKP) sensor wire from the rib portion of the chain cover.
   c. Disengage the clamp and disconnect the Crankshaft Position (CKP) sensor connector from the connector clamp bracket.
   d. Disconnect the Crankshaft Position (CKP) sensor wire from the wire harness clamp bracket.
   e. Disengage the clamp and disconnect the Crankshaft Position (CKP) sensor wire from the water pump.
   f. Remove the bolt and Crankshaft Position (CKP) sensor.

➡If the Crankshaft Position (CKP) sensor has been dropped or subjected to a strong shock, replace it with a new one.

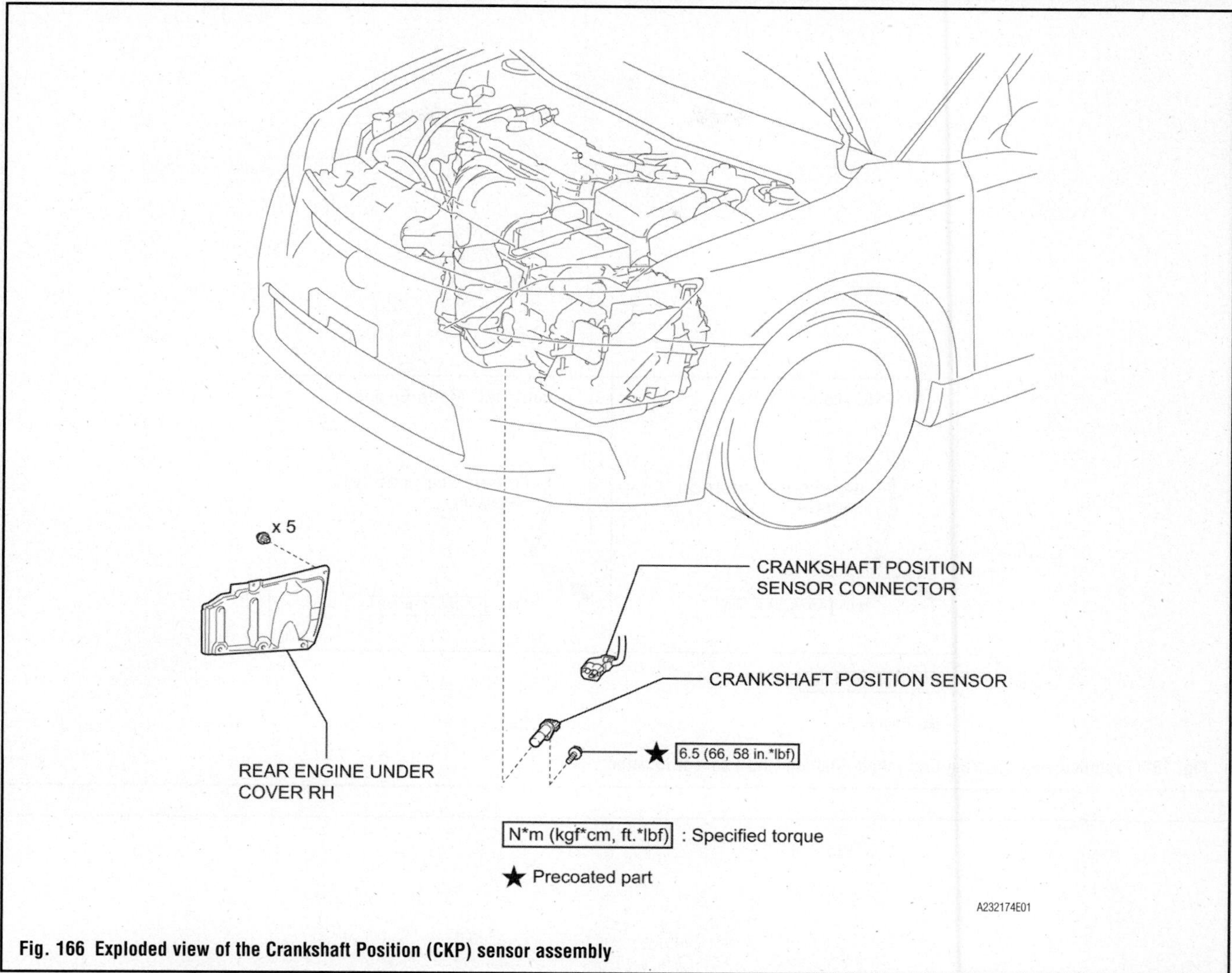

Fig. 166 Exploded view of the Crankshaft Position (CKP) sensor assembly

4. Installation is the reverse order of removal.

xD

1. Remove engine under cover RH.
2. Remove Crankshaft Position (CKP) sensor.
   a. Disconnect the Crankshaft Position (CKP) sensor connector.
   b. Remove the bolt and remove the Crankshaft Position (CKP) sensor.
3. Installation is the reverse order of removal.

ELECTRONIC CONTROL MODULE (ECM)

LOCATION

tC

See Figure 170.

xB

See Figure 171.

xD

See Figure 172.

REMOVAL & INSTALLATION

tC

See Figure 173.

1. Disconnect cable from negative battery terminal.
2. Remove air cleaner cap sub-assembly.
   a. Disconnect the Mass Air Flow (MAF) meter connector.
   b. Disconnect the purge VSV connector.
   c. Detach the 2 wire harness clamps.
   d. Disconnect the 2 purge line hoses.
   e. Disconnect the No. 2 ventilation hose from the cylinder head cover.
   f. Disconnect the purge line hose from the clamp.
   g. Lock the No. 1 air cleaner hose clamp, and then disconnect the No. 1 air cleaner hose from the throttle body.
   h. Detach the 2 hook clamps, and then remove the air cleaner cap.
   i. Remove the air cleaner filter element from the air cleaner case.
3. Remove air cleaner case.
   a. Detach the harness clamp.
   b. Remove the 3 bolts and air cleaner case.
4. Remove the bolt and disconnect the air cleaner inlet.
5. Remove ECM.
   a. Disconnect the 2 ECM connectors.
   b. Raise the 2 levers while pushing the locks on the 2 levers and disconnect the 2 ECM connectors.
   c. Detach the clamp.
6. Remove the 2 bolts and ECM with bracket.

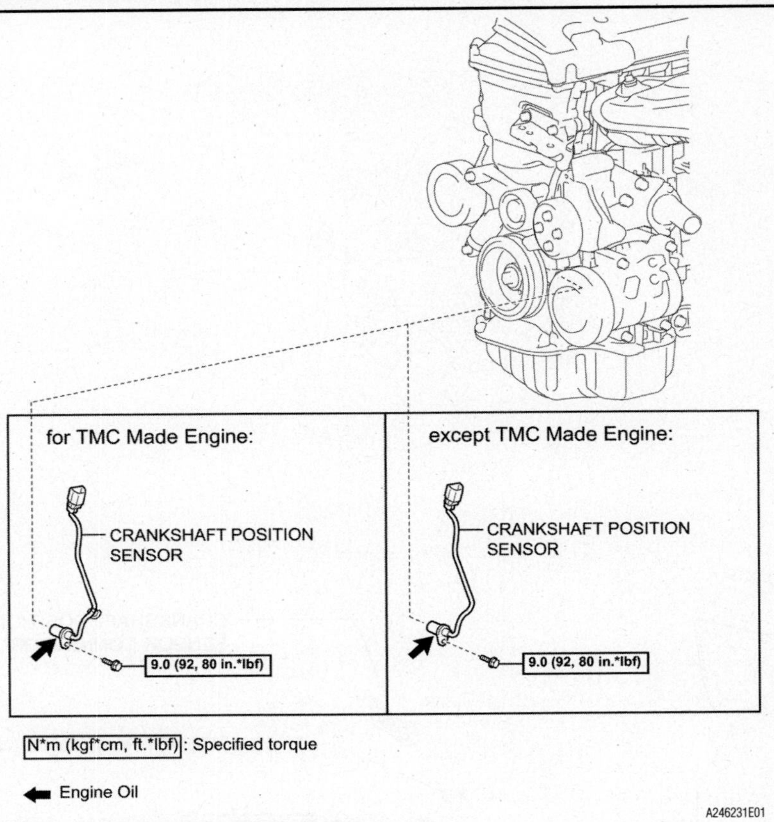

Fig. 167 Exploded view showing Crankshaft Position (CKP) sensor location

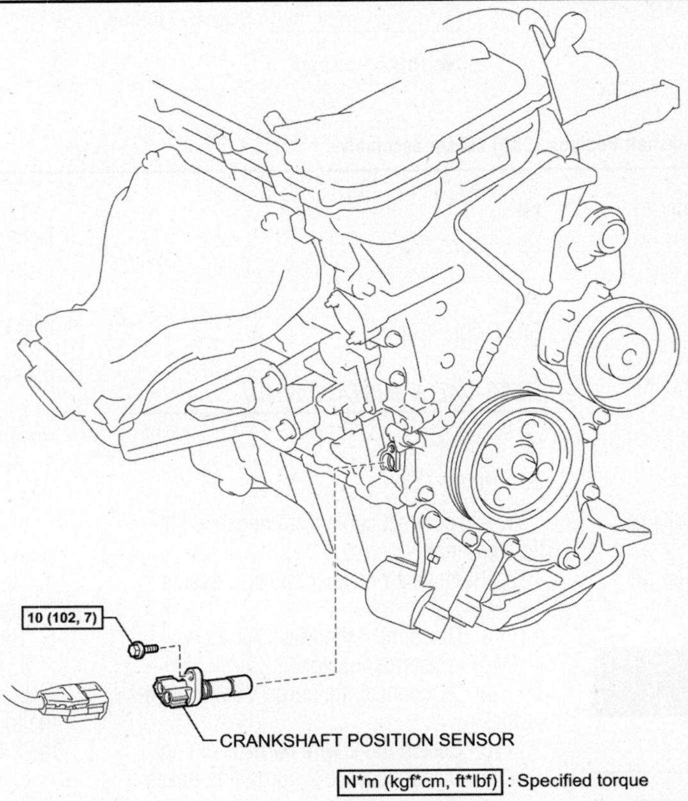

Fig. 168 Exploded view showing Crankshaft Position (CKP) sensor location

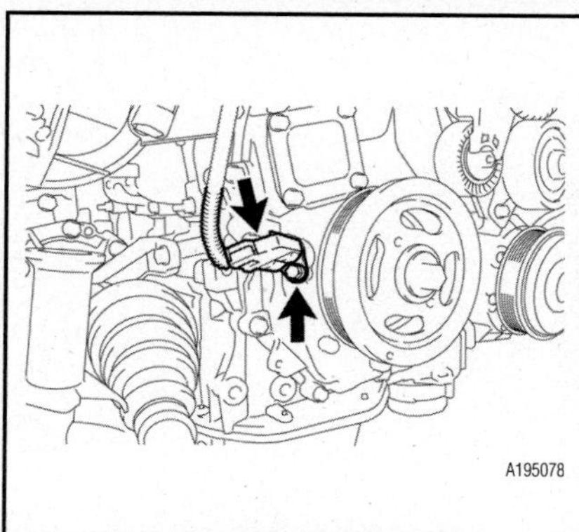

Fig. 169 Remove Crankshaft Position (CKP) sensor

a. Remove the 4 screws and 2 ECM brackets.

7. Installation is the reverse order of removal.

xB

See Figure 174.

1. Remove air cleaner assembly.
   a. Disconnect the Mass Air Flow (MAF) meter connector.
   b. Separate the wire harness.
   c. Loosen the No. 1 air cleaner hose clamp, unlock the air cleaner assembly clamp and remove the air cleaner cap sub-assembly.
   d. Remove the air cleaner filter element sub-assembly.
   e. Separate the wire harness.
   f. Remove the 3 bolts and remove the air cleaner case sub-assembly.
2. Remove ECM.
   a. Remove the 2 lock knobs and harness clamp.
   b. Disconnect the 2 ECM connectors.
   c. Separate the engine wire.
   d. Remove the 2 bolts from the ECM.
   e. Remove the 4 screws and ECM bracket.
3. Installation is the reverse order of removal.

3.0 (31, 27 in.*lbf)
x 2
NO. 1 ECM BRACKET
3.0 (31, 27 in.*lbf)
x 2
NO. 2 ECM BRACKET
ECM
ECM CONNECTOR
8.0 (82, 71 in.*lbf)
x 2
ECM WITH BRACKET
N*m (kgf*cm, ft.*lbf) : Specified torque
A232169E01

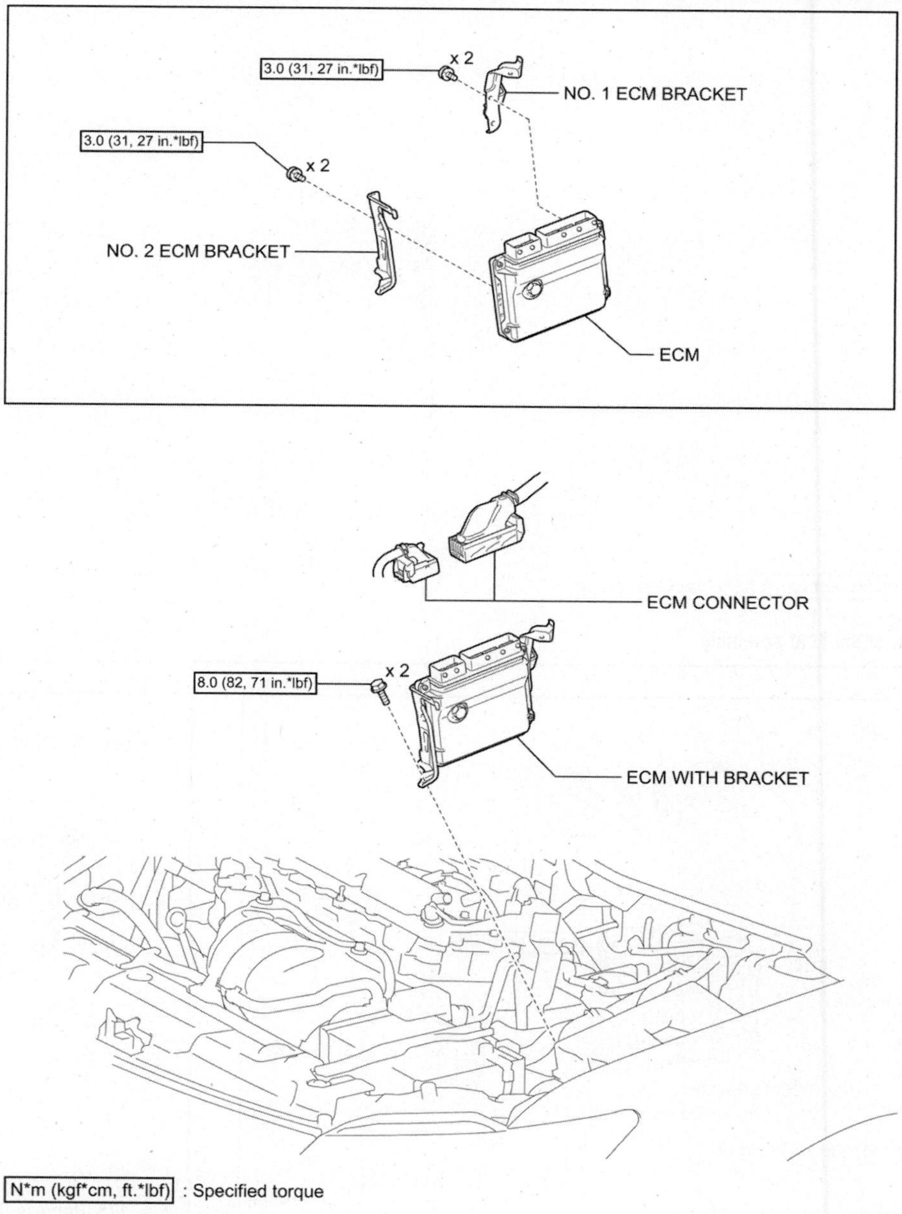

Fig. 170 Exploded view of the ECM assembly

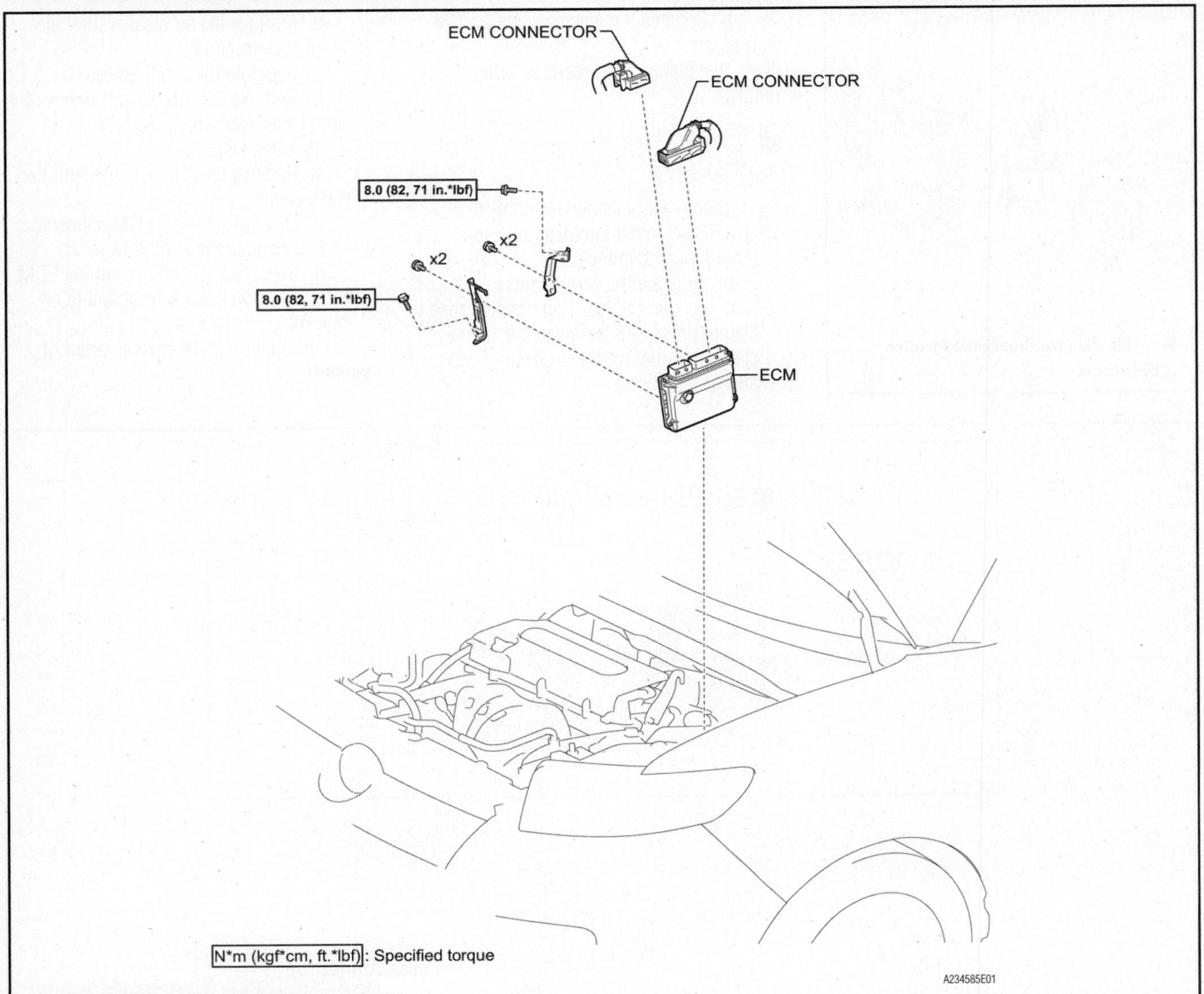

Fig. 171 Exploded view of the ECM assembly

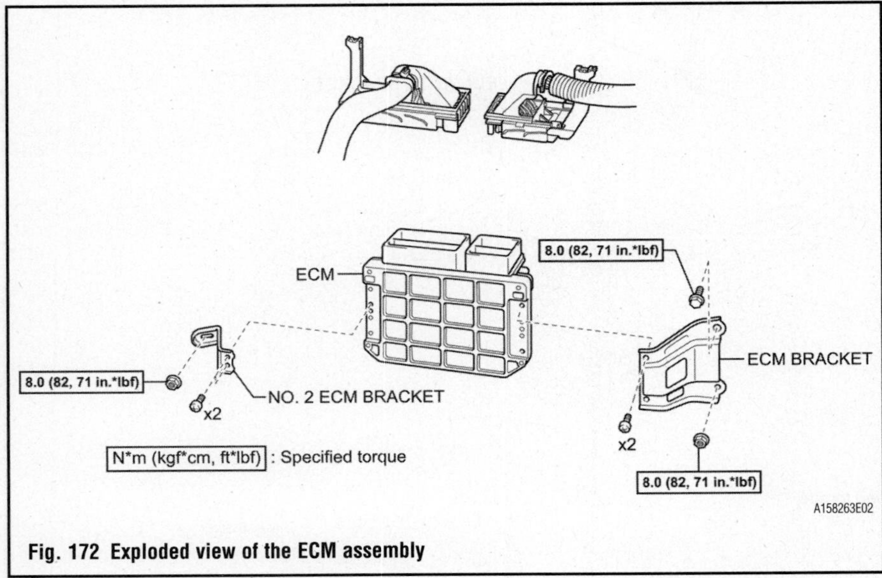

Fig. 172 Exploded view of the ECM assembly

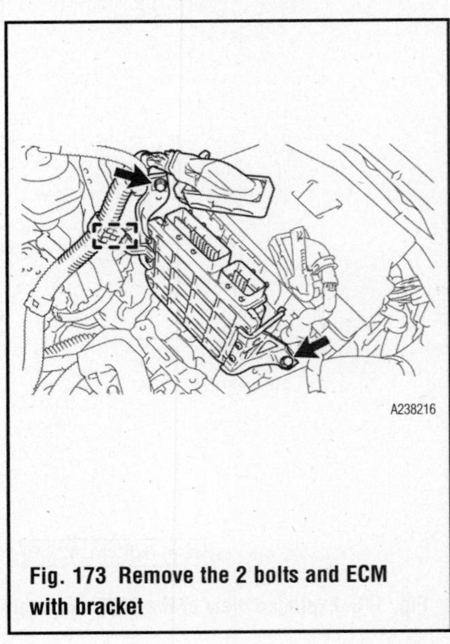

Fig. 173 Remove the 2 bolts and ECM with bracket

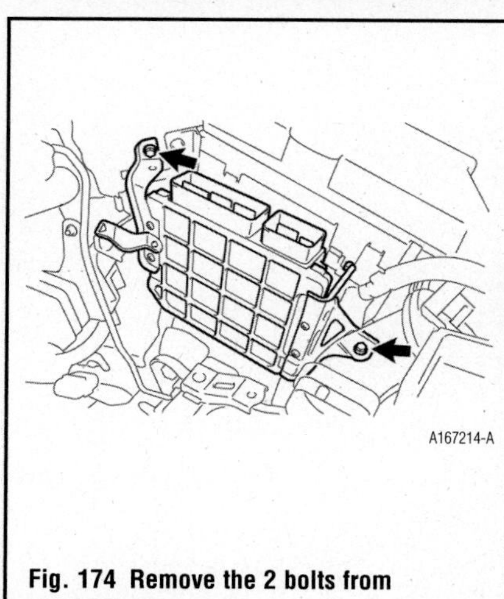

Fig. 174 Remove the 2 bolts from the ECM

xD

See Figure 175.

1. Remove front wiper arm head cap.
2. Remove front wiper arm and blade assembly LH.
3. Remove front wiper arm and blade assembly RH.
4. Remove hood to cowl top seal.
5. Remove cowl top ventilator louver RH.
6. Remove cowl top ventilator louver LH.
7. Remove front wiper motor and link.
8. Remove front air shutter seal RH.
9. Remove outer cowl top panel.
10. Remove ECM.
    a. Remove the 2 lock knobs and harness clamp.
    b. Disconnect the 2 ECM connectors.
    c. Remove the bolt and 2 nuts and remove the ECM.

Fig. 175 Remove the bolt and 2 nuts and remove the ECM

11. Remove the 2 screws and the ECM bracket.
12. Remove the 2 screws and the No. 2 ECM bracket.
13. Installation is the reverse order of removal.

ENGINE COOLANT TEMPERATURE (ECT) SENSOR

LOCATION

tC

See Figure 176.

xB

See Figure 177.

xD

See Figure 178.

REMOVAL & INSTALLATION

tC

See Figure 179.

1. Remove engine under cover.
2. Drain engine coolant.
3. Remove engine coolant temperature sensor.
    a. Disconnect the sensor connector.
    b. Using SST, remove the sensor and gasket.
4. Installation is the reverse order of removal.

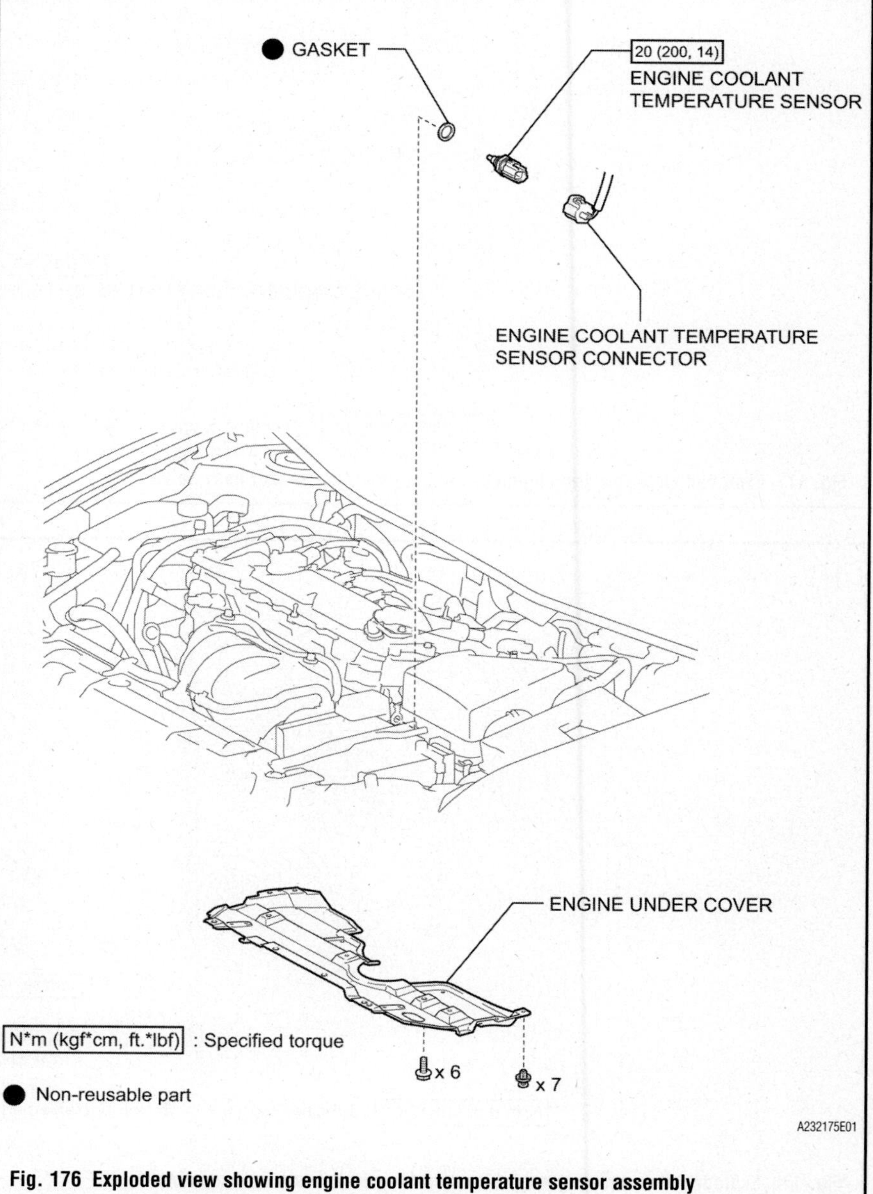

Fig. 176 Exploded view showing engine coolant temperature sensor assembly

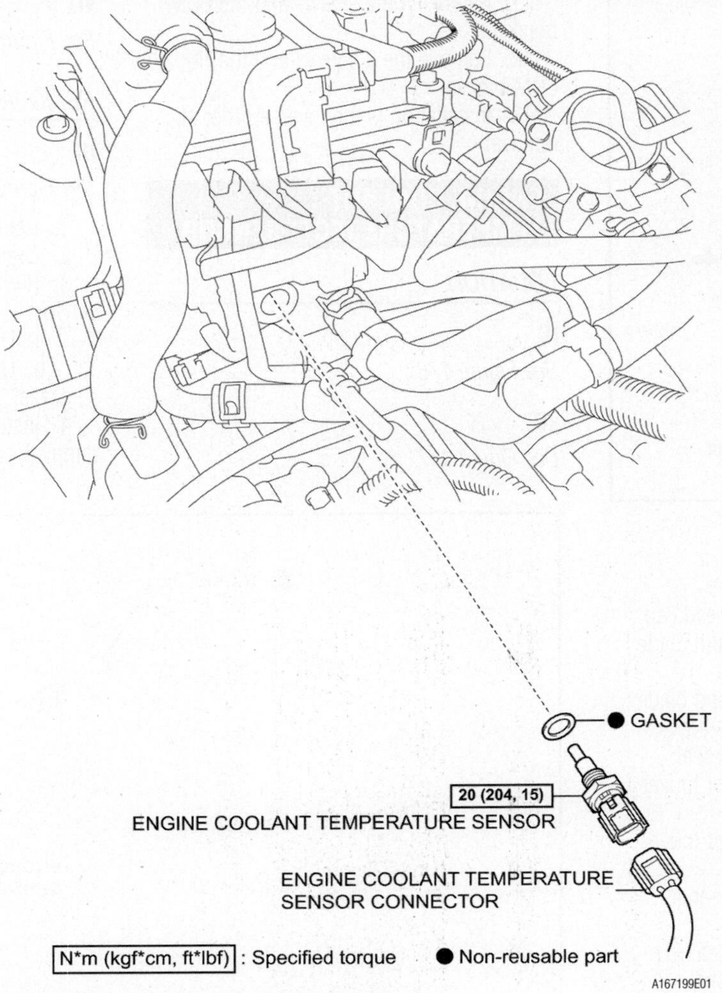

GASKET
20 (204, 15)
ENGINE COOLANT TEMPERATURE SENSOR
ENGINE COOLANT TEMPERATURE
SENSOR CONNECTOR
N*m (kgf*cm, ft*lbf) : Specified torque
Non-reusable part
A167199E01

Fig. 177 Exploded view showing engine coolant temperature sensor assembly

GASKET
20 (200, 14)
ENGINE COOLANT
TEMPERATURE SENSOR
N*m (kgf*cm, ft*lbf) : Specified torque
Non-reusable part
A158228E02

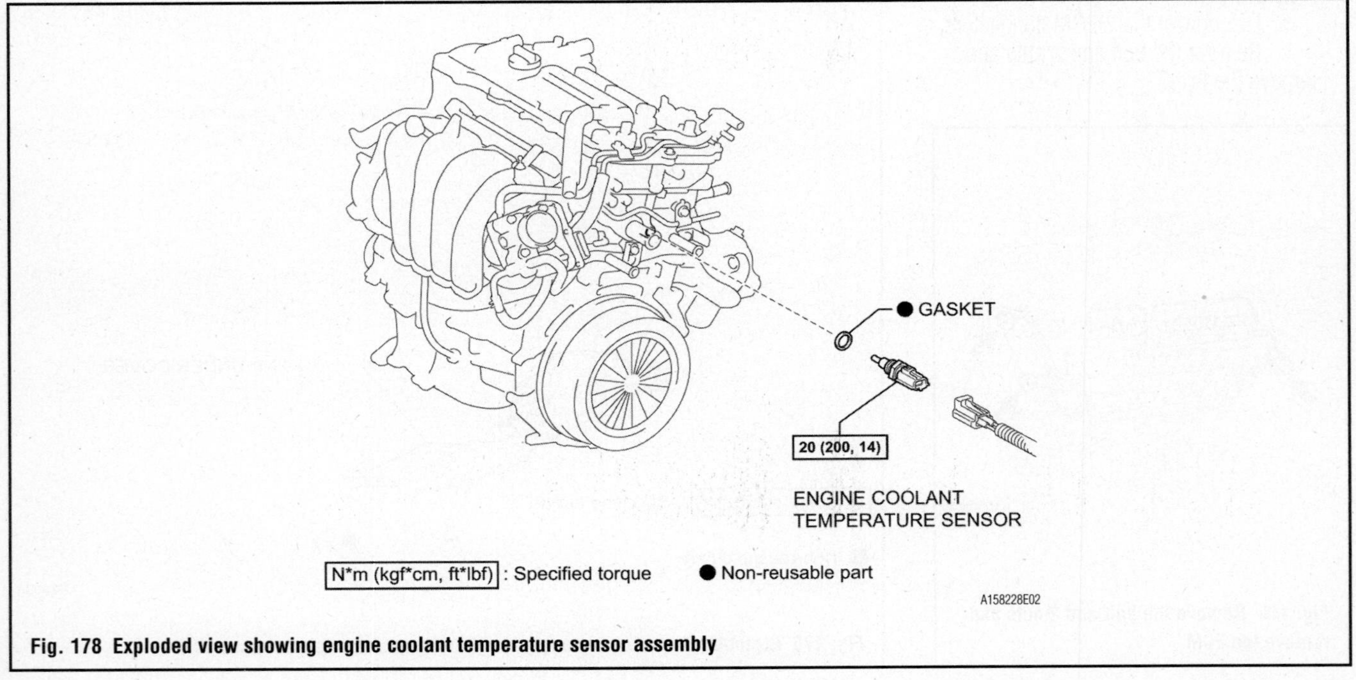

Fig. 178 Exploded view showing engine coolant temperature sensor assembly

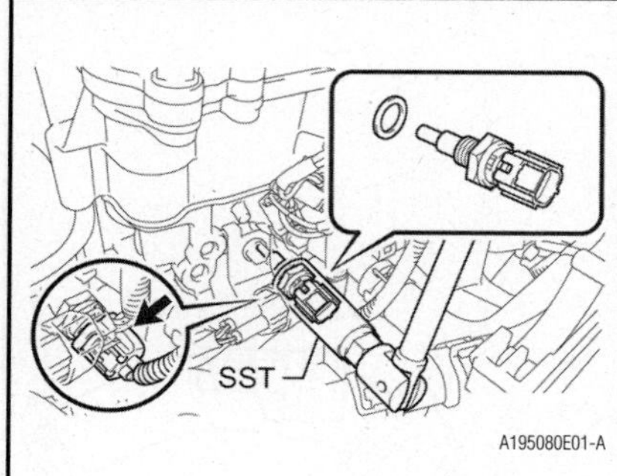

Fig. 179 Using SST (09817-33190), remove the sensor and gasket

xB

See Figure 180.

1. Drain engine coolant.
2. Remove air cleaner assembly.
   a. Disconnect the Mass Air Flow (MAF) meter connector.
   b. Separate the wire harness.
   c. Separate the No. 1 vacuum switching valve.
   d. Disconnect the ventilation hose.
   e. Separate the No. 2 fuel vapor feed hose.
   f. Loosen the No. 1 air cleaner hose clamp, unlock the air cleaner assembly clamp and remove the air cleaner cap sub-assembly with No. 1 air cleaner hose.
   g. Remove the air cleaner filter element sub-assembly.
   h. Separate the wire harness.
   i. Remove the 3 bolts and remove the air cleaner case sub-assembly.
3. Remove engine coolant temperature sensor.
   a. Disconnect the engine coolant temperature sensor connector.

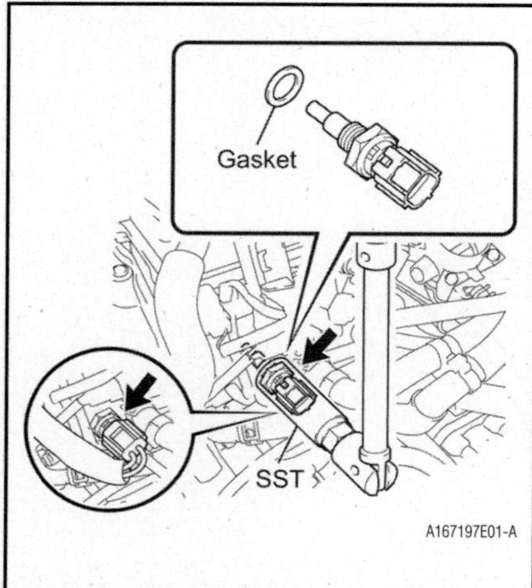

Fig. 180 Using SST (09817-33190), remove the sensor and gasket

   b. Using SST (09817-33190), remove the engine coolant temperature sensor and gasket.
4. Installation is the reverse order of removal.

xD

See Figure 181.

1. Drain engine coolant
2. Remove the No. 2 cylinder head cover.
3. Remove the Engine Coolant Temperature (ECT) sensor connector and remove the Engine Coolant Temperature (ECT) sensor.
4. Installation is the reverse of the removal procedure.
5. Tighten the sensor to 14 ft. lbs. (20 Nm).

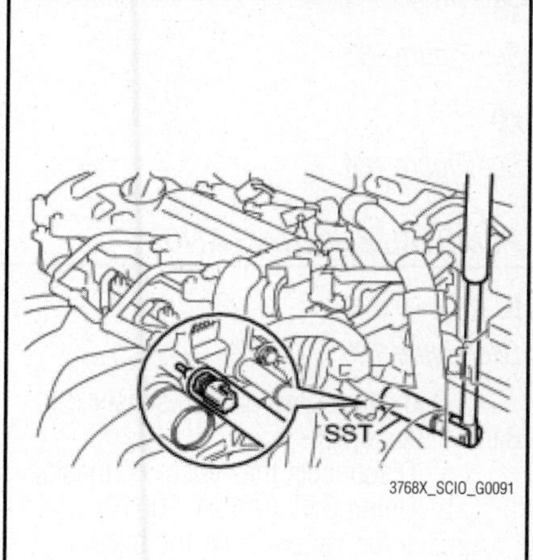

Fig. 181 Remove the Engine Coolant Temperature (ECT) sensor connector and remove the Engine Coolant Temperature (ECT) sensor.

HEATED OXYGEN SENSOR (HO2S)

LOCATION

tC

See Figure 182.

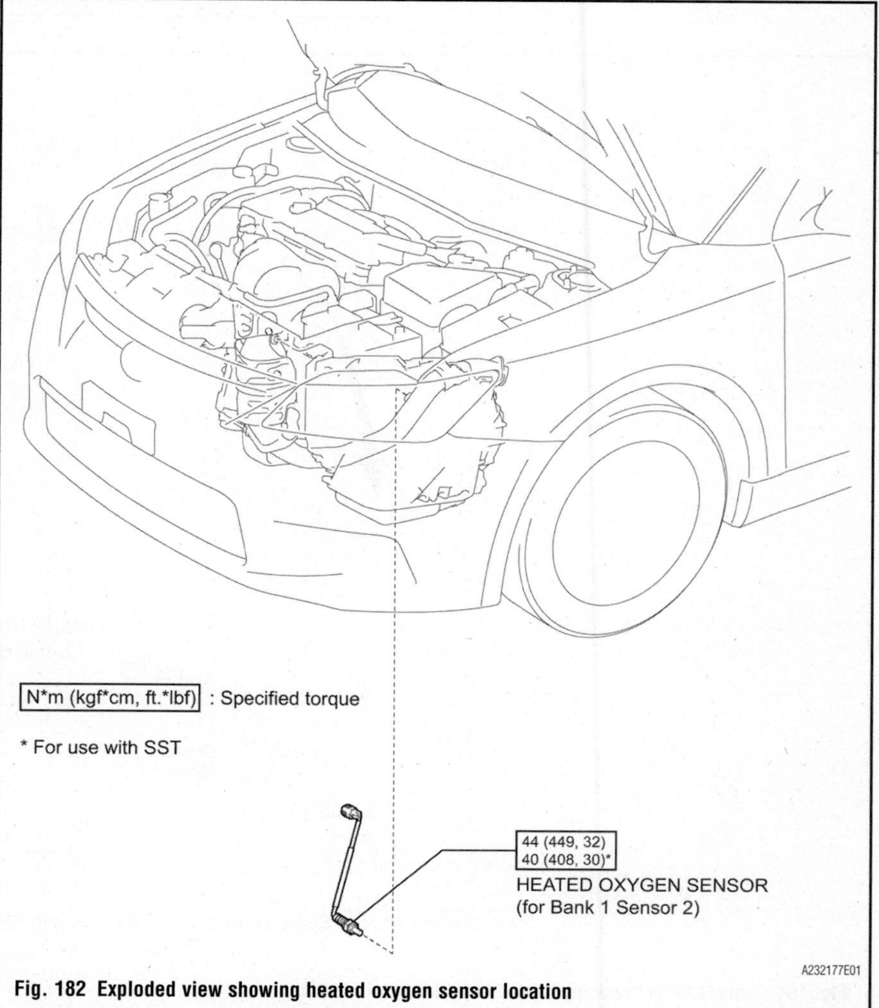

Fig. 182 Exploded view showing heated oxygen sensor location

xB

See Figure 183

xD

See Figure 184.

REMOVAL & INSTALLATION

tC

See Figure 185.

1. Remove heated oxygen sensor (for Bank 1 Sensor 2).
   a. Disconnect the sensor connector.
   b. Using SST (09224-00010), remove the sensor from the front exhaust pipe.
2. Installation is the reverse order of removal.

xB

1. Remove heated oxygen sensor
   a. Disconnect the heated oxygen sensor connector.
   b. Using SST (09224-00010), remove the heated oxygen sensor.

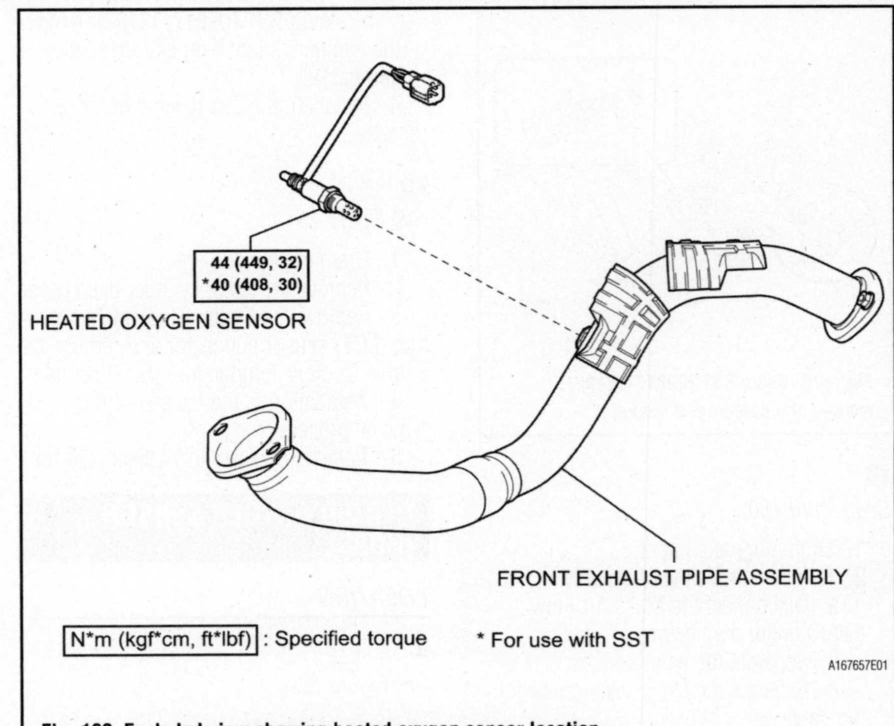

Fig. 183 Exploded view showing heated oxygen sensor location

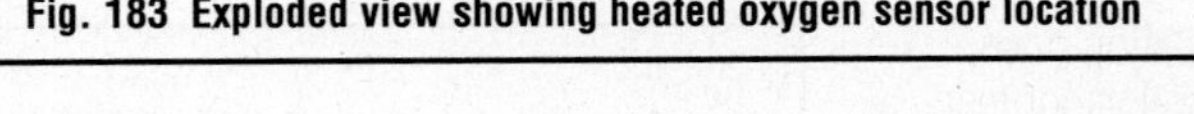

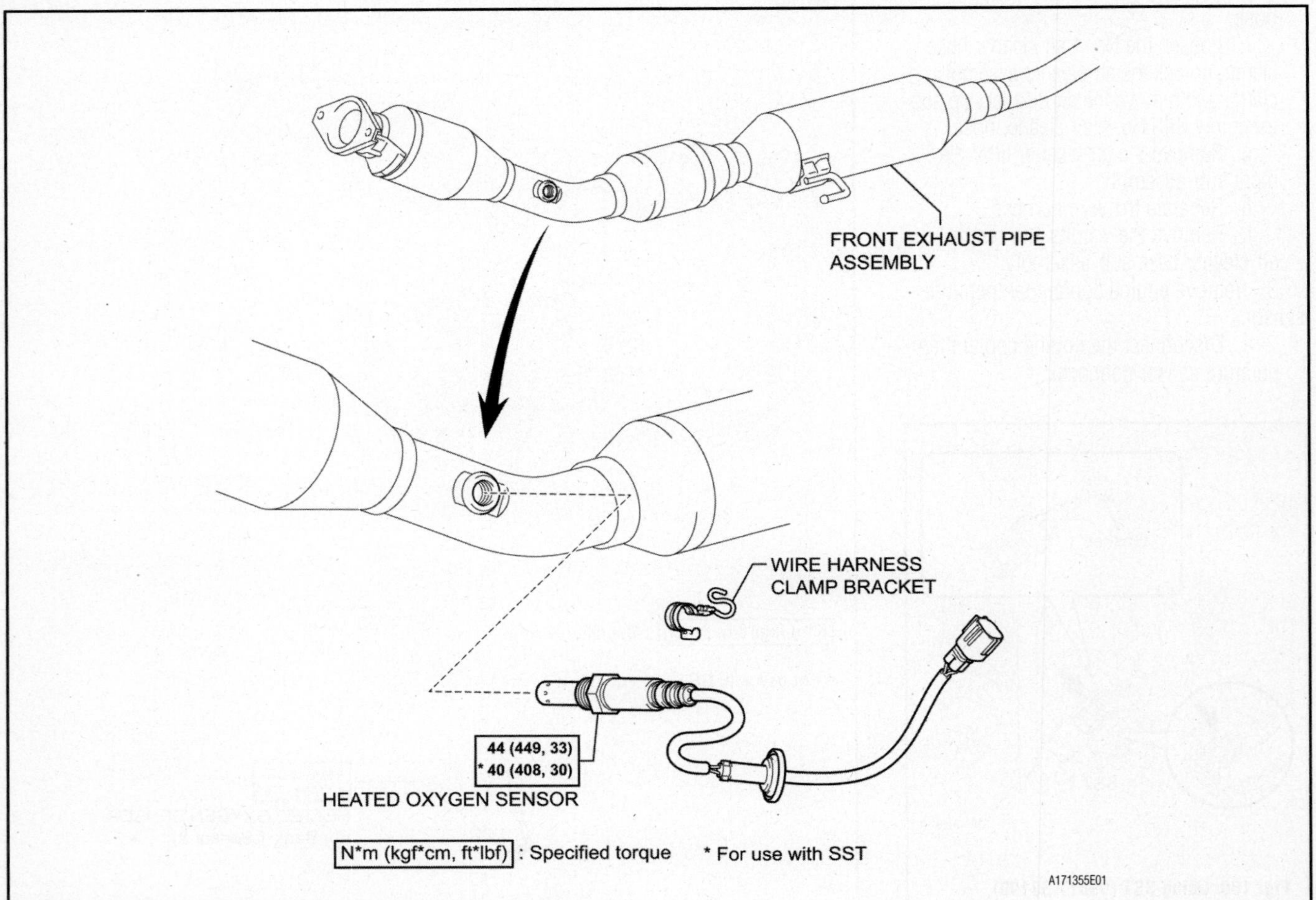

Fig. 184 Exploded view showing heated oxygen sensor location

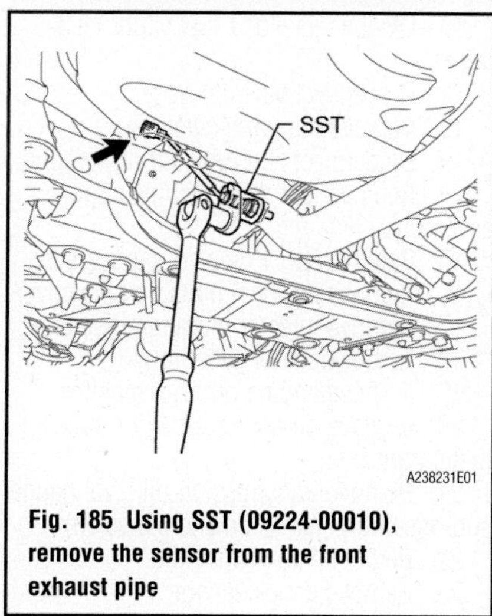

Fig. 185 Using SST (09224-00010), remove the sensor from the front exhaust pipe

2. Installation is the reverse order of removal.

xD

1. Remove shift lever knob sub-assembly (M/T).
2. Remove upper console panel.
3. Remove rear console box cover.
4. Remove console box carpet.
5. Remove rear console box sub-assembly.
6. Remove front console box.
7. Remove heated oxygen sensor.
   a. Disconnect the heated oxygen sensor connector.
   b. Remove the grommet and pull the sensor connector out of the cabin through the floor panel.
   c. Remove the wire harness clamp bracket.
   d. Using SST (09224-00010), remove heated oxygen sensor.
8. Installation is the reverse of removal.
9. Tighten to 33 ft. lbs. (44 Nm).

KNOCK SENSOR (KS)

LOCATION

tC

See Figure 186.

xB

See Figure 187.

xD

See Figure 188.

REMOVAL & INSTALLATION

tC

1. Remove intake manifold.

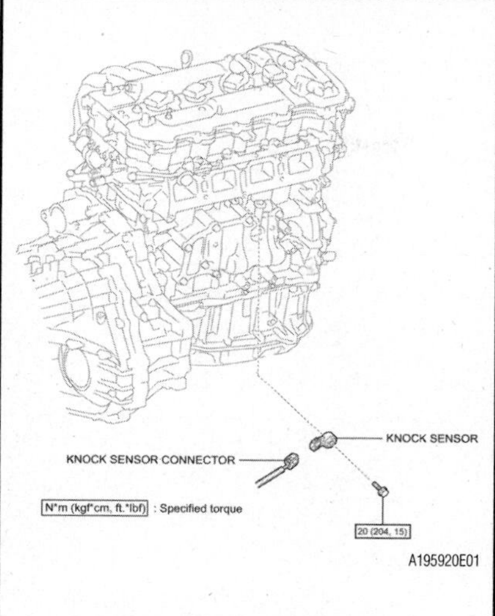

Fig. 186 Exploded view showing knock sensor location

2. Remove knock sensor.
   a. Disconnect the sensor connector.
   b. Remove the bolt and sensor.
3. Installation is the reverse order of removal.

xB

1. Discharge fuel system pressure.
2. Drain coolant.
3. Remove windshield wiper arm cover.
4. Remove front wiper arm and blade assembly LH.
5. Remove front wiper arm and blade assembly RH.
6. Remove hood to cowl top seal.
7. Remove cowl top ventilator louver RH.
8. Remove cowl top ventilator louver LH.
9. Remove front wiper motor and link
10. Remove outer cowl top panel.

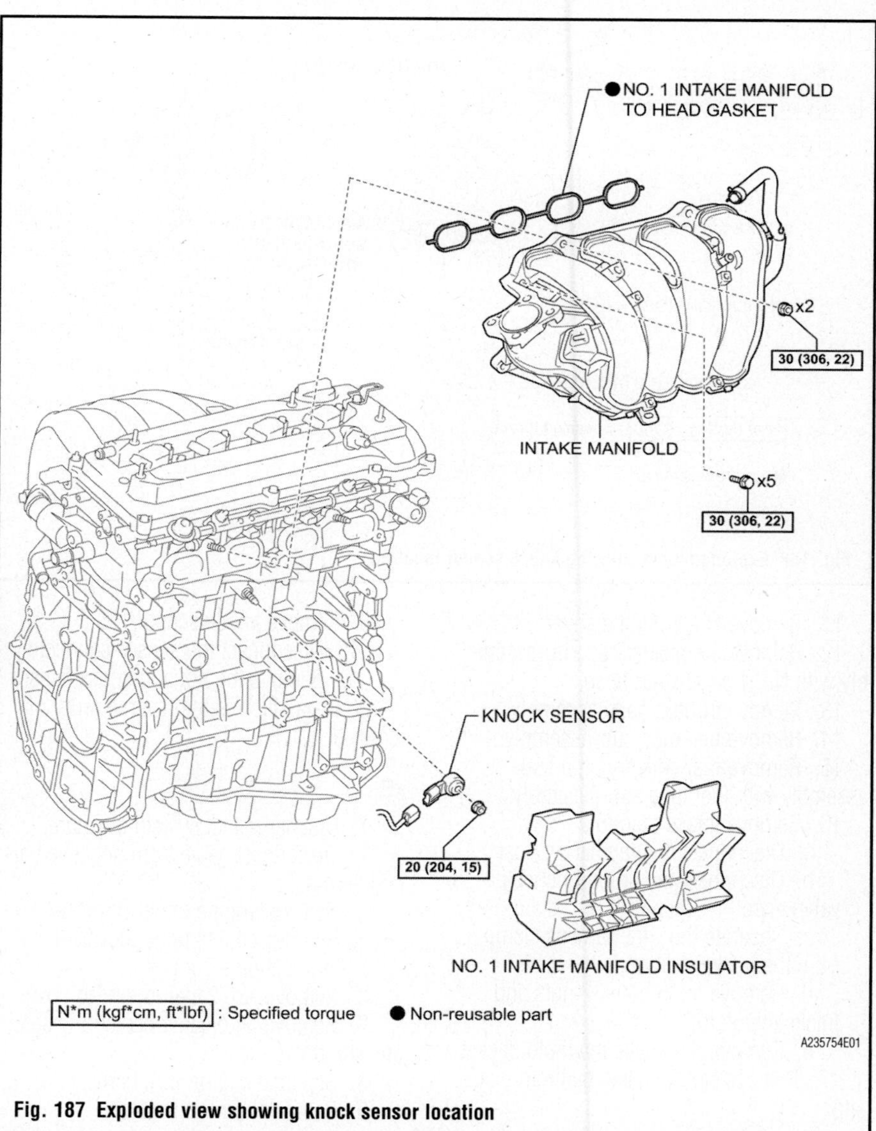

Fig. 187 Exploded view showing knock sensor location

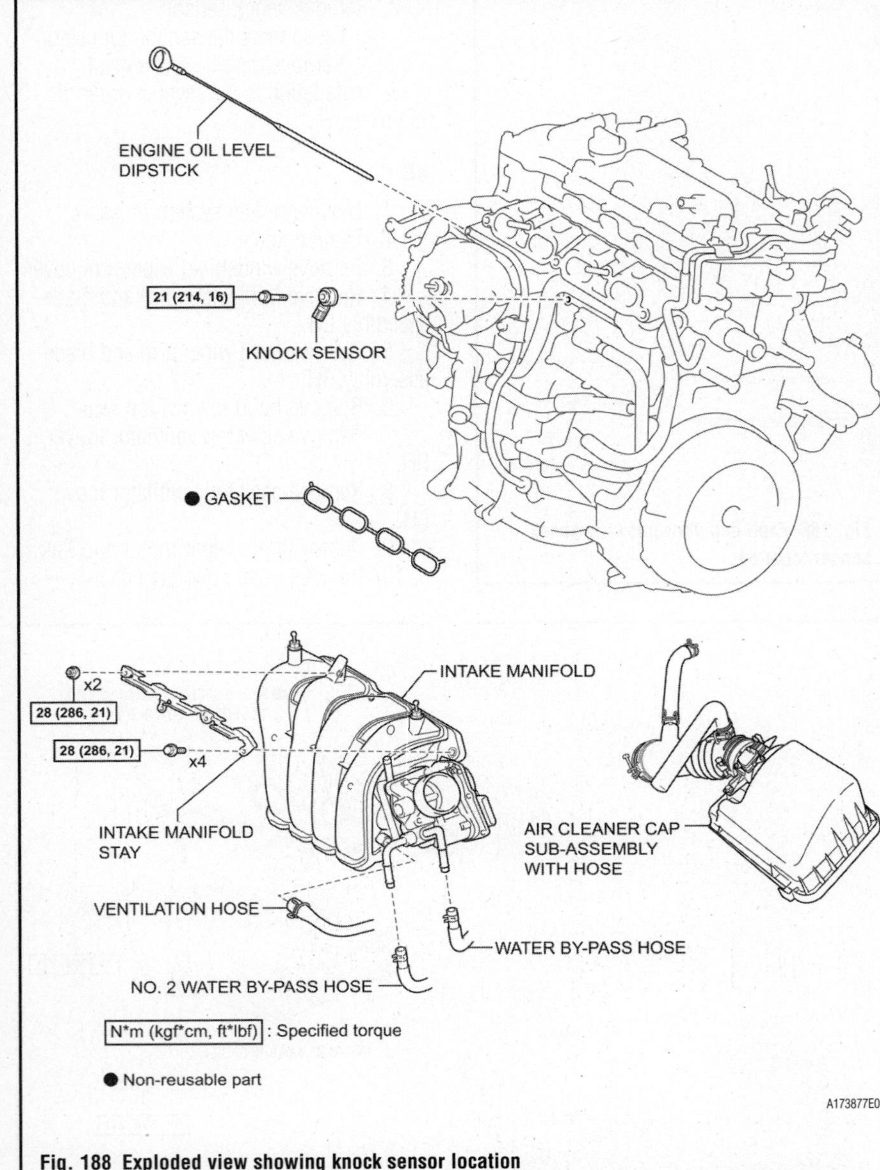

Fig. 188 Exploded view showing knock sensor location

11. Remove No. 1 engine cover.
12. Remove air cleaner cap sub-assembly with No. 1 air cleaner hose.
13. Remove throttle body assembly.
14. Remove fuel tube sub-assembly.
15. Remove fuel delivery pipe sub-assembly with fuel tube sub-assembly.
16. Remove intake manifold.
    a. Disconnect the ventilation hose.
    b. Disconnect the union to check valve hose.
    c. Separate the wire harness clamp from the intake manifold.
    d. Remove the 5 bolts, 2 nuts and intake manifold.
    e. Remove the intake manifold gasket.
17. Remove No. 1 intake manifold insulator.
18. Remove knock sensor.
    a. Disconnect the sensor connector.
    b. Remove the nut and knock sensor.
19. Installation is the reverse order of removal.

xD

1. Discharge fuel system pressure.
2. Disconnect cable from negative battery terminal.
3. Remove engine under cover LH.
4. Remove engine under cover RH.
5. Drain engine coolant.
6. Remove No. 2 cylinder head cover.
7. Remove air cleaner cap sub-assembly with hose.
8. Separate engine wire harness.
9. Remove harness bracket.
10. Disconnect No. 1 fuel vapor feed hose.
11. Disconnect vacuum hose.
12. Remove fuel pipe clamp.
13. Disconnect fuel tube sub-assembly.
14. Remove fuel delivery pipe sub-assembly.
15. Remove No. 1 delivery pipe spacer.
16. Remove flywheel housing side cover.
17. Remove injector vibration insulator.
18. Remove starter assembly.
19. Remove engine oil level dipstick.
20. Remove oil cooler tube (for automatic transaxle).
21. Remove transmission oil level gauge sub-assembly (for automatic transaxle).
22. Remove intake manifold.
23. Remove knock sensor.
    a. Disconnect the knock sensor connector.
    b. Remove the bolt and remove the knock sensor.
24. Installation is the reverse order of removal.

MASS AIR FLOW (MAF) METER

LOCATION

tC

See Figure 189.

xB

See Figure 190.

xD

See Figure 191.

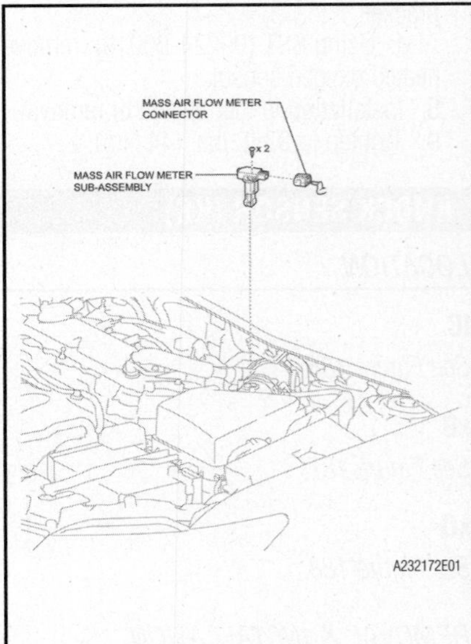

Fig. 189 Exploded view showing the Mass Air Flow (MAF) meter location

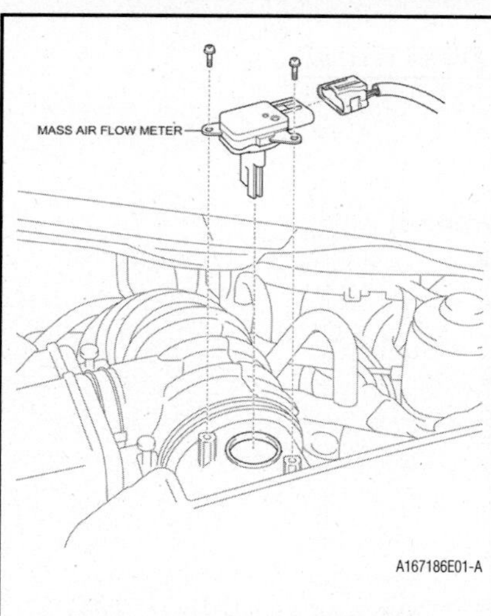

Fig. 190 Exploded view showing the Mass Air Flow (MAF) meter location

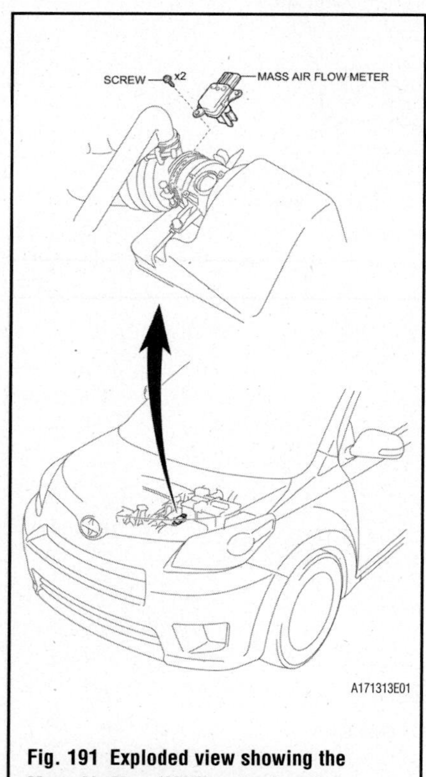

Fig. 191 Exploded view showing the Mass Air Flow (MAF) meter location

REMOVAL & INSTALLATION

tC and xB

1. Remove Mass Air Flow (MAF) meter sub-assembly.
   a. Disconnect the Mass Air Flow (MAF) meter connector.
   b. Remove the 2 screws and MAF meter.
2. Installation is the reverse order of removal.

xD

1. Remove Mass Air Flow (MAF) meter:
   a. Disconnect the Mass Air Flow (MAF) meter connector.
   b. Remove the 2 screws and the mass air flow meter.
2. Installation is the reverse order of removal.

THROTTLE POSITION SENSOR (TPS)

LOCATION

The Throttle Position Sensor (TPS) is an integral part of the throttle body. Refer to Throttle Body Removal and Installation when servicing this component.

VEHICLE SPEED SENSOR (VSS)

LOCATION

tC (U760E Transaxle)

See Figure 192.

xB (U241E Transaxle)

See Figure 193.

xD (U341E Transaxle)

See Figure 194.

REMOVAL & INSTALLATION

tC (U760E Transaxle)

1. Remove transmission valve body assembly.
2. Remove speed sensor.

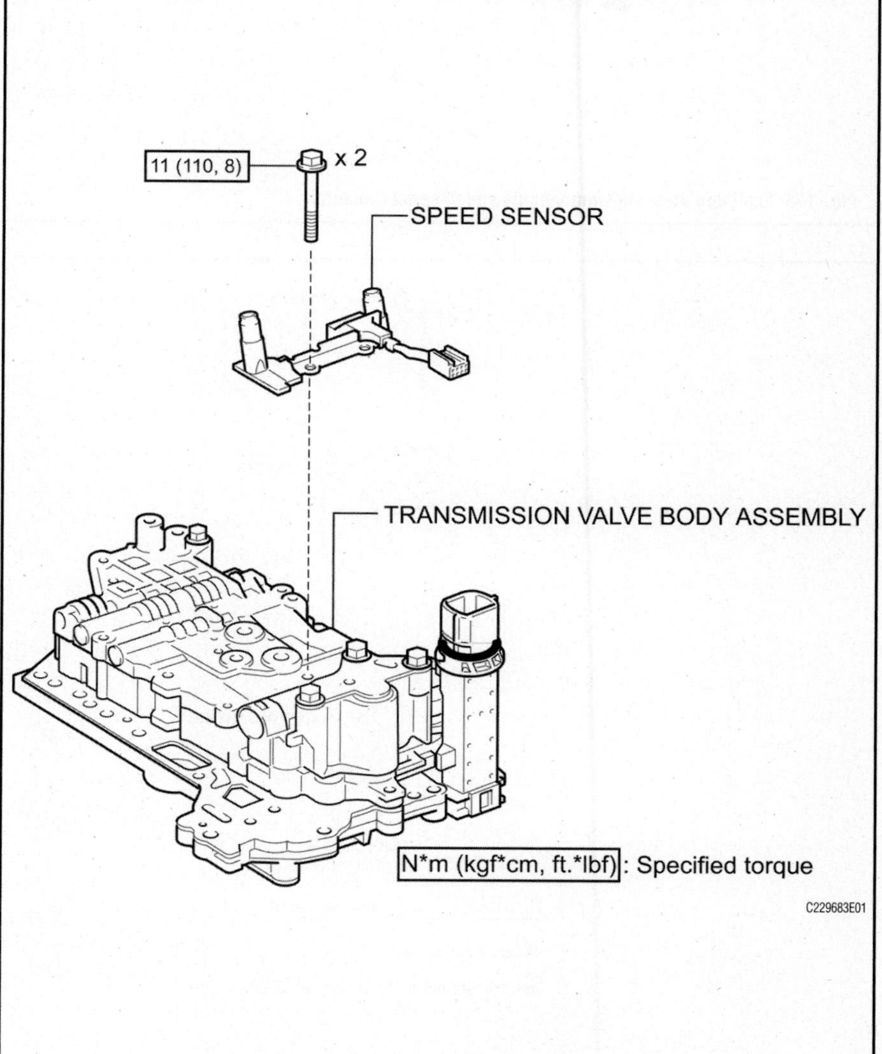

Fig. 192 Exploded view showing vehicle speed sensor location

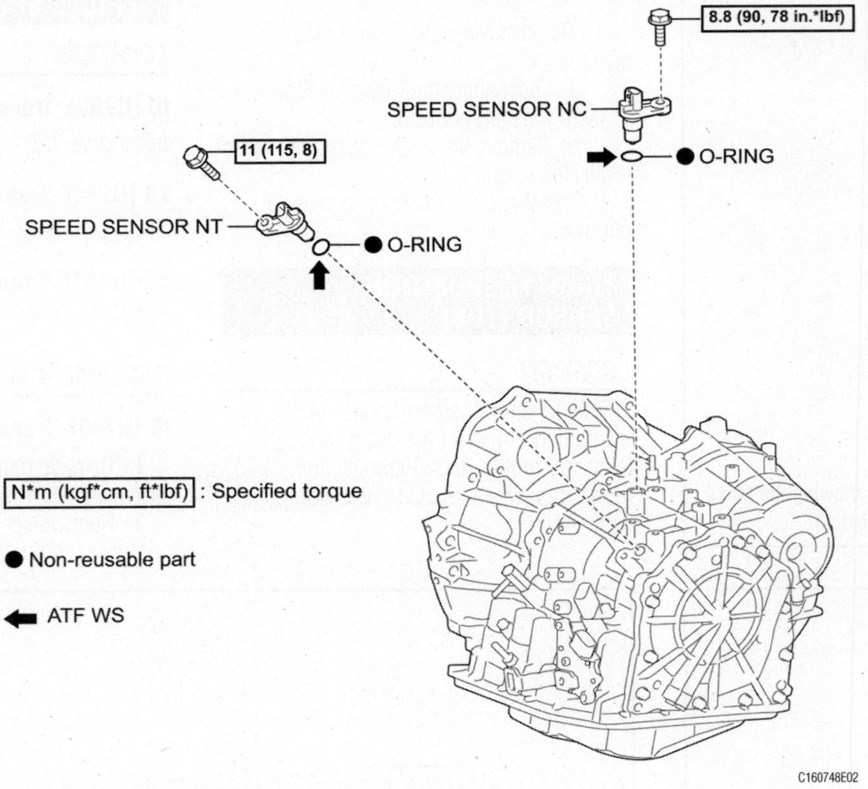

Fig. 193 Exploded view showing vehicle speed sensor location

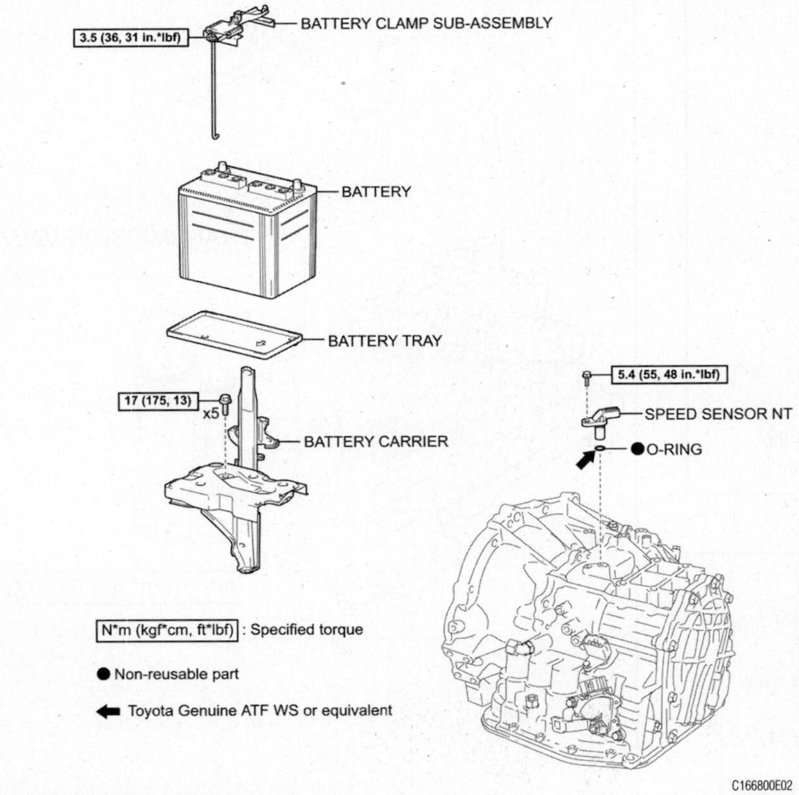

Fig. 194 Exploded view showing vehicle speed sensor location

a. Disconnect the connector of the speed sensor.

b. Remove the 2 bolts and speed sensor from the transmission valve body assembly.

3. Installation is the reverse order of removal.

xB (U241E Transaxle)

1. Remove battery.
2. Remove No. 2 battery carrier.
3. Remove air cleaner cap sub-assembly with No. 1 air cleaner hose.
4. Remove air cleaner filter element sub-assembly.
5. Remove air cleaner case sub-assembly.
6. Remove speed sensor NT.

a. Disconnect the speed sensor NT connector.

b. Remove the bolt and remove the speed sensor NT.

c. Remove the O-ring from the speed sensor NT.

7. Remove speed sensor NC.

a. Disconnect the speed sensor NC connector.

b. Remove the bolt and remove the speed sensor NC.

c. Remove the O-ring from the speed sensor NC.

8. Installation is the reverse order of removal.

xD (U341E Transaxle)

1. Remove battery.
2. Remove battery tray.
3. Remove battery carrier.
4. Remove speed sensor NT.

a. Disconnect the speed sensor NT connector.

b. Remove the bolt and speed sensor NT.

c. Remove the O-ring from the speed sensor NT.

5. Installation is the reverse order of removal.

FUEL — GASOLINE FUEL INJECTION SYSTEM

FUEL SYSTEM SERVICE PRECAUTIONS

Safety is the most important factor when performing not only fuel system maintenance but any type of maintenance. Failure to conduct maintenance and repairs in a safe manner may result in serious personal injury or death. Maintenance and testing of the vehicle's fuel system components can be accomplished safely and effectively by adhering to the following rules and guidelines.

- To avoid the possibility of fire and personal injury, always disconnect the negative battery cable unless the repair or test procedure requires that battery voltage be applied.
- Always relieve the fuel system pressure prior to disconnecting any fuel system component (injector, fuel rail, pressure regulator, etc.), fitting or fuel line connection. Exercise extreme caution whenever relieving fuel system pressure to avoid exposing skin, face and eyes to fuel spray. Please be advised that fuel under pressure may penetrate the skin or any part of the body that it contacts.
- Always place a shop towel or cloth around the fitting or connection prior to loosening to absorb any excess fuel due to spillage. Ensure that all fuel spillage (should it occur) is quickly removed from engine surfaces. Ensure that all fuel soaked cloths or towels are deposited into a suitable waste container.
- Always keep a dry chemical (Class B) fire extinguisher near the work area.
- Do not allow fuel spray or fuel vapors to come into contact with a spark or open flame.
- Always use a back-up wrench when loosening and tightening fuel line connection fittings. This will prevent unnecessary stress and torsion to fuel line piping.
- Always replace worn fuel fitting O-rings with new Do not substitute fuel hose or equivalent where fuel pipe is installed.

Before servicing the vehicle, make sure to also refer to the precautions in the beginning of this section as well.

RELIEVING FUEL SYSTEM PRESSURE

tC

1. Remove the rear seat cushion.
2. Remove the rear floor service hole cover.
3. Disconnect the fuel pump and sender gauge connector.
4. Start the engine. After the engine has stopped, turn the ignition switch off.
5. Check that the engine does not start.
6. Remove the fuel tank cap and discharge the pressure in the fuel tank completely.
7. Connect the fuel pump and sender gauge connector.
8. Install the rear floor service hole cover.
9. Install the rear seat cushion.

xB

1. Remove the rear seat cushion with cover pad subassembly.
2. Remove the rear seat under tray cover.
3. Remove the bench type rear seat cushion assembly.
4. Remove the rear door scuff plate RH.
5. Remove the rear door scuff plate LH.
6. Remove the rear floor service hose cover.
7. Disconnect the connector.
8. Start the engine. After the engine stops naturally, turn the ignition switch off.
9. Start the engine again and make sure that engine does not start.
10. Remove the fuel tank cap, and let the air out of the fuel tank.
11. Connect the connector.
12. Install the rear floor service hole cover with new butyl tape.
13. Install the rear door scuff plate RH.
14. Install the rear door scuff plate LH.
15. Install the bench type rear seat cushion assembly.
16. Install the rear seat under tray cover.
17. Install the rear seat cushion with cover pad subassembly.

xD

1. Remove the rear seat cushion assembly LH.
2. Remove the rear floor service hose cover.
3. Disconnect the connector from the fuel pump assembly.
4. Start the engine. After the engine stops naturally, turn the ignition switch OFF.
5. Start the engine again and make sure that engine does not start.
6. Remove the fuel tank cap, and let the air out of the fuel tank.
7. Connect the connector.
8. Install the rear floor service hole cover with new butyl tape.
9. Install the rear seat cushion assembly LH.

FUEL FILTER

REMOVAL & INSTALLATION

tC

See Figure 195.

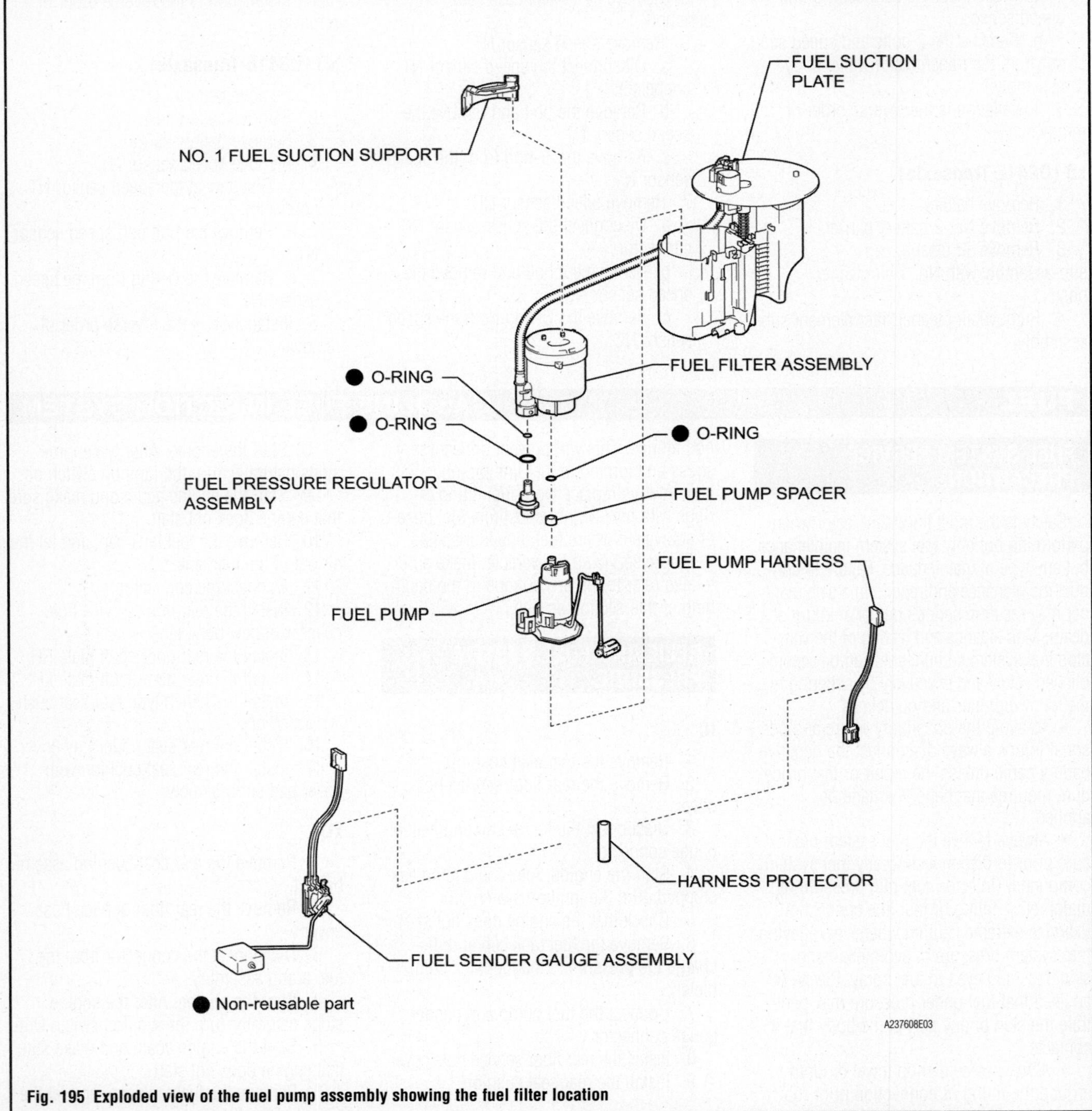

Fig. 195 Exploded view of the fuel pump assembly showing the fuel filter location

➡Do not try to remove the black nylon tube as it is welded to the fuel suction tube assembly.

1. Remove fuel suction tube assembly with pump and gauge.
2. Remove fuel sender gauge assembly.
3. Remove fuel pump.
4. Remove fuel filter assembly.
   a. Remove the fuel pressure regulator from the fuel filter.
   b. Remove the 2 O-rings.

To install:

5. Install fuel filter assembly.
   a. Apply a light coat of gasoline to 2 new O-rings and install them to the fuel pressure regulator.
   b. Install the fuel pressure regulator to the fuel filter.
6. Install fuel pump.
7. Install fuel sender gauge assembly.
8. Install fuel suction tube assembly with pump and gauge.

xB and xD

The fuel filter is an integral part of the fuel pump assembly. Refer to Fuel Pump Removal and Installation when servicing this component.

FUEL INJECTORS

REMOVAL & INSTALLATION

tC

See Figure 196.

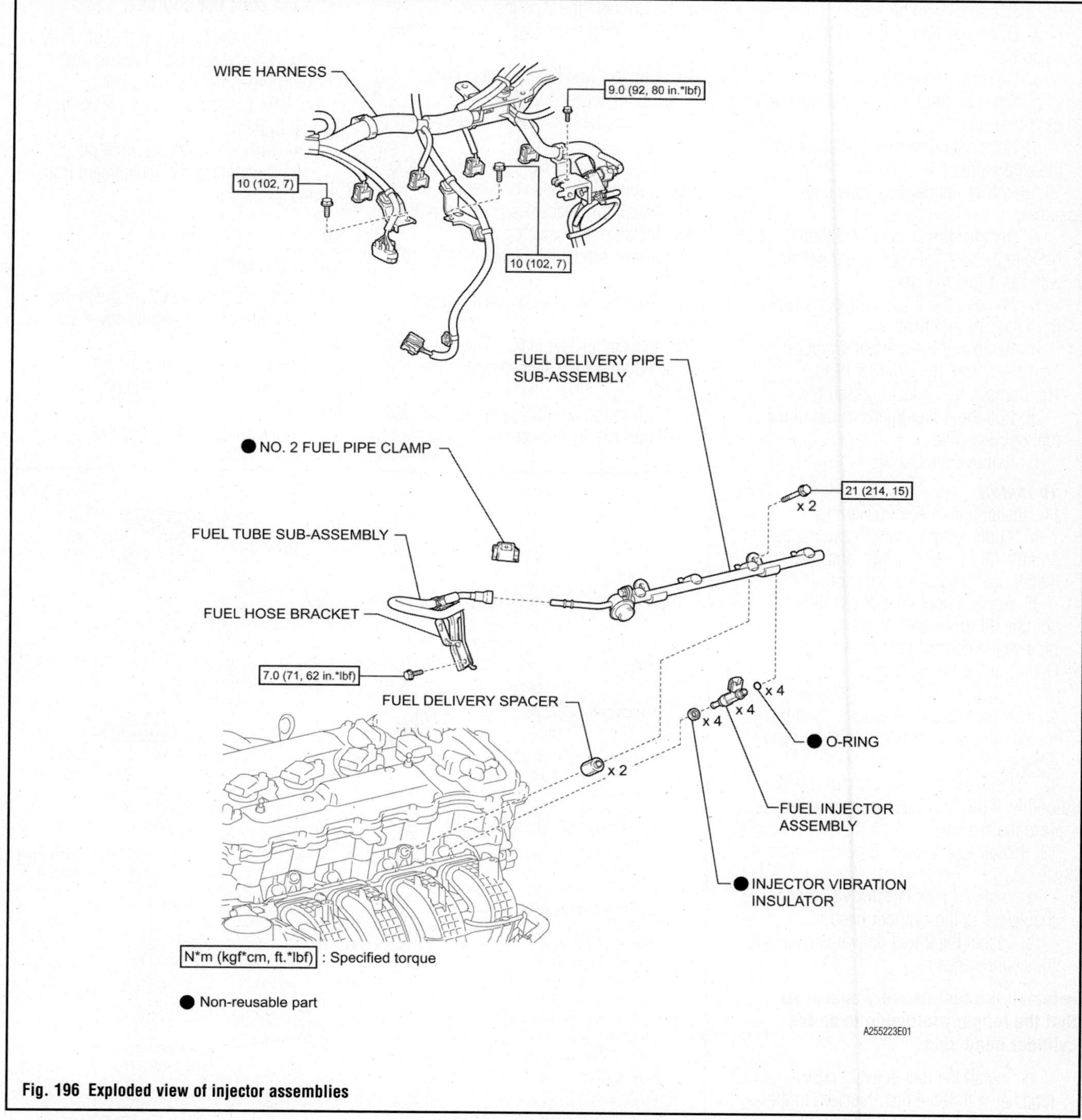

Fig. 196 Exploded view of injector assemblies

1. Discharge fuel system pressure.
2. Disconnect cable from negative battery terminal.
3. Remove windshield wiper motor and link.
4. Remove outer cowl top panel.
5. Remove air cleaner cap sub-assembly.
6. Disconnect fuel hose bracket.
7. Disconnect fuel tube sub-assembly.
   a. Remove the No. 2 fuel pipe clamp.
   b. Wipe off any dirt on the fuel tube connector.
   c. Hold the fuel tube connector, and then install SST 09268-21011.
   d. Turn SST to align the retainer inside the fuel tube connector with the chamfered part of SST.
   e. Insert SST into the fuel tube and hold it. Then push the fuel tube connector toward SST.
   f. Mount the retainer of the fuel tube connector onto the chamfered part of SST.
   g. Slide SST and the fuel tube connector together towards the fuel tube until they make a "click" sound, and then disconnect the fuel tube.
   h. Drain the fuel remaining inside the fuel tube.
   i. Cover the fuel tube and fuel pipe with a plastic bag to protect the disconnected parts.

8. Disconnect wire harness.
   a. Disconnect the 4 fuel injector connectors.
   b. Disconnect the 2 connectors.
   c. Remove the 3 bolts and 3 wire harness brackets.
   d. Detach the clamp and disconnect the wire harness.
9. Remove fuel delivery pipe sub-assembly.
   a. Remove the 2 bolts, and then remove the fuel delivery pipe together with the 4 fuel injectors.
   b. Remove the 2 fuel delivery spacers from the cylinder head.
   c. Remove the 4 injector vibration insulators from the cylinder head.
10. Remove fuel injector assembly.
   a. Pull the 4 fuel injectors out of the fuel delivery pipe.
   b. Remove the O-ring.

To install:

11. Install fuel injector assembly.
   a. Apply a light coat of gasoline or spindle oil to new O-rings, and then install one to each fuel injector.
   b. Apply a light coat of gasoline or spindle oil to the part of the fuel delivery pipe which comes into contact with the O-ring of the fuel injector.
   c. Apply a light coat of gasoline or spindle oil to the O-ring again, and then install the fuel injectors to the fuel delivery pipe.
12. Check that the fuel injector rotates smoothly. If the fuel injector does not rotate, replace the O-ring.
13. Install fuel delivery pipe sub-assembly.
   a. Install 4 new injector vibration insulators to the cylinder head.
   b. Install the 2 fuel delivery spacers to the cylinder head.

➡Install the fuel delivery spacer so that the longer protrusion is on the cylinder head side.

   c. Install the fuel delivery pipe together with the 4 fuel injectors to the cylinder head, and then temporarily install the 2 bolts.
   d. Check that the fuel injector rotates smoothly.

➡If the fuel injector does not rotate, replace the O-ring.

   e. Tighten the 2 bolts to 15 ft. lbs. (21 Nm).
14. Connect wire harness.
   a. Attach the clamp to connect the wire harness.
   b. Connect the 2 connectors.
   c. Connect the 4 fuel injector connectors.
15. Connect fuel tube sub-assembly.
   a. Connect the fuel tube.
   b. Push the fuel tube connector until it makes a "click" sound.
   c. Install a new No. 2 fuel pipe clamp.
16. Connect fuel hose bracket.
17. Install air cleaner cap sub-assembly.
18. Install outer cowl top panel.
19. Install windshield wiper motor and link.
20. Connect cable to negative battery terminal.
21. Inspect for fuel leak.
   a. Connect the Techstream to the DLC3.
   b. Turn the ignition switch to ON, and then turn the Techstream on.

➡Do not start the engine.

   c. Enter the following menus: Powertrain / Engine and ECT / Active Test / Control the Fuel Pump / Speed.
22. Check that there are no leaks from the fuel system.
   a. Turn the ignition switch off.
   b. Disconnect the Techstream from the DLC3.

xB

See Figure 197.

1. Discharge fuel system pressure.
2. Remove No. 1 engine cover sub-assembly.
3. Remove air cleaner cap sub-assembly with No. 1 air cleaner hose.
4. Disconnect fuel main tube.

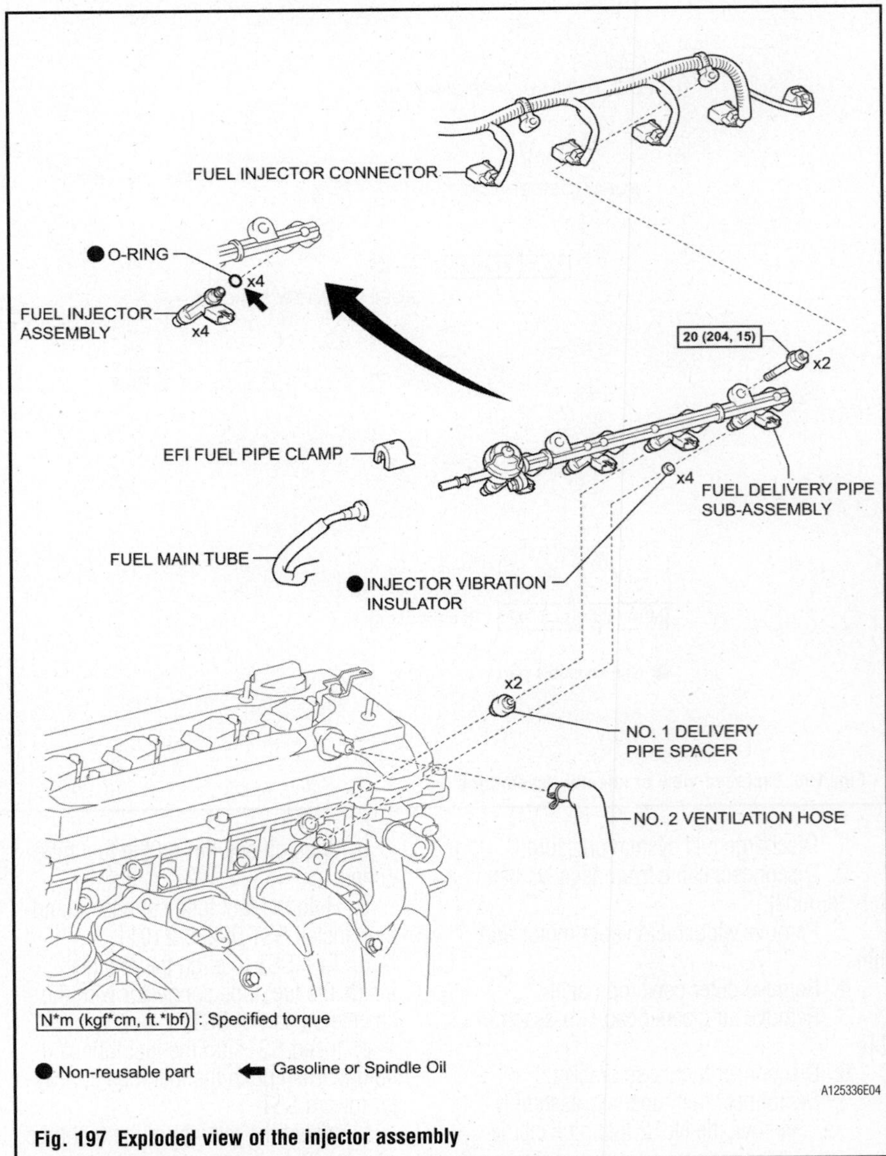

Fig. 197 Exploded view of the injector assembly

a. Remove the fuel tube from the fuel hose clamp.

b. Remove the EFI fuel pipe clamp.

c. Wipe off any dirt on the fuel tube connector.

d. Hold the fuel tube connector, and then install SST (09268-21011).

e. Turn SST to align the retainer inside the fuel tube connector with the chamfered part of SST.

5. Insert SST into the fuel tube and hold it. Then push the fuel tube connector toward SST.

a. Mount the retainer of the fuel tube connector onto the chamfered part of SST.

b. Slide SST and fuel tube connector together towards the fuel tube until they make a "click" sound, and then disconnect the fuel tube.

c. Drain the fuel remaining inside the fuel tube.

d. Cover the fuel tube and fuel pipe with a plastic bag to protect the disconnected part.

6. Disconnect No. 2 ventilation hose.

7. Remove fuel delivery pipe sub-assembly.

a. Disconnect the 2 wire harness clamps.

b. Disconnect the 4 fuel injector connectors.

c. Remove the 2 bolts, then remove the fuel delivery pipe together with the 4 fuel injectors.

d. Remove the 2 No. 1 delivery pipe spacers from the cylinder head.

e. Remove the 4 injector vibration insulators from the cylinder head.

8. Remove fuel injector assembly.

To install:

9. Install fuel injector assembly.

a. Apply a light coat of gasoline or spindle oil to new O-rings, then install one onto each fuel injector.

b. Apply a light coat of gasoline or spindle oil to the part of the fuel delivery pipe which comes into contact with the O-ring of the fuel injector.

c. Apply a light coat of gasoline or spindle oil to the O-ring again, then install the fuel injectors into the fuel delivery pipe.

10. Check that the fuel injector rotates smoothly. If the fuel injector does not rotate, replace the O-ring.

11. Install fuel delivery pipe sub-assembly.

a. Install 4 new injector vibration insulators onto the cylinder head.

b. Install the 2 No. 1 delivery pipe spacers onto the cylinder head.

c. Install the fuel delivery pipe together with the 4 fuel injectors, then temporarily tighten the 2 bolts.

d. Check that the fuel injector rotates smoothly.

If the fuel injector does not rotate, replace the O-ring.

e. Fully tighten the 2 bolts to the specified torque.

f. Connect the 4 fuel injector connectors.

g. Connect the 2 wire harness clamps.

12. Connect No. 2 ventilation hose.

13. Connect fuel main tube.

a. Connect the fuel main tube.

b. Push the fuel tube connector until it makes a "click" sound.

c. Install the EFI fuel pipe clamp.

d. Install the fuel tube into the fuel hose clamp.

14. Install air cleaner cap sub-assembly with No. 1 air cleaner hose.

15. Inspect for fuel leak.

16. Install No. 1 engine cover sub-assembly.

xD

See Figure 198.

1. Discharge fuel system pressure.
2. Remove No. 2 cylinder head cover.
3. Remove No. 2 ventilation hose.
4. Separate engine wire harness.

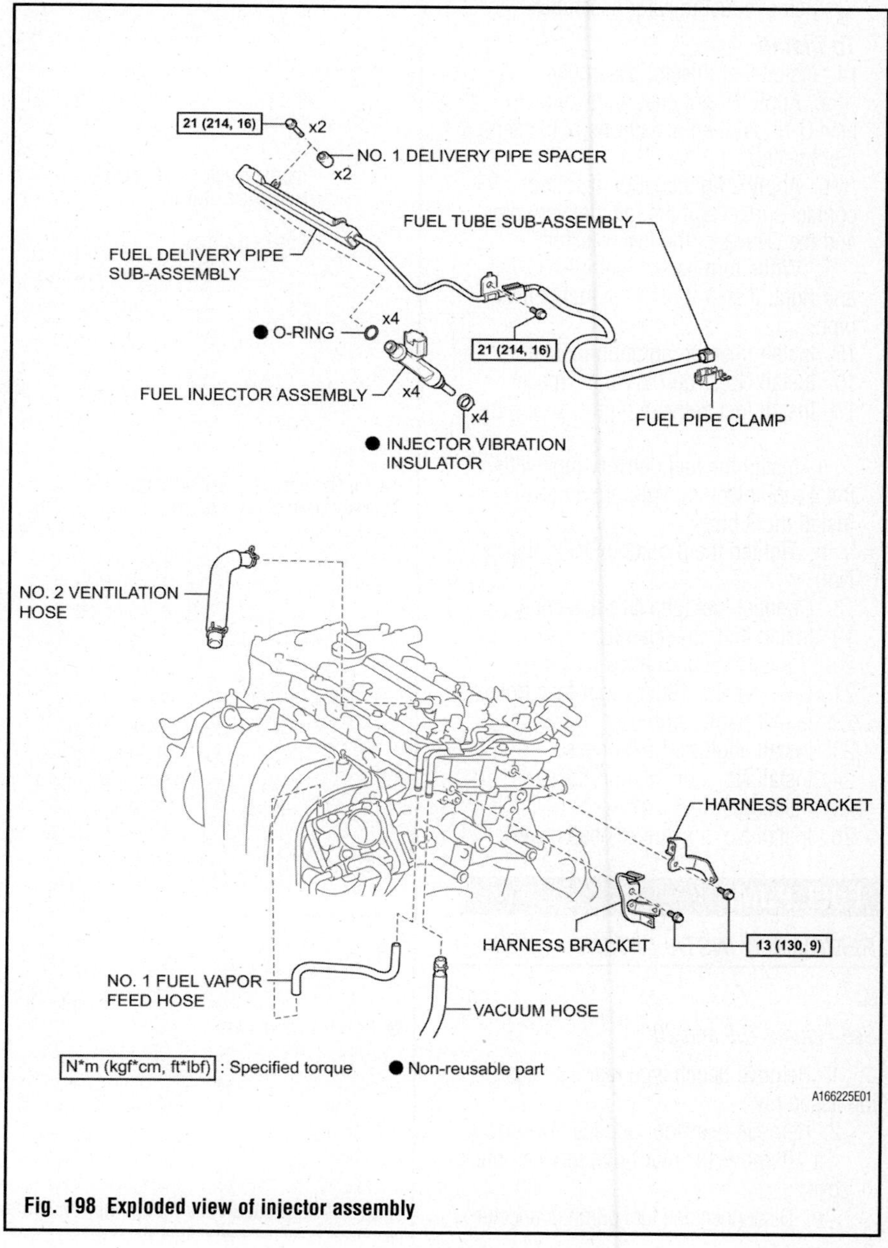

Fig. 198 Exploded view of injector assembly

5. Remove harness bracket.
   a. Disconnect the 5 wire harness clamps.
   b. Remove the 2 bolts and remove the 2 harness brackets.
6. Disconnect No. 1 fuel vapor feed hose.
7. Disconnect vacuum hose.
8. Remove fuel pipe clamp.
9. Disconnect fuel tube sub-assembly.
10. Remove fuel delivery pipe sub-assembly.
    a. Remove the bolt.
    b. Remove the 2 bolts and remove the fuel delivery pipe with the 4 fuel injectors.
11. Remove No. 1 delivery pipe spacer.
12. Remove injector vibration insulator.
13. Remove fuel injector assembly.

To install:

14. Install fuel injector assembly.
    a. Apply a light coat of gasoline to new O-rings, then install one onto each fuel injector.
    b. Apply a light coat of gasoline to the contact surfaces of the fuel delivery pipe and the O-ring of the fuel injector.
    c. While turning the fuel injector left and right, install it onto the fuel delivery pipe.
15. Install injector vibration insulator.
16. Install No. 1 delivery pipe spacer.
17. Install fuel delivery pipe sub-assembly.
    a. Install the fuel delivery pipe with the 4 fuel injectors, then provisionally install the 3 bolts.
    b. Tighten the 3 bolts to 16 ft. lbs. (21 Nm).
18. Connect fuel tube sub-assembly.
19. Install fuel pipe clamp.
20. Connect vacuum hose.
21. Connect No. 1 fuel vapor feed hose.
22. Install harness bracket.
23. Install engine wire harness.
24. Install No. 2 ventilation hose.
25. Inspect for fuel leak.
26. Install No. 2 cylinder head cover.

FUEL PUMP

REMOVAL & INSTALLATION

tC

See Figures 199 and 200.

1. Remove bench type rear seat cushion assembly.
2. Remove rear floor service hole cover.
   a. Remove the rear floor service hole cover.
   b. Disconnect the fuel pump connector.
3. Discharge fuel system pressure.
4. Disconnect cable from negative battery terminal.
5. Disconnect fuel tank main tube sub-assembly.
6. Remove fuel pump gauge retainer.
   a. Using 6 mm socket hexagon wrench, install SST to the fuel pump gauge retainer.

➡Securely attach the claws of SST to the protrusions of the fuel pump gauge retainer and fix SST in place. Install SST while pressing the claws of SST against the fuel pump gauge retainer (towards the center of SST).

   b. Using SST, loosen the fuel pump gauge retainer.
   c. Remove the fuel pump gauge retainer while holding the fuel suction tube assembly by hand.
7. Remove fuel suction tube assembly with pump and gauge.
   a. Remove the fuel suction tube assembly with pump and gauge from the fuel tank.
   b. Remove the gasket from the fuel tank.

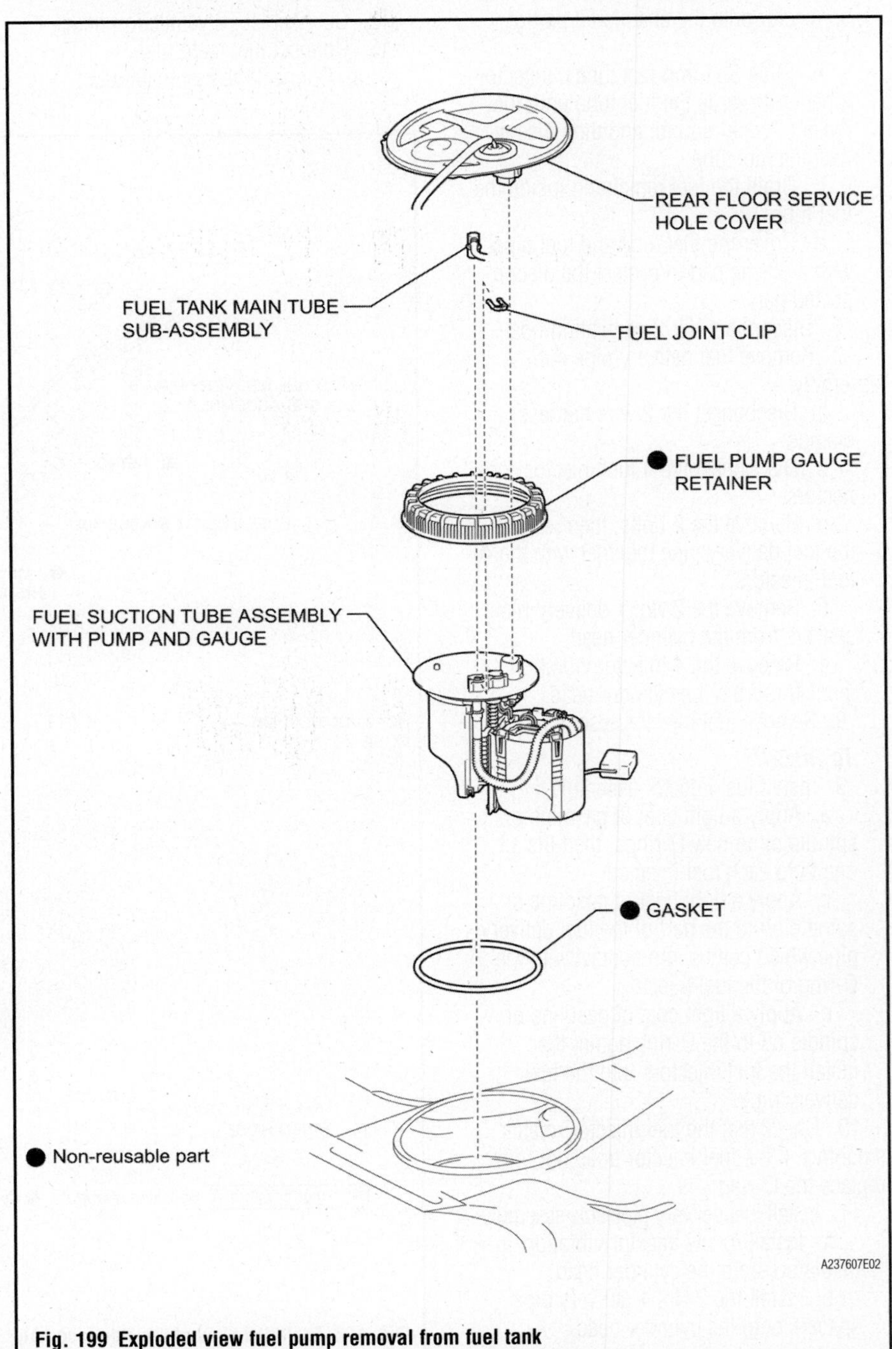

Fig. 199 Exploded view fuel pump removal from fuel tank

To install:

8. Install fuel suction tube assembly with pump and gauge.
   a. Install a new gasket to the fuel tank.
   b. Install the fuel suction tube assembly with pump and gauge to the fuel tank.
   c. Align the protrusions of the fuel suction tube assembly with pump and gauge with the groove of the fuel tank.
9. Install fuel pump gauge retainer.
   a. Put a new retainer on the fuel tank. While holding the fuel suction tube assembly with pump and gauge, tighten the retainer one complete turn by hand.
   b. Using a 6 mm socket hexagon wrench, install SST to the fuel pump gauge retainer.
   c. Using SST, tighten the retainer until the mark on the retainer is within range A on the fuel tank.
10. Connect fuel tank main tube sub-assembly.
   a. Install the fuel tank main tube with the fuel joint clip.
   b. Connect the fuel pump connector.

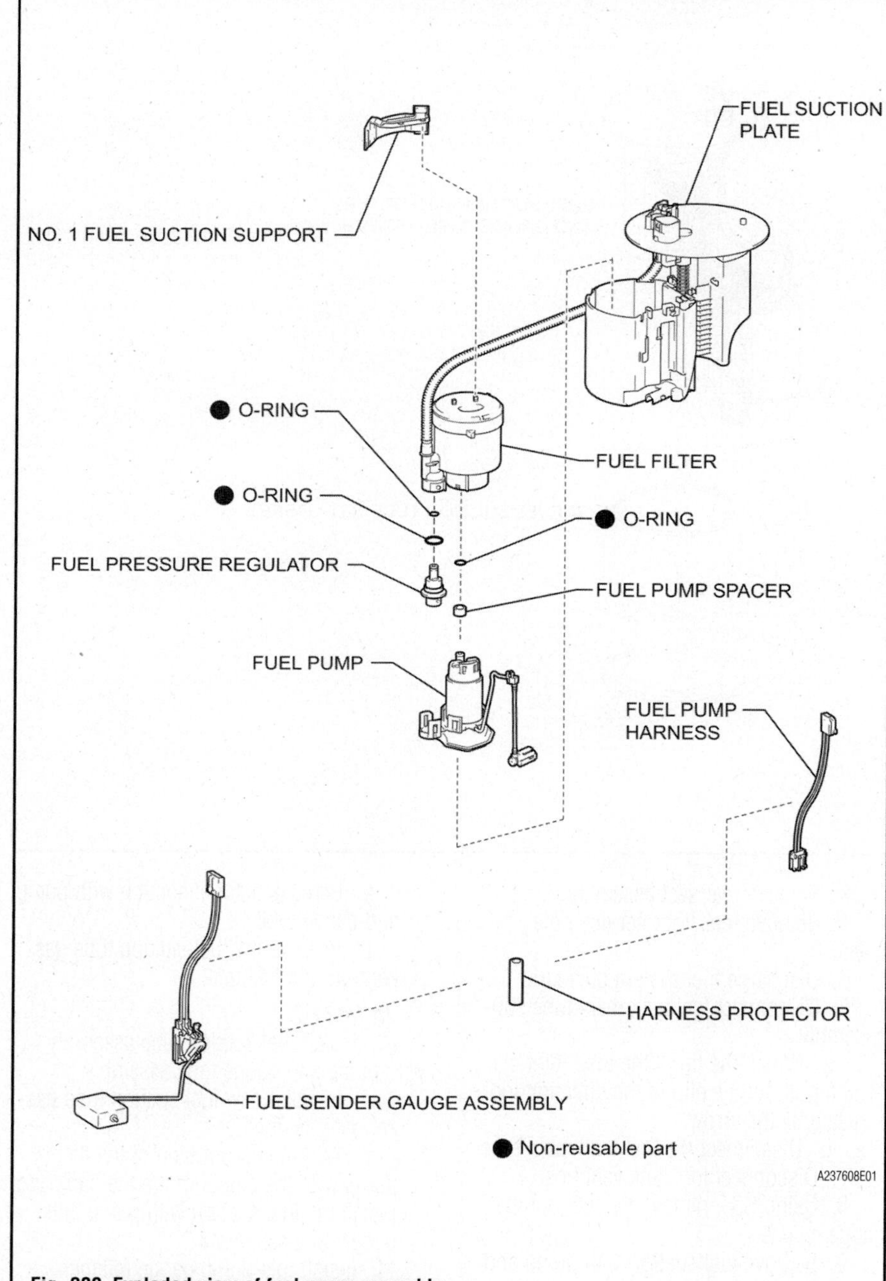

Fig. 200 Exploded view of fuel pump assembly

11. Connect cable to negative battery terminal.
12. Inspect for fuel leak.
13. Install rear floor service hole cover.
14. Install bench type rear seat cushion assembly.

xB

See Figure 201.

1. Remove rear seat cushion with cover pad sub-assembly.
2. Remove rear seat under tray cover.
3. Remove bench type rear seat cushion assembly.
4. Remove rear door scuff plate RH
5. Remove rear door scuff plate LH.
6. Remove rear floor service hole cover.
   a. Remove the rear floor service hole cover.
   b. Disconnect the connector.
7. Discharge fuel system pressure.
8. Remove fuel tank main tube sub-assembly.
   a. Widen the tip of the tube joint clip and pull out the clip.
   b. Pull out the fuel tank main tube to disconnect it.
9. Remove fuel pump gauge retainer.
10. Remove fuel suction with pump and gauge tube assembly.
   a. Remove the fuel suction with pump and gauge tube assembly.
   b. Remove the fuel suction tube set gasket from the fuel tank.

To install:

11. Install fuel suction with pump and gauge tube assembly.
   a. Install a new fuel suction tube set gasket onto the fuel tank.
   b. Align the protrusion on the fuel pump with the cutout on the fuel tank and install the fuel suction with pump and gauge tube assembly.
12. Install fuel pump gauge retainer.
   a. While holding the fuel suction with pump and gauge tube assembly by hand to prevent it from tilting, align the mark on the fuel pump gauge retainer with the start mark on the fuel tank and tighten the fuel pump gauge retainer 180° by hand.
   b. Using SST, tighten the fuel pump gauge retainer approximately 360° so that the mark on the retainer comes within the range.
13. Connect fuel tank main tube sub-assembly.
   a. Insert the fuel tank main tube.
   b. Install the tube joint clip.
14. Install rear floor service hole cover.
   a. Connect the fuel pump connector.

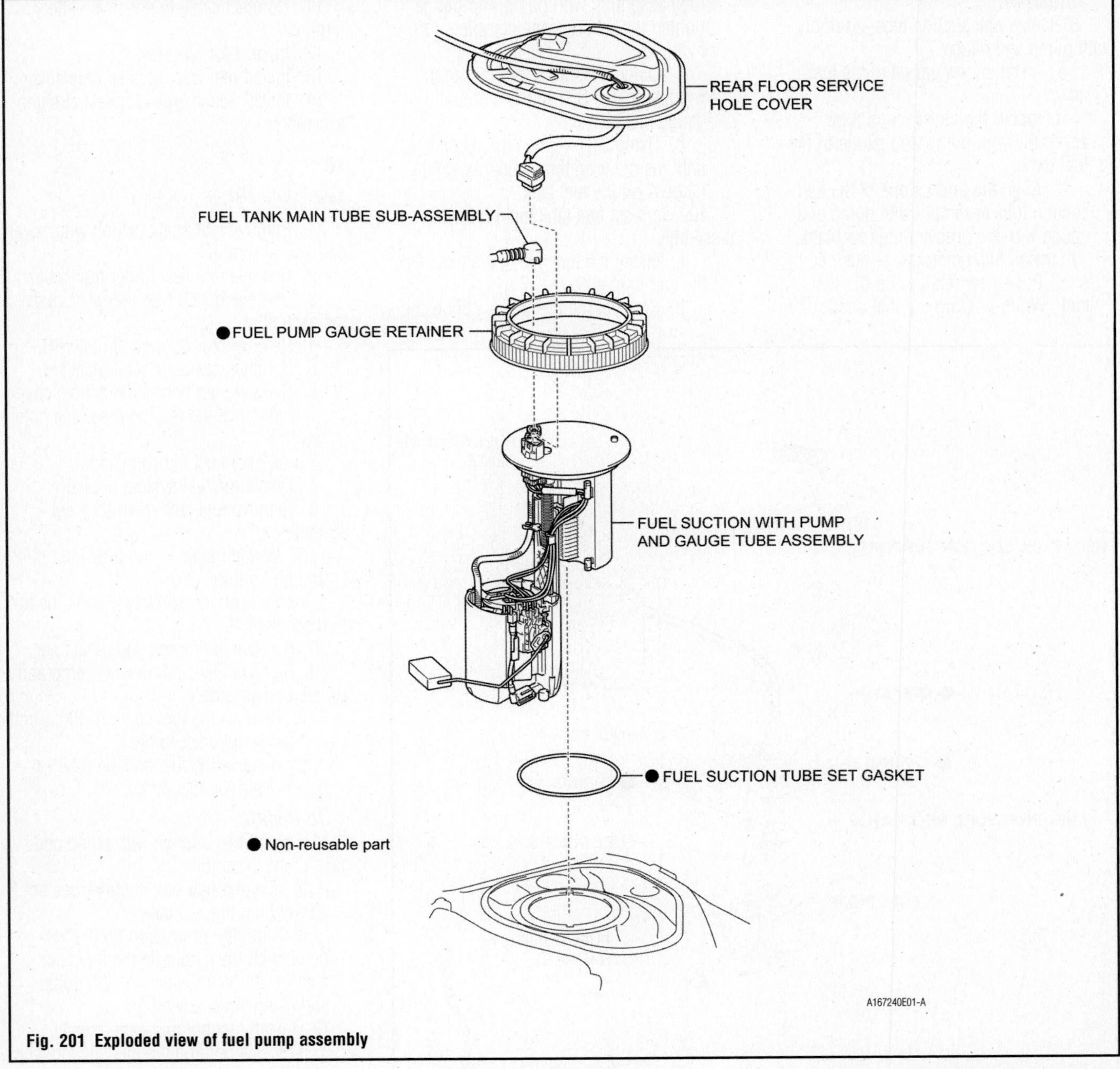

Fig. 201 Exploded view of fuel pump assembly

b. Install the rear floor service hole cover with new butyl tape.

15. Inspect for fuel leak.
16. Install rear door scuff plate RH.
17. Install rear door scuff plate LH.
18. Install bench type rear seat cushion assembly.
19. Install rear seat under tray cover.
20. Install rear seat cushion with cover pad sub-assembly.

xD

See Figure 202.

1. Remove No. 1 rear seat leg cover.
2. Remove No. 2 seat leg cover.
3. Remove rear seat assembly.
4. Remove rear floor service hole cover.
5. Discharge fuel system pressure.
6. Disconnect fuel tank main tube sub-assembly.
   a. Widen the tip of the tube joint clip and pull out the clip in the direction indicated by the arrow.
   b. Disconnect the fuel tank main tube.
7. Disconnect fuel tank vent hose.
8. Using SST, remove the fuel pump gauge retainer.
9. Remove fuel suction with pump and gauge tube assembly.
   a. Remove the fuel suction with pump and gauge tube.
   b. Remove the fuel suction tube gasket from the fuel tank.

To install:

10. Install fuel suction tube assembly with pump and gauge tube assembly.
    a. Install a new fuel suction tube gasket onto the fuel tank.
    b. Align the protrusion of the fuel pump with the cutout of the fuel tank and install the fuel suction with pump and gauge tube assembly.
11. Install fuel pump gauge retainer.
12. Connect fuel tank vent hose.

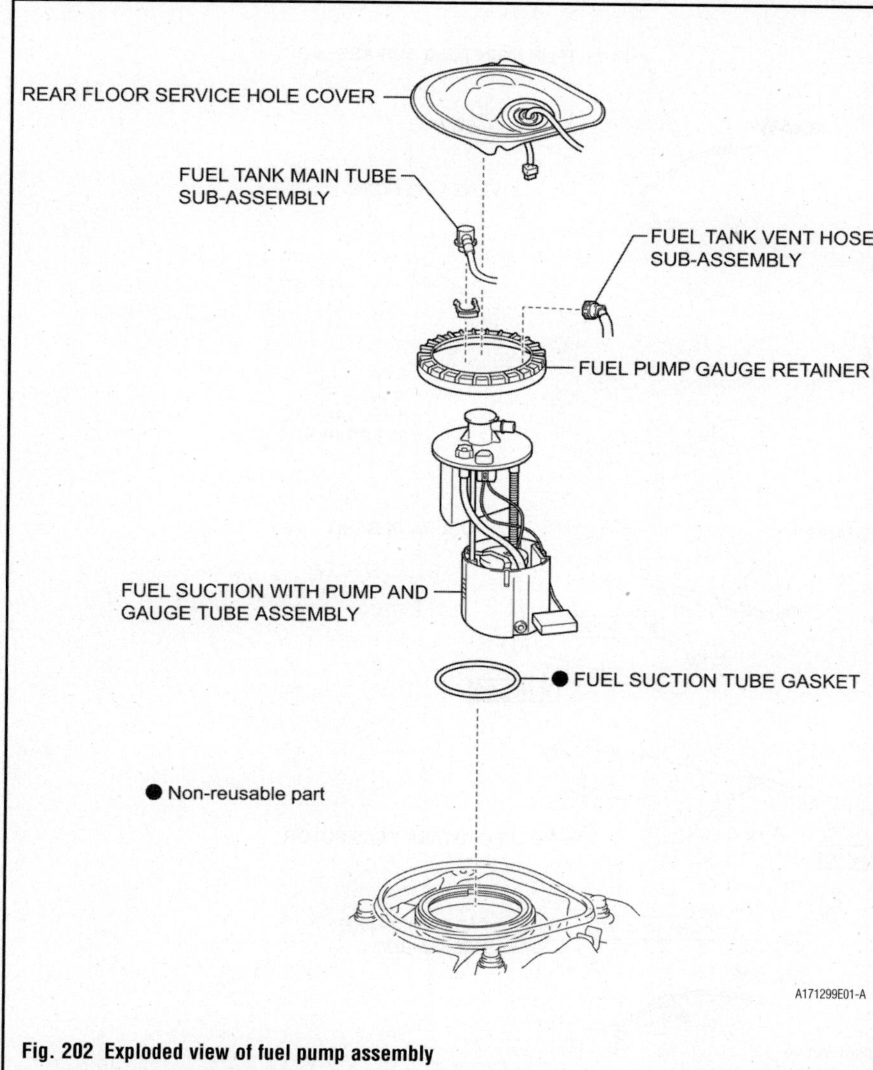

Fig. 202 Exploded view of fuel pump assembly

13. Connect fuel tank main tube sub-assembly.
    a. Insert the fuel tank main tube.
    b. Install the tube joint clip.
14. Install rear floor service hole cover.
    a. Connect the fuel pump connector.
    b. Install the rear service hole cover with new butyl tape.
15. Install rear seat cushion with cover pad sub-assembly LH.
16. Remove No. 2 seat leg cover.
17. Install No. 1 seat leg cover.
18. Connect cable to negative battery terminal.
19. Check for fuel leak.

FUEL TANK

REMOVAL & INSTALLATION

tC

See Figure 203.

1. Disconnect cable from negative battery terminal.
2. Remove fuel suction tube assembly with pump and gauge.
3. Drain fuel.
4. Remove center exhaust pipe assembly.
5. Remove No. 1 fuel tank protector.
6. Disconnect fuel tank main tube sub-assembly.
7. Remove fuel tank assembly.
    a. Remove the 2 bolts and disconnect the No. 2 parking brake cable assembly and No. 3 parking brake cable assembly.
    b. Disconnect the vent line hose.
    c. Loosen the hose clamp bolt, and then disconnect the fuel tank to filler pipe hose.
    d. Hold the fuel tank using an engine lifter.
    e. Remove the 4 bolts and 2 fuel tank bands.
    f. Operate the engine lifter, and then remove the fuel tank.
8. Remove fuel tank main tube sub-assembly.

To install:

9. Install fuel tank main tube sub-assembly.
10. Install fuel tank assembly.
    a. Set the fuel tank onto an engine lifter.
    b. Operate the engine lifter, and then install the fuel tank to the vehicle.
    c. Install the 2 fuel tank bands with the 4 bolts. Tighten to 29 ft. lbs. (39 Nm).
    d. Connect the fuel tank to filler pipe hose to the fuel tank, and then tighten the hose clamp.
    e. Connect the vent line hose.
    f. Connect the No. 2 parking brake cable assembly and No. 3 parking brake cable assembly with the 2 bolts.
11. Connect fuel tank main tube sub-assembly.
12. Install No. 1 fuel tank protector.
13. Install center exhaust pipe assembly.
14. Add fuel.
15. Install fuel suction tube assembly with pump and gauge.
16. Connect cable to negative battery terminal.
17. Inspect for fuel leak.

xB

See Figures 204 and 205.

1. Remove fuel suction with pump and gauge tube assembly.
2. Drain fuel.
3. Disconnect fuel tank vent hose sub-assembly.
4. Remove No. 2 center exhaust pipe assembly.
5. Remove rear floor side member cover RH.
6. Remove rear floor side member cover LH.
7. Remove rear floor side member brace sub-assembly.
8. Remove No. 1 fuel tank protector sub-assembly.
9. Disconnect fuel tank main tube sub-assembly.
10. Disconnect fuel tank to filler pipe hose.
11. Disconnect No. 6 fuel tank breather tube.
12. Remove fuel tank assembly.
    a. Remove the 2 bolts and separate the parking brake cable assembly.
    b. Set the mission jack under the fuel tank.

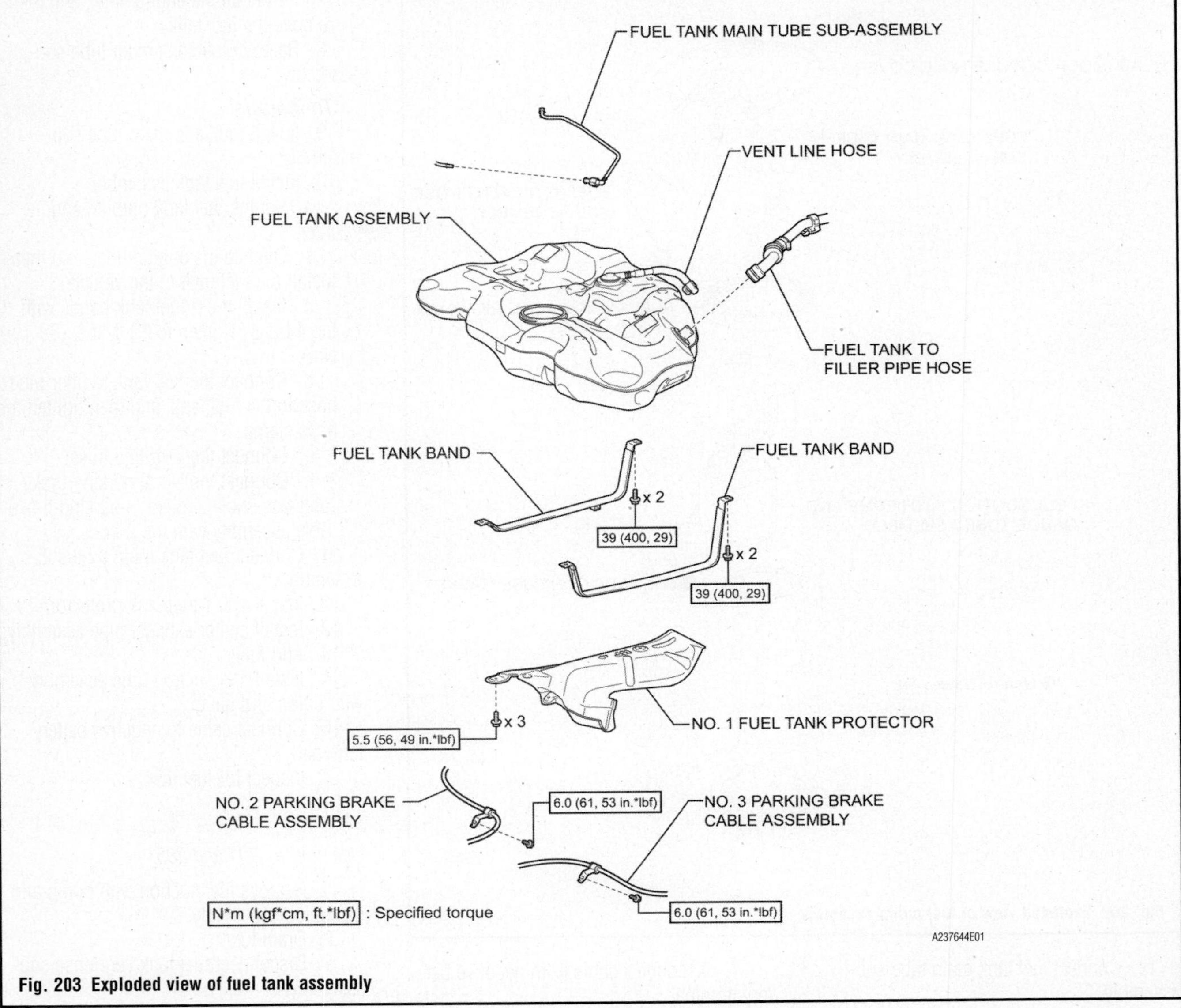

Fig. 203 Exploded view of fuel tank assembly

c. Remove the 4 bolts and remove the fuel tank assembly.

13. Remove fuel tank cushion.
14. Remove fuel tank main tube sub-assembly.

To install:

15. Install fuel tank main tube sub-assembly.
16. Install fuel tank cushion.
17. Install fuel tank assembly.
    a. Support the fuel tank using an mission jack.
    b. Temporary install the fuel tank with 4 new bolts.
    c. Tighten the 4 bolts in order to 29 ft. lbs. (39 Nm).
    d. Connect the parking brake cable assembly with the 2 bolts.
18. Connect fuel tank main tube sub-assembly.
19. Connect No. 6 fuel tank breather tube.
20. Connect fuel tank to filler pipe hose.
21. Install No. 1 fuel tank protector sub-assembly.
22. Install No. 2 center exhaust pipe assembly.
23. Install rear floor side member brace sub-assembly.
24. Install rear floor side member cover LH.
25. Install rear floor side member cover RH.
26. Connect fuel tank vent hose sub-assembly.
27. Install fuel suction with pump and gauge tube assembly.
28. Add fuel.

xD

See Figure 206.

1. Remove No. 1 seat leg cover.
2. Remove No. 2 seat leg cover.
3. Remove rear seat cushion with cover pad sub-assembly LH.
4. Remove rear floor service hole cover.
5. Discharge fuel system pressure.
6. Disconnect fuel tank main tube sub-assembly.
7. Disconnect fuel tank vent hose.
8. Remove fuel pump gauge retainer.
9. Remove fuel suction with pump and gauge tube assembly.
10. Drain fuel.
11. Remove No. 4 front floor heat insulator.
12. Remove No. 1 fuel tank protector.

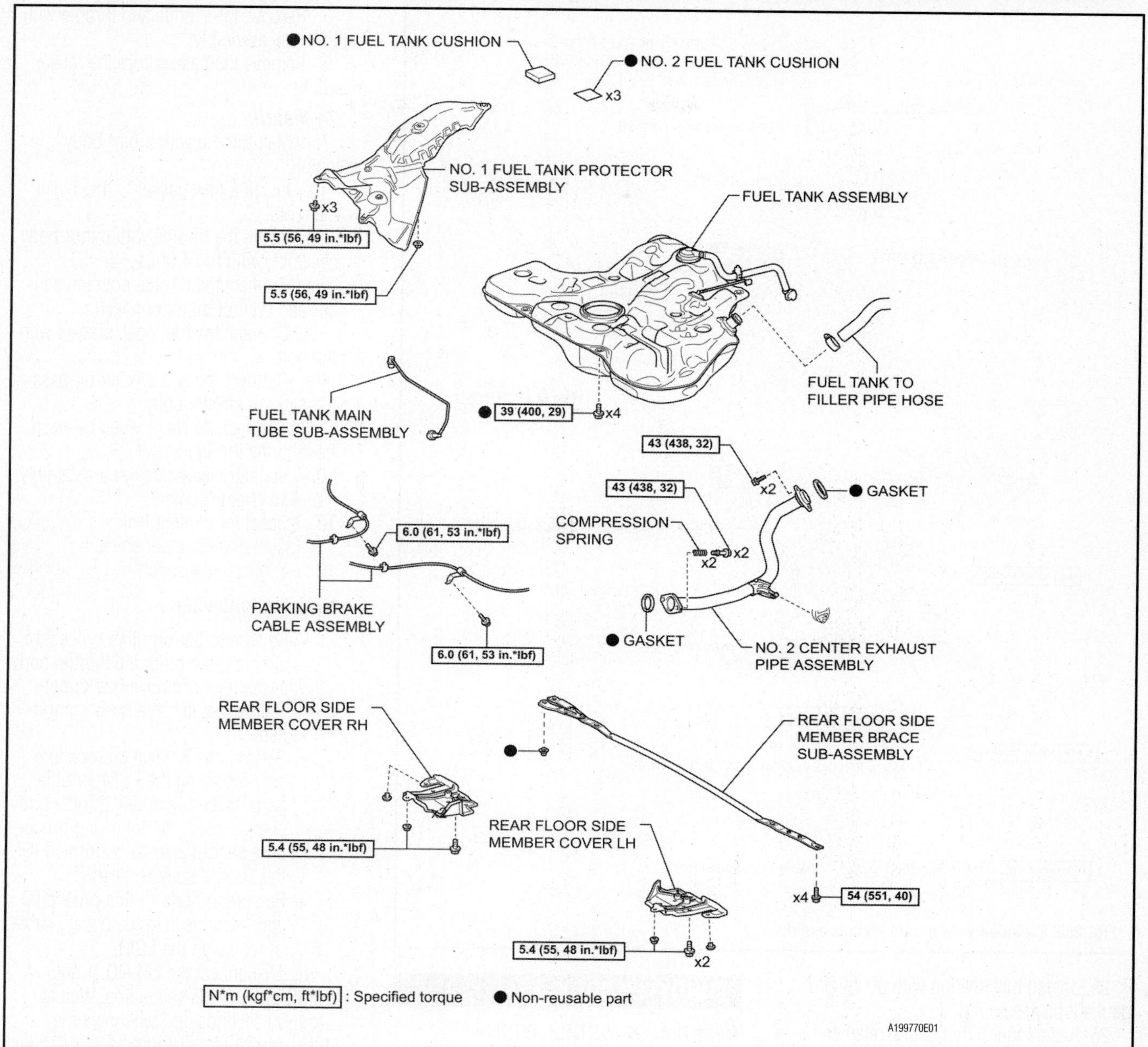

NO. 1 FUEL TANK CUSHION
NO. 2 FUEL TANK CUSHION
x3
NO. 1 FUEL TANK PROTECTOR SUB-ASSEMBLY
x3
5.5 (56, 49 in.*lbf)
5.5 (56, 49 in.*lbf)
FUEL TANK ASSEMBLY
FUEL TANK TO FILLER PIPE HOSE
FUEL TANK MAIN TUBE SUB-ASSEMBLY
39 (400, 29) x4
43 (438, 32)
43 (438, 32) x2
GASKET
COMPRESSION SPRING
x2 x2
6.0 (61, 53 in.*lbf)
PARKING BRAKE CABLE ASSEMBLY
6.0 (61, 53 in.*lbf)
GASKET
NO. 2 CENTER EXHAUST PIPE ASSEMBLY
REAR FLOOR SIDE MEMBER COVER RH
REAR FLOOR SIDE MEMBER BRACE SUB-ASSEMBLY
5.4 (55, 48 in.*lbf)
REAR FLOOR SIDE MEMBER COVER LH
x4 54 (551, 40)
5.4 (55, 48 in.*lbf) x2
N*m (kgf*cm, ft*lbf) : Specified torque ● Non-reusable part
A199770E01

Fig. 204 Exploded view of fuel tank assembly

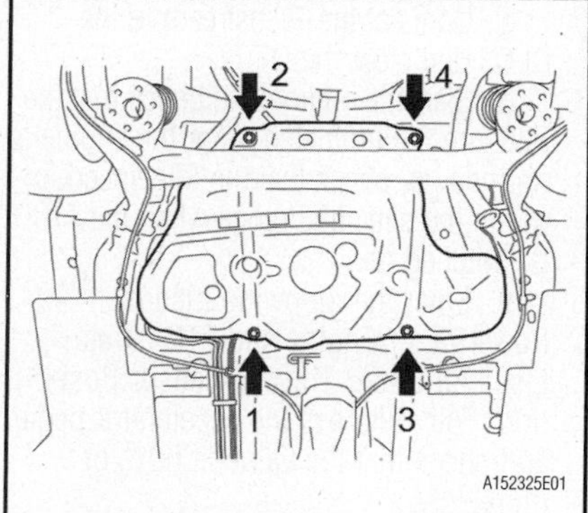

Fig. 205 Tighten the 4 bolts in order to 29 ft. lbs. (39 Nm)

13. Disconnect fuel tank main tube sub-assembly.
14. Disconnect fuel tank vent hose.
15. Disconnect fuel tank breather hose.
16. Disconnect lower fuel tank filler pipe sub-assembly.
17. Remove fuel tank assembly.
    a. Remove the 4 bolts and remove the fuel tank.
    b. Remove the fuel tank main tube from the fuel tank.
    c. Remove the fuel tank vent hose from the fuel tank.

To install:

18. Install fuel tank assembly;
    a. Install the fuel vent hose onto the fuel tank.
    b. Install the fuel tank main tube onto the fuel tank.
    c. Clean the bolt hole and remove any grease.
    d. Install the fuel tank with 4 new bolts. Tighten to 10 ft. lbs. (14 Nm).
19. Connect lower fuel tank filler pipe sub-assembly;
20. Connect fuel tank breather tube.
21. Connect fuel tank vent hose.
22. Connect fuel tank main tube sub-assembly.
23. Install No. 1 fuel tank protector.
24. Install No. 4 front floor heat insulator.

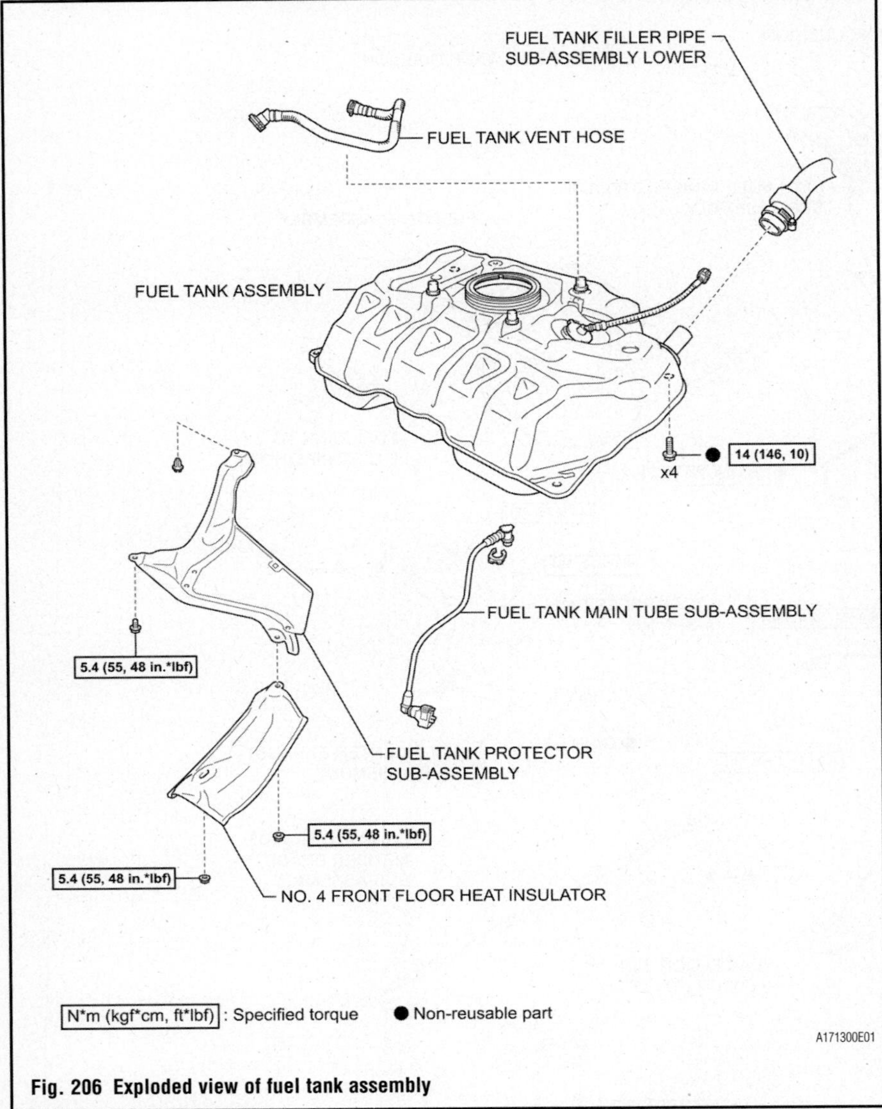

Fig. 206 Exploded view of fuel tank assembly

25. Install fuel suction with pump and gauge tube assembly.
26. Install fuel pump gauge retainer.
27. Connect fuel tank vent hose sub-assembly.
28. Connect fuel tank main tube sub-assembly.
29. Connect rear floor service hole cover.
30. Install rear seat cushion with cover pad sub-assembly LH.
31. Install No. 2 seat leg cover.
32. Install No. 1 seat leg cover.
33. Add fuel.
34. Inspect for fuel leak.

IDLE SPEED

ADJUSTMENT

Idle speed is controlled by the Engine Control Module (ECM). No adjustment is necessary or possible.

THROTTLE BODY

REMOVAL & INSTALLATION

tC

See Figure 207.

1. Remove engine under cover.
2. Drain engine coolant.
3. Remove air cleaner cap sub-assembly.
4. Remove throttle with motor body assembly.
   a. Disconnect the No. 1 water by-pass hose from the throttle body.
   b. Disconnect the No. 2 water by-pass hose from the throttle body.
   c. Remove the bolt and disconnect the fuel hose bracket.
   d. Disconnect the throttle position sensor and control motor connector.
5. Remove the 4 bolts and throttle with motor body assembly.
6. Remove the gasket from the intake manifold.

To install:

7. Install throttle with motor body assembly.
   a. Install a new gasket to the intake manifold.
   b. Install the throttle with motor body assembly with the 4 bolts.
   c. Connect the throttle position sensor and control motor connector.
   d. Connect the fuel hose bracket with the bolt.
   e. Connect the No. 2 water by-pass hose to the throttle body.
   f. Connect the No. 1 water by-pass hose to the throttle body.
8. Install air cleaner cap sub-assembly.
9. Add engine coolant.
10. Inspect for coolant leak.
11. Install engine under cover.
12. Perform initialization.

➡Note the following:

- Be sure to perform this procedure after reassembling the throttle body assembly or removing and reinstalling any throttle body component.
- Perform the following procedure after replacing the ECM, throttle body assembly or any throttle body components. The following procedure should also be performed if the throttle body is cleaned.
- Be sure to perform this procedure after reconnecting the battery cable or replacing the ECM.

   a. Disconnect the EFI NO. 1 and ETCS fuses at the same time. Wait at least 60 seconds and reconnect the fuses.
   b. Turn the ignition switch to ON without operating the accelerator pedal.
   c. Connect the Techstream to the DLC3 and clear the DTCs.
   d. Start the engine and check that the MIL is not illuminated. After the engine is warmed up, check that the idle speed is within the specified range when the A/C is switched off.
   e. Enter the following menus: Powertrain / Engine and ETC / Data List / All Data / Throttle Sensor Position. Fully depress the accelerator pedal and check that the value is 60% or more.
13. Perform a road test and confirm that there are no abnormalities.

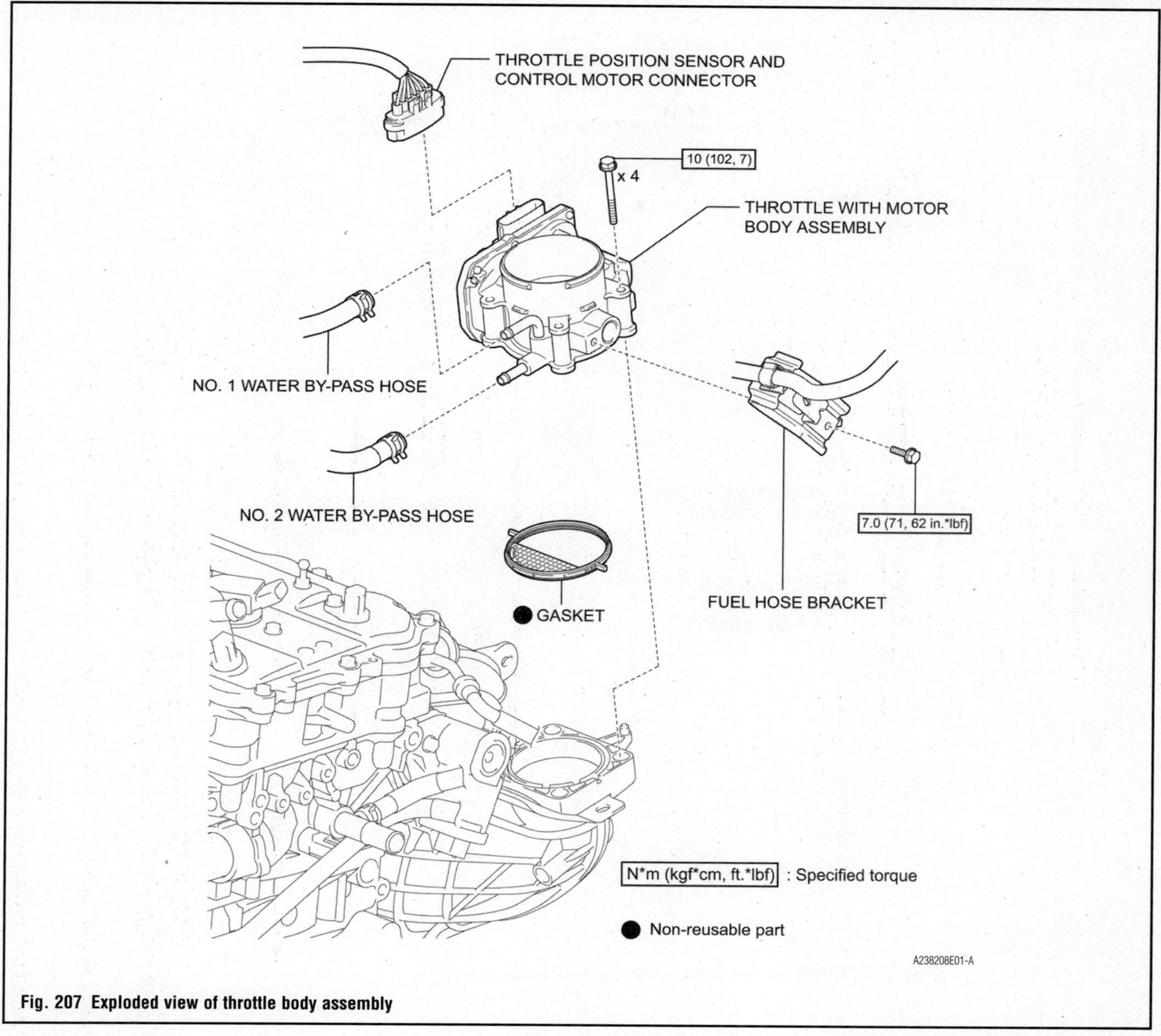

Fig. 207 Exploded view of throttle body assembly

xB

See Figure 208.

1. Drain engine coolant.
2. Remove No. 1 engine cover sub-assembly.
3. Remove air cleaner cap sub-assembly with No. 1 air cleaner hose.
    a. Disconnect the Mass Air Flow (MAF) meter connector.
    b. Disengage the wire harness clamp.
    c. Separate the No. 1 vacuum switching valve assembly.
    d. Disconnect the ventilation hose.
    e. Disengage the hose clamp and separate the No. 2 fuel vapor feed hose.
    f. Loosen the No. 1 air cleaner hose clamp, unlock the air cleaner assembly clamp and remove the air cleaner cap sub-assembly with No. 1 air cleaner hose.
    g. Remove the air cleaner filter element sub-assembly.
4. Remove throttle with motor body assembly.
    a. Disconnect the No. 2 fuel vapor feed hose.
    b. Disconnect the No. 1 water by-pass hose.
    c. Disconnect the No. 2 water by-pass hose.
    d. Disconnect the throttle body connector.
    e. Remove the 4 bolts and the throttle with motor body assembly.
    f. Remove the gasket.

To install:

5. Install throttle with motor body assembly
    a. Install a new gasket onto the intake manifold.
    b. Install the throttle with motor body assembly with the 4 bolts. Tighten to 22 ft. lbs. (30 Nm).
    c. Connect the throttle body connector.
    d. Connect the No. 2 water by-pass hose to the throttle with motor body assembly.
    e. Connect the No. 1 water by-pass hose to the throttle with motor body assembly.
    f. Connect the No. 2 fuel vapor feed hose to the throttle with motor body assembly.

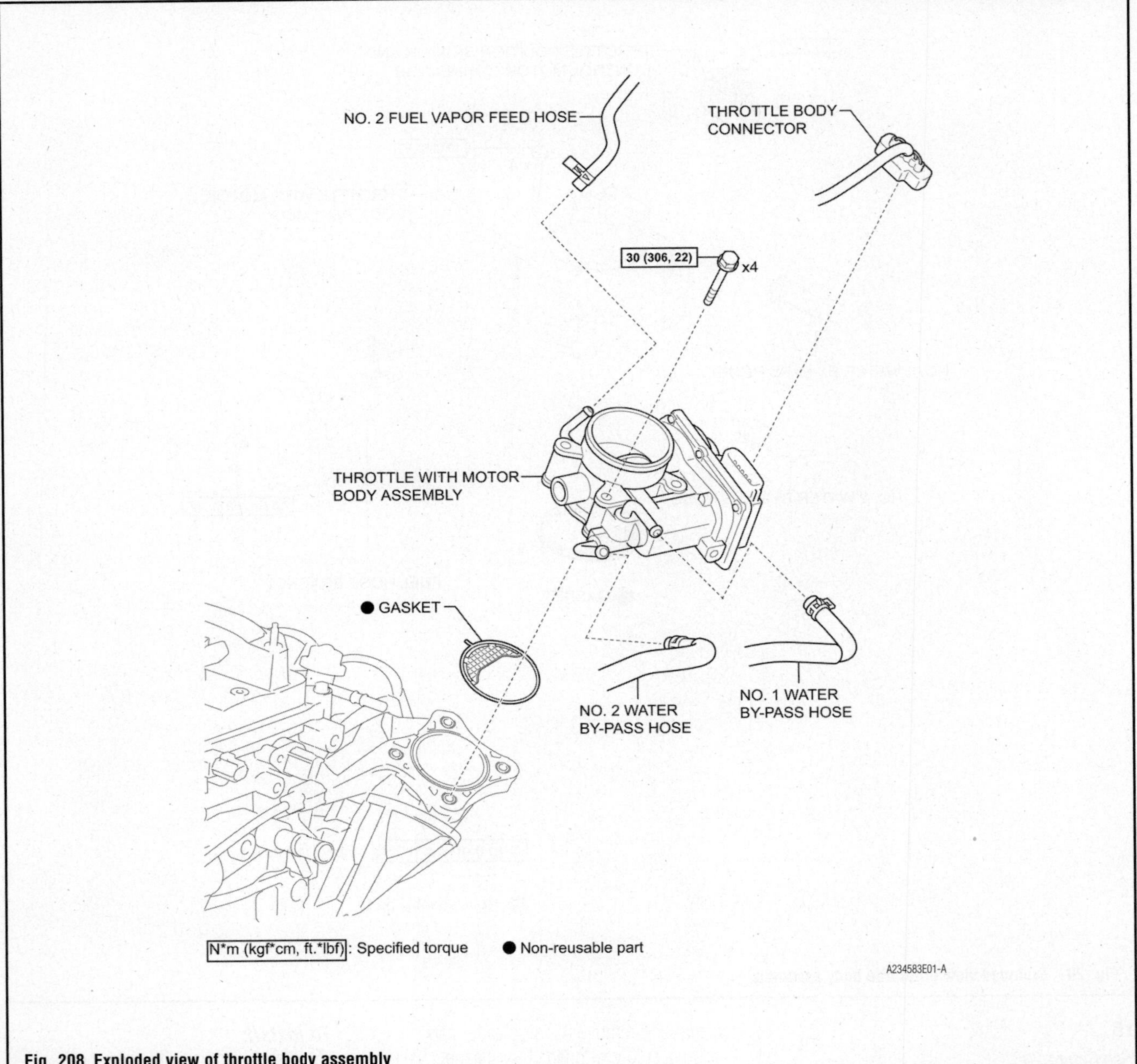

Fig. 208 Exploded view of throttle body assembly

6. Install air cleaner cap sub-assembly with No. 1 air cleaner hose.
7. Add engine coolant.
8. Inspect for coolant leak.
9. Install No. 1 engine cover sub-assembly.
10. Perform initialization.
   a. Disconnect the cable from the negative (-) battery terminal. Wait at least 60 seconds and reconnect the cable.
   b. Turn the ignition switch to ON without operating the accelerator pedal.
   c. Connect the Techstream to the DLC3 and clear the DTCs.
   d. Start the engine and check that the MIL is not illuminated and that the idle speed is within the specified range when the A/C is switched off after the engine is warmed up.
   e. Enter the following menus: Powertrain/ Engine / Data List/ Throttle Sensor Position. Fully depress the accelerator pedal and check that the value is 60% or more.
   f. Perform a road test and confirm that there are no abnormalities.

xD

See Figure 209.

1. Drain engine coolant.
2. Remove No. 2 cylinder head cover.
3. Remove air cleaner cap sub-assembly with hose.
   a. Disconnect the wire harness clamp and Mass Air Flow (MAF) meter connector.
   b. Disconnect the No. 2 ventilation hose.
   c. Loosen the hose clamp, unlock the 2 clamps and remove air cleaner cap sub-assembly with hose.
4. Remove throttle body assembly.
   a. Disconnect the throttle body connector.

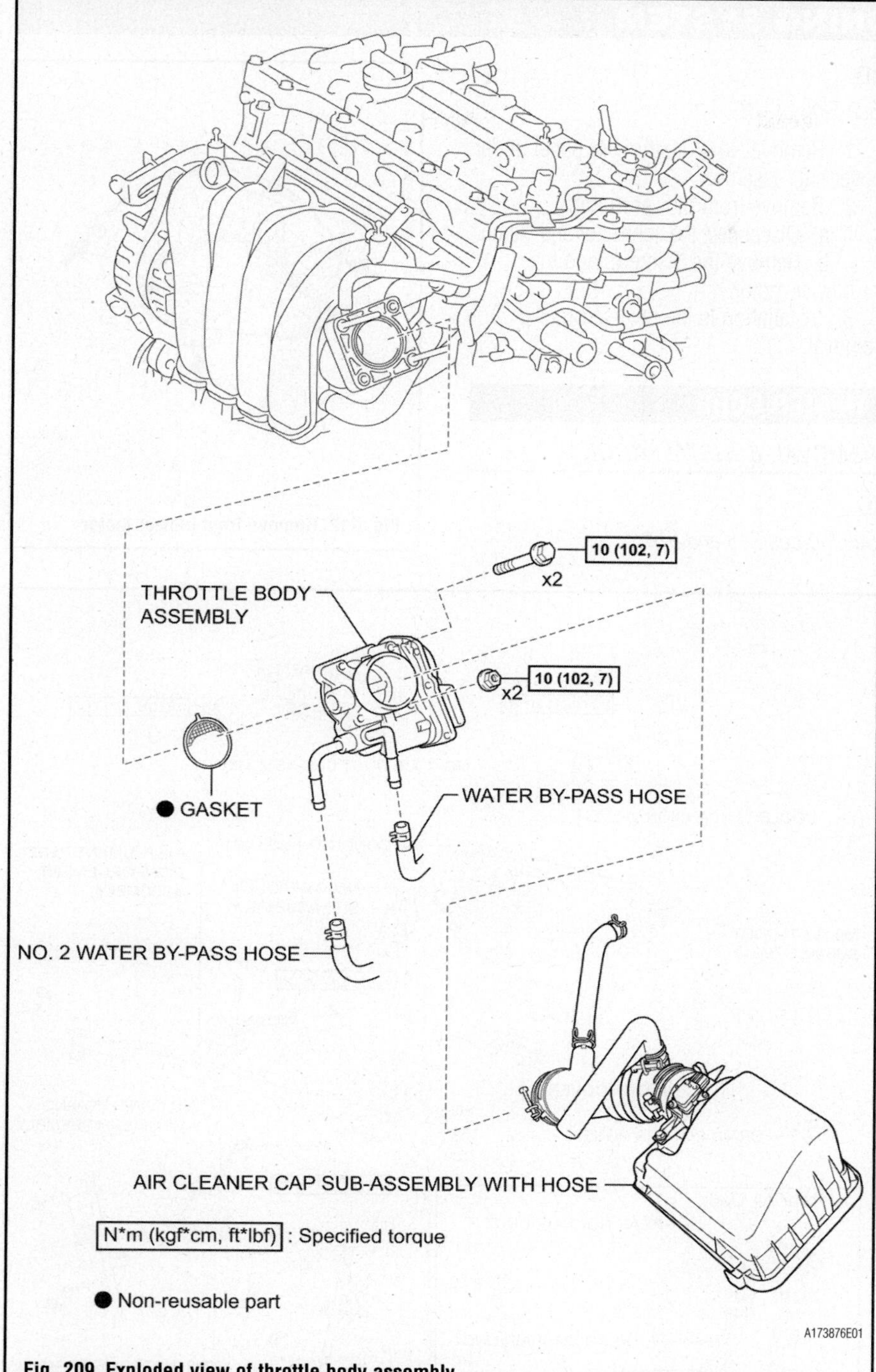

Fig. 209 Exploded view of throttle body assembly

b. Disconnect the water by-pass hose and the No. 2 water by-pass hose.

c. Remove the 2 bolts and 2 nuts and remove the throttle body assembly.

d. Remove the gasket from the intake manifold.

To install:

5. Install throttle body assembly.
   a. Install a new gasket onto the intake manifold.
   b. Install the throttle body assembly with the 2 bolts and 2 nuts.
   c. Connect the water by-pass hose and the No. 2 water by-pass hose.
   d. Connect the throttle body assembly connector.
6. Install air cleaner cap sub-assembly with hose.
   a. Install air cleaner cap sub-assembly with hose and lock the 2 clamps.
   b. Tighten the hose clamp to the specified torque.
   c. Connect the No. 2 ventilation hose.
   d. Connect the wire harness clamp and Mass Air Flow (MAF) meter connector.
7. Add engine coolant.
8. Inspect for coolant leak.
9. Install No. 2 cylinder head cover.
10. Perform initialization.
    a. Disconnect the cable from the negative (-) battery terminal. Wait at least 60 seconds and reconnect the cable.
    b. Turn the ignition switch to ON without operating the accelerator pedal.
    c. Connect the Techstream to the DLC3 and clear the DTCs.
    d. Start the engine and check that the MIL is not illuminated and that the idle speed is within the specified range when the A/C is switched off after the engine is warmed up.
    e. Enter the following menus: Powertrain/ Engine and ECT/ Data List/ Throttle Pos. Sensor Output. Fully depress the accelerator pedal and check that the value is 60% or more.
    f. Perform a road test and confirm that there are no abnormalities.

HEATING & AIR CONDITIONING SYSTEM

BLOWER MOTOR

REMOVAL & INSTALLATION

tC

See Figure 210.

1. Remove air conditioning unit
2. Remove the 3 screws and blower assembly.
3. Installation is the reverse order of removal.

xB

See Figure 211.

1. Remove blower motor sub-assembly.
2. Disconnect the connector and clamp.
3. Remove the 3 screws and the blower motor.
4. Installation is the reverse order of removal.

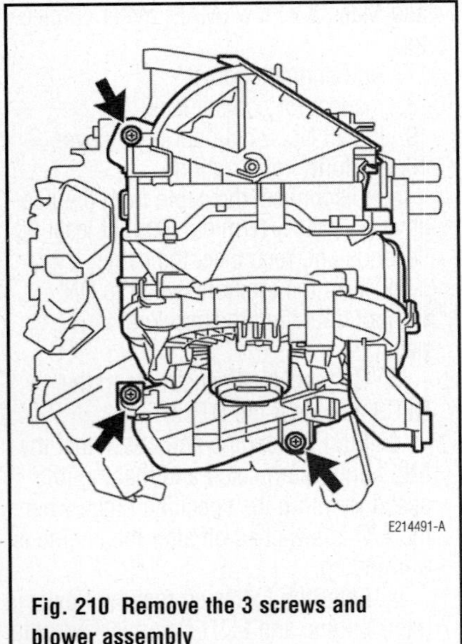

Fig. 210 Remove the 3 screws and blower assembly

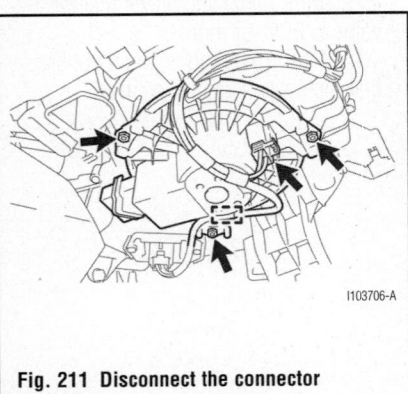

Fig. 211 Disconnect the connector and clamp

xD

See Figure 212.

1. Remove No. 2 instrument panel under cover sub-assembly.
2. Remove front blower motor.
   a. Disconnect the connector and clamp.
   b. Remove the 3 screws and the blower motor.
3. Installation is the reverse order of removal.

HEATER CORE

REMOVAL & INSTALLATION

tC

See Figures 213 and 214.

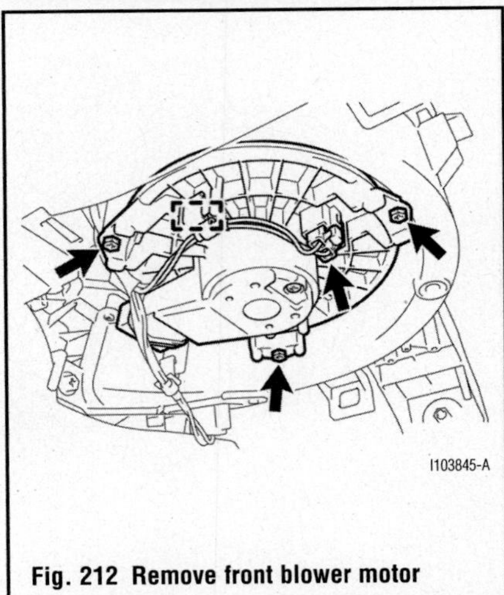

Fig. 212 Remove front blower motor

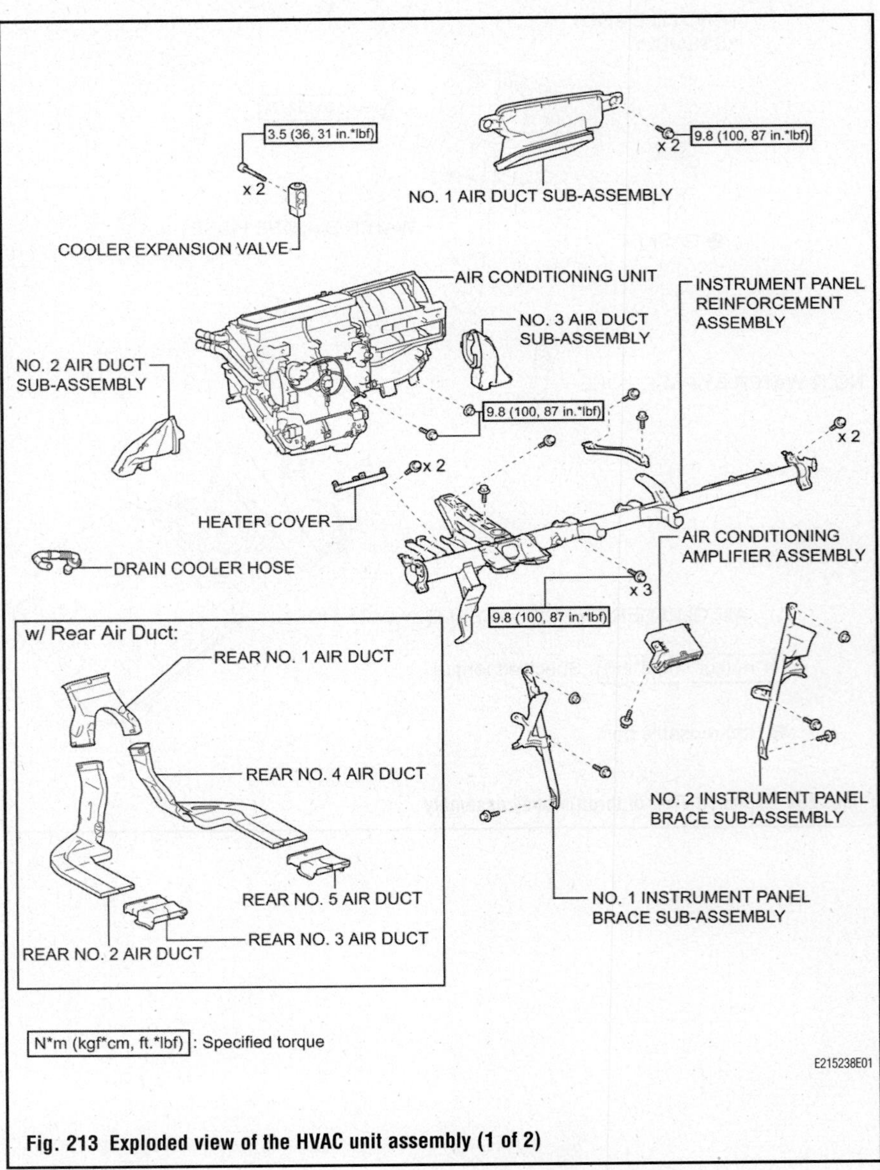

Fig. 213 Exploded view of the HVAC unit assembly (1 of 2)

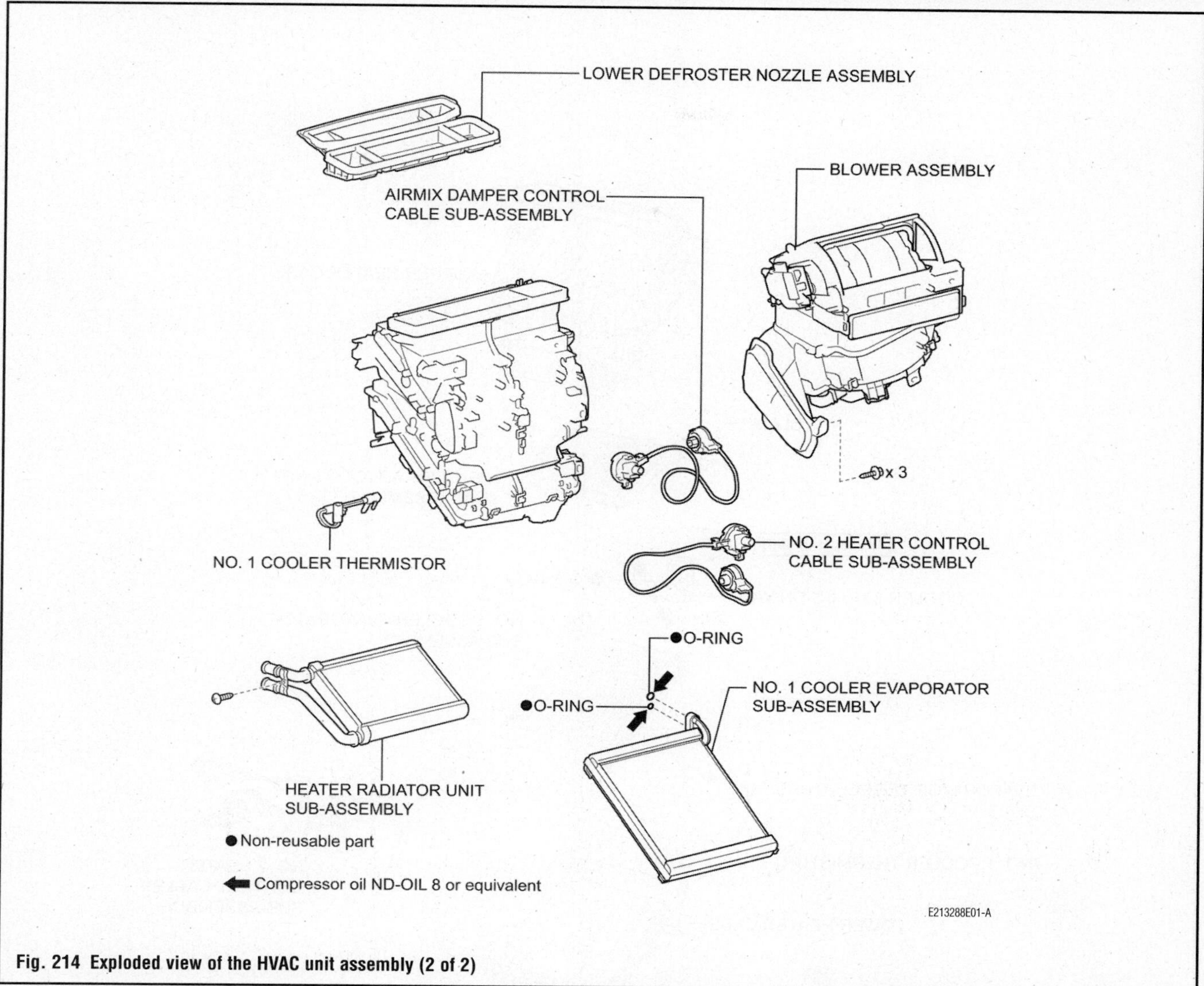

Fig. 214 Exploded view of the HVAC unit assembly (2 of 2)

➡The manufacturer does not provide a specific Removal and Installation procedure for this component. Refer to the graphic(s) when servicing this component.

xB

See Figure 215.

➡The manufacturer does not provide a specific Removal and Installation procedure for this component. Refer to the graphic(s) when servicing this component.

xD

See Figure 216.

➡The manufacturer does not provide a specific Removal and Installation procedure for this component. Refer to the graphic(s) when servicing this component.

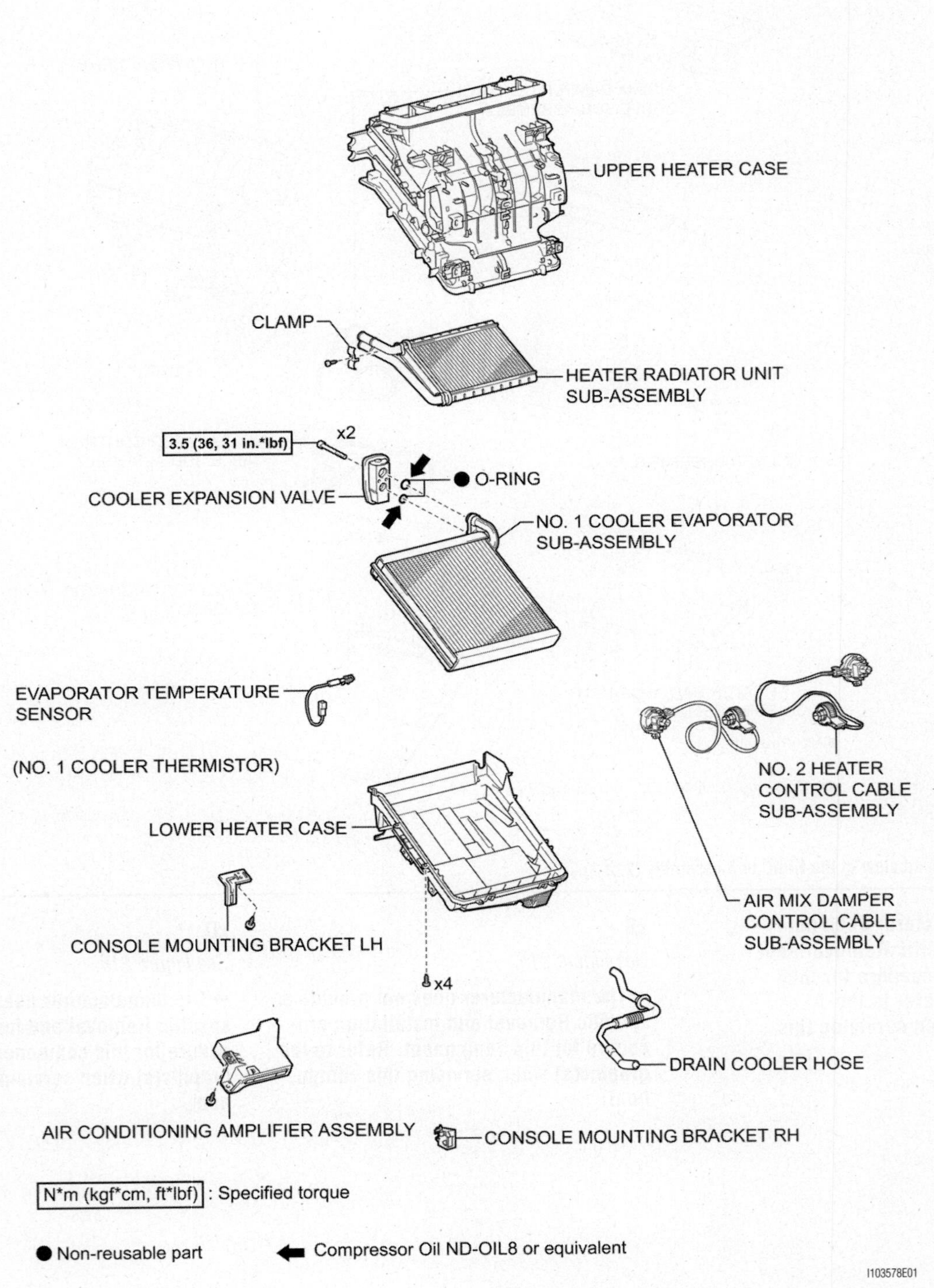
UPPER HEATER CASE
CLAMP
HEATER RADIATOR UNIT SUB-ASSEMBLY
3.5 (36, 31 in.*lbf)
x2
O-RING
COOLER EXPANSION VALVE
NO. 1 COOLER EVAPORATOR SUB-ASSEMBLY
EVAPORATOR TEMPERATURE SENSOR
(NO. 1 COOLER THERMISTOR)
NO. 2 HEATER CONTROL CABLE SUB-ASSEMBLY
LOWER HEATER CASE
AIR MIX DAMPER CONTROL CABLE SUB-ASSEMBLY
CONSOLE MOUNTING BRACKET LH
x4
DRAIN COOLER HOSE
AIR CONDITIONING AMPLIFIER ASSEMBLY
CONSOLE MOUNTING BRACKET RH
N*m (kgf*cm, ft*lbf) : Specified torque
Non-reusable part
Compressor Oil ND-OIL8 or equivalent
I103578E01

Fig. 215 Exploded view of HVAC unit assembly

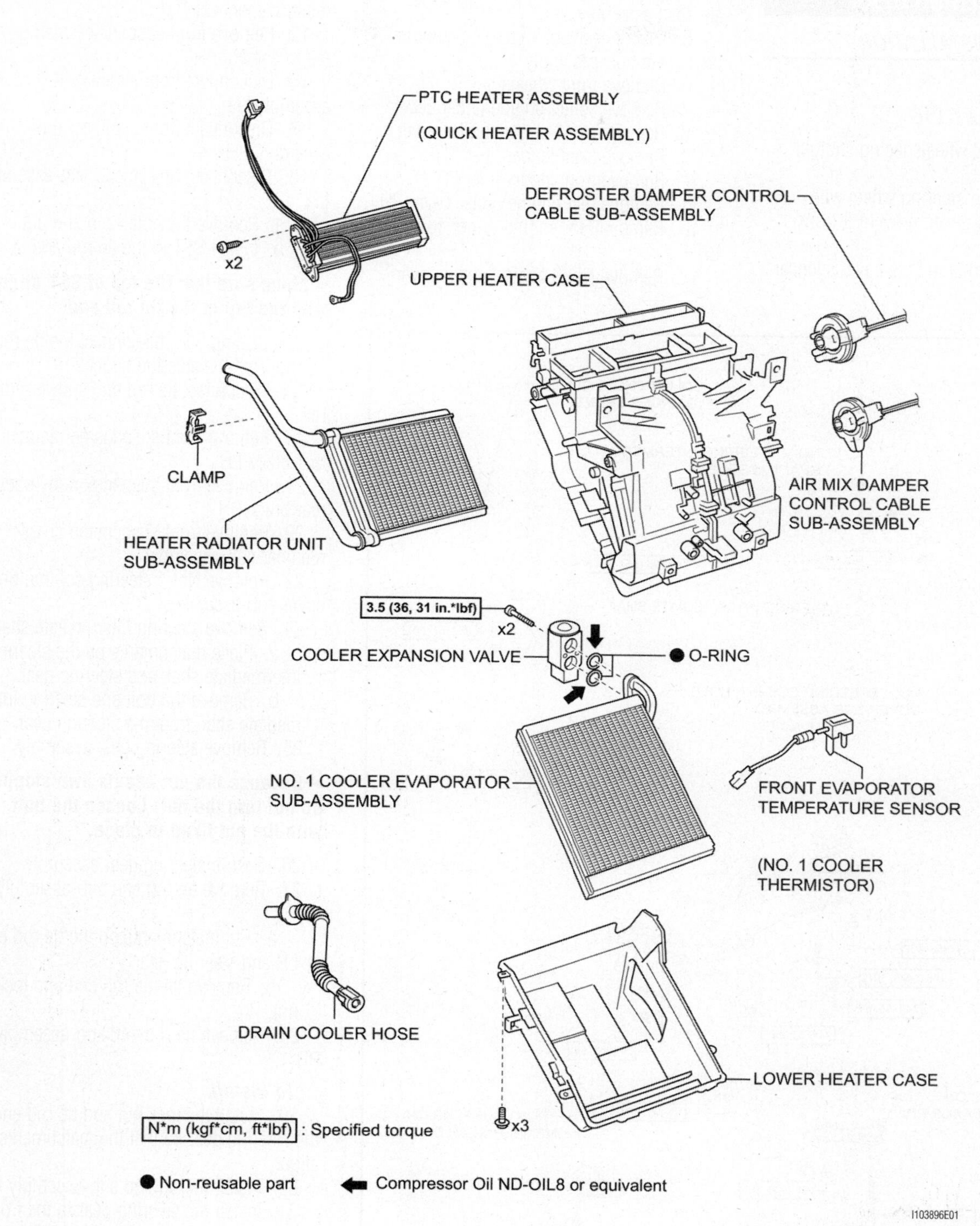
PTC HEATER ASSEMBLY
(QUICK HEATER ASSEMBLY)
x2
DEFROSTER DAMPER CONTROL CABLE SUB-ASSEMBLY
UPPER HEATER CASE
CLAMP
AIR MIX DAMPER CONTROL CABLE SUB-ASSEMBLY
HEATER RADIATOR UNIT SUB-ASSEMBLY
3.5 (36, 31 in.*lbf)
x2
COOLER EXPANSION VALVE
O-RING
NO. 1 COOLER EVAPORATOR SUB-ASSEMBLY
FRONT EVAPORATOR TEMPERATURE SENSOR
(NO. 1 COOLER THERMISTOR)
DRAIN COOLER HOSE
LOWER HEATER CASE
x3
N*m (kgf*cm, ft*lbf) : Specified torque
Non-reusable part
Compressor Oil ND-OIL8 or equivalent
I103896E01

Fig. 216 Exploded view of HVAC unit assembly

STEERING

POWER STEERING GEAR

REMOVAL & INSTALLATION

tC and xB

See Figures 217 and 218.

1. Place front wheels facing straight ahead
2. Secure the steering wheel with the seat belt in order to prevent it from rotating.
3. Remove column hole cover silencer sheet
4. Disconnect No. 2 steering intermediate shaft assembly
5. Disconnect No. 1 steering column hole cover sub-assembly
6. Remove front wheels
7. Remove center engine under cover
8. Remove No. 1 engine under cover
9. Remove front fender liner RH
10. Remove front fender liner LH
11. Remove rear engine under cover RH
12. Remove rear engine under cover LH
13. Remove No. 2 engine under cover
14. Remove front suspension member reinforcement LH
15. Remove front suspension member reinforcement RH
16. Disconnect front stabilizer link assembly LH
17. Disconnect front stabilizer link assembly RH
18. Disconnect tie rod end sub-assembly LH
    a. Remove the cotter pin and nut.
    b. Install SST to the tie rod end.

➡Make sure that the top of SST aligns with the top of the tie rod end.

    c. Using SST, disconnect the tie rod end from the steering knuckle.
19. Disconnect tie rod end sub-assembly RH
20. Remove front suspension member rear brace LH
21. Remove front suspension member rear brace RH
22. Remove front suspension cross-member sub-assembly
23. Remove No. 1 steering column hole cover sub-assembly
24. Remove steering intermediate shaft
    a. Place matchmarks on the steering intermediate shaft and steering gear.
    b. Remove the bolt and steering intermediate shaft from the steering gear.
25. Remove steering gear assembly

➡Because the nut has its own stopper, do not turn the nut. Loosen the bolt with the nut fixed in place.

26. Secure steering gear assembly
27. Remove tie rod end sub-assembly LH
    a. Put matchmarks on the tie rod end LH and steering gear.
    b. Remove the tie rod end and lock nut.
28. Remove tie rod end sub-assembly RH

To install:

29. Install the lock nut and tie rod end to the steering gear so that the matchmarks align.
30. Install tie rod end sub-assembly RH
31. Install the steering gear to the front suspension crossmember with the 2 bolts and 2 nuts. Tighten to 102 ft. lbs. (138 nm).

➡Make sure to tighten the bolts starting from the left side of the vehicle. Because the nut has its own stopper, do not turn the nut. Tighten the bolt with the nut fixed in place.

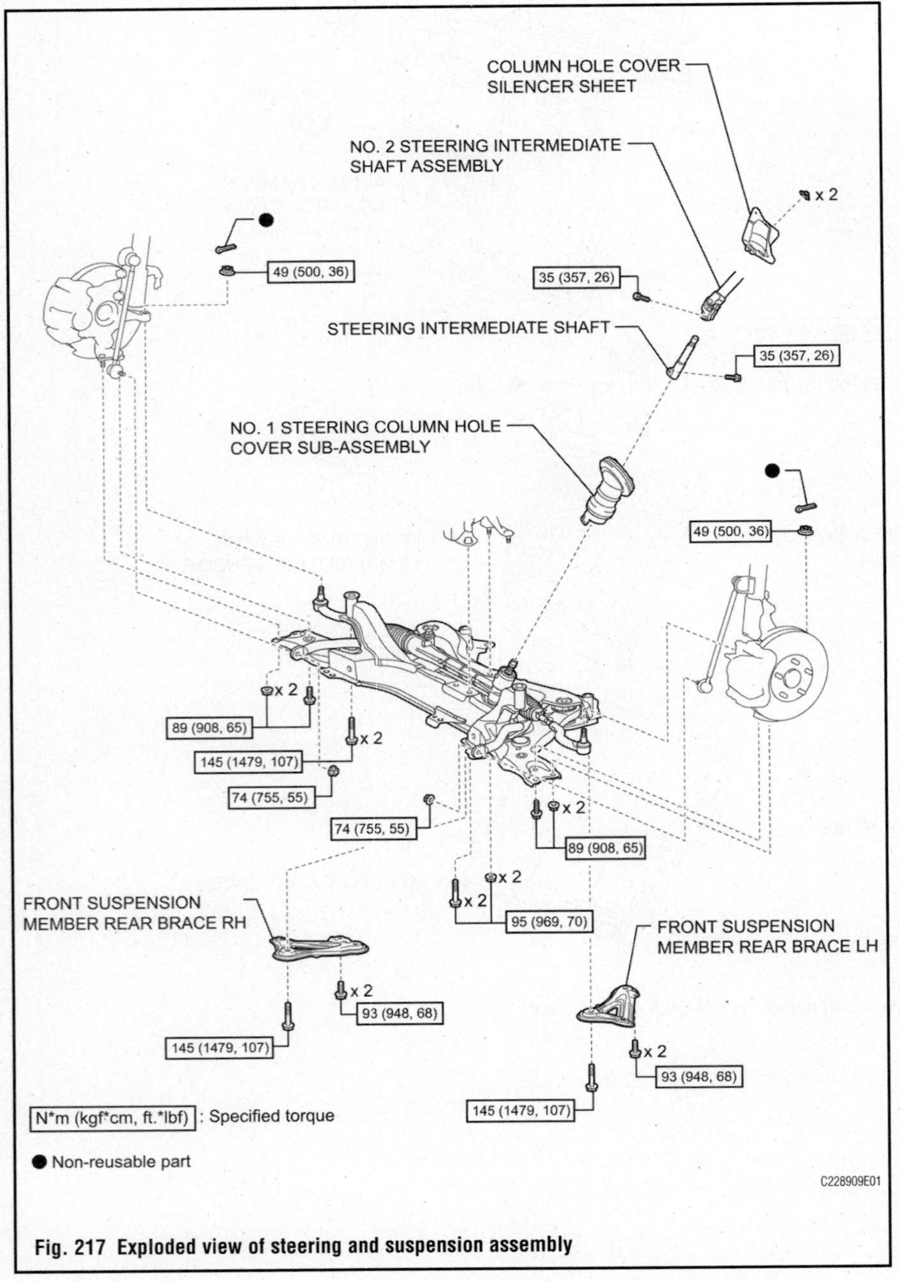

Fig. 217 Exploded view of steering and suspension assembly

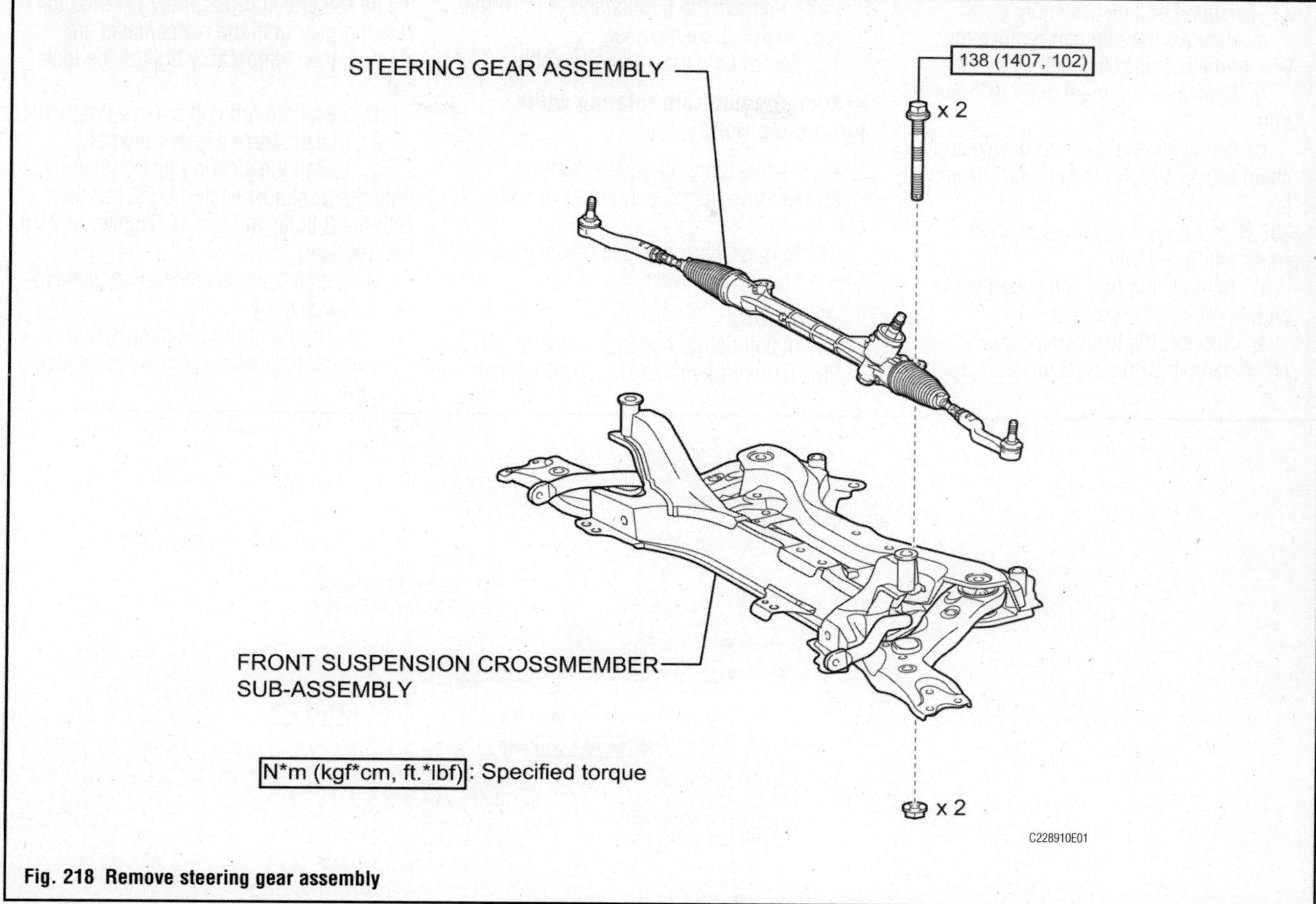

Fig. 218 Remove steering gear assembly

32. Install steering intermediate shaft
    a. Align the matchmarks and install the steering intermediate shaft to the steering gear.
    b. Install the bolt. Tighten to 26 ft. lbs. (35 Nm).
33. Install No. 1 steering column hole cover sub-assembly
34. Install front suspension crossmember sub-assembly
35. Install front suspension member rear brace LH
36. Install front suspension member rear brace RH
37. Connect tie rod end sub-assembly LH

➡Tighten the nut up to an additional 60° if the holes for the cotter pin are not aligned.

38. Install a new cotter pin.
39. Connect tie rod end sub-assembly RH
40. Connect front stabilizer link assembly LH
41. Connect front stabilizer link assembly RH
42. Install front suspension member reinforcement LH
43. Install front suspension member reinforcement RH
44. Install No. 2 engine under cover
45. Install rear engine under cover LH
46. Install rear engine under cover RH
47. Install front fender liner LH
48. Install front fender liner RH
49. Install No. 1 engine under cover
50. Install center engine under cover
51. Connect No. 1 steering column hole cover sub-assembly
52. Connect No. 2 steering intermediate shaft assembly
53. Install column hole cover silencer sheet
54. Install front wheels
55. Adjust front wheel alignment

xD

See Figures 219 through 222.

1. Position front wheels facing straight ahead.
2. Disconnect cable from negative battery terminal.
3. Remove front wiper arm head cap.
4. Remove front wiper arm and blade assembly LH.
5. Remove front wiper arm and blade assembly RH.
6. Remove hood to cowl top seal.
7. Remove cowl top ventilator louver sub-assembly.
8. Remove cowl top ventilator louver LH.
9. Remove front wiper motor and link.
10. Remove front air shutter seal RH.
11. Remove outer cowl top panel.
12. Remove column hole cover silencer sheet.
13. Remove steering sliding yoke sub-assembly.
14. Remove No. 1 steering column hole cover sub-assembly.
15. Remove front wheel.
16. Separate tie rod end sub-assembly LH.
17. Separate tie rod end sub-assembly RH.
18. Separate front stabilizer link assembly LH.
19. Separate front stabilizer link assembly RH.
20. Remove front suspension lower arm LH.
21. Remove front suspension lower arm RH.

22. Suspend engine assembly.
a. Remove the bolt and remove the wire harness clamp bracket.
b. Install the engine hanger with the bolt.
c. Using an engine sling device and a chain block, support the engine assembly.
23. Remove front suspension crossmember sub-assembly
a. Remove the bolt and separate the engine moving control rod.
b. Support the front suspension crossmember with a transmission jack.
c. Remove the 6 bolts and remove the suspension crossmember.
24. Remove steering gear assembly.

➡Keep the nut from rotating while turning the bolt.

25. Secure steering gear assembly.
26. Remove tie rod end sub-assembly LH.
27. Remove tie rod end sub-assembly RH.

To install:

28. Install tie rod end sub-assembly LH.
29. Temporarily install the lock nut and the tie rod end sub-assembly LH onto the steering gear until the matchmarks are aligned, then temporarily tighten the lock nut.
30. Install tie rod end sub-assembly RH.
31. Install steering gear assembly.
32. Install the steering gear assembly onto the front suspension crossmember with the 2 bolts and 2 nuts. Tighten to 71 ft. lbs. (96 Nm).
33. Install front suspension crossmember sub-assembly.
a. Support the front suspension crossmember with a transmission jack.

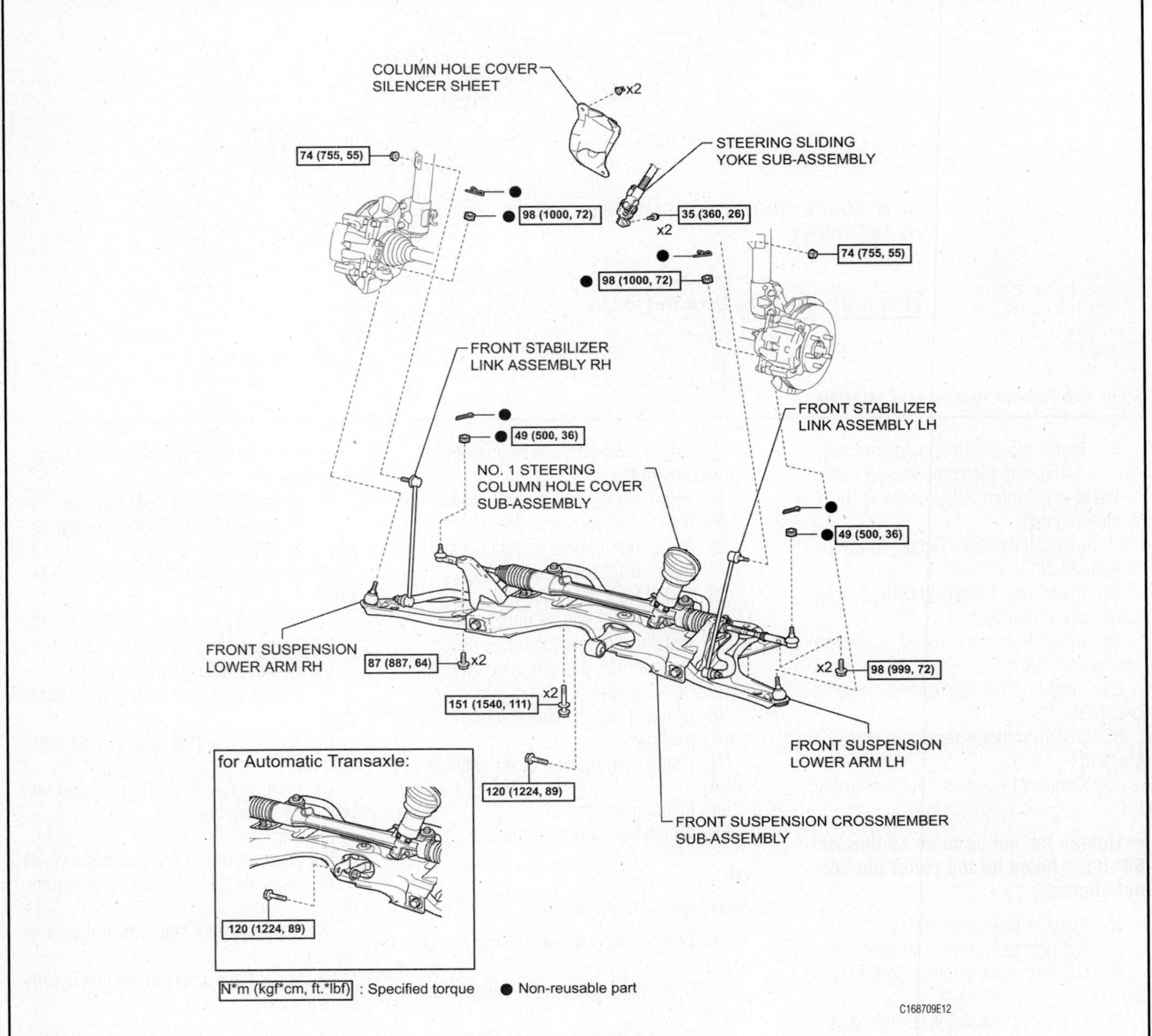

Fig. 219 Exploded view of the steering and suspension assembly

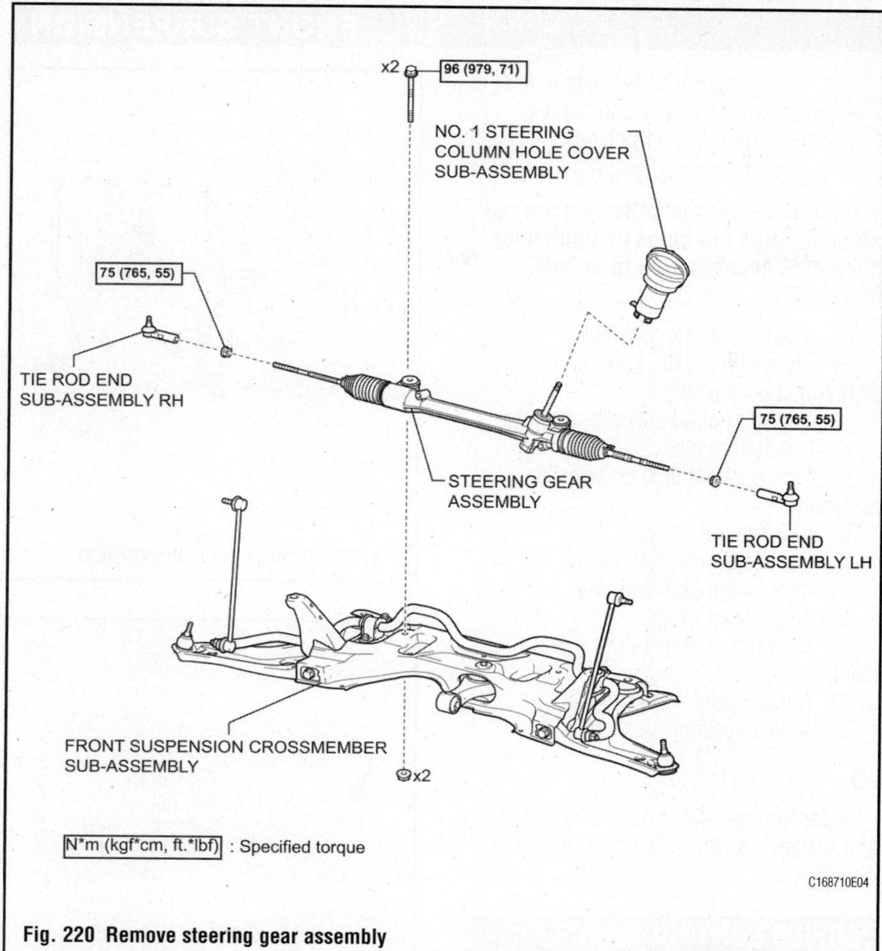

Fig. 220 Remove steering gear assembly

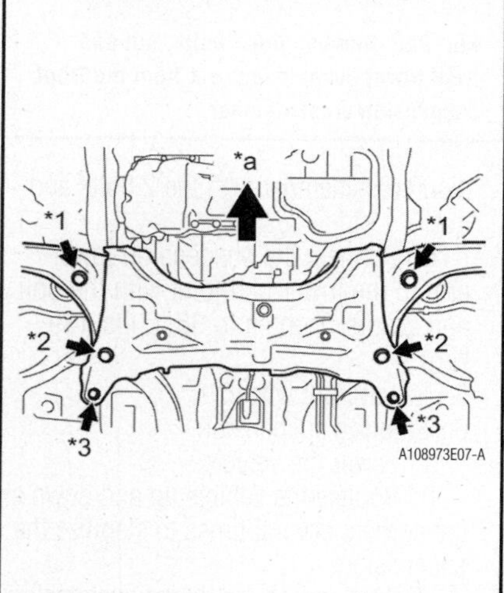

Fig. 221 Install the front suspension crossmember onto the body with the 6 bolts

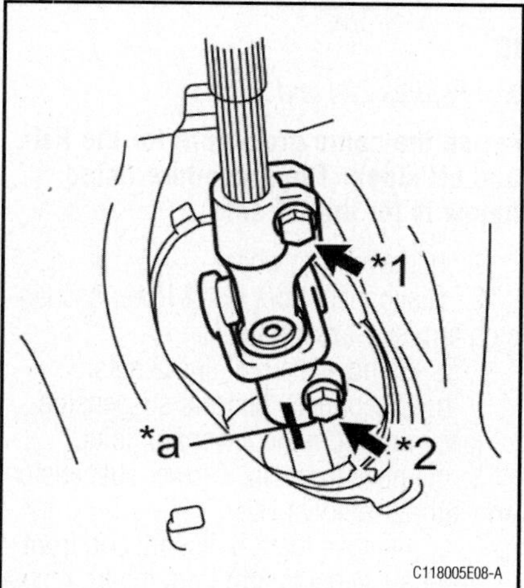

Fig. 222 Align the matchmarks and install the sliding yoke onto the intermediate shaft with bolt B (2), Bolt A (1); matchmark (a)

b. Provisionally install the front suspension crossmember onto the body with the 6 bolts.

c. By inserting SST into the datum holes in the front suspension crossmembers RH and LH alternately, tighten bolts A, B and C on both sides to the specified torque, in several steps.

- Bolt A: 64 ft. lbs. (87 Nm)
- Bolt B: 111 ft. lbs. (1514 Nm)
- Bolt C: 72 ft. lbs. (98 Nm)

d. Install the engine moving control rod with the bolt. Tighten to 89 ft. lbs. (120 Nm).

34. Install front suspension lower arm LH.
35. Install front suspension lower arm RH.
36. Install front stabilizer link assembly LH.
37. Install front stabilizer link assembly RH.
38. Install tie rod end sub-assembly LH.
39. Install tie rod end sub-assembly RH.
40. Install front wheel.
41. Install No. 1 steering column hole cover sub-assembly.
42. Install steering sliding yoke sub-assembly.

a. Align the matchmarks and install the sliding yoke onto the intermediate shaft with bolt B. Tighten to 26 ft. lbs. (35 Nm).

b. Tighten bolt A. Tighten to 26 ft. lbs. (35 Nm).

c. Install column hole cover silencer sheet.

d. Install the column hole cover silencer sheet with the 2 clips.

e. Install the floor carpet.

43. Install outer cowl top panel.
44. Install front air shutter seal RH.
45. Install front wiper motor and link.
46. Install cowl top ventilator louver LH.
47. Install cowl top ventilator louver sub-assembly.
48. Install hood to cowl top seal.
49. Install front wiper arm and blade assembly LH.
50. Install front wiper arm and blade assembly RH.
51. Install front wiper arm head cap.
52. Connect cable to negative battery terminal.
53. Position front wheels facing straight ahead.
54. Inspect and adjust front wheel alignment.
55. Check VSC sensor signal.

LOWER BALL JOINT

REMOVAL & INSTALLATION

tC and xB

See Figure 223.

➡Use the same procedure for the RH and LH sides. The procedure listed below is for the LH side.

1. Remove front wheel.
2. Remove front axle shaft nut LH.
3. Disconnect front speed sensor LH.
4. Disconnect front flexible hose.
5. Disconnect tie rod end sub-assembly LH.
6. Disconnect front disc brake caliper assembly LH.
7. Remove front disc.
8. Disconnect front No. 1 lower suspension arm sub-assembly LH.
9. Remove front axle assembly LH.
10. Remove front lower ball joint assembly LH.
    a. Secure the front axle assembly between aluminum plates in a vise.
    b. Remove the cotter pin and nut.
    c. Install SST to the front lower ball joint.

➡Make sure that the clearance measurement (1) between SST and the front axle assembly is 0.03937 inches (1 mm).

d. Using SST, remove the front lower ball joint from the front axle assembly.

To install:

11. Install front lower ball joint assembly.
    a. Secure the front axle assembly between aluminum plates in a vise.

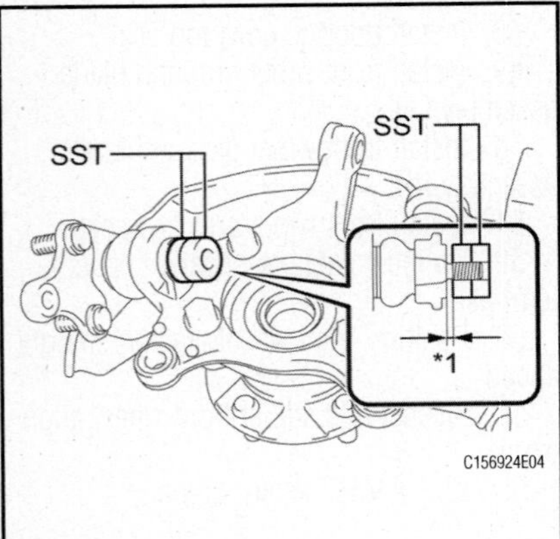

Fig. 223 Install SST to the front lower ball joint

b. Install the front lower ball joint to the front axle assembly with the nut. Tighten to 98 ft. lbs. (133 Nm).
c. Install a new cotter pin.

➡If the holes for the cotter pin are not aligned, align the holes by tightening the nut as necessary up to an additional 60°.

12. Install front axle assembly.
13. Connect front No. 1 lower suspension arm sub-assembly.
14. Connect tie rod end sub-assembly.
15. Install front disc.
16. Connect front disc brake caliper assembly.
17. Connect front flexible hose.
18. Connect front speed sensor.
19. Install front axle shaft nut.
20. Install front wheel.
21. Inspect and adjust front wheel alignment.
22. Check speed sensor signal.
23. Adjust headlight assembly.

xD

Refer to Lower Control Arm Removal and Installation when servicing this component.

LOWER CONTROL ARM

REMOVAL & INSTALLATION

tC

See Figures 224 and 225.

➡Use the same procedure for the RH and LH sides. The procedure listed below is for the LH side.

1. Remove front wheel.
2. Disconnect front No. 1 lower suspension arm sub-assembly LH.
    a. Remove the bolt and 2 nuts.
    b. Disconnect the front suspension lower arm from the lower ball joint.
3. Remove front No. 1 lower suspension arm sub-assembly LH.
    a. Remove the 2 bolts, nut and front lower suspension arm from the front suspension crossmember.

➡Because the nut has its own stopper, do not turn the nut. Loosen the bolt with the nut fixed in place.

To install:

4. Temporarily install front No. 1 lower suspension arm sub-assembly LH.
    a. Temporarily install the front lower suspension arm LH to the front suspension crossmember with the 2 bolts and nut.
    b. Connect the suspension lower arm to the lower ball joint with the bolt and 2 nuts. Tighten to 66 ft. lbs. (89 Nm).
5. Install front wheel.
6. Stabilize suspension.
    a. Lower the vehicle.
    b. Bounce the vehicle up and down at the corners several times to stabilize the suspension.
7. Tighten front No. 1 lower suspension arm sub-assembly LH.
    a. Using SST, tighten bolt A to 172 ft. lbs. (233 Nm).
    b. Tighten bolt B to 172 ft. lbs. (233 Nm).
8. Inspect and adjust front wheel alignment.
9. Adjust headlight assembly.

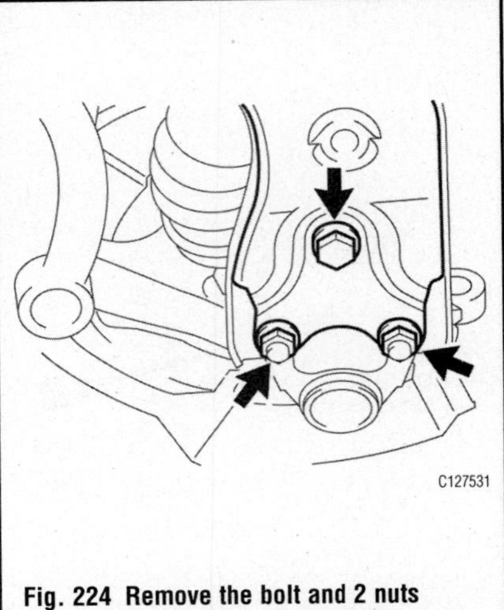

Fig. 224 Remove the bolt and 2 nuts

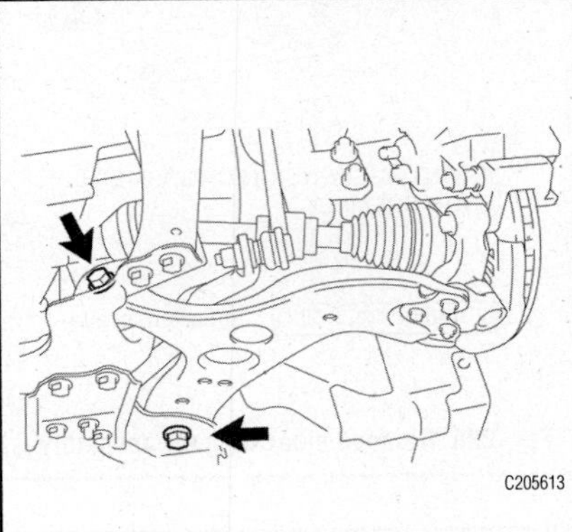

Fig. 225 Remove the 2 bolts, nut and front lower suspension arm from the front suspension crossmember

xB

See Figures 226 through 231.

1. Remove front wheels
2. Remove No. 1 engine under cover
3. Remove No. 2 engine under cover
4. Remove engine under cover rear LH
5. Remove engine under cover rear RH
6. Loosen front lower No. 1 suspension arm sub-assembly LH

➡Because the nut has its own stopper, do not turn the nut. Loosen the bolt with the nut fixed.

7. Loosen front lower No. 1 suspension arm sub-assembly RH
8. Separate front lower No. 1 suspension arm sub-assembly LH
9. Separate front lower No. 1 suspension arm sub-assembly RH
10. Remove front lower No. 1 suspension arm sub-assembly LH (for manual transaxle)
11. Place front wheels facing straight ahead
12. Secure steering wheel
13. Remove column hole cover silencer sheet
14. Separate No. 2 steering intermediate shaft assembly
15. Separate No. 1 steering column hole cover sub-assembly
16. Separate front stabilizer link assembly LH
17. Separate front stabilizer link assembly RH
18. Separate tie rod end sub-assembly LH
19. Separate tie rod end sub-assembly RH
20. Remove front suspension member reinforcement LH
21. Remove front suspension member reinforcement RH
22. Remove front suspension member brace rear LH
23. Remove front suspension member brace rear RH
24. Remove front suspension cross-member sub-assembly (for manual transaxle)
    a. Support the front suspension crossmember with a transmission jack.
    b. Remove the 4 bolts, 2 nuts, and front suspension crossmember sub-assembly.
25. Remove front suspension cross-member sub-assembly (for Automatic Transaxle)
    a. Support the front suspension crossmember with a transmission jack.
    b. Remove the 4 bolts, 2 nuts, and front suspension crossmember sub-assembly.
26. Remove front lower No. 1 suspension arm sub-assembly LH (for Automatic Transaxle)
27. Remove front lower No. 1 suspension arm sub-assembly RH

To install:

28. Temporarily install front lower No. 1 suspension arm sub-assembly RH.
    a. Position the front suspension crossmember on a level surface.
    b. Temporarily install the front lower No. 1 suspension arm sub-assembly RH onto the front suspension crossmember with the 2 bolts and nut.
29. Temporarily install front lower No. 1 suspension arm sub-assembly LH (for Automatic Transaxle).
    a. Position the front suspension crossmember on a level surface.
    b. Temporarily install the front lower No. 1 suspension arm sub-assembly LH onto the front suspension crossmember with the 2 bolts and nut.

➡Because the nut has its own stopper, do not turn the nut. Tighten the bolt with the nut fixed.

30. Install front suspension crossmember sub-assembly (for Manual Transaxle).

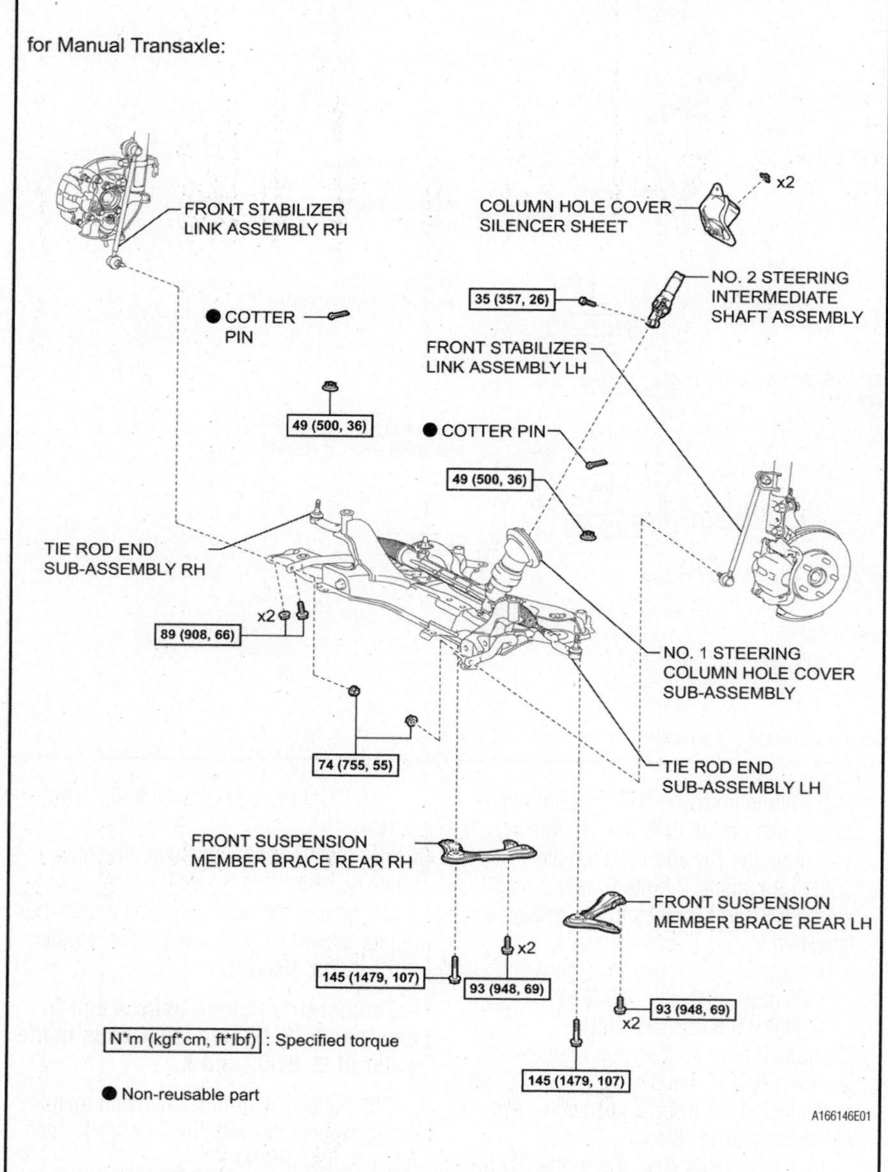

Fig. 226 Exploded view of front suspension assembly for Manual Transaxle

for Automatic Transaxle:

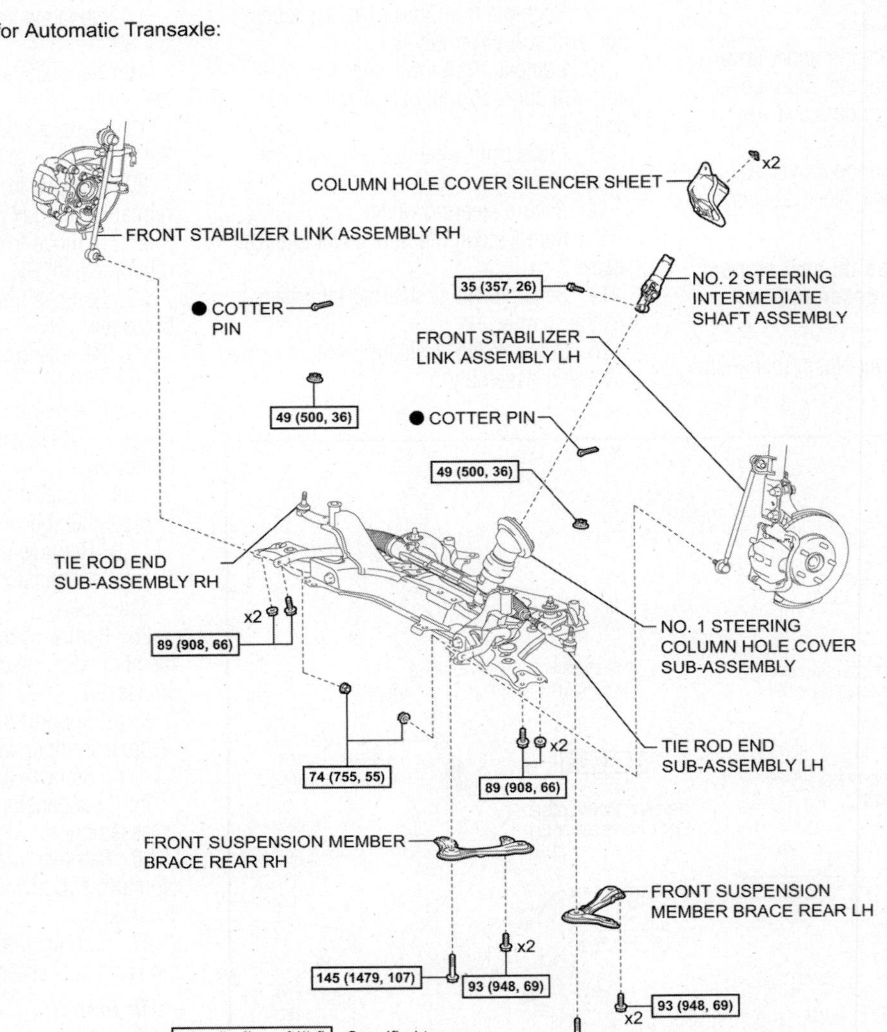

Fig. 227 Exploded view of front suspension assembly for Automatic Transaxle

a. Support the front suspension crossmember with a transmission jack.

b. While inserting SST into the reference holes on the front suspension crossmember RH and LH alternately, tighten 2 bolts A, 2 bolts B, and 2 nuts on the RH and LH sides to the respective specified torque in several steps.

- Bolt A: 107 ft. lbs. (145 Nm)
- Bolt B: 70 ft. lbs. (95 Nm)
- Nut: 70 ft. lbs. (95 Nm)

31. Install front suspension crossmember sub-assembly (for Automatic Transaxle).

a. Support the front suspension crossmember with a transmission jack.

b. While inserting SST into the reference holes on the front suspension crossmember RH and LH alternately, tighten 2 bolts A, 2 bolts B, and 2 nuts on the RH and LH sides to the respective specified torque in several steps.

- Bolt A: 107 ft. lbs. (145 Nm)
- Bolt B: 70 ft. lbs. (95 Nm)
- Nut: 70 ft. lbs. (95 Nm)

32. Install front suspension member brace rear LH.

a. Install the front suspension member brace rear LH with the 3 bolts.

b. Tighten Bolt A to 107 ft. lbs. (145 Nm).

c. Tighten Bolt B to 69 ft. lbs. (93 Nm).

33. Install front suspension member brace rear RH.

34. Install the front suspension member reinforcement LH with the 4 bolts. Tighten to 71 ft. lbs. (96 Nm).

➡Temporarily tighten bolts A and B, and then fully tighten the 4 bolts in the order of C, B, D, and A.

35. Install the front suspension member reinforcement RH with the 4 bolts. Tighten to 71 ft. lbs. (96 Nm).

➡Temporarily tighten bolts A and B,

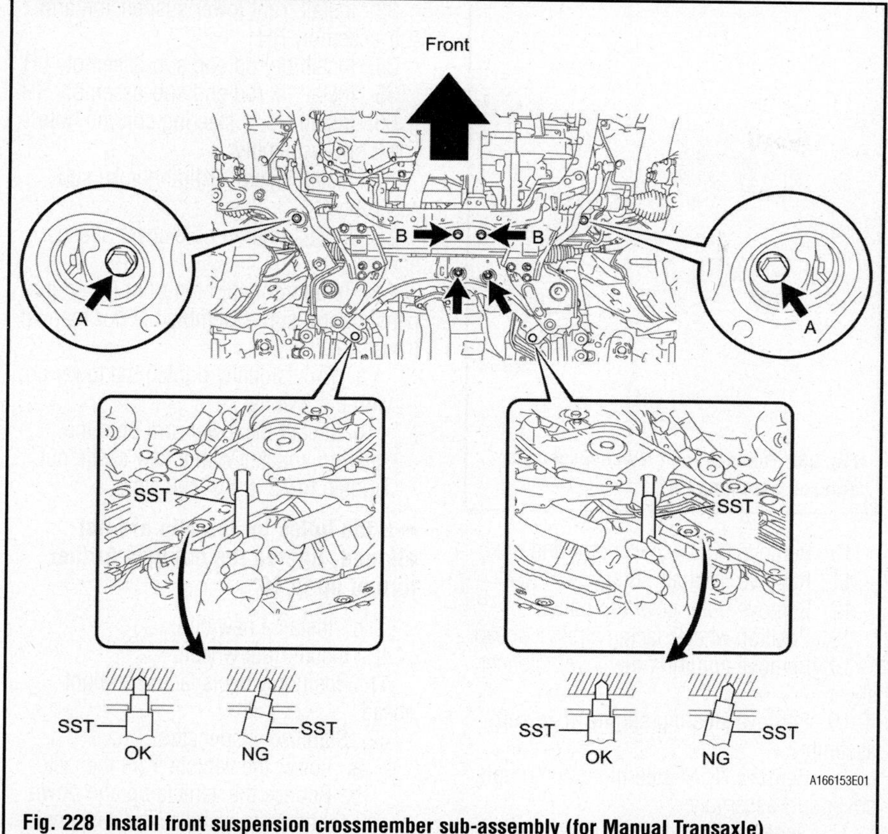

Fig. 228 Install front suspension crossmember sub-assembly (for Manual Transaxle)

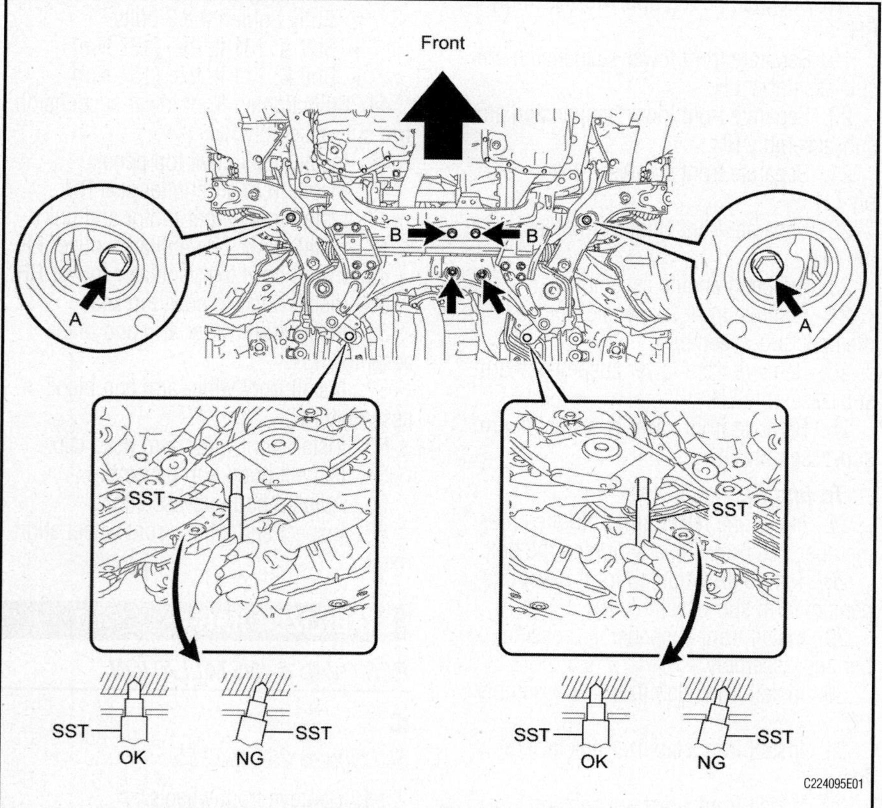

Fig. 229 Install front suspension crossmember sub-assembly (for Automatic Transaxle)

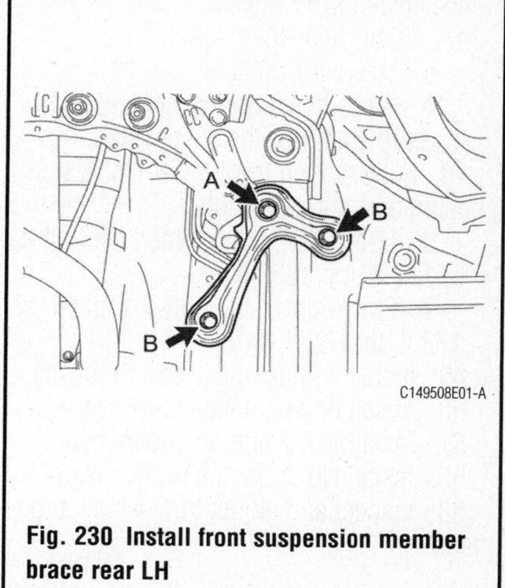

Fig. 230 Install front suspension member brace rear LH

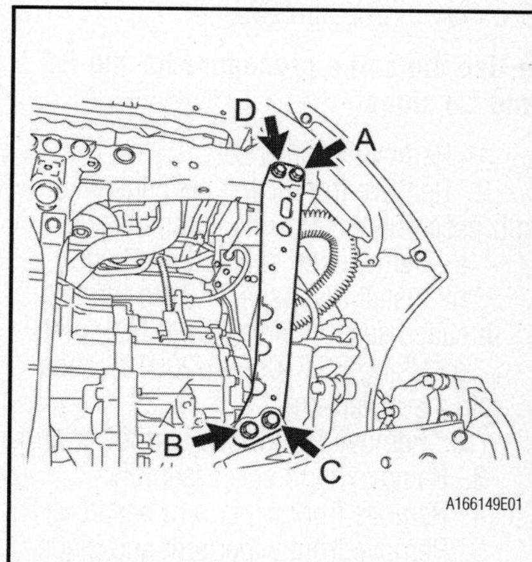

Fig. 231 Install the front suspension member reinforcement LH with the 4 bolts

and then fully tighten the 4 bolts in the order of C, B, D, and A.

36. Temporarily install front lower No. 1 suspension arm sub-assembly LH (for Manual Transaxle).
37. Connect front lower No. 1 suspension arm sub-assembly LH.
38. Connect front lower No. 1 suspension arm sub-assembly RH.
39. Connect tie rod end sub-assembly LH.
40. Connect tie rod end sub-assembly RH.
41. Install the front stabilizer link assembly LH to the front stabilizer bar with the nut. Tighten to 55 ft. lbs. (74 Nm).
42. Install front stabilizer link assembly RH.
43. Connect No. 1 steering column hole cover sub-assembly.
44. Connect No. 2 steering intermediate shaft assembly.
45. Install column hole cover silencer sheet.

46. Install front wheels.
47. Stabilize suspension.
    a. Lower the vehicle.
    b. Press down on the vehicle several times to stabilizer the suspension.
48. Fully tighten front lower No. 1 suspension arm sub-assembly.
    a. Using SST, fully tighten the bolt A to 172 ft. lbs. (233 Nm).
    b. Fully tighten the bolt B. Tighten to 172 ft. lbs. (233 Nm).
49. Install engine under cover rear RH.
50. Install engine under cover rear LH.
51. Install No. 2 engine under cover.
52. Install No. 1 engine under cover.
53. Inspect and adjust front wheel alignment.

xD

See Figures 232 and 233.

➡Use the same procedure for the RH and LH sides.

1. Remove front wheel.
2. Remove front lower suspension arm sub-assembly LH (for manual transaxle).
    a. Remove the clip and castle nut.
    b. Install SST (spacer B) to the threaded section of the lower ball joint.
    c. Using SST (09960-20010), separate the lower arm.
    d. Remove the 2 bolts and lower arm.
3. Remove hood sub-assembly.
4. Remove front wiper arm head cap.
5. Remove front wiper arm and blade assembly LH.
6. Remove front wiper arm and blade assembly RH.
7. Remove hood to cowl top seal.
8. Remove cowl top ventilator louver RH.
9. Remove cowl top ventilator louver LH.

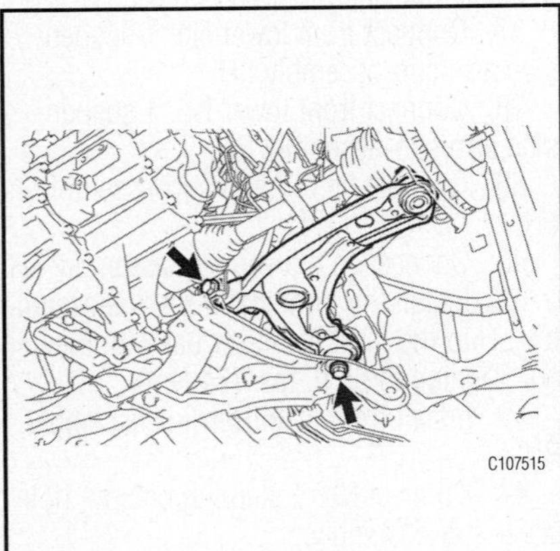

Fig. 232 Remove the 2 bolts and lower arm

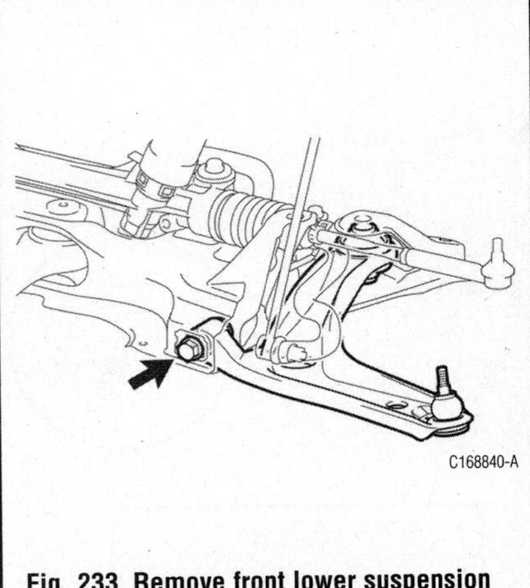

Fig. 233 Remove front lower suspension arm sub-assembly LH

10. Remove front wiper motor and link.
11. Remove front air shutter seal RH.
12. Remove outer cowl top panel.
13. Position wheels facing straight ahead.
14. Remove column hole cover silencer sheet.
15. Remove steering sliding yoke sub-assembly.
16. Remove No. 1 steering column hole cover sub-assembly.
17. Separate tie rod end sub-assembly LH.
18. Separate tie rod end sub-assembly RH.
19. Separate front lower suspension arm sub-assembly LH.
20. Separate front lower suspension arm sub-assembly RH.
21. Separate front stabilizer link assembly LH.
22. Separate front stabilizer link assembly RH.
23. Suspend engine assembly.
24. Remove front suspension cross-member sub-assembly.
25. Remove front lower suspension arm sub-assembly LH.
26. Remove front lower suspension arm sub-assembly RH.

To install:

27. Install the lower arm onto the cross-member and provisionally tighten the bolt.
28. Temporarily tighten front lower suspension arm sub-assembly RH.
29. Install front suspension crossmember sub-assembly.
30. Install front stabilizer link assembly LH.
31. Install front stabilizer link assembly RH.
32. Install front lower suspension arm sub-assembly LH.
33. Install front lower suspension arm sub-assembly RH.
34. Install tie rod end sub-assembly LH.
35. Install tie rod end sub-assembly RH.
36. Install No. 1 steering column hole cover sub-assembly.
37. Install steering sliding yoke sub-assembly.
38. Install column hole cover silencer sheet.
39. Temporarily tighten front lower suspension arm sub-assembly LH (for manual transaxle).
    a. Provisionally tighten the lower arm with the 2 bolts.
    b. Install the lower arm onto the steering knuckle with a new castle nut. Tighten to 72 ft. lbs. (98 Nm).

➡If the holes for the clip are not aligned, tighten the nut by a further turn of up to 60°.

    c. Install a new clip.
40. Install front wheel.
41. Position wheels facing straight ahead.
42. Stabilize suspension.
    a. Lower the vehicle from the jack.
    b. Bounce the vehicle up and down several times to stabilize the suspension.
43. Fully tighten front lower suspension arm sub-assembly LH.
    a. Fully tighten the 2 bolts.
        - Bolt A: 101 ft. lbs. (137 Nm)
        - Bolt B: 111 ft. lbs. (151 Nm)
44. Fully tighten front lower suspension arm sub-assembly RH.
45. Install outer cowl top panel.
46. Install front air shutter seal RH.
47. Install front wiper motor and link.
48. Install cowl top ventilator louver LH.
49. Install cowl top ventilator louver RH.
50. Install hood to cowl top seal.
51. Install front wiper arm and blade assembly LH.
52. Install front wiper arm and blade assembly RH.
53. Install front wiper arm head cap.
54. Inspect hood sub-assembly.
55. Adjust hood sub-assembly.
56. Inspect and adjust front wheel alignment.

STABILIZER BAR

REMOVAL & INSTALLATION

tC

See Figures 234 and 235.

1. Remove front wheels.
2. Remove No. 2 engine under cover.

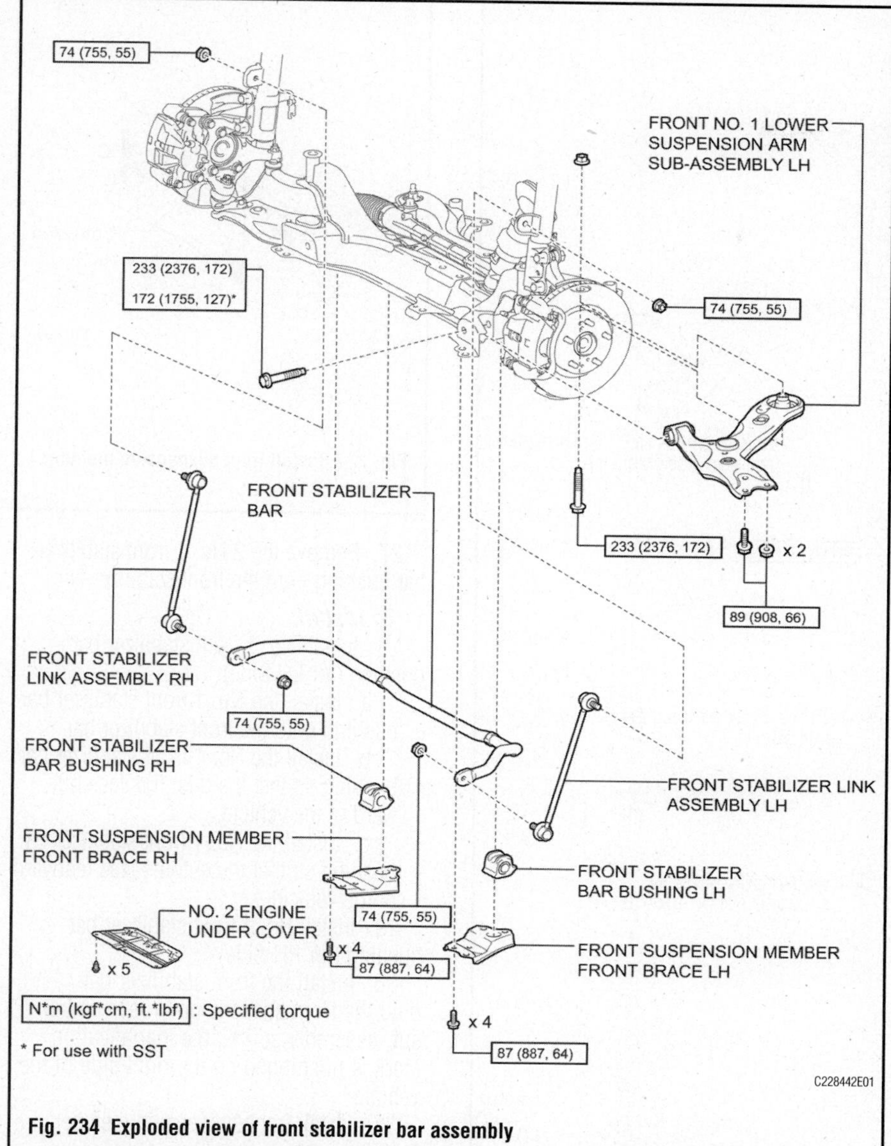

Fig. 234 Exploded view of front stabilizer bar assembly

3. Remove front stabilizer link assembly LH.
4. Remove front stabilizer link assembly RH.
5. Disconnect front No. 1 lower suspension arm sub-assembly LH.
6. Remove front No. 1 lower suspension arm sub-assembly LH.
7. Remove front suspension member front brace LH.
8. Remove front suspension member front brace RH.
9. Remove front stabilizer bar.
10. Remove front stabilizer bar bushing LH.
11. Remove front stabilizer bar bushing RH.

To install:

12. Install front stabilizer bar bushing LH.
    a. Install the front stabilizer bar bushing to the front stabilizer bar.
    b. Install the front stabilizer bar bushing so that the dust lips face the outside of the vehicle. Install the front stabilizer bar bushing so that the cutouts face the rear of the vehicle.
13. Install front stabilizer bar bushing RH.
14. Install the front stabilizer bar to the front suspension crossmember so that the identification mark is positioned on the right side of the vehicle.
15. Install front suspension member front brace LH.
    a. Install the front brace with the 4 bolts. Tighten to 64 ft. lbs. (87 Nm).
    b. Tighten the bolts in the order of B, C, D and A. Make sure that the protrusion of the front stabilizer bar bushing protrudes from the hole of the front suspension member front brace LH when installing the front suspension member front brace LH.

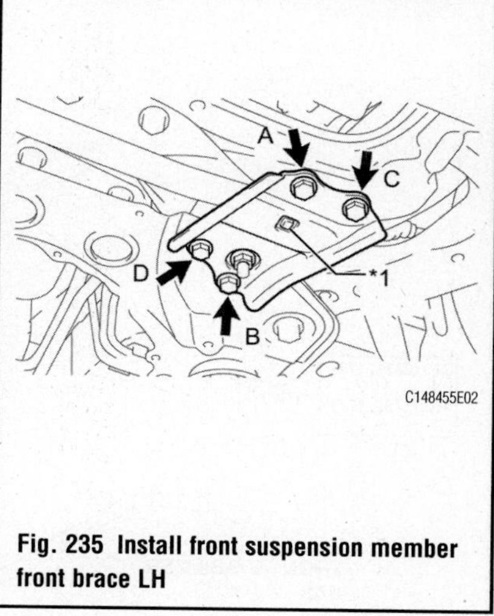

Fig. 235 Install front suspension member front brace LH

16. Install front suspension member front brace RH.
    a. Install the front brace with the 4 bolts. Tighten to 64 ft. lbs. (87 Nm).
    b. Tighten the bolts in the order of B, C, D and A. Make sure that the protrusion of the front stabilizer bar bushing protrudes from the hole of the front suspension member front brace RH when installing the front suspension member front brace RH.
17. Temporarily install front No. 1 lower suspension arm sub-assembly LH.
18. Connect front No. 1 lower suspension arm sub-assembly LH.
19. Install the front stabilizer link assembly LH with the 2 nuts. Tighten to 55 ft. lbs. (74 Nm).
20. Install front stabilizer link assembly RH.
21. Install No. 2 engine under cover.
22. Install front wheels.
23. Stabilize suspension.
24. Tighten front No. 1 lower suspension arm sub-assembly LH.
25. Inspect and adjust front wheel alignment.

xB

See Figures 236 and 237.

1. Remove No. 1 engine under cover.
2. Remove No. 2 engine under cover.
3. Remove engine under cover rear LH.
4. Remove engine under cover rear RH.
5. Place front wheels facing straight ahead.
6. Secure steering wheel.
7. Remove column hole cover silencer sheet.

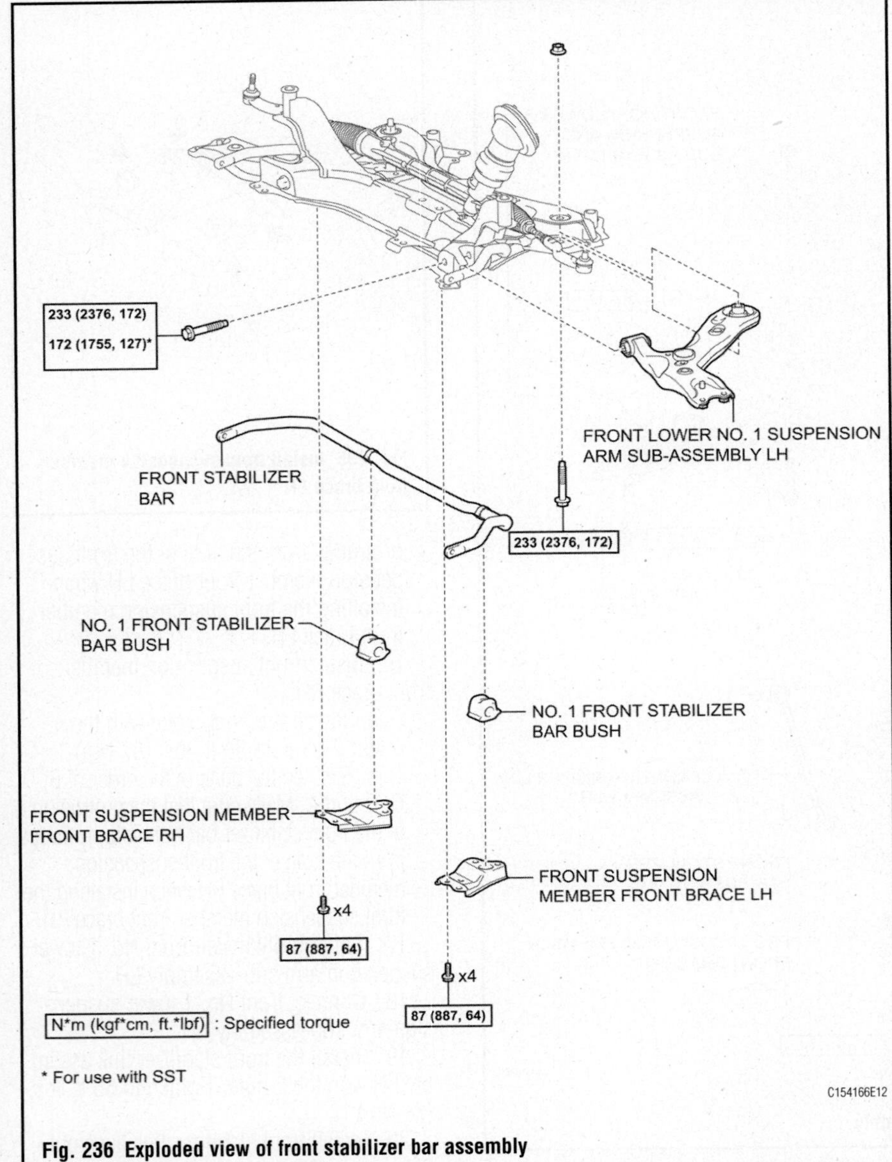

Fig. 236 Exploded view of front stabilizer bar assembly

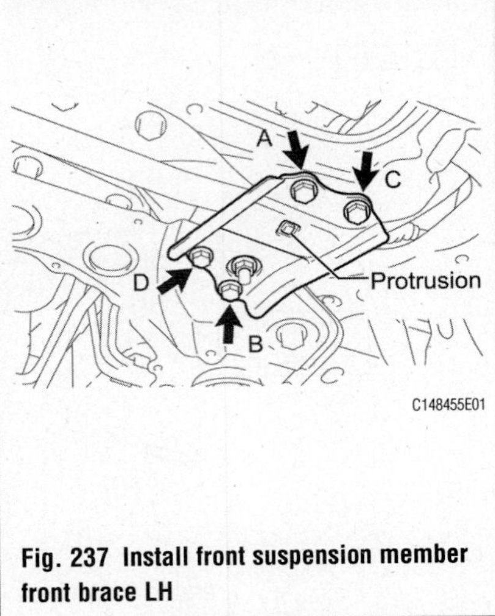

Fig. 237 Install front suspension member front brace LH

8. Separate No. 2 steering intermediate shaft assembly.
9. Separate No. 1 steering column hole cover sub-assembly.
10. Remove front wheels.
11. Loosen front lower No. 1 suspension arm sub-assembly LH.
12. Separate front lower No. 1 suspension arm sub-assembly LH.
13. Separate front lower No. 1 suspension arm sub-assembly RH.
14. Separate front stabilizer link assembly LH.
15. Separate front stabilizer link assembly RH.
16. Separate tie rod end sub-assembly LH.
17. Separate tie rod end sub-assembly RH.
18. Remove front suspension member reinforcement LH.
19. Remove front suspension member reinforcement RH.
20. Remove front suspension member brace rear LH.
21. Remove front suspension member brace rear RH.
22. Remove front suspension crossmember sub-assembly.
23. Remove front lower No. 1 suspension arm sub-assembly LH.
24. Remove the 4 bolts and front suspension member front brace LH.
25. Remove front suspension member front brace RH.
26. Remove the front stabilizer bar from the front suspension crossmember sub-assembly.
27. Remove the 2 No. 1 front stabilizer bar bushing from the front stabilizer bar.

To install:

28. Install No. 1 front stabilizer bar bushing (for LH Side).
    a. Install the No. 1 front stabilizer bar bushing onto the front stabilizer bar.
    b. Install the No. 1 front stabilizer bar bushing so that the dust lips face outward of the vehicle.
    c. Install the No. 1 front stabilizer bar bushing so that the cutouts face rearward of the vehicle.
29. Install No. 1 front stabilizer bar bushing (for RH Side).
30. Install the front stabilizer bar onto the front suspension crossmember sub-assembly so that the identification mark is positioned on the right side of the vehicle.
31. Install front suspension member front brace LH.
    a. Install the front suspension member front brace LH with the 4 bolts. Tighten to 64 ft. lbs. (87 Nm).
    b. Temporarily tighten bolt A, and then fully tighten the 4 bolts in the order of B, C, D, and A.
    c. After installing the front suspension member front brace LH, make sure that the protrusion of the No. 1 front stabilizer bar bushing comes out.
32. Install front suspension member front brace RH.

Install the front suspension member front brace RH with the 4 bolts. Tighten to 64 ft. lbs. (87 Nm).
    a. Temporarily tighten bolt A, and then fully tighten the 4 bolts in the order of B, C, D, and A.
    b. After installing the front suspension member front brace RH, make sure

that the protrusion of the No. 1 front stabilizer bar bushing comes out.

33. Temporarily install front lower No. 1 suspension arm sub-assembly LH.
34. Install front suspension crossmember sub-assembly.
35. Install front suspension member brace rear LH.
36. Install front suspension member brace rear RH.
37. Install front suspension member reinforcement LH.
38. Install front suspension member reinforcement RH.
39. Connect front lower No. 1 suspension arm sub-assembly LH.
40. Connect front lower No. 1 suspension arm sub-assembly RH.
41. Connect tie rod end sub-assembly LH.
42. Connect tie rod end sub-assembly RH.
43. Install front stabilizer link assembly LH.
44. Install front stabilizer link assembly RH.
45. Connect No. 1 steering column hole cover sub-assembly.
46. Connect No. 2 steering intermediate shaft assembly.
47. Install column hole cover silencer sheet.
48. Install front wheels.
49. Stabilize suspension.
50. Fully tighten front lower No. 1 suspension arm sub-assembly LH.
51. Install engine under cover rear RH.
52. Install engine under cover rear LH.
53. Install No. 2 engine under cover.
54. Install No. 1 engine under cover.
55. Inspect and adjust front wheel alignment.

xD

See Figures 238 and 239.

1. Remove hood sub-assembly
2. Remove front wiper arm head cap
3. Remove front wiper arm and blade assembly LH
4. Remove front wiper arm and blade assembly RH
5. Remove hood to cowl top seal
6. Remove cowl top ventilator louver RH
7. Remove cowl top ventilator louver LH
8. Remove front wiper motor and link
9. Remove front air shutter seal RH
10. Remove outer cowl top panel
11. Position wheels facing straight ahead

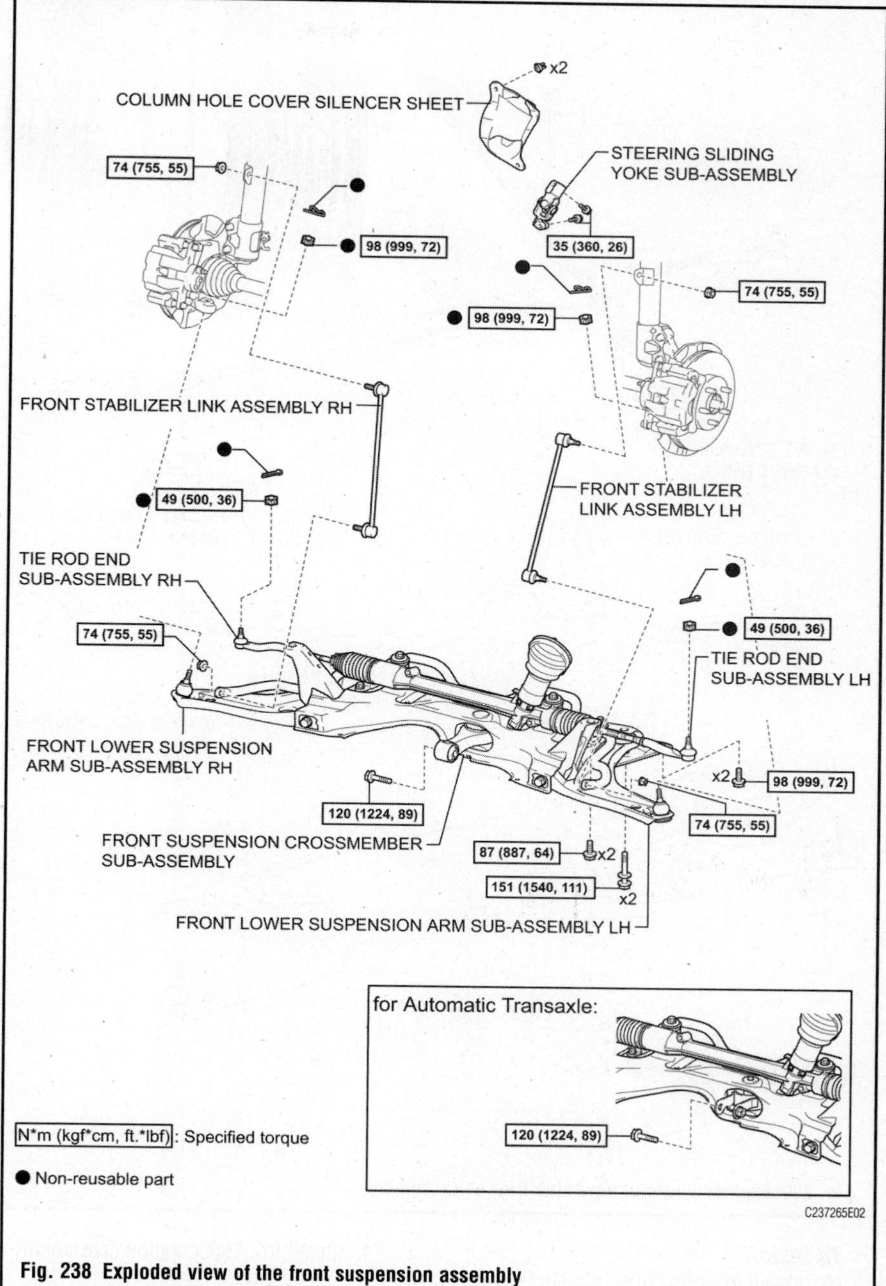

Fig. 238 Exploded view of the front suspension assembly

12. Remove front wheel
13. Remove column hole cover silencer sheet
14. Remove steering sliding yoke sub-assembly
15. Remove No. 1 steering column hole cover sub-assembly
16. Separate tie rod end sub-assembly LH
17. Separate tie rod end sub-assembly RH
18. Separate front lower suspension arm sub-assembly LH
19. Separate front lower suspension arm sub-assembly RH
20. Remove front stabilizer link assembly LH
21. Remove front stabilizer link assembly RH
22. Suspend engine assembly
23. Remove front suspension crossmember sub-assembly
24. Remove power steering gear
25. Remove front stabilizer No. 1 bracket LH
26. Remove front stabilizer No. 1 bracket RH
27. Remove front stabilizer bar
28. Remove front stabilizer bar No. 1 bushing

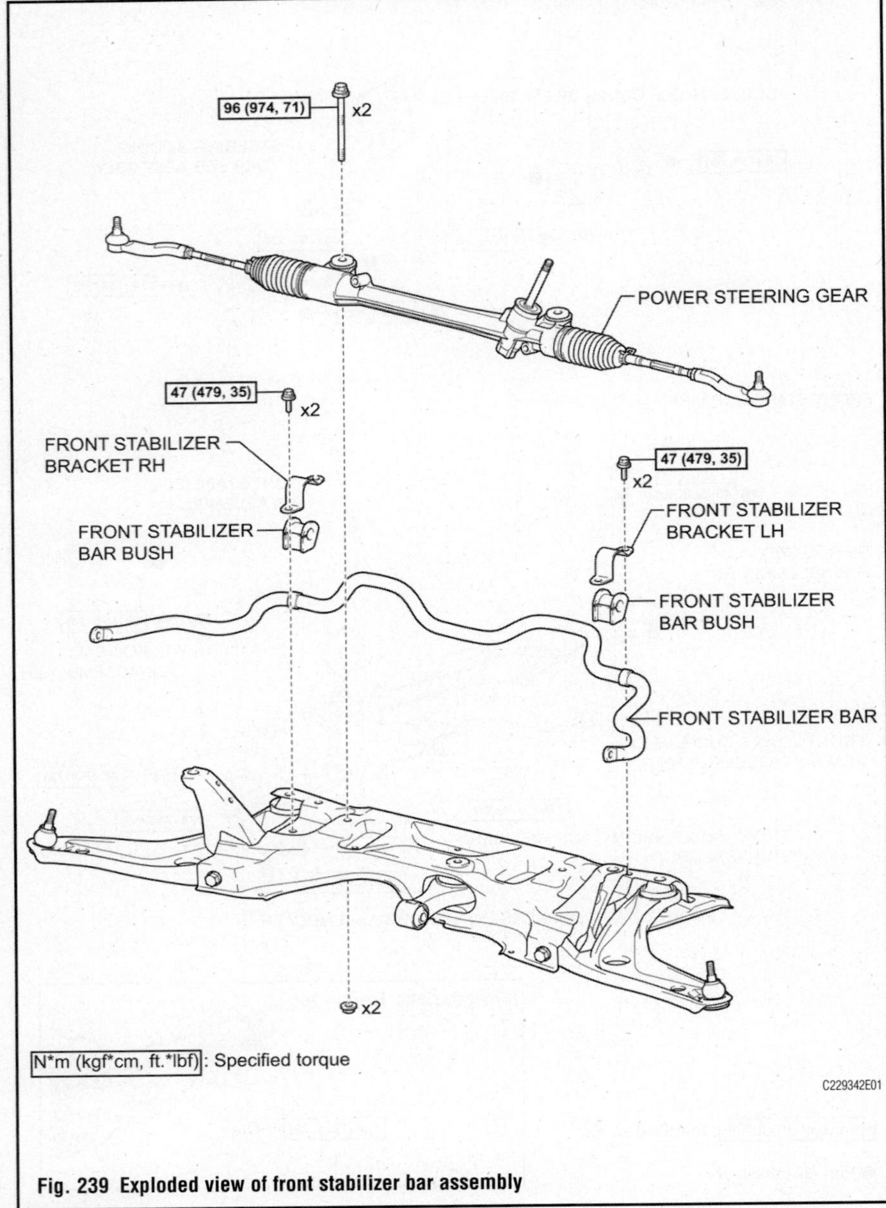

Fig. 239 Exploded view of front stabilizer bar assembly

To install:

29. Install front stabilizer bar No. 1 bushing
    a. Install the stabilizer bar bushing onto the stabilizer bar.

➡Install the bushing onto the stabilizer so that the bushing stopper of the stabilizer bar faces the outside of the vehicle.

    b. Install the bushing with its cutout facing the front of the vehicle.
30. Install front stabilizer bar
31. Install front stabilizer No. 1 bracket LH. Tighten to 35 ft. lbs. (47 Nm).
32. Install front stabilizer No. 1 bracket RH
33. Install power steering gear
34. Install front suspension crossmember sub-assembly
35. Install front stabilizer link assembly LH. Tighten to 55 ft. lbs. (74 Nm).
36. Install front stabilizer link assembly RH
37. Install front lower suspension arm sub-assembly LH
38. Install front lower suspension arm sub-assembly RH
39. Install tie rod end sub-assembly LH
40. Install tie rod end sub-assembly RH
41. Install steering column No. 1 hole cover sub-assembly
42. Install steering sliding yoke sub-assembly
43. Install column hole cover silencer sheet
44. Install front wheel
45. Position wheels facing straight ahead
46. Install outer cowl top panel
47. Install front air shutter seal RH
48. Install front wiper motor and link
49. Install cowl top ventilator louver LH
50. Install cowl top ventilator louver RH
51. Install hood to cowl top seal
52. Install front wiper arm and blade assembly LH
53. Install front wiper arm and blade assembly RH
54. Install front wiper arm head cap
55. Inspect hood sub-assembly
56. Adjust hood sub-assembly
57. Inspect and adjust front wheel alignment

STEERING KNUCKLE

REMOVAL & INSTALLATION

tC

See Figure 240.

➡Use the same procedure for the RH and LH sides. The procedure listed below is for the LH side.

1. Remove front axle hub sub-assembly LH.
2. Remove front lower ball joint assembly LH.
3. Remove steering knuckle LH.
    a. Secure the front axle assembly in a vise.
    b. Remove the 4 bolts and front axle hub sub-assembly from the steering knuckle.
    c. Remove the front brake dust cover from the steering knuckle.

To install:

4. Install steering knuckle LH.
    a. Secure the steering knuckle in a vise.
    b. Install the front brake dust cover to the steering knuckle.
    c. Install the front axle hub sub-assembly with the 4 bolts. Tighten to 71 ft. lbs. (96 Nm).
5. Install front lower ball joint assembly LH.
6. Install front axle hub sub-assembly LH.

xB

See Figure 241.

➡Use the same procedure for the RH and LH sides. The procedure listed below is for the LH side.

1. Remove front wheel.

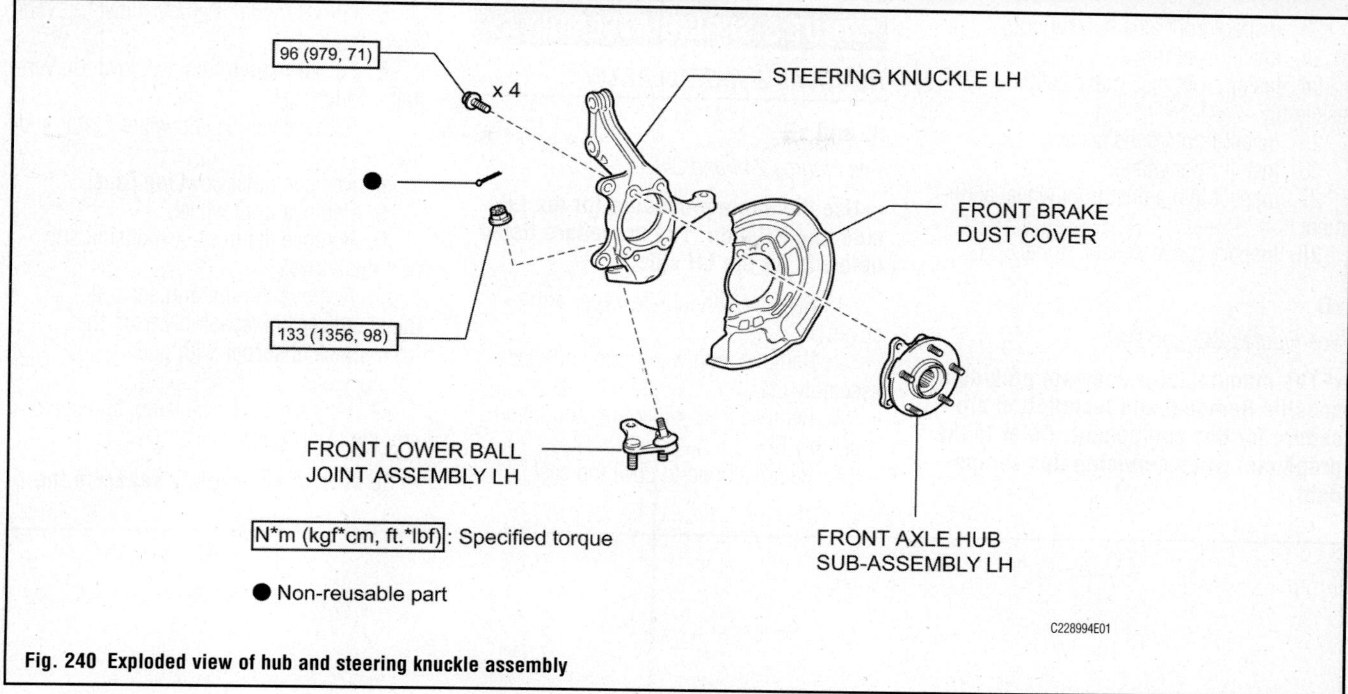

Fig. 240 Exploded view of hub and steering knuckle assembly

2. Remove front axle hub nut.
3. Separate front speed sensor.
4. Separate tie rod end sub-assembly.
5. Separate front disc brake caliper assembly.
6. Remove front disc.
7. Separate front lower No. 1 suspension arm sub-assembly.
8. Remove front axle assembly.
9. Remove front lower ball joint assembly.
10. Remove steering knuckle.
    a. Remove the 4 bolts and axle hub from the steering knuckle.
    b. Remove the front disc brake dust cover from the steering knuckle.

To install:

11. Install steering knuckle
    a. Install the front disc brake dust cover onto the steering knuckle.
    b. Install the axle hub with the 4 bolts. Tighten to 71 ft. lbs. (96 Nm).
12. Install front lower ball joint assembly.
13. Install front axle assembly.
14. Install front lower No. 1 suspension arm sub-assembly.
15. Install front disc.
16. Install front disc brake caliper assembly.
17. Install tie rod end sub-assembly.
18. Install front axle hub nut.
19. Separate front disc brake caliper assembly.
20. Remove front disc.
21. Inspect front axle hub bearing looseness.

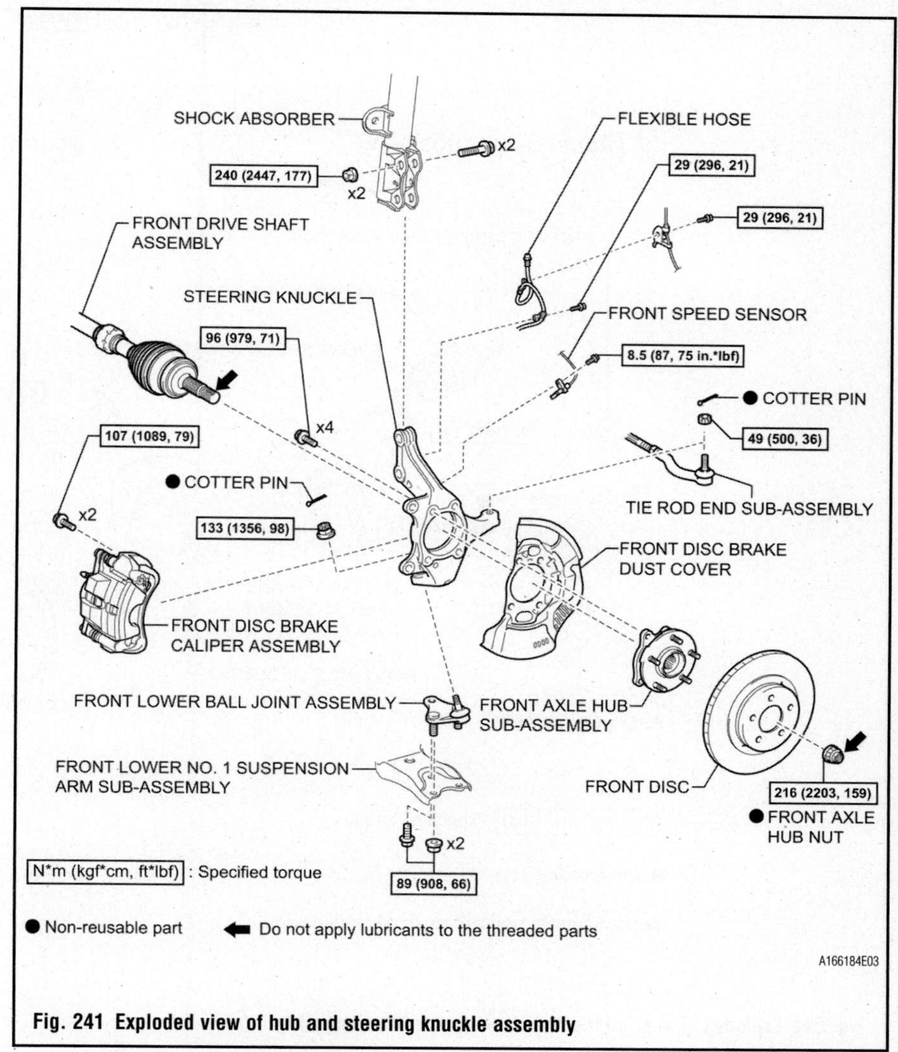

Fig. 241 Exploded view of hub and steering knuckle assembly

22. Inspect front axle hub runout.
23. Install front disc.
24. Install front disc brake caliper assembly.
25. Install front speed sensor.
26. Install front wheel.
27. Inspect and adjust front wheel alignment.
28. Inspect speed sensor signal.

xD

See Figures 242 and 243.

➡The manufacturer does not provide a specific Removal and Installation procedure for this component. Refer to the graphic(s) when servicing this component.

STRUT & SPRING ASSEMBLY

REMOVAL & INSTALLATION

tC and xB

See Figures 244 and 245.

➡Use the same procedure for the LH side and RH side. The procedure listed below is for the LH side.

1. Remove windshield wiper arm cover.
2. Remove front wiper arm and blade assembly LH.
3. Remove front wiper arm and blade assembly RH.
4. Remove hood to cowl top seal.
5. For xB model: Remove cowl top ventilator louver RH.
6. For xB model: Remove cowl top ventilator louver LH.
7. Remove windshield wiper motor and link.
8. Remove outer cowl top panel.
9. Remove front wheel.
10. Remove the front suspension support dust cover.
11. Remove the nut and separate the stabilizer link assembly from the front shock absorber with coil spring.
12. Remove the bolt, clamp, and separate the front speed sensor.

➡Be sure to completely separate the

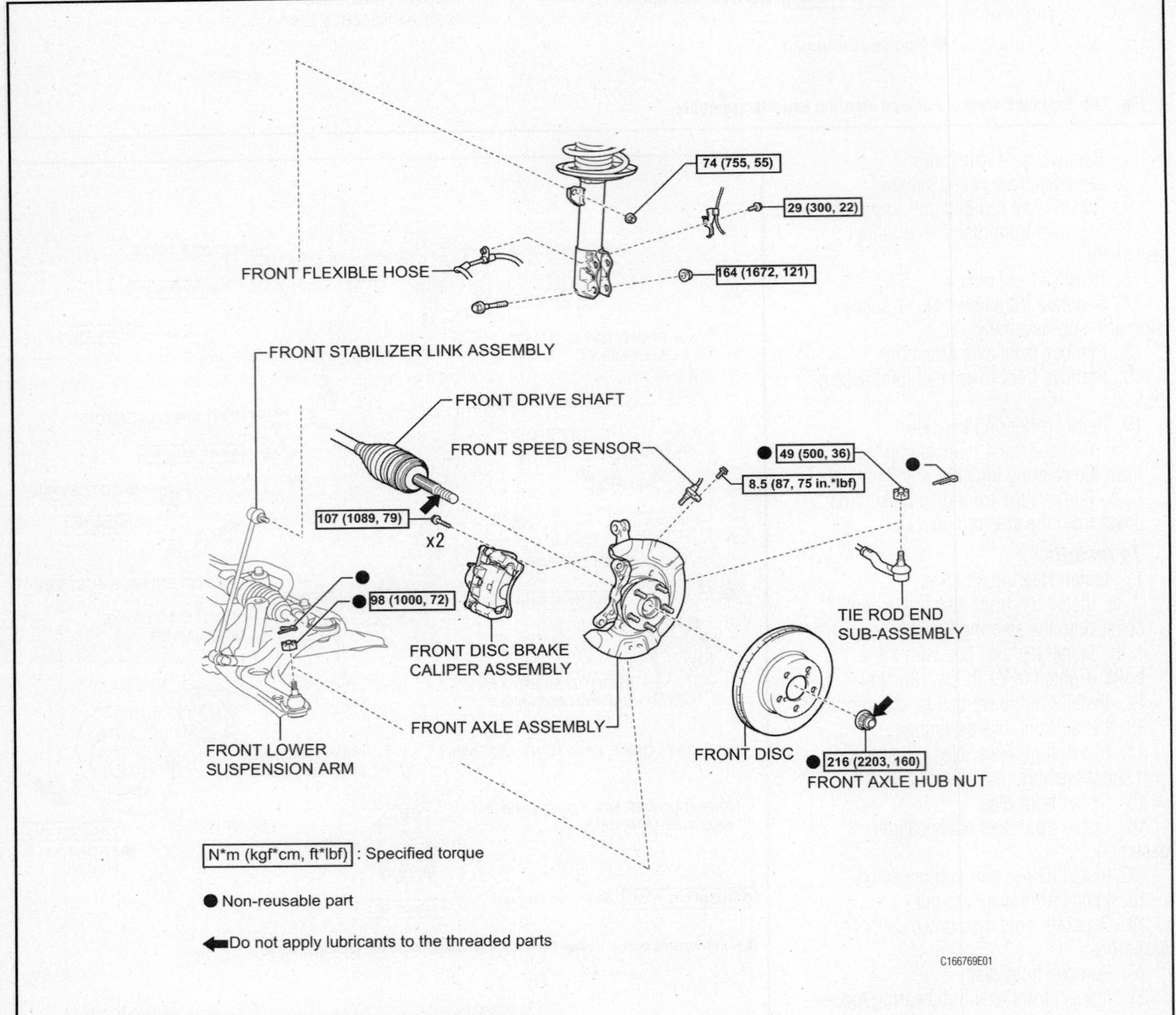

Fig. 242 Exploded view of front axle hub and suspension assembly

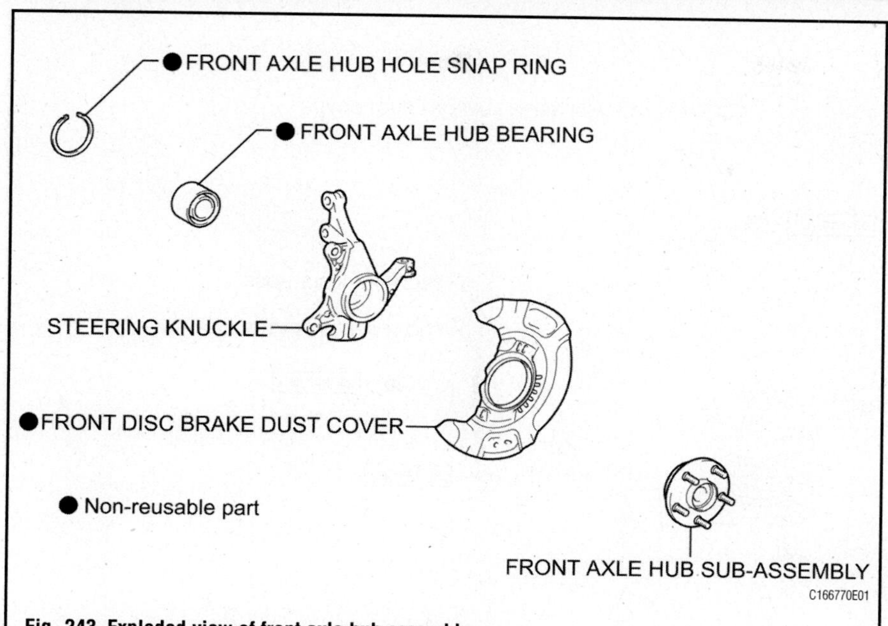

Fig. 243 Exploded view of front axle hub assembly

front speed sensor from the front shock absorber with coil spring.

13. Remove the bolt and separate the front flexible hose.
14. Remove front shock absorber with coil spring.
    a. Loosen the front support to front shock absorber nut of the front shock absorber.

➡Do not remove the front support to front shock absorber nut. Loosen the nut only when the front shock absorber with coil spring needs to be disassembled.

    b. Support the front axle using a jack and wooden blocks.
    c. Remove the 2 bolts and 2 nuts, and separate the front shock absorber with coil spring (lower side) from the steering knuckle.
    d. Remove the 3 nuts and front shock absorber with coil spring.

To install:

15. Install front shock absorber with coil spring.
    a. Install the front shock absorber with coil spring (upper side) with the 3 nuts. Tighten to 37 ft. lbs. (50 Nm).
    b. Install the front shock absorber with coil spring (lower side) to the steering knuckle and insert the 2 bolts and 2 nuts. Tighten to 177 ft. lbs. (240 Nm).
    c. Fully tighten the front support to front shock absorber nut. Tighten to 35 ft. lbs. (47 Nm).
16. Install the front flexible hose to the steering knuckle with the bolt. Tighten to 21 ft. lbs. (29 Nm).
17. Install the front speed sensor to the front shock absorber with coil spring with the bolt and clamp. Tighten to 21 ft. lbs. (29 Nm).
18. Install the front stabilizer link assembly to the front shock absorber with coil spring with the nut. Tighten to 55 ft. lbs. (74 Nm).
19. Install the front suspension support dust cover.
20. Install front wheel.
21. Install outer cowl top panel.
22. Install windshield wiper motor and link.
23. Install cowl top ventilator louver LH.
24. Install cowl top ventilator louver RH.
25. Install front wiper arm and blade assembly RH.
26. Install front wiper arm and blade assembly LH.

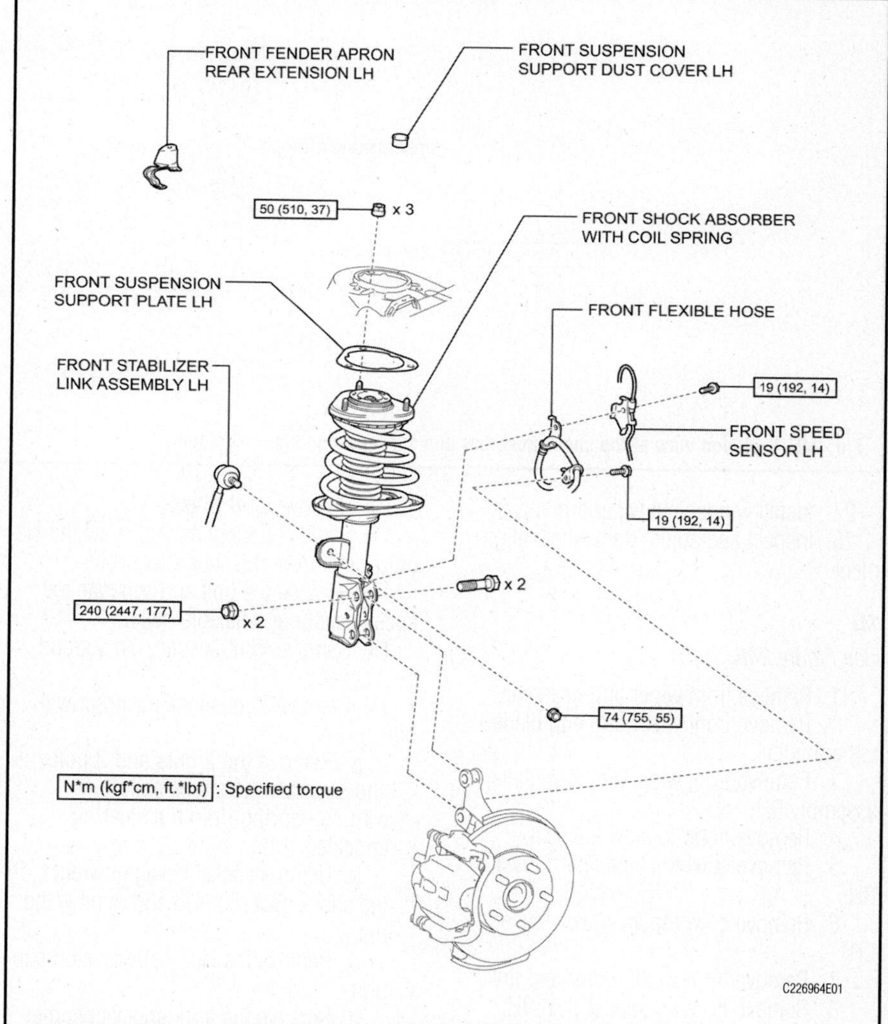

Fig. 244 Exploded view of the shock absorber and spring assembly—tC model

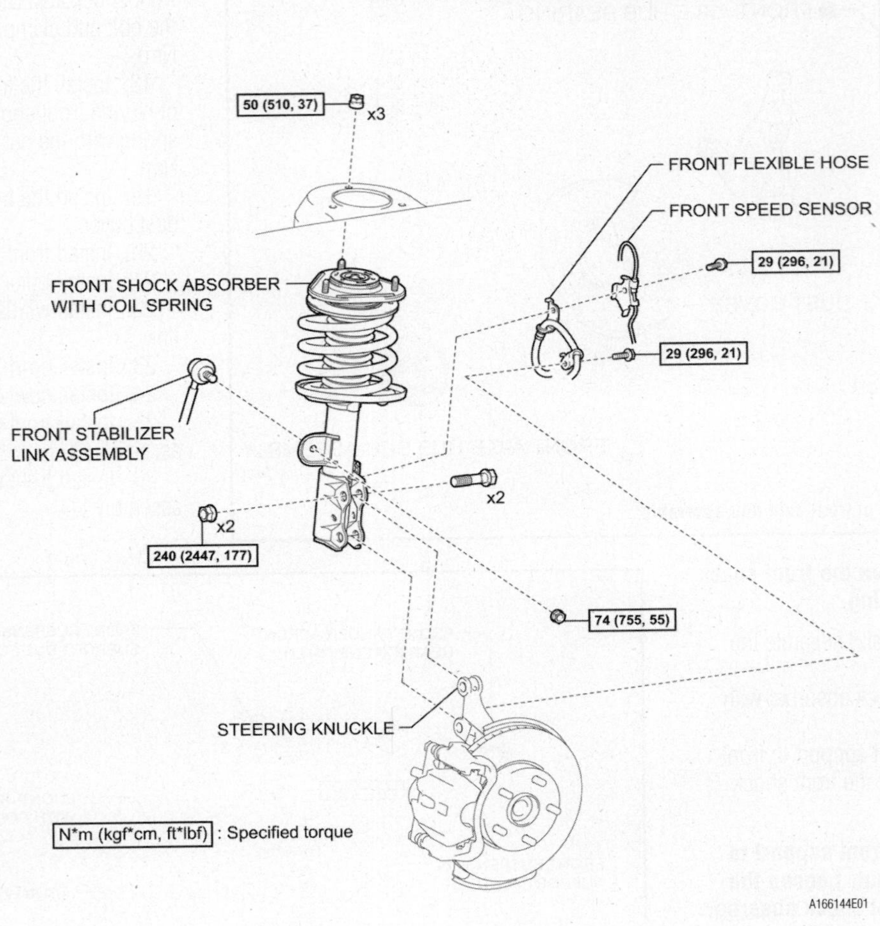

Fig. 245 Exploded view of the shock absorber and spring assembly—xB model

27. Install windshield wiper arm cover.
28. Inspect and adjust front wheel alignment.

xD

See Figure 246.

1. Remove front wiper arm head cap.
2. Remove front wiper arm and blade assembly LH.
3. Remove front wiper arm and blade assembly RH.
4. Remove hood to cowl top seal.
5. Remove cowl top ventilator louver RH.
6. Remove cowl top ventilator louver LH.
7. Remove front wiper motor and link.
8. Remove front air shutter seal RH.
9. Remove outer cowl top panel.
10. Remove front wheel.
11. Remove the nut and separate the stabilizer link from the shock absorber.
12. Remove the bolt and separate the speed sensor and flexible hose.
13. Remove front suspension support dust cover.
14. Remove front shock absorber with coil spring.
    a. Remove the 2 nuts and 2 bolts and separate the shock absorber with coil spring from the steering knuckle.
    b. Using a socket hexagon wrench, fix the shock absorber rod and remove the nut.
    c. Remove the No. 2 suspension support.
    d. Remove the front shock absorber with coil spring from the vehicle.

To install:

15. Temporarily tighten front shock absorber with coil spring.
    a. Provisionally tighten a new nut through No. 2 suspension support.
    b. Install the front shock absorber with coil spring onto the steering knuckle.
    c. Install the 2 bolts and 2 nuts. Tighten to 121 ft. lbs. (164 Nm).
16. Install the flexible hose and speed sensor with the bolt. Tighten to 22 ft. lbs. (29 Nm).
17. Install the stabilizer link with the nut. Tighten to 55 ft. lbs. (74 Nm).
18. Install front wheel.
19. Fully tighten front shock absorber with coil spring. Tighten to 41 ft. lbs. (55 Nm).

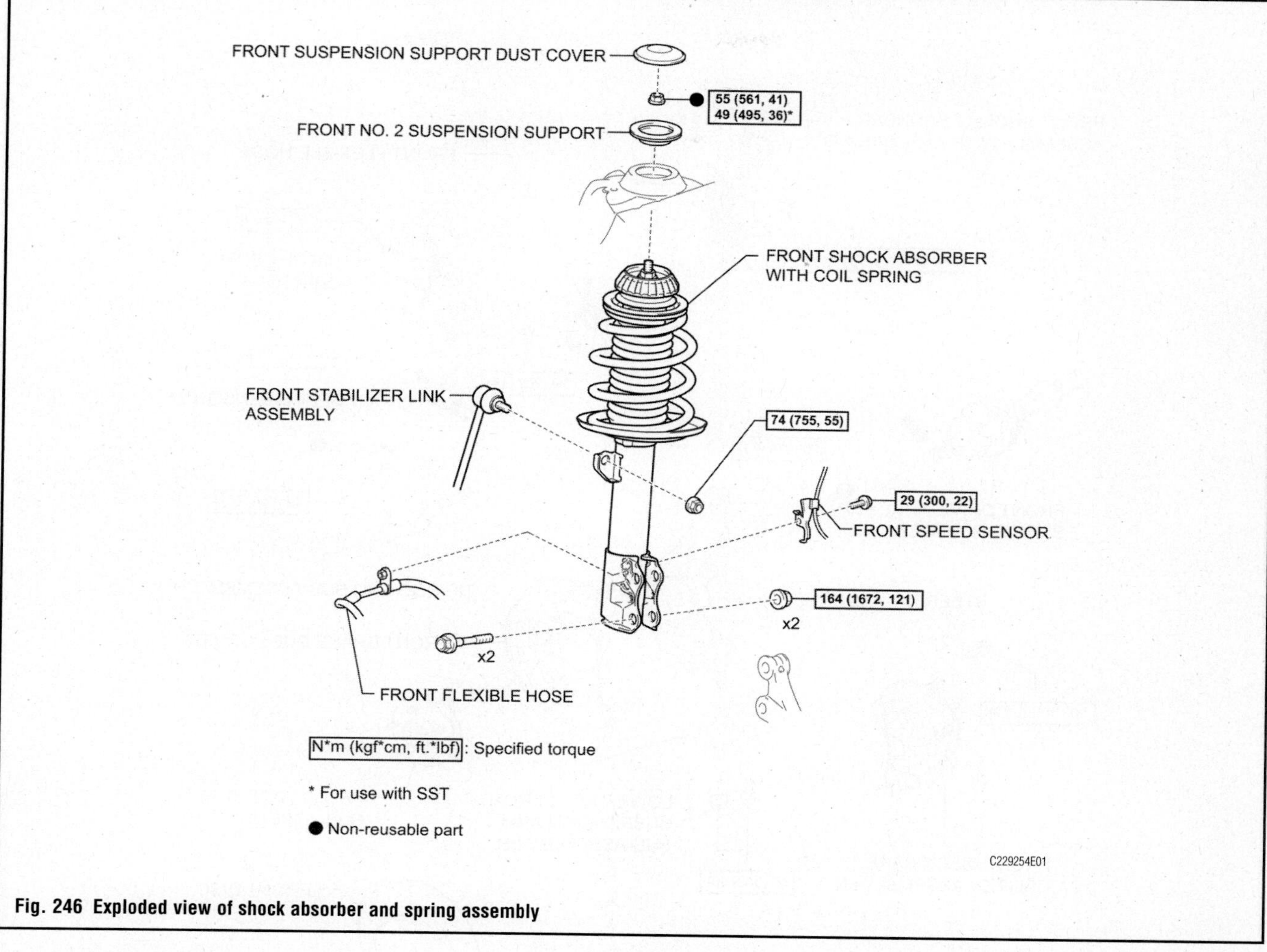

Fig. 246 Exploded view of shock absorber and spring assembly

20. Install front suspension support dust cover.
21. Install outer cowl top panel.
22. Install front air shutter seal RH.
23. Install front wiper motor and link.
24. Install cowl top ventilator louver LH.
25. Install cowl top ventilator louver RH.
26. Install hood to cowl top seal.
27. Install front wiper arm and blade assembly LH.
28. Install front wiper arm and blade assembly RH.
29. Install front wiper arm head cap.
30. Inspect and adjust front wheel alignment.

WHEEL BEARINGS

REMOVAL & INSTALLATION

tC and xB

See Figures 247 and 248.

➡Use the same procedure for the RH side and LH side. The procedure listed below is for the LH side.

1. Remove front wheel.
2. Remove front axle shaft nut.
3. Disconnect front speed sensor.
   a. Remove the bolt and clamp, and disconnect the front speed sensor.
   b. Remove the bolt and disconnect the front speed sensor from the steering knuckle.
4. Disconnect front flexible hose.
5. Remove the 2 bolts and disconnect the front disc brake caliper assembly from the steering knuckle.
6. Remove front disc.
7. Disconnect tie rod end sub-assembly.
8. Disconnect lower No. 1 front suspension arm sub-assembly.
9. Remove front axle assembly.
   a. Remove the 2 bolts and 2 nuts, and disconnect the front shock absorber assembly with coil spring from the steering knuckle.
   b. Put matchmarks on the drive shaft and front axle hub sub-assembly.
   c. Using a plastic-faced hammer, remove the steering knuckle assembly.
10. Remove front axle hub sub-assembly.

To install:

11. Install front axle hub sub-assembly.
    a. Secure the steering knuckle in a vise.
    b. Install the front brake dust cover to the steering knuckle.
    c. Install the front axle hub sub-assembly with the 4 bolts. Tighten to 71 ft. lbs. (96 Nm).
12. Install front axle assembly.
    a. Align the matchmarks and install the front drive shaft assembly to the front axle hub sub-assembly.
    b. Install the front axle assembly to the front shock absorber assembly with coil spring with the 2 bolts and 2 nuts. Tighten to 177 ft. lbs. (240 Nm).
13. Connect lower No. 1 front suspension arm sub-assembly.
14. Connect tie rod end sub-assembly.
15. Install front disc.
16. Connect the front disc brake caliper assembly to the steering knuckle with the 2 bolts. Tighten to 79 ft. lbs. (107 Nm).
17. Connect front flexible hose.

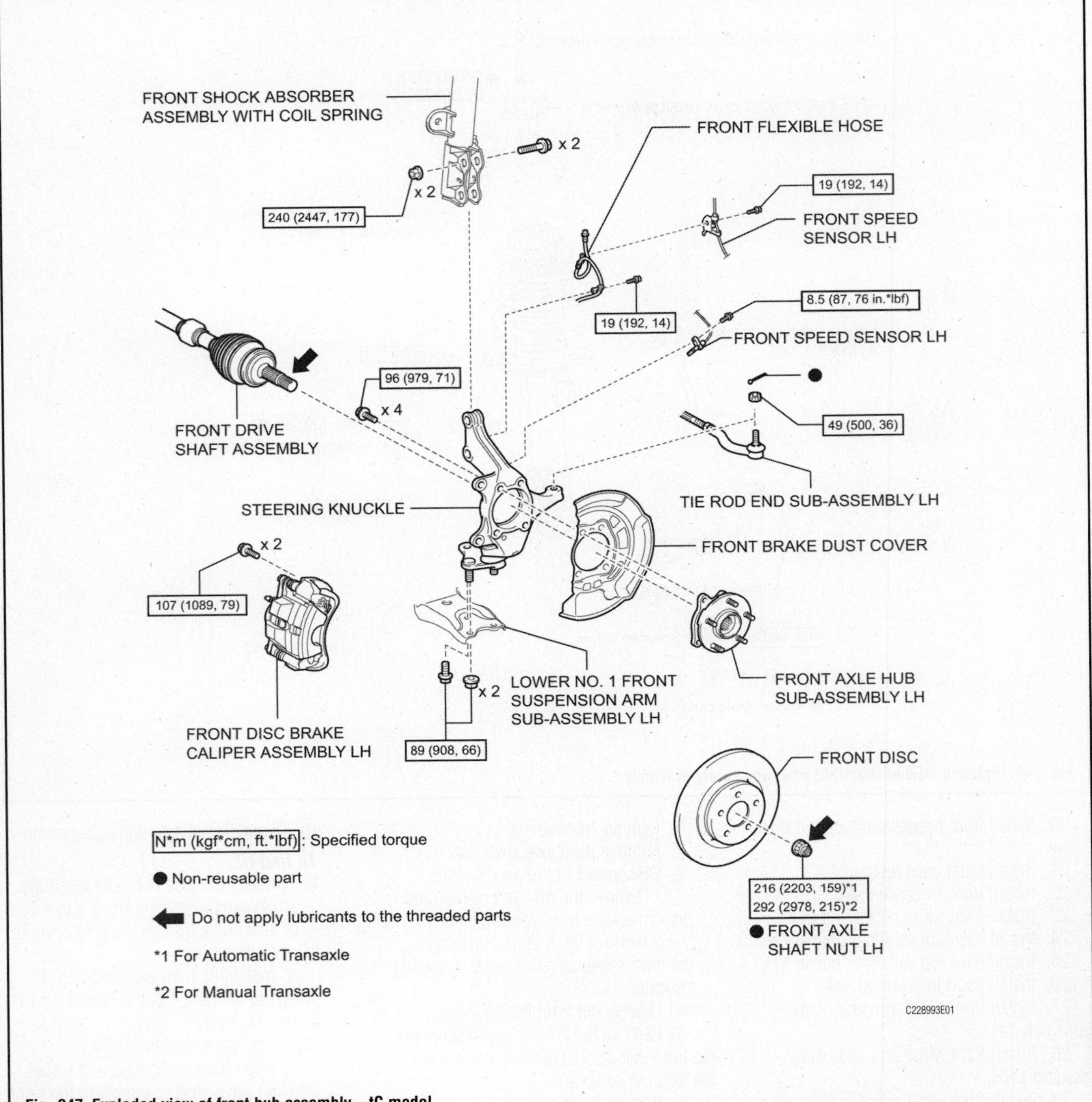

Fig. 247 Exploded view of front hub assembly—tC model

18. Connect front speed sensor.
    a. Connect the front speed sensor and front flexible hose to the front shock absorber with the bolt and clamp.
    b. Install the front speed sensor to the steering knuckle with the bolt.
19. Install front axle shaft nut.
20. Install front wheel.
21. Inspect and adjust front wheel alignment.
22. Check for speed sensor signal.

xD

See Figure 249.

1. Remove front wheel.
2. Remove front axle hub nut.
3. Remove the 2 bolts and separate the disc brake caliper from the steering knuckle.
4. Remove front disc.
5. Separate front speed sensor.
6. Separate front stabilizer link assembly.
7. Separate tie rod end sub-assembly.
8. Separate front lower suspension arm.
9. Remove front axle assembly.
    a. Using a plastic hammer, tap the end of the drive shaft and disengage the fitting between the drive shaft and front axle.
    b. Push the front axle out of the vehicle to remove the drive shaft from the front axle.

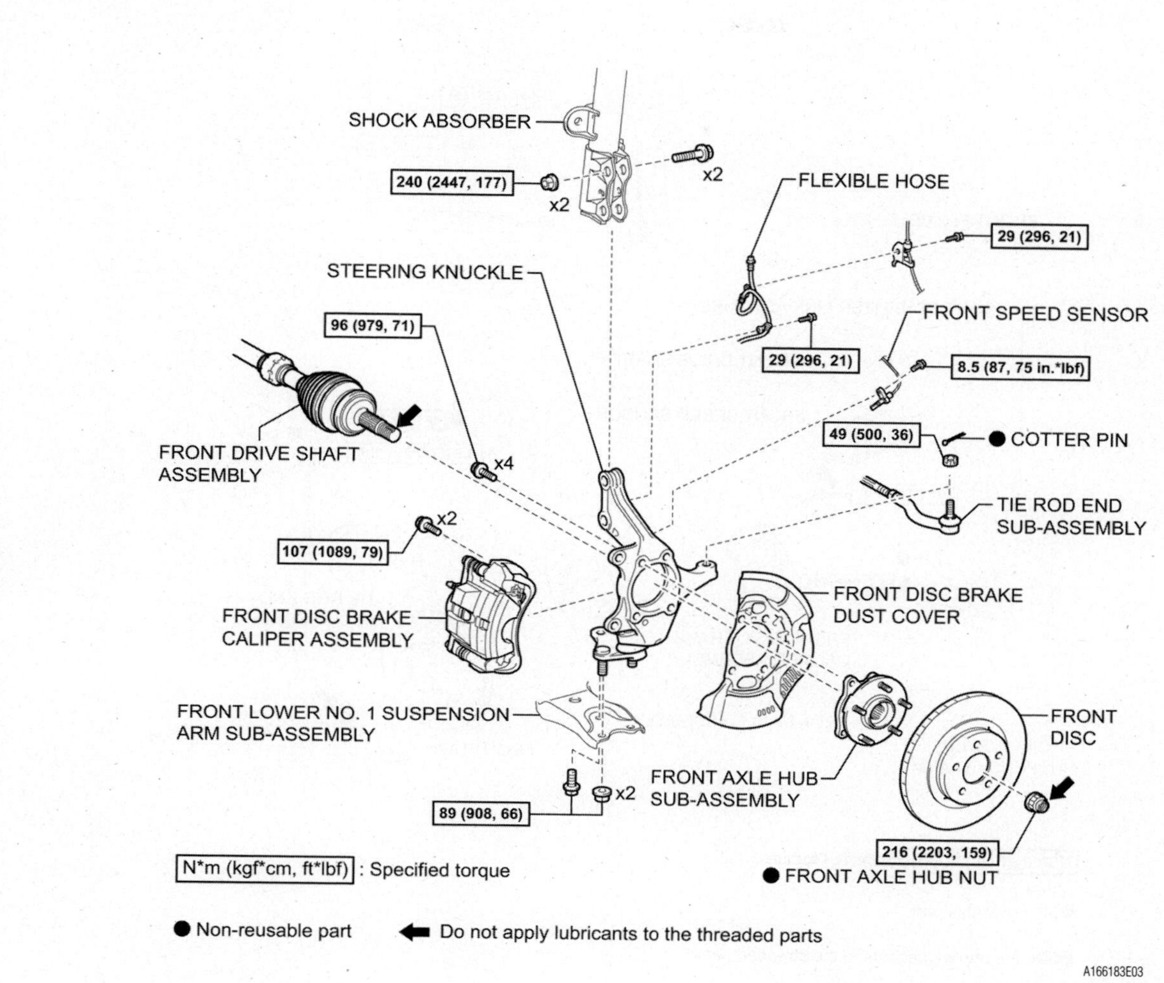

Fig. 248 Exploded view of front hub assembly—xB model

c. Remove the 2 nuts and 2 bolts and remove the front axle assembly.

10. Using snap ring pliers, remove the hole snap ring.

11. Remove front axle hub sub-assembly.

a. Fix the steering knuckle in a vise between aluminum plates.

b. Using SST, remove the axle hub.

c. Using SST, remove the hub bearing inner race from the axle hub.

12. Remove the dust cover by tapping the back side with a brass bar and hammer.

13. Remove front axle hub bearing.

To install:

14. Install front axle hub bearing.

15. Install front disc brake dust cover.

a. Provisionally install a new disc brake dust cover.

b. Using SST and a hammer, install the disc brake dust cover.

c. Using a chisel, fix the 3 points on the circumference.

16. Install front axle hub sub-assembly.

17. Install front axle hub hole snap ring.

18. Install front axle assembly.

a. Install the front axle assembly onto the shock absorber.

b. Install the 2 bolts and 2 nuts. Tighten to 121 ft. lbs. (164 Nm).

c. Push the front axle out of the vehicle to align the spline of the drive shaft with the front axle and insert the front axle.

19. Install front lower suspension arm.

20. Install tie rod end sub-assembly.

21. Install front stabilizer link assembly.

22. Install front speed sensor.

23. Install front disc.

24. Install the disc brake caliper onto the steering knuckle. Tighten to 79 ft. lbs. (107 Nm).

25. Install front axle hub nut.

26. Inspect front axle hub bearing.

27. Install front wheel.

28. Inspect and adjust front wheel alignment.

29. Check VSC sensor signal.

ADJUSTMENT

➡Wheel bearings are an integral, sealed unit in the hub assembly, with no allowance for adjustment.

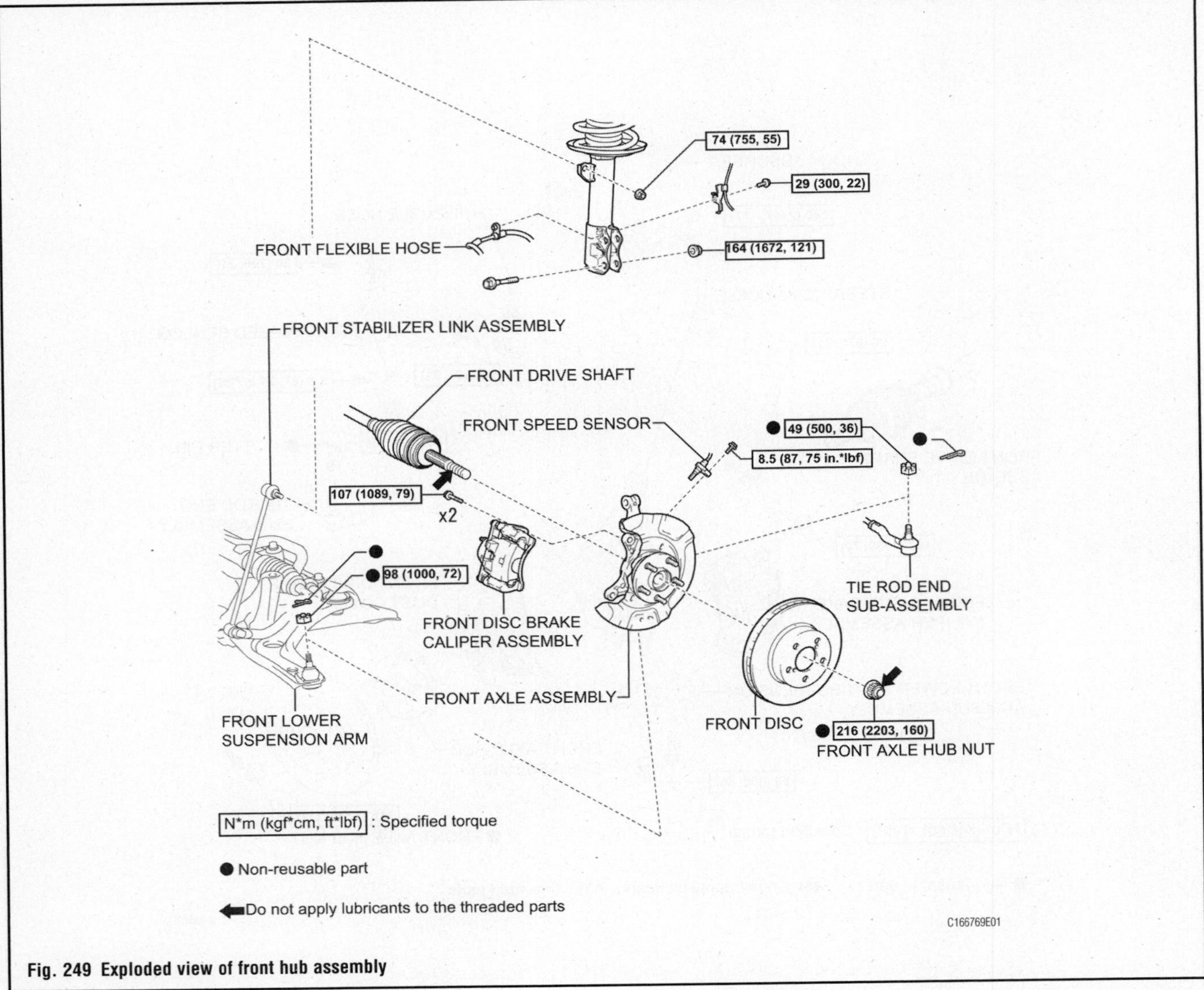

Fig. 249 Exploded view of front hub assembly

AXLE BEAM

REMOVAL & INSTALLATION

xB

See Figures 250 through 252.

1. Remove rear wheels.
2. Drain brake fluid.
3. Remove upper console panel sub-assembly.
4. Loosen wire adjusting nut.
5. Remove rear floor side member cover LH.
6. Remove rear floor side member cover RH.
7. Remove the 4 bolts, clamp, and rear floor side member brace sub-assembly.
8. Separate skid control sensor wire (for LH Side).
   a. Using a screwdriver, disconnect the connector from the rear speed sensor.
   b. Remove the nut and separate the 2 clamps and skid control sensor wire.
9. Separate skid control sensor wire (for RH Side).
10. Separate rear flexible hose LH.
   a. Using union nut wrench, disconnect the brake line while holding the rear flexible hose LH with a wrench.
   b. Remove the bolt and separate the rear flexible hose LH.
11. Separate rear flexible hose RH.
   a. Using union nut wrench, disconnect the brake line while holding the rear flexible hose RH with a wrench.
   b. Remove the clip and separate the rear flexible hose RH.
12. Separate rear brake tube flexible hose (for LH Side).
   a. Using union nut wrench, disconnect the brake line while holding the rear brake flexible hose with a wrench.
   b. Remove the clip and separate the rear brake flexible hose.
13. Separate rear brake tube flexible hose (for RH Side).
14. Remove the nut, and remove the No. 4 rear brake tube from the rear axle beam assembly.
15. Remove No. 3 rear brake tube.
16. Separate No. 3 parking brake cable assembly.
   a. Remove the bolt.

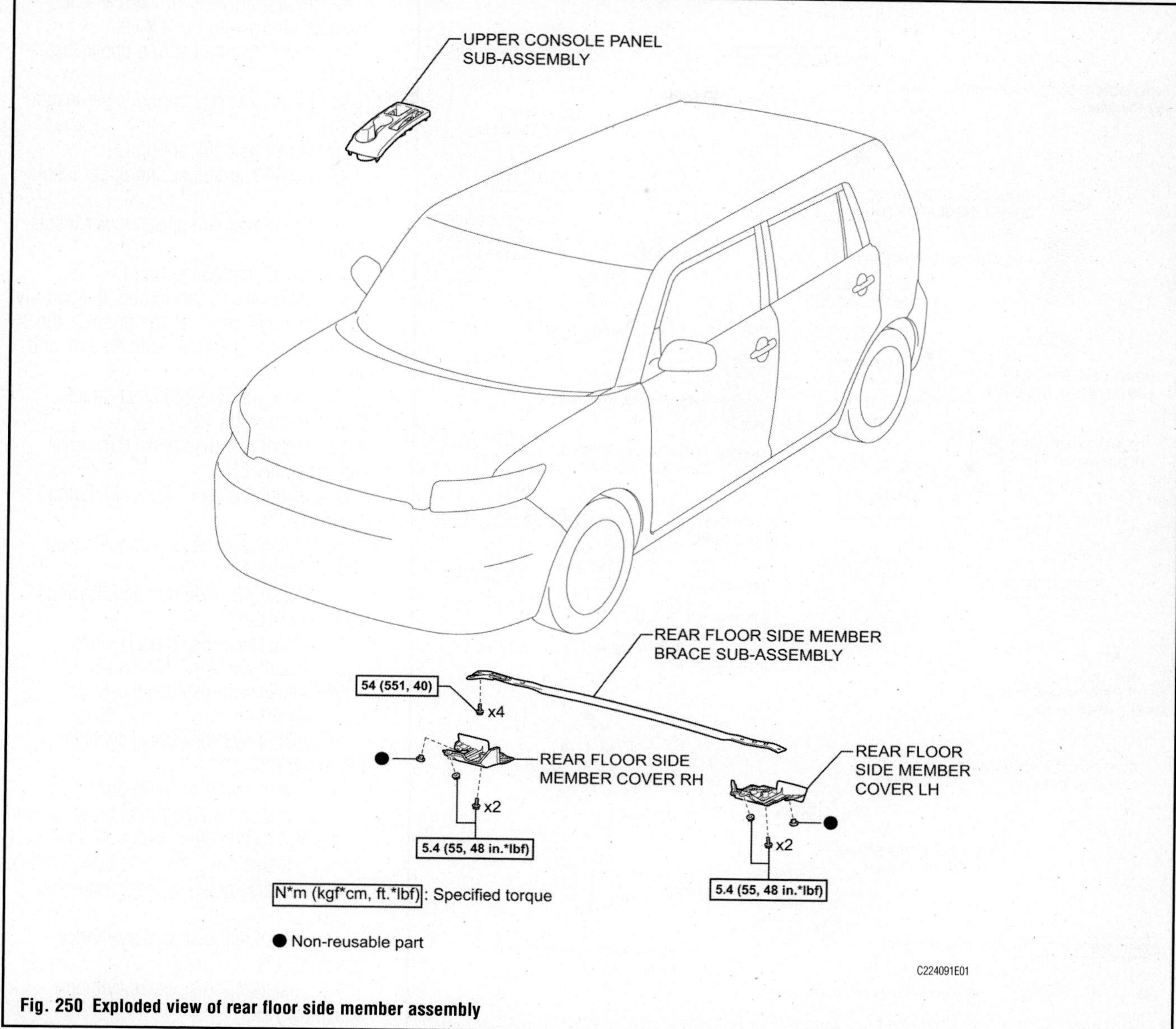

Fig. 250 Exploded view of rear floor side member assembly

b. Disconnect the No. 3 parking brake cable assembly from the rear disc brake cylinder operation lever.

c. Insert an offset wrench all the way into the clip of the No. 3 parking brake cable assembly to release the clip claws and separate the No. 3 parking brake cable assembly from the rear disc brake cylinder assembly LH.

17. Separate No. 2 parking brake cable assembly.

18. Remove rear disc brake cylinder assembly LH.

19. Remove rear disc brake cylinder assembly RH.

20. Remove rear disc brake pad.

a. Remove the 2 disc brake pads from the disc brake cylinder mounting LH.

b. Remove the 2 disc brake pads from the disc brake cylinder mounting RH.

21. Remove rear disc brake cylinder mounting LH.

22. Remove rear disc brake cylinder mounting RH.

23. Remove rear disc (for LH side).

24. Remove rear disc (for RH side).

25. Remove rear axle hub and bearing assembly LH.

26. Remove rear axle hub and bearing assembly RH.

27. Remove rear disc brake dust cover sub-assembly LH.

28. Remove rear disc brake dust cover sub-assembly RH.

29. Loosen the 2 bolts of the rear axle beam assembly.

➡Do not remove the bolts.

30. Separate rear shock absorber assembly LH.

a. Support the rear axle beam assembly using a jack and wooden block.

b. Remove the bolt while holding the nut and separate the rear shock absorber LH.

31. Separate rear shock absorber assembly RH.

32. Remove rear coil spring LH.

33. Remove rear coil spring RH.

34. Remove rear coil spring upper insulator LH.

35. Remove rear coil spring upper insulator RH.

36. Remove rear coil spring lower insulator LH.

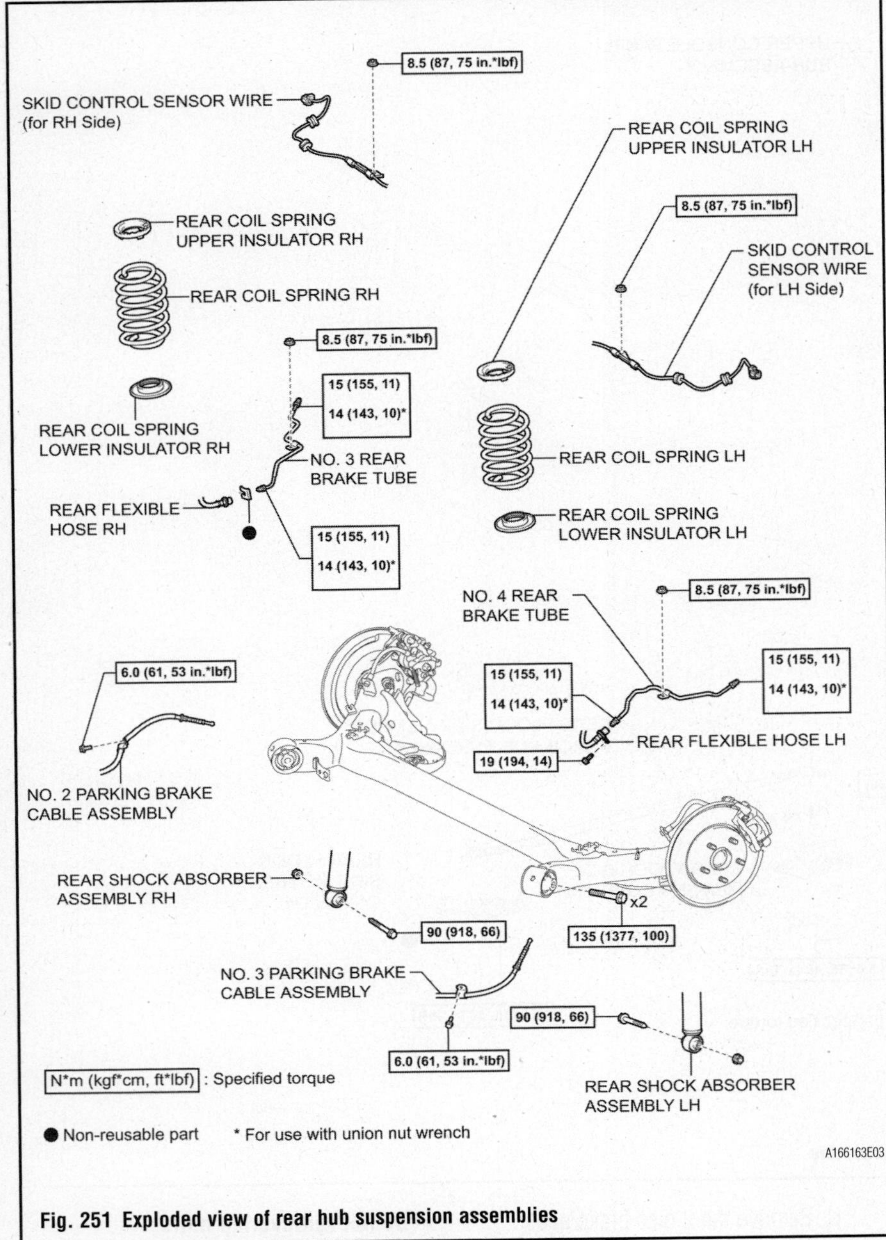

Fig. 251 Exploded view of rear hub suspension assemblies

37. Remove rear coil spring lower insulator RH.
38. Remove the 2 bolts and the rear axle beam assembly.
39. Remove rear trailing bushing LH.
   a. Put a matchmark on the rear axle beam assembly so that the mark aligns with the arrow mark on the rear trailing bushing LH. (If the rear axle beam assembly is reused.)
   b. Using a chisel and hammer, bend the 2 ribs on the bushing.

➡When removing the bushing, do not erase the matchmark on the rear axle beam assembly.

   c. Using SST, remove the bushing from the rear axle beam assembly.
40. Remove rear trailing bushing RH.

To install:

41. Install rear trailing bushing LH.
   a. Align the arrow mark on a new rear trailing bushing LH with the matchmark on the rear axle beam assembly and temporarily install the rear trailing bushing to the rear axle beam assembly. (If the rear axle beam assembly is reused.)
   b. Temporarily install the new rear trailing bushing LH.
   c. Using SST, install the bushing to the rear axle beam assembly.
42. Install rear trailing bushing RH.
43. Temporarily install the rear axle beam assembly with the 2 bolts.
44. Install rear coil spring upper insulator LH.
45. Install rear coil spring lower insulator LH.
46. Install rear coil spring LH.
47. Install rear coil spring upper insulator RH.
48. Install rear coil spring lower insulator RH.
49. Install rear coil spring RH.
50. While slowly jacking up, temporarily tighten the rear shock absorber LH to the rear axle beam assembly with the bolt and nut.
51. Temporarily tighten rear shock absorber assembly RH.
52. Install rear disc brake dust cover sub-assembly LH.
53. Install rear axle hub and bearing assembly LH.
54. Install rear disc brake dust cover sub-assembly RH.
55. Install rear axle hub and bearing assembly RH.
56. Install rear disc (for LH side).
57. Install rear disc (for RH side).
58. Install rear disc brake cylinder mounting LH.
59. Install rear disc brake cylinder mounting RH.
60. Install rear disc brake pad.
   a. Install the 2 rear disc brake pads to the rear disc brake cylinder mounting LH.
   b. Install the 2 rear disc brake pads to the rear disc brake cylinder mounting RH.
61. Install rear disc brake cylinder assembly LH.
62. Install rear disc brake cylinder assembly RH.
63. Install No. 3 parking brake cable assembly.
   a. Insert the No. 3 parking brake cable assembly into the rear disc brake cylinder assembly and engage the clip claws of the No. 3 parking brake cable to the rear disc brake cylinder guide.
   b. Connect the No. 3 parking brake cable end to the rear disc brake cylinder operation lever.
   c. Install the No. 3 parking brake cable with the bolt.
64. Install No. 2 parking brake cable assembly.
65. Install the No. 4 rear brake tube onto the axle beam with the nut.
66. Install No. 3 rear brake tube.
67. Install rear flexible hose LH.
   a. Install the rear flexible hose LH with the bolt.

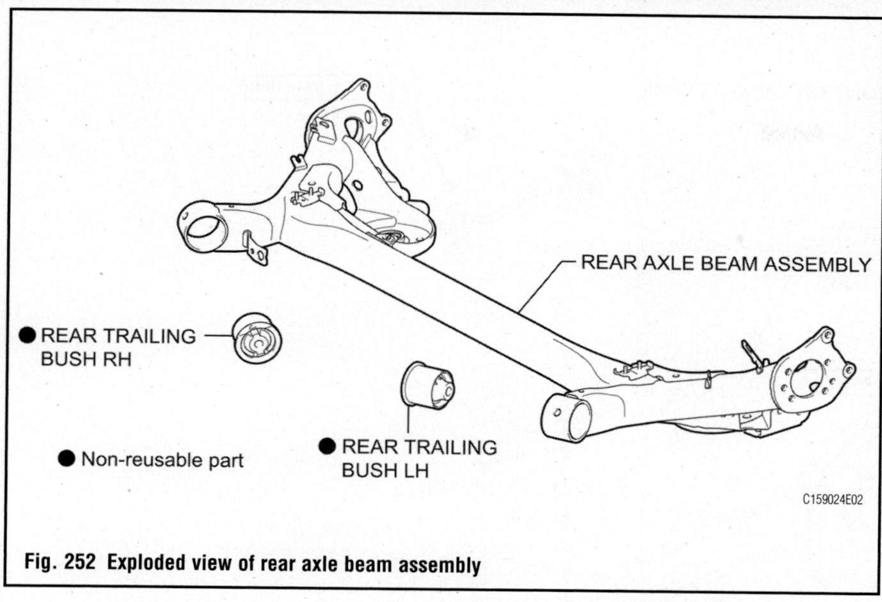

Fig. 252 Exploded view of rear axle beam assembly

b. Using union nut wrench, connect the No. 4 rear brake tube to the rear flexible hose LH while holding the rear flexible hose LH with a wrench.

68. Install rear flexible hose RH.

a. Install the rear flexible hose RH with a new clip.

➡Install the clip as far as it will go.

b. Using union nut wrench, connect the No. 3 rear brake tube to the rear flexible hose RH while holding the rear flexible hose RH with a wrench.

69. Connect rear brake tube flexible hose (for LH Side).

a. Install a new clip.

➡Install the clip as far as it will go.

b. Using union nut wrench, connect the No. 4 rear brake tube to the rear brake flexible hose while holding the rear brake flexible hose with a wrench.

70. Connect rear brake tube flexible hose (for RH Side).

71. Connect skid control sensor wire (for LH Side).

a. Install the nut and connect the 2 clamps.

b. Connect the connector.

72. Connect skid control sensor wire (for RH Side).

73. Install the rear floor side member brace sub-assembly with the 4 bolts and clamp. Tighten to 40 ft. lbs. (54 Nm).

74. Fill reservoir with brake fluid.

75. Bleed brake master cylinder.

76. Bleed brake line.

77. Bleed brake actuator.

78. Inspect fluid level in reservoir.

79. Inspect for brake fluid leak.

80. Inspect parking brake lever travel.

81. Adjust parking brake lever travel.

82. Install upper console panel sub-assembly.

83. Install rear wheels.

84. Jack up the rear axle carrier, placing a wooden block to avoid damage. Apply load to the suspension so that the rear drive shaft assembly is horizontally positioned.

85. Fully tighten the rear axle beam assembly with the left and right bolts to 100 ft. lbs. (135 Nm).

86. Fully tighten rear shock absorber assembly LH to 66 ft. Lbs. (90 Nm).

87. Fully tighten rear shock absorber assembly RH to 66 ft. Lbs. (90 Nm).

88. Install rear floor side member cover LH.

89. Install rear floor side member cover RH.

90. Inspect rear wheel alignment.

91. Inspect speed sensor signal.

xD

See Figures 253 through 255.

1. Remove rear wheel.
2. Drain brake fluid.
3. Separate skid control sensor wire.
4. Separate rear brake No. 4 tube.

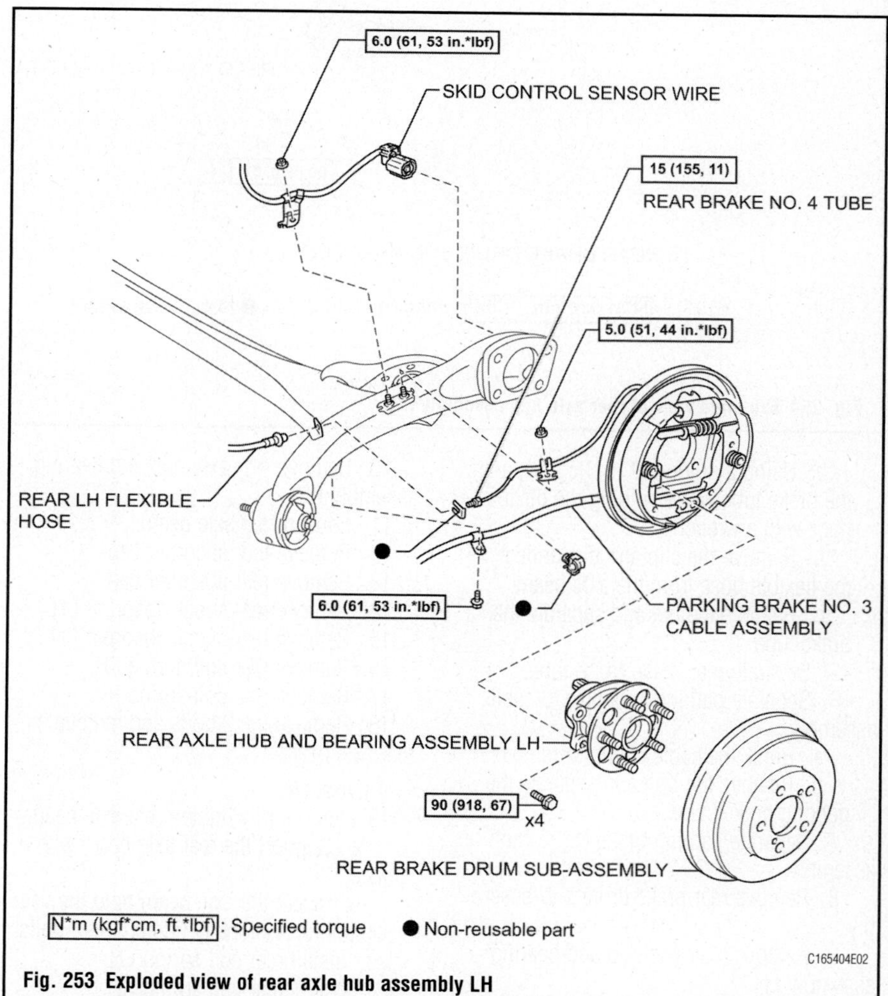

Fig. 253 Exploded view of rear axle hub assembly LH

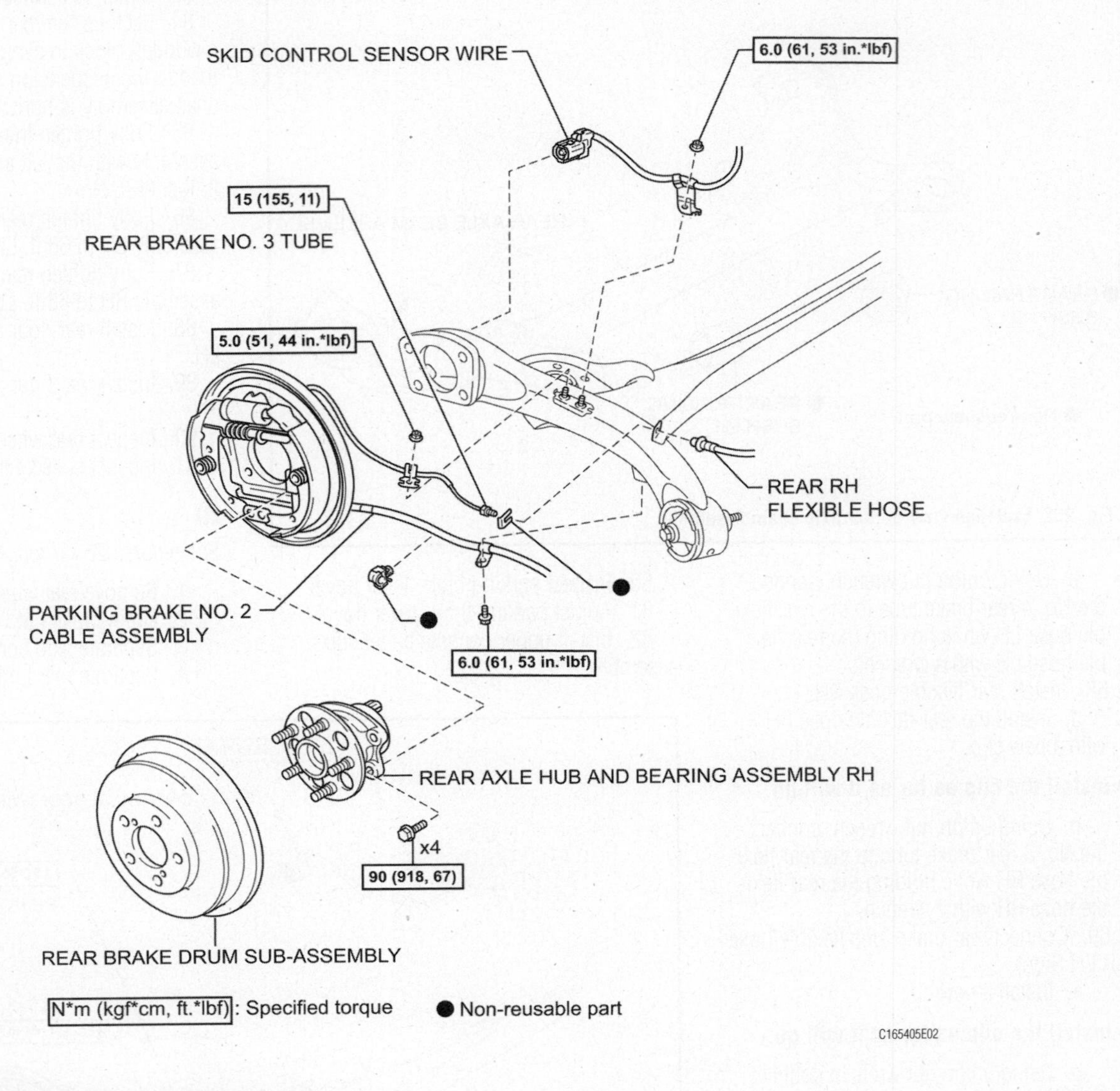

Fig. 254 Exploded view of rear axle hub assembly RH

a. Using a union nut wrench, separate the brake tube while holding the flexible hose with a wrench.

b. Remove the clip and disconnect the flexible hose from the axle beam.

c. Remove the nut and separate the brake tube.

5. Separate rear brake No. 3 tube.
6. Separate parking brake No. 3 cable assembly.

a. Break the cable clamp to remove it.

b. Remove the bolt and separate the parking brake cable.

7. Separate parking brake No. 2 cable assembly
8. Remove rear brake drum sub-assembly..
9. Remove rear axle hub and bearing assembly LH.
10. Remove rear axle hub and bearing assembly RH.
11. Loosen rear axle beam.
12. Remove rear absorber cap.
13. Remove rear absorber cap.
14. Remove rear shock absorber LH.
15. Remove rear shock absorber RH.
16. Remove rear coil spring LH.
17. Remove rear coil spring RH.
18. Remove the 2 bolts and remove the rear axle beam.

To install:

19. Temporarily tighten rear axle beam.

a. Support the rear axle beam with a jack.

b. Install the axle beam onto the vehicle and provisionally tighten the 2 bolts.

20. Install rear coil spring LH.
21. Install rear coil spring RH.
22. Temporarily tighten rear shock absorber LH.
23. Temporarily tighten rear shock absorber RH.
24. Install rear absorber cap.
25. Install rear absorber cap.
26. Install rear axle hub and bearing assembly LH.
27. Install rear axle hub and bearing assembly RH.
28. Inspect rear axle hub bearing.
29. Install rear brake drum sub-assembly.
30. Adjust rear drum brake shoe clearance.
31. Install the parking brake cable with the bolt and a new clamp.
32. Install parking brake No. 2 cable assembly.

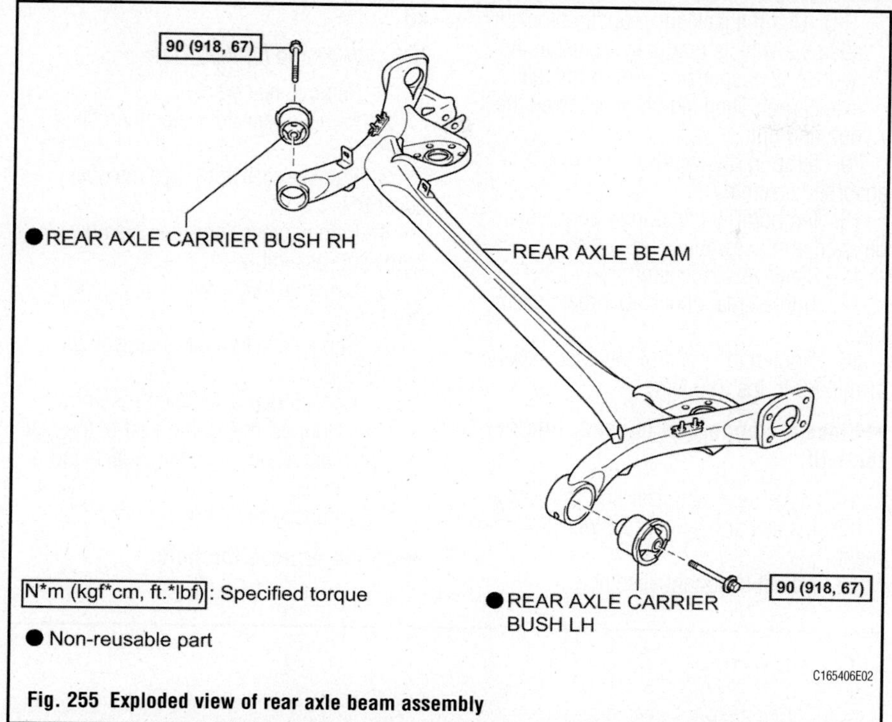

Fig. 255 Exploded view of rear axle beam assembly

33. Install rear brake No. 4 tube.
    a. Install the brake tube onto the axle beam with the nut.
    b. Connect the flexible hose to the axle beam with a new clip.
    c. Using a union nut wrench, separate the brake tube while holding the flexible hose with a wrench.
34. Install rear brake No. 3 tube.
35. Install skid control sensor wire.
36. Install rear wheel.
37. Stabilize suspension.
    a. Lower the vehicle from the jack.
    b. Bounce the vehicle up and down several times to stabilize the suspension.
38. Fully tighten rear axle beam.
    a. Suspend the jack on the rear axle spring seat and adjust the length of the shock absorber.
    b. Fully tighten the 2 bolts to 67 ft. lbs. (90 Nm).
39. Fully tighten rear shock absorber LH.
40. Fully tighten rear shock absorber RH.
41. Fill reservoir with brake fluid.
42. Bleed master cylinder.
43. Bleed brake line.
44. Bleed brake actuator.
45. Inspect fluid level in reservoir.
46. Inspect for brake fluid leak.
47. Inspect rear wheel alignment.
48. Check VSC sensor signal.

COIL SPRING

REMOVAL & INSTALLATION

tC

See Figures 256 and 257.

➡Use the same procedure for the RH and LH sides. The procedure listed below is for the LH side.

1. Remove rear wheel.
2. Remove rear stabilizer link assembly.
3. Loosen the 2 bolts of the suspension arm.

➡Do not remove the bolts and nuts, only loosen them. Since a stopper nut is used, loosen the bolt.

4. Disconnect rear shock absorber assembly.
    a. Support the rear No. 2 suspension arm with a jack using a wooden block to avoid damage.
    b. Remove the bolt and nut, and then

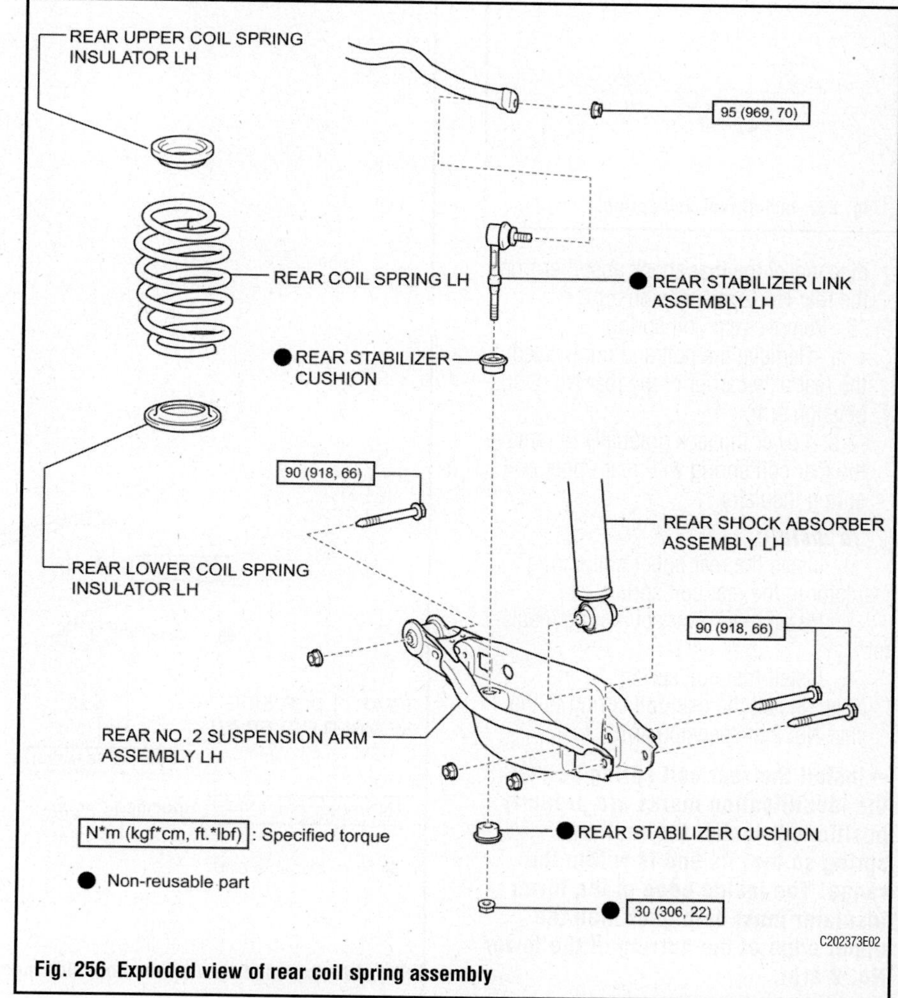

Fig. 256 Exploded view of rear coil spring assembly

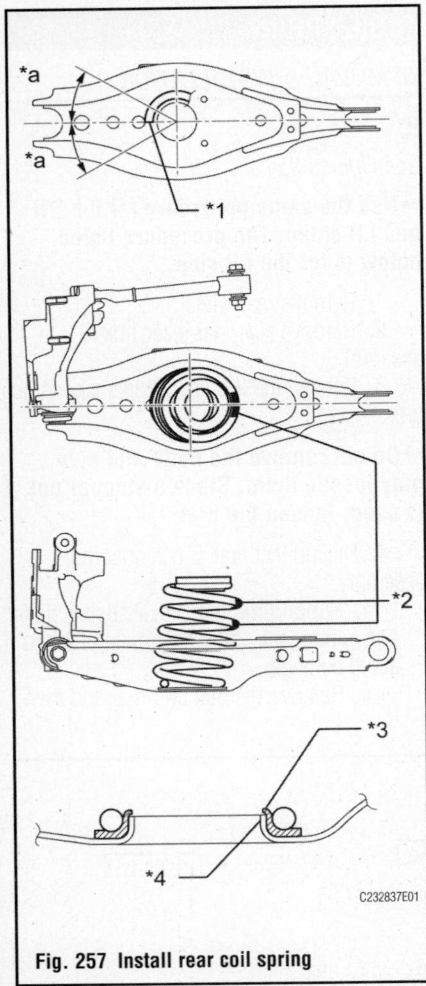

Fig. 257 Install rear coil spring

disconnect the rear shock absorber from the rear No. 2 suspension arm.

5. Remove rear coil spring.

a. Remove the bolt and nut located on the rear axle carrier of the rear No. 2 suspension arm.

b. Lower the jack gradually to remove the rear coil spring with rear upper coil spring insulator.

To install:

6. Install the rear upper coil spring insulator to the rear coil spring.

7. Install rear lower coil spring insulator.

8. Install rear coil spring.

a. Install the rear coil spring to the rear No. 2 suspension arm.

➡Install the rear coil spring so that the identification marks are properly positioned. Install the rear coil spring so that its end is within the range. The inside edge of the lower insulator must be higher than the upper edge of the burring of the lower No. 2 arm.

b. Using a jack and wooden block, raise the vehicle gradually to install the rear No. 2 suspension arm to the rear axle carrier. Then temporarily install the bolt and nut.

9. Temporarily install rear shock absorber assembly.

10. Temporarily install rear no. 2 suspension arm assembly.

11. Stabilize suspension.

12. Tighten rear shock absorber assembly.

13. Tighten the bolts of the suspension arm to 66 ft. lbs. (90 Nm).

➡Since a stopper nut is used, tighten the bolt.

14. Install rear stabilizer link assembly.

15. Inspect and adjust rear wheel alignment.

16. Adjust headlight assembly.

xB

See Figures 258 through 260.

1. Remove rear wheels.
2. Remove rear floor side member cover LH.
3. Remove rear floor side member cover RH.
4. Remove rear floor side member brace sub-assembly.
5. Separate skid control sensor wire (for LH side).
6. Separate skid control sensor wire (for RH side).
7. Loosen rear axle beam assembly.

a. Support the spring seat of the rear axle beam assembly using a jack and wooden block.

b. Loosen the 2 bolts.

➡Do not remove the bolts.

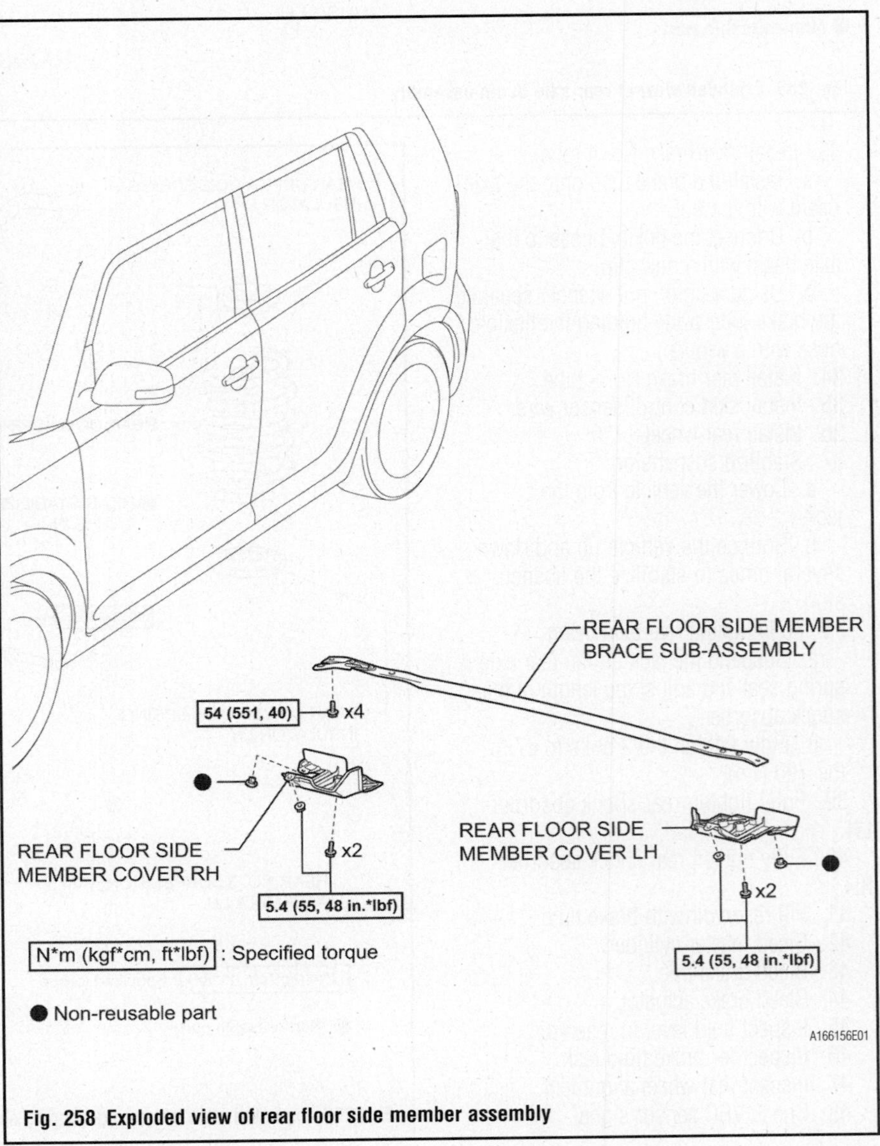

Fig. 258 Exploded view of rear floor side member assembly

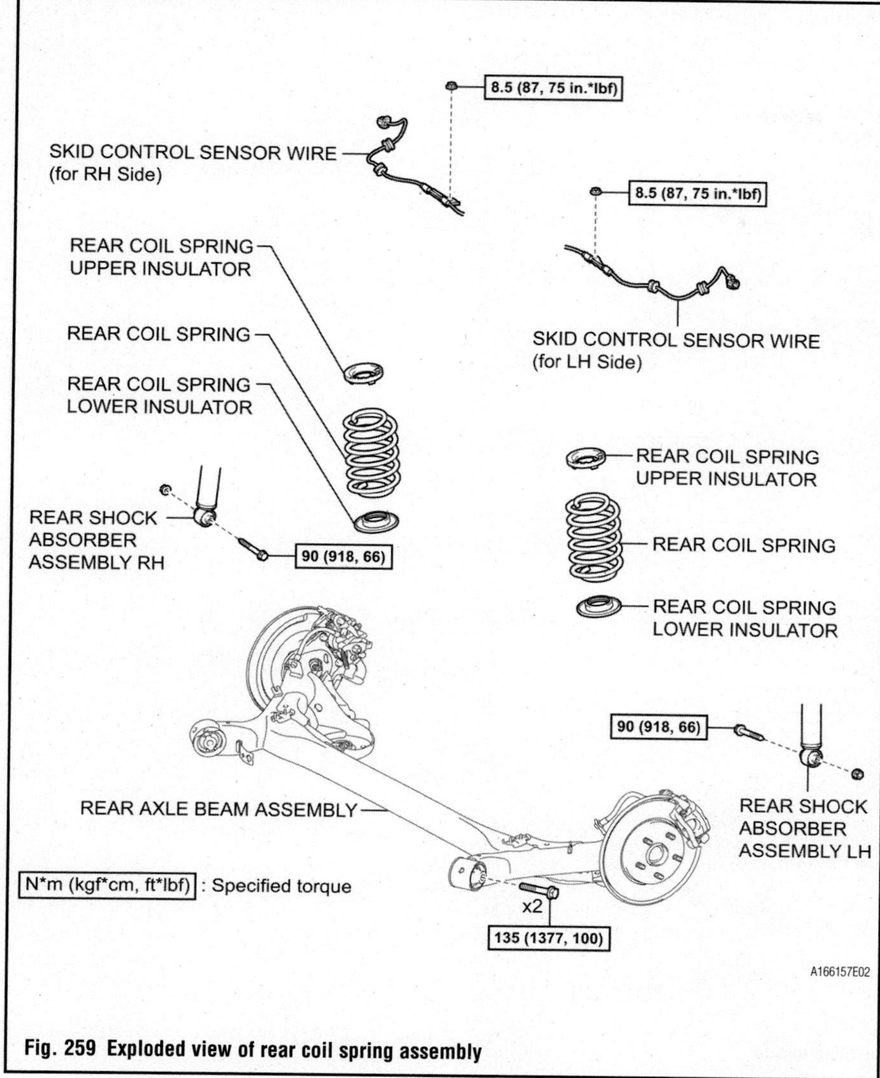

Fig. 259 Exploded view of rear coil spring assembly

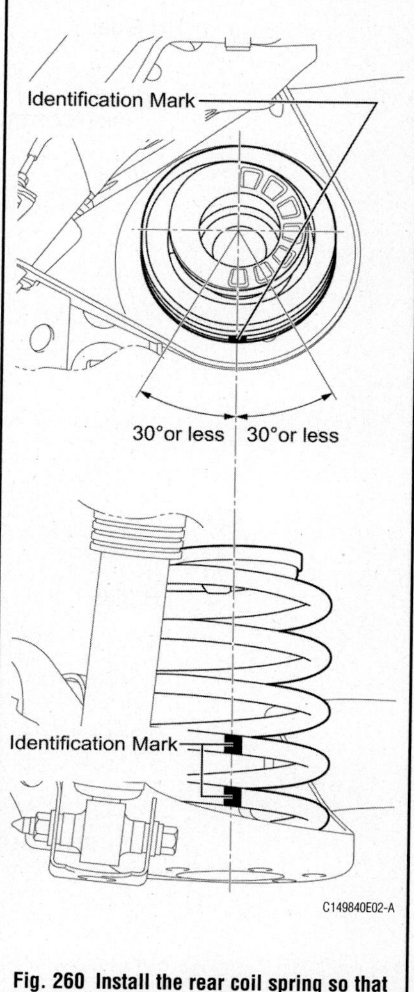

Fig. 260 Install the rear coil spring so that the identification mark is positioned properly

8. Separate rear shock absorber assembly LH.
9. Separate rear shock absorber assembly RH.
10. Slowly lower the jack, and remove the rear coil spring.

To install:

11. Install the rear coil spring upper insulator onto the rear coil spring.
12. Install rear coil spring lower insulator.
13. Install rear coil spring.
    a. Install the rear coil spring onto the rear axle beam.
    b. Install the rear coil spring so that the identification mark is positioned properly.
14. Temporarily tighten rear shock absorber assembly LH.
15. Temporarily tighten rear shock absorber assembly RH.
16. Connect skid control sensor wire (for LH side).
17. Connect skid control sensor wire (for RH side).
18. Install rear wheels.
19. Install rear floor side member brace sub-assembly.
20. Install rear floor side member cover RH.
21. Install rear floor side member cover LH.
22. Stabilize suspension.
23. Fully tighten rear axle beam assembly.
24. Fully tighten rear shock absorber assembly LH.
25. Fully tighten rear shock absorber assembly RH.
26. Inspect rear wheel alignment.
27. Inspect speed sensor signal.

xD

See Figures 261 and 262.

1. Remove rear wheel.
2. Drain brake fluid.
3. Separate skid control sensor wire.
    a. Using a screwdriver, remove the claw of the connector lock portion and disconnect the skid control sensor wire connector.

➡Do not remove the connector cover from the connector because the skid control sensor wire may be damaged.

    b. Remove the nut and separate the skid control sensor wire.
4. Separate rear LH flexible hose.
    a. Using a union nut wrench, separate the brake tube while holding the flexible hose with a wrench.
    b. Remove the clip and disconnect the flexible hose from the axle beam.
5. Separate rear RH flexible hose.
6. Loosen rear axle beam. Loosen the 2 bolts.

➡Do not remove the bolts.

7. Remove rear absorber cap.

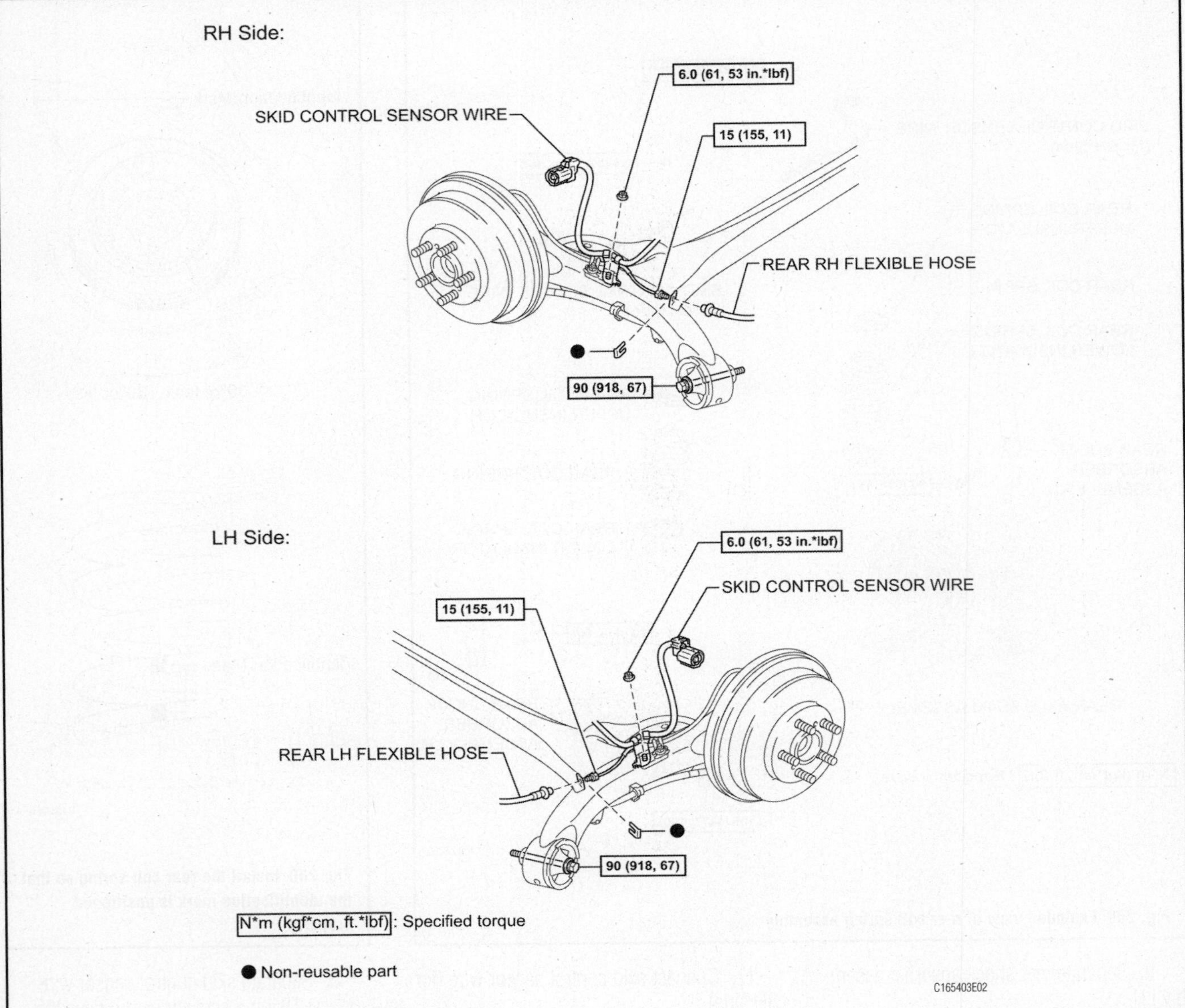

Fig. 261 Exploded view of skid control sensor wire assemblies

8. Remove rear absorber cap.
9. Remove rear shock absorber LH.
10. Remove rear shock absorber RH.
11. Remove rear coil spring LH.
    a. Lower the jacks slowly.
    b. Remove the coil spring, coil spring insulator upper and coil spring insulator lower.
12. Remove rear coil spring RH.

To install:

13. Install rear coil spring LH.
    a. Install the coil spring insulator lower onto the axle beam.
    b. Install the coil spring insulator upper so that its gap fits onto the end of coil spring.
    c. Install the coil spring onto the axle beam.

➡The paint mark of the coil spring should be towards the underside and rear side of the vehicle.

14. Install rear coil spring RH.
15. Temporarily tighten rear shock absorber LH.
16. Temporarily tighten rear shock absorber RH.
17. Install rear absorber cap.
18. Install rear absorber cap.
19. Install rear LH flexible hose.
    a. Connect the flexible hose onto the axle beam with a new clip.
    b. Using a union nut wrench, install the brake tube onto the flexible hose while holding the flexible hose with a wrench.
20. Install rear RH flexible hose.
21. Install skid control sensor wire.
    a. Install the skid control sensor wire onto the axle beam with the nut.
    b. Connect the skid control sensor wire connector.
22. Install rear wheel.
23. Stabilize suspension.
    a. Lower the vehicle from the jack.
    b. Bounce the vehicle up and down several times to stabilize the suspension.
24. Fully tighten rear axle beam.
25. Fully tighten rear shock absorber LH.
26. Fully tighten rear shock absorber RH.
27. Fill reservoir with brake fluid.
28. Bleed master cylinder.
29. Bleed brake line.
30. Bleed brake actuator.

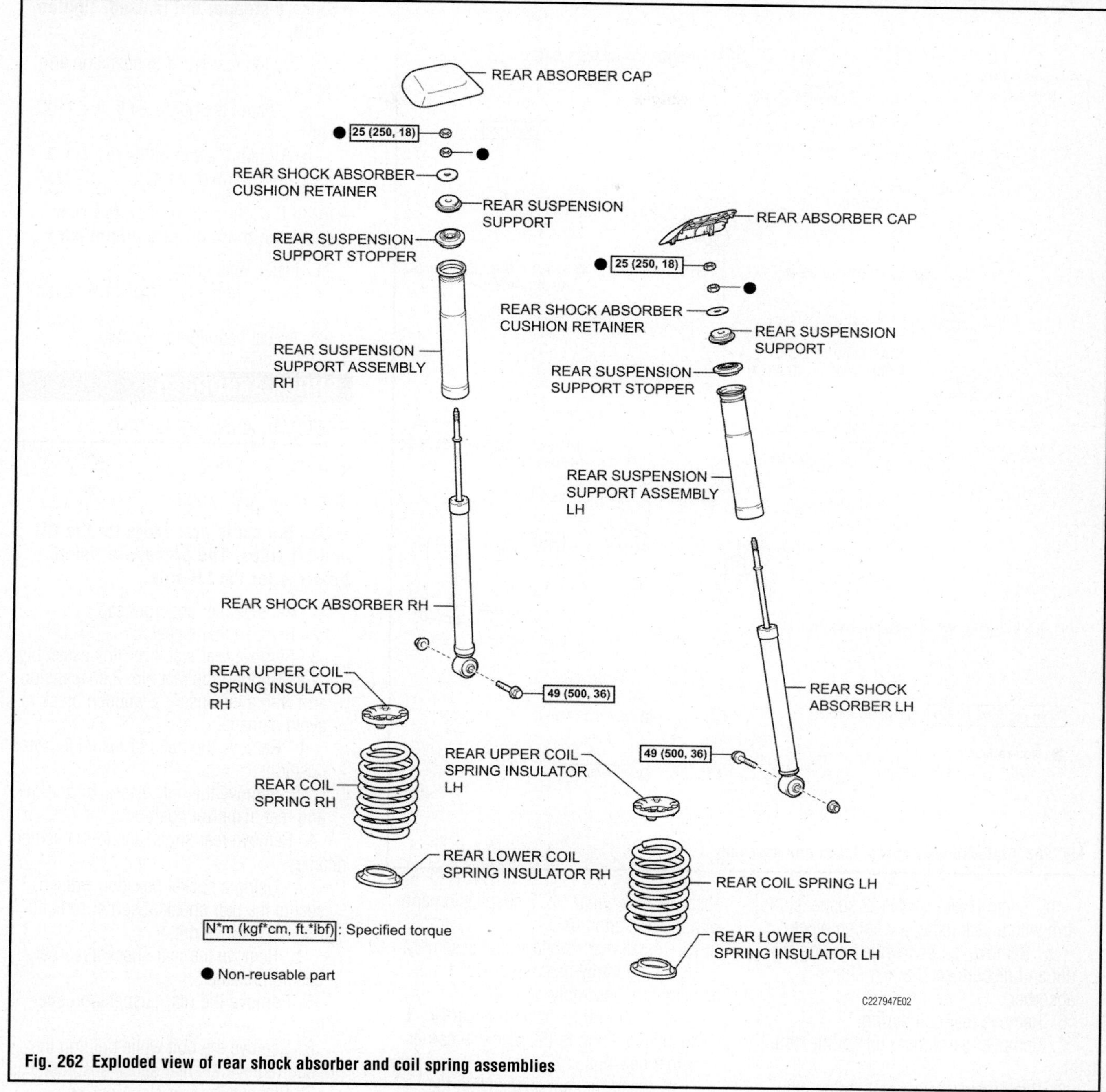

Fig. 262 Exploded view of rear shock absorber and coil spring assemblies

31. Inspect fluid level in reservoir.
32. Inspect for brake fluid leak.
33. Inspect rear wheel alignment.

LOWER CONTROL ARM

REMOVAL & INSTALLATION

tC

See Figure 263.

➡**Use the same procedure for the RH and LH sides. The procedure listed below is for the LH side.**

1. Remove rear wheel.
2. Remove rear No. 1 suspension arm assembly.
   a. Remove the nut from the axle carrier.
   b. Install SST to the rear No. 1 suspension arm.
   c. Using SST, disconnect the rear No. 1 suspension arm from the rear axle carrier.
   d. Put matchmarks on the rear suspension toe adjust cam, No. 2 camber adjust cam and rear suspension member.
   e. Remove the nut, No. 2 camber adjust cam, rear suspension toe adjust cam and rear No. 1 suspension arm assembly.
3. Remove rear stabilizer link assembly.
4. Loosen the 2 bolts of the rear No. 2 suspension arm.

➡**Do not remove the bolts and nuts, only loosen them. Since a stopper nut is used, loosen the bolt.**

5. Disconnect rear shock absorber assembly.

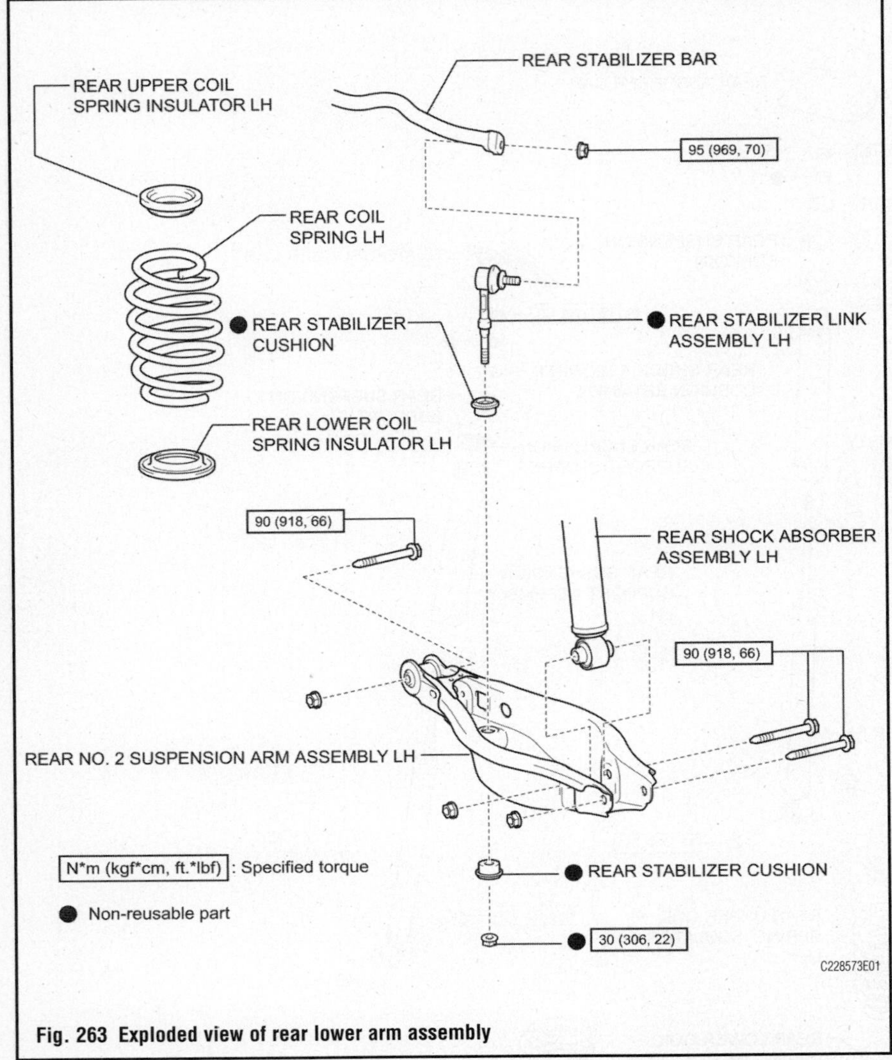

Fig. 263 Exploded view of rear lower arm assembly

a. Support the rear No. 2 suspension arm with a jack using a wooden block.

b. Remove the bolt while holding the nut and disconnect the rear shock absorber.

6. Remove rear coil spring.
7. Remove rear upper coil spring insulator.
8. Remove rear lower coil spring insulator.
9. Remove the bolt, nut and rear No. 2 suspension arm from the suspension member.

To install:

10. Temporarily install the rear No. 2 suspension arm to the suspension member with the bolt and nut.
11. Install rear upper coil spring insulator.
12. Install rear lower coil spring insulator.
13. Install rear coil spring.
14. Temporarily install the rear shock absorber to the rear No. 2 suspension arm with the bolt and nut.
15. Install rear stabilizer link assembly.
16. Temporarily install rear No. 1 suspension arm assembly.

a. Temporarily install the rear No. 1 suspension arm to the rear axle carrier with a new nut.

b. Insert the rear suspension toe adjust cam sub-assembly from the rear of the vehicle to install the rear No. 1 suspension arm assembly, and then temporarily install the No. 2 camber adjust cam with the nut.

17. Stabilize suspension.
18. Tighten the 2 bolts and 2 nuts of the rear No. 2 suspension arm assembly to 66 ft. lbs. (90 Nm).

➡Since a stopper nut is used, tighten the bolt.

19. Tighten the bolt of the rear shock absorber to 66 ft. lbs. (90 Nm).

➡Since a stopper nut is used, tighten the bolt.

20. Tighten rear No. 1 suspension arm assembly.

a. Tighten the nut to 74 ft. lbs. (100 Nm).

b. Tighten the nut of the rear No. 1 suspension arm to 74 ft. lbs. (100 Nm).

➡Align the matchmarks on the rear suspension member and adjust cam.

21. Install rear wheel.
22. Inspect and adjust rear wheel alignment.
23. Adjust headlight assembly.

SHOCK ABSORBER

REMOVAL & INSTALLATION

tC

See Figure 264.

➡Use the same procedure for the RH and LH sides. The procedure listed below is for the LH side.

1. Remove rear absorber cap.
2. Remove rear wheel.
3. Remove rear stabilizer link assembly.

a. Support the rear No. 2 suspension arm with a jack using a wooden block to avoid damage.

b. Remove the nut and rear stabilizer cushion.

c. Remove the nut, rear stabilizer link and rear stabilizer cushion.

4. Remove rear shock absorber cushion retainer.

a. Using a socket hexagon wrench, secure the rear shock absorber rod and remove the lock nut.

b. Remove the rear shock absorber cushion retainer.

5. Remove the rear suspension support.
6. Remove the bolt while holding the nut and remove the rear shock absorber.
7. Remove the rear No. 1 spring bumper from the rear shock absorber assembly.

To install:

8. Install the rear No. 1 spring bumper to the rear shock absorber assembly.
9. Temporarily install rear shock absorber assembly.

a. Support the rear No. 2 suspension arm assembly LH with a jack using a wooden block to avoid damage.

b. Temporarily install the rear shock absorber assembly to the rear No. 2 suspension arm with the bolt and nut.

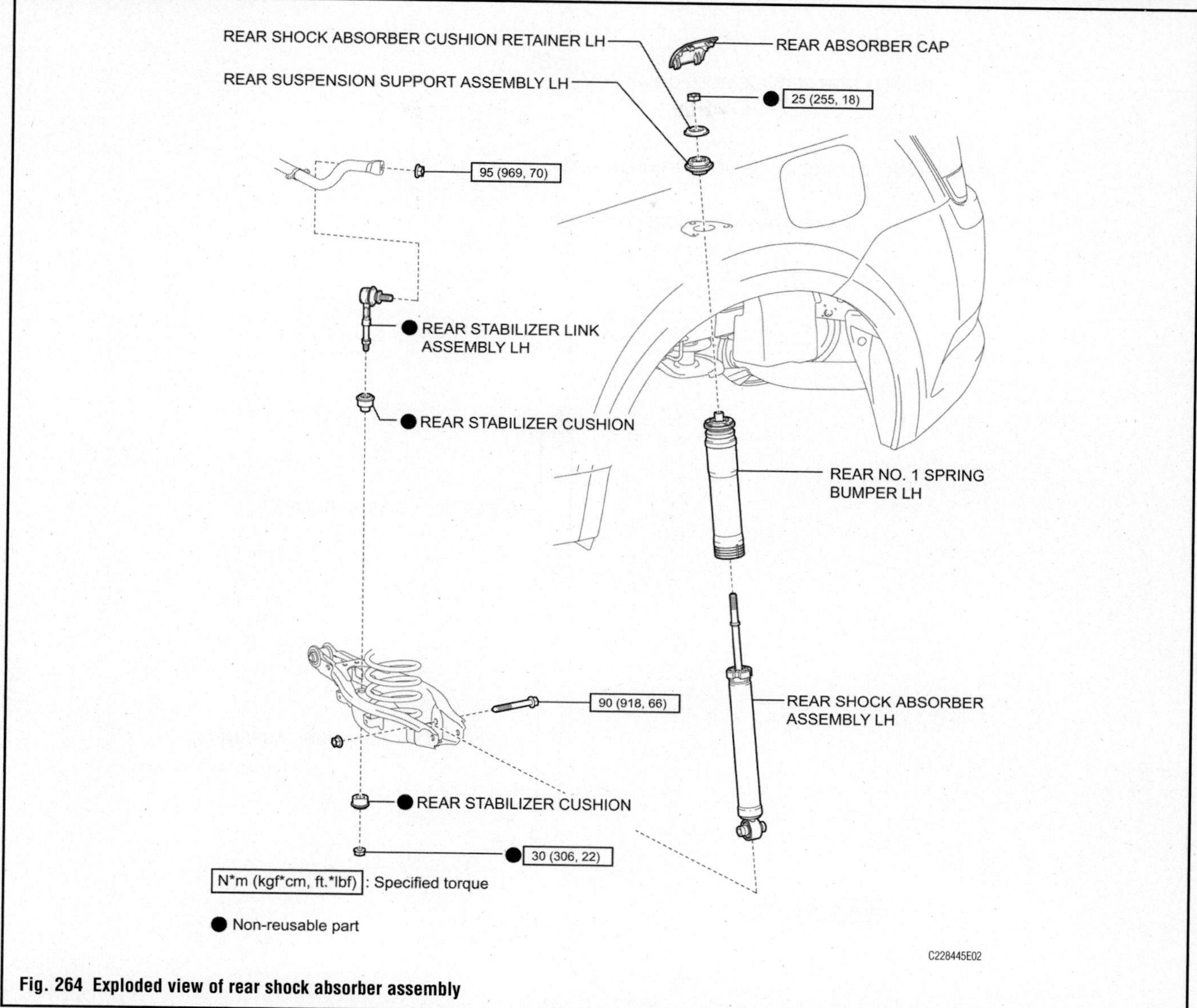

Fig. 264 Exploded view of rear shock absorber assembly

10. Install rear suspension support assembly.
11. Install rear shock absorber cushion retainer.
    a. Install the rear shock absorber cushion retainer.
    b. Using a socket hexagon wrench, secure the rear shock absorber assembly and install a new lock nut. Tighten to 18 ft. lbs. (25 Nm).
12. Jack up the rear No. 2 suspension arm assembly, placing a wooden block underneath to avoid damage. Apply load to the suspension so that the rear upper control arm assembly is properly positioned.
13. Tighten the bolt on the rear shock absorber (lower side) to 66 ft. lbs. (90 Nm).

➡Since a stopper nut is used, tighten the bolt.

14. Install rear stabilizer link assembly.
    a. Temporarily install a new rear stabilizer link and 2 new rear stabilizer cushions to the rear No. 2 suspension arm with a new nut.
    b. Install the upper nut. Tighten to 70 ft. lbs. (95 Nm).
    c. Tighten the lower nut to 22 ft. lbs. (30 Nm)..
15. Install rear absorber cap.
16. Install rear wheel.
17. Inspect and adjust rear wheel alignment.
18. Adjust headlight assembly.

xB

See Figure 265.

➡Use the same procedure for the LH side and RH side. The procedure listed below is for the LH side.

1. Remove rear wheels.
2. Remove rear floor service hole cover.
3. Remove rear shock absorber cushion retainer.
    a. Support the spring seat of the rear axle beam assembly using a jack and wooden block.
    b. Using a socket hexagon wrench, secure the rear shock absorber rod and remove the lock nut.
    c. Remove the rear shock absorber cushion retainer.
4. Remove the rear suspension support.
5. Remove the bolt while holding the nut and remove the rear shock absorber.
6. Remove rear No. 1 spring bumper.

To install:

7. Install the rear No. 1 spring bumper onto the rear shock absorber.
8. Install rear shock absorber assembly.

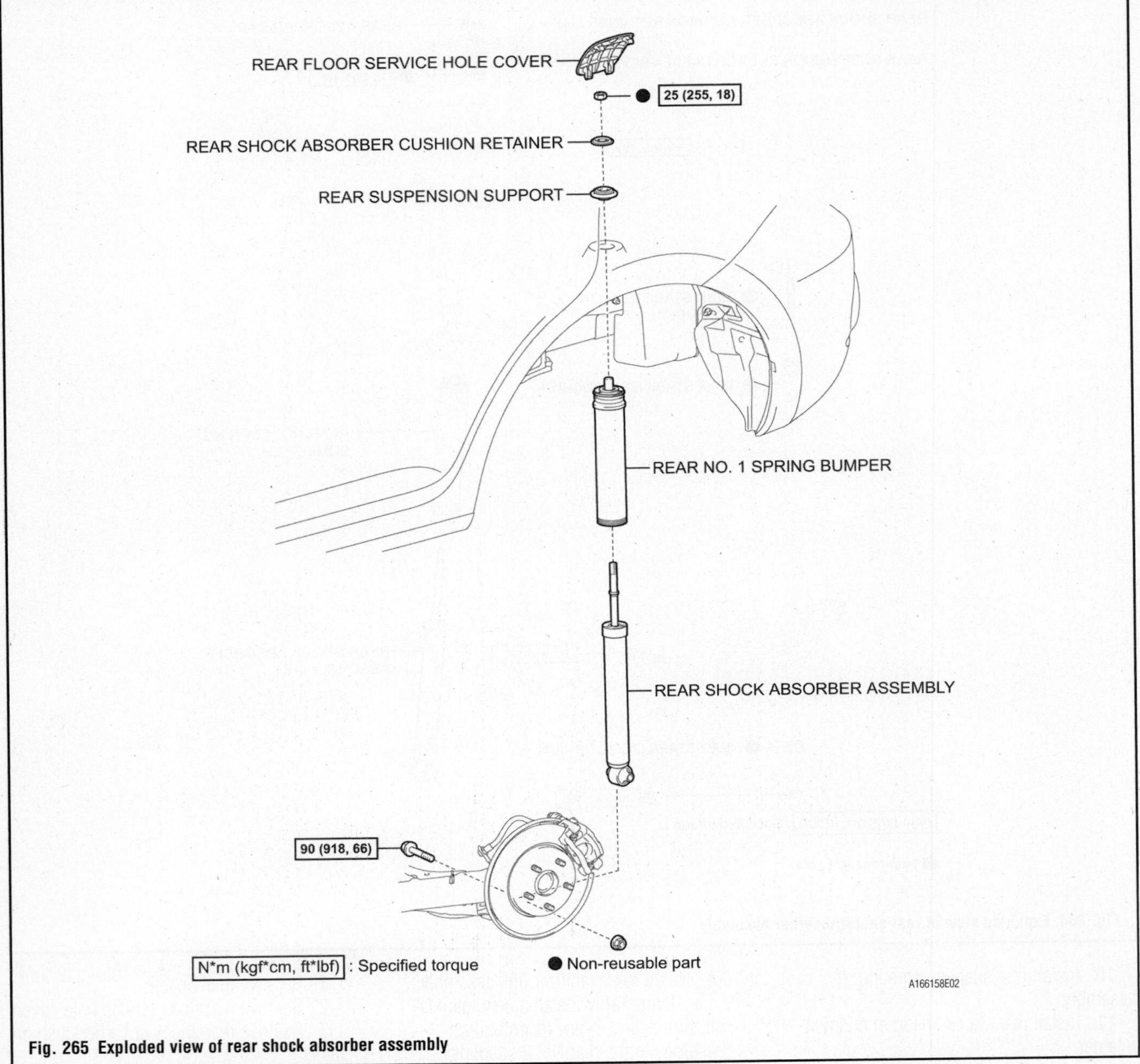

Fig. 265 Exploded view of rear shock absorber assembly

a. Support the spring seat of the rear axle beam assembly using a jack and wooden block.

b. Temporarily tighten the rear shock absorber assembly to the rear axle beam assembly with the bolt and nut.

c. Slowly raise the jack and insert the upper part of the rear shock absorber to the installation hole.

9. Install the rear suspension support.

10. Install rear shock absorber cushion retainer.

a. Install the rear shock absorber cushion retainer.

b. Using a union nut wrench, fully tighten the lock nut while holding the rod of the rear shock absorber assembly with a socket hexagon wrench. Tighten to 18 ft. lbs. (25 Nm).

11. Install rear wheels.

12. Lower the vehicle and bounce it up and down several times to stabilize the rear suspension.

13. Tighten the bolt on the rear shock absorber (lower side) to 66 ft. lbs. (90 Nm).

➡Since the stopper nut is used, tighten the bolt.

14. Install rear floor service hole cover.

xD

See Figure 266.

1. Remove rear wheel.
2. Remove rear absorber cap.
3. Remove rear shock absorber.

a. Support the axle beam with a jack. Insert a wooden block between the jack and the rear axle spring seat to prevent damage.

b. Remove the 2 nuts while keeping the piston rod from rotating.

c. Remove the cushion retainer and suspension support.

d. Remove the bolt while keeping the nut from rotating and remove the shock absorber.

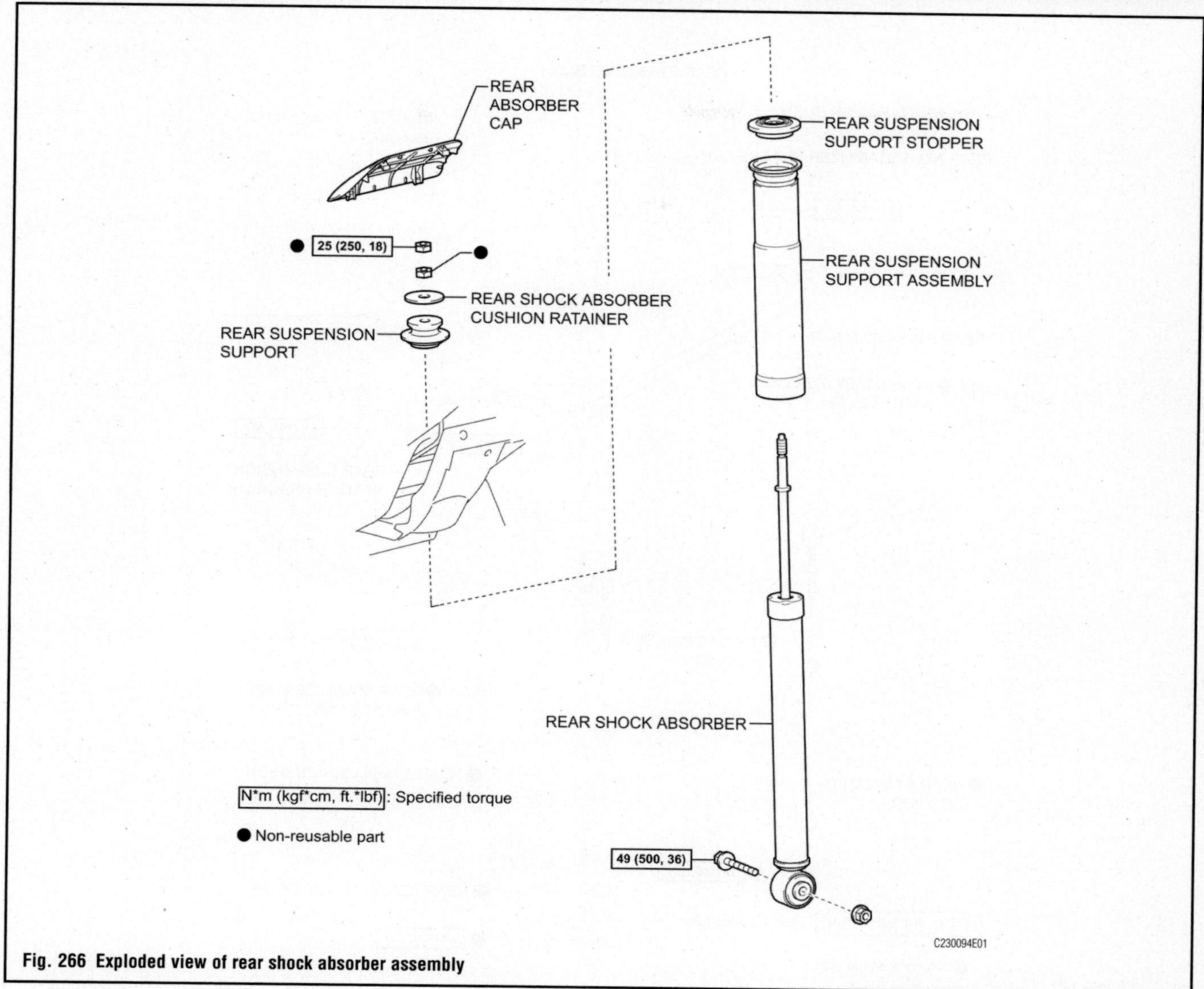

Fig. 266 Exploded view of rear shock absorber assembly

➡Remove the nut from the bolt side because the one on the lower side is a jam nut.

4. Remove rear suspension support stopper.
5. Remove rear suspension support assembly.

To install:

6. Install rear suspension support assembly.
7. Install rear suspension support stopper.
8. Temporarily tighten rear shock absorber.
    a. Support the axle beam with a jack. Insert a wooden block between the jack and the rear axle spring seat to prevent damage.
    b. Jack up the axle beam slowly, and provisionally install the shock absorber (lower side) with the bolt and nut onto the axle beam.
    c. Install the suspension support and cushion retainer.
    d. While holding the piston rod, install a new nut (lower nut).
    e. Hold the lower nut and tighten the upper nut against it. Tighten to 18 ft. lbs. (25 Nm).
9. Install rear absorber cap.
10. Install rear wheel.
11. Stabilize suspension.
    a. Lower the vehicle from the jack.
    b. Bounce the vehicle up and down several times to stabilize the suspension.
12. Fully tighten the shock absorber (lower side) with the bolt to 36 ft. lbs. (49 Nm).
13. Inspect rear wheel alignment.

STABILIZER BAR

REMOVAL & INSTALLATION

tC

See Figure 267.

1. Remove the 3 bolts and member brace from the suspension member.
2. Remove rear stabilizer link assembly LH.
    a. Remove the nut and rear stabilizer cushion.
    b. Remove the nut, rear stabilizer link and rear stabilizer cushion.
    c. Remove the rear stabilizer cushion from the stabilizer link.
3. Remove rear stabilizer link assembly RH.
4. Remove rear stabilizer bar
    a. Remove the 4 bolts and 2 stabilizer brackets from the suspension member.

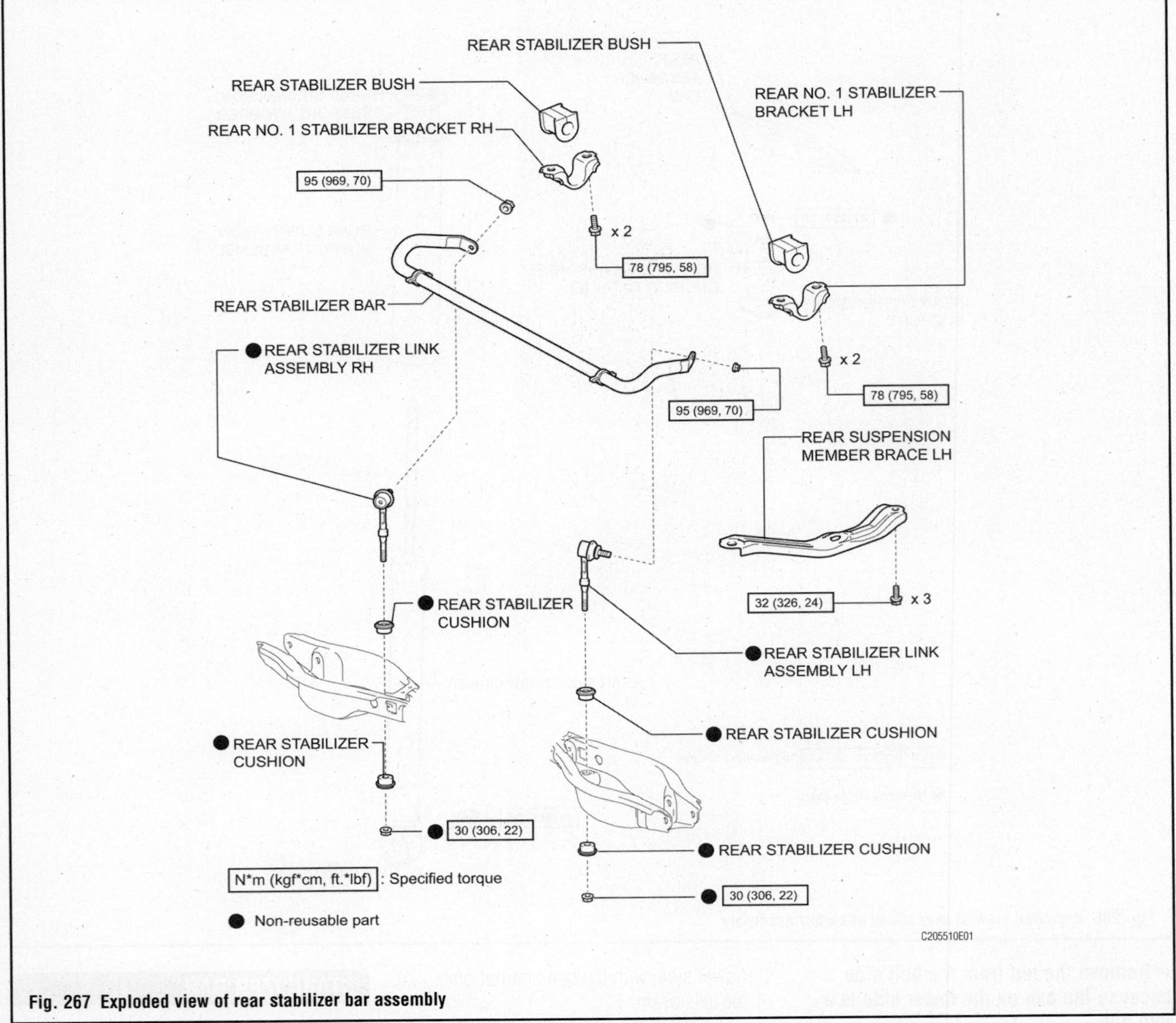

Fig. 267 Exploded view of rear stabilizer bar assembly

b. Remove the stabilizer bar from the suspension member.

5. Remove the 2 rear stabilizer bushings from the rear stabilizer bar.

To install:

6. Install the 2 rear stabilizer bushings to the rear stabilizer bar on the outside of the bushing stoppers.

7. Install the rear stabilizer bar and rear No. 1 stabilizer bracket LH and RH to the rear suspension member with the 4 bolts. Tighten to 58 ft. lbs. (78 Nm).

8. Install rear stabilizer link assembly LH.

a. Temporarily install a new rear stabilizer link and 2 new rear stabilizer cushions to the rear No. 2 suspension arm with 2 new nuts.

b. Tighten the upper nut to 70 ft. lbs. (95 Nm).

c. Tighten the lower lock nut to 22 ft. lbs. (30 Nm).

9. Install rear stabilizer link assembly RH.

10. Install the rear suspension member brace to the rear suspension member with the 3 bolts. Tighten to 24 ft. lbs. (32 Nm).

TRAILING ARM

REMOVAL & INSTALLATION

tC

See Figures 268 and 269.

➡Use the same procedure for the RH and LH sides. The procedure listed below is for the LH side.

1. Remove rear wheel.
2. Disconnect parking brake cable
3. Disconnect skid control sensor wire.
4. Disconnect rear No. 1 suspension arm assembly.

a. Put matchmarks on the rear suspension toe adjust cam, No. 2 camber adjust cam and rear suspension member.

b. Remove the nut, No. 2 camber adjust cam and rear suspension toe adjust cam and disconnect the rear No. 1 suspension arm assembly LH from the rear suspension member.

5. Remove the 4 bolts and trailing arm from the axle carrier and suspension member.

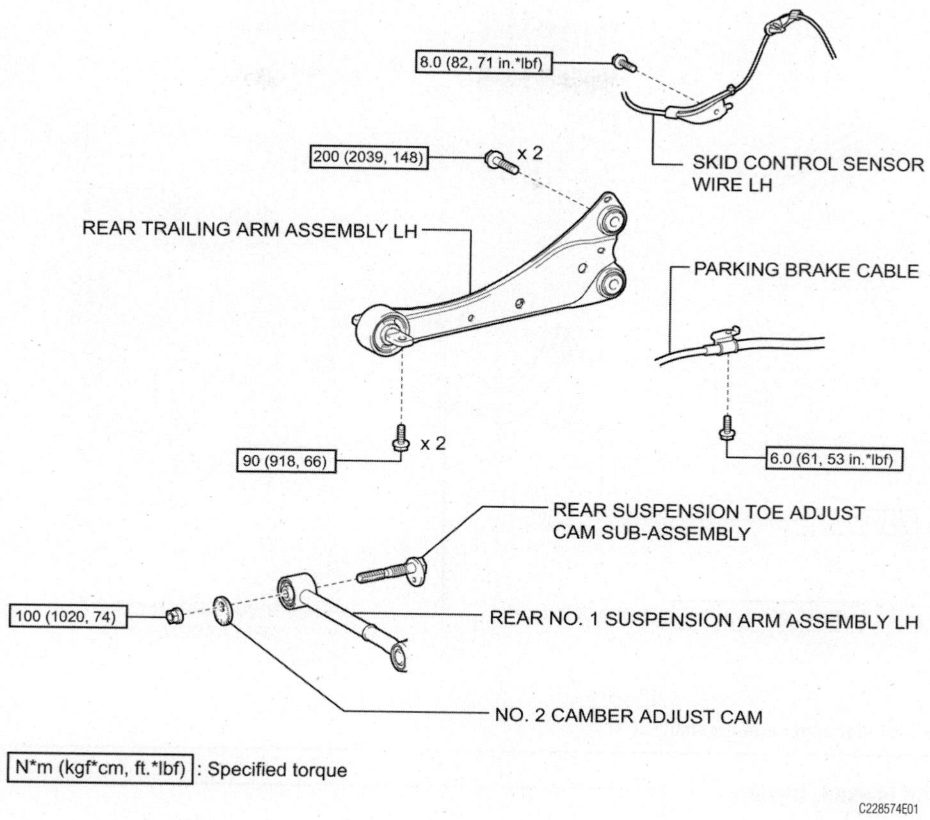

Fig. 268 Exploded view of rear trailing arm assembly

To install:

6. Install the trailing arm to the axle carrier and suspension member with the 4 bolts.
   - Bolt A: 148 ft. lbs. (200 Nm)
   - Bolt B: 66 ft. lbs. (90 Nm)
7. Insert the rear suspension toe adjust cam sub-assembly from the rear of the vehicle to install the rear No. 1 suspension arm assembly LH, and then temporarily install the No. 2 camber adjust cam with the nut.

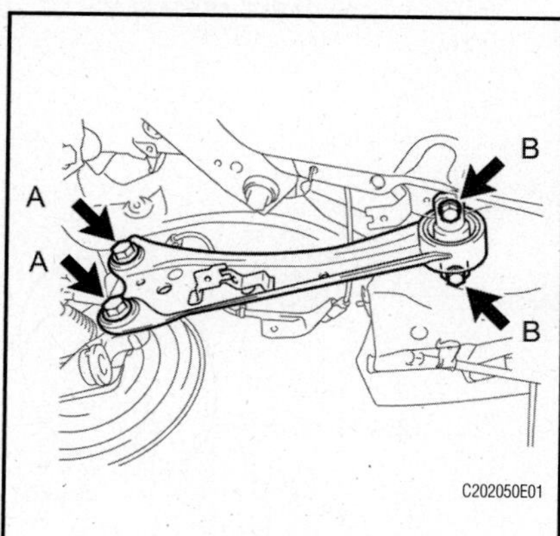

Fig. 269 Install the trailing arm to the axle carrier and suspension member with the 4 bolts

8. Connect parking brake cable.
9. Connect skid control sensor wire.
10. Stabilize suspension.
11. Tighten rear No. 1 suspension arm assembly. Tighten the nut to 74 ft. lbs. (100 Nm).

➡Align the matchmarks on the rear suspension member and adjust cam.

12. Install rear wheel.
13. Inspect and adjust rear wheel alignment.
14. Adjust headlight assembly.

UPPER CONTROL ARM

REMOVAL & INSTALLATION

tC

See Figure 270.

➡Use the same procedure for the RH and LH sides. The procedure listed below is for the LH side.

1. Remove rear wheel.
2. Remove rear upper control arm assembly.
   a. Using a jack and wooden block, raise the vehicle and hold the rear No. 2 suspension arm assembly LH with the rear axle carrier at the standard vehicle height.
   b. Remove the 2 bolts, 2 nuts and upper control arm.

➡Since a stopper nut is used, remove the bolt.

To install:

3. Temporarily install the rear upper control arm to the rear suspension member and rear axle carrier with the 2 bolts and 2 nuts.
4. Stabilize suspension.
5. Tighten rear upper control arm assembly.
   a. Using a jack and wooden block, raise the vehicle so that the rear No. 2 suspension arm assembly is at the standard vehicle height to apply load to the rear suspension.
   b. Tighten the 2 bolts and 2 nuts of the upper control arm to 66 ft. lbs. (90 Nm).

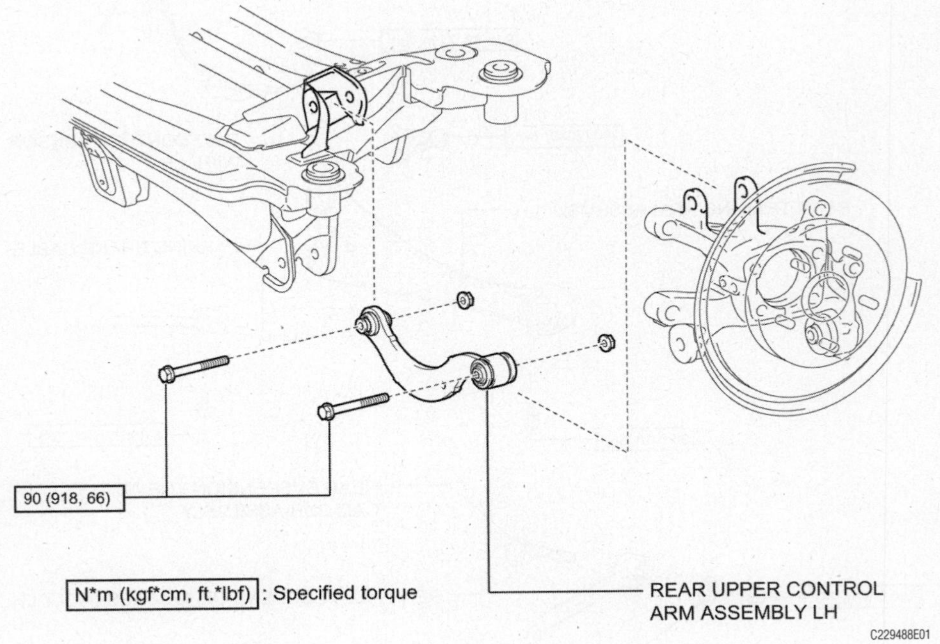

90 (918, 66)

N*m (kgf*cm, ft.*lbf) : Specified torque

REAR UPPER CONTROL ARM ASSEMBLY LH

C229488E01

Fig. 270 Exploded view of rear upper arm assembly

➡Since a stopper nut is used, tighten the bolt.

6. Install rear wheel.
7. Inspect and adjust rear wheel alignment.

WHEEL BEARINGS

REMOVAL & INSTALLATION

tC

See Figure 271.

➡Use the same procedure for the RH side and LH side. The procedure listed below is for the LH side.

1. Remove rear wheel.
2. Remove rear console box sub-assembly.
3. Loosen parking brake cable.
4. Remove parking brake lever protector.
5. Disconnect No. 3 parking brake cable assembly.
6. Remove the 2 bolts, and disconnect the rear disc brake caliper assembly.
7. Remove rear disc.
8. Disconnect skid control sensor wire.
9. Remove the 4 bolts and the rear axle hub and bearing assembly.

To install:

10. Install the rear axle hub and bearing assembly with 4 bolts. Tighten to 66 ft. lbs. (90 Nm).

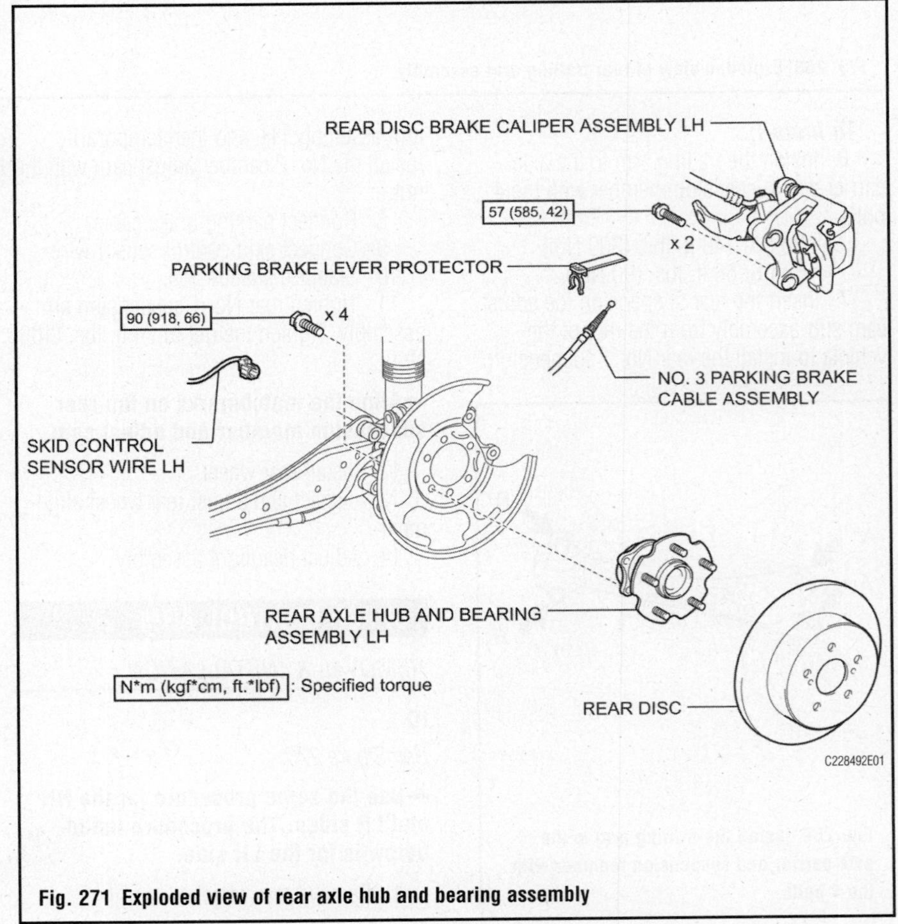

Fig. 271 Exploded view of rear axle hub and bearing assembly

11. Connect skid control sensor wire.
12. Inspect rear axle hub bearing looseness.
13. Install rear disc.
14. Connect the rear disc brake caliper assembly with the 2 bolts. Tighten to 42 ft. lbs. (57 Nm).
15. Connect No. 3 parking brake cable assembly.
16. Install parking brake lever protector.
17. Inspect and adjust parking brake.
18. Install rear console box sub-assembly.
19. Inspect rear wheel alignment.
20. Check for speed sensor signal.

xB

See Figure 272.

➡Use the same procedures for the RH side and LH side. The procedures listed below are for the LH side.

1. Remove rear wheel.
2. Remove upper console panel sub-assembly.
3. Loosen wire adjusting nut.
4. Remove rear floor side member cover.
5. Separate No. 3 parking brake cable assembly.
6. Remove the 2 bolts, and separate the rear disc brake caliper assembly.
7. Remove rear disc.
8. Separate skid control sensor wire.
9. Remove the 4 bolts and the hub and bearing assembly.
10. Remove rear disc brake dust cover sub-assembly.

To install:

11. Install rear disc brake dust cover sub-assembly.
12. Install the hub and bearing assembly with the 4 bolts. Tighten to 66 ft. lbs. (90 Nm).
13. Install skid control sensor wire.
14. Inspect rear axle hub bearing looseness.
15. Inspect rear axle hub runout.
16. Install rear disc.
17. Install the rear disc brake caliper with the 2 bolts. Tighten to 42 ft. lbs. (57 Nm).
18. Install No. 3 parking brake cable assembly.
19. Inspect parking brake lever travel.
20. Adjust parking brake lever travel.
21. Install rear floor side member cover.
22. Install upper console panel sub-assembly.
23. Install rear wheel.
24. Inspect rear wheel alignment.
25. Inspect speed sensor signal.

xD

See Figure 273.

1. Remove rear wheel.
2. Remove rear brake drum sub-assembly.
3. Disconnect speed sensor wire.
4. Using a screwdriver, remove the claw of the connector lock portion and disconnect the skid control sensor wire connector.

➡Do not remove the connector cover from the connector because the skid control sensor wire may be damaged.

5. Remove the 4 bolts and remove the axle hub and bearing from the axle beam.

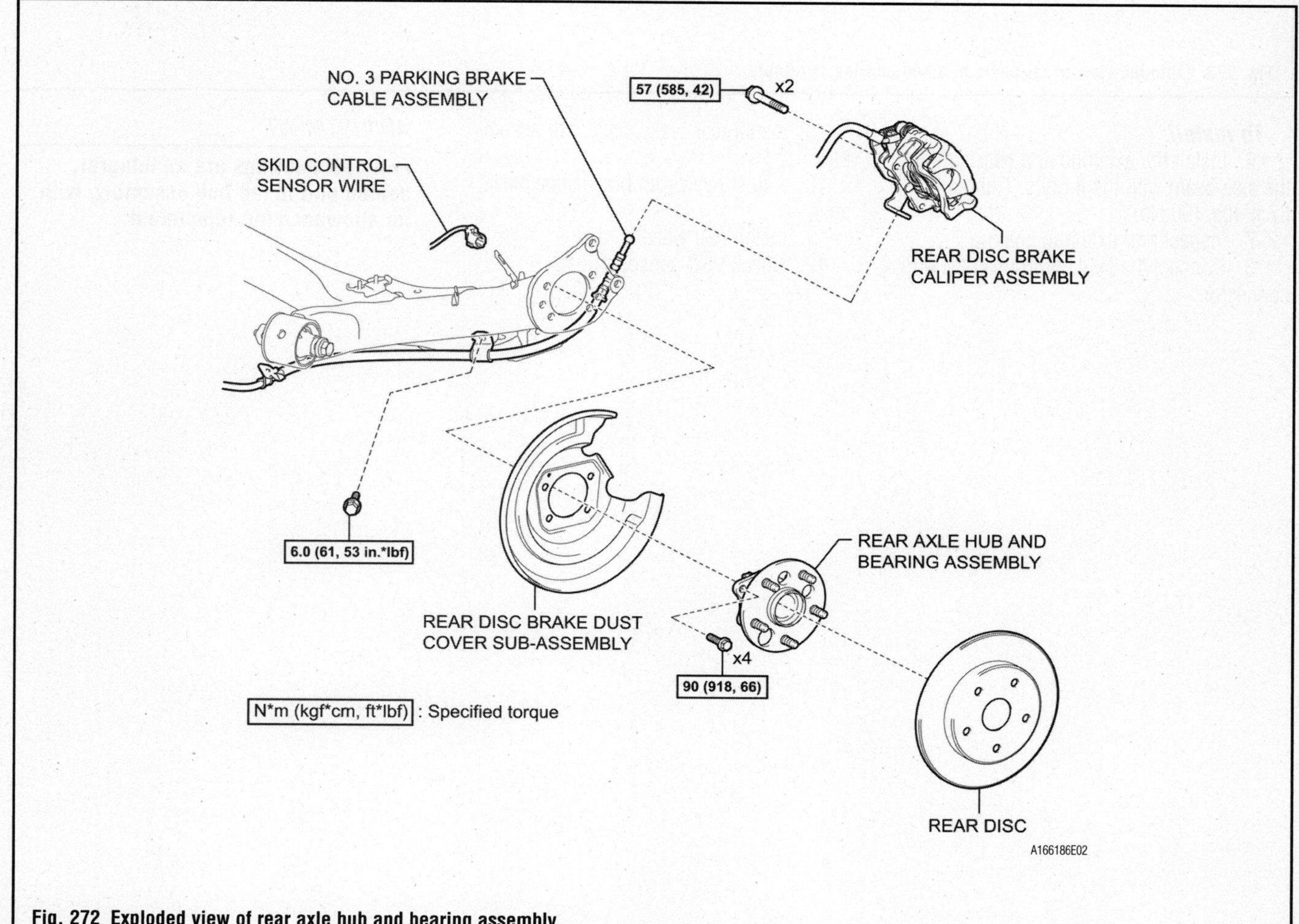

Fig. 272 Exploded view of rear axle hub and bearing assembly

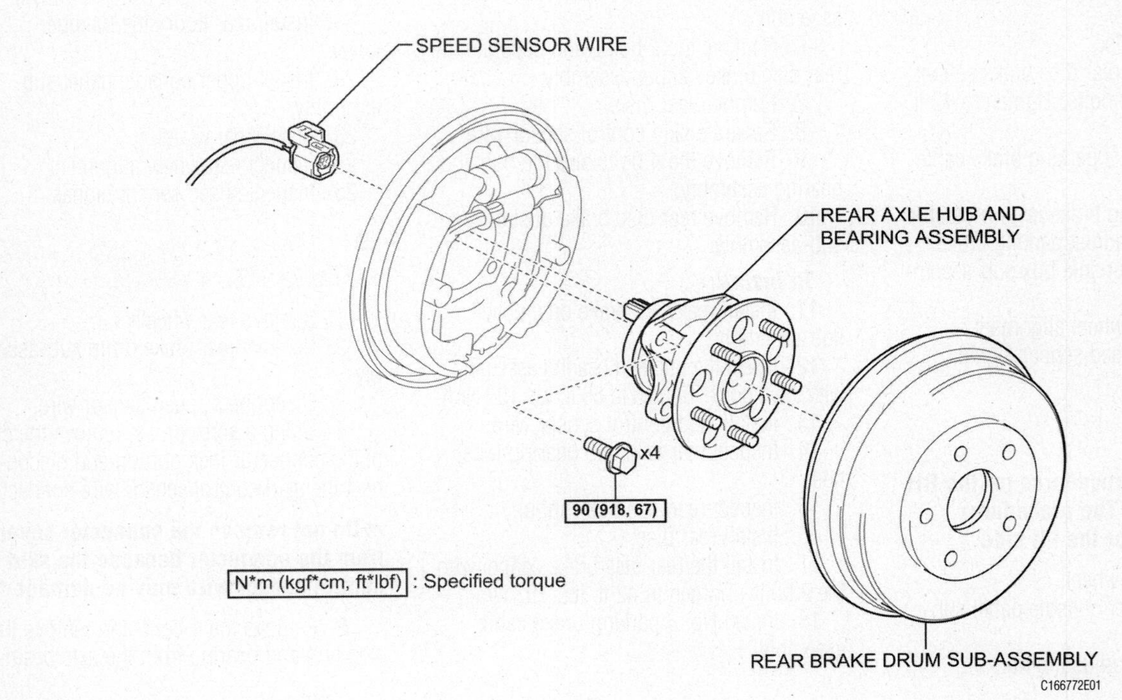

Fig. 273 Exploded view of rear axle hub and bearing assembly

To install:

6. Install the axle hub and bearing onto the axle beam with the 4 bolts. Tighten to 67 ft. lbs. (90 Nm).
7. Inspect rear axle hub bearing.
8. Connect the skid control sensor wire connector.
9. Install rear brake drum sub-assembly.
10. Adjust rear drum brake shoe clearance.
11. Install rear wheel.
12. Check VSC sensor signal.

ADJUSTMENT

➡Wheel bearings are an integral, sealed unit in the hub assembly, with no allowance for adjustment.

SCION

Diagnostic Trouble Codes

DIAGNOSTIC TROUBLE CODES

OBD II VEHICLE APPLICATIONS

SCION

iQ
2012
- 1.3L EFI . . . . . . Engine Code: 1NR-FE

tC
2011–2012
- 2.5L V6 SEFI . . Engine Code: 2AR-FE

xB
2011–2012
- 2.4L SEFI . . . . . Engine Code: 2AZ-FE

xD
2011–2012
- 1.8L SEFI . . . . . Engine Code: 2ZR-FE

OBD II Trouble Code List (P0XXX Codes)

DTC	Trouble Code Title and Conditions
DTC: P0010 **1T ECM, MIL: Yes** **Year:** 2010, 2011 **Model:** tC, xD **Engine:** 1.8L L4 VIN U, 2.4L L4, 2.5L L4	**Camshaft Position "A" Actuator Circuit (Bank 1):** Starter: OFF Engine Switch: On (IG) Time after turning engine switch off to on (IG): 0.5 seconds or more
DTC: P0010 **2T ECM, MIL: Yes** **Year:** 2010, 2011 **Model:** xB **Engine:** 2.4L L4	**VVT Oil Control Circuit Malfunction (Bank 1):** Key on or engine running; and the PCM detected an unexpected voltage condition on the VVT Oil Control Valve Bank 1 circuit. The VVT system controls the intake camshaft in order to provide optimal valve timing during all conditions based signals from the ECT, IAT and TP sensor. The VVT regulates the intake camshaft angle using oil pressure through the Oil Control Valve. This results in the relative position between the camshaft and crankshaft to become optimal. The result is higher torque, better fuel economy and low emissions.
DTC: P0011 **1T ECM, MIL: Yes** **Year:** 2010, 2011 **Model:** tC, xD **Engine:** 1.8L L4 VIN U, 2.4L L4, 2.5L L4	**Camshaft Position "A" - Timing Over-Advanced or System Performance (Bank 1):** Valve timing is not adjusted in valve timing advance. Battery Voltage: 11V or more Engine: 500-4000 rpm ECT: 167-212 degrees F (75-100 degrees C)
DTC: P0011 **2T** **Year:** 2010, 2011 **Model:** xB **Engine:** 2.4L L4	**Camshaft Position 'A' Over-Advanced Or System Performance (Bank 1):** Engine started, ECT sensor more than 158°F, vehicle driven at an engine speed of 400-4000 rpm, and the PCM detected the valve timing did not change from the "current" valve timing, or the valve timing remain fixed during testing. The VVT system controls the intake camshaft in order to provide optimal valve timing during all conditions based signals from the ECT, IAT and TP sensor. The VVT regulates the intake camshaft angle using oil pressure through the Oil Control Valve. This results in the relative position between the camshaft and crankshaft to become optimal. The result is better engine torque, fuel economy and lower emissions.
DTC: P0012 **2T ECM, MIL: Yes** **Year:** 2010, 2011 **Model:** tC, xD **Engine:** 1.8L L4 VIN U, 2.4L L4, 2.5L L4	**Camshaft Position "A" - Timing Over-Retarded (Bank 1):** Valve timing is not adjusted in valve timing retard range. Battery Voltage: 11V or more Engine: 500-4000 rpm ECT: 167-212 degrees F (75-100 degrees C)
DTC: P0013 **1T ECM, MIL: Yes** **Year:** 2010, 2011 **Model:** xD **Engine:** 1.8L L4 VIN U	**Camshaft Position "B" Actuator Circuit / Open (Bank 1):** All of following conditions met, starter OFF, ignition switch ON. Time after ignition switch OFF to ON 0.5 seconds or more. One of following conditions met 1). All of following conditions met: Battery voltage 11 to 13 V, CPU commanded duty less than 70%, current cut status Not Cut, 2). All of following conditions met: Battery voltage 13 V or more, Target duty ratio less than 80%, current cut status Not Cut
DTC: P0013 **1T ECM, MIL: Yes** **Year:** 2011 **Model:** tC **Engine:** 2.5L L4	**Camshaft Position "B" Actuator Circuit / Open (Bank 1):** Starter: OFF Engine Switch: On (IG) Time after turning engine switch off to on (IG): 0.5 seconds or more
DTC: P0014 **2T ECM, MIL: Yes** **Year:** 2011 **Model:** tC **Engine:** 2.5L L4	**Camshaft Position "B" - Timing Over-Advanced or System Performance (Bank 1):** Battery Voltage: 11V or more Engine: 500-4000 rpm Engine Coolant Temperature: 167-212 degrees F (75-100 degrees C)
DTC: P0014 **2T ECM, MIL: Yes** **Year:** 2010, 2011 **Model:** xD **Engine:** 1.8L L4 VIN U	**Camshaft Position "B" - Timing Over-Advanced or System Performance (Bank 1):** Battery voltage 11 V or more, engine RPM 500 to 4000 rpm, engine coolant temperature 75 to 100°C (167 to 212°F). The valve timing is not adjusted in exhaust valve timing advance range.

DTC	Trouble Code Title and Conditions
DTC: P0015 **1T ECM, MIL: Yes** **Year:** 2011 **Model:** tC **Engine:** 2.5L L4	**Camshaft Position "B" - Timing Over-Retarded (Bank 1):** Battery Voltage: 11V or more Engine: 500-4000 rpm Engine coolant temprature: 167-212 degrees F (75-100 degrees C)
DTC: P0015 **1T ECM, MIL: Yes** **Year:** 2010, 2011 **Model:** xD **Engine:** 1.8L L4 VIN U	**Camshaft Position "B" - Timing Over-Retarded (Bank 1):** Battery voltage 11 V or more, engine RPM 500 to 4000 rpm, engine coolant temperature 75 to 100°C (167 to 212°F). The valve timing is not adjusted in exhaust valve timing retard range.
DTC: P0016 **2T ECM, MIL: Yes** **Year:** 2010, 2011 **Model:** tC, xD **Engine:** 1.8L L4 VIN U, 2.4L L4, 2.5L L4	**Crankshaft Position - Camshaft Position Correlation (Bank 1 Sensor A):** Deviations in crankshaft and camshaft position sensor signals. Engine RPM: 500-1000 RPM
DTC: P0017 **2T ECM, MIL: Yes** **Year:** 2010, 2011 **Model:** xD **Engine:** 1.8L L4 VIN U	**Crankshaft Position - Camshaft Position Correlation (Bank 1 Sensor B):** Engine RPM 500 to 1000 rpm. A deviation in crankshaft position sensor signal and camshaft position sensor (for exhaust camshaft) signal is detected.
DTC: P0017 **2T ECM, MIL: Yes** **Year:** 2011 **Model:** tC **Engine:** 2.5L L4	**Crankshaft Position - Camshaft Position Correlation (Bank 1 Sensor B):** Engine RPM: 500-1000 RPM
DTC: P0031 **1T ECM, MIL: Yes** **Year:** 2010, 2011 **Model:** xD **Engine:** 1.8L L4 VIN U	**Oxygen (A/F) Sensor Heater Control Circuit Low (Bank 1 Sensor 1):** Battery voltage 10.5 V or more, time after heater ON 5 seconds or more. Heater output duty 10 % or more. The Air Fuel Ratio (A/F) sensor heater current is less than 0.8 A.
DTC: P0031 **1T ECM, MIL: Yes** **Year:** 2011 **Model:** tC **Engine:** 2.5L L4	**Oxygen (A/F) Sensor Heater Control Circuit Low (Bank 1 Sensor 1):** Battery voltage: 10.5 V or more Heater output duty: 50% or more Time after engine start: 10 seconds or more Active heater off control: Not operating Active heater on control: Not operating
DTC: P0031 **1T ECM, MIL: Yes** **Year:** 2010, 2011 **Model:** xB **Engine:** 2.4L L4	**Oxygen (A/F) Sensor Heater Control Circuit Low (Bank 1 Sensor 1):** The battery voltageis 10.5 V or more, A/F sensor heater duty-cycle ratio is 50% or more. Time after engine start 10 seconds or more. The Air Fuel Ratio (A/F) sensor heater current less than 0.8 A.
DTC: P0032 **1T ECM, MIL: Yes** **Year:** 2010, 2011 **Model:** tC, xB, xD **Engine:** 1.8L L4 VIN U, 2.4L L4, 2.5L L4	**Oxygen (A/F) Sensor Heater Control Circuit High (Bank 1 Sensor 1):** Battery voltage: 10.5 V or more Heater output duty: More than 50% Time after engine start: 10 seconds or more Active heater off control: Not operating Active heater on control: Not operating
DTC: P0037 **1T ECM, MIL: Yes** **Year:** 2010, 2011 **Model:** tC, xB, xD **Engine:** 1.8L L4 VIN U, 2.4L L4, 2.5L L4	**Oxygen Sensor Heater Control Circuit Low (Bank 1 Sensor 2):** Battery voltage: 10.5-20V. Heated Oxygen (HO2) sensor (sensor 2) heater current less than 0.3 A.

DTC	Trouble Code Title and Conditions
DTC: P0038 **1T ECM, MIL: Yes** **Year:** 2010, 2011 **Model:** tC, xB, xD **Engine:** 1.8L L4 VIN U, 2.4L L4, 2.5L L4	**Oxygen Sensor Heater Control Circuit High (Bank 1 Sensor 2):** (Case 1) Battery voltage: 10.5 V or more Engine: Running Starter: OFF (Case 2) Battery voltage: 10.5-20V
DTC: P0100 **1T ECM, MIL: Yes** **Year:** 2010 **Model:** tC **Engine:** 2.4L L4	**Mass or Volume Air Flow Circuit:** Monitor runs whenever following DTCs are not present: None MAF meter voltage: Less than 0.2 V, or more than 4.9 V
DTC: P0101 **2T ECM, MIL: Yes** **Year:** 2010, 2011 **Model:** tC, xB, xD **Engine:** 1.8L L4 VIN U, 2.4L L4, 2.5L L4	**Mass Air Flow Circuit Range / Performance Problem:** TP (Throttle position) sensor voltage: 0.24 to 2 V Engine: Running Battery voltage: 10.5 V or more ECT: 158°F (70°C) or more Estimated load: 30 to 70%
DTC: P0102 **1T ECM, MIL: Yes** **Year:** 2010, 2011 **Model:** tC, xB, xD **Engine:** 1.8L L4 VIN U, 2.4L L4, 2.5L L4	**Mass or Volume Air Flow Circuit Low Input:** Monitor runs whenever following DTCs are not present Mass air flow meter voltage: Less than 0.2 V for 3 seconds.
DTC: P0103 **1T ECM, MIL: Yes** **Year:** 2010, 2011 **Model:** tC, xB, xD **Engine:** 1.8L L4 VIN U, 2.4L L4	**Mass or Volume Air Flow Circuit High Input:** MAF meter voltage is more than 4.9 V for 3 seconds.
DTC: P0103 **1T ECM, MIL: Yes** **Year:** 2011 **Model:** tC **Engine:** 2.5L L4	**Mass or Volume Air Flow Circuit High Input:** Monitor runs whenever following DTCs are not present: None Mass air flow meter voltage more than 4.9 V for 3 seconds.
DTC: P0110 **1T ECM, MIL: Yes** **Year:** 2010 **Model:** tC **Engine:** 2.4L L4	**Intake Air Temperature Circuit Malfunction:** Open or short in IAT sensor circuit for 0.5 seconds.
DTC: P0111 **2T ECM, MIL: Yes** **Year:** 2010, 2011 **Model:** tC, xB, xD **Engine:** 1.8L L4 VIN U, 2.4L L4, 2.5L L4	**Intake Air Temperature Sensor Gradient Too High:** After Engine Stop: Time after engine start: 10 seconds or more Battery voltage: 10.5 V or more ECT change since engine stop: -40°F (-40°C) or more Accumulated MAF amount before engine stop: 2033 g or more Key-off duration: 30 minutes After Cold Engine Start: Key-off duration:5 hours Time after engine start: 10 seconds or more ECT: 158° F (70° C) or more Accumulated MAF amount: 2033 g or more One of the following conditions 1 or 2 is met: 1. Duration while engine load is low: 120 seconds or more 2. Duration while engine load is high: 10 seconds or more

DTC	Trouble Code Title and Conditions
DTC: P0112 **1T ECM, MIL: Yes** **Year:** 2010, 2011 **Model:** tC, xB, xD **Engine:** 1.8L L4 VIN U, 2.4L L4, 2.5L L4	**Intake Air Temperature Circuit Low Input:** Monitor runs whenever following DTCs are not present: None Short in intake air temperature sensor circuit for 0.5 seconds. Battery voltage: 8 V or more Engine switch: On (IG) Starter: OFF
DTC: P0113 **1T ECM, MIL: Yes** **Year:** 2010, 2011 **Model:** tC, xB, xD **Engine:** 1.8L L4 VIN U, 2.4L L4	**Intake Air Temperature Circuit High Input:** Open in IAT sensor circuit for 0.5 seconds.
DTC: P0113 **1T ECM, MIL: Yes** **Year:** 2011 **Model:** tC **Engine:** 2.5L L4	**Intake Air Temperature Circuit High Input:** Monitor runs whenever following DTCs are not present: None
DTC: P0115 **1T ECM, MIL: Yes** **Year:** 2010, 2011 **Model:** tC, xB, xD **Engine:** 1.8L L4 VIN U, 2.4L L4, 2.5L L4	**Engine Coolant Temperature Circuit Malfunction:** Monitor runs whenever following DTCs are not present: None Engine coolant temperature sensor voltage: Less than 0.14 V, or more than 4.91 V
DTC: P0116 **2T ECM, MIL: Yes** **Year:** 2010, 2011 **Model:** tC, xB, xD **Engine:** 1.8L L4 VIN U, 2.4L L4, 2.5L L4	**Engine Coolant Temperature Circuit Range / Performance Problem:** Monitor runs whenever following DTCs are not present: None ECT sensor cold start monitor: Battery voltage: 10.5 V or more Time after engine start: 1 second or more ECT at engine start: Less than 140°F (60°C) Soak time: 0 second Accumulated MAF:1241 g or more Fuel cut: OFF Difference between ECT at engine start and IAT: Less than 104°F (40°C) ECT sensor soak monitor: Battery voltage: 10.5 V or more Engine: Running Soak time: 5 hours or more ECT at engine start: 140°F (60°C) or more Accumulated MAF: 2334 g or more
DTC: P0117 **1T ECM, MIL: Yes** **Year:** 2010, 2011 **Model:** tC, xB, xD **Engine:** 1.8L L4 VIN U, 2.4L L4, 2.5L L4	**Engine Coolant Temperature Circuit Low Input:** Monitor runs whenever following DTCs are not present: None Engine coolant temperature sensor voltage: Less than 0.14 V
DTC: P0118 **1T ECM, MIL: Yes** **Year:** 2010, 2011 **Model:** tC, xB, xD **Engine:** 1.8L L4 VIN U, 2.4L L4, 2.5L L4	**Engine Coolant Temperature Circuit High Input:** Monitor runs whenever following DTCs are not present: None Engine coolant temperature sensor voltage: More than 4.91 V

DTC	Trouble Code Title and Conditions
DTC: P011B **2T ECM, MIL: Yes** **Year:** 2010, 2011 **Model:** tC, xB, xD **Engine:** 1.8L L4 VIN U, 2.4L L4, 2.5L L4	**Engine Coolant Temperature / Intake Air Temperature Correlation:** Monitor runs whenever following DTCs are not present: None All of following conditions met: Conditions 1 and 2 1. All of the following conditions are met: Conditions (a), (b), (c) and (d) (a) After engine switch on (IG) and engine not running time: Less than 20 seconds (b) Soak Time: 7 hours or more (c) Battery voltage: 10.5 V or more (d) Time after engine start: 15 seconds or more 2. Either of the following conditions are met: Condition (a) and (b) (a) Minium IAT after engine start: 14° F (-10° C) or more (b) ECT before engine start: 14° F (-10° C) or more
DTC: P0120 **1T ECM, MIL: Yes** **Year:** 2010, 2011 **Model:** tC, xB, xD **Engine:** 1.8L L4 VIN U, 2.4L L4, 2.5L L4	**Throttle / Pedal Position Sensor / Switch "A" Circuit Malfunction:** Monitor runs whenever following DTCs are not present: None Either of the following conditions A or B is met: A. Engine switch on (IG): 0.012 seconds or more B. Electronic throttle actuator power: ON
DTC: P0121 **1T ECM, MIL: Yes** **Year:** 2010, 2011 **Model:** tC, xB, xD **Engine:** 1.8L L4 VIN U, 2.4L L4, 2.5L L4	**Throttle / Pedal Position Sensor / Switch "A" Circuit Range / Performance Problem:** Monitor runs whenever following DTCs are not present: None Either of the following conditions A or B is met: A. Engine switch: On (IG) B. Electric throttle motor power: ON Throttle position sensor malfunction (P0120, P0122, P0123, P0220, P0222, P0223, P2135): not detected
DTC: P0122 **1T ECM, MIL: Yes** **Year:** 2010, 2011 **Model:** tC, xB, xD **Engine:** 1.8L L4 VIN U, 2.4L L4, 2.5L L4	**Throttle / Pedal Position Sensor / Switch "A" Circuit Low Input:** Monitor runs whenever following DTCs are not present: None Either of the following conditions A or B is met: A. Engine switch on (IG): 0.012 seconds or more B. Electronic throttle actuator power: ON
DTC: P0123 **1T ECM, MIL: Yes** **Year:** 2010, 2011 **Model:** tC, xB, xD **Engine:** 1.8L L4 VIN U, 2.4L L4, 2.5L L4	**Throttle / Pedal Position Sensor / Switch "A" Circuit High Input:** Monitor runs whenever following DTCs are not present: None Either of the following conditions A or B is met: A. Engine switch on (IG): 0.012 seconds or more B. Electronic throttle actuator power: ON
DTC: P0125 **T ECM, MIL: Yes** **Year:** 2010, 2011 **Model:** tC, xB, xD **Engine:** 1.8L L4 VIN U, 2.4L L4, 2.5L L4	**Insufficient Coolant Temperature for Closed Loop Fuel Control:** Monitor runs whenever following DTCs are not present: MAF sensor circuit fail (P0102, P0103) IAT sensor circuit fail (P0112, P0113) ECT sensor circuit fail (P0115, P0117, P0118) Thermostat fail (P0128)
DTC: P0128 **2T ECM, MIL: Yes** **Year:** 2010, 2011 **Model:** tC, xB, xD **Engine:** 1.8L L4 VIN U, 2.4L L4, 2.5L L4	**Coolant Thermostat (Coolant Temperature Below Thermostat Regulating Temperature):** Battery voltage: 11V or more Either following condition 1 or 2 is met: 1. All of the following conditions are met: *ECT at engine start - IAT at engine start: 5-44.6°F (-15-7°C) *ECT at engine start: 14-132.8°F (-10-56°C) *IAT at engine start: 14-132.8°F (-10-56°C) 2. All of following conditions are met: *ECT at engine start - IAT at engine start: More than 44.6°F (7°C) *ECT at engine start: 132.8°F (56°C) or less *IAT at engine start: 14°F (-10°C) or more Accumulated time that vehicle speed is 80 mph (128 km/h) or more speed: Less than 20 seconds

DTC	Trouble Code Title and Conditions
DTC: P0136 **2T ECM, MIL: Yes** **Year:** 2010, 2011 **Model:** tC, xB, xD **Engine:** 1.8L L4 VIN U, 2.4L L4, 2.5L L4	**Oxygen Sensor Circuit Malfunction (Bank 1 Sensor 2):** Active A/F control: Performing Battery voltage: 11 V or more ECT: 167°F (75°C) or more Idle: OFF Engine rpm: Less than 3200 rpm A/F sensor status: Activatted Fuel system status: Closed loop Fuel cut: OFF Engineload: 10 to 70% Shift position: 3rd or more Estimated rear HO2S temperature: Less than 1292°F (700°C) ECM monitor: Completed P0607: Not set
DTC: P0137 **2T ECM, MIL: Yes** **Year:** 2010, 2011 **Model:** tC, xB, xD **Engine:** 1.8L L4 VIN U, 2.4L L4, 2.5L L4	**Oxygen Sensor Circuit Low Voltage (Bank 1 Sensor 2):** Active A/F control: Performing Battery voltage: 11 V or more ECT: 167°F (75°C) or more Idle: OFF Engine rpm: Less than 3200 rpm A/F sensor status: Activatted Fuel system status: Closed loop Fuel cut: OFF Engineload: 10 to 70% Shift position: 4rd or more Battery voltage: 11 V or more Estimated rear HO2S temperature: 842-1382°F (450-750°C) P0607: Not preset
DTC: P0138 **2T ECM, MIL: Yes** **Year:** 2010, 2011 **Model:** tC, xB, xD **Engine:** 1.8L L4 VIN U, 2.4L L4, 2.5L L4	**Oxygen Sensor Circuit High Voltage (Bank 1 Sensor 2):** Active A/F control: Performing Battery voltage: 11 V or more ECT: 167°F (75°C) or more Idle: OFF Engine rpm: Less than 3200 rpm A/F sensor status: Activatted Fuel system status: Closed loop Fuel cut: OFF Engineload: 10 to 70% Shift position: 4rd or more Battery voltage: 11 V or more Time after engine start: 2 seconds or more
DTC: P0139 **2T ECM, MIL: Yes** **Year:** 2010, 2011 **Model:** tC, xB, xD **Engine:** 1.8L L4 VIN U, 2.4L L4, 2.5L L4	**Oxygen Sensor Circuit Slow Response (Bank 1 Sensor 2):** ECT: 167°F (75°C) or more Estimated catalyst temperature: 752°F (400°C) or more Fuel cut: ON
DTC: P0141 **2T ECM, MIL: Yes** **Year:** 2010, 2011 **Model:** tC, xB, xD **Engine:** 1.8L L4 VIN U, 2.4L L4, 2.5L L4	**Oxygen Sensor Heater Circuit Malfunction (Bank 1 Sensor 2):** One of the following conditions is met: Condition A or B A. All of the following conditions are met: Conditions 1, 2, 3, 4 and 5 1. Battery voltage: 10.5 V or more 2. Fuel cut: OFF 3. Time after fuel cut ON to OFF: 30 seconds or more 4. Accumulated heater ON time: 100 seconds or more 5. Learned heater OFF current operation: Completed B. Duration that rear heated oxygen sensor impedance is less than 15 kΩ: 2 seconds or more

DTC	Trouble Code Title and Conditions
DTC: P014C **2T ECM, MIL: Yes** **Year:** 2011 **Model:** tC **Engine:** 2.5L L4	**A/F Sensor Slow Response - Rich to Lean Bank 1 Sensor 1:** Monitor runs whenever following DTCs are not stored: None Active air fuel ratio control: Performing Active air fuel ratio control is performed when the following conditions are met: Battery voltage: 11 V or higher Engine coolant temperature: 167°F (75°C) or more Idle: OFF Engine speed: 1000-4000 rpm Air fuel ratio sensor status: Activated Fuel cut: OFF Engine load: 10-70% Shift position: 2nd or higher Catalyst monitor: Not yet Mass air flow: 5-15 g/sec. Rich to lean response rate deterioration leve: 0.035 V or less
DTC: P014D **2T , MIL: Yes** **Year:** 2011 **Model:** tC **Engine:** 2.5L L4	**A/F Sensor Slow Response - Lean to Rich Bank 1 Sensor 1:** Monitor runs whenever following DTCs are not stored: None Active air fuel ratio control: Performing Active air fuel ratio control is performed when the following conditions are met: Battery voltage: 11 V or higher Engine coolant temperature: 167°F (75°C) or more Idle: OFF Engine speed: 1000-4000 rpm Air fuel ratio sensor status: Activated Fuel cut: OFF Engine load: 10-70% Shift position: 2nd or higher Catalyst monitor: Not yet Mass air flow: 5-15 g/sec. Lean to Rich Response rate deterioration level: -0.038 V or higher
DTC: P015A **2T ECM, MIL: Yes** **Year:** 2011 **Model:** tC **Engine:** 2.5L L4	**A/F Sensor Delayed Response - Rich to Lean Bank 1 Sensor 1:** Monitor runs whenever following DTCs are not stored: None Active air fuel ratio control: Performing Active air fuel ratio control is performed when the following conditions are met: Battery voltage: 11 V or higher Engine coolant temperature: 167°F (75°C) or more Idle: OFF Engine speed: 1000-4000 rpm Air fuel ratio sensor status: Activated Fuel cut: OFF Engine load: 10-70% Shift position: 2nd or higher Catalyst monitor: Not yet Mass air flow: 5-15 g/sec. Rich to Lean delay level: 230 msec or more

DTC	Trouble Code Title and Conditions
DTC: P015B **2T , MIL: Yes** **Year:** 2011 **Model:** tC **Engine:** 2.5L L4	**A/F Sensor Delayed Response - Lean to Rich Bank 1 Sensor 1:** Monitor runs whenever following DTCs are not stored: None Active air fuel ratio control: Performing Active air fuel ratio control is performed when the following conditions are met: Battery voltage: 11 V or higher Engine coolant temperature: 167°F (75°C) or more Idle: OFF Engine speed: 1000-4000 rpm Air fuel ratio sensor status: Activated Fuel cut: OFF Engine load: 10-70% Shift position: 2nd or higher Catalyst monitor: Not yet Mass air flow: 5-15 g/sec. Lean to Rich delay level: 230 msec or more
DTC: P0171 **2T ECM, MIL: Yes** **Year:** 2011 **Model:** tC **Engine:** 2.5L L4	**System Too Lean (Bank 1):** Monitor runs whenever following DTCs not stored: P0010 (VVT Oil Control Valve) P0011 (VVT System - Advance) P0012 (VVT System - Retard) P0016 (VVT System - Misalignment) P0013 (Exhaust VVT Oil Control Valve) P0014 (Exhaust VVT System - Advance) P0015 (Exhaust VVT System - Retard) P0017 (Exhaust VVT System - Misalignment) P0031, P0032, P101D (Air Fuel Ratio Sensor Heater) P0102, P0103 (Mass Air Flow Sensor) P0115, P0117, P0118 (Engine Coolant Temperature Sensor) P0125 (Insufficient Coolant Temperature for Closed Loop Fuel Control) P0120, P0121, P0122, P0123, P0220, P0222, P0223, P2135 (Throttle Position Sensor) P0335 (Crankshaft Position Sensor) P0340 (Camshaft Position Sensor) P0365, P0367, P0368 (Exhaust VVT Sensor) P0351 - P0354 (Igniter) P0500 (Vehicle Speed Sensor) Fuel system status: Closed loop Battery voltage: 11 V or higher Either of the following conditions is met: 1. Engine speed: Less than 800 rpm 2. Engine load: 11.52% or more Catalyst monitor: Not executed

DTC	Trouble Code Title and Conditions
DTC: P0171 **2T ECM, MIL: Yes** **Year:** 2010, 2011 **Model:** tC, xB, xD **Engine:** 1.8L L4 VIN U, 2.4L L4	**System Too Lean (Bank 1):** Monitor runs whenever following DTCs are not present: P0010, P0020 (OCV bank 1, 2) P0011, P0021 (VVT system bank 1, 2 - advance) P0012, P0022 (VVT system bank 1, 2 - retard) P0013, P0023 (Exhaust OCV bank 1, 2) P0014, P0024 (Exhaust VVT system bank 1, 2 - advance) P0015, P0025 (Exhaust VVT system bank 1, 2 - retard) P0016, P0018 (VVT system bank 1, 2 - misalignment) P0017, P0019 (Exhaust VVT system bank 1, 2 - misalignment) P0031, P0032, P0051, P0052 (A/F sensor heater) P0102, P0103 (MAF meter) P0115, P0117, P0118 (ECT sensor) P0120, P0121, P0122, P0123, P0220, P0222, P0223, P2135 (TP sensor) P0125 (Insufficient ECT for closed loop) P0335 (CKP sensor) P0340, P0342, P0343, P0345, P0347, P0348 (VVT sensor) P0351, P0352, P0353, P0354, P0355, P0356 (Igniter) P0365, P0367, P0368, P0390, P0392, P0393 (Exhaust VVT sensor) P0500 (Vehicle speed sensor) Fuel system status: Closed loop Battery voltage: 11 V or more Either following condition is met: 1. Engine RPM: less than 1100 rpm 2. Intake air amount per revolution: 0.22 g/rev or more Catalyst monitor: Not executed
DTC: P0172 **2T ECM, MIL: Yes** **Year:** 2011 **Model:** tC **Engine:** 2.5L L4	**System Too Rich (Bank 1):** Monitor runs whenever following DTCs not stored: P0010 (VVT Oil Control Valve) P0011 (VVT System - Advance) P0012 (VVT System - Retard) P0016 (VVT System - Misalignment) P0013 (Exhaust VVT Oil Control Valve) P0014 (Exhaust VVT System - Advance) P0015 (Exhaust VVT System - Retard) P0017 (Exhaust VVT System - Misalignment) P0031, P0032, P101D (Air Fuel Ratio Sensor Heater) P0102, P0103 (Mass Air Flow Sensor) P0115, P0117, P0118 (Engine Coolant Temperature Sensor) P0125 (Insufficient Coolant Temperature for Closed Loop Fuel Control) P0120, P0121, P0122, P0123, P0220, P0222, P0223, P2135 (Throttle Position Sensor) P0335 (Crankshaft Position Sensor) P0340 (Camshaft Position Sensor) P0365, P0367, P0368 (Exhaust VVT Sensor) P0351 - P0354 (Igniter) P0500 (Vehicle Speed Sensor) Fuel system status: Closed loop Battery voltage: 11 V or higher Either of the following conditions is met: 1. Engine speed: Less than 800 rpm 2. Engine load: 11.52% or more Catalyst monitor: Not executed

DTC	Trouble Code Title and Conditions
DTC: P0172 **2T ECM, MIL: Yes** **Year:** 2010, 2011 **Model:** tC, xB, xD **Engine:** 1.8L L4 VIN U, 2.4L L4	**System Too Rich (Bank 1):** Monitor runs whenever following DTCs are not present: P0010, P0020 (OCV bank 1, 2) P0011, P0021 (VVT system bank 1, 2 - advance) P0012, P0022 (VVT system bank 1, 2 - retard) P0013, P0023 (Exhaust OCV bank 1, 2) P0014, P0024 (Exhaust VVT system bank 1, 2 - advance) P0015, P0025 (Exhaust VVT system bank 1, 2 - retard) P0016, P0018 (VVT system bank 1, 2 - misalignment) P0017, P0019 (Exhaust VVT system bank 1, 2 - misalignment) P0031, P0032, P0051, P0052 (A/F sensor heater) P0102, P0103 (MAF meter) P0115, P0117, P0118 (ECT sensor) P0120, P0121, P0122, P0123, P0220, P0222, P0223, P2135 (TP sensor) P0125 (Insufficient ECT for closed loop) P0335 (CKP sensor) P0340, P0342, P0343, P0345, P0347, P0348 (VVT sensor) P0351, P0352, P0353, P0354, P0355, P0356 (Igniter) P0365, P0367, P0368, P0390, P0392, P0393 (Exhaust VVT sensor) P0500 (Vehicle speed sensor) Fuel system status: Closed loop Battery voltage: 11 V or more Either following condition is met: 1. Engine RPM: less than 1100 rpm 2. Intake air amount per revolution: 0.22 g/rev or more Catalyst monitor: Not executed
DTC: P0220 **1T ECM, MIL: Yes** **Year:** 2010, 2011 **Model:** tC, xB, xD **Engine:** 1.8L L4 VIN U, 2.4L L4, 2.5L L4	**Throttle / Pedal Position Sensor / Switch "B" Circuit:** Monitor runs whenever following DTCs are not present: None Either of the following conditions A or B is met: A. Engine switch on (IG): 0.012 seconds or more B. Electronic throttle actuator power: ON
DTC: P0222 **1T ECM, MIL: Yes** **Year:** 2010, 2011 **Model:** tC, xB **Engine:** 2.4L L4, 2.5L L4	**Throttle / Pedal Position Sensor / Switch "B" Circuit Low Input:** Monitor runs whenever following DTCs are not present: None Either of the following conditions A or B is met: A. Engine switch on (IG): 0.012 seconds or more B. Electronic throttle actuator power: ON
DTC: P0223 **1T ECM, MIL: Yes** **Year:** 2010, 2011 **Model:** tC, xB, xD **Engine:** 1.8L L4 VIN U, 2.4L L4, 2.5L L4	**Throttle / Pedal Position Sensor / Switch "B" Circuit High Input:** Monitor runs whenever following DTCs are not present: None Either of the following conditions A or B is met: A. Engine switch on (IG): 0.012 seconds or more B. Electronic throttle actuator power: ON

DTC	Trouble Code Title and Conditions
DTC: P0300 **2T ECM, MIL: Yes** **Year:** 2010, 2011 **Model:** tC, xB, xD **Engine:** 1.8L L4 VIN U, 2.4L L4, 2.5L L4	**Random / Multiple Cylinder Misfire Detected:** Battery voltage: 8 V or more VVT system: Not operated by scan tool Engine RPM: 450-6500 rpm Either of the following conditions (a) or (b) is met: (a) Engine coolant temperature at engine start: More than 19.4°F (-7°C) (b) Engine coolant temperature: More than 68°F (20°C) Fuel cut: OFF Monitor period of emission-related misfire: First 1000 revolutions after engine start, or Check Mode: Crankshaft 1000 revolutions Except above: Crankshaft 1000 revolutions X 4 Monitor period of catalyst-damaged misfire (MIL blinks): All of the following conditions 1, 2 and 3 are met: Crankshaft 200 revolutions 1. Driving cycles: 1st 2. Check mode: OFF 3. Enine RPM: Less than 2300 rpm Except above: Crankshaft 200 revolutions X 3
DTC: P0301 **2T ECM, MIL: Yes** **Year:** 2010, 2011 **Model:** tC, xB, xD **Engine:** 1.8L L4 VIN U, 2.4L L4, 2.5L L4	**Cylinder 1 Misfire Detected:** Battery voltage: 8 V or more VVT system: Not operated by scan tool Engine RPM: 450-6500 rpm Either of the following conditions (a) or (b) is met: (a) Engine coolant temperature at engine start: More than 19.4°F (-7°C) (b) Engine coolant temperature: More than 68°F (20°C) Fuel cut: OFF Monitor period of emission-related misfire: First 1000 revolutions after engine start, or Check Mode: Crankshaft 1000 revolutions Except above: Crankshaft 1000 revolutions X 4 Monitor period of catalyst-damaged misfire (MIL blinks): All of the following conditions 1, 2 and 3 are met: Crankshaft 200 revolutions 1. Driving cycles: 1st 2. Check mode: OFF 3. Enine RPM: Less than 2300 rpm Except above: Crankshaft 200 revolutions X 3
DTC: P0302 **2T ECM, MIL: Yes** **Year:** 2010, 2011 **Model:** tC, xB, xD **Engine:** 1.8L L4 VIN U, 2.4L L4, 2.5L L4	**Cylinder 2 Misfire Detected:** Battery voltage: 8 V or more VVT system: Not operated by scan tool Engine RPM: 450-6500 rpm Either of the following conditions (a) or (b) is met: (a) Engine coolant temperature at engine start: More than 19.4°F (-7°C) (b) Engine coolant temperature: More than 68°F (20°C) Fuel cut: OFF Monitor period of emission-related misfire: First 1000 revolutions after engine start, or Check Mode: Crankshaft 1000 revolutions Except above: Crankshaft 1000 revolutions X 4 Monitor period of catalyst-damaged misfire (MIL blinks): All of the following conditions 1, 2 and 3 are met: Crankshaft 200 revolutions 1. Driving cycles: 1st 2. Check mode: OFF 3. Enine RPM: Less than 2300 rpm Except above: Crankshaft 200 revolutions X 3

DTC	Trouble Code Title and Conditions
DTC: P0303 **2T ECM, MIL: Yes** **Year:** 2010, 2011 **Model:** tC, xB, xD **Engine:** 1.8L L4 VIN U, 2.4L L4, 2.5L L4	**Cylinder 3 Misfire Detected:** Battery voltage: 8 V or more VVT system: Not operated by scan tool Engine RPM: 450-6500 rpm Either of the following conditions (a) or (b) is met: (a) Engine coolant temperature at engine start: More than 19.4°F (-7°C) (b) Engine coolant temperature: More than 68°F (20°C) Fuel cut: OFF Monitor period of emission-related misfire: First 1000 revolutions after engine start, or Check Mode: Crankshaft 1000 revolutions Except above: Crankshaft 1000 revolutions X 4 Monitor period of catalyst-damaged misfire (MIL blinks): All of the following conditions 1, 2 and 3 are met: Crankshaft 200 revolutions 1. Driving cycles: 1st 2. Check mode: OFF 3. Enine RPM: Less than 2300 rpm Except above: Crankshaft 200 revolutions X 3
DTC: P0304 **2T ECM, MIL: Yes** **Year:** 2010, 2011 **Model:** tC, xB, xD **Engine:** 1.8L L4 VIN U, 2.4L L4, 2.5L L4	**Cylinder 4 Misfire Detected:** Battery voltage: 8 V or more VVT system: Not operated by scan tool Engine RPM: 450-6500 rpm Either of the following conditions (a) or (b) is met: (a) Engine coolant temperature at engine start: More than 19.4°F (-7°C) (b) Engine coolant temperature: More than 68°F (20°C) Fuel cut: OFF Monitor period of emission-related misfire: First 1000 revolutions after engine start, or Check Mode: Crankshaft 1000 revolutions Except above: Crankshaft 1000 revolutions X 4 Monitor period of catalyst-damaged misfire (MIL blinks): All of the following conditions 1, 2 and 3 are met: Crankshaft 200 revolutions 1. Driving cycles: 1st 2. Check mode: OFF 3. Enine RPM: Less than 2300 rpm Except above: Crankshaft 200 revolutions X 3
DTC: P0327 **1T ECM, MIL: Yes** **Year:** 2010, 2011 **Model:** tC, xB, xD **Engine:** 1.8L L4 VIN U, 2.4L L4, 2.5L L4	**Knock Sensor 1 Circuit Low Input (Bank 1 or Single Sensor):** Monitor runs whenever following DTCs are not present: None Battery voltage: 10.5 V or more Time after engine start: 5 seconds or more Engine switch: On (IG) Starter: OFF
DTC: P0328 **1T ECM, MIL: Yes** **Year:** 2010, 2011 **Model:** tC, xB, xD **Engine:** 1.8L L4 VIN U, 2.4L L4, 2.5L L4	**Knock Sensor 1 Circuit High Input (Bank 1 or Single Sensor):** Monitor runs whenever following DTCs are not present: None Battery voltage: 10.5 V or more Time after engine start: 5 seconds or more Engine switch: On (IG) Starter: OFF

DTC	Trouble Code Title and Conditions
DTC: P0335 **1T ECM, MIL: Yes** **Year:** 2010, 2011 **Model:** tC, xB, xD **Engine:** 1.8L L4 VIN U, 2.4L L4, 2.5L L4	**Crankshaft Position Sensor "A" Circuit:** Monitor runs whenever following DTCs are not present: None CKP Sensor Range Check/Rationality: Time after starter OFF to ON: 3 seconds or more Battery voltage: 7 V or more Minimum battery voltage while starter ON: Less than 11 V Number of VVT sensor signal pulse: 6 times Camshaft position sensor circuit fail (P0340,P0342,P0343): Not detected CKP Sensor Verify Pulse Input (Case 1): Engine speed: 600 rpm or less Starter: OFF Time after starter ON to OFF: 3 seconds or more CKP Sensor Verify Pulse Input (Case 2): Starter: ON Minimum battery voltage starter ON: Below 11 V
DTC: P0339 **1T ECM, MIL: Yes** **Year:** 2010, 2011 **Model:** tC, xB, xD **Engine:** 1.8L L4 VIN U, 2.4L L4, 2.5L L4	**Crankshaft Position Sensor "A" Circuit Intermittent:** Monitor runs whenever following conditions are met: Engine speed 1,000 rpm or more. No Crankshaft Position (CKP) sensor signal for 0.05 seconds or more. 3 seconds or more have elapsed since starter signal switched from ON to OFF.
DTC: P0340 **2T ECM, MIL: Yes** **Year:** 2011 **Model:** tC **Engine:** 2.5L L4	**Camshaft Position Sensor Circuit Malfunction:** VVT Sensor (Bank 1) Verify Pulse Input (Case 1): Engine Speed: 600 rpm or more Starter: OFF VVT system misalignment: Not detected VVT sensor range check fail (P0342, P0343): Not detected VVT sensor voltage: 0.3 V or more, and 4.7 V or less VVT Sensor (Bank 1) Verify Pulse Input (Case 2): Starter: ON Minimal battery voltage while starter ON: Below 11 V
DTC: P0340 **2T ECM, MIL: Yes** **Year:** 2010, 2011 **Model:** tC, xB, xD **Engine:** 1.8L L4 VIN U, 2.4L L4	**Camshaft Position Sensor "A" Circuit (Bank 1 or Single Sensor):** Monitor runs whenever following DTCs are not present: None Camshaft Position Sensor Range Check: Stater: ON Minimal battery voltage while starter ON: Less than 11 V Camshaft Position/Crankshaft Position Misalignment: Engine speed: 600 rpm or more Starter: OFF
DTC: P0342 **1T ECM, MIL: Yes** **Year:** 2011 **Model:** tC **Engine:** 2.5L L4	**Camshaft Position Sensor "A" Circuit Low Input (Bank 1 or Single Sensor):** Starter: OFF Engine switch: On (IG) Time after engine switch off to on (IG): 2 seconds or more VVT sensor verify pulse input fail (P0340): Not detected Battery voltage: 8 V or more
DTC: P0342 **1T ECM, MIL: Yes** **Year:** 2010, 2011 **Model:** xD **Engine:** 1.8L L4 VIN U	**Camshaft Position Sensor "A" Circuit Low Input (Bank 1 or Single Sensor):** Starter OFF, ignition switch ON. Time after ignition switch OFF to ON 2 seconds. The output voltage of CMP sensor is less than 0.3 V for 4 seconds.
DTC: P0343 **1T ECM, MIL: Yes** **Year:** 2010, 2011 **Model:** xD **Engine:** 1.8L L4 VIN U	**Camshaft Position Sensor "A" Circuit High Input (Bank 1 or Single Sensor):** Starter OFF, ignition switch ON. Time after ignition switch OFF to ON 2 seconds. The output voltage of the CMP sensor is more than 4.7 V for 4 seconds.

DTC	Trouble Code Title and Conditions
DTC: P0343 **1T ECM, MIL: Yes** **Year:** 2011 **Model:** tC **Engine:** 2.5L L4	**Camshaft Position Sensor "A" Circuit High Input (Bank 1 or Single Sensor):** Starter: OFF Engine switch: On (IG) Time after engine switch off to on (IG): 2 seconds or more VVT sensor verify pulse input fail (P0340): Not detected Battery voltage: 8 V or more
DTC: P0351 **1T ECM, MIL: Yes** **Year:** 2010, 2011 **Model:** xB **Engine:** 2.4L L4	**Ignition Coil "A" Primary / Secondary Circuit:** No IGF signal to ECM while engine running .
DTC: P0351 **1T ECM, MIL: Yes** **Year:** 2011 **Model:** tC **Engine:** 2.5L L4	**Ignition Coil "A" Primary / Secondary Circuit:** Monitor runs whenever following DTCs are not present: None Either of the following condition A or B is met: A. Engine RPM: 1500 rpm or less B. Starter: OFF Either of the following condition C or D is met: C. Both of the following conditions are met: (a) Engine speed: 500 rpm or less (b) Battery voltage: 6 V or more D. All of the following conditions are met: (a) Engine speed: More than 500 rpm (b) Battery voltage: 10 V or more (c) Number of sparks after CPU reset: 5 sparks or more
DTC: P0351 **1T ECM, MIL: Yes** **Year:** 2010, 2011 **Model:** xD **Engine:** 1.8L L4 VIN U	**Ignition Coil "A" Primary / Secondary Circuit:** No IGF signal to ECM while engine is running, the ECM does not receive any IGF signal despite ECM sending IGT signal to igniter.
DTC: P0352 **1T ECM, MIL: Yes** **Year:** 2010, 2011 **Model:** tC, xB, xD **Engine:** 1.8L L4 VIN U, 2.4L L4, 2.5L L4	**Ignition Coil "B" Primary / Secondary Circuit:** Monitor runs whenever following DTCs are not present: None Either of the following condition A or B is met: A. Engine RPM: 1500 rpm or less B. Starter: OFF Either of the following condition C or D is met: C. Both of the following conditions are met: (a) Engine speed: 500 rpm or less (b) Battery voltage: 6 V or more D. All of the following conditions are met: (a) Engine speed: More than 500 rpm (b) Battery voltage: 10 V or more (c) Number of sparks after CPU reset: 5 sparks or more
DTC: P0353 **1T ECM, MIL: Yes** **Year:** 2010, 2011 **Model:** tC, xB, xD **Engine:** 1.8L L4 VIN U, 2.4L L4, 2.5L L4	**Ignition Coil "C" Primary / Secondary Circuit:** Monitor runs whenever following DTCs are not present: None Either of the following condition A or B is met: A. Engine RPM: 1500 rpm or less B. Starter: OFF Either of the following condition C or D is met: C. Both of the following conditions are met: (a) Engine speed: 500 rpm or less (b) Battery voltage: 6 V or more D. All of the following conditions are met: (a) Engine speed: More than 500 rpm (b) Battery voltage: 10 V or more (c) Number of sparks after CPU reset: 5 sparks or more

DTC	Trouble Code Title and Conditions
DTC: P0354 **1T ECM, MIL: Yes** **Year:** 2010, 2011 **Model:** tC, xB, xD **Engine:** 1.8L L4 VIN U, 2.4L L4, 2.5L L4	**Ignition Coil "D" Primary / Secondary Circuit:** Monitor runs whenever following DTCs are not present: None Either of the following condition A or B is met: A. Engine RPM: 1500 rpm or less B. Starter: OFF Either of the following condition C or D is met: C. Both of the following conditions are met: (a) Engine speed: 500 rpm or less (b) Battery voltage: 6 V or more D. All of the following conditions are met: (a) Engine speed: More than 500 rpm (b) Battery voltage: 10 V or more (c) Number of sparks after CPU reset: 5 sparks or more
DTC: P0365 **1T ECM, MIL: Yes** **Year:** 2011 **Model:** tC **Engine:** 2.5L L4	**Camshaft Position Sensor "B" Circuit (Bank 1):** Monitor runs whenever following DTCs are not present: None Engine speed: 600 rpm or more Starter: OFF Exhaust VVT sensor range check fail (P0367, P0368, P0392, P0393): Not detected Exhaust VVT sensor voltage: 0.3 V or more, and 4.7 V or less Battery voltage: 8 V or more Engine switch: On (IG) Time after engine switch off to on (IG): 0.5 seconds or more
DTC: P0365 **1T ECM, MIL: Yes** **Year:** 2010, 2011 **Model:** xD **Engine:** 1.8L L4 VIN U	**Camshaft Position Sensor "B" Circuit (Bank 1):** When one of following conditions is met: Misfiring exhaust camshaft position sensor signal signal for 5 seconds at engine speed of 600 rpm or more (1 trip detection logic). Input voltage to ECM remains less than 0.3 V, or more than 4.7 V for 4 seconds when 2 or more seconds have elapsed after turning ignition switch ON (1 trip detection logic).
DTC: P0367 **1T ECM, MIL: Yes** **Year:** 2010, 2011 **Model:** xD **Engine:** 1.8L L4 VIN U	**Camshaft Position Sensor "B" Circuit Low Input (Bank 1):** Starter OFF, ignition switch ON. Time after ignition switch OFF to ON 2 seconds. The output voltage of exhaust camshaft position sensor is less than 0.3 V for 4 seconds.
DTC: P0367 **1T ECM, MIL: Yes** **Year:** 2011 **Model:** tC **Engine:** 2.5L L4	**Camshaft Position Sensor "B" Circuit Low Input (Bank 1):** Monitor runs whenever following DTCs are not present: None Starter: OFF Engine switch: On (IG) Time after engine switch off to on (IG): 2 seconds or more Exhaust VVT sensor verify pulse input fail (P0365, P0390): Not detected Battery voltage: 8 V or more
DTC: P0368 **1T ECM, MIL: Yes** **Year:** 2010, 2011 **Model:** xD **Engine:** 1.8L L4 VIN U	**Camshaft Position Sensor "B" Circuit High Input (Bank 1):** Starter OFF, ignition switch ON. Time after ignition switch OFF to ON 2 seconds. The output voltage of exhaust camshaft position sensor is more than 4.7 V for 4 seconds.
DTC: P0368 **1T ECM, MIL: Yes** **Year:** 2011 **Model:** tC **Engine:** 2.5L L4	**Camshaft Position Sensor "B" Circuit High Input (Bank 1):** Monitor runs whenever following DTCs are not present: None Starter: OFF Engine switch: On (IG) Time after engine switch off to on (IG): 2 seconds or more Exhaust VVT sensor verify pulse input fail (P0365, P0390): Not detected Battery voltage: 8 V or more

DTC	Trouble Code Title and Conditions
DTC: P0420 **2T ECM, MIL: Yes** **Year:** 2010, 2011 **Model:** tC, xB, xD **Engine:** 1.8L L4 VIN U, 2.4L L4, 2.5L L4	**Catalyst System Efficiency Below Threshold (Bank 1):** Monitor runs whenever following DTCs are not present: P0010, P0020 (OCV bank 1, 2) P0011, P0021 (VVT system bank 1, 2 - advance) P0012, P0022 (VVT system bank 1, 2 - retard) P0013, P0023 (Exhaust OCV bank 1, 2) P0014, P0024 (Exhaust VVT system bank 1, 2 - advance) P0015, P0025 (Exhaust VVT system bank 1, 2 - retard) P0016, P0018 (VVT system bank 1, 2 - misalignment) P0017, P0019 (Exhaust VVT system bank 1, 2 - misalignment) P0031, P0032, P0051, P0052 (A/F sensor heater) P0037, P0038, P0057, P0058 (HO2 sensor heater) P0102, P0103 (MAF meter) P0115, P0117, P0118 (ECT sensor) P0120, P0121, P0122, P0123, P0220, P0222, P0223, P2135 (TP sensor) P0125 (Insufficient ECT for closed loop) P0136, P0137, P0138, P0139 (HO2 sensor bank 1) P014C, P014D, P014E, P014F (Air fuel ratio sensor - slow response) P015A, P015B, P015C, P015D (Air fuel ratio sensor - delayed response) P0156, P0157, P0158, P0159 (HO2 sensor bank 2) P0171, P0172, P0174, P0175 (Fuel system) P0301, P0302, P0303, P0304, P0305, P0306 (Misfire) P0335 (CKP sensor) P0340, P0342, P0343, P0345, P0347, P0348 (VVT sensor) P0351, P0352, P0353, P0354, P0355, P0356 (Igniter) P0365, P0367, P0368, P0390, P0392, P0393 (Exhaust VVT sensor) P0500 (Vehicle speed sensor) P0607 (HO2 sensor - sensor 2) P2195, P2196, P2197, P2198 (A/F sensor - rationality) P2237, P2240 (A./F sensor - open) P2238, P2241, P2252, P2255 (A/F sensor - low impedance) P2239, P2242, P2253, P2256 (A/F sensor - high impedance) Battery voltage: 11 V or more IAT: 14°F (-10°C) or more ECT: 167°F (75°C) or more Atmospheric pressure: 570 mmHg (76 kPa) or more Idle: OFF Enigne RPM: Less than 3200 rpm A/F Sensor: Activated Fuel system status: Closed loop Engine load: 10-70% All of the following conditions 1, 2 and 3 are met: 1. MAF: 5-60 g/sec. 2. Front catalyst temperature (estimated): 1112-1472°F (600-800°C) 3. Rear catalyst temperature (estimated): 212-1652°F (100-900°C) Shift position: 4rd or higher
DTC: P043E **1T ECM, MIL: Yes** **Year:** 2010, 2011 **Model:** tC, xB, xD **Engine:** 1.8L L4 VIN U, 2.4L L4, 2.5L L4	**Evaporative Emission System Reference Orifice Clog Up:** Reference orifice clogged. Reference orifice high-flow. Leak detection pump OFF malfunction. Leak detection pump ON malfunction. Vent valve ON (close) malfunction.
DTC: P043F **1T ECM, MIL: Yes** **Year:** 2010, 2011 **Model:** tC, xB, xD **Engine:** 1.8L L4 VIN U, 2.4L L4, 2.5L L4	**Evaporative Emission System Reference Orifice High Flow:** Reference orifice clogged. Reference orifice high-flow. Leak detection pump OFF malfunction. Leak detection pump ON malfunction. Vent valve ON (close) malfunction.

DTC	Trouble Code Title and Conditions
DTC: P0441 **2T ECM, MIL: Yes** **Year:** 2010, 2011 **Model:** tC, xB, xD **Engine:** 1.8L L4 VIN U, 2.4L L4, 2.5L L4	**Evaporative Emission Control System Incorrect Purge Flow:** Purge VSV (Vacuum Switching Valve) stuck open: Leak detection pump creates negative pressure (vacuum) in EVAP system and EVAP system pressure is measured. 0.02 inch leak pressure standard is measured at start and at end of leak check. If stabilized pressure is higher than [second 0.02 inch leak pressure standard x 0.2], ECM determines that Purge VSV is stuck open. Purge VSV stuck closed: After EVAP leak check is performed, Purge VSV is turned ON (open), and atmospheric air is introduced into EVAP system. 0.02 inch leak pressure standard is measured at start and at end of the check. If pressure does not return to near atmospheric pressure, ECM determines that purge valve is stuck closed. Purge flow: While engine is running, following conditions are successively met: * Negative pressure is not created in EVAP system when Purge VSV is turned ON (open) * EVAP system pressure change is less than 0.5 kPa (3.75 mmHg) when vent valve is turned ON (closed) * Atmospheric pressure change before and after purge flow monitor is less than 0.1 kPa (0.75 mmHg)
DTC: P0443 **1T ECM, MIL: Yes** **Year:** 2011 **Model:** tC **Engine:** 2.5L L4	**Evaporative Emission Control System Purge Control Valve Circuit:** NA
DTC: P0450 **1T , MIL: Yes** **Year:** 2010 **Model:** tC **Engine:** 2.4L L4	**Evaporative Emission Control System Pressure Sensor / Switch:** Monitor runs whenever following DTCs are not present: None Either of following conditions is met: (a) or (b) (a) Ignition switch: ON (b) Soak timer: ON Battery voltage: 8 V or more Starter: OFF
DTC: P0451 **2T ECM, MIL: Yes** **Year:** 2010, 2011 **Model:** tC, xB, xD **Engine:** 1.8L L4 VIN U, 2.4L L4, 2.5L L4	**Evaporative Emission Control System Pressure Sensor Range / Performance:** Canister pressure sensor noise: Atmospheric pressure: 70 kPa (525 mmHg) to 110 kPa (825 mmHg) Battery voltage: 10.5 V or more Intake air temperature: 40 to 95°F (4.4 to 35°C) EVAP pressure sensor malfunction (P0452, P0453): Not detected Either of following conditions is met: 1. Engine: Running 2. Time after key off: 5, 7 or 9.5 hours
DTC: P0452 **1T ECM, MIL: Yes** **Year:** 2011 **Model:** tC **Engine:** 2.5L L4	**Evaporative Emission Control System Pressure Sensor / Switch Low Input:** Monitor runs whenever following DTCs are not present: None Battery voltage: 8 V or more Starter: OFF Engine switch: ON
DTC: P0452 **1T ECM, MIL: Yes** **Year:** 2010, 2011 **Model:** xD **Engine:** 1.8L L4 VIN U	**Evaporative Emission Control System Pressure Sensor / Switch Low Input:** Ignition switch ON, EVAP monitoring (ignition switch OFF). The EVAP pressure is less than 42.1 kPa for 0.5 seconds.
DTC: P0452 **1T ECM, MIL: Yes** **Year:** 2010, 2011 **Model:** xB **Engine:** 2.4L L4	**Evaporative Emission Control System Pressure Sensor / Switch Low Input:** With the ignition switch ON, EVAP monitoring. (ignition switch OFF) The EVAP pressure is less than 42.1 kPa for 0.5 seconds.
DTC: P0453 **1T ECM, MIL: Yes** **Year:** 2010, 2011 **Model:** tC, xB, xD **Engine:** 1.8L L4 VIN U, 2.4L L4, 2.5L L4	**Evaporative Emission Control System Pressure Sensor / Switch High Input:** Monitor runs whenever following DTCs are not present: None Battery voltage: 8 V or more Starter: OFF Engine switch: ON

DTC	Trouble Code Title and Conditions
DTC: P0455 **2T ECM, MIL: Yes** **Year:** 2010, 2011 **Model:** tC, xB, xD **Engine:** 1.8L L4 VIN U, 2.4L L4, 2.5L L4	**Evaporative Emission Control System Leak Detected (Gross Leak):** EVAP key-off monitor runs when all of following conditions are met: Atmospheric pressure: 70 to 110 kPa (525 to 825 mmHg) Battery voltage: 10.5 V or more Vehicle speed: Below 2.5 mph (4 km/h) Engine switch: OFF Time after key off: 5, 7 or 9.5 hours EVAP pressure sensor malfunction (P0451, P0452 and P0453): Not detected Purge VSV: Not operated by scan tool Vent valve: Not operated by scan tool Leak detection pump: Not operated by scan tool Both of following conditions are met before key OFF: Conditions 1 and 2 1. Duration that vehicle is being driven: 5 minutes or more 2. EVAP purge operation: Performed ECT: 4.4 to 35°C (40 to 95°F): P0456 4.4 to 50°C (40 to 122°F): P0455 IAT:4.4 to 35°C (40 to 95°F): P0456 4.4 to 50°C (40 to 122°F): P0455
DTC: P0456 **2T ECM, MIL: Yes** **Year:** 2010, 2011 **Model:** tC, xB, xD **Engine:** 1.8L L4 VIN U, 2.4L L4, 2.5L L4	**Evaporative Emission Control System Leak Detected (Very Small Leak):** EVAP key-off monitor runs when all of following conditions are met: Atmospheric pressure: 70 to 110 kPa (525 to 825 mmHg) Battery voltage: 10.5 V or more Vehicle speed: Below 2.5 mph (4 km/h) Engine switch: OFF Time after key off: 5, 7 or 9.5 hours EVAP pressure sensor malfunction (P0451, P0452 and P0453): Not detected Purge VSV: Not operated by scan tool Vent valve: Not operated by scan tool Leak detection pump: Not operated by scan tool Both of following conditions are met before key OFF: Conditions 1 and 2 1. Duration that vehicle is being driven: 5 minutes or more 2. EVAP purge operation: Performed ECT: 4.4 to 35°C (40 to 95°F): P0456 IAT:4.4 to 35°C (40 to 95°F): P0456
DTC: P0500 **1T ECM, MIL: Yes** **Year:** 2010, 2011 **Model:** tC, xB, xD **Engine:** 1.8L L4 VIN U, 2.4L L4, 2.5L L4	**Vehicle Speed Sensor "A":** The monitor will run whenever these DTCs are not present: None Time after engine switch off to on: 3 seconds or more Engine: Running Battery voltage: 8 V or more Engine switch: ON Starter: OFF Either of the following conditions are met: Condition 1 or 2 1. All fo the following conditions are met: Condition (a), (b) or (c) (a) ECT: 60°F (20°C) or more (b) ECT fail: Not detected (c) Time after park/neutral position switch ON to OFF: 10 seconds or more 2. All of the following conditions are met: Either (a) or (b) is set (a) ECT: Less than 68°F (20°C) (b) ECT fail: Detected (c) Time after park/neutral position switch ON to OFF: 30 seconds or more
DTC: P0503 **T ECM** **Year:** 2010, 2011 **Model:** tC, xB, xD **Engine:** 1.8L L4 VIN U, 2.4L L4	**Vehicle Speed Sensor "A" Intermittent / Erratic / High:** Momentary interruption and noise are detected when a rapid change of vehicle speed occurs while cruise control is in operation.

DTC	Trouble Code Title and Conditions
DTC: P0503 **1T ECM, MIL: Yes** **Year:** 2011 **Model:** tC **Engine:** 2.5L L4	**Vehicle Speed Sensor Circuit Malfunction:** Momentary interruption and noise malfunction codes are detected when a rapid change of vehicle speed occurs while the cruise control is in operation.
DTC: P0504 **1T ECM, MIL: Yes** **Year:** 2010, 2011 **Model:** tC, xB, xD **Engine:** 1.8L L4 VIN U, 2.4L L4, 2.5L L4	**Brake Switch "A" / "B" Correlation:** The monitor will run whenever these DTCs are not present: None Engine switch: ON Starter: OFF Battery voltage: 8 V or more GO (Vehicle speed is 18.65 mph (30 km/h) or more): Once STOP (Vehicle speed is less than 1.86 mph (3 km/h)): Once
DTC: P0505 **2T ECM, MIL: Yes** **Year:** 2010, 2011 **Model:** tC, xB, xD **Engine:** 1.8L L4 VIN U, 2.4L L4, 2.5L L4	**Idle Control System Malfunction:** Monitor will run whenever these DTCs are not present: P0010, P0020 (OCV bank 1, 2) P0011, P0021 (VVT system bank 1, 2 - advance) P0012, P0022 (VVT system bank 1, 2 - retard) P0013, P0023 (Exhaust OCV bank 1, 2) P0014, P0024 (Exhaust VVT system bank 1, 2 - advance) P0015, P0025 (Exhaust VVT system bank 1, 2 - retard) P0016, P0018 (VVT system bank 1, 2 - misalignment) P0017, P0019 (Exhaust VVT system bank 1, 2 - misalignment) P0031, P0032, P0051, P0052 (A/F sensor heater) P0102, P0103 (MAF meter) P0115, P0117, P0118 (ECT sensor) P0120, P0121, P0122, P0123, P0220, P0222, P0223, P2135 (TP sensor) P0125 (Insufficient ECT for closed loop) P0171, P0172, P0174, P0175 (Fuel system) P0301, P0302, P0303, P0304, P0305, P0306 (Misfire) P0335 (CKP sensor) P0340, P0342, P0343, P0345, P0347, P0348 (VVT sensor) P0351, P0352, P0353, P0354, P0355, P0356 (Igniter) P0365, P0367, P0368, P0390, P0392, P0393 (Exhaust VVT sensor) P0451, P0452, P0452 (EVAP system) P0500 (Vehicle speed sensor) P2195, P2196, P2197, P2198 (A/F sensor - rationality) P2237, P2240 (A/F sensor - open) P2238, P2241, P2252, P2255 (A/F sensor - low impedance) P2239, P2242, P2253, P2256 (A/F sensor - high impedance) P2A00, P2A03 (A/F sensor - slow response) Engine: Running

DTC	Trouble Code Title and Conditions
DTC: P050A **2T ECM, MIL: Yes** **Year:** 2010, 2011 **Model:** tC, xB, xD **Engine:** 1.8L L4 VIN U, 2.4L L4, 2.5L L4	**Cold Start Idle Air Control System Performance:** Monitor runs whenever following DTCs are not present: P0010, P0020 (OCV bank 1, 2) P0011, P0021 (VVT system bank 1, 2 - advance) P0012, P0022 (VVT system bank 1, 2 - retard) P0013, P0023 (Exhaust OCV bank 1, 2) P0014, P0024 (Exhaust VVT system bank 1, 2 - advance) P0015, P0025 (Exhaust VVT system bank 1, 2 - retard) P0016, P0018 (VVT system bank 1, 2 - misalignment) P0017, P0019 (Exhaust VVT system bank 1, 2 - misalignment) P0102, P0103 (MAF meter) P0115, P0117, P0118 (ECT sensor) P0120, P0121, P0122, P0123, P0220, P0222, P0223, P2135 (TP sensor) P0125 (Insufficient ECT for closed loop) P0171, P0172, P0174, P0175 (Fuel system) P0301, P0302, P0303, P0304, P0305, P0306 (Misfire) P0335 (CKP sensor) P0340, P0342, P0343, P0345, P0347, P0348 (VVT sensor) P0351, P0352, P0353, P0354, P0355, P0356 (Igniter) P0365, P0367, P0368, P0390, P0392, P0393 (Exhaust VVT sensor) P0500 (Vehicle speed sensor) P2195, P2196, P2197, P2198 (A/F sensor - rationality) P2237, P2240 (A/F sensor - open) P2238, P2241, P2252, P2255 (A/F sensor - low impedance) P2239, P2242, P2253, P2256 (A/F sensor - high impedance) P2A00, P2A03 (A/F sensor - slow response) Battery voltage: 8 V or more Time after engine start: 3 seconds or more Starter: OFF ECT at engine start: -10°C (14°F) or more ECT: -10°C to 50°C (14°F to 122°F) Engine idling time: 3 seconds or more Fuel-cut: OFF Vehicle speed: 1.875 mph (3.01 km/h) or less Time after shift position changed: 1 second or more Atmospheric pressure: 76 kPa (570 mmHg) or more
DTC: P050B **2T ECM, MIL: Yes** **Year:** 2010, 2011 **Model:** xD **Engine:** 1.8L L4 VIN U	**Cold Start Ignition Timing Performance:** Battery voltage 8 V or more, time after engine start 3 seconds or more. Starter OFF, ECT at engine start -10°C (14°F) or more, ECT -10 to 50°C (14 to 122°F). Engine idling time 3 seconds or more, Fuel-cut OFF. Vehicle speed less than 3 km/h (1.875 mph). Insufficient ignition timing retard at cold start.
DTC: P050B **2T ECM, MIL: Yes** **Year:** 2011 **Model:** tC **Engine:** 2.5L L4	**Cold Start Ignition Timing Performance:** Battery voltage: 8 V or more Time after engine start: 3 seconds or more Starter: OFF ECT at engine start: -10°C (14°F) or more ECT: -10 to 50°C (14 to 122°F) Engine idling time: 3 seconds or more Fuel cut: OFF Vehicle speed: 1.875 mph (3.01 km/h) or less Atmospheric pressure: 76 kPa (570 mmHg) or more
DTC: P0560 **1T TCM, MIL: Yes** **Year:** 2011 **Model:** tC **Engine:** 2.5L L4	**System Voltage:** Monitor runs whenever following DTCs are not present: None Stand-by RAM: Initialized

DTC	Trouble Code Title and Conditions
DTC: P0560 **1T ECM, MIL: Yes** **Year:** 2010, 2011 **Model:** tC, xB, xD **Engine:** 1.8L L4 VIN U, 2.4L L4, 2.5L L4	**System Voltage:** Monitor runs whenever following DTCs are not present: None Stand by RAM: Initialize
DTC: P0571 **1T ECM, MIL: Yes** **Year:** 2010, 2011 **Model:** tC, xB, xD **Engine:** 1.8L L4 VIN U, 2.4L L4, 2.5L L4	**Brake Switch "A" Circuit:** When voltage of STP terminal and that of ST1- terminal of ECM are less than 1 V for 0.5 seconds or more.
DTC: P0575 **1T ECM, MIL: Yes** **Year:** 2010, 2011 **Model:** xB, xD **Engine:** 1.8L L4 VIN U, 2.4L L4	**Cruise Control Input Circuit:** When both of the following conditions are met: STP signals input to the ECM supervisory CPU and control ECU are different for 0.15 seconds or more. 0.4 seconds have passed after cruise cancel input signal (STP input) is input to the ECM.
DTC: P0575 **T , MIL: Yes** **Year:** 2010, 2011 **Model:** tC **Engine:** 2.4L L4, 2.5L L4	**Cruise Control Input Circuit:** When both of the following conditions are met: * STP signals input to the hybrid vehicle control ECU supervisory CPU and control ECU are different for 0.15 seconds or more * 0.4 seconds have passed after cruise cancel input signal (STP input) is input to the hybrid vehicle control ECU
DTC: P0604 **1T ECM, MIL: Yes** **Year:** 2010, 2011 **Model:** tC, xB, xD **Engine:** 1.8L L4 VIN U, 2.4L L4, 2.5L L4	**Internal Control Module Random Access Memory (RAM) Error:** The ECM continuously monitors its internal memory status. This self-check ensures that the ECM is functioning properly. It is diagnosed by internal "mirroring" of the main CPU and sub CPU to detect the Random Access Memory (RAM) errors. If outputs from these CPUs are different and deviate from the standards, the ECM will illuminate the MIL and set a DTC immediately. Monitor will run whenever this DTC is not present: None
DTC: P0606 **1T ECM, MIL: Yes** **Year:** 2010, 2011 **Model:** tC, xB, xD **Engine:** 1.8L L4 VIN U, 2.4L L4, 2.5L L4	**ECM/PCM Processor:** Monitor will run whenever this DTC is not present: None With the engine running. Estimated A/F sensor temperature 450 to 800°C (842 to 1,472°F). Estimated HO2S temperature 450 to 80. ECM CPUs malfunction. A/F sensor transistors malfunction. HO2S transistors malfunction.
DTC: P0607 **T ECM, MIL: Yes** **Year:** 2010, 2011 **Model:** tC, xB, xD **Engine:** 1.8L L4 VIN U, 2.4L L4, 2.5L L4	**Control Module Performance:** Monitor runs whenever the following DTCs are not present: None Engine: Running
DTC: P060A **T ECM, MIL: Yes** **Year:** 2011 **Model:** tC **Engine:** 2.5L L4	**Internal Control Module Monitoring Processor Performance:** CPU reset: Occurred
DTC: P060A **T ECM, MIL: Yes** **Year:** 2010, 2011 **Model:** xB **Engine:** 2.4L L4	**Internal Control Module Monitoring Processor Performance:** When either of the following conditions below are met: Condition: 1. CPU reset 1 time or more, Learned TP - Learned APP 0.4 V or more, Electronic throttle actuator is OFF Condition: 2. CPU reset 2 times or more.
DTC: P060A **T ECM, MIL: Yes** **Year:** 2010, 2011 **Model:** xD **Engine:** 1.8L L4 VIN U	**Internal Control Module Monitoring Processor Performance:** ECM sub CPU error, the MIL operation is immediate.

DTC	Trouble Code Title and Conditions
DTC: P060B **1T ECM, MIL: Yes** **Year:** 2011 **Model:** tC **Engine:** 2.5L L4	**Internal Control Module A/D Processing Performance:** Monitor runs whenever following DTCs not stored: None
DTC: P060D **T ECM, MIL: Yes** **Year:** 2010, 2011 **Model:** xB, xD **Engine:** 1.8L L4 VIN U, 2.4L L4	**Internal Control Module Accelerator Pedal Position Performance:** When the difference of the main APP and sub APP are 0.3 V or more. for mor than 1 second. MIL operation is immediate.
DTC: P060D **T ECM, MIL: Yes** **Year:** 2011 **Model:** tC **Engine:** 2.5L L4	**Internal Control Module Accelerator Pedal Position Performance:** None provided
DTC: P060E **T ECM, MIL: Yes** **Year:** 2011 **Model:** tC **Engine:** 2.5L L4	**Internal Control Module Throttle Position Performance:** DMA communication error: Not detected
DTC: P060E **T ECM, MIL: Yes** **Year:** 2010, 2011 **Model:** xB, xD **Engine:** 1.8L L4 VIN U, 2.4L L4	**Internal Control Module Throttle Position Performance:** When one of following conditions is met: Condition 1 or 2 1. Difference of main TP and sub TP 0.3 V or more. 2. Difference of main brake switch signal and sub brake switch signal. The MIL operation is immediate.
DTC: P0617 **1T ECM, MIL: Yes** **Year:** 2010, 2011 **Model:** tC, xB, xD **Engine:** 1.8L L4 VIN U, 2.4L L4, 2.5L L4	**Starter Relay Circuit High:** Monitor runs whenever this DTC is not present: None Battery voltage: 10.5 V or more Vehicle speed: 12.43 mph (20 km/h) or more Engine speed: 1000 rpm or more
DTC: P0617 **1T TCM, MIL: Yes** **Year:** 2011 **Model:** tC **Engine:** 2.5L L4	**Starter Relay Circuit High:** Monitor runs whenever these DTCs are not present: None Battery voltage: 10.5 V or more Vehicle speed: 12.43 mph (20 km/h) or more Engine speed: 1000 rpm or more
DTC: P062F **T ECM, MIL: Yes** **Year:** 2010, 2011 **Model:** xB, xD **Engine:** 1.8L L4 VIN U, 2.4L L4	**Internal Control Module EEPROM Error:** Time after engine start 10 seconds or more, battery voltage 8 V or higher, ignition switch ON, starter OFF. MIL operation is immediate. Mismatch (3 times or more)
DTC: P062F **1T ECM, MIL: Yes** **Year:** 2011 **Model:** tC **Engine:** 2.5L L4	**Internal Control Module EEPROM Error:** Monitor runs whenever following DTCs not stored: None Time after engine start: 10 seconds or more Battery voltage: 8 V or more Ignition switch: ON Starter: OFF
DTC: P062F **T TCM, MIL: Yes** **Year:** 2011 **Model:** tC **Engine:** 2.5L L4	**Internal Control Module EEPROM Error:** Monitor runs whenever following DTCs not present: None Time after engine start: 10 seconds or more Battery voltage:8 V or higher Ignition switch: ON Starter: OFF

DTC	Trouble Code Title and Conditions
DTC: P0630 **T ECM, MIL: Yes** **Year:** 2010, 2011 **Model:** tC, xB, xD **Engine:** 1.8L L4 VIN U, 2.4L L4, 2.5L L4	**VIN not Programmed or Mismatch - ECM / PCM:** Battery voltage: 8 V or more Engine switch: ON Starter: OFF
DTC: P0657 **T ECM, MIL: Yes** **Year:** 2010, 2011 **Model:** tC, xB, xD **Engine:** 1.8L L4 VIN U, 2.4L L4, 2.5L L4	**Actuator Supply Voltage Circuit / Open:** Monitor will run whenever this DTC is not present: None Engine switch: Front ON to OFF
DTC: P0705 **2T ECM, MIL: Yes** **Year:** 2011 **Model:** tC **Engine:** 2.5L L4	**Transmission Range Sensor Circuit Malfunction (PRNDL Input):** The monitor will run whenever this DTC is not present: None Engine switch: ON Battery voltage: 10.5 V or more
DTC: P0705 **2T ECM, MIL: Yes** **Year:** 2010, 2011 **Model:** tC, xB, xD **Engine:** 1.8L L4 VIN U, 2.4L L4, 2.5L L4	**Transmission Range Sensor Circuit Malfunction (PRNDL Input):** The monitor will run whenever this DTC is not present: None Engine switch: ON Battery voltage: 10.5 V or more Condition (C) One of the following conditions is met: Condition (a) or (b) (a) Park Neutral position switch: ON (b) R range positions switch: ON
DTC: P0711 **2T ECM, MIL: Yes** **Year:** 2010, 2011 **Model:** tC, xB, xD **Engine:** 1.8L L4 VIN U, 2.4L L4, 2.5L L4	**Transmission Fluid Temperature Sensor "A" Performance:** The monitor will run whenever this DTC is not present (Not circuit malfunction) P0712, P0713 (ATF temperature sensor circuit (TFT sensor)) P0115, P0117, P0118 (ECT sensor circuit) P0112, P0113 (IAT sensor circuit) P0715, P0717 (Turbine speed sensor circuit) P0791, P0793 (Intermediate shaft speed sensor circuit) P0748 (Shift solenoid valve SL1 circuit) P0778 (Shift solenoid valve SL2 circuit) P0798 (Shift solenoid valve SL3 circuit) P2810 (Shift solenoid valve SL4 circuit) P0327, P0328, P0332, P0333 (KCS sensor circuit) P0120, P0121, P0122, P0123, P0220, P0222, P0223, P0604, P0606, P060A, P060B, P060D, P060E, P0657, P1607, P2102, P2103, P2111, P2112, P2118, P2119, P2135 ((ETCS) Electronic throttle control system) U0100 (CAN communication system) TFT (Transmission fluid temperature) sensor circuit: No circuit malfunction ECT (Engine coolant temperature) sensor circuit: No circuit malfunction IAT (Intake air temperature) sensor circuit: No circuit malfunction Turbine speed sensor circuit: No circuit malfunction Intermediate shaft speed sensor: No circuit malfunction Intermediate shaft speed sensor: No circuit malfunction Shift solenoid valve SL1 circuit: No circuit malfunction Shift solenoid valve SL2 circuit: No circuit malfunction Shift solennoid valve SL3 circuit: No circuit malfunction Shift solenoid valve SL4 circuit: No circuit malfunction (KCS) Knock control sensor circuit: No circuit malfunction (ETCS) Electronic throttle control system: No system down CAN communication system: Not system down Time after engine start: 16 min. and 40 sec. or more ECT (Engine Coolant Temperature): 5°F (-15°C) or more

DTC	Trouble Code Title and Conditions
DTC: P0712 **1T ECM, MIL: Yes** **Year:** 2010, 2011 **Model:** tC, xB **Engine:** 2.4L L4, 2.5L L4	**Transmission Fluid Temperature Sensor "A" Circuit Low Input:** The monitor will run whenever this DTC is not present: None The typical enabling condition is not available.
DTC: P0713 **1T ECM, MIL: Yes** **Year:** 2010, 2011 **Model:** tC, xB, xD **Engine:** 1.8L L4 VIN U, 2.4L L4, 2.5L L4	**Transmission Fluid Temperature Sensor "A" Circuit High Input:** The monitor will run whenever this DTC is not present: None The typical enabling condition is not available.
DTC: P0715 **T TCM, MIL: Yes** **Year:** 2011 **Model:** tC **Engine:** 2.5L L4	**Input / Turbine Speed Sensor Circuit Malfunction:** Battery voltage: 8 V or more Engine switch: ON Starter: OFF
DTC: P0717 **1T TCM, MIL: Yes** **Year:** 2011 **Model:** tC **Engine:** 2.5L L4	**Input Speed Sensor Circuit No Signal:** The monitor will run whenever this DTC is not present:P0500 (VSS), P0748, P0778, P0798 (Shift solenoid valve (range)) Shift change: Shift change is completed before starting next shift change operation TCM selected gear: 2nd, 3rd, 4th, 5th or 6th Output shaft rpm: 1,000 rpm or more STAR switch: OFF R switch: OFF Engine: Running Battery voltage: 8 V or more Engine switch: ON Starter: OFF
DTC: P0717 **1T ECM, MIL: Yes** **Year:** 2010, 2011 **Model:** xB, xD **Engine:** 1.8L L4 VIN U, 2.4L L4	**Input Speed Sensor Circuit No Signal:** ECM detects conditions (a), (b) and (c) continuously for 5 seconds or more. (a) Output shaft speed: 1000 rpm or more (b) Park/Neutral position switch (R) is OFF (c) Speed sensor NT: Less than 300 rpm
DTC: P0724 **2T ECM, MIL: Yes** **Year:** 2011 **Model:** tC **Engine:** 2.5L L4	**Brake Switch "B" Circuit High:** The monitor will run whenever these DTCs are not present: None Engine switch: ON Starter: OFF Battery voltage: 8 V or more GO (Vehicle speed is 18.65 mph (30 km/h) or more): Once STOP (Vehicle speed is less than 1.86 mph (3 km/h)): Once
DTC: P0724 **2T TCM, MIL: Yes** **Year:** 2010, 2011 **Model:** tC, xB, xD **Engine:** 1.8L L4 VIN U, 2.4L L4, 2.5L L4	**Brake Switch "B" Circuit High:** The monitor will run whenever this DTC is not present: None GO: (Vehicle speed is 18.63 mph (30 km/h) or more): 18.7 mph (30 km/h) or more STOP: (Vehicle speed is less than 1.86 mph (3 km/h)): Less than 1.86 mph (3 km/h) Starter: OFF Battery voltage: 8 V or more Engine switch: ON

DTC	Trouble Code Title and Conditions
DTC: P0741 **2T TCM, MIL: Yes** **Year:** 2011 **Model:** tC **Engine:** 2.5L L4	**Torque Converter Clutch Solenoid Performance (Shift Solenoid Valve SL):** The monitor will run whenever this DTC is not present (Not circuit malfunction): P0712, P0713 (ATF temperature sensor circuit (TFT sensor)) P0115, P0117, P0118 (ECT sensor circuit) P0715, P0717 (Turbine speed sensor circuit) P0791, P0793 (Intermediate shaft speed sensor circuit) P0748 (Shift solenoid valve SL1 circuit) P0778 (Shift solenoid valve SL2 circuit) P0798 (Shift solenoid valve SL3 circuit) P2810 (Shift solenoid valve SL4 circuit) P2759 (Shift solenoid valve SLU circuit) P2769, P2770 (Shift solenoid valve SL circuit) P0327, P0328, P0332, P0333 (KCS sensor circuit) P0120, P0121, P0122, P0123, P0220, P0222, P0223, P0604, P0606, P060A, P060B, P060D, P060E, P0657, P1607, P2102, P2103, P2111, P2112, P2118, P2119, P2135 ((ETCS) Electronic throttle control system) U0100 (CAN communication system) ECT (Engine coolant temperature): 104°F (40°C) or more Spark advance from Max. retard timing by KCS control: 0°CA or more Transmission range: "D" TFT (Transmission fluid temperature): 14°F (-10°C) or more TFT (Transmission fluid temperature): -10°C (14°F) or more TFT (Transmission fluid temperature) sensor circuit: No circuit malfunction ECT (Engine coolant temperature) sensor circuit:No circuit malfunction Turbine speed sensor circuit:No circuit malfunction Intermediate shaft speed sensor circuit: No circuit malfunction Shift solenoid valve SL1 circuit: No circuit malfunction Shift solenoid valve SL2 circuit: No circuit malfunction Shift solenoid valve SL3 circuit:No circuit malfunction Shift solenoid valve SL4 circuit: No circuit malfunction Shift solenoid valve SLU circuit: No circuit malfunction Shift solenoid valve SL circuit: No circuit malfunction (KCS) Knock control sensor circuit: No circuit malfunction (ETCS) Electronic throttle control system: Not system down CAN communication system: Not system down TCM selected gear:Not 1st Vehicle speed: 15.5 mph (25 km/h) or more Turbine speed/Output speed (NT/NO) with 1st: 3.304 to 7.724 Turbine speed/Output speed (NT/NO) with 2nd:1.901 to 2.340 Turbine speed/Output speed (NT/NO) with 3rd: 1.399 to 1.649 Turbine speed/Output speed (NT/NO) with 4th: 0.998 to 1.138 Turbine speed/Output speed (NT/NO) with 5th: 0.705 to 0.836 Turbine speed/Output speed (NT/NO) with 6th: 0.568 to 0.695
DTC: P0741 **2T ECM, MIL: Yes** **Year:** 2010, 2011 **Model:** tC, xB **Engine:** 2.4L L4	**Torque Converter Clutch Solenoid Performance (Shift Solenoid Valve DSL):** Lock-up does not occur when driving in lock-up range (normal driving at 80 km/h [50 mph]), or lock-up remains ON in lock-up OFF range (2 trip detection logic)

DTC	Trouble Code Title and Conditions
DTC: P0746 **2T TCM, MIL: Yes** **Year:** 2010, 2011 **Model:** tC, xB **Engine:** 2.4L L4, 2.5L L4	**Pressure Control Solenoid "A" Performance (Shift Solenoid Valve SL1):** The monitor will run whenever this DTC is not present. (Not circuit malfunction): P0115, P0117, P0118 (ECT sensor circuit) P0715, P0717 (Turbine speed sensor circuit) P0791, P0793 (Intermediate shaft speed sensor circuit) P0748 (Shift solenoid valve SL1 circuit) P0778 (Shift solenoid valve SL2 circuit) P0798 (Shift solenoid valve SL3 circuit) P2810 (Shift solenoid valve SL4 circuit) P0327, P0328, P0332, P0333 (KCS sensor circuit) P0120, P0121, P0122, P0123, P0220, P0222, P0223, P0604, P0606, P060A, P060B, P060D, P060E, P0657, P1607, P2102, P2103, P2111, P2112, P2118, P2119, P2135 ((ETCS) Electronic throttle control system) U0100 (CAN communication system) Transmission range: "D" TFT (Transmission fluid temperature): -10°C (14°F) or more TFT (Transmission fluid temperature) sensor circuit: No circuit malfunction ECT (Engine coolant temperature) sensor circuit: No circuit malfunction Turbine speed sensor circuit: No circuit malfunction Intermediate shaft speed sensor circuit: No circuit malfunction Shift solenoid valve SL1 circuit: No circuit malfunction Shift solenoid valve SL2 circuit: No circuit malfunction Shift solenoid valve SL3 circuit: No circuit malfunction Shift solenoid valve SL4 circuit: No circuit malfunction (KCS) Knock control sensor circuit: No circuit malfunction (ETCS) Electronic throttle control system: Not system down CAN communication system: Not system down
DTC: P0748 **1T ECM, MIL: Yes** **Year:** 2010, 2011 **Model:** tC, xB **Engine:** 2.4L L4, 2.5L L4	**Pressure Control Solenoid "A" Electrical (Shift Solenoid Valve SL1):** The monitor will run whenever this DTC is not present: None Engine switch: ON Starter: OFF Condition (A): Battery voltage: 12 V or more Condition (B): Battery voltage: 10 V or more and less than 12 V Target current: Less than 0.75 A Condition (C): Battery voltage: 8 V or more Target current: 0.25 A or more
DTC: P0751 **2T ECM, MIL: Yes** **Year:** 2010, 2011 **Model:** xD **Engine:** 1.8L L4 VIN U	**Shift Solenoid "A" Performance (Shift Solenoid Valve S1) :** Transmission shift position D, time after shifting N to D 4.5 seconds or more ECT (Engine Coolant Temperature) 60°C (140°F) or more. The gear required by the ECM does not match the actual gear when driving.
DTC: P0756 **2T ECM, MIL: Yes** **Year:** 2010, 2011 **Model:** xD **Engine:** 1.8L L4 VIN U	**Shift Solenoid "B" Performance (Shift Solenoid Valve S2):** Transmission shift position D, time after shifting N to D 4.5 seconds or more. ECT (Engine Coolant Temperature) 60°C (140°F) or more. The gear required by the ECM does not match the actual gear when driving.
DTC: P0776 **2T ECM, MIL: Yes** **Year:** 2010, 2011 **Model:** tC, xB **Engine:** 2.4L L4	**Pressure Control Solenoid "B" Performance (Shift Solenoid Valve SL2) :** The gear required by the ECM does not match the actual gear when driving.

DTC	Trouble Code Title and Conditions
DTC: P0776 **2T TCM, MIL: Yes** **Year:** 2011 **Model:** tC **Engine:** 2.5L L4	**Pressure Control Solenoid "B" Performance (Shift Solenoid Valve SL2):** The monitor will run whenever this DTC is not present. (Not circuit malfunction): P0115, P0117, P0118 (ECT sensor circuit) P0715, P0717 (Turbine speed sensor circuit) P0791, P0793 (Intermediate shaft speed sensor circuit) P0748 (Shift solenoid valve SL1 circuit) P0778 (Shift solenoid valve SL2 circuit) P0798 (Shift solenoid valve SL3 circuit) P2810 (Shift solenoid valve SL4 circuit) P0327, P0328, P0332, P0333 (KCS sensor circuit) P0120, P0121, P0122, P0123, P0220, P0222, P0223, P0604, P0606, P060A, P060B, P060D, P060E, P0657, P1607, P2102, P2103, P2111, P2112, P2118, P2119, P2135 ((ETCS) Electronic throttle control system) U0100 (CAN communication system) Transmission range: "D" TFT (Transmission fluid temperature): -10°C (14°F) or more TFT (Transmission fluid temperature) sensor circuit: No circuit malfunction ECT (Engine coolant temperature) sensor circuit: No circuit malfunction Turbine speed sensor circuit: No circuit malfunction Intermediate shaft speed sensor circuit: No circuit malfunction Shift solenoid valve SL1 circuit: No circuit malfunction Shift solenoid valve SL2 circuit: No circuit malfunction Shift solenoid valve SL3 circuit: No circuit malfunction Shift solenoid valve SL4 circuit: No circuit malfunction (KCS) Knock control sensor circuit: No circuit malfunction (ETCS) Electronic throttle control system: Not system down CAN communication system: Not system down
DTC: P0778 **1T ECM, MIL: Yes** **Year:** 2010, 2011 **Model:** tC, xB **Engine:** 2.4L L4	**Pressure Control Solenoid "B" Electrical (Shift Solenoid Valve SL2):** ECM checks for an open or short circuit in shift solenoid valves SL2 (1-trip detection logic) Hybrid IC for solenoid indicates fail.
DTC: P0778 **1T TCM, MIL: Yes** **Year:** 2011 **Model:** tC **Engine:** 2.5L L4	**Pressure Control Solenoid "B" Electrical (Shift Solenoid Valve SL2):** The monitor will run whenever this DTC is not present: None Engine switch: ON Starter: OFF Condition (A): Battery voltage: 12 V or more Condition (B): Battery voltage: 10 V or more and less than 12 V Target current: Less than 0.75 A Condition (C): Battery voltage: 8 V or more Target current: 0.25 A or more
DTC: P0787 **1T ECM, MIL: Yes** **Year:** 2010, 2011 **Model:** xD **Engine:** 1.8L L4 VIN U	**Shift / Timing Solenoid Low (Shift Solenoid Valve ST):** Solenoid ON, time after solenoid OFF to ON more than 0.008 seconds. Battery voltage 8 V or more, starter OFF. The ECM detects short in solenoid valve ST circuit 2-4 times when solenoid valve ST is operated.
DTC: P0788 **1T ECM, MIL: Yes** **Year:** 2010, 2011 **Model:** xD **Engine:** 1.8L L4 VIN U	**Shift / Timing Solenoid High (Shift Solenoid Valve ST):** Solenoid OFF, time after solenoid OFF to ON more than 0.008 seconds. Battery voltage 8 V or more. Starter OFF. The ECM detects open in solenoid valve ST circuit 2 times when solenoid valve ST is not operated.
DTC: P0791 **T TCM, MIL: Yes** **Year:** 2011 **Model:** tC **Engine:** 2.5L L4	**Intermediate Shaft Speed Sensor "A" Circuit:** The monitor will run whenever this DTC is not present: P0500 (VSS), P0748, P0778, P0798 (Shift solenoid valve (range)) Vehicle speed: 15.5 mph (25 km/h) or more Battery voltage: 8 V or more Engine switch: ON Starter: OFF

DTC	Trouble Code Title and Conditions
DTC: P0793 **1T ECM, MIL: Yes** **Year:** 2010, 2011 **Model:** tC, xB **Engine:** 2.4L L4	**Intermediate Shaft Speed Sensor "A":** ECM detects conditions (a), (b) and (c) continuously for 5 sec. or more: (1-trip detection logic) (a) Vehicle speed: 50 km/h (31 mph) or more (b) Park/neutral position switch (STAR) is OFF (c) Speed sensor (NC): less than 300 rpm
DTC: P0793 **T TCM, MIL: Yes** **Year:** 2011 **Model:** tC **Engine:** 2.5L L4	**Intermediate Shaft Speed Sensor "A":** The monitor will run whenever this DTC is not present: P0500 (VSS), P0748, P0778, P0798 (Shift solenoid valve (range)) Vehicle speed: 15.5 mph (25 km/h) or more Battery voltage: 8 V or more Engine switch: ON Starter: OFF
DTC: P0796 **2T TCM, MIL: Yes** **Year:** 2011 **Model:** tC **Engine:** 2.5L L4	**Pressure Control Solenoid "C" Performance (Shift Solenoid Valve SL3):** The monitor will run whenever this DTC is not present. (Not circuit malfunction): P0115, P0117, P0118 (ECT sensor circuit) P0715, P0717 (Turbine speed sensor circuit) P0791, P0793 (Intermediate shaft speed sensor circuit) P0748 (Shift solenoid valve SL1 circuit) P0778 (Shift solenoid valve SL2 circuit) P0798 (Shift solenoid valve SL3 circuit) P2810 (Shift solenoid valve SL4 circuit) P0327, P0328, P0332, P0333 (KCS sensor circuit) P0120, P0121, P0122, P0123, P0220, P0222, P0223, P0604, P0606, P060A, P060B, P060D, P060E, P0657, P1607, P2102, P2103, P2111, P2112, P2118, P2119, P2135 ((ETCS) Electronic throttle control system) U0100 (CAN communication system) Transmission range:"D" TFT (Transmission fluid temperature): -10°C (14°F) or more TFT (Transmission fluid temperature) sensor circuit: No circuit malfunction ECT (Engine coolant temperature) sensor circuit: No circuit malfunction Turbine speed sensor circuit: No circuit malfunction Intermediate shaft speed sensor circuit: No circuit malfunction Shift solenoid valve SL1 circuit: No circuit malfunction Shift solenoid valve SL2 circuit: No circuit malfunction Shift solenoid valve SL3 circuit: No circuit malfunction Shift solenoid valve SL4 circuit: No circuit malfunction (KCS) Knock control sensor circuit: No circuit malfunction (ETCS) Electronic throttle control system: Not system down CAN communication system: Not system down
DTC: P0798 **1T TCM, MIL: Yes** **Year:** 2011 **Model:** tC **Engine:** 2.5L L4	**Pressure Control Solenoid "C" Electrical (Shift Solenoid Valve SL3):** The monitor will run whenever this DTC is not present:None Engine switch: ON Starter: OFF Condition (A): Battery voltage: 12 V or more Condition (B): Battery voltage: 10 V or more and less than 12 V Target current: Less than 0.75 A Condition (C): Battery voltage: 8 V or more Target current: 0.25 A or more
DTC: P0973 **1T ECM, MIL: Yes** **Year:** 2010, 2011 **Model:** xD **Engine:** 1.8L L4 VIN U	**Shift Solenoid "A" Control Circuit Low (Shift Solenoid Valve S1):** Solenoid ON, time after solenoid OFF to ON more than 0.008 seconds. The ECM detects short in solenoid valve S1 circuit 2 times when solenoid valve S1 is operated.

DTC	Trouble Code Title and Conditions
DTC: P0974 **1T ECM, MIL: Yes** **Year:** 2010, 2011 **Model:** xD **Engine:** 1.8L L4 VIN U	**Shift Solenoid "A" Control Circuit High (Shift Solenoid Valve S1):** Solenoid OFF, time after solenoid OFF to ON more than 0.008 seconds. The ECM detects open in solenoid valve S1 circuit 2 times when solenoid valve S1 is not operated.
DTC: P0976 **1T TCM, MIL: Yes** **Year:** 2010, 2011 **Model:** xD **Engine:** 1.8L L4 VIN U	**Shift Solenoid "B" Control Circuit Low (Shift Solenoid Valve S2):** The monitor will run whenever the following DTCs are not stored: None Battery voltage: 8 V or higher Ignition switch: ON Starter: OFF Shift solenoid valve S2: ON
DTC: P0977 **1T TCM, MIL: Yes** **Year:** 2010, 2011 **Model:** xD **Engine:** 1.8L L4 VIN U	**Shift Solenoid "B" Control Circuit High (Shift Solenoid Valve S2):** The monitor will run whenever the following DTCs are not stored: None Battery voltage: 8 V or higher Ignition switch: ON Starter: OFF Shift solenoid valve S2: OFF
DTC: P0982 **1T ECM, MIL: Yes** **Year:** 2010, 2011 **Model:** tC, xB **Engine:** 2.4L L4	**Shift Solenoid "D" Control Circuit Low (Shift Solenoid Valve S4):** ECM detects short in solenoid valve S4 circuit 2 times when solenoid valve S4 is operated.
DTC: P0983 **1T ECM, MIL: Yes** **Year:** 2010, 2011 **Model:** tC, xB **Engine:** 2.4L L4	**Shift Solenoid "D" Control Circuit High (Shift Solenoid Valve S4):** ECM detects open in solenoid valve S4 circuit 2 times when solenoid valve S4 is not operated.

OBD II Trouble Code List (P0XXX Codes)

DTC	Trouble Code Title and Conditions
DTC: P101D **1T ECM, MIL: Yes** **Year:** 2010, 2011 **Model:** xD **Engine:** 1.8L L4 VIN U	**A/F Sensor Heater Circuit Performance Bank 1 Sensor 1 Stuck ON:** Battery voltage 10.5 V or higher, time after heater ON 5 seconds or more. A/F sensor heater duty-cycle 10 to 60%, A/F sensor heater current 0.8 A or higher. The heater current is higher than the specified value while the heater is not operating.
DTC: P101D **1T ECM, MIL: Yes** **Year:** 2011 **Model:** tC **Engine:** 2.5L L4	**A/F Sensor Heater Circuit Performance Bank 1 Sensor 1 Stuck ON:** Battery voltage: 10.5 V or more Time after engine start: 10 seconds or more Air fuel ration sensor heater range check low current (P0031): OK Active heater off control: Not operating Active heater on control: Not operating Air fuel ratio sensor heater duty-cycle ratio: 10 to 60% Air fuel ratio sensor heater on current: 0.8 A or more
DTC: P102D **1T ECM, MIL: Yes** **Year:** 2010, 2011 **Model:** xD **Engine:** 1.8L L4 VIN U	**O2 Sensor Heater Circuit Performance Bank 1 Sensor 2 Stuck ON:** Monitor runs whenever following DTCs are not present: None Battery voltage: 10.5 V or more Engine: Running Starter: OFF Catalyst active air fuel ratio control: Not operating Time after heater ON: 10 seconds or more Learned heater OFF current operation completed flag: ON Heated oxygen sensor heater OFF current: More than 3.5 A Hybrid IC high current limiter monitor input: Fail Heated oxygen sensor heater high current fail (P0038): Not detected

DTC	Trouble Code Title and Conditions
DTC: P102D **1T ECM, MIL: Yes** **Year:** 2011 **Model:** tC **Engine:** 2.5L L4	**O2 Sensor Heater Circuit Performance Bank 1 Sensor 2 Stuck ON:** Battery voltage: 10.5 V or more Engine: Running Starter: OFF Catalyst active air fuel ratio control: Not operating Time after heater on: 10 seconds or more Learned heater off current operation: Complete Hybrid IC high current limiter monitor input: Fail
DTC: P1603 **1T ECM, MIL: Yes** **Year:** 2010, 2011 **Model:** xB, xD **Engine:** 1.8L L4 VIN U, 2.4L L4	**Engine Stall History:** After 5 seconds or more elapse after starting the engine, with the engine running, the engine stops (the engine speed drops to 200 rpm or less) for 0.5 seconds or more, without the ignition switch or engine switch being operated.
DTC: P1604 **1T ECM, MIL: Yes** **Year:** 2010, 2011 **Model:** xD **Engine:** 1.8L L4 VIN U	**Startability Malfunction:** The engine speed is below 500 rpm with the STA signal on for a certain amount of time. After the engine starts (engine speed is 500 rpm or more), the engine speed drops to 200 rpm or less within approximately 2 seconds.
DTC: P1605 **T ECM, MIL: Yes** **Year:** 2010, 2011 **Model:** xB, xD **Engine:** 1.8L L4 VIN U, 2.4L L4	**Rough Idling:** 5 seconds or more elapse after starting the engine and the engine is running. The engine speed drops to 400 rpm or less.
DTC: P1607 **T ECM, MIL: Yes** **Year:** 2010, 2011 **Model:** xB, xD **Engine:** 1.8L L4 VIN U, 2.4L L4	**Cruise Control Input Processor:** The ECM continuously monitors its main and sub CPUs for the cruise control. This self-check ensures that the ECM is functioning properly. If outputs from the CPUs are different and deviate from the standards, the ECM will illuminate the MIL and set the DTC immediately.
DTC: P1607 **T ECM, MIL: Yes** **Year:** 2011 **Model:** tC **Engine:** 2.5L L4	**Cruise Control Input Processor:** Monitor runs whenever the following DTCs are not present: None

OBD II Trouble Code List (P0XXX Codes)

DTC	Trouble Code Title and Conditions
DTC: P2004 **2T ECM, MIL: Yes** **Year:** 2011 **Model:** tC **Engine:** 2.5L L4	**Intake Manifold Runner Control Stuck Open (Bank 1):** Monitor runs whenever following DTCs are not present: P0110 - P0113 (IAT sensor) P0115 - P0118 (ECT sensor) Battery voltage: 8 V or more ECT: -10°C (14°F) or more IAT: -10°C (14°F) or more Ignition switch: ON Command IMRC valve: Closed
DTC: P2006 **2T ECM, MIL: Yes** **Year:** 2011 **Model:** tC **Engine:** 2.5L L4	**Intake Manifold Runner Control Stuck Closed (Bank 1):** Monitor runs whenever following DTCs are not present: P0110 - P0113 (IAT sensor) P0115 - P0118 (ECT sensor) Battery voltage: 8 V or more Command IMRC valve: Closed
DTC: P2014 **1T ECM, MIL: Yes** **Year:** 2011 **Model:** tC **Engine:** 2.5L L4	**Intake Manifold Runner Position Sensor / Switch Circuit (Bank 1):** Monitor runs whenever following DTCs are not present: None IMRC valve position sensor voltage: Less than 0.2 V, or more than 4.8 V

DTC	Trouble Code Title and Conditions
DTC: P2016 **1T ECM, MIL: Yes** **Year:** 2011 **Model:** tC **Engine:** 2.5L L4	**Intake Manifold Runner Position Sensor / Switch Circuit Low (Bank 1):** Monitor runs whenever following DTCs are not present: None IMRC valve position sensor voltage: Less than 0.2 V
DTC: P2017 **1T ECM, MIL: Yes** **Year:** 2011 **Model:** tC **Engine:** 2.5L L4	**Intake Manifold Runner Position Sensor / Switch Circuit High (Bank 1):** Monitor runs whenever following DTCs are not present: None IMRC valve position sensor voltage: More than 4.8 V
DTC: P2102 **1T ECM, MIL: Yes** **Year:** 2010, 2011 **Model:** tC, xB, xD **Engine:** 1.8L L4 VIN U, 2.4L L4, 2.5L L4	**Throttle Actuator Control Motor Circuit Low:** Monitor runs whenever following DTCs are not present: None Throttle motor: 80 % or more Throttle actuator power supply: 8 V or more Current motor current-Motor current before 0.016 sec.: Less than 0.2A
DTC: P2103 **T ECM, MIL: Yes** **Year:** 2010, 2011 **Model:** tC, xB, xD **Engine:** 1.8L L4 VIN U, 2.4L L4, 2.5L L4	**Throttle Actuator Control Motor Circuit High:** Monitor runs whenever following DTCs are not present: None Throttle motor: ON Either following condition 1 or 2 is met: 1. Throttle actuator power supply: 8 V or more 2. Throttle actuator power: ON Battery voltage: 8 V or more Starter: OFF
DTC: P2109 **5T ECM, MIL: Yes** **Year:** 2010, 2011 **Model:** xB, xD **Engine:** 1.8L L4 VIN U, 2.4L L4	**Throttle / Pedal Position Sensor "A" Minimum Stop Performance:** The engine is started with a coolant temperature of 45°C (113°F) or lower, and after the engine has warmed up, the ISC learning conditions are met, or the engine has been running for at least an hour after ignition switch is ON, the engine is warmed up, and the ISC learning conditions are met. Vehicle has been driven at a speed of 30 km/h (18.65 mph) or more. Mass air flow meter is operating correctly. Atmospheric Pressure is 85 kPa (637.5 mmHg) or more (typically when altitude is 1400 m or lower).
DTC: P2111 **1T ECM, MIL: Yes** **Year:** 2010, 2011 **Model:** tC, xB, xD **Engine:** 1.8L L4 VIN U, 2.4L L4, 2.5L L4	**Throttle Actuator Control System - Stuck Open:** Monitor runs whenever following DTCs are not present: None All of the following conditions are met: System guard*:ON Throttle actuator current: 2 A or more Duty cycle to close throttle: 80% or more * System guard is ON when the following conditions are met Throttle actuator: ON Throttle actuator duty calculation: Executing Throttle position sensor: Fail determined Throttle actuator current-cut operation: Not executing Throttle actuator power supply: 4 V or more Throttle actuator: Fail determined
DTC: P2112 **1T ECM, MIL: Yes** **Year:** 2010, 2011 **Model:** tC, xB, xD **Engine:** 1.8L L4 VIN U, 2.4L L4, 2.5L L4	**Throttle Actuator Control System - Stuck Closed:** Monitor runs whenever following DTCs are not present: None All of the following conditions are met: System guard*:ON Throttle actuator current: 2 A or more Duty cycle to open throttle: 80% or more * System guard is ON when the following conditions are met Throttle actuator: ON Throttle actuator duty calculation: Executing Throttle position sensor: Fail determined Throttle actuator current-cut operation: Not executing Throttle actuator power supply: 4 V or more Throttle actuator: Fail determined

DTC	Trouble Code Title and Conditions
DTC: P2118 **1T ECM, MIL: Yes** **Year:** 2010, 2011 **Model:** tC, xB, xD **Engine:** 1.8L L4 VIN U, 2.4L L4, 2.5L L4	**Throttle Actuator Control Motor Current Range / Performance:** Monitor runs whenever following DTCs are not present: None Battery voltage: 8 V or more Throttle actuator power: ON
DTC: P2119 **1T ECM, MIL: Yes** **Year:** 2010, 2011 **Model:** tC, xB, xD **Engine:** 1.8L L4 VIN U, 2.4L L4, 2.5L L4	**Throttle Actuator Control Throttle Body Range / Performance:** Monitor runs whenever following DTCs are not present: None System guard*: ON *System guard is set when following conditions are met: Throttle actuator: ON Throttle actuator duty calculation: Executing TP sensor: Fail determined Throttle actuator current cut operation: Not executing Throttle actuator power supply: 4 V or more Throttle actuator: Fail determined
DTC: P2120 **1T ECM, MIL: Yes** **Year:** 2010, 2011 **Model:** tC, xB, xD **Engine:** 1.8L L4 VIN U, 2.4L L4, 2.5L L4	**Throttle / Pedal Position Sensor / Switch "D" Circuit:** VPA fluctuates rapidly beyond upper and lower malfunction thresholds for 0.5 seconds or more, No other DTC's are present.
DTC: P2121 **1T ECM, MIL: Yes** **Year:** 2010, 2011 **Model:** tC, xB, xD **Engine:** 1.8L L4 VIN U, 2.4L L4, 2.5L L4	**Throttle / Pedal Position Sensor / Switch "D" Circuit Range / Performance:** Monitor runs whenever following DTCs are not present: None Either of the following conditions met: Condition (a) or (b) (a) Engine switch: ON (b) Throttle actuator power: ON
DTC: P2122 **1T ECM, MIL: Yes** **Year:** 2010, 2011 **Model:** tC, xB, xD **Engine:** 1.8L L4 VIN U, 2.4L L4, 2.5L L4	**Throttle / Pedal Position Sensor / Switch "D" Circuit Low Input:** Monitor runs whenever following DTCs are not present: None VPA 0.4 V or less for 0.5 seconds or more when accelerator pedal depressed.
DTC: P2123 **1T ECM, MIL: Yes** **Year:** 2010, 2011 **Model:** tC, xB, xD **Engine:** 1.8L L4 VIN U, 2.4L L4, 2.5L L4	**Throttle / Pedal Position Sensor / Switch "D" Circuit High Input:** Monitor runs whenever following DTCs are not present: None VPA 4.8 V or more for 2.0 seconds or more.
DTC: P2125 **1T ECM, MIL: Yes** **Year:** 2010, 2011 **Model:** tC, xB, xD **Engine:** 1.8L L4 VIN U, 2.4L L4, 2.5L L4	**Throttle / Pedal Position Sensor / Switch "E" Circuit:** Monitor runs whenever following DTCs are not present: None VPA2 1.2 V or less for 0.5 seconds or more when accelerator pedal depressed.
DTC: P2127 **1T ECM, MIL: Yes** **Year:** 2010, 2011 **Model:** tC, xB, xD **Engine:** 1.8L L4 VIN U, 2.4L L4, 2.5L L4	**Throttle / Pedal Position Sensor / Switch "E" Circuit Low Input:** Monitor runs whenever following DTCs are not present: None
DTC: P2128 **1T ECM, MIL: Yes** **Year:** 2010, 2011 **Model:** tC, xB, xD **Engine:** 1.8L L4 VIN U, 2.4L L4, 2.5L L4	**Throttle / Pedal Position Sensor / Switch "E" Circuit High Input:** Monitor runs whenever following DTCs are not present: None With the ignition switch ON for 0.012 seconds or more and throttle actuator power ON. Conditions (a) and (b) continue for 2.0 seconds or more: (a) VPA2 4.8 V or more (b) VPA between 0.4 V and 3.45 V

DTC	Trouble Code Title and Conditions
DTC: P2135 **1T ECM, MIL: Yes** **Year:** 2010, 2011 **Model:** tC, xB, xD **Engine:** 1.8L L4 VIN U, 2.4L L4, 2.5L L4	**Throttle / Pedal Position Sensor / Switch "A" / "B" Voltage Correlation:** Monitor runs whenever following DTCs are not present: None Either of the following conditions A or B is met: A. Engine switch on (IG): 0.012 seconds or more B. Electronic throttle actuator power: ON
DTC: P2138 **1T ECM, MIL: Yes** **Year:** 2010, 2011 **Model:** tC, xB, xD **Engine:** 1.8L L4 VIN U, 2.4L L4, 2.5L L4	**Throttle / Pedal Position Sensor / Switch "D" / "E" Voltage Correlation:** Monitor runs whenever following DTCs are not present: None
DTC: P2195 **2T ECM, MIL: Yes** **Year:** 2010, 2011 **Model:** tC, xB, xD **Engine:** 1.8L L4 VIN U, 2.4L L4, 2.5L L4	**Oxygen (A/F) Sensor Signal Stuck Lean (Bank 1 Sensor 1):** Monitor runs whenever following DTCs are not present: P0016, P0018 (VVT system bank 1, 2 - misalignment) P0017, P0019 (Exhaust VVT system bank 1, 2 - misalignment) P0031, P0032, P0051, P0052 (A/F sensor heater) P0102, P0103 (MAF meter) P0110, P0112, P0113 (IAT sensor) P0115, P0117, P0118 (ECT sensor) P0120, P0121, P0122, P0123, P0220, P0222, P0223, P2135 (TP sensor) P0125 (Insufficient ECT for closed loop) P0128 (Thermostat) P0171, P0172, P0174, P0175 (Fuel system) P0301, P0302, P0303, P0304, P0305, P0306 (Misfire) P0335 (CKP sensor) P0451, P0452, P0452 (EVAP system) P0500 (Vehicle speed sensor) P0505 (Idle speed control) Sensor voltage detection monitor (Lean side malfunction) Time after engine start: 30 seconds or more Fuel system status: Closed loop Sensor current detection monitor (High and low side malfunction): Battery voltage: 11 V or more Atmospheric pressure: 76 kPa (570 mmHg) or more A/F sensor status: Activated Continuous time of fuel cut: 3-10 seconds ECT: 167°F (75°C) or more

DTC	Trouble Code Title and Conditions
DTC: P2196 **2T ECM, MIL: Yes** **Year:** 2010, 2011 **Model:** tC, xB, xD **Engine:** 1.8L L4 VIN U, 2.4L L4, 2.5L L4	**Oxygen (A/F) Sensor Signal Stuck Rich (Bank 1 Sensor 1):** Monitor runs whenever following DTCs are not present: P0016, P0018 (VVT system bank 1, 2 - misalignment) P0017, P0019 (Exhaust VVT system bank 1, 2 - misalignment) P0031, P0032, P0051, P0052 (A/F sensor heater) P0102, P0103 (MAF meter) P0110, P0112, P0113 (IAT sensor) P0115, P0117, P0118 (ECT sensor) P0120, P0121, P0122, P0123, P0220, P0222, P0223, P2135 (TP sensor) P0125 (Insufficient ECT for closed loop) P0128 (Thermostat) P0171, P0172, P0174, P0175 (Fuel system) P0301, P0302, P0303, P0304, P0305, P0306 (Misfire) P0335 (CKP sensor) P0451, P0452, P0452 (EVAP system) P0500 (Vehicle speed sensor) P0505 (Idle speed control) Sensor voltage detection monitor (Rich side malfunction): Time after engine start: 30 seconds or more Fuel system status: Closed loop Sensor current detection monitor (High and low side malfunction): Battery voltage: 11 V or more Atmospheric pressure: 76 kPa (570 mmHg) or more A/F sensor status: Activated Continuous time of fuel cut: 3-10 seconds ECT: 167°F (75°C) or more
DTC: P219A **1T ECM, MIL: Yes** **Year:** 2011 **Model:** tC **Engine:** 2.5L L4	**Bank 1 Air-Fuel Ratio Imbalance:** The difference in air-fuel ratios between the cylinders exceeds the threshold.
DTC: P2237 **2T ECM, MIL: Yes** **Year:** 2010, 2011 **Model:** tC, xB, xD **Engine:** 1.8L L4 VIN U, 2.4L L4, 2.5L L4	**Oxygen (A/F) Sensor Pumping Current Circuit / Open (Bank 1 Sensor 1):** Monitor runs whenever following DTCs are not present: P0016, P0018 (VVT system bank 1, 2 - misalignment) P0017, P0019 (Exhaust VVT system bank 1, 2 - misalignment) P0031, P0032, P0051, P0052 (A/F sensor heater) P0102, P0103 (MAF meter) P0110, P0112, P0113 (IAT sensor) P0115, P0117, P0118 (ECT sensor) P0120, P0121, P0122, P0123, P0220, P0222, P0223, P2135 (TP sensor) P0125 (Insufficient ECT for closed loop) P0128 (Thermostat) P0171, P0172, P0174, P0175 (Fuel system) P0301, P0302, P0303, P0304, P0305, P0306 (Misfire) P0335 (CKP sensor) P0340 (CMP sensor) P0451, P0452, P0452 (EVAP system) P0500 (Vehicle speed sensor) P0505 (Idle speed control) Open circuit between AF+ and AF-: Estimated sensor temperature: 842-1022°F (450-550°C)
DTC: P2238 **2T ECM, MIL: Yes** **Year:** 2010, 2011 **Model:** xB **Engine:** 2.4L L4	**Oxygen (A/F) Sensor Pumping Current Circuit Low (Bank 1 Sensor 1):** Air Fuel Ratio (A/F) sensor output drops while engine is running. Voltage at terminal A1A+ is 0.5 V or less. Voltage difference between terminals A1A+ and A1A- is 0.1 V or less.

DTC	Trouble Code Title and Conditions
DTC: P2238 **2T ECM, MIL: Yes** **Year:** 2011 **Model:** tC **Engine:** 2.5L L4	**Oxygen (A/F) Sensor Pumping Current Circuit Low (Bank 1 Sensor 1):** Monitor runs whenever following DTCs are not present: P0016, P0018 (VVT system bank 1, 2 - misalignment) P0017, P0019 (Exhaust VVT system bank 1, 2 - misalignment) P0031, P0032, P0051, P0052 (A/F sensor heater) P0102, P0103 (MAF meter) P0110, P0112, P0113 (IAT sensor) P0115, P0117, P0118 (ECT sensor) P0120, P0121, P0122, P0123, P0220, P0222, P0223, P2135 (TP sensor) P0125 (Insufficient ECT for closed loop) P0128 (Thermostat) P0171, P0172, P0174, P0175 (Fuel system) P0301, P0302, P0303, P0304, P0305, P0306 (Misfire) P0335 (CKP sensor) P0340 (CMP sensor) P0451, P0452, P0452 (EVAP system) P0500 (Vehicle speed sensor) P0505 (Idle speed control) Open circuit between AF+ and AF-: Estimated sensor temperature: 1292-1472°F (700-800°C) ECT: 41°F (5°C) or more (varies with ECT at engine start) Fuel cut: Not executed
DTC: P2239 **2T ECM, MIL: Yes** **Year:** 2010, 2011 **Model:** tC, xB, xD **Engine:** 1.8L L4 VIN U, 2.4L L4, 2.5L L4	**Oxygen (A/F) Sensor Pumping Current Circuit High (Bank 1 Sensor 1):** Monitor runs whenever following DTCs are not present: P0016, P0018 (VVT system bank 1, 2 - misalignment) P0017, P0019 (Exhaust VVT system bank 1, 2 - misalignment) P0031, P0032, P0051, P0052 (A/F sensor heater) P0102, P0103 (MAF meter) P0110, P0112, P0113 (IAT sensor) P0115, P0117, P0118 (ECT sensor) P0120, P0121, P0122, P0123, P0220, P0222, P0223, P2135 (TP sensor) P0125 (Insufficient ECT for closed loop) P0128 (Thermostat) P0171, P0172, P0174, P0175 (Fuel system) P0301, P0302, P0303, P0304, P0305, P0306 (Misfire) P0335 (CKP sensor) P0340 (CMP sensor) P0451, P0452, P0452 (EVAP system) P0500 (Vehicle speed sensor) P0505 (Idle speed control) Battery voltage: 11 V or more Engine switch: ON Time after engine switch OFF to ON: 5 seconds or more

DTC	Trouble Code Title and Conditions
DTC: P2252 **2T ECM, MIL: Yes** **Year:** 2010, 2011 **Model:** tC, xB, xD **Engine:** 1.8L L4 VIN U, 2.4L L4, 2.5L L4	**Oxygen (A/F) Sensor Reference Ground Circuit Low (Bank 1 Sensor 1):** Monitor runs whenever following DTCs are not present: P0016, P0018 (VVT system bank 1, 2 - misalignment) P0017, P0019 (Exhaust VVT system bank 1, 2 - misalignment) P0031, P0032, P0051, P0052 (A/F sensor heater) P0102, P0103 (MAF meter) P0110, P0112, P0113 (IAT sensor) P0115, P0117, P0118 (ECT sensor) P0120, P0121, P0122, P0123, P0220, P0222, P0223, P2135 (TP sensor) P0125 (Insufficient ECT for closed loop) P0128 (Thermostat) P0171, P0172, P0174, P0175 (Fuel system) P0301, P0302, P0303, P0304, P0305, P0306 (Misfire) P0335 (CKP sensor) P0340 (CMP sensor) P0451, P0452, P0452 (EVAP system) P0500 (Vehicle speed sensor) P0505 (Idle speed control) Battery voltage: 11 V or more Engine switch: ON Time after engine switch OFF to ON: 5 seconds or more
DTC: P2253 **2T ECM, MIL: Yes** **Year:** 2010, 2011 **Model:** tC, xB, xD **Engine:** 1.8L L4 VIN U, 2.4L L4, 2.5L L4	**Oxygen (A/F) Sensor Reference Ground Circuit High (Bank 1 Sensor 1):** Monitor runs whenever following DTCs are not present: P0016, P0018 (VVT system bank 1, 2 - misalignment) P0017, P0019 (Exhaust VVT system bank 1, 2 - misalignment) P0031, P0032, P0051, P0052 (A/F sensor heater) P0102, P0103 (MAF meter) P0110, P0112, P0113 (IAT sensor) P0115, P0117, P0118 (ECT sensor) P0120, P0121, P0122, P0123, P0220, P0222, P0223, P2135 (TP sensor) P0125 (Insufficient ECT for closed loop) P0128 (Thermostat) P0171, P0172, P0174, P0175 (Fuel system) P0301, P0302, P0303, P0304, P0305, P0306 (Misfire) P0335 (CKP sensor) P0340 (CMP sensor) P0451, P0452, P0452 (EVAP system) P0500 (Vehicle speed sensor) P0505 (Idle speed control) Battery voltage: 11 V or more Engine switch: ON Time after engine switch OFF to ON: 5 seconds or more
DTC: P2401 **2T ECM, MIL: Yes** **Year:** 2010, 2011 **Model:** tC, xB, xD **Engine:** 1.8L L4 VIN U, 2.4L L4, 2.5L L4	**Evaporative Emission System Leak Detection Pump Control Circuit Low:** Reference orifice clogged. Reference orifice high-flow. Leak detection pump OFF malfunction. Leak detection pump ON malfunction. Vent valve ON (close) malfunction.
DTC: P2402 **2T ECM, MIL: Yes** **Year:** 2010, 2011 **Model:** tC, xB, xD **Engine:** 1.8L L4 VIN U, 2.4L L4, 2.5L L4	**Evaporative Emission System Leak Detection Pump Control Circuit High:** Reference orifice clogged. Reference orifice high-flow. Leak detection pump OFF malfunction. Leak detection pump ON malfunction. Vent valve ON (close) malfunction.

DTC	Trouble Code Title and Conditions
DTC: P2419 **2T ECM, MIL: Yes** **Year:** 2010, 2011 **Model:** tC, xB, xD **Engine:** 1.8L L4 VIN U, 2.4L L4, 2.5L L4	**Evaporative Emission System Switching Valve Control Circuit Low:** Reference orifice clogged. Reference orifice high-flow. Leak detection pump OFF malfunction. Leak detection pump ON malfunction. Vent valve ON (close) malfunction.
DTC: P2420 **2T ECM, MIL: Yes** **Year:** 2010, 2011 **Model:** tC, xB, xD **Engine:** 1.8L L4 VIN U, 2.4L L4, 2.5L L4	**Evaporative Emission System Switching Valve Control Circuit High:** The monitor will run whenever these DTCs are not present: None Key-off monitor is run when all of the following conditions are met: Atmospheric pressure: 70 to 110 kPa (525 to 825 mmHg) Battery voltage: 10.5 V or more Vehicle speed: 2.5 mph (4 km/h) or less Engine switch: OFF Time after key-off: 5 or 7 or 9.5 hours EVAP pressure sensor malfunction (P0452, P0453): Not detected EVAP canister purge valve: Not operated by scan tool EVAP canister vent valve: Not operated by scan tool EVAP leak detection pump: Not operated by scan tool Both of the following conditions 1 and 2 are met before key-off: 1. Duration that vehicle has been driven: 5 minutes or more 2. EVAP purge operation: Performed ECT: 40-95°F (4.4-35°C) IAT: 40-95°F (4.4-35°C)
DTC: P2610 **2T ECM, MIL: Yes** **Year:** 2010, 2011 **Model:** tC, xB, xD **Engine:** 1.8L L4 VIN U, 2.4L L4, 2.5L L4	**ECM / PCM Internal Engine Off Timer Performance:** Case 1: Engine switch: ON Engine: Running Battery voltage: 8 V or more Starter: OFF Case 2: Internal engine OFF timer (elapsed time from engine stop): 10-30 minutes Battery voltage: 8 V or more Engine switch: ON Starter: OFF Case 3: Internal engine OFF timer (elapsed time from engine stop): 40 minutes Battery voltage: 8 V or more Engine switch: ON Starter: OFF

DTC	Trouble Code Title and Conditions
DTC: P2714 **2T , MIL: Yes** **Year:** 2010, 2011 **Model:** tC, xB, xD **Engine:** 1.8L L4 VIN U, 2.4L L4, 2.5L L4	**Pressure Control Solenoid "D" Performance (Shift Solenoid Valve SLT):** The monitor will run whenever this DTC is not present. (Not circuit malfunction): P0712, P0713 (ATF temperature sensor circuit (TFT sensor)) P0115, P0117, P0118 (ECT sensor circuit) P0715, P0717 (Turbine speed sensor circuit) P0791, P0793 (Intermediate shaft speed sensor circuit) P0748 (Shift solenoid valve SL1 circuit) P0778 (Shift solenoid valve SL2 circuit) P0798 (Shift solenoid valve SL3 circuit) P2810 (Shift solenoid valve SL4 circuit) P2716 (Shift solenoid valve SLT circuit) P0327, P0328, P0332, P0333 (KCS sensor circuit) P0120, P0121, P0122, P0123, P0220, P0222, P0223, P0604, P0606, P060A, P060B, P060D, P060E, P0657, P1607, P2102, P2103, P2111, P2112, P2118, P2119, P2135 ((ETCS) Electronic throttle control system) U0100 (CAN communication system) Transmission range: "D" Duration time from shifting "N" to "D": 4 sec. or more TFT (Transmission fluid temperature): -10°C (14°F) or more TFT (Transmission fluid temperature) sensor circuit: No circuit malfunction ECT (Engine coolant temperature) sensor circuit: No circuit malfunction Turbine speed sensor circuit: No circuit malfunction Intermediate shaft speed sensor circuit: No circuit malfunction Shift solenoid valve SL1 circuit: No circuit malfunction Shift solenoid valve SL2 circuit: No circuit malfunction Shift solenoid valve SL3 circuit: No circuit malfunction Shift solenoid valve SL4 circuit: No circuit malfunction Shift solenoid valve SLT circuit: No circuit malfunction (KCS) Knock control sensor circuit: No circuit malfunction (ETCS) Electronic throttle control system: Not system down CAN communication system: Not system down Engine: Starting Input turbine torque: 100 N*m or more Turbine speed: 250 rpm or more Output speed: 250 rpm or more
DTC: P2716 **1T TCM, MIL: Yes** **Year:** 2011 **Model:** tC **Engine:** 2.5L L4	**Pressure Control Solenoid "D" Electrical (Shift Solenoid Valve SLT):** The monitor will run whenever this DTC is not present: None Solenoid current cut status: Not cut Engine switch: ON Starter: OFF Malfunction (A): Battery voltage: 10-12 V Malfunction (B): Battery voltage: 10 V or more and less than 12 V Target current: 0.75 A Malfunction (C): Battery voltage: 8 V or more Target current: 1 A Malfunction (D): Battery voltage: 11 V or more Target current: 1 A or more Malfunction (E): Battery voltage: 11 V or more Target current: 0.1 A or more Commanded voltage - last commanded voltage: Less than 0.000019 V

DTC	Trouble Code Title and Conditions
DTC: P2716 **1T ECM, MIL: Yes** **Year:** 2010, 2011 **Model:** tC, xB, xD **Engine:** 1.8L L4 VIN U, 2.4L L4	**Pressure Control Solenoid "D" Electrical (Shift Solenoid Valve SLT):** Open or short is detected in shift solenoid valve SLT circuit for 1 second or more while driving (1-trip detecting logic). Solenoid current cut status: Not cut Battery voltage 11 V or more, ignition switch ON, starter OFF, CPU commanded duty ratio to SLT 19% or more.
DTC: P2757 **2T TCM, MIL: Yes** **Year:** 2010, 2011 **Model:** tC, xD **Engine:** 1.8L L4 VIN U, 2.5L L4	**Torque Converter Clutch Pressure Control Solenoid Performance (Shift Solenoid Valve SLU):** The monitor will run whenever this DTC is not present. (Not circuit malfunction): P0712, P0713 (ATF temperature sensor circuit (TFT sensor)) P0115, P0117, P0118 (ECT sensor circuit) P0715, P0717 (Turbine speed sensor circuit) P0791, P0793 (Intermediate shaft speed sensor circuit) P0748 (Shift solenoid valve SL1 circuit) P0778 (Shift solenoid valve SL2 circuit) P0798 (Shift solenoid valve SL3 circuit) P2810 (Shift solenoid valve SL4 circuit) P2759 (Shift solenoid valve SLU circuit) P2769, P2770 (Shift solenoid valve SL circuit) P0327, P0328, P0332, P0333 (KCS sensor circuit) P0120, P0121, P0122, P0123, P0220, P0222, P0223, P0604, P0606, P060A, P060B, P060D, P060E, P0657, P1607, P2102, P2103, P2111, P2112, P2118, P2119, P2135 ((ETCS) Electronic throttle control system) U0100 (CAN communication system) ECT (Engine coolant temperature): 40°C (104°F) or more Spark advance from Max. retard timing by KCS control: 0°CA or more Transmission range: "D" TFT (Transmission fluid temperature): -10°C (14°F) or more TFT (Transmission fluid temperature) sensor circuitL: No circuit malfunction ECT (Engine coolant temperature) sensor circuit: No circuit malfunction Turbine speed sensor circuit: No circuit malfunction Intermediate shaft speed sensor circuit: No circuit malfunction Shift solenoid valve SL1 circuit: No circuit malfunction Shift solenoid valve SL2 circuit: No circuit malfunction Shift solenoid valve SL3 circuit:n: No circuit malfunction Shift solenoid valve SL4 circuit: No circuit malfunction Shift solenoid valve SLU circuit: No circuit malfunction Shift solenoid valve SL circuit: No circuit malfunction (KCS) Knock control sensor circuit: No circuit malfunction (ETCS) Electronic throttle control system: Not system down CAN communication system: Not system down TCM selected gear: Not 1st Vehicle speed: 5.5 mph (25 km/h) or more Turbine speed/Output speed (NT/NO) with 1st: 3.304 to 7.724 Turbine speed/Output speed (NT/NO) with 2nd: 1.901 to 2.340 Turbine speed/Output speed (NT/NO) with 3rd: 1.399 to 1.649 Turbine speed/Output speed (NT/NO) with 4th: 0.998 to 1.138 Turbine speed/Output speed (NT/NO) with 5th: 0.705 to 0.836 Turbine speed/Output speed (NT/NO) with 6th: 0.568 to 0.695
DTC: P2759 **1T TCM, MIL: Yes** **Year:** 2011 **Model:** tC **Engine:** 2.5L L4	**Torque Converter Clutch Pressure Control Solenoid Control Circuit Electrical (Shift Solenoid Valve SLU):** The monitor will run whenever this DTC is not present: None Engine switch: ON Starter: OFF Condition (A): Battery voltage: 12 V or more Condition (B): Battery voltage: 10 V or more and less than 12 V Target current: Less than 0.75 A Condition (C): Battery voltage: 8 V or more Target current: 0.25 A or more

DTC	Trouble Code Title and Conditions
DTC: P2759 **1T ECM, MIL: Yes** **Year:** 2010, 2011 **Model:** xD **Engine:** 1.8L L4 VIN U	**Torque Converter Clutch Pressure Control Solenoid Control Circuit Electrical (Shift Solenoid Valve SLU):** The monitor will run whenever this DTC is not present: None Solenoid current cut status: Not cut CPU commanded duty: 19% or more Battery voltage: 8 V or more Engine switch: ON Starter: OFF
DTC: P2769 **2T TCM, MIL: Yes** **Year:** 2011 **Model:** tC **Engine:** 2.5L L4	**Short in Torque Converter Clutch Solenoid Circuit (Shift Solenoid Valve SL):** The monitor will run whenever the following DTCs are not present: None Solenoid: ON Time after solenoid OFF to ON: More than 0.008 sec.
DTC: P2769 **2T ECM** **Year:** 2010, 2011 **Model:** tC, xB **Engine:** 2.4L L4	**Torque Converter Clutch Solenoid Circuit Low (Shift Solenoid Valve DSL):** The ECM detects a short in the solenoid valve DSL circuit (0.1 sec.) when solenoid valve DSL is operated. Shift solenoid valve DSL: ON, Solenoid current cut status: Not cut, battery voltage 8 V or more, ignition switch ON, starter OFF.
DTC: P2770 **2T ECM, MIL: Yes** **Year:** 2010, 2011 **Model:** tC, xB **Engine:** 2.4L L4	**Torque Converter Clutch Solenoid Circuit High (Shift Solenoid Valve DSL):** The ECM detects a short in the solenoid valve DSL circuit (0.1 sec.) when solenoid valve DSL is operated. Shift solenoid valve DSL: ON, Solenoid current cut status: Not cut, battery voltage 8 V or more, ignition switch ON, starter OFF.
DTC: P2770 **2T TCM, MIL: Yes** **Year:** 2011 **Model:** tC **Engine:** 2.5L L4	**Open in Torque Converter Clutch Solenoid Circuit (Shift Solenoid Valve SL):** The monitor will run whenever the following DTCs are not present: None Solenoid: ON Time after solenoid ON to OFF: More than 0.008 sec.
DTC: P2808 **2T TCM, MIL: Yes** **Year:** 2011 **Model:** tC **Engine:** 2.5L L4	**Pressure Control Solenoid "G" Performance (Shift Solenoid Valve SL4):** The monitor will run whenever this DTC is not present. (Not circuit malfunction): P0115, P0117, P0118 (ECT sensor circuit) P0715, P0717 (Turbine speed sensor circuit) P0791, P0793 (Intermediate shaft speed sensor circuit) P0748 (Shift solenoid valve SL1 circuit) P0778 (Shift solenoid valve SL2 circuit) P0798 (Shift solenoid valve SL3 circuit) P2810 (Shift solenoid valve SL4 circuit) P0327, P0328, P0332, P0333 (KCS sensor circuit) P0120, P0121, P0122, P0123, P0220, P0222, P0223, P0604, P0606, P060A, P060B, P060D, P060E, P0657, P1607, P2102, P2103, P2111, P2112, P2118, P2119, P2135 ((ETCS) Electronic throttle control system) U0100 (CAN communication system) Transmission range: "D" TFT (Transmission fluid temperature): -10°C (14°F) or more TFT (Transmission fluid temperature) sensor circuit: No circuit malfunction ECT (Engine coolant temperature) sensor circuit: No circuit malfunction Turbine speed sensor circuit: No circuit malfunction Intermediate shaft speed sensor circuit: No circuit malfunction Shift solenoid valve SL1 circuit: No circuit malfunction Shift solenoid valve SL2 circuit: No circuit malfunction Shift solenoid valve SL3 circuit: No circuit malfunction Shift solenoid valve SL4 circuit: No circuit malfunction (KCS) Knock control sensor circuit: No circuit malfunction (ETCS) Electronic throttle control system: Not system down CAN communication system: Not system down

DTC	Trouble Code Title and Conditions
DTC: P2810 **1T TCM, MIL: Yes** **Year:** 2011 **Model:** tC **Engine:** 2.5L L4	**Pressure Control Solenoid "G" Electrical (Shift Solenoid Valve SL4):** The monitor will run whenever this DTC is not present: None Engine switch: ON Starter: OFF Condition (A): Battery voltage: 12 V or more Condition (B): Battery voltage: 10 V or more and less than 12 V Target condition: Less than 0.75 A Condition (C): Battery voltage: 8 V or more Target condition: 0.25 A or more
DTC: P2A00 **2T ECM, MIL: Yes** **Year:** 2010, 2011 **Model:** tC, xB, xD **Engine:** 1.8L L4 VIN U, 2.4L L4, 2.5L L4	**A/F Sensor Circuit Slow Response (Bank 1 Sensor 1):** Monitor runs whenever following DTCs are not present: P0016, P0018 (VVT system bank 1, 2 - misalignment) P0017, P0019 (Exhaust VVT system bank 1, 2 - misalignment) P0031, P0032, P0051, P0052 (A/F sensor heater) P0102, P0103 (MAF meter) P0110, P0112, P0113 (IAT sensor) P0115, P0117, P0118 (ECT sensor) P0120, P0121, P0122, P0123, P0220, P0222, P0223, P2135 (TP sensor) P0125 (Insufficient ECT for closed loop) P0128 (Thermostat) P0171, P0172, P0174, P0175 (Fuel system) P0301, P0302, P0303, P0304, P0305, P0306 (Misfire) P0335 (CKP sensor) P0340 (CMP sensor) P0451, P0452, P0452 (EVAP system) P0500 (Vehicle speed sensor) P0505 (Idle speed control) Active A/F control: Performing Battery voltage: 11 V or more ECT: 167°F (75°C) or more Idle: OFF Engine rpm: Less than 4000 rpm A/F sensor status: Activated Fuel cut: OFF Engine load: 10-70% Shift position: 2 or more Catalyst monitor: Not yet MAF: 2.5-15 g/sec

TOYOTA

.4Runner

3

SPECIFICATIONS AND MAINTENANCE CHARTS

ENGINE AND VEHICLE IDENTIFICATION

Engine							Model Year	
Code ①	Liters (cc)	Cu. In.	Cyl.	Fuel Sys.	Engine Type	Eng. Mfg.	Code ②	Year
2TR-FE	2.7 (2697)	164.4	4	SFI	DOHC	Toyota	B	2011
1GR-FE	4.0 (3956)	241	6	SFI	DOHC	Toyota	C	2012

SFI: Sequential Fuel Injection

DOHC: Double Overhead Camshaft

① Stamped on the left side of the engine block

② 10th digit of the Vehicle Identification Number (VIN)

71099_4RUN_C0001

GENERAL ENGINE SPECIFICATIONS

Year	Model	Engine Displacement Liters	Engine Series ID	Net Horsepower @ rpm	Net Torque @ rpm (ft. lbs.)	Bore x Stroke (in.)	Com-pression Ratio	Oil Pressure @ rpm
2011	4Runner	4.0	1GR-FE	270@5600	278@4400	3.70x3.74	10.4:01	43-85@3000
	4Runner	2.7	2TR-FE	157@5200	178@3800	3.74x3.74	9.6:01	23-71@3000
2012	4Runner	4.0	1GR-FE	270@5600	278@4400	3.70x3.74	10.4:01	43-85@3000

71099_4RUN_C0002

ENGINE TUNE-UP SPECIFICATIONS

Year	Engine Displacement Liters	Engine ID	Spark Plug Gap (in.)	Ignition Timing (deg.)*	Fuel Pump (psi)	Idle Speed (rpm)		Valve Clearance	
						MT	AT	Intake	Exhaust
2011	4.0	1GR-FE	0.043	N/A	46.5-47.4	—	690-790	NA	NA
	2.7	2TR-FE	0.043	N/A	41-42	—	600-700	NA	NA
2012	4.0	1GR-FE	0.043	N/A	46.5-47.4	—	690-790	NA	NA

NOTE: The Vehicle Emission Control Information label often reflects specification changes made during production. The label figures must be used if they differ from those in this chart.

* With terminals TC and E1 connected to DLC1 or for 4.7L Terminal TC and CG of DLC3 connected

71099_4RUN_C0003

CAPACITIES

Year	Model	Engine Displacement Liters	Engine ID	Engine Oil with Filter (qts.)	Transmission (qts.) 5-Spd	Transmission (qts.) Auto.*	Transfer Case (pts.)	Drive Axle Front (pts.)	Drive Axle Rear (pts.)	Fuel Tank (gal.)	Cooling System (qts.)
2011	4Runner	4.0	1GR-FE	5.5	NA	11.3	3.0	3.2	6.4	23.0	11.1
	4Runner	2.7	2TR-FE	5.1	NA	10.5	2.2	3.0	6.4	23.0	8.6
2012	4Runner	4.0	1GR-FE	5.5	NA	11.3	3.0	3.2	6.4	23.0	11.1

*After draining, add the following amounts, then fill to the cold full line

NA: Not available

71099_4RUN_C0004

FLUID SPECIFICATIONS

Year	Model	Engine Displacement Liters	Engine ID/VIN	Engine Oil	Auto. Trans.	Power Steering Fluid	Brake Master Cylinder
2011	4Runner	4.0	1GR-FE	5W-30	Toyota Genuine ATF WS	ATF Dexron® II or III	DOT 3
		2.7	2TR-FE	5W-30	Toyota Genuine ATF WS	ATF Dexron® II or III	DOT 3
2012	4Runner	4.0	1GR-FE	5W-30	Toyota Genuine ATF WS	ATF Dexron® II or III	DOT 3

DOT: Department Of Transpotation

NA: Not Available

71099_4RUN_C0014

VALVE SPECIFICATIONS

Year	Engine Displacement Liters	Engine ID	Seat Angle (deg.)	Face Angle (deg.)	Spring Test Pressure (lbs. @ in.)	Spring Installed Height (in.)	Stem-to-Guide Clearance (in.) Intake	Stem-to-Guide Clearance (in.) Exhaust	Stem Diameter (in.) Intake	Stem Diameter (in.) Exhaust
2011	4.0	1GR-FE	NA	NA	53-58.3@ 1.45	1.311	0.0010-0.0024	0.0012-0.0026	0.2154-0.2159	0.2152-0.0158
	2.7	2TR-FE	NA	NA	47.2-50.7@ 1.378	1.380	0.0010-0.0024	0.0012-0.0026	0.2154-0.2159	0.2152-0.2157
2012	4.0	1GR-FE	NA	NA	53-58.3@ 1.45	1.311	0.0010-0.0024	0.0012-0.0026	0.2154-0.2159	0.2152-0.0158

71099_4RUN_C0005

CAMSHAFT AND BEARING SPECIFICATIONS CHART

All measurements are given in inches.

Year	Engine Displacement Liters	Engine ID/VIN	Journal Dia.	Brg. Oil Clearance	Shaft End-play	Runout	Journal Bore	Lobe Height	
								Intake	Exhaust
2011	4.0	1GR-FE	①	②	NA	NA	NA	1.728-1.7320	1.743-1.7470
	2.7	2TR-FE	1.415-1.4160	NA	NA	0.00118	NA	1.687-1.6910	1.687-1.6910
2012	4.0	1GR-FE	①	②	NA	NA	NA	1.728-1.7320	1.743-1.7470

NA: Not Available

① No. 1 journal: 1.4152-1.4157

Other Journals: 1.0221-1.0226 in.

② No. 1 journal: 0.00126-0.00248 in.

Other journal: 0.000984-0.00244 in.

Other Maximum Journals: 0.00276 in.

71099_4RUN_C0013

CRANKSHAFT AND CONNECTING ROD SPECIFICATIONS

All measurements are given in inches.

Year	Engine Displacement Liters	Engine ID	Crankshaft				Connecting Rod		
			Main Brg. Journal Dia.	Main Brg. Clearance	Shaft End-play	Thrust on No.	Journal Diameter	Oil Clearance	Side Clearance
2011	4.0	1GR-FE	2.8342-2.8346	0.0010-0.0018	–	–	2.2044-2.2047	0.0016 0.0026	0.0059 0.0138
	2.7	2TR-FE	2.361-2.3620	②	–	–	2.086-2.0870	0.0015-0.0026	0.00591-0.0138
2012	4.0	1GR-FE	2.8342-2.8346	0.0010-0.0018	–	–	2.2044-2.2047	0.0016 0.0026	0.0059 0.0138

71099_4RUN_C0006

PISTON AND RING SPECIFICATIONS

All measurements are given in inches.

Year	Engine Displacement Liters	Engine ID	Piston Clearance	Ring Gap			Ring Side Clearance		
				Top Comp.	Bottom Comp.	Oil Control	Top Comp.	Bottom Comp.	Oil Control
2011	4.0	1GR-FE	0.000354-0.0020	0.00866 0.0126	0.0138-0.0177	0.0394-0.0157	0.0008 0.0028	0.0008 0.0024	0.0028 0.0059
	2.7	2TR-FE	0.0035-0.0044	0.0102-0.0150	0.0232-0.0280	0.0039-0.0157	0.00079-0.0030	0.00079-0.0026	0.00079-0.0028
2012	4.0	1GR-FE	0.000354-0.0020	0.00866 0.0126	0.0138-0.0177	0.0394-0.0157	0.0008 0.0028	0.0008 0.0024	0.0028 0.0059

71099_4RUN_C0007

TORQUE SPECIFICATIONS

All readings in ft. lbs.

Year	Engine Displacement Liters	Engine ID	Cylinder Head Bolts	Main Bearing Bolts	Rod Bearing Bolts	Crankshaft Damper Bolts	Flywheel Bolts	Manifold Intake	Manifold Exhaust	Spark Plugs	Oil Pan Drain Plug
2011	4.0	1GR-FE	①	⑤	②	184	61	15	15	13	30
	2.7	2TR-FE	⑥	⑦	③	181	④	18	27	13	28
2012	4.0	1GR-FE	①	⑤	②	184	61	15	15	13	30

① Step 1: 45 ft. lbs.
Step 2: Plus 180 degrees

② Step 1: 18 ft. lbs.
Step 2: Plus 90 degre

③ Step 1: 18 ft. lbs.
Step 2: Plus 90 degrees

④ Step 1: 22 ft. lbs.
Step 2: Plus 90 degrees

⑤ Step 1: 45 ft. lbs.
Step 2: Plus 90 degrees

⑥ Step 1: 29 ft. lbs.
Step 2: Plus 90 degrees
Step 3: Plus 90 degrees

⑦ Bolt A: 9 ft. lbs.
Bolt B: 11 ft. lbs.

71099_4RUN_C0008

WHEEL ALIGNMENT

Year	Model	Caster Range (+/-Deg.)	Caster Preferred Setting (Deg.)	Camber Range (+/-Deg.)	Camber Preferred Setting (Deg.)	Toe-in (in.)	Steering Axis Inclination (Deg.)
2011	4Runner	0.75	①	0.75	②	0.08+/-0.16	③
2012	4Runner	0.75	①	0.75	②	0.08+/-0.16	③

Note: All alignment specifications are based on nominal ride height and standard tires

① 2WD except air suspension +3.38
2WD with air suspension +3.55
4WD except air suspension +3.22
4WD with air suspension +3.37

② 2WD except air suspension -0.47
2WD with air suspension -0.50
4WD except air suspension -0.
4WD with air suspension -0.17

③ 2WD except air suspension 12.97+/-.075
2WD with air suspension 13.00+/-.075
4WD except air suspension 12.65+/-.075
4WD with air suspension 12.67+/-.075

71099_4RUN_C0009

TIRE, WHEEL AND BALL JOINT SPECIFICATIONS

Year	Model	OEM Tires		Tire Pressures (psi)		Wheel	Ball Joint	Lug Nut Torque
		Standard	Optional	Front	Rear	Size	Inspection	(ft. lbs.)
2011	4Runner	P265/70R16	P265/65R17	①	①	7-J/7.5J	②	83
2012	4Runner	P265/70R16	P265/65R17	①	①	7-J/7.5J	②	83

OEM: Original Equipment Manufacturer

① See placard on vehicle

② Upper arm ball joint turning torque: 39 inch lbs.

Lower arm ball joint turning torque: 26 inch lbs.

71099_4RUN_C0010

BRAKE SPECIFICATIONS

All measurements in inches unless noted

Year	Model		Brake Disc			Minimum	Brake Caliper	
			Original Thickness	Minimum Thickness	Maximum Runout	Lining Thickness	Bracket Bolts (ft. lbs.)	Mounting Bolts (ft. lbs.)
2011	4Runner	F	1.102	1.024	0.0020	0.039	—	91
		R	0.709	0.630	0.0079	0.039	77	65
2012	4Runner	F	1.260	1.150	0.0020	0.039	—	91
		R	0.709	0.630	0.0079	0.039	77	65

F: Front

R: Rear

71099_4RUN_C0011

SCHEDULED MAINTENANCE INTERVALS

4Runner

TO BE SERVICED	TYPE OF SERVICE	VEHICLE MILEAGE INTERVAL (x1000)													
		5	10	15	20	25	30	35	40	45	50	55	60	90	120
Engine oil & filter	R	✓	✓	✓	✓	✓	✓	✓	✓	✓	✓	✓	✓	✓	✓
Automatic transmission fluid	S/I			✓			✓			✓			✓	✓	✓
Ball joints & dust covers	S/I			✓			✓			✓			✓	✓	✓
Bolts & nuts on chassis & body	S/I			✓			✓			✓			✓	✓	✓
Brake linings & drums	S/I	✓	✓	✓	✓	✓	✓	✓	✓	✓	✓	✓	✓	✓	✓
Brake line pipes & hoses	S/I			✓			✓			✓			✓	✓	✓
Brake pads & discs (front & rear)	S/I	✓	✓	✓	✓	✓	✓	✓	✓	✓	✓	✓	✓	✓	✓
Brake fluid	R						✓						✓	✓	✓
Rack and pinion assembly	S/I			✓			✓			✓			✓	✓	✓
Steering linkage & boots	S/I			✓			✓			✓			✓	✓	✓
Air cleaner filter	R						✓						✓	✓	✓
Spark plugs ①	R														✓
Drive belts	S/I												✓	✓	✓
Exhaust pipes & mountings	S/I			✓			✓			✓			✓	✓	✓
Fuel lines & connections	S/I						✓						✓	✓	✓
Engine coolant ②	S/I			✓			✓			✓			✓	✓	
Rear differential & transfer case oil	S/I			✓			✓			✓			✓	✓	✓
Fuel tank cap gasket	S/I						✓						✓	✓	✓
Rotate tires	S/I			✓			✓			✓			✓		✓
Clean air conditioning filter ③	S/I			✓			✓			✓			✓		✓
Axle shaft bolts	S/I			✓			✓			✓			✓	✓	✓
Brake pad thickness and rotor runout	S/I						✓						✓	✓	✓

R: Replace S/I: Service or Inspect

① Spark plugs are replaced at 120,000 miles

② Replace engine coolant at 100,000 miles and then inspect every 15,000 miles

③ Replace air conditioning filter every 30,000 miles

FREQUENT OPERATION MAINTENANCE (SEVERE SERVICE)

If a vehicle is operated under any of the following conditions it is considered severe service:

- Extremely dusty areas.

- 50% or more of the vehicle operation is in 32°C (90°F) or higher temperatures, or constant temperatures below 0°C (32°F).

- Prolonged idling (vehicle operation in stop and go traffic).

- Frequent short running periods (engine does not warm to normal operating temperatures).

- Police, taxi, delivery usage or trailer towing usage.

Air cleaner filter: service or inspect every 5000 miles

Rear differential & transfer case oil: replace every 15,000 miles.

Ball joints & dust covers: service or inspect every 5000 miles.

Bolts & nuts on chassis & body: service or inspect every 5000 miles.

Axle shaft bolts: service or inspect every 5000 miles.

Steering linkage: service or inspect every 5000 miles.

71099_4RUN_C0012

PRECAUTIONS

Before servicing any vehicle, please be sure to read all of the following precautions, which deal with personal safety, prevention of component damage, and important points to take into consideration when servicing a motor vehicle:

- Never open, service or drain the radiator or cooling system when the engine is hot; serious burns can occur from the steam and hot coolant.
- Observe all applicable safety precautions when working around fuel. Whenever servicing the fuel system, always work in a well-ventilated area. Do not allow fuel spray or vapors to come in contact with a spark, open flame, or excessive heat (a hot drop light, for example). Keep a dry chemical fire extinguisher near the work area. Always keep fuel in a container specifically designed for fuel storage; also, always properly seal fuel containers to avoid the possibility of fire or explosion. Refer to the additional fuel system precautions later in this section.
- Fuel injection systems often remain pressurized, even after the engine has been turned **OFF**. The fuel system pressure must be relieved before disconnecting any fuel lines. Failure to do so may result in fire and/or personal injury.
- Brake fluid often contains polyglycol ethers and polyglycols. Avoid contact with the eyes and wash your hands thoroughly after handling brake fluid. If you do get brake fluid in your eyes, flush your eyes with clean, running water for 15 minutes. If eye irritation persists, or if you have taken brake fluid internally, IMMEDIATELY seek medical assistance.
- The EPA warns that prolonged contact with used engine oil may cause a number of skin disorders, including cancer. You should make every effort to minimize your exposure to used engine oil. Protective gloves should be worn when changing oil. Wash your hands and any other exposed skin areas as soon as possible after exposure to used engine oil. Soap and water, or waterless hand cleaner should be used.
- All new vehicles are now equipped with an air bag system, often referred to as a Supplemental Restraint System (SRS) or Supplemental Inflatable Restraint (SIR) system. The system must be disabled before performing service on or around system components, steering column, instrument panel components, wiring and sensors. Failure to follow safety and disabling procedures could result in accidental air bag deployment, possible personal injury and unnecessary system repairs.
- Always wear safety goggles when working with, or around, the air bag system. When carrying a non-deployed air bag, be sure the bag and trim cover are pointed away from your body. When placing a non-deployed air bag on a work surface, always face the bag and trim cover upward, away from the surface. This will reduce the motion of the module if it is accidentally deployed. Refer to the additional air bag system precautions later in this section.
- Clean, high quality brake fluid from a sealed container is essential to the safe and proper operation of the brake system. You should always buy the correct type of brake fluid for your vehicle. If the brake fluid becomes contaminated, completely flush the system with new fluid. Never reuse any brake fluid. Any brake fluid that is removed from the system should be discarded. Also, do not allow any brake fluid to come in contact with a painted surface; it will damage the paint.
- Never operate the engine without the proper amount and type of engine oil; doing so WILL result in severe engine damage.
- Timing belt maintenance is extremely important. Many models utilize an interference-type, non-freewheeling engine. If the timing belt breaks, the valves in the cylinder head may strike the pistons, causing potentially serious (also time-consuming and expensive) engine damage. Refer to the maintenance interval charts for the recommended replacement interval for the timing belt, and to the timing belt section for belt replacement and inspection.
- Disconnecting the negative battery cable on some vehicles may interfere with the functions of the on-board computer system(s) and may require the computer to undergo a relearning process once the negative battery cable is reconnected.
- When servicing drum brakes, only disassemble and assemble one side at a time, leaving the remaining side intact for reference.
- Only an MVAC-trained, EPA-certified automotive technician should service the air conditioning system or its components.

BRAKES

ANTI-LOCK BRAKE SYSTEM (ABS)

GENERAL INFORMATION

PRECAUTIONS

- Certain components within the ABS system are not intended to be serviced or repaired individually.
- Do not use rubber hoses or other parts not specifically specified for and ABS system. When using repair kits, replace all parts included in the kit. Partial or incorrect repair may lead to functional problems and require the replacement of components.
- Lubricate rubber parts with clean, fresh brake fluid to ease assembly. Do not use shop air to clean parts; damage to rubber components may result.
- Use only DOT 3 brake fluid from an unopened container.
- If any hydraulic component or line is removed or replaced, it may be necessary to bleed the entire system.
- A clean repair area is essential. Always clean the reservoir and cap thoroughly before removing the cap. The slightest amount of dirt in the fluid may plug an orifice and impair the system function. Perform repairs after components have been thoroughly cleaned; use only denatured alcohol to clean components. Do not allow ABS components to come into contact with any substance containing mineral oil; this includes used shop rags.
- The Anti-Lock control unit is a microprocessor similar to other computer units in the vehicle. Ensure that the ignition switch is **OFF** before removing or installing controller harnesses. Avoid static electricity discharge at or near the controller.
- If any arc welding is to be done on the vehicle, the control unit should be unplugged before welding operations begin.

SPEED SENSORS

REMOVAL & INSTALLATION

Front

See Figure 1.

➡The procedure listed below is for the LH side.

1. Remove the front wheel.
2. Remove the front skid control sensor wire.

a. Disconnect the connector from the front speed sensor.

b. Remove the 2 bolts and 2 harness clamps.

c. Detach the clip.

d. Remove the 2 bolts and the 2 harness clamps.

e. Disconnect the connector as follows.

LH

a. Disconnect the connector.

b. Detach the connector.

RH

a. Disconnect the connector.

b. Detach the connector from the skid control sensor clamp.

c. Detach the clip.

d. Remove the bolt and skid control sensor clamp.

3. Remove the skid control sensor clamp.

a. Remove the bolt and skid control sensor clamp from the knuckle.

4. Remove the front speed sensor from the knuckle.

➡Pull out the sensor while trying as much as possible not to rotate it.

To install:

5. Install the front speed sensor. Tighten to 75 inch lbs. (8.5 Nm).

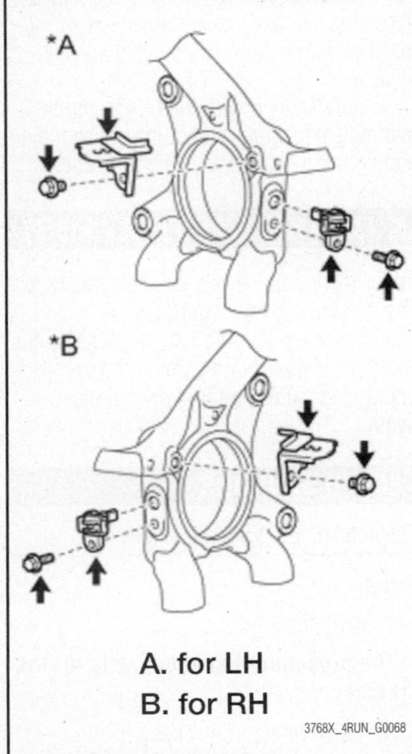

Fig. 1 Removing the skid control sensor clamp

➡Make sure there are no pieces of iron or other foreign matter attached to the sensor tip.

➡While inserting the speed sensor into the knuckle hole, do not strike or damage the sensor tip.

➡After installing the speed sensor, make sure there is no clearance or foreign matter between the sensor stay part and the knuckle.

➡Make sure there is no foreign matter attached to the speed sensor rotor.

6. Install the skid control sensor clamp. Tighten the bolt to 9 ft. lbs. (13 Nm).

➡Install the clamp so that the rotation stopper touches the knuckle.

7. Install the front skid control sensor wire.

a. For the LH, attach the connector, and then connector the connector.

b. For the RH, install the skid control sensor clamp with the bolt. Tighten to 44 inch lbs. (5 Nm).

➡Make sure the clamp rotation stopper touches the installation position.

c. For the RH, attach the connector, and then connect the connector.

➡Securely connect the connector.

➡When connecting the connector, do not twist the wire harness.

d. For the RH, attach the clip.

e. Install the 2 harness clamps with the 2 bolts. Tighten to 9 ft. lbs. (13 Nm).

➡When installing the clamps, do not twist the wire harness.

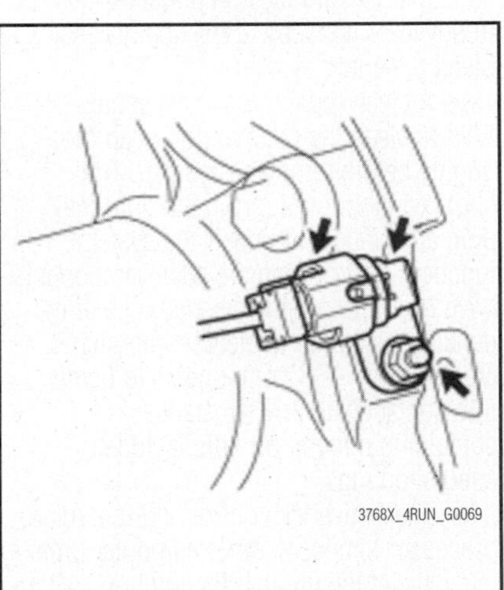

Fig. 2 Remove the rear speed sensor LH

➡Make sure the clamp rotation stopper touches the installation position.

f. Install the 2 harness clamps with the 2 bolts. Tighten to 9 ft. lbs. (13 Nm).

➡When installing the clamps, do not twist the wire harness.

➡Make sure the clamp rotation stopper touches the installation position.

g. Attach the clip.

h. Connect the connector.

➡Securely connect the connector.

8. Install the front wheel.

9. Check the speed sensor signal.

Rear

See Figures 2 and 3.

1. Remove the rear wheel.

2. Remove the rear speed sensor LH.

a. Disconnect the speed sensor connector.

b. Remove the nut and speed sensor.

➡Pull out the sensor while trying as much as possible not to rotate it.

3. Remove the rear speed sensor RH.

4. Remove the skid control sensor wire.

a. Disconnect the connector.

b. Detach the connector.

c. Detach the 8 clamps.

d. Remove the 2 bolts and 2 sensor clamps.

To install:

5. Install the skid control sensor wire.

a. Install the 2 sensor clamps with the 2 bolts. Tighten to 9 ft. lbs. (13 Nm).

➡Make sure the clamp rotation stopper touches the installation position.

Fig. 3 Removing the skid control sensor wire

b. Connect the connector.

➡**Securely connect the connector.**

c. Attach the connector.
d. Attach the 8 clamps.

➡**When attaching the clamps, do not twist the wire harness.**

6. Install the rear speed sensor LH.
a. Install the speed sensor with the nut. Tighten to 69 inch lbs. (7.8 Nm).

➡**Make sure there are no pieces of iron or other foreign matter attached to the sensor tip.**

➡**While inserting the speed sensor into the axle hole, do not strike or damage the sensor tip.**

➡**After installing the speed sensor, make sure there is no clearance or foreign matter between the sensor stay part and the axle.**

➡**Make sure there is no foreign matter attached to the speed sensor rotor.**

b. Connect the speed sensor connector.

➡**Securely connect the connector.**

7. Install the rear speed sensor RH.
8. Install the rear wheel.
9. Check the speed sensor signal.

BRAKES

BLEEDING THE BRAKE SYSTEM

BLEEDING PROCEDURE

BLEEDING PROCEDURE

➡**If any work is done on the brake system or if air is suspected in the brake lines, bleed the air from the system.**

➡**Do not let brake fluid remain on a painted surface. Wash it off immediately.**

1. Before servicing the vehicle, refer to the Precautions section.
2. Check the fluid level in the reservoir after bleeding each wheel. Add DOT3 fluid, if necessary.
3. If the hydraulic brake booster was disassembled or if the reservoir becomes empty, bleed the air from the hydraulic brake booster as follows:

➡**Perform this step only if the brake booster with accumulator pump assembly is removed and/or installed.**

a. Turn the ignition switch OFF, depress the brake pedal 20 times or more to release the pressure from the accumulator.
b. Fully depress the brake pedal 10 times.
c. Turn the ignition switch to the ON position and start the brake booster pump.
d. Make sure the pump operates for 8 to 14 seconds.

➡**If the pump does not operate as specified, repeat the above and recheck the operating time.**

MASTER CYLINDER BLEEDING

1. Bleeding Master Cylinder Solenoid is only possible with a Toyota proprietary scan system.

BRAKE LINE BLEEDING

Front Brake Lines

a. Turn the ignition switch to the ON position and wait until the pump motor has stopped.
b. Connect the vinyl tube to the brake caliper.
c. Depress the brake pedal several times, then loosen the bleeder plug with the pedal held down.
d. At the point when the fluid stops coming out, tighten the bleeder plug, 8 ft. lbs. (11 Nm) then release the brake pedal.
e. Repeat procedure until all the air in the fluid has been bled out.
f. Repeat the above procedures to bleed the other brake line.

Rear Brake Lines

g. Turn the ignition switch to the ON position and depress the brake pedal.
h. Connect the vinyl tube to the brake caliper.
i. Loosen the bleeder plug and release air.

➡**Brake fluid is sent through the pump, so keep the brake pedal depressed until the air is completely bled out.**

j. When the air is completely bled out of the brake fluid through the bleeder plug, tighten the bleeder plug to 8 ft. lbs. (11 Nm) then release.
k. Repeat the above procedures to bleed the other brake line.

CAUTION

Dust and dirt accumulating on brake parts during normal use may contain asbestos fibers from production or aftermarket brake linings. Breathing excessive concentrations of asbestos fibers can cause serious bodily harm. Exercise care when servicing brake parts. Do not sand or grind brake lining unless equipment used is designed to contain the dust residue. Do not clean brake parts with compressed air or by dry brushing. Cleaning should be done by dampening the brake components with a fine mist of water, then wiping the brake components clean with a dampened cloth. Dispose of cloth and all residue containing asbestos fibers in an impermeable container with the appropriate label. Follow practices prescribed by the Occupational Safety and Health Administration (OSHA) and the Environmental Protection Agency (EPA) for the handling, processing, and disposing of dust or debris that may contain asbestos fibers.

BRAKE CALIPER

REMOVAL & INSTALLATION

See Figures 4 and 5.

➡Use the same procedure for the RH and LH sides.

➡The procedure listed below is for the LH side.

1. Remove front wheel.
2. Drain the brake fluid.

➡Wash off the brake fluid immediately if it comes into contact with a painted surface.

3. Remove the front disc brake pad.
4. Remove the disc brake caliper assembly LH.
   a. Using a union nut wrench, disconnect the brake tube from the disc brake caliper assembly.

➡Use a container to catch the brake fluid as it drains out.

   b. Remove the 2 bolts and the disc brake caliper assembly.

To install:

5. Install the disc brake caliper assembly LH.
   a. Install the disc brake caliper assembly with the 2 bolts. Tighten to 91 ft. lbs. (123 Nm).
   b. Using a union nut wrench, connect the brake tube to the disc brake caliper assembly. Tighten to 11 ft. lbs. (15 Nm).
6. Install the front No. 1 anti-squeal shim to the front disc brake pads.

➡If necessary, replace the anti-squeal shim when replacing the brake pad.

➡There should be no oil or grease on the friction surfaces of the front disc pads and the front disc.

7. Install the front disc brake pad.
8. Bleed the brake line.
9. Install the front wheel.

DISC BRAKE PADS

REMOVAL & INSTALLATION

See Figures 6 and 7.

➡Use the same procedure for the RH and LH sides.

➡The procedure listed below is for the LH side.

1. Remove front wheel.
2. Drain the brake fluid.

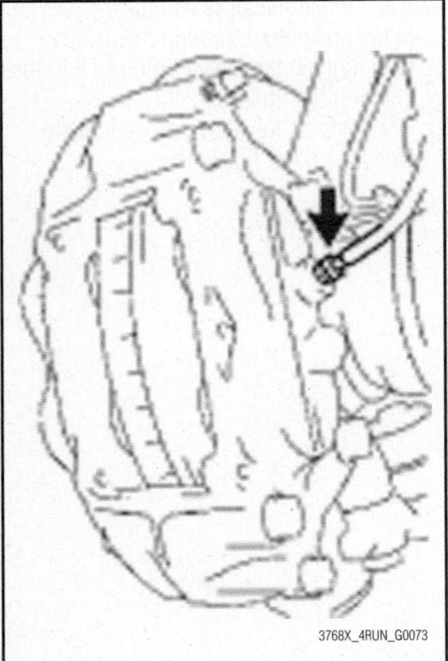

Fig. 4 Disconnecting the brake tube from the disc brake caliper

Fig. 5 Removing the disc brake caliper assembly

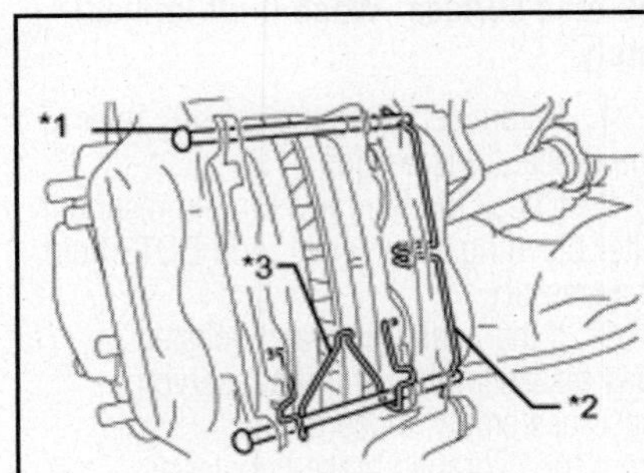

Fig. 6 Removing the pin hold clip

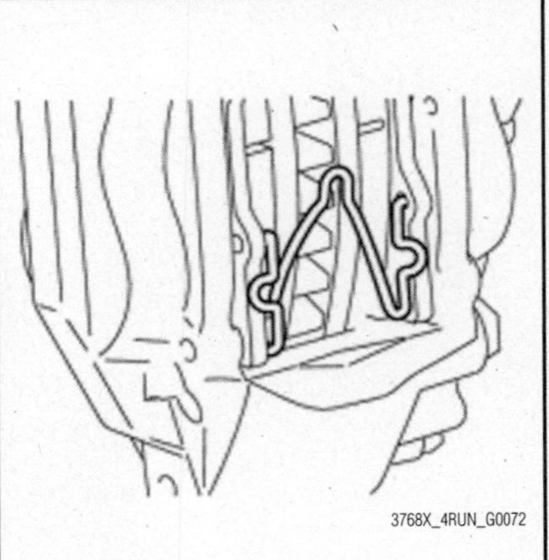

Fig. 7 Removing the front disc anti-rattle spring

➡Wash off the brake fluid immediately if it comes into contact with a painted surface.

3. Remove the pin hold clip.

➡The pin hold clip can be reused if it has sufficient rebound; no deformation or wear; and has had all rust, dirt and foreign matter cleaned off.

4. Remove the 2 hole pins.
5. Remove the front disc brake anti-rattle spring.

➡The anti-rattle spring can be reused if it has sufficient rebound; no deformation, cracks or wear; and has had rust, dirt and foreign matter cleaned.

6. Remove the 2 front disc brake pads from the disc brake caliper.
7. Remove the front No. 1 anti-squeal shims from each pad.

To install:

8. Install the 2 front disc brake pads to the disc brake caliper.
9. Install the anti-rattle spring between the 2 front brake pads.

➡The anti-rattle spring can be reused if it has sufficient rebound; no deformation, cracks or wear, and has had all rust, dirt and foreign matter cleaned off.

10. Install the 2 hole pins.
11. Install the pin hold clip.

➡The anti-rattle spring can be reused if it has sufficient rebound; no deformation, cracks or wear, and has had all rust, dirt and foreign matter cleaned off.

12. Bleed the brake line.
13. Install the front wheel.

BRAKES

REAR DISC BRAKES

⁂ CAUTION

Dust and dirt accumulating on brake parts during normal use may contain asbestos fibers from production or aftermarket brake linings. Breathing excessive concentrations of asbestos fibers can cause serious bodily harm. Exercise care when servicing brake parts. Do not sand or grind brake lining unless equipment used is designed to contain the dust residue. Do not clean brake parts with compressed air or by dry brushing. Cleaning should be done by dampening the brake components with a fine mist of water, then wiping the brake components clean with a dampened cloth. Dispose of cloth and all residue containing asbestos fibers in an impermeable container with the appropriate label. Follow practices prescribed by the Occupational Safety and Health Administration (OSHA) and the Environmental Protection Agency (EPA) for the handling, processing, and disposing of dust or debris that may contain asbestos fibers.

BRAKE CALIPER

REMOVAL & INSTALLATION

See Figure 8.

1. Remove the rear wheel.
2. Drain the brake fluid.

➡Wash the brake fluid off immediately if immediately if it adheres to any painted surfaces.

3. Disconnect the rear flexible hose LH.
   a. Remove the union bolt and gasket from the rear disc brake caliper, and then disconnect the rear flexible hose from the rear disc brake caliper.

➡Use a container to catch brake fluid as it drains out.

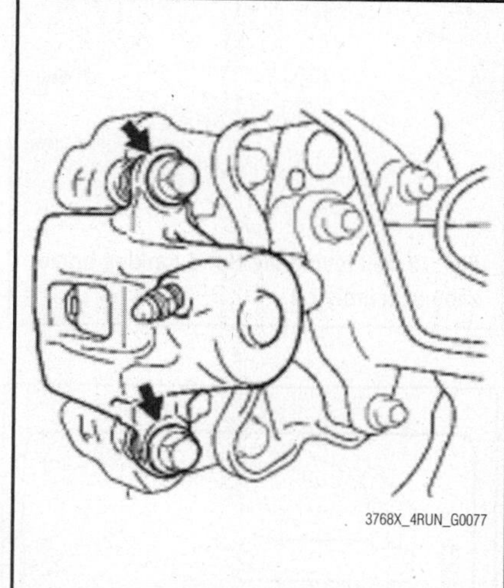

Fig. 8 Removing the rear brake cylinder

4. Remove the rear disc brake caliper assembly LH.
   a. Remove the 2 caliper slide pins from the rear disc brake caliper.
   b. Remove the rear disc brake caliper from the rear disc brake caliper mounting.

To install:

5. To install, reverse the removal procedure. Apply a light coat of lithium soap base glycol grease to the sliding surfaces of the rear disc brake caliper slide pins. Tighten the bolts to 65 ft. lbs. (88 Nm).

DISC BRAKE PADS

REMOVAL & INSTALLATION

1. Remove the rear wheel.
2. Drain the brake fluid.
3. Remove the rear brake cylinder assembly LH.
4. Remove the rear disc brake pad.
   a. Remove the 2 rear disc brake pads together with the rear disc brake anti-squeal shims from the rear disc brake caliper mounting.

To install:

5. To install, reverse the removal procedure.

PARKING BRAKE CABLES

ADJUSTMENT

1. Remove rear wheel.
2. Adjust parking brake shoe clearance.
3. Install rear wheel an tighten to 82 ft. lbs. (112 Nm)
4. Inspect parking brake lever travel.
5. Slowly depress the parking brake lever all the way, and count the number of clicks.
   a. Parking brake lever travel at 66 ft. lbs. (294 Nm) 5 to 7 clicks.
6. Adjust parking brake lever travel by removing the console panel upper.
7. Turn the adjusting nut until the parking brake lever travel becomes correct.
8. Check whether parking brake drags or not.
9. When operating the parking brake lever, check that the parking brake lever indicator light comes on.
10. Install the console panel upper.

PARKING BRAKE SHOES

REMOVAL & INSTALLATION

See Figures 9 through 16.

➡Use the same procedure for the RH and LH sides.

➡The procedure listed below is for the LH side.

1. Remove the rear wheel.
2. Disconnect the rear disc brake caliper assembly LH.
   a. Remove the 2 bolts and disconnect the rear disc brake caliper.

➡Do not disconnect the flexible hose from the disc brake caliper.

➡Do not twist or bend the flexible hose.

3. Remove the rear disc.
4. Remove the parking brake shoe return tension spring.
   a. Remove the 2 parking brake shoe return tension springs.
5. Disconnect the No. 1 parking brake shoe assembly LH.
   a. Remove the parking brake shoe hold down spring cup and parking brake shoe hold down spring to disconnect the No. 1 parking brake shoe assembly from the backing plate.
6. Remove the parking brake shoe strut and parking brake shoe strut compression spring.
7. Disconnect the No. 2 parking brake shoe assembly LH.

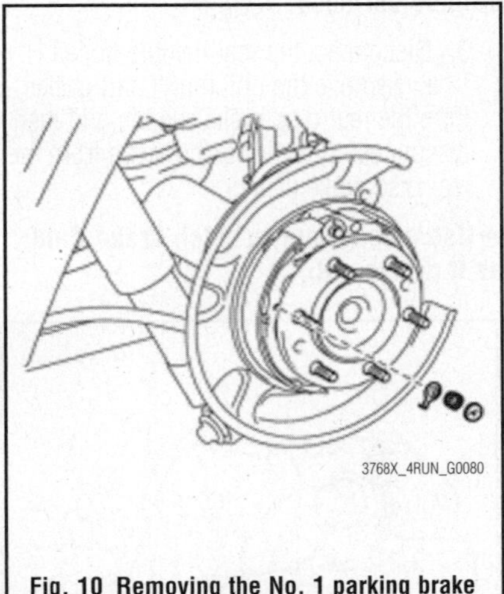

Fig. 10 Removing the No. 1 parking brake shoe assembly LH

   a. Remove the parking brake shoe hold down spring cup and parking brake shoe hold down spring to disconnect the No. 2 parking brake shoe assembly from the backing plate.
8. Remove the parking brake shoe adjusting screw set.
9. Remove the No. 1 parking brake shoe assembly LH.
   a. Disconnect the parking brake shoe return tension spring to remove the No. 1 parking brake shoe assembly.
10. Remove the parking brake shoe return tension spring from the No. 2 parking brake shoe assembly.
11. Remove the No. 2 parking brake shoe assembly with parking brake shoe lever.
   a. Using needle-nose pliers, disconnect the No. 3 parking brake cable assembly from the parking brake shoe lever.

➡Be careful not to damage the No. 3 parking brake cable assembly.

12. Remove the parking brake shoe lever.
   a. Remove the C-washer, shim and parking brake shoe lever from the No. 2 parking brake shoe assembly.
13. Remove the parking brake shoe hold down spring pin.
   a. Remove the parking brake shoe hold down spring pin (for front side).
   b. Remove the parking brake shoe hold down spring pin (for rear side).

To install:

14. Install the parking brake shoe hold down spring.
15. Apply high temperature grease.

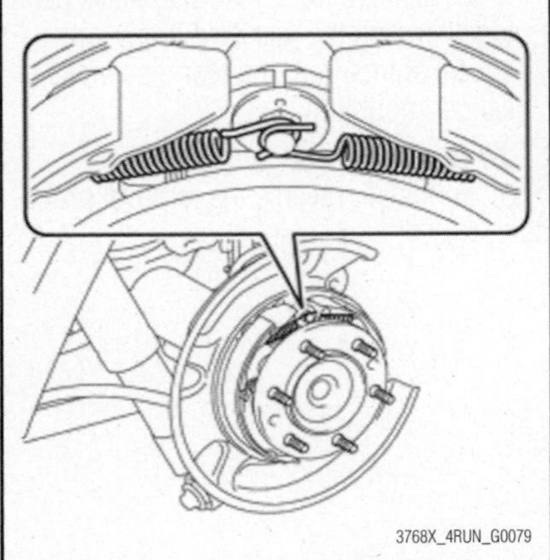

Fig. 9 Removing the parking brake shoe return tension spring

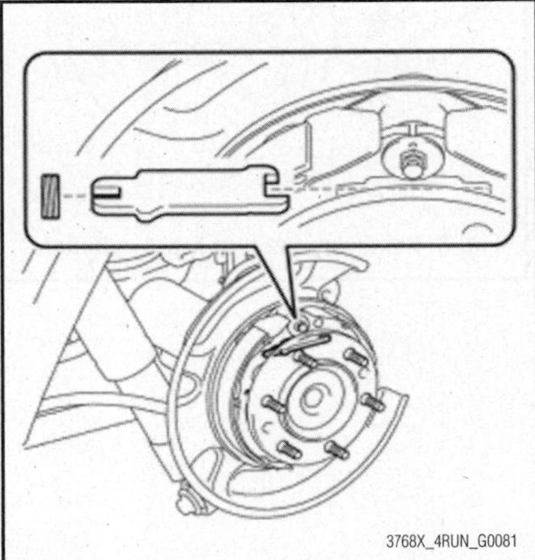

Fig. 11 Removing the parking brake shoe strut LH

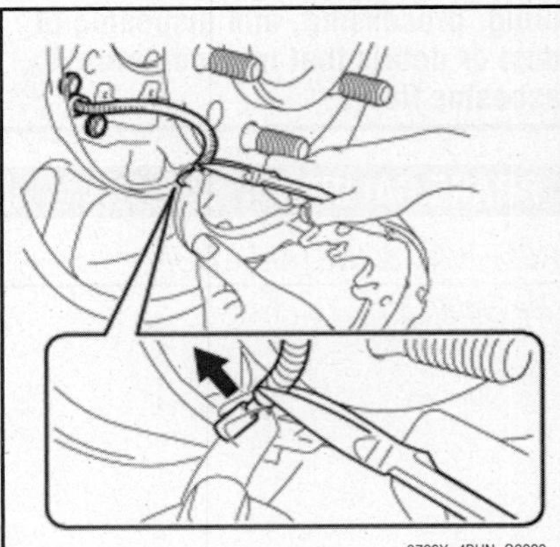

Fig. 12 Removing the No. 2 parking brake shoe assembly with the parking brake shoe lever

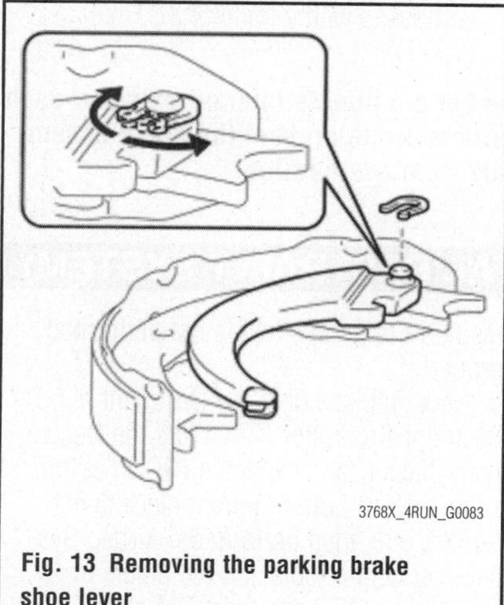

Fig. 13 Removing the parking brake shoe lever

a. Apply a light coat of high-temperature grease to the areas of the backing plate which make contact with the shoe.

16. Install the parking brake shoe lever.

a. Apply a light coat of high-temperature grease to the areas of the parking brake shoe lever which make contact with the No. 2 parking brake shoe assembly.

b. Install the parking brake shoe lever and shim to the No. 2 parking brake shoe assembly with a new C-washer.

c. Using a feeler gauge, measure the clearance between the No. 2 parking brake shoe assembly and the parking brake lever. Standard clearance is 0.00984 inch (0.25 mm). If the clearance is not within the specification, replace the shim with one of a different thickness so that the clearance is within the specification.

17. Install the No. 2 parking brake shoe assembly with the parking brake shoe lever.

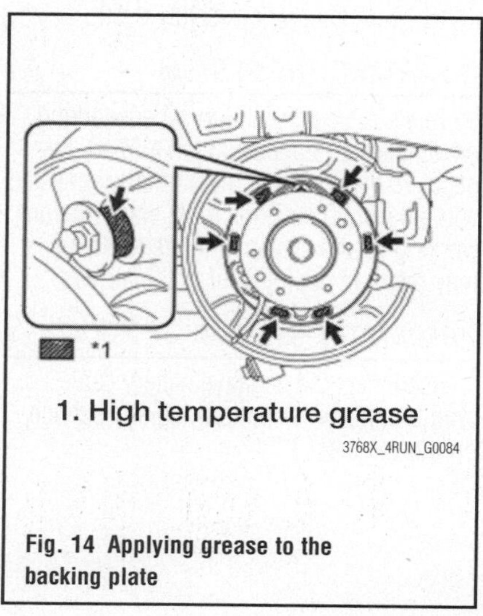

Fig. 14 Applying grease to the backing plate

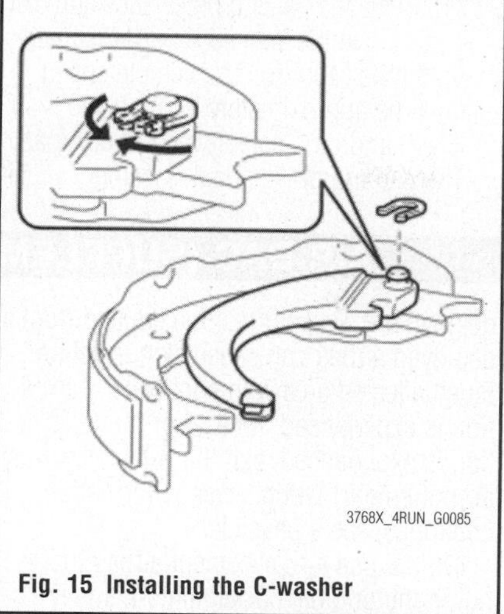

Fig. 15 Installing the C-washer

a. Using needle-nose pliers, connect the No. 3 parking brake cable assembly to the parking brake shoe lever.

➡Be careful not to damage the No. 3 parking brake cable assembly.

18. Install the parking brake shoe return tension spring.

a. Install the parking brake shoe return tension spring to the No. 2 parking b rake shoe assembly.

19. Install the No. 1 parking brake shoe assembly LH.

a. Connect the parking brake shoe return tension spring to install the No. 1 parking brake shoe assembly.

20. Install the parking brake shoe adjusting screw set.

a. Apply a light coat of high-temperature grease to the areas of the shoe adjusting screw set.

b. Install the parking brake shoe adjusting screw set.

21. Install the No. 2 parking brake shoe assembly LH.

a. Install the No. 2 parking brake shoe assembly to the backing plate with the parking brake shoe hold down spring cup and parking brake shoe hold down spring.

22. Install the parking brake shoe strut and parking brake shoe strut compression spring.

23. Install the parking brake shoe assembly LH.

a. Install the No. 1 parking brake shoe assembly to the backing plate with the parking brake shoe hold down spring cup and parking brake shoe hold down spring.

24. Install the parking brake shoe return tension spring.

➡First install the front side spring, and then the rear side spring.

25. Check the parking brake installation.

a. Check that each part is installed properly.

➡There should be no oil or grease adhering to the friction surfaces of the shoe lining and disc.

26. Install the rear disc.

27. Connect the rear disc brake caliper assembly LH. Tighten the 2 bolts to 77 ft. lbs. (105 Nm).

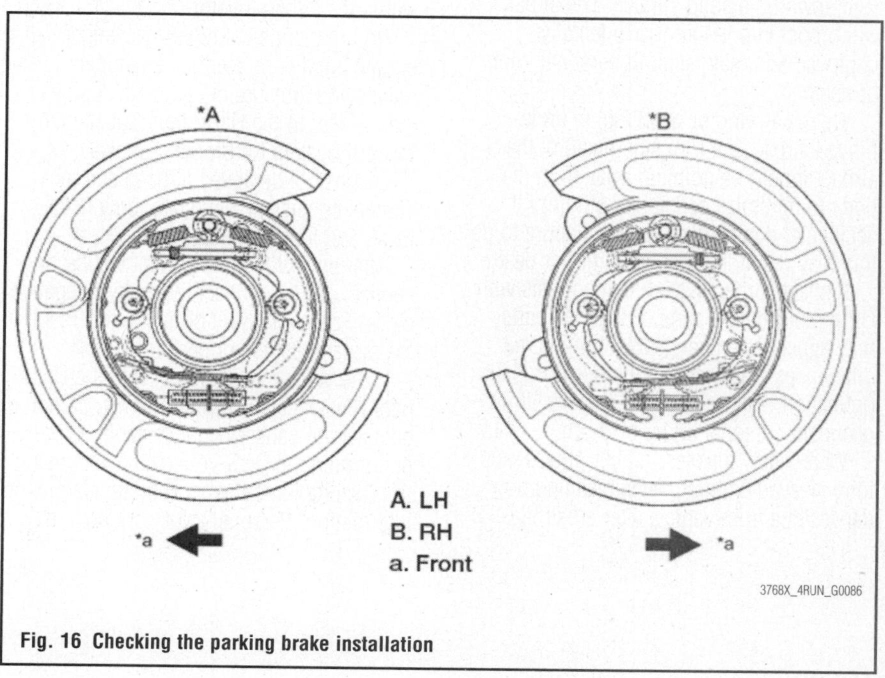

Fig. 16 Checking the parking brake installation

28. Adjust the parking brake shoe clearance and parking brake pedal travel.
29. Inspect the parking brake pedal travel.
30. Install the rear wheel.
31. Settle the parking brake shoe and disc.
   a. Drive the vehicle for approximately 0.25 mile (400 m). The vehicle speed must be approximately 31 mph (50 km/h) and on a safe, level and dry road. The parking brake pedal is being depressed with a force of 37.7 lbs. (150 N).

➡Set a 5 minute interval between each procedure to prevent the brake assembly from overheating.

CHASSIS ELECTRICAL

AIR BAG (SUPPLEMENTAL RESTRAINT SYSTEM)

GENERAL INFORMATION

⁂ CAUTION

These vehicles are equipped with an air bag system. The system must be disarmed before performing service on, or around, system components, the steering column, instrument panel components, wiring and sensors. Failure to follow the safety precautions and the disarming procedure could result in accidental air bag deployment, possible injury and unnecessary system repairs.

SERVICE PRECAUTIONS

Disconnect and isolate the battery negative cable before beginning any airbag system component diagnosis, testing, removal, or installation procedures. Allow system capacitor to discharge for two minutes before beginning any component service. This will disable the airbag system. Failure to disable the airbag system may result in accidental airbag deployment, personal injury, or death.

Do not place an intact undeployed airbag face down on a solid surface. The airbag will propel into the air if accidentally deployed and may result in personal injury or death.

When carrying or handling an undeployed airbag, the trim side (face) of the airbag should be pointing away from the body to minimize possibility of injury if accidental deployment occurs. Failure to do this may result in personal injury or death.

Replace airbag system components with OEM replacement parts. Substitute parts may appear interchangeable, but internal differences may result in inferior occupant protection. Failure to do so may result in occupant personal injury or death.

Wear safety glasses, rubber gloves, and long sleeved clothing when cleaning powder residue from vehicle after an airbag deployment. Powder residue emitted from a deployed airbag can cause skin irritation. Flush affected area with cool water if irritation is experienced. If nasal or throat irritation is experienced, exit the vehicle for fresh air until the irritation ceases. If irritation continues, see a physician.

Do not use a replacement airbag that is not in the original packaging. This may result in improper deployment, personal injury, or death.

The factory installed fasteners, screws and bolts used to fasten airbag components have a special coating and are specifically designed for the airbag system. Do not use substitute fasteners. Use only original equipment fasteners listed in the parts catalog when fastener replacement is required.

During, and following, any child restraint anchor service, due to impact event or vehicle repair, carefully inspect all mounting hardware, tether straps, and anchors for proper installation, operation, or damage. If a child restraint anchor is found damaged in any way, the anchor must be replaced. Failure to do this may result in personal injury or death.

Deployed and non-deployed airbags may or may not have live pyrotechnic material within the airbag inflator.

Do not dispose of driver/passenger/curtain airbags or seat belt tensioners unless you are sure of complete deployment. Refer to the Hazardous Substance Control System for proper disposal.

Dispose of deployed airbags and tensioners consistent with state, provincial, local, and federal regulations.

After any airbag component testing or service, do not connect the battery negative cable. Personal injury or death may result if the system test is not performed first.

If the vehicle is equipped with the Occupant Classification System (OCS), do not connect the battery negative cable before performing the OCS Verification Test using the scan tool and the appropriate diagnostic information. Personal injury or death may result if the system test is not performed properly.

Never replace both the Occupant Restraint Controller (ORC) and the Occupant Classification Module (OCM) at the same time. If both require replacement, replace one, then perform the Airbag System test before replacing the other.

Both the ORC and the OCM store Occupant Classification System (OCS) calibration data, which they transfer to one another when one of them is replaced. If both are replaced at the same time, an irreversible fault will be set in both modules and the OCS may malfunction and cause personal injury or death.

If equipped with OCS, the Seat Weight Sensor is a sensitive, calibrated unit and must be handled carefully. Do not drop or handle roughly. If dropped or damaged, replace with another sensor. Failure to do so may result in occupant injury or death.

If equipped with OCS, the front passenger seat must be handled carefully as well. When removing the seat, be careful when setting on floor not to drop. If dropped, the sensor may be inoperative, could result in occupant injury, or possibly death.

If equipped with OCS, when the passenger front seat is on the floor, no one should sit in the front passenger seat. This uneven force may damage the sensing ability of the seat weight sensors. If sat on and damaged, the sensor may be inoperative, could result in occupant injury, or possibly death.

DISARMING THE SYSTEM

To avoid personal injury when working on vehicles equipped with an air bag, the negative battery cable must be disconnected and at least 90 seconds must elapse before working on the system. Failure to do so may result in deployment of the air bag.

ARMING THE SYSTEM

Reconnect the negative battery cable. Wait 2 minutes for performing any service on the vehicle.

DRIVE TRAIN

TRANSFER CASE ASSEMBLY

REMOVAL & INSTALLATION

1. Drain the transfer oil.
2. Remove the automatic transmission assembly.
3. Remove the transfer assembly.
   a. Remove the 8 bolts.
   b. Remove the transfer from the transmission.

To install:

4. Install the transfer case.
   a. Install the transfer to the transmission.
   b. Install the 8 bolts. Tighten to 18 ft. lbs. (24 Nm).
5. Install the automatic transmission assembly.
6. Add transfer oil.
7. Check for transfer oil leaks.

FRONT DRIVESHAFT

REMOVAL & INSTALLATION

See Figure 17.

➡Use the same procedure for the RH and LH sides.

➡The procedures listed below is for the LH side.

1. Remove the front wheel.
2. Drain the differential oil.
3. Remove the front axle hub grease cap.
4. Remove the front axle shaft nut.
5. Remove the front speed sensor.
6. Disconnect the tie rod end sub assembly LH.
7. Disconnect the front lower ball joint attachment LH.
   a. Remove the 2 bolts and disconnect the lower ball joint attachment from the steering knuckle.
8. Remove the front drive shaft assembly LH.

➡Be careful not to damage the dust cover or oil seal.

9. Keep the drive shaft level while handling it.

To install:

10. Install the front drive shaft assembly.
    a. Coat the spline of the inboard joint shaft assembly with ATF.
    b. Align the shaft splines and install the drive shaft with a brass bar and hammer.

➡Set the snap ring with the opening facing downward.

➡Be careful not to damage the oil seal, boot or dust cover.

➡Whether the inboard joint shaft is in contact with the pinion shaft or not can be confirmed from the sound or feeling when tapping in the shaft.

11. Install the front speed sensor.
12. Install the lower ball joint attachment LH.
    a. Install the lower ball joint attachment with the 2 bolts. Tighten to 118 ft. lbs. (160 Nm).
13. Connect the tie rod end sub assembly LH.
14. Install the front axle shaft nut.
15. Install the front grease hub cap.
16. Add differential oil.
17. Install the front wheel. Tighten to 83 ft. lbs. (112 Nm).
18. Check the front speed sensor signal.

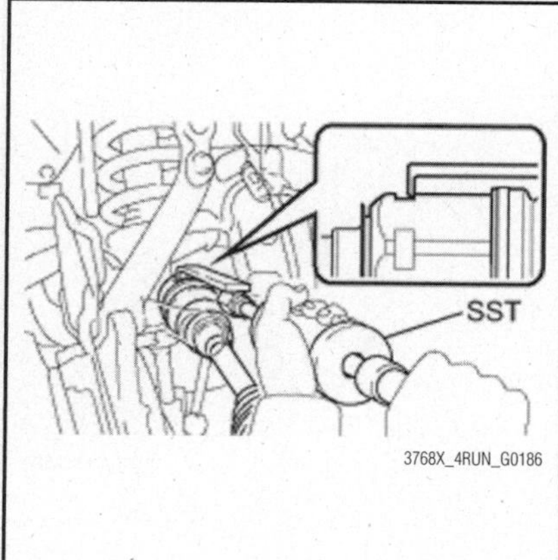

Fig. 17 Removing the front drive shaft assembly LH

FRONT PINION SEAL

REMOVAL & INSTALLATION

See Figures 18 through 20.

1. Remove the differential vacuum actuator assembly (w/A.D.D.).
   a. Remove the 4 bolts.
   b. Using a hammer handle, pry out the actuator from the differential tube.
2. Remove the front differential tube assembly.
   a. Using a E14 TORX® socket wrench, remove the 4 bolts.
   b. Using a plastic faced hammer, tap out the differential tube.
3. Remove the differential side gear shaft oil seal.
4. Remove the differential side gear inner shaft sub assembly (w/A.D.D.).
   a. Remove the snap ring from the side gear inter shaft.
5. Remove the front differential side bearing retainer deflector (w/A.D.D.).
   a. Using a screwdriver pry out the bearing retainer deflector.

➡Tape the screwdriver tip before use.

6. Remove the front drive pinion companion flange nut.
   a. Using a special tool and a hammer, unstake the nut.
   b. Using the special tool to hold the companion flange, remove the nut.
7. Remove the front drive pinion companion flange sub assembly.

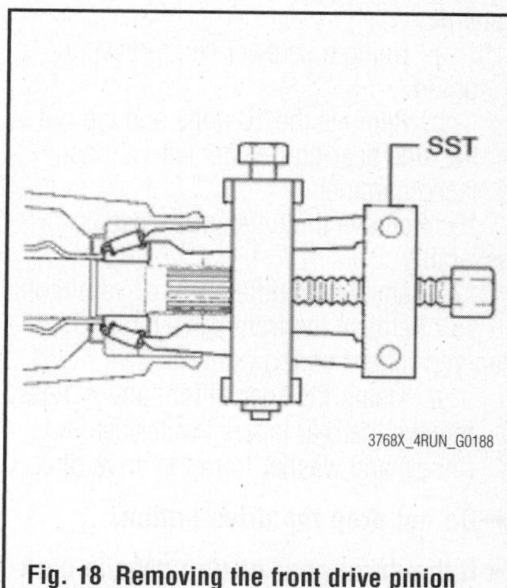

Fig. 18 Removing the front drive pinion front tapered roller bearing (inner)

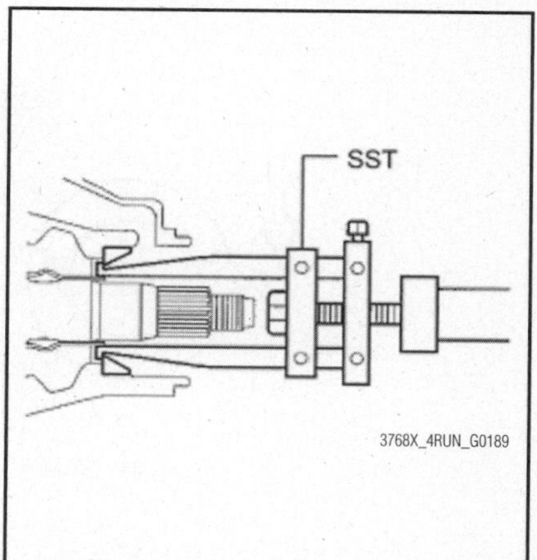

Fig. 19 Removing the front drive pinion front tapered roller bearing (outer)

8. Remove the front differential dust deflector.
9. Remove the front differential carrier oil seal.
10. Remove the front differential drive pinion oil slinger.
   a. Remove the oil slinger from the drive pinion.
11. Remove the front drive pinion front tapered roller bearing (inner).
   a. Using the special tool, remove the front tapered roller bearing (inner) from the drive pinion.
12. Remove the front drive pinion front tapered roller bearing (outer).
   a. Using the special tool, remove the front tapered roller bearing (outer).
13. Remove the front differential oil storage ring.
   a. Using a screwdriver and hammer, tap out the oil storage ring.
14. Remove the front differential drive pinion bearing spacer.
15. Remove the differential side bearing retainer.
   a. Using a screwdriver, remove the union.
   b. Remove the 10 bolts and tap out the side bearing retainer with a plastic faced hammer.
16. Remove the differential case assembly.
17. Remove the differential drive pinion.
18. Remove the front drive pinion rear tapered roller bearing (inner).
   a. Using the special tool and a press, remove the rear tapered roller bearing (inner) and washer from the drive pinion.

➡Do not drop the drive pinion.

➡If the drive gear or ring gear is damaged, replace them as a set.

19. Remove the front drive pinion tapered roller bearing (outer).
20. Using a brass bar and hammer, remove the rear tapered roller bearing (outer).

To install:

21. To install, reverse the removal procedure.

REAR AXLE SHAFT, BEARING & SEAL

REMOVAL & INSTALLATION

See Figures 21 and 22.

➡Use the same procedure for the RH and LH sides.

➡The procedure listed below is for the LH side.

1. Disconnect the cable from the negative battery terminal.

➡When disconnecting the cable, some systems need to be initialized after the cable is reconnected.

2. Remove the rear wheel.
3. Drain the brake fluid.
4. Disconnect the rear flexible hose LH.
5. Remove the rear speed sensor LH.
6. Remove the parking brake assembly.
7. Remove the rear axle shaft with the parking brake plate LH.
   a. Remove the 4 nuts and rear axle shaft with parking brake plate.
   b. Remove the O-ring.
8. Remove the rear axle shaft oil seal LH.

To install:

9. To install, reverse the removal procedure. Take note to tighten the rear axle shaft with the parking brake plate nuts to 44 ft. lbs. (60 Nm). Tighten the rear wheel to 82 ft. lbs. (112 Nm).

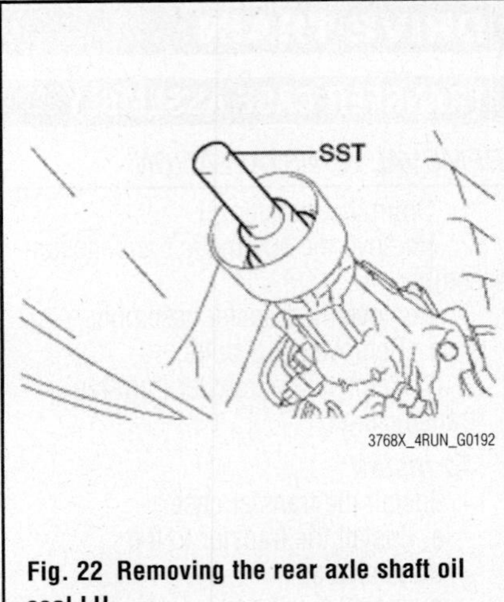

Fig. 22 Removing the rear axle shaft oil seal LH

REAR DRIVESHAFT

REMOVAL & INSTALLATION

2WD Vehicles

1. Remove the propeller shaft assembly.
   a. Place the matchmarks on the propeller shaft flange and differential flange.
   b. Remove the 4 nuts, 4 bolts and 4 washers.
   c. Remove the propeller shaft.
   d. Insert the special tool into the transmission to prevent oil leakage.

To install:

2. To install, reverse the removal procedure. Tighten the propeller shaft assembly bolts and nuts to 65 ft. lbs. (88 Nm).

4WD Vehicles

See Figure 23 and 24.

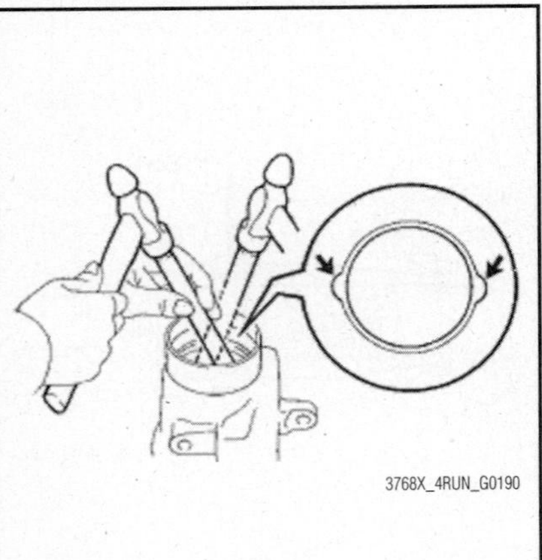

Fig. 20 Removing the rear tapered roller bearing (outer)

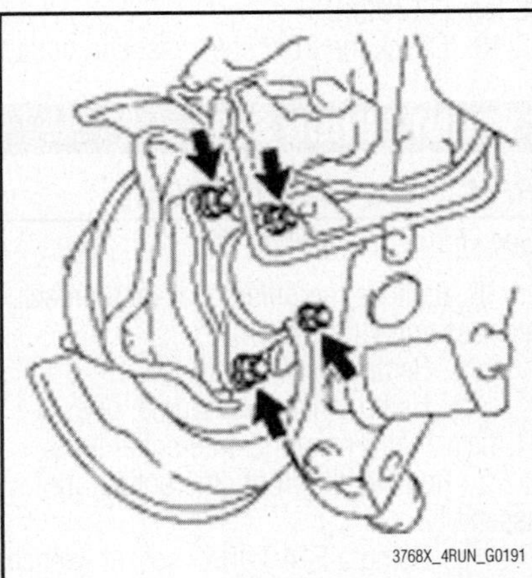

Fig. 21 Removing the rear axle shaft with the parking brake plate LH

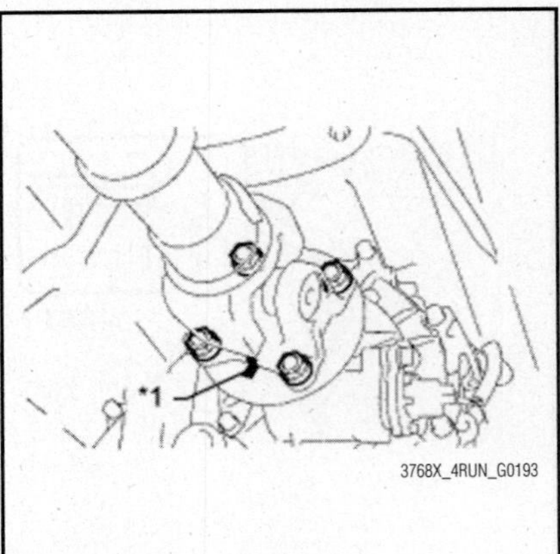

Fig. 23 Placing matchmarks on the propeller shaft flange and transfer flange

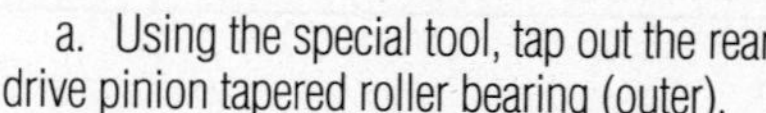

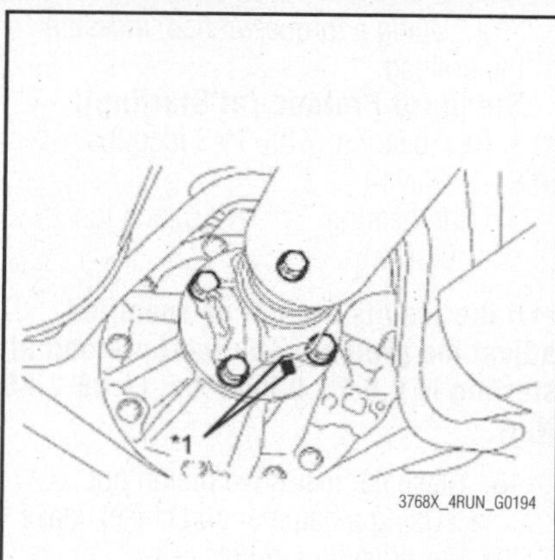

Fig. 24 Placing matchmarks on the propeller shaft flange and differential flange

1. Remove the propeller shaft assembly.
   a. Place matchmarks on the propeller shaft flange and transfer flange.
   b. Remove the 4 nuts, 4 washers and propeller shaft assembly.
   c. Place matchmarks on the propeller shaft flange and differential flange.
   d. Remove the 4 nuts, 4 bolts and 4 washers.

To install:

2. To install, reverse the removal procedure. Tighten the propeller shaft bolts and nuts to 65 ft. lbs. (88 Nm).

REAR PINION SEAL

REMOVAL & INSTALLATION

See Figures 25 through 29.

1. Remove the rear propeller shaft assembly.
2. Remove the rear drive pinion nut.
   a. Using the special tool and a hammer, loosen the staked part of the rear drive pinion nut.

➡Be sure to use special tool with the tapered surface facing the shaft.

➡Do not grind the tip of the SST with a grinder, etc.

➡Completely loosen the staked part of the nut when removing it.

➡Do not damage the threads of the drive pinion nut.

   b. Use the special tool to hold the companion flange.
   c. Using a 30 mm socket wrench, remove the rear drive pinion nut.
3. Remove the rear drive pinion companion flange sub assembly.
   a. Using a special tool, remove the rear drive pinion companion flange sub assembly.

➡Before using the special tool (center bolt), apply hypoid gear oil to its threads and tip.

4. Remove the rear differential carrier oil seal.
5. Remove the rear differential drive pinion oil slinger.

➡Apply grease to the threads and tip of the special tool center bolt before use.

6. Remove the rear drive pinion from tapered roller bearing.
7. Using the special tool, remove the rear drive pinion tapered roller bearing (inner).

➡Apply grease to the threads and tip of the special tool center bolt before use.

   a. Using the special tool, tap out the rear drive pinion tapered roller bearing (outer).
8. Remove the differential drive pinion bearing spacer.

To install:

9. Install the rear differential drive pinion bearing spacer.

➡Install the spacer so that it is facing in the correct direction.

10. Install the differential oil storage ring using a hammer.

➡Be careful not to damage the oil storage ring.

11. Install the rear drive pinion front tapered roller bearing.
   a. Using the special tool and a hammer, tap in the rear drive pinion front roller bearing (outer).
   b. Install the rear drive pinion front roller bearing (inner).

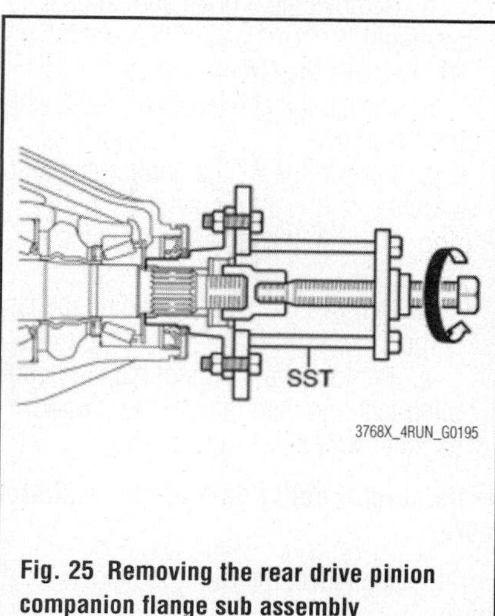

Fig. 25 Removing the rear drive pinion companion flange sub assembly

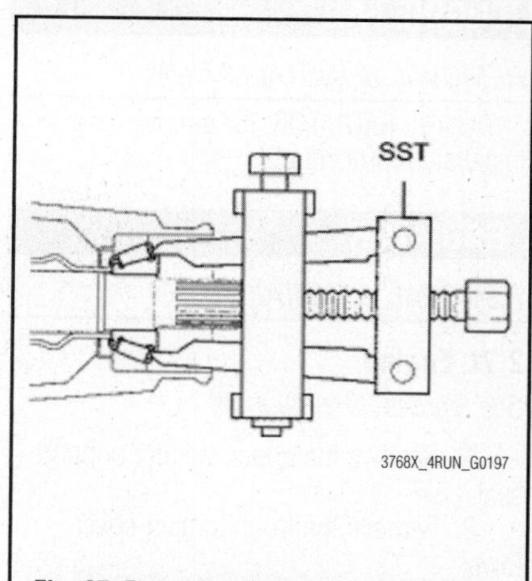

Fig. 27 Removing the rear drive pinion tapered roller bearing (inner)

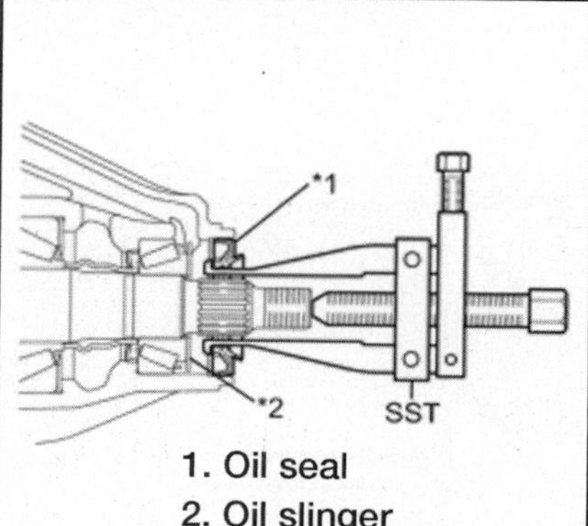

Fig. 26 Removing the rear differential carrier oil seal

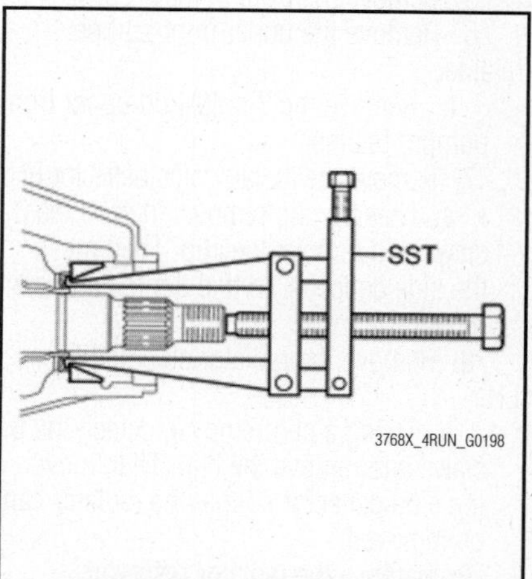

Fig. 28 Tapping out the rear drive pinion tapered roller bearing (outer)

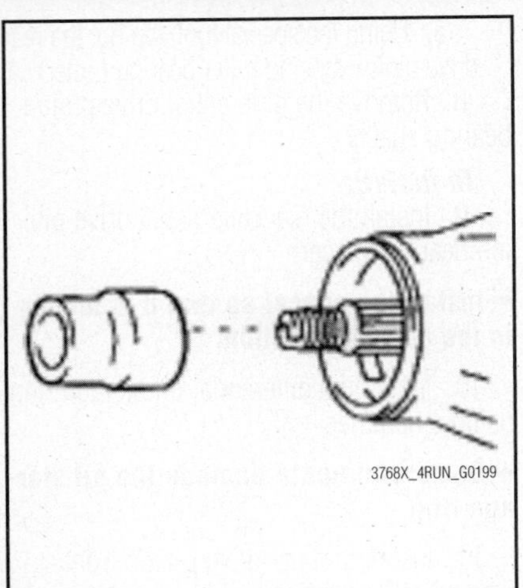

Fig. 29 Installing the rear differential drive pinion bearing spacer

12. Install the front differential drive pinion oil slinger.
13. Install the rear differential carrier oil seal.
    a. Apply MP grease to the lip of a new oil seal.
    b. Using the special tool and a hammer, tap in the rear differential carrier oil seal.
14. Install the rear drive pinion companion flange sub assembly.

➡**Before using the special tool (center bolt), apply hypoid gear oil to its threads and tip.**

    a. Using the special tool to hold the companion flange in place, install the rear drive pinion nut. Tighten to 337 ft. lbs. (457 Nm).
15. Inspect the differential drive pinion preload.
    a. Using a torque wrench, measure the preload.

Standard Preload (at Starting):

- New bearing: 7.35-19.3 inch lbs. (0.83-2.18 Nm)
- Used bearing: 7.79-17.5 inch lbs. (0.88-1.98 Nm)

➡**If the results are not as specified, adjust the preload. The total preload at starting is 9.6-21.0 inch lbs. (1.08-2.38 Nm).**

16. Stake the rear drive pinion nut.
    a. Using a hammer and chisel, stake the rear drive pinion nut.
17. Install the rear propeller shaft assembly.
18. Install the differential oil.
19. Check for differential oil leakage.

ENGINE COOLING

ENGINE FAN

REMOVAL & INSTALLATION

Refer to RADIATOR, for removal and installation procedures.

RADIATOR

REMOVAL & INSTALLATION

2.7L Engine

See Figures 30 through 39.

1. Remove the upper radiator support seal.
2. Remove the front bumper cover lower.
3. Remove the No. 1 engine under cover sub assembly.
4. Drain the engine coolant.
5. Remove the front bumper cover.
6. Remove the upper front bumper retainer.
    a. Remove the 3 bolts and upper front bumper retainer.
7. Remove the radiator side deflector RH.
    a. Using a clip remover, detach the 3 claws and remove the clip. Then move the side deflector so that the radiator can be removed.
8. Remove the radiator side deflector LH.
    a. Using a clip remover, detach the 3 claws and remove the clip. Then move the side deflector so that the radiator can be removed.
9. Remove the radiator reservoir.
    a. Disconnect the reservoir hose from the radiator.
    b. Remove the 3 bolts and radiator reservoir.
10. Remove the fan shroud.
    a. Detach the claw to open the flexible hose clamp.
    b. Loosen the 4 nuts holding the fluid coupling fan.
    c. Remove the fan and alternator v-belt.
    d. Remove the 2 bolts holding the fan shroud.
    e. Remove the 4 nuts of the fluid coupling fan, and then remove the shroud together with the coupling fan.

➡**Be careful not to damage the radiator core.**

11. Disconnect the No. 1 radiator hose.
    a. Disconnect the No. 1 radiator hose from the radiator.
12. Disconnect the No. 2 radiator hose from the radiator.
13. Disconnect the oil cooler inlet hose from the radiator.
14. Disconnect the oil cooler outlet hose from the radiator.
15. Remove the radiator assembly.
    a. Remove the 4 bolts and the radiator.
16. Remove the No. 1 radiator support.
    a. Remove the 2 radiator supports and 2 No. 1 radiator support bushes.
17. Remove the No. 2 radiator support.
    a. Remove the 2 radiator supports and 2 No. 2 radiator support bushes.
18. Remove the No. 1 radiator to support seal from the radiator.

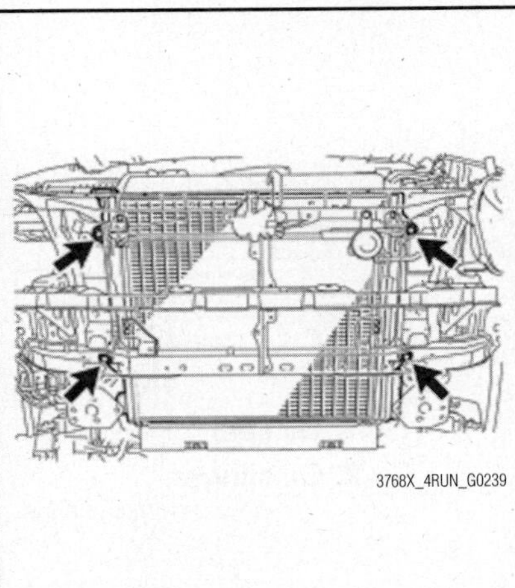

Fig. 30 Removing the radiator

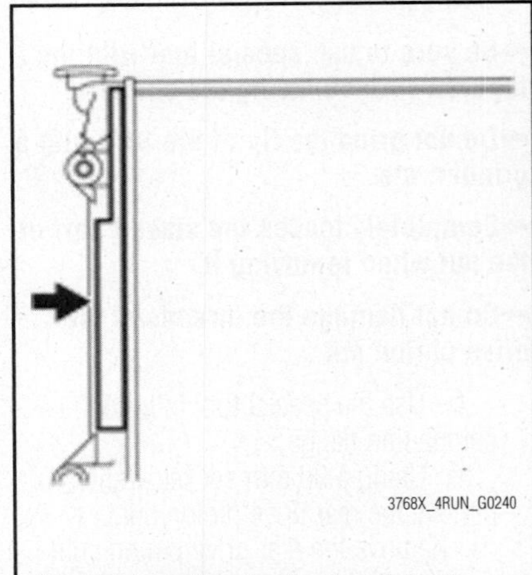

Fig. 31 Installing the No. 1 radiator to support seal

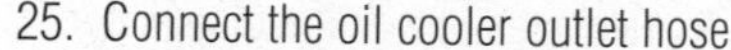

19. Remove the No. 2 radiator to support seal from the radiator.

To install:

20. Install the No. 1 radiator to support seal.
    a. Install a new seal to the radiator as shown.
21. Install the No. 2 radiator to support seal.
22. Install the No. 2 radiator support.
    a. Install the 2 radiator supports and 2 No. 1 radiator support bushes.
23. Install the No. 1 radiator support.
    a. Install the 2 radiator supports and 2 No. 1 radiator support bushes.
24. Install the radiator assembly.
    a. Insert the No. 1 radiator support hooks into the radiator support holes.
    b. Install the radiator with the 4 bolts. Tighten the bolts to 13 ft. lbs. (18 Nm).
25. Connect the oil cooler outlet hose.
    a. Connect the oil cooler outlet hose to the radiator.

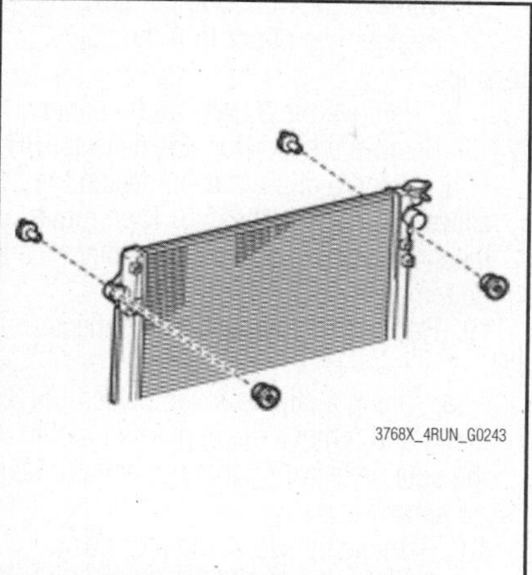

Fig. 34 Installing the No. 1 radiator support

➡Make sure the direction of the hose clamp is as shown.

26. Connect the oil cooler inlet hose to the radiator.
27. Connect the No. 2 radiator hose to the radiator.

➡Make sure the direction of the hose clamp is as shown.

28. Connect the No. 1 radiator hose to the radiator.

➡Make sure the direction of the hose clamp is as shown.

29. Install the fan shroud.
    a. Install the fan pulley to the water pump.
    b. Place the shroud together with the coupling fan between the radiator and engine.

➡Be careful not to damage the radiator core.

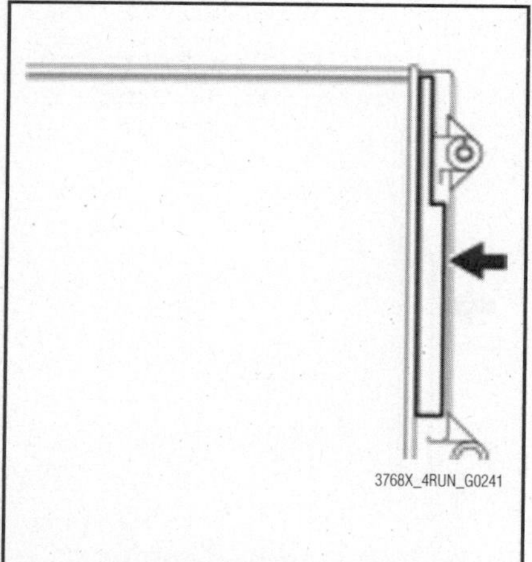

Fig. 32 Installing the No. 2 radiator to support seal

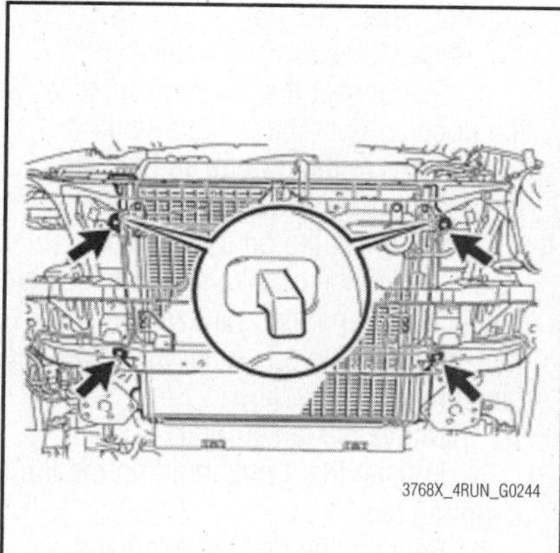

Fig. 35 Inserting the No. 1 radiator support hooks into the radiator support holes

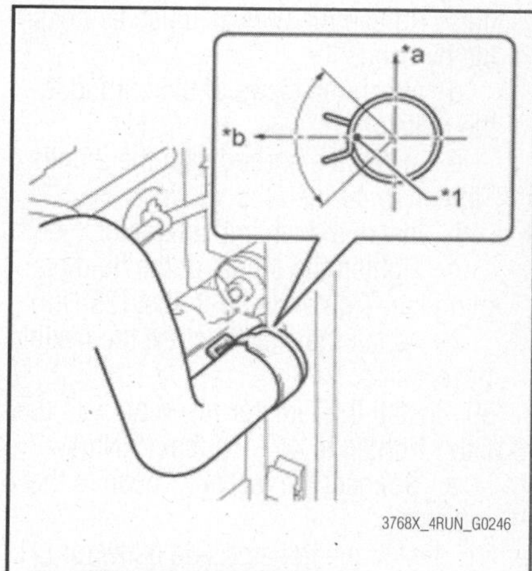

Fig. 37 Connecting the No. 2 radiator hose

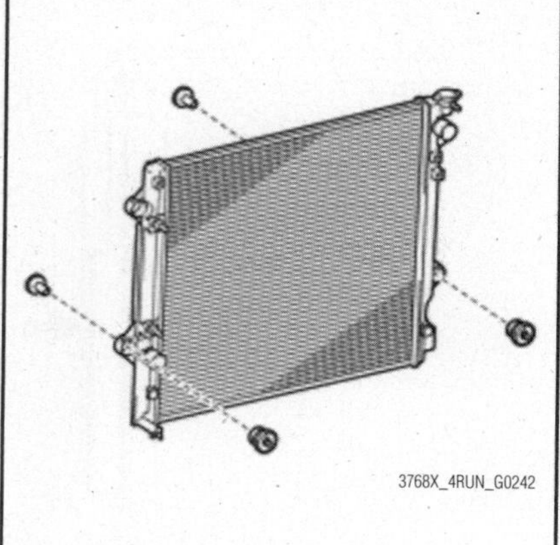

Fig. 33 Installing the 2 radiator supports and 2 No. 1 radiator support bushes

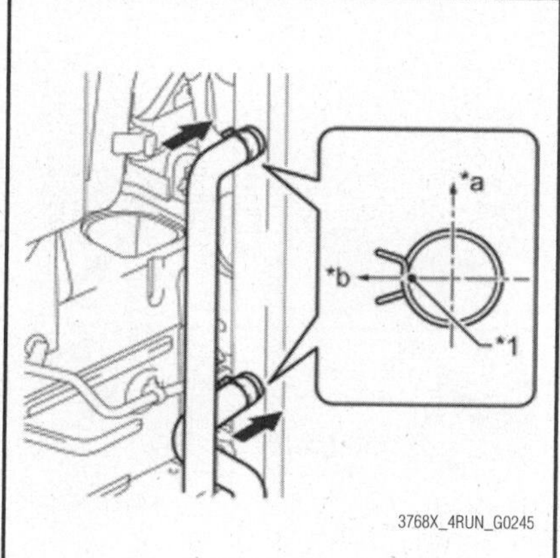

Fig. 36 Connecting the oil cooler outlet hose

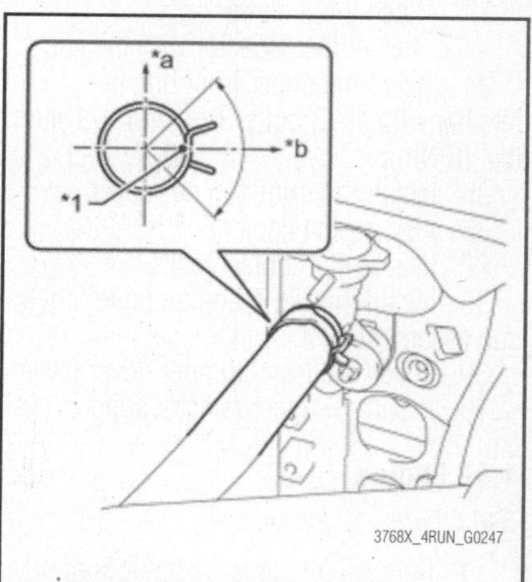

Fig. 38 Connecting the No. 1 radiator hose

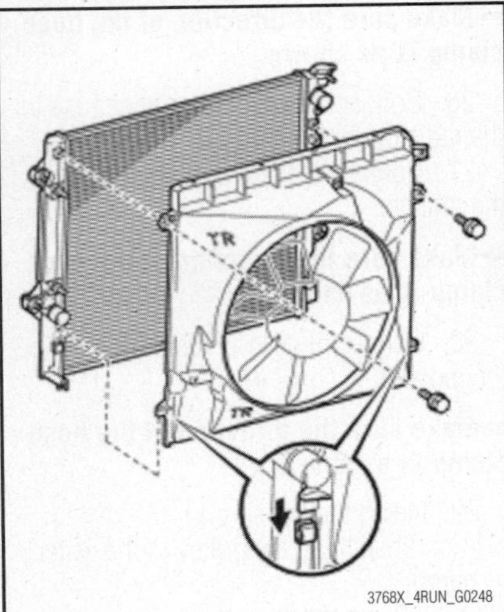

Fig. 39 Attaching the claws of the shroud to the radiator

c. Temporarily install the fluid coupling fan to the water pump with the 4 nuts. Tighten the nuts as much as possible by hand.

d. Attach the claws of the shroud to the radiator.

e. Install the shroud with the 2 bolts. Tighten to 44 inch lbs. (5 Nm).

f. Install the fan and alternator v-belt.

g. Tighten the 4 nuts of the fluid coupling fan. Tighten to 18 ft. lbs. (25 Nm).

h. Attach the claw to close the flexible hose clamp.

30. Install the radiator reservoir with the 3 bolts. Tighten to 44 inch lbs. (5 Nm).

a. Connect the reservoir hose to the radiator.

31. Install the radiator side deflector LH.

a. Attach the 3 claws.

b. Install the deflector with the clip.

32. Install the radiator side deflector RH.

33. Attach the 3 claws.

a. Install the deflector with the clip.

34. Install the upper front bumper retainer with the 3 bolts. Tighten to 71 inch lbs. (8 Nm).

35. Install the front bumper cover.

36. Add engine coolant.

37. Inspect for coolant leaks.

38. Install the No. 1 engine under cover sub assembly.

39. Install the front bumper cover lower.

40. Install the upper radiator support seal.

4.0L Engine

See Figures 40 through 48.

1. Remove the upper radiator support seal.

2. Remove the front bumper cover lower.

3. Remove the No. 1 engine under cover sub assembly.

4. Drain the engine coolant.

5. Remove the V-bank cover.

6. Remove the front bumper cover.

7. Remove the upper front bumper retainer.

a. Remove the 3 bolts and retainer.

8. Remove the radiator side deflector RH.

a. Using a clip remover, detach the 3 claws and remove the clip. Then move the side deflector so that the radiator can be removed.

9. Remove the radiator side deflector LH.

a. Using a clip remover, detach the 3 claws and remove the clip. Then move the side deflector so that the radiator can be removed.

10. Remove the No. 1 radiator hose.

11. Remove the No. 2 radiator hose.

a. Disconnect the radiator hose from the water inlet.

b. Detach the clamp and remove the radiator hose.

12. Remove the radiator reservoir.

a. Disconnect the reservoir hose from the upper side of the radiator tank.

b. Remove the 3 bolts and radiator reservoir.

13. Disconnect the oil cooler tube.

a. Detach the claw to open the flexible hose clamp, and then remove the 2 bolts to disconnect the oil cooler tube from the fan shroud.

14. Remove the fan shroud.

a. Loosen the 4 nuts holding the fluid coupling fan.

b. Remove the fan and alternator v-belt.

c. Remove the 2 bolts holding the fan shroud.

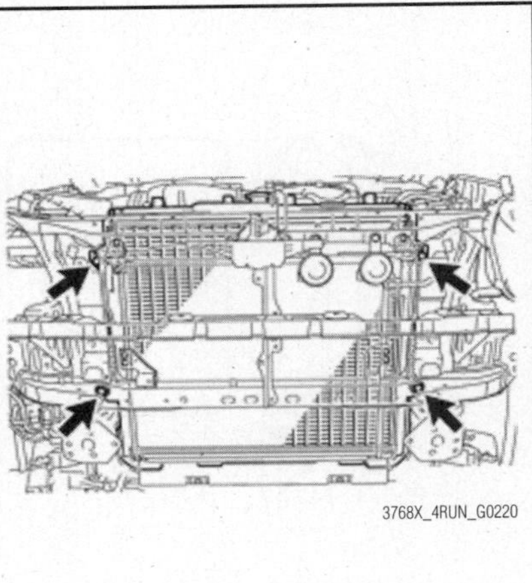

Fig. 40 Removing the radiator

d. Remove the 4 nuts of the fluid coupling fan, and then remove the shroud together with the coupling fan.

➡Be careful not to damage the radiator core.

e. Remove the fan pulley from the water pump.

15. Remove the radiator assembly.

a. Disconnect the 2 oil cooler hoses.

b. Remove the 4 bolts and radiator.

16. Remove the No. 1 radiator to support seal.

a. Remove the seal from the radiator assembly.

17. Remove the No. 2 radiator to support seal.

a. Remove the seal from the radiator assembly.

To install:

18. Install the No. 1 radiator to support seal.

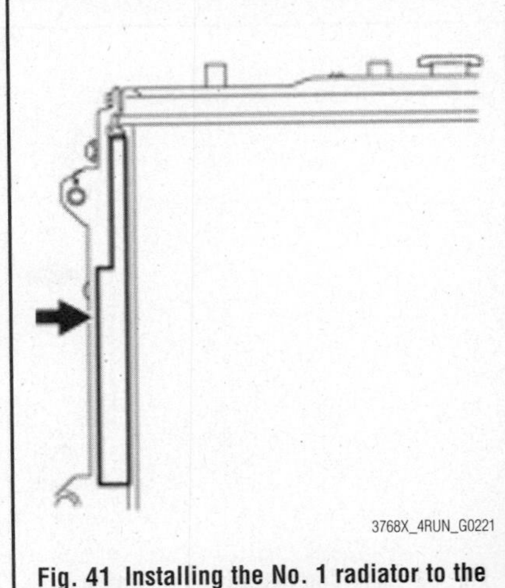

Fig. 41 Installing the No. 1 radiator to the support seal

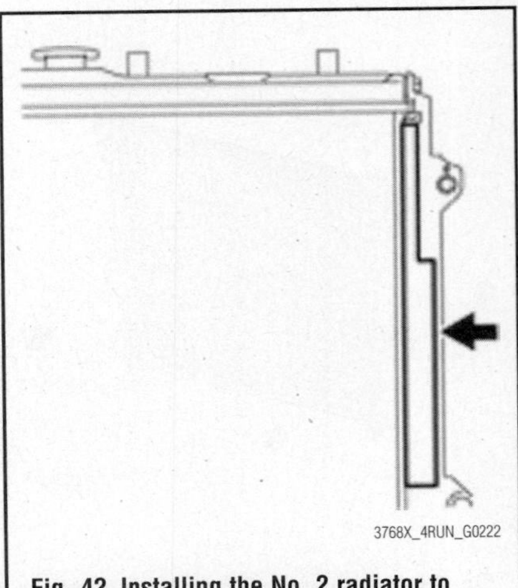

Fig. 42 Installing the No. 2 radiator to support seal

19. Install the No. 2 radiator to support seal.

20. Install the radiator assembly.

a. Insert the radiator bracket hooks into the radiator support holes.

b. Install the radiator with the 4 bolts. Tighten the bolts to 13 ft. lbs. (18 Nm).

c. Connect the 2 oil cooler hoses.

➡Make sure the direction of the hose clamp is as shown.

21. Install the fan shroud.

a. Install the fan pulley to the water pump.

b. Place the shroud together with the coupling fan between the radiator and engine.

➡Be careful not to damage the radiator core.

c. Align the pain marks on the heads of the water pump stud bolts with the paint marks of the same color on the outer edge of the fluid coupling flange and install the fluid coupling to the water pump.

d. Temporarily install the fluid coupling fan to the water pump with the 4 nuts. Tighten the nuts as much as possible by hand.

e. Attach the claws of the shroud to the radiator.

f. Install the shroud with the 2 bolts. Tighten the bolts to 44 inch lbs. (5 Nm).

g. Install the fan and alternator v-belt.

h. Tighten the 4 nuts of the fluid coupling fan to 15 ft. lbs. (21 Nm).

22. Connect the oil cooler tube with the 2 bolts, and attach the claw to close the flexible hose clamp. Tighten to 49 inch lbs. (5.5 Nm).

23. Install the radiator reservoir with the 3 bolts. Tighten to 44 inch lbs. (5 Nm). Connect the reservoir hose to the upper side of the radiator tank upper.

24. Install the No. 2 radiator hose.

a. Connect the No. 2 radiator hose so that its paint mark aligns with the radiator protrusion.

➡Make sure the direction of the hose clamp is as shown.

b. Connect the No. 2 radiator hose so that its paint mark aligns with the water inlet housing protrusion.

➡Make sure the direction of the hose clamp is as shown.

25. Install the No. 1 radiator hose.

Fig. 43 Inserting the radiator bracket hooks into the radiator support holes

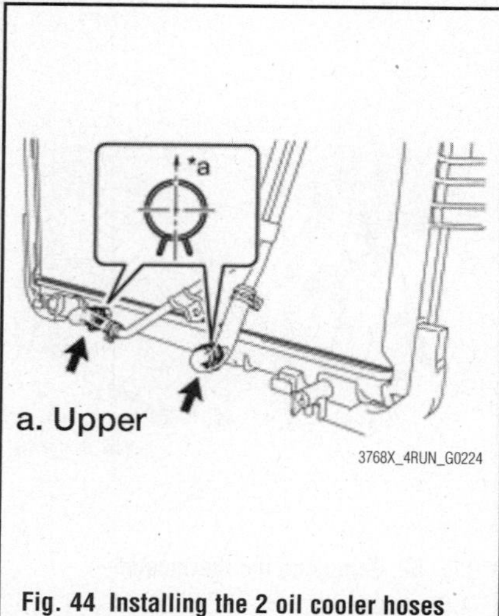

Fig. 44 Installing the 2 oil cooler hoses

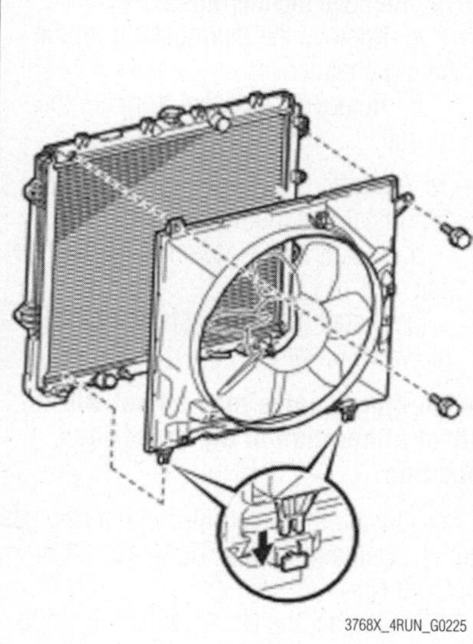

Fig. 45 Attaching the claws of the shroud to the radiator

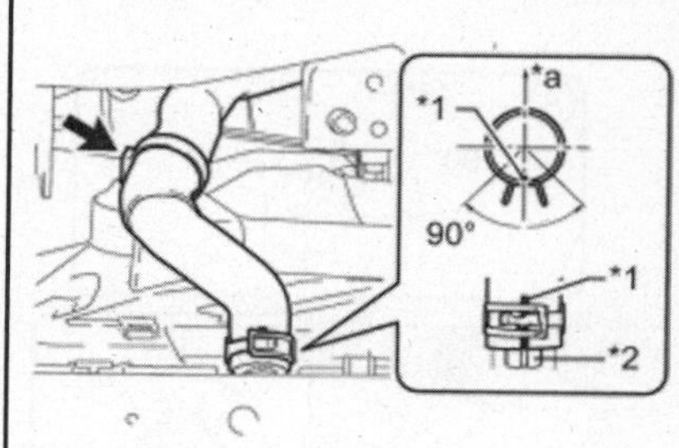

Fig. 46 Connecting the No. 2 radiator hose so that the paint marks align with the radiator protrusion

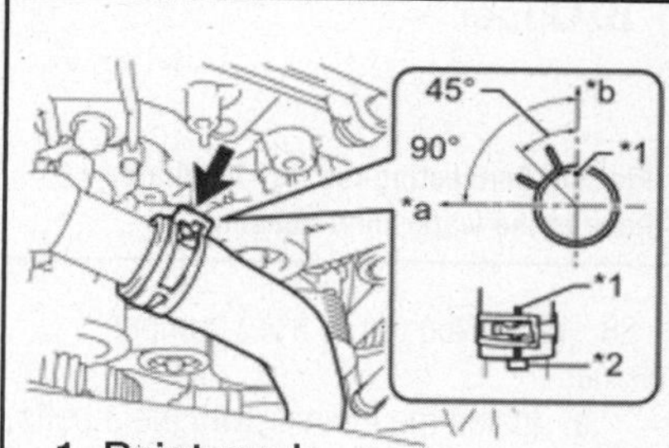

Fig. 47 Connecting the No. 2 radiator hose so that its paint mark aligns with the water inlet housing protrusion

a. Connect the No. 1 radiator hose to the water inlet housing.

➡Make sure the paint mark of the No. 1 radiator hose is facing upward.

➡Make sure the direction of the hose clamp is as shown in the illustration.

b. Connect the No. 1 radiator hose so that its paint mark aligns with the radiator protrusion as shown in the illustration labeled B.

➡Make sure the paint mark of the No. 1 radiator hose is facing upward.

➡Make sure the direction of the hose clamp is as shown

26. Install the radiator side deflector RH.

a. Attach the 3 screws.

b. Install the deflector with the clip.

27. Install the radiator side deflector RH.

a. Attach the 3 claws.

b. Install the deflector with the clip.

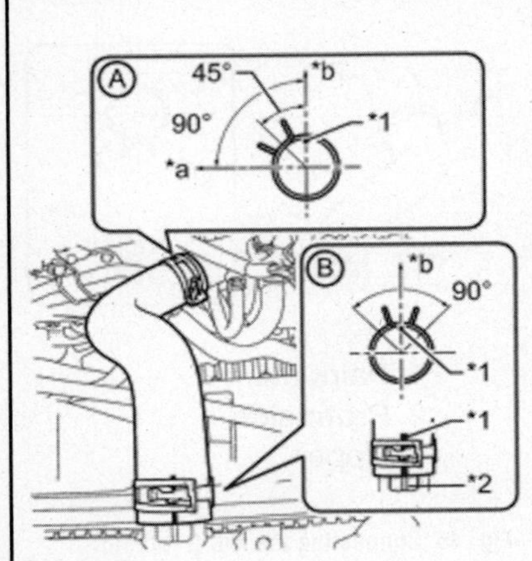

Fig. 48 Connecting the No. 1 radiator hose to the water inlet housing

28. Install the upper front bumper retainer.
    a. Install the retainer with the 3 bolts. Tighten to 71 inch lbs. (8 Nm).
29. Install the front bumper cover.
30. Add engine coolant.
31. Inspect for engine coolant leaks.
32. Install the No. 1 engine under cover sub assembly.
33. Install the front bumper cover lower.
34. Install the v-bank cover.
35. Install the upper radiator support seal.

THERMOSTAT

REMOVAL & INSTALLATION

2.7L Engine

See Figures 49 through 51.

1. Remove the upper radiator support seal.
2. Remove the front bumper cover lower.
3. Remove the No. 1 engine under cover sub assembly.
4. Drain the engine coolant.
5. Remove the fan and alternator v-belt.
6. Disconnect the vane pump assembly.
7. Disconnect the No. 2 radiator hose from the water inlet.
8. Remove the water inlet.
    a. Remove the gasket from the timing chain cover.

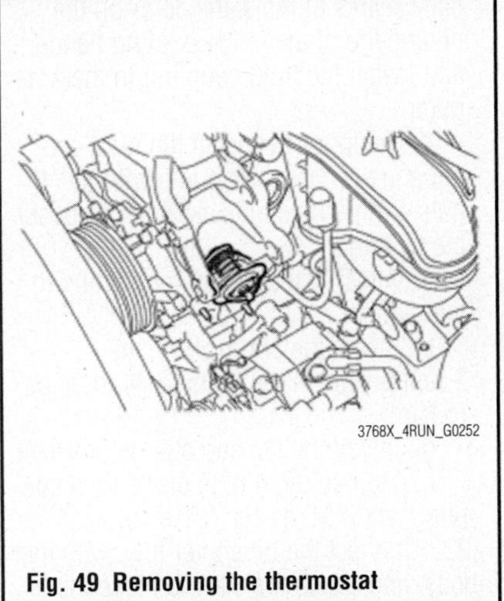

Fig. 49 Removing the thermostat

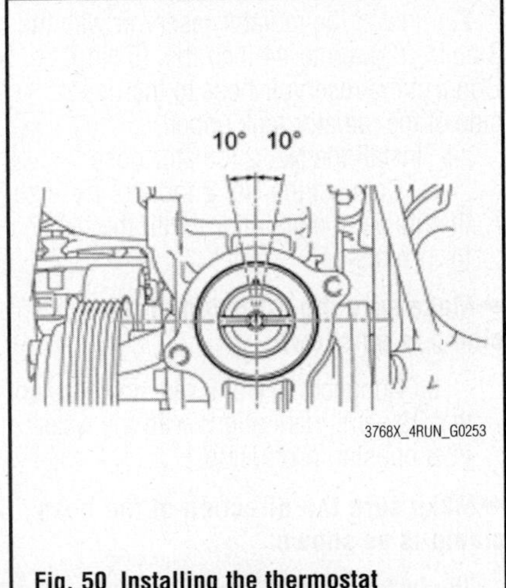

Fig. 50 Installing the thermostat

9. Remove the thermostat.
    a. Remove the thermostat from the timing chain cover.
    b. Remove the gasket from the thermostat.

To install:

10. Install the thermostat.
    a. Install a new gasket to the thermostat.
    b. Install the thermostat with the jiggle valve upward.

➡The jiggle valve may be set within 10°of either side of the prescribed position.

11. Install the water inlet with a new gasket, the bolt and 2 nuts. Tighten to 21 ft. lbs. (28 Nm).
12. Connect the No. 2 radiator hose to the water inlet.

➡Make sure the direction of the hose clamp is as shown.

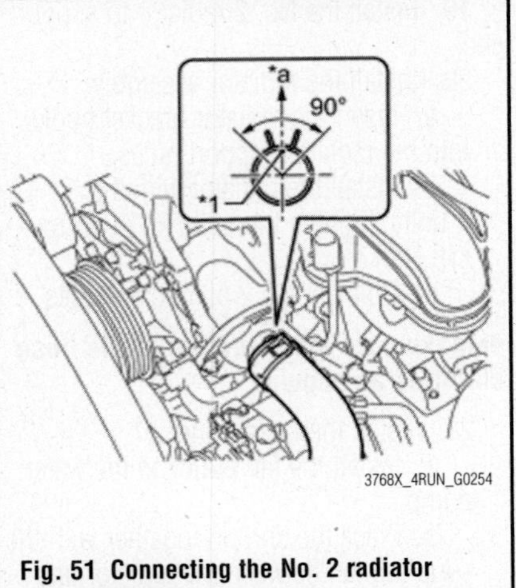

Fig. 51 Connecting the No. 2 radiator hose to the water inlet

13. Connect the vane pump assembly.
14. Install the fan and alternator v-belt.
15. Add engine coolant.
16. Inspect for coolant leaks.
17. Install the No. 1 engine under cover sub assembly.
18. Install the front bumper cover lower.
19. Install the upper radiator support seal.

4.0L Engine

See Figure 52.

1. Remove the upper radiator support seal.
2. Remove the front bumper cover lower.
3. Remove the No. 1 engine under cover sub assembly.
4. Drain the engine coolant.
5. Remove the v-bank cover.
6. Disconnect the No. 2 radiator hose.
    a. Disconnect the No. 2 radiator hose.
7. Remove the water inlet with the thermostat.

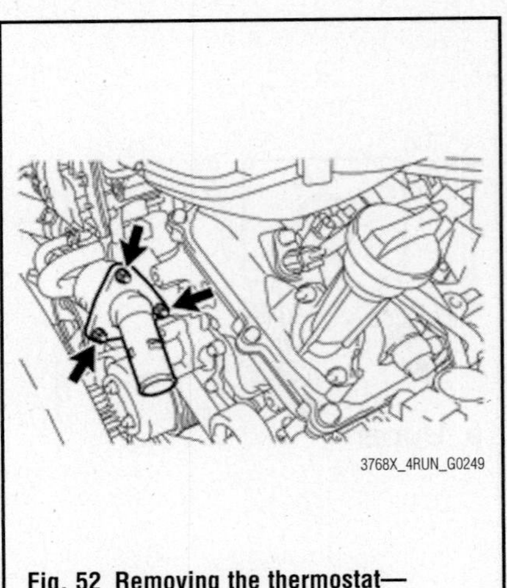

Fig. 52 Removing the thermostat—4.0L engine

➡If the thermostat was not installed, cooling efficiency would decrease. Even if the engine tends to overheat, do not remove the thermostat.

a. Remove the 3 nuts, water inlet with the thermostat and gasket.

To install:

8. Install a new gasket and the water inlet with thermostat with the 3 nuts. Tighten to 80 inch lbs. (9 Nm). To complete installation, reverse the removal procedure.

WATER PUMP

REMOVAL & INSTALLATION

2.7L Engine

See Figure 53.

1. Disconnect the cable from the negative battery terminal.

➡When disconnecting the cable some systems need to be initialized after the cable is reconnected.

2. Remove the upper radiator support seal.
3. Remove the front bumper cover lower.
4. Remove the No. 1 engine under cover sub assembly.
5. Drain the engine coolant.
6. Remove the radiator reservoir.
7. Remove the fan shroud.
8. Remove the air cleaner and hose.
9. Remove the alternator assembly.
10. Remove the No. 1 idler pulley sub assembly.
11. Remove the v-ribbed belt tensioner assembly.
12. Remove the water pump assembly.
    a. Remove the 10 bolts, water pump and gasket.

To install:

13. Install the water pump assembly.
    a. Install a new gasket and the water pump with the 10 bolts. Tighten the A bolts (black arrows) to 19 ft. lbs. (26 Nm). Tighten the B bolts (white arrows) to 79 inch lbs. (8.9 Nm).
14. To complete installation, reverse the installation procedure.

4.0L Engine

See Figures 54 and 55.

1. Remove the upper radiator support seal.
2. Remove the front bumper cover lower.
3. Remove the No. 1 engine under cover sub assembly.
4. Drain the engine coolant.
5. Remove the v-bank cover.
6. Remove the No. 1 radiator hose.
7. Remove the No. 2 radiator hose.
8. Remove the radiator reservoir.
9. Disconnect the oil cooler tube.
10. Remove the fan shroud.
11. Remove the water inlet housing.
    a. Disconnect the throttle body connector.
    b. Disconnect the 5 water by pass hoses.
    c. Disconnect the oil cooler hose.
    d. Disconnect the No. 2 oil cooler hose.
    e. Remove the 5 bolts and water inlet.
    f. Remove the O-ring from the water outlet pipe.
    g. Remove the gasket from the water pump.
12. Remove the No. 2 idler pulley sub assembly.
13. Disconnect the vane pump assembly.
    a. Disconnect the 2 connectors.
    b. Detach the 2 wire harness clamps.
    c. Remove the 2 bolts and disconnect the vane pump.
14. Remove the alternator assembly.
15. Remove the cooler compressor assembly.
16. Remove the v-ribbed belt tensioner assembly.
17. Remove the water pump assembly.
    a. Remove the 17 bolts, water pump and gasket.

To install:

18. Install the water pump assembly.
    a. Install a new gasket and the water pump with the 17 bolts. Tighten bolt A to 35 ft. lbs. (47 Nm). Tighten bolt B to 8 ft. lbs. (11 Nm). Tighten bolt C to 17 ft. lbs. (23 Nm).
19. Install the v-ribbed belt tensioner assembly.
20. Install the cooler compressor assembly.
21. Install the alternator assembly.
22. Connect the vane pump assembly. Tighten the 2 bolts to 32 ft. lbs. (43 Nm).
    a. Attach the 2 wire harness clamps.
    b. Connect the 2 connectors.
23. Install the No. 2 idler puller sub assembly. Tighten to 40 ft. lbs. (54 Nm).
24. Install the water inlet housing.
    a. Install a new O-ring to the water outlet pipe.
    b. Install a new gasket to the water pump.
    c. Apply soapy water to the gasket of the water outlet pipe.
    d. Install the water inlet with the 5 bolts. Tighten to 7 ft. lbs. (10 Nm).
    e. Connect the 5 water by-pass hoses.

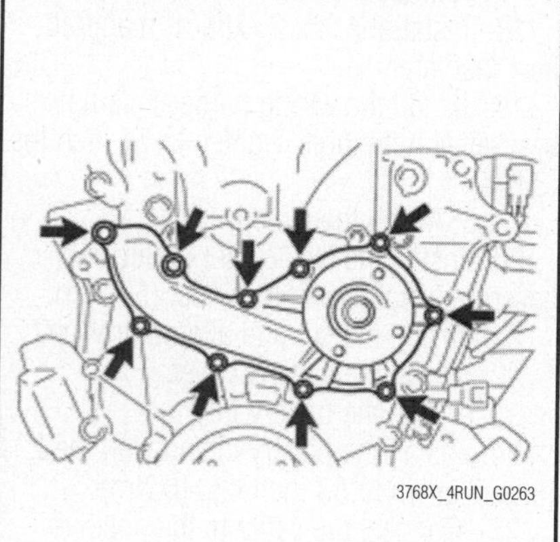

Fig. 53 Removing the water pump—2.7L engine

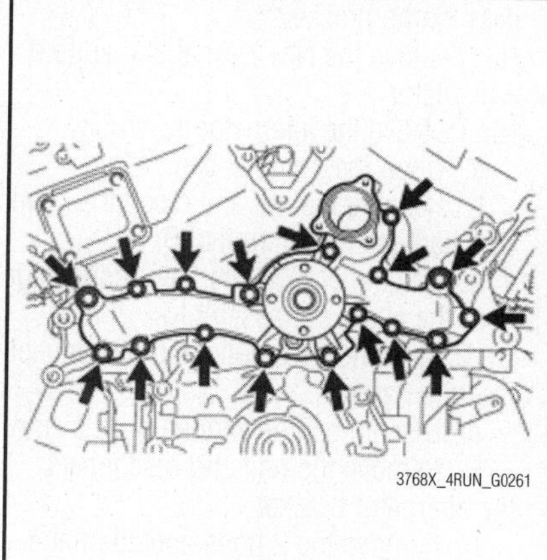

Fig. 54 Removing the water pump

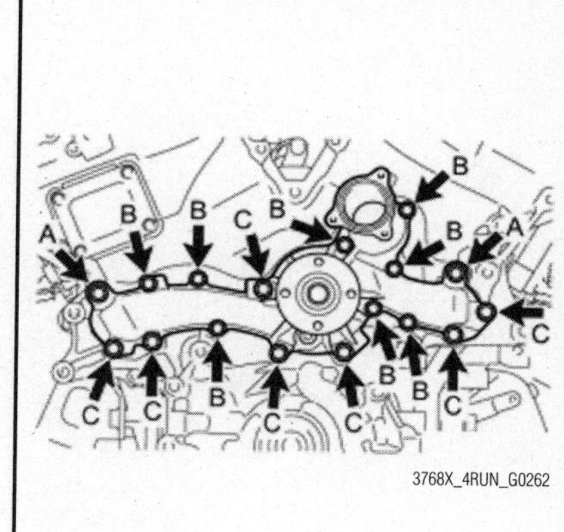

Fig. 55 Installing the water pump

f. Connect the No. 2 oil cooler hose.
g. Connect the oil cooler hose.
h. Connect the throttle body connector.
25. Install the fan shroud.
26. Connect the oil cooler tube.
27. Install the radiator reservoir.
28. Install the No. 2 radiator hose.
29. Install the No. 1 radiator hose.
30. Install the v-bank cover.
31. Add engine coolant.
32. Inspect for engine coolant leak.
33. Install the No. 1 engine under cover sub assembly.
34. Install the front bumper cover lower.
35. Install the upper radiator support seal.

ENGINE ELECTRICAL

CHARGING SYSTEM

ALTERNATOR

REMOVAL & INSTALLATION

2.7L Engine

See Figure 56.

1. Disconnect the cable from the negative battery terminal.

➡**When disconnecting the cable, some systems need to be initialized after the cable is reconnected.**

2. Remove the air cleaner and hose.
3. Remove the fan and alternator v-belt.
4. Remove the alternator.
   a. Disconnect the alternator connector.
   b. Remove the terminal cap.
   c. Remove the nut and disconnect the alternator wire from terminal B.
   d. Remove the 3 bolts and the alternator.

To install:

5. Install the alternator assembly.
   a. Install the alternator with the 3 bolts. Tighten to 32 ft. lbs. (43 Nm).
   b. Connect the alternator wire to terminal B with the nut. Tighten to 87 inch lbs. (9.8 Nm).
   c. Install the terminal cap.
   d. Connect the alternator connector.
6. Install the fan and alternator v-belt.
7. Install the air cleaner and hose.

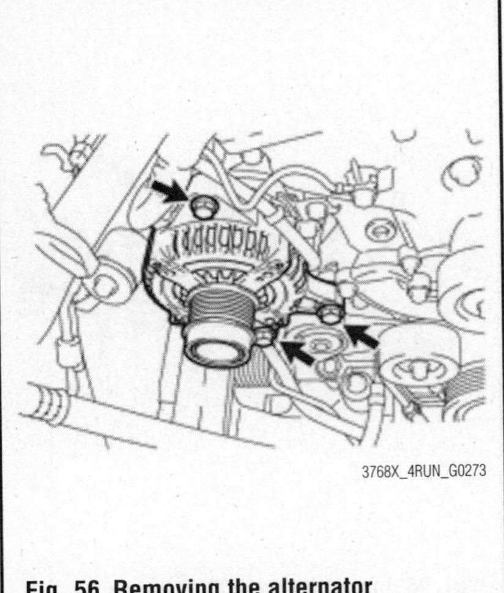

Fig. 56 Removing the alternator

8. Connect the cable to the negative battery terminal.

➡**When disconnecting the cable, some systems need to be initialized after the cable is reconnected.**

4.0L Engine

See Figure 57.

1. Disconnect the cable from the negative battery terminal.

➡**When disconnecting the cable, some systems need to be initialized after the cable is reconnected.**

2. Disconnect the cable from the positive battery terminal.
3. Remove the battery clamp.
   a. Loosen the 2 nuts and remove the battery clamp.
4. Remove the battery.
5. Remove the battery tray.
6. Remove the v-bank cover.
7. Remove the fan and alternator v-belt.
8. Remove the No. 2 idler pulley sub assembly.
9. Remove the battery.
10. Remove the battery tray.
11. Remove the v-bank cover.
12. Remove the fan and alternator v-belt.
13. Remove the No. 2 idler pulley sub assembly.
14. Remove the wiring harness clamp bracket.
    a. Detach the clamp.
    b. Remove the bolt and wiring harness clamp bracket.
15. Remove the No. 2 exhaust manifold heat insulator.
16. Remove the alternator assembly.
    a. Open the terminal cap.
    b. Remove the nut and disconnect the wire harness from terminal B.
    c. Disconnect the alternator connector from the alternator assembly.
    d. Remove the 2 bolts and disconnect the wire harness.
    e. Disconnect the wire harness clamp.
    f. Remove the bolt and disconnect the alternator bracket.
    g. Remove the 2 bolts and alternator assembly.
    h. Remove the bolt and alternator bracket.

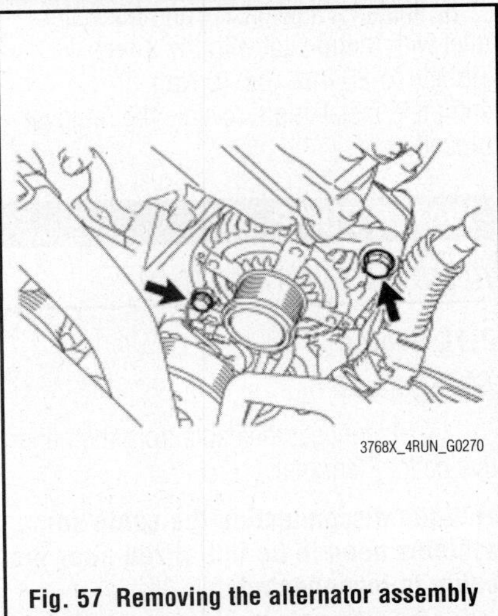

Fig. 57 Removing the alternator assembly

To install:

17. Install the alternator assembly.
    a. Install the alternator bracket to the alternator with the bolt. Tighten to 15 ft. lbs. (20 Nm).
    b. Install the alternator with the 2 bolts. Tighten to 32 ft. lbs. (43 Nm).
    c. Install the alternator bracket with the bolt. Tighten to 15 ft. lbs. (20 Nm).
    d. Attach the wire harness clamp.
    e. Install the wire harness with the 2 bolts. Tighten to 71 inch lbs. (8 Nm).
    f. Connect the alternator connector to the alternator.
    g. Install the alternator wire with the nut. Tighten to 87 inch lbs. (9.8 Nm).
    h. Close the terminal cap.
18. Install the No. 2 exhaust manifold heat insulator.
19. Install the wiring harness clamp bracket with the bolt. Tighten to 71 inch lbs. (8 Nm).
    a. Attach the clamp.
20. Install the No. 2 idler pulley sub assembly. Tighten to 40 ft. lbs. (54 Nm).
21. Install the fan and alternator v-belt.
22. Install the v-bank cover.
23. Install the battery tray.
24. Install the battery clamp with the 2 nuts. Tighten to 53 inch lbs. (6 Nm).
25. Connect the cable to the positive battery terminal.
26. Connect the cable to the negative battery terminal.

FIRING ORDER

2.7L Engine Firing order: 1–3–4–2
4.0L Engine Firing order: 1–2–3–4–5–6

IGNITION COIL

REMOVAL & INSTALLATION

2.7L Engine

See Figure 58.

1. Remove the air cleaner.
2. Remove the intake air connector.
3. Remove the ignition coil assembly.
   a. Disconnect the 4 ignition coil connectors.
   b. Remove the 4 bolts and 4 ignition coils.
4. Remove the spark plugs using a 16 mm spark plug wrench.

To install:

5. Install the spark plugs using a 16 mm spark plug wrench. Tighten to 13 ft. lbs. (18 Nm).
6. Install the ignition coil assembly.
   a. Install the 4 ignition coils with the 4 bolts. Tighten to 80 inch lbs. (9 Nm).
   b. Connect the 4 ignition coil connectors.
7. Install the intake air connector.
8. Install the air cleaner and hose.

4.0L Engine

See Figures 59 and 60.

1. Remove the v-bank cover.
2. Remove the air cleaner cap and hose.
   a. Disconnect the mass air flow meter connector, vacuum hose and ventilation hose and detach the 4 clamps.
   b. Loosen the clamp.
   c. Unfasten the 4 hook clamps, and then remove the bolt and air cleaner cap and hose.
3. Remove the ignition coil assembly.
   a. Disconnect the 6 ignition coil connectors.
   b. Remove the 6 bolts and 6 ignition coils.
4. Remove the 6 spark plugs.

To install:

5. Install the spark plugs and tighten to 13 ft. lbs. (18 Nm).
6. Install the ignition coil assembly with the bolts. Tighten to 7 ft. lbs. (10 Nm). Connect the connectors.
7. Install the air cleaner cap and hose.

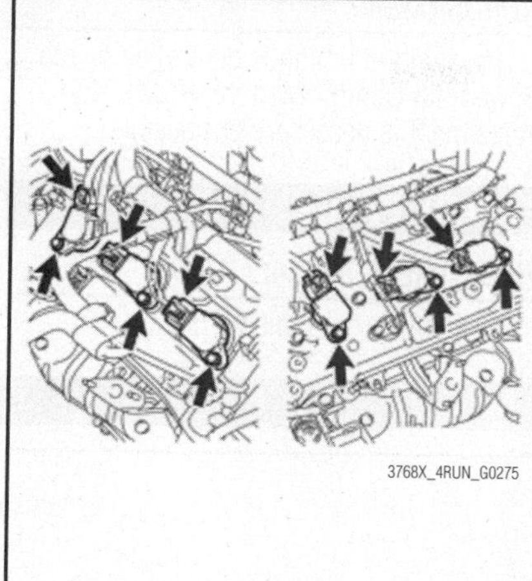

Fig. 59 Removing the ignition coil

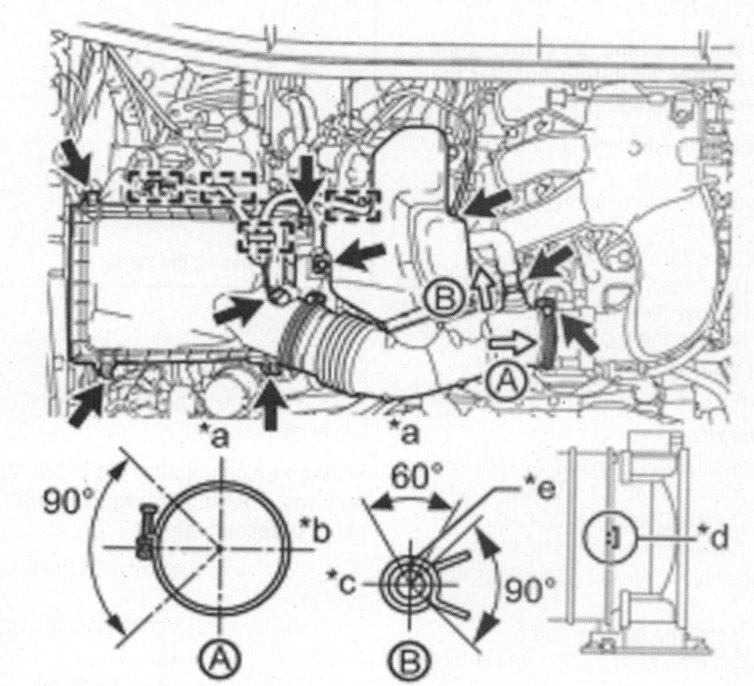

a. Top
b. Front
c. RH
d. Align cutout portion of the hose with the protrusion of the throttle
e. Paint mark

3768X_4RUN_G0276

Fig. 60 Installing the air cleaner cap and hose

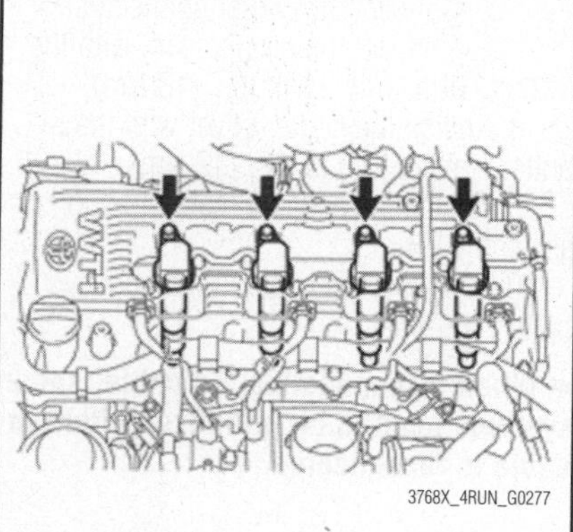

Fig. 58 Removing the ignition coils

a. Install the air cleaner cap and hose with the bolt and fasten the 4 hook clamps. Tighten to 44 inch lbs. (5 Nm).

b. Tighten the clamp to 44 inch lbs. (5 Nm).

c. Attach the 4 clamps and connect the ventilation hose, vacuum hose and mass air flow meter connector.

➡The direction of the hose clamp is illustrated.

IGNITION TIMING

ADJUSTMENT

The ignition timing is controlled by the Powertrain Control Module (PCM). No adjustment is necessary or possible.

SPARK PLUGS

REMOVAL & INSTALLATION

1. Remove the ignition coils.
2. Using a 16 mm plug wrench, remove the spark plugs.
3. Clean the spark plugs.

To install:

4. Adjust the spark plug electrode gap. Electrode gap for new spark plug is 0.039 to 0.043 inch (1.0 to 1.1 mm).
5. Using a 16 mm plug wrench, install the spark plugs and tighten to 13 ft. lbs. (17.5 Nm.).
6. Reinstall the ignition coils.

ENGINE ELECTRICAL

STARTING SYSTEM

STARTER

REMOVAL & INSTALLATION

2.7L Engine

See Figure 61.

1. Disconnect the cable from the negative battery terminal.

➡When disconnecting the cable, some systems need to be initialized after the cable is reconnected.

2. Remove the rear engine under cover assembly.
3. Remove the transmission oil filter tube sub assembly.
4. Remove the starter assembly.
    a. Disconnect the starter connector.
    b. Remove the terminal cap.
    c. Remove the nut and disconnect the starter wire.
    d. Remove the 2 bolts and the starter.

To install:

5. Install the starter assembly with the 2 bolts. Tighten to 27 ft. lbs. (37 Nm).
    a. Connect the starter connector.
    b. Connect the starter wire harness with the nut. Tighten to 87 inch lbs. (9.8 Nm).
    c. install the terminal cap.
6. Install the transmission oil filler tube sub assembly.
7. Install the rear engine under cover assembly.
8. Connect the cable to the negative battery terminal.

➡When disconnecting the cable, some systems need to be initialized after the cable is reconnected.

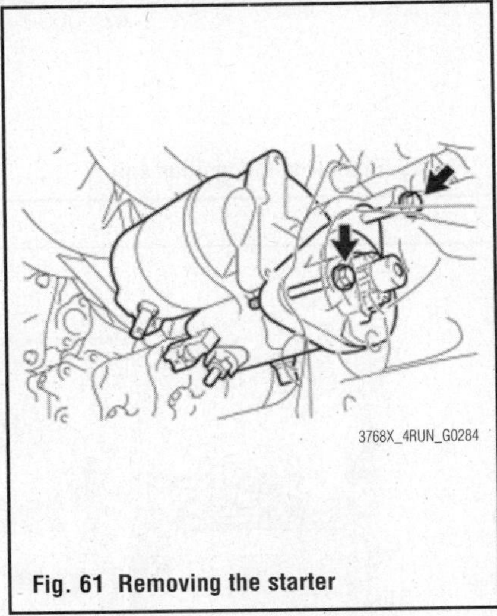

Fig. 61 Removing the starter

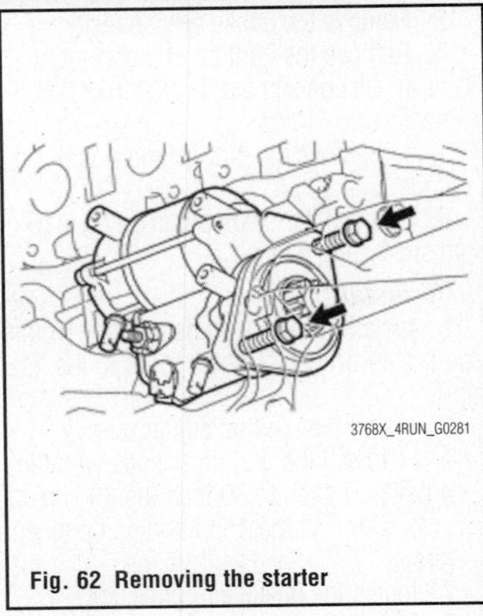

Fig. 62 Removing the starter

4.0L Engine

See Figure 62.

1. Disconnect the cable from the negative battery terminal.

➡When disconnecting the cable, some systems need to be initialized after the cable is reconnected.

2. Remove the exhaust manifold sub assembly LH.
3. Remove the 3 bolts and the starter cover.
4. Remove the starter assembly.
    a. Remove the bolt and disconnect the ground wire.
    b. Disconnect the starter connector.
    c. Open the terminal cap.
    d. Remove the 2 bolts and starter.
5. Remove the flywheel housing side cover.

To install:

6. Install the flywheel housing side cover.
7. Install the starter assembly with the 2 bolts. Tighten the bolts to 27 ft. lbs. (37 Nm).
    a. Connect the starter wire with the bolt and nut. Tighten the bolts to 71 inch lbs. (8 Nm). Tighten the nut to 87 inch lbs. (9.8 Nm).
    b. Close the terminal cap.
    c. Connect the starter connector.
    d. Connect the ground wire with the bolt. Tighten to 10 ft. lbs. (13 Nm).
8. Install the starter cover with the 3 bolts. Tighten to 8 ft. lbs. (12 Nm).
9. Install the exhaust manifold sub assembly LH.
10. Connect the cable to the negative battery terminal.

➡When disconnecting the cable, some systems need to be initialized after the cable is reconnected.

ENGINE MECHANICAL

➡Disconnecting the negative battery cable may interfere with the functions of the on board computer systems and may require the computer to undergo a relearning process, once the negative battery cable is reconnected.

ACCESSORY DRIVE BELTS

ACCESSORY BELT ROUTING

See Figure 63.

INSPECTION

Inspect the drive belt for signs of glazing or cracking. A glazed belt will be perfectly smooth from slippage, while a good belt will have a slight texture of fabric visible. Cracks will usually start at the inner edge of the belt and run outward. All worn or damaged drive belts should be replaced immediately.

ADJUSTMENT

Belt adjustment is automatic and non-adjustable.

REMOVAL & INSTALLATION

See Figure 64.

1. Before servicing the vehicle, refer to the Precautions section.
2. Loosen the drive belt tension by turning the drive belt tensioner counterclockwise, and remove the drive belt.

To install:

3. Installation is the reverse of removal.

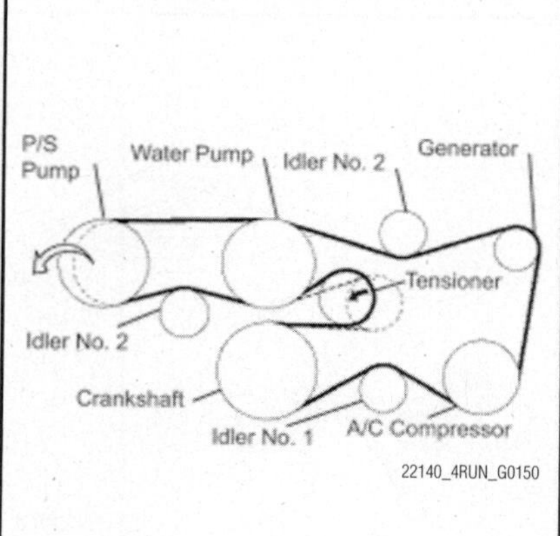

Fig. 63 Accessory drive belt routing—4.0L engine

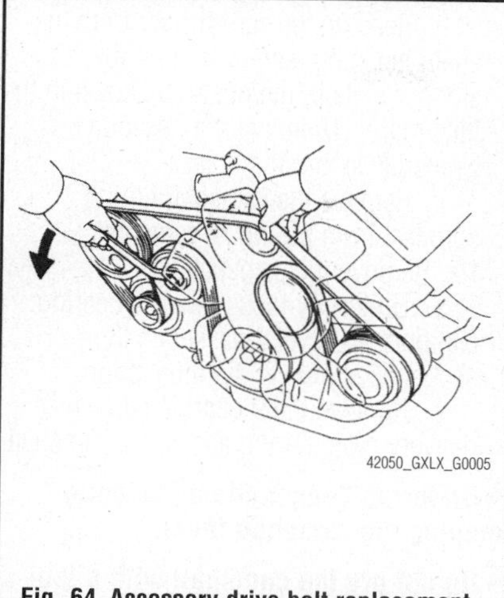

Fig. 64 Accessory drive belt replacement

CAMSHAFT AND VALVE LIFTERS

REMOVAL & INSTALLATION

2.7L Engine

See Figures 65 through 80.

1. Disconnect the cable from the negative battery terminal.

➡When disconnecting the cable, some systems need to be initialized after the cable is reconnected.

2. Remove the front bumper cover lower.
3. Remove the No. 1 engine under cover sub assembly.
4. Remove the upper radiator support seal.
5. Drain the engine oil.
6. Remove the air cleaner cap sub assembly.
7. Remove the intake air connector.
8. Remove the radiator reservoir.
9. Remove the fan shroud.
10. Remove the ignition coil assembly.
11. Remove the Camshaft Position (CMP) sensor.
12. Remove the cylinder head cover sub assembly.
13. Remove the timing chain.
    a. Remove the 2 bolts, chain guide and O-ring.
14. Remove the camshaft timing sprocket.
    a. Turn the crankshaft pulley, and align its groove with the "0" timing mark of the timing chain cover.
    b. Check the timing marks of the camshaft timing gear and sprocket are aligned with the timing marks of the No. 1 bearing cap.

➡If the timing marks do not align, rotate the crankshaft clockwise again and align the timing marks.

    c. Place paint marks on the timing chain, camshaft timing gear and sprocket.
    d. Hold the camshaft with a wrench and loosen the sprocket bolt.

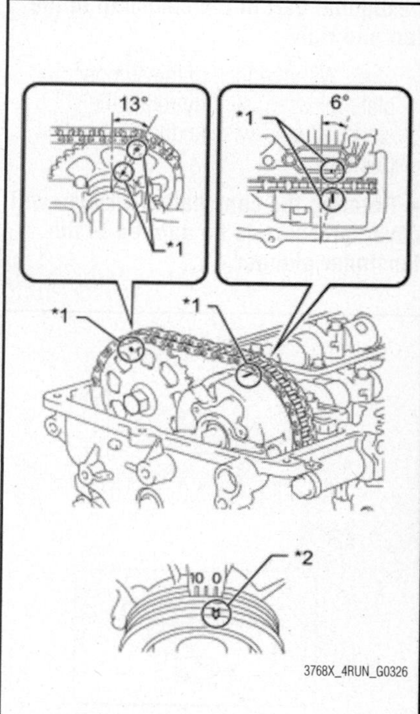

Fig. 65 Aligning the timing mark of the timing chain

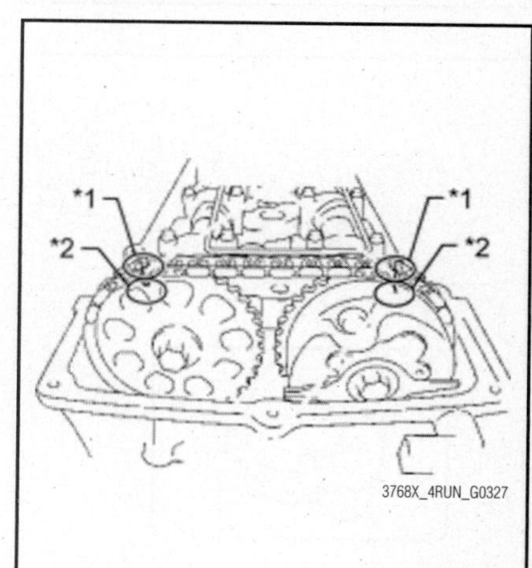

Fig. 66 Marking the timing chain, camshaft timing gear and sprocket

➡**Be careful not to damage the oil delivery pipe.**

e. Using a 10 mm socket hexagon wrench, remove the straight screw plug.

f. Using a screwdriver, access the tensioner stopper plate through the chain tensioner service hole. Move the stopper plate (*1) upward to release the lock. Then hold the plate in that position as shown in the illustration.

➡**If the lock of the stopper plate is difficult to release, slightly rotate the hexagonal part of the camshaft to the left and right.**

g. With the lock of the stopper plate released, slightly rotate the camshaft clockwise and keep it in that position.

➡**Rotating the camshaft clockwise will cause pressure to be applied to the tensioner plunger.**

➡**Be careful not to damage the oil delivery pipe.**

h. Remove the screwdriver from the chain tensioner service hole. Move the stopper plate to the position shown in the illustration. Then insert a hexagon wrench (*1) into the hole.

i. Remove the camshaft timing sprocket from the No. 2 camshaft.

15. Remove the camshaft.

a. Uniformly loosen the 21 bearing cap bolts in several passes in the sequence shown in the illustration.

b. Remove the 9 bearing caps, oil delivery pipe, O-ring and No. 2 camshaft.

➡**Uniformly loosen the bolts while keeping the camshaft level.**

➡**Do not pry the camshaft with a tool or apply excessive force to it.**

c. Remove the camshaft while holding the timing chain.

d. Secure the timing chain with a string.

16. Remove the No. 1 valve rocket arm sub assembly.

a. Remove the 16 valve rocket arms from the cylinder head.

➡**Arrange the removed parts in the correct order.**

17. Remove the valve lash adjuster assembly.

a. Remove the 16 valve lash adjusters from the cylinder head.

➡**Arrange the removed parts in the correct order.**

18. Remove the camshaft timing gear assembly.

a. Remove the flange bolt and camshaft timing gear.

➡**Be sure not to remove the other 3 bolts.**

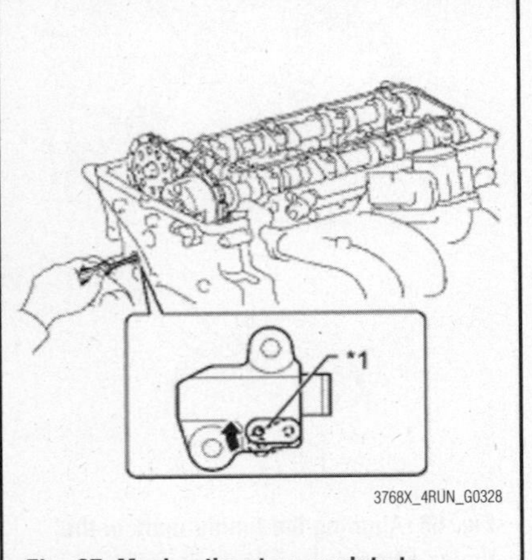

Fig. 67 Moving the stopper plate to release the lock

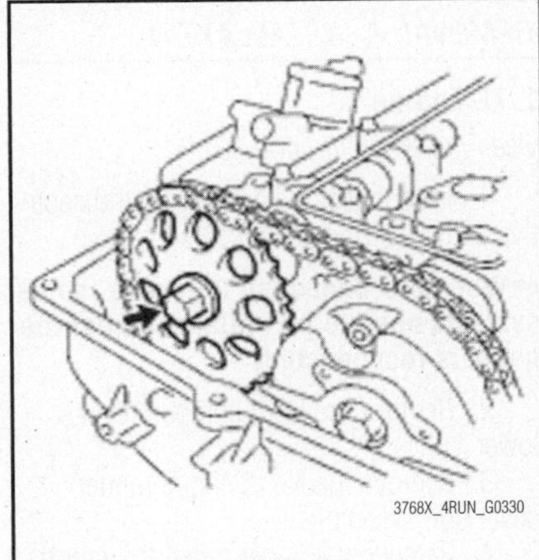

Fig. 69 Removing the camshaft timing sprocket from the No. 2 camshaft

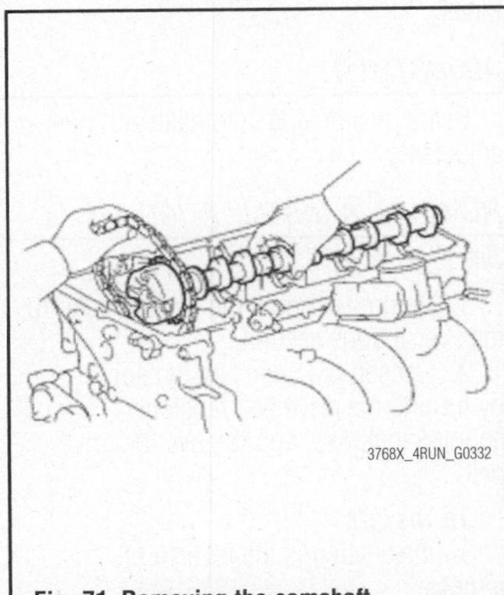

Fig. 71 Removing the camshaft

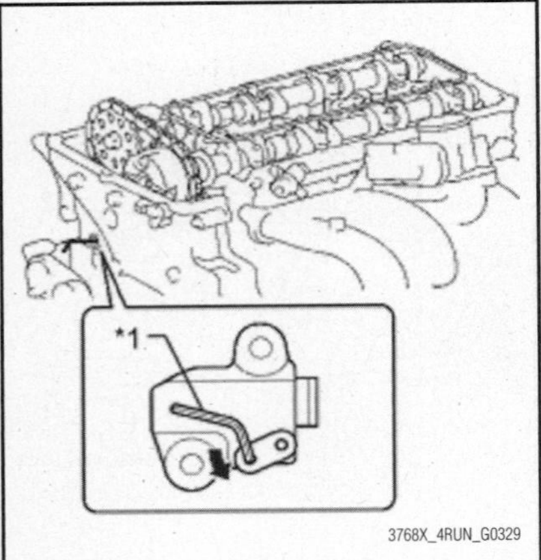

Fig. 68 Moving the stopper plate to a specified position

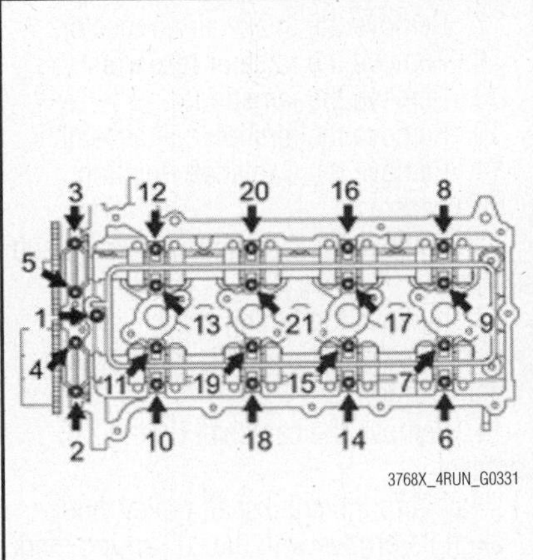

Fig. 70 Bearing cap loosening sequence

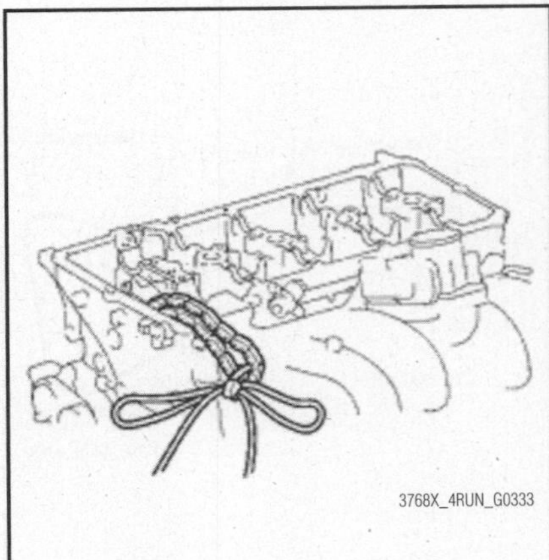

Fig. 72 Securing the timing chain with string

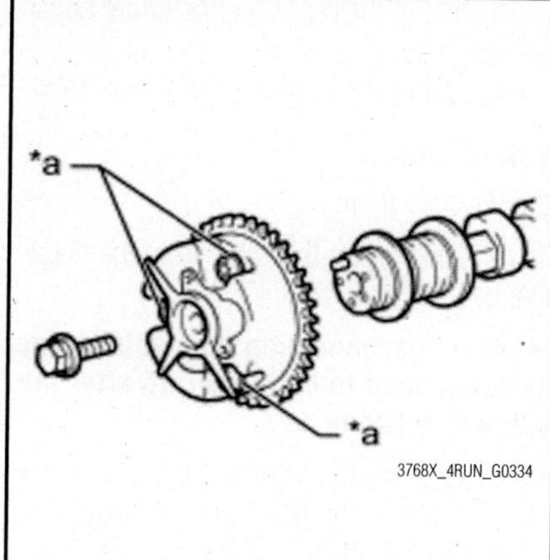

Fig. 73 Removing the camshaft timing gear assembly

➡If planning to reuse the gear, be sure to release the straight pin lock before installing the gear.

19. Inspect the valve lash adjuster assembly.
20. Inspect the camshaft timing gear assembly.

To install:

21. Install the camshaft timing gear assembly.
    a. Align the pin hole and straight pin and install the camshaft timing gear to the camshaft.
    b. Lightly press the gear against the camshaft and turn the gear. Push further at the position where the pin enters the groove.
    c. Check that there is no gap between the flange of the gear and the camshaft.
    d. With the camshaft timing gear fixed in place, install the flange bolt. Tighten to 58 ft. lbs. (78 Nm).
    e. Check that the camshaft timing gear can move in the retard direction and becomes locked at the most retarded position.
22. Install valve lash adjuster assembly.
    a. Inspect each valve lash adjuster before installing it.
    b. Install the 16 valve lash adjusters to the cylinder head.

➡Install each lash adjuster to the same place it was removed from.

23. Install the No. 1 valve rocker arm sub assembly.
    a. Apply clean engine oil to the valve lash adjuster tips and valve stem cap surfaces.
    b. Install the 16 valve rocker arms as shown.

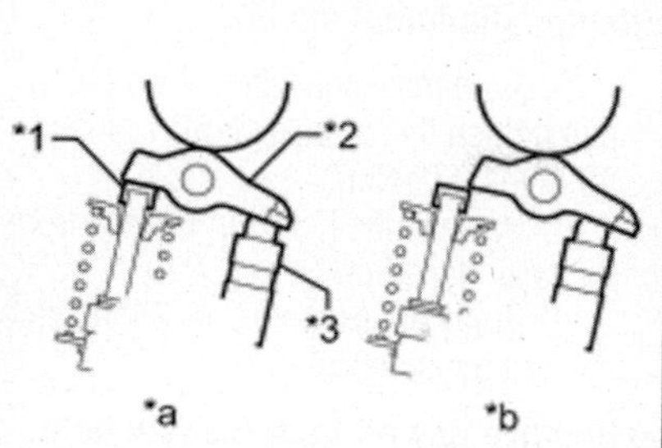

Fig. 74 Installing the No. 1 valve rocker arm sub assembly

➡install each valve rocker arm to the same place it was removed from.

24. Install the camshaft.
    a. Apply clean engine oil to the camshaft cams and cylinder head journals.
    b. Install the timing chain to the camshaft timing gear with the painted mark of the link aligned with the timing mark of the camshaft timing gear.
    c. Position the 2 camshafts as shown in the illustration

➡Align the paint mark and timing mark before positioning the camshaft.

➡Before and after positioning the camshaft and No. 2 camshaft, check that the rocker arm is firmly set on the lash adjuster.

    d. Temporarily install the No. 1 camshaft bearing cap.
    e. Check the proper location of each camshaft bearing cap and install each one.
    f. Install a new O-ring to the No. 1 camshaft bearing cap.
    g. Temporarily install the oil delivery pipe.
    h. Install the 21 bolts and tighten them in the order shown in the illustration. Tighten bolt A to 9 ft. lbs. (12 Nm). Tighten all other bolts to 11 ft. lbs. (16 Nm).
25. Install the camshaft timing sprocket.
    a. Rotate the camshaft so that the camshaft timing mark and No. 2 camshaft knock pin are as shown in the illustration.

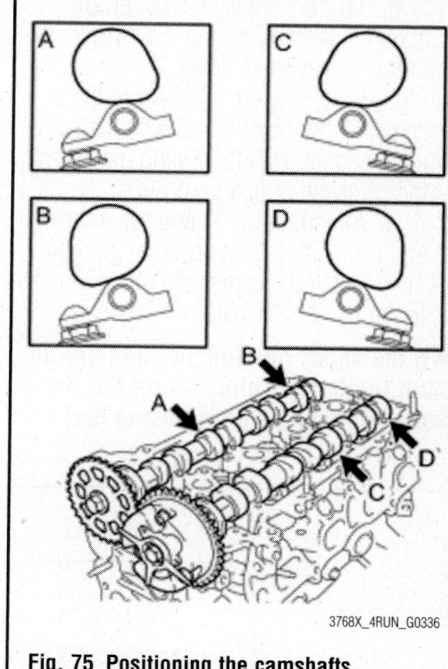

Fig. 75 Positioning the camshafts

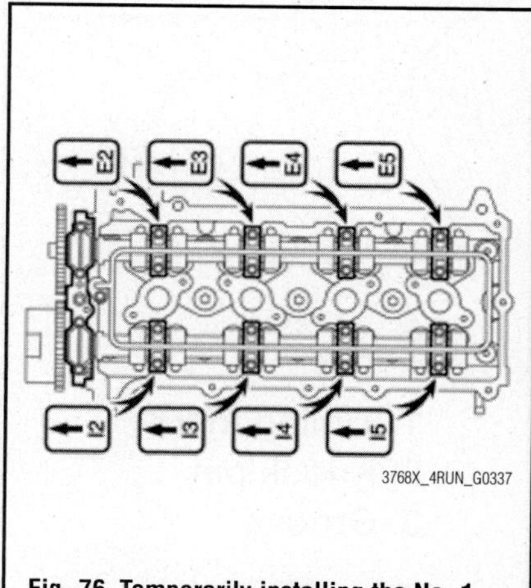

Fig. 76 Temporarily installing the No. 1 camshaft bearing cap

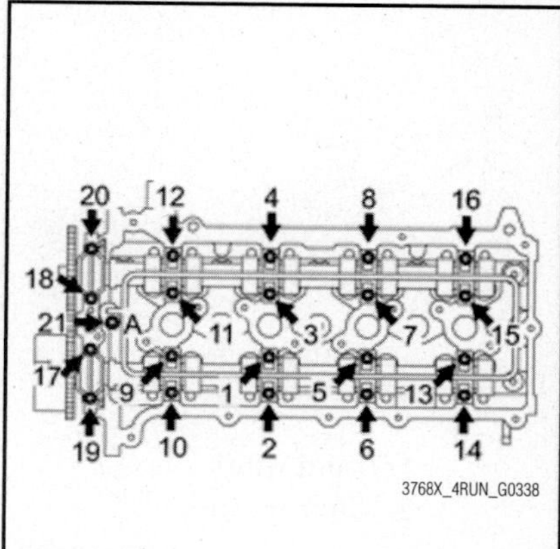

Fig. 77 Identifying the No. 1 camshaft bearing cap bolt torque sequence

b. Turn the crankshaft pulley and align its groove with the "0" timing mark of the timing chain cover.

c. Install the timing chain to the camshaft timing sprocket with the paint mark aligned with the timing marks on the camshaft timing sprocket.

d. Align the No. 2 camshaft knock pin and camshaft timing sprocket pin hole. Then install the camshaft timing sprocket to the No. 2 camshaft.

➡If the knock pin and pin hole are difficult to align, slightly rotate the No. 2 camshaft back and forth using the hexagonal part of the camshaft. Then attempt alignment again.

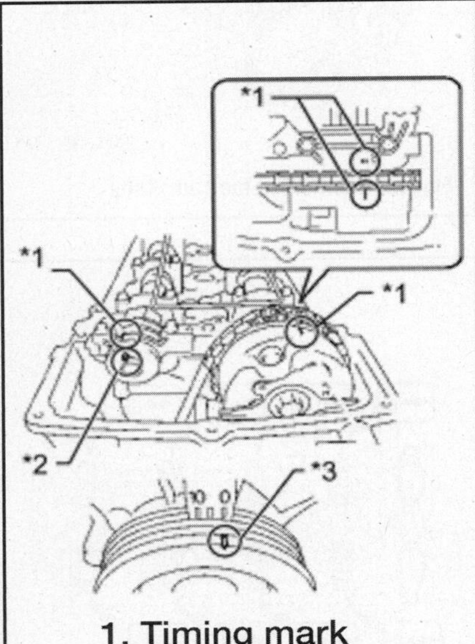

Fig. 78 Rotating the camshaft

e. Hold the camshaft with a wrench and tighten the sprocket bolt. Tighten to 58 ft. lbs. (78 Nm).

f. Remove the hexagon wrench from the chain tensioner.

g. Apply adhesive to 2 or 3 threads of the straight screw plug.

➡Remove any oil from the bolt hole.

h. Using a 10 mm socket hexagon wrench, install the straight screw plug. Tighten to 12 ft. lbs. (17 Nm).

26. Install the timing chain guide.

a. Install a new O-ring to the camshaft bearing cap.

b. Install the timing chain guide with the 2 bolts. Tighten to 7 ft. lbs. (10 Nm).

27. Install the cylinder head cover sub assembly.

28. Install the Camshaft Position (CMP) sensor.

29. Install the ignition coil assembly.

30. Install the fan shroud.

31. Install the radiator reservoir.

32. Install the intake air connector.

33. Install the air cleaner cap sub assembly.

34. Connect the cable to the negative battery terminal.

➡When disconnecting the cable, some systems need to be initialized after the cable is reconnected.

35. Add engine oil.

36. Inspect the engine oil level.

37. Inspect for oil leaks.

38. Inspect the ignition timing.

39. Inspect the engine idle speed.

40. Install the upper radiator support seal.

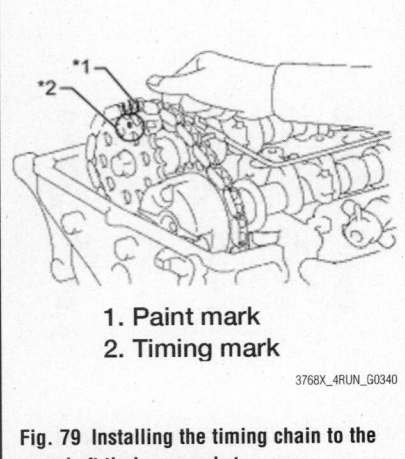

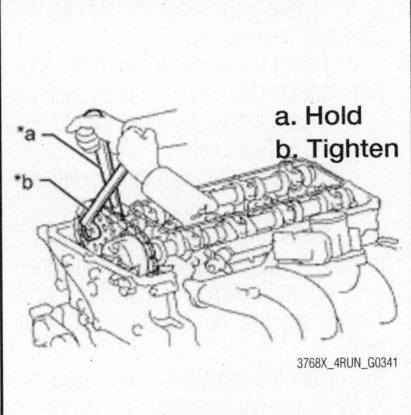

Fig. 79 Installing the timing chain to the camshaft timing sprocket

Fig. 80 Tightening the camshaft sprocket bolt

41. Install the No. 1 engine under cover sub assembly.

42. Install the front bumper cover lower.

4.0L Engine

See Figures 81 through 114.

1. Disconnect the cable from the negative battery terminal.

➡When disconnecting the cable, some systems need to be initialized after the cable is reconnected.

2. Remove the front bumper.

3. Remove the No. 1 engine under cover sub assembly.

4. Drain the engine oil.

5. Drain the coolant.

6. Remove the upper radiator support seal.

7. Disconnect the cable from the positive battery terminal.

8. Remove the battery hold down clamp.

9. Remove the battery.

10. Remove the battery tray.

11. Remove the v-bank cover.

12. Remove the air cleaner cap and hose.

13. Remove the air cleaner case sub assembly.

14. Remove the No. 1 radiator hose.

15. Remove the No. 2 radiator hose.

16. Remove the radiator reservoir.

17. Disconnect the oil cooler tube.

18. Remove the fan shroud.

19. Remove the intake air surge tank.

20. Remove the ignition coil assembly.

21. Disconnect the vane pump assembly.

22. Remove the No. 2 idler pulley sub assembly.

23. Remove the wiring harness clamp bracket.

24. Remove the No. 2 exhaust manifold heat insulator.

25. Remove the alternator assembly.

26. Remove the engine oil level dipstick guide.

27. Remove the water by pass pipe sub assembly.

28. Remove the No. 1 oil pipe.

29. Remove the No. 2 oil pipe.

30. Remove the rear cylinder head cover.

31. Disconnect the fuel pipe sub assembly.

32. Remove the cylinder head cover sub assembly LH.

33. Remove the cylinder head cover sub assembly.

34. Remove the timing chain cover plate.

a. Remove the 4 bolts, timing chain cover plate and gasket.

35. Set the No. 1 cylinder to TDC/compression.

a. Turn the crankshaft pulley and align the notch with the "0" timing mark of the timing chain cover.

b. Check that the timing marks of the camshaft timing gears are aligned with the timing marks of the bearing caps as shown in the illustration.

➡If the marks are not aligned, turn the crankshaft again to align the marks.

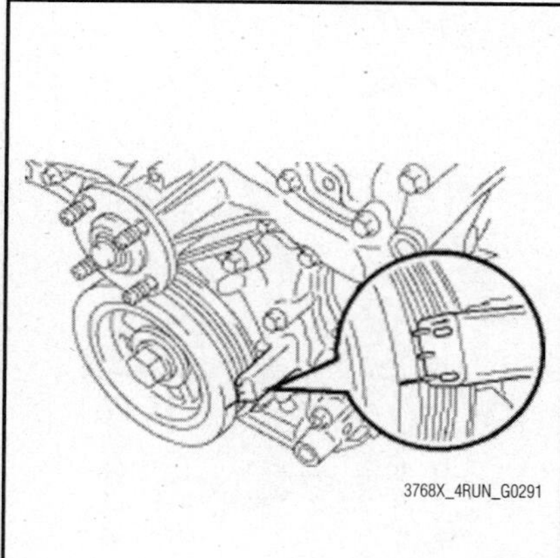

Fig. 81 Aligning the notch with the "0" timing mark of the timing chain cover

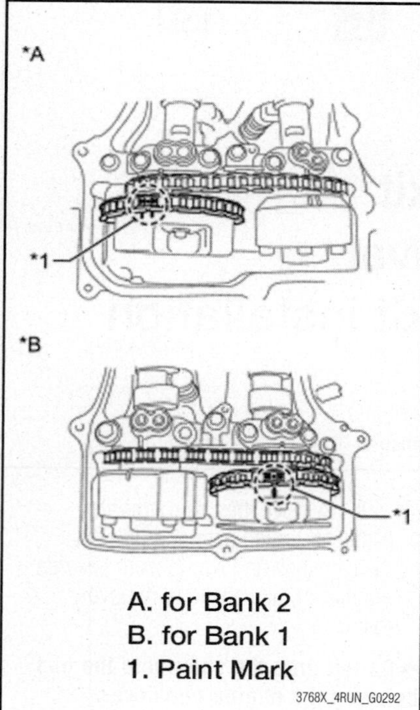

Fig. 82 Checking the timing marks of the camshaft timing gears are aligned with the timing marks of the bearing caps

c. Place paint marks on the timing marks and sprockets of each camshaft timing gear and on the links of the No. 1 chain.

➡Be sure to place the paint marks on 2 links of the chain and on the sprockets of the camshaft timing gears at the locations of the timing marks of the camshaft timing gears.

36. Remove the No. 1 chain tensioner assembly.

a. Turn the crankshaft approximately 30° counterclockwise so that there is some slack in the chain.

➡This prevents the valves and pistons from interfering with each other.

b. Align the hole in the lever of the tensioner with the hole in the tensioner body as shown in the illustration, and then insert a pin with a diameter of 0.0500 inch (1.27 mm) into the hole.

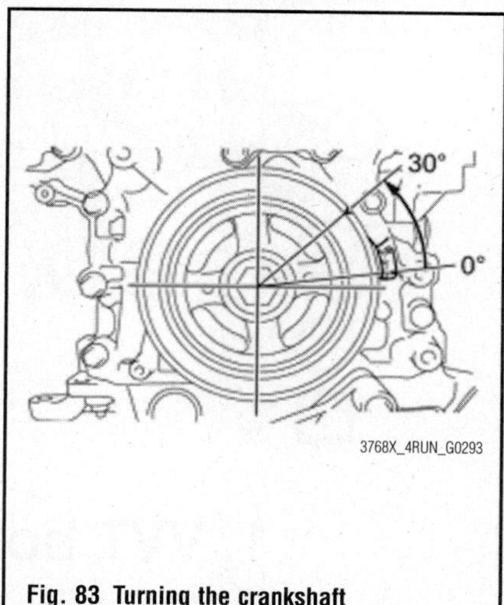

Fig. 83 Turning the crankshaft

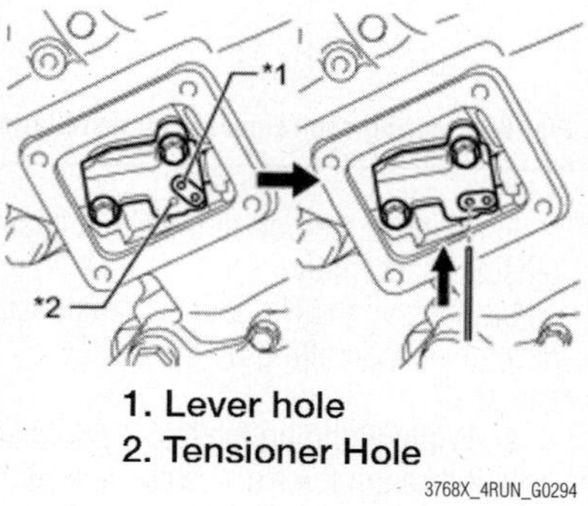

Fig. 84 Aligning the hole in the lever of the tensioner with the hole in the tensioner body

c. Turn the crankshaft clockwise and align the notch with the "0" timing mark of the timing chain cover.

d. Remove the 2 bolts and chain tensioner.

➡Do not drop the No. 1 chain tensioner assembly or bolts into the timing chain cover.

37. Disconnect the chain sub assembly (for Bank 1).

a. Turn the crankshaft clockwise until it is in the position shown in the illustration so that there is some slack in the chain between the banks.

➡When turning the crankshaft, engine oil may spray out of the oil holes.

⁂ CAUTION

As the camshafts turn suddenly, do not touch the camshafts or camshaft timing gears.

b. Turn the crankshaft clockwise until it is in the position shown in the illustration so that the chain can be removed easily.

➡When turning the crankshaft, engine oil may spray out of the oil holes.

c. Remove the chain from the sprocket of the camshaft timing gear and set it on the gear.

⁂ CAUTION

As the camshaft may turn suddenly and pinch your fingers when the chain is removed, pinch the chain and lift it upward to remove it from the sprocket.

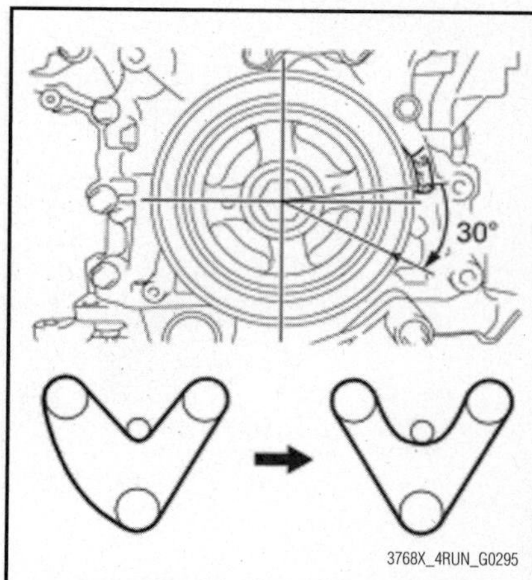

Fig. 85 Turning the crankshaft clockwise to the correct position

38. Remove the camshaft bearing cap (for Bank 1).

a. Remove the bolts and bearing caps in the order shown in the illustration. Immediately after removing a bearing cap, install VVT bolt kit in the order shown in the illustration. Tighten to 21 ft. lbs. (28 Nm).

➡Arrange the removed parts so that they can be reinstalled in their original locations.

➡Do not install the bearing caps when installing VVT bolt kit.

➡Be sure to follow the numerical order when performing this procedure.

➡Do not allow the VVT bolt kit to contact the camshaft.

➡Do not drop the VVT bolt kit into the cylinder head.

39. Remove the No. 2 camshaft.

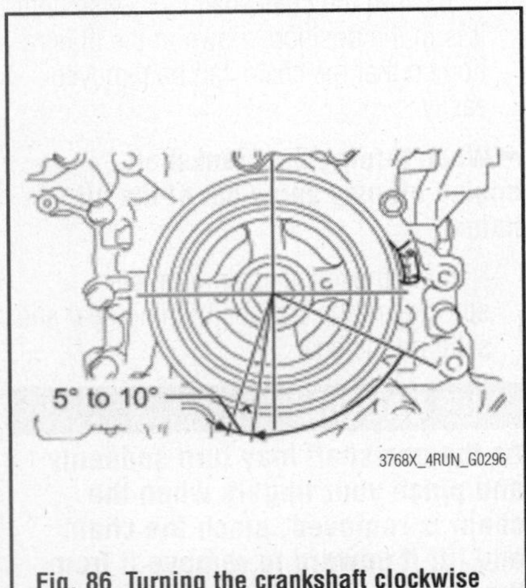

Fig. 86 Turning the crankshaft clockwise to remove the chain easily

Fig. 87 Removing the chain from the sprocket of the camshaft timing gear

1. VVT bolt kit
a. Part removal
b. VVT bolt kit installation

3768X_4RUN_G0298

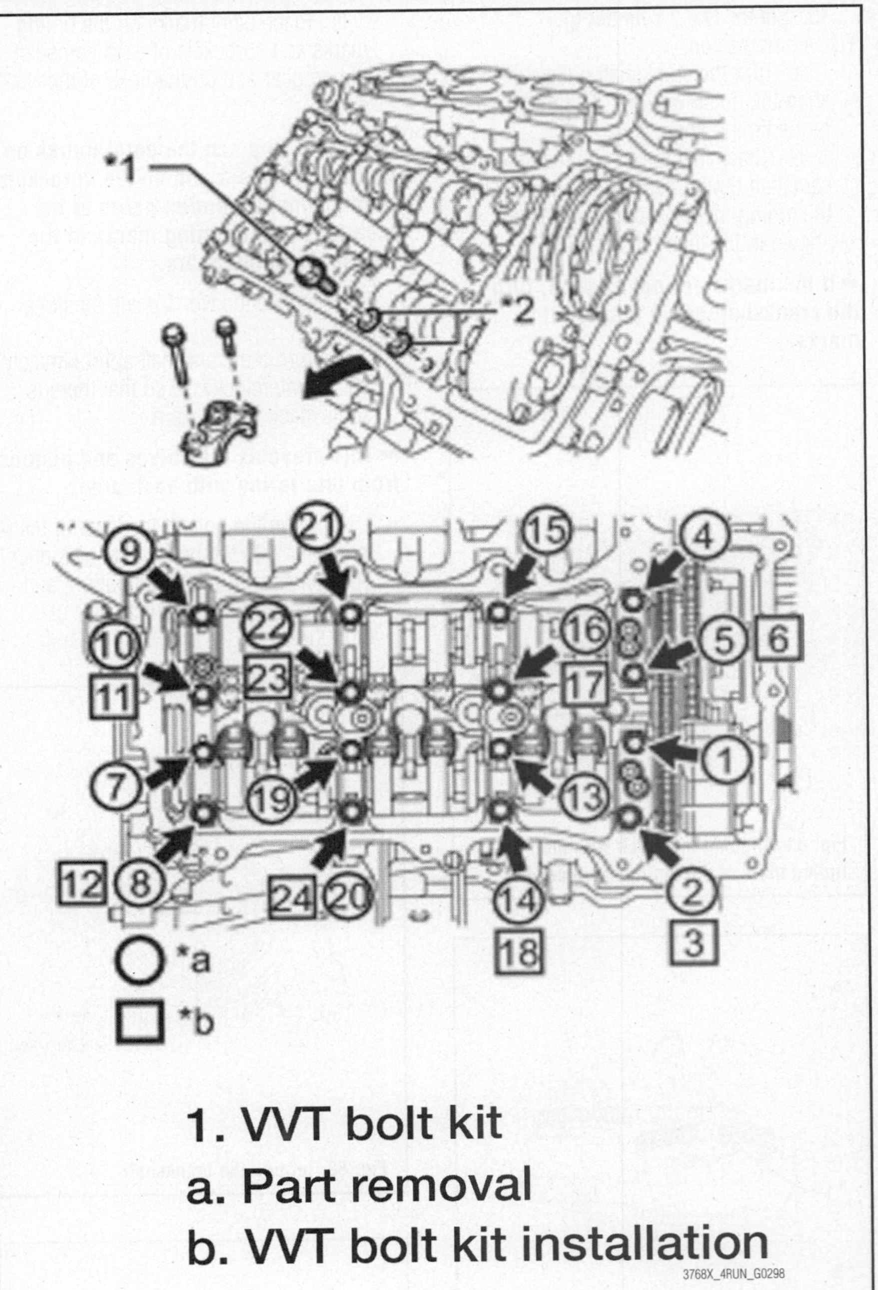

Fig. 88 Bearing cap removal and installation sequence

a. Remove the bolt of the No. 2 chain tensioner assembly.

b. Remove the No. 2 chain tensioner assembly while lifting up the No. 2 camshaft.

c. While lifting up the No. 2 camshaft, pass it through the No. 2 chain and pull it out towards the front of the vehicle to remove it.

40. Remove the camshaft.

a. Lift up the rear of the camshaft so that it is at an angle.

b. Remove the chain from the camshaft timing gear and pull out the camshaft and No. 2 chain towards the rear of the vehicle to remove them.

➡Do not drop the chain into the gap between the engine and cover.

c. Suspend the chain with a string or equivalent.

41. Disconnect the chain sub assembly (for Bank 2).

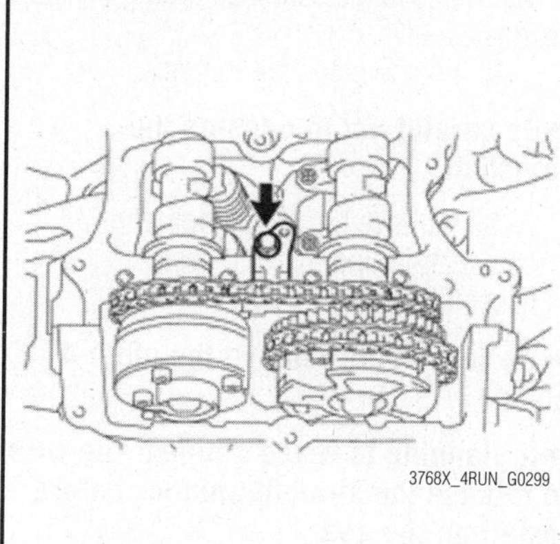

Fig. 89 Removing the No. 2 chain tensioner assembly bolt

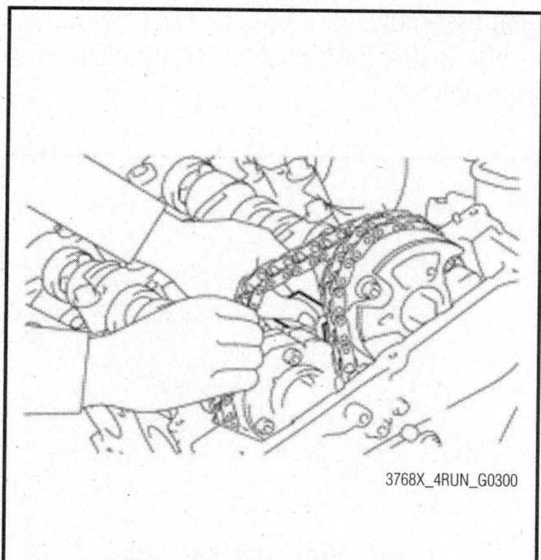

Fig. 90 Removing the No. 2 chain tensioner assembly

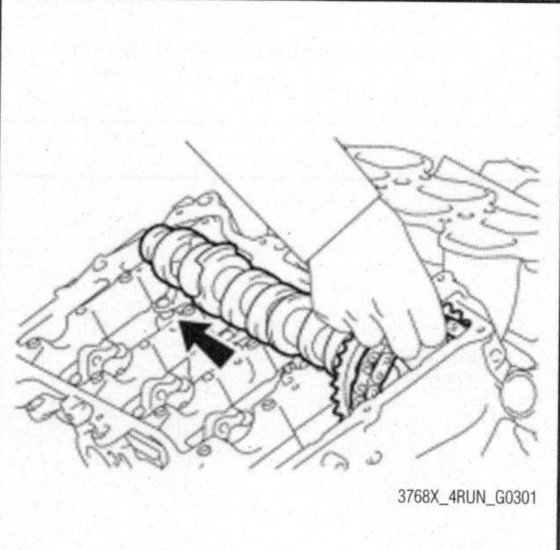

Fig. 91 Removing the chain from the camshaft timing gear

a. Turn the crankshaft counterclockwise and align the notch with the "0" timing mark of the timing chain cover.

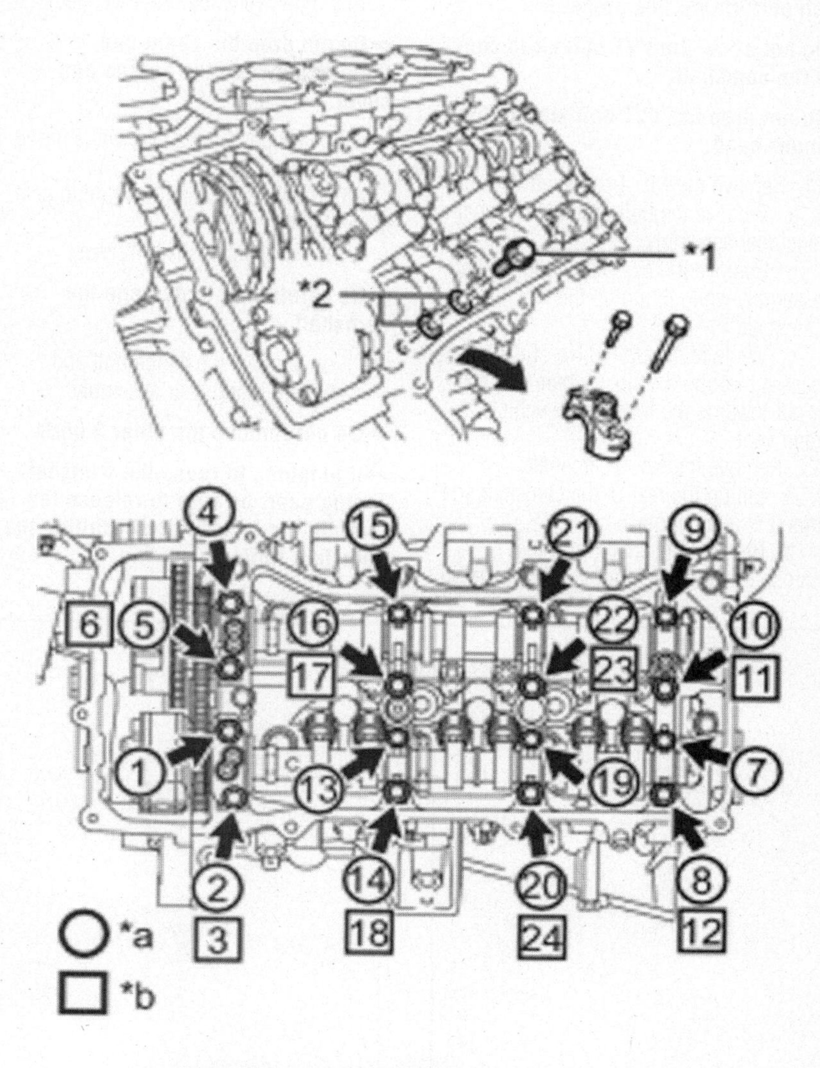

Fig. 92 Identifying the bearing cap (for Bank 2) removal and installation sequence

42. Remove the chain from the sprocket of the camshaft timing gear and set it on the gear.

CAUTION

As the camshaft may turn suddenly and pinch your fingers when the chain is removed, pinch the chain and lift it upward to remove it from the sprocket.

43. Remove the camshaft bearing cap (for Bank 2).

a. Remove the bolts and bearing caps in the order shown in the illustration. Immediately after removing a bearing cap, install VVT bolt kit in the order shown in the illustration. Tighten to 21 ft. lbs. (28 Nm).

➡Arrange the removed parts so that they can be reinstalled in their original locations.

➡Do not install the bearing caps when installing VVT bolt kit.

➡Be sure to follow the numerical order when performing this procedure.

➡Do not allow the VVT bolt kit to contact the camshaft.

➡Do not drop the VVT bolt kit into the cylinder head.

44. Remove the No. 4 camshaft.
   a. Remove the bolt of the No. 3 chain tensioner assembly.
   b. Remove the No. 3 chain tensioner assembly while lifting up the No. 4 camshaft.
   c. While lifting up the No. 4 camshaft, pass it through the No. 2 chain and pull it out towards the front of the vehicle to remove it.
45. Remove the No. 3 camshaft.
   a. Lift up the rear of the camshaft so that it is at an angle.
   b. Remove the chain from the camshaft timing gear and pull out the No. 3 camshaft and No. 2 chain towards the rear of the vehicle to remove them.

➡Do not drop the chain into the gap between the engine and cover.

   c. Suspend the chain with a string or equivalent.
46. Remove the camshaft timing gear assembly.
   a. Fix the camshaft in place.

➡Be careful not to damage the camshaft.

   b. Remove the flange bolt and camshaft timing gear assembly.

➡Do not remove the other 3 bolts.

➡If planning to reuse the camshaft timing gear, be sure to release the straight pin lock before installing the camshaft timing gear.

47. Remove the camshaft timing exhaust gear assembly.
   a. Fix the camshaft in place.

➡Be careful not to damage the camshaft.

   b. Remove the flange bolt and camshaft timing exhaust gear assembly.

➡Be sure not to remove the other 4 bolts.

➡If planning to reuse the gear, be sure to release the straight pin lock before installing the gear.

To install:

48. Inspect the camshaft timing gear assembly.
49. Inspect the camshaft timing exhaust gear assembly.
50. Install the camshaft timing gear assembly.

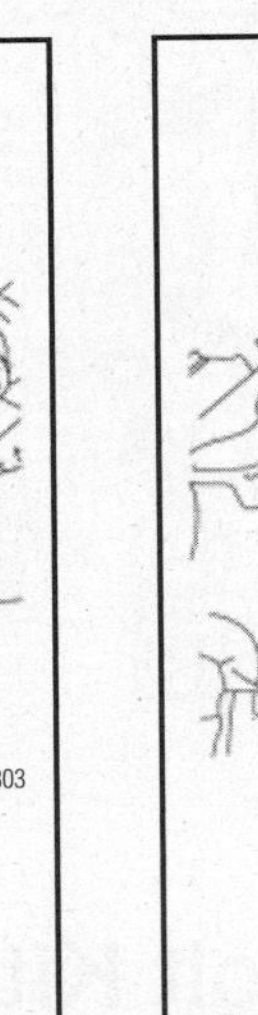

Fig. 93 Removing the No. 3 chain tensioner

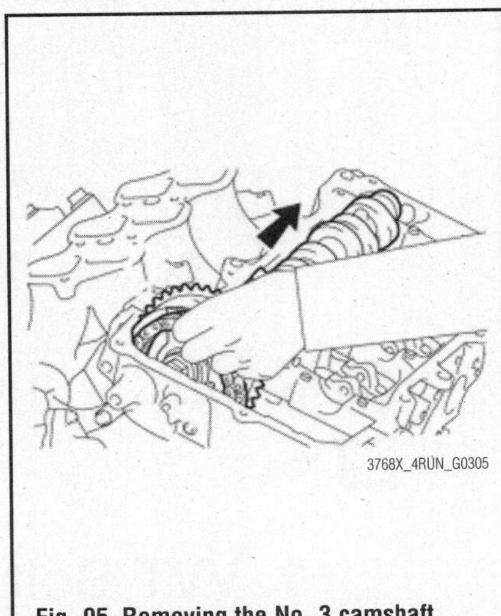

Fig. 95 Removing the No. 3 camshaft

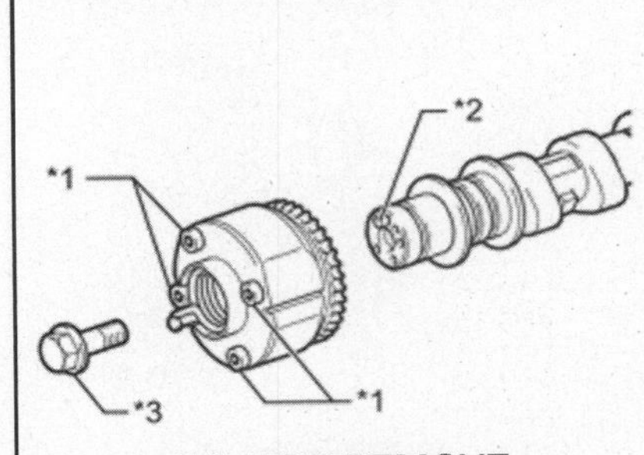

Fig. 97 Removing the camshaft timing exhaust gear assembly

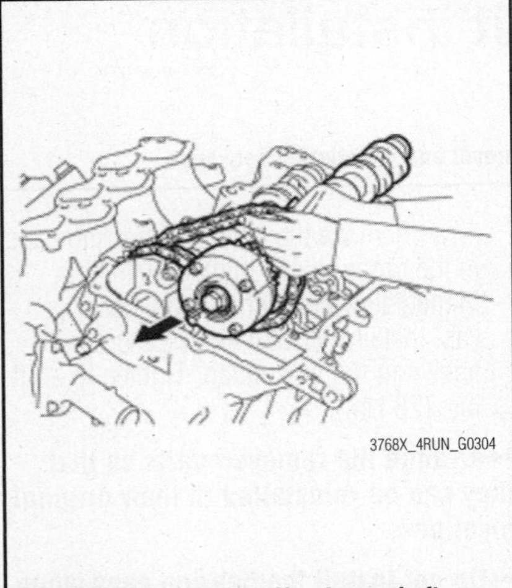

Fig. 94 Removing the No. 4 camshaft

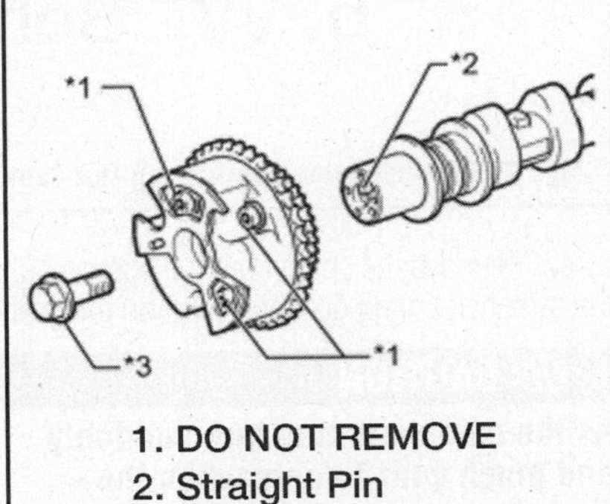

Fig. 96 Removing the camshaft timing gear assembly

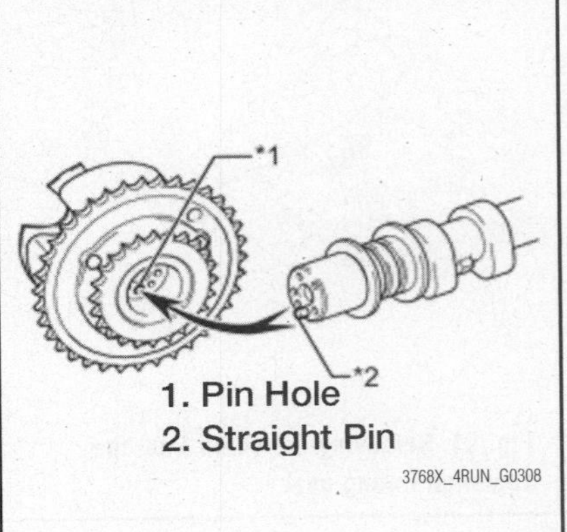

Fig. 98 Putting the camshaft timing gear and camshaft together

a. Fix the camshaft in place.

➡**Be careful not to damage the camshaft.**

b. Put the camshaft timing gear assembly and camshaft together by aligning the pin hole and straight pin.

c. Lightly press and turn the camshaft timing gear assembly against the camshaft, and press harder after the pin enters the hole.

➡**Be sure not to turn the camshaft timing gear assembly in the retard direction.**

d. Check that there is no clearance between the camshaft timing gear assembly flange and camshaft.

e. Install the flange bolt while holding the camshaft. Tighten to 74 ft. lbs. (100 Nm).

f. Check the lock of the camshaft timing gear assembly.

g. Fix the camshaft in place and confirm that the camshaft timing gear assembly is locked.

➡**Be careful not to damage the camshaft.**

51. Install the camshaft timing exhaust gear assembly.

a. Fix the camshaft in place.

➡**Be careful not to damage the camshaft.**

b. Put the camshaft timing exhaust gear assembly and camshaft together by aligning the pin hole and straight pin.

c. Lightly press and turn the camshaft timing gear assembly against the camshaft, and press harder after the pin enters the hole.

➡**Be sure not to turn the camshaft timing exhaust gear in the advanced direction.**

d. Check that there is no clearance between the gear flange and camshaft.

e. Install the flange bolt while holding the camshaft. Tighten to 74 ft. lbs. (100 Nm).

f. Check the camshaft timing exhaust gear lock.

g. Make sure that the camshaft timing exhaust gear assembly locks.

52. Install the No. 3 camshaft.

a. Check that the notch is aligned with the "0" timing mark of the timing chain cover.

b. Align the mark plate (yellow) with the timing mark of the camshaft timing gear as shown in the illustration and install the No. 2 chain to the camshaft timing gear.

c. Clean the camshaft housing LH and camshaft journals and apply engine oil to them.

d. Make sure that the No. 1 valve rocker arm sub-assembly is installed as shown in the illustration.

e. Install the chain to the No. 3 camshaft, and then install the camshaft to the camshaft housing LH.

➡**Place the chain on the camshaft timing gear but do not engage the teeth of the sprocket and the chain.**

➡**Install the camshaft so that the timing mark is facing upward.**

53. Install the No. 4 camshaft.

a. Clean the camshaft housing LH and camshaft journals and apply engine oil to them.

b. Pass the No. 4 camshaft through the No. 2 chain from the front of the vehicle, align the mark plate (yellow) with the timing mark and install the No. 2 chain to the camshaft timing exhaust gear.

54. While lifting up the No. 4 camshaft, pass the No. 3 chain tensioner assembly through the No. 2 chain and set it in place.

a. Install the No. 4 camshaft to the camshaft housing LH, and then install the No. 3 chain tensioner assembly with the bolt. Tighten to 15 ft. lbs. (21 Nm).

55. Install the camshaft bearing cap (for Bank 2).

a. Clean the camshaft bearing caps and apply engine oil to them.

b. Make sure that the No. 1 valve rocker arm sub assembly.

c. Check the marks and numbers on the camshaft bearing caps, and then remove VVT bolt kit in the order shown in the illustration. Immediately after removing VVT bolt kit in the location for a bearing cap, install the bearing cap with the bolts in the order shown in the illustration. Tighten bolt A to 21 ft. lbs. (28 Nm). Tighten bolt B to 12 ft. lbs. (16 Nm).

➡**Be sure to follow the numerical order when performing this procedure.**

➡**Do not drop the VVT bolt kit into the cylinder head.**

d. Check the torque of each bolt again.

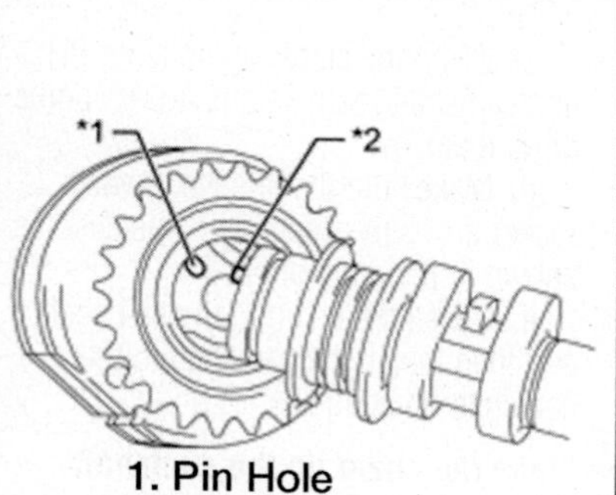

Fig. 99 Putting the camshaft timing exhaust gear assembly and camshaft together

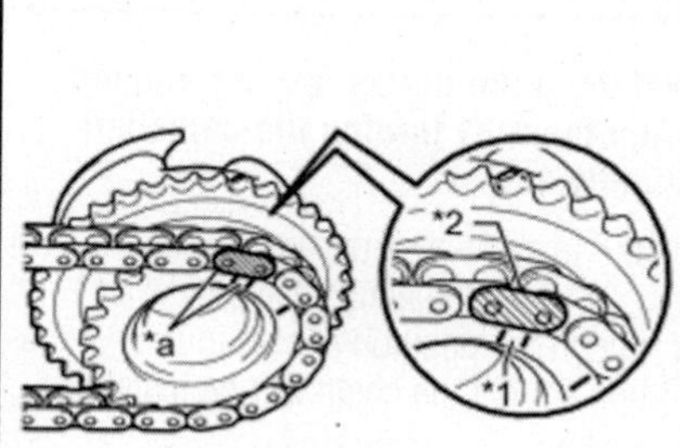

Fig. 100 Aligning the mark plate with the timing mark of the camshaft timing gear and installing the No. 2 chain to the camshaft timing gear

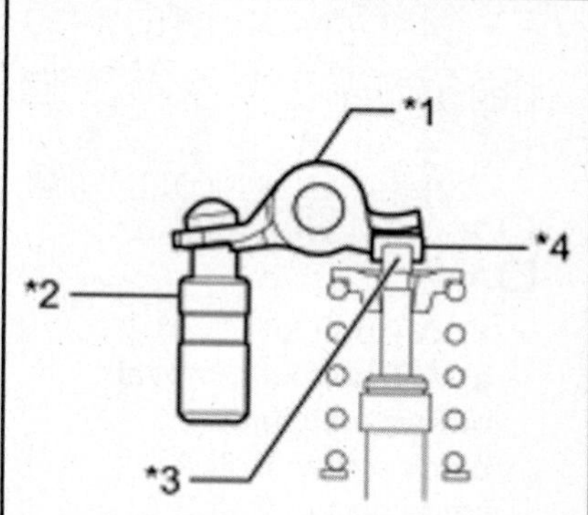

Fig. 101 Checking the No. 1 valve rocker arm sub assembly

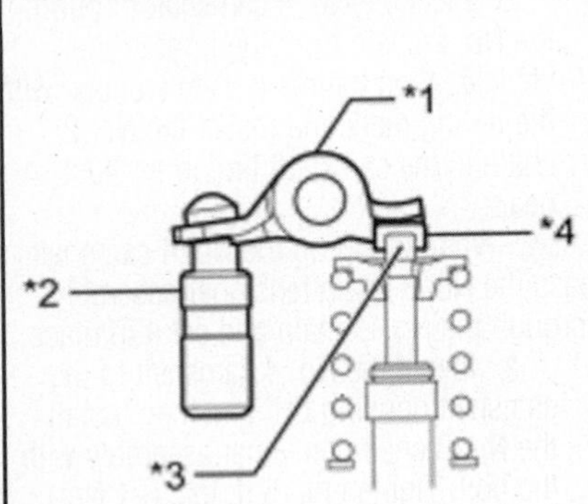

Fig. 102 Installing the No. 1 valve rocker arm sub assembly

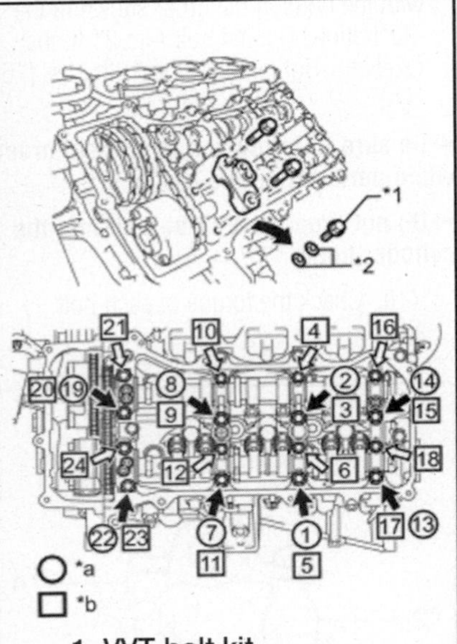

Fig. 103 Checking the bearing caps

56. Connect the chain sub assembly (for Bank 2).

a. Align the paint marks (*1) on the camshaft timing gear and No. 1 chain and install the No. 1 chain to the camshaft timing gear.

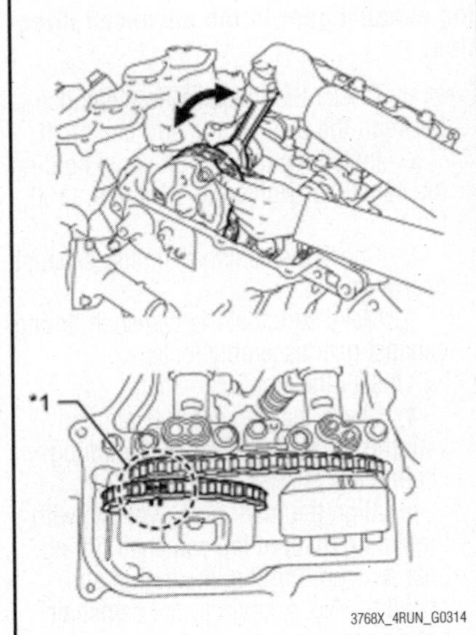

Fig. 104 Connecting the chain sub assembly (for Bank 2)

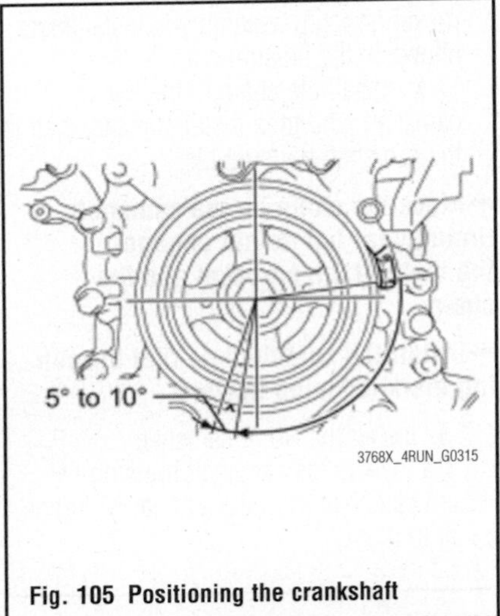

Fig. 105 Positioning the crankshaft

➡If the paint marks are not aligned, align them by turning the camshaft slightly.

57. Install the camshaft.

a. Turn the crankshaft clockwise until it is in the position shown in the illustration so that the chain can be installed easily.

➡When turning the crankshaft, engine oil may spray out of the oil holes.

b. Align the mark plate (yellow) with the timing mark of the camshaft timing gear as shown in the illustration and install the No. 2 chain to the camshaft timing gear.

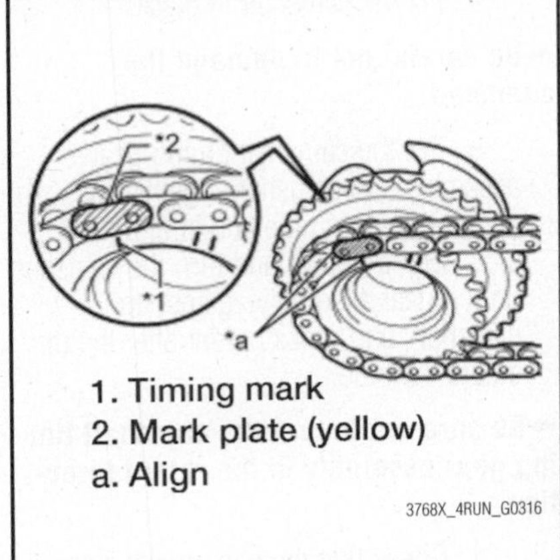

Fig. 106 Installing the No. 2 chain to the camshaft timing gear

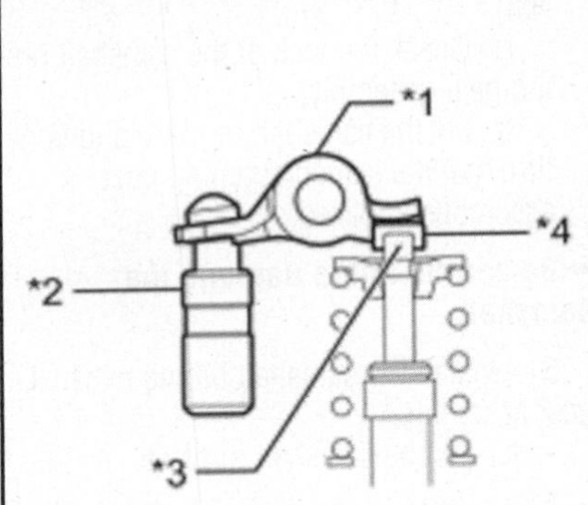

Fig. 107 Installing the No. 1 valve rocker arm

c. Clean the camshaft housing RH and camshaft journals and apply engine oil to them.

d. Make sure that the No. 1 valve rocker arm sub-assembly is installed as shown in the illustration.

e. Install the chain to the camshaft, and then install the camshaft to the camshaft housing RH.

➡Place the chain on the camshaft timing gear but do not engage the teeth of the sprocket and the chain.

➡Install the camshaft so that the timing mark is facing upward.

58. Install the No. 2 camshaft.

a. Clean the camshaft housing RH and camshaft journals and apply engine oil to them.

b. Pass the No. 2 camshaft through the No. 2 chain from the front of the vehicle, align the mark plate (yellow) with the timing mark and install the No. 2 chain to the camshaft timing exhaust gear.

c. While lifting up the No. 2 camshaft, pass the No. 2 chain tensioner assembly through the No. 2 chain and set it in place.

d. Install the No. 2 camshaft to the camshaft housing RH, and then install the No. 2 chain tensioner assembly with the bolt. Tighten to 15 ft. lbs. (21 Nm).

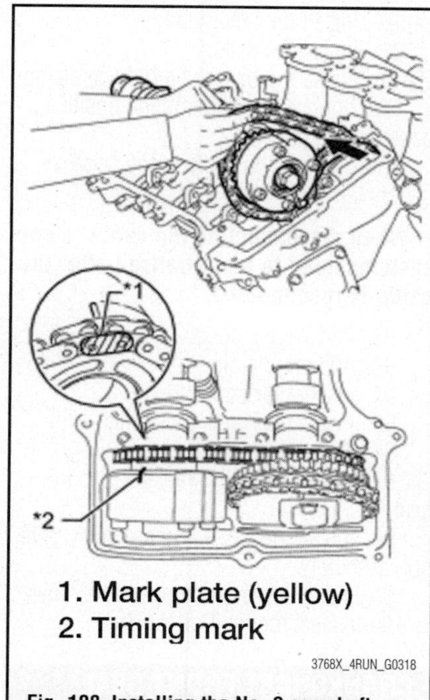

Fig. 108 Installing the No. 2 camshaft

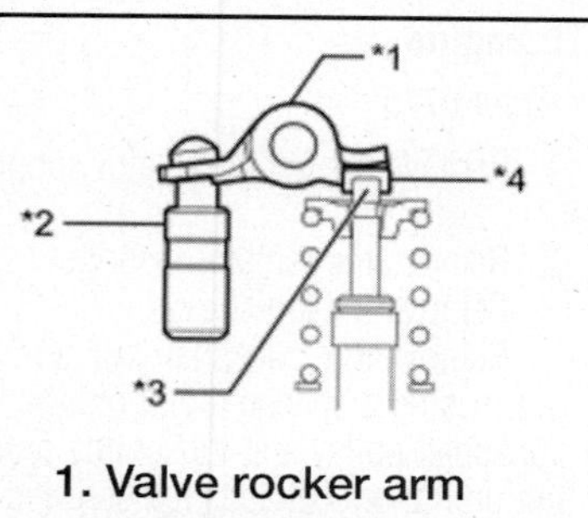

Fig. 109 Installing the No. 1 valve rocker arm sub assembly

59. Install the camshaft bearing cap (for Bank 1).

a. Clean the camshaft bearing caps and apply engine oil to them.

b. Make sure that the No. 1 valve rocker arm sub assembly is installed as shown in the illustration.

c. Check the marks and numbers on the camshaft bearing caps, and then remove VVT bolt kit in the order shown in the illustration. Immediately after removing VVT bolt kit in the location for a bearing cap, install the bearing cap with the bolts in the order shown in the illustration. Tighten bolt A to 21 ft. lbs. (28 Nm). Tighten bolt B to 12 ft. lbs. (16 Nm).

➡**Be sure to follow the numerical order when performing this procedure.**

➡**Do not drop the VVT bolt kit into the cylinder head.**

d. Check the torque of each bolt again.

60. Connect the chain sub assembly (for Bank 1).

a. Align the paint marks on the camshaft timing gear and No. 1 chain and install the No. 1 chain to the camshaft timing gear.

➡**If the paint marks are not aligned, align them by turning the camshaft slightly.**

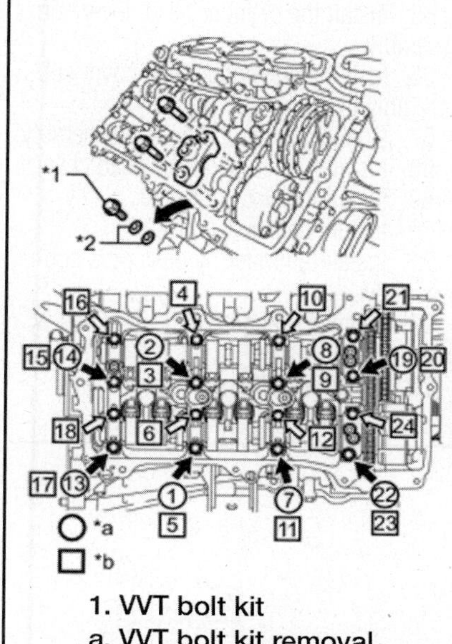

Fig. 110 Checking the bearing cap installation

61. Install the No. 1 chain tensioner assembly.

a. Turn the crankshaft counterclockwise 30° past the "0" timing mark, and then turn it clockwise to align the notch with the "0" timing mark.

b. Turn the crankshaft slightly to eliminate the slack in the chain.

➡**Make sure there is some slack in the chain around the area where the chain tensioner is installed.**

c. While turning the stopper plate of the tensioner clockwise, push in the

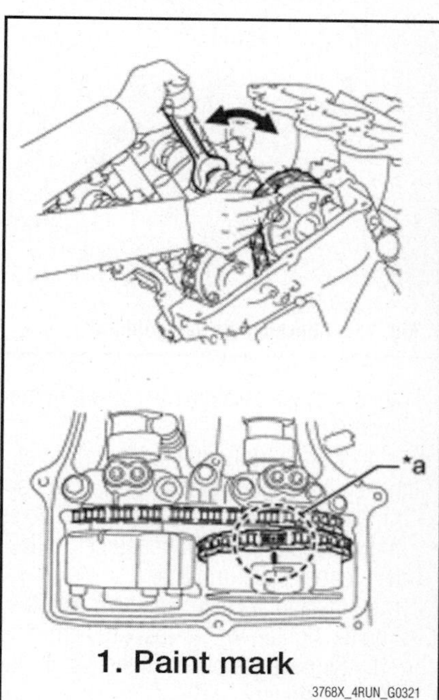

Fig. 111 Aligning the paint marks on the camshaft timing gear and No. 1 chain to the camshaft timing gear

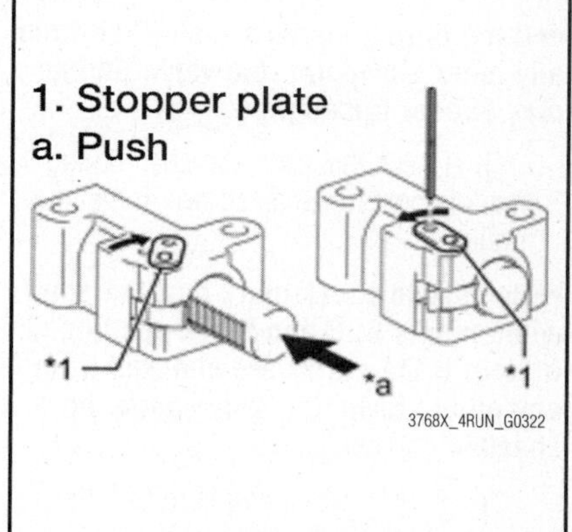

Fig. 112 Pushing the plunger of the tensioner

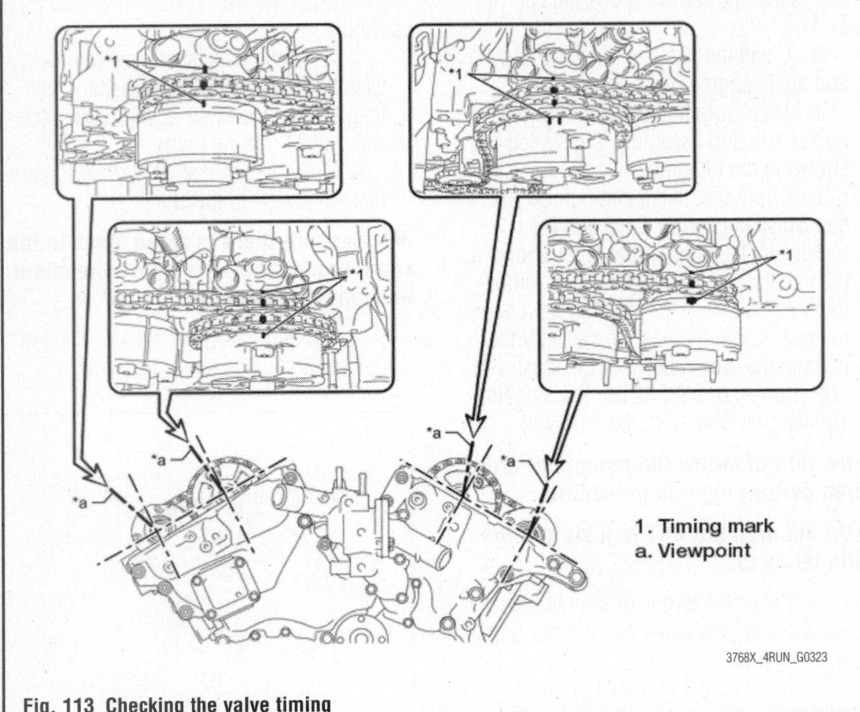

Fig. 113 Checking the valve timing

plunger of the tensioner as shown in the illustration.

d. While turning the stopper plate of the tensioner counterclockwise, insert a pin 0.0500 inch (1.27 mm) into the holes in the stopper plate and tensioner to fix the stopper plate in place.

e. Install the chain tensioner with the 2 bolts. Tighten to 7 ft. lbs. (10 Nm).

f. Remove the pin from the No. 1 chain tensioner.

62. Inspect the valve timing.

a. Check the camshaft timing marks.

➡Check each timing mark from a viewpoint directly in line with the center of the camshaft and the timing mark on each camshaft timing gear.

➡If the timing marks are checked from any other viewpoint, the valve timing may appear misaligned.

b. Check that each camshaft timing mark is positioned as shown in the illustration.

➡Be sure to check mark A at the point when marks B, C and D are positioned in line. If the marks are checked from any other viewpoint, they cannot be checked correctly.

c. If the valve timing is misaligned, reinstall the timing chain.

d. Turn the crankshaft 2 revolutions, set the No. 1 cylinder to TDC/compression and check the timing marks again.

63. Install the timing chain cover plate with a new gasket and the 4 bolts. Tighten to 80 inch lbs. (9 Nm).

64. Pour engine oil.

65. Install the cylinder head cover sub assembly.

66. Install the cylinder head cover sub assembly LH.

67. Connect the fuel pipe sub assembly.

68. Install the rear cylinder head cover.

69. Install the No. 2 oil pipe.

70. Install the No. 1 oil pipe.

71. Install the water by pass pipe sub assembly.

72. Install the engine oil level dipstick guide.

73. Install the alternator assembly.

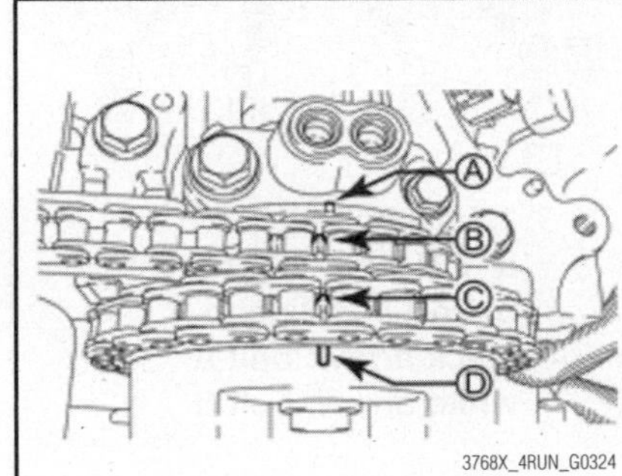

Fig. 114 Checking the intake camshaft

74. Install the No. 2 exhaust manifold heat insulator.

75. Install the wiring harness clamp bracket.

76. Install the No. 2 idler pulley sub assembly.

77. Connect the vane pump assembly.

78. Install the ignition coil assembly.

79. Install the intake air surge tank.

80. Install the fan shroud.

81. Connect the oil cooler tube.

82. Install the radiator reservoir.

83. Install the No. 2 radiator hose.

84. Install the No. 1 radiator hose.

85. Install the air cleaner case sub assembly.

86. Install the air cleaner cap and hose.

87. Install the battery tray.

88. Install the battery.

89. Install the battery hold down clamp.

90. Connect the cable to the positive battery terminal.

91. Connect the cable to the negative battery terminal.

➡When disconnecting the cable, some systems need to be initialized after the cable is reconnected.

92. Add engine coolant.

93. Add engine oil.

94. Inspect for coolant leaks.

95. Inspect for engine oil leaks.

96. Install the v-bank cover.

97. Install the upper radiator support seal.

98. Install the No. 1 engine under cover sub assembly.

99. Install the front bumper cover lower.

100. Inspect the ignition timing.

CRANKSHAFT FRONT SEAL

REMOVAL & INSTALLATION

2.7L Engine

See Figure 115.

1. Remove the upper radiator support seal.

2. Remove the radiator reservoir.

3. Remove the fan shroud.

4. Remove the crankshaft pulley.

a. Using a special tool, hold the crankshaft pulley and loosen the pulley bolt until 2 or 3 threads are screwed into the crankshaft.

b. Using a special tool and pulley bolt, remove the crankshaft pulley.

5. Remove the front crankshaft oil seal.

a. Using a screwdriver, pry out the oil seal.

➡Tape the screwdriver tip before use.

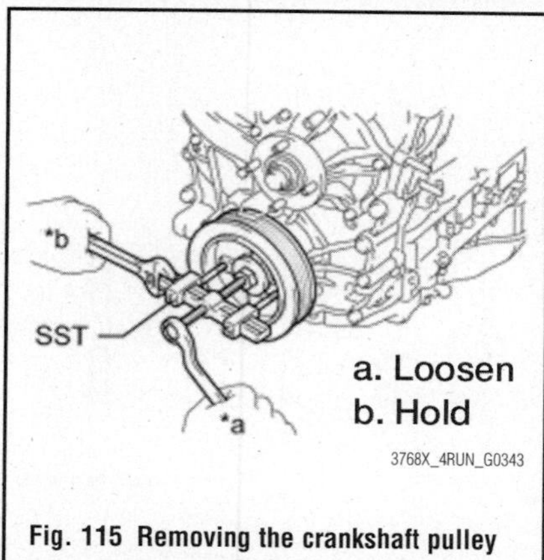

Fig. 115 Removing the crankshaft pulley

➡**Do not damage the surface of the oil seal press fit hole of the crankshaft.**

➡**Check the crankshaft for damage after removing the oil seal. If the crankshaft is damaged, smooth the surface with 400 grit sandpaper.**

To install:

6. Install the front crankshaft oil seal.
   a. Apply MP grease to the lip of a new oil seal.

➡**Do not allow foreign matter to contact the lip of the oil seal.**

➡**Do not allow MP grease to contact the dust seal.**

   b. Temporarily install the oil seal to the timing chain cover.
   c. Using the special tool and a hammer, tap in the oil seal until its surface is flush with the chin cover edge.

➡**Keep the lip free from foreign matter.**

➡**Do not tap the oil seal at an angle.**

7. Install the crankshaft pulley.
   a. Align the key groove of the pulley with the pulley set key and slide on the pulley.
   b. Using the special tool, install a new crankshaft pulley bolt. Tighten the bolt to 192 ft. lbs. (260 Nm).

➡**Do not reuse the pulley bolt.**

8. Install the fan shroud.
9. Install the radiator reservoir.
10. Install the upper radiator support seal.
11. Inspect for oil leaks.
12. Inspect the engine oil level.

4.0L Engine

See Figure 116.

1. Remove the radiator assembly.
2. Disconnect the water by pass pipe sub assembly.

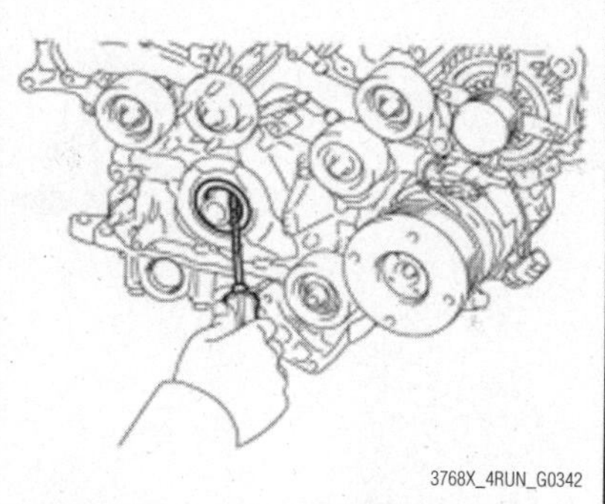

Fig. 116 Removing the front crankshaft oil seal

3. Remove the oil filter bracket.
4. Remove the crankshaft pulley.
5. Remove the front crankshaft oil seal.
   a. Using a screwdriver, pry out the oil seal.

➡**Tape the tip of the screwdriver before use.**

➡**Do not damage the surfaces of the oil seal press fit hole or crankshaft.**

To install:

6. Install the front crankshaft oil seal.
   a. Apply MP grease to the lip of a new oil seal.
   b. Using a special tool and hammer, tap in the oil seal until its surface is flush with the timing chain cover edge.

➡**Keep the lip free from foreign matter.**

➡**Do not tap the oil seal at an angle.**

7. Install the crankshaft pulley.
8. Install the oil filter bracket.
9. Connect the water by pass pipe sub assembly.
10. Install the radiator assembly.

CYLINDER HEAD

REMOVAL & INSTALLATION

2.7L Engine

See Figures 117 through 131.

1. Remove the engine assembly.
2. Remove the No. 1 exhaust manifold heat insulator.
3. Remove the No. 4 intake pipe.
4. Remove the air switching valve assembly.
5. Remove the exhaust manifold.
6. Remove the timing chain cover sub assembly.
7. Remove the No. 1 cylinder to TDC/compression.
   a. Temporarily install the crankshaft pulley bolt.
   b. Rotate the crankshaft clockwise so that the timing marks on the crankshaft timing gear and camshaft timing gears are as shown in the illustration.

➡**If the timing marks do not align, rotate the crankshaft clockwise again and align the timing marks.**

   c. Remove the crankshaft pulley bolt.
8. Removing the timing chain guide.
9. Remove the No. 1 chain tensioner assembly.

➡**When the chain tensioner is removed, do not rotate the crankshaft.**

➡**When the chain is removed and the camshaft needs to be rotated, rotate the crankshaft 90° to the right.**

   a. Move the stopper plate upward to release the lock and push the plunger deep into the tensioner.
   b. Move the stopper plate downward to set the lock and insert a 3.0 mm (0.118 in.) diameter bar into the stopper plate hole.

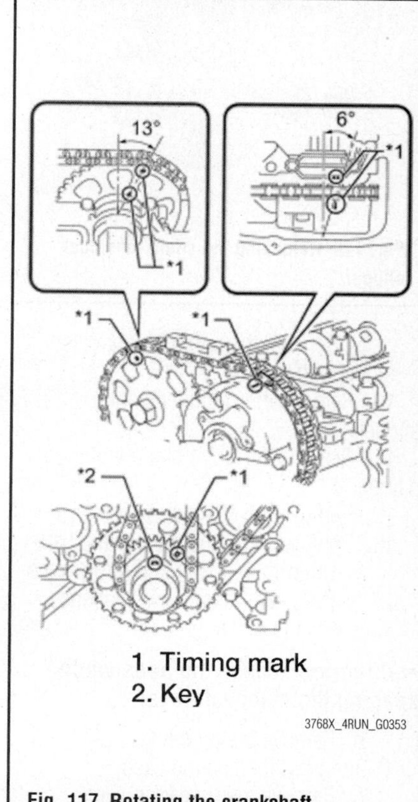

Fig. 117 Rotating the crankshaft

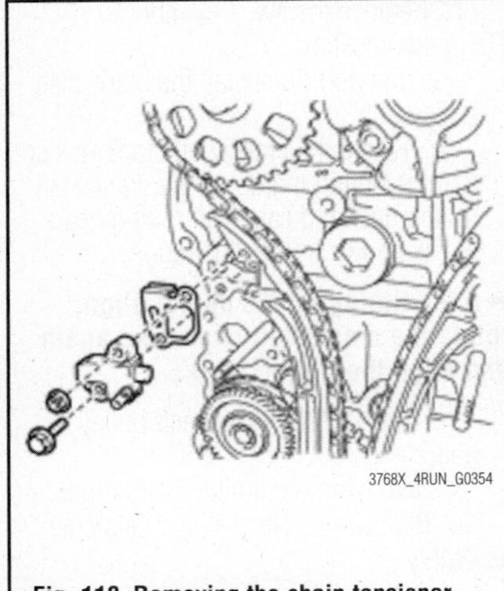
Fig. 118 Removing the chain tensioner

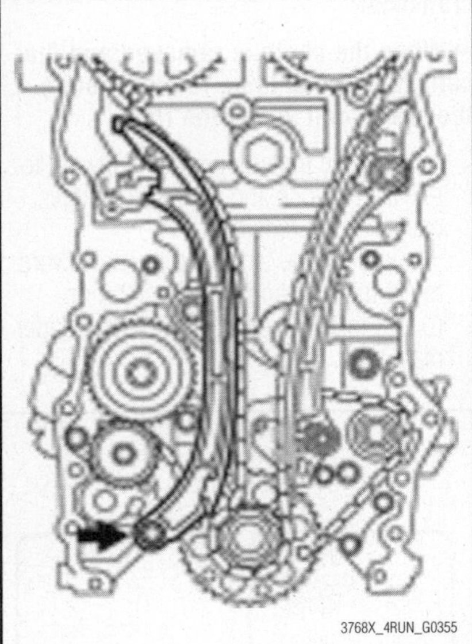
Fig. 119 Removing the chain tensioner slipper

c. Remove the bolt, nut, chain tensioner and gasket.

10. Remove the chain tensioner slipper.

11. Remove the No. 1 chain vibration damper.

12. Remove the chain sub assembly.

13. Remove the camshaft bearing cap.

a. Uniformly loosen and remove the 21 bearing cap bolts in the sequence shown in the illustration.

➡Uniformly loosen the bolts while keeping the camshaft level.

b. Remove the oil delivery pipe and O-ring from the bearing caps.

c. Remove the 9 bearing caps.

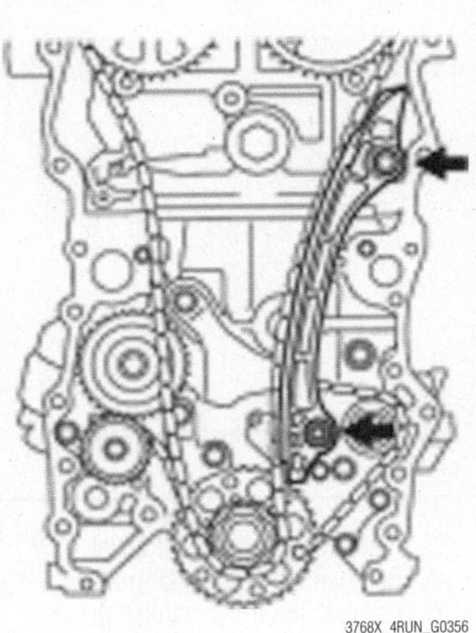
Fig. 120 Removing the No. 1 chain vibration damper

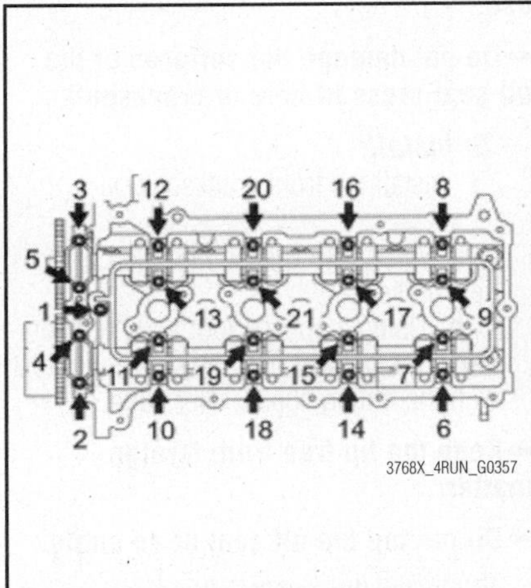

Fig. 121 Removing the camshaft bearing cap bolts in sequence

➡Arrange the removed parts in the correct order.

14. Remove the camshaft.

15. Remove the No. 2 camshaft.

16. Remove the No. 1 valve rocker arm sub assembly.

17. Remove the valve lash adjuster assembly.

18. Remove the valve step cap.

➡Arrange the removed parts in the correct order.

19. Remove the cylinder head sub assembly.

a. Uniformly loosen the 10 bolts in the sequence shown in the illustration. Remove the 10 cylinder head bolts and plate washers.

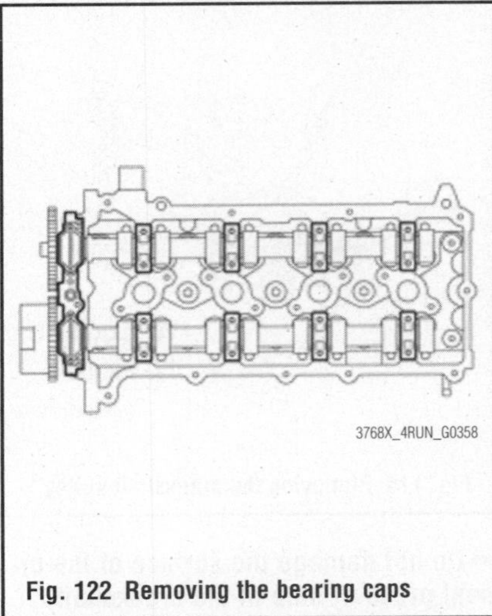
Fig. 122 Removing the bearing caps

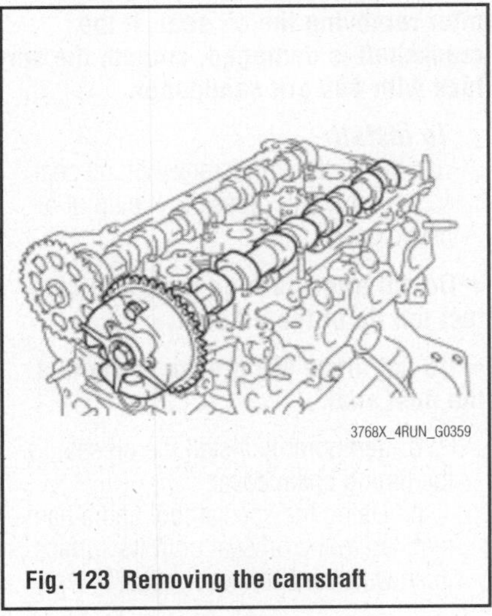
Fig. 123 Removing the camshaft

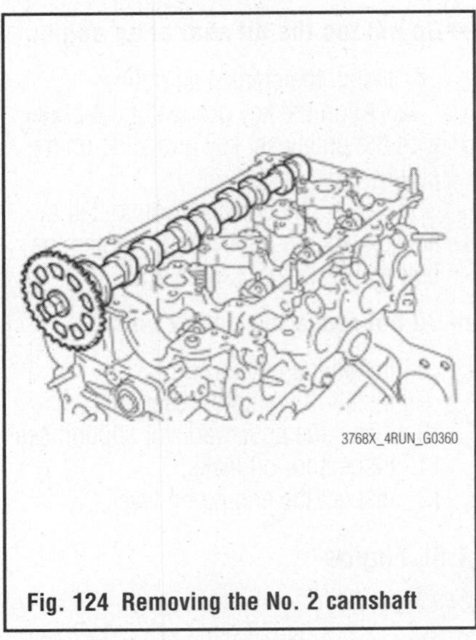
Fig. 124 Removing the No. 2 camshaft

➡Be careful not to drop the washers into the cylinder head.

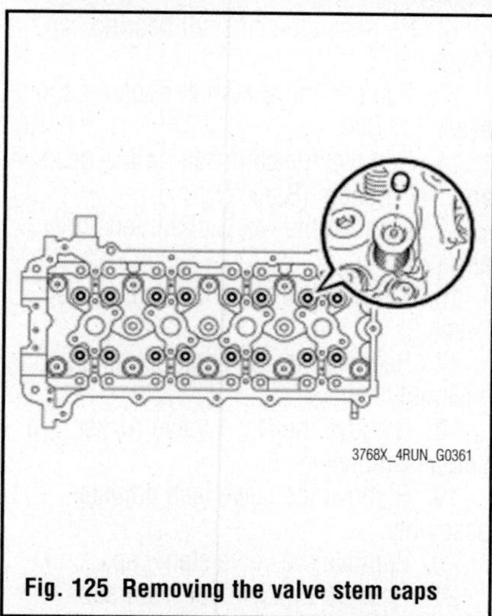

Fig. 125 Removing the valve stem caps

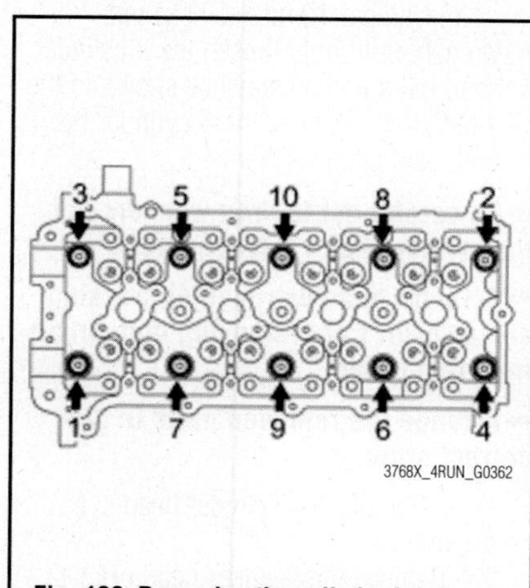

Fig. 126 Removing the cylinder head sub assembly

➡Head warpage or cracking could result from removing the bolts in the wrong order.

20. Remove the cylinder head gasket.
21. Inspect the cylinder head set bolt.
22. Inspect the cylinder head sub assembly.

To install:

23. Install the cylinder head gasket.
   a. Place a new cylinder head gasket on the cylinder block surface with the lot No. stamp facing upward.

➡Make sure that the cylinder head gasket is installed so that it is facing in the correct direction.

24. Install the cylinder head sub assembly.
   a. Place the cylinder head on the cylinder block.

➡Make sure that no oil is on the mounting surface of the cylinder head.

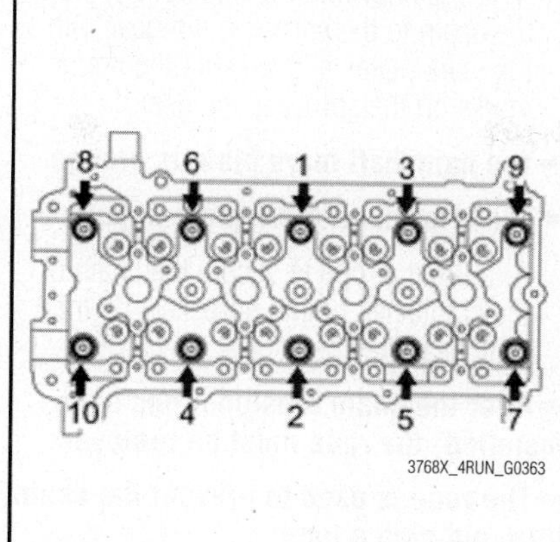

Fig. 127 Installing the cylinder head bolts in sequence

➡Place the cylinder head on the cylinder block gently in order not to damage the gasket with the bottom part of the head.

   b. Install the cylinder head bolts.

➡The cylinder head bolts are tightened in 3 successive steps.

   c. Install the plate washers to the cylinder head bolts.
   d. Apply a light coat of engine oil to the threads and under the heads of the cylinder head bolts.

Step 1:

   a. Using several steps, install and uniformly tighten the 10 cylinder head bolts with plate washers in the sequence shown in the illustration. Tighten to 29 ft. lbs. (39 Nm).
   b. Mark the front of each cylinder head bolt head with paint.

Step 2:

   a. Tighten the cylinder head bolts 90°in the sequence shown in step 1.

Step 3:

   a. Tighten the cylinder head bolts another 90°in the sequence shown in step 1.
   b. Check that the paint marks are now at a 180c. angle to the front.

25. Install the valve step cap.
   a. Apply a light coat of engine oil to the valve stem ends.
   b. Install the 16 valve stem caps to the cylinder head.

➡Do not drop the valve stem caps into the cylinder head.

26. Install the valve lash adjuster assembly.
27. Install the No. 1 valve rocker arm sub assembly.
28. Install the camshaft.
   a. Apply clean engine oil to the camshaft cams and cylinder head journals.
   b. Position the camshaft and No. 2 camshaft as shown in the illustration.
29. Install the camshaft bearing cap.
   a. Temporarily install the No. 1 camshaft bearing cap.
   b. Check the proper location of each No. 2 camshaft bearing cap and install them.
   c. Install a new O-ring to the No. 1 camshaft bearing cap.
   d. Temporarily install the oil delivery pipe.
   e. Install the 21 bolts and tighten them in the order shown. Tighten

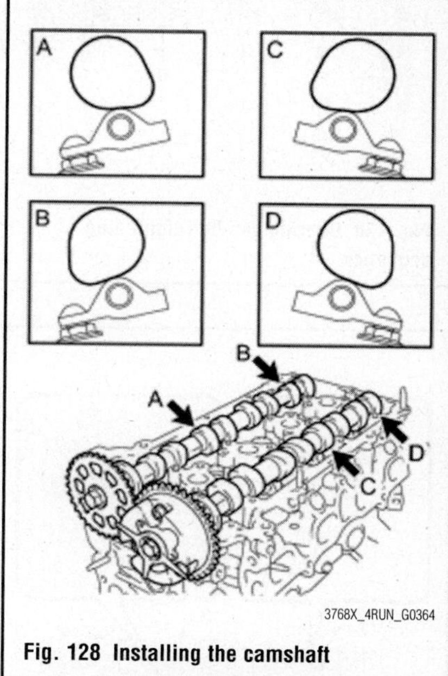

Fig. 128 Installing the camshaft

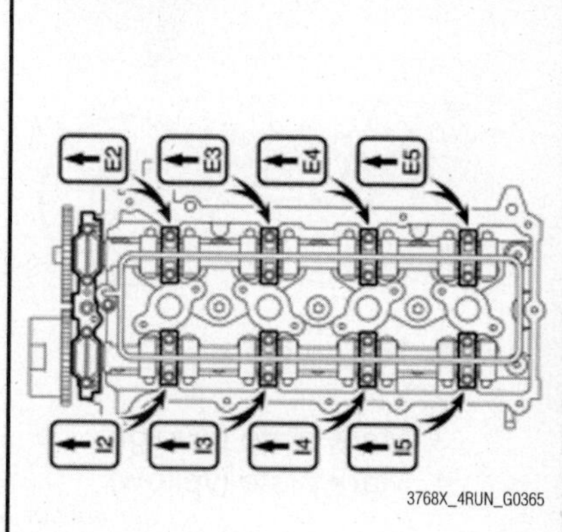

Fig. 129 Installing the camshaft bearing caps

bolt A to 9 ft. lbs. (12 Nm). Tighten all the remaining bolts to 11 ft. lbs. (16 Nm).

30. Install the No. 1 chain vibration damper.

a. Install the vibration damper with the bolt and nut. Tighten the bolt to 15 ft. lbs. (21 Nm). Tighten the nut to 13 ft. lbs. (18 Nm).

31. Install the chain sub assembly.

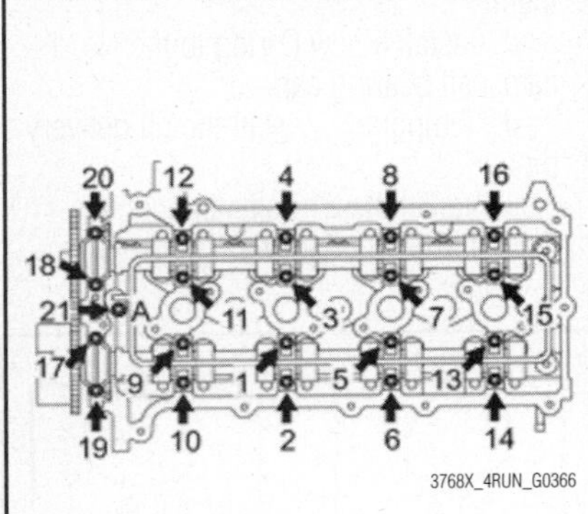

Fig. 130 Bearing cap bolt tightening sequence

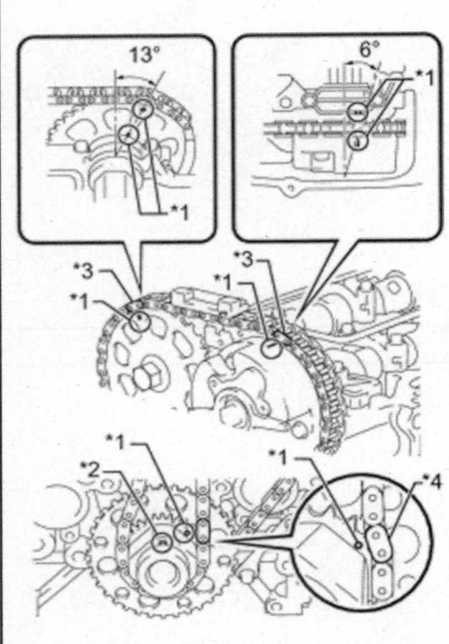

Fig. 131 Installing the chain to the sprocket and gear

a. As shown in the illustration, install the chain to the sprocket and gear with the mark plates aligned with the timing marks on the sprocket and gear.

➡The camshaft mark plate is orange.

➡The crankshaft mark plate is yellow.

b. Use a rope to secure the chain of the crankshaft timing sprocket. Tie the rope near the sprocket.

➡After the chain tensioner has been installed, the rope must be removed.

➡The rope is used to prevent the chain from jumping a tooth.

32. Install the chain tensioner slipper with the bolt. Tighten to 15 ft. lbs. (21 Nm).

33. Install the No. 1 chain tensioner assembly.

a. Move the stopper plate upward to release the lock and push the plunger deep into the tensioner.

b. Move the stopper plate downward to set the lock and insert a hexagon wrench into the hole of the stopper plate.

c. Install a new gasket and the chain tensioner with the bolt and nut. Tighten to 7 ft. lbs. (10 Nm).

34. Install the timing chain guide.

35. Install the timing chain sub assembly.

36. Install the exhaust manifold.

37. Install the air switching valve assembly.

38. Install the No. 4 intake pipe.

39. Install the No. 1 exhaust manifold heat insulator.

40. Install the engine assembly.

4.0L Engine

See Figures 132 through 140.

1. Remove the timing chain cover sub assembly.
2. Set the No. 1 cylinder to TDC/compression.
3. Remove the No. 1 chain tensioner assembly.
4. Remove the chain tensioner slipper.
5. Remove the chain sub assembly.
6. Remove the No. 1 idle gear shaft.
7. Remove the No. 1 chain vibration damper.
8. Remove the No. 2 chain vibration damper.
9. Remove the crankshaft timing sprocket.
10. Remove the camshaft timing gears and No. 2 chain (Bank 1).
11. Remove the No. 2 chain tensioner assembly.
12. Remove the camshaft bearing cap (Bank 1).
13. Remove the camshaft housing sub assembly RH.
14. Remove the camshaft timing gears and No. 2 chain (Bank 2).
15. Remove the No. 3 chain tensioner assembly.
16. Remove the camshaft bearing cap (Bank 2).
17. Remove the camshaft housing sub assembly LH.
18. Remove the No. 1 valve rocker arm sub assembly.
19. Remove the valve lash adjuster assembly.
20. Remove the valve stem cap.
21. Remove the cylinder head sub assembly.

a. Using a 10 mm bi-hexagon wrench, uniformly loosen the 8 cylinder head bolts in the sequence shown in the illustration. Remove the 8 cylinder head bolts and plate washers.

➡Be careful not to drop washers into the cylinder head sub assembly.

➡Cylinder head warpage or cracking could result from removing bolts in an incorrect order.

➡Arrange the removed parts in the correct order.

b. Remove the cylinder head sub assembly.

22. Remove the cylinder head LH.

a. Uniformly loosen and remove the 2 cylinder head set bolts in several steps in the sequence shown in the illustration.

b. Using a 10mm bi-hexagon wrench, uniformly loosen the 8 bolts in

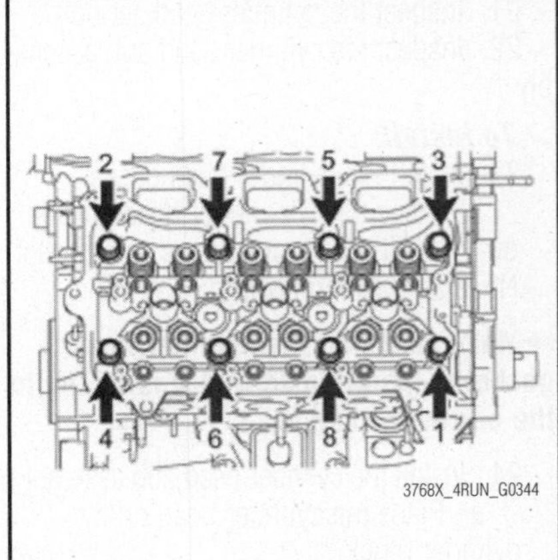

Fig. 132 Removing the cylinder head sub assembly

the sequence shown in the illustration. Remove the 8 cylinder head bolts and plate washers.

➡Be careful not to drop washers into the cylinder head sub-assembly.

➡Cylinder head warpage or cracking could result from removing bolts in an incorrect order.

➡Be sure to keep the removed parts for each installation position separate.

c. Remove the cylinder head LH.

To install:

23. Inspect the cylinder head set bolt.
24. Inspect the cylinder head sub assembly.
25. Install the cylinder head gasket.

a. Remove any old packing (FIPG) material and be careful not to drop any oil on the contact surfaces of the cylinder head or cylinder block.

b. Apply seal packing to a new cylinder head gasket as shown in the illustration.

➡Remove any oil from the contact surface.

➡Install the cylinder head gasket within 3 minutes and tighten the bolts within 15 minutes after applying seal packing.

➡Do not add engine oil within 2 hours of installation.

c. Place the cylinder head gasket on the cylinder block surface with the front face of the Lot No. stamp upward.

➡Make sure that the gasket is installed facing the proper direction.

26. Install the cylinder head sub assembly.

a. Place the cylinder head on the cylinder block.

➡Gently place the cylinder head in order not to damage the gasket with the bottom part of the head.

➡Make sure that no oil is on the mounting surface of the cylinder head.

➡The cylinder head bolts are tightened in 3 progressive steps.

b. Apply a light coat of engine oil to the threads and under the heads of the cylinder head bolts.

Step 1:

a. Using a 10 mm bi-hexagon wrench, install and uniformly tighten the 8 cylinder head bolts with the plate washers in several steps in the sequence shown in the illustration. Tighten to 27 ft. lbs. (36 Nm).

Step 2:

a. Mark the front side of each cylinder head bolt head with paint.

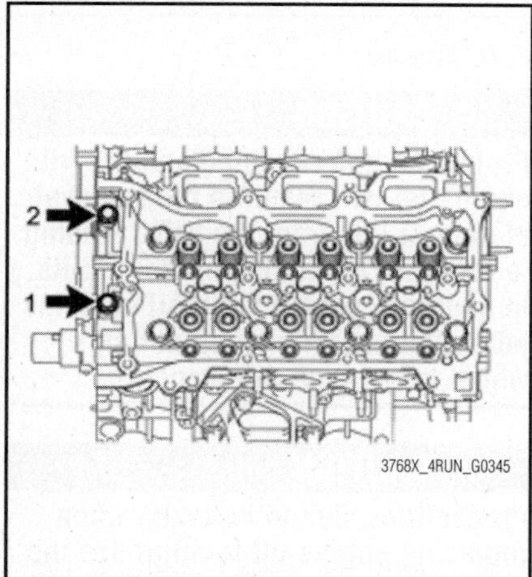

Fig. 133 Loosening the 2 cylinder head set bolts

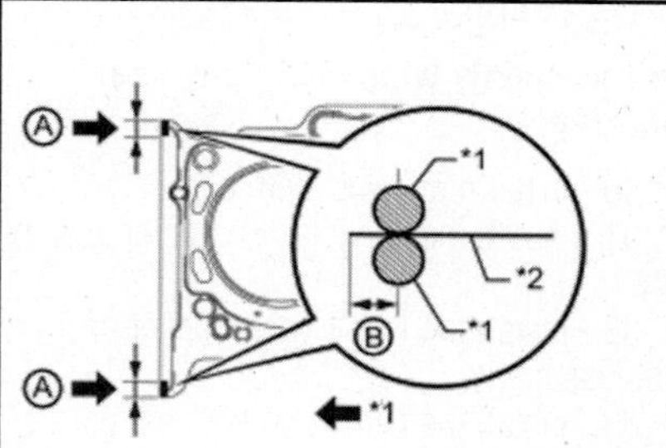

A. 0.394-0.591 inch (10-15 mm)
B. 0.0492-0.0591 inch (1.25-1.5 mm)
1. Seal packing
2. Gasket

3768X_4RUN_G0347

Fig. 135 Applying seal packing to a new cylinder head gasket

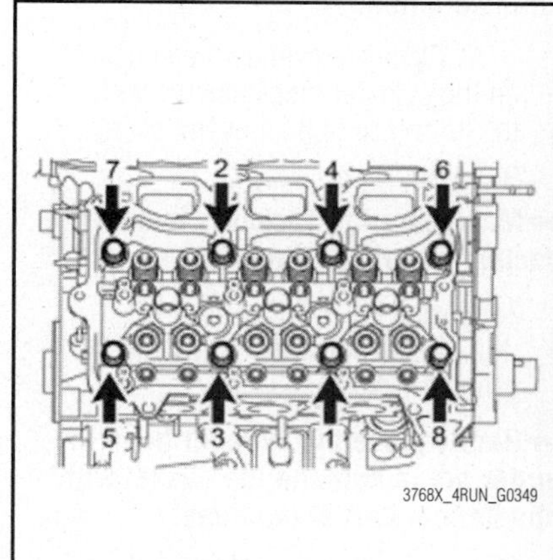

Fig. 137 Cylinder head sub assembly bolt installation sequence

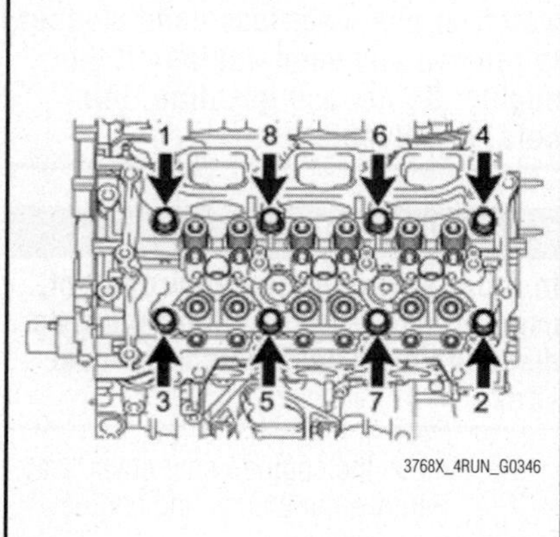

Fig. 134 Removing the 8 cylinder head bolts and plate washers

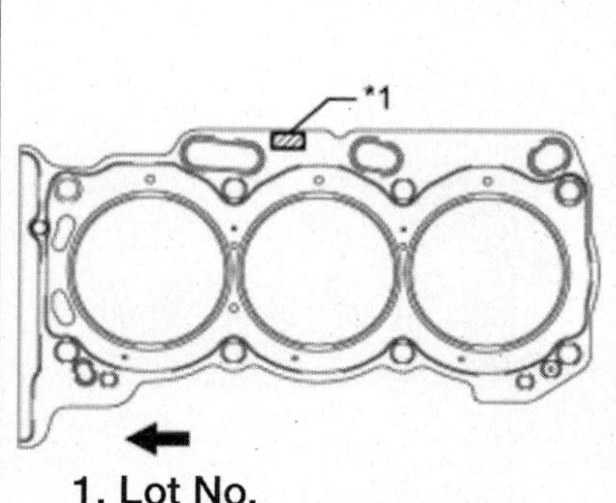

3768X_4RUN_G0348

Fig. 136 Installing the gasket

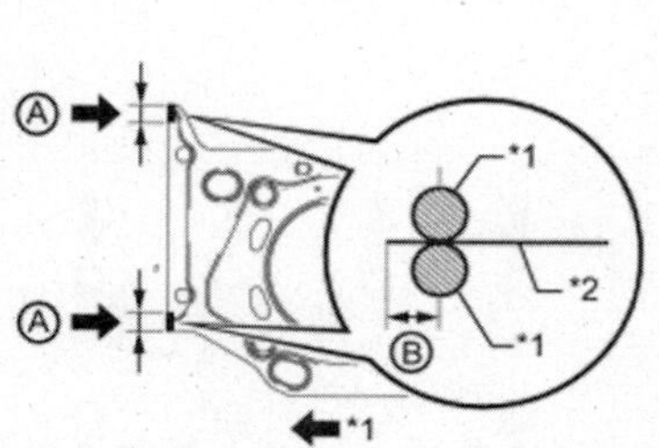

A. 0.394-0.591 inch (10-15 mm)
B. 0.0492-0.0591 inch (1.25-1.5 mm)
1. Seal packing
2. Gasket

3768X_4RUN_G0350

Fig. 138 Installing the No. 2 cylinder head gasket

b. Tighten the cylinder head bolts another 90°.

Step 3:

a. Tighten the cylinder head bolts an additional 90°.

b. Check that the paint mark is now at a 180° angle to the front.

➡**Thoroughly wipe clean any seal packing.**

27. Install No. 2 cylinder head gasket.

a. Remove any old packing (FIPG) material and be careful not to drop any oil on the contact surfaces of the cylinder head or cylinder block.

➡**Remove any oil from the contact surface.**

➡**Install the cylinder head gasket within 3 minutes and tighten the bolts within 15 minutes after applying seal packing.**

➡**Do not add engine oil within 2 hours of installation.**

b. Place the cylinder head gasket on the cylinder block surface with the front face of the Lot No. stamp upward.

➡**Make sure the gasket is installed facing the proper direction.**

28. Install the cylinder head LH.

a. Place the cylinder head on the cylinder block.

➡**Gently place the cylinder head in order not to damage the gasket with the bottom part of the head.**

➡**Make sure that no oil is on the mounting surface of the cylinder head.**

➡**The cylinder head bolts are tightened in 3 progressive steps.**

b. Apply a light coat of engine oil to the threads and under the heads of the cylinder head bolts.

Step 1:

a. Using a 10 mm bi-hexagon wrench, install and uniformly tighten the 8 cylinder head bolts with plate washers in several steps in the sequence shown. Tighten to 27 ft. lbs. (36 Nm).

Step 2:

a. Mark the front side of each cylinder head bolt head with paint.

b. Tighten the cylinder head bolts another 90°.

Step 3:

29. Tighten the cylinder head bolts and additional 90°.

a. Check that the paint mark is now at a 180° angle to the front.

b. Tighten the 2 bolts in the order shown in the illustration. Tighten to 22 ft. lbs. (30 Nm).

➡**Thoroughly wipe clean any seal packing.**

30. Install the valve stem cap.
31. Install the valve lash adjuster assembly.
32. Install the No. 1 valve rocker arm sub assembly.
33. Install the camshaft bearing cap (Bank 2).
34. Install the camshaft housing sub assembly LH.
35. Install the camshaft bearing cap (Bank 1).
36. Install the camshaft housing sub assembly RH.
37. Install the No. 3 chain tensioner assembly.
38. Install the camshaft timing gears and No. 2 chain (Bank 2).
39. Install the No. 2 chain tensioner assembly.
40. Install the camshaft timing gears and No. 2 chain (Bank 1).
41. Install the No. 1 chain vibration damper.
42. Install the No. 2 chain vibration damper.
43. Install the crankshaft timing sprocket.
44. Install No. 1 idle gear shaft.
45. Install the chain sub assembly.
46. Install the chain tensioner slipper
47. Install the No. 1 chain tensioner assembly.
48. Inspect the valve timing.
49. Install the timing chain cover sub assembly.

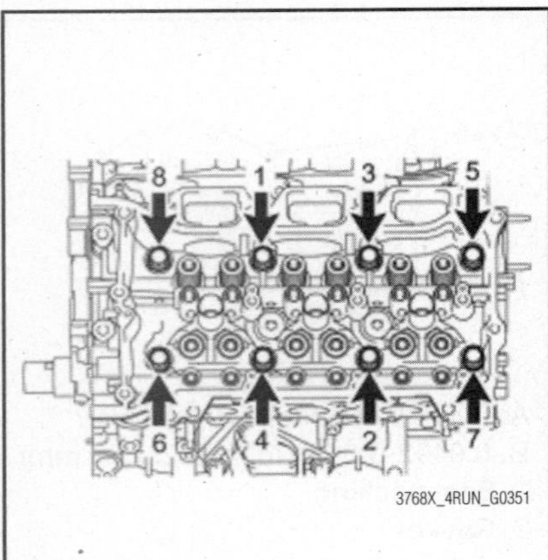

Fig. 139 LH cylinder head bolt tightening sequence

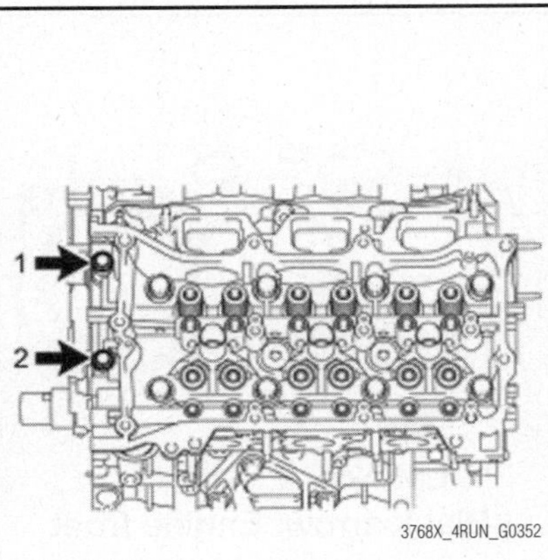

Fig. 140 Tightening the 2 remaining bolts

ENGINE OIL & FILTER

REPLACEMENT

2.7L Engine

CAUTION

Prolonged and repeated contact with engine oil will result in the removal of natural oils from the skin, leading to dryness, irritation and dermatitis. In addition, used engine oil contains potentially harmful contaminants which may cause skin cancer.

CAUTION

Precautions should be taken when replacing engine oil to minimize the risk of your skin making contact with used engine oil. Protective clothing and gloves that cannot be penetrated by oil should be worn. The skin should be washed with soap and water, or use waterless hand cleaner, to remove any used engine oil thoroughly. Do not use gasoline, thinners, or solvents.

WARNING

In order to protect the environment, used oil and used oil filters must be disposed of at designated disposal sites.

1. Remove the engine under cover seal.

a. Remove the 2 bolts and engine under cover seal.

2. Drain the engine oil.

a. Remove the oil filler cap.

b. Remove the oil drain plug and gasket, and drain the oil into a container.

3. Remove the rear engine under cover assembly.

4. Remove the oil filter sub assembly.

a. Using SST (09228-07501), remove the oil filter.

5. Install the oil filter sub assembly.

a. Check and clean the oil filter installation surface.

b. Apply clean engine oil to the gasket of a new oil filter.

c. Lightly screw the oil filter into place by hand. Tighten it until the gasket contacts the seat.

d. Using SST (09228-07501), tighten the filter. Tighten to 13 ft. lbs. (17 Nm). If enough space is not available to use a torque wrench, tighten the oil filter ¾ turn by hand or use a common wrench.

6. Add engine oil.

a. Clean and install the oil drain plug and new gasket. Tighten to 28 ft. lbs. (38 Nm).

b. Add fresh engine oil.

c. Install the oil filter cap.

7. Inspect for oil leak.

a. Start the engine. Make sure that there are no oil leaks from the areas that were worked on.

8. Inspect the engine oil level.

9. Install the rear engine unde4r cover seal with the 2 bolts. Tighten to 21 ft. lbs. (29 Nm).

4.0L Engine

See Figures 141 and 142.

✲✲ CAUTION

Prolonged and repeated contact with engine oil will result in the removal of natural oils from the skin, leading to dryness, irritation and dermatitis. In addition, used engine oil contains potentially harmful contaminants which may cause skin cancer.

✲✲ CAUTION

Precautions should be taken when replacing engine oil to minimize the risk of your skin making contact with used engine oil. Protective clothing and gloves that cannot be penetrated by oil should be worn. The skin should be washed with soap and water, or use waterless hand cleaner, to remove any used engine oil thoroughly. Do not use gasoline, thinners, or solvents.

✲✲ WARNING

In order to protect the environment, used oil and used oil filters must be disposed of at designated disposal sites.

1. Remove the 2 bolts and engine under cover seal.

2. Disconnect the front bumper cover lower.

a. Remove the 2 bolts and clip and disconnect the front bumper cover lower/

3. Disconnect the No. 1 engine under cover sub assembly.

a. Remove the 4 bolts and open the No. 1 engine under cover.

4. Drain the engine oil.

a. Remove the oil filler cap.

b. Remove the oil pan drain plug and drain the engine oil into a container.

c. Install a new gasket and the oil pan drain plug. Tighten to 30 ft. lbs. (40 Nm).

5. Remove the oil filter element.

a. Connect a hose with an inside diameter of 15 mm (0.591 in.) to the pipe.

b. Remove the oil filter drain plug.

c. Install the pipe to the oil filter cap.

➡Use a container to catch the draining oil.

➡If the O-ring is removed with the drain plug, install the O-ring together with the pipe.

d. Check that oil is drained from the oil filter. Then, disconnect the pipe and remove the O-ring.

e. Using SST (09228-06501), remove the filler cap.

➡Do not remove the oil filter bracket clip.

f. Remove the oil filter element and O-ring from the oil filter cap.

➡Be sure to remove the O-ring (for the cap) by hand, without using any tools, to prevent damage to the groove for the O-ring in the cap.

6. Install the oil filter element.

a. Clean the inside of the oil filter cap, the threads and O-ring groove.

b. Apply a small amount of engine oil to a new O-ring and install the O-ring to the oil filter cap.

c. Set a new oil filter element on the oil filter cap.

d. Remove any dirt or foreign matter from the installation surface of the engine.

e. Apply a small amount of engine oil to the O-ring again and temporarily install the oil filter cap.

f. Using SST (09228-06501), tighten the oil filter cap. Tighten to 18 ft. lbs. (25 Nm).

✲✲ WARNING

When tightening the oil filter cap, do not remove the oil filter bracket clip.

➡Make sure that the oil filter is installed securely.

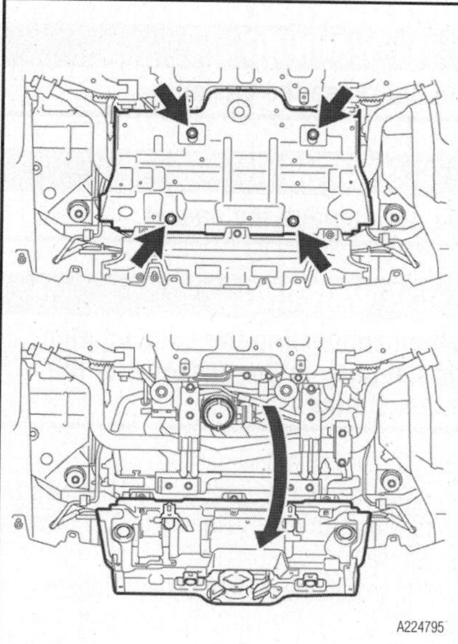

Fig. 141 Removing the 4 bolts and opening the No. 1 engine under cover

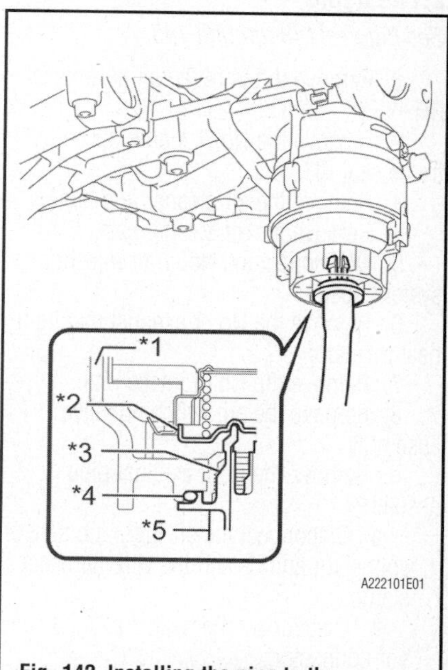

Fig. 142 Installing the pipe to the oil filter cap

✲✲ WARNING

Make sure that the O-ring does not get caught between the parts.

g. Apply a small amount of engine oil to a new drain plug O-ring and install the O-ring to the oil filter cap.

➡Before installing the O-ring, remove any dirt or foreign matter from the installation surface of the oil filter cap.

h. Install the oil filter drain plug. Tighten to 10 ft. lbs. (13 Nm).

➡Make sure that the O-ring does not get caught between the parts.

7. Add engine oil.
8. Check the engine oil level.
9. Inspect for engine oil leaks.
a. Start the engine. Make sure that there are no oil leaks from the area that was worked on.
10. Connect the No. 1 engine under cover sub assembly.
a. Install the 4 bolts to close the No. 1 engine under cover. Tighten to 21 ft. lbs. (29 Nm).
11. Connect the front bumper cover lower with the 2 bolts and clip. Tighten to 71 inch lbs. (8 Nm).
12. Install the engine under cover seal with the 2 bolts. Tighten to 21 ft. lbs. (29 Nm).

EXHAUST MANIFOLD

REMOVAL & INSTALLATION

2.7L Engine

See Figures 143 through 145.

1. Remove the front fender apron seal RH.
2. Remove the No. 1 fender apron to frame seal RH.
3. Remove the air cleaner and hose.
4. Remove the air cleaner case.
5. Disconnect the No. 1 air injection system hose.
6. Remove the No. 1 exhaust manifold heat insulator.
7. Remove the No. 4 intake pipe.
8. Remove the air switching valve assembly.
9. Remove the front exhaust pipe assembly.
a. Disconnect the air fuel ratio sensor connector and detach the wire harness clamp.
b. Disconnect the heated oxygen sensor connector.
c. Remove the 4 bolts and 4 compression springs.

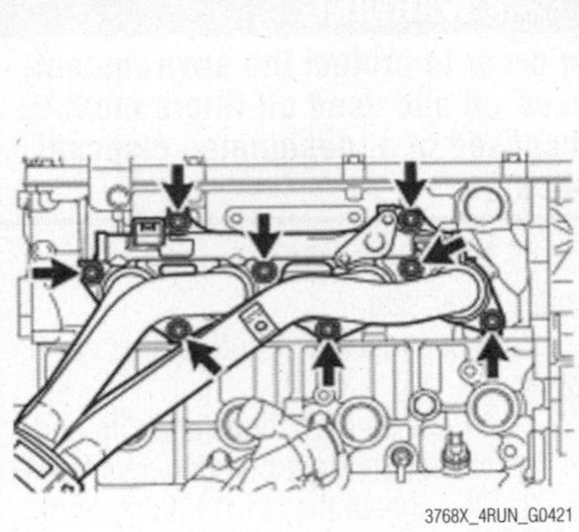

Fig. 143 Removing the exhaust manifold

d. Remove the front exhaust pipe from the pipe support.
e. Remove the gasket.
10. Remove the manifold stay.
11. Remove the exhaust manifold.
a. Remove the 8 nuts and exhaust manifold.
b. Remove the 2 gaskets.

To install:

12. Install the exhaust manifold.
a. Using a plastic-faced hammer and wooden block, tap in a new gasket until its surface is flush with the exhaust manifold.

✲✲ WARNING

Be sure to install the gasket so that it faces the correct direction.

✲✲ WARNING

Do not reuse the gasket.

✲✲ WARNING

Do not damage the gasket.

✲✲ WARNING

When connecting the exhaust pipe, do not push in the gasket with the exhaust pipe.

b. Install a new gasket.
c. Install the exhaust manifold with 8 new nuts in order. Tighten to 27 ft. lbs. (36 Nm).
13. Install the manifold stay. Tighten bolt A to 52 ft. lbs. (71 Nm). Tighten bolt B to 32 ft. lbs. (44 Nm). Tighten bolt C to 22 ft. lbs. (30 Nm).
14. Install the front exhaust pipe assembly.
a. Using a plastic-faced hammer and wooden block, tap in a new gasket until its surface is flush with the front exhaust pipe.

✲✲ WARNING

Be sure to install the gasket so that it faces the correct direction.

✲✲ WARNING

Do not reuse the gasket.

✲✲ WARNING

Do not damage the gasket.

✲✲ WARNING

When connecting the exhaust pipe, do not push in the gasket with the exhaust pipe.

b. Using a vernier caliper, measure the free length of the compression spring. Minimum free length is 1.65 inch (42 mm).

✲✲ WARNING

If the free length is less than the minimum, replace the compression spring.

c. Install the front exhaust pipe to the pipe support.
d. Connect the front exhaust pipe to the exhaust manifold with the 2 compression springs and 2 bolts. Alternately tighten the bolts in several passes. Tighten to 32 ft. lbs. (43 Nm).
e. Connect the front exhaust pipe to the center exhaust pipe with the 2 compression springs and 2 bolts. Alternately tighten the bolts in several passes. Tighten to 32 ft. lbs. (43 Nm).
f. Connect the air fuel ratio sensor connector and attach the wire harness clamp.
g. Connect the heated oxygen sensor connector.
15. Install the air switching valve assembly.
16. Install the No. 4 intake pipe.
17. Install the No. 1 exhaust manifold heat insulator.
18. Connect the No. 1 air injection system hose.
19. Install the air cleaner case.
20. Install the air cleaner and hose.

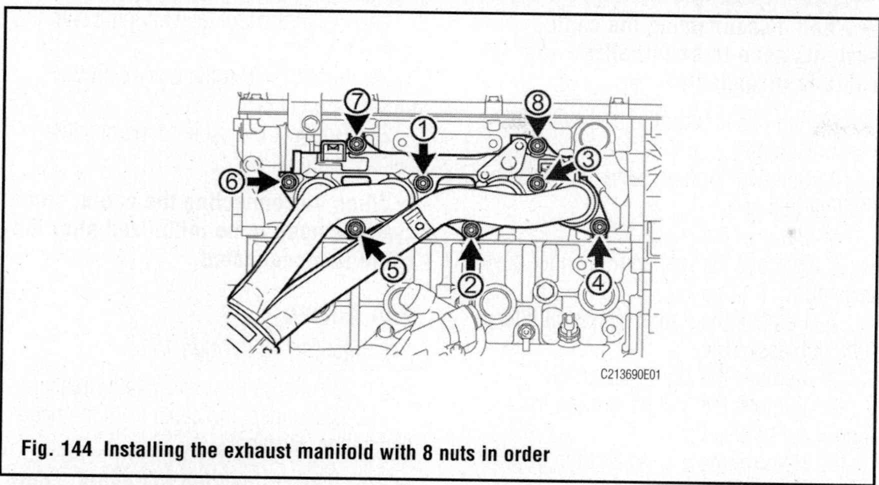

Fig. 144 Installing the exhaust manifold with 8 nuts in order

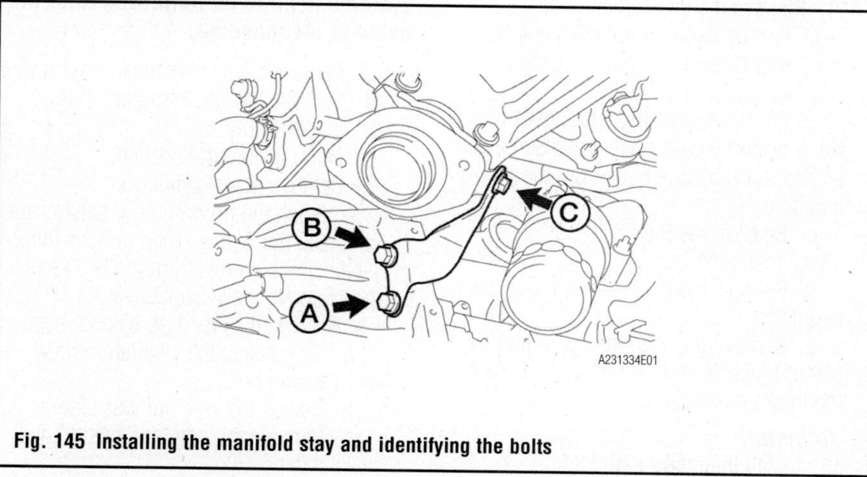

Fig. 145 Installing the manifold stay and identifying the bolts

21. Check for exhaust gas leak.
22. Install the front No. 1 fender apron to frame seal RH with the 5 clips.
23. Install the front fender apron seal RH with the 5 clips.
24. Inspect for exhaust gas leak.

4.0L Engine

See Figure 146.

1. Remove the upper radiator support seal.
2. Remove the v-bank cover.
3. Remove the air cleaner cap and hose.
4. Remove the air cleaner case sub assembly.
5. Remove the front fender apron seal LH.
6. Remove the front fender apron seal RH.
7. Remove the front No. 1 fender apron to frame seal LH.
8. Remove the front No. 1 fender apron to frame seal RH.
9. Remove the front exhaust pipe assembly.
10. Remove the manifold stay.
   a. Remove the 3 bolts and manifold stay.
11. Remove the No. 1 exhaust manifold heat insulator.
12. Remove the exhaust manifold sub assembly RH.
   a. Disconnect the air fuel ratio sensor connector.
   b. Remove the 6 nuts, manifold and gasket.
13. Remove the air fuel ratio sensor (Bank 1, Sensor 1).
14. Remove the No. 2 manifold stay.
   a. Remove the 3 bolts and the No. 2 manifold stay.
15. Remove the No. 1 exhaust manifold heat insulator.
   a. Remove the 3 bolts and heat insulator.
16. Remove the exhaust manifold sub assembly LH.
   a. Disconnect the air fuel ratio sensor connector.
   b. Remove the 6 nuts, manifold and gasket.
17. Remove the air fuel ratio sensor (Bank 2, Sensor 1).

To install:

18. Install the air fuel ratio sensor (Bank 2, Sensor 1).
19. Install the exhaust manifold sub assembly LH.

➡Be careful of the installation direction.

   a. Temporarily install the manifold with 6 new nuts.
   b. Tighten the 6 nuts in the sequence shown in the illustration. Tighten to 15 ft. lbs. (21 Nm).
   c. Connect the air fuel ratio sensor connector.
20. Install the No. 2 exhaust manifold heat in the illustration. Tighten to 15 ft. lbs. (21 Nm).
   a. Connect the air fuel ratio sensor connector.
21. Install the No. 2 exhaust manifold heat insulator with the 3 bolts. Tighten to 10 ft. lbs. (13 Nm).
22. Install the manifold stay with the 3 bolts. Tighten to 30 ft. lbs. (40 Nm).

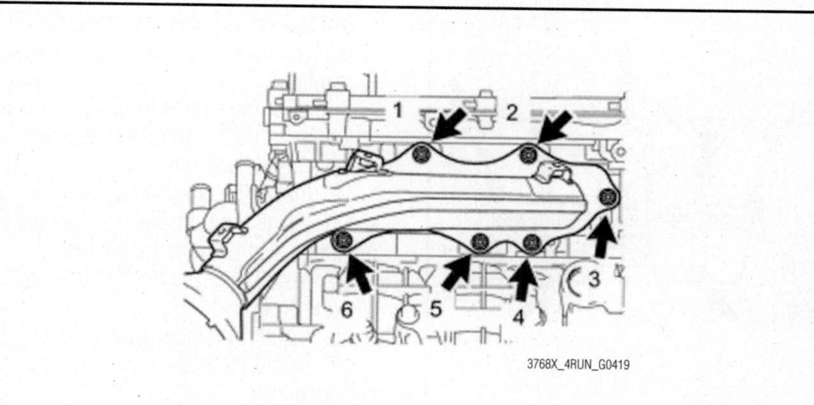

Fig. 146 Identifying the exhaust manifold sub assembly RH nut tightening sequence

23. Install the air fuel ratio sensor (Bank 1, Sensor 1).
24. Install the exhaust manifold sub assembly RH.
   a. Install a new gasket onto the cylinder head.

➡**Be careful of the installation direction.**

   b. Temporarily install the manifold with 6 new nuts.
   c. Tighten the 6 nuts in the sequence shown.
   d. Connect the air fuel ratio sensor connector.
25. Install the No. 1 exhaust manifold heat insulator with the 3 bolts. Tighten to 10 ft. lbs. (13 Nm).
26. Install the No. 2 manifold stay with the 3 bolts. Tighten to 30 ft. lbs. (40 Nm).
27. Install the front exhaust pipe assembly.
28. Install the air cleaner case sub assembly.
29. install the air cleaner cap and hose.
30. Install the v-bank cover.
31. Install the radiator support seal upper.
32. Inspect for exhaust gas leaks.
33. Install the front fender apron seal LH.
34. Install the front fender apron seal RH.
35. Install the front No. 1 fender apron to frame seal LH.
36. Install the front No. 1 fender apron to frame seal RH.

INTAKE MANIFOLD

REMOVAL & INSTALLATION

2.7L Engine

See Figure 147.

1. Discharge the fuel system pressure.
2. Disconnect the cable from the negative battery terminal.

➡**When disconnecting the cable, some systems need to be initialized after the cable is reconnected.**

3. Remove the throttle body with the motor assembly.
4. Remove the fuel delivery pipe with the fuel injectors.
5. Remove the purge VSV.
6. Remove the rear engine under cover assembly.
7. Remove the transmission oil filter tube sub assembly.
8. Remove the starter assembly.
9. Remove the 5 clips and the front fender apron seal LH.
10. Remove the 5 clips and the front No. 1 fender apron to frame seal LH.
11. Remove the intake manifold.
   a. For the engine front side, detach the 2 wire harness clamps from the 2 wire harness clamp brackets.
   b. For the engine rear side, detach the No. 2 water by pass hose and disconnect the No. 3 ventilation hose from the intake manifold.
   c. Remove the 5 bolts, 2 nuts and intake manifold.
   d. Remove the gasket from the intake manifold.
   e. Remove the 2 bolts, 2 wire harness clamp brackets and purge line hose from the intake manifold.

To install:

12. Install the intake manifold.
   a. Install the purge line hose to the intake manifold.
   b. Install the 2 wire harness clamp brackets to the intake manifold with the 2 bolts. Tighten to 71 inch lbs. (8 Nm).
   c. Install a new gasket to the intake manifold.
   d. Install the intake manifold with the 5 bolts and 2 nuts. Tighten to 18 ft. lbs. (25 Nm).
   e. For the engine rear side, attach the No. 2 water by pass hose and connect the No. 3 ventilation hose to the intake manifold.
   f. For the engine front side, attach the 2 wire harness clamps to the 2 wire harness clamp brackets.
13. Install the purge VSV.
14. Install the front No. 1 fender apron to the frame seal LH with the 5 clips.
15. Install the front fender apron seal LH with the 5 clips.
16. Install the starter assembly.
17. Install the transmission oil filler tube sub assembly.
18. Install the rear engine under cover assembly.
19. Install the fuel delivery pipe with the fuel injectors.
20. Install the throttle body with the motor assembly.
21. Connect the cable to the negative battery terminal.

➡**When disconnecting the cable, some systems need to be initialized after the cable is reconnected.**

4.0L Engine

See Figures 148 through 151.

1. Discharge the fuel system pressure.
2. Disconnect the cable from the negative battery terminal.

➡**When disconnecting the cable, some systems need to be initialized after the cable is reconnected.**

3. Remove the front bumper cover lower.
4. Remove the No. 1 engine under cover sub assembly.
5. Drain the engine coolant.
6. Remove the v-bank cover.
   a. Raise the front of the v-bank cover to detach the 2 pins. Then remove the 2 v-bank cover hooks from the bracket, then remove the v-bank cover.
7. Remove the No. 1 air cleaner hose.
   a. Disconnect the ventilation hose and vacuum hose.
   b. Detach the wire harness clamp.
   c. Remove the bolt and loosen the 2 hose clamps.
   d. remove the No. 1 air cleaner hose.
8. Remove the intake air surge tank.
   a. Disconnect the throttle body connector.
   b. Disconnect the No. 4 water by pass hose.
   c. Disconnect the No. 5 water by pass hose.

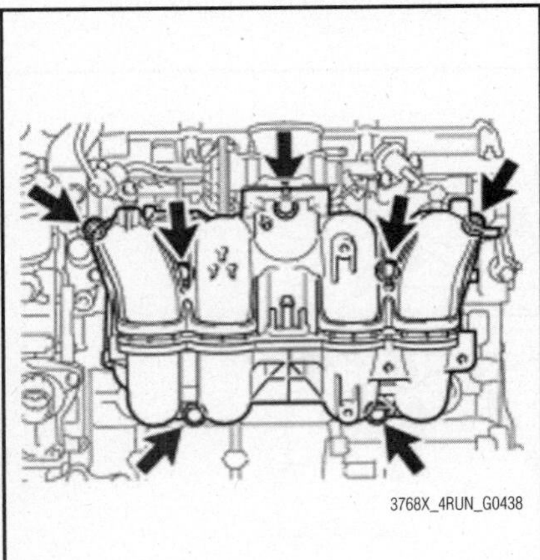

Fig. 147 Removing the intake manifold

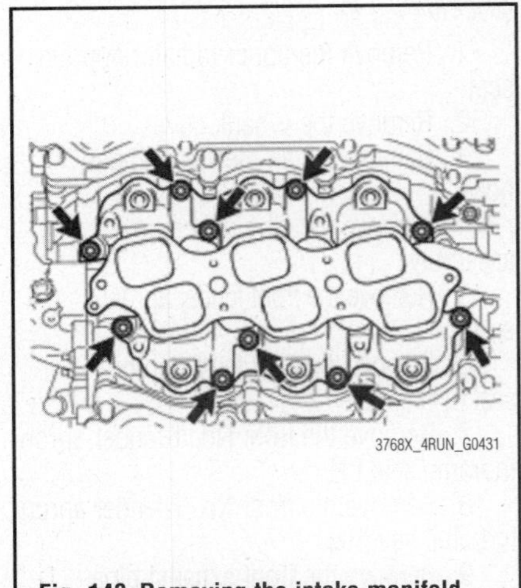

Fig. 148 Removing the intake manifold

d. Disconnect the No. 1 fuel vapor feed hose.

e. Disconnect the No. 1 vacuum switching valve connector.

f. Disconnect the No. 1 ventilation hose.

g. Detach the 2 heater hose clamps.

h. Remove the 2 bolts and throttle body bracket.

i. Using a clip remover, detach the wire harness clamp.

j. Remove the 2 bolts and No. 1 surge tank stay.

k. Remove the 2 bolts and No. 2 surge tank stay.

l. Remove the 2 nuts, 4 bolts and intake air surge tank.

m. Remove the gasket.

9. Remove the fuel delivery pipe sub assembly.

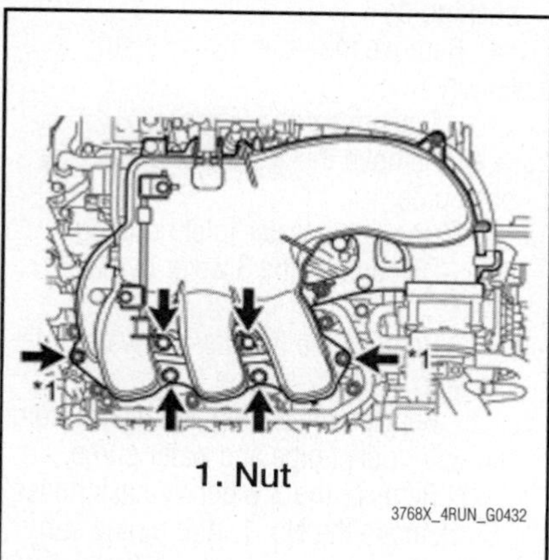

Fig. 149 Identifying the intake air surge tank nut and bolt installation sequence

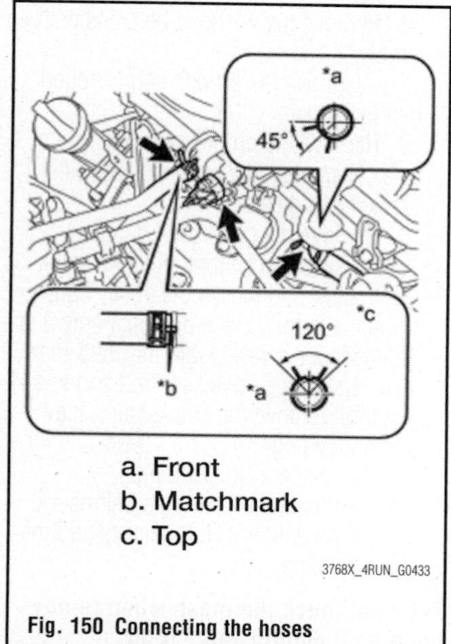

Fig. 150 Connecting the hoses

10. Remove the intake manifold.

a. Remove the 4 nuts, 6 bolts and 2 gaskets.

To install:

11. Install the intake manifold.

a. Set a new gasket on each cylinder head.

➡Align the port holes of the gasket and cylinder head.

➡Be careful of the installation direction.

b. Set the intake manifold on the cylinder heads.

c. Install and uniformly tighten the 6 bolts and 4 nuts in several passes. Tighten to 15 ft. lbs. (21 Nm).

➡Tighten the inner installation bolts of the intake manifold before tightening the outer bolts.

12. Install the fuel delivery pipe sub assembly.

13. Install the intake air surge tank.

a. Install a new gasket to the intake air surge tank.

b. Install the intake air surge tank with the 4 bolts and 2 nuts in the order shown. Tighten to 21 ft. lbs. (28 Nm).

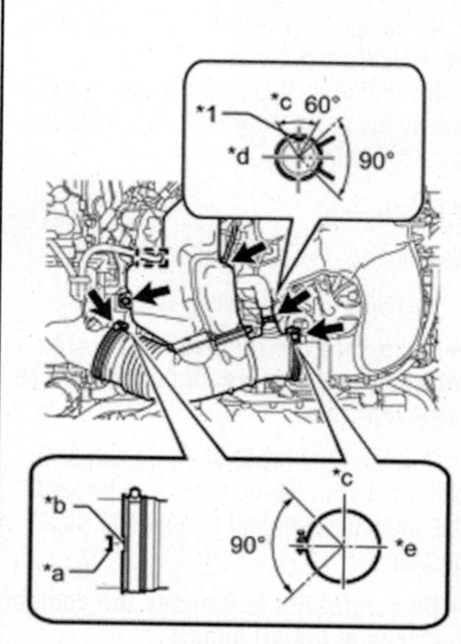

Fig. 151 Installing the No. 1 air cleaner hose

c. Install the No. 1 surge tank stay with the 2 bolts. Tighten to 15 ft. lbs. (21 Nm).

d. Attach the wire harness clamp.

e. Install the No. 2 surge tank stay with the 2 bolts. Tighten to 15 ft. lbs. (21 Nm).

f. Install the throttle body bracket with the 2 bolts. Tighten to 15 ft. lbs. (21 Nm).

g. Connect the No. 1 ventilation hose.

h. Connect the No. 1 vacuum switching valve connector.

i. Connect the No.1 fuel feed hose.

➡Connect the hose so that the direction of the hose clamp is as indicated in the illustration.

j. Connect the throttle body connector.

k. Connect the No. 4 water by-pass hose.

l. Connect the No. 5 water by-pass hose.

➡Connect the hose so that the direction of the hose clamp is as indicated in the illustration.

m. Connect the 2 heater hose clamps.

14. Install the No. 1 air cleaner hose with the 2 clamps. Tighten to 44 inch lbs. (5 Nm).

a. Install the bolt. Tighten to 44 inch lbs. (5 Nm).

b. Connect the vacuum hose and ventilation hose.

➡The direction of the hose clamp is indicated in the illustration.

15. Connect the cable to the negative battery terminal.

➡When disconnecting the cable, some systems need to be initialized after the cable is reconnected.

16. Add engine coolant.

17. Inspect for coolant leaks.

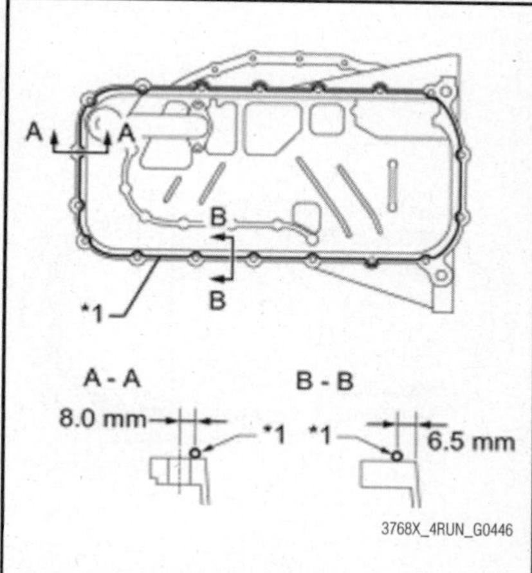

Fig. 152 Applying seal packing to the oil pan

18. Inspect for fuel leaks.
19. Install the v-bank cover.
    a. Attach the 2 V-bank cover hooks to the bracket. Then align the 2 V-bank cover grommets with the 2 pins and press down on the V-bank cover to attach the pins.
20. Install the No. 1 engine under cover sub assembly.
21. Install the front bumper cover lower.

OIL PAN

REMOVAL & INSTALLATION

2.7L Engine

See Figures 152 and 153.

1. Remove the drain plug and gasket.
    a. Remove the 18 bolts and 2 nuts.
2. Insert the blade of an oil pan seal cutter between the oil pans. Cut through the applied sealer and remove the oil pan.

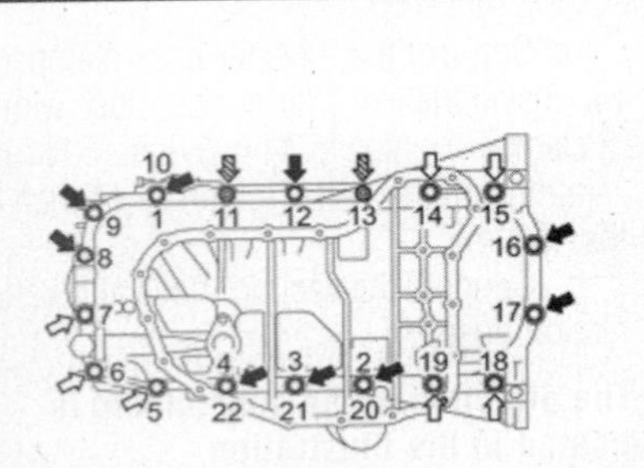

Fig. 153 Identifying the oil pan bolt and nut tightening sequence

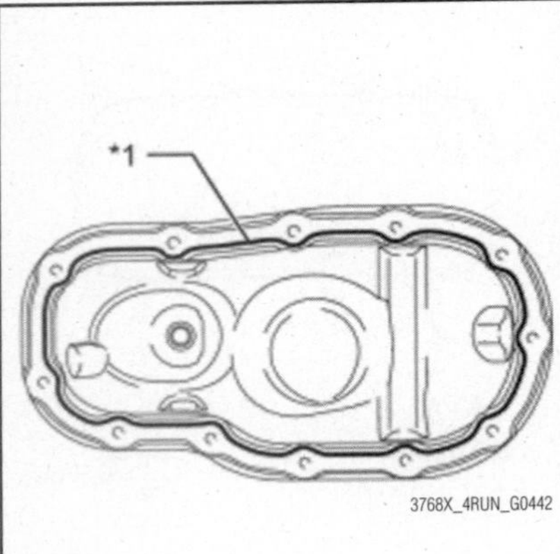

Fig. 154 Applying seal packing to the oil pan

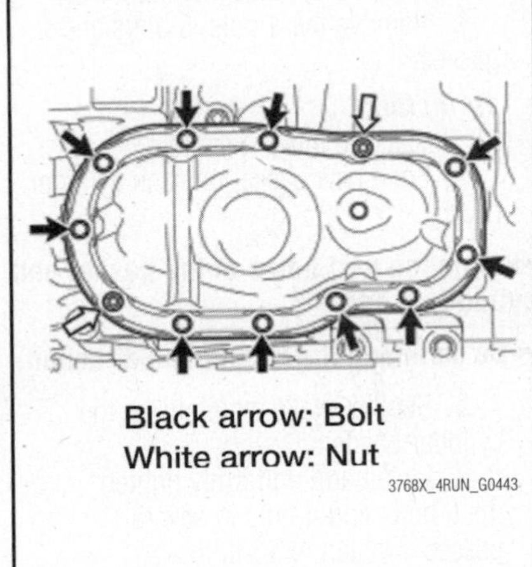

Fig. 155 Installing the oil pan

➡Be careful not to damage the contact surfaces of the oil pans.

To install:

3. Apply seal packing in a continuous line.

➡Remove any oil from the contact surface.

➡Install the oil pan within 3 minutes after applying seal packing.

➡Do not start the engine for at least 4 hours after the installation.

4. Temporarily install the oil pan with the 16 bolts and 2 nuts.
5. Uniformly tighten the 16 bolts and 2 nuts in the order shown. Tighten to 19 ft. lbs. (26 Nm).

4.0L Engine

See Figures 154 and 155.

1. Remove the 10 bolts and 2 nuts.

➡Make sure the removed parts are returned to the same places they were removed from.

2. Insert the blade of an oil pan seal cutter between the oil pans. Cut through the applied sealer and remove the No. 2 oil pan.

➡Be careful not to damage the contact surfaces of the oil pans.

To install:

3. Apply seal packing (* 1) in a continuous line.

➡Remove any oil from the contact surface.

➡Install the No. 2 oil pan within 3 minutes after applying seal packing.

    a. Install the No. 2 oil pan with the 10 bolts and 2 nuts. Tighten the bolts and nuts uniformly in several steps. Tighten to 7 ft. lbs. (10 Nm).

➡Tighten the nuts first. After tightening the bolts, check that the nuts and bolts are tightened to the specified torque.

➡Do not start the engine for at least 2 hours after the installation.

OIL PUMP

REMOVAL & INSTALLATION

See Figures 156 through 165.

1. Remove the engine assembly.
2. Remove the ignition coil assembly.
3. Remove the engine oil level dipstick guide.
    a. Remove the dipstick.
    b. Remove the bolt and dipstick guide.
    c. Remove the O-ring from the dipstick guide.
4. Remove the water by pass sub assembly.
    a. Disconnect the 2 hoses.
    b. Remove the 3 bolts and water by pass pipe.
5. Remove the water inlet housing.
    a. Disconnect the 3 water by pass hoses.
    b. Remove the 5 bolts and water inlet housing.
    c. Remove the O-ring and gasket from the water outlet pipe and water pump.
    d. Remove the 3 water by pass hoses.
6. Remove the No. 1 idler pulley sub assembly.
    a. Remove the bolt and No. 1 idler pulley.
7. Remove the No. 2 idler pulley sub assembly.
8. Remove the v-ribbed belt tensioner assembly.
    a. Remove the 5 bolts and v-ribbed belt tensioner.
9. Remove the oil filter bracket.
    a. Remove the 2 nuts, bolt, oil filter bracket and gasket.
10. Remove the crankshaft pulley.
    a. Using special tool, hold the crankshaft pulley and loosen the pulley bolt. Continue to loosen the bolt until only 2 or 3 threads are screwed into the crankshaft.
    b. Using the pulley set bolt and special tool, remove the crankshaft pulley and pulley bolt.
11. Remove the No. 1 oil pipe.
    a. Remove the 2 oil pipe unions, oil control valve filter LH, 3 gaskets and the No. 1 oil pipe.

➡Do not touch the mesh when removing the oil control valve filter.

12. Remove the rear cylinder head cover.
13. Disconnect the fuel pipe sub assembly.
    a. Remove the 2 bolts and disconnect the fuel pipe.
14. Remove the cylinder head cover sub assembly LH.
    a. Remove the 12 bolts, seal washer, cylinder head cover and gasket.

➡**Make sure the removed parts are returned to the same places they were removed from.**

    b. Remove the 3 gaskets.
15. Remove the cylinder head cover sub assembly.
    a. Remove the 12 bolts, seal washer, cylinder head cover and gasket.

➡**Make sure the removed parts are returned to the same places they were removed from.**

    b. Remove the 3 gaskets.
16. Remove the No. 2 oil pan sub assembly.
    a. Remove the 10 bolts and 2 nuts.

➡**Make sure the removed parts are returned to the same places they were removed from.**

    b. Insert the blade of an oil pan seal cutter between the oil pans. Cut through the applied sealer and remove the No. 2 oil pan.

➡**Be careful not to damage the contact surfaces of the oil pans.**

17. Remove the oil strainer sub assembly.
    a. Remove the 2 nuts, oil strainer and gasket.
18. Remove the oil pan sub assembly.
    a. Remove the 17 bolts and 2 nuts.

➡**Make sure the removed parts are returned to the same places they were removed from.**

    b. Using the screwdriver, remove the oil pan by prying between the oil pan and cylinder block.

➡**Tape the screwdriver tip before use.**

➡**Be careful not to damage the contact surfaces of the cylinder block and oil pan.**

    c. Remove the 3 O-rings from the timing chain cover.
19. Remove the timing chain cover sub assembly.
    a. Remove the 26 bolts and the 2 nuts.
    b. Remove the timing chain cover by prying between the timing chain cover and cylinder head or cylinder block with a screwdriver.

➡**Tape the screwdriver tip before use.**

➡**Be careful not to damage the contact surfaces of the timing chain cover, cylinder block and cylinder head.**

    c. Remove the oil pump gasket from the cylinder block.
20. Remove the front crankshaft oil seal.
    a. Using a screwdriver and wooden block, pry out the oil seal.

➡**Tape the screwdriver tip before use.**

➡**Do not damage the surface of the oil seal press fit hole.**

To install:

21. Install the front crankshaft oil seal.
    a. Using SST and a hammer, tap in a new oil seal until its surface is flush with the timing chain cover edge.

➡**Keep the lip free from foreign matter.**

➡**Do not tap on the oil seal at an angle.**

➡**Make sure that the oil seal edge does not protrude from the timing chain case.**

22. Install the timing chain cover sub assembly.
    a. Remove any old packing (FIPG material) and be careful not to drop any oil on the contact surfaces of the timing chain cover, cylinder head and cylinder block.

➡**Be sure to clean and degrease the contact surfaces, especially the surfaces indicated in the illustration.**

    b. Apply a light coat of engine oil to a new oil pump gasket.
    c. Install the oil pump gasket.
    d. Apply seal packing as shown in the illustration.
    e. Apply seal packing to the timing chain cover in a continuous line as shown in the following illustration.

➡**When the contact surfaces are wet, wipe them with an oil-free cloth before applying seal packing.**

➡**Install the chain cover within 3 minutes and tighten the bolts within 10 minutes after applying seal packing.**

    f. Align the oil pump drive rotor spline and crankshaft as shown in the illustration. Install the drive rotor and chain cover to the crankshaft.
    g. Install the timing chain cover with the 26 bolts, labeled A, B, C and D, and the 2 nuts. Tighten the bolts and nuts uniformly in several steps.

Standard bolt:

- Bolt A: 0.984 inch (25 mm)
- Bolt B: 2.17 inch (55 mm)

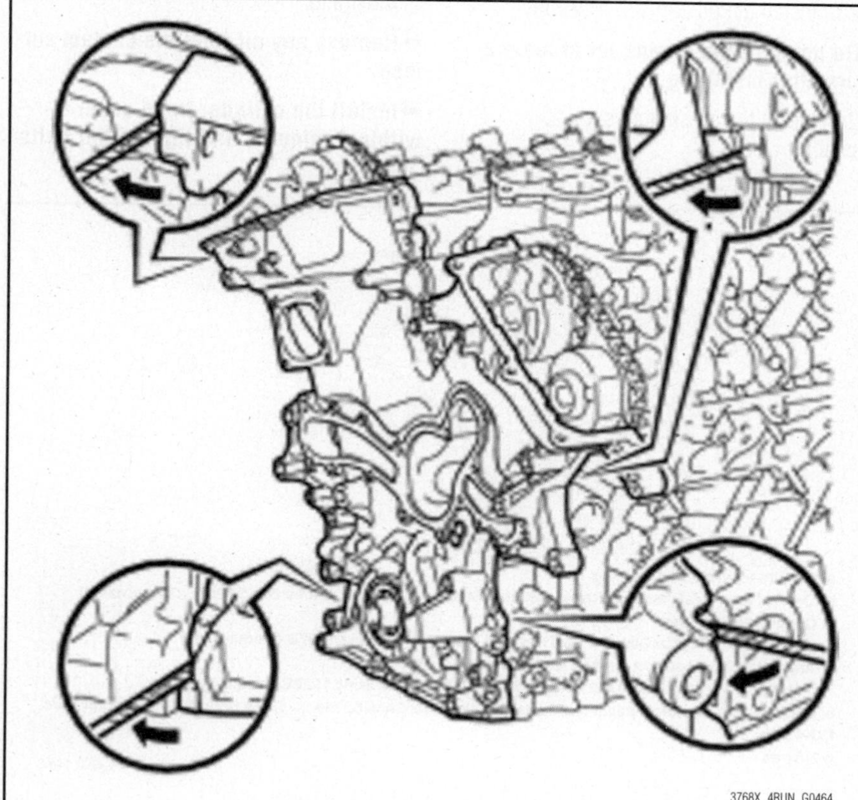

Fig. 156 Locating the pry points

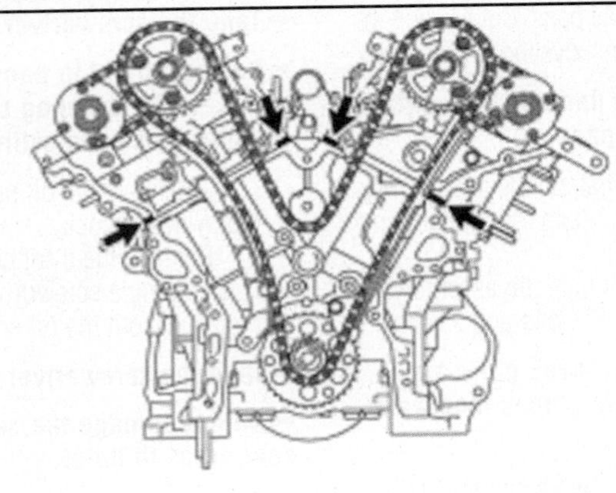

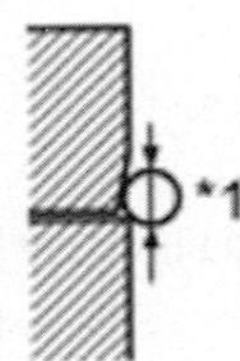

1. Seal diameter
Black arrow: Seal packing 0.118-0.157 inch (3-4 mm) diameter

3768X_4RUN_G0465

Fig. 157 Applying seal packing

- Bolt C: 1.38 inch (35 mm)
- Bolt D: 2.56 inch (65 mm)

➡Make sure that there is no oil on the bolt threads.

a. Tighten the bolts in area 1 to 35 ft. lbs. (47 Nm).

➡Tighten the bolts from bottom to top as shown.

b. Tighten the bolts in area 2 to 17 ft. lbs. (23 Nm).

c. Tighten the bolts in area 3 to 17 ft. lbs. (23 Nm).

➡Tighten the bolts and nuts from top to bottom.

d. Tighten the bolts in area 4 to 17 ft. lbs. (23 Nm).

➡Tighten the bolts from bottom to top as shown in the illustration.

➡Do not start the engine for at least 2 hours after installation.

➡Wipe off any seal packing that seeped out around the surfaces of the oil pan and head cover and make sure that there is no seal packing seeping out around the edges.

23. Install the oil pan sub assembly.

➡Do not start the engine for at least 2 hours after installing.

24. Install the oil strainer sub assembly.

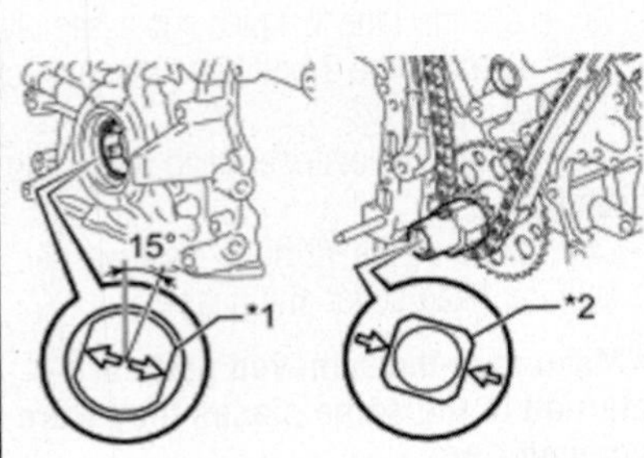

1. Drive rotor spline
2. Crankshaft

3768X_4RUN_G0467

Fig. 159 Aligning the oil pump drive rotor spline and crankshaft

a. Install a new gasket and the oil strainer with the 2 nuts. Tighten to 80 inch lbs. (9 Nm).

25. Install the No. 2 oil pan sub assembly.

26. Install the cylinder head cover sub assembly.

a. Remove any old packing (FIPG material) and be careful not to drop any oil on the contact surfaces of the cylinder head, timing chain cover and cylinder head cover.

b. Apply seal packing as shown in the illustration.

➡Remove any oil from the contact surface.

➡Install the cylinder head cover within 3 minutes and tighten the bolts

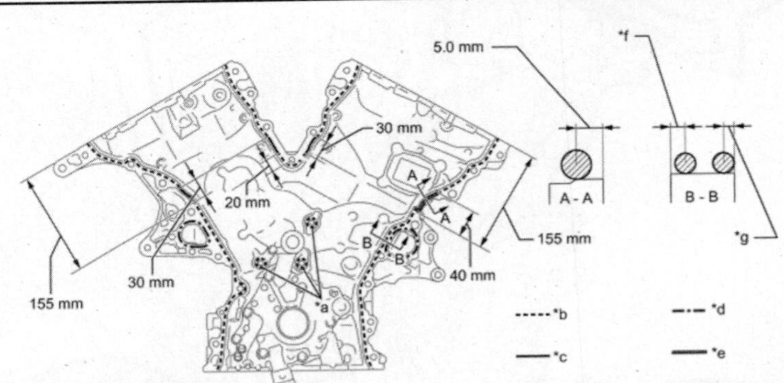

a. Be sure to apply seal packing
b. Dashed area (seal packing: Toyota Genuine Seal Packing Black, Three Bond 1207B or equivalent)
c. Continuous line area
(seal packing: Toyota Genuine Seal Packing Black, Three Bond 1207B or equivalent)
d. Alternate long and short dashed line area
(seal packing: Toyota Genuine Seal Packing 1282B, Three Bond 1282B or equivalent)
e. Diagonal line area (seal packing: Toyota Genuine Seal Packing Black, Three Bond 1207B or equivalent
f. 3-4 mm
g 2-3 mm

3768X_4RUN_G0466

Fig. 158 Applying seal packing to the timing chain cover

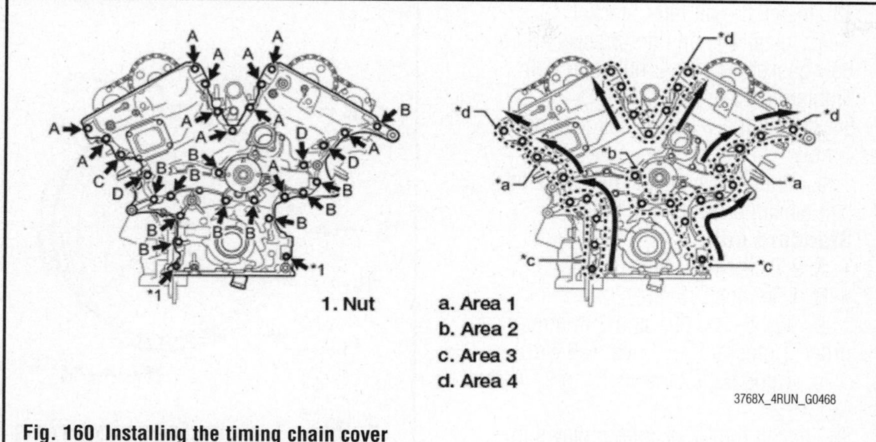

Fig. 160 Installing the timing chain cover

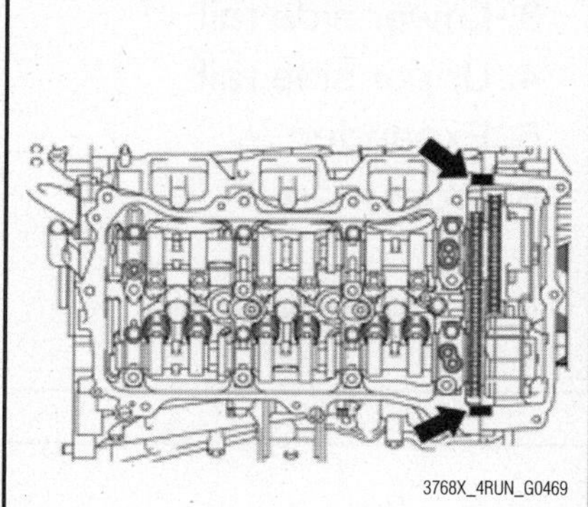

Fig. 161 Applying the seal packing to the cylinder head cover sub assembly

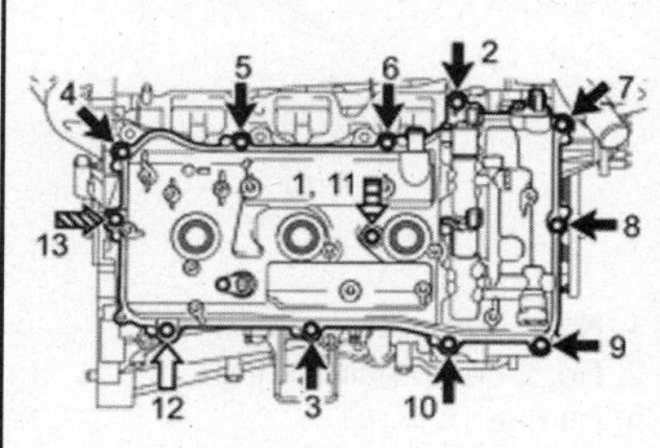

Fig. 162 Tightening the cylinder head bolts

within 15 minutes after applying seal packing.

c. Install 3 new gaskets.

d. Install a new gasket to the cylinder head cover.

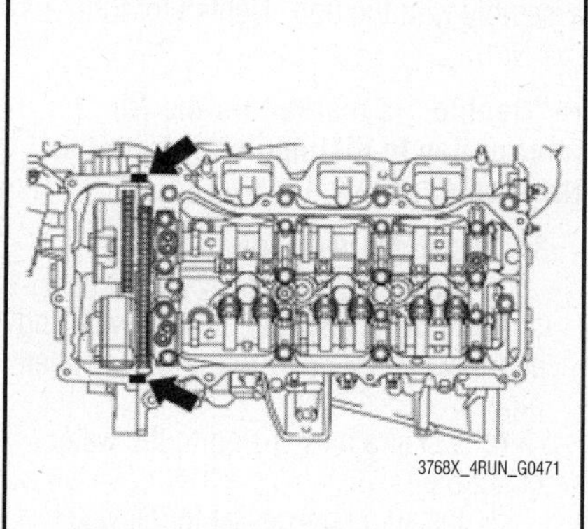

Fig. 163 Applying seal packing to the cylinder head cover sub assembly LH

e. Install the seal washers to the bolts.

f. Temporarily install the cylinder head cover with the 12 bolts. Tighten the bolts uniformly in several steps. Tighten bolts A and D to 7 ft. lbs. (10 Nm). Tighten bolts B and C to 15 ft. lbs. (21 Nm).

Standard Bolts:

- A: 0.984 inch (25 mm)
- B: 1.38 inch (35 mm)
- C: 2.56 inch (65 mm)
- D: 2.36 inch (60 mm)

➡Do not start the engine for at least 2 hours after installation.

27. Install the cylinder head cover sub assembly LH.

a. Remove any old packing (FIPG material) and be careful not to drop any oil on the contact surfaces of the cylinder head, timing chain cover and cylinder head cover.

b. Apply seal packing as shown in the illustration.

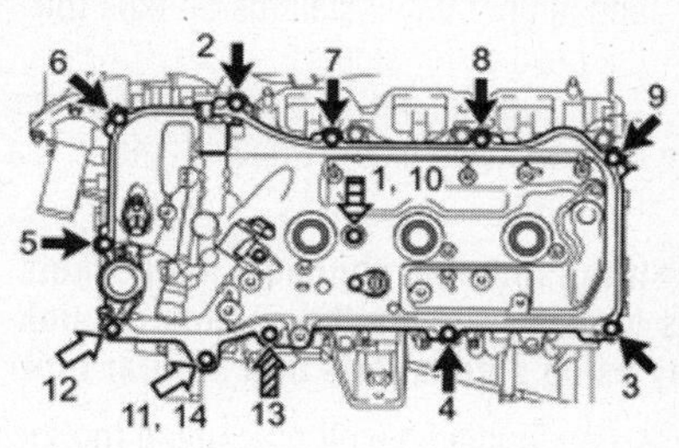

Fig. 164 Installing the cylinder head cover LH

➡Remove any oil from the contact surface.

➡Install the cylinder head cover within 3 minutes and tighten the bolts within 15 minutes after applying seal packing.

c. Install 3 new gaskets.

d. Install a new gasket to the cylinder head cover.

e. Install the seal washers to the bolts.

f. Temporarily install the cylinder head cover with the 12 bolts. Tighten the bolts uniformly in several steps. Tighten bolts A and D to 7 ft. lbs. (10 Nm). Tighten bolts B and C to 15 ft. lbs. (21 Nm).

Standard Bolts:

- A: 0.984 inch (25 mm)
- B: 1.38 inch (35 mm)
- C: 2.76 inch (70 mm)
- D: 2.36 inch (60 mm)

➡Do not start the engine for at least 2 hours after installation.

28. Connect the fuel pipe sub assembly with the 2 bolts. Tighten to 80 inch lbs. (9 Nm).

29. Install the rear cylinder head cover.

30. Install the No. 2 oil pipe.

a. Make sure that there is no foreign matter on the mesh of the oil control valve filter RH.

➡Do not touch the mesh when installing the oil control valve filter.

b. Install a new gasket and temporarily install the oil pipe (on the cylinder head side) with the oil check valve bolt.

c. Install the oil control valve filter RH to the oil pipe union. Install new gaskets and temporarily install the oil pipe (on the head cover side).

d. Tighten the oil pipe union (on the cylinder head side). Tighten to 48 ft. lbs. (65 Nm).

➡If the link that connects the gaskets is broken, remove the connecting link by using side cutters or a similar tool.

e. Tighten the oil pipe union (on the head cover side). Tighten to 48 ft. lbs. (65 Nm).

31. Install the No. 1 oil pipe.

a. Make sure that there is no foreign matter on the mesh of the oil control valve filter LH.

➡Do not touch the mesh when installing the oil control valve filter.

b. Install a new gasket and temporarily install the oil pipe (on the cylinder head side) with the oil check valve bolt.

c. Install the oil control valve filter LH to the oil pipe union. Install new gaskets and temporarily install the oil pipe (on the head cover side).

d. Tighten the oil pipe union (on the cylinder head side). Tighten to 48 ft. lbs. (65 Nm).

➡If the link that connects the gaskets is broken, remove the connecting link by using side cutters or a similar tool.

e. Tighten the oil pipe union (on the head cover side). Tighten to 48 ft. lbs. (65 Nm).

32. Install the crankshaft pulley.

a. Using the special tool, install the crankshaft pulley with the pulley set bolt. Tighten to 192 ft. lbs. (260 Nm).

Fig. 165 Installing the v-ribbed belt tensioner assembly

33. Install the oil filter bracket.

a. Install the oil filter bracket and a new gasket with the 2 nuts and bolt. Tighten to 15 ft. lbs. (21 Nm).

34. Install the v-ribbed belt tensioner assembly.

a. Temporarily install the v-ribbed belt tensioner with the 5 bolts.

Standard bolt:

- A: 2.76 inch (70 mm)
- B: 1.30 inch (33 mm)

a. Tighten bolts 1 and 2 in numerical order. Tighten to 27 ft. lbs. (36 Nm).

b. Tighten the other bolts to 27 ft. lbs. (36 Nm).

35. Install the No. 2 idler pulley sub assembly.

36. Install the No. 1 idler pulley sub assembly with the bolt. Tighten to 40 ft. lbs. (54 Nm).

➡"Double" is marked on the No. 1 idler pulley to distinguish it from the No. 2 idler pulley.

37. Install the water inlet housing.

a. Install the 3 water by-pass hoses.

b. Apply soapy water to a new O-ring and install the O-ring to the water outlet pipe.

c. Install a new O-ring to the water outlet pipe.

d. Install a new gasket to the water pump.

e. Install the water inlet with the 5 bolts. Tighten to 7 ft. lbs. (10 Nm).

f. Connect the 3 water by pass hoses.

38. Install the water by pass pipe sub assembly with the 3 bolts. Tighten to 7 ft. lbs. (10 Nm).

a. Connect the 2 hoses.

➡The direction of the hose clamp is indicated in the illustration.

39. Install the engine oil level dipstick guide.

a. Install a new O-ring to the dipstick guide.

b. Apply a light coat of engine oil to the O-ring.

c. Push the dipstick guide end into the guide hole.

d. Install the dipstick guide with the bolt. Tighten to 7 ft. lbs. (10 Nm).

e. Install the dipstick.

40. Install the ignition coil assembly.

41. Install the engine assembly.

PISTON AND RING

POSITIONING

See Figures 166 and 167.

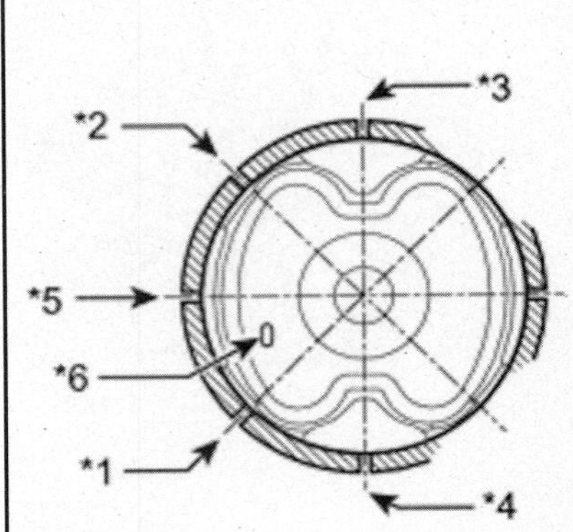

Fig. 166 Piston ring positioning—4.0L engine

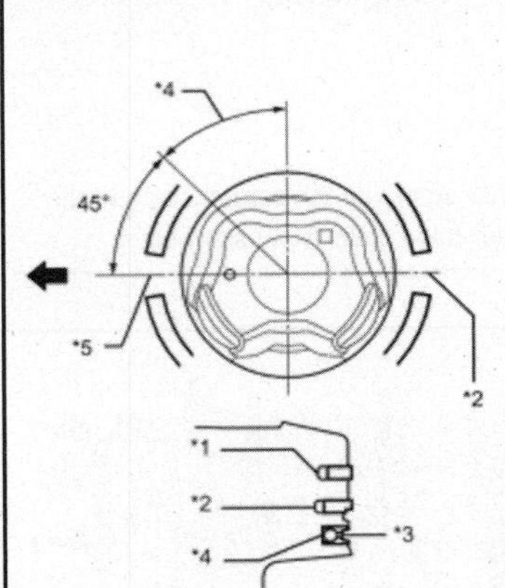

Fig. 167 Piston ring positioning—2.7L engine

TIMING CHAIN & SPROCKETS

REMOVAL & INSTALLATION

Refer to CAMSHAFT for removal and installation procedures.

ENGINE PERFORMANCE & EMISSION CONTROLS

CAMSHAFT POSITION (CMP) SENSOR

LOCATION

See Figures 168 and 169.

REMOVAL & INSTALLATION

2.7L Engine

See Figure 170.

1. Remove the Camshaft Position (CMP) sensor.
    a. Disconnect the Camshaft Position (CMP) sensor connector.
    b. Remove the bolt and Camshaft Position (CMP) sensor.

To install:

2. Install the Camshaft Position (CMP) sensor.
    a. Apply a light coat of engine oil to the O-ring of the Camshaft Position (CMP) sensor.

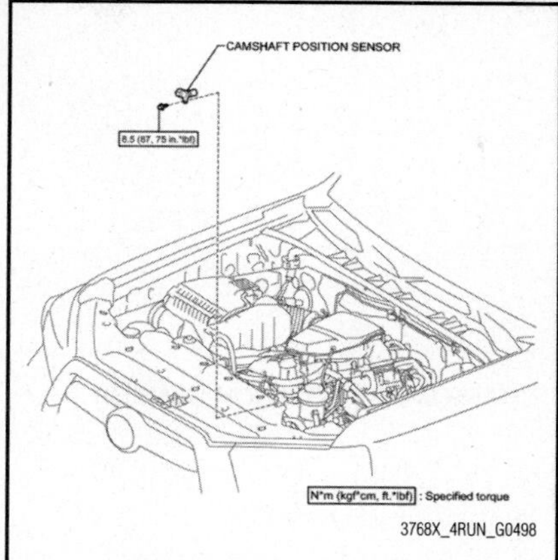

Fig. 169 Camshaft Position (CMP) sensor—2.7L engine

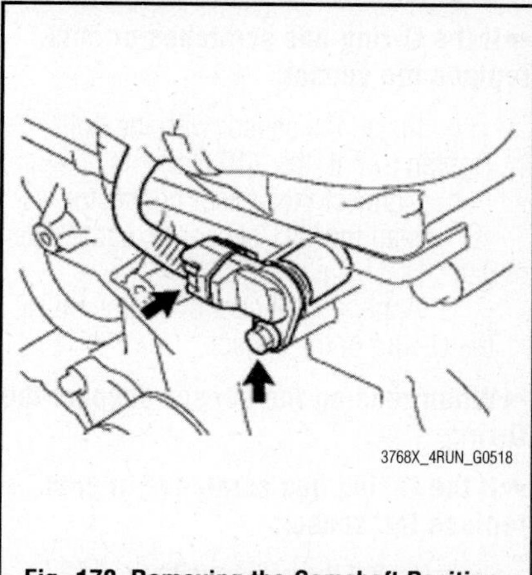

Fig. 170 Removing the Camshaft Position (CMP) sensor

    b. Install the Camshaft Position (CMP) sensor with the bolt. Tighten to 75 inch lbs. (8.5 Nm).

➡Make sure that the O-ring is not cracked or jammed when installing it.

    c. Connect the Camshaft Position (CMP) sensor connector.

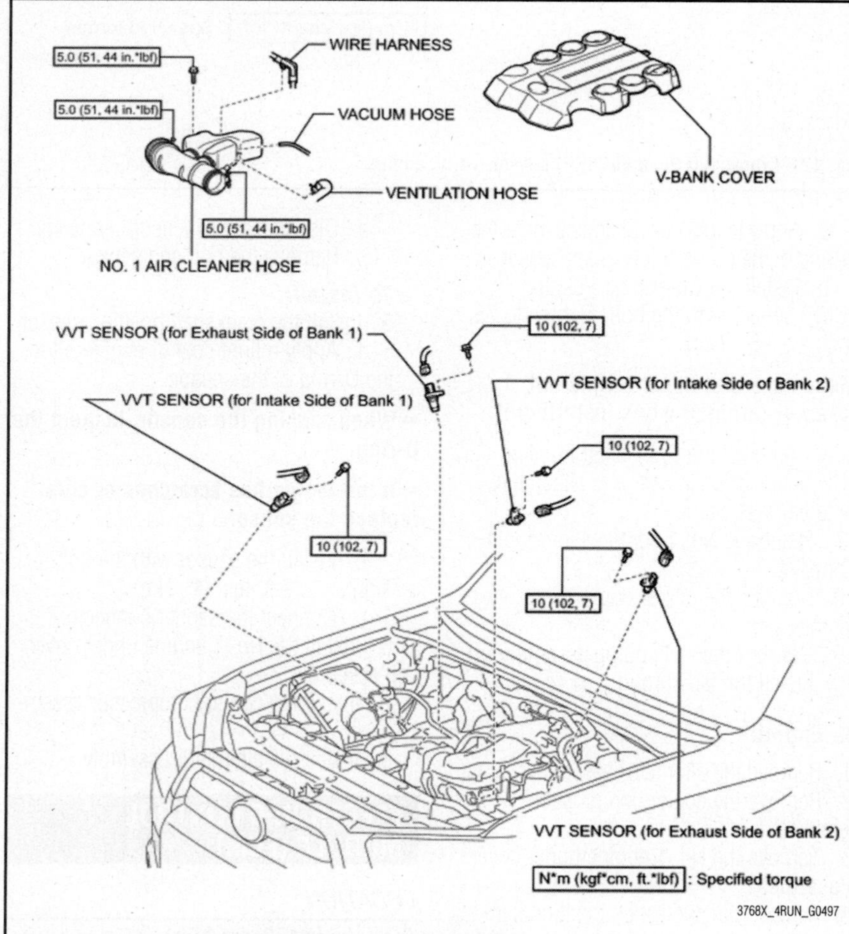

Fig. 168 Camshaft Position (CMP) sensor—4.0L engine

4.0L Engine

1. Remove the v-bank cover.
2. Remove the No. 1 air cleaner hose.
3. Remove the VVT sensor (for intake side of Bank 1).
    a. Disconnect the sensor connector.
    b. Remove the bolt and sensor.
4. Remove the VVT sensor (exhaust side of Bank 1).
    a. Disconnect the sensor connector.
    b. Remove the bolt and sensor.
5. Remove the VVT sensor (intake side of Bank 2).
    a. Disconnect the sensor connector.
    b. Remove the bolt and sensor.
6. Remove the VVT sensor (exhaust side of Bank 2).
    a. Disconnect the sensor connector.
    b. Remove the bolt and sensor.

To install:

7. Install the VVT sensor (exhaust side of Bank 2).
    a. Apply a light coat of engine oil to the O-ring of the VVT sensor.

➡When reusing the sensor, inspect the O-ring.

➡If the O-ring has scratches or cuts, replace the sensor.

    b. Install the VVT sensor with the bolt. Tighten to 7 ft. lbs. (10 Nm).
    c. Connect the sensor connector.
8. Install the VVT sensor (intake side of Bank 2).
    a. Apply a light coat of engine oil to the O-ring of the sensor.

➡When reusing the sensor, inspect the O-ring.

➡If the O-ring has scratches or cuts, replace the sensor.

b. Install the sensor with the bolt. Tighten to 7 ft. lbs. (10 Nm).
c. Connect the sensor connector.
9. Install the VVT sensor (exhaust side of Bank 1).
a. Apply a light coat of engine oil to the O-ring of the sensor.

➡When reusing the sensor, inspect the O-ring.

➡If the O-ring has scratches or cuts, replace the sensor.

b. Install the sensor with the bolt. Tighten to 7 ft. lbs. (10 Nm).
c. Connect the sensor connector.
10. Install the VVT sensor (intake side of Bank 1).
a. Apply a light coat of engine oil to the O-ring of the sensor.

➡When reusing the sensor, inspect the O-ring.

➡If the O-ring has scratches or cuts, replace the sensor.

b. Install the sensor with the bolt. Tighten to 7 ft. lbs. (10 Nm).
11. Install the No. 1 air cleaner hose.
12. Install the v-bank cover.

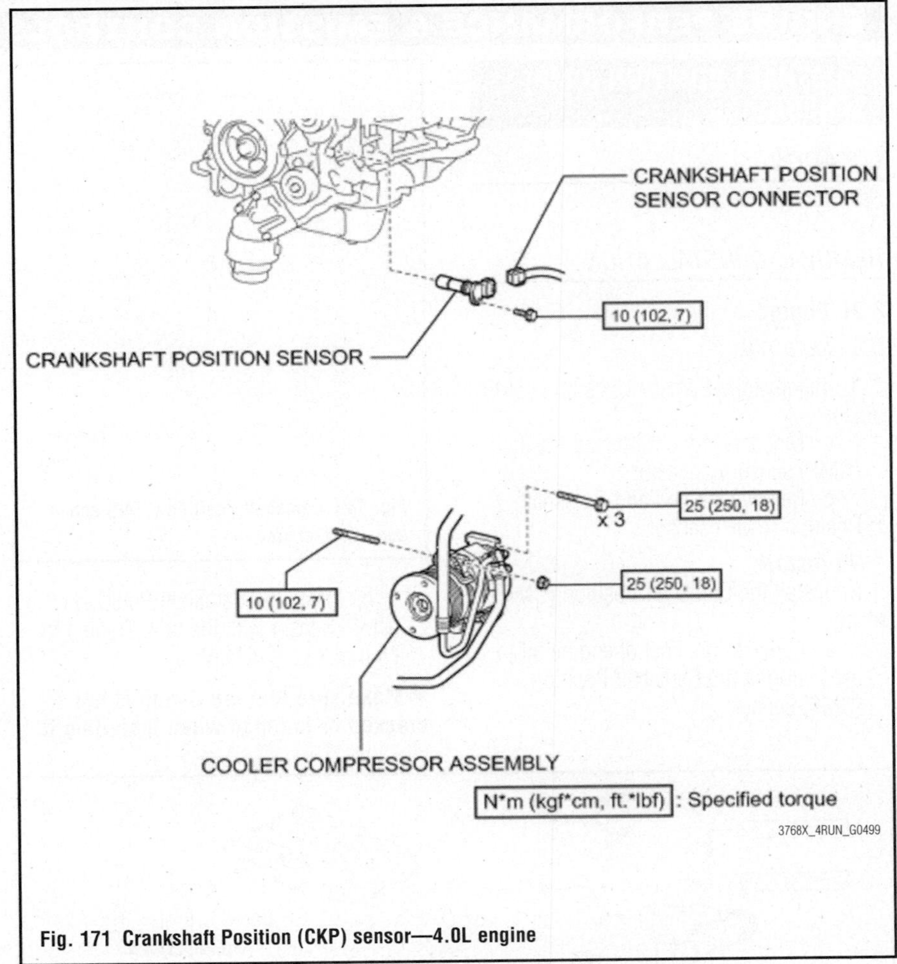

Fig. 171 Crankshaft Position (CKP) sensor—4.0L engine

CRANKSHAFT POSITION (CKP) SENSOR

LOCATION

See Figures 171 and 172.

REMOVAL & INSTALLATION

2.7L Engine

1. Remove the fan shroud.
2. Disconnect the vane pump assembly.
3. Disconnect the cooler compressor assembly.
4. Remove the No. 1 compressor mounting bracket.
5. Remove the crankshaft position sensor.
a. Disconnect the Crankshaft Position (CKP) sensor connector and detach the wire harness clamp.
b. Remove the bolt and crankshaft position sensor.

To install:

6. Install the crankshaft position sensor.
a. Apply a light coat of engine oil to the O-ring of the crankshaft position sensor.
b. Install the Crankshaft Position (CKP) sensor with the bolt. Tighten to 75 inch lbs. (8.5 Nm).

➡Make sure that the O-ring is not cracked or jammed when installing it.

c. Connect the Crankshaft Position (CKP) sensor connector and attach the wire harness clamp.
7. Install the No. 1 compressor mounting bracket.
8. Connect the cooler compressor assembly.
9. Connect the vane pump assembly.
10. Install the fan shroud.

4.0L Engine

1. Remove the alternator assembly.
2. Remove the cooler compressor assembly.
3. Remove the No. 1 engine under cover sub assembly.
4. Remove the crankshaft position sensor.
a. Disconnect the sensor connector.
b. Remove the bolt and sensor.

To install:

5. Install the crankshaft position sensor.
a. Apply a light coat of engine oil to the O-ring of the sensor.

➡When reusing the sensor, inspect the O-ring.

➡If the O-ring has scratches or cuts, replace the sensor.

b. Install the sensor with the bolt. Tighten to 7 ft. lbs. (10 Nm).
c. Connect the sensor connector.
6. Install the No. 1 engine under cover sub assembly.
7. Install the cooler compressor assembly.
8. Install the alternator assembly.

ELECTRONIC CONTROL MODULE (ECM)

LOCATION

See Figures 173 and 174.

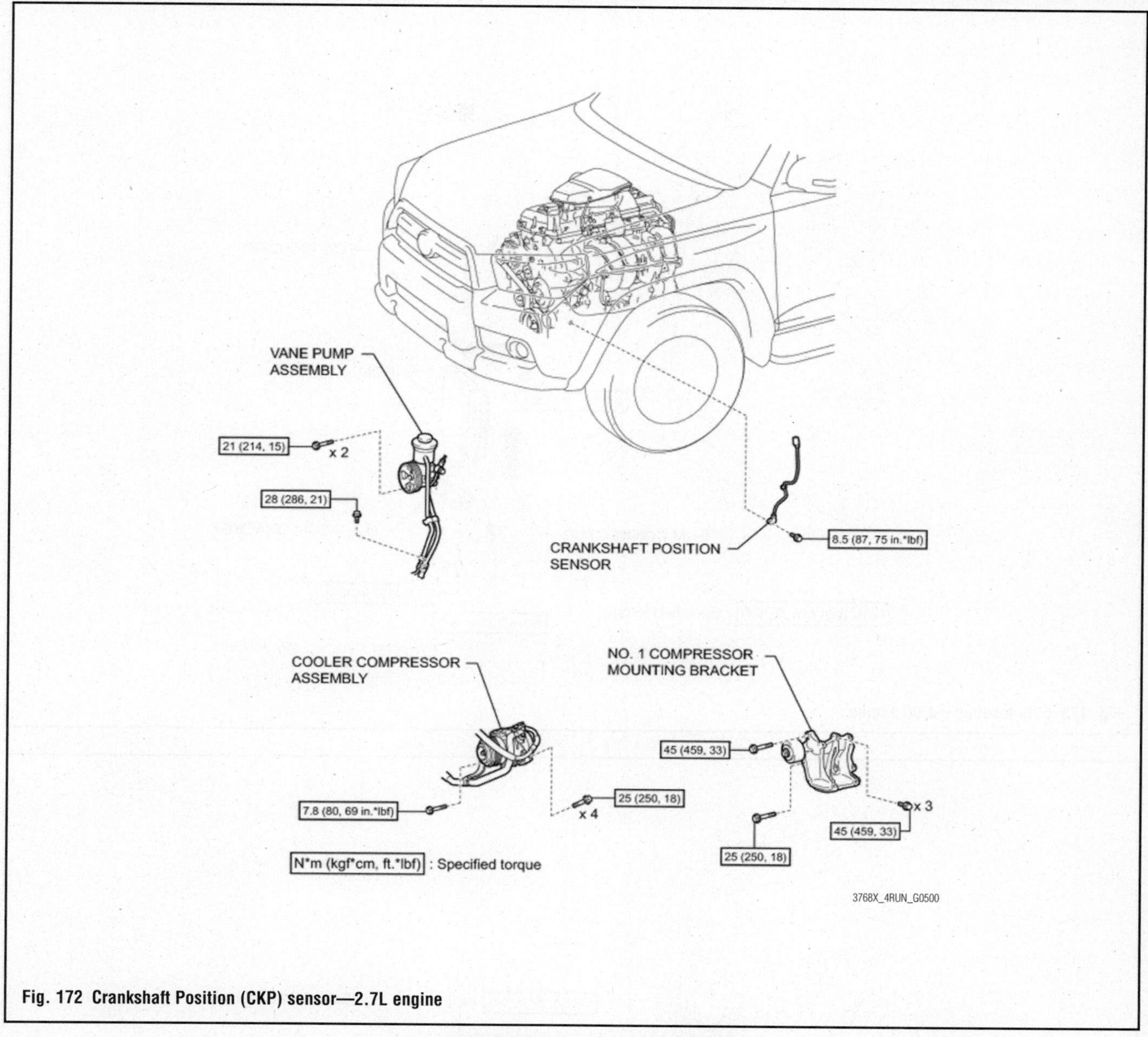

Fig. 172 Crankshaft Position (CKP) sensor—2.7L engine

REMOVAL & INSTALLATION

1. Remove the lower instrument panel sub assembly.
2. Remove the ECM.
   a. Disconnect the 6 connectors.
   b. Remove the bolt, nut and ECM.
3. Remove the 2 screws and the ECM bracket.
4. Remove the 2 screws and the No. 2 ECM bracket.

To install:

5. To install, reverse the removal procedure. Tighten the ECM bracket bolts to 27 inch lbs. (3 Nm). Tighten the ECM to 71 inch lbs. (8 Nm).

ENGINE COOLANT TEMPERATURE (ECT) SENSOR

LOCATION

See Figures 175 and 176.

REMOVAL & INSTALLATION

2.7L Engine

1. Remove the intake manifold.
2. Remove the Engine Coolant Temperature (ECT) sensor.
   a. Disconnect the Engine Coolant Temperature (ECT) sensor connector.
   b. Using a 19 mm deep socket wrench, remove the Engine Coolant Temperature (ECT) sensor and gasket.

To install:

3. To install, reverse the removal procedure. Tighten the sensor to 14 ft. lbs. (20 Nm).

4.0L Engine

1. Remove the intake manifold.
2. Remove the Engine Coolant Temperature (ECT) sensor.
   a. Disconnect the sensor connector.
   b. Using a 19 mm deep socket wrench, remove the sensor.
   c. Remove the gasket from the sensor.

To install:

3. To install, reverse the removal procedure. Tighten the sensor to 14 ft. lbs. (20 Nm).

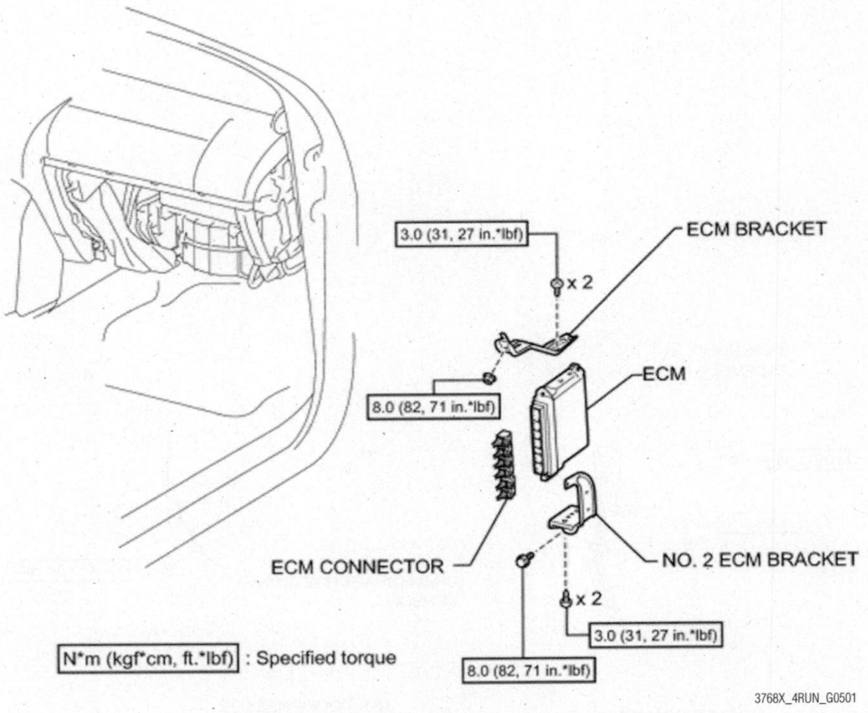

Fig. 173 ECM location—4.0L engine

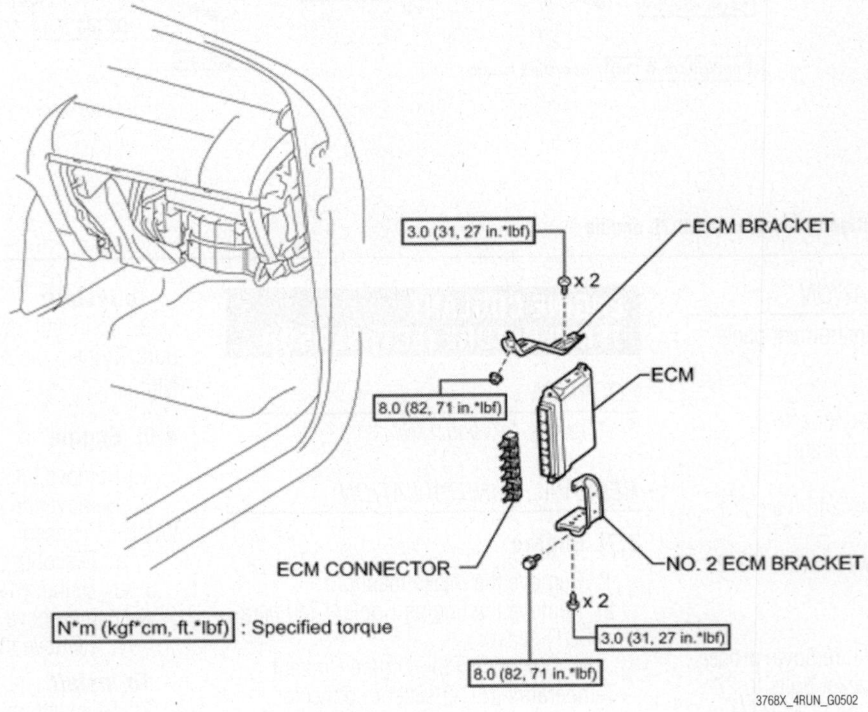

Fig. 174 ECM location—2.7L engine

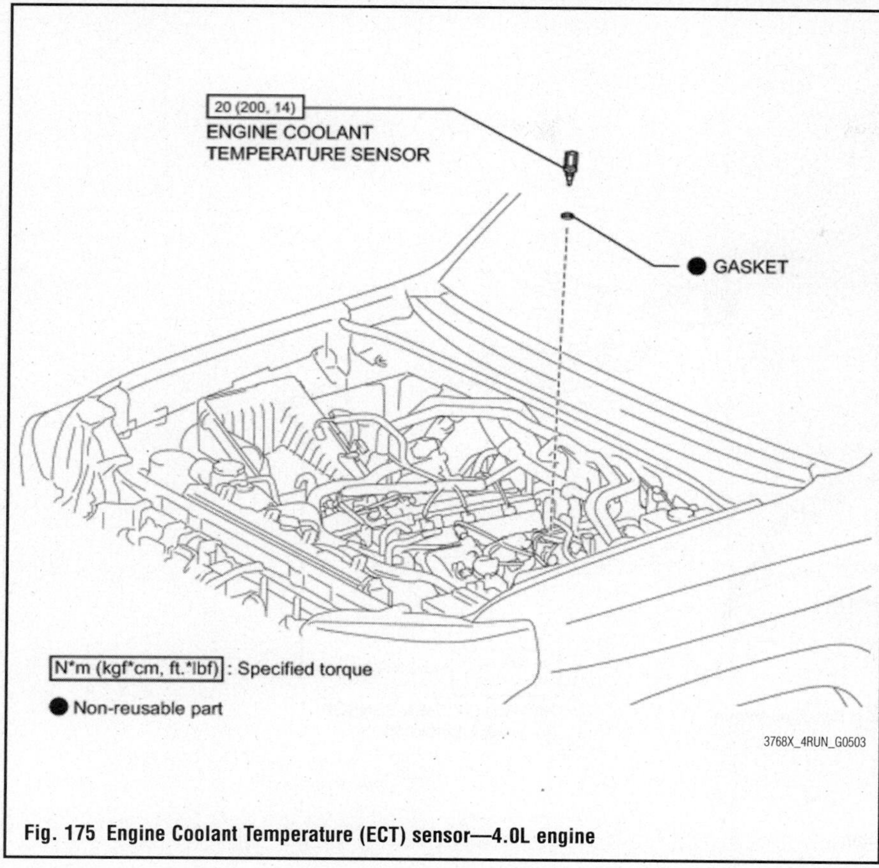

Fig. 175 Engine Coolant Temperature (ECT) sensor—4.0L engine

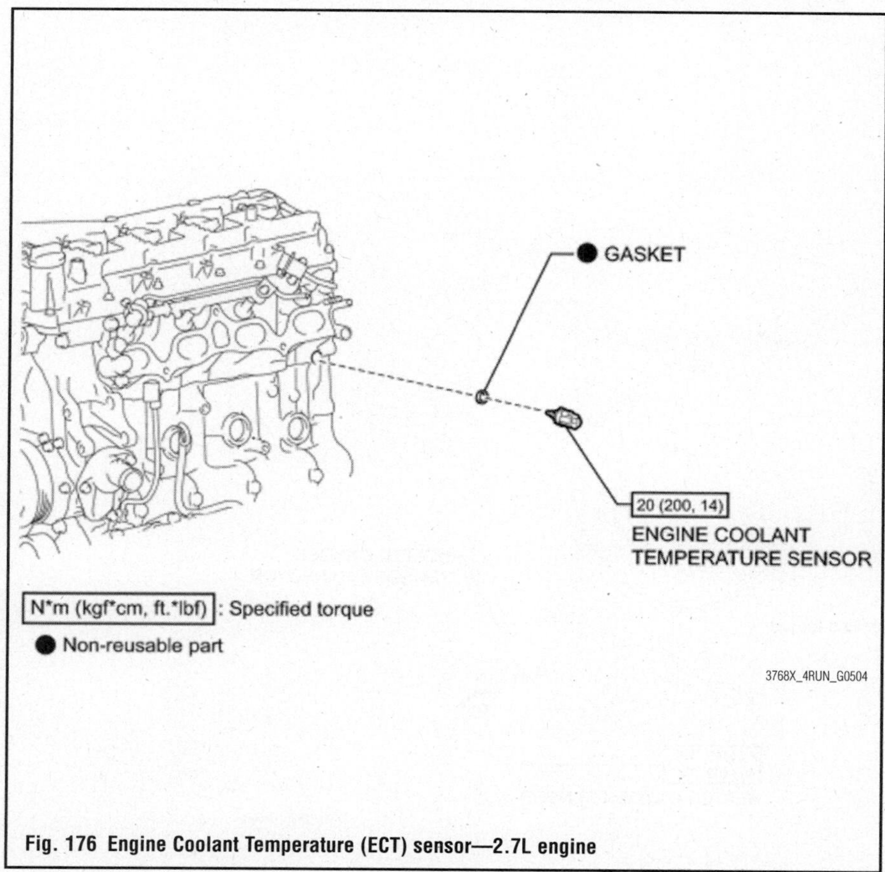

Fig. 176 Engine Coolant Temperature (ECT) sensor—2.7L engine

HEATED OXYGEN SENSOR (HO2S)

LOCATION

See Figures 177 and 178.

REMOVAL & INSTALLATION

2.7L Engine

See Figure 179.

1. Remove the heated oxygen sensor.
   a. Disconnect the heated oxygen sensor connector.
   b. Using the special tool (09224-00010), remove the heated oxygen sensor.

To install:

2. To install, reverse the removal procedure. Tighten the heated oxygen sensor to 32 ft. lbs. (44 Nm) if using the special tool. If tightening with the special tool, tighten to 30 ft. lbs. (40 Nm).

4.0L Engine

See Figures 180 and 181.

1. Remove the heated oxygen sensor (Bank 1 Sensor 2).
   a. Using the special tool (09224-00010), remove the sensor.
2. Remove the heated oxygen sensor (Bank 2 Sensor 2).
   a. Using the special tool (09224-00010), remove the sensor.

To install:

3. Install the heated oxygen sensor (Bank 1 Sensor 2).
   a. Temporarily install the sensor to the exhaust pipe by hand.
   b. Using the special tool (09224-00010). If tightening without the special tool, tighten to 32 ft. lbs. (44 Nm). If tightening with the special tool, tighten to 30 ft. lbs. (40 Nm).

➡Use a torque wrench with a fulcrum length of 11.8 inch (30 cm). When using a torque wrench with a fulcrum length that is not 11.8 inch (30 cm), calculate the torque specification for the torque wrench and SST based on the "without SST" torque specification.

➡Make sure SST and the wrench are connected in a straight line.

   c. Connect the sensor connector.
4. Install the heated oxygen sensor (Bank 2 Sensor 2).
   a. Temporarily install the sensor to the exhaust pipe by hand.

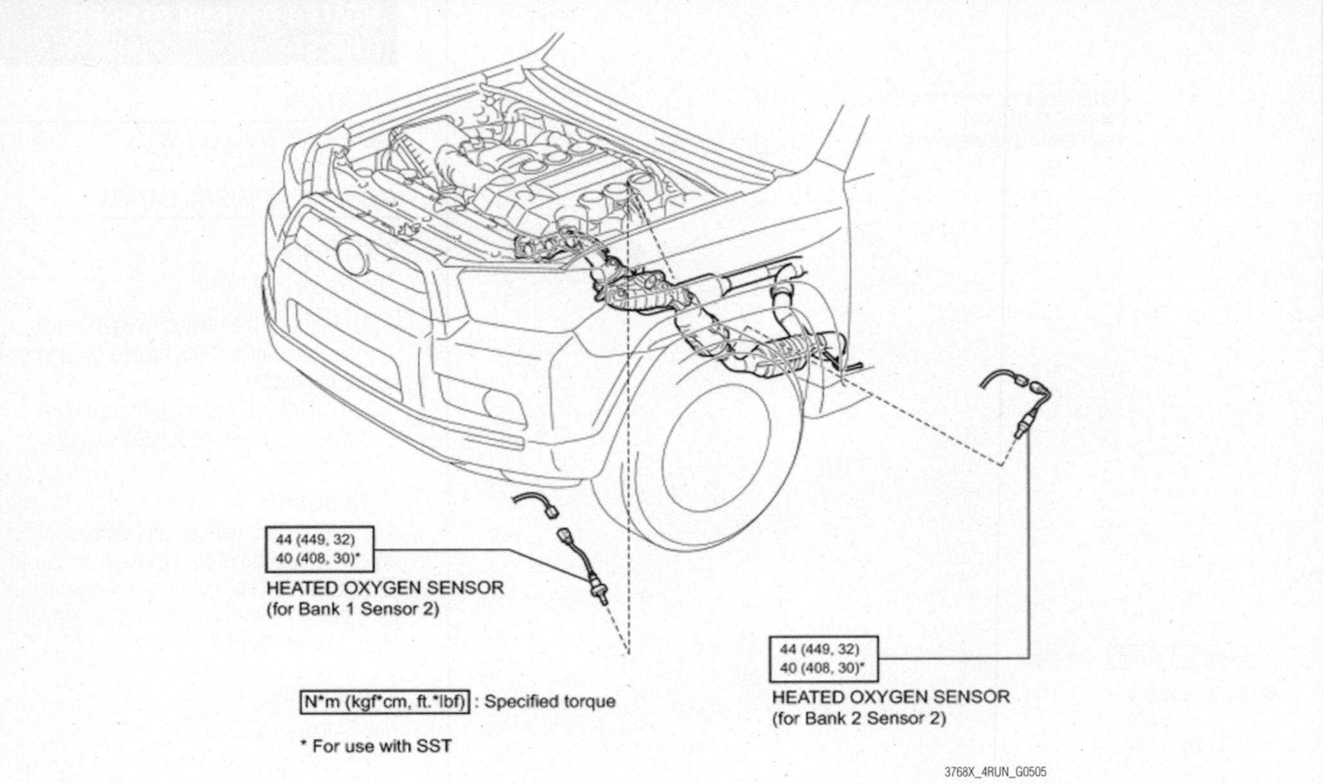

Fig. 177 Heated Oxygen Sensor (HO2S)—4.0L engine

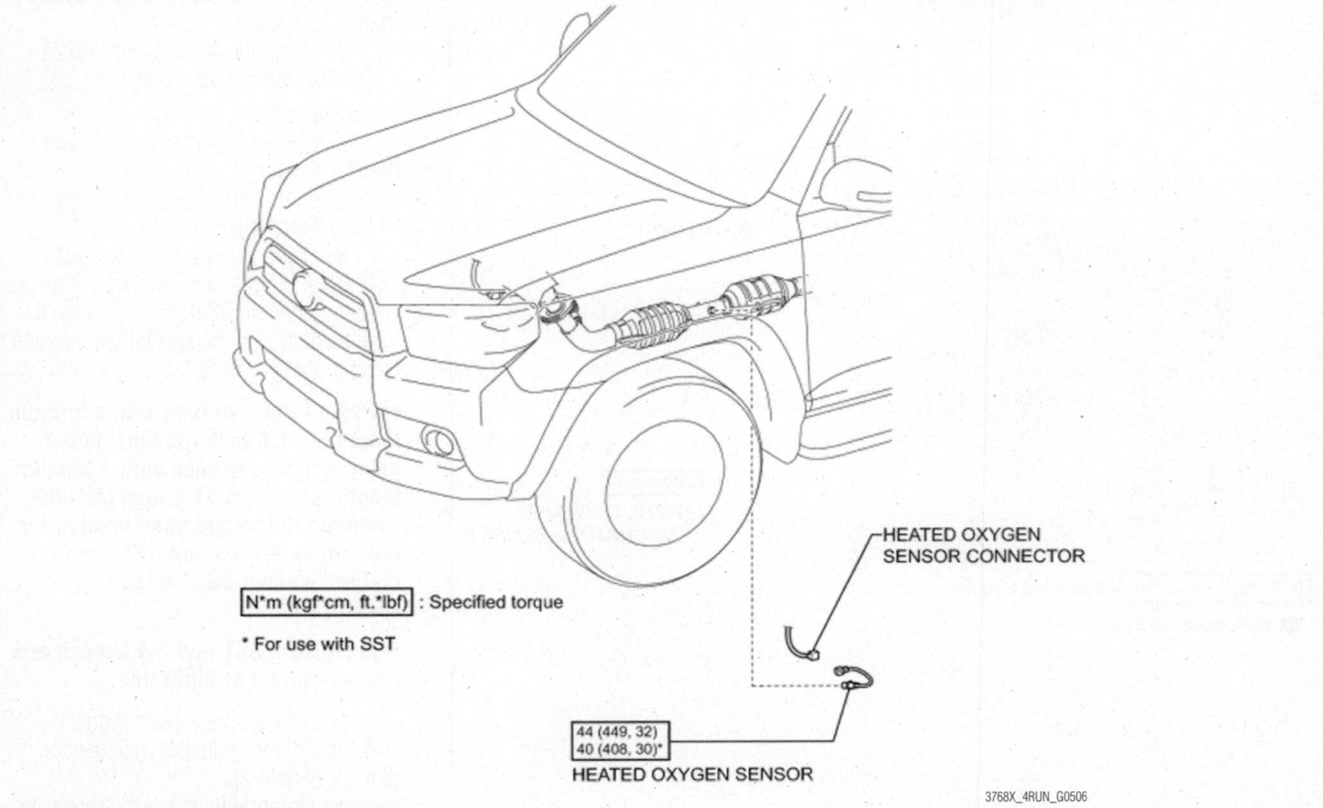

Fig. 178 Heated oxygen sensor—2.7L engine

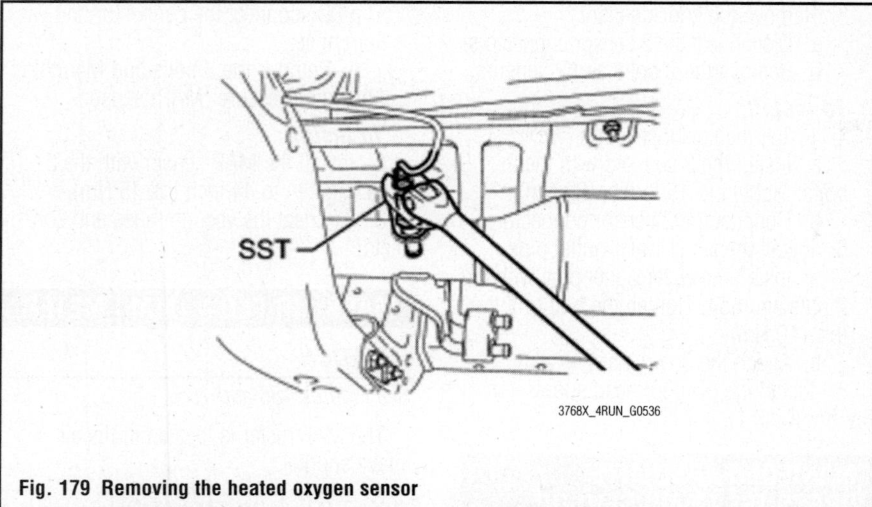

Fig. 179 Removing the heated oxygen sensor

b. Using the special tool (09224-00010). If tightening without the special tool, tighten to 32 ft. lbs. (44 Nm). If tightening with the special tool, tighten to 30 ft. lbs. (40 Nm).

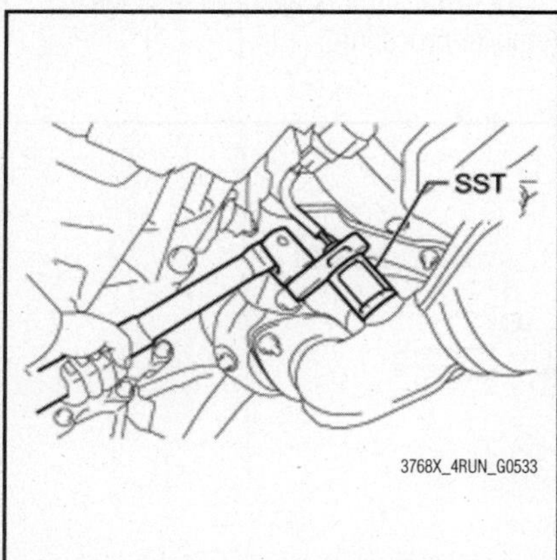

Fig. 180 Removing the sensor (Bank 1 Sensor 2)

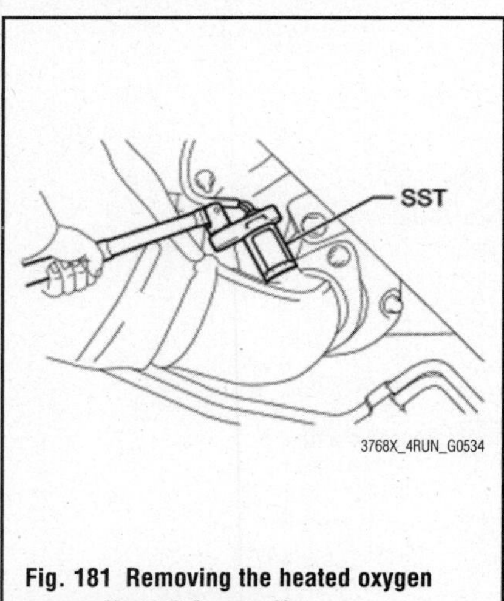

Fig. 181 Removing the heated oxygen sensor (Bank 2 Sensor 2)

➡Use a torque wrench with a fulcrum length of 11.8 inch (30 cm). When using a torque wrench with a fulcrum length that is not 11.8 inch (30 cm), calculate the torque specification for the torque wrench and SST based on the "without SST" torque specification.

➡Make sure SST and the wrench are connected in a straight line.

c. Connect the sensor connector.

INTAKE AIR TEMPERATURE (IAT) SENSOR

LOCATION

The Intake Air Temperature (IAT) sensor, built into the Mass Air Flow (MAF) meter.

REMOVAL & INSTALLATION

See Mass Air Flow Meter.

KNOCK SENSOR (KS)

LOCATION

See Figures 182 and 183.

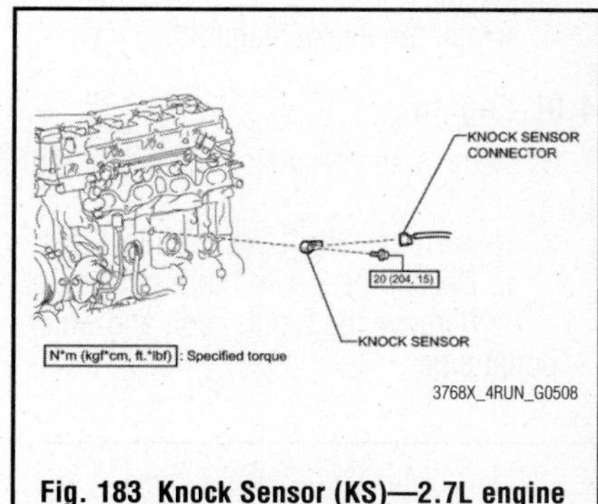

Fig. 183 Knock Sensor (KS)—2.7L engine

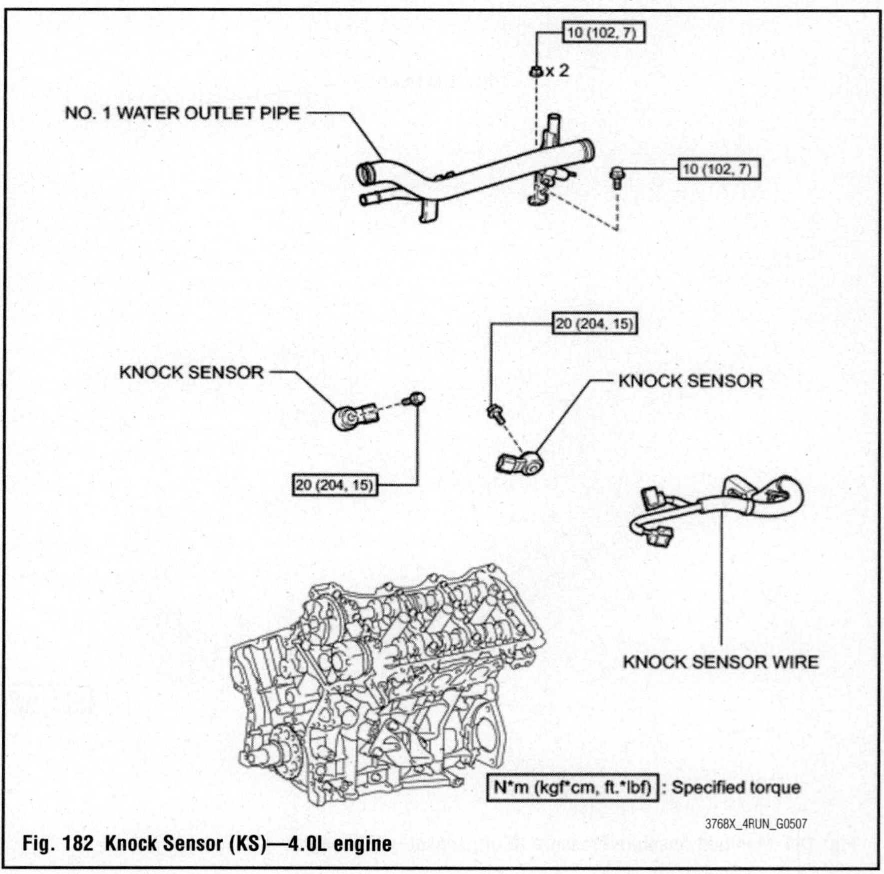

Fig. 182 Knock Sensor (KS)—4.0L engine

REMOVAL & INSTALLATION

2.7L Engine

1. Remove the intake manifold.
2. Remove the knock sensor.
   a. Disconnect the knock sensor connector.
   b. Remove the bolt and knock sensor.

To install:

3. Install the knock sensor.
   a. Install the knock sensor with the bolt. Tighten to 15 ft. lbs. (20 Nm).

➡Make sure that the knock sensor is at the correct angle when installing it.

   b. Connect the knock sensor connector.
4. Install the intake manifold.

4.0L Engine

1. Remove the cylinder head sub assembly (Bank 1).
2. Remove the No. 1 water outlet pipe.
   a. Detach the 3 wire harness clamps.
   b. Remove the 2 nuts, bolt and water outlet pipe.
3. Remove the knock sensor.
   a. Disconnect the 2 sensor connectors.
   b. Remove the 2 bolts and 2 sensors.

To install:

4. Install the knock sensor.
   a. Install the 2 sensors with the 2 bolts. Tighten to 15 ft. lbs. (20 Nm).
   b. Connect the 2 sensor connectors.
5. Install the No. 1 water outlet pipe.
   a. Install the water outlet pipe with the 2 nuts and bolt. Tighten the bolt to 7 ft. lbs. (10 Nm).
   b. Attach the 3 wire harness clamps.
6. Install the cylinder head sub assembly (for Bank 1).

MANIFOLD ABSOLUTE PRESSURE (MAP) SENSOR

LOCATION

See Figure 184.

REMOVAL & INSTALLATION

1. Remove the manifold absolute pressure sensor.
   a. Disconnect the connector and vacuum hose.
   b. Remove the 2 bolts and Manifold Absolute Pressure (MAP) sensor.

To install:

2. Install the MAP sensor with the 2 bolts. Tighten to 44 inch lbs. (5 Nm).
3. Connect the vacuum hose and connector.

MASS AIR FLOW (MAF) METER

LOCATION

See Figures 185 and 186.

The MAF meter is located in the air intake snorkel.

REMOVAL & INSTALLATION

1. Disconnect the electrical connector.
2. Remove attaching screws and remove MAF meter.

To install:

3. Installation is the reverse of the removal procedure.

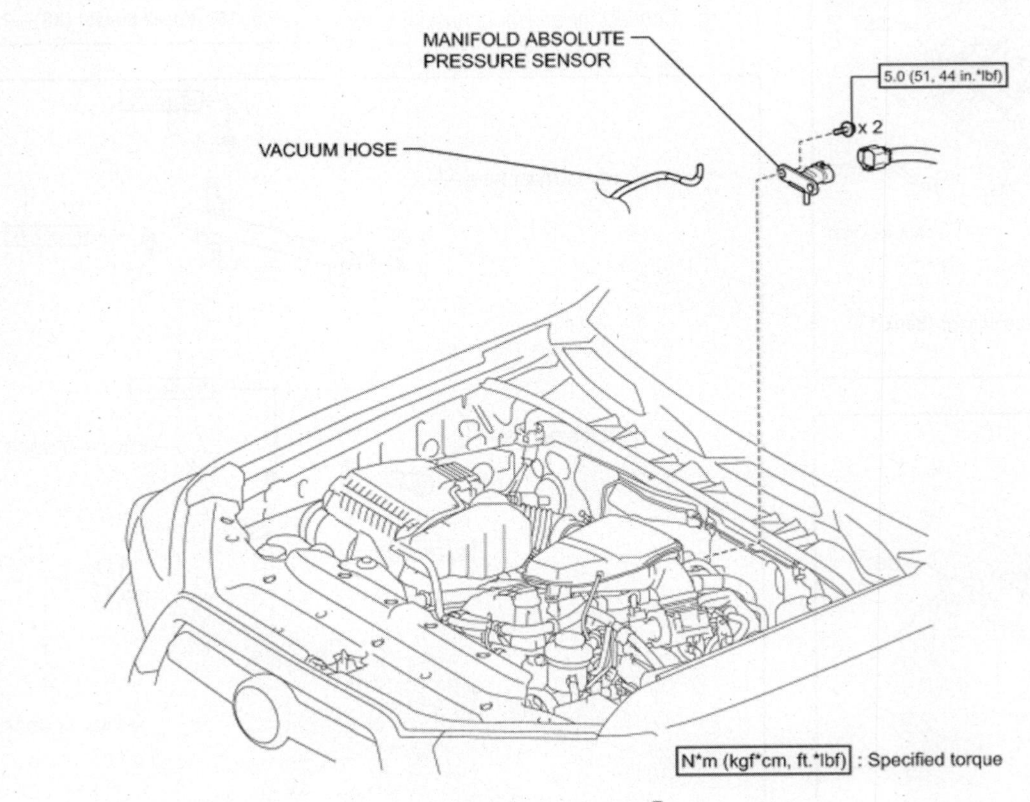

Fig. 184 Manifold Absolute Pressure (MAP) sensor—2.7L engine

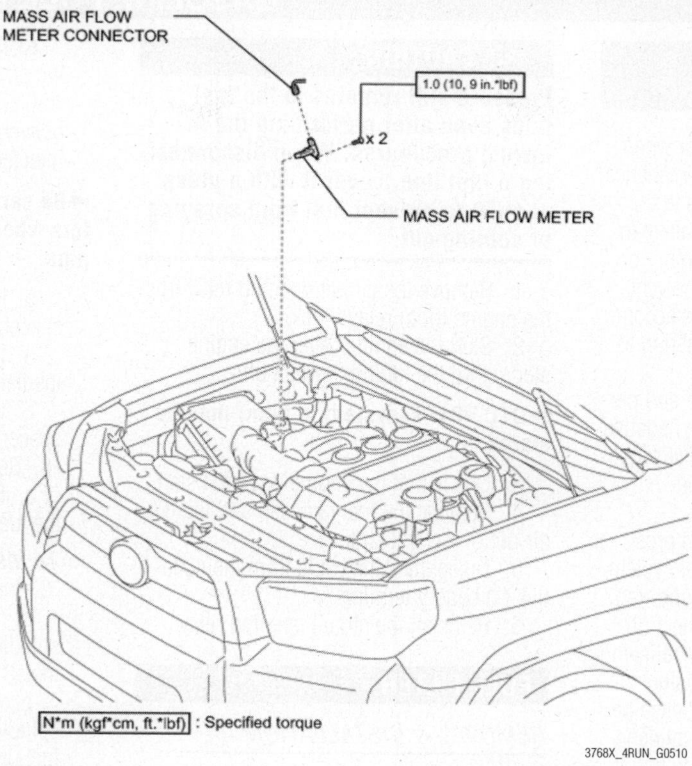

Fig. 185 Mass Air Flow (MAF) meter—4.0L engine

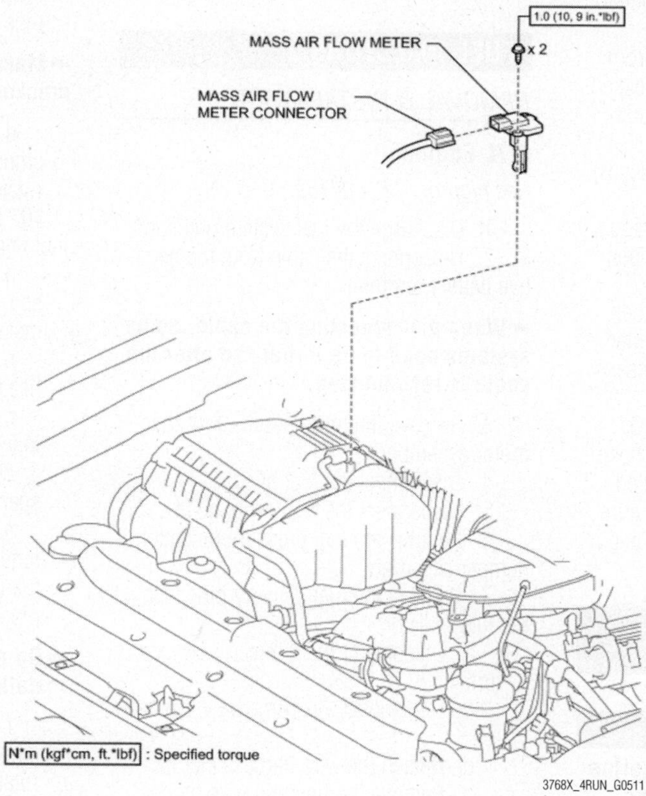

Fig. 186 Mass Air Flow (MAF) meter—2.7L engine

FUEL SYSTEM SERVICE PRECAUTIONS

Safety is the most important factor when performing not only fuel system maintenance but any type of maintenance. Failure to conduct maintenance and repairs in a safe manner may result in serious personal injury or death. Maintenance and testing of the vehicle's fuel system components can be accomplished safely and effectively by adhering to the following rules and guidelines.

- To avoid the possibility of fire and personal injury, always disconnect the negative battery cable unless the repair or test procedure requires that battery voltage be applied.
- Always relieve the fuel system pressure prior to disconnecting any fuel system component (injector, fuel rail, pressure regulator, etc.), fitting or fuel line connection. Exercise extreme caution whenever relieving fuel system pressure to avoid exposing skin, face and eyes to fuel spray. Please be advised that fuel under pressure may penetrate the skin or any part of the body that it contacts.
- Always place a shop towel or cloth around the fitting or connection prior to loosening to absorb any excess fuel due to spillage. Ensure that all fuel spillage (should it occur) is quickly removed from engine surfaces. Ensure that all fuel soaked cloths or towels are deposited into a suitable waste container.
- Always keep a dry chemical (Class B) fire extinguisher near the work area.
- Do not allow fuel spray or fuel vapors to come into contact with a spark or open flame.
- Always use a back-up wrench when loosening and tightening fuel line connection fittings. This will prevent unnecessary stress and torsion to fuel line piping.
- Always replace worn fuel fitting O-rings with new Do not substitute fuel hose or equivalent where fuel pipe is installed.

Before servicing the vehicle, make sure to also refer to the precautions in the beginning of this section as well.

RELIEVING FUEL SYSTEM PRESSURE

CAUTION

Take precautions to prevent gasoline from spilling out before removing fuel system parts.

CAUTION

Pressure still remains in the fuel lines even after performing the following procedures. When disconnecting a fuel line, cover it with a piece of cloth to prevent fuel from spraying or coming out.

1. Remove the circuit opening relay from the engine room relay block.
2. Start the engine. After the engine stops, turn the ignition switch off.

➡DTC P0171 (system to lean) may be stored.

3. Check that the engine does not start.
4. Remove the fuel tank cap and let the air out of the fuel tank.
5. Disconnect the cable from the negative (-) battery terminal.
6. Reinstall the circuit opening relay.

FUEL FILTER

REMOVAL & INSTALLATION

See Figure 187.

The fuel filter is part of the fuel pump module unit and is not a normally replaced item.

FUEL INJECTORS

REMOVAL & INSTALLATION

2.7L Engine

See Figures 188 and 189.

1. Discharge the fuel system pressure.
2. Disconnect the cable from the negative battery terminal.

➡When disconnecting the cable, some systems need to be initialized after the cable is reconnected.

3. Remove the throttle body with the motor assembly.
4. Disconnect the fuel hose.
5. Disconnect the No. 2 fuel hose.
6. Remove the fuel pressure pulsation damper assembly.
7. Remove the fuel delivery pipe with the fuel injector.
   a. Disconnect the 4 injector connectors.
   b. Disconnect the purge VSV connector.
   c. Detach the wire harness clamp.
   d. Remove the bolt and wire harness clamp bracket.
   e. Remove the bolt and disconnect the purge VSV.
   f. Remove the 2 bolts and fuel delivery pipe together with the fuel injectors.

➡Be careful not to drop the fuel injectors when removing the fuel delivery pipe.

   g. Remove the 2 No. 1 delivery pipe spacers.
   h. Remove the 4 injector vibration insulators.
   i. Using a screwdriver, remove the 4 spacers.
8. Remove the fuel injector assembly.
   a. Remove the 4 fuel injectors from the fuel delivery pipe.

To install:

9. Install the fuel injector assembly.
   a. Apply a light coat of gasoline or spindle oil to new O-rings, and then install one to each fuel injector.
   b. Apply a light coat of gasoline or spindle oil to the contact surfaces of the fuel delivery pipe and the O-ring of the fuel injector.
   c. Apply a light coat of gasoline or spindle oil to the O-ring again, and then install the fuel injector to the fuel delivery pipe by turning it right and left.

➡Make sure that the O-ring is not cracked or jammed when installing.

   d. Check that the fuel injector rotates smoothly. If the fuel injector does not rotate smoothly, replace the O-ring.
10. Install the fuel delivery pipe with the fuel injector.
   a. Apply a light coat of gasoline or spindle oil to new O-rings, and then install one to each spacer.
   b. Install the 4 spacers to the cylinder head.
   c. Install 4 new injector vibration insulators to the cylinder head.
   d. Install the 2 No. 1 delivery pipe spacers to the cylinder head.
   e. Install the fuel delivery pipe together with the 4 fuel injectors, and then temporarily install the fuel delivery pipe with the 2 bolts.

➡Do not drop the fuel injectors when installing the fuel delivery pipe.

   f. Check that the fuel injectors rotate smoothly. If any fuel injector does not rotate smoothly, replace its O-ring.
   g. Tighten the 2 bolts to 9 ft. lbs. (12 Nm).

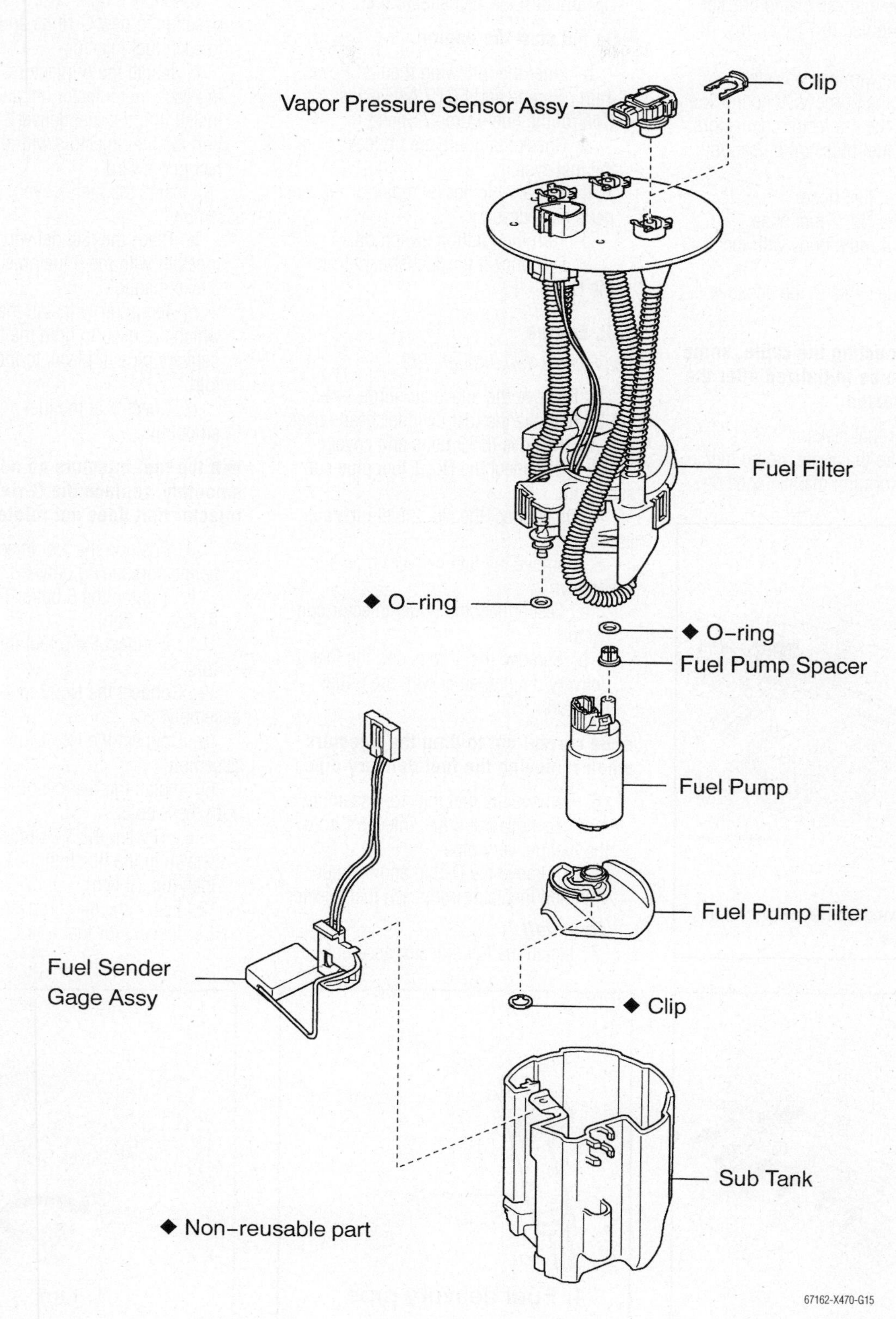

Fig. 187 Fuel pump components

h. Install the purge VSV with the bolt. Tighten to 80 inch lbs. (9 Nm).

i. Install the harness clamp bracket with the bolt. Tighten to 71 inch lbs. (8 Nm).

j. Attach the wire harness clamp.

k. Connect the purge VSV connector.

l. Connect the 4 injector connectors.

11. Install the fuel pressure pulsation damper assembly.
12. Connect the fuel hose.
13. Connect the No. 2 fuel hose.
14. Install the throttle body with the motor assembly.
15. Connect the cable to the negative battery terminal.

➡**When disconnecting the cable, some systems need to be initialized after the cable is reconnected.**

16. Inspect for fuel leaks.

a. Make sure that there are no fuel leaks after performing maintenance on the fuel system. Connect the Techstream to the DLC3. Turn the ignition switch to ON and turn the Techstream on.

➡**Do not start the engine.**

b. Enter the following menus: Powertrain / Engine and ECT / Active Test / Control the Fuel Pump / Speed.

c. Check that there are no leaks from the fuel system.

If there are fuel leaks, repair or replace parts as necessary.

d. Turn the ignition switch off.

e. Disconnect the Techstream from the DLC3.

Fig. 188 Removing the fuel delivery pipe with the injectors

4.0L Engine

See Figures 190 through 192.

1. Remove the intake air surge tank.
2. Remove the rear cylinder head cover:

a. Remove the 3 bolts and cover.

3. Disconnect the No. 1 fuel pipe sub assembly.
4. Disconnect the No. 2 fuel pipe sub assembly.
5. Remove the fuel delivery pipe sub assembly.

a. Disconnect the 6 fuel injector connectors.

b. Remove the 4 bolts and the fuel delivery pipe together with the 6 fuel injectors.

➡**Be careful not to drop the injectors when removing the fuel delivery pipe.**

6. Remove the fuel injector assembly.

a. Remove the 6 fuel injectors from the fuel delivery pipe.

b. Remove the O-ring and injector vibration insulator from each fuel injector.

To install:

7. Install the fuel injector assembly.

a. Install a new insulator to each fuel injector.

b. Apply a light coat of spindle oil or gasoline to new O-rings and install one to each fuel injector.

c. Install the 6 injectors. While turning each fuel injector left and right, install it to the fuel delivery pipe. Position the fuel injectors with the connectors facing outward.

8. Install the fuel delivery pipe sub assembly.

a. Place the fuel delivery pipe together with the 6 fuel injectors on the intake manifold.

b. Temporarily install the 4 bolts, which are used to hold the fuel delivery pipe in place, to the intake manifold.

c. Check that the fuel injectors rotate smoothly.

➡**If the fuel injectors do not rotate smoothly, replace the O-ring of any injector that does not rotate smoothly.**

d. Position the fuel injectors with the connectors facing outward.

e. Tighten the 6 bolts. Tighten to 15 ft. lbs. (21 Nm).

f. Connect the 6 fuel injector connectors.

9. Connect the No. 2 fuel pipe sub assembly.
10. Connect the No. 1 fuel pipe sub assembly.
11. Install the rear cylinder head cover with the 3 bolts.

a. Tighten the 3 bolts in the sequence shown in the illustration. Tighten to 80 inch lbs. (9 Nm).

12. Install the air surge tank.
13. Inspect for fuel leaks.

Fig. 189 Removing the delivery pipe spacers, injector vibration insulators and spacers

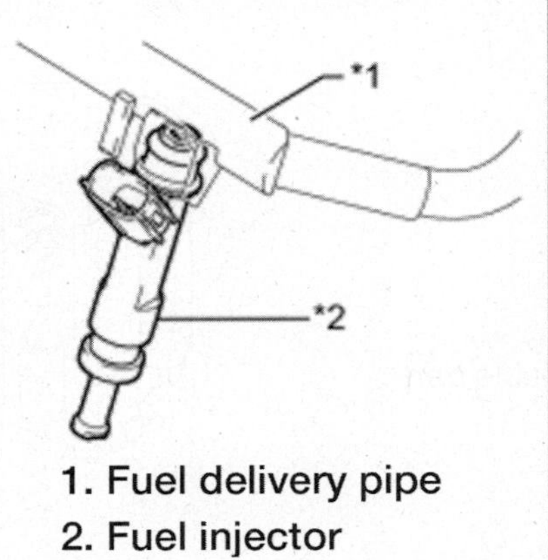

Fig. 190 Removing the fuel injector assembly

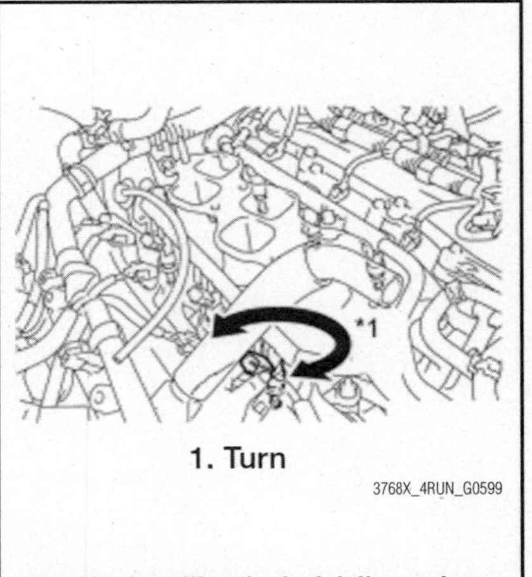

Fig. 191 Installing the fuel delivery pipe sub assembly

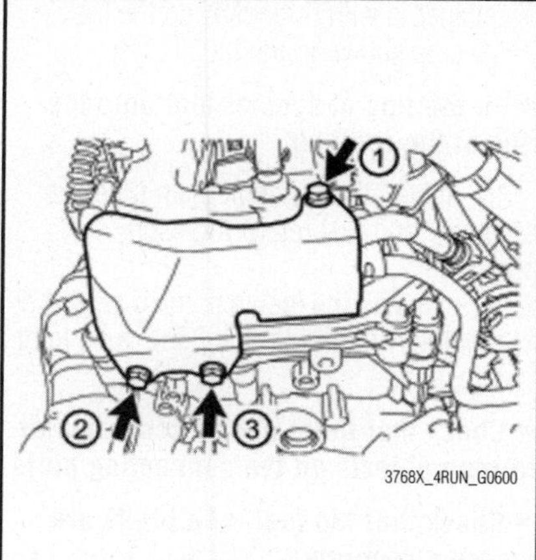

Fig. 192 Identifying the rear cylinder head cover bolt tightening sequence

a. Make sure that there are no fuel leaks after performing maintenance on the fuel system. Connect the Techstream to the DLC3. Turn the ignition switch to ON and turn the Techstream ON.

➡**Do not start the engine.**

b. Enter the following menus: Powertrain / Engine and ECT / Active Test / Control the Fuel Pump / Speed.

c. Check that there are no leaks from the fuel system.

If there are fuel leaks, repair or replace parts as necessary.

d. Turn the ignition switch off.

e. Disconnect the Techstream from the DLC3.

FUEL PUMP

REMOVAL & INSTALLATION

See Figures 193 through 198.

1. Discharge the fuel system pressure.
2. Remove the rear seat assembly LH.
   a. For a 60/40 split double folding seat type LH side, remove the rear seat assembly.
   b. For a 60/40 split slide walk in seat type LH side, remove the rear seat assembly LH.
3. Remove the rear floor service hole cover.
   a. Remove the 3 screws and rear floor service hole cover.
4. Remove the No. 1 fuel tank protector sub assembly.
   a. For standard type, remove the 6 bolts and No. 1 fuel tank protector.
   b. For half cover types, remove the 4 bolts and the No. 1 fuel tank protector.
5. Disconnect the fuel tank main tube sub assembly.

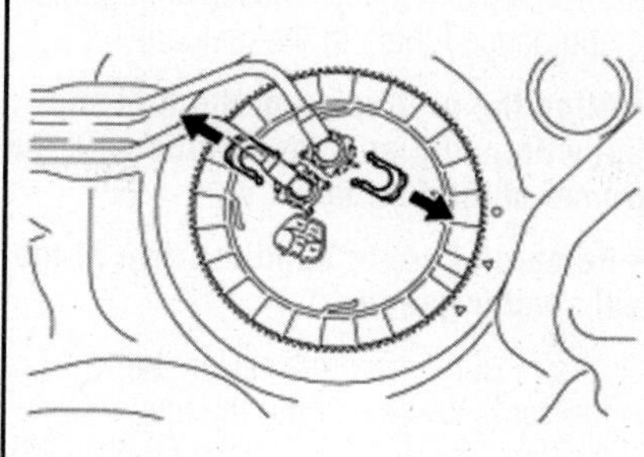

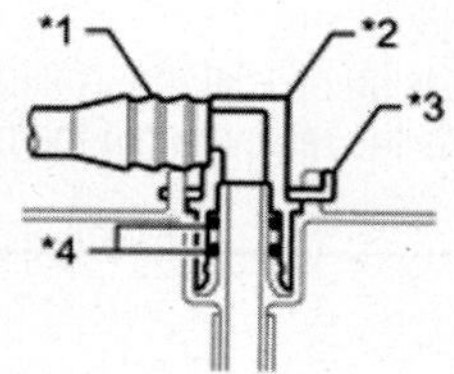

Fig. 193 Removing the 2 fuel tube joint clips and pulling out the fuel tank main tube and fuel return tube

6. Drain the fuel.
   a. Connect the Techstream to the DLC3.
   b. Turn the ignition switch to ON.

➡**Do not start the engine.**

   c. Turn the Techstream ON.
   d. Enter the following menus: Powertrain/Engine and ECT/Active Test/ Control the Fuel Pump/Speed.
   e. Operate the fuel pump and drain the fuel from the fuel tank.

⁂ CAUTION

Do not smoke or be near an open flame when working on the fuel system.

⁂ CAUTION

Secure good ventilation.

⁂ CAUTION

Keep gasoline away from rubber or leather parts.

   f. Disconnect the fuel pump and sender gauge connector.
7. Disconnect the cable from the negative battery terminal.

➡**When disconnecting the cable, some systems need to be initialized after the cable is reconnected.**

8. Disconnect the fuel return tube sub assembly.
   a. Disconnect the fuel return tube.
9. Disconnect the fuel tank vent hose sub assembly.
   a. Disconnect the fuel tank vent hose.
10. Disconnect the fuel tank breather tube sub assembly.
    a. Disconnect the fuel tank breather tube.
11. Disconnect the fuel tank to the filler pipe hose.
12. Remove the fuel tank sub assembly.
    a. Place a transmission jack under the fuel tank.
    b. Remove the 2 bolts, 2 clips, 2 pins and 2 fuel tank bands.
    c. Slowly lower the transmission jack slightly.
13. Remove the fuel tank cushion.
    a. Remove the No. 1, No. 2 and No. 3 fuel tank cushions from the fuel tank.
14. Remove the No. 1 fuel tank protector.
15. Remove the fuel tank breather tube sub assembly.
16. Remove the fuel tank main tube sub assembly and fuel return tube sub assembly.
    a. Remove the 2 fuel tube joint clips and pull out the fuel tank main tube and fuel return tube.

➡**Remove any dirt and foreign matter on the fuel tube joint before performing this work.**

➡**Do not allow any scratches or foreign matter on the parts when disconnecting them, as the fuel tube joint contains the O-rings that seal the plug.**

➡**Perform this work by hand. Do not use any tools.**

➡**Do not forcibly bend, twist or turn the nylon tube.**

➡**Protect the disconnected part by covering it with a plastic bag and tape after disconnecting the fuel tubes.**

    b. Remove the fuel tank main tube and fuel return tube from the fuel tank.
17. Remove the fuel suction with pump and gauge tube assembly.
    a. Using the special tool (09808-14020), loosen the retainer.

➡**Fit the tips of the special tool onto the ribs of the retainer.**

➡**When the retainer is loosened, be careful as the pump and gauge tube will spring upward from the force of the spring.**

➡**Remove any foreign matter around the fuel suction with pump and gauge tube before this operation.**

b. Remove the retainer.

c. Remove the fuel suction with pump and gauge tube assembly from the fuel tank.

➡**Be careful not to bend the arm of the fuel sender gauge.**

d. Remove the gasket from the fuel tank.

To install:

18. Install the fuel suction with the pump and gauge tube assembly.

a. Apply a light coat of gasoline or grease to a new gasket and install the gasket to the fuel tank.

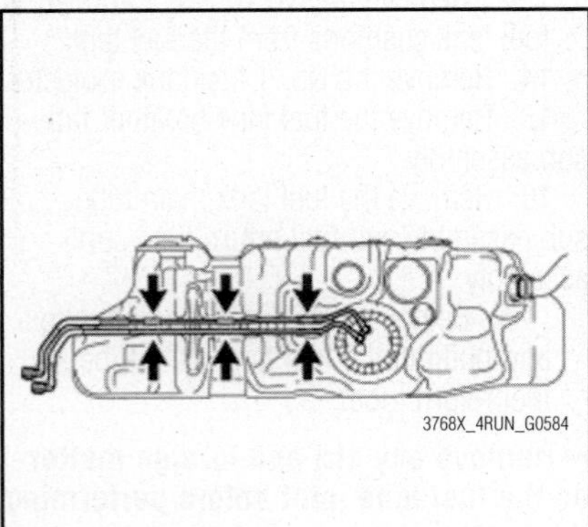

Fig. 194 Removing the fuel tank main tube and fuel return tube from the fuel tank

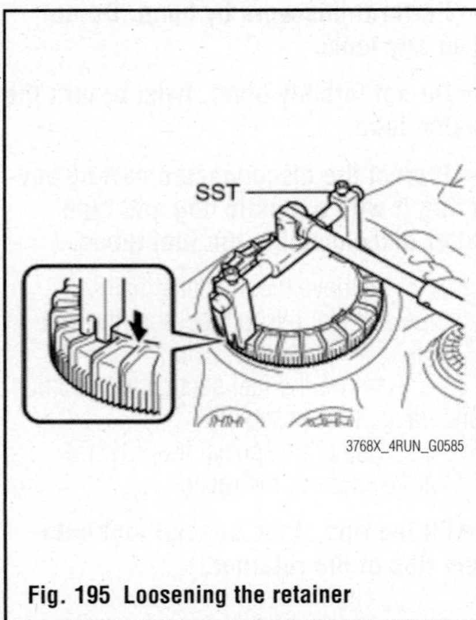

Fig. 195 Loosening the retainer

b. Install the fuel suction with pump and gauge tube into the fuel tank.

➡**Align the protrusion of the fuel suction with pump and gauge tube with the groove of the fuel tank.**

➡**Be careful not to bend the arm of the fuel sender gauge.**

c. Put the new retainer on the fuel tank. While holding the fuel suction with pump and gauge tube, tighten the retainer one complete turn by hand.

d. Using special tool (09808-14020), tightens the retainer until the mark on the retainer is within range A on the fuel tank, as shown in the illustration.

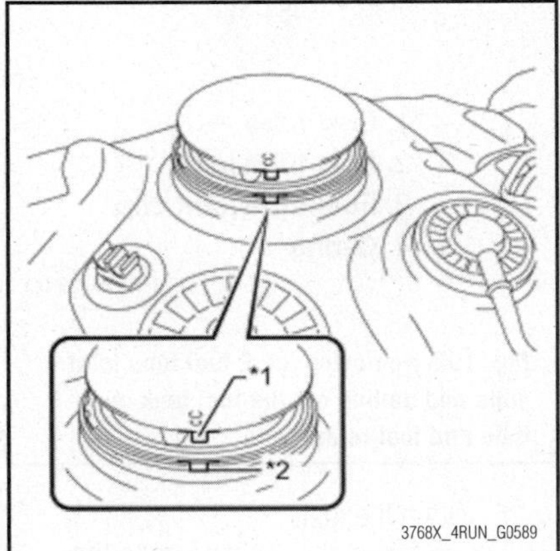

Fig. 196 Installing the fuel suction with the pump and gauge tube assembly

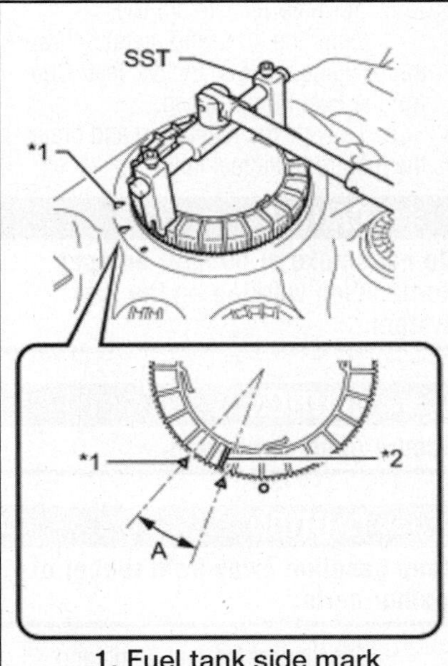

Fig. 197 Tightening the retainer

➡**Fit the tips of special tool onto the ribs of the retainer.**

19. Install the fuel tank main tube sub assembly and fuel return tube sub assembly.

a. Install the fuel tank main tube and fuel return tube with the 2 fuel tube joint clips.

➡**Check that there are no scratches or foreign objects on the connecting parts.**

➡**Check that the fuel tube joints are inserted securely.**

➡**Check that the fuel tube joint clips are on the collars of the fuel tube joints.**

➡**After installing the fuel tube joint clips, check that the fuel tube joints cannot be pulled off.**

b. Install the fuel tank main tube and fuel return tube to the fuel tank.

20. Install the fuel tank breather tube sub assembly.

21. Install the No. 1 fuel tank protector to the fuel tank.

22. Install the fuel tank cushion.

a. Install the 3 new fuel tank cushions to the fuel tank.

23. Install the fuel tank sub assembly.

a. Set the fuel tank on a transmission jack and lift up the transmission jack.

➡**Do not allow the fuel tank to contact the vehicle, especially the differential.**

b. Connect the 2 fuel tank bands with the 2 bolts. Tighten to 30 ft. lbs. (40 Nm).

24. Connect the fuel tank to filler pipe hose to the filler pipe.

25. Connect the fuel tank breather tube sub assembly.

26. Connect the fuel tank vent hose sub assembly.

27. Connect the fuel return tube sub assembly.

28. Connect the fuel tank main tube sub assembly.

29. Install the No. 1 fuel tank protector sub assembly. Tighten to 15 ft. lbs. (20 Nm).

30. Install the rear floor service hole cover.

a. Connect the fuel pump and fuel sender gauge connector.

b. Install the rear floor service hole cover with the 3 screws.

31. Install the rear seat assembly LH.

32. Connect the cable to the negative battery terminal.

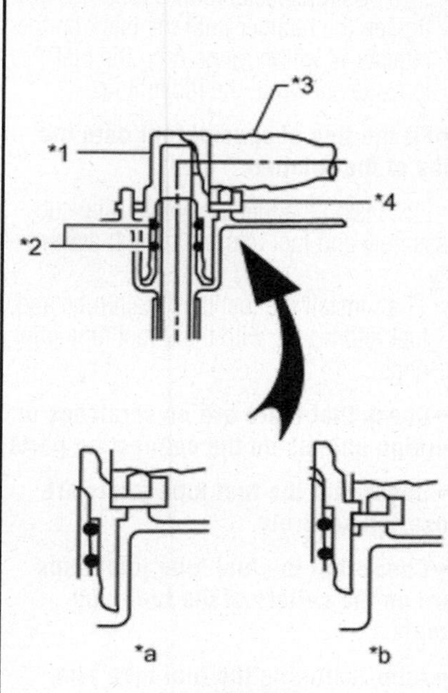

Fig. 198 Installing the fuel tank main tube sub assembly and fuel return tube sub assembly

➡When disconnecting the cable, some systems need to be initialized after the cable is reconnected.

33. Inspect for fuel leaks.

FUEL TANK

DRAINING

1. Connect the Techstream to the DLC3.
2. Turn the ignition switch to ON.

➡Do not start the engine.

3. Turn the Techstream on.
4. Enter the following menus: Powertrain / Engine and ECT / Active Test / Control the Fuel Pump / Speed.
5. Operate the fuel pump and drain the fuel from the fuel tank main tube.

✱✱ CAUTION

Do not smoke or be near an open flame when working on the fuel system.

✱✱ CAUTION

Secure good ventilation.

✱✱ CAUTION

Keep gasoline away from rubber or leather parts.

6. Disconnect the fuel pump and sender gauge connector.

REMOVAL & INSTALLATION

See Figures 199 through 201.

1. Discharge the fuel system pressure.
2. Remove the rear seat assembly LH.
   a. For a 60/40 split double folding seat type LH side, remove the rear seat assembly.
   b. For a 60/40 split slide walk in seat type LH side, remove the rear seat assembly LH.
3. Remove the rear floor service hole cover.
   a. Remove the 3 screws and rear floor service hole cover.
4. Remove the No. 1 fuel tank protector sub assembly.
   a. For standard type, remove the 6 bolts and No. 1 fuel tank protector.
   b. For half cover types, remove the 4 bolts and the No. 1 fuel tank protector.
5. Disconnect the fuel tank main tube sub assembly.
6. Drain the fuel.
   a. Connect the Techstream to the DLC3.
   b. Turn the ignition switch to ON.

➡Do not start the engine.

   c. Turn the Techstream ON.
   d. Enter the following menus: Powertrain/Engine and ECT/Active Test/ Control the Fuel Pump/Speed.
   e. Operate the fuel pump and drain the fuel from the fuel tank.

✱✱ CAUTION

Do not smoke or be near an open flame when working on the fuel system.

✱✱ CAUTION

Secure good ventilation.

✱✱ CAUTION

Keep gasoline away from rubber or leather parts.

   f. Disconnect the fuel pump and sender gauge connector.
7. Disconnect the cable from the negative battery terminal.

➡When disconnecting the cable, some systems need to be initialized after the cable is reconnected.

8. Disconnect the fuel return tube sub assembly.
   a. Disconnect the fuel return tube.
9. Disconnect the fuel tank vent hose sub assembly.
   a. Disconnect the fuel tank vent hose.
10. Disconnect the fuel tank breather tube sub assembly.
   a. Disconnect the fuel tank breather tube.
11. Disconnect the fuel tank to the filler pipe hose.
12. Remove the fuel tank sub assembly.
   a. Place a transmission jack under the fuel tank.
   b. Remove the 2 bolts, 2 clips, 2 pins and 2 fuel tank bands.
   c. Slowly lower the transmission jack slightly.
13. Remove the fuel tank cushion.
   a. Remove the No. 1, No. 2 and No. 3 fuel tank cushions from the fuel tank.
14. Remove the No. 1 fuel tank protector.
15. Remove the fuel tank breather tube sub assembly.
16. Remove the fuel tank main tube sub assembly and fuel return tube sub assembly.
   a. Remove the 2 fuel tube joint clips and pull out the fuel tank main tube and fuel return tube.

➡Remove any dirt and foreign matter on the fuel tube joint before performing this work.

➡Do not allow any scratches or foreign matter on the parts when disconnecting them, as the fuel tube joint contains the O-rings that seal the plug.

➡Perform this work by hand. Do not use any tools.

➡Do not forcibly bend, twist or turn the nylon tube.

➡Protect the disconnected part by covering it with a plastic bag and tape after disconnecting the fuel tubes.

   b. Remove the fuel tank main tube and fuel return tube from the fuel tank.
17. Remove the fuel suction with pump and gauge tube assembly.
   a. Using the special tool (09808-14020), loosen the retainer.

➡Fit the tips of the special tool onto the ribs of the retainer.

➡When the retainer is loosened, be careful as the pump and gauge tube will spring upward from the force of the spring.

➡Remove any foreign matter around the fuel suction with pump and gauge tube before this operation.

b. Remove the retainer.

c. Remove the fuel suction with pump and gauge tube assembly from the fuel tank.

➡Be careful not to bend the arm of the fuel sender gauge.

d. Remove the gasket from the fuel tank.

18. Remove the No. 3 fuel tank protector.

a. Remove the 2 bolts.

b. Detach the 4 clamps and remove the No. 3 fuel tank protector.

19. Remove the fuel tank to filler pipe hose.

a. Remove the fuel tank to filler pipe hose from the fuel tank.

To install:

20. Install the fuel tank to filler pipe hose.

➡Align the fuel tank side mark with the hose side mark when installing the hose.

a. Install the 2 bolts and tighten to 44 inch lbs. (5 Nm).

21. Install the fuel suction with the pump and gauge tube assembly.

a. Apply a light coat of gasoline or grease to a new gasket and install the gasket to the fuel tank.

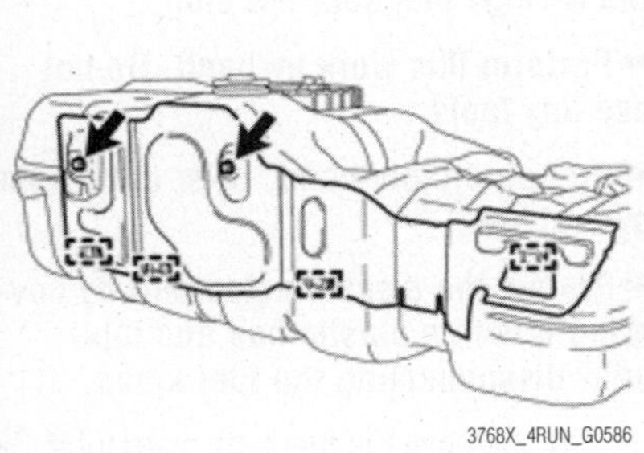

Fig. 199 Removing the No. 3 fuel tank protector

b. Install the fuel suction with pump and gauge tube into the fuel tank.

➡Align the protrusion of the fuel suction with pump and gauge tube with the groove of the fuel tank.

➡Be careful not to bend the arm of the fuel sender gauge.

c. Put the new retainer on the fuel tank. While holding the fuel suction with pump and gauge tube, tighten the retainer one complete turn by hand.

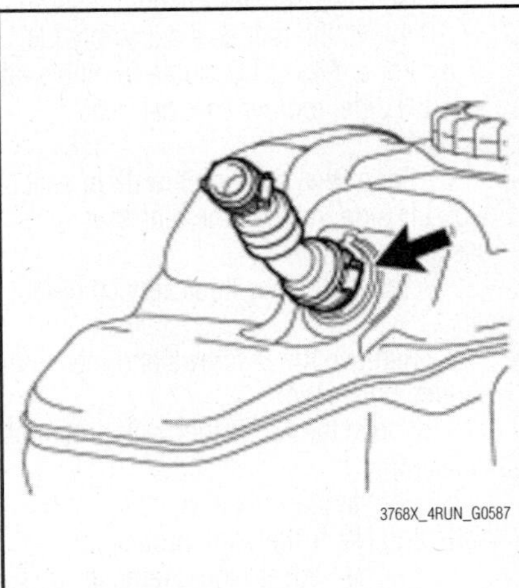

Fig. 200 Removing the fuel tank to filler pipe hose

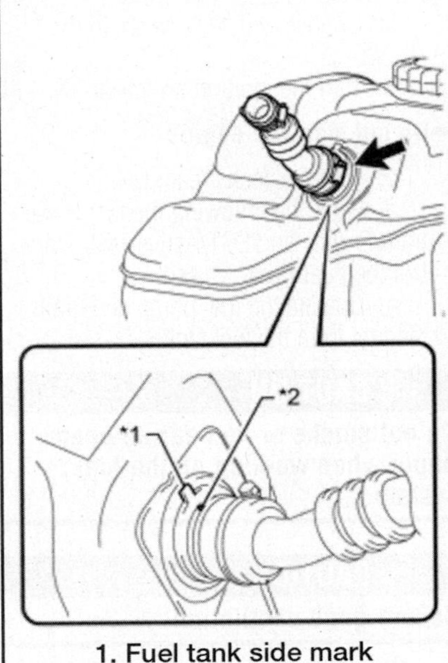

Fig. 201 Installing the fuel tank to the filler pipe hose

d. Using special tool (09808-14020), tighten the retainer until the mark on the retainer is within range A on the fuel tank, as shown in the illustration.

➡Fit the tips of special tool onto the ribs of the retainer.

22. Install the fuel tank main tube sub assembly and fuel return tube sub assembly.

a. Install the fuel tank main tube and fuel return tube with the 2 fuel tube joint clips.

➡Check that there are no scratches or foreign objects on the connecting parts.

➡Check that the fuel tube joints are inserted securely.

➡Check that the fuel tube joint clips are on the collars of the fuel tube joints.

➡After installing the fuel tube joint clips, check that the fuel tube joints cannot be pulled off.

b. Install the fuel tank main tube and fuel return tube to the fuel tank.

23. Install the fuel tank breather tube sub assembly.

24. Install the No. 1 fuel tank protector to the fuel tank.

25. Install the fuel tank cushion.

a. Install the 3 new fuel tank cushions to the fuel tank.

26. Install the fuel tank sub assembly.

a. Set the fuel tank on a transmission jack and lift up the transmission jack.

➡Do not allow the fuel tank to contact the vehicle, especially the differential.

b. Connect the 2 fuel tank bands with the 2 bolts. Tighten to 30 ft. lbs. (40 Nm).

27. Connect the fuel tank to filler pipe hose to the filler pipe.

28. Connect the fuel tank breather tube sub assembly.

29. Connect the fuel tank vent hose sub assembly.

30. Connect the fuel return tube sub assembly.

31. Connect the fuel tank main tube sub assembly.

32. Install the No. 1 fuel tank protector sub assembly. Tighten to 15 ft. lbs. (20 Nm).

33. Install the rear floor service hole cover.

a. Connect the fuel pump and fuel sender gauge connector.

b. Install the rear floor service hole cover with the 3 screws.

34. Install the rear seat assembly LH.
35. Connect the cable to the negative battery terminal.

➡**When disconnecting the cable, some systems need to be initialized after the cable is reconnected.**

36. Inspect for fuel leaks.

IDLE SPEED

ADJUSTMENT

Idle speed is maintained by the Powertrain Control Module (PCM). No adjustment is necessary or possible.

THROTTLE BODY

REMOVAL & INSTALLATION

2.7L Engine

See Figures 202 and 203.

1. Remove the front bumper cover lower.
2. Remove the No. engine under cover sub assembly.
3. Drain the engine coolant.
4. Remove the air cleaner and hose.
   a. Detach the 3 clamps and disconnect the mass air flow meter connector.
   b. Detach the 4 clamps.
   c. Loosen the hose clamp and remove the air cleaner and hose.
5. Remove the intake air connector.
   a. Disconnect the vacuum hose from the fuel pressure regulator.
   b. Disconnect the No. 2 ventilation hose.
   c. Detach the wire harness clamp.
   d. Disconnect the connector.
   e. Disconnect the vacuum hose from the manifold absolute pressure sensor.
   f. Detach the vacuum hose.
   g. Loosen the clamp.
   h. Remove the 3 bolts and intake air connector.
6. Remove the throttle body with the motor assembly.
   a. Disconnect the water by pass hose.
   b. Disconnect the No. 2 water by-pass hose.
   c. Disconnect the throttle position sensor and throttle control motor connector.
   d. Remove the 2 bolts, 2 nuts and throttle body with motor.
   e. Remove the gasket from the intake manifold.

To install:

7. Install the throttle body with the motor assembly.
   a. Align the protrusion of a new gasket with the groove of the intake manifold.
   b. Install a new gasket to the intake manifold.
   c. Install the throttle body with motor with the 2 bolts and 2 nuts. Tighten to 80 inch lbs. (9 Nm).
   d. Connect the water by-pass hose.
   e. Connect the No. 2 water by-pass hose.
   f. Connect the throttle position sensor and throttle control motor connector.
8. Install the intake air connector.
   a. Install the intake air connector with the 3 bolts. Tighten to 71 inch lbs. (8 Nm).
   b. Tighten the hose clamp. Tighten to 44 inch lbs. (5 Nm).
   c. Attach the vacuum hose.
   d. Connect the vacuum hose to the manifold absolute pressure sensor.
   e. Connect the connector.
   f. Attach the wire harness clamp.
   g. Connect the No. 2 ventilation hose.
   h. Connect the vacuum hose to the fuel pressure regulator.
9. Install the air cleaner and hose.
   a. Install the air cleaner and hose, align its matchmark with the matchmark of the air cleaner cap as shown in the illustration.
   b. Tighten the hose clamp. Tighten to 44 inch lbs. (5 Nm).
   c. Attach the 4 clamps.
   d. Attach the 3 clamps and connect the mass air flow meter connector.
10. Add engine coolant.
11. Inspect for coolant leak.
12. Install the No. 1 engine under cover sub assembly.
13. Install the front bumper cover lower.

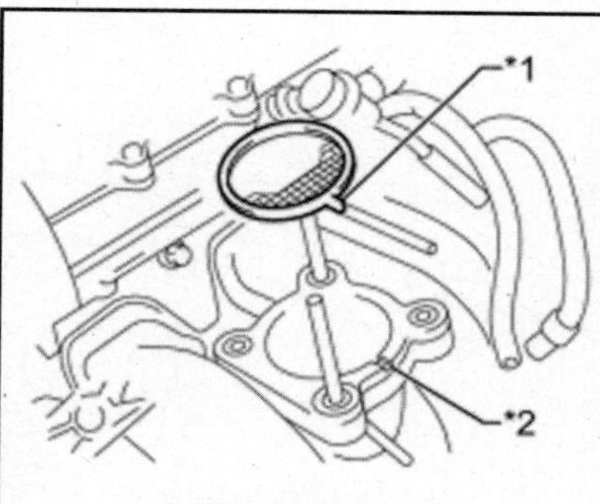

Fig. 202 Aligning the protrusion of the new gasket with the groove of the intake manifold

14. Perform the initialization.

➡**Be sure to perform this procedure after reassembling the throttle body assembly or removing and reinstalling any throttle body component.**

➡**Perform the following procedure after replacing the ECM, throttle body assembly or any throttle body components. The following procedure should also be performed if the throttle body is cleaned.**

➡**Be sure to perform this procedure after reconnecting the battery cable and after replacing the ECM.**

   a. Disconnect the cable from the negative (-) battery terminal. Wait at least 60 seconds and reconnect the cable.
   b. Turn the ignition switch to ON without operating the accelerator pedal.

➡**If the accelerator pedal is operated, perform the above steps again.**

   c. Connect the Techstream to the DLC3 and clear the DTCs.
   d. Start the engine, and check that the MIL is not illuminated and that the idle speed is within the specified range when the A/C is switched off after the engine is warmed up. Engine idle speed with the A/C switched off should be 600-700 rpm.

➡**Be sure to perform this step with all accessories off.**

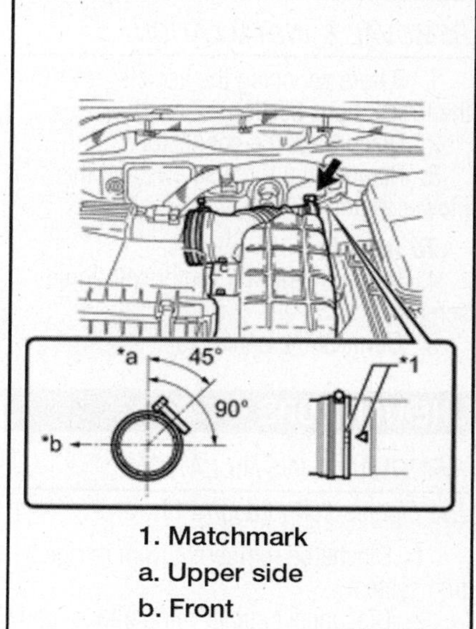

Fig. 203 Installing the air cleaner

➡Make sure that the shift lever is in neutral.

e. Enter the following menus: Powertrain / Engine and ECT / Data List / Throttle Sensor Position. Fully depress the accelerator pedal and check that the value is 60% or more.

f. Perform a road test and confirm that there are no abnormalities.

4.0L Engine

1. Remove the upper radiator support seal.
2. Remove the front bumper cover lower.
3. Remove the No. 1 engine under cover sub assembly.
4. Drain the engine coolant.
5. Remove the v-bank cover.
6. Remove the No. 1 air cleaner hose.
7. Remove the throttle body with the motor assembly.

a. Disconnect the No. 5 water by pass hose.

b. Disconnect the No. 4 water by pass hose.

c. Disconnect the throttle position sensor and throttle control motor connector.

d. Remove the 4 bolts, throttle body with motor and gasket.

To install:

8. Install the throttle body with the motor assembly with the 4 bolts. Tighten to 7 ft. lbs. (10 Nm).

a. Connect the throttle position sensor and throttle control motor connector.

b. Connect the No. 4 water by pass hose.

c. Connect the No. 5 water by pass hose.

9. Install the No. 1 air cleaner hose.
10. Add engine coolant.
11. Inspect for engine coolant leak.
12. Install the No. 1 engine under cover sub assembly.
13. Install the front bumper cover lower.
14. Install the v-bank cover.
15. Install the upper radiator support seal.
16. Perform the initialization.

➡Be sure to perform this procedure after reassembling the throttle body or removing and reinstalling any throttle body component.

➡Perform the following procedure after replacing the ECM, throttle body or any throttle body components. The following procedure should also be performed if the throttle body is cleaned.

➡Be sure to perform this procedure after replacing the ECM and reconnecting the battery cable.

a. Disconnect the EFI and ETCS fuses at the same time. Wait at least 60 seconds, and then reconnect the fuses.

b. Turn the ignition switch to ON without operating the accelerator pedal.

➡If the accelerator pedal is operated, perform the above steps again.

c. Connect the Techstream to the DLC3 and clear the DTCs.

d. Start the engine and check that the MIL is not illuminated and that the idle speed is within the specified range when the A/C is switched off after the engine is warmed up.

The standard idle speed is 690 to 790 rpm.

➡Be sure to perform this step with all accessories off.

➡Make sure that the shift lever is in neutral.

e. Enter the following menus: Powertrain / Engine and ECT / Data List / All Data / Throttle Sensor Position. Fully depress the accelerator pedal and check that the value is 60% or more.

f. Perform a road test and confirm that there are no abnormalities.

HEATING & AIR CONDITIONING SYSTEM

BLOWER MOTOR

REMOVAL & INSTALLATION

1. Before servicing the vehicle, refer to the Precautions section.
2. Disconnect the connector.
3. Remove the three screws and the blower motor.

To install:

4. Install the blower motor with three screws.
5. Connect the connector.

HEATER CORE

REMOVAL & INSTALLATION

See Figures 204 through 211.

1. Discharge refrigerant from refrigeration system.
2. Disconnect cooler refrigerant suction pipe c.
3. Disconnect cooler refrigerant liquid pipe c.
4. Using pliers, grip the claws of the clip and slide the clip and disconnect the heater water outlet hose.
5. Disconnect heater water outlet hose.
6. Remove instrument panel safety pad sub-assembly.
7. Remove air conditioning amplifier assembly.
8. Remove the 2 clips.
9. Release the 2 claw fittings and remove the side defroster nozzle duct No. 1.

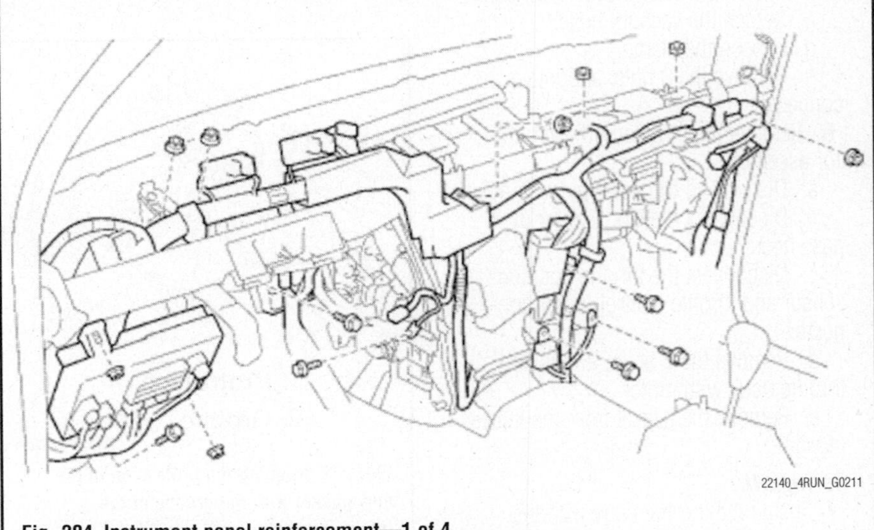

Fig. 204 Instrument panel reinforcement—1 of 4

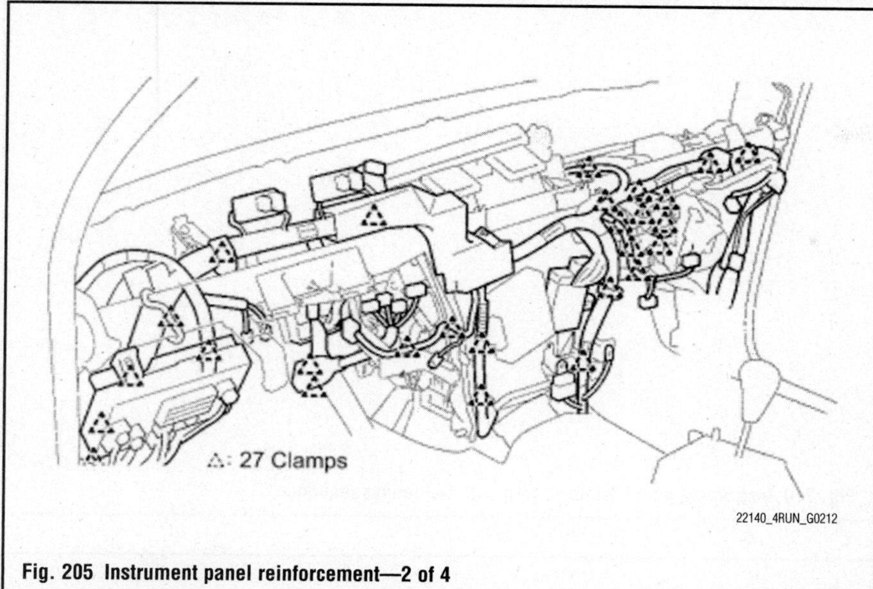

Fig. 205 Instrument panel reinforcement—2 of 4

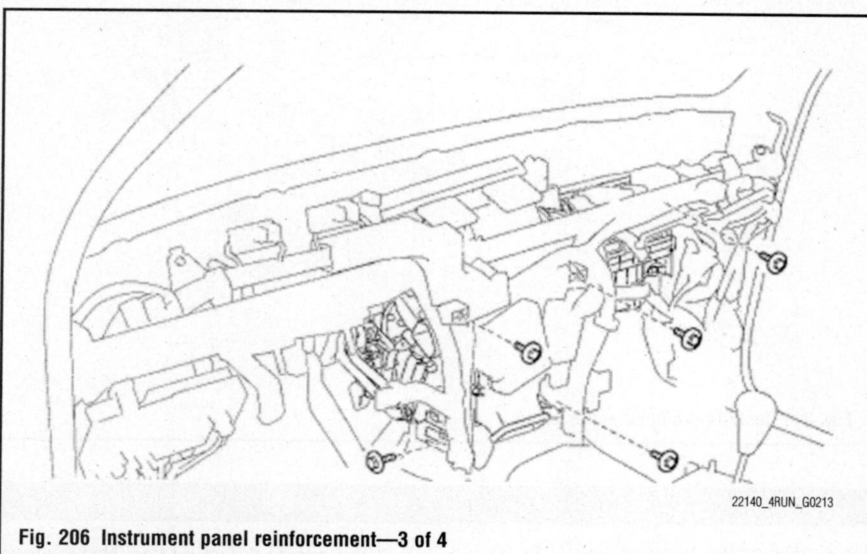

Fig. 206 Instrument panel reinforcement—3 of 4

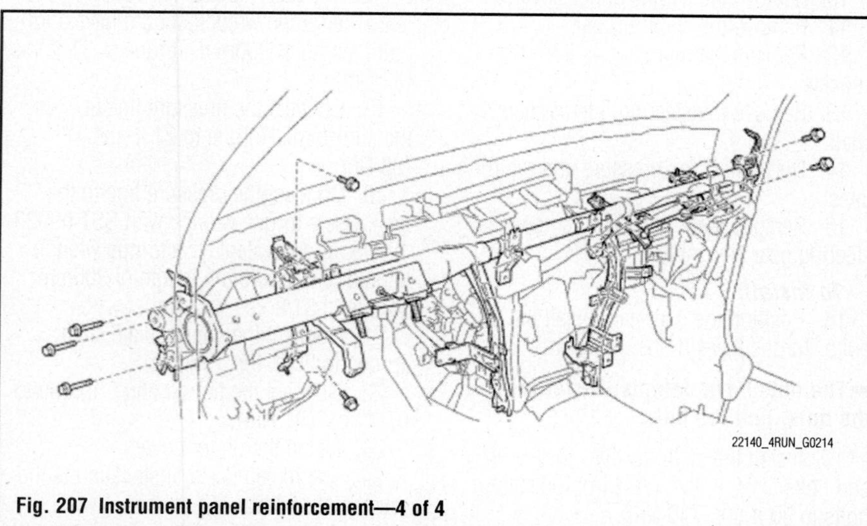

Fig. 207 Instrument panel reinforcement—4 of 4

10. Remove the 2 clips.
11. Release the 2 claw fittings and remove the side defroster nozzle duct No. 2.
12. Remove the screw.
13. Release the 2 pin fittings and remove the heater to register duct No. 1.
14. Remove the screw.
15. Release the 2 pin fittings and remove the heater to register duct No. 3.
16. Release the 6 claw fittings and 3 clamps and remove the air duct rear No. 2.
17. Release the 6 claw fittings and 3 clamps and remove the air duct rear No. 1.
18. Remove the clip and disconnect the console box duct No. 1.
19. Release the 3 claw fittings and remove the air duct No. 1.
20. Release the 3 claw fittings and remove the air duct No. 2.
21. Remove the bolt, nut and instrument panel brace mounting brackets.
22. Release the 3 clamps and disconnect the connector.
23. Remove the 4 nuts and disconnect the steering column assembly.
24. Remove instrument panel reinforcement.
    a. Remove the 6 bolts and 8 nuts.
    b. Release the 27 clamps.
    c. Disconnect the connectors.
    d. Remove the 5 bolts.
    e. Remove the 7 bolts and instrument panel reinforcement.
25. Release the 4 claw fittings and remove the heater to register duct center.
26. Release the 4 claw fittings and remove the defroster nozzle assembly lower.
27. Disconnect the connectors.
28. Remove the 2 nut and air conditioner unit assembly.
29. Remove the 2 screws and air conditioning radiator assembly.

To install:

30. Installation is reverse of removal.
31. Tighten bolts and nuts to the following torque:
    - A/C unit nuts 48 inch lbs. (5.4 Nm)
    - (7) Instrument panel reinforcement bolts 87 inch lbs. (9.8 Nm)
    - (5) Instrument panel reinforcement bolts 87 inch lbs. (9.8 Nm) in the order of the illustration
    - 6 bolts and 8 nuts as shown in illustration
    - Steering column and tighten to 19 ft. lbs. (26 Nm)

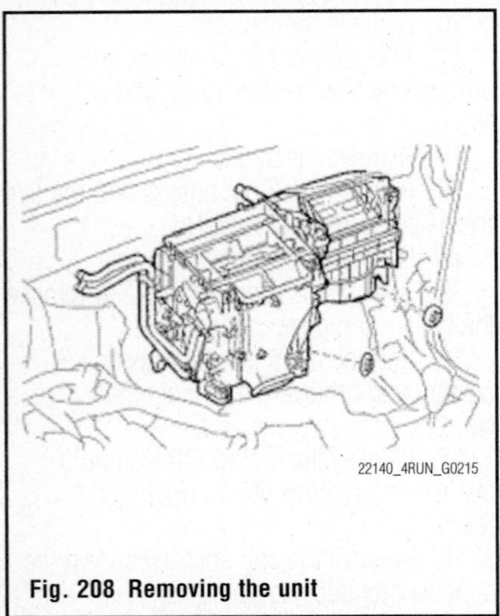

Fig. 208 Removing the unit

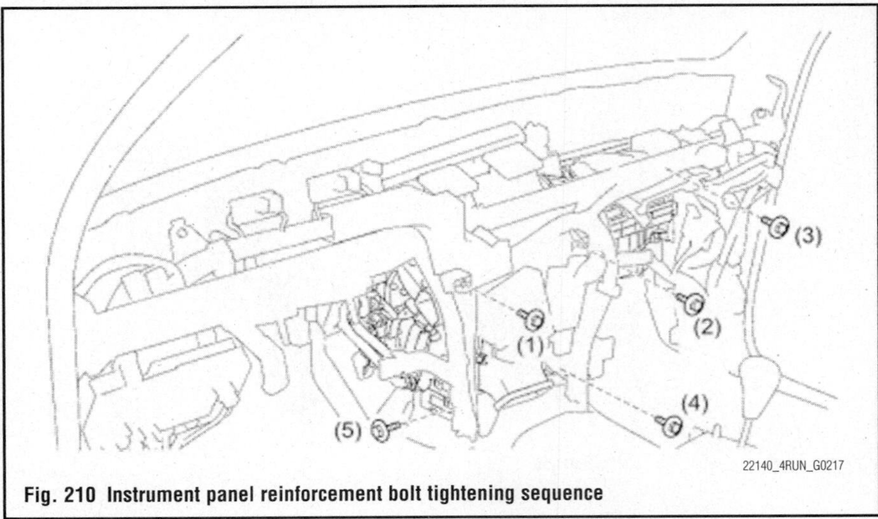

Fig. 210 Instrument panel reinforcement bolt tightening sequence

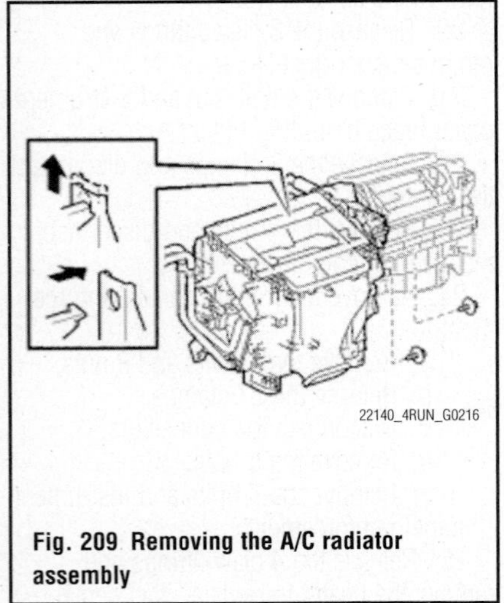

Fig. 209 Removing the A/C radiator assembly

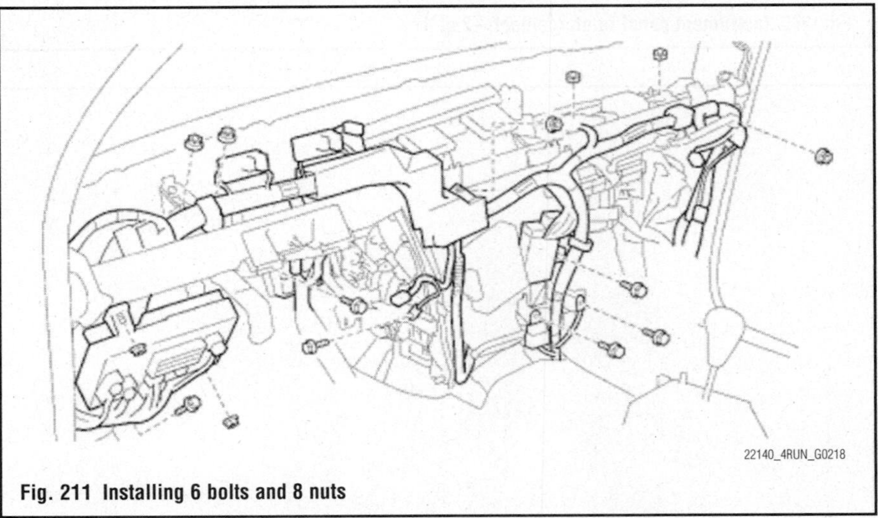

Fig. 211 Installing 6 bolts and 8 nuts

STEERING

POWER RACK & PINION STEERING GEAR

REMOVAL & INSTALLATION

See Figure 212.

1. Before servicing the vehicle, refer to the precautions section.
2. Disconnect the battery ground cable.
3. Place the front wheels in the straight ahead position.
4. Remove the horn pad.
5. Remove the steering wheel.
6. Remove the lower steering column cover.
7. Remove the turn signal switch.
8. Remove the spiral cable assembly.
9. Remove the front wheels.
10. Remove the engine under-covers.
11. Remove the stabilizer bar.
12. Remove the tie rod ends from the knuckle.
13. Remove the steering intermediate shaft.
14. Disconnect the pressure and return lines.
15. Remove the 2 bolts and remove the steering gear assembly.

To install:

16. Position the gear and install the 2 bolts. Torque to 74 ft. lbs. (100 Nm).

➡The nuts have detents. Never turn the nuts, just the bolts.

17. Install the stabilizer bar. Torque the end links to 52 ft. lbs. (70 Nm); the clamp bolts to 30 ft. lbs. (40 Nm).
18. Connect the return line. Use a torque wrench with SST 09023-12700, or equivalent. The torque wrench should have a fulcrum length of 300mm. Torque to 31 ft. lbs. (42 Nm).
19. Connect the pressure line at the sub-frame. Torque to 21 ft. lbs. (28 Nm).
20. Connect the pressure line to the gear. Use a torque wrench with SST 09023-12700, or equivalent. The torque wrench should have a fulcrum length of 300mm. Torque to 31 ft. lbs. (42 Nm).
21. Connect the intermediate shaft. Torque to 26 ft. lbs. (36 Nm).
22. Connect the tie rod ends. Torque to 67 ft. lbs. (91 Nm).
23. Install the under-covers.
24. The remainder of installation is the reverse of removal.

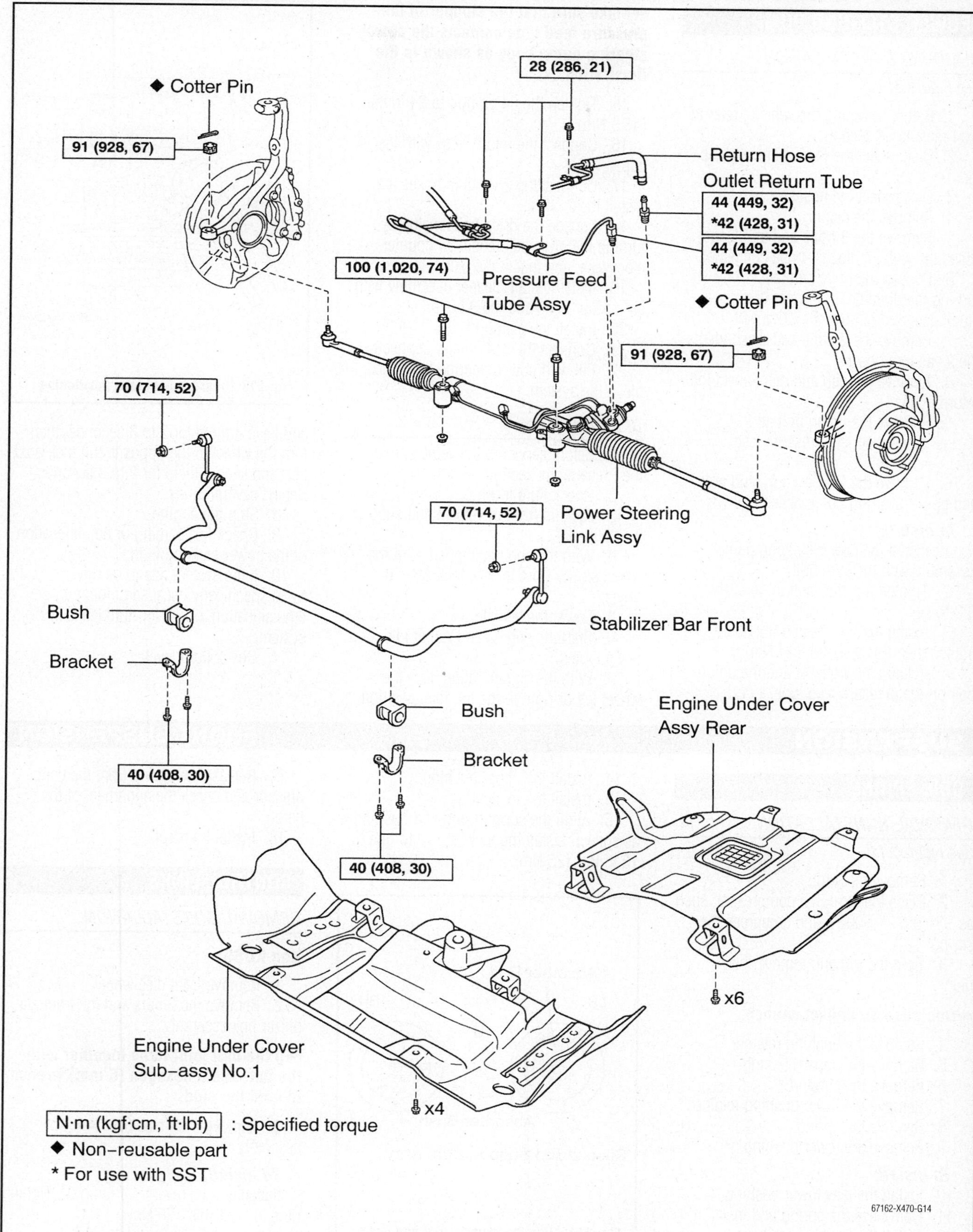

28 (286, 21)
◆ Cotter Pin
91 (928, 67)
Return Hose
Outlet Return Tube
44 (449, 32)
*42 (428, 31)
44 (449, 32)
*42 (428, 31)
100 (1,020, 74)
Pressure Feed
Tube Assy
◆ Cotter Pin
91 (928, 67)
70 (714, 52)
70 (714, 52)
Power Steering
Link Assy
Bush
Stabilizer Bar Front
Bracket
Bush
Engine Under Cover
Assy Rear
Bracket
40 (408, 30)
40 (408, 30)
x6
Engine Under Cover
Sub–assy No.1
x4
N·m (kgf·cm, ft·lbf) : Specified torque
◆ Non–reusable part
* For use with SST
67162-X470-G14

Fig. 212 Steering gear and related parts

POWER STEERING PUMP

REMOVAL & INSTALLATION

See Figure 213.

1. Before servicing the vehicle, refer to the Precautions section.
2. Disconnect the MAF meter connector.
3. Disconnect the hoses.
4. Remove the clamp.
5. Remove the 3 bolts and air cleaner assembly with air cleaner hose connected.
6. Loosen the drive belt tension by turning the drive belt tensioner counterclockwise, and remove the drive belt.
7. Remove the 2 clips and disconnect the 2 vacuum hoses.
8. Remove the clip and disconnect the return hose.
9. Remove the union bolt and gasket, disconnect the pressure feed tube.
10. Remove the 2 bolts, nut, stud bolt and power steering pump assembly.

To install:

11. Install the power steering pump assembly with the stud bolt.
12. Tighten the stud bolt to 16 ft. lbs. (22 Nm).
13. Install the 2 bolts and nut and tighten them to 33 ft. lbs. (44 Nm).
14. Install a new gasket and the union bolt on the pressure feed tube.

➡Make sure that the stopper of the pressure feed tube contacts the power steering pump body as shown in the illustration.

15. Tighten the union bolt to 34 ft. lbs. (46.5 Nm).
16. Connect the return hose with the clip.
17. Connect the 2 vacuum hoses and install the 2 clips.
18. Loosen the drive belt tension by turning the drive belt tensioner counterclockwise, and install the belt.
19. Install the air cleaner assembly with air cleaner hose and the 3 bolts.
20. Install the clamp.
21. Connect the MAF meter connector.
22. Fill with power steering fluid and bleed the system.

BLEEDING

1. Before servicing the vehicle, refer to the Precautions section.
2. Check fluid level.
3. Jack up front of vehicle and support it with stands.
4. With the engine stopped, turn the wheel slowly from lock to lock several times.
5. Lower the vehicle.
6. Start the engine and run at idle for a few minutes.
7. With the engine idling, turn the wheel left or right to the full lock position and keep it there for 2 to 3 seconds, then turn the wheel to the opposite full lock position and keep it there for 2 to 3 seconds. Repeat several times.
8. Stop the engine.
9. Check for foaming or emulsification of the power steering fluid.
10. If the system has to be bled twice specifically because of foaming or emulsification, check for fluid leaks in the system.
11. Check fluid level.

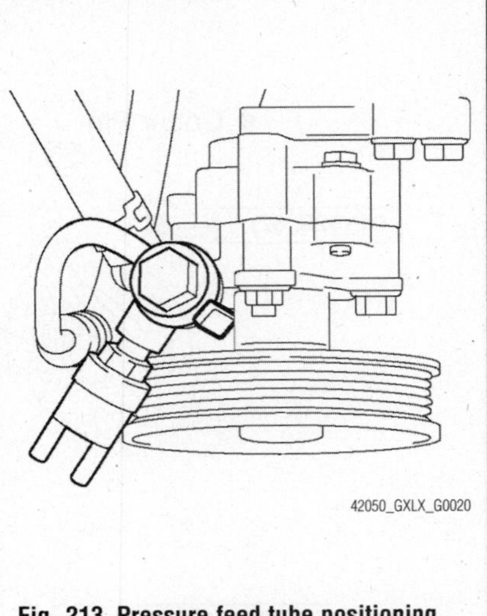

Fig. 213 Pressure feed tube positioning

SUSPENSION

FRONT SUSPENSION

COIL SPRING

REMOVAL & INSTALLATION

See Figure 214.

1. Remove the strut.
2. Place the strut in a compressor, such as SST 09727-30021, and compress the spring.
3. Hold the rod and remove the nut.

➡Don't use an impact wrench.

4. Remove the bushing retainer
5. Remove the upper bushing.
6. Remove the support.
7. Remove the lower bushing retainer.
8. Remove the spring.
9. Remove the lower bushing.

To install:

10. Install the new lower bushing.
11. Compress the spring and install it.
12. Install the bushing retainer.
13. Install the suspension support.
14. Install the upper bushing.
15. Install the retainer.
16. Align the support, rod and bushing as shown. Install the locknut and torque to 18 ft. lbs. (25 Nm).

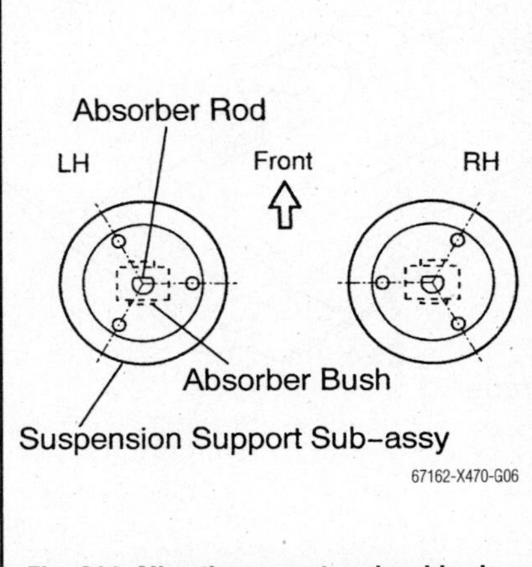

Fig. 214 Align the support, rod and bushing as shown

17. Release the spring from the compressor and check the alignment of the parts.
18. Install the strut.

CONTROL LINKS

REMOVAL & INSTALLATION

Non KDSS

1. Remove front disc wheel.
2. Remove the 2 nuts and the front stabilizer link assembly.

➡If the ball joint turns together with the nut, use a hexagon (6 mm) wrench to hold the stud.

3. Remove front stabilizer link assembly.

To install:

Installation is reverse of removal. Tighten nuts to 52 ft. lbs. (70 Nm).

LOWER BALL JOINT

REMOVAL & INSTALLATION

The lower ball joint is serviced with the lower control arm as an assembly.

LOWER CONTROL ARM

REMOVAL & INSTALLATION

See Figure 215.

1. Before servicing the vehicle, refer to the precautions section.
2. Remove the wheel.
3. Support the lower arm with a jack.
4. Remove the lower strut bolt.
5. Remove the 2 bolts and separate the lower ball joint attachment from the knuckle.
6. Place matchmarks on the camber adjusting cam and toe adjusting cam.
7. Remove the 2 nuts and remove the arm along with the cams.

To install:

8. Installation is the reverse of removal. Align all matchmarks. Use new nuts and cotter pins. Don't fully tighten the control arm bolts until the vehicle is on the ground and the suspension jounced a few times. Observe the following torques:
   - Lower ball joint stud: 103 ft. lbs. (140 Nm)
   - Lower ball joint attachment bolts: 166 ft. lbs. (225 Nm)
   - Lower arm bolts: 100 ft. lbs. (135 Nm)

MACPHERSON STRUT

REMOVAL & INSTALLATION

Non REAS Suspension

See Figure 216.

1. Before servicing the vehicle, refer to the precautions section.
2. Remove the wheel.
3. Remove the stabilizer bar.
4. Remove the clamps and connector.
5. Remove the wire bracket.
6. Remove the lower strut bolt.
7. Remove the 3 upper strut nuts.
8. Remove the strut.

To install:

9. Installation is the reverse of removal. Do not fully tighten the lower strut bolt until the vehicle is resting on the ground and the suspension has been jounced a few times. Observe the following torques:
   - Upper nuts: 47 ft. lbs. (64 Nm)
   - Bracket nut: 11 ft. lbs. (15 Nm)
   - Stabilizer bar links: 52 ft. lbs. (70 Nm)
   - Wheel: 83 ft. lbs. (112 Nm)
   - Lower strut bolt: 100 ft. lbs. (135 Nm)

Rear Suspension

See Figures 217 and 218.

1. For REAS suspension follow the instructions above and install as follows:
   a. Install the bolt.

➡Be sure to fit the detents attached to the bracket into a hole on the frame side.

2. As shown in the illustration, tighten the nut of clearance to standard value. Tighten to 18 ft. lbs. (25 Nm).

OVERHAUL

Coil spring removal and installation and any other strut disassembly should go here

STABILIZER BAR

REMOVAL & INSTALLATION

Without KDSS

1. Remove front disc wheel.
2. Remove the 2 nuts and the front stabilizer link assembly.
3. Remove front stabilizer link assembly.
4. Remove the 2 bolts and front stabilizer bracket
5. Remove the 2 front stabilizer bar bush.
6. Remove stabilizer bar front.

To install:

7. Install the 2 front stabilizer bar bush.

➡Install the bushing to the inner side of the bushing stopper on the stabilizer bar. Install the stabilizer bush No. 1 as the protrusion to be on the inner side of the vehicle.

8. Install the front stabilizer bracket and tighten to 30 ft. lbs. (40 Nm).
9. Install the front stabilizer link assemblies.
10. Install front wheels.

UPPER BALL JOINT

REMOVAL & INSTALLATION

The upper ball joint is serviced with the upper control arm as an assembly.

UPPER CONTROL ARM

REMOVAL & INSTALLATION

See Figure 219.

1. Before servicing the vehicle, refer to the precautions section.
2. Remove the wheel.
3. Disconnect the skid control wire.
4. Support the lower arm with a jack.
5. Remove the cable bracket.
6. Disconnect the ball joint from the knuckle.
7. Remove the through-bolt, washers and nut.
8. Remove the arm.

To install:

9. Installation is the reverse of removal. Don't fully tighten the through-bolt until the vehicle is on the ground and the suspension is jounced a few times.
   - Ball joint nut: 81 ft. lbs. (110 Nm)
   - Through-bolt: 85 ft. lbs. (115 Nm)

WHEEL BEARINGS

REMOVAL & INSTALLATION

1. Remove the wheel.
2. Remove the caliper.
3. Remove the hub grease cap.
4. Remove the cotter pin.
5. Remove the hub nut.
6. Remove the speed sensor.
7. Remove the stabilizer links from the knuckles.
8. Remove the tie rod end from the knuckle.
9. Remove the lower arm from the knuckle.
10. Remove the upper arm from the knuckle.
11. Remove the hub/knuckle assembly from the shaft.
12. Mount the assembly in a vise.
13. Remove the knuckle oil seal.
14. Remove the 4 bolts and remove the hub assembly from the knuckle.
15. Using SST 09710-30021 and its components, remove the bearing from the hub.
16. Remove the oil seal.

To install:

17. Using a seal driver, install a new seal.

➡Take care to avoid damage to the spacer.

18. Press a new bearing into the hub.
19. Coat a new O-ring with MP grease and install it in the hub.
20. Attach the hub to the knuckle. Torque to 59 ft. lbs. (80 Nm).
21. Install a new knuckle oil seal.
22. The remainder of installation is the reverse of removal. Observe the following torques:

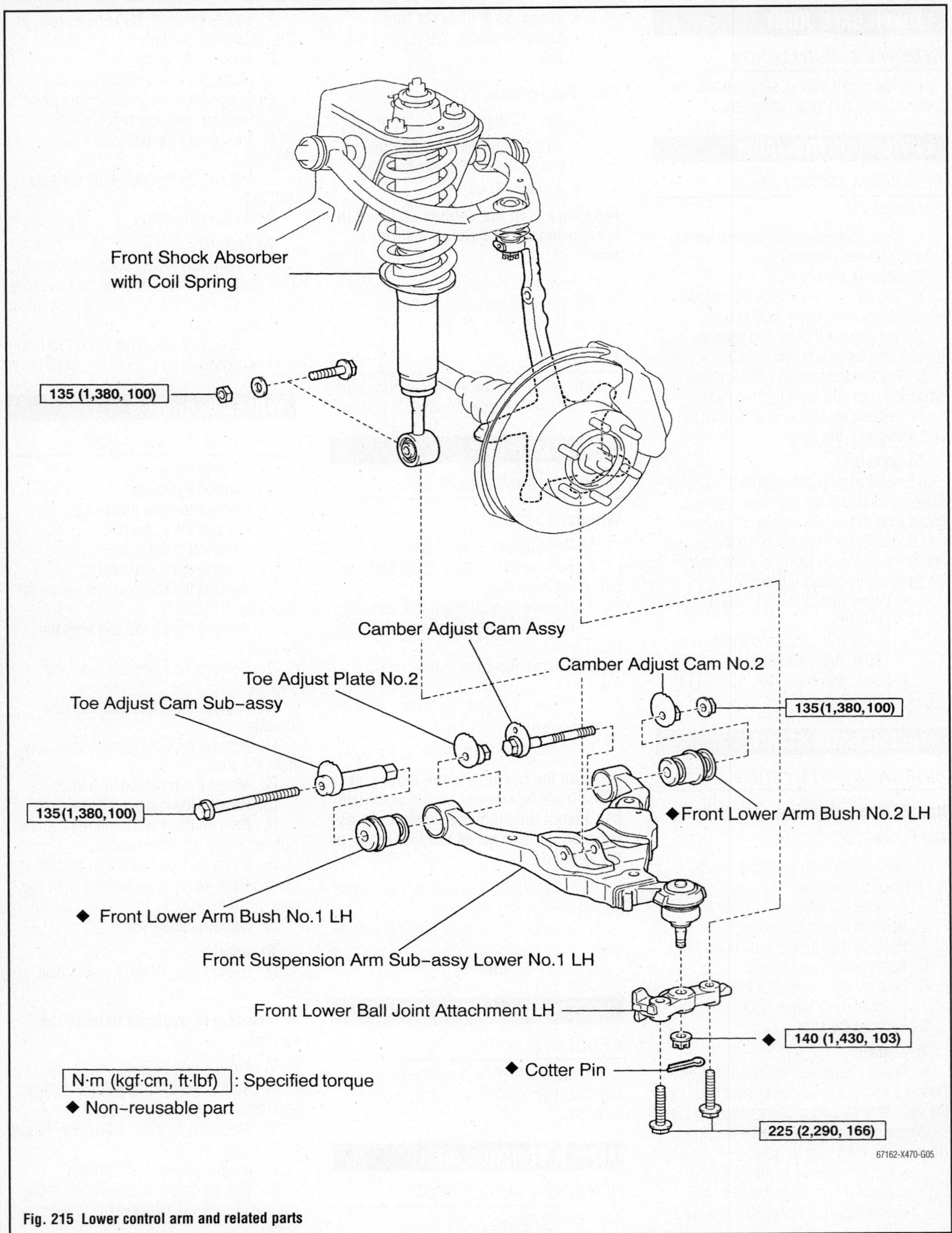
Front Shock Absorber with Coil Spring
135 (1,380, 100)
Camber Adjust Cam Assy
Camber Adjust Cam No.2
Toe Adjust Plate No.2
Toe Adjust Cam Sub-assy
135(1,380,100)
135(1,380,100)
◆Front Lower Arm Bush No.2 LH
◆ Front Lower Arm Bush No.1 LH
Front Suspension Arm Sub-assy Lower No.1 LH
Front Lower Ball Joint Attachment LH
◆ 140 (1,430, 103)
◆ Cotter Pin
N·m (kgf·cm, ft·lbf) : Specified torque
◆ Non-reusable part
225 (2,290, 166)
67162-X470-G05

Fig. 215 Lower control arm and related parts

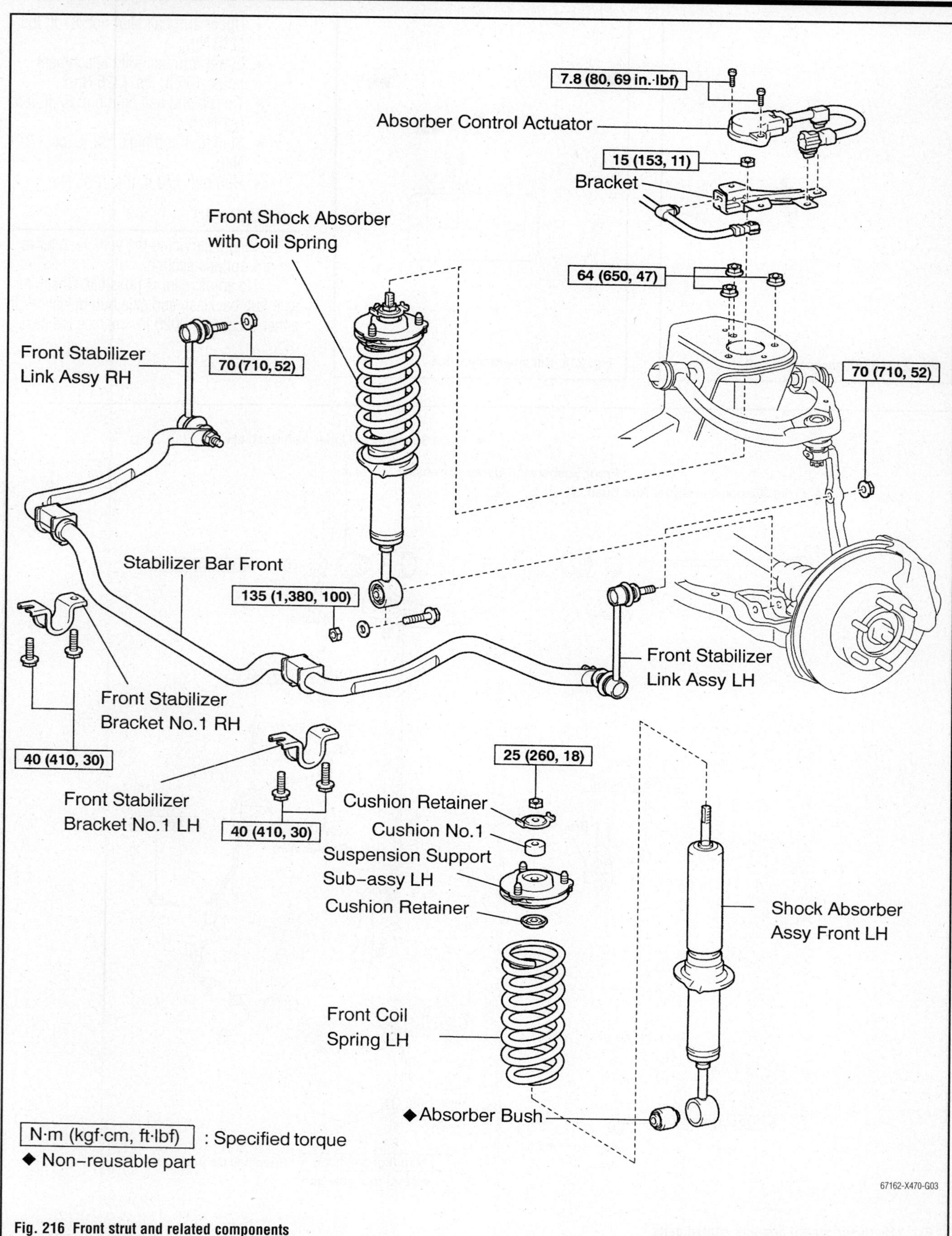
7.8 (80, 69 in.·lbf)
Absorber Control Actuator
15 (153, 11)
Bracket
Front Shock Absorber with Coil Spring
64 (650, 47)
Front Stabilizer Link Assy RH
70 (710, 52)
70 (710, 52)
Stabilizer Bar Front
135 (1,380, 100)
Front Stabilizer Link Assy LH
Front Stabilizer Bracket No.1 RH
40 (410, 30)
25 (260, 18)
Front Stabilizer Bracket No.1 LH
40 (410, 30)
Cushion Retainer
Cushion No.1
Suspension Support Sub-assy LH
Cushion Retainer
Shock Absorber Assy Front LH
Front Coil Spring LH
◆ Absorber Bush
N·m (kgf·cm, ft·lbf) : Specified torque
◆ Non-reusable part
67162-X470-G03

Fig. 216 Front strut and related components

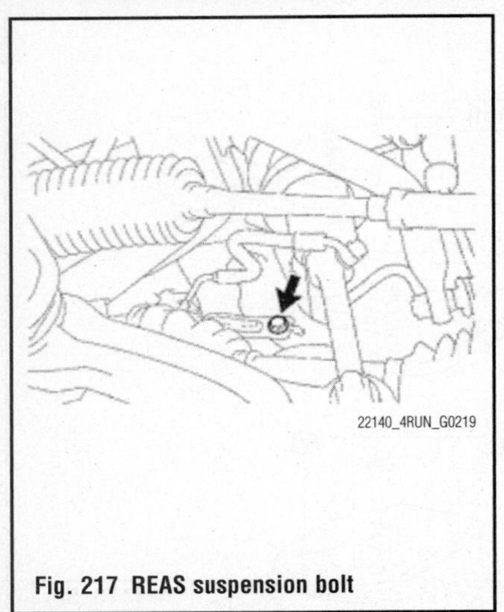

Fig. 217 REAS suspension bolt

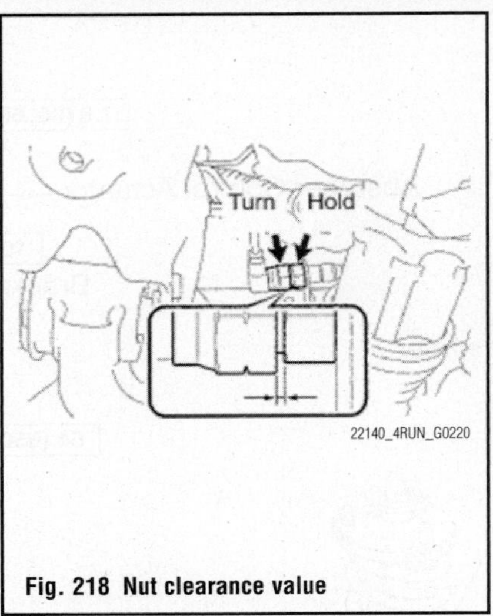

Fig. 218 Nut clearance value

- Upper arm ball stud nut: 81 ft. lbs. (110 Nm)
- Lower arm ball joint attachment bolts: 166 ft. lbs. (225 Nm)
- Tie rod end ball stud nut: 67 ft. lbs. (91 Nm)
- Stabilizer end links: 52 ft. lbs. (70 Nm)
- Hub nut: 173 ft. lbs. (235 Nm)

ADJUSTMENT

1. Before servicing the vehicle, refer to the precautions section.
2. No adjustment is possible. Check for axle hub backlash and axle hub deviation. If either exceeds 0.0020 in., replace the bearing.

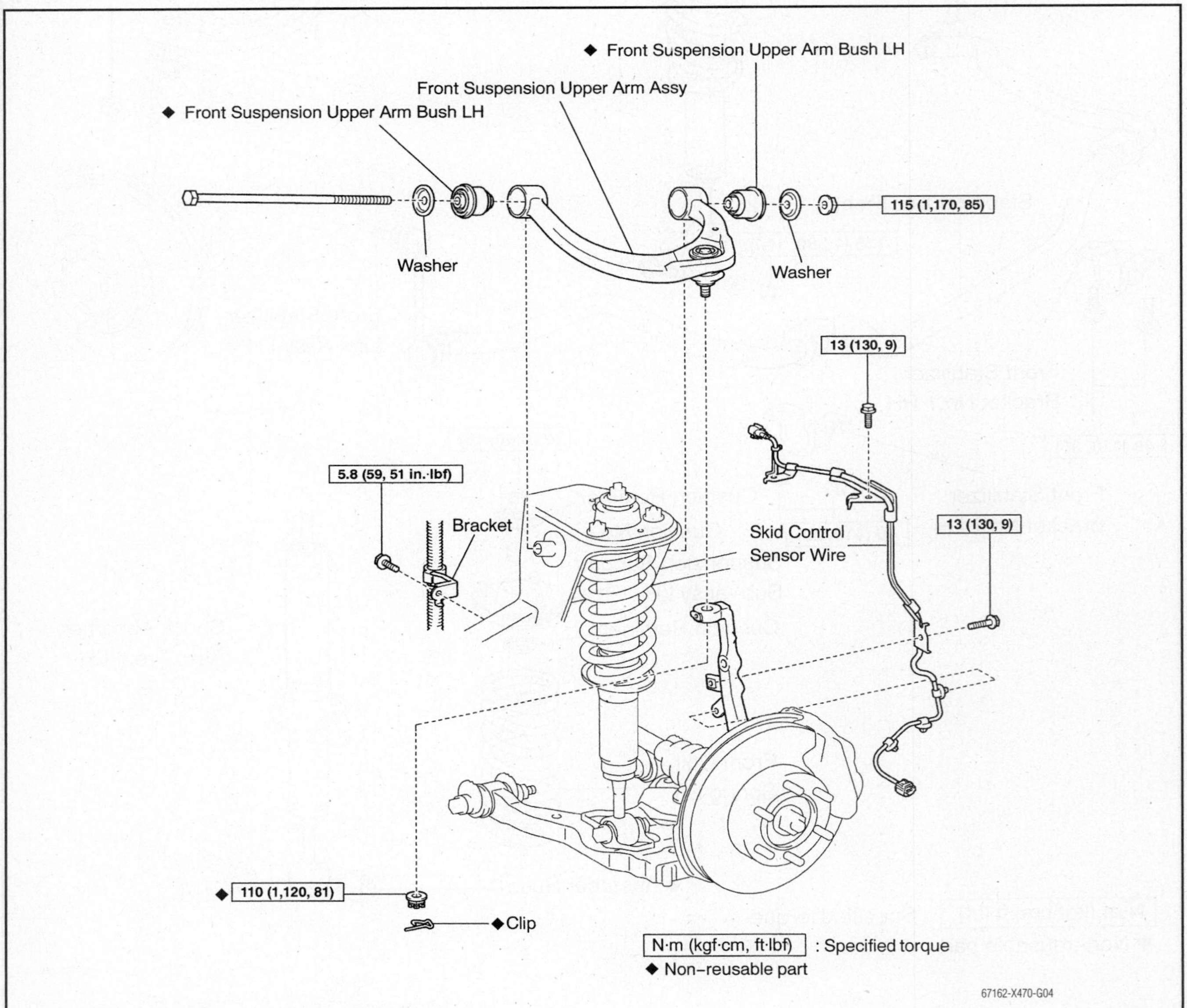

Fig. 219 Upper control arm and related parts

SHOCK ABSORBER

REMOVAL & INSTALLATION

1. Before servicing the vehicle, refer to the precautions section.
2. Support the axle with a jackstand.
3. Disconnect the actuator at the shock absorber.

➡Don't over-extend the pneumatic shock.

4. Remove the lower shock bolt.
5. Remove the upper nut and remove the shock.

To install:

6. Installation is the reverse of removal. Don't fully tighten the lower bolt until the vehicle is on the ground and the suspension jounced a few times. Torque the upper nut to 18 ft. lbs. (25 Nm); the lower bolt to 72 ft. lbs. (98 Nm).

WHEEL BEARINGS

REMOVAL & INSTALLATION

See Figures 220 through 226.

1. Remove rear wheel.
2. Remove rear speed sensor (w/ ABS).
3. Separate rear disc brake caliper assembly.
4. Remove rear disc.
5. Remove parking brake shoe return tension spring.
6. Remove parking brake shoe strut compression spring.
7. Remove parking brake shoe strut.
8. Remove parking brake shoe.
9. Remove rear axle shaft with backing plate.
10. Remove the 4 nuts and rear axle shaft with backing plate.
11. Remove the O-ring.
12. Using a snap ring expander, remove the snap ring.
13. Remove the rear axle shaft from bearing.
14. Remove the rear axle bearing retainer inner from the rear axle bearing assembly.
15. Remove the rear axle shaft washer from the rear axle bearing assembly.
16. Attach the 4 nuts to the parking brake plate to rear axle housing bolts.
17. Using a hammer, remove the 4 parking brake plate to rear axle housing bolts and rear axle bearing assembly.

➡Do not reuse the nuts previously removed from the vehicle.

18. Remove the 6 hub bolts.

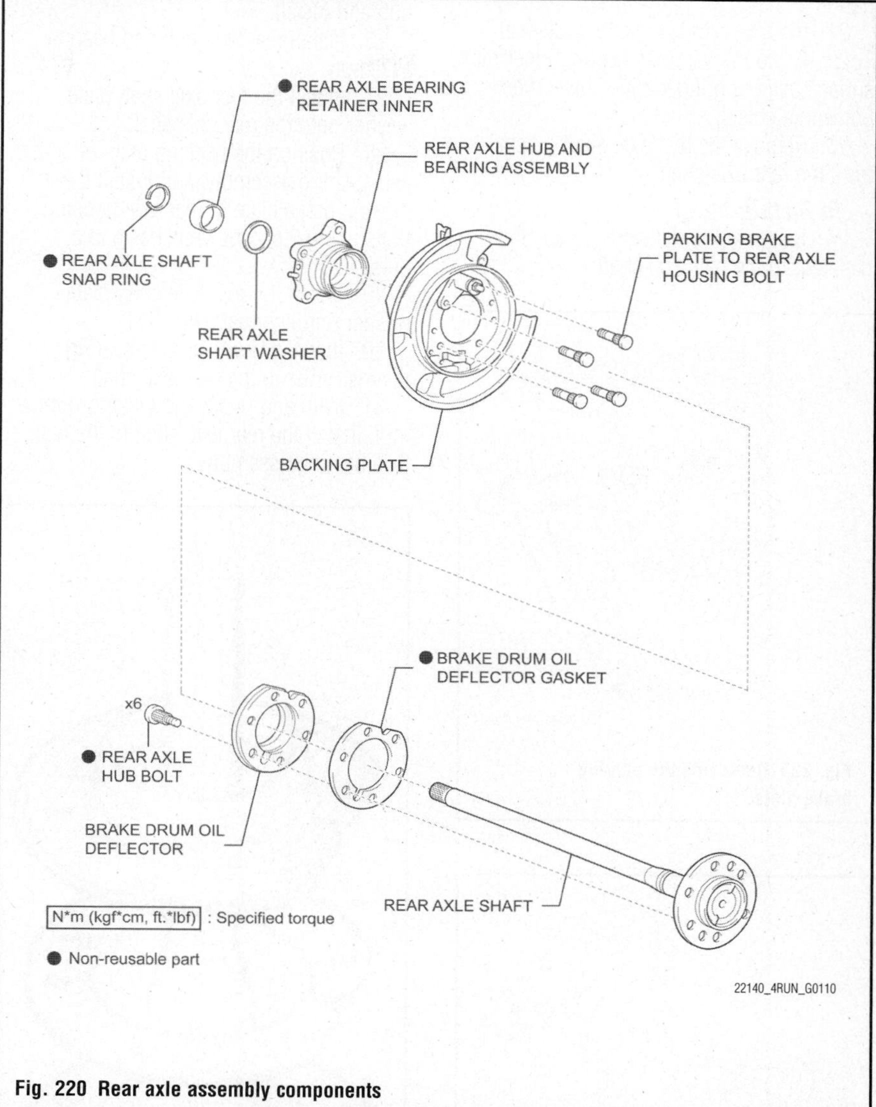

Fig. 220 Rear axle assembly components

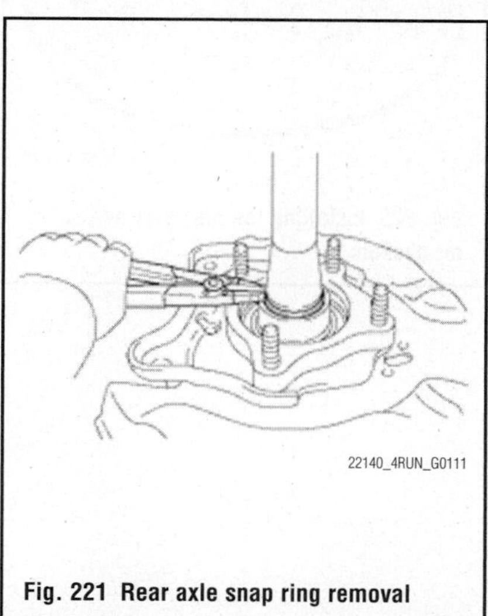

Fig. 221 Rear axle snap ring removal

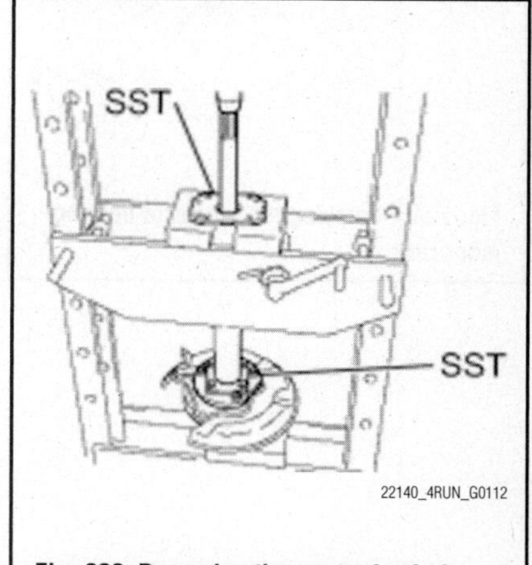

Fig. 222 Removing the rear axle shaft from the bearing

19. Remove brake drum oil deflector.
20. Remove brake drum oil deflector gasket.
21. Remove rear axle bearing oil seal.
22. Grind the rear axle bearing inner race surface using a grinder, then chisel them out with a chisel.
23. Remove the rear axle shaft oil seal from the rear axle shaft.

To install:

24. Install a new deflector gasket and deflector to the rear axle shaft.

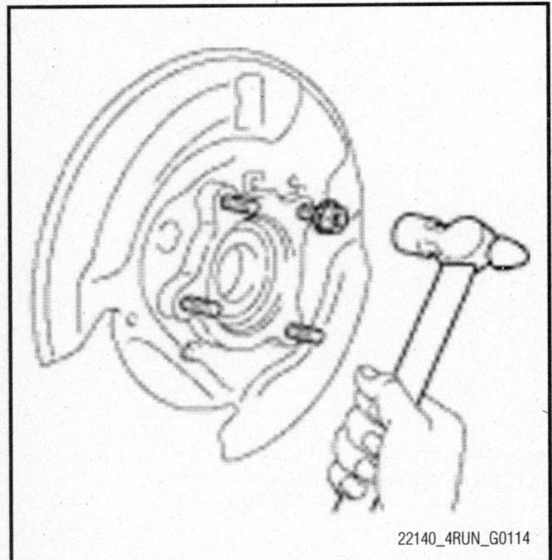

Fig. 223 Removing the parking brake plate

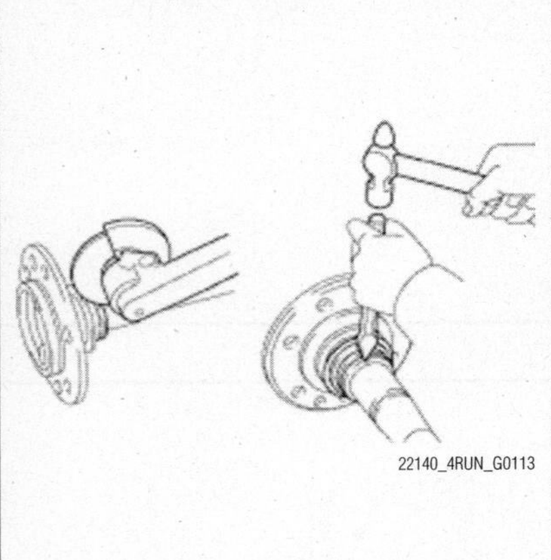

Fig. 224 Removing the rear axle bearing inner race

➡Align the 2 notches.

25. Pass the 6 bolts through the axle hub and install.
26. Install rear axle hub and bearing assembly.
27. Install the rear axle shaft plate washer onto the rear axle shaft.
28. Position the backing plate on a rear axle bearing assembly, and install the 4 parking brake plate to rear axle housing bolts using 2 socket wrenches and a press.
29. Install the rear axle shaft plate washer onto the rear axle shaft.
30. Install a new rear axle bearing retainer inner to the rear axle shaft.
31. With and press and the appropriate tool, install the rear axle shaft to the rear axle bearing assembly.

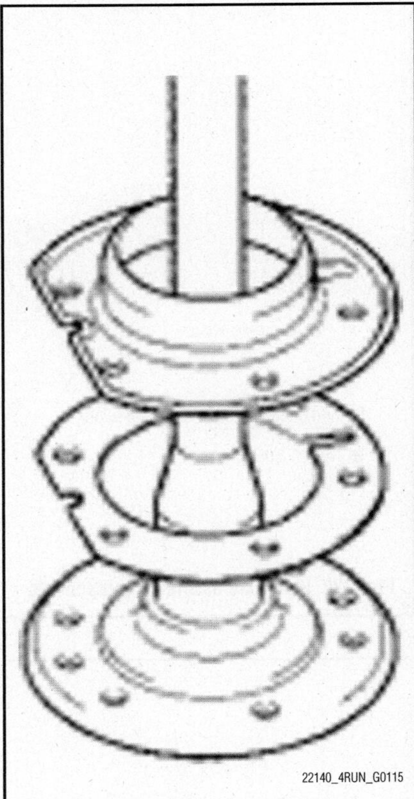

Fig. 225 Installing the rear axle deflector gaskets

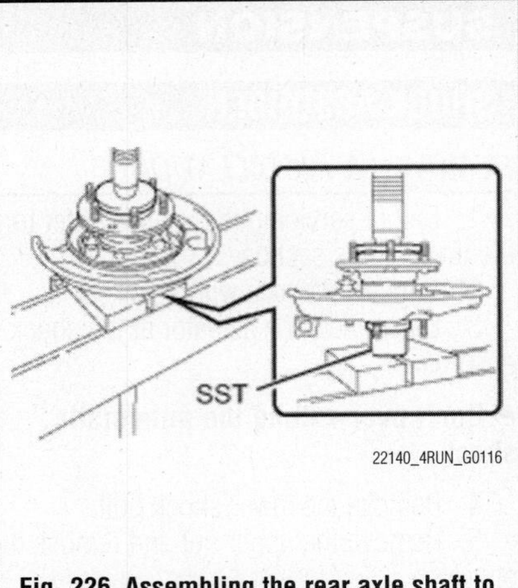

Fig. 226 Assembling the rear axle shaft to axle bearing assembly

➡Do not damage the speed sensor rotor.

32. Using a snap ring expander, install a new rear axle shaft snap ring.
33. Install a new O-ring.
34. Install the rear axle shaft with backing plate with the 4 nuts, tighten to 89 ft. lbs. (120 Nm).

➡Do not damage the speed sensor rotor. Inspect no damage and no foreign matter at the speed sensor rotor.

35. Install parking brake shoe.
36. Install parking brake shoe strut.
37. Install parking brake shoe strut compression spring.
38. Install parking brake shoe return tension spring.
39. Install rear disc.
40. Connect rear disc brake caliper assembly.
41. Install rear speed sensor (w/ ABS).
42. Fill up differential oil as necessary.
43. Inspect brake fluid level in reservoir.
44. Inspect brake fluid leakage.
45. Install rear wheel tighten to 83 ft. lbs. (112 Nm).
46. Inspect and adjust parking brake lever travel.
47. Inspect ABS speed sensor signal (w/ ABS).

TOYOTA

Avalon

4

SPECIFICATIONS AND MAINTENANCE CHARTS

ENGINE AND VEHICLE IDENTIFICATION

Engine						
Code ①	Liters (cc)	Cu. In.	Cyl.	Fuel Sys.	Engine Type	Eng. Mfg.
2GR-FE	3.5 (3456)	210	6	SFI	DOHC	Toyota

Model Year	
Code ②	Year
B	2011
C	2012

SFI: Sequential Fuel Injection

DOHC: Double Overhead Camshaft

NA: Information not available

① Stamped on the left side of the engine block

② 10th digit of the Vehicle Identification Number (VIN)

71099_AVAL_C0001

GENERAL ENGINE SPECIFICATIONS

Year	Model	Engine Displacement Liters	Engine Series ID	Net Horsepower @ rpm	Net Torque @ rpm (ft. lbs.)	Bore x Stroke (in.)	Com-pression Ratio	Oil Pressure (psi) @ rpm
2011	Avalon	3.5	2GR-FE	270@6200	251@4700	3.70x3.27	10.8:1	36-78@3000
2012	Avalon	3.5	2GR-FE	270@6200	251@4700	3.70x3.27	10.8:1	36-78@3000

71099_AVAL_C0002

ENGINE TUNE-UP SPECIFICATIONS

Year	Engine Displacement Liters	Engine ID	Spark Plug Gap (in.)	Ignition Timing (deg.)	Fuel Pump (psi)	Idle Speed (rpm)	Valve Clearance	
							Intake	Exhaust
2011	3.5	2GR-FE	0.039-0.043	NA	44-50	650-750	NA	NA
2012	3.5	2GR-FE	0.039-0.043	NA	44-50	650-750	NA	NA

NOTE: The Vehicle Emission Control Information label often reflects specification changes made during production.

The label figures must be used if they differ from those in this chart.

NA: Not available

71099_AVAL_C0003

CAPACITIES

Year	Model	Engine Displacement Liters	Engine ID	Engine Oil with Filter (qts.)	Transmission (qts.) 5-Spd	Transmission (qts.) Auto.*	Transfer Case (pts.)	Drive Axle Front (pts.)	Drive Axle Rear (pts.)	Fuel Tank (gal.)	Cooling System (qts.)
2011	Avalon	3.5	2GR-FE	6.4	—	6.9	—	—	—	18.5	8.8
2012	Avalon	3.5	2GR-FE	6.4	—	6.9	—	—	—	18.5	8.8

71099_AVAL_C0004

FLUID SPECIFICATIONS

Year	Model	Engine Displacement Liters	Engine ID/VIN	Engine Oil	Auto. Trans.	Drive Axle	Power Steering Fluid	Brake Master Cylinder
2011	Avalon	3.5	2GR-FE	5W-30	NA	—	ATF Dexron® II or III	DOT 3
2012	Avalon	3.5	2GR-FE	5W-30	NA	—	ATF Dexron® II or III	DOT 3

DOT: Department Of Transportation

NA: Not Available

71099_AVAL_C0005

VALVE SPECIFICATIONS

Year	Engine Displacement Liters	Engine ID	Seat Angle (deg.)	Face Angle (deg.)	Spring Test Pressure (lbs. @ in.)	Spring Installed Height (in.)	Stem-to-Guide Clearance (in.) Intake	Stem-to-Guide Clearance (in.) Exhaust	Stem Diameter (in.) Intake	Stem Diameter (in.) Exhaust
2011	3.5	2GR-FE	45	44.5	NA	NA	0.0010-0.0024	0.0012-0.0026	0.2154-0.2159	0.2152-0.2158
2012	3.5	2GR-FE	45	44.5	NA	NA	0.0010-0.0024	0.0012-0.0026	0.2154-0.2159	0.2152-0.2158

NA: Information not available

71099_AVAL_C0006

CAMSHAFT AND BEARING SPECIFICATIONS CHART

All measurements are given in inches.

Year	Engine Displ. Liters	Engine ID/VIN	Journal Dia.	Brg. Oil Clearance	Shaft End-play	Runout	Journal Bore	Lobe Height Intake	Lobe Height Exhaust
2011	3.5	2GR-FE	①	②	NA	0.0016	NA	1.7447-1.7487	1.7426-1.7465
2012	3.5	2GR-FE	①	②	NA	0.0016	NA	1.7447-1.7487	1.7426-1.7465

NA: Not Available

① No. 1 journal: 1.4152-1.4157

Other Journals: 1.0220-1.0226 in.

② No. 1 journal: 0.0016-0.0031

Other Journals: 1.0010-1.0024 in.

Maximum No.1 Journal: 0.0039 in.

Other Maximum Journals: 0.0035 in.

71099_AVAL_C0007

CRANKSHAFT AND CONNECTING ROD SPECIFICATIONS

All measurements are given in inches.

Year	Engine Displacement Liters	Engine ID	Crankshaft Main Brg. Journal Dia.	Crankshaft Main Brg. Oil Clearance	Crankshaft Shaft End-play	Crankshaft Thrust on No.	Connecting Rod Journal Diameter	Connecting Rod Oil Clearance	Connecting Rod Side Clearance
2011	3.5	2GR-FE	2.4011-2.4016	0.0010-0.0019	0.0016-0.0094	2	2.0863-2.0866	0.0018-0.0026	0.0059-0.0157
2012	3.5	2GR-FE	2.4011-2.4016	0.0010-0.0019	0.0016-0.0094	2	2.0863-2.0866	0.0018-0.0026	0.0059-0.0157

71099_AVAL_C0008

PISTON AND RING SPECIFICATIONS

All measurements are given in inches.

Year	Engine Displ. Liters	Engine ID	Piston Clearance	Ring Gap Top Comp.	Ring Gap Bottom Comp.	Ring Gap Oil Control	Ring Side Clearance Top Comp.	Ring Side Clearance Bottom Comp.	Ring Side Clearance Oil Control
2011	3.5	2GR-FE	0.0018-0.0020	0.0098-0.0138	0.0197-0.0236	0.0039-0.0157	0.0008-0.0028	0.0008-0.0024	0.0028-0.0059
2012	3.5	2GR-FE	0.0018-0.0020	0.0098-0.0138	0.0197-0.0236	0.0039-0.0157	0.0008-0.0028	0.0008-0.0024	0.0028-0.0059

71099_AVAL_C0009

TORQUE SPECIFICATIONS

All readings in ft. lbs.

Year	Engine Displacement Liters	Engine ID	Cylinder Head Bolts	Main Bearing Bolts	Rod Bearing Bolts	Crankshaft Damper Bolts	Flywheel Bolts	Manifold Intake	Manifold Exhaust	Spark Plugs	Oil Pan Drain Plug
2011	3.5	2GR-FE	①	②	③	184	61	15	15	13	30
2012	3.5	2GR-FE	①	②	③	184	61	15	15	13	30

① Step 1: 10 mm bolts to 27 ft. lbs.
Step 2: 10mm point cap bolts plus 90 degrees
Step 3: 10mm point cap bolts plus 90 degrees
Step 4: Front bolts to 22 ft. lbs.

③ Step 1: 16 cap bolts to 45 ft. lbs.
Step 2: 16 cap bolts plus 90 degrees
Step 3: 8 side bolts to 38 ft. lbs.

③ Step 1: 18 ft. lbs.
Step 2: Plus 90 degrees

71099_AVAL_C0010

WHEEL ALIGNMENT

Year	Model		Caster Range (+/-Deg.)	Caster Preferred Setting (Deg.)	Camber Range (+/-Deg.)	Camber Preferred Setting (Deg.)	Toe-in (in.)	Steering Axis Inclination (Deg.)
2011	Avalon XLS	Front	0.75	2.70	0.75	-0.72	0+/-0.04	12.33+/-0.75
		Rear	—	—	0.75	-1.22	0.16+/-0.08	—
	Avalon Limited	Front	0.75	2.80	0.75	-0.72	0+/-0.04	12.33+/-0.75
		Rear	—	—	0.75	-1.25	0.16+/-0.08	—
2012	Avalon XLS	Front	0.75	2.70	0.75	-0.72	0+/-0.04	12.33+/-0.75
		Rear	—	—	0.75	-1.22	0.16+/-0.08	—
	Avalon Limited	Front	0.75	2.80	0.75	-0.72	0+/-0.04	12.33+/-0.75
		Rear	—	—	0.75	-1.25	0.16+/-0.08	—

71099_AVAL_C0011

TIRE, WHEEL AND BALL JOINT SPECIFICATIONS

Year	Model	OEM Tires Standard	OEM Tires Optional	Tire Pressures (psi) Front	Tire Pressures (psi) Rear	Wheel Size	Ball Joint Inspection	Lug Nut Torque (ft. lbs.)
2011	Avalon	P215/60R16	P215/55R17	29	29	6.5-JJ	①	76
2012	Avalon	P215/60R16	P215/55R17	29	29	6.5-JJ	①	76

OEM: Original Equipment Manufacturer

PSI: Pounds Per Square Inch

STD: Standard

OPT: Optional

① Replace if any measurable movement is found.

71099_AVAL_C0012

BRAKE SPECIFICATIONS

All measurements in inches unless noted

Year	Model		Brake Disc Original Thickness	Brake Disc Minimum Thickness	Brake Disc Maximum Runout	Minimum Lining Thickness	Brake Caliper Bracket Bolts (ft. lbs.)	Brake Caliper Mounting Bolts (ft. lbs.)
2011	Avalon	F	1.102	0.983	0.0020	0.039	79	25
		R	0.390	0.334	0.0059	0.039	46	20
2012	Avalon	F	1.102	0.983	0.0020	0.039	79	25
		R	0.390	0.334	0.0059	0.039	46	20

F: Front

R: Rear

71099_AVAL_C0013

SCHEDULED MAINTENANCE INTERVALS

TOYOTA—AVALON

TO BE SERVICED	TYPE OF SERVICE	VEHICLE MILEAGE INTERVAL (x1000)													
		5	10	15	20	25	30	35	40	45	50	55	60	90	120
Engine oil & filter	R	✓	✓	✓	✓	✓	✓	✓	✓	✓	✓	✓	✓	✓	✓
Automatic transmission fluid	S/I			✓			✓			✓			✓	✓	✓
Ball joints & dust covers	S/I			✓			✓			✓			✓	✓	✓
Bolts & nuts on chassis & body	S/I			✓			✓			✓			✓	✓	✓
Brake linings & drums	S/I	✓	✓	✓	✓	✓	✓	✓	✓	✓	✓	✓	✓	✓	✓
Brake line pipes & hoses	S/I			✓			✓			✓			✓	✓	✓
Brake pads & discs (front & rear)	S/I	✓	✓	✓	✓	✓	✓	✓	✓	✓	✓	✓	✓	✓	✓
Brake fluid	R						✓						✓	✓	✓
Rack and pinion assembly	S/I			✓			✓			✓			✓	✓	✓
Steering linkage & boots	S/I			✓			✓			✓			✓	✓	✓
Air cleaner filter	R						✓						✓	✓	✓
Spark plugs ①	R														✓
Drive belts	S/I												✓	✓	✓
Exhaust pipes & mountings	S/I			✓			✓			✓			✓	✓	✓
Fuel lines & connections	S/I						✓						✓	✓	✓
Engine coolant ②	S/I			✓			✓			✓			✓	✓	
Fuel tank cap gasket	S/I						✓						✓	✓	✓
Rotate tires	S/I			✓			✓			✓			✓		✓
Clean air conditioning filter ③	S/I			✓			✓			✓			✓		✓
Axle shaft bolts	S/I			✓			✓			✓			✓	✓	✓
Brake pad thickness and rotor runout	S/I						✓						✓	✓	✓

R: Replace   S/I: Service or Inspect

① Spark plugs are replaced at 120,000 miles

② Replace engine coolant at 100,000 miles and then inspect every 15,000 miles

③ Replace air conditioning filter every 30,000 miles

FREQUENT OPERATION MAINTENANCE (SEVERE SERVICE)

If a vehicle is operated under any of the following conditions it is considered severe service:

- Extremely dusty areas.

- 50% or more of the vehicle operation is in 32°C (90°F) or higher temperatures, or constant temperatures below 0°C (32°F).

- Prolonged idling (vehicle operation in stop and go traffic).

- Frequent short running periods (engine does not warm to normal operating temperatures).

- Police, taxi, delivery usage or trailer towing usage.

Air cleaner filter: service or inspect every 5000 miles

Rear differential & transfer case oil: replace every 15,000 miles.

Ball joints & dust covers: service or inspect every 5000 miles.

Bolts & nuts on chassis & body: service or inspect every 5000 miles.

Axle shaft bolts: service or inspect every 5000 miles.

Steering linkage: service or inspect every 5000 miles.

71099_AVAL_C0014

PRECAUTIONS

Before servicing any vehicle, please be sure to read all of the following precautions, which deal with personal safety, prevention of component damage, and important points to take into consideration when servicing a motor vehicle:

- Never open, service or drain the radiator or cooling system when the engine is hot; serious burns can occur from the steam and hot coolant.
- Observe all applicable safety precautions when working around fuel. Whenever servicing the fuel system, always work in a well-ventilated area. Do not allow fuel spray or vapors to come in contact with a spark, open flame, or excessive heat (a hot drop light, for example). Keep a dry chemical fire extinguisher near the work area. Always keep fuel in a container specifically designed for fuel storage; also, always properly seal fuel containers to avoid the possibility of fire or explosion. Refer to the additional fuel system precautions later in this section.
- Fuel injection systems often remain pressurized, even after the engine has been turned **OFF**. The fuel system pressure must be relieved before disconnecting any fuel lines. Failure to do so may result in fire and/or personal injury.
- Brake fluid often contains polyglycol ethers and polyglycols. Avoid contact with the eyes and wash your hands thoroughly after handling brake fluid. If you do get brake fluid in your eyes, flush your eyes with clean, running water for 15 minutes. If eye irritation persists, or if you have taken brake fluid internally, IMMEDIATELY seek medical assistance.
- The EPA warns that prolonged contact with used engine oil may cause a number of skin disorders, including cancer. You should make every effort to minimize your exposure to used engine oil. Protective gloves should be worn when changing oil. Wash your hands and any other exposed skin areas as soon as possible after exposure to used engine oil. Soap and water, or waterless hand cleaner should be used.
- All new vehicles are now equipped with an air bag system, often referred to as a Supplemental Restraint System (SRS) or Supplemental Inflatable Restraint (SIR) system. The system must be disabled before performing service on or around system components, steering column, instrument panel components, wiring and sensors. Failure to follow safety and disabling procedures could result in accidental air bag deployment, possible personal injury and unnecessary system repairs.
- Always wear safety goggles when working with, or around, the air bag system. When carrying a non-deployed air bag, be sure the bag and trim cover are pointed away from your body. When placing a non-deployed air bag on a work surface, always face the bag and trim cover upward, away from the surface. This will reduce the motion of the module if it is accidentally deployed. Refer to the additional air bag system precautions later in this section.
- Clean, high quality brake fluid from a sealed container is essential to the safe and proper operation of the brake system. You should always buy the correct type of brake fluid for your vehicle. If the brake fluid becomes contaminated, completely flush the system with new fluid. Never reuse any brake fluid. Any brake fluid that is removed from the system should be discarded. Also, do not allow any brake fluid to come in contact with a painted surface; it will damage the paint.
- Never operate the engine without the proper amount and type of engine oil; doing so WILL result in severe engine damage.
- Timing belt maintenance is extremely important. Many models utilize an interference-type, non-freewheeling engine. If the timing belt breaks, the valves in the cylinder head may strike the pistons, causing potentially serious (also time-consuming and expensive) engine damage. Refer to the maintenance interval charts for the recommended replacement interval for the timing belt, and to the timing belt section for belt replacement and inspection.
- Disconnecting the negative battery cable on some vehicles may interfere with the functions of the on-board computer system(s) and may require the computer to undergo a relearning process once the negative battery cable is reconnected.
- When servicing drum brakes, only disassemble and assemble one side at a time, leaving the remaining side intact for reference.
- Only an MVAC-trained, EPA-certified automotive technician should service the air conditioning system or its components.

BRAKES

ANTI-LOCK BRAKE SYSTEM (ABS)

GENERAL INFORMATION

PRECAUTIONS

- Certain components within the ABS system are not intended to be serviced or repaired individually.
- Do not use rubber hoses or other parts not specifically specified for and ABS system. When using repair kits, replace all parts included in the kit. Partial or incorrect repair may lead to functional problems and require the replacement of components.
- Lubricate rubber parts with clean, fresh brake fluid to ease assembly. Do not use shop air to clean parts; damage to rubber components may result.
- Use only DOT 3 brake fluid from an unopened container.
- If any hydraulic component or line is removed or replaced, it may be necessary to bleed the entire system.
- A clean repair area is essential. Always clean the reservoir and cap thoroughly before removing the cap. The slightest amount of dirt in the fluid may plug an orifice and impair the system function. Perform repairs after components have been thoroughly cleaned; use only denatured alcohol to clean components. Do not allow ABS components to come into contact with any substance containing mineral oil; this includes used shop rags.
- The Anti-Lock control unit is a microprocessor similar to other computer units in the vehicle. Ensure that the ignition switch is **OFF** before removing or installing controller harnesses. Avoid static electricity discharge at or near the controller.
- If any arc welding is to be done on the vehicle, the control unit should be unplugged before welding operations begin.

SPEED SENSORS

REMOVAL & INSTALLATION

Front

1. Remove the front wheel.
2. Remove the front fender liner.
3. Remove the front speed sensor, as follows:
   a. Disconnect the speed sensor connector and clamp.
   b. Remove the 2 bolts and separate the speed sensor harness from the body and shock absorber assembly.

c. Remove the clamp from the steering knuckle.

d. Remove the bolt and the front speed sensor.

➡**Do not allow foreign matter to attach to the sensor tip.**

➡**Clean the installation hole and surface for the speed sensor every time the speed sensor is removed.**

To install:

4. Install the front speed sensor with the bolt and tighten to 71 inch lbs. (8 Nm).

➡**Do not allow foreign matter to attach to the sensor tip.**

5. Install the sensor harness brackets with the 2 bolts to the body and shock absorber assembly and tighten to 44 inch lbs. (5 Nm), and 14 ft. lbs. (19 Nm).

a. Connect the clamp to the steering knuckle.

➡**Do not twist the sensor wire when installing the sensor.**

6. Connect the speed sensor connector and clamp.
7. Install the front fender liner.
8. Install the front wheel and tighten the lug nuts to 76 ft. lbs. (103 Nm).
9. Check ABS speed sensor signal.

Rear

See Figure 1.

1. Remove the rear wheel.
2. Disconnect the connector from the skid control sensor.
3. Separate the rear disc brake caliper assembly.

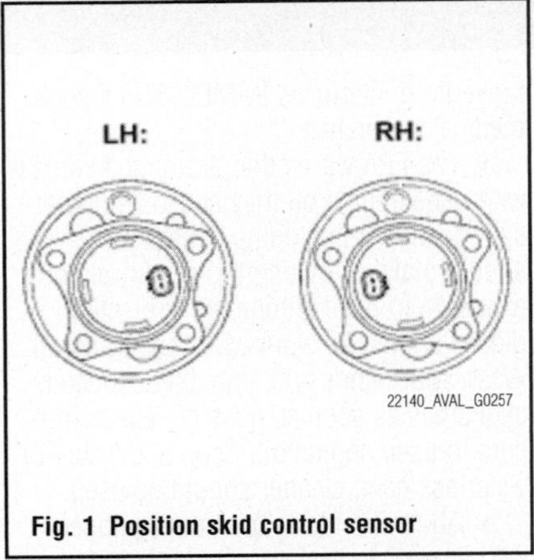

Fig. 1 Position skid control sensor

4. Remove the rear disc.
5. Remove the 4 bolts and rear axle hub and bearing assembly.
6. Remove the rear speed sensor, as follows:

a. Mount the rear axle hub in a vise between aluminum plates.

WARNING

Replace the hub and bearing assembly if it is dropped or receives a strong shock.

b. Using a pin punch and hammer, drive out the 2 pins and remove the 2 attachments from SST.

c. Using SST (SST: 09520-00031, SST: 09521-00020, SST: 09950-00020 or equivalent) and 2 bolts (Diameter: 12 mm, pitch: 1.5 mm), remove the rear speed sensor from the rear axle hub.

➡**If the sensor rotor is damaged, replace the hub and bearing assembly.**

➡**Do not scratch the contact surface of the axle hub and speed sensor.**

To install:

7. Install the rear speed sensor, as follows:

a. Clean the contact surface of the axle hub and a new skid control sensor.

➡**Make sure the sensor rotor is clean.**

b. Place the skid control sensor on the rear axle hub so that the connector is positioned as shown in the illustration.

c. Using SST (SST: 09830-36010, 09950-60010, 09950-70010 or equivalent) and a press, install the skid control sensor to the axle hub.

➡**Do not tap the skid control sensor directly with a hammer.**

➡**Check that the skid control sensor detection part is clean.**

➡**Press in the skid control sensor straight and slowly.**

8. Install the rear axle hub and bearing assembly with the 4 bolts and tighten to 59 ft. lbs. (80 Nm).
9. Install the rear disc.
10. Install the rear disc brake caliper assembly.
11. Connect the connector to the skid control sensor.
12. Install the rear wheel and tighten the lug nuts to 76 ft. lbs. (103 Nm).
13. Inspect and adjust rear wheel alignment.
14. Check ABS speed sensor signal.

BRAKES

BLEEDING THE BRAKE SYSTEM

BLEEDING PROCEDURE

MASTER CYLINDER BLEEDING

If the master cylinder is reinstalled or if the reservoir becomes empty, bleed the air from the master cylinder. To prevent brake fluid from adhering, cover nearly painted surfaces with a shop rag or a piece of cloth.

1. Using a union nut wrench (10 mm), disconnect the 2 brake lines from the master cylinder, using a suitable brake line wrench.
2. Have an assistant slowly depress the brake pedal and hold it.
3. Cover the 2 outer holes with your fingers, and have your assistant release the brake pedal.
4. Repeat the previous 2 steps 3 or 4 times.
5. Using a union nut wrench (10 mm), connect the 2 the brake lines to the master cylinder and tighten to 11 ft. lbs. (15 Nm).

➡**Use a torque wrench with a fulcrum length of 250 mm (9.84 in.).**

➡**This torque value is effective when the union nut wrench is parallel to the torque wrench.**

BRAKE LINE BLEEDING

➡**Bleed air from the brake line of the wheel farthest from the master cylinder.**

1. Raise and safely support the vehicle.
2. Connect a vinyl tube to the bleeder plug.
3. Have an assistant depress the brake pedal several times, then loosen the bleeder plug while the pedal is depressed.
4. When fluid stops coming out, tighten the bleeder plug, then release the brake pedal.
5. Repeat steps 3 and 4 until all the air in the fluid has been bled out.
6. Tighten the brake bleeder plug to 73 inch lbs. (8.3 Nm).
7. Repeat the above steps to bleed the air out of the brake line for each wheel.

BLEEDING THE ABS SYSTEM

➡**After bleeding the air from the brake system, if the height or feel of the brake pedal cannot be obtained, perform air bleeding in the brake actuator**

assembly with a hand-held tester by following the procedures below.

1. Depress the brake pedal more than 20 times with the engine off.
2. Connect the hand-held tester to the DLC3, then turn the ignition switch to the **ON** position, but do NOT start the engine.
3. Select "AIR BLEEDING" on the hand-held tester.

➡Refer to the hand-held tester operator's manual for more details.

4. Bleed the air out of the regular brake line when "Step 1: Increase" appears on the hand-held tester display, as follows:

➡Bleed the air by following the steps displayed on the hand-held tester. Make sure that the brake fluid in the master cylinder reservoir tank does not become empty.

a. Connect the vinyl tube to either one of the bleeder plugs.
b. Have an assistant depress the brake pedal several times, and then loosen the bleeder plug connected to the vinyl tube with the pedal depressed.
c. When fluid stops coming out, tighten the bleeder plug and release the brake pedal.
d. Repeat the previous 2 steps until all the air in the fluid is completely bled out.
e. Tighten the bleeder plug completely to 73 inch lbs. (8.3 Nm).
f. Repeat the above procedures for each wheel to bleed the air out of the brake line.

5. Bleed the air out of the suction line when "Step 2: Inhalation" appears on the hand-held tester display, as follows:

➡Bleed the air by following the steps displayed on the hand-held tester. Make sure that the brake fluid in the master cylinder reservoir tank does not become empty.

a. Connect the vinyl tube to the bleeder plug at the right front wheel or the right rear wheel and loosen the bleeder plug.
b. Operate the brake actuator assembly to perform air bleeding from the suction line using the hand-held tester.

➡This operation stops automatically after 4 seconds. At this time, be sure to release the brake pedal.

c. Check if the operation has stopped by referring to the hand-held tester display and tighten the bleeder plug.
d. Repeat the previous 2 steps until all air in the fluid is completely bled out.
e. Tighten the bleeder plug completely to 73 inch lbs. (8.3 Nm).
f. Repeat the above procedures to bleed the air out of the brake line for each wheel.

6. Bleed the air out of the pressure reduction line when "Step 3: Decrease" appears on the hand-held tester display, as follows:

➡Bleed the air by following the steps displayed on the hand-held tester. Make sure that the brake fluid in the master cylinder reservoir tank does not become empty.

a. Connect a vinyl tube to either one of the bleeder plugs.
b. Loosen the bleeder plug.
c. Using the hand-held tester, operate the brake actuator assembly, completely depress the brake pedal and keep it depressed.

➡The operation stops automatically after 4 seconds. When performing this procedure continuously, set an interval of at least 20 seconds. When the operation is complete, the brake pedal goes down slightly. This is a normal phenomenon caused when the solenoid opens. During this procedure, the pedal will feel heavy, but completely depress it so that the brake fluid comes out from the bleeder plug. Be sure to keep depressing the brake pedal. Do not depress and release the pedal repeatedly.

d. Tighten the bleeder plug, then release the brake pedal.
e. Repeat the previous 2 steps until all the air in the fluid is completely bled out.
f. Tighten the bleeder plug completely to 73 inch lbs. (8.3 Nm).
g. Repeat the above procedures for each wheel to bleed the air out of the brake line.

7. Bleed the air out of the brake line again when "Step 4: Increase" appears on the hand-held tester display, as follows:

➡Bleed the air by following the steps displayed on the hand-held tester. Make sure that the brake fluid in the master cylinder reservoir tank does not become empty.

a. Connect the vinyl tube to either one of the bleeder plugs.
b. Depress the brake pedal several times, then loosen the bleeder plug connected to the vinyl tube with the pedal depressed.

8. When fluid stops coming out, tighten the bleeder plug, then release the brake pedal.

a. Repeat the previous 2 steps until all the air in the fluid is completely bled out.
b. Tighten the bleeder plug completely to 73 inch lbs. (8.3 Nm).
c. Repeat the above procedures for each wheel to bleed the air out of the brake line.
d. Finish "AIR BLEEDING" on the hand-held tester and turn the hand-held tester off.
e. Disconnect the hand-held tester from the DLC3 from the DLC3.
f. Turn the ignition switch off.
g. Inspect for fluid leak.

9. Check the fluid level and add fluid if necessary. Use SAE J1703 or FMVSS No. 116 DOT3 Brake fluid.

FLUID FILL PROCEDURE

1. With the vehicle on a flat level surface, check the fluid level and add fluid if necessary. Use SAE J1703 or FMVSS No. 116 DOT3 Brake fluid.
2. Add brake fluid to the level between the MIN and MAX lines of the reservoir.

CAUTION

When filling the reservoir. Take care because brake fluid can harm your hands or eyes and damage painted surfaces. If fluid gets in your eyes, flush your eyes with clean water immediately. If you still experience discomfort, see a doctor.

➡If the fluid level is low or high. It is normal for the brake fluid level to go down slightly as the brake pads wear or when the fluid level in the accumulator is high. If the reservoir needs frequent refilling, it may indicate a serious problem.

CAUTION

Dust and dirt accumulating on brake parts during normal use may contain asbestos fibers from production or aftermarket brake linings. Breathing excessive concentrations of asbestos fibers can cause serious bodily harm. Exercise care when servicing brake parts. Do not sand or grind brake lining unless equipment used is designed to contain the dust residue. Do not clean brake parts with compressed air or by dry brushing. Cleaning should be done by dampening the brake components with a fine mist of water, then wiping the brake components clean with a dampened cloth. Dispose of cloth and all residue containing asbestos fibers in an impermeable container with the appropriate label. Follow practices prescribed by the Occupational Safety and Health Administration (OSHA) and the Environmental Protection Agency (EPA) for the handling, processing, and disposing of dust or debris that may contain asbestos fibers.

BRAKE CALIPER

REMOVAL & INSTALLATION

See Figure 2.

1. Before servicing the vehicle, refer to the Precautions Section.
2. Remove the front wheel.
3. Drain brake fluid.

WARNING

Do not let brake fluid sit on painted surfaces, as it will eat through the paint. Wash it off immediately.

4. Remove the union bolt and gasket from the disc brake cylinder assembly, then disconnect the front brake flexible hose.
5. Hold the front disc brake cylinder slide pin and remove the 2 bolts and disc brake cylinder assembly.

➡Remove the disc brake cylinder assembly while holding both of the brake pads or the anti-squeal springs may fall off the brake pads.

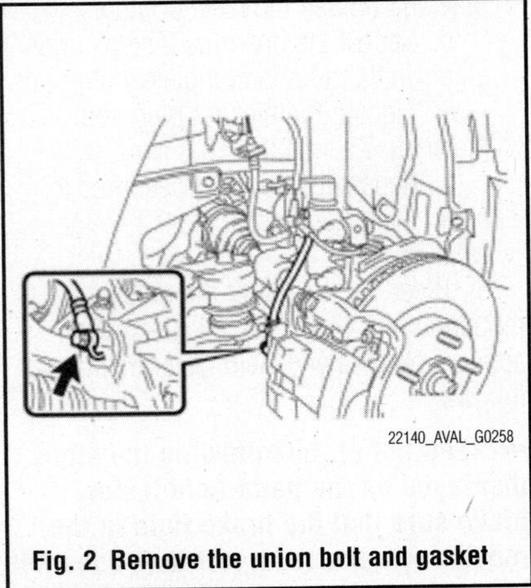

Fig. 2 Remove the union bolt and gasket

To install:

6. Install the disc brake cylinder assembly with the 2 bolts and tighten to 25 ft. lbs. (34 Nm).

➡Be sure that the anti-squeal springs are installed to the front disc brake pads.

7. Connect the flexible hose with the union bolt and a new gasket and tighten to 21 ft. lbs. (29 Nm).

➡Install the front brake flexible hose lock securely in the lock hole in the disc brake cylinder.

8. Fill reservoir with brake fluid.
9. Bleed brake line.
10. Inspect for brake fluid leak.
11. Inspect brake fluid level in reservoir.
12. Install the front wheel and tighten the lug nuts to 76 ft. lbs. (103 Nm).

DISC BRAKE PADS

REMOVAL & INSTALLATION

See Figures 3 and 4.

1. Before servicing the vehicle, refer to the Precautions Section.
2. Remove the 2 front disc brake cylinder slide pins (upper and lower) from the front disc brake cylinder mounting.
3. Remove brake cylinder.
4. Remove the 2 anti-squeal springs.
5. Remove the 2 brake pads from the front disc brake cylinder mounting.

To install:

6. Install the 2 brake pads with front anti-squeal shims to the front disc brake cylinder mounting.

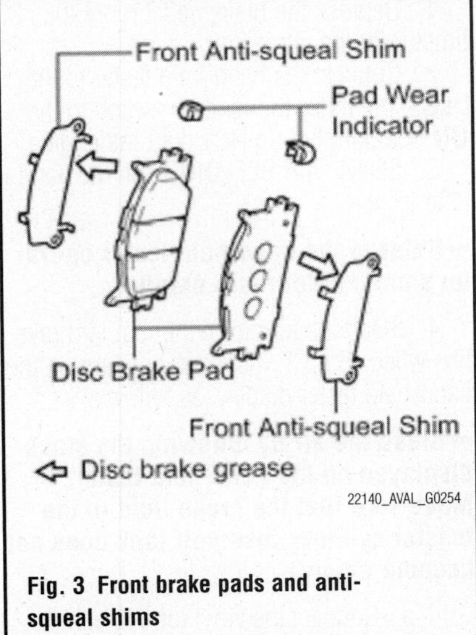

Fig. 3 Front brake pads and anti-squeal shims

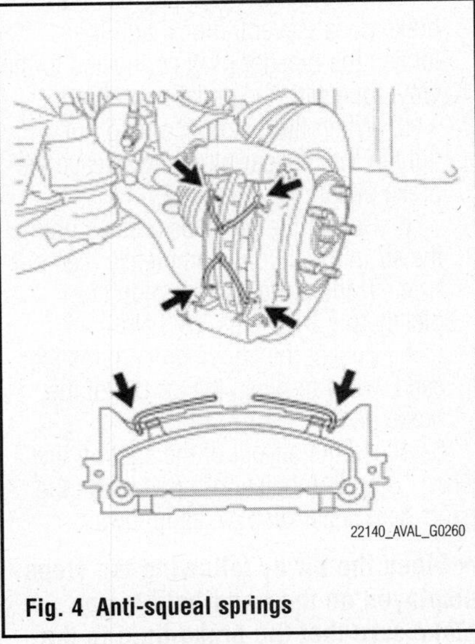

Fig. 4 Anti-squeal springs

➡When replacing worn pads, the front anti-squeal springs must be replaced at the same time.

➡Be sure to install the anti-squeal springs into the front disc brake pad installation holes as far as they will go.

7. Install the 2 front disc brake cylinder slide pins (upper and lower) from the front disc brake cylinder mounting.
8. Install the brake cylinder.

** CAUTION

Dust and dirt accumulating on brake parts during normal use may contain asbestos fibers from production or aftermarket brake linings. Breathing excessive concentrations of asbestos fibers can cause serious bodily harm. Exercise care when servicing brake parts. Do not sand or grind brake lining unless equipment used is designed to contain the dust residue. Do not clean brake parts with compressed air or by dry brushing. Cleaning should be done by dampening the brake components with a fine mist of water, then wiping the brake components clean with a dampened cloth. Dispose of cloth and all residue containing asbestos fibers in an impermeable container with the appropriate label. Follow practices prescribed by the Occupational Safety and Health Administration (OSHA) and the Environmental Protection Agency (EPA) for the handling, processing, and disposing of dust or debris that may contain asbestos fibers.

BRAKE CALIPER

REMOVAL & INSTALLATION

See Figure 5.

1. Before servicing the vehicle, refer to the Precautions Section.
2. Remove the rear wheel.
3. Drain brake fluid.

** WARNING

Do not let brake fluid sit on painted surfaces, as it will eat through the paint. Wash it off immediately.

4. Remove the union bolt and the gasket from the rear disc brake cylinder assembly, then disconnect the rear brake flexible hose.
5. Hold the 2 rear disc brake cylinder slide pins and remove the 2 bolts and rear disc brake cylinder assembly.

To install:

6. Install the rear disc brake cylinder assembly with the 2 bolts and tighten to 20 ft. lbs. (27 Nm).

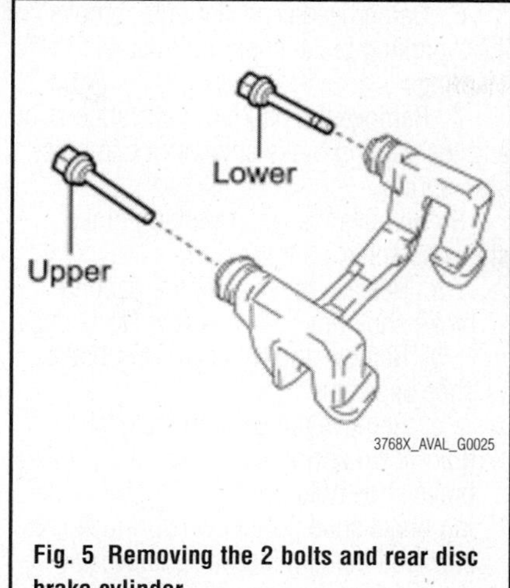

Fig. 5 Removing the 2 bolts and rear disc brake cylinder

7. Connect the rear brake flexible hose with the union bolt and a new gasket and tighten to 24 ft. lbs. (33 Nm).
8. Fill reservoir with brake fluid.
9. Bleed brake line.
10. Inspect for brake fluid leak.
11. Inspect brake fluid level in reservoir.
12. Install the rear wheel and tighten the lug nuts to 76 ft. lbs. (103 Nm).

DISC BRAKE PADS

REMOVAL & INSTALLATION

See Figure 6.

1. Before servicing the vehicle, refer to the Precautions Section.
2. Remove the 2 rear disc brake cylinder slide pins (upper and lower) from the rear disc brake cylinder mounting.
3. Remove brake cylinder.
4. Remove the 2 brake pads with the rear anti-squeal shims.

To install:

5. Installation is the reverse of removal procedure.

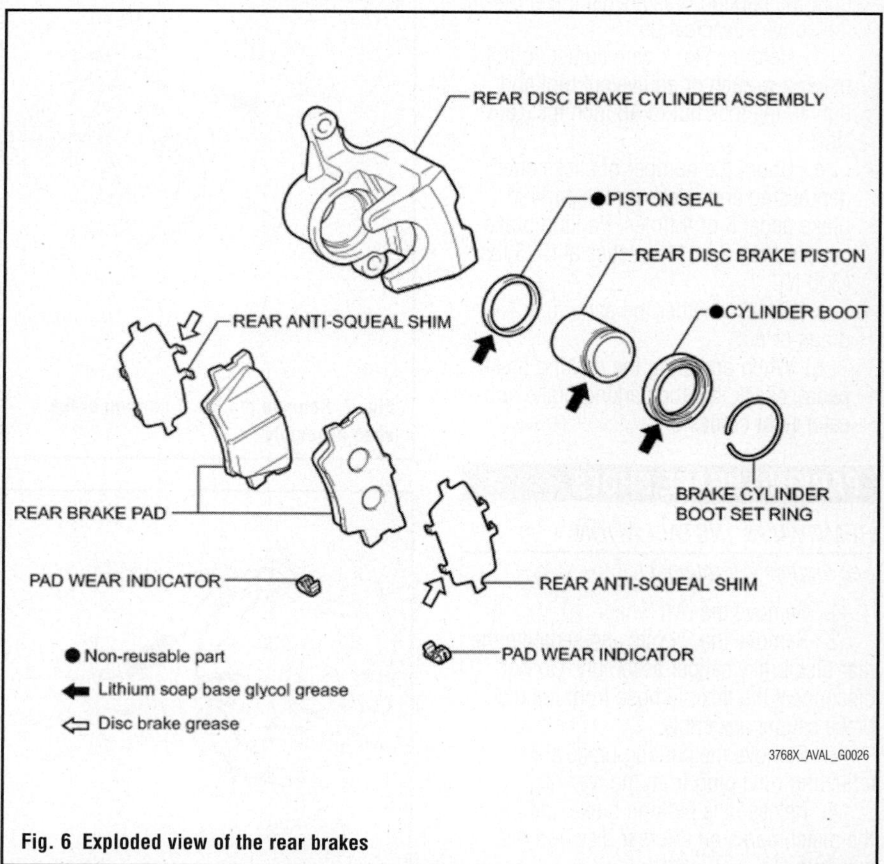

Fig. 6 Exploded view of the rear brakes

PARKING BRAKE CABLES

ADJUSTMENT

1. Inspect the parking brake pedal travel, as follows:
   a. Fully depress the parking brake pedal and release it to engage the parking brake.
   b. Depress the pedal to the floor again, and release it to disengage the parking brake.
   c. Slowly depress the parking brake pedal to the floor, and count the number of clicks. Parking brake pedal travel: 9 to 11 notches at 67.5 lbs. (300 N).
2. Adjust the parking brake pedal travel, as follows:
   a. Depress the parking brake pedal. Hold the wire adjusting nut No. 1 using a wrench and loosen the lock nut.
   b. Release the parking brake pedal.
   c. Turn the No. 1 wire adjusting nut until the parking brake pedal travel meets the above specification.
   d. Hold the No. 1 wire adjusting nut using a wrench or equivalent tool and tighten the lock nut to 48 inch lbs. (5.4 Nm).
   e. Count the number of clicks after depressing and releasing the parking brake pedal 3 or 4 times. Parking brake pedal travel: 9 to 11 notches at 67.5 lbs. (300 N).
   f. Check whether the parking brake drags or not.
   g. When operating the parking brake pedal, check that the parking brake indicator light comes on.

PARKING BRAKE SHOES

REMOVAL & INSTALLATION

See Figures 7 through 11.

1. Remove the rear wheel.
2. Remove the 2 bolts and separate the rear disc brake caliper assembly. Do not disconnect the flexible hose from the disc brake caliper assembly.
3. Remove the parking brake shoe adjusting hole plug from the rear disc.
4. Release the parking brake and place the matchmarks on the rear disc and the axle hub.
5. Remove the rear disc.

➡**If the disc cannot be removed easily, turn the shoe adjuster until the disc turns freely.**

6. Using needle-nose pliers, remove the 2 parking brake shoe return tension No. 1 springs.
7. Remove the parking brake shoe strut and the parking brake shoe strut compression spring.
8. Remove the No. 1 parking brake shoe assembly, as follows:
   a. Release the claw of the parking brake shoe hold down spring No. 2 cup.
   b. Remove the No. 1 parking brake shoe assembly.
   c. Remove the parking brake shoe hold down spring No. 1 cup, the parking brake shoe hold down spring, the parking brake shoe hold down spring No. 2 cup, and the parking brake shoe hold down spring No. 1 pin.
9. Remove the parking brake shoe adjusting screw set.
10. Remove the parking brake shoe return tension No. 2 spring.

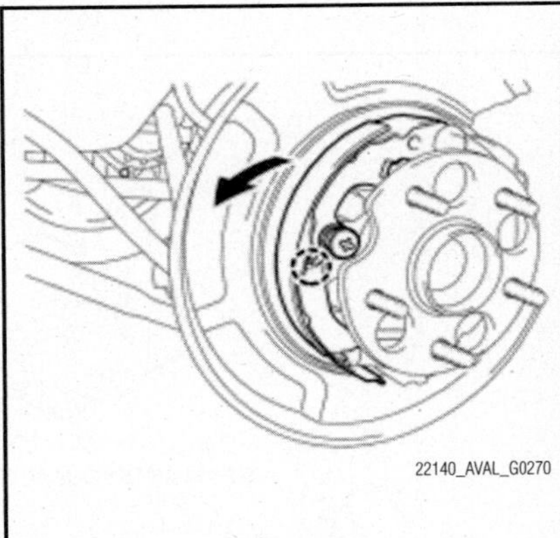

Fig. 7 Remove the No. 1 parking brake shoe assembly

Fig. 8 Remove the No. 2 parking brake shoe assembly

11. Remove the No. 2 parking brake shoe assembly, as follows:
    a. Release the claw of the parking brake shoe hold down spring No. 2 cup.
    b. Remove the No. 2 parking brake shoe assembly.
    c. Remove the parking brake shoe hold down spring No. 1 cup, the parking brake shoe hold down spring, the parking brake shoe hold down spring No. 2 cup, and the parking brake shoe hold down spring No. 2 pin.
    d. Using needle-nose pliers, disconnect the No. 3 parking brake cable assembly from the parking brake shoe lever.

➡**Be careful not to damage the No. 3 parking brake cable assembly.**

12. Using a screwdriver, remove the C-washer, shim and the parking brake shoe lever.
13. Remove the parking brake shoe guide plate set bolt and the parking brake shoe guide plate.

To install:

14. Apply high temperature grease to the backing plate where it contacts the shoe.
15. Apply adhesive (Toyota Genuine Adhesive 1344, Three Bond® 1344 or equivalent) to the threads of the parking brake shoe guide plate set bolt.
16. Install the parking brake shoe guide plate with the parking brake shoe guide plate set bolt and tighten to 13 ft. lbs. (18 Nm).
17. Install the parking brake shoe lever and shim to the No. 2 parking brake shoe assembly with a new C-washer.
18. Using a feeler gauge, measure the clearance between the No. 2 parking brake shoe assembly and parking brake shoe lever. Standard clearance: Less than 0.014 inch (0.35 mm). If the clearance is not as specified, replace the shim with one of the correct size.
19. Install the No. 2 parking brake shoe assembly as follows:
    a. Using needle-nose pliers, connect the No. 3 parking brake cable assembly to the parking brake shoe lever.
    b. Install the No. 2 parking brake shoe assembly with the parking brake shoe hold down spring No. 2 pin, the parking brake shoe hold down spring No. 2 cup, the parking brake shoe hold down spring and the parking brake shoe hold down spring No. 1 cup.
    c. Engage the claw of the parking

Shim Thickness	Shim Thickness
0.3 mm (0.012 in.)	0.9 mm (0.035 in.)
0.6 mm (0.024 in.)	-

22140_AVAL_G0272

Fig. 9 Shim thickness

brake shoe hold down spring No. 2 cup to the No. 2 parking brake shoe assembly.

20. Install the parking brake shoe adjusting screw set, as follows:

a. Apply high temperature grease to the parking brake shoe adjusting screw set as shown in the illustration.

b. Install the parking brake shoe return tension No. 2 spring to the No. 1 parking brake shoe assembly and the No. 2 parking brake shoe assembly.

c. Install the parking brake shoe adjusting screw set to the No. 1 parking brake shoe assembly and the No. 2 parking brake shoe assembly.

21. Install the No. 1 parking brake shoe assembly as follows:

a. Install the No. 1 parking brake shoe assembly with the parking brake shoe hold down spring No. 1 pin, parking brake shoe hold down spring No. 2 cup, parking brake shoe hold down spring and parking brake shoe hold down spring No. 1 cup.

b. Engage the claw of the parking brake shoe hold down spring No. 2 cup to the No. 1 parking brake shoe assembly.

22. Attach the parking brake shoe strut and the parking brake shoe strut compression spring to the No. 1 parking brake shoe assembly and No. 2 parking brake shoe assembly.

23. Using needle-nose pliers, install the 2 parking brake shoe return tension No. 1 springs. First install the front side spring and then the rear side spring.

24. Inspect parking brake installation and check that each part is installed properly.

➡There should be no oil or grease on the friction surfaces of the shoe linings and discs.

25. Install the rear disc.

26. Install the parking brake shoe adjusting hole plug.

27. Adjust parking brake shoe clearance.

28. Install the rear disc brake caliper assembly with the 2 bolts and tighten to 46 ft. lbs. (62 Nm).

29. Install the rear wheel.

30. Adjust the parking brake pedal travel.

31. Bed in parking brake shoes to discs, as follows:

a. Drive the vehicle at about 31 mph (50 km/h) on a safe, level and dry road.

b. Depress the parking brake pedal with 34 lbs. (150 N) of force.

32. Drive the vehicle about 0.25 miles (400 m) in this condition.

a. Repeat this procedure 3 times using 5-minute intervals between each procedure to prevent the parking brake assembly from overheating.

33. Remove the rear wheel.

34. Adjust parking brake shoe clearance.

35. Adjust the parking brake pedal travel.

36. Install the rear wheel and tighten the lug nuts to 76 ft. lbs. (103 Nm).

ADJUSTMENT

1. Adjust parking brake shoe clearance, as follows:

a. Temporarily install the hub nuts.

b. Remove the shoe adjusting hole plug, turn the adjuster and expand the shoes until the disc locks.

c. Contract the shoe adjuster until the disc rotates smoothly. Standard: returns 8 notches

d. Check that the disc has no brake drag.

e. Install the shoe adjusting hole plug.

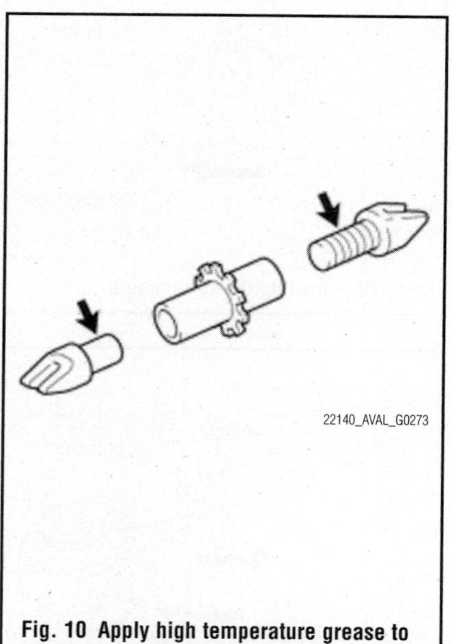

Fig. 10 Apply high temperature grease to the parking brake shoe adjusting screw set

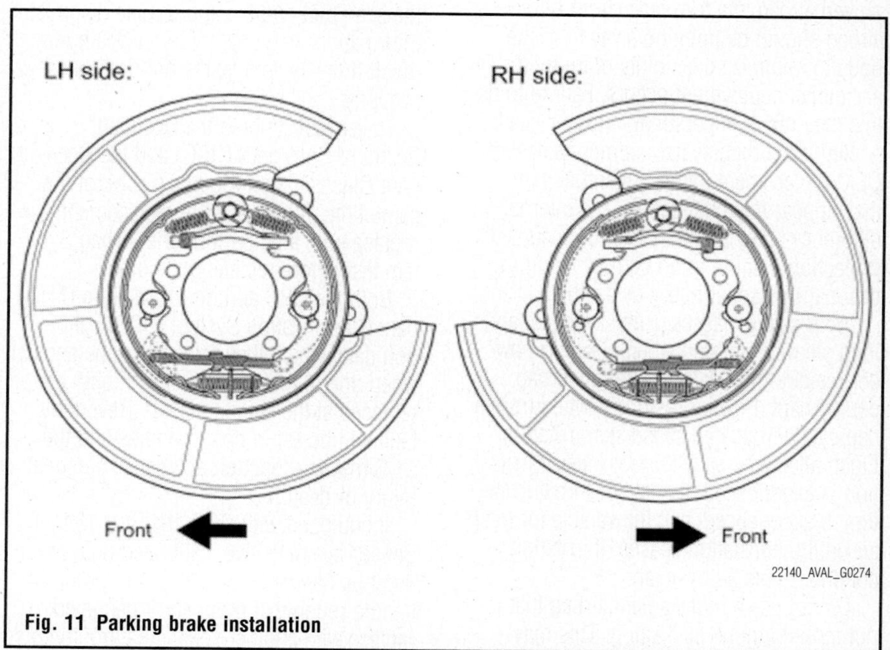

Fig. 11 Parking brake installation

GENERAL INFORMATION

CAUTION

These vehicles are equipped with an air bag system. The system must be disarmed before performing service on, or around, system components, the steering column, instrument panel components, wiring and sensors. Failure to follow the safety precautions and the disarming procedure could result in accidental air bag deployment, possible injury and unnecessary system repairs.

SERVICE PRECAUTIONS

Disconnect and isolate the battery negative cable before beginning any airbag system component diagnosis, testing, removal, or installation procedures. Allow system capacitor to discharge for two minutes before beginning any component service. This will disable the airbag system. Failure to disable the airbag system may result in accidental airbag deployment, personal injury, or death.

Do not place an intact undeployed airbag face down on a solid surface. The airbag will propel into the air if accidentally deployed and may result in personal injury or death.

When carrying or handling an undeployed airbag, the trim side (face) of the airbag should be pointing away from the body to minimize possibility of injury if accidental deployment occurs. Failure to do this may result in personal injury or death.

Replace airbag system components with OEM replacement parts. Substitute parts may appear interchangeable, but internal differences may result in inferior occupant protection. Failure to do so may result in occupant personal injury or death.

Wear safety glasses, rubber gloves, and long sleeved clothing when cleaning powder residue from vehicle after an airbag deployment. Powder residue emitted from a deployed airbag can cause skin irritation. Flush affected area with cool water if irritation is experienced. If nasal or throat irritation is experienced, exit the vehicle for fresh air until the irritation ceases. If irritation continues, see a physician.

Do not use a replacement airbag that is not in the original packaging. This may result in improper deployment, personal injury, or death.

The factory installed fasteners, screws and bolts used to fasten airbag components have a special coating and are specifically designed for the airbag system. Do not use substitute fasteners. Use only original equipment fasteners listed in the parts catalog when fastener replacement is required.

During, and following, any child restraint anchor service, due to impact event or vehicle repair, carefully inspect all mounting hardware, tether straps, and anchors for proper installation, operation, or damage. If a child restraint anchor is found damaged in any way, the anchor must be replaced. Failure to do this may result in personal injury or death.

Deployed and non-deployed airbags may or may not have live pyrotechnic material within the airbag inflator.

Do not dispose of driver/passenger/curtain airbags or seat belt tensioners unless you are sure of complete deployment. Refer to the Hazardous Substance Control System for proper disposal.

Dispose of deployed airbags and tensioners consistent with state, provincial, local, and federal regulations.

After any airbag component testing or service, do not connect the battery negative cable. Personal injury or death may result if the system test is not performed first.

If the vehicle is equipped with the Occupant Classification System (OCS), do not connect the battery negative cable before performing the OCS Verification Test using the scan tool and the appropriate diagnostic information. Personal injury or death may result if the system test is not performed properly.

Never replace both the Occupant Restraint Controller (ORC) and the Occupant Classification Module (OCM) at the same time. If both require replacement, replace one, then perform the Airbag System test before replacing the other.

Both the ORC and the OCM store Occupant Classification System (OCS) calibration data, which they transfer to one another when one of them is replaced. If both are replaced at the same time, an irreversible fault will be set in both modules and the OCS may malfunction and cause personal injury or death.

If equipped with OCS, the Seat Weight Sensor is a sensitive, calibrated unit and must be handled carefully. Do not drop or handle roughly. If dropped or damaged, replace with another sensor. Failure to do so may result in occupant injury or death.

If equipped with OCS, the front passenger seat must be handled carefully as well. When removing the seat, be careful when setting on floor not to drop. If dropped, the sensor may be inoperative, could result in occupant injury, or possibly death.

If equipped with OCS, when the passenger front seat is on the floor, no one should sit in the front passenger seat. This uneven force may damage the sensing ability of the seat weight sensors. If sat on and damaged, the sensor may be inoperative, could result in occupant injury, or possibly death.

DISARMING THE SYSTEM

To avoid personal injury when working on vehicles equipped with an air bag, the negative battery cable must be disconnected and at least 90 seconds must elapse before working on the system. Failure to do so may result in deployment of the air bag.

ARMING THE SYSTEM

To arm the system after service is finished, connect the negative battery cable.

CLOCKSPRING CENTERING

See Figures 12 and 13.

1. Before servicing the vehicle, refer to the Precautions Section.
2. Check that the ignition switch is **OFF**.
3. Check that the battery negative (-) terminal is disconnected.

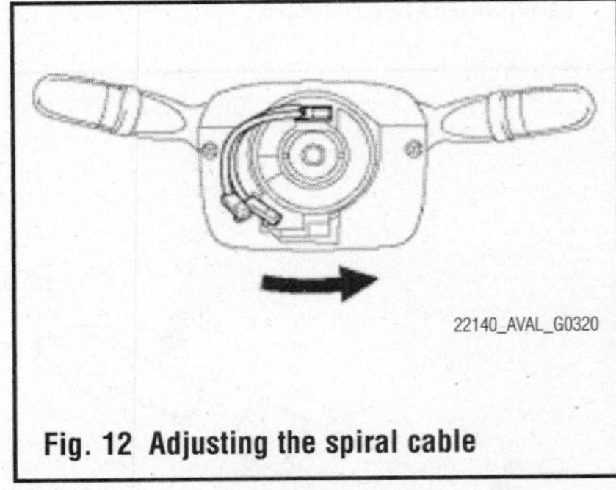

Fig. 12 Adjusting the spiral cable

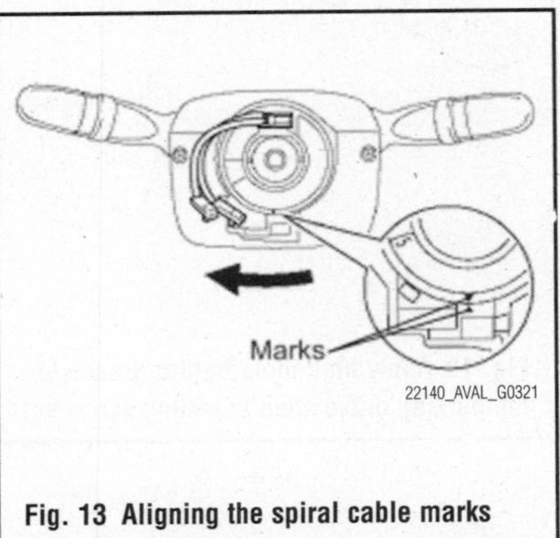

Fig. 13 Aligning the spiral cable marks

⁂ CAUTION

After removing the terminal, wait for at least 90 seconds before starting the operation.

4. Rotate the spiral cable counterclockwise slowly by hand until it feels firm.
5. Rotate the spiral cable clockwise approximately 2.5 turns to align the marks.

➡Do not turn the spiral cable by the airbag wire harness.

➡The spiral cable will rotate approximately 2.5 turns to both the left and right from the center.

DRIVE TRAIN

AUTOMATIC TRANSAXLE FLUID

DRAIN AND REFILL

See Figures 14 through 20

1. Review the precautions and work description.
   a. The U660E automatic transaxle does not have an oil filler tube and oil level gauge. When adding fluid, add fluid through the refill hole on the transaxle case. The fluid level can be adjusted by draining excess fluid (allowing excess fluid to overflow) through the overflow tube of the oil pan.

➡"Overflow" indicates the condition under which fluid comes out of the overflow plug hole.

   b. Before adjusting the fluid level, add the specified amount of fluid when the engine is cold and warm up the engine to circulate the fluid in the transaxle. Ensure that the fluid temperature is as specified and the engine is idling.
   c. The U660E automatic transaxle requires Toyota Genuine ATF WS.
   d. The adjustment should be per-

Fluid Filling Procedure:

1. PERFORM INITIAL FILLING

Add fluid to the oil pan to the specified level.

2. ADD SPECIFIED AMOUNT

Add the correct amount of fluid specified for the operation that was performed.

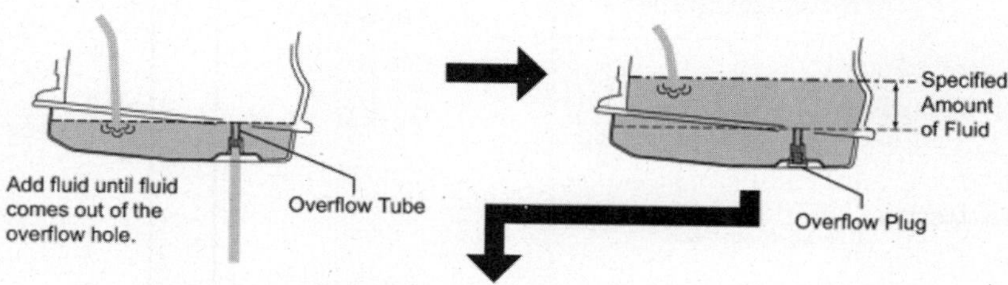

Specified Amount of Fluid
Add fluid until fluid comes out of the overflow hole.
Overflow Tube
Overflow Plug

3. ADJUST FLUID TEMPERATURE

Start the engine to circulate the fluid. Activate the fluid temperature detection mode and engine idle speed control mode and adjust the fluid temperature to the specified value.

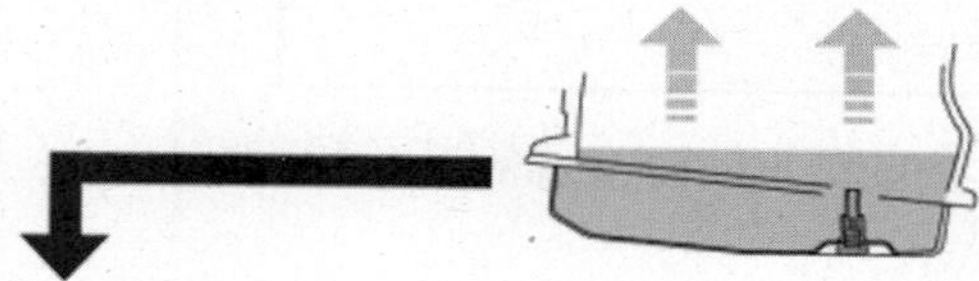

4. ADJUST FLUID LEVEL

Drain excess fluid at the specified fluid temperature.

If no fluid comes out, add fluid until fluid comes out of the overflow hole.

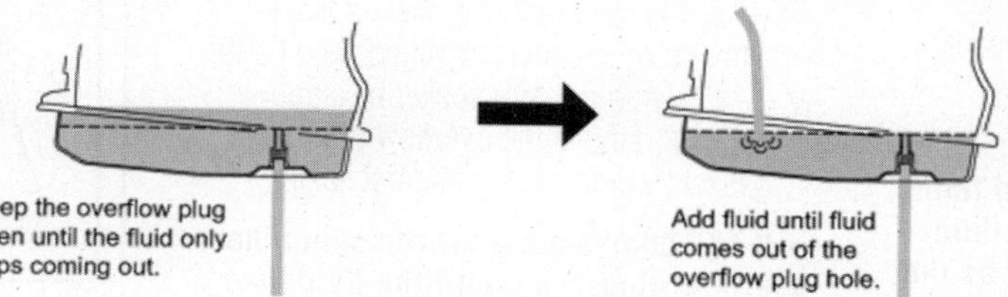

Keep the overflow plug open until the fluid only drips coming out.
Add fluid until fluid comes out of the overflow plug hole.

3768X_AVAL_G0001

Fig. 14 Fluid filling procedure

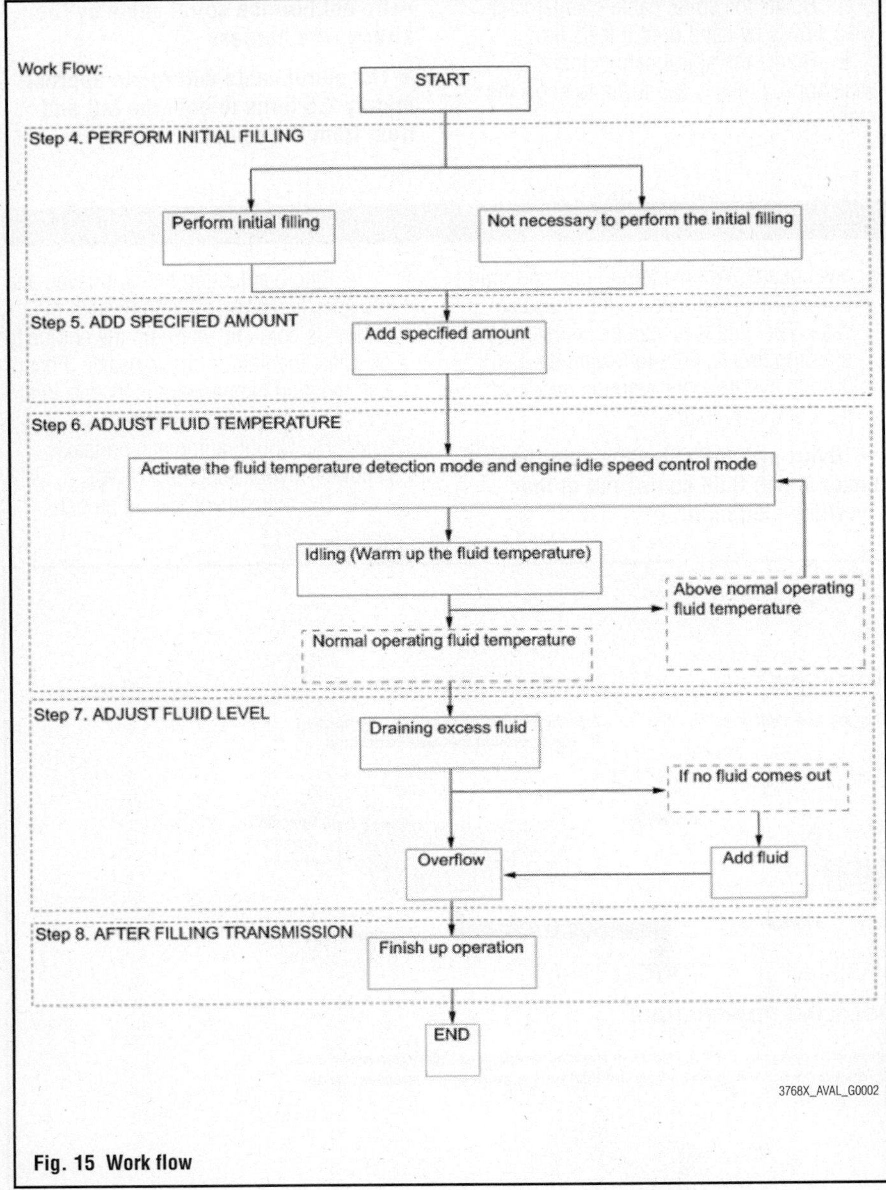

Fig. 15 Work flow

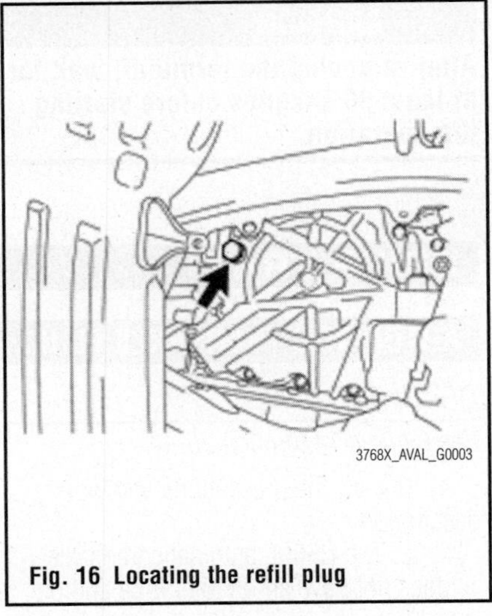

Fig. 16 Locating the refill plug

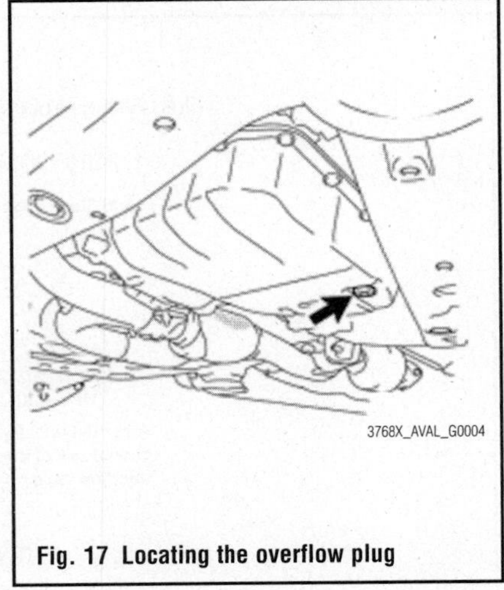

Fig. 17 Locating the overflow plug

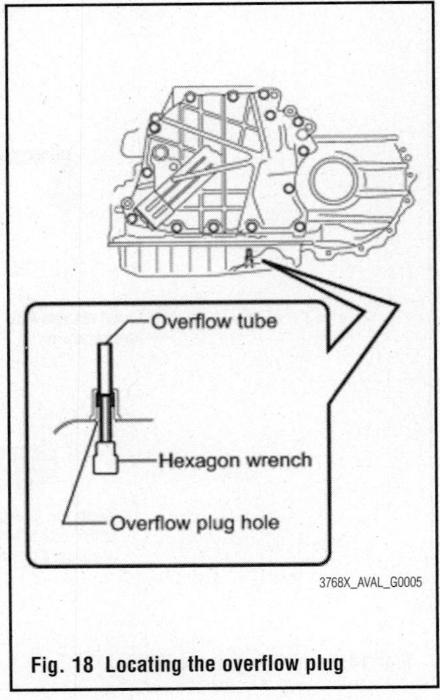

Fig. 18 Locating the overflow plug

formed according to the procedures and notes.

2. Follow the work flow.

a. The adjustment should be performed according to the procedures referring to the work flow below.

3. Before filling the transaxle with fluid, lift the vehicle, remove the engine under covers RH and LH and remove the fender apron seal LH.

4. Perform the initial filling.

➡If the transaxle is hot (AT fluid temperature is high), wait until the fluid temperature becomes the same as the ambient temperature before starting the following procedure. (Recommended ATF temperature: around 68°F [20°C]).

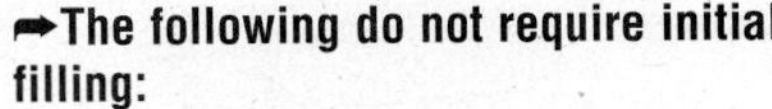

➡The following do not require initial filling:

a. Disconnection of the oil cooler tube or oil cooler hose

b. Repair of fluid leakage due to a loose case plug, or a faulty plug gasket or O-ring

c. Replacement of a new transaxle with a torque converter (filled fluid parts)

d. Using a 6 mm socket hexagon wrench, remove the overflow plug and gasket from the automatic transaxle.

➡If ATF comes out after removing the overflow plug, wait until the fluid flow slows and only drips come out.

➡If ATF comes out, it is not necessary to perform the initial filling procedure.

After checking the tightening torque of the overflow tube, temporarily tighten the overflow plug.

e. Using a 6 mm socket hexagon wrench, check that the overflow tube is tightened to the specified torque. Tighten to 15 inch lbs. (1.7 Nm).

➡If the transmission overflow tube is not tightened to the specified torque, the amount of fluid cannot be precisely adjusted.

➡To check the torque of the overflow tube, insert the hexagon socket into the overflow hole.

f. Perform the initial filling. Fill the transaxle through the refill hole until fluid begins to trickle out of the overflow plug hole.

➡Use Toyota Genuine ATF WS.

g. Wait until the fluid flow slows and only drips come out.

h. Temporarily install the overflow plug.

➡Reuse the old gasket. The plug will be removed again to adjust the fluid level.

5. Add the specified amount of fluid.

➡Refill amount differs depending on the operation that was performed.

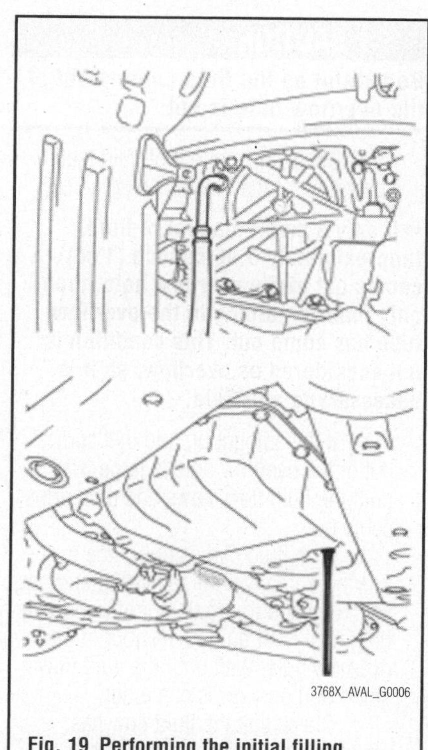

3768X_AVAL_G0006

Fig. 19 Performing the initial filling

Standard capacity:

a. Disconnection of the oil cooler tube or oil cooler hose: 0.5 qts. (0.5 L)

b. Repair of fluid leakage due to a loose case plug, or a faulty plug gasket or O-ring: 0.5 qts. (0.5 L)

c. Replacement of a new transaxle with torque converter (fluid filled parts): 0.5 qts. (0.5 L)

d. Removal and installation of the drive shaft: 3 qts. (2.8 L)

e. Removal and installation of the transaxle case oil seal: 3 qts. (2.8 L)

f. Removal and installation of the oil pan: 3 qts. (2.8 L)

g. Removal and installation of the valve body: 3.4 qts. (3.2 L)

h. Removal and installation of the transaxle (torque converter not removed): 3.6 qts. (3.4L)

i. Removal and installation of the torque converter (torque converter reused): 5.1 qts. (4.8 L)

j. Replacement of a new torque converter: 5.6 qts. (5.3 L)

k. Temporarily install the refill plug to avoid fluid splash.

➡Reuse the old gasket. The plug will be removed again to adjust the fluid level.

l. Lower the vehicle.

6. Adjust the fluid temperature.

Using the Techstream

➡The actual ATF temperature can be checked on the data list using the Techstream.

a. Connect the Techs ream to the DLC3 with the ignition switch off.

b. Turn the ignition switch on (IG) and turn the Techstream main switch ON.

➡Check that electrical systems such as the air conditioning system, audio system and lighting system are off.

c. Enter the following items: "Powertrain / ECT / Active Test / Connect the TC and TE1".

d. Select the data list menu: "A/T Oil Temperature 1".

e. Check "A/T Oil Temperature 1".

➡If the fluid temperature is below 45°C (113°F), proceed to the next step. (Recommended ATF temperature: 104°F [40°C] or less).

➡If the fluid temperature is 113°F (45°C) or more, turn the ignition switch off and wait until the fluid temperature drops below 113°F (45°C).

f. Push the active test menu: "Connect the TC and TE1 / ON".

➡Indicator lights in the meter will blink to indicate a DTC when "Connect the TC and TE1 / ON" is selected.

g. Depress and hold down the brake pedal.

h. Start the engine.

i. Slowly move the shift lever from the P position to the D position, then back to the P position.

➡Slowly move the shift lever to circulate the fluid through each part of the transaxle.

j. While observing the D shift indicator on the combination meter, move the shift lever back and forth between the N position and the D position at an interval of 1.5 seconds for 6 seconds or more.

➡Do not pause for more than 1.5 seconds.

➡Performing this operation will cause the vehicle to enter the fluid temperature detection mode.

k. Check that the D shift indicator comes on for 2 seconds.

➡When the fluid temperature detection mode is activated, the D shift indicator on the combination meter comes on for 2 seconds.

➡If the D shift indicator does not come on for 2 seconds, return to the step where terminal TC is first turned on and perform the procedure again.

l. Move the shift lever from the N position to the P position.

m. Release the brake pedal.

n. Push the active test menu: "Connect the TC and TE1 / OFF".

➡Be sure to select "Connect the TC and TE1 / OFF". If the active test continues, the fluid level cannot be precisely adjusted due to fluctuations in idle speed.

➡Selecting "Connect the TC and TE1 / OFF" activates the engine idle speed control mode.

➡In the engine idle speed control mode, engine idle speed control starts when the fluid temperature becomes 95°F (35°C) or more and the engine speed is maintained at approximately 800 rpm.

➡Even after selecting "Connect the TC and TE1 / OFF", the fluid temperature

detection mode is active until the ignition switch is turned off.

o. Warm up the engine with the engine idling until the fluid temperature reaches the normal operating temperature (104 to 113°F [40 to 45°C]).

Operating temperatures

a. Below normal: 104°F or less (40°C or less)

b. Normal: 104 to 113°F (40 to 45°C

c. Above normal: 113°F or more (45°C or more

➡**If the fluid temperature is within the normal operating temperature range, immediately proceed to the "ADJUST FLUID LEVEL" procedure.**

➡**If the fluid temperature is 113°F (45°C) or more, stop the engine and wait until the fluid temperature drops to 104°F (40°C) or less. Then perform the "ADJUST FLUID TEMPERATURE" procedure again from the beginning.**

➡**In the fluid temperature detection mode, the D shift indicator comes on, goes off, or blinks depending on the fluid temperature.**

D shift indicator

a. Off: below normal operating temperature

b. On: normal operating temperature

c. Blinking: above normal operating temperature

Not using the Techstream

a. Using SST (09843-18040), connect terminals 13 (TC) and 4 (CG) of the DLC3 with the ignition switch off.

b. Depress and hold down the brake pedal.

c. Start the engine.

➡**Check that electrical systems such as the air conditioning system, audio system and lighting system are off.**

➡**Indicator lights in the meter will blink to indicate a DTC when terminals TC and CG are connected.**

d. Slowly move the shift lever from the P position to the D position, then back to the P position.

➡**Slowly move the shift lever to circulate the fluid through each part of the transaxle.**

e. While observing the D shift indicator on the combination meter, move the shift lever back and forth between the N position and the D position at an interval of 1.5 seconds for 6 seconds or more.

➡**Do not pause for more than 1.5 seconds.**

➡**Performing this operation will cause the vehicle to enter the fluid temperature detection mode.**

f. Check that the D shift indicator comes on for 2 seconds.

➡**When the fluid temperature detection mode is activated, the D shift indicator on the combination meter comes on for 2 seconds.**

➡**If the D shift indicator does not come on for 2 seconds, return to the step where terminal TC is first turned on and perform the procedure again.**

g. Move the shift lever from the N position to the P position.

h. Release the brake pedal.

i. Remove SST (09843-18040) from terminals 13 (TC) and 4 (CG).

➡**Be sure that terminals TC and CG are not connected. If the terminals are connected, the fluid level cannot be precisely adjusted due to fluctuations in engine speed.**

➡**If SST is removed from terminals TC and CG, the engine idle control mode is activated.**

➡**In the engine idle speed control mode, engine idle speed control starts when the fluid temperature becomes 95°F (35°C) or more and the engine speed is maintained at approximately 800 rpm.**

➡**Even after SST is removed from terminals TC and CG, the fluid temperature detection mode is active until the ignition switch is turned off.**

j. Allow the engine to idle until the D shift indicator comes on again.

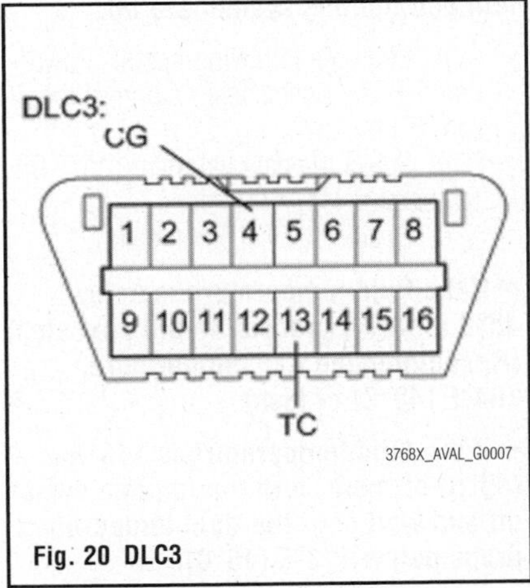

Fig. 20 DLC3

D shift indicator

a. Off: below normal operating temperature

b. On: normal operating temperature

c. Blinking: above normal operating temperature

d. If the D shift indicator is on, immediately proceed to the "ADJUST FLUID LEVEL" procedure.

e. If the D shift indicator blinks, stop the engine and wait until the fluid temperature drops to 104°F (40°C) or less (the indicator goes off). Then perform the "ADJUST FLUID TEMPERATURE" procedure again from the beginning.

f. In the fluid temperature detection mode, the D shift indicator comes on, goes off, or blinks depending on the fluid temperature.

g. Fluid filling procedure should be performed when the D shift indicator is on (the fluid temperature is within the normal operating temperature range).

7. Adjust the fluid level.

⁂ CAUTION

To work while the engine is idling and the radiator fan is rotating, care must be taken.

a. Lift the vehicle.

b. Adjust the fluid level.

c. Using a 6 mm socket hexagon wrench, remove the overflow plug and gasket.

⁂ CAUTION

Be careful as the fluid coming out of the overflow hole is hot.

d. Check the amount of fluid that comes out of the overflow plug hole.

➡**If only a small amount of fluid (approximately 0.06 cu inch (1 cc)) comes out of the overflow hole, then only fluid remaining in the overflow tube has come out. This condition is not considered as overflow, so it is necessary to add fluid.**

e. If the amount of fluid that comes out of the overflow hole is large, wait until the fluid flow slows and only drips come out.

f. If no fluid comes out of the overflow hole, remove the refill plug and gasket. Then add transaxle fluid through the refill hole until fluid comes out of the overflow hole. Wait until the fluid flow slows and only drips come out.

g. Check that the fluid flow has slowed and only drips come out.

➡**The fluid will not completely stop because the fluid expands as its temperature increases.**

h. Install the overflow plug with a new gasket. Tighten to 30 ft. lbs. (40 Nm).
i. Install the refill plug with a new gasket. Tighten to 36 ft. lbs. (49 Nm).
j. Lower the vehicle.
k. Turn the ignition switch off.

➡**Turning the ignition switch off exits the fluid temperature detection mode.**

l. Remove the Techstream from the DLC3 (when using the Techstream).
8. After filling the transmission perform the following:
a. Lift the vehicle.
b. Clean each part.
c. Install the fender apron seal LH, and engine under covers RH and LH.
d. Lower the vehicle.

FRONT HALFSHAFT

REMOVAL & INSTALLATION

See Figures 21 through 26.

1. Remove the engine under cover.
2. Remove the drain plug and gasket, and then drain the automatic transaxle fluid.
3. Install a new gasket and drain plug and tighten to 36 ft. lbs. (49 Nm).
4. Remove front wheel.
5. Using SST (SST: 09930-00010) and hammer, release the staked part of the front axle hub nut.

➡**Loosen the staked part of the nut completely, otherwise the screw of the driveshaft may be damaged.**

6. While applying the brakes, remove the front axle hub nut.
7. Remove the nut and separate the front stabilizer link assembly.

➡**If the ball joint turns together with the nut, use a hexagon wrench (6mm) to hold the stud.**

8. Remove the bolt and clip, and separate the speed sensor wire and flexible hose from the shock absorber.
9. Remove the bolt and separate the front speed sensor from the steering knuckle.

➡**Do not allow foreign matter to adhere to the speed sensor. Be careful not to damage the speed sensor.**

10. Separate tie rod end sub-assembly, as follows:
a. Remove the cotter pin and nut.
b. Using SST (SST: 09628-62011) or equivalent, separate the tie rod end sub-assembly from the steering knuckle.

➡**Do not damage the ball joint dust cover.**

11. Remove the bolt and 2 nuts, and separate the lower No. 1 front suspension arm sub-assembly from the lower ball joint.
12. Put matchmarks on the front driveshaft assembly and the axle hub.
13. Using a plastic hammer, separate the front driveshaft assembly from the front axle hub sub-assembly.

➡**Be careful not to damage the driveshaft boot and speed sensor rotor.**

14. Remove the front driveshaft assembly(s), as follows:
a. For left front driveshaft, use SST (SST: 09520-01010, SST: 09520-24010) or equivalent, and remove the left front driveshaft assembly.

➡**Be careful not to damage the driveshaft dust cover, boot and oil seal. Be careful not to drop the driveshaft assembly.**

b. For right front driveshaft, use a screwdriver and remove the bearing bracket hole snap ring.
c. Remove the bolt and right front driveshaft assembly from the driveshaft bearing bracket.

➡**Do not damage the boot and oil seal.**

15. Secure the front axle hub bearing. The hub bearing could be damaged if it is subjected to the vehicle's full weight, such as moving the vehicle with the driveshaft removed. If it is necessary to place the vehicle's weight on the hub bearing, first support it with SST (SST: 09608-16042).

To install:

16. Install the front driveshaft assembly(s), as follows:

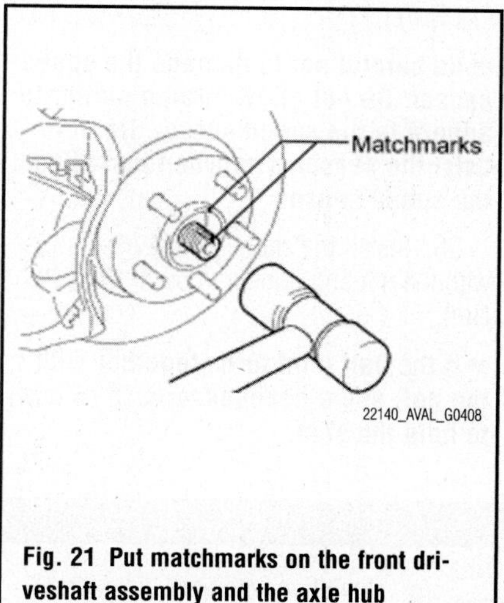

Fig. 21 Put matchmarks on the front driveshaft assembly and the axle hub

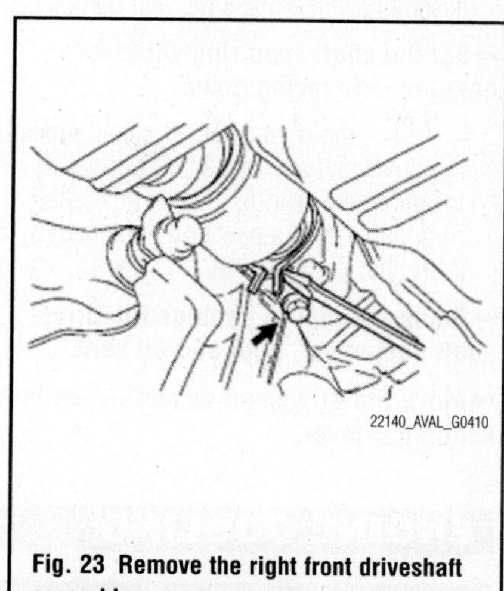

Fig. 23 Remove the right front driveshaft assembly

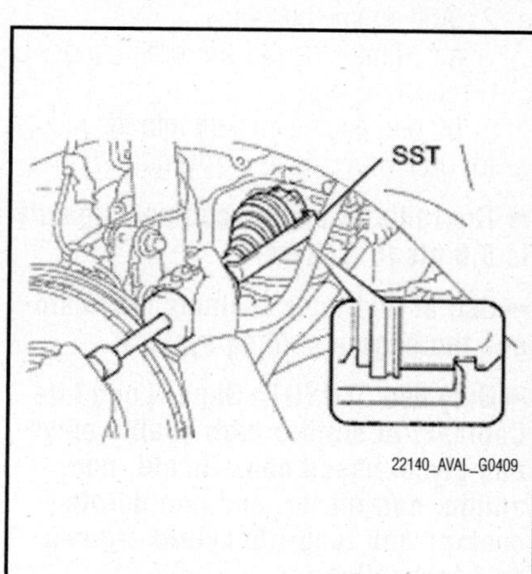

Fig. 22 Remove the left front driveshaft assembly

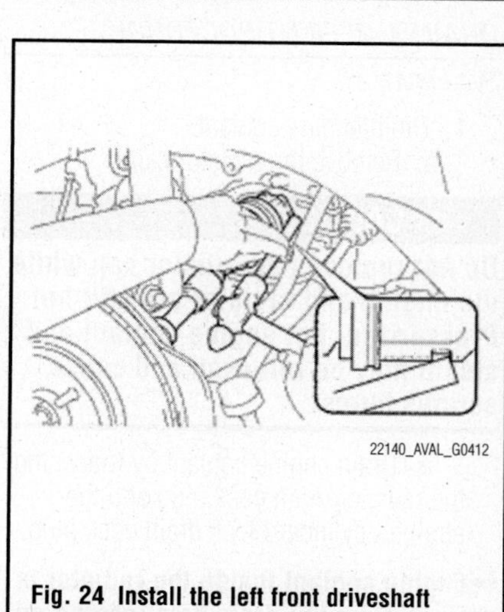

Fig. 24 Install the left front driveshaft assembly

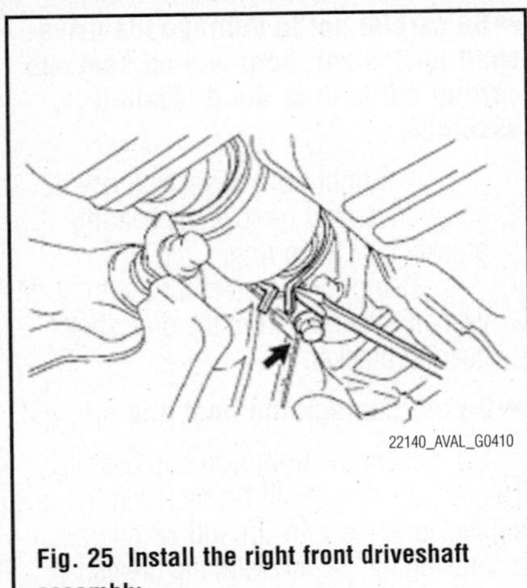

Fig. 25 Install the right front driveshaft assembly

a. Coat the spline of the inboard joint shaft assembly with automatic transaxle fluid.

b. For the left front driveshaft, align the shaft splines and install the driveshaft assembly with a brass bar and hammer.

➡**Set the shaft snap ring with the opening side facing down.**

c. For the right front driveshaft, install the driveshaft and use a screwdriver to install a new bearing bracket hole snap ring, and install a new bolt tightened to 24 ft. lbs. (32 Nm).

➡**Be careful not to damage the driveshaft dust cover, boot and oil seal.**

➡**Move the driveshaft assembly while keeping it level.**

17. Align the matchmarks and install the front driveshaft assembly to the front axle hub sub-assembly.

➡**Be careful not to damage the driveshaft boot and speed sensor rotor.**

18. Install the lower ball joint to the lower No. 1 front suspension arm sub-assembly with the bolt and 2 nuts and tighten to 55 ft. lbs. (75 Nm).
19. Install the tie rod end sub-assembly to the steering knuckle with the nut and tighten to 36 ft. lbs. (49 Nm).
20. Install a new cotter pin. If the holes for the cotter pin are not aligned, tighten the nut up to 60° further.
21. Install the front speed sensor to the steering knuckle with the bolt and tighten to 71 inch lbs. (8 Nm).
22. Install the flexible hose and the speed sensor to the shock absorber with the bolt and set the sensor clip on the knuckle and tighten to 14 ft. lbs. (19 Nm).

➡**Be careful not to damage the speed sensor. Do not allow foreign matter to adhere to the speed sensor. Do not twist the sensor wire when installing the speed sensor.**

23. Install the stabilizer link assembly with the nut and tighten to 55 ft. lbs. (74 Nm).

➡**If the ball joint turns together with the nut, use a hexagon wrench (6 mm) to hold the stud.**

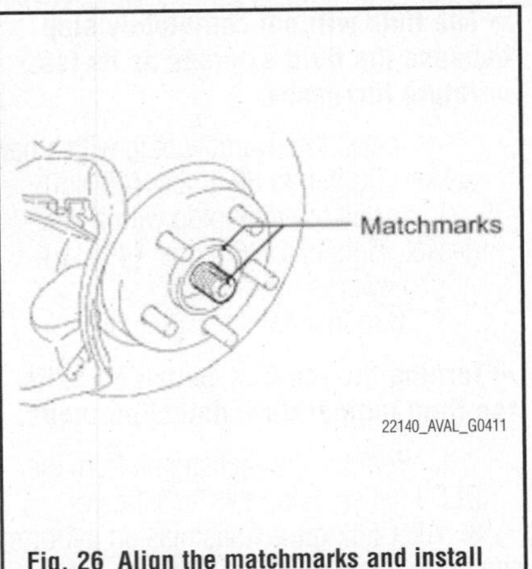

Fig. 26 Align the matchmarks and install the front driveshaft assembly

24. Clean the threaded parts on the driveshaft and axle hub nut using a non-residue solvent.

➡**Be sure to perform this work for a new driveshaft. Keep the threaded parts free of oil and foreign objects.**

25. Using a socket wrench (30 mm), install a new axle hub nut and tighten to 217 ft. lbs. (294 Nm).
26. Using a chisel and hammer, stake the front axle hub nut.
27. Install front wheel.
28. Add automatic transaxle fluid.
29. Inspect automatic transaxle fluid.
30. Inspect and adjust front wheel alignment.
31. Install the engine under cover.
32. Check the ABS speed sensor signal.

ENGINE COOLING

ENGINE COOLANT

DRAIN & REFILL PROCEDURE

See Figure 27.

1. Drain engine coolant.
   a. Remove the radiator cap.

CAUTION

Do not remove the radiator cap while the engine and radiator are still hot. Pressurized, hot engine coolant and steam may be released and cause serious burns.

b. Drain engine coolant by loosening the radiator drain cock plug and the engine's cylinder block drain cock plug.

➡**Engine coolant inside the radiator is drained from the drain hole located on the bottom of the engine under cover.**

c. Tighten the cylinder block drain cock plugs to 10 ft. lbs. (13 Nm).

2. Add engine coolant.
   a. Tighten the radiator drain plug (Procedure "A").
   b. Add engine coolant into the radiator until it overflows (Procedure "B").

➡**The radiator engine coolant capacity is 8.8 qts (8.3 liters).**

➡**Use of improper coolants may damage the engine cooling system.**

➡**Only use "TOYOTA Super Long Life Coolant" or similar high quality ethylene glycol based non-silicate, non-amine, non-nitrite, and non-borate coolant with long-life hybrid organic acid technology.**

➡**New TOYOTA vehicles are filled with TOYOTA Super Long Life Coolant (color is pink, premixed ethylene-glycol concentration is approximately 50% and freezing temperature is -35%°C (-31°F)). When replacing the coolant, TOYOTA Super Long Life Coolant is recommended.**

➡**Observe the coolant level inside the radiator by pressing the inlet and outlet radiator hoses several times by hand. If the coolant level goes down, add the coolant.**

CAUTION

Do not use plain water alone.

c. Pour coolant into the radiator reservoir tank until the coolant reaches the full line (Procedure "C").

d. Install the radiator cap (Procedure "D").

e. Warm up the engine (Procedure "E").

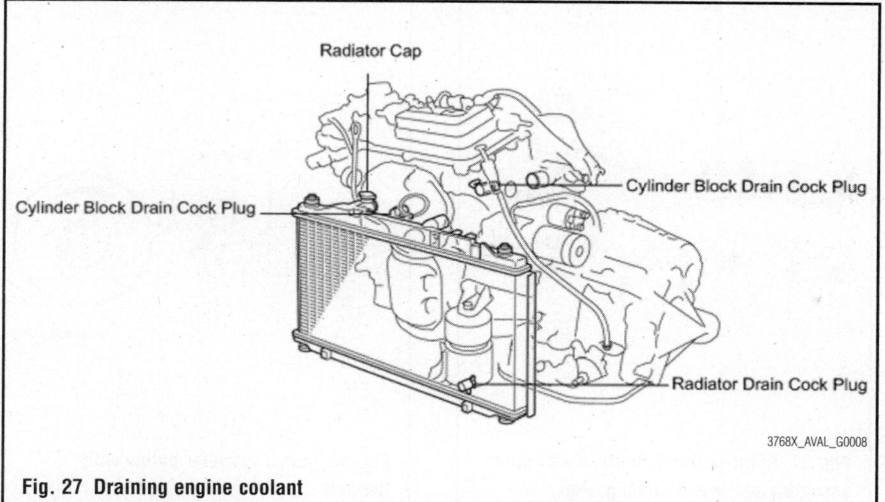

Fig. 27 Draining engine coolant

➡As the engine warms up, press the inlet and outlet radiator hoses several times by hand.

f. Stop the engine and wait until the coolant cools down to room temperature (Procedure "F").

g. Remove the radiator cap and check the coolant level inside the radiator (Procedure "G"). If the coolant level is below the full level, repeat procedure "C" to "G" until the coolant level. If it is below the full line, add coolant.

3. Inspect for coolant leak.

a. Fill the radiator with coolant and attach a radiator cap tester.

b. Pump it to 17.1 psi (118 kPa) and check leakage.

ENGINE FAN

REMOVAL & INSTALLATION

Refer to the Radiator removal and installation procedure.

RADIATOR

REMOVAL & INSTALLATION

See Figures 28 and 29.

1. Drain engine coolant.
2. Remove battery.
3. Remove the No. 2 air cleaner inlet.
4. Remove the No. 1 air cleaner inlet.
5. Separate the radiator reserve tank hose from the radiator assembly.
6. Separate the radiator hose inlet from the radiator assembly.
7. Separate the radiator hose outlet from the radiator assembly.
8. Separate the No. 1 oil cooler inlet tube from the radiator assembly.
9. Separate the No. 1 oil cooler outlet tube from the radiator assembly.
10. Remove the radiator support upper, as follows:

a. Remove the hood lock nut cap from the hood lock assembly.

b. Remove the 3 bolts and separate the hood lock assembly from the upper radiator support.

c. Remove the 5 bolts and radiator support upper.

11. Remove the radiator assembly as follows:

a. Disconnect the fan motor connector.

b. Remove the 4 bolts and separate the condenser assembly from the radiator assembly.

c. Remove the radiator assembly from the body.

12. Remove the 2 radiator support cushions from the radiator assembly.

Fig. 28 Separate the hood lock assembly from the upper radiator support

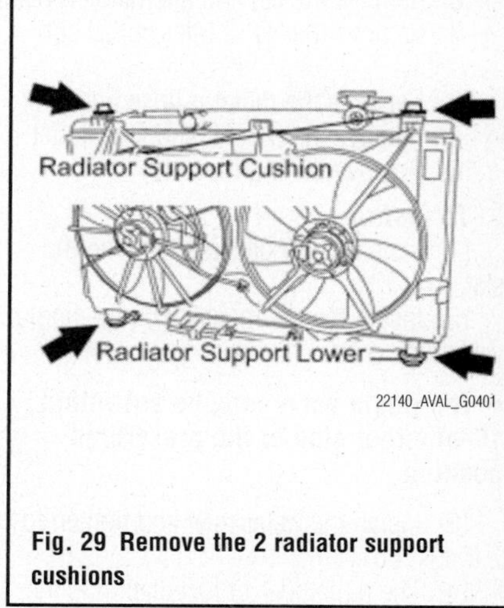

Fig. 29 Remove the 2 radiator support cushions

13. Remove the 2 radiator support lowers from the radiator assembly.
14. Disassemble the 3 snap fits and lift the fan assembly w/motor from the radiator.

To install:

15. Installation is the reverse of removal.
16. After installation, inspect for coolant leak.

THERMOSTAT

REMOVAL & INSTALLATION

See Figure 30.

1. Drain engine coolant.
2. Remove the V-bank cover sub-assembly.
3. Remove the engine moving control rod.
4. Remove the engine mounting control bracket.
5. Remove the left front No. 1 engine mounting bracket.

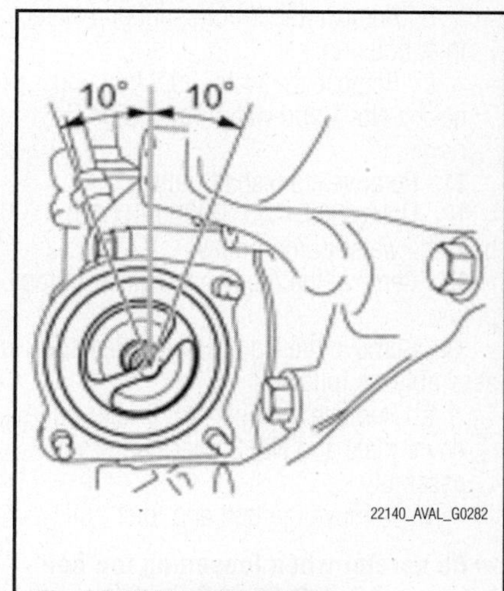

Fig. 30 Radiator jiggle valve

6. Remove the fan and alternator V-belt.
7. Remove the No. 2 idler pulley sub-assembly.
8. Separate the radiator hose outlet.
9. Remove the 2 bolts and water inlet.
10. Remove the thermostat.

To install:

11. Install a new gasket to the thermostat.
12. Install the thermostat with the jiggle valve facing up.

➡The jiggle valve may be set within 10°of either side of the prescribed position.

13. Install the water inlet and tighten to 7 ft. lbs. (10 Nm).
14. The remainder of installation is the reverse of removal.
15. After installation, inspect for coolant leak.

WATER PUMP

REMOVAL & INSTALLATION

See Figures 31 through 34.

1. Drain engine coolant.
2. Remove the right front wheel.
3. Remove the right engine under cover.
4. Remove the V-bank cover sub-assembly.
5. Remove the engine moving control rod.
6. Remove the engine mounting control bracket.
7. Remove the left front No. 1 engine mounting bracket.
8. Remove the fan and alternator V-belt.
9. Separate the radiator hose outlet.
10. Remove water inlet housing, as follows:
    a. Separate the water hose.
    b. Remove the 2 bolts, nut and water inlet housing.
    c. Remove the water inlet housing gasket No. 1 and water outlet pipe O-ring.
11. Remove crankshaft pulley.
12. Using SST (SST: 09960-10010), hold the water pump pulley.
13. Remove the 4 bolts and water pump pulley.
14. Remove the No. 2 idler pulley sub-assembly, as follows:
    a. Remove the bolt, No. 2 idler pulley cover plate and No. 2 idler pulley sub-assembly.
    b. Remove the bolt and idler pulley.

➡Be careful when loosening the bolt because it is left-hand threaded.

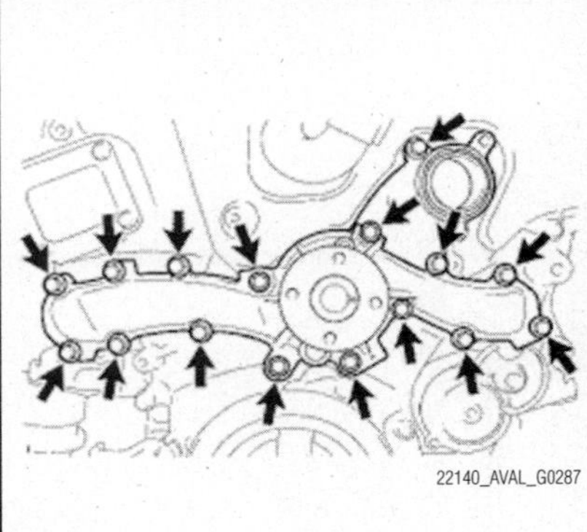

Fig. 31 Remove the 16 bolts, water pump assembly and water pump gasket

Fig. 32 Water pump assembly bolt position

15. Separate the vane pump assembly.
16. Remove the water pump assembly, as follows:
    a. Remove the 16 bolts, water pump assembly and water pump gasket.

To install:

17. Install a new water pump gasket and the water pump assembly with the 16 bolts and tighten to 15 ft. lbs. (21 Nm), and 81 inch lbs. (9.1 Nm).

➡Make sure that there is no oil on the threads of the A bolts.

➡Be sure to replace the 2 C bolts with new ones or reuse them after applying adhesive (Part No. 08833-00080, Three Bond® 1344 or equivalent).

18. Install vane pump assembly.
19. Install the idler pulley with the bolt and tighten to 32 ft. lbs. (43 Nm).

➡Be careful when tightening the bolt because it is left-hand threaded.

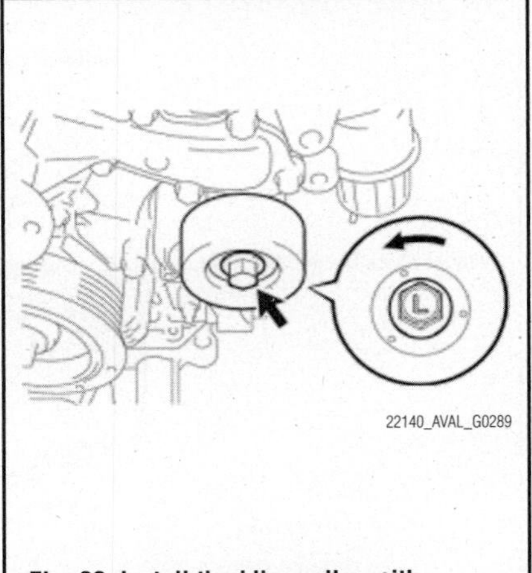

Fig. 33 Install the idler pulley with the bolt

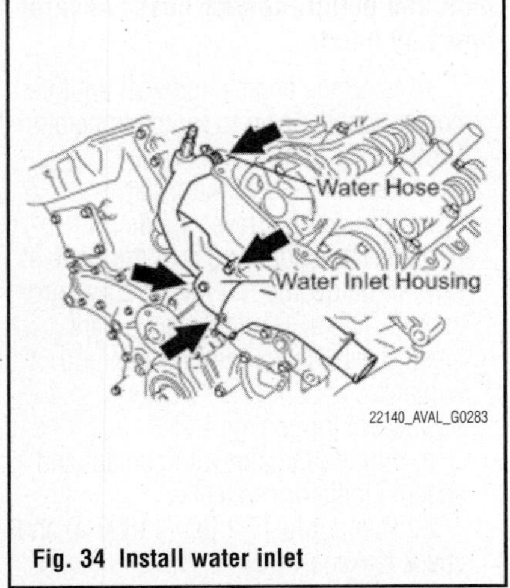

Fig. 34 Install water inlet

20. Install the No. 2 idler pulley cover plate and No. 2 idler pulley sub-assembly with the bolt tighten to 32 ft. lbs. (43 Nm).
21. Install the water pump pulley, as follows:
22. Temporarily install the water pump pulley with the 4 bolts.
    a. Using SST (SST: 09960-10010), hold the water pump pulley.
    b. Tighten the 4 bolts to 15 ft. lbs. (21 Nm).
23. Install crankshaft pulley.
24. Install water inlet housing, as follows:
    a. Install a new No. 1 water inlet housing gasket and water outlet pipe O-ring.
    b. Install the water inlet with the 2 bolts and nut and tighten to 7 ft. lbs. (10 Nm).

➡Be careful not to allow the O-ring to get caught between the parts.

c. Install the water hose.
25. Install the radiator hose outlet.
26. Install the fan and alternator V-belt.
27. Install the left front No. 1 engine mounting bracket.
28. Install the engine mounting control bracket.
29. Install the engine moving control rod.
30. Install the V-bank cover sub-assembly.
31. Add engine coolant.
32. Inspect for coolant leak.
33. Install the right engine under cover.
34. Install the right front wheel.

ENGINE ELECTRICAL — BATTERY SYSTEM

BATTERY

REMOVAL & INSTALLATION

1. Disconnect negative battery cable.
2. Disconnect positive battery cable.
3. Remove battery hold down bracket bolt from body.
 a. Pivot hold down bracket out of way of battery.
4. Remove battery.

To install

5. Reverse removal procedure to install.

➡**Clean battery cable ends and apply battery terminal protector to the cable ends.**

6. Perform initialization.

ENGINE ELECTRICAL — CHARGING SYSTEM

ALTERNATOR

REMOVAL & INSTALLATION

See Figure 35.

1. Disconnect the negative battery cable.
2. Remove the V-bank cover sub-assembly.
3. Remove the V-ribbed belt.
4. Remove the alternator assembly, as follows:
 a. Disconnect the wire harness clamp.
 b. Remove the terminal cap.
 c. Remove the nut and disconnect the wire harness from terminal B.
 d. Disconnect the alternator connector from the alternator assembly.
 e. Remove the nut from the cylinder block.
 f. Remove the 2 bolts and alternator assembly.

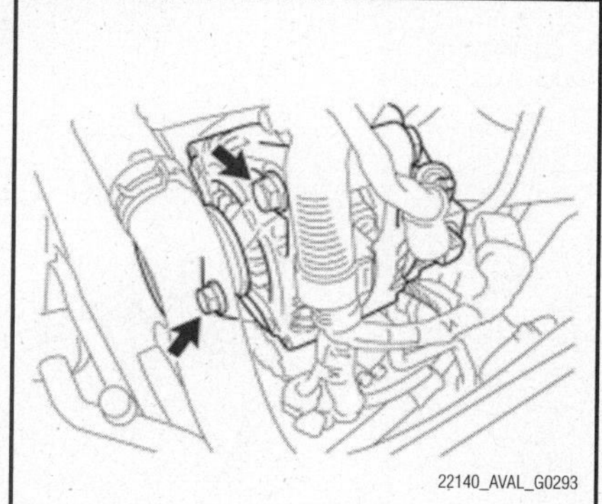

Fig. 35 Remove the 2 bolts and alternator assembly

 g. Remove the bolt and wire harness clamp stay.
 h. Remove the bolt and bracket.

To install:

5. Install the alternator assembly, as follows:
 a. Install the bracket with the bolt and tighten to 15 ft. lbs. (20 Nm).
 b. Install the wire harness clamp stay and tighten to 74 inch lbs. (8.4 Nm).
 c. Install the alternator assembly with the 2 bolts and tighten to 32 ft. lbs. (43 Nm).
 d. Install the nut to the cylinder block and tighten to 15 ft. lbs. (20 Nm).
 e. Connect the alternator connector to the alternator assembly.
 f. Install the alternator wire with the nut and tighten to 87 inch lbs. (9.8 Nm).
 g. Install the terminal cap.
 h. Connect the wire harness clamp.
6. Install the V-ribbed belt.
7. Install the V-bank cover sub-assembly.
8. Connect the negative battery cable.
9. Perform initialization.

ENGINE ELECTRICAL — IGNITION SYSTEM

FIRING ORDER

Firing order for 3.5L engine: 1–2–3–4–5–6

IGNITION COIL

REMOVAL & INSTALLATION

See Figure 36.

1. Disconnect the negative battery cable.
2. Remove the V-bank cover.
3. Remove the intake air surge tank.
4. Disconnect the 6 ignition coil connectors.
5. Remove the 6 bolts and 6 ignition coils.

To install:

6. Installation is the reverse of removal. Torque the ignition coils to 66 inch lbs. (7.5 Nm).

IGNITION TIMING

ADJUSTMENT

This vehicle is equipped with a Distributorless Ignition System (DIS). No timing adjustment is possible.

SPARK PLUGS

REMOVAL & INSTALLATION

1. Disconnect the negative battery cable.
2. Remove the V-bank cover.
3. Remove the intake air surge tank.
4. Disconnect the 6 ignition coil connectors.
5. Remove the 6 bolts and 6 ignition coils.
6. Using a 16 mm (0.63 in.) plug wrench, remove the spark plugs.

To install:

7. Installation is the reverse of removal, noting the following:
 a. Torque the ignition coils to 66 inch lbs. (7.5 Nm) and the spark plugs to 13 ft. lbs (18 Nm).

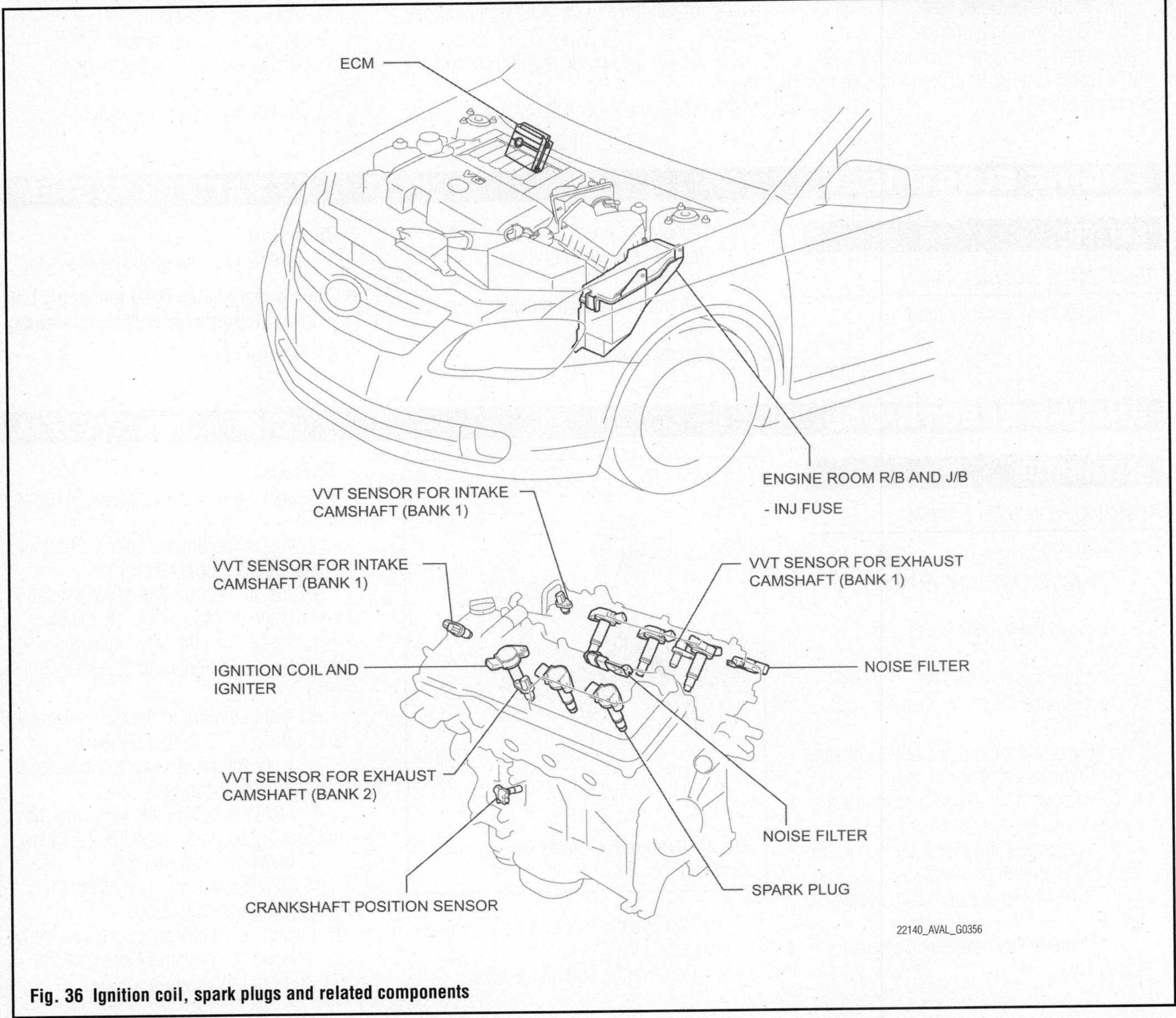

Fig. 36 Ignition coil, spark plugs and related components

ENGINE ELECTRICAL

STARTING SYSTEM

STARTER

REMOVAL & INSTALLATION

See Figure 37.

1. Disconnect the negative battery cable.
2. Remove the air cleaner assembly with hose.
3. Remove the No. 1 air cleaner inlet.
4. Remove the starter assembly, as follows:
    a. Disconnect terminal 50 of the connector from the starter assembly.
    b. Open the terminal cap, remove the nut and disconnect the wire harness from terminal 30.
    c. Remove the 2 bolts and the starter assembly.

To install:

5. Install the starter assembly with the 2 bolts and tighten to 26 ft. lbs. (37 Nm).
6. Connect the wire harness to terminal 30 and install the nut and tighten to 87 inch lbs. (9.8 Nm).
7. Cover the nut with the cap.
8. Connect terminal 50 to the starter assembly.
9. Install the No. 1 air cleaner inlet.
10. Install the air cleaner assembly with hose.
11. Connect the negative battery cable.
12. Perform initialization.

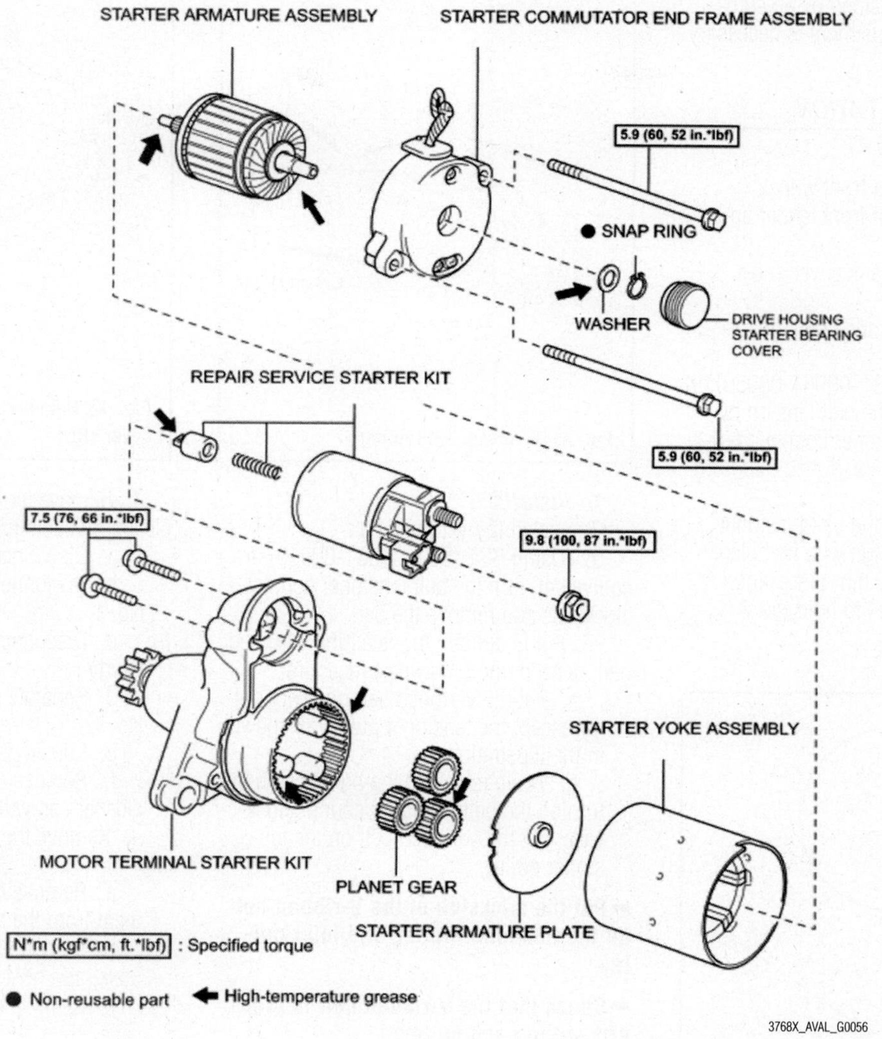

Fig. 37 Starter and related component locations

ENGINE MECHANICAL

➡Disconnecting the negative battery cable may interfere with the functions of the on board computer systems and may require the computer to undergo a relearning process, once the negative battery cable is reconnected.

ACCESSORY DRIVE BELTS

ACCESSORY BELT ROUTING

See Figure 38.

Refer to the accompanying illustration for Drive Belt routing.

INSPECTION

Visually check the V-ribbed belt for excessive wear, frayed cords, etc. All worn or damaged drive belts should be replaced immediately. Cracks on the rib side of a V-ribbed belt are considered acceptable, If the drive belt has chunks missing from its ribs, it should be replaced. After installing the V-ribbed belt, check that it fits properly in the ribbed grooves. Check to confirm that the belt has not slipped out of the grooves on the bottom of the crank pulley by hand.

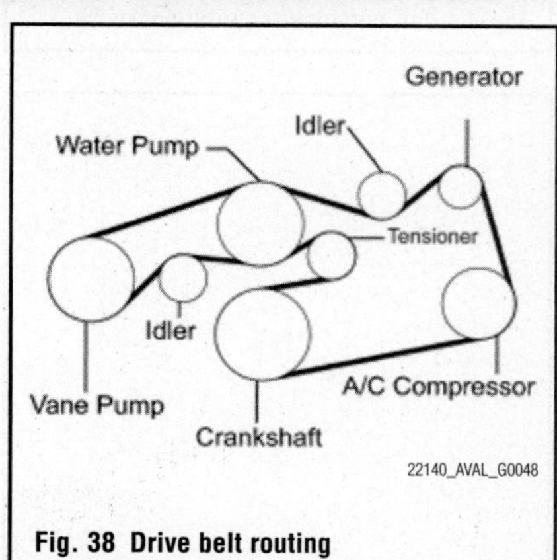

Fig. 38 Drive belt routing

ADJUSTMENT

Belt tension is maintained by an automatic tensioner. No adjustment is necessary or possible.

REMOVAL & INSTALLATION

See Figures 39 through 41.

1. Remove the right front wheel.
2. Remove the right front fender apron seal.
3. Remove the V-bank cover sub-assembly.
4. Remove the V-ribbed belt, as follows:
   a. Using SST (SST: 09961-00950) or equivalent, release the belt tension by turning the belt tensioner counterclockwise, and remove the V-ribbed belt from the belt tensioner.
   b. While turning the belt tensioner counterclockwise, align with its holes, and then insert the 5 mm bi-hexagon wrench into the holes to hold the V-ribbed belt tensioner.

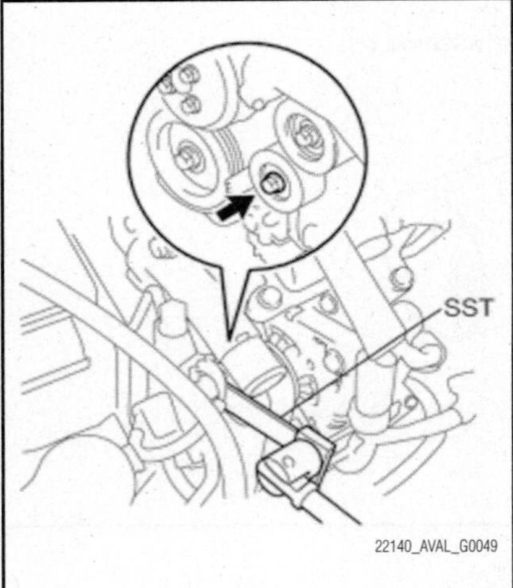

Fig. 39 Release the belt tension by turning the belt tensioner counterclockwise

Fig. 40 Align the tensioner holes

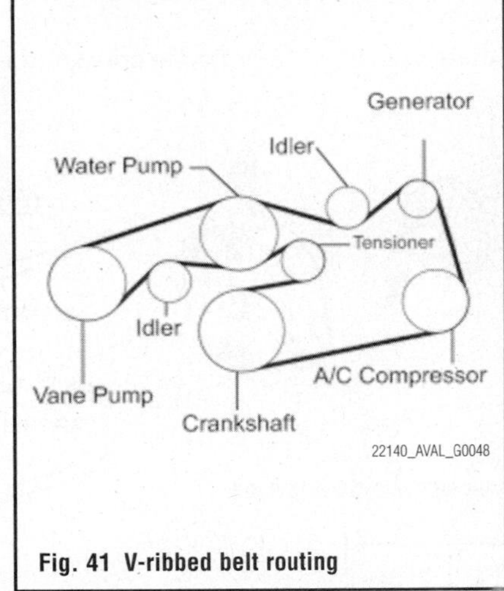

Fig. 41 V-ribbed belt routing

To install:

5. Install the V-ribbed belt.
6. Using SST (SST: 09961-00950) or equivalent, turn the belt tensioner counterclockwise and remove the bar.
7. If it is difficult to install the V-ribbed belt, perform the following procedure:
   a. Put the V-ribbed belt on every pulley except the tensioner pulley as shown in the illustration.
   b. While releasing the belt tension by turning the belt tensioner counterclockwise, put the V-ribbed belt on the tensioner pulley.

➡Put the backside of the V-ribbed belt on the tensioner pulley and idler pulley.

➡Check that the V-ribbed belt is properly set to each pulley.

8. Install the V-bank cover sub-assembly.
9. Install the right front fender apron seal.
10. Install the right front wheel and tighten the lug nuts to 76 ft. lbs. (103 Nm).

AIR CLEANER

REMOVAL & INSTALLATION

See Figures 42 and 43.

1. Remove v-bank cover sub-assembly.
   a. Disengage the 3 clips and remove the V-bank cover.

➡Be sure to disengage the clips in the order from A to B.

2. Remove no. 2 air cleaner inlet.
   a. Disconnect the clamp and 2 vacuum hoses.
   b. Remove the 2 bolts and No. 2 air cleaner inlet.

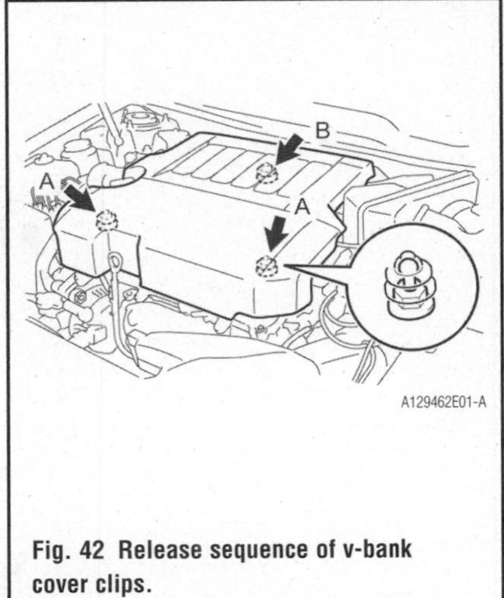

Fig. 42 Release sequence of v-bank cover clips.

3. Remove air cleaner cap with air cleaner hose as follows:
   a. Disconnect the 3 vacuum hoses.
   b. Disconnect the ventilation hose No. 1.
   c. Disconnect the air flow meter connector.
   d. Separate the fuel vapor feed hose No. 1.
   e. Loosen the hose clamp bolt.
   f. Remove the 3 clamps and air cleaner cap with air cleaner hose.
4. Remove the lower air cleaner housing.
   a. Remove the air cleaner filter element from the air cleaner case.
   b. Disconnect the connector, vacuum hose, and hose clamp.
   c. Remove the 3 bolts and air cleaner case.

To install

5. Reverse removal procedure to install the air cleaner housing assembly

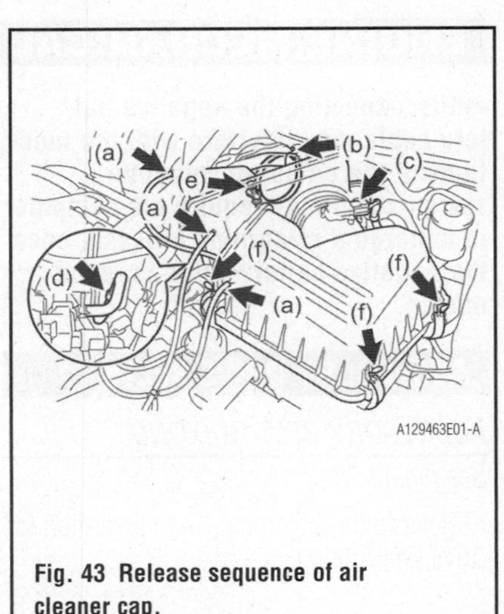

Fig. 43 Release sequence of air cleaner cap.

FILTER/ELEMENT REPLACEMENT

1. Remove the 3 clamps and pull upward the air cleaner cap to gain access to the element.
2. Remove air filter element.

To install

➡Clean out any debris in lower housing prior to installing the new element.

3. Reverse removal procedure to install the air cleaner housing assembly

CAMSHAFT AND VALVE LIFTERS

INSPECTION

See Figures 44 through 49.

1. To inspect the camshaft, place the camshaft on V-blocks.
   a. Using a dial indicator, measure the circle runout at the center journal.
   b. Maximum runout: 0.0016 inches

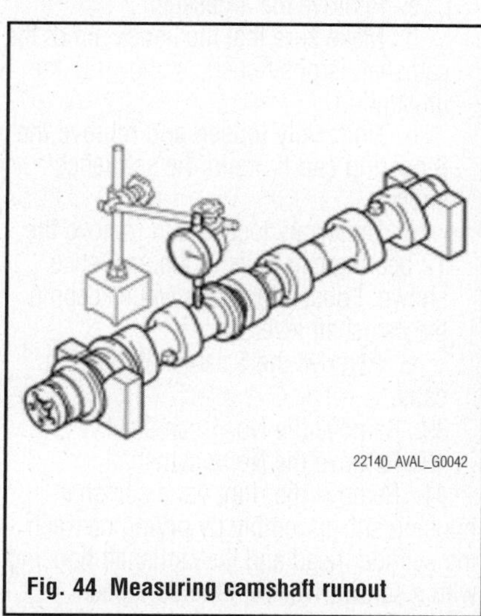

Fig. 44 Measuring camshaft runout

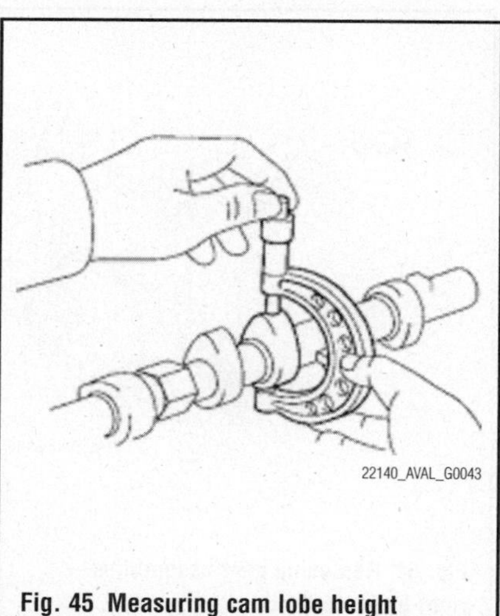

Fig. 45 Measuring cam lobe height

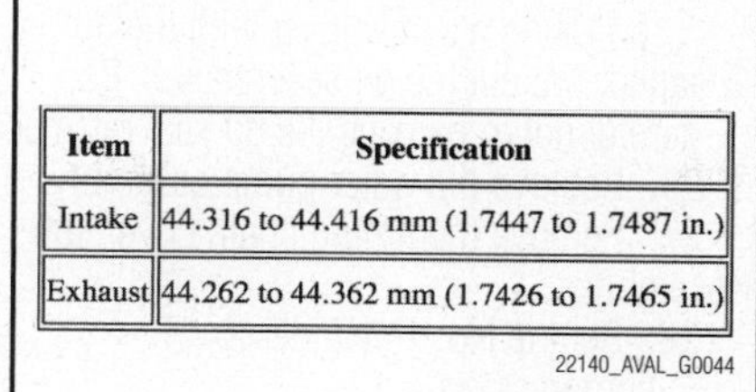

Item	Specification
Intake	44.316 to 44.416 mm (1.7447 to 1.7487 in.)
Exhaust	44.262 to 44.362 mm (1.7426 to 1.7465 in.)

22140_AVAL_G0044

Fig. 46 Standard cam lobe height

Item	Specification
Intake	44.166 mm (1.7388 in.)
Exhaust	44.112 mm (1.7367 in.)

22140_AVAL_G0045

Fig. 47 Maximum cam lobe height

Item	Specification
No. 1 journal	35.946 to 35.960 mm (1.4152 to 1.4157 in.)
Other journal	25.959 to 25. 975 mm (1.0220 to 1.0226 in.)

22140_AVAL_G0046

Fig. 48 Standard journal diameter

(0.04 mm). If the runout is greater than the maximum, replace the camshaft.

➡Check the oil clearance after replacing the camshaft.

2. Using a micrometer, measure the cam lobe height.
3. Using a micrometer, measure the journal diameter.
   a. If the journal diameter is not as specified, check the oil clearance.
4. The valve lash adjuster assembly can be inspected, as follows:

➡Keep the lash adjuster free of dirt and foreign objects. Only use clean engine oil.

   a. Place the lash adjuster into a container filled with engine oil.

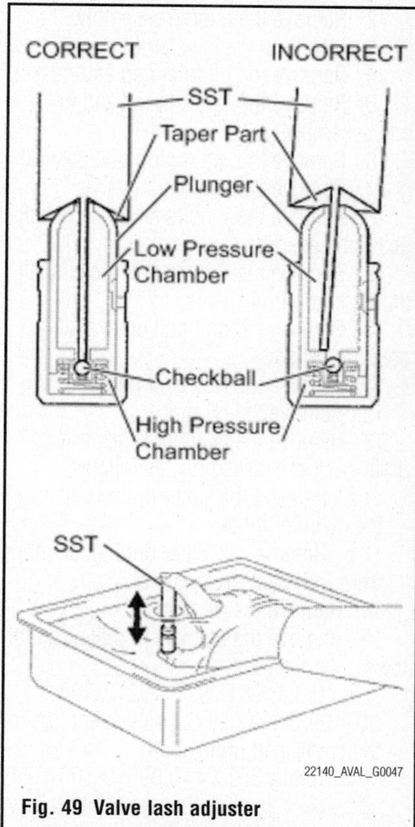

Fig. 49 Valve lash adjuster

   b. Insert the SST's tip (SST: 09276-75010) into the lash adjuster's plunger and use the tip to press down on the check ball inside the plunger.
   c. Squeeze the SST and lash adjuster together to move the plunger up and down 5 to 6 times.
   d. Check the movement of the plunger and bleed the air. OK: Plunger moves up and down.

➡When bleeding air from the high-pressure chamber, make sure that the tip of the SST is actually pressing the check ball as shown in the illustration. If the check ball is not pressed, air will not bleed.

   e. After bleeding the air, remove the SST. Then, try to press the plunger quickly and firmly with a finger. OK: Plunger is very difficult to move. If the result is not as specified, replace the lash adjuster.
   f. Install the lash adjusters. Install the lash adjuster to the same place where it was removed from.

REMOVAL & INSTALLATION

See Figures 50 through 76.

1. Before servicing the vehicle, refer to the Precautions Section.

2. Remove the engine assembly.
3. Install on engine stand.
4. Remove the oil filler cap and gasket.
5. Remove the spark plugs and ignition coil assembly.
6. Remove the drain plug and gasket.
7. Remove the ventilation valve.
8. Remove the 4 bolts and 4 camshaft position sensors.
9. Remove the 4 bolts and 4 camshaft timing oil control valves.
10. Remove the bolt and crankshaft position sensor.
11. Remove the No. 1 oil pipe.
12. Remove the oil pipe.
13. Remove the cylinder block water drain cock sub-assembly, as follows:
   a. Remove the water drain cocks from the cylinder block.
   b. Remove the water drain cock plugs from the water drain cocks.
14. Remove the oil filter.
15. Remove the crankshaft pulley, as follows:
   a. Using SST (SST: 09213-70011, SST: 09330-00021) or equivalent, loosen the crankshaft pulley bolt.
   b. Using SST (SST: 09950-50013) or equivalent, remove the crankshaft pulley bolt and crankshaft pulley.
16. Remove the 6 bolts and the left hand No. 1 front engine mounting bracket.
17. Remove the water inlet housing, as follows:
   a. Remove the 2 nuts, water inlet and thermostat.
   b. Remove the gasket.
   c. Remove the drain cock plug.
   d. Remove the drain cock.
   e. Remove the 2 stud bolts.
   f. Remove the 2 bolts, nut, and water inlet housing.
   g. Remove the 2 O-rings.
18. Remove the water outlet, as follows:
   a. Remove the 2 bolts, 4 nuts and water outlet.
   b. Remove the 2 gaskets and O-ring.
19. Remove the 12 bolts, cylinder head cover sub-assembly (for Bank 1), and gasket.
20. Remove the 12 bolts, cylinder head cover sub-assembly (for Bank 2), and gasket.
21. Remove the No. 2 oil pan sub-assembly.
22. Remove the oil strainer sub-assembly.
23. Remove the oil pan sub-assembly.
24. Remove the No. 1 oil pan baffle plate.
25. Remove the engine rear oil seal, as follows:
   a. Remove the 6 bolts.
   b. Using a screwdriver with the tip taped, pry out the oil seal retainer. Be careful not to damage the oil seal retainer.
26. Remove the water pump assembly.
27. Remove the timing chain cover sub-assembly.
28. Set the No. 1 cylinder to TDC/compression.
29. Remove the No. 1 chain tensioner assembly.
30. Remove the chain tensioner slipper.
31. Remove the chain sub-assembly.
32. Remove the idle sprocket assembly.
33. Remove the No. 1 chain vibration damper.
34. Remove the No. 2 chain vibration damper.
35. Remove the crankshaft timing gear, as follows:
   a. Remove the pulley set bolt.
   b. Remove the crankshaft timing gear from the crankshaft.

Fig. 50 Crankshaft timing gear

   c. Remove the 2 pulley set keys from the crankshaft.
36. Remove the camshaft timing gears and No. 2 chain (for right-hand bank), as follows:
   a. While raising the No. 2 chain tensioner, insert a pin of 0.039 inch (1.0 mm) into the hole to hold the No. 2 chain tensioner.
   b. Hold the hexagonal portion of the camshaft with a wrench, and remove the 2 bolts and 2 camshaft timing gears.

➡Be careful not to damage the cylinder head with the wrench.

➡Do not disassemble the camshaft timing gear assemblies.

   c. Remove the No. 2 chain.
37. Remove the bolt and No. 2 chain tensioner assembly.
38. Remove the camshaft bearing cap, as follows:
   a. Remove the 3 gaskets.
   b. Make sure that the knock pin of the camshaft is positioned as shown in the illustration.
   c. Uniformly loosen and remove the 8 bearing cap bolts in the sequence shown.
   d. Uniformly loosen and remove the 12 bearing cap bolts in the sequence shown. Loosen the bolts while keeping the camshaft level
   e. Remove the 5 camshaft bearing caps.
39. Remove the No. 1 camshaft.
40. Remove the No. 2 camshaft.
41. Remove the right hand camshaft housing sub-assembly by prying between the cylinder head and the camshaft housing with a screwdriver with the tip taped.

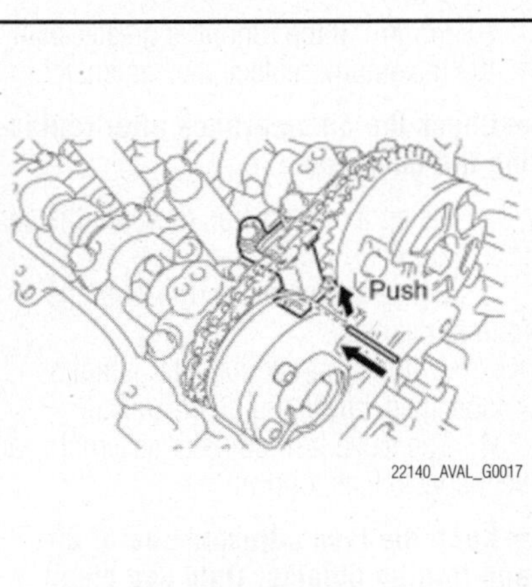

Fig. 51 Pinning No. 2 tensioner—right-hand bank

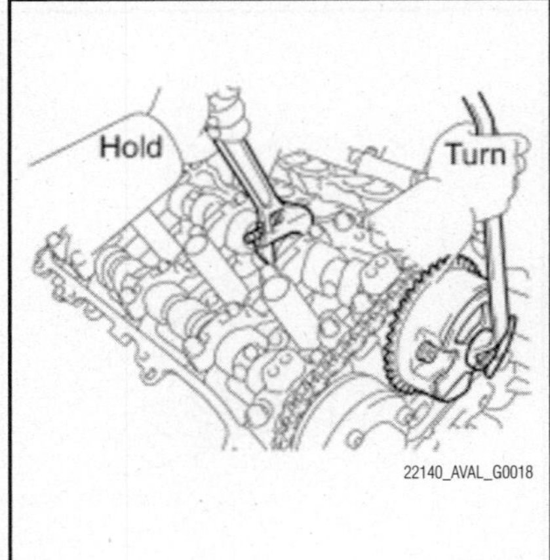

Fig. 52 Removing gear assemblies—right-hand bank

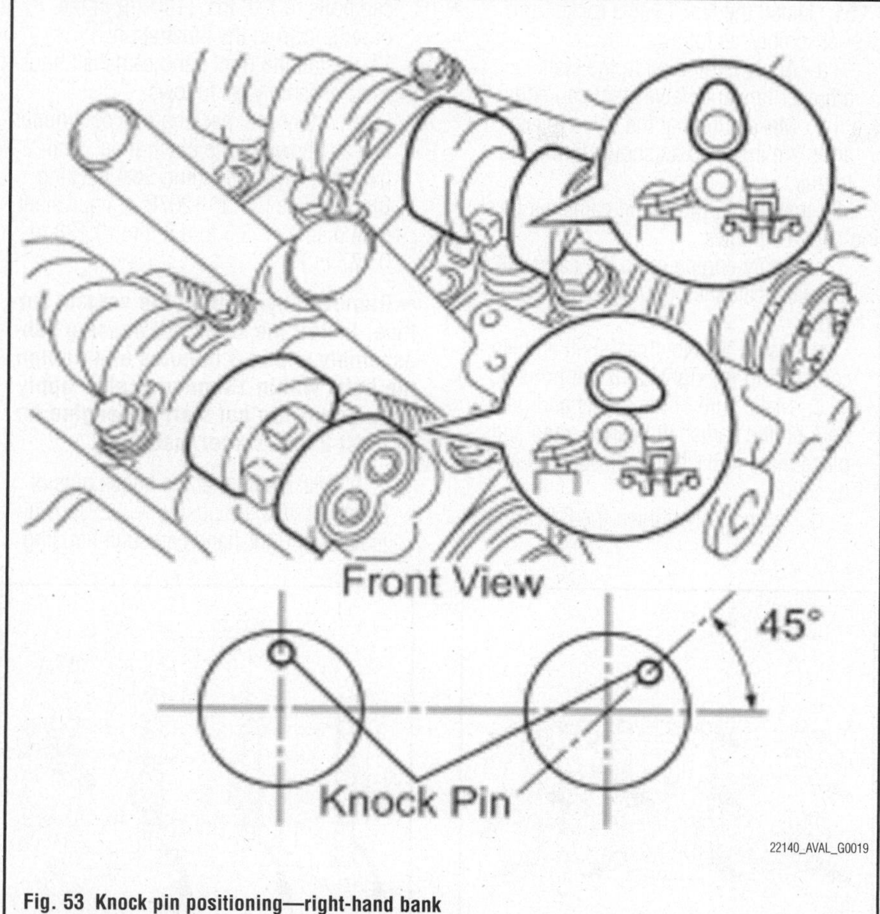

Fig. 53 Knock pin positioning—right-hand bank

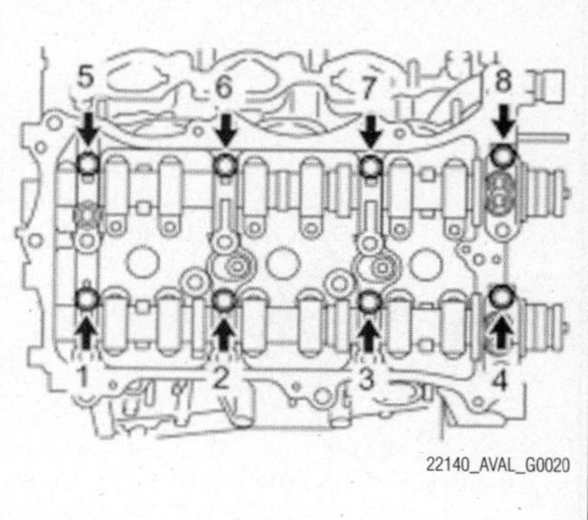

Fig. 54 Bearing cap 8 bolt removal sequence—right-hand bank

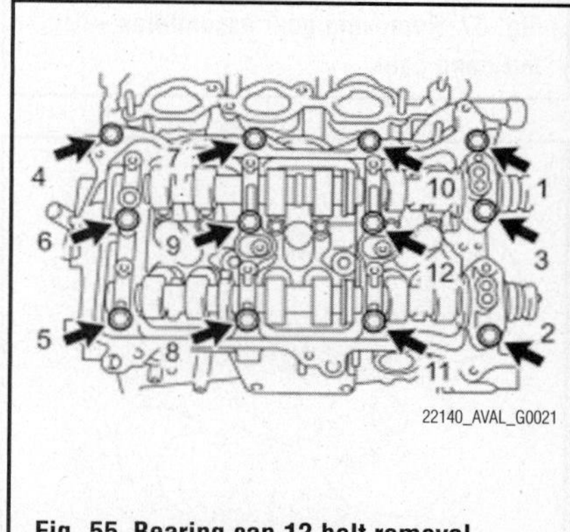

Fig. 55 Bearing cap 12 bolt removal sequence—right-hand bank

➡Be careful not to damage the contact surfaces of the cylinder head and the camshaft housing.

42. Remove the camshaft timing gears and No. 2 chain (for left-hand bank), as follows:
   a. While pushing down the No. 3 chain tensioner, insert a pin of 1.0 mm (0.039 in.) into the hole to hold the No. 3 chain tensioner.
   b. Hold the hexagonal portion of the camshaft with a wrench, and remove the 2 bolts and 2 camshaft timing gears.

➡Be careful not to damage the cylinder head with the wrench.

➡Do not disassemble the camshaft timing gear assemblies.

   c. Remove the No. 2 chain.
43. Remove the bolt and No. 3 chain tensioner.
44. Remove the camshaft bearing cap, as follows:
   a. Remove the 3 gaskets.
   b. Make sure that the knock pin of the camshaft is positioned as shown in the illustration.
   c. Uniformly loosen and remove the 8 bearing cap bolts in the sequence shown in the illustration.
   d. Uniformly loosen and remove the 13 bearing cap bolts in the sequence shown in the illustration. Loosen the bolts while keeping the camshaft level
   e. Remove the 5 camshaft bearing caps.
45. Remove the No. 4 camshaft.
46. Remove the No. 3 camshaft.
47. Remove the left hand camshaft housing sub-assembly by prying between the cylinder head and the camshaft housing with a screwdriver with the tip taped.

➡Be careful not to damage the contact surfaces of the cylinder head and the camshaft housing.

48. Remove the No. 1 valve rocker arm sub-assembly, as follows:
   a. Remove the 24 valve rocker arms.

➡Arrange the removed parts in the correct order.

49. Remove the valve lash adjuster assembly, as follows:
   a. Remove the 24 valve lash adjusters from the cylinder head.

➡Arrange the removed parts in the correct order.

To install:

50. Install the valve lash adjuster assembly.

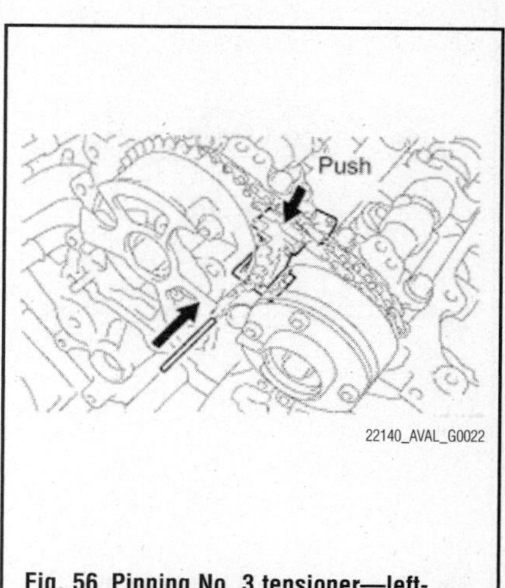

Fig. 56 Pinning No. 3 tensioner—left-hand bank

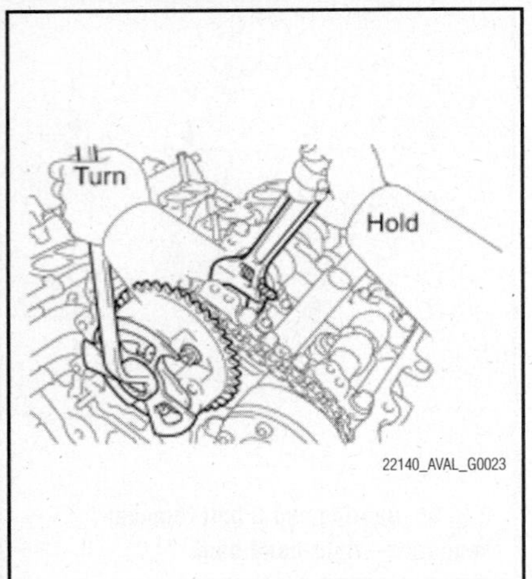

Fig. 57 Removing gear assemblies—left-hand bank

51. Install the No. 1 valve rocker arm sub-assembly, as follows:
   a. Apply engine oil to the lash adjuster tips and valve stem cap ends.
   b. Make sure that the valve rocker arms are installed as shown in the illustration.
52. Install the right-hand camshaft bearing cap, as follows:
   a. Apply engine oil to the camshaft journals, camshaft housing and bearing caps.
   b. Install the No. 1camshaft and No. 2 camshaft to the right camshaft housing.
   c. Make sure of the marks and numbers on the camshaft bearing caps and place them in each proper position and direction.
   d. Temporarily tighten the 8 bearing cap bolts to 7 ft. lbs. (10 Nm) in the order shown in the illustration.
53. Install the right hand camshaft housing sub-assembly, as follows:
   a. Apply seal packing in a continuous line as shown in the illustration. Seal packing: Toyota Genuine Seal Packing Black, Three Bond® 1207B or equivalent. Seal diameter: 3.5 to 4.5 mm (0.138 to 0.177 in.).

➡Remove any oil from the contact surface. Install the camshaft housing sub-assembly within 3 minutes and tighten the bolts within 15 minutes after applying sealant. Do not start the engine for at least 2 hours after installing.

   b. Make sure that the knock pins of the camshafts are positioned as shown. Install the right-hand camshaft housing

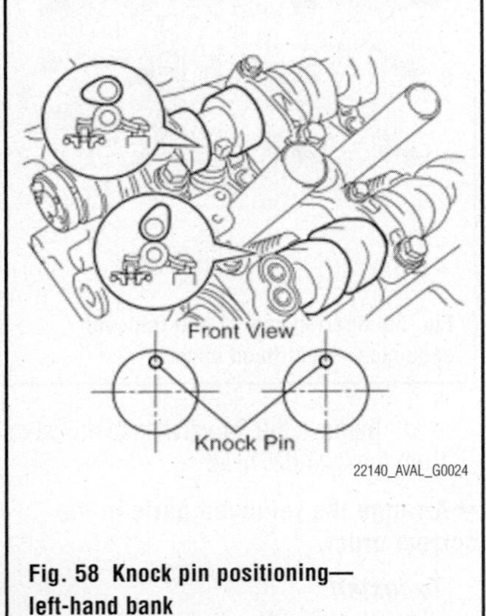

Fig. 58 Knock pin positioning—left-hand bank

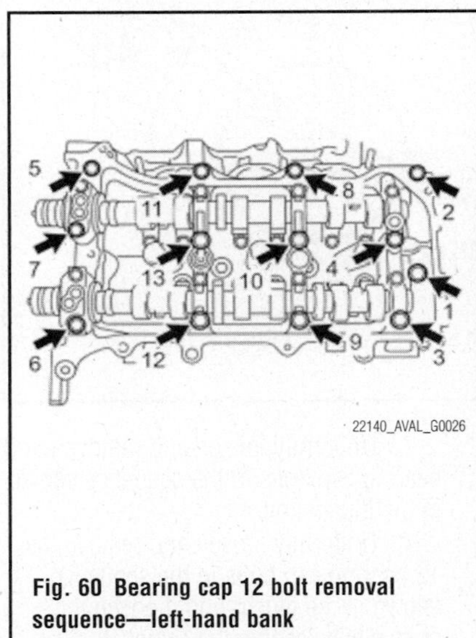

Fig. 60 Bearing cap 12 bolt removal sequence—left-hand bank

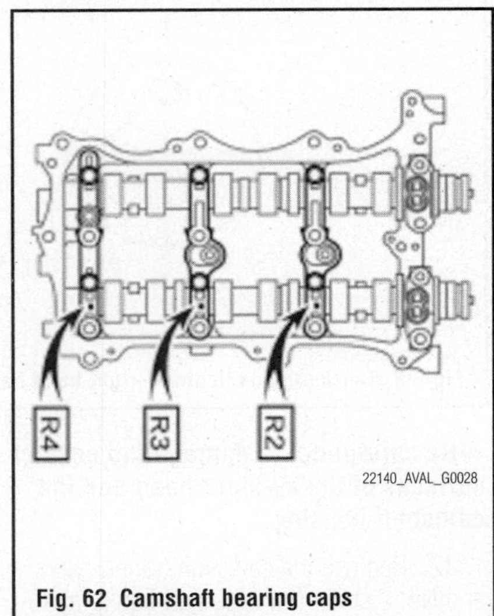

Fig. 62 Camshaft bearing caps placement—right-hand

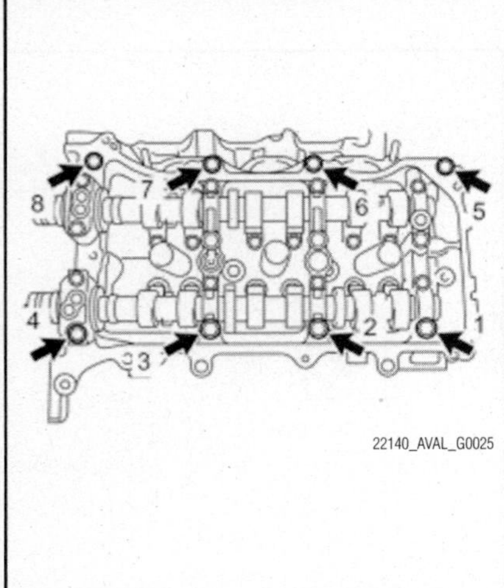

Fig. 59 Bearing cap 8 bolt removal sequence—left-hand bank

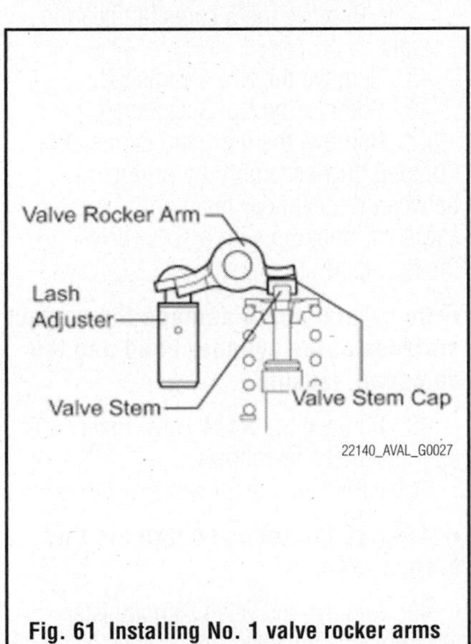

Fig. 61 Installing No. 1 valve rocker arms

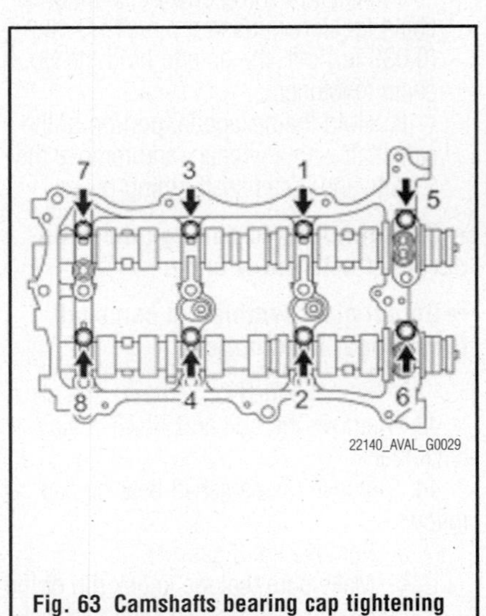

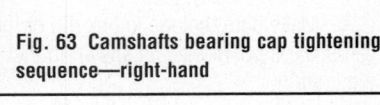

Fig. 63 Camshafts bearing cap tightening sequence—right-hand

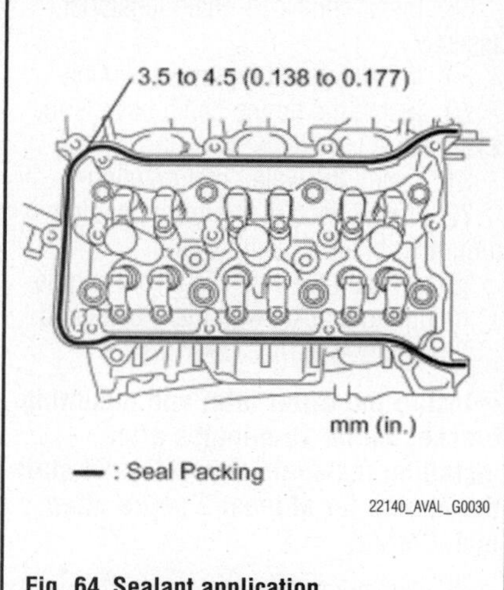

Fig. 64 Sealant application

and tighten the 12 bolts in the order shown in the illustration to 21 ft. lbs. (28 Nm).

c. Tighten the 8 bolts to 12 ft. lbs. (16 Nm) in the order shown in the illustration.

➡Thoroughly wipe clean any seal packing.

d. Install 3 new gaskets.

54. Install the camshaft bearing cap, as follows:

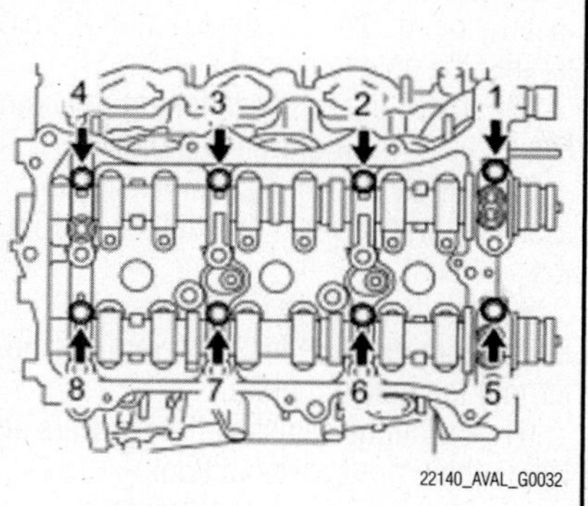

Fig. 66 Camshaft 8 bolt tightening sequence—right-hand

a. Apply engine oil to the camshaft journals, camshaft housing and bearing caps.

b. Install the No. 3 camshaft and No. 4 camshaft to the left hand camshaft housing.

c. Make sure of the marks and numbers on the camshaft bearing caps and place them in each proper position and direction.

d. Temporarily tighten the 8 bolts in the order shown in the illustration to 7 ft. lbs. (10 Nm).

55. Install the left camshaft housing sub-assembly, as follows:

a. Apply seal packing in a continuous line as shown in the illustration. Seal packing: Toyota Genuine Seal Packing Black, Three Bond® 1207B or equivalent.

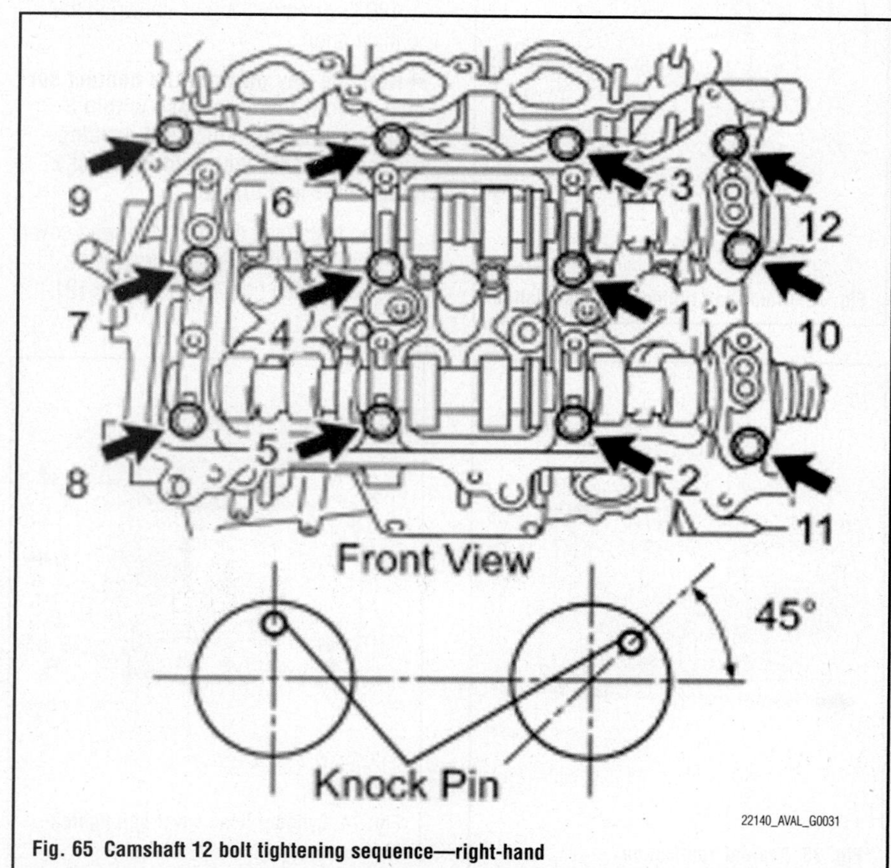

Fig. 65 Camshaft 12 bolt tightening sequence—right-hand

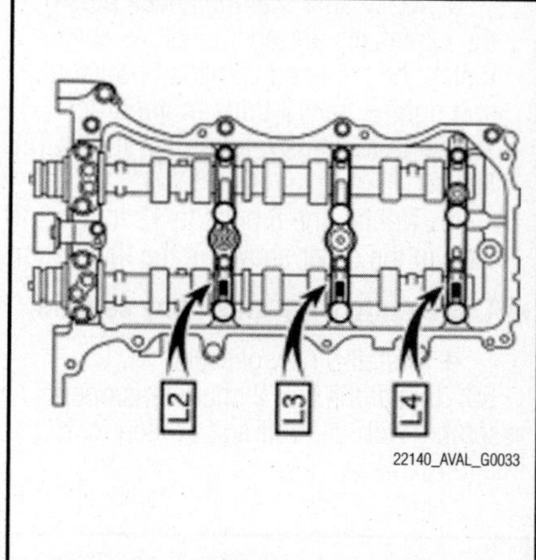

Fig. 67 Camshaft bearing cap positioning—left-hand

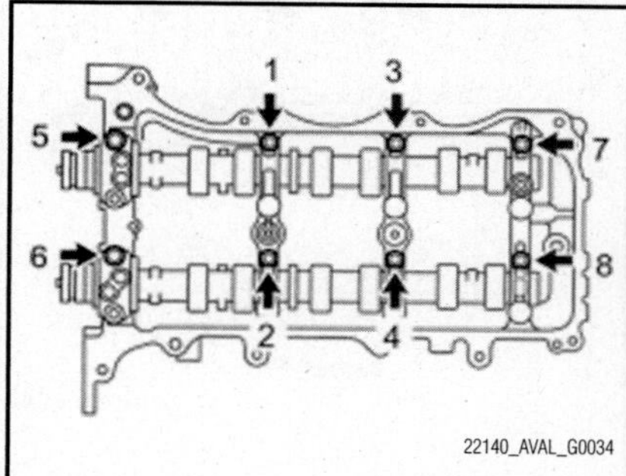

Fig. 68 Camshaft 8 bolt tightening sequence—left-hand

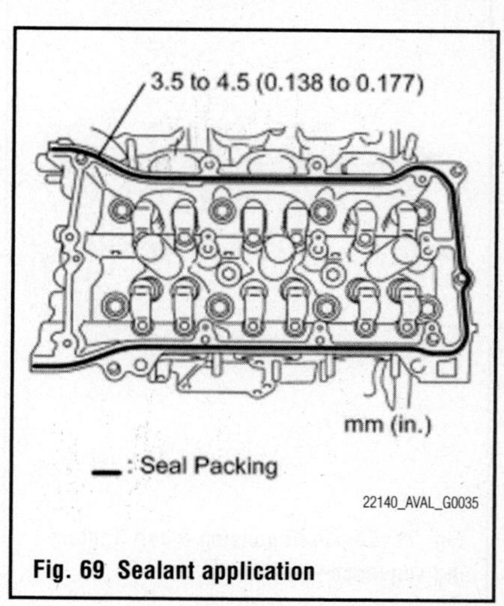

Fig. 69 Sealant application

Seal diameter: 3.5 to 4.5 mm (0.138 to 0.177 in.).

➡Remove any oil from the contact surface. Install the camshaft housing sub-assembly within 3 minutes and tighten the bolts within 15 minutes after applying sealant. Do not start the engine for at least 2 hours after installing.

b. Make sure that the knock pins of the camshafts are positioned as shown. Install the left hand camshaft housing and tighten the 13 bolts in the order shown in the illustration to 21 ft. lbs. (28 Nm).

c. Tighten the 8 bolts to 12 ft. lbs. (16 Nm) in the order shown in the illustration.

➡Thoroughly wipe clean any sealant.

d. Install 3 new gaskets.

56. Install the No. 2 chain tensioner assembly with the bolt and tighten to 15 ft. lbs. (21 Nm).

57. While pushing in the tensioner, insert a pin of 0.039 inch (1.0 mm) diameter into the hole to hold it.

58. Install the camshaft timing gears and No. 2 chain (for Right-hand Bank).

59. Install the No. 3 chain tensioner assembly with the bolt and tighten to 15 ft. lbs. (21 Nm).

60. While pushing in the tensioner, insert a pin of 0.039 inch (1.0 mm) diameter into the hole to hold it.

61. Install the camshaft timing gears and No. 2 chain (for Left-hand Bank).

62. Install the No 1 chain vibration damper with the 2 bolts and tighten to 17 ft. lbs. (23 Nm).

63. Install the No 2 chain vibration damper.

64. Install the timing gear set keys and timing gear as shown in the illustration.

65. Install the idle sprocket assembly.

66. Install the chain sub-assembly.

67. Install the chain tensioner slipper.

68. Install the No. 1 chain tensioner assembly.

69. Install the water pump assembly.

70. Install the timing chain cover sub-assembly.

71. Install the water inlet housing.

72. Install the No. 1 left front engine mounting bracket, as follows:

a. Install the No. 1 left front engine mounting bracket with the 6 bolts and tighten to 40 ft. lbs. (54 Nm).

➡Install the water inlet and mounting bracket within 15 minutes after installing the chain cover. Do not start the engine for at least 2 hours after installation.

73. Install the No. 1 oil pan baffle plate with the 7 bolts and tighten to 7 ft. lbs. (10 Nm).

74. Install the oil pan sub-assembly.

75. Install the oil strainer sub-assembly.

76. Install the No. 2 oil pan sub-assembly.

77. Install a new gasket and oil pan drain plug and tighten to 30 ft. lbs. (40 Nm).

78. Install the cylinder head cover sub-assembly, as follows:

a. Apply seal packing (Toyota Genuine Seal Packing Black, Three Bond® 1207B or equivalent) as shown in the illustration.

➡Remove any oil from the contact surface. Install the crankcase within 3 minutes after applying seal packing. Do not start the engine for at least 2 hours after installation.

b. Install the gasket to the head cover.

c. Install the head cover with the 12 bolts. Tighten bolt A to 15 ft. lbs. (21

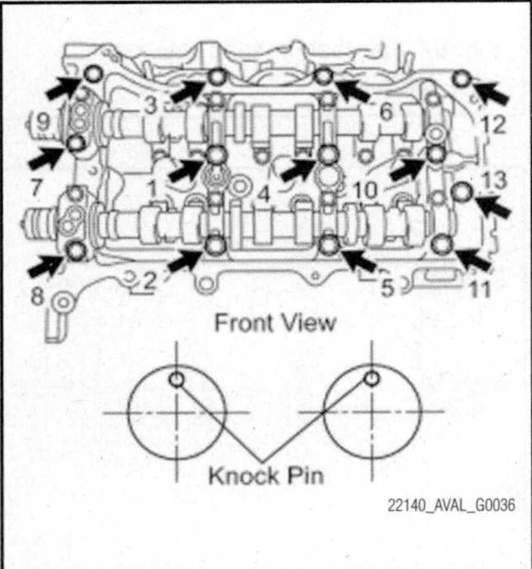

Fig. 70 Camshaft housing 13 bolt tightening sequence—left-hand

Fig. 72 Crankshaft timing gear installation

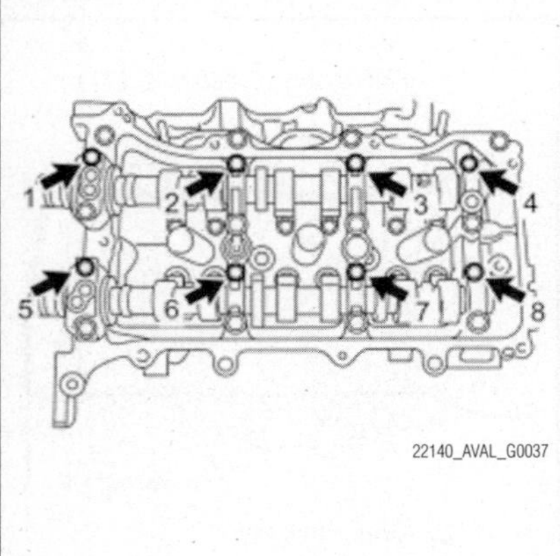

Fig. 71 Camshaft housing 8 bolt tightening sequence—left-hand

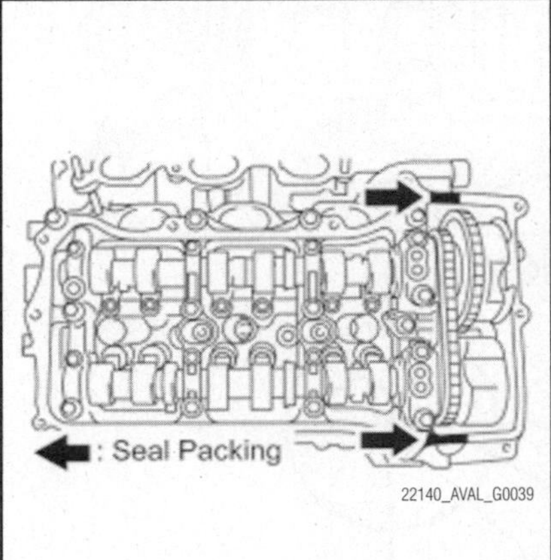

Fig. 73 Sealant application

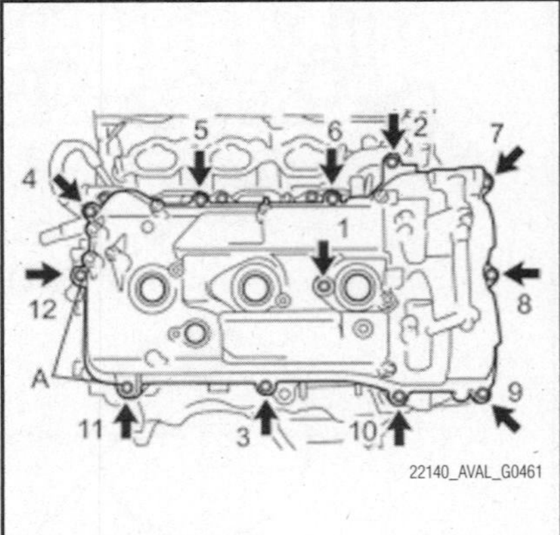

Fig. 74 Cylinder head cover bolt tightening sequence

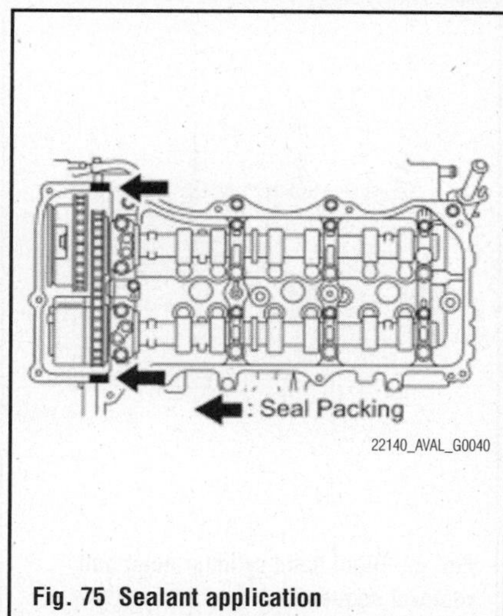

Fig. 75 Sealant application

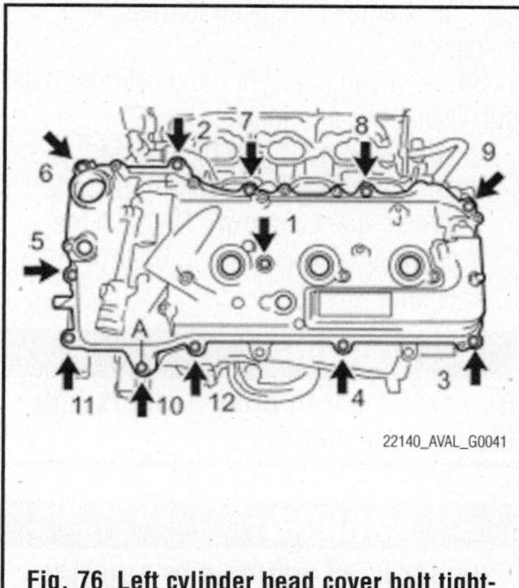

Fig. 76 Left cylinder head cover bolt tightening sequence

Nm), and other bolts to 7 ft. lbs. (10 Nm). Be certain to tighten bolt 1.

79. Install the left-hand cylinder head cover sub-assembly, as follows:

a. Apply seal packing (Toyota Genuine Seal Packing Black, Three Bond® 1207B or equivalent) as shown.

➡**Remove any oil from the contact surface. Install the crankcase within 3 minutes after applying seal packing. Do not start the engine for at least 2 hours after installation.**

b. Install the gasket to the head cover.

c. Install the head cover with the 14 bolts. Tighten bolt A to 15 ft. lbs. (21 Nm), and other bolts to 7 ft. lbs. (10 Nm). Be certain to tighten bolts 1 and 10.

80. Install water outlet.

81. Install the crankshaft pulley.

82. Install the oil filter element.

83. Install the cylinder block water drain cock sub-assembly, as follows:

a. Apply adhesive around the drain cocks. Adhesive: Toyota Genuine Adhesive 1324, Three Bond® 1324 or Equivalent.

b. Install the water drain cocks and tighten to 18 ft. lbs. (25 Nm). Do not rotate the drain cocks more than 1 revolution (360°) after tightening the drain cocks with the specified torque. Do not loosen after setting correctly.

c. Install the water drain cock plug to the water drain cocks and tighten to 9 ft. lbs. (13 Nm).

84. Install the No. 1 oil pipe.

85. Install the oil pipe.

86. Install the crankshaft position sensor with the bolt and tighten to 7 ft. lbs. (10 Nm).

87. Install the 4 camshaft timing oil control valves with the 4 bolts and tighten to 7 ft. lbs. (10 Nm).

88. Install the 4 camshaft position sensors with the 4 bolts and tighten to 7 ft. lbs. (10 Nm).

89. Install the ventilation valve sub-assembly, as follows:

a. Apply adhesive (Toyota Genuine Adhesive 1324, Three Bond® 1324 or equivalent) around the ventilation valve.

b. Install the ventilation valve and tighten to 20 ft. lbs. (27 Nm).

90. Install the 6 spark plugs and the ignition coil assembly.

91. Install the oil filler cap sub-assembly.

92. Remove the engine stand.

93. Install the engine assembly.

CRANKSHAFT FRONT SEAL

REMOVAL & INSTALLATION

See Figures 77 through 79.

1. Remove the RH front wheel.
2. Remove RH front fender apron seal.
3. Remove V-bank cover sub-assembly.
4. Remove drive belt.
5. Remove crankshaft pulley.

a. Using SST (SST: 09213-70011, SST: 09330-00021) or equivalent, loosen the crankshaft pulley bolt.

b. Using SST (SST: 09950-50013) or equivalent, remove the crankshaft pulley bolt and crankshaft pulley.

6. Using a medium screwdriver or small pry bar carefully on the seal lip to remove from cover.

➡**Be careful not to damage the crankshaft by applying tape to the prying tip of the pry tool.**

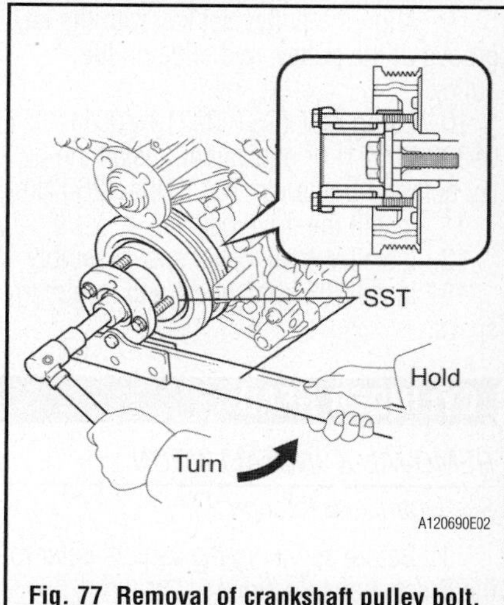

Fig. 77 Removal of crankshaft pulley bolt.

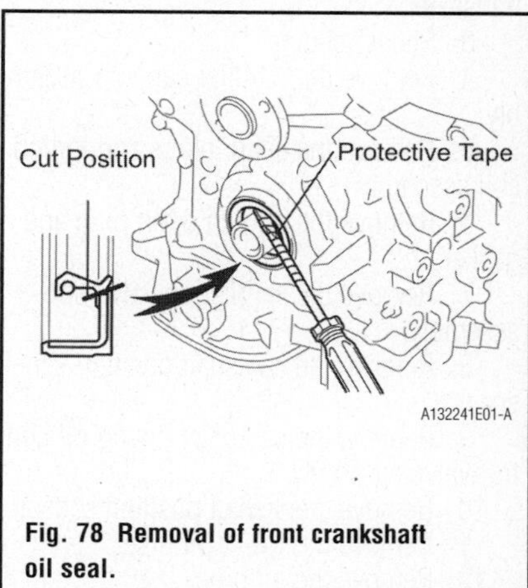

Fig. 78 Removal of front crankshaft oil seal.

To install:

7. Apply MP grease to a new timing chain case oil seal lip.

8. Using SST (SST: 09223-22010) or equivalent, tap in the seal until its flush with the timing chain cover edge.

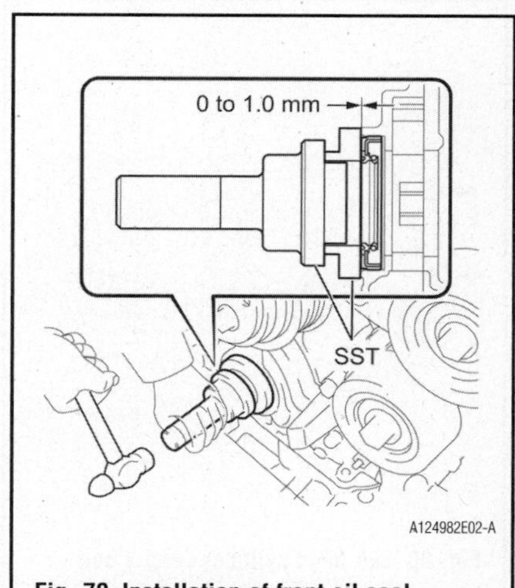

Fig. 79 Installation of front oil seal.

9. Align the pulley set key with the key groove of the pulley, and slide on the pulley.
10. Using SST (SST: 09213-70011, SST: 09330-00021) or equivalent, install the pulley bolt and tighten to 184 ft. lbs. (250 Nm).
11. Install the drive belt.
12. Install V-bank cover sub-assembly.
13. Install RH front wheel and torque to specs.

CYLINDER HEAD

REMOVAL & INSTALLATION

See Figures 80 through 88.

1. Before servicing the vehicle, refer to the Precautions Section.
2. Remove the engine assembly with transaxle.
3. Secure engine.
4. Remove the oil filler cap sub-assembly.
5. Remove the spark plugs and ignition coil assembly.
6. Remove the oil pan drain plug and gasket.
7. Remove the ventilation valve sub-assembly.
8. Remove the camshaft position sensor.
9. Remove the camshaft timing oil control valve assembly.
10. Remove crankshaft position sensor.
11. Remove the No. 1 oil pipe.
12. Remove the oil pipe.
13. Remove the cylinder block water drain cock sub-assembly.
14. Remove the oil filter.
15. Remove the crankshaft pulley.
16. Remove the left hand No. 1 front engine mounting bracket.
17. Remove the water inlet housing.

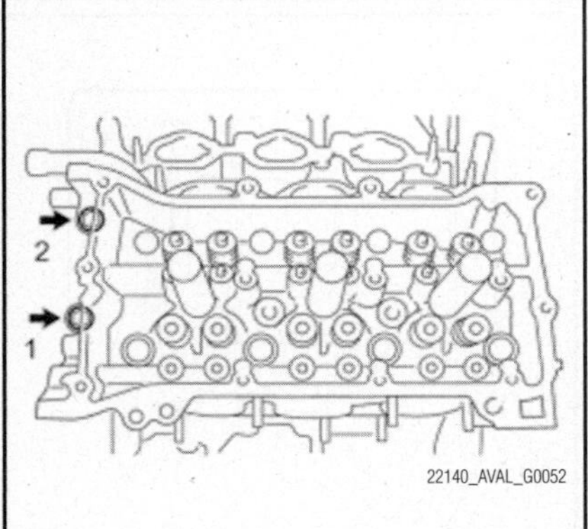

Fig. 80 Left hand cylinder head 2 bolt removal sequence

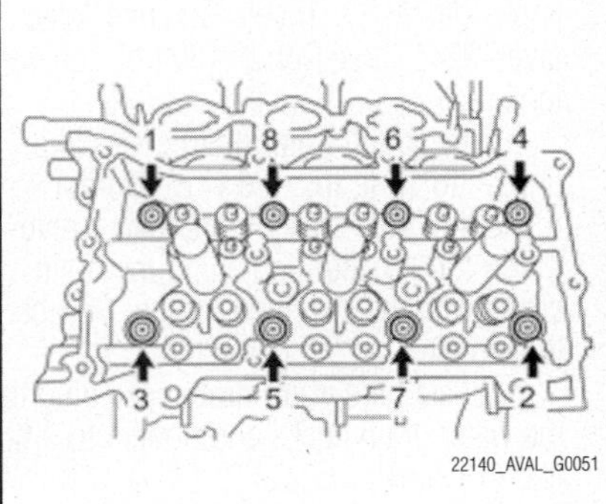

Fig. 81 Left hand cylinder head 8 bolt removal sequence

18. Remove the water outlet.
19. Remove the cylinder head covers and gaskets.
20. Remove the No. 2 oil pan sub-assembly.
21. Remove the oil strainer sub-assembly.
22. Remove the oil pan sub-assembly.
23. Remove the No. 1 oil pan baffle plate.
24. Remove the engine rear oil seal.
25. Remove the water pump assembly.
26. Remove the timing chain cover.
27. Set the No. 1 cylinder to TDC/compression.
28. Remove the No. 1 chain tensioner assembly.
29. Remove the chain tensioner slipper.
30. Remove the chain sub-assembly.
31. Remove the idle sprocket assembly.
32. Remove the No. 1 and 2 chain vibration damper.
33. Remove the left hand cylinder head sub-assembly, as follows:
    a. Uniformly loosen and remove the 2 bolts in the sequence shown.
    b. Using a 10 mm bi-hexagon wrench, uniformly loosen the 8 bolts in the sequence shown. Remove the 8 cylinder head bolts and plate washers.

WARNING

Be careful not to drop washers into the cylinder head.

WARNING

Cylinder head warpage or cracking could result from removing bolts in an incorrect order.

➡Be sure to keep separate the removed parts for each installation position.

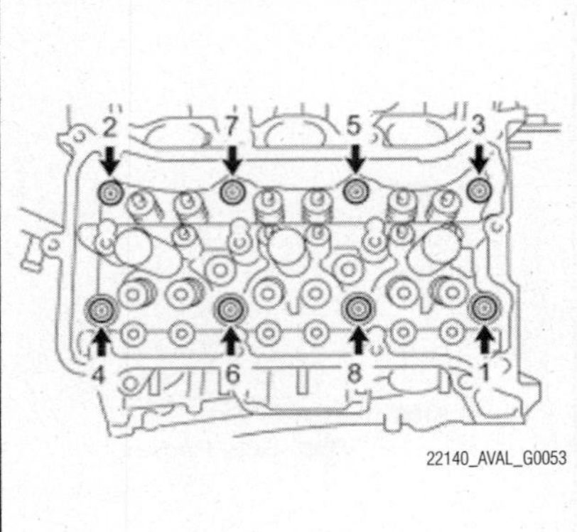

Fig. 82 Right hand cylinder head bolt removal sequence

    c. Remove the cylinder head and gasket.
34. Remove the right hand cylinder head sub-assembly, as follows:
    a. Using a 10 mm bi-hexagon wrench, uniformly loosen the 8 bolts in the sequence shown. Remove the 8 cylinder head bolts and plate washers.

WARNING

Be careful not to drop washers into the cylinder head.

WARNING

Cylinder head warpage or cracking could result from removing bolts in an incorrect order.

➡Be sure to keep separate the removed parts for each installation position.

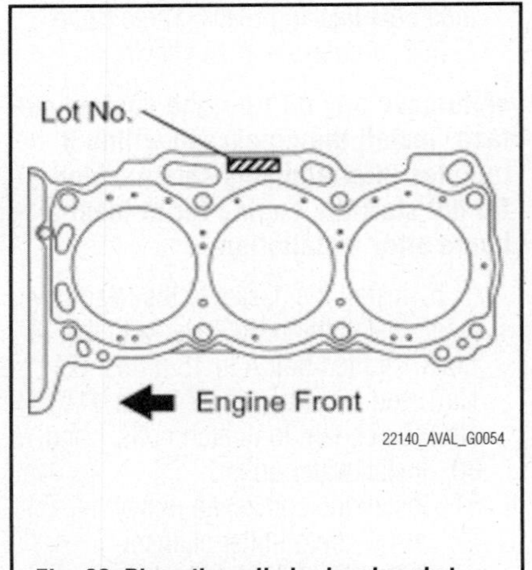

Fig. 83 Place the cylinder head gasket with Lot No. stamp upward

b. Remove the cylinder head and gasket.

To install:

35. Place the right hand cylinder head gasket on the cylinder block surface with the front face of the Lot No. stamp upward.

➡**Be careful of the installation direction.**

⁂ WARNING

Gently place the cylinder head in order not to damage the gasket with the bottom part of the head.

36. Place the cylinder head on the cylinder block.

➡**Do not allow oil to adhere to the mounting surface of the cylinder head.**

37. Apply a light coat of engine oil to the threads and under the heads of the cylinder head bolts.

38. The cylinder head bolts are tightened in 3 progressive steps:

a. Step 1: Using a 10 mm bi-hexagon wrench, install and uniformly tighten the 8 cylinder head bolts with the plate washers in several steps and in the sequence shown in the illustration. Tighten to 27 ft. lbs. (36 Nm).

b. Step 2: Mark the cylinder head bolt head with paint as shown in the illustration. Tighten the cylinder head bolts another 90°.

c. Step 3: Tighten the cylinder head bolts an additional 90°. Check that the painted mark is now facing rearward.

d. Seal packing will seep out on the engine's front side. Thoroughly wipe clean any seal packing.

39. Place the left hand cylinder head gasket on the cylinder block surface with the front face of the Lot No. stamp upward.

➡**Be careful of the installation direction.**

➡**Gently place the cylinder head in order not to damage the gasket with the bottom part of the head.**

40. Place the cylinder head on the cylinder block.

➡**Do not allow oil to adhere to the mounting surface of the cylinder head.**

41. Apply a light coat of engine oil to the threads and under the heads of the cylinder head bolts.

42. The cylinder head bolts are tightened in 3 progressive steps:

a. Step 1: Using a 10 mm bi-hexagon wrench, install and uniformly tighten the 8 cylinder head bolts with the plate washers in several steps and in the sequence shown in the illustration. Tighten to 27 ft. lbs. (36 Nm).

b. Step 2: Mark the cylinder head bolt head with paint as shown in the illustration. Tighten the cylinder head bolts another 90°.

c. Step 3: Tighten the cylinder head bolts an additional 90°. Check that the painted mark is now facing rearward.

d. Tighten the 2 bolts in the order shown in the illustration to 22 ft. lbs. (30 Nm). Only use the specifications stated above when tightening the bolts 1 and 2 shown in the illustration.

e. Seal packing will seep out on the engine's front side. Thoroughly wipe clean any seal packing.

43. Install the No. 2 chain tensioner assembly.

44. Install the No. 3 chain tensioner.

45. Install the No. 1 and 2 chain vibration damper.

46. Install the idle sprocket assembly.

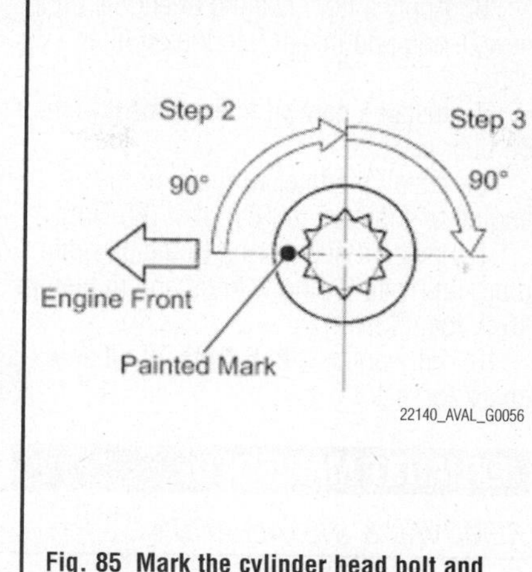

Fig. 85 Mark the cylinder head bolt and tighten another 90°

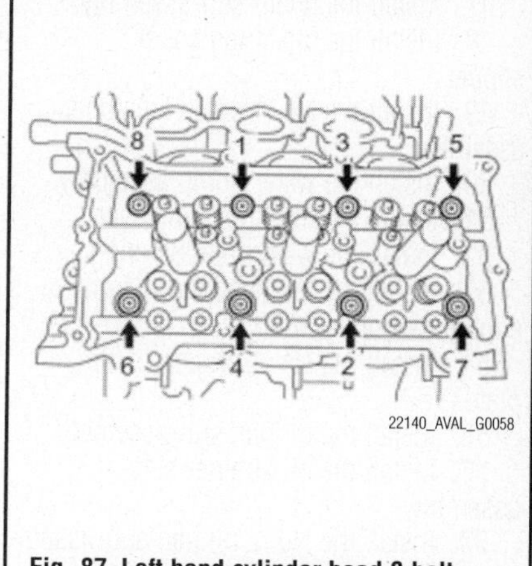

Fig. 87 Left hand cylinder head 8 bolt tightening sequence

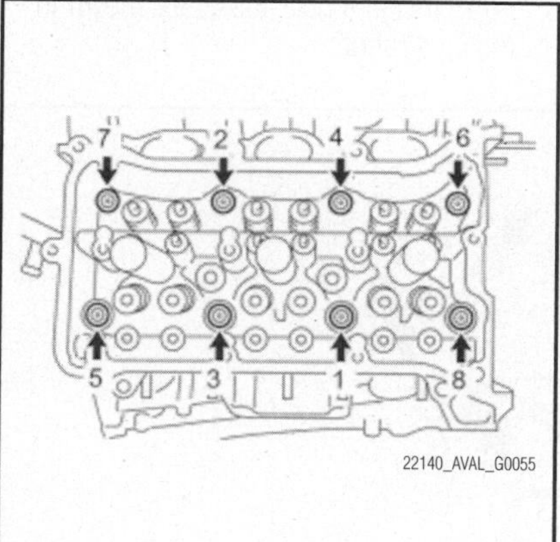

Fig. 84 Right hand cylinder head bolt tightening sequence

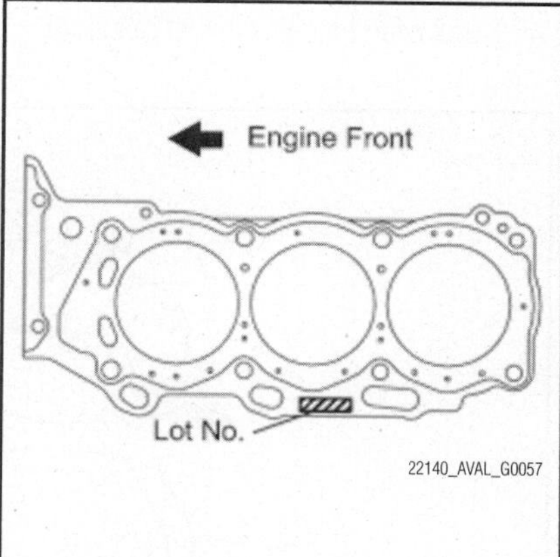

Fig. 86 Place the cylinder head with Lot No. stamp upward

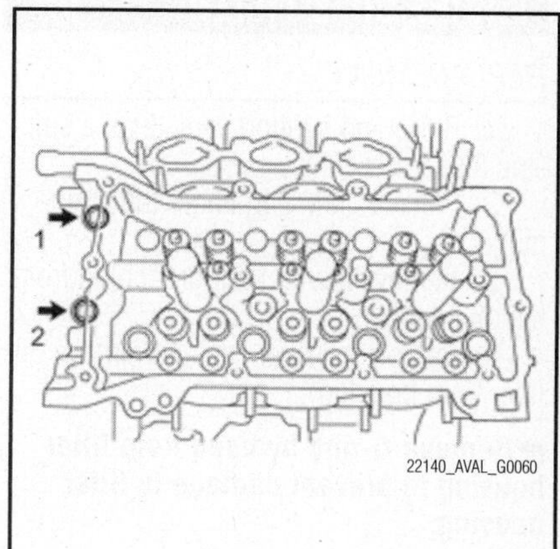

Fig. 88 Left hand cylinder head 2 bolt tightening sequence

47. Install the chain sub-assembly.
48. Install the chain tensioner slipper.
49. Install the No. 1 chain tensioner assembly.
50. Install the water pump assembly.
51. Install the timing chain cover.
52. Install the water inlet housing.
53. Install the left hand No. 1 front engine mounting bracket.
54. Install the No. 1 oil pan baffle plate.
55. Install the oil pan sub-assembly.
56. Install the oil strainer sub-assembly.
57. Install the No. 2 oil pan sub-assembly.
58. Install the oil pan drain plug and gasket.
59. Install the cylinder head covers and gaskets
60. Install the water outlet.
61. Install the crankshaft pulley.
62. Install the oil filter.
63. Install the cylinder block water drain cock sub-assembly.
64. Install the No. 1 oil pipe.
65. Install the oil pipe.
66. Install the crankshaft position sensor.
67. Install the camshaft timing oil control valve assembly.
68. Install the camshaft position sensor.
69. Install the ventilation valve sub-assembly.
70. Install the spark plugs and ignition coil assembly.
71. Install the oil filler cap sub-assembly.
72. Install the engine assembly with transaxle.

ENGINE OIL & FILTER

REPLACEMENT

1. Raise and support vehicle on a suitable lift.
2. Place a drain pan under the drain plug.
3. Remove the oil filter drain plug from the oil filter cap.
4. Remove oil filter cap and remove oil filter from housing.

➡Remove O-ring by hand from filter housing to prevent damage to filter housing.

To install:

5. Clean the inside of the oil filter cap, threads and O-ring groove.
6. Apply a light coating of engine oil to new O-ring and install it to the oil filter cap.
7. Install a new oil filter element to the oil filter cap.
8. Install the filter with cap to the engine and tighten to 18 ft. lbs. (25 Nm)
9. Install a new O-ring to the oil drain plug, lubricate O-ring with oil and tighten to 10 ft. lbs. (13Nm) .
10. Fill engine with 6.4 qts. of oil and check for leaks.

EXHAUST MANIFOLD

REMOVAL & INSTALLATION

See Figures 89 through 94.

1. Before servicing the vehicle, refer to the Precautions Section.
2. Remove the right No. 2 engine mounting stay.
3. Remove the right exhaust manifold sub-assembly, as follows:
   a. Uniformly loosen and remove the 6 nuts.
   b. Remove the manifold and gasket.
4. Remove the left exhaust manifold sub-assembly, as follows:
   a. Remove the oil level gauge guide sub-assembly.
   b. Remove the bolt, nut and No. 2 manifold stay.
   c. Remove the 3 bolts and No. 2 exhaust manifold heat insulator.
   d. Uniformly loosen and remove the 6 nuts.
   e. Remove the manifold and gasket.

To install:

5. Install the left exhaust manifold sub-assembly, as follows:
   a. Install a new gasket as shown in the illustration.
   b. Install the left exhaust manifold sub-assembly with the 6 nuts and tighten to 15 ft. lbs. (21 Nm).
6. Install the No. 2 exhaust manifold heat insulator with the 3 bolts and tighten to 75 inch lbs. (8.5 Nm).
7. Install the No. 2 manifold stay with the bolt and nut and tighten to 25 ft. lbs. (34 Nm).
8. Install the oil level gauge guide sub-assembly.
9. Install the right exhaust manifold sub-assembly, as follows:
   a. Install a new gasket as shown in the illustration.

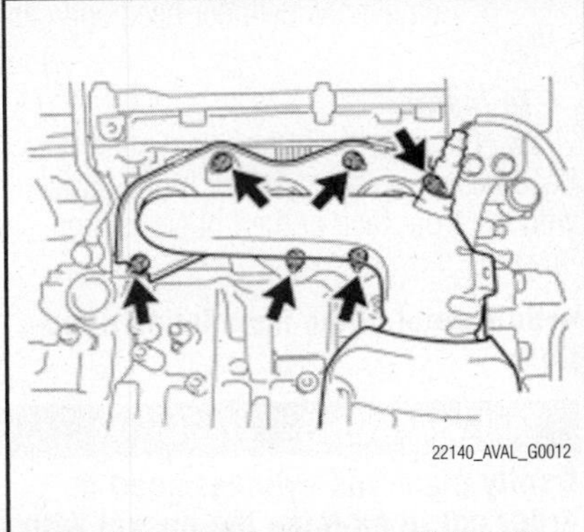

Fig. 90 Left-hand exhaust manifold nuts

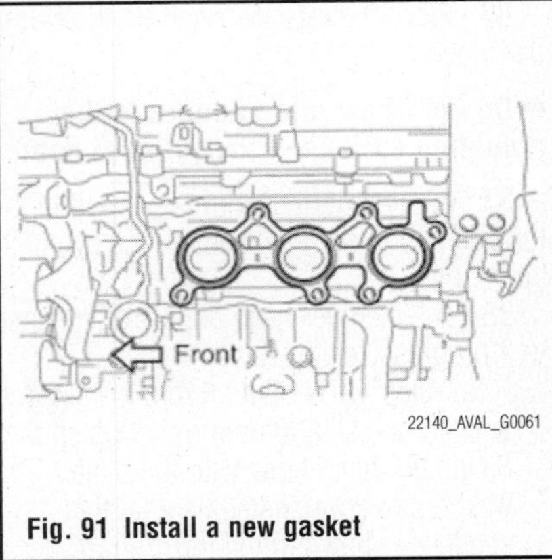

Fig. 91 Install a new gasket

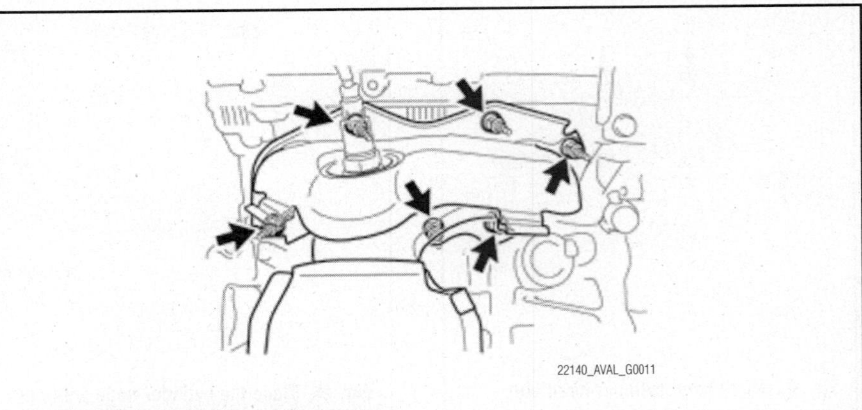

Fig. 89 Right-hand exhaust manifold nuts

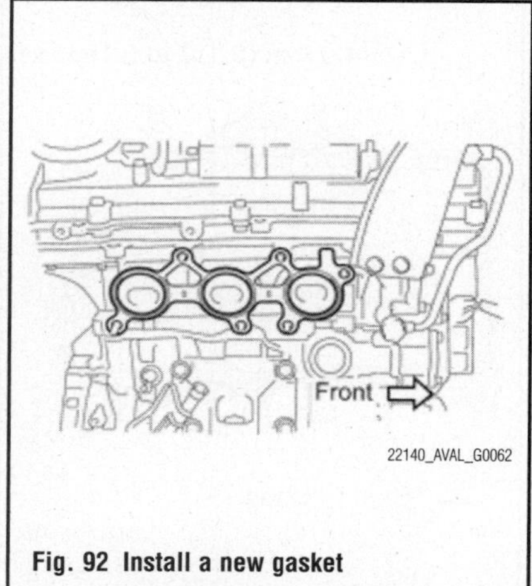

Fig. 92 Install a new gasket

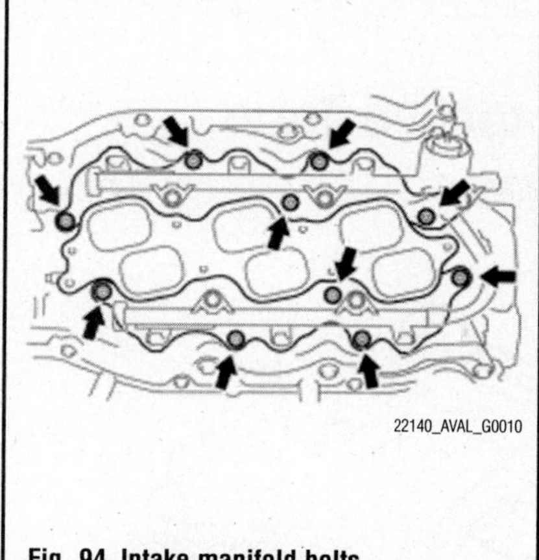

Fig. 94 Intake manifold bolts

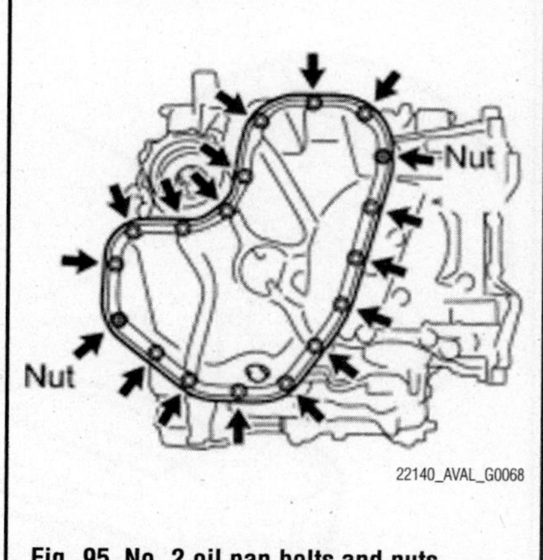

Fig. 95 No. 2 oil pan bolts and nuts

b. Install the right exhaust manifold sub-assembly with the 6 nuts and tighten to 15 ft. lbs. (21 Nm).

10. Install the right No. 2 engine mounting stay.

INTAKE MANIFOLD

REMOVAL & INSTALLATION

See Figures 94 and 95.

1. Before servicing the vehicle, refer to the Precautions Section.
2. Relieve the fuel system pressure.
3. Drain the cooling system.
4. Remove or disconnect the following:
   a. Negative battery cable.
   b. Wiper link assembly from both wipers.
   c. Outer top cowl panel.
   d. Wiper motor and linkage assembly.
   e. Engine v-bank cover.
   f. Air cleaner cap with air cleaner hoses.
   g. All hoses from the throttle body and intake air surge tanks.
   h. Intake air surge tank.
   i. Wire connectors from fuel injectors.
   j. Fuel hose from the fuel rail.
   k. Intake manifold with gaskets.

To install

5. Installation is the reverse order of removal. Note the following:
   a. Install the intake manifold with new gaskets.
   b. Tighten the intake manifold mounting bolts to 15 ft. lbs. (21 Nm).

OIL PAN

REMOVAL & INSTALLATION

See Figures 96 through 102.

1. Before servicing the vehicle, refer to the Precautions Section.
2. Drain the engine oil.
3. Remove the engine assembly with transaxle.
4. Secure the engine.
5. Remove the oil filler cap and gasket.
6. Remove the oil pan drain plug and gasket.
7. Remove the No. 1 oil pipe, as follows:
   a. Remove the 2 oil pipe unions and oil pipe.
   b. Remove the left hand oil control valve filter and gaskets.
8. Remove the oil pipe, as follows:
   a. Remove the bolt.
   b. Remove the 2 oil pipe unions and oil pipe.
   c. Remove the right oil control valve filter and gaskets.
9. Remove the oil filter element, as follows:
   a. Remove the drain plug. Do not remove the O-ring.
   b. Connect the hose to the pipe.
   c. Insert the pipe with the hose into the oil filter cap.
   d. Make sure that the oil is completely drained and remove the pipe and O-ring.
   e. Using SST (SST: 09228-06501) or equivalent, remove the oil filter cap.
   f. Remove the oil filter element and O-ring from the oil filter cap. Do not use any tools when removing the O-ring to prevent the O-ring groove from being damaged.

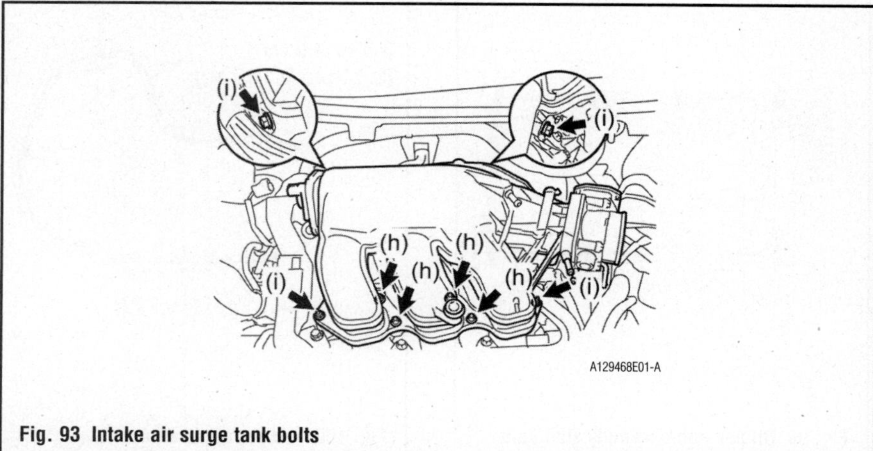

Fig. 93 Intake air surge tank bolts

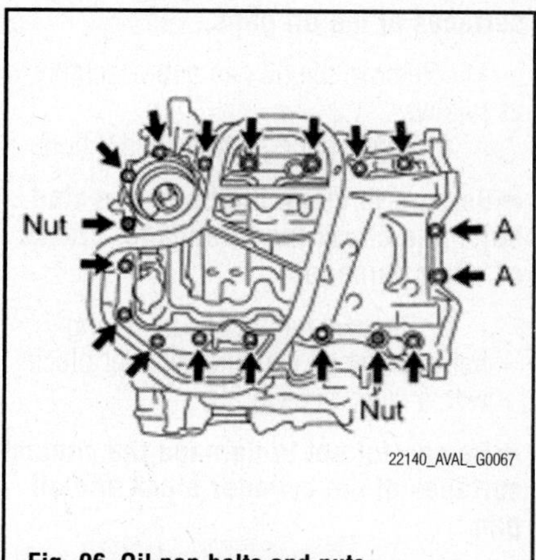

Fig. 96 Oil pan bolts and nuts

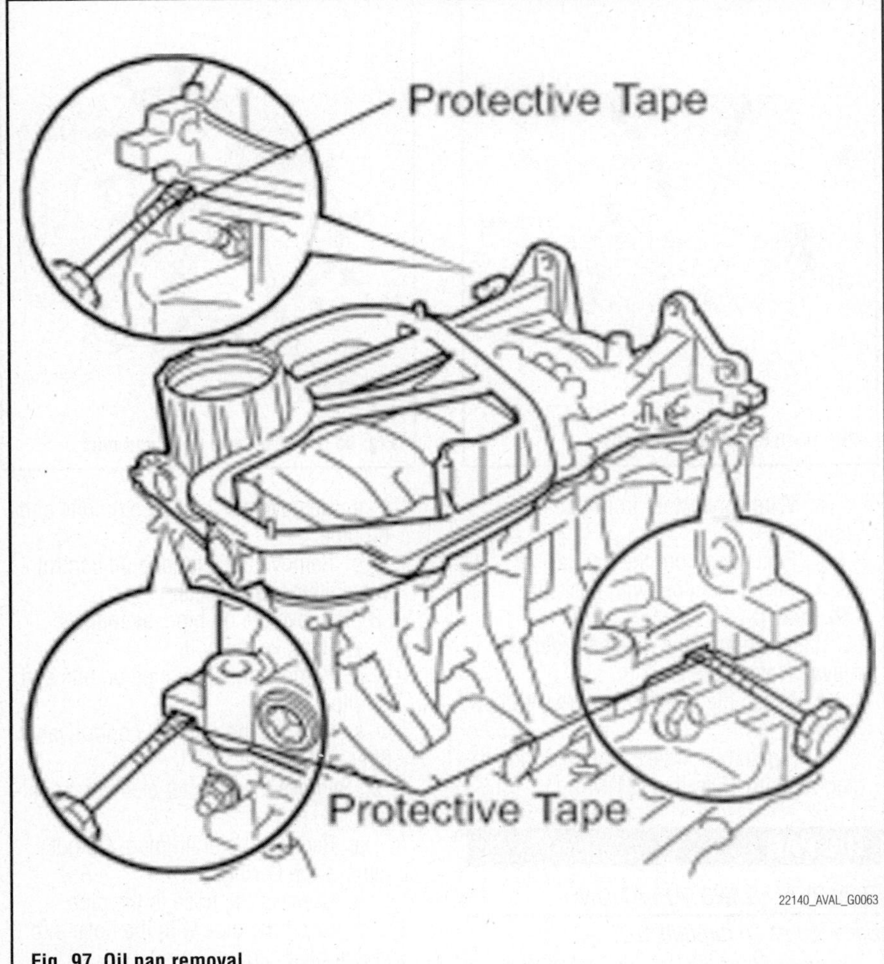

Fig. 97 Oil pan removal

10. Remove the No. 2 oil pan sub-assembly, as follows:
    a. Remove the 16 bolts and 2 nuts.
    b. Insert the blade of SST (SST: 09032-00100) or equivalent tool between the oil pans. Cut through the applied sealer and remove the No. 2 oil pan sub-assembly.

➡Be careful not to damage the contact surfaces of the oil pans.

11. Remove the oil pan sub-assembly, as follows:
    a. Remove the 16 bolts and 2 nuts.

➡Be sure to clean the bolts and stud bolts and check the threads for cracks or other damage.

    b. Remove the oil pan by prying between the oil pan and cylinder block with a taped screwdriver.

➡Be careful not to damage the contact surfaces of the cylinder block and oil pan.

    c. Remove the 2 O-rings.

To install:

12. Install the oil pan sub-assembly, as follows:
    a. Using an E8 Torx® socket wrench, install the stud bolts as shown in the illustration. Tighten to 7 ft. lbs (10 Nm).

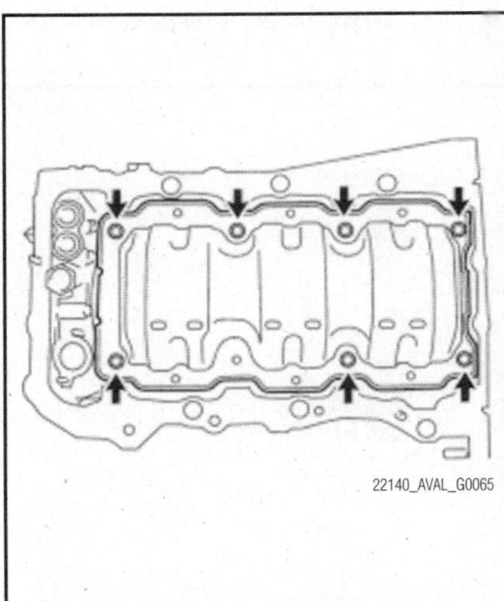

Fig. 98 Oil pan sub-assembly stud bolts

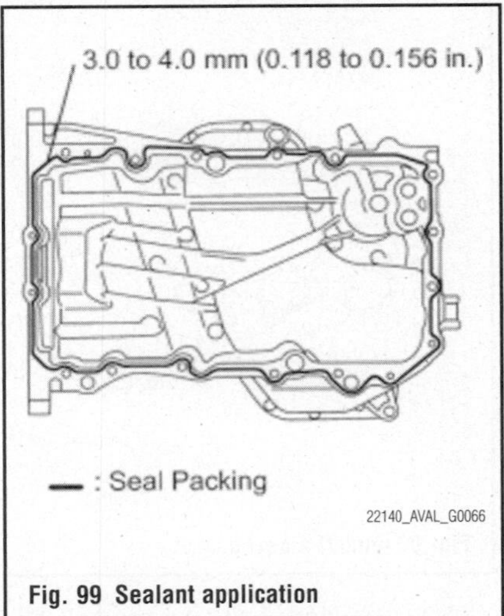

Fig. 99 Sealant application

    b. Apply seal packing (Toyota Genuine Seal Packing Black, Three Bond® 1207B or equivalent) in a continuous line as shown in the illustration. Seal diameter: 0.118 to 0.156 inch (3.0 to 4.0 mm).

➡Remove any oil from the contact surface.

➡Install the oil pan within 3 minutes after applying seal packing.

➡Do not start the engine for at least 2 hours after installing.

    c. Install 2 new O-rings.
    d. Install the oil pan with the 16 bolts and 2 nuts and tighten to 7 ft. lbs (10 Nm), and 15 ft. lbs (21 Nm).

13. Install the No. 2 oil pan sub-assembly, as follows:
    a. Apply seal packing (Toyota Genuine Seal Packing Black, Three Bond® 1207B or equivalent) in a continuous line

Fig. 100 Sealant application

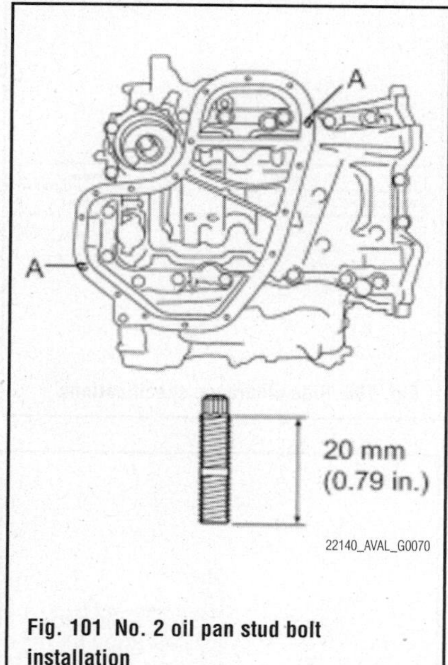

Fig. 101 No. 2 oil pan stud bolt installation

as shown in the illustration. Seal diameter: 0.118 to 0.156 inch (3.0 to 4.0 mm).

➡**Remove any oil from the contact surface.**

➡**Install the No. 2 oil pan within 3 minutes after applying seal packing.**

➡**Do not start the engine for at least 2 hours after installing.**

b. Using an E6 Torx® socket wrench, install the stud bolts as shown in the illustration and tighten to 35 inch lbs (4 Nm).

c. Install the No. 2 oil pan with the 16 bolts and 2 nuts and tighten to 7 ft. lbs (10 Nm).

14. Install the oil pan drain plug and a new gasket. Tighten to 30 ft. lbs (40 Nm).

15. Install the oil filter element, as follows:

a. Clean the inside of the oil filter cap, the threads and O-ring groove.

b. Apply a small amount of engine oil to a new O-ring and install it to the oil filter cap.

c. Set a new oil filter element to the oil filter cap.

d. Remove dirt or foreign matter from the installation surface and inside of the engine.

e. Apply a small amount of engine oil to the O-ring again and install the oil filter cap.

➡**Be careful that the O-ring does not get caught between the parts. The O-ring must not be twisted on the groove.**

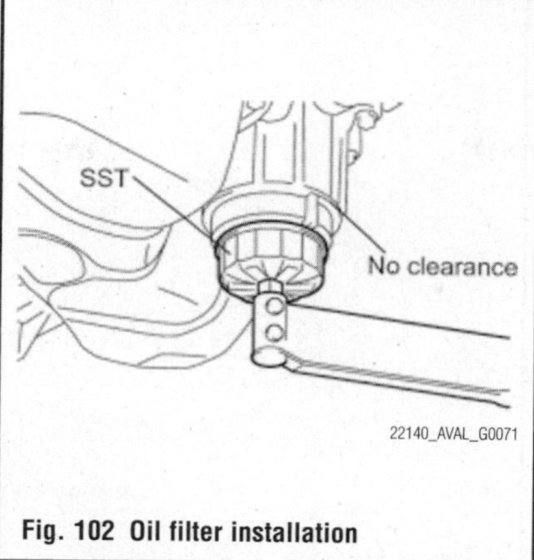

Fig. 102 Oil filter installation

f. Using SST (SST: 09228-06501) or equivalent, install the oil filter cap and tighten to 18 ft. lbs (25 Nm). Make sure that the oil filter is installed securely as shown.

g. Apply a light coat of engine oil to a new O-ring and install it to the oil filter cap. Remove all dirt and foreign matter from the installation surface.

h. Install the oil filter drain plug to the oil filter cap and tighten to 9 ft. lbs (13 Nm). Make sure that the O-ring does not get caught between the parts.

16. Install the oil filler cap sub-assembly.

17. Install the engine assembly with transaxle.

OIL PUMP

REMOVAL & INSTALLATION

See Figures 103 and 104.

1. Before servicing the vehicle, refer to the Precautions Section.
2. Remove the engine assembly with transaxle.
3. Secure engine.
4. Remove the engine wire.
5. Remove the front frame assembly.
6. Remove the starter assembly.
7. Remove the automatic transaxle assembly.
8. Remove the oil level gauge guide sub-assembly.
9. Remove the right and left exhaust manifold sub-assemblies.
10. Remove the drive plate and ring gear sub-assembly.
11. Remove the No. 2 idler pulley sub-assembly.
12. Remove the V-ribbed belt tensioner assembly.
13. Remove the water pump pulley.
14. Remove the water inlet housing.
15. Remove the crankshaft pulley.
16. Remove the No. 2 oil pan sub-assembly.
17. Remove the oil strainer sub-assembly.
18. Remove the oil pan sub-assembly.
19. Remove the intake air surge tank assembly.
20. Remove the ignition coil assembly.
21. Remove the No. 1 and 2 oil pipes.
22. Remove the right and left cylinder head cover sub-assemblies.
23. Remove the timing chain cover sub-assembly. Refer to Timing Chain Cover and Seal procedures for instructions.
24. Using a screwdriver with the tip taped, pry out the timing gear case or timing chain case oil seal.
25. Using a 27 mm socket wrench, remove the relief valve plug.
26. Remove the valve spring and oil pump relief valve.
27. Remove the 8 bolts, oil pump cover, drive rotor and driven rotor.

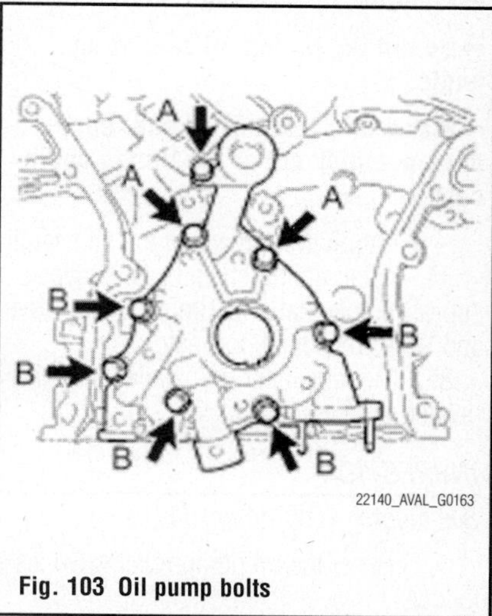

Fig. 103 Oil pump bolts

Fig. 104 Oil pump gears

To install:

28. Coat the drive and driven rotors with engine oil and place them into the timing chain cover with the marks facing outward (oil pump cover side). Check that the rotors revolve smoothly.
29. Install the oil pump cover with the 8 bolts and tighten to 81 inch lbs. (9.1 Nm). Bolt length: 0.87 in. (22 mm) for bolt A, 1.58 in. (40 mm) for bolt B.
30. Coat the oil pump relief valve with engine oil.
31. Insert the relief valve and relief valve spring into the oil pump cover hole.
32. Using a 27 mm socket wrench, install the plug and tighten to 36 ft. lbs. (49 Nm).
33. Install timing gear case or timing chain case oil seal, as follows:
    a. Using SST (SST: 09316-60011) or equivalent tool, tap in a new oil seal until its surface is flush with the timing chain case edge.

➡Keep the lip free from foreign matter.

➡Do not tap on the oil seal at an angle.

➡Make sure that the oil seal edge does not stick out of the timing chain case.

    b. Apply MP grease to the oil seal lip.
34. Install timing chain or belt cover sub-assembly. Refer to Timing Chain Cover and Seal procedures for instructions.
35. The remainder of installation is the reverse of removal.

INSPECTION

See Figures 105 through 111.

1. Inspect the oil pump relief valve, as follows:
    a. Coat the relief valve with engine oil and check that it falls smoothly into the valve hole by its own weight. If the valve does not fall smoothly, replace the relief valve. If necessary, replace the oil pump assembly.
2. Inspect the oil pump rotor set, as follows:
    a. Install the rotors to the timing chain cover with the rotors' marks outward. Check that the rotors rotate smoothly.
    b. Check the tip clearance: using a feeler gauge, measure the clearance between the drive and driven rotor tips, as shown. If the clearance is greater than the maximum, replace the drive and driven rotors.
    c. Check the side clearance: using a feeler gauge and precision straightedge, measure the clearance between the rotors and precision straightedge, as shown in the illustration. If the side clearance is greater than the maximum, replace the timing chain cover sub-assembly.

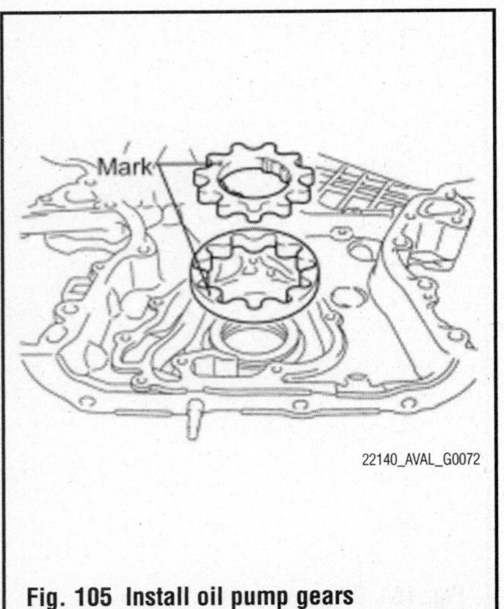

Fig. 105 Install oil pump gears

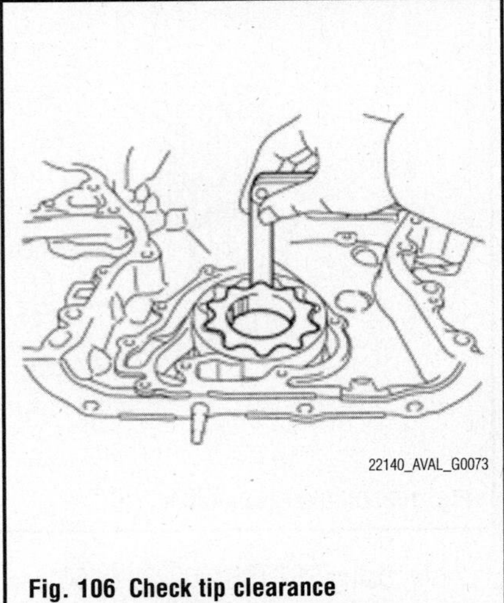

Fig. 106 Check tip clearance

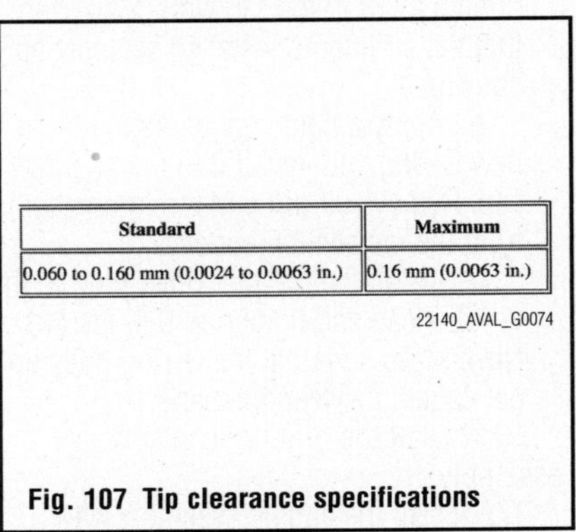

Standard	Maximum
0.060 to 0.160 mm (0.0024 to 0.0063 in.)	0.16 mm (0.0063 in.)

22140_AVAL_G0074

Fig. 107 Tip clearance specifications

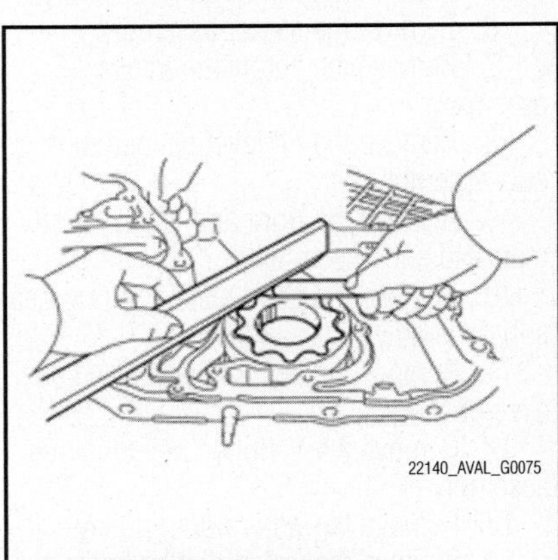

Fig. 108 Check side clearance

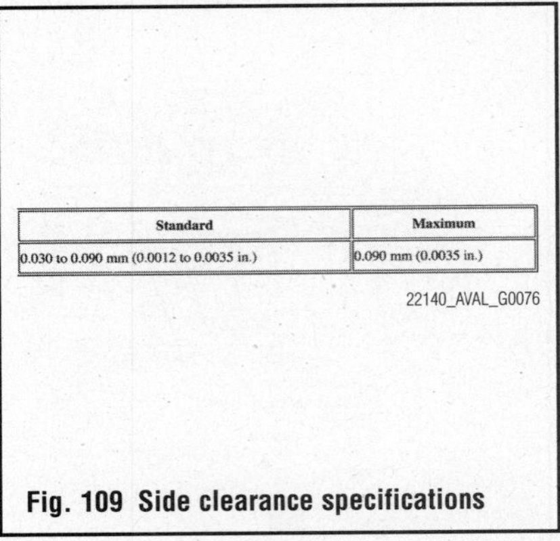

Standard	Maximum
0.030 to 0.090 mm (0.0012 to 0.0035 in.)	0.090 mm (0.0035 in.)

22140_AVAL_G0076

Fig. 109 Side clearance specifications

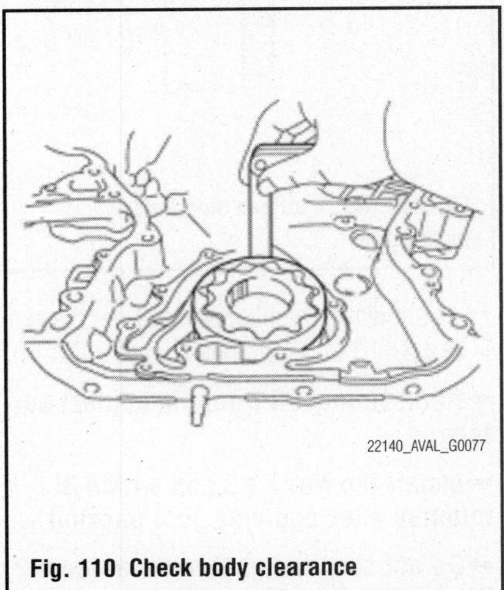

Fig. 110 Check body clearance

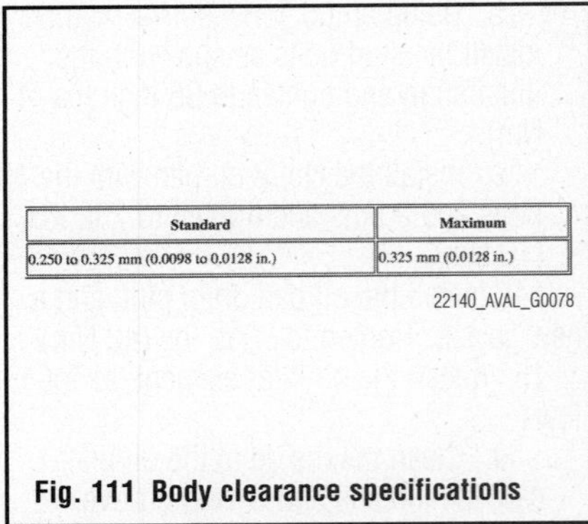

Standard	Maximum
0.250 to 0.325 mm (0.0098 to 0.0128 in.)	0.325 mm (0.0128 in.)

22140_AVAL_G0078

Fig. 111 Body clearance specifications

d. Check the body clearance: using a feeler gauge, measure the clearance between the timing chain cover and driven rotor, as shown in the illustration. If the body clearance is greater than the maximum, replace the timing chain cover sub-assembly.

PISTON AND RING

POSITIONING

See Figure 112.

Refer to the graphic provided for piston ring positioning.

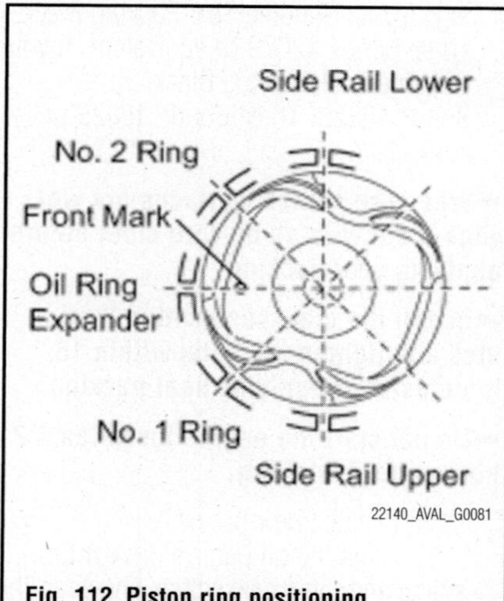

Fig. 112 Piston ring positioning

REAR MAIN SEAL

REMOVAL & INSTALLATION

See Figures 113 and 114.

1. Before servicing the vehicle, refer to the Precautions Section.
2. Remove the automatic transaxle assembly.
3. Remove the drive plate and ring gear sub-assembly.
4. Remove the rear main seal, as follows:
   a. Using a knife, cut off the oil seal lip.
   b. Using a screwdriver with the tip taped, pry out the oil seal.

➡Be careful not to damage the crankshaft.

To install:

5. Apply MP grease to a new oil seal lip.
6. Using SST (SST: 09223-15030, SST: 09950-70010) or equivalent and a hammer, tap in the oil seal. Oil seal tap in depth: -0.020 to 0.020 inch (-0.5 to 0.5 mm).

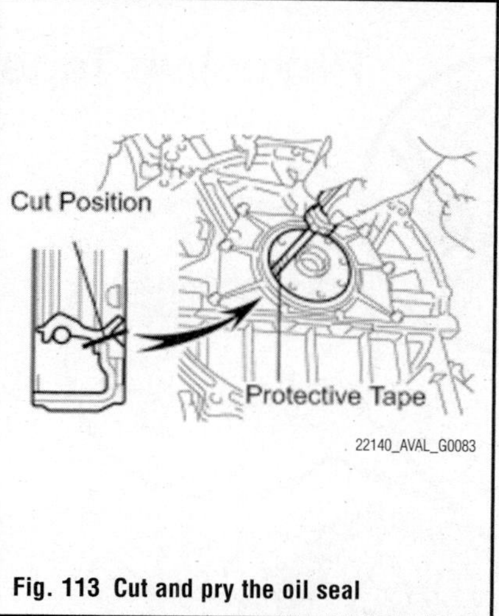

Fig. 113 Cut and pry the oil seal

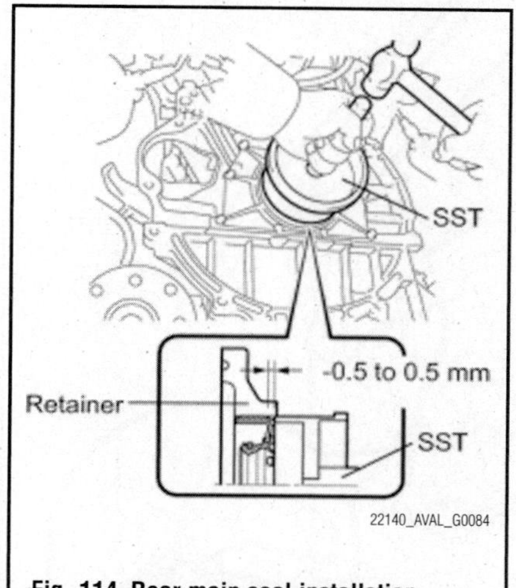

Fig. 114 Rear main seal installation

7. Install the drive plate and ring gear sub-assembly.
8. Install automatic transaxle assembly.

TIMING CHAIN FRONT COVER

REMOVAL & INSTALLATION

See Figures 115 through 124.

1. Before servicing the vehicle, refer to the Precautions Section.
2. Remove the engine assembly with transaxle.
3. Secure engine.
4. Remove the oil filler cap sub-assembly.
5. Remove the spark plugs and ignition coil assembly.
6. Remove the oil pan drain plug and gasket.
7. Remove the ventilation valve sub-assembly.
8. Remove the camshaft position sensor.

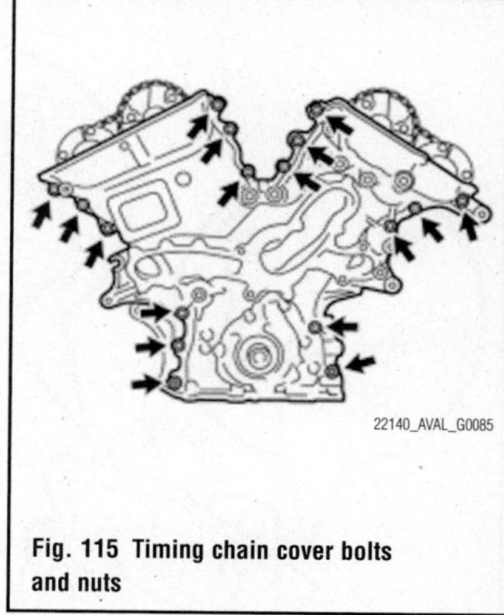

Fig. 115 Timing chain cover bolts and nuts

9. Remove the camshaft timing oil control valve assembly.
10. Remove crankshaft position sensor.
11. Remove the No. 1 oil pipe.
12. Remove the oil pipe.
13. Remove the cylinder block water drain cock sub-assembly.
14. Remove the oil filter.
15. Remove the crankshaft pulley.
16. Remove the left hand No. 1 front engine mounting bracket.
17. Remove the water inlet housing.
18. Remove the water outlet.
19. Remove the left-hand cylinder head cover sub-assembly and gasket.
20. Remove the cylinder head cover sub-assembly and gasket.
21. Remove the No. 2 oil pan sub-assembly.
22. Remove the oil strainer sub-assembly.
23. Remove the oil pan sub-assembly.
24. Remove the water pump assembly.
25. Remove the timing chain cover sub-assembly, as follows:
    a. Remove the 15 bolts and 2 nuts as shown in the illustration.
    b. Remove the timing chain cover by prying between the timing chain cover and cylinder head or cylinder block with a screwdriver with the tip taped.

➡Be careful not to damage the contact surfaces of the cylinder head, cylinder block and chain cover.

    c. Remove the 4 bolts, chain cover plate and gasket.
    d. Remove the gasket.
26. Remove the timing chain case oil seal, as follows:

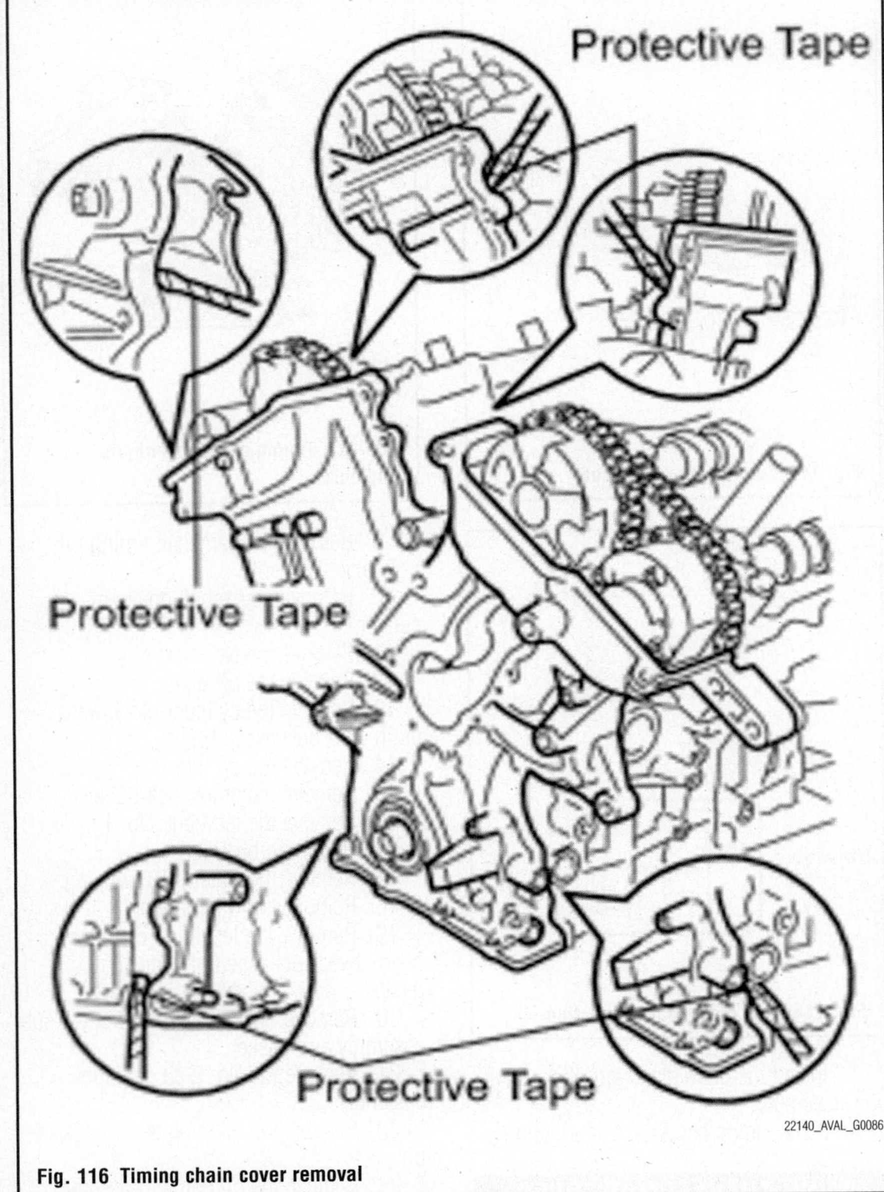

Fig. 116 Timing chain cover removal

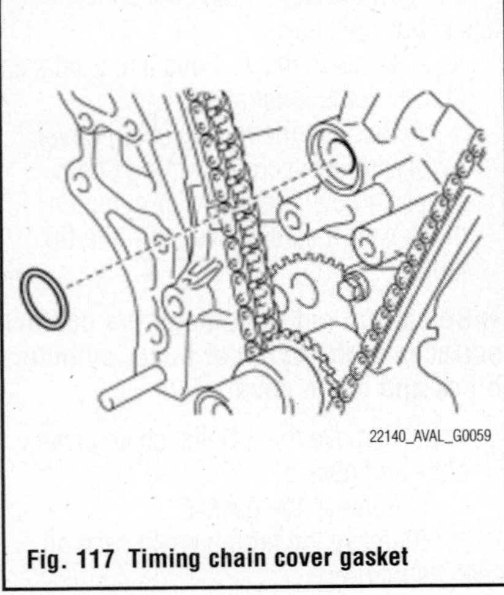

Fig. 117 Timing chain cover gasket

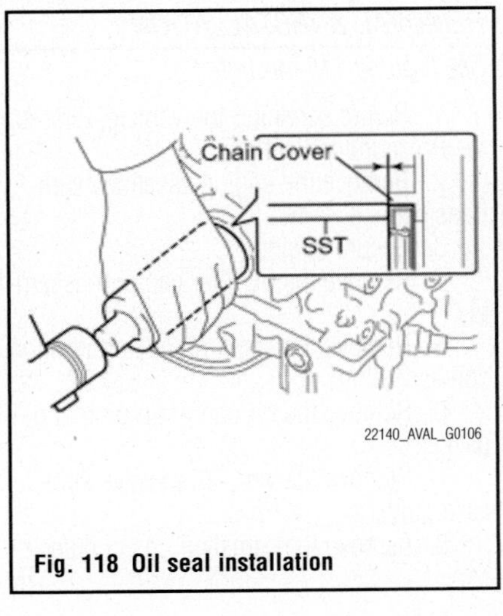

Fig. 118 Oil seal installation

a. Using a screwdriver with the tip taped, pry out the oil seal.

To install:

27. Install timing gear case or timing chain cover oil seal, as follows:

a. Apply MP grease to a new oil seal lip.

b. Using SST (SST: 09316-60011) and a hammer, tap in the oil seal until its surface is flush with the timing chain cover edge.

➡Keep the lip free from foreign matter.

➡Do not tap on the oil seal at an angle.

➡Make sure that the oil seal edge does not stick out of the timing chain cover.

28. Install the timing chain cover sub-assembly, as follows:

a. Install a new gasket and the chain cover plate with the 4 bolts and tighten to 81 inch lbs. (9.1 Nm).

b. Apply seal packing (Toyota Genuine Seal Packing Black, Three Bond® 1207B or equivalent) in a continuous line to the engine unit as shown in the illustration. Seal diameter: 3.0 mm (0.118 in.).

➡Be sure to clean and degrease the contact surfaces, especially the surfaces indicated by C in the illustration.

➡When the contact surfaces are wet, wipe them with an oil-free cloth before applying seal packing.

➡Install the chain cover within 3 minutes after applying seal packing.

➡Do not start the engine for at least 2 hours after installing.

c. Apply seal packing in a continuous line to the timing chain cover as shown in the following illustration. Seal packing: Toyota Genuine Seal Packing Black, Three Bond® 1207B or equivalent, Toyota Genuine Seal Packing Black, Three Bond® 1282B, Three Bond® 1282B or equivalent.

➡When the contact surfaces are wet, wipe them with an oil-free cloth before applying seal packing.

➡Install the crankcase within 3 minutes and tighten the bolts within 15 minutes after applying seal packing.

➡Do not start the engine for at least 2 hours after installing.

d. Install a new gasket.

e. Align the oil pump's drive rotor spline and the crankshaft as shown in the

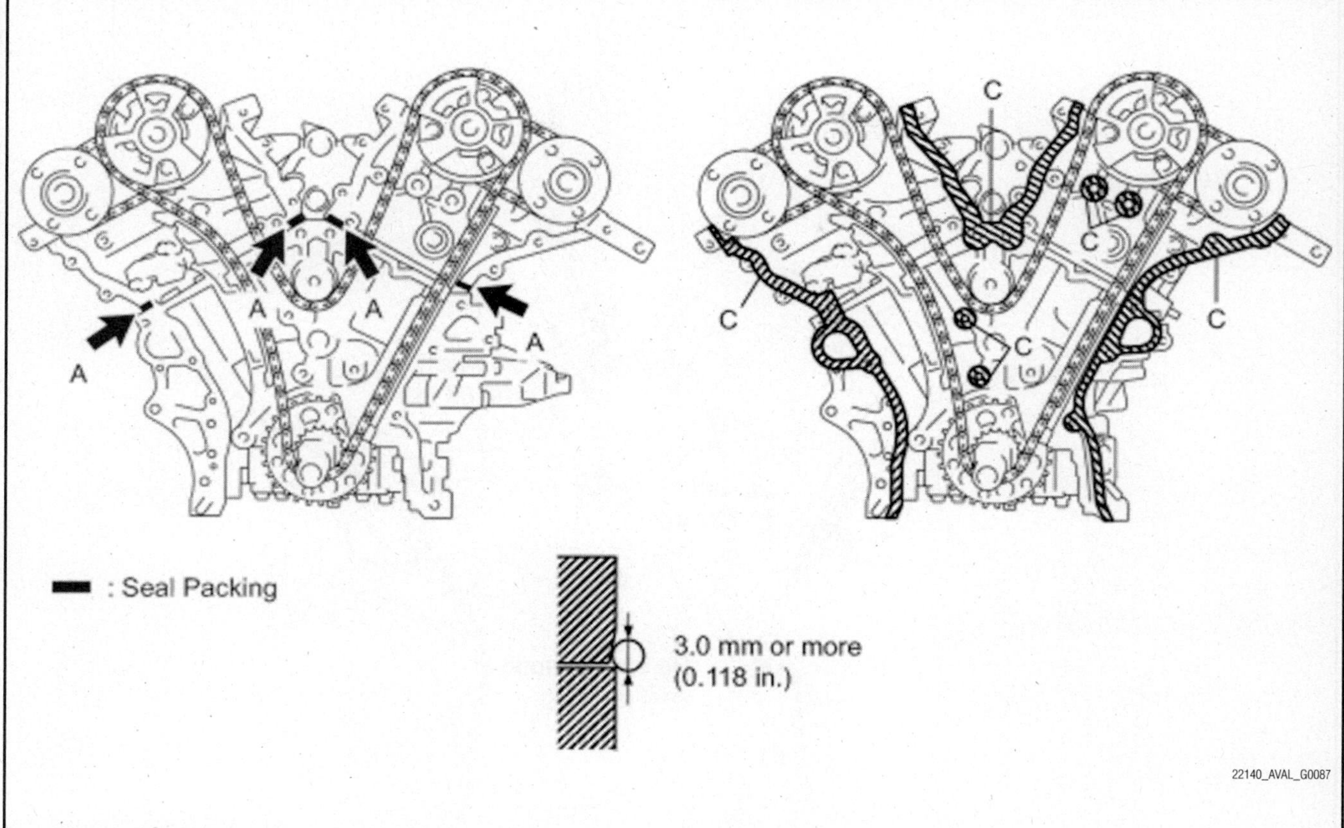

Fig. 119 Sealant application

illustration. Install the spline and chain cover to the crankshaft.

f. Loosely install the timing chain cover with the 23 bolts and 2 nuts, but do not tighten the bolts and 2 nuts yet.

WARNING

Make sure that there is no oil on the bolt and nut threads.

g. Fully tighten the bolts in this order: Area 1 and Area 2, tighten to 15 ft. lbs. (21 Nm).

h. Fully tighten the bolts in Area 3 to 15 ft. lbs. (21 Nm). Tighten the bolts and nuts in the order of upper to lower as shown in the illustration.

i. Fully tighten the bolts in Area 4 to 32 ft. lbs. (43 Nm), and to 15 ft. lbs. (21 Nm). Tighten the bolts and nuts in the order of lower to upper as shown in the illustration.

29. Install the water pump assembly.
30. Install the water inlet housing.
31. Install the left hand No. 1 front engine mounting bracket.
32. Install the oil pan sub-assembly.
33. Install the oil strainer sub-assembly.
34. Install the No. 2 oil pan sub-assembly.
35. Install the oil pan drain plug and gasket.
36. Install the cylinder head cover sub-assembly.
37. Install the left-hand cylinder head cover sub-assembly.
38. Install the water outlet.
39. Install the crankshaft pulley.
40. Install the oil filter.
41. Install the cylinder block water drain cock sub-assembly.
42. Install the No. 1 oil pipe.
43. Install the oil pipe.
44. Install the crankshaft position sensor.
45. Install the camshaft timing oil control valve assembly.
46. Install the camshaft position sensor.
47. Install the ventilation valve sub-assembly.
48. Install the spark plugs and ignition coil assembly.
49. Install the oil filler cap sub-assembly.
50. Install the water pump assembly.
51. Install the engine assembly with transaxle.

TIMING CHAIN & SPROCKETS

REMOVAL & INSTALLATION

See Figures 125 through 137.

1. Before servicing the vehicle, refer to the Precautions Section.
2. Remove the engine assembly with transaxle.
3. Secure engine.
4. Remove the oil filler cap sub-assembly.
5. Remove the spark plugs and ignition coil assembly.
6. Remove the oil pan drain plug and gasket.
7. Remove the ventilation valve sub-assembly.
8. Remove the camshaft position sensor.
9. Remove the camshaft timing oil control valve assembly.
10. Remove crankshaft position sensor.
11. Remove the No. 1 oil pipe.
12. Remove the oil pipe.
13. Remove the cylinder block water drain cock sub-assembly.
14. Remove the oil filter.
15. Remove the crankshaft pulley.
16. Remove the left hand No. 1 front engine mounting bracket.

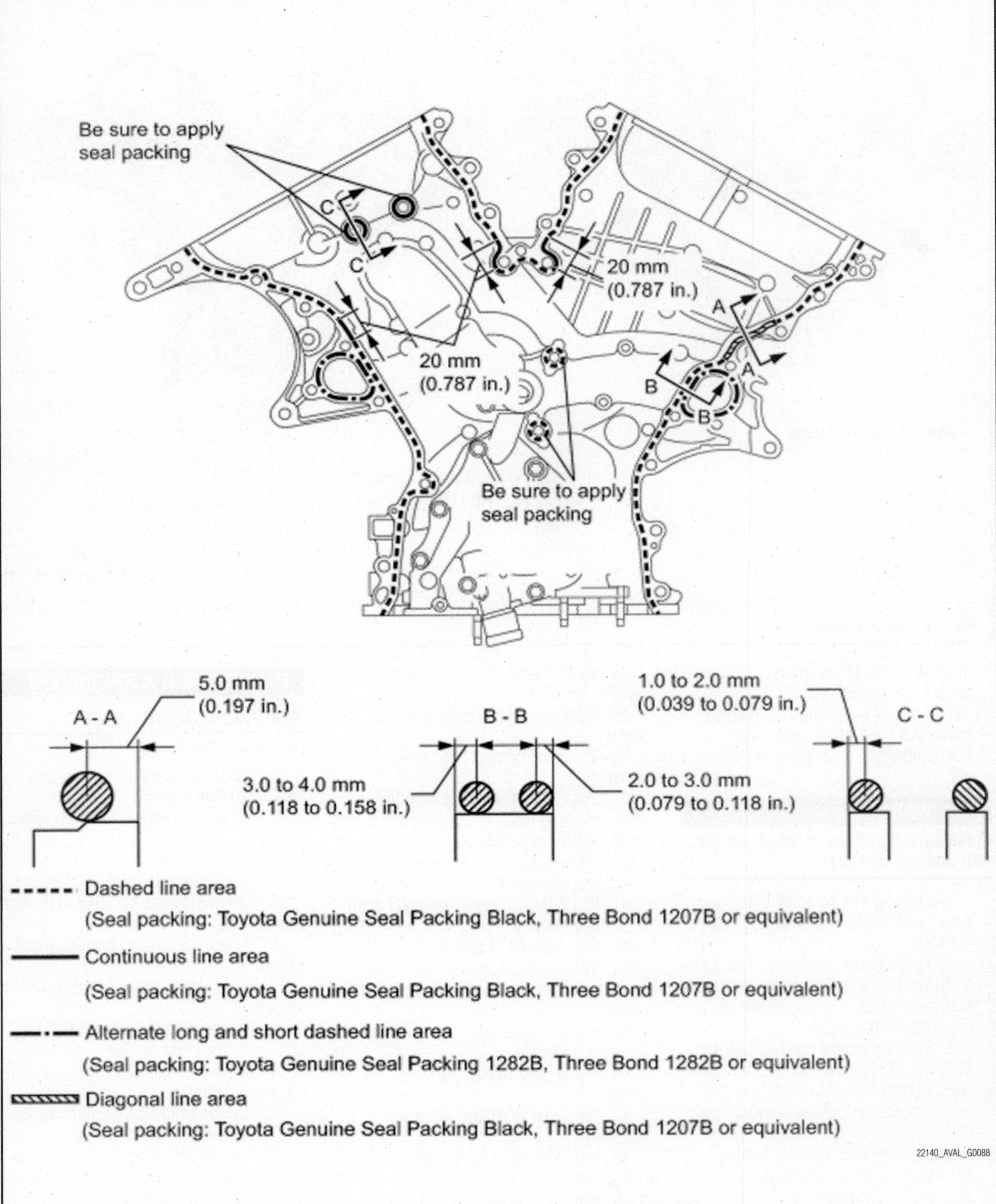

Fig. 120 Timing chain cover sealant application

Area	Seal Packing Diameter	Application Position from Inside Seal Line
Dashed line area	3.5 mm or more (0.138 in.)	3.0 to 4.0 mm (0.118 to 0.158 in.)
Continuous line area	4.5 mm or more (0.177 in.)	3.0 to 4.0 mm (0.118 to 0.158 in.)
Alternate long and short dashed line area	3.5 mm or more (0.138 in.)	2.0 to 3.0 mm (0.079 to 0.118 in.)
Diagonal line area	6.0 mm or more (0.236 in.)	5.0 mm (0.197 in.)

22140_AVAL_G0089

Fig. 121 Sealant application diameter and position

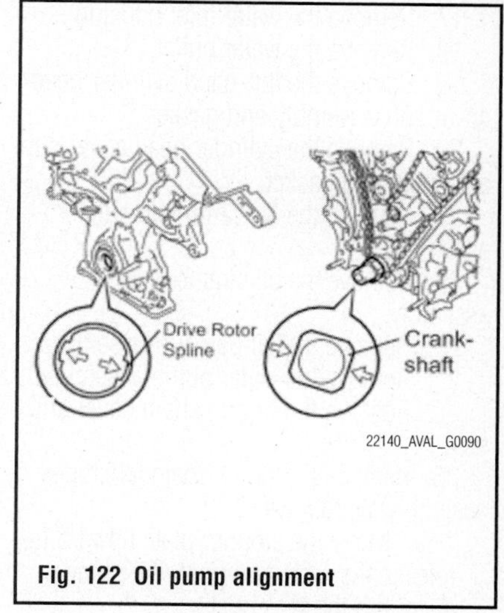

Fig. 122 Oil pump alignment

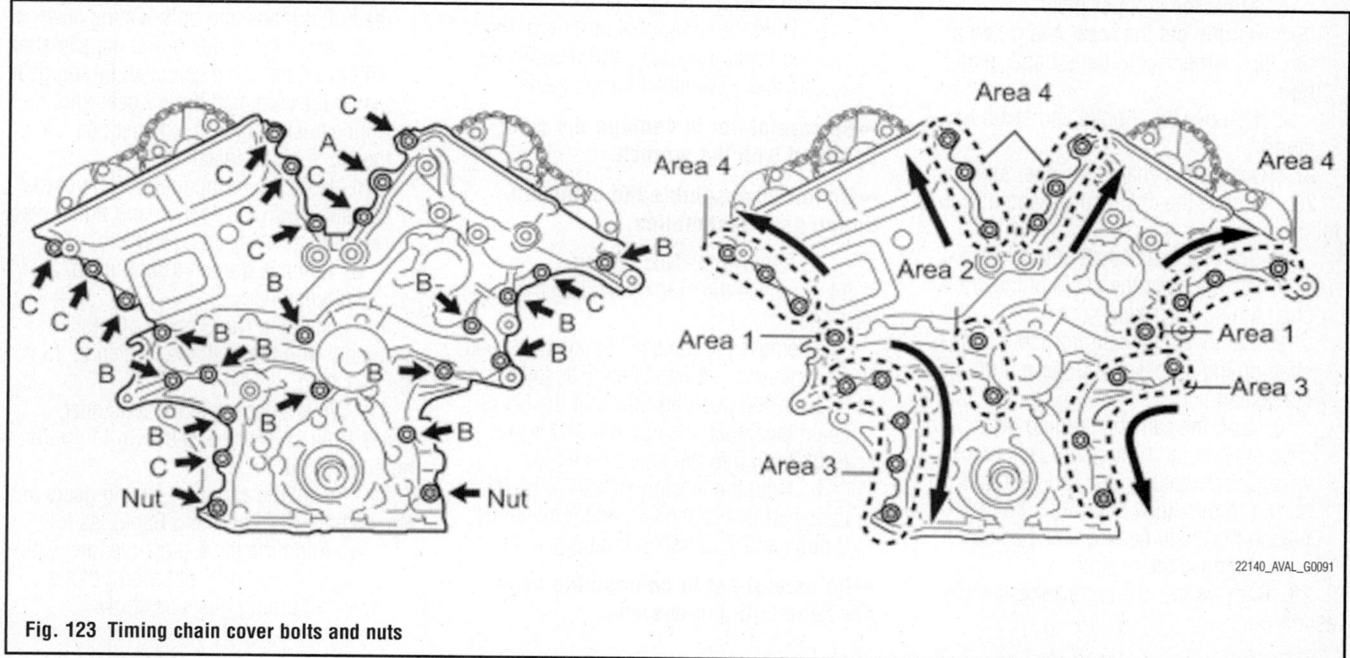

Fig. 123 Timing chain cover bolts and nuts

Item	Length
Bolt A	40 mm (1.57 in.)
Bolt B	55 mm (2.17 in.)
Bolt C	25 mm (0.98 in.)

22140_AVAL_G0092

Fig. 124 Bolt length

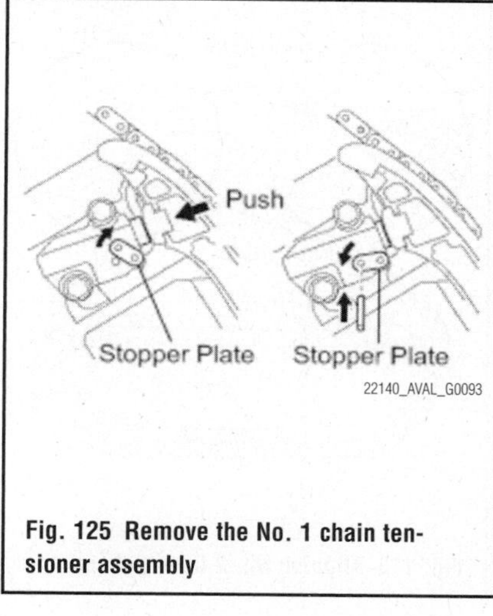

Fig. 125 Remove the No. 1 chain tensioner assembly

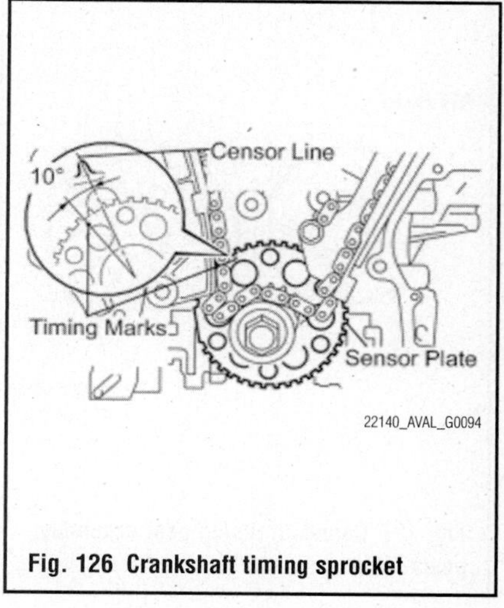

Fig. 126 Crankshaft timing sprocket

17. Remove the water inlet housing.
18. Remove the water outlet.
19. Remove the left-hand cylinder head cover sub-assembly and gasket.
20. Remove the cylinder head cover sub-assembly and gasket.
21. Remove the No. 2 oil pan sub-assembly.
22. Remove the oil strainer sub-assembly.
23. Remove the oil pan sub-assembly.
24. Remove the water pump assembly.
25. Remove the timing chain cover and seal.
26. Remove the No. 1 chain tensioner assembly, as follows:
    a. Move the stopper plate upward to release the lock, and push the plunger deep into the tensioner.
    b. Move the stopper plate downward to set the lock, and insert a hexagon wrench into the stopper plate's hole.
    c. Remove the 2 bolts and chain tensioner.
27. Remove the chain tensioner slipper.
28. Remove the chain sub-assembly, as follows:
    a. Turn the crankshaft counterclockwise 10°to loosen the chain of the crankshaft timing sprocket.
    b. Remove the chain from the crankshaft timing sprocket and place it on the crankshaft.
    c. Turn the camshaft timing gear assembly on the right hand bank clockwise (approximately 60°) and set it as shown in the illustration. Be sure to loosen the chain between the banks.
    d. Remove the chain.
29. Remove the idle sprocket assembly, as follows:
    a. Using a 10 mm hexagon wrench, remove the No. 2 idle gear shaft, sprocket and No. 1 idle gear shaft.
30. Remove the 2 bolts and the No. 1 chain vibration damper.
31. Remove the No. 2 chain vibration damper.
32. Remove the crankshaft timing sprocket, as follows:
    a. Remove the pulley set bolt.
    b. Remove the crankshaft timing gear from the crankshaft.
    c. Remove the 2 pulley set keys from the crankshaft.
33. Remove the camshaft timing gears and No. 2 chain (for Right-hand Bank), as follows:
    a. While raising up the No. 2 chain tensioner, insert a pin of 1.0 mm (0.039 in.) into the hole to hold it.
    b. Hold the hexagonal portion of the camshaft with a wrench, and remove the 2 bolts and 2 camshaft timing gears.

➡Be careful not to damage the cylinder head with the wrench.

➡Do not disassemble the camshaft timing gear assemblies.

    c. Remove the No. 2 chain.
34. Remove the bolt and No. 2 chain tensioner.
35. Remove the camshaft timing gears and No. 2 chain (for Left-hand Bank), as follows:
    a. While pushing down on the No. 3 chain tensioner, insert a pin of 1.0 mm (0.039 in.) into the hole to hold it.
    b. Hold the hexagonal portion of the camshaft with a wrench, and remove the 2 bolts and 2 camshaft timing gears.

➡Be careful not to damage the cylinder head with the wrench.

➡Do not disassemble the camshaft timing gear assemblies.

    c. Remove the No. 2 chain.
36. Remove the bolt and the No. 3 chain tensioner.

To install:

37. Install the No. 2 chain tensioner assembly with the bolt and tighten to 15 ft. lbs. (21 Nm).
38. While pushing in the tensioner, insert a pin of 1.0 mm (0.039 in.) into the hole to hold it.
39. Install the camshaft timing gears and No. 2 chain (for Right-hand Bank), as follows:
    a. Align the mark plate with the timing marks (1-dot mark) of the camshaft timing gears as shown.
    b. Apply a light coat of engine oil to the bolt threads and bolt-seating surface.
    c. Align the knock pin of the camshaft with pin hole of the camshaft timing gear. Install the camshaft timing gear and the right camshaft timing exhaust gear with the No. 2 chain installed.
    d. Hold the hexagonal portion of the camshaft with the wrench and tighten the two bolts to 74 ft. lbs. (100 Nm).
    e. Remove the pin from the No. 2 chain tensioner.
40. Install the No. 3 chain tensioner assembly with the bolt and tighten to 15 ft. lbs. (21 Nm).
41. While pushing in the tensioner, insert a pin of 1.0 mm (0.039 in.) into the hole to hold it.
42. Install the camshaft timing gears and No. 2 chain (for Left-hand Bank), as follows:
    a. Align the mark plate (yellow) with the timing marks (2-dot mark) of the camshaft timing gears as shown.

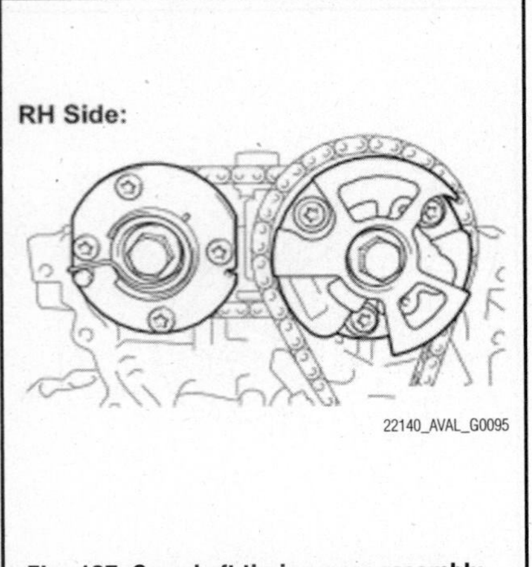

Fig. 127 Camshaft timing gear assembly positioning

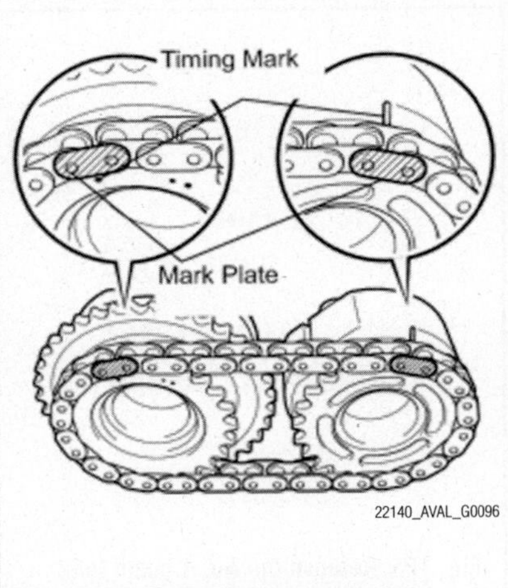

Fig. 128 Aligning No. 2 timing chain

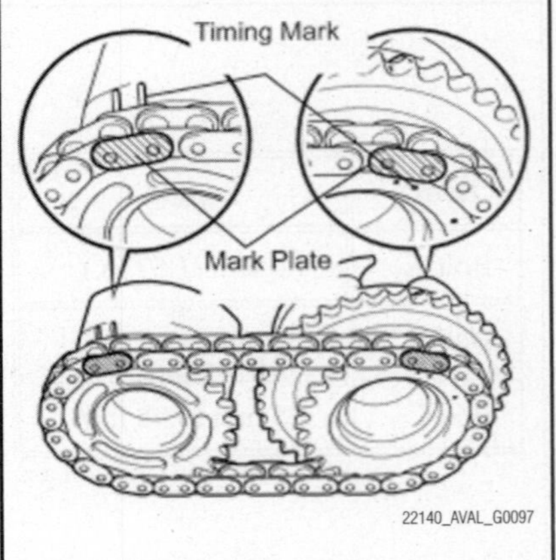

Fig. 129 Aligning left-hand No. 2 timing chain

b. Apply a light coat of engine oil to the bolt threads and bolts seating surface.

c. Align the knock pin of the camshaft with pin hole of the camshaft timing gear. Install the camshaft timing gear and the left camshaft timing exhaust gear with the No. 2 chain installed.

d. Hold the hexagonal portion of the camshaft with the wrench and tighten the two bolts to 74 ft. lbs. (100 Nm).

e. Remove the pin from the No 2 chain tensioner.

43. Install the No. 1 and 2 chain vibration dampers.

44. Install the timing gear set keys and crankshaft timing sprocket as shown in the illustration.

45. Install the idle sprocket assembly, as follows:

a. Apply a light coat of engine oil to the rotating surface of the No. 1 idle gear shaft.

b. Temporarily install the No. 1 idle gear shaft and idle sprocket with the No. 2 idle gear shaft while aligning the knock pin of the No. 1 idle gear with the knock pin groove of the cylinder block. Be careful of the idle gear direction.

c. Using a 10 mm hexagon wrench, tighten the No. 2 idle gear shaft to 44 ft. lbs. (60 Nm).

d. After installing the idle sprocket assembly, check that the idle sprocket turns smoothly.

46. Install the chain sub-assembly, as follows:

a. Align the mark plate and timing marks as shown in the illustration and install the chain. The camshaft mark plate is orange.

Fig. 130 Crankshaft timing sprocket

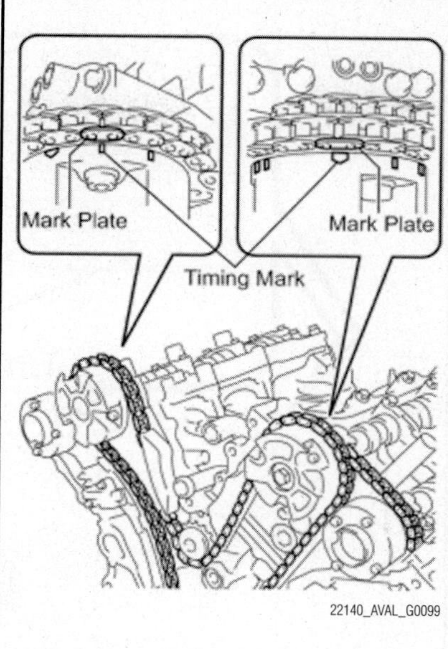

Fig. 131 Aligning timing chain sub-assembly

b. Do not pass the chain over the crankshaft, just put it on.

c. Turn the camshaft timing gear assembly on the right bank counterclockwise to tighten the chain between the banks.

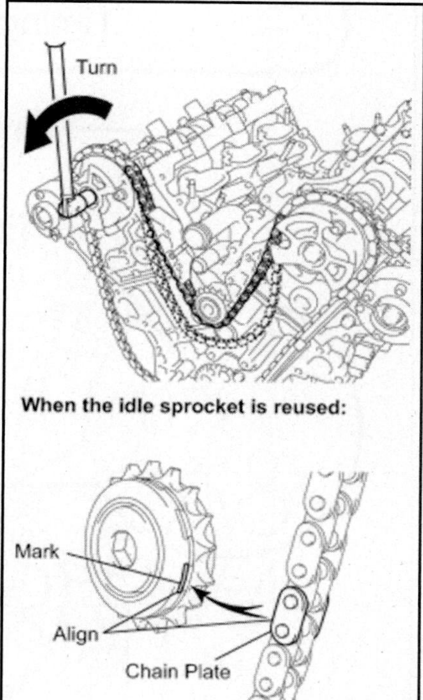

Fig. 132 Timing chain and idle sprocket alignment

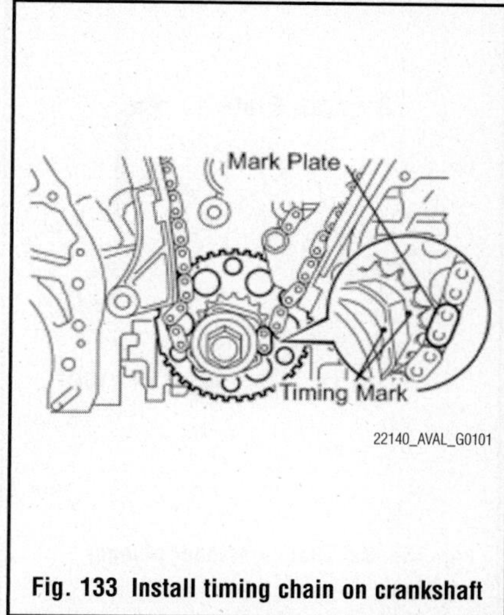

Fig. 133 Install timing chain on crankshaft

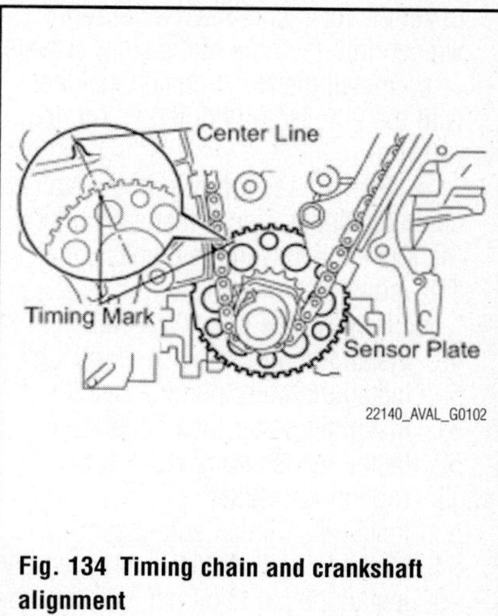

Fig. 134 Timing chain and crankshaft alignment

➡When the idle sprocket assembly is reused, align the timing chain plate with the mark on the sprocket in order to tighten the chain between the banks.

d. Align the mark plate and timing marks as shown in the illustration and install the chain onto the crankshaft timing sprocket. The crankshaft to mark plate is yellow.

e. Temporarily tighten the pulley set bolt.

f. Turn the crankshaft clockwise to set it to the right-hand block bore more centerline. (TDC/compression).

47. Install the chain tensioner slipper.

48. Install the No. 1 chain tensioner assembly, as follows:

a. Move the stopper plate upward to release the lock, and push the plunger deep into the tensioner.

b. Move the stopper plate downward

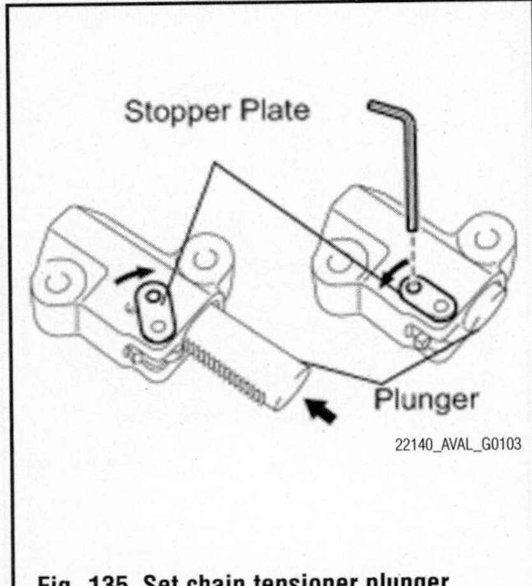

Fig. 135 Set chain tensioner plunger position

to set the lock, and insert a hexagon wrench into the hole of the stopper plate.

c. Install the No. 1 chain tensioner with the 2 bolts and tighten to 7 ft. lbs. (10 Nm).

d. Remove the lock pin of the No. 1 chain tensioner. Check that each timing mark is aligned with the crankshaft at TDC/compression.

e. Remove the pulley set bolt.

49. Install timing chain cover and seal.
50. Install the water pump assembly.
51. Install the water inlet housing.
52. Install the left hand No. 1 front engine mounting bracket.
53. Install the oil pan sub-assembly.
54. Install the oil strainer sub-assembly.
55. Install the No. 2 oil pan sub-assembly.
56. Install the oil pan drain plug and gasket.
57. Install the cylinder head cover sub-assembly.

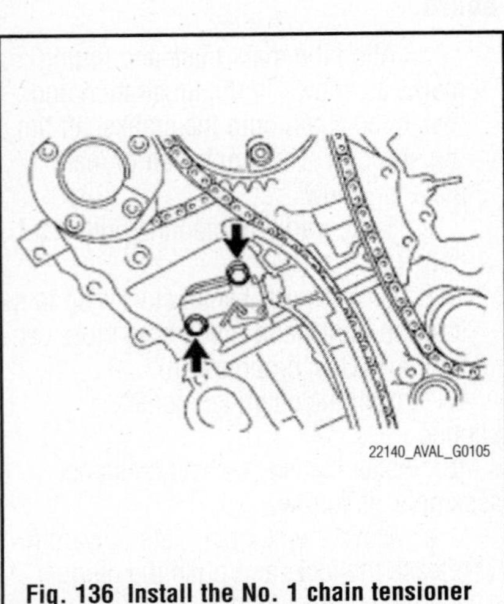

Fig. 136 Install the No. 1 chain tensioner with the 2 bolts

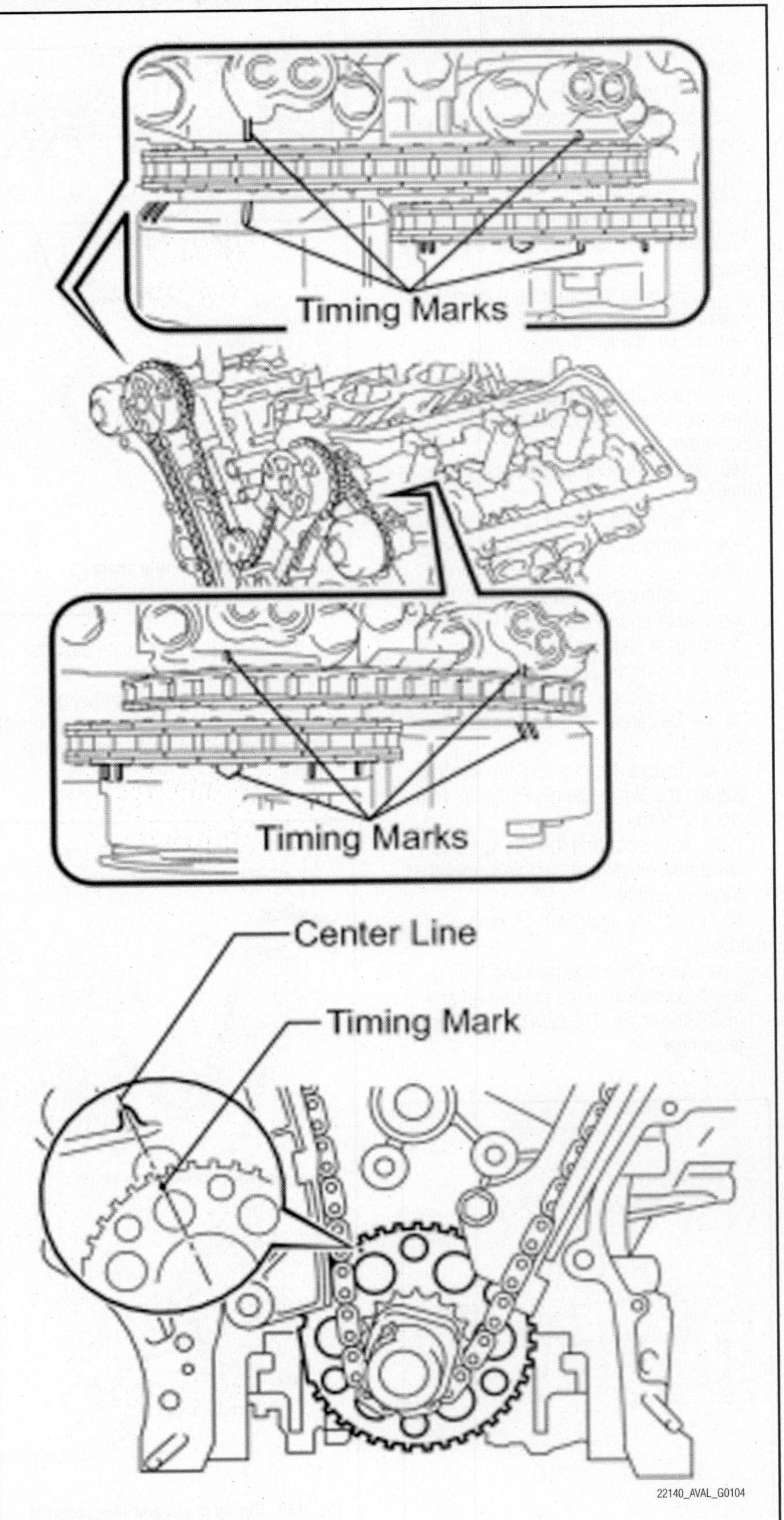

Fig. 137 Aligning timing marks

58. Install the left-hand cylinder head cover sub-assembly.
59. Install the water outlet.
60. Install the crankshaft pulley.
61. Install the oil filter.
62. Install the cylinder block water drain cock sub-assembly.
63. Install the No. 1 oil pipe.
64. Install the oil pipe.
65. Install the crankshaft position sensor.
66. Install the camshaft timing oil control valve assembly.
67. Install the camshaft position sensor.
68. Install the ventilation valve sub-assembly.
69. Install the spark plugs and ignition coil assembly.
70. Install the oil filler cap sub-assembly.
71. Install the engine assembly with transaxle.

VALVE COVERS

REMOVAL & INSTALLATION

See Figures 138 through 145.

1. Before servicing the vehicle, refer to the Precautions Section.

RH or Rear Bank

1. Remove or disconnect the following:
   a. Negative battery cable.
   b. Wiper link assembly from both wipers.
   c. Outer top cowl panel.
   d. Wiper motor and linkage assembly.
   e. Engine v-bank cover.
   f. Air cleaner cap with air cleaner hoses.
   g. All hoses from the throttle body and intake air surge tanks.

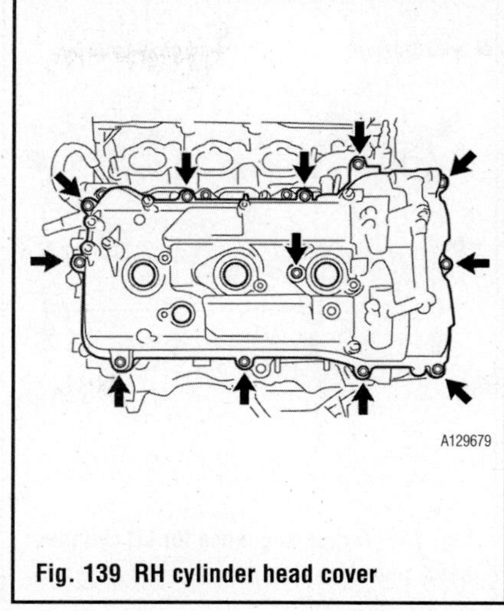

Fig. 139 RH cylinder head cover

   h. Intake air surge tank.
   i. Remove ignition coils.
   j. Disconnect cam sensor connector.
   k. Disconnect camshaft timing control valve connectors.
   l. Remove oil pipe
2. Remove cylinder head cover sub-assembly.
   a. Remove the 12 bolts, seal washer, cylinder head cover sub-assembly and cylinder head cover gasket.

LH or Front Bank

1. Remove or disconnect the following:
   a. Negative battery cable.
   b. Engine v—bank cover.
   c. Air cleaner cap with air cleaner hoses.
   d. Remove ignition coils.
   e. Disconnect cam sensor connector.
   f. Disconnect camshaft timing control valve connectors.
   g. Disconnect ventilation valve hose.

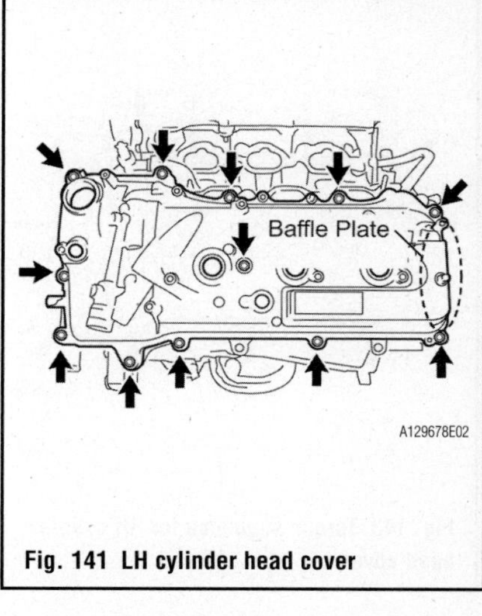

Fig. 141 LH cylinder head cover

   h. Remove oil pipe
2. Remove LH cylinder head cover sub-assembly.

➡The baffle plate is located on the back of the portion shown in the illustration. Do not damage the baffle plate when removing the head cover.

To install

RH or Rear Bank

1. Installation is the reverse order of removal. Note the following:
   a. Apply seal packing as shown in the illustration.

➡Remove any oil from the contact surface.

➡Install the head cover within 3 minutes after applying seal packing.

➡Do not start the engine for at least 2 hours after installing.

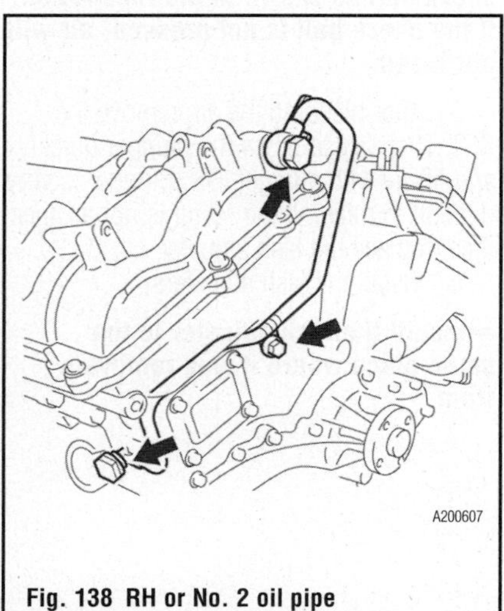

Fig. 138 RH or No. 2 oil pipe

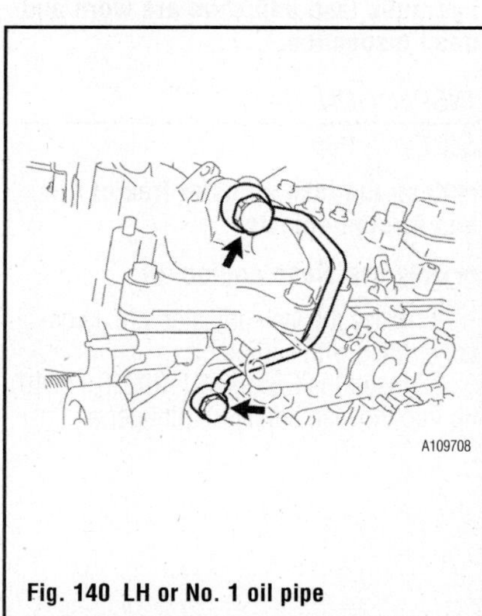

Fig. 140 LH or No. 1 oil pipe

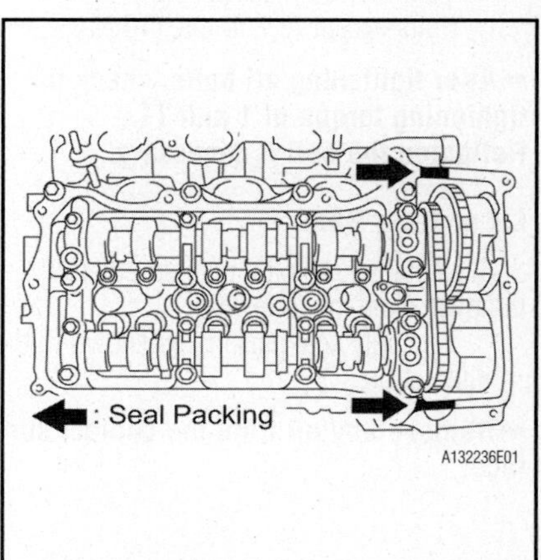

Fig. 142 Seal packing locations for RH cylinder head

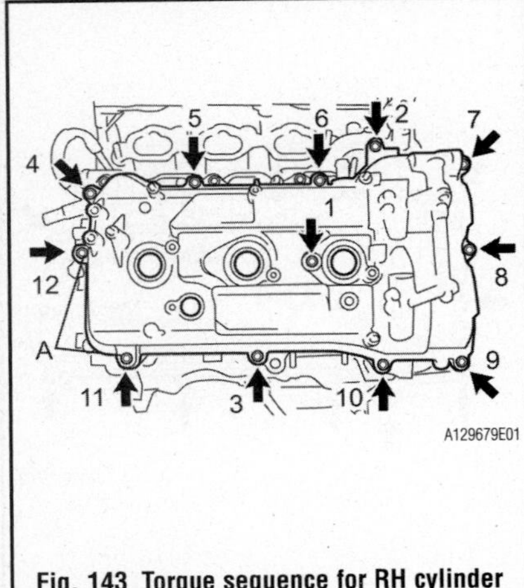

Fig. 143 Torque sequence for RH cylinder head cover

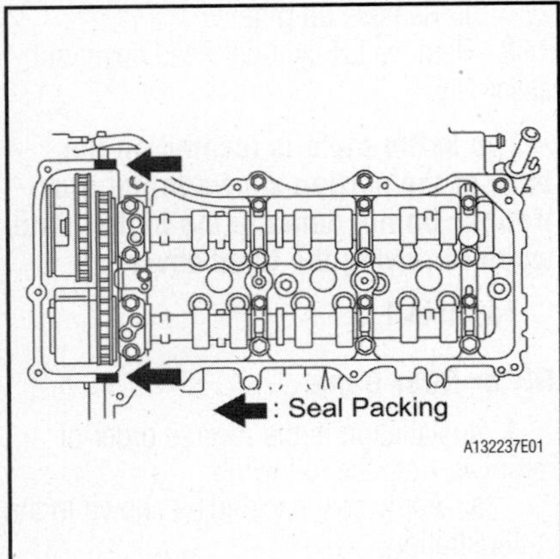

Fig. 144 Seal packing locations for LH cylinder head

b. Install a new gasket to the head cover.

c. Install a head cover with the 12 bolts and a new washer.

Bolts A: 15 ft. lbs. (21 Nm)

Bolts except A: 7 ft. lbs. (10 Nm)

➡After tightening all bolts, check the tightening torque of 1 and 11. Retighten the bolt if necessary.

LH or Rear Bank

1. Installation is the reverse order of removal. Note the following:

a. Apply seal packing as shown in the illustration.

➡Remove any oil from the contact surface.

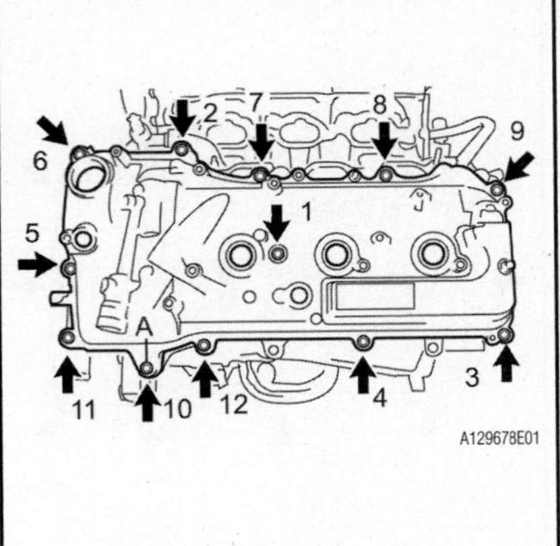

Fig. 145 Torque sequence for LH cylinder head cover

➡Install the head cover within 3 minutes after applying seal packing.

➡Do not start the engine for at least 2 hours after installing.

b. Install a new gasket to the head cover.

c. Install a head cover with the 12 bolts and a new washer.

Bolts A: 15 ft. lbs. (21 Nm)

Bolts except A: 7 ft. lbs. (10 Nm)

➡After tightening all bolts, check the tightening torque of 1 and 10. Retighten the bolt if necessary.

VALVE LASH

ADJUSTMENT

➡This vehicle is equipped with hydraulic lash adjusters and they are not adjustable. If the vehicle has valve train noise it is possible that the hydraulic lash adjusters are worn and need inspection.

INSPECTION

See Figure 146.

➡Keep the lash adjuster free of dirt and foreign objects.

➡Only use clean engine oil.

1. Place the lash adjuster into a container filled with engine oil.
2. Insert the SST's (SST: 09276-75010) tip into the lash adjuster's plunger and use the tip to press down on the check ball inside the plunger.

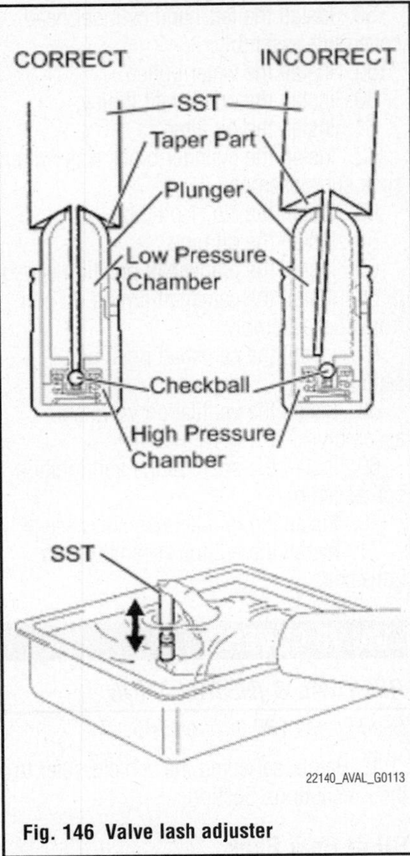

Fig. 146 Valve lash adjuster

3. Squeeze the SST and lash adjuster together to move the plunger up and down 5 to 6 times.
4. Check the movement of the plunger and bleed the air. OK: Plunger moves up and down.

➡When bleeding air from the high-pressure chamber, make sure that the tip of the SST is actually pressing the check ball as shown in the illustration. If the check ball is not pressed, air will not bleed.

5. After bleeding the air, remove the SST. Then, try to press the plunger quickly and firmly with a finger. OK: Plunger is very difficult to move. If the result is not as specified, replace the lash adjuster.
6. Install the lash adjusters.

➡Install the lash adjuster to the same place where it was removed from.

ENGINE PERFORMANCE & EMISSION CONTROLS

ACCELERATOR PEDAL POSITION (APP) SENSOR

REMOVAL & INSTALLATION

1. Remove left center floor carpet cover.
2. Disconnect the accelerator pedal connector.
3. Remove the 2 nuts and accelerator pedal assembly.

➡Avoid physical shock to the accelerator pedal assembly.

➡Do not disassemble the accelerator pedal assembly.

To install:

Installation is the reverse of the removal procedure. Tighten the accelerator pedal rod nuts to 43 inch lbs. (5.4 Nm).

CAMSHAFT POSITION (CMP) SENSOR

LOCATION

See Figure 147.

Refer to the accompanying illustration for sensor location.

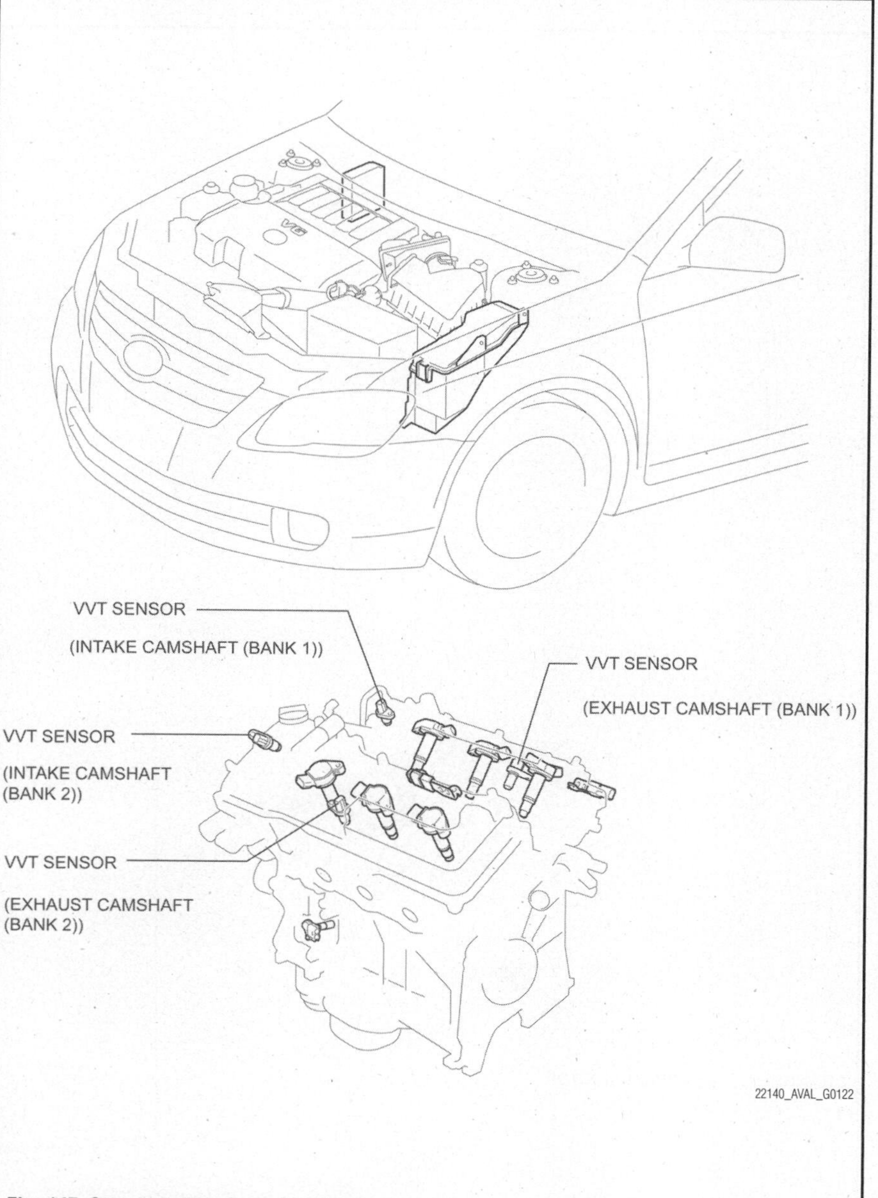

Fig. 147 Camshaft VVT sensor location

REMOVAL & INSTALLATION

See Figures 148 through 151.

1. Remove the V-bank cover sub-assembly.
2. Remove the Intake camshaft VVT sensor (Bank 1), as follows:
   a. Disconnect the VVT sensor connector.

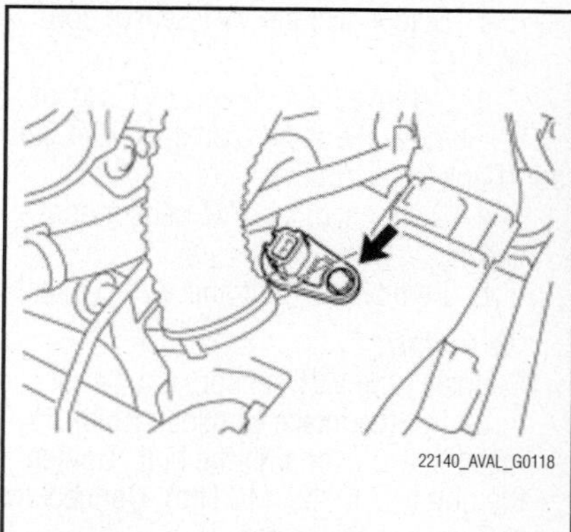

Fig. 148 Bank 1 intake camshaft VVT sensor

Fig. 149 Bank 1 exhaust camshaft VVT sensor

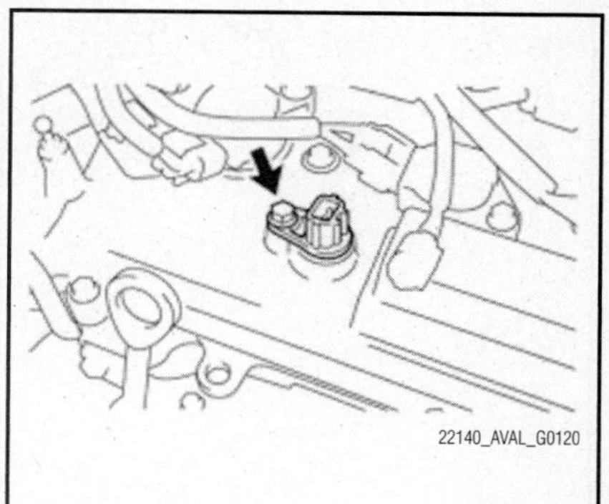

Fig. 150 Bank 2 intake camshaft VVT sensor

b. Remove the bolt and VVT sensor.

3. Remove the Exhaust camshaft VVT sensor (Bank 1), as follows:

a. Remove the windshield wiper link assembly.

b. Remove the front cowl top outside panel.

c. Disconnect the VVT sensor connector.

d. Remove the bolt and VVT sensor.

4. Remove the Exhaust camshaft VVT sensor (Bank 2), as follows:

a. Disconnect the VVT sensor connector.

b. Remove the bolt and VVT sensor.

5. Remove the Intake camshaft VVT sensor (Bank 2), as follows:

a. Disconnect the VVT sensor connector.

b. Remove the bolt and VVT sensor.

To install:

6. Install the VVT sensor.

a. For the intake camshaft (Bank 1), install the sensor with the bolt. Tighten the bolt to 7 ft. lbs. (10 Nm). Connect the VVT sensor connector.

b. For the exhaust camshaft (Bank 1), install the VVT sensor with the bolt. Tighten the bolt to 7 ft. lbs. (10 Nm). Connect the connector. Install the cowl top outside panel front. Install the windshield wiper link assembly.

c. For the exhaust camshaft (Bank 2), install the VVT sensor with the bolt. Tighten the bolt to 7 ft. lbs. (10 Nm). Connect the VVT sensor connector.

d. For the intake camshaft (Bank 2), install the VVT sensor with the bolt. Tighten the bolt to 7 ft. lbs. (10 Nm). Connect the VVT sensor connector.

7. Install the V-bank cover sub assembly.

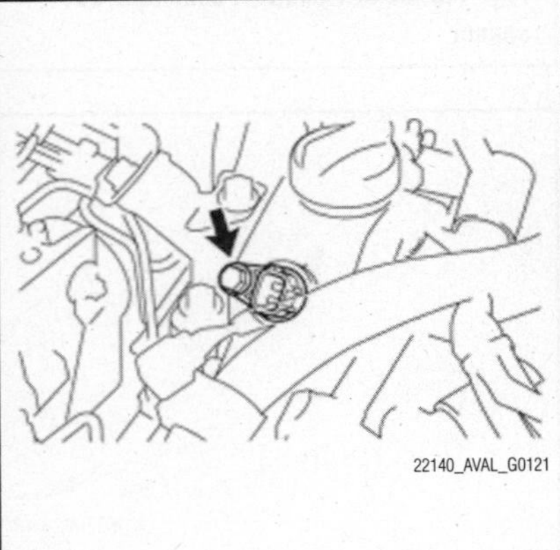

Fig. 151 Bank 2 exhaust camshaft VVT sensor

CRANKSHAFT POSITION (CKP) SENSOR

LOCATION

See Figure 152.

Refer to the accompanying illustration for sensor location.

REMOVAL & INSTALLATION

1. Disconnect the negative battery terminal.
2. Remove the V-ribbed belt.
3. Remove alternator assembly.
4. Disconnect the cooler compressor assembly.
5. Remove the crankshaft position sensor connector.
6. Remove the bolt, and then remove the crankshaft position sensor.

To install:

7. Apply a light coat of engine oil to the O-ring on the crankshaft position sensor.
8. Install the crankshaft position sensor with the bolt and tighten to 7 ft. lbs. (10 Nm).
9. Connect the crankshaft position sensor connector.
10. The remainder of installation is the reverse of the removal procedure.

ELECTRONIC CONTROL MODULE (ECM)

LOCATION

See Figure 153.

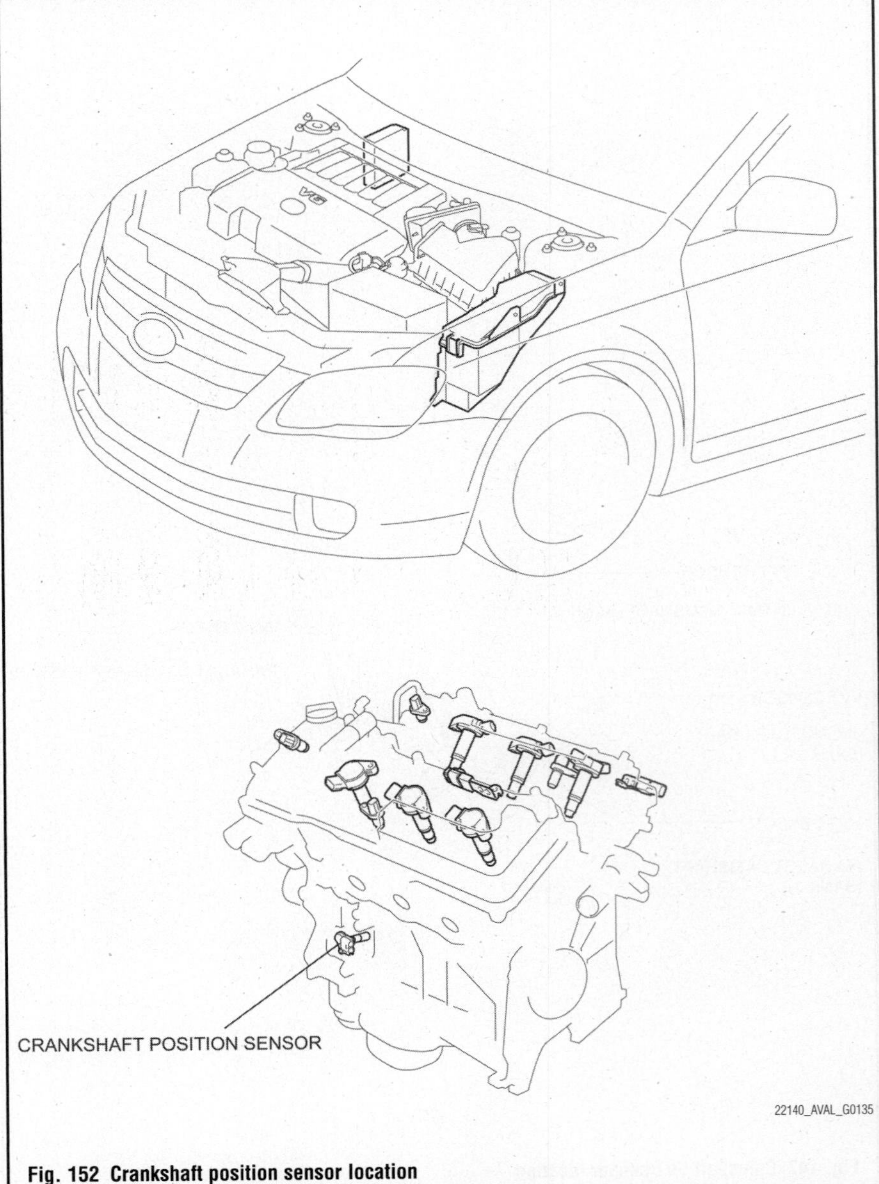

Fig. 152 Crankshaft position sensor location

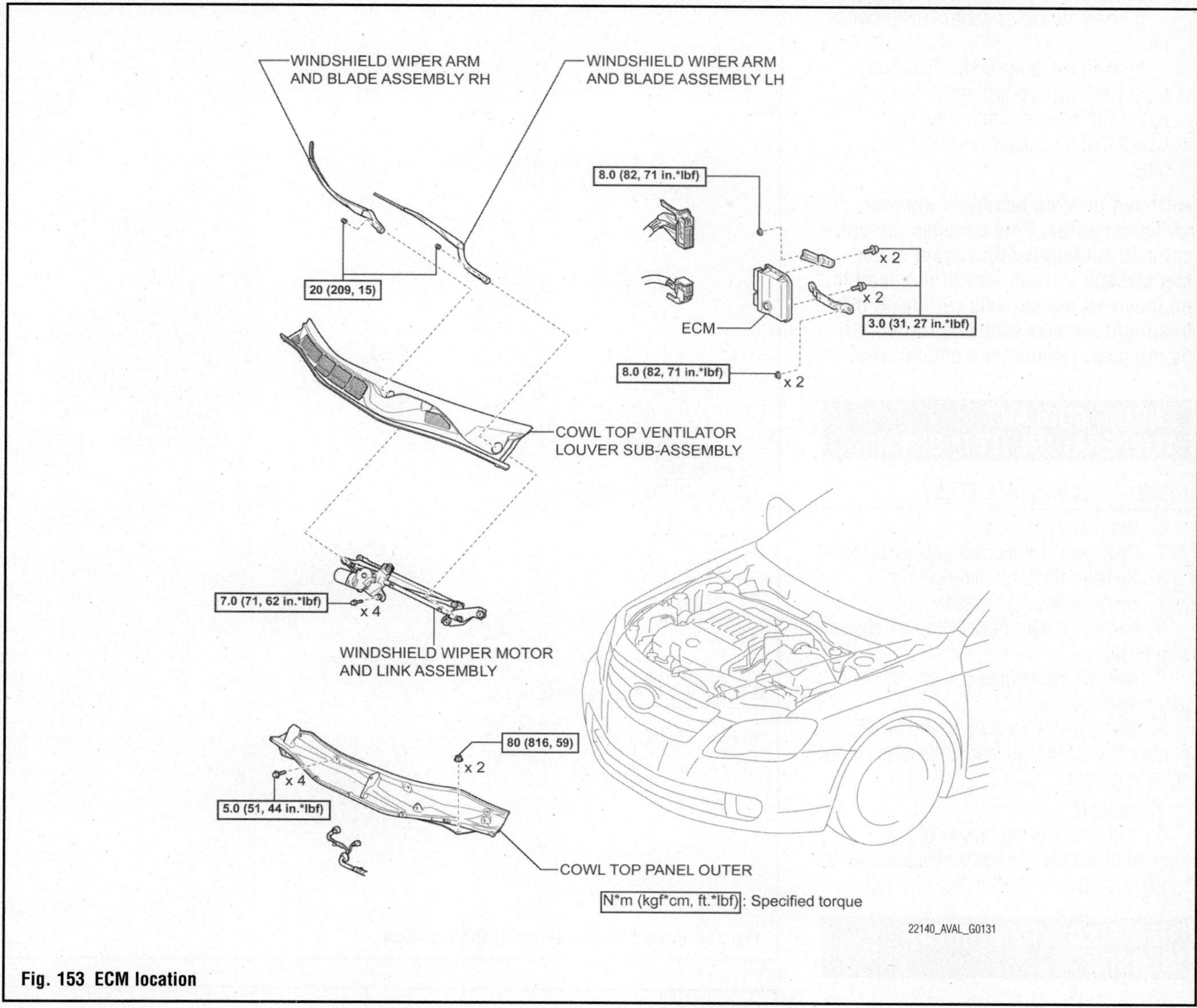

Fig. 153 ECM location

Refer to the accompanying illustration for ECM location.

REMOVAL & INSTALLATION

1. Disconnect the negative battery cable.
2. Remove both windshield wiper arm and blade assemblies.
3. Remove the cowl top ventilator louver sub-assembly.
4. Remove the windshield wiper motor and link assembly.
5. Remove the outer cowl top panel.
6. Remove the ECM, as follows:
   a. Remove the 3 nuts.
   b. Separate the ECM from the body. When separating the ECM, do not apply excessive force to the wire harness.
   c. Raise the 2 levers while pushing the locks on the 2 levers, and disconnect the 2 ECM connectors.

➡After disconnecting the connectors, make sure that dirt, water or other foreign matter does not contact the connections of the connectors.

   d. Remove the ECM.
   e. Remove the 4 screws and 2 ECM brackets.

To install:

7. Install the 2 ECM brackets with the 4 screws, and tighten to 27 inch lbs. (3 Nm).
8. Connect the 2 ECM connectors and lower the 2 levers.

➡Make sure that dirt, water or other foreign matter does not contact the connections of the connectors.

9. Install the ECM to the body.
10. Attach the ECM with the 3 nuts and tighten to 71 inch lbs. (8 Nm).
11. Install the outer cowl top panel.
12. Install the windshield wiper motor and link assembly.
13. Install the cowl top ventilator louver sub-assembly.
14. Install both windshield wiper arm and blade assemblies.
15. Connect the negative battery cable.
16. Register the immobilizer communication ID. If the ECM is replaced, register the ECM communication ID for the immobilizer system (refer to the Service Bulletin for registration).
17. Perform initialization. After replacing the ECM on vehicles with a dynamic laser cruise control system, it is necessary to initialize the ECM so that the ECM can recognize the dynamic laser cruise control system.
18. Be sure to perform the following procedure after replacing the ECM:
   a. Turn the ignition switch on (IG).

b. Turn the cruise control main switch on.

c. With the brake pedal depressed, push the cruise control main switch to RES/ACC 3 times within 3 seconds. Check that the buzzer sounds at this time.

➡Do not turn the headlight dimmer switch on at this time because the optical axis automatic adjustment mode has already started, which may lead to an incorrect optical axis setting. If the headlight dimmer switch is turned on by mistake, readjust the optical axis.

ENGINE COOLANT TEMPERATURE (ECT) SENSOR

REMOVAL & INSTALLATION

1. Drain engine coolant.
2. Remove V-bank cover sub-assembly.
3. Remove No. 2 air cleaner inlet.
4. Remove No. 1 air cleaner inlet.
5. Remove the air cleaner cap air cleaner with hose.
6. Remove the engine coolant temperature sensor connector.
7. Using a 19 mm deep socket wrench, remove the engine coolant temperature sensor and gasket.

To install:

8. Installation is the reverse of the removal procedure. Torque the engine coolant temperature sensor to 15 ft. lbs. (20 Nm).

HEATED OXYGEN (HO2S) SENSOR

LOCATION

See Figure 154.

Refer to the accompanying illustration for sensor location.

REMOVAL & INSTALLATION

See Figures 155 and 156.

1. Disconnect the 2 Heated Oxygen Sensor (HO2S) connectors.

   a. Using SST (SST: 09224-00010) or equivalent, remove the 2 oxygen sensors from the front pipe assembly.

To install:

2. Install the 2 oxygen sensors to the front pipe assembly. Tighten to 32 ft. lbs. (44 Nm) and 30 ft. lbs. (40 Nm). Use a torque wrench with a fulcrum length of 11.81 inch (300 mm).
3. Connect the 2 oxygen sensor connectors.

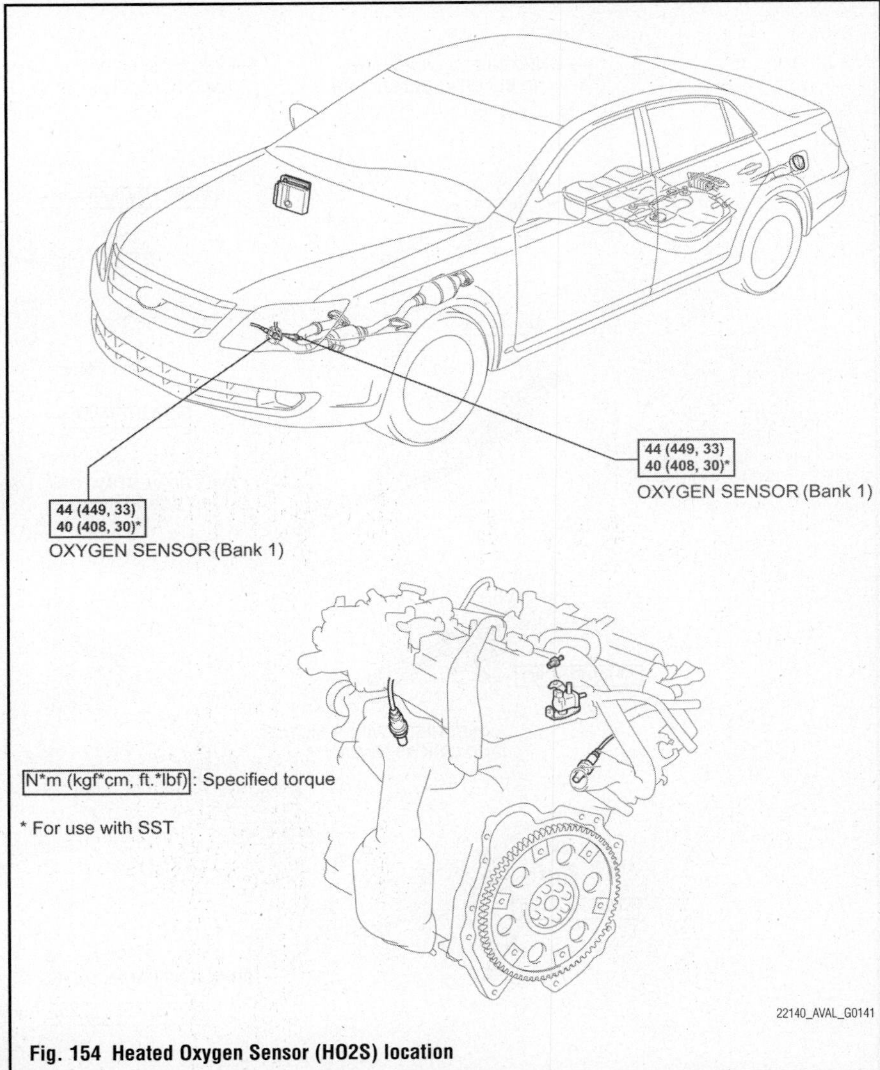

Fig. 154 Heated Oxygen Sensor (HO2S) location

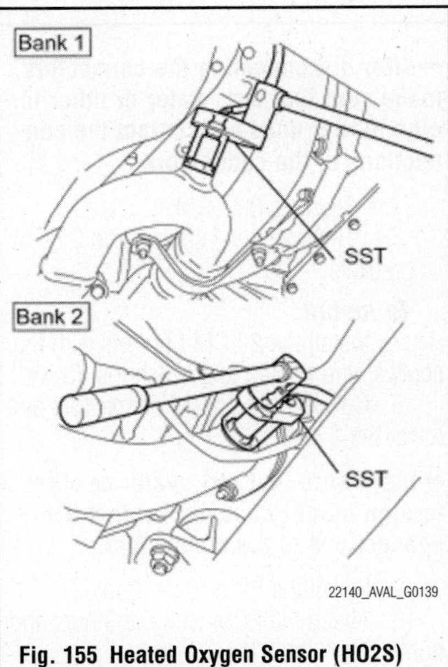

Fig. 155 Heated Oxygen Sensor (HO2S) removal

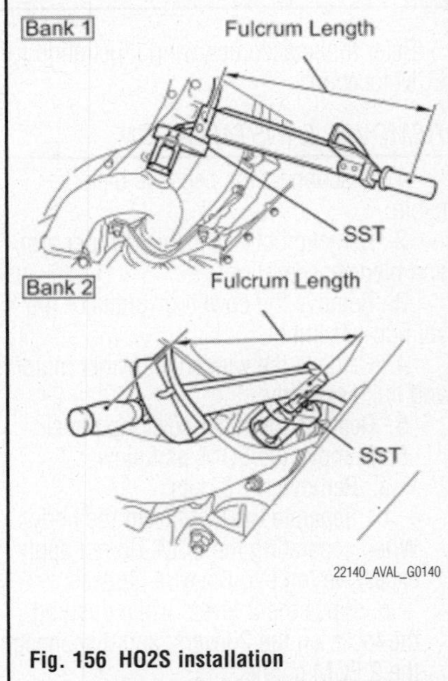

Fig. 156 HO2S installation

INTAKE AIR TEMPERATURE (IAT) SENSOR

LOCATION

The Intake Air Temperature (IAT) sensor is an integral component of the Mass Air Flow (MAF) meter.

REMOVAL & INSTALLATION

See Mass Air Flow (MAF) meter.

KNOCK SENSOR (KS)

LOCATION

See Figure 157.

Refer to the accompanying illustration for sensor location.

REMOVAL & INSTALLATION

1. Properly discharge the fuel system pressure.
2. Disconnect battery negative cable.
3. Drain engine coolant.
4. Remove the windshield wiper motor and link assembly.
5. Remove the outer cowl top panel.
6. Remove the V-bank cover sub-assembly.
7. Remove the air cleaner cap with air cleaner hose.
8. Remove the intake air surge tank.
9. Remove the intake manifold.

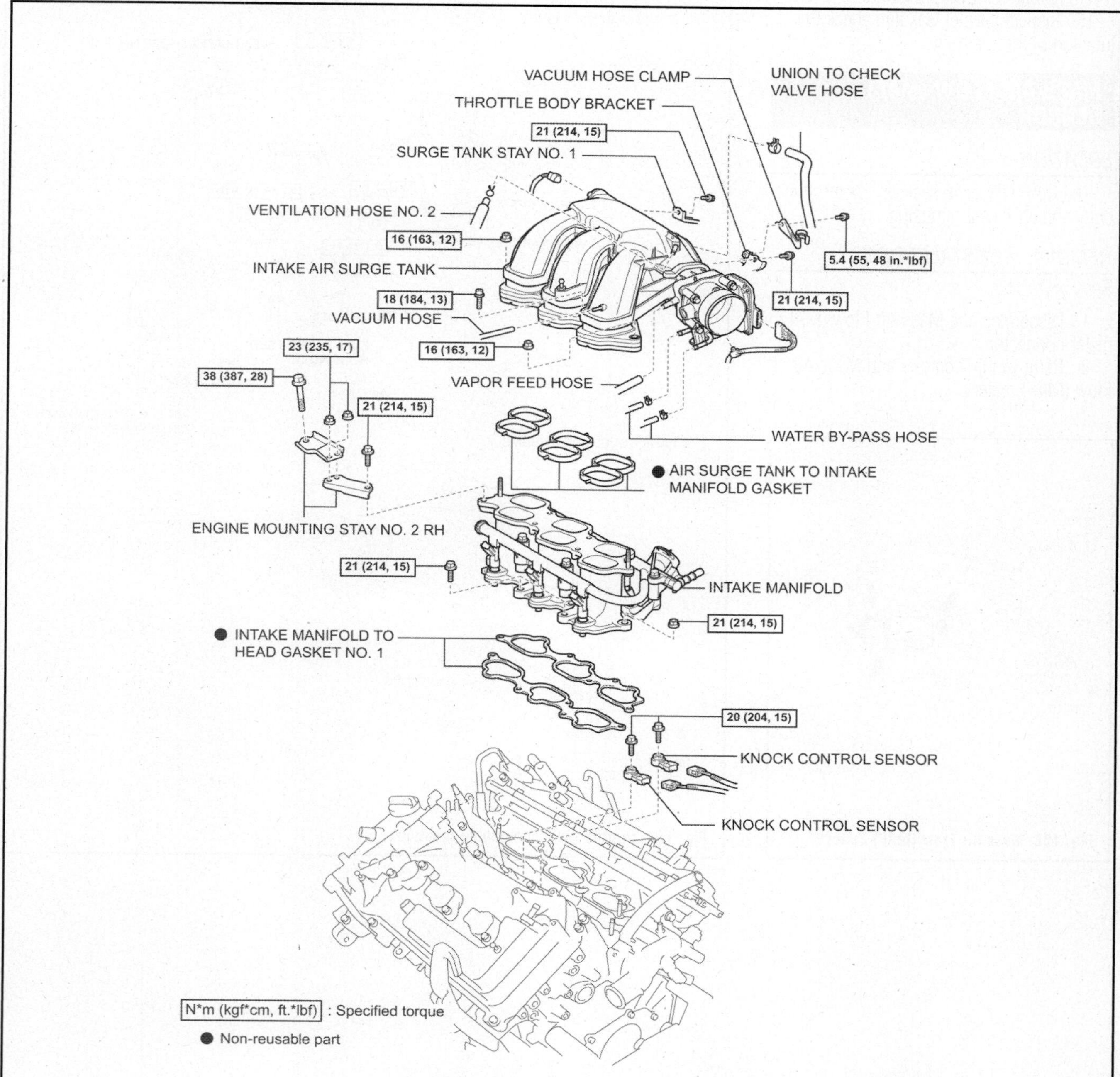

Fig. 157 Knock sensor location

10. Disconnect the 2 knock control sensor connectors.
11. Remove the 2 bolts and 2 knock control sensors.

To install:

12. Install the 2 knock control sensors with the 2 bolts as shown in the illustration and tighten to 15 ft. lbs. (20 Nm).
13. Connect the 2 knock control sensor connectors.
14. The remainder of installation is the reverse of the removal procedure.
15. Inspect for fuel leak and check the function of throttle body.

MASS AIR FLOW (MAF) SENSOR

LOCATION

The MAF sensor is between the throttle body and air cleaner housing.

REMOVAL & INSTALLATION

See Figure 158.

1. Disconnect the Mass Air Flow (MAF) meter connector.
2. Remove the 2 screws and Mass Air Flow (MAF) meter.

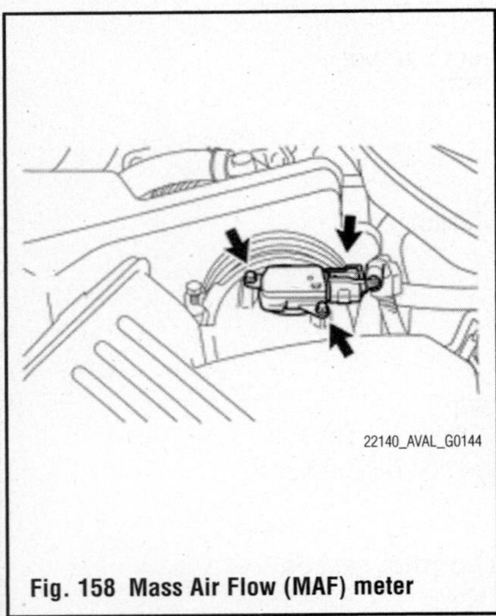

Fig. 158 Mass Air Flow (MAF) meter

To install:

3. Installation is the reverse of the removal procedure.

THROTTLE POSITION SENSOR (TPS)

LOCATION

See Figure 159.

The Throttle Position Sensor (TPS) is located on the throttle body.

REMOVAL & INSTALLATION

Refer to the Throttle Body removal and installation procedures.

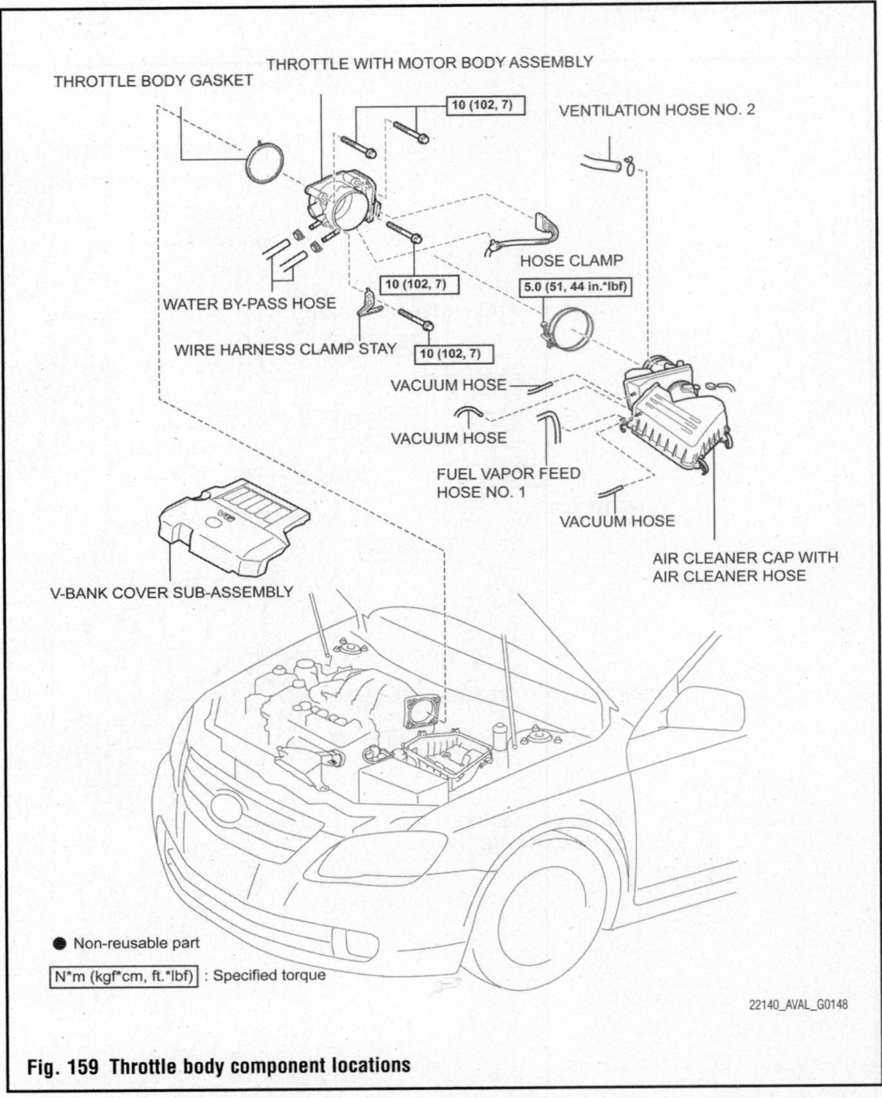

Fig. 159 Throttle body component locations

FUEL SYSTEM SERVICE PRECAUTIONS

Safety is the most important factor when performing not only fuel system maintenance but any type of maintenance. Failure to conduct maintenance and repairs in a safe manner may result in serious personal injury or death. Maintenance and testing of the vehicle's fuel system components can be accomplished safely and effectively by adhering to the following rules and guidelines.

- To avoid the possibility of fire and personal injury, always disconnect the negative battery cable unless the repair or test procedure requires that battery voltage be applied.
- Always relieve the fuel system pressure prior to disconnecting any fuel system component (injector, fuel rail, pressure regulator, etc.), fitting or fuel line connection. Exercise extreme caution whenever relieving fuel system pressure to avoid exposing skin, face and eyes to fuel spray. Please be advised that fuel under pressure may penetrate the skin or any part of the body that it contacts.
- Always place a shop towel or cloth around the fitting or connection prior to loosening to absorb any excess fuel due to spillage. Ensure that all fuel spillage (should it occur) is quickly removed from engine surfaces. Ensure that all fuel soaked cloths or towels are deposited into a suitable waste container.
- Always keep a dry chemical (Class B) fire extinguisher near the work area.
- Do not allow fuel spray or fuel vapors to come into contact with a spark or open flame.
- Always use a back-up wrench when loosening and tightening fuel line connection fittings. This will prevent unnecessary stress and torsion to fuel line piping.
- Always replace worn fuel fitting O-rings with new Do not substitute fuel hose or equivalent where fuel pipe is installed.

Before servicing the vehicle, make sure to also refer to the precautions in the beginning of this section as well.

RELIEVING FUEL SYSTEM PRESSURE

CAUTION

Perform the following procedures to prevent fuel from spilling out before removing any fuel system parts.

CAUTION

Pressure will still remain in the fuel line even after performing the following procedures. When disconnecting the fuel line, cover with a shop rag or a piece of cloth to prevent fuel from spraying or spewing out.

1. Disconnect the fuel pump connector.
   a. Remove the rear seat cushion assembly.
   b. Remove the rear floor service hole cover.
   c. Disconnect the fuel pump connector.
   d. Start the engine.
   e. After the engine stops, turn the ignition switch off.

➡DTC P0171/25 (fuel problem) may be detected.

   f. Crank the engine again. Check that the engine does not start.
   g. Remove the fuel tank cap to discharge pressure from the fuel tank.
   h. Disconnect the cable from the negative (-) battery terminal.
   i. Reconnect the fuel pump connector.
   j. Install the rear floor service hole cover.
   k. Install the rear seat.

FUEL FILTER

REMOVAL & INSTALLATION

See Figure 160.

1. Before servicing the vehicle, refer to the precautions section.

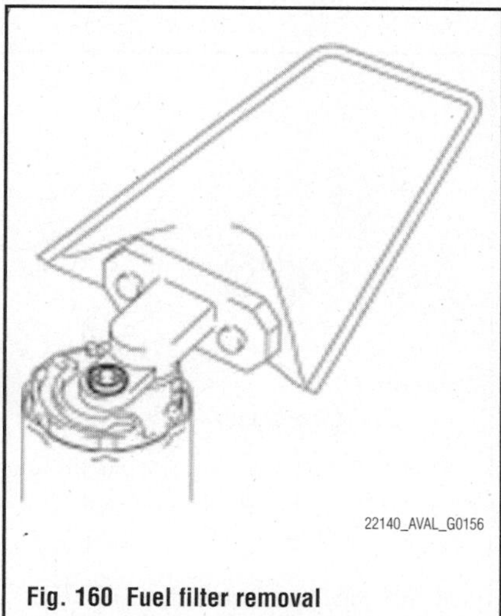

Fig. 160 Fuel filter removal

2. Remove the fuel pump from the vehicle.
3. Remove the fuel pump filter, as follows:
   a. Using a screwdriver, pry out the clip.
   b. Pull out the fuel pump filter from the fuel pump.

To install:

4. Install the fuel pump filter with a new clip.
5. Install the fuel pump.

FUEL INJECTORS

REMOVAL & INSTALLATION

See Figures 161 through 163.

1. Properly discharge the fuel system pressure.
2. Disconnect battery negative cable.
3. Drain engine coolant.
4. Remove both windshield wiper arm and blade assemblies.
5. Remove the right cowl top ventilator louver.
6. Remove the windshield wiper motor and link assembly.
7. Remove the outer cowl top panel.
8. Remove the V-bank cover sub-assembly.
9. Remove the air cleaner cap with air cleaner hose.
10. Remove the intake air surge tank.
11. Disconnect the fuel tube sub-assembly, as follows:

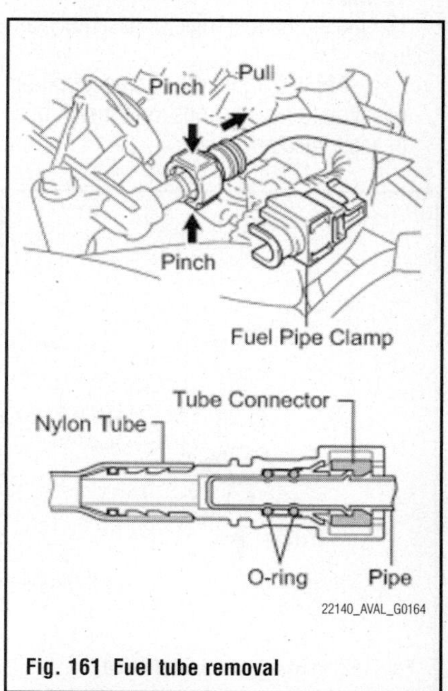

Fig. 161 Fuel tube removal

a. Remove the No. 2 fuel pipe clamp.

b. Pinch the tube connector and then pull out the fuel pipe.

➡**Check that there is no dirt or other foreign objects around the connector before removing fuel tube, and clean the connector as necessary.**

➡**It is necessary to prevent mud or dirt from entering the connector. If mud or dirt gets in the connector, the O-rings may not seal properly.**

➡**Do not use any tools in this operation.**

➡**Do not bend, kink or twist the nylon tube. Protect the connector by covering it with a plastic bag.**

➡**When the pipe and connector are stuck, push and pull the connector to release and pull the connector out carefully.**

12. Remove the fuel injector assembly, as follows:

a. Disconnect the 6 fuel injector connectors.

b. Remove the 5 bolts and fuel delivery pipe together with the 6 fuel injectors.

➡**Be careful not to drop the fuel injectors when removing the fuel delivery pipe.**

c. Remove the 6 insulators from the intake manifold.

d. Pull out the fuel injector from the fuel delivery pipe.

To install:

13. Install the fuel injector assembly, as follows:

a. Apply a light coat of spindle oil or gasoline to new O-rings, and install one to each injector.

b. Apply a light coat of spindle oil or gasoline where the fuel delivery pipe contacts the O-ring.

c. Push the fuel injector while twisting it back and forth to install it in the fuel delivery pipe.

d. Position the fuel injector connector outward.

➡**Be careful not to twist the O-ring.**

➡**After installing the fuel injector, check that it turns smoothly. If not, reinstall it with a new O-ring.**

e. Install 6 new insulators to the intake manifold.

f. Place the fuel delivery pipe and the 6 fuel injectors together to the intake manifold.

➡**Be careful not to drop the fuel injectors when installing the fuel delivery pipe.**

g. Temporarily install the 6 bolts which are used to hold the fuel delivery pipe to the intake manifold.

➡**After installing the fuel injector, check that it turns smoothly. If not, reinstall it with a new O-ring.**

h. Tighten the 5 bolts which are used to hold the fuel delivery pipe to the intake manifold to 15 ft. lbs. (21 Nm).

14. Push in the tube connector to the pipe until the tube connector makes a "click" sound.

➡**Before connecting the tube, make sure that it is not damaged. Make sure that there is no dirt present on the connecting surfaces.**

➡**After connecting, check if the fuel tube connector and the pipe are securely connected by pulling on them.**

15. Install the No. 2 fuel pipe clamp.

16. The remainder of installation is the reverse of the removal procedure.

17. Check for coolant leak and fuel leak.

FUEL PUMP

REMOVAL & INSTALLATION

See Figures 164 and 165.

1. Before servicing the vehicle, refer to the precautions section.

2. Discharge fuel system pressure.

3. Disconnect battery negative cable.

4. Remove the rear seat cushion assembly.

a. Remove the rear floor service hole cover.

b. Disconnect the fuel pump connector.

5. Separate the fuel pump tube sub-assembly, as follows:

a. Remove the tube joint clip, and pull out the fuel pump tube.

➡**Check if there is any dirt or mud around the connector before this operation and remove the dirt as necessary.**

➡**Be careful of mud because the quick connector has an O-ring which seals the pipe and connector that can be contaminated.**

➡**Do not use any tools in this operation.**

➡**Do not bend or twist the nylon tube. Cover the fuel tube joint with a plastic bag.**

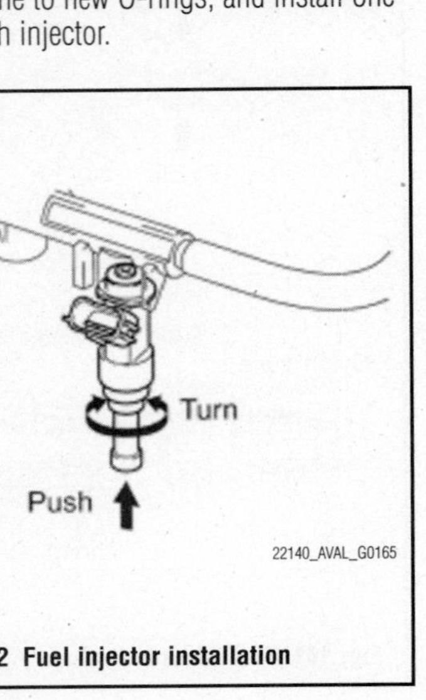

Fig. 162 Fuel injector installation

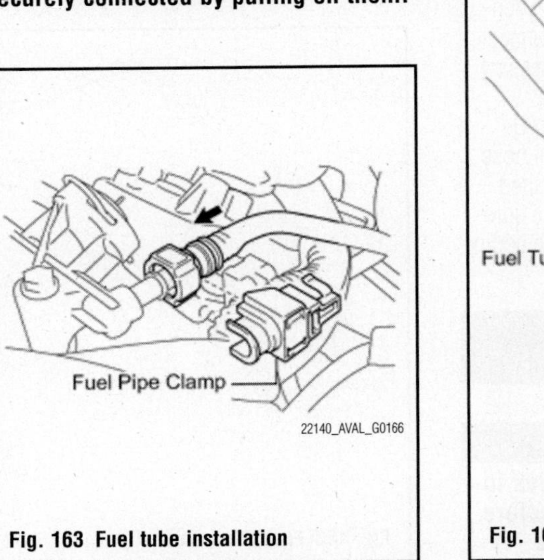

Fig. 163 Fuel tube installation

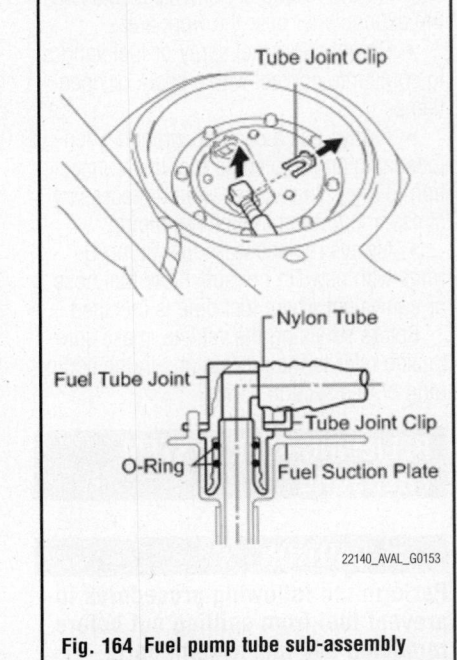

Fig. 164 Fuel pump tube sub-assembly

➡When the fuel tube joint and fuel suction plate are stuck, pinch the fuel tank tube between fingers, and turn it carefully to release it. Disconnect the fuel tank tube.

6. Remove fuel tank vent tube set plate, as follows:
   a. Remove the 8 bolts and set plate.
7. Remove fuel suction tube assembly with pump and gauge, as follows:
   a. Pull out the fuel suction tube from the fuel tank.

➡Do not damage the fuel pump filter.

➡Be careful not to bend the arm of the fuel sender gauge.

   b. Remove the gasket from the fuel suction tube.

To install:

8. Install a new gasket to the fuel suction tube.
9. Install the fuel suction tube.

➡Do not damage the fuel pump filter.

➡Be careful not to bend the arm of the fuel sender gauge.

10. Install the fuel tank vent tube set plate, as follows:
    a. Align the mark of the set plate with the fuel suction tube.
    b. Install the set plate with the 8 bolts and tighten to 52 inch lbs. (5.9 Nm).
11. Install the fuel pump tube with the tube joint clip.

➡Check that there is no scratches or foreign objects on the connecting part.

➡Check that the fuel tube joint is inserted securely.

➡Check that the tube joint clip is on the collar of the fuel tube joint.

➡After installing the tube joint clip, check that the fuel tube joint is pulled off.

12. Connect battery negative cable.
13. Inspect for fuel leak.
14. Install the rear floor service hole cover.
15. Install the rear seat cushion assembly.

FUEL TANK

REMOVAL & INSTALLATION

See Figures 166 through 168.

1. Before servicing the vehicle, refer to the precautions section.
2. Discharge fuel system pressure.
3. Disconnect battery negative cable.
4. Remove rear seat cushion assembly.
5. Remove rear floor service hole cover.
6. Separate fuel pump tube sub-assembly.
7. Remove fuel tank vent tube set plate.
8. Remove fuel suction tube assembly with pump and gauge.
9. Drain fuel.
10. Remove center exhaust pipe assembly.
11. Disconnect no. 2 parking brake cable assembly.
12. Disconnect no. 3 parking brake cable assembly.
13. Remove the 4 bolts and the lower center fuel tank protector.
14. Disconnect the fuel pump tube, as follows:

➡Check that there is no dirt or other foreign objects around the connector before removing fuel tubes, and clean the connector as necessary.

➡It is necessary to prevent mud or dirt from entering the connector. If mud or dirt gets in the connector, the O-rings may not seal properly.

➡Do not use any tools in these operations.

➡Do not bend, kink or twist the nylon tubes. Protect the connector by covering it with a plastic bag.

➡When the pipe and connector are stuck, push and pull the connector to release and pull the connector out carefully.

   a. Pinch the tabs of the retainer to remove the lock claws and pull it down as shown in the illustration.
   b. Pull out the fuel tank main tube.
15. Pinch the tube connector and then pull out the No. 1 fuel tube.
16. Set up a transmission jack underneath the fuel tank.
17. Remove the 2 set bolts of the fuel tank bands.
18. Remove the hose clamp and disconnect the fuel tank to filter pipe hose.
19. Disconnect the fuel tank vent hose from the charcoal canister, as follows:
    a. Push the connector deep into the charcoal canister to release the locking pin.
    b. Pinch portion A.

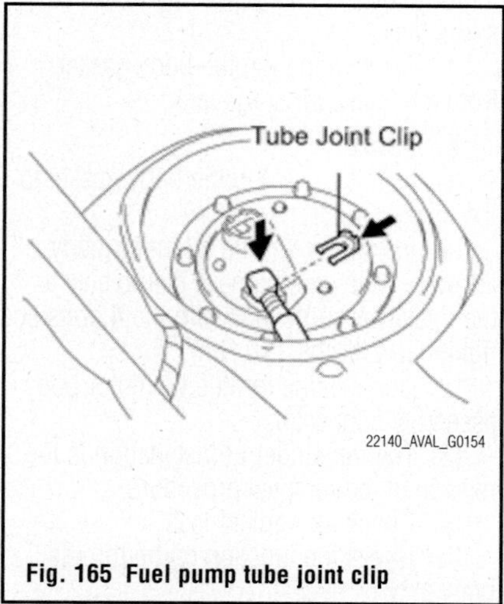

Fig. 165 Fuel pump tube joint clip

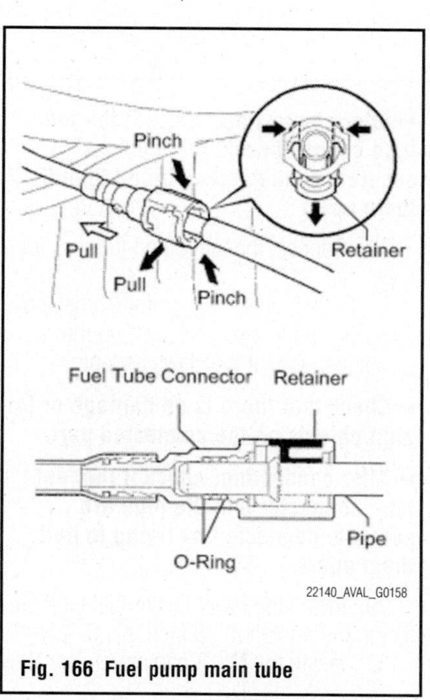

Fig. 166 Fuel pump main tube

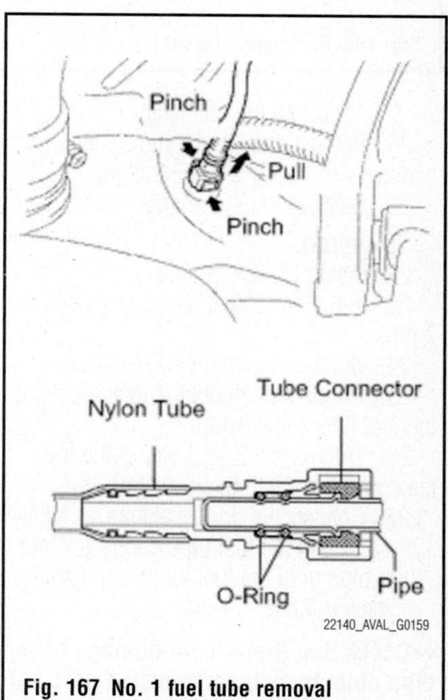

Fig. 167 No. 1 fuel tube removal

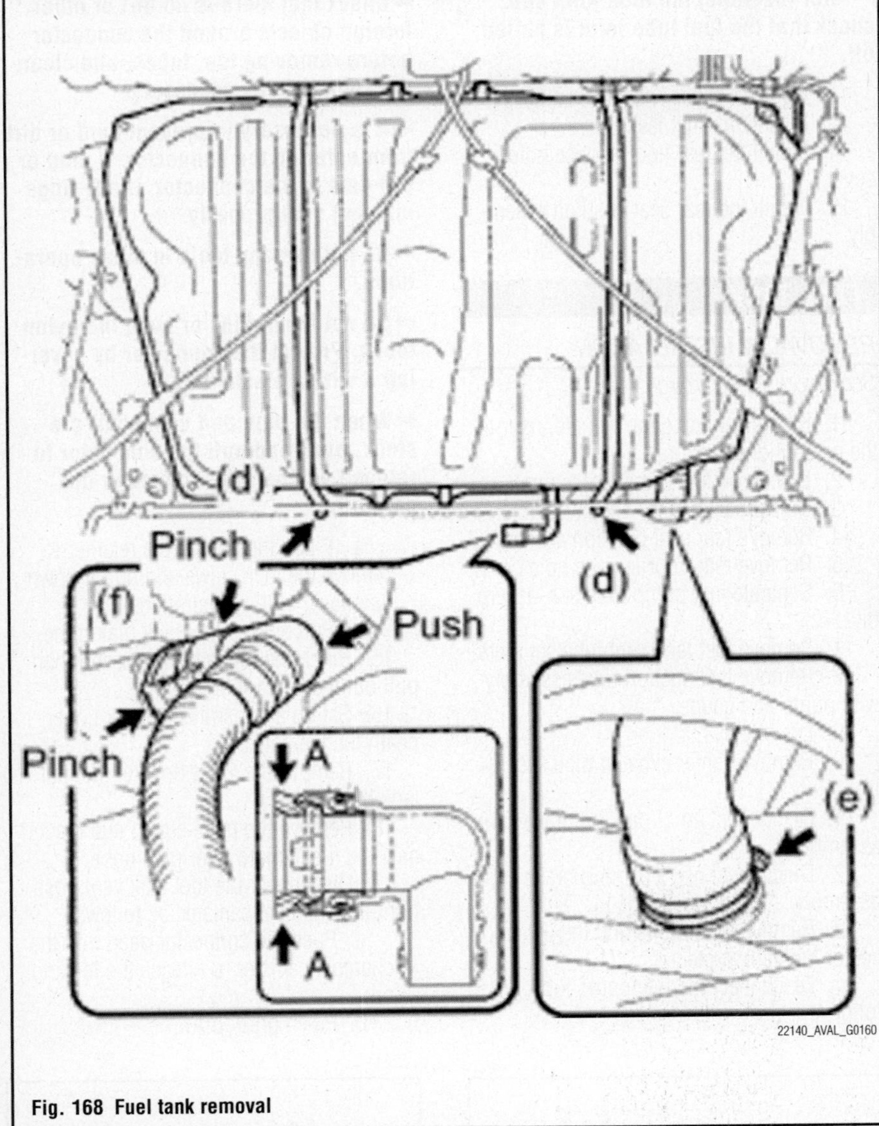

Fig. 168 Fuel tank removal

c. Pull out the connector.

20. Remove the 2 pins and 2 fuel tank bands as shown in the illustration.
21. Remove the 4 clip nuts.

To install:

22. Install the 4 clip nuts.
23. Install the 2 fuel tank bands with the 2 pins.
24. Connect the fuel tank vent hose.
25. Connect the fuel tank inlet pipe with the fuel filter pipe clamp.
26. Tighten the 2 set bolts of the fuel tank bands to 29 ft. lbs. (39 Nm).
27. Connect the No. 1 fuel tube, as follows:
    a. Push the fuel tube connector into the pipe until the fuel tube connector makes a "click" sound.

➡Check that there is no damage or foreign objects on the connected part.

➡After connecting, check if the fuel tube connector and the pipe are securely connected by trying to pull them apart.

28. Connect the fuel pump tube, as follows:
    a. Push in the fuel pumps tube connector to the pipe and push up the retainer so that the claws engage.

➡Check that there is no damage or foreign objects on the connected part.

➡After connecting, check if the fuel tube connector and the pipe are securely connected by trying to pull them apart.

29. Install the lower center fuel tank protector and tighten to 48 inch lbs. (5.4 Nm).
30. Install the No. 3 parking brake cable assembly with the bolt and nut and tighten to 53 inch lbs. (6 Nm), and 75 inch lbs. (8.5 Nm).
31. Install the No. 2 parking brake cable assembly with the bolt and nut and tighten to 53 inch lbs. (6 Nm), and 75 inch lbs. (8.5 Nm).
32. Install the center exhaust pipe assembly.
33. Install the fuel suction tube assembly with pump and gauge.
34. Install the fuel tank vent tube set plate.
35. Connect the fuel pump tube sub-assembly.
36. Add fuel.
37. Connect battery negative cable.
38. Inspect for fuel leak and exhaust gas leak.
39. Install the rear floor service hole cover.
40. Install the rear seat cushion assembly.

IDLE SPEED

ADJUSTMENT

Idle speed is maintained by the ECM. No adjustment is necessary or possible.

THROTTLE BODY

REMOVAL & INSTALLATION

1. Before servicing the vehicle, refer to the precautions section.
2. Disconnect battery negative cable.
3. Drain engine coolant.
4. Remove the V-bank cover sub-assembly.
5. Remove the air cleaner cap with air cleaner hose.
6. Disconnect the 2 water by-pass hoses from the throttle w/ motor body assembly.
7. Disconnect the throttle w/ motor body assembly connector.
8. Remove the 4 bolts and throttle w/ motor body assembly from the intake air surge tank.
9. Remove the throttle body gasket from the intake air surge tank.

To install:

10. Install a new throttle body gasket to the intake air surge tank.
11. Install the throttle w/ motor body assembly and wire harness clamp stay to the intake air surge tank with the 4 bolts and tighten to 7 ft. lbs. (10 Nm).
12. Connect the throttle w/ motor body assembly connector.
13. The remainder of installation is the reverse of the removal procedure.
14. Check for coolant leak.
15. Check the function of the throttle body.

HEATING & AIR CONDITIONING SYSTEM

BLOWER MOTOR

REMOVAL & INSTALLATION

1. Remove the glove box assembly.
2. Disconnect the connector from the blower motor.
3. Remove the 3 screws and the blower motor.

To install:

4. Replace the blower motor and the three screws.
5. Connect the electrical connector.
6. Replace the glove box assembly.

HEATER CORE

REMOVAL & INSTALLATION

See Figures 169 through 179.

1. Before servicing the vehicle, refer to the precautions section.
2. Disconnect battery negative terminal.

CAUTION

Wait for 90 seconds after disconnecting the cable to prevent the airbag from deploying.

3. Drain engine coolant.
4. Remove both windshield wiper arm and blade assemblies.
5. Remove the right cowl top ventilator louver.
6. Remove the windshield wiper motor and link.
7. Remove the outer front cowl top panel.
8. Discharge refrigerant from refrigeration system.
9. Disconnect the suction hose sub-assembly, as follows:
   a. Remove the bolt, and slide the hook connector.
   b. Disconnect the suction hose sub-assembly.
   c. Remove the O-ring from the suction hose sub-assembly.

➡Seal the openings of the disconnected parts using vinyl tape to prevent moisture and foreign matter from entering.

10. Disconnect the air conditioner tube and accessory.
11. Remove the O-ring from the air conditioner tube and accessory.
12. Slide the clip and disconnect the heater water outlet hose A.
13. Slide the clip and disconnect the heater water inlet hose A.

Fig. 170 No. 2 heater to register duct

➡Do not apply excessive force to the heater water hoses

➡Prepare a drain pan or cloth in case the cooling water leaks.

14. Remove the instrument panel safety pad sub-assembly.

➡Refer to the removal procedures for the instrument panel safety pad sub-assembly w/ front passenger airbag assembly.

15. Remove the 3 clips and No. 2 heater to register duct.
16. Disengage the 4 claws and then remove the No. 6 heater to register duct.
17. Remove the clip and No. 1 console box duct.
18. Remove the left and right hand floor carpet brackets, as follows:
   a. Release the clamps (2 from the left and 3 from the right).
   b. Turn back the floor carpet.
   c. Remove the 3 clips.
   d. Remove the floor carpet brackets.
19. Release the 2 claws and remove the rear No. 2 air duct.
20. Release the 2 claws and remove the rear No. 1 air duct.
21. Remove the No. 1 air duct sub-assembly.
22. Separate the steering intermediate shaft assembly.
23. Remove the steering column assembly.
24. Remove instrument panel reinforcement assembly, as follows:
   a. Disconnect each connector and remove each clamp. Disconnect the wire harness.

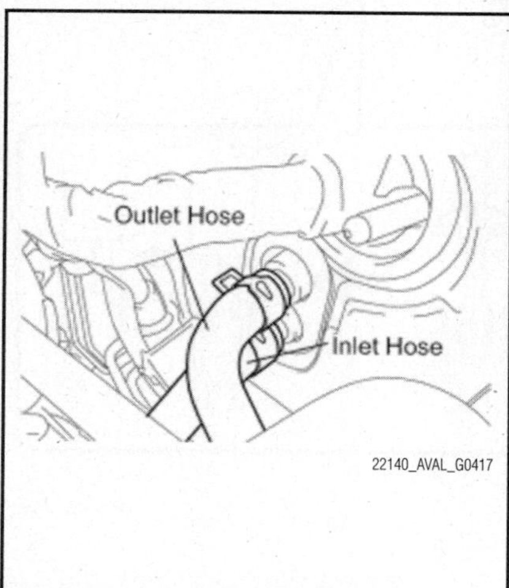

Fig. 169 Heater water outlet hoses

Fig. 171 No. 6 heater to register duct

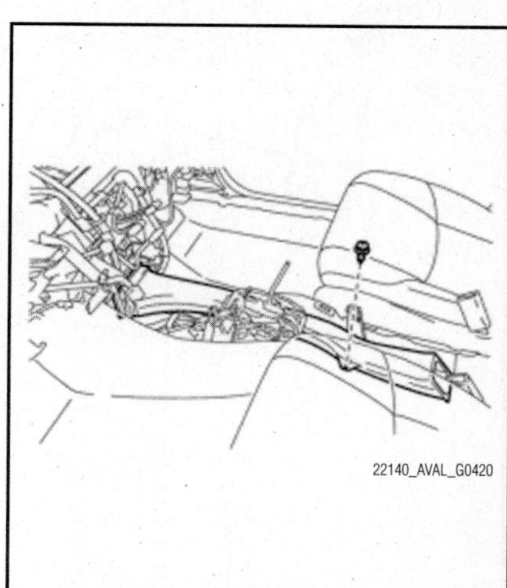

Fig. 172 No. 1 console box duct

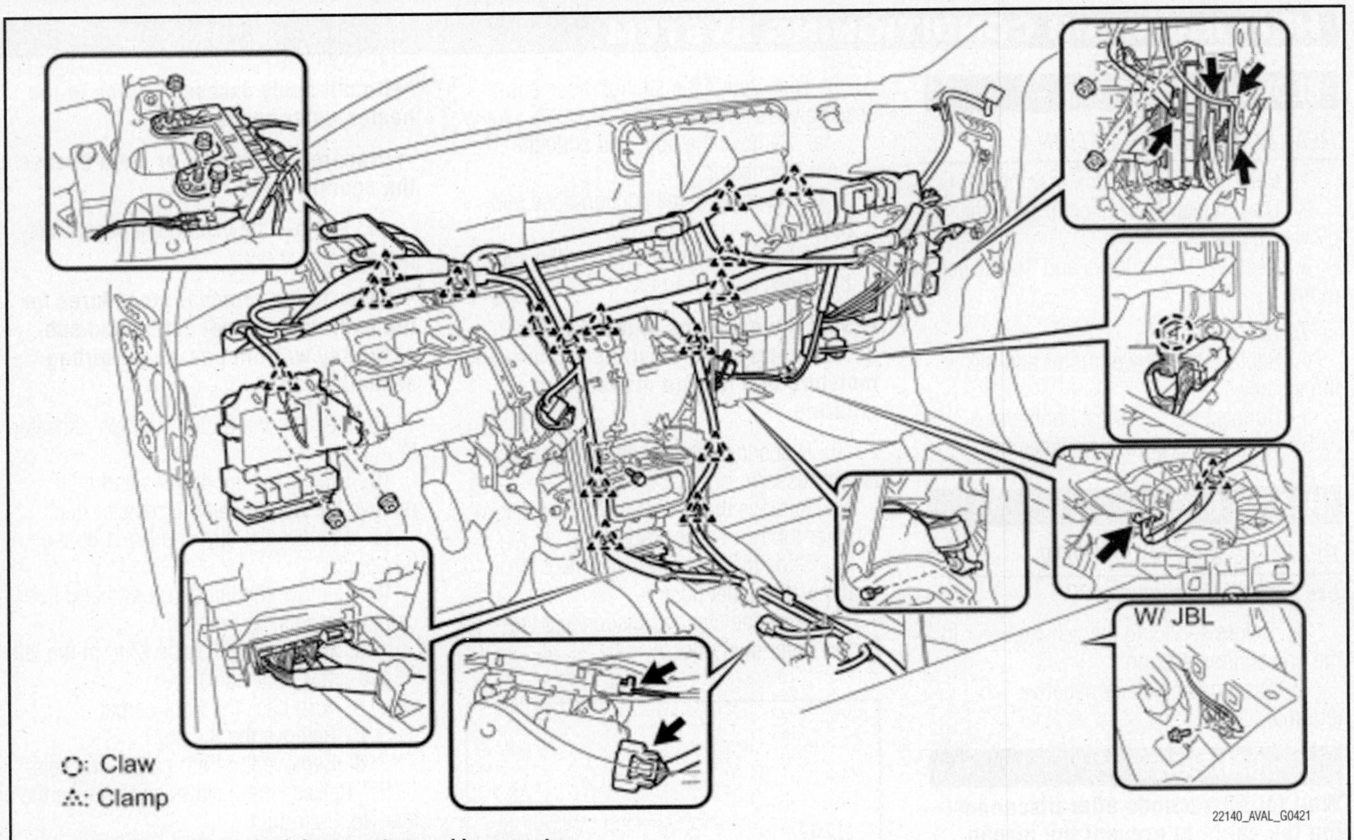

Fig. 173 Instrument panel reinforcement assembly connectors

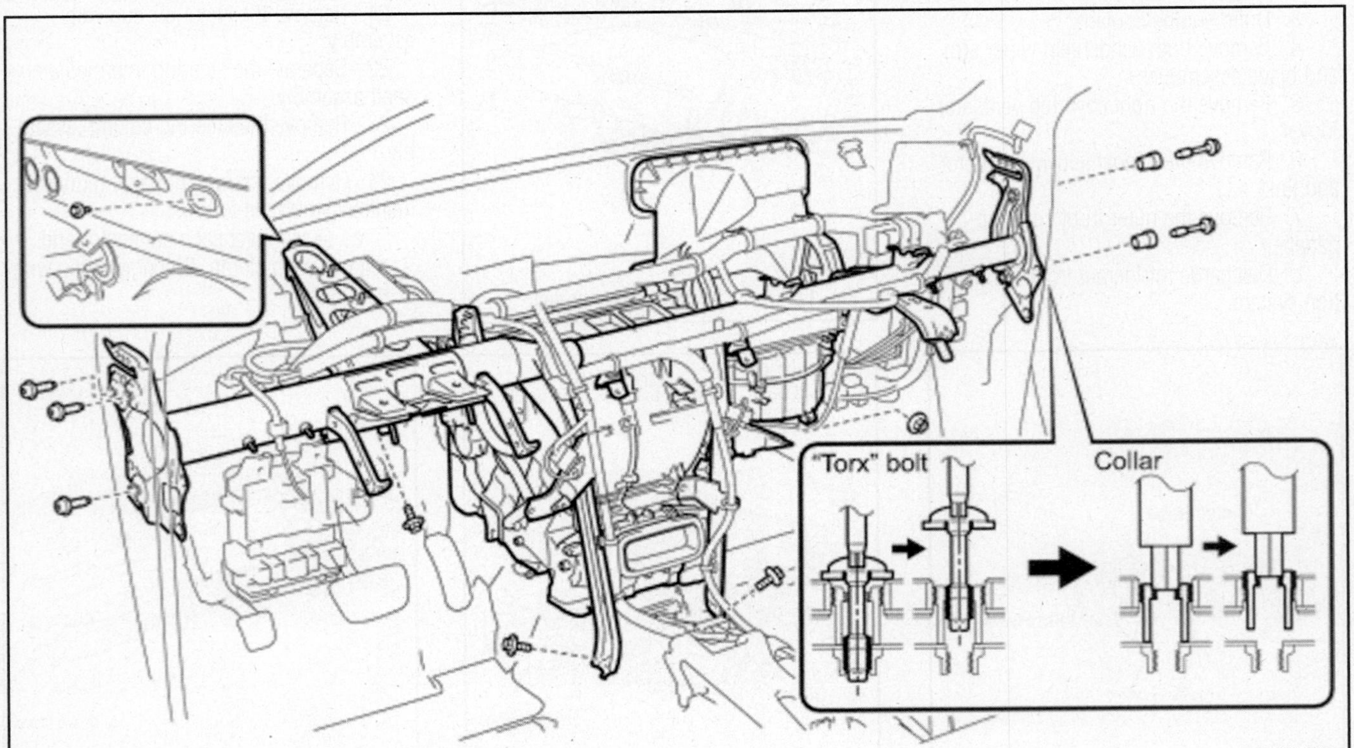

Fig. 174 Instrument panel reinforcement assembly with the air conditioner unit assembly

b. Remove the 6 nuts and 3 bolts.

c. Remove the 3 bolts and nut.

d. Using a Torx® socket wrench (T40), remove the 5 Torx® bolts.

➡The Torx® bolts on the passenger side can be removed with the collar or for adjustment.

e. Using a hexagon wrench 12 mm, remove the 2 collars and instrument panel reinforcement assembly with the air conditioner unit assembly.

f. Remove the 3 bolts, 2 screws and instrument panel reinforcement assembly.

25. Remove the No. 2 air duct sub-assembly.

26. Remove the blower assembly.

27. Remove the air outlet control servo motor.

28. Remove the air mix control servo motor.

29. Remove the heater radiator unit sub-assembly, as follows:

a. Remove the screw and clamp.

b. Release the 4 claws and remove the clamp.

c. Remove the heater radiator unit sub-assembly from the air conditioning radiator assembly.

➡Prepare a drain pan or cloth in case the cooling water leaks.

To install:

30. Install the heater radiator unit sub-assembly.

31. Install the air mix control servo motor.

32. Install the air outlet control servo motor.

33. Install the blower assembly.

34. Install the No. 2 air duct sub-assembly.

35. Install the air conditioner unit assembly to the instrument panel reinforcement assembly with the 2 screws and 3 bolts and tighten in the order shown to 87 inch lbs. (9.8 Nm).

36. Driver seat:

a. Using a Torx® socket wrench (T40), install the instrument panel reinforcement assembly with the 3 Torx® bolts and tighten to 13 ft. lbs. (17 Nm).

37. Passenger seat:

a. Using a hexagon wrench 12 mm, install the instrument panel reinforcement assembly with the 2 bolts and tighten to 53 inch lbs. (6 Nm).

b. Using a Torx® socket wrench (T40), install the instrument panel reinforcement assembly with the 2 Torx® bolts and tighten to 15 inch lbs. (20 Nm).

c. Install the 3 bolts and nut and tighten to 87 inch lbs. (9.8 Nm) and 15 ft. lbs. (20 Nm).

d. Connect the connectors and clamps.

e. Install the 6 nuts and 3 bolts.

38. The remainder of installation is the reverse of removal.

39. Initialize system.

40. After adding engine coolant and charging refrigerant, check for coolant and refrigerant leaks.

22140_AVAL_G0423

Fig. 175 Instrument panel reinforcement assembly with the air conditioner unit assembly

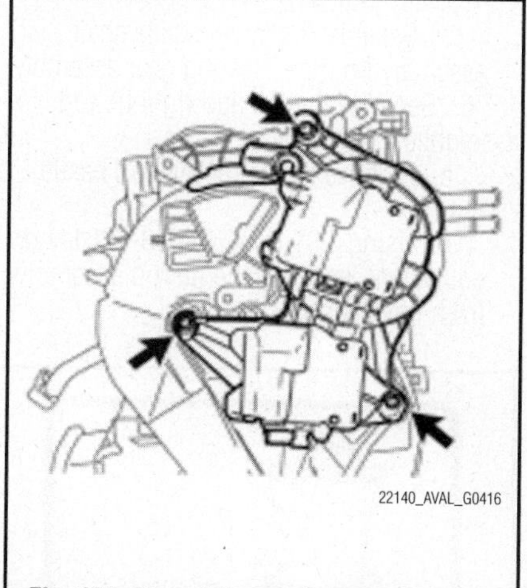

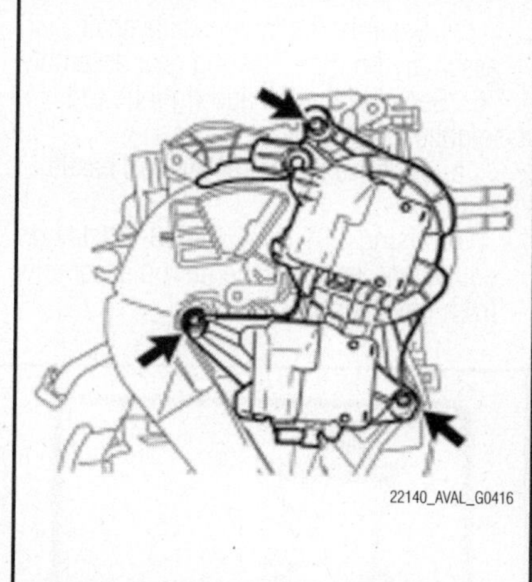

Fig. 177 Air outlet control servomotor

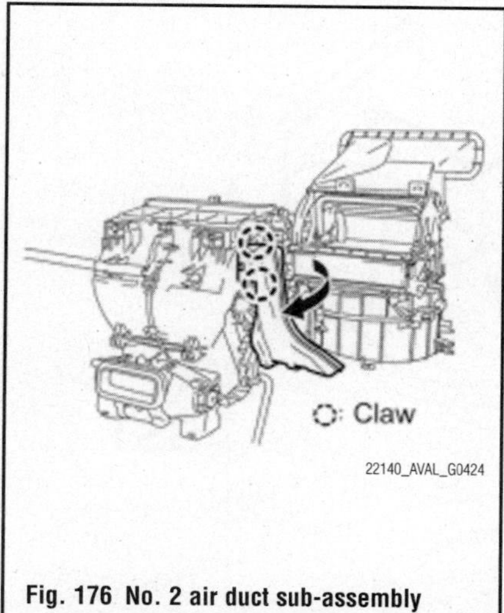

Fig. 176 No. 2 air duct sub-assembly

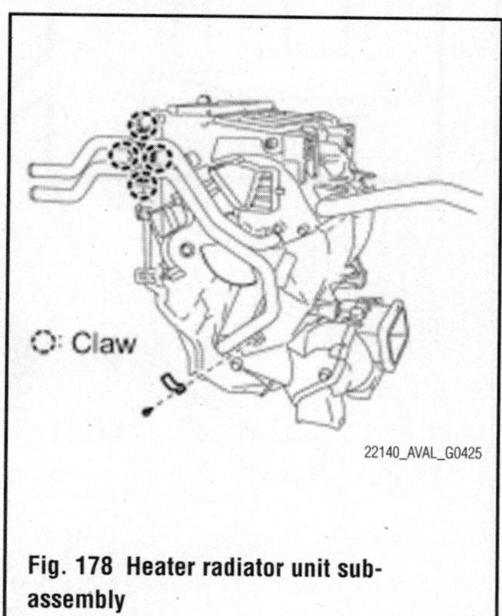

Fig. 178 Heater radiator unit sub-assembly

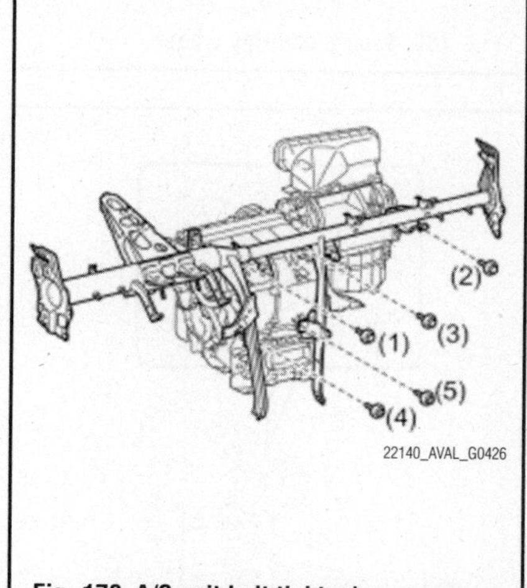

Fig. 179 A/C unit bolt tightening sequence

STEERING

POWER STEERING GEAR

REMOVAL & INSTALLATION

See Figures 180 through 188.

1. Before servicing the vehicle, refer to the precautions section.
2. Position the front wheels straight ahead.
3. Remove front wheel.
4. Remove right and left front fender apron seals.
5. Separate steering intermediate shaft assembly, as follows:
   a. Secure the steering wheel with the seat belt in order to prevent rotation. This will help prevent damage to the spiral cable.
   b. Remove the bolt.
   c. Put matchmarks on the steering intermediate shaft assembly and the power steering gear assembly.

Fig. 180 Secure steering wheel

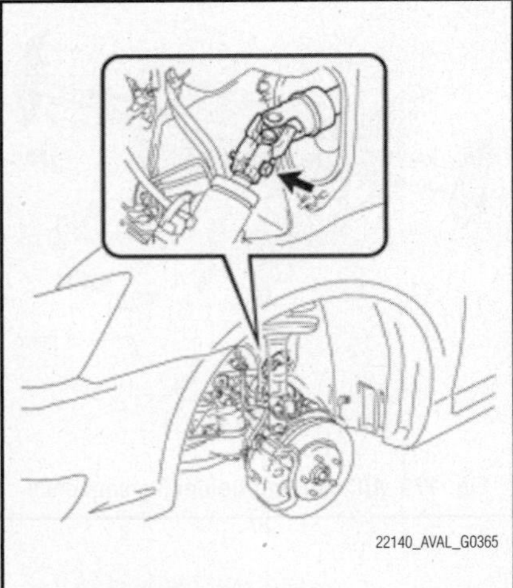

Fig. 181 Remove the bolt

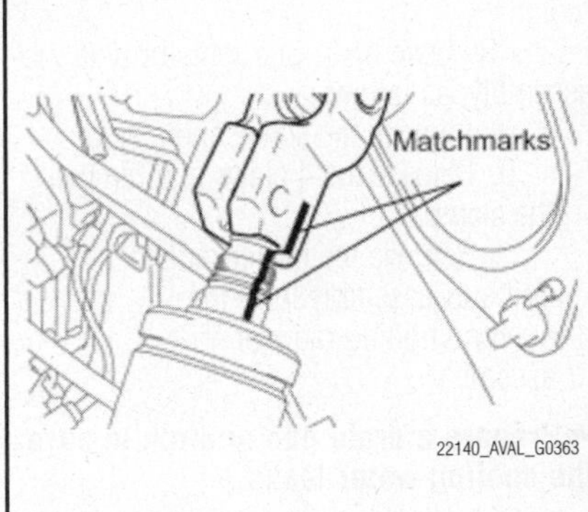

Fig. 182 Matchmarks on the steering intermediate shaft assembly and the power steering gear assembly

   d. Separate the intermediate shaft assembly from the steering gear assembly.
6. Separate the left and right tie rod assemblies, as follows:
   a. Remove the cotter pin and castle nut.
   b. Using SST (SST: 09628-62011) or equivalent, separate the tie rod assembly from the steering knuckle.
7. Remove the right and left front stabilizer link assemblies.
8. Remove the 2 bolts and the left and right No. 1 stabilizer bracket and the No. 1 stabilizer bar bush.
9. Disconnect pressure feed tube assembly, as follows:
   a. Using SST (09023-12701) or equivalent, disconnect the return tube assembly from the power steering gear assembly.
   b. Using SST (09023-12701), disconnect the pressure feed tube assembly from the power steering gear assembly.
   c. Remove the 2 bolts and separate the pressure feed tube clamp from the power steering gear assembly.
10. Remove the 2 bolts, nuts and the power steering gear assembly.

➡Do not turn the nut because the nut has its own stopper. Loosen the bolt with the nut secured.

To install:

11. Install the power steering gear assembly with the 2 bolts and 2 nuts and tighten to 52 ft. lbs. (70 Nm).

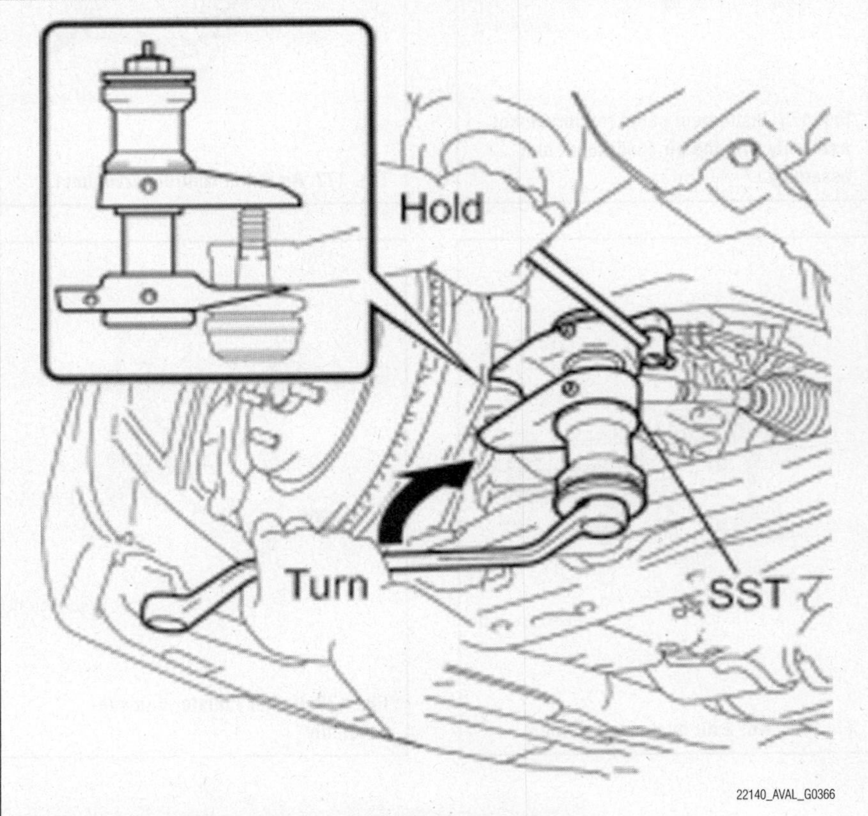

Fig. 183 Separate the tie rod assembly from the steering knuckle

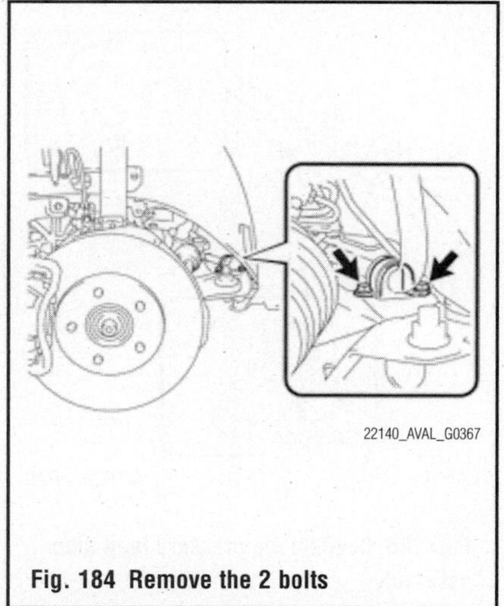

Fig. 184 Remove the 2 bolts

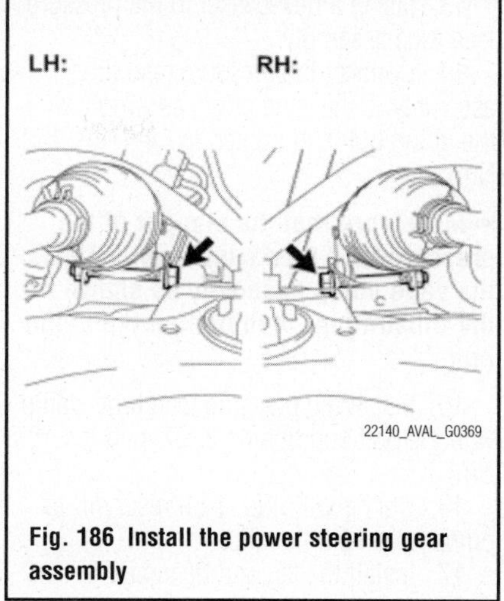

Fig. 186 Install the power steering gear assembly

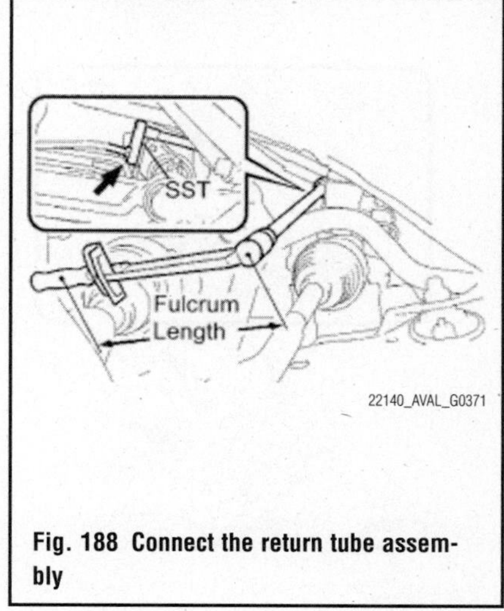

Fig. 188 Connect the return tube assembly

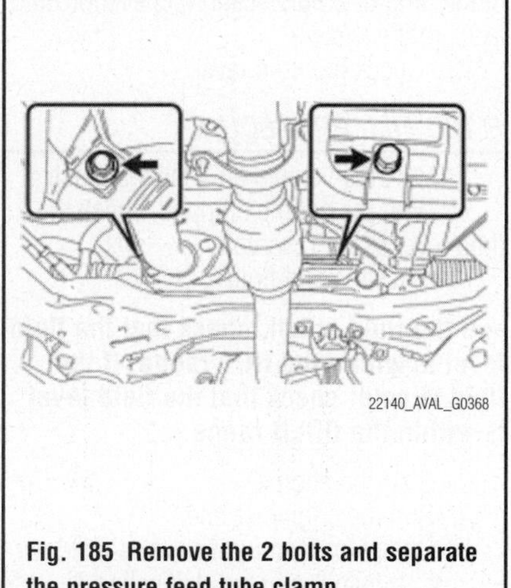

Fig. 185 Remove the 2 bolts and separate the pressure feed tube clamp

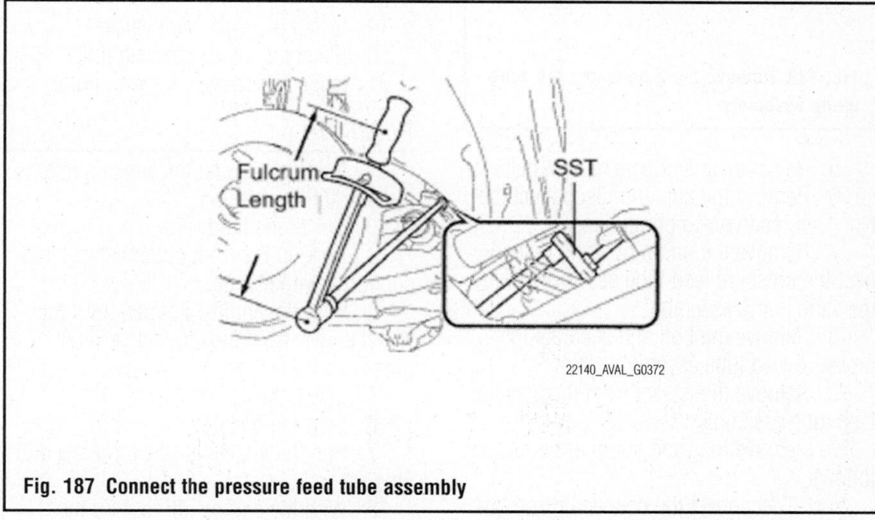

Fig. 187 Connect the pressure feed tube assembly

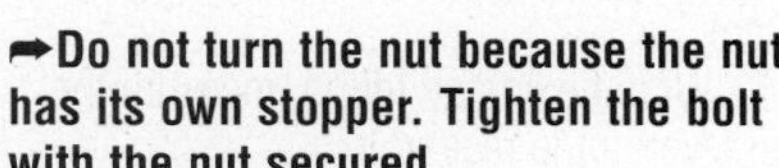

➡Do not turn the nut because the nut has its own stopper. Tighten the bolt with the nut secured.

➡For the next 2 steps, use a torque wrench with a fulcrum length of 300 mm (11.81 in.). These torque values are effective when SST is parallel to the torque wrench.

12. Using SST (09023-12701) or equivalent, connect the pressure feed tube assembly to the power steering gear assembly and tighten to 16 ft. lbs. (22 Nm).
13. Using SST (09023-12701) or equivalent, connect the return tube assembly to the power steering gear assembly and tighten to 16 ft. lbs. (22 Nm).
14. Install the pressure feed tube clamp with the 2 bolts and tighten to 87 inch lbs. (9.8 Nm).
15. Install the left and right front stabilizer brackets, as follows:
   a. Install the No. 1 stabilizer bar bush to the stabilizer bar.
   b. Install the No. 1 front stabilizer bracket with the 2 bolts and tighten to 20 ft. lbs. (27 Nm).
16. Install the right and left front stabilizer link assemblies.
   a. Connect the left and right tie rod assemblies, as follows:
   b. Connect the tie rod assembly to the steering knuckle with the castle nut and tighten to 36 ft. lbs. (49 Nm).
   c. Install a new cotter pin. If the holes for the cotter pin are not aligned, tighten the nut up to 60° further.
17. Align the matchmarks on the steering intermediate shaft assembly and the power steering gear assembly.
   a. Install the bolt and tighten to 26 ft. lbs. (35 Nm).
18. Bleed power steering fluid.
19. Inspect for power steering fluid.
20. Install the left and right front fender apron seals.
21. Install the front wheel.
22. Inspect the front wheels.
23. Inspect the front wheel alignment.

POWER STEERING PUMP

REMOVAL & INSTALLATION

See Figures 189 and 190.

1. Before servicing the vehicle, refer to the precautions section.

➡Do not over tighten when using a vise.

➡When installing, coat the parts indicated by arrows with power steering fluid. Take care not to spill power steering fluid on the V-belt.

2. Remove the right front wheel.
3. Drain power steering fluid.
4. Remove the right front fender apron seal.

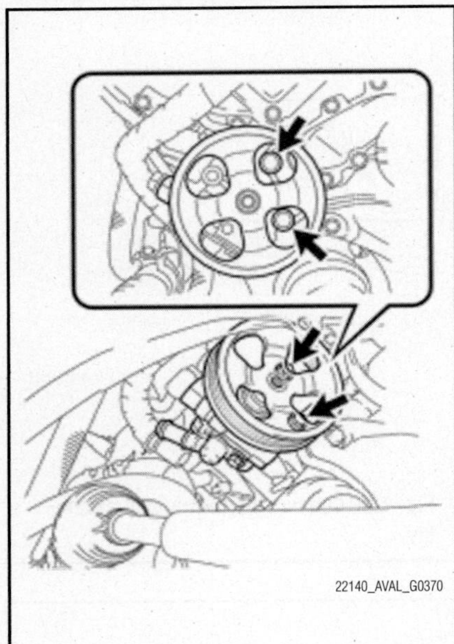

Fig. 189 Remove the 2 bolts and the vane pump assembly

5. Remove fan and generator V-belt.
6. Remove the clip and disconnect the No. 1 oil reservoir to pump hose.
7. Remove the union bolt and disconnect the pressure feed tube assembly from the vane pump assembly.
8. Remove the bolt and separate the pressure feed tube assembly clamp.
9. Remove the gasket from the pressure feed tube assembly.
10. Remove the vane pump assembly, as follows:
    a. Disconnect the connector from the power steering oil pressure switch.
    b. Remove the 2 bolts and the vane pump assembly.

To install:

11. Install the vane pump assembly with the 2 bolts and tighten to 32 ft. lbs. (43 Nm).
12. Connect the connector to the power steering oil pressure switch.
13. Install a new gasket to the pressure feed tube assembly.
14. Connect the pressure feed tube assembly to the vane pump assembly with the union bolt and tighten to 38 ft. lbs. (52 Nm).

➡Make sure that the stopper of the pressure feed tube assembly contacts the vane pump assembly as shown in the illustration, then tighten the union bolt.

15. Install the pressure feed tube clamp with the bolt and tighten to 87 inch lbs. (9.8 Nm).
16. Connect the No. 1 oil reservoir to pump hose with the clip.
17. Install the fan and generator V-belt.
18. Install the right front fender apron seal.
19. Install the right front wheel.
20. Bleed the power steering fluid.
21. Inspect for power steering fluid.

BLEEDING

1. Before servicing the vehicle, refer to the precautions section.
2. Check the fluid level.
3. Jack up the front of the vehicle and support it with stands.
4. With the engine stopped, turn the wheel slowly from lock to lock several times.
5. Lower the vehicle.
6. Start the engine.
7. Run the engine at idle for a few minutes.
8. With the engine idling, turn the wheel left or right to the full lock position and keep it there for 2 to 3 seconds, then turn the wheel to the opposite full lock position and keep it there for 2 to 3 seconds.
9. Repeat the above steps several times.
10. Stop the engine.
11. Check for foaming or emulsification. If the system has to be bled twice because of foaming or emulsification, check for fluid leaks in the system.
12. Check the fluid level.

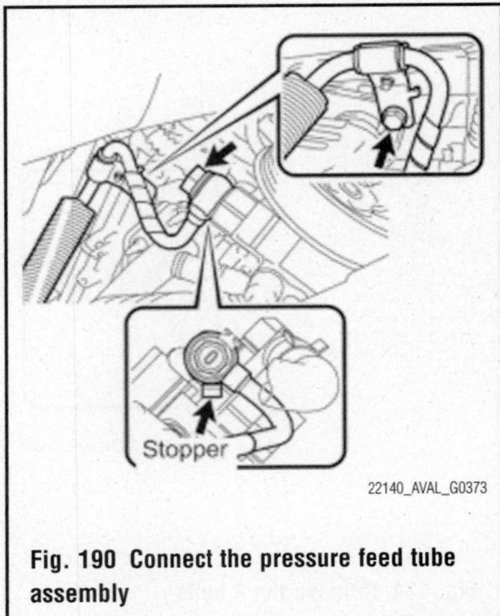

Fig. 190 Connect the pressure feed tube assembly

FLUID FILL PROCEDURE

1. Park the vehicle on a level surface.
2. With the engine stopped, check the fluid level in the reservoir.

If necessary, add fluid.

➡If the fluid is hot, check that the fluid level is within the HOT range. If the fluid is cold, check that the fluid level is within the COLD range.

3. Start the engine.
4. Run the engine at idle.
5. Check for foaming or emulsification.
    a. If foaming or emulsification is identified, bleed the power steering system.
6. With the engine idling, measure the fluid level in the reservoir.
7. Stop the engine.
8. Wait a few minutes and remeasure the fluid level in the reservoir.
    a. If a problem is found, bleed the power steering system
9. Check the fluid level.

SUSPENSION

CONTROL LINKS

REMOVAL & INSTALLATION

1. Raise and support the front of the vehicle.
2. Remove the front LH or RH wheel.
3. Remove the 2 nuts and front stabilizer link assembly.

To install:

4. Install the stabilizer link assembly with the 2 nuts and tighten to 55 ft. lbs. (74 Nm).

➡Use a hexagon (6 mm) wrench to hold the stud if the ball joint turns together with the nut.

5. Install the front wheel.

LOWER BALL JOINT

REMOVAL & INSTALLATION

See Figure 191.

1. Remove the front wheel.
2. Remove the front axle hub nut.
3. Separate the front speed sensor.
4. Separate the front disc the brake caliper assembly.
5. Remove front disc.
6. Separate the tie rod assembly.
7. Separate the front lower suspension arm.
8. Remove the front axle assembly.
9. Remove the front lower ball joint assembly, as follows:
   a. Remove the cotter pin and castle nut.
   b. Using SST (SST: 09628-62011) or equivalent, remove the front lower ball joint assembly.

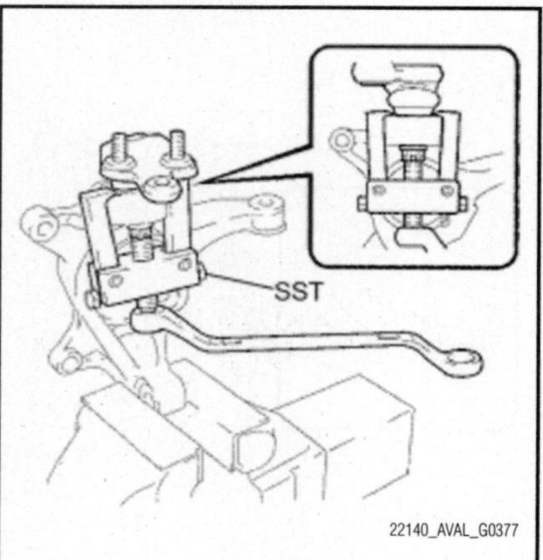

Fig. 191 Remove the front lower ball joint assembly

To install:

10. Installation is the reverse of the removal procedure, noting the following:
    a. Install the front lower ball joint assembly to the steering knuckle with the castle nut and tighten to 91 ft. lbs. (123 Nm). Further tighten the nut up to 60°if the holes for the cotter pin are not aligned.
    b. Inspect and adjust the front wheel alignment.
    c. Inspect the ABS speed sensor signal.

LOWER CONTROL ARM

REMOVAL & INSTALLATION

See Figures 192 and 193.

1. Remove the engine assembly with transaxle.

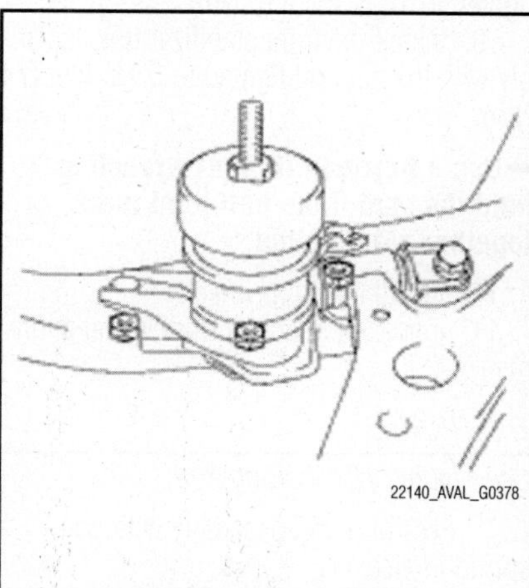

Fig. 192 Transverse engine mounting insulator with 3 nuts

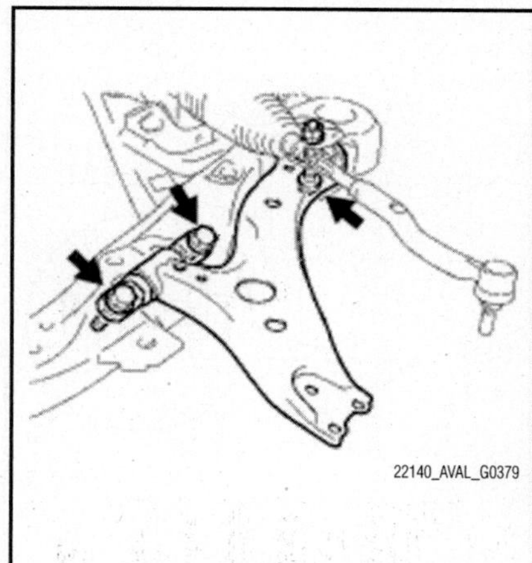

Fig. 193 Lower No. 1 front suspension arm sub-assembly

2. Remove the 3 nuts and transverse engine mounting insulator.
3. Remove the 2 bolts on the front side of the lower No. 1 front suspension arm sub-assembly.
4. Remove the bolt and nut on the rear side of the lower No. 1 front suspension arm sub-assembly.
5. Remove the lower No. 1 front suspension arm sub-assembly.
6. Remove the front lower arm bush stopper from the lower No. 1 front suspension arm sub-assembly.

To install:

7. Install the front lower arm bush stopper to the lower No. 1 front suspension arm sub-assembly.
8. Install the 2 bolts on the front side of the lower No. 1 front suspension arm sub-assembly and tighten to 148 ft. lbs. (200 Nm).
9. Install the bolt and nut on the rear side of the lower No. 1 front suspension arm sub-assembly.
10. Install the transverse engine mounting insulator with the 3 nuts.
11. Install the engine assembly with transaxle.

MACPHERSON STRUT

REMOVAL & INSTALLATION

See Figures 194 through 196.

1. Remove the front wheel.
2. Remove the nut and disconnect the front stabilizer link assembly from the front shock absorber assembly.

➡Use a hexagon (6 mm) wrench to hold the stud if the ball joint turns together with the nut.

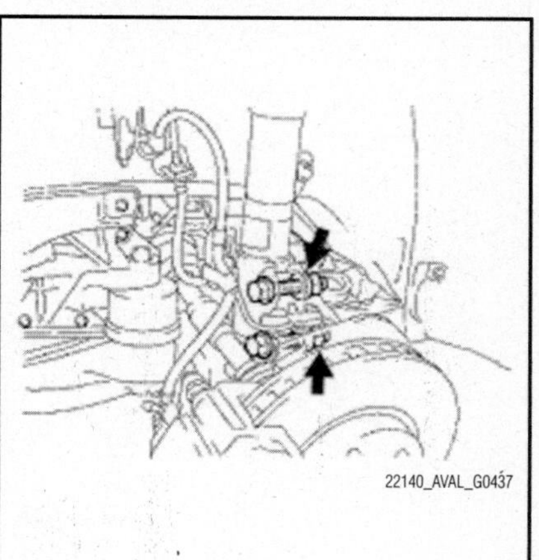

Fig. 194 Remove the 2 nuts on the lower side of the front shock absorber

3. Remove the front shock absorber with coil spring, as follows:

a. Loosen the lock nut. If not disassembling the shock absorber it is not necessary to loosen the nut.

b. Remove the bolt and disconnect the No. 1 front flexible hose and front speed sensor wire harness.

c. Remove the 2 nuts on the lower side of the front shock absorber with coil spring. Keep the bolts inserted.

d. Remove the 3 nuts on the upper side of the front shock absorber with coil spring.

e. Remove the 2 bolts on the lower side of the front shock absorber and front shock absorber with coil spring.

➡Be careful not to drop the collar in the case that there is front suspension upper brace center.

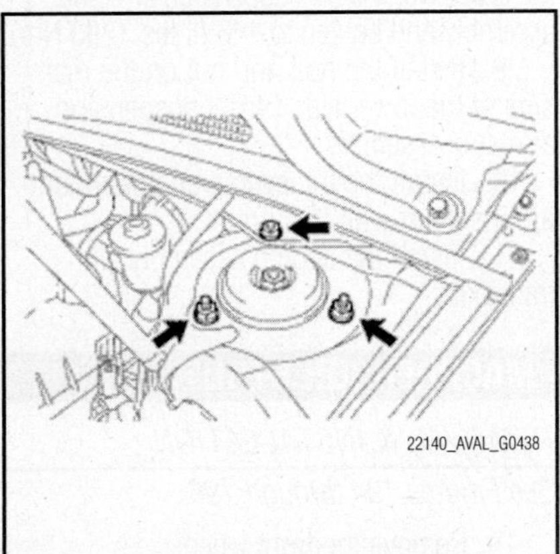

Fig. 195 Remove the 3 nuts on the upper side of the front shock absorber

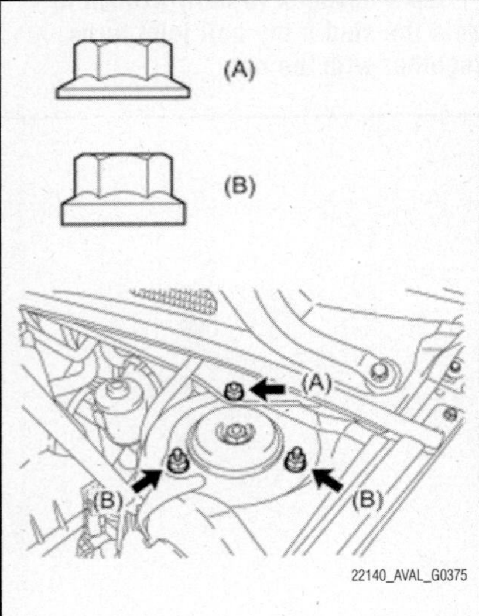

Fig. 196 Install the 3 nuts to the upper side of the front shock absorber

To install:

4. Install the front shock absorber with coil spring.

5. Install the 3 nuts to the upper side of the front shock absorber with coil spring and tighten to 63 ft. lbs. (85 Nm).

➡Be careful not to drop the collar in the case that there is front suspension upper brace center.

6. Install the 2 bolts and 2 nuts to the lower side of the front shock absorber with coil spring and tighten to 155 ft. lbs. (210 Nm).

➡Keep the bolts from rotating and torque the 2 nuts when installing the 2 nuts.

➡Insert the bolts from the front side of the vehicle.

7. Fully tighten the lock nut to 51 ft. lbs. (70 Nm).

8. Install the No. 1 front flexible hose and front speed sensor with the bolt and tighten to 14 ft. lbs. (19 Nm).

9. Install the front stabilizer link assembly with the nut and tighten to 55 ft. lbs. (74 Nm).

➡Use a hexagon (6 mm) wrench to hold the stud if the ball joint turns together with the nut.

10. Install the front wheel.

11. Inspect and adjust front wheel alignment.

OVERHAUL

See Figures 197 through 202

1. Fix front strut assembly with coil spring in vise.

a. As shown in the illustration, secure the front shock absorber with coil spring in a vise by clamping onto a double nutted bolt affixed to the bracket at the bottom of the absorber.

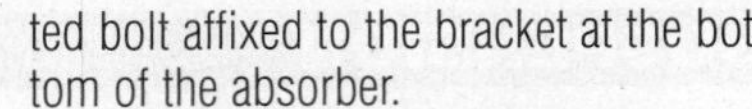

2. Using SST: 09727-30021 (09727-00010, 09727-00021), compress the front coil spring.

➡Do not use an impact wrench. It will damage the SST.

3. Disassemble as follows:

a. Remove the front suspension support sub-assembly.

b. Remove front suspension support bearing.

c. Remove front coil spring seat upper.

d. Remove front coil spring insulator upper.

e. Remove front coil spring.

f. Remove front spring bumper.

g. Remove front coil spring insulator lower from the strut assembly.

To reassemble

4. Place new strut in vise as when disassembling.

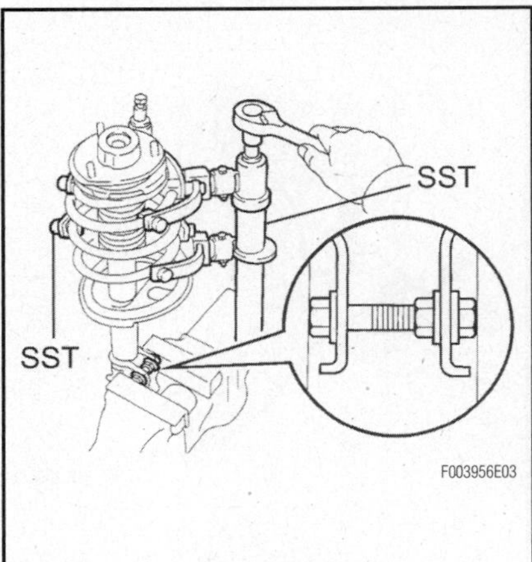

Fig. 197 Mounting strut in vise for service.

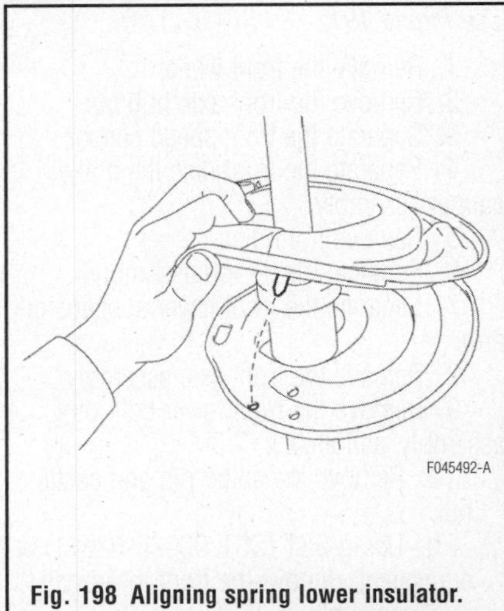

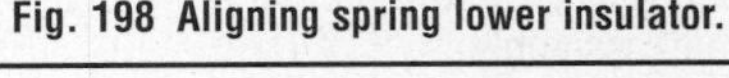

Fig. 198 Aligning spring lower insulator.

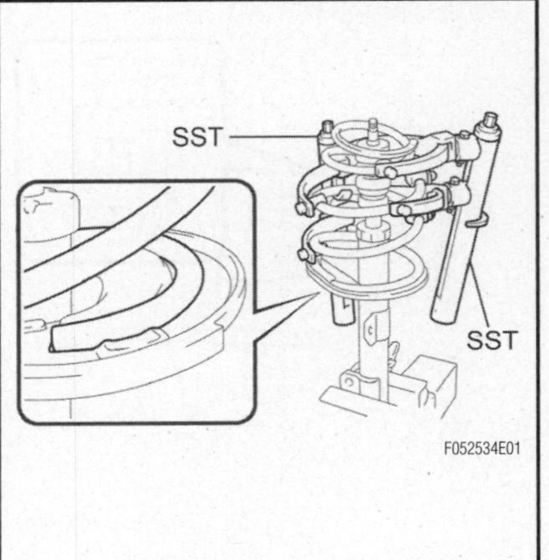

Fig. 199 Aligning coil spring to lower insulator.

a. Install the front coil spring insulator lower onto the strut assembly.

b. Install front spring bumper to the piston rod.

c. Using SST: 09727-30021 (09727-00010, 09727-00021), compress the front coil spring.

➡Do not use an impact wrench. It will damage the SST.

d. Install front coil spring, aligning coil as shown in the illustration..

e. Install the front coil spring upper insulator as shown in the illustration.

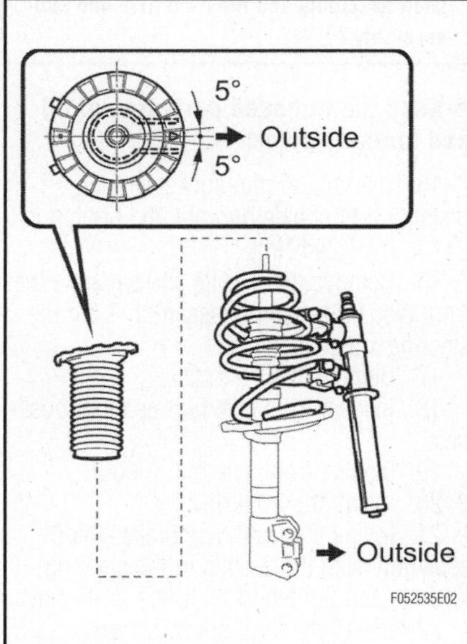

Fig. 200 Aligning coil spring upper insulator.

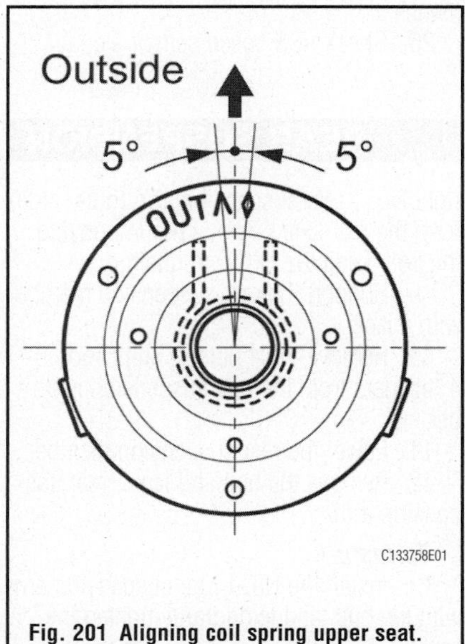

Fig. 201 Aligning coil spring upper seat.

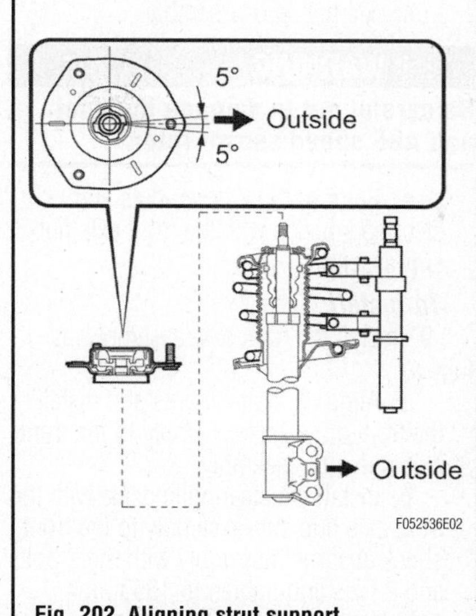

Fig. 202 Aligning strut support.

f. Install the front coil spring upper seat to the strut assembly with the mark facing the outside of the vehicle.

g. Install a new front suspension support bearing.

h. Install the front suspension support sub-assembly.

i. Install and tighten a new lock nut.

STABILIZER BAR

REMOVAL & INSTALLATION

See Figures 203 through 205

1. Remove the front wheel.
2. Remove the nuts and the right and left front stabilizer link assemblies.

➡Use a hexagon (6 mm) wrench to hold the stud if the ball joint turns together with the nut.

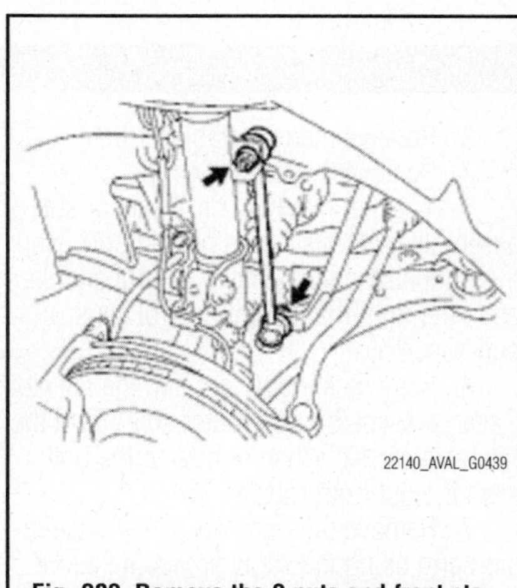

Fig. 203 Remove the 2 nuts and front stabilizer link assembly (left hand shown)

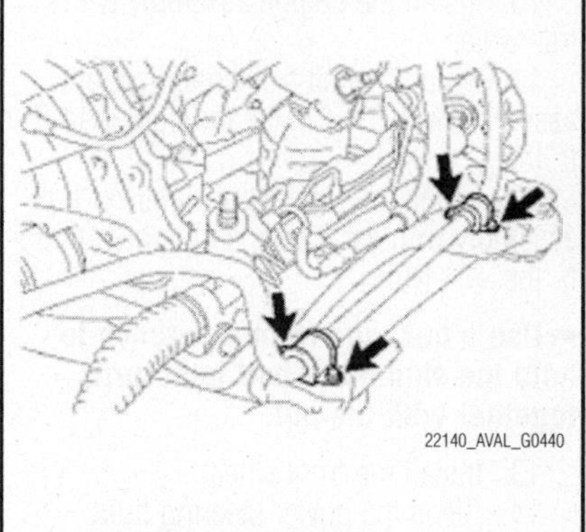

Fig. 204 Remove the 2 bolts and No. 1 stabilizer bracket (left hand shown)

3. Remove the engine assembly with transaxle.
4. Remove the bolts and the left and right No. 1 stabilizer brackets.
5. Remove the 2 bushes from the stabilizer.
6. Remove the front stabilizer bar from the vehicle.

To install:

7. Install the stabilizer bar front to the vehicle.

➡Install the bushes as to the outer side of each bush stopper on the stabilizer bar.

➡Place the cutout of the stabilizer bushes as facing the rear side as shown in the illustration.

8. Install the No. 1 left front stabilizer bracket with the 2 bolts and tighten to 20 ft. lbs. (27 Nm).
9. Install the No. 1 right front stabilizer bracket with the 2 bolts and tighten to 20 ft. lbs. (27 Nm).

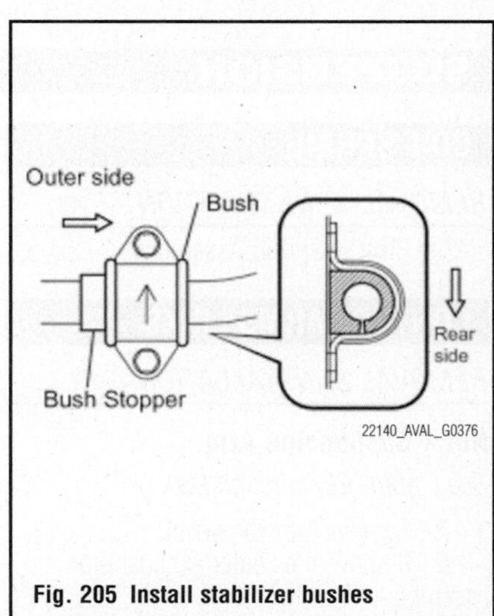

Fig. 205 Install stabilizer bushes

10. Install the engine assembly with transaxle.

11. Install the left front stabilizer link assembly with the 2 nuts and tighten to 55 ft. lbs. (74 Nm).

12. Install the right front stabilizer link assembly with the 2 nuts and tighten to 55 ft. lbs. (74 Nm).

➡Use a hexagon (6 mm) wrench to hold the stud if the ball joint turns together with the nut.

13. Install the front wheel.

14. Bleed the power steering fluid.

15. Inspect and adjust the front wheel alignment.

STEERING KNUCKLE

REMOVAL & INSTALLATION

See Wheel Hub and Bearing.

WHEEL HUB & BEARING

REMOVAL & INSTALLATION

See Figure 206.

1. Remove the front wheel.
2. Remove the front axle shaft nut.
3. Separate the front speed sensor.
4. Remove the 2 bolts and separate the front disc brake caliper assembly from the steering knuckle.
5. Remove the front disc.
6. Separate the tie rod end sub-assembly.
7. Separate the No. 1 front suspension arm sub-assembly.
8. Remove the front axle assembly, as follows:
   a. Put matchmarks on the front driveshaft assembly and the front axle hub sub-assembly.
   b. Using a plastic hammer, separate the front driveshaft assembly from the front axle hub sub-assembly.

⁂ WARNING

Be careful not to damage the boot and ABS speed sensor rotor.

   c. Remove the 2 bolts, nuts and steering knuckle with the front axle hub sub-assembly.

To install:

9. Install the front axle assembly, as follows:
   a. Align the matchmarks and install the front driveshaft assembly to the front axle hub sub-assembly.
   b. Install the steering knuckle with the front axle hub sub-assembly to the front shock absorber assembly with the 2 bolts and 2 nuts and tighten to 155 ft. lbs. (210 Nm).

➡Only when reusing the bolts and nuts, apply the small amount of engine oil to the screw part of the nuts.

⁂ WARNING

Be careful not to damage the driveshaft boot or speed sensor rotor.

10. Install the lower No. 1 front suspension arm sub-assembly.
11. Install the tie rod end sub-assembly.
12. Install the front disc.
13. Install the front disc brake caliper assembly with the 2 bolts to the steering knuckle and tighten to 79 ft. lbs. (107 Nm).
14. Clean the threaded parts on the driveshaft and axle hub nut using a non-residue solvent.

➡Be sure to perform this work for a new driveshaft.

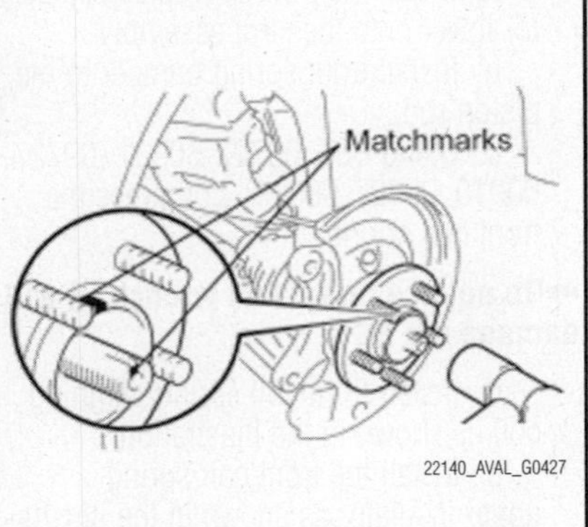

Fig. 206 Matchmarks on the front driveshaft assembly and the front axle hub sub-assembly

➡Keep the threaded parts free of oil and foreign objects.

15. Using a 30 mm socket wrench, install the front axle hub nut and tighten to 217 ft. lbs. (294 Nm).
16. Remove the 2 bolts and separate the front disc brake caliper assembly from the steering knuckle.
17. Remove the front disc.
18. Inspect front axle hub bearing looseness.
19. Inspect front axle hub runout.
20. Install the front disc.
21. Install the front disc brake caliper assembly with the 2 bolts to the steering knuckle and tighten to 79 ft. lbs. (107 Nm).
22. Install the front speed sensor.
23. Using a chisel and hammer, stake the axle hub nut.
24. Install the front wheel.
25. Inspect and adjust front wheel alignment.
26. Check ABS speed sensor signal.

SUSPENSION

COIL SPRING

REMOVAL & INSTALLATION

See Strut & Spring Assembly.

CONTROL ARMS/LINKS

REMOVAL & INSTALLATION

No. 1 Suspension Arm

See Figures 207 through 220.

1. Remove the rear wheel.
2. Remove the center exhaust pipe assembly.
3. Remove the rear stabilizer bar.
4. Remove the rear strut rod.
5. The rear height control sensor sub-assembly (with discharge heard light) is on the right side only. Remove the nut and separate the right rear height control sensor sub-assembly.
6. Remove the bolt, nut and the left No. 2 rear suspension arm (outer side) from the rear axle carrier. When removing the bolt, keep the nut from rotating.
7. Remove the right No. 2 rear suspension arm using the same procedure as for the left side.
8. Remove the bolt, nut and the left and right No. 1 rear suspension arm (outer side) from the rear axle carrier. When removing the bolt, keep the nut from rotating.
9. Support the rear suspension member with a jack.
10. Remove the 4 nuts, 2 bolts and 4 retainers from the rear suspension member.
11. Lower the rear suspension member.
12. Remove the bolt and No. 1 rear suspension arm.

To install:

13. Install the No. 1 rear suspension arm with the bolt, and temporarily tighten the bolt.

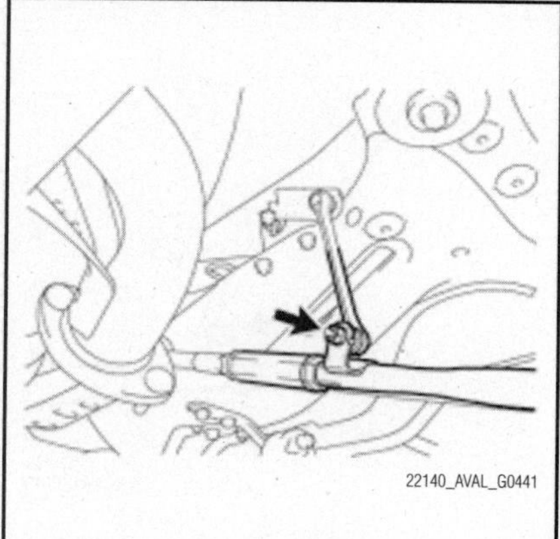

Fig. 207 Remove the nut and separate the right rear height control sensor sub-assembly

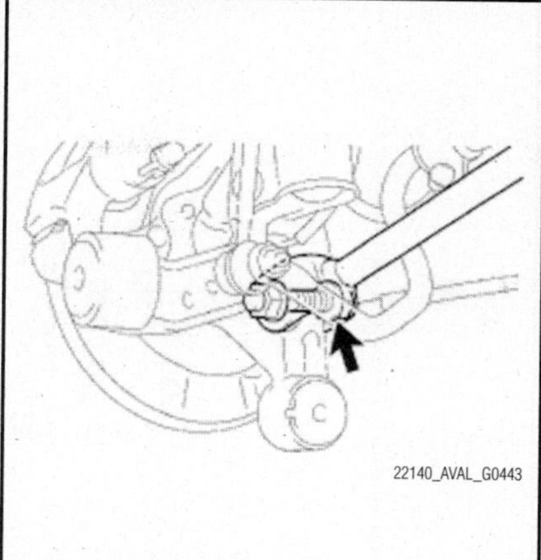

Fig. 209 Remove the bolt, nut and No. 1 rear suspension arm

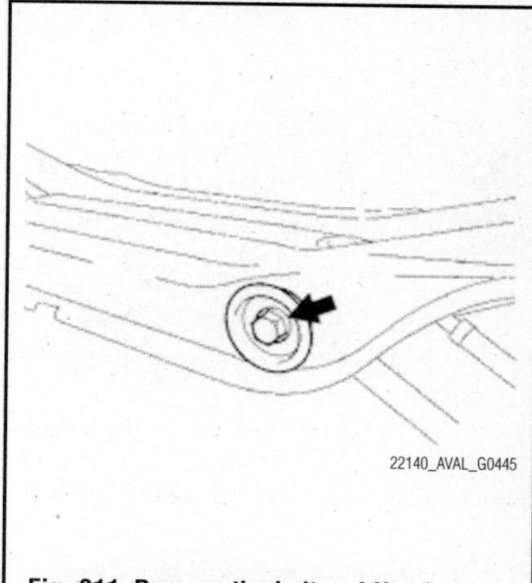

Fig. 211 Remove the bolt and No. 1 rear suspension arm

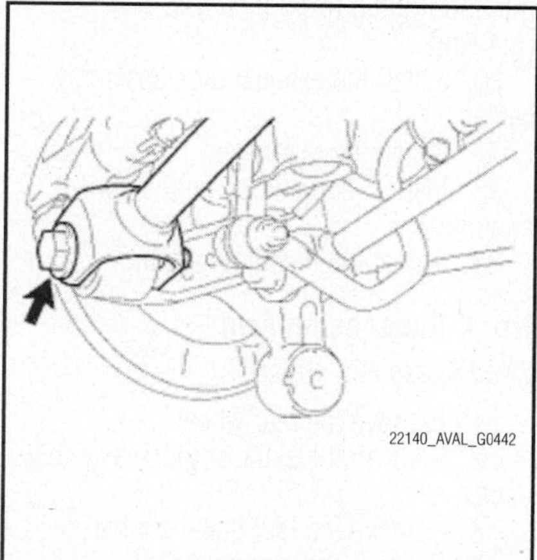

Fig. 208 Remove the bolt, nut and the left No. 2 rear suspension arm

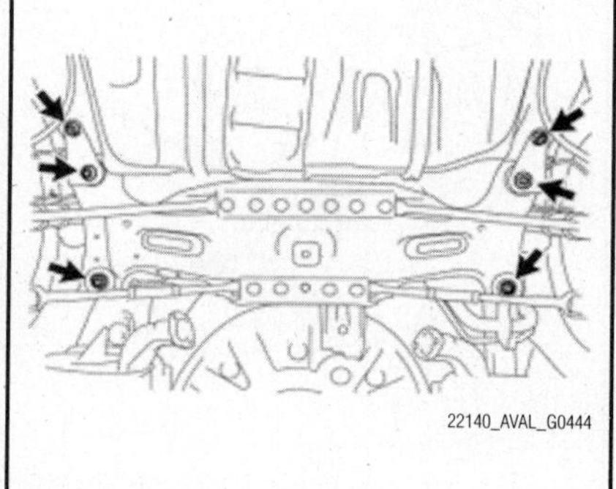

Fig. 210 Remove the 4 nuts, 2 bolts and 4 retainers from the rear suspension member

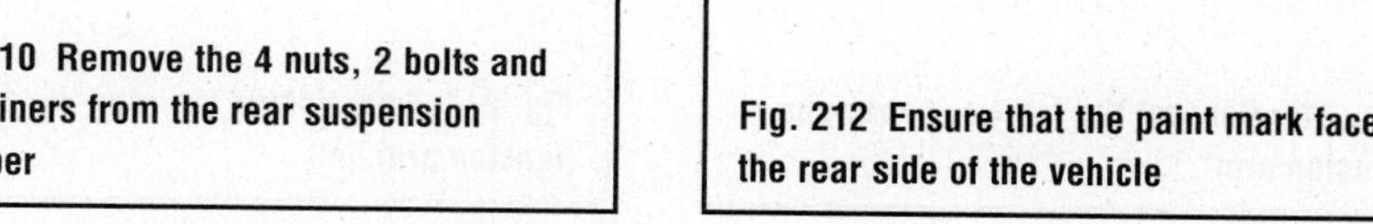

Fig. 212 Ensure that the paint mark faces the rear side of the vehicle

➡Install the No. 1 rear suspension arm so that the bracket leans toward the front side of the vehicle, as shown in the illustration.

➡Ensure that the paint mark faces the rear side of the vehicle.

14. Set the No. 1 rear suspension arm in the position shown in the illustration, and fully tighten the bolt to 74 ft. lbs. (100 Nm).

15. Raise the rear suspension member with a jack.

16. Install the rear suspension member with the 4 nuts, 2 bolts and 4 retainers and tighten to 41 ft. lbs. (55 Nm), and 28 ft. lbs. (38 Nm).

17. Connect the left No. 1 rear suspension arm (outer side) to the rear axle carrier with the bolt and nut and temporarily tighten. Insert the bolt from the front side of the vehicle and temporarily install it.

18. Temporarily tighten the right rear No. 1 suspension arm using the same procedure as for the left side.

19. Connect the left No. 2 rear suspension arm (outer side) to the rear axle carrier with the bolt and nut and temporarily tighten the bolt.

20. Temporarily tighten the right rear No. 2 suspension arm using the same procedure as for the left side.

21. Temporarily tighten the rear strut rod.

22. Jack up the rear axle carrier, placing a wooden block to avoid damage. Apply a load to the suspension so that the installed bolt of the No. 1 suspension arm assembly (inner side of the vehicle) is horizontally aligned with the center of the rear axle hub.

23. Fully tighten the left rear No. 1 suspension arm bolt to 74 ft. lbs. (100 Nm). When installing the bolt, hold the nut and tighten the bolt.

24. Fully tighten the right rear No. 1 suspension arm using the same procedure as for the left side.

25. Fully tighten the left rear No. 2 suspension arm bolt to 74 ft. lbs. (100 Nm). When installing the bolt, hold the nut and tighten the bolt.

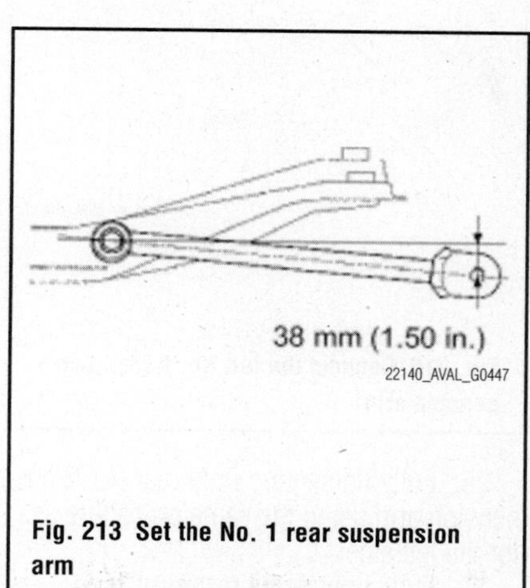

Fig. 213 Set the No. 1 rear suspension arm

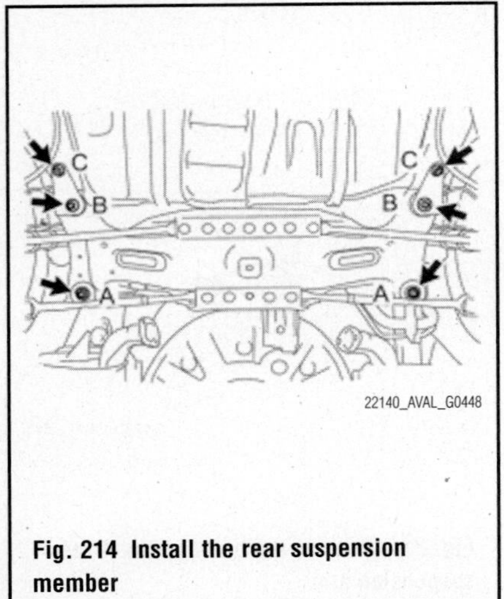

Fig. 214 Install the rear suspension member

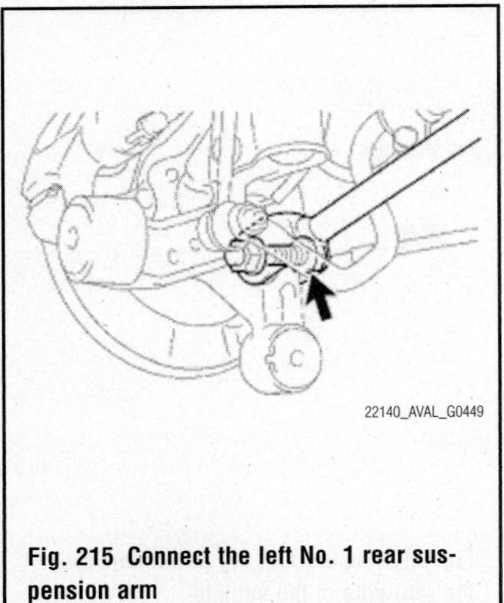

Fig. 215 Connect the left No. 1 rear suspension arm

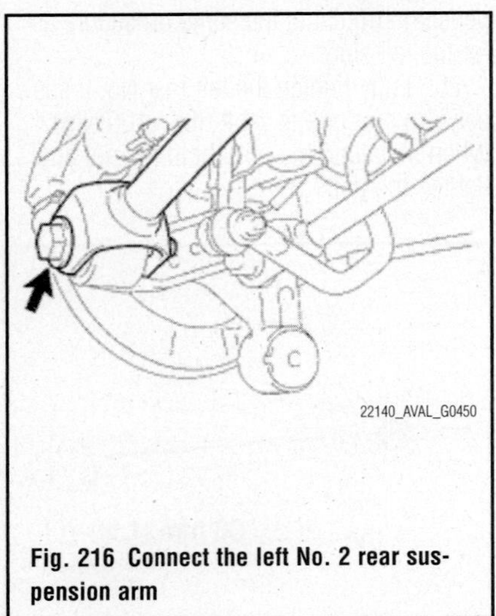

Fig. 216 Connect the left No. 2 rear suspension arm

26. Fully tighten the right rear No. 2 suspension arm using the same procedure as for the left side.
27. Fully tighten the rear strut rod.

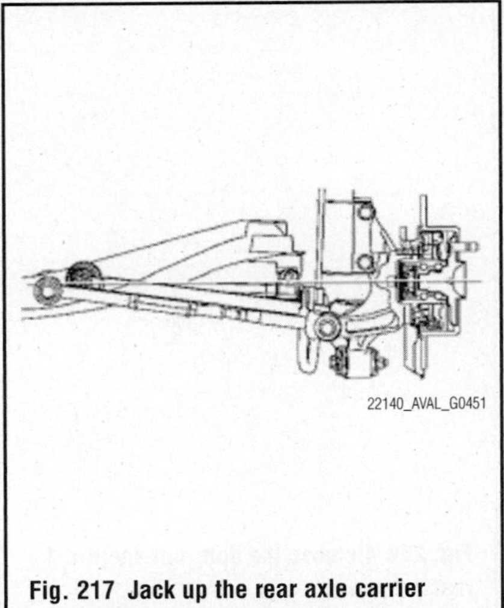

Fig. 217 Jack up the rear axle carrier

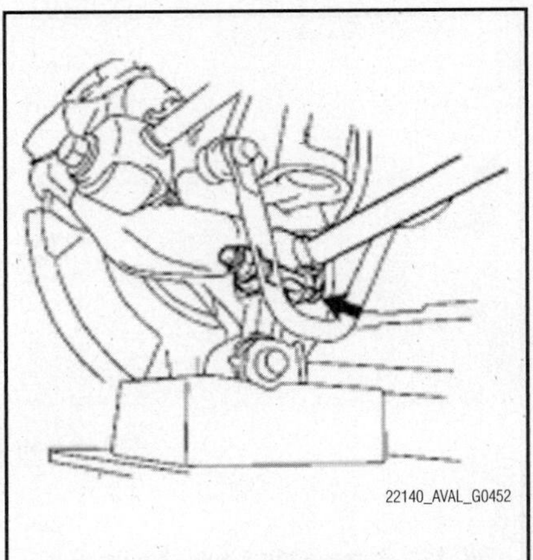

Fig. 218 Fully tighten the rear No. 1 suspension arm bolt

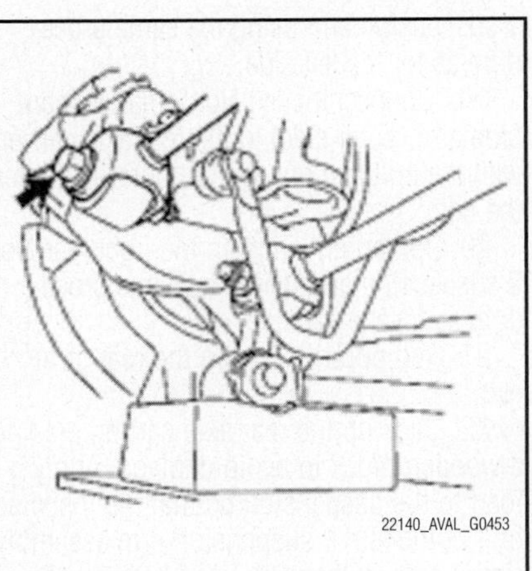

Fig. 219 Fully tighten the rear No. 2 suspension arm bolt

28. Install the rear stabilizer bar.
29. Install the right rear height control sensor sub-assembly (with discharge head light) with the

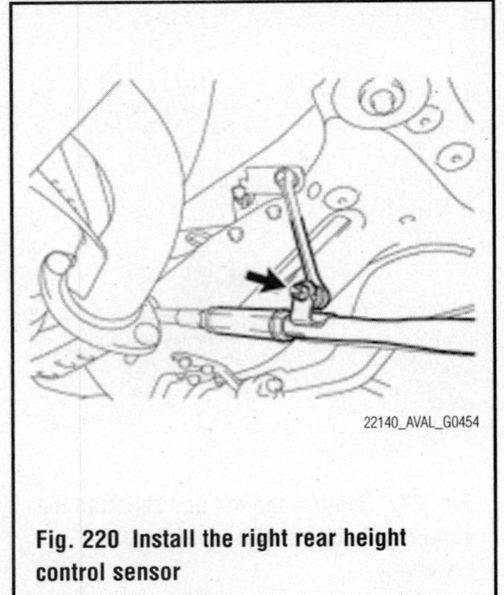

Fig. 220 Install the right rear height control sensor

nut and tighten to 48 inch lbs. (5.4 Nm).
30. Install the exhaust pipe assembly center.
31. Install the rear wheel.
32. Inspect and adjust the rear wheel alignment.
33. Adjust the headlight aiming.

No. 2 Suspension Arm

See Figures 221 and 222.

1. Remove the rear wheel.
2. Remove the exhaust pipe assembly center.
3. Remove the rear stabilizer bar.
4. Remove the rear strut rod.
5. Separate the rear height control sensor sub-assembly.
6. Separate the left and right rear No. 1 suspension arms.
7. Separate the left and right rear No. 2 suspension arms.
8. Remove the rear suspension member sub-assembly.
9. Remove the bolt, and disconnect the rear No. 2 suspension arm (inner side).

To install:

10. Install the rear No. 2 suspension arm (inner side) with the bolt and tighten to 74 ft. lbs. (100 Nm).

➡Ensure that the paint marks face the rear side of the vehicle.

11. Install the rear suspension member sub-assembly.
12. Temporarily tighten the left and right rear No. 1 suspension arms.
13. Temporarily tighten the left and right rear No. 2 suspension arms.
14. Temporarily tighten the rear strut rod.
15. Stabilize the suspension.

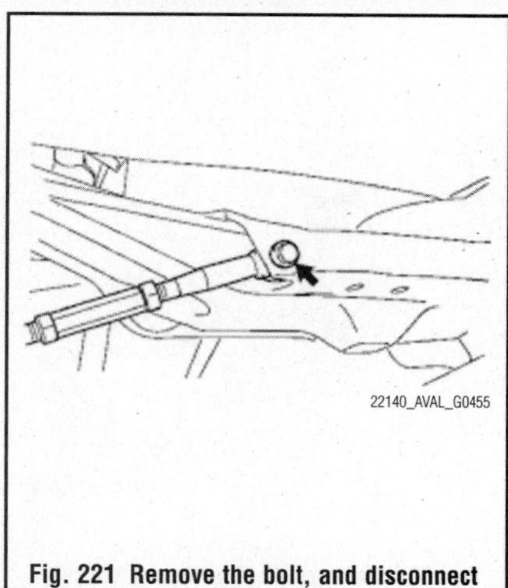
Fig. 221 Remove the bolt, and disconnect the rear No. 2 suspension arm

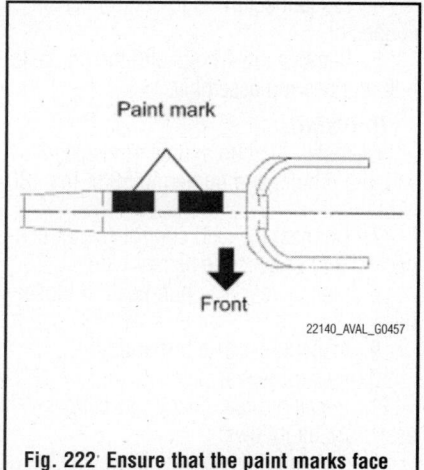

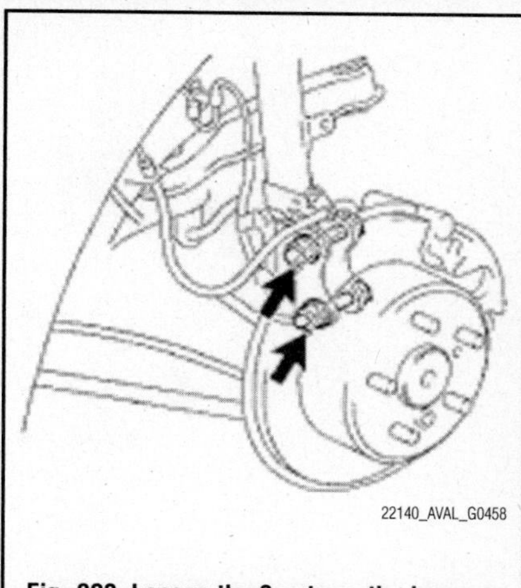
Fig. 223 Loosen the 2 nuts on the lower side of the shock absorber

Fig. 222 Ensure that the paint marks face the rear side of the vehicle

16. Fully tighten the left and right rear No. 1 suspension arms.
17. Fully tighten the left and right rear No. 2 suspension arms.
18. Fully tighten the rear strut rod.
19. Install the rear stabilizer bar.
20. Install the rear height control sensor sub-assembly.
21. Install the exhaust pipe assembly center.
22. Install the rear wheel.
23. Inspect and adjust the rear wheel alignment.
24. Adjust the headlight aiming.

STABILIZER BAR

REMOVAL & INSTALLATION

➡If an old gasket still remains on the pipe, remove it. Also, if any bolts or nuts are rusted, replace them.

1. Remove rear wheels.
2. Remove center exhaust pipe assembly.
3. Separate rear stabilizer links from stabilizer bar assembly.
4. Remove 2 bolts for LH rear stabilizer bar bracket.
5. Remove 2 bolts for RH rear stabilizer bar bracket.
6. Remove rear stabilizer bar with bushings.
7. Remove the 2 bushings from the bar.

To install:

8. Reverse removal procedure to install.
 a. Torque the bushing brackets to 23 ft. lbs. (31 Nm).
 b. Torque the link nuts to 29 ft. lbs. (39 Nm).
 c. Torque the wheel nuts to 76 ft. lbs. (103 Nm).

STRUT & SPRING ASSEMBLY

REMOVAL & INSTALLATION

See Figures 223 through 226

1. Remove the rear seat cushion assembly.
2. Remove rear seat headrest plate cover.
3. Remove rear seat headrest assembly.
4. Remove the rear seatback assembly.
5. Remove the rear wheel.
6. Separate rear stabilizer link assembly, as follows:
 a. Support the rear axle carrier with a jack.
 b. Remove the nut, and disconnect the stabilizer link from the shock absorber.

➡If the ball joint turns together with the nut, use a hexagon wrench (5 mm) to hold the stud.

7. Remove the 2 bolts, and disconnect the flexible hose and skid control sensor wire from the shock absorber.
8. Remove the rear shock absorber with coil spring, as follows:
 a. Loosen the 2 nuts on the lower side of the shock absorber. Do not remove the 2 bolts and 2 nuts.
 b. Remove the No. 1 rear suspension support cover.
 c. Loosen the support suspension center nut. Do not remove the nut. It is not necessary to loosen the nut if the shock absorber is not being disassembled.
 d. Remove the 3 nuts.
 e. Lower the rear axle carrier, and remove the 2 nuts and 2 bolts on the lower side of the shock absorber.
 f. Remove the shock absorber with coil spring.

Fig. 224 Remove the 3 nuts

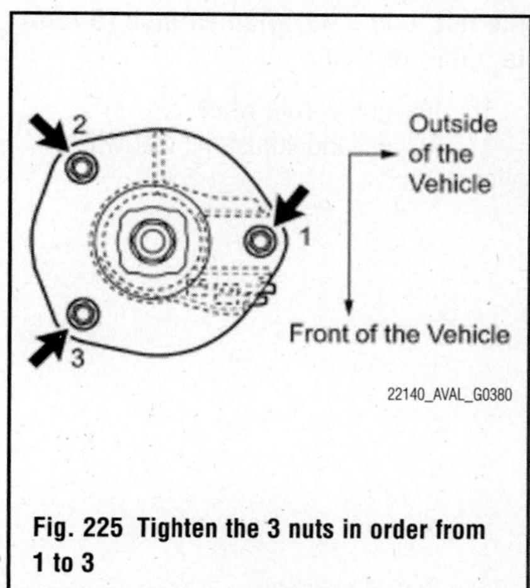

Fig. 225 Tighten the 3 nuts in order from 1 to 3

To install:

9. Temporarily install the 3 nuts to the upper side of the rear shock absorber with coil spring.
10. Fully tighten the 3 nuts in order from 1 to 3 to 29 ft. lbs. (39 Nm).

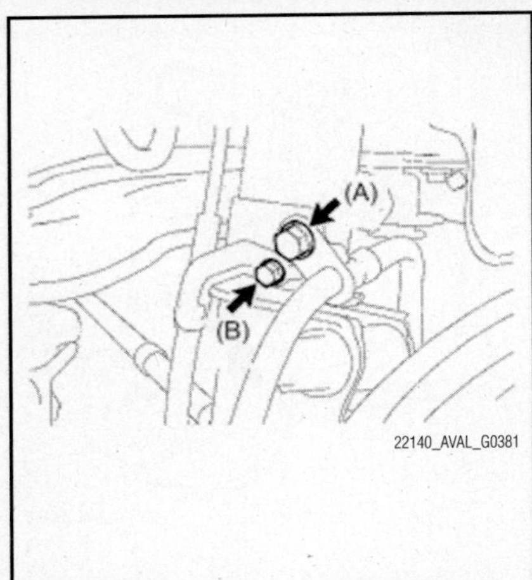

Fig. 226 Flexible hose and skid control sensor wire bolt tightening sequence

11. Install the 2 bolts and 2 nuts to the shock absorber with coil spring tighten to 133 ft. lbs. (180 Nm).
12. Fully tighten the nut installed on the top of the shock absorber with coil spring and tighten to 41 ft. lbs. (55 Nm). If the shock absorber has not been disassembled, it is not necessary to torque the nut.
13. Install the No. 1 rear suspension support cover.
14. Install the flexible hose and skid control sensor wire with the 2 bolts and tighten to 14 ft. lbs. (19 Nm), 49 inch lbs. (5.5 Nm).
15. Install the stabilizer link to the shock absorber with the nut and tighten to 29 ft. lbs. (39 Nm).

➡If the ball joint turns together with the nut, use a hexagon wrench (5 mm) to hold the stud.

16. Install the front wheel.
17. Inspect and adjust the rear wheel alignment.

STRUT ROD

REMOVAL & INSTALLATION

1. Remove rear wheel.
2. Remove strut rod as follows.
   a. Remove parking brake cable bolt and position the cable out of the way.
   b. Remove the strut rod front bolt.
   c. Remove the strut rod rear bolt from the axle carrier and remove the strut rod.

To install:

3. Install the strut rod (rear side), bolt and nut, and temporarily tighten the bolt.
4. Connect the strut rod (front side) with the bolt and nut and temporarily tighten.
5. Stabilize suspension as follows:
   a. Jack up the rear axle carrier, placing a wooden block to avoid damage. Apply a load to the suspension so that the installed bolt of the suspension arm assembly No. 1 (inner side of the vehicle) is horizontally aligned with the center of the rear axle hub.b. Torque the link nuts to 29 ft. lbs. (39 Nm).
   c. Torque the strut rod fasteners 83 ft. lbs. (113 Nm).
6. Install the parking brake cable with the bolt torque to 53 ft. lbs. (60Nm).
7. Install rear wheel.
8. Inspect and Adjust rear wheel alignment.

WHEEL BEARINGS

REMOVAL & INSTALLATION

See Figure 227.

1. Remove the rear wheel.
2. Separate the rear disc brake caliper assembly, as follows:
   a. Remove the bolt and separate the flexible hose from the shock absorber.
   b. Remove the 2 bolts and separate the rear disc brake caliper assembly.
3. Remove the rear disc.

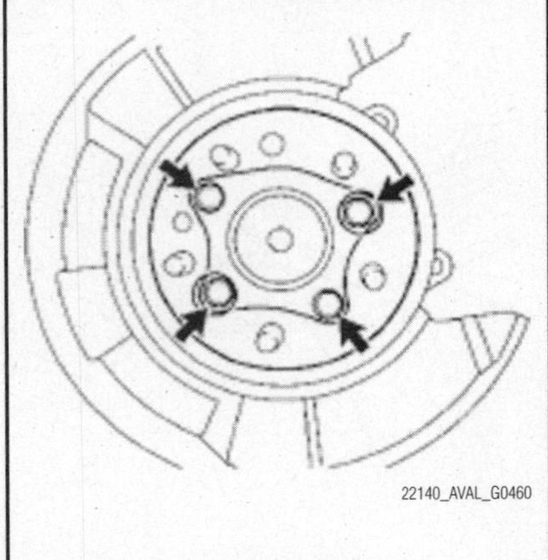

Fig. 227 Remove the 4 bolts and the rear axle hub and bearing assembly

4. Disconnect the skid control sensor connector.
5. Remove the 4 bolts and the rear axle hub and bearing assembly.

To install:

6. Install the hub and bearing assembly with the 4 bolts and tighten to 59 ft. lbs. (80 Nm).
7. Connect the skid control sensor connector. Do not twist the sensor wire.
8. Inspect rear axle hub bearing looseness.
9. Inspect rear axle hub runout.
10. Install the rear disc.
11. Install the rear disc brake caliper assembly, as follows:
    a. Install the rear disc brake caliper with the 2 bolts and tighten to 46 ft. lbs. (62 Nm).
    b. Install the flexible hose with the bolt and tighten to 14 ft. lbs. (19 Nm).
12. Install the rear wheel.
13. Inspect and adjust the rear wheel alignment.
14. Check ABS speed sensor signal.

TOYOTA

Camry • Camry Hybrid

5

SPECIFICATIONS AND MAINTENANCE CHARTS

ENGINE AND VEHICLE IDENTIFICATION

Engine						
Code ①	Liters (cc)	Cu. In.	Cyl.	Fuel Sys.	Engine Type	Eng. Mfg.
2AZ-FE	2.4 (2362)	144	4	EFI	DOHC	Toyota
2AZ-FXE	2.4 (2362)	144	4	EFI	DOHC	Toyota
2AR-FE	2.5 (2494)	152	4	EFI	DOHC	Toyota
2AR-FXE	2.5 (2494)	152	4	EFI	DOHC	Toyota
2GR-FE	3.5 (3456)	210	6	EFI	DOHC	Toyota

Model Year	
Code ②	Year
B	2011
C	2012

EFI: Electronic Fuel Injection

DOHC: Double Overhead Camshaft

① Stamped on the left side of the engine block

② 10th digit of the Vehicle Identification Number (VIN)

71099_CAMR_C0001

GENERAL ENGINE SPECIFICATIONS

All measurements are given in inches.

Year	Model	Engine Displacement Liters	Engine Series ID	Net Horsepower @ rpm	Net Torque @ rpm (ft. lbs.)	Bore x Stroke (in.)	Com-pression Ratio	Oil Pressure @ rpm
2011	Camry/ Camry HV	2.5	2AR-FE	179@6000	171@4000	3.54x3.86	10.4:1	55@3000
		2.4	2AZ-FXE	155@6000	158@4000	3.48x3.78	9.8:1	55@3000
		3.5	2GR-FE	268@6200	248@4700	3.70x3.27	10.8:1	36-78@3000
2012	Camry/ Camry HV	2.5	2AR-FE	179@6000	171@4000	3.54x3.86	10.4:1	55@3000
		2.5	2AR-FXE	179@6000	171@4000	3.54x3.86	10.4:1	55@3000
		3.5	2GR-FE	268@6200	248@4700	3.70x3.27	10.8:1	36-78@3000

71099_CAMR_C0002

GASOLINE ENGINE TUNE-UP SPECIFICATIONS

Year	Engine Displacement Liters	Engine ID	Spark Plug Gap (in.)	Ignition Timing (deg.)	Fuel Pump (psi)	Idle Speed (rpm)	Valve Clearance (in.)	
							Intake	Exhaust
2011	2.4	2AZ-FXE	0.039-0.043	N/A	44-50	①	0.0075-0.0114	0.0150-0.0189
	2.5	2AR-FE	0.039-0.043	N/A	44-50	①	0.0075-0.0114	0.0150-0.0189
	3.5	2GR-FE	0.039-0.043	N/A	44-50	650-750	N/A	N/A
2012	2.5	2AR-FE	0.039-0.043	N/A	44-50	①	0.0075-0.0114	0.0150-0.0189
	2.5	2AR-FXE	0.039-0.043	N/A	44-50	①	0.0075-0.0114	0.0150-0.0189
	3.5	2GR-FE	0.039-0.043	N/A	44-50	650-750	N/A	N/A

NOTE: The Vehicle Emission Control Information label often reflects specification changes made during production. The label figures must be used if they differ from those in this chart.

NA: Not available

① Manual transmission: 650 to 750 rpm, Automatic transmission: 610 to 710 rpm

71099_CAMR_C0003

CAPACITIES

Year	Model	Engine Displacement Liters	Engine ID	Engine Oil with Filter (qts.)	Transmission (pts.)		Fuel Tank (gal.)	Cooling System (qts.)
					5-Spd	Auto.		
2011	Camry	2.4	2AZ-FE	4.5	—	3.7	18.5	6.6
		3.5	2GR-FE	6.4	—	6.8	18.5	8.8
		2.5	2AR-FE	4.6	—	3.7	17.0	7.5
	Camry HV	2.4	2AZ-FXE	4.5	—	3.7	18.5	6.6
2012	Camry	2.5	2AR-FE	4.6	—	3.7	17.0	7.5
		3.5	2GR-FE	6.4	—	6.8	17.0	9.5
	Camry HV	2.5	2AR-FXE	4.6	—	3.7	17.0	7.5

NOTE: All capacities are approximate. Add fluid gradually and check to be sure a proper fluid level is obtained.

71099_CAMR_C0004

FLUID SPECIFICATIONS

Year	Model	Engine Disp. Liters	Engine ID	Engine Oil	Auto. Trans.	Power Steering Fluid	Engine Coolant	Brake Master Cylinder
2011	Camry	2.5	2AR-FE	0W-20	N/A	ATF Dexron® II or III	①	DOT 3
		3.5	2GR-FE	5W-30	N/A	ATF Dexron® II or III	①	DOT 3
	Camry HV	2.4	2AZ-FXE	5W-20	H V fluid ②	N/S	①	DOT 3
2012	Camry	2.5	2AR-FE	0W-20	N/A	ATF Dexron® II or III	①	DOT 3
		3.5	2GR-FE	5W-30	N/A	ATF Dexron® II or III	①	DOT 3
	Camry HV	2.5	2AR-FXE	0W-20	H V fluid ②	N/S	①	DOT 3

DOT: Department Of Transpotation

N/A: Not Available

N/S: Not Specified

① Toyota Super long life Coolant (SLLC)

② Hybrid Transaxle fluid.

71099_CAMR_C0005

VALVE SPECIFICATIONS

Year	Engine Displacement Liters	Engine ID	Seat Angle (deg.)	Face Angle (deg.)	Spring Test Pressure (lbs. @ in.)	Spring Installed Height (in.)	Stem-to-Guide Clearance (in.)		Stem Diameter (in.)	
							Intake	Exhaust	Intake	Exhaust
2011	2.5	2AR-FE	45	44.5	NA	NA	0.0010-0.0031	0.0012-0.0039	0.2154-0.2159	0.2151-0.2157
	2.4	2AZ-FXE	45	44.5	NA	NA	0.0010-0.0031	0.0012-0.0039	0.2154-0.2159	0.2151-0.2157
	3.5	2GR-FE	45	44.5	NA	NA	0.0010-0.0024	0.0012-0.0026	0.2154-0.2159	0.2151-0.2157
2012	2.5	2AR-FE	45	44.5	NA	NA	0.0010-0.0031	0.0012-0.0039	0.2154-0.2159	0.2151-0.2157
	2.5	2AR-FXE	45	44.5	NA	NA	0.0010-0.0031	0.0012-0.0039	0.2154-0.2159	0.2151-0.2157
	3.5	2GR-FE	45	44.5	NA	NA	0.0010-0.0024	0.0012-0.0026	0.2154-0.2159	0.2151-0.2157

NA: Not Available

71099_CAMR_C0006

CAMSHAFT AND BEARING SPECIFICATIONS CHART

All measurements are given in inches.

Year	Engine Displacement Liters	Engine ID	Journal Dia.	Brg. Oil Clearance	Shaft End-play	Runout	Lobe Height	
							Intake	Exhaust
2011	2.5	2AR-FE	④	NA	NA	0.0012	1.7387-1.7443	1.7379-1.7435
	2.4	2AZ-FXE	①	NA	NA	NA	1.8624-1.8664	1.8104-1.1843
	3.5	2GR-FE	②	③	NA	0.0016	1.7447-1.7487	1.7426-1.7465
2012	2.5	2AR-FE	④	NA	NA	0.0012	1.7387-1.7443	1.7379-1.7435
	2.5	2AR-FXE	④	NA	NA	0.0012	1.725-1.7290	1.691-1.6940
	3.5	2GR-FE	②	③	NA	0.0016	1.7447-1.7487	1.7426-1.7465

NA: Not Available

① Mark 1, 2 and 3: 1.4162-1.4167

② No. 1 journal: 1.4152-1.4157
Other Journals: 1.0220-1.0226 in.

③ No. 1 journal: 0.0016-0.0031
Other Journals: 1.0010-1.0024 in.
Maximum No.1 Journal: 0.0039 in.
Other Maximum Journals: 0.0035 in.

④ No 1 journal: 1.3562-1.3568
Other journals: 0.9039-0.9045

71099_CAMR_C0007

CRANKSHAFT AND CONNECTING ROD SPECIFICATIONS

All measurements are given in inches.

	Engine		Crankshaft				Connecting Rod		
Year	Displacement Liters	Engine ID	Main Brg. Journal Dia.	Main Brg. Oil Clearance	Shaft End-play	Thrust on No.	Journal Diameter	Oil Clearance	Side Clearance
2011	2.5	2AR-FE	2.1649-2.1654	0.0006-0.0015	0.0016-0.0095	3	1.8894-1.8898	0.0012-0.0025	0.0063-0.0202
	2.4	2AZ-FXE	2.0863-2.0866	0.0007-0.0016	0.0016-0.0095	3	1.8894-1.8898	0.0013-0.0025	0.0063-0.0143
	3.5	2GR-FE	2.4011-2.4016	0.0010-0.0019	0.0016-0.0095	2	2.0863-2.0866	0.0018-0.0026	0.0059-0.0157
2012	2.5	2AR-FE	2.1649-2.1654	0.0006-0.0015	0.0016-0.0095	3	1.8894-1.8898	0.0012-0.0025	0.0063-0.0202
	2.5	2AR-FXE	2.1649-2.1654	0.0006-0.0015	0.0016-0.0095	3	1.8894-1.8898	0.0012-0.0025	0.0063-0.0202
	3.5	2GR-FE	2.4011-2.4016	0.0010-0.0019	0.0016-0.0095	2	2.0863-2.0866	0.0018-0.0026	0.0059-0.0157

71099_CAMR_C0008

PISTON AND RING SPECIFICATIONS

All measurements are given in inches.

	Engine			Ring Gap			Ring Side Clearance		
Year	Displ. Liters	Engine ID	Piston Clearance	Top Comp.	Bottom Comp.	Oil Control	Top Comp.	Bottom Comp.	Oil Control
2011	2.5	2AR-FE	0.0020-0.0029	0.0094-0.0122	0.0130-0.0169	0.0040-0.0119	0.0008-0.0028	0.0021-0.0037	0.0023-0.0085
	2.4	2AZ-FXE	0.0020-0.0029	0.0094-0.0122	0.0130-0.0169	0.0040-0.0119	0.0008-0.0028	0.0008-0.0024	0.0008-0.0028
	3.5	2GR-FE	0.0018-0.0020	0.0098-0.0138	0.0197-0.0413	0.0039-0.0157	0.0008-0.0028	0.0008-0.0024	0.0028-0.0059
2012	2.5	2AR-FE	0.0020-0.0029	0.0094-0.0122	0.0130-0.0169	0.0040-0.0119	0.0008-0.0028	0.0021-0.0037	0.0023-0.0085
	2.5	2AR-FXE	0.0020-0.0029	0.0094-0.0122	0.0130-0.0169	0.0040-0.0119	0.0008-0.0028	0.0021-0.0037	0.0023-0.0085
	3.5	2GR-FE	0.0018-0.0020	0.0098-0.0138	0.0197-0.0413	0.0039-0.0157	0.0008-0.0028	0.0008-0.0024	0.0028-0.0059

71099_CAMR_C0009

TORQUE SPECIFICATIONS

All readings in ft. lbs.

Year	Engine Disp. Liters	Engine VIN	Cylinder Head Bolts	Main Bearing Bolts	Rod Bearing Bolts	Crankshaft Damper Bolts	Flywheel Bolts	Manifold Intake	Manifold Exhaust	Spark Plugs	Oil Pan Drain Plug
2011	2.4	2AZ-FE	①	②	③	125	④	22	27	13	18
	2.4	2AZ-FXE	①	②	③	125	④	22	27	13	18
	2.5	2AR-FE	⑧	②	⑨	125	④	22	27	13	18
	3.5	2GR-FE	⑤	⑥	⑦	184	61	15	15	13	30
2012	2.5	2AR-FE	⑧	②	⑨	125	④	22	27	13	18
	2.5	2AR-FXE	⑧	②	⑨	125	④	22	27	13	18
	3.5	2GR-FE	⑤	⑥	⑦	184	61	15	15	13	30

① Step 1: 52 ft. lbs.
Step 2: plus 90 degrees

② Step 1: 15 ft. lbs.
Step 2: 30 ft. lbs.
Step 3: +90 degrees

③ Step 1: 18 ft. lbs.
Step 2: plus 90 degrees

④ Auto driveplate: 72 ft. lbs.
Manual Flywheel : 96 ft. lbs.

⑤ Step 1: 10mm point cap bolts to 27 ft. lbs.
Step 2: 10mm point cap bolts plus 90 degrees
Step 3: 10mm point cap bolts plus 90 degrees
Step 4: Front bolts to 22 ft. lbs.

⑥ Step 1: 18 ft. lbs.
Step 2: Plus 90 degrees

⑦ Step 1: 16 cap bolts to 45 ft. lbs.
Step 2: 16 cap bolts plus 90 degrees
Step 3: 8 side bolts to 38 ft. lbs.

⑧ Step 1: 27 ft. lbs.
Step 2: 27 ft. lbs.
Step 3: Plus 90 degrees
Step 3: Plus 90 degrees

⑨ Step 1: 30 ft. lbs.
Step 2: Plus 90 degrees

71099_CAMR_C0010

WHEEL ALIGNMENT

Year	Model		Caster Range (+/-Deg.)	Caster Preferred Setting (Deg.)	Camber Range (+/-Deg.)	Camber Preferred Setting (Deg.)	Toe-in (in.)	Steering Axis Inclination (Deg.)
2011	Camry/ Camry HV	Front	0.75	2.92	0.75	-0.67	0+/-0.08	12.25+/-0.75
		Rear	—	—	0.75	-1.33	0.16+/-0.08	—
2012	Camry	Front	NA	NA	NA	NA	NA	NA
		Rear	NA	NA	NA	NA	NA	NA
	Camry HV	Front	NA	NA	NA	NA	0+/-0.08	NA
		Rear	NA	NA	NA	NA	0.09+/-0.08	NA

71099_CAMR_C0011

TIRE, WHEEL AND BALL JOINT SPECIFICATIONS

Year	Model	OEM Tires		Tire Pressures (psi)		Wheel	Ball Joint	Lug Nut Torque
		Standard	Optional	Front	Rear	Size	Inspection	(ft. lbs.)
2011	Camry	P215/60R16	P215/55R17	31	31	NA	①	76
	Camry HV	P215/60R16	P215/55R17	31	31	NA	①	76
2012	Camry	P215/60R16	P215/55R17	31	31	NA	①	76
	Camry HV	P215/60R16	P215/55R17	31	31	NA	①	76

NA: Not Available

OEM: Original Equipment Manufacturer

OPT: Optional

PSI: Pounds Per Square Inch

STD: Standard

① Replace if any measurable movement is found.

71099_CAMR_C0012

BRAKE SPECIFICATIONS

All measurements in inches unless noted

Year	Model		Brake Disc			Minimum	Brake Caliper	
			Original Thickness	Minimum Thickness	Maximum Runout	Lining Thickness	Bracket Bolts (ft. lbs.)	Mounting Bolts (ft. lbs.)
2011	Camry &	F	1.102	0.983	0.0020	0.039	79	25
	Camry HV	R	0.390	0.334	0.0059	0.039	46	32
2012	Camry &	F	1.102	0.983	0.0020	0.039	79	25
	Camry HV	R	0.390	0.334	0.0059	0.039	46	32

71099_CAMR_C0013

SCHEDULED MAINTENANCE INTERVALS

TOYOTA—CAMRY & CAMRY HV

TO BE SERVICED	TYPE OF SERVICE	VEHICLE MILEAGE INTERVAL (x1000)													
		5	10	15	20	25	30	35	40	45	50	55	60	90	120
Engine oil & filter	R	✓	✓	✓	✓	✓	✓	✓	✓	✓	✓	✓	✓	✓	✓
Automatic transmission fluid	S/I			✓			✓			✓			✓	✓	✓
Ball joints & dust covers	S/I			✓			✓			✓			✓	✓	✓
Bolts & nuts on chassis & body	S/I			✓			✓			✓			✓	✓	✓
Brake linings & drums	S/I	✓	✓	✓	✓	✓	✓	✓	✓	✓	✓	✓	✓	✓	✓
Brake line pipes & hoses	S/I			✓			✓			✓			✓	✓	✓
Brake pads & discs (front & rear)	S/I	✓	✓	✓	✓	✓	✓	✓	✓	✓	✓	✓	✓	✓	✓
Brake fluid	R						✓						✓	✓	✓
Rack and pinion assembly	S/I			✓			✓			✓			✓	✓	✓
Steering linkage & boots	S/I			✓			✓			✓			✓	✓	✓
Air cleaner filter	R						✓						✓	✓	✓
Spark plugs ①	R														✓
Drive belts	S/I												✓	✓	✓
Exhaust pipes & mountings	S/I			✓			✓			✓			✓	✓	✓
Fuel lines & connections	S/I						✓						✓	✓	✓
Engine coolant ②	S/I			✓			✓			✓			✓	✓	
Fuel tank cap gasket	S/I						✓						✓	✓	✓
Rotate tires	S/I	✓	✓	✓	✓	✓	✓	✓	✓	✓	✓	✓	✓	✓	✓
Clean air conditioning filter ③	S/I			✓			✓			✓			✓		✓
Axle shaft bolts	S/I			✓			✓			✓			✓	✓	✓
Brake pad thickness and rotor runout	S/I						✓						✓	✓	✓

R: Replace S/I: Service or Inspect

① Spark plugs are replaced at 120,000 miles

② Replace engine coolant at 100,000 miles and then inspect every 15,000 miles

③ Replace air conditioning filter every 30,000 miles

FREQUENT OPERATION MAINTENANCE (SEVERE SERVICE)

If a vehicle is operated under any of the following conditions it is considered severe service:

- **Extremely dusty areas.**
- **50% or more of the vehicle operation is in 32°C (90°F) or higher temperatures, or constant temperatures below 0°C (32°F).**
- **Prolonged idling (vehicle operation in stop and go traffic).**
- **Frequent short running periods (engine does not warm to normal operating temperatures).**
- **Police, taxi, delivery usage or trailer towing usage.**

Air cleaner filter: service or inspect every 5000 miles

Rear differential & transfer case oil: replace every 15,000 miles.

Ball joints & dust covers: service or inspect every 5000 miles.

Bolts & nuts on chassis & body: service or inspect every 5000 miles.

Axle shaft bolts: service or inspect every 5000 miles.

Steering linkage: service or inspect every 5000 miles.

71099_CAMR_C0014

PRECAUTIONS

Before servicing any vehicle, please be sure to read all of the following precautions, which deal with personal safety, prevention of component damage, and important points to take into consideration when servicing a motor vehicle:

- Never open, service or drain the radiator or cooling system when the engine is hot; serious burns can occur from the steam and hot coolant.
- Observe all applicable safety precautions when working around fuel. Whenever servicing the fuel system, always work in a well-ventilated area. Do not allow fuel spray or vapors to come in contact with a spark, open flame, or excessive heat (a hot drop light, for example). Keep a dry chemical fire extinguisher near the work area. Always keep fuel in a container specifically designed for fuel storage; also, always properly seal fuel containers to avoid the possibility of fire or explosion. Refer to the additional fuel system precautions later in this section.
- Fuel injection systems often remain pressurized, even after the engine has been turned **OFF**. The fuel system pressure must be relieved before disconnecting any fuel lines. Failure to do so may result in fire and/or personal injury.
- Brake fluid often contains polyglycol ethers and polyglycols. Avoid contact with the eyes and wash your hands thoroughly after handling brake fluid. If you do get brake fluid in your eyes, flush your eyes with clean, running water for 15 minutes. If eye irritation persists, or if you have taken brake fluid internally, IMMEDIATELY seek medical assistance.
- The EPA warns that prolonged contact with used engine oil may cause a number of skin disorders, including cancer. You should make every effort to minimize your exposure to used engine oil. Protective gloves should be worn when changing oil. Wash your hands and any other exposed skin areas as soon as possible after exposure to used engine oil. Soap and water, or waterless hand cleaner should be used.
- All new vehicles are now equipped with an air bag system, often referred to as a Supplemental Restraint System (SRS) or Supplemental Inflatable Restraint (SIR) system. The system must be disabled before performing service on or around system components, steering column, instrument panel components, wiring and sensors. Failure to follow safety and disabling procedures could result in accidental air bag deployment, possible personal injury and unnecessary system repairs.
- Always wear safety goggles when working with, or around, the air bag system. When carrying a non-deployed air bag, be sure the bag and trim cover are pointed away from your body. When placing a non-deployed air bag on a work surface, always face the bag and trim cover upward, away from the surface. This will reduce the motion of the module if it is accidentally deployed. Refer to the additional air bag system precautions later in this section.
- Clean, high quality brake fluid from a sealed container is essential to the safe and proper operation of the brake system. You should always buy the correct type of brake fluid for your vehicle. If the brake fluid becomes contaminated, completely flush the system with new fluid. Never reuse any brake fluid. Any brake fluid that is removed from the system should be discarded. Also, do not allow any brake fluid to come in contact with a painted surface; it will damage the paint.
- Never operate the engine without the proper amount and type of engine oil; doing so WILL result in severe engine damage.
- Timing belt maintenance is extremely important. Many models utilize an interference-type, non-freewheeling engine. If the timing belt breaks, the valves in the cylinder head may strike the pistons, causing potentially serious (also time-consuming and expensive) engine damage. Refer to the maintenance interval charts for the recommended replacement interval for the timing belt, and to the timing belt section for belt replacement and inspection.
- Disconnecting the negative battery cable on some vehicles may interfere with the functions of the on-board computer system(s) and may require the computer to undergo a relearning process once the negative battery cable is reconnected.
- When servicing drum brakes, only disassemble and assemble one side at a time, leaving the remaining side intact for reference.
- Only an MVAC-trained, EPA-certified automotive technician should service the air conditioning system or its components.

BRAKES

ANTI-LOCK BRAKE SYSTEM (ABS)

GENERAL INFORMATION

PRECAUTIONS

- Certain components within the ABS system are not intended to be serviced or repaired individually.
- Do not use rubber hoses or other parts not specifically specified for and ABS system. When using repair kits, replace all parts included in the kit. Partial or incorrect repair may lead to functional problems and require the replacement of components.
- Lubricate rubber parts with clean, fresh brake fluid to ease assembly. Do not use shop air to clean parts; damage to rubber components may result.
- Use only DOT 3 brake fluid from an unopened container.
- If any hydraulic component or line is removed or replaced, it may be necessary to bleed the entire system.
- A clean repair area is essential. Always clean the reservoir and cap thoroughly before removing the cap. The slightest amount of dirt in the fluid may plug an orifice and impair the system function. Perform repairs after components have been thoroughly cleaned; use only denatured alcohol to clean components. Do not allow ABS components to come into contact with any substance containing mineral oil; this includes used shop rags.
- The Anti-Lock control unit is a microprocessor similar to other computer units in the vehicle. Ensure that the ignition switch is **OFF** before removing or installing controller harnesses. Avoid static electricity discharge at or near the controller.
- If any arc welding is to be done on the vehicle, the control unit should be unplugged before welding operations begin.

SPEED SENSORS

REMOVAL & INSTALLATION

Front

See Figure 1.

1. Remove front wheel.
2. Remove 3 screws and front wheel opening extension pad.
3. Remove front fender liner.
    a. Remove the 5 clips and bolt.
    b. Remove the 4 screws, grommet, and the front fender splash shield sub assembly.
4. Remove front speed sensor.

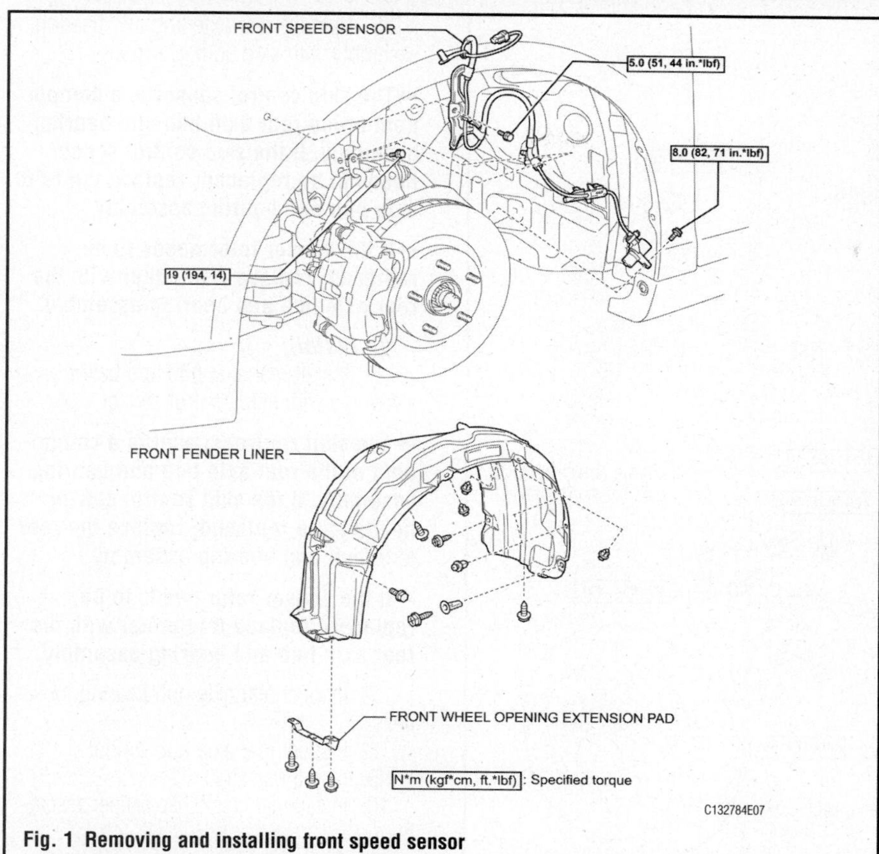

Fig. 1 Removing and installing front speed sensor

a. Disconnect the front speed sensor connector and clamp.

b. Remove the 2 bolts from the body and the shock absorber assembly.

c. Disengage the 2 claws from the steering knuckle.

d. Remove the bolt and the front speed sensor.

To install:

⁂ WARNING

Prevent foreign matter from attaching to the sensor tip.

➡Clean the installation hole and the surface for the speed sensor every time the speed sensor is removed.

5. Install the front speed sensor with bolt. Tighten to 71 ft. lbs. (96 Nm).
6. Install the front speed sensor with the 2 bolts and engage the 2 claws.

a. Tighten the body bolts to 44 inch. lbs. (5 Nm).

b. Tighten the bolts to the shock absorber to 14 ft. lbs. (19 Nm).

c. Connect the front speed sensor connector and clamp.

7. Install front fender liner.
8. Install front wheel opening extension pad.
9. Install front wheel. Tighten to 76 ft. lbs. (103 Nm).
10. Check speed sensor signal.

Rear

See Figures 2 and 3.

➡Use the same procedure for the RH side and LH side.

➡The procedure listed below is for the LH side.

➡The skid control sensor is a component of the rear axle hub and bearing assembly. If the skid control sensor needs to be replaced, replace the rear axle hub and bearing assembly.

➡If the sensor rotor needs to be replaced, replace it together with the rear axle hub and bearing assembly.

1. Remove rear wheel.
2. Disconnect skid control sensor wire.

a. Disconnect the connector from the skid control sensor.

➡Be careful not to damage the speed sensor.

3. Separate rear disc brake caliper assembly.
4. Remove rear disc.

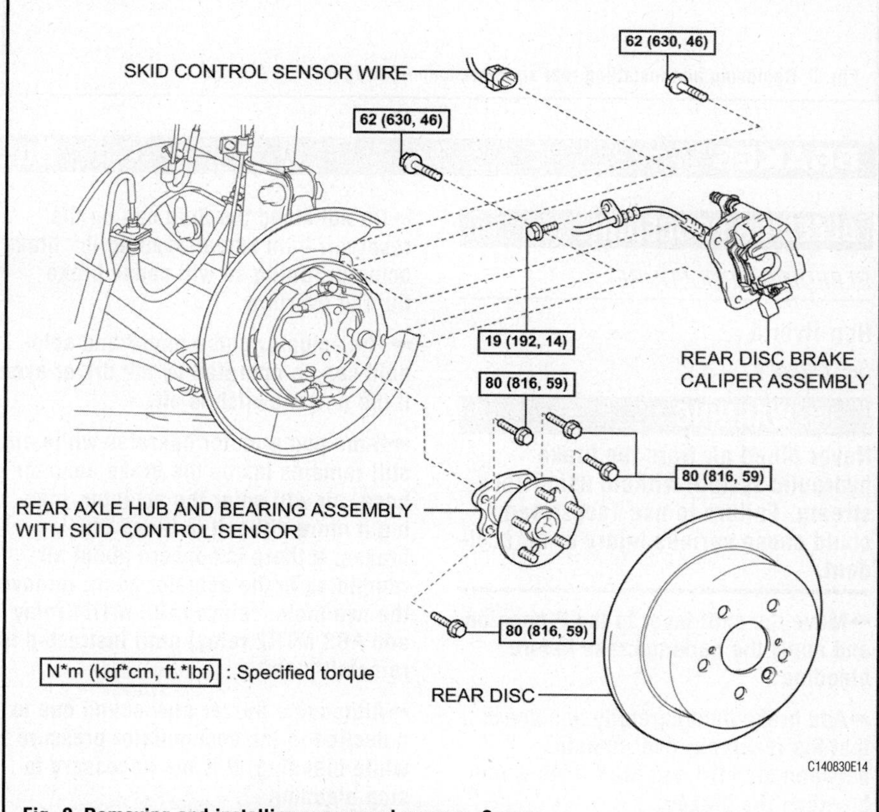

Fig. 2 Removing and installing rear speed sensor—Camry

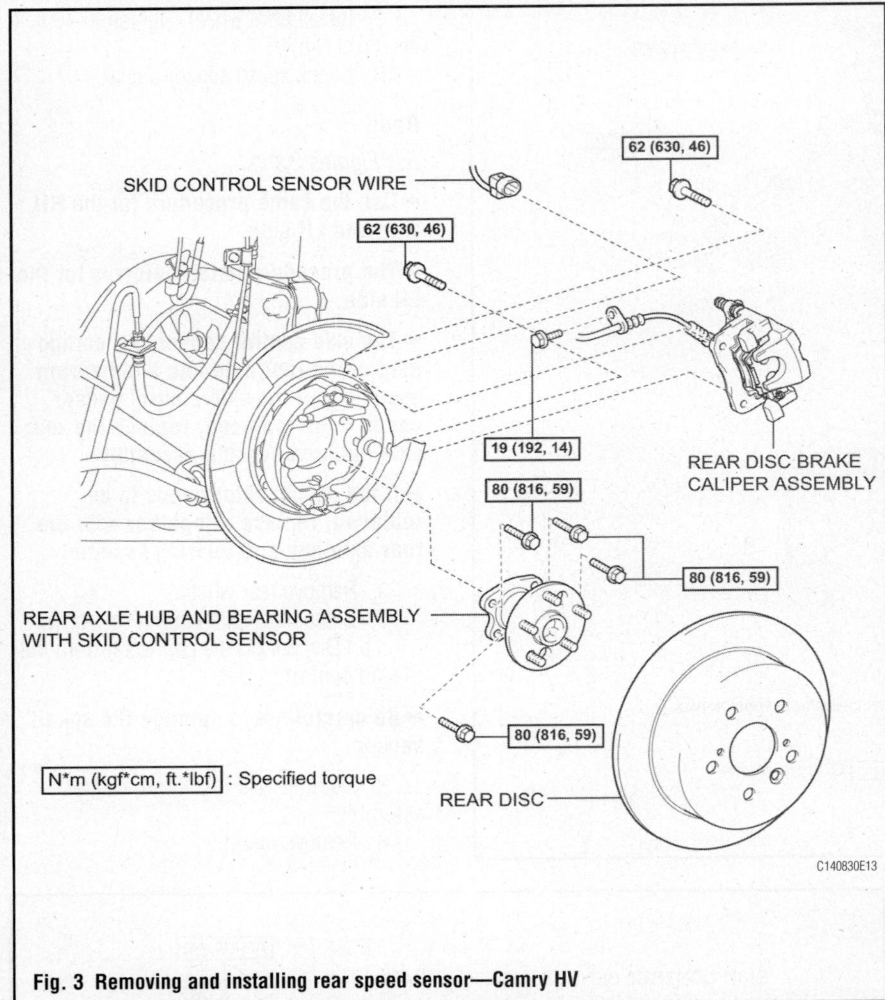

Fig. 3 Removing and installing rear speed sensor—Camry HV

5. Remove rear axle hub and bearing assembly with skid control sensor.

➡The skid control sensor is a component of the rear axle hub and bearing assembly. If the skid control sensor needs to be replaced, replace the rear axle hub and bearing assembly.

➡If the sensor rotor needs to be replaced, replace it together with the rear axle hub and bearing assembly.

To install:

6. Install rear axle hub and bearing assembly with skid control sensor.

➡The skid control sensor is a component of the rear axle hub and bearing assembly. If the skid control sensor needs to be replaced, replace the rear axle hub and bearing assembly.

➡If the sensor rotor needs to be replaced, replace it together with the rear axle hub and bearing assembly.

7. Inspect rear axle hub bearing looseness.
8. Inspect rear axle hub runout.
9. Install rear disc.
10. Install rear disc brake caliper assembly.
11. Connect skid control sensor wire.
12. Install rear wheel. Tighten to 76 ft. lbs. (103 Nm).
13. Inspect and adjust rear wheel alignment.
14. Check speed sensor signal.

BRAKES

BLEEDING THE BRAKE SYSTEM

BLEEDING PROCEDURE

BLEEDING PROCEDURE

Non-Hybrid

See Figure 4.

⁂ WARNING

Never bleed air from the brake hydraulic system without using Techstream. Failure to use Techstream could cause serious injury or an accident.

➡Move the shift lever to the P position and apply the parking brake before bleeding.

➡Add brake fluid carefully and check that the reservoir level remains between the MIN and MAX lines while bleeding the brakes.

➡Do not stand the fluid can on the reservoir inlet when bleeding the brake actuator. Doing so will cause brake fluid to overflow.

➡The actuator pump motor and solenoid can be operated by the driver even if the power switch is off.

➡If the pump motor operates while air still remains inside the brake actuator hose, air will enter the actuator, making it more difficult to bleed the brakes. If there is concern about air remaining in the actuator hose, remove the two motor relays (ABS MTR1 relay and ABS MTR2 relay) until instructed to reinstall them.

➡Although a buzzer may sound due to a decline in the accumulator pressure while bleeding, it is not necessary to stop bleeding.

➡DTCs indicating a malfunction in the motor relays (ABS MTR1 relay and ABS MTR2 relay) or the pressure sensor are stored after bleeding. Clear the DTCs when instructed during or after bleeding.

⁂ WARNING

Be sure to perform the bleeding operation in the order described.

➡If air enters the brake actuator hose because the fluid level in the reservoir is low, be sure to bleed the brake actuator hose.

1. Add brake fluid to the MAX line in the reservoir. (SAE J1703 or FMVSS No. 116 DOT3).

⁂ WARNING

Add brake fluid carefully and check that the reservoir level remains

between the MIN and MAX lines while bleeding the brakes.

WARNING

Do not stand the fluid can on the reservoir inlet when bleeding the brake actuator. Doing so will cause brake fluid to overflow.

2. Disable brake control.

When using Techstream:

➡Refer to the Techstream operator's manual for further details.

➡Bleed the air by following the steps displayed on Techstream.

a. Move the shift lever to the P position and apply the parking brake.
b. Connect Techstream to the DLC3 with the power switch off.
c. Turn the power switch on (IG).

➡Do not start the engine.

d. Turn Techstream on and select "Chassis / ABS/VSC/TRC / Utility / ECB (Electronically Controlled Brake system) Utility / ECB (Electronically Controlled Brake system) Invalid".

WARNING

If the pump motor operates while air remains inside the brake actuator hose, air will enter the actuator, and this will make bleeding the brakes more difficult.

When removing the ABS relay:

a. Remove the 2 ABS motor relays (ABS MTR1 relay and ABS MTR2 relay) with the power switch off in order to disable brake control.

WARNING

If the pump motor operates while air remains inside the brake actuator hose, air will enter the actuator, and this will make bleeding the brakes more difficult.

➡After the brake actuator assembly has been replaced, remove the ABS motor relay before bleeding the brakes.

➡Before removing the brake actuator assembly, make sure to perform accumulator pressure zero down.

3. Bleed brake actuator hose.
a. Connect SST to the reservoir with the brake reservoir pressure adapter.
b. Loosen the bleeder plug of the actuator.
c. Connect a vinyl tube to the bleeder plug of the actuator.
d. Use the SST to boost pressure in the reservoir.

➡Standard pressure is 11.6 psi (50-80 kPa).

e. Drain approximately 100 cc (3.4 fl. oz.) of fluid.
f. Tighten the bleeder plug and boost the pressure in the reservoir again (50 to 80 kPa (0.5 to 0.8 kgf/cm2)). Then, loosen the bleeder plug and bleed the brake actuator hose.

➡Repeat this procedure at least 5 times.

g. When air is completely bled out from the hose between the reservoir and the actuator, tighten the bleeder plug. Tighten to 74 inch lbs. (8.3 Nm).
4. Bleed master cylinder.
5. Bleed front brake system.
a. Depress the brake pedal several times and bleed the front brake system from the bleeder plugs on the front brake cylinders RH and LH.

➡Repeat the procedure until air is completely bled from the front brake system.

b. Tighten the bleeder plugs after bleeding. Tighten to 74 inch lbs. (8.3 Nm).
6. Cancel brake control disable.

WARNING

Install the 2 ABS motor relays (ABS MTR1 relay and ABS MTR2 relay). (If they have been removed.)

a. Cancel brake control prevention following the prompts on the Techstream screen. (If brake control has been prevented using Techstream.)
7. Clear DTC for brake control system.
8. Bleed rear brake system.

WARNING

Never bleed air from the brake hydraulic system without using Techstream. Failure to use Techstream could cause serious injury or an accident.

➡Bleed air by following the steps displayed on Techstream.

a. Connect Techstream to the DLC3 with the power switch off.
b. Check that the parking brake is applied and turn the power switch on (IG).
c. Turn Techstream on and select "Chassis / ABS/VSC/TRC / Utility / ECB (Electronically Controlled Brake system) Utility / ECB (Electronically Controlled Brake system) Invalid".
d. With the brake pedal depressed, bleed the rear brake system from the bleeder plug on the rear disc brake cylinder LH while the pump motor and solenoid are operating.

WARNING

Keep the fluid inside the reservoir above the LOW level by replenishing.

➡Depress and hold the brake pedal.

➡After the solenoid operates for approximately 30 seconds, release the brake pedal to stop the solenoid.

➡Repeat the procedures until air is completely bled from the rear brake system.

➡The brake control warning light comes on and the buzzer sounds while bleeding, but they do not indicate a malfunction.

e. Tighten the bleeder plug after bleeding. Tighten to 74 inch lbs. (8.3 Nm).
9. Perform accumulator pressure zero down.

WARNING

Never bleed air from the brake hydraulic system without using Techstream. Failure to use Techstream could cause serious injury or an accident.

WARNING

Be sure to perform this procedure before replacement, removal, or installation of the actuator.

➡Perform accumulator zero down by following the steps displayed on Techstream.

a. Drain the brake fluid in the reservoir tank near the MIN line.
b. Connect Techstream to the DLC3 with the power switch off.
c. Depressurize the accumulator.
d. Check that the parking brake is applied and turn the power switch on (IG).
e. Turn Techstream on and select "Chassis / ABS/VSC/TRC / Utility / ECB (Electronically Controlled Brake system) Utility / Zero Down" on Techstream.

f. When the buzzer sounds, turn the power switch off.

g. Circulate the fluid in the accumulator.

h. Depressurize the accumulator 5 times.

➡Accumulator pressure is released and accumulated repeatedly, which circulates the fluid inside the accumulator, when repeating accumulator zero down.

➡The pump motor rotates and the accumulator is pressurized every time the power switch is turned from off to on (IG).

10. Check brake fluid level in reservoir.

a. Adjust the brake fluid to the MAX line with the power switch on (IG).

➡The brake fluid level must always be adjusted to the MAX line.

11. Clear DTC for brake control system.
12. Perform linear valve offset learning.

a. When the brake actuator assembly is replaced, perform linear valve offset learning after bleeding is completed.

➡If any work is performed on the brake system or if air in the brake line is suspected, bleed the air from the brake system.

➡Move the shift lever to the P position and apply the parking brake before bleeding the brakes.

➡Add brake fluid to keep the level between MIN and MAX lines of the reservoir while bleeding the brakes.

➡If brake fluid leaks onto any painted surface, wash or remove it completely.

Hybrid

See Figure 4.

1. Fill the reservoir with brake fluid. (SAE J1703 or FMVSS No. 116 DOT3).
2. Bleed master cylinder.

➡If the master cylinder is reinstalled or if the reservoir becomes empty, bleed the air from the master cylinder.

➡To avoid brake fluid from adhering, cover the painted surface with a shop rag or piece of cloth.

a. Using SST, disconnect the 2 brake lines from the master cylinder.

3. Slowly depress the brake pedal and hold it.
4. Cover the 2 outer holes with fingers, and release the brake pedal.
5. Repeat steps 3 and 4 four times.
6. Using SST, connect the brake lines to the master cylinder.

a. For TMC made without special tool, tighten to 11 ft. lbs. (15 Nm).

b. For TMC made with special tool, tighten to 10 ft. lbs. (14 Nm).

c. For TMMK made without special tool, tighten to 14 ft. lbs. (19 Nm).

d. For TMMK made with special tool, tighten to 13 ft. lbs. (17 Nm).

➡Use a torque wrench with a fulcrum length of 250 mm (9.84 in.).

➡This torque value is effective when SST is parallel to the torque wrench.

7. Bleed brake line.
8. Bleed brake actuator assembly (TMC made only).

➡After bleeding the air from the brake system, if the height or feel of the brake pedal cannot be obtained, bleed the air from the brake actuator assembly with the Techstream by following the procedures listed below.

a. Depress the brake pedal more than 20 times with the ignition switch off.

b. Connect the Techstream to the DLC3, then turn the ignition switch on (IG).

➡Do not start the engine.

c. Turn the Techstream on and select "AIR BLEEDING" on the screen.

d. Bleed the air according to "Step 1: Increase" on the Techstream display.

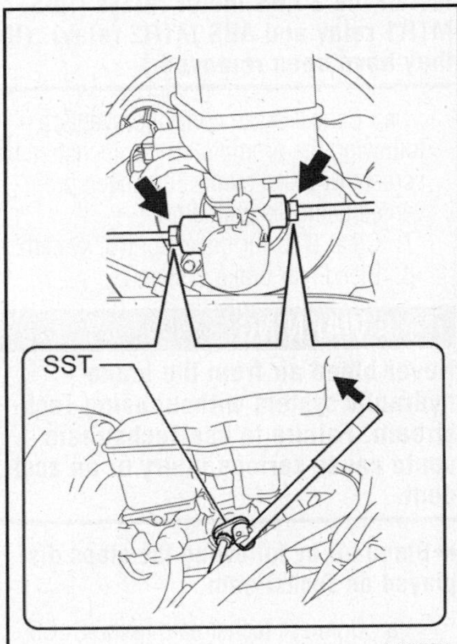

Fig. 4 Disconnecting brake lines from master cylinder

➡Make sure that the master cylinder reservoir tank does not become empty of brake fluid.

e. Connect a vinyl tube to either one of the bleeder plugs.

f. Depress the brake pedal several times, then loosen the bleeder plug connected to the vinyl tube with the pedal depressed.

g. When fluid stops coming out, tighten the bleeder plug, then release the brake pedal.

h. Repeat until all the air in the fluid is completely bled out.

i. Tighten the bleeder plug completely. Tighten to 71 inch lbs. (8 Nm).

j. Bleed the air from the suction line according to "Step 2: Inhalation" on the Techstream display.

k. Connect a vinyl tube to the bleeder plug at the right front wheel or the right rear wheel and loosen the bleeder plug.

l. Operate the brake actuator assembly to bleed the air using the Techstream.

➡The operation stops auto9matically in 4 seconds.

➡At this time, be sure to release the brake pedal.

m. Check that the operation has stopped by referring to the Techstream display and tighten the bleeder plug.

n. Repeat until all the air in the fluid is completely bled out.

o. Tighten the bleeder plug completely. Tighten to 71 inch lbs. (8 Nm).

➡For the rest of the wheels, bleed the air in the same way as stated in the above procedures.

p. Bleed the air from the pressure reduction line according to "Step 3: Decrease" on the Techstream display.

q. Connect a vinyl tube to either one of the bleeder plugs.

r. Loosen the bleeder plug.

s. Using the Techstream, operate the brake actuator assembly, completely depress the brake pedal, and hold it.

➡The operation stops automatically in 4 seconds. When performing this procedure continuously, an interval of at least 20 seconds is required.

➡When the operation is completed, the brake pedal slightly goes down. This is a normal phenomenon when the solenoid opens.

➡During this procedure, the pedal seems heavy, but completely depress it so that the brake fluid comes out from the bleeder plug.

➡Be sure to keep the brake pedal depressed. Never depress and release the pedal repeatedly.

t. Tighten the bleeder plug, then release the brake pedal.

u. Repeat until all the air in the fluid is completely bled out.

9. Tighten the bleeder plug to 71 inch lbs. (8 Nm).

➡Repeat the above process for the rest of the brakes to bleed the air from the brake line.

a. Bleed the air from the brake line again according to "Step 4: Increase" on the Techstream display.

b. Connect a vinyl tube to either one of the bleeder plugs.

c. Depress the brake pedal several times, then loosen the bleeder plug connected to the vinyl tube with the pedal depressed.

d. When fluid stops coming out, tighten the bleeder plug, then release the brake pedal.

e. Repeat until all the air in the fluid is completely bled out.

10. Tighten the bleeder plug to 71 inch lbs. (8 Nm).

➡Repeat the above process for the rest of the brakes to bleed the air from the brake line.

a. Finish "AIR BLEEDING" on the Techstream, and turn off the Techstream.

b. Disconnect the Techstream from the DLC3.

c. Turn the ignition switch off.

11. Inspect fluid level in reservoir.

a. Add fluid if necessary. (SAE J1703 or FMVSS No. 116 DOT3.

➡If fluid leaks, tighten or replace leaking part.

MASTER CYLINDER BLEEDING

See Figure 5.

➡If the master cylinder has been disassembled or if the reservoir becomes empty, bleed the air from the master cylinder.

1. Using SST 09023-00101, disconnect the brake stroke simulator line from the master cylinder.
2. Using SST 09023-00101, disconnect the brake line from the master cylinder.
3. Slowly depress and hold the brake pedal.
4. Cover the outer holes with fingers, and release the brake pedal.
5. Repeat steps 3 and 4 four times.
6. Use the SST 09023-00101 to fully tighten the brake line.

a. With special tool tighten to 10 ft. lbs. (14 Nm). If no special tool, tighten to 11 ft. lbs. (15 Nm).

➡If brake fluid leaks onto any painted surface of the vehicle, wash or remove it completely.

➡Bleed air by following the steps displayed on Techstream.

➡Do not bend or damage the brake line.

➡Do not allow any foreign matter such as dirt and dust to enter the brake line from the connecting points.

➡Use a torque wrench with a fulcrum length of 250 mm (9.84 in.).

➡This torque value is effective when SST is parallel to a torque wrench.

➡Air can be easily bled from the front brake system if air has been bled from the master cylinder when replacing the brake master cylinder assembly.

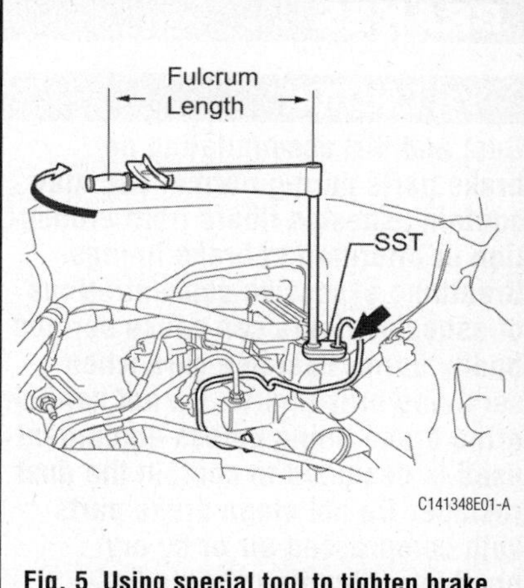

Fig. 5 Using special tool to tighten brake stroke simulator line

7. Use the SST to fully tighten the brake stroke simulator line.

a. With special tool tighten to 10 ft. lbs. (14 Nm). If no special tool, tighten to 11 ft. lbs. (15 Nm).

BRAKE LINE BLEEDING

➡Bleed the air from the brake line of the wheel farthest from the master cylinder.

1. Connect a vinyl tube to the bleeder plug.
2. Depress the brake pedal several times, then loosen the bleeder plug with the pedal depressed.
3. When fluid stops coming out, tighten the bleeder plug, then release the brake pedal.
4. Repeat steps 2 and 3 until all the air in the fluid is completely bled out.
5. Tighten the bleeder plug completely. Tighten to 71 inch lbs. (8 Nm).
6. Repeat the above procedure for each wheel to bleed the air from the brake line.

CAUTION

Dust and dirt accumulating on brake parts during normal use may contain asbestos fibers from production or aftermarket brake linings. Breathing excessive concentrations of asbestos fibers can cause serious bodily harm. Exercise care when servicing brake parts. Do not sand or grind brake lining unless equipment used is designed to contain the dust residue. Do not clean brake parts with compressed air or by dry brushing. Cleaning should be done by dampening the brake components with a fine mist of water, then wiping the brake components clean with a dampened cloth. Dispose of cloth and all residue containing asbestos fibers in an impermeable container with the appropriate label. Follow practices prescribed by the Occupational Safety and Health Administration (OSHA) and the Environmental Protection Agency (EPA) for the handling, processing, and disposing of dust or debris that may contain asbestos fibers.

BRAKE CALIPER

REMOVAL & INSTALLATION

Hybrid

See Figures 6 through 8.

WARNING

When the brake caliper is removed from the brake disc, do not start the hybrid system. If the hybrid system is started, the brake fluid pressure may increase.

➡Use the same procedures for the RH side and LH side.

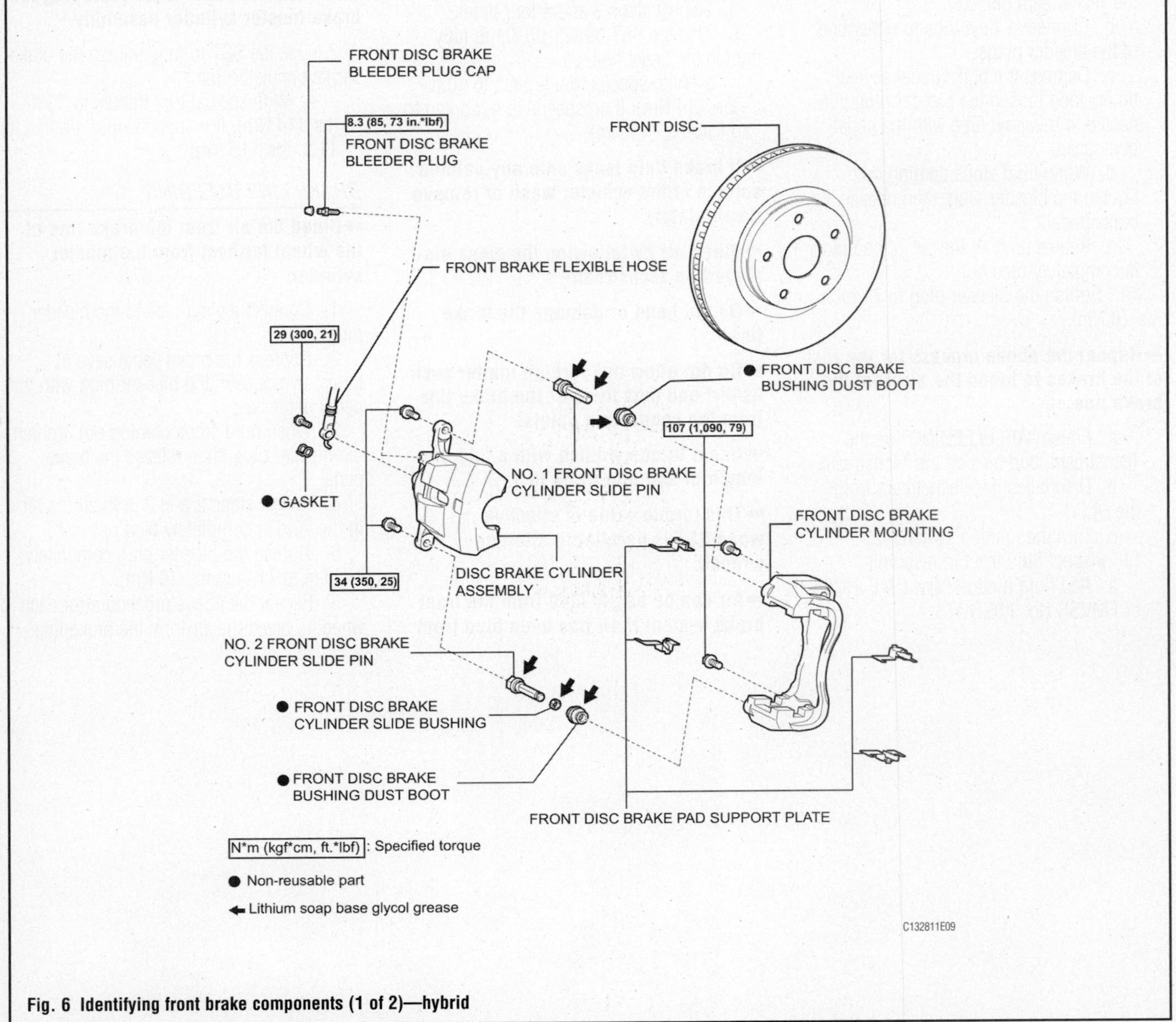

Fig. 6 Identifying front brake components (1 of 2)—hybrid

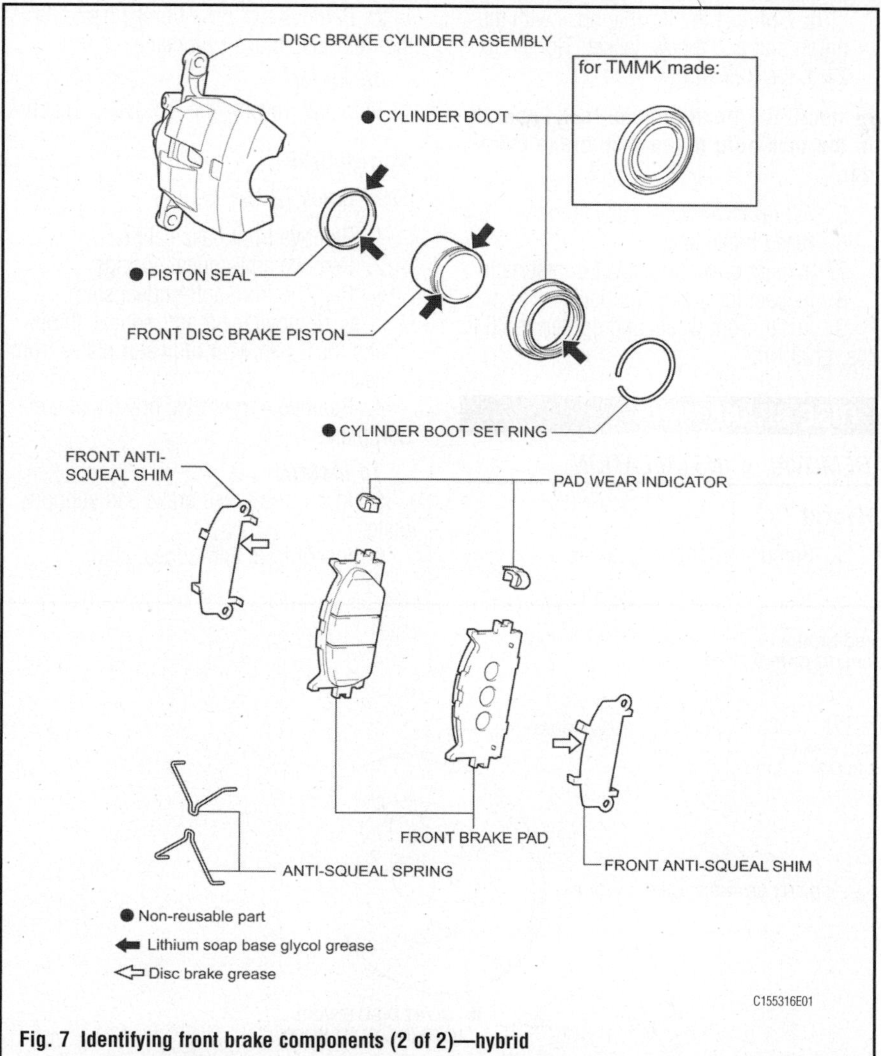

Fig. 7 Identifying front brake components (2 of 2)—hybrid

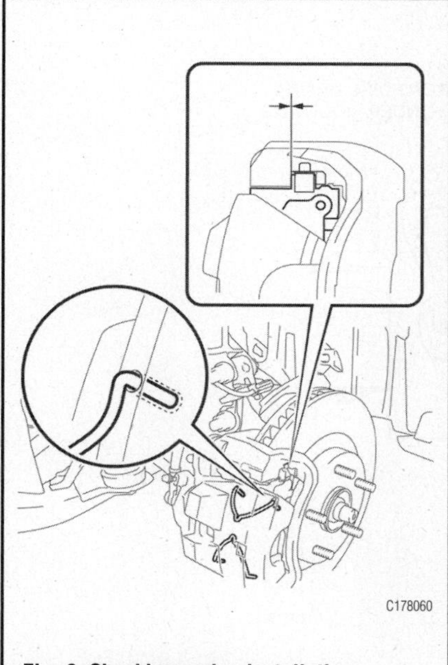

Fig. 8 Checking spring installation

➡The procedures listed below are for the LH side.

1. Remove the 2 ABS motor relays with the power switch off.
2. Remove front wheel.
3. Drain brake fluid.

➡If brake fluid leaks onto any painted surface of the vehicle, wash or remove it completely.

4. Remove disc brake cylinder assembly.
   a. Remove the union bolt and gasket from the disc brake cylinder assembly, then disconnect the front brake flexible hose.
   b. Hold the front disc brake cylinder slide pin and remove the 2 bolts and disc brake cylinder assembly.

➡Remove the disc brake cylinder assembly while holding both of the brake pads or the anti-squeal springs may fall off the brake pads.

To install:

5. Install disc brake cylinder assembly.
   a. Install the disc brake cylinder assembly with the 2 bolts. Tighten to 25 ft. lbs. (34 Nm).
   b. Check the installation of the anti-squeal springs.
   c. Visually check for any clearance between the brake pad and front disc brake pad support plates.

➡If the anti-squeal springs are installed correctly, there will be no clearance between the brake pad and the front disc brake pad support plates. If there is a clearance, the anti-squeal springs may not be installed properly.

➡Check all 4 contact surfaces between the brake pad and the front disc brake pad support plates.

   d. Connect the front flexible hose with the union bolt and a new gasket. Tighten to 21 ft. lbs. (29 Nm).

➡Install the flexible hose lock securely in the lock hole in the disc brake cylinder.

6. Add brake fluid to reservoir.
7. Disable brake control.
8. Bleed front brake system.
9. Cancel brake control disable.
10. Clear DTC for brake control system.
11. Perform accumulator pressure zero down.
12. Inspect for brake fluid leak.
13. Check brake fluid leak.
14. Check brake fluid level in reservoir.
15. Install front wheel and tighten to 76 ft. lbs. (103 Nm).

Non-Hybrid

See Figures 9 through 11.

➡Use the same procedures for the RH side and LH side.

➡The procedures listed below are for the LH side.

1. Remove front wheel.
2. Drain brake fluid.
3. Remove disc brake cylinder assembly.
   a. Remove the union bolt and gasket from the disc brake cylinder assembly, then disconnect the flexible hose.
   b. Hold the front disc brake cylinder slide pin and remove the 2 bolts and disc brake cylinder assembly.

➡Remove the disc brake cylinder assembly while holding both of the brake pads or the anti-squeal springs may fall off the brake pads.

To install:

4. Install disc brake cylinder assembly.
   a. Install the disc brake cylinder assembly with the 2 bolts. Tighten to 25 ft. lbs. (34 Nm).
   b. Check the installation of the anti-squeal springs.
   c. Visually check for any clearance between the brake pad and front disc brake pad support plates.

➡If the anti-squeal springs are installed correctly, there will be no clearance between the brake pad and the front disc brake pad support plates. If there is a clearance, the anti-squeal springs may not be installed properly.

➡Check all 4 contact surfaces between the brake pad and the front disc brake pad support plates.

   d. Connect the flexible hose with the union bolt and a new gasket. Tighten to 21 ft. lbs. (29 Nm).

➡Install the flexible hose lock securely in the lock hole in the disc brake cylinder.

5. Fill reservoir with brake fluid.
6. Bleed brake line.
7. Inspect brake fluid level in reservoir.
8. Inspect for brake fluid leak.
9. Install front wheel and tighten to 76 ft. lbs. (103 Nm).

DISC BRAKE PADS

REMOVAL & INSTALLATION

Hybrid

1. Remove front brake caliper.
2. Remove the 2 front brake pads from the front disc brake cylinder.

To install:

To install, reverse the removal procedure.

Non-Hybrid

See Figures 12 and 13.

1. Remove front bake caliper.
2. Remove anti-squeal springs.
3. Remove front anti-squeal shim.
   a. Remove the 2 anti-squeal shims and the 2 pad wear indicator plates from each pad.
4. Remove 4 front disc brake pad support plates.

To install:

5. Install front disc brake pad support plate.
6. Install from anti-squeal shim.

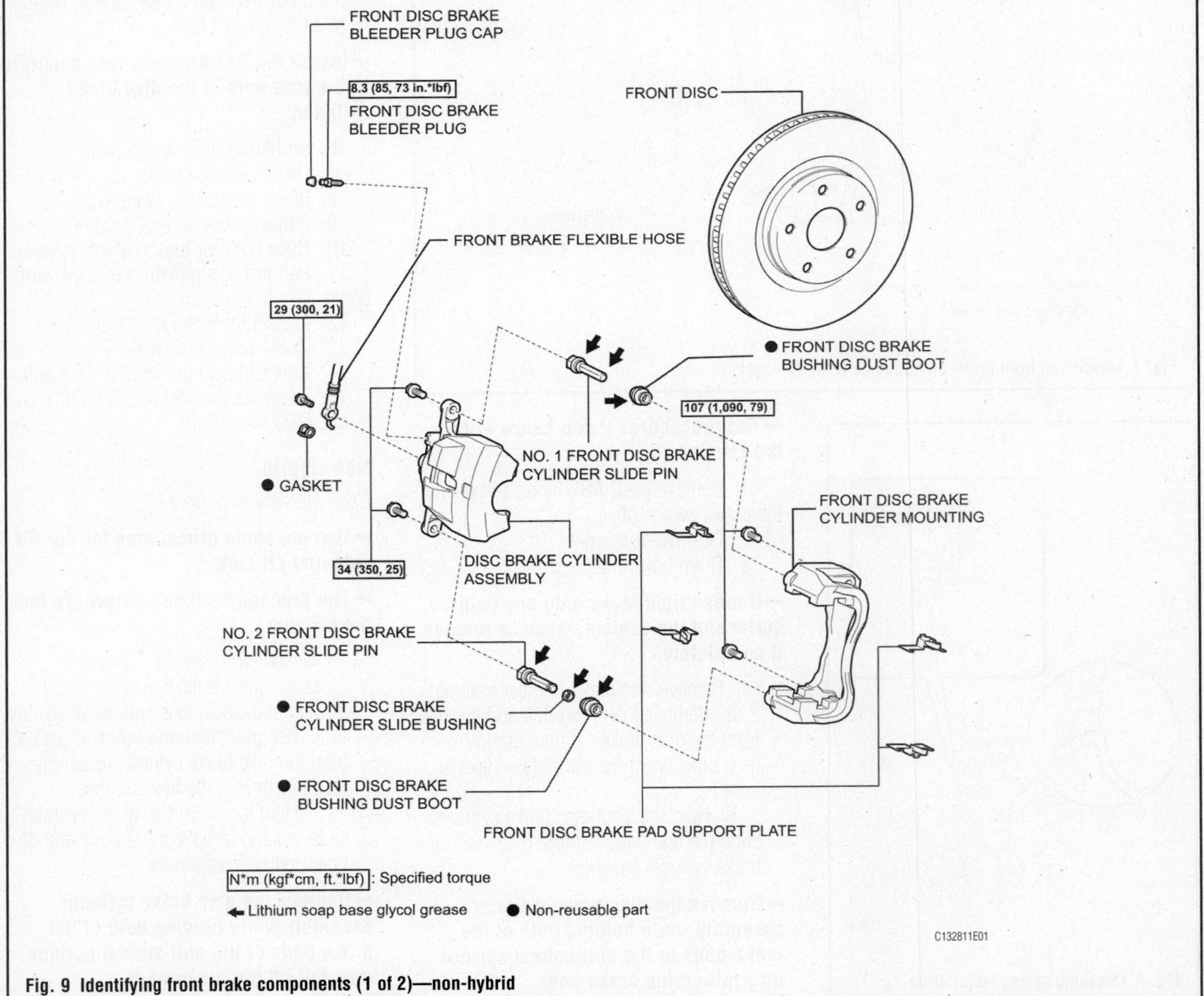

Fig. 9 Identifying front brake components (1 of 2)—non-hybrid

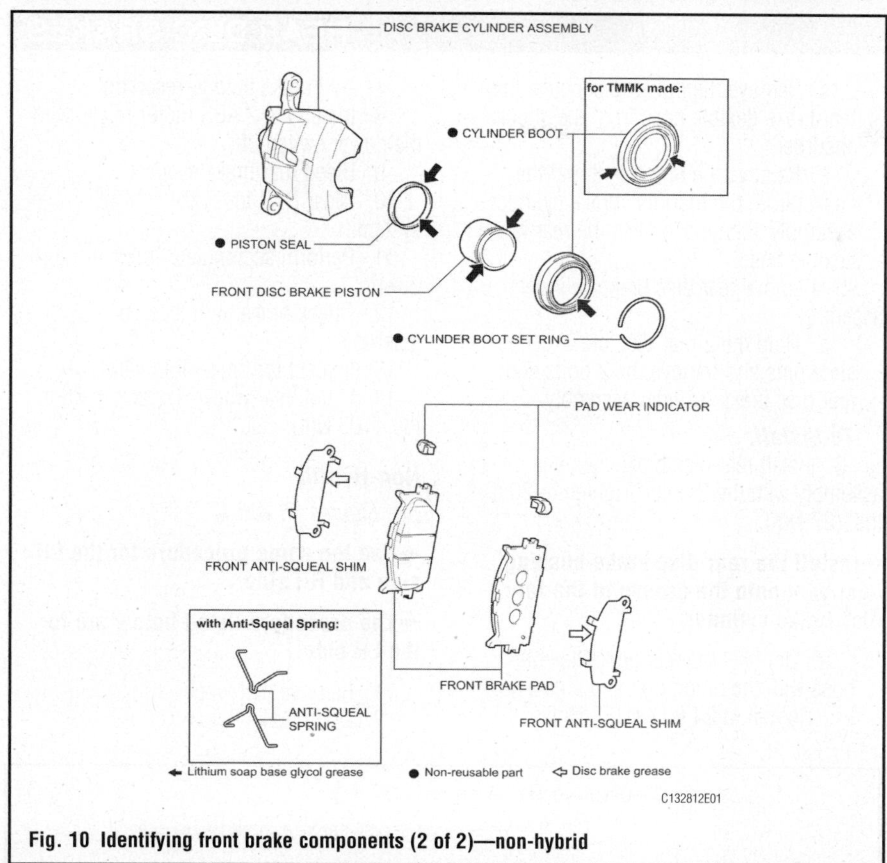

Fig. 10 Identifying front brake components (2 of 2)—non-hybrid

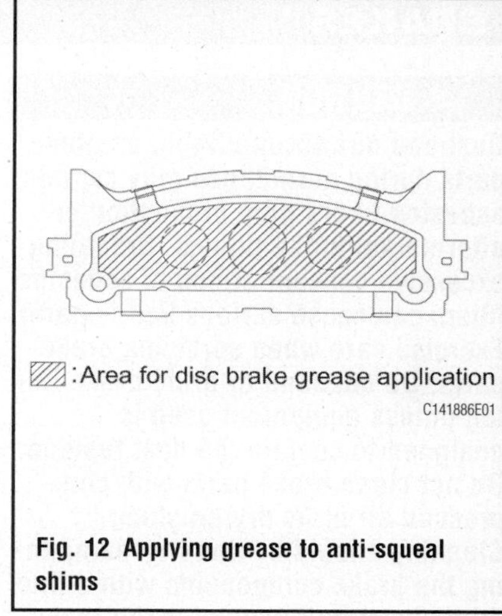

Fig. 12 Applying grease to anti-squeal shims

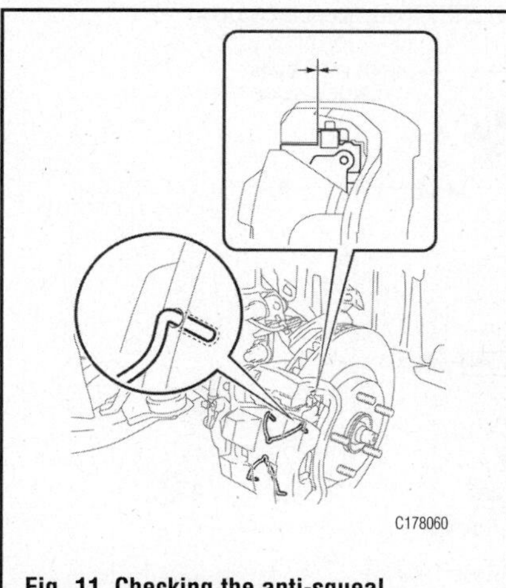

Fig. 11 Checking the anti-squeal spring installation

a. Apply disc brake grease o the anti-squeal shims.

➡When replacing worn pads, the anti-squeal shims must be replaced together with the pads.

➡Install the shims in the correct positions and directions.

➡Apply disc brake grease to the area that contacts the anti-squeal shim.

➡Disc brake grease can come out slightly from the area where the anti-squeal shim is installed.

➡Make sure that disc brake grease is not applied onto the lining surface.

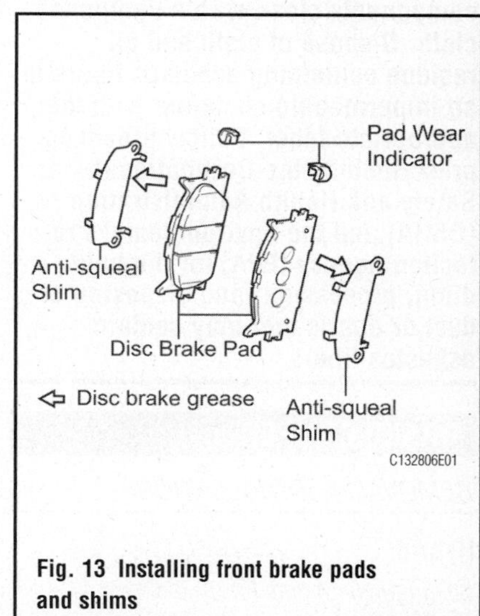

Fig. 13 Installing front brake pads and shims

b. Install the 2 anti-squeal shims and the 2 pad wear indicators to the pads.

➡Install the pad wear indicators in the correct positions and directions.

7. Install 2 brake pads with anti-squeal shims to the disc brake cylinder mounting.
8. Install the brake caliper.

⁂ CAUTION

Dust and dirt accumulating on brake parts during normal use may contain asbestos fibers from production or aftermarket brake linings. Breathing excessive concentrations of asbestos fibers can cause serious bodily harm. Exercise care when servicing brake parts. Do not sand or grind brake lining unless equipment used is designed to contain the dust residue. Do not clean brake parts with compressed air or by dry brushing. Cleaning should be done by dampening the brake components with a fine mist of water, then wiping the brake components clean with a dampened cloth. Dispose of cloth and all residue containing asbestos fibers in an impermeable container with the appropriate label. Follow practices prescribed by the Occupational Safety and Health Administration (OSHA) and the Environmental Protection Agency (EPA) for the handling, processing, and disposing of dust or debris that may contain asbestos fibers.

BRAKE CALIPER

REMOVAL & INSTALLATION

Hybrid

See Figures 14 and 15.

⁂ WARNING

When the brake caliper is removed from the brake disc, do not start the hybrid system. If the hybrid system is started, the brake fluid pressure may increase.

➡Use the same procedures for the LH side and RH side.

➡The procedure listed below are for the LH side.

1. Remove the 2 ABS motor relays with the power switch off.
2. Remove rear wheel.
3. Drain brake fluid.

⁂ WARNING

If brake fluid leaks onto any painted surface, wash or remove it completely.

4. Disconnect rear brake flexible hose.
   a. Remove the bolt and separate the rear brake flexible hose from the shock absorber.
   b. Remove the union bolt and the gasket from the rear disc brake cylinder assembly, then disconnect the rear brake flexible hose.
5. Remove rear disc brake cylinder assembly.
   a. Hold the 2 rear disc brake cylinder slide pins and remove the 2 bolts and rear disc brake cylinder assembly.

To install:

6. Install rear disc brake cylinder assembly with the 2 bolts. Tighten to 20 ft. lbs. (27 Nm).

➡Install the rear disc brake bushing dust boot onto the groove of the rear disc brake cylinder.

   a. Connect the rear brake flexible hose with the union bolt and a new gasket. Tighten to 24 ft. lbs. (33 Nm).
7. Add brake fluid to reservoir.
8. Install the 2 ABS motor relays with the power switch off.
9. Bleed rear brake system.
10. Clear DTC for brake control system.
11. Perform accumulator pressure zero down.
12. Check brake fluid level in reservoir.
13. Inspect for brake fluid leak.
14. Install rear wheel. Tighten to 76 ft. lbs. (103 Nm).

Non-Hybrid

See Figures 16 and 17.

➡Use the same procedure for the LH side and RH side.

➡The procedure listed below are for the LH side.

1. Remove rear wheel.
2. Drain brake fluid.

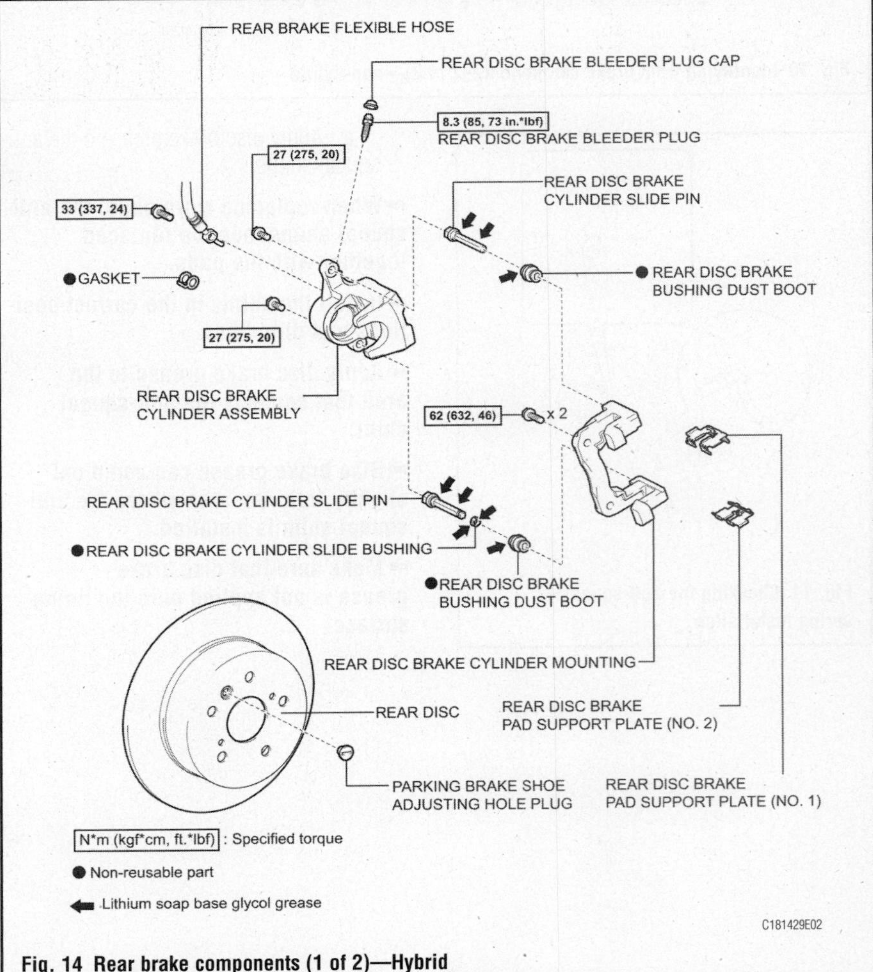

Fig. 14 Rear brake components (1 of 2)—Hybrid

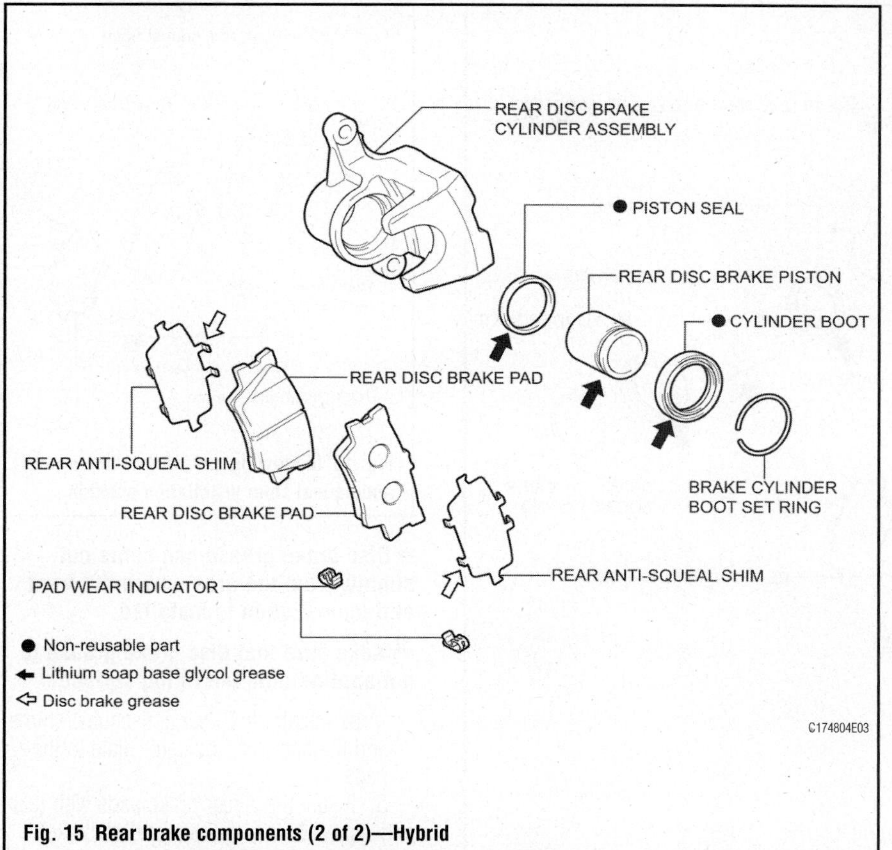

Fig. 15 Rear brake components (2 of 2)—Hybrid

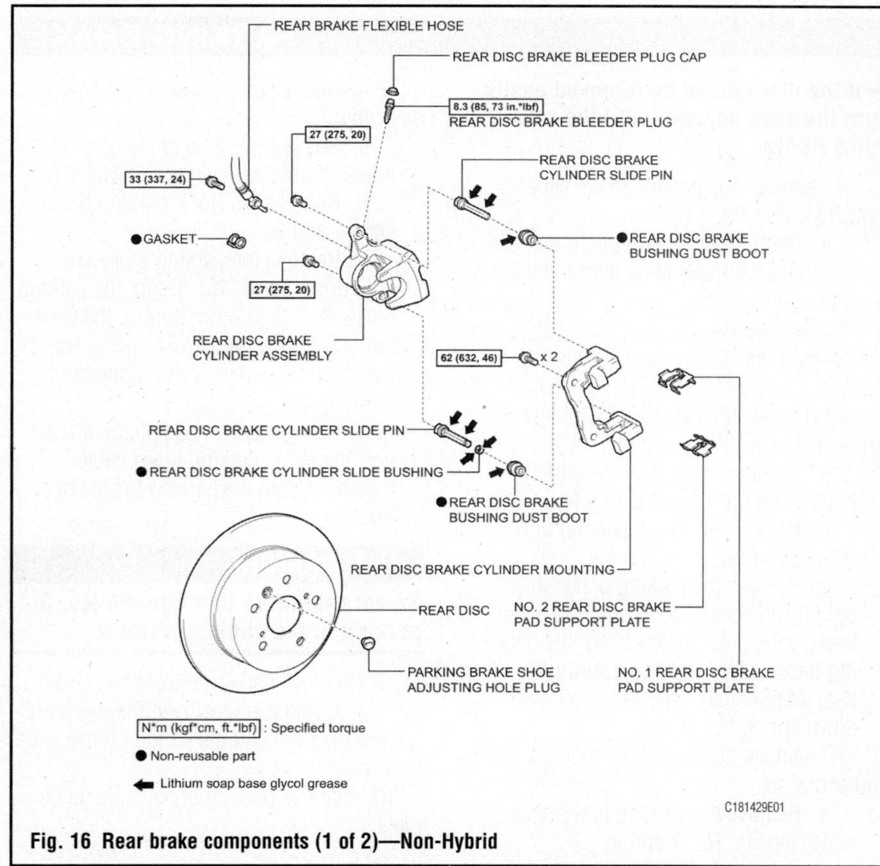

Fig. 16 Rear brake components (1 of 2)—Non-Hybrid

⁂ WARNING

If brake fluid leaks onto any painted surface, wash or remove it completely.

3. Disconnect rear brake flexible hose.
   a. Remove the union bolt and the gasket from the rear disc brake cylinder assembly, then disconnect the flexible hose.
4. Remove rear disc brake cylinder assembly.
   a. Hold the 2 rear disc brake cylinder slide pins and remove the 2 bolts and rear disc brake cylinder assembly.

To install:

5. Install the disc brake cylinder assembly with the 2 bolts. Tighten to 20 ft. lbs. (27 Nm).

➡Install the rear disc brake bushing dust boot onto the groove of the rear disc brake cylinder.

6. Connect the flexible hose with the union bolt and a new gasket. Tighten to 24 ft. lbs. (33 Nm).
7. Fill reservoir with brake fluid.
8. Bleed brake line.
9. Inspect fluid level in reservoir.
10. Inspect for brake fluid leak.
11. Install rear wheel. Tighten to 76 ft. lbs. (103 Nm).

DISC BRAKE PADS

REMOVAL & INSTALLATION

See Figure 18.

1. Remove rear brake caliper.
2. Remove rear disc brake pad.
   a. Remove the 2 rear disc brake pads with the rear anti-squeal shims.
3. Remove rear anti-squeal shim.
   a. Remove the 2 rear anti-squeal shims and the 2 pad wear indicators from each pad.

To install:

4. Install rear anti-squeal shim.
   a. Apply disc brake grease to the inside of the rear anti-squeal shims and install them and the pad wear indicators to each of the 2 rear disc brake pads.

➡When replacing worn pads, the anti-squeal shims must be replaced together with the pads.

➡Apply disc brake grease to the area that contacts the front anti-squeal shim.

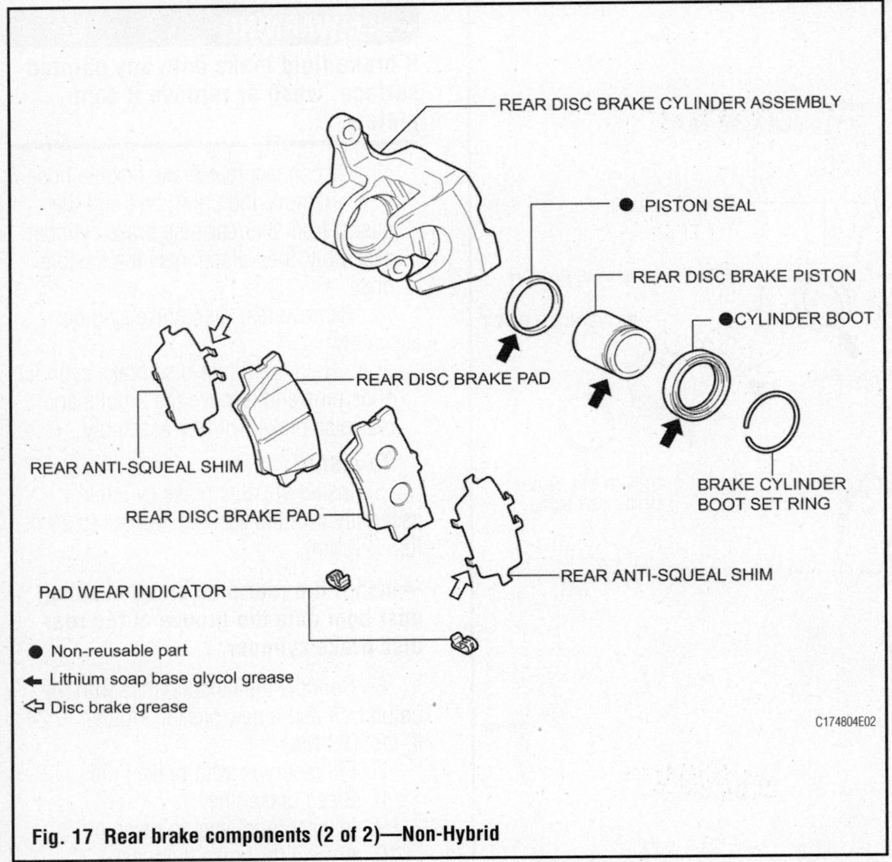

Fig. 17 Rear brake components (2 of 2)—Non-Hybrid

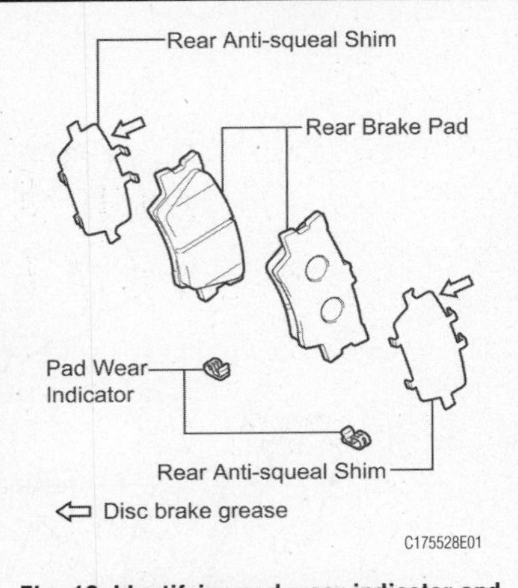

Fig. 18 Identifying pad wear indicator and anti-squeal shim installation position

➡Disc brake grease can come out slightly from the area where the front anti-squeal shim is installed.

➡Make sure that disc brake grease is not applied onto the lining surface.

b. Install the 2 rear anti-squeal shims and the pad wear indicator plate to the pads.

5. Install the 2 rear brake pads with rear anti-squeal shims to the rear disc brake cylinder mounting.

PARKING BRAKE SHOES

REMOVAL & INSTALLATION

Hybrid

See Figures 19 through 23.

➡Use the same procedures for the RH side and the LH side.

➡The procedures listed below are for the LH side.

1. Remove rear wheel.
2. Separate rear disc brake caliper assembly.
   a. Remove the 2 bolts and separate the rear disc brake caliper assembly.

⁂ WARNING

Do not disconnect the flexible hose from the disc brake caliper assembly.

3. Remove the rear disc.
   a. Release the parking brake and put matchmarks on the rear disc and axle hub.
   b. Remove the rear disc.

➡If the disc cannot be removed easily, turn the shoe adjuster until the wheel turns freely.

4. Remove the parking brake shoe return tension No. 1 spring.
   a. Using needle-nose pliers, remove the 2 parking brake shoe return tension No. 1 springs.
5. Remove parking brake shoe strut and the parking brake shoe strut compression spring.
6. Remove the No. 1 parking brake shoe assembly.
   a. Release the claw of the parking brake shoe hold down spring No. 1 cup.
   b. Remove the No. 1 parking brake shoe assembly.
   c. Remove the parking brake shoe hold down spring No. 1 cup, the parking brake shoe hold down spring, the parking brake shoe hold down spring No. 2 cup, and the parking brake shoe hold down spring No. 1 pin.
7. Remove parking brake shoe adjusting screw set.
   a. Remove the parking brake shoe return tension No. 2 spring.
8. Remove No. 2 parking brake shoe assembly.
   a. Release the claw of the parking brake shoe hold down spring No. 1 cup.
   b. Remove the No. 2 parking brake shoe assembly.
   c. Remove the parking brake shoe hold down spring No. 1 cup, the parking brake shoe hold down spring, the parking brake shoe hold down spring No. 2 cup, and the parking brake shoe hold down spring No. 2 pin.
   d. Using needle-nose pliers, disconnect the No. 3 parking brake cable assembly from the parking brake shoe lever.

⁂ WARNING

Be careful not to damage the No. 3 parking brake cable assembly.

9. Remove parking brake shoe lever.
   a. Using a screwdriver, remove the C-washer, shim and the parking brake shoe lever.
10. Remove parking brake shoe guide plate.

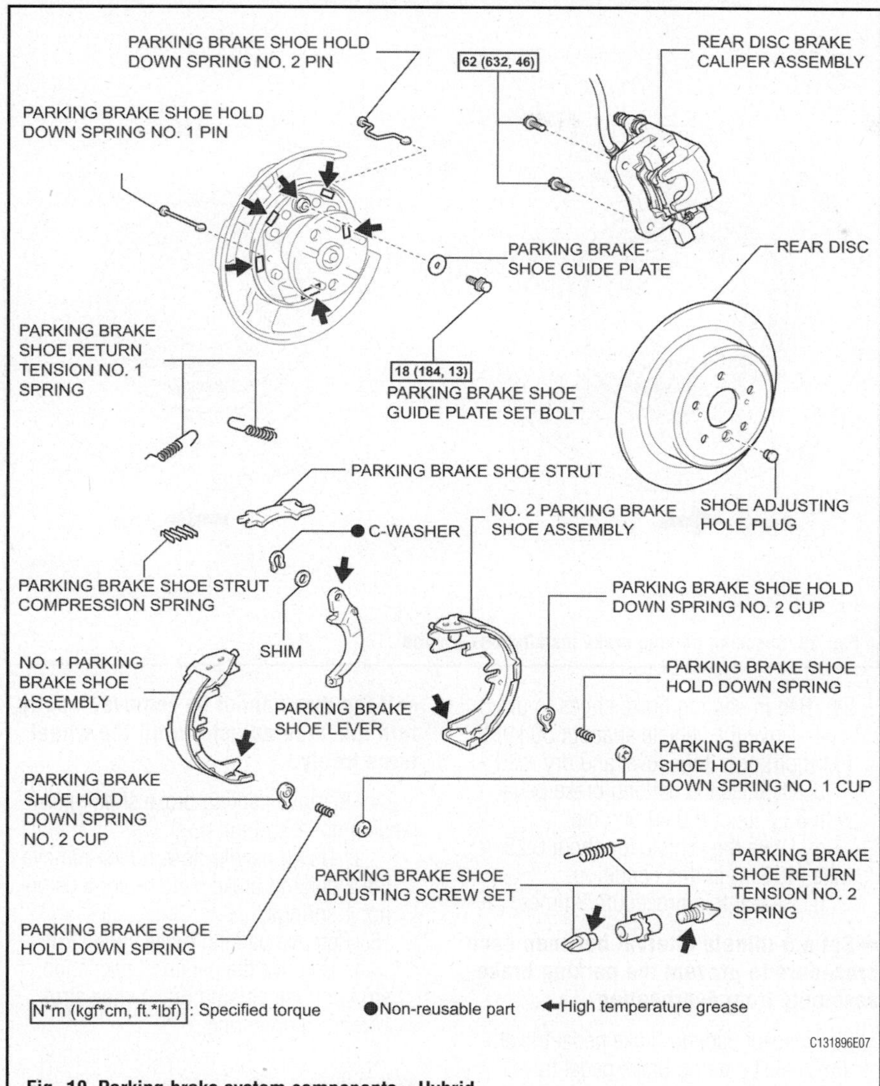

Fig. 19 Parking brake system components—Hybrid

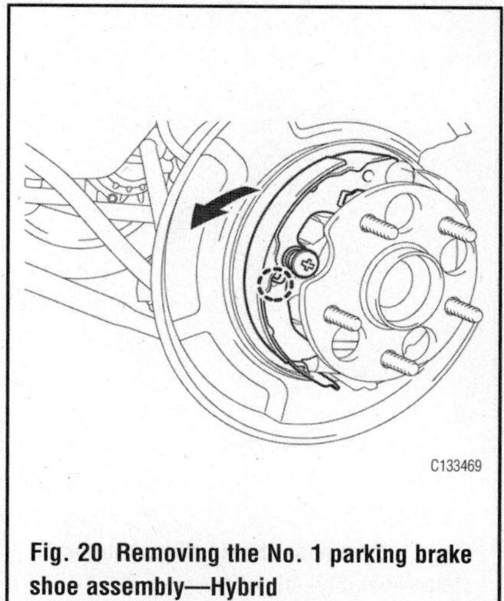

Fig. 20 Removing the No. 1 parking brake shoe assembly—Hybrid

a. Remove the parking brake shoe guide plate set bolt and the parking brake shoe guide plate.

To install:

11. Apply high temperature grease

a. Apply high temperature grease to the backing plate which makes contact with the shoe.

12. Install parking brake shoe guide plate

a. Apply adhesive to the threads of the parking brake shoe guide plate set bolt.

➡Toyota genuine adhesive 1344, three bond 1344 or equivalent

b. Install the parking brake shoe guide plate with the parking brake shoe guide plate set bolt. Tighten to 13 ft. lbs. (18 Nm).

13. Install parking brake shoe lever

a. Install the parking brake shoe lever and shim to the no. 2 parking brake shoe assembly with a new c-washer.

b. Using a feeler gauge, measure the clearance between the No. 2 parking brake shoe assembly and parking brake shoe lever.

➡Standard clearance is less than 0.014 inch (0.35 mm).

➡If the clearance is not as specified, replace the shim with one of the correct size.

14. Install no. 2 parking brake shoe assembly

a. Using needle-nose pliers, connect the no. 3 parking brake cable assembly to the parking brake shoe lever.

b. Install the No. 2 parking brake shoe assembly with the parking brake shoe hold down spring No. 2 pin, the parking brake shoe hold down spring No. 2 cup, the parking brake shoe hold down spring and the parking brake shoe hold down spring No. 1 cup.

c. Engage the claw of the parking brake shoe hold down spring No. 1 cup to the No. 2 parking brake shoe assembly.

15. Install parking brake shoe adjusting screw set

a. Apply high temperature grease to the parking brake adjusting screw.

b. Install the parking brake shoe return tension No. 2 spring to the No. 1 parking brake shoe assembly and the No. 2 parking brake shoe assembly.

c. Install the parking brake shoe adjusting screw set to the No. 1 parking brake shoe assembly and the No. 2 parking brake shoe assembly.

16. Install no. 1 parking brake shoe assembly

a. Install the No. 1 parking brake shoe assembly with the parking brake shoe hold down spring No. 1 pin, parking brake shoe hold down spring No. 2 cup, parking brake shoe hold down spring and parking brake shoe hold down spring No. 1 cup.

b. Engage the claw of the parking brake shoe hold down spring No. 1 cup to the No. 1 parking brake shoe assembly.

17. Install parking brake shoe strut

a. Attach the parking brake shoe strut and the parking brake shoe strut compression spring to the No. 1 parking brake shoe assembly and No. 2 parking brake shoe assembly.

18. Install parking brake shoe return tension no. 1 spring

a. Using needle-nose pliers, install

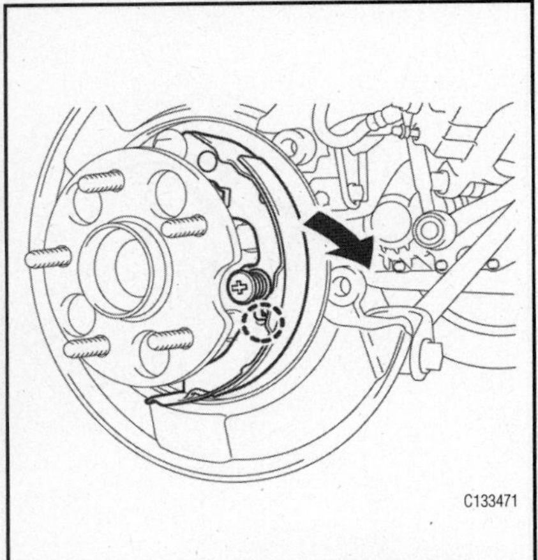

Fig. 21 Removing the No. 2 parking brake shoe assembly—Hybrid

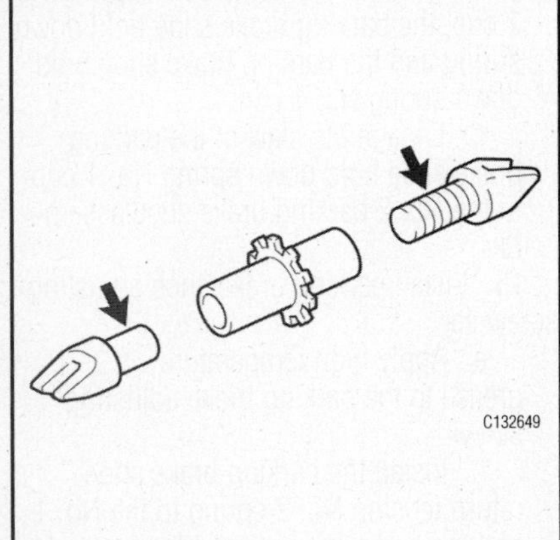

Fig. 22 Installing parking brake shoe adjusting screw set—Hybrid

the 2 parking brake shoe return tension No. 1 springs.

➡First install the front side spring and then the rear side spring.

19. Check that each part is installed properly.

➡There should be no oil or grease on the friction surface of the shoe lining and disc.

20. Install rear disc.
   a. Align the matchmarks and install the rear disc.
21. Adjust the parking brake shoe clearance.
22. Install rear disc brake caliper assembly.
   a. Install the rear brake caliper assembly with the 2 bolts. Tighten to 46 ft. lbs. (62 Nm).
23. Install rear wheel. Tighten to 76 ft. lbs. (103 Nm).

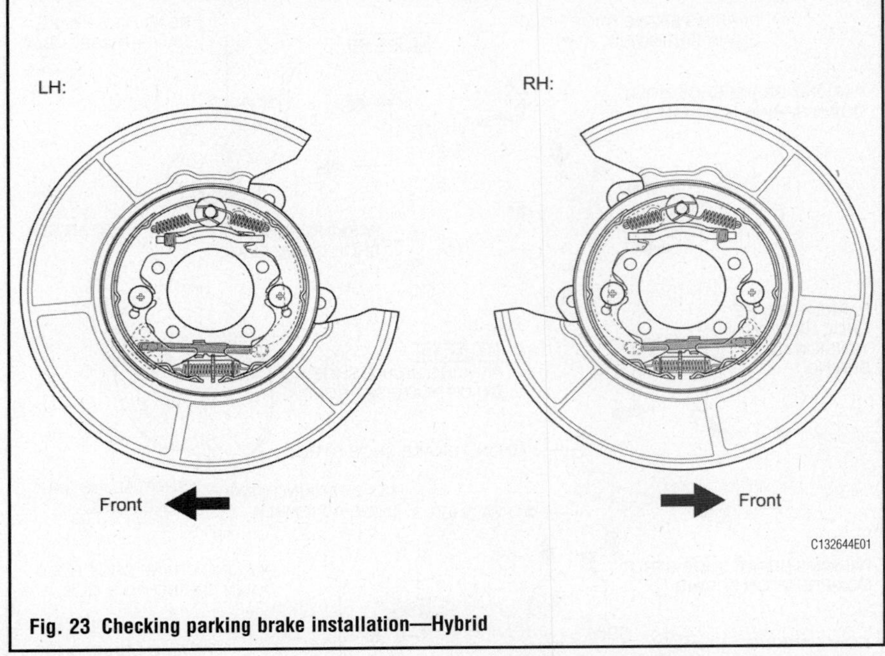

Fig. 23 Checking parking brake installation—Hybrid

24. Bed in parking brake shoes to disc.
   a. Drive the vehicle at about 50 km/h (31 mph) on a safe, level and dry road.
   b. Depress the parking brake pedal with 33.7 lbs. (150 N) of force.
   c. Drive the vehicle for about 0.25 mile (400 m) in this condition.
   d. Repeat this procedure 3 times.

➡Set a 5-minute interval between each procedure to prevent the parking brake assembly from overheating.

25. Inspect parking brake pedal travel.
26. Adjust parking brake pedal travel.

Non-Hybrid

See Figures 24 through 27

➡Use the same procedures for the RH side and the LH side.

➡The procedures listed below are for the LH side.

1. Remove rear wheel
2. Separate rear disc brake caliper assembly
   a. Remove the 2 bolts and separate the rear disc brake caliper assembly.

WARNING

Do not disconnect the flexible hose from the disc brake caliper assembly.

3. Remove rear disc.
   a. Release the parking brake and place the matchmarks on the rear disc and the axle hub.
   b. Remove the rear disc.

➡If the disc cannot be removed easily, turn the shoe adjuster until the wheel turns freely.

4. Remove parking brake shoe return tension no. 1 spring
   a. Using needle-nose pliers, remove the 2 parking brake shoe return tension no. 1 springs.
5. Remove parking brake shoe strut
   a. Remove the parking brake shoe strut and the parking brake shoe strut compression spring.
6. Remove no. 1 parking brake shoe assembly
   a. Release the claw of the parking brake shoe hold down spring no. 1 cup.
   b. Remove the No. 1 parking brake shoe assembly as shown in the illustration.
   c. Remove the parking brake shoe hold down spring No. 1 cup, the parking brake shoe hold down spring, the parking brake shoe hold down spring No. 2 cup, and the parking brake shoe hold down spring No. 1 pin.
7. Remove parking brake shoe adjusting screw set
   a. Remove the parking brake shoe adjusting screw set.
   b. Remove the parking brake shoe return tension No. 2 spring.
8. Remove no. 2 parking brake shoe assembly
   a. Release the claw of the parking brake shoe hold down spring No. 1 cup.
   b. Remove the No. 2 parking brake shoe assembly.

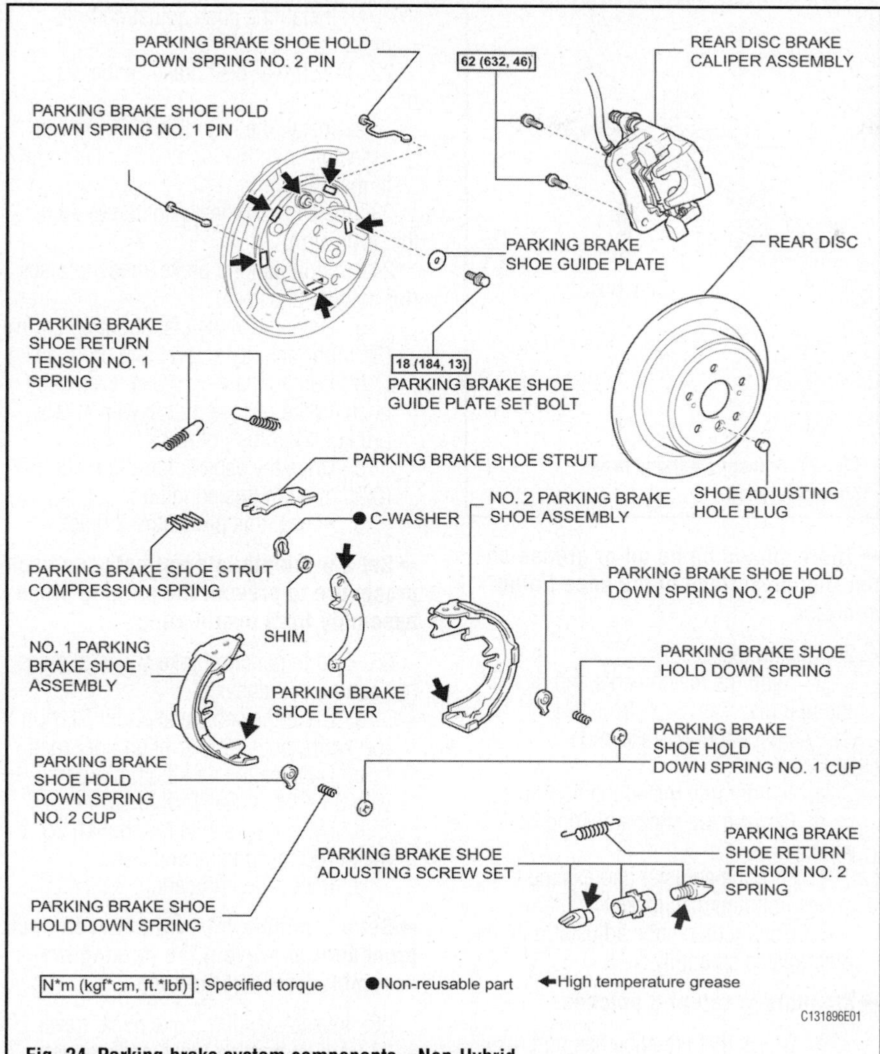

Fig. 24 Parking brake system components—Non-Hybrid

c. Remove the parking brake shoe hold down spring No. 1 cup, the parking brake shoe hold down spring, the parking brake shoe hold down spring No. 2 cup, and the parking brake shoe hold down spring No. 2 pin.

d. Using needle-nose pliers, disconnect the No. 3 parking brake cable assembly from the parking brake shoe lever.

➡**Be careful not to damage the No. 3 parking brake cable assembly.**

9. Remove parking brake shoe lever.

a. Using a screwdriver, remove the C-washer, shim and the parking brake shoe lever.

10. Remove the parking brake shoe guide plate.

a. Remove the parking brake shoe guide plate set bolt and the parking brake shoe guide plate.

To install:

11. Apply high temperature grease to the backing plate which makes contact with the shoe.

12. Install parking brake shoe guide plate.

➡**Apply adhesive to the threads of the parking brake shoe guide plate set bolt.**

➡**Toyota Genuine Adhesive 1344, Three Bond 1344 or equivalent**

a. Install the parking brake shoe guide plate with the parking brake shoe guide plate set bolt. Tighten to 13 ft. lbs. (18 Nm).

13. Install parking brake shoe lever

a. Install the parking brake shoe lever and shim to the No. 2 parking brake shoe assembly with a new C-washer.

b. Using a feeler gauge, measure the clearance between the No. 2 parking

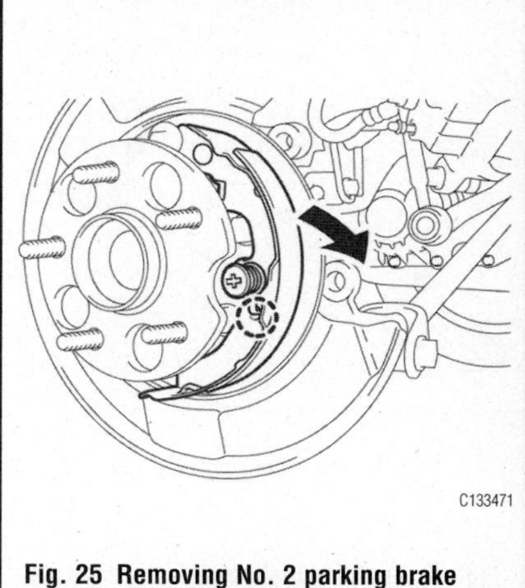

Fig. 25 Removing No. 2 parking brake shoe assembly—Non-Hybrid

brake shoe assembly and parking brake shoe lever.

➡**Standard clearance is less than 0.014 inch (0.35 mm).**

➡**If the clearance is not as specified, replace the shim with one of the correct size.**

14. Install no. 2 parking brake shoe assembly

a. Using needle-nose pliers, connect the No. 3 parking brake cable assembly to the parking brake shoe lever.

b. Install the No. 2 parking brake shoe assembly with the parking brake shoe hold down spring No. 2 pin, the parking brake shoe hold down spring No. 2 cup, the parking brake shoe hold down spring and the parking brake shoe hold down spring No. 1 cup.

c. Engage the claw of the parking brake shoe hold down spring No. 1 cup to the No. 2 parking brake shoe assembly.

15. Install parking brake shoe adjusting screw set

a. Apply high temperature grease to the parking brake adjusting screw.

b. Install the parking brake shoe return tension No. 2 spring to the No. 1 parking brake shoe assembly and the No. 2 parking brake shoe assembly.

c. Install the parking brake shoe adjusting screw set to the No. 1 parking brake shoe assembly and the No. 2 parking brake shoe assembly.

16. Install no. 1 parking brake shoe assembly

a. Install the No. 1 parking brake shoe assembly with the parking brake shoe hold down spring No. 1 pin, parking brake shoe hold down spring No. 2 cup, parking brake shoe hold down

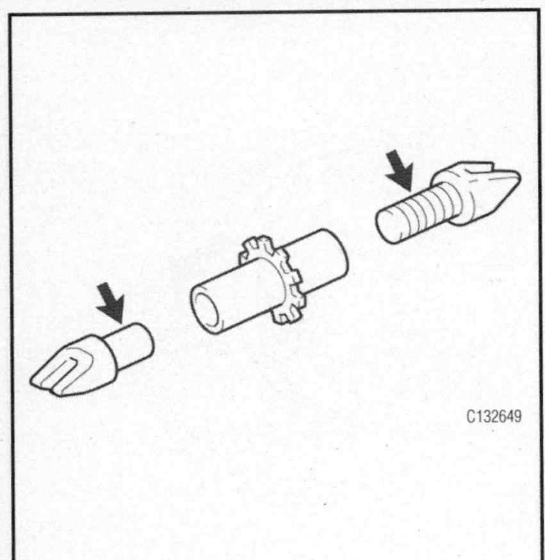

Fig. 26 Applying high temperature grease to the parking brake adjusting screw

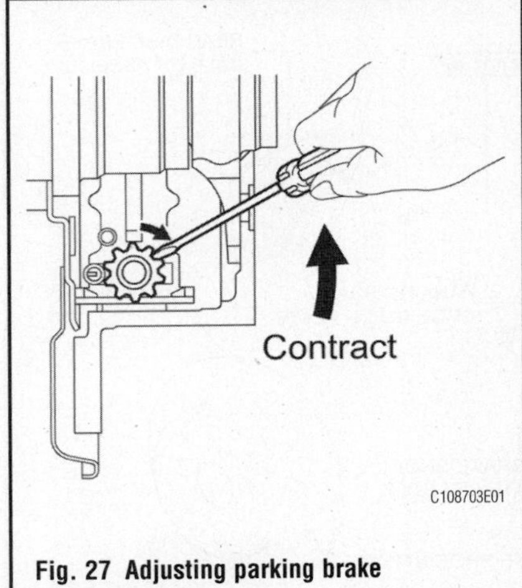

Fig. 27 Adjusting parking brake shoe clearance

spring and parking brake shoe hold down spring No. 1 cup.

17. Install parking brake shoe strut
    a. Attach the parking brake shoe strut and the parking brake shoe strut compression spring to the No. 1 parking brake shoe assembly and No. 2 parking brake shoe assembly.
18. Install parking brake shoe return tension no. 1 spring
    a. Using needle-nose pliers, install the 2 parking brake shoe return tension No. 1 springs.

➡First install the front side spring and then the rear side spring.

19. Check parking brake installation
    a. Check that each part is installed properly.

➡There should be no oil or grease on the friction surface of the shoe lining and disc.

20. Install rear disc
    a. Align the matchmarks and install the rear disc.
21. Adjust parking brake shoe clearance
    a. Temporarily install the hub nuts.
    b. Remove the shoe adjusting hole plug.
    c. Turn the adjuster and expand the shoes until the disc locks.
    d. Contract the shoe adjuster until the disc rotates smoothly.

➡Standard is return 8 notches.

    e. Check that the shoe has no brake drag.
    f. Install the shoe adjusting hole plug.
22. Install rear disc brake caliper assembly
    a. Install the rear disc brake caliper assembly with the 2 bolts. Tighten to 46 ft. lbs. (62 Nm).
23. Install rear wheel. Tighten to 76 ft. lbs. (103 Nm).
24. Bed in parking brake shoes to discs (for manual transaxle).
    a. Drive the vehicle at about 50 km/h (31 mph) on a safe, level and dry road.
    b. Pull the parking brake lever with pushing the release button with 100 N (10 kgf, 22.5 lbs.) of force.
    c. Drive the vehicle for about 400 m (0.25 mile) in this condition.
    d. Repeat this procedure 3 times.

➡Set a 5-minute interval between each procedure to prevent the parking brake assembly from overheating.

25. Bed in parking brake shoes to discs (for automatic transaxle).
    a. Drive the vehicle at about 31 mph (50 km/h) on a safe, level and dry road.
    b. Depress the parking brake pedal with 33.7 lbs. (150 N) of force.
    c. Drive the vehicle for about 0.25 mile (400 m) in this condition.
    d. Repeat this procedure 3 times.

➡Set a 5-minute interval between each procedure to prevent the parking brake assembly from overheating.

26. Inspect parking brake pedal travel.
27. Adjust parking brake pedal travel.

CHASSIS ELECTRICAL

AIR BAG (SUPPLEMENTAL RESTRAINT SYSTEM)

GENERAL INFORMATION

** CAUTION

These vehicles are equipped with an air bag system. The system must be disarmed before performing service on, or around, system components, the steering column, instrument panel components, wiring and sensors. Failure to follow the safety precautions and the disarming procedure could result in accidental air bag deployment, possible injury and unnecessary system repairs.

SERVICE PRECAUTIONS

Disconnect and isolate the battery negative cable before beginning any airbag system component diagnosis, testing, removal, or installation procedures. Allow system capacitor to discharge for two minutes before beginning any component service. This will disable the airbag system. Failure to disable the airbag system may result in accidental airbag deployment, personal injury, or death.

Do not place an intact undeployed airbag face down on a solid surface. The airbag will propel into the air if accidentally deployed and may result in personal injury or death.

When carrying or handling an undeployed airbag, the trim side (face) of the airbag should be pointing away from the body to minimize possibility of injury if accidental deployment occurs. Failure to do this may result in personal injury or death.

Replace airbag system components with OEM replacement parts. Substitute parts may appear interchangeable, but internal differences may result in inferior occupant protection. Failure to do so may result in occupant personal injury or death.

Wear safety glasses, rubber gloves, and long sleeved clothing when cleaning powder residue from vehicle after an airbag deployment. Powder residue emitted from a deployed airbag can cause skin irritation. Flush affected area with cool water if irritation is experienced. If nasal or throat irritation is experienced, exit the vehicle for fresh air until the irritation ceases. If irritation continues, see a physician.

Do not use a replacement airbag that is not in the original packaging. This may result in improper deployment, personal injury, or death.

The factory installed fasteners, screws

and bolts used to fasten airbag components have a special coating and are specifically designed for the airbag system. Do not use substitute fasteners. Use only original equipment fasteners listed in the parts catalog when fastener replacement is required.

During, and following, any child restraint anchor service, due to impact event or vehicle repair, carefully inspect all mounting hardware, tether straps, and anchors for proper installation, operation, or damage. If a child restraint anchor is found damaged in any way, the anchor must be replaced. Failure to do this may result in personal injury or death.

Deployed and non-deployed airbags may or may not have live pyrotechnic material within the airbag inflator.

Do not dispose of driver/passenger/curtain airbags or seat belt tensioners unless you are sure of complete deployment. Refer to the Hazardous Substance Control System for proper disposal.

Dispose of deployed airbags and tensioners consistent with state, provincial, local, and federal regulations.

After any airbag component testing or service, do not connect the battery negative cable. Personal injury or death may result if the system test is not performed first.

If the vehicle is equipped with the Occupant Classification System (OCS), do not connect the battery negative cable before performing the OCS Verification Test using the scan tool and the appropriate diagnostic information. Personal injury or death may result if the system test is not performed properly.

Never replace both the Occupant Restraint Controller (ORC) and the Occupant Classification Module (OCM) at the same time. If both require replacement, replace one, then perform the Airbag System test before replacing the other.

Both the ORC and the OCM store Occupant Classification System (OCS) calibration data, which they transfer to one another when one of them is replaced. If both are replaced at the same time, an irreversible fault will be set in both modules and the OCS may malfunction and cause personal injury or death.

If equipped with OCS, the Seat Weight Sensor is a sensitive, calibrated unit and must be handled carefully. Do not drop or handle roughly. If dropped or damaged, replace with another sensor. Failure to do so may result in occupant injury or death.

If equipped with OCS, the front passenger seat must be handled carefully as well. When removing the seat, be careful when setting on floor not to drop. If dropped, the sensor may be inoperative, could result in occupant injury, or possibly death.

If equipped with OCS, when the passenger front seat is on the floor, no one should sit in the front passenger seat. This uneven force may damage the sensing ability of the seat weight sensors. If sat on and damaged, the sensor may be inoperative, could result in occupant injury, or possibly death.

DISARMING THE SYSTEM

1. Disconnect cable from negative battery terminal.

⁂ WARNING

Wait for 90 seconds after disconnecting the cable to prevent airbag deployment.

ARMING THE SYSTEM

1. Connect cable to negative battery terminal.
2. Perform and necessary initializations.

DRIVE TRAIN

AUTOMATIC TRANSAXLE FLUID

DRAIN AND REFILL

Non-Hybrid

See Figure 28.

⁂ CAUTION

The U660E and U760E automatic transaxle does not have an oil filler tube and oil level gauge. When adding fluid, add fluid through the refill hole on the transaxle case. The fluid level can be adjusted by draining excess fluid (allowing excess fluid to overflow) through the overflow tube of the oil pan.

Hybrid

See Figure 29.

1. Using a 10 mm socket hexagon wrench, remove the filler plug and gasket.
2. Using a 10 mm socket hexagon wrench, remove the drain plug and gasket.
3. Using a 10 mm socket hexagon wrench, install the drain plug and a new gasket. Tighten to 29 ft. lbs. (39 Nm).
4. Add hybrid transaxle fluid. Fill capacity is 4.3 qts (4.1 L).
5. Check that the fluid surface is within 0.20 inch (5 mm) from the inner surface of the differential filler plug opening.

➡Stop the vehicle on a level surface.

➡Recheck the transaxle fluid level after driving following fluid replacement.

➡Insufficient or excessive amounts of transaxle fluid may damage the hybrid transaxle.

6. Check the oil level.
7. Using a 10 mm socket hexagon wrench, install the filler plug and a new gasket. Tighten to 29 ft. lbs. (39 Nm).
8. Using a 10 mm socket hexagon wrench, remove the filler plug and gasket.
9. Check that the fluid surface is within 5 mm (0.20 in.) from the inner surface of the differential filler plug opening.

➡Stop the vehicle on a level surface.

➡Recheck the transaxle fluid level after driving following fluid replacement.

➡Insufficient or excessive amounts of transaxle fluid may damage the hybrid transaxle.

10. Check for leaks if the fluid level is low.
11. Using a 10 mm socket hexagon wrench, install the filler plug and a new gasket. Tighten to 29 ft. lbs. (39 Nm).

MANUAL TRANSAXLE FLUID

DRAIN AND REFILL

See Figure 30.

1. Remove the filler plug and gasket.
   a. Remove the drain plug and gasket to drain the manual transaxle oil.
2. Install the drain plug with a new gasket. Tighten to 29 ft. lbs. (39 Nm).
3. Add oil until the oil level is within 5 mm (0.197 in.) from the bottom of the transmission filler plug opening.
4. Install the transmission filler plug

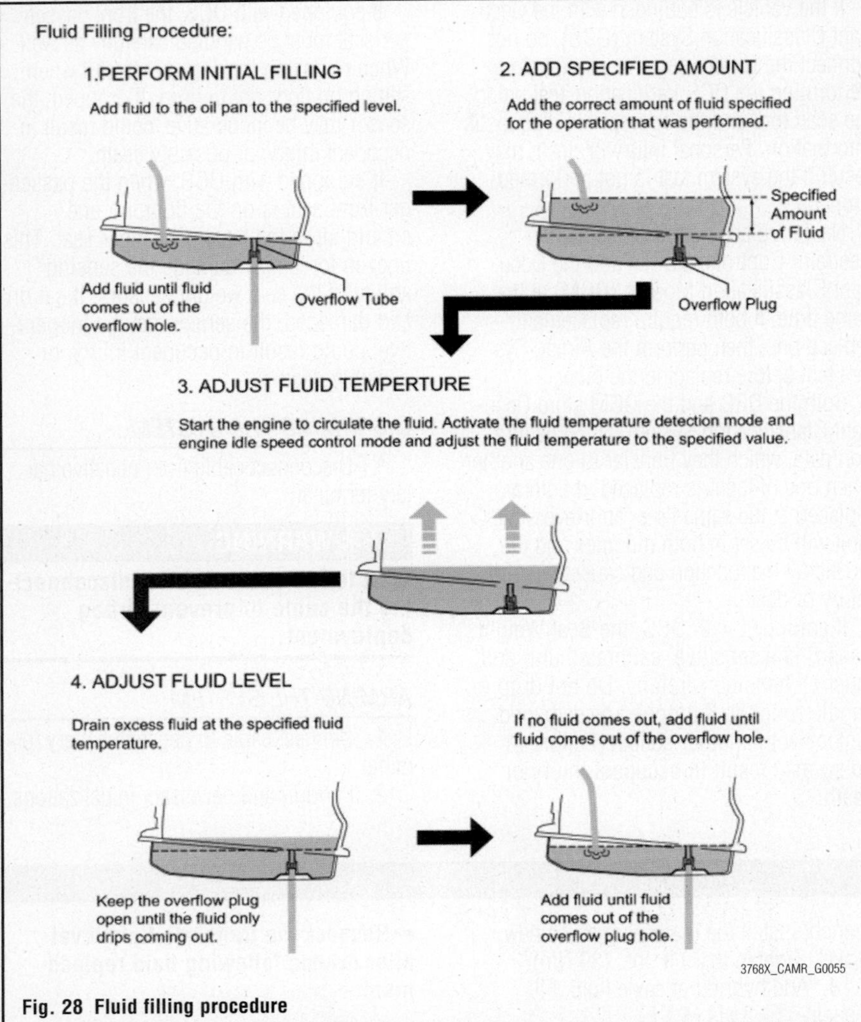

Fluid Filling Procedure:
1.PERFORM INITIAL FILLING
Add fluid to the oil pan to the specified level.
2. ADD SPECIFIED AMOUNT
Add the correct amount of fluid specified for the operation that was performed.
Specified Amount of Fluid
Add fluid until fluid comes out of the overflow hole.
Overflow Tube
Overflow Plug
3. ADJUST FLUID TEMPERTURE
Start the engine to circulate the fluid. Activate the fluid temperature detection mode and engine idle speed control mode and adjust the fluid temperature to the specified value.
4. ADJUST FLUID LEVEL
Drain excess fluid at the specified fluid temperature.
If no fluid comes out, add fluid until fluid comes out of the overflow hole.
Keep the overflow plug open until the fluid only drips coming out.
Add fluid until fluid comes out of the overflow plug hole.
3768X_CAMR_G0055

Fig. 28 Fluid filling procedure

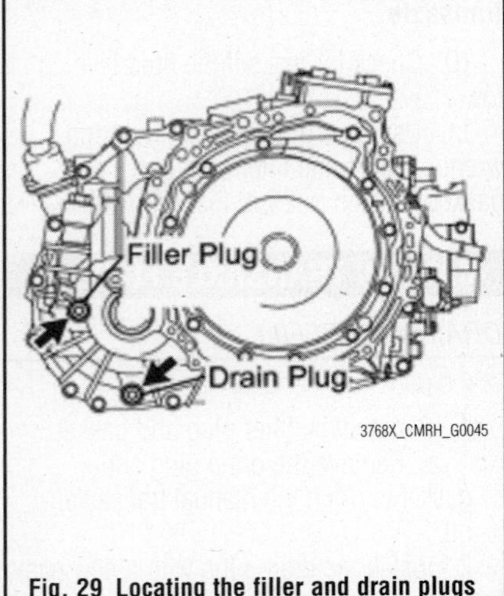

Fig. 29 Locating the filler and drain plugs

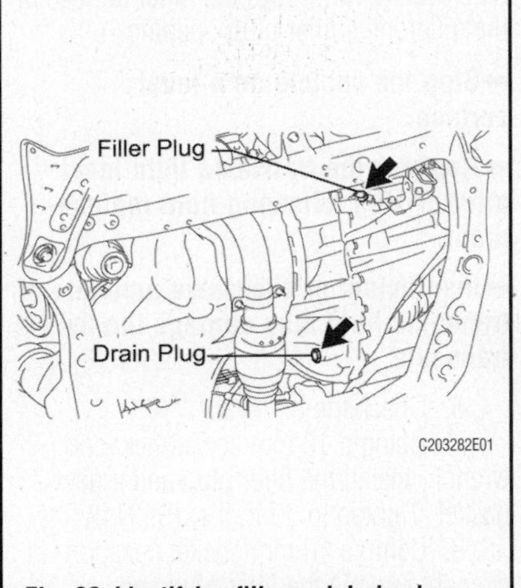

Fig. 30 Identifying filler and drain plug

with a new gasket. Tighten to 29 ft. lbs. (39 Nm).

➡When adding transaxle oil, make sure that the vehicle is level.

➡An excessively large or small amount of oil may cause problems.

➡After adding oil, drive the vehicle and recheck the oil level.

5. Inspect for oil leaks.

FRONT DRIVESHAFT

REMOVAL & INSTALLATION

Hybrid

See Figure 31.

➡Use the same procedure for the RH side and LH side.

➡The procedure listed below is for the LH side.

1. Remove front wheels
2. Remove front wheel opening extension pad LH.
3. Remove engine under cover LH.
4. Remove front wheel opening extension pad RH.
5. Remove engine under cover RH.
6. Remove front fender apron seal LH.
7. Remove front fender apron seal RH.
8. Drain hybrid transaxle fluid.
9. Remove front axle shaft nut.
   a. Using SST (09930-00010) and a hammer, release the staked part of the front axle shaft nut.

⁂ WARNING

Loosen the staked part of the nut completely, otherwise the threads of the drive shaft may be damaged.

   b. While applying the brakes, remove the front axle shaft nut.
10. Separate front stabilizer link assembly.
11. Separate front speed sensor.
12. Separate tie rod assembly.
13. Separate front lower No. 1 suspension arm sub assembly.
14. Separate front drive shaft assembly.
15. Remove front drive shaft assembly LH.
   a. Using SST (09520-00031, 09520-01010), remove the front drive shaft assembly LH.

⁂ WARNING

Do not damage the hybrid vehicle transaxle assembly type T oil seal.

⁂ WARNING

Do not damage the front axle inboard joint boot.

⁂ WARNING

Do not drop the front drive shaft assembly LH.

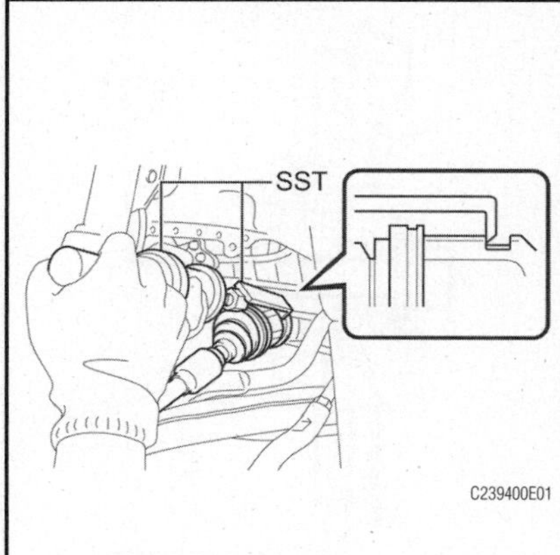

Fig. 31 Removing front drive shaft assembly LH

16. Remove front drive shaft assembly RH.
   a. Separate the bearing bracket hole snap ring from the drive shaft bearing bracket.
   b. Remove the bolt and front drive shaft assembly RH from the drive shaft bearing bracket.

WARNING

Do not damage the hybrid vehicle transaxle assembly type T oil seal.

WARNING

Do not damage the front axle inboard joint boot.

WARNING

Do not drop the front drive shaft assembly RH.

   c. Remove the bearing bracket hole snap ring from the front drive shaft assembly RH.
17. Remove front drive shaft hole snap ring LH.
   a. Using a screwdriver, remove the front drive shaft hole snap ring LH.

To install:

18. Install front drive shaft assembly LH.
   a. Coat the spline of the front drive inboard joint assembly with ATF.
   b. Coat the lip of the hybrid vehicle transaxle assembly type T oil seal with MP grease.
   c. Align the inboard joint splines, and using a brass bar and a hammer, install the front drive shaft assembly LH.

WARNING

Face the end gap of the front drive shaft hole snap ring LH downward.

WARNING

Do not damage the hybrid vehicle transaxle assembly type T oil seal.

WARNING

Do not damage the front axle inboard joint boot.

WARNING

Make sure to center the front drive shaft assembly LH during installation to prevent damage to the front drive shaft hole snap ring LH.

➡Confirm whether the drive shaft is securely driven in by checking the reaction force and sound.

   d. Apply mineral oil base grease to the areas on the front drive shaft assembly LH contact surface of the front axle hub bearing.

➡Apply 0.1 to 0.3 g (0.00353 to 0.0106 oz.) of mineral oil base grease to each area.

   e. Align the matchmarks and install the front drive shaft assembly LH to the front axle hub sub-assembly.

WARNING

Be careful not to damage the front axle outboard joint boot or speed sensor rotor.

19. Install front drive shaft assembly RH.
   a. Coat the spline of the front drive inboard joint assembly with ATF.
   b. Coat the lip of the hybrid vehicle transaxle assembly type T oil seal with MP grease.
   c. Install a new bearing bracket hole snap ring to the front drive shaft assembly RH.
   d. Install the front drive shaft assembly RH.

WARNING

Do not damage the hybrid vehicle transaxle assembly type T oil seal.

WARNING

Do not damage the front axle inboard joint boot.

WARNING

When inserting the front drive shaft assembly RH, keep it level.

   e. Install the bearing bracket hole snap ring and a new bolt. Tighten to 24 ft. lbs. (32 Nm).
   f. Apply mineral oil base grease to the front drive shaft assembly RH contact surface of the front axle hub bearing.

➡Apply 0.1 to 0.3 g (0.00353 to 0.0106 oz.) of mineral oil base grease to each area.

   g. Align the matchmarks and install the front drive shaft assembly RH to the front axle hub sub-assembly.

WARNING

Be careful not to damage the front axle outboard joint boot or speed sensor rotor.

20. Connect front lower No. 1 suspension arm sub assembly.
21. Connect tie rod assembly.
22. Install front speed sensor.
23. Install front stabilizer link assembly.
24. Install front axle shaft nut.
   a. Clean the threaded parts on the front drive shaft assembly and a new axle shaft nut using a non-residue solvent.

WARNING

Be sure to perform this work even when using a new drive shaft.

WARNING

Keep the threaded parts free of oil and foreign matter.

   b. Using a socket wrench (30 mm), install the front axle shaft nut. Tighten to 217 ft. lbs. (294 Nm).
   c. Using a chisel and a hammer, stake the front axle shaft nut.
25. Install front wheels. Tighten to 76 ft. lbs. (103 Nm).
26. Add hybrid transaxle fluid.
27. Inspect hybrid transaxle fluid.
28. Install front fender apron seal LH.
29. Install front fender apron seal RH.
30. Install engine under cover LH.
31. Install front wheel opening extension pad LH.
32. Install engine under cover RH.
33. Install front wheel opening extension pad RH.
34. Adjust front wheel alignment.
35. Check for speed sensor signal.

TMC Made

See Figures 32 through 34.

➡Use the same procedures for the RH side and LH side.

➡The procedures listed below are for the LH side.

1. Drain the automatic transaxle fluid.
2. Remove the front wheel.
3. Remove the front axle hub nut.
   a. Using the special tool and a hammer, release the staked part of the front axle hub nut.

➡Loosen the staked part of the nut completely, otherwise the thread of the drive shaft may be damaged.

   b. While applying the brakes, remove the front axle hub nut.
4. Separate the front stabilizer link assembly.
   a. Remove the nut and separate the front stabilizer link assembly.
5. Separate the front speed sensor.
   a. Remove the bolt and clip, and separate the speed sensor wire and flexible hose from the shock absorber.
   b. Remove the bolt and separate the speed sensor from the steering knuckle.

➡Prevent foreign matter from adhering to the speed sensor.

➡Be careful not to damage the speed sensor.

6. Separate the tie rod end sub assembly.
   a. Remove the cotter pin and nut.
   b. Using the special tool, separate the tie rod end sub assembly from the steering knuckle.

➡Make sure that the string of the SST is securely tied to the vehicle.

➡Be careful not to damage the ball joint dust cover.

➡Be careful not to damage the steering knuckle.

➡Be careful not to damage the front disc brake dust cover.

7. Separate the front suspension lower No. 1 arm.
   a. Remove the bolt and 2 nuts, and separate the front suspension lower No. 1 arm from the lower ball joint.
8. Separating the front axle assembly.
   a. Put matchmarks on the front drive shaft assembly and the axle hub.
   b. Using a plastic hammer, separate the front drive shaft assembly from the front axle assembly.

➡Be careful not to damage the drive shaft boot and speed sensor rotor.

9. Remove the front drive shaft assembly LH.
   a. Using the special tool, remove the front drive shaft assembly LH.

➡Be careful not to damage the drive shaft dust cover, boot or seal.

➡Be careful not to drop the drive shaft assembly.

10. Remove the front drive shaft assembly RH.
   a. Using a screwdriver, remove the bearing bracket hole snap ring.
   b. Remove the bolt and the front drive shaft assembly RH from the drive shaft bearing bracket.

➡Do not damage the boot or oil seal.

11. Secure the front axle hub sub assembly.
   a. Secure the front axle hub bearing.

➡The hub bearing may be damaged if it is subjected to the vehicle's full weight, such as moving the vehicle with the drive shaft removed. If it is necessary to place the vehicle's weight on the hub bearing, first support it with SST.

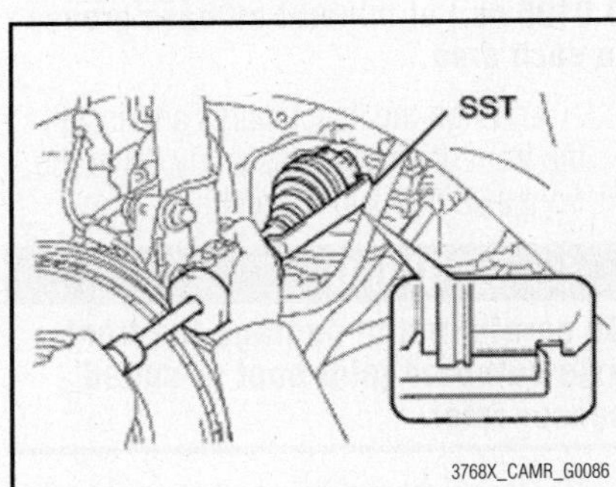

Fig. 32 Removing the front drive shaft assembly LH

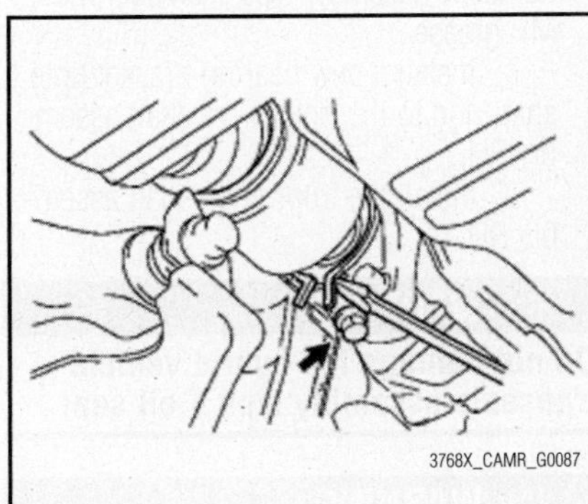

Fig. 33 Removing the front drive shaft assembly RH

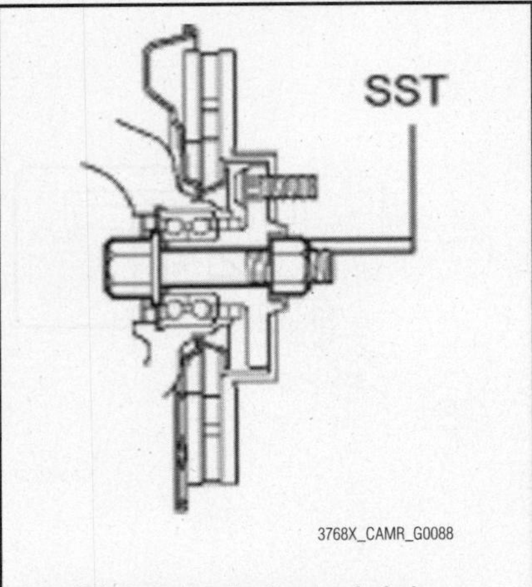

Fig. 34 Securing the front axle hub sub assembly

To install:

12. Install the front drive shaft assembly LH.
   a. Coat the spline of the inboard joint shaft assembly with ATF.
   b. Align the shaft splines and install the drive shaft assembly LH with a brass bar and a hammer.

➡Set the shaft snap ring with the opening side facing down.

➡Be careful not to damage the drive shaft dust cover, boot, or oil seal.

➡Move the drive shaft assembly while keeping it level.

13. Install the front drive shaft assembly RH.
   a. Coat the spline of the inboard joint shaft assembly with ATF.
   b. Install the front drive shaft assembly RH.
   c. Using a screwdriver, install a new bearing bracket hole snap ring.

➡Do not damage the boot and oil seal.

➡Move the drive shaft assembly while keeping it level.

   d. Install a new bolt and tighten to 24 ft. lbs. (32 Nm).
14. Install the front axle assembly.
   a. Align the matchmarks and install the front drive shaft assembly to the front axle hub sub assembly.

➡Be careful not to damage the drive shaft boot or speed sensor rotor.

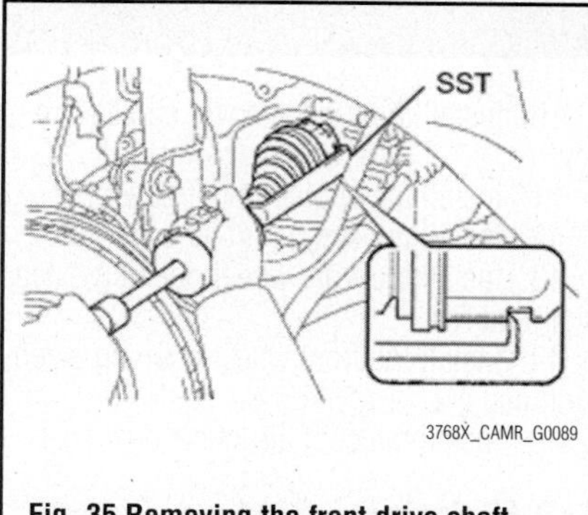

Fig. 35 Removing the front drive shaft assembly LH

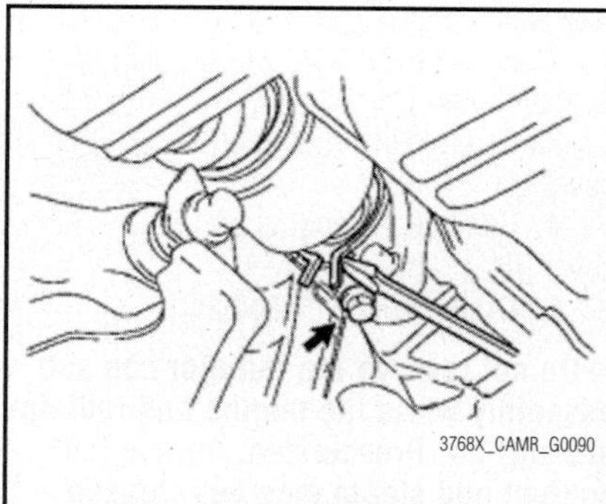

Fig. 36 Removing the front drive shaft assembly RH

15. Install the front suspension lower No. 1 arm.
    a. Install the lower ball joint to the front suspension lower No. 1 arm with the bolt and 2 nuts. Tighten to 55 ft. lbs. (75 Nm).
16. Install the tie rod end sub assembly.
    a. Install the tie rod end sub assembly to the steering knuckle with the nut. Tighten to 36 ft. lbs. (49 Nm).
    b. Install a new cotter pin.

➡If the holes for the cotter pin are not aligned, tighten the nut up to 60°further.

17. Install the front speed sensor.
    a. Install the front speed sensor to the steering knuckle with the bolt. Tighten to 71 inch lbs. (8 Nm).

➡Prevent foreign matter from adhering to the speed sensor.

➡Be careful not to damage the speed sensor.

    b. Install the flexible hose and the speed sensor to the shock absorber with the bolt and the set sensor clip on the knuckle. Tighten to 14 ft. lbs. (19 Nm).

➡Be careful not to damage the speed sensor.

➡Prevent foreign matter from adhering to the speed sensor.

➡Do not twist the sensor wire when installing the speed sensor.

18. Install the front stabilizer link assembly.
    a. Install the stabilizer link assembly with the nut. Tighten to 55 ft. lbs. (74 Nm).
19. Install the front axle hub nut.
    a. Clean the threaded parts on the drive shaft and axle hub nut using a non-residue solvent.

➡Be sure to perform this work for a new drive shaft.

➡Keep the threaded parts free of oil and foreign matter.

    b. Using a socket wrench (30 mm), install a new axle hub. Tighten to 217 ft. lbs. (294 Nm).
    c. Using a chisel and hammer, stake the front axle hub nut.
20. Install the front wheel. Tighten to 76 ft. lbs. (103 Nm).
21. Add automatic transaxle fluid.
22. Inspect the automatic transaxle fluid.
23. Adjust the front wheel alignment.
24. Check the ABS speed sensor signal.

TMMK Made

See Figures 35 and 36.

➡Use the same procedures for the RH side and the LH side.

➡The procedures listed below are for the LH side.

1. Drain the automatic transaxle fluid.
2. Drain the manual transaxle oil (MT).
3. Remove the front wheel.
4. Remove the front axle hub nut.
5. Separate the front stabilizer link assembly.
6. Separate the front speed sensor.
7. Separate the front suspension lower No. 1 arm.
8. Separate the front axle assembly.
9. Remove the front drive shaft assembly LH.

➡Be careful not to damage the drive shaft dust cover, boot and oil seal.

➡Be careful not to drop the drive shaft assembly.

10. Remove the front drive shaft assembly RH.
    a. Using a screwdriver, remove the bearing bracket hole snap ring.
    b. Remove the bolt and front drive shaft assembly RH front the drive shaft bearing bracket.

➡Do not damage the boot and oil seal.

11. Secure the front axle hub sub assembly.
12. Inspect the front drive shaft assembly.

To install:

13. Install the front drive shaft assembly LH.
    a. Coat the spline of the inboard joint shaft assembly with ATF.
    b. Align the shaft splines and install the drive shaft assembly LH with a brass bar and hammer.

➡Set the shaft snap ring with the opening side facing down.

➡Be careful not to damage the drive shaft dust cover, boot, and oil seal.

➡Move the drive shaft assembly while keeping it level.

14. Install the front drive shaft assembly RH.
    a. Coat the spline of the inboard joint shaft assembly with ATF.
    b. Install the front drive shaft assembly RH.
    c. Using a screwdriver, install a new bearing bracket hole snap ring.

➡Do not damage the boot and oil seal.

➡Move the drive shaft assembly while keeping it level.

    d. Install a new bolt. Tighten to 24 ft. lbs. (32 Nm).
15. Install the front axle assembly.
16. Install the front suspension lower No. 1 arm.
17. Install the tie rod end sub assembly.
18. Install the front speed sensor.
19. Install the front stabilizer link assembly.
20. Install the front axle hub nut.
21. Install the front wheel. Tighten to 76 ft. lbs. (103 Nm).
22. Add automatic transaxle fluid (AT).
23. Add manual transaxle oil (MT).
24. Inspect the automatic transaxle fluid.
25. Inspect the manual transaxle oil.
26. Adjust the front wheel alignment.
27. Check the ABS speed sensor signal.

ENGINE COOLING

ENGINE COOLANT

DRAIN & REFILL PROCEDURE

2GR-FE Engine

See Figure 37.

1. Remove the front wheel opening extension pad RH.
2. Remove the front wheel opening extension pad LH.
3. Remove the engine under cover RH.
4. Remove the engine under cover LH.
5. Remove the v-bank cover sub assembly.
6. Remove the cool air intake duct seal.
7. Drain the engine coolant.

➡Do not remove the radiator cap sub assembly while the engine and radiator are still hot. Pressurized, hot engine coolant and steam may be released and cause serious burns.

a. Remove the radiator cap sub assembly from the radiator assembly.
b. Loosen the radiator drain cock plug and 2 cylinder block drain cock plugs, then drain the coolant.

➡Collect the coolant in a container and dispose of it according to the regulations in your area.

8. Add engine coolant.
a. Close the radiator drain cock plug and 2 cylinder block drain cock plugs. Tighten to 9 ft. lbs. (13 Nm) for the cylinder drain cock plug.
b. Slowly fill the radiator with TOYOTA SLLC. The specified capacity is 9.5 qts. (9 L).

➡Toyota vehicle are filled with TOYOTA SLLC at the factory. In order to avoid damage to the engine cooling system and other technical problems, only use TOYOTA SLLC or similar high quality ethylene glycol based non-silicate, non-amine, non-nitrite, non-borate coolant with long life hybrid organic acid technology (coolant with long life hybrid organic acid technology consists of a combination of low phosphates and organic acids).

➡Contact your TOYOTA dealer for further details.

c. Slowly pour coolant into the radiator reservoir tank until it reaches the FULL line.
d. Press the inlet and outlet radiator hoses several times by hand, and then check the level of the coolant. If the coolant is low, add coolant.
e. Install the radiator cap sub assembly and reservoir tank cap.
f. Start the engine, and warm it up.

➡Adjust the air conditioner set temperature to MAX (hot).

g. Stop the engine, and wait until the engine coolant cools down.
h. Add engine coolant to the FULL line on the radiator reservoir.
9. Inspect for coolant leak.

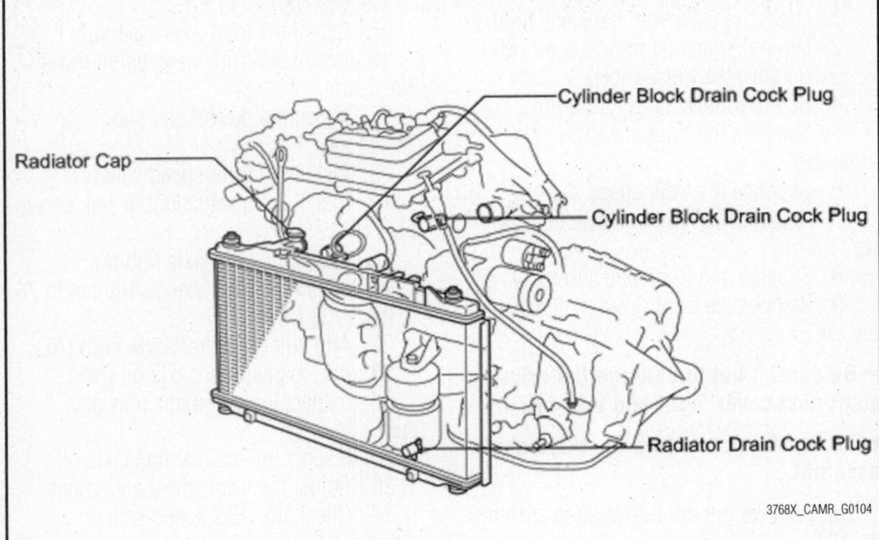

Fig. 37 Draining engine coolant (2GR-FE)

10. Install the v-bank cover sub assembly.
11. Install the engine under cover RH.
12. Install the engine under cover LH.
13. Install the front wheel opening extension pad RH.
14. Install the front wheel opening extension pad LH.
15. Install the cool air intake duct seal.

2AR-FE Engine

See Figure 38.

1. Remove the front wheel opening extension pad RH.
2. Remove the front wheel opening extension pad LH.
3. Remove the engine under cover LH.
4. Remove the engine under cover RH.
5. Drain the engine coolant.

➡Do not remove the radiator cap sub assembly while the engine and radiator are still ho. Pressurized, hot engine coolant and steam may be released and cause serious burns.

a. Remove the radiator cap sub assembly from the radiator assembly.
b. Loosen the radiator drain cock plug and cylinder block drain cock plug, then drain the coolant.

➡Collect the coolant in a container and dispose of it according to the regulations in your area.

6. Add engine coolant.
a. Tighten the radiator drain cock plug by hand.
b. Tighten the cylinder block drain cock plug to 9 ft. lbs. (13 Nm).
c. Slowly fill the radiator with TOYOTA Super Long Life Coolant (SLLC). Add 7.5 qts. (7.1 L).

➡Never use water as a substitute for engine coolant.

➡Toyota vehicle are filled with TOYOTA SLLC at the factory. In order to avoid damage to the engine cooling system and other technical problems, only use TOYOTA SLLC or similar high quality ethylene glycol based non-silicate, non-amine, non-nitrite, non-borate coolant with long life hybrid organic acid technology (coolant with long life hybrid organic acid technology consists of a combination of low phosphates and organic acids).

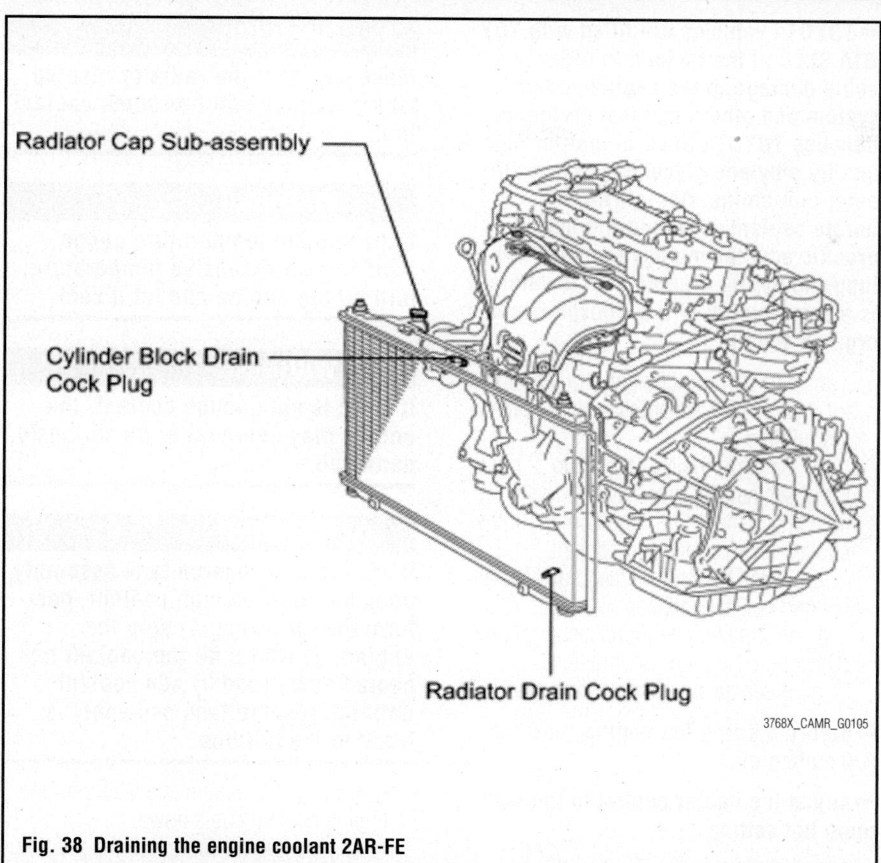

Fig. 38 Draining the engine coolant 2AR-FE

1. Remove the front wheel opening extension pad RH and LH.
2. Remove the engine under cover LH and RH.
3. Drain the coolant.

➡Do not remove the radiator cap sub-assembly while the engine and radiator are still hot. Pressurized, hot coolant and steam may be released and cause serious burns.

a. Remove the radiator cap sub-assembly from the radiator assembly.
b. Loosen the radiator drain cock plug and cylinder block drain cock plug, then drain the coolant.

➡Collect the coolant in a container and dispose of it according to the regulations in your area.

4. Add coolant.
a. Close the radiator drain cock plug and cylinder block cock plug. Tighten the cylinder block drain cock plug to 9 ft. lbs. (13 Nm).
b. Slowly fill the radiator with Toyota Super Long Life Coolant (SLLC). Fill capacity is 6.6 qts (6.2 L).

➡TOYOTA vehicles are filled with TOYOTA SLLC at the factory. In order to

➡Contact your TOYOTA dealer for further details.

d. Slowly pour coolant into the radiator reservoir tank until it reaches the FULL line.
e. Press the inlet and outlet radiator hoses several times by hand, and then check the level of the coolant. If the coolant is low, add coolant.
f. Install the radiator cap sub assembly and reservoir tank cap.
g. Start the engine, and warm it up.

➡Adjust the air conditioner set temperature to MAX (Hot).

h. Stop the engine, and wait until the engine coolant cools down.
i. Add engine coolant to the FULL line on the radiator reservoir.

7. Inspect for coolant leaks.
8. Install the engine under cover RH.
9. Install the engine under cover LH.
10. Install the front wheel opening extension pad LH.
11. Install the front wheel opening extension pad RH.

2AZ-FXE Hybrid Engine

See Figure 39.

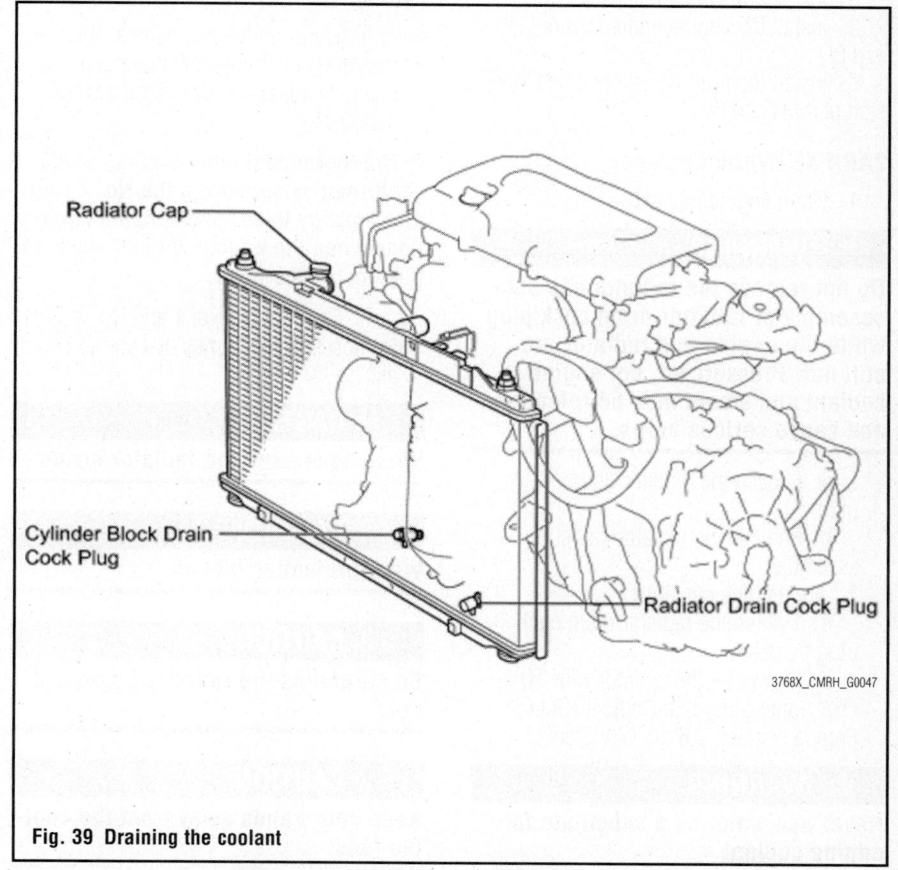

Fig. 39 Draining the coolant

avoid damage to the engine cooling system and other technical problems, only use TOYOTA SLLC or similar high quality ethylene glycol based non-silicate, non-amine, non-nitrite, non-borate coolant with long-life hybrid organic acid technology (coolant with long-life hybrid organic acid technology consists of a combination of low phosphates and organic acids).

➡Contact your TOYOTA dealer for further details.

c. Slowly pour coolant into the radiator reservoir tank until it reaches the FULL line.

d. Press the inlet and outlet radiator hoses several times by hand, and then check the level of the coolant. If the coolant level is low, add coolant.

e. Install the radiator cap sub-assembly and reservoir tank cap.

f. Set the vehicle to inspection mode.

g. Start the engine, and warm it up.

➡Adjust the air conditioner set temperature to MAX (HOT).

h. Stop the engine, and wait until the coolant cools down.

i. Add engine coolant to the FULL line on the radiator reservoir.

5. Inspect for coolant leaks.
6. Install the engine under covers LH and RH.
7. Install the front wheel opening extension pad LH and RH.

2AR-FXE Hybrid Engine

1. Drain engine coolant.

⁂ WARNING

Do not remove the radiator cap sub-assembly or radiator drain cock plug while the engine and radiator are still hot. Pressurized, hot engine coolant and steam may be released and cause serious burns.

a. Loosen the radiator drain cock plug.

b. Remove the radiator cap sub-assembly.

2. Add engine coolant.

a. Tighten the radiator drain cock plug by hand.

b. Slowly fill the radiator with TOYOTA Super Long Life Coolant (SLLC). Engine coolant: 7.6 qts (7.2 liters).

⁂ WARNING

Never use water as a substitute for engine coolant.

➡TOYOTA vehicles are filled with TOYOTA SLLC at the factory. In order to avoid damage to the engine cooling system and other technical problems, only use TOYOTA SLLC or similar high quality ethylene glycol based non-silicate, non-amine, non-nitrite, non-borate coolant with long-life hybrid organic acid technology (coolant with long-life hybrid organic acid technology is a combination of low phosphates and organic acids).

c. Slowly pour coolant into the radiator reserve tank assembly until it reaches the full line.

d. Squeeze the No. 1 and No. 2 radiator hoses several times by hand, and then check the level of the coolant. If the coolant level is low, add coolant.

e. Install the radiator cap sub-assembly and reserve tank cap.

f. Put the engine in inspection mode (Click here for more information).

g. Bleed air from the cooling system.

➡Before starting the engine, turn the A/C switch off.

➡Adjust the heater control to the maximum hot setting.

➡Adjust the blower speed to the low setting.

h. Warm up the engine until the thermostat opens. While the thermostat is open, circulate the coolant for several minutes.

➡The thermostat open timing can be confirmed by squeezing the No. 2 radiator hose by hand, and sensing vibrations when the engine coolant starts to flow inside the hose.

i. Squeeze the No. 1 and No. 2 radiator hoses several times by hand to bleed air.

⁂ CAUTION

When squeezing the radiator hoses:

⁂ CAUTION

Wear protective gloves.

⁂ CAUTION

Be careful as the radiator hoses are hot.

⁂ CAUTION

Keep your hands away from the cooling fans.

⁂ WARNING

Make sure that the radiator reserve tank assembly still has some coolant in it.

⁂ WARNING

If the coolant temperature gauge indicates an excessive temperature, turn off the engine and let it cool.

⁂ WARNING

If there is not enough coolant, the engine may overheat or be seriously damaged.

⁂ WARNING

If the radiator reserve tank assembly does not have enough coolant, perform the following: 1) stop the engine, 2) wait until the coolant has cooled down, and 3) add coolant until the reserve tank assembly is filled to the full line.

j. Stop the engine and wait until the engine coolant cools down.

k. Add engine coolant to the full line on the radiator reserve tank assembly.

3. Inspect for coolant leak.

Inspection Mode Procedure

See Figure 40.

Inspection Mode

1. Activate inspection mode (not using Techstream).

➡Perform the following 4 steps (Procedure "A" through "D") within 60 seconds.

a. Turn the power switch on (IG) with the shift lever in the P position (Procedure "A").

b. Fully depress the accelerator pedal twice with the shift P position (Procedure "B").

c. Move the shift lever to the N position and fully depress the accelerator pedal twice (Procedure "C").

d. Move the shift lever to the P position and fully depress the accelerator pedal twice (Procedure "D").

e. Check that the "FWD MAINTENANCE MODE" is displayed on the multi-information display of the combination meter. (Procedure "E").

➡If the "FWD MAINTENANCE MODE" message appears on the multi-information display do not come on in proce-

dure "E", repeat the procedures from "A" to "E".

2. Activate inspection mode (Using Techstream).
   a. Connect Techstream to the DLC3 (Procedure "A").
   b. Turn the power switch on (IG) (Procedure "B").
   c. Turn Techstream on (Procedure "C").
   d. Select following menu items: Powertrain / Hybrid Control / Utility / Inspection Mode -2WD Inspection (Procedure "D").
   e. Check that the "FWD MAINTENANCE MODE" is displayed on the multi-information display of the combination meter (Procedure "E").
   f. If the "FWD MAINTENANCE MODE" message appears on the multi-information display do not come on in procedure "E", repeat the procedures from "A" to "E".

➡If the "FWD MAINTENANCE MODE" message appears on the multi-information display do not come on in procedure "E", repeat the procedures from "A" to "E".

3. Deactivate the inspection mode.
   a. Turn the power switch off. The HV main system turns off simultaneously.

✲✲ CAUTION

The idling speed in inspection mode is approximately 900 rpm. The engine speed increases to 1,500 rpm if the accelerator pedal is depressed by less than 60 %. If the accelerator pedal is depressed by more than 60%, the engine speed increases to 2,500 rpm.

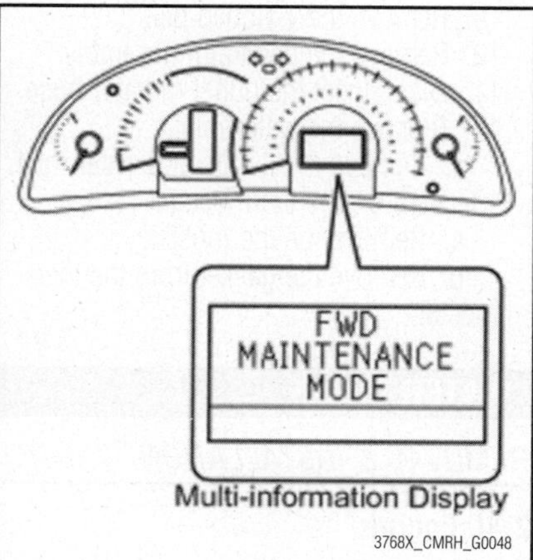

Fig. 40 Checking the "FWD MAINTENANCE MODE" message

✲✲ CAUTION

If a DTC is set during inspection mode, the master warning light and the warning massage illuminate on the multi-information display.

✲✲ CAUTION

When the master warning light illuminates during the inspection mode, deactivate the inspection mode, and check DTC(s).

✲✲ CAUTION

Driving the vehicle without deactivating the inspection mode may damage the transaxle.

ENGINE FAN

REMOVAL & INSTALLATION

See radiator removal and installation.

RADIATOR

REMOVAL & INSTALLATION

See Figure 41.

1. Before servicing the vehicle, refer to the Precautions Section.

✲✲ WARNING

After turning the power switch off, waiting time may be required before disconnecting the cable from the negative (-) auxiliary battery terminal. Therefore, make sure to read the disconnecting the cable from the negative (-) auxiliary battery terminal notice before proceeding with work.

2. Drain engine coolant.
3. Remove both front wheel opening extension pads.
4. Remove both engine under cover.
5. For 2.4L and 2.5L engines, perform the following:
   a. Remove air cleaner cap sub-assembly.
   b. Remove air cleaner inlet assembly.
6. For 3.5L engines, perform the following:
   a. Remove v-bank cover sub-assembly.
   b. Remove cool air intake duct seal.
   c. Remove air cleaner cap sub-assembly.
   d. Remove air cleaner inlet sub-assembly.
   e. Remove no. 1 air cleaner inlet.
7. For all vehicles, remove front bumper assembly.
8. Remove front bumper energy absorber.
9. Separate the radiator reserve tank hose from the radiator assembly.
10. Disconnect the inlet and outlet radiator hose from the radiator assembly.
11. Disconnect the oil from the radiator assembly.
12. For vehicles with A/T, disconnect the oil cooler inlet hose and outlet hose from the radiator assembly.
13. For 2.5L engines, remove the hood lock assembly.
14. Remove the upper radiator support, as follows:
   a. Disconnect the horn connector.
   b. Remove the 3 bolts and separate the hood lock assembly from the radiator support upper.
   c. Remove the clamp and separate the hood lock control cable from the radiator support upper.
   d. Remove the 5 bolts and radiator support upper.
15. Remove the radiator assembly, as follows:
   a. Remove the 3 clamps and 2 connectors.
   b. Remove the 4 bolts and separate the condenser assembly from the radiator assembly.
   c. Remove the radiator assembly from the body by pulling upwards.
16. Release the 3 snap fits and lift the fan assembly with motor from the radiator.
17. Remove the 2 radiator support cushions from the radiator assembly.
18. Remove the 2 radiator support lowers from the radiator assembly.

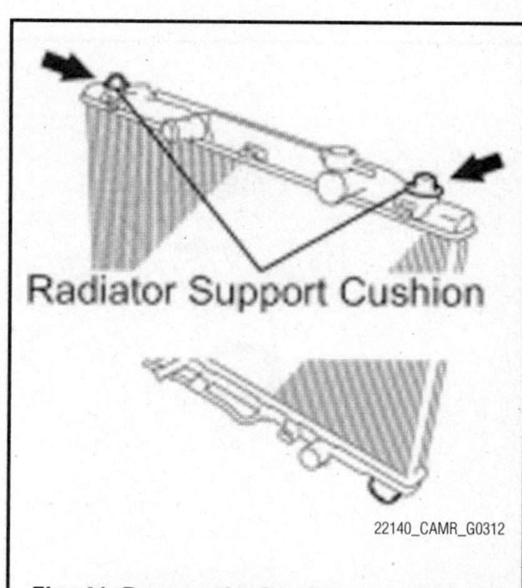

Fig. 41 Remove the 2 radiator support cushions

To install:

19. Installation is the reverse of removal.
20. After installation, inspect for coolant leak.

THERMOSTAT

REMOVAL & INSTALLATION

2.4L & 3.5L Engines

See Figures 42 through 45.

1. Before servicing the vehicle, refer to the Precautions Section.
2. For 2.4L engines, perform the following:
    a. Remove both front wheel opening extension pads.
    b. Remove both engine under cover.
    c. Drain engine coolant.
3. For 3.5L engines, perform the following:
    a. Drain engine coolant.
    b. Remove the V-bank cover sub-assembly.
    c. Remove the RH front fender apron seal.
    d. Remove the RH No. 2 engine mounting stay.
    e. Remove the drive belt.
    f. Remove No. 2 idler pulley sub-assembly.
4. Separate the radiator hose outlet.
5. Remove the 2 nuts and disconnect the water inlet from the cylinder block.
6. Remove the thermostat and the gasket from the thermostat.

To install:

7. Install a new gasket to the thermostat.
8. Install a new gasket to the thermostat.

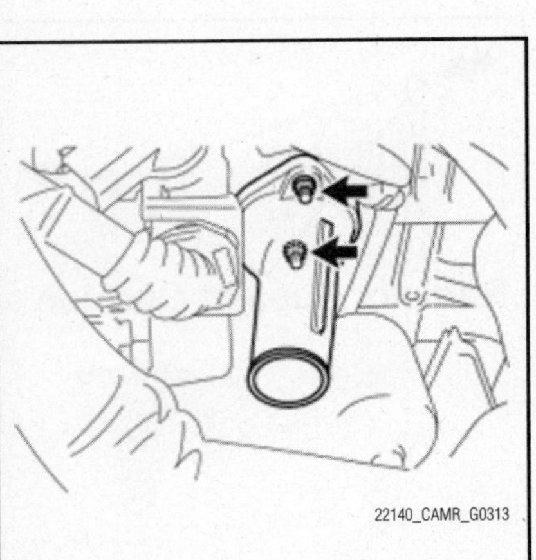

Fig. 42 Removing the 2 thermostat nuts—2.4L engine

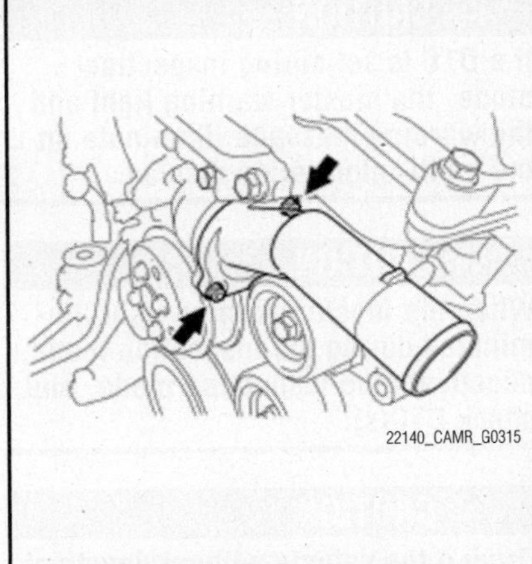

Fig. 43 Removing the 2 thermostat nuts—3.5L engine

➡The jiggle valve may be set within 10° of either side of the prescribed position.

9. Install the thermostat with the jiggle valve facing up.
10. Install the water inlet and tighten to 7 ft. lbs. (10 Nm).
11. The remainder of installation is the reverse of removal.
12. After installation, inspect for coolant leak.

2.5L Engine

See Figure 46.

1. Disconnect the cable from the negative battery terminal.
2. Remove the front wheel RH.
3. Remove the front wheel opening extension pad RH.
4. Remove the front wheel opening extension pad LH.
5. Remove the engine under cover RH and LH.

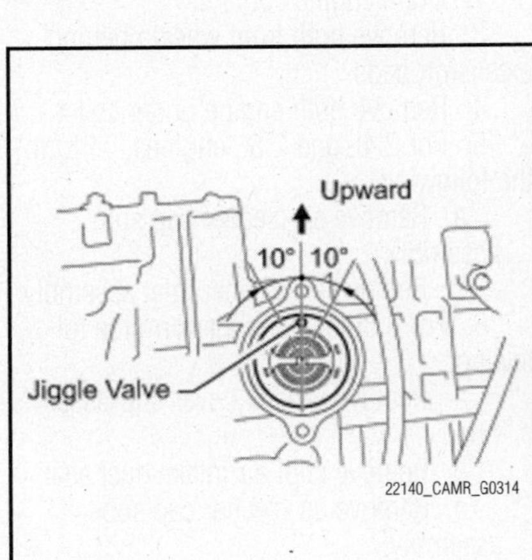

Fig. 44 Radiator jiggle valve—2.4L engine

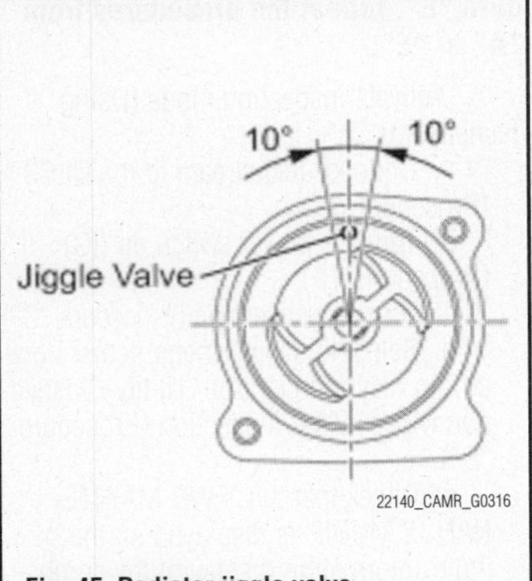

Fig. 45 Radiator jiggle valve—3.5L engine

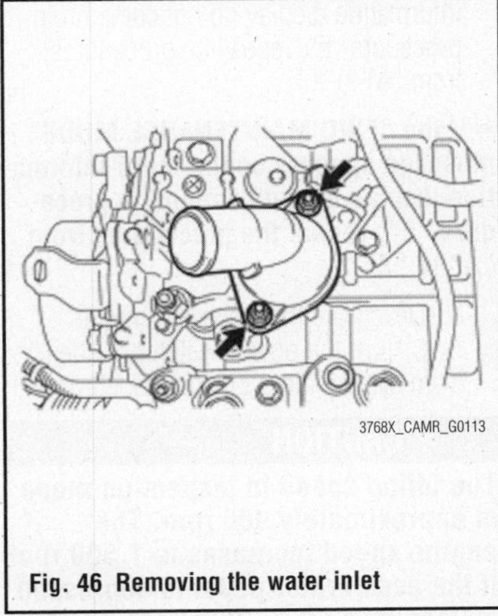

Fig. 46 Removing the water inlet

6. Remove the front fender apron seal RH.
7. Remove the No. 1 engine cover sub assembly.
8. Drain the engine coolant.
9. Remove the v-ribbed belt.
10. Remove the generator assembly.
11. Disconnect the outlet radiator hose.
12. Remove the water inlet.
    a. Remove the 2 nuts and water inlet.
13. Remove the thermostat.
    a. Remove the thermostat.
    b. Remove the gasket from the thermostat.

WATER PUMP

REMOVAL & INSTALLATION

2.4L Engine

See Figures 47 and 48.

1. Before servicing the vehicle, refer to the Precautions Section.

2. Disconnect the negative battery cable.
3. Remove both front wheel opening extension pads.
4. Remove both engine under covers.
5. Drain and recycle the engine coolant.
6. Remove RH front fender apron seal.
7. Remove RH No. 2 engine mounting stay.
8. Remove engine moving control rod sub-assembly.
9. Remove RH No. 2 engine mounting bracket
10. Remove the drive belt.
11. Remove generator assembly.
12. Using SST: 09960-10010, remove the 4 bolts and water pump pulley.
13. Remove water pump assembly by performing the following:
a. Remove the clamp of the crankshaft position sensor from the water pump.
b. Disconnect the wire of the crankshaft position sensor from the clamp bracket.
c. Remove the 4 bolts, 2 nuts and clamp bracket.

➡Be careful not to damage the contact surfaces of the water pump and cylinder block.

d. Using a screwdriver, pry between the water pump and cylinder block, and then remove the water pump. Tape the screwdriver tip before use.

To install:

14. Install water pump assembly:
a. Remove any old seal packing material from the water pump assembly contact surface.

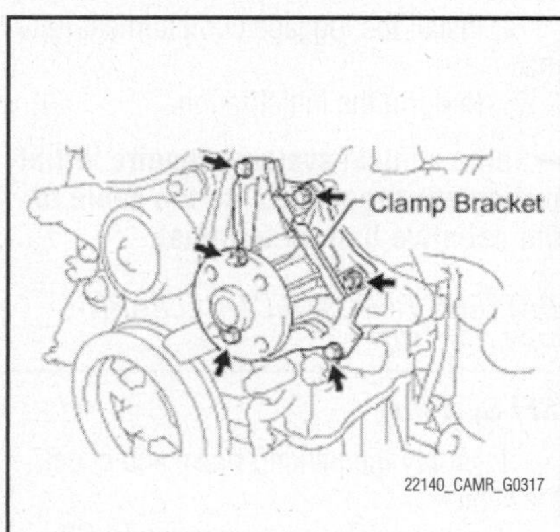

Fig. 47 Removing the 4 bolts, 2 nuts and clamp bracket—2.4L engine

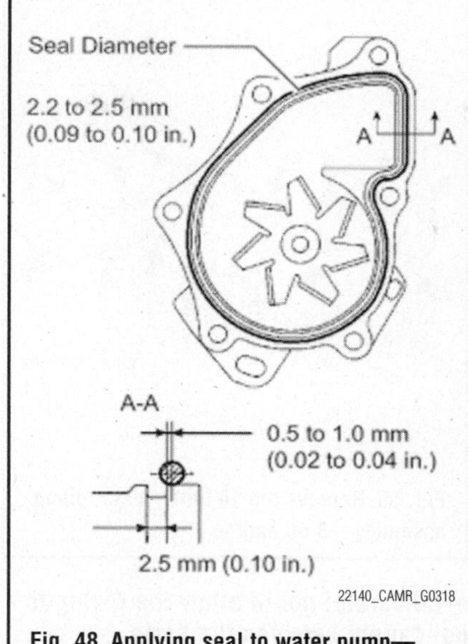

Fig. 48 Applying seal to water pump—2.4L engine

➡Remove any oil from the contact surface. The parts must be set within 3 minutes after applying seal packing. Otherwise, the material must be removed and reapplied.

b. Apply a continuous line of seal packing. Seal packing: Toyota Genuine Seal Packing Black, Three Bond 1207B or Equivalent. Standard seal diameter: 0.09 to 0.10 in. (2.2 to 2.5 mm)
15. Install the water pump and clamp bracket with the 4 bolts and 2 nuts and tighten to 80 inch lbs. (9 Nm).
16. To complete installation, reverse removal procedure.
17. Tighten the water pump bolts to 19 ft. lbs. (26 Nm).
18. Add engine coolant.
19. Inspect for coolant leak.

2.5L Engine

See Figure 49.

1. Disconnect the cable from the negative battery terminal.
2. Remove the front wheel RH.
3. Remove the front wheel opening extension pad RH.
4. Remove the front wheel opening extension pad LH.
5. Remove the engine under cover RH.
6. Remove the engine under cover LH.
7. Remove the front fender apron seal RH.
8. Remove the No. 1 engine cover sub assembly.
9. Drain the engine coolant.
10. Remove the v-ribbed belt.

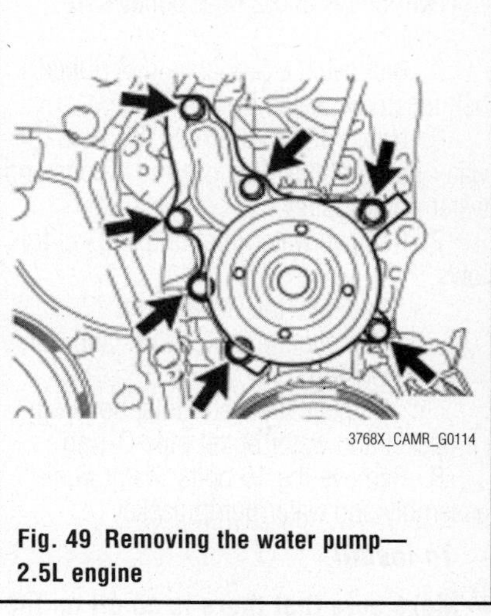

Fig. 49 Removing the water pump—2.5L engine

11. Remove the generator assembly.
12. Remove the v-ribbed belt tensioner assembly (for compressor and generator).
13. Remove the water pump assembly.
a. Remove the 7 bolts, water pump and water pump gasket.

To install:

14. Install the water pump assembly.
a. Install a new gasket and the water pump with the 7 bolts. Tighten to 15 ft. lbs. (21 Nm).
15. Install the v-ribbed belt tensioner assembly (for compressor and generator).
16. Install the generator assembly.
17. Install the v-ribbed belt.
18. Connect the cable to the negative battery terminal.
19. Add engine coolant.
20. Inspect for coolant leaks.
21. Install the No. 1 engine cover sub assembly.
22. Install the front fender apron seal RH.
23. Install the engine under cover RH.
24. Install the engine under cover LH.
25. Install the front wheel opening extension pad RH.
26. Install the front wheel opening extension pad LH.
27. Install the front wheel RH.

3.5L Engine

See Figures 50 and 51.

1. Before servicing the vehicle, refer to the Precautions Section.
2. Remove engine assembly and transaxle. Secure the engine stand.
3. Remove RH front No. 1 engine mounting bracket.
4. Remove the No. 2 idler pulley sub-assembly, as follows:
a. Remove the 2 bolts, 2 idler pulley

cover plates and 2 idler pulley sub-assemblies.

5. Remove the 5 bolts and V-ribbed belt tensioner assembly.

6. Using SST: 09960-10010, hold the water pump pulley. Remove the 4 bolts and water pump pulley.

7. Remove water inlet housing, as follows:

a. Separate the water hose.

b. Remove the 2 bolts, nut and water inlet housing.

c. Remove the water inlet housing gasket and water outlet pipe O-ring.

8. Remove the 16 bolts, water pump assembly and water pump gasket.

To install:

➡Make sure that there is no oil on the threads of the A bolts.

➡Be sure to replace the 2 C bolts with new ones or reuse them after applying adhesive (Part No. 08833-00080, three bond 1344 or equivalent).

9. Install a new water pump gasket and the water pump assembly with the 16 bolts and tighten to:

a. Bolt A: 15 ft. lbs. (21 Nm).

b. Bolt B: 81 inch lbs. (9.1 Nm).

c. Bolt C: 81 inch lbs. (9.1 Nm).

10. Install a new water inlet housing No. 1 gasket and water outlet pipe O-ring.

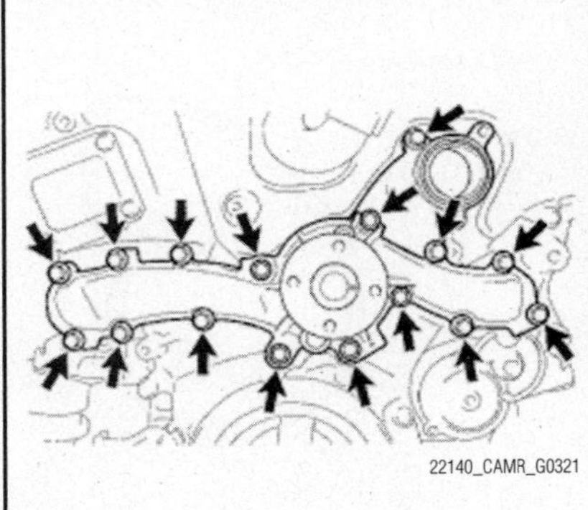

Fig. 50 Remove the 16 bolts, water pump assembly—3.5L engine

➡Be careful not to allow the O-ring to get caught between the parts.

11. Install water inlet housing, as follows:

a. Install a new No. 1 water inlet housing gasket and water outlet pipe O-ring.

b. Install the water inlet with the 2 bolts and nut and tighten to 7 ft. lbs. (10 Nm).

12. Temporarily install the water pump pulley with the 4 bolts.

a. Using SST (SST: 09960-10010) or equivalent, hold the water pump pulley.

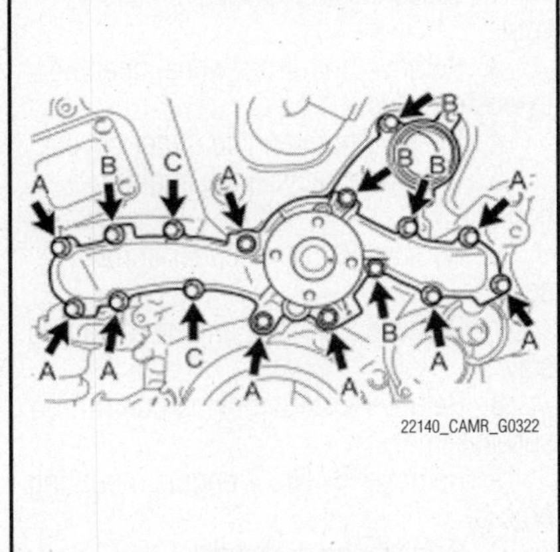

Fig. 51 Water pump tightening sequence—3.5L engine

b. Tighten the 4 bolts to 15 ft. lbs. (21 Nm).

13. Install the V-ribbed belt tensioner assembly with the 5 bolts and tighten to 32 ft. lbs. (43 Nm).

14. Install the 2 idler pulley cover plates and idler pulley sub-assemblies with the 2 bolts and tighten to 32 ft. lbs. (43 Nm).

15. To complete installation, reverse the remaining removal procedure.

16. Add engine coolant.

17. Inspect for coolant leak.

ENGINE ELECTRICAL

BATTERY SYSTEM

BATTERY

REMOVAL & INSTALLATION

See Figure 52.

➡The auxiliary battery is exclusive to the hybrid model. Be sure to replace the auxiliary battery at a dealer.

1. Remove the luggage compartment floor mat.

2. Remove the luggage trim service hole cover.

3. Remove the auxiliary battery.

a. Disconnect the cable from the negative (-) terminal of the auxiliary battery.

⁂ CAUTION

Wait at least 90 seconds after disconnecting the cable from the negative battery terminal to prevent airbag and seat belt pretensioner activation.

b. Disconnect the cable from the positive (+) terminal of the auxiliary battery.

c. Disconnect the thermistor sensor connector.

d. Disconnect the battery room ventilation hose.

e. Remove the nut, bolt and battery clamp.

f. Remove the auxiliary battery.

To install:

4. Install the auxiliary battery.

a. Install the battery.

b. Install the battery clamp with the nut and bolt. Torque to 48 inch lbs. (5.4 Nm).

c. Connect the thermistor sensor connector.

d. Connect the battery room ventilation hose.

e. Connect the cables to the positive (+) and negative (-) terminals of the auxiliary battery. Torque to 48 inch lbs. (5.4 Nm).

➡When the auxiliary battery has been disconnected and reconnected, attempting to turn the power switch on (READY) may not start the system (the system may not enter the READY-on state) on the first attempt. If so, turn the power switch off and reattempt to turn the power switch on (READY).

5. Install the luggage trim service hole cover.

6. Install the luggage compartment floor mat.

7. Perform the initialization.

➡Some vehicle systems require initialization after reconnecting the cable to the negative battery terminal.

BATTERY RECONNECT/RELEARN PROCEDURE

SFI System

1. Apply the parking brake and chock the vehicle.

2. Connect Techstream to the DLC3.

3. Put the engine in the inspection mode.

4. Select the following menu items:

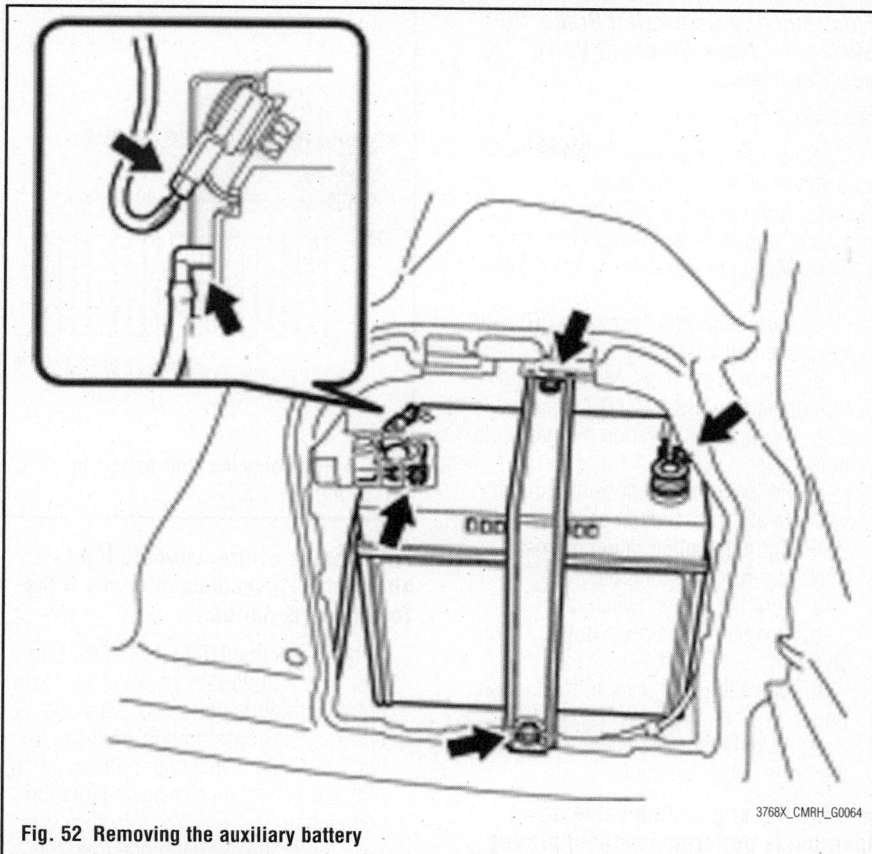

Fig. 52 Removing the auxiliary battery

Powertrain / Engine / Data List / Coolant Temp.

5. Warm up the engine with the A/C switch off until the coolant temperature becomes 181°F (83°C) or more.

6. Select the following menu items: Powertrain / Hybrid Control / Data List / SOC.

7. Move the shift lever to the N position, set the A/C switch to MAX cool or turn on the front defroster, and leave the vehicle until the SOC value becomes below 40°

➡If the SOC is substantially below 40%, the HV battery may become fully discharged. Therefore, do not allow the SOC to reach substantially below 40%.

8. Move the shift lever to the P position and turn the power switch off.

9. Disconnect the negative terminal from the auxiliary battery and leave it for 1 minute or more.

10. Reconnect the negative terminal to the auxiliary battery and turn the power switch on (IG).

11. Select the following menu items: Powertrain / Engine / Data List / Coolant Temp.

12. Check that the engine coolant temperature is 181°F (83°C) or more.

➡If the engine coolant temperature is below 181°F (83°C), start the engine by depressing the accelerator pedal with the shift lever in the P position to increase the coolant temperature.

13. Select the following menu items: Powertrain / Hybrid Control / Data List / Calculate Load.

14. Move the shift lever to the D position while depressing the brake pedal with your left foot.

15. Depress the accelerator pedal with your right foot while firmly depressing the brake pedal with your left foot, and maintain an engine load value of 45% or more (accelerator pedal depressed 60 to 70%) for approximately 30 seconds.

➡Do not perform this step for 40 seconds or more.

16. Shift the shift lever to the P position. Turn the power switch off and then wait for 5 seconds before putting the vehicle into the READY-on state again.

17. Select the following menu items: Powertrain / Engine / Data List / ISC Learning.

18. Check that the air conditioner is off. Lightly depress the accelerator pedal and release it when the engine starts.

19. Check that "Complete" is displayed on the Techstream screen.

➡If "ISC Learning" is not completed within 1 minute, repeat steps (o) through®.

➡The engine will usually stop when "ISC Learning" is completed. However, the engine will not stop even when "ISC Learning" is completed in such cases when the charge level of the auxiliary battery is decreased.

Tire Pressure Warning System

See Figure 53.

1. Turn the power switch to on (IG).
2. Press and hold the tire pressure warning reset switch for 3 seconds or more so that the tire pressure warning light blinks 3 times.
3. Turn the power switch off.
4. Connect Techstream to the DLC3.
5. Turn the power switch on (IG) and turn the tester on.
6. Enter the following menus: Chassis / Tire Pressure Monitor / Data List.
7. Check that the initialization has been completed.
8. Confirm that the tire pressure data of all tires are displayed on the tester screen.

➡The initialization is normally completed within 2 to 3 minutes.

➡If the initialization has not been completed successfully, DTC C2177/77 is

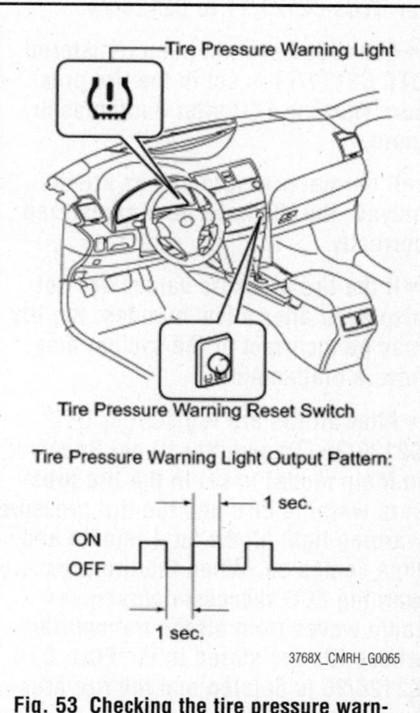

Fig. 53 Checking the tire pressure warning lights

set after a vehicle speed of 8 km/h (5 mph) or more continues for 20 minutes or more.

➡The initialization can be terminated by connecting terminals TC and CG of the DLC3 connector.

Tire Pressure Warning System—Register Transmitter ID

1. Set the air pressure of all wheels to the specified value.
2. Turn the power switch off.
3. Connect Techstream to the DLC3.
4. Turn the power switch on (IG) and Techstream on.
5. Enter the menu items in this order: Chassis / Tire Pressure Monitor / Utility / ID Registration
6. Perform the following procedures displayed on the screen.
7. Confirmation of transmitter ID registration.
    a. Enter the menu items in this order: Chassis / Tire Pressure Monitor / Data List
    b. Read the "ID Tire Inflation Pressure" values.
    c. Confirm that the data of tire pressure of all tires are displayed on the screen.

➡It may take up to about 2 or 3 minutes to update the tire pressure data. If the values are not displayed after a few minutes, perform troubleshooting according to the inspection procedure for DTCs C2121/21 to C2124/24.

➡If the IDs have not been registered, DTC C2171/71 is set in the tire pressure warning ECU after 3 minutes or more.

➡If normal pressure values are displayed, the IDs have been registered correctly.

➡If the tire pressure values are not displayed after a few minutes, the IDs may be incorrect or the system may have a malfunction.

➡After all IDs are registered, DTC C2126/26 (Transmitter ID not Received in Main Mode) is set in the tire pressure warning ECU and the tire pressure warning light blinks for 1 minute and then comes on. When the tire pressure warning ECU successfully receives radio waves from all the transmitters whose IDs are stored in the ECU, DTC C2126/26 is deleted and the tire pressure warning light goes off.

Electronically Controlled Brake System—Linear Solenoid Valve and Calibration

See Figure 54.

1. Clear the stored value of linear solenoid valve and calibration data.
    a. Turn the power switch off.
    b. Connect Techstream to the DLC3.
    c. Move the shift lever to the P position.
    d. Turn the power switch on (IG) with the brake pedal released.
    e. Perform the Reset Memory function under the ABS/VSC/TRAC menu.
    f. Perform initialization of linear solenoid valve.
    g. Perform zero point calibration of yaw rate and acceleration sensor.
2. Perform the initialization of linear solenoid valve and calibration.
    a. Turn the power switch off.
    b. Connect Techstream to the DLC3.
    c. Move the shift lever to the P position.
    d. Turn the power switch on (IG) with the brake pedal released.

➡If the linear solenoid valve offset learning is performed without turning the power switch on (IG), the learning process may not be completed properly because of insufficient auxiliary battery voltage.

➡When the linear solenoid valve offset learning is interrupted, or the learning process is performed with the shift lever not in the P position, DTC C1345/66 will be stored.

    e. Set Techstream to Test Mode (select "ECB Utility").

➡Refer to the Techstream operator's manual for further details.

    f. Leave the vehicle stationary without depressing the brake pedal for 1 or 2 minutes.
    g. Check that the interval between blinks of the brake control warning light changes from 1 second to 0.25 seconds.

➡The time needed to complete initialization of linear solenoid valve and calibration varies depending on auxiliary battery voltage.

➡The brake control warning light blinks at 1 second intervals during initialization of linear solenoid valve and calibration and changes to the Test Mode display.

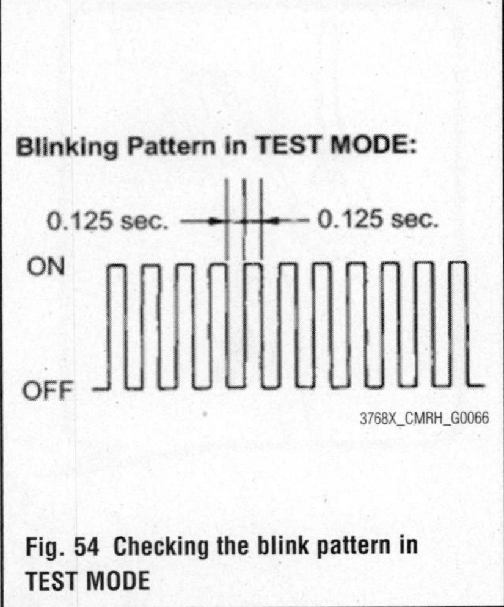

Fig. 54 Checking the blink pattern in TEST MODE

➡The brake control warning light blinks at 0.25 seconds intervals if the Test Mode is normal.

    h. Check that DTC C1345/66 which indicates trouble with stroke sensor zero point learning is not output when the brake control warning light changes to the Test Mode display upon initialization of linear solenoid valve and calibration completion.
    i. Enter the normal mode from the Test Mode following Techstream prompts.

➡Refer to the Techstream operator's manual for further details.

Electronically Controled Brake System—Linear Solenoid Valve and Calibration

1. Clear the stored value of linear solenoid valve and calibration data.
    a. Turn the power switch off.
    b. Move the shift lever to the P position.
    c. Turn the power switch on (IG) with the brake pedal released.
    d. Using SST (09843-18040), connect and disconnect terminals TS and CG of the DLC3 4 times or more within 8 seconds.
    e. Check that no codes other than ABS code 42, VSC code 45 and Electronically Controlled Brake code 48, 66, or 95 are stored in the diagnostic system.

➡The ABS warning, brake control warning lights and multi information display do not indicate the normal system code.

➡Illustrations may differ from the actual screen displayed depending on

the specifications of the vehicle and customized settings.

f. Remove the SST from the terminals of the DLC3.

g. Using the check wire, perform initialization of linear solenoid valve and calibration.

2. Perform initialization of linear solenoid valve and calibration.

a. Turn the power switch off.

b. Using SST (09843-18040), connect terminals TS and CG of the DLC3.

c. Move the shift lever to the P position.

d. Turn the power switch on (IG) with the brake pedal released.

➡If the linear solenoid valve offset learning is performed without turning the power switch on (IG), the learning process may not be completed properly because of insufficient auxiliary battery voltage.

➡When the linear solenoid valve offset learning is interrupted, or the learning process is performed with the shift lever not in the P position, DTC C1345/66 will be stored.

e. Leave the vehicle stationary without depressing the brake pedal for 1 or 2 minutes.

f. Check that the interval between blinks of the brake control warning light changes from 1 second to 0.25 seconds.

➡The time needed to complete initialization of linear solenoid valve and calibration varies depending on the auxiliary battery voltage.

➡The brake control warning light blinks at 1 second intervals during initialization of linear solenoid valve and calibration and changes to the Test Mode display.

➡The brake control warning light blinks at 0.25 seconds intervals if the Test Mode is normal.

g. Check that DTC C1345/66 which indicates trouble with stroke sensor zero point learning is not output when the brake control warning light changes to the Test Mode display upon initialization of linear solenoid valve and calibration completion.

h. Turn the power switch off and disconnect the SST from the DLC3.

ENGINE ELECTRICAL

CHARGING SYSTEM

ALTERNATOR

REMOVAL & INSTALLATION

2.4L Engine

See Figure 55.

1. Disconnect the negative battery cable.
2. Remove the front wheel RH.
3. Remove the engine front cover RH.
4. Remove the front fender apron seal RH.
5. Remove the V-bank cover sub-assembly.
6. Remove the V-ribbed belt.
7. Remove the alternator assembly, as follows:

a. Disconnect the generator connector.

b. Remove the nut and disconnect the wire harness from terminal B.

c. Remove the bolt and wire harness clamp bracket.

d. Remove the wire harness clamps.

e. Remove the 2 bolts and generator assembly.

To install:

8. Install the alternator assembly, as follows:

a. Install the bracket with the bolt and tighten to 15 ft. lbs. (20 Nm).

b. Install the wire harness clamp stay and tighten to 74 inch lbs. (8.4 Nm).

c. Install the alternator assembly with the 2 bolts and tighten to 32 ft. lbs. (43 Nm).

d. Install the nut to the cylinder block and tighten to 15 ft. lbs. (20 Nm).

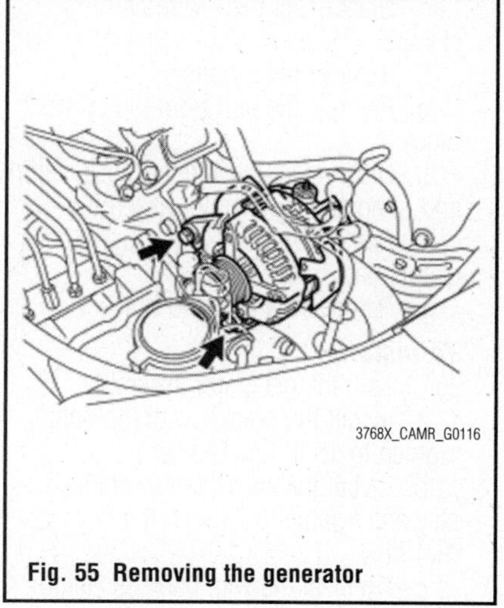

Fig. 55 Removing the generator

e. Connect the alternator connector to the alternator assembly.

f. Install the alternator wire with the nut and tighten to 87 inch lbs. (9.8 Nm).

g. Install the terminal cap.

h. Connect the wire harness clamp.

9. Install the V-ribbed belt.
10. Install the V-bank cover sub-assembly.
11. Connect the negative battery cable.
12. Perform initialization.

2.5L Engine

See Figure 56.

1. Disconnect the cable from the negative battery terminal.
2. Remove the No. 1 engine cover sub assembly.
3. Remove the v-ribbed belt.
4. Remove the generator assembly.

a. Disconnect the generator connector.

b. Turn back the terminal cap.

c. Remove the nut and disconnect the generator wire.

d. Remove the bolt and wire harness clamp bracket.

e. Remove the 2 bolts and generator.

To install:

5. Install the generator assembly.

a. Install the generator with the 2 bolts. Tighten to 38 ft. lbs. (52 Nm).

b. Install the wire harness clamp bracket with the bolt. Tighten to 74 inch lbs. (8.4 Nm).

c. Connect the generator wire with the nut. Tighten to 87 inch lbs. (9.8 Nm).

d. Install the terminal cap.

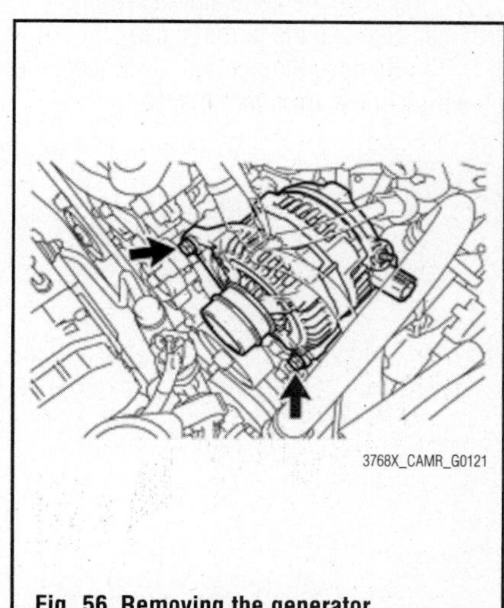

Fig. 56 Removing the generator

e. Connect the generator connector.
6. Install the v-ribbed belt.
7. Install the No. 1 engine cover sub assembly.
8. Connect the cable to the negative battery terminal.

3.5L Engine

See Figure 57.

1. Disconnect the cable from the negative battery terminal.
2. Remove the front wheel.
3. Remove the front fender apron seal RH.
4. Remove the front wheel opening extension pad RH.
5. Remove the front wheel opening extension pad LH.
6. Remove the engine under cover RH.
7. Remove the engine under cover LH.
8. Drain the engine coolant.
9. Remove the v-bank cover sub assembly.
10. Remove the cool air intake duct seal.
11. Remove the air cleaner inlet assembly.
12. Remove the No. 1 air cleaner inlet.
13. Remove the front bumper assembly.
14. Remove the front bumper energy absorber.
15. Separate the radiator reserve tank hose.
16. Separate the radiator inlet hose.
17. Separate the radiator outlet hose.
18. Separate the No. 1 oil cooler inlet hose.
19. Separate the No. 1 oil cooler outlet hose.
20. Remove the radiator support upper.
21. Remove the fan shroud.
22. Remove the radiator assembly.
23. Remove the v-ribbed belt.
24. Remove the generator assembly.
a. Remove the terminal cap.
b. Remove the nut and disconnect the wire harness from terminal B.

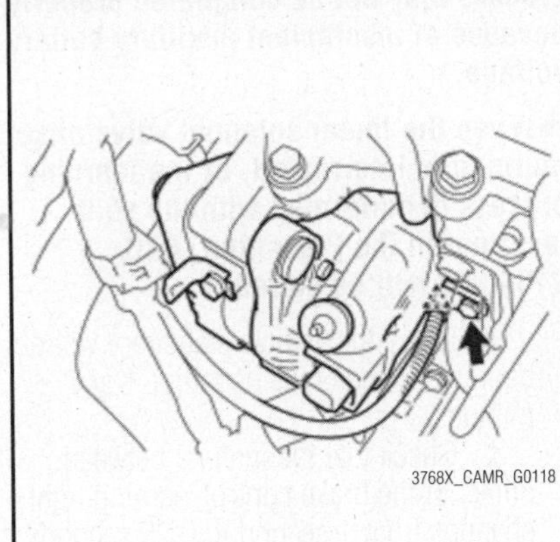

Fig. 57 Removing the bolt from the cylinder block

c. Disconnect the generator connector from the generator assembly.
d. Disconnect the connector from the compressor and magnetic clutch.
e. Disconnect the 2 wire harness clamps.
f. Remove the 2 bolts.
g. Remove the bolt from the cylinder block.
h. Disconnect the wire harness clamp and remove the generator assembly.
i. Remove the bolt and wire harness clamp stay.
j. Remove the bolt and bracket.

To install:

25. Install the generator assembly.
a. Install the bracket with the bolt. Tighten to 15 ft. lbs. (20 Nm).
b. Install the wire harness clamp stay and tighten to 74 inch lbs. (8.4 Nm).
c. Connect the wire harness clamp.
d. Install the generator assembly to the cylinder block with the bolt. Tighten to 15 ft. lbs. (20 Nm).
e. Install the 2 bolts. Tighten to 32 ft. lbs. (43 Nm).
f. Connect the generator connector to the generator assembly.
g. Install the generator wire with the nut and tighten to 87 inch lbs. (9.8 Nm).
h. Install the terminal cap.
i. Connect the 2 wire harness clamps.
j. Connect the magnetic clutch connector to the compressor and magnetic clutch.
26. Install the v-ribbed belt.
27. Install the radiator assembly.
28. Install the fan shroud.
29. Install the radiator support upper.
30. Connect the No. 1 oil cooler outlet tube.
31. Connect the No. 1 oil cooler inlet tube.
32. Connect the radiator reserve tank hose.
33. Install the front bumper energy absorber.
34. Install the front bumper assembly.
35. Install the No. 1 air cleaner inlet.
36. Install the air cleaner cap sub assembly.
37. Install the air cleaner inlet assembly.
38. Connect the cable to the negative battery terminal.
39. Add engine coolant.
40. Check for engine coolant leaks.
41. Install the v-bank cover sub assembly.
42. Install the cool air intake duct seal.
43. Install the front fender apron seal RH.
44. Install the engine under cover RH.
45. Install the engine under cover LH.
46. Install the front wheel opening extension pad RH.
47. Install the front wheel opening extension pad LH.
48. Install the front wheel and tighten to 76 ft. lbs. (103 Nm).

PRECAUTIONS

Before working on any part of the Hybrid high voltage system, observe the following precautions:

⁂ CAUTION

The nominal high voltage traction battery voltage is 330 volts DC. The buffer zone must be set up and insulated rubber gloves and a face shield must be worn. Failure to follow these instructions may result in severe injury or death.

⁂ CAUTION

The high voltage traction battery and charging system contains high voltage components and wiring. High voltage insulated safety gloves and a face shield must be worn when carrying out any diagnostics on this vehicle. Failure to follow these instructions may result in severe personal injury or death.

⁂ CAUTION

Before carrying out any removal and installation procedures of the high voltage traction battery system, the high voltage traction battery must be Disarmed. Failure to follow these instructions may result in severe personal injury or death.

⁂ CAUTION

The rubber insulating gloves that are to be worn while working on the high voltage system should be of the appropriate safety and protection rating for use on the high voltage system. They must be inspected before use and must always be worn in conjunction with the leather outer gloves. Any hole in the rubber insulating glove is a potential entry point for high voltage. Failure to follow these instructions may result in severe personal injury or death.

➡The high voltage insulated safety gloves must be re-certified every 6 months to remain within Occupational Safety and Health Administration (OSHA) guidelines:

- Roll the glove up from the open end until the lower portion of the glove begins to balloon from the resulting air pressure. If the glove leaks any air, it must not be used.
- The gloves should not be used if they exhibit any signs of wear and tear.
- The leather gloves must always be worn over the rubber insulating gloves in order to protect them.
- The rubber insulating gloves must be class "00" and meet all of the American Society for Testing and Materials (ASTM) standards

⁂ CAUTION

High voltage insulated safety gloves and a face shield must be worn when working with high voltage cables. The ignition switch must be OFF for a minimum of 5 minutes before removing high voltage cables. Failure to follow these instructions may result in severe personal injury or death.

⁂ CAUTION

Establish a buffer zone before servicing the high voltage system. The buffer zone is required only when working with the high voltage system. See the text for buffer zone establishment. Failure to follow these instructions may result in severe personal injury or death. Do not allow any unauthorized personnel into the buffer zone during repairs involving the high voltage system. Only personnel trained for repair on the high voltage system are to be permitted in the buffer zone.

⁂ CAUTION

Disarm the high voltage traction battery (HVTB) before working on the high voltage system. See the text for the Disarming procedure. Failure to follow these instructions may result in severe personal injury or death.

BUFFER ZONE

1. Before servicing the vehicle, refer to the Precautions Section.

⁂ CAUTION

Before proceeding, read and observe all of the High Voltage System Precautions.

2. Establish a buffer zone around the vehicle:
   a. Position the vehicle in the repair bay.
   b. Position 4 orange cones at the corners of the vehicle to mark off a 1 m (3 ft.) perimeter around the vehicle.
   c. Do not allow any unauthorized personnel into the buffer zone during repairs involving the high voltage system. Only personnel trained for repair on the high voltage system are to be permitted in the buffer zone.

HIGH VOLTAGE TRACTION BATTERY

REMOVAL & INSTALLATION

See Figures 58 through 65.

1. Check for DTCs.

⁂ WARNING

Confirm that P0AA6 (Hybrid Battery Voltage System Isolation Fault) is not output before doing removal or installation work HV battery. If this DTC is output, perform troubleshooting for this DTC first.

2. Remove service plug grip.
3. Remove connector cover assembly.
4. Check terminal voltage.
5. Install connector cover assembly.
6. Remove auxiliary battery.
   a. Open the auxiliary battery terminal cap.
   b. Loosen the nut, and disconnect the positive (+) auxiliary battery cable.
   c. Disconnect the auxiliary battery hose.
   d. Loosen the nut, and remove the bolt.
   e. Remove the auxiliary battery clamp.
   f. Remove the auxiliary battery.
7. Remove rear floor finish plate.
8. Remove No. 1 luggage compartment trim hook.
9. Remove luggage compartment inner trim LH and RH.
10. Remove rear seat cushion assembly.
11. Remove rear seat cushion lock hook.
12. Remove rear seatback assembly RH.
13. Remove No. 2 room partition cover.
14. Remove rear seatback assembly LH.
15. Remove rear side seatback assembly LH and RH.
16. Disconnect rear door opening trim weatherstrip LH and RH.

17. Disconnect rear seat outer belt assembly LH and RH.
18. Remove child restraint seat anchor bracket sub assembly LH.
19. Remove rear seat inner with center belt assembly LH.
20. Remove roof side inner garnish LH and RH.
21. Remove center stop light set.
22. Remove child restraint seat tether anchor cover.
23. Remove rear seat shoulder belt cover.
24. Remove package tray trim panel assembly.
25. Remove No. 1 room partition board.
26. Remove No. 1 hybrid battery intake duct.
   a. Remove the 3 clips and No. 1 hybrid battery intake duct.
27. Remove No. 4 hybrid battery shield panel.
28. Disconnect No. 4 floor wire.
29. Disconnect low voltage connector.
   a. Disconnect the connector and clamp.
   b. Disconnect the 2 connectors.
30. Remove HV battery.

CAUTION

Wear insulated gloves.

   a. Remove the 6 bolts from the HV battery.
   b. Install the spare wheel cover assembly upside down.
   c. Cut and shape a piece of cardboard so that it fits the area between the HV battery bolt attachment positions, and then insert it.
   d. Using a tire lever to hold up the HV battery, insert the cardboard until it cannot be inserted any further.

➡Secure the No. 4 floor wire with electrical tape to prevent it from getting caught when moving the battery or other parts.

➡Attach tape to the feet and edges of the battery to protect tools and the vehicle body.

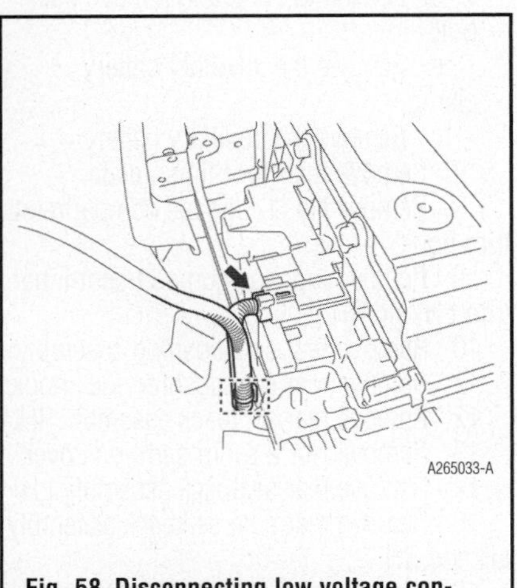

Fig. 58 Disconnecting low voltage connector and clamp

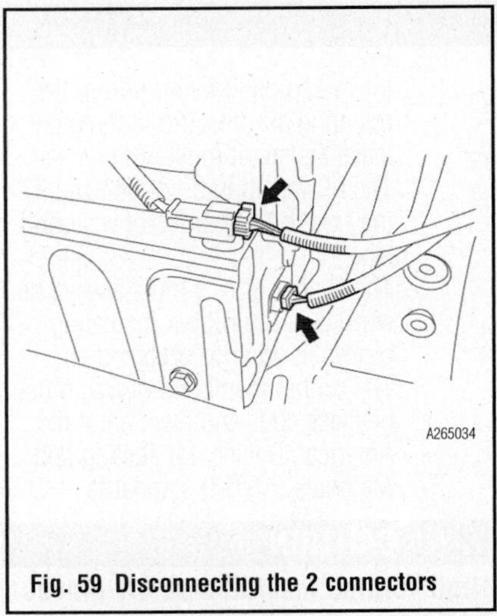

Fig. 59 Disconnecting the 2 connectors

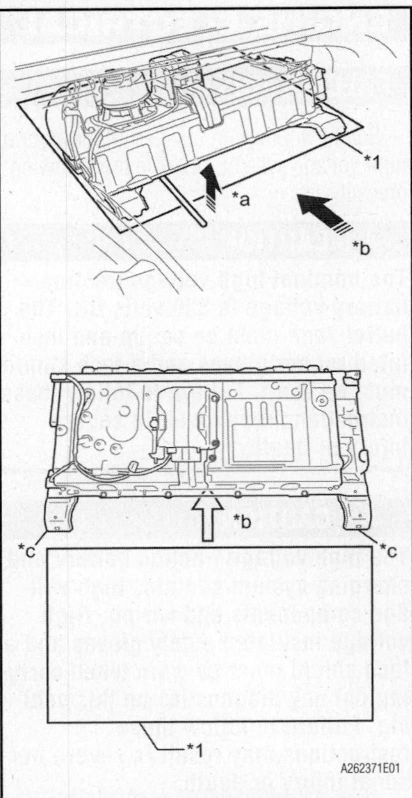

Fig. 61 Placing cardboard between HV battery bolt attachment positions

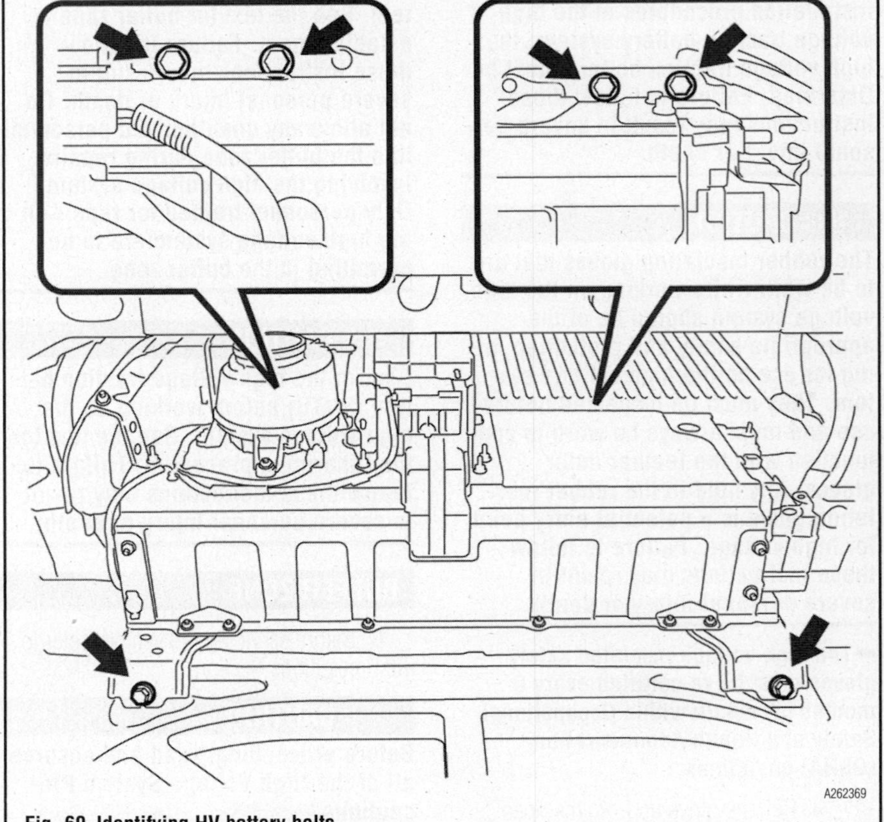

Fig. 60 Identifying HV battery bolts

e. Pull the cardboard and HV battery together to the middle of the luggage compartment.

➡Use cardboard or other similar material to protect the HV battery and vehicle body from damage.

f. Remove the 2 nuts and No. 2 hybrid vehicle battery upper cover bracket.

g. Remove the clip.

h. Disengage the 2 claws and remove the No. 2 hybrid battery intake duct.

i. Disconnect the connector from the battery cooling blower assembly.

j. Remove the 3 nuts and battery cooling blower assembly.

WARNING

Be sure not to touch the fan part of the battery cooling blower assembly.

WARNING

Do not lift the battery cooling blower assembly using the wire harness.

k. Remove the 2 clips.

l. Turn the cardboard and HV battery 180° together.

m. Pull the HV battery together with the cardboard toward the rear of the vehicle.

n. Using a suitable adaptor such as a rope, remove the HV battery while tilting it.

WARNING

While lowering the HV battery from the vehicle, do not allow it to contact the vehicle.

31. Remove the No. 2 hybrid vehicle battery shield panel.

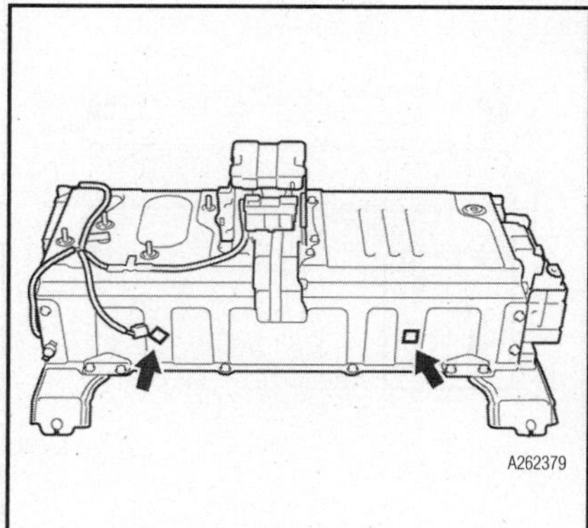

Fig. 62 Removing the 2 clips

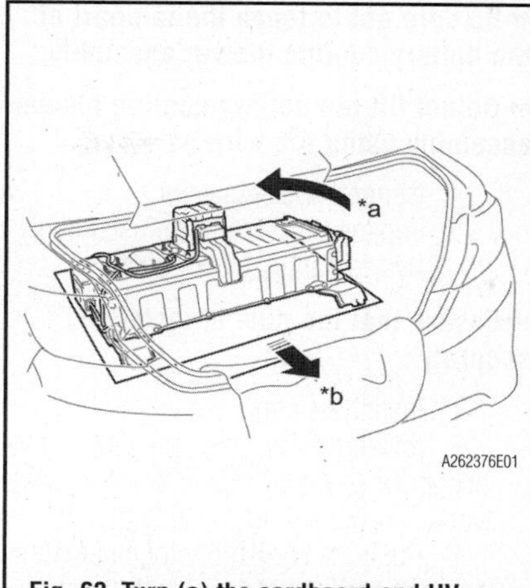

Fig. 63 Turn (a) the cardboard and HV battery 180° together and pull (b)

CAUTION

Wear insulated gloves.

a. Using the service plug grip, remove the battery cover lock striker.

➡Insert the projection part of the service plug grip, and turn the button of the battery cover lock striker counterclockwise, and release the lock.

b. Disconnect the connector.

c. Remove the 2 bolts, nut and No. 2 hybrid vehicle battery shield panel.

d. Disconnect the 3 wire harness clamps.

32. Remove the battery smart unit.

a. Disconnect the connector.

➡Insulate the removed connector with insulating tape.

b. Remove the bolt and battery smart unit.

33. Remove No. 1 hybrid battery shield sub assembly.

CAUTION

Wear insulated gloves.

a. Remove the bolt, 3 nuts and No. 1 hybrid battery shield sub assembly.

34. Remove hybrid battery junction block assembly.

35. Separate No. 2 hybrid battery pack wire.

CAUTION

Wear insulated gloves.

a. Disconnect the connector and 5 clamps, and remove the No. 2 hybrid battery pack wire.

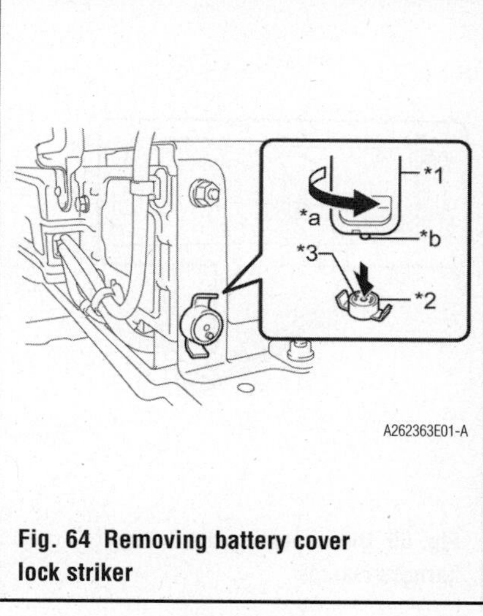

Fig. 64 Removing battery cover lock striker

36. Remove No. 1 hybrid vehicle battery carrier bracket sub assembly.

a. Remove the 2 bolts and No. 1 hybrid vehicle battery carrier bracket sub assembly.

37. Remove No. 2 hybrid vehicle battery carrier bracket sub assembly.

a. Remove the 2 bolts and No. 2 hybrid vehicle battery carrier bracket sub assembly.

To install:

38. Install the No. 2 hybrid vehicle battery carrier bracket with the 2 bolts. Tighten to 14 ft. lbs. (19 Nm).

39. Install the No. 1 hybrid vehicle battery carrier bracket with the 2 bolts. Tighten to 14 ft. lbs. (19 Nm).

➡Be sure to align the claw of the No. 1 hybrid vehicle battery carrier bracket with the hole.

40. Install No. 2 hybrid battery pack wire.

CAUTION

Wear insulated gloves.

a. Connect the connector and 5 clamps and install the No. 2 hybrid battery pack wire.

41. Install hybrid battery junction block assembly.

42. Install the No. 1 hybrid battery shield sub assembly with the bolt and 3 nuts. Tighten to 66 inch lbs. (7.5 Nm).

CAUTION

Wear insulated gloves.

43. Install the battery smart unit with the bolt. Tighten to 66 inch lbs. (7.5 Nm).

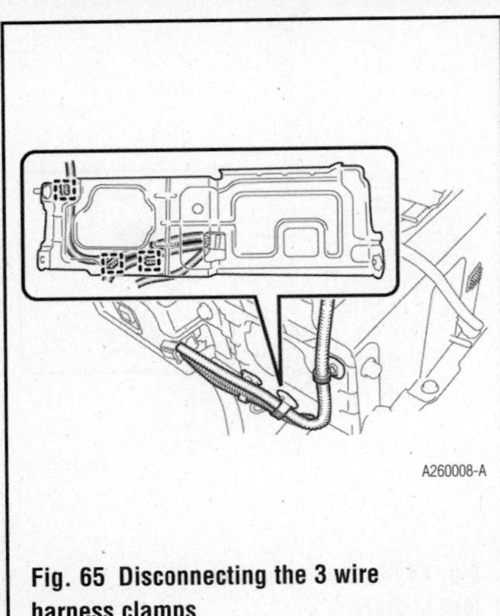

Fig. 65 Disconnecting the 3 wire harness clamps

CAUTION

Wear insulated gloves.

a. Connect the connector.

➡The connectors should be connected securely.

44. Install the No. 2 hybrid vehicle battery shield panel with the 2 bolts and nut. Tighten to 66 inch lbs. (7.5 Nm).

CAUTION

Wear insulated gloves.

a. Connect the 3 wire harness clamps.
b. Connect the connector.
c. Install the battery cover lock striker, then push the button to lock it.

45. Install HV battery.

CAUTION

Wear insulated gloves.

a. Place a piece of cardboard in the luggage compartment.
b. Using a suitable adaptor such as a rope, install the HV battery to the vehicle in the same direction it was facing during removal.

➡Use cardboard or other similar material to protect the HV battery and vehicle body from damage.

c. Push the cardboard and HV battery together to the middle of the luggage compartment.
d. Turn the cardboard and HV battery 180° together.
e. Install the 2 clips.
f. Install the battery cooling blower assembly with the 3 nuts. Tighten to 66 inch lbs. (7.5 Nm).

➡Be sure not to touch the fan part of the battery cooling blower assembly.

➡Do not lift the battery cooling blower assembly using the wire harness.

g. Connect the connector.
h. Engage the 2 claws to install the No. 2 hybrid battery intake duct.

➡Ensure that the duct is installed securely.

i. Install the clip.
j. Install the No. 2 hybrid vehicle battery upper cover bracket with the 2 nuts. Tighten to 66 inch lbs. (7.5 Nm).
k. Push the HV battery together with the cardboard toward the front of the vehicle.

➡Align the holes for the HV battery holding bolts.

l. Use a tire lever to hold up the HV battery and pull out the cardboard.
m. Remove the spare wheel cover assembly.
n. Install the HV battery with the 6 bolts. Tighten to 14 ft. lbs. (19 Nm).

➡Install the ground bolts to the appropriate locations.

46. Connect the low voltage connector.
    a. Connect the 3 conn3ctors and wire harness clamp.
47. Connect No. 4 floor wire.
48. Install No. 4 hybrid battery shield panel.
49. Install No. 1 hybrid battery intake duct.
50. Install No. 1 room partition board.
51. Install package tray trim panel assembly.
52. Install rear sweat shoulder belt cover.
53. Install child restraint seat tether anchor cover.
54. Install center stop light set.
55. Install roof side inner garnish LH and RH.
56. Connect rear seat inner with center belt assembly LH.
57. Connect rear seat outer belt assembly LH and RH.
58. Install rear door opening trim weatherstrip LH and RH.
59. Install rear side seatback assembly LH and RH.
60. Install rear seatback assembly LH.
61. Install No. 2 room partition cover.
62. Install rear seatback assembly RH.
63. Install rear seat cushion lock hook.
64. Install rear seat cushion assembly.
65. Install luggage compartment inner trim cover LH and RH.
66. Install No. 1 luggage compartment trim hook.
67. Install rear floor finish plate.
68. Install auxiliary battery.
69. Install service plug grip.

MOTOR ELECTRONICS RADIATOR

REMOVAL & INSTALLATION

See Figures 66 and 67.

1. Drain coolant (for inverter).
2. Remove front bumper assembly.
3. Remove front bumper energy absorber.
4. Remove front bumper reinforcement sub assembly.
5. Remove inlet air cleaner assembly.
6. Remove No. 5 inverter bracket.
7. Remove radiator reserve tank assembly.
8. Remove hood lock assembly.
9. Remove upper radiator support.
10. Remove radiator side deflector RH.
    a. Disengage the 3 claws and remove the radiator side deflector RH.
11. Remove radiator assembly.
    a. Disconnect the 2 water hoses.

➡Put pieces of cloth into the pipes and in the disconnected hoses or cover the pipes and hoses with plastic bags to prevent entry of foreign matter.

b. Remove the 4 bolts and radiator assembly.

To install:

12. Install radiator assembly.
    a. Install the radiator assembly with the 4 bolts. Tighten bolt A to 44 inch lbs. (5 Nm). Tighten bolt B to 80 inch lbs. (9 Nm).

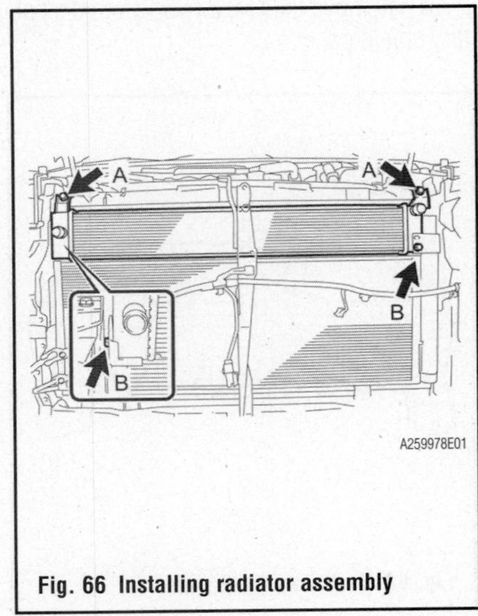

Fig. 66 Installing radiator assembly

b. Connect the 2 water hoses.

➡**Make sure that the clip is positioned as shown.**

➡**Make sure to align the alignment marks of the hoses with the ribs of the radiator assembly.**

➡**Do not remove the pieces of cloth or plastic bags from the pipes and disconnected hoses until installation.**

13. Engage the 3 claws to install the radiator side deflector RH.
14. Install upper radiator support.
15. Install hood lock assembly.
16. Install radiator reserve tank assembly.
17. Install No. 5 inverter bracket.
18. Install inlet air cleaner assembly.
19. Install front bumper reinforcement sub assembly.
20. Install front bumper energy absorber.
21. Install from bumper assembly.
22. Add coolant (for inverter).
23. Inspect for coolant leak (for inverter).

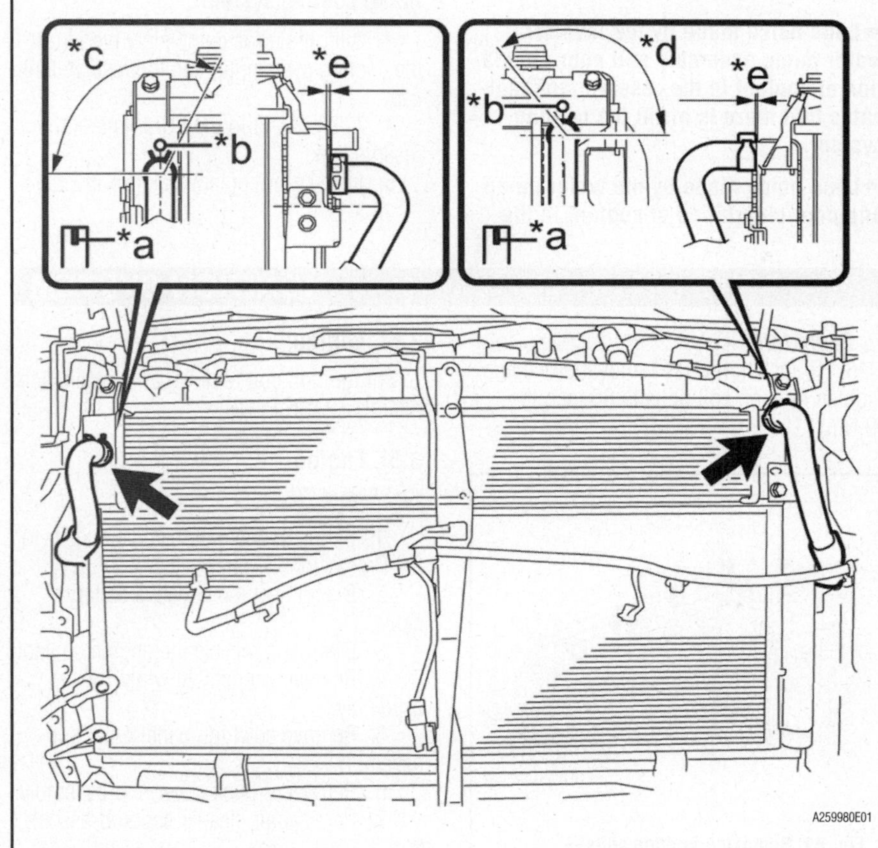

Fig. 67 Connecting the water hoses

MOTOR ELECTRONICS COOLING SYSTEM DRAINING AND FILLING

DRAINING

See Figure 68.

⁂ WARNING

Do not reuse the drained coolant because it may contain foreign objects.

⁂ WARNING

Collect the drained coolant and measure its volume to establish a benchmark. When adding coolant, make sure to add more coolant than the measured amount.

1. Remove the reserve tank cap.

⁂ CAUTION

To avoid the danger of being burned, do not remove the reserve tank cap while the coolant for the inverter is still hot.

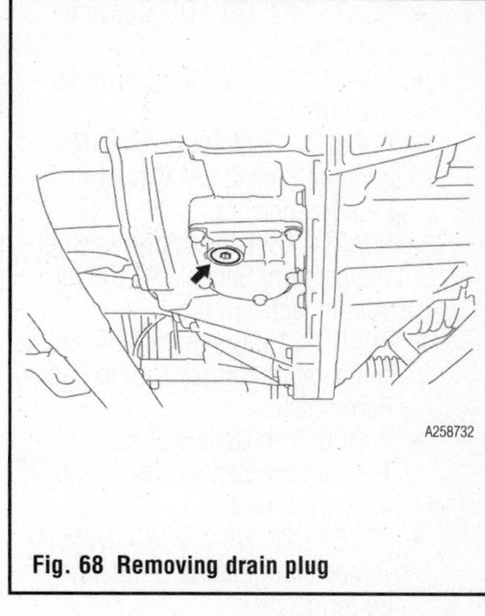

Fig. 68 Removing drain plug

2. Using a hexagon wrench (10 mm), remove the drain plug and drain the coolant.

⁂ CAUTION

Use caution when handling coolant immediately after driving or in summer because it may be hot.

3. Install the plug with a new gasket. Tighten to 29 ft. lbs. (39 Nm).

FILLING

⁂ WARNING

Do not reuse the drained coolant because it may contain foreign objects.

⁂ WARNING

If the vehicle is driven with air in the inverter cooling system, damage may occur and the following DTCs may be set.

- P0A01-726: Motor Electronics Coolant Temperature Sensor Circuit Range / Performance
- P0A04-725: Motor Electronics Coolant Temperature Sensor Circuit Intermittent
- P0A08-264: DC / DC Converter Status Circuit
- P0A78-284: Drive Motor "A" Inverter Performance
- P0A78-286: Drive Motor "A" Inverter Performance
- P0A7A-322: Generator Inverter Performance
- P0A7A-324: Generator Inverter Performance
- P0A93-346: Inverter Cooling System Performance

- P0A94-553: DC / DC Converter Performance
- P0A94-557: DC / DC Converter Performance
- P0AEE-277: Motor Inverter Temperature Sensor "A" Circuit Range / Performance
- P0AF1-276: Drive Motor Inverter Temperature Sensor "A" Circuit Intermittent / Erratic
- P0BCD-315: Generator Inverter Temperature Sensor Circuit Range / Performance
- P0BD0-314: Generator Inverter Temperature Sensor Circuit Intermittent / Erratic
- P0C39-626: DC / DC Converter Temperature Sensor "A" Range / Performance
- P0C3C-625: DC / DC Converter Temperature Sensor "A" Intermittent / Erratic
- P0C3E-628: DC / DC Converter Temperature Sensor "B" Range / Performance
- P0C41-627: DC / DC Converter Temperature Sensor "B" Intermittent / Erratic
- P0C73-776: Motor Electronics Coolant Pump "A" Control Performance

a. Slowly pour coolant into the reserve tank until it reaches the FULL line. (Coolant quantity: 3.4 qts (3.2 liters)).

➡To prevent foreign matter such as dust or dirt from entering the cooling system, make sure to confirm that the container used to add coolant is clean and free of foreign matter such as dust or dirt.

1. When using the Techstream:
   a. Connect the Techstream to the DLC3.
   b. Turn the power switch on (IG).
   c. Enter the following menus: Powertrain / Hybrid Control / Active Test / Activate the Water Pump.
   d. Keep the coolant at the FULL line in the reserve tank to compensate for the drop in coolant level when the air bleeds.

➡Air bleeding from the inverter cooling system is completed when the noise made by the inverter water pump assembly becomes smaller and the circulation of coolant in the reserve tank improves.

➡If free spinning of the inverter water pump is detected for approximately 5 seconds, failsafe control will be activated to suspend the operation of the pump for approximately 15 seconds and resume operation for approximately 4 seconds repeatedly. Operation of the inverter water pump will return to normal if coolant is added.

➡Loud noise made by the inverter water pump assembly and poor circulation of coolant in the reserve tank indicates that there is air in the cooling system.

➡Loud noise made by the water pump and poor circulation of coolant in the reserve tank indicates that there is air in the cooling system.

2. When not using the Techstream:
   a. Turn the power switch on (READY).[*1]
   b. Turn the power switch off and add coolant to the FULL line because the coolant level drops as the air bleeds.[*2]

➡Be sure to turn the power switch off before adding SLLC.

➡Do not work on the components in the engine compartment while the vehicle is in the READY-on state because the engine is in intermittent operation.

   c. Repeat steps [*1] and [*2] until air bleeding from the cooling system is completed.

➡Air bleeding from the inverter cooling system is completed when the noise made by the inverter water pump assembly becomes smaller and the circulation of coolant in the reserve tank improves.

➡Loud noise made by the water pump and poor circulation of coolant in the reserve tank indicates that there is air in the cooling system.

3. After the air is completely bled from the cooling system, tighten the reserve tank cap.
4. Add coolant to the FULL line of the reserve tank.
5. Inspect for coolant leak (for inverter).

ENGINE ELECTRICAL

IGNITION SYSTEM

FIRING ORDER

Firing order for the 2.4L and 2.5L engines are 1–3–4–2.

Firing order for 3.5L engine: 1–2–3–4–5–6.

IGNITION COIL

REMOVAL & INSTALLATION

2.4L Engine

See Figure 69.

1. Before servicing the vehicle, refer to the Precautions Section.
2. Disconnect the negative battery cable.
3. Remove engine cover(s).
4. Disconnect the 4 ignition coil connectors. Remove the 4 bolts and 4 ignition coils.

To install:

5. To install, reverse removal procedure. Tighten ignition coil bolts to 80 inch lbs. (9 Nm).

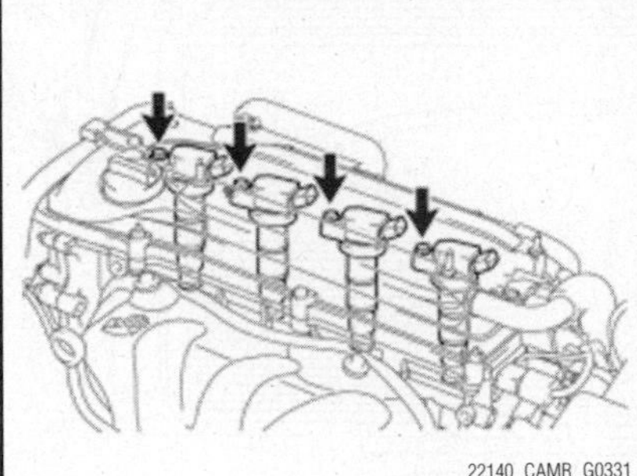

Fig. 69 Removing ignition coils—2.4L engines

2.5L Engine

For ignition coil removal and installation refer to SPARK PLUG.

3.5L Engine

See Figure 70.

1. Before servicing the vehicle, refer to the Precautions Section.
2. Disconnect the negative battery cable.
3. Drain and recycle the engine coolant.
4. Remove windshield wiper link assembly.
5. Remove cowl top panel outer sub-assembly.
6. Remove v-bank cover sub-assembly.
7. Remove air cleaner cap sub-assembly.

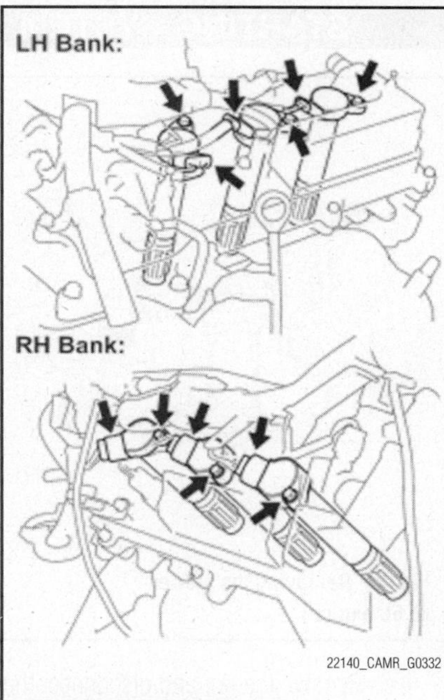

Fig. 70 Removing ignition coils—3.5L engines

8. Remove intake air surge tank assembly.
9. Remove No. 1 surge tank stay by performing the following:
   a. Remove the bolt and disconnect the harness clamp.
   b. Remove the bolt and No. 1 surge tank stay.
10. Disconnect the 6 ignition coil connectors.
11. Remove the 6 bolts and 6 ignition coils.

To install:

12. To install, reverse removal procedure.
13. Tighten the following to specification:
   a. 6 ignition coil bolt: 10 ft. lbs. (10 Nm).
   b. No. 1 surge tank stay bolt: 15 ft. lbs. (21 Nm).
   c. No. 1 surge tank stay bolt and clamp: 62 inch lbs. (7 Nm).

IGNITION TIMING

ADJUSTMENT

All engines are equipped with a Distributorless Ignition System (DIS). No timing adjustment is possible.

SPARK PLUGS

REMOVAL & INSTALLATION

2.4L Engine

See Figure 71.

1. Before servicing the vehicle, refer to the Precautions Section.
2. Remove the plastic engine cover.
3. Disconnect the 4 ignition coil connectors and remove the 4 bolts and ignition coils.
4. Using a 16 mm (0.63 in.) spark plug wrench, remove the 4 spark plugs.

To install:

5. Installation is the reverse of removal.
   a. Torque the ignition coils to 66 inch lbs. (7.5 Nm) and the spark plugs to 13 ft. lbs (18 Nm).

2.5L Engine

See Figure 72.

1. Remove the No. 1 engine cover sub assembly.
2. Remove the ignition coil assembly.
   a. Disconnect the 4 ignition coil assembly connectors.
   b. Remove the 4 bolts and 4 ignition coil assemblies.
3. Remove the spark plug.
   a. Using a spark plug wrench, remove the 4 spark plugs.

To install:

4. Install the spark plugs.
   a. Using a spark plug wrench, install the 4 spark plugs. Tighten to 18 ft. lbs. (25 Nm).

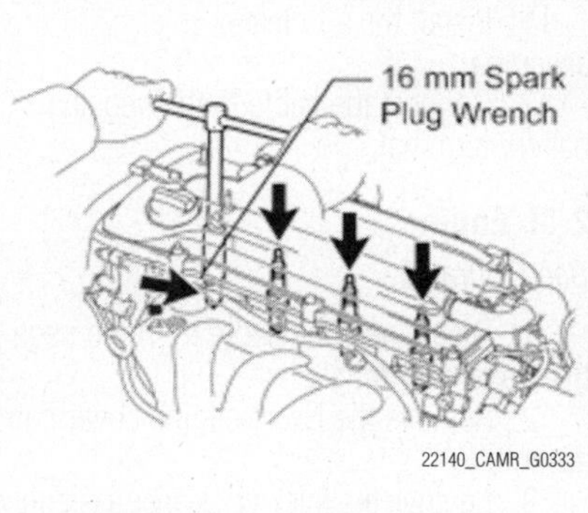

Fig. 71 Removing spark plugs—2.4L engines

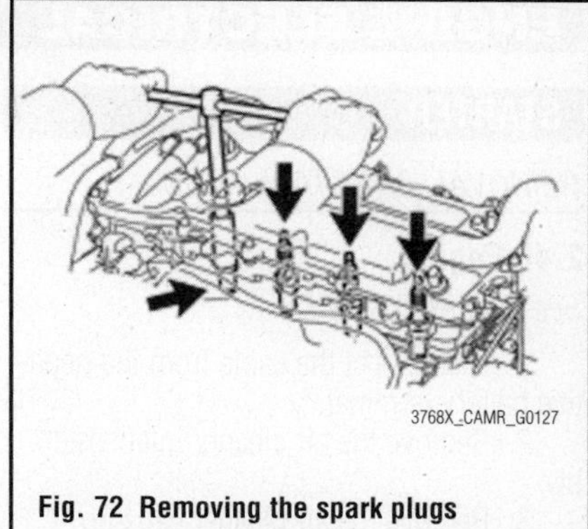

Fig. 72 Removing the spark plugs

5. Install the ignition coil assembly.
   a. Install the 4 ignition coil assemblies with the 4 bolts. Tighten to 7 ft. lbs. (10 Nm).

3.5L Engine

See Figure 73.

1. Before servicing the vehicle, refer to the Precautions Section.
2. Remove the V-bank cover.
3. Remove the intake air surge tank.
4. Disconnect the 6 ignition coil connectors.
5. Remove the 6 bolts and 6 ignition coils.
6. Using a 16 mm (0.63 in.) plug wrench, remove the spark plugs.

To install:

7. Installation is the reverse of removal, noting the following:
   a. Torque the ignition coils to 66 inch lbs. (7.5 Nm) and the spark plugs to 13 ft. lbs. (18 Nm).

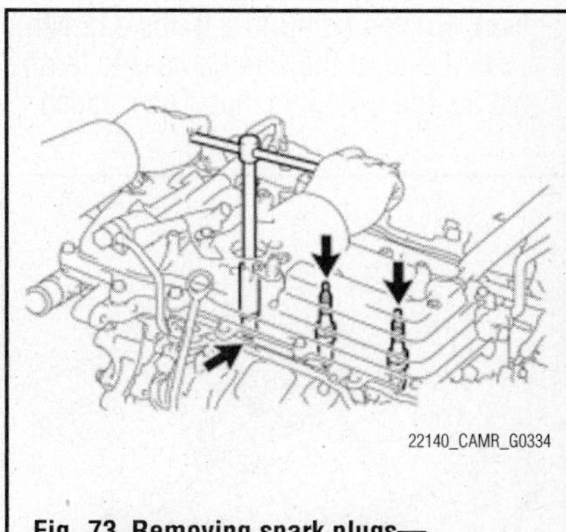

Fig. 73 Removing spark plugs—3.5L engines

STARTER

REMOVAL & INSTALLATION

2.4L Engine

See Figures 74 and 75.

1. Disconnect the cable from the negative battery terminal.
2. Remove the air cleaner inlet assembly.
3. Remove the air cleaner cap sub assembly.
4. Remove the air cleaner case sub assembly.
5. Remove the starter assembly (manual transaxle).
   a. Disconnect the terminal 50 connector from the starter assembly.
   b. Remove the nut and disconnect the wire harness from terminal 30.
   c. Remove the 3 bolts, clutch flexible hose bracket and starter assembly.
6. Remove the starter assembly (automatic transaxle).
   a. Disconnect the terminal 50 connector from the starter assembly.
   b. Remove the nut and disconnect the wire harness from terminal 30.
   c. Remove the 2 bolts and starter assembly.

To install:

7. Install the starter assembly (manual transaxle).
   a. Install the starter assembly and clutch flexible hose bracket with the 3 bolts. Tighten bolt A to 28 ft. lbs. (37 Nm). Tighten bolt B to 9 ft. lbs. (12 Nm).
   b. Connect the wire harness to terminal 30 and install the nut. Then, attach the terminal cap. Tighten to 87 inch lbs. (9.8 Nm).
   c. Connect the terminal 50 connector to the starter assembly.
8. Install the starter assembly (automatic transaxle).
   a. Install the starter assembly with the 2 bolts. Tighten to 28 ft. lbs. (37 Nm).
   b. Connect the wire harness to terminal 30 and install the nut. Then, attach the terminal cap. Tighten to 87 inch lbs. (9.8 Nm).
   c. Connect the terminal 50 connector to the starter assembly.
9. Install the air cleaner case sub assembly.
10. Install the air cleaner cap sub assembly.
11. Install the air cleaner inlet assembly.
12. Connect the cable to the negative battery terminal.

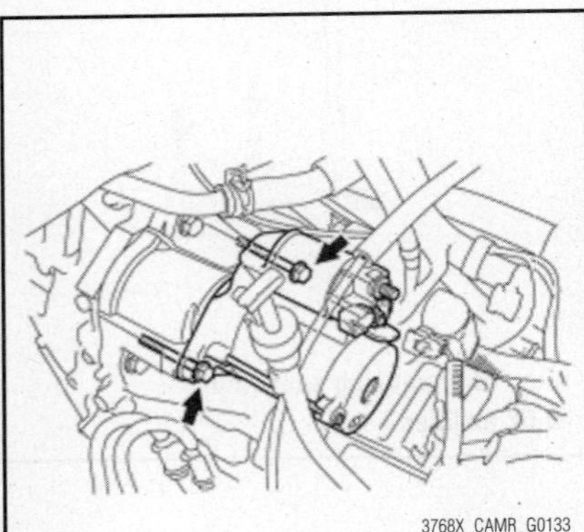

Fig. 74 Removing the starter assembly (2.4L engine, automatic transaxle)

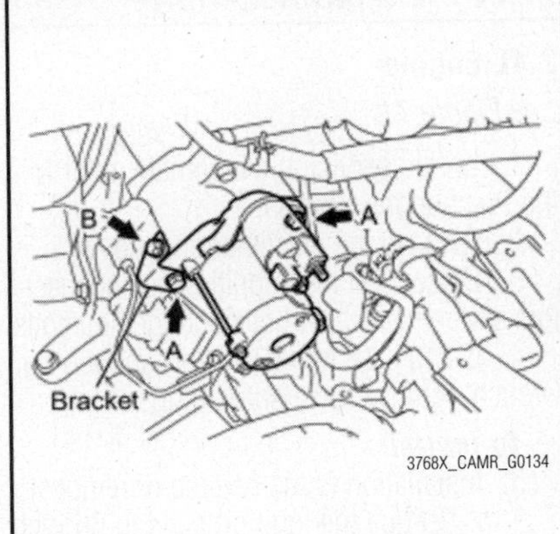

Fig. 75 Installing the starter assembly (manual transaxle)

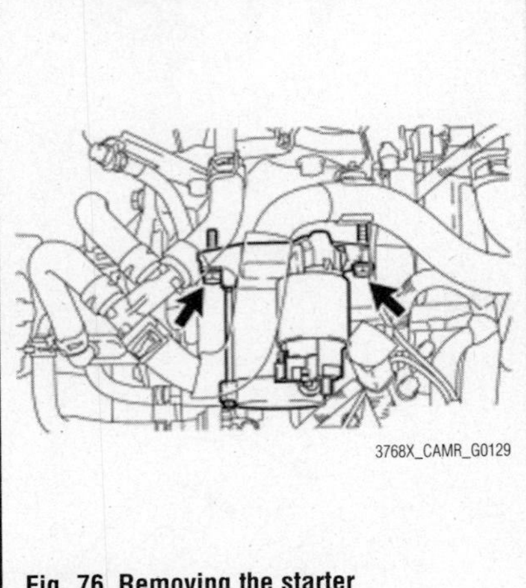

Fig. 76 Removing the starter (2.5L engine)

2.5L Engine

See Figure 76.

1. Disconnect the cable from the negative battery terminal.
2. Remove the No. 1 engine cover sub assembly.
3. Remove the inlet air cleaner assembly.
4. Remove the air cleaner cap sub assembly.
5. Remove the air cleaner filter element.
6. Remove the air cleaner case sub assembly.
7. Remove the starter assembly.
   a. Disconnect the terminal 50 connector from the starter assembly.
   b. Turn back the terminal cap.
   c. Remove the nut and disconnect the wire harness from terminal 30.
   d. Remove the 2 bolts and starter assembly

To install:

8. Install the starter assembly.
   a. Install the starter assembly with the 2 bolts. Tighten to 27 ft. lbs. (37 Nm).
   b. Connect the wire harness to terminal 30 and install the nut. Then, attach the terminal cap. Tighten to 87 inch lbs. (9.8 Nm).
   c. Connect the terminal 50 connector to the starter assembly.
9. Install the air cleaner case sub assembly.
10. Install the air cleaner filter element.
11. Install the air cleaner cap sub assembly.
12. Install the air cleaner assembly.
13. Install the No. 1 engine cover sub assembly.
14. Connect the cable to the negative battery terminal.

3.5L Engine

See Figure 77.

1. Before servicing the vehicle, refer to the Precautions Section.
2. Disconnect the negative battery cable.
3. Remove cool air intake duct seal.
4. Remove v-bank cover sub-assembly.
5. Remove air cleaner inlet assembly.
6. Remove air cleaner cap sub-assembly.
7. Remove air cleaner case sub-assembly.
8. Remove No. 1 air cleaner inlet.
9. Disconnect the terminal 50 connector from the starter assembly.

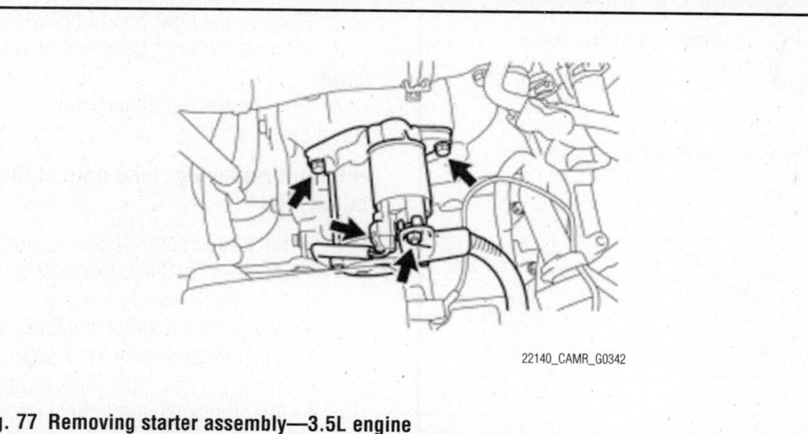

Fig. 77 Removing starter assembly—3.5L engine

10. Remove the nut and disconnect the wire harness from terminal 30.
11. Remove the 2 bolts and starter assembly.

To install:

12. Install the starter assembly with the 2 bolts and tighten to 26 ft. lbs. (37 Nm).
13. Connect the wire harness to terminal 30 and install the nut and tighten to 87 inch lbs. (9.8 Nm).
14. Cover the nut with the cap.
15. Connect terminal 50 to the starter assembly.
16. To complete installation, reverse removal procedure.

ENGINE MECHANICAL

➡Disconnecting the negative battery cable may interfere with the functions of the on board computer systems and may require the computer to undergo a relearning process, once the negative battery cable is reconnected.

ACCESSORY DRIVE BELTS

ACCESSORY BELT ROUTING

See Figures 78 through 80.

INSPECTION

See Figure 81.

Visually check the V-ribbed belt for excessive wear, frayed cords, etc. If any defect has been found, replace the V-ribbed belt.

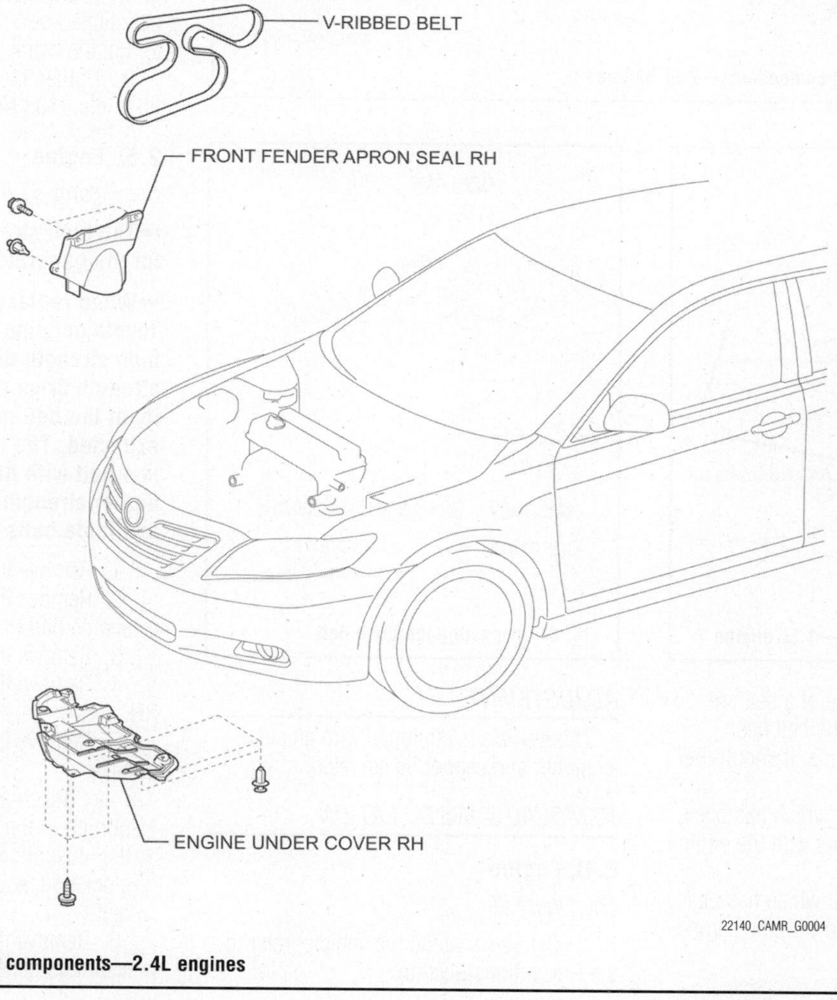

V-RIBBED BELT
FRONT FENDER APRON SEAL RH
ENGINE UNDER COVER RH
22140_CAMR_G0004

Fig. 78 Locating drive belt components—2.4L engines

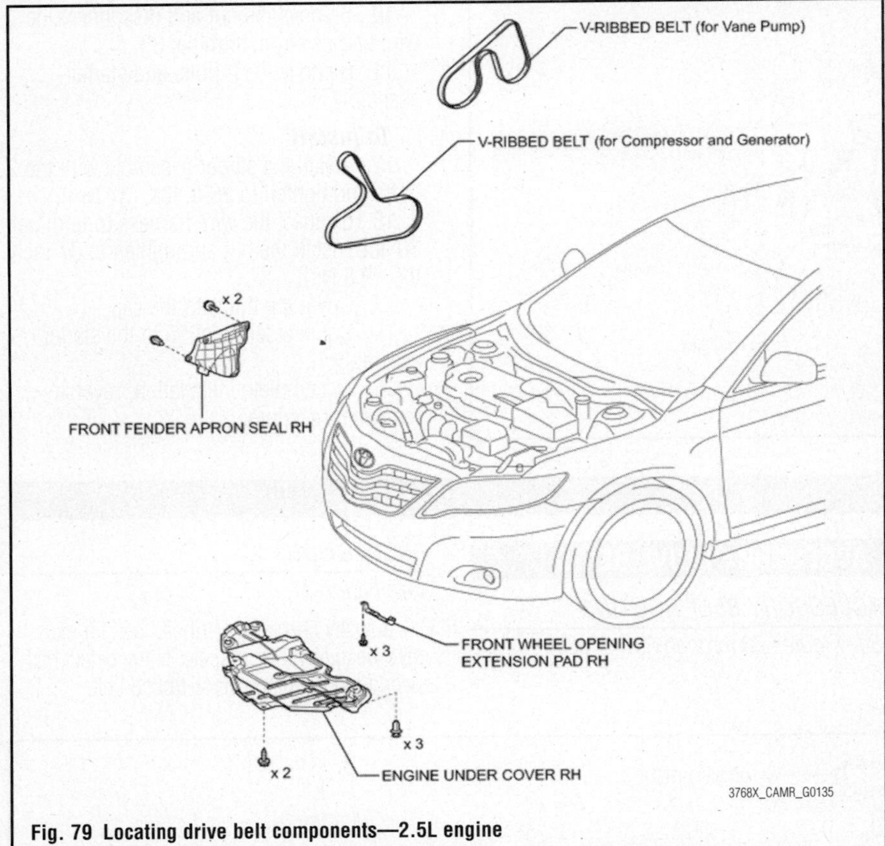

Fig. 79 Locating drive belt components—2.5L engine

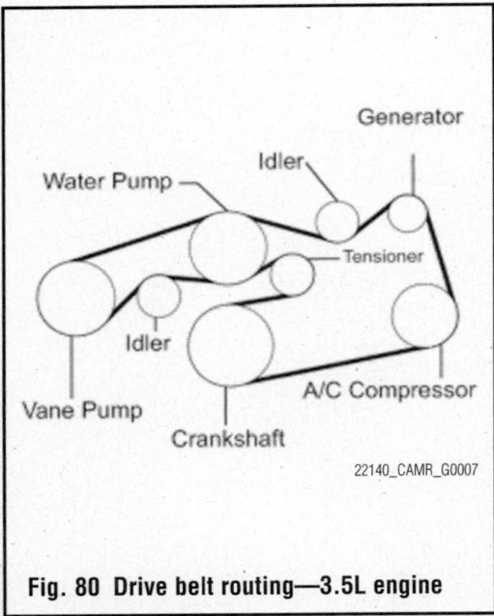

Fig. 80 Drive belt routing—3.5L engine

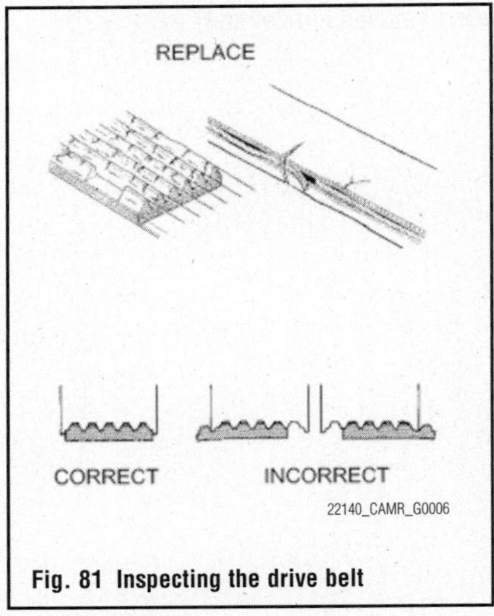

Fig. 81 Inspecting the drive belt

- Cracks on the rib side of a belt are considered acceptable. If the belt has chunks missing from the ribs, it should be replaced
- A "new belt" is a belt which has been used for less than 5 minutes with the engine running
- A "used belt" is a belt which has been used for 5 minutes or more with the engine running

ADJUSTMENT

This vehicle is equipped with an auto-tensioner and cannot be adjusted.

REMOVAL & INSTALLATION

2.4L Engine

See Figure 78

1. Before servicing the vehicle, refer to the Precautions Section.
2. Remove the right hand front wheel.
3. Remove the right hand engine under cover.
4. Remove the right hand front fender apron seal.

➡Before removing, take note of the following:

- Be sure to connect Special Tool: 09216-42010 and the tools so that they are in line during use
- When retracting the tensioner, turn it clockwise slowly for 3 seconds or more. Do not apply force rapidly
- After the tensioner is fully retracted, do not apply force any more than necessary

5. Using the Special Tool and a 19 mm socket wrench, loosen the v-ribbed belt tensioner arm clockwise, then remove the v-ribbed belt.
6. Remove the v-ribbed belt.

To install:

7. To install, reverse removal procedure.
8. After installing the V-ribbed belt, check that it fits properly in the ribbed grooves. Check to confirm that the belt has not slipped out of the grooves on the bottom of the crank pulley by hand.
9. Tighten the right hand front wheel to: 76 ft. lbs. (103 Nm).

2.5L Engine

See Figures 82 through 85.

➡The high strength drive belt is used for the generator side drive belt.

➡When replacing the drive belt, use Toyota genuine drive belt or equivalent high strength drive belt. If the high strength drive belt is not used, durability of the belt may become less than expected. The high strength drive belt is a belt with Aramid core which has higher strength compared to usually available belts with PET or PEN core.

1. Remove the front wheel RH.
2. Remove the front wheel opening extension pad RH.
3. Remove the engine under cover RH.
4. Remove the front fender apron seal RH.
5. Remove the v-ribbed belt (for vane pump).
    a. Attach a wrench to the hexagonal portion of the belt tensioner as shown in the illustration, rotate the belt tensioner clockwise, and remove the V-ribbed belt.
6. Remove the v-ribbed belt (for compressor and generator).

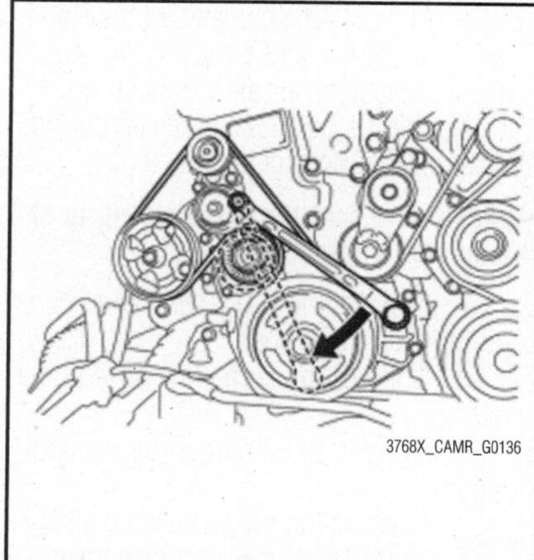

Fig. 82 Removing the v-ribbed belt (for vane pump)

a. Attach a wrench to the hexagonal portion of the belt tensioner as shown in the illustration, rotate the belt tensioner clockwise, and remove the V-ribbed belt.

To install:

7. Install the v-ribbed belt (for compressor and generator).
   a. Set the v-ribbed belt onto each part as shown, except the water pump pulley.
   b. Loosen the V-ribbed belt by turning the belt tensioner clockwise.
   c. Set the V-ribbed belt onto the water pump pulley.

➡Make sure that the belt is attached to each pulley. In particular, make sure that the belt is securely fitted into the grooves of the crankshaft pulley.

8. Install the v-ribbed belt (for vane pump).
   a. Set the V-ribbed belt onto each part as shown in the illustration except the tensioner pulley.

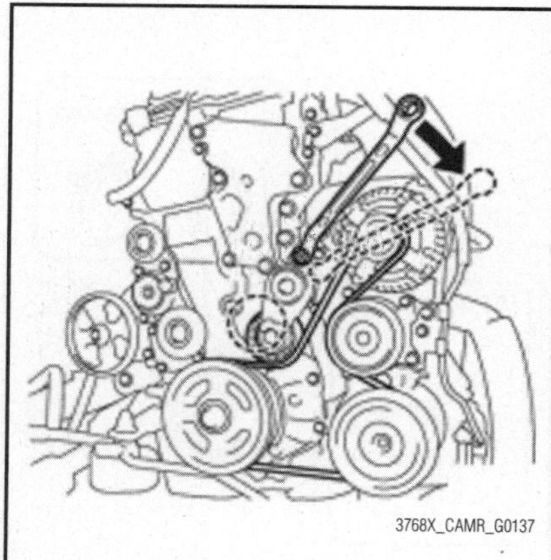

Fig. 83 Removing the v-ribbed belt (for compressor and generator)

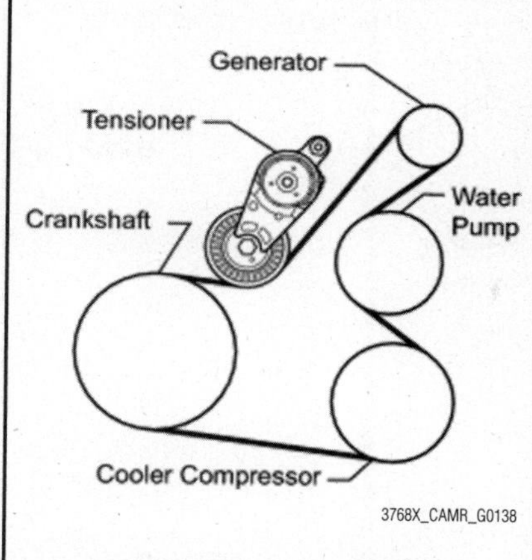

Fig. 84 Installing the v-ribbed belt (for compressor and generator)

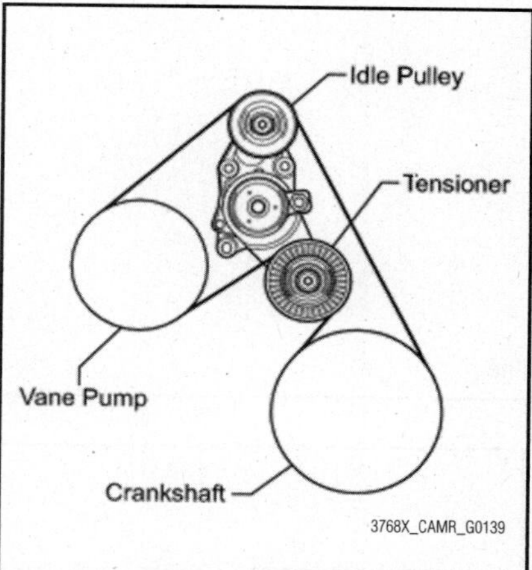

Fig. 85 Installing the v-ribbed belt (for vane pump)

   b. Loosen the V-ribbed belt by turning the belt tensioner clockwise.
   c. Set the V-ribbed belt onto the tensioner pulley.

➡Make sure that the belt is attached to each pulley. In particular, make sure that the belt is securely fitted into the grooves of the crankshaft pulley.

9. Install the front fender apron seal RH.
10. Install the engine under cover RH.
11. Install the front wheel opening extension pad RH.
12. Install the front wheel RH. Tighten to 76 ft. lbs. (103 Nm).

3.5L Engine

See Figures 86 and 87.

1. Before servicing the vehicle, refer to the Precautions Section.
2. Remove the right hand front wheel.
3. Remove the right hand front fender apron seal.
4. Remove the V-bank cover sub-assembly.
5. Using Special Tool: 09249-63010, release the belt tension by turning the belt tensioner counterclockwise, and remove the V-ribbed belt from the belt tensioner.
6. While turning the belt tensioner counterclockwise, align with its holes and then insert the 5 mm bi-hexagon wrench into the holes to fix the V-ribbed belt tensioner.
7. Remove the v-ribbed belt.

To install:

8. To install, reverse removal procedure.
9. If it is difficult to install the V-ribbed belt, perform the following procedure:
   a. Put the V-ribbed belt on every pulley except the tensioner pulley.
   b. While releasing the belt tension by turning the belt tensioner counterclockwise, put the V-ribbed belt on the tensioner pulley.

➡Put the backside of the V-ribbed belt on the tensioner pulley and idler pulley. Check that the V-ribbed belt is properly set to each pulley.

10. After installing the V-ribbed belt, check that it fits properly in the ribbed grooves. Check to confirm that the belt has not slipped out of the grooves on the bottom of the crank pulley by hand.
11. Tighten the right hand front wheel to: 76 ft. lbs. (103 Nm).

AIR CLEANER

REMOVAL & INSTALLATION

Non-Hybrid Models

See Figures 88 and 89.

1. Remove the air cleaner inlet assembly.
   a. Remove the 2 bolts, clamp ad air cleaner inlet.
2. Remove the air cleaner cap sub assembly.
   a. Disconnect the Mass Air Flow meter connector.
   b. Disconnect the purge VSV connector.
   c. Disconnect the 2 purge VSV vacuum hoses.
   d. Disconnect the purge line hose from the clamp.
   e. Disconnect the No. 2 ventilation hose from the air cleaner hose.
   f. Lock the No. 1 air cleaner hose

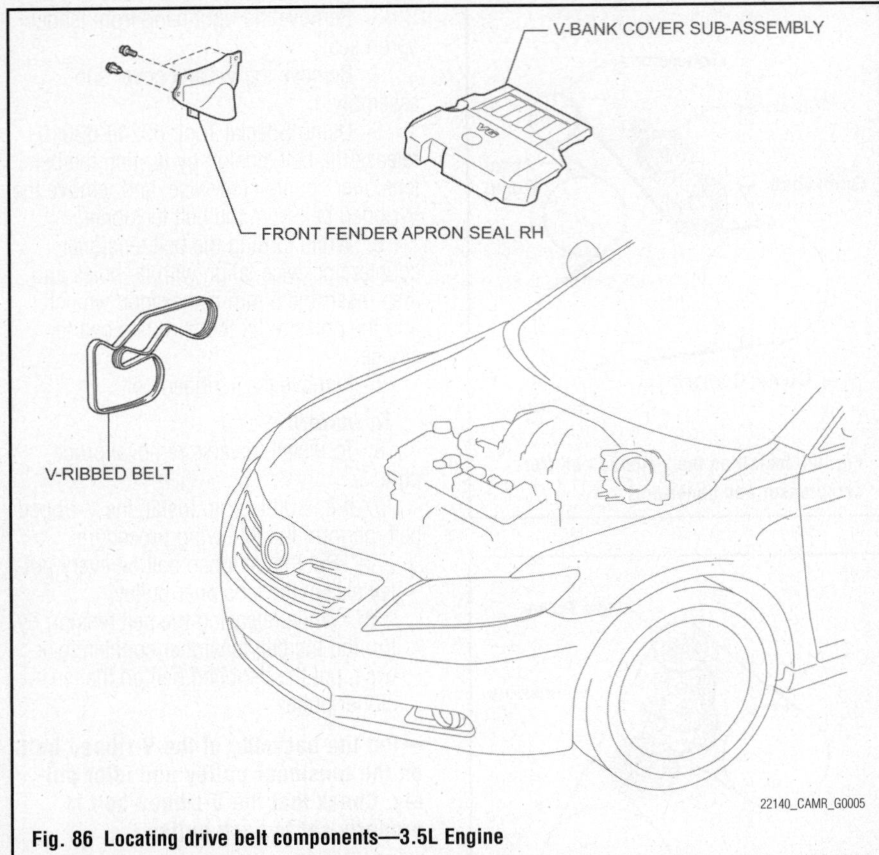

Fig. 86 Locating drive belt components—3.5L Engine

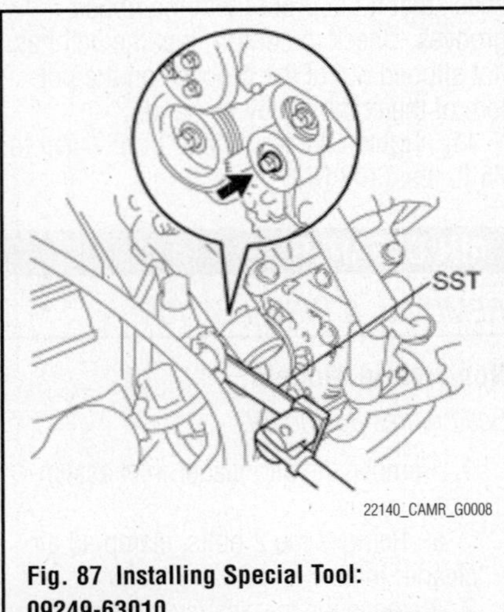

Fig. 87 Installing Special Tool: 09249-63010

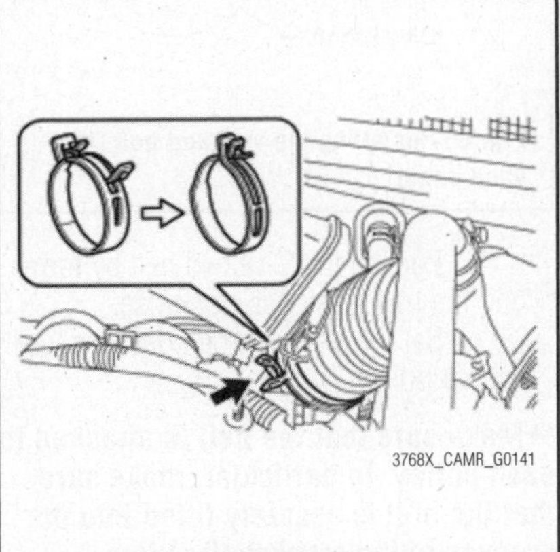

Fig. 88 Locking the No. 1 air cleaner hose clamp, and then disconnecting the No. 1 air cleaner hose from the throttle body

clamp, and then disconnect the No. 1 air cleaner hose from the throttle body.

g. Remove the 2 bolts and the air cleaner cap.

h. Remove the air cleaner filter element from the air cleaner case.

To install:

3. Install the air cleaner case sub assembly.

a. Install the air cleaner case with the 3 bolts. Tighten to 44 inch lbs. (5 Nm).

b. Connect the hose clamp.

4. Install the air cleaner cap sub assembly.

a. Install the air cleaner filter element onto the air cleaner case.

b. Insert the hinges. Install the air cleaner cap sub assembly with the 2 bolts.

c. Align the matchmarks of the No. 1 air cleaner hose and throttle body, and then connect the air cleaner hose No. 1 to the throttle body and unfasten the No. 1 air cleaner hose clamp.

➡Make sure that the hose clamp is at the correct angle.

d. Connect the No. 2 ventilation hose to the air cleaner hose.

e. Connect the purge line hose to the clamp.

f. Connect the 2 purge VSV vacuum hoses.

g. Connect the purge VSV connector.

h. Connect the mass air flow meter connector.

5. Install the air cleaner inlet assembly with 2 bolts. Tighten to 44 inch lbs. (5 Nm).

Hybrid Models

See Figure 90.

1. Remove the air cleaner inlet assembly.

a. Remove the 2 bolts, clamp and air cleaner inlet.

2. Remove the air cleaner.

To install:

To install, reverse the removal procedure.

FILTER/ELEMENT REPLACEMENT

For filter/element replacement, see AIR CLEANER.

CAMSHAFT AND VALVE LIFTERS

REMOVAL & INSTALLATION

⁂ CAUTION

All models are equipped with a Supplemental Restraint System (SRS),

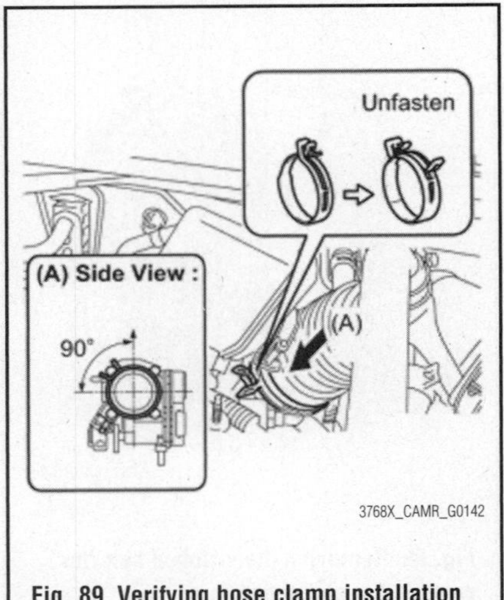

Fig. 89 Verifying hose clamp installation

Fig. 90 Removing the air cleaner inlet assembly

which uses an air bag. Whenever working near any of the SRS components, such as the impact sensors, the air bag module, steering column and instrument panel, disable the SRS.

2.4L Engine

See Figures 91 through 102.

1. Before servicing the vehicle, refer to the Precautions Section.
2. Disconnect the negative battery cable.
3. Loosen the lug nuts on the front right hand wheel.
4. Apply the parking brake, block the rear wheels, then raise and safely support the front of the vehicle securely on jackstands.
5. Remove the front right hand wheel.
6. Remove the left and right hand under cover.

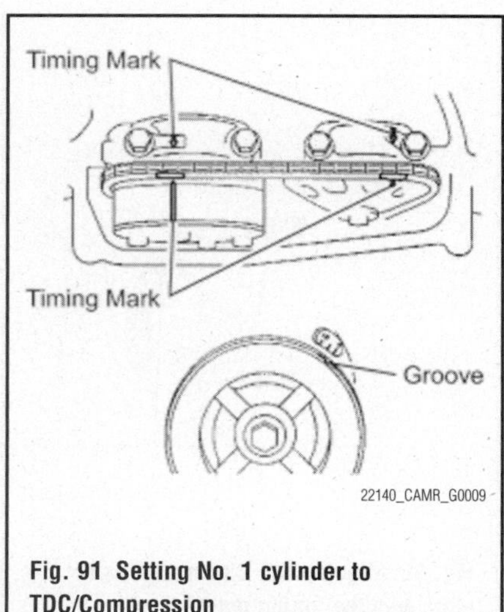

Fig. 91 Setting No. 1 cylinder to TDC/Compression

7. Remove the No.1 engine cover subassembly and 2 nuts.
8. Remove the ignition coil assembly.
9. Remove the cylinder head cover subassembly
10. Set No.1 Cylinder to TDC/Compression by performing the following:
    - Turn the crankshaft pulley until its groove and the timing mark "0" of the timing chain cover are aligned
    - Check that each timing mark of the camshaft timing gear and sprocket is aligned with each timing mark located on the No. 1 and No. 2 bearing caps as shown in the illustration. If not, turn the crankshaft by 1 revolution (360°) to align the timing marks

➡Do not turn the crankshaft without the chain tensioner.

11. Remove No.1 Chain tensioner assembly. Remove the 2 nuts, tensioner and gasket.
12. Remove the No. 2 camshaft by performing the following:
    a. While holding the camshaft with a wrench, loosen the camshaft timing set bolt.
    b. Using several steps, uniformly loosen and remove the 10 bearing cap bolts in the sequence shown in the illustration.
    c. While holding the No. 2 camshaft by hand, remove the camshaft timing sprocket set bolt.
    d. Remove the camshaft timing sprocket from the No. 2 camshaft with the timing chain wrapped on the sprocket.
    e. Remove the camshaft timing sprocket from the timing chain.

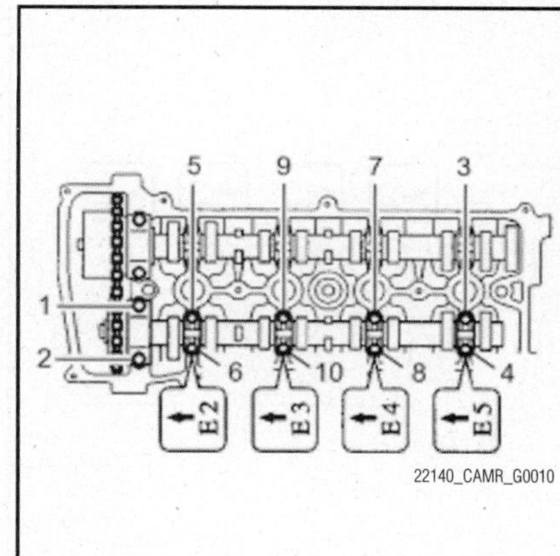

Fig. 92 No. 2 camshaft bearing cap bolt removal sequence

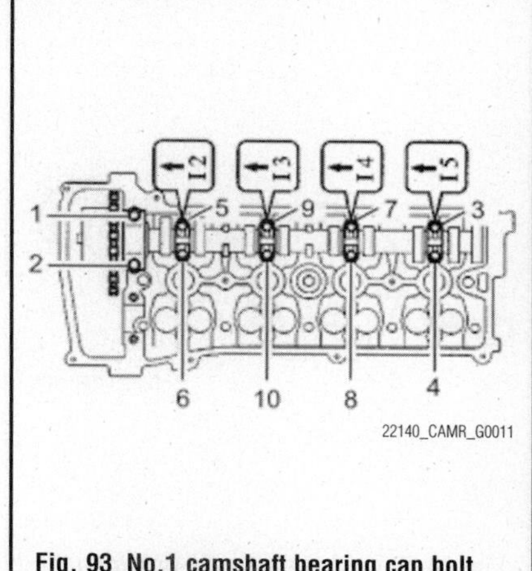

Fig. 93 No.1 camshaft bearing cap bolt removal sequence

13. Remove the No. 1 camshaft by performing the following:
    a. Using several steps, uniformly loosen and remove the 10 bearing cap bolts in the sequence shown in the illustration.
    b. Remove the 5 bearing caps.
    c. Remove the camshaft and camshaft timing gear while holding the timing chain by hand.

➡Be careful not to drop anything inside the timing chain cover.

    d. Tie the timing chain with a string.
14. Remove the camshaft timing gear assembly by performing the following:
    a. Clamp the camshaft in a vise, and make sure that the camshaft timing gear does not rotate.
    b. Cover all the oil ports except the advance side port shown in the illustration with vinyl tape

➡Cover the paths with a shop rag or piece of cloth to avoid oil splashes.

➡Depending on the air pressure, the camshaft timing gear will turn to the advance angle side without applying force by hand. Also, if the pressure is difficult to apply because of air leakage from the port, the lock may be difficult to release.

    c. Apply air pressure of 14 psi to the oil path, then turn the camshaft timing gear in the advance direction (counterclockwise) by hand.
    d. Remove the flange bolt of the camshaft timing gear. Be sure not to remove the other four bolts. If planning to reuse the gear, be sure to release the straight pin lock before installing the gear.

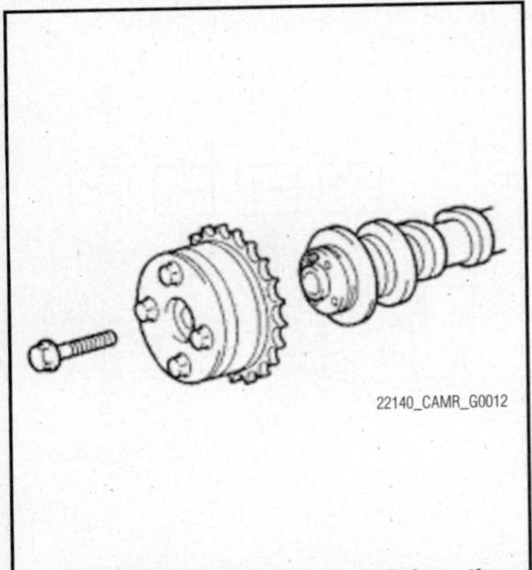

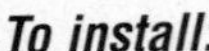

Fig. 94 Removing the flange bolt from the camshaft timing gear

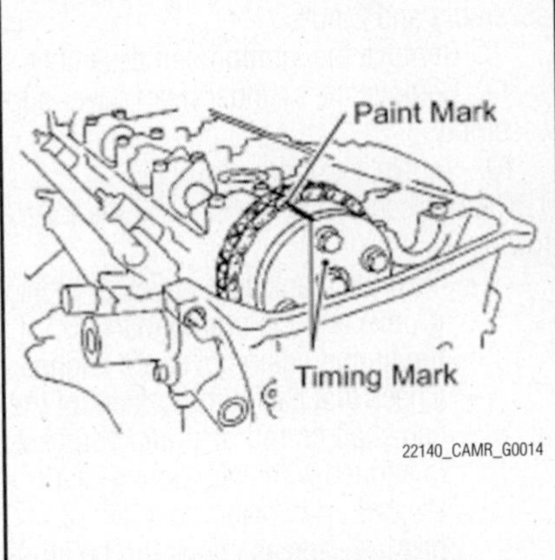

Fig. 96 Aligning No. 1 camshaft paint mark with the timing mark

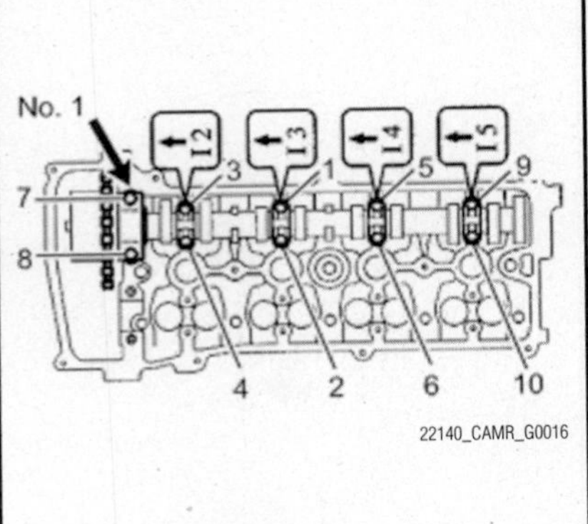

Fig. 98 No. 1 camshaft bearing cap tightening sequence

To install:

15. Put the camshaft timing gear and camshaft together with the straight pin and key groove misaligned.

➡Be sure not to turn the camshaft timing gear to the retard angle side (the right angle).

16. Turn the camshaft timing gear as shown in the illustration while pushing it gently against the camshaft. Push further at the position where the pin fits into the groove.
17. Check that there is no clearance between the gear and camshaft.
18. Tighten the flange bolt with the camshaft timing gear fixed in place and tighten to 40 ft. lbs. (54 Nm).
19. Check that the camshaft timing gear can move to the retard angle side (the right direction) and is locked in the most retarded position.
20. To install the No.1 camshaft, perform the following:
    a. Apply a light coat of engine oil to the journal portion of the camshaft.
    b. Install the timing chain onto the camshaft timing gear with the paint mark aligned with the timing mark in the camshaft timing gear.
    c. Examine the front marks and numbers, and check that the order is as shown in the illustration below. Then install the bearing caps into the cylinder head.
    d. Apply a light coat of engine oil to the threads and under the heads of the bearing cap bolts.
    e. Using several steps, uniformly tighten the 10 bearing cap bolts in the sequence shown in the illustration. Tighten the following to specification:
    - No. 1 Bearing cap: 22 ft. lbs. (30 Nm)
    - No. 3 Bearing cap: 80 inch lbs. (9 Nm)
21. To install camshaft No. 2, perform the following:
    a. Apply a light coat of engine oil to the journal portion of the No. 2 camshaft.
    b. Put the No. 2 camshaft on the cylinder head with the paint mark of the chain aligned with the timing mark on the camshaft timing sprocket.
    c. While holding the No. 2 camshaft by hand, temporarily tighten the camshaft timing sprocket set bolt.
    d. Examine the front marks and numbers, and check that the order is as shown in the illustration. Then install the bearing caps onto the cylinder head.
    e. Apply a light coat of engine oil to the threads and under the heads of the bearing cap bolts.
    f. Using several steps, uniformly tighten the 10 bearing cap bolts in the sequence shown in the illustration. Tighten the following to specification:

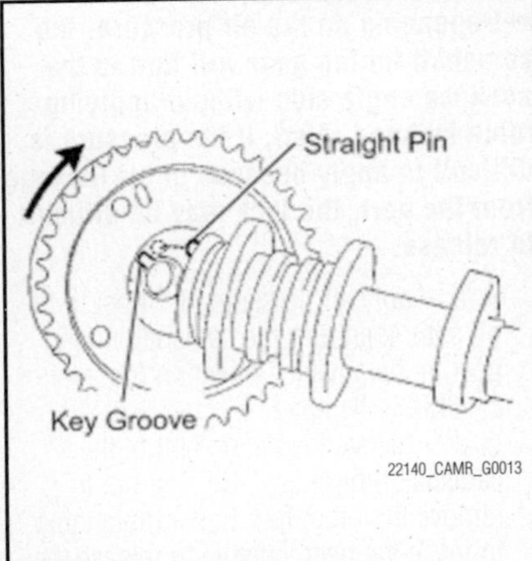

Fig. 95 No. 1 camshaft straight pin and key groove misaligned

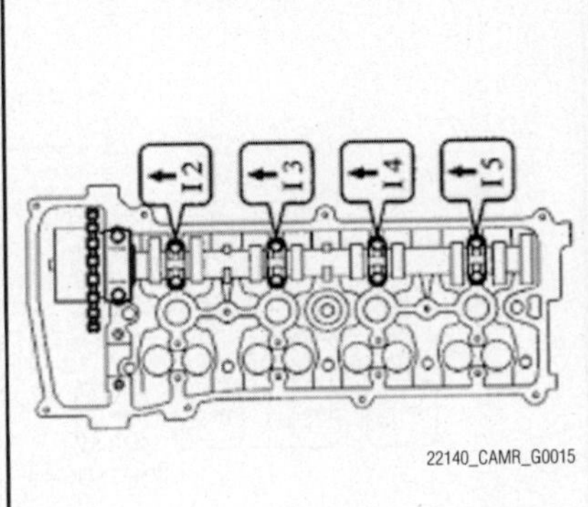

Fig. 97 Checking the No. 1 camshaft bearing cap front marks, numbers and order

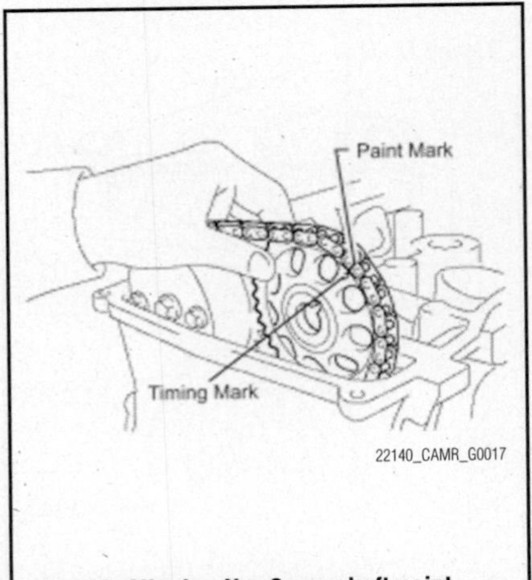

Fig. 99 Aligning No. 2 camshaft paint mark with the timing mark

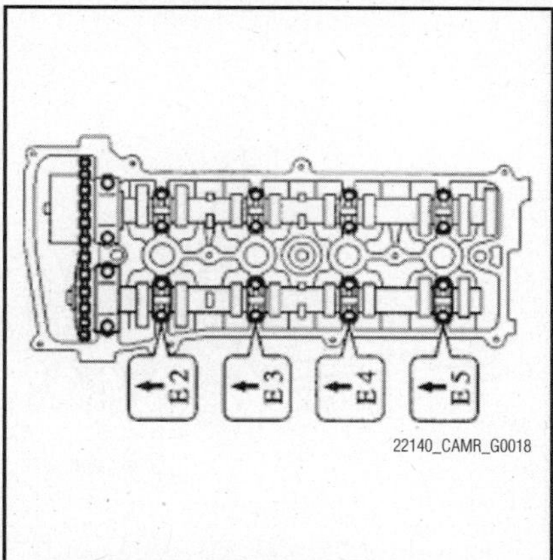

Fig. 100 Checking the No. 2 camshaft bearing cap front marks, numbers and order

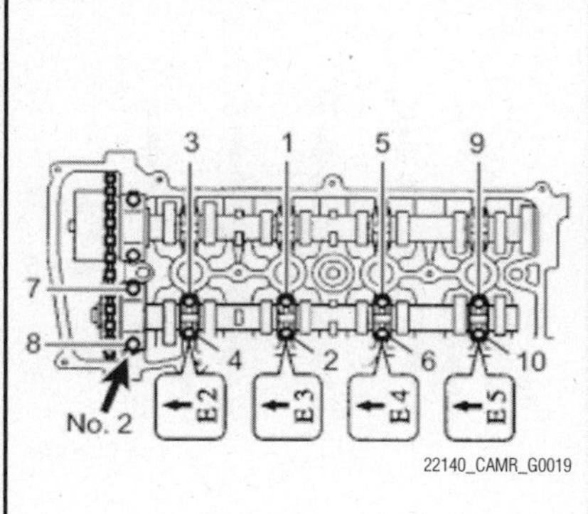

Fig. 101 No. 2 camshaft bearing cap tightening sequence

- No. 1 Bearing cap: 22 ft. lbs. (30 Nm)
- No. 3 Bearing cap: 80 inch lbs. (9 Nm)

g. While holding the camshaft with a wrench, tighten the camshaft timing sprocket set bolt and tighten to 40 ft. lbs. (54 Nm).

h. Check that the paint marks on the chain are aligned with the timing marks on the camshaft timing gear and camshaft timing sprocket. Also, check that the crankshaft pulley groove is aligned with the timing mark "0" of the timing chain cover.

22. To complete installation, reverse remaining removal procedure.

23. Check engine for oil leaks.

24. Connect the negative battery cable.

2.5L Engine

See Figures 103 through 129.

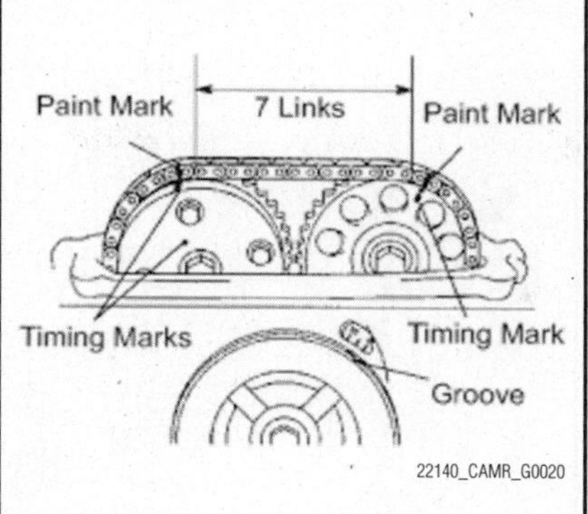

Fig. 102 Aligning paint marks on the chain with the timing marks on the camshaft timing gear and camshaft timing sprocket

1. Remove the timing chain cover sub assembly.

2. Set the No. 1 cylinder to TDC/Compression.

a. Temporarily install the crankshaft pulley bolt.

➡"A" is not a timing mark.

b. Rotate the crankshaft clockwise so that the timing marks on the crankshaft timing gear and camshaft timing gears are as shown in the illustration.

➡If the timing marks do not align, rotate the crankshaft clockwise again and align the timing marks.

c. Remove the crankshaft pulley bolt.

3. Remove the timing chain guide.

4. Remove the No. 1 chain tensioner assembly.

a. Allow the plunger to extend slightly, and then rotate the stopper plate counterclockwise to release the lock. Once the lock is released, push the plunger into the tensioner.

b. Move the stopper plate clockwise to set the lock, and insert a pin into the stopper plate hole.

c. Remove the 2 bolts, chain tensioner and the gasket.

5. Remove the chain tensioner slipper.

6. Remove the chain sub assembly.

7. Remove the 2 bolts and No. 1 chain vibration damper.

8. Remove the camshaft timing gear assembly.

a. Hold the hexagonal portion of the camshaft with a wrench and remove the bolt and camshaft timing gear.

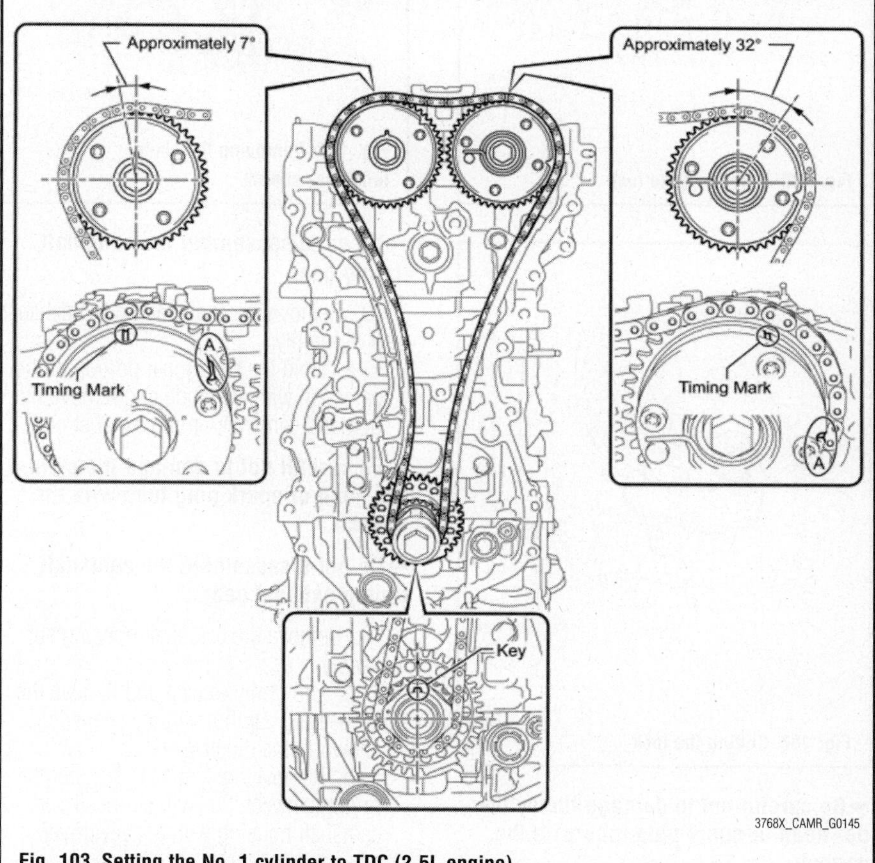

Fig. 103 Setting the No. 1 cylinder to TDC (2.5L engine)

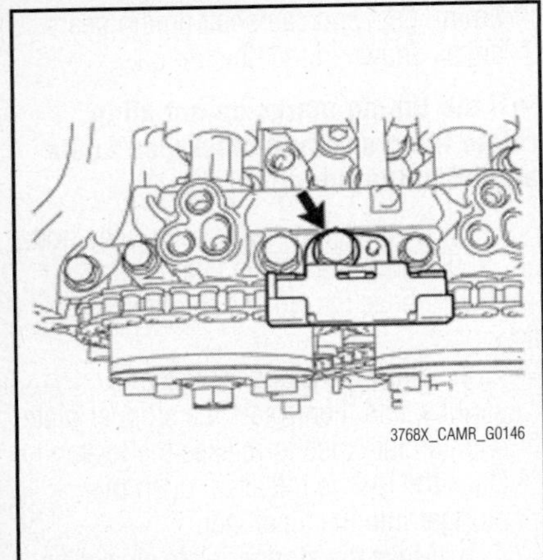

Fig. 104 Removing the timing chain bolt and guide (2.5L engine)

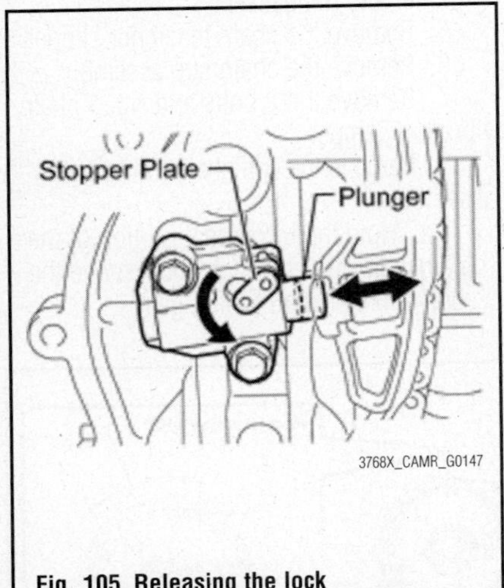

Fig. 105 Releasing the lock

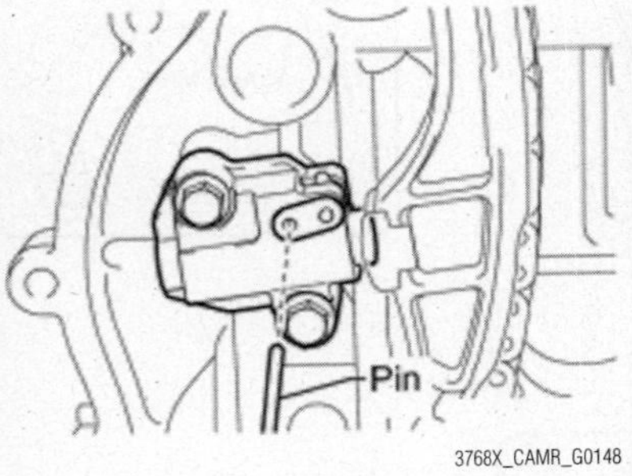

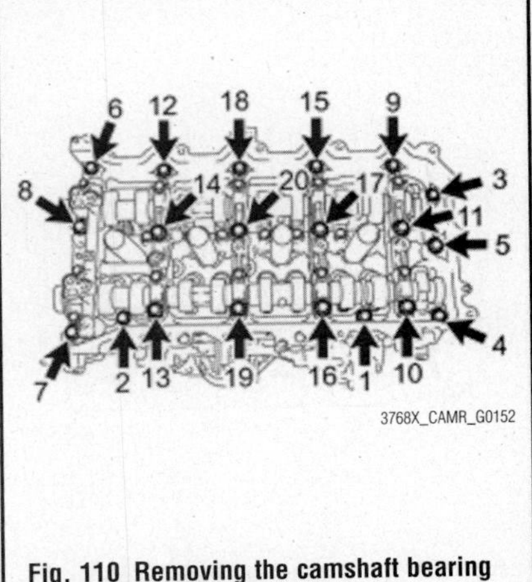

Fig. 106 Setting the lock

➡**Be careful not to damage the cylinder head or spark plug tube with the wrench.**

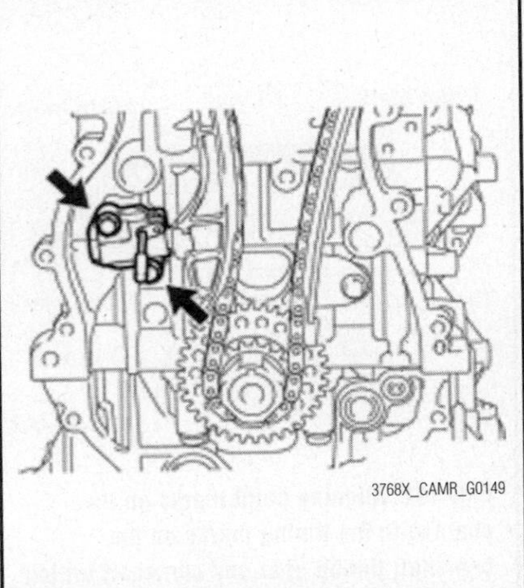

Fig. 107 Removing the chain tensioner

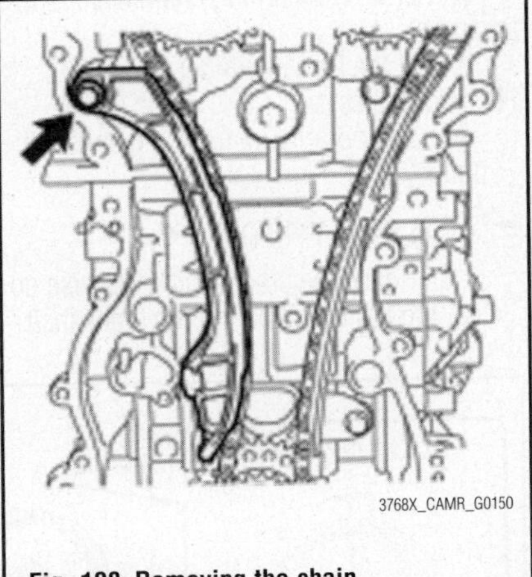

Fig. 108 Removing the chain tensioner slipper

➡**Do not disassemble the camshaft timing gear.**

9. Remove the camshaft timing exhaust gear assembly.
   a. Hold the hexagonal portion of the camshaft with a wrench and remove the bolt and camshaft timing exhaust gear.

➡**Be careful not to damage the cylinder head or spark plug tube with the wrench.**

➡**Do not disassemble the camshaft timing exhaust gear.**

10. Remove the camshaft housing sub assembly.
    a. Uniformly loosen and remove the 20 bearing cap bolts in the sequence shown in the illustration.
    b. Remove the camshaft housing by prying between the cylinder head and camshaft housing with a screwdriver.

➡**Tape the screwdriver tip before use.**

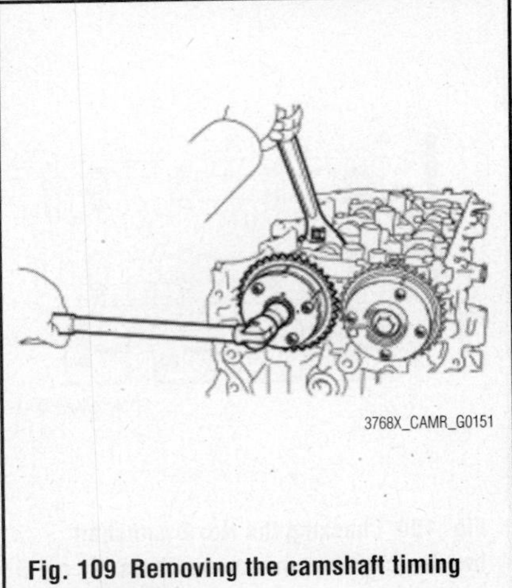

Fig. 109 Removing the camshaft timing gear assembly

Fig. 110 Removing the camshaft bearing cap bolts in sequence

➡**Be careful not to damage the contact surfaces of the cylinder head and camshaft housing.**

11. Remove the camshaft bearing cap.
    a. Remove the 11 bearing cap bolts in the sequence shown in the illustration.
    b. Removing the 5 bearing caps.

➡**Arrange the removed parts in the correct order.**

12. Remove the oil control valve filter.
13. Remove the No. 1 camshaft bearing.
14. Remove the camshafts.
15. Remove the No. 2 camshaft bearing.
16. Remove the No. 1 valve rocker arm sub assembly.
    a. Remove the 16 valve rocker arms from the cylinder head.

➡**Arrange the removed parts in the correct order.**

17. Remove the valve lash adjuster assembly.

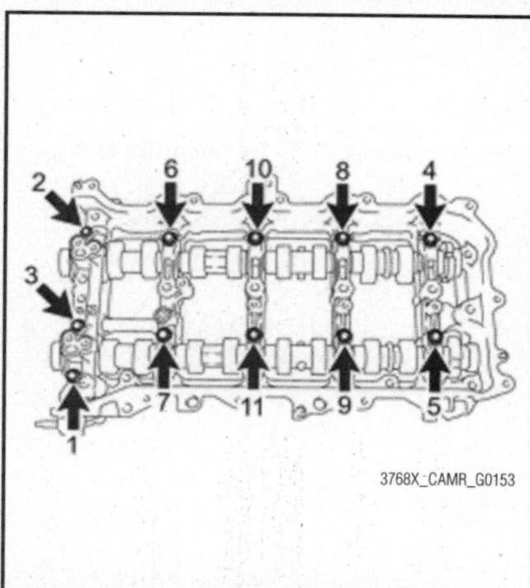

Fig. 111 Removing the 11 bearing cap bolts in sequence

a. Remove the 16 valve lash adjusters from the cylinder head.

➡Arrange the removed parts in the correct order.

To install:

18. Set the camshaft timing gear assembly.

➡When installing the camshaft timing gear, release the lock pin and set the camshaft timing gear to the advanced position before installation.

a. Check the camshaft timing gear position.

➡If the camshaft timing gear is set to the advanced position, do not let the camshaft timing gear rotate clockwise during installation.

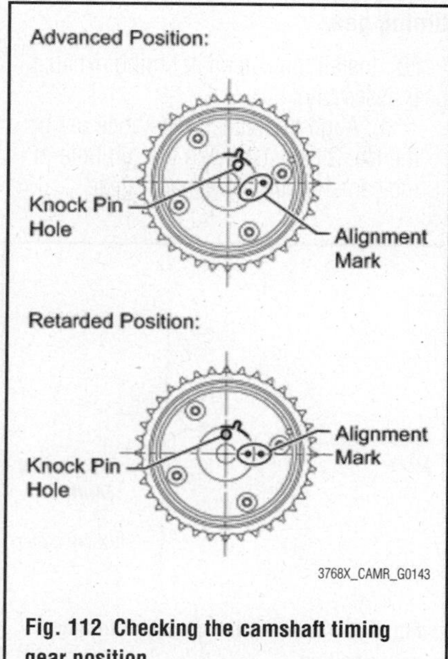

Fig. 112 Checking the camshaft timing gear position

➡If the camshaft timing gear rotates to the retarded position, release the lock pin and set the camshaft timing gear to the advanced position.

b. Align and attach the knock pin of the No. 1 camshaft with the pin hole of the camshaft timing gear.

c. Check that there is no clearance between the camshaft timing gear and camshaft flange.

d. Secure the camshaft in place by hand, and then install the installation bolt of the camshaft timing gear by hand.

➡Do not use any tools to install the bolt. If the bolt is installed using a tool, the lock pin will be damaged.

e. Release the lock pin. Clean the camshaft journal with non-residue solvent. Cover the 4 oil paths of the cam journal with vinyl tape as shown in the illustration.

➡There are 4 oil paths in the grooves of the camshaft. Plug three of the paths with pieces of rubber.

f. Open a hole at port A shown in the illustration.

g. While applying approximately 29 psi (200 kPa) of air pressure to the oil path, forcibly turn the camshaft timing gear assembly in the advance direction (counterclockwise).

⁂ CAUTION

Cover the paths with a piece of cloth when applying pressure to keep oil from splashing.

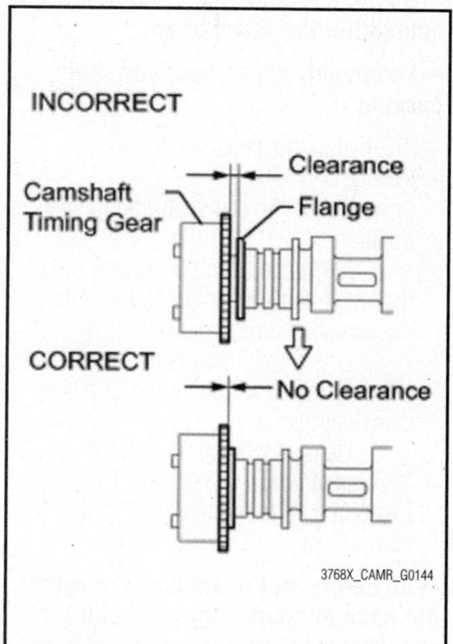

Fig. 113 Checking camshaft clearance

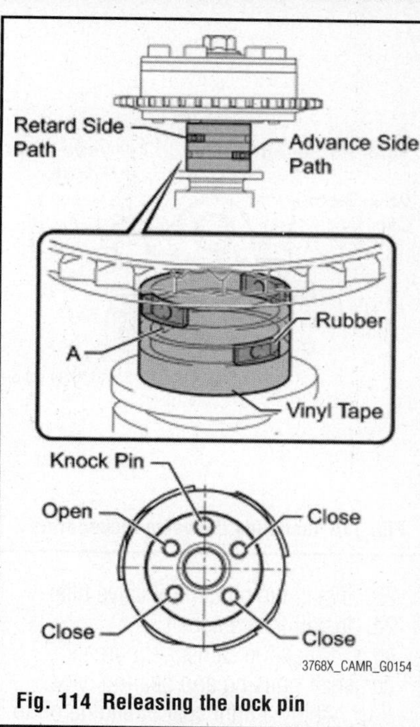

Fig. 114 Releasing the lock pin

➡Do not allow the camshaft timing gear assembly to lock. If it locks, release the lock pin again.

➡The camshaft timing gear assembly may be turned in the advance direction without applying any force.

➡If enough air pressure cannot be applied because of air leakage from the port, releasing the lock pin may be difficult.

h. Remove the vinyl tape and rubber pieces from the camshaft.

i. Remove the bolt and camshaft timing gear.

➡Do not allow the camshaft timing gear assembly to lock. If it locks, release the lock pin again.

19. Install the valve lash adjuster assembly.

a. Inspect the valve lash adjuster before installing it.

b. Inspect the 16 lash adjusters to the cylinder head.

➡Install the lash adjuster to the same place it was removed from.

20. Install the No. 1 valve rocker arm sub assembly.

a. Apply engine oil to the lash adjuster tips and valve stem caps.

b. Install the 16 valve rocker arms as shown.

21. Install the No. 2 camshaft bearing.

22. Install the No. 1 camshaft bearing.

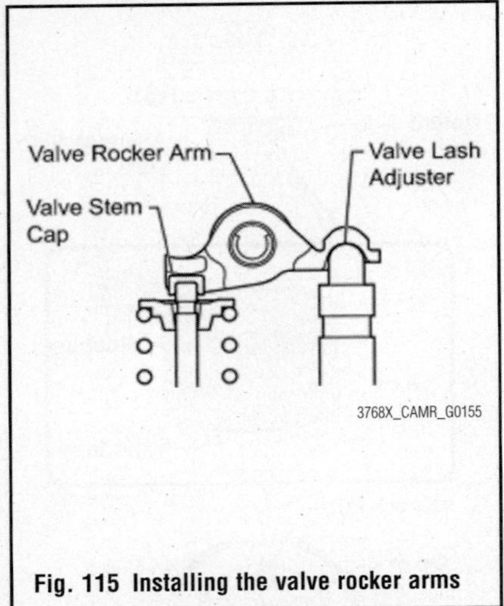

Fig. 115 Installing the valve rocker arms

23. Install the oil control valve filter.
24. Install the camshaft.
   a. Clean the camshaft journals, camshaft housing and bearing caps.
   b. Apply a light coat of engine oil to the camshaft journal, camshaft housing and bearing caps.
   c. Install the No. 1 and No. 2 camshafts to the camshaft housing.
25. Install the camshaft bearing cap.
   a. Confirm the marks and numbers on the camshaft bearing caps and place them in their proper positions and directions.
   b. Install the 11 bolts in the order shown in the illustration. Tighten to 12 ft. lbs. (16 Nm).

➡Make sure that the camshaft rotates smoothly after installing the bearing caps.

26. Install the camshaft housing sub assembly.
   a. Check that the valve rocker arms are installed as shown in the illustration.
   b. Apply seal packing in a continuous line. The line diameter should be 0.118-0.157 inch (3-4 mm).

➡Remove any oil from the contact surface.

➡Install the camshaft housing within 3 minutes and tighten the bolts within 10 minutes after applying seal packing.

   c. Position the knock pin of the No. 1 and No. 2 camshafts as shown in the illustration.
   d. Install the camshaft housing, and then install the 20 bolts in the order shown in the illustration. Tighten to 20 ft. lbs. (27 Nm).

➡Do not apply oil for at least 4 hours after the installation.

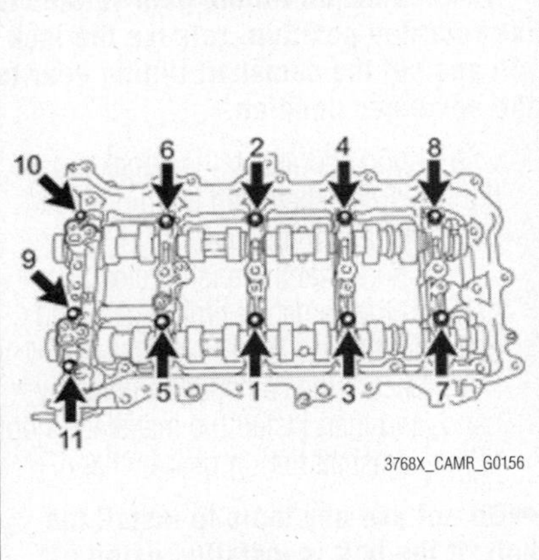

Fig. 116 Identifying the camshaft bearing cap order of installation

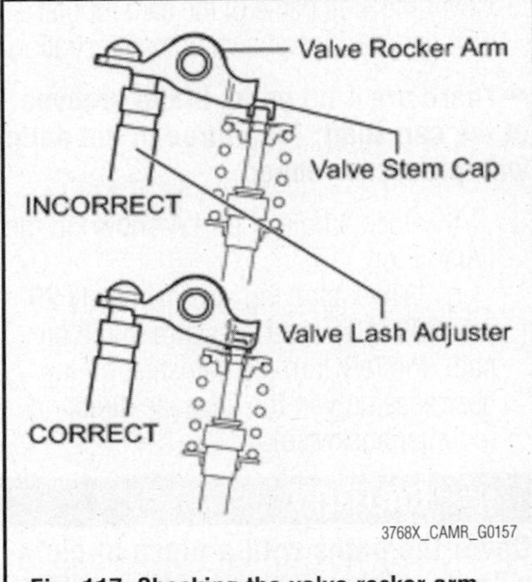

Fig. 117 Checking the valve rocker arm installation

➡Do not start the engine for at least 4 hours after the installation.

➡Thoroughly wipe clean any seal packing.

27. Install the camshaft timing gear assembly.
   a. Check the camshaft timing gear position.
   b. Align and attach the knock pin of the No. 1 camshaft with the pin hole of the camshaft timing gear.
   c. Check that there is no clearance between the camshaft timing gear and camshaft flange.
   d. Using a wrench to hold the hexagonal portion of the No. 1 camshaft, install the bolt. Tighten to 63 ft. lbs. (85 Nm).

➡Be careful not to damage the cylinder head or spark plug tube with the wrench.

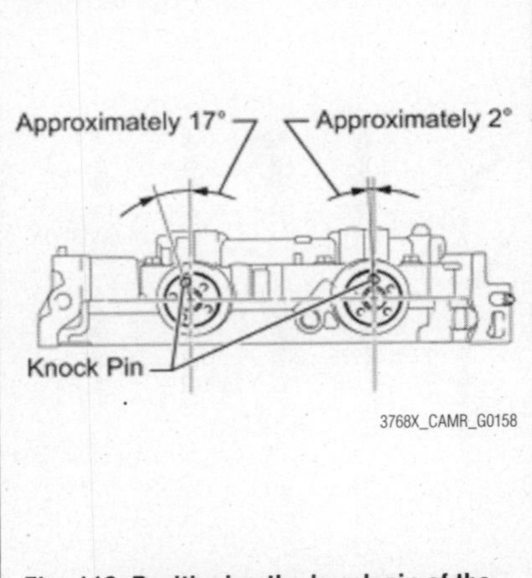

Fig. 118 Positioning the knock pin of the No. 1 and No. 2 camshafts

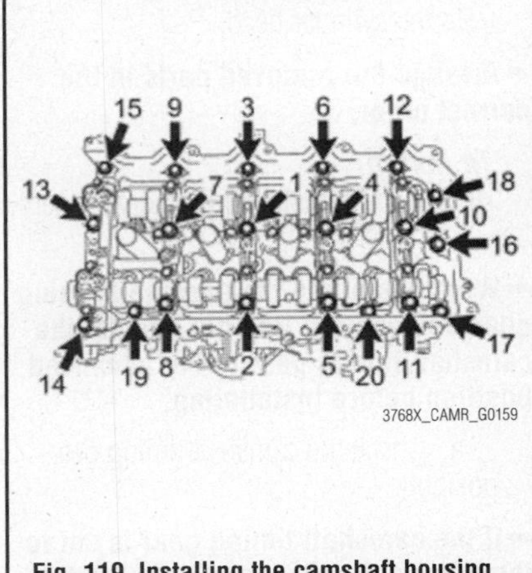

Fig. 119 Installing the camshaft housing bolts in order

➡Do not disassemble the camshaft timing gear

28. Install the camshaft timing exhaust gear assembly.
   a. Align and attach the knock pin of the No. 2 camshaft with the pin hole of the camshaft timing exhaust gear.

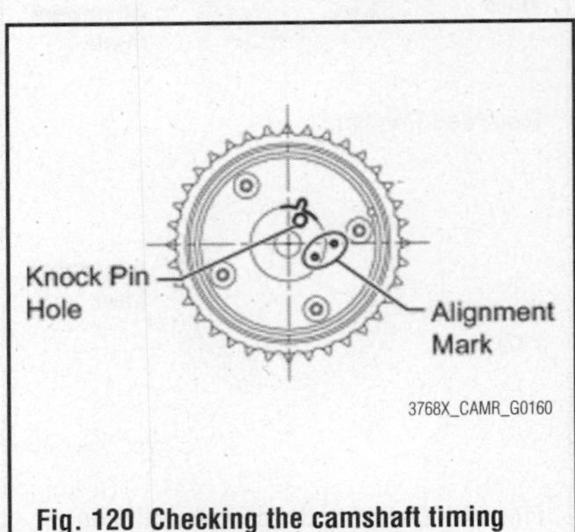

Fig. 120 Checking the camshaft timing gear position

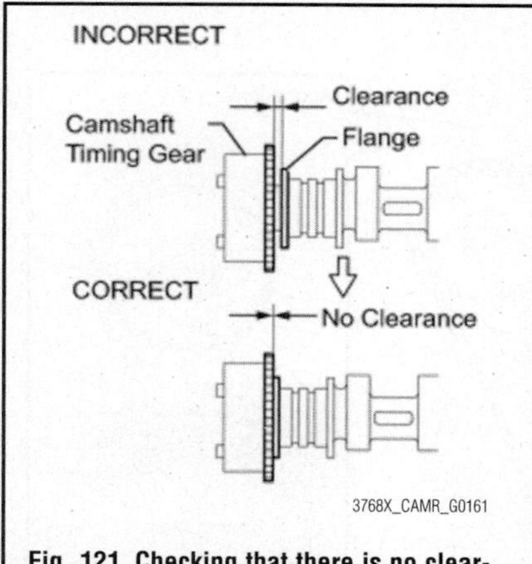

Fig. 121 Checking that there is no clearance between the camshaft timing gear and camshaft flange

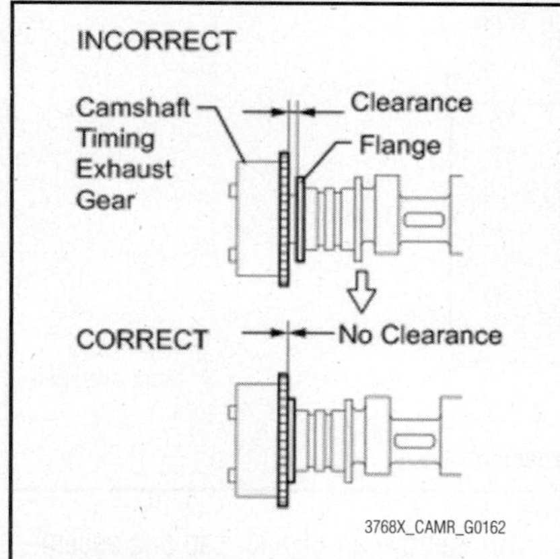

Fig. 122 Checking that there is no clearance between the camshaft timing exhaust gear and the camshaft flange

b. Check that there is no clearance between the camshaft timing exhaust gear and camshaft flange.

c. Using a wrench to hold the hexagonal portion of the No. 2 camshaft, install the bolt. Tighten to 63 ft. lbs. (85 Nm).

➡Be careful not to damage the cylinder head or spark plug tube with the wrench.

➡Do not disassemble the camshaft timing exhaust gear.

29. Add engine oil.

a. Add 1.7 fl. oz. (50 cc) of engine oil into the oil hole.

➡Oil must be added if the lash adjusters were removed.

➡Make sure that the low pressure chamber and oil paths of the lash adjusters are full of engine oil.

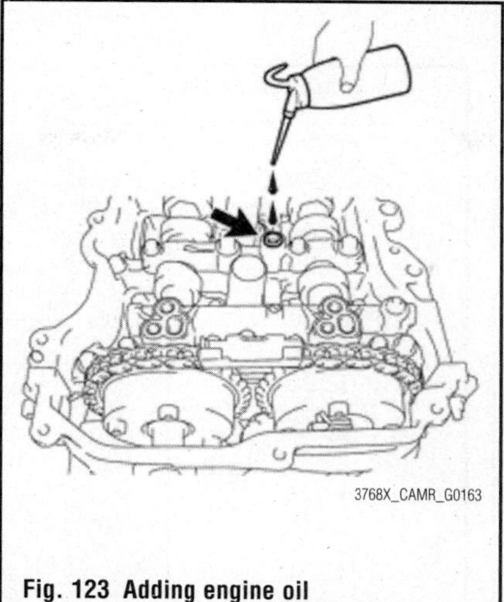

Fig. 123 Adding engine oil

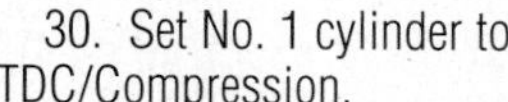

30. Set No. 1 cylinder to TDC/Compression.

a. Temporarily install the crankshaft pulley bolt.

b. Rotate the crankshaft 40° counterclockwise to position the crankshaft pulley key as shown in the illustration.

c. Check that the timing marks of the camshaft timing gears are as shown in the illustration.

➡"A" is not a timing mark.

31. Install the No. 1 chain vibration damper with the 2 bolts. Tighten to 15 ft. lbs. (21 Nm).

32. Install the chain sub assembly.

➡Make sure the mark plate of the chain faces away from the engine.

➡It is not necessary to install the chain to the teeth of the gears and sprocket.

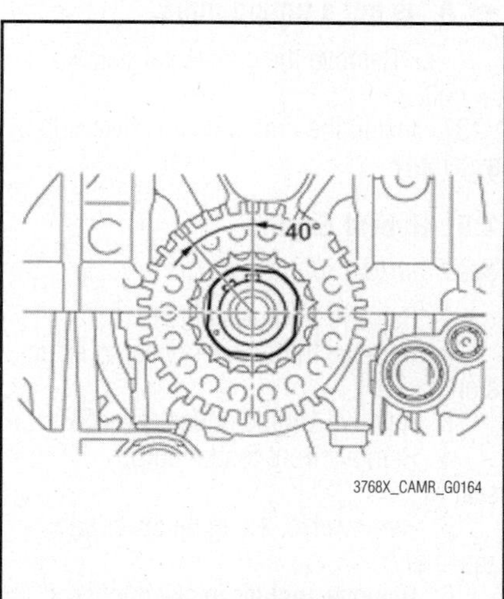

Fig. 124 Rotating the crankshaft 40°

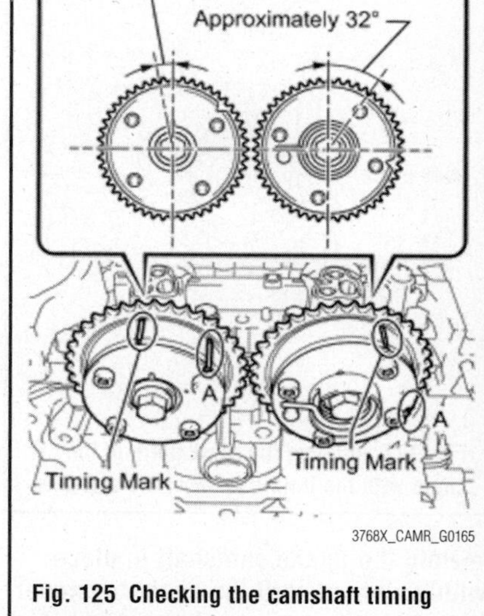

Fig. 125 Checking the camshaft timing gear timing marks

a. Align the mark plate (yellow or gold) of the chain with the timing mark of the camshaft timing exhaust gear and install the chain to the camshaft timing exhaust gear.

b. Align the mark plate (pink or gold) of the chain with the timing mark of the crankshaft timing sprocket and install the chain to the crankshaft timing sprocket.

c. Tie a string above the crankshaft timing sprocket so that the chain is secure.

d. Using the hexagonal portion of the intake camshaft, rotate the intake camshaft counterclockwise with a wrench, align the timing mark of the camshaft timing gear with the mark plate (yellow or gold) of the chain and install the chain to the camshaft timing gear.

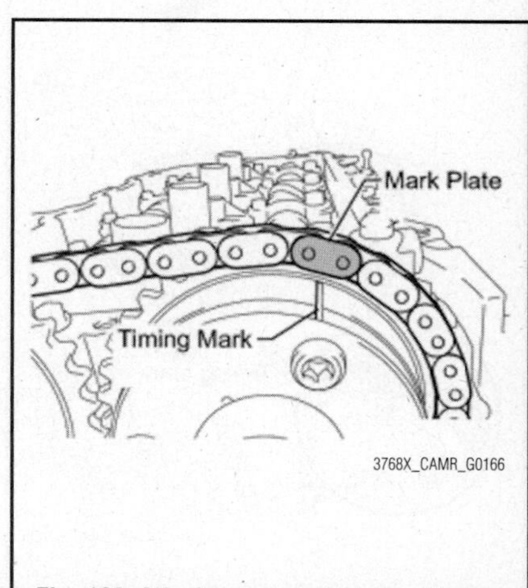

Fig. 126 Aligning the mark plate with the timing mark

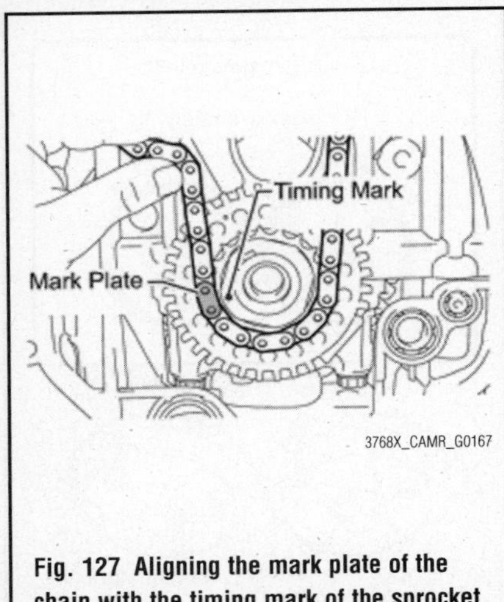

Fig. 127 Aligning the mark plate of the chain with the timing mark of the sprocket

➡Hold the intake camshaft in place with a wrench until the chain tensioner is installed.

e. Remove the string above the crankshaft timing sprocket, rotate the crankshaft clockwise, and loosen the chain so that the chain tensioner slipper can be installed.

➡Make sure the chain is secure.

33. Install the chain tensioner slipper with the bolt. Tighten to 15 ft. lbs. (21 Nm).
34. Install the No. 1 chain tensioner assembly with 2 bolts. Tighten to 7 ft. lbs. (10 Nm).
   a. Remove the pin from the stopper plate.
35. Install the timing chain guide with the bolt. Tighten to 15 ft. lbs. (21 Nm).

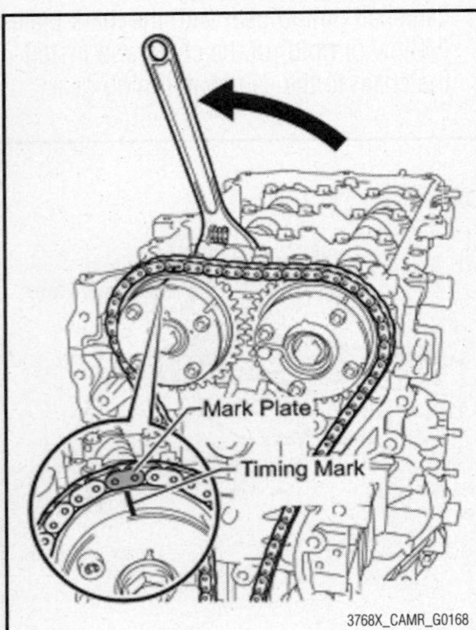

Fig. 128 Rotating the intake camshaft to align the timing marks

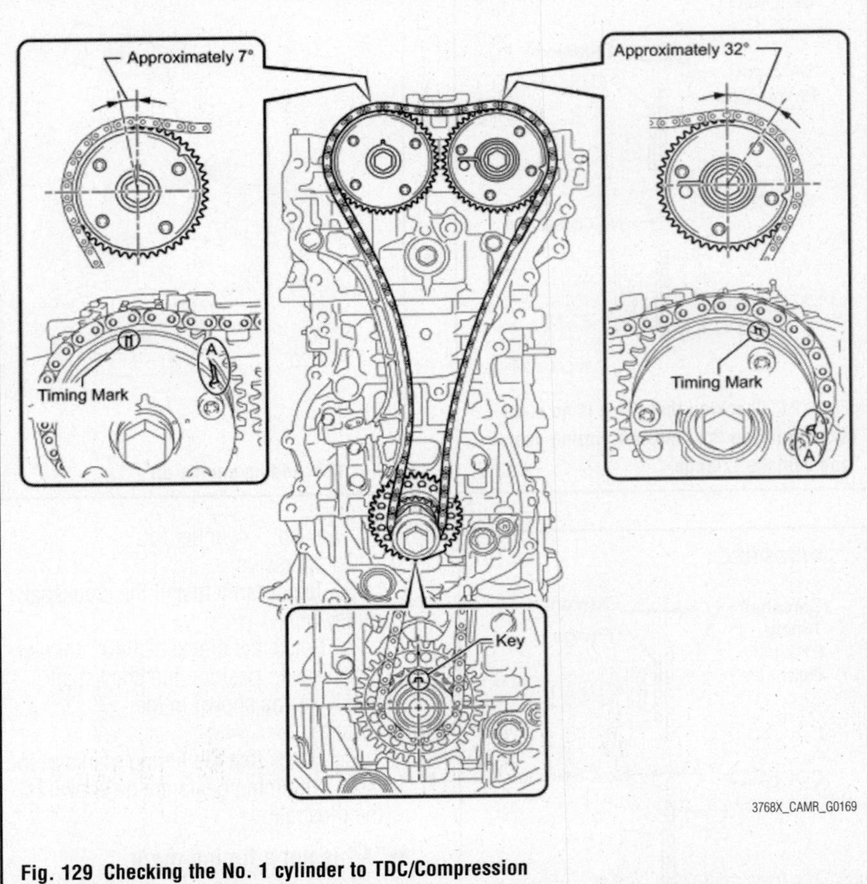

Fig. 129 Checking the No. 1 cylinder to TDC/Compression

36. Check the No. 1 cylinder to TDC/Compression.
   a. Temporarily install the crankshaft pulley bolt.
   b. Rotate the crankshaft clockwise, and check that the timing marks on the crankshaft timing sprocket and camshaft timing gears are as shown in the illustration.

➡"A" is not a timing mark.

   c. Remove the crankshaft pulley bolt.
37. Install the timing chain cover sub assembly.

2.5L Hybrid Engine

See Figures 130 through 139

1. Remove front wheel RH.
2. Remove front wheel opening extension pad RH.
3. Remove engine under cover RH.
4. Remove front fender apron seal RH.
5. Remove No. 1 engine cover sub assembly.
6. Remove cool air intake duct seal.
7. Remove inlet air cleaner assembly.
8. Remove air cleaner cap sub assembly.
9. Remove air cleaner filter element.
10. Remove air cleaner case sub assembly.
11. Remove No. 2 engine mounting stay RH.
12. Disconnect earth wire.
13. Remove engine moving control rod sub assembly.
14. Disconnect inverter reserve tank assembly (for LHD).
   a. Remove the 2 bolts and disconnect the inverter reserve tank assembly.
15. Disconnect No. 2 engine room relay block (for LHD).
   a. Remove the screw and disconnect the front fender liner RH.
   b. Remove the bolt, harness clamp and disconnect the No. 2 engine room relay block.
16. Disconnect the engine wire.
   a. Disconnect the 8 connectors and 2 harness clamps.
   b. Remove the 5 bolts and 3 nuts and disconnect the engine wire from the engine.
17. Remove the ignition coil assembly.

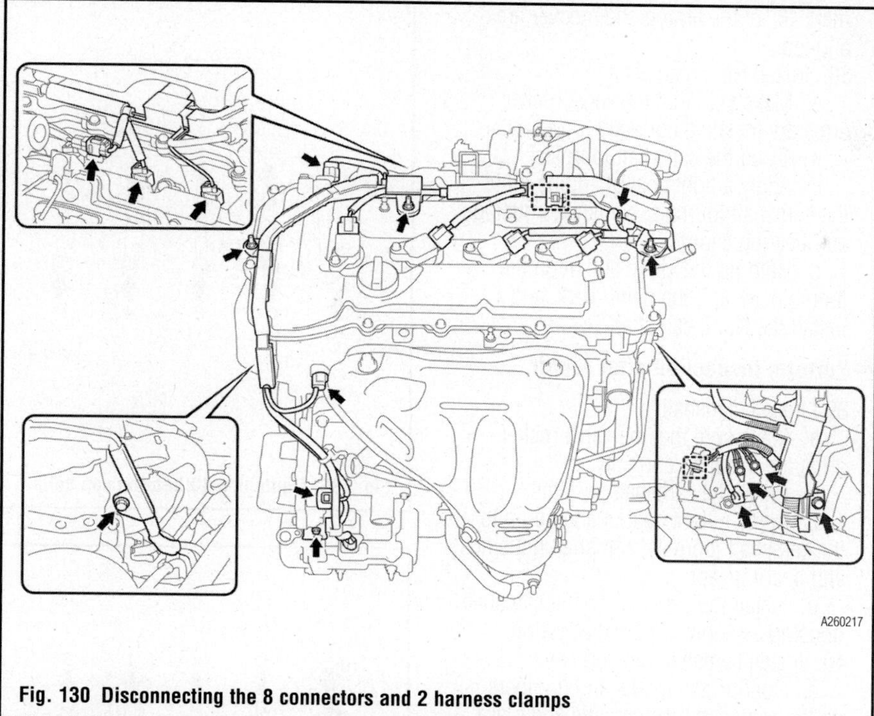

Fig. 130 Disconnecting the 8 connectors and 2 harness clamps

18. Remove the cylinder head cover sub assembly.

19. Set No. 1 cylinder to TDC/COMPRESSION.

a. Turn the crankshaft pulley until its timing notch (groove) and the timing mark "0" of the timing chain cover are aligned.

b. Check that each timing mark of the camshaft timing gear assembly and camshaft timing sprocket are aligned with each timing mark located as shown in the illustration. If not, turn the crankshaft 1 revolution (360°) to align the timing marks as shown in the illustration.

c. Place paint marks on the chain in alignment with the timing marks on the camshaft timing gear and camshaft timing sprocket.

20. Remove timing chain cover plate.

a. Remove the 4 bolts, timing chain cover plate and gasket.

21. Remove the No. 1 chain tensioner assembly.

a. Turn the crankshaft approximately 10° clockwise.

b. Turn the crankshaft approximately 10° counterclockwise.

c. Align the holes of the stopper plate and tensioner, and insert a pin into the stopper plate hole to lock the tensioner.

d. Turn the crankshaft approximately 10° clockwise.

e. Remove the 2 bolts, chain tensioner and gasket.

WARNING

Make sure not to drop the gasket inside the timing chain cover.

f. Turn the crankshaft approximately 10° counterclockwise.

22. Remove timing chain guide.

23. Remove oil pump relief valve plug.

24. Remove camshaft timing gear assembly.

a. Hold the hexagonal portion of the camshaft with a wrench and remove the bolt from the camshaft.

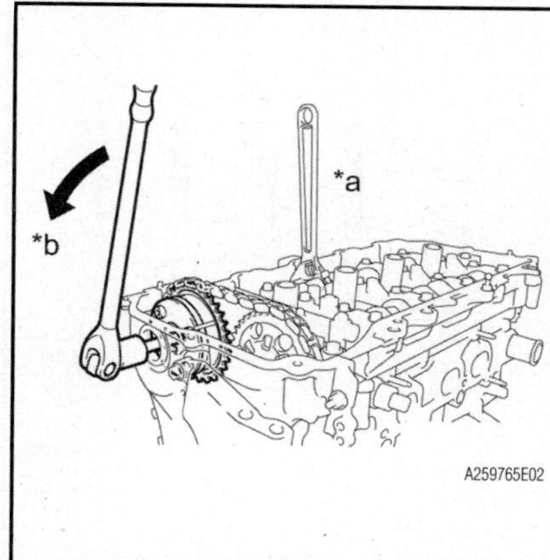

Fig. 131 Removing camshaft timing gear assembly (a: hold, b: turn)

WARNING

Be careful not to damage the cylinder head or spark plug tube with the wrench.

b. Separate the camshaft timing gear assembly from the camshaft.

c. Remove the timing chain from the camshaft timing gear assembly, and turn the camshaft timing gear assembly approximately 180°.

d. Remove the camshaft timing gear assembly.

WARNING

Do not disassemble the camshaft timing gear.

25. Remove camshaft bearing cap.

a. Using several steps, remove the 11 bearing cap bolts in sequence.

b. Using several steps, remove the 10 bearing cap bolts in sequence.

c. Remove the 5 bearing caps.

➡Arrange the removed parts in the correct order.

26. Remove the camshaft from the camshaft housing.

27. Remove the No. 2 camshaft.

a. Hold up the chain and remove the No. 2 camshaft from the camshaft housing.

b. Suspend the chain with a string or equivalent.

WARNING

Be careful not to drop the chain inside the timing chain cover.

28. Remove the camshaft timing sprocket.

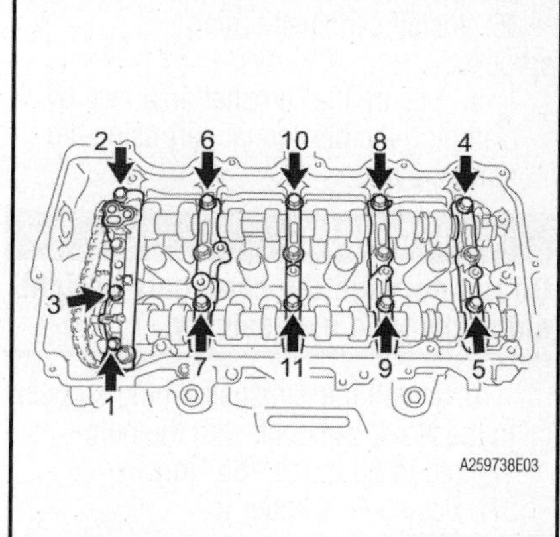

Fig. 132 Removing the 11 bearing cap bolts in sequence

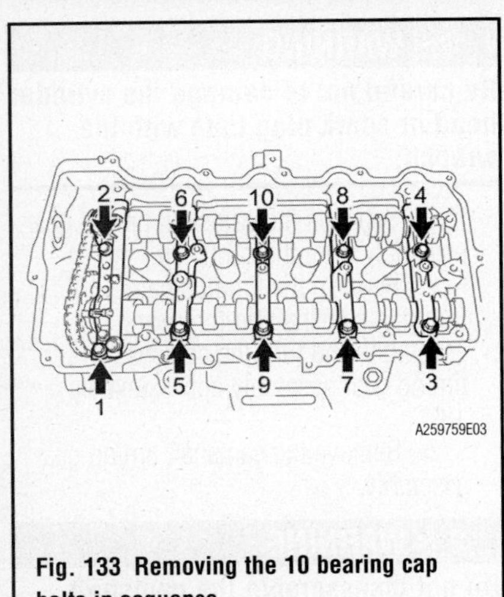

Fig. 133 Removing the 10 bearing cap bolts in sequence

a. Secure the camshaft in a vice by clamping the hexagonal part using aluminum plates.

⁂ WARNING

Do not damage the camshaft by tightening the vice excessively.

b. Remove the flange bolt and camshaft timing sprocket.

⁂ WARNING

Do not damage the No. 2 camshaft or camshaft timing sprocket.

29. Remove oil control valve filter.
30. Remove No. 1 camshaft bearing.
31. Remove No. 2 camshaft bearing.

To install:

➡Perform inspection after repair.

32. Inspect camshaft timing gear assembly.
33. Install No. 2 camshaft bearing.
34. Install No. 1 camshaft bearing.
35. Install oil control valve filter.
36. Install camshaft timing sprocket.
    a. Secure the camshaft in a vice by clamping the hexagonal part using aluminum plates.

⁂ WARNING

Do not damage the camshaft by tightening the vice excessively.

b. Install the camshaft timing sprocket to the No. 2 camshaft with the bolt. Tighten to 63 ft. lbs. (85 Nm).

37. Set No. 1 cylinder to TDC/COMPRESSION.
    a. Turn the crankshaft pulley until its timing notch (groove) and the timing mark "0" of the timing chain cover are aligned.
38. Install No. 2 camshaft.
    a. Make sure that the valve rocker arms are installed correctly.
    b. Clean the camshaft journals.
    c. Apply a light coat of engine oil to the camshaft journals, camshaft housings and bearing caps.
    d. Hold up the chain and align the timing mark and the paint mark, and install the No. 2 camshaft.

➡Perform inspection after repair.

39. Install camshaft.
    a. Make sure that the valve rocker arms are installed correctly.
    b. Clean the camshaft journals.
    c. Apply a light coat of engine oil to the camshaft journals, camshaft housings and bearing caps.
    d. Install the camshaft to the camshaft housing as shown in the illustration.
40. Install camshaft bearing cap.
    a. Confirm the marks and numbers on the camshaft bearing caps and place them in their proper positions and directions.
    b. Using several steps, uniformly tighten the 10 bolts in sequence. Tighten to 20 ft. lbs. (27 Nm).
    c. Using several steps, uniformly tighten the 11 bolts in sequence. Tighten to 12 ft. lbs. (16 Nm).
    d. Check the torque of each bolt again.
41. Install camshaft timing gear assembly.
    a. Check the camshaft timing gear position.

⁂ WARNING

If the camshaft timing gear is set to the advanced position, do not let the camshaft timing gear rotate clockwise during installation.

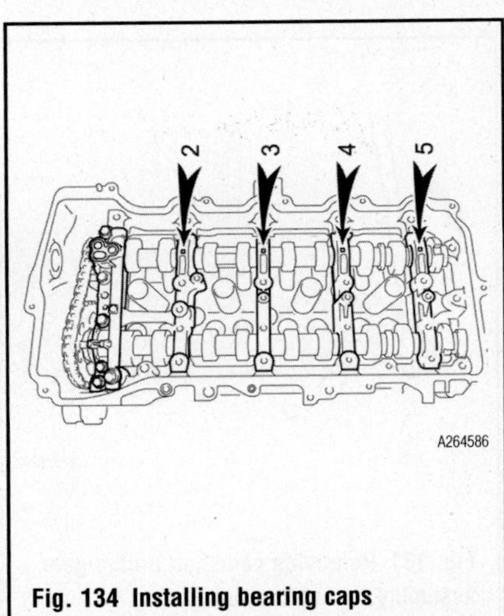

Fig. 134 Installing bearing caps

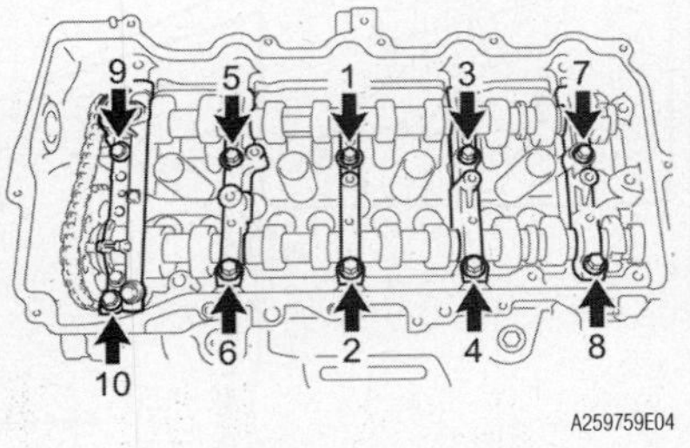

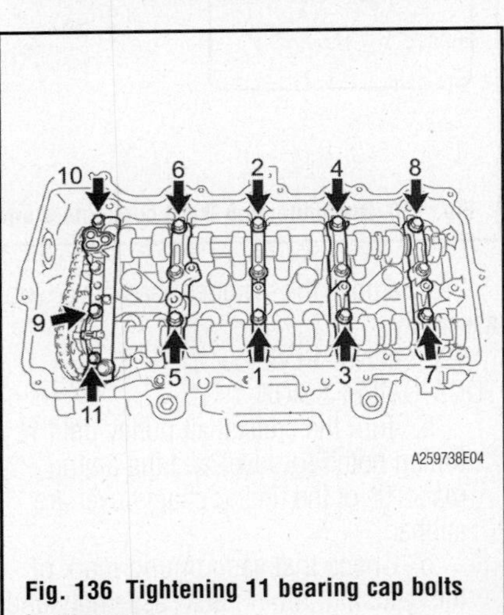

Fig. 136 Tightening 11 bearing cap bolts in sequence

⁂ WARNING

If the camshaft timing gear has rotated to the most retarded position, make sure to release the lock pin and set the camshaft timing gear to the most advanced position before tightening the camshaft timing gear.

b. Install the camshaft timing gear.
c. Turn the camshaft timing gear approximately 180° counterclockwise.
d. Align the paint mark with the timing mark to install the chain.
e. Align and attach the knock pin of the camshaft with the pin hole of the camshaft timing gear.
f. Check that there is no clearance between the camshaft timing gear and camshaft flange.

Fig. 135 Tightening 10 bearing cap bolts

g. Secure the camshaft in place by hand, and then install the installation bolt of the camshaft timing gear by hand.

⁂ WARNING

Do not use any tools to install the bolt. If the bolt is installed using a tool, the lock pin will be damaged.

➡If the lock pin has not been released, release it.

h. After cleaning and degreasing the intake side VVT oil hole on the No. 1 camshaft bearing cap, completely seal the oil hole with adhesive tape or equivalent as shown in the illustration to prevent air from leaking.

⁂ WARNING

Be sure to seal the oil hole completely because air leaks due to insufficient sealing will prevent the lock pin from being released.

i. Make a hole in the adhesive tape covering the oil hole.

j. Apply approximately 29 psi (200 kPa) of air pressure to the hole made in procedure A to release the lock pin.

➡If air leaks out, reattach the adhesive tape.

➡Cover the oil hole with a piece of cloth when applying air pressure to prevent oil from spraying.

k. Forcibly turn the camshaft timing gear in the advance direction (counterclockwise).

➡Depending on the air pressure applied, the camshaft timing gear may turn in the advance direction without assistance by hand.

l. Remove the adhesive tape from the No. 1 camshaft bearing cap.

m. Using a wrench to hold the hexagonal portion of the camshaft, install the bolt. Tighten to 63 ft. lbs. (85 Nm).

⁂ WARNING

Be careful not to damage the cylinder head or spark plug tube with the wrench.

n. Check that each timing mark of the camshaft timing gear and camshaft timing sprocket are aligned with each timing mark.

42. Perform inspection after repair.
43. Add engine oil.
44. Install timing chain guide.
45. Install No. 1 chain tensioner assembly.

a. Turn the crankshaft approximately 10° clockwise.

b. Install a new gasket and the No. 1 chain tensioner assembly with the 2 bolts. Tighten to 7 ft. lbs. (10 Nm).

c. Remove the pin from the stopper plate.

46. Set No. 1 cylinder to TDC/COMPRESSION.

a. Turn the crankshaft pulley until its timing notch (groove) and the timing mark "0" of the timing chain cover are aligned.

b. Check that the timing marks of the camshaft timing gears are as shown in the illustration. If not, turn the crankshaft 1 revolution (360°) to align the timing marks.

➡"A" is not a timing mark.

47. Install oil pump relief valve plug.
48. Install cylinder head cover sub assembly.
49. Install ignition coil assembly.
50. Connect engine wire.

a. Install the engine wire to the engine with the 5 bolts and 3 nuts. Tighten A bolts to 71 inch lbs. (8 Nm). Tighten B bolts to 7 ft. lbs. (10 Nm). Tighten the nuts to 71 inch lbs. (8 Nm).

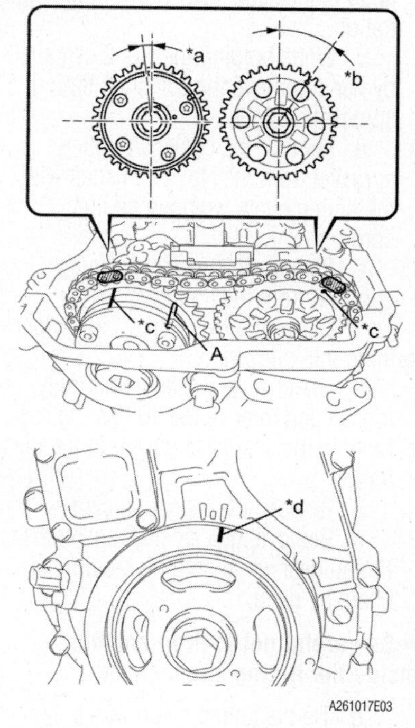

Fig. 137 Checking timing marks (c) and timing notch (d) camshaft timing gears (a: approx. 7°, b: approx. 32°)

b. Connect the 8 connectors and 2 har4ness clamps.

51. Install the No. 2 engine room relay block (for LHD).

a. Install the No. 2 engine room relay bl9ock with the bolt and harness clamp. Tighten to 71 inch lbs. (8 Nm).

52. Install inverter reserve tank assembly (for LHD).

a. Temporarily install the inverter reserve tank assembly with the 2 bolts.

b. Tighten the 2 bolts to the inverter reserve tank assembly in order of bolt A and then bolt B. Tighten to 7 ft. lbs. (10 Nm).

53. Install engine moving control rod sub assembly.
54. Connect earth wire.
55. Install No. 2 engine mounting stay RH.
56. Install air cleaner case sub assembly.
57. Install air cleaner filter element.
58. Install air cleaner cap sub assembly.
59. Install inlet air cleaner assembly.
60. Inspect for oil leak.
61. Install front fender apron seal RH.
62. Install engine under cover RH.
63. Install front wheel opening extension pad RH.
64. Install No. 1 engine cover sub assembly.
65. Install cool air intake duct seal.
66. Install front wheel. Tighten to 76 ft. lbs. (103 Nm).

CRANKSHAFT FRONT SEAL

REMOVAL & INSTALLATION

⁂ CAUTION

All models are equipped with a Supplemental Restraint System (SRS), which uses an air bag. Whenever working near any of the SRS components, such as the impact sensors, the air bag module, steering column and instrument panel, disable the SRS.

2.4L Engine

See Figures 91 through 102.

1. Before servicing the vehicle, refer to the Precautions Section.
2. Disconnect the negative battery cable.
3. Loosen the lug nuts on the front right hand wheel.

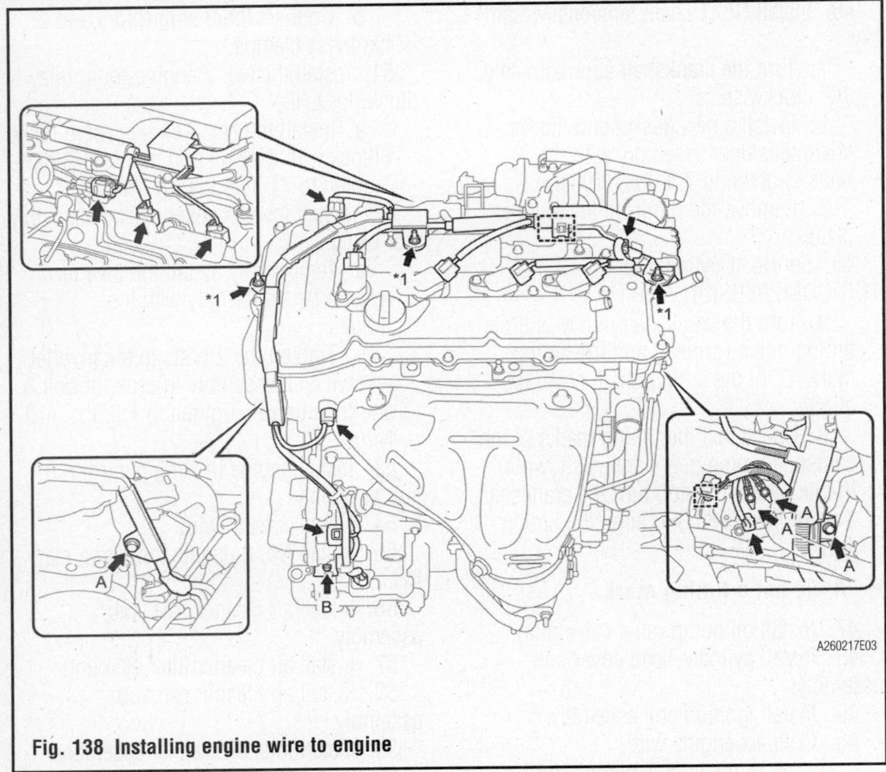

Fig. 138 Installing engine wire to engine

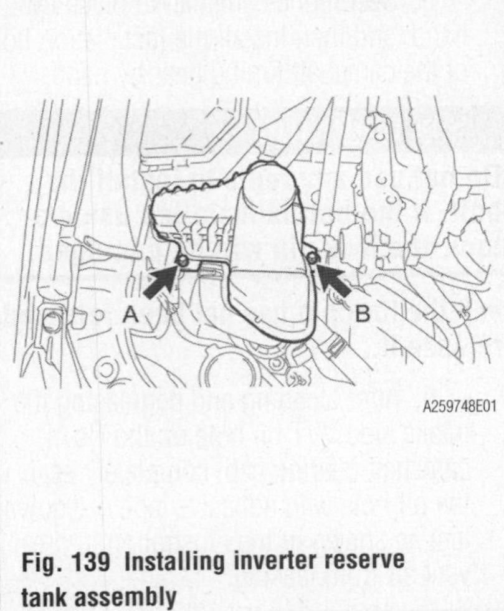

Fig. 139 Installing inverter reserve tank assembly

4. Apply the parking brake, block the rear wheels, then raise and safely support the front of the vehicle securely on jackstands.
5. Remove the front right hand wheel.
6. Remove the left and right hand under cover.
7. Remove the No.1 engine cover sub-assembly and 2 nuts.
8. Remove the ignition coil assembly.
9. Remove the cylinder head cover sub-assembly
10. Set No.1 Cylinder to TDC/ Compression by performing the following:
- Turn the crankshaft pulley until its groove and the timing mark "0" of the timing chain cover are aligned
- Check that each timing mark of the camshaft timing gear and sprocket is aligned with each timing mark located on the No. 1 and No. 2 bearing caps as shown in the illustration. If not, turn the crankshaft by 1 revolution (360°) to align the timing marks

➡Do not turn the crankshaft without the chain tensioner.

11. Remove No.1 Chain tensioner assembly. Remove the 2 nuts, tensioner and gasket.
12. Remove the No. 2 camshaft by performing the following:
a. While holding the camshaft with a wrench, loosen the camshaft timing set bolt.
b. Using several steps, uniformly loosen and remove the 10 bearing cap bolts in the sequence shown in the illustration.
c. While holding the No. 2 camshaft by hand, remove the camshaft timing sprocket set bolt.
d. Remove the camshaft timing sprocket from the No. 2 camshaft with the timing chain wrapped on the sprocket.
e. Remove the camshaft timing sprocket from the timing chain.
13. Remove the No. 1 camshaft by performing the following:
a. Using several steps, uniformly loosen and remove the 10 bearing cap bolts in the sequence shown in the illustration.
b. Remove the 5 bearing caps.
c. Remove the camshaft and camshaft timing gear while holding the timing chain by hand.

➡Be careful not to drop anything inside the timing chain cover.

d. Tie the timing chain with a string.
14. Remove the camshaft timing gear assembly by performing the following:
a. Clamp the camshaft in a vise, and make sure that the camshaft timing gear does not rotate.
b. Cover all the oil ports except the advance side port shown in the illustration with vinyl tape

➡Cover the paths with a shop rag or piece of cloth to avoid oil splashes.

➡Depending on the air pressure, the camshaft timing gear will turn to the advance angle side without applying force by hand. Also, if the pressure is difficult to apply because of air leakage from the port, the lock may be difficult to release.

c. Apply air pressure of 14 psi to the oil path, then turn the camshaft timing gear in the advance direction (counterclockwise) by hand.
d. Remove the flange bolt of the camshaft timing gear. Be sure not to remove the other four bolts. If planning to reuse the gear, be sure to release the straight pin lock before installing the gear.

To install:

15. Put the camshaft timing gear and camshaft together with the straight pin and key groove misaligned.

➡Be sure not to turn the camshaft timing gear to the retard angle side (the right angle).

16. Turn the camshaft timing gear as shown in the illustration while pushing it gently against the camshaft. Push further at the position where the pin fits into the groove.
17. Check that there is no clearance between the gear and camshaft.
18. Tighten the flange bolt with the camshaft timing gear fixed in place and tighten to 40 ft. lbs. (54 Nm).

19. Check that the camshaft timing gear can move to the retard angle side (the right direction) and is locked in the most retarded position.

20. To install the No.1 camshaft, perform the following:

a. Apply a light coat of engine oil to the journal portion of the camshaft.

b. Install the timing chain onto the camshaft timing gear with the paint mark aligned with the timing mark in the camshaft timing gear.

c. Examine the front marks and numbers, and check that the order is as shown in the illustration below. Then install the bearing caps into the cylinder head.

d. Apply a light coat of engine oil to the threads and under the heads of the bearing cap bolts.

e. Using several steps, uniformly tighten the 10 bearing cap bolts in the sequence shown in the illustration. Tighten the following to specification:

- No. 1 Bearing cap: 22 ft. lbs. (30 Nm)
- No. 3 Bearing cap: 80 inch lbs. (9 Nm)

21. To install camshaft No. 2, perform the following:

a. Apply a light coat of engine oil to the journal portion of the No. 2 camshaft.

b. Put the No. 2 camshaft on the cylinder head with the paint mark of the chain aligned with the timing mark on the camshaft timing sprocket.

c. While holding the No. 2 camshaft by hand, temporarily tighten the camshaft timing sprocket set bolt.

d. Examine the front marks and numbers, and check that the order is as shown in the illustration. Then install the bearing caps onto the cylinder head.

e. Apply a light coat of engine oil to the threads and under the heads of the bearing cap bolts.

f. Using several steps, uniformly tighten the 10 bearing cap bolts in the sequence shown in the illustration. Tighten the following to specification:

- No. 1 Bearing cap: 22 ft. lbs. (30 Nm)
- No. 3 Bearing cap: 80 inch lbs. (9 Nm)

g. While holding the camshaft with a wrench, tighten the camshaft timing sprocket set bolt and tighten to 40 ft. lbs. (54 Nm).

h. Check that the paint marks on the chain are aligned with the timing marks on the camshaft timing gear and camshaft timing sprocket. Also, check that the crankshaft pulley groove is aligned with the timing mark "0" of the timing chain cover.

22. To complete installation, reverse remaining removal procedure.

23. Check engine for oil leaks.

24. Connect the negative battery cable.

2.5L Engine

See Figures 103 through 121, 123 through 125 and 127 through 129.

1. Remove the timing chain cover sub assembly.

2. Set the No. 1 cylinder to TDC/Compression.

a. Temporarily install the crankshaft pulley bolt.

➡"A" is not a timing mark.

b. Rotate the crankshaft clockwise so that the timing marks on the crankshaft timing gear and camshaft timing gears are as shown in the illustration.

➡If the timing marks do not align, rotate the crankshaft clockwise again and align the timing marks.

c. Remove the crankshaft pulley bolt.

3. Remove the timing chain guide.

4. Remove the No. 1 chain tensioner assembly.

a. Allow the plunger to extend slightly, and then rotate the stopper plate counterclockwise to release the lock. Once the lock is released, push the plunger into the tensioner.

b. Move the stopper plate clockwise to set the lock, and insert a pin into the stopper plate hole.

c. Remove the 2 bolts, chain tensioner and the gasket.

5. Remove the chain tensioner slipper.

6. Remove the chain sub assembly.

7. Remove the 2 bolts and No. 1 chain vibration damper.

8. Remove the camshaft timing gear assembly.

a. Hold the hexagonal portion of the camshaft with a wrench and remove the bolt and camshaft timing gear.

➡Be careful not to damage the cylinder head or spark plug tube with the wrench.

➡Do not disassemble the camshaft timing gear.

9. Remove the camshaft timing exhaust gear assembly.

a. Hold the hexagonal portion of the camshaft with a wrench and remove the bolt and camshaft timing exhaust gear.

➡Be careful not to damage the cylinder head or spark plug tube with the wrench.

➡Do not disassemble the camshaft timing exhaust gear.

10. Remove the camshaft housing sub assembly.

a. Uniformly loosen and remove the 20 bearing cap bolts in the sequence shown in the illustration.

b. Remove the camshaft housing by prying between the cylinder head and camshaft housing with a screwdriver.

➡Tape the screwdriver tip before use.

➡Be careful not to damage the contact surfaces of the cylinder head and camshaft housing.

11. Remove the camshaft bearing cap.

a. Remove the 11 bearing cap bolts in the sequence shown in the illustration.

b. Removing the 5 bearing caps.

➡Arrange the removed parts in the correct order.

12. Remove the oil control valve filter.

13. Remove the No. 1 camshaft bearing.

14. Remove the camshafts.

15. Remove the No. 2 camshaft bearing.

16. Remove the No. 1 valve rocker arm sub assembly.

a. Remove the 16 valve rocker arms from the cylinder head.

➡Arrange the removed parts in the correct order.

17. Remove the valve lash adjuster assembly.

a. Remove the 16 valve lash adjusters from the cylinder head.

➡Arrange the removed parts in the correct order.

To install:

18. Set the camshaft timing gear assembly.

➡When installing the camshaft timing gear, release the lock pin and set the camshaft timing gear to the advanced position before installation.

a. Check the camshaft timing gear position.

➡If the camshaft timing gear is set to the advanced position, do not let the camshaft timing gear rotate clockwise during installation.

➡If the camshaft timing gear rotates to the retarded position, release the lock pin and set the camshaft timing gear to the advanced position.

b. Align and attach the knock pin of the No. 1 camshaft with the pin hole of the camshaft timing gear.

c. Check that there is no clearance between the camshaft timing gear and camshaft flange.

d. Secure the camshaft in place by hand, and then install the installation bolt of the camshaft timing gear by hand.

➡Do not use any tools to install the bolt. If the bolt is installed using a tool, the lock pin will be damaged.

e. Release the lock pin. Clean the camshaft journal with non-residue solvent. Cover the 4 oil paths of the cam journal with vinyl tape as shown in the illustration.

➡There are 4 oil paths in the grooves of the camshaft. Plug three of the paths with pieces of rubber.

f. Open a hole at port A shown in the illustration.

g. While applying approximately 29 psi (200 kPa) of air pressure to the oil path, forcibly turn the camshaft timing gear assembly in the advance direction (counterclockwise).

⁂ CAUTION

Cover the paths with a piece of cloth when applying pressure to keep oil from splashing.

➡Do not allow the camshaft timing gear assembly to lock. If it locks, release the lock pin again.

➡The camshaft timing gear assembly may be turned in the advance direction without applying any force.

➡If enough air pressure cannot be applied because of air leakage from the port, releasing the lock pin may be difficult.

h. Remove the vinyl tape and rubber pieces from the camshaft.

i. Remove the bolt and camshaft timing gear.

➡Do not allow the camshaft timing gear assembly to lock. If it locks, release the lock pin again.

19. Install the valve lash adjuster assembly.

a. Inspect the valve lash adjuster before installing it.

b. Inspect the 16 lash adjusters to the cylinder head.

➡Install the lash adjuster to the same place it was removed from.

20. Install the No. 1 valve rocker arm sub assembly.

a. Apply engine oil to the lash adjuster tips and valve stem caps.

b. Install the 16 valve rocker arms as shown.

21. Install the No. 2 camshaft bearing.
22. Install the No. 1 camshaft bearing.
23. Install the oil control valve filter.
24. Install the camshaft.

a. Clean the camshaft journals, camshaft housing and bearing caps.

b. Apply a light coat of engine oil to the camshaft journal, camshaft housing and bearing caps.

c. Install the No. 1 and No. 2 camshafts to the camshaft housing.

25. Install the camshaft bearing cap.

a. Confirm the marks and numbers on the camshaft bearing caps and place them in their proper positions and directions.

b. Install the 11 bolts in the order shown in the illustration. Tighten to 12 ft. lbs. (16 Nm).

➡Make sure that the camshaft rotates smoothly after installing the bearing caps.

26. Install the camshaft housing sub assembly.

a. Check that the valve rocker arms are installed as shown in the illustration.

b. Apply seal packing in a continuous line. The line diameter should be 0.118-0.157 inch (3-4 mm).

➡Remove any oil from the contact surface.

➡Install the camshaft housing within 3 minutes and tighten the bolts within 10 minutes after applying seal packing.

c. Position the knock pin of the No. 1 and No. 2 camshafts as shown in the illustration.

d. Install the camshaft housing, and then install the 20 bolts in the order shown in the illustration. Tighten to 20 ft. lbs. (27 Nm).

➡Do not apply oil for at least 4 hours after the installation.

➡Do not start the engine for at least 4 hours after the installation.

➡Thoroughly wipe clean any seal packing.

27. Install the camshaft timing gear assembly.

a. Check the camshaft timing gear position.

b. Align and attach the knock pin of the No. 1 camshaft with the pin hole of the camshaft timing gear.

c. Check that there is no clearance between the camshaft timing gear and camshaft flange.

d. Using a wrench to hold the hexagonal portion of the No. 1 camshaft, install the bolt. Tighten to 63 ft. lbs. (85 Nm).

➡Be careful not to damage the cylinder head or spark plug tube with the wrench.

➡Do not disassemble the camshaft timing gear

28. Install the camshaft timing exhaust gear assembly.

a. Align and attach the knock pin of the No. 2 camshaft with the pin hole of the camshaft timing exhaust gear.

b. Check that there is no clearance between the camshaft timing exhaust gear and camshaft flange.

c. Using a wrench to hold the hexagonal portion of the No. 2 camshaft, install the bolt. Tighten to 63 ft. lbs. (85 Nm).

➡Be careful not to damage the cylinder head or spark plug tube with the wrench.

➡Do not disassemble the camshaft timing exhaust gear.

29. Add engine oil.

a. Add 1.7 fl. oz. (50 cc) of engine oil into the oil hole.

➡Oil must be added if the lash adjusters were removed.

➡Make sure that the low pressure chamber and oil paths of the lash adjusters are full of engine oil.

30. Set No. 1 cylinder to TDC/Compression.

a. Temporarily install the crankshaft pulley bolt.

b. Rotate the crankshaft 40° counterclockwise to position the crankshaft pulley key as shown in the illustration.

c. Check that the timing marks of the camshaft timing gears are as shown in the illustration.

➡"A" is not a timing mark.

31. Install the No. 1 chain vibration

damper with the 2 bolts. Tighten to 15 ft. lbs. (21 Nm).

32. Install the chain sub assembly.

➡**Make sure the mark plate of the chain faces away from the engine.**

➡**It is not necessary to install the chain to the teeth of the gears and sprocket.**

a. Align the mark plate (yellow or gold) of the chain with the timing mark of the camshaft timing exhaust gear and install the chain to the camshaft timing exhaust gear.

b. Align the mark plate (pink or gold) of the chain with the timing mark of the crankshaft timing sprocket and install the chain to the crankshaft timing sprocket.

c. Tie a string above the crankshaft timing sprocket so that the chain is secure.

d. Using the hexagonal portion of the intake camshaft, rotate the intake camshaft counterclockwise with a wrench, align the timing mark of the camshaft timing gear with the mark plate (yellow or gold) of the chain and install the chain to the camshaft timing gear.

➡**Hold the intake camshaft in place with a wrench until the chain tensioner is installed.**

e. Remove the string above the crankshaft timing sprocket, rotate the crankshaft clockwise, and loosen the chain so that the chain tensioner slipper can be installed.

➡**Make sure the chain is secure.**

33. Install the chain tensioner slipper with the bolt. Tighten to 15 ft. lbs. (21 Nm).

34. Install the No. 1 chain tensioner assembly with 2 bolts. Tighten to 7 ft. lbs. (10 Nm).

a. Remove the pin from the stopper plate.

35. Install the timing chain guide with the bolt. Tighten to 15 ft. lbs. (21 Nm).

36. Check the No. 1 cylinder to TDC/Compression.

a. Temporarily install the crankshaft pulley bolt.

b. Rotate the crankshaft clockwise, and check that the timing marks on the crankshaft timing sprocket and camshaft timing gears are as shown in the illustration.

➡**"A" is not a timing mark.**

c. Remove the crankshaft pulley bolt.

37. Install the timing chain cover sub assembly.

2.5L Hybrid Engine

See Figures 140 and 141.

1. Remove front wheel RH.
2. Remove front wheel opening extension pad RH.
3. Remove engine under cover RH.
4. Remove front fender apron seal RH.
5. Remove crankshaft pulley.

a. Using SST, hold the crankshaft pulley and loosen the pulley bolt. Further loosen the bolt until 2 or 3 threads are screwed into the crankshaft.

b. Using SST and the pulley bolt, remove the crankshaft pulley.

➡**Apply a lubricant to the threads and end of SST.**

6. Remove timing chain cover oil seal.

a. Using a screwdriver, pry out the oil seal.

➡**Tape the screwdriver tip before use.**

⁂ WARNING

Do not damage the surface of the oil seal press fit hole or the crankshaft.

To install:

7. Install timing chain cover oil seal.

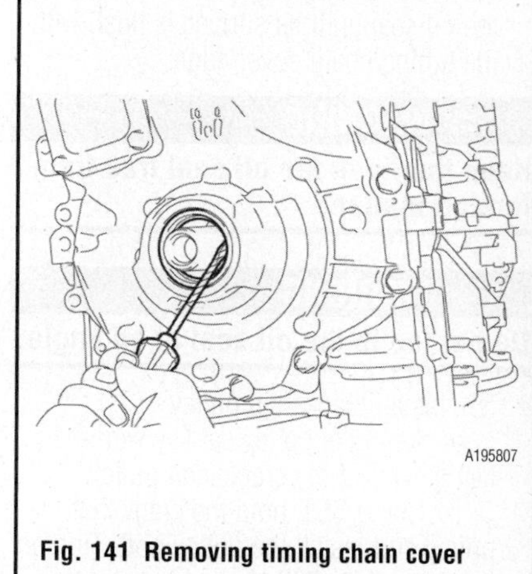

Fig. 141 Removing timing chain cover oil seal

a. Apply MP grease to the lip of a new oil seal.

⁂ WARNING

Do not allow foreign matter to contact the lip of the oil seal.

⁂ WARNING

Do not allow MP grease to contact the dust seal.

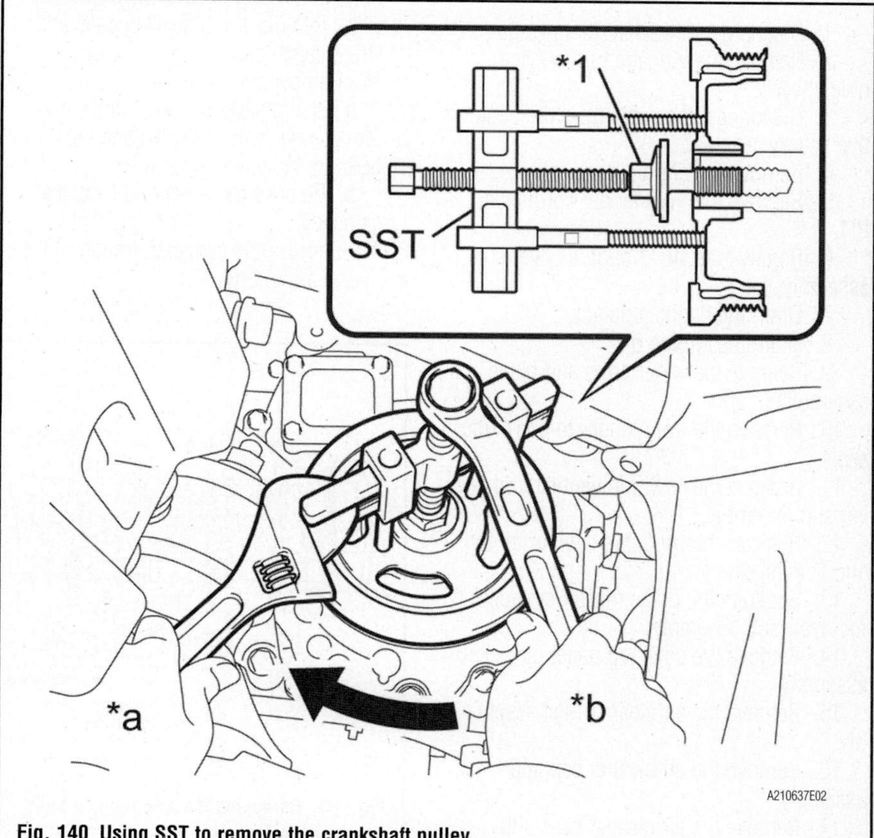

Fig. 140 Using SST to remove the crankshaft pulley

b. Using SST and a hammer, tap in the oil seal until its surface is flush with the timing chain cover edge.

WARNING

Keep the lip of the oil seal free from foreign matter.

WARNING

Do not tap in the oil seal at an angle.

8. Install crankshaft pulley.
   a. Align the pulley set key with the key groove of the crankshaft pulley.
   b. Using SST, hold the crankshaft pulley and install the pulley bolt. Tighten to 192 ft. lbs. (260 Nm).
9. Inspect for oil leak.
10. Install front fender apron seal RH.
11. Install engine under cover RH.
12. Install front wheel opening extension pad RH.
13. Install front wheel RH. Tighten to 76 ft. lbs. (103 Nm).

CYLINDER HEAD

REMOVAL & INSTALLATION

2.4L Hybrid Engine

See Figures 142 through 152.

1. Discharge the fuel system pressure.
2. Remove the luggage trim service hole cover.
3. Disconnect the cable from the negative battery terminal.
4. Remove the engine under covers.
5. Remove the front fender apron seal RH.
6. Remove the No. 1 engine cover sub assembly.
7. Drain the engine coolant.
8. Drain the engine oil.
9. Remove the wiper arms and blade assemblies.
10. Remove the front fender to cowl side seals.
11. Remove the cowl top ventilator louver sub assembly.
12. Remove the windshield wiper motor and link assembly.
13. Separate the brake master cylinder reservoir sub assembly.
14. Remove the cowl top panel outer sub assembly.
15. Remove the air cleaner inlet assembly.
16. Remove the air cleaner cap sub assembly.
17. Remove the air cleaner case sub assembly.
18. Remove the throttle body assembly.
19. Disconnect the fuel tube sub assembly.
20. Remove the fuel delivery pipe with the injector.
21. Remove the intake manifold.
22. Remove the No. 1 intake manifold insulator.
23. Remove the front exhaust pipe assembly.
24. Remove the No. 2 engine mounting stay RH.
25. Remove the engine moving control rod sub assembly.
26. Remove the No. 2 engine mounting bracket RH.
27. Remove the V-ribbed belt tensioner cover sub assembly.
28. Remove the V-ribbed belt.
29. Remove the oil level gage sub assembly.
30. Remove the oil level gauge guide.
31. Remove the manifold stay.
32. Remove the No. 2 manifold stay.
33. Remove the exhaust manifold converter sub assembly.
34. Remove the chain sub assembly.
35. Remove the No. 2 camshaft.
   a. Using several steps, uniformly loosen and remove the 10 bearing cap bolts in the sequence shown.
   b. Remove the 5 bearing caps and No. 2 camshaft.
36. Remove the camshaft.
   a. Using several steps, uniformly loosen and remove the 10 bearing cap bolts in the sequence shown.
   b. Remove the 5 bearing caps and camshaft.
37. Remove the camshaft timing oil control valve assembly.

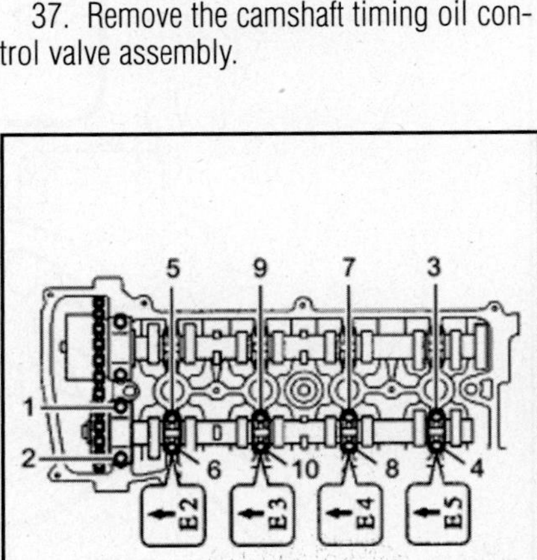

Fig. 142 Removing the bearing cap bolts in sequence

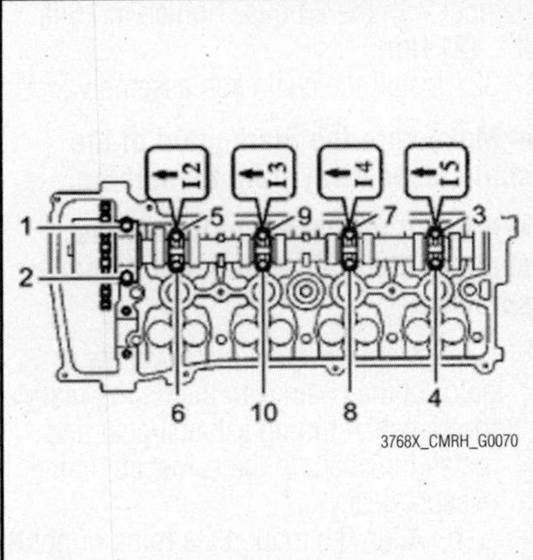

Fig. 143 Removing the 10 bearing cap bolts

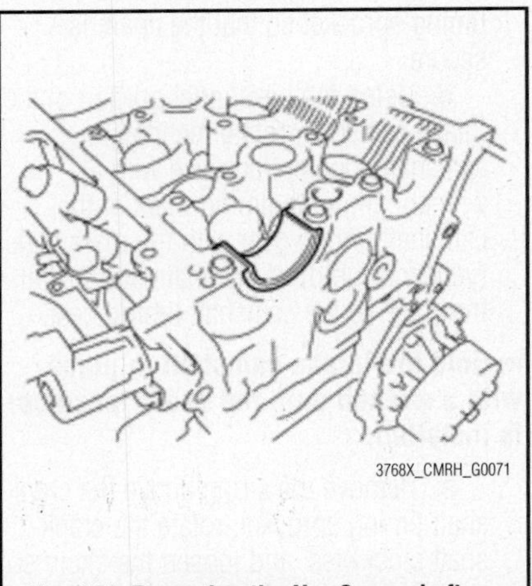

Fig. 144 Removing the No. 2 camshaft bearing

38. Disconnect the radiator hose inlet.
39. Disconnect the engine wire.
   a. Disconnect the radio setting condenser connector.

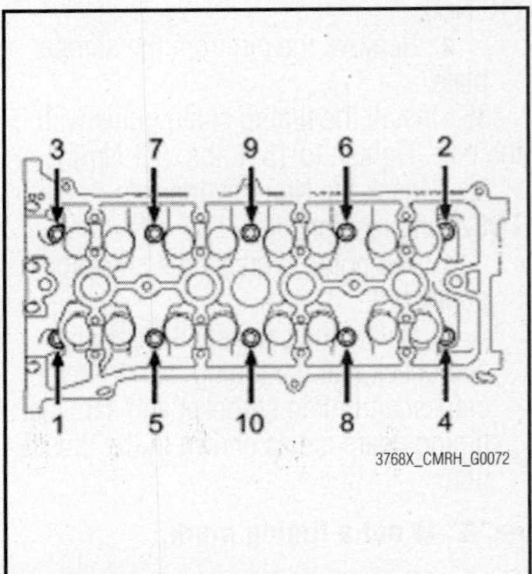

Fig. 145 Removing the 10 cylinder head bolts

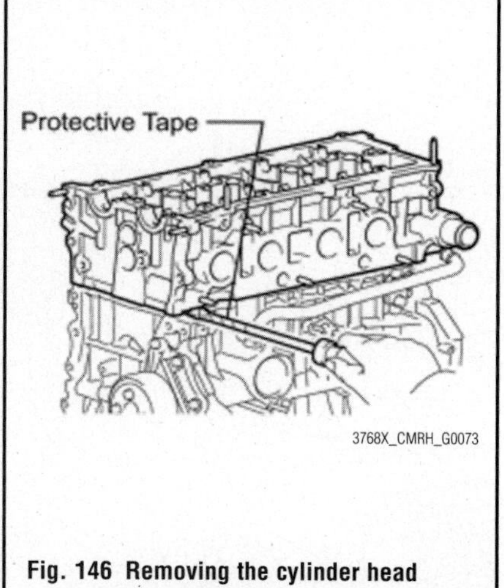

Fig. 146 Removing the cylinder head

b. Disconnect the engine oil pressure switch connector.

c. Disconnect the engine coolant temperature sensor connector.

d. Disconnect the camshaft position sensor connector.

e. Remove the bolt and disconnect the ground cable.

40. Remove the No. 2 camshaft bearing.

41. Remove the cylinder head sub assembly.

a. Using several steps, uniformly loosen and remove the 10 cylinder head bolts and 10 plate washers with a 10 mm bi-hexagon wrench in the sequence shown.

➡Head warpage or cracking could result from removing the bolts in the wrong order.

b. Using a screwdriver with its tip wrapped with protective tape, pry between the cylinder head and cylinder block, and remove the cylinder head.

➡Be careful not to damage the contact surfaces between the cylinder head and cylinder block.

42. Remove the cylinder head gasket.

To install:

43. Install the cylinder head gasket.

a. Place a new gasket on the cylinder block surface with the Lot No. stamp facing upward.

➡Remove any oil from the contact surfaces.

➡Make sure that the gasket is installed in the correct direction.

44. Install the cylinder head sub assembly.

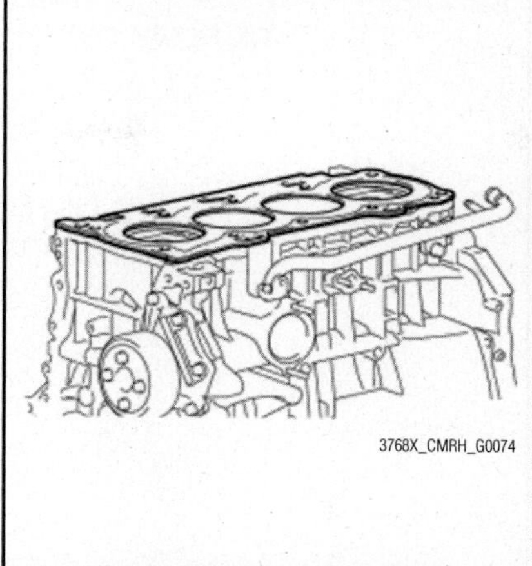

Fig. 147 Removing the cylinder head gasket

a. Install the cylinder head on the cylinder block.

➡The cylinder head bolts are tightened in 2 progressive steps.

b. Apply a light coat of engine oil to the bolt threads and the area beneath the bolt heads that come in contact with the washers.

c. Install the bolts and plate washers to the cylinder head.

➡Do not drop the washers into the cylinder head.

d. Using several steps, uniformly install and tighten the 10 cylinder head set bolts and plate washers with a 10 mm bi-hexagon wrench in the order shown. Tighten to 52 ft. lbs. (70 Nm).

e. Mark the front side of the cylinder head bolts with paint.

f. Retighten the cylinder head bolts 90°in the sequence shown.

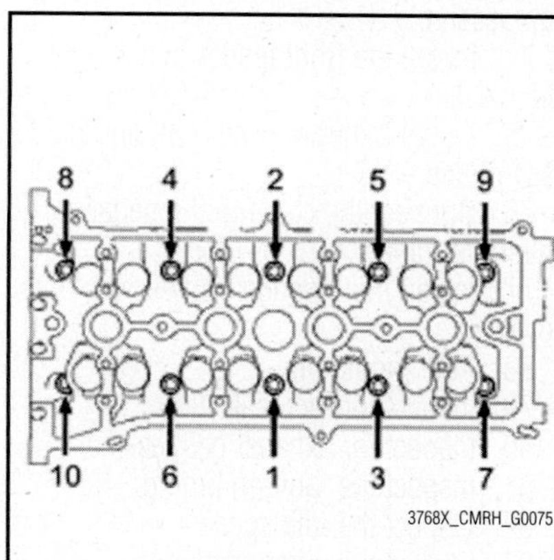

Fig. 148 Installing the cylinder head bolts

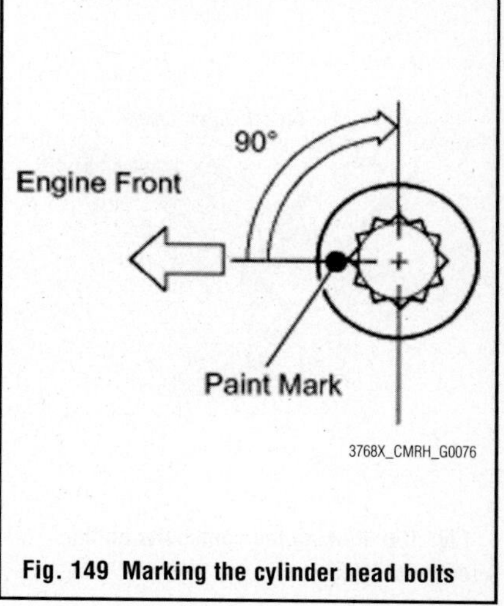

Fig. 149 Marking the cylinder head bolts

g. Check that the paint mark is now at a 90° angle from the front.

45. Connect the engine wire.

a. Connect the ground cable with the bolt. Tighten the bolt to 74 inch lbs. (8.4 Nm).

b. Connect the camshaft position sensor connector.

c. Connect the engine coolant temperature sensor connector.

d. Connect the engine oil pressure switch connector.

e. Connect the radio setting condenser connector.

46. Connect the radiator hose inlet.

47. Install the No. 2 camshaft bearing.

a. Install the No. 2 camshaft bearing.

48. Install the camshaft timing oil control valve assembly.

49. Install the camshaft.

a. Apply a light coat of engine oil to the journal portion of the camshaft.

b. Place the 2 camshafts on the cylinder head with the No. 1 cam lobes facing the directions shown.

c. Examine the front marks and numbers, and check that the order is as shown in the illustration. Then install the bearing caps onto the cylinder head.

d. Apply a light coat of engine oil to the threads and under the heads of the bearing cap bolts.

e. Using several steps, uniformly tighten the 20 bearing cap bolts in the sequence shown. Tighten the No.1 and No. 2 bearing cap to 22 ft. lbs. (30 Nm). Tighten the No. 3 bearing cap to 80 inch lbs. (9 Nm).

50. Install the chain sub assembly.

51. Install the exhaust manifold converter sub assembly.

52. Install the No. 2 manifold stay.

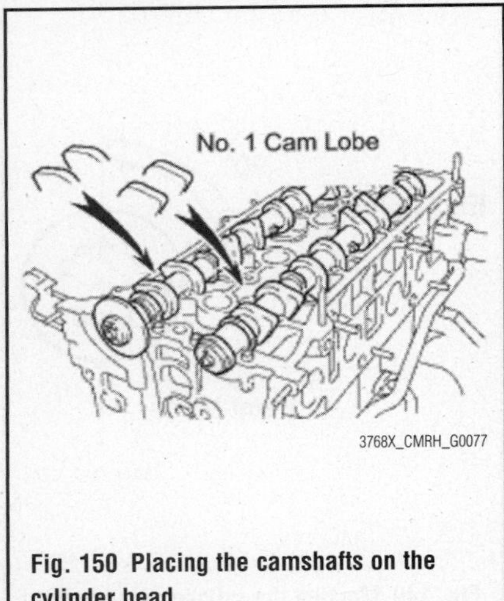

Fig. 150 Placing the camshafts on the cylinder head

Fig. 151 Examining the front marks and numbers

53. Install the manifold stay.
54. Install the oil level gauge guide.
55. Install the oil level gauge sub assembly.
56. Install the V-ribbed belt.
57. Install the V-ribbed belt tensioner cover sub assembly.
58. Install the No. 2 engine mounting bracket RH.
59. Install the engine moving control rod sub assembly.
60. Install the No. 2 engine mounting stay RH.
61. Install the front exhaust pipe assembly.
62. Install the No. 1 intake manifold insulator.
63. Install the intake manifold.
64. Install the fuel delivery pipe with the injector.
65. Connect the fuel tube sub assembly.
66. Install the throttle body assembly.

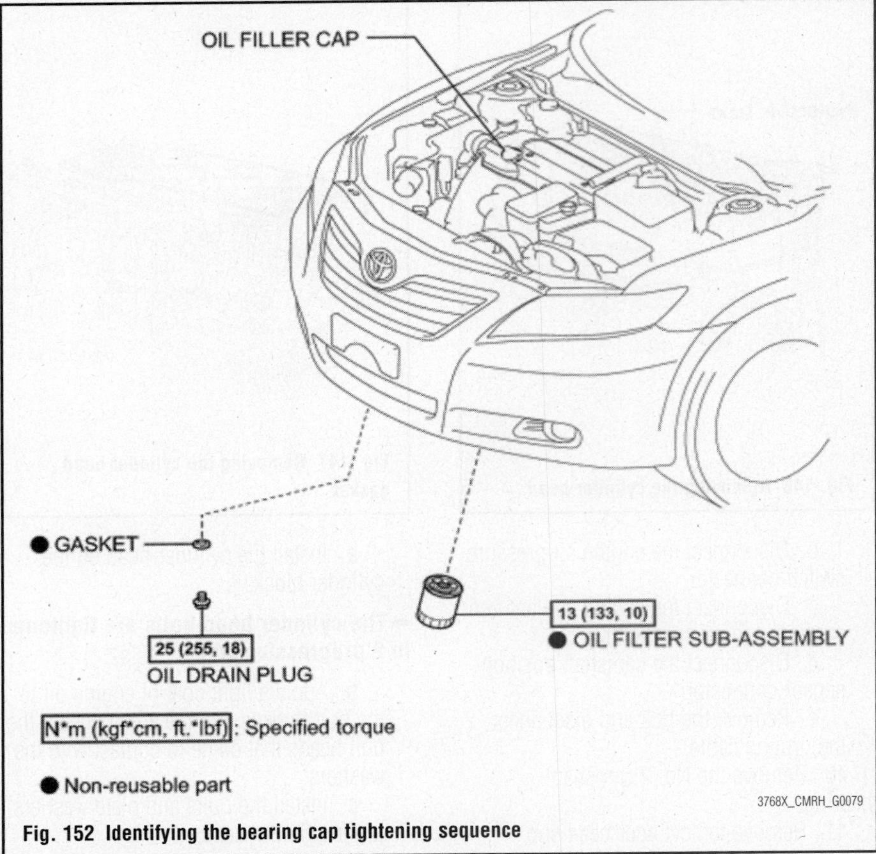

Fig. 152 Identifying the bearing cap tightening sequence

67. Install the air cleaner case sub assembly.
68. Install the air cleaner cap sub assembly.
69. Install the air cleaner inlet assembly.
70. Install the cowl top panel outer sub assembly.
71. Install the brake master cylinder reservoir sub assembly.
72. Install the windshield wiper motor and link assembly.
73. Install the cowl top ventilator louver sub assembly.
74. Install the front fender to the cowl side seals.
75. Install the front wiper arm and blade assemblies.
76. Connect the cable to the negative battery terminal.
77. Install the luggage trim service hole cover.
78. Add engine oil.
79. Inspect for fuel leak.
80. Inspect for exhaust gas leak.
81. Inspect the ignition timing.
82. Inspect the idle speed.
83. Inspect the compression.
84. Inspect CO/HC.
85. Install the No. 1 engine cover sub assembly.
86. Install the front fender apron seal RH.
87. Install the engine under covers.
88. Install the front wheel RH.
89. Perform initialization.

2.5L Engine

See Figures 153 through 173.

1. Remove timing chain cover sub-assembly.
2. Remove manifold stay.
3. Remove no. 2 manifold stay.
4. Remove EGR inlet exhaust manifold plate (for PZEV).
5. Remove No. 1 exhaust manifold heat insulator (except PZEV).
6. Remove exhaust manifold converter sub-assembly.
7. Remove throttle with motor body assembly.
8. Remove vacuum switching valve assembly (for ACIS).
9. Remove fuel delivery pipe sub-assembly.
10. Disconnect No. 2 ventilation hose.
11. Remove intake manifold.
12. Set No. 1 cylinder to TDC/compression.
    a. Temporarily install the crankshaft pulley bolt.

b. Rotate the crankshaft clockwise so that the timing marks on the crankshaft timing gear and camshaft timing gears are as shown in the illustration.

➡If the timing marks do not align, rotate the crankshaft clockwise again and align the timing marks.

c. Remove the crankshaft pulley bolt.

13. Remove timing chain guide.

a. Remove the bolt and timing chain guide.

14. Remove no. 1 chain tensioner assembly.

a. Allow the plunger to extend slightly, and then rotate the stopper plate counterclockwise to release the lock. Once the lock is released, push the plunger into the tensioner.

b. Move the stopper plate clockwise to set the lock, and insert a pin into the stopper plate hole.

c. Remove the 2 bolts, chain tensioner and gasket.

15. Remove chain tensioner slipper.

a. Remove the bolt and chain tensioner slipper.

16. Remove chain sub-assembly.

a. Remove the chain sub-assembly.

17. Remove no. 1 chain vibration damper.

a. Remove the 2 bolts and chain vibration damper.

18. Remove camshaft timing gear assembly.

a. Hold the hexagonal portion of the camshaft with a wrench and remove the bolt and camshaft timing gear assembly.

⁂ WARNING

Be careful not to damage the cylinder head sub-assembly or spark plug tube with the wrench.

⁂ WARNING

Do not disassemble the camshaft timing gear assembly.

19. Remove camshaft timing exhaust gear assembly.

a. Hold the hexagonal portion of the camshaft with a wrench and remove the bolt and camshaft timing exhaust gear assembly.

⁂ WARNING

Be careful not to damage the cylinder head sub-assembly or spark plug tube with the wrench.

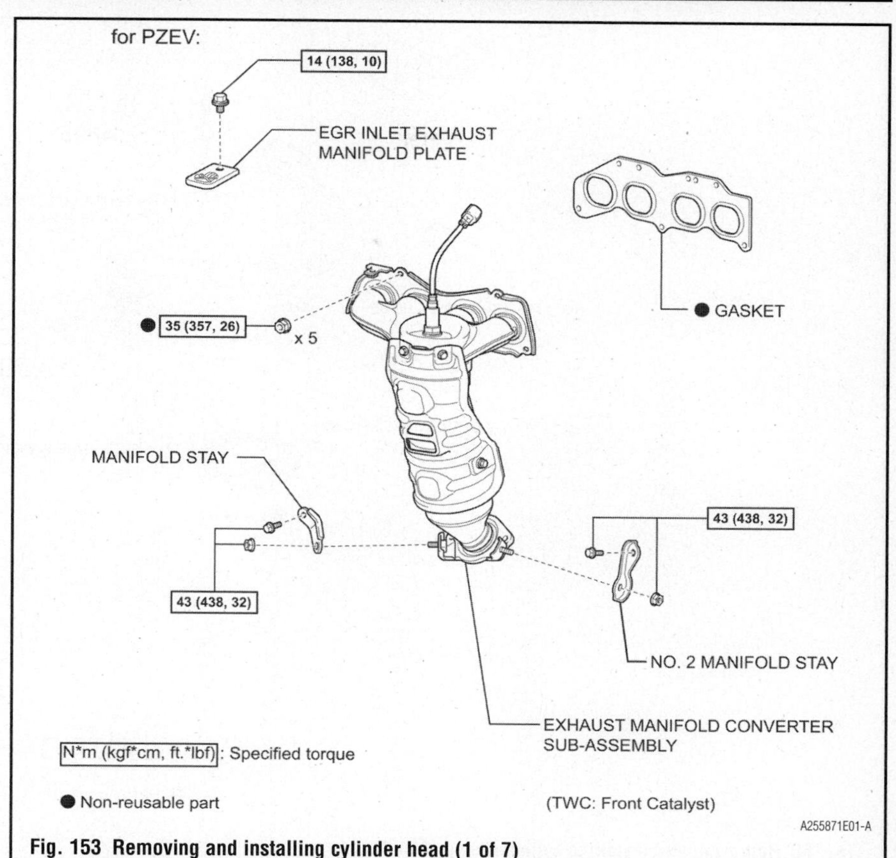

Fig. 153 Removing and installing cylinder head (1 of 7)

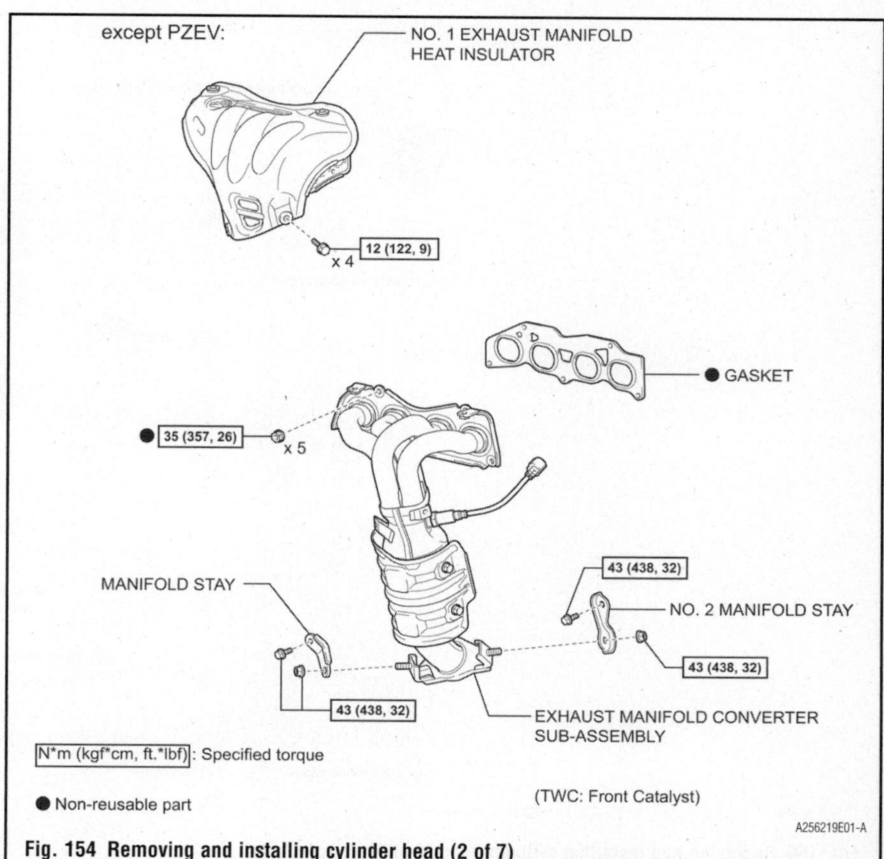

Fig. 154 Removing and installing cylinder head (2 of 7)

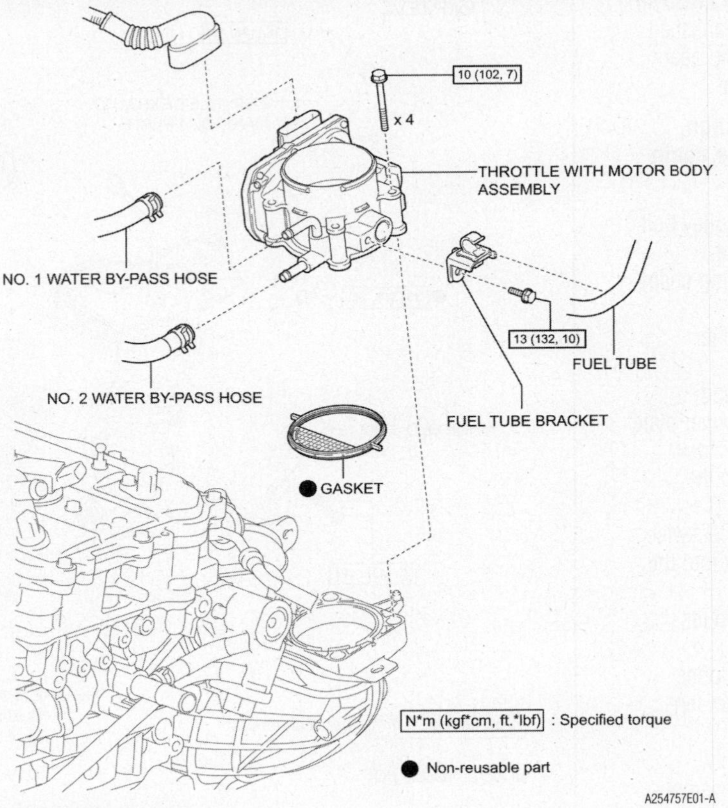

Fig. 155 Removing and installing cylinder head (3 of 7)

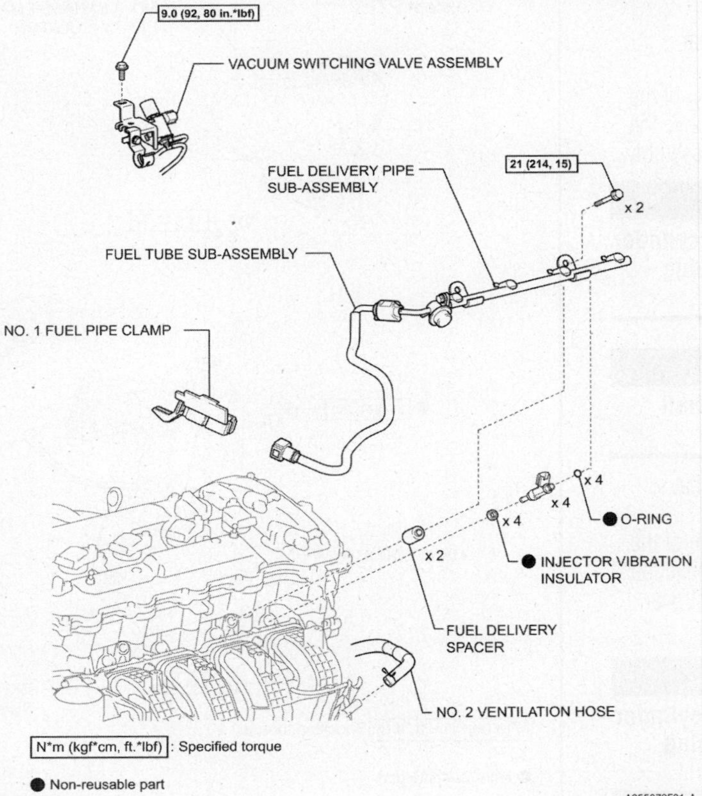

Fig. 156 Removing and installing cylinder head (4 of 7)

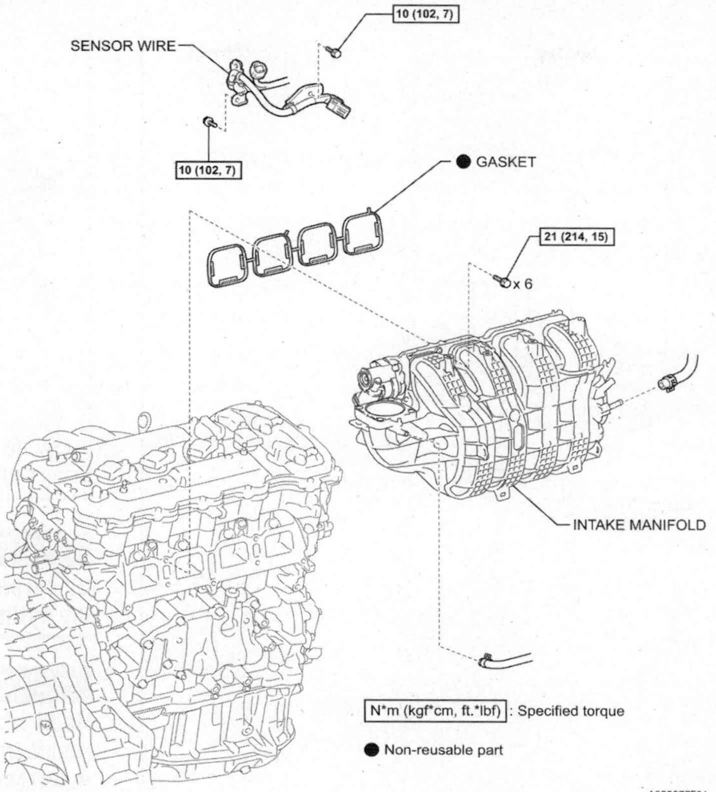

Fig. 157 Removing and installing cylinder head (5 of 7)

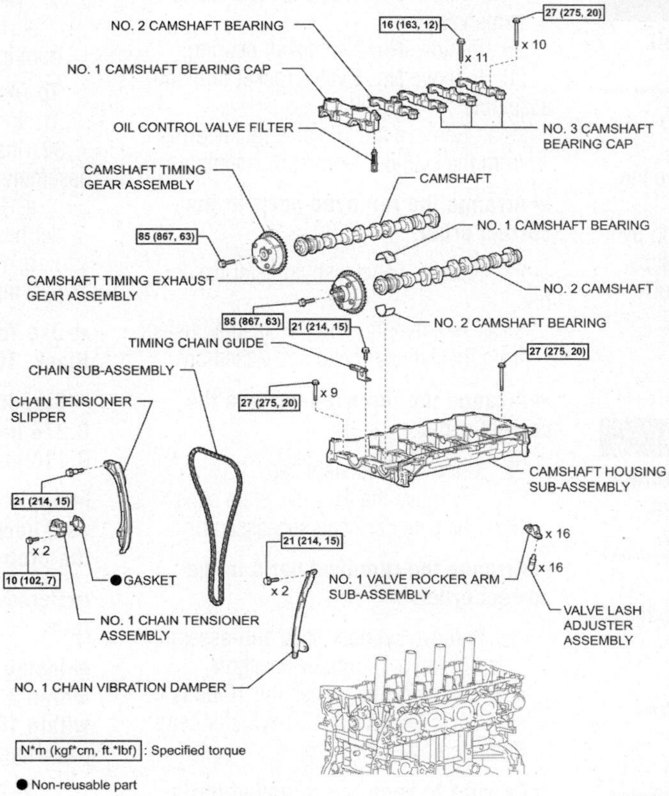

Fig. 158 Removing and installing cylinder head (6 of 7)

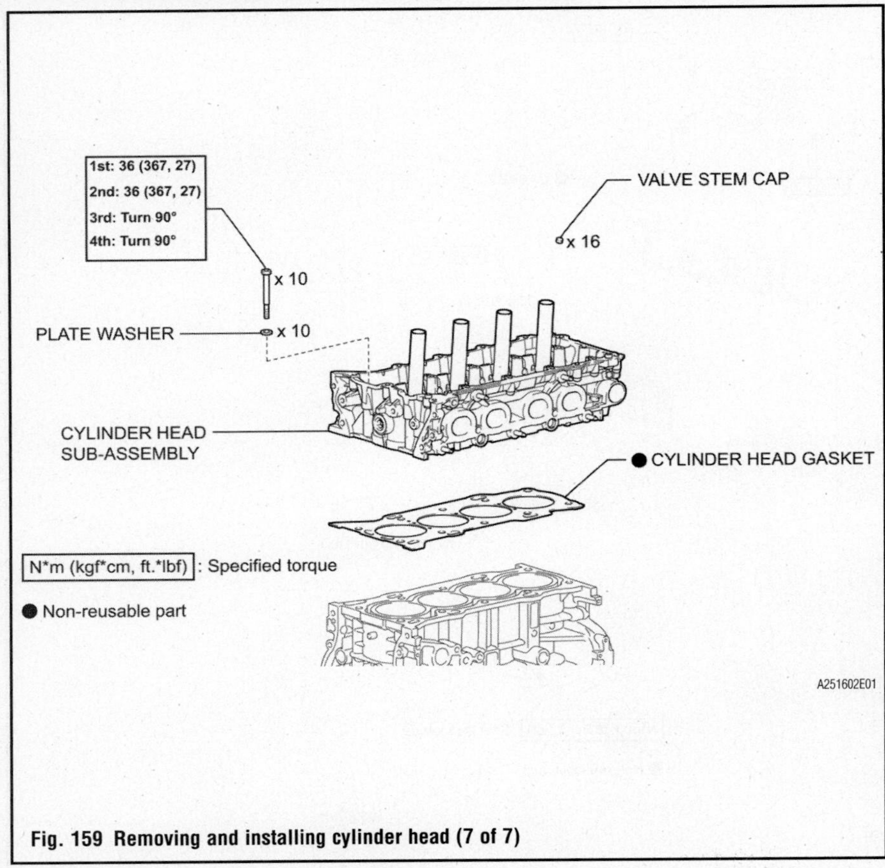

Fig. 159 Removing and installing cylinder head (7 of 7)

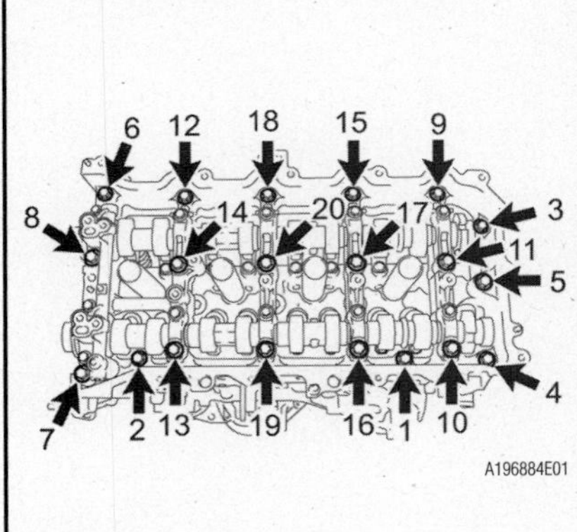

Fig. 160 Loosening and removing bearing cap bolts in sequence

WARNING

Be careful not to drop washers into the cylinder head sub-assembly.

WARNING

Head warpage or cracking could result from removing bolts in the incorrect order.

WARNING

Do not disassemble the camshaft timing exhaust gear assembly.

20. Remove camshaft housing sub-assembly.
   a. Uniformly loosen and remove the 20 bearing cap bolts in sequence.
   b. Remove the camshaft housing by prying between the cylinder head sub-assembly and camshaft housing sub-assembly with a screwdriver.

➡Tape the screwdriver tip before use.

WARNING

Be careful not to damage the contact surfaces of the cylinder head and camshaft housing sub-assembly.

21. Remove camshaft bearing cap.
   a. Remove the 11 bearing cap bolts in sequence.
   b. Remove the 5 bearing caps.

➡Arrange the removed parts in the correct order.

22. Remove oil control valve filter.
23. Remove No. 1 camshaft bearing.
24. Remove camshaft.
   a. Remove the camshaft and No. 2 camshaft.
25. Remove No. 2 camshaft bearing.
26. Remove No. 1 valve rocker arm sub-assembly
   a. Remove the 16 valve rocker arms from the cylinder head sub-assembly.

➡Arrange the removed parts in the correct order.

27. Remove valve lash adjuster assembly.
   a. Remove the 16 valve lash adjusters from the cylinder head sub-assembly.

➡Arrange the removed parts in the correct order.

28. Remove valve stem cap.
   a. Remove the 16 valve stem caps from the cylinder head sub-assembly.

➡Arrange the removed parts in the correct order.

29. Remove cylinder head sub-assembly.
   a. Using a 10 mm bi-hexagon wrench, uniformly loosen the 10 bolts in sequence. Remove the 10 cylinder head bolts and plate washers.

➡Be sure to keep the removed parts separate for each installation position.

   b. Remove the cylinder head sub-assembly.
30. Remove cylinder head gasket.
   a. Remove the cylinder head gasket from the cylinder block.

To install:

31. Inspect cylinder head bolt.
32. Inspect cylinder head sub assembly.
   a. Clean the cylinder block and cylinder head with solvent.
   b. Apply a continuous line of seal packing to a new cylinder head gasket.

➡Use Toyota Genuine Seal Packing Black, Three Bond 1207B or equivalent

➡Standard seal dimension: 0.118 to 0.276 inch (3.0 to 7.0 mm) wide and 0.118 inch (3.0 mm) thick

➡Apply at least 0.787 inch (20 mm) of seal packing from the inside edge of the protrusion of the cylinder block.

➡Remove any oil from the contact surface.

➡Install the cylinder head gasket within 3 minutes and tighten the bolts within 15 minutes after applying seal packing.

   c. Place a new cylinder head gasket on the cylinder block surface with the

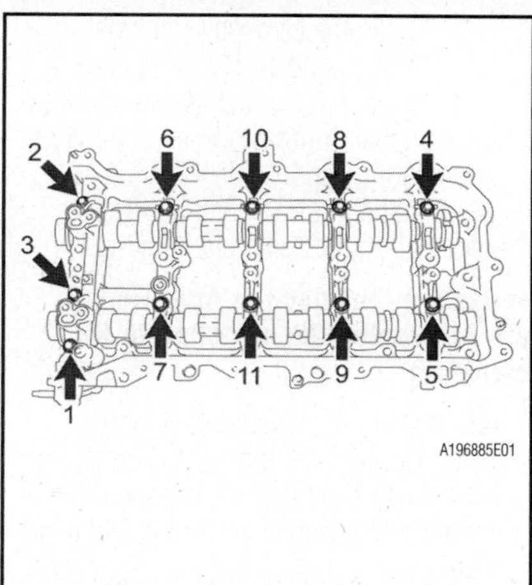

Fig. 161 Removing 11 bearing cap bolts in sequence

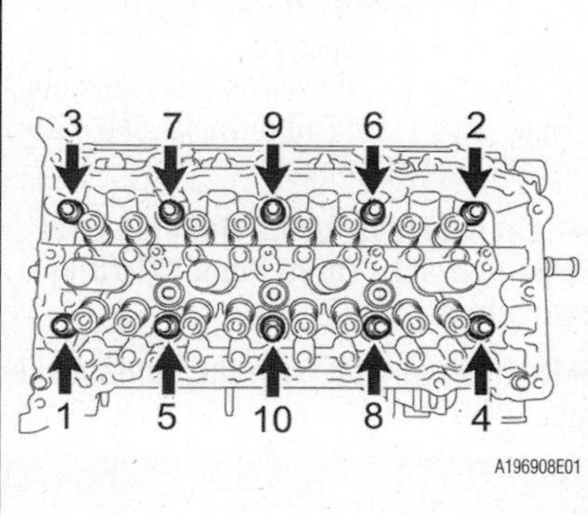

Fig. 162 Loosening and removing bolts in sequence

front face of the Lot No. stamp facing upward.

➡**Pay attention to the installation direction.**

33. Install cylinder head sub assembly.

➡**The cylinder head bolts are tightened in 4 progressive steps.**

a. Place the cylinder head on the cylinder block.

➡**Ensure that no oil is on the mounting surface of the cylinder head sub assembly.**

➡**Place the cylinder head on the cylinder block gently in order not to damage the gasket with the bottom part of the head.**

b. Install the plate washers to the cylinder head bolts.

c. Apply a light coat of engine oil to the threads and under the heads of the cylinder head bolts.

d. Step 1: Using a 10 mm bi-hexagon wrench, install and uniformly tighten the 10 cylinder head bolts in several steps, in sequence. Tighten to 27 ft. lbs. (36 Nm).

✲✲ WARNING

Do not drop the plate washer for the cylinder head bolt into the cylinder head sub assembly.

e. Step 2: Tighten the cylinder head bolts again in the same sequence as the previous step. Make sure that they are tightened to 27 ft. lbs. (36 Nm).

f. Step 3: Mark each cylinder head bolt head with paint.

g. Tighten the cylinder head bolts 90° in sequence.

h. Step 4: Tighten the cylinder head bolts another 90° in sequence.

i. Check that the paint marks are now facing rearward.

✲✲ WARNING

Do not apply oil for at least 2 hours after installation.

✲✲ WARNING

Do not start the engine for at least 2 hours after installation.

✲✲ WARNING

After installation, if the seal packing has seeped out, wipe it off.

➡**Perform inspection after repair.**

34. Install valve stem cap.

a. Apply a light coat of engine oil to the valve stem ends.

b. Install the 16 valve stem caps to the cylinder head.

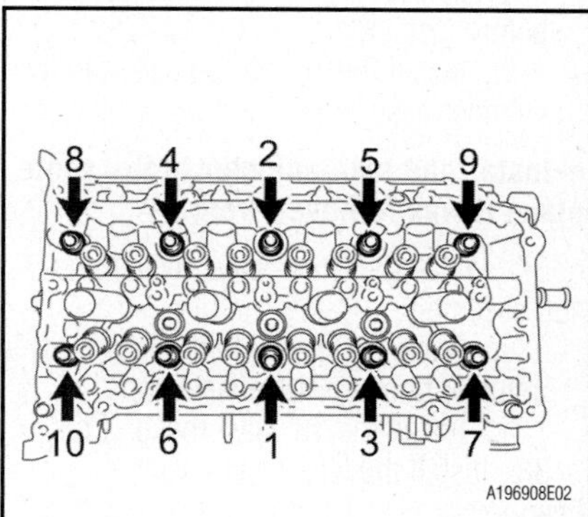

Fig. 163 Tightening cylinder head bolts in sequence

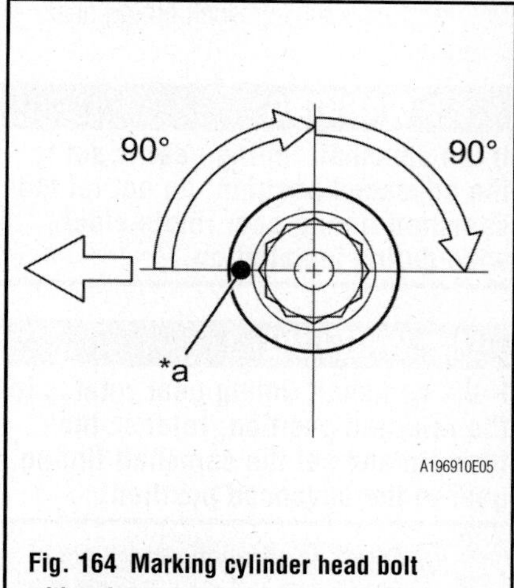

Fig. 164 Marking cylinder head bolt with paint

✲✲ WARNING

Do not drop the valve stem caps into the cylinder head.

35. Set camshaft timing gear assembly.

➡**When installing camshaft timing gear, release the lock pin and set the camshaft timing gear to the advanced position before installation.**

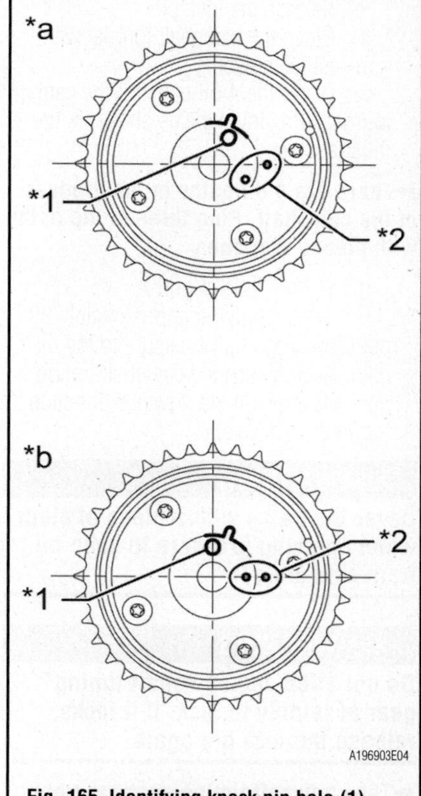

Fig. 165 Identifying knock pin hole (1), Alignment mark (2), Advanced position (a) and Retarded position (b)

a. Check the camshaft timing gear position.

WARNING

If the camshaft timing gear is set to the advanced position, do not let the camshaft timing gear rotate clockwise during installation.

WARNING

If the camshaft timing gear rotates to the retarded position, release the lock pin and set the camshaft timing gear to the advanced position.

b. Align and attach the knock pin of the camshaft with the pin hole of the camshaft timing gear assembly.

c. Check that there is no clearance between the camshaft timing gear assembly and camshaft flange.

d. Secure the camshaft in place by hand, and then install the installation bolt of the camshaft timing gear by hand.

WARNING

Do not use any tools to install the bolt. If the bolt is installed using a tool, the lock pin will be damaged.

e. Release the lock pin.

f. Clean the camshaft journal with non-residue solvent.

g. Cover the 4 oil paths of the cam journal with vinyl tape as shown in the illustration.

➡There are 4 oil paths in the grooves of the camshaft. Plug three of the paths with pieces of rubber.

h. Open a hole at port A.

i. While applying approximately 29 psi (200 kPa) of air pressure to the oil path, forcibly turn the camshaft timing gear assembly in the advance direction (counterclockwise).

WARNING

Cover the paths with a piece of cloth when applying pressure to keep oil from splashing.

WARNING

Do not allow the camshaft timing gear assembly to lock. If it locks, release the lock pin again.

➡The camshaft timing gear assembly may be turned in the advance direction without applying any force.

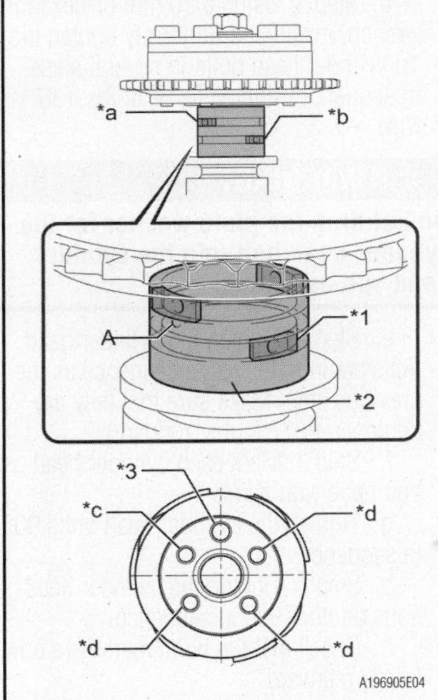

Fig. 166 Identifying rubber (1), vinyl tape (2), knock pin (3), retard side path (a), advance side path (b), open (c) and close (d) positioning

➡If enough air pressure cannot be applied because of air leakage from the port, releasing the lock pin may be difficult.

j. Remove the vinyl tape and rubber pieces from the camshaft.

k. Remove the bolt and camshaft timing gear.

WARNING

Do not allow the camshaft timing gear assembly to lock. If it locks, release the lock pin again.

36. Install valve lash adjuster assembly.

a. Inspect the valve lash adjuster before installing.

b. Install the 16 lash adjusters to the cylinder head.

➡Install the lash adjuster to the same place it was removed from.

37. Install the No. 1 valve rocker arm sub assembly.

a. Apply engine oil to the lash adjuster tips and valve stem caps.

b. Install the 16 valve rocker arms.

38. Install the No. 2 camshaft bearing.

39. Install the No. 1 camshaft bearing.

40. Install the oil control valve filter.

41. Install the camshaft.

a. Clean the camshaft journals, camshaft housing and bearing caps.

b. Apply a light coat of engine oil to the camshaft journal, camshaft housing and bearing caps.

c. Install the camshaft and No. 2 camshaft to the camshaft housing.

➡Perform "Inspection After Repair" after replacing the camshaft or No. 2 camshaft.

42. Install the camshaft bearing cap.

a. Confirm the marks and numbers on the camshaft bearing caps and place them in their proper positions and directions.

b. Install the 11 bolts in order. Tighten to 12 ft. lbs. (16 Nm).

43. Install camshaft housing sub assembly.

a. Check that the valve rocker arms are installed properly.

b. Apply seal packing in a continuous line, with a standard diameter of 0.118-0.157 inch (3-4 mm).

➡Seal packing: Toyota Genuine Seal Packing Black, Three Bond 1207B or equivalent.

➡Remove any oil from the contact surface.

WARNING

Install the camshaft housing within 3 minutes and tighten the bolts within 10 minutes after applying seal packing.

c. Position the knock pin of the camshaft and No. 2 camshaft.

d. Install the camshaft housing, and then install the 20 bolts in the order. Tighten to 20 ft. lbs. (27 Nm).

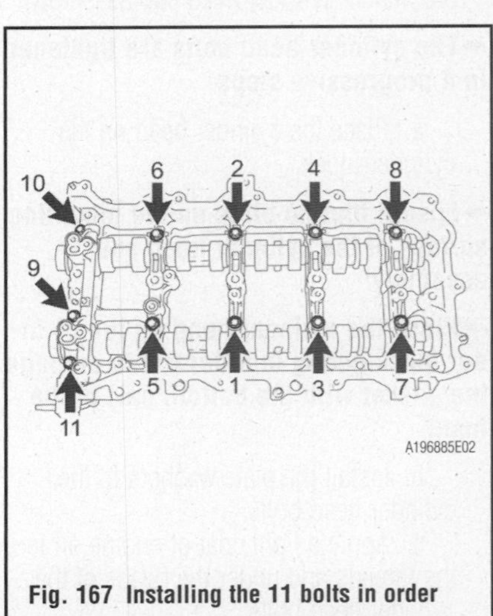

Fig. 167 Installing the 11 bolts in order

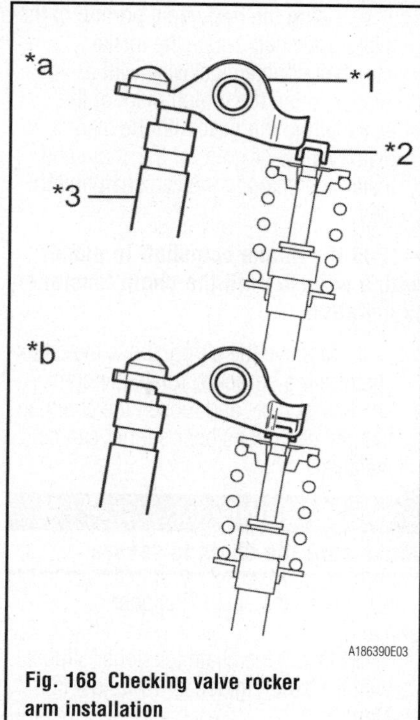

Fig. 168 Checking valve rocker arm installation

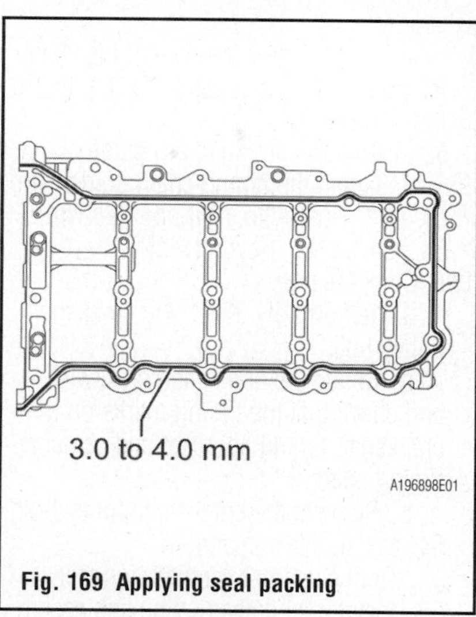

Fig. 169 Applying seal packing

⁂ WARNING

Do not apply oil for at least 2 hours after the installation.

⁂ WARNING

Do not start the engine for at least 2 hours after the installation.

⁂ WARNING

Thoroughly wipe clean any seal packing.

44. Install camshaft timing gear assembly.

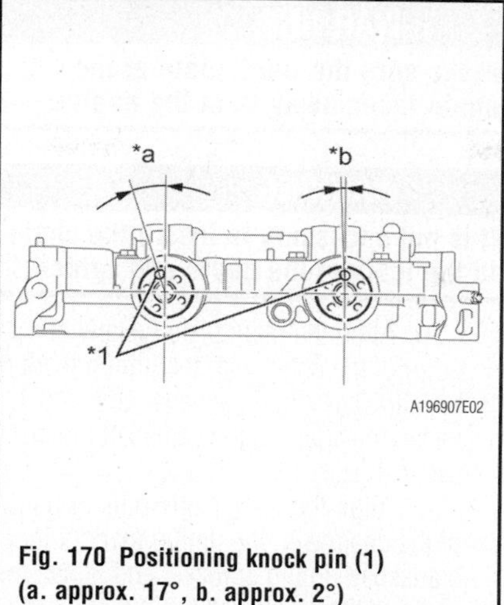

Fig. 170 Positioning knock pin (1) (a. approx. 17°, b. approx. 2°)

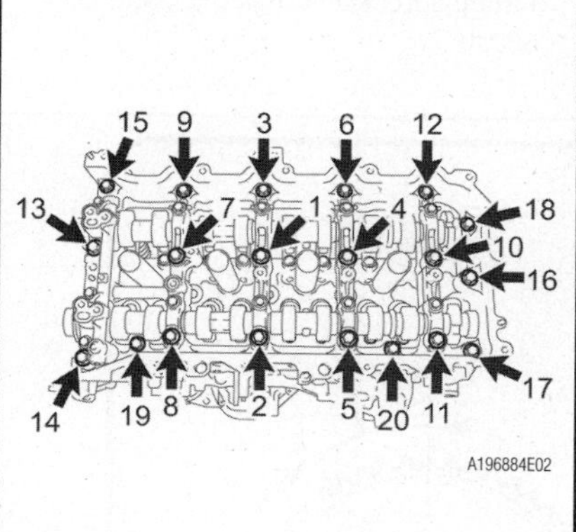

Fig. 171 Installing camshaft housing and tightening bolts in order

a. Check the camshaft timing gear position.

➡If the camshaft timing gear is not set to the advanced position, release the lock pin and reset the camshaft timing gear.

b. Align and attach the knock pin of the camshaft with the pin hole of the camshaft timing gear assembly.

c. Check that there is no clearance between the camshaft timing gear and camshaft flange.

d. Using a wrench to hold the hexagonal portion of the camshaft, install the bolt. Tighten to 63 ft. lbs. (85 Nm).

⁂ WARNING

Be careful not to damage the cylinder head or spark plug tube with the wrench.

⁂ WARNING

Do not disassemble the camshaft timing gear assembly.

➡Perform "Inspection After Repair" after replacing the camshaft timing gear assembly.

45. Install camshaft timing exhaust gear assembly.

a. Align and attach the knock pin of the No. 2 camshaft with the pin hole of the camshaft timing exhaust gear assembly.

b. Check that there is no clearance between the camshaft timing exhaust gear and camshaft flange.

c. Using a wrench to hold the hexagonal portion of the No. 2 camshaft, install the bolt. Tighten to 63 ft. lbs. (85 Nm).

⁂ WARNING

Be careful not to damage the cylinder head or spark plug tube with the wrench.

⁂ WARNING

Do not disassemble the camshaft timing exhaust gear assembly.

➡Perform "Inspection After Repair" after replacing the camshaft timing exhaust gear assembly.

46. Add engine oil.

a. Add 1.7 fl. oz. (50 cc) of engine oil into the oil hole.

⁂ WARNING

Oil must be added if the lash adjusters were removed.

Fig. 172 Identifying oil hole

⁂ WARNING

Make sure that the low pressure chamber and oil paths of the lash adjusters are full of engine oil.

47. Set No. 1 cylinder to TDC/Compression.
    a. Temporarily install the crankshaft pulley bolt.
    b. Rotate the crankshaft 40° counterclockwise to position the crankshaft pulley key.
    c. Check the timing marks of the camshaft timing gears.
48. Install No. 1 chain vibration damper with the 2 bolts. Tighten to 15 ft. lbs. (21 Nm).
49. Install chain sub assembly.
    a. Place the chain onto the camshaft timing gears and crankshaft timing sprocket.

⁂ WARNING

Make sure the mark plate of the chain faces away from the engine.

⁂ WARNING

It is not necessary to install the chain to the teeth of the gears and sprocket.

    b. Align the mark plate (yellow or gold) of the chain with the timing mark of the camshaft timing exhaust gear and install the chain to the camshaft timing exhaust gear.
    c. Align the mark plate (pink or gold) of the chain with the timing mark of the crankshaft timing sprocket and install the chain to the crankshaft timing sprocket.
    d. Tie a string above the crankshaft timing sprocket so that the chain is secure.
    e. Using the hexagonal portion of the intake camshaft, rotate the intake camshaft counterclockwise with a wrench, align the timing mark of the camshaft timing gear with the mark plate (yellow or gold) of the chain and install the chain to the camshaft timing gear.

➡Hold the intake camshaft in place with a wrench until the chain tensioner is installed.

    f. Remove the string above the crankshaft timing sprocket, rotate the crankshaft clockwise, and loosen the chain so that the chain tensioner slipper can be installed.

⁂ WARNING

Make sure the chain is secure.

50. Install the chain tensioner slipper.
    a. Install the chain tensioner slipper with the bolt. Tighten to 15 ft. lbs. (21 Nm).
51. Install the No. 1 chain tensioner assembly.
    a. Install a new gasket and the chain tensioner with the 2 bolts. Tighten to 7 ft. lbs. (10 Nm).
52. Install the timing chain guide.
    a. Install the timing chain guide with the bolt. Tighten to 15 ft. lbs. (21 Nm).
53. Check No. 1 cylinder at TDC/Compression.
    a. Temporarily install the crankshaft pulley bolt.
    b. Rotate the crankshaft clockwise, and check that the timing marks on the crankshaft timing sprocket and camshaft timing gears.
    c. Remove the crankshaft pulley bolt.
54. Install intake manifold.
55. Connect No. 2 ventilation hose.
56. Install fuel delivery pipe sub assembly.
57. Install vacuum switching valve assembly (for ACIS).
58. Install throttle with motor body assembly.
59. Install exhaust manifold converter sub assembly.
60. Install EGR inlet exhaust manifold plate (for PZEV).
61. Install No. 1 exhaust manifold heat insulator.
62. Install No. 2 manifold stay.
63. Install manifold stay.
64. Install timing chain cover sub assembly.

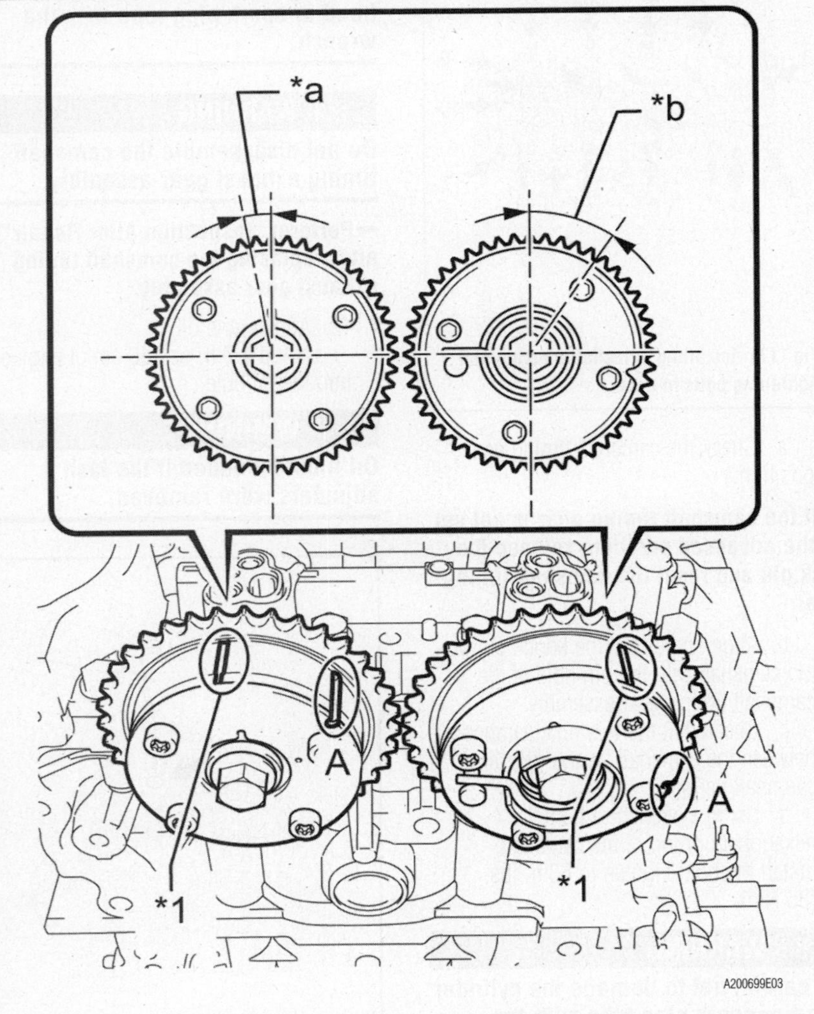

Fig. 173 Rotating crankshaft 40° (a: approx. 7°, b: approx. 32°) and checking timing marks of camshaft timing gears (A: is not a timing mark)

3.5L Engine

See Figures 174 through 182.

1. Remove camshaft.
2. Remove No. 2 camshaft.
3. Remove camshaft housing sub assembly RH.
4. Remove camshaft timing gears and No. 2 chain (for bank 2).
5. Remove No. 3 chain tensioner assembly.
6. Remove camshaft bearing cap (for bank 2).
7. Remove No. 3 camshaft.
8. Remove No. 4 camshaft.
9. Remove camshaft housing sub assembly LH.
10. Remove No. 1 valve rocker arm sub assembly.
    a. Remove the 24 valve rocker arms.

➡Arrange the removed parts in the correct order.

11. Remove valve lash adjuster assembly.
    a. Remove the 24 valve lash adjuster assemblies from the cylinder head.

➡Arrange the removed parts in the correct order.

12. Remove the 24 valve step caps.
13. Remove water outlet.
14. Remove cylinder head sub assembly RH.
    a. Using a 10 mm bi-hexagon wrench, uniformly loosen the 8 cylinder head bolts in sequence. Remove the 8 cylinder head bolts and plate washers.

⁂ WARNING

Be careful not to drop washers into the cylinder head sub-assembly.

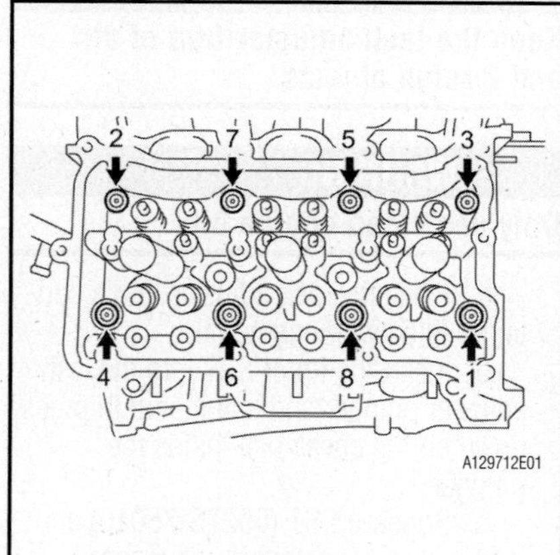

Fig. 174 Removing cylinder head bolts and plate washers in order

⁂ WARNING

Cylinder head warpage or cracking could result from removing bolts in the incorrect order.

➡Arrange the removed parts in the correct order.

    b. Remove the cylinder head sub-assembly RH.
15. Remove the cylinder head gasket RH.
16. Remove the cylinder head sub assembly LH.
    a. Uniformly loosen and remove the 2 cylinder head set bolts in several steps and in sequence.
    b. Using a 10 mm bi-hexagon wrench, uniformly loosen the 8 bolts in sequence. Remove the 8 cylinder head bolts and plate washers.

⁂ WARNING

Be careful not to drop washers into the cylinder head sub-assembly.

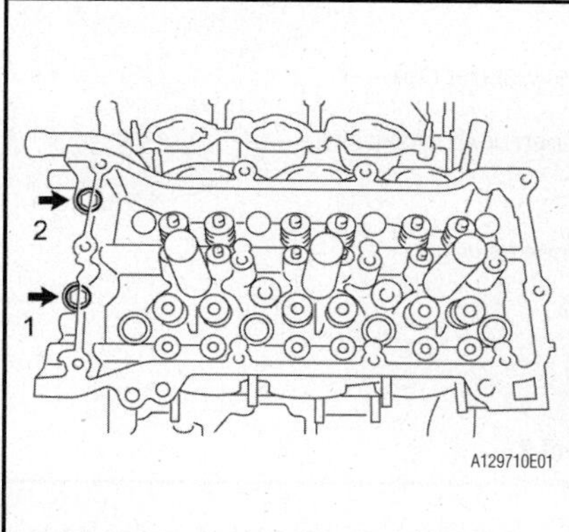

Fig. 175 Removing cylinder head set bolts

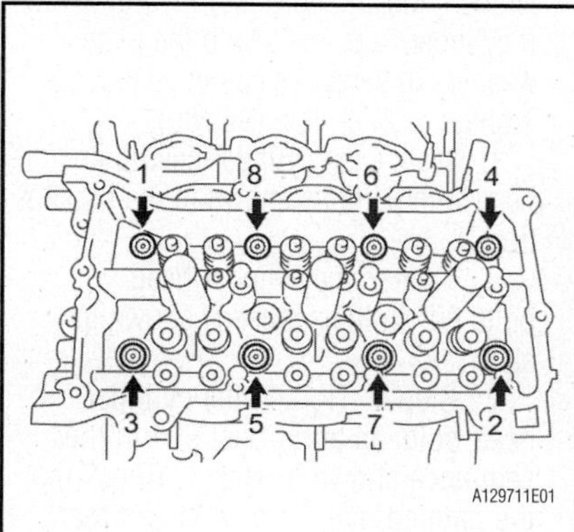

Fig. 176 Removing the 8 cylinder head bolts and plate washers

⁂ WARNING

Cylinder head warpage or cracking could result from removing bolts in the incorrect order.

➡Be sure to keep separate the removed parts for each installation position.

    c. Remove the cylinder head sub-assembly LH.
17. Remove the cylinder head gasket LH.

To install:

⁂ WARNING

Perform inspection after repair.

18. Install cylinder head gasket RH.
    a. Place a new cylinder head gasket RH on the cylinder block surface with the Lot No. stamp facing upward.

⁂ WARNING

Be careful of installation direction.

19. Install cylinder head sub assembly RH.
    a. Place the cylinder head on the cylinder block.

⁂ WARNING

Be careful not to allow oil to adhere to the bottom part of the cylinder head.

⁂ WARNING

Gently lower the cylinder head in order not to damage the gasket with the bottom part of the head.

➡The cylinder head bolts are tightened in 3 progressive steps.

    b. Apply a light coat of engine oil to the threads and under the heads of the cylinder head bolts.
    c. Step 1: Using a 10 mm bi-hexagon wrench, install and uniformly tighten the 8 cylinder head bolts with the plate washers in several steps and in sequence. Tighten to 27 ft. lbs. (36 Nm).
    d. Step 2: Mark each cylinder head bolt head with paint. Tighten the cylinder head bolts 90° in sequence from Step 1.
    e. Step 3: Tighten the cylinder head bolts an additional 90° in the sequence from Step 1. Check that the painted mark is now facing rearward.
20. Perform inspection after repair.
21. Install cylinder head gasket LH.

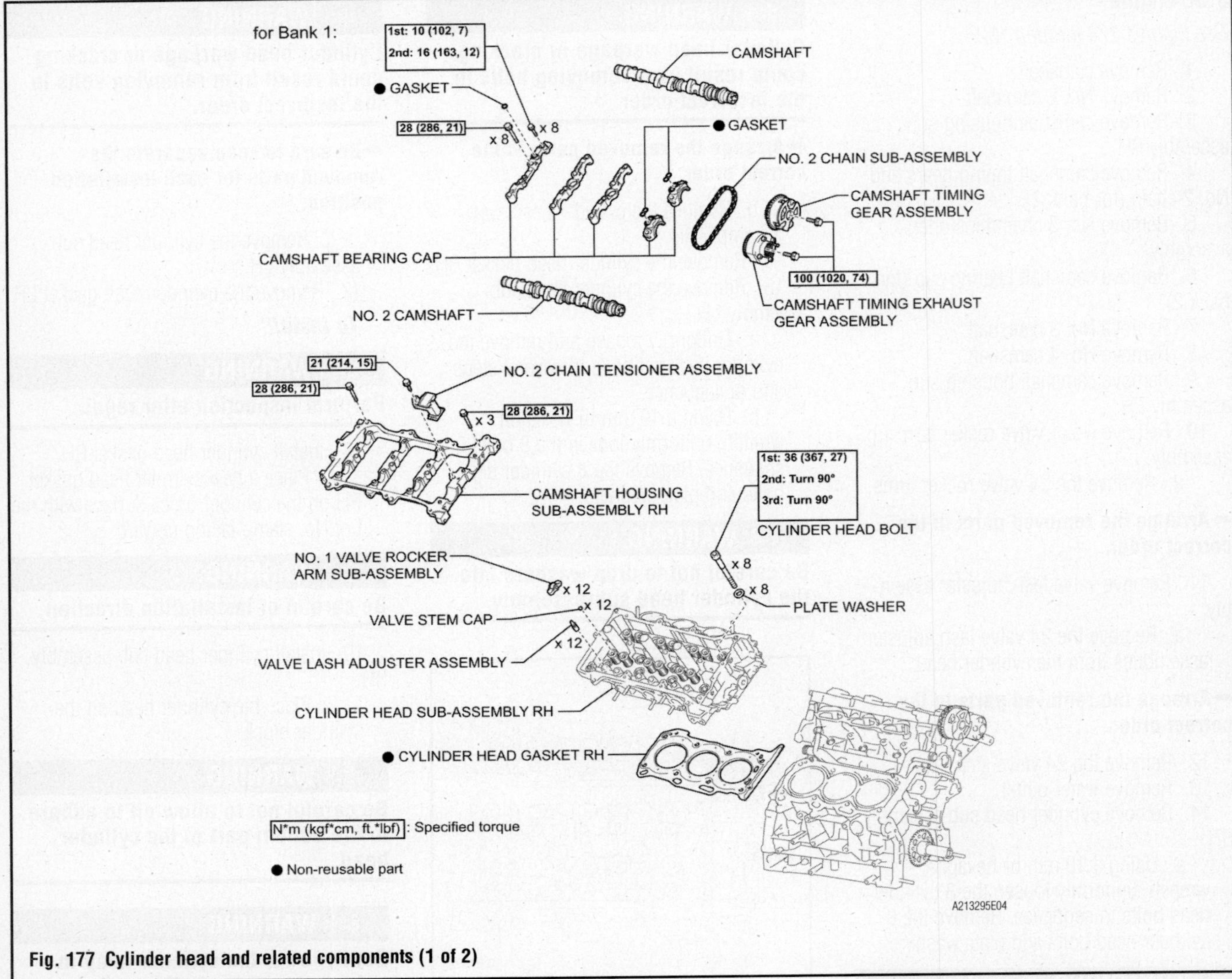

Fig. 177 Cylinder head and related components (1 of 2)

a. Place a new cylinder head gasket LH on the cylinder block surface with the Lot No. stamp facing upward.

➡Be careful of installation direction.

22. Install cylinder head sub assembly LH.

a. Place the cylinder head on the cylinder block.

⁂ WARNING

Be careful not to allow oil to adhere to the bottom part of the cylinder head.

⁂ WARNING

Gently lower the cylinder head in order not to damage the gasket with the bottom part of the head.

➡The cylinder head bolts are tightened in 3 progressive steps.

b. Apply a light coat of engine oil to the threads and under the heads of the cylinder head bolts.

c. Step 1: Using a 10 mm bi-hexagon wrench, install and uniformly tighten the 8 cylinder head bolts with the plate washers in several steps in sequence. Tighten to 27 ft. lbs. (36 Nm).

d. Mark each cylinder head bolt head with paint as shown in the illustration.

e. Tighten the cylinder head bolts 90° in the sequence shown in step 1.

f. Step 3: Tighten the cylinder head bolts an additional 90° in the sequence shown in step 1. Check that the painted mark is now facing rearward.

g. Tighten the 2 bolts in order. Tighten to 22 ft. lbs. (30 Nm).

➡Perform inspection after repair.

23. Install water outlet.
24. Install 24 valve stem caps.
25. Install valve lash adjuster assembly.

⁂ WARNING

Keep the lash adjuster free of dirt and foreign objects.

⁂ WARNING

Only use clean engine oil.

a. Place the lash adjuster into a container filled with engine oil.

b. Insert the tip of SST into the lash adjuster plunger and use the tip to press down on the check ball inside the plunger.

c. Squeeze SST (09276-75010) and lash adjuster together to move the plunger up and down 5 to 6 times.

d. Check the movement of the plunger and bleed the air.

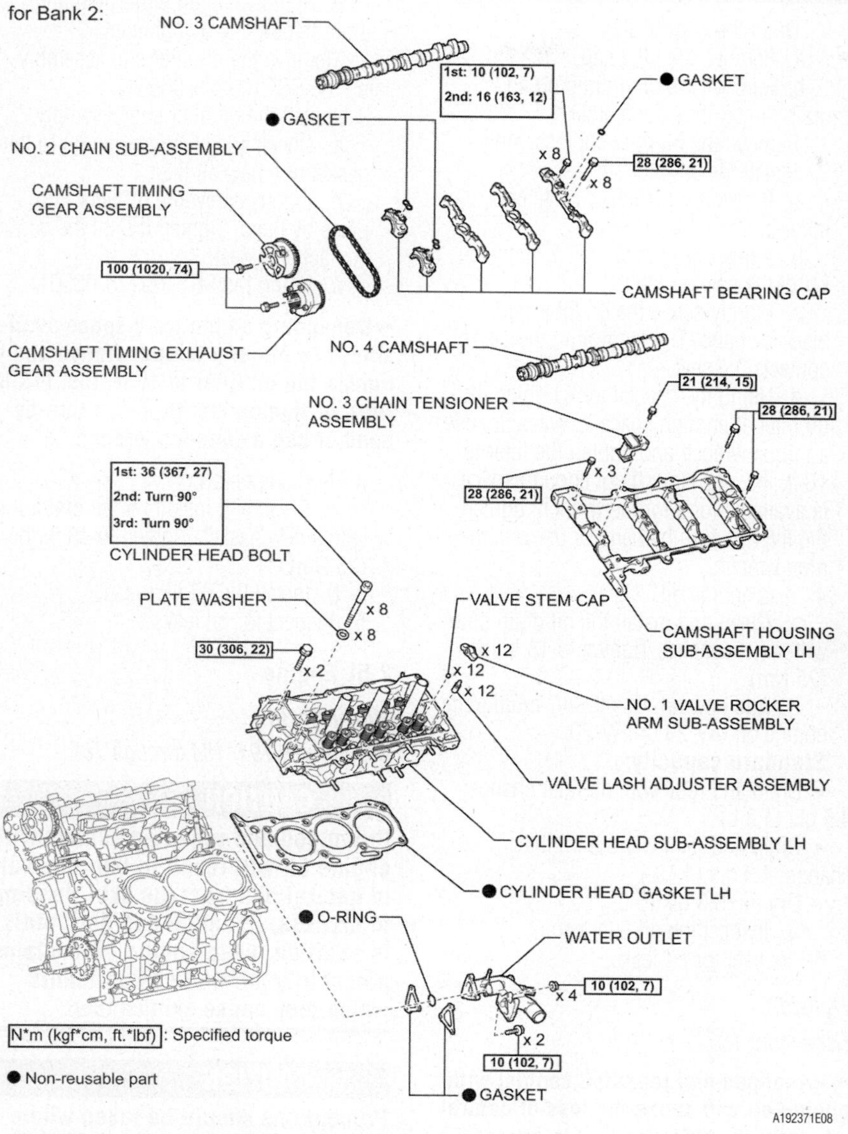

Fig. 178 Cylinder head and related components (2 of 2)

➡It is OK, if plunger moves up and down.

⁂ WARNING

When bleeding air from the high pressure chamber, make sure that the tip of the SST is actually pressing the check ball. If the check ball is not pressed, air will not bleed.

e. After bleeding the air, remove SST. Then, try to press the plunger quickly and firmly by hand.

➡It is OK, if plunger is very difficult to move. If the result is not as specified, replace the valve lash adjuster.

f. Install the valve lash adjusters.

➡Install each valve lash adjuster to the same place it was removed from.

26. Install No. 1 valve rocker arm sub assembly.

a. Apply engine oil to the lash adjuster tip and valve stem cap end.

b. Install the valve rocker arms.

27. Install No. 3 camshaft.

28. Install No. 4 camshaft.

29. Install camshaft bearing cap (for bank 2).

30. Install camshaft housing sub assembly LH.

31. Install camshaft.

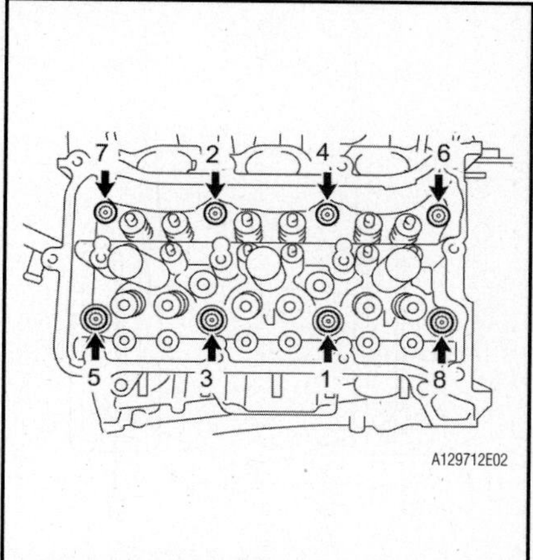

Fig. 179 Tightening the 8 cylinder head bolts with the plate washers in sequence

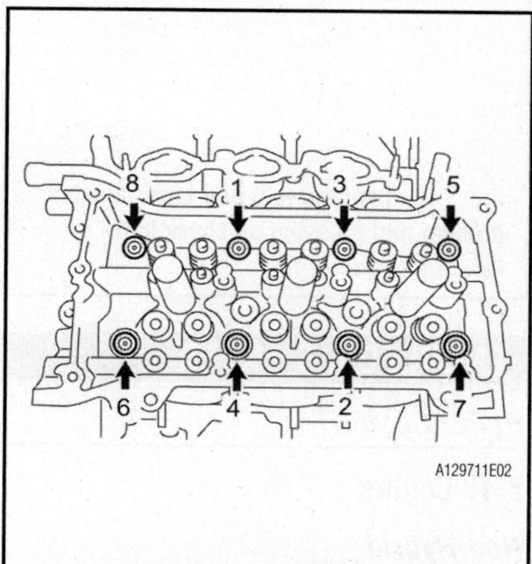

Fig. 180 Installing cylinder head bolts with plate washers in sequence

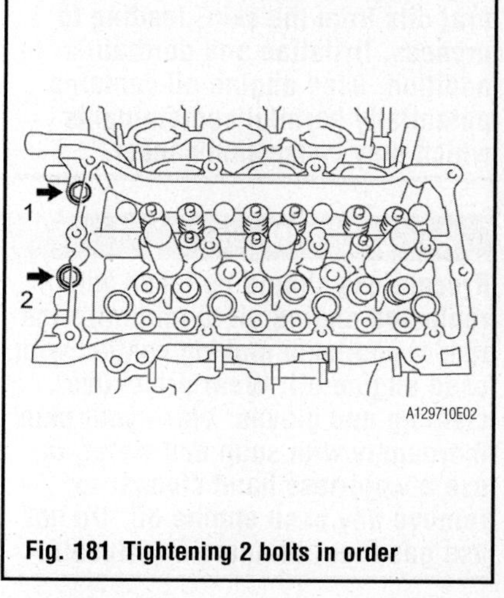

Fig. 181 Tightening 2 bolts in order

32. Install camshaft bearing cap (for bank 1).

33. Install camshaft housing sub assembly RH.

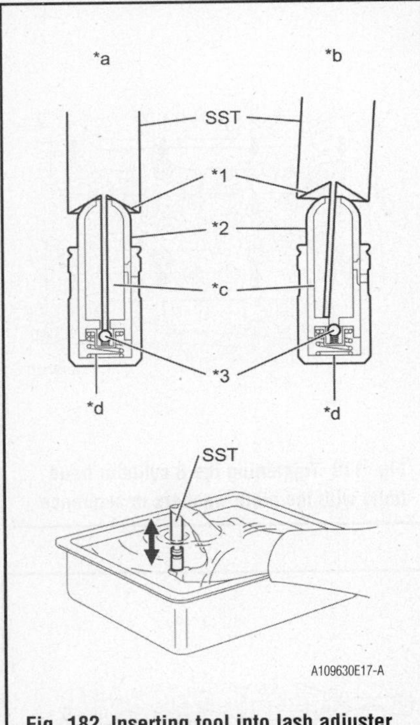

Fig. 182 Inserting tool into lash adjuster plunger and pressing on check ball inside plunger

ENGINE OIL & FILTER

REPLACEMENT

2.4L Engine

Non-Hybrid

CAUTION

Prolonged and repeated contact with engine oil will cause the loss of natural oils from the skin, leading to dryness, irritation and dermatitis. In addition, used engine oil contains potentially harmful contaminants which may cause skin cancer.

CAUTION

Precautions should be taken when replacing engine oil to minimize the risk of your skin making contact with used engine oil. Wear protective clothing and gloves. Wash your skin thoroughly with soap and water, or use a waterless hand cleaner to remove any used engine oil. Do not use gasoline, thinners or solvents.

CAUTION

For environmental protection, used oil and used oil filters must be disposed of at designated disposal sites.

1. Drain the engine oil.
   a. Remove the filler cap.
   b. Remove the oil drain plug and drain the oil into a container.
2. Remove the oil filter sub assembly.
3. install the oil filter sub assembly.
   a. Check and clean the oil filter installation surface.
   b. Apply clean engine oil to the gasket of the new oil filter.
   c. Lightly screw the oil filter into place by hand. Tighten it until the gasket contacts the seat.
   d. Using the special tool, tighten the oil filter. If enough space is available, use a torque wrench and tighten the filter to 10 ft. lbs. (13 Nm). If not enough space is available for a torque wrench, tighten the filter ¾ turn by hand or use a common wrench.
4. Add engine oil.
   a. Clean and install the oil drain plug with a new gasket. Tighten to 18 ft. lbs. (25 Nm).
   b. Add new oil. Use ILSAC multigrade engine oil 0W-20 or 5W-20.

Standard capacity:

- Drain and refill with oil filter change: 4.5 qts (4.3 L)
- Drain and refill without oil filter change: 4.3 qts (4.1 L)
- Dry fill: 5.3 qts (5 L)

   a. Inspect the oil filler cap.
5. Inspect for oil leaks.

Hybrid

See Figure 183.

➡**Prolonged and repeated contact with engine oil will cause the loss of natural oils from the skin, leading to dryness, irritation and dermatitis. In addition, used engine oil contains potentially harmful contaminants which may cause skin cancer.**

➡**Precautions should be taken when replacing engine oil to minimize the risk of your skin making contact with used engine oil. Wear protective clothing and gloves. Wash your skin thoroughly with soap and water, or use a waterless hand cleaner to remove any used engine oil. Do not use gasoline, thinners or solvents.**

➡**For environmental protection, used oil and used oil filters must be disposed of at designated disposal sites.**

1. Drain the engine oil.
   a. Remove the oil filler cap.
   b. Remove the oil drain plug and drain the oil into a container.
2. Remove the oil filter sub assembly using the SST (09228-06501).
3. Install the oil filter sub assembly.
   a. Check and clean engine oil to the gasket of a new oil filter.
   b. Lightly screw the oil filter into place by hand. Tighten it until the gasket contacts the seat.
   c. Using the SST (09228-06501).

➡**Depending on the work space available, use either a torque wrench and tighten the oil filter to 10 ft. lbs. (13 ft. Nm), or tighten the filter ¾ a turn by hand or use a common wrench.**

4. Add engine oil.
   a. Clean and install the oil drain plug with a new gasket. Tighten to 18 ft. lbs. (25 Nm).
   b. Install the oil filler cap.
5. Inspect for oil leaks.

2.5L Engine

Non-Hybrid

See Figures 194, 184 through 186.

CAUTION

Prolonged and repeated contact with engine oil will result in the removal of natural oils from the skin, leading to dryness, irritation and dermatitis. In addition, used engine oil contains potentially harmful contaminants which may cause skin cancer.

CAUTION

Precautions should be taken when replacing engine oil to minimize the risk of your skin making contact with used engine oil. Protective clothing and gloves that cannot be penetrated by oil should be worn. The skin should be washed with soap and water, or use waterless hand cleaner, to remove any used engine oil thoroughly. Do not use gasoline, thinners, or solvents.

CAUTION

In order to protect the environment, used oil and used oil filters must be disposed of at designated disposal sites.

1. Drain the engine oil.
   a. Remove the oil filler cap.

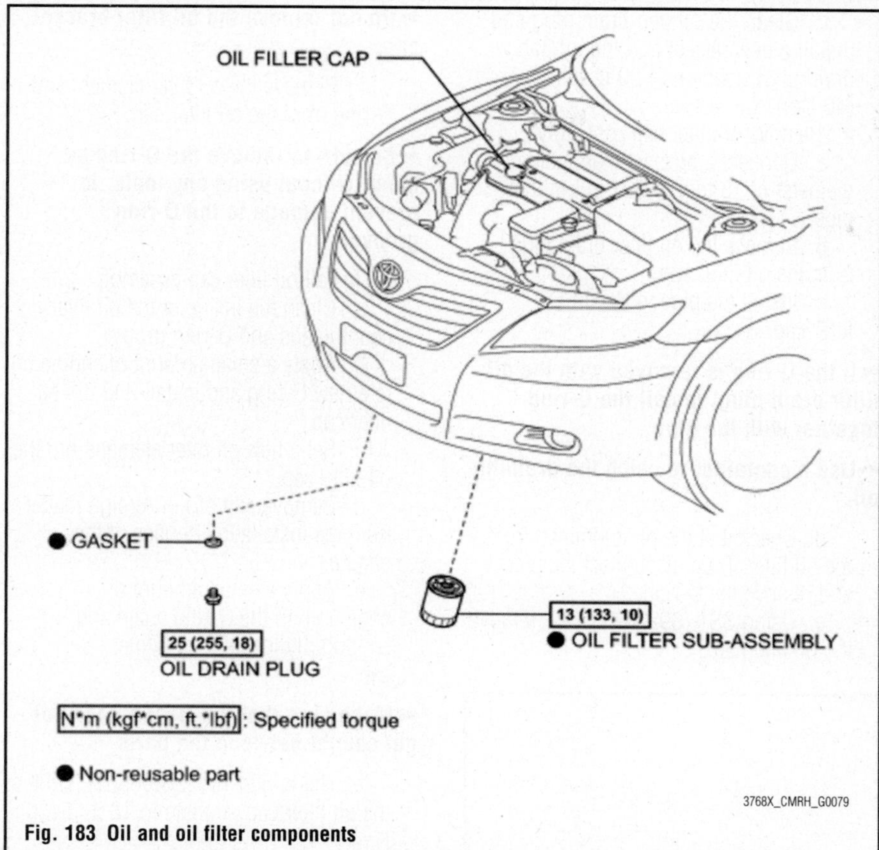

Fig. 183 Oil and oil filter components

b. Remove the oil pan drain plug and gasket, and drain the engine oil into a container.

c. Clean and install a new gasket and the oil pan drain plug. Tighten to 30 ft. lbs. (40 Nm).

2. Remove the oil filter cap assembly.

a. Connect a hose with an inside diameter of 0.591 inch (15 mm) to the pipe.

b. Remove the oil filter drain plug from the oil filter cap.

c. Install the pipe to the oil filter cap.

➡If the O-ring is removed with the drain plug, install the O-ring together with the pipe.

➡Use a container to catch the draining oil.

d. Check that the oil is drained from the oil filter. Then disconnect the pipe as shown in the illustration and remove the O-ring.

e. Remove the oil filter cap with the special tool.

➡Do not remove the oil filter bracket clip.

f. Remove the oil filter element and the O-ring from the oil filter cap.

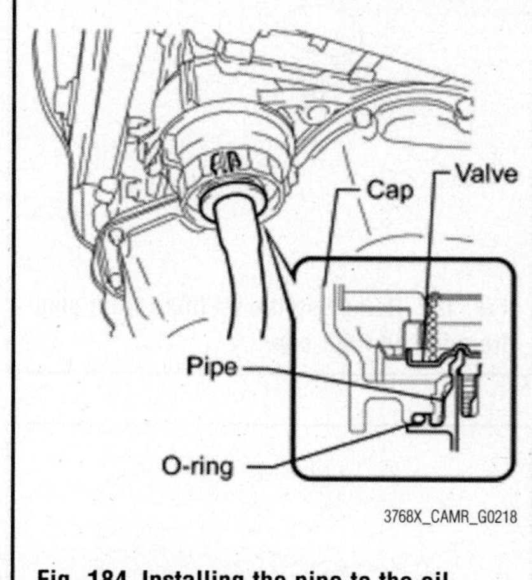

Fig. 184 Installing the pipe to the oil filter cap

➡Be sure to remove the cap O-ring by hand, without using any tools, to prevent damage to the cap O-ring groove.

3. Install the oil filter cap assembly.

a. Clean the inside of the oil filter cap, its threads and its O-ring groove.

b. Apply a small amount of engine oil to a new O-ring and install it to the oil filter cap.

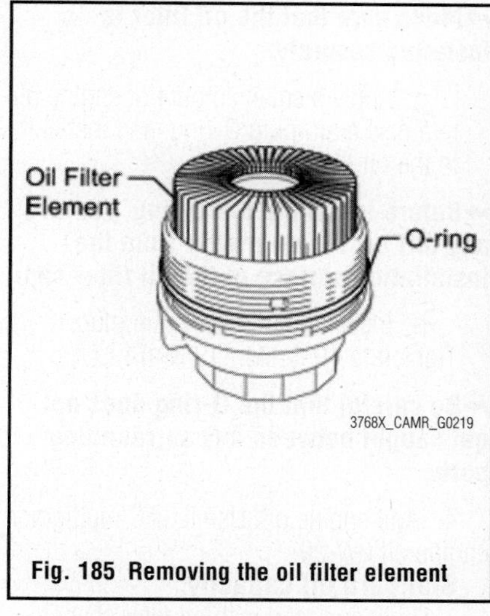

Fig. 185 Removing the oil filter element

c. Set a new oil filter element into the oil filter cap.

d. Remove any dirt or foreign matter from the installation surface of the engine.

e. Apply a small amount of engine oil to the O-ring again and temporarily install the oil filter cap.

➡Make sure that the O-ring does not get caught between the parts.

f. Using the special tool, tighten the oil filter cap. Tighten to 18 ft. lbs. (25 Nm).

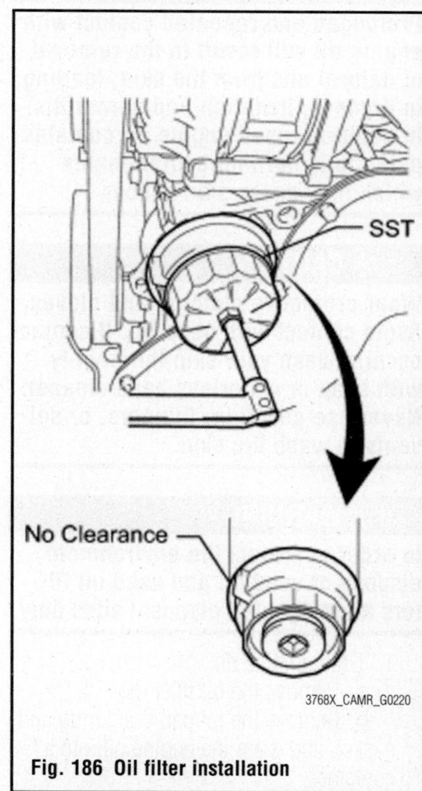

Fig. 186 Oil filter installation

➡**Make sure that the oil filter is installed securely.**

g. Apply a small amount of engine oil to a new drain plug O-ring, and install it to the oil filter cap.

➡**Before installing the O-ring, remove any dirt or foreign matter from the installation surface of the oil filter cap.**

h. Install the oil filter drain plug. Tighten to 10 ft. lbs. (13 Nm).

➡**Be careful that the O-ring does not get caught between any surrounding parts.**

4. Add engine oil. Use ILSAC multigrade engine oil 0W-20.

Standard oil capacity:

- Drain and refill with oil filter change: 4.6 qts (4.4 L)
- Drain and refill without oil filter change: 4.2 qts (4 L)
- Dry fill: 6 qts (5.7 L)

a. Install the oil filler cap.

5. Inspect for oil leaks.

a. Start the engine. Make sure that there are no oil leaks from the areas that were worked on

6. Inspect the engine oil level.

Hybrid

See Figures 187 through 190.

⁂ CAUTION

Prolonged and repeated contact with engine oil will result in the removal of natural oils from the skin, leading to dryness, irritation and dermatitis. In addition, used engine oil contains potentially harmful contaminants which may cause skin cancer.

⁂ CAUTION

Wear protective clothing and gloves. Avoid contact with used oil. If contact occurs, wash your skin thoroughly with soap or waterless hand cleaner. Never use gasoline, thinners, or solvents to wash the skin.

⁂ WARNING

In order to protect the environment, dispose of used oil and used oil filters at designated disposal sites only.

1. Drain engine oil.

a. Remove the oil filler cap.

b. Remove the oil pan drain plug and gasket, and drain the engine oil into a container.

c. Clean the oil pan drain plug and install a new gasket and the oil pan drain plug. Tighten to 30 ft. lbs. (40 Nm).

2. Remove oil filter cap assembly.

a. Connect a hose with an inside diameter of 15 mm (0.591 in.) to the pipe.

b. Remove the oil filter drain plug from the oil filter cap.

c. Install the pipe to the oil filter cap.

➡**If the O-ring is removed with the oil filter drain plug, install the O-ring together with the pipe.**

➡**Use a container to catch the draining oil.**

d. Check that the oil is drained from the oil filter. Then disconnect the pipe and remove the O-ring.

e. Using SST (09228-06501), remove the oil filter cap.

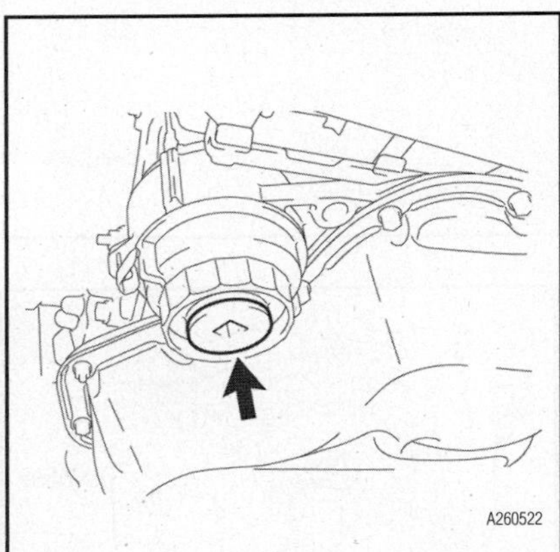

Fig. 187 Removing the oil filter drain plug from the oil filter cap

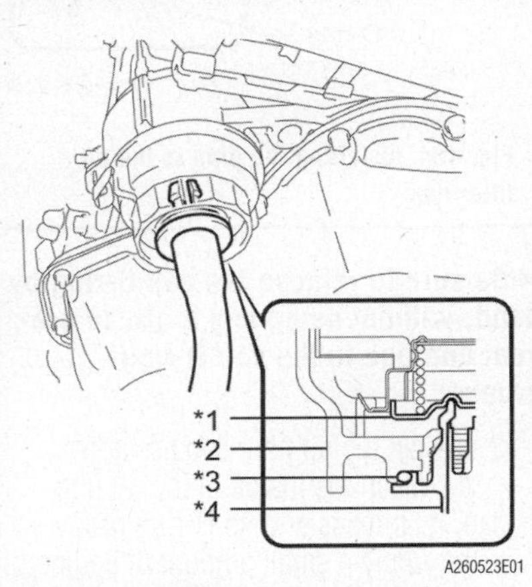

Fig. 188 Installing pipe to the oil filter cap

➡**Do not remove the oil filter bracket clip.**

f. Remove the oil filter element and O-ring from the oil filter cap.

➡**Be sure to remove the O-ring by hand, without using any tools, to prevent damage to the O-ring groove.**

3. Install oil filter cap assembly.

a. Clean the inside of the oil filter cap, threads and O-ring groove.

b. Apply a small amount of engine oil to a new O-ring and install it to the oil filter cap.

c. Set a new oil filter element into the oil filter cap.

d. Remove any dirt or foreign matter from the installation surface of the engine.

e. Apply a small amount of engine oil to the O-ring again and temporarily install the oil filter cap.

➡**Make sure that the O-ring does not get caught between the parts.**

f. Using SST (09228-06501), tighten the oil filter cap. Tighten to 18 ft. lbs. (25 Nm).

➡**Make sure that the oil filter is installed securely.**

g. Apply a small amount of engine oil to a new O-ring, and install it to the oil filter cap.

➡**Before installing the O-ring, remove any dirt or foreign matter from the installation surface of the oil filter cap.**

h. Install the oil filter drain plug. Tighten to 10 ft. lbs. (13 Nm).

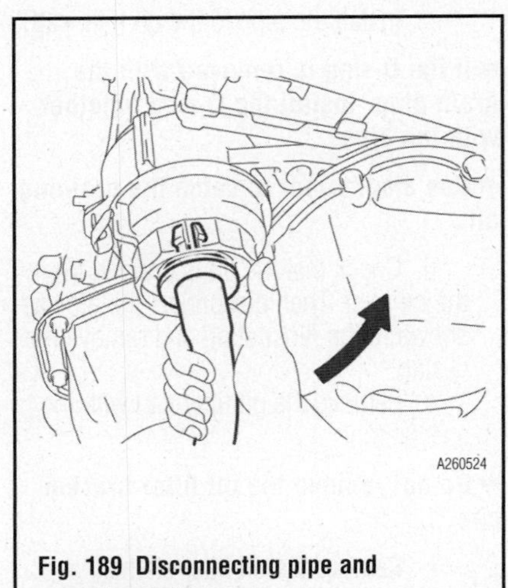

Fig. 189 Disconnecting pipe and removing O-ring

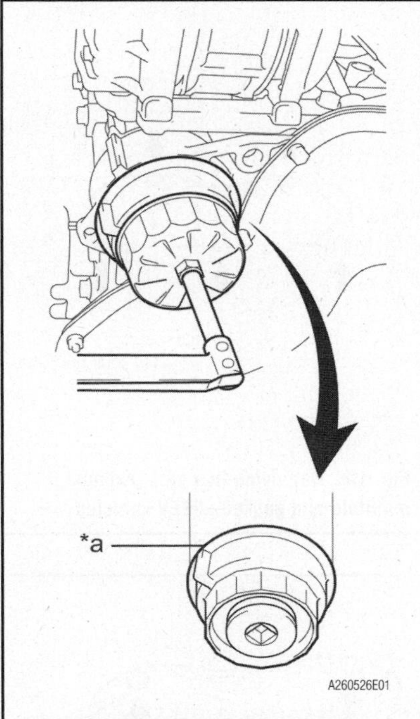

Fig. 190 Checking oil filter installation (*a: no clearance)

✱✱ WARNING

Be careful that the O-ring does not get caught between any surrounding parts.

4. Add engine oil.
   a. Add new oil.
   b. Install the oil filler cap.
5. Inspect for oil leak.
6. Inspect engine oil level.

3.5L Engine

See Figures 191 through 194.

✱✱ CAUTION

Prolonged and repeated contact with engine oil will result in the removal of natural oils from the skin, leading to dryness, irritation and dermatitis. In addition, used engine oil contains potentially harmful contaminants which may cause skin cancer.

✱✱ CAUTION

Precautions should be taken when replacing engine oil to minimize the risk of your skin making contact with used engine oil. Protective clothing and gloves that cannot be penetrated by oil should be worn. The skin should be washed with soap and water, or use water-less hand cleaner, to remove any used engine oil thoroughly. Do not use gasoline, thinners, or solvents.

✱✱ CAUTION

In order to preserve the environment, used oil and used oil filters must be disposed of at designated disposal sites.

1. Drain the engine oil.
   a. Remove the oil filler cap.
   b. Remove the oil drain plug and drain the oil into a container.
2. Remove the oil filter element.
   a. Connect the hose with an inside diameter of 0.59 inch (15 mm) to the pipe.
   b. Remove the oil filter drain plug from the oil filter cap.
   c. insert the pipe with the hose into the oil filter cap.

➡Be sure to insert the pipe with the O-ring installed on the oil filter cap side.

➡Place the hose end into a container before draining the oil from the hose.

   d. Make sure that the oil is completely drained, and remove the pipe O-ring.

➡Be sure to turn the pipe in the direction or the arrow to remove it.

   e. Using the special tool, remove the oil filter cap.
   f. Remove the oil filter element and O-ring from the oil filter cap.

➡Do not use any tools to remove the O-ring in order to prevent the cap from being damaged. Be sure to remove it by hand.

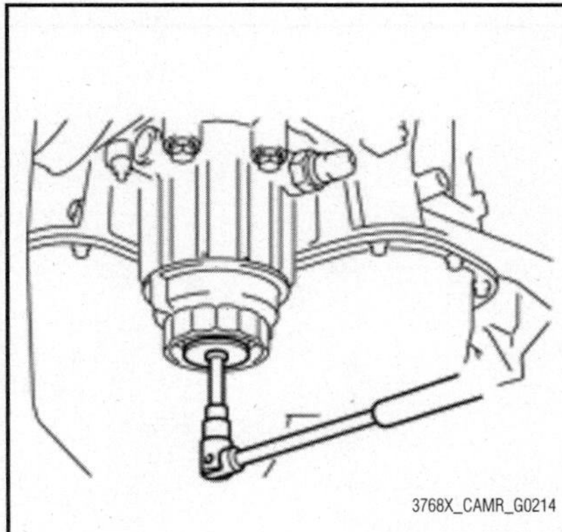

Fig. 191 Removing the oil filter drain plug from the oil filter cap

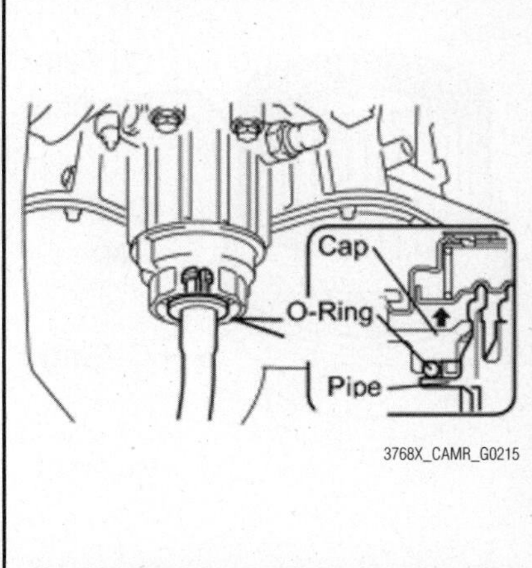

Fig. 192 Inserting the pipe with the hose into the oil filter cap

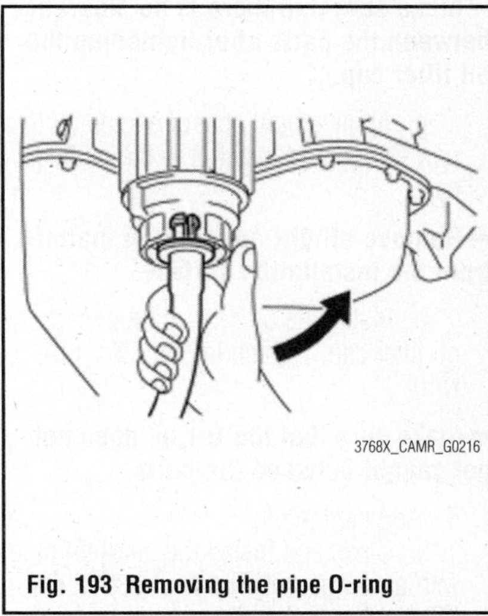

Fig. 193 Removing the pipe O-ring

3. Install the oil filter element.
   a. Clean the inside of the oil filter cap, threads, and O-ring groove.
   b. Apply a light coat of engine oil to a mew O-ring and install it to the oil filter cap.

➡Make sure that the O-ring does not get twisted on the groove.

   c. Install a new oil filter element to the oil filter cap.
   d. Remove all dirt and foreign matter from the installation surface and the inside of the cap on the engine side.
   e. Apply a light coat of engine oil to the O-ring again and install the oil filter cap.

➡Make sure that the O-ring does not get caught between the parts.

   f. Install the oil filter cap with the special tool. Tighten to 18 ft. lbs. (25 Nm).

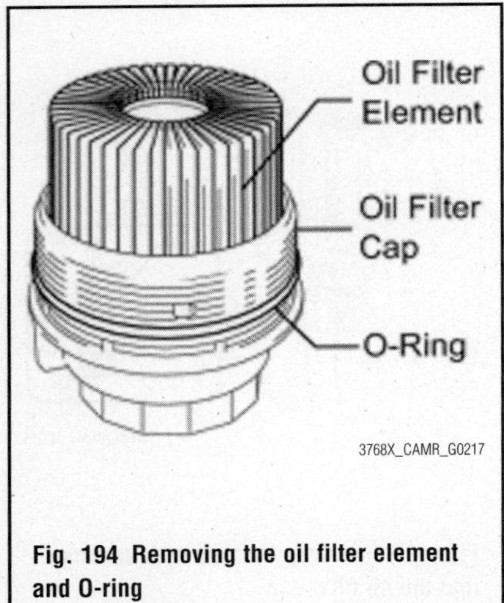

Fig. 194 Removing the oil filter element and O-ring

➡Make sure that there is no clearance between the parts after tightening the oil filter cap.

g. Apply a light coat of engine oil to a new O-ring and install it to the oil filter cap.

➡Remove all dirt and foreign matter from the installation surface.

h. Install the oil filter drain plug to the oil filter cap. Tighten to 10 ft. lbs. (13 Nm).

➡make sure that the O-ring does not get caught between the parts.

4. Add engine oil.
   a. Clean and install the oil drain plug with a new gasket. Tighten to 30 ft. lbs. (40 Nm).
   b. Add new oil. Use ILSAC multigrade engine oil 5W-30.

Standard capacity:

- Drain and refill with oil filter change: 6.4 qts (6.1 L)
- Drain and refill without oil filter change: 6 qts (5.7L)
- Dry fill: 6.9 qts (6.5 L)

   a. Install the oil filler cap.
5. Inspect for oil leaks.
   a. Start the engine. Check for engine oil leaks from the connected parts of the oil filter cap and oil filter drain plug.

EXHAUST MANIFOLD

REMOVAL & INSTALLATION

2.4L Engine

See Figures 195 through 202.

1. Before servicing the vehicle, refer to the Precautions Section.

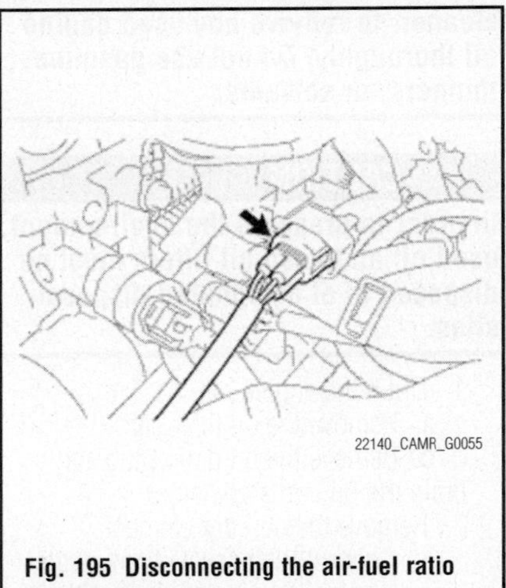

Fig. 195 Disconnecting the air-fuel ratio sensor connector

Fig. 196 Removing exhaust manifold stay—1 of 2

2. Disconnect the negative battery cable.
3. Remove engine cover.
4. Remove air cleaner assembly.

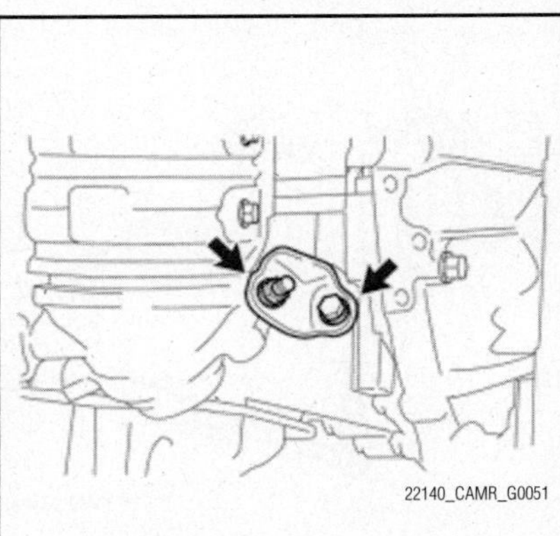

Fig. 197 Removing exhaust manifold stay—2 of 2

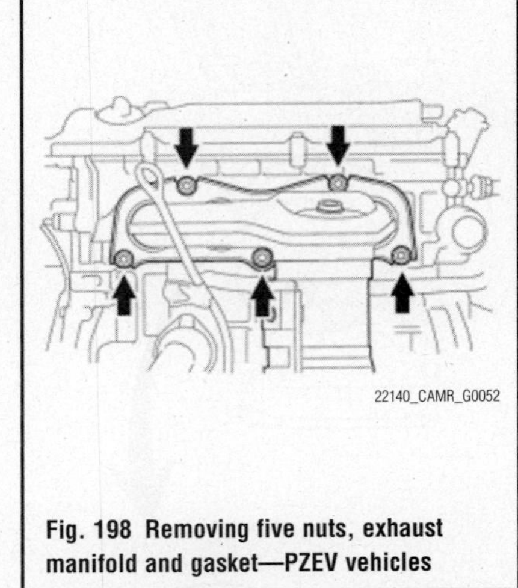

Fig. 198 Removing five nuts, exhaust manifold and gasket—PZEV vehicles

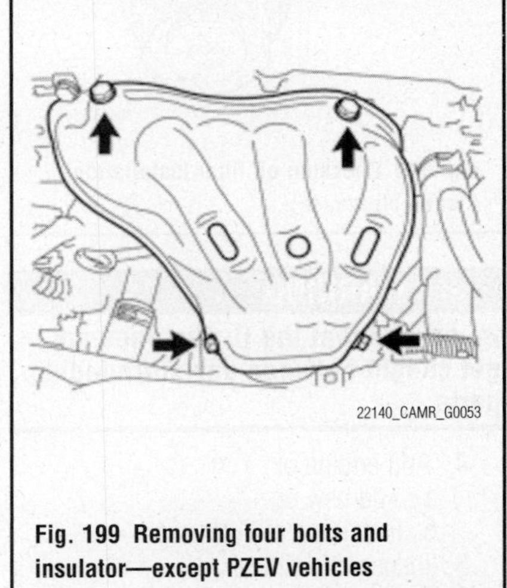

Fig. 199 Removing four bolts and insulator—except PZEV vehicles

5. Remove manifold stays.
6. Disconnect the air-fuel ratio sensor connector.
7. Remove or disconnect remaining components from the exhaust manifold.
8. For PZEV vehicles, perform the following:
   a. Remove the five nuts, manifold converter and gasket.
9. For non-PZEV vehicles, perform the following:
   a. Remove the four bolts and insulator.
   b. Remove the five nuts, manifold converter and gasket.
10. Remove exhaust manifold from catalytic converter.

To install:

11. To install, reverse removal procedure.
12. Install new gaskets for exhaust manifolds.
13. For non-PZEV vehicles, tighten

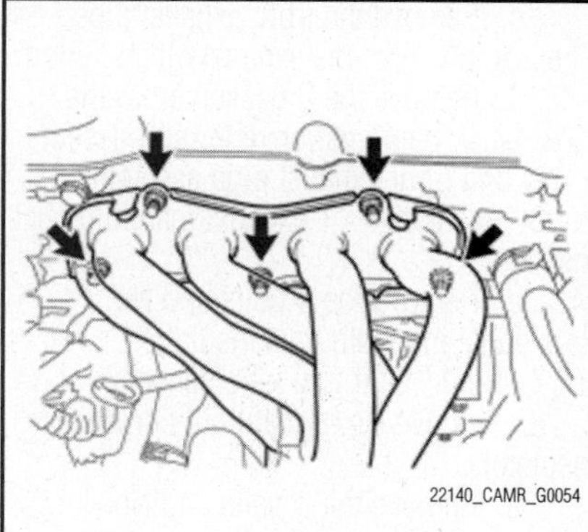

Fig. 200 Removing the five nuts, exhaust manifold and gasket—except PZEV vehicles

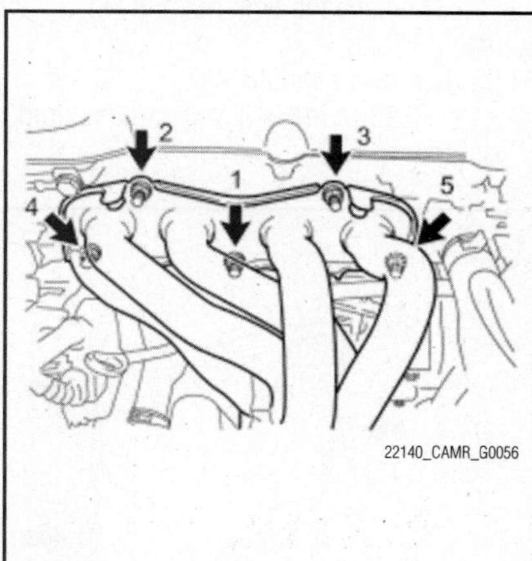

Fig. 201 Exhaust manifold tightening sequence—except PZEV vehicles

exhaust manifold bolts in sequence to: 27 ft. lbs. (37 Nm). Tighten the exhaust manifold heat insulator bolts to: 9 ft. lbs. (12 Nm).

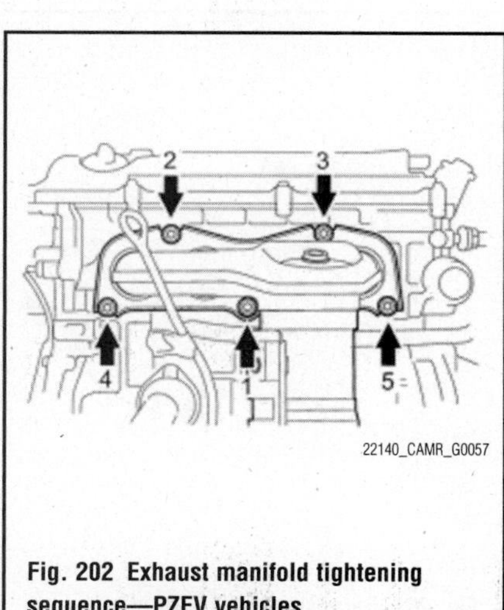

Fig. 202 Exhaust manifold tightening sequence—PZEV vehicles

14. For PZEV vehicles, tighten exhaust manifold bolts in sequence to: 27 ft. lbs. (37 Nm).

15. Tighten the exhaust manifold stays to 32 ft. lbs. (44 Nm).

2.5L Engine

NON-HYBRID

See Figures 203 through 208.

1. Remove the front wheel opening extension pad RH.
2. Remove the front wheel opening extension pad LH.
3. Remove the engine under cover LH.
4. Remove the engine under cover RH.
5. Remove the front exhaust pipe assembly.
   a. Disconnect the connector.
   b. Remove the 2 bolts and support bracket.
   c. Remove the 2 nuts and bracket.
   d. Remove the 4 nuts, 2 bolts and front exhaust pipe assembly.
   e. Remove the 2 gaskets from the front exhaust pipe assembly and center exhaust pipe assembly.
6. Remove the air fuel ratio sensor (bank 1, sensor 1).
7. Remove the manifold stay.
8. Remove the No. 2 manifold stay.
9. Remove the No. 1 exhaust manifold heat insulator.
   a. Remove the 4 bolts and No. 1 exhaust manifold heat insulator.
10. Remove the exhaust manifold converter sub assembly.
    a. Remove the 5 nuts and exhaust manifold converter sub assembly.
11. Remove the No. 2 exhaust manifold heat insulator.
    a. Remove the 2 bolts and No. 2 exhaust manifold heat insulator.

Fig. 203 Removing the front exhaust pipe support bracket

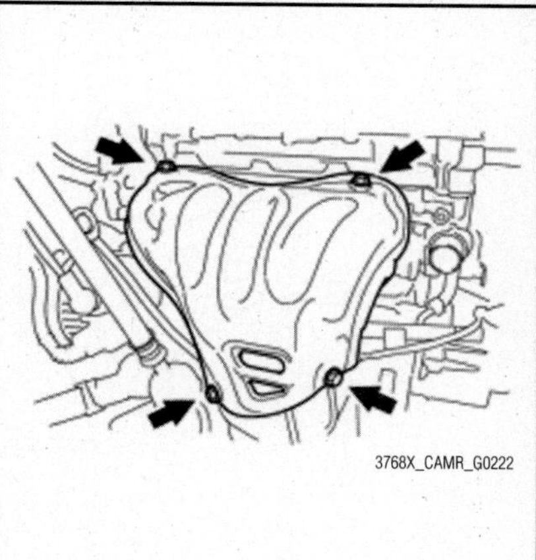

Fig. 204 Removing the exhaust manifold heat insulator

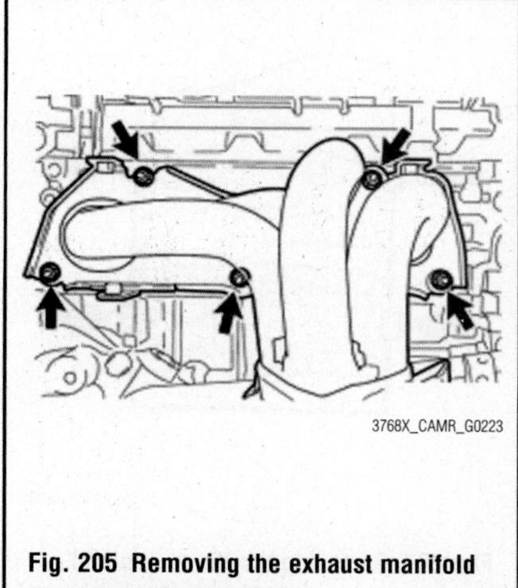

Fig. 205 Removing the exhaust manifold converter sub assembly

12. Remove the No. 1 manifold converter insulator.
    a. Remove the 4 bolts and the No. 1 manifold converter insulator.

To install:

13. Install the No. 1 manifold converter insulator.
    a. Install the No. 1 manifold converter insulator with the 4 bolts. Tighten to 9 ft. lbs. (12 Nm).
14. Install the No. 2 exhaust manifold heat insulator with the 2 bolts. Tighten to 9 ft. lbs. (12 Nm).
15. Install the exhaust manifold converter sub assembly.
    a. Install a new gasket onto the cylinder head.
    b. Temporarily install the exhaust manifold converter sub assembly with the 5 nuts.
    c. Tighten the 5 nuts in the sequence shown. Tighten to 26 ft. lbs. (35 Nm).

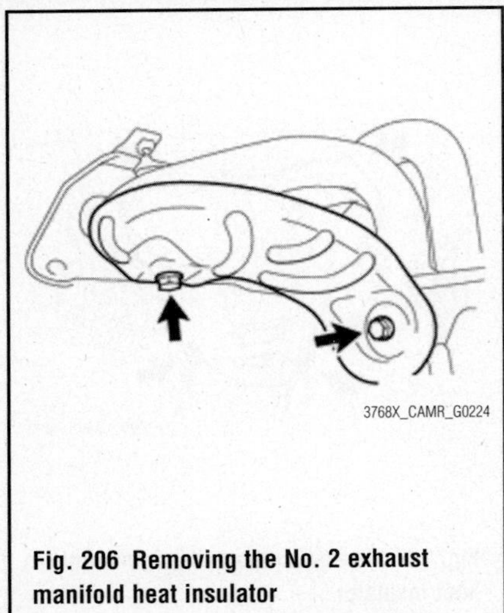

Fig. 206 Removing the No. 2 exhaust manifold heat insulator

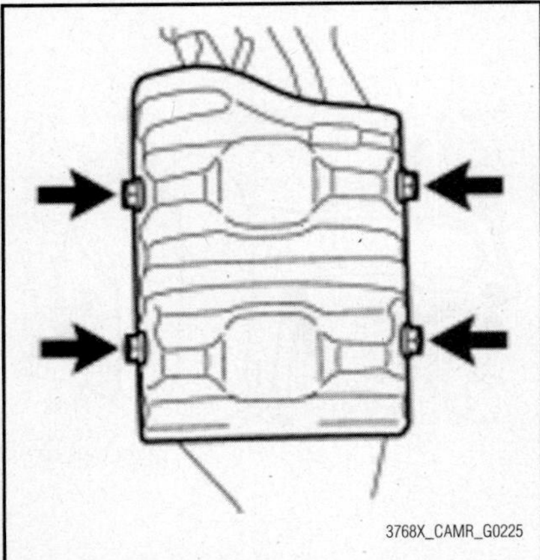

Fig. 207 Removing the No. 1 manifold converter insulator

16. Install the No. 1 exhaust manifold heat insulator with the 4 bolts. Tighten to 9 ft. lbs. (12 Nm).
17. Install the No. 2 manifold stay with the bolt and nut. Tighten to 32 ft. lbs. (43 Nm).
18. Install the manifold stay with the bolt and nut. Tighten to 32 ft. lbs. (43 Nm).
19. Install the air fuel ratio sensor (bank 1, sensor 1).
20. Install the front exhaust pipe assembly.
    a. Install the 2 new gaskets to the front exhaust pipe assembly and center exhaust pipe assembly.
    b. Install the front exhaust pipe with the 2 nuts to the exhaust manifold converter sub assembly. Tighten to 40 ft. lbs. (55 Nm).
    c. Install the front exhaust pipe with the 2 nuts and 2 bolts to the center exhaust pipe assembly. Tighten to 36 ft. lbs. (49 Nm).

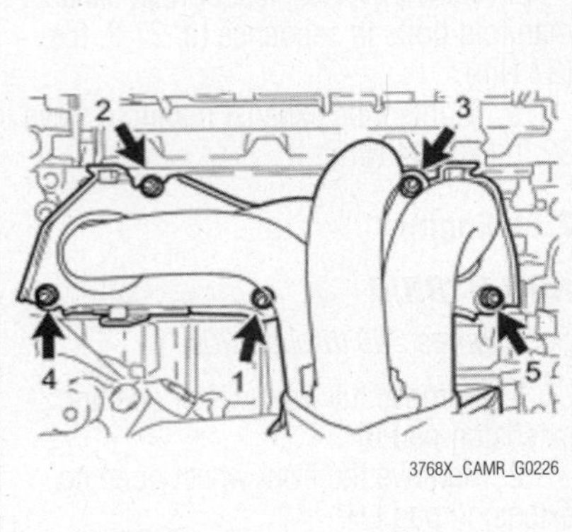

Fig. 208 Installing the exhaust manifold converter sub assembly

    d. Install the bracket with the 2 nuts. Tighten to 24 ft. lbs. (33 Nm).
    e. Install the support bracket with the 2 bolts. Tighten to 24 ft. lbs. (33 Nm).
    f. Connect the connector.
21. Inspect for exhaust gas leaks.
22. Install the engine under covers.
23. Install the front wheel opening extension pads.

HYBRID

See Figures 209 through 217.

1. Remove front wheel opening extension pad RH.
2. Remove engine under cover RH.
3. Remove front wheel opening extension pad LH.
4. Remove engine under cover LH.
5. Remove front exhaust pipe assembly.
    a. Disconnect the heated oxygen sensor connector.
    b. Remove the 2 bolts, 2 compression springs and 2 nuts

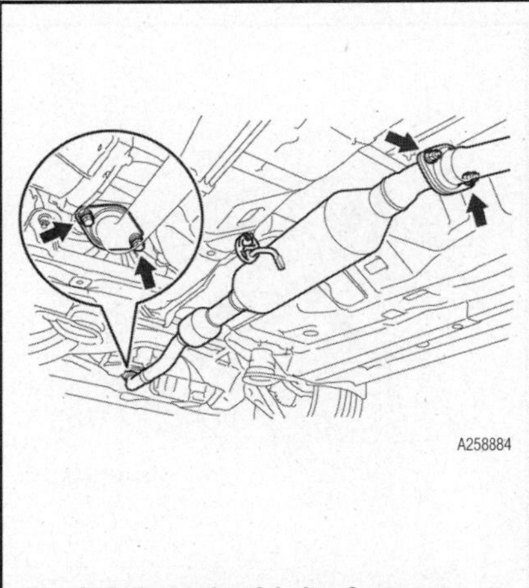

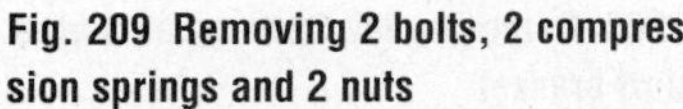

Fig. 209 Removing 2 bolts, 2 compression springs and 2 nuts

    c. Remove the front exhaust pipe assembly from the exhaust pipe support.
    d. Remove the 2 gaskets from the exhaust manifold converter sub-assembly and front exhaust pipe assembly.
6. Remove no. 1 exhaust manifold heat insulator.
    a. Remove the 4 bolts and No. 1 exhaust manifold heat insulator.
7. Remove air fuel ratio sensor.
8. Remove No. 1 manifold converter insulator.
    a. Remove the 3 bolts and No. 1 manifold converter insulator.
9. Remove No. 2 EGR pipe.
    a. Remove the 2 bolts, 2 nuts and No. 2 EGR pipe.
    b. Remove the 2 gaskets.
10. Remove no. 2 manifold stay.
    a. Remove the bolt, nut and No. 2 manifold stay.
11. Remove manifold stay.
    a. Remove the bolt, nut and manifold stay.

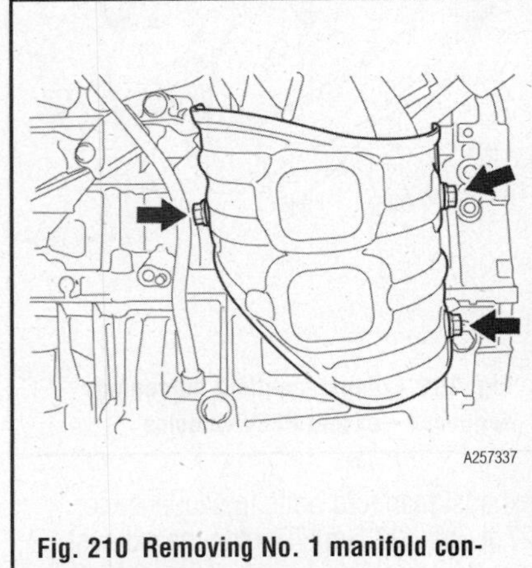

Fig. 210 Removing No. 1 manifold converter insulator

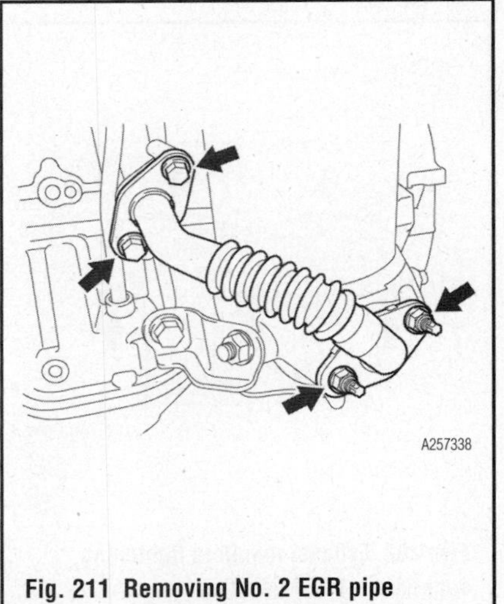

Fig. 211 Removing No. 2 EGR pipe

12. Remove exhaust manifold converter sub assembly.

a. Remove the 5 nuts and exhaust manifold converter sub assembly.

b. Remove the gasket.

13. Remove No. 2 exhaust manifold heat insulator.

a. Remove the 2 bolts and No. 2 exhaust manifold heat insulator.

To install:

14. Install the No. 2 exhaust manifold heat insulator with the 2 bolts. Tighten to 10 ft. lbs. (14 Nm).

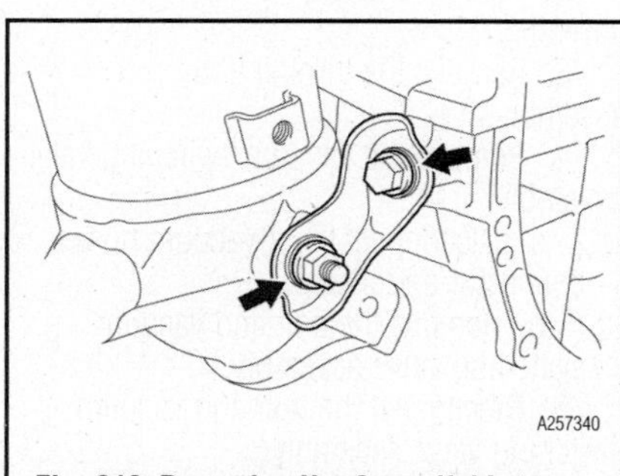

Fig. 212 Removing No. 2 manifold stay

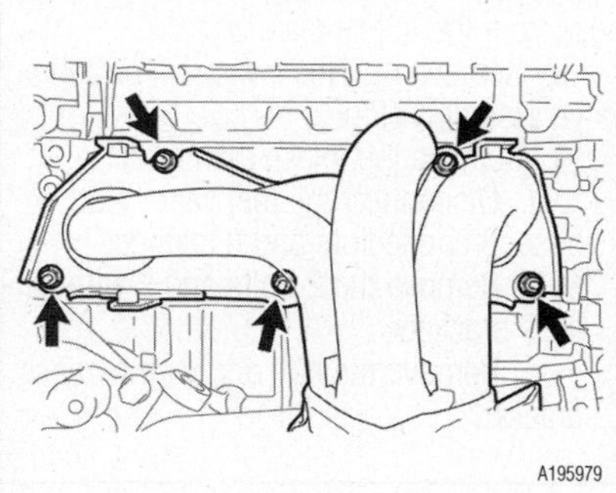

Fig. 213 Removing exhaust manifold converter sub assembly

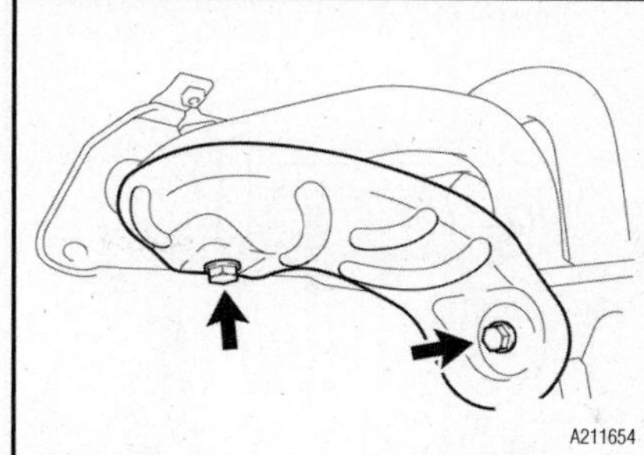

Fig. 214 Removing No. 2 exhaust manifold heat insulator

15. Install exhaust manifold converter sub assembly.

a. Set a new gasket to the cylinder head sub-assembly.

b. Temporarily install the exhaust manifold converter sub-assembly with the 5 nuts.

c. Temporarily install the No. 2 manifold stay and manifold stay with the 2 bolts and 2 nuts.

d. Tighten the 5 nuts in order. Tighten to 26 ft. lbs. (35 Nm).

16. Install manifold stay.

a. Tighten the bolt and nut in the order. Tighten to 32 ft. lbs. (43 Nm).

17. Install No. 2 manifold stay.

a. Tighten the bolt and nut in order. Tighten to 32 ft. lbs. (43 Nm).

18. Install No. 2 EGR pipe.

a. Install 2 new gaskets to the No. 2 EGR pipe.

b. Install the No. 2 EGR pipe with the 2 bolts and 2 nuts. Tighten to 26 ft. lbs. (36 Nm).

19. Install No. 1 manifold converter insulator with the 3 bolts. Tighten to 10 ft. lbs. (14 Nm).

20. Install air fuel ratio sensor.

21. Install No. 1 exhaust manifold heat insulator with the 4 bolts. Tighten to 9 ft. lbs. (12 Nm).

22. Install front exhaust pipe assembly.

a. Using a vernier caliper, measure the free length of the compression spring.

- Standard length: 1.65 inch (42 mm)
- Minimum length: 1.6 inch (40.5 mm)

➡If the length is less than the minimum, replace the compression spring.

b. Fully insert a new gasket to the exhaust manifold converter sub-assembly.

c. Using a plastic hammer and wooden block, tap in the new gasket until its surface is flush with the exhaust manifold converter sub-assembly.

⁂ WARNING

Be sure to install the gasket in the correct direction.

⁂ WARNING

Do not reuse the gasket.

⁂ WARNING

Do not damage the gasket.

⁂ WARNING

Do not push in the gasket by using the exhaust pipe when connecting it.

d. Install a new gasket to the front exhaust pipe assembly.

e. Connect the front exhaust pipe assembly to the exhaust pipe support.

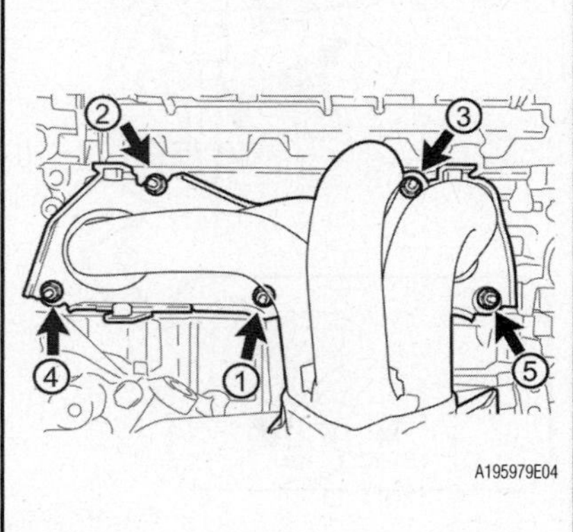

Fig. 215 Identifying exhaust manifold converter sub assembly bolt tightening sequence

Fig. 216 Installing manifold stay bolt and nut in order

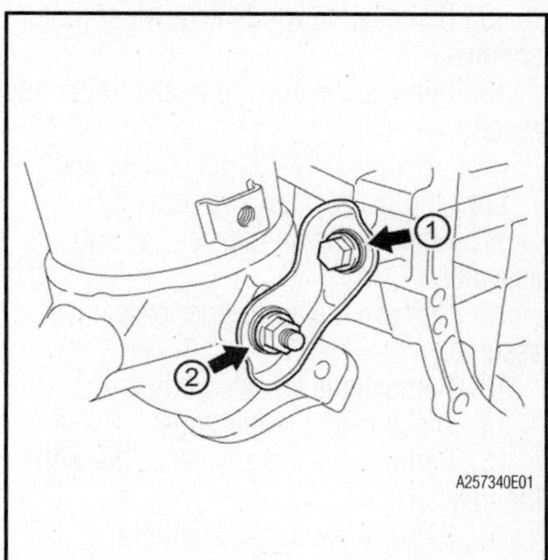

Fig. 217 Installing No. 2 manifold stay bolt and nut in order

f. Install the front exhaust pipe assembly with the 2 compression springs, 2 bolts and 2 nuts. Tighten to 32 ft. lbs. (43 Nm).

WARNING

After the installation, check that the gaps between the flanges of the exhaust manifold converter sub-assembly and front exhaust pipe assembly are consistent front-to-rear and left-to-right.

g. Connect the heated oxygen sensor connector.
23. Install engine under cover LH.
24. Install front wheel opening extension pad LH.
25. Install engine under cover RH.
26. Install front wheel opening extension pad RH.
27. Inspect for exhaust gas leak.

INTAKE MANIFOLD

REMOVAL & INSTALLATION

2.4L Hybrid Engine

See Figure 218.

1. Disconnect the fuel system pressure.
2. Disconnect the cable from the negative battery terminal.
3. Remove the No. 1 engine cover sub assembly.
4. Drain the engine coolant.
5. Remove the windshield wiper arm and blade assemblies.
6. Remove the front fender to cowl side seals.
7. Remove the cowl top ventilator louver sub assembly.
8. Separate the brake master cylinder reservoir sub assembly.
9. Remove the windshield wiper motor and link.
10. Remove the cowl top panel outer sub assembly.
a. Remove the 4 bolts, 4 nuts and the cowl top panel outer sub assembly.
11. Remove the air cleaner cap sub assembly.
12. Remove the air cleaner case sub assembly.
13. Remove the throttle body.
14. Disconnect the fuel tube.
15. Remove the fuel delivery pipe with the injector.
16. Remove the intake manifold.
a. Disconnect the union to check valve hose from the brake booster.
b. Disconnect the camshaft timing oil control valve connector.

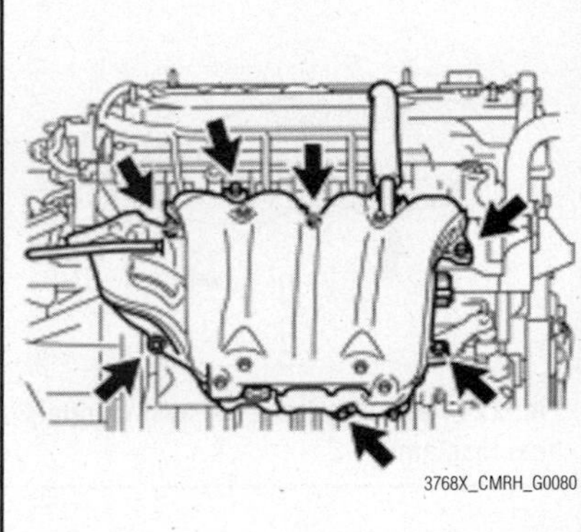

Fig. 218 Removing and installing the intake manifold

c. Remove the wire harness clamp.
d. Remove the union to check valve hose from the vacuum hose clamp.
e. Remove the 5 bolts, 2 nuts and the intake manifold.

To install:

17. Install the intake manifold.
a. Install a new gasket into the intake manifold.
b. Install the intake manifold with the 5 bolts and 2 nuts. Tighten to 22 ft. lbs. (30 Nm).
c. Fit the union to check valve hose into the vacuum hose clamp.
d. Install the wire harness clamp.
e. Connect the camshaft timing oil control valve connector.
f. Connect the union to check valve hose to the brake booster.
18. Install the fuel delivery pipe with the injector.
19. Install the fuel tube.
20. Install the throttle body.
21. Install the air cleaner case sub assembly.
22. Install the air cleaner cap sub assembly.
23. Install the air cleaner inlet assembly.
24. Connect the cable to the negative battery terminal.
25. Add engine coolant.
26. Inspect for coolant leaks.
27. Inspect for fuel leaks.
28. Install the No. 1 engine cover sub assembly.
29. Install the cowl top panel outer sub assembly. Tighten the bolts to 44 inch lbs. (5 Nm). Tighten the nuts to 63 ft. lbs. (85 Nm).
30. Install the brake master cylinder reservoir sub assembly.
31. Install the windshield wiper motor and link.
32. Install the cowl top ventilator louver sub assembly.
33. Install the front fender to cowl side seals.
34. Install the windshield wiper arm and blade assemblies.
35. Perform initialization.

2.5L Engine

NON-HYBRID

See Figures 219 through 227.

1. Drain engine coolant.
2. Remove the fuel delivery pipe sub assembly.
3. Remove the throttle body assembly.
4. Remove the vacuum switching valve assembly.
a. Disconnect the 2 vacuum hoses, connector and clamp.
b. Remove the bolt and vacuum switching valve assembly.
5. Disconnect the bolt and vacuum switching valve assembly.
6. Disconnect the No. 2 ventilation hose from the intake manifold.
7. Remove the union to connector tube hose from the intake manifold.
8. Remove the No. 1 vacuum switching valve assembly (PZEV).
9. Remove the intake manifold.
a. Disconnect the fuel vapor feed hose, 3 connectors and 8 clamps.
b. Remove the 2 bolts and 2 wire harness brackets.
c. Remove the bolt and wire harness bracket.

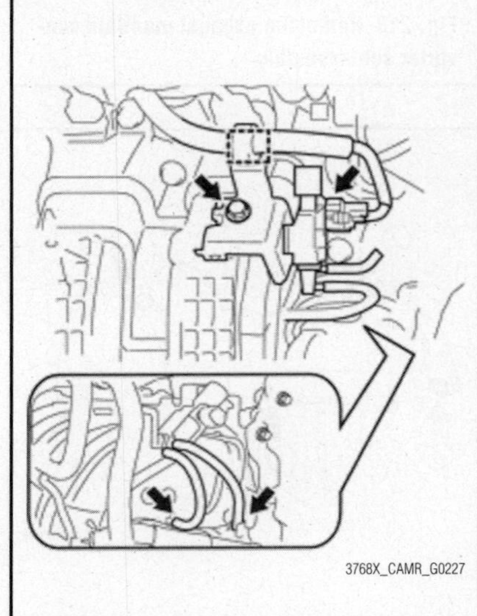

Fig. 219 Removing the vacuum switching valve assembly

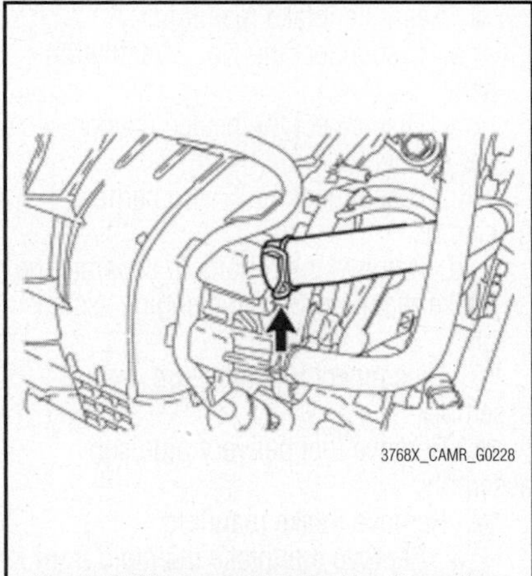

Fig. 220 Disconnecting the No. 2 ventilation hose from the intake manifold

d. Apply battery voltage to the terminals of the connector to close the tumble control valves.

Positive (+) battery voltage applied to terminal 8 (M-), and negative (-) battery voltage applied to terminal 4 (M+): Open → Closed

➡If this procedure is not performed, the valves may be damaged when the intake manifold is removed.

➡Apply battery voltage for 1 to 3 seconds.

➡If battery voltage is applied for more than 3 seconds, the actuator may be damaged.

➡Do not allow the lead wires to contact the other terminals.

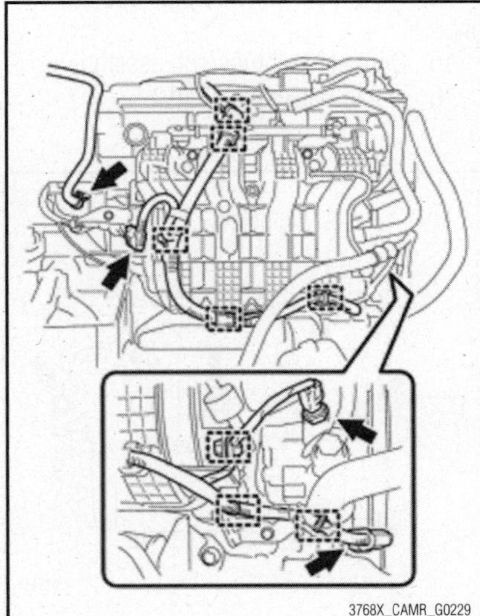

Fig. 221 Disconnecting the fuel vapor feed hose, connectors and clamps from the intake manifold

Fig. 222 Removing the intake manifold 2 wire harness brackets

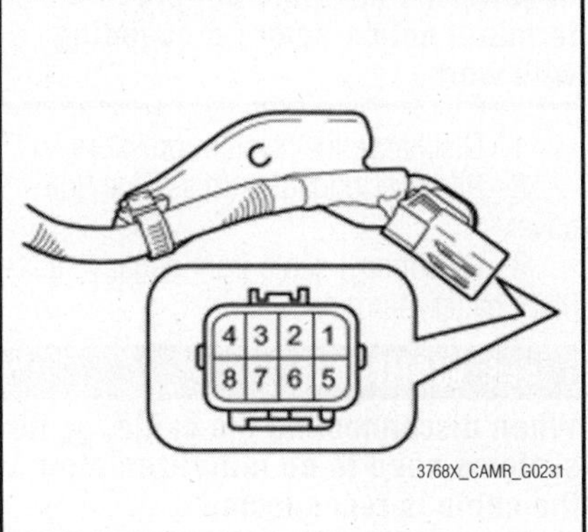

Fig. 223 Applying battery voltage to the terminals of the connector to close the tumble control valves

a. Remove the bolt and separate the wire harness.

b. Detach the 2 clamps from the intake manifold and bracket.

c. Disconnect the intake air control valve actuator connector.

d. Remove the 6 bolts and intake manifold.

➡The tumble control valves may be damaged if they are not closed before installing the intake manifold.

➡Connect the battery to the terminals of the actuators to operate the motor and close the valves.

e. Remove the intake manifold gasket from the intake manifold.

10. Remove the check valve.

a. Disconnect the 2 vacuum hoses from the intake manifold and remove the check valve.

11. Remove the wiring harness clamp bracket.

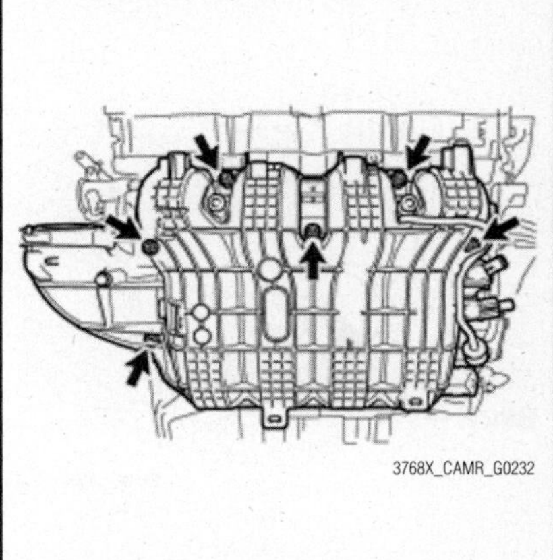

Fig. 224 Removing the intake manifold bolts

To install:

12. Install the wiring harness clamp bracket with the bolt. Tighten to 7 ft. lbs. (10 Nm).

13. Install the check valve.

a. Connect the 2 vacuum hoses to the intake manifold to install the check valve.

b. Check that the check valve is installed properly.

14. Install the intake manifold.

a. Close the tumble control valves.

➡The tumble control valves may be damaged if they are not closed before installing the intake manifold.

➡Connect the battery to the terminals of the actuator to operate the motor and close the valves.

b. Install a new gasket to the intake manifold.

c. Install the intake manifold by tightening the 6 bolts in the sequence shown. Tighten to 15 ft. lbs. (21 Nm).

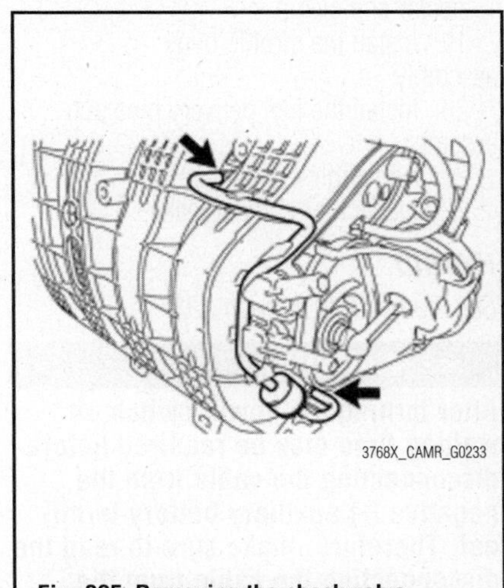

Fig. 225 Removing the check valve

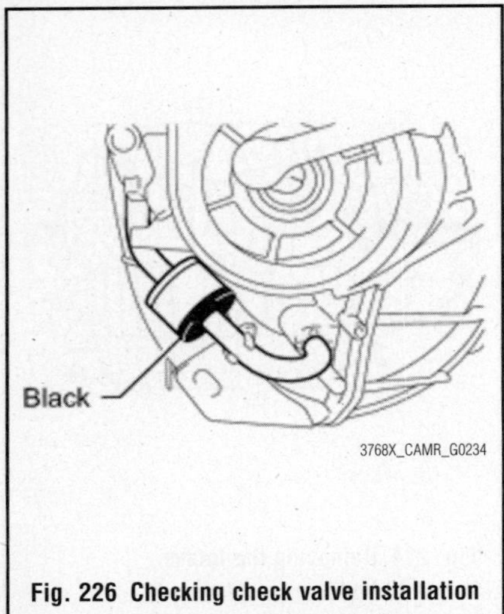

Fig. 226 Checking check valve installation

d. Connect the intake air control actuator connector.

e. Attach the 2 clamps to the intake manifold and bracket.

f. Install the wire harness with the bolt. Tighten the bolt to 7 ft. lbs. (10 Nm).

g. Install the 2 wire harness brackets with the 2 bolts. Tighten the bolts to 7 ft. lbs. (10 Nm).

h. Connect the fuel vapor feed hose, 3 connectors and 8 clamps.

15. Install the No. 1 vacuum switching valve assembly (PZEV).
16. Install the union to the connector tube hose to the intake manifold.
17. Connect the No. 2 ventilation hose to the intake manifold.
18. Install the vacuum switching valve assembly (ACIS).

a. Install the vacuum switching valve with the bolt. Tighten to 80 inch lbs. (9 Nm).

b. Connect the 2 vacuum hoses, connector and clamp.

19. Install the throttle body assembly.
20. Install the fuel delivery pipe sub assembly.
21. Add engine coolant.
22. Inspect for coolant leaks.

HYBRID

See Figures 228 through 232.

⁂ WARNING

After turning the power switch off, waiting time may be required before disconnecting the cable from the negative (-) auxiliary battery terminal. Therefore, make sure to read the disconnecting the cable from the negative (-) auxiliary battery terminal notice before proceeding with work.

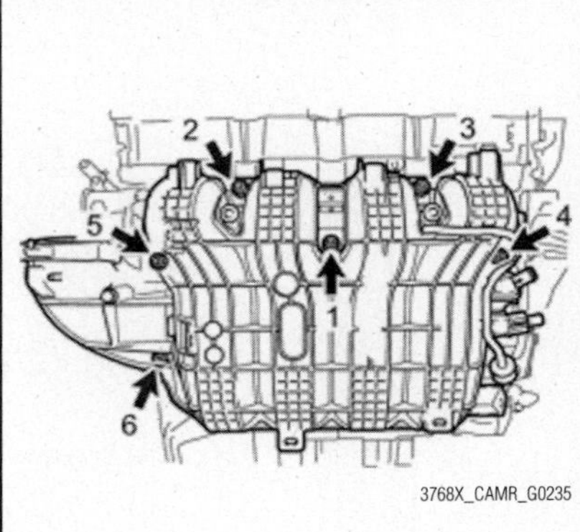

Fig. 227 Identifying the intake manifold bolt tightening sequence

1. Discharge fuel system pressure.
2. Remove luggage trim service hole cover.
3. Disconnect cable from negative auxiliary battery terminal.

⁂ WARNING

When disconnecting the cable, some systems need to be initialized after the cable is reconnected.

4. Remove the throttle with motor body assembly.
5. Remove the windshield wiper motor and link assembly.
6. Remove the front outer cowl top panel sub assembly.
7. Remove EGR valve assembly.
8. Remove manifold absolute pressure sensor.
9. Separate intake manifold.

a. Disconnect the No. 2 ventilation hose.

b. Disconnect the heated oxygen sensor connector.

c. Disconnect the 6 wire harness clamps.

d. Remove the 6 bolts to separate the intake manifold from the engine assembly.

10. Disconnect fuel tube sub assembly.
11. Remove fuel delivery pipe sub assembly.
12. Remove intake manifold.

a. Remove the intake manifold from the vehicle.

b. Remove the gasket from the intake manifold.

c. Remove the bolt and wire harness clamp bracket.

d. Remove the vacuum hose from the intake manifold.

e. Remove the fuel vapor feed hose from the intake manifold.

To install:

13. Set intake manifold.

a. Install the fuel vapor feed hose to the intake manifold.

b. Install the vacuum hose to the intake manifold.

c. Install the wire harness clamp bracket with the bolt. Tighten to 7 ft. lbs. (10 Nm).

d. Install a new gasket to the intake manifold.

e. Set the intake manifold on the engine assembly.

14. Install fuel delivery pipe sub assembly.
15. Connect fuel tube sub assembly.
16. Install intake manifold.

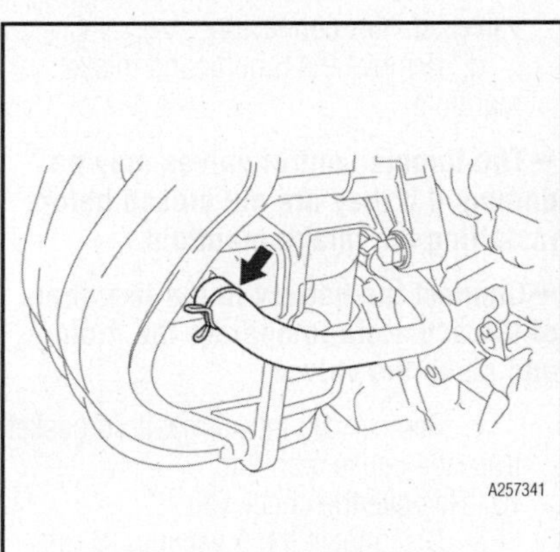

Fig. 228 Disconnecting the No. 2 ventilation hose

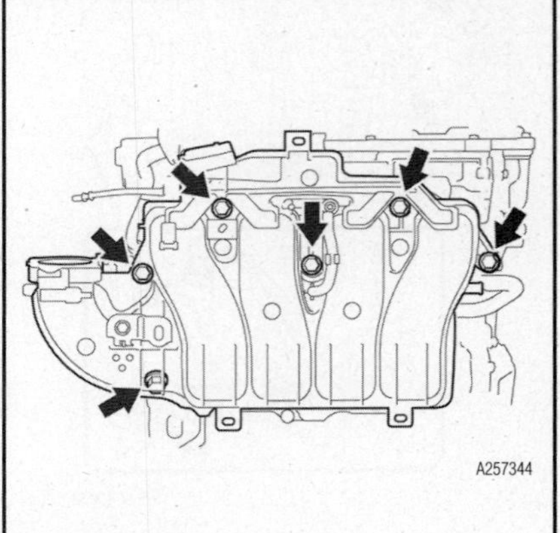

Fig. 229 Removing 6 bolts to separate the intake manifold from the engine assembly

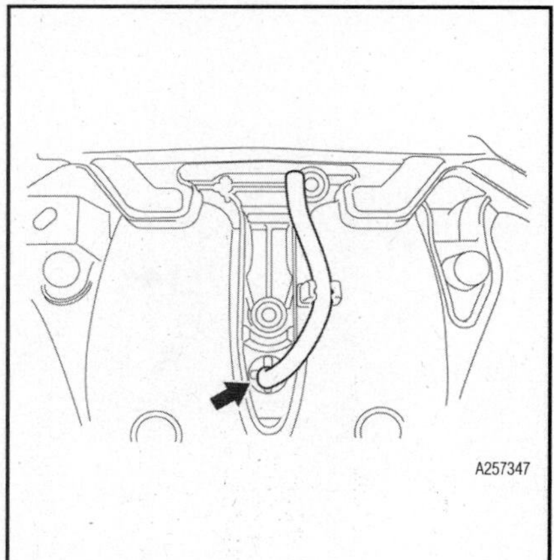

Fig. 230 Removing the vacuum hose from the intake manifold

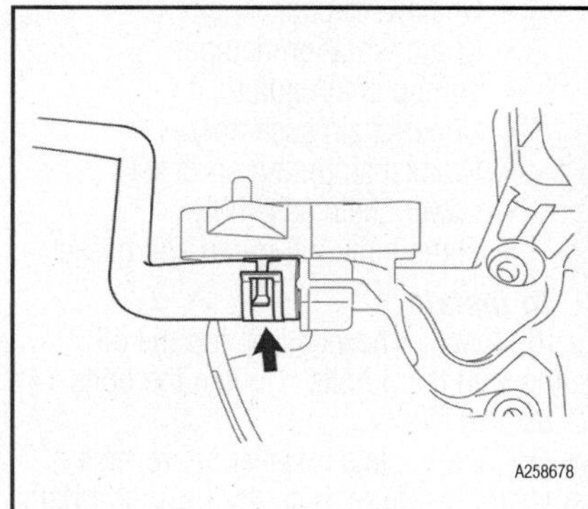

Fig. 231 Removing the fuel vapor feed hose from the intake manifold

a. Temporarily install the intake manifold with the 6 bolts.

b. Tighten the 6 bolts in order. Tighten to 15 ft. lbs. (21 Nm).

c. Connect the 6 wire harness clamps.

d. Connect the heated oxygen sensor connector.

e. Connect the No. 2 ventilation hose.

17. Install manifold absolute pressure sensor.

18. Install EGR valve assembly.

19. Install front outer cowl top panel sub assembly.

20. Install windshield wiper motor and link assembly.

21. Install throttle with motor body assembly.

22. Connect cable from negative auxiliary battery terminal.

⁂ WARNING

When disconnecting the cable, some systems need to be initialized after the cable is reconnected.

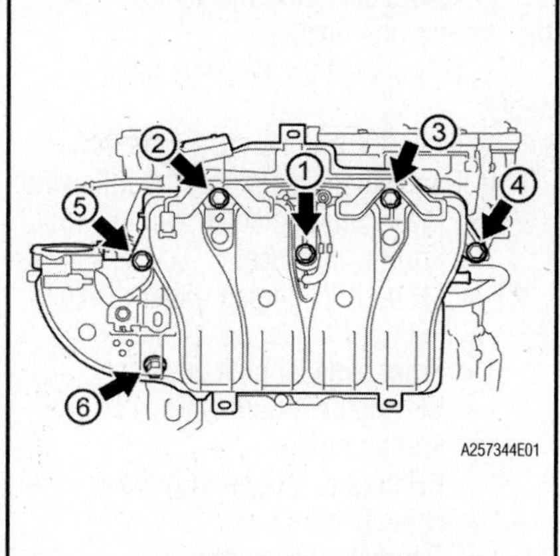

Fig. 232 Tightening the intake manifold bolts in order

23. Install luggage trim service hole cover.

24. Inspect for fuel leak.

OIL PAN

REMOVAL & INSTALLATION

2.4L Engine

See Figures 233 and 234.

1. Before servicing the vehicle, refer to the Precautions Section.
2. Disconnect the negative battery cable.
3. Drain the engine oil.
4. Remove the oil pan drain plug and gasket.
5. Remove the oil pan 12 bolts and 2 nuts.

➡Be careful not to damage the contact surfaces of the crankcase, chain cover and oil pan.

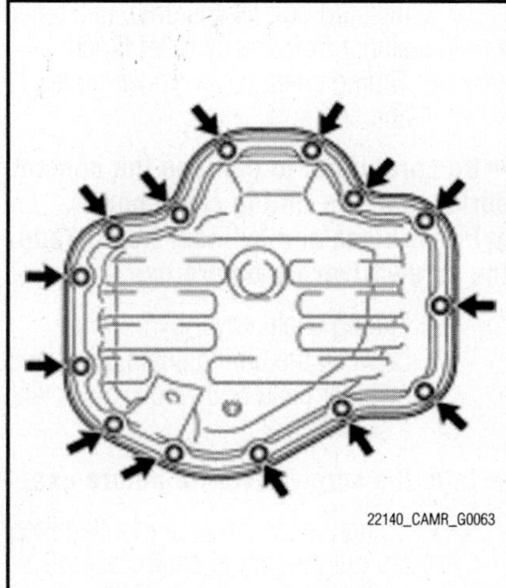

Fig. 233 Removing oil pan bolts

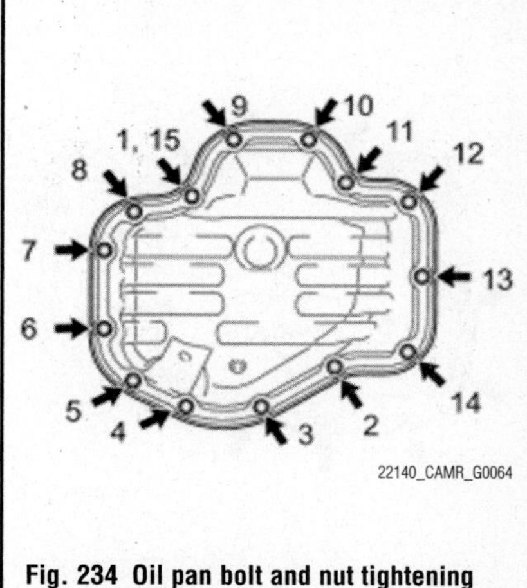

Fig. 234 Oil pan bolt and nut tightening sequence

6. Insert the blade of Special Tool: 09032-00100 between the crankcase and oil pan. Cut through the sealer and remove the oil pan.

To install:

7. Remove any old packing material and be careful not to drop any oil on the contact surfaces of the cylinder block and oil pan.
8. Apply a continuous bead of seal packing (Diameter 0.118 to 0.157 inches (3.0 to 4.0 mm)). Use Toyota Genuine Seal Packing Block, Three Bond 1207B or Equivalent.

➡Remove any oil from the contact surfaces. Install the oil pan within 3 minutes after applying seal packing. Do not start the engine for at least 2 hours after installing.

9. Uniformly tighten the 12 bolts and 2 nuts in sequence. Tighten the bolts and nuts to 80 inch lbs. (9 Nm).
10. To complete installation, reverse remaining removal procedure.
11. Add oil and check for leaks.

2.5L Engine

See Figures 235 and 236.

1. Remove the oil filter cap assembly.
2. Remove the oil pan sub assembly.
   a. Remove the 11 bolts and 2 nuts.
   b. Insert the blade of an oil pan seal cutter between the oil and stiffening crankcase, cut off the applied sealer and remove the oil pan.

To install:

3. Install the oil pan sub assembly.
   a. Apply seal packing in a continuous line. Use Toyota Genuine Seal Packing Black, Three Bond 1207B or equivalent. The seal diameter should ne 0.0984-0.138 inch (2.5-3.5 mm).

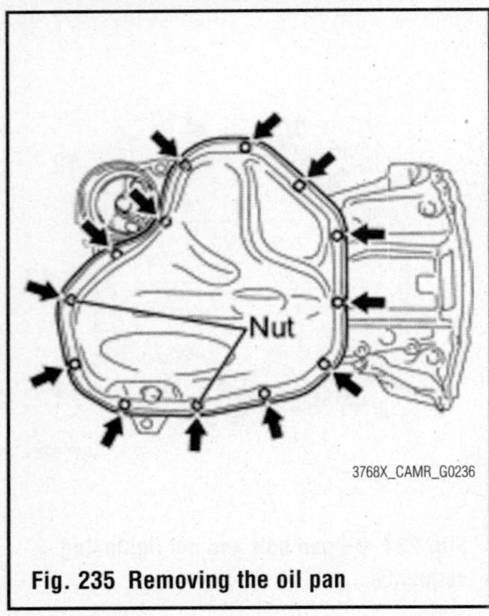

Fig. 235 Removing the oil pan

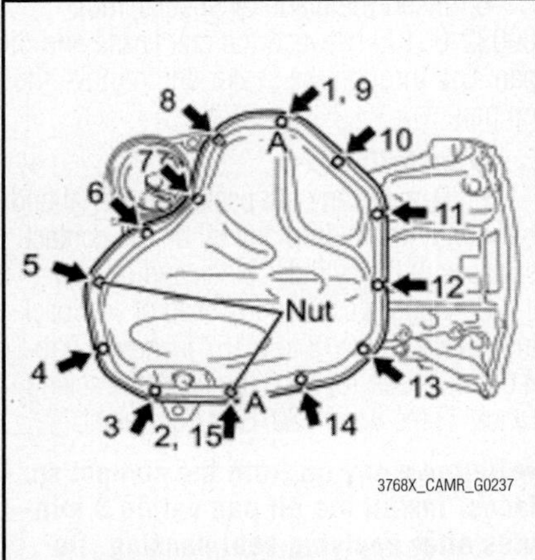

Fig. 236 Identifying the oil pan bolt installation pattern

➡Remove any foreign oil from the contact surface.

➡Install the oil pan within 3 minutes and tighten the bolts and nuts within 10 minutes after applying seal packing.

➡Do not apply oil for at least 4 hours after the installation.

b. Install the oil pan with the 11 bolts and 2 nuts in several steps, in the sequence shown in the illustration. Tighten to 7 ft. lbs. (10 Nm).

➡Bolt A and nut A are tightened twice.

4. Install the oil filter cap assembly.

OIL PUMP

REMOVAL & INSTALLATION

2.4L Engine

See Figure 237.

1. Before servicing the vehicle, refer to the Precautions Section.
2. Disconnect the negative battery cable.
3. Drain the engine oil.
4. Remove or disconnect the following:
   - Plastic engine cover and two nuts
   - Front right wheel
   - LH and RH engine under cover
   - RH front fender apron seal
   - Front exhaust pipe assembly
   - RH engine mount stay
   - Engine mount
   - RH engine mounting bracket
   - Drive belt
   - Generator assembly
   - Vane pump
   - Ignition coil assembly
   - Ventilation hoses
   - Valve cover
5. Turn the crankshaft pulley until its groove and the timing mark "0" of the timing chain cover are aligned.
6. Check that each timing mark of the camshaft timing gear and sprocket is aligned with each timing mark located on the intake and exhaust bearing caps. If not, turn the crankshaft by 1 revolution (360°) to align the timing marks.
7. Remove or disconnect the following:
   - Crankshaft pulley
   - Crankshaft position sensor
   - Oil pan
   - Chain upper tensioner assembly
8. Install the No. 1 engine hanger (12281-28010) and No. 2 engine hanger (12282-28010) with the bolts (91512-61020) and tighten to: 28 ft. lbs. (38 Nm).
9. Remove or disconnect the following:
   - Drive belt tensioner
   - Engine mount insulator
   - RH engine mount bracket
   - Using an E10 Torx® socket, remove the stud bolt for the drive belt tensioner from the cylinder block
   - Timing chain cover twelve bolts and two nuts

➡Be careful not to damage the contact surfaces of the timing chain cover, cylinder block and cylinder head. Tape the screwdriver tip before use.

   - Timing chain cover by prying between the timing chain cover and cylinder head or cylinder block with a screwdriver

➡Tape the screwdriver tip before use.

   - Using a screwdriver and a hammer, tap out the timing chain case oil seal
   - Crankshaft position sensor plate

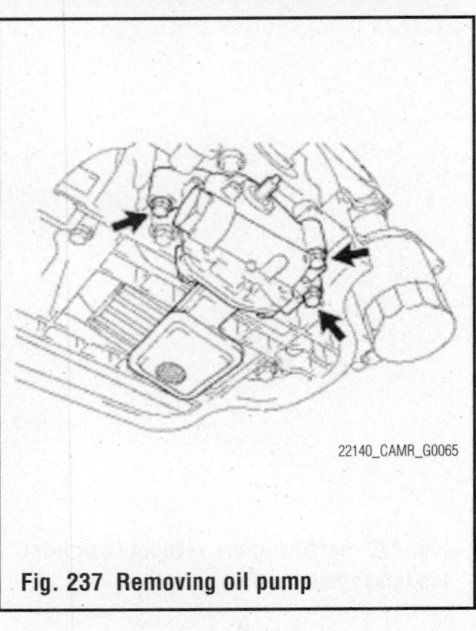

Fig. 237 Removing oil pump

   - Chain tensioner slipper
   - Chain vibration damper
   - Timing chain guide
   - Upper chain assembly
   - Crankshaft timing sprocket
   - Lower chain assembly
   - Three bolts, oil pump and gasket

To install:

10. Install a new gasket and the oil pump with the 3 bolts. Tighten the bolts 14 ft. lbs. (19 Nm).
11. To complete installation, reverse removal procedure. Refer to the appropriate sections to install components correctly.

2.5L Engine

Non-Hybrid

See Figures 238 through 241.

➡Do not remove the oil pump or oil pump relief valve from the timing chain cover sub assembly.

1. Remove the engine and transaxle.
2. Remove the engine wire.
3. Remove the ignition coil assembly.
4. Remove the cylinder head cover sub assembly.
5. Remove the crankshaft position sensor.
6. Remove the crankshaft pulley.
7. Remove the No. 2 timing chain cover.
   a. Remove the 2 bolts and No. 2 timing chain cover.
8. Remove the 5 bolts engine mounting bracket RH.
9. Remove the v-ribbed belt tensioner assembly (compressor and generator).
10. Remove the v-ribbed belt tensioner assembly (vane pump).
11. Remove the timing chain cover sub assembly.

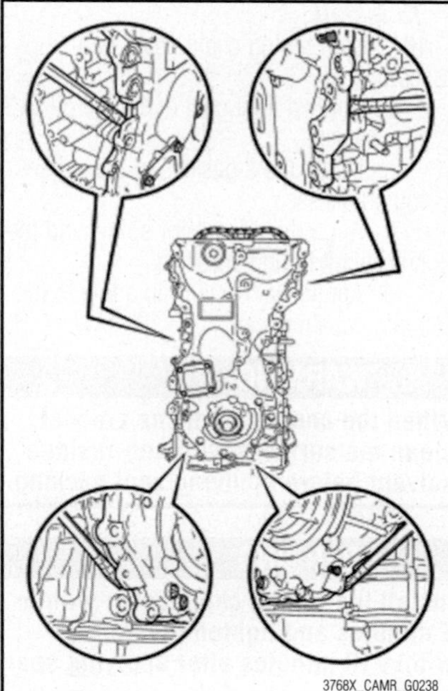

3768X_CAMR_G0238

Fig. 238 Removing the timing chain cover

a. Remove the 17 bolts and 2 nuts.

b. Remove the timing chain cover by prying between the timing chain cover and the cylinder head, camshaft housing, cylinder block and stiffening crankcase with a screwdriver.

➡Be careful not to damage the contact surfaces of the cylinder head, camshaft housing, cylinder block, stiffening crankcase or chain cover.

➡Tape the screwdriver tip before use.

c. Remove the 3 gaskets from the stiffening crankcase.

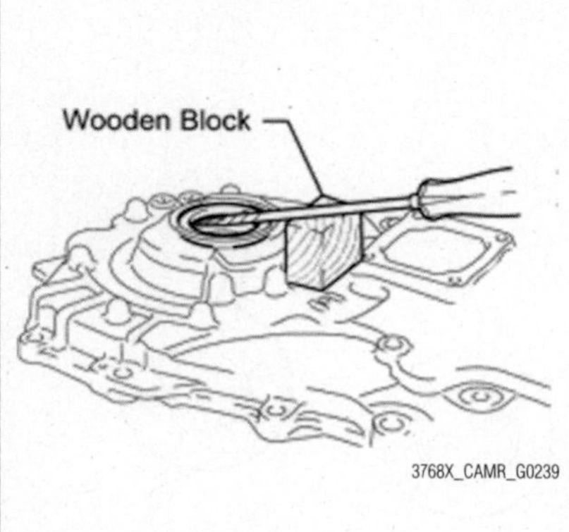

3768X_CAMR_G0239

Fig. 239 Removing the timing chain cover oil seal

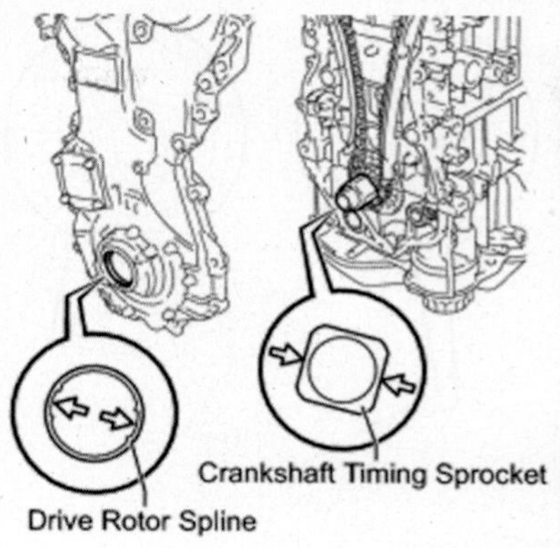

3768X_CAMR_G0240

Fig. 240 Aligning the drive rotor spline and the crankshaft timing sprocket

12. Remove the timing chain cover oil seal.

a. Using a screwdriver and wooden block, pry out the oil seal.

➡Do not damage the surface of the oil seal press fit hole.

13. Tape the screwdriver tip before use.

To install:

14. Install the timing chain cover sub assembly.

a. Apply a light coat of engine oil to 3 new gaskets.

b. Install the 3 gaskets to the stiffening crankcase.

c. Align the drive rotor spline and the crankshaft timing sprocket.

d. Apply seal packing in a line to the timing chain cover.

Hybrid

See Figures 242 through 251.

⁂ WARNING

Do not remove the oil pump or oil pump relief valve from the timing chain cover sub-assembly.

1. Remove engine and transaxle.
2. Remove engine wire.
3. Remove ignition coil assembly.
4. Remove cylinder head cover sub assembly.

a. Remove the 16 bolts, 3 seal washers, cylinder head cover and cylinder head cover gasket.

b. Remove the 2 gaskets from the camshaft bearing caps.

5. Remove crankshaft position sensor.
6. Remove crankshaft pulley.

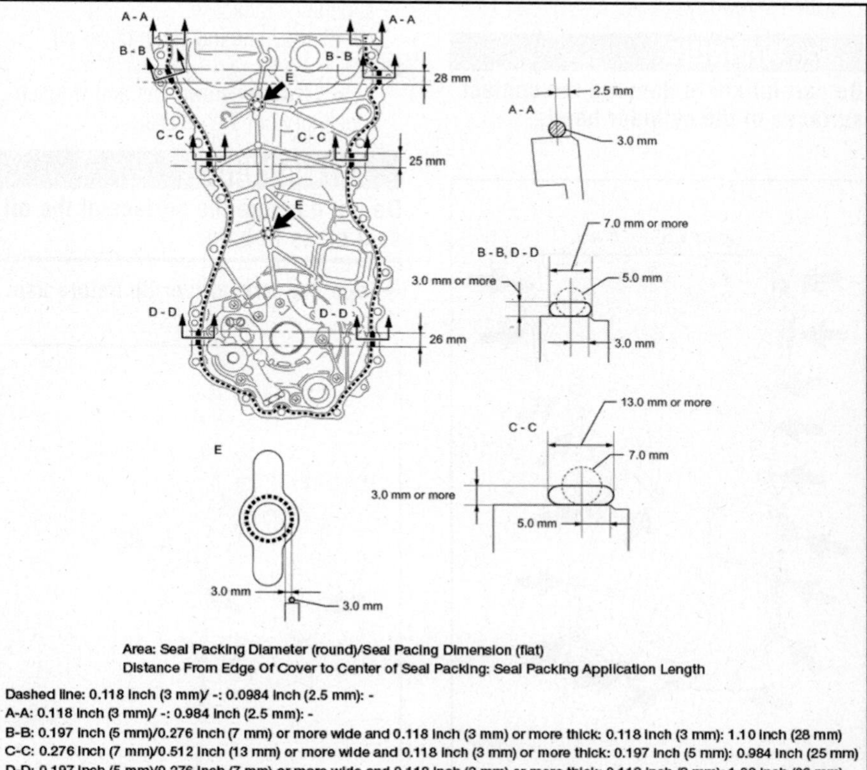

Area: Seal Packing Diameter (round)/Seal Pacing Dimension (flat)
Distance From Edge Of Cover to Center of Seal Packing: Seal Packing Application Length

Dashed line: 0.118 inch (3 mm)/ -: 0.0984 inch (2.5 mm): -
A-A: 0.118 inch (3 mm)/ -: 0.984 inch (2.5 mm): -
B-B: 0.197 inch (5 mm)/0.276 inch (7 mm) or more wide and 0.118 inch (3 mm) or more thick: 0.118 inch (3 mm): 1.10 inch (28 mm)
C-C: 0.276 inch (7 mm)/0.512 inch (13 mm) or more wide and 0.118 inch (3 mm) or more thick: 0.197 inch (5 mm): 0.984 inch (25 mm)
D-D: 0.197 inch (5 mm)/0.276 inch (7 mm) or more wide and 0.118 inch (3 mm) or more thick: 0.118 inch (3 mm): 1.02 inch (26 mm)
E: 0.118 inch (3 mm)/ -: 0.118 inch (3 mm): -

3768X_CAMR_G0241

Fig. 241 Applying seal packing to the timing chain cover

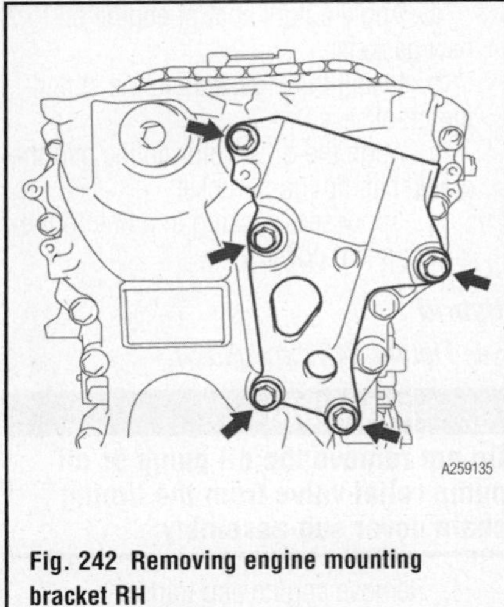
Fig. 242 Removing engine mounting bracket RH

7. Remove engine mounting bracket RH.
   a. Remove the 5 bolts and engine mounting bracket RH.
8. Remove timing chain cover sub assembly.
   a. Remove the 17 bolts and 2 nuts.
   b. Remove the timing chain cover by prying between the timing chain cover and cylinder head, camshaft housing, cylinder block and stiffening crankcase with a screwdriver.

WARNING

Be careful not to damage the contact surfaces of the cylinder head, camshaft housing, cylinder block, stiffening crankcase or chain cover.

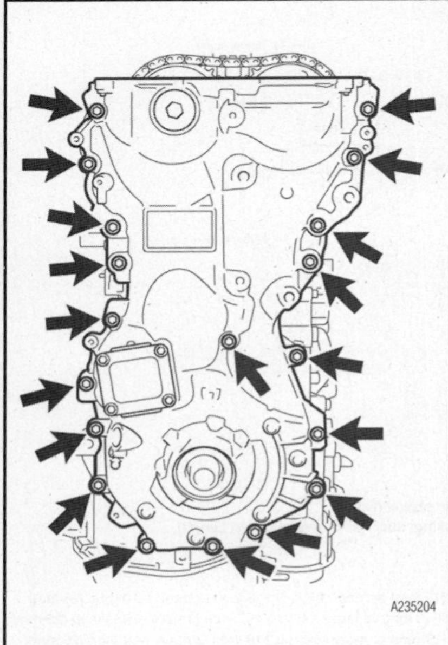
Fig. 243 Removing timing chain cover sub assembly

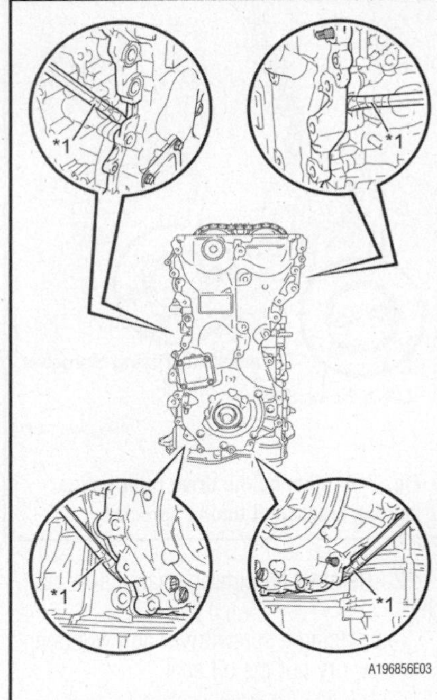

Fig. 244 Removing timing chain

➡Tape the screwdriver tip before use.

   c. Remove the 3 gaskets from the stiffening crankcase.
9. Remove timing chain cover oil seal.
   a. Using a screwdriver and wooden block, pry out the oil seal.

WARNING

Do not damage the surface of the oil seal press fit hole.

➡Tape the screwdriver tip before use.

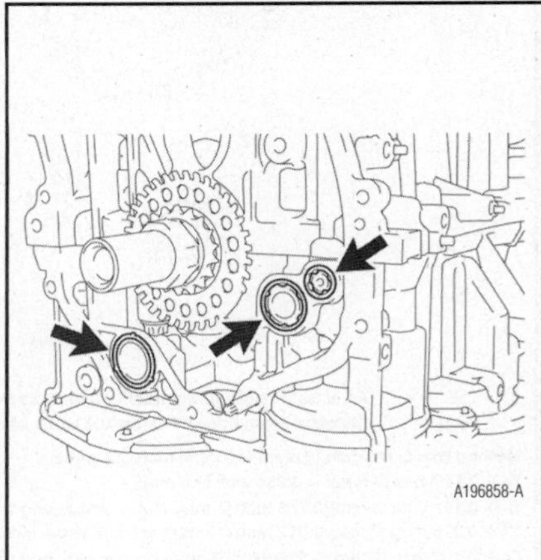
Fig. 245 Removing gaskets from stiffening crankcase

To install:

10. Install timing chain cover sub assembly.
   a. Apply a light coat of engine oil to 3 new gaskets.
   b. Install the 3 gaskets to the stiffening crankcase.
   c. Align the drive rotor spline and the crankshaft timing sprocket.
   d. Apply seal packing in a line to the timing chain cover.

WARNING

When the contact surfaces are wet, clean the surfaces with non-residue solvent before applying seal packing.

WARNING

Install the timing chain cover within 3 minutes and tighten the bolts within 10 minutes after applying seal packing.

WARNING

After applying seal packing to the timing chain cover, install the engine mounting bracket within 10 minutes.

WARNING

Do not apply oil for at least 2 hours after the installation.

WARNING

Do not start the engine for at least 2 hours after the installation.

   e. Temporarily install the timing chain cover with the 17 bolts and 2 nuts.

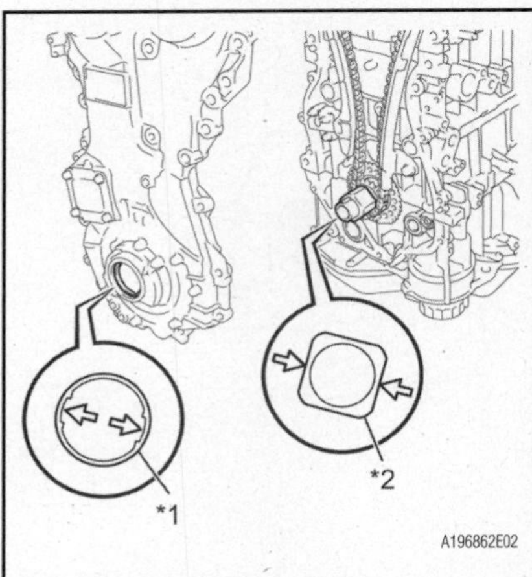

Fig. 246 Aligning the drive rotor spline (1) and crankshaft timing sprocket (2)

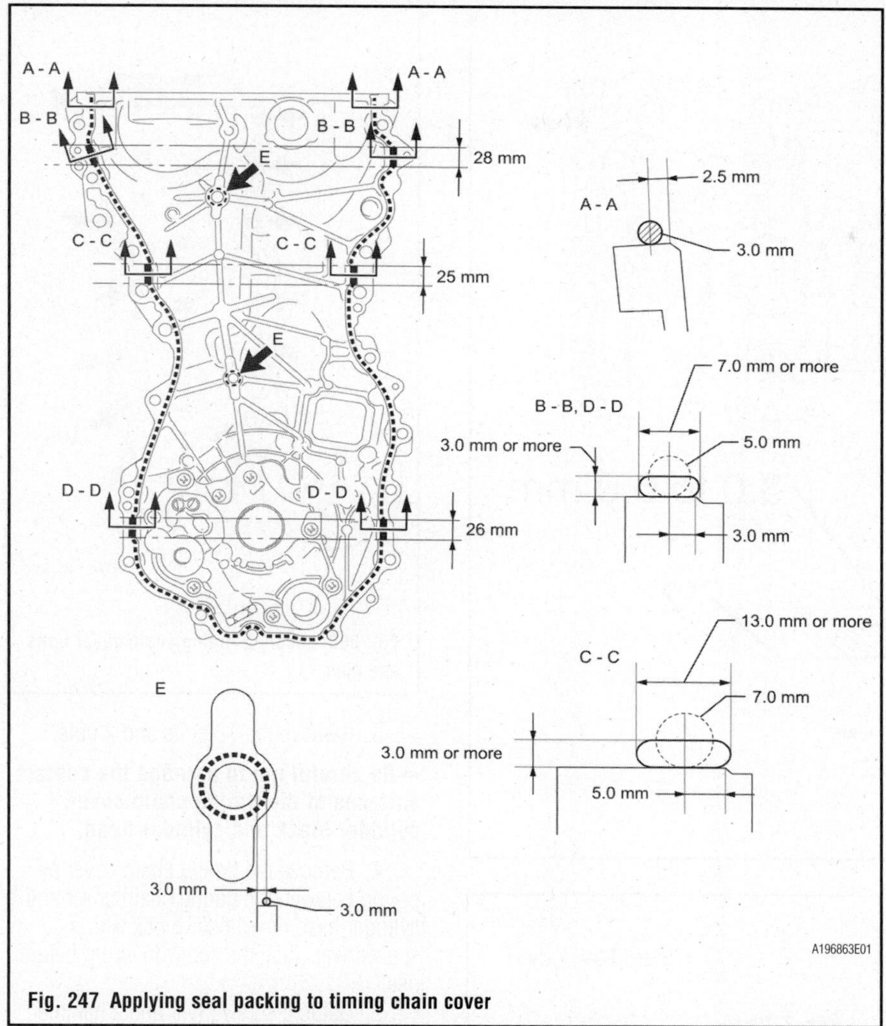

Fig. 247 Applying seal packing to timing chain cover

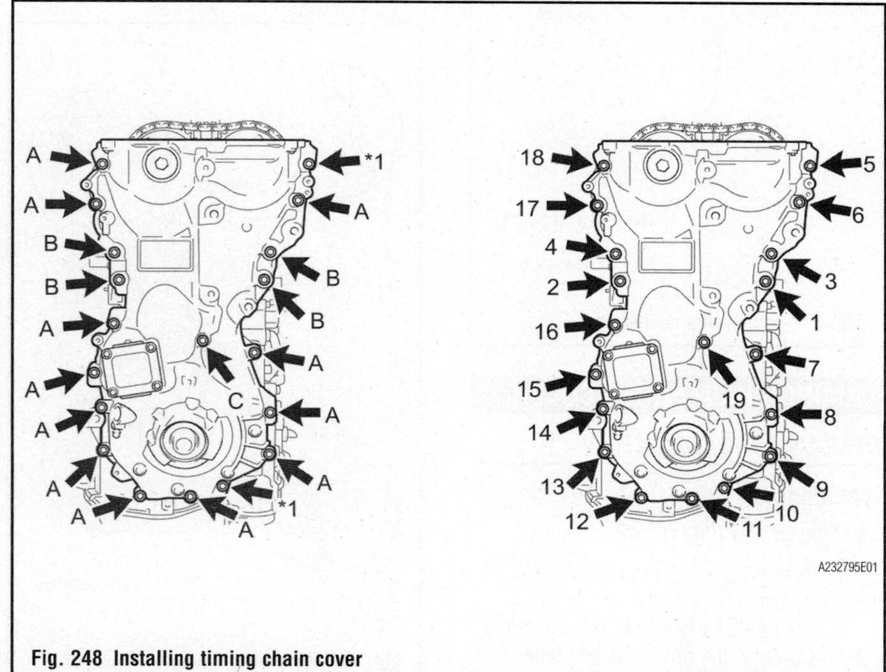

Fig. 248 Installing timing chain cover

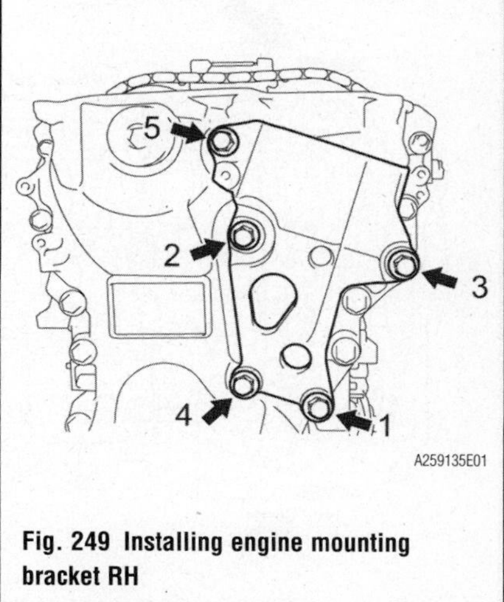

Fig. 249 Installing engine mounting bracket RH

f. Tighten the 17 bolts and 2 nuts in several steps, in sequence. Tighten bolts A, C and Nut to 15 ft. lbs. (21 Nm). Tighten B bolts to 41 ft. lbs. (55 Nm).

11. Install engine mounting bracket RH with the 5 bolts in sequence. Tighten bolts 1, 2 and 3 to 41 ft. lbs. (55 Nm). Tighten bolts 4 and 5 to 15 ft. lbs. (21 Nm).

➡After applying seal packing to the timing chain cover, install the engine mounting bracket within 10 minutes.

12. Install timing chain cover oil seal.
13. Install crankshaft pulley.
14. Install crankshaft position sensor.
15. Install cylinder head cover sub assembly.

a. Apply a light coat of engine oil to 2 new gaskets.

b. Install the 2 gaskets to the camshaft bearing caps.

c. Visually check the spark plug tube gasket.

d. Install a new cylinder head cover gasket to the cylinder head cover.

➡Remove any oil from the contact surface.

e. Apply seal packing (diameter: 0.118-0.236 inch).

f. Remove any oil from the contact surface.

g. Install the cylinder head cover within 3 minutes and tighten the bolts within 15 minutes of applying seal packing.

h. Align the cylinder head cover with pin A. Then align the cylinder head cover with pin B and install the cylinder head cover.

i. Install 3 new seal washers and the 16 bolts, and then tighten the bolts in the order. Tighten to 9 ft. lbs. (12 Nm).

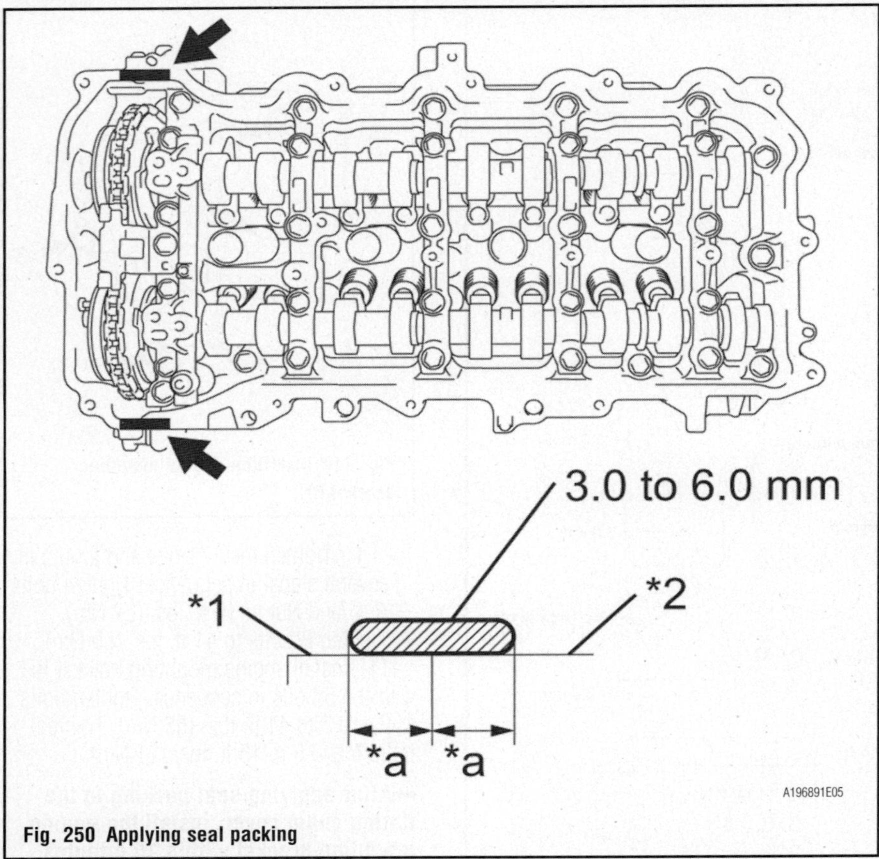

Fig. 250 Applying seal packing

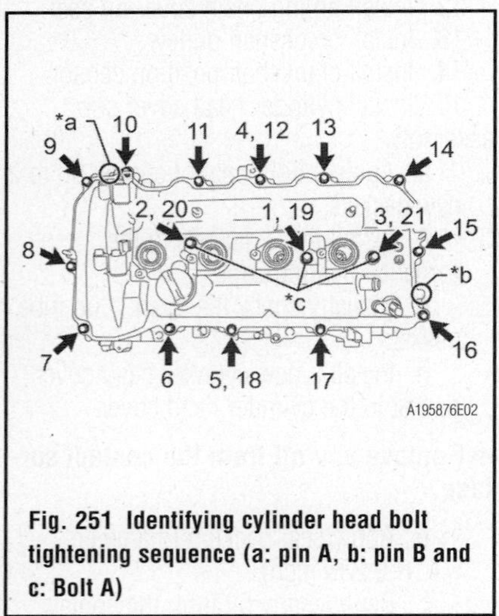

Fig. 251 Identifying cylinder head bolt tightening sequence (a: pin A, b: pin B and c: Bolt A)

➡**Do not apply oil for at least 2 hours after the installation.**

16. Install ignition coil assembly.
17. Install engine wire.
18. Install engine and transaxle.

PISTON AND RING

POSITIONING

See Figure 252.

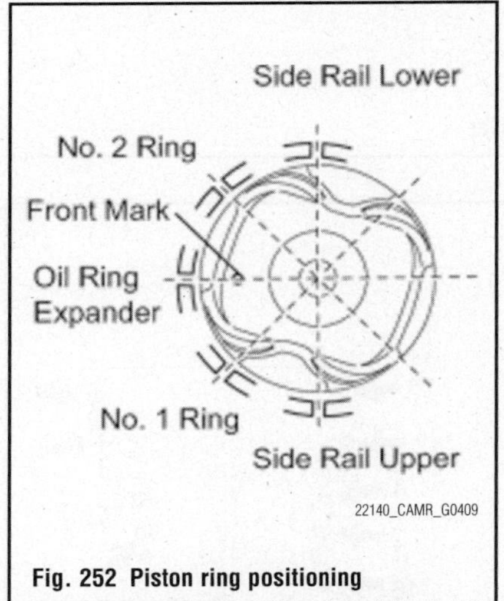

Fig. 252 Piston ring positioning

TIMING CHAIN FRONT COVER

REMOVAL & INSTALLATION

2.4L Engine

See Figures 253 through 258.

1. Before servicing the vehicle, refer to the Precautions Section.
2. Using a E10 Torx® socket, remove the stud bolt for the drive belt tensioner from the cylinder block.

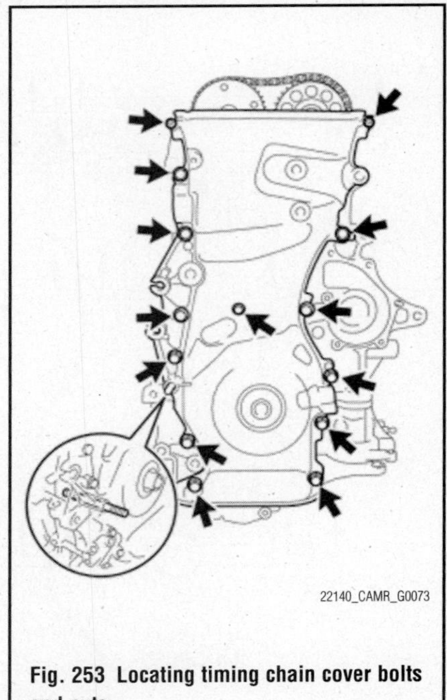

Fig. 253 Locating timing chain cover bolts and nuts

3. Remove the 12 bolts and 2 nuts.

➡**Be careful not to damage the contact surfaces of the timing chain cover, cylinder block and cylinder head.**

4. Remove the timing chain cover by prying between the timing chain cover and cylinder head or cylinder block with a screwdriver. Tape the screwdriver tip before use.
5. Using a screwdriver and a hammer, tap out the oil seal.

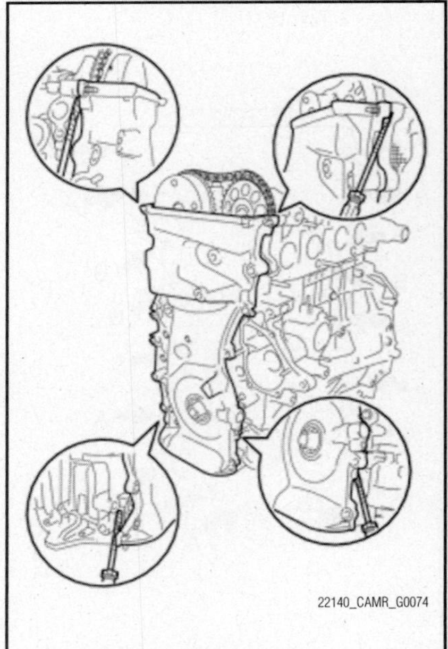

Fig. 254 Removing timing chain cover

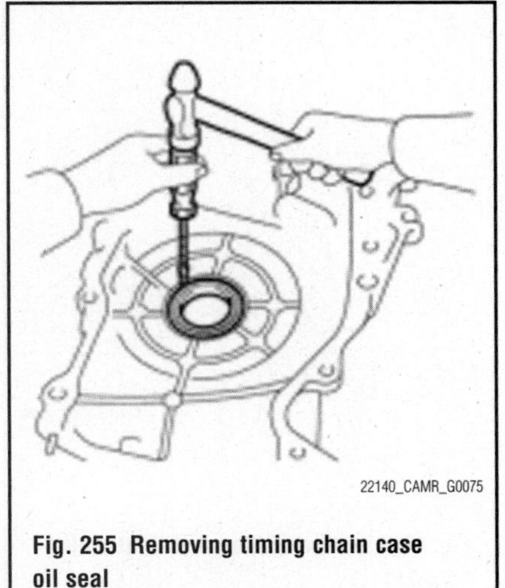

Fig. 255 Removing timing chain case oil seal

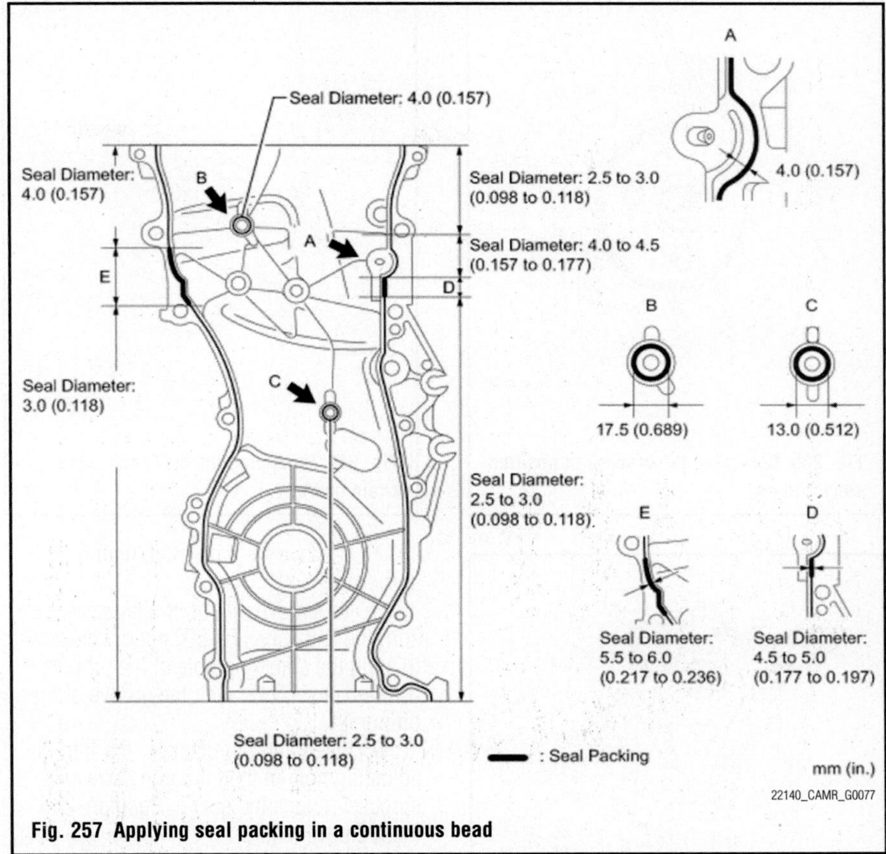

Fig. 257 Applying seal packing in a continuous bead

To install:

➡Keep the gap between the timing chain cover edge and the oil seal free of foreign matter.

6. Using Special Tool: 09223-22010, tap in a new oil seal until its surface is flush with the timing chain cover edge. Apply a light coat of MP grease to the lip of the oil seal.
7. Remove any old packing (FIPG) material and be careful not to drop any oil on the contact surfaces of the timing chain cover, cylinder head and cylinder block.
8. Apply Toyota Genuine Seal Packing Black, Three Bond 1207B or Equivalent seal packing in a diameter of 0.157 to 0.177 inches (4.0 to 4.5 mm).
9. Apply seal packing in a continuous bead as shown in the illustration below.

➡Remove any oil from the contact surface. Install the chain cover within 3 minutes after applying seal packing.

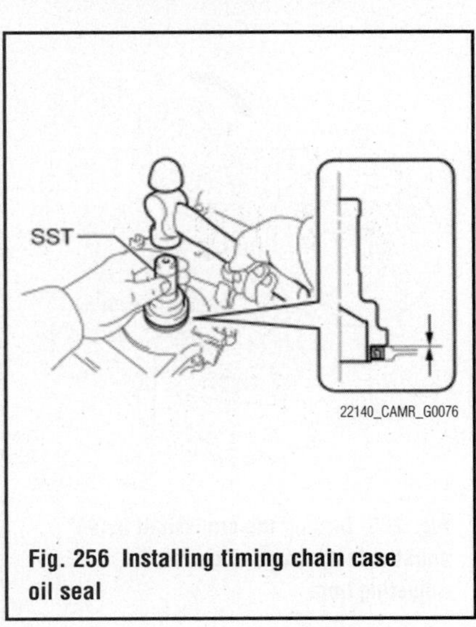

Fig. 256 Installing timing chain case oil seal

Do not start the engine for at least 2 hours after installing.

10. Install the timing chain cover with the twelve bolts and two nuts in sequence to the following torque specification:

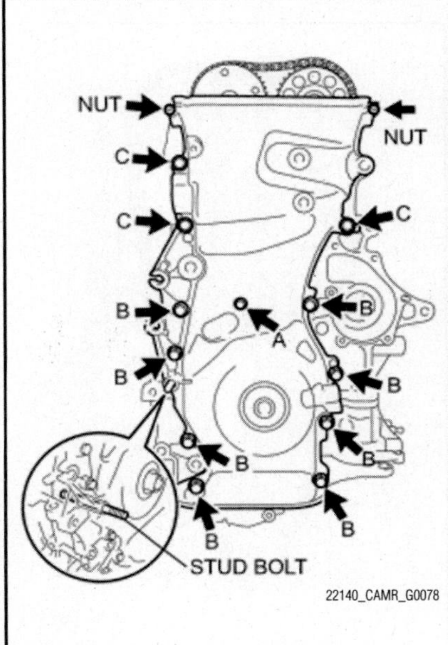

Fig. 258 Installing the twelve bolts and two nuts

- Bolt A length: 1.18 inches (30 mm) for 10 mm head: 80 inch lbs. (9 Nm)
- Bolt B length: 1.18 inches (30 mm) for 12 mm head: 18 ft. lbs. (25 Nm)
- Bolt C length: 1.57 inches (40 mm) for 14 mm head: 41 ft. lbs. (55 Nm)
- Nut: 8 ft. lbs. (11 Nm)

11. Using a E10 Torx® socket, install the stud bolt to the drive belt tensioner and tighten to 16 ft. lbs. (22 Nm).

2.5L Engine

For removal and installation procedures, refer to TIMING CHAIN & SPROCKETS.

TIMING CHAIN & SPROCKETS

REMOVAL & INSTALLATION

2.4L Engine

See Figures 259 through 281.

1. Before servicing the vehicle, refer to the Precautions Section.
2. Remove the crankshaft position sensor plate.
3. Remove the bolt and chain tensioner slipper.
4. Remove the two bolts and chain vibration damper.

Fig. 259 Removing the crankshaft position sensor plate

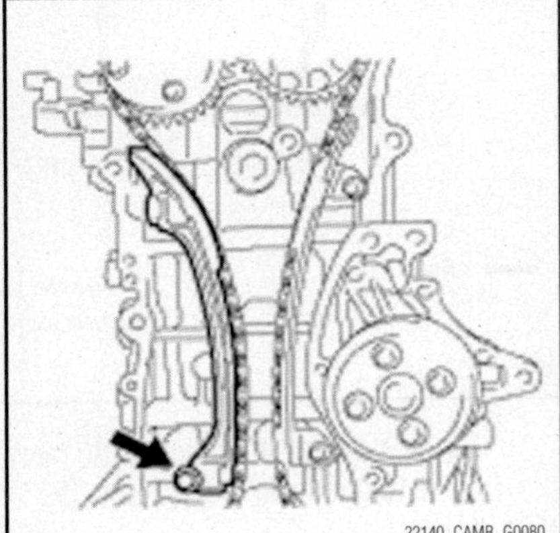

Fig. 260 Removing the bolt and chain tensioner slipper

5. Remove the bolt and timing chain guide.

6. Remove the upper chain assembly.

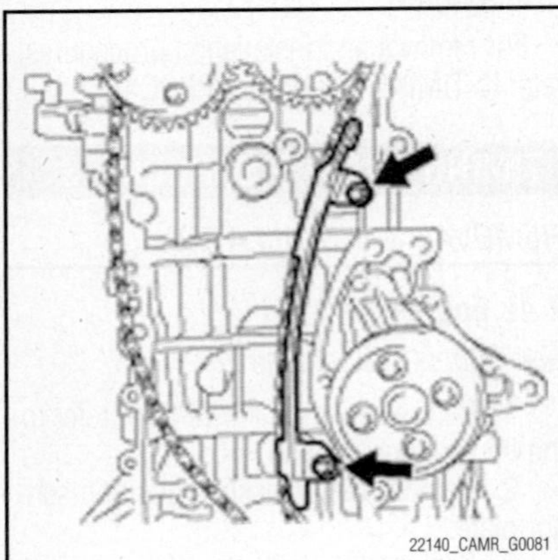

Fig. 261 Removing the two bolts and upper chain vibration damper

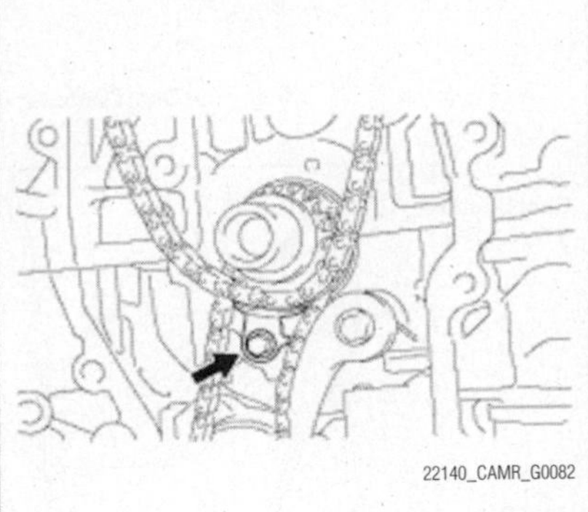

Fig. 262 Removing the bolt and timing chain guide

7. Remove the crankshaft timing sprocket.

8. To remove the lower chain assembly. Turn the crankshaft by 90° counterclockwise to align the adjusting hole of the oil pump drive shaft sprocket with the groove of the oil pump.

9. Insert a 4 mm diameter bar into the adjusting hole of the oil pump drive shaft sprocket to lock the gear in position, and then remove the nut.

10. Remove the bolt, chain tensioner plate and spring.

11. Remove the chain tensioner, oil pump driven sprocket and chain.

To install:

12. Set the crankshaft key into the left horizontal position.

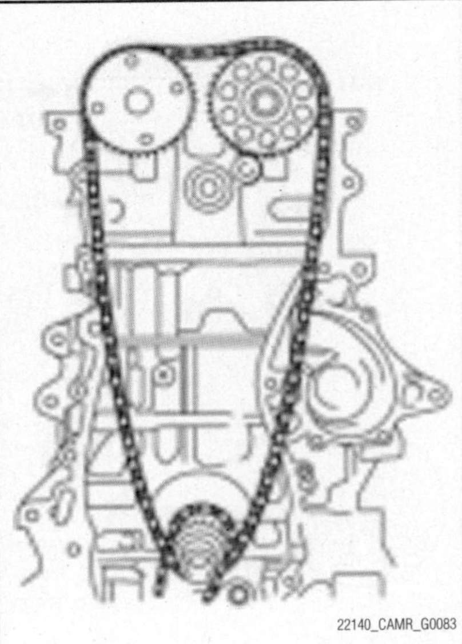

Fig. 263 Removing the upper chain assembly

Fig. 264 Removing the crankshaft timing sprocket

13. Turn the drive shaft so that the cutout faces upward.

14. Align the yellow mark links with the timing marks of each gear.

15. Install the sprockets onto the crankshaft and oil pump shaft with the chain wrapped on the gears

16. Temporarily tighten the oil pump drive shaft sprocket with the nut.

17. Insert the damper spring into the adjusting hole, and then install the chain tensioner plate with the bolt and tighten to 9 ft. lbs. (12 Nm).

18. Align the adjusting hole of the oil pump drive shaft sprocket with the groove of the oil pump.

19. Insert a 4 mm diameter bar into the adjusting hole of the oil pump drive shaft gear to lock the gear in position, and then tighten the nut to 22 ft. lbs. (30 Nm).

20. Rotate the crankshaft clockwise by 90°, and align the crankshaft key to the top.

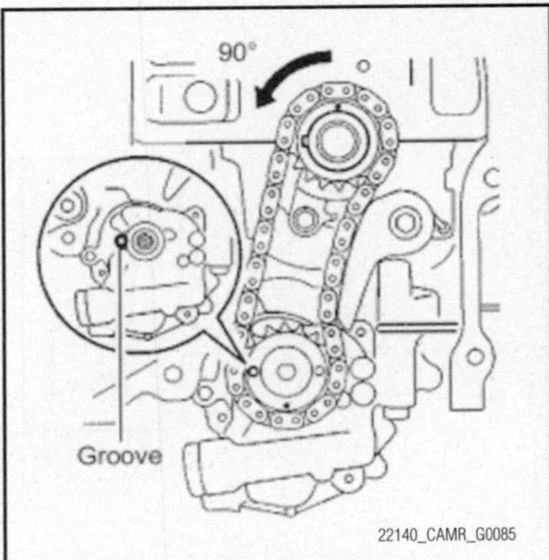

Fig. 265 Turning the crankshaft by 90° counterclockwise to align the adjusting hole

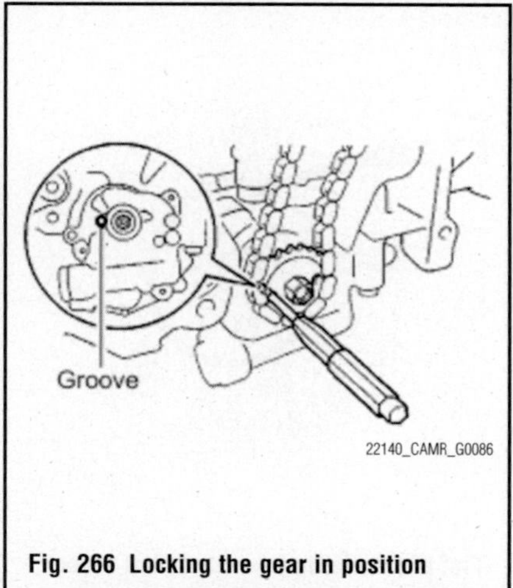

Fig. 266 Locking the gear in position

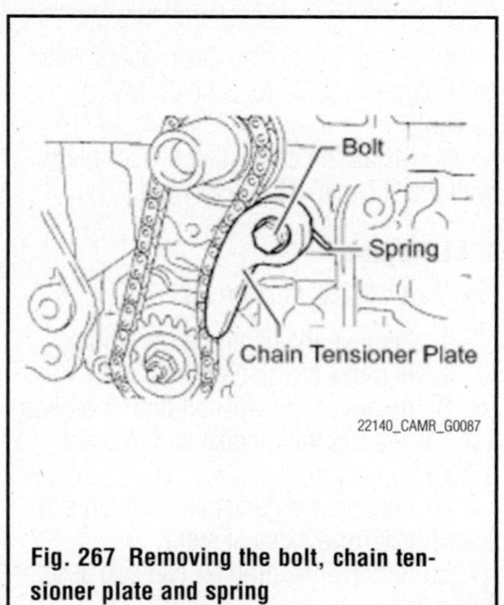

Fig. 267 Removing the bolt, chain tensioner plate and spring

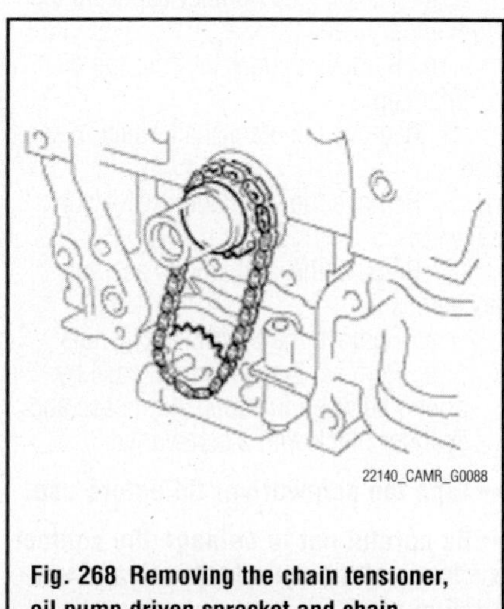

Fig. 268 Removing the chain tensioner, oil pump driven sprocket and chain

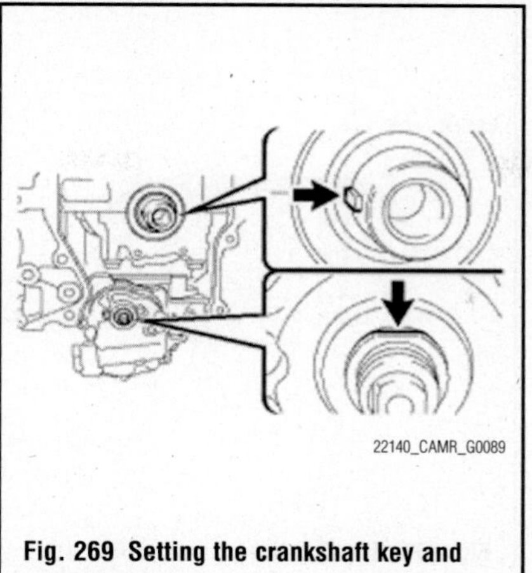

Fig. 269 Setting the crankshaft key and cutout face

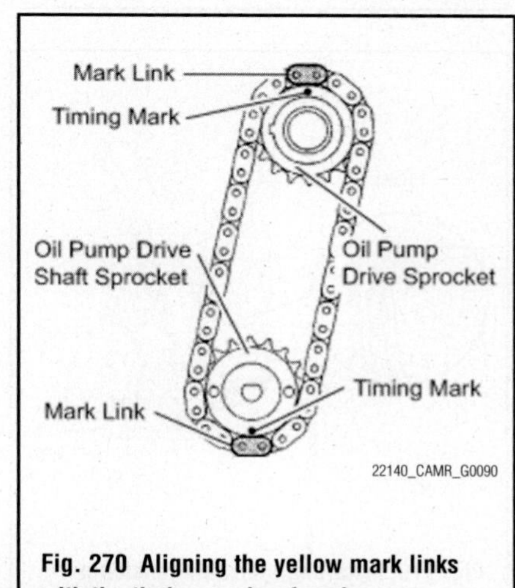

Fig. 270 Aligning the yellow mark links with the timing marks of each gear

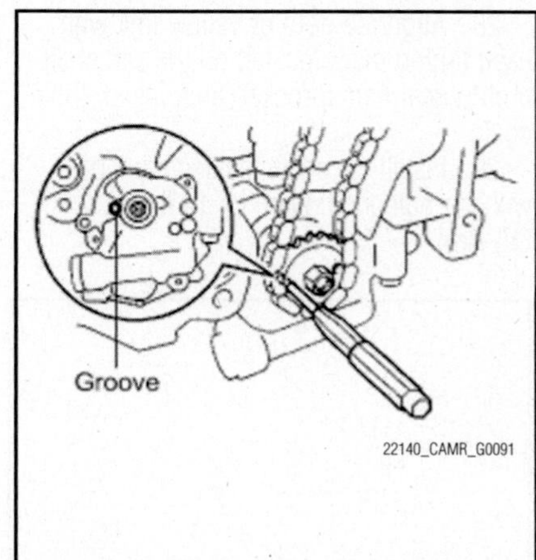

Fig. 271 Aligning the adjusting hole of the oil pump drive shaft sprocket

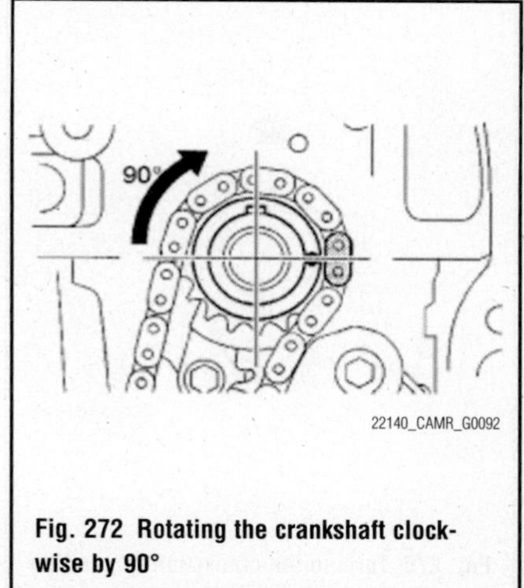

Fig. 272 Rotating the crankshaft clockwise by 90°

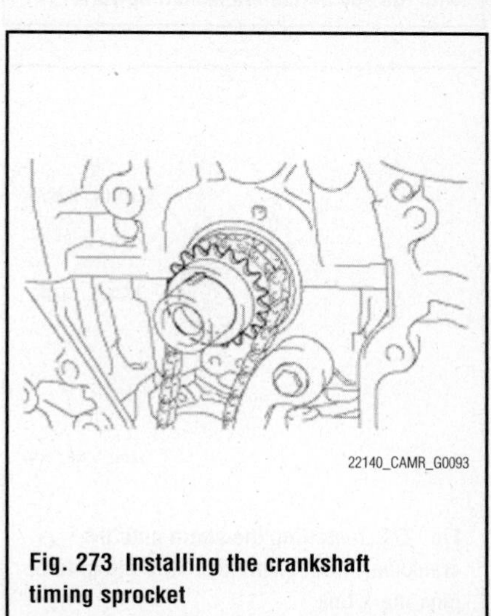

Fig. 273 Installing the crankshaft timing sprocket

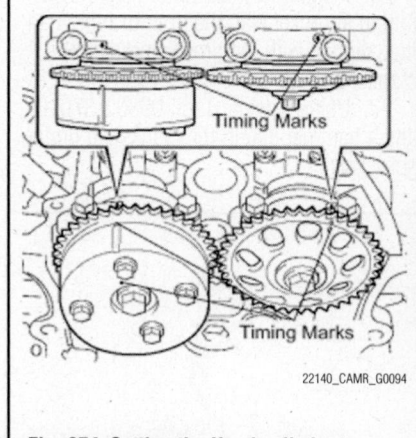

Fig. 274 Setting the No. 1 cylinder to TDC/compression

21. Install the crankshaft timing sprocket.

22. Install the chain vibration damper with the 2 bolts and tighten to 80 inch lbs. (9 Nm).

23. Set the No. 1 cylinder to TDC/compression.

24. Turn the camshafts with a wrench (using the hexagonal lobe) to align the timing marks of the camshaft timing gear with each timing mark located on the No. 1 and No. 2 bearing caps.

25. Using the crankshaft pulley bolt, turn the crankshaft to position with the key on the crankshaft upward.

Fig. 275 Turning the crankshaft to position with the key on the crankshaft upward

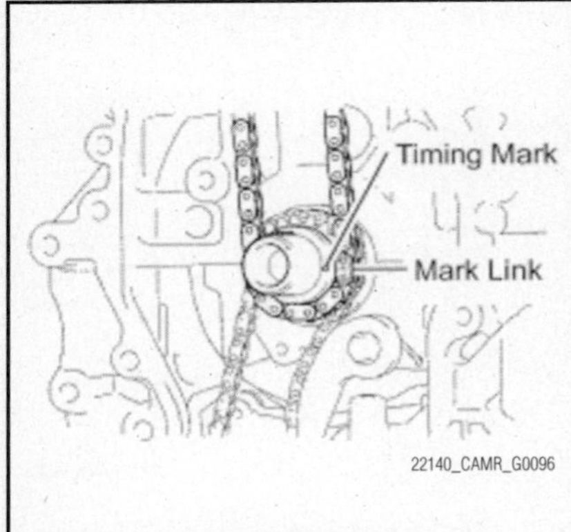

Fig. 276 Installing the chain onto the crankshaft timing sprocket with the gold or pink mark link

26. Install the chain onto the crankshaft timing sprocket with the gold or pink mark link aligned with the timing mark on the crankshaft.
27. Using Special Tool: 09309-37010 and a hammer, tap in the crankshaft timing sprocket.

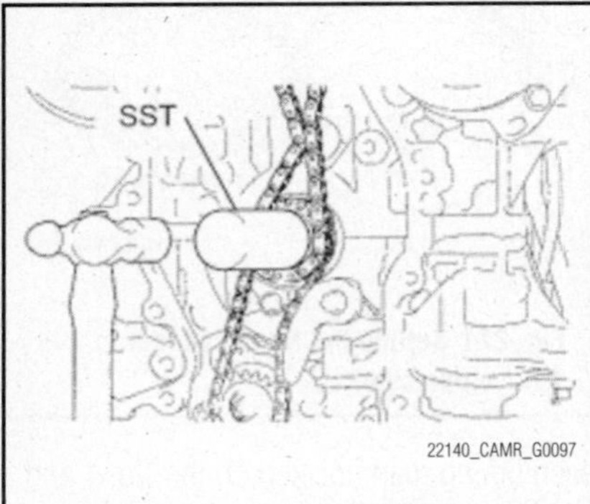

Fig. 277 Using Special Tool: 09309-37010 and a hammer to install the crankshaft timing sprocket

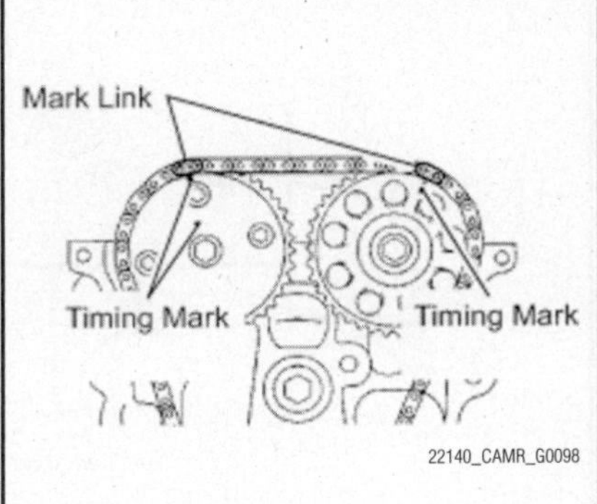

Fig. 278 Aligning the gold or yellow link with each timing mark located on the camshaft timing gear and sprocket

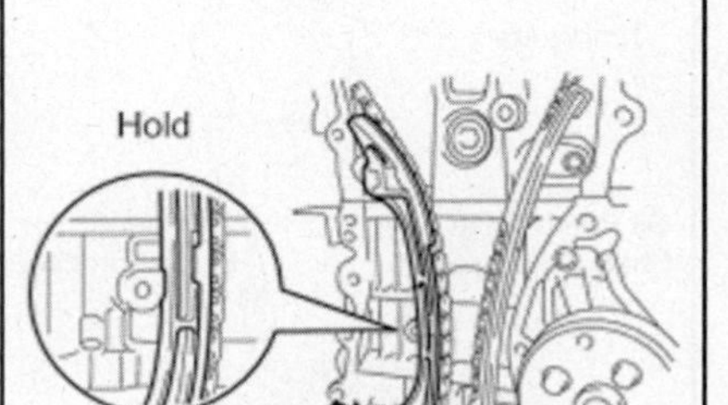

Fig. 279 Installing the chain tensioner slipper and bolt

28. Align the gold or yellow link with each timing mark located on the camshaft timing gear and sprocket, then install the chain.
29. Install the chain tensioner slipper with the bolt and tighten to: 14 ft. lbs. (19 Nm).

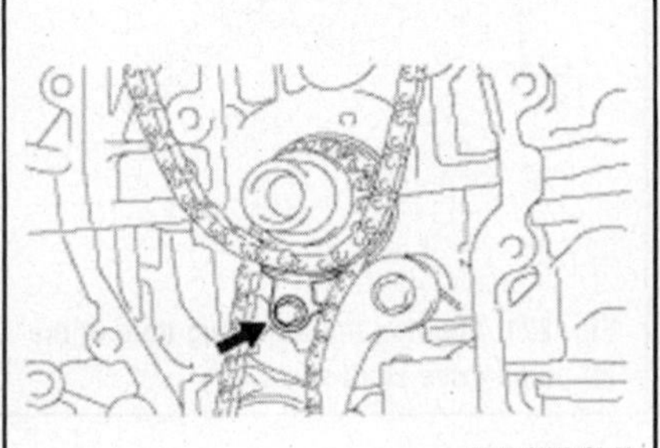

Fig. 280 Installing the timing chain guide and bolt

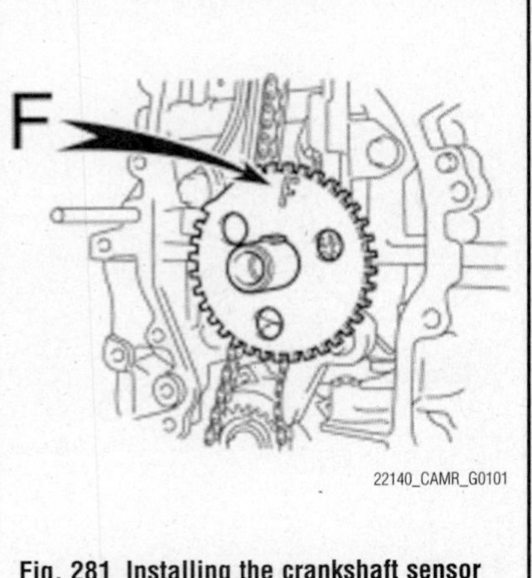

Fig. 281 Installing the crankshaft sensor plate with the "F" mark facing forward

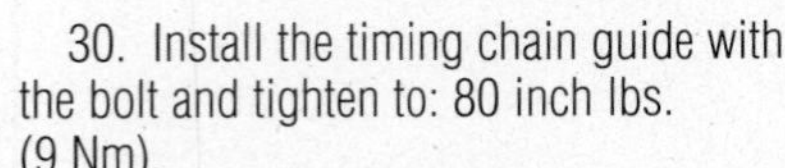

30. Install the timing chain guide with the bolt and tighten to: 80 inch lbs. (9 Nm).
31. Install the crankshaft sensor plate with the "F" mark facing forward.

2.5L Engine

See Figures 282 through 293.

1. Remove the engine cover joints.
2. Remove the spark plug.
3. Remove the camshaft timing oil control valve assembly (intake and exhaust sides).
4. Remove the camshaft position sensors (intake and exhaust side).
5. Remove the oil filler cap sub assembly.
    a. Remove the oil filler cap from the cylinder head.
    b. Remove the gasket from the oil filler cap.
6. Remove the crankshaft position sensor.
7. Remove the ventilation valve sub assembly.
8. Remove the ventilation case sub assembly.
    a. Remove the 8 bolts and 2 nuts.
    b. Remove the ventilation case by prying between the ventilation case and cylinder block with a screwdriver.

➡Tape the screwdriver tip before use.

➡Be careful not to damage the contact surfaces of the cylinder block and ventilation case.

9. Remove the separator case.
    a. Remove the 2 bolts, separator case and gasket.
10. Remove the No. 1 water by pass pipe and gasket.
11. Remove the inlet water.

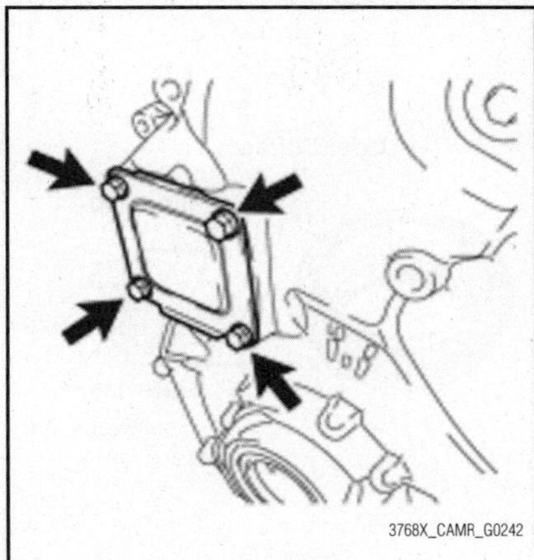

Fig. 282 Removing the timing chain cover plate

12. Remove the thermostat.
13. Remove the v-ribbed belt tensioner assembly (compressor and generator).
14. Remove the water pump assembly.
15. Remove the water drain cock plug from the water drain cock.
   a. Remove the water drain cock from the cylinder block.
16. Remove the oil cooler assembly (with the oil cooler).
17. Remove the inlet water housing.
   a. Remove the 4 bolts, nut, inlet water housing and gasket.
18. Remove the crankshaft pulley.
   a. Using the special tool, hold the crankshaft pulley ad loosen the pulley bolt. Further loosen the bolt until 2 or 3 threads are screwed into the crankshaft.
   b. Using the special tool and the pulley bolt, remove the crankshaft pulley.

➡Apply a lubricant to the threads and end of the special tool.

19. Remove the cylinder head cover sub assembly and 3 gaskets from the camshaft bearing caps.
20. Remove the spark plug tube gasket.
   a. Using a screwdriver, pry out the 4 plug tube gaskets.

➡Be careful not to damage the cylinder head cover.

➡Tape the screwdriver tip before use.

21. Remove the No. 2 timing chain cover.
22. Remove the engine mounting bracket RH.
23. Remove the timing chain cover sub assembly.
24. Remove the timing chain cover tight plug.
   a. Using a 14 mm hexagon wrench, remove the plug and gasket.
25. Remove the timing chain cover plate.
   a. Remove the 4 bolts, timing chain cover plate and gasket.
26. Remove the timing chain cover oil seal.
27. Set the No. 1 cylinder to TDC/Compression.
28. Remove the timing chain guide.
29. Remove the No. 1 chain tensioner assembly.
   a. Allow the plunger to extend slightly, and then rotate the stopper plate counterclockwise to release the lock. Once the lock is released, push the plunger into the tensioner.
   b. Move the stopper plate clockwise to set the lock, and insert a pin into the stopper plate hole.
   c. Remove the 2 bolts, chain tensioner and gasket.
30. Remove the chain tensioner slipper.
31. Remove the chain sub assembly.
32. Remove the No. 1 crankshaft timing sprocket.

To install:

33. Install the crankshaft timing sprocket to the crankshaft.
34. Add engine oil.
35. Set the No. 1 cylinder to TDC/Compression.
36. Install the No. 1 chain vibration damper with the 2 bolts. Tighten to 15 ft. lbs. (21 Nm).
37. Install the chain sub assembly.
38. Install the chain tensioner slipper. Tighten to 15 ft. lbs. (21 Nm).
39. Install the No. 1 chain tensioner assembly.

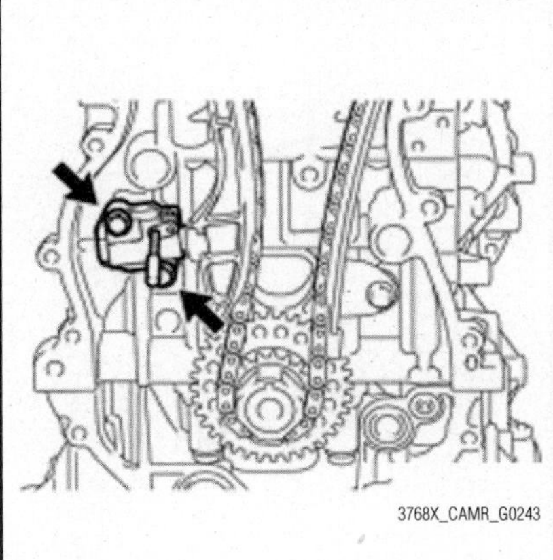

Fig. 283 Removing the chain tensioner

Fig. 284 Removing the No. 1 crankshaft timing sprocket

   a. Install a new gasket and the chain tensioner with the 2 bolts. Tighten to 7 ft. lbs. (10 Nm).
   b. Remove the pin from the stopper plate.
40. Install the timing chain guide. Tighten the bolt to 15 ft. lbs. (21 Nm).
41. Check the No. 1 cylinder to TDC/Compression.
   a. Temporarily install the crankshaft pulley bolt.
   b. Rotate the crankshaft clockwise, and check that the timing marks on the crankshaft timing sprocket and camshaft timing gears are as shown in the illustration.

➡"A" is not a timing mark.

   c. Remove the crankshaft pulley bolt.
42. Install the timing chain cover sub assembly.
43. Install the timing chain cover plate. Tighten the 4 bolts to 7 ft. lbs. (10 Nm).
44. Install the timing chain cover tight plug.
   a. Using a 14 mm hexagon wrench, install a new gasket and the plug. Tighten to 15 ft. lbs. (20 Nm).
45. Install the timing chain cover sub assembly.
   a. Apply a light coat of engine oil to the 3 new gaskets.
   b. Install the 3 gaskets to the stiffening crankcase.
   c. Align the drive rotor spline and the crankshaft timing sprocket.
46. Install the engine mounting bracket RH.
   a. Install the engine mounting bracket, and install the 5 bolts in the order shown. Tighten bolts 1, 2 and 3 to 41 ft. lbs. (55 Nm). Tighten bolts 4 and 5 to 16 ft. lbs. (21 Nm).

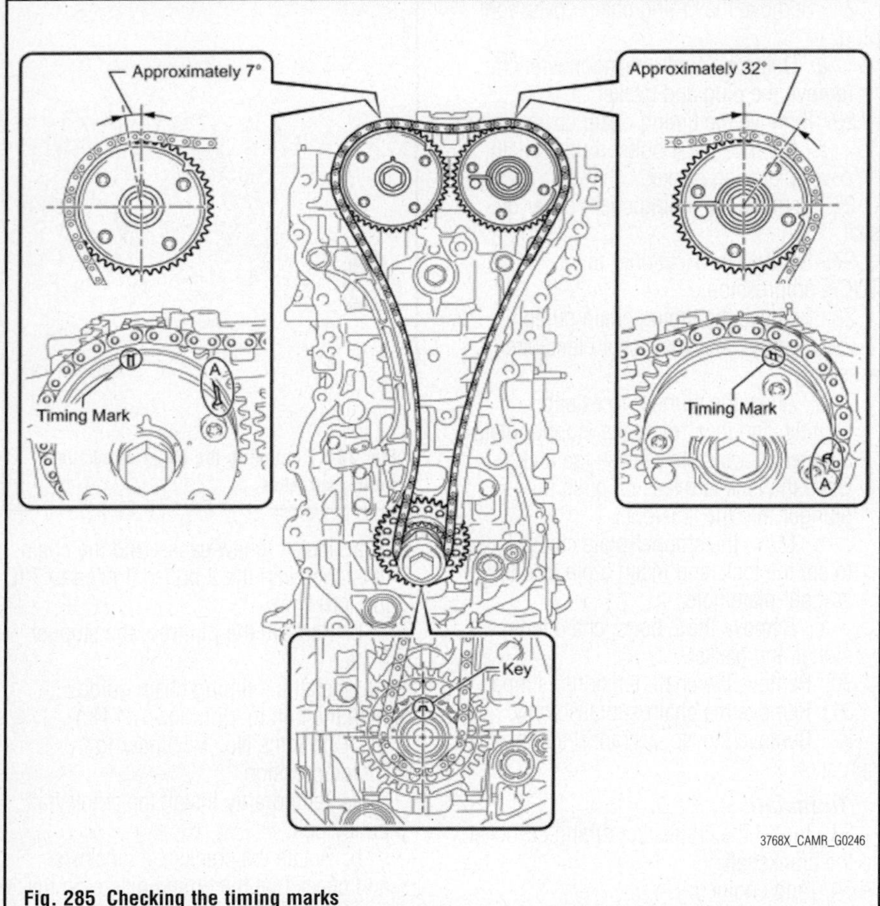

Fig. 285 Checking the timing marks

➡After applying seal packing to the timing chain cover install the engine mounting bracket within 10 minutes.

47. Install the No. 2 timing chain cover with the 2 bolts. Tighten to 7 ft. lbs. (10 Nm).

48. Install the timing chain cover oil seal.

49. Install the spark plug tube gasket.

a. Visually check the spark plug tube gasket. If the results are not as specified, replace the spark plug tube gasket.

- Upper surface: no scratches or deformation
- Outer lip: no scratches or deformation

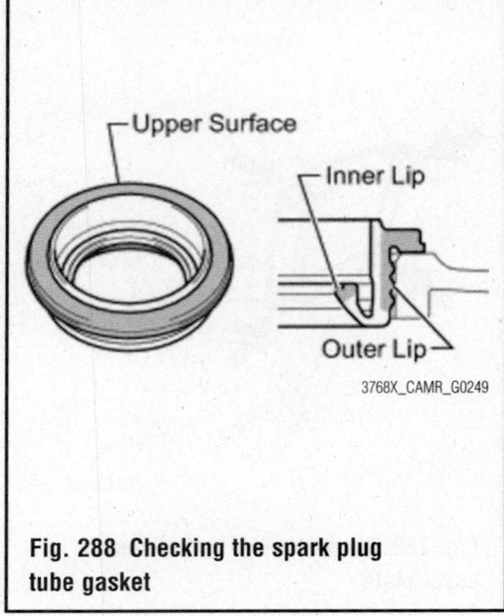

Fig. 288 Checking the spark plug tube gasket

- Inner lip: No scratches

b. Install the 4 plug tube gaskets to the cylinder head cover.

➡After pressing in the spark plug tube gasket, make sure the gasket protrudes 0.0394 inch (1 mm) or less from the cylinder head cover.

50. Install the cylinder head cover sub assembly.

a. Apply a light coat of engine oil to 3 new gaskets.

b. Install the 3 gaskets to the camshaft bearing caps.

c. Install a new gasket to the cylinder head cover.

➡Remove any oil from the contact surface.

d. Apply seal packing as shown.

➡Remove any oil from the contact surface.

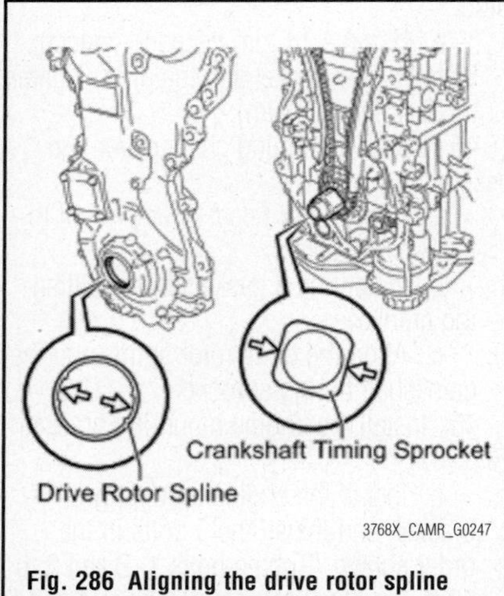

Fig. 286 Aligning the drive rotor spline and the crankshaft timing sprocket

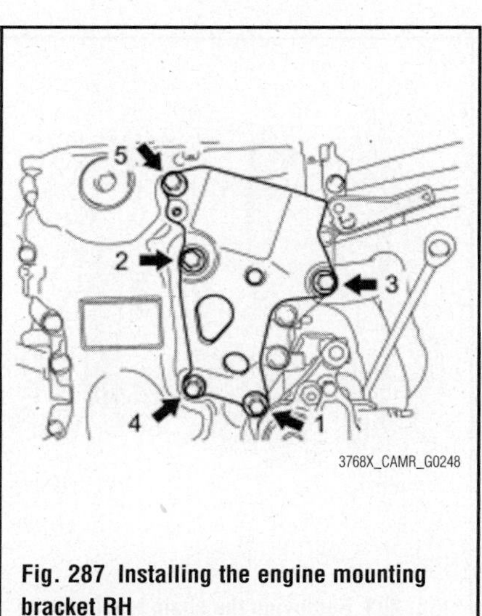

Fig. 287 Installing the engine mounting bracket RH

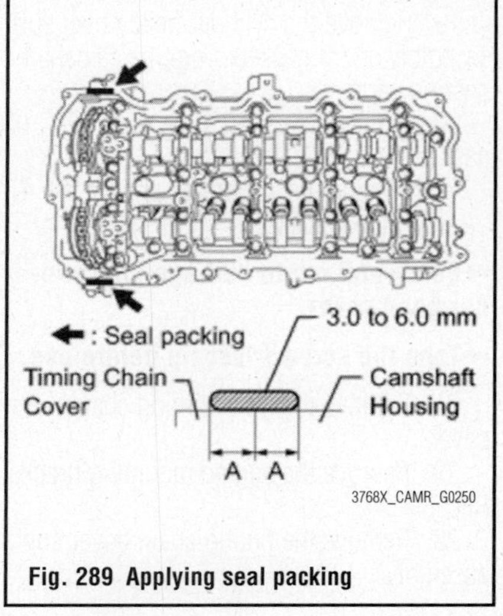

Fig. 289 Applying seal packing

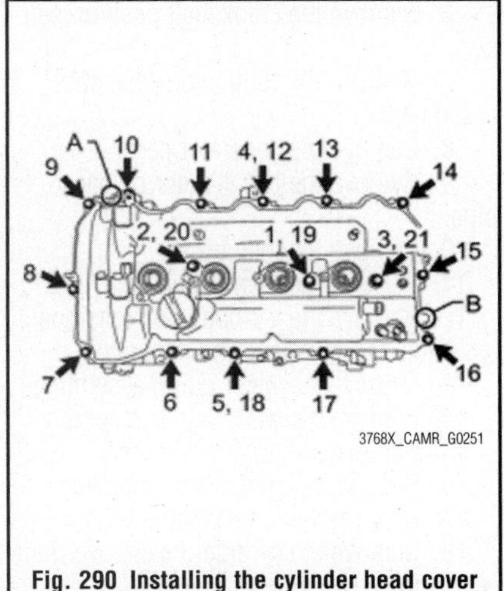

Fig. 290 Installing the cylinder head cover

➡Install the cylinder head cover within 3 minutes and tighten the bolts within 15 minutes after applying seal packing.

e. Align the cylinder head cover with pin A. Then align the cylinder head cover with pin B and install the cylinder head cover.

f. Install 3 new seal washers and the 16 bolts, and then tighten the bolts in the order shown in the illustration. Tighten to 9 ft. lbs. (12 Nm).

➡Do not apply oil for at least 4 hours after the installation.

51. Install the crankshaft pulley.

a. Align the pulley set key with the key groove of the crankshaft pulley.

b. Using the special tool, hold the crankshaft pulley and install the pulley bolt. Tighten the bolt to 192 ft. lbs. (260 Nm).

52. Install the crankshaft position sensor.

53. Install the inlet water housing with the 4 bolts. Tighten to 32 ft. lbs. (43 Nm).

54. Install the oil cooler assembly (with oil cooler).

55. Install the water pump assembly.

56. Install the v-ribbed belt tensioner assembly (compressor and generator). Tighten to 15 ft. lbs. (21 Nm).

57. Install the thermostat.

58. Install the inlet water.

59. Install the water drain cock.

a. Apply adhesive to 2 or 3 threads of the drain cock.

b. Install the water drain cock as shown.

➡Do not rotate the drain cock more than 1 revolution (360°) after tightening the drain cock to the specified torque.

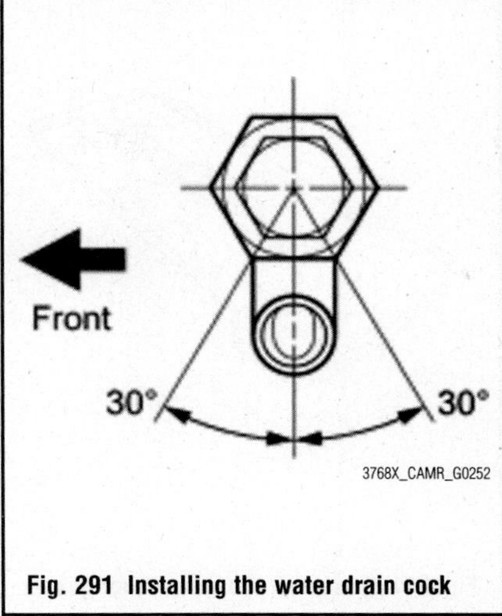

Fig. 291 Installing the water drain cock

➡Do not loosen the drain cock to adjust it. If an adjustment is necessary, remove the drain cock and reinstall it.

c. Install the water drain cock plug to the water drain cock. Tighten to 9 ft. lbs. (13 Nm).

60. Install the No. 1 water by pass pipe.

a. Install a new gasket and the water by-pass pipe with the 2 nuts and bolt. Tighten to 7 ft. lbs. (10 Nm).

61. Install the separator case.

a. Apply a light coat of engine oil to a new gasket.

b. Install the gasket to the separator case.

c. Install the separator case with the 2 bolts. Tighten to 7 ft. lbs. (10 Nm).

62. Install the ventilation case sub assembly.

a. Apply seal packing in a continuous line as shown.

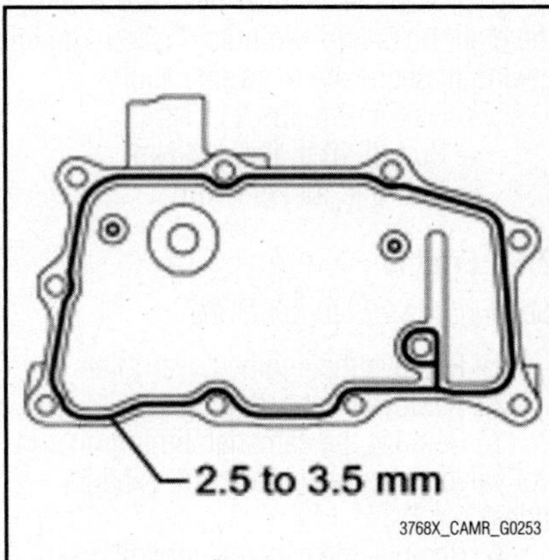

Fig. 292 Applying seal packing to the separator case

Fig. 293 Installing the ventilation case

➡Remove any oil from the contact surface.

➡Install the ventilation case within 3 minutes and tighten the bolts and nuts within 15 minutes after applying seal packing.

b. Install the ventilation case, and install the 8 bolts and 2 nuts in the order shown. Tighten to 15 ft. lbs. (21 Nm).

➡Bolt A is tightened twice.

c. Install the ventilation valve sub assembly.

63. Install the camshaft position sensor (intake and exhaust side).

64. Install the camshaft timing oil control valve assembly (intake and exhaust side).

65. Install the oil filler cap sub assembly.

a. Install a new gasket to the filler cap.

b. Install the oil filler cap to the cylinder head.

66. Install the spark plug.

67. Install the engine cover joints. Tighten to 7 ft. lbs. (10 Nm).

VALVE COVERS

REMOVAL & INSTALLATION

2.4L Engine

See Figures 294 through 298.

1. Before servicing the vehicle, refer to the Precautions Section.
2. Remove the oil filler cap.
3. Remove the ventilation valve.
4. Remove the spark plugs.
5. Drain the engine oil and remove the oil filter.

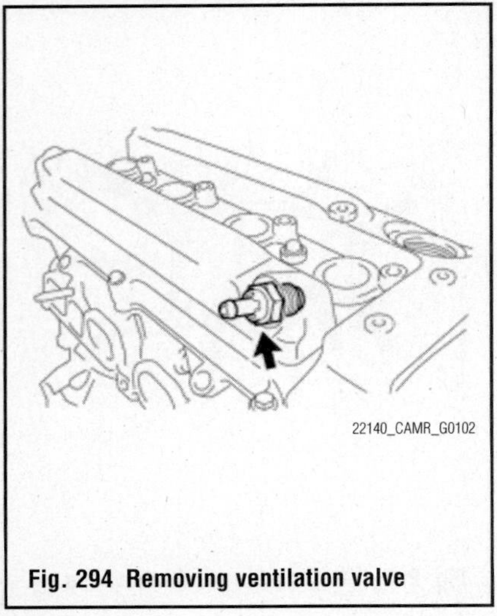
Fig. 294 Removing ventilation valve

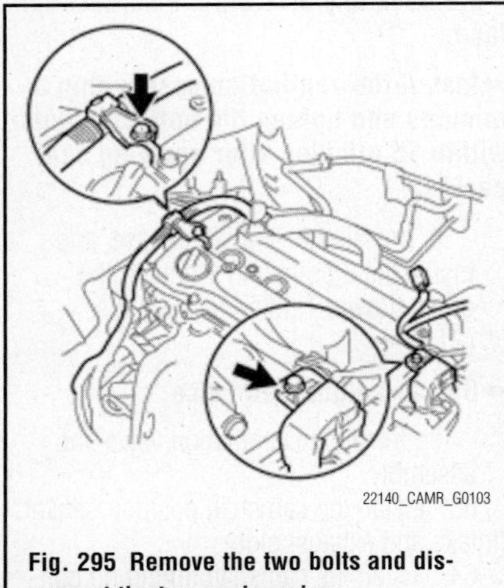
Fig. 295 Remove the two bolts and disconnect the two engine wires

6. Remove the two bolts and disconnect the two engine wires.
7. Remove the eight bolts, two nuts and the valve cover.

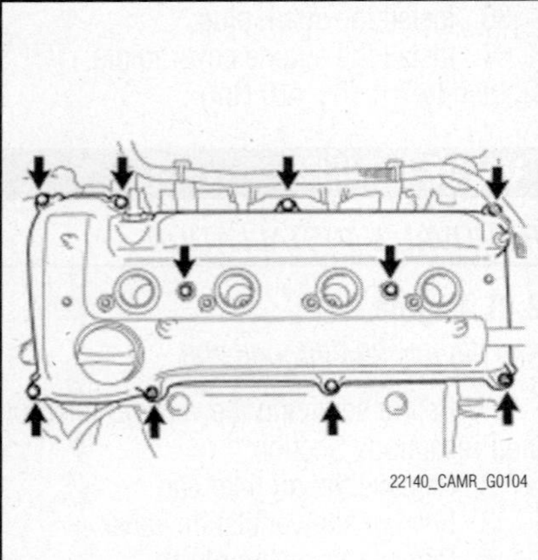
Fig. 296 Removing the eight bolts, two nuts and the valve cover

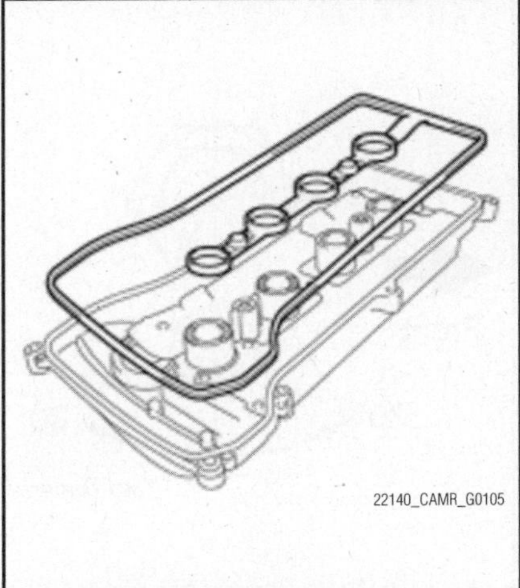
Fig. 297 Removing the valve cover gasket

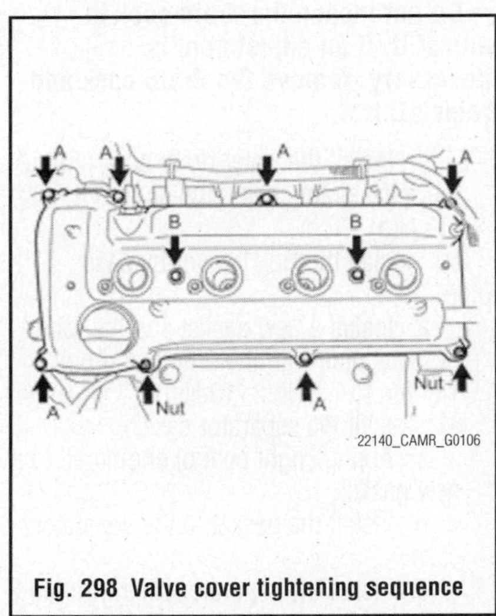

Fig. 298 Valve cover tightening sequence

8. Remove the cylinder head cover gasket.

To install:

9. To install, reverse removal procedure.
10. Install the cylinder head cover with the eight bolts and two nuts. Tighten the following in sequence to specification:
- Bolt A: 8 ft. lbs. (11 Nm)
- Bolt B: 10 ft. lbs. (14 Nm)
- Nut: 8 ft. lbs. (11 Nm)

2.5L Engine

See Figures 299 through 302.

1. Remove the engine cover joints.
2. Remove the spark plugs.
3. Remove the camshaft timing oil control valve assembly (intake and exhaust sides).
4. Remove the camshaft position sensors (intake and exhaust sides).
5. Remove the oil filler cap sub assembly.
6. Remove the crankshaft position sensor.
7. Remove the ventilation case sub assembly.
8. Remove the separator case.
9. Remove the No. 1 water by pass pipe.
10. Remove the inlet water.
11. Remove the thermostat.
12. Remove the v-ribbed belt tensioner assembly (compressor and generator).
13. Remove the water pump assembly.
14. Remove the water pump assembly.
15. Remove the water drain cock.
16. Remove the inlet water housing;
17. Remove the crankshaft pulley.
18. Remove the cylinder head cover sub assembly.
   a. Remove the 16 bolts, 3 seal washers, cylinder head cover and gasket.

To install:

19. Install the cylinder head cover sub assembly.
   a. Apply a light coat of engine oil to the 3 new gaskets.
   b. Install the 3 gaskets to the camshaft bearing caps.
   c. Install a new gasket to the cylinder head cover.

➡Remove any oil from the contact surface.

   d. Apply seal packing as shown.

➡Remove any oil from the contact surface.

➡Install the cylinder head cover within 3 minutes and tighten the bolts within 15 minutes after applying seal packing.

   e. Align the cylinder head cover with pin A. Then align the cylinder head cover

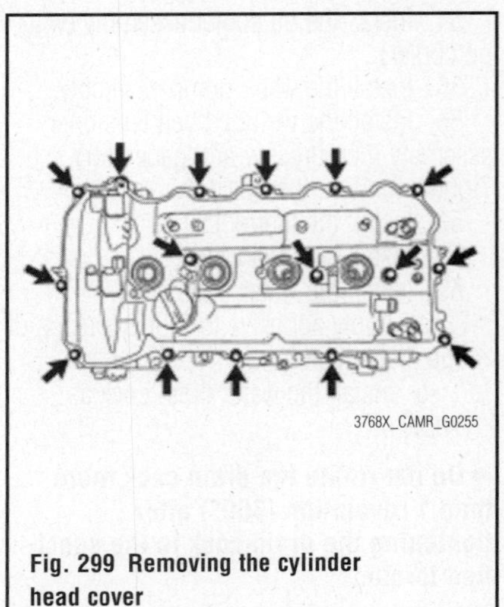
Fig. 299 Removing the cylinder head cover

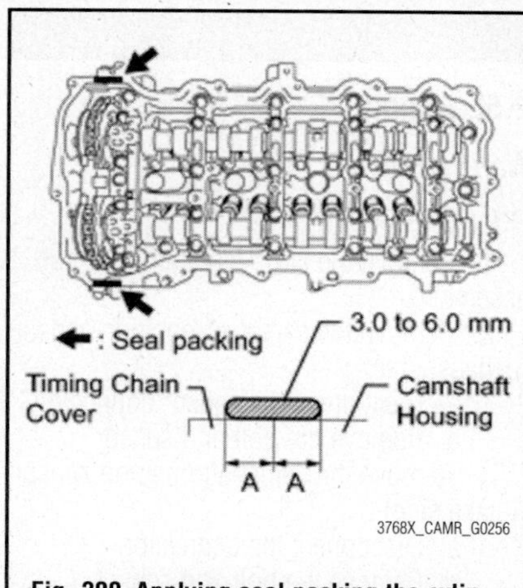

Fig. 300 Applying seal packing the cylinder head cover

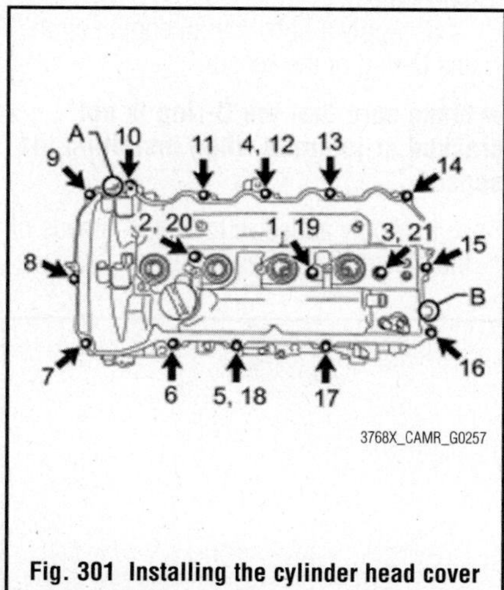

Fig. 301 Installing the cylinder head cover

with pin B and install the cylinder head cover.

f. Install 3 new seal washers and the 16 bolts, and then tighten the bolts in the order shown in the illustration. Tighten to 9 ft. lbs. (12 Nm).

➡Do not apply oil for at least 4 hours after installation.

20. Install the crankshaft pulley. Tighten to 192 ft. lbs. (260 Nm).
21. Install the inlet water housing. Tighten to 32 ft. lbs. (43 Nm).
22. Install the oil cooler assembly (with oil cooler).
23. Install the water pump assembly.
24. Install the v-ribbed belt tensioner assembly (compressor and generator). Tighten to 15 ft. lbs. (21 Nm).
25. Install the thermostat.
26. Install the inlet water.
27. Install the water drain cock. Tighten to 15 ft. lbs. (20 Nm).

➡Do not rotate the drain cock more than 1 revolution (360°) after tightening the drain cock to the specified torque.

➡Do not loosen the drain cock to adjust it. If an adjustment is necessary, remove the drain cock and reinstall it.

a. Install the water drain cock plug to the water drain cock. Tighten to 9 ft. lbs. (13 Nm).

28. Install the No. 1 water by pass pipe. Tighten to 7 ft. lbs. (10 Nm).
29. Install the separator case. Tighten to 7 ft. lbs. (10 Nm).
30. Install the ventilation case sub assembly.

➡Remove any oil from the contact surface.

➡Install the ventilation case within 3 minutes and tighten the bolts and nuts within 15 minutes after applying seal packing.

a. Install the ventilation case, and install the 8 bolts and 2 nuts in the order shown. Tighten to 15 ft. lbs. (21 Nm).

➡Bolt A is tightened twice.

31. Install the ventilation valve sub assembly.
32. Install the camshaft position sensor (intake and exhaust sides).
33. Install the camshaft timing oil control valve assembly (intake and exhaust sides).
34. Install the oil filler cap sub assembly.
35. Install the spark plug.
36. Install the engine cover joints.

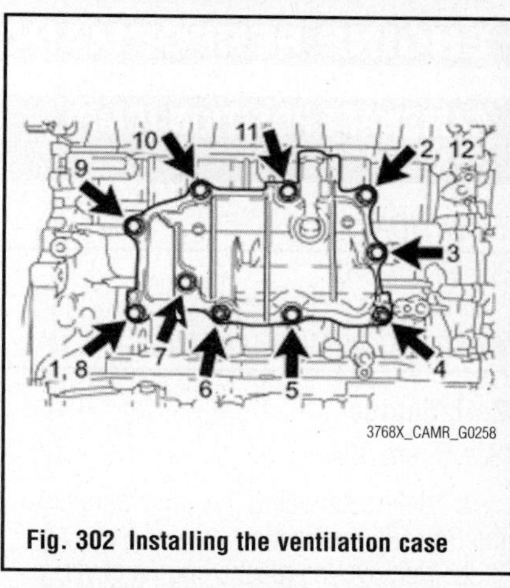

Fig. 302 Installing the ventilation case

VALVE LASH

ADJUSTMENT

➡Keep the lash adjuster free of dirt and foreign objects.

➡Only use clean engine oil.

1. Place the lash adjuster into a container filled with engine oil.
2. Insert the SST's (SST: 09276-75010) tip into the lash adjuster's plunger and use the tip to press down on the check ball inside the plunger.
3. Squeeze the SST and lash adjuster together to move the plunger up and down 5 to 6 times.
4. Check the movement of the plunger and bleed the air. OK: Plunger moves up and down.

➡When bleeding air from the high-pressure chamber, make sure that the tip of the SST is actually pressing the check ball as shown in the illustration. If the check ball is not pressed, air will not bleed.

5. After bleeding the air, remove the SST. Then, try to press the plunger quickly and firmly with a finger. OK: Plunger is very difficult to move. If the result is not as specified, replace the lash adjuster.
6. Install the lash adjusters.

➡Install the lash adjuster to the same place where it was removed from.

ENGINE PERFORMANCE & EMISSION CONTROLS

CAMSHAFT POSITION (CMP) SENSOR

LOCATION

See Figure 303.

REMOVAL & INSTALLATION

2.4L Engine

See Figure 304.

1. Before servicing the vehicle, refer to the Precautions Section.
2. Remove the plastic engine cover.
3. Remove air cleaner cap sub-assembly.
4. Disconnect the camshaft position sensor connector.
5. Remove the bolt and camshaft position sensor.

To install:

➡Make sure that the O-ring is not cracked or jammed when installing it.

6. Apply a light coat of engine oil to the O-ring of the sensor.

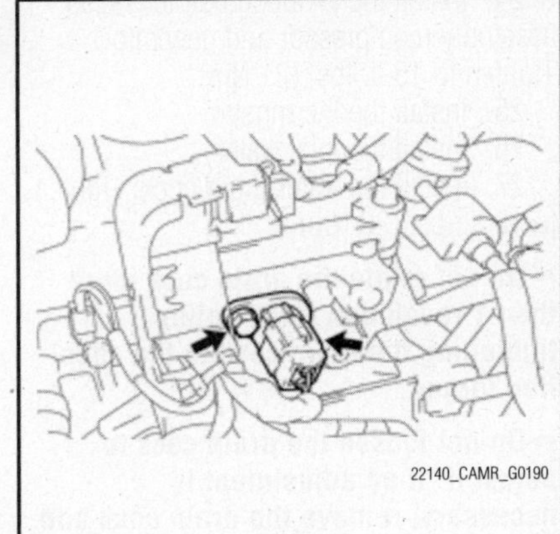

Fig. 304 Remove the bolt and camshaft position sensor—2.4L Engine

7. Install the camshaft position sensor with the bolt and tighten to: 80 inch lbs. (9 Nm).
8. To complete installation, reverse removal procedure.

2.5L Engine

Non-Hybrid

See Figures 305 and 306.

1. Remove the No. 1 engine cover sub assembly.
2. Remove the camshaft position sensor (exhaust side).
   a. Disconnect the sensor connector.
   b. Remove the bolt and sensor.
3. Remove the camshaft position sensor (intake side).
   a. Disconnect the connector.
   b. Remove the bolt and sensor.

To install:

4. Install the camshaft position sensor (exhaust side).
   a. Apply a light coat of engine oil to the O-ring of the sensor.

➡Make sure that the O-ring is not cracked or jammed when installing the sensor.

   b. Apply adhesive to 2 or 3 threads of the bolt.

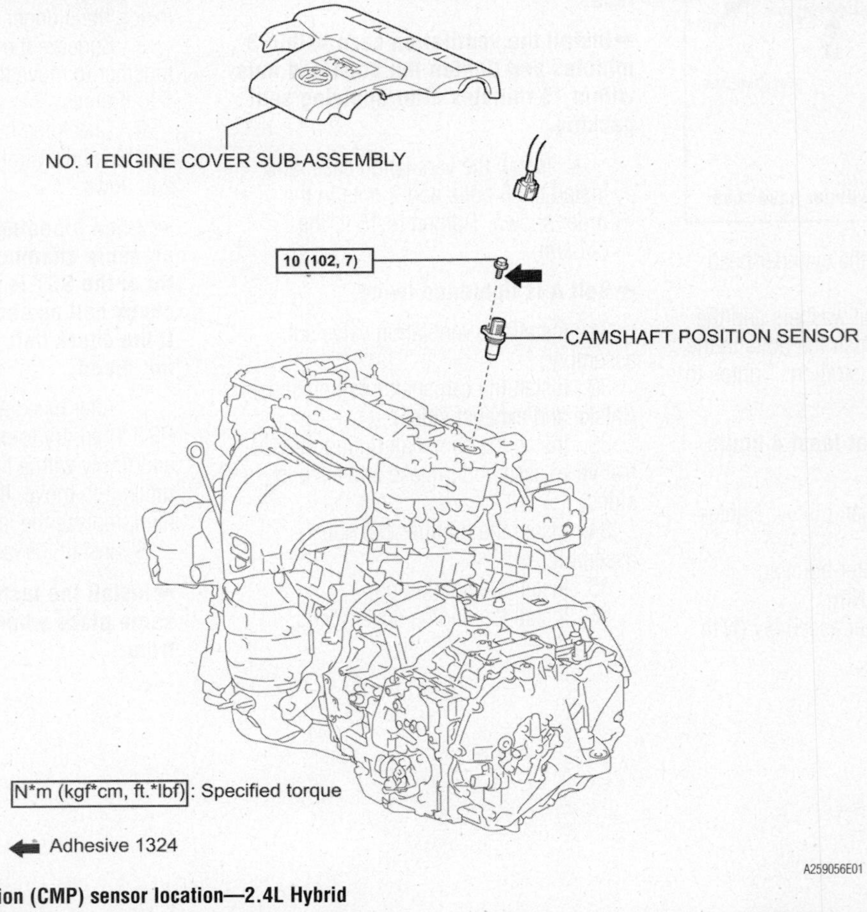

NO. 1 ENGINE COVER SUB-ASSEMBLY
10 (102, 7)
CAMSHAFT POSITION SENSOR
N*m (kgf*cm, ft.*lbf): Specified torque
Adhesive 1324
A259056E01

Fig. 303 Camshaft Position (CMP) sensor location—2.4L Hybrid

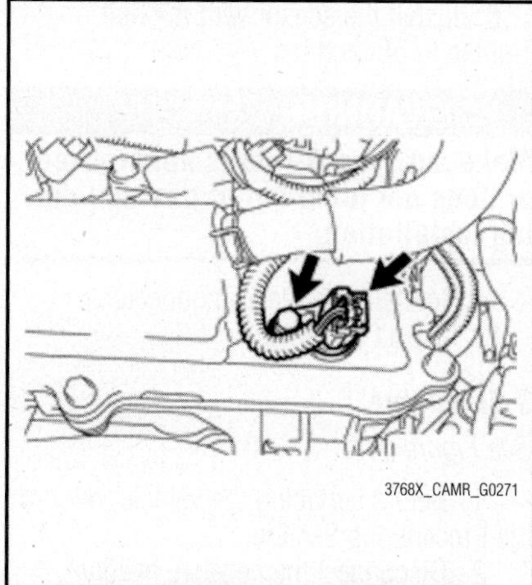

Fig. 305 Removing the camshaft position sensor (exhaust side)

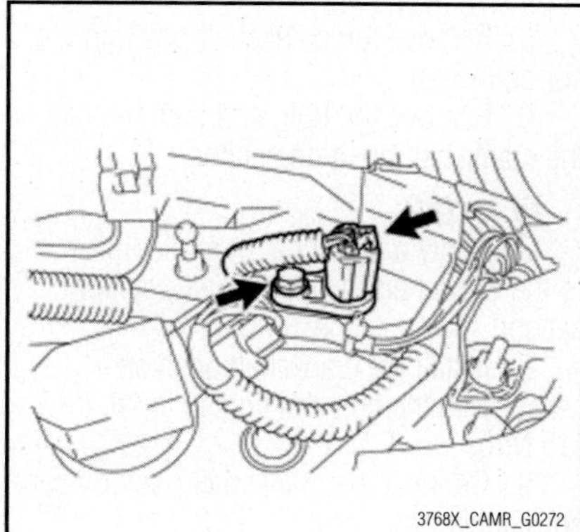

Fig. 306 Removing the camshaft position sensor (intake side)

c. Install the sensor with the bolt. Tighten to 7 ft. lbs. (10 Nm).

d. Connect the sensor connector.

5. Repeat the above steps for the camshaft position sensor (intake side).

Hybrid

See Figure 307.

1. Remove No. 1 engine cover sub assembly.
2. Remove camshaft position sensor.
   a. Disconnect the sensor connector.
   b. Remove the bolt and sensor.

To install:

3. Install camshaft position sensor.
   a. Clean and degrease the threads of the bolt for the camshaft position sensor.
   b. Apply a light coat of engine oil to the O-ring of the sensor.

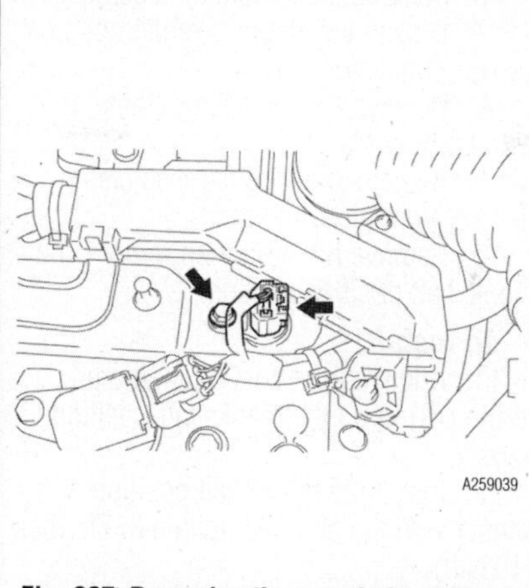

Fig. 307 Removing the camshaft position sensor

➡Make sure that the O-ring is not cracked or does not jump out of position during installation.

   c. Apply adhesive to 2 or 3 threads of the bolt.
4. Install the sensor with the bolt. Tighten to 7 ft. lbs. (10 Nm).
   a. Connect the sensor connector.
5. Inspect for oil leak.
6. Install No. 1 engine cover sub assembly.

3.5L Engine

See Figures 308 through 311.

1. Before servicing the vehicle, refer to the Precautions Section.
2. Drain and recycle the engine coolant.
3. Disconnect the negative battery cable.
4. Remove the V-bank cover sub-assembly.
5. Remove the windshield wiper link assembly.
6. Remove the front cowl top outside panel.
7. Remove the Intake camshaft VVT sensor (Bank 1), as follows:
   a. Disconnect the VVT sensor connector.
   b. Remove the bolt and VVT sensor.
8. Remove the Exhaust camshaft VVT sensor (Bank 1), as follows:
   a. Disconnect the VVT sensor connector.
   b. Remove the bolt and VVT sensor.
9. Remove the Exhaust camshaft VVT sensor (Bank 2), as follows:
   a. Disconnect the VVT sensor connector.
   b. Remove the bolt and VVT sensor.
10. Remove the Intake camshaft VVT sensor (Bank 2), as follows:

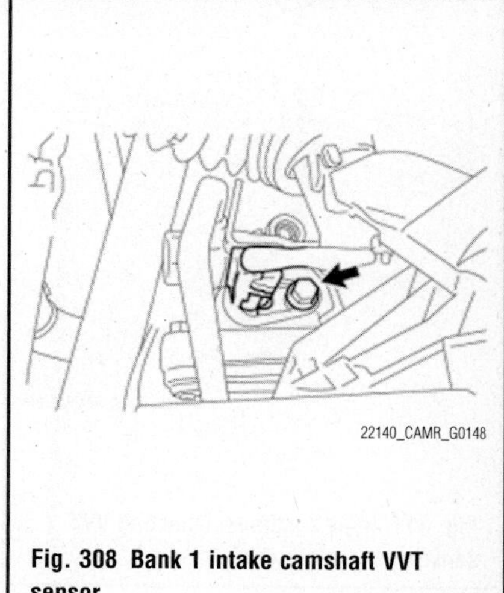

Fig. 308 Bank 1 intake camshaft VVT sensor

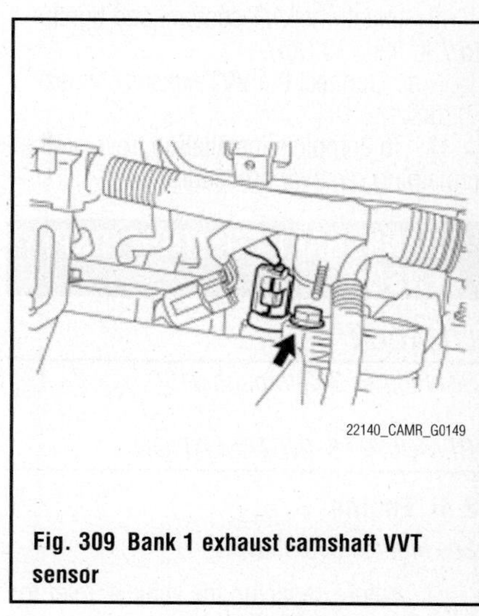

Fig. 309 Bank 1 exhaust camshaft VVT sensor

   a. Disconnect the VVT sensor connector.
   b. Remove the bolt and VVT sensor.

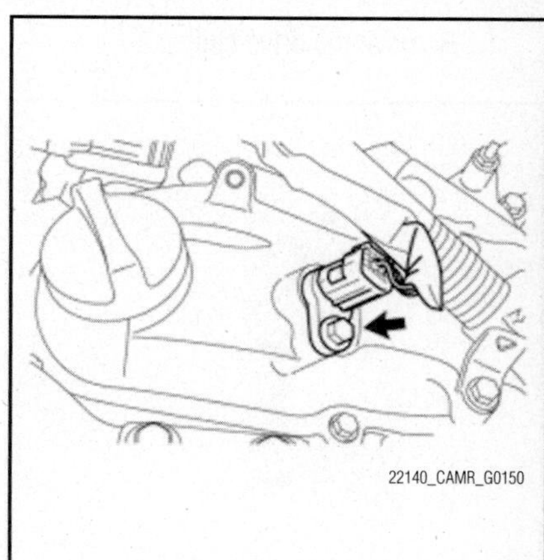

Fig. 310 Bank 2 intake camshaft VVT sensor

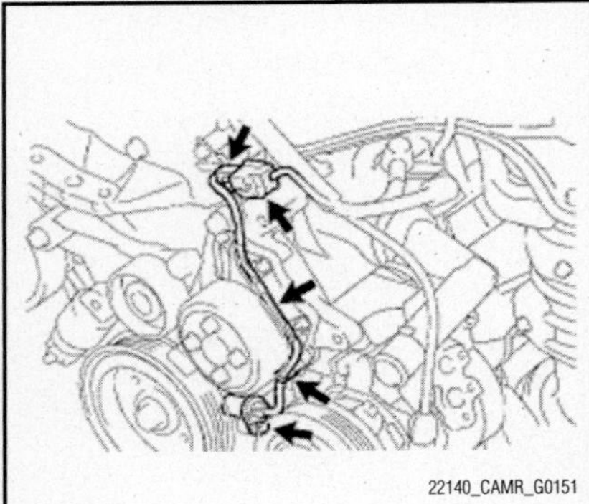

Fig. 311 Bank 2 exhaust camshaft VVT sensor

To install:

11. Install the VVT sensors and tighten to 7 ft. lbs. (10 Nm).
    a. Connect the VVT sensor connectors.
12. To complete installation, reverse the remaining removal procedure.

CRANKSHAFT POSITION (CKP) SENSOR

LOCATION

See Figures 312 through 315.

REMOVAL & INSTALLATION

2.4L Engine

See Figures 316 and 317.

1. Before servicing the vehicle, refer to the Precautions Section.
2. Disconnect the negative battery cable.
3. Remove the front right wheel.
4. Remove the right hand front fender apron.
5. Remove the drive belt.

Fig. 312 Crankshaft position sensor location—2.4L engine

6. Remove the generator assembly.
7. Disconnect the crankshaft position sensor connector.
8. Remove the connector clamp and wire harness clamp.
9. Remove the wire harness clamp bracket from the wire harness.
10. Remove the bolt, then remove the crankshaft position sensor.

To install:

11. Apply a light coat of engine oil to the O-ring on the crankshaft position sensor.
12. Install the crankshaft position sensor with the bolt and tighten to 7ft. lbs. (10 Nm).
13. Connect the crankshaft position sensor connector.
14. The remainder of installation is the reverse of the removal procedure.

2.5L Engine

Non-Hybrid

See Figure 318.

1. Remove the front apron seal RH.
2. Remove the crankshaft position sensor.
    a. Disconnect the sensor connector.
    b. Remove the bolt and sensor.

To install:

3. Install the crankshaft position sensor.
    a. Apply a light coat of engine oil to the O-ring of the sensor.

➡Make sure that the O-ring is not cracked or jammed when installing the sensor.

    b. Apply adhesive to 2 or 3 threads of the bolt.
    c. Install the sensor with the bolt and tighten to 7 ft. lbs. (10 Nm).
    d. Connect the sensor connector.
4. Inspect for oil leaks.
5. Install the front fender apron seal RH.

Hybrid

1. Disconnect the sensor connector.
2. Remove the bolt and sensor.

To install:

3. Clean and degrease the threads of the bolt for the crankshaft position sensor.
4. Apply a light coat of engine oil to the O-ring of the sensor.
5. Apply adhesive to 2 or 3 threads of the bolt.

➡Use Toyota Genuine Adhesive 1324, Three Bond 1324 or equivalent.

6. Install the sensor with the bolt. Tighten to 58 inch lbs. (6.5 Nm).

⁂ WARNING

Make sure the O-ring is not cracked or does not jump out of position during installation.

7. Connect the sensor connector.
8. Inspect for leaks.

3.5L Engine

See Figure 319.

1. Before servicing the vehicle, refer to the Precautions Section.
2. Disconnect the negative battery cable.
3. Remove alternator assembly.
4. Disconnect the cooler compressor assembly.
5. Remove the crankshaft position sensor connector.
6. Remove the bolt, and then remove the crankshaft position sensor.

To install:

7. Apply a light coat of engine oil to the O-ring on the crankshaft position sensor.
8. Install the crankshaft position sensor with the bolt and tighten to 7ft. lbs. (10 Nm).
9. Connect the crankshaft position sensor connector.
10. The remainder of installation is the reverse of the removal procedure.

ELECTRONIC CONTROL MODULE (ECM)

LOCATION

See Figure 320.

REMOVAL & INSTALLATION

2.4L Engine

See Figure 321.

1. Before servicing the vehicle, refer to the Precautions Section.
2. Disconnect the negative battery cable.
3. Remove the plastic engine cover.
4. Remove the air cleaner inlet assembly.
5. Remove the air cleaner cap sub-assembly.
6. Remove the air cleaner case sub-assembly.
7. Remove the two bolts and air cleaner bracket.

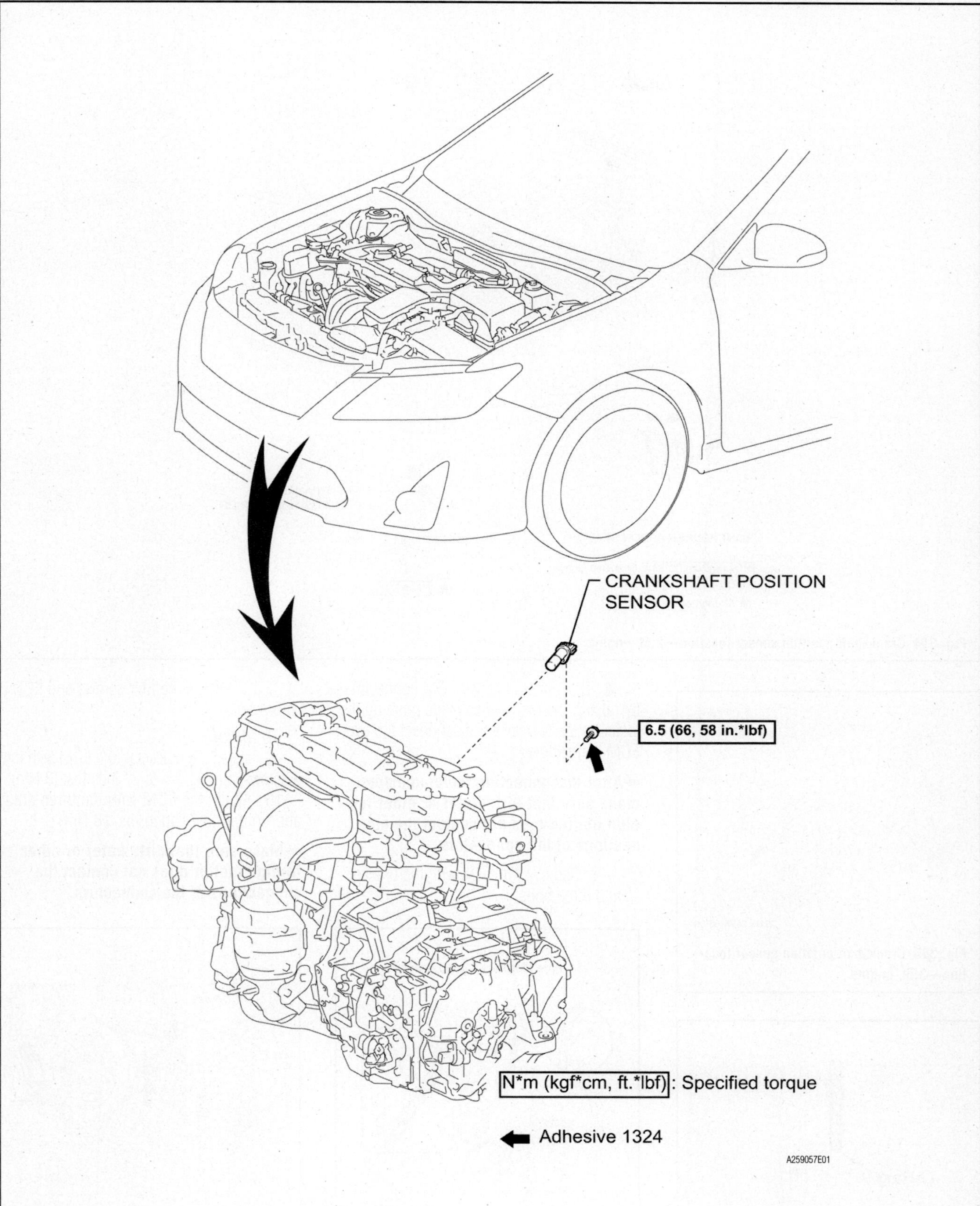

Fig. 313 Crankshaft position sensor location–2.5L engine Hybrid (2012)

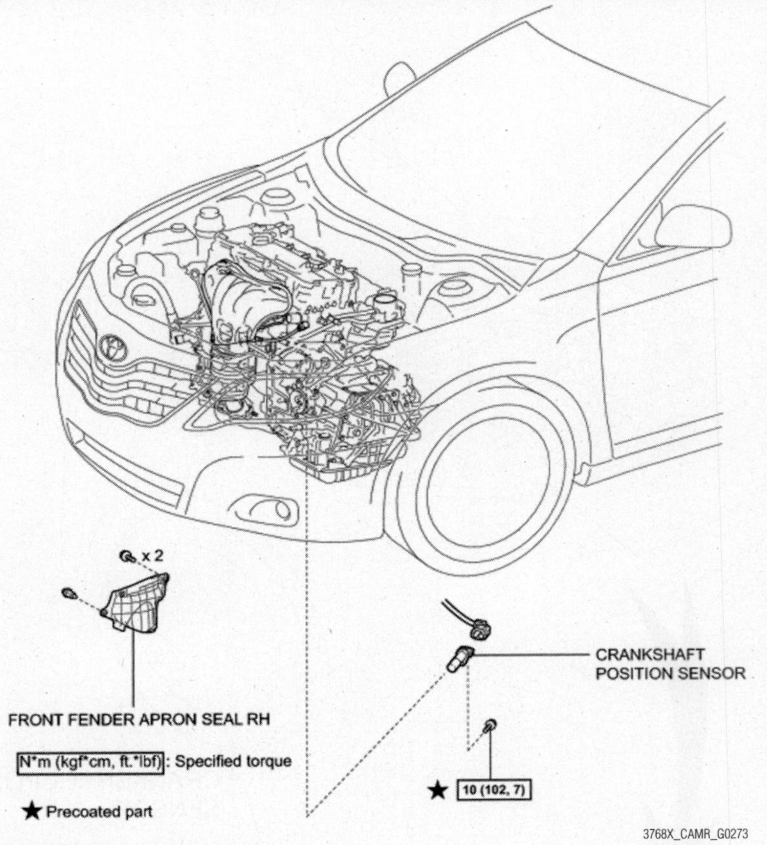

x 2

FRONT FENDER APRON SEAL RH

N*m (kgf*cm, ft.*lbf): Specified torque

★ Precoated part

CRANKSHAFT POSITION SENSOR

★ 10 (102, 7)

3768X_CAMR_G0273

Fig. 314 Crankshaft position sensor location—2.5L engine

Fig. 315 Crankshaft position sensor location—3.5L engine

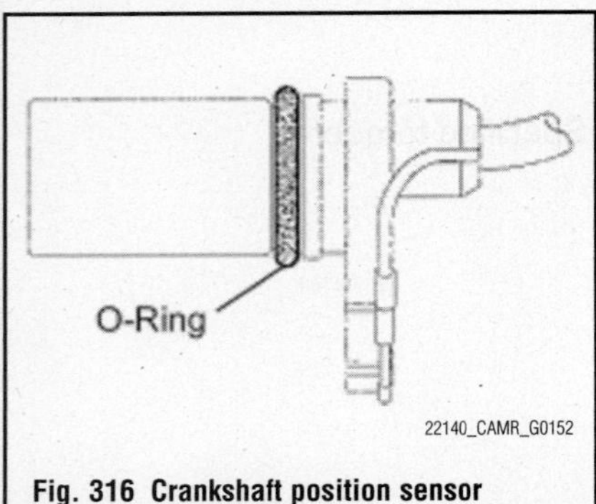

Fig. 316 Crankshaft position sensor

8. Disconnect the two ECM connectors by raising the two levers. While pushing the locks on the two levers, disconnect the two ECM connectors.

➡After disconnecting the connectors, make sure that dirt, water or other foreign matter does not contact the connections of the connectors.

a. Remove the ECM with the bracket and three bolts.

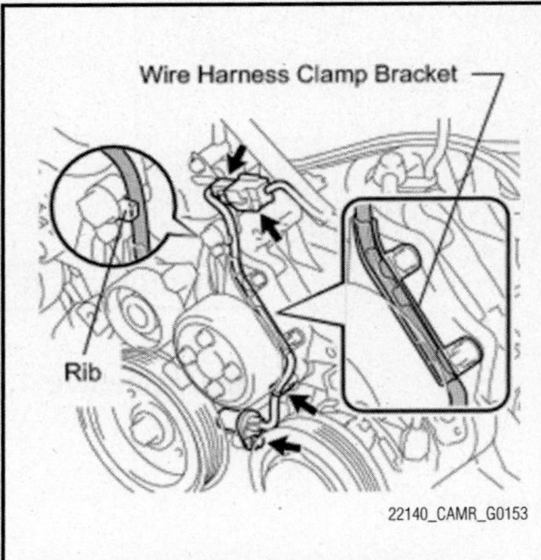

Fig. 317 Installing crankshaft position sensor—2.4L Engine

b. Remove the four screws and ECM brackets.

To install:

9. Install the bracket to the ECM with the 4 screws, and tighten to 27 inch lbs. (3 Nm).

10. Attach the ECM with the three nuts and tighten to 71 inch lbs. (8 Nm).

➡Make sure that dirt, water or other foreign matter does not contact the connections of the connectors.

Fig. 318 Removing and installing the crankshaft position sensor—2.5L engine

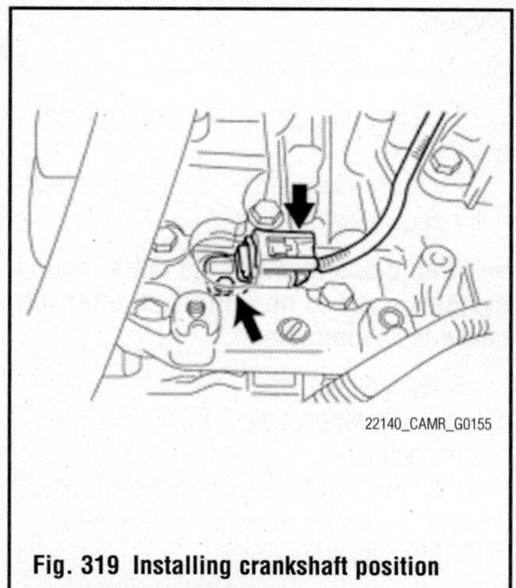

Fig. 319 Installing crankshaft position sensor—3.5L Engine

11. Connect the two ECM connectors and lower the two levers.
12. Install the ECM to the body.
13. Install the two bolts and air cleaner bracket.
14. Install the air cleaner case sub-assembly.
15. Install the air cleaner cap sub-assembly.
16. Install the air cleaner inlet assembly.
17. Remove the plastic engine cover.
18. Connect the negative battery cable.
19. Register the immobilizer communication ID. If the ECM is replaced, register the ECM communication ID for the immobilizer system (refer to the Service Bulletin for registration).
20. Perform initialization. After replacing the ECM on vehicles with a dynamic laser cruise control system, it is necessary to initialize the ECM so that the ECM can recognize the dynamic laser cruise control system.
21. Be sure to perform the following procedure after replacing the ECM:
    a. Turn the ignition switch on (IG).
    b. Turn the cruise control main switch on.

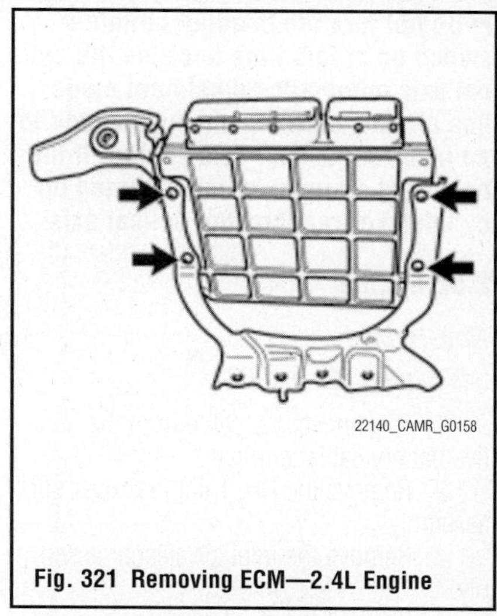

Fig. 321 Removing ECM—2.4L Engine

    c. With the brake pedal depressed, push the cruise control main switch to RES/ACC 3 times within 3 seconds. Check that the buzzer sounds at this time.

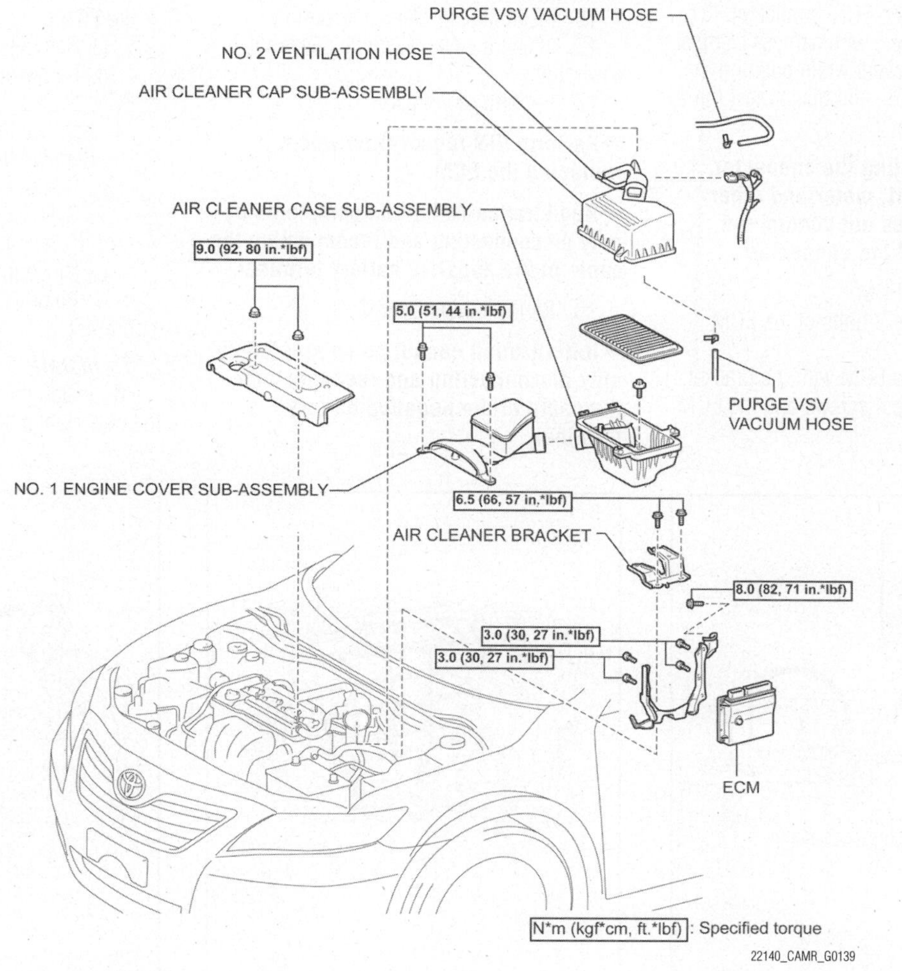

Fig. 320 ECM location

➡Do not turn the headlight dimmer switch on at this time because the optical axis automatic adjustment mode has already started, which may lead to an incorrect optical axis setting. If the headlight dimmer switch is turned on by mistake, readjust the optical axis.

2.5L Engine

Non-Hybrid

See Figures 322 and 323.

1. Disconnect the cable from the negative battery cable terminal.
2. Remove the No. 1 engine cover sub assembly.
3. Remove the inlet air cleaner assembly.
4. Remove the air cleaner cap sub assembly.
5. Remove the air cleaner filter element.
6. Remove the air cleaner case sub assembly.
7. Remove the air cleaner bracket.
   a. Remove the 2 bolts and the air cleaner bracket.
8. Disconnect the ECM connector.
   a. Separate the 2 wire harness clamps.
   b. Raise the levers while pushing the locks on the levers, and disconnect the 2 ECM connectors.

➡After disconnecting the connector, make sure that dirt, water and other foreign matter does not contact the connecting part of the connector.

9. Remove the ECM.
   a. Remove the 3 bolts of the ECM bracket.
   b. Remove the ECM with the bracket.
   c. Remove the 4 screws and the ECM bracket.

To install:

10. Install the ECM.
    a. Install the bracket to the ECM with the 4 screws.
    b. Install the ECM with the 3 bolts. Tighten to 71 inch lbs. (8 Nm).
11. Connect the connector.
    a. Connect the 2 ECM connectors and lower the 2 levers.

➡When connecting the connectors, make sure that dirt, water and other foreign matter does not become stuck between the connectors and other parts.

➡Make sure that the 2 levers are securely lowered.

    b. Install the 2 wire harness clamps.
12. Install the air cleaner bracket with the 2 bolts. Tighten to 69 inch lbs. (7.8 Nm). Instll the air cleaner case sub assembly.
13. Install the air cleaner filter element.
14. Install the air cleaner cap sub assembly.
15. Install the air cleaner assembly,
16. Install the No. 1 engine cover sub assembly.
17. Perform registration.

➡Perform VIN registration when replacing the ECM.

➡Registration cannot be completed by only disconnecting and reconnecting the cable of the negative battery terminal.

18. Perform initialization.

➡Initialization cannot be completed by only disconnecting and reconnecting the cable of the negative battery terminal.

Hybrid

See Figures 324 through 326.

1. Remove luggage trim service hole cover.
2. Disconnect cable from negative auxiliary battery terminal.

➡When disconnecting the cable, some systems need to be initialized after the cable is reconnected.

3. Remove ECM.
   a. Disconnect the 2 ECM connectors.
   b. Push in the locks on the 2 levers, raise the levers, and disconnect the 2 ECM connectors.

⁂ WARNING

After disconnecting the connectors, make sure that dirt, water or other foreign matter does not contact the connecting part of the connectors.

   c. Remove the nut.
   d. Remove the screw (A).
   e. Loosen the screw (B) and remove the ECM.
   f. Remove the screw (B).
   g. Remove the 2 screws and No. 2 ECM bracket
4. Remove the No. 1 relay block cover.
5. Remove the No. 1 ECM bracket.
   a. Remove the nut.
   b. Release the 2 clamps and disconnect the wire harness.
   c. Disconnect the wire harness clamp.
   d. Remove the nut and No. 1 ECM bracket.

To install:

6. Install the No. 1 ECM bracket with the nut. Tighten to 9 ft. lbs. (12 Nm).

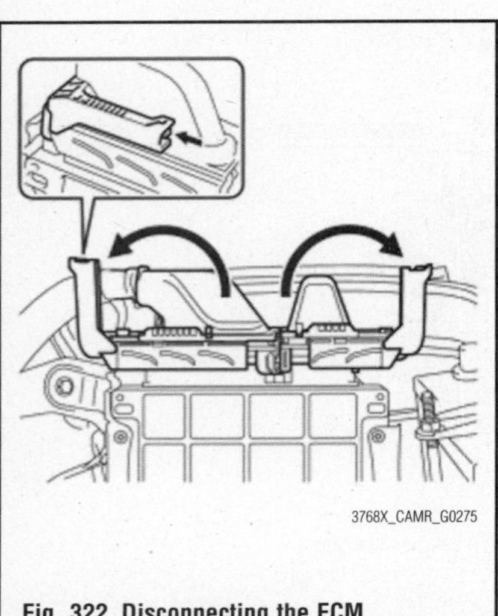

Fig. 322 Disconnecting the ECM

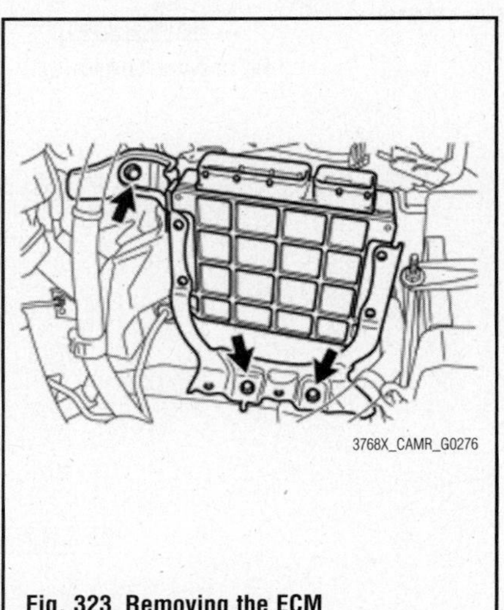

Fig. 323 Removing the ECM

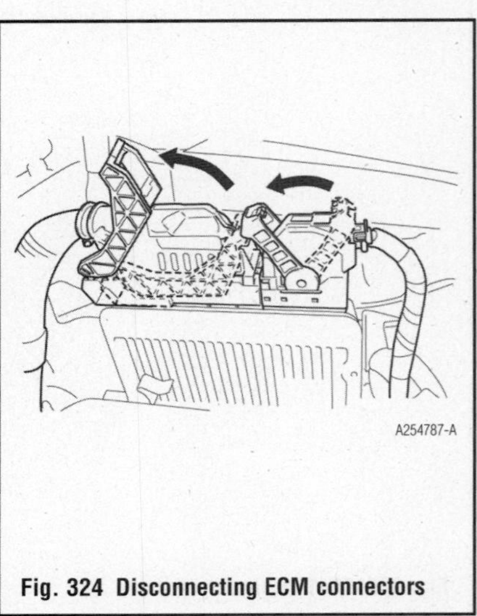

Fig. 324 Disconnecting ECM connectors

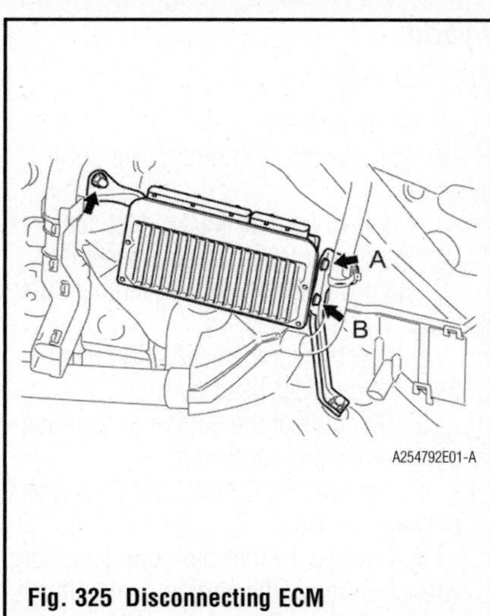

Fig. 325 Disconnecting ECM

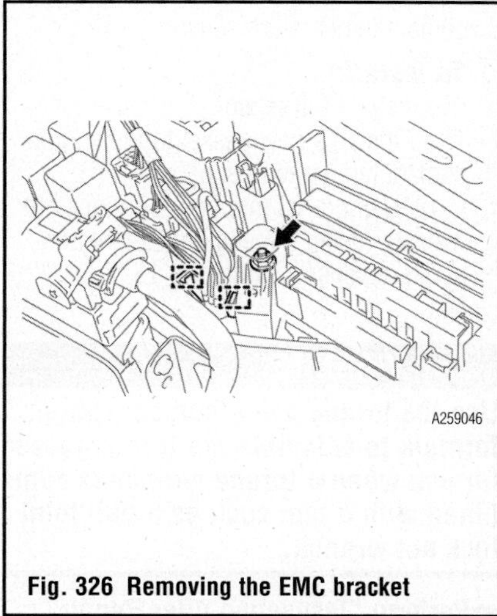

Fig. 326 Removing the EMC bracket

a. Connect the wire harness clamp.

b. Install the wire harness and 2 clamps.

c. Install the nut.

7. Install the No. 1 relay block cover.

8. Install the ECM.

a. Install the No. 2 ECM brackets to the ECM with the 2 screws. Tighten to 27 inch lbs. (3 Nm).

b. Temporarily install the screw (B).

c. Install the ECM with the nut. Tighten to 9 ft. lbs. (12 Nm).

d. Install screw (A). Tighten to 27 inch lbs. (3 Nm).

e. Install screw (B). Tighten screw (B) to 27 inch lbs. (3 Nm).

f. Connect the 2 ECM connectors and lower the 2 levers.

WARNING

When connecting the connectors, make sure that dirt, water or other foreign matter does not become stuck between the connectors and other part.

WARNING

Make sure that the 2 levers are securely locked.

g. Connect the 2 wire harness clamps.

9. Connect cable to negative auxiliary battery terminal.

WARNING

When disconnecting the cable, some systems need to be initialized after the cable is reconnected.

10. Install luggage trim service hole cover.

11. Perform initialization.

3.5L Engine

See Figure 327.

1. Before servicing the vehicle, refer to the Precautions Section.

2. Disconnect the negative battery cable.

3. Remove both windshield wiper arm and blade assemblies.

4. Remove the cowl top ventilator louver sub-assembly.

5. Remove the windshield wiper motor and link assembly.

6. Remove the outer cowl top panel.

7. Remove the ECM, as follows:

a. Remove the 3 nuts.

b. Separate the ECM from the body. When separating the ECM, do not apply excessive force to the wire harness.

c. Raise the 2 levers while pushing the locks on the 2 levers, and disconnect the 2 ECM connectors.

➡After disconnecting the connectors, make sure that dirt, water or other foreign matter does not contact the connections of the connectors.

d. Remove the ECM.

e. Remove the 4 screws and 2 ECM brackets.

To install:

8. Install the 2 ECM brackets with the 4 screws, and tighten to 27 inch lbs. (3 Nm).

9. Connect the 2 ECM connectors and lower the 2 levers.

➡Make sure that dirt, water or other foreign matter does not contact the connections of the connectors.

10. Install the ECM to the body.

11. Attach the ECM with the 3 nuts and tighten to 71 inch lbs. (8 Nm).

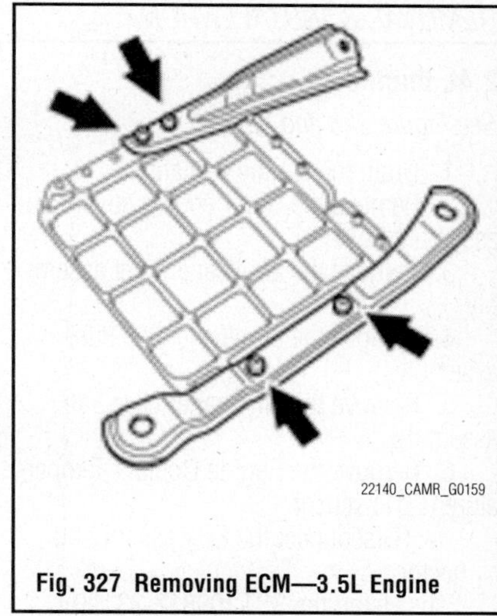

Fig. 327 Removing ECM—3.5L Engine

12. Install the outer cowl top panel.

13. Install the windshield wiper motor and link assembly.

14. Install the cowl top ventilator louver sub-assembly.

15. Install both windshield wiper arm and blade assemblies.

16. Connect the negative battery cable.

17. Register the immobilizer communication ID. If the ECM is replaced, register the ECM communication ID for the immobilizer system (refer to the Service Bulletin for registration).

18. Perform initialization. After replacing the ECM on vehicles with a dynamic laser cruise control system, it is necessary to initialize the ECM so that the ECM can recognize the dynamic laser cruise control system.

19. Be sure to perform the following procedure after replacing the ECM:

a. Turn the ignition switch on (IG).

b. Turn the cruise control main switch on.

c. With the brake pedal depressed, push the cruise control main switch to RES/ACC 3 times within 3 seconds. Check that the buzzer sounds at this time.

➡Do not turn the headlight dimmer switch on at this time because the optical axis automatic adjustment mode has already started, which may lead to an incorrect optical axis setting. If the headlight dimmer switch is turned on by mistake, readjust the optical axis.

ENGINE COOLANT TEMPERATURE (ECT) SENSOR

LOCATION

The ECT sensor is connected to the air cleaner case sub assembly.

REMOVAL & INSTALLATION

2.4L Engine

See Figure 328 and 329.

1. Drain the engine coolant.
2. Remove the No. 1 engine cover sub assembly.
3. Remove the air cleaner inlet assembly.
4. Remove the air cleaner cap sub assembly.
5. Remove the air cleaner case sub assembly.
6. Remove the Engine Coolant Temperature (ECT) sensor.
    a. Disconnect the ECT sensor connector.
    b. Using the SST (09817-33190), remove the ECT sensor and gasket.

To install:

7. Install the ECT sensor.
    a. Install a new gasket onto the ECT.
    b. Using the SST (09817-33190), install the ECT. Tighten to 15 ft. lbs. (20 Nm).
8. Install the air cleaner case sub assembly.
9. Install the air cleaner cap sub assembly.
10. Install the air cleaner inlet assembly.
11. Add engine coolant.
12. Inspect for coolant leaks.
13. Install the No. 1 engine cover sub assembly.

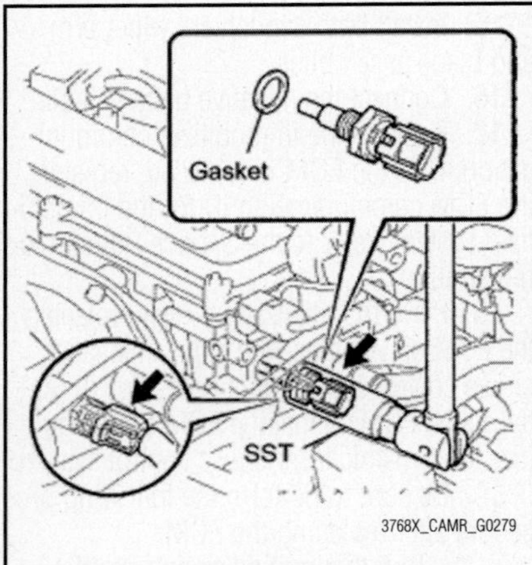

Fig. 328 Removing the ECT sensor—2.4L non-hybrid engine

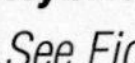

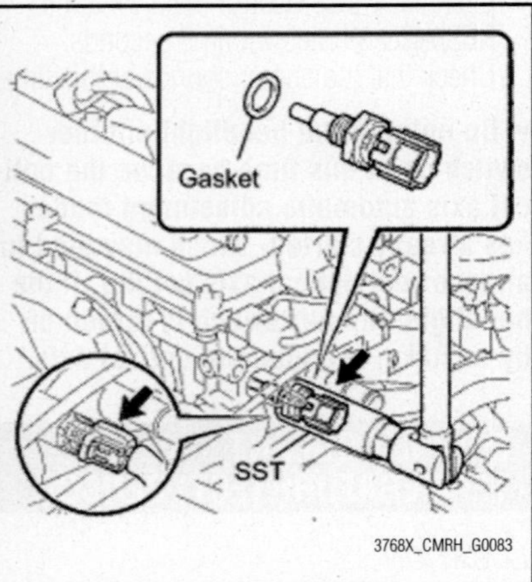

Fig. 329 Removing the ECT sensor—2.4L hybrid engine

2.5L Engine

Non-Hybrid

See Figure 330.

1. Remove the front wheel opening extension pads.
2. Remove the engine under covers.
3. Drain the engine coolant.
4. Remove the No. 1engine cover sub assembly.
5. Remove the inlet air cleaner assembly.
6. Remove the air cleaner cap sub assembly.
7. Remove the air cleaner filter element.
8. Remove the air cleaner case sub assembly.
9. Remove the engine coolant temperature sensor.
    a. Disconnect the engine coolant temperature sensor connector.
    b. Remove the engine coolant temperature sensor and gasket.

To install:

To install, reverse the removal procedure. Tighten the ECT sensor to 15 ft. lbs. (20 Nm).

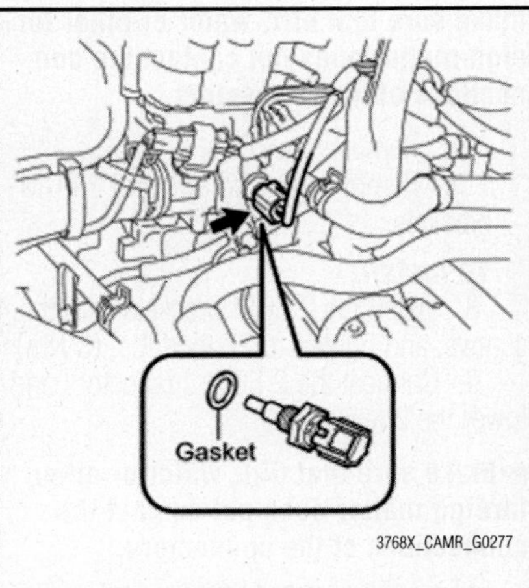

Fig. 330 Removing the Engine Coolant Temperature (ECT) sensor—2.5L engine

Hybrid

See Figure 331.

1. Drain coolant.
2. Remove No. 1 engine cover sub-assembly.
3. Remove cool air intake duct seal.
4. Remove inlet air cleaner assembly.
5. Remove engine coolant temperature sensor.
    a. Disconnect the ventilation hose from the cylinder head cover.
    b. Disconnect the engine coolant temperature sensor connector.
    c. Remove the 2 bolts and disconnect the wire harness.
    d. Using a 19 mm ball joint lock nut wrench, remove the engine coolant temperature sensor.
    e. Remove the gasket from the engine coolant temperature sensor.

To install:

6. Install ECT sensor.
    a. Install a new gasket to the engine coolant temperature sensor.
    b. Using a 19 mm ball joint lock nut wrench, install the engine coolant temperature sensor. Tighten to 15 ft. lbs. (20 Nm).

✱✱ WARNING

Use the torque value compensation formula to calculate the torque value for use when a torque wrench is combined with a tool such as a ball joint lock nut wrench.

➡Perform "Inspection After Repair" after replacing the engine coolant temperature sensor.

    c. Connect the engine coolant temperature sensor connector.

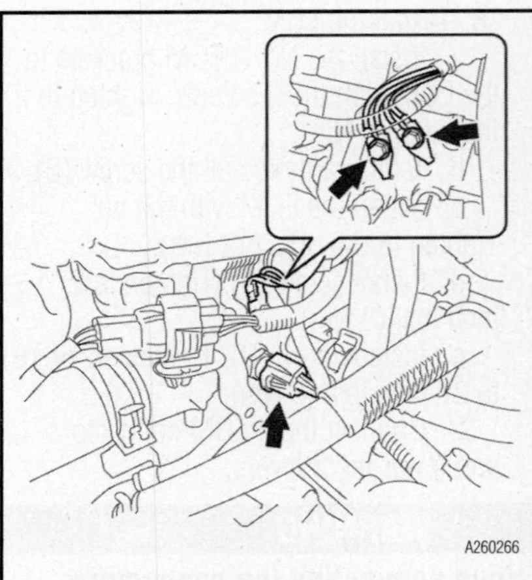

Fig. 331 Disconnecting ECT sensor connector

d. Install the wire harness with the 2 bolts. Tighten to 71 inch lbs. (8 Nm).

e. Connect the ventilation hose to the cylinder head cover.

7. Install inlet air cleaner assembly.
8. Install cool air intake duct seal.
9. Add coolant (for engine).
10. Inspect for coolant leak (for engine).
11. Install No. 1 engine cover sub assembly.

3.5L Engine

See Figure 332.

1. Remove the engine under covers.
2. Drain the engine coolant.
3. Remove the v-bank cover sub assembly.
4. Remove the air cleaner inlet assembly.
5. Remove the air cleaner cap sub assembly.
6. Remove the air cleaner case sub assembly.
7. Remove the No. 1 air cleaner inlet.
8. Remove the Engine Coolant Temperature (ECT) sensor.
   a. Remove the ECT sensor connector.
   b. Remove the engine coolant temperature sensor.

To install:

To install, reverse the removal procedure. Tighten the ECT sensor to 15 ft. lbs. (20 Nm).

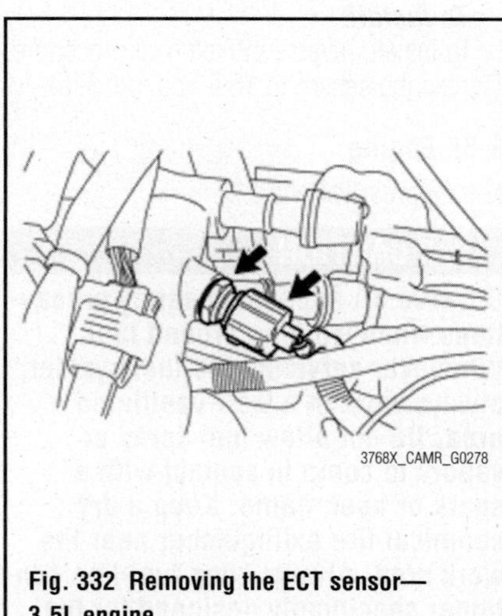

Fig. 332 Removing the ECT sensor—3.5L engine

HEATED OXYGEN (HO2S) SENSOR

REMOVAL & INSTALLATION

2.4L Engine

See Figures 333 and 334.

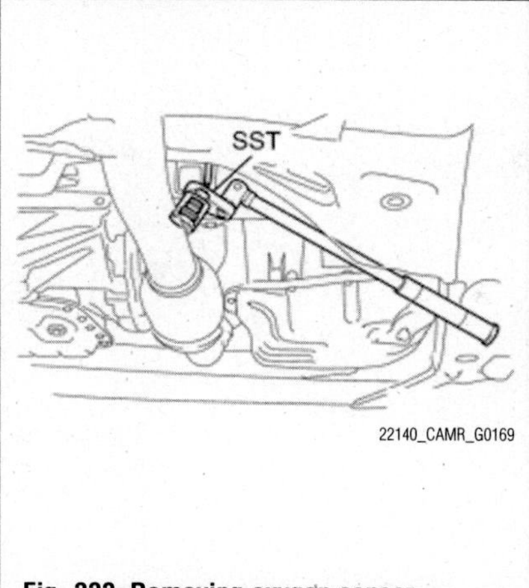

Fig. 333 Removing oxygen sensor—2.4L Engine

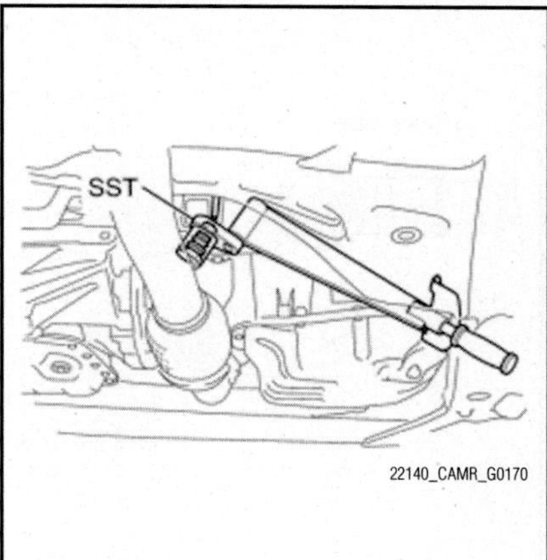

Fig. 334 Installing oxygen sensor—2.4L Engine

1. Before servicing the vehicle, refer to the Precautions Section.
2. Disconnect the oxygen sensor connectors.

➡Do not damage the heated oxygen sensor.

3. Using Special Tool: 09224-00010 or equivalent, remove the two oxygen sensors from the front pipe assembly.

To install:

4. Install the oxygen sensor to the front pipe assembly. Tighten to 32 ft. lbs. (44 Nm). Use a torque wrench with a fulcrum length of 300 mm (11.81 in.).
5. Connect the oxygen sensor connector.

2.5L Engine

See Figure 335.

1. Remove the heated oxygen sensor.
   a. Disconnect the heated oxygen sensor connector.

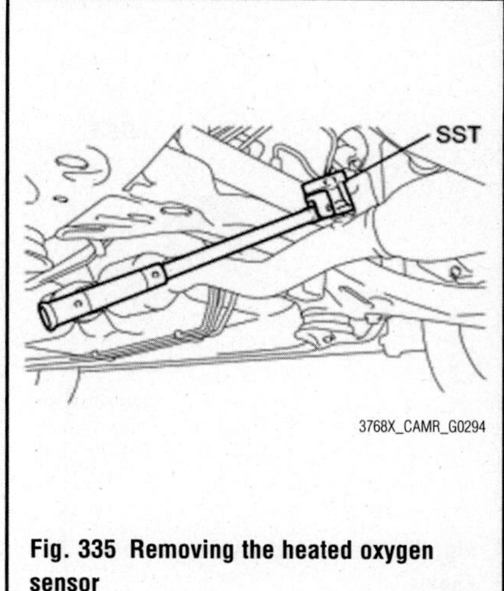

Fig. 335 Removing the heated oxygen sensor

   b. Using the special tool, remove the heated oxygen sensor from the front exhaust pipe.

To install:

To install, reverse the removal procedure. If using the special tool tighten the heated oxygen sensor to 30 ft. lbs. (40 Nm). If tightening without the special tool, tighten to 32 ft. lbs. (44 Nm). Inspect for exhaust gas leaks.

➡The "with special tool" torque value is effective when using SST with a fulcrum length of 1.18 inch (30 mm).

➡The "with special tool" torque value is effective when using a torque wrench with a fulcrum length of 11.81 inch (300 mm).

3.5L Engine

See Figures 336 and 337.

1. Before servicing the vehicle, refer to the Precautions Section.
2. Remove the front exhaust pipe assembly.
3. Disconnect the 2 oxygen sensor connectors.
4. Using Special Tool: 09224-00010 or equivalent, remove the 2 oxygen sensors from the front pipe assembly.

To install:

5. Install the 2 oxygen sensors to the front pipe assembly. Tighten to 32 ft. lbs. (44 Nm) and 30 ft. lbs. (40 Nm). Use a torque wrench with a fulcrum length of 300 mm (11.81 in.).
6. Connect the 2 oxygen sensor connectors.
7. Remove the front exhaust pipe assembly.

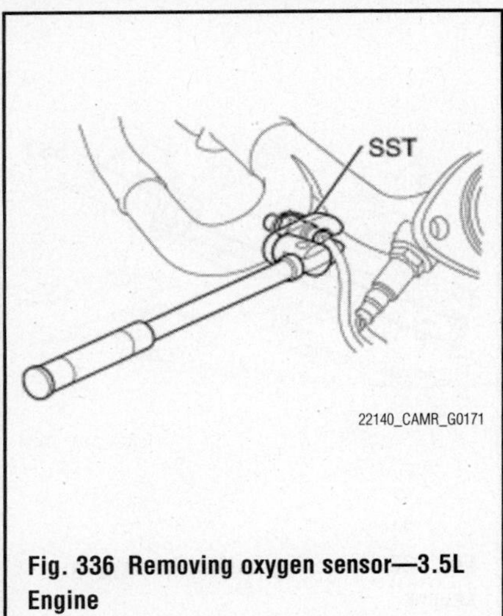

Fig. 336 Removing oxygen sensor—3.5L Engine

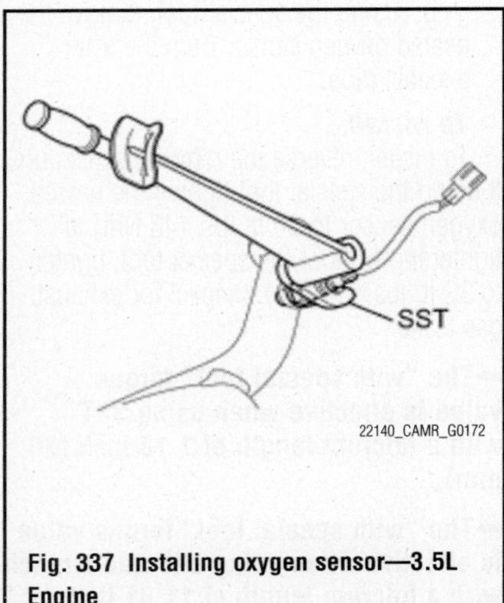

Fig. 337 Installing oxygen sensor—3.5L Engine

KNOCK SENSOR (KS)

REMOVAL & INSTALLATION

2.4L Engine

See Figures 338 and 339.

CAUTION

Observe all applicable safety precautions when working around fuel. Whenever servicing the fuel system, always work in a well-ventilated area. Do not allow fuel spray or vapors to come in contact with a spark or open flame. Keep a dry chemical fire extinguisher near the work area. Always keep fuel in a container specifically designed for fuel storage; also, always properly seal fuel containers to avoid the possibility of fire or explosion.

1. Before servicing the vehicle, refer to the Precautions Section.

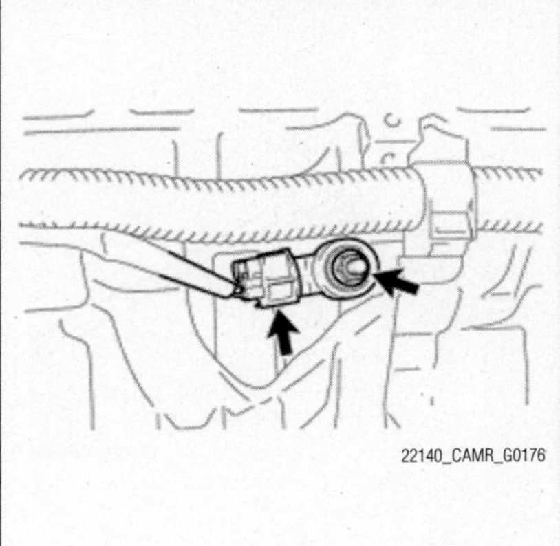

Fig. 338 Removing the nut and knock sensor—2.4L Engine

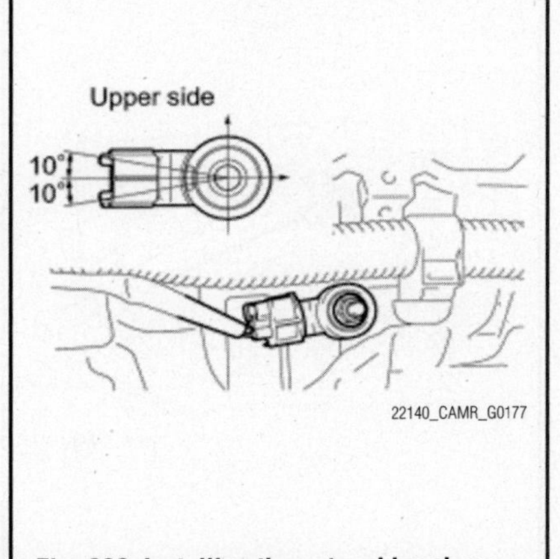

Fig. 339 Installing the nut and knock sensor—2.4L Engine

2. Properly discharge the fuel system pressure.
3. Disconnect battery negative cable.
4. Remove plastic engine cover.
5. Drain and recycle the engine coolant.
6. Remove windshield wiper arms and blade assemblies.
7. Remove both front fender to cowl side seals.
8. Remove cowl top ventilator louver sub-assembly.
9. Remove windshield wiper motor and link.
10. Remove the four bolts, four nuts and cowl top panel outer sub-assembly.
11. Remove air cleaner cap sub-assembly.
12. Remove air cleaner case sub-assembly.
13. Remove throttle body.
14. Disconnect fuel tube.
15. Remove fuel delivery pipe with injector.
16. Disconnect the union to check valve hose from the brake booster.
17. Disconnect the camshaft timing oil control valve connector.
18. Remove the wire harness clamp.
19. Remove the union to check valve hose from the vacuum hose clamp.
20. Remove the 5 bolts, 2 nuts and intake manifold. Remove the gasket from the intake manifold.
21. Disconnect the knock sensor connector. Remove the nut and knock sensor.

To install:

➡Make sure that the knock sensor is in the correct position.

22. Install the knock control sensor with the nut and tighten to 15 ft. lbs. (20 Nm).
23. Connect the knock control sensor connector.
24. The remainder of installation is the reverse of the removal procedure.
25. Inspect for fuel leak and check the function of throttle body.

2.5L Engine

See Figure 340.

1. Remove the intake manifold.
2. Remove the knock sensor.
   a. Disconnect the sensor connector.
   b. Remove the bolt and sensor.

To install:

To install, reverse the removal procedure. Tighten the sensor to 15 ft. lbs. (20 Nm).

3.5L Engine

See Figures 341 and 342.

CAUTION

Observe all applicable safety precautions when working around fuel. Whenever servicing the fuel system, always work in a well-ventilated area. Do not allow fuel spray or vapors to come in contact with a spark or open flame. Keep a dry chemical fire extinguisher near the work area. Always keep fuel in a container specifically designed for fuel storage; also, always properly seal fuel containers to avoid the possibility of fire or explosion.

1. Before servicing the vehicle, refer to the Precautions Section.
2. Properly discharge the fuel system pressure.
3. Disconnect battery negative cable.

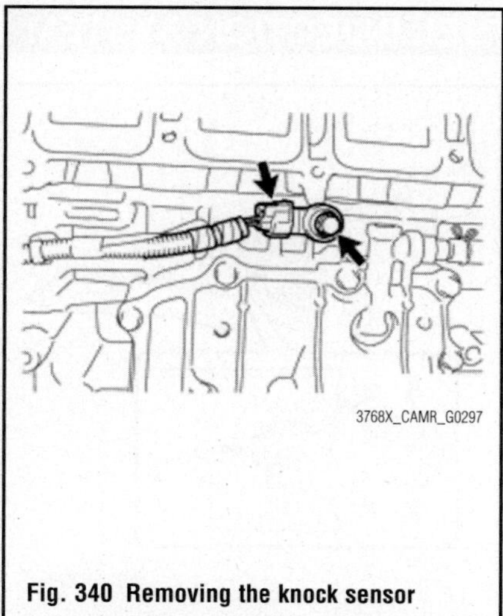

Fig. 340 Removing the knock sensor

4. Drain and recycle the engine coolant.
5. Remove both plastic engine under covers.
6. Remove windshield wiper arms and blade assemblies.
7. Remove both of the front fender to cowl side seals.
8. Remove cowl top ventilator louver sub-assembly.
9. Remove the windshield wiper motor and link assembly.
10. Remove the outer cowl top panel.
11. Remove the cool air intake duct seal.
12. Remove the V-bank cover sub-assembly.
13. Remove air cleaner inlet assembly.
14. Remove the air cleaner cap sub-assembly.
15. Remove air cleaner case sub-assembly.
16. Remove the intake air surge tank.
17. Remove no. 1 air cleaner inlet.
18. Separate fuel tube sub-assembly.
19. Remove the intake manifold.
20. Disconnect the 2 knock control sensor connectors.
21. Remove the 2 bolts and 2 knock control sensors.

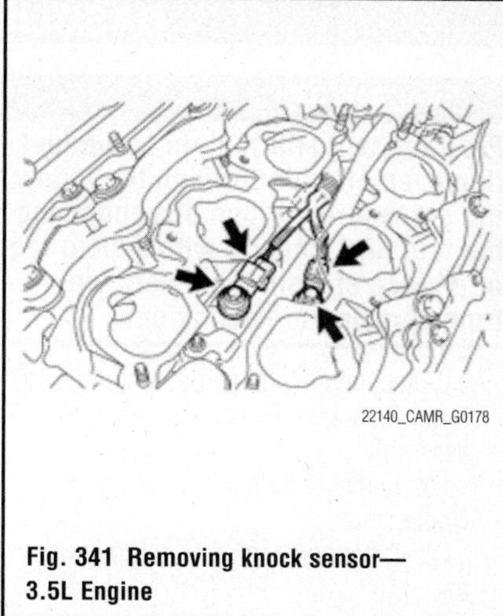

Fig. 341 Removing knock sensor—3.5L Engine

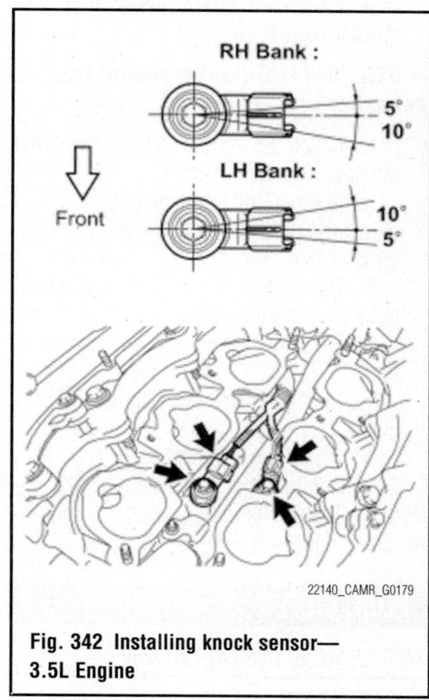

Fig. 342 Installing knock sensor—3.5L Engine

To install:

22. Install the 2 knock control sensors with the 2 bolts as shown in the illustration and tighten to 15 ft. lbs. (20 Nm).
23. Connect the 2 knock control sensor connectors.
24. The remainder of installation is the reverse of the removal procedure.
25. Inspect for fuel leak and check the function of throttle body.

MASS AIR FLOW (MAF) SENSOR

REMOVAL & INSTALLATION

See Figure 343.

1. Before servicing the vehicle, refer to the Precautions Section.
2. Disconnect the mass air flow meter connector.
3. Remove the 2 screws and mass air flow meter.

To install:

4. Installation is the reverse of the removal procedure.

THROTTLE POSITION SENSOR (TPS)

REMOVAL & INSTALLATION

Refer to the Throttle Body removal and installation procedures in the Fuel System Section.

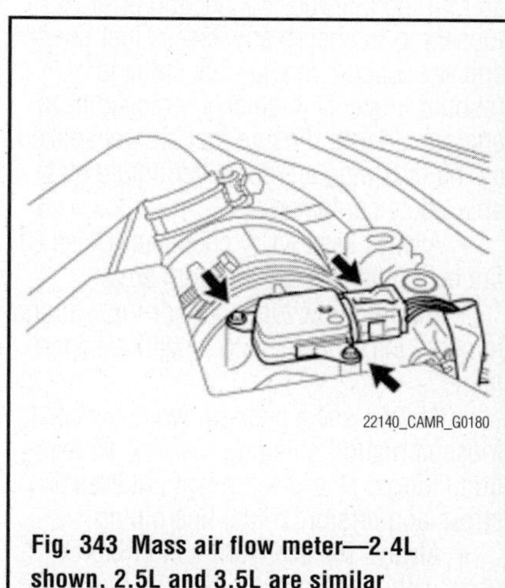

Fig. 343 Mass air flow meter—2.4L shown, 2.5L and 3.5L are similar

FUEL SYSTEM SERVICE PRECAUTIONS

Safety is the most important factor when performing not only fuel system maintenance but any type of maintenance. Failure to conduct maintenance and repairs in a safe manner may result in serious personal injury or death. Maintenance and testing of the vehicle's fuel system components can be accomplished safely and effectively by adhering to the following rules and guidelines.

- To avoid the possibility of fire and personal injury, always disconnect the negative battery cable unless the repair or test procedure requires that battery voltage be applied.
- Always relieve the fuel system pressure prior to disconnecting any fuel system component (injector, fuel rail, pressure regulator, etc.), fitting or fuel line connection. Exercise extreme caution whenever relieving fuel system pressure to avoid exposing skin, face and eyes to fuel spray. Please be advised that fuel under pressure may penetrate the skin or any part of the body that it contacts.
- Always place a shop towel or cloth around the fitting or connection prior to loosening to absorb any excess fuel due to spillage. Ensure that all fuel spillage (should it occur) is quickly removed from engine surfaces. Ensure that all fuel soaked cloths or towels are deposited into a suitable waste container.
- Always keep a dry chemical (Class B) fire extinguisher near the work area.
- Do not allow fuel spray or fuel vapors to come into contact with a spark or open flame.
- Always use a back-up wrench when loosening and tightening fuel line connection fittings. This will prevent unnecessary stress and torsion to fuel line piping.
- Always replace worn fuel fitting O-rings with new Do not substitute fuel hose or equivalent where fuel pipe is installed.

Before servicing the vehicle, make sure to also refer to the precautions in the beginning of this section as well.

RELIEVING FUEL SYSTEM PRESSURE

CAUTION

Perform the following procedures to prevent fuel from spilling out before removing any fuel system parts.

CAUTION

Pressure will still remain in the fuel line even after performing the following procedures. When disconnecting the fuel line, cover it with a shop rag or a piece of cloth to prevent fuel from spraying or coming out.

1. Disconnect the fuel pump connector:
   a. Remove the rear seat cushion assembly.
   b. Remove the rear floor service whole cover.
   c. Disconnect the fuel pump connector.
   d. Start the engine.
   e. After the engine stops, turn the ignition switch off.

➡DTC P0171/25 (fuel problem) may be detected.

   f. Crank the engine again. Check that the engine does not start.
   g. Remove the fuel tank cap to discharge pressure from the fuel tank.
   h. Disconnect the cable from the negative (-) battery terminal.
   i. Reconnect the fuel pump connector.
   j. Install the rear floor service hole cover.
   k. Install the rear seat.
2. Check that there are no fuel leaks from the fuel system after doing any maintenance or repairs.

FUEL FILTER

REMOVAL & INSTALLATION

See Figure 344.

1. Before servicing the vehicle, refer to the precautions section.
2. Remove the fuel pump from the vehicle.
3. Remove the fuel pump filter, as follows:

➡Do not damage the fuel pump filter. Do not remove the suction filter.

   a. Using a screwdriver, pry out the clips.
   b. Pull out the fuel pump filter from the fuel pump.

To install:

4. Install the fuel pump filter with a new clip.
5. Install the fuel pump.

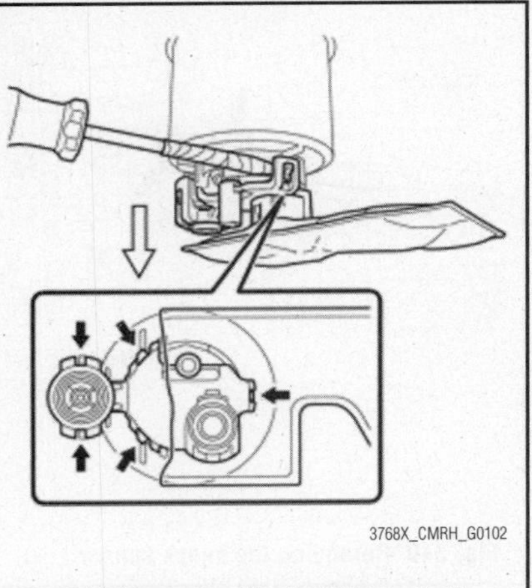

Fig. 344 Removing the fuel filter—2.4L hybrid engine shown

FUEL INJECTORS

REMOVAL & INSTALLATION

2.4L Engine

See Figures 345 through 348.

1. Before servicing the vehicle, refer to the precautions section.
2. Discharge fuel system pressure.
3. Disconnect battery negative cable.
4. Remove air cleaner cap sub-assembly.
5. Remove No. 1 engine cover sub-assembly.

➡Check for foreign matter on the pipe and around the connector before disconnecting the quick connector. Clean the connector if necessary.

6. Disconnect fuel tube sub-assembly, as follows:
   a. Remove the No. 1 fuel pipe clamp.

CAUTION

Do not use any tools in this following procedure. Check for foreign matter on the sealing surface of the disconnected pipe. Clean it if necessary.

   b. If the connector and pipe are stuck, pinch the connector, and push and pull the pipe to disconnect them.
   c. Separate the fuel tube from the fuel hose clamp.
7. Disconnect the No. 2 ventilation hose from the ventilation valve.
8. Remove fuel delivery pipe with injector as follows:
   a. Remove the 2 wire harness clamps.

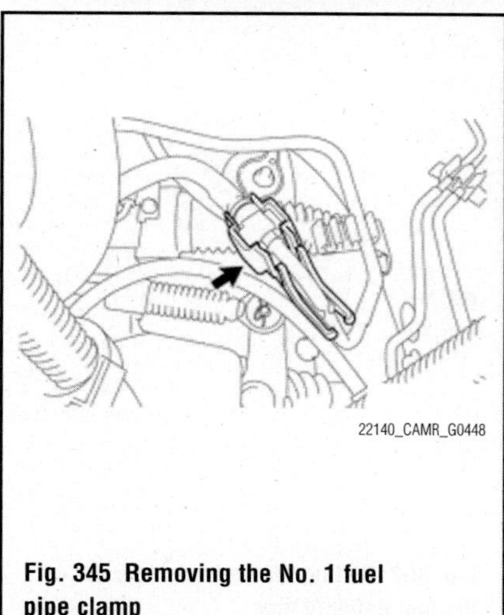

Fig. 345 Removing the No. 1 fuel pipe clamp

CAUTION

Be careful not to drop the fuel injectors when removing the fuel delivery pipe.

b. Remove the 2 bolts, then remove the fuel delivery pipe together with the 4 fuel injectors.

c. Remove the 2 delivery pipe spacers from the cylinder head.

d. Remove the 4 insulators from the cylinder head.

9. Pull out the 4 injectors from the delivery pipe.

10. Remove the 4 O-rings from the injectors.

To install:

11. Install the fuel injector assembly, as follows:

a. Apply a light coat of spindle oil or gasoline to new O-rings, and install one to each injector.

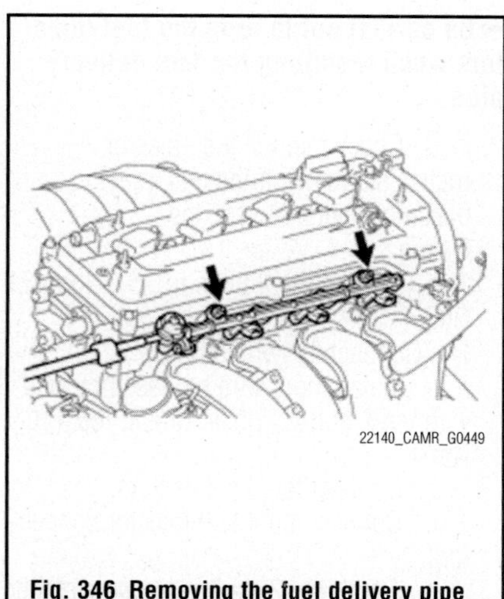

Fig. 346 Removing the fuel delivery pipe together with the 4 fuel injectors

b. Apply a light coat of spindle oil or gasoline where the fuel delivery pipe contacts the O-ring.

c. Apply a light coat of gasoline or spindle oil to the O-ring again, then install the right and left fuel injectors onto the fuel delivery pipe.

➡Make sure that the O-ring is not cracked or jammed before installing the injector.

d. Check that the fuel injector rotates smoothly. If the fuel injector does not rotate, replace the O-ring.

12. Install fuel delivery pipe with injector as follows:

a. Install 4 new insulators into the cylinder head.

b. Install the 2 delivery pipe spacers onto the cylinder head.

c. Install the fuel delivery pipe together with the 4 fuel injectors, then temporarily tighten the 2 bolts.

➡Be careful not to drop the fuel injectors when installing the fuel delivery pipe.

d. Check that the fuel injector rotates smoothly. If the fuel injector does not rotate smoothly, replace the O-ring.

e. Tighten the 2 bolts to the specified torque and tighten to 15 ft. lbs. (20 Nm).

f. Connect the 4 fuel injector connectors.

g. Install the 2 wire harness clamps.

13. Connect the No. 2 ventilation hose to the ventilation valve.

14. Install the fuel tube to the fuel hose clamp.

15. Push the fuel tube connector until it makes a "click" sound.

16. Install the No. 1 fuel pipe clamp.

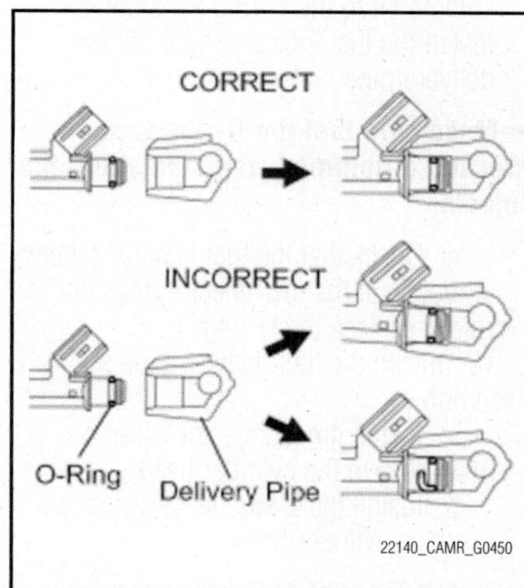

Fig. 347 Installing fuel injector to fuel tube

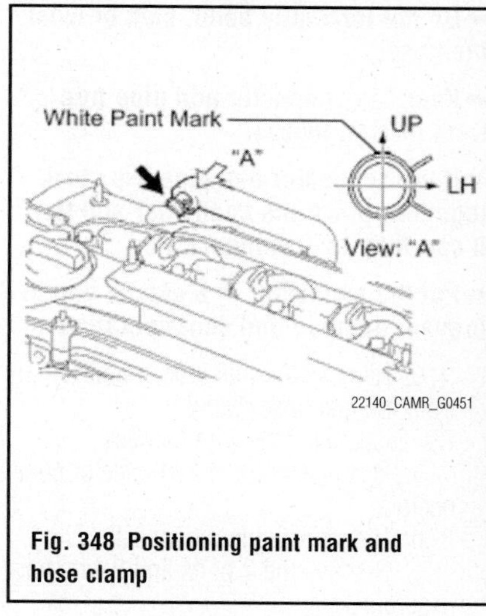

Fig. 348 Positioning paint mark and hose clamp

17. Install air cleaner cap sub-assembly.

18. Connect cable to negative battery terminal.

19. Check for fuel leaks.

20. Install No. 1 engine cover sub-assembly.

2.5L Engine

See Figures 349 through 353.

1. Discharge the fuel system pressure.
2. Disconnect the cable from the negative battery terminal.
3. Remove the No. 1 engine cover sub assembly.
4. Remove the air cleaner cap sub assembly.
5. Remove the air cleaner filter element sub assembly.
6. Remove the front wiper arm and blade assemblies.
7. Remove the front fender to cowl side seals.
8. Remove the cowl top ventilator louver sub assembly.
9. Remove the windshield wiper motor and link assembly.
10. Remove the cowl top outer front panel sub assembly.
11. Disconnect the fuel tube sub assembly.

a. Remove the No. 1 fuel pipe clamp.

b. Pinch the tube connector, and then pull the tube connector off the pipe.

➡Check for foreign matter in the fuel tube around the fuel tube connector. Clean it if necessary. Foreign matter can affect the ability of the O-ring to seal the connector and fuel pipe.

➡Do not use any tools to separate the connector and pipe.

➡**Do not forcefully bend, kink or twist the hose.**

➡**Keep the connector and pipe free from foreign matter.**

➡**If the connector and pipe are stuck together, pinch the connector and turn it carefully to disconnect it.**

➡**Put the connector in a plastic bag to prevent damage and contamination.**

c. Remove the fuel tube sub assembly from the fuel hose clamp.

12. Disconnect the wire harness.

a. Disconnect the 4 fuel injector connectors.

b. Disconnect the 3 connectors.

c. Remove the 2 bolts and 2 wire harness brackets.

d. Detach the 2 clamps to disconnect the wire harness.

13. Remove the vacuum switching valve assembly (ACIS).

a. Remove the 2 bolts, and then remove the fuel delivery pipe together with the 4 fuel injectors.

➡**Be careful not to drop the fuel injectors when removing the fuel delivery pipe.**

b. Remove the 2 fuel delivery spacers from the cylinder head.

c. Remove the 4 injector vibration insulators from the cylinder head.

14. Remove the fuel injector assembly.

a. Pull the 4 fuel injectors out of the fuel delivery pipe.

To install:

15. Install the fuel injector assembly.

a. Apply a light coat of gasoline or spindle oil to the new O-rings, and then install one onto each fuel injector.

b. Apply a light coat of gasoline or spindle oil to the part of the fuel delivery pipe which comes into contact with the O-ring of the fuel injector.

c. Apply a light coat of gasoline or spindle oil to the O-ring again, and then install the fuel injectors onto the fuel delivery pipe.

➡**Make sure that the O-ring is not cracked or jammed when installing the injector.**

d. Check that the fuel injector rotates smoothly. If the fuel injector does not rotate, replace the O-ring.

16. Install the fuel delivery pipe sub assembly.

a. Install 4 new injector vibration insulators to the cylinder head.

b. Install the 2 fuel delivery spacers onto the cylinder head.

➡**Install the fuel delivery spacer so that the longer protrusion is on the cylinder head side.**

c. Install the fuel delivery pipe together with the 4 fuel injectors to the cylinder head, and then temporarily install the 2 bolts.

➡**Be careful not to drop the fuel injectors when installing the fuel delivery pipe.**

d. Check that the fuel injector rotates smoothly. If the fuel injector does not rotate, replace the O-ring.

e. Tighten the 2 bolts to 15 ft. lbs. (21 Nm).

17. Connect the wire harness.

a. Install the 2 wire harness brackets with the 2 bolts. Tighten to 7 ft. lbs. (10 Nm).

b. Connect the 3 connectors.

c. Connect the 4 fuel injector connectors.

d. Attach the 2 clamps to connect the wire harness.

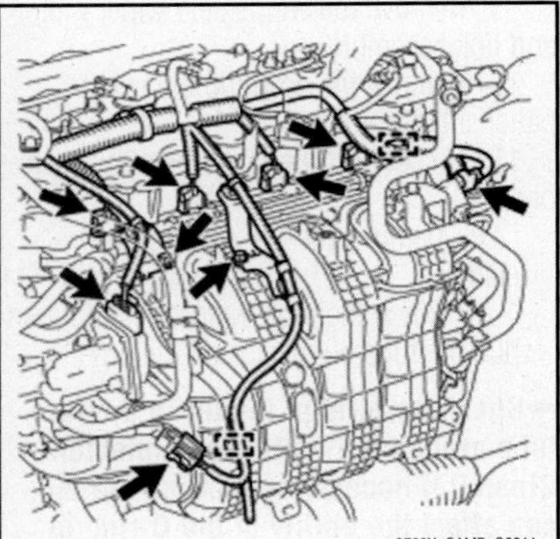

Fig. 349 Disconnecting the wire harness and injector connectors

Fig. 350 Removing the fuel delivery pipe with the injectors

Fig. 351 Removing the fuel delivery spacers and injector vibration insulators from the cylinder head

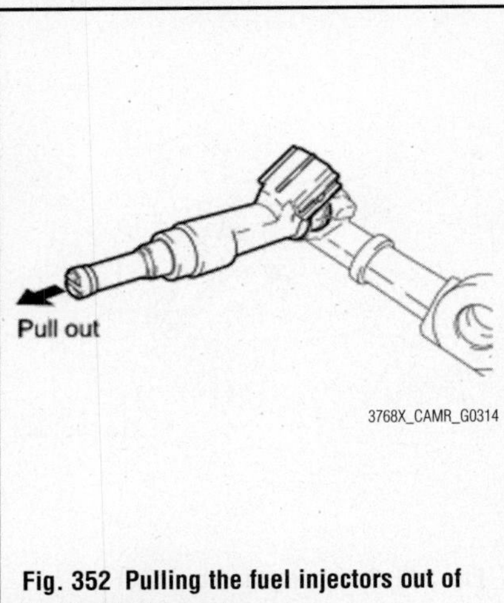

Fig. 352 Pulling the fuel injectors out of the fuel delivery pipe

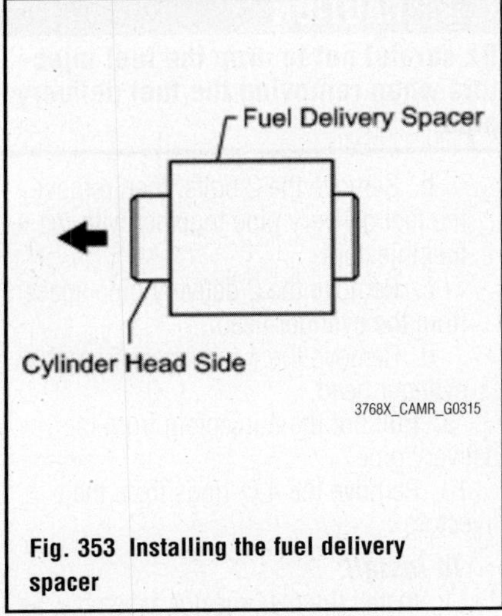

Fig. 353 Installing the fuel delivery spacer

18. Install the vacuum switching valve assembly (ACIS).
19. Connect the fuel tube sub assembly.
a. Push the tube connector to the pipe until the tube connector makes a "click" sound.

➡Before connecting the connector and fuel pipe, check that there is no damage or foreign matter on the connecting part of the fuel pipe.

➡After connecting the fuel tube connector and pipe, check that they are securely connected by trying to pull them apart.

b. Install the No. 1 fuel pipe clamp.
c. Install the fuel tube sub-assembly to the fuel hose clamp.
20. Install the cowl top outer front panel sub assembly.
21. Install the windshield wiper motor and link assembly.
22. Install the cowl top ventilator louver sub assembly.
23. Install the front fender to cowl side seals.
24. Install the front wiper arm and blade assemblies.
25. Install the air cleaner cap sub assembly.
26. Install the No. 1 engine cover sub assembly.
27. Connect the cable to the negative battery terminal.
28. Inspect for fuel leaks.

3.5L Engine

See Figures 354 through 356.

1. Properly discharge the fuel system pressure.
2. Disconnect battery negative cable.
3. Drain engine coolant.
4. Remove both windshield wiper arm and blade assemblies.
5. Remove the right cowl top ventilator louver.
6. Remove the windshield wiper motor and link assembly.
7. Remove front fender to cowl side seal.
8. Remove the cowl top ventilator louver sub-assembly.
9. Remove windshield wiper motor and link assembly.
10. Remove cowl top panel outer sub-assembly.
11. Remove the V-bank cover sub-assembly.
12. Remove the air cleaner cap with air cleaner hose.
13. Remove the intake air surge tank.
14. Disconnect the fuel tube sub-assembly, as follows:
a. Remove the No. 2 fuel pipe clamp.
b. Pinch the tube connector and then pull out the fuel pipe.

➡Check that there is no dirt or other foreign objects around the connector before removing fuel tube, and clean the connector as necessary.

➡It is necessary to prevent mud or dirt from entering the connector. If mud or dirt gets in the connector, the O-rings may not seal properly.

➡Do not use any tools in this operation.

➡Do not bend, kink or twist the nylon tube. Protect the connector by covering it with a plastic bag.

➡When the pipe and connector are stuck, push and pull the connector to release and pull the connector out carefully.

15. Remove the fuel injector assembly, as follows:
a. Disconnect the 6 fuel injector connectors.
b. Remove the 5 bolts and fuel delivery pipe together with the 6 fuel injectors.

➡Be careful not to drop the fuel injectors when removing the fuel delivery pipe.

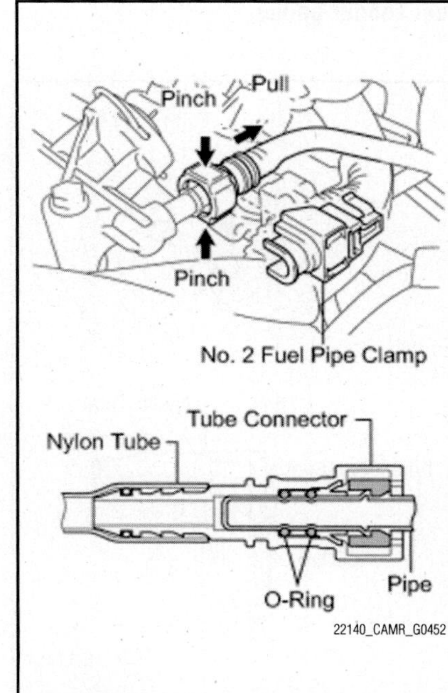

Fig. 354 Remove the No. 2 fuel pipe clamp

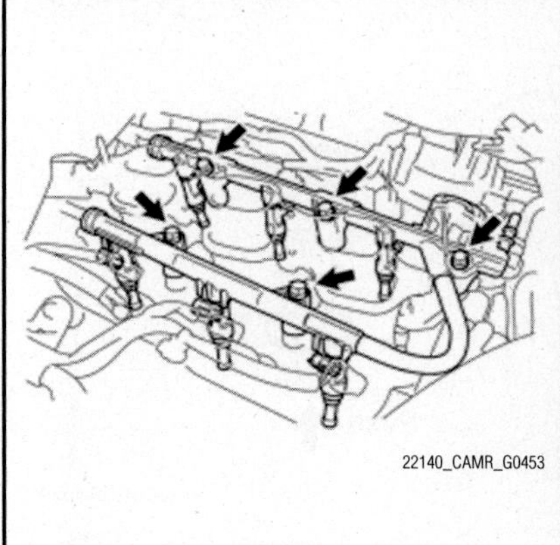

Fig. 355 Removing the 5 bolts and fuel delivery pipe together with the 6 fuel injectors

c. Remove the 6 insulators from the intake manifold.
d. Pull out the fuel injector from the fuel delivery pipe.
e. Remove the 6 O-rings from the injectors.

To install:

16. Install the fuel injector assembly, as follows:
a. Apply a light coat of spindle oil or gasoline to new O-rings, and install one to each injector.
b. Apply a light coat of spindle oil or gasoline where the fuel delivery pipe contacts the O-ring.

➡Be careful not to twist the O-ring.

➡After installing the fuel injector, check that it turns smoothly. If not, reinstall it with a new O-ring.

c. Push the fuel injector while twisting it back and forth to install it in the fuel delivery pipe.
d. Position the fuel injector connector outward.
e. Install 6 new insulators to the intake manifold.
f. Place the fuel delivery pipe and the 6 fuel injectors together to the intake manifold.

➡Be careful not to drop the fuel injectors when installing the fuel delivery pipe.

g. Temporarily install the 6 bolts which are used to hold the fuel delivery pipe to the intake manifold.

➡After installing the fuel injector, check that it turns smoothly. If not, reinstall it with a new O-ring.

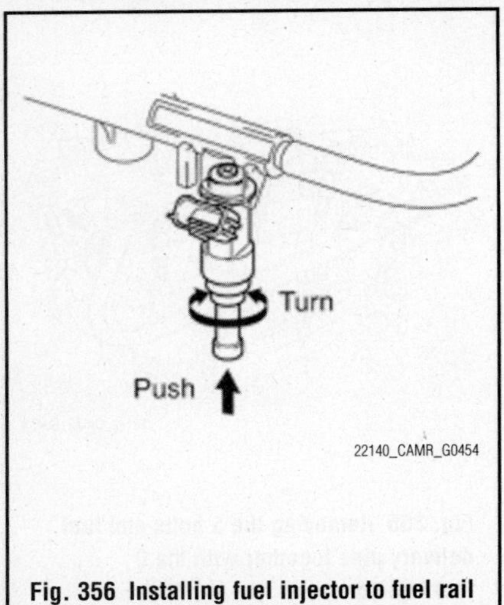

Fig. 356 Installing fuel injector to fuel rail

h. Tighten the 5 bolts which are used to hold the fuel delivery pipe to the intake manifold to 15 ft. lbs. (21 Nm).

17. Push in the tube connector to the pipe until the tube connector makes a "click" sound.

➡Before connecting the tube, make sure that it is not damaged. Make sure that there is no dirt present on the connecting surfaces.

➡After connecting, check if the fuel tube connector and the pipe are securely connected by pulling on them.

18. Install the No. 2 fuel pipe clamp.
19. The remainder of installation is the reverse of the removal procedure.
20. Check for coolant leak and fuel leak.

FUEL PUMP

REMOVAL & INSTALLATION

See Figures 357 and 358.

1. Before servicing the vehicle, refer to the precautions section.
2. Discharge fuel system pressure.
3. Disconnect battery negative cable.
4. Remove the rear seat cushion assembly.
5. Remove the rear floor service hole cover.
6. Disconnect the fuel pump connector.
7. Separate the fuel pump tube sub-assembly, as follows:

➡Check if there is any dirt or mud around the connector before this operation and remove the dirt as necessary.

➡Be careful of mud because the quick connector has an O-ring which seals the pipe and connector that can be contaminated.

➡Do not use any tools in this operation.

➡Do not bend or twist the nylon tube. Cover the fuel tube joint with a plastic bag.

➡When the fuel tube joint and fuel suction plate are stuck, pinch the fuel tank tube between fingers, and turn it carefully to release it. Disconnect the fuel tank tube.

a. Remove the tube joint clip, and pull out the fuel pump tube.

8. Remove fuel tank vent tube set plate, as follows:

a. Remove the 8 bolts and set plate.

9. Remove fuel suction tube assembly with pump and gauge, as follows:

a. Pull out the fuel suction tube from the fuel tank.

➡Do not damage the fuel pump filter.

➡Be careful not to bend the arm of the fuel sender gauge.

b. Remove the gasket from the fuel suction tube.

To install:

10. Install a new gasket to the fuel suction tube.
11. Install the fuel suction tube.

➡Do not damage the fuel pump filter.

➡Be careful not to bend the arm of the fuel sender gauge.

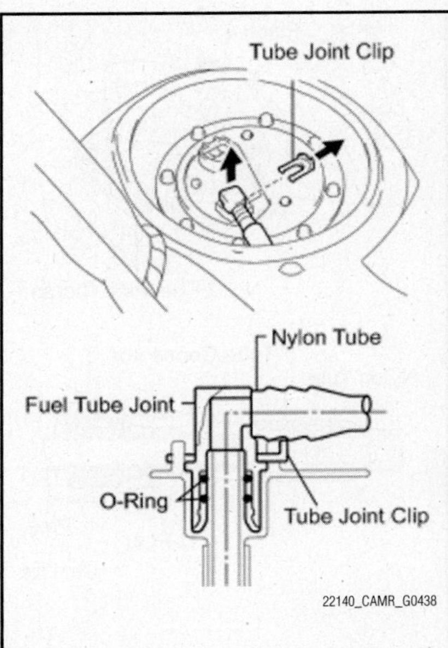

Fig. 357 Fuel pump tube sub-assembly

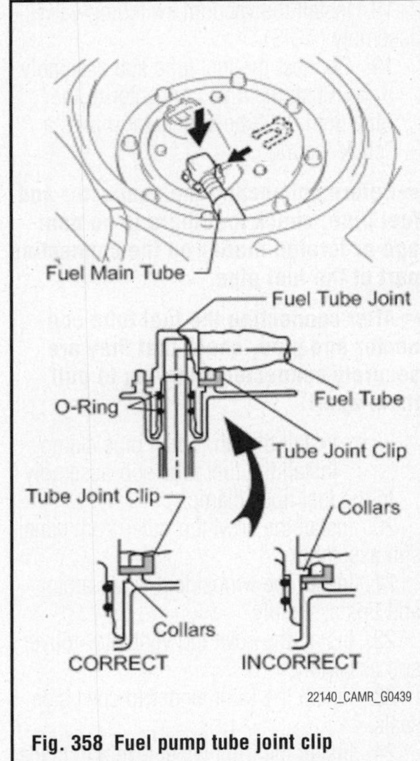

Fig. 358 Fuel pump tube joint clip

12. Install the fuel tank vent tube set plate, as follows:

a. Align the mark of the set plate with the fuel suction tube.

b. Install the set plate with the 8 bolts and tighten to 52 inch lbs. (5.9 Nm).

13. Install the fuel pump tube with the tube joint clip.

➡Check that there is no scratches or foreign objects on the connecting part.

➡Check that the fuel tube joint is inserted securely.

➡Check that the tube joint clip is on the collar of the fuel tube joint.

➡After installing the tube joint clip, check that the fuel tube joint is pulled off.

14. Connect battery negative cable.
15. Inspect for fuel leak.
16. Install the rear floor service hole cover.
17. Install the rear seat cushion assembly.

FUEL TANK

REMOVAL & INSTALLATION

See Figures 359 through 364.

1. Before servicing the vehicle, refer to the precautions section.
2. Discharge fuel system pressure.

3. Disconnect battery negative cable.
4. Remove rear seat cushion assembly.
5. Remove rear floor service whole cover.
6. Separate fuel pump tube sub-assembly.
7. Remove fuel tank vent tube set plate.
8. Remove fuel suction tube assembly with pump and gauge.
9. Drain fuel.
10. Remove center exhaust pipe assembly.
11. Disconnect no. 2 parking brake cable assembly.
12. Disconnect no. 3 parking brake cable assembly.
13. Remove rear stabilizer BAR No. 1 bracket.
14. Remove lower center fuel tank protector, as follows:
  a. Remove the 4 bolts and the (Except SE grade).
  b. Remove the 4 bolts and 2 clips (for SE Grade).
  c. Remove the fuel tank protector (for SE Grade).

➡Check that there is no dirt or other foreign objects around the connector before removing fuel tubes, and clean the connector as necessary.

➡It is necessary to prevent mud or dirt from entering the connector. If mud or dirt gets in the connector, the O-rings may not seal properly.

➡Do not use any tools in these operations.

➡Do not bend, kink or twist the nylon tubes. Protect the connector by covering it with a plastic bag.

➡When the pipe and connector are stuck, push and pull the connector to release and pull the connector out carefully.

15. Disconnect the fuel pump tube, as follows:
  a. Pinch the tabs of the retainer to remove the lock claws and pull it down as shown in the illustration.
  b. Pull out the fuel tank main tube.
16. Pinch the tube connector and then pull out the No. 1 fuel tube.
17. Set up a transmission jack underneath the fuel tank.
18. Remove the 2 set bolts of the fuel tank bands.

➡Check that there is no dirt or other foreign objects around the connector before removing fuel tubes, and clean the connector as necessary.

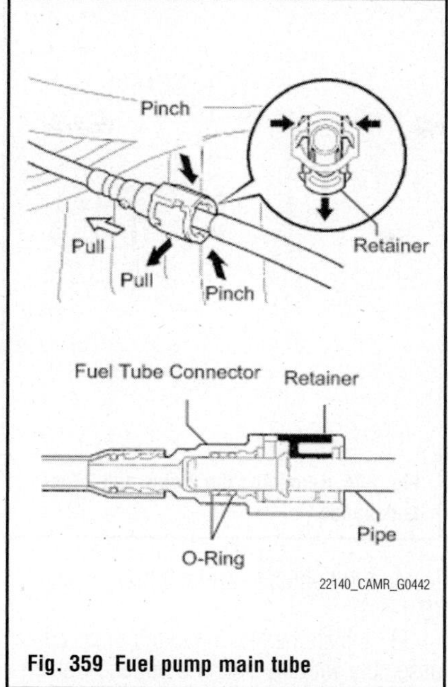

Fig. 359 Fuel pump main tube

➡It is necessary to prevent mud or dirt from entering the connector. If mud or dirt gets in the connector, the O-rings may not seal properly.

➡Do not use any tools in these operations.

➡Do not bend, kink or twist the nylon tubes. Protect the connector by covering it with a plastic bag.

➡When the pipe and connector are stuck, push and pull the connector to release and pull the connector out carefully.

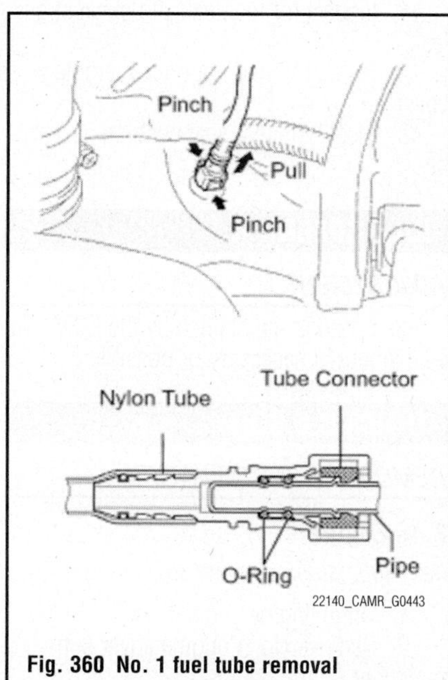

Fig. 360 No. 1 fuel tube removal

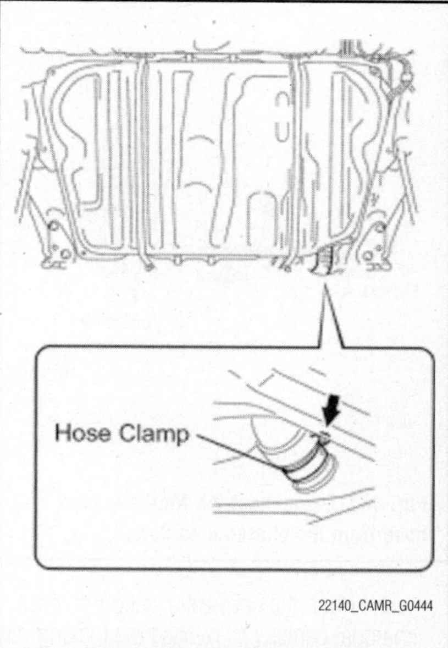

Fig. 361 Disconnecting the fuel tank to filter pipe hose—except PZEV

19. Remove the hose clamp and disconnect the fuel tank to filter pipe hose (except PZEV).
20. Remove the clamp and disconnect the fuel tank to filter pipe hose (for PZEV).
21. Slightly lower the transmission jack.
22. Disconnect the fuel tank vent hose from the charcoal canister, as follows:

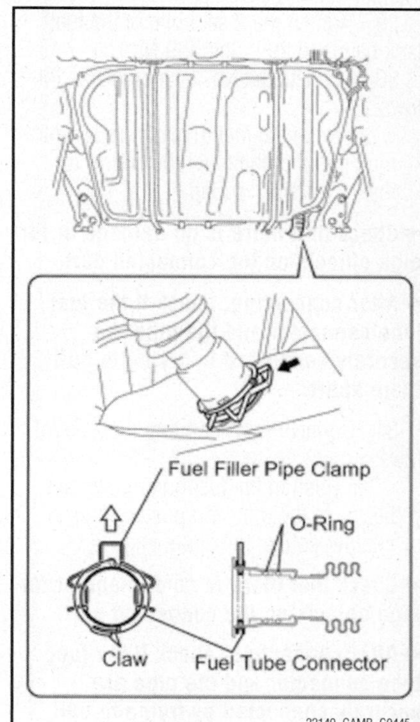

Fig. 362 Disconnecting the fuel tank to filter pipe hose—PZEV

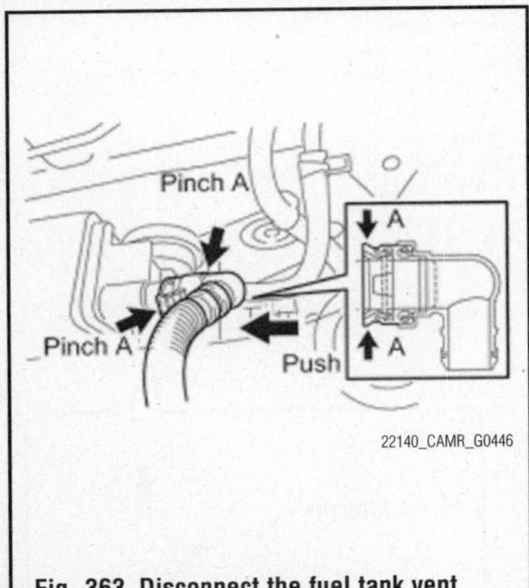

Fig. 363 Disconnect the fuel tank vent hose from the charcoal canister

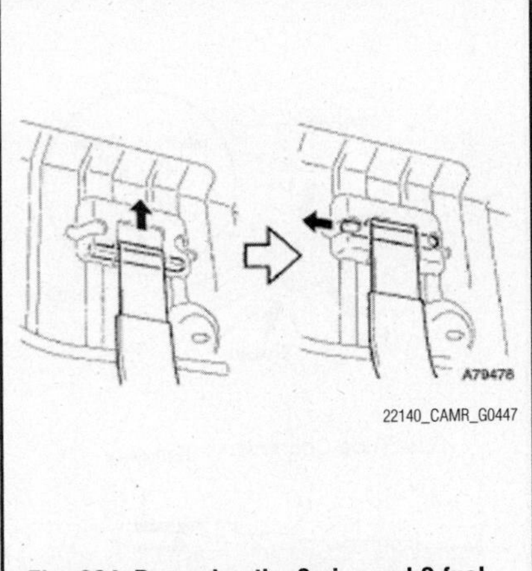

Fig. 364 Removing the 2 pins and 2 fuel tank bands

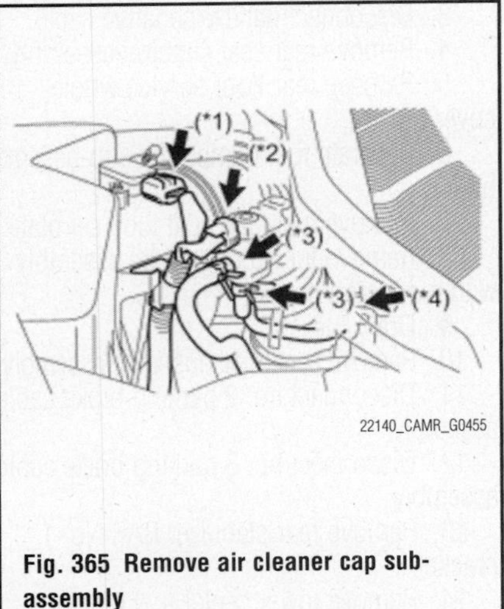

Fig. 365 Remove air cleaner cap sub-assembly

a. Push the connector deep into the charcoal canister to release the locking pin.

b. Pinch portion A.

c. Pull out the connector.

23. Remove the 2 pins and 2 fuel tank bands as shown in the illustration.

24. Remove the 4 clip nuts.

To install:

25. Install the 4 clip nuts.

26. Install the 2 fuel tank bands with the 2 pins.

27. Connect the fuel tank vent hose.

28. Connect the fuel tank inlet pipe with the fuel filter pipe clamp.

29. Tighten the 2 set bolts of the fuel tank bands to 29 ft. lbs. (39 Nm).

30. Connect the No. 1 fuel tube, as follows:

a. Push the fuel tube connector into the pipe until the fuel tube connector makes a "click" sound.

➡Check that there is no damage or foreign objects on the connected part.

➡After connecting, check if the fuel tube connector and the pipe are securely connected by trying to pull them apart.

31. Connect the fuel pump tube, as follows:

a. Push in the fuel pump tube connector to the pipe and push up the retainer so that the claws engage.

➡Check that there is no damage or foreign objects on the connected part.

➡After connecting, check if the fuel tube connector and the pipe are securely connected by trying to pull them apart.

32. Install the lower center fuel tank protector and tighten to 48 inch lbs. (5.4 Nm).

33. Install the No. 3 parking brake cable assembly with the bolt and nut and tighten to 53 inch lbs. (6 Nm), and 75 inch lbs. (8.5 Nm).

34. Install the No. 2 parking brake cable assembly with the bolt and nut and tighten to 53 inch lbs. (6 Nm), and 75 inch lbs. (8.5 Nm).

35. Install the center exhaust pipe assembly.

36. Install the fuel suction tube assembly with pump and gauge.

37. Install the fuel tank vent tube set plate.

38. Connect the fuel pump tube sub-assembly.

39. Add fuel.

40. Connect battery negative cable.

41. Inspect for fuel leak and exhaust gas leak.

42. Install the rear floor service hole cover.

43. Install the rear seat cushion assembly.

IDLE SPEED

ADJUSTMENT

Idle speed is maintained by the ECM. No adjustment is necessary or possible.

THROTTLE BODY

REMOVAL & INSTALLATION

2.4L Engine

See Figures 365 through 367.

1. Drain engine coolant.

2. Remove No. 1 engine cover sub-assembly.

3. Remove air cleaner cap sub-assembly, as follows:

a. Disconnect the mass air flow meter connector.

b. Disconnect the purge VSV connector.

c. Disconnect the 2 purge VSV vacuum hoses.

d. Disconnect the purge line hose from the clamp.

e. Disconnect the No. 2 ventilation hose from the air cleaner hose.

f. Lock the No. 1 air cleaner hose clamp, and then disconnect the No. 1 air cleaner hose from the throttle body.

g. Remove the 2 bolts and air cleaner cap.

h. Remove the air cleaner filter element from the air cleaner case.

4. Remove air cleaner case sub-assembly.

5. Disconnect the throttle position sensor connector and wire harness clamp.

6. Remove the 4 bolts, and then remove the fuel pipe support and throttle body.

7. Disconnect the purge line hose from the throttle body.

8. Disconnect the water by-pass hose from the throttle body.

9. Disconnect the No. 2 water by-pass hose from the throttle body.

10. Disconnect the No. 1 throttle body hose from the throttle body.

11. Remove the gasket from the intake manifold.

To install:

12. Install a new gasket onto the intake manifold.

13. Connect the purge line hose to the throttle body.

14. Connect the water by-pass hose to the throttle body.

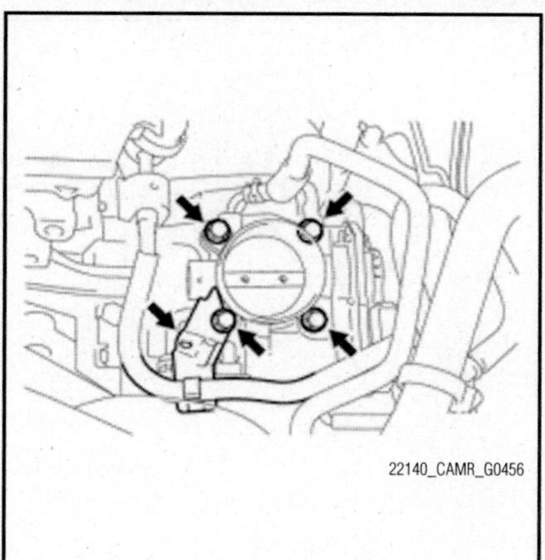
Fig. 366 Removing the 4 bolts and fuel pipe support

Fig. 367 Disconnecting the purge line hose

15. Connect the No. 2 water by-pass hose to the throttle body.
16. Connect the No. 1 throttle body hose to the throttle body.
17. Install the throttle body and fuel pipe clamp with the 4 bolts and tighten to 22 ft. lbs. (30 Nm).
18. Connect the fuel tube into the clamp.
19. Connect the throttle position sensor connector.
20. Connect the wire harness clamp.
21. Install air cleaner case sub-assembly.
22. Install air cleaner cap sub-assembly, as follows:
   a. Install the air cleaner filter element onto the air cleaner case.
   b. Insert the hinges. Install the air cleaner cap sub-assembly with the 2 bolts.
   c. Align the matchmarks of the No. 1 air cleaner hose and throttle body, and then connect the air cleaner hose No. 1 to the throttle body and unfasten the No. 1 air cleaner hose clamp.

➡Make sure that the hose clamp is at the correct angle.

   d. Connect the No. 2 ventilation hose to the air cleaner hose.
   e. Connect the purge line hose to the clamp.
   f. Connect the 2 purge VSV vacuum hoses.
   g. Connect the purge VSV connector.
   h. Connect the mass air flow meter connector.
23. Install air cleaner inlet assembly.
24. Add engine coolant.
25. Check for engine coolant leaks.
26. Install No. 1 engine cover sub-assembly.

2.5L Engine

See Figures 368 through 370.

1. Remove the front wheel opening extension pads.
2. Remove the engine under covers.
3. Drain the engine coolant.
4. Remove the No. 1 engine cover sub assembly.
5. Remove the air cleaner cap sub assembly (PZEV).
   a. Disconnect the vacuum hose and separate it from the air cleaner hose.
   b. Disconnect the vacuum switching valve connector and 2 hoses.
   c. Separate the hose from the air cleaner cap sub-assembly
   d. Disconnect the mass air flow meter connector and separate the wire harness clamp from the air cleaner cap.
   e. Loosen the hose clamp and disconnect the hose.
   f. Loosen the hose clamp and disconnect the air cleaner hose.
   g. Remove the 2 bolts and the air cleaner cap sub-assembly.
6. Remove the air cleaner cap sub assembly (except PZEV).
   a. Disconnect the mass air flow meter connector and separate the wire harness clamp from the air cleaner cap.
   b. Disconnect the vacuum switching valve connector and 2 hoses.
   c. Separate the hose from the air cleaner cap sub-assembly.
   d. Loosen the hose clamp and disconnect the hose.
   e. Loosen the hose clamp and disconnect the air cleaner hose.
   f. Remove the 2 bolts and the air cleaner cap sub-assembly.
7. Remove the throttle body.

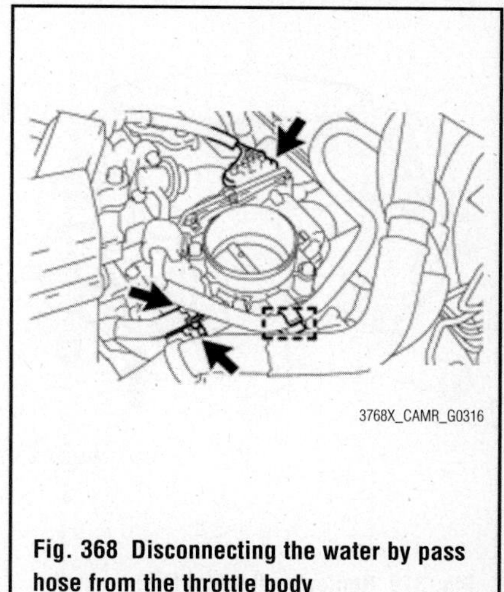
Fig. 368 Disconnecting the water by pass hose from the throttle body

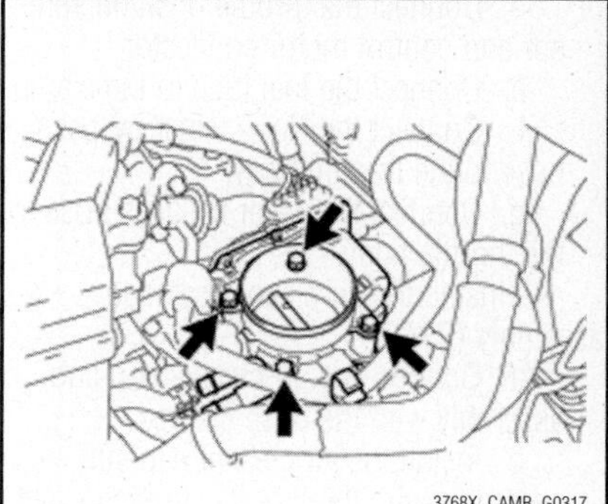
Fig. 369 Removing the throttle body with the fuel tube bracket

   a. Disconnect the water by pass hose from the throttle body.
   b. Disconnect the No. 2 water by-pass hose from the throttle body.
   c. Disconnect the fuel tube from the clamp.
   d. Disconnect the throttle position sensor and control motor connector.
   e. Remove the 4 bolts and the throttle body with fuel tube bracket.
   f. Remove the bolt and fuel tube bracket.
   g. Remove the gasket from the intake manifold.

To install:

8. Install the throttle body assembly.
   a. Install a new gasket to the intake manifold.
   b. Install the fuel tube bracket with the bolt. Tighten to 10 ft. lbs. (13 Nm).
   c. Install the throttle body (with the fuel tube bracket) with the 4 bolts. Tighten to 7 ft. lbs. (10 Nm).

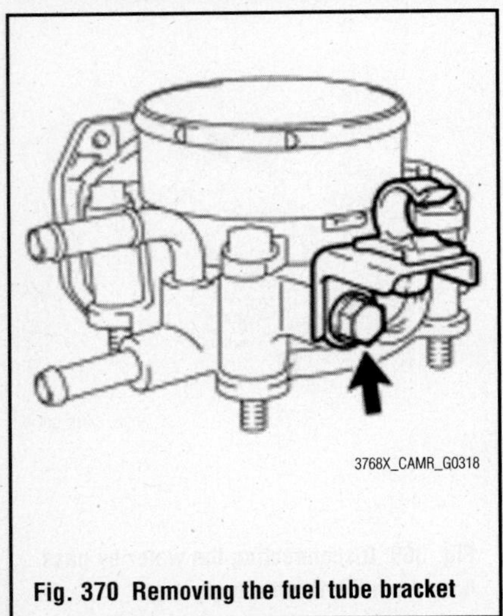

Fig. 370 Removing the fuel tube bracket

d. Connect the throttle position sensor and control motor connector.
e. Connect the fuel tube to the clamp.
f. Connect the No. 2 water by-pass hose to the throttle body.
g. Connect the water by-pass hose to the throttle body.
9. Install the air cleaner cap sub assembly (PZEV).
a. Connect the air cleaner cap sub assembly with the hose clamp.
b. Install the air cleaner cap sub assembly with the 2 bolts. Tighten to 44 inch lbs. (5 Nm).
c. Connect the hose with the hose clamp.
d. Engage the hose with the hose clamp.
e. Connect the vacuum switching valve connector and 2 hoses.
f. Connect the mass air flow meter connector and wire harness clamp to the air cleaner cap sub assembly.
g. Connect the hose and engage the hose to the clips on the air cleaner hose.
10. Install the air cleaner cap sub assembly (except PZEV).
a. Connect the air cleaner cap sub assembly with the hose clamp.
b. Install the air cleaner cap sub assembly with the 2 bolts. Tighten to 44 inch lbs. (5 Nm).
c. Connect the hose with the hose clamp.
d. Engage the hose with the hose clamp.
e. Connect the vacuum switching valve connector and 2 hoses.
f. Connect the mass air flow meter connector and wire harness clamp to the air cleaner cap sub assembly.
11. Add engine coolant.
12. Inspect for coolant leaks.
13. Install the No. 1 engine cover sub assembly.
14. Install the engine under covers.
15. Install the front wheel opening extension pads.
16. Perform the initialization.

➡Be sure to perform this procedure after reassembling the throttle body assembly or removing and reinstalling any throttle body component.

➡Perform the following procedure after replacing the ECM, throttle body assembly or any throttle body components. The following procedure should also be performed if the throttle body is cleaned.

➡Be sure to perform this procedure after reconnecting the battery cable or replacing the ECM.

a. Disconnect the cable of the negative (-) battery terminal. Wait at least 60 seconds and reconnect the cable of the negative (-) battery terminal.
b. Turn the ignition switch to ON without operating the accelerator pedal.

➡If the accelerator pedal is operated, perform the above steps again.

c. Connect the Techstream to the DLC3 and clear the DTCs.
d. Start the engine and check that the MIL is not illuminated. After the engine is warmed up, check that the idle speed is within the specified range when the A/C is switched off.

➡Standard idle speed for an automatic transaxle with the A/C switched off is 610-710 rpm. For a manual transaxle the standard idle speed with the A/C switched off is 650-750 rpm.

➡Be sure to perform this step with all accessories off.

➡Make sure that the shift lever is in neutral.

e. Enter the following menus: Powertrain / Engine / Data List / Throttle Sensor Position. Fully depress the accelerator pedal and check that the value is 60% or more.
f. Perform a road test and confirm that there are no abnormalities.

3.5L Engine

See Figures 371 through 374.

1. Before servicing the vehicle, refer to the precautions section.
2. Disconnect battery negative cable.

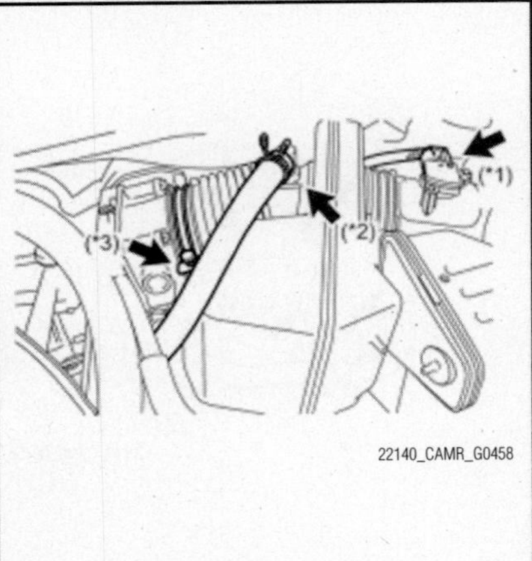

Fig. 371 Removing air cleaner cap sub-assembly

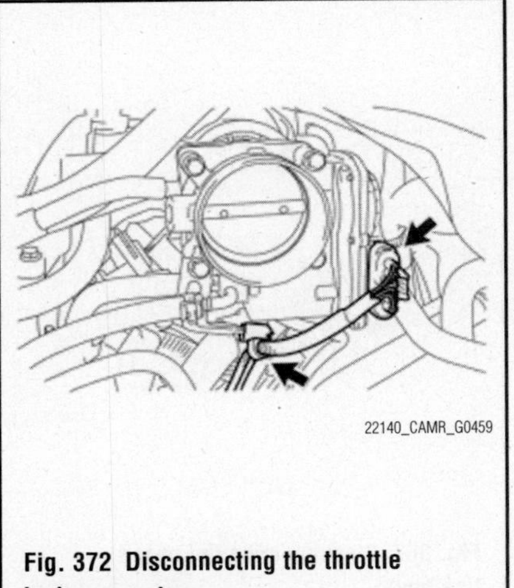

Fig. 372 Disconnecting the throttle body connector

3. Drain engine coolant.
4. Remove cool air intake duct seal.
5. Remove the V-bank cover sub-assembly.
6. Remove air cleaner inlet assembly.
7. Remove air cleaner cap sub-assembly, as follows:
a. Disconnect the 3 vacuum hoses.
b. Disconnect the mass air flow meter connector (1).
c. Disconnect the No. 2 ventilation hose (2).
d. Disconnect the hose band (3).
e. Disconnect the 3 bands, and remove the air cleaner cap sub-assembly.
8. Remove air cleaner case sub-assembly.
9. Remove No. 1 air cleaner inlet.
10. Disconnect the throttle body connector and clamp.
11. Disconnect the 2 water by-pass hoses from the throttle body.

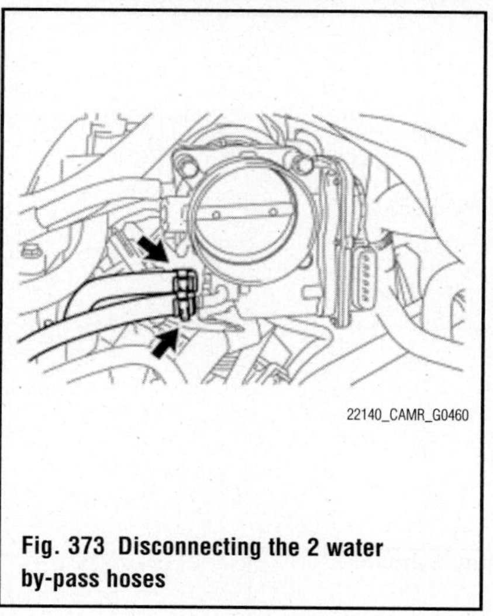
Fig. 373 Disconnecting the 2 water by-pass hoses

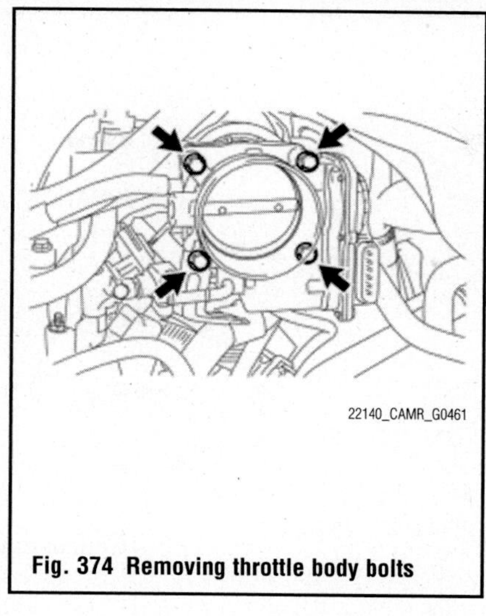
Fig. 374 Removing throttle body bolts

12. Remove the 4 bolts and throttle body.
13. Remove the throttle body gasket from the intake air surge tank.

To install:

14. Install a new throttle body gasket to the intake air surge tank.
15. Install the throttle w/ motor body assembly and wire harness clamp stay to the intake air surge tank with the 4 bolts and tighten to 7 ft. lbs. (10 Nm).
16. Connect the throttle w/ motor body assembly connector.
17. The remainder of installation is the reverse of the removal procedure.
18. Check for coolant leak.
19. Check the function of the throttle body.

HEATING & AIR CONDITIONING SYSTEM

BLOWER MOTOR

REMOVAL & INSTALLATION

See Figures 375 and 376.

1. Drain and recycle the engine coolant.
2. Disconnect the negative battery cable.
3. Remove instrument panel.
4. For TMC made:
   a. Disconnect the connector.
   b. Remove the 2 screws and blower assembly.
   c. Remove the 3 screws and blower with fan motor sub-assembly.
5. For TMMK made:
   a. Remove cooler expansion valve.
   b. Remove the connector and clamp, and disconnect the wire harness.

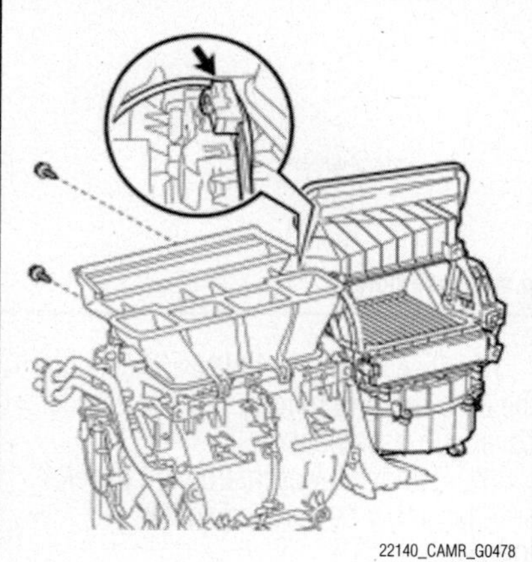
Fig. 375 Removing the connector, clamp and wire harness—TMMK

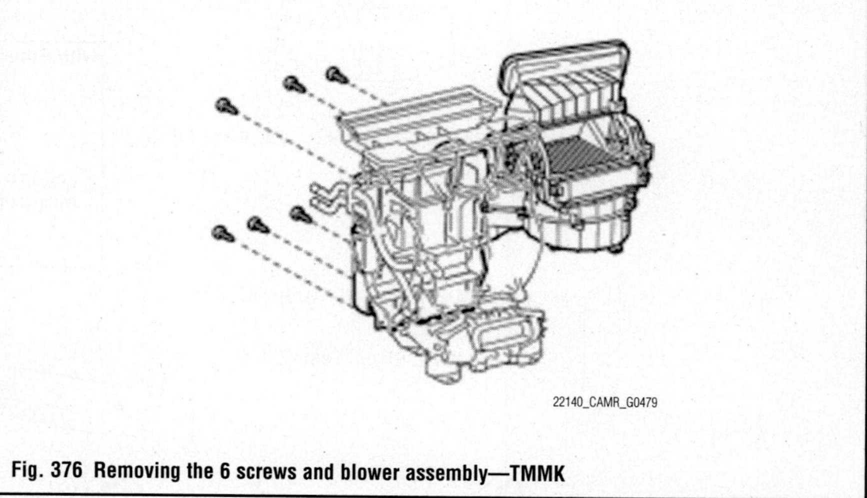
Fig. 376 Removing the 6 screws and blower assembly—TMMK

   c. Remove the 6 screws and then the blower assembly with the cooler evaporator sub-assembly.
   d. Remove the 3 screws and blower with fan motor sub-assembly.

To install:

6. To install, reverse removal procedure.

HEATER CORE

REMOVAL & INSTALLATION

See Figures 377 through 384.

Whenever working near any of the SRS components, such as the impact sensors, the air bag module, steering column and instrument panel, disable the SRS.

1. Before servicing the vehicle, refer to the Precautions Section.

⁂ CAUTION

Wait for 90 seconds after disconnecting the cable to prevent airbag deployment.

2. Disconnect battery negative terminal.
3. Remove the lower No. 2 and No. 3 steering wheel covers.
4. Remove the steering pad.
5. Remove the steering wheel assembly.
6. Remove the LH front door scuff plate.
7. Remove the LH cowl side trim sub-assembly.
8. Remove steering column cover.
9. Remove the turn signal switch assembly.
10. For vehicles without Smart Key System, disengage the 2 claws and 2 clips and

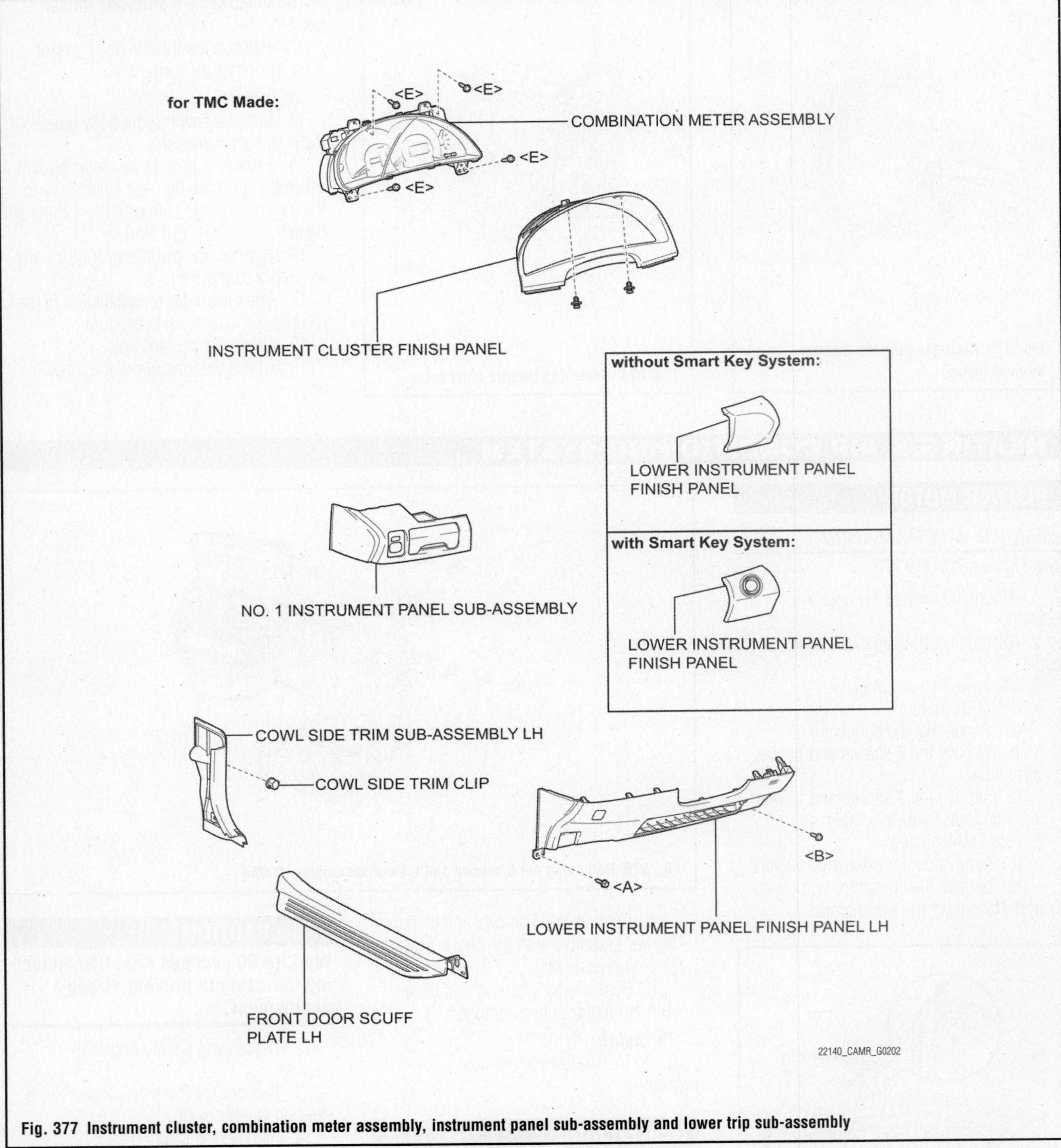

Fig. 377 Instrument cluster, combination meter assembly, instrument panel sub-assembly and lower trip sub-assembly

then remove the lower instrument panel finish panel.

11. For vehicles with Smart Key System, disengage the 2 claws and 2 clips. Disconnect the connector and remove the lower instrument panel finish panel.

12. Using a molding remover, disengage the 2 clips. Disengage the guide and 4 claws, and then remove the No. 1 instrument cluster finish panel.

13. Remove the 4 screws. Disconnect each connector and remove the combination meter assembly.

14. Remove the RH front door scuff plate.

15. Remove cowl side trim sub-assembly.

16. Disengage the 4 claws. Disengage the 2 guides and remove the No. 2 under cover sub-assembly.

17. Remove lower instrument panel sub-assembly by performing the following:

a. Remove the 4 screws.

b. Disengage the 3 claws and the 3 clips.

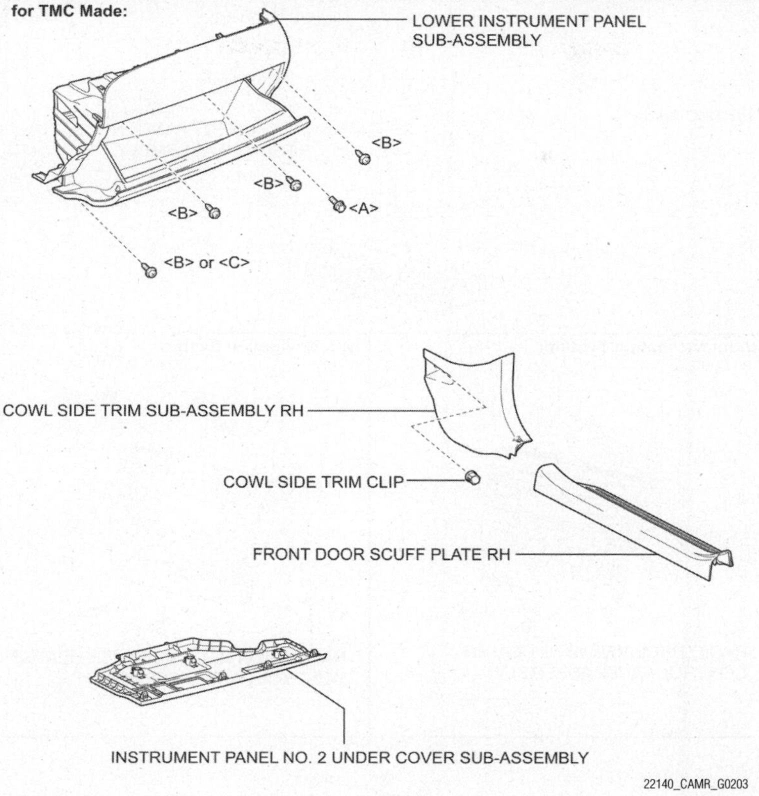

Fig. 378 Lower instrument panel sub-assembly and No. 2 under cover sub-assembly

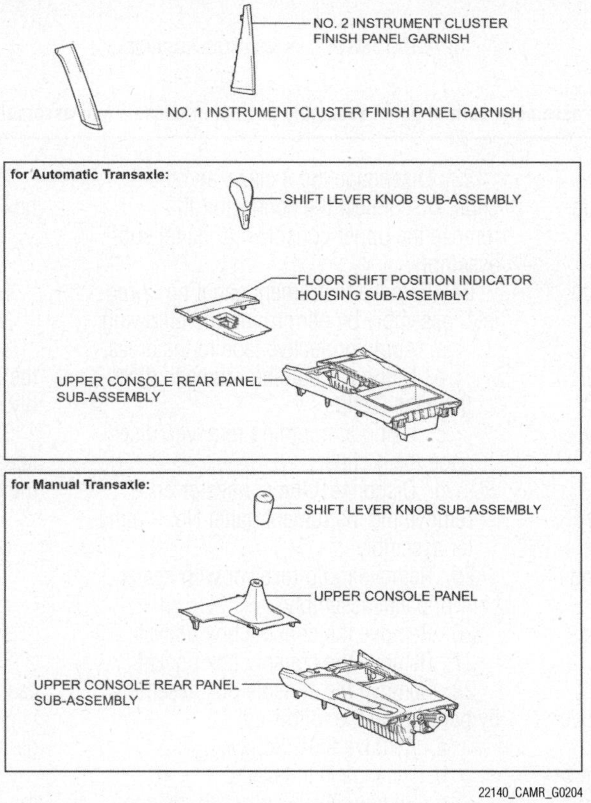

Fig. 379 Instrument cluster panel garnish, floor housing sub-assembly and upper console rear panel sub-assembly

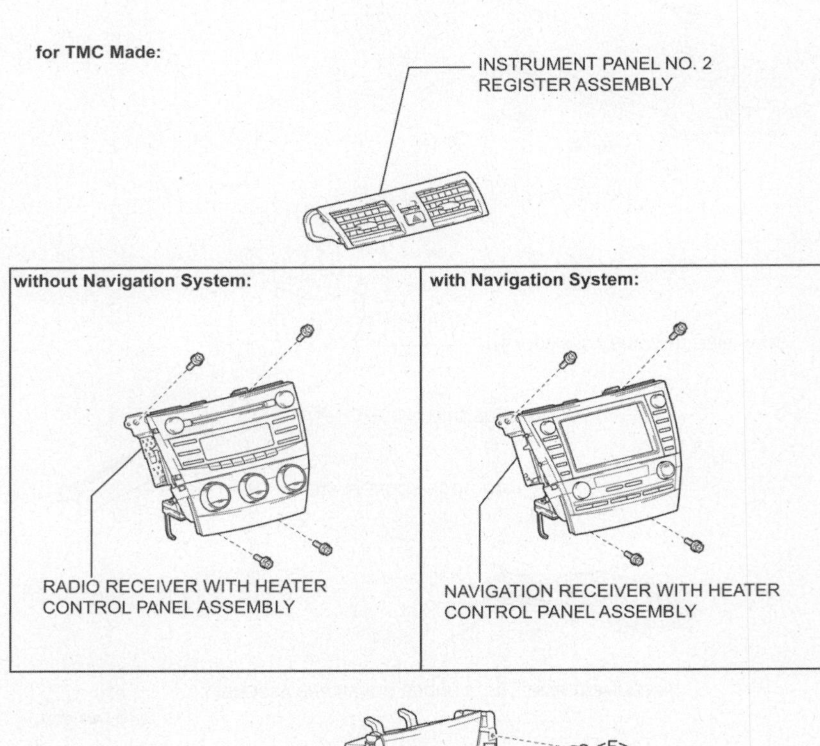

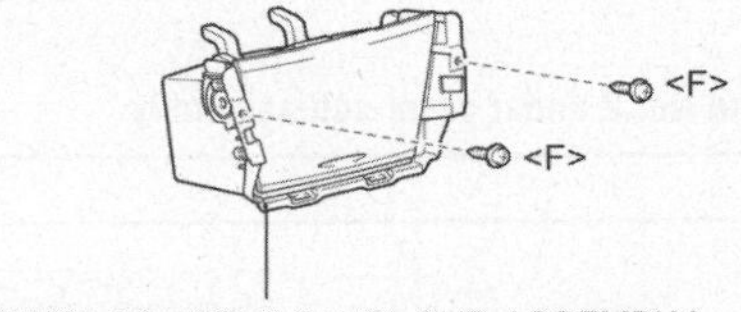

Fig. 380 Instrument panel no. 2 register assembly, control panel assembly and upper console sub-assembly

c. Disconnect the connector and remove the lower instrument panel sub-assembly.

18. Turn the shift lever knob counter-clockwise and remove the shift lever knob sub-assembly.

19. Disengage the 2 clips and remove the No. 1 instrument cluster finish panel garnish.

20. Disengage the 2 clips and remove the No. 2 instrument cluster finish panel garnish.

21. For A/T vehicles, Disengage the 6 claws and the 3 clips, and then remove the floor shift position indicator housing sub-assembly. If equipped with Seat Heater System, disconnect each connector.

22. For M/T vehicles, open the lid of the upper console panel. Apply protective tape to the area. Using a moulding remover, disengage the 2 claws and the 5 clips, and then remove the upper console panel.

23. Disengage the 3 claws and the 5 clips. Disconnect the connector and remove the upper console rear panel sub-assembly.

24. Remove instrument panel no. 2 register assembly by performing the following:

a. Apply protective tape to the areas.

b. Using a moulding remover, disengage the 3 clips.

c. Using a moulding remover, disengage the 4 clips.

d. Disconnect the connector and remove the instrument panel No. 2 register assembly.

25. Remove radio receiver with heater control panel assembly.

26. Remove the console box pocket

27. Remove the console box pocket.

28. Remove the console box assembly by performing the following:

a. Remove the 2 screws.

b. Disengage the clamp.

c. Remove the 2 bolts and the console box assembly.

29. Remove both of the front console box inserts by performing the following:

a. Remove the 3 screws.

b. Disengage the clip and remove the front console box insert.

30. Remove the LH front pillar garnish.

31. Disengage the 4 clips and remove the instrument panel No. 1 register assembly.

32. Remove instrument panel no. 1 speaker panel sub-assembly by performing the following:

a. Disengage the 6 claws and the 2 clips.

b. Disengage the 2 guides and remove the instrument panel No. 1 speaker panel sub-assembly.

33. Remove RH front no. 2 speaker assembly.

34. Remove the RH front pillar garnish.

35. Disengage the 4 clips and remove the instrument panel No. 3 register assembly.

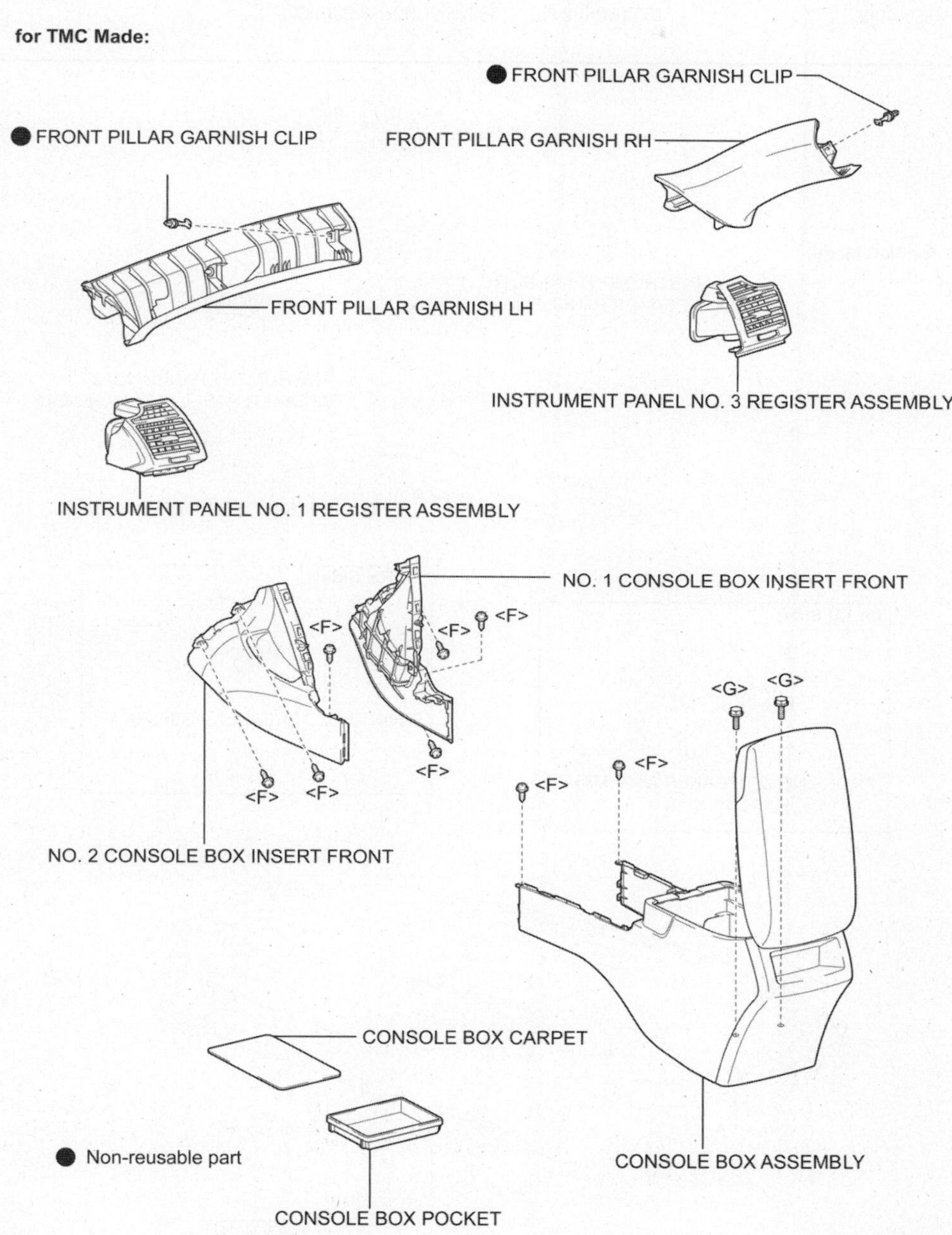

Fig. 381 Console box assembly, front console box inserts, LH front pillar garnish and instrument panel No. 1 and No. 3 register assembly

36. Remove instrument panel no. 2 speaker panel sub-assembly by performing the following:
   a. Disengage the 6 claws and the 2 clips.
   b. Disengage the 2 guides and remove the instrument panel No. 2 speaker panel sub-assembly.

37. Remove LH and RH front no. 2 speaker assemblies.

38. Remove no. 1 defroster nozzle garnish by performing the following:
   a. Disengage the 8 clips and the 4 guides.
   b. Disconnect each connector and remove the No. 1 defroster nozzle garnish.

39. Disconnect instrument panel wire assembly.

40. Remove instrument panel safety pad assembly by performing the following:
   a. Disengage each clamp.
   b. Disconnect each connector.
   c. Remove the bolt (J).

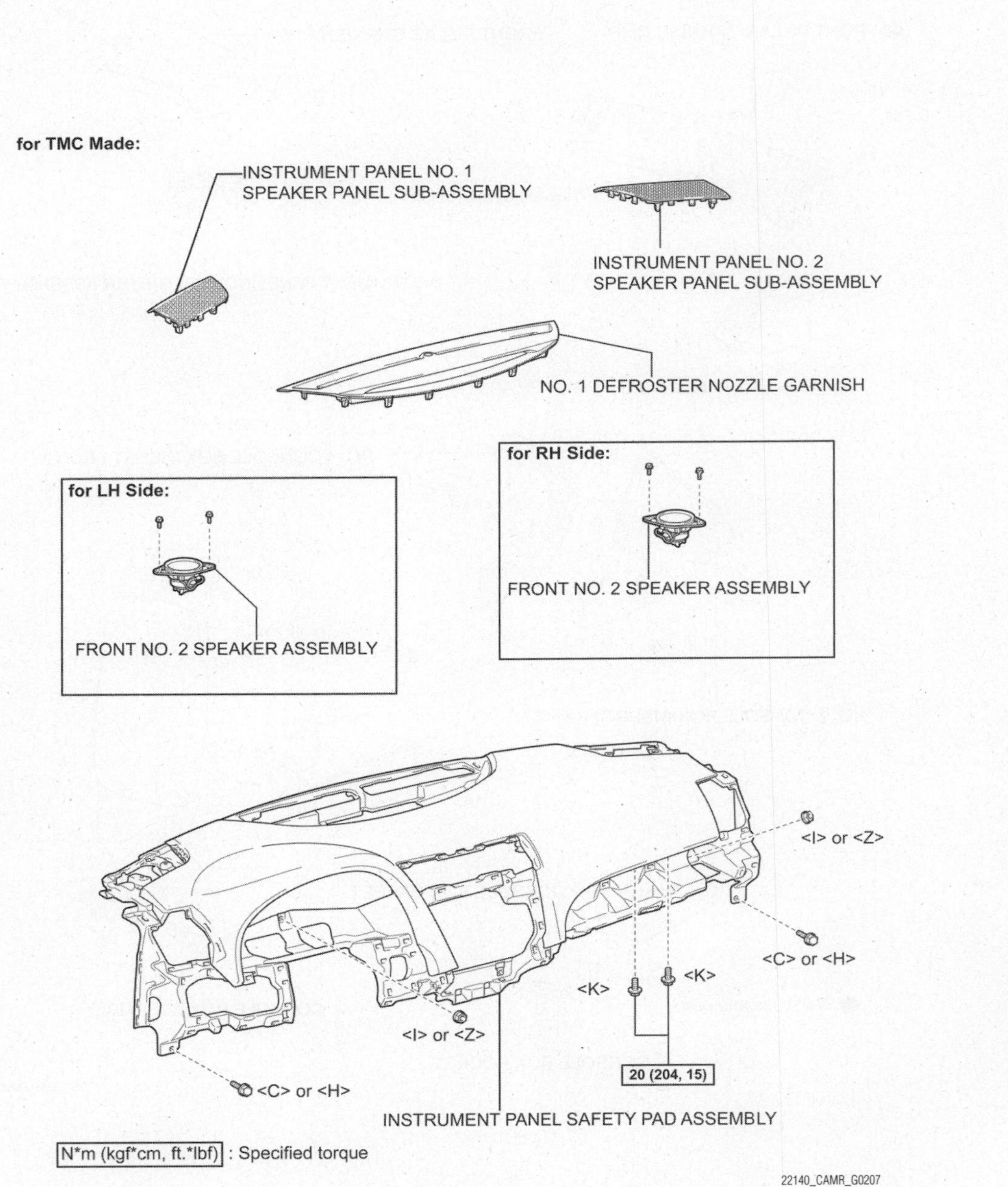

Fig. 382 Speakers, speaker panel sub-assembly and No. 1 defroster nozzle garnish

⁂ CAUTION

Models are equipped with a Supplemental Restraint System (SRS), which uses an air bag. Whenever working near any of the SRS components, such as the impact sensors, the air bag module, steering column and instrument panel, disable the SRS.

d. Remove the 2 passenger airbag bolts (K).

e. If equipped with Plasmacluster, disconnect the connector

f. Disconnect the connector

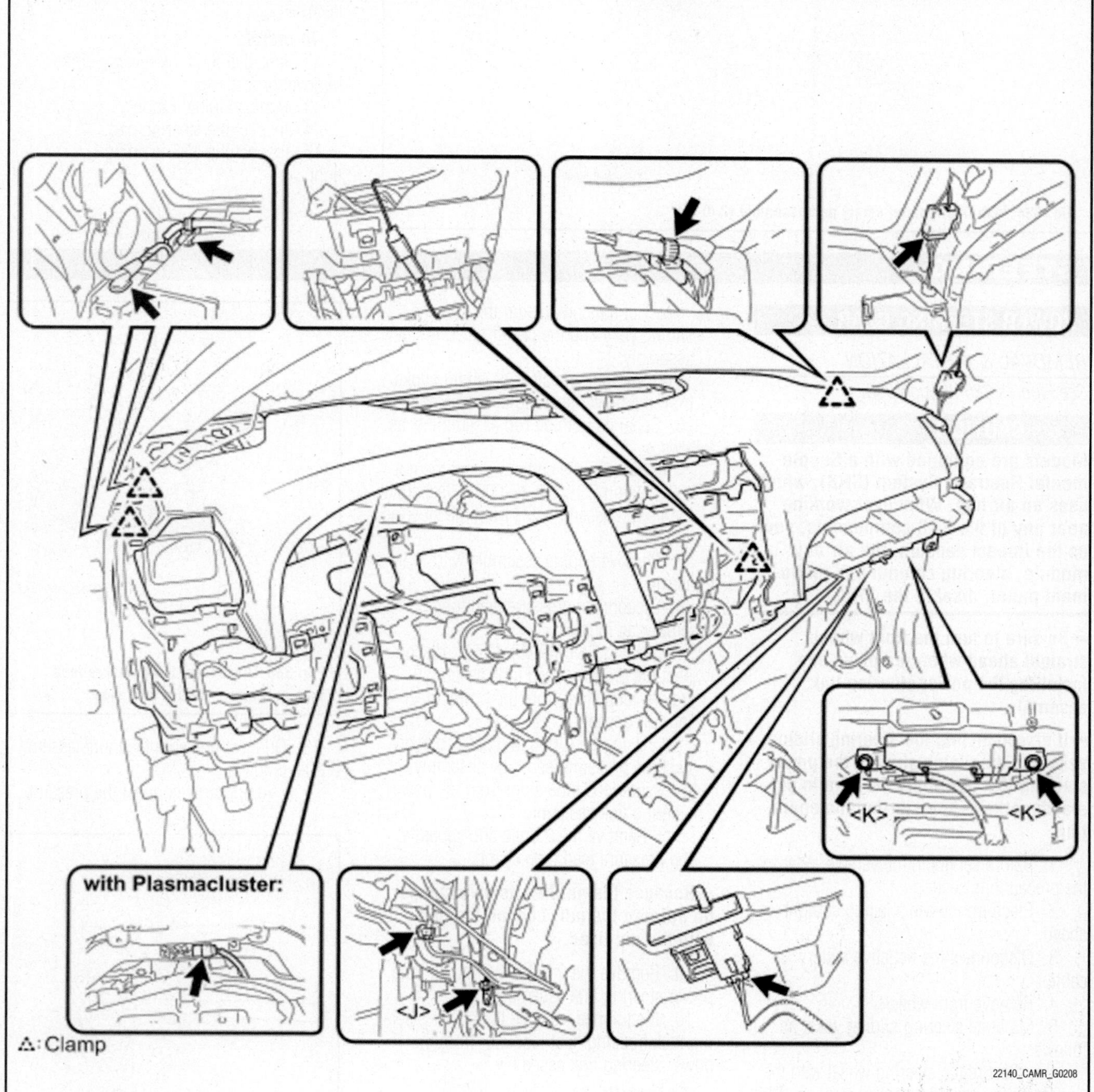

Fig. 383 Instrument panel safety pad assembly (1 of 2)

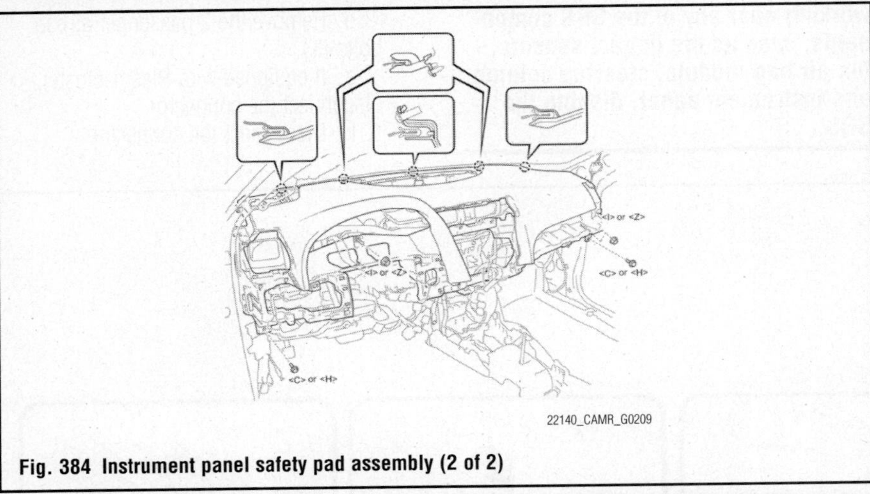

Fig. 384 Instrument panel safety pad assembly (2 of 2)

g. Remove the 2 bolts (C) or (H).

h. Disengage the 5 claws and remove the instrument panel safety pad assembly.

i. Disengage the claw and remove the 5 instrument panel stays.

41. Remove heater core as necessary.

To install:

42. Installation is the reverse of the removal procedure.

43. Perform initialization.

44. Inspect the steering pad.

45. Inspect the SRS warning light.

STEERING

POWER STEERING GEAR

REMOVAL & INSTALLATION

See Figures 385 through 388.

⁂ CAUTION

Models are equipped with a Supplemental Restraint System (SRS), which uses an air bag. Whenever working near any of the SRS components, such as the impact sensors, the air bag module, steering column and instrument panel, disable the SRS.

➡Be sure to turn the front wheels straight ahead when removing and installing the power steering link assembly.

➡If disconnecting the steering sliding yoke and the pinion shaft of the power steering link assembly, be sure to put matchmarks before starting the operation.

1. Before servicing the vehicle, refer to the precautions section.

2. Place front wheels facing straight ahead.

3. Disconnect the negative battery cable.

4. Remove front wheels.

5. Separate steering sliding yoke, as follows:

a. Secure the steering wheel with the seat belt in order to prevent rotation. This operation is useful to prevent damage to the spiral cable.

b. Remove the bolt and slide the steering sliding yoke. Do not separate the steering sliding yoke from the power steering link assembly.

c. Put matchmarks on the steering sliding yoke and the power steering link assembly.

d. Separate the steering sliding yoke from the power steering link assembly.

6. Separate both tie rod assemblies, as follows:

a. Remove the cotter pin and the nut.

b. Using SST: 09628-00011, separate the tie rod assembly LH from the steering knuckle.

7. Remove engine assembly with transaxle.

8. Disconnect pressure feed tube assembly, as follows:

a. Using SST: 09023-12701, disconnect the pressure feed tube assembly (return tube side) from the power steering link assembly.

b. Using SST: 09023-12701, disconnect the pressure feed tube assembly (pressure feed tube side) from the power steering link assembly.

c. Remove the 2 bolts and separate the pressure feed tube clamp.

➡Because the nut has its own stopper, do not turn the nut. Loosen the bolt with the nut fixed.

9. Remove the 2 bolts, 2 nuts, and the power steering link assembly.

10. If equipped, remove the power steering rack housing heat insulator from the power steering link assembly.

To install:

11. If equipped, install the power steering rack housing heat insulator to the power steering link assembly.

12. Install the power steering link assembly with the 2 bolts and 2 nuts. Tighten to 52 ft. lbs. (70 Nm).

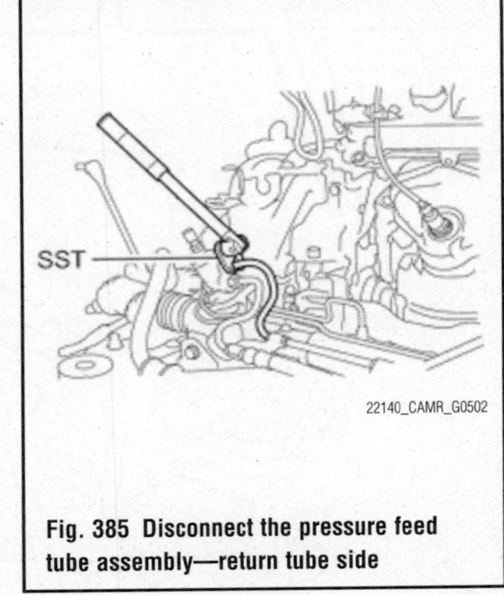

Fig. 385 Disconnect the pressure feed tube assembly—return tube side

13. Connect pressure feed tube assembly, as follows:

a. Temporarily connect the pressure

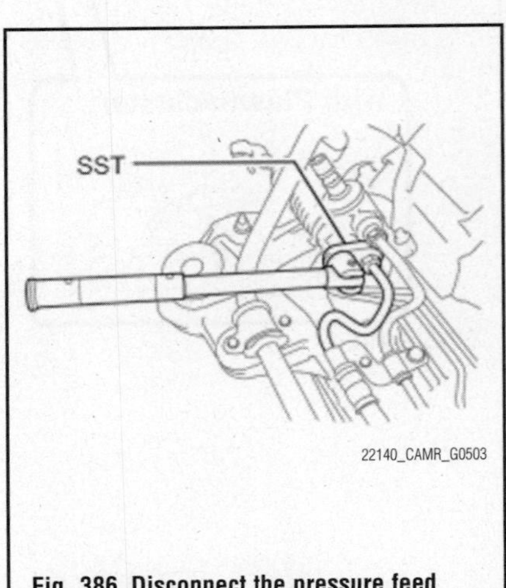

Fig. 386 Disconnect the pressure feed tube assembly—pressure feed tube side

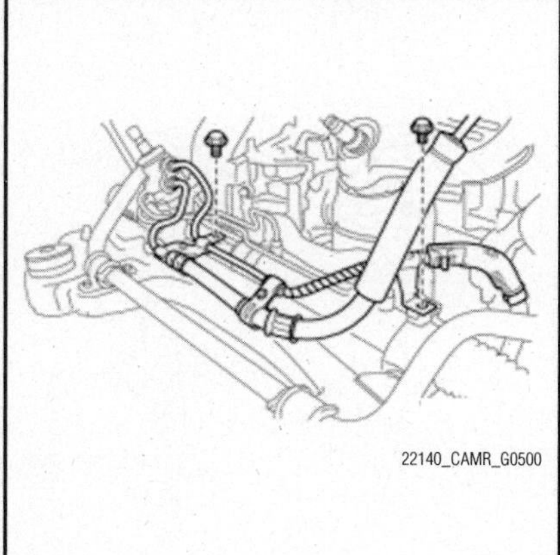

Fig. 387 Removing the 2 bolts and pressure feed tube clamp

Fig. 388 Removing power steering link assembly

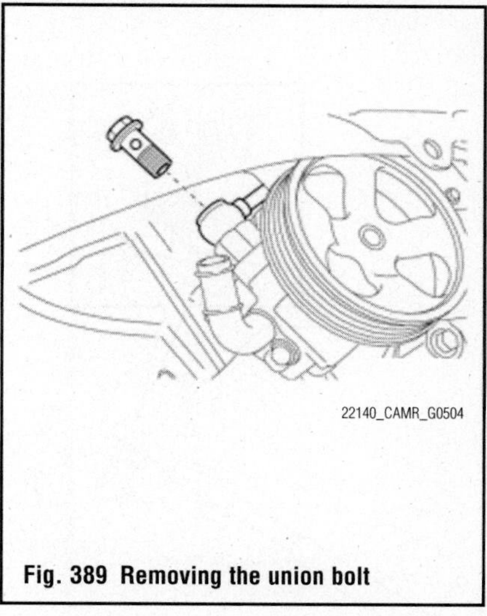

Fig. 389 Removing the union bolt

feed tube assembly to the power steering link assembly.

b. Install the pressure feed tube assembly clamp with the 2 bolts and tighten to 87 inch lbs. (9.8 Nm).

➡Use a torque wrench with a fulcrum length of 11.81 inches (300 mm).

➡This torque value is effective when SST is parallel to the torque wrench.

c. Using SST: 09023-12701, tighten the pressure feed tube assembly (pressure feed tube side) to 16 ft. lbs. (22 Nm).

d. Using SST: 09023-12701, tighten the pressure feed tube assembly (return tube side) to 16 ft. lbs. (22 Nm).

14. Install engine assembly with transaxle.

15. Connect both of the tie rod assemblies to the steering knuckle with the nut. Tighten to 36 ft. lbs. (49 Nm). Install a new cotter pin. Further tighten the nut up to 60°if the holes for the cotter pin are not aligned.

16. Connect steering sliding yoke, as follows:

a. Align the matchmarks on the steering sliding yoke and the steering link assembly.

b. Install the bolt to 26 ft. lbs. (35 Nm).

17. Install front wheels and tighten to 76 ft. lbs. (103 Nm).

18. Connect cable to negative battery terminal.

19. Bleed power steering fluid.

20. Check power steering fluid level.

21. Check for power steering fluid leakage.

22. Check for exhaust gas leaks.

23. Place front wheels facing straight ahead.

24. Inspect and adjust front wheel alignment.

POWER STEERING PUMP

REMOVAL & INSTALLATION

2.4L Engine

See Figures 389 through 392.

1. Before servicing the vehicle, refer to the precautions section.
2. Drain power steering fluid.
3. Remove RH engine under cover.
4. Remove RH front fender apron seal.
5. Remove fan and generator v belt.
6. Slide the clip and disconnect the No. 1 fluid reservoir to pump hose from the vane pump assembly.
7. Disconnect pressure feed tube assembly, as follows:

a. Remove the union bolt and disconnect the pressure feed tube assembly from the vane pump assembly.

b. Remove the gasket from the pressure feed tube assembly.

8. Disconnect the power steering fluid pressure switch connector.
9. Remove vane pump assembly, as follows:

a. Using SST: 09249-63010, loosen the 2 bolts and remove the vane pump assembly.

b. Remove the 2 bolts from the vane pump assembly.

To install:

10. Install vane pump assembly, as follows:

a. Temporarily install the 2 bolts to the vane pump assembly.

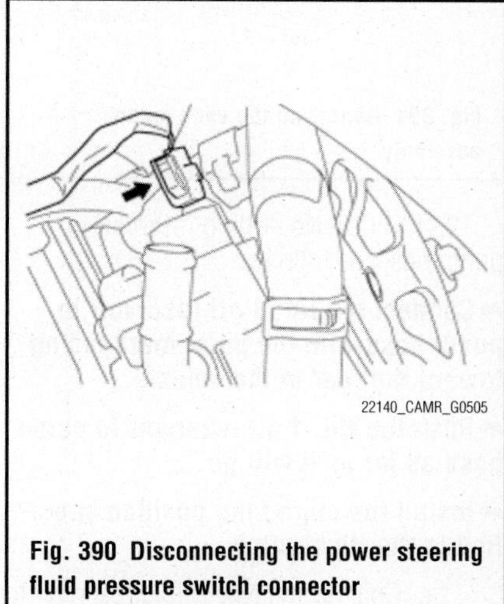

Fig. 390 Disconnecting the power steering fluid pressure switch connector

b. Install the vane pump assembly.

➡Use a torque wrench with a fulcrum length of 11.81 inches (300 mm).

➡This torque value is effective when SST is parallel to the torque wrench.

c. Using SST: 09249-63010, tighten the 2 bolts to 32 ft. lbs. (43 Nm).

11. Connect the connector to the power steering fluid pressure switch.

12. Connect pressure feed tube assembly, as follows:

a. Install a new gasket to the pressure feed tube assembly.

➡Make sure that the stopper of the pressure feed tube assembly contacts the vane pump assembly securely as shown in the illustration.

b. Connect the pressure feed tube assembly to the vane pump assembly with the union bolt. Tighten to 37 ft. lbs. (50 Nm).

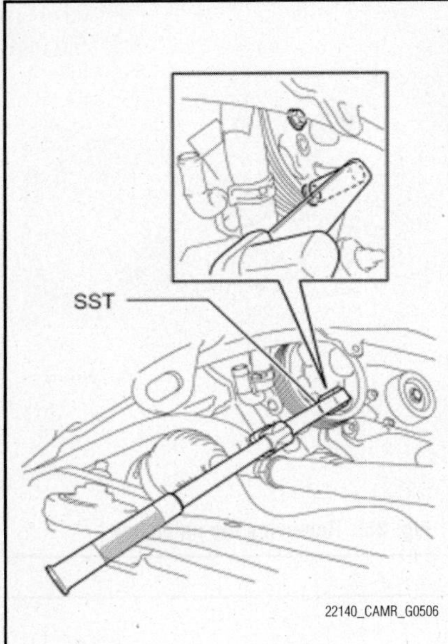

Fig. 391 Removing the vane pump assembly

13. Connect No. 1 fluid reservoir to pump hose, as follows:

➡Connect the No. 1 oil reservoir to pump hose with the paint mark facing toward the rear of the vehicle.

➡Push the No. 1 oil reservoir to pump hose as far as it will go.

➡Install the clip at the position specified in the illustration.

a. Connect the No. 1 fluid reservoir to pump hose to the vane pump assembly with the clip.

14. To complete installation, reverse removal procedure.

2.5L Engine

See Figures 393 through 396.

1. Drain the power steering fluid.
2. Remove the front wheel RH.
3. Remove the front fender apron seal RH.
4. Remove the v-ribbed belt.
   a. Attach a wrench to the hexagonal portion of the belt tensioner as shown in the illustration, rotate the belt tensioner clockwise, and remove the V-ribbed belt.
5. Disconnect the No. 1 fluid reservoir to pump hose.
   a. Slide the clip and disconnect the No. 1 fluid reservoir to pump hose from the vane pump assembly.
6. Disconnect the pressure feed tube assembly.
   a. Remove the union bolt and disconnect the pressure feed tube assembly from the vane pump assembly.
   b. Remove the union bolt and disconnect the pressure feed tube assembly from the vane pump assembly.
7. Disconnect the power steering fluid pressure switch connector.
8. Remove the vane pump assembly.
   a. Using the special tool, loosen the 2 bolts and remove the vane pump assembly.
   b. Remove the 2 bolts from the vane pump assembly.

To install:

9. Install the vane pump.
   a. Temporarily install the 2 bolts to the vane pump assembly.
   b. Install the vane pump assembly.
   c. Using the special tool, tighten to 2 bolts. Using the special tool, tighten to 21 ft. lbs. (29 Nm). Tightening without the special tool, tighten to 32 ft. lbs. (43 Nm).
10. Connect the power steering fluid pressure switch connector.
11. Connect the pressure feed tube assembly and a new gasket. Tighten to 37 ft. lbs. (50 Nm).

➡Make sure that the stopper of the pressure feed tube assembly contacts the vane pump assembly securely.

12. Connect the No. 1 fluid reservoir to pump hose with the clip.

➡Connect the No. 1 fluid reservoir to pump hose with the paint mark facing the rear of the vehicle.

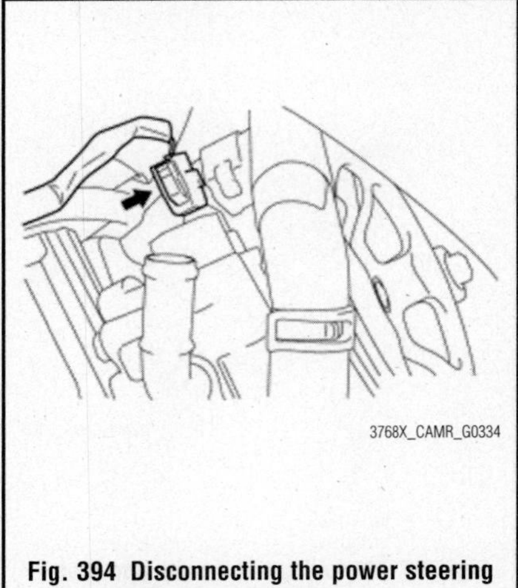

Fig. 394 Disconnecting the power steering fluid pressure switch connector

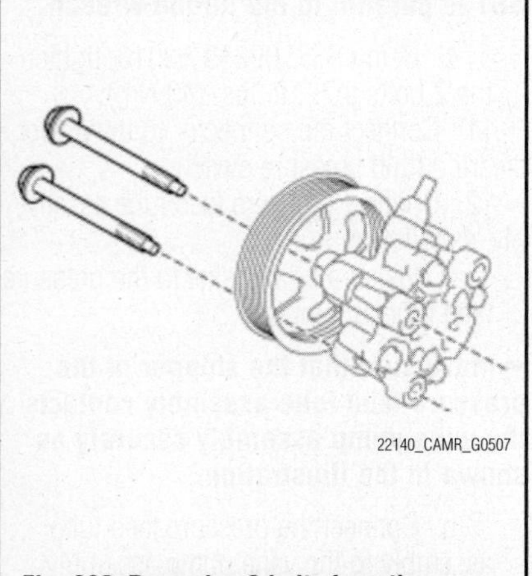

Fig. 392 Removing 2 bolts from the vane pump assembly

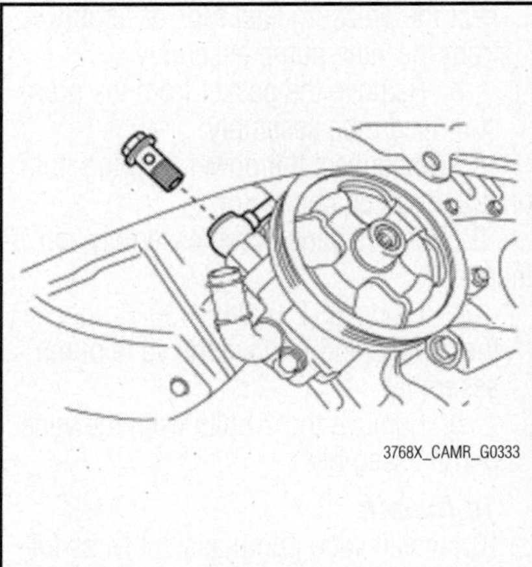

Fig. 393 Disconnecting the pressure feed tube assembly

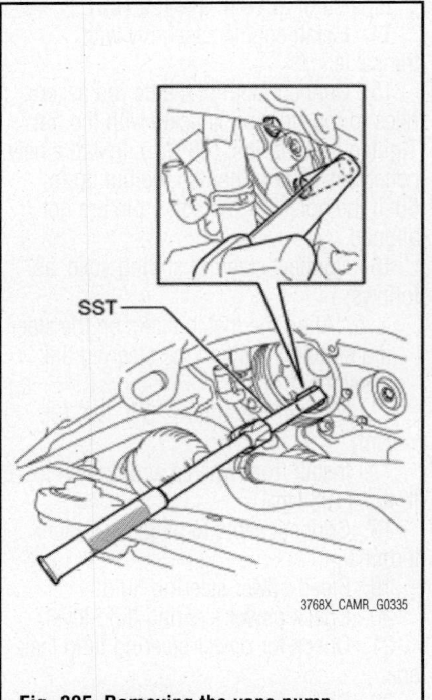

Fig. 395 Removing the vane pump

➡**Push the No. 1 fluid reservoir to pump hose as far as it will go as shown in the illustration.**

➡**Install the clip at the position specified in the illustration.**

13. Install the v-ribbed belt.
    a. Set the v-ribbed belt onto each part.
    b. Loosen the v-ribbed belt by turning the belt tensioner clockwise.
    c. Set the v-ribbed belt onto the tensioner pulley.

➡**Make sure that the belt is attached to each pulley. In particular, make sure that the belt is securely fitted into the grooves of the crankshaft pulley.**

14. Add power steering fluid.
15. Bleed the power steering fluid.
16. Check the power steering fluid level.
17. Inspect for power steering fluid leaks.
18. Install the front fender apron seal RH.
19. Install the front wheel RH. Tighten to 76 ft. lbs. (103 Nm).

3.5L Engine

See Figures 397 through 400.

1. Before servicing the vehicle, refer to the precautions section.
2. Drain power steering fluid.
3. Remove RH engine under cover.
4. Remove RH front fender apron seal.
5. Remove v-bank cover sub-assembly.
6. Remove fan and generator v belt.
7. Slide the clip and disconnect the No. 1 fluid reservoir to pump hose from the vane pump assembly.
8. Disconnect pressure feed tube assembly, as follows:
    a. Remove the union bolt and disconnect the pressure feed tube assembly from the vane pump assembly.
    b. Remove the bolt and separate the pressure feed tube clamp.

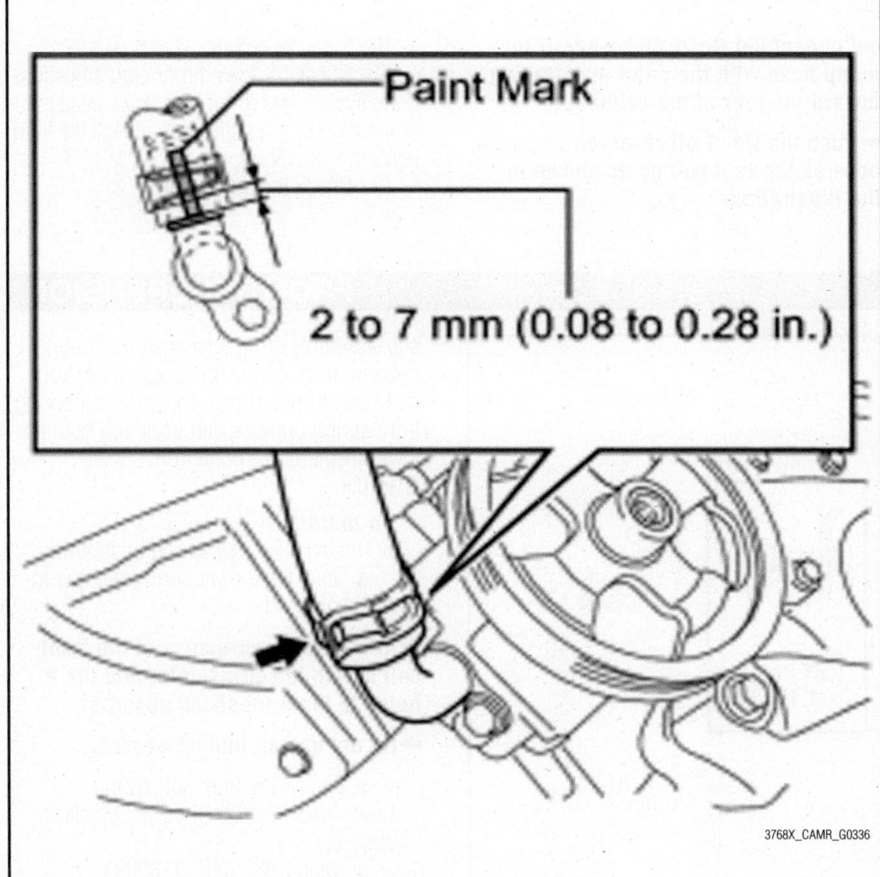

Fig. 396 Connecting the No. 1 fluid reservoir to the pump hose

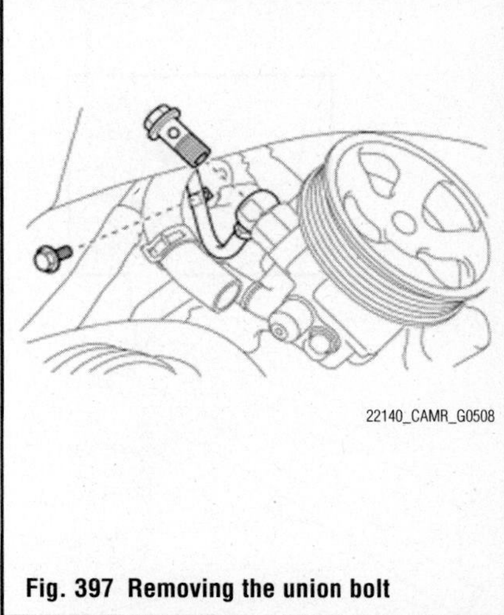

Fig. 397 Removing the union bolt

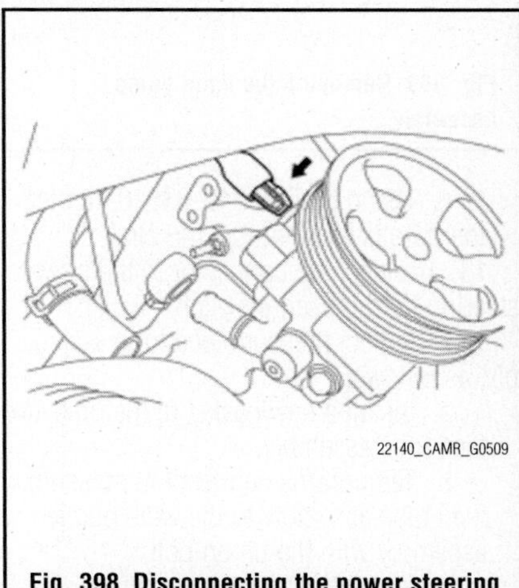

Fig. 398 Disconnecting the power steering fluid pressure switch connector

    c. Remove the gasket from the pressure feed tube assembly.
9. Disconnect the power steering fluid pressure switch connector.
10. Using SST: 09249-63010, loosen bolt (A) and remove bolt (B), and then remove the vane pump assembly.
11. Remove the bolt from the vane pump assembly.

To install:

12. Install vane pump assembly, as follows:
    a. Temporarily install the bolt to the vane pump assembly.
    b. Install the vane pump assembly.

➡**Use a torque wrench with a fulcrum length of 11.81 inches (300 mm).**

➡**This torque value is effective when SST is parallel to the torque wrench.**

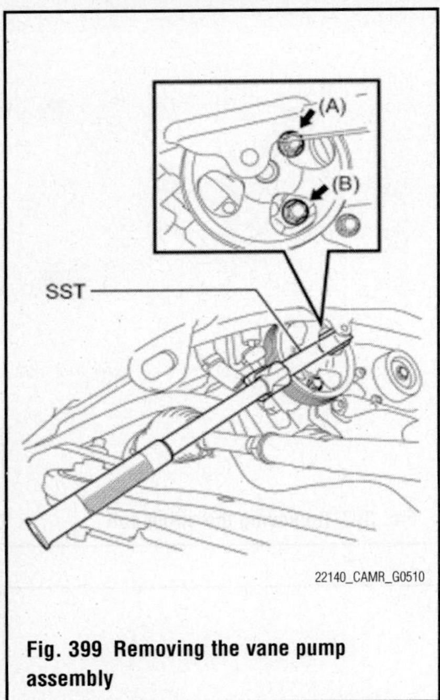

Fig. 399 Removing the vane pump assembly

c. Using SST: 09249-63010, tighten the 2 bolts to 32 ft. lbs. (43 Nm).

13. Connect the connector to the power steering fluid pressure switch.

14. Connect pressure feed tube assembly, as follows:

a. Install a new gasket to the pressure feed tube assembly.

b. Temporarily connect the pressure feed tube assembly to the vane pump assembly with the union bolt.

c. Install the pressure feed tube

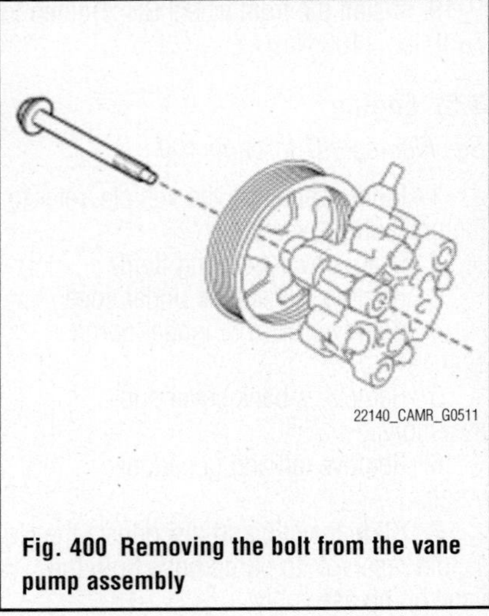

Fig. 400 Removing the bolt from the vane pump assembly

assembly clamp with the bolt. Tighten to 87 ft. lbs. (10 Nm).

d. Fully tighten the union bolt and tighten to 37 ft. lbs. (50 Nm).

➡Make sure that the stopper of the pressure feed tube assembly contacts the vane pump assembly securely.

15. Connect No. 1 fluid reservoir to pump hose, as follows:

➡Connect the No. 1 oil reservoir to pump hose with the paint mark facing toward the rear of the vehicle.

➡Push the No. 1 oil reservoir to pump hose as far as it will go as shown in the illustration.

➡Install the clip at the position specified in the illustration.

a. Connect the No. 1 fluid reservoir to pump hose to the vane pump assembly with the clip.

16. To complete installation, reverse removal procedure.

BLEEDING

1. Before servicing the vehicle, refer to the precautions section.
2. Check the fluid level.
3. Jack up the front of the vehicle and support it with stands.
4. With the engine stopped, turn the wheel slowly from lock to lock several times.
5. Lower the vehicle.
6. Start the engine.
7. Run the engine at idle for a few minutes.
8. With the engine idling, turn the wheel left or right to the full lock position and keep it there for 2 to 3 seconds, then turn the wheel to the opposite full lock position and keep it there for 2 to 3 seconds.
9. Repeat the above steps several times.
10. Stop the engine.
11. Check for foaming or emulsification. If the system has to be bled twice because of foaming or emulsification, check for fluid leaks in the system.
12. Check the fluid level.

SUSPENSION

FRONT SUSPENSION

COIL SPRING

REMOVAL & INSTALLATION

See Figures 401 through 406.

1. Before servicing the vehicle, refer to the precautions section.
2. Remove the front shock absorber.
3. As shown in the illustration, secure the front shock absorber with coil spring in a vise using aluminum plates by clamping onto a double nutted bolt affixed to the bracket at the bottom of the absorber.

➡Do not use an impact wrench.

➡If the front coil spring is compressed at an angle, using 2 SST will make the work easier.

4. Using SST: 09727-30021, compress the front coil spring.
5. Remove the front suspension support sub-assembly, front suspension support bearing, front coil spring upper seat, front coil spring upper insulator, front coil spring, front spring bumper, and front coil spring lower insulator from the front shock absorber.

28 mm (1.1 in.)

22140_CAMR_G0525

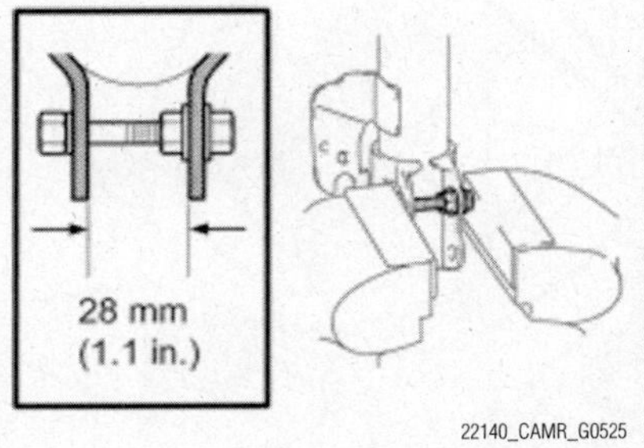

Fig. 401 Secure the front shock absorber

To install:

6. Install front coil spring as follows:

a. Install the front spring bumper to the piston rod.

➡Align the 2 protrusions of the front coil spring lower insulator and the 2 holes in the front shock absorber.

➡Do not use an impact wrench.

b. Install the front coil spring lower insulator onto the front shock absorber.

c. Using SST: 09727-30021, compress the front coil spring.

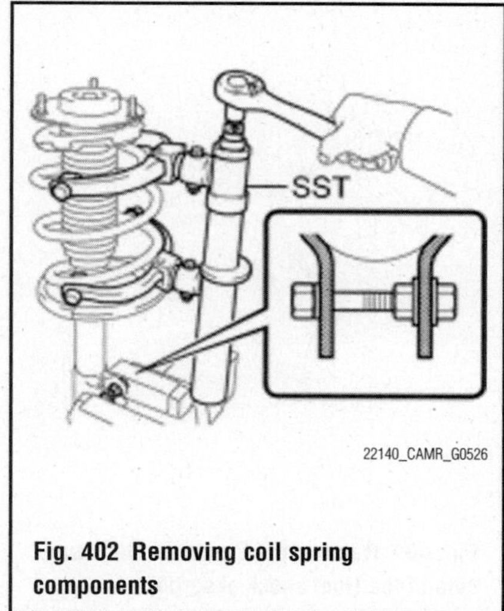

Fig. 402 Removing coil spring components

➡The smaller diameter end of the front coil spring must face upward.

➡Fit the lower end of the front coil spring into the gap of the insulator.

d. Install the front coil spring to the front shock absorber.

➡Any misalignment between the front shock absorber lower bracket and the matchmark must be +/-5°.

e. Install the front coil spring upper insulator as shown in the illustration.

➡Any misalignment between the front shock absorber lower bracket and the matchmark must be +/-5°.

7. Install the front coil spring upper seat with the mark facing to the outside of the vehicle.

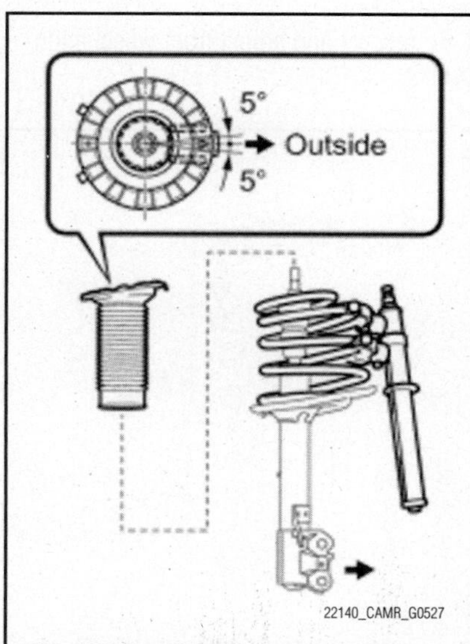

Fig. 403 Installing the front coil spring upper insulator

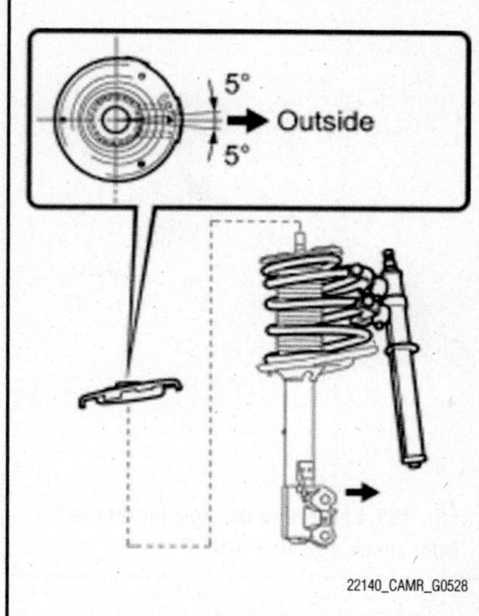

Fig. 404 Installing the front coil spring upper seat

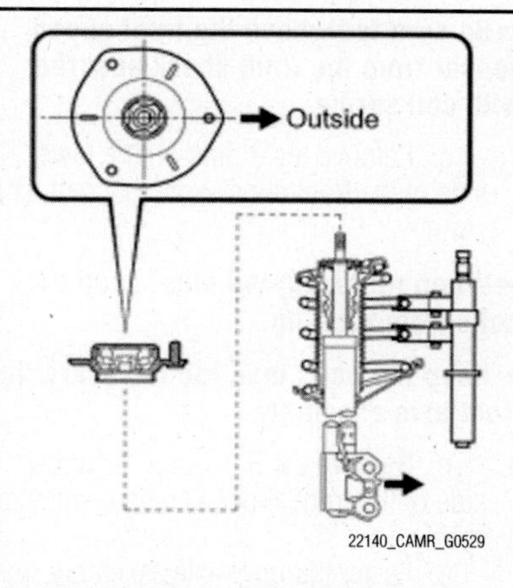

Fig. 405 the front suspension support sub-assembly

➡If there is foreign matter inside the front suspension support bearing, replace it with a new one.

a. Install a new front suspension support bearing.

➡Check that the flats on the piston rod and the flats on the front suspension support sub-assembly are aligned.

b. Install the front suspension support sub-assembly. Temporarily tighten a new lock nut.

➡Do not use an impact wrench.

➡Any misalignment between the front shock absorber lower bracket and the matchmark must be +/-5°.

c. Remove the SST slowly in order to release the coil spring.

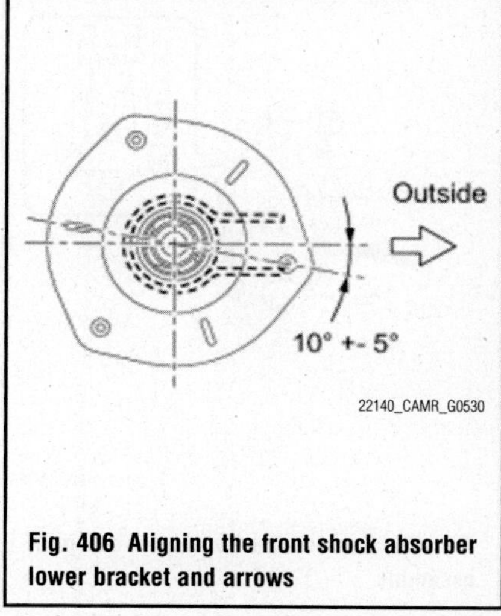

Fig. 406 Aligning the front shock absorber lower bracket and arrows

CONTROL LINKS

REMOVAL & INSTALLATION

See Front Stabilizer Bar.

LOWER BALL JOINT

REMOVAL & INSTALLATION

See Figure 407.

1. Before servicing the vehicle, refer to the precautions section.
2. Remove the front wheel.
3. Remove the front axle hub nut.
4. Separate the front speed sensor.
5. Separate the front disc the brake caliper assembly.
6. Remove front disc.
7. Separate the tie rod assembly.
8. Separate the No. 1 front lower suspension arm.
9. Remove the front axle assembly.
10. Remove front wheel No. 1 bearing dust deflector.
11. Remove front axle hub hole snap ring.
12. Remove front axle hub.
13. Remove front disc brake dust cover.
14. Remove the front lower ball joint assembly, as follows:

a. Secure the steering knuckle in a vise using aluminum plates.

b. Remove the cotter pin and castle nut.

➡Do not damage the dust cover of the ball joint.

➡Do not damage the steering knuckle.

c. Using SST (SST: 09628-62011) or equivalent, remove the front lower ball joint assembly.

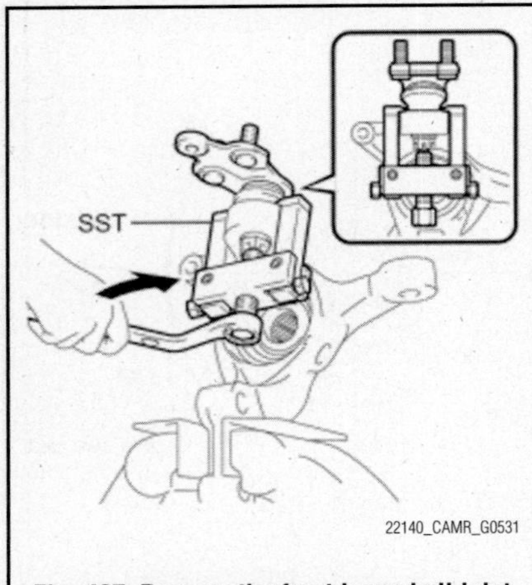

Fig. 407 Remove the front lower ball joint assembly

To install:

15. Installation is the reverse of the removal procedure, noting the following:
   a. Install the front lower ball joint assembly to the steering knuckle with the castle nut and tighten to 91 ft. lbs. (123 Nm). Further tighten the nut up to 60° if the holes for the cotter pin are not aligned.
   b. Inspect and adjust the front wheel alignment.
   c. Inspect the ABS speed sensor signal.

STEERING KNUCKLE

REMOVAL & INSTALLATION

See Wheel Hub and Bearing.

STRUT & SPRING ASSEMBLY

REMOVAL & INSTALLATION

See Figures 408 through 410.

➡Use the same procedures for the RH side and the LH side. The procedures listed below are for the LH side.

1. Before servicing the vehicle, refer to the precautions section.
2. Remove the front wheel.
3. Remove the nut and disconnect the front stabilizer link assembly from the front shock absorber assembly.
4. Remove front shock absorber with coil spring, as follows:
   a. Loosen the lock nut of the front shock absorber with coil spring.

➡Do not remove the lock nut.

➡Only loosen the nut when disassembling the front shock absorber with coil spring.

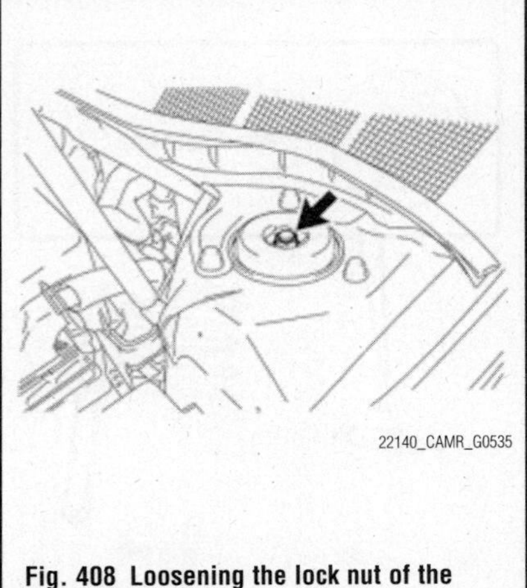

Fig. 408 Loosening the lock nut of the front shock absorber arm

   b. Remove the bolt and disconnect the front flexible hose and front speed sensor wire harness from the front shock absorber with coil spring.

➡Be sure to remove the front speed sensor from the front shock absorber with coil spring.

   c. Remove the 2 nuts on the lower side of the front shock absorber with coil spring.

➡When removing the nuts, keep the bolts from rotating.

➡Keep the bolts inserted to secure the front axle assembly.

   d. Remove the 3 nuts on the upper side of the front shock absorber with coil spring.
   e. Lower the front axle assembly, and remove the 2 bolts on the lower side of the front shock absorber.

➡Make sure that the front speed sensor is disconnected from the front shock absorber with coil spring.

   f. Remove the front shock absorber with coil spring.

To install:

5. Install front shock absorber with coil spring, as follows:
   a. Install the front shock absorber with coil spring to the front axle assembly and insert the 2 bolts from the front side of the vehicle.
   b. Slowly jack up the vehicle using a wooden block and install the front shock absorber with coil spring (upper side) to the vehicle.
   c. Install the 3 nuts to the upper side of the front shock absorber with coil spring and tighten to 63 ft. lbs. (85 Nm).

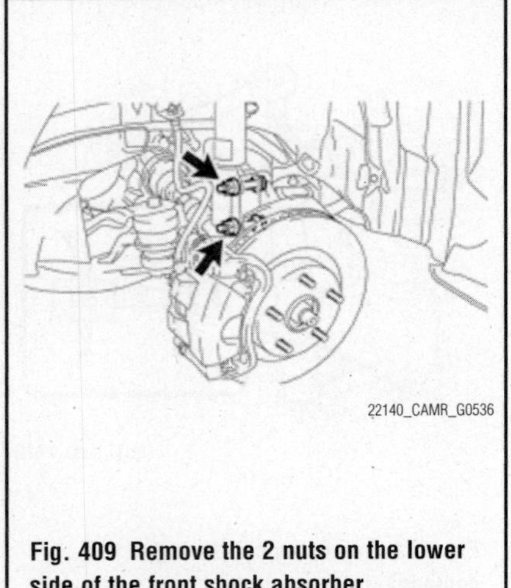

Fig. 409 Remove the 2 nuts on the lower side of the front shock absorber

➡When installing the nuts, keep the bolts from rotating.

   d. Install the 2 nuts to the lower side of the front shock absorber with coil spring and tighten to 155 ft. lbs. (210 Nm).
   e. Install the front flexible hose and front speed sensor wire harness with the bolt and tighten to 14 ft. lbs. (19 Nm).
   f. Fully tighten the lock nut and tighten to 52 ft. lbs. (70 Nm).

➡If the ball joint turns together with the nut, use a hexagon wrench (6 mm) to hold the stud.

6. Install the front stabilizer link assembly with the nut and tighten to 55 ft. lbs. (74 Nm).
7. Install front wheel and tighten to 76 ft. lbs. (103 Nm).
8. Inspect and adjust front wheel alignment.

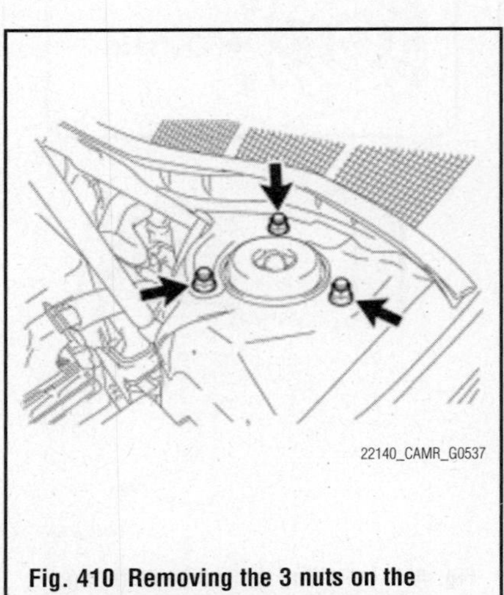

Fig. 410 Removing the 3 nuts on the upper side of the front shock absorber

WHEEL BEARINGS

REMOVAL & INSTALLATION

See Figures 411 through 417.

1. Before servicing the vehicle, refer to the Precautions Section.
2. Remove front wheel.
3. Remove front axle hub nut.
4. Separate front speed sensor.
5. Remove the 2 bolts and separate the front disc brake caliper assembly from the steering knuckle. Use wire or an equivalent tool to keep the brake caliper from hanging down by the flexible hose.
6. Remove front disc.
7. Separate tie rod end sub-assembly.
8. Separate front suspension lower no. 1 arm.
9. Remove front axle assembly.
10. Using a screwdriver with its tip wrapped with vinyl tape, remove the No. 1 front wheel bearing dust deflector. Be careful not to damage the steering knuckle.
11. Using snap ring pliers, remove the front axle hub hole snap ring.
12. Remove front axle hub sub-assembly by performing the following:
    a. Hold the front axle assembly between aluminum plates in a vise.

➡Do not overtighten the vise.

    b. Using SST 09520-00031, remove the front axle hub sub-assembly.

➡Be careful not to drop the front axle hub sub-assembly.

    c. Using SST 09555-55010, SST: 09950-60010 and SST: 09950-70010 and a press, remove the bearing inner race (outside) from the front axle hub sub-assembly.

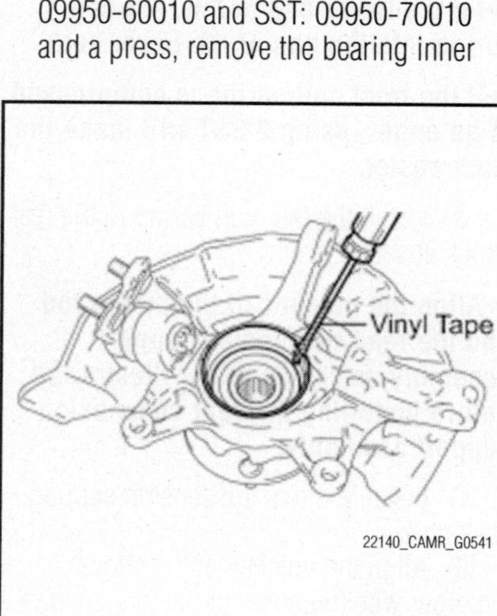

Fig. 411 Remove the No. 1 front wheel bearing dust deflector

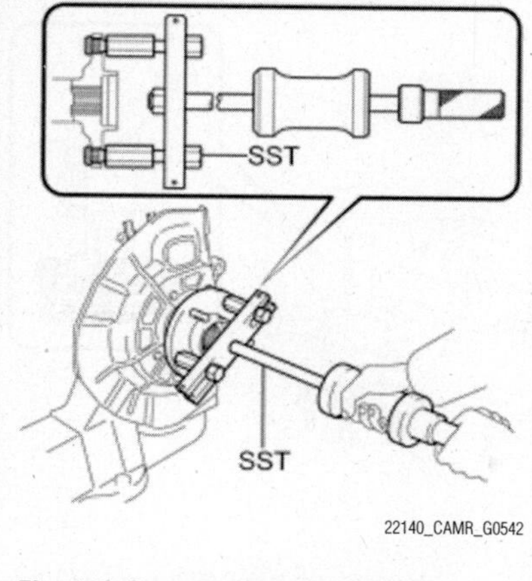

Fig. 412 Remove the front axle hub sub-assembly

13. Remove the 4 bolts and disc brake dust cover from the steering knuckle.
14. Remove front lower ball joint assembly.
15. Remove front axle hub bearing by performing the following:
    a. Place the bearing inner race (outside) on the front axle hub bearing.
    b. Using SST 09527-17011, SST: 09950-60010 and a press, press the front axle hub bearing until it contacts the SST: 09950-70010.
    c. Using SST: 09527-20011, SST: 09950-60010 to make the steering knuckle horizontal, fix it to the V-block.
16. Using SST: 09950-70010 and a press, remove the front axle hub bearing from the steering knuckle.

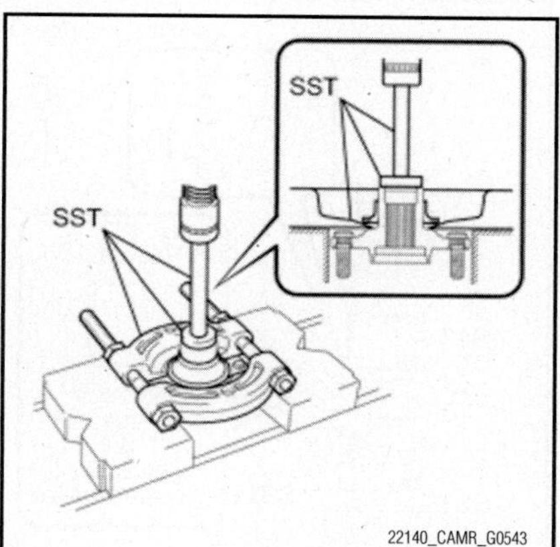

Fig. 413 Remove the bearing inner race (outside) from the front axle hub sub-assembly

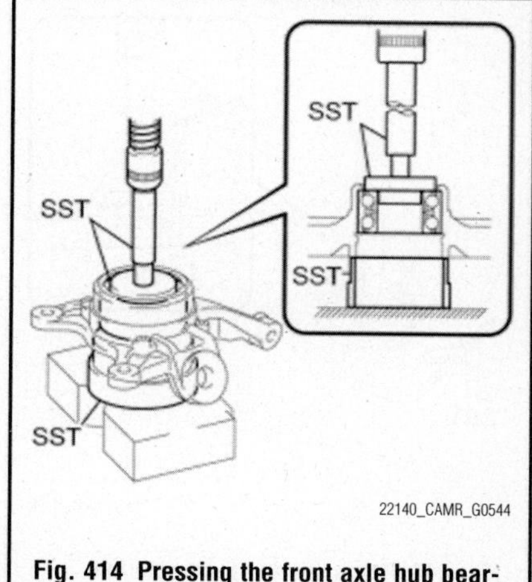

Fig. 414 Pressing the front axle hub bearing

To install:

17. Using SST's: 09950-60020, 09950-70010 and a press, install a new front axle hub bearing to the steering knuckle.
18. Install front lower ball joint assembly.
19. Install the disc brake dust cover to the steering knuckle with the 4 bolts and tighten to 73 inch lbs. (8.3 Nm).
20. Using SST's: 09608-32010, 09950-60020, 09950-70010 and a press, install the front axle hub sub-assembly.
21. Using snap ring pliers, install a new front axle hub hole snap ring.

➡Align the hole for the speed sensor in the No. 1 front wheel bearing dust deflector with the steering knuckle.

22. Using SST's: 09316-60011, 09608-32010 and a hammer, install a new No. 1 front wheel bearing dust deflector.

➡Only when reusing the bolts and nuts, apply the small amount of engine oil to the screw part of the nuts.

➡Be careful not to damage the drive shaft boot or speed sensor rotor.

23. Align the matchmarks and install the front drive shaft assembly to the front axle hub sub-assembly.
24. Install the steering knuckle with the front axle hub sub-assembly to the front shock absorber assembly with the 2 bolts and 2 nuts and tighten to 155 ft. lbs. (210 Nm).
25. Install the lower No. 1 front suspension arm sub-assembly.
26. Install the tie rod end sub-assembly.
27. Install the front disc.

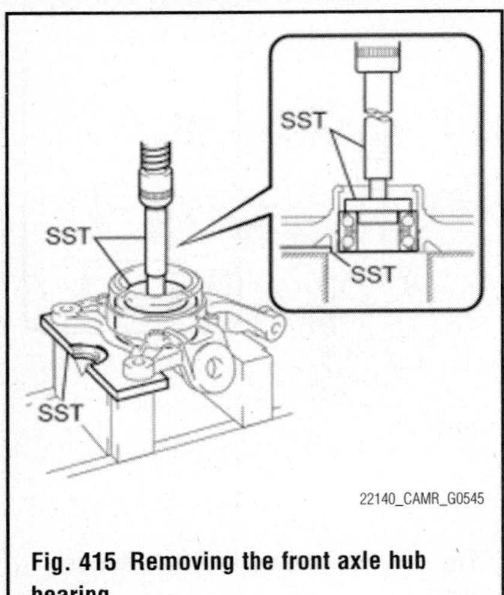

Fig. 415 Removing the front axle hub bearing

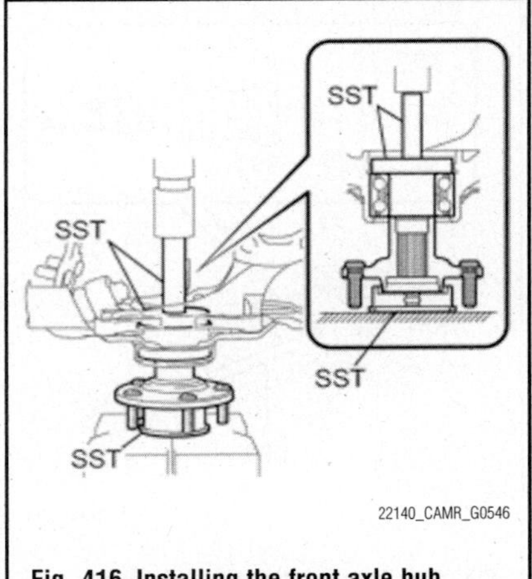

Fig. 416 Installing the front axle hub sub-assembly

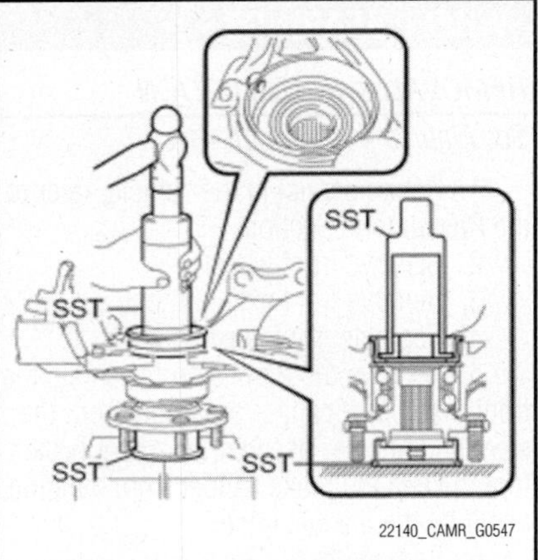

Fig. 417 Installing No. 1 front wheel bearing dust deflector

28. Install the front disc brake caliper assembly with the 2 bolts to the steering knuckle and tighten to 79 ft. lbs. (107 Nm).

29. Clean the threaded parts on the drive shaft and axle hub nut using a non-residue solvent.

➡Be sure to perform this work for a new drive shaft.

➡Keep the threaded parts free of oil and foreign objects.

30. Using a 30 mm socket wrench, install the front axle hub nut and tighten to 217 ft. lbs. (294 Nm).

31. Remove the 2 bolts and separate the front disc brake caliper assembly from the steering knuckle.

32. Remove the front disc.

33. Inspect front axle hub bearing looseness.

34. Inspect front axle hub runout.

35. Install the front disc.

36. Install the front disc brake caliper assembly with the 2 bolts to the steering knuckle and tighten to 79 ft. lbs. (107 Nm).

37. Install the front speed sensor.

38. Using a chisel and hammer, stake the axle hub nut.

39. Install the front wheel.

40. Inspect and adjust front wheel alignment.

41. Check ABS speed sensor signal.

SUSPENSION

REAR SUSPENSION

COIL SPRING

REMOVAL & INSTALLATION

See Figures 418 through 420.

1. Before servicing the vehicle, refer to the precautions section.

2. Secure the rear shock absorber with coil spring in a vise using aluminum plates by closing the vise onto the double nutted bolt affixed to the bracket at the bottom of the absorber.

➡Do not use an impact wrench.

➡If the rear coil spring is compressed at an angle, using 2 SST will make the work easier.

3. Using SST: 09727-30021, compress the rear coil spring

4. Remove the nut, rear shock absorber collar and rear suspension support assembly.

5. Remove the rear coil spring, rear No. 1 spring bumper, and rear coil spring lower insulator.

To install:

6. Install the rear No. 1 spring bumper to the piston rod.

7. Install the rear coil spring lower insulator onto the rear shock absorber.

➡Do not use an impact wrench.

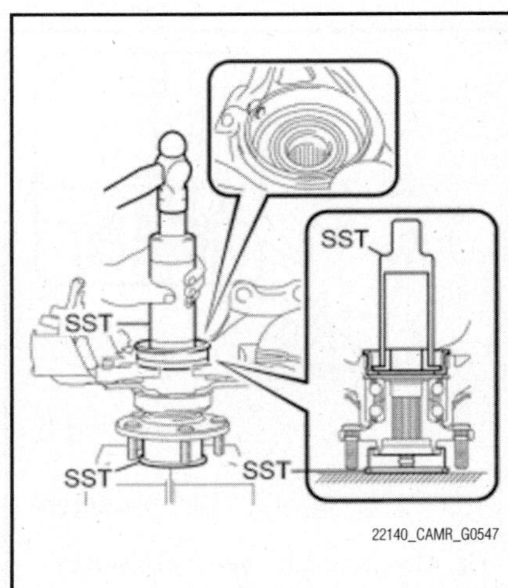

Fig. 418 Remove the rear coil spring

8. Using SST: 09727-30021, compress the rear coil spring.

➡The smaller diameter end must face upward.

➡Fit the lower end of the rear coil spring into the gap of the lower seat.

➡If the front coil spring is compressed at an angle, using 2 SST will make the work easier.

9. Install the rear coil spring to the rear shock absorber.

➡Align the notches of the piston rod and the rear suspension support assembly as shown in the illustration before installing the rear suspension support assembly.

10. Install the rear suspension support assembly.

11. Align the notches of the shock absorber with the notch of the rear suspension support assembly so that the notches face the outside of the vehicle.

12. Install the rear shock absorber collar.

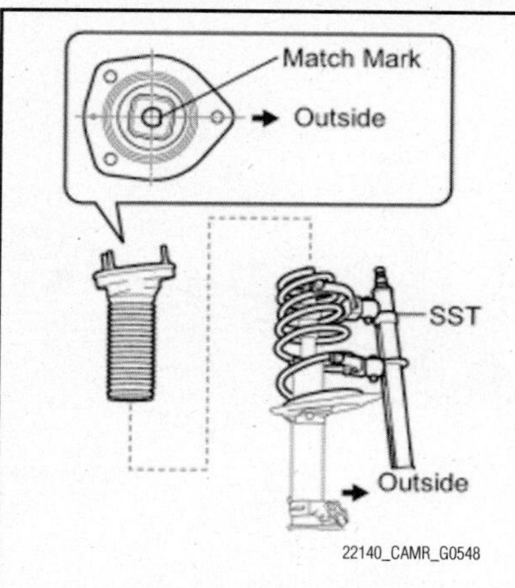

Fig. 419 Align the notches of the shock absorber

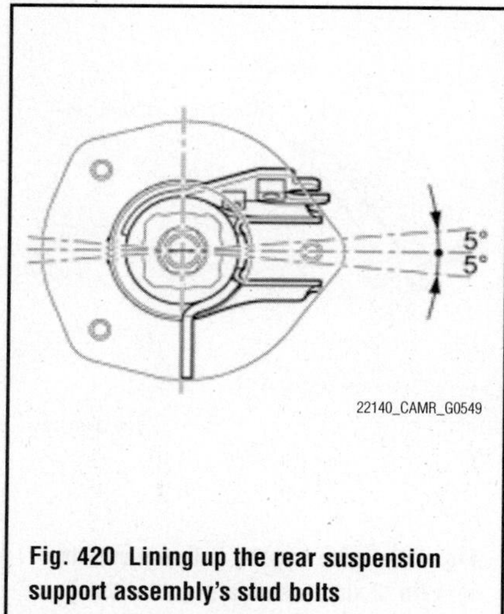

Fig. 420 Lining up the rear suspension support assembly's stud bolts

13. Loosely tighten a new lock nut to the rear suspension piston rod.

➡Do not use an impact wrench.

➡When lining up the rear suspension support assembly's stud bolts at the middle point between the two sides of the bracket, the maximum permissible degree of error is plus or minus 5°.

14. Release the spring while adjusting the rear suspension support assembly to the position shown in the illustration, and remove the SST from the rear coil spring.

CONTROL ARM & LINKS

REMOVAL & INSTALLATION

No. 1 Suspension Arm

See Figures 421 through 426.

1. Before servicing the vehicle, refer to the precautions section.

➡Check if an old gasket still remains on the pipe. If so, remove it. Also, check if any bolts or nuts are rusted. If so, replace them.

2. Remove rear wheel.
3. Remove center exhaust pipe assembly.
4. Remove tail exhaust pipe assembly.
5. Separate both rear stabilizer link assemblies.
6. Remove rear stabilizer bar no. 2 and no. 1 bracket.
7. Remove rear stabilizer bar.
8. Remove rear stabilizer bushing.
9. Separate rear strut rod.

➡When removing the bolt, keep the nut from rotating.

10. Remove the bolt, nut and separate the rear suspension No. 2 arm (outer side) from the rear axle carrier.

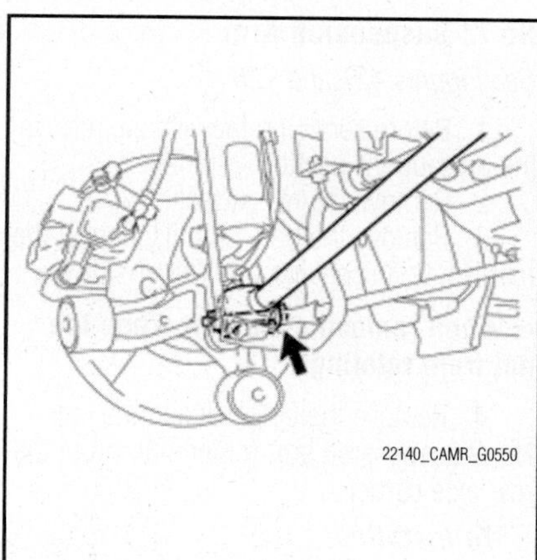

Fig. 421 Removing the bolt, nut and the rear No. 1 suspension arm—LH shown

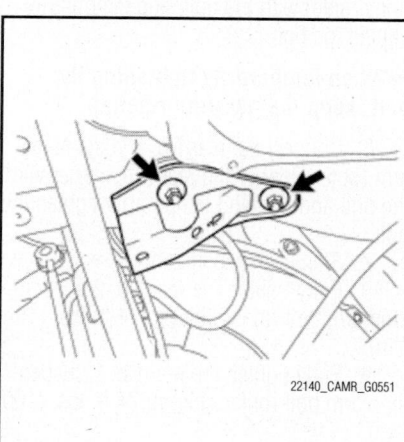

Fig. 422 Removing the 2 nuts and the LH rear suspension member lower stopper

➡When removing the bolt, keep the nut from rotating.

11. Remove the bolt, nut and the rear No. 1 suspension arm (outer side) from the rear axle carrier.
12. Remove the 2 nuts and the LH rear suspension member lower stopper.
13. Remove the 2 nuts and the RH rear suspension member lower stopper.
14. Support the rear suspension member with a jack.
15. Remove the 2 bolts, and the rear suspension member sub-assembly.
16. Remove the bolt and rear No. 1 suspension arm assembly.

To install:

17. Install the No. 1 rear suspension arm (inner side) with the bolt, and temporarily tighten the bolt.
18. Install the rear No. 1 suspension arm so that the bracket leans toward the front side of the vehicle.
19. Ensure that the paint mark faces the rear side of the vehicle.
20. Set the rear No.1 suspension arm in the position shown in the illustration, and fully tighten the bolt to 74 ft. lbs. (100 Nm).
21. Raise the rear suspension member with a jack. Install the rear suspension member with the 2 bolts and tighten to 41 ft. lbs. (56 Nm).
22. Install both the rear suspension member lower stoppers with the 2 nuts and tighten to:
    a. Nut A: 41 ft. lbs. (55 Nm).
    b. Nut B: 28 ft. lbs. (38 Nm).

➡Insert the bolt from the front of the vehicle and temporarily install the bolt.

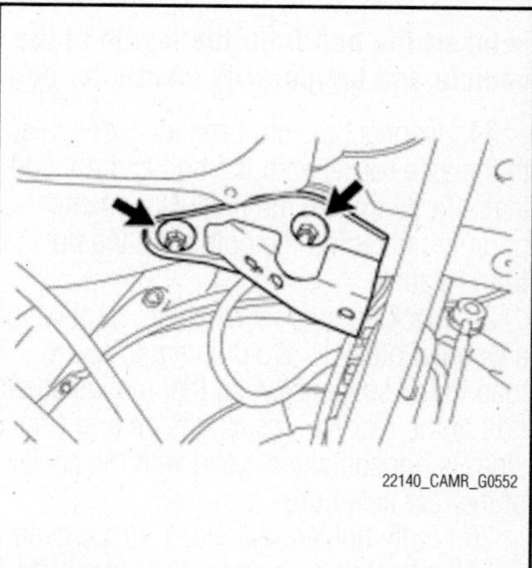

Fig. 423 Removing the 2 nuts and the RH rear suspension member lower stopper

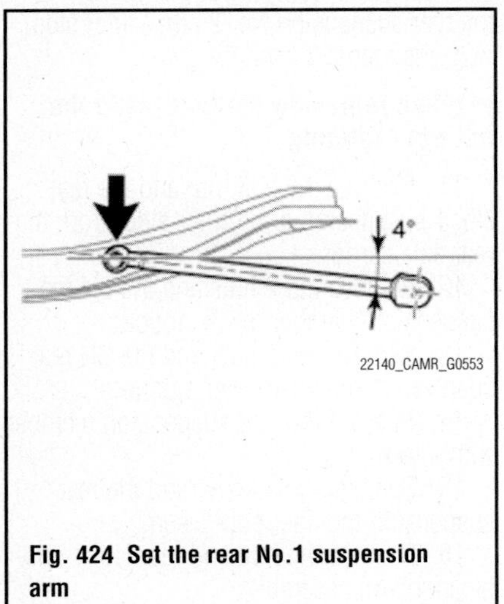

Fig. 424 Set the rear No.1 suspension arm

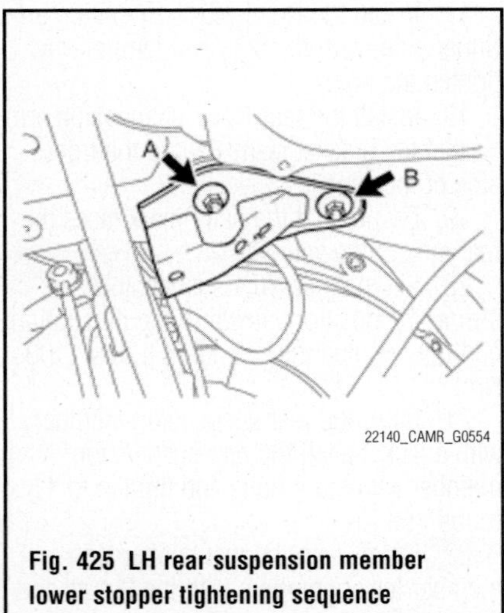

Fig. 425 LH rear suspension member lower stopper tightening sequence

23. Connect the rear No.1 suspension arm (outer side) to the rear axle carrier with the bolt and nut and temporarily tighten the bolt and nut. When temporarily tightening the bolt, keep the nut from rotating.

➡Insert the bolt from the inside of the vehicle and temporarily install the bolt.

24. Connect the strut rod assembly rear to the axle carrier with the bolt and nut and temporarily tighten the bolt. When temporarily tightening the bolt, keep the nut from rotating.
25. Jack up the rear axle carrier, placing a wooden block to avoid damage. Apply load to the suspension so that the installed bolt of the rear No. 1 suspension arm (inner side) is horizontally aligned with the center of the rear axle hub.
26. Fully tighten rear No. 1 suspension arm and tighten the bolt to 74 ft. lbs. (100 Nm).
27. Fully tighten rear No. 2 suspension arm and tighten the bolt to 74 ft. lbs. (100 Nm).
28. To complete installation, reverse removal procedure.

Fig. 426 LH rear suspension member lower stopper tightening sequence

No. 2 Suspension Arm

See Figures 427 and 428.

1. Before servicing the vehicle, refer to the precautions section.
2. Remove the rear wheel.
3. Remove the bolt, and disconnect the rear No. 2 suspension arm (inner side).

➡When removing the bolt, keep the nut from rotating.

4. Remove the bolt, nut and the rear No. 2 suspension arm (outer side) from the rear axle carrier.

To install:

➡Ensure that the paint mark faces to the rear of the vehicle.

5. Install the rear No. 2 suspension arm (inner side) with the bolt, and temporarily tighten the bolt.

➡When temporarily tightening the bolt, keep the nut from rotating.

6. Connect the rear No. 2 suspension arm (outer side) to the rear axle carrier with the bolt and nut, and temporarily tighten the bolt.
7. Stabilize suspension.
8. Fully tighten the rear No. 2 suspension arm bolt (inner side) to 74 ft. lbs. (100 Nm).
9. Fully tighten the rear No. 2 suspension arm bolt (outer side) to 74 ft. lbs. (100 Nm).
10. Install the rear wheel.
11. Inspect and adjust the rear wheel alignment.

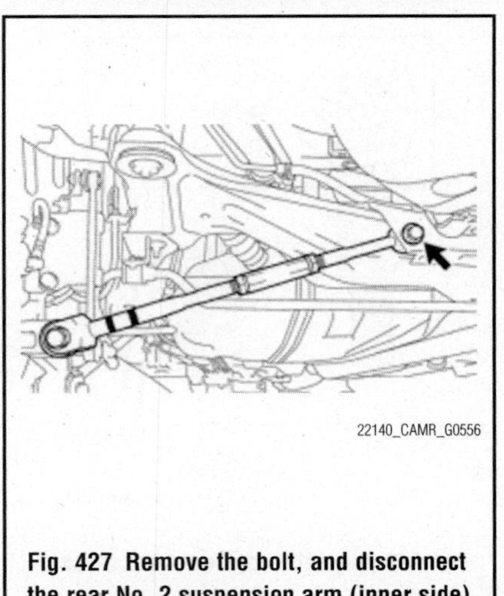

Fig. 427 Remove the bolt, and disconnect the rear No. 2 suspension arm (inner side)

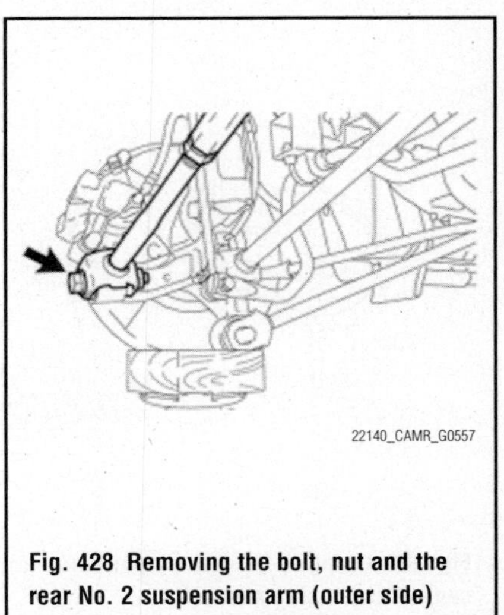

Fig. 428 Removing the bolt, nut and the rear No. 2 suspension arm (outer side)

STABILIZER BAR

REMOVAL & INSTALLATION

See Figures 429 and 430.

1. Before servicing the vehicle, refer to the precautions section.
2. Remove rear wheels.
3. For 2.4L and 2.5L engines, remove center exhaust pipe assembly.
4. For 3.5L engines, remove tail exhaust pipe assembly.
5. For 3.5L engines, center exhaust pipe assembly.
6. Remove rear stabilizer link assembly.
7. Remove the 2 bolts and rear stabilizer bar No. 2 bracket.
8. Remove the 2 bolts and rear stabilizer bar No. 1 bracket.
9. Remove the 2 rear stabilizer bushings from the rear stabilizer bar.
10. Remove rear stabilizer bar.

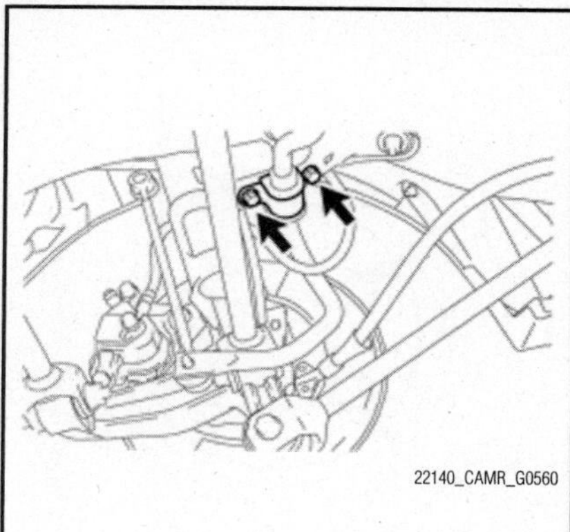

Fig. 429 Removing the 2 bolts and No. 2 bracket

To install:

11. Install the 2 rear stabilizer bushings to the outside of the stopper ring on the stabilizer bar.
12. Install the rear stabilizer bar No. 2 and No. 1 bracket.
13. Install the rear stabilizer bar with the 2 bolts and tighten to 23 ft. lbs. (31 Nm).
14. Install rear stabilizer link assembly.
15. To complete installation, reverse remaining removal.
16. Check abs speed sensor signal.
17. Inspect and adjust rear wheel alignment.

STRUT & SPRING ASSEMBLY

REMOVAL & INSTALLATION

See Figures 431 and 432.

1. Before servicing the vehicle, refer to the precautions section.
2. Remove the rear seat cushion assembly.

Fig. 430 Removing the 2 bolts and No. 1 bracket

3. Remove rear seat headrest plate cover.
4. Remove rear seat headrest assembly.
5. Remove the rear seatback assembly.
6. Remove the rear wheel.
7. Separate LH rear stabilizer link assembly.
8. Remove the 2 bolts, and disconnect the rear brake flexible hose and rear speed sensor from the rear shock absorber with coil spring and rear axle carrier.
9. Remove the 4 claws and the rear suspension support No. 1 cover.
10. Remove the rear shock absorber with coil spring, as follows:

➡Do not remove the lock nut.

➡Only loosen the nut when disassembling the rear shock absorber with coil spring.

a. Loosen the lock nut of the rear shock absorber with coil spring.

➡When removing the nuts, keep the bolts from rotating.

➡Keep one bolt inserted to secure the hub and disc rotor.

b. Remove the 2 nuts and 2 bolts on the lower side of the rear shock absorber with coil spring.
c. Remove the 3 nuts on the upper side of the rear shock absorber with coil spring.

➡Make sure that the rear speed sensor is disconnected from the rear shock absorber with coil spring.

d. Lower the rear axle carrier, and remove the 2 bolts on the lower side of the rear shock absorber with coil spring.

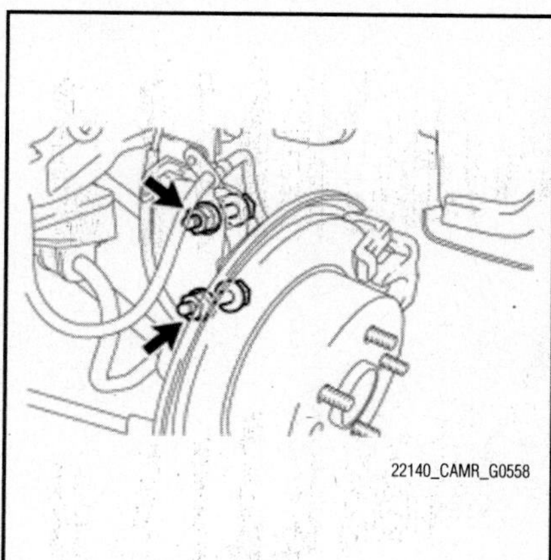

Fig. 431 Loosen the 2 nuts on the lower side of the shock absorber

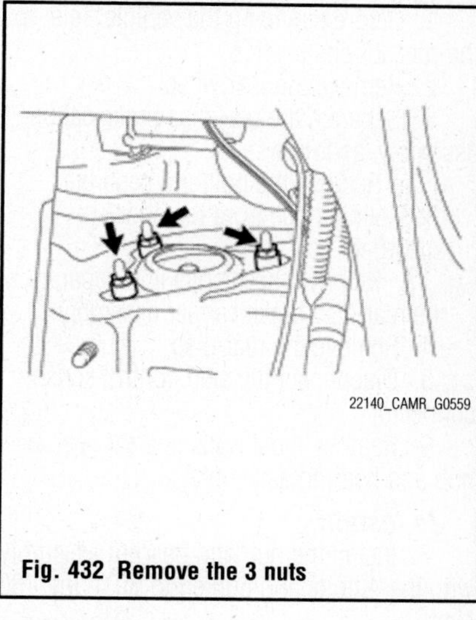

Fig. 432 Remove the 3 nuts

To install:

11. Install the rear shock absorber with coil spring to the rear axle carrier assembly and insert the 2 bolts from the rear of the vehicle.
12. Slowly jack up the vehicle using a wooden block and install the rear shock absorber with coil spring (upper side) to the vehicle.
13. Install the 3 nuts to the upper side of the rear shock absorber with coil spring and tighten to 29 ft. lbs. (39 Nm).

➡When installing the nuts, keep the bolts from rotating.

14. Install the 2 nuts and 2 bolts to the lower side of the rear shock absorber with coil spring and tighten 133 ft. lbs. (180 Nm).
15. Fully tighten the lock nut to 41 ft. lbs. (55 Nm).
16. Connect rear speed sensor.
17. Install the LH rear stabilizer link assembly.
18. Engage the 4 claws and install the rear suspension support No. 1 cover.
19. To complete installation, reverse remaining removal.
20. Check abs speed sensor signal.
21. Inspect and adjust rear wheel alignment.

WHEEL BEARINGS

REMOVAL & INSTALLATION

See Figure 433.

➡Use the same procedures for the RH side and LH side.

➡The procedures listed below are for the LH side.

1. Before servicing the vehicle, refer to the precautions section.
2. Remove the rear wheel.
3. Separate the rear disc brake caliper assembly, as follows:
    a. Remove the bolt and separate the flexible hose from the shock absorber.
    b. Remove the 2 bolts and separate the rear disc brake caliper assembly.
4. Remove the rear disc.
5. Disconnect the skid control sensor connector.
6. Remove the 4 bolts and the rear axle hub and bearing assembly.

To install:

7. Install the hub and bearing assembly with the 4 bolts and tighten to 59 ft. lbs. (80 Nm).
8. Connect the skid control sensor connector. Do not twist the sensor wire.
9. Inspect rear axle hub bearing looseness.
10. Inspect rear axle hub runout.
11. Install the rear disc.
12. Install the rear disc brake caliper assembly, as follows:
    a. Install the rear disc brake caliper with the 2 bolts and tighten to 46 ft. lbs. (62 Nm).
    b. Install the flexible hose with the bolt and tighten to 14 ft. lbs. (19 Nm).
13. Install the rear wheel.
14. Inspect and adjust the rear wheel alignment.
15. Check ABS speed sensor signal.

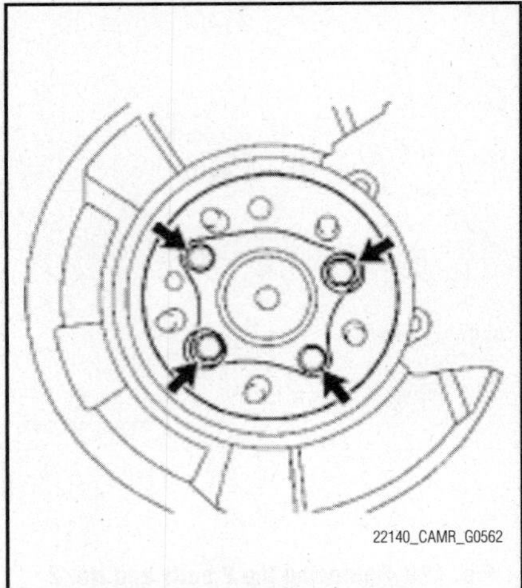

Fig. 433 Remove the 4 bolts and the rear axle hub and bearing assembly

TOYOTA

Corolla

6

SPECIFICATIONS AND MAINTENANCE CHARTS

ENGINE AND VEHICLE IDENTIFICATION

Engine							Model Year	
Code ①	Liters (cc)	Cu. In.	Cyl.	Fuel Sys.	Engine Type	Eng. Mfg.	Code ②	Year
2ZR-FE	1.8 (1794)	109.5	4	EFI	DOHC	Toyota	B	2011
2AZ-FE	2.4 (2400)	146	4	EFI	DOHC	Toyota	C	2012

EFI: Electronic Fuel Injection

DOHC: Double Overhead Camshaft

① 5th digit of VIN

②10th digit of VIN

71099_CORO_C0001

GENERAL ENGINE SPECIFICATIONS

Year	Model	Engine Displacement Liters (cc)	Engine Series (ID/VIN)	Fuel System	Net Horsepower @ rpm	Net Torque @ rpm (ft. lbs.)	Bore x Stroke (in.)	Com-pression Ratio	Oil Pressure @ rpm
2011	Corolla	1.8 (1794)	2ZR-FE	EFI	132@6000	128@4400	3.16x3.48	10.0:1	4.3
		2.4 (2400)	2AZ-FE	EFI	158@6000	162@4000	3.48x3.78	9.8	NA
2012	Corolla	1.8 (1794)	2ZR-FE	EFI	132@6000	128@4400	3.16x3.48	10.0:1	4.3
		2.4 (2400)	2AZ-FE	EFI	158@6000	162@4000	3.48x3.78	9.8	NA

NA: Not Available

EFI: Electronic Fuel Injection

71099_CORO_C0002

ENGINE TUNE-UP SPECIFICATIONS

Year	Engine Displacement Liters	Engine ID/VIN	Spark Plug Gap (in.)	Ignition Timing (deg.)	Fuel Pump (psi)	Idle Speed (rpm) MT	Idle Speed (rpm) AT	Valve Clearance In.	Valve Clearance Ex.
2011	1.8	2ZR-FE	0.043	①	44-50	600-700	600-700	0.007-0.014	0.015-0.018
	2.4	2AZ-FE	0.043	②	44-50	600-700	600-700	NA	NA
2012	1.8	2ZR-FE	0.043	①	44-50	600-700	600-700	0.007-0.014	0.015-0.018
	2.4	2AZ-FE	0.043	②	44-50	600-700	600-700	NA	NA

Note: The Vehicle Emission Control Information label often reflects specification changes made during production.

NA: Not Available

BTDC: Before Top Dead Center

① With Techstream (ODB-II scanner or equivalent): BTDC 5-15 at idle

Without Techstream: BTDC 8-12 at idle. Connect terminals 13 (TC) and 4 (CG)

② BTDC 8-12 with terminal TC and CG of DLC3 connected

71099_CORO_C0003

CAPACITIES

Year	Model	Engine Displacement Liters	Engine ID/VIN	Engine Oil with Filter	Transmission (pts.) 5-Spd	Transmission (pts.) Auto.	Drive Axle Front (pts.)	Drive Axle Rear (pts.)	Fuel Tank (gal.)	Cooling System (qts.)
2011	Corolla	1.8	2ZR-FE	4.4	4.0	16.0	NA	NE	13.2	6.0
		2.4	2AZ-FE	4.4	4.0	6.0	NA	NE	13.2	6.0
2012	Corolla	1.8	2ZR-FE	4.4	4.0	16.0	NA	NE	13.2	6.0
		2.4	2AZ-FE	4.4	4.0	6.0	NA	NE	13.2	6.0

Note: All capacities are approximate. Add fluid gradually and check to be sure a proper fluid level is obtained. Auto trans is a drain and fill capacity.

NE: Not Equipped

NA: Not Available

71099_CORO_C0004

FLUID SPECIFICATIONS

Year	Model	Engine Displacement Liters	Engine Oil	Man. Trans.	Auto. Trans.	Drive Axle	Power Steering Fluid	Brake Master Cylinder
2011	Corolla	1.8	5W-30	SAE 75W	ATF Fluid	NA	ATF Fluid	DOT 3
		2.4	5W-30	SAE 75W	ATF Fluid	NA	ATF Fluid	DOT 3
2012	Corolla	1.8	5W-30	SAE 75W	ATF Fluid	NA	ATF Fluid	DOT 3
		2.4	5W-30	SAE 75W	ATF Fluid	NA	ATF Fluid	DOT 3

NA: Not Available

DOT: Department Of Transpotation

71099_CORO_C0005

VALVE SPECIFICATIONS

Year	Engine Displacement Liters	Engine ID/VIN	Seat Angle (deg.)	Face Angle (deg.)	Spring Test Pressure (lbs. @ in.)	Spring Installed Height (in.)	Stem-to-Guide Clearance (in.) Intake	Stem-to-Guide Clearance (in.) Exhaust	Stem Diameter (in.) Intake	Stem Diameter (in.) Exhaust
2011	1.8	2ZR-FE	45	44.5	35.7-39.5@ 1.323	1.323	0.0010-0.0024	0.0012-0.0025	0.2154-0.2159	0.2152-0.2158
	2.4	2AZ-FE	45	44.5	35.7-39.5@ 1.323	1.323	0.0010-0.0024	0.0012-0.0025	0.2154-0.2159	0.2152-0.2158
2012	1.8	2ZR-FE	45	44.5	35.7-39.5@ 1.323	1.323	0.0010-0.0024	0.0012-0.0025	0.2154-0.2159	0.2152-0.2158
	2.4	2AZ-FE	45	44.5	35.7-39.5@ 1.323	1.323	0.0010-0.0024	0.0012-0.0025	0.2154-0.2159	0.2152-0.2158

71099_CORO_C0007

CAMSHAFT AND BEARING SPECIFICATIONS

All measurements are given in inches.

Year	Engine Displacement Liters	Engine VIN	Journal Diameter	Brg. Oil Clearance	Shaft End-play	Runout	Journal Bore	Lobe Lift Intake	Lobe Lift Exhaust
2011	1.8	2ZR-FE	①	0.0012-0.0025	NA	0.0016	NA	1.685-1.6890	1.745-1.7490
	2.4	2AZ-FE	②	③	NA	0.00118	NA	1.865-1.8660	1.813-1.8170
2012	1.8	2ZR-FE	①	0.0012-0.0025	NA	0.0016	NA	1.685-1.6890	1.745-1.7490
	2.4	2AZ-FE	②	③	NA	0.00118	NA	1.865-1.8660	1.813-1.8170

NA: Not Available

① No. 1 journal diameter: 1.356
Other: 0.903 to 0.904

② No. 1 journal diamter: 1.416
Other: 0.903 to 0.904

③ Intake: 0.00276 (Max.)
Exhaust: 0.00394 (Max.)

71099_CORO_C0006

CRANKSHAFT AND CONNECTING ROD SPECIFICATIONS

All measurements are given in inches.

Year	Engine Displacement Liters	Engine ID/VIN	Crankshaft Main Brg. Journal Dia.	Crankshaft Main Brg. Oil Clearance	Crankshaft Shaft End-play	Crankshaft Thrust on No.	Connecting Rod Journal Diameter	Connecting Rod Oil Clearance	Connecting Rod Side Clearance
2011	1.8	2ZR-FE	1.8893-1.8898	0.0006-0.0015	0.0008-0.0087	3	1.7320-1.7323	0.0012-0.0024	0.0063-0.0135
	2.4	2AZ-FE	2.164-2.165	0.00236	NA	3	0.283-0.2870	0.00248	0.0063-0.0143
2012	1.8	2ZR-FE	1.8893-1.8898	0.0006-0.0015	0.0008-0.0087	3	1.7320-1.7323	0.0012-0.0024	0.0063-0.0135
	2.4	2AZ-FE	2.164-2.165	0.00236	NA	3	0.283-0.2870	0.00248	0.0063-0.0143

NA: Not Available

71099_CORO_C0010

PISTON AND RING SPECIFICATIONS

All measurements are given in inches.

Year	Engine Displacement Liters	Engine ID/VIN	Piston Clearance	Ring Gap Top Compression	Ring Gap Bottom Compression	Ring Gap Oil Control	Ring Side Clearance Top Compression	Ring Side Clearance Bottom Compression	Ring Side Clearance Oil Control
2011	1.8	2ZR-FE	0.0011-0.0020	0.0079-0.0118	0.0118-0.0197	0.0039-0.0157	0.0008-0.0028	0.0008-0.0024	0.0008-0.0026
	2.4	2AZ-FE	0.00082-0.0017	0.0094-0.0122	0.0130-0.0169	0.0039-0.0118	0.0007-0.0027	0.0007-0.0023	0.0007-0.0027
2012	1.8	2ZR-FE	0.0011-0.0020	0.0079-0.0118	0.0118-0.0197	0.0039-0.0157	0.0008-0.0028	0.0008-0.0024	0.0008-0.0026
	2.4	2AZ-FE	0.00082-0.0017	0.0094-0.0122	0.0130-0.0169	0.0039-0.0118	0.0007-0.0027	0.0007-0.0023	0.0007-0.0027

71099_CORO_C0009

TORQUE SPECIFICATIONS

All readings in ft. lbs.

Year	Engine Displacement Liters	Engine ID/VIN	Cylinder Head Bolts	Main Bearing Bolts	Rod Bearing Bolts	Crankshaft Damper Bolts	Flywheel Bolts	Manifold Intake	Manifold Exhaust	Spark Plugs	Oil Pan Drain Plug
2011	1.8	2ZR-FE	①	②	③	102	65	21	16	15	27
	2.4	2AZ-FE	④	NA	⑤	NA	⑥	22	27	14	27
2012	1.8	2ZR-FE	①	②	③	102	65	21	16	15	27
	2.4	2AZ-FE	④	NA	⑤	NA	⑥	22	27	14	27

NA: Not Available

① Step 1: 36 ft. lbs.
Step 2: 90 degree turn
Step 3: 45 degree turn

② Inner 12 point bolts:
Step 1: 16 ft. lbs.
Step 2: 32 ft. lbs.
Step 3: 45 degree turn
Step 4: 45 degree turn
Outer cap bolts: 13 ft. lbs.

③ Step 1: 15 ft. lbs.
Step 2: 90 degree turn

④ Step 1: 52 ft. lbs.
Step 2: 90 degree turn

⑤ Step 1: 18 ft. lbs.
Step 2: 90 degree turn

⑥ A/T: 72 ft. lbs.
M/T: 96 ft. lbs.

71099_CORO_C0008

WHEEL ALIGNMENT

Year	Model		Caster Range (+/-Deg.)	Caster Preferred Setting (Deg.)	Camber Range (+/-Deg.)	Camber Preferred Setting (Deg.)	Toe-in (Deg.)	Steering Axis Inclination (Deg.)
2011	Corolla	Front	0.75	+2.50	0.75	-0.35	0+/-0.2	12.53+/-0.75
		Rear	—	—	0.5	-1.45	①	—
2012	Corolla	Front	0.75	+2.50	0.75	-0.35	0+/-0.2	12.53+/-0.75
		Rear	—	—	0.5	-1.45	①	—

① For P185/65R15: 0.25+/-0.25

For P195/65R15: 0.26+/-0.26 for 195/55R16: 0.34+/-0.25

For 195/55R16: 0.34+/-0.25

71099_CORO_C0011

TIRE, WHEEL AND BALL JOINT SPECIFICATIONS

Year	Model	OEM Tires Standard	OEM Tires Optional	Tire Pressures (psi) Front	Tire Pressures (psi) Rear	Wheel Size	Ball Joint Inspection	Lug Nut (ft. lbs.)
2011	Corolla	P195/65R15 89S	None	30	30	NA	NA	76
		P195/65R15 91H	None	30	30	NA	NA	76
		P205/55R16 89H	None	32	32	NA	NA	76
		P215/45R17 87W	None	32	32	NA	NA	76
		P205/55R16 91V	None	32	32	NA	NA	76
2012	Corolla	P195/65R15 89S	None	30	30	NA	NA	76
		P195/65R15 91H	None	30	30	NA	NA	76
		P205/55R16 89H	None	32	32	NA	NA	76
		P215/45R17 87W	None	32	32	NA	NA	76
		P205/55R16 91V	None	32	32	NA	NA	76

PSI: Pounds Per Square Inch

NA: Not Available

71099_CORO_C0012

BRAKE SPECIFICATIONS

All measurements in inches unless noted

Year	Model		Brake Disc Original Thickness	Brake Disc Minimum Thickness	Brake Disc Maximum Runout	Brake Drum Diameter Original Inside Diameter	Brake Drum Diameter Max. Wear Limit	Brake Drum Diameter Maximum Machine Diameter	Minimum Lining Thickness	Brake Caliper Bracket Bolts (ft. lbs.)	Brake Caliper Mounting Bolts (ft. lbs.)
2011	Corolla	F	0.866	0.748	0.0019	NE	NE	NE	0.039	NA	25
		R	0.35	0.30	0.00591	9.00	9.04	NA	0.039	NA	26
2012	Corolla	F	0.866	0.748	0.0019	NE	NE	NE	0.039	NA	25
		R	0.35	0.30	0.00591	9.00	9.04	NA	0.039	NA	26

NE: Not Equipment

NA: Not Available

71099_CORO_C0013

SCHEDULED MAINTENANCE INTERVALS
TOYOTA—COROLLA

TO BE SERVICED	TYPE OF SERVICE	VEHICLE MILEAGE INTERVAL (x1000)													
		5	10	15	20	25	30	35	40	45	50	55	60	90	120
Engine oil & filter	R	✓	✓	✓	✓	✓	✓	✓	✓	✓	✓	✓	✓	✓	✓
Drive belts	S/I						✓						✓	✓	✓
Automatic transaxle fluid & filter	S/I						✓						✓	✓	✓
Brake line pipes & hoses	S/I	✓	✓	✓	✓	✓	✓	✓	✓	✓	✓	✓	✓	✓	✓
Brake linings & drums	S/I	✓	✓	✓	✓	✓	✓	✓	✓	✓	✓	✓	✓	✓	✓
Brake pads & discs (front & rear if equipped)	S/I	✓	✓	✓	✓	✓	✓	✓	✓	✓	✓	✓	✓	✓	✓
Cabin air filter	R				✓				✓				✓		✓
Differential oil	S/I						✓						✓	✓	✓
Drive shaft boots	S/I	✓	✓	✓	✓	✓	✓	✓	✓	✓	✓	✓	✓	✓	✓
Drive shaft bolt (tighten)	S/I	✓	✓	✓	✓	✓	✓	✓	✓	✓	✓	✓	✓	✓	✓
Engine coolant	S/I			✓			✓			✓			✓	✓	✓
Manual transaxle oil	S/I						✓						✓	✓	✓
Steering gear housing oil	S/I	✓	✓	✓	✓	✓	✓	✓	✓	✓	✓	✓	✓	✓	✓
Steering linkage	S/I	✓	✓	✓	✓	✓	✓	✓	✓	✓	✓	✓	✓	✓	✓
Air filter	R						✓						✓	✓	✓
Rotate tires	S/I	✓	✓	✓	✓	✓	✓	✓	✓	✓	✓	✓	✓	✓	✓
Spark plugs	R									✓				✓	
Fuel lines & connections	S/I						✓			✓			✓	✓	✓
Fuel tank cap gasket	R									✓				✓	
Charcoal canister	S/I									✓				✓	

R: Replace S/I: Service or Inspect

FREQUENT OPERATION MAINTENANCE (SEVERE SERVICE)

If a vehicle is operated under any of the following conditions it is considered severe service:

- Extremely dusty areas.

- 50% or more of the vehicle operation is in 32°C (90°F) or higher temperatures, or constant operation in temperatures below 0°C (32°F).

- Prolonged idling (vehicle operation in stop and go traffic).

- Frequent short running periods (engine does not warm to normal operating temperatures).

- Police, taxi, delivery usage or trailer towing usage.

Oil & oil filter: change every 5000 miles.

Bolts & nuts on chassis & body: tighten every 5000 miles.

Ball joints & dust covers: service or inspect every 5,000 miles.

Drive shaft boots & except Supra): service or inspect every 12,000 miles.

Steering linkage: service or inspect every 12,000 miles.

Air filter: service or inspect every 5,000 miles.

Exhaust system: service or inspect every 15,000 miles.

Timing belt: replace every 60,000 miles.

71099_CORO_C0014

PRECAUTIONS

Before servicing any vehicle, please be sure to read all of the following precautions, which deal with personal safety, prevention of component damage, and important points to take into consideration when servicing a motor vehicle:

- Never open, service or drain the radiator or cooling system when the engine is hot; serious burns can occur from the steam and hot coolant.
- Observe all applicable safety precautions when working around fuel. Whenever servicing the fuel system, always work in a well-ventilated area. Do not allow fuel spray or vapors to come in contact with a spark, open flame, or excessive heat (a hot drop light, for example). Keep a dry chemical fire extinguisher near the work area. Always keep fuel in a container specifically designed for fuel storage; also, always properly seal fuel containers to avoid the possibility of fire or explosion. Refer to the additional fuel system precautions later in this section.
- Fuel injection systems often remain pressurized, even after the engine has been turned **OFF**. The fuel system pressure must be relieved before disconnecting any fuel lines. Failure to do so may result in fire and/or personal injury.
- Brake fluid often contains polyglycol ethers and polyglycols. Avoid contact with the eyes and wash your hands thoroughly after handling brake fluid. If you do get brake fluid in your eyes, flush your eyes with clean, running water for 15 minutes. If eye irritation persists, or if you have taken brake fluid internally, IMMEDIATELY seek medical assistance.
- The EPA warns that prolonged contact with used engine oil may cause a number of skin disorders, including cancer. You should make every effort to minimize your exposure to used engine oil. Protective gloves should be worn when changing oil. Wash your hands and any other exposed skin areas as soon as possible after exposure to used engine oil. Soap and water, or waterless hand cleaner should be used.
- All new vehicles are now equipped with an air bag system, often referred to as a Supplemental Restraint System (SRS) or Supplemental Inflatable Restraint (SIR) system. The system must be disabled before performing service on or around system components, steering column, instrument panel components, wiring and sensors. Failure to follow safety and disabling procedures could result in accidental air bag deployment, possible personal injury and unnecessary system repairs.
- Always wear safety goggles when working with, or around, the air bag system. When carrying a non-deployed air bag, be sure the bag and trim cover are pointed away from your body. When placing a non-deployed air bag on a work surface, always face the bag and trim cover upward, away from the surface. This will reduce the motion of the module if it is accidentally deployed. Refer to the additional air bag system precautions later in this section.
- Clean, high quality brake fluid from a sealed container is essential to the safe and proper operation of the brake system. You should always buy the correct type of brake fluid for your vehicle. If the brake fluid becomes contaminated, completely flush the system with new fluid. Never reuse any brake fluid. Any brake fluid that is removed from the system should be discarded. Also, do not allow any brake fluid to come in contact with a painted surface; it will damage the paint.
- Never operate the engine without the proper amount and type of engine oil; doing so WILL result in severe engine damage.
- Timing belt maintenance is extremely important. Many models utilize an interference-type, non-freewheeling engine. If the timing belt breaks, the valves in the cylinder head may strike the pistons, causing potentially serious (also time-consuming and expensive) engine damage. Refer to the maintenance interval charts for the recommended replacement interval for the timing belt, and to the timing belt section for belt replacement and inspection.
- Disconnecting the negative battery cable on some vehicles may interfere with the functions of the on-board computer system(s) and may require the computer to undergo a relearning process once the negative battery cable is reconnected.
- When servicing drum brakes, only disassemble and assemble one side at a time, leaving the remaining side intact for reference.
- Only an MVAC-trained, EPA-certified automotive technician should service the air conditioning system or its components.

BRAKES

ANTI-LOCK BRAKE SYSTEM (ABS)

GENERAL INFORMATION

PRECAUTIONS

- Certain components within the ABS system are not intended to be serviced or repaired individually.
- Do not use rubber hoses or other parts not specifically specified for and ABS system. When using repair kits, replace all parts included in the kit. Partial or incorrect repair may lead to functional problems and require the replacement of components.
- Lubricate rubber parts with clean, fresh brake fluid to ease assembly. Do not use shop air to clean parts; damage to rubber components may result.
- Use only DOT 3 brake fluid from an unopened container.
- If any hydraulic component or line is removed or replaced, it may be necessary to bleed the entire system.
- A clean repair area is essential. Always clean the reservoir and cap thoroughly before removing the cap. The slightest amount of dirt in the fluid may plug an orifice and impair the system function. Perform repairs after components have been thoroughly cleaned; use only denatured alcohol to clean components. Do not allow ABS components to come into contact with any substance containing mineral oil; this includes used shop rags.
- The Anti-Lock control unit is a microprocessor similar to other computer units in the vehicle. Ensure that the ignition switch is **OFF** before removing or installing controller harnesses. Avoid static electricity discharge at or near the controller.
- If any arc welding is to be done on the vehicle, the control unit should be unplugged before welding operations begin.

SPEED SENSORS

REMOVAL & INSTALLATION

Front

See Figure 1.

1. The right side and the left side procedures are the same.
2. Remove the wheel.
3. Remove the fender liner.
4. Disengage the speed sensor wire harness clamp from the body.

5. Disconnect the speed sensor connector.
6. Remove the 2 clamp bolts holding the sensor harness from the body and shock absorber.
7. Remove the bolt and speed sensor.

WARNING

Keep the tip of the sensor clean.

To install:

8. Install the speed sensor with the bolt. Tighten to 71 inch. lbs. (8 Nm) of torque.
9. Install the sensor harness clamp with the 2 bolts to the body and shock absorber.
10. Connect the speed sensor connector.
11. Engage the speed sensor wire harness clamp to the body.
12. Install the fender liner.
13. Install the wheel.

Rear

See Figures 2 through 5.

➡Use the same procedure for the RH side and LH side. The procedure listed below is for the LH side.

➡If the sensor rotor needs to be replaced, replace it together with the rear axle hub and bearing assembly with rear speed sensor.

➡Special tools needed for this procedure: SST: 09520-00031, 09521-00010, 09520-00040) (SST: 09521-00020)

1. Disconnect cable from negative battery terminal.
2. For vehicles equipped with disc brakes, remove lower LH and RH instrument panel finish panel LH. Refer to Instrument Panel in Interior under Body.
3. For vehicles equipped with disc brakes, remove shift lever knob sub-assembly.
4. For vehicles equipped with disc brakes, remove center instrument cluster finish panel assembly. Refer to Instrument Panel in Interior under Body.
5. For vehicles equipped with disc brakes, remove upper console panel sub-assembly. Refer to Console in Interior under Body.
6. Remove rear wheel.
7. Loosen parking brake cable (for disc type). Refer to Parking Brake in this section.
8. Disconnect rear speed sensor wire (for Drum Type). Using a screwdriver, disconnect the connector from the rear speed sensor. Be careful not to damage the rear speed sensor.

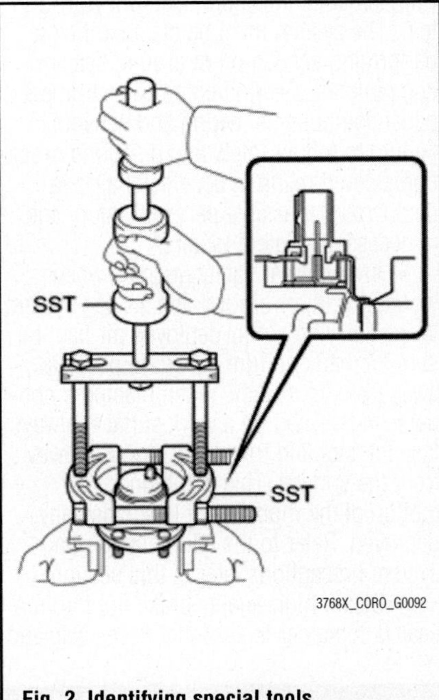

Fig. 2 Identifying special tools

9. Disconnect rear speed sensor wire (for Disc Type). Using a screwdriver, disconnect the connector from the rear speed sensor. Be careful not to damage the rear speed sensor.
10. Remove parking brake lever protector.
11. Separate no. 3 parking brake cable assembly (for Disc Type). Refer to Parking Brake in this section.
12. Separate rear disc brake caliper assembly (for Disc Type). Refer to Brake Caliper under Rear Disc Brake in this section.
13. Remove rear disc (for disc type). Refer to Disc under Rear Disc Brake in this section.
14. Remove rear brake drum (for drum type). Refer to Brake Drum under Rear Drum Brake in this section.
15. Remove rear axle hub and bearing assembly with rear speed sensor (for drum type). Refer to Wheel Hub and Bearing Assembly.
16. Remove rear axle hub and bearing assembly with rear speed sensor (for disc type). Refer to Wheel Hub and Bearing Assembly.
17. Remove rear speed sensor:
    a. Install the 3 hub nuts and mount the rear axle hub and bearing assembly in a vise using aluminum plates.
    b. Replace the rear axle hub and bearing assembly if it is dropped or receives a strong shock.
    c. Using a pin punch and a hammer, drive out the 2 pins and remove the 2 attachments from SST.
    d. Using SST and 2 bolts (diameter: 12 mm, pitch: 1.5 mm), remove the rear

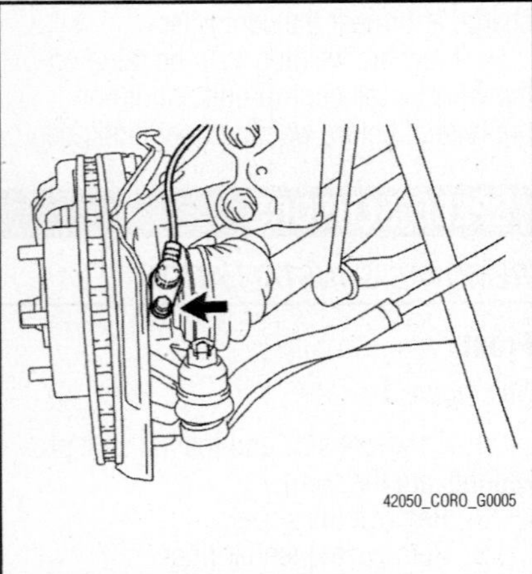

Fig. 1 Removal and installation of the speed sensor

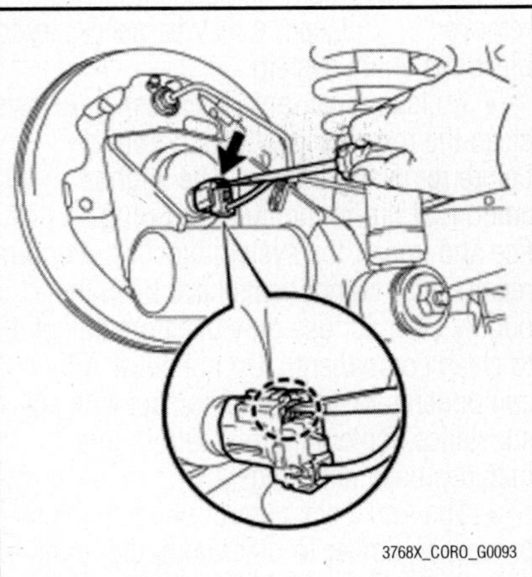

Fig. 3 Disconnecting the rear speed sensor connector—Drum

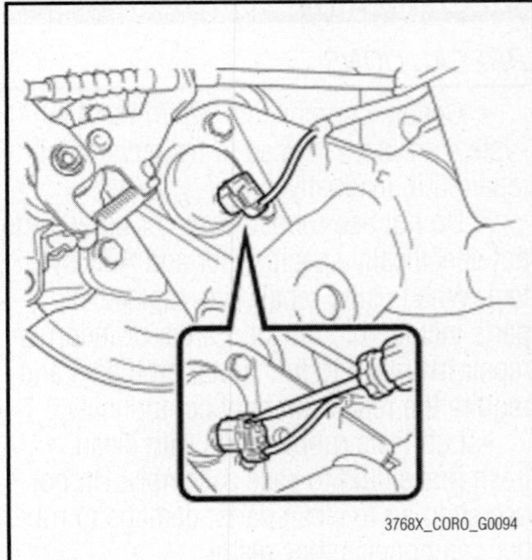

Fig. 4 Disconnecting the rear speed sensor connector—Drum

speed sensor from the rear axle hub and bearing assembly.

18. Keep the rear speed sensor away from magnets.

19. Pull the rear speed sensor off straight, taking care not to allow it to contact with the rear speed sensor rotor.

20. If the rear speed sensor rotor is damaged or deformed, replace the rear axle hub and bearing assembly.

21. Do not scratch the contact surface between the rear axle hub and bearing assembly and the rear speed sensor.

22. Prevent foreign matter from attaching to the speed sensor rotor or tip.

➡If the sensor rotor needs to be replaced, replace it together with the rear axle hub and bearing assembly with rear speed sensor.

To install:

➡If the sensor rotor needs to be replaced, replace it together with the rear axle hub and bearing assembly with rear speed sensor.

23. Clean the contact surface between the rear axle hub and bearing assembly and a new rear speed sensor. Prevent foreign matter from attaching to the sensor rotor.

24. Place the rear speed sensor on the rear axle hub and bearing assembly so that the connector is positioned at the bottom.

25. Using SST: 09214-76011, steel plates and a press, install a new speed sensor to the rear axle hub and bearing assembly.

➡Keep the rear speed sensor away from magnets. Do not use a hammer to install the rear speed sensor. Check that there is no foreign matter such as iron chips on the detecting portion of the rear speed sensor. Slowly press in the rear speed sensor straight.

26. To complete the installation, reverse remaining removal procedure.

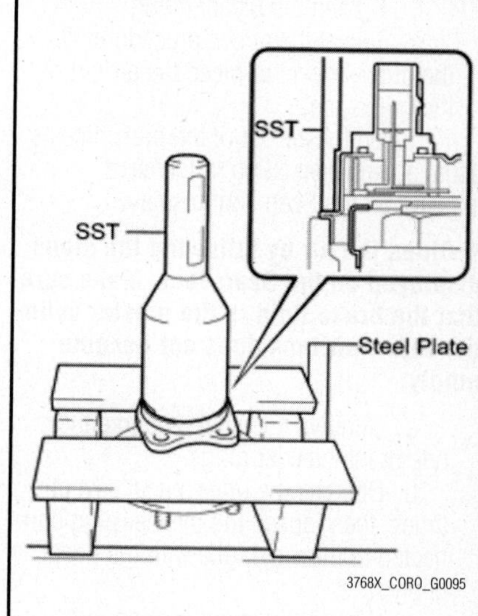

Fig. 5 Installing the new speed sensor to the rear axle hub and bearing assembly using SST: 09214-76011

BRAKES

BLEEDING THE BRAKE SYSTEM

BLEEDING PROCEDURE

BLEEDING PROCEDURE

➡After bleeding the air from the brake system, if the height or feel of the brake pedal cannot be obtained, perform air bleeding in the brake actuator assembly with the Scan Tool by following the procedures below.

1. Depress the brake pedal more than 20 times with the engine off.
2. Connect the Scan Tool to the DLC3, and turn the ignition switch to the ON position.
3. Select "AIR BLEEDING" on the Scan Tool.
4. Bleed the air out of the brake line as usual when "Step 1: Increase" appears on the Scan Tool display.

➡Bleed the air by following the steps displayed on the Scan Tool. Make sure that the brake fluid in the master cylinder reservoir tank does not become empty.

a. Connect the vinyl tube to either one of the bleeder plugs.
b. Depress the brake pedal several times, then loosen the bleeder plug connected to the vinyl tube with the pedal depressed.
c. When fluid stops coming out, tighten the bleeder plug and release the brake pedal.
d. Repeat this procedure until all air in the fluid is completely bled out.
e. Tighten the bleeder plug completely.
f. Repeat the above procedures for each wheel to bleed the air out of the brake line.

5. Bleed the air out of the suction line when "Step 2: Inhalation" appears on the Scan Tool display.

➡Bleed the air by following the steps displayed on the Scan Tool. Make sure that the brake fluid in the master cylinder reservoir tank does not become empty.

a. Connect the vinyl tube to the bleeder plug at the right front wheel or the right rear wheel and loosen the bleeder plug.
b. Operate the brake actuator assembly to bleed the air using the Scan Tool.

➡At this time, be sure to release the brake pedal.

➡This operation stops automatically after 4 seconds.

c. Check if the operation has stopped by referring to the Scan Tool display.
d. Repeat this until all air in the fluid is completely bled out.
e. Tighten the bleeder plug.
f. Repeat the above procedures for the other wheels to bleed the air out of the brake line.

6. Bleed the air out of the pressure reduction line when "Step 3: Decrease" appears on the Scan Tool display.

➡Bleed the air by following the steps displayed on the Scan Tool. Make sure that the brake fluid in the master cylinder reservoir tank does not become empty.

a. Connect a vinyl tube to either one of the bleeder plugs.
b. Loosen the bleeder plug.
c. Using the Scan Tool, operate the brake actuator assembly, completely depress the brake pedal and keep it.

➡During this procedure, the pedal will feel heavy, but completely depress it so that the brake fluid comes out from the bleeder plug. Be sure to keep depressing the brake pedal. Do not depress and release the pedal repeatedly.

➡The operation stops automatically after 4 seconds. When performing this procedure continuously, set an interval of at least 20 seconds. When the operation is complete, the brake pedal goes down slightly. This is a normal phenomenon caused when the solenoid opens.

d. Tighten the bleeder plug, then release the brake pedal.
e. Repeat this until all the air in the fluid is completely bled out.

f. Tighten the bleeder plug.

g. Repeat the above procedures for the other wheels to bleed the air out of the brake line.

7. Bleed the air out of the brake line as usual again when "Step 4: Increase" appears on the Scan Tool display.

➡**Bleed the air by following the steps displayed on the Scan Tool. Make sure that the brake fluid in the master cylinder reservoir tank does not become empty.**

a. Connect the vinyl tube to either one of the bleeder plugs.

b. Depress the brake pedal several times, then loosen the bleeder plug connected to the vinyl tube with the pedal depressed.

c. When fluid stops coming out, tighten the bleeder plug, then release the brake pedal.

d. Repeat this until all the air in the fluid is completely bled out.

e. Tighten the bleeder plug.

f. Repeat the above procedures for the other wheels to bleed the air out of the brake line.

g. Make sure that the air bleeding is complete by referring to the Scan Tool display.

h. Check the fluid level and add fluid if necessary.

MASTER CYLINDER BLEEDING

Refer to BLEEDING PROCEDURE.

BRAKE LINE BLEEDING

⁂ WARNING

Move the shift lever to P and apply the parking brake before bleeding the brakes.

⁂ WARNING

Add brake fluid to keep the level between the MIN and MAX lines of the reservoir while bleeding the brakes.

⁂ WARNING

If brake fluid leaks onto any painted surface, immediately wash it off.

1. Remove center cowl top ventilator louver.

a. Slide the hood to cowl top seal and disengage the clip.

b. Disengage the 5 claws and remove the center cowl top ventilator louver.

2. Fill the reservoir with brake fluid. (Standard fluid: SAE J1703 or FMVSS No. 116 DOT 3).

➡**Add brake fluid to keep the level between the MIN and MAX lines of the reservoir while bleeding the brakes.**

3. Bleed the brake line.

➡**Bleed the brake line of the wheel farthest from the master cylinder first.**

➡**Add brake fluid to keep the level between the MIN and MAX lines of the reservoir while bleeding the brakes.**

a. Connect a vinyl tube to the bleeder plug.

b. Depress the brake pedal several times, and then loosen the bleeder plug with the pedal depressed.

c. When fluid stops coming out, tighten the bleeder plug, and then release the brake pedal.

d. Repeat steps b. and c. until all the air in the fluid is completely bled out.

e. Tighten the bleeder plug completely. Tighten front bleeder plug to 73 inch lbs. (8.3 Nm). Tighten rear bleeder plug (drum brake) to 75 inch lbs. (8.5 Nm). Tighten rear bleeder plug (disc brake) to 7 ft. lbs. (10 Nm).

f. Repeat the above procedure for each wheel to bleed the brake line.

4. Inspect fluid level.

5. Inspect for brake fluid leak.

6. Install center cowl top ventilator louver.

a. Engage the 5 claws to install the center cowl top ventilator louver.

b. Push and engage the clip, and slide and engage the claw of the hood to cowl top seal

FLUID FILL PROCEDURE

See Figures 6 through 8.

⁂ WARNING

Clean, high quality brake fluid is essential to the safe and proper operation of the brake system. You should always buy the highest quality brake fluid that is available. If the brake fluid becomes contaminated, drain and flush the system, then refill the master cylinder with new fluid. Never reuse any brake fluid. Any brake fluid that is removed from the system should be discarded. Also, do not allow any brake fluid to come in contact with a painted surface; it will damage the paint.

⁂ CAUTION

Brake fluid contains polyglycol ethers and polyglycols. Avoid contact with

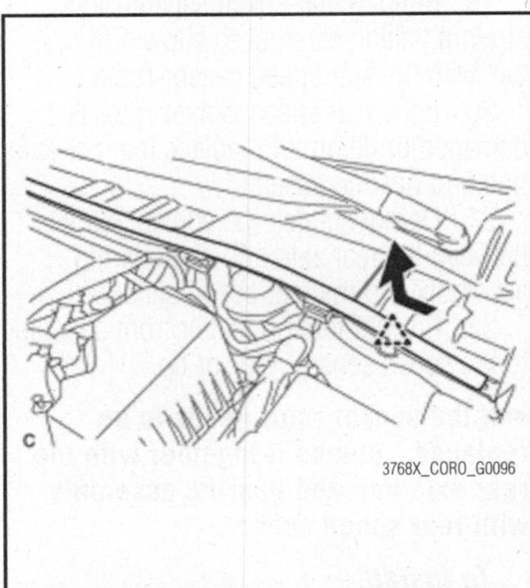

Fig. 6 Sliding the hood cowl top seal and disengaging the clip

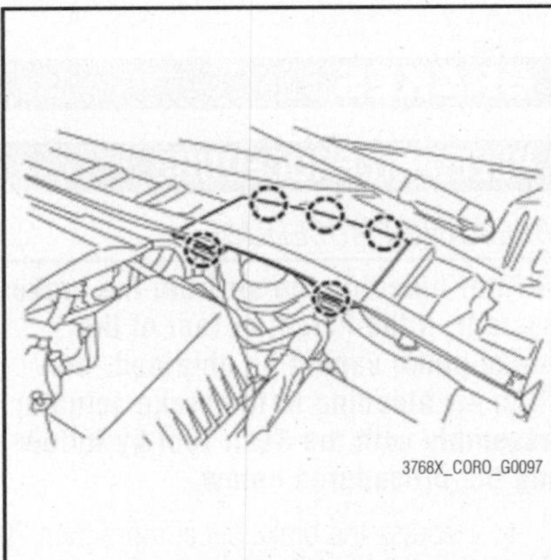

Fig. 7 Disengaging the 5 claws and removing the center cowl top ventilator louver

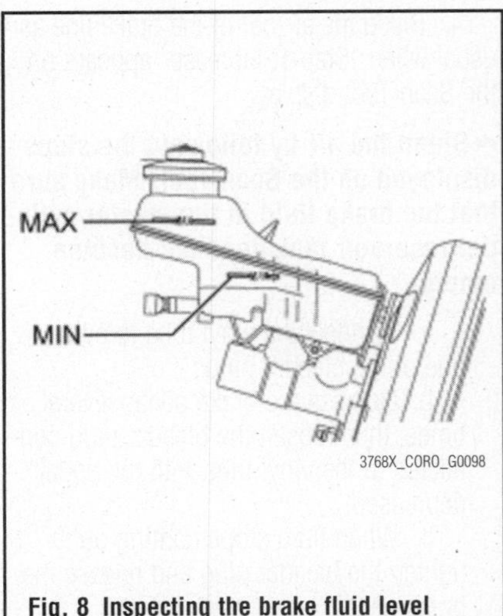

Fig. 8 Inspecting the brake fluid level

the eyes and wash your hands thoroughly after handling brake fluid. If you do get brake fluid in your eyes, flush your eyes with clean, running water for 15 minutes. If eye irritation persists, or if you have taken brake fluid internally, IMMEDIATELY seek medical assistance.

Note the following:

- If any work is performed on the brake system or if air in the brake lines is suspected, bleed the brake system.
- Move the shift lever to P and apply the parking brake before bleeding the brakes.
- Add brake fluid to keep the level between the MIN and MAX lines of the reservoir while bleeding the brakes.
- If brake fluid leaks onto any painted surface, immediately wash it off.

1. Remove center cowl top ventilator louver:
   a. Slide the hood to cowl top seal and disengage the clip as shown in the illustration.
   b. Disengage the 5 claws and remove the center cowl top ventilator louver.
2. Fill the reservoir with brake fluid (SAE DOT 3).
3. Add brake fluid between the MIN and MAX lines of the reservoir, especially if you are bleeding the brakes.
4. If brake fluid level is lower than the MIN line, check for leaks and inspect the disc brake pads. If necessary, refill the reservoir with brake fluid to the MAX line after repair or replacement.

BRAKES

FRONT DISC BRAKES

CAUTION

Dust and dirt accumulating on brake parts during normal use may contain asbestos fibers from production or aftermarket brake linings. Breathing excessive concentrations of asbestos fibers can cause serious bodily harm. Exercise care when servicing brake parts. Do not sand or grind brake lining unless equipment used is designed to contain the dust residue. Do not clean brake parts with compressed air or by dry brushing. Cleaning should be done by dampening the brake components with a fine mist of water, then wiping the brake components clean with a dampened cloth. Dispose of cloth and all residue containing asbestos fibers in an impermeable container with the appropriate label. Follow practices prescribed by the Occupational Safety and Health Administration (OSHA) and the Environmental Protection Agency (EPA) for the handling, processing, and disposing of dust or debris that may contain asbestos fibers.

BRAKE CALIPER

REMOVAL & INSTALLATION

See Figure 9.

1. Before servicing the vehicle, refer to the Precautions section.
2. Remove the wheels.
3. Disconnect the brake hose from the caliper.
4. Remove the bolts that attach the caliper to the torque plate. If applicable, hold the flats of the sliding pin with a wrench while loosening the caliper attaching bolts.

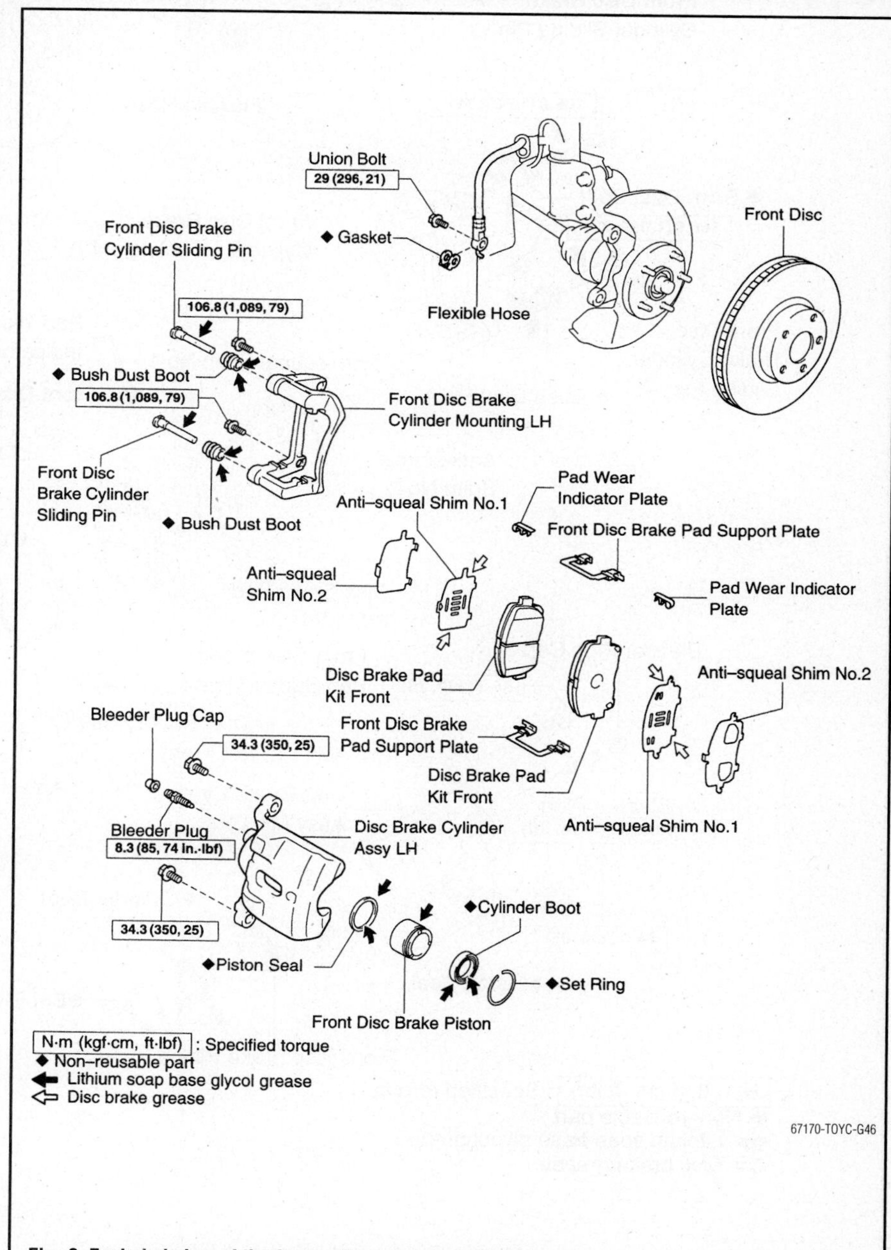

Fig. 9 Exploded view of the front caliper and components

5. Lift up and remove the caliper assembly.

To install:

6. Install the caliper and loosely install the bolts.

7. Hold the flats of the sliding pin with a wrench, then tighten the bolts. Tighten to 25 ft. lbs. (34 Nm).

8. Connect the brake hose to the caliper, using 2 new washers.

9. Fill the brake system to the proper level and bleed the brake system.

10. Add brake fluid to the reservoir to fill to the correct level.

11. Lower the vehicle to the ground.

DISC BRAKE PADS

REMOVAL & INSTALLATION

See Figure 10.

1. Before servicing the vehicle, refer to the Precautions section.

2. Remove the wheels.

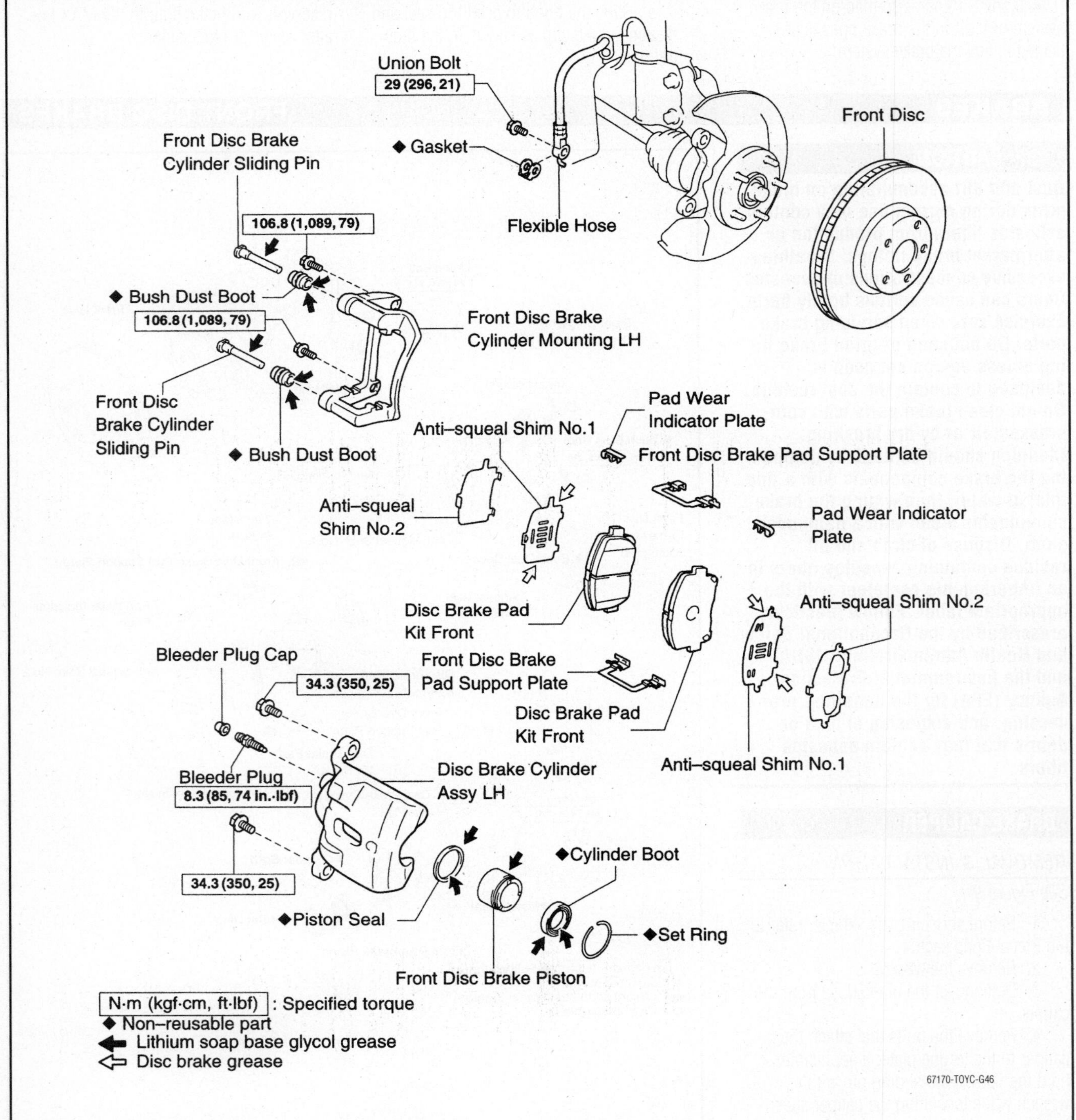

Fig. 10 Exploded view of the front brake pads and components

3. Loosen and remove the caliper mounting bolts, then remove the caliper assembly, without disconnecting the brake line. Position it aside.

4. Slide out the old brake pads along with any anti-squeal shims, springs, pad wear indicators and pad support plates.

To install:

5. Install the pad support plates into the torque plate.

6. Install the pad wear indicators onto the pads. Be sure the arrow on the indicator plate is pointing in the direction of rotation.

7. Install the anti-squeal shims on the outside of each pad and then install the pad assemblies into the torque plate.

8. Compress the caliper piston into the bore.

9. Position the caliper back down over the pads.

10. Install and tighten the caliper mounting bolts.

11. Install the wheels. Check the brake fluid level.

BRAKES

REAR DISC BRAKES

⁂ CAUTION

Dust and dirt accumulating on brake parts during normal use may contain asbestos fibers from production or aftermarket brake linings. Breathing excessive concentrations of asbestos fibers can cause serious bodily harm. Exercise care when servicing brake parts. Do not sand or grind brake lining unless equipment used is designed to contain the dust residue. Do not clean brake parts with compressed air or by dry brushing. Cleaning should be done by dampening the brake components with a fine mist of water, then wiping the brake components clean with a dampened cloth. Dispose of cloth and all residue containing asbestos fibers in an impermeable container with the appropriate label. Follow practices prescribed by the Occupational Safety and Health Administration (OSHA) and the Environmental Protection Agency (EPA) for the handling, processing, and disposing of dust or debris that may contain asbestos fibers.

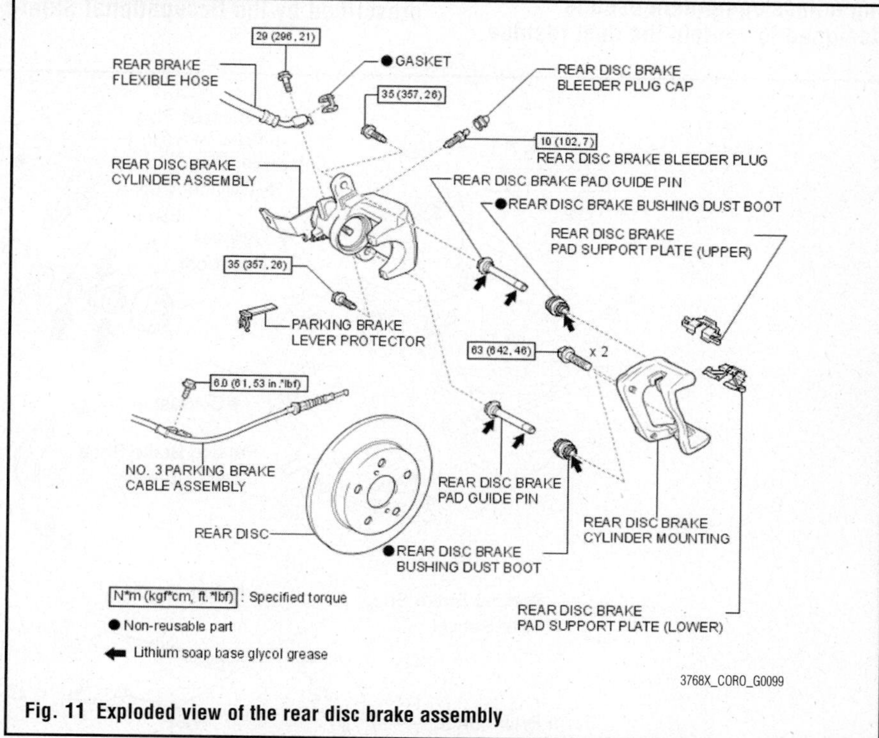

Fig. 11 Exploded view of the rear disc brake assembly

BRAKE CALIPER

REMOVAL & INSTALLATION

See Figure 11.

1. Before servicing the vehicle, refer to the Precautions section.

2. Remove the wheels.

3. Disconnect the brake hose from the caliper.

4. Remove the bolts that attach the caliper to the torque plate. If applicable, hold the flats of the sliding pin with a wrench while loosening the caliper attaching bolts.

5. Lift up and remove the caliper assembly.

To install:

6. Install the caliper and loosely install the bolts.

7. Hold the flats of the sliding pin with a wrench, then tighten the bolts. Tighten the bolts on rear calipers to 34 ft. lbs. (46 Nm).

8. Connect the brake hose to the caliper, using 2 new washers.

9. Fill the brake system to the proper level and bleed the brake system.

10. Add brake fluid to the reservoir to fill to the correct level.

11. Lower the vehicle to the ground.

DISC BRAKE PADS

REMOVAL & INSTALLATION

See Figure 11.

1. Before servicing the vehicle, refer to the Precautions section.

2. Remove the wheels.

3. Loosen and remove the caliper mounting bolts, then remove the caliper assembly, without disconnecting the brake line. Position it aside.

4. Slide out the old brake pads along with any anti-squeal shims, springs, pad wear indicators and pad support plates.

To install:

5. Install the pad support plates into the torque plate.

6. Install the pad wear indicators onto the pads. Be sure the arrow on the indicator plate is pointing in the direction of rotation.

7. Install the anti-squeal shims on the outside of each pad and then install the pad assemblies into the torque plate.

8. Compress the caliper piston into the bore. Use tool SST 09719-14020, to rotate the piston clockwise while pressing it into the bore until it locks.

9. Position the caliper back down over the pads.

10. Install and tighten the caliper mounting bolts.

11. Install the wheels. Check the brake fluid level.

⁂ CAUTION

Dust and dirt accumulating on brake parts during normal use may contain asbestos fibers from production or aftermarket brake linings. Breathing excessive concentrations of asbestos fibers can cause serious bodily harm. Exercise care when servicing brake parts. Do not sand or grind brake lining unless equipment used is designed to contain the dust residue. Do not clean brake parts with compressed air or by dry brushing. Cleaning should be done by dampening the brake components with a fine mist of water, then wiping the brake components clean with a dampened cloth. Dispose of cloth and all residue containing asbestos fibers in an impermeable container with the appropriate label. Follow practices prescribed by the Occupational Safety and Health Administration (OSHA) and the Environmental Protection Agency (EPA) for the handling, processing, and disposing of dust or debris that may contain asbestos fibers.

BRAKE DRUM

REMOVAL & INSTALLATION

See Figure 12.

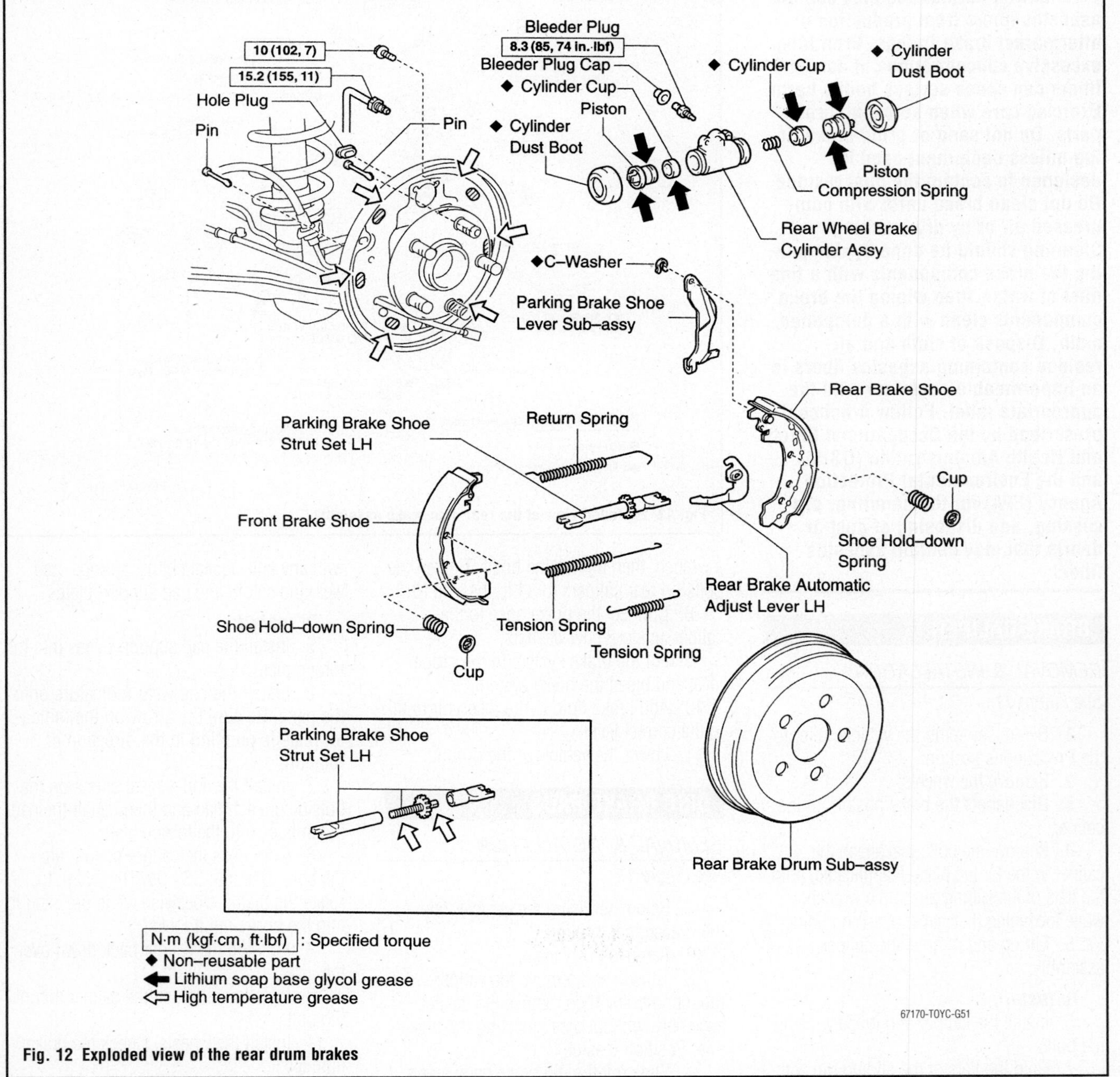

Fig. 12 Exploded view of the rear drum brakes

1. Before servicing the vehicle, refer to the Precautions section.
2. Remove the wheels.
3. Remove the brake drum from the axle hub.

To install:

4. Install the brake drum.
5. Install the rear wheels, tighten the wheel lug nuts.

BRAKE SHOES

REMOVAL & INSTALLATION

See Figure 12.

1. Before servicing the vehicle, refer to the Precautions section.
2. Remove the wheels.
3. Remove the brake drum.
4. Unhook the return spring from the leading (front) brake shoe. Remove the hold-down spring and the pin. Pull out the brake shoe and unhook the anchor spring from the lower edge.
5. Remove the hold-down spring from the trailing (rear) shoe. Pull the shoe out with the adjuster strut, automatic adjuster assembly and springs attached and disconnect the parking brake cable. Unhook the return spring and then remove the adjusting strut. Remove the anchor spring.
6. Remove the adjusting strut. Unhook the adjusting lever spring from the rear shoe and then remove the automatic adjuster assembly by popping out the C-clip.

To install:

7. Mount the automatic adjuster assembly onto a new rear brake shoe. Make sure the C-clip fits properly. Connect the adjusting strut/return spring and then install the adjusting spring.
8. Connect the parking brake cable to the rear shoe and then position the shoe so the lower end rides in the anchor plate and the upper end is against the boot in the wheel cylinder. Install the pin and the hold-down spring.
9. Install the anchor spring between the front and rear shoes. Install the hold-down spring and pin.
10. Connect the return spring/adjusting strut between the 2 shoes so it rides freely.
11. Install the drum.
12. Install the wheel.

ADJUSTMENT

1. Temporarily install the hub nuts.
2. Remove the hole plug, and turn the adjuster and expand the shoe until the drum locks.
3. Using a screwdriver, back off the adjuster 8 notches.
4. Install the hole plug.

BRAKES — PARKING BRAKE

PARKING BRAKE CABLES

ADJUSTMENT

See Figure 13.

1. Remove the rear wheel.
2. Adjust the brake shoe clearance.
3. Install the rear wheel.

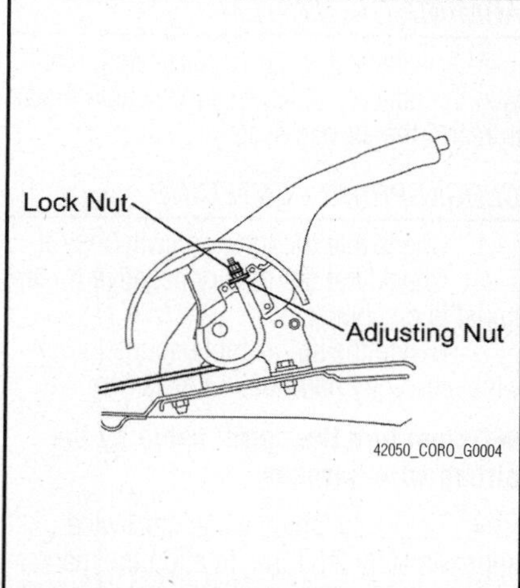

Fig. 13 Adjusting the parking brake cable

4. Pull the parking brake lever to the fully applied position, and count the number of clicks (6 to 9 is optimal).
5. Remove the rear console box sub-assembly.
6. Loosen the lock nut and turn the adjusting nut until the lever travel turns correct (6-9 Clicks).
7. Tighten the lock nut to 44 in. lbs. (5 Nm) of torque.
8. Install the rear console box assembly.

PARKING BRAKE SHOES

REMOVAL & INSTALLATION

Drum

1. Before servicing the vehicle, refer to the Precautions section.
2. Remove the wheels.
3. Remove the brake drum.
4. Unhook the return spring from the leading (front) brake shoe. Remove the hold-down spring and the pin. Pull out the brake shoe and unhook the anchor spring from the lower edge.
5. Remove the hold-down spring from the trailing (rear) shoe. Pull the shoe out with the adjuster strut, automatic adjuster assembly and springs attached and disconnect the parking brake cable. Unhook the return spring and then remove the adjusting strut. Remove the anchor spring.
6. Remove the adjusting strut. Unhook the adjusting lever spring from the rear shoe and then remove the automatic adjuster assembly by popping out the C-clip.

To install:

7. Mount the automatic adjuster assembly onto a new rear brake shoe. Make sure the C-clip fits properly. Connect the adjusting strut/return spring and then install the adjusting spring.
8. Connect the parking brake cable to the rear shoe and then position the shoe so the lower end rides in the anchor plate and the upper end is against the boot in the wheel cylinder. Install the pin and the hold-down spring.
9. Install the anchor spring between the front and rear shoes. Install the hold-down spring and pin.
10. Connect the return spring/adjusting strut between the 2 shoes so it rides freely.
11. Install the drum.
12. Install the wheel.

GENERAL INFORMATION

⁂ CAUTION

These vehicles are equipped with an air bag system. The system must be disarmed before performing service on, or around, system components, the steering column, instrument panel components, wiring and sensors. Failure to follow the safety precautions and the disarming procedure could result in accidental air bag deployment, possible injury and unnecessary system repairs.

SERVICE PRECAUTIONS

Disconnect and isolate the battery negative cable before beginning any airbag system component diagnosis, testing, removal, or installation procedures. Allow system capacitor to discharge for two minutes before beginning any component service. This will disable the airbag system. Failure to disable the airbag system may result in accidental airbag deployment, personal injury, or death.

Do not place an intact undeployed airbag face down on a solid surface. The airbag will propel into the air if accidentally deployed and may result in personal injury or death.

When carrying or handling an undeployed airbag, the trim side (face) of the airbag should be pointing away from the body to minimize possibility of injury if accidental deployment occurs. Failure to do this may result in personal injury or death.

Replace airbag system components with OEM replacement parts. Substitute parts may appear interchangeable, but internal differences may result in inferior occupant protection. Failure to do so may result in occupant personal injury or death.

Wear safety glasses, rubber gloves, and long sleeved clothing when cleaning powder residue from vehicle after an airbag deployment. Powder residue emitted from a deployed airbag can cause skin irritation. Flush affected area with cool water if irritation is experienced. If nasal or throat irritation is experienced, exit the vehicle for fresh air until the irritation ceases. If irritation continues, see a physician.

Do not use a replacement airbag that is not in the original packaging. This may result in improper deployment, personal injury, or death.

The factory installed fasteners, screws and bolts used to fasten airbag components have a special coating and are specifically designed for the airbag system. Do not use substitute fasteners. Use only original equipment fasteners listed in the parts catalog when fastener replacement is required.

During, and following, any child restraint anchor service, due to impact event or vehicle repair, carefully inspect all mounting hardware, tether straps, and anchors for proper installation, operation, or damage. If a child restraint anchor is found damaged in any way, the anchor must be replaced. Failure to do this may result in personal injury or death.

Deployed and non-deployed airbags may or may not have live pyrotechnic material within the airbag inflator.

Do not dispose of driver/passenger/curtain airbags or seat belt tensioners unless you are sure of complete deployment. Refer to the Hazardous Substance Control System for proper disposal.

Dispose of deployed airbags and tensioners consistent with state, provincial, local, and federal regulations.

After any airbag component testing or service, do not connect the battery negative cable. Personal injury or death may result if the system test is not performed first.

If the vehicle is equipped with the Occupant Classification System (OCS), do not connect the battery negative cable before performing the OCS Verification Test using the scan tool and the appropriate diagnostic information. Personal injury or death may result if the system test is not performed properly.

Never replace both the Occupant Restraint Controller (ORC) and the Occupant Classification Module (OCM) at the same time. If both require replacement, replace one, then perform the Airbag System test before replacing the other.

Both the ORC and the OCM store Occupant Classification System (OCS) calibration data, which they transfer to one another when one of them is replaced. If both are replaced at the same time, an irreversible fault will be set in both modules and the OCS may malfunction and cause personal injury or death.

If equipped with OCS, the Seat Weight Sensor is a sensitive, calibrated unit and must be handled carefully. Do not drop or handle roughly. If dropped or damaged, replace with another sensor. Failure to do so may result in occupant injury or death.

If equipped with OCS, the front passenger seat must be handled carefully as well. When removing the seat, be careful when setting on floor not to drop. If dropped, the sensor may be inoperative, could result in occupant injury, or possibly death.

If equipped with OCS, when the passenger front seat is on the floor, no one should sit in the front passenger seat. This uneven force may damage the sensing ability of the seat weight sensors. If sat on and damaged, the sensor may be inoperative, could result in occupant injury, or possibly death.

DISARMING THE SYSTEM

To avoid personal injury when working on vehicles equipped with an air bag, the negative battery cable must be disconnected and at least 90 seconds must elapse before working on the system. Failure to do so may result in deployment of the air bag.

ARMING THE SYSTEM

After vehicle service is completed, reattach the battery cables (positive cable first!) to rearm the air bag system.

CLOCKSPRING CENTERING

1. Check that the ignition switch is off.
2. Check that the battery negative (-) terminal is disconnected.
3. Rotate the spiral cable counterclockwise slowly by hand until it feels firm.

➡Do not turn the spiral cable by the airbag wire harness.

4. Rotate the spiral cable clockwise approximately 2.5 turns to align the marks.

➡Do not turn the spiral cable by the airbag wire harness. The spiral cable will rotate approximately 2.5 turns to both the left and right from the center.

DRIVE TRAIN

AUTOMATIC TRANSAXLE FLUID

DRAIN AND REFILL

1. Remove the drain plug and gasket to drain the ATF.
2. Install a new gasket and the drain plug. Tighten to 36 ft. lbs. (49 Nm).
3. Add the correct fluid amount via the dipstick tube.
4. Check the fluid level:
   a. Drive the vehicle so that the engine and transaxle are at normal operating temperature.
   b. Park the vehicle on a level surface and set the parking brake.
   c. With the engine idling and the brake pedal depressed, move the shift lever to all positions from P to S. Then return it to P.
   d. Pull out the dipstick and wipe it clean.
   e. Push the dipstick back fully into the pipe.
   f. Pull the dipstick out again and check that the fluid level is within the HOT range. If the fluid level is below the HOT range, add new fluid and recheck the fluid level. If the fluid level exceeds the HOT range, drain the fluid once, add a proper amount of new fluid and recheck the fluid level.

FILTER REPLACEMENT

See Figures 14 through 17.

1. Remove the LH engine under cover.
2. Drain the ATF fluid.
3. For the U250E transaxles, remove the 18 bolts, oil pan and gasket.

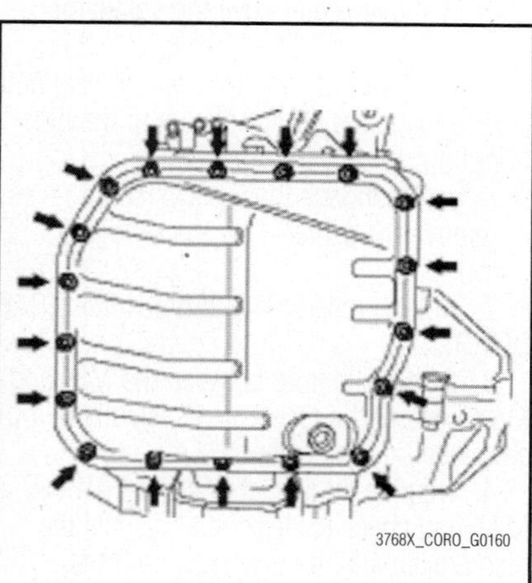

Fig. 14 Removing the 18 bolts—U250E transaxles

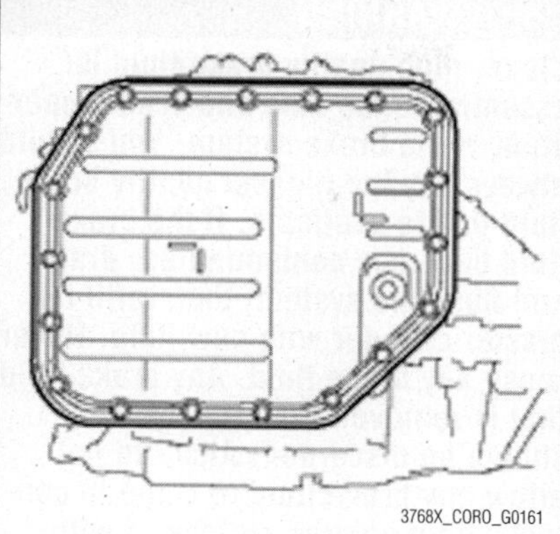

Fig. 15 Removing the 19 bolts—U341E transaxles

4. For the U341E transaxles, remove the 19 bolts, oil pan and oil pan gasket.
5. Remove the 2 magnets from the oil pan.
6. Examine particles in the pan:
   a. Collect any steel chips with the removed magnets. Look carefully at the chips and particles in the pan and on the magnets to see the type of wear which might be found in the transaxle.
   b. Result:
      - Steel (magnetic): bearing, gear and plate wear
      - Brass (non-magnetic): bushing wear
7. Remove the 3 bolts, oil strainer and O-ring.

To install:

8. Coat a new O-ring with ATF, and install it to the oil strainer.

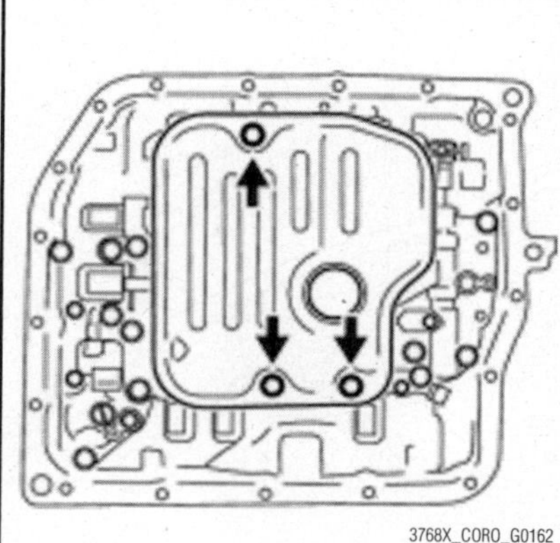

Fig. 16 Removing the 3 bolts and oil strainer—U250E transaxles

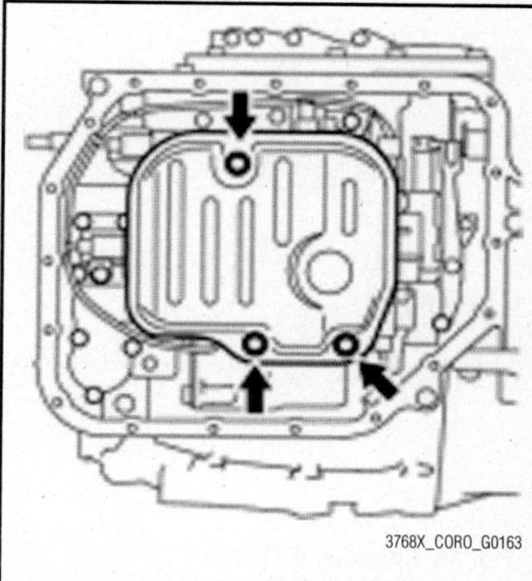

Fig. 17 Removing the 3 bolts and oil strainer—U341E transaxles

➡Install the O-ring carefully to avoid twists or pinches. Apply sufficient ATF to the O-ring prior to installation.

➡Apply ATF to the bolts prior to installation.

9. Install the oil strainer to the valve body with the 3 bolts.
10. Tighten the bolts to 8 ft. lbs. (11 Nm).
11. Install the 2 magnets to the oil pan.
12. Install a new oil pan gasket and the oil pan to the transaxle case.
13. Tighten the bolts to 69 inch. lbs. (8 Nm).

MANUAL TRANSAXLE FLUID

DRAIN AND REFILL

1. Remove the filler plug and the gasket.
2. Remove the drain plug and gasket, and then drain the manual transaxle oil in to a suitable container.
3. Install a new gasket to the drain plug. Add manual transaxle oil.
4. Install the transmission filler plug and a new gasket. Tighten to 29 ft. lbs. (39 Nm).

Inspection

See Figure 18.

1. Inspect transaxle oil:
   a. Stop the vehicle in a level place.
   b. Remove the transmission filler plug and the gasket.
   c. Check that the oil surface is within 0.197 inch (5 mm) of the bottom of the transmission filler plug opening.

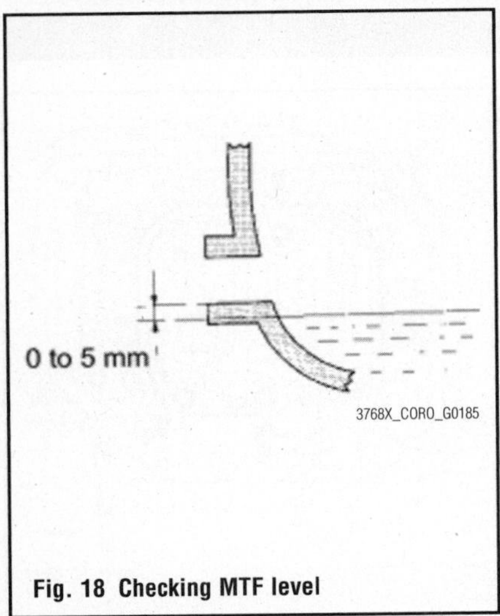

Fig. 18 Checking MTF level

➡Excessively large or small amounts of oil may cause problems. After replacing the oil, drive the vehicle and check the oil level again.

d. Check for oil leakage when the oil level is low.

CLUTCH

BLEEDING

⁂ WARNING

Wash off clutch fluid immediately if it comes in contact with any painted surface.

➡If any work is performed on the clutch system or if any air in the clutch lines is suspected, bleed air from the clutch hydraulic system.

1. Fill brake fluid reservoir.
2. Bleed clutch line.
   a. Remove the bleeder plug cap.
   b. Connect a vinyl tube to the bleeder plug.
   c. Depress the clutch pedal several times, and then loosen the bleeder plug while the pedal is depressed.
   d. When fluid no longer comes out, tighten the bleeder plug, and then release the clutch pedal.
   e. Repeat the previous 2 steps until all the air in the fluid is completely bled.
   f. Tighten the bleeder plug. Tighten to 74 inch lbs. (8.4 Nm).
   g. Install the bleeder plug cap.
   h. Check that all the air has been bled from the clutch line.
3. Inspect the brake fluid level in the reservoir.

FLUID FILL PROCEDURE

See Figures 19 through 21.

⁂ WARNING

Clean, high quality brake fluid is essential to the safe and proper operation of the brake system. You should always buy the highest quality brake fluid that is available. If the brake fluid becomes contaminated, drain and flush the system, then refill the master cylinder with new fluid. Never reuse any brake fluid. Any brake fluid that is removed from the system should be discarded. Also, do not allow any brake fluid to come in contact with a painted surface; it will damage the paint.

⁂ CAUTION

Brake fluid contains polyglycol ethers and polyglycols. Avoid contact with the eyes and wash your hands thoroughly after handling brake fluid. If you do get brake fluid in your eyes, flush your eyes with clean, running water for 15 minutes. If eye irritation persists, or if you have taken brake fluid internally, IMMEDIATELY seek medical assistance.

➡Note the following:

- If any work is performed on the brake system or if air in the brake lines is suspected, bleed the brake system.
- Move the shift lever to P and apply the parking brake before bleeding the brakes.
- Add brake fluid to keep the level between the MIN and MAX lines of the reservoir while bleeding the brakes.
- If brake fluid leaks onto any painted surface, immediately wash it off.

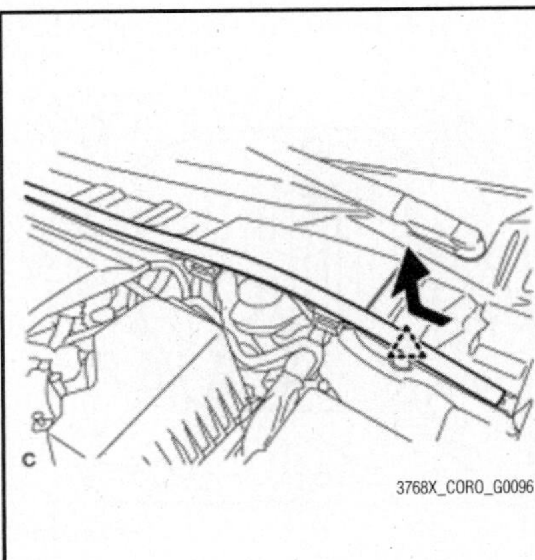

Fig. 19 Sliding the hood cowl top seal and disengaging the clip

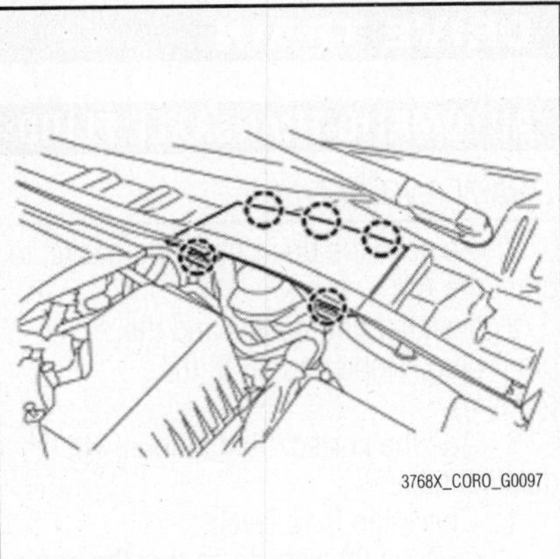

Fig. 20 Disengaging the 5 claws and removing the center cowl top ventilator louver

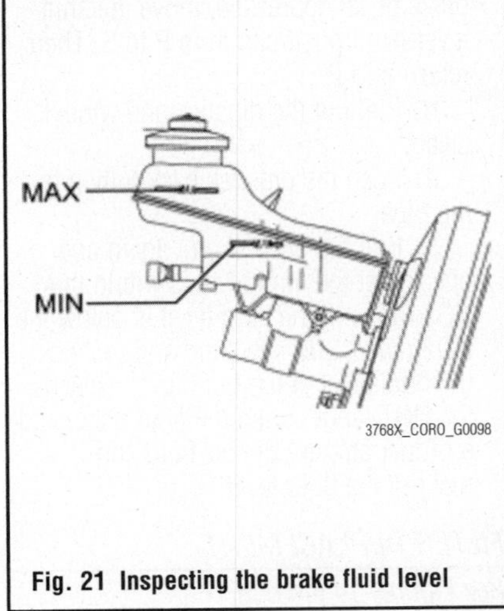

Fig. 21 Inspecting the brake fluid level

1. Remove center cowl top ventilator louver:
   a. Slide the hood to cowl top seal and disengage the clip as shown in the illustration.
   b. Disengage the 5 claws and remove the center cowl top ventilator louver.
2. Fill the reservoir with brake fluid (SAE DOT 3).
3. Add brake fluid between the MIN and MAX lines of the reservoir, especially if you are bleeding the brakes.
4. If brake fluid level is lower than the MIN line, check for leaks and inspect the disc brake pads. If necessary, refill the reservoir with brake fluid to the MAX line after repair or replacement.

FRONT HALFSHAFT

REMOVAL & INSTALLATION

See Figures 22 and 23.

➡The hub bearing could be damaged if subjected to the full weight of the vehicle, such as if the vehicle is moved without the halfshafts. If it is absolutely necessary to place the full vehicle weight on the hub bearing, first support the bearing with SST No. 09608–16042.

➡This procedure is for removing both LH and RH halfshafts.

1. Remove the LH and RH engine under covers.
2. Drain transmission fluid.
3. Raise and support the vehicle.
4. Remove the front wheels.
5. Remove front axle shaft nut:
   a. Using SST: 09930-00010 and a hammer, release the staked part of the front axle shaft nut.
   b. Insert SST into the groove with the flat surface facing up.
   c. Do not damage the tip of SST using grinders.
   d. Completely unstake the staked part before removing the axle hub nut.
   e. Do not damage the threads of the drive shaft.
   f. Using a socket wrench (30 mm), remove the front axle shaft nut.
6. Separate the wheel speed sensors.
7. Separate the front stabilizer link assembly. Refer to Stabilizer in Front Suspension.
8. Separate the tie rod end sub-assembly. Refer to Tie Rod in Front Suspension.
9. Separate front lower suspension. Refer to Front Suspension.
10. Separate front axle assembly:
    a. Using a plastic hammer, tap the end of the drive shaft and disengage the fitting between the drive shaft and front axle assembly.
    b. If it is difficult to disengage the fitting, tap the end of the drive shaft with a brass bar and hammer.
    c. Push the front axle assembly out of the vehicle to remove the drive shaft from the front axle assembly.

➡Do not push out the front axle further than necessary. Do not damage the outboard joint boot. Do not damage the speed sensor rotor.

11. If equipped, remove the 2 bolts and the manual transmission case protector.
12. Using a halfshaft puller, remove the LH halfshaft assembly.
13. Remove front RH drive shaft assembly:
    a. For 1.8L Engines:
    b. Using a brass bar and hammer, remove the front drive shaft assembly.

➡Do not damage the oil seal. Do not damage the inboard joint boot. Do not drop the drive shaft.

    c. For 2.4L Engines:
    d. Remove the bearing bracket hole snap ring.

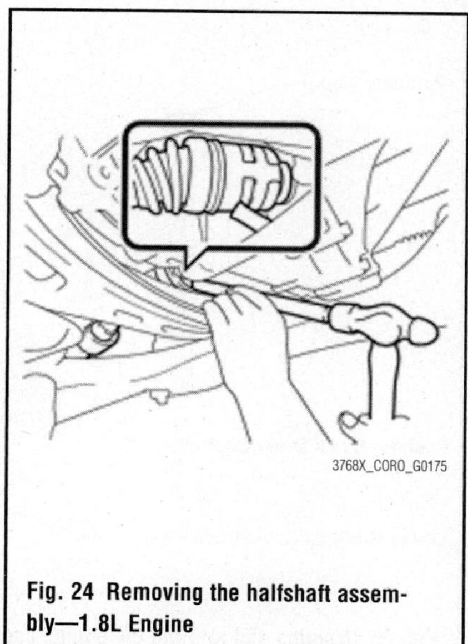

Fig. 24 Removing the halfshaft assembly—1.8L Engine

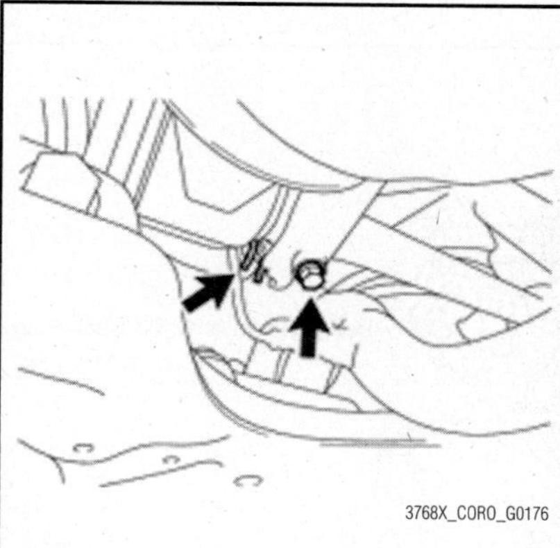

Fig. 25 Removing the bearing bracket hole snap ring and halfshaft assembly—2.4L Engine

    e. Remove the bolt and front drive shaft assembly RH from the drive shaft bearing bracket.

➡Do not damage the boot or oil seal.

14. Using a screwdriver, remove the snap ring from the front LH and RH drive inboard joint assembly.

To install:

15. Install a new front drive shaft hole snap ring to the front drive inboard joint assembly.
16. Align the inboard joint splines, and using a brass bar and a hammer, install the front drive shaft assembly.

➡Face the end gap of the front drive inboard joint hole snap ring downward. Do not damage the oil seal. Do not damage the inboard joint boot.

➡Confirm whether the drive shaft is securely driven in by checking the reaction force and sound.

17. For 2.4L engines:
    a. Using a screwdriver, install a new bearing bracket hole snap ring.

➡Do not damage the boot or oil seal. Move the drive shaft assembly while keeping it level.

    b. Install a new bolt. Tighten to 24 ft. lbs. (32 Nm).
18. If equipped, install the manual transmission case protector with the 2 bolts. Tighten to 13 ft. lbs. (18 Nm).
19. To complete installation, reverser remaining removal procedure.
20. Install front axle shaft nut:
    a. Clean the threaded parts on the drive shaft and axle shaft nut using a non-residue solvent.

➡Be sure to perform this work for a new drive shaft. Keep the threaded parts free of oil and foreign matter.

    b. Using a socket wrench (30 mm), install a new axle shaft nut. Tighten to 160 ft. lbs. (216 Nm).
    c. Using a chisel and hammer, caulk the axle shaft nut.
21. Tighten wheel lug nuts to 76 ft. lbs. (103 Nm).
22. Add transmission fluid.
23. Inspect for fluid leaks.
24. Inspect and adjust the wheel alignment.

ENGINE COOLING

ENGINE COOLANT

DRAIN & REFILL PROCEDURE

See Figures 24 and 25.

CAUTION

Do not loosen the lower radiator drain cock plug while the engine and radiator are still hot. Pressurized, hot engine coolant and steam may be released and cause serious burns.

➡Collect the coolant in a container and dispose of it according to the regulations in your area.

1. Loosen the lower radiator drain cock plug.
2. Remove the radiator cap.
3. Loosen the cylinder block drain cock plug.
4. To refill, perform the following:
   a. Tighten the lower radiator drain cock plug.
   b. Tighten the cylinder block drain cock plug. Tighten to 9 ft. lbs. (13 Nm).
   c. Loosen the upper radiator drain cock plug.
   d. Slowly fill the radiator with TOYOTA Super Long Life Coolant (SLLC).
   e. Standard Capacity: 6 quarts (5.7 liters).

➡TOYOTA vehicles are filled with TOYOTA SLLC at the factory. In order to avoid damage to the engine cooling system and other technical problems, only use TOYOTA SLLC or similar high quality ethylene glycol based non-silicate, non-amine, non-nitrite, non-borate coolant with long-life hybrid organic acid technology (coolant with long-life hybrid organic acid technology consists of a combination of low phosphates and organic acids).

➡Never use water as a substitute for engine coolant.

   f. Squeeze the inlet and outlet radiator hoses several times by hand, and then check the level of the coolant. If the coolant level is low, add coolant.
   g. Tighten the upper radiator drain cock plug.
   h. Slowly pour coolant into the radiator reservoir tank until it reaches the FULL line.
   i. Install the radiator cap sub-assembly and reservoir tank cap.

➡Before starting the engine, turn the A/C switch off. Adjust the air conditioning temperature setting to MAX (HOT). Adjust the air conditioning blower setting to LO.

   j. Start the engine and warm it up.
   k. Bleed air from the cooling system.

➡Thermostat opening timing can be determined by squeezing the inlet radiator hose, and sensing vibrations when the engine coolant starts to flow inside the hose.

CAUTION

When squeezing the radiator hoses: Wear protective gloves. Be careful as the radiator hoses are hot. Keep your hands away from the radiator fan.

   l. Stop the engine, and wait until the engine coolant cools down.
   m. Add engine coolant to the FULL line on the radiator reservoir.
5. Inspect for coolant leaks:

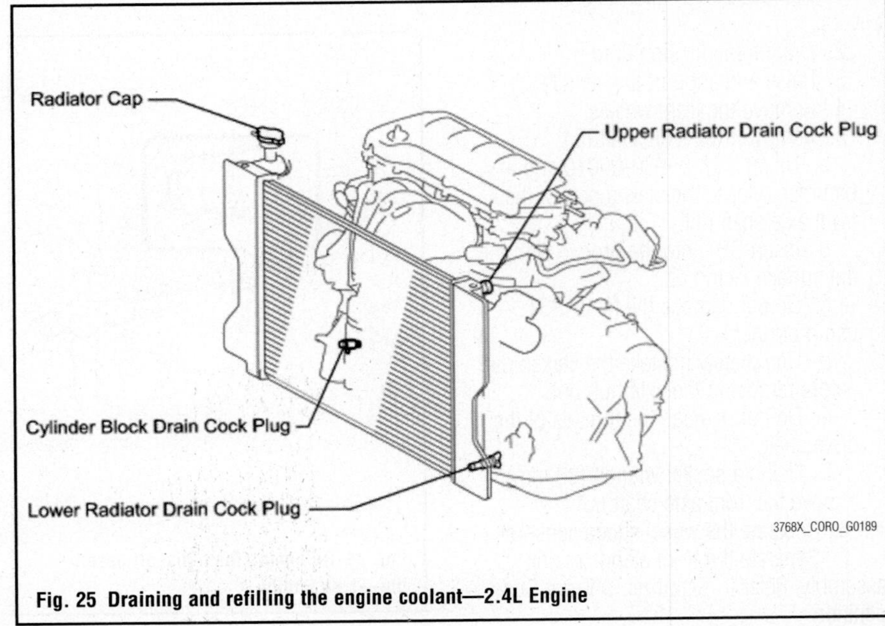

Fig. 25 Draining and refilling the engine coolant—2.4L Engine

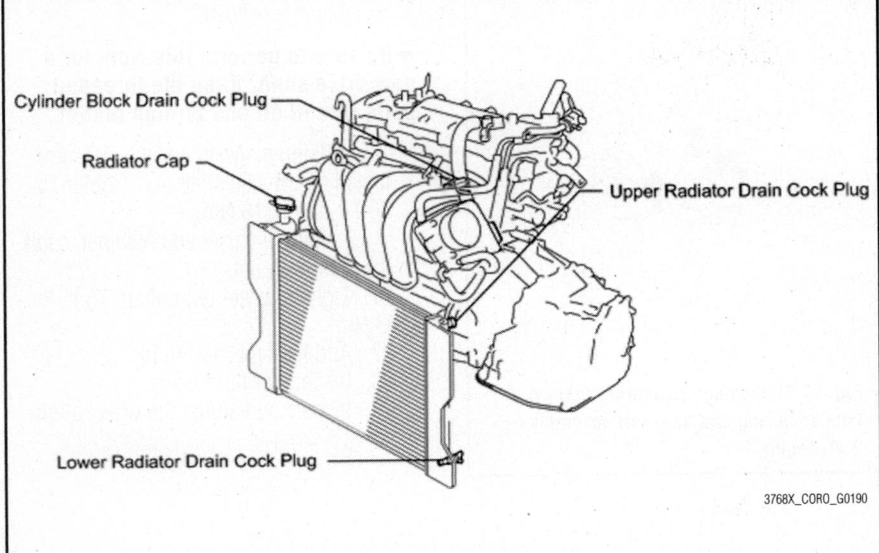

Fig. 24 Draining and refilling the engine coolant—1.8L Engine

a. Fill the radiator assembly with engine coolant, then attach a radiator cap tester.

b. Pump the tester to 17.1 psi, then check that the pressure does not drop.

c. If the pressure drops, check the hoses, radiator assembly and water pump assembly for leakage. If there are no signs or traces of external engine coolant leakage, check the heater core, cylinder block and head.

ENGINE FAN

REMOVAL & INSTALLATION

1.8L Engine

See Figure 26.

1. Disconnect the negative battery cable.
2. Remove the radiator assembly. Refer to Radiator in this section.
3. Remove the nut, then remove the fan.

To install:

4. To install, reverse removal procedure.

2.4L Engine

See Figure 27.

1. Disconnect the negative battery cable.
2. Remove the radiator assembly. Refer to Radiator in this section.
3. Remove the 2 nuts and 2 fans.

To install:

4. To install, reverse removal procedure.

RADIATOR

REMOVAL & INSTALLATION

1.8L Engine

See Figures 28 and 29.

1. Disconnect the negative battery cable.
2. Drain the coolant.
3. Remove battery.
4. Remove the 2 radiator grille protectors.
5. Remove front bumper assembly:
   a. Remove the clip.
   b. Using a screwdriver, turn the pin 90 degrees and remove the pin hold clip.
   c. Put protective tape around the front bumper assembly.
   d. Remove the 2 screws and 3 clips.
   e. Remove the 4 screws.
   f. Remove the 2 clips.
   g. Disengage the 6 claws and remove the front bumper assembly.
   h. Disconnect the fog light connector. (w/ Fog Light).
6. Disconnect the radiator reservoir tank hose from the radiator assembly.
7. Disconnect the both radiator hose from the radiator assembly.
8. Remove the 2 bolts and 2 upper radiator supports.

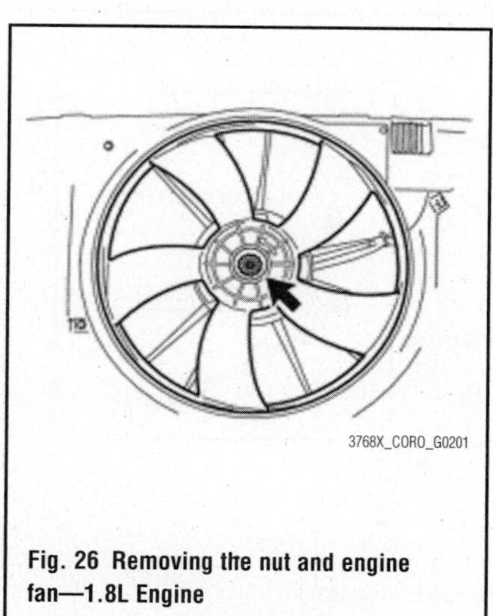

Fig. 26 Removing the nut and engine fan—1.8L Engine

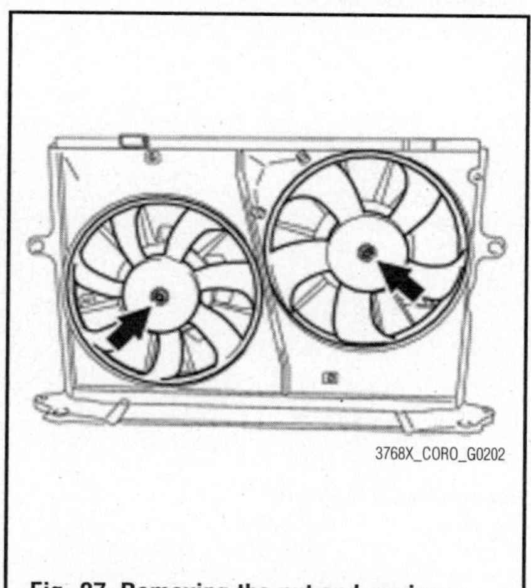

Fig. 27 Removing the nut and engine fan—2.4L Engine

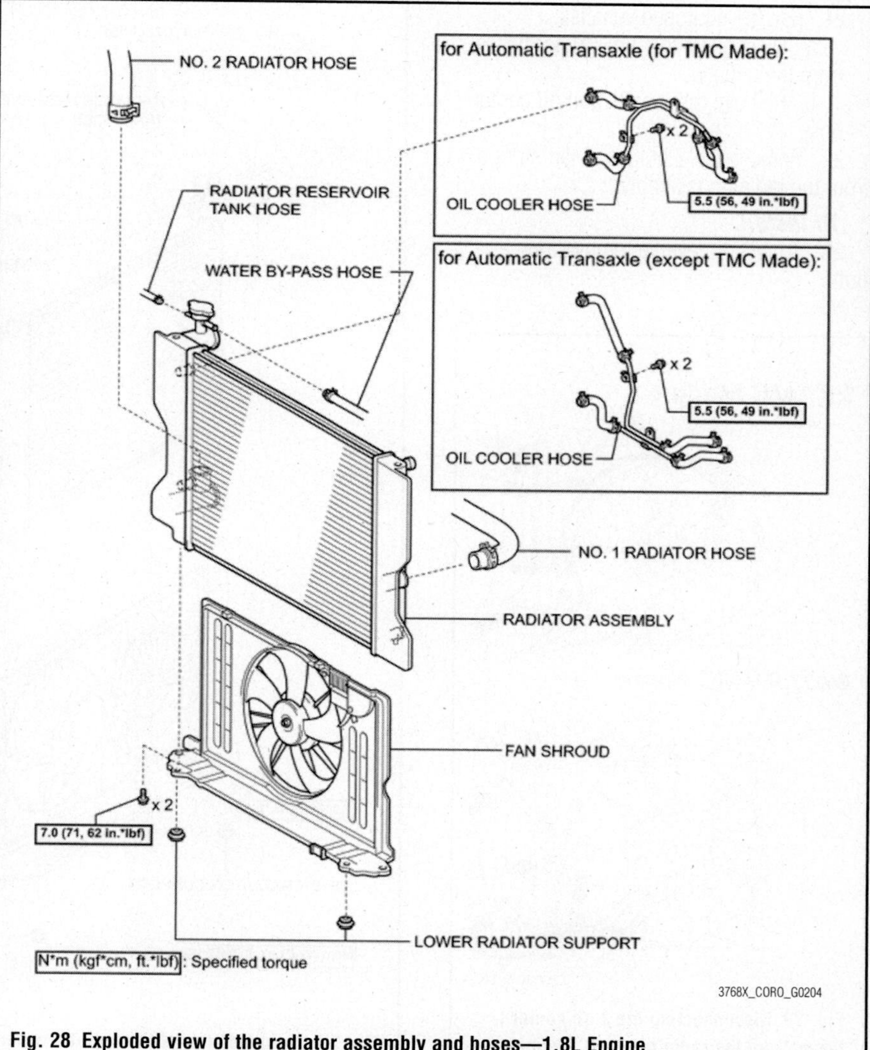

Fig. 28 Exploded view of the radiator assembly and hoses—1.8L Engine

9. Remove the 2 support cushions from the 2 upper radiator supports.
10. Remove the 3 bolts from the hood lock assembly.
11. Disconnect the hood lock control cable and remove the hood lock assembly.
12. Separate the water by-pass hose from the 3 clamps.
13. Disconnect the water by-pass hose from the radiator assembly.
14. Remove the 2 bolts and hood lock support sub-assembly.
15. Disconnect the horn connector.
16. Remove the 4 bolts and upper radiator support sub-assembly.
17. Remove the 2 bolts, disengage the 2 claws, and remove the No. 2 fan shroud from the radiator assembly.
18. Disconnect the cooling fan ECU connector and wire harness clamp.
19. Remove the radiator assembly with the fan shroud.
20. Remove the 2 lower radiator supports.
21. For A/T equipped vehicles:
   a. Disconnect the 2 oil cooler hoses from the radiator.
   b. Remove the 2 bolts and oil cooler hose.
22. Remove the 2 bolts and fan shroud from the radiator assembly.

To install:

23. To install, reverse removal procedure.

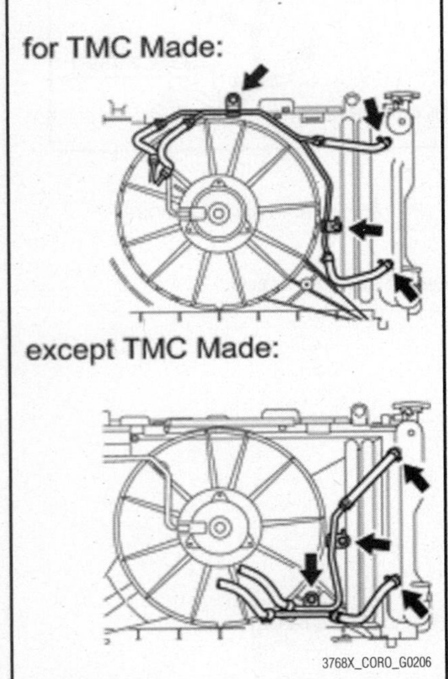

Fig. 29 Disconnecting the 2 oil cooler hoses from the radiator—1.8L Engine

➡Do not apply any excessive force to the cooler condenser assembly or pipe when installing the radiator assembly.

2.4L Engine

See Figures 30 and 31.

1. Disconnect the negative battery cable.
2. Drain the coolant.
3. Remove battery.
4. Remove the 2 radiator grille protectors.
5. Remove front bumper assembly:
   a. Remove the clip.
   b. Using a screwdriver, turn the pin 90 degrees and remove the pin hold clip.
   c. Put protective tape around the front bumper assembly.
   d. Remove the 2 screws and 3 clips.
   e. Remove the 4 screws.
   f. Remove the 2 clips.
   g. Disengage the 6 claws and remove the front bumper assembly.
   h. Disconnect the fog light connector. (w/ Fog Light).
6. Disconnect the radiator reserve tank hose from the radiator assembly.
7. Disconnect both radiator hoses from the radiator assembly.
8. Disconnect oil cooler hose (for automatic transaxle).
9. Remove the 2 bolts and 2 upper radiator supports.
10. Remove the 2 support cushions from the 2 upper radiator supports.
11. Remove the 3 bolts from the hood lock assembly.
12. Disconnect the hood lock control cable and remove the hood lock assembly.
13. Separate the water by-pass hose from the 2 clamps.

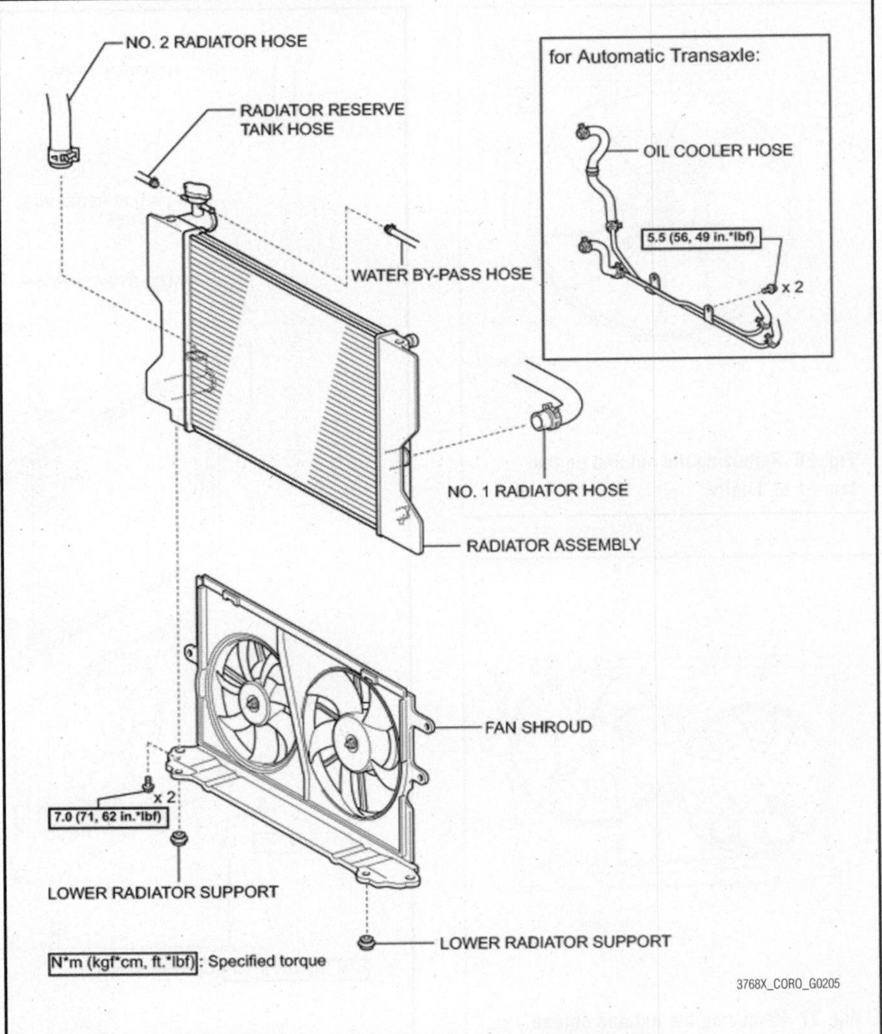

Fig. 30 Exploded view of the radiator assembly and hoses—2.4L Engine

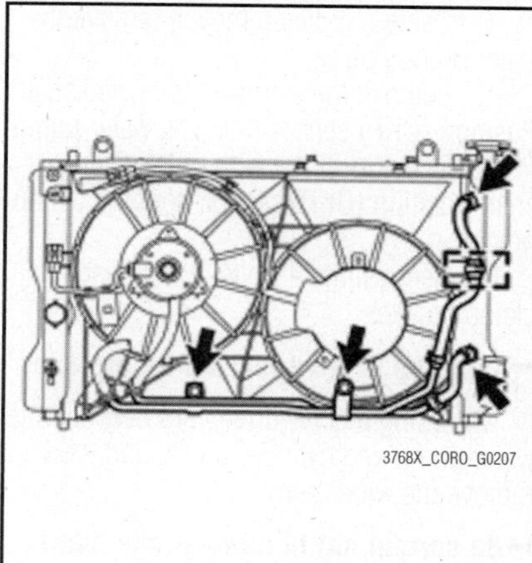

Fig. 31 Disconnecting the 2 oil cooler hoses from the radiator—2.4L Engine

14. Disconnect the water by-pass hose from the radiator assembly.
15. Remove the 2 bolts and hood lock support sub-assembly.
16. Disconnect the horn connector.
17. Remove the 4 bolts and upper radiator support sub-assembly.
18. Remove the 2 bolts, disengage the 2 claws, and remove the No. 2 fan shroud from the radiator assembly.
19. Disconnect the 2 cooling fan motor connectors and wire harness clamp.
20. Remove the radiator assembly with the fan shroud.

➡Do not apply any excessive force to the cooler condenser assembly or pipe when removing the radiator assembly.

21. Disconnect the 2 oil cooler hoses from the radiator.
22. Disconnect the clamp from the fan shroud.
23. Remove the 2 bolts and oil cooler hose.
24. Remove the 2 bolts and fan shroud from the radiator assembly.

To install:

25. To install, reverse removal procedure.

THERMOSTAT

REMOVAL & INSTALLATION

See Figures 32 through 34.

1. Disconnect the negative battery cable.
2. Drain the engine coolant.
3. For 2.4L engines, remove the no. 2 coolant hose.
4. Remove the 2 nuts and water inlet.
5. Remove the thermostat.

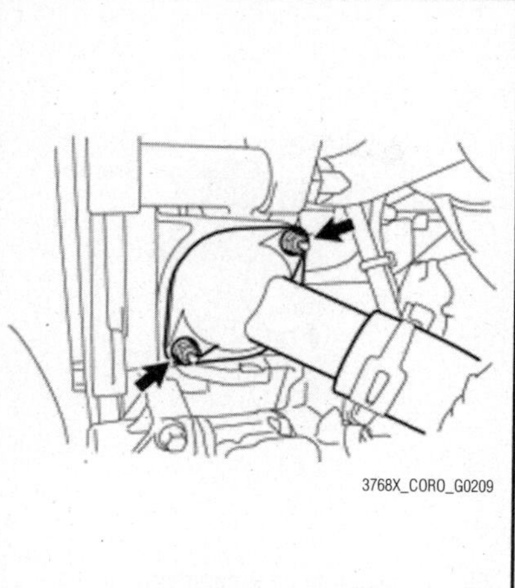

Fig. 32 Removing the 2 nuts and water inlet—1.8L Engine

Fig. 33 Removing the 2 nuts and water inlet—2.4L Engine

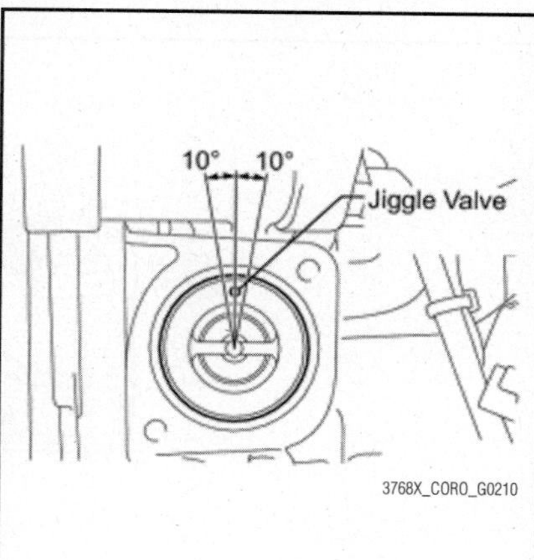

Fig. 34 Installing the thermostat and jiggle valve alignment

6. Remove the gasket from the thermostat.

To install:

7. Install a new gasket on the thermostat.
8. Install the thermostat to the water inlet with the jiggle valve upward.

➡The jiggle valve may be set to within 10° on either side of the indicated position.

9. Install the water inlet with the 2 nuts. Tighten to 7 ft. lbs. (10 Nm).
10. For 2.4L engines, install the no. 2 coolant hose
11. Add coolant and inspect for leaks.
12. Inspect reservoir tank engine coolant level.

WATER PUMP

REMOVAL & INSTALLATION

1.8L Engine

See Figures 35 and 36.

1. Disconnect the negative battery cable.
2. Remove the engine cover.
3. Remove the RH engine under cover.
4. Drain the engine coolant.
5. Remove the accessory drive belt.
6. Remove the alternator. Refer to Alternator in Charging System under Engine Electrical.
7. Remove the 5 bolts and water pump assembly from the timing chain cover.
8. Remove the water pump gasket from the timing chain cover.

To install:

9. Align the protrusion of a new water pump gasket with the cutout in the timing

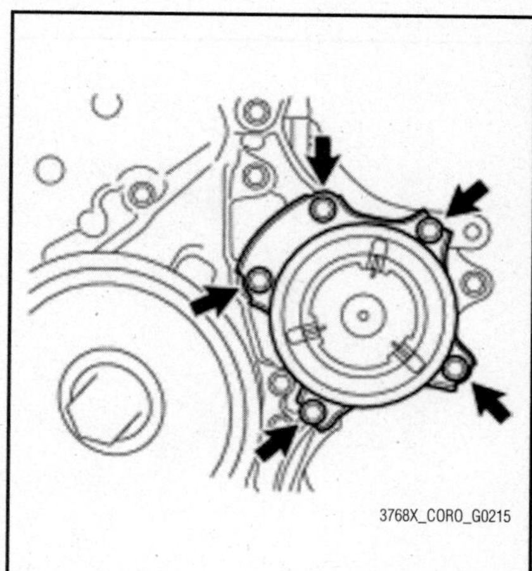

Fig. 35 Locating the 5 water pump bolts—1.8L Engine

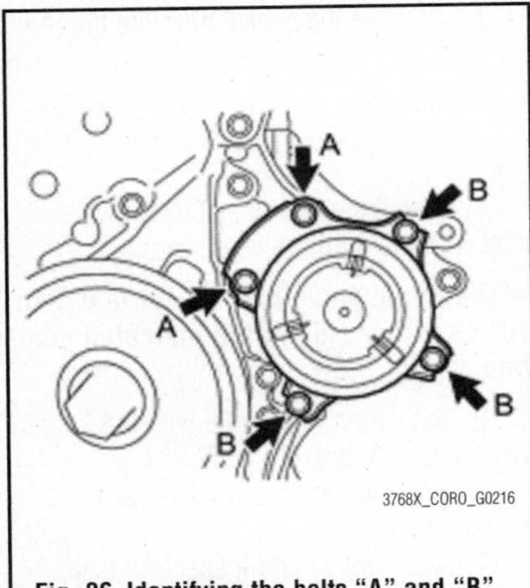

Fig. 36 Identifying the bolts "A" and "B" for installation—1.8L Engine

chain cover and install the gasket to the groove of the timing chain cover.

➡Be sure to clean the contact surfaces.

10. Install the water pump assembly to the timing chain cover with the 5 bolts. Tighten to:
 a. Bolt A (1.38 inches): 19 ft. lbs. (26 Nm).
 b. Bolt B: (0.709 inches): 18 ft. lbs. (24 Nm).
11. To complete installation, reverse remaining removal procedure.
12. Inspect for coolant leaks.

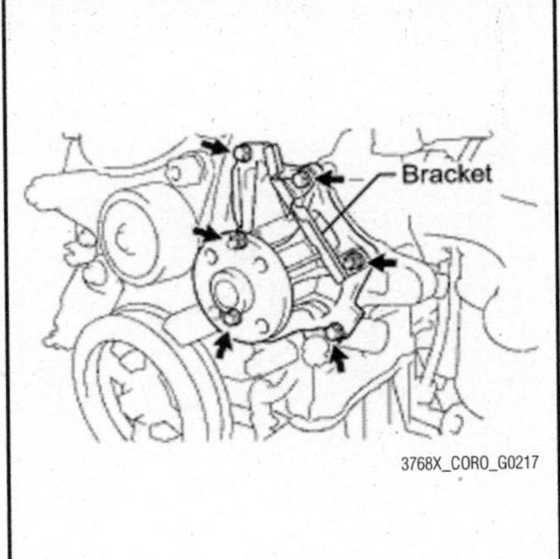

Fig. 37 Locating and identifying fasteners—2.4L Engine

2.4L Engine

See Figure 37.

1. Disconnect the negative battery cable.
2. Remove the engine cover.
3. Remove the LH and RH engine under cover.
4. Drain the engine coolant.
5. Remove the accessory drive belt.
6. Remove the alternator. Refer to Alternator in Charging System under Engine Electrical.
7. Remove the RH engine mounting insulator sub-assembly. Refer to Engine under Engine Mechanical.
8. Using a pulley remover, remove the water pump pulley.
9. Remove the clamp of the Crankshaft Position (CKP) sensor from the water pump.
10. Disconnect the wire of the Crankshaft Position (CKP) sensor from the clamp bracket.
11. Remove the 4 bolts, 2 nuts and clamp bracket.

➡Tape the screwdriver tip before use.

12. Using a screwdriver, pry between the water pump and cylinder block, and then remove the water pump.

➡Be careful not to damage the contact surfaces of the water pump and cylinder block.

➡Remove any oil from the contact surface. The parts must be set within 3 minutes after applying seal packing. Otherwise, the material must be removed and reapplied.

To install:

13. To install, reverse the removal procedure and pay attention to the following:
 a. Apply a continuous line of seal packing. Use Toyota Genuine Seal Packing Black, Three Bond® 1207B or equivalent.
 b. Tighten the water pump bolts to 80 inch lbs. (9 Nm).
 c. Tighten the water pump pulley to 19 ft. lbs. (26 Nm).

ENGINE ELECTRICAL

BATTERY SYSTEM

BATTERY

REMOVAL & INSTALLATION

See Figures 38 and 39.

1. Remove the bolt, nut and battery clamp.
2. Remove the battery and battery insulator.
3. If necessary, separate the 2 wire harness clamps from the battery carrier, remove the 4 bolts and battery carrier.

To install:

4. To install, reverse removal procedure.

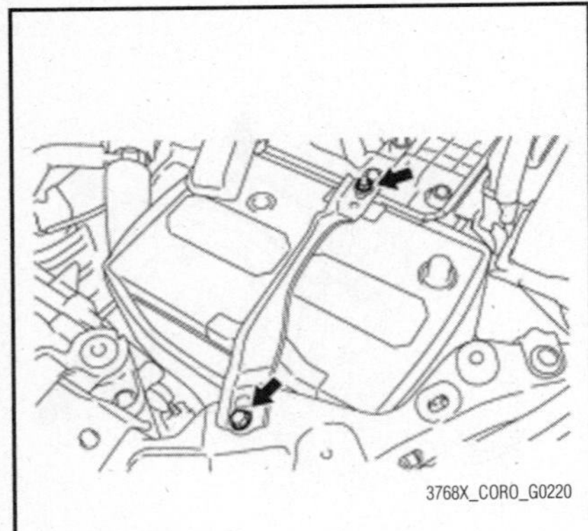

Fig. 38 Removing the bolt, nut and battery clamp

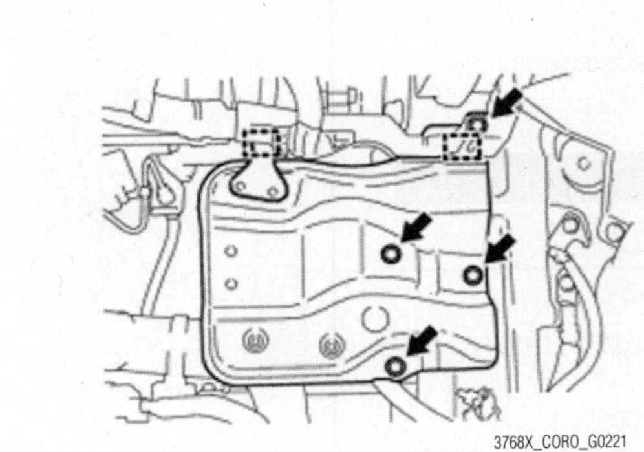

Fig. 39 Separating the 2 wire harness clamps from the battery carrier, removing the 4 bolts and battery carrier

ENGINE ELECTRICAL — CHARGING SYSTEM

ALTERNATOR

REMOVAL & INSTALLATION

1.8L Engine

See Figure 40.

1. Disconnect the negative battery cable.
2. Remove the RH engine under cover.
3. Remove the accessory drive belt.
4. Remove generator assembly:
   a. Remove the terminal cap.

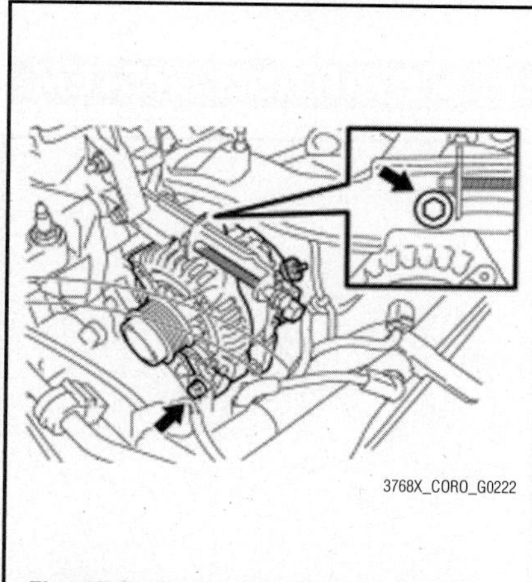

Fig. 40 Removing the alternator assembly—1.8L Engine

   b. Remove the nut and disconnect the wire harness from terminal B.
   c. Disconnect the connector and harness clamp.
   d. Remove the 2 bolts and generator assembly.
   e. Remove the bolt and wire harness clamp bracket.

To install:

5. To install, reverse removal procedure. Pay attention to the following:
   a. Alternator upper mounting bolt: Tighten to 32 ft. lbs. (43 Nm).
   b. Alternator lower mounting bolt: Tighten to 14 ft. lbs. (19 Nm).

2.4L Engine

See Figure 41.

1. Disconnect the negative battery cable.
2. Remove the RH engine under cover.
3. Remove the accessory drive belt.
4. Remove generator assembly:
   a. Disconnect the generator connector.
   b. Remove the nut and disconnect the wire harness from terminal B.
   c. Separate the 2 wire harness clamps.
   d. Remove the 2 bolts and generator assembly.

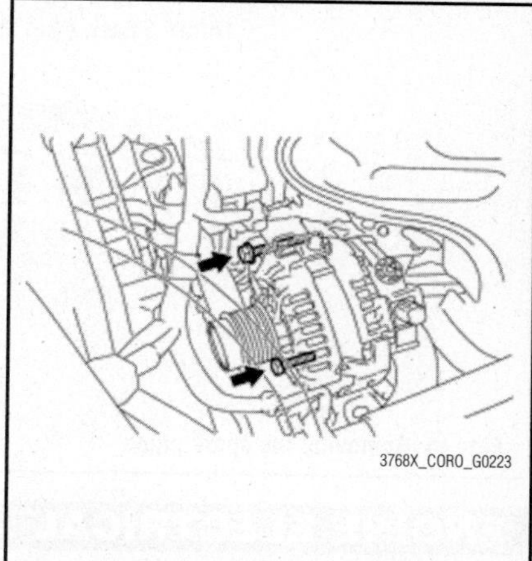

Fig. 41 Removing the alternator assembly—2.4L Engine

   e. Remove the bolt and wire harness clamp bracket.

To install:

5. To install, reverse removal procedure. Pay attention to the following:
   a. Alternator upper mounting bolt: Tighten to 16 ft. lbs. (21 Nm).
   b. Alternator lower mounting bolt: Tighten to 38 ft. lbs. (52 Nm).

ENGINE ELECTRICAL — IGNITION SYSTEM

FIRING ORDER

See Figure 42.

②③④①
4
3
2
1
Front of the Vehicle
79233G64

Fig. 42 Firing order: 1–3–4–2 Distributorless ignition system

IGNITION COIL

REMOVAL & INSTALLATION

See Figure 43.

1. Disconnect the battery's negative terminal.
2. Remove the two nuts and clips and remove the cylinder head cover.
3. Disconnect the four ignition coil connectors.
4. Remove the four bolts that retain the coils.

To install:

5. Install the coils and the four bolts.
6. Connect the coil connections.
7. Install the cylinder head cover.

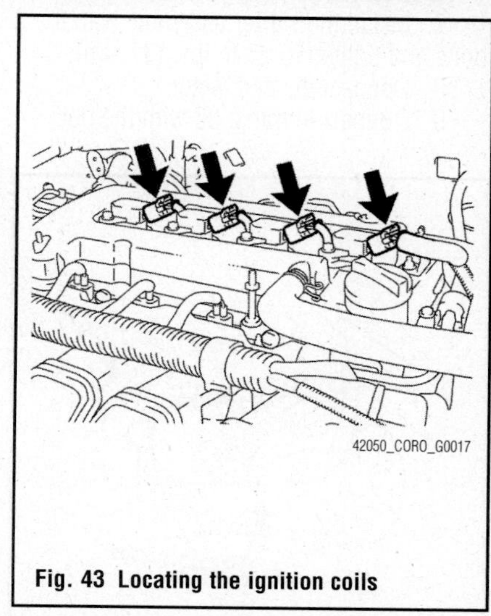

Fig. 43 Locating the ignition coils

IGNITION TIMING

ADJUSTMENT

The ignition timing is controlled by the Powertrain Control Module (PCM). No adjustment is necessary or possible.

SPARK PLUGS

REMOVAL & INSTALLATION

See Figure 44.

1. Disconnect the negative battery cable.

Fig. 44 Removing the spark plugs

2. Remove the ignition coil pack. Refer to Ignition Coil Pack under Distributorless Ignition System.
3. Remove the 4 spark plugs.

To install:

4. To install, use a 14 mm spark plug wrench, install the 4 spark plugs and tighten to 15 ft. lbs. (20 Nm).
5. To complete installation, reverse remaining removal procedure.

ENGINE ELECTRICAL

STARTER

REMOVAL & INSTALLATION

1.8L Engine

See Figure 45.

1. Disconnect the negative battery cable.
2. Remove transmission oil filler tube sub-assembly (for automatic transaxle).
3. Remove the terminal cap.
4. Remove the nut and disconnect terminal 30.
5. Disconnect the connector.
6. Remove the 2 bolts and starter assembly.

To install:

7. Install the starter assembly with the 2 bolts and tighten to 27 ft. lbs. (37 Nm).
8. Connect the connector.
9. Connect terminal 30 with the nut.
10. Close the terminal cap.
11. To complete the installation, reverse remaining removal procedure.

2.4L Engine

See Figures 46 and 47.

1. Disconnect the negative battery cable.
2. Remove no. 1 engine cover sub-assembly.
3. Remove air cleaner cap sub-assembly with hose and air cleaner case. Refer to Air Cleaner in Engine Mechanical.
4. Remove battery and carrier if necessary.
5. Release the claw, and disconnect the wire harness.
6. For M/T vehicles, perform the following:
   a. Disconnect the terminal 50 connector from the starter assembly.
   b. Remove the nut and disconnect the wire harness from terminal 30.
   c. Remove the 3 bolts, clutch accumulator bracket, wire harness clamp bracket and starter assembly.
7. For A/T vehicles:
   a. Disconnect the terminal 50 connector from the starter assembly.
   b. Remove the nut and disconnect the wire harness from terminal 30.
   c. Remove the 2 bolts, wire harness clamp bracket and starter assembly.

To install:

8. To install, reverse removal procedure.
9. For M/T vehicles, tighten the starter bolts to:
   a. Bolt A: 27 ft. lbs. (27 Nm).
   b. Bolt B: 9 ft. lbs. (12 Nm).
10. For A/T vehicles, tighten the bolts to 27 ft. lbs. (37 Nm).

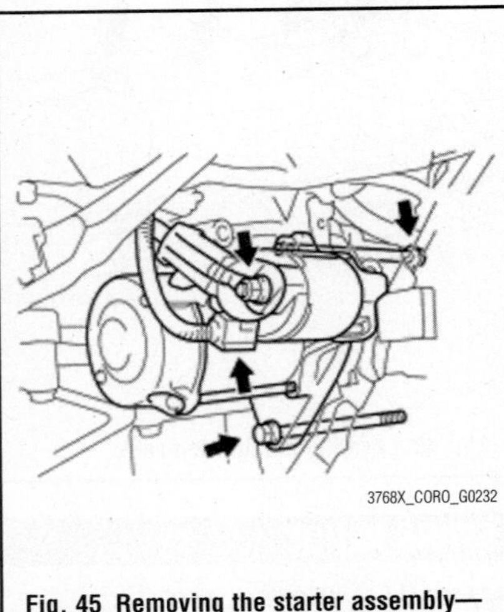

Fig. 45 Removing the starter assembly—1.8L Engine

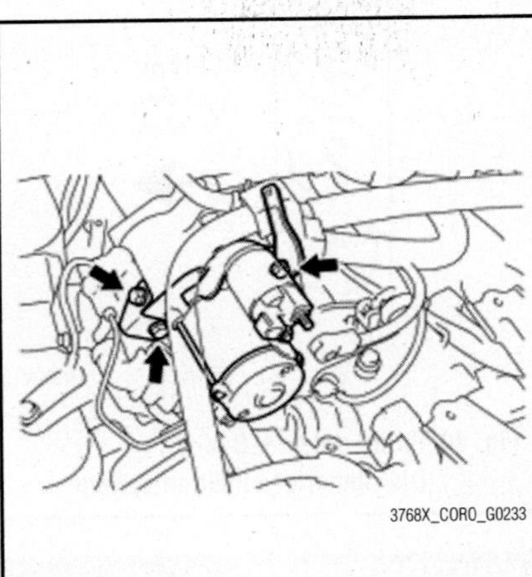

Fig. 46 Removing the starter—M/T vehicles

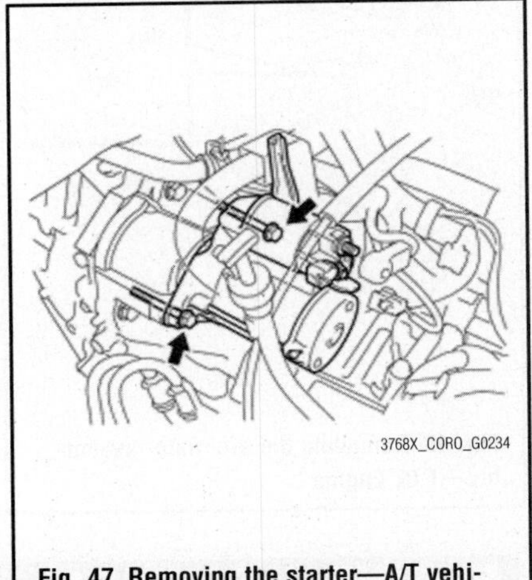

Fig. 47 Removing the starter—A/T vehicles

ENGINE MECHANICAL

➡**Disconnecting the negative battery cable may interfere with the functions of the on board computer systems and may require the computer to undergo a relearning process, once the negative battery cable is reconnected.**

ACCESSORY DRIVE BELTS

ACCESSORY BELT ROUTING

See Figures 48 and 49.

INSPECTION

Inspect the drive belt for signs of glazing or cracking. A glazed belt will be perfectly smooth from slippage, while a good belt will have a slight texture of fabric visible. Cracks will usually start at the inner edge of the belt and run outward. All worn or damaged drive belts should be replaced immediately.

ADJUSTMENT

Only the 1.8L engine belt requires adjustment. Refer to the removal procedure for adjustment procedure.

REMOVAL & INSTALLATION

1.8L Engine

See Figure 50.

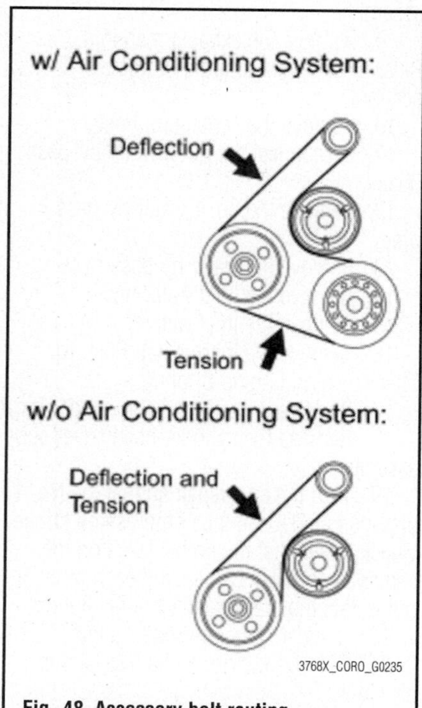

Fig. 48 Accessory belt routing—1.8L Engine

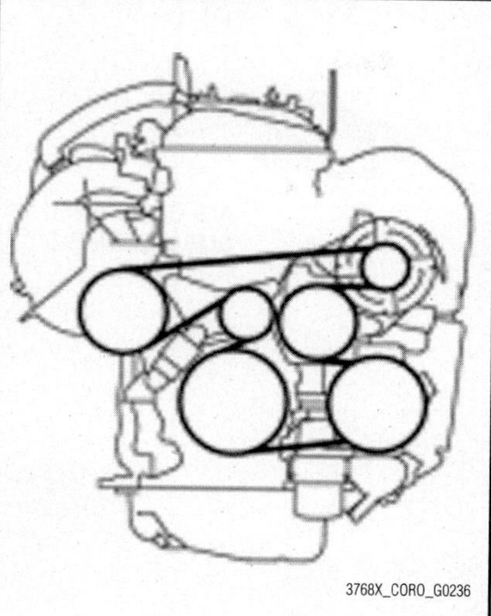

Fig. 49 Accessory belt routing—2.4L Engine

➡**Refer to the illustration for bolt identification.**

1. Remove the engine cover.
2. Remove the RH engine under cover.
3. Loosen bolts A and B.
4. Loosen bolt C, then remove the V-ribbed belt.

➡**Do not loosen bolt D.**

To install:

5. Install the belt.
6. Turn bolt C to adjust the tension of the V-ribbed belt.
7. Tighten bolts:
   a. A: 14 ft. lbs. (19 Nm).
   b. B: 32 ft. lbs. (43 Nm).

➡**Confirm that bolt D is not loosened.**

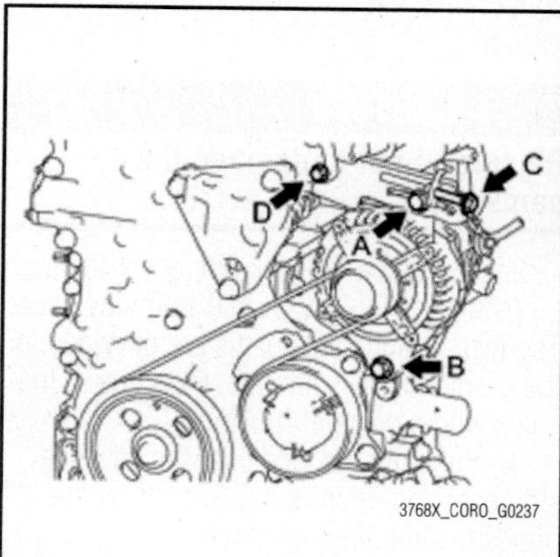

Fig. 50 Removing the accessory belt—1.8L Engine

8. Inspect the belt.
9. To complete installation, reverse remaining removal procedure.

2.4L Engine

See Figure 51.

➡**Adjustment is not possible or necessary with the auto-tensioner.**

1. Remove the right hand cover under the engine.
2. Turn the drive belt tensioner clockwise to relieve tension on the belt.
3. Remove the fan and generator V belt.
4. Return the tensioner to the unloaded position.

➡**When retracting the tensioner, turn it clockwise slowly in 3 sec. or more. Be sure not to apply force rapidly. After the tensioner is retracted all the way, do not apply force any more than necessary.**

To install:

5. Turn the drive belt tensioner clockwise.
6. Install the belt.
7. Install the right hand cover under the engine.

AIR CLEANER

REMOVAL & INSTALLATION

1.8L Engine

See Figure 52.

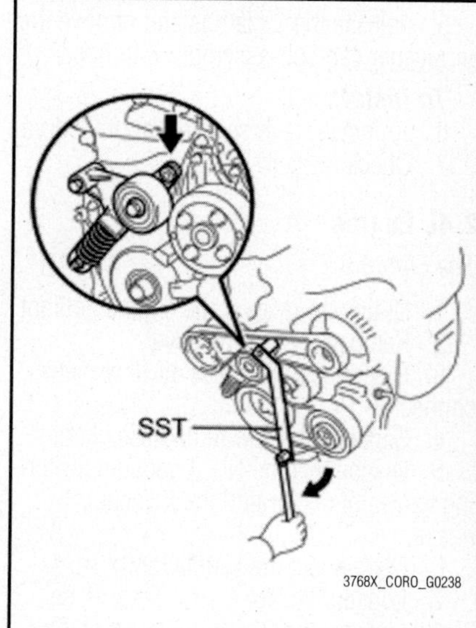

Fig. 51 Removing the serpentine belt—2.4L Engine

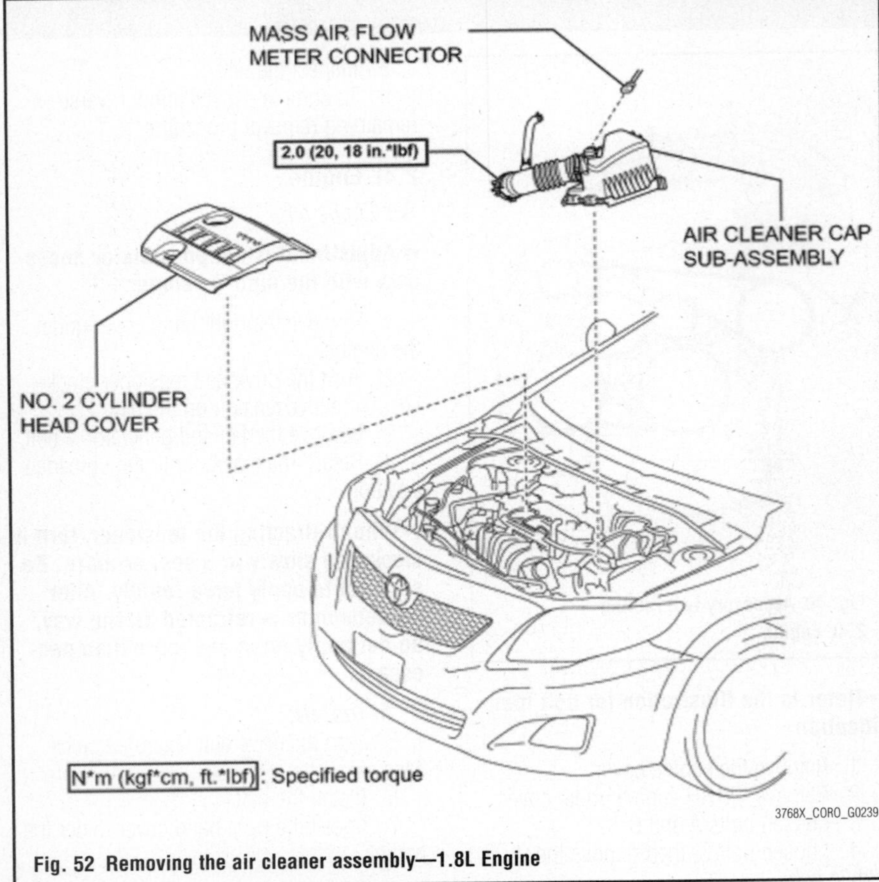

Fig. 52 Removing the air cleaner assembly—1.8L Engine

1. Drain and recover the engine coolant.
2. Remove the engine cover.
3. Disconnect the mass air flow meter connector and the 2 wire harness clamps.
4. Disconnect the ventilation hose and loosen the hose clamp.
5. Release the 2 clamps and remove the air cleaner cap sub-assembly with hose.

To install:

6. To install, reverse removal procedure.
7. Check for leaks.

2.4L Engine

See Figure 53.

1. Drain and recover the engine coolant.
2. Remove the engine cover.
3. Disconnect the mass air flow meter connector.
4. Separate the 2 wire harness clamps.
5. Disconnect the No. 1 vacuum switching valve connector and the 2 vacuum hoses.
6. Disconnect the ventilation hose.
7. Loosen the No. 1 air cleaner hose clamp, release the 3 air cleaner assembly clamps and remove the air cleaner cap sub-assembly with No. 1 hose.

To install:

8. To install, reverse removal procedure.
9. Check for leaks.

CAMSHAFT AND VALVE LIFTERS

INSPECTION

1. To check the lock of the camshaft timing gear, clamp the camshaft in a vice and confirm that the camshaft timing gear is locked.

⁂ CAUTION

Be careful not to damage the camshaft.

2. To release the lock pin, cover the four oil paths on the cam journal with vinyl tape. The two advance side paths are provided in the groove of the camshaft. Plug one of the paths with a piece of rubber.
3. Puncture the tape over the advance side path and retard the side path on the opposite side of the groove.
4. Apply air pressure to the two broken paths (the advance side path and the retard side path).

⁂ WARNING

Cover the paths with a shop rag to avoid oil splashes.

5. Confirm that the camshaft timing gear revolves in the timing advance direction when weakening the air pressure of the timing retard path.
6. When the camshaft timing gear comes to the most advanced position, release the air pressure of the timing retard side path, then release that of the timing advance side path.
7. There should be a smooth revolution of the gear.

REMOVAL & INSTALLATION

1.8L Engine

See Figures 54 through 61.

1. Disconnect the negative battery cable.

remove engine assembly with transaxle.

2. Install the engine stand.
3. Remove the intake manifold. Refer to Intake Manifold in this section.
4. Remove the fuel tube sub-assembly. Refer to Fuel System.
5. Remove the fuel delivery pipe sub-assembly. Refer to Fuel System.
6. Remove the fuel injector assembly. Refer to Fuel System.
7. Remove the ignition coil assembly.
8. Remove the oil level dipstick sub-assembly.
9. Remove the exhaust manifold. Refer to Exhaust Manifold in this section.
10. Remove the ventilation hose.
11. Disconnect the no. 3 water by-pass hose.
12. Remove the no. 1 water by-pass pipe.
13. Remove the water by-pass hose.
14. Remove the inlet water hose.
15. Remove the inlet water.
16. Remove the thermostat. Refer to Thermostat in Engine Cooling.
17. Remove the radio setting condenser.
18. Remove the cylinder head cover sub-assembly.
19. Turn the crankshaft so that the No. 1 piston is at TDC on the compression stroke. Check to see that the point marks on the camshaft sprockets are facing each other, if not, rotate the crankshaft 1 full revolution.
20. Remove the crankshaft pulley.
21. Remove the no. 1 chain tensioner assembly.
22. Remove the timing chain cover sub-assembly.

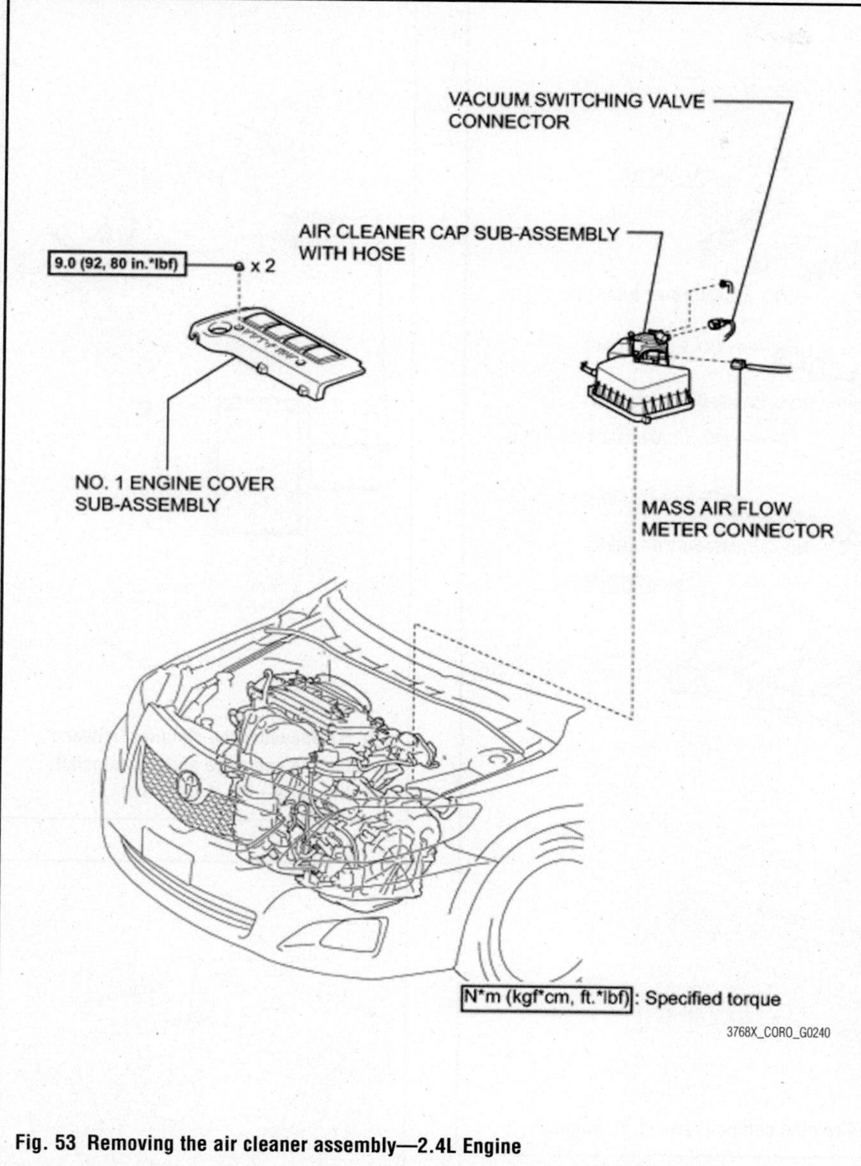

Fig. 53 Removing the air cleaner assembly—2.4L Engine

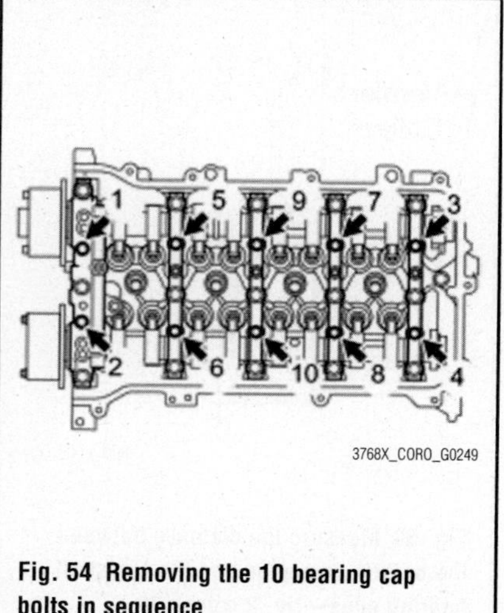

Fig. 54 Removing the 10 bearing cap bolts in sequence

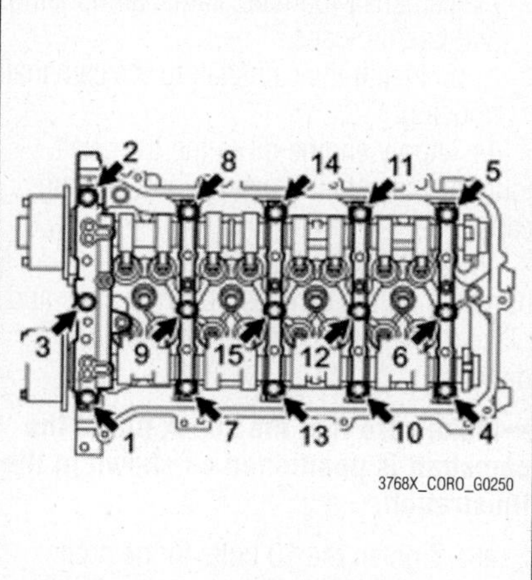

Fig. 55 Remove the 15 bearing cap bolts in sequence

23. Remove the timing chain cover oil seal.
24. Remove the chain tensioner slipper.
25. Remove the no. 1 chain vibration damper.
26. Remove the chain sub-assembly.
27. Remove the no. 2 chain vibration damper.
28. Uniformly loosen and remove the 10 bearing cap bolts in the sequence shown in the illustration.
29. Uniformly loosen and remove the 15 bearing cap bolts in the sequence shown in the illustration.

➡Uniformly loosen the bolts while keeping the camshaft level.

30. Remove the 5 bearing caps.
31. Remove the camshaft.
32. Remove the No. 2 camshaft.
33. Remove the rocker arm. Refer to Rocker Arm in this section.
34. Remove the valve lash adjuster assembly. Refer to Valve Lash Adjuster in this section.
35. Remove the 2 No. 1 camshaft bearings.
36. Remove the 2 No. 2 camshaft bearings.
37. If necessary, remove camshaft housing sub-assembly:
    a. Remove the 2 bolts.
    b. Remove the camshaft housing by prying between the cylinder head and camshaft housing with a screwdriver.

➡Be careful not to damage the contact surfaces of the cylinder head and camshaft housing. Tape the screwdriver tip before use.

To install:

38. Remove the valve lash adjuster assembly. Refer to Valve Lash Adjuster in this section.
39. Install the rocker arm. Refer to Rocker Arm in this section.
40. Install the no. 1 camshaft bearing:
    a. Clean both surfaces of the bearings.
    b. Install the 2 No. 1 camshaft bearings.
    c. Using a vernier caliper, measure the distance between the bearing cap edge and the camshaft bearing edge. Dimension A-B: 0.7 mm (0.0276 in.) or less.

➡Position the bearings to the center of the bearing cap by measuring dimensions A.

41. Install the no. 2 camshaft bearing:

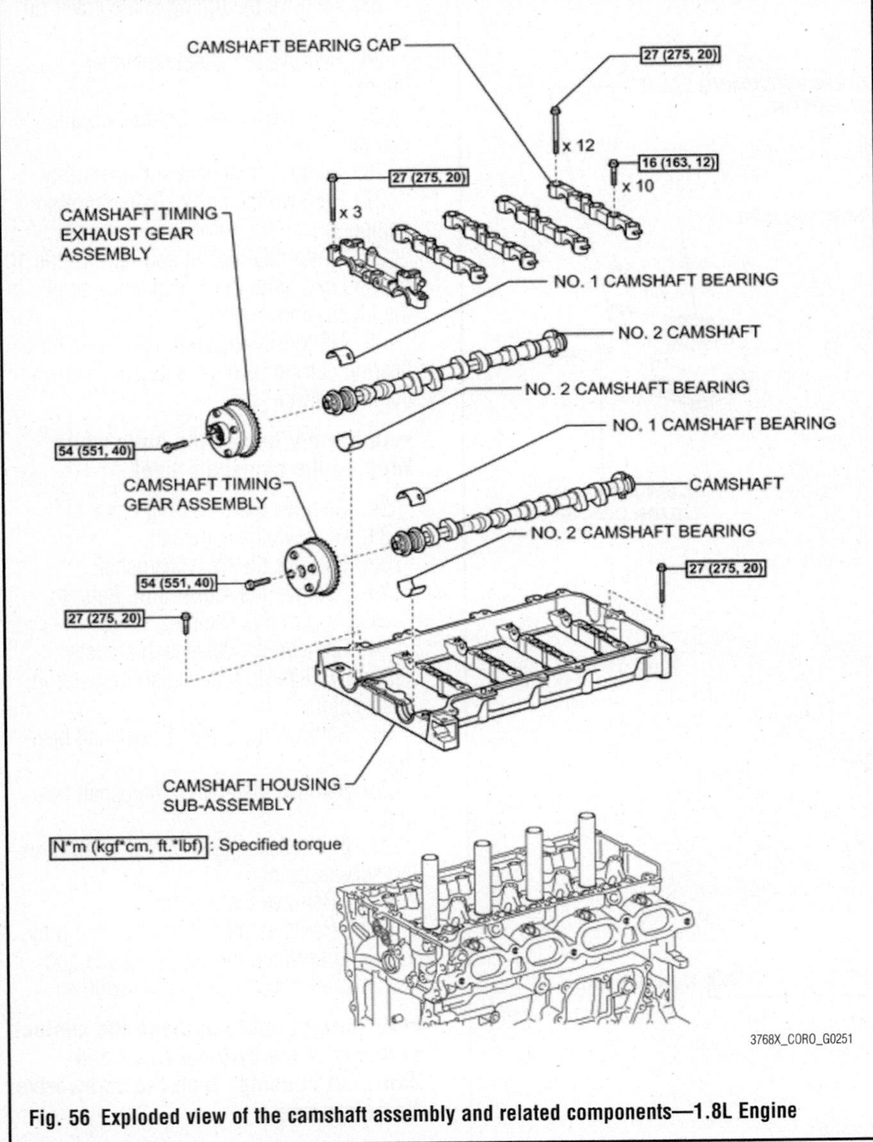

Fig. 56 Exploded view of the camshaft assembly and related components—1.8L Engine

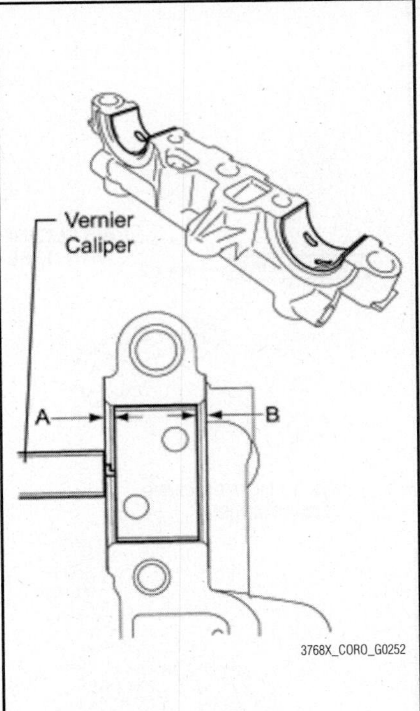

Fig. 57 Measure the distance between the bearing cap edge and the camshaft bearing edge—No. 1 camshaft

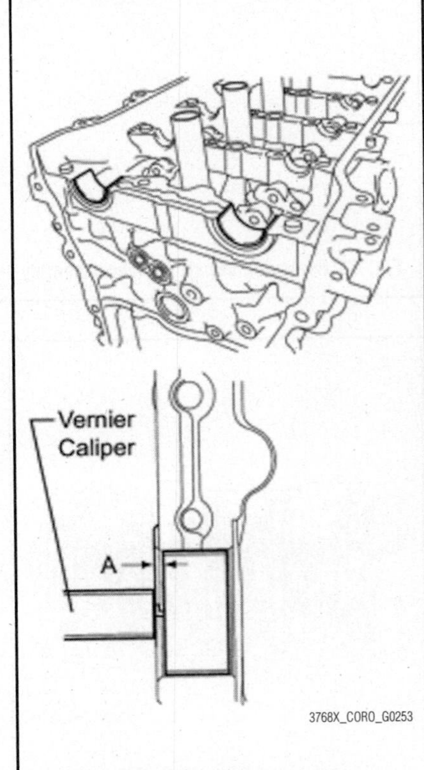

Fig. 58 Measure the distance between the bearing cap edge and the camshaft bearing edge—No. 2 camshaft

a. Clean both surfaces of the bearings.

b. Install the 2 No. 2 camshaft bearings.

c. Using a vernier caliper, measure the distance between the bearing cap edge and the camshaft bearing edge. Dimension: 1.05 to 1.75 mm (0.0413 to 0.0689 in.)

➡Position the bearings to the center of the bearing cap by measuring dimensions A.

42. Install the no. 2 camshaft:

a. Clean the camshaft journals.

b. Apply a light coat of engine oil to the camshaft journals, camshaft housings and bearing caps.

c. Install the No. 2 camshaft to the camshaft housing.

43. Install the camshaft:

a. Apply a light coat of engine oil to the camshaft journals, camshaft housings and bearing caps.

b. Install the camshaft to the camshaft housing.

44. Apply engine oil to the camshaft journals, camshaft housings and bearing caps.

45. Make sure of the marks and numbers on the camshaft bearing caps and place them in each proper position and direction.

➡Make sure that the knock pin of the camshaft is positioned as shown in the illustration.

46. Tighten the 10 bolts in the order shown in the illustration.

47. If removed, install the camshaft housing sub-assembly:

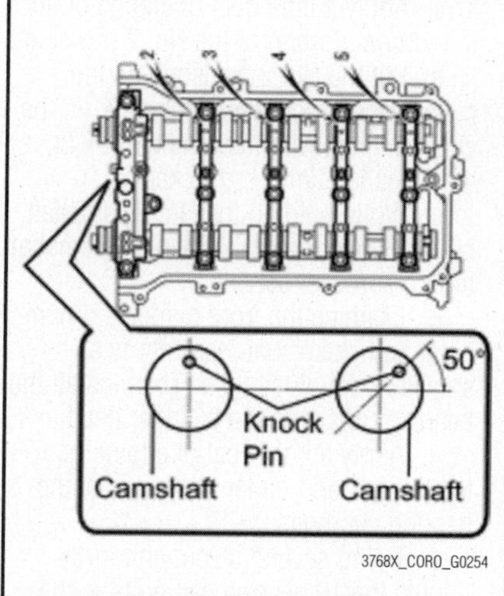

Fig. 59 Making sure of the marks and numbers on the camshaft bearing caps

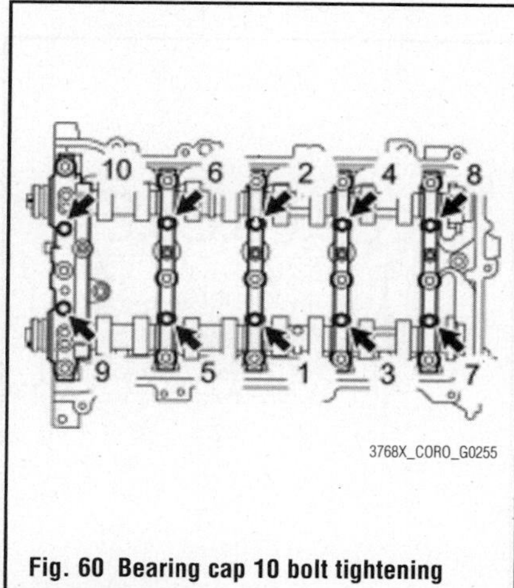

Fig. 60 Bearing cap 10 bolt tightening sequence

a. Make sure that the valve rocker arm is installed correctly. Refer to Rocker Arms.

b. Apply seal packing in a continuous bead. Use Toyota Genuine Seal Packing Black, Three Bond® 1207B or equivalent.

➡Remove any oil from the contact surface. Install the camshaft housing sub-assembly within 3 minutes and tighten the bolts within 15 minutes after applying seal packing. Do not start the engine for at least 2 hours after installing.

48. Set the camshaft and No. 2 camshaft as shown in the illustration.

49. Install the camshaft housing and tighten the 17 bolts in the order shown in the illustration. Tighten to 20 ft. lbs. (27 Nm).

➡After installing the camshaft housing, make sure that the cam lobes are positioned as shown in the illustration.

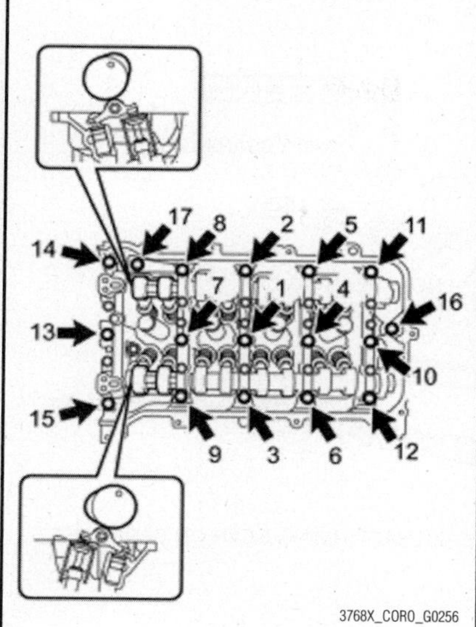

Fig. 61 Camshaft housing 17 bolt tightening sequence

- If any of the bolts are loosened during installation, remove the camshaft housing, clean the installation surfaces, and reapply seal packing.
- If the camshaft housing is removed because any of the bolts are loosened during installation, make sure that the previously applied seal packing does not enter any oil passages.
- After installing the camshaft housing, wipe off any seal packing that seeped out from between the housing and the cylinder head.

50. To complete the installation, reverse remaining removal procedure.

2.4L Engine

See Figures 62 through 68.

1. Disconnect the negative battery cable.

2. Remove the RH engine under cover.

3. Remove the No. 1 engine cover sub-assembly. Refer to Valve Cover in this section.

4. Remove the ignition coil assembly. Refer to Ignition Coil in Engine Electrical.

5. Remove the spark plugs. Refer to Spark Plugs in Engine Electrical.

6. Remove the valve cover. Refer to Valve Cover in this section.

7. Set no. 1 cylinder to TDC/compression:

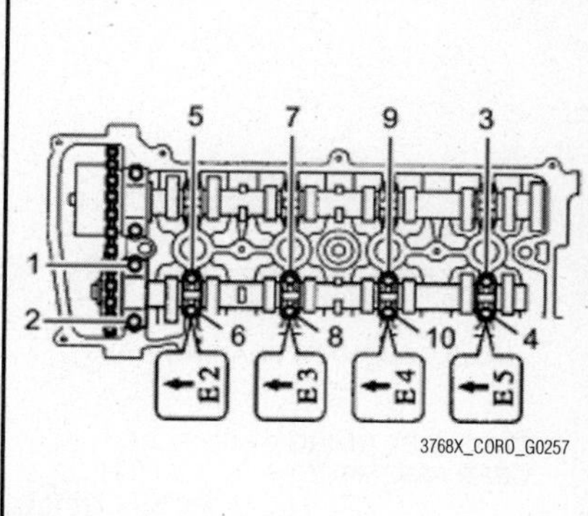

Fig. 62 10 bearing cap bolts removal sequence and identifying the 5 bearing caps

a. Turn the crankshaft pulley until the groove and the timing mark "0" on the timing chain cover are aligned.

b. Check that each timing mark on the camshaft timing gear and sprocket is aligned with each timing mark located on the No. 1 and No. 2 bearing caps.

c. If not, turn the crankshaft pulley 1 revolution (360°) to align the timing marks.

d. Place paint marks on the chain in alignment with the timing marks on the camshaft timing gear and camshaft timing sprocket.

8. While holding the No. 2 camshaft with a wrench, loosen the No. 2 camshaft timing set bolt.

9. Using several steps, uniformly loosen and remove the 10 bearing cap bolts in the sequence shown in the illustration.

10. Remove the 5 bearing caps.

11. While holding the No. 2 camshaft by hand, remove the camshaft timing sprocket set bolt.

12. Remove the camshaft timing sprocket from the No. 2 camshaft with the timing chain wrapped on the sprocket.

13. Remove the camshaft timing sprocket from the timing chain.

14. In several steps, uniformly loosen and remove the 10 bearing caps bolts in the sequence shown in the illustration.

15. Remove the 5 bearing caps.

16. Remove the camshaft and camshaft timing gear assembly while holding the timing chain by hand.

17. Support the timing chain with a string to prevent it from slipping off the crankshaft sprocket.

➡Be careful not to drop anything inside the timing chain cover.

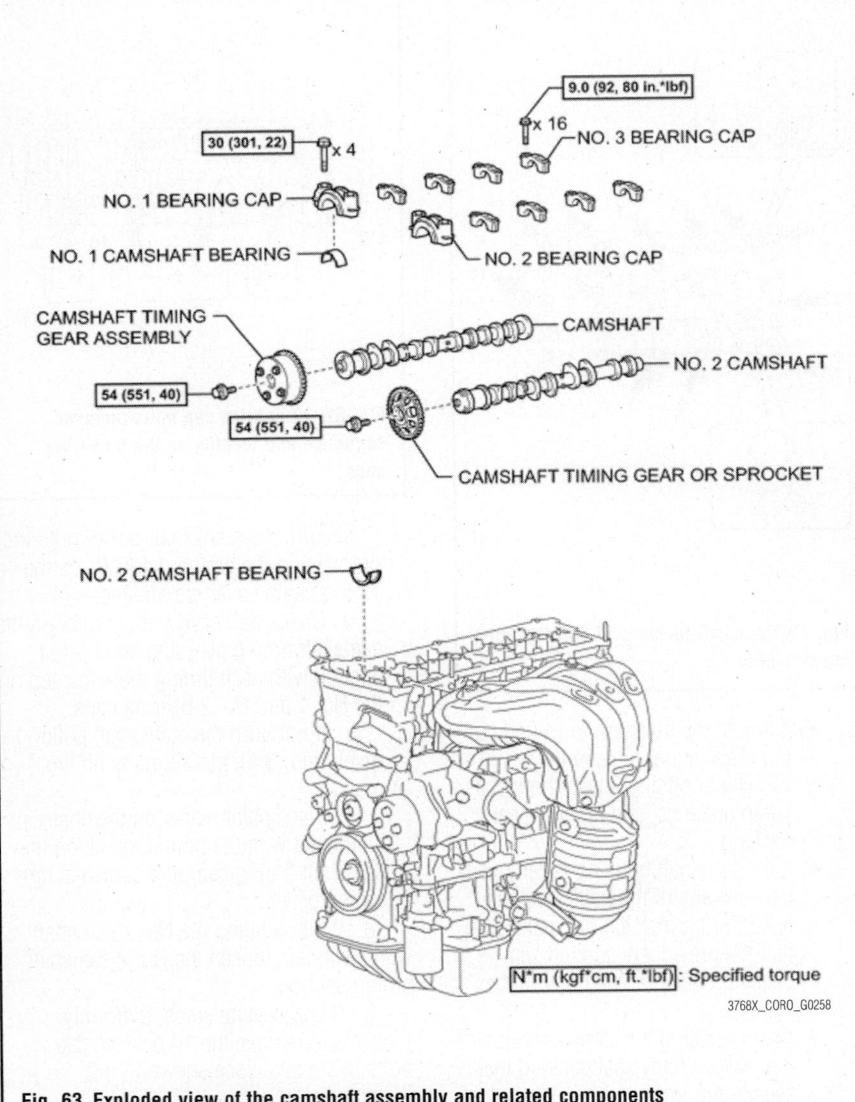

Fig. 63 Exploded view of the camshaft assembly and related components

To install:

18. Apply a light coat of engine oil to the journal portion of the camshaft.
19. Install the timing chain onto the camshaft timing gear with the paint mark aligned with the timing mark on the camshaft timing gear.
20. Examine the front marks and numbers, and check that the order is as shown in the illustration. Then install the bearing caps into the cylinder head.
21. Apply a light coat of engine oil on the threads and under the heads of the bearing cap bolts.
22. Using several steps, uniformly tighten the 10 bearing cap bolts in the sequence shown in the illustration.
    a. No. 1 bearing: 22 ft. lbs. (30 Nm).
    b. No. 3 bearing: 80 inch lbs. (9 Nm).
23. Install the no. 2 camshaft:
    a. Apply a light coat of engine oil to the journal portion of the No. 2 camshaft.
    b. Put the No. 2 camshaft on the cylinder head with the paint mark on the chain aligned with the timing mark on the camshaft timing sprocket.
    c. While holding the No. 2 camshaft by hand, temporarily tighten the camshaft timing sprocket set bolt.
    d. Examine the front marks and numbers, and check that the order is as shown in the illustration. Then install the bearing caps onto the cylinder head.
    e. Apply a light coat of engine oil to the threads and under the heads of the bearing cap bolts.
    f. Using several steps, uniformly tighten the 10 bearing cap bolts in the sequence shown in the illustration.
    g. No. 2 bearing: 22 ft. lbs. (30 Nm).
    h. No. 3 bearing: 80 inch lbs. (9 Nm).

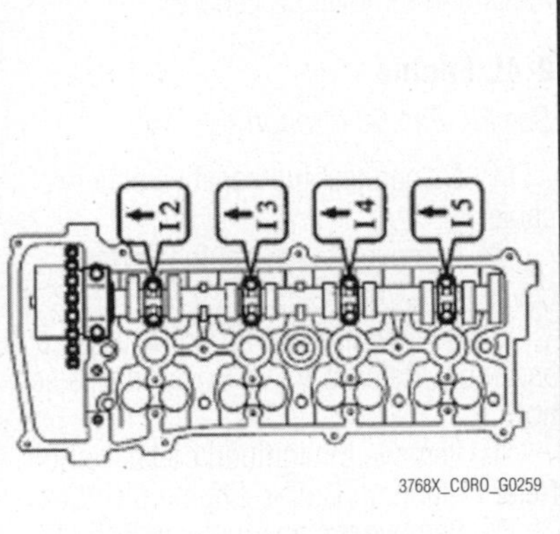

Fig. 64 Checking camshaft no. 1 bearing cap orientation

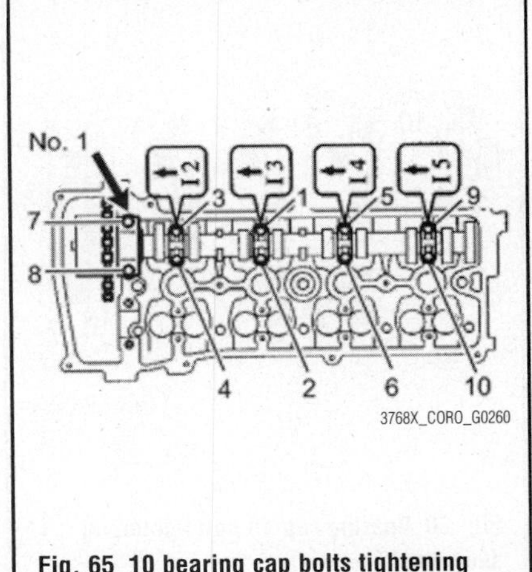

Fig. 65 10 bearing cap bolts tightening sequence

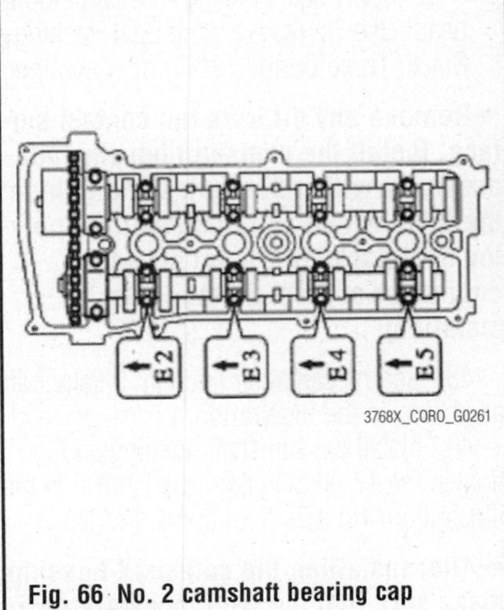

Fig. 66 No. 2 camshaft bearing cap orientation

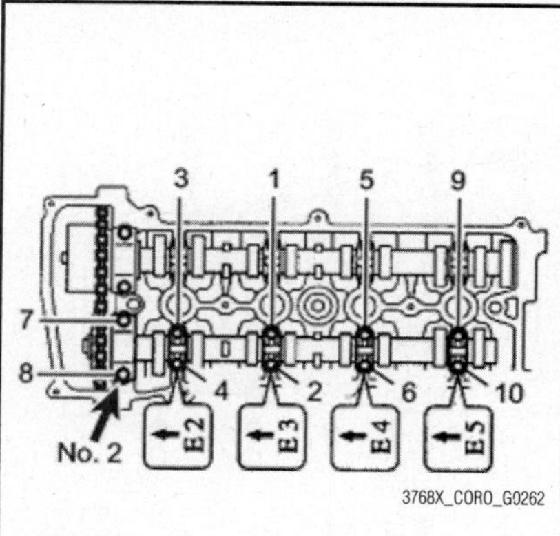

Fig. 67 No. 2 camshaft bearing cap Tightening sequence

i. While holding the camshaft with a wrench, tighten the camshaft timing sprocket set bolt. Tighten to 40 ft. lbs. (54 Nm.)

➡Be careful not to damage the valve lifter.

j. Check that the paint marks on the chain are aligned with the timing marks on the camshaft timing gear and camshaft timing sprocket. Also, check that the crankshaft pulley groove is aligned with the timing mark "0" on the timing mark chain cover.

24. Install the no. 1 chain tensioner assembly:

a. Release the ratchet pawl, then fully push in the plunger and set the hook to the pin so that the plunger is in the position shown in the illustration.

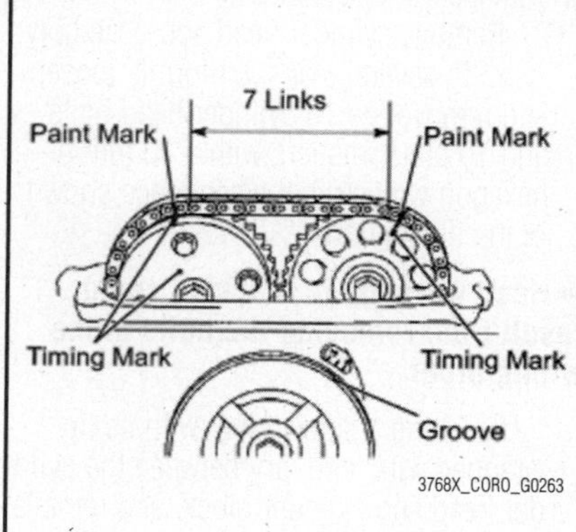

Fig. 68 Check that the paint marks on the chain are aligned with the timing marks on the camshaft timing gear and camshaft timing sprocket

b. Install a new gasket and the chain tensioner with the 2 nuts. Tighten the nuts to 80 inch. lbs. (9 Nm).

➡When installing the chain tensioner, set the hook again if the hook releases the plunger.

25. Turn the crankshaft counterclockwise, then disconnect the hook from the pin.

26. Turn the crankshaft clockwise, then check that the plunger is extended.

27. Set no. 1 cylinder to TDC/compression:

a. Turn the crankshaft pulley until the groove and the timing mark "0" on the timing chain cover are aligned.

b. Check that each timing mark on the camshaft timing gear and sprocket is aligned with each timing mark located on the No. 1 and No. 2 bearing caps.

c. If not, turn the crankshaft pulley 1 revolution (360°) to align the timing marks.

d. Place paint marks on the chain in alignment with the timing marks on the camshaft timing gear and camshaft timing sprocket.

28. Check valve clearance.

29. Adjust valve clearance.

30. To complete installation, reverse remaining removal procedure.

CATALYTIC CONVERTER

REMOVAL & INSTALLATION

The catalytic converter is integrated with the exhaust pipe and cannot be removed separately. The catalytic converter must be replaced with the exhaust pipe as a unit. Refer to Exhaust Manifold in this section.

CYLINDER HEAD

REMOVAL & INSTALLATION

1.8L Engine

See Figures 69 through 72.

1. Remove the camshaft assembly. Refer to Camshaft in this section.

2. Remove cylinder head sub-assembly:

a. Using several steps, uniformly loosen and remove the 10 cylinder head bolts and 10 plate washers with a 10 mm bi-hexagon wrench in the sequence shown in the illustration.

➡Head warpage or cracking could result from removing the bolts in the wrong order.

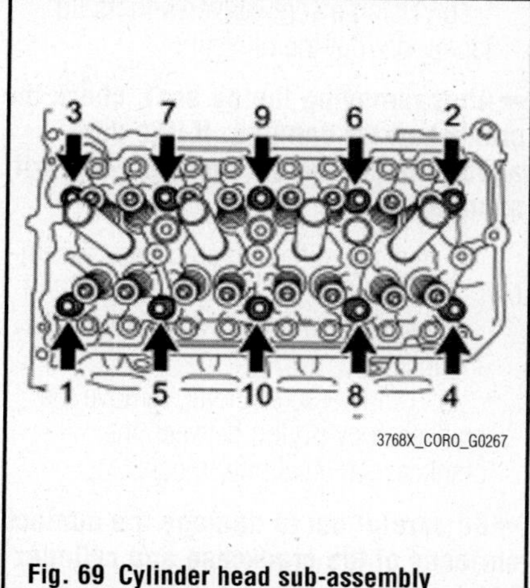

Fig. 69 Cylinder head sub-assembly removal sequence

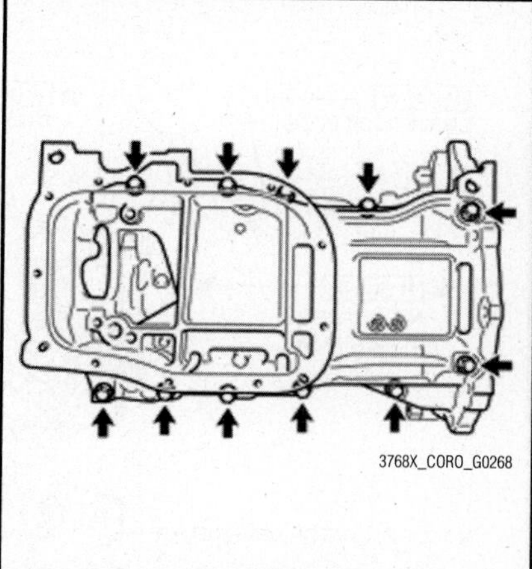

Fig. 70 Locating the 11 stiffening crankcase assembly bolts

b. Using a screwdriver with its tip wrapped with tape, pry between the cylinder head and cylinder block, and remove the cylinder head.

➡Be careful not to damage the contact surfaces of the cylinder head and cylinder block.

3. Remove the cylinder head gasket.

4. Remove the water drain cock plug from the water drain cock sub-assembly.

5. Remove the cylinder block water drain cock sub-assembly from the cylinder block.

6. Remove the ventilation valve.

7. Remove the oil pan drain plug and gasket. Remove the oil pan. Refer to Oil Pan in this section.

8. Remove oil pump assembly. Refer to Oil Pump in this section.

9. Remove rear engine oil seal:

a. Using a knife, cut off the oil seal lip.

b. Using a screwdriver with its tip taped, pry out the oil seal.

➡After removing the oil seal, check the crankshaft for damage. If it is damaged, smooth the surface with 400-grit sandpaper.

10. Remove stiffening crankcase assembly:

a. Uniformly loosen and remove the 11 bolts.

b. Using a screwdriver, remove the crankcase by prying between the crankcase and cylinder block.

➡Be careful not to damage the contact surfaces of the crankcase and cylinder block.

To install:

11. Install the stiffening crankcase with the 11 bolts. Tighten to 16 ft. lbs. (21 Nm).

a. Bolt A: 5.43 inches.

b. Bolt B: 1.38 inches.

c. Bolt C: 2.76 inches.

12. Recheck the torque for bolts 1 and 2. Tighten to 16 ft. lbs. (21 Nm).

13. Wipe off any excess seal packing with a clean piece of cloth.

14. Using a seal installer (SST 09223-15030 and 09950-70010) and a hammer, evenly tap the oil seal until its surface is flush with the rear oil seal retainer edge.

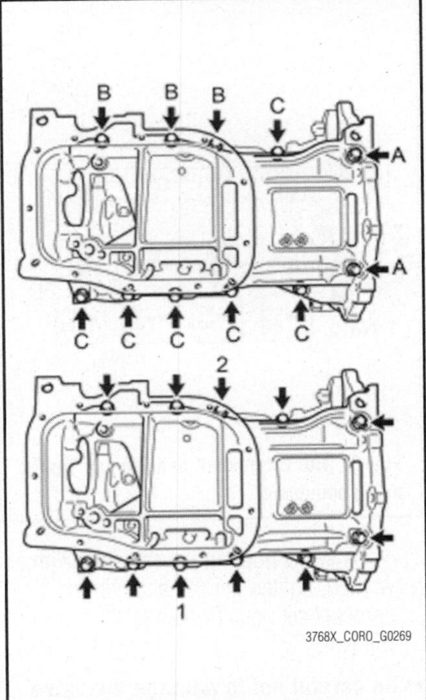

Fig. 72 Identifying and locating the 11 stiffening crankcase bolts

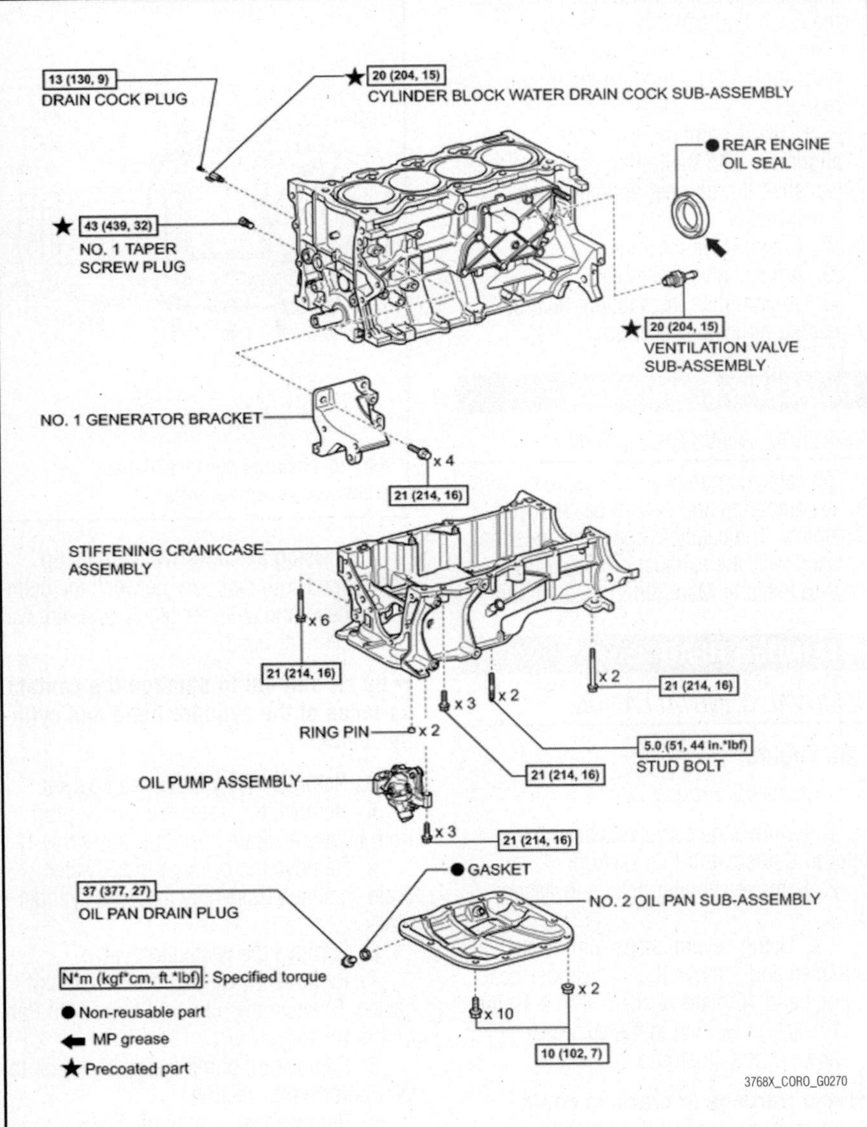

Fig. 71 Exploded view of the cylinder head and related components—1.8L Engine

➡Keep the lip free of foreign matter. Do not tap on the oil seal at an angle.

15. Apply MP grease to a new oil seal lip.

➡Wipe off extra grease on the crankshaft.

16. To complete the installation, reverse remaining removal procedure.

2.4L Engine

See Figures 73 through 75.

1. Remove the camshaft assembly. Refer to Camshaft in this section.

2. Remove cylinder head sub-assembly:

a. In several steps, uniformly loosen and remove the 10 cylinder head bolts and 10 plate washers with a 10 mm bi-hexagon wrench in the sequence shown in the illustration.

➡Head warpage or cracking could result from removing the bolts in the wrong order.

b. Using a screwdriver with its tip wrapped with tape, pry between the cylinder head and cylinder block, and remove the cylinder head.

➡Be careful not to damage the contact surfaces between the cylinder head and cylinder block.

3. Remove cylinder head gasket.

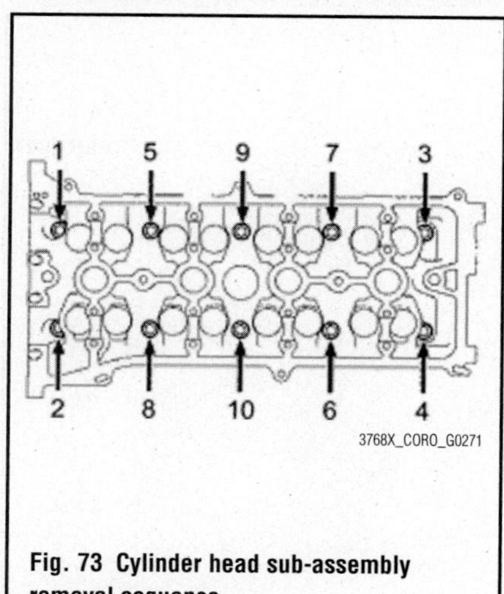

Fig. 73 Cylinder head sub-assembly removal sequence

To install:

4. Place a new cylinder head gasket on the cylinder block surface with the Lot No. stamp facing upward.

➡Remove any oil from the contact surface. Be careful of the installation direction.

5. Place the cylinder head on the cylinder head gasket.

➡Place the cylinder head gently in order to avoid damaging the cylinder head gasket.

➡The cylinder head bolts are tightened in 2 successive steps.

6. Install the cylinder head bolts:

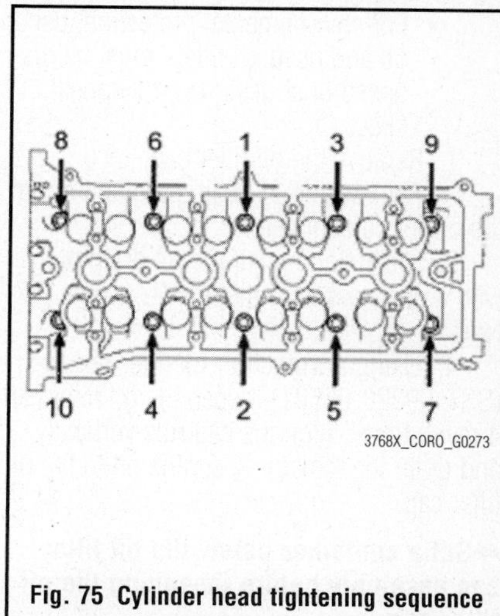

Fig. 75 Cylinder head tightening sequence

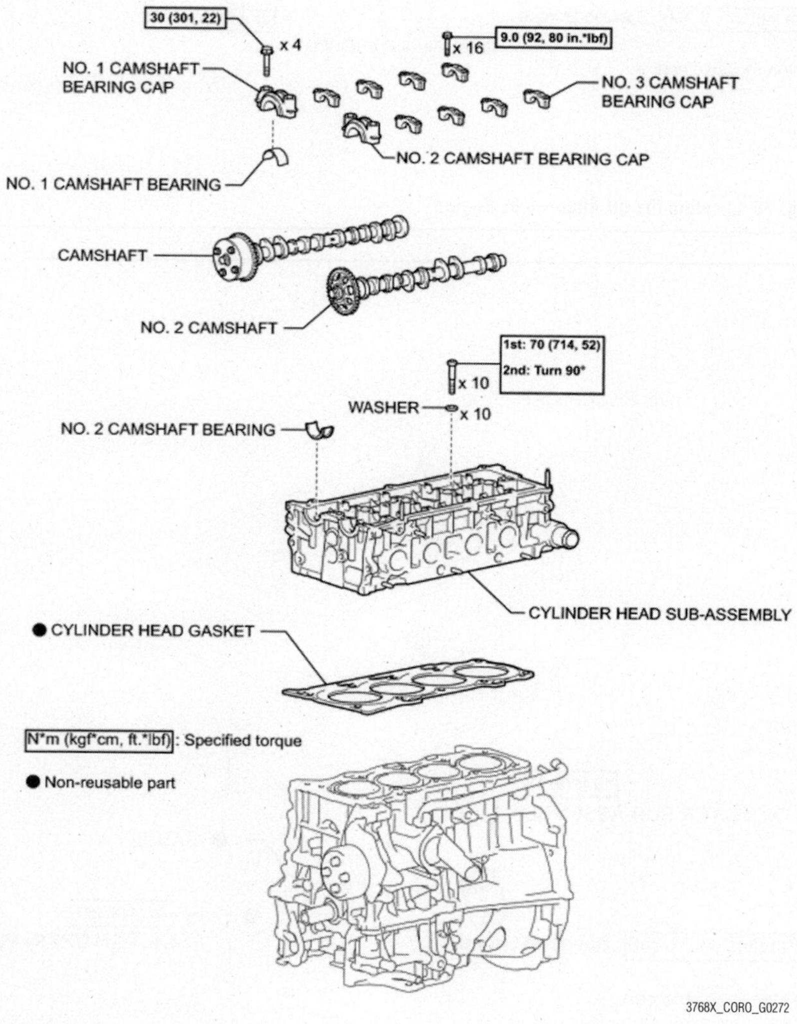

Fig. 74 Exploded view of the cylinder head and related components—2.4L Engine

a. Apply a light coat of engine oil to the threads and under the heads of the cylinder head set bolts.

b. Using several steps, uniformly install and tighten the 10 cylinder head set bolts and plate washers with a 10 mm bi-hexagon wrench in the order shown in the illustration. Tighten to 52 ft. lbs. (70 Nm).

c. Mark the front of the cylinder head bolts with paint.

d. Further tighten the cylinder head bolts 90°.

e. Check that the paint mark is now at a 90° angle to the front.

7. To complete the installation, reverse remaining removal procedure.

ENGINE OIL & FILTER

REPLACEMENT

See Figures 76 through 79.

⁂ CAUTION

Prolonged and repeated contact with engine oil will cause removal of natural oils from the skin, leading to dryness, irritation and dermatitis. In addition, used engine oil contains potentially harmful contaminants which may cause skin cancer.

- Precautions should be taken when replacing engine oil to minimize the risk of your skin making contact with used engine oil. Wear protective clothing and gloves. Wash your skin thoroughly with soap and water, or use a waterless hand cleaner to remove any used engine oil. Do not use gasoline, thinners or solvents.

- For environmental protection, used oil and used oil filters must be disposed of at designated disposal sites.

1. Remove the oil filler cap.
2. Remove the oil drain plug and drain the oil into a container.
3. Clean and install the oil drain plug with a new gasket. Tighten to 27 ft. lbs. (37 Nm).
4. Using a wrench or oil filter remover (SST: 09228-06501) loosen the oil filter cap 4 revolutions, align the cap ribs vertically, and drain the remaining engine oil in the oil filter cap.

➡Set a container below the oil filter cap assembly before loosening the oil filter cap.

5. Remove the oil filter cap assembly.
6. Remove oil filter element and O-ring from the oil filter cap.

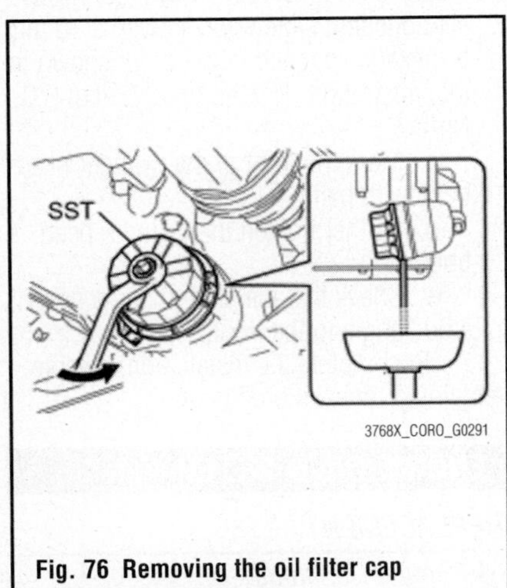

Fig. 76 Removing the oil filter cap assembly—1.8L Engine

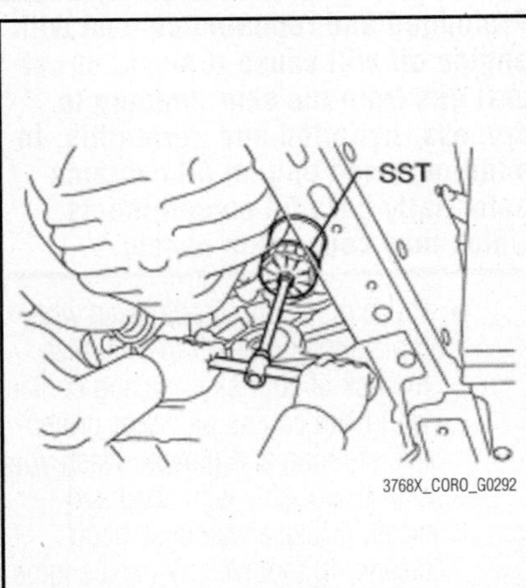

Fig. 77 Removing the oil filter cap assembly—2.4L Engine

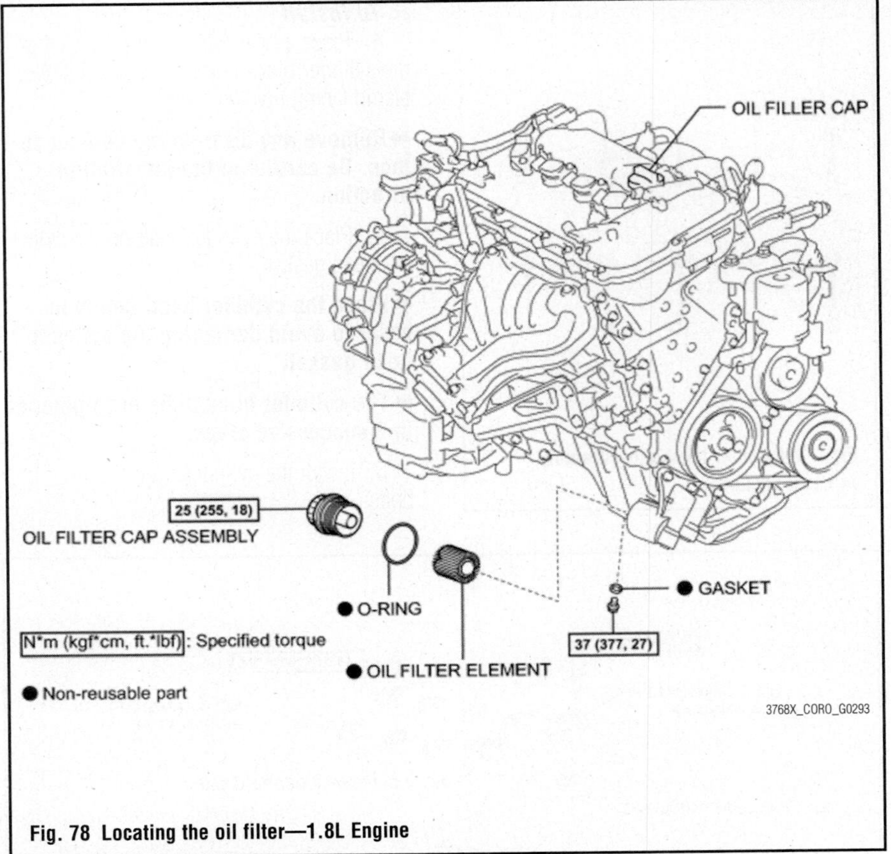

Fig. 78 Locating the oil filter—1.8L Engine

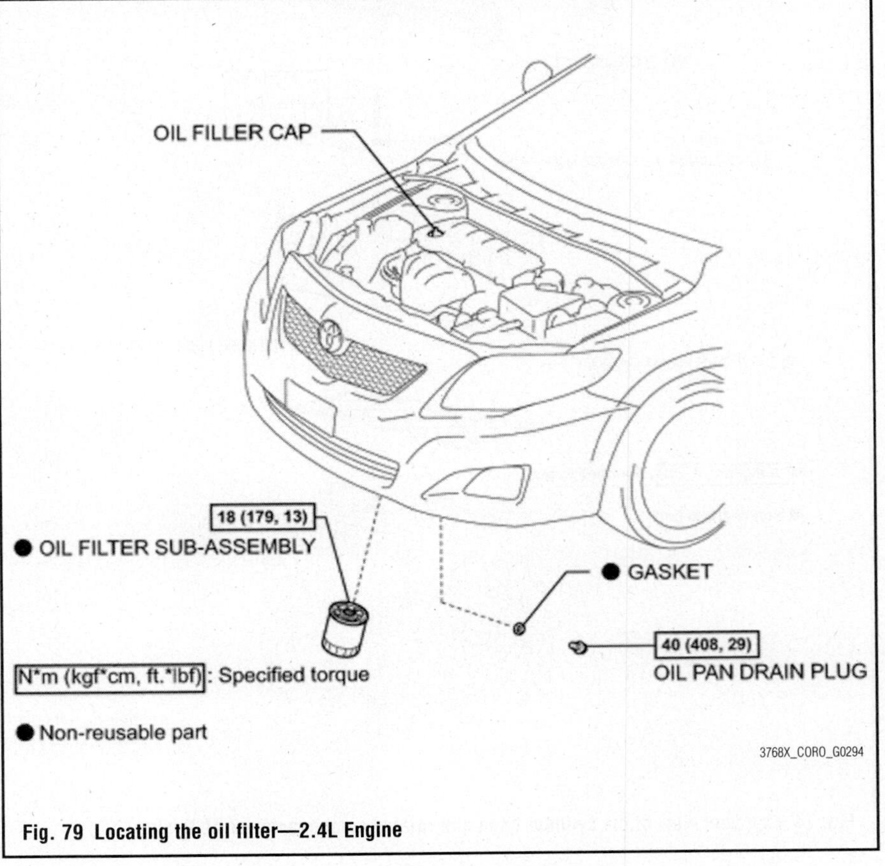

Fig. 79 Locating the oil filter—2.4L Engine

➡Be sure to remove the O-ring (for the cap) by hand, without using any tools, to prevent damage to the groove for the O-ring on the cap.

To install:

7. Clean the oil filter cap threads and O-ring groove.
8. Apply a small amount of engine oil to a new O-ring and install it to the oil filter cap.
9. Set a new oil filter element in the oil filter cap.
10. Remove any dirt or foreign matter from the installation surface and inside of the engine.
11. Reapply a small amount of engine oil to the O-ring of the oil filter cap assembly. Align the cutout in the oil filter cap threads 90°to the grooves in the oil filter bracket and temporarily tighten the cap.

➡Make sure that the O-ring does not get caught between the parts.

12. Tighten the oil filter cap assembly to 18 ft. lbs. (25 Nm).

➡After tightening the oil filter cap assembly, check for gaps between the installation surfaces.

- Do not remove the oil filter bracket clip when installing the oil filter cap assembly.
- Do not cross thread the oil filter cap assembly.

13. Add 5 quarts of fresh engine oil and install the oil filler cap.
14. Start the engine for 30 seconds, turn off and check the fluid level.
15. Inspect for oil leaks.

EXHAUST MANIFOLD

REMOVAL & INSTALLATION

1.8L Engine

See Figures 80 and 81.

1. Disconnect the negative battery cable.
2. Remove front wiper arm motor and assembly.
3. Remove the valve cover.
4. Remove air fuel ratio sensor. Refer to Air Fuel Ratio Sensor in Engine Performance and Emission Controls.
5. Remove the 4 bolts and the exhaust manifold heat insulator.
6. Disconnect the heated oxygen sensor connector. Refer to Heated Oxygen Sensor in Engine Performance and Emission control.
7. Remove the 4 bolts and 4 compression springs.
8. Remove the front exhaust pipe assembly from the 2 exhaust pipe supports.
9. Remove the 3 bolts and manifold stay.
10. Remove the 5 nuts and exhaust manifold.
11. Remove the exhaust manifold gasket.
12. Remove the 3 bolts and exhaust manifold heat insulator.

To install:

13. Install the No. 2 exhaust manifold heat insulator with the 3 bolts. Tighten to 9 ft. lbs. (12 Nm).
14. Install a new exhaust manifold gasket.
15. Install the exhaust manifold with the 5 nuts. Tighten to 16 ft. lbs. (21 Nm).
16. Install the manifold stay with the 3 bolts. Tighten to 32 ft. lbs. (43 Nm).
17. Using a vernier caliper, measure the free length of the compression springs.

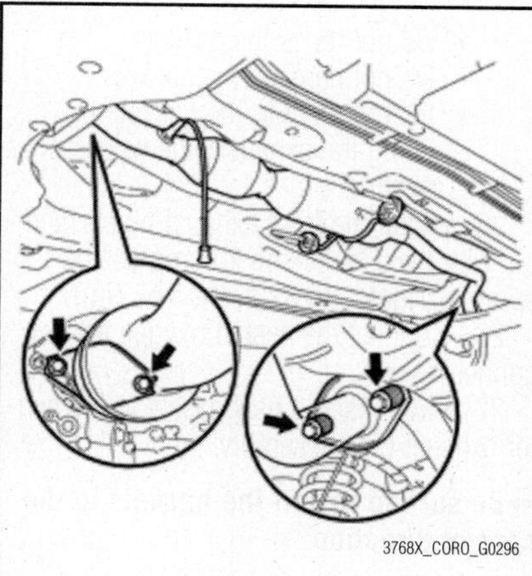

Fig. 80 Removing the 4 bolts and 4 compression springs

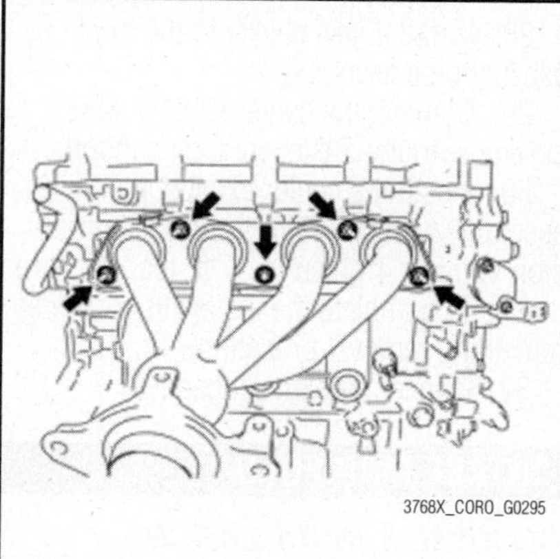

Fig. 81 Removing the 5 nuts and exhaust manifold

a. Front springs: Minimum: 1.63 inches.
b. Rear springs: Minimum: 1.52 inches.

➡If the free length is less than the minimum, replace the compression spring.

18. Fully insert new gaskets to the exhaust manifold and front exhaust pipe assembly.
19. Using a plastic hammer and wooden block, tap in the 2 new gaskets until their surface are flush with the exhaust manifold and front exhaust pipe assembly.

➡Be sure to install the gaskets in the correct direction.

- Do not reuse the gaskets.
- Do not damage the gaskets.
- Do not push in the gaskets by using the exhaust pipe when connecting it.

20. Connect the front exhaust pipe assembly to the 2 exhaust pipe supports.
21. Install the front exhaust pipe assembly with the 4 bolts and 4 compression springs. Tighten to 32 ft. lbs. (43 Nm).
22. Connect the heated oxygen sensor connector.
23. Install the outer exhaust manifold heat insulator with the 4 bolts. Tighten to 9 ft. lbs. (12 Nm).
24. To complete the installation, reverse remaining removal procedure.
25. Inspect for exhaust gas leak.

2.4L Engine

See Figures 80 through 83.

1. Disconnect the negative battery cable.
2. Remove the drive belt. Refer to Accessory Drive Belts in the section.
3. Remove the alternator. Refer to Alternator in Engine Electrical.
4. Remove the 4 bolts and 2 compression springs.
5. Remove the center exhaust pipe assembly from the 2 exhaust pipe supports.
6. Remove the 2 gaskets.
7. Remove front exhaust pipe assembly:
   a. Disconnect the heated oxygen sensor connector.
   b. Remove the 2 bolts, 2 compression springs and front exhaust pipe assembly.
   c. Remove the gasket from the exhaust manifold.
8. Remove the bolt, nut and LH manifold stay.
9. Remove the bolt, nut and RH manifold stay.

10. Remove the 4 bolts and upper exhaust manifold heat insulator.
11. Remove air fuel ratio sensor. Refer to Air Fuel Ratio Sensor in Engine Performance and Emission Controls.
12. Remove the 5 nuts and exhaust manifold converter sub-assembly.
13. Remove the gasket.
14. Remove the 2 bolts and underside exhaust manifold heat insulator.
15. Remove the 4 bolts and catalytic converter insulator.

To install:

16. Install the 4 bolts and catalytic converter insulator. Tighten to 9 ft. lbs. (12 Nm).
17. Install the underside exhaust manifold heat insulator with the 2 bolts. Tighten to 9 ft. lbs. (12 Nm).
18. Install the exhaust manifold converter sub-assembly:
    a. Install a new gasket.

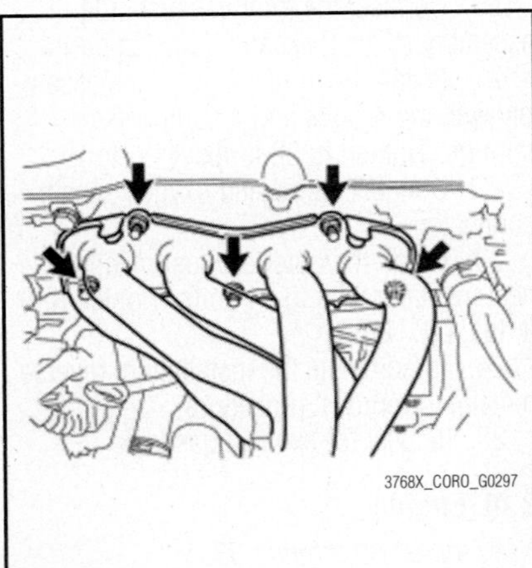

Fig. 82 Remove the 5 nuts and exhaust manifold converter sub-assembly

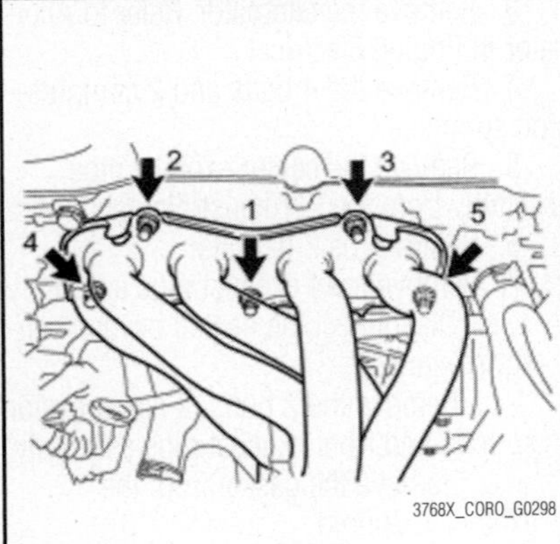

Fig. 83 Exhaust manifold tightening sequence

    b. Install the exhaust manifold converter sub-assembly with the 5 nuts in the order shown in the illustration. Tighten to 27 ft. lbs. (37 Nm).
19. Install the RH and LH manifold stay with the bolt and nut. Tighten to 33 ft. lbs. (44 Nm).
20. Install the air fuel ratio sensor.
21. Install the outer exhaust manifold heat insulator with the 4 bolts. 9 ft. lbs. (12 Nm).
22. Using a vernier caliper, measure the free length of the compression springs.
    a. Front springs: Minimum: 1.63 inches.
    b. Rear springs: Minimum: 1.52 inches.
23. Remove the remains of exhaust manifold converter with wire brash.
24. Fully insert a new gasket to the exhaust manifold.

➡Be sure to install the gaskets in the correct direction.

- Do not reuse the gaskets.
- Do not damage the gaskets.
- Do not push in the gaskets by using the exhaust pipe when connecting it.

25. Install the front exhaust pipe assembly with the 2 bolts and 2 compression springs. Tighten to 32 ft. lbs. (43 Nm).
26. Connect the heated oxygen sensor connector.
27. Fully insert a new gasket to the center exhaust pipe assembly.

➡Be sure to install the gaskets in the correct direction.

- Do not reuse the gaskets.
- Do not damage the gaskets.
- Do not push in the gaskets by using the exhaust pipe when connecting it.

28. Install a new gasket to the front exhaust pipe assembly.
29. Connect the center exhaust pipe assembly to the 2 exhaust pipe supports.
30. Install the center exhaust pipe assembly with the 4 bolts and 2 compression springs. Tighten to 32 ft. lbs. (43 Nm).
31. To complete the installation, reverse remaining removal procedure.
32. Inspect for exhaust gas leak.

INTAKE MANIFOLD

REMOVAL & INSTALLATION

1.8L Engine

See Figures 84 and 85.

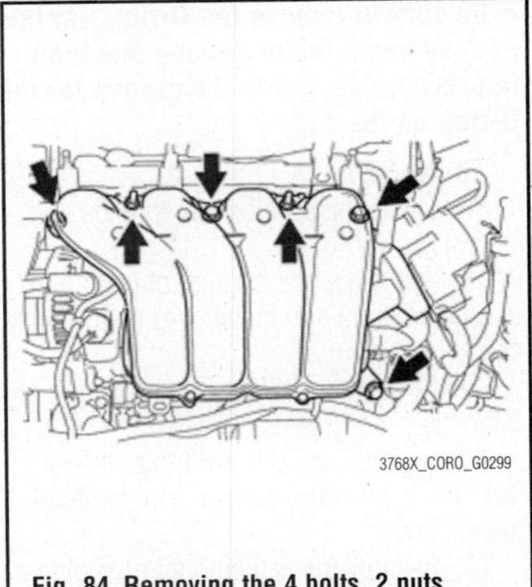

Fig. 84 Removing the 4 bolts, 2 nuts, intake manifold stay and intake manifold

1. Disconnect the negative battery cable.
2. Drain and recycle the engine coolant.
3. Remove the engine cover.
4. Remove the air cleaner assembly. Refer to Air Cleaner in this section.
5. Remove the throttle body. Refer to Throttle Body in Fuel System.
6. Remove the intake manifold:
    a. Remove the bolt and disconnect the wire harness bracket.
    b. Disconnect the 3 hoses.
    c. Remove the 4 bolts, 2 nuts, intake manifold stay and intake manifold.
    d. Remove the gasket from the intake manifold.
    e. Remove the bolt and bracket.
    f. Using a TORX® socket E6, remove the 2 stud bolts from the intake manifold.

To install:

7. Using a TORX® socket E6, install the 2 stud bolts to the intake manifold. Tighten to 44 inch lbs. (5 Nm).
8. Install the bracket with the bolt. Tighten to 8 ft. lbs. (10 Nm).
9. Install a new gasket into the intake manifold.
10. Install the intake manifold and intake manifold stay with the 4 bolts and 2 nuts. Tighten to 21 ft. lbs. (28 Nm).
11. To complete the installation, reverse remaining removal procedure.

2.4L Engine

See Figures 86 and 87.

✲✲ CAUTION

Observe all applicable safety precautions when working around fuel. Whenever servicing the fuel system, always work in a well-ventilated area. Do not allow fuel spray or

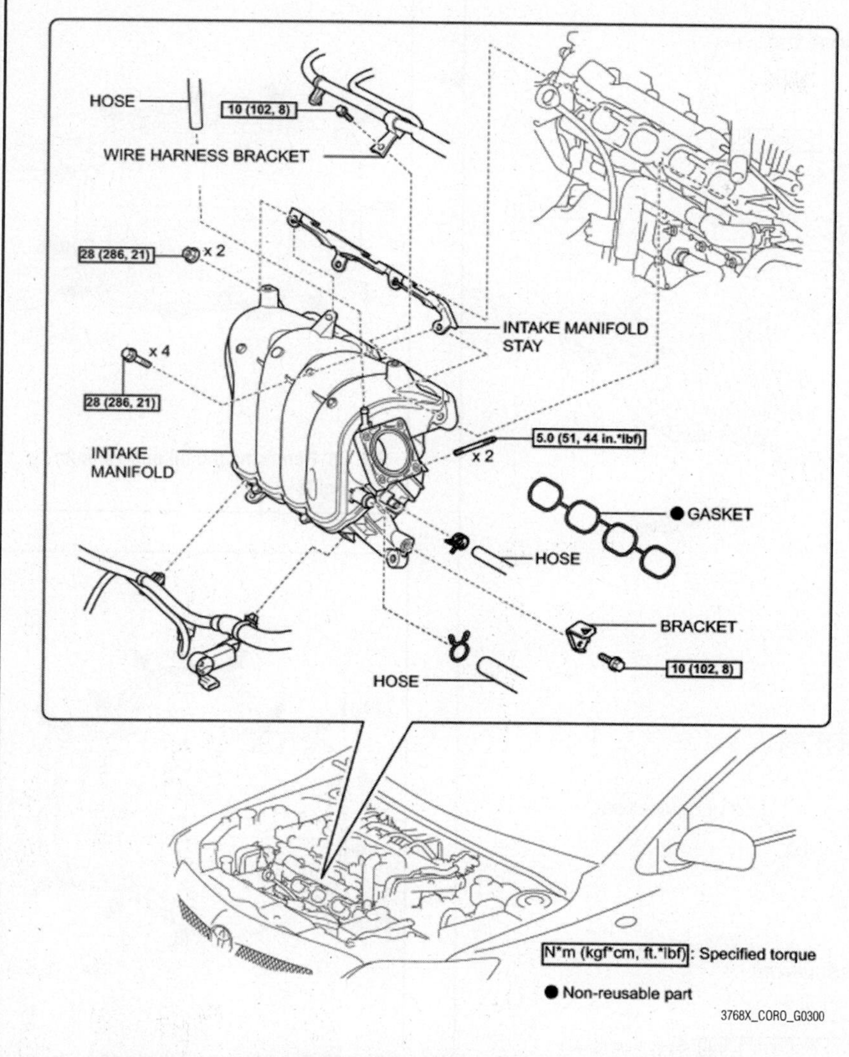

Fig. 85 Exploded view of the intake manifold and related components—1.8L Engine

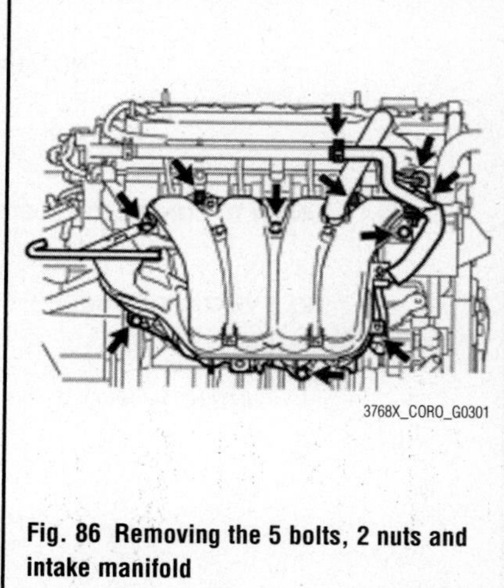

Fig. 86 Removing the 5 bolts, 2 nuts and intake manifold

vapors to come in contact with a spark or open flame. Keep a dry chemical fire extinguisher near the work area. Always keep fuel in a container specifically designed for fuel storage; also, always properly seal fuel containers to avoid the possibility of fire or explosion.

1. Discharge fuel system pressure.
2. Disconnect the cable from negative battery terminal.
3. Remove the engine cover.
4. Drain and recycle the engine coolant.
5. Remove the windshield wiper motor and link assembly.
6. Remove suspension tower damper assembly (with front strut bar). Refer to Suspension Tower Damper Assembly in Front Suspension.
7. Remove the air cleaner assembly. Refer to Air Cleaner in this section.
8. Remove the throttle body. Refer to Throttle Body in Fuel System.
9. Disconnect the main fuel tube.
10. Disconnect the ventilation hose.
11. Remove fuel delivery pipe with injector. Refer to Fuel Rail and Injectors in Fuel System.
12. Remove intake manifold:
    a. Disconnect the union to connector tube hose from the No. 2 hose to hose tube.
    b. Disconnect the camshaft timing oil control valve connector.
    c. Remove the wire harness clamp.
    d. Remove the union to connector tube hose from the vacuum hose clamp.
    e. Remove the 5 bolts, 2 nuts and intake manifold.
    f. Remove the gasket from the intake manifold.

To install:

13. To install, reverse removal procedure. Refer to exploded view illustration for torque values.

OIL PAN

REMOVAL & INSTALLATION

1.8L Engine

See Figure 88.

1. Remove the oil pan drain plug and gasket and drain the engine oil.
2. Remove the 10 bolts and 2 nuts.
3. Insert the blade of oil pan seal cutter between the crankcase and oil pan. Cut through the sealer and remove the oil pan.

➡Be careful not to damage the contact surfaces of the crankcase and oil pan.

To install:

4. Remove any old packing material and be careful not to drop any oil on the contact surfaces of the cylinder block and oil pan.
5. Apply a continuous bead of seal packing (Diameter 4.0 mm (0.157 in.)). Seal packing: Toyota Genuine Seal Packing Black, Three Bond 1207B or equivalent.

➡Remove any oil from the contact surfaces.

- Install the oil pan within 3 minutes after applying seal packing.
- Do not start the engine for at least 2 hours after installing the oil pan.

6. Install the oil pan with the 10 bolts and 2 nuts. Tighten to 7 ft. lbs. (10 Nm).

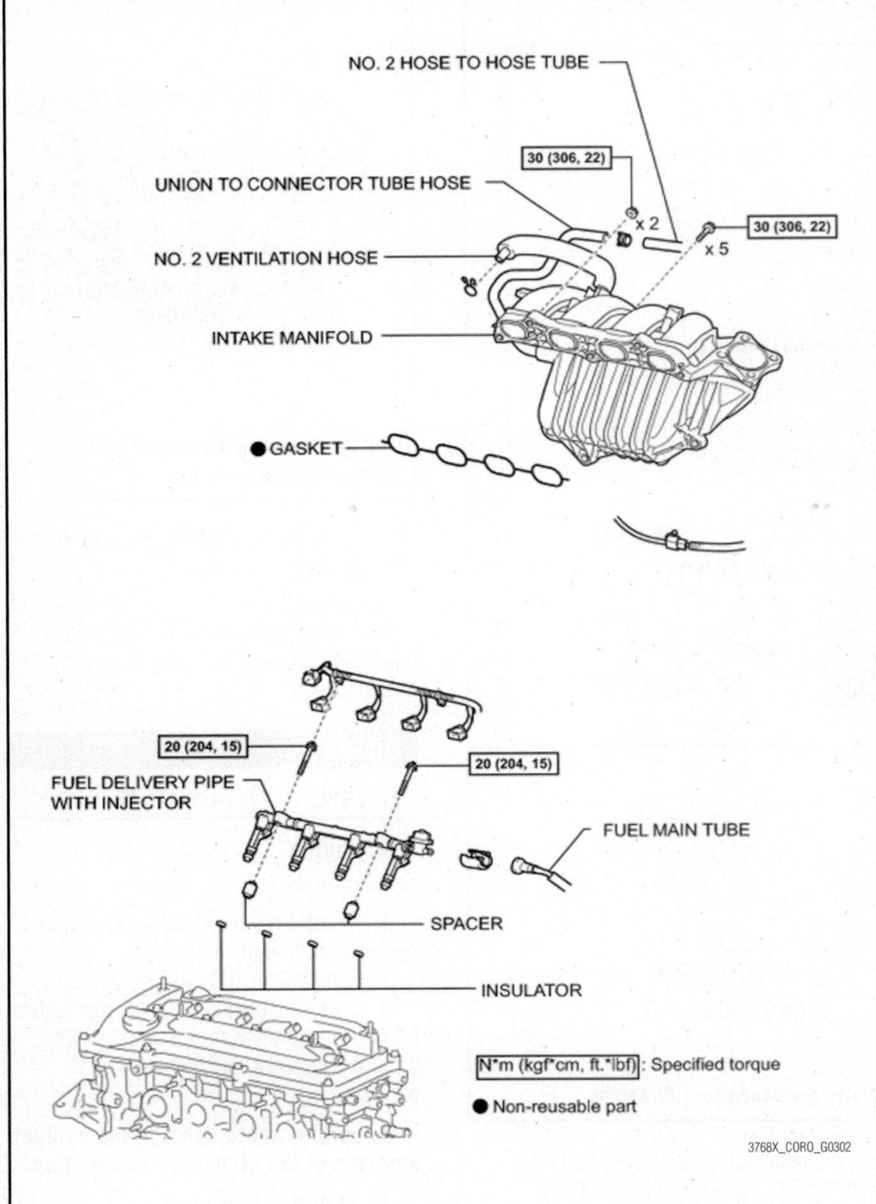

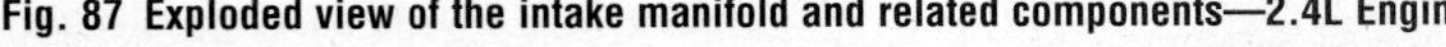

Fig. 87 Exploded view of the intake manifold and related components—2.4L Engine

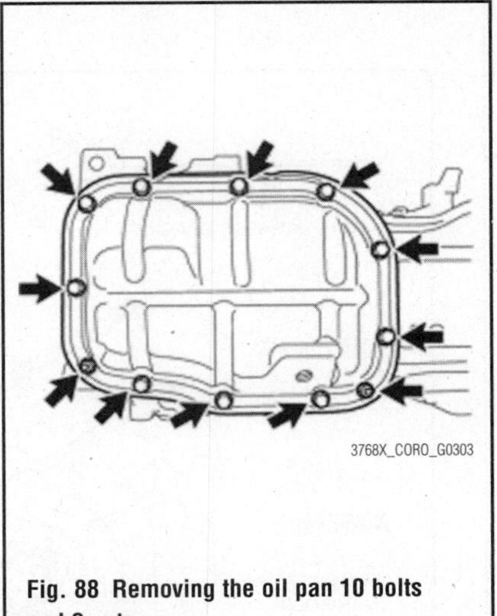

Fig. 88 Removing the oil pan 10 bolts and 2 nuts

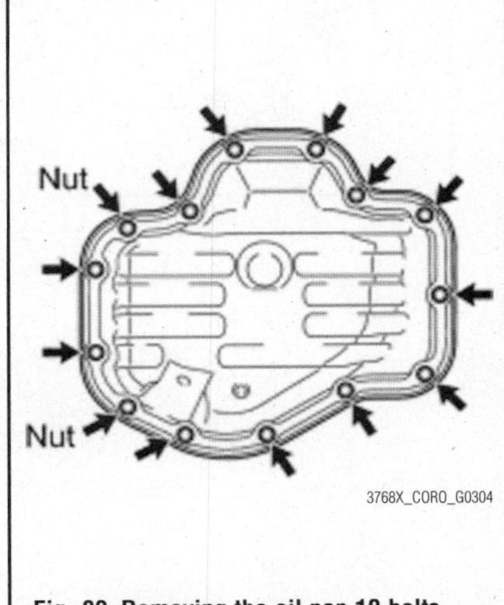

Fig. 89 Removing the oil pan 10 bolts and 2 nuts

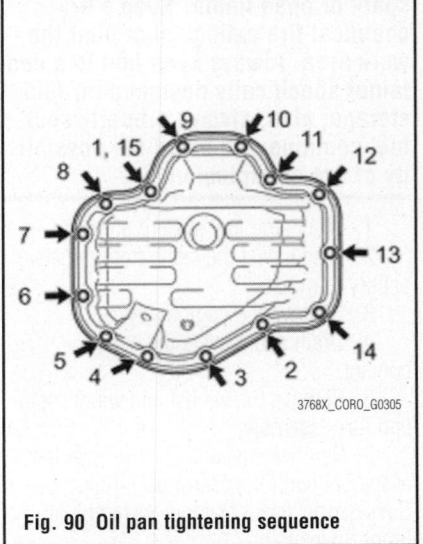

Fig. 90 Oil pan tightening sequence

2.4L Engine

See Figures 89 and 90.

1. Remove the oil pan drain plug and gasket and drain the engine oil.
2. Remove the 12 bolts and 2 nuts.
3. Insert the blade of oil pan seal cutter between the crankcase and oil pan. Cut through the sealer and remove the oil pan.

➡Be careful not to damage the contact surfaces of the crankcase and oil pan.

To install:

4. Remove any old packing material and be careful not to drop any oil on the contact surfaces of the cylinder block and oil pan.
5. Apply a continuous bead of seal packing (Diameter 4.0 mm (0.157 in.)). Seal packing: Toyota Genuine Seal Packing Black, Three Bond 1207B or equivalent.

➡Remove any oil from the contact surfaces.

- Install the oil pan within 3 minutes after applying seal packing.
- Do not start the engine for at least 2 hours after installing the oil pan.

6. Uniformly tighten the 12 bolts and 2 nuts in the sequence shown in the illustration. Tighten to 80 inch lbs. (9 Nm).

OIL PUMP

REMOVAL & INSTALLATION

See Figures 91 and 92.

1. Remove the engine assembly with the transaxle.
2. Remove the timing chain front cover. Refer to Timing Chain Front Cover in this section.
3. Remove the oil pan. Refer to Oil Pan in this section.
4. Remove the 3 bolts and oil pump.
5. For 2.4L engines, remove the gasket.

To install:

6. To install, reverse the removal procedure. Tighten the oil pump assembly bolts to 16 ft. lbs. (21 Nm).

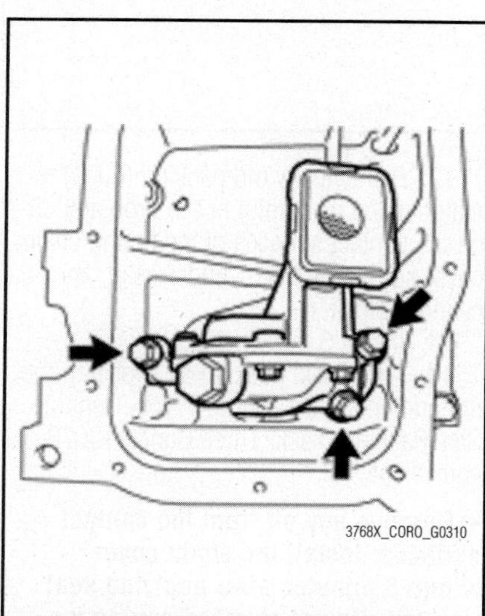

Fig. 91 Removing the 3 bolts and oil pump—1.8L Engine

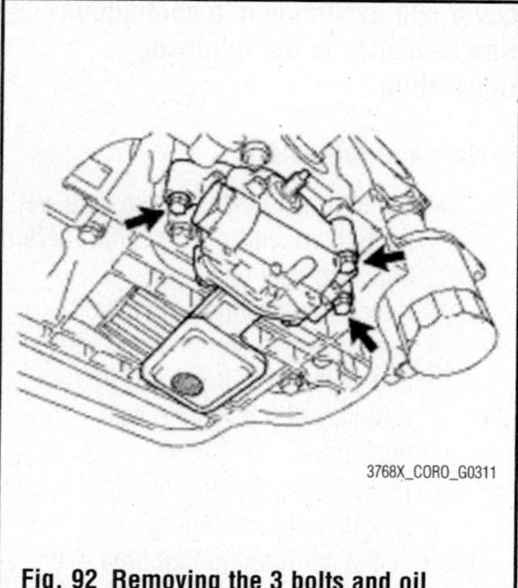

Fig. 92 Removing the 3 bolts and oil pump—2.4L Engine

PISTON AND RING

POSITIONING

See Figures 93 through 95.

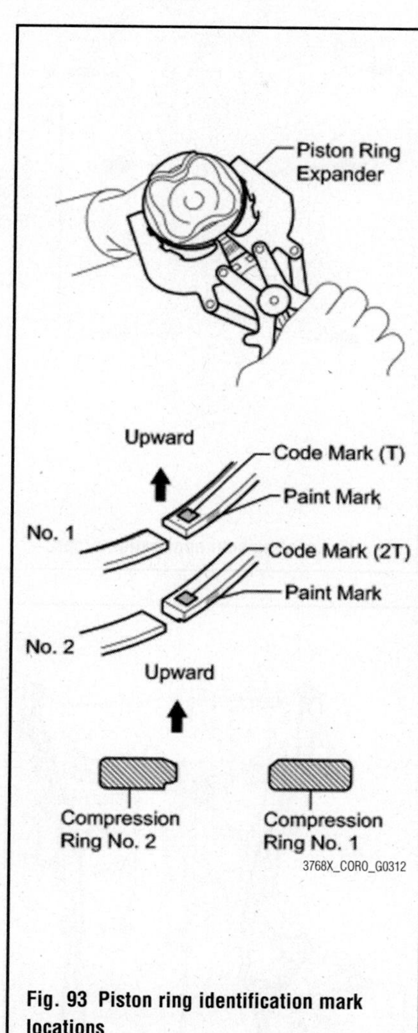

Fig. 93 Piston ring identification mark locations

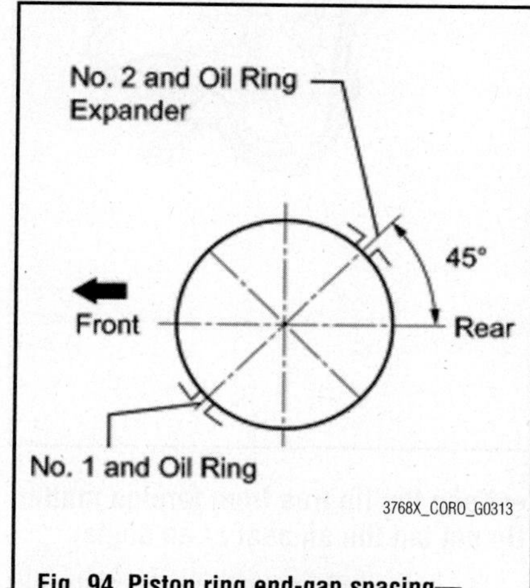

Fig. 94 Piston ring end-gap spacing—1.8L Engine

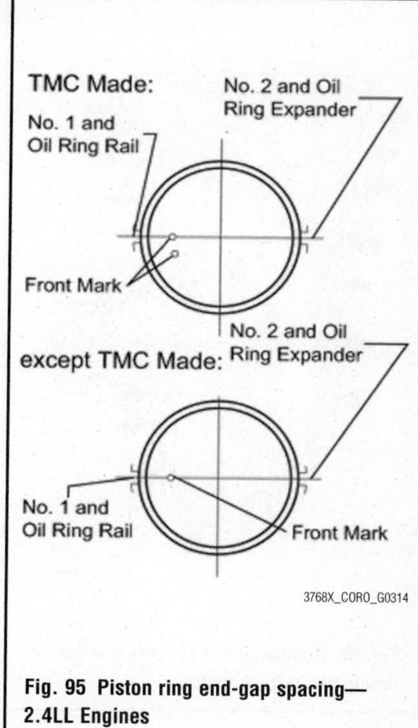

Fig. 95 Piston ring end-gap spacing—2.4LL Engines

TIMING CHAIN FRONT COVER

REMOVAL & INSTALLATION

1.8L Engine

See Figures 96 through 101.

1. Remove the 3 bolts and engine mounting bracket.
2. Remove the 4 bolts and oil filter bracket.
3. Remove the 2 O-rings.
4. Remove the 19 bolts.
5. Remove the timing chain cover by prying between the timing chain cover and cylinder head or cylinder block with a screwdriver.

➡Be careful not to damage the contact surfaces of the timing chain cover, cylinder block, and cylinder head.

➡Tape the screwdriver tip before use.

6. Place the timing chain cover on wooden blocks.
7. Using a screwdriver and a hammer, pry out the oil seal.

➡Do not damage the surface of the oil seal press fit hole.

8. Remove the 3 O-rings.
9. Remove the 3 bolts and water pump. Remove the gasket.

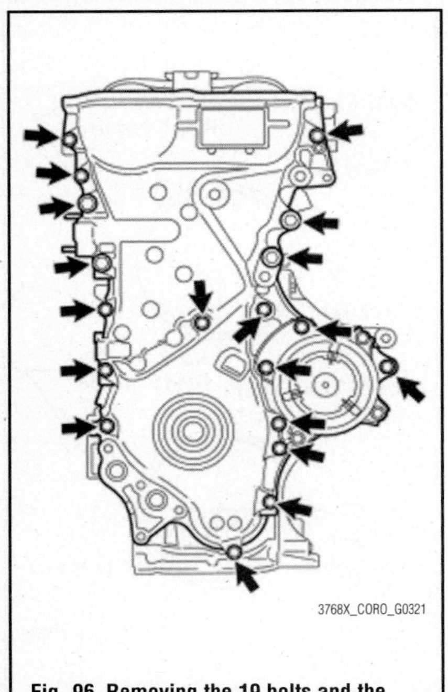

Fig. 96 Removing the 19 bolts and the timing chain front cover

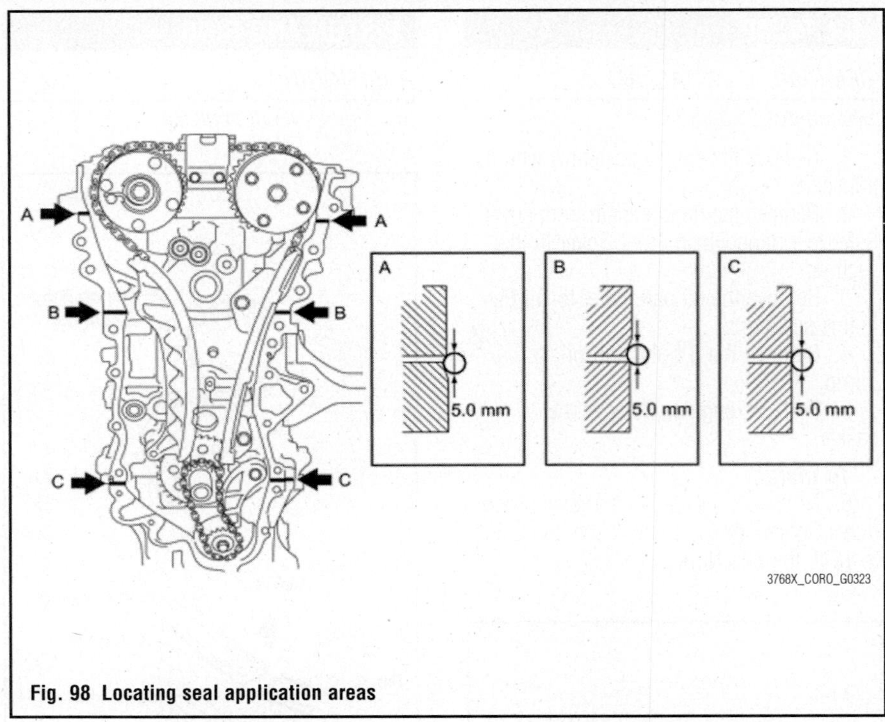

Fig. 98 Locating seal application areas

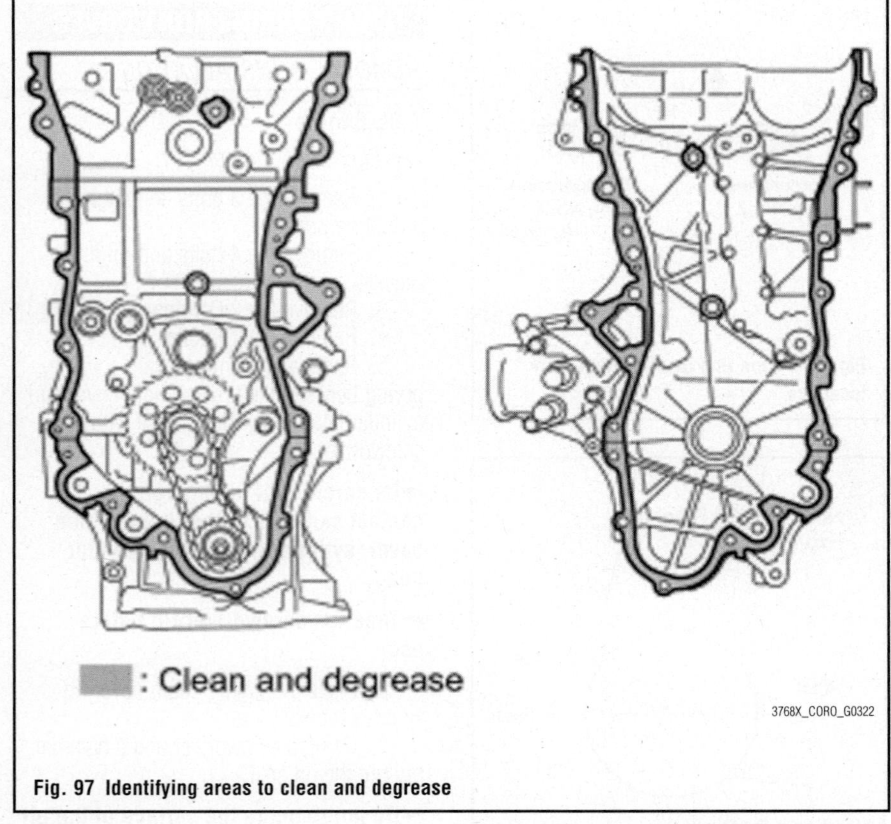

Fig. 97 Identifying areas to clean and degrease

To install:

10. Using a seal installer (SST 09223-15030) and a hammer, evenly tap the oil seal until its surface is flush with the timing chain cover sub-assembly edge.

➡Keep the lip free from foreign matter. Do not tap the oil seal at an angle.

11. Apply MP grease to the lip of the oil seal.

12. Remove any old packing (FIPG) material and be careful not to drop any oil on the contact surfaces of the timing chain cover sub-assembly, cylinder head, and cylinder block.

13. Install 3 new O-rings.

14. Apply seal packing as shown in the illustration. Seal packing: Toyota Genuine Seal Packing Black, Three Bond® 1207B or equivalent.

➡Remove any oil from the contact surfaces. Install the chain cover within 3 minutes after applying seal packing. Do not start the engine for at least 2 hours after installing the timing chain cover sub-assembly. Apply seal packing to the timing chain cover sub-assembly in a continuous line as shown in the following illustration.

➡Note the following:

- When the contact surfaces are wet, wipe them with oil-free cloth before applying seal packing.
- Install the timing chain cover sub-assembly within 3 minutes and tighten the bolts within 15 minutes after applying seal packing.
- Do not start the engine for at least 2 hours after installing.

15. For the continuous line area, use: Toyota Genuine Seal Packing Black, Three Bond® 1207B or equivalent.

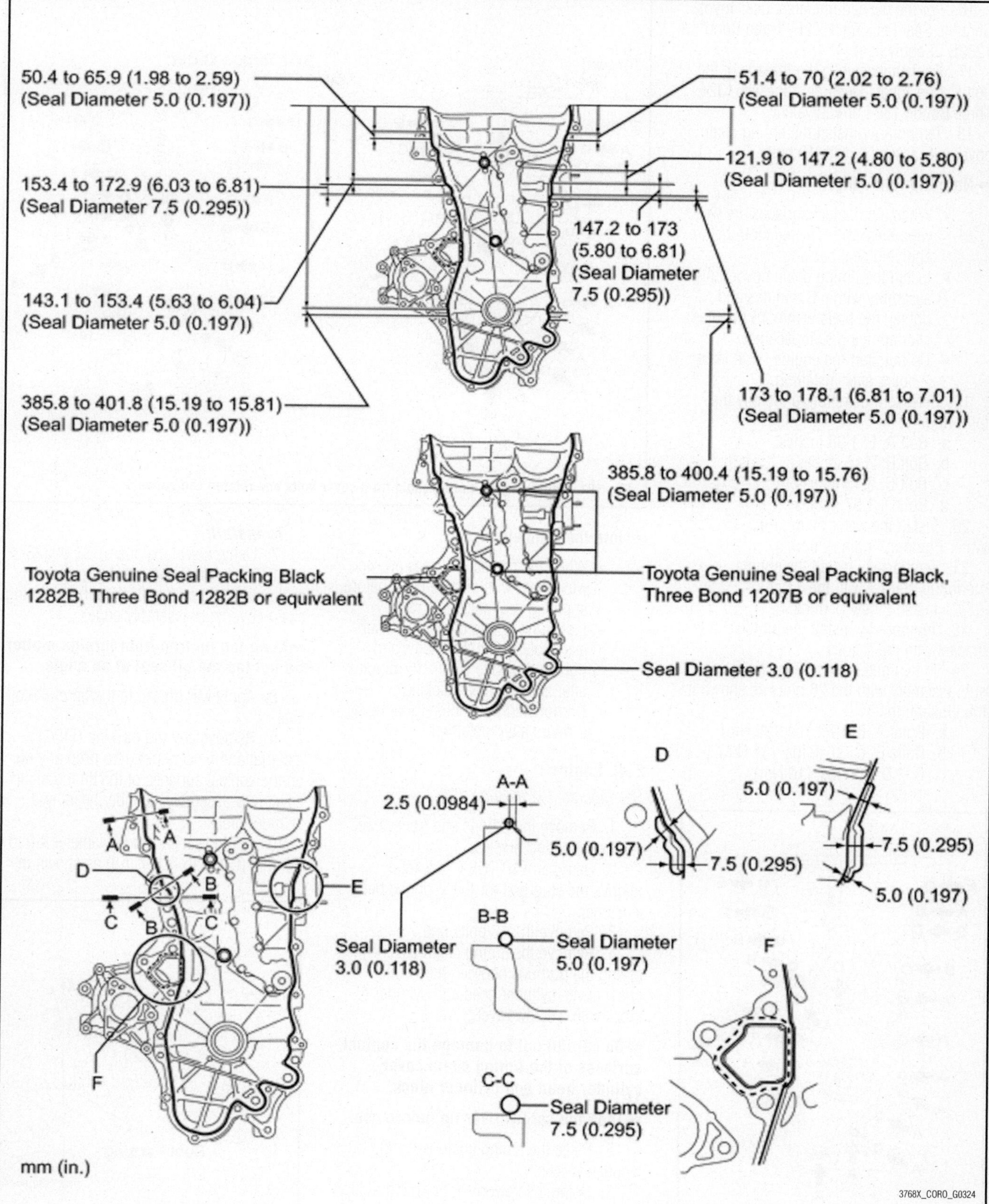

Fig. 99 Applying seal packing to the timing chain cover sub-assembly in a continuous line

16. For the dashed line area, use: Toyota Genuine Seal Packing 1282B, Three Bond® 1282B or equivalent.

17. Apply adhesive to the threads of the bolt E. Use Toyota Genuine Adhesive 1324, Three Bond® 1324 or equivalent.

18. Temporarily install the timing chain cover sub-assembly with 19 bolts.

➡Note the following:

- When the contact surfaces are wet, wipe them with oil-free cloth before applying seal packing.
- Install the timing chain cover sub-assembly within 3 minutes and tighten the bolts within 15 minutes after applying seal packing.
- Do not start the engine for at least 2 hours after installing.

19. Refer to illustration to identify the following:

  a. Bolt A, E: 1.38 inches.
  b. Bolt B: 2.16 inches.
  c. Bolt C: 3.15 inches.
  d. Bolt D: 1.57 inches

20. Install the water pump. Refer to Water Pump in Engine Cooling.

21. Temporarily install the engine mounting bracket with the 3 bolts.

22. Install 2 new O-rings.

23. Temporarily install the oil filter bracket with the 4 bolts.

24. Fully tighten the timing chain cover sub-assembly with the 26 bolts as shown in the illustration.

  a. Bolts A, E: 19 ft. lbs. (26 Nm).
  b. Bolts B, C: 37 ft. lbs. (51 Nm).
  c. Bolt D: 7 ft. lbs. (10 Nm).

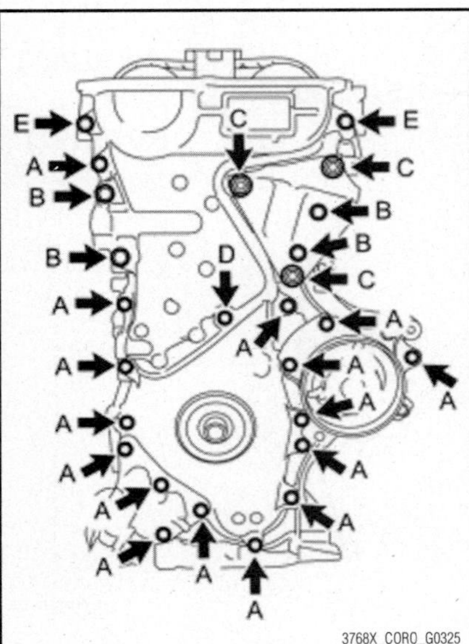

Fig. 100 Identifying the timing chain front cover bolts

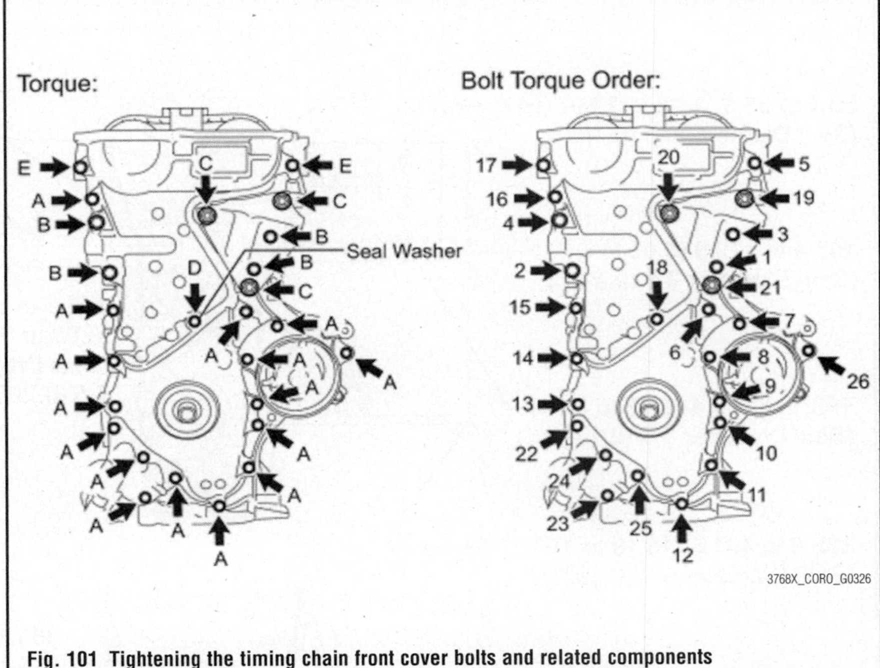

Fig. 101 Tightening the timing chain front cover bolts and related components

➡Note the following:

- When the contact surfaces are wet, wipe them with oil-free cloth before applying seal packing.
- Install the timing chain cover sub-assembly within 3 minutes and tighten the bolts within 15 minutes after applying seal packing.
- Do not start the engine for at least 2 hours after installing.

2.4L Engine

See Figures 102 through 105.

1. Remove the 3 bolts and transverse engine mounting bracket.

2. Using an E10 TORX® socket, remove the stud bolt for the V-ribbed belt tensioner.

3. Remove the 12 bolts and 2 nuts.

4. Remove the timing chain cover by prying the portions between the timing chain cover, cylinder head and cylinder block with a screwdriver.

➡Be careful not to damage the contact surfaces of the timing chain cover, cylinder head and cylinder block.

➡Tape the screwdriver tip before use.

5. Place the timing chain cover on wooden blocks.

6. Using a screwdriver, pry out the oil seal.

➡Do not damage the surface of the oil seal press fit hole.

To install:

7. Using a seal installer (SST 09223-22010) and a hammer, evenly tap the oil seal until its surface is flush with the timing chain cover sub-assembly edge.

➡Keep the lip free from foreign matter. Do not tap the oil seal at an angle.

8. Apply MP grease to the lip of the oil seal.

9. Remove any old packing (FIPG) material and be careful not to drop any oil on the contact surfaces of the timing chain cover sub-assembly, cylinder head, and cylinder block.

10. Apply seal packing (Diameter 4.0 to 4.5 mm (0.157 to 0.177 in.)) as shown in

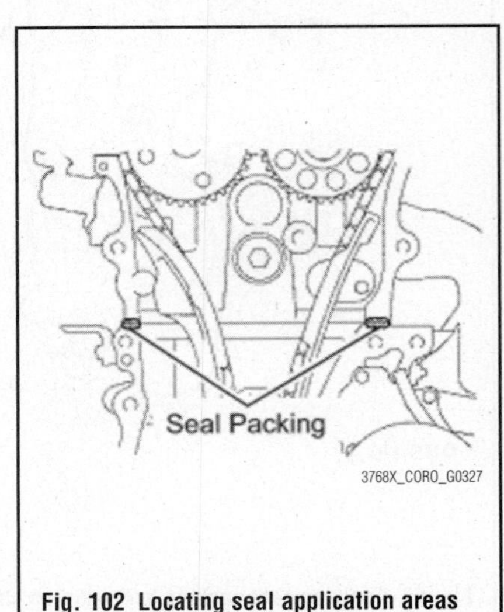

Fig. 102 Locating seal application areas

the illustration. Use Toyota Genuine Seal Packing Black, Three Bond 1207B or equivalent.

➡Remove any oil from the contact surfaces. Install the chain cover within 3 minutes of applying seal packing. Do not add engine oil for at least 2 hours after installing the chain cover.

11. Apply a continuous bead of seal packing as shown in the illustration. Use Toyota Genuine Seal Packing Black, Three Bond 1207B or equivalent.

➡Remove any oil from the contact surfaces. Install the chain cover within 3 minutes of applying seal packing. Do not add engine oil for at least 2 hours after installing the chain cover.

12. Apply adhesive to the threads of the bolt "A". Use Toyota Genuine Adhesive 1324, Three Bond 1324 or equivalent.

13. Temporarily install the timing chain cover with the 12 bolts and 2 nuts.:
   a. Bolt A: 1.18 inches (10 mm head)
   b. Bolt B: 1.18 inches (12 mm head
   c. Bolt C: 1.57 inches (14 mm head)

14. Temporarily install the engine transverse engine mounting bracket with the 3 bolts.

15. Fully tighten the timing chain cover with the 15 bolts and 2 nuts as shown in the illustration.

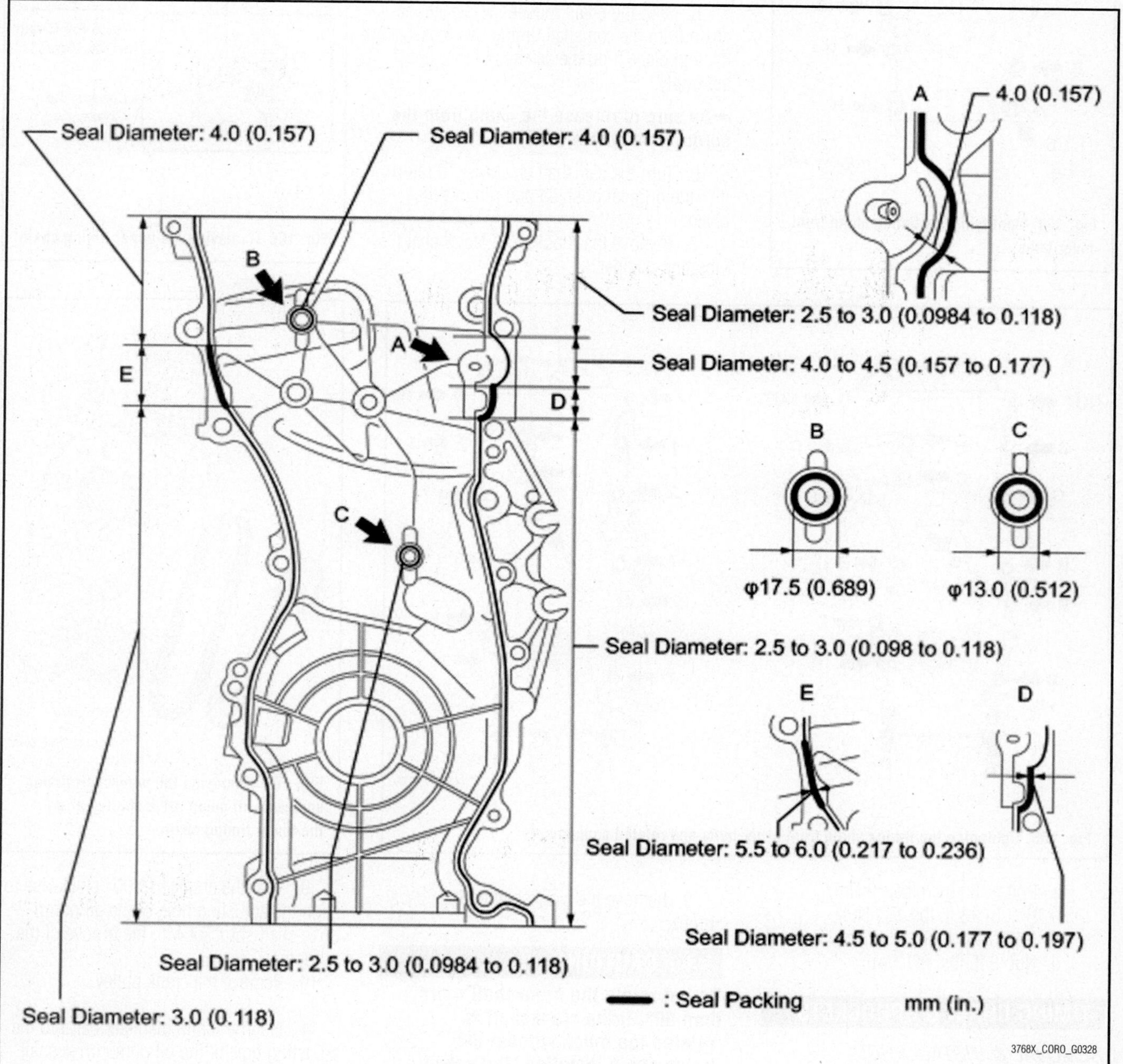

Fig. 103 Applying seal packing to the timing chain cover sub-assembly in a continuous line

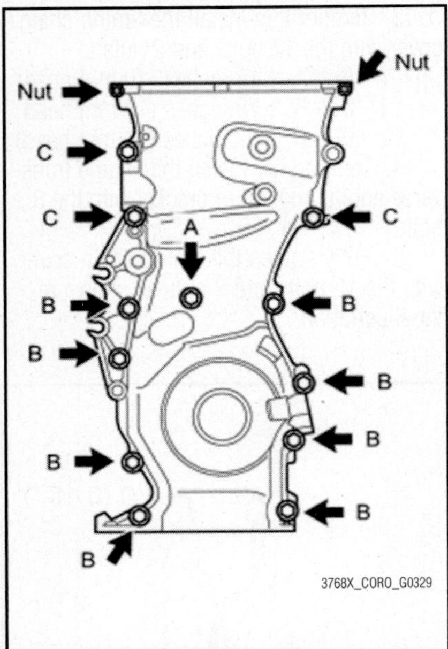

Fig. 104 Identifying the timing chain front cover bolts

➡Do not turn the crankshaft without the chain tensioner installed.

1. Remove the 2 nuts, bracket, tensioner and gasket.
2. Remove the chain tensioner slipper.
3. Remove the 2 bolts and LH chain vibration damper.
4. Hold the hexagonal portion of the camshaft with a wrench and turn the camshaft timing gear assembly counterclockwise to loosen the chain between the camshaft timing gears.
5. With the chain loosened, release the chain from the camshaft timing gear assembly and place it on the camshaft timing gear assembly.

➡Be sure to release the chain from the sprocket completely.

6. Turn the camshaft clockwise to return it to the original position and remove the chain.
7. Remove the 2 bolts and No. 2 chain vibration damper.

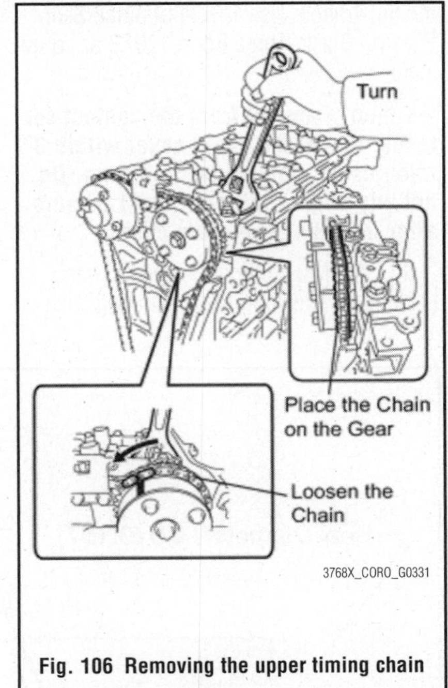

Fig. 106 Removing the upper timing chain

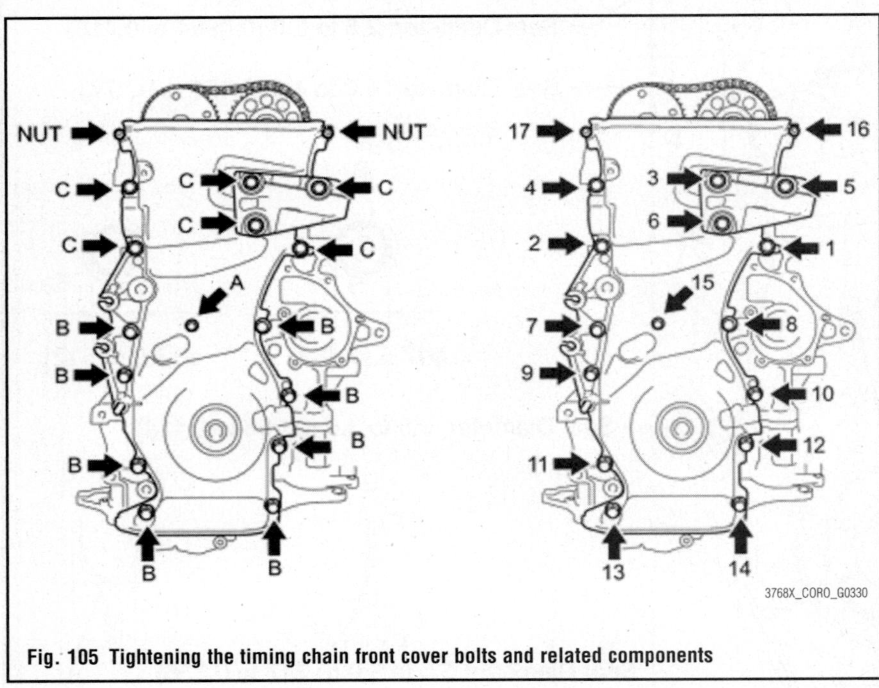

Fig. 105 Tightening the timing chain front cover bolts and related components

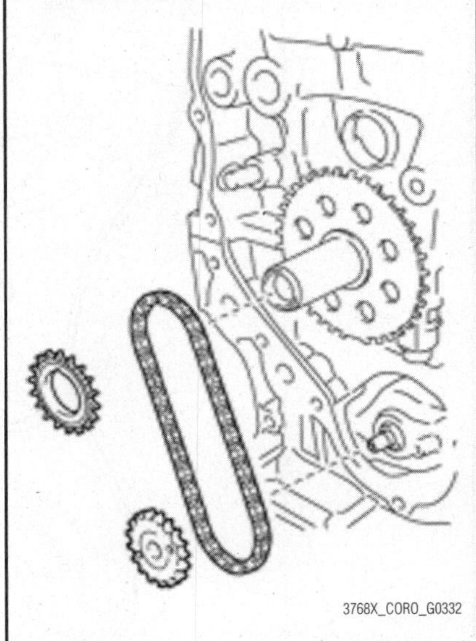

Fig. 107 Removing the crankshaft timing sprocket, oil pump drive shaft gear and the lower timing chain

a. Bolt A: 80 inch lbs. (9 Nm).
b. Bolt B: 18 ft. lbs. (25 Nm).
c. Bolt C: 41 ft. lbs. (55 Nm).
d. Nut: 8 ft. lbs. (11 Nm).

TIMING CHAIN & SPROCKETS

REMOVAL & INSTALLATION

1.8L Engine

See Figures 106 through 114.

8. Remove the crankshaft timing sprocket.

CAUTION

Do not rotate the crankshaft more than 90°. If the crankshaft is rotated too much without the timing chain installed, the valves may hit the pistons and cause damage.

9. Turn the crankshaft 90° clockwise to align the adjusting hole of the oil pump drive shaft sprocket with the groove of the oil pump.
10. Remove the crank pulley bolt.
11. Insert a 3 mm diameter bar into the adjusting hole of the oil pump drive shaft sprocket to lock the gear in position, and then remove the nut.
12. Remove the bolt, chain tensioner plate, and spring.

13. Remove the crankshaft timing sprocket, oil pump drive shaft gear, and No. 2 chain sub-assembly.

To install:

14. Set the crankshaft key as shown in the illustration.
15. Turn the drive shaft so that the cutout faces the right horizontal position.
16. Align the yellow mark links with the timing marks of each gear as shown in the illustration.
17. Install the sprockets onto the crankshaft and oil pump shaft with the chain on the gears.
18. Temporarily tighten the oil pump drive shaft sprocket with the nut.
19. Insert the damper spring into the adjusting hole, and then install the chain tensioner plate with the bolt. Tighten to 7 ft. lbs. (10 Nm).
20. Align the adjusting hole of the oil pump drive shaft sprocket with the groove of the oil pump.

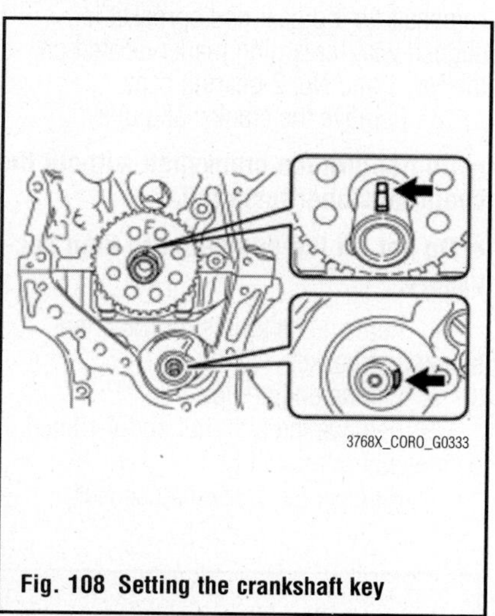

Fig. 108 Setting the crankshaft key

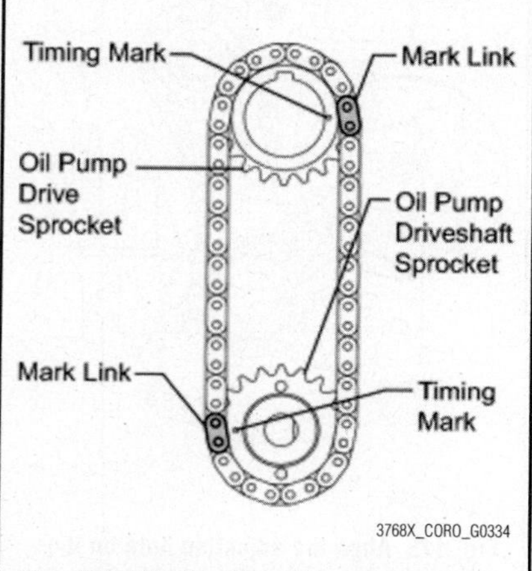

Fig. 109 Aligning the yellow mark links with the timing marks of each gear

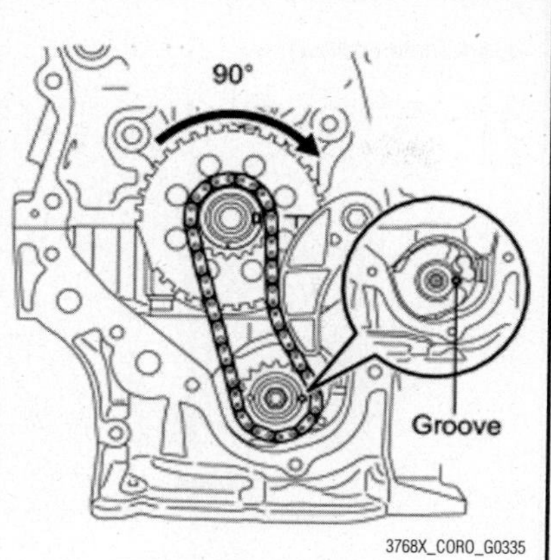

Fig. 110 Aligning the adjusting hole of the oil pump drive shaft sprocket with the groove of the oil pump

21. Insert a 3 mm diameter bar into the adjusting hole of the oil pump drive shaft gear to lock the gear in position, and then tighten the nut. Tighten to 21 ft. lbs. (28 Nm).
22. Install the crankshaft timing sprocket.
23. Install the lower chain vibration damper with the 2 bolts. Tighten to 16 ft. lbs. (21 Nm).
24. Install the upper chain vibration damper with the 2 bolts. Tighten to 7 ft. lbs. (10 Nm).
25. Check the No. 1 cylinder TDC/compression.
26. Temporarily tighten the crankshaft pulley bolt.
27. Turn the crankshaft counterclockwise to position the timing gear key to the top.

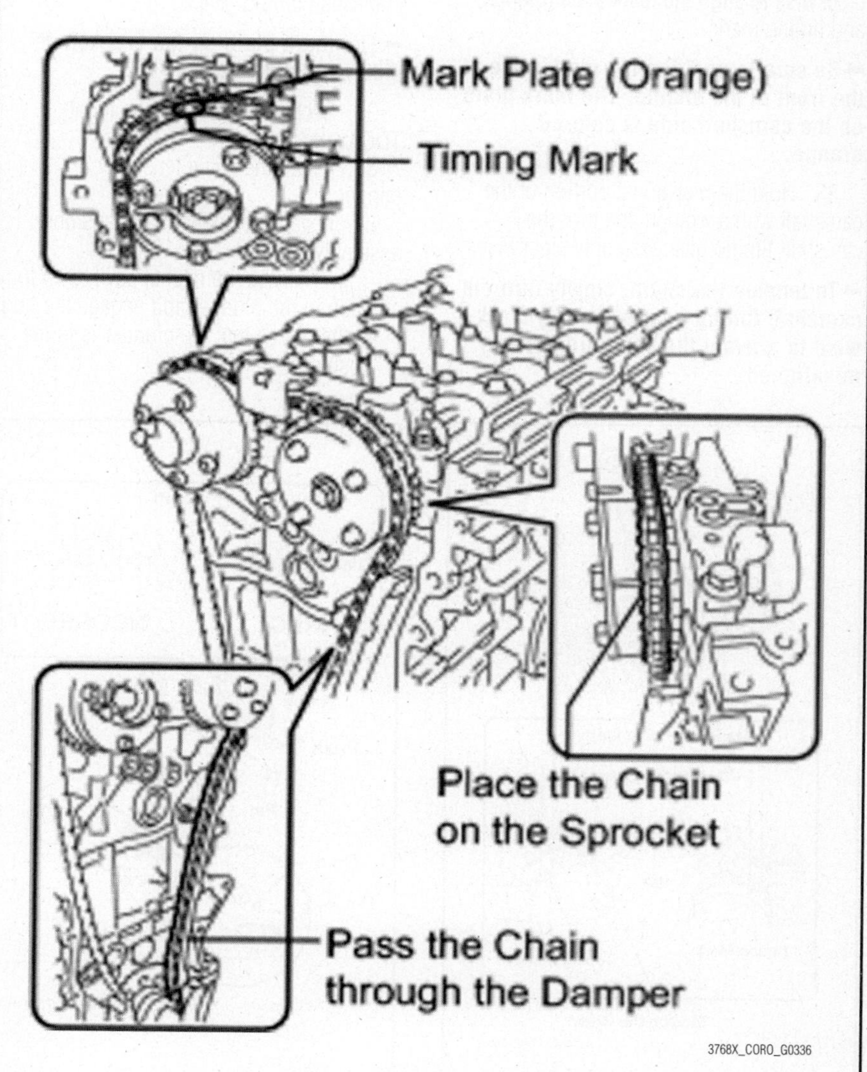

Fig. 111 Align the mark plate (orange) with the timing mark of the No. 2 camshaft

28. Remove the crankshaft pulley bolt.
29. Check the timing marks on each camshaft timing gear.
30. Align the mark plate (orange) with the timing mark of the No. 2 camshaft as shown in the illustration and install the chain.

➡Be sure to position the mark plate at the front of the engine. The mark plate on the camshaft side is colored orange. Do not pass the chain around the sprocket of the camshaft timing gear assembly. Only place it on the sprocket. Pass the chain through the No. 1 vibration damper.

31. Place the chain on the crankshaft without passing it around the shaft.
32. Hold the hexagonal portion of the camshaft with a wrench and turn the camshaft timing gear assembly counter-clockwise to align the mark plate (orange) and timing mark.

➡Be sure to position the mark plate at the front of the engine. The mark plate on the camshaft side is colored orange.

33. Hold the hexagonal portion of the camshaft with a wrench and turn the camshaft timing gear assembly clockwise.

➡To tension the chain, slowly turn the camshaft timing gear assembly clockwise to prevent the chain from being misaligned.

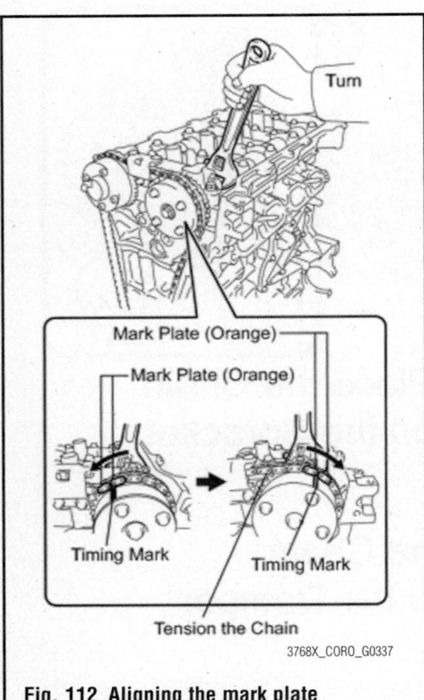

Fig. 112 Aligning the mark plate (orange) and timing mark

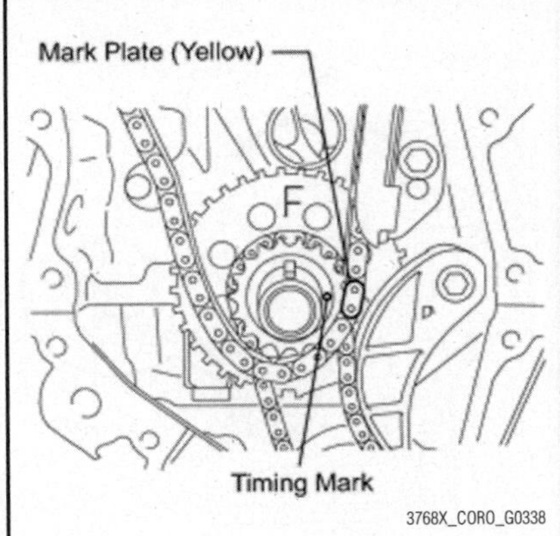

Fig. 113 Aligning the mark plate (yellow) and timing mark on the crankshaft timing gear

34. Align the mark plate (yellow) and timing mark and install the chain to the crankshaft timing gear.

➡The mark plate on the crankshaft side is colored yellow.

35. Recheck each timing mark at TDC/compression.
36. Install the chain tensioner slipper.
37. Install the upper chain tensioner assembly:
   a. Release the ratchet pawl, then fully push in the plunger and engage the hook to the pin so that the plunger is in the position.

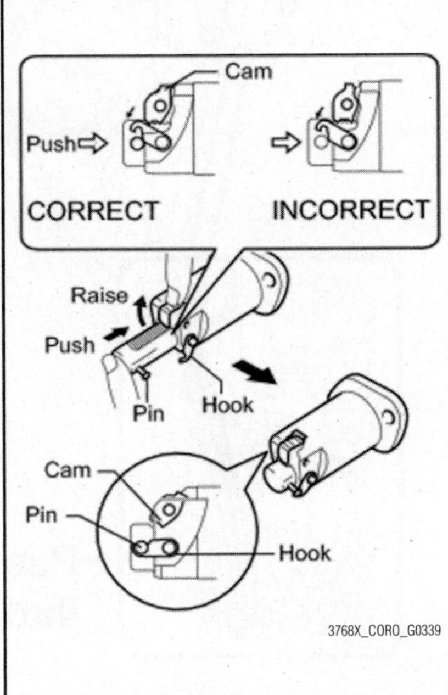

Fig. 114 Correct plunger position and orientation.

➡Make sure that the cam engages the first tooth of the plunger to allow the hook to pass over the pin.

   b. Install a new gasket, bracket and No. 1 chain tensioner with the 2 nuts.

➡If the hook releases the plunger while the chain tensioner is being installed, engage the hook again.

   c. Turn the crankshaft counterclockwise, then disconnect the hook from plunger knock pin.
   d. Turn the crankshaft clockwise, then check that the plunger is extended.

38. To complete installation, reverse remaining removal procedure.

2.4L Engine

See Figures 115 through 121.

1. Turn the crankshaft pulley until the groove and the timing mark "0" on the timing chain cover are aligned.
2. Check that each timing mark on the camshaft timing gear and sprocket is aligned with the timing marks located on the No. 1 and No. 2 bearing caps.
3. Remove the crankshaft pulley.

➡Do not turn the crankshaft without the chain tensioner installed.

➡Do not lift the engine more than necessary.

4. Remove the 2 nuts, bracket, tensioner and gasket.
5. Lift the engine upward.
6. Remove the bolt, nut and V-ribbed belt tensioner assembly.
7. Remove the crankshaft sensor.

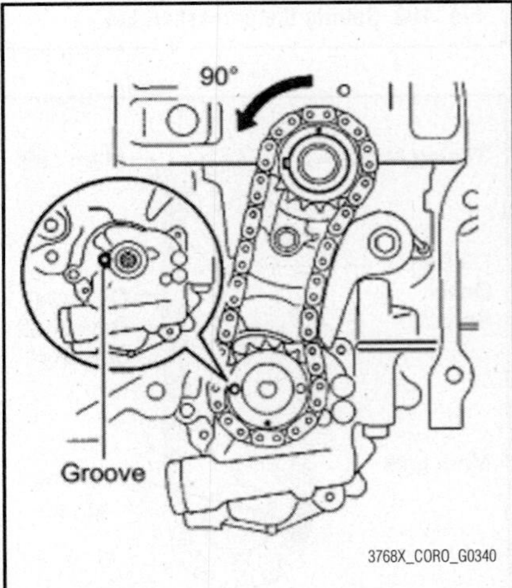

Fig. 115 Align the adjusting hole on the oil pump drive shaft sprocket with the groove on the oil pump

8. Remove the timing chain and sprocket cover.
9. Remove the No. 1 Crankshaft Position (CKP) sensor plate.
10. Remove the bolt and timing chain guide.
11. Remove the bolt and chain tensioner slipper.
12. Remove the 2 bolts and No. 1 chain vibration damper.
13. Remove the chain sub-assembly.
14. Remove the crankshaft timing gear or sprocket from the crankshaft.
15. Turn the crankshaft 90°counterclockwise to align the adjusting hole on the oil pump drive shaft sprocket with the groove on the oil pump.
16. Insert a 4 mm diameter bar into the adjusting hole of the oil pump drive shaft sprocket to lock the gear in position, and then remove the nut.
17. Remove the bolt, chain tensioner plate and spring.

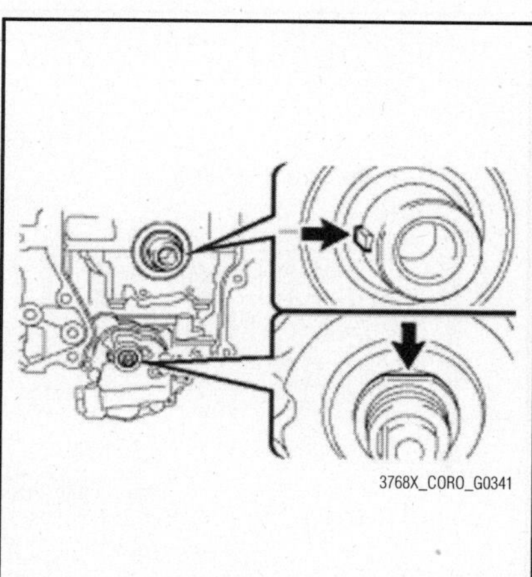

Fig. 116 Setting the crankshaft key in the left horizontal position

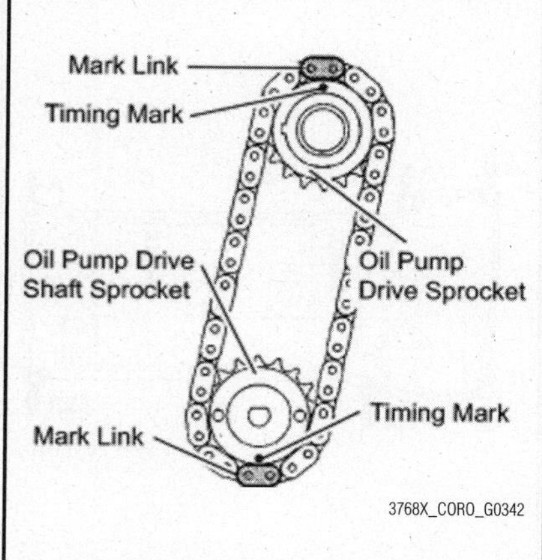

Fig. 117 Aligning the yellow mark links with the timing marks of each gear

18. Remove the oil pump drive sprocket, oil pump drive shaft sprocket and No. 2 chain.

To install:

19. Set the crankshaft key in the left horizontal position.
20. Turn the cutout of the drive shaft so that it faces upward.
21. Align the yellow mark links with the timing marks of each gear as shown in the illustration.
22. Install the sprockets onto the crankshaft and oil pump shaft with the chain wrapped on the gears.
23. Temporarily tighten the oil pump drive shaft sprocket with the nut.
24. Insert the damper spring into the adjusting hole, and then install the chain tensioner plate with the bolt. Tighten to 9 ft. lbs. (12 Nm).
25. Align the adjusting hole on the oil pump drive shaft sprocket with the groove on the oil pump.
26. Insert a 4 mm diameter bar into the adjusting hole on the oil pump drive shaft gear to lock the gear in position, and then tighten the nut. Tighten to 22 ft. lbs. (30 Nm).
27. Install the crankshaft timing gear or sprocket to the crankshaft.
28. Install the No. 1 chain vibration damper with the 2 bolts. Tighten to 80 inch lbs. (9 Nm).
29. Set the No. 1 cylinder to TDC/compression.
30. Turn the camshafts with a wrench (using the hexagonal lobe) to align the timing marks on the camshaft timing gear with the timing marks located on the No. 1 and No. 2 bearing caps
31. Using the crankshaft pulley bolt, turn the crankshaft to position the key on the crankshaft upward.
32. Install the chain onto the crankshaft timing sprocket with the gold or orange mark link aligned with the timing mark on the crankshaft.
33. Using SST 09309-37010 and a hammer, tap in the crankshaft timing sprocket.
34. Align the gold or yellow links with the timing marks located on the camshaft timing gear and sprocket, then install the chain.
35. Install the chain tensioner slipper with the bolt. Tighten to 14 ft. lbs. (19 Nm).
36. Install the timing chain guide with the bolt. Tighten to 80 inch lbs. (9 Nm).
37. Install the sensor plate with the "F" mark facing forward.
38. Install the timing chain cover.
39. Release the ratchet pawl, then fully push in the plunger and set the hook to the pin so that the plunger is in the position.

Fig. 118 Turning the crankshaft to position the key on the crankshaft upward

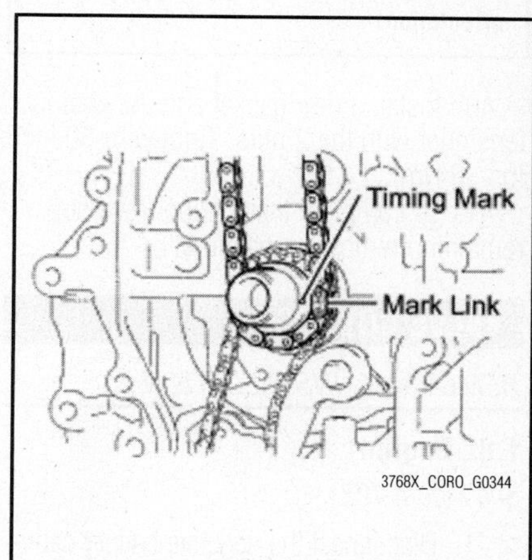

Fig. 119 Install the chain onto the crankshaft timing sprocket

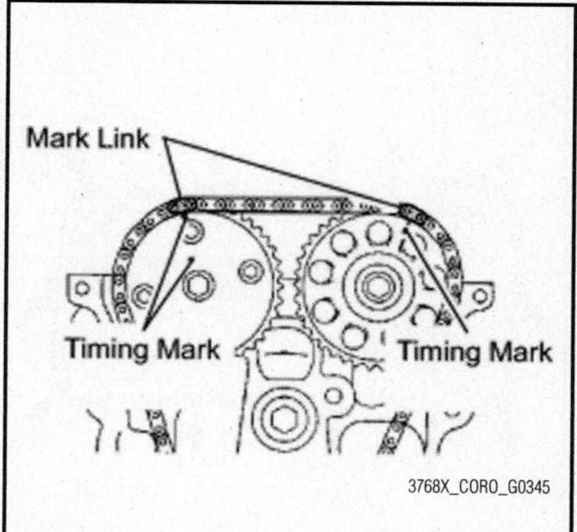

Fig. 120 Aligning the gold or yellow links with the timing marks located on the camshaft timing gear and sprocket

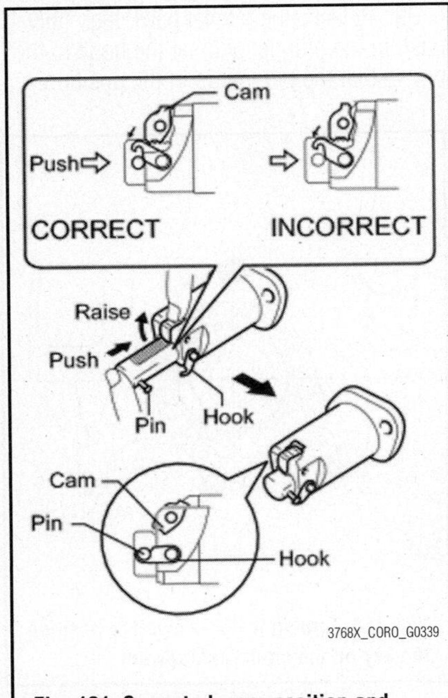

Fig. 121 Correct plunger position and orientation.

40. Install a new gasket and the chain tensioner with the 2 nuts. Tighten to 80 inch lbs. (9 Nm).
41. To complete installation, reverse remaining removal procedure.

VALVE COVERS

REMOVAL & INSTALLATION

1.8L Engine

See Figure 122.

1. Disconnect the negative battery cable.
2. Remove the oil filler cap.
3. Remove the gasket from the oil filler cap.

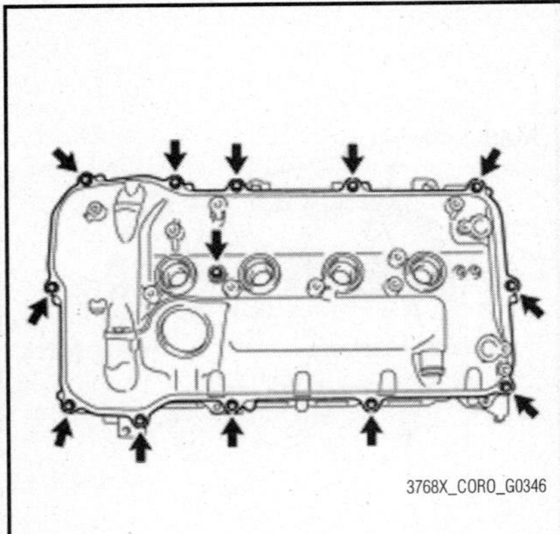

Fig. 122 Removing the valve cover and 13 bolts

4. Remove the 2 engine cover joints.
5. Using a 14 mm spark plug wrench, remove the 4 spark plugs.
6. Remove the 2 bolts and 2 sensors.
7. Remove the 2 bolts, O-rings, bracket and 2 oil control valves.
8. Remove the 13 bolts, seal washer and cylinder head cover.
9. Be careful not to drop any of the gaskets into the engine when removing the cylinder head cover because the gaskets may stick to the cylinder head cover.
10. Remove the cylinder head cover gasket.

To install:

➡Remove any oil from the contact surfaces.

11. Install a new gasket to the cylinder head cover.
12. Install 3 new gaskets to the No. 1 camshaft bearing cap.
13. Apply seal packing as shown the illustration. Use Toyota Genuine Seal Packing Black, Three Bond® 1207B or equivalent.

➡Remove any oil from the contact surfaces. Install the cylinder head cover within 3 minutes and tighten the bolts within 15 minutes after applying seal packing. Do not start the engine for at least 2 hours after the installation.

14. Install the cylinder head cover with a new seal washer and the 13 bolts. Tighten to 7 ft. lbs. (10 Nm).
15. To complete installation, reverse remaining removal procedure.

2.4L Engine

See Figures 123 through 125.

1. Disconnect the negative battery cable.
2. Remove the 2 ventilation hoses from the cylinder head cover sub-assembly.
3. Remove the 2 bolts and separate the 2 wire harness brackets.
4. Remove the ignition coil assembly.
5. Remove the oil filler cap from the cylinder head cover.
6. Remove the oil filler cap gasket from the oil filler cap.
7. Using a 22 mm deep socket wrench, remove the ventilation valve from the cylinder head cover.
8. Remove the spark plugs.
9. Remove the 8 bolts and 2 nuts, and remove the cylinder head cover.
10. Remove cylinder head cover gasket from the cylinder head cover.

To install:

11. Install the cylinder head cover gasket onto the cylinder head cover.

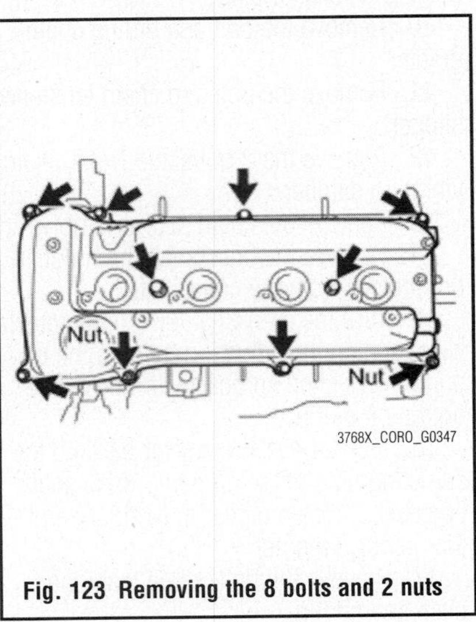

Fig. 123 Removing the 8 bolts and 2 nuts

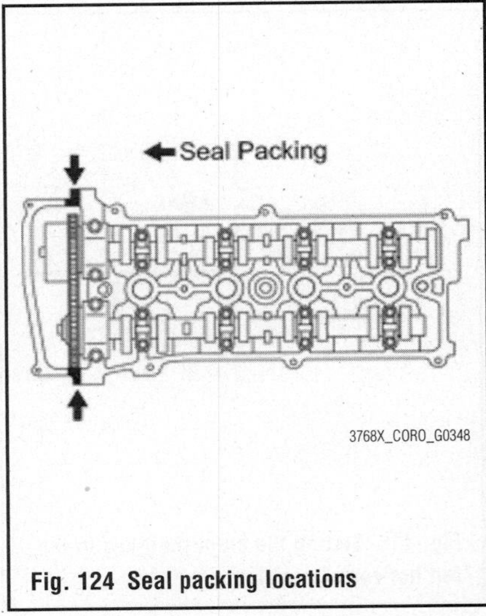

Fig. 124 Seal packing locations

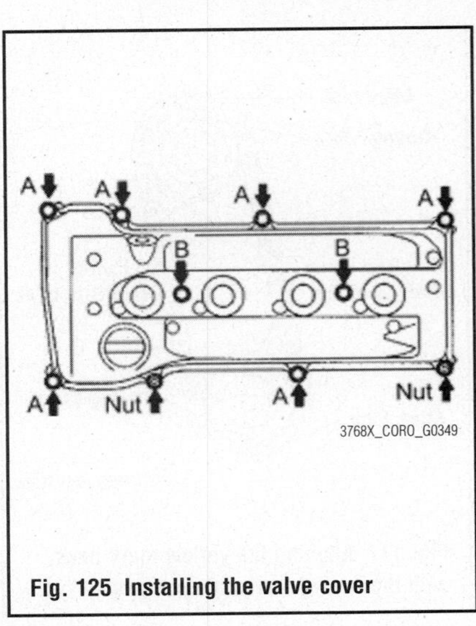

Fig. 125 Installing the valve cover

12. Remove any old packing material from the contact surface.

13. Apply seal packing to the 2 locations shown in the illustration. Use Toyota Genuine Seal Packing Black, Three Bond® 1207B or equivalent.

➡Remove any oil from the contact surfaces. Install the cylinder head cover within 3 minutes of applying seal packing. Do not start the engine for at least 2 hours after installing the cylinder head cover.

14. Install the cylinder head cover with the 8 bolts and 2 nuts.
    a. Bolt A: 8 ft. lbs. (11 Nm).
    b. Bolt B: 10 ft. lbs. (14 Nm).
    c. Bolt C: 8 ft. lbs. (11 Nm).

15. To complete installation, reverse remaining removal procedure.

ENGINE PERFORMANCE & EMISSION CONTROLS

COMPONENT LOCATIONS

See Figures 126 through 128.

CAMSHAFT POSITION (CMP) SENSOR

LOCATION

See Figures 129 and 130.

REMOVAL & INSTALLATION

1.8L Engine

See Figures 131 through 133

➡Although the part name refers to the No. 1 crank position sensors, this procedure is for the Camshaft Position (CMP) sensors.

1. Disconnect the negative battery cable.
2. Exhaust camshaft side:
    a. Disconnect the duty vacuum switching valve connector and 3 engine wire harness clamps.
    b. Disconnect the No. 1 crank position sensor connector.
    c. Remove the bolt and No. 1 crank position sensor.
3. Intake camshaft side:
    a. Disconnect the No. 1 crank position sensor connector.
    b. Remove the bolt and No. 1 crank position sensor.

To install:

4. To install, reverse removal procedure.
5. Apply a light coat of engine oil to the O-rings on the No. 1 crank position sensors.
6. Tighten the CMP sensor bolts to 7 ft. lbs. (10 Nm).
7. Inspect for oil leaks.

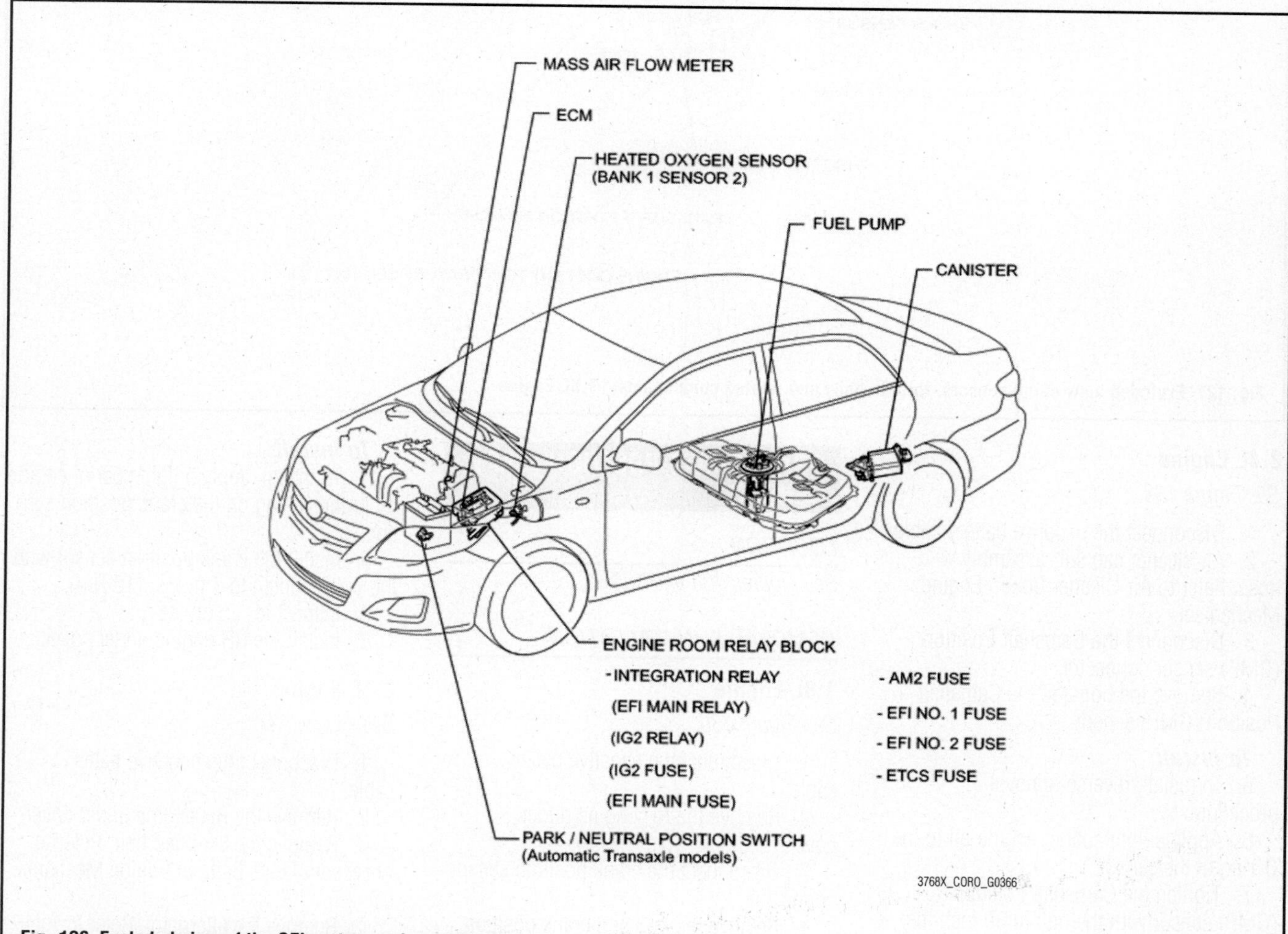

Fig. 126 Exploded view of the SFI system and components

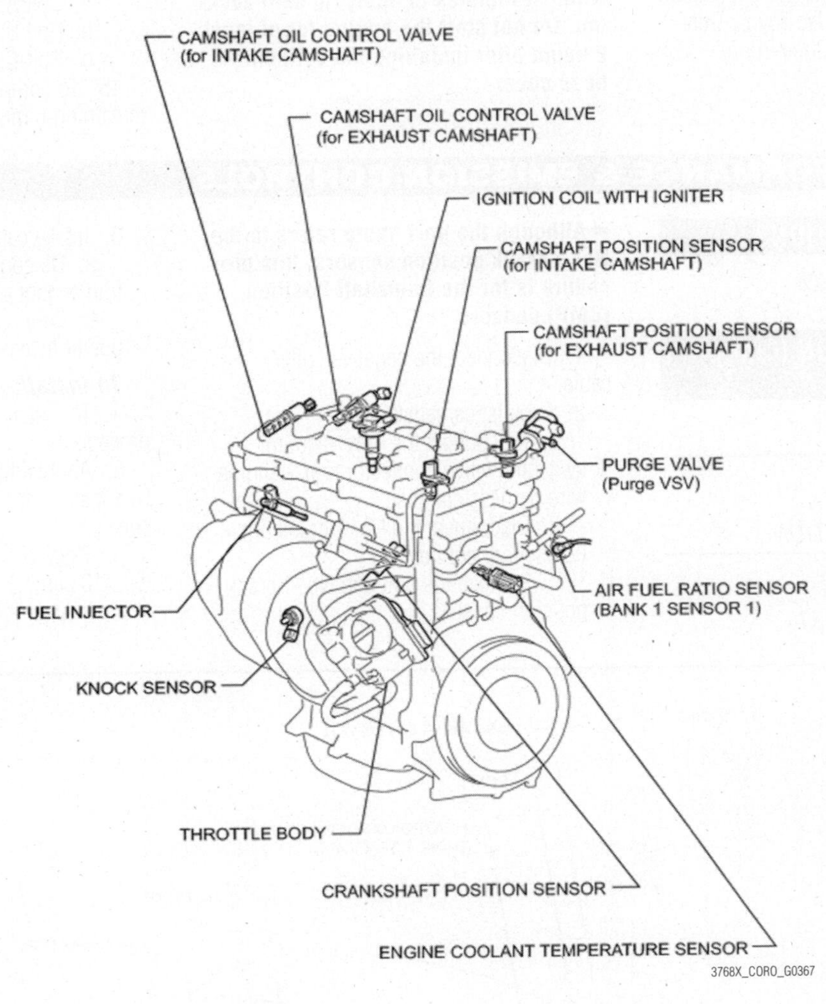

Fig. 127 Exploded view of the sensors, throttle body and related components—1.8L Engine

2.4L Engine

See Figure 133.

1. Disconnect the negative battery cable.
2. Air cleaner cap sub-assembly with hose. Refer to Air Cleaner Box in Engine Mechanical.
3. Disconnect the Camshaft Position (CMP) sensor connector.
4. Remove the bolt and the Camshaft Position (CMP) sensor.

To install:

5. To install, reverse removal procedure.
6. Apply a light coat of engine oil to the O-ring of the sensor.
7. Tighten the Camshaft Position (CMP) sensor with the bolt to 80 inch lbs. (9 Nm).

CRANKSHAFT POSITION (CKP) SENSOR

LOCATION

See Figures 134 and 135.

REMOVAL & INSTALLATION

1.8L Engine

See Figure 136.

1. Disconnect the negative battery cable.
2. Remove the RH engine under cover.
3. Disconnect the crank position sensor connector.
4. Remove the bolt and crank position sensor.

To install:

5. To install, apply a light coat of engine oil to the O-ring on the crank position sensor.
6. Install the crank position sensor with the bolt. Tighten to 7 ft. lbs. (10 Nm).
7. Inspect for an oil leak.
8. Install the RH engine under cover.

2.4L Engine

See Figure 137.

1. Disconnect the negative battery cable.
2. Remove the RH engine under cover.
3. Remove the v-ribbed belt. Refer to Accessory Drive Belts in Engine Mechanical.
4. Remove the alternator. Refer to Alternator in Engine Electrical.

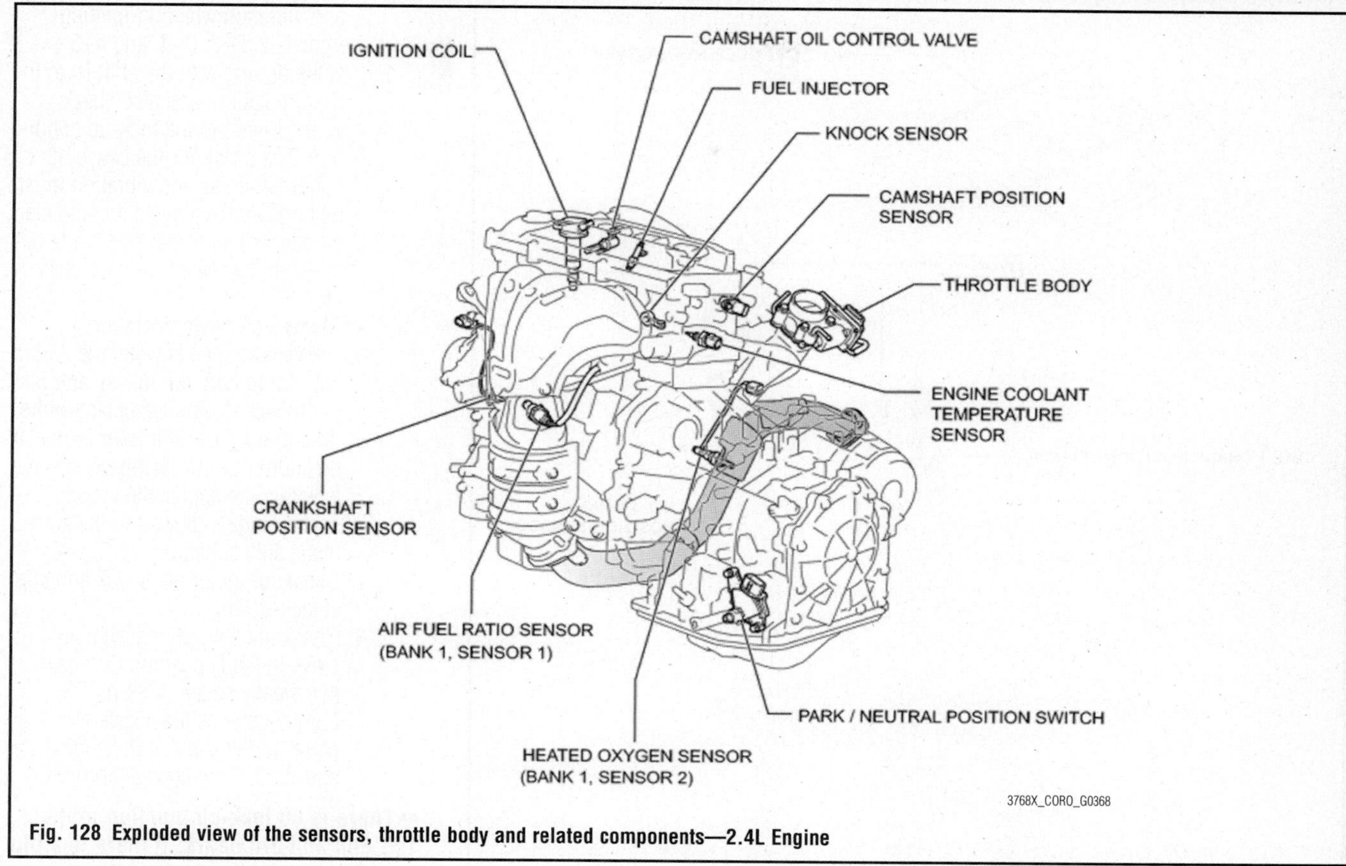

Fig. 128 Exploded view of the sensors, throttle body and related components—2.4L Engine

5. Disconnect the crank position sensor connector.
6. Separate the crank position sensor connector clamp and wire harness.
7. Remove the bolt and crank position sensor.

To install:

To install, reverse the removal procedure.

ELECTRONIC CONTROL MODULE (ECM)

LOCATION

See Figures 138 and 139.

REMOVAL & INSTALLATION

See Figure 140.

1. Disconnect the negative battery cable.
2. Remove the air cleaner assembly. Refer to Air Cleaner Assembly in Engine Mechanical.
3. Separate the wire harness clamp.
4. Disconnect the 2 ECM connectors. Push the locks on the 2 levers, then raise the levers, and disconnect the 2 ECM connectors.

➡After disconnecting the connectors, make sure that dirt, water or other foreign matter does not contact the connecting parts of the connectors.

5. Remove the 3 bolts and the ECM.
6. Remove the 4 screws and 2 ECM brackets.

To install:

7. To install, reverse removal procedure.
8. Perform the VIN registration procedure and/or Automatic Transaxle Initialization.

RESET AND REGISTRATION

Automatic Transaxle Initialization

➡The ECM memorizes the condition that the ECT controls the automatic transaxle assembly and engine assembly according to those characteristics. Therefore, when the automatic transaxle assembly, engine assembly, or ECM has been replaced, it is necessary to reset the memory so that the ECM can memorize the new information.

1. Turn the ignition switch off.
2. Connect the Techstream to the DLC3.
3. Turn the ignition switch to ON and the Techstream main switch on.
4. Enter the following menus: Powertrain / Engine and ECT / Utility / Reset Memory. Then, press "Next".

⁂ CAUTION

After performing the reset memory, be sure to perform the road test.

5. The ECM is learned by performing the road test.

VIN Registration

➡The Vehicle Identification Number (VIN) must be input into a replacement ECM.

➡The VIN is a 17-digit alphanumeric vehicle identification number. The Techstream is required to register the VIN.

➡This registration section consists of two parts: Read VIN and Write VIN.

1. Read VIN: This process allows the VIN stored in the ECM to be read in order to confirm that the two VINs, provided with the vehicle and stored in the vehicle's ECM, are the same.
2. Write VIN: This process allows the VIN to be input into the ECM. If the ECM is changed, or the ECM VIN and vehicle VIN

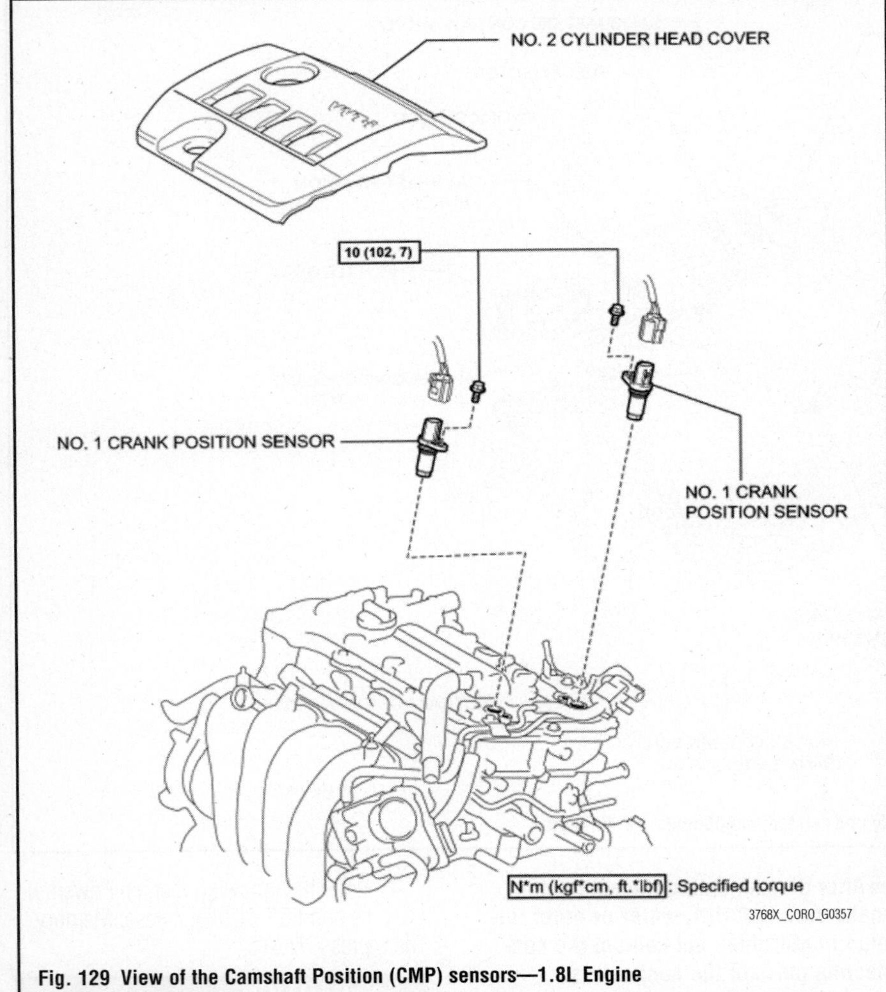

Fig. 129 View of the Camshaft Position (CMP) sensors—1.8L Engine

do not match, the VIN can be registered, or overwritten in the ECM by following this procedure.

3. Read the VIN:
   a. Confirm the vehicle VIN.
   b. Connect the Techstream to the DLC3.
   c. Turn the ignition switch to ON.
   d. Turn the Techstream on.
   e. Enter the following menus: Powertrain / Engine and ECT / Utility / VIN / VIN Read.
4. Write the VIN:
   a. Confirm the vehicle VIN.
   b. Connect the Techstream to the DLC3.
   c. Turn the ignition switch to ON.
   d. Turn the Techstream on.
   e. Enter the following menus: Powertrain / Engine and ECT / Utility / VIN / VIN Write.

ROAD TEST

1. Perform the test at the ATF (Automatic Transmission Fluid) temperature 50 to 80 °C (122 to 176 °F) in the normal operation.

a. D position test: Shift into the D position and fully depress the accelerator pedal and check the following points.

- Check up-shift operation.

Check that 1–2, 2–3, 3–4 and 4–5th up-shifts take place, and that the shift points conform to the automatic shift schedule.

➡5th Gear Up-shift Prohibition Control: Engine coolant temperature is 55°C (131°F) or less and vehicle speed is at 70km/h (43 mph) or less. ATF temperature is -2°C (28°F) or less.

➡5th and 4th Gear Lock-up Prohibition Control: Brake pedal is depressed. Accelerator pedal is released. Engine coolant temperature is 60°C (140°F) or less.

- Check for shift shock and slip. Check for shock and slip at the 1–2, 2–3, 3–4, and 4–5th up-shifts.
- Check for abnormal noise and vibration. Check for abnormal noise and vibration when up-shifting from 1–2, 2–3, 3–4, and 4–5 while driving with the shift lever in the D position, and also check while driving in the lock-up condition. The check for the cause of abnormal noise and vibration must be done thoroughly as it could also be due to loss of balance in the differential, torque converter clutch, etc.
- Check kick-down operation.

Check vehicle speeds when the 2nd to 1st, 3rd to 2nd, 4th to 3rd, and 5th to 4th kick-downs take place while driving with the shift lever in the D position. Confirm that each speed is within the applicable vehicle speed range indicated in the automatic shift schedule.

- Check for abnormal shock and slip at kick-down.
- Check the lock-up mechanism: Drive in the D position (5th gear), at a steady speed (lock-up ON). Lightly depress the accelerator pedal and check that the engine speed does not change abruptly.

➡There is no lock-up function in the 1st, 2nd and 3rd gears. If there is a big jump in engine speed, there is no lock-up.

b. S position test: Shift to the S position, depress the accelerator pedal and check the following points:

- Check shift operation.
- While driving in the D position and 5th gear, shift into the S position and back to the D position. Check that the gear change 5–4 down-shift and 4–5 up-shift can be performed.
- With the shift lever in the S position (while the vehicle is stopped), shift into the "+" position to check that the shift position on the combination meter changes as follows: 1–2, 2–3, 3–4, and 4–5.
- While driving in the 4(S) position and 4th gear (at a vehicle speed of approximately 40 to 50 km/h (25 to 31mph)), shift into the "-" position and check if the 3rd gear down-shift occurs and the engine brake performs properly.
- While driving in the 3(S) position and 3rd gear (at a vehicle speed of approximately 30 to 40 km/h (19 to 25 mph)), shift into the "-" position and check if the 2nd gear down-shift occurs and the engine brake performs properly.

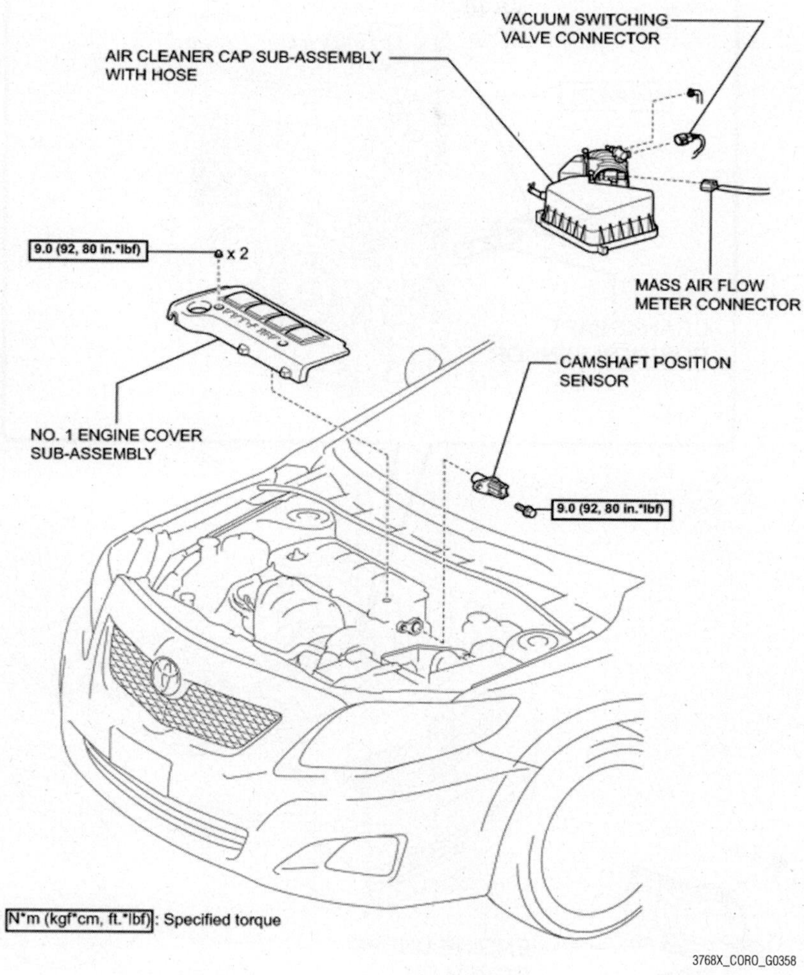

VACUUM SWITCHING VALVE CONNECTOR
AIR CLEANER CAP SUB-ASSEMBLY WITH HOSE
9.0 (92, 80 in.*lbf)
x 2
MASS AIR FLOW METER CONNECTOR
CAMSHAFT POSITION SENSOR
NO. 1 ENGINE COVER SUB-ASSEMBLY
9.0 (92, 80 in.*lbf)
N*m (kgf*cm, ft.*lbf): Specified torque
3768X_CORO_G0358

Fig. 130 View of the Camshaft Position (CMP) sensors—2.4L Engine

3768X_CORO_G0359

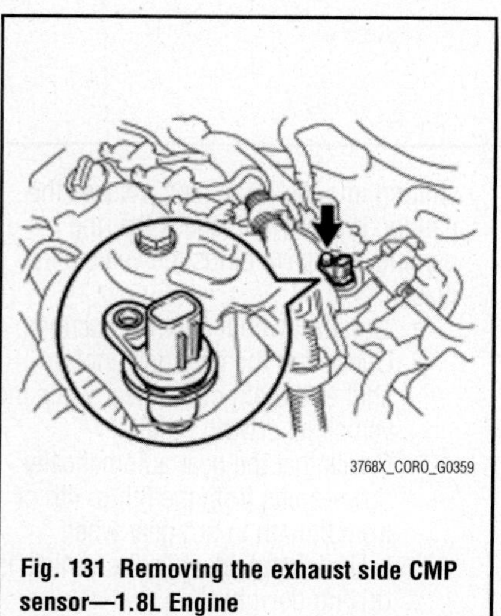

Fig. 131 Removing the exhaust side CMP sensor—1.8L Engine

3768X_CORO_G0360

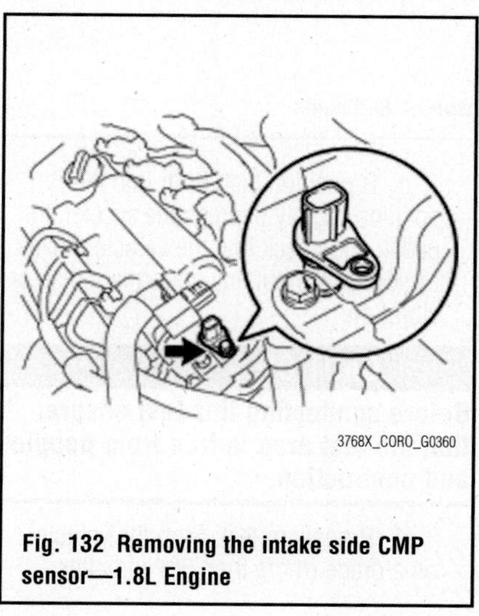

Fig. 132 Removing the intake side CMP sensor—1.8L Engine

3768X_CORO_G0361

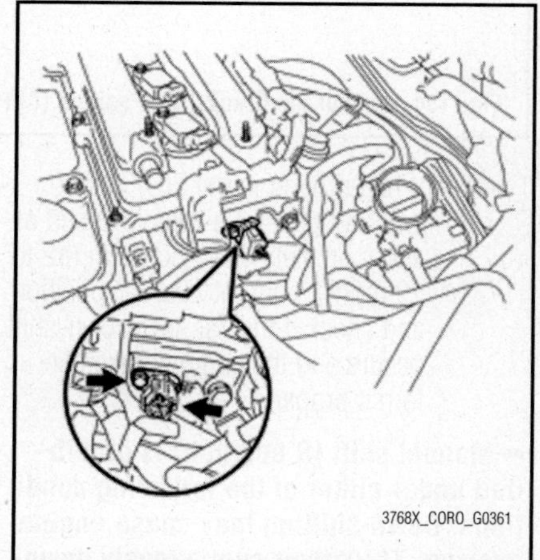

Fig. 133 Removing the Camshaft Position (CMP) sensor—2.4L Engine

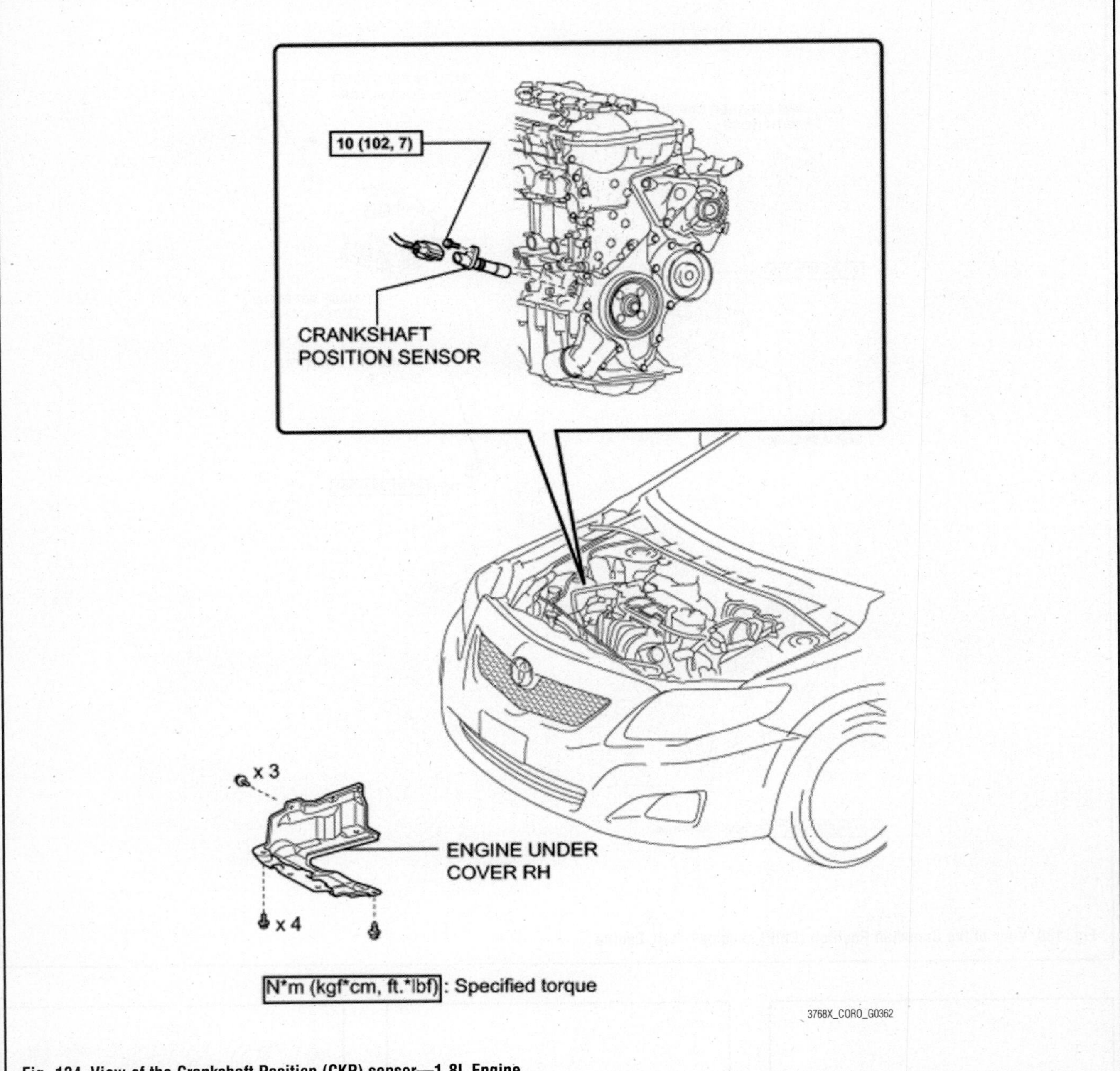

Fig. 134 View of the Crankshaft Position (CKP) sensor—1.8L Engine

- While driving in the 2(S) position and 2nd gear (at a vehicle speed of approximately 20 to 30 km/h (12 to 19 mph)), shift into the "-" position and check if the 1st gear down-shift occurs and the engine brake performs properly.

➡Manual shift (S position) is prohibited under either of the following conditions: Down-shifting may cause engine overrun. The driver continuously down-shifts. (Down-shifting to 1st gear may not be performed.)

c. R position test: Shift into the R position, lightly depress the accelerator pedal, and check that the vehicle moves backward without any abnormal noise or vibration.

CAUTION

Before conducting this test ensure that the test area is free from people and obstruction.

d. P position test: Stop the vehicle on a grade (more than 5°) and after shifting into the P position, release the parking brake. Then, check that the parking lock pawl holds the vehicle in place.

e. Uphill/downhill control function:

- Check that the gear does not up-shift to the 4th or 5th gear while the vehicle is driving uphill.
- Check that the gear automatically down-shifts from the 5th to 4th or from the 4th to 3rd gear when brake is applied while the vehicle is driving downhill.

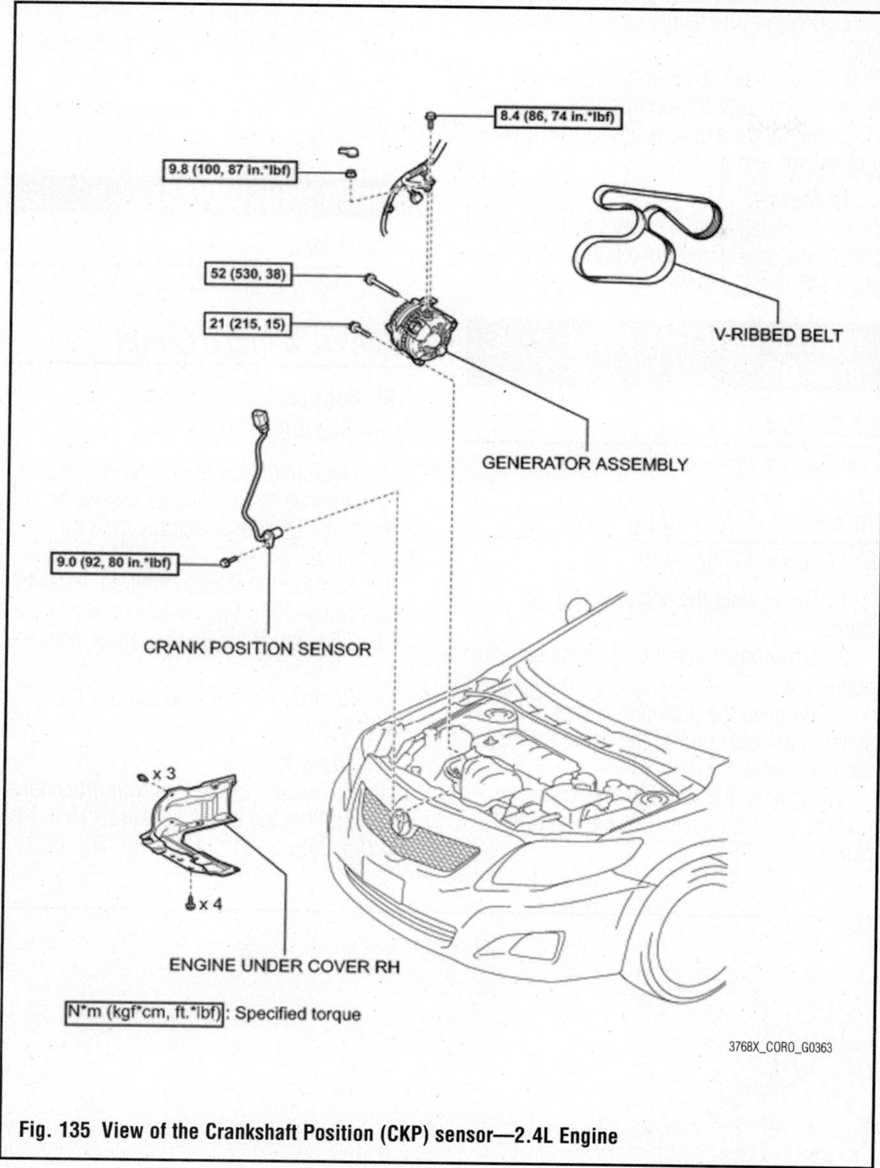

Fig. 135 View of the Crankshaft Position (CKP) sensor—2.4L Engine

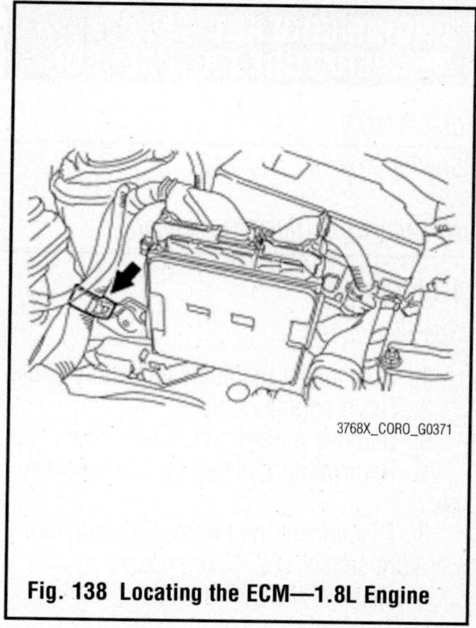

Fig. 138 Locating the ECM—1.8L Engine

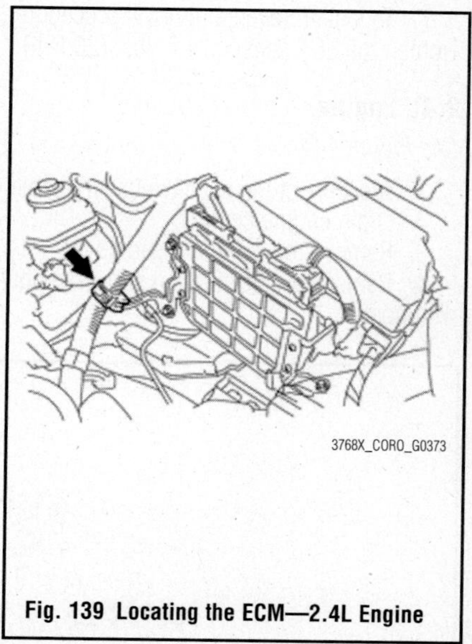

Fig. 139 Locating the ECM—2.4L Engine

Fig. 136 Removing the CKP sensor—1.8L Engine

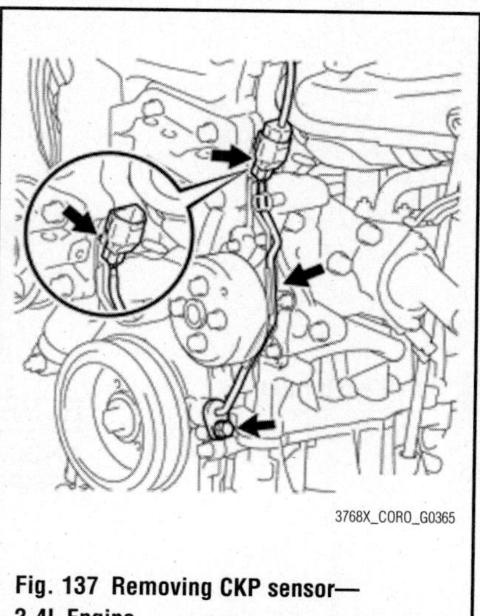

Fig. 137 Removing CKP sensor—2.4L Engine

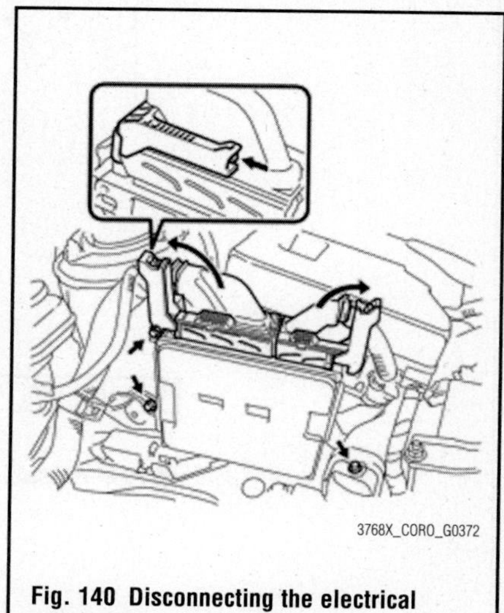

Fig. 140 Disconnecting the electrical connectors

ENGINE COOLANT TEMPERATURE (ECT) SENSOR

LOCATION

See Figures 141 and 142.

REMOVAL & INSTALLATION

1.8L Engine

See Figure 143.

1. Disconnect the negative battery cable.
2. Drain engine coolant.
3. Remove the cylinder head cover.
4. Remove air cleaner cap sub-assembly.
5. Disconnect the Engine Coolant Temperature sensor (ECT) connector.
6. Remove the ECT sensor.

To install:

7. To install, reverse removal procedure. Tighten the ECT sensor 14 ft. lbs. (20 Nm).

2.4L Engine

See Figure 144.

1. Disconnect the negative battery cable.
2. Drain engine coolant.
3. Remove the cylinder head cover.
4. Remove air cleaner cap sub-assembly with hose.
5. Remove air cleaner case sub-assembly.
6. Disconnect the Engine Coolant Temperature sensor (ECT) connector.
7. Remove the engine coolant temperature sensor and gasket.

To install:

8. To install, reverse removal procedure. Use a new gasket and tighten the ECT sensor 14 ft. lbs. (20 Nm).

HEATED OXYGEN (HO2S) SENSOR

LOCATION

See Figures 145 and 146.

REMOVAL & INSTALLATION

See Figures 147 and 148.

1. Disconnect the negative battery cable.
2. Disconnect the heated oxygen sensor connector.
3. Remove the grommet and pull the sensor connector out of the cabin through the floor panel.
4. Remove the wire harness clamp bracket and disconnect the wire harness clamp.
5. Using a wrench or SST: 09224-00010, remove the heated oxygen sensor.

To install:

6. To install, reverse removal procedure.
7. Tighten to 32 ft. lbs. (44 Nm).

KNOCK SENSOR (KS)

LOCATION

See Figures 149 and 150.

REMOVAL & INSTALLATION

1.8L Engine

See Figure 149.

1. Drain and recycle the engine coolant.
2. Remove the air cleaner assembly. Refer to Air Cleaner Assembly in Engine Mechanical.
3. Remove the intake manifold. Refer to Intake Manifold in Engine Mechanical.
4. Disconnect the knock sensor connector.
5. Remove the bolt and remove the knock sensor.

To install:

6. To install, reverse removal procedure.
7. Tighten the knock sensor to 15 ft. lbs. (20 Nm) with +/- 20°.

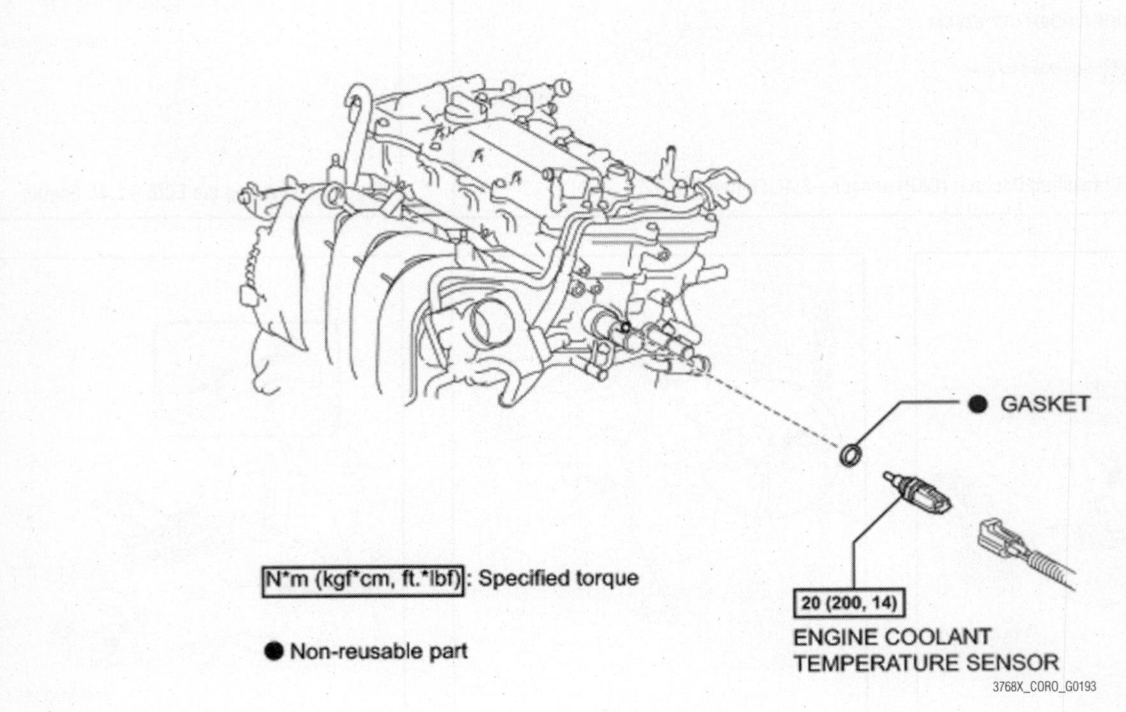

Fig. 141 Locating the Engine Coolant Temperature (ECT) sensor—1.8L Engine

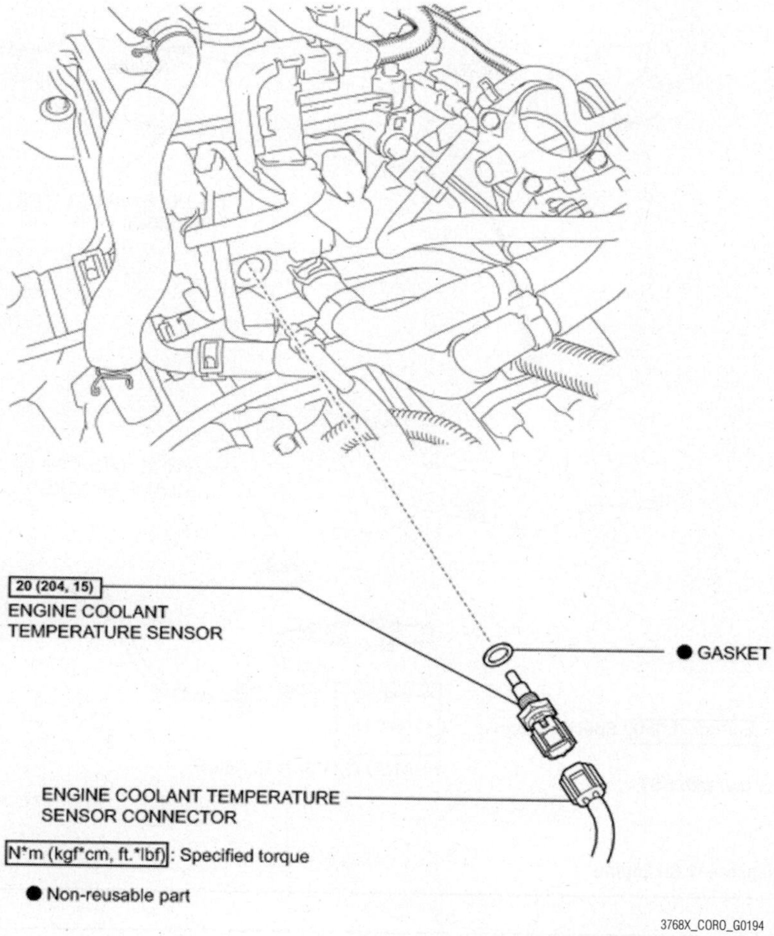

20 (204, 15)
ENGINE COOLANT TEMPERATURE SENSOR
GASKET
ENGINE COOLANT TEMPERATURE SENSOR CONNECTOR
N*m (kgf*cm, ft.*lbf) : Specified torque
Non-reusable part
3768X_CORO_G0194

Fig. 142 Locating the Engine Coolant Temperature (ECT) sensor—2.4L Engine

3768X_CORO_G0195

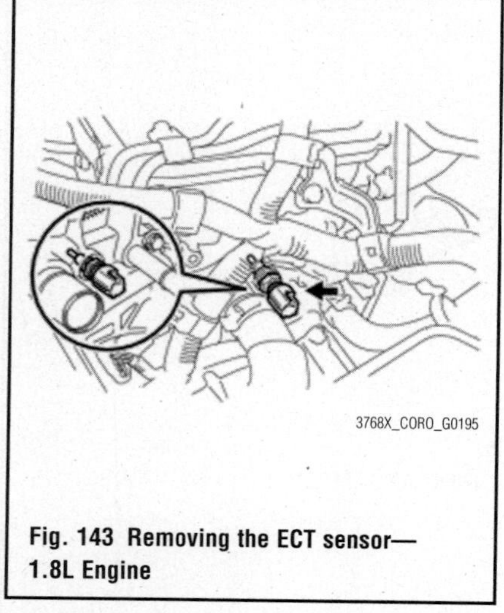

Fig. 143 Removing the ECT sensor—1.8L Engine

Gasket
3768X_CORO_G0196

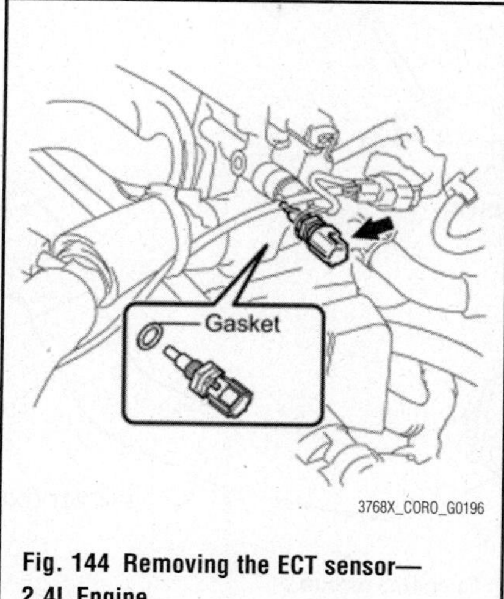

Fig. 144 Removing the ECT sensor—2.4L Engine

2.4L Engine

See Figure 150.

1. Discharge fuel system pressure.
2. Drain and recycle the engine coolant.
3. Remove the wiper motor assembly.
4. Remove suspension tower damper assembly (with front strut bar). Refer to Suspension Tower Damper Assembly in Front Suspension.
5. Remove the air cleaner assembly. Refer to Air Cleaner Assembly in Engine Mechanical.
6. Remove the throttle body. Refer to Throttle Body in this section.
7. Disconnect fuel tube sub-assembly.
8. Remove fuel delivery pipe sub-assembly with fuel tube sub-assembly.
9. Remove the intake manifold. Refer to Intake Manifold in Engine Mechanical.
10. Disconnect the sensor connector.
11. Remove the nut and knock sensor.

To install:

12. To install, reverse removal procedure.
13. Tighten the knock sensor to 15 ft. lbs. (20 Nm) with +/- 10°.

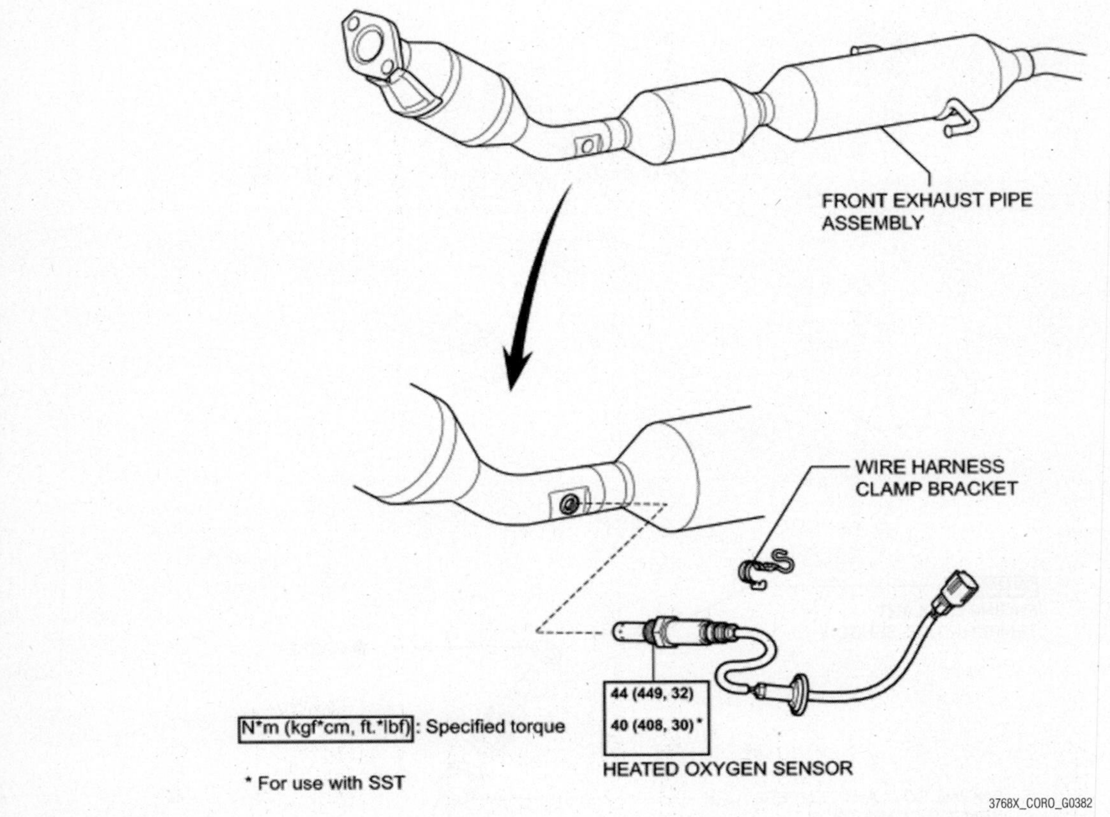

FRONT EXHAUST PIPE ASSEMBLY
WIRE HARNESS CLAMP BRACKET
44 (449, 32)
40 (408, 30)*
N*m (kgf*cm, ft.*lbf): Specified torque
* For use with SST
HEATED OXYGEN SENSOR
3768X_CORO_G0382

Fig. 145 Locating the HO2S sensor—1.8L Engine

44 (449, 32)
40 (408, 30)*
HEATED OXYGEN SENSOR
FRONT EXHAUST PIPE ASSEMBLY
N*m (kgf*cm, ft.*lbf): Specified torque
* For use with SST
3768X_CORO_G0383

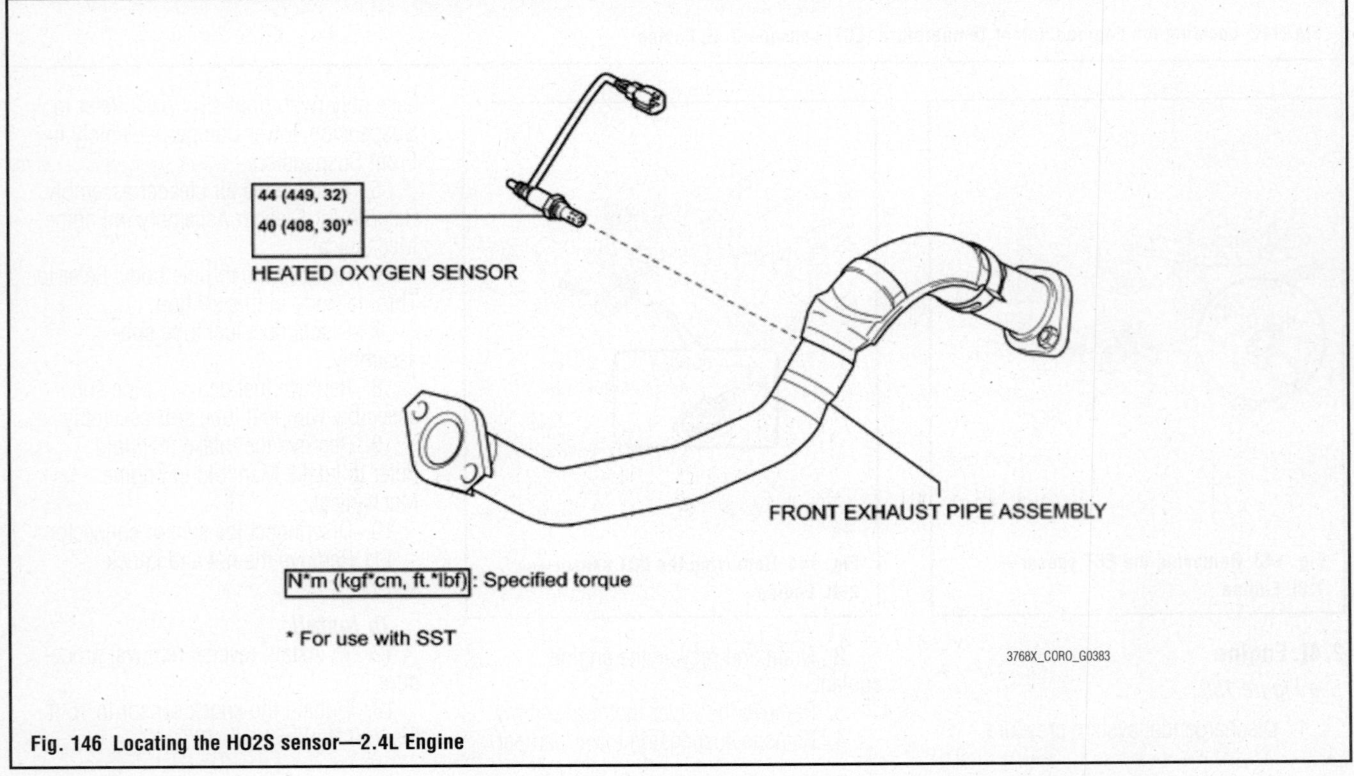

Fig. 146 Locating the HO2S sensor—2.4L Engine

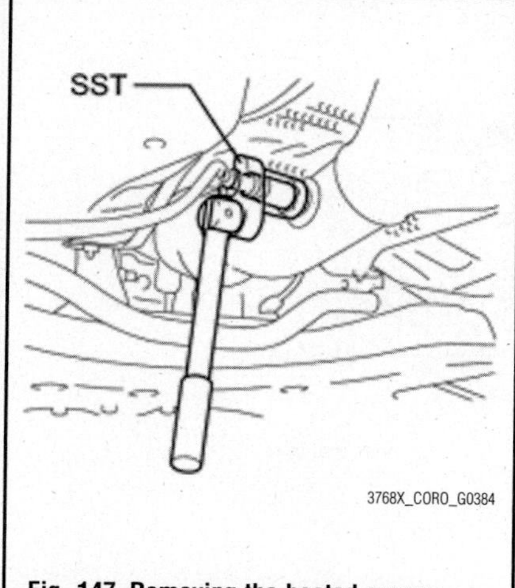

Fig. 147 Removing the heated oxygen sensor—1.8L Engine

Fig. 150 Locating the knock sensor—2.4L Engine

MASS AIR FLOW (MAF) METER

LOCATION

See Figures 151 and 152.

REMOVAL & INSTALLATION

See Figures 151 and 152.

1. Disconnect the negative battery cable.
2. Disconnect the mass air flow meter connector.
3. Remove the 2 screws and the mass air flow meter.

➡Make sure that the O-ring is not cracked or does not jump out of position during installation.

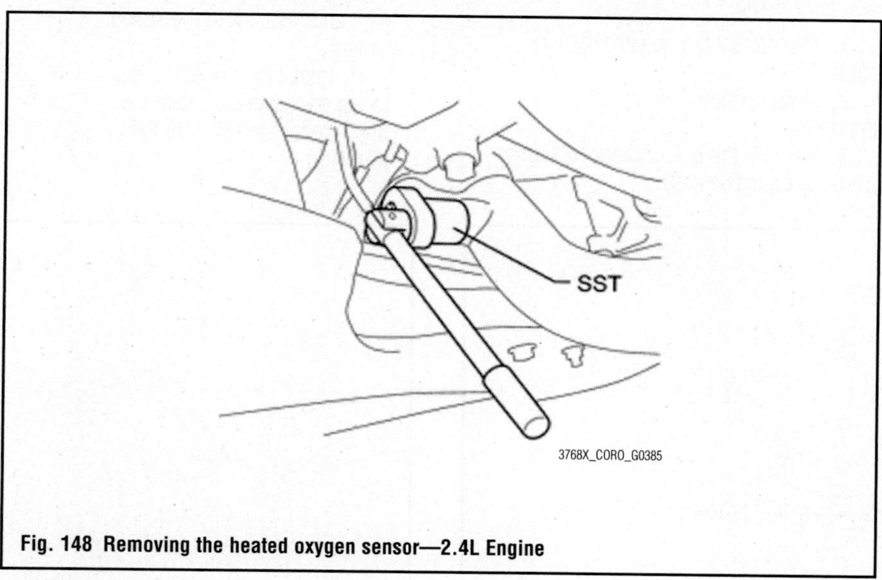

Fig. 148 Removing the heated oxygen sensor—2.4L Engine

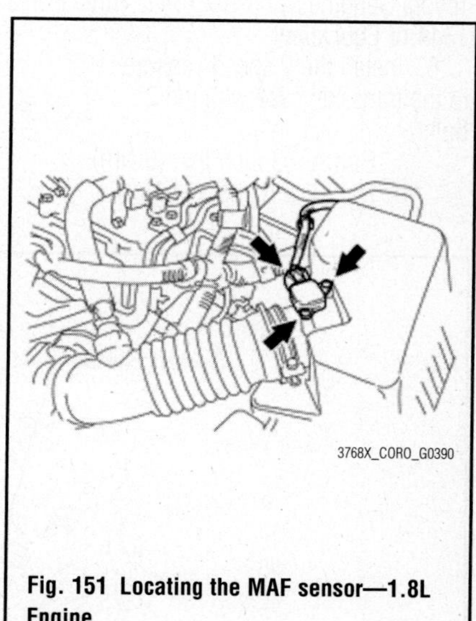

Fig. 151 Locating the MAF sensor—1.8L Engine

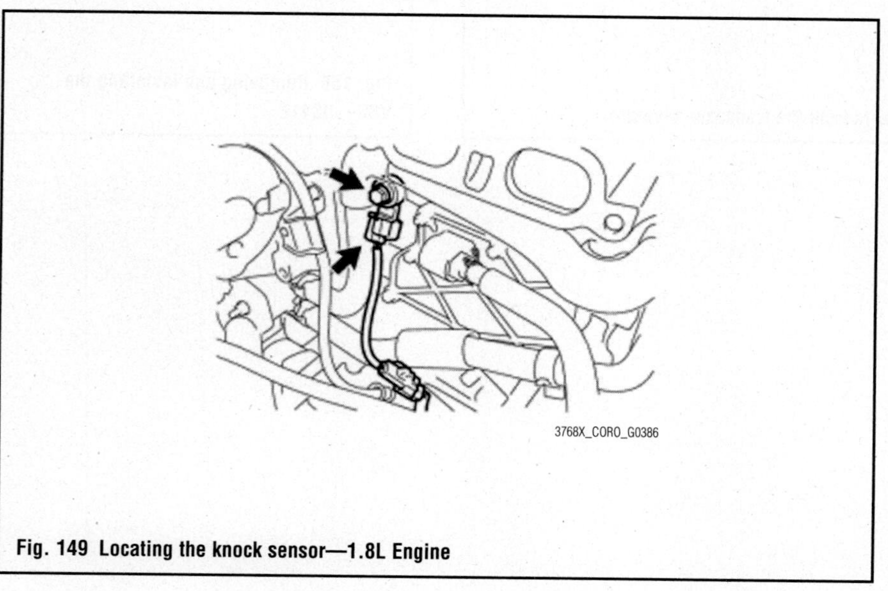

Fig. 149 Locating the knock sensor—1.8L Engine

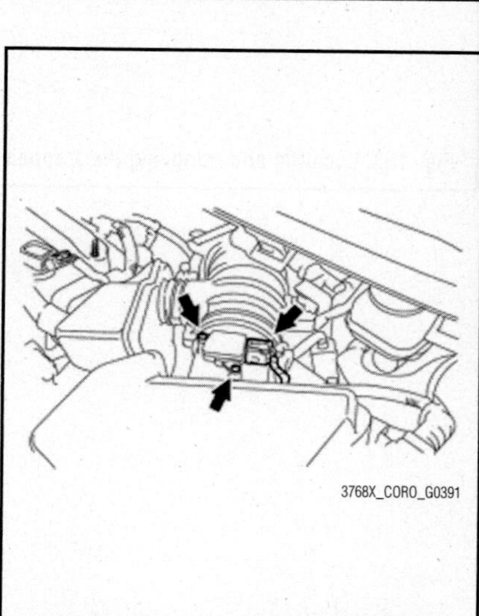

Fig. 152 Locating the MAF sensor—2.4L Engine

To install:

4. To install, reverse removal procedure.

VEHICLE SPEED SENSOR (VSS)

REMOVAL & INSTALLATION

U250e Automatic Transaxle

See Figures 153 and 154.

1. Disconnect the negative battery cable.
2. Remove the 2 bolts and the 2 speed sensors from the transaxle assembly.
3. Remove the 2 O-rings from the 2 speed sensors.

To install:

4. Coat 2 new O-rings with ATF, and install them to the 2 speed sensors.
5. Apply adhesive to the bolts threads. Toyota Genuine Adhesive 1344, Three Bond 1344 or Equivalent.
6. Install the 2 speed sensors to the transaxle case with the 2 bolts:
   a. Bolt A: 78 inch lbs. (9 Nm).
   b. Bolt B: 8 ft. lbs. (11 Nm).

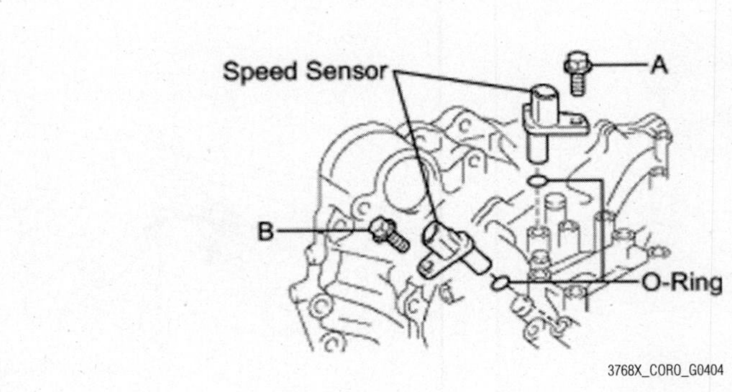

Fig. 154 Installing the VSS fasteners—U250E

U341e Automatic Transaxle

See Figure 155.

1. Disconnect the negative battery cable.
2. Disconnect the electrical connector
3. Remove the bolt and speed sensor from the transaxle case.

To install:

a. Coat a new O-ring with ATF, and install it to the speed sensor.
b. Install the speed sensor to the transaxle case with the bolt. Tighten to 48 inch. lbs. (5.4 Nm).

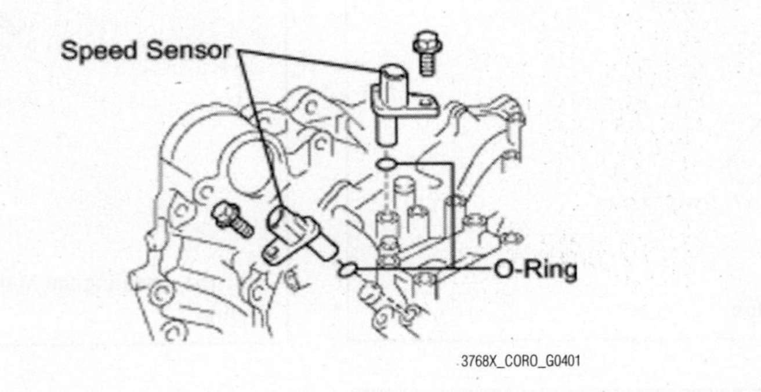

Fig. 153 Locating and removing the 2 speed sensors from the transaxle assembly

Fig. 155 Removing and installing the VSS—U341E

FUEL SYSTEM SERVICE PRECAUTIONS

Safety is the most important factor when performing not only fuel system maintenance but any type of maintenance. Failure to conduct maintenance and repairs in a safe manner may result in serious personal injury or death. Maintenance and testing of the vehicle's fuel system components can be accomplished safely and effectively by adhering to the following rules and guidelines.

- To avoid the possibility of fire and personal injury, always disconnect the negative battery cable unless the repair or test procedure requires that battery voltage be applied.
- Always relieve the fuel system pressure prior to disconnecting any fuel system component (injector, fuel rail, pressure regulator, etc.), fitting or fuel line connection. Exercise extreme caution whenever relieving fuel system pressure to avoid exposing skin, face and eyes to fuel spray. Please be advised that fuel under pressure may penetrate the skin or any part of the body that it contacts.
- Always place a shop towel or cloth around the fitting or connection prior to loosening to absorb any excess fuel due to spillage. Ensure that all fuel spillage (should it occur) is quickly removed from engine surfaces. Ensure that all fuel soaked cloths or towels are deposited into a suitable waste container.
- Always keep a dry chemical (Class B) fire extinguisher near the work area.
- Do not allow fuel spray or fuel vapors to come into contact with a spark or open flame.
- Always use a back-up wrench when loosening and tightening fuel line connection fittings. This will prevent unnecessary stress and torsion to fuel line piping.
- Always replace worn fuel fitting O-rings with new Do not substitute fuel hose or equivalent where fuel pipe is installed.

Before servicing the vehicle, make sure to also refer to the precautions in the beginning of this section as well.

RELIEVING FUEL SYSTEM PRESSURE

✲✲ CAUTION

Observe all applicable safety precautions when working around fuel. Whenever servicing the fuel system, always work in a well-ventilated area. Do not allow fuel spray or vapors to come in contact with a spark or open flame. Keep a dry chemical fire extinguisher near the work area. Always keep fuel in a container specifically designed for fuel storage; also, always properly seal fuel containers to avoid the possibility of fire or explosion.

✲✲ CAUTION

Perform the following procedure to prevent fuel from spilling out before removing any fuel system parts. Pressure will still remain in the fuel lines even after performing the following procedure. When disconnecting a fuel line, cover it with a piece of cloth to prevent fuel from spraying or coming out.

1. Remove the rear seat cushion assembly. Refer to Seats in Body Interior.
2. Remove the rear floor service hole cover.
3. Disconnect the connector from the fuel pump assembly.
4. Start the engine. After the engine stops naturally, turn the ignition switch off.

➡Do not increase engine speed or drive the vehicle while waiting for the engine to stop naturally.

➡DTC P0171/25 (system too lean) may be set.

5. Crank the engine again and make sure that the engine does not start.
6. Remove the fuel tank cap and discharge the pressure from the fuel tank.
7. Disconnect the cable from the negative (-) battery terminal.
8. Connect the connector of the fuel pump assembly.
9. Install the rear floor service hole cover.
10. Install the rear seat cushion assembly.

FUEL FILTER

REMOVAL & INSTALLATION

The fuel filter is integrated in the fuel pump and must be replaced as a unit. Refer to Fuel Pump in this section.

FUEL INJECTORS

REMOVAL & INSTALLATION

1.8L Engine

See Figures 156 through 159.

✲✲ CAUTION

Observe all applicable safety precautions when working around fuel. Whenever servicing the fuel system, always work in a well-ventilated area. Do not allow fuel spray or vapors to come in contact with a spark or open flame. Keep a dry chemical fire extinguisher near the work area. Always keep fuel in a container specifically designed for fuel storage; also, always properly seal fuel containers to avoid the possibility of fire or explosion.

1. Properly relieve the fuel system pressure.
2. Disconnect the negative battery cable.
3. Disconnect the No. 2 ventilation hose.
4. Remove the 2 bolts and disconnect the ground wires.
5. Disconnect the 4 fuel injector assembly connectors.
6. Disconnect the 2 wire harness clamps.
7. Disconnect the 4 wire harness clamps.
8. Remove the 2 bolts and 2 wire harness brackets.
9. Remove the No. 2 fuel pipe clamp (Type A).

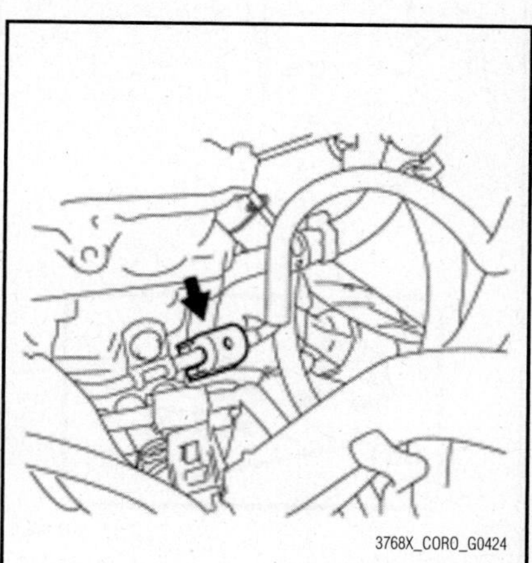

Fig. 156 Removing the No. 2 fuel pipe clamp (Type A)

10. Remove the No. 2 fuel pipe clamp (Type B).

11. Using SST 09268-21010, disconnect the fuel tube sub-assembly.

12. Remove the fuel delivery pipe sub-assembly:

a. Remove the bolt and wire harness bracket.

b. Remove the 2 bolts.

c. Remove the bolt and the fuel delivery pipe sub-assembly.

d. Remove the 2 No. 1 delivery pipe spacers.

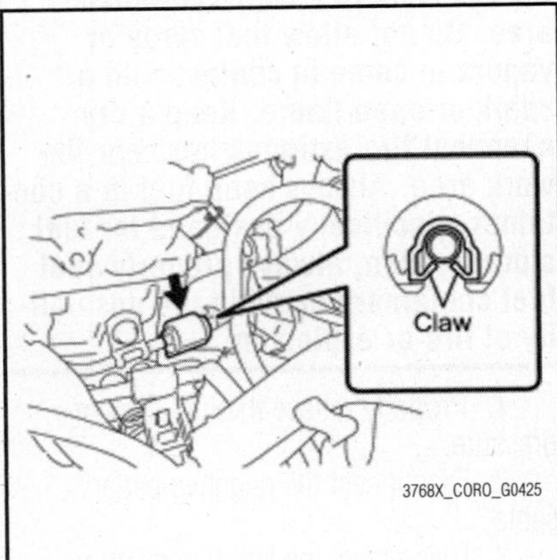

Fig. 157 Removing the No. 2 fuel pipe clamp (Type B)

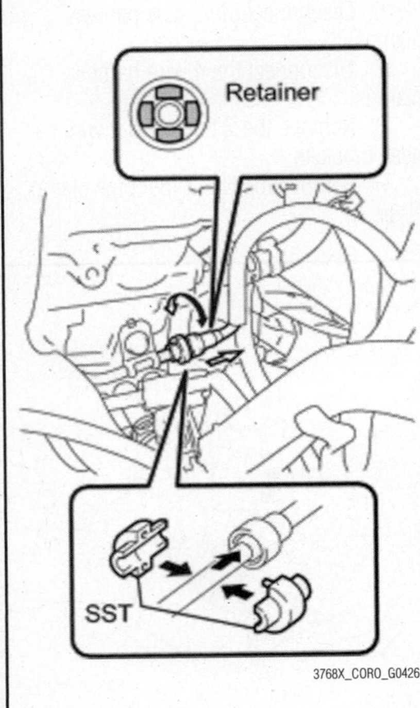

Fig. 158 Using SST 09268-21010, disconnect the fuel tube sub-assembly

13. Pull the 4 fuel injector assemblies out of the fuel delivery pipe sub-assembly.

14. For reinstallation, attach a tag or label to the injector shaft.

➡Prevent entry of foreign objects by covering the fuel injector with a plastic bag.

15. Remove the O-rings from the fuel injector assemblies.

16. Remove the 4 injector vibration insulators.

To install:

17. Install 4 new injector vibration insulators to the 4 fuel injector assemblies.

18. Apply a light coat of gasoline or spindle oil to the contact surfaces of the O-rings of the fuel injector assemblies.

19. While turning the fuel injector assembly left and right, install it onto the fuel delivery pipe sub-assembly.

➡Do not twist the O-ring. After installing the fuel injectors, check that they turn smoothly.

20. Install the 2 No. 1 delivery pipe spacers onto the cylinder head.

➡Install the No. 1 delivery pipe spacers in the correct direction.

21. Install the fuel delivery pipe sub-assembly with the 4 fuel injector assemblies, then temporarily install the 2 bolts. Tighten the bolts to 15 ft. lbs. (21 Nm).

➡Do not drop the fuel injectors when installing the fuel delivery pipe sub-assembly. Check that the fuel injector assemblies rotate smoothly after installing the fuel delivery pipe sub-assembly.

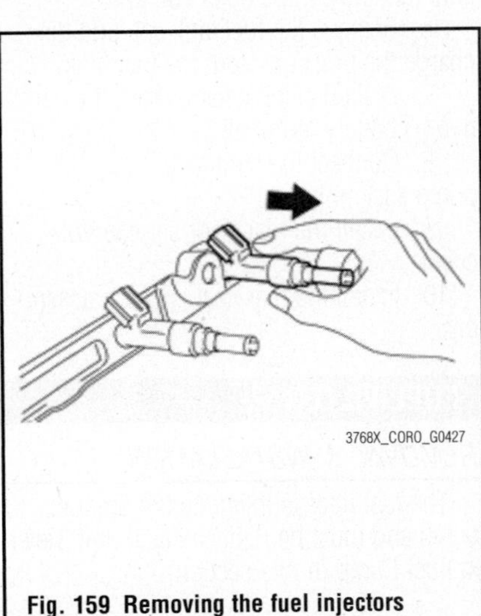

Fig. 159 Removing the fuel injectors

22. Install the bolt to secure the fuel delivery pipe sub-assembly. Tighten the bolt to 15 ft. lbs. (21 Nm).

23. Install the wire harness bracket with the bolt. Tighten to 44 inch lbs. (5 Nm).

24. Insert the fuel tube sub-assembly connector into the fuel delivery pipe until a "click" sound can be heard.

➡Check that there are no scratches or foreign matter around the contact surfaces of the fuel tube connector and pipe before performing this work. After connecting the fuel tube, check that the fuel tube connector and pipe are securely connected by pulling on them.

25. Install a new No. 2 fuel pipe clamp (Type B).

26. Install a new No. 2 fuel pump clamp (Type A).

27. Install the 2 wire harness brackets with the 2 bolts. Tighten to 10 ft. lbs. (13 Nm).

28. Connect the 4 wire harness clamps.

29. Connect the 4 fuel injector assembly connectors.

30. Connect the 2 wire harness clamps.

31. Connect the ground wires with the 2 bolts.

32. Install the air cleaner assembly.

33. Connect the No. 2 ventilation hose.

34. Connect cable to negative battery terminal.

35. Inspect for fuel leaks.

2.4L Engine

See Figures 160 through 166.

⁂ CAUTION

Observe all applicable safety precautions when working around fuel. Whenever servicing the fuel system, always work in a well-ventilated area. Do not allow fuel spray or vapors to come in contact with a spark or open flame. Keep a dry chemical fire extinguisher near the work area. Always keep fuel in a container specifically designed for fuel storage; also, always properly seal fuel containers to avoid the possibility of fire or explosion.

1. Properly relieve the fuel system pressure.

2. Disconnect the negative battery cable.

3. Remove the air cleaner assembly. Refer to Air Cleaner in Engine Mechanical.

➡Do not forcibly bend, kink or twist the fuel main tube.

4. Remove the fuel tube from the fuel hose clamp.
5. Remove the fuel pipe clamp.
6. Wipe off any dirt on the fuel tube connector.
7. Hold the fuel tube connector, and then install SST 09268-21010.
8. Turn SST to align the retainer inside the fuel tube connector with the chamfered part of SST.
9. Insert SST into the fuel tube and hold it. Then push the fuel tube connector toward SST.
10. Mount the retainer of the fuel tube connector onto the chamfered part of SST.

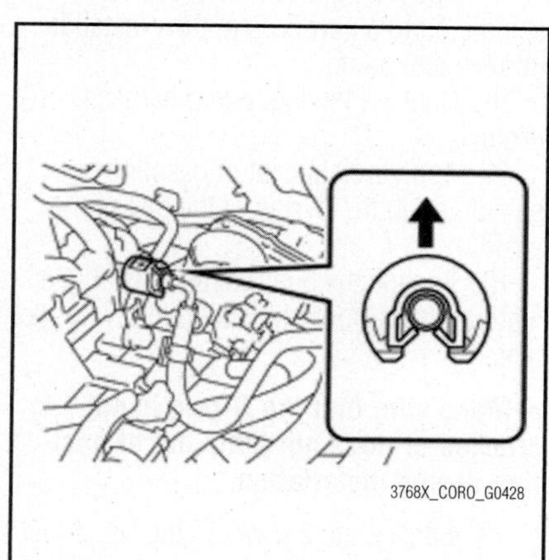

Fig. 160 Removing the fuel tube from the fuel hose clamp

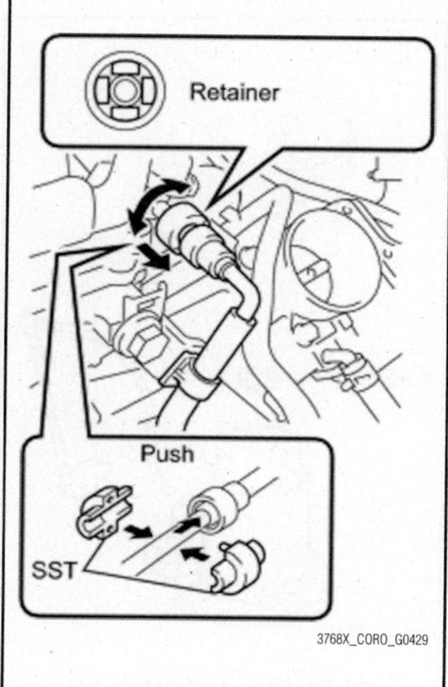

Fig. 161 Installing SST: 09268-21010 to the fuel tube connector

11. Slide SST and fuel tube connector together towards the fuel tube until they make a "click" sound, and then disconnect the fuel tube.
12. Drain the fuel remaining inside the fuel tube.
13. Cover the fuel tube and fuel pipe with a plastic bag to protect the disconnected part.
14. Disconnect the No. 2 ventilation hose from the ventilation valve.
15. Remove the 2 wire harness clamps.
16. Disconnect the 4 fuel injector connectors.
17. Remove the 2 bolts, then remove the fuel delivery pipe together with the 4 fuel injectors.

➡Be careful not to drop the fuel injectors when removing the fuel delivery pipe.

18. Remove the 2 delivery pipe spacers from the cylinder head.

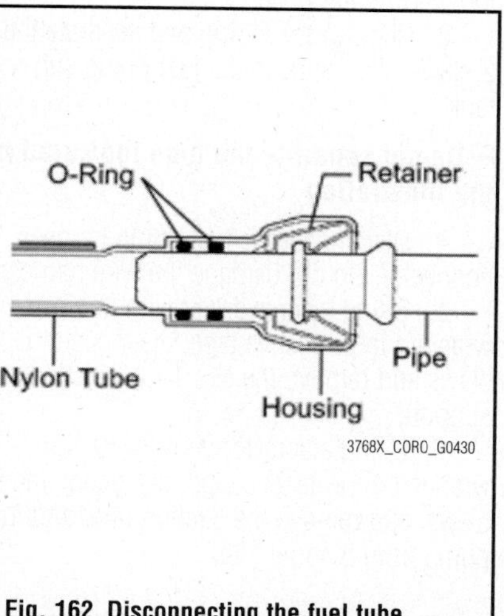
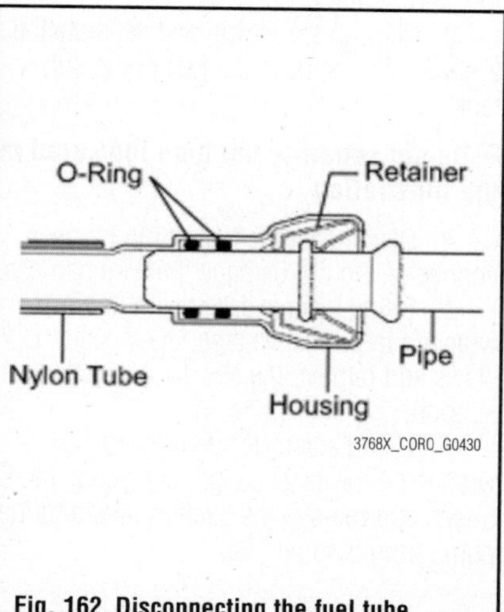

Fig. 162 Disconnecting the fuel tube

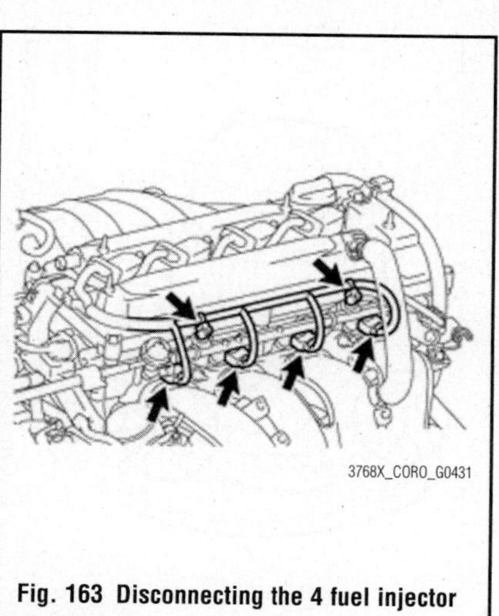

Fig. 163 Disconnecting the 4 fuel injector connectors

19. Remove the 4 insulators from the cylinder head.
20. Pull the 4 fuel injectors out of the fuel delivery pipe.
21. Remove the 4 O-rings from the 4 fuel injectors.

To install:

22. Apply a light coat of gasoline or spindle oil to new O-rings, then install them to the fuel injectors.
23. Apply a light coat of gasoline or spindle oil to the part of the fuel delivery pipe which comes into contact with the O-ring of the fuel injector.
24. Apply a light coat of gasoline or spindle oil to the O-ring again, then install the right and left fuel injectors onto the fuel delivery pipe.

➡Make sure that the O-rings are not cracked or does not jump out of position during installation.

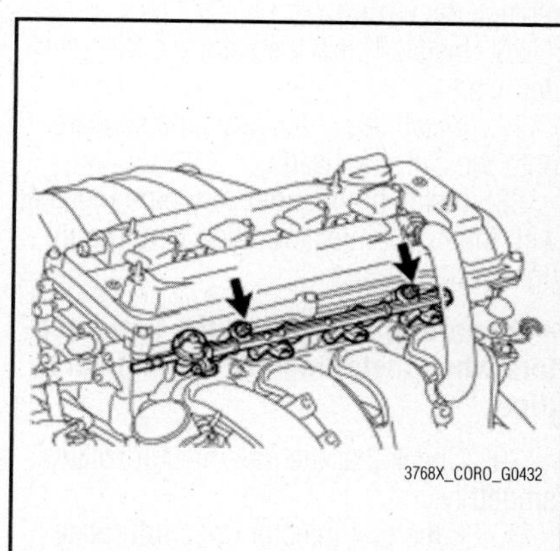

Fig. 164 Removing the 2 bolts, then remove the fuel delivery pipe together with the 4 fuel injectors

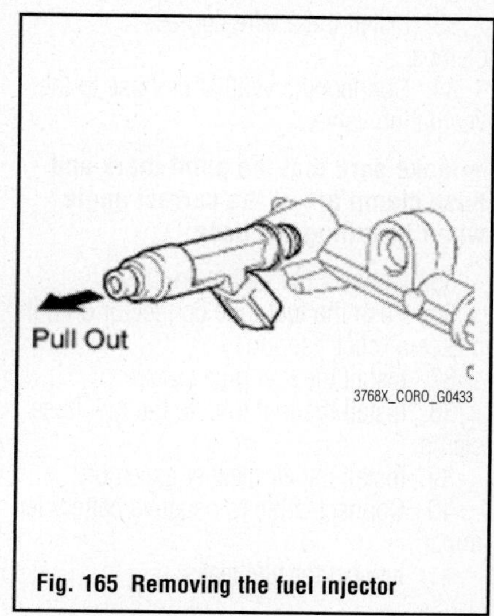

Fig. 165 Removing the fuel injector

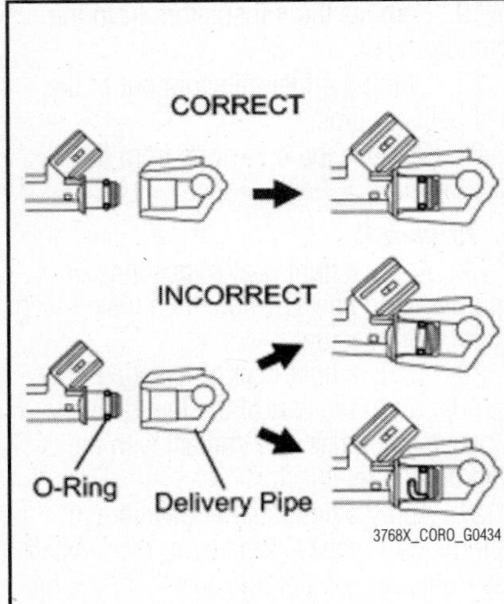

Fig. 166 Correct fuel injector installation position

25. Check that the fuel injectors rotate smoothly. If the fuel injector does not rotate, replace the O-ring.
26. Install 4 new insulators to the cylinder head.
27. Install the 2 delivery pipe spacers onto the cylinder head.
28. Install the fuel delivery pipe together with the 4 fuel injectors, then temporarily tighten the 2 bolts.

➡Be careful not to drop the fuel injectors when installing the fuel delivery pipe.

29. Check that the fuel injector rotates smoothly.
30. If the fuel injector does not rotate smoothly, replace the O-ring.
31. Tighten the 2 fuel rail bolts to 15 ft. lbs. (20 Nm).
32. Connect the 4 fuel injector connectors.
33. Install the 2 wire harness clamps.
34. Connect the ventilation hose to the ventilation valve.

➡Make sure that the paint mark and hose clamp are at the correct angle when installing the hose.

35. Connect the fuel main tube.
36. Push the fuel tube connector until it makes a "click" sound.
37. Install the fuel pipe clamp.
38. Install the fuel tube to the fuel hose clamp.
39. Install the air cleaner assembly.
40. Connect cable to negative battery terminal.
41. Inspect for fuel leaks.

FUEL PUMP

REMOVAL & INSTALLATION

See Figures 161 and 168.

⁂ CAUTION

Observe all applicable safety precautions when working around fuel. Whenever servicing the fuel system, always work in a well-ventilated area. Do not allow fuel spray or vapors to come in contact with a spark or open flame. Keep a dry chemical fire extinguisher near the work area. Always keep fuel in a container specifically designed for fuel storage; also, always properly seal fuel containers to avoid the possibility of fire or explosion.

1. Remove the fuel pump module. Refer to Fuel Pump Module in this section.
2. Release the claw and disconnect the fuel pump filter hose.
3. Remove the E-ring and separate the 2 claws, then remove the fuel pump sub-tank.

➡Do not separate the tube indicated in the illustration.

4. Disconnect the fuel pump harness connector. Do not damage the wire harness.
5. Using a screwdriver with its tip wrapped in protective tape, disengage the 2 claws and remove the No. 1 fuel suction support.
6. Using a screwdriver with its tip wrapped in protective tape, disengage the 5 claws, and remove the suction filter and fuel pump from the fuel filter.

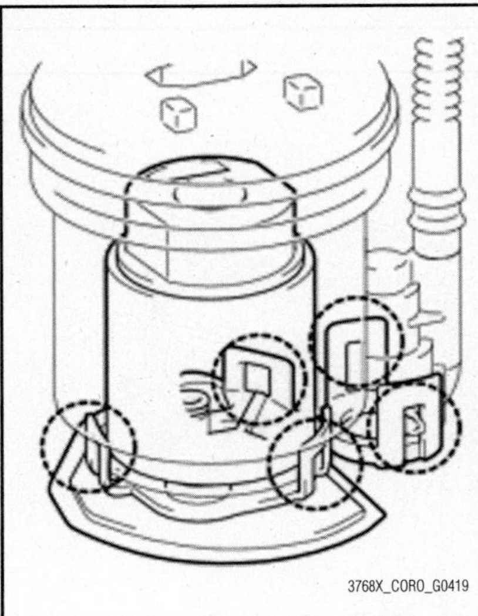

Fig. 167 Locating the 5 claws

➡Do not damage the fuel pump filter or fuel filter. Do not remove the suction filter. Do not use either the fuel pump or the suction filter if the suction filter is removed from the fuel pump.

7. Disconnect the fuel pump connector.
8. Remove the O-ring from the fuel pump.

➡Do not disassemble the fuel pump and the fuel filter because they are non-reusable parts.

To install:

9. Apply a light coat of gasoline or spindle oil to a new O-ring, then install it into the fuel pump.
10. Connect the fuel pump harness connector.
11. Apply a light coat of gasoline or spindle oil to the O-ring of the fuel pump again.
12. Engage the 5 claws, and install the fuel pump filter onto the fuel pump with fuel filter.

➡Make sure that the O-ring is not cracked or does not jump out of position during installation.

13. Engage the 2 claws of the No. 1 fuel suction support.
14. Connect the fuel pump harness connector.
15. Engage the 2 claws and install a new E-ring and fuel pump sub-tank.

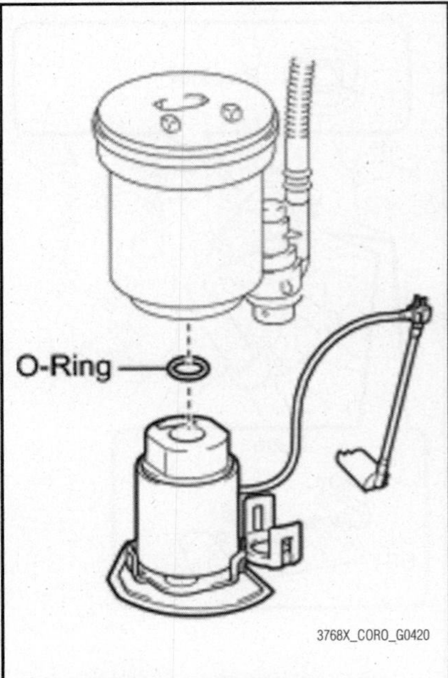

Fig. 168 Removing the fuel pump and O-ring

16. Align the groove of the fuel pump filter hose with the cutout of the fuel sub-tank and install the hose.

17. To complete installation reverse remaining removal procedure.

FUEL TANK

REMOVAL & INSTALLATION

See Figures 169 through 172.

CAUTION

Observe all applicable safety precautions when working around fuel. Whenever servicing the fuel system, always work in a well-ventilated area. Do not allow fuel spray or vapors to come in contact with a spark or open flame. Keep a dry chemical fire extinguisher near the work area. Always keep fuel in a container specifically designed for fuel storage; also, always properly seal fuel containers to avoid the possibility of fire or explosion.

1. Properly relieve the fuel system pressure.
2. Disconnect the negative battery cable.
3. Remove the fuel pump module. Refer to Fuel Pump Module in this section.
4. Drain fuel.
5. Remove front exhaust pipe assembly. Refer to Exhaust Manifold in Engine Mechanical.
6. Remove the 4 bolts and the No. 1 fuel tank protector.
7. Remove the 4 bolts, and separate the parking brake cables.
8. Disconnect the fuel tank vent hose.
9. Pull the fuel tank vent hose out of the pipe.

➡Check that there is no dirt or other foreign objects around the connector before disconnecting it. Clean the connector if necessary.

- It is necessary to prevent mud or dirt from entering the connector. If mud or dirt gets in the connector, the O-rings may not seal properly.
- Only disconnect the quick connector by hand.
- Do not bend, kink or twist the nylon tubes.
- Protect the connector by covering it with a plastic bag.

10. Disconnect breather tube fuel hose:
    a. Pinch the retainer of the fuel tube connector, then pull the fuel tube connector out of the pipe.

➡Check that there is no dirt or other foreign objects around the connector before disconnecting it. Clean the connector if necessary.

- It is necessary to prevent mud or dirt from entering the connector. If mud or dirt gets in the connector, the O-rings may not seal properly.
- Only disconnect the quick connector by hand.
- Do not bend, kink or twist the nylon tubes.
- Protect the connector by covering it with a plastic bag.
- If the pipe and connector are stuck, carefully try wiggling or pushing and pulling on the connector to release it. Pull the connector off the pipe carefully.

    b. Separate the fuel breather tube fuel hose.

11. Disconnect the fuel tank main tube sub-assembly:
    a. Pinch the tabs of the retainer of the fuel tube connector to disengage the lock claws and push it down as shown in the illustration.
    b. Pull the fuel tank main tube out of the pipe.

➡Check that there is no dirt or other foreign objects around the connector before disconnecting it. Clean the connector if necessary.

- It is necessary to prevent mud or dirt from entering the connector. If mud or dirt gets in the connector, the O-rings may not seal properly.
- Only disconnect the quick connector by hand.
- Do not bend, kink or twist the nylon tubes.
- Protect the connector by covering it with a plastic bag.
- If the pipe and connector are stuck, carefully try wiggling or pushing and pulling on the connector to release it. Pull the connector off the pipe carefully.

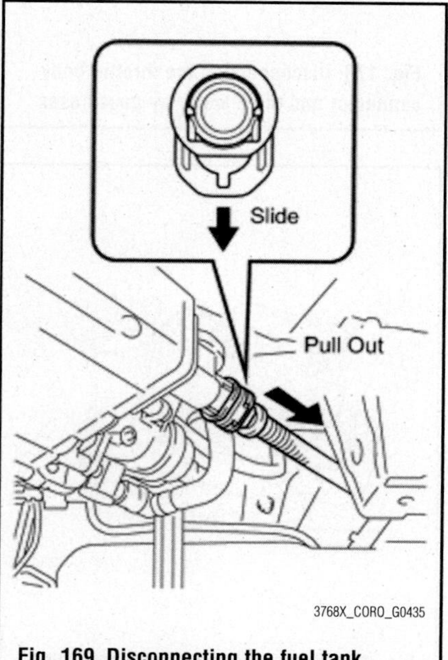

Fig. 169 Disconnecting the fuel tank vent hose

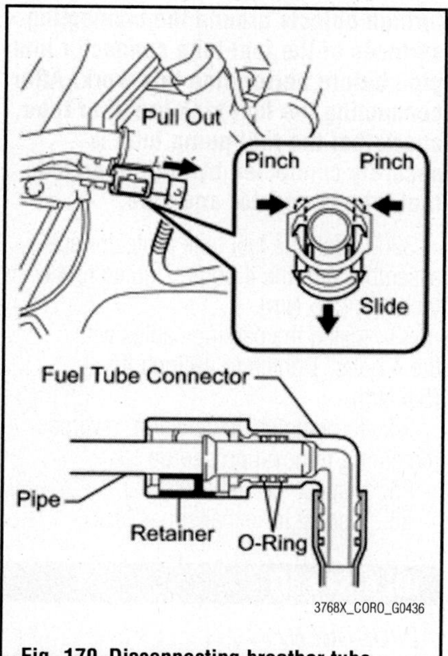

Fig. 170 Disconnecting breather tube fuel hose

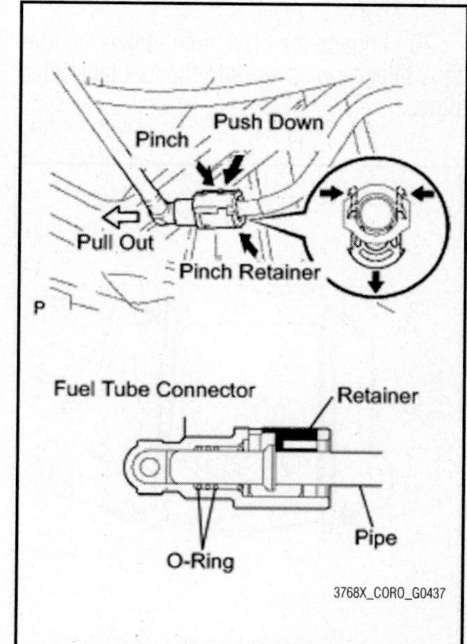

Fig. 171 Disconnecting the fuel tank main tube sub-assembly

12. Disconnect fuel tank filler pipe sub-assembly:
    a. Using a screwdriver, unfasten the claw. Then remove the fuel tank filler pipe cover from the fuel tank filler pipe.
    b. Loosen the hose clamp bolt, then disconnect the fuel tank filler pipe hose from the fuel tank.
13. Hold the fuel tank using a transmission jack.
14. Remove the 4 bolts, then remove the 2 No. 1 fuel tank bands.
15. Operate the transmission jack, then remove the fuel tank.
16. Remove the fuel tank main tube from the fuel tank.
17. Remove the fuel tank vent hose from the fuel tank clamp.
18. Remove the fuel tank cushions from the fuel tank.

To install:

19. Install new fuel tank cushions onto the fuel tank.
20. Install the fuel tank vent hose onto the fuel tank clamp.
21. Install the fuel tank main tube onto the fuel tank.
22. Set the fuel tank on a transmission jack.
23. Operate the transmission jack, then install the fuel tank into the vehicle.
24. Install the 2 No. 1 fuel tank bands with the 4 bolts. Tighten to 29 ft. lbs. (39 Nm).
25. Connect the fuel tank filler pipe to the fuel tank.

➡Make sure that the hose clamp is facing in the correct direction when installing.

26. Engage the claw, then install the fuel tank filler pipe cover onto the fuel tank filler pipe.

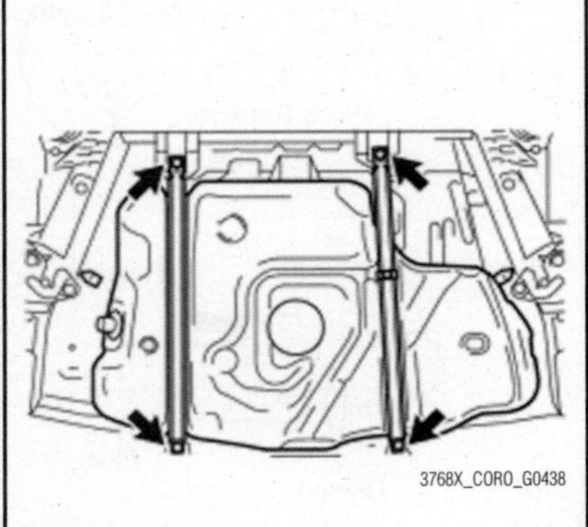

Fig. 172 Removing the 4 bolts, then remove the 2 No. 1 fuel tank bands

27. Align the fuel tube connector with the pipe, then push the fuel tube connector in until it comes into contact with the seat to connect the fuel tank main tube to the pipe, then push the retainer up until the claws lock.

➡Check that there are no scratches or foreign objects around the connecting surfaces of the fuel tube connector and pipe before performing this work. After connecting the fuel tank main tube, check that the fuel tank main tube is securely connected by pulling on the fuel tube connector and pipe.

28. Connect the breather tube fuel hose to the clamp.
29. Align the fuel hose connector with the pipe, then push the fuel hose connector in until the retainer makes a "click" sound to connect the fuel tank breather tube fuel hose to the pipe.

➡Check that there are no scratches or foreign objects around the connecting surfaces of the fuel tube connector and pipe before performing this work. After connecting the fuel tank breather tube, check that the fuel pump tube is securely connected by pulling on the fuel tube connector and pipe.

30. Align the fuel tube connector with the pipe, then push the fuel tube connector in until it comes into contact with the seat to connect the fuel tank vent hose to the pipe.
31. Slide the retainer of the fuel tube connector to lock the claws

➡Check that there are no scratches or foreign objects around the connecting surfaces of the fuel tube connector and pipe before performing this work. After connecting the fuel tank breather tube, check that the fuel pump tube is securely connected by pulling on the fuel tube connector and pipe.

32. Install the fuel tank protector sub-assembly with the 4 bolts. Tighten to 49 inch lbs. (5.5 Nm).
33. Install the parking cables with the 4 bolts. Tighten to 49 inch lbs. (5.5 Nm).
34. To complete installation, reverse remaining removal procedure.
35. Inspect for fuel leaks.
36. Inspect for exhaust gas leak.

IDLE SPEED

ADJUSTMENT

Adjustment is not available or necessary.

THROTTLE BODY

REMOVAL & INSTALLATION

1.8L Engine

See Figures 173 and 174.

1. Drain and recycle the engine coolant.
2. Disconnect the negative battery cable.
3. Disconnect the mass air flow meter connector and the 2 wire harness clamps.
4. Remove the air cleaner assembly. Refer to Air Cleaner Assembly in Engine Mechanical.
5. Disconnect the throttle body connector and the 2 water by-pass hoses.
6. Remove the 2 bolts and 2 nuts, and remove the throttle body assembly.
7. Remove the gasket from the intake manifold.

To install:

8. Install a new gasket onto the intake manifold.

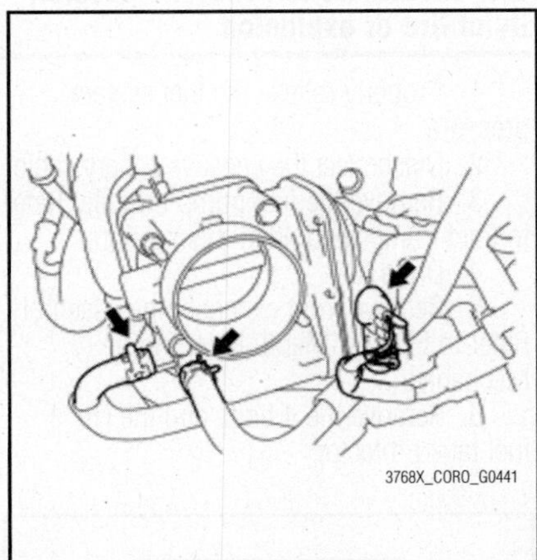

Fig. 173 Disconnecting the throttle body connector and the 2 water by-pass hoses

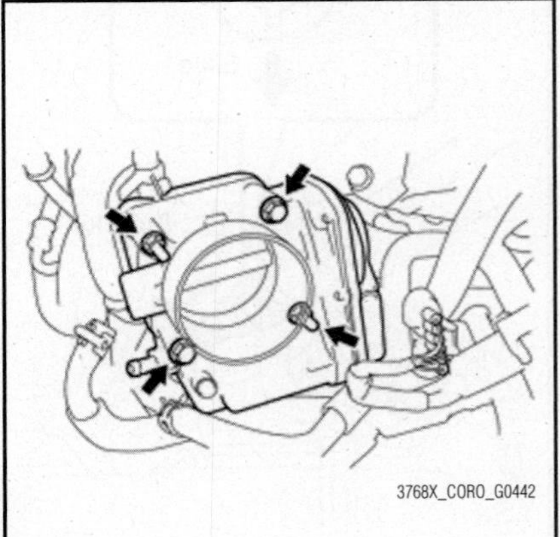

Fig. 174 Removing the 2 bolts and 2 nuts, and removing the throttle body assembly

9. Install the throttle body assembly with the 2 bolts and 2 nuts. Tighten to 7 ft. lbs. (10 Nm).

10. Connect the 2 water by-pass hoses and the throttle body connector.

11. To complete installation, reverse remaining removal procedure.

12. Perform the initialization procedure.

2.4L Engine

See Figures 175 and 176.

1. Drain and recycle the engine coolant.
2. Disconnect the negative battery cable.
3. Disconnect the mass air flow meter connector and the 2 wire harness clamps.
4. Remove the air cleaner assembly. Refer to Air Cleaner Assembly in Engine Mechanical.
5. Disconnect the 2 water by-pass hoses.
6. Disconnect the throttle body assembly connector.
7. Disconnect the throttle body hose.
8. Remove the 4 bolts and throttle body assembly.
9. Remove the gasket from the intake manifold.

To install:

10. Install a new gasket onto the intake manifold.
11. Install the throttle body assembly with the 4 bolts. Tighten to 22 ft. lbs. (30 Nm).
12. Connect the 2 water by-pass hoses and the throttle body connector.
13. To complete installation, reverse remaining removal procedure.

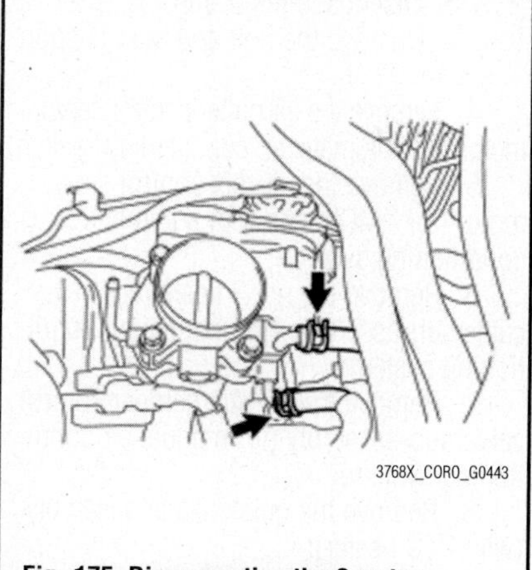

Fig. 175 Disconnecting the 2 water by-pass hoses

14. Perform the initialization procedure.

INITIALIZATION

➡Be sure to perform this procedure after reassembling the throttle body assembly, removing and reinstalling any throttle body component or replacing the ECM.

1. Disconnect the cable from the negative (-) battery terminal. Wait at least 60 seconds and reconnect the cable.
2. Turn the ignition switch to ON without operating the accelerator pedal.

➡If the accelerator pedal is operated, perform the above steps again.

3. Connect the Techstream to the DLC3 and clear the DTC's.

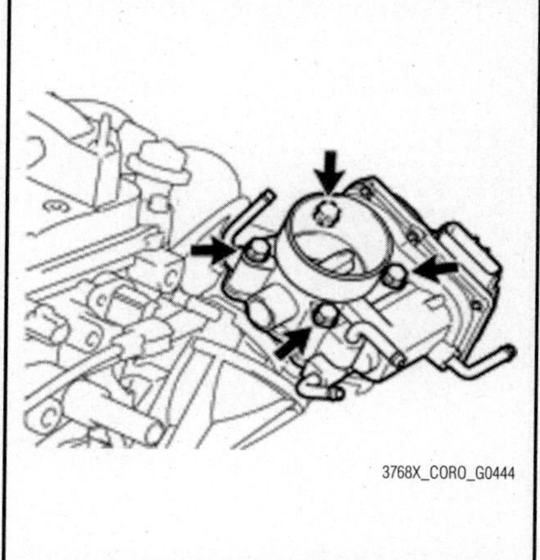

Fig. 176 Removing the 4 bolts and throttle body assembly

4. Start the engine and check that the MIL is not illuminated and that the idle speed is within the specified range when the air conditioning is switched off after the engine is warmed up.
   a. Standard:
   - Condition: A/C switched off
   - Engine idle speed: 600 to 700 rpm

➡Be sure to perform this step with all accessories off. Make sure that the shift lever is in N or P.

5. Enter the following menus: Powertrain/ Engine and ECT/ Data List/ Throttle Pos. Sensor Output. Fully depress the accelerator pedal and check that the value is 60% or more.
6. Perform a road test and confirm that there are no abnormalities. Refer to Drive Train.

HEATING & AIR CONDITIONING SYSTEM

BLOWER MOTOR

REMOVAL & INSTALLATION

See Figures 177 through 179.

1. Disconnect the negative battery cable.
2. Remove the no. 2 instrument panel under cover sub-assembly (w/ instrument panel under cover)
   a. Disengage the 3 claws.
   b. Disengage the guide and remove the No. 2 instrument panel under cover sub-assembly.
3. Remove blower motor sub-assembly (without PTC heater):
   a. Disconnect the connector.
   b. Remove the 3 screws and the blower motor sub-assembly.
4. Remove blower motor sub-assembly (with PTC heater):

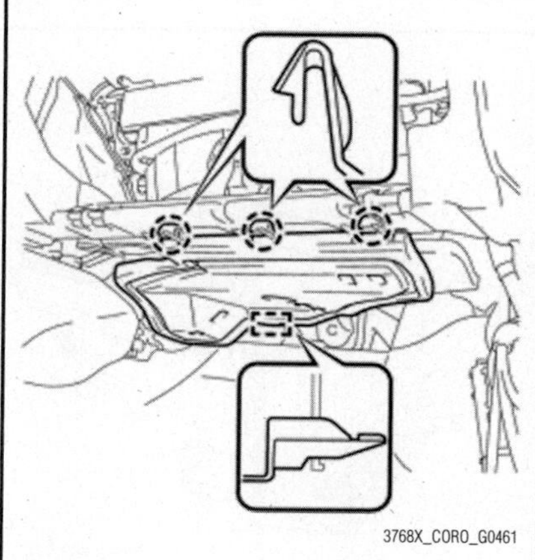

Fig. 177 Removing the No. 2 instrument panel under cover sub-assembly

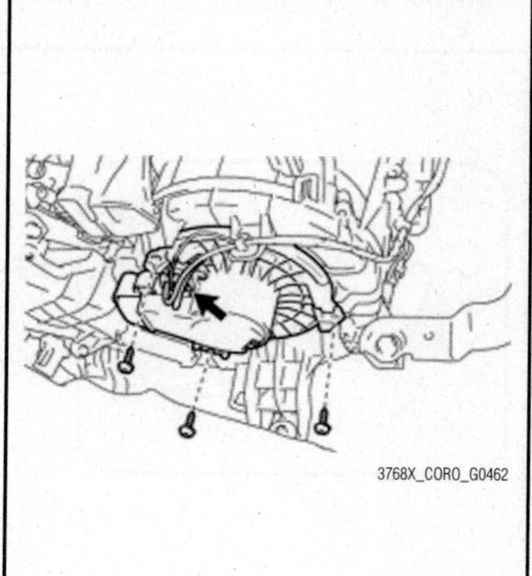

Fig. 178 Removing blower motor sub-assembly (without PTC heater)

Fig. 179 Remove blower motor sub-assembly (with PTC heater)

a. Remove the quick heater connector screw.

b. Disconnect the connector.

c. Remove the 4 screws and the blower motor sub-assembly.

To install:

5. To install, reverse removal procedure.

HEATER CORE

REMOVAL & INSTALLATION

See Figures 180 through 185.

1. Remove Heater and A/C Unit.
2. Remove immobilizer code ECU (with smart key system):
   a. Disconnect the connector.
   b. Disengage the clamp.
   c. Remove the bolt and the immobilizer code ECU.
3. Remove transponder key ECU (without smart key system):
   a. Disconnect the connector.
   b. Disengage the clamp.
   c. Remove the bolt and the transponder key ECU.
4. Remove the air outlet control servo motor (for automatic air conditioning system).
5. Remove the air mix control servo motor (for TMC made with automatic air conditioning system).
6. Remove the no. 2 heater control cable sub-assembly (for manual air conditioning system).
7. Remove the air mix damper control cable sub-assembly (for manual air conditioning system).
8. Remove the quick heater assembly (with PTC heater):
   a. Disengage each clamp.
   b. Remove the 2 screws.
   c. Remove the quick heater assembly as shown in the illustration.
9. Remove the air conditioning harness assembly (for automatic air conditioning system):

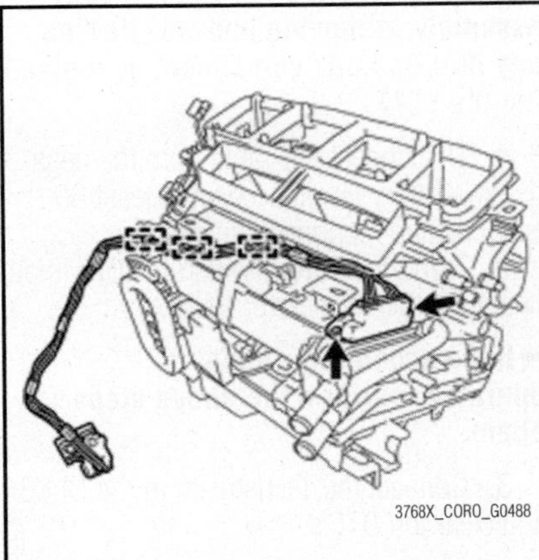

Fig. 181 Disengaging the clamps and removing the 2 quick heater screws

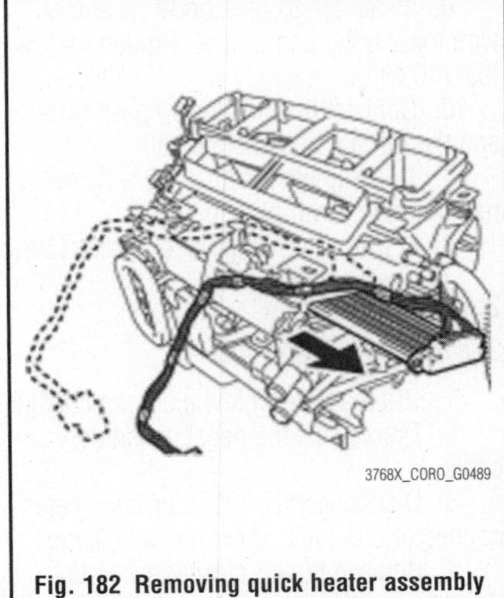

Fig. 182 Removing quick heater assembly

   a. Disconnect the connector.
   b. Disengage each clamp and remove the air conditioning harness.
10. Disengage the 2 claws and remove the RH console mounting bracket.
11. Remove the screw and the console LH mounting bracket.
12. Remove air conditioning amplifier assembly:
    a. Remove the screw.
    b. Disengage the claw and remove the air conditioning amplifier assembly.
13. Remove the air conditioning duct sub-assembly (for automatic air conditioning system).
14. Remove the drain cooler hose from the air conditioning radiator assembly.
15. Remove heater radiator unit sub-assembly:
    a. Remove the screw and the clamp.
    b. Remove the heater radiator unit sub-assembly from the air conditioning unit assembly.

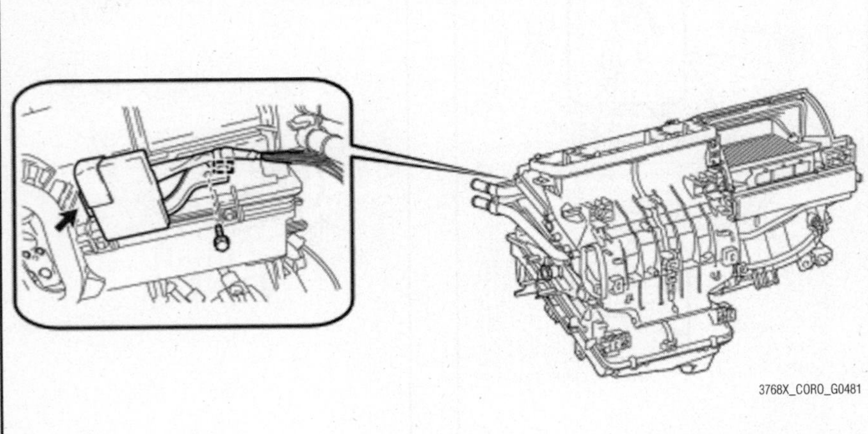

Fig. 180 Removing the immobilizer code ECU/ transponder key ECU

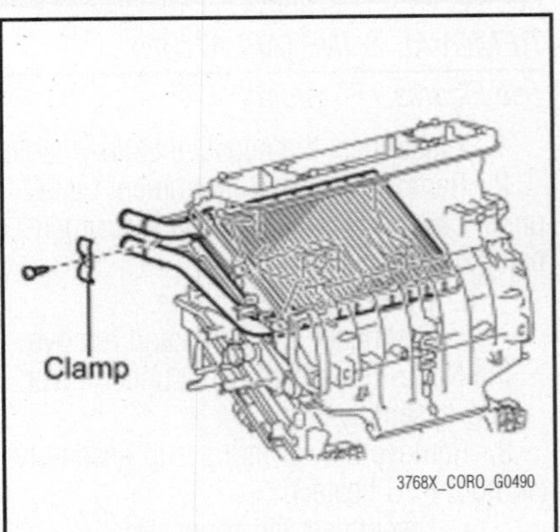

Fig. 183 Removing the heater radiator unit sub-assembly from the air conditioning unit assembly

16. Using a 4 mm hexagon wrench, remove the 2 hexagon bolts and cooler expansion valve.

17. Remove no. 1 cooler evaporator sub-assembly:

a. Remove the 4 screws.

b. Disengage the 4 claws and remove the lower heater case.

c. Disengage the clamp, and remove the No. 1 cooler evaporator sub-assembly together with the No. 1 cooler thermistor.

d. Remove the 2 O-rings.

To install:

18. Sufficiently apply compressor oil to 2 new O-rings and the fitting surfaces. Install the 2 O-rings to the No. 1 cooler evaporator sub-assembly. Compressor oil: ND-OIL 8 or equivalent.

➡Keep the O-rings and O-ring fitting surfaces clean from dirt or foreign matter.

19. Install the No. 1 cooler evaporator sub-assembly together with the No. 1 cooler thermistor as a set.

20. Engage the clamp.

21. Engage the 4 claws

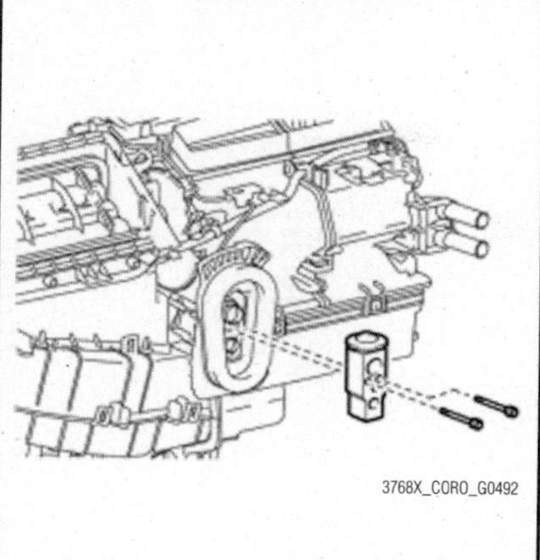

Fig. 184 Remove the 2 hexagon bolts and cooler expansion valve

22. Install the lower heater case with the 4 screws.

23. Using a 4 mm hexagon wrench, install the cooler expansion valve with the 2 hexagon bolts. Tighten to 31 inch lbs. (4 Nm).

24. To complete installation, reverse remaining removal procedure.

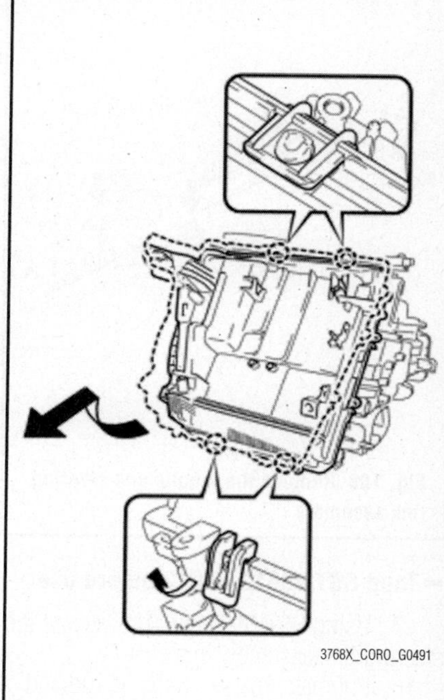

Fig. 185 Disengaging the 4 claws and remove the lower heater case.

STEERING

POWER STEERING GEAR

REMOVAL & INSTALLATION

See Figures 186 through 190.

✻✻ CAUTION

Some models covered by this manual may be equipped with a Supplemental Restraint System (SRS), which uses an air bag. Whenever working near any of the SRS components, such as the impact sensors, the air bag module, steering column and instrument panel, disable the SRS, as described in the Chassis Electrical Section.

1. Place front wheels facing straight ahead.

2. Secure the steering wheel with the seat belt in order to prevent rotation.

➡This operation is useful to prevent damage to the spiral cable.

3. Disconnect the negative battery cable.

4. Remove column hole cover silencer sheet.

5. Separate no. 2 steering intermediate shaft assembly.

6. Remove clips A and the No. 1 steering column hole cover sub-assembly and disengage clip B from the body.

7. Remove the front wheels.

8. Remove the engine under covers.

9. Separate the tie rod end sub-assembly.

10. Separate front stabilizer link assembly.

11. Separate front lower suspension arms.

12. Remove front suspension cross-member sub-assembly. Refer to Front Suspension.

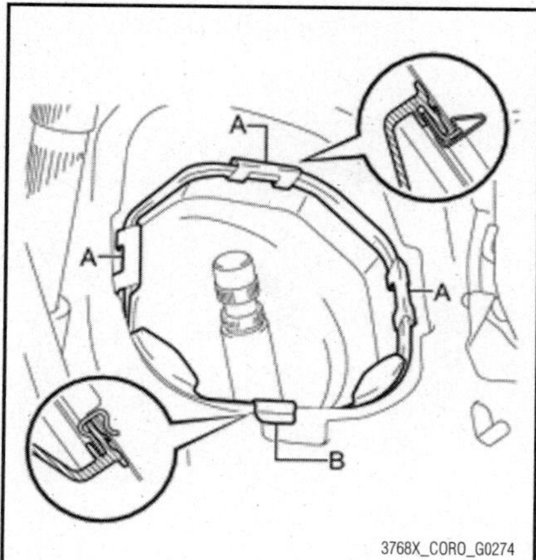

Fig. 186 Remove clips A and the No. 1 steering column hole cover sub-assembly

13. Remove the No. 1 steering column hole cover sub-assembly from the steering link assembly.

14. Put match marks on the steering intermediate shaft and the steering link assembly.

15. Remove the bolt and the steering intermediate shaft from the steering link assembly.

16. Remove the 4 bolts and steering link assembly from the front suspension cross-member sub-assembly.

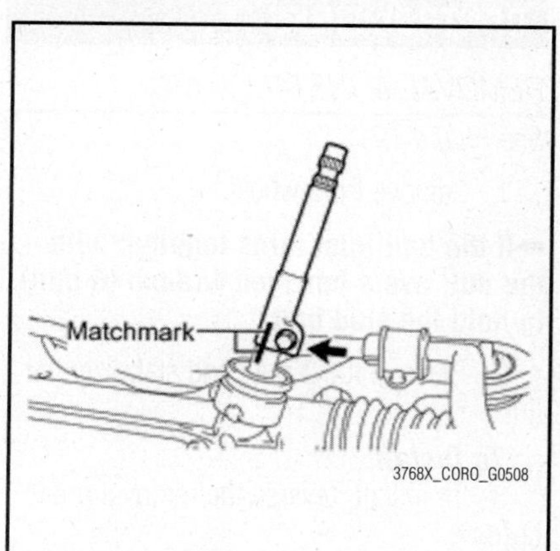

Fig. 187 Putting match marks on the steering intermediate shaft and the steering link assembly

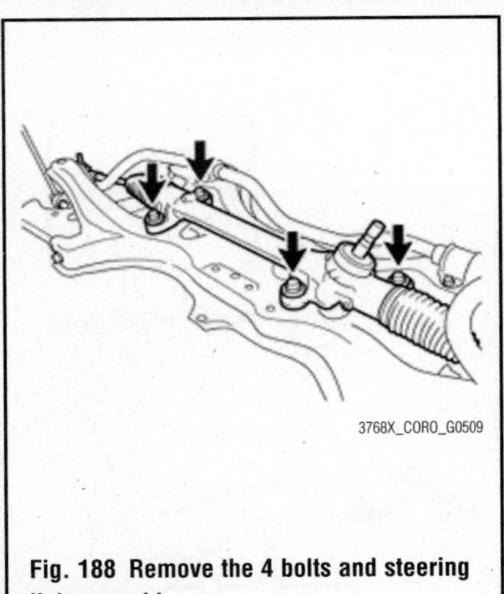

Fig. 188 Remove the 4 bolts and steering link assembly

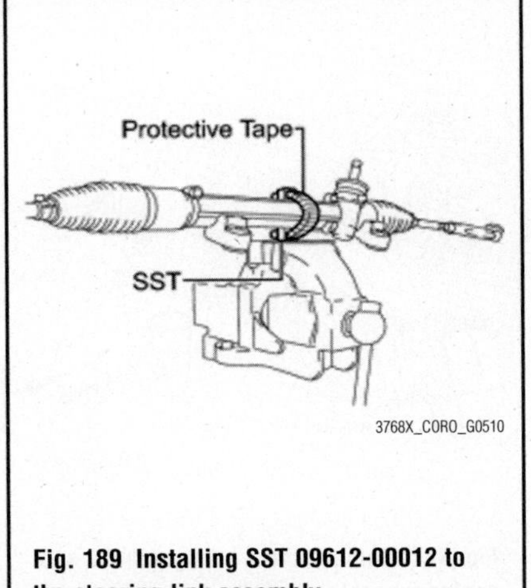

Fig. 189 Installing SST 09612-00012 to the steering link assembly

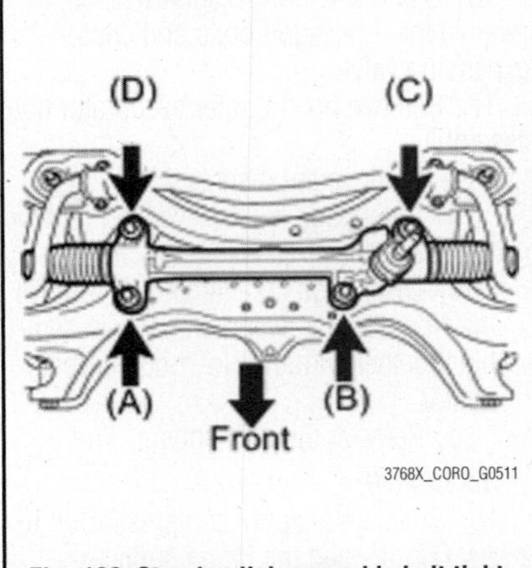

Fig. 190 Steering link assembly bolt tightening sequence

➡Tape SST 09612-00012 before use.

17. Using SST 09612-00012, secure the steering link assembly in a vise.
18. Put matchmarks on the tie rod end sub-assembly LH and RH and steering gear assembly.
19. Remove the tie rod end sub-assembly and lock nuts.

To install:

20. Install the lock nut and the tie rod end sub-assembly to the steering gear assembly until the match marks are aligned.

➡After adjusting the toe-in, tighten the lock nut.

21. Install the steering link assembly to the front suspension crossmember sub-assembly with the 4 bolts:
    a. Temporarily tighten the bolts in order of (A), (B), (C), and then (D).
    b. Tighten the bolts in order of (A), (B), (C), and then (D).
    c. Tighten the bolts to 43 ft. lbs. (58 Nm).
22. Align the matchmarks and install the steering intermediate shaft to the steering link assembly.
23. Install the bolt. Tighten to 26 ft. lbs. (35 Nm).
24. To complete installation, reverse remaining removal procedure.

POWER STEERING PUMP

REMOVAL & INSTALLATION

➡The Corolla does not use a power steering pump. The Corolla utilizes a Power Steering Motor that works in conjunction with a power steering ECU. The power steering motor is integrated in to the steering column assembly and must be replaced as a unit.

SUSPENSION

FRONT SUSPENSION

COIL SPRING

REMOVAL & INSTALLATION

Refer to MacPherson Strut.

CONTROL LINKS

REMOVAL & INSTALLATION

See Figure 191.

1. Remove front wheel.

➡If the ball joint turns together with the nut, use a hexagon wrench (6 mm) to hold the stud bolt.

2. Remove the 2 nuts and stabilizer bar link.

To install:

3. To install, reverse the removal procedure..
4. Tighten the link assembly nuts to 55 ft. lbs. (74 Nm).

LOWER BALL JOINT

REMOVAL & INSTALLATION

See Figure 192.

1. Before servicing the vehicle, refer to the Precautions section.
2. Remove or disconnect the following:
    - Negative battery cable. On vehicles equipped with an air bag, wait at

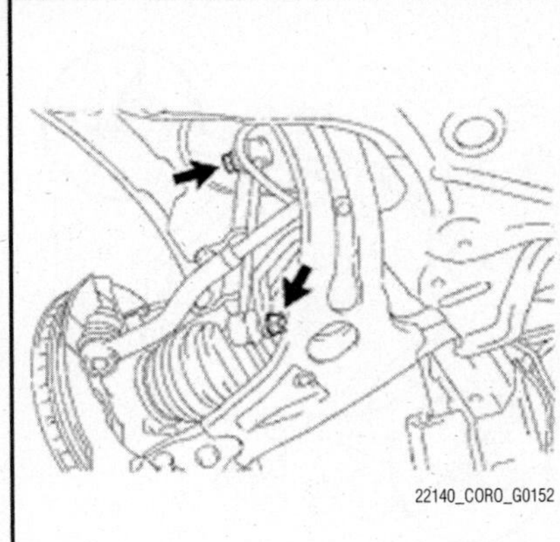

Fig. 191 Removing front stabilizer link assembly

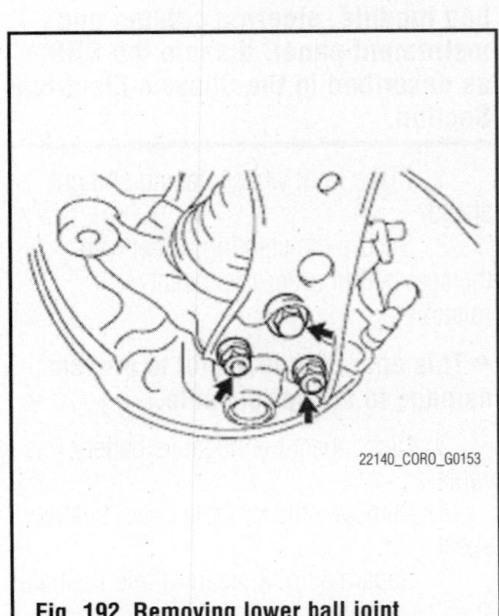

Fig. 192 Removing lower ball joint

least 90 seconds before proceeding
- Front wheels
- Cotter pin from the bearing locknut cap, then remove the cap

3. Depress the brake pedal and loosen the axle nut
4. Remove or disconnect the following:
   - Brake caliper attaching hardware, position the caliper aside with the hydraulic line still attached and suspend it with a wire
   - ABS speed sensor, if equipped
   - Rotor
5. Loosen the 2 nuts holding the strut to the steering knuckle assembly. Do not remove at this time.
6. Remove or disconnect the following:
   - Cotter pin and nut from the tie rod end. Using a tie rod end removal tool, separate the tie rod end from the steering knuckle
   - Steering knuckle from the strut assembly
   - Axle nut and grasp the hub and knuckle assembly. With a plastic hammer tap the axle shaft to remove knuckle and hub

➡Cover the halfshaft boot with a shop rag to protect it from any damage.

7. Clamp the steering knuckle in a vise and remove the dust deflector. Remove the nut holding the steering knuckle to the ball joint. Press the ball joint out of the steering knuckle.
8. Remove the ball joint from the arm.

To install:

9. Install the Lower ball joint to the lower arm. Tighten the fasteners to: 66 ft. lbs. (89 Nm).
10. Install the ball joint to the steering knuckle. Tighten the ball joint-to-steering knuckle nut to: 76 ft. lbs. (103 Nm).
11. Install or connect the following:
    - New cotter pin. Drive the deflector shield onto the knuckle
    - Knuckle and hub assembly to the axle and temporarily tighten the axle nut
    - Knuckle assembly to the lower strut bracket. Temporarily insert the mounting bolts from the rear and install the nuts
    - Tie rod end to the knuckle
12. Tighten the bolts on the lower side of the strut assembly.
13. Install or connect the following:
    - ABS speed sensor
    - Brake disc and the caliper
14. Tighten the axle nut.
15. Connect the negative battery cable.
16. Check and adjust the alignment, if needed.

LOWER CONTROL ARM

REMOVAL & INSTALLATION

See Figures 193 through 197.

1. Remove front wheel.
2. Disconnect front stabilizer link assembly.
3. Separate front suspension arm sub assembly lower left and right:
   a. Remove the bolt and 2 nuts, and separate the front suspension arm sub-assembly lower No. 1 LH from the lower ball joint assembly front LH.
4. Separate rack and pinion power steering gear assembly:
   a. Remove the 4 bolts, separate the rack & pinion power steering gear assembly.

➡Loosen the bolt since the nut cannot be rotated.

   b. Suspend the rack & pinion power steering gear assembly.
5. Suspended engine assembly:
6. Separate front suspension cross member subassembly:
   a. Remove the 3 bolts and 3 nuts, disconnect the transverse engine mounting insulator and engine mounting member sub-assembly center from the front suspension cross member sub-assembly.
   b. Remove the 4 bolts.
   c. Lower the transmission jack, remove the front suspension cross member sub-assembly.
7. Remove front suspension arm sub assembly lower:
   a. Remove the 2 bolts, nut and front suspension arm sub-assembly lower No. 1 LH from the front suspension cross member sub-assembly.

To install:

8. Temporarily tighten front suspension arm subassembly lower.
   a. Install the front suspension arm sub-assembly lower, temporarily tighten the 2 bolts and nut.
9. Install front suspension cross member subassembly:
   a. Lift the front suspension cross member sub-assembly up with a transmission jack.
   b. Insert service tool to the base hole of the RH side cross member and RH side of the vehicle.
   c. Tighten the bolts temporarily.
   d. Insert service tool to the base hole of the LH side of cross member and LH side of the vehicle.

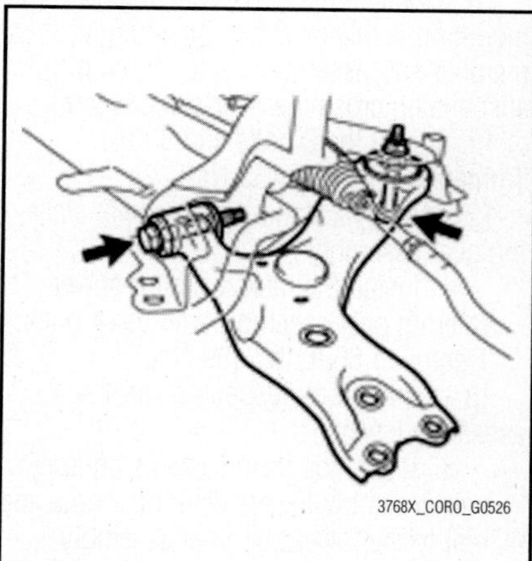

Fig. 194 Remove the 2 bolts, nut and front lower control arm

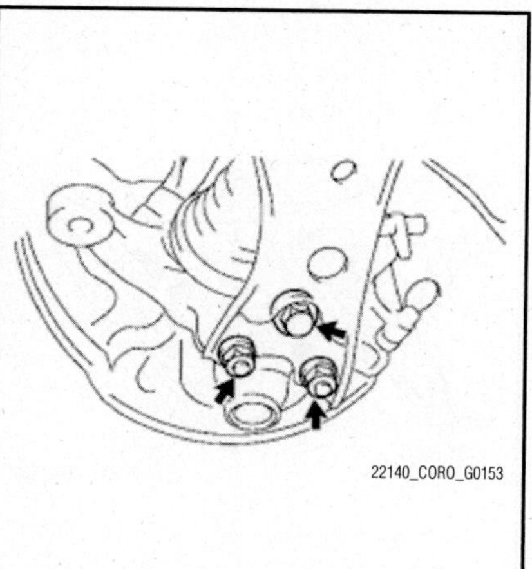

Fig. 193 Removing front suspension arm lower

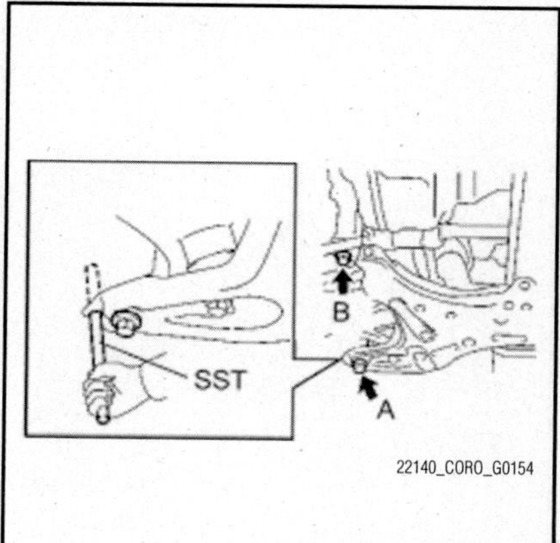

Fig. 195 Installing front suspension arm lower

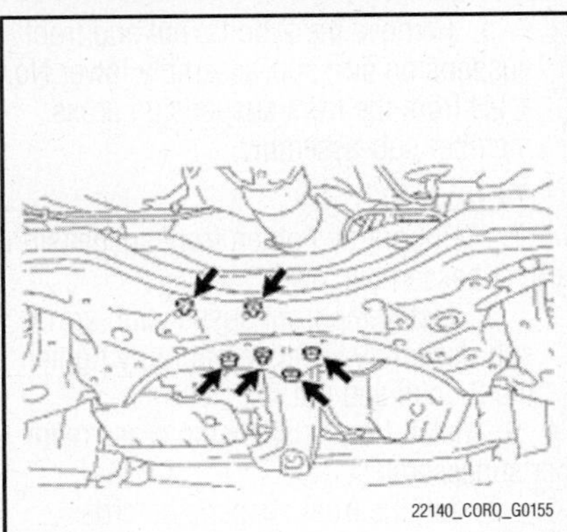

Fig. 196 Installing engine mounting member subassembly

e. Tighten the bolts temporarily.

f. Then tighten the bolt A and B by the specified torque A: 116 ft. lbs. (157 Nm) B: 83 ft. lbs. (113 Nm)

10. Connect the transverse engine mounting insulator and engine mounting member sub-assembly center to the front suspension cross member sub-assembly.

11. Install the 3 bolts and 3 nuts. Tighten to 38 ft. lbs. (52 Nm)

12. Install rack and pinion power steering gear assembly:

a. Install the rack & pinion power steering gear assembly with the 4 bolts. Tighten to 66 ft. lbs. (89 Nm)

13. Install front suspension arm sub assembly lowers:

a. Install the front suspension arm sub-assembly lowers with the 2 nuts and bolt to the lower ball joint assembly front.

14. Install front stabilizer link assemblies.

15. Stabilizes suspension:

a. Install the front wheel and jack down the vehicle.

b. Bounce the vehicle up and down several times to stabilize the suspension.

16. Fully tighten front suspension arm subassembly lower:

a. Fully tighten the 2 bolts and nut.

17. Inspect and adjust front wheel alignment.

➡Tighten the bolt since the nut cannot be rotated.

MACPHERSON STRUT

REMOVAL & INSTALLATION

See Figure 198.

1. Before servicing the vehicle, refer to the Precautions section.
2. Remove front wheel.
3. Remove front wiper arm head cap.
4. Remove front LH and RH wiper arm and blade assembly.
5. Remove hood to cowl top seal.
6. Remove center no.1 cowl top ventilator louver.
7. Remove LH cowl top ventilator louver.
8. Remove windshield wiper motor and link assembly
9. Remove outer cowl top panel.
10. Separate front stabilizer link assembly.
11. Remove the bolt and separate the front flexible hose and the front speed sensor from the front shock absorber with coil spring.

➡Be sure to separate the front speed sensor from the front shock absorber with coil spring completely.

12. Remove the front suspension support dust cover.
13. Loosen the front support to front shock absorber nut of the front shock absorber.

➡Do not remove the front support to front shock absorber nut. Loosen the nut only when the front shock absorber with coil spring needs to be disassembled.

14. Support the front axle using a jack and wooden block.
15. Remove the 2 bolts and 2 nuts, and separate the front shock absorber with coil spring (lower side) from the steering knuckle.
16. Remove the 3 nuts and front shock absorber with coil spring.

➡Make sure that the front speed sensor is completely separated from the front shock absorber with coil spring.

To install:

17. Install or connect the following:
    - Nuts holding the strut to the strut tower. Nuts to 29 ft. lbs. (39 Nm)
    - Steering knuckle to the strut lower bracket
18. Insert the 2 bolts from the rear side and tighten the strut-to-steering knuckle arm bolts. Tighten as follows: 113 ft. lbs. (153 Nm).
19. Install or connect the following:
    - Brake line to the steering knuckle
    - If equipped with ABS, secure the wiring harness
    - Wheel
    - Negative battery cable
20. Check and adjust the alignment, if needed.

OVERHAUL

See Figure 199.

1. To disassemble the strut:

a. Install a bolt and 2 nuts to the bracket at the lower portion of the strut shell and secure it in a vise.

b. Compress the coil spring.

c. Remove the dust cover and hold the spring seat so that it will not turn. Remove the nut on the top of the strut.

d. Remove the suspension support, bearing, dust seal, spring seat, spring, insulators and bumper.

e. Remove front coil spring lower insulator and front shock absorber

To assemble:

2. To assemble the strut:

a. Install the spring bumper to piston.

Fig. 197 Tightening front suspension arm lower

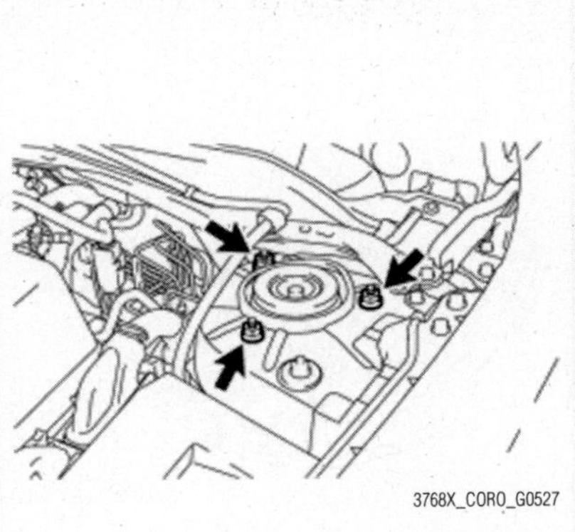

Fig. 198 Removing the 3 nuts and front shock absorber with coil spring

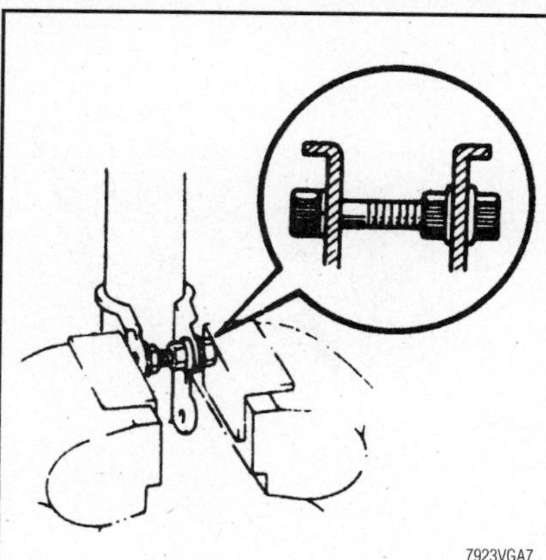
7923VGA7

Fig. 199 Proper method of supporting the strut in a vise

b. Using a spring compressor, compress the spring.

c. Install the coil spring to the strut. Fit the lower end of the coil spring into the gap of the lower seat.

d. Install the spring seat with the insulator.

e. Install the dust seal on the spring seat.

f. Install the suspension support and tighten 35 ft. lbs. (47 Nm). After the nut has been tighten, release the compressor tool tension.

g. Pack multipurpose grease into the suspension support. Install the dust cover.

➡Do not use an impact wrench to tighten the nut. Also, check that the bearing fits into the recess in the suspension support.

STABILIZER BAR

REMOVAL & INSTALLATION

See Figure 200.

1. Remove the column hole cover silencer sheet.
2. Separate no. 2 steering intermediate shaft assembly.
3. Separate no. 1 steering column hole cover sub-assembly.
4. Remove front wheels.
5. Separate tie rod end sub-assembly.
6. Remove the front stabilizer link assembly.
7. Separate front lower suspension arm.
8. Remove front suspension cross-member sub-assembly.
9. Remove the 4 bolts, 2 No. 1 front stabilizer brackets and front stabilizer bar from the front suspension crossmember.
10. Remove the 2 No. 1 front stabilizer bar bushings from the front stabilizer bar.

To install:

11. To install, reverse the removal procedure.
12. Tighten the four bolts to 14 ft. lbs. (19 Nm).
13. Inspect and adjust front wheel alignment.

STEERING KNUCKLE

REMOVAL & INSTALLATION

See Figure 201.

1. Remove front wheel.
2. Remove front axle hub nut.
3. Separate front speed sensor.
4. Separate front disc brake caliper assembly.
5. Remove front disc.
6. Separate tie rod end sub-assembly.
7. Separate front lower suspension arm.
8. Remove front axle assembly.
9. Remove front lower ball joint
10. Remove front axle hub hole snap ring.
11. Remove front axle hub sub-assembly.
12. Remove front disc brake dust cover.
13. Place the front axle hub bearing inner race (outside) on the front axle hub bearing.
14. Using SST's 09223-15020, 09387-02010, 09950-60010, 09950-70010 and a press, remove the front axle hub bearing from the steering knuckle.

To install:

15. Using SST's and a press, install a new front axle hub bearing to the steering knuckle.
16. To compete the installation, reverse the removal procedure.
17. Check the wheel alignment and for speed sensor signal.

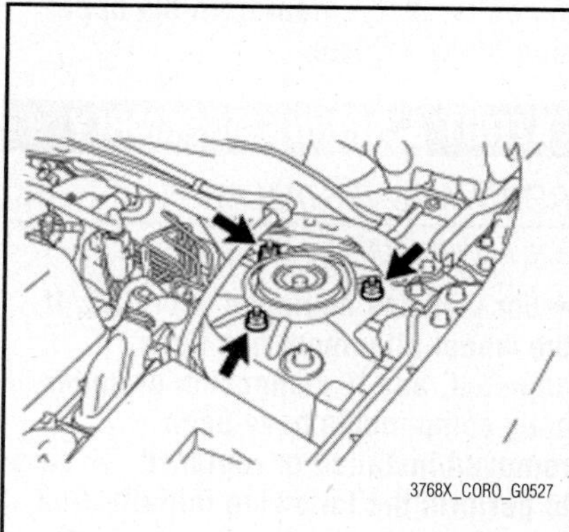
3768X_CORO_G0527

Fig. 200 Removing the 4 bolts, 2 No. 1 front stabilizer brackets and front stabilizer bar from the front suspension

WHEEL BEARINGS

REMOVAL & INSTALLATION

See Figures 202 through 204.

1. Remove front wheel.
2. Remove front axle hub nut. Refer to Front Brakes.
3. Separate front speed sensor. Refer to Front Brakes.
4. Separate front disc brake caliper assembly. Refer to Front Brakes.
5. Remove front disc.
6. Separate tie rod end sub-assembly.
7. Separate front lower suspension arm.
8. Put match marks on the front drive shaft assembly and the front axle hub sub-assembly.
9. Using a plastic hammer, separate the front drive shaft assembly from the front axle assembly.

➡Be careful not to damage the drive shaft boot or speed sensor rotor.

10. Remove the 2 bolts, 2 nuts, and front axle assembly.
11. Remove the lower ball joint.
12. Using snap ring pliers, remove the front axle hub hole snap ring.

➡Do not over tighten the vise.

13. Remove front axle hub sub-assembly:

a. Using a hub puller (SST 09520-00031), remove the front axle hub sub-assembly.

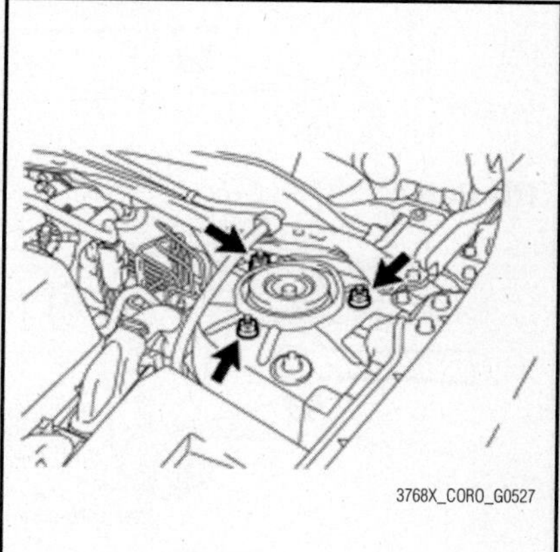
3768X_CORO_G0527

Fig. 201 Identifying the special tools and removing the steering knuckle

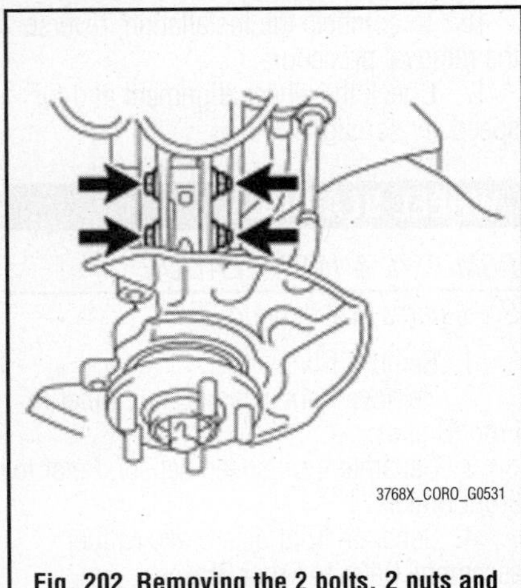

Fig. 202 Removing the 2 bolts, 2 nuts and front axle assembly

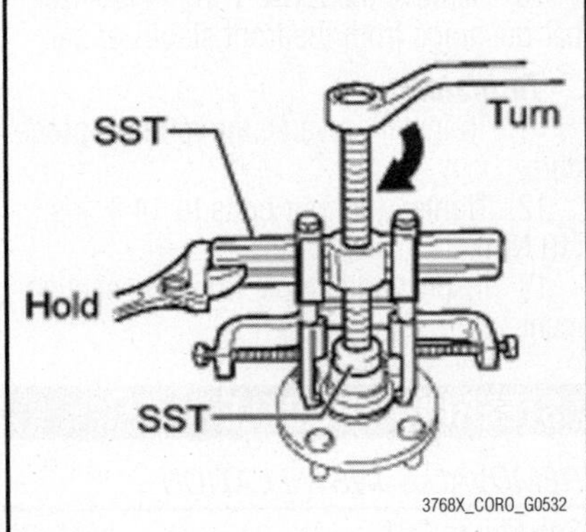

Fig. 203 Identifying the special tools and removing the bearing inner race from the front axle hub sub-assembly

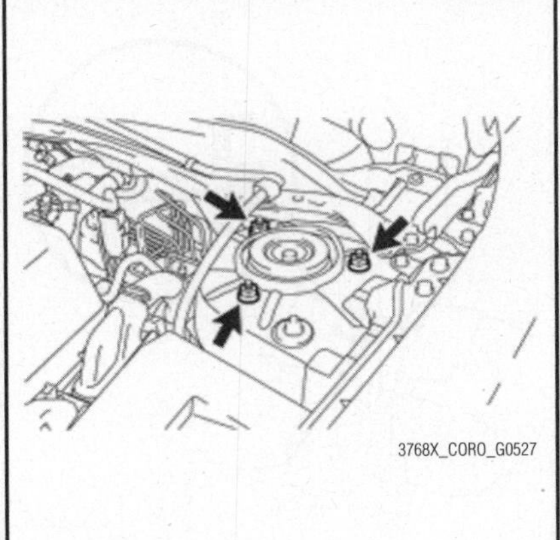

Fig. 204 Identifying the special tools and removing the steering knuckle

➡Apply a small amount of grease to the threads and tip of SST (09953-04020) before use.

b. Using SST 09950-40011 and 09950-60010, remove the bearing inner race (outside) from the front axle hub sub-assembly.

14. Remove the 3 bolts and the front disc brake dust cover from the steering knuckle.

15. Remove front axle hub bearing:

a. Place the front axle hub bearing inner race (outside) on the front axle hub bearing.

b. Using SST and a press, remove the front axle hub bearing from the steering knuckle.

16. Using SST's 09223-15020, 09387-02010, 09950-60010, 09950-70010 and a press, remove the front axle hub bearing from the steering knuckle.

SUSPENSION

REAR SUSPENSION

STABILIZER BAR

REMOVAL & INSTALLATION

See Figure 205.

1. Remove the rear axle beam damper.

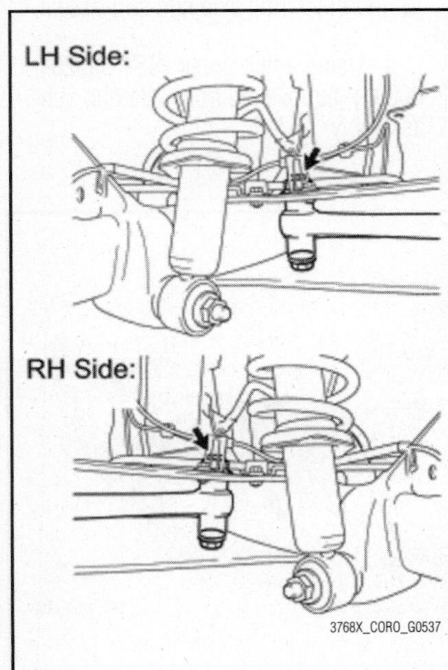

Fig. 205 Removing and installing the rear stabilizer bar

2. Remove the 2 bolts, 2 nuts and rear stabilizer bar.

➡Be sure to loosen the nuts.

To install:

3. To install, reverse removal procedure.
4. Check that the identification mark of the rear stabilizer bar is positioned on the right side of the vehicle.
5. Install the rear stabilizer bar with the 2 bolts and 2 nuts. Tighten to 184 ft. lbs. (250 Nm).

➡Be sure to tighten the nuts. If reusing the bolts, insert them from the upper side of the vehicle.

STRUT & SPRING ASSEMBLY

REMOVAL & INSTALLATION

See Figures 206 and 207.

➡For vehicles equipped with VSC, if the wheel alignment has been adjusted, and if suspension or underbody components have been removed/installed or replaced, be sure to perform the following initialization procedure in order for the system to function normally:

- Disconnect the cable from the negative (-) battery terminal for more than 2 seconds.
- Reconnect the cable to the negative (-) battery terminal.
- Perform zero point calibration of the yaw rate and acceleration sensor and test mode inspection.

1. Before servicing the vehicle, refer to the Precautions section.
2. Remove spare wheel cover.
3. Remove rear floor finish plate:

a. Using a clip remover, remove the 2 clips.

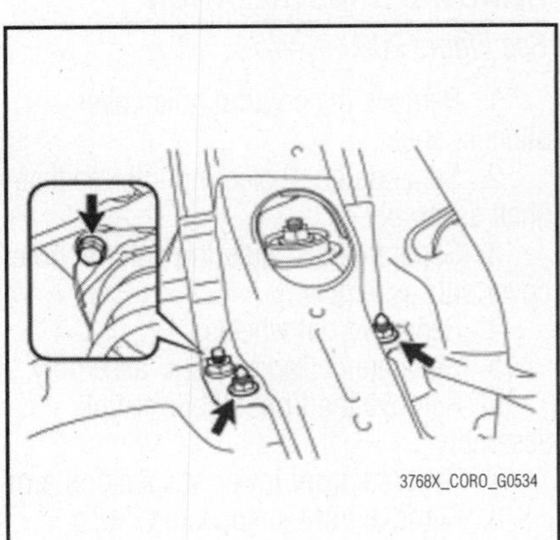

Fig. 206 Removing the 2 nuts from the rear shock absorber with coil spring (upper side)

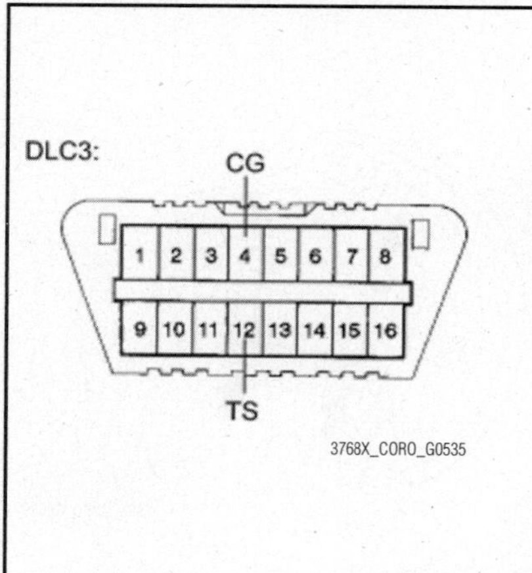

Fig. 207 Rear shock absorber and coil spring positioning

b. Disengage the 8 claws and remove the rear floor finish plate.

4. Remove the 4 clips and inner luggage compartment trim cover .

5. Remove rear wheel.

6. Support the rear axle beam assembly using a jack and wooden block.

7. Remove the nut and rear shock absorber cushion retainer.

8. Remove the 2 nuts from the rear shock absorber with coil spring (upper side).

9. Remove the bolt (lower side) from the rear shock absorber with coil spring.

10. Slowly lower the jack and remove the rear shock absorber with coil spring.

To install:

11. Install the rear shock absorber with coil spring with the bolt (lower side). Tighten to 59 ft. lbs. (80 Nm).

12. Install the rear shock absorber with coil spring with the 2 nuts (upper side). Tighten to Tighten to 59 ft. lbs. (80 Nm).

13. Support the rear axle beam using a jack and wooden block.

14. Temporarily install the rear shock absorber with coil spring (lower side) and rear shock absorber cushion retainer with the nut.

15. Install the rear wheel.

16. Lower the vehicle.

17. Bounce the vehicle up and down several times to stabilize the suspension.

18. Jack up the rear axle beam, placing a wooden block underneath to avoid damage. Apply load to the suspension so that the rear shock absorber with coil spring is positioned as shown in the illustration.

a. Length A: 8.11 inches

b. If the rear shock absorber with coil spring cannot be positioned as shown in the illustration even when the rear axle beam is jacked up, apply additional load to the vehicle such as by having a person sit in the rear seat.

CAUTION

Do not jack up the rear axle beam too high as the vehicle may fall.

19. Fully tighten the nut on the rear shock absorber with coil spring (lower side). Tighten to Tighten to 59 ft. lbs. (80 Nm).

➡The final torque must be applied under standard vehicle height conditions.

20. To complete the installation, reverse the removal procedure.

21. Perform yaw rate sensor zero point calibration.

22. Check steering angle sensor zero point calibration:

a. Drive the vehicle straight ahead at 35 km/h (22 mph) or more for at least 5 seconds.

b. Check that the center position of the steering wheel is correctly set while driving straight ahead.

➡If front wheel alignment and steering position are adjusted as a result of an off-centered position of the steering wheel, acquire yaw rate and acceleration sensor zero point again after the adjustments are completed.

c. If the center position of the steering wheel is correctly set, reconfirm the DTC.

d. If the center position of the steering wheel is not correctly set, adjust the front wheel alignment or steering position.

OVERHAUL

1. Place the strut assembly in a pipe vise or strut vise.

➡Do not attempt to clamp the strut assembly in a flat jaw vise as this will result in damage to the strut tube.

2. Compress the spring until the upper suspension support is free of any spring tension. Do not over-compress the spring.

3. Hold the upper support, then remove the nut on the end of the shock piston rod.

4. Remove the support, coil spring, insulator, and bumper.

5. Inspect the strut as follows:

a. Check the shock absorber by moving the piston shaft through its full range of travel. It should move smoothly and evenly throughout its entire travel without any trace of binding or notching.

b. Use a small straightedge to check the piston shaft for any bending or deformation.

c. Inspect the spring for any sign of deterioration or cracking. The waterproof coating on the coils should be intact to prevent rusting.

YAW RATE SENSOR ZERO POINT CALIBRATION (WITH VSC)

See Figure 208.

➡While obtaining the zero point, keep the vehicle stationary and do not vibrate, tilt, move, or shake it. (Do not start the engine.) Be sure to perform this procedure on a level surface (with an inclination of less than 1 degree).

1. Clear the zero point calibration data:

a. Turn the ignition switch off.

b. Check that the steering wheel is centered.

c. Check that the shift lever is in P (for automatic transaxle model) or the parking brake is applied (for manual transaxle model).

d. Turn the ignition switch to ON.

e. The warning and indicator light come on for 3 seconds to indicate that the initial check is completed.

f. Connect and disconnect terminals TS and CG of the DLC3 (ODB II connector) 4 times or more within 8 seconds.

g. Check that the VSC OFF indicator light comes on.

➡If the ignition switch is turned to ON for more than 15 seconds with the shift lever in P (for automatic transaxle model) or the parking brake applied (for manual transaxle model) after zero

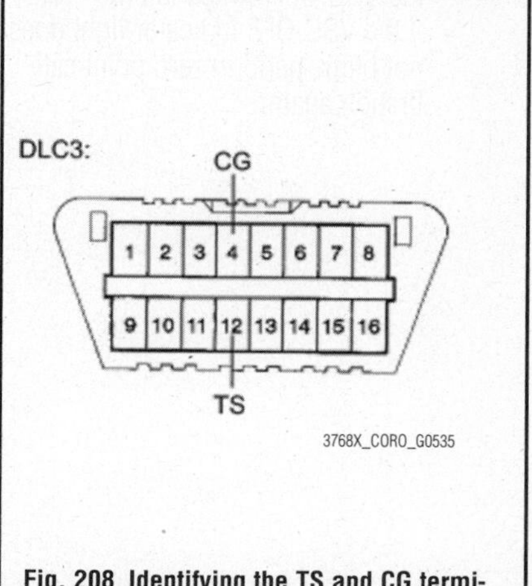

Fig. 208 Identifying the TS and CG terminals on the DLC3 (ODB II connector)

point of the yaw rate and acceleration sensor has been cleared, only the zero point of the yaw rate sensor will be stored. If the vehicle is driven under these conditions, the skid control ECU will store the zero point calibration for the acceleration sensor as not being completed. The skid control ECU will then also indicate this as a malfunction of the VSC system using the indicator lights.

2. Perform zero point calibration of the yaw rate and acceleration sensor:
    a. Turn the ignition switch off.
    b. Check that the steering wheel is centered.
    c. Check that the shift lever is in P (for automatic transaxle model) or the parking brake is applied (for manual transaxle model).

➡DTCs C1210/36 and C1336/39 will be recorded if the shift lever is not in P (for automatic transaxle model) or the parking brake is not applied (for manual transaxle model)

    d. Connect terminals TS and CG of the DLC3 (ODB II connector).
    e. Turn the ignition switch to ON.
    f. Keep the vehicle stationary on a level surface for 2 seconds or more.
    g. Check that the VSC OFF indicator light comes on for several seconds and then blinks in Test Mode.

➡

- The slip indicator light remains on during Test Mode because traction control operation is prohibited (The slip indicator light goes off when the VSC OFF switch is on).
- If the VSC OFF indicator light does not blink, perform zero point calibration again.
- The zero point calibration is performed only once after the system enters Test Mode.
- Calibration cannot be performed again until the stored data is cleared.

    h. Turn the ignition switch off and disconnect SST from the DLC3 (ODB II connector).

WHEEL BEARINGS

REMOVAL & INSTALLATION

See Figures 209 and 210.

1. Disconnect cable from negative battery terminal.
2. Remove lower instrument panel finish panel.
3. Remove shift lever knob sub-assembly.
4. Remove center instrument cluster finish panel assembly.
5. Remove upper console panel sub-assembly.
6. Loosen parking brake cable (for rear disc brake).
7. Remove rear wheel.
8. Disconnect rear speed sensor wire (for rear drum brake).
9. Disconnect rear speed sensor wire (for rear disc brake).
10. Remove parking brake lever protector (for rear disc brake).
11. Separate no. 3 parking brake cable assembly (for rear disc brake).
12. Separate rear disc brake caliper assembly (for rear disc brake).
13. Remove rear disc (for rear disc brake).
14. Remove rear brake drum (for rear drum brake).
15. Remove the 4 bolts and the rear axle hub and bearing assembly.

To install:

16. Install the rear axle hub and bearing assembly with the 4 bolts. Tighten to 74 ft. lbs. (100 Nm).

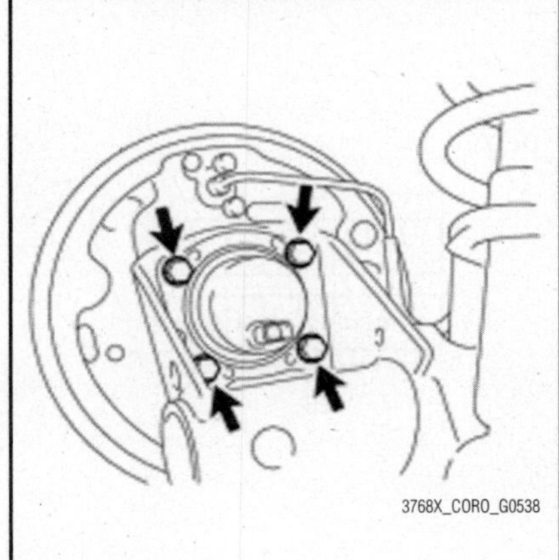

Fig. 209 Removing the 4 bolts and the rear axle hub and bearing assembly—Drum brakes

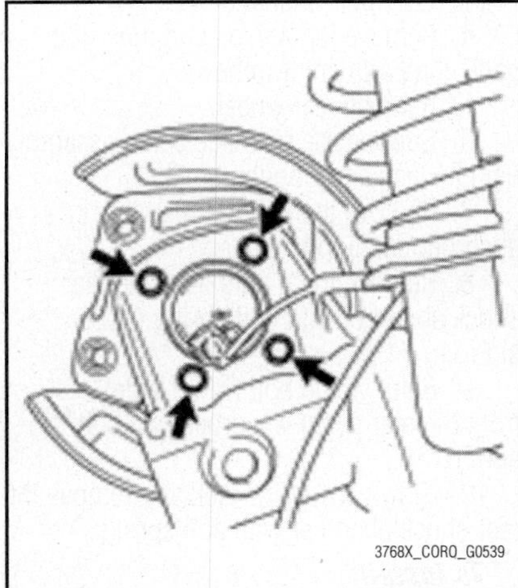

Fig. 210 Removing the 4 bolts and the rear axle hub and bearing assembly—Disc brakes

17. To complete installation, reverse remaining removal procedure.
18. Inspect rear wheel alignment.
19. Check for speed sensor signal.

TOYOTA

FJ Cruiser

7

SPECIFICATIONS AND MAINTENANCE CHARTS

ENGINE AND VEHICLE IDENTIFICATION

Engine						
Code/ID ①	Liters (cc)	Cu. In.	Cyl.	Fuel Sys.	Engine Type	Eng. Mfg.
1GR-FE	4.0 (3956)	241	6	SFI	DOHC	Toyota

Model Year	
Code ②	Year
B	2011
C	2012

DOHC: Double Overhead Camshaft

① 1GR-FE engine: stamped on the right side of the engine block. Engine ID is the fifth character of the VIN number.

② 10th digit of the VIN number

71099_FJCR_C0001

GENERAL ENGINE SPECIFICATIONS

Year	Model	Engine Displacement Liters	Engine Series Code/ID	Net Horsepower @ rpm	Net Torque @ rpm (ft. lbs.)	Bore x Stroke (in.)	Com-pression Ratio	Oil Pressure @ rpm
2011	FJ Cruiser	4.0	1GR-FE	260@5600	271@4400	3.70x3.74	10.4:1	43-85@3000
2012	FJ Cruiser	4.0	1GR-FE	260@5600	271@4400	3.70x3.74	10.4:1	43-85@3000

71099_FJCR_C0002

TUNE-UP SPECIFICATIONS

Year	Engine Displacement Liters	Engine Code/ID	Spark Plug Gap (in.)	Ignition Timing (deg.)	Fuel Pump (psi)	Idle Speed (rpm)		Valve Clearance	
						MT	AT	Intake	Exhaust
2011	4.0	1GR-FE	0.039-0.043	7-24B ①	41-42	650-750	650-750	NA	NA
2012	4.0	1GR-FE	0.039-0.043	7-24B ①	41-42	650-750	650-750	NA	NA

NOTE: The Vehicle Emission Control Information label often reflects specification changes made during production.

The label figures must be used if they differ from those in this chart.

B: Before top dead center idle

① With terminals TC and CG of the DLC3 (ODB II connector) disconnected

71099_FJCR_C0003

CAPACITIES

Year	Model	Engine Displacement Liters	Engine Code/ID	Engine Oil with Filter (qts.)	Transmission (pts.) 6-Spd	Transmission (pts.) Auto.	Transfer Case (pts.)	Drive Axle Front (pts.)	Drive Axle Rear (pts.)	Fuel Tank (gal.)	Cooling System (qts.)
2011	FJ Cruiser	4.0	1GR-FE	6.4	3	3.6	①	②	6.4 ③	21.1	⑤
2012	FJ Cruiser	4.0	1GR-FE	6.4	3	3.6	①	④	6.4 ③	21.1	⑤

① MT: 3.0 pts
AT: 2.2 pts
② Full Time 4WD: 3.0 pts
Part Time 4WD: 3.2 pts
③ W/ Locking Differential: 6.2 pts
④ MT: 3.0 pts
AT: 3.4 pts
⑤ MT: 9.9
AT: 10.4

71099_FJCR_C0004

FLUID SPECIFICATIONS

Year	Model	Engine Displ. Liters	Engine Oil	Man. Trans.	Auto. Trans.	Drive Axle Front	Drive Axle Rear	Transfer Case	Power Steering Fluid	Brake Master Cylinder	Cooling System
2011	FJ Cruiser	4.0	0W-20	75W-90	ATF WS	①	80W-90	75W-90	Dexron® II or III	DOT 3	S-LLC
2012	FJ Cruiser	4.0	0W-20	75W-90	ATF WS	②	80W-90	75W-90	Dexron® II or III	DOT 3	S-LLC

DOT: Department Of Transportation
S-LLC: Toyota Super Long Life Coolant
① Full Time 4WD: 80W-90
Part Time 4WD: 75W-85
② MT: 80W-90
AT: 75W-85

71099_FJCR_C0014

VALVE SPECIFICATIONS

Year	Engine Displacement Liters	Engine Code/ID	Seat Angle (deg.)	Face Angle (deg.)	Spring Test Pressure (lbs. @ in.)	Spring Installed Height (in.)	Stem-to-Guide Clearance (in.) Intake	Stem-to-Guide Clearance (in.) Exhaust	Stem Diameter (in.) Intake	Stem Diameter (in.) Exhaust
2011	4.0	1GR-FE	NA	45.5	41.9-46.3@ 1.311	1.910	0.0010-0.0024	0.0012-0.0026	0.2154-0.2159	0.2152-0.2158
2012	4.0	1GR-FE	NA	45.5	41.9-46.3@ 1.311	1.910	0.0010-0.0024	0.0012-0.0026	0.2154-0.2160	0.2152-0.2160

NA: Not Available

71099_FJCR_C0005

CAMSHAFT SPECIFICATIONS

All measurements in inches unless noted

Year	Engine Displacement Liters	Engine Code/ID	Journal Dia.	Brg. Oil Clearance	Shaft End-play	Circle Runout	Lobe Height	
							Intake	Exhaust
2011	4.0	1GR-FE	①	②	0.00315-0.0512	0.00157	1.728-1.7320	1.743-1.7470
2012	4.0	1GR-FE	①	②	0.00315-0.0512	0.00157	1.728-1.7320	1.743-1.7470

① No. 1: 1.4152-1.4157
All others: 0.9039-0.9045

② No. 1: 0.00157-0.00311
All others: 0.00984-0.00244

71099_FJCR_C0006

CRANKSHAFT AND CONNECTING ROD SPECIFICATIONS

All measurements are given in inches.

Year	Engine Displacement Liters	Engine Code/ID	Crankshaft				Connecting Rod		
			Main Brg. Journal Dia.	Main Brg. Oil Clearance	Shaft End-play	Thrust on No.	Journal Diameter	Oil Clearance	Side Clearance
2011	4.0	1GR-FE	2.8342-2.8346	0.0007-0.0012	0.0016-0.0094	NA	2.2044-2.2047	0.0010-0.0018	0.0059-0.0118
2012	4.0	1GR-FE	2.8342-2.8346	0.0007-0.0012	0.0016-0.0094	NA	2.2044-2.2047	0.0010-0.0018	0.0059-0.0118

NA: Not Available

71099_FJCR_C0007

PISTON AND RING SPECIFICATIONS

All measurements are given in inches.

Year	Engine Displacement Liters	Engine Code/ID	Piston Clearance	Ring Gap			Ring Side Clearance		
				Top Compression	Bottom Compression	Oil Control	Top Compression	Bottom Compression	Oil Control
2011	4.0	1GR-FE	0.0035-0.0020	0.008-0.0142	0.0138-0.0193	0.0039-0.0173	0.0007-0.0027	0.0007-0.0023	0.0027-0.0059
2012	4.0	1GR-FE	0.0035-0.0020	0.008-0.0142	0.0138-0.0193	0.0039-0.0173	0.0007-0.0027	0.0007-0.0023	0.0027-0.0059

71099_FJCR_C0008

TORQUE SPECIFICATIONS

All readings in ft. lbs.

Year	Engine Displacement Liters	Engine Code/ID	Cylinder Head Bolts	Main Bearing Bolts	Rod Bearing Bolts	Crankshaft Damper Bolts	Flywheel Bolts	Manifold Intake	Manifold Exhaust	Spark Plugs	Oil Pan Drain Plug
2011	4.0	1GR-FE	①	②	③	185	61	19	22	15	30
2012	4.0	1GR-FE	①	②	③	185	61	19	22	15	30

① Right side: 27 ft. lbs. plus 180 degrees

Left side (recessed head): 27 ft. lbs. plus 180 degrees

Left side (0.55 inch head): 22 ft. lbs.

② 12 pointed head: 45 ft. lbs. plus 90 degrees

12mm head: 18 ft. lbs.

③ Step 1: 18 ft. lbs.

Step 2: Plus 90 degrees

71099_FJCR_C0009

WHEEL ALIGNMENT

Year	Model	Caster Range (+/-Deg.)	Caster Preferred Setting (Deg.)	Camber Range (+/-Deg.)	Camber Preferred Setting (Deg.)	Toe-in (in.)	Steering Axis Inclination (Deg.)
2011	FJ Cruiser 2WD	0.50	3.57	0.50	-0.57	0.04+/-0.08	12.92+/-0.50
	FJ Cruiser 4WD	0.50	2.82	0.50	0.15	0.04+/-0.08	12.35+/-0.50
2012	FJ Cruiser 2WD	0.50	3.57	0.50	-0.57	0.04+/-0.08	12.92+/-0.50
	FJ Cruiser 4WD	0.50	2.82	0.50	0.15	0.04+/-0.08	12.35+/-0.50

NOTE: All alignment figures based on the following nominal ride heights:

2WD Front: 4.56 in.

2WD Rear: 3.20 in.

4WD Front: 3.43 in.

4WD Rear: 2.43 in.

71099_FJCR_C0010

TIRE, WHEEL AND BALL JOINT SPECIFICATIONS

Year	Model	OEM Tires		Tire Pressures (psi)		Wheel	Ball Joint	Lug Nut Torque
		Standard	Optional	Front	Rear	Size	Inspection	(ft. lbs.)
2011	FJ Cruiser	P265/70R17	NA	32	32	NA	①	82
2012	FJ Cruiser	P265/70R17	NA	32	32	NA	①	82

NA: Not available

① Lower ball joint excessive play, all models: 0.020 inch

71099_FJCR_C0011

BRAKE SPECIFICATIONS

All measurements in inches unless noted

Year	Model		Brake Disc Original Thickness	Brake Disc Minimum Thickness	Brake Disc Maximum Runout	Minimum Lining Thickness Front	Minimum Lining Thickness Rear	Brake Caliper Bracket Bolts (ft. lbs.)	Brake Caliper Mounting Bolts (ft. lbs.)
2011	FJ Cruiser	Front	1.102	1.024	0.0020	0.039	—	NA	91
		Rear	0.709	0.630	0.0079	—	0.039	NA	65
2012	FJ Cruiser	Front	1.102	1.024	0.0020	0.039	—	NA	91
		Rear	0.709	0.630	0.0079	—	0.039	NA	65

NA: Not Available

71099_FJCR_C0012

SCHEDULED MAINTENANCE INTERVALS

TOYOTA—FJ Cruiser

TO BE SERVICED	TYPE OF SERVICE	VEHICLE MILEAGE INTERVAL (x1000)																		
		5	10	15	20	25	30	35	40	45	50	55	60	65	70	75	80	85	90	95
Automatic transmission and differential fluid	S/I			✓			✓			✓			✓			✓			✓	
Ball joints and boots	S/I			✓			✓			✓			✓			✓			✓	
Brake system	S/I	✓	✓	✓	✓	✓	✓	✓	✓	✓	✓	✓	✓	✓	✓	✓	✓	✓	✓	✓
Charcoal canister	S/I												✓							
Drive belts	S/I						✓						✓						✓	
Driveshaft bushing	L						✓						✓						✓	
Engine coolant	R			✓			✓			✓			✓			✓			✓	
Engine oil & filter	R		✓		✓		✓		✓		✓		✓		✓		✓		✓	
Exhaust pipes & mounts	S/I			✓			✓			✓			✓			✓			✓	
Fuel tank cap gasket	S/I						✓						✓						✓	
Halfshaft boots & flange bolts	S/I			✓			✓			✓			✓			✓			✓	
Limited slip differential fluid	R						✓						✓						✓	
Manual transmission and differential fluid	S/I						✓						✓						✓	
Platinum spark plugs	R												✓							
Propeller shaft bolts	S/I			✓			✓			✓			✓			✓			✓	
Rack and pinion assembly	S/I			✓			✓			✓			✓			✓			✓	
Tires (rotate)	S/I	✓	✓	✓	✓	✓	✓	✓	✓	✓	✓	✓	✓	✓	✓	✓	✓	✓	✓	✓
Transfer case and differential fluid	S/I			✓			✓			✓			✓			✓			✓	
Valves	S/I												✓							

R: Replace S/I: Service or Inspect L: Lubricate

FREQUENT OPERATION MAINTENANCE (SEVERE SERVICE)

If a vehicle is operated under any of the following conditions it is considered severe service:

- Towing a trailer or using a camper or car-top carrier.

- Repeated short trips of less than 5 miles in temperatures below freezing.

- Excessive idling or low-speed driving for long distances as in heavy commercial use, such as delivery, taxi or police cars.

- Operating on rough, muddy or salt-covered roads.

- Operating on unpaved or dusty roads.

Oil filter: service or inspect every 5000 miles or 4 months, whichever occurs first.

Brake linings and discs or drums: service or inspect every 5000 miles or 4 months, whichever occurs first.

Steering linkage: service or inspect every 5000 miles or 4 months, whichever occurs first.

Ball joints and boots: service or inspect every 5000 miles or 4 months, whichever occurs first.

Brake discs & pads (front): service or inspect every 6000 miles.

Halfshaft boots: service or inspect every 5000 miles or 4 months. Retighten the flange bolts, whichever occurs first.

Body chassis bolts and nuts: service or inspect every 5000 miles or 4 months, whichever occurs first.

Transmission and differential fluid: replace every 15,000 miles or 12 months, whichever occurs first.

Transfer case and differential fluid: replace every 15,000 miles or 12 months, whichever occurs first.

71099_FJCR_C0013

PRECAUTIONS

Before servicing any vehicle, please be sure to read all of the following precautions, which deal with personal safety, prevention of component damage, and important points to take into consideration when servicing a motor vehicle:

- Never open, service or drain the radiator or cooling system when the engine is hot; serious burns can occur from the steam and hot coolant.
- Observe all applicable safety precautions when working around fuel. Whenever servicing the fuel system, always work in a well-ventilated area. Do not allow fuel spray or vapors to come in contact with a spark, open flame, or excessive heat (a hot drop light, for example). Keep a dry chemical fire extinguisher near the work area. Always keep fuel in a container specifically designed for fuel storage; also, always properly seal fuel containers to avoid the possibility of fire or explosion. Refer to the additional fuel system precautions later in this section.
- Fuel injection systems often remain pressurized, even after the engine has been turned **OFF**. The fuel system pressure must be relieved before disconnecting any fuel lines. Failure to do so may result in fire and/or personal injury.
- Brake fluid often contains polyglycol ethers and polyglycols. Avoid contact with the eyes and wash your hands thoroughly after handling brake fluid. If you do get brake fluid in your eyes, flush your eyes with clean, running water for 15 minutes. If eye irritation persists, or if you have taken brake fluid internally, IMMEDIATELY seek medical assistance.
- The EPA warns that prolonged contact with used engine oil may cause a number of skin disorders, including cancer. You should make every effort to minimize your exposure to used engine oil. Protective gloves should be worn when changing oil. Wash your hands and any other exposed skin areas as soon as possible after exposure to used engine oil. Soap and water, or waterless hand cleaner should be used.
- All new vehicles are now equipped with an air bag system, often referred to as a Supplemental Restraint System (SRS) or Supplemental Inflatable Restraint (SIR) system. The system must be disabled before performing service on or around system components, steering column, instrument panel components, wiring and sensors. Failure to follow safety and disabling procedures could result in accidental air bag deployment, possible personal injury and unnecessary system repairs.
- Always wear safety goggles when working with, or around, the air bag system. When carrying a non-deployed air bag, be sure the bag and trim cover are pointed away from your body. When placing a non-deployed air bag on a work surface, always face the bag and trim cover upward, away from the surface. This will reduce the motion of the module if it is accidentally deployed. Refer to the additional air bag system precautions later in this section.
- Clean, high quality brake fluid from a sealed container is essential to the safe and proper operation of the brake system. You should always buy the correct type of brake fluid for your vehicle. If the brake fluid becomes contaminated, completely flush the system with new fluid. Never reuse any brake fluid. Any brake fluid that is removed from the system should be discarded. Also, do not allow any brake fluid to come in contact with a painted surface; it will damage the paint.
- Never operate the engine without the proper amount and type of engine oil; doing so WILL result in severe engine damage.
- Timing belt maintenance is extremely important. Many models utilize an interference-type, non-freewheeling engine. If the timing belt breaks, the valves in the cylinder head may strike the pistons, causing potentially serious (also time-consuming and expensive) engine damage. Refer to the maintenance interval charts for the recommended replacement interval for the timing belt, and to the timing belt section for belt replacement and inspection.
- Disconnecting the negative battery cable on some vehicles may interfere with the functions of the on-board computer system(s) and may require the computer to undergo a relearning process once the negative battery cable is reconnected.
- When servicing drum brakes, only disassemble and assemble one side at a time, leaving the remaining side intact for reference.
- Only an MVAC-trained, EPA-certified automotive technician should service the air conditioning system or its components.

BRAKES

ANTI-LOCK BRAKE SYSTEM (ABS)

GENERAL INFORMATION

PRECAUTIONS

- Certain components within the ABS system are not intended to be serviced or repaired individually.
- Do not use rubber hoses or other parts not specifically specified for and ABS system. When using repair kits, replace all parts included in the kit. Partial or incorrect repair may lead to functional problems and require the replacement of components.
- Lubricate rubber parts with clean, fresh brake fluid to ease assembly. Do not use shop air to clean parts; damage to rubber components may result.
- Use only DOT 3 brake fluid from an unopened container.
- If any hydraulic component or line is removed or replaced, it may be necessary to bleed the entire system.
- A clean repair area is essential. Always clean the reservoir and cap thoroughly before removing the cap. The slightest amount of dirt in the fluid may plug an orifice and impair the system function. Perform repairs after components have been thoroughly cleaned; use only denatured alcohol to clean components. Do not allow ABS components to come into contact with any substance containing mineral oil; this includes used shop rags.
- The Anti-Lock control unit is a microprocessor similar to other computer units in the vehicle. Ensure that the ignition switch is **OFF** before removing or installing controller harnesses. Avoid static electricity discharge at or near the controller.
- If any arc welding is to be done on the vehicle, the control unit should be unplugged before welding operations begin.

WHEEL SPEED SENSORS

REMOVAL & INSTALLATION

Front

See Figure 1.

1. Before servicing the vehicle, refer to the precautions.
2. Disconnect the negative battery cable.
3. Raise and safely support the vehicle securely on jackstands.
4. Remove front wheel
5. Disconnect the skid control sensor wire.

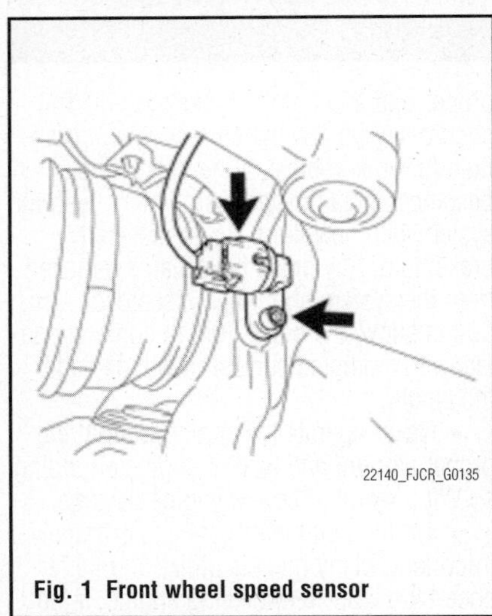

Fig. 1 Front wheel speed sensor

6. Using a 5 mm hexagon wrench, remove the bolt and the front speed sensor.

WARNING

Do not let any foreign matter attach to the sensor tip or enters the sensor bore.

To install:

WARNING

Make sure that the sensor tip is clean.

7. Install the wheel speed sensor and using a 5 mm hexagon wrench, tighten the bolt to 73 inch lbs. (98 Nm).
8. Connect the skid control sensor wire.
9. Install front wheel and lower the vehicle.
10. Connect the negative battery cable.
11. Using a scan tool, check VSC sensor signal.

Rear

See Figure 2.

1. Before servicing the vehicle, refer to the precautions.
2. Disconnect the negative battery cable.
3. Raise and safely support the vehicle securely on jackstands.
4. Remove rear wheel
5. Disconnect the skid control sensor wire.
6. Remove the nut and the rear speed sensor.

WARNING

Do not let any foreign matter attach to the sensor tip or enters the sensor bore.

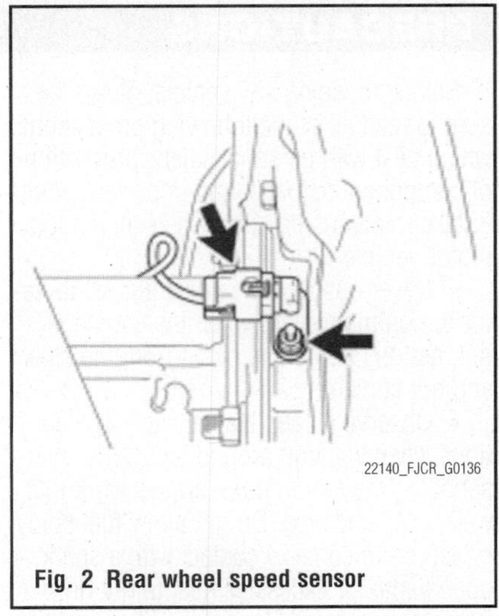

Fig. 2 Rear wheel speed sensor

To install:

WARNING

Make sure that the sensor tip is clean.

7. Install the wheel speed sensor and tighten the bolt to 73 inch lbs. (98 Nm).
8. Connect the skid control sensor wire.
9. Install rear wheel and lower the vehicle.
10. Connect the negative battery cable.
11. Using a scan tool, check VSC sensor signal.

BRAKES — BLEEDING THE BRAKE SYSTEM

BLEEDING & FILLING PROCEDURE

See Figure 3.

1. Before servicing the vehicle, refer to the precautions.

WARNING

Depressing the brake pedal with the reservoir cap removed will cause the fluid to spray. When bleeding, maintain the amount of fluid in the reservoir between the Min. and Max. lines.

2. Fill reservoir with brake fluid.
3. Bleed brake booster with accumulator pump assembly.
4. If the brake master cylinder is disassembled, the brake line is disconnected from the brake master cylinder or if the reservoir becomes empty, bleed the brake master cylinder.
   a. Turn the ignition switch to **ON**, and wait until the pump motor has stopped. Pump operating sound can be heard.
   b. Turn the ignition switch to **OFF**, and depress the brake pedal more than 20 times.
   c. When pressure in the accumulator is released, the reaction force becomes lighter and the stroke becomes longer.
   d. Repeat the first two steps 5 times.
   e. Turn the ignition switch to **ON**, and check that the pump stops after approximately 8 to 14 seconds.

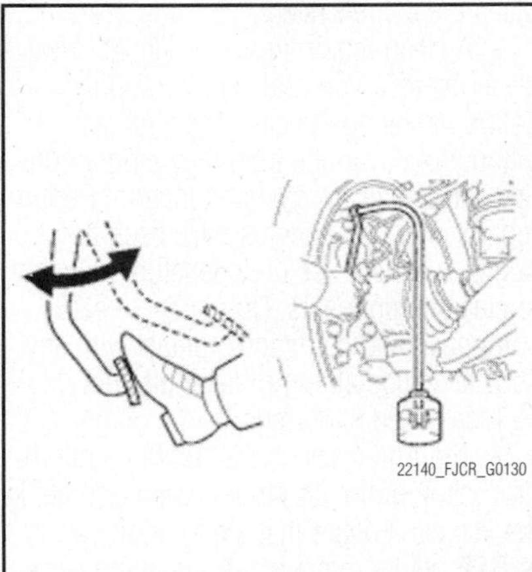

Fig. 3 Connect the vinyl tube to the brake caliper

WARNING

If the pump does not stop, repeat the procedure again.

5. Bleed brake line.
   a. Turn the ignition switch to **ON**, and wait until the pump motor has stopped. Pump operating sound can be heard.
   b. For the front brake line, connect the vinyl tube to the brake caliper. Depress the brake pedal several times, then loosen the bleeder plug with the pedal held down.
   c. At the point when the fluid stops coming out, tighten the bleeder plug, then release the brake pedal.
   d. Repeat the steps until all the air in the fluid has been bled out. Tighten the bleeder plug to 8 ft. lbs. (11 Nm).
   e. Repeat the above procedures to bleed the other brake line.
6. For the rear brake line, connect the vinyl tube to the brake caliper. Depress the brake pedal, hold it, and then loosen the bleeder plug. Brake fluid is pumped out automatically.

a. Loosen the bleeder plug and release the air. Keep the brake fluid in the reservoir tank above the **MIN** line during the above procedures.

b. When the air is completely bled out of the brake fluid through the bleeder plug, tighten the bleeder plug. Tighten the bleeder plug to 8 ft. lbs. (11 Nm).

c. Repeat the above procedures to bleed the other brake line.

7. Bleed master cylinder solenoid.

8. If the brake master cylinder is disassembled, the brake line is disconnected from the brake master cylinder or if the reservoir becomes empty, bleed the brake master cylinder.

a. Connect the scan tool to the diagnostic link connector.

b. Turn the ignition switch to **ON**.

c. Select **ACTIVE TEST** mode on the scan tool.

d. Connect the vinyl tube to the rear brake caliper.

e. Loosen the bleeder plug.

f. Select **SRMF** to drive the solenoids and bleed air from the rear brake caliper.

⁂ WARNING

Do not depress the brake pedal. Keep the brake fluid in the reservoir tank above the MIN line during the above procedures. Brake fluid is sent through the pump.

➡To protect the solenoids, the Techstream turns OFF automatically 2 seconds after every solenoid has been turned ON.

g. Repeat steps until all the air in the brake fluid is bled out.

h. When the air is completely bled out of the brake fluid through the bleeder plug, tighten the bleeder plug to 8 ft. lbs. (11 Nm).

i. Repeat the above procedures to bleed the other brake line.

j. Turn the ignition switch to **OFF**. Turn the ignition switch to **ON**.

k. Clear DTC.

9. Check fluid level in reservoir

a. Turn the ignition switch to **OFF**, and depress the brake pedal more than 20 times (until the pedal reaction feels light and pedal stroke becomes longer), and adjust the fluid level to the MAX level.

b. When the ignition switch is turned to **ON**, brake fluid is sent to the accumulator and the fluid level decreases by approximately 5 mm from the level when the ignition switch is **OFF** (normal).

BRAKES — FRONT DISC BRAKES

BRAKE CALIPER

REMOVAL & INSTALLATION

See Figures 4 and 5.

1. Before servicing the vehicle, refer to the precautions.
2. Raise and safely support the vehicle securely on jackstands.
3. Remove the front wheels.
4. Drain brake fluid.
5. Remove front disc brake anti-rattle with hole pin.

a. Remove the 2 pin hold clips, then remove the 2 hole pins from the disc brake caliper.

b. Remove the anti-rattle spring from the disc brake pad.

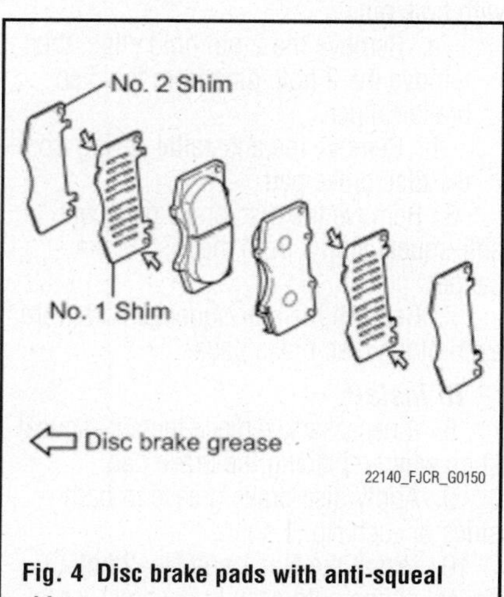

Fig. 4 Disc brake pads with anti-squeal shims

6. Remove the disc brake pads with anti-squeal shims from the disc brake caliper.
7. Remove the anti-squeal shims from each of the disc brake pads.
8. Using a union nut wrench, separate the brake tube from the disc brake caliper.

➡Use a container to collect the brake fluid as it drains out.

9. Remove the disc brake caliper.

To install:

10. Install the disc brake caliper and tighten the two bolts to 91 ft. lbs. (123 Nm).
11. Install the brake tube onto the disc brake caliper and tighten to 10 ft. lbs. (14 Nm).

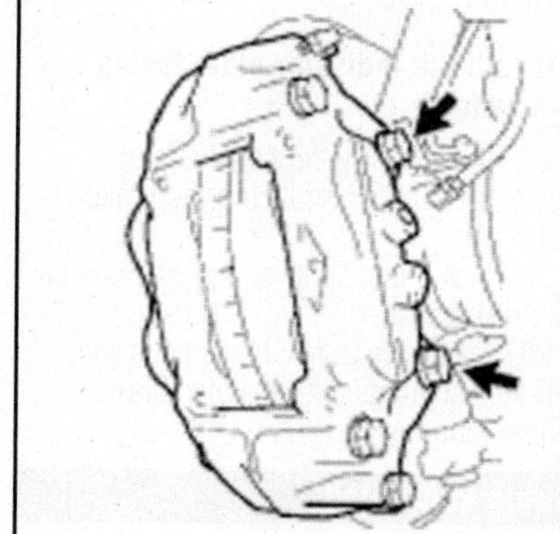

Fig. 5 Disc brake caliper retaining bolts

12. If necessary, replace the anti-squeal shim when replacing the brake pad.
13. Apply disc brake grease to both sides of each No. 1 shim.
14. Install the No. 1 and No. 2 anti-squeal shims onto each brake pad.
15. Install the disc brake pads with anti-squeal shims onto the disc brake caliper.

⁂ WARNING

There should be no oil or grease on the friction surfaces of the disc brake pads or the front disc.

16. Install the anti-rattle spring and hole pins onto the disc brake caliper.

➡The anti-rattle spring is installed onto the lower hole pin.

17. Install the pin hold clip with its handle oriented in the center of the vehicle.
18. Fill reservoir with brake fluid.
19. Bleed the brakes.
20. Install the front wheel and lower the vehicle.

DISC BRAKE PADS

REMOVAL & INSTALLATION

See Figure 4.

1. Before servicing the vehicle, refer to the precautions.
2. Raise and safely support the vehicle securely on jackstands.
3. Remove the front wheels.

4. Drain brake fluid.
5. Remove front disc brake anti-rattle with hole pin.
   a. Remove the 2 pin hold clips, then remove the 2 hole pins from the disc brake caliper.
   b. Remove the anti-rattle spring from the disc brake pad.
6. Remove the disc brake pads with anti-squeal shims from the disc brake caliper.
7. Remove the anti-squeal shims from each of the disc brake pads.

To install:

8. If necessary, replace the anti-squeal shim when replacing the brake pad.
9. Apply disc brake grease to both sides of each No. 1 shim.
10. Install the No. 1 and No. 2 anti-squeal shims onto each brake pad.
11. Install the disc brake pads with anti-squeal shims onto the disc brake caliper.

**** WARNING**

There should be no oil or grease on the friction surfaces of the disc brake pads or the front disc.

12. Install the anti-rattle spring and hole pins onto the disc brake caliper.

➡The anti-rattle spring is installed onto the lower hole pin.

13. Install the pin hold clip with its handle oriented in the center of the vehicle.
14. Fill reservoir with brake fluid.
15. Bleed the brakes.
16. Install the front wheel and lower the vehicle.

BRAKES

REAR DISC BRAKES

BRAKE CALIPER

REMOVAL & INSTALLATION

See Figure 6.

1. Before servicing the vehicle, refer to the precautions.
2. Raise and safely support the vehicle securely on jackstands.
3. Remove the rear wheels.
4. Drain brake fluid.
5. Using a union nut wrench, separate the brake tube from the disc brake caliper.

➡Use a container to collect the brake fluid as it drains out.

6. Remove the disc brake caliper by remove the 2 slide pins.
7. Remove the 2 disc brake pads with anti-squeal shim from the disc brake caliper mounting.
8. Remove the 2 anti-squeal shims from each brake.
9. Remove the indicator plate from the inner side of the brake.

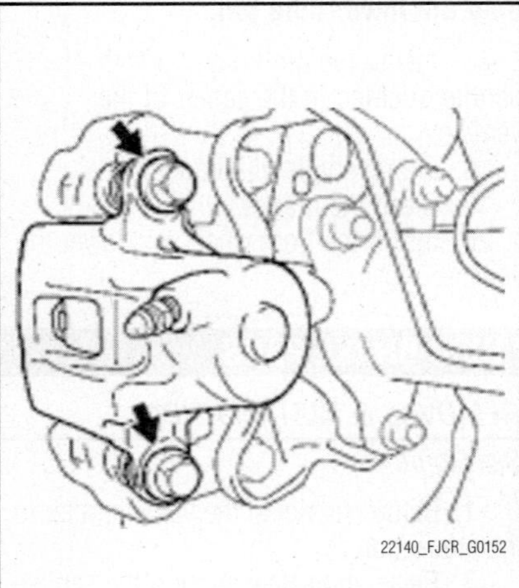

Fig. 6 Disc brake caliper retaining bolts

10. Remove the 4 pad support plates from the disc brake caliper mounting.
11. Remove the disc brake caliper mounting.
12. Remove the slide bushing from the disc brake caliper mounting.
13. Remove the dust boot from the disc brake caliper mounting.
14. Remove the hole plug from the disc brake caliper mounting.

To install:

15. Install a new hole plug onto the disc brake caliper mounting.
16. Apply lithium soap base glycol grease to a new dust boot.
17. Install the dust boot onto the disc brake caliper mounting.
18. Apply lithium soap base glycol grease to a new slide bushing.
19. Install the slide bushing onto the disc brake caliper mounting.
20. Install the disc brake caliper mounting with the 2 bolts and 2 washers. Tighten to 78 ft. lbs. (105 Nm).
21. Install the 4 pad support plates onto the disc brake caliper mounting.
22. Install the indicator plate onto the inner side brake pad.

➡Install the indicator plate facing downward.

23. If necessary, replace the anti-squeal shim when replacing the brake pad.
24. Install the 2 anti-squeal shims onto the brake pads.
25. Install the 2 disc brake pads with anti-squeal shims onto the disc brake caliper mounting.

**** WARNING**

There should be no oil or grease on the friction surfaces of the disc brake pads or the rear disc.

26. Apply lithium soap base glycol grease to the 2 slide pins.
27. Install the disc brake caliper onto the disc brake caliper mounting with the 2 slide pins. Tighten to 65 ft. lbs. (88 Nm).
28. Install the flexible hose with the union bolt and a new gasket. Tighten to 23 ft. lbs. (31 Nm).
29. Fill reservoir with brake fluid.
30. Bleed brake system.
31. Check fluid level in reservoir.
32. Check for brake fluid leakage.
33. Install rear wheel.

DISC BRAKE PADS

REMOVAL & INSTALLATION

See Figure 4

1. Before servicing the vehicle, refer to the precautions.
2. Raise and safely support the vehicle securely on jackstands.
3. Remove the front wheels.
4. Drain brake fluid.
5. Remove front disc brake anti-rattle with hole pin.
   a. Remove the 2 pin hold clips, then remove the 2 hole pins from the disc brake caliper.
   b. Remove the anti-rattle spring from the disc brake pad.
6. Remove the disc brake pads with anti-squeal shims from the disc brake caliper.
7. Remove the anti-squeal shims from each of the disc brake pads.

To install:

8. If necessary, replace the anti-squeal shim when replacing the brake pad.
9. Apply disc brake grease to both sides of each No. 1 shim.
10. Install the No. 1 and No. 2 anti-squeal shims onto each brake pad.

11. Install the disc brake pads with anti-squeal shims onto the disc brake caliper.

WARNING

There should be no oil or grease on the friction surfaces of the disc brake pads or the front disc.

12. Install the anti-rattle spring and hole pins onto the disc brake caliper.

➡The anti-rattle spring is installed onto the lower hole pin.

13. Install the pin hold clip with its handle oriented in the center of the vehicle.
14. Fill reservoir with brake fluid.
15. Bleed the brakes.
16. Install the front wheel and lower the vehicle.

BRAKES

PARKING BRAKE

PARKING BRAKE CABLES

ADJUSTMENT

See Figure 7.

1. Before servicing the vehicle, refer to the precautions.
2. Inspect the parking brake lever travel.
   a. Slowly pull the parking brake lever to the fully applied position, counting the number of clicks.
   b. The parking brake lever should travel 5–7 clicks when pulled at 45 lbs. of force.
3. Turn the adjusting nut until the parking brake lever travel is corrected to within the specified range.
4. Operate the parking brake lever 3 to 4 times, and check the parking brake lever travel.
5. Check whether the parking brake drags or not.
6. When operating the parking brake lever, check that the brake warning light illuminates. The brake warning light always illuminates at the first click.

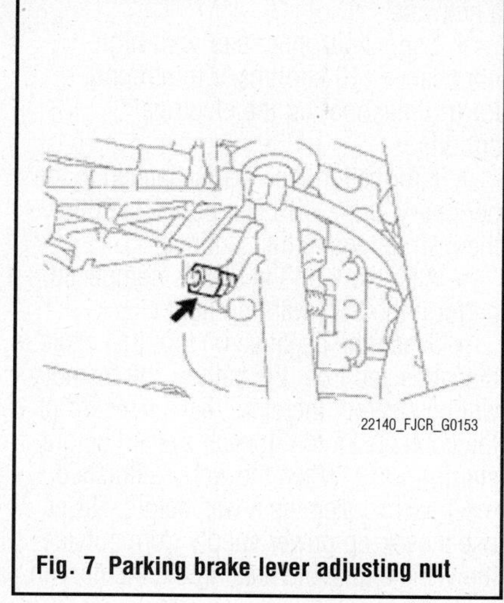

Fig. 7 Parking brake lever adjusting nut

PARKING BRAKE SHOES

REMOVAL & INSTALLATION

See Figure 8.

1. Before servicing the vehicle, refer to the precautions.
2. Raise and safely support the vehicle securely on jackstands.
3. Remove rear wheel.
4. Remove the disc brake caliper assembly and suspend it out of the way.
5. Release the parking brake lever.
6. Place matchmarks on the disc and axle hub and remove the disc.

➡If the disc cannot be removed easily, turn the shoe adjuster until the wheel turns freely.

To install:

7. Align the matchmarks of the disc and axle hub and install the disc.
8. Temporarily install 2 hub nuts.
9. Remove the hole plug, and turn the adjuster to expand the shoe until the disc locks.
10. Contract the shoe adjuster until the disc rotates smoothly. Then return 8 notches.
11. Install the hole plug.
12. Install the disc brake caliper.
13. Install rear wheel and lower the vehicle.
14. Inspect parking brake lever travel.
15. Adjust parking brake lever travel.

ADJUSTMENT

See Figure 8

1. Before servicing the vehicle, refer to the precautions.
2. Raise and safely support the vehicle securely on jackstands.
3. Remove rear wheel.
4. Temporarily install 2 hub nuts.
5. Remove the hole plug, and turn the adjuster to expand the shoe until the disc locks.
6. Contract the shoe adjuster until the disc rotates smoothly. Then return 8 notches.
7. Install rear wheel and lower the vehicle.
8. Inspect parking brake lever travel.
9. Adjust parking brake lever travel.

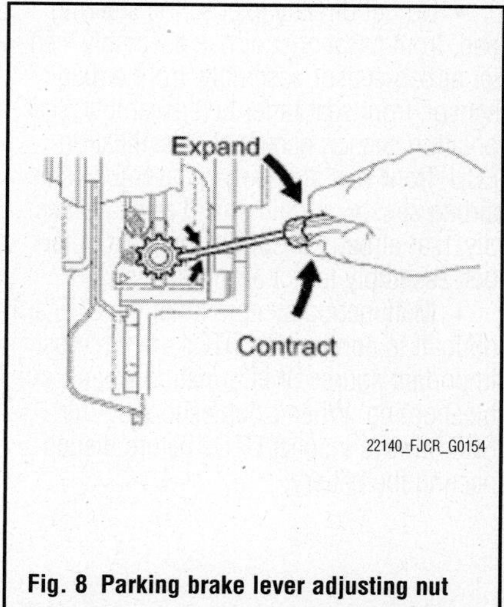

Fig. 8 Parking brake lever adjusting nut

GENERAL INFORMATION

PRECAUTIONS

Note the following:

- The vehicle is equipped with a Supplemental Restraint System (SRS), which consists of a driver airbag, front passenger airbag, side airbags, curtain shield airbags and front seat belt pretensioner. Failure to carry out service operations in the correct sequence could cause the SRS to unexpectedly deploy during servicing, possibly leading to a serious accident. Furthermore, if a mistake is made in servicing the SRS, it is possible that the SRS may fail to operate when required. Before performing servicing (including removal or installation of parts, inspection or replacement), be sure to read the following items carefully, then follow the correct procedures as indicated in the repair manual.
- Wait at least 90 seconds after the ignition switch is turned off and the negative (-) terminal cable is disconnected from the battery before starting the operation. The SRS is equipped with a back-up power source, so if work is started within 90 seconds of disconnecting the negative (-) terminal cable of the battery, the SRS may be deployed.
- Do not directly expose the steering pad, front passenger airbag assembly, center airbag sensor assembly, front airbag sensor, front seat inner belt assembly, seat position sensor, occupant classification ECU, front seat side airbag assembly, side airbag sensor, curtain shield airbag assembly, rear airbag sensor or front seat outer belt assembly to hot air or flames.
- Malfunction symptoms of the SRS are difficult to confirm, so DTCs are the most important source of information when troubleshooting. When troubleshooting the SRS, always inspect DTCs before disconnecting the battery.
- Never use SRS parts from another vehicle. When replacing parts, replace them with new ones.
- Never disassemble or repair any of the following parts in order to reuse them. If any of these parts have been dropped, or a defect is found (e.g. cracks, dents or any other defects) in any of the housings, brackets or connectors, then replace the part with a new one.
- Use a volt/ohmmeter with high impedance (10 kilohms/V minimum) for troubleshooting the electrical circuits.
- Information labels are attached to the periphery of the SRS components. Follow the instructions in the cautions.
- After work on the SRS is completed, perform the SRS warning light check.
- When the negative (-) terminal cable is disconnected from the battery, the memory will be cleared. therefore make a record of the contents stored in each system before starting work. When the work is finished, reset each system as it was before. Never use a back-up power supply from outside the vehicle to avoid erasing the memory in any system.
- When disconnecting the cable from the negative (-) battery terminal, initialize the meter / gauge system after the cable is reconnected.
- In the airbag system, the center airbag sensor assembly, front airbag sensor LH and RH, side airbag sensor LH and RH, rear airbag sensor LH and RH are collectively referred to as the airbag sensors.

DISARMING THE SYSTEM

Wait at least 90 seconds after the ignition switch is turned off and the negative (-) terminal cable is disconnected from the battery before starting the operation. The SRS is equipped with a back-up power source, so if work is started within 90 seconds of disconnecting the negative (-) terminal cable of the battery, the SRS may be deployed.

ARMING THE SYSTEM

Reconnect the negative battery cable and perform an SRS warning light check.

CLOCKSPRING CENTERING

See Figure 9.

1. Check that the ignition switch is turned to **OFF**.
2. Rotate the spiral cable counterclockwise slowly by hand until it feels firm.

WARNING

Do not turn the spiral cable with the airbag wire harness or connector.

3. Rotate the spiral cable clockwise approximately 2.5 turns to align the marks.
4. The spiral cable will rotate approximately 2.5 turns both ways from the center.

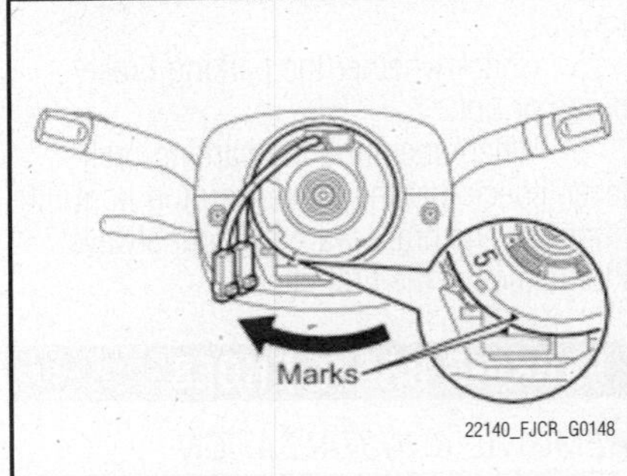

Fig. 9 Rotate the spiral cable counterclockwise slowly by hand until it feels firm, then rotate the spiral cable clockwise approximately 2.5 turns to align the marks

DRIVE TRAIN

HYDRAULIC SYSTEM BLEEDING

BLEEDING PROCEDURE

➡If any maintenance on the clutch system was performed or the system is suspected of containing air, bleed the system. Use care; brake fluid will remove the paint from any surface. If the brake fluid spills onto any painted surface, wash it off immediately with soap and water.

1. Before servicing the vehicle, refer to the precautions section.
2. Fill the clutch reservoir with brake fluid. Check the reservoir level frequently and add fluid as needed.
3. Connect one end of a vinyl tube to the bleeder plug on the slave cylinder and submerge the other end into a clear container half-filled with brake fluid.
4. Slowly pump the clutch pedal several times.
5. Have an assistant hold the clutch pedal down and loosen the bleeder plug until fluid and/or air starts to run out of the bleeder plug. Close the bleeder plug while the pedal is held to the floor.

➡Do not allow the pedal to rise back-up while the bleeder is still open. If this happens, it will allow air to re-enter the slave cylinder and cause the clutch system not to work properly.

6. Repeat Steps 2 and 3 until all the air bubbles are removed from the system.
7. Tighten the bleeder plug when all the air is gone.
8. Refill the master cylinder to the proper level as required.
9. Check the system for leaks.

TRANSFER CASE ASSEMBLY

REMOVAL & INSTALLATION

See Figure 10.

1. Before servicing the vehicle, refer to the precautions.
2. Disconnect the negative battery cable.
3. Drain transfer oil.
4. Remove lower transfer case protector.
5. Remove the transmission assembly.
6. Remove transfer case assembly.

To install:

7. Install the transfer case onto the transmission. Tighten the bolts to 17 ft. lbs. (24 Nm).

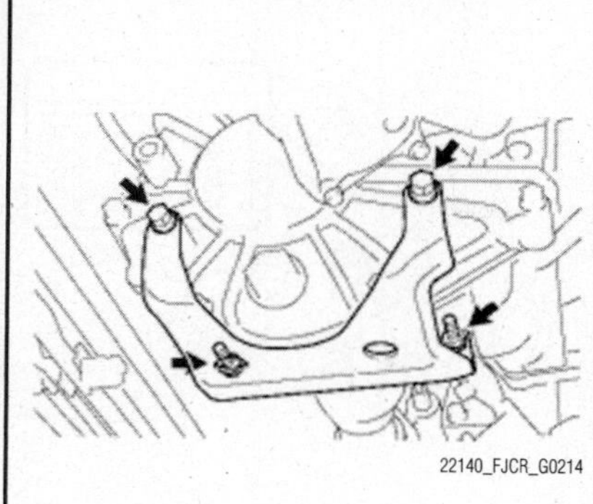

Fig. 10 Lower transfer case protector mounting bolts

8. Install the transmission.
9. Install the lower transfer case protector with the 4 bolts. Tighten the bolts to 13 ft. lbs. (18 Nm).
10. Inspect and adjust the transfer case oil.
11. Connect the negative battery cable.
12. Using a scan tool, perform the reset memory procedure to initialize the transmission.
13. Check for transfer oil leakage.

FRONT DIFFERENTIAL CARRIER

REMOVAL & INSTALLATION

1. Before servicing the vehicle, refer to the precautions.
2. Disconnect the negative battery cable.
3. Remove the front wheel.
4. Remove the No.1 engine under cover.
5. Remove the rear engine under cover assembly.
6. Drain the differential oil.
7. Remove the front halfshaft assembly.
8. Disconnect the front speed sensors.
9. Remove front axle hub grease cap.
10. Separate tie rod end.
11. Separate front lower ball joint attachment.
12. Remove front axle hub nut.
13. Remove front drive shaft assembly.
14. Remove the bolt and disconnect the differential breather tube bracket.
15. Support the differential with a jack.
16. Remove the No. 1 differential mounting nut.
17. Remove the 2 mounting bolts and 2 nuts.
18. Lower the jack and remove the front differential.
19. Remove the 3 bolts and front No. 1 differential support.
20. Remove the 2 bolts and front No. 2 differential support.
21. Remove the 2 bolts and front No. 3 differential support.

To install:

22. Install the front No. 3 differential support with the 2 bolts. Tighten to 80 ft. lbs. (108 Nm).
23. Install the front No. 2 differential support with the 2 bolts. Tighten to 118 ft. lbs. (160 Nm).
24. Install the front No. 1 differential support with the 3 bolts. Tighten to 137 ft. lbs. (186 Nm).
25. Support the front differential with a jack.
26. Install the 2 front mounting bolts and 2 nuts. Tighten to 101 ft. lbs. (137 Nm).
27. Install front differential mounting nut No. 1. Tighten to 64 ft. lbs. (87 Nm).
28. Install the front differential breather tube bracket with the bolt. Tighten to 10 ft. lbs. (13 Nm).
29. Lower the jack.
30. Install front halfshaft assembly.
31. Attach front lower ball joint
32. Install tie rod end.
33. Install front axle hub nut.
34. Install front axle hub grease cap.
35. Install front speed sensor.
36. Install front wheel.
37. Inspect differential oil level and correct as necessary.
38. Check for differential oil leakage.
39. Install rear engine under cover assembly.
40. Install No. 1 engine under cover.
41. Connect the negative battery cable.
42. Using a scan tool, check VSC sensor signal.
43. Inspect and adjust front wheel alignment

FRONT DRIVESHAFT

REMOVAL & INSTALLATION

See Figure 11.

1. Before servicing the vehicle, refer to the precautions.
2. Place matchmarks on the driveshaft flange and differential flange.
3. Remove the 4 nuts, 4bolts and 4 washers.
4. Place matchmarks on the driveshaft flange and transfer flange.

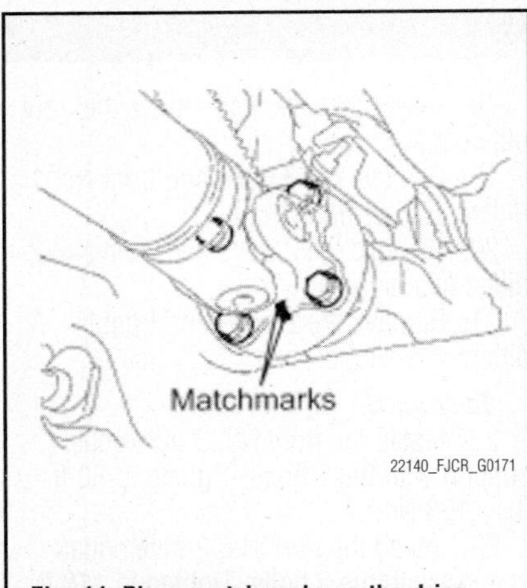

Fig. 11 Place matchmarks on the driveshaft flange for installation reference

5. Remove the 4 nuts, 4 washers and the driveshaft assembly.

To install:

6. Align the matchmarks on the yoke and differential flange.
7. Install the driveshaft assembly with the 4 bolts, 4 nuts and 4 washers. Tighten to 65 ft. lbs. (88 Nm).
8. Align the matchmarks on the yoke and transfer flange.
9. Install the driveshaft assembly with the 4 nuts and 4 washers. Tighten to 65 ft. lbs. (88 Nm).
10. Before servicing the vehicle, refer to the precautions.
11. Place matchmarks on the driveshaft flange and differential flange.
12. Remove the 4 nuts, 4bolts and 4 washers.
13. Place matchmarks on the driveshaft flange and transfer flange.
14. Remove the 4 nuts, 4 washers and the driveshaft assembly.

To install:

15. Align the matchmarks on the yoke and differential flange.
16. Install the driveshaft assembly with the 4 bolts, 4 nuts and 4 washers. Tighten to 65 ft. lbs. (88 Nm).
17. Align the matchmarks on the yoke and transfer flange.
18. Install the driveshaft assembly with the 4 nuts and 4 washers. Tighten to 65 ft. lbs. (88 Nm).

FRONT HALFSHAFT

REMOVAL & INSTALLATION

See Figure 12.

1. Disconnect the negative battery cable.

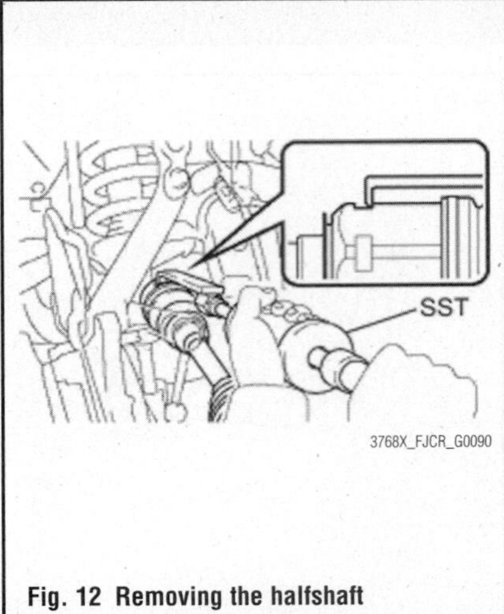

Fig. 12 Removing the halfshaft

2. Remove the front wheel.
3. Drain differential oil.
4. Separate the front speed sensor:
   a. Using a 5 mm hexagon wrench, remove the bolt and separate the front speed sensor.
   b. Disengage the 3 clamps.
   c. Remove the bolt and separate the skid control sensor wire from the steering knuckle.
5. Using a screwdriver and a hammer, remove the front axle hub grease cap.
6. Remove the cotter pin and adjusting lock cap.
7. Remove the front axle hub nut.
8. Separate tie rod end sub-assembly. Refer to Tie Rod End in Steering.
9. Separate front lower ball joint attachment. Refer to Lower Ball Joint in Front Suspension.
10. Using SST 09520-01010 and 09520-24010, remove the halfshaft.

To install:

11. Coat the spline of the inboard joint shaft with ATF.
12. Align the shaft splines and install the drive shaft with a brass bar and hammer.

➡Set the snap-ring with the opening side facing downward. Do not damage the oil seal.

➡Whether the inboard joint shaft is in contact with the pinion shaft or not can be confirmed from the sound or feeling when driving it.

13. To complete installation, reverse remaining removal procedure.
14. Inspect differential oil.
15. Check VSC sensor signal.
16. Check for differential oil leakage.
17. Inspect and adjust the front wheel alignment.

FRONT PINION SEAL

REMOVAL & INSTALLATION

See Figures 13 through 22.

1. Before servicing the vehicle, refer to the precautions.
2. Remove No. 1 engine under cover.
3. Remove the 4 bolts, then remove the rear engine under cover.
4. Remove front halfshaft assembly.

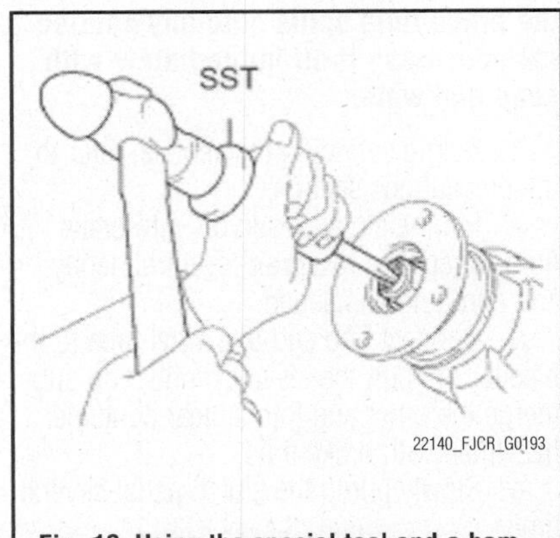

Fig. 13 Using the special tool and a hammer, loosen the staked part of the nut

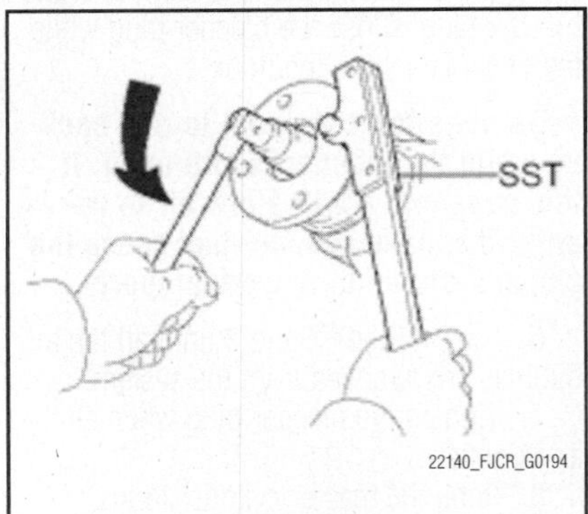

Fig. 14 Using the special tool hold the companion flange and remove the nut

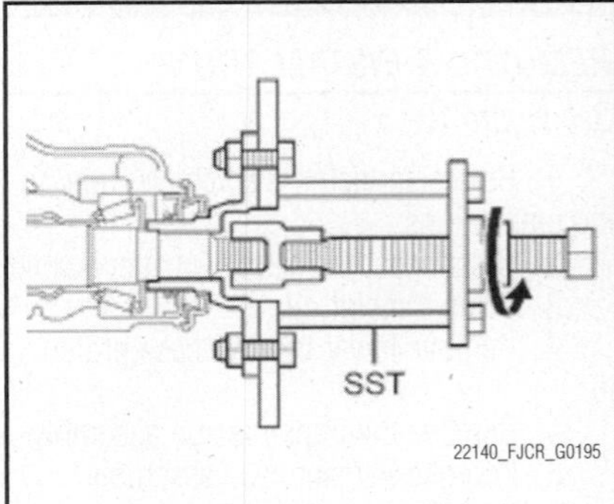

Fig. 15 Using the special tool remove the companion flange

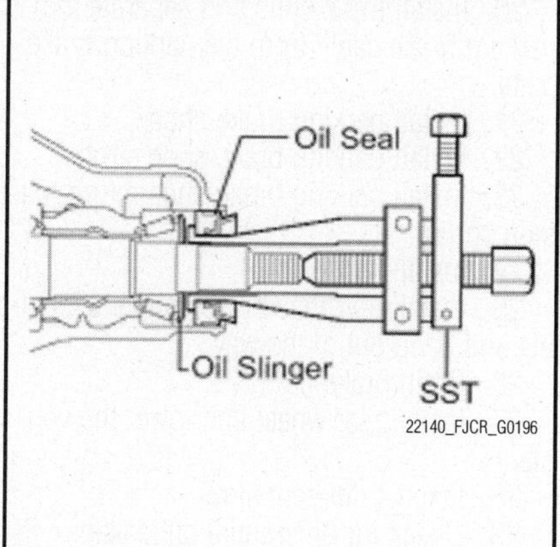

Fig. 16 Using the special tool remove the oil seal and oil slinger

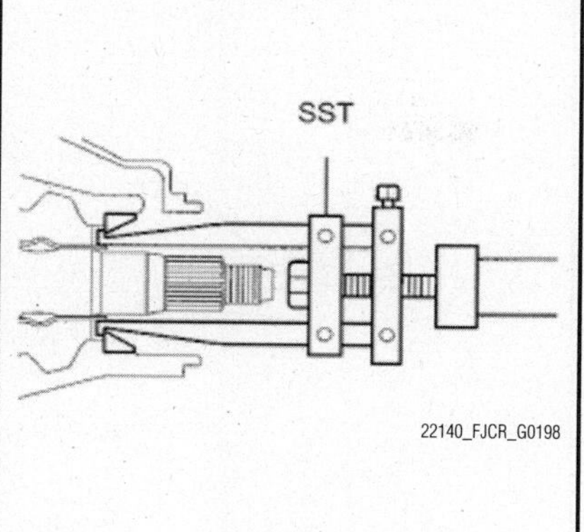

Fig. 18 Using the special tool remove the outer roller bearing

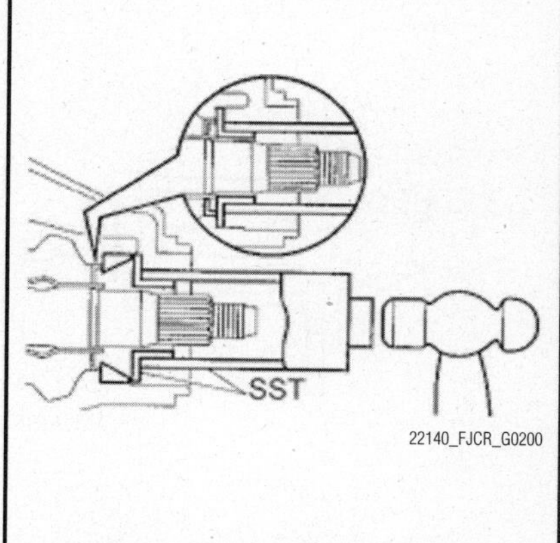

Fig. 20 Using the special tool and a hammer, install the outer roller bearing

5. Remove front drive pinion companion flange nut.
   a. Using SST: 09930-00010 and a hammer, loosen the staked part of the nut.
   b. Using SST: 09330-00021 to hold the companion flange, remove the nut.
6. Remove front drive pinion companion flange.
   a. Before using the special tool (center bolt), apply hypoid gear oil to its threads and tip.
   b. Using SST: 09950-30012, remove the companion flange.
7. Remove front differential carrier oil seal.
   a. Using SST: 09308-10010, remove the oil seal.
8. Remove front differential drive pinion oil slinger.
9. Remove front drive pinion rear tapered roller bearing.
   a. Using SST: 09556-22010, remove the inner roller bearing.
   b. Using SST: 09308-00010, tap out the outer roller bearing.
10. Using a screwdriver and hammer, tap out the oil storage ring.
11. Remove front differential drive pinion bearing spacer.

To install:

12. Install front differential drive pinion bearing spacer.

➡Install the spacer in the correct direction.

13. Using a brass bar and hammer, tap in a new oil storage ring.

WARNING

Be careful not to damage the oil storage ring.

14. Install front drive pinion rear tapered roller bearing.
   a. Using SST: 09316-60011 and a hammer, install the outer roller bearing.
   b. Install the inner roller bearing.
15. Install front differential drive pinion oil slinger.
16. Install front differential carrier oil seal.
   a. Apply MP grease to the lip of a new oil seal.
   b. Using SST: 09554-22010 and a hammer, tap in the oil seal to a depth of 0.154 to 0.188 in. (3.9 to 4.8 mm).
17. Install front drive pinion companion flange.
   a. Before using the special tool (center bolt), apply hypoid gear oil to its threads and tip.
   b. Using SST: 09950-30012, install the companion flange.
   c. Using SST: 09330-00021 to hold the companion flange, install the nut. Tighten to 273 ft. lbs. (370 Nm).
18. Using a torque wench, measure the differential drive pinion preload.
   a. Preload for a new bearing should be 8.7–13.9 inch lbs. (0.98–1.57 Nm).

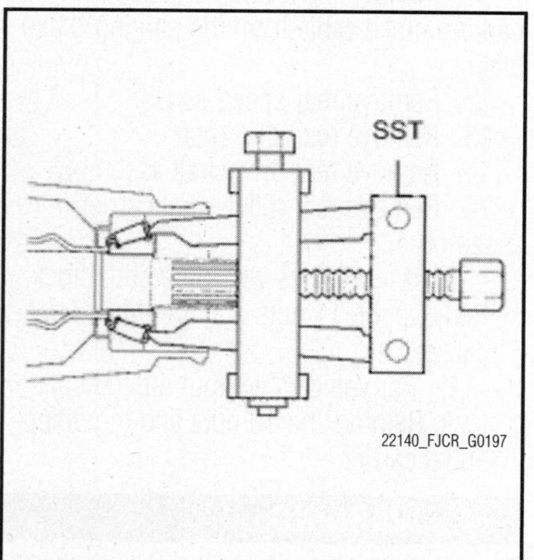

Fig. 17 Using the special tool remove the inner roller bearing

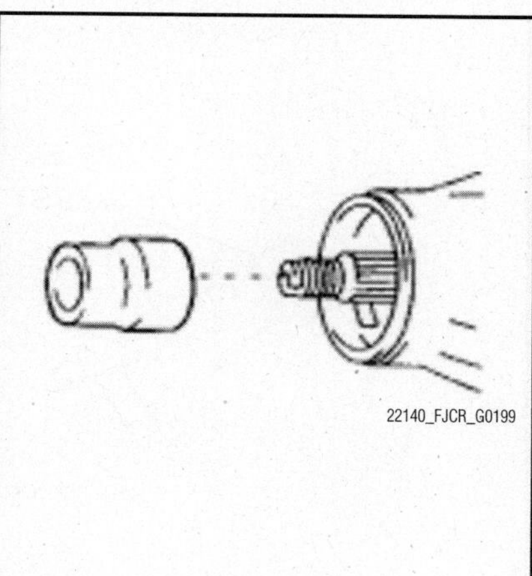

Fig. 19 Install the spacer in the correct direction

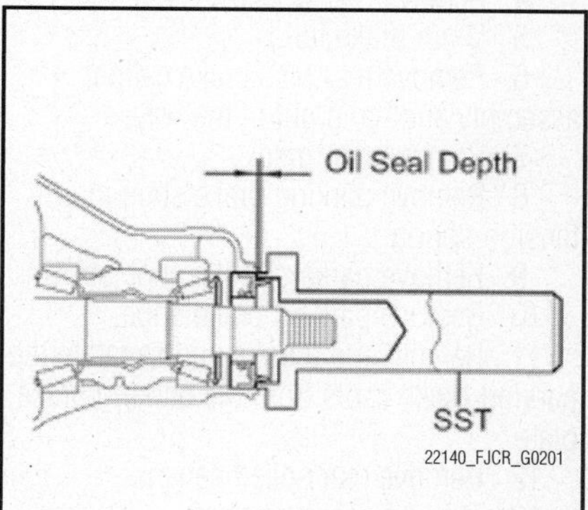

Fig. 21 Using the special tool and a hammer, tap in the oil seal to a depth of 0.154 to 0.188 in. (3.9 to 4.8 mm)

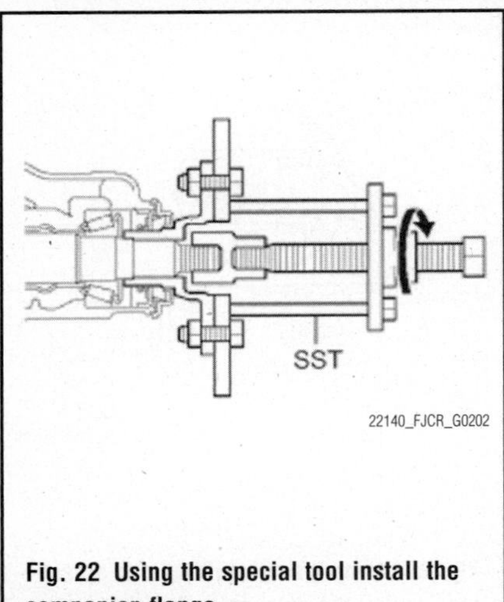

Fig. 22 Using the special tool install the companion flange

b. Preload for a used bearing should be 4.3–6.9 inch lbs. (0.49–0.78 Nm).

c. If the result is not as specified, adjust the preload.

19. Using a chisel and hammer, stake the nut.
20. Install front halfshaft assembly.
21. Add differential oil.
22. Check for differential oil leakage.
23. Install rear engine under cover assembly.
24. Install No. 1 engine under cover.

REAR AXLE SHAFT, BEARING & SEAL

REMOVAL & INSTALLATION

See Figures 23 and 24.

1. Before servicing the vehicle, refer to the precautions.
2. Disconnect the negative battery cable.
3. Raise and safely support the vehicle securely on jackstands.
4. Remove rear wheel.
5. Drain brake fluid.
6. Remove rear disc brake caliper assembly and hang out of the way.
7. Remove rear disc.
8. Remove parking brake shoe return tension spring.
9. Remove parking brake shoe strut.
10. Remove parking brake shoe.
11. Remove the 2 bolts and separate the parking brake cable from the parking brake plate.
12. Remove rear speed sensor.
13. Remove rear axle shaft.
    a. Remove the 4 nuts and rear axle shaft with parking brake plate.
    b. Remove the O-ring.
14. Using SST: 09308-00010, remove rear axle shaft oil seal.

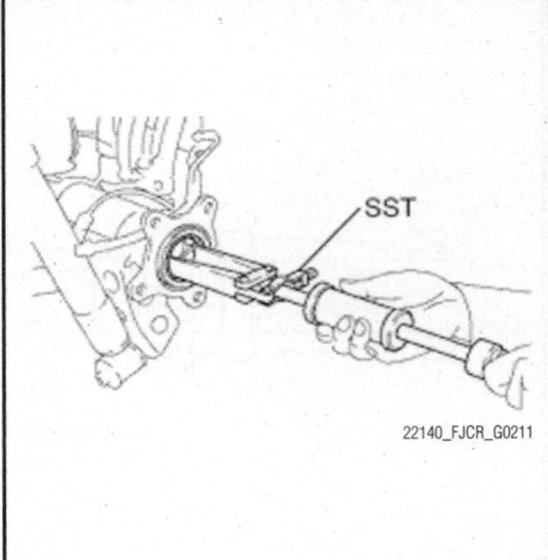

Fig. 23 Using the special tool, remove rear axle shaft oil seal

To install:

15. Using SST: 09950-60020, install rear axle shaft oil seal.
    a. Install the O-ring.
    b. Install the 4 nuts and rear axle shaft with parking brake plate. Tighten to 89 ft. lbs. (120 Nm).
16. Install rear axle shaft.
17. Inspect the axle shaft backlash.
    a. Using a dial indicator, check the backlash near the center of the axle shaft. Maximum backlash is 0.0020 in. (0.05 mm).
    b. If the backlash is greater than the maximum, replace the bearing.
18. Inspect the axle shaft runout.
    a. Using a dial indicator, check the runout of the surface of the axle shaft. Maximum runout is 0.0020 in. (0.05 mm).
    b. If the runout is greater than the maximum, replace the bearing.
19. Install rear speed sensor. Tighten to 71 inch lbs. (8 Nm).

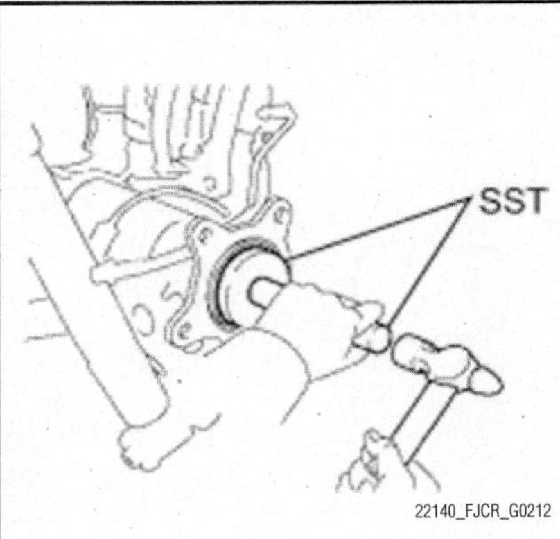

Fig. 24 Using the special tool, install rear axle shaft oil seal

20. Install the 2 bolts and separate the parking brake cable from the parking brake plate.
21. Install parking brake shoe.
22. Install parking brake shoe strut.
23. Install parking brake shoe return tension spring.
24. Install rear disc.
25. Install rear disc brake caliper assembly and hang out of the way.
26. Drain brake fluid.
27. Install rear wheel and lower the vehicle.
28. Inspect differential oil.
29. Check for differential oil leakage.
30. Check and adjust fluid level in reservoir.
31. Check for brake fluid leakage.
32. Inspect and adjust parking brake lever travel.
33. Using a scan tool, check VSC sensor signal.

REAR DIFFERENTIAL HOUSING

REMOVAL & INSTALLATION

See Figures 25 through 27.

1. Before servicing the vehicle, refer to the precautions.
2. Disconnect the negative battery cable.
3. Raise and safely support the vehicle securely on jackstands.
4. Remove rear wheel.
5. Drain brake fluid.
6. Remove rear disc brake caliper assembly and hang out of the way.
7. Remove rear disc.
8. Remove parking brake shoe return tension spring.
9. Remove parking brake shoe strut.
10. Remove parking brake shoe.
11. Remove the 2 bolts and separate the parking brake cable from the parking brake plate.
12. Remove rear speed sensor.
13. Remove rear axle shaft.
14. Remove rear driveshaft assembly.
15. Remove rear differential carrier assembly:
    a. For vehicles with differential lock: Remove the 11 nuts and the differential carrier.
    b. For vehicles without differential lock: Remove the 10 nuts and the differential carrier.

⁂ WARNING

Be careful not to damage the contact surface.

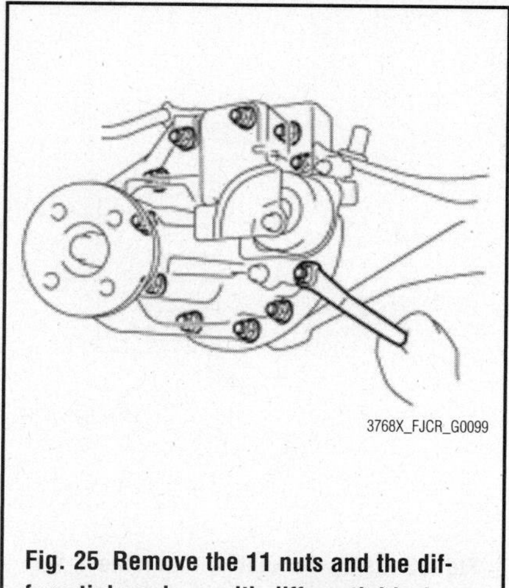

Fig. 25 Remove the 11 nuts and the differential carrier—with differential lock

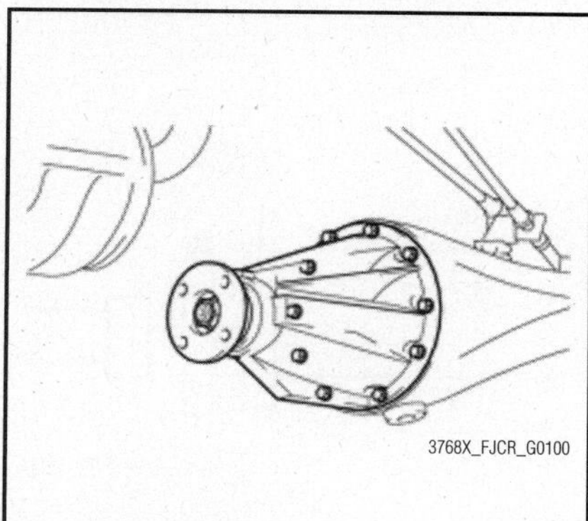

Fig. 26 Remove the 10 nuts and the differential carrier—without differential lock

16. Remove rear differential carrier gasket.

To install:

17. Remove any dust and oil from the differential carrier assembly and the contact surfaces of the axle housing.
18. Apply liquid gasket to both sides of a new gasket.

➡Do not apply the liquid gasket to the stud bolt.

19. Install a new gasket and the differential carrier assembly with the 10 and/or 11 nuts and 10 and/or 11 washers. Tighten to 38 ft. lbs. (52 Nm).
20. Install rear axle shaft.
21. Inspect the axle shaft backlash.
    a. Using a dial indicator, check the backlash near the center of the axle shaft. Maximum backlash is 0.0020 in. (0.05 mm).
    b. If the backlash is greater than the maximum, replace the bearing.
22. Inspect the axle shaft runout.
    a. Using a dial indicator, check the runout of the surface of the axle shaft. Maximum runout is 0.0020 in. (0.05 mm).
    b. If the runout is greater than the maximum, replace the bearing.
23. Install rear speed sensor. Tighten to 71 inch lbs. (8 Nm).
24. Install the 2 bolts and separate the parking brake cable from the parking brake plate.
25. Install parking brake shoe.
26. Install parking brake shoe strut.
27. Install parking brake shoe return tension spring.
28. Install rear disc.
29. Install rear disc brake caliper assembly and hang out of the way.
30. Drain brake fluid.
31. Install rear wheel and lower the vehicle.
32. Inspect differential oil.
33. Check for differential oil leakage.
34. Check and adjust fluid level in reservoir.
35. Check for brake fluid leakage.
36. Inspect and adjust parking brake lever travel.
37. Using a scan tool, check VSC sensor signal.

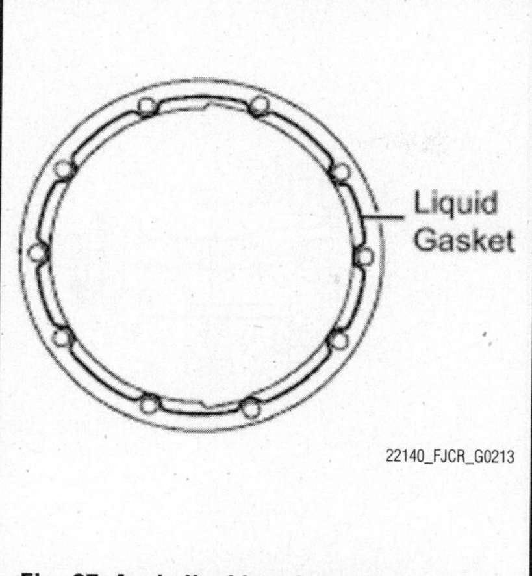

Fig. 27 Apply liquid gasket to both sides of a new gasket

REAR DRIVESHAFT

REMOVAL & INSTALLATION

See Figure 28.

1. Before servicing the vehicle, refer to the precautions.
2. Place matchmarks on the driveshaft flange and differential flange.
3. Remove the 4 nuts, 4bolts and 4 washers.
4. Place matchmarks on the driveshaft flange and transfer flange.

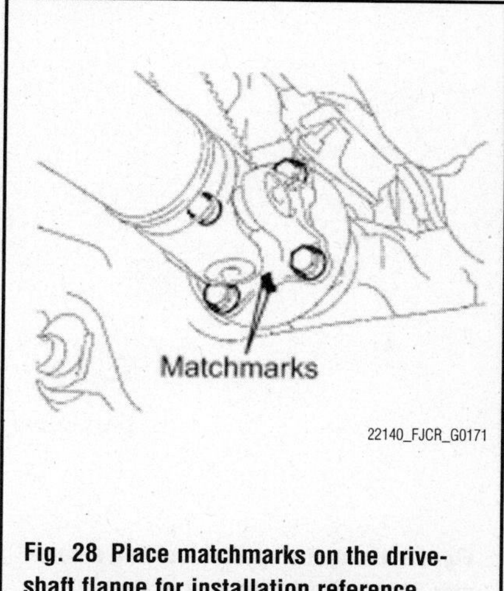

Fig. 28 Place matchmarks on the driveshaft flange for installation reference

5. Remove the 4 nuts, 4 washers and the driveshaft assembly.

To install:

6. Align the matchmarks on the yoke and differential flange.
7. Install the driveshaft assembly with the 4 bolts, 4 nuts and 4 washers. Tighten to 65 ft. lbs. (88 Nm).
8. Align the matchmarks on the yoke and transfer flange.
9. Install the driveshaft assembly with the 4 nuts and 4 washers. Tighten to 65 ft. lbs. (88 Nm).
10. Before servicing the vehicle, refer to the precautions.
11. Place matchmarks on the driveshaft flange and differential flange.
12. Remove the 4 nuts, 4bolts and 4 washers.
13. Place matchmarks on the driveshaft flange and transfer flange.
14. Remove the 4 nuts, 4 washers and the driveshaft assembly.

To install:

15. Align the matchmarks on the yoke and differential flange.
16. Install the driveshaft assembly with the 4 bolts, 4 nuts and 4 washers. Tighten to 65 ft. lbs. (88 Nm).
17. Align the matchmarks on the yoke and transfer flange.
18. Install the driveshaft assembly with the 4 nuts and 4 washers. Tighten to 65 ft. lbs. (88 Nm).

REAR PINION SEAL

REMOVAL & INSTALLATION

See Figures 29 through 38.

1. Before servicing the vehicle, refer to the precautions.

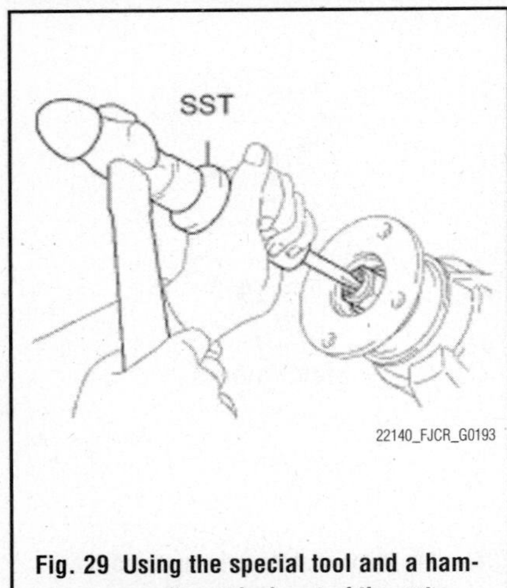

Fig. 29 Using the special tool and a hammer, loosen the staked part of the nut

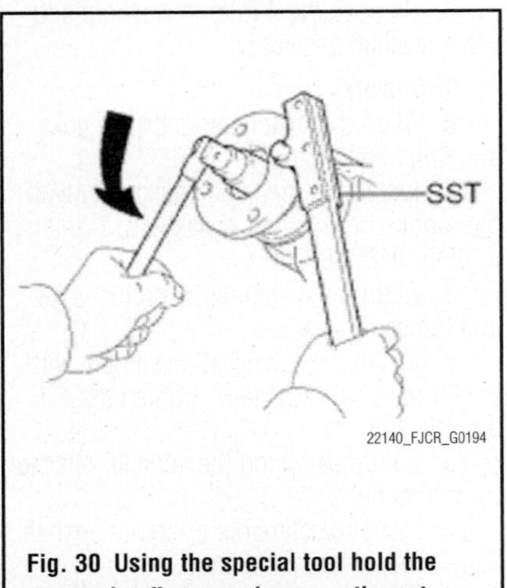

Fig. 30 Using the special tool hold the companion flange and remove the nut

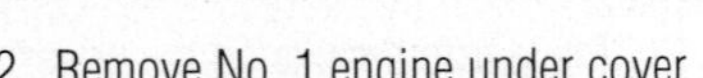

2. Remove No. 1 engine under cover.
3. Remove the 4 bolts, then remove the rear engine under cover.
4. Remove rear halfshaft assembly.
5. Remove rear drive pinion companion flange nut.
   a. Using SST: 09930-00010 and a hammer, loosen the staked part of the nut.
   b. Using SST: 09330-00021 to hold the companion flange, remove the nut.
6. Remove rear drive pinion companion flange.
   a. Before using the special tool (center bolt), apply hypoid gear oil to its threads and tip.
   b. Using SST: 09950-30012, remove the companion flange.
7. Remove rear differential carrier oil seal.
   a. Using SST: 09308-10010, remove the oil seal.
8. Remove rear differential drive pinion oil slinger.

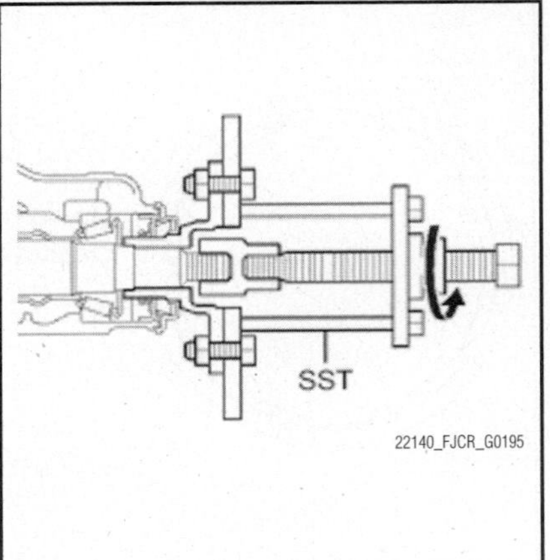

Fig. 31 Using the special tool remove the companion flange

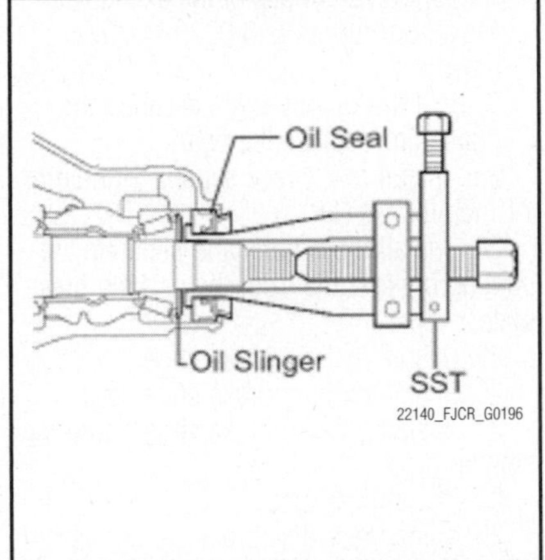

Fig. 32 Using the special tool remove the oil seal and oil slinger

9. Remove rear drive pinion rear tapered roller bearing.
   a. Using SST: 09556-22010, remove the inner roller bearing.
   b. Using SST: 09308-00010, tap out the outer roller bearing.
10. Using a screwdriver and hammer, tap out the oil storage ring.
11. Remove rear differential drive pinion bearing spacer.

To install:

12. Install rear differential drive pinion bearing spacer.

➡Install the spacer in the correct direction.

13. Using a brass bar and hammer, tap in a new oil storage ring.

WARNING

Be careful not to damage the oil storage ring.

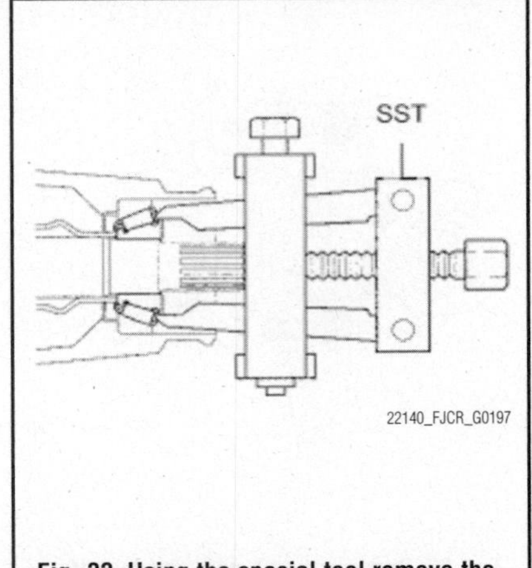

Fig. 33 Using the special tool remove the inner roller bearing

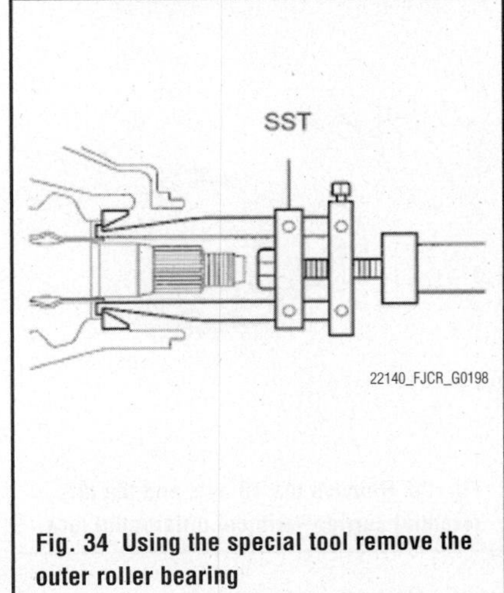

Fig. 34 Using the special tool remove the outer roller bearing

14. Install rear drive pinion rear tapered roller bearing.
    a. Using SST: 09316-60011 and a hammer, install the outer roller bearing.
    b. Install the inner roller bearing.
15. Install rear differential drive pinion oil slinger.
16. Install rear differential carrier oil seal.

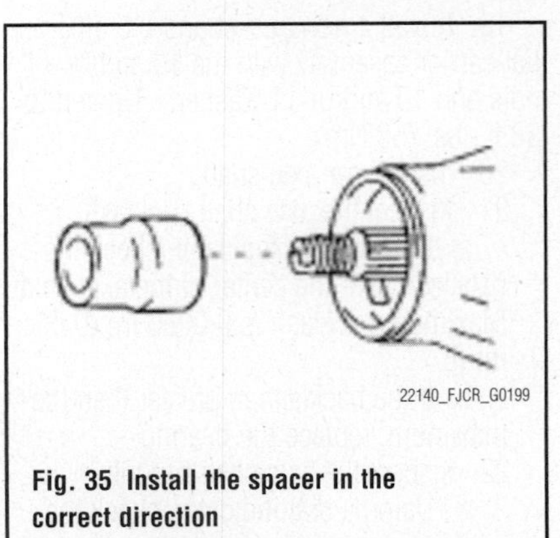

Fig. 35 Install the spacer in the correct direction

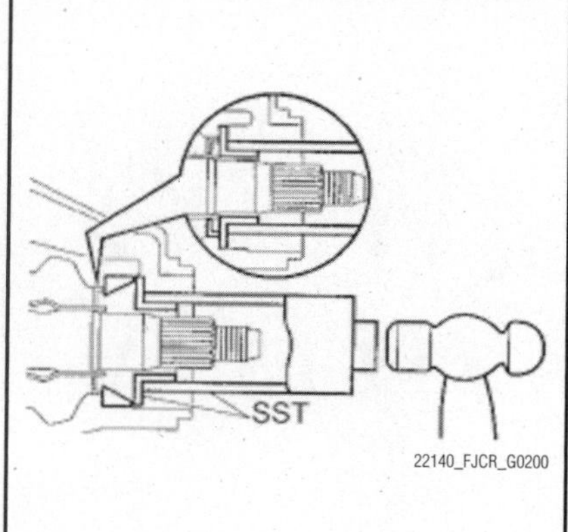

Fig. 36 Using the special tool and a hammer, install the outer roller bearing

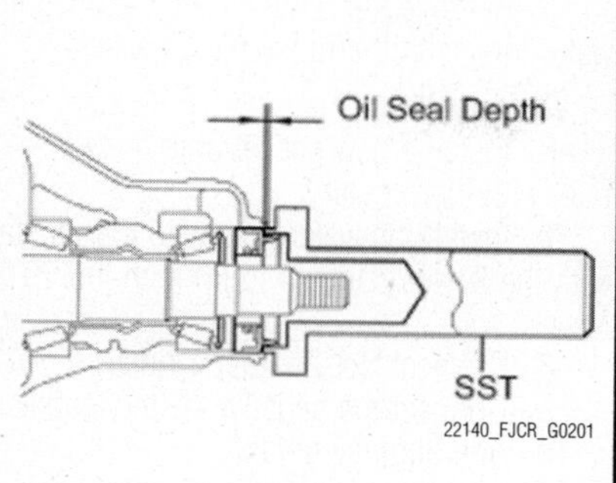

Fig. 37 Using the special tool and a hammer, tap in the oil seal to a depth of 0.154 to 0.188 in. (3.9 to 4.8 mm)

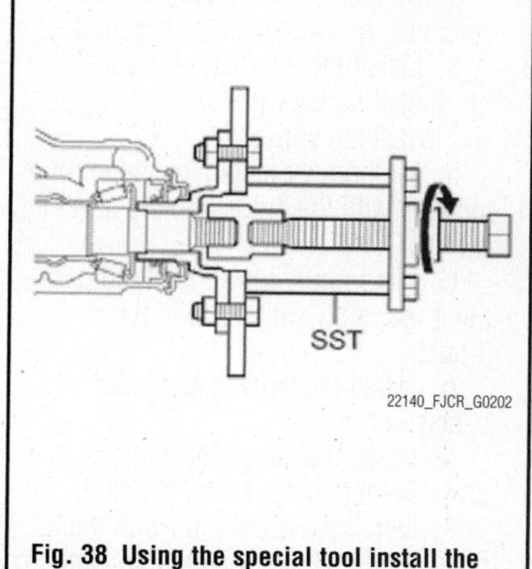

Fig. 38 Using the special tool install the companion flange

a. Apply MP grease to the lip of a new oil seal.

b. Using SST: 09554-22010 and a hammer, tap in the oil seal to a depth of 0.154 to 0.188 in. (3.9 to 4.8 mm).

17. Install rear drive pinion companion flange.

a. Before using the special tool (center bolt), apply hypoid gear oil to its threads and tip.

b. Using SST: 09950-30012, install the companion flange.

c. Using SST: 09330-00021 to hold the companion flange, install the nut. Tighten to 273 ft. lbs. (370 Nm).

18. Using a torque wench, measure the differential drive pinion preload.

a. Preload for a new bearing should be 8.7–13.9 inch lbs. (0.98–1.57 Nm).

b. Preload for a used bearing should be 4.3–6.9 inch lbs. (0.49–0.78 Nm).

c. If the result is not as specified, adjust the preload.

19. Using a chisel and hammer, stake the nut.

20. Install rear halfshaft assembly.

21. Add differential oil.

22. Check for differential oil leakage.

23. Install rear engine under cover assembly.

24. Install No. 1 engine under cover.

ENGINE COOLING

ENGINE FAN

REMOVAL & INSTALLATION

See Figure 39.

1. Before servicing the vehicle, refer to the precautions.
2. Remove V-bank cover.
3. Remove no.1 engine under cover.

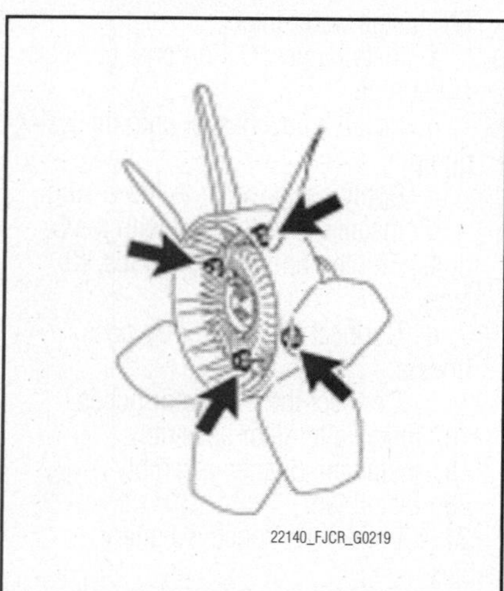

Fig. 39 Fan coupling fastener locations

4. Remove radiator support seal upper.
5. Remove fan shroud.
6. Remove the 4 nuts and the fan from the fluid coupling.

To install:

7. Install the 4 nuts and the fan from the fluid coupling. Tighten to 80 inch lbs. (9 Nm).
8. Install fan shroud.
9. Install radiator support seal upper.
10. Install no.1 engine under cover.
11. Install V-bank cover.

RADIATOR

REMOVAL & INSTALLATION

1. Before servicing the vehicle, refer to the precautions.
2. Remove V-bank cover.
3. Remove No. 1 engine under cover.
4. Drain engine coolant.
5. Disengage the 7 clips, then remove the radiator support seal.
6. Remove fan shroud.

a. Disconnect the hose from the radiator.

b. Loosen the 4 nuts from the fan pulley.

c. Remove the fan and alternator V-belt.

d. Remove the 4 nuts and the fluid coupling assembly with fan.

e. Remove the bolt from the reserve tank.

f. Remove the 2 bolts from the fan shroud.

g. Remove the fluid coupling assembly with fan and fan shroud.

h. Remove the fan pulley.

7. Remove radiator grille.
8. Remove radiator assembly.

a. Disconnect the 2 radiator hoses.

b. Disconnect the 2 oil cooler hoses for the automatic transmission, as required.

c. Remove the 4 bolts, then remove the radiator.

9. Remove radiator hoses.
10. Remove inlet oil cooler hose for automatic transmission, as required.

To install:

11. Install inlet oil cooler hose for automatic transmission, as required.
12. Install the radiator hoses.
13. Install the radiator with the 4 bolts. Tighten to 13 ft. lbs. (18 Nm).

a. Connect the 2 oil cooler hoses automatic transmission, as required.
b. Install the 2 radiator hoses.
14. Install radiator grille.
15. Install fan shroud.
a. Temporarily install the fan shroud together with the fluid coupling assembly with fan.
b. Temporarily install the fluid coupling assembly with fan with the 4 nuts.
c. Install the fan and alternator V-belt.
d. Install the fan pulley with the 4 nuts. Tighten to 15 ft. lbs. (21 Nm).
f. Install the fan shroud with the 2 bolts. Tighten to 44 inch lbs. (5 Nm).
g. Install the bolt to the reserve tank. Tighten to 44 inch lbs. (5 Nm).
h. Connect the 2 oil cooler hoses with the 2 clamps to the automatic transmission, as required.
i. Connect the hose to the radiator reserve tank.
16. Engage the 7 clips, then install the radiator support seal.
17. Connect the negative battery cable.
18. Check and adjust the engine coolant level.
19. Start the engine and check engine for coolant leaks.
20. Install No. 1 engine under cover.
21. Install V-bank cover

THERMOSTAT

REMOVAL & INSTALLATION

See Figure 40.

1. Before servicing the vehicle, refer to the precautions.
2. Drain engine coolant.
3. Remove V-bank cover.
4. Remove No. 2 radiator hose.
5. Remove the 3 nuts, then remove the water inlet with thermostat and gasket.

To install:

6. Install a new gasket onto the water inlet with thermostat.
7. Install the water inlet with thermostat with the 3 nuts. Tighten to 80 inch lbs. (9 Nm)
8. Connect No. 2 radiator hose.
9. Connect the negative battery cable.
10. Add engine coolant.
11. Check for engine coolant leakage.
12. Install V-bank cover.

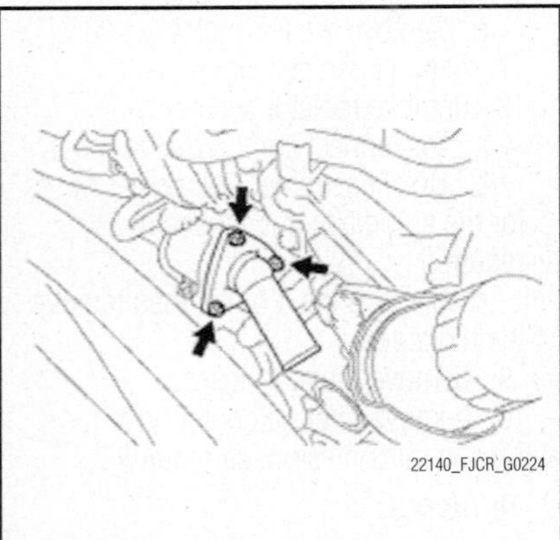

Fig. 40 Remove the 3 nuts, then remove the water inlet

WATER PUMP

REMOVAL & INSTALLATION

See Figures 41 and 42.

1. Before servicing the vehicle, refer to the precautions.
2. Remove fan.
3. Remove V-bank cover.
4. Remove air cleaner assembly.
5. Remove alternator assembly.
6. Remove water inlet.
a. Disconnect the 2 radiator hoses.
b. Disconnect the 5 water by-pass hoses.
c. Remove the 5 bolts and water inlet.
d. Remove the O-ring from the water outlet pipe.
e. Remove the gasket from the water pump.
7. Remove No. 2 idler pulley.
8. Remove the air conditioning air conditioning compressor assembly and lay it aside.

➡Do not disconnect the air conditioning hoses from the compressor.

Fig. 41 Water inlet fastener locations

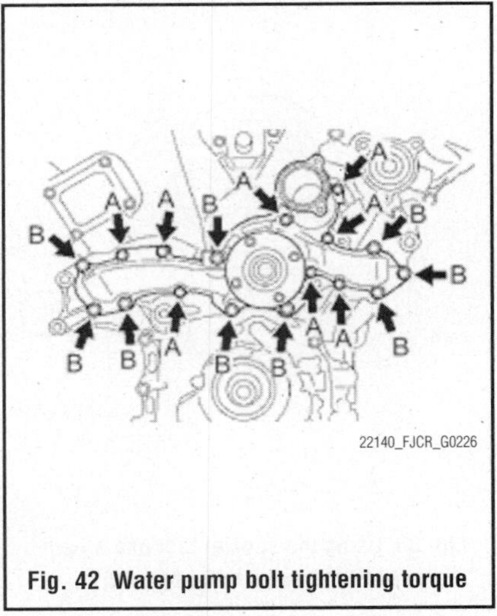

Fig. 42 Water pump bolt tightening torque

9. Remove V-ribbed belt tensioner assembly.
10. Separate power steering pump assembly and set it aside.

➡Do not disconnect the power steering hoses.

11. Remove the 17 bolts, then remove the water pump and gasket.

To install:

12. Install a new gasket and the water pump with the 17 bolts. Tighten bolts **A** to 80 inch lbs. (9 Nm) and bolts **B** to 17 ft. lbs. (23 Nm).
13. Install power steering pump assembly and set it aside.

➡Do not disconnect the power steering hoses.

14. Install V-ribbed belt tensioner assembly.
15. Install air conditioning compressor assembly.
16. Install No. 2 idler pulley.
17. Install water inlet.
a. Install a new O-ring onto the water outlet pipe.
b. Install a new gasket onto the water pump.
c. Apply soapy water to the O-ring.
d. Install the water inlet with the 5 bolts. Tighten bolt to 80 inch lbs. (9 Nm).
e. Connect the 5 water by-pass hoses.
f. Connect the 2 radiator hoses.
18. Install alternator assembly.
19. Install air cleaner assembly.
20. Install fan.
21. Connect the negative battery cable.
22. Install V-bank cover.

ENGINE ELECTRICAL

CHARGING SYSTEM

ALTERNATOR

REMOVAL & INSTALLATION

See Figure 43.

1. Before servicing the vehicle, refer to the precautions.
2. Remove battery.
3. Remove V-bank cover.
4. Remove No. 1 engine under cover.
5. Remove fan and alternator V-belt.
6. Disconnect the wire harness.
7. Remove the bolt and wire harness bracket.
8. Disconnect the connector from the alternator assembly.
9. Remove the terminal cap and nut.
10. Disconnect the wire harness from terminal B.
11. Remove the 2 bolts, then separate the wire harness clamp bracket from the alternator assembly.
12. Remove the 2 bolts, then remove the alternator assembly.

To install:

13. Install the alternator assembly with the 2 bolts. Tighten to 32 ft. lbs. (43 Nm).
14. Install the wire harness clamp bracket with the 2 bolts. Tighten to 71 inch lbs. (8 Nm).
15. Connect the wire harness.
16. Connect the wire harness to terminal B and install the nut. Tighten to 7 ft. lbs. (10 Nm).
17. Connect the connector to the alternator assembly.
18. Install the wire harness bracket with the bolt. Tighten to 51 inch lbs. (6 Nm).
19. Install fan and alternator V-belt.
20. Install No. 1 engine under cover.
21. Install V-bank cover.
22. Install battery

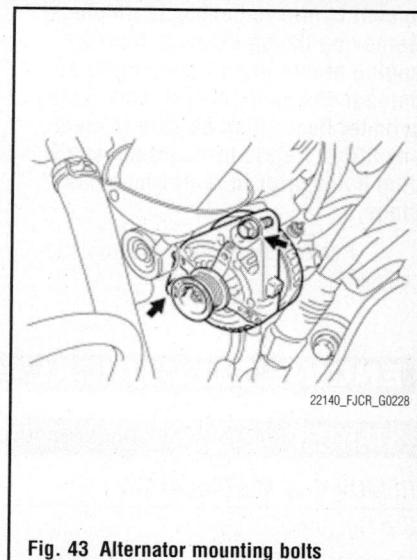

Fig. 43 Alternator mounting bolts

ENGINE ELECTRICAL

IGNITION SYSTEM

FIRING ORDER

See Figure 44.

IGNITION COIL

REMOVAL & INSTALLATION

See Figure 45.

1. Before servicing the vehicle, refer to the precautions.
2. Disconnect the negative battery cable.
3. Remove V-bank cover.
4. Remove air cleaner assembly.
5. Disconnect the 6 ignition coil connectors.
6. Remove the 6 bolts and 6 ignition coils.
7. To install, reverse removal procedure.
8. Tighten the bolts to 7 ft. lbs. (10 Nm).

IGNITION TIMING

ADJUSTMENT

The ignition timing is controlled by the Electronic Control Module (ECM). No adjustment is possible.

SPARK PLUGS

REMOVAL & INSTALLATION

See Figure 46.

1. Before servicing the vehicle, refer to the precautions.
2. Remove the V-bank cover.
3. Disconnect the electrical connector from each ignition coil.
4. Remove the bolt securing each ignition coil from the cylinder head cover.
5. Remove each ignition coil from the cylinder head cover.

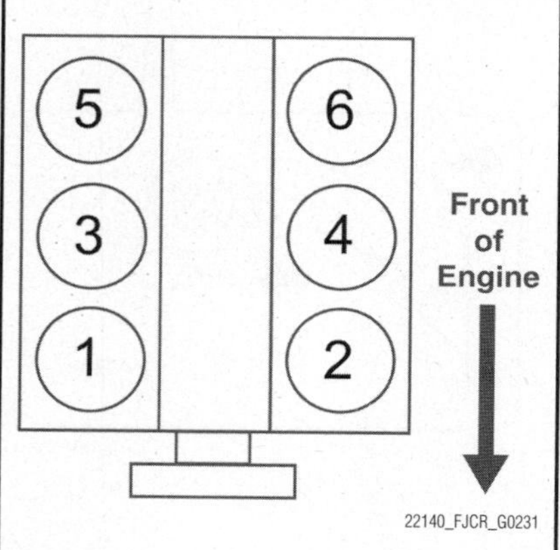

Fig. 44 4.0L V6 DOHC engine
Firing order: 1–2–3–4–5–6
Distributorless ignition system

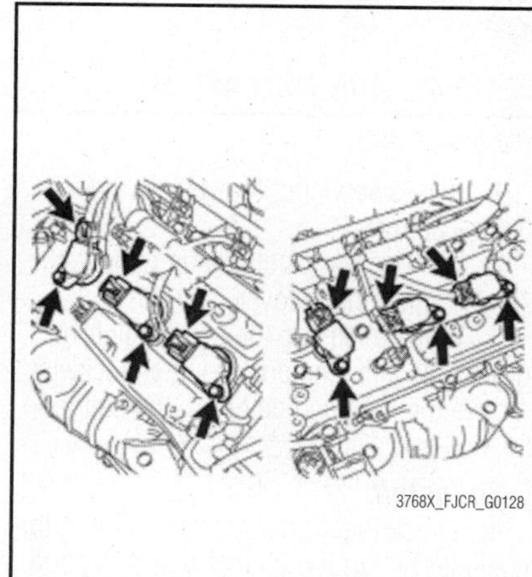

Fig. 45 Removing the 6 ignition coils

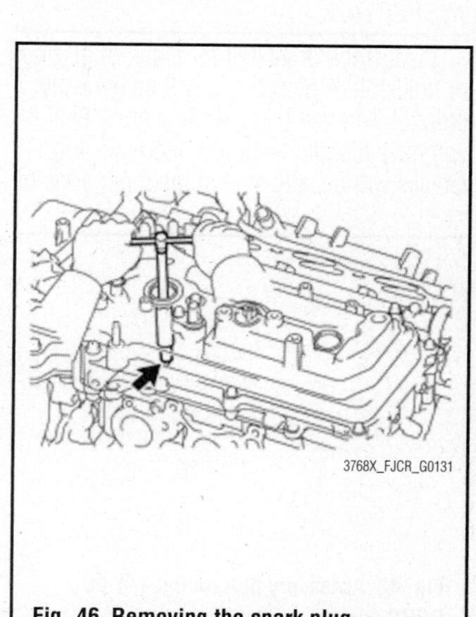

Fig. 46 Removing the spark plug

➡This engine is equipped with an aluminum cylinder head. Allow the engine to cool before removing spark plugs. Removing the spark plugs from an engine at operating temperature may damage the spark plug threads in the cylinder head. Also be sure to clean any dirt or debris from around spark plug holes prior to removing spark plugs.

6. Remove the spark plugs from the cylinder head.
7. Inspect the spark plugs for electrode wear, carbon deposits and insulator damage.

To install:

⁂ WARNING

Do not touch the tip of the spark plug. Do not damage the iridium surface of the electrode when gapping the plug.

➡Do not adjust the gap on used spark plugs. Replace the spark plug if the gap is greater than specification.

8. Apply a small quantity of anti-seize compound to the plug threads, and screw the plugs into the cylinder head finger-tight.
9. Tighten the spark plugs to 15 ft. lbs. (20 Nm).
10. Install each ignition coil to the cylinder head cover.
11. Secure the ignition coils using the bolts. Tighten the bolts to 80 inch lbs (9 Nm).
12. Connect each ignition coil electrical connector.
13. Install the V-bank cover.

ENGINE ELECTRICAL

STARTER

REMOVAL & INSTALLATION

See Figure 47.

1. Before servicing the vehicle, refer to the precautions.
2. Disconnect the negative battery cable.
3. Remove the LH exhaust manifold.
4. Remove the 3 bolts and starter cover.
5. Remove the bolt and ground wire.
6. Disconnect the starter connector.
7. Open the terminal cap.
8. Remove the 2 bolts and starter.
9. Remove flywheel housing side cover.

Fig. 47 Removing the starter assembly

ENGINE MECHANICAL

ACCESSORY DRIVE BELTS

ACCESSORY BELT ROUTING

See Figure 48.

INSPECTION

Inspect the drive belt for signs of glazing or cracking. A glazed belt will be perfectly smooth from slippage, while a good belt will have a slight texture of fabric visible. Cracks will usually start at the inner edge of the belt and run outward. All worn or damaged drive belts should be replaced immediately.

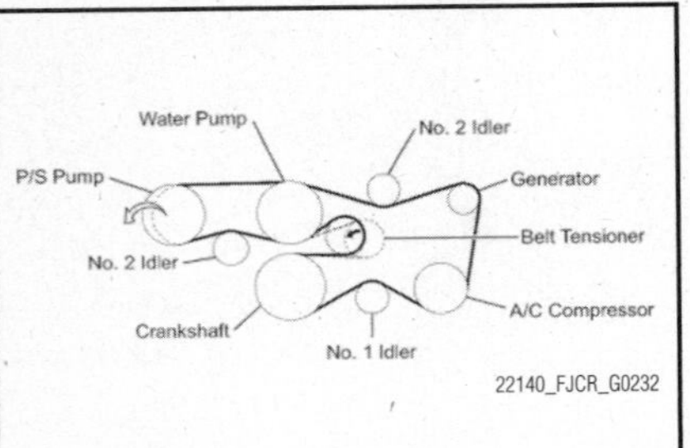

Fig. 48 Accessory belt routing—4.0L DOHC engine

ADJUSTMENT

The belt tension is maintained by an automatic tensioner. No adjustment is possible.

REMOVAL & INSTALLATION

See Figure 49.

1. Before servicing the vehicle, refer to the precautions.
2. Remove the 4 bolts, then remove the No. 1 engine under cover.
3. While releasing the belt tension by turning the belt tensioner counterclockwise, remove the V-ribbed belt from the belt tensioner.

To install:

4. Check that nothing gets caught in the tensioner by turning it clockwise and counterclockwise. If a malfunction exists, replace the tensioner.
5. While turning the belt tensioner coun-

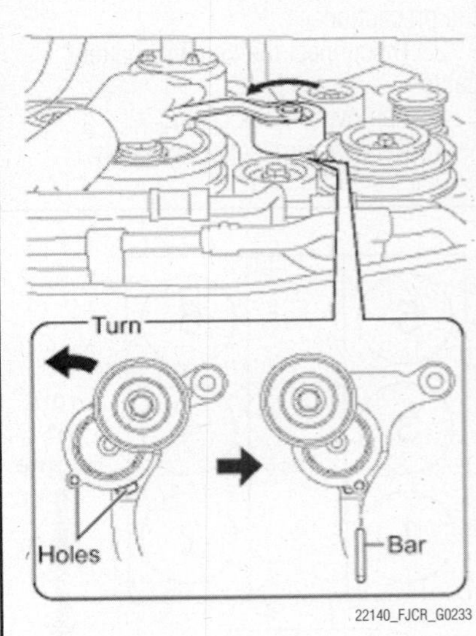

Fig. 49 While turning the belt tensioner counterclockwise, align the holes as shown, and then insert a bar of 0.24 in. (6 mm) into the holes to fix the belt tensioner

terclockwise, align the holes as shown, and then insert a bar of 0.24 in. (6 mm) into the holes to fix the belt tensioner.

6. Install the V-ribbed belt.
7. While turning the belt tensioner counterclockwise, remove the bar.
8. If it is hard to install the V-ribbed belt, perform the following procedure:
    a. Put the V-ribbed belt on all parts except the P/S pump, as shown in the illustration.
    b. While releasing the belt tension by turning the belt tensioner counterclockwise, put the V-ribbed belt on the P/S pump.
9. Install the No. 1 engine under cover with the 4 bolts. Tighten to 21 ft. lbs. (29 Nm).

CAMSHAFT AND VALVE LIFTERS

INSPECTION

See Figure 50.

1. Inspect the camshaft for runout.
    a. Place the camshaft on V-blocks.
    b. Using a dial indicator, measure the circle runout at the center journal. Maximum runout is0.0024 in. (0.06 mm).
    c. If the circle runout is greater than the maximum, replace the camshaft.
2. Inspect the cam lobes.
    a. Using a micrometer, measure the cam lobe height. Minimum cam lobe height is 1.7330 in. (44.018 mm) for the intake and 1.7492 in. (44.430 mm) for the exhaust lobes.
    b. If the cam lobe height is less than the minimum, replace the camshaft.
3. Using a micrometer, measure the journal diameter. No. 1 journal diameter is 1.4162 to 1.4167 in. (35.971 to 35.985 mm) and all other journal diameters are 0.9039 to 0.9045 in. (22.959 to 22.975 mm)
    a. If the journal diameter is not as specified, check the oil clearance.
4. Inspect camshaft timing gear assembly.
    a. Fix the intake camshaft in a vise.

⁂ WARNING

Be careful not to damage the camshaft.

b. Align the knock pin hole in the camshaft timing gear assembly with the knock pin of the camshaft, and install the camshaft timing gear assembly with the bolt. Tighten to 74 ft. lbs. (100 Nm).

c. Confirm that the camshaft timing gear assembly is locked.

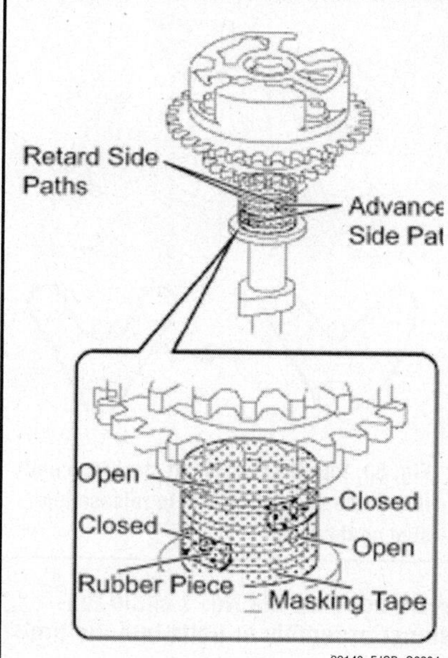

Fig. 50 Cover the 4 oil paths of the cam journal with masking tape as shown

d. Release that lock pin.

e. Cover the 4 oil paths of the cam journal with masking tape as shown in the illustration.

f. One of the 2 grooves on the cam journal is for the retard side path (upper) and the other is for the advance side path (lower). Each groove has 2 oil paths. Plug one of the oil paths for each groove with a piece of rubber before wrapping the cam journal with tape.

g. Puncture the tape covering the advance oil path and retard oil path on the opposite side of the groove as shown in the illustration.

h. Apply compressed air at 29 PSI (200 kPa) into the 2 broken paths (the advance side path and the retard side path).

i. Cover the paths with a shop rag or piece of cloth to avoid oil splashes.

j. Confirm that the camshaft timing gear assembly rotates in the timing advance direction when reducing the compressed air on the timing retard path.

k. When the lock pin is released, the camshaft timing gear rotates in the advance direction.

l. When the camshaft timing gear comes to the most advanced position, release the compressed air on the timing retard side path, and release that on the timing advance side path.

m. Camshaft timing assembly gear occasionally shifts to the retard side abruptly if the air compression on the advanced side path is released first. This often causes breakage of the lock pin.

n. Check the smooth revolution.

o. Rotate the camshaft timing gear several times within the movable range except for the most retarded position and check it rotates smoothly. The gear should moves smoothly in a range of about 31°.

p. Be sure to perform this check by hand, instead of using compressed air.

q. Check that the lock is in the most retarded position.

r. Confirm that the camshaft timing gear assembly is locked in the most retarded position.

s. Remove the set bolt, then remove the camshaft timing gear assembly.

➡Do not remove the other 3 bolts.

REMOVAL & INSTALLATION

See Figures 51 through 79.

1. Remove no. 1 engine under cover sub-assembly.
2. Drain engine oil.
3. Drain engine coolant.
4. Remove v-bank cover.
5. Remove air cleaner assembly.
6. Remove intake manifold.
7. Remove fan shroud.
8. Remove ignition coil assembly.
9. Separate vane pump assembly.
10. Remove generator assembly.

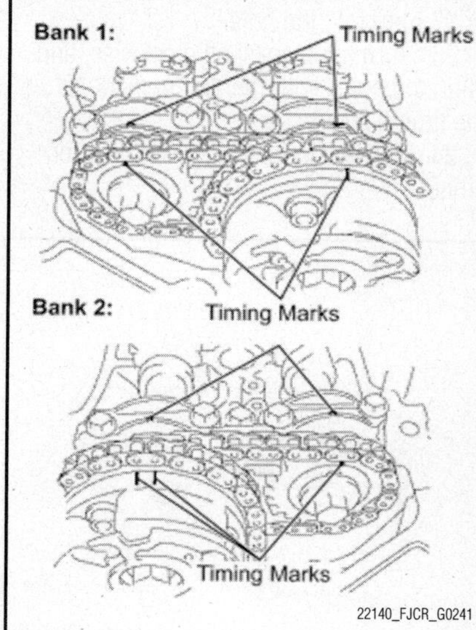

Fig. 51 Check that the timing marks of the camshaft timing gears are aligned with the timing marks of the bearing caps as shown

11. Remove engine oil level dipstick guide.
12. Remove no. 1 oil pipe.
13. Remove no. 2 oil pipe.
14. Remove rear cylinder head cover.
15. Disconnect fuel pipe sub-assembly.
16. Remove the valve covers.
17. Remove the 4 bolts, timing chain cover plate and gasket.
18. Turn the crankshaft pulley and align the notch with the "0" timing mark of the timing chain cover.

➡If the marks are not aligned, turn the crankshaft again to align the marks.

19. Check that the timing marks of the camshaft timing gears are aligned with the timing marks of the bearing cap.

➡Be sure to place the paint marks on 2 links of the chain and on the sprockets of the camshaft timing gears at the locations of the timing marks of the camshaft timing gears.

20. Place paint marks on the timing marks and sprockets of each camshaft timing gear and on the links of the No. 1 chain.

➡This prevents the valves and pistons from interfering with each other.

21. Turn the crankshaft approximately 30° counterclockwise so that there is some slack in the chain.
22. Align the hole in the lever of the tensioner with the hole in the tensioner body as shown in the illustration, and then insert a pin with a diameter of 0.0500 inches. (1.27 mm) into the hole.
23. Turn the crankshaft clockwise, and align the notch with the timing mark "0" of the timing chain cover.
24. Remove the 2 bolts and chain tensioner.

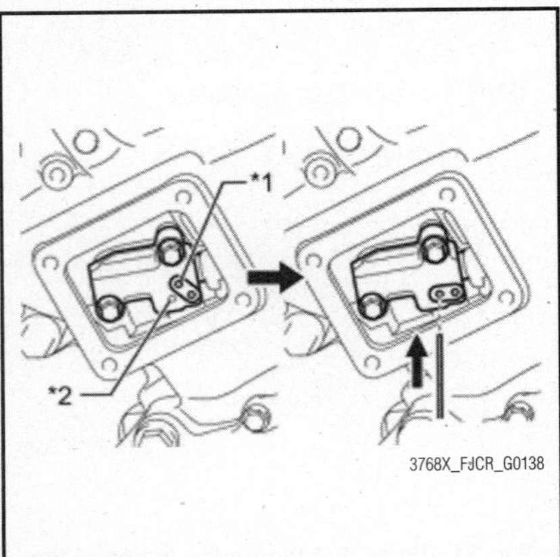

Fig. 52 Lever hole (1) and tensioner hole (2)

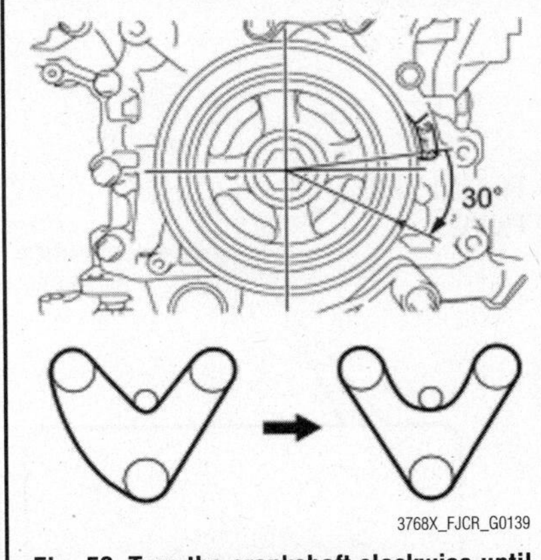

Fig. 53 Turn the crankshaft clockwise until it is in the position shown to release tension on the chain

➡Do not drop the No. 1 chain tensioner assembly or bolts into the timing chain cover.

25. Disconnect the chain sub-assembly (for bank 1):

➡When turning the crankshaft, engine oil may spray out of the oil holes.

a. Turn the crankshaft clockwise until it is in the position shown in the illustration so that there is some slack in the chain between the banks.

CAUTION

As the camshafts turn suddenly, do not touch the camshafts or camshaft timing gears.

b. Turn the crankshaft clockwise until it is in the position shown in the illustration so that the chain can be removed easily.

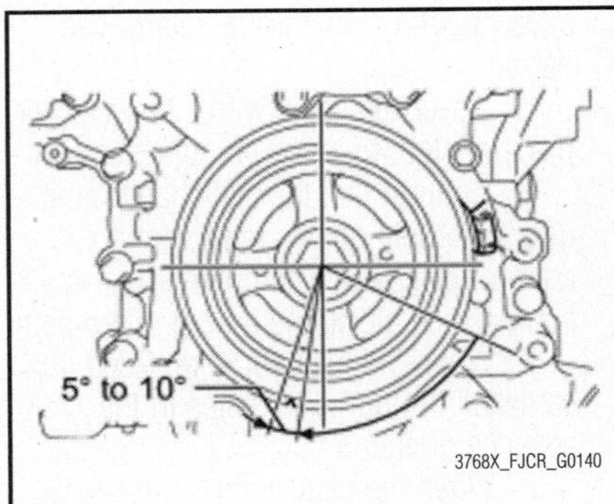

Fig. 54 Turn the crankshaft clockwise until it is in the position shown in the illustration so that the chain can be removed easily

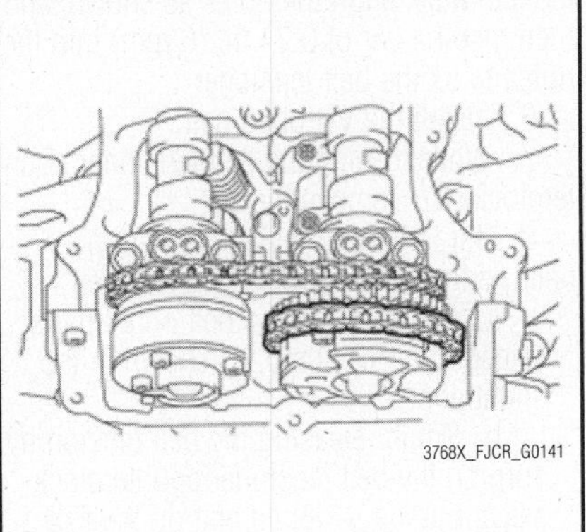

Fig. 55 Remove the chain from the sprocket of the camshaft timing gear and set it on the gear

CAUTION

As the camshaft may turn suddenly and pinch your fingers when the chain is removed, pinch the chain and lift it upward to remove it from the sprocket.

26. Remove the chain from the sprocket of the camshaft timing gear and set it on the gear.

➡Arrange the removed parts so that they can be reinstalled in their original locations:

- Do not install the bearing caps when installing VVT bolt kit.

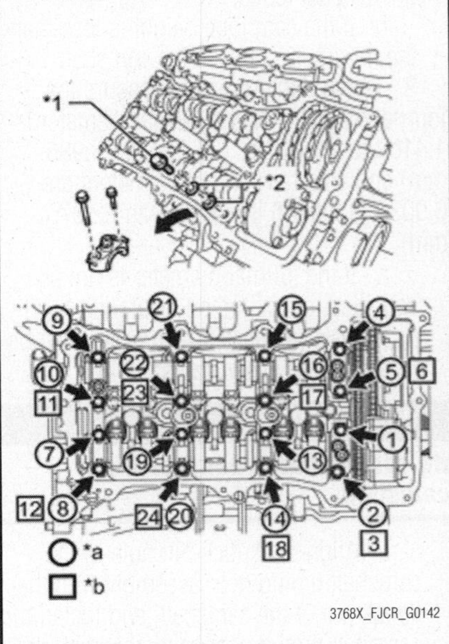

Fig. 56 Bank 1 bearing cap removal sequence—VVT bolt kit (1), part removal (a), VVT bolt kit installation (b)

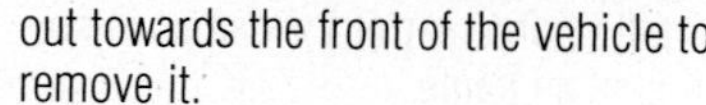

- Be sure to follow the numerical order when performing this procedure.
- Do not allow the VVT bolt kit to contact the camshaft.
- Do not drop the VVT bolt kit into the cylinder head.

27. Remove camshaft bearing cap (for bank 1):

a. Remove the bolts and bearing caps in the order shown in the illustration. Immediately after removing a bearing cap, install VVT bolt kit in the order shown in the illustration.

28. Remove no. 2 camshaft:

a. Remove the bolt of the No. 2 chain tensioner assembly.

29. Remove the No. 2 chain tensioner assembly while lifting up the No. 2 camshaft.

30. While lifting up the No. 2 camshaft, pass it through the No. 2 chain and pull it out towards the front of the vehicle to remove it.

31. Remove camshaft:

a. Lift up the rear of the camshaft so that it is at an angle.

b. Remove the chain from the camshaft timing gear and pull out the camshaft and No. 2 chain towards the rear of the vehicle to remove them.

➡Do not drop the chain into the gap between the engine and cover.

32. Suspend the chain with a string or equivalent.

33. Disconnect chain sub-assembly (for bank 2):

a. Turn the crankshaft counterclockwise, and align the notch with the "0" timing mark of the timing chain cover.

b. Remove the chain from the sprocket of the camshaft timing gear and set it on the gear.

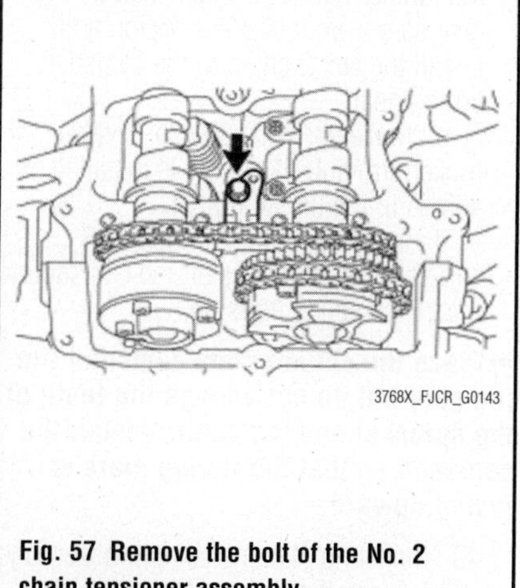

Fig. 57 Remove the bolt of the No. 2 chain tensioner assembly

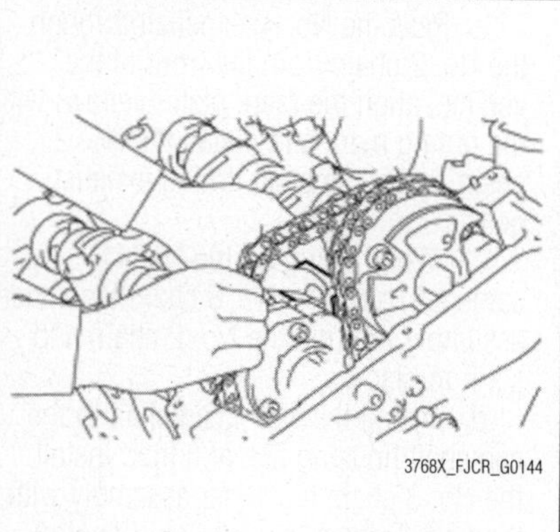

Fig. 58 Remove the No. 2 chain tensioner assembly while lifting up the No. 2 camshaft

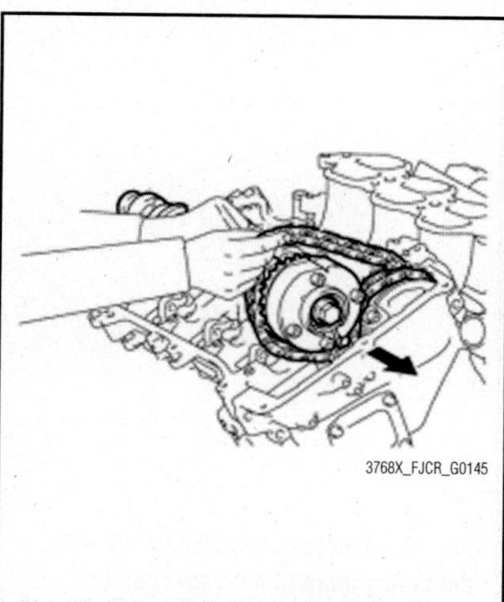

Fig. 59 Removing the No. 2 camshaft

Fig. 60 Removing the camshaft

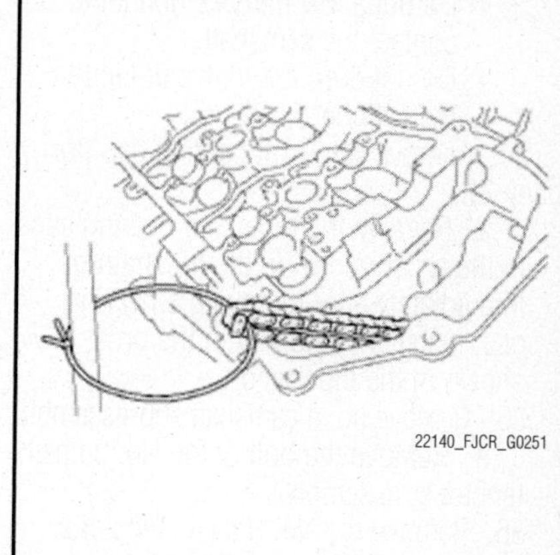

Fig. 61 Tie the No. 1 chain with a piece of string as shown

CAUTION

As the camshaft may turn suddenly and pinch your fingers when the chain is removed, pinch the chain and lift it upward to remove it from the sprocket.

➡Arrange the removed parts so that they can be reinstalled in their original locations.

- Do not install the bearing caps when installing VVT bolt kit.
- Be sure to follow the numerical order when performing this procedure.

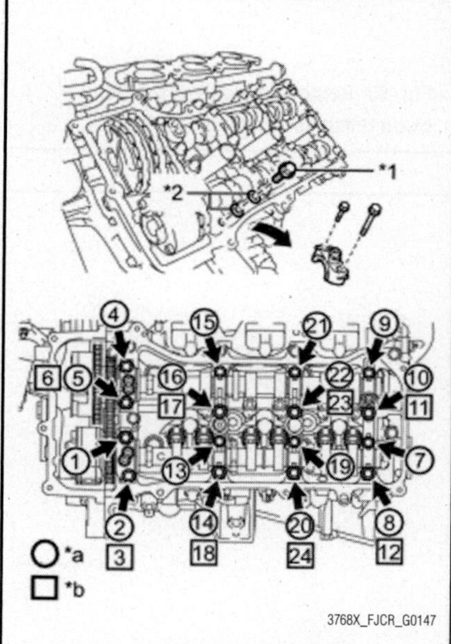

Fig. 62 Bank 2 bearing cap removal sequence—VVT bolt kit (1), part removal (a), VVT bolt kit installation (b)

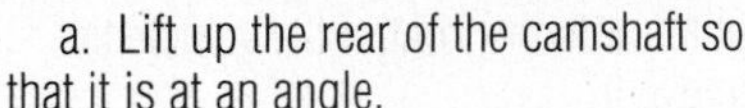

- Do not allow the VVT bolt kit to contact the camshaft.
- Do not drop the VVT bolt kit into the cylinder head.

34. Remove camshaft bearing cap (for bank 2):

a. Remove the bolts and bearing caps in the order shown in the illustration. Immediately after removing a bearing cap, install VVT bolt kit in the order shown in the illustration.

35. Remove no. 4 camshaft sub-assembly:

a. Remove the bolt of the No. 3 chain tensioner assembly.

36. Remove the No. 3 chain tensioner assembly while lifting up the No. 4 camshaft.

37. While lifting up the No. 4 camshaft, pass it through the No. 2 chain and pull it out towards the front of the vehicle to remove it.

38. Remove no. 3 camshaft sub-assembly:

a. Lift up the rear of the camshaft so that it is at an angle.

b. Remove the chain from the camshaft timing gear and pull out the No. 3 camshaft and No. 2 chain towards the rear of the vehicle to remove them.

➡Do not drop the chain into the gap between the engine and cover.

c. Suspend the chain with a string or equivalent.

➡Do not remove the other 3 bolts. If planning to reuse the camshaft timing gear, be sure to release the straight pin lock before installing the camshaft timing gear.

To install:

39. Install no. 3 camshaft sub-assembly:

a. Check that the notch is aligned with the "0" timing mark of the timing chain cover.

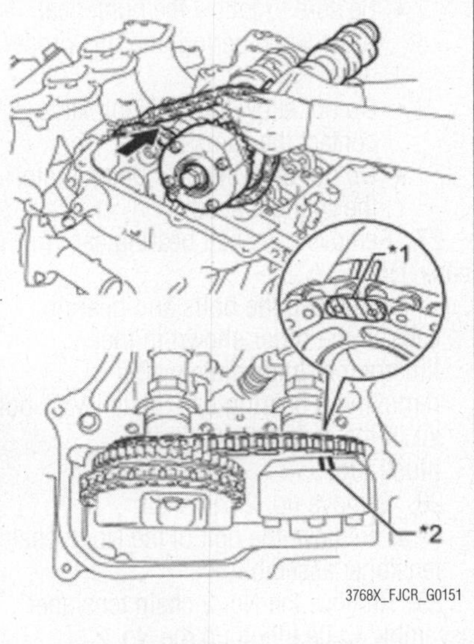

Fig. 67 Installing the no. 4 camshaft—Mark plate (yellow) (1), timing mark (2)

b. Align the mark plate (yellow) with the timing mark of the camshaft timing gear as shown in the illustration and install the No. 2 chain to the camshaft timing gear.

c. Make sure that the No. 1 valve rocker arm sub-assembly is installed as shown in the illustration.

d. Install the chain to the No. 3 camshaft, and then install the camshaft to the LH camshaft housing.

➡Place the chain on the camshaft timing gear but do not engage the teeth of the sprocket and the chain. Install the camshaft so that the timing mark is facing upward.

40. Install no. 4 camshaft sub-assembly:

a. Clean the camshaft housing LH and camshaft journals and apply engine oil to them.

b. Pass the No. 4 camshaft through the No. 2 chain from the front of the vehicle, align the mark plate (yellow) with the timing mark and install the No. 2 chain to the camshaft timing exhaust gear.

c. While lifting up the No. 4 camshaft, pass the No. 3 chain tensioner assembly through the No. 2 chain and set it in place.

d. Install the No. 4 camshaft to the camshaft housing LH, and then install the No. 3 chain tensioner assembly with the bolt. Tighten to 15 ft. lbs. (21 Nm).

41. Install camshaft bearing cap (for bank 2):

a. Clean the camshaft bearing caps and apply engine oil to them.

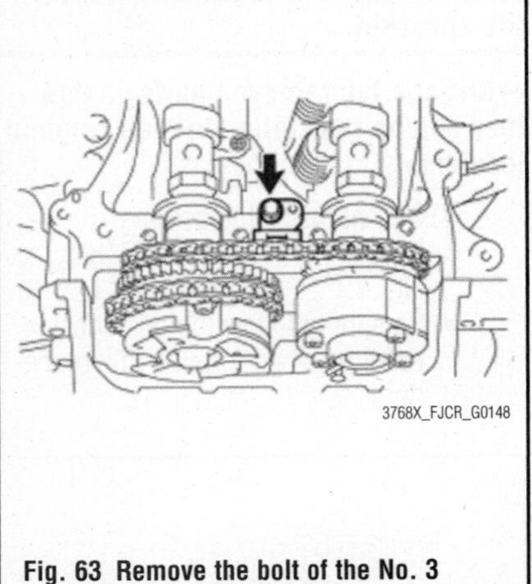

Fig. 63 Remove the bolt of the No. 3 chain tensioner assembly

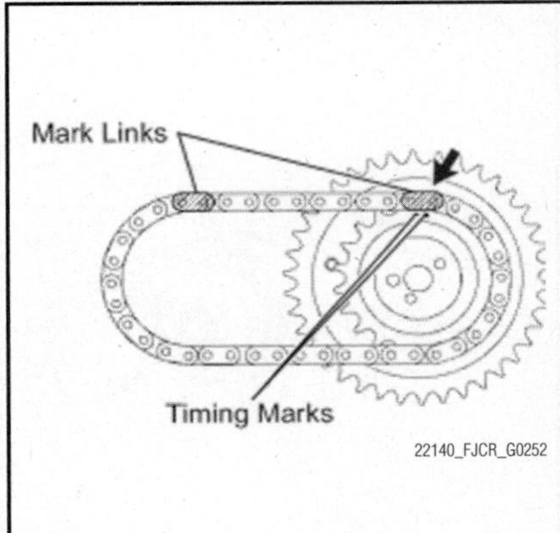

Fig. 65 Align the yellow mark link with the timing mark (2 dot marks) of the camshaft timing gear as shown

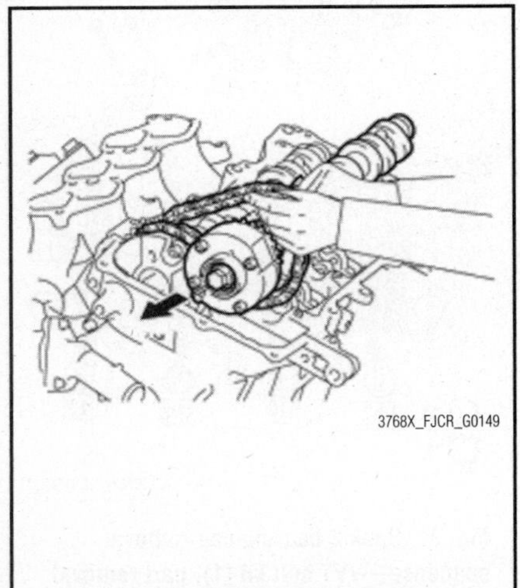

Fig. 64 Removing the No. 4 camshaft

*1
*2
*3
*4
3768X_FJCR_G0150

Fig. 66 No. 1 valve rocker arm sub-assembly installation orientation

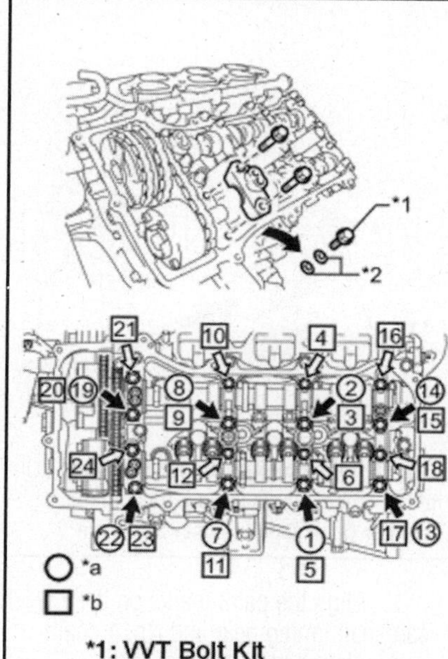

Fig. 68 Install the bank 2 bearing cap with the bolts in the order shown in the illustration

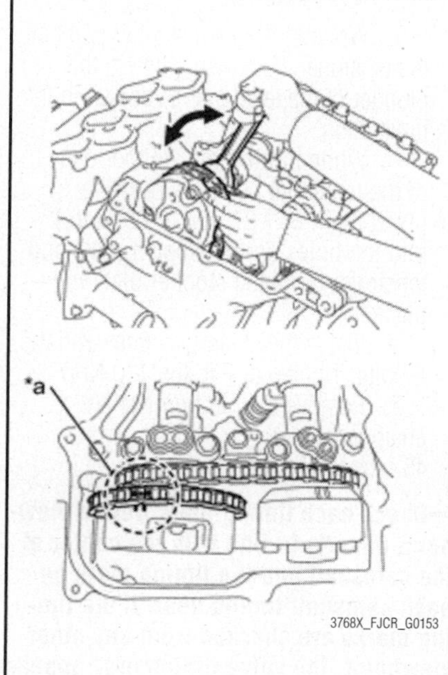

Fig. 69 Align the paint marks on the camshaft timing gear and No. 1 chain and install the No. 1 chain to the camshaft timing gear

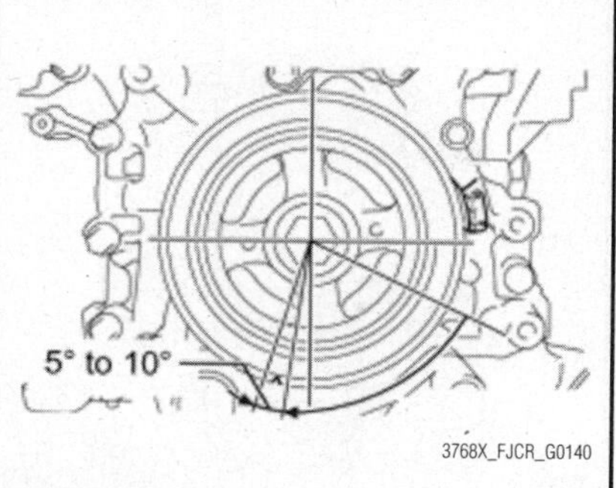

Fig. 70 Turn the crankshaft clockwise until it is in the position shown in the illustration so that the chain can be removed easily

b. Make sure that the No. 1 valve rocker arm sub-assembly is installed as shown in the illustration.

➡Be sure to follow the numerical order when performing this procedure. Do not drop the VVT bolt kit into the cylinder head.

c. Check the marks and numbers on the camshaft bearing caps, and then remove VVT bolt kit in the order shown in the illustration. Immediately after removing VVT bolt kit in the location for a bearing cap, install the bearing cap with the bolts in the order shown in the illustration.

- Bolt A: 21 ft. lbs. (28 Nm).
- Bolt B: 12 ft. lbs. (16 Nm).

d. Check the torque of each bolt again.

➡If the paint marks are not aligned, align them by turning the camshaft slightly.

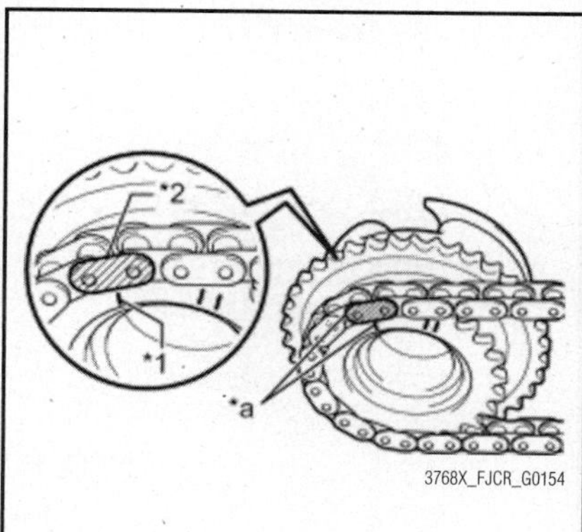

Fig. 71 Align the mark plate (yellow) with the timing mark of the camshaft timing gear as shown

42. Connect chain sub-assembly (for bank 2):

a. Align the paint marks on the camshaft timing gear and No. 1 chain and install the No. 1 chain to the camshaft timing gear.

43. Install camshaft:

a. Turn the crankshaft clockwise until it is in the position shown in the illustration so that the chain can be removed easily.

➡When turning the crankshaft, engine oil may spray out of the oil holes.

b. Align the mark plate (yellow) with the timing mark of the camshaft timing gear as shown in the illustration and install the No. 2 chain to the camshaft timing gear.

➡The mark plate is yellow.

c. Clean the camshaft housing RH and camshaft journals and apply engine oil to them.

d. Make sure that the No. 1 valve rocker arm sub-assembly is installed as shown in the illustration.

e. Install the chain to the camshaft, and then install the camshaft to the RH camshaft housing.

44. Install no. 2 camshaft:

a. Clean the camshaft housing RH and camshaft journals and apply engine oil to them.

➡The mark plate is yellow.

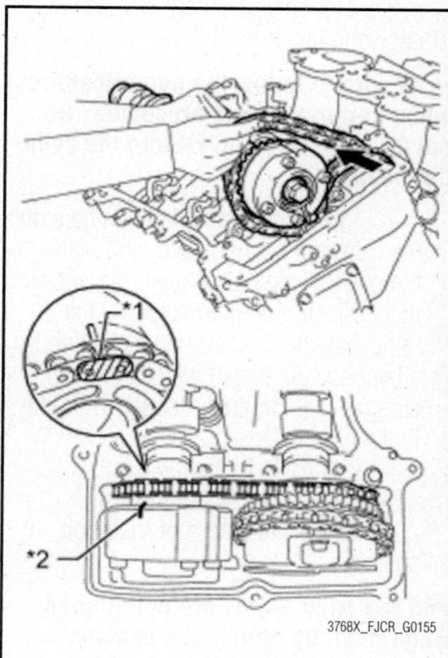

Fig. 72 Installing the No. 2 camshaft—Mark plate (yellow) (1), Timing mark (2)

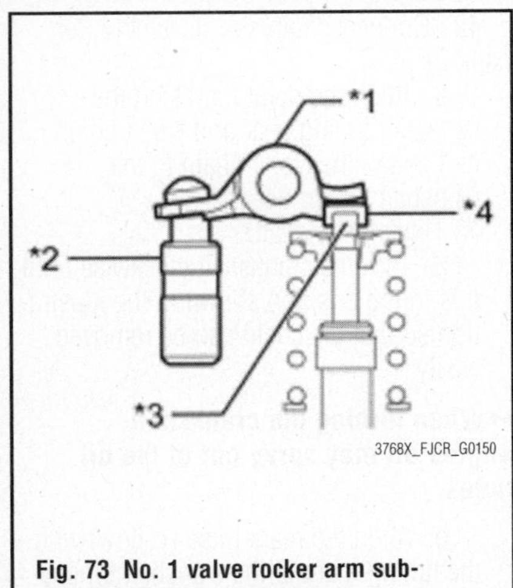

Fig. 73 No. 1 valve rocker arm sub-assembly installation orientation

b. Pass the No. 2 camshaft through the No. 2 chain from the front of the vehicle, align the mark plate (yellow) with the timing mark and install the No. 2 chain to the camshaft timing exhaust gear.

c. While lifting up the No. 2 camshaft, pass the No. 2 chain tensioner assembly through the No. 2 chain and set it in place.

d. Install the No. 2 camshaft to the camshaft housing RH, and then install the No. 2 chain tensioner assembly with the bolt. Tighten to 15 ft. lbs. (21 Nm).

45. Install camshaft bearing cap (for bank 1):

a. Clean the camshaft bearing caps and apply engine oil to them.

b. Make sure that the No. 1 valve rocker arm sub-assembly is installed as shown in the illustration.

➡Be sure to follow the numerical order when performing this procedure. Do not drop the VVT bolt kit into the cylinder head.

c. Check the marks and numbers on the camshaft bearing caps, and then remove VVT bolt kit in the order shown in the illustration. Immediately after removing VVT bolt kit in the location for a bearing cap, install the bearing cap with the bolts in the order shown in the illustration.

- Bolt A: 21 ft. lbs. (28 Nm).
- Bolt B: 12 ft. lbs. (16 Nm).

d. Check the torque of each bolt again.

➡If the paint marks are not aligned, align them by turning the camshaft slightly.

46. Connect chain sub-assembly (for bank 1):

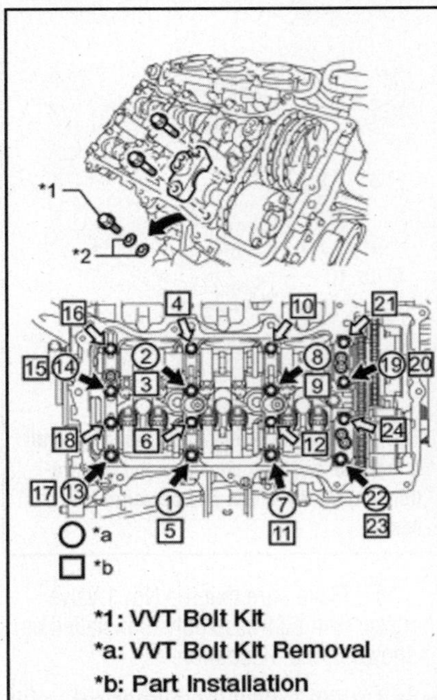

Fig. 74 Install the bank 1 bearing cap with the bolts in the order shown in the illustration

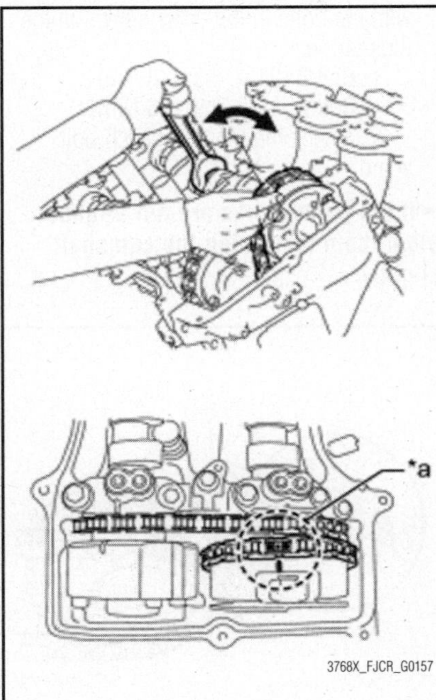

Fig. 75 Align the paint marks on the camshaft timing gear and No. 1 chain and install the No. 1 chain to the camshaft timing gear

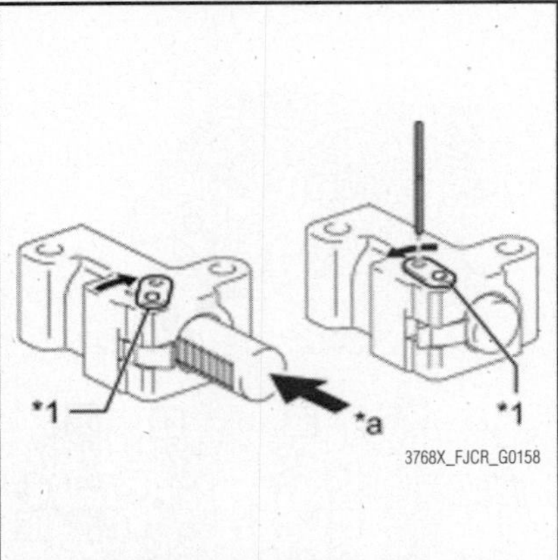

Fig. 76 While turning the stopper plate (1) of the tensioner clockwise, push in the plunger (a) of the tensioner as shown

a. Align the paint marks on the camshaft timing gear and No. 1 chain and install the No. 1 chain to the camshaft timing gear.

47. Install no. 1 chain tensioner assembly:

a. Turn the crankshaft counterclockwise 30° past the "0" timing mark, and then turn it clockwise to align the notch with the "0" timing mark.

b. Turn the crankshaft slightly to eliminate the slack in the chain.

➡Make sure there is some slack in the chain around the area where the chain tensioner is installed.

c. While turning the stopper plate of the tensioner clockwise, push in the plunger of the tensioner as shown in the illustration.

d. While turning the stopper plate of the tensioner counterclockwise, insert a pin of 1.27 mm (0.0500 in.) into the holes on the stopper plate and tensioner to fix the stopper plate in place.

e. Install the chain tensioner with the 2 bolts. Tighten to 7 ft. lbs. (10 Nm).

f. Remove the pin from the No. 1 chain tensioner.

48. Inspect valve timing:

➡Check each timing mark from a viewpoint directly in-line with the center of the camshaft and the timing mark on each camshaft timing gear. If the timing marks are checked from any other viewpoint, the valve timing may appear misaligned.

a. Check the camshaft timing marks.

b. Check that each camshaft timing mark is positioned as shown in the illustration.

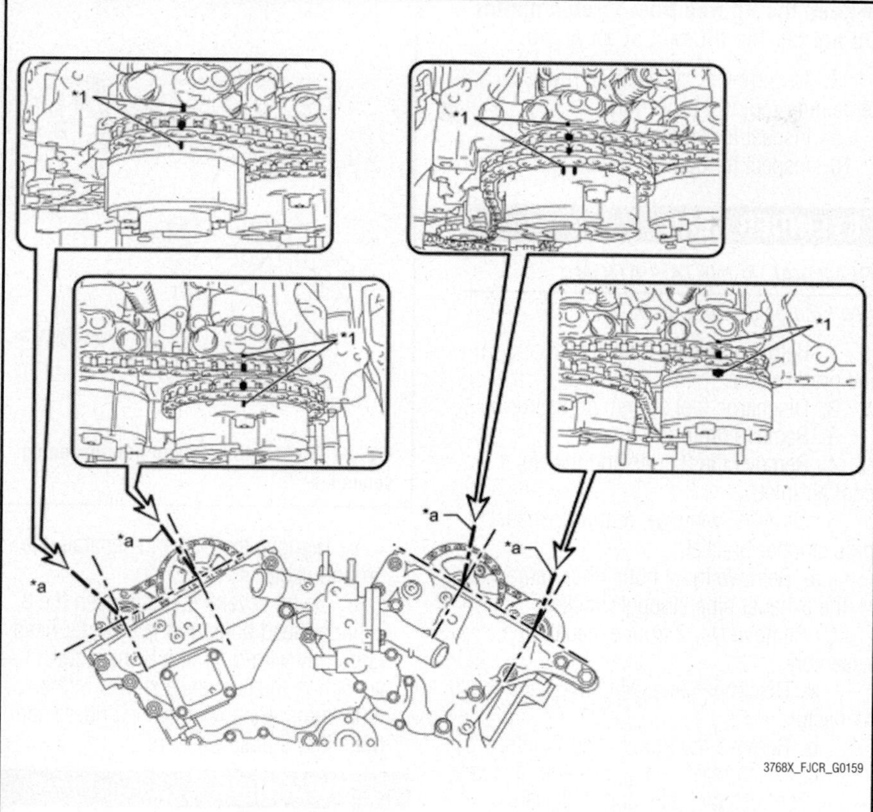

Fig. 77 Check that each camshaft timing mark is positioned as shown—Timing mark (1), viewpoint (a)

➡For the intake camshaft: Be sure to check mark A at the point when marks B, C, and D are positioned in line. If the marks are checked from any other viewpoint, they cannot be checked correctly.

c. If the valve timing is misaligned, reinstall the timing chain.

d. Turn the crankshaft 2 revolutions, set the No. 1 cylinder to TDC/compression and check the timing marks again.

49. Install a new gasket and the timing chain cover plate with the 4 bolts.

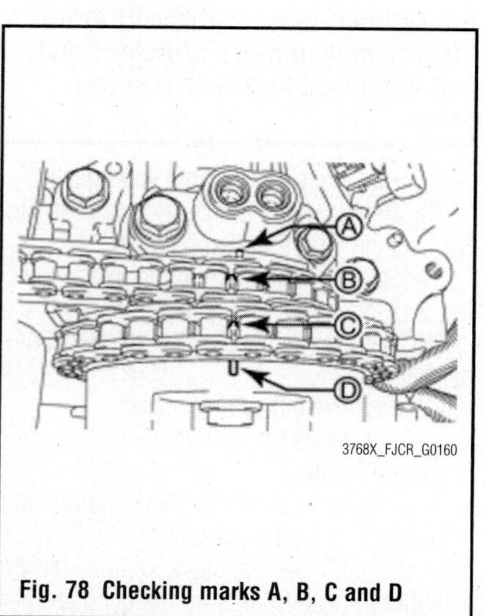

Fig. 78 Checking marks A, B, C and D

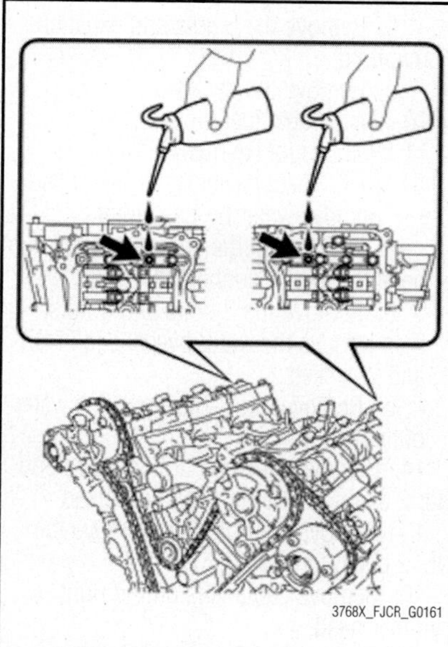

Fig. 79 Oil lubrication points

50. Before installing the cylinder head cover, pour engine oil into the locations shown in the illustration.

51. To complete installation, reverse remaining removal procedure.

CRANKSHAFT DAMPER

REMOVAL & INSTALLATION

See Figures 80 and 81.

1. Before servicing the vehicle, refer to the precautions.
2. Remove the accessory drive belt.
3. Using SST: 09213-54015, hold the crankshaft pulley and loosen the pulley set bolt.
4. Using the pulley set bolt and SST: 09950-50013, remove the crankshaft pulley.

To install:

5. Install the crankshaft pulley.
6. Using SST: 09213-54015, install the pulley set bolt. Tighten to 184 ft. lbs. (250 Nm).

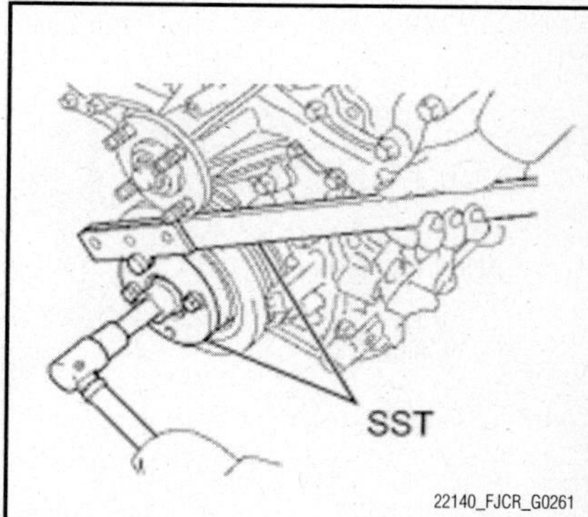

Fig. 80 Using SST: 09213-54015, hold the crankshaft pulley and loosen the pulley set bolt

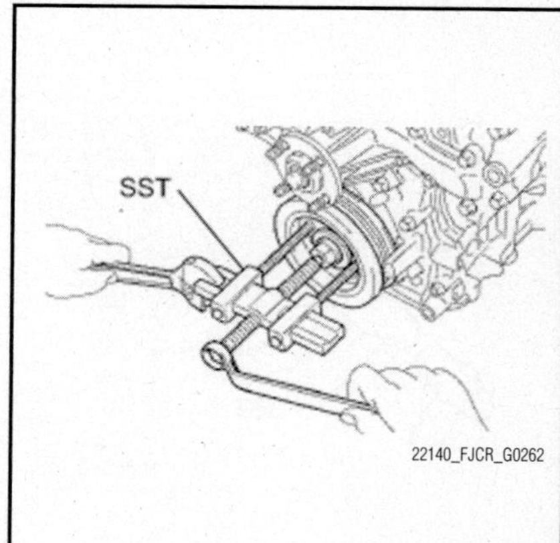

Fig. 81 Using the pulley set bolt and SST: 09950-50013, remove the crankshaft pulley

CRANKSHAFT FRONT SEAL

REMOVAL & INSTALLATION

See Figures 82 and 83.

1. Disconnect the negative battery cable.
2. Remove the fan shroud.
3. Remove oil filter bracket.
4. Remove the crankshaft damper.

➡Do not damage the surface of the oil seal press fit hole and crankshaft.

➡Tape the screwdriver tip before use.

5. Using a screwdriver, pry out the oil seal.

To install:

6. Apply MP grease to the lip of a new oil seal.
7. Using SST: 09226-10010 and a hammer, tap in the oil seal until its surface is flush with the timing chain cover edge.

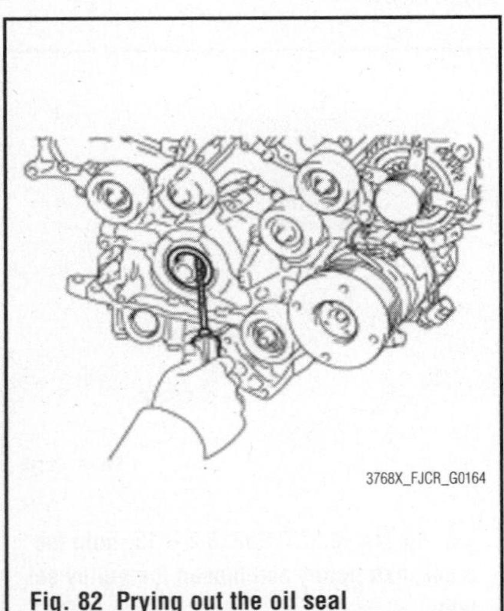

Fig. 82 Prying out the oil seal

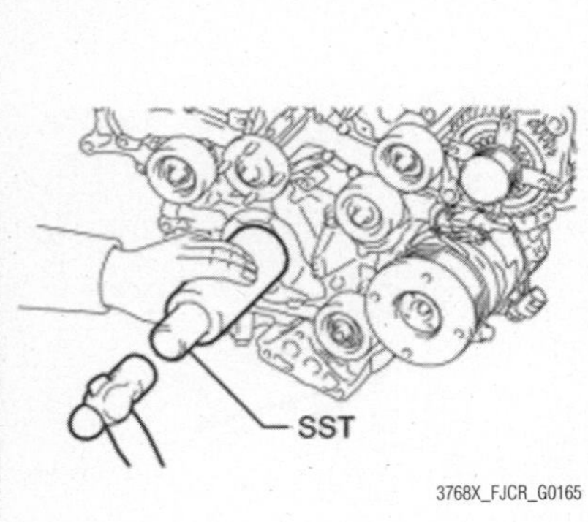

Fig. 83 Using SST: 09226-10010 and a hammer, tap in the oil seal

➡Keep the lip free from foreign matter. Do not tap the oil seal at an angle.

8. To complete installation, reverse remaining removal procedure.
9. Inspect for engine oil leaks.
10. Inspect for coolant leaks.

CYLINDER HEAD

REMOVAL & INSTALLATION

See Figures 84 through 91.

1. Before servicing the vehicle, refer to the precautions.
2. Discharge fuel system pressure.
3. Remove timing chain.
4. Remove the 2 bolts and the No. 1 cool air inlet.
5. On 4WD vehicles, remove exhaust pipe stopper bracket.
    a. Remove the 2 bolts, then remove the exhaust pipe stopper bracket.
6. Remove No. 2 front exhaust pipe assembly.
    a. Disconnect the oxygen sensor connector.
    b. Remove the 2 bolts and 2 nuts.
    c. Disengage the support and remove the front exhaust pipe and 2 gaskets.
7. Remove front exhaust pipe assembly.
    a. Disconnect the oxygen sensor connector.
    b. Remove the 2 bolts, 2 springs and 2 nuts, then separate the front exhaust pipe from the exhaust manifold.
8. Remove the 3 bolts and manifold bracket.
9. Remove exhaust manifolds.
    a. Disconnect the oxygen sensor connector.
    b. Remove the 6 nuts and exhaust manifold.
    c. Remove the gasket.
10. Disconnect fuel pipes.
11. Remove intake manifold.
12. Disconnect heater water inlet hose.
13. Remove water by-pass joint.
    a. Disconnect the engine coolant temperature sensor connector.
    b. Remove the 2 bolts and 4 nuts, then remove the water by-pass joint RR and 2 gaskets.
    c. Remove the O-ring from the water outlet hose.
14. Remove camshaft timing gears and No. 2 chain on the right cylinder head.
15. Remove the bolt, then remove the No. 2 chain tensioner.
16. Remove camshafts on the right cylinder head.
17. Remove No. 2 camshaft bearing.
18. Remove right cylinder head.

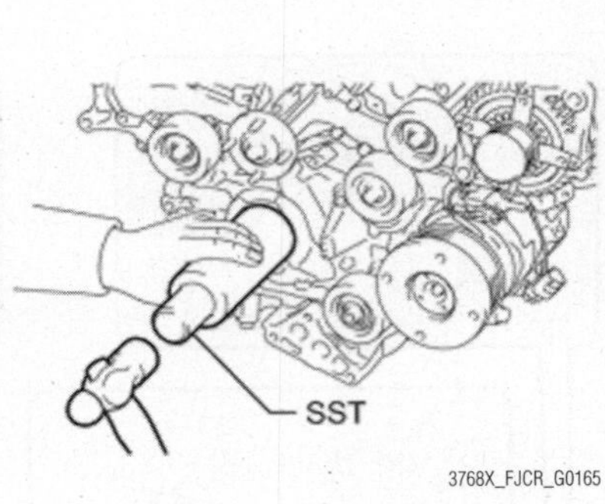

Fig. 84 Right cylinder head bolt loosening sequence

    a. Remove the bolt and separate the ground cable.
    b. Using several steps, loosen the 8 cylinder head bolts on the cylinder head uniformly with a 10 mm bi-hexagon wrench in the sequence shown in the illustration. Remove the 8 cylinder head bolts and 8 plate washers.

WARNING

Be careful not to drop the plate washers into the cylinder head. Cylinder head warpage or cracking could result from removing the bolts in the wrong order.

    c. Lift the cylinder head from the dowels on the cylinder block, and place the cylinder head on wooden blocks on a bench.

WARNING

Be careful not to drop the plate washers into the cylinder head. If the cylinder head is difficult to lift off, pry between the cylinder head and cylinder block with a screwdriver.

19. Remove right cylinder head gasket.
20. Remove the 2 bolts, then remove the No. 1 chain vibration damper.
21. Remove camshaft timing gears and No. 2 chain on the left cylinder head.
22. Remove camshafts on the left cylinder head.
23. Remove left cylinder head.
    a. Remove the bolt, then separate the ground cable.
    b. Remove the bolt, then separate the oxygen connector bracket.
    c. Using several steps, remove the 2 cylinder head bolts from the cylinder

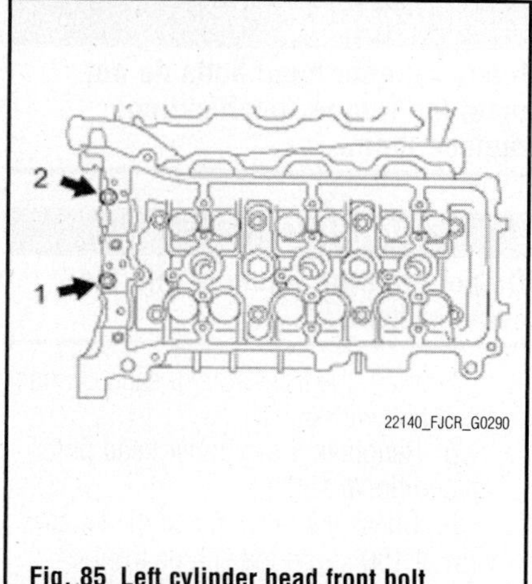

Fig. 85 Left cylinder head front bolt loosening sequence

head in the sequence shown in the illustration.

d. Using several steps, uniformly loosen the 8 cylinder head bolts on the cylinder head with a 10 mm bi-hexagon wrench in the sequence shown in the illustration. Remove the 8 cylinder head bolts and 8 plate washers.

WARNING

Be careful not to drop the plate washers into the cylinder head. Cylinder head warpage or cracking could result from removing the bolts in the wrong order.

e. Lift the cylinder head from the dowels on the cylinder block, and place the cylinder head on wooden blocks on a bench.

WARNING

Be careful not to drop the plate washers into the cylinder head. If the cylinder head is difficult to remove, pry between the cylinder head and cylinder block with a screwdriver.

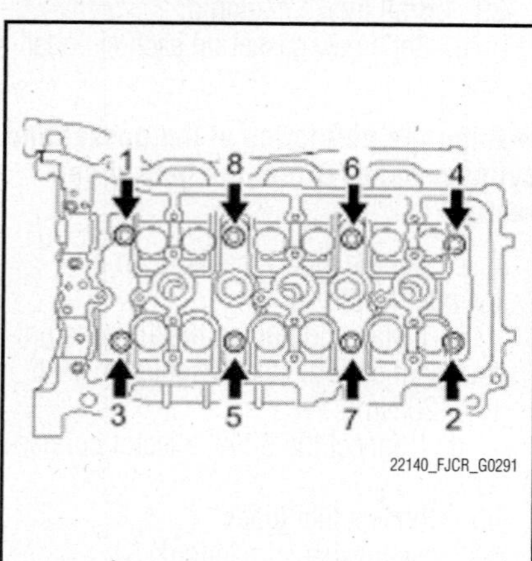

Fig. 86 Left cylinder head bolt loosening sequence

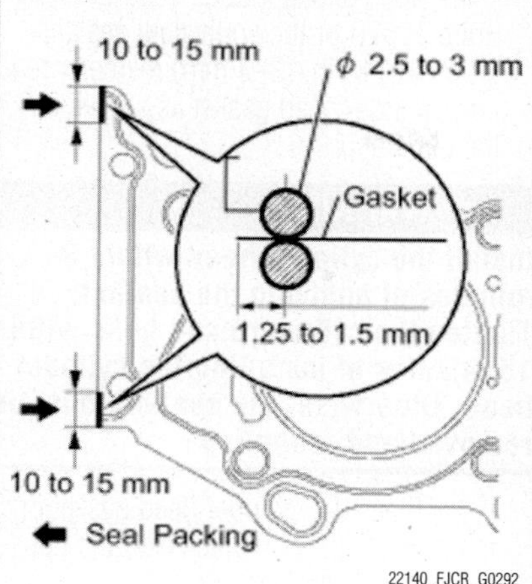

Fig. 87 Apply a continuous bead of Three Bond 1207B or the equivalent sealant 0.098–0.118 in. (2–3 mm) diameter

24. Remove right cylinder head gasket.

To install:

25. Install No. 2 cylinder head gasket.
   a. Remove any old sealant.

WARNING

Do not drop any oil on the contact surface of the cylinder head and cylinder block.

b. Apply a continuous bead of Three Bond 1207B or the equivalent sealant 0.098–0.118 in. (2–3 mm) diameter to a new cylinder head gasket as shown in the illustration.

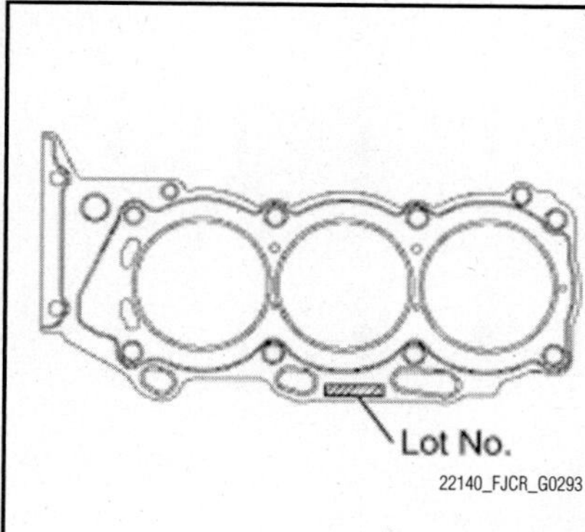

Fig. 88 Apply a continuous bead sealant in the pattern shown

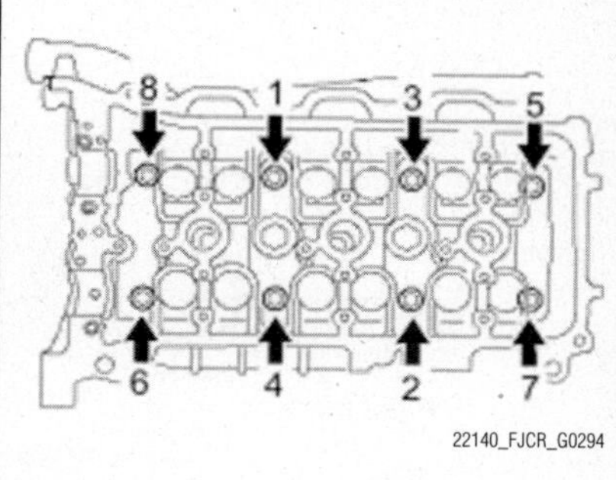

Fig. 89 Using several steps, uniformly tighten each bolt with a 10 mm bi-hexagon wrench in the sequence shown—left cylinder head

WARNING

Install the cylinder head within 3 minutes of applying the sealant. Tighten the cylinder head bolts within 15 minutes of installing the cylinder head. Otherwise, the sealant must be removed and reapplied.

c. Place the cylinder head gasket on the cylinder block surface with the Lot No. stamp facing upward.

WARNING

Orient the cylinder head gasket correctly. Place the cylinder head carefully in order not to damage the gasket with the bottom part of the head.

26. Install the left cylinder head.
   a. Place the cylinder head on the cylinder head gasket.
   b. Install the 8 cylinder head bolts.

➡The cylinder head bolts are tightened in 2 successive steps. If any cylinder head bolts are broken or deformed, replace them.

c. Apply a light coat of engine oil to the threads of the cylinder head bolts.

d. Install the plate washer onto the cylinder head bolt.

e. Using several steps, uniformly tighten each bolt with a 10 mm bi-hexagon wrench in the sequence shown in the illustration. Tighten to 27 ft. lbs. (36 Nm).

WARNING

If any cylinder head bolts do not meet the torque specification, replace them.

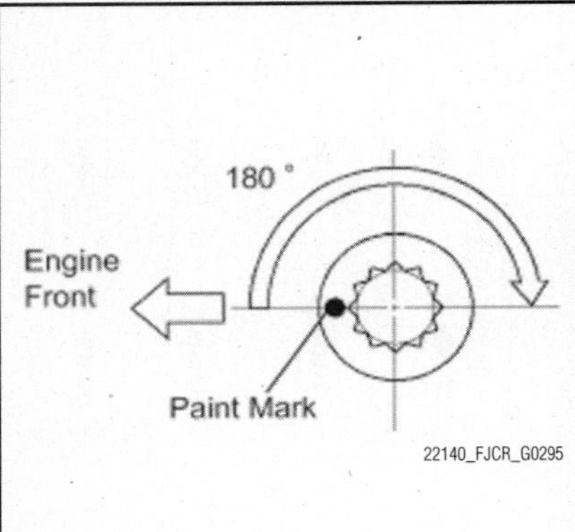

Fig. 90 Check that the painted marks are now at 180°from the engine front

⁂ WARNING

Do not drop the washers into the cylinder head.

f. Mark the front side of each cylinder head bolt with paint.

g. Retighten the cylinder head bolts an additional 180°.

h. Check that the painted marks are now at 180°from the engine front.

i. Install the front 2 cylinder head bolts.

j. Apply a light coat of engine oil to the threads of the cylinder head bolts.

k. Using several steps, uniformly install and tighten the 2 cylinder head bolts in the sequence shown in the illustration. Tighten to 22 ft. lbs. (30 Nm).

l. Install the ground cable with the bolt. Tighten to 71 ft. lbs. (96 Nm).

m. Install the oxygen sensor connector bracket with bolt. Tighten to 14 ft. lbs. (19 Nm).

27. Install camshafts on the left cylinder head.

28. Install No. 3 chain tensioner assembly.

29. Install camshaft timing gears and No. 2 chain on the left cylinder head.

30. Install No. 1 chain vibration damper.

a. Install the chain vibration damper with the 2 bolts. Tighten to 14 ft. lbs. (19 Nm).

31. Install the right cylinder head gasket.

a. Remove any old sealant.

⁂ WARNING

Do not drop any oil on the contact surfaces of the cylinder head and cylinder block.

b. Apply a continuous bead of Three Bond 1207B or the equivalent sealant 0.098–0.118 in. (2–3 mm) diameter to a new cylinder head gasket as shown in the illustration.

⁂ WARNING

Install the cylinder head within 3 minutes of applying the sealant. Tighten the cylinder head bolts within 15 minutes of installing the cylinder head. Otherwise, the sealant must be removed and reapplied.

c. Place the cylinder head gasket on the cylinder block surface with the Lot No. stamp facing upward.

⁂ WARNING

Orient the cylinder head gasket correctly. Place the cylinder head carefully in order not to damage the gasket.

32. Install the right cylinder head.

a. Place the cylinder head on the cylinder head gasket.

b. Install the 8 cylinder head bolts.

➡The cylinder head bolts are tightened in 2 successive steps. If any cylinder head bolts are broken or deformed, replace them.

c. Apply a light coat of engine oil to the threads of the cylinder head bolts.

d. Install the plate washer onto the cylinder head bolt.

e. Using several steps, uniformly tighten each bolt with a 10 mm bi-hexagon wrench in the sequence shown in the illustration. Tighten to 27 ft. lbs. (36 Nm).

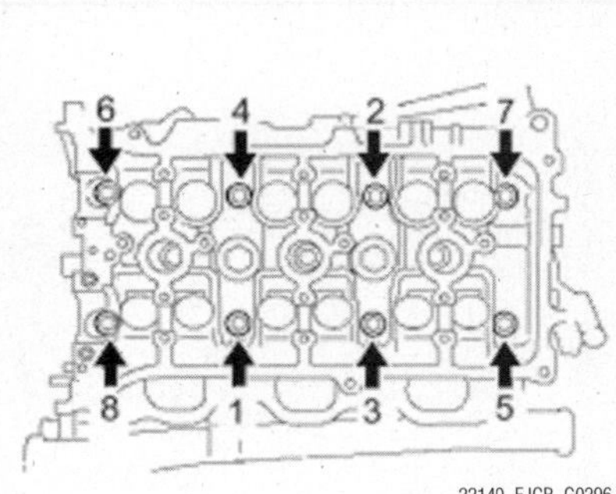

Fig. 91 Using several steps, uniformly tighten each bolt with a 10 mm bi-hexagon wrench in the sequence shown—right cylinder head

⁂ WARNING

If any cylinder head bolts do not meet the torque specification, replace them.

⁂ WARNING

Do not drop the washers into the cylinder head.

f. Mark the front side of each cylinder head bolt with paint.

g. Retighten the cylinder head bolts an additional 180°.

h. Check that the painted marks are now at 180° from the engine front.

i. Install the ground cable with the bolt. Tighten to 71 ft. lbs. (96 Nm).

33. Install No. 2 camshaft bearing.

a. Install the No. 2 camshaft bearing onto the cylinder head.

⁂ WARNING

Clean the installation planes of the back side of the bearing and cylinder head and keep them free of oil.

34. Install camshafts on the right cylinder head.

35. Install No. 2 chain tensioner assembly.

36. Install camshaft timing gears and No. 2 chain on the right cylinder head.

37. Install water by-pass joint.

a. Install a new O-ring onto the water outlet pipe.

b. Apply soapy water to the O-ring.

c. Install 2 new gaskets and water bypass joint rear with the 2 bolts and 4 nuts. Tighten to 80 inch lbs. (9 Nm).

d. Connect the engine coolant temperature sensor connector.

38. Connect heater water inlet hose.

39. Install intake manifold.

a. Set a new gasket on each cylinder head.

➡Align the port holes of the gasket and cylinder head. Orient the gasket correctly.

b. Set the intake manifold on the cylinder heads.

c. Install and tighten the 10 bolts uniformly in several steps. Tighten to 19 ft. lbs. (26 Nm).

d. Connect the 6 fuel injector connectors.

40. Connect fuel pipes.

41. Install exhaust manifolds.

a. Set a new gasket to the cylinder head with the oval shape facing forward.

➡**Orient the new gasket correctly.**

b. Install the exhaust manifold with the 6 nuts. Tighten to 22 ft. lbs. (30 Nm).

c. Connect the oxygen sensor connector.

42. Install the manifold bracket with the 3 bolts. Tighten to 30 ft. lbs. (40 Nm).

43. Install front exhaust pipe assembly.

a. Check the free length.

b. Using Vernier calipers, measure the free length of the compression spring. Minimum length should be 0.594 in. (5mm). If the free length is less than the minimum, replace the compression spring.

c. Install a new gasket onto the right exhaust manifold.

d. Install a new gasket onto the exhaust front pipe.

e. Using a wooden block and plastic-faced hammer, tap in the new gasket until it is flush with the exhaust front pipe.

➡**Make sure that the gasket is in the correct direction when installing. Do not reuse the removed gasket.**

f. Install the exhaust front pipe with new 2 nuts. Tighten to 40 ft. lbs. (54 Nm).

g. Install the exhaust center pipe onto the front exhaust pipe with the 2 bolts and 2 springs. Tighten to 32 ft. lbs. (43 Nm).

h. Connect the oxygen sensor connector.

44. Install No. 2 front exhaust pipe assembly.

a. Install a new gasket onto the left exhaust manifold.

b. Install a new gasket onto the No. 2 front exhaust pipe.

c. Install the No. 2 front exhaust pipe with the 2 new nuts and 2 bolts. Tighten the nut to 40 ft. lbs. (54 Nm) and the bolt to 35 ft. lbs. (48 Nm).

d. Connect the oxygen sensor connector.

45. On 4WD vehicles, install the exhaust pipe stopper bracket with the 2 bolts. Tighten the nut to 14 ft. lbs. (19 Nm).

46. Install the cool air inlet with the 2 bolts. Tighten to 9 ft. lbs. (12 Nm).

47. Install the timing chain.

48. Check for fuel leakage.

49. Check for exhaust gas leakage.

50. Inspect ignition timing.

51. Inspect engine idling speed.

52. Inspect the cylinder compression.

53. Inspect CO/HC.

54. Inspect and adjust front wheel alignment

EXHAUST MANIFOLD & FRONT CATALYTIC CONVERTER

REMOVAL & INSTALLATION

See Figures 92 through 95.

1. Remove the v-bank cover.
2. Remove the air cleaner assembly.
3. Remove the 5 clips and LH fender apron seal.
4. Remove the 4 clips and RH fender apron seal.
5. Remove the 5 clips and LH frame seal.
6. Remove the 5 clips and RH fender apron seal.
7. Remove the 2 bolts and the exhaust pipe stopper bracket.
8. Remove the exhaust pipe assembly.
9. Remove the 3 bolts and manifold stay.
10. Remove the 3 bolts and heat insulator.
11. Remove RH exhaust manifold:

a. Disconnect the air fuel ratio sensor connector.

b. Remove the 6 nuts, manifold and gasket.

12. Remove air fuel ratio sensor (for Bank 1 Sensor 1).
13. Remove the 3 bolts and No. 2 manifold stay.
14. Remove the 3 bolts and heat insulator.
15. Remove LH exhaust manifold:

a. Disconnect the air fuel ratio sensor connector.

b. Remove the 6 nuts, manifold and gasket.

16. Remove air fuel ratio sensor (for Bank 2 Sensor 1).

To install:

17. Install air fuel ratio sensor (for Bank 2 Sensor 1).
18. Install the LH exhaust manifold:

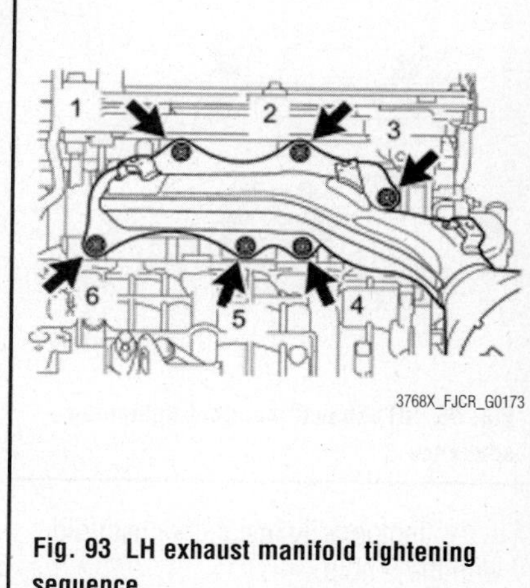

Fig. 93 LH exhaust manifold tightening sequence

➡**Be careful of the installation direction.**

a. Install a new gasket onto the cylinder head.

b. Temporarily install the manifold with the 6 nuts.

c. Tighten the 6 nuts in the sequence shown in the illustration. Tighten to 15 ft. lbs. (21 Nm).

d. Connect the air fuel ratio sensor connector.

19. Install no. 2 exhaust manifold heat insulator. Tighten the three bolts to 10 ft. lbs. (13 Nm).

20. Install the manifold stay with the 3 bolts. Tighten to 30 ft. lbs. (13 Nm).

21. Install air fuel ratio sensor (for Bank 1 Sensor 1).

22. Install the RH exhaust manifold:

➡**Be careful of the installation direction.**

a. Install a new gasket onto the cylinder head.

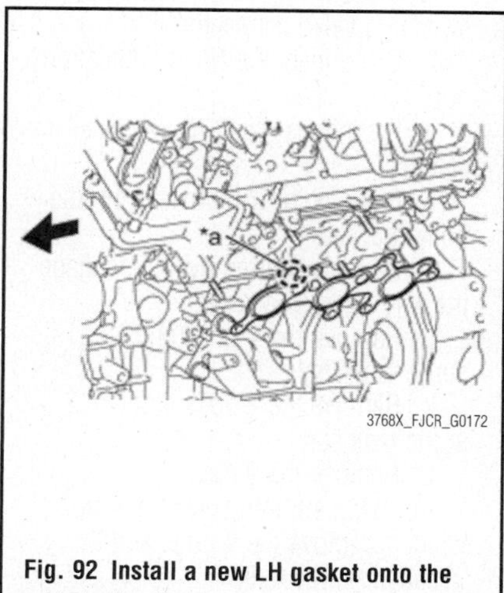

Fig. 92 Install a new LH gasket onto the cylinder head

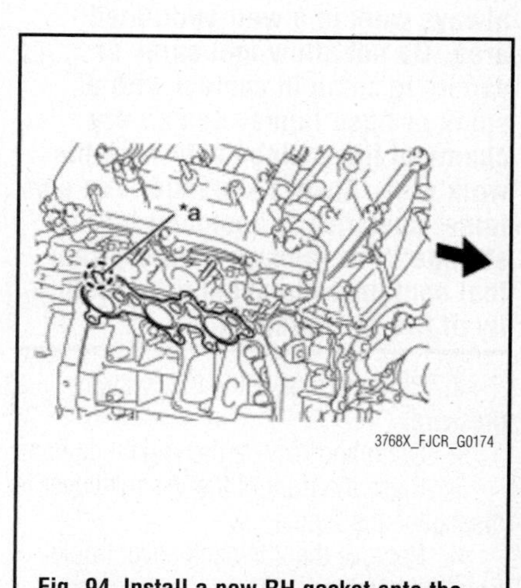

Fig. 94 Install a new RH gasket onto the cylinder head

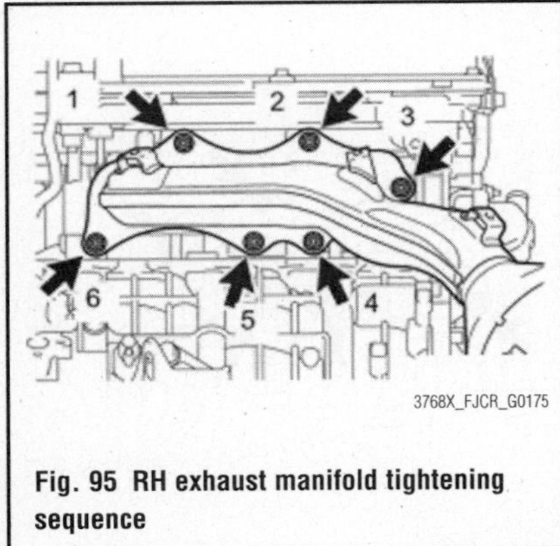

Fig. 95 RH exhaust manifold tightening sequence

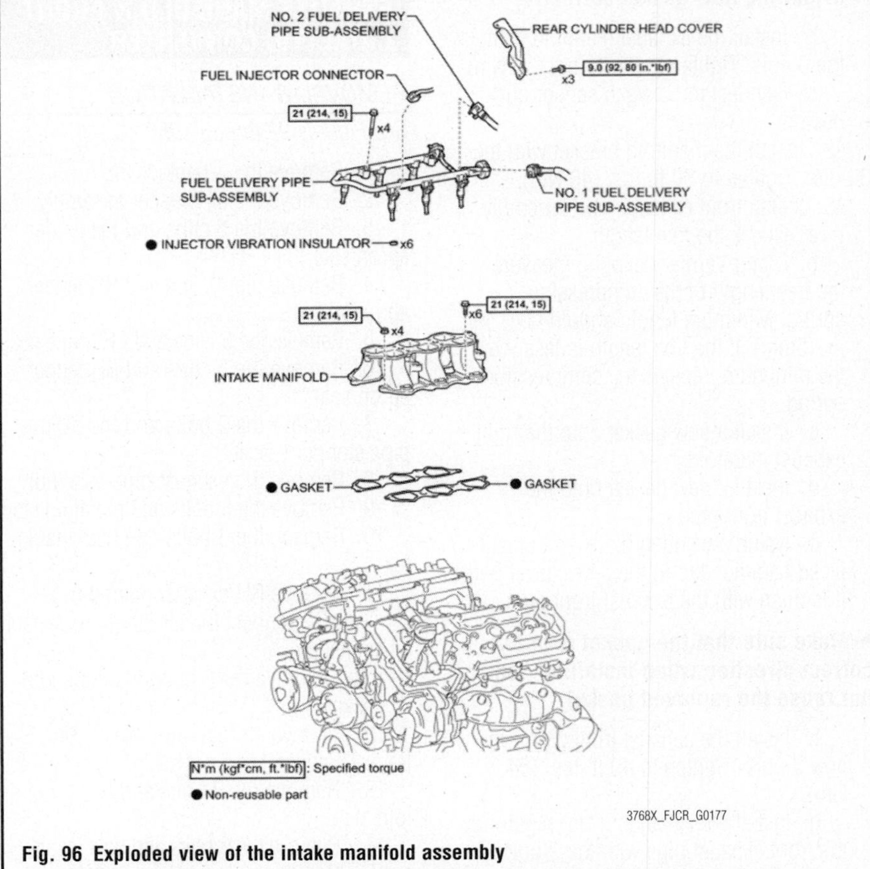

Fig. 96 Exploded view of the intake manifold assembly

b. Temporarily install the manifold with the 6 nuts.

c. Tighten the 6 nuts in the sequence shown in the illustration. Tighten to 15 ft. lbs. (21 Nm).

d. Connect the air fuel ratio sensor connector.

23. Install no. 1 exhaust manifold heat insulator. Tighten the three bolts to 10 ft. lbs. (13 Nm).

24. Install the no. 2 manifold stay with the 3 bolts. Tighten to 30 ft. lbs. (13 Nm).

25. To complete the installation, reverse remaining removal procedure.

26. Install the exhaust pipe stopper bracket with the 2 bolts. Tighten to 14 ft. lbs. (19 Nm).

INTAKE MANIFOLD

REMOVAL & INSTALLATION

See Figures 96 and 97.

⁂ CAUTION

Observe all applicable safety precautions when working around fuel. Whenever servicing the fuel system, always work in a well-ventilated area. Do not allow fuel spray or vapors to come in contact with a spark or open flame. Keep a dry chemical fire extinguisher near the work area. Always keep fuel in a container specifically designed for fuel storage; also, always properly seal fuel containers to avoid the possibility of fire or explosion.

1. Properly relieve the fuel system pressure.
2. Drain and recycle the engine coolant.
3. Raise the front of the V-bank cover to disengage the 2 pins.
4. Remove the 2 V-bank cover hooks from the bracket.
5. Remove the V-bank cover.
6. Remove the air cleaner assembly.
7. Remove intake air surge tank:

a. Disconnect the No. 4 water by-pass hose.

b. Disconnect the No. 5 water by-pass hose.

c. Disconnect the throttle body connector.

d. Disconnect the No. 1 fuel vapor feed hose.

e. Disconnect the No. 1 vacuum switching valve connector.

f. Disconnect the No. 1 ventilation hose.

g. Disconnect the 2 heater water hose clamps.

h. Remove the 2 bolts and throttle body bracket.

i. Using a clip remover, disengage the wire harness clamp.

j. Remove the 2 bolts and No. 1 surge tank stay.

k. Remove the 2 bolts and No. 2 surge tank stay.

l. Remove the 2 nuts.

m. Using a 8 mm socket hexagon wrench, remove the 4 bolts and the intake air surge tank.

n. Remove the gasket.

8. Remove the 3 bolts and the rear cylinder head cover.
9. Disconnect fuel pipes.

a. Remove the fuel pipe clamp.

b. Pinch the tube connector, and then pull the fuel pipe out of the delivery pipe.

10. Remove fuel delivery pipe assembly.

a. Disconnect the 6 fuel injector connectors.

b. Remove the 6 bolts and remove the fuel delivery pipe together with the 6 fuel injectors.

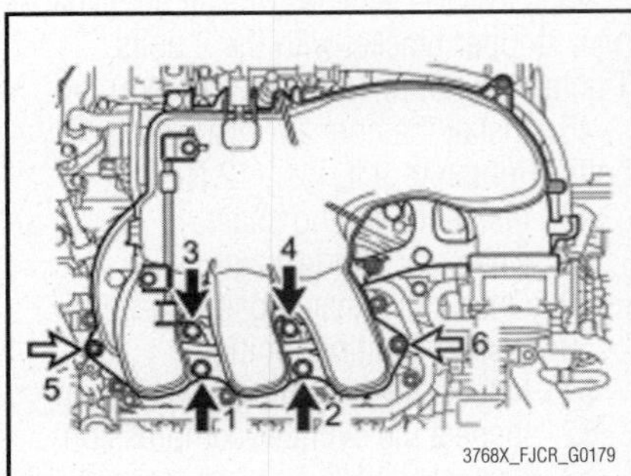

Fig. 97 Install the intake air surge tank in the order shown

✲✲ WARNING

Do not drop the injectors when removing the fuel delivery pipe.

11. Remove the 4 nuts, 6 bolts and intake manifold.
12. Remove the 2 gaskets.

To install:

13. Set a new gasket on each cylinder head. Align the port holes of the gasket and cylinder head. Be careful of the installation direction.
14. Set the intake manifold on the cylinder heads.
15. Install and uniformly tighten the 6 bolts and 4 nuts in several passes. Tighten to 15 ft. lbs. (21 Nm).

➡Tighten the inner installation bolts of the intake manifold before tightening the outer bolts.

16. Install and connect the fuel pipe delivery system. Refer to Fuel Rail and Injectors in Fuel System.
17. Install rear cylinder head cover. Loosely install the 3 bolts and tighten to 80 inch lbs. (9 Nm).
18. Install intake air surge tank:
    a. Install a new gasket to the intake air surge tank.
    b. Install the intake air surge tank in the order shown in the illustration.
    - Using a 8 mm socket hexagon wrench, temporarily tighten the 4 bolts to the intake air surge tank.
    - Temporarily tighten the 2 nuts to the intake air surge tank.
    - Tighten the 4 bolts and 2 nuts to 21 ft. lbs. (28 Nm).
19. To complete installation, reverse remaining removal procedure.
20. Refer to the exploded view illustration for torque values.
21. Add engine coolant. Inspect for coolant leak .
22. Inspect for fuel leaks.
23. Engage the 2 V-bank cover hooks to the bracket.
24. Align the 2 V-bank cover grommets with the 2 pins and press down on the V-bank cover to engage the pins.

OIL PAN

REMOVAL & INSTALLATION

See Figures 98 through 101.

1. Before servicing the vehicle, refer to the precautions.
2. Drain engine oil.

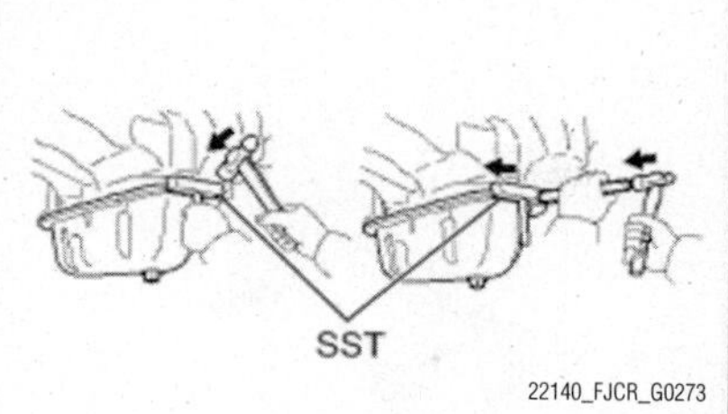

Fig. 98 Insert the blade of SST: 09032-00100 between the oil pan and No. 2 oil pan, cut off applied sealer

3. Remove No. 2 oil pan.
    a. Remove the 10 bolts and 2 nuts.
    b. Insert the blade of SST: 09032-00100 between the oil pan and No. 2 oil pan, cut off applied sealer and remove the No. 2 oil pan.

✲✲ WARNING

Be careful not to damage the contact surfaces of the oil pan and No. 2 oil pan. Be careful not to damage the No. 2 oil pan flange.

4. Remove the 2 nuts, then remove the oil strainer and gasket.
5. Remove oil pan.

To install:

6. Install oil pan.
    a. Remove any old sealant.

✲✲ WARNING

Do not drop any oil on the contact surfaces of the cylinder block, rear oil seal and oil pan.

b. Install a new O-ring onto the oil pump.

c. Apply a continuous bead of Three Bond or the equivalent sealant

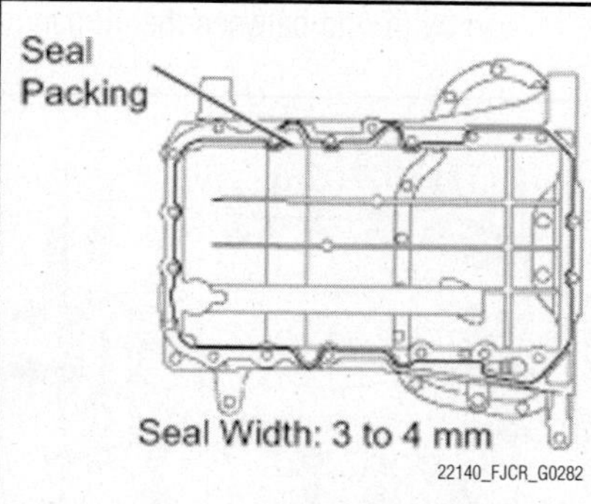

Fig. 99 Apply a continuous bead of Three Bond or the equivalent sealant 0.12–0.16 in. (3–4 mm) diameter to the oil pan as shown

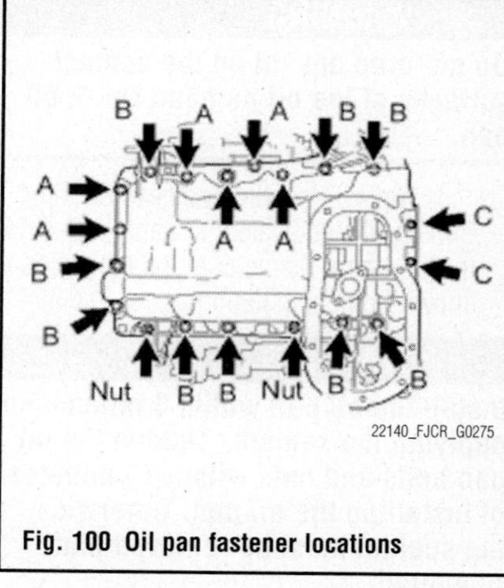

Fig. 100 Oil pan fastener locations

0.12–0.16 in. (3–4 mm) diameter to the oil pan as shown in the illustration.

✲✲ WARNING

Install the oil pan within 3 minutes of applying the sealant. Tighten the oil pan bolts and nuts within 15 minutes of installing the oil pan. Otherwise, the sealant must be removed and reapplied.

d. Install the oil pan with the 17 bolts and 2 nuts, and tighten the bolts and nuts uniformly in several steps. Tighten the 0.39 in. (10 mm) head oil pan bolt to 7 ft. lbs. (10 Nm), the 0.47 in. (12 mm) head oil pan bolt to 16 ft. lbs. (21 Nm) and the nuts to 16 ft. lbs. (21 Nm).

✲✲ WARNING

Install bolts in their proper locations. Bolt lengths are as follows: A is 0.98 in. (25 mm) B 1.77 in. (45 mm) and C 0.55 in. (14 mm).

To install:

e. Remove any old sealant.

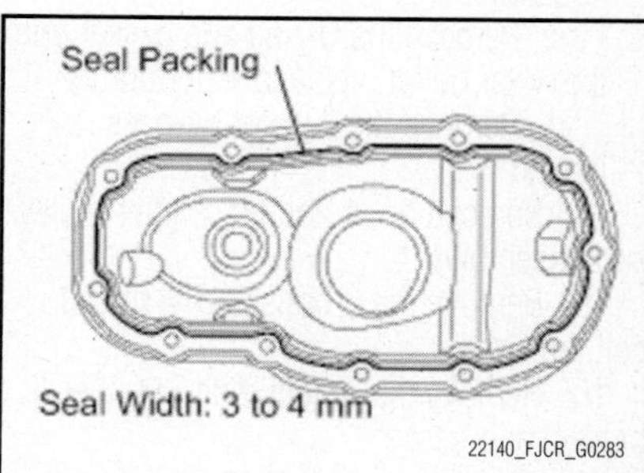

Fig. 101 Apply a continuous bead of Three Bond or the equivalent sealant 0.12–0.16 in. (3–4 mm) diameter to the oil pan

WARNING

Do not drop any oil on the contact surfaces of the oil pan and No. 2 oil pan.

f. Apply a continuous bead of Three Bond or the equivalent sealant 0.12–0.16 in. (3–4 mm) diameter to the oil pan as shown in the illustration.

WARNING

Install the oil pan within 3 minutes of applying the sealant. Tighten the oil pan bolts and nuts within 15 minutes of installing the oil pan. Otherwise, the sealant must be removed and reapplied.

g. Install the oil pan with the 17 bolts and 2 nuts, and tighten the bolts and nuts uniformly in several steps. Tighten the bolts to 80 inch lbs. (9 Nm) and the nuts to 7 ft. lbs. (10 Nm).

7. Add engine oil.
8. Add power steering fluid and bleed the system, as necessary.
9. On 4WD vehicles, inspect and adjust the differential oil.
10. Check for engine oil leakage.

OIL PUMP

REMOVAL & INSTALLATION

See Figures 102 through 107.

1. Before servicing the vehicle, refer to the precautions.
2. Remove the engine.
3. Remove the ignition coil assembly.
4. Remove the dipstick.
5. Remove the bolt and dipstick guide.
6. Remove the O-ring from the dipstick guide.
7. Remove water inlet housing:

a. Disconnect the 3 water by-pass hoses.

b. Remove the 5 bolts and water inlet housing.

c. Remove the O-ring and gasket from the water outlet pipe and water pump.

d. Remove the 3 water by-pass hoses.

8. Remove no. 1 and no. 2 idler pulley sub-assembly.
9. Remove the 5 bolts and V-ribbed belt tensioner.
10. Remove the 2 nuts, bolt, oil filter bracket and gasket.
11. Remove crankshaft pulley.
12. Remove no. 1 oil pipe:

a. Remove the 2 oil pipe unions, gaskets and No. 1 oil pipe.

Fig. 102 Remove the 2 nuts, oil strainer and gasket

b. Remove the oil control valve filter LH and gaskets.

➡ **Do not touch the mesh when removing the oil control valve filter.**

13. Remove no. 2 oil pipe:

a. Remove the 2 oil pipe unions, gaskets and No. 2 oil pipe.

b. Remove the oil control valve filter RH and gaskets.

➡ **Do not touch the mesh when removing the oil control valve filter.**

14. Remove rear cylinder head cover.
15. Remove the 2 bolts and disconnect the fuel pipe.
16. Remove the LH and RH valve covers.

➡ **Make sure the removed parts are returned to the same places they were removed from.**

17. Remove the oil pan.
18. Remove the 2 nuts, oil strainer and gasket.
19. Remove oil pan sub-assembly:

a. Remove the 17 bolts and 2 nuts.

➡ **Make sure the removed parts are returned to the same places they were removed from.**

b. Using a screwdriver, remove the oil pan by prying between the oil pan and cylinder block as shown in the illustration.

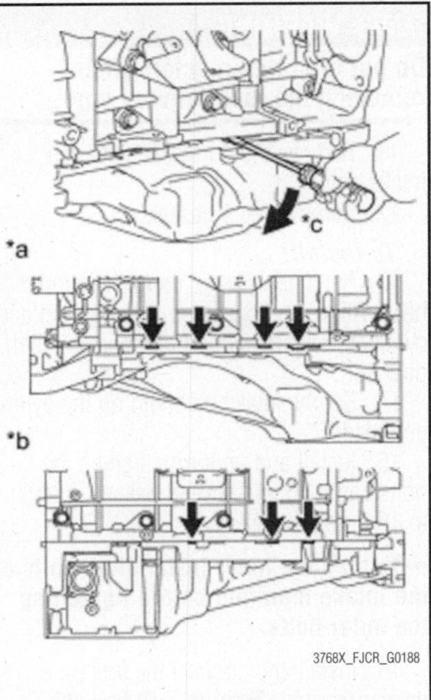

Fig. 104 Remove the oil pan by prying at the indicated locations between the oil pan and cylinder block as shown

➡ **Be careful not to damage the contact surfaces of the cylinder block or oil pan.**

➡ **Tape the screwdriver tip before use.**

c. Remove the 3 O-rings from the timing chain cover.

d. Remove the 3 O-rings from the timing chain cover.

20. Remove timing chain cover sub-assembly.
21. Remove front crankshaft oil seal.
22. Remove the 8 bolts, screw, oil pump cover, drive rotor and driven rotor.
23. Remove oil pump relief valve:

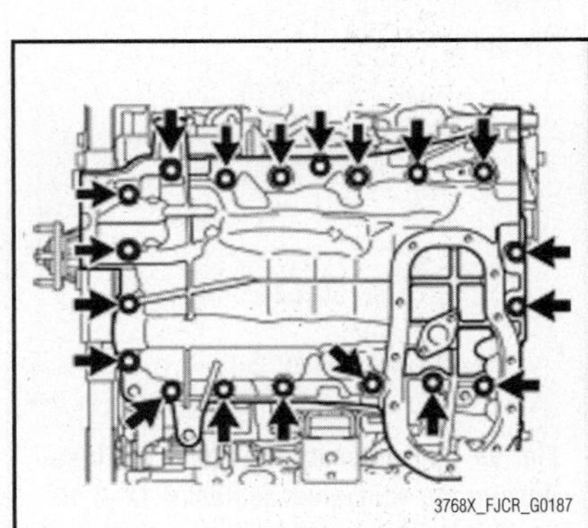

Fig. 103 Remove the 17 bolts and 2 nuts

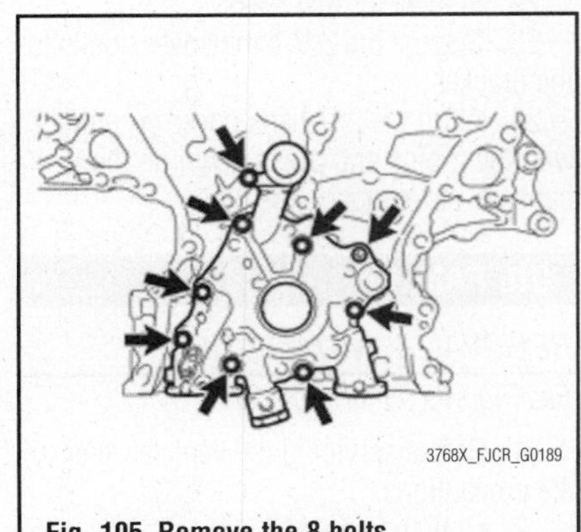

Fig. 105 Remove the 8 bolts

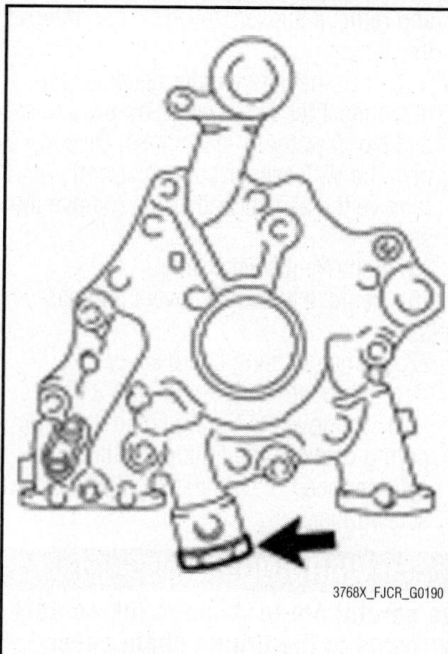

Fig. 106 Remove oil pump relief valve

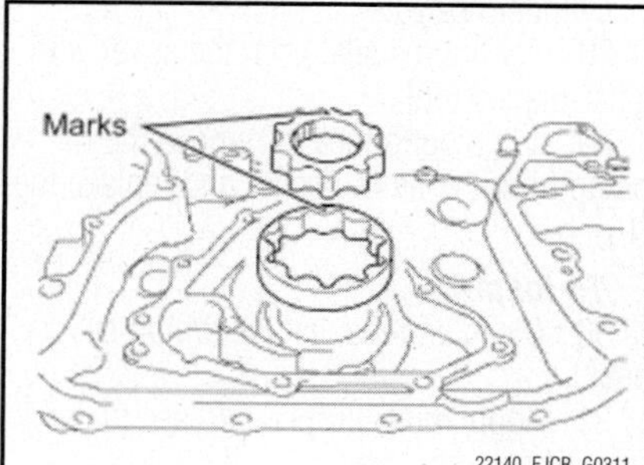

Fig. 107 Place the drive and driven rotors into the timing chain cover with the marks facing the oil pump cover side

a. Remove the relief valve plug.
b. Remove the spring and relief valve.

To install:

24. Install oil pump relief valve
a. Coat the relief valve with engine oil and insert the relief valve and spring into the valve hole.
b. Install the relief valve plug. Tighten to 36 ft. lbs. (49 Nm).
25. Install oil pump cover:
a. Apply fresh engine oil to the drive and driven rotors.
b. Place the drive and driven rotors into the timing chain cover with the matchmarks facing the oil pump cover side.
26. Install the oil pump cover with the 8 bolts and screw:
a. Screw: 7 ft. lbs. (10 Nm).
b. Bolts: 80 inch. lbs. (9 Nm).

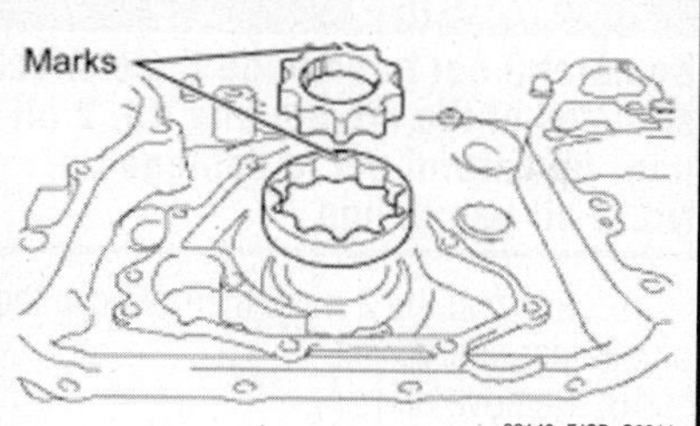

Fig. 108 Place the drive and driven rotors into the timing chain cover with the marks facing upward

27. To complete installation, reverse remaining removal procedure.

INSPECTION

See Figures 108 through 111.

1. Inspect oil pump relief valve.
a. Coat the valve with engine oil, then check that it falls smoothly into the valve hole by its own weight. If not, replace the relief valve. If necessary, replace the oil pump assembly.
2. Inspect oil pump rotor set.
a. Place the drive and driven rotors into the timing chain cover with the marks facing upward.
b. Check the rotor tip clearance.
c. Using a feeler gauge, measure the clearance between the drive and driven rotor tips.
d. Standard clearance is 0.0024–0.0063 in. (0.06–0.16 mm). Maximum tip clearance is 0.0063 in. (0.16 mm).
e. If the clearance is greater than the maximum, replace the drive and driven rotors together.
3. Check the rotor side clearance.
a. Using a feeler gauge and precision straight edge, measure the clearance between the rotors and precision straight edge.
b. Standard clearance is 0.0012–0.0035 in. (0.03–0.09 mm). Maximum side clearance is 0.0035 in. (0.09 mm).
c. If the clearance is greater than the maximum, replace the drive and driven rotors. If necessary, replace the timing chain cover assembly.
4. Check the rotor body clearance.
a. Using a feeler gauge, measure the clearance between the driven rotor and body.
b. Standard clearance is 0.0098–0.0128 in. (0.250–0.325 mm). Maximum body clearance is 0.0128 in. (0.325 mm)
5. If the clearance is greater than the maximum, replace the drive and driven rotors. If necessary, replace the timing chain cover assembly.

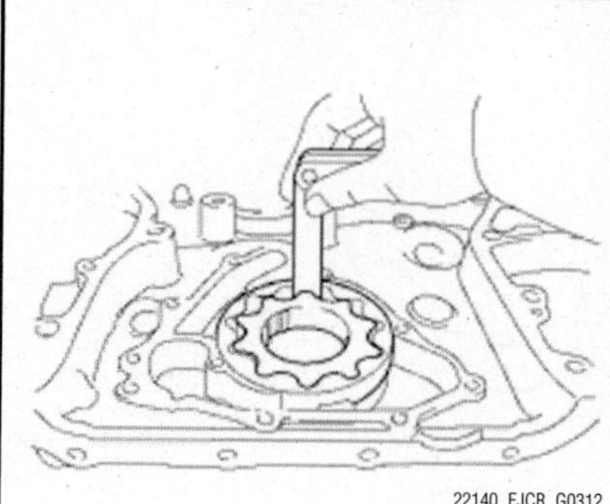

Fig. 109 Using a feeler gauge, measure the clearance between the drive and driven rotor tips

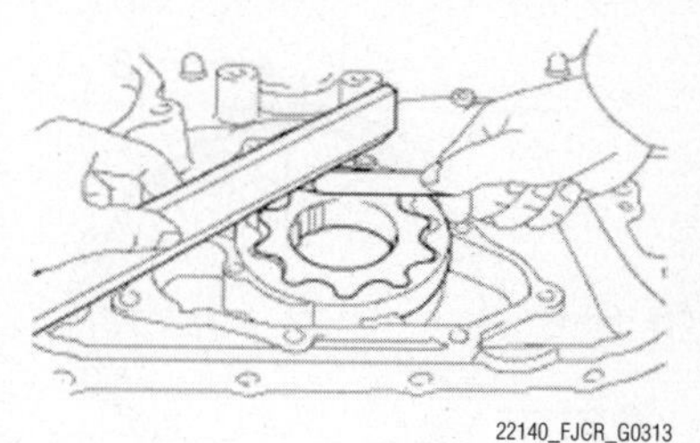

Fig. 110 Using a feeler gauge and precision straight edge, measure the clearance between the rotors and precision straight edge

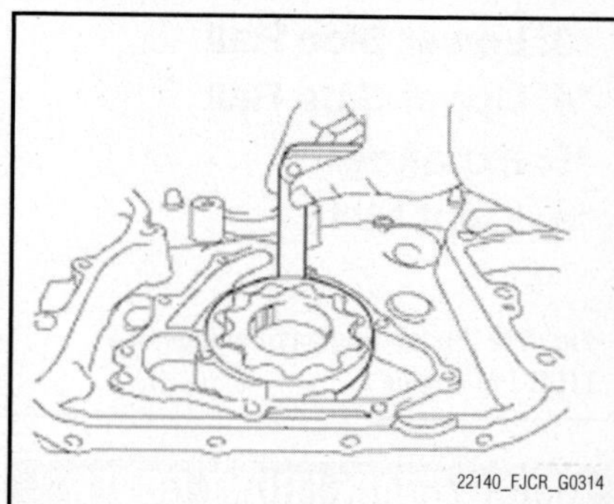

Fig. 111 Using a feeler gauge, measure the clearance between the driven rotor and body

PISTON AND RING

POSITIONING

See Figures 112

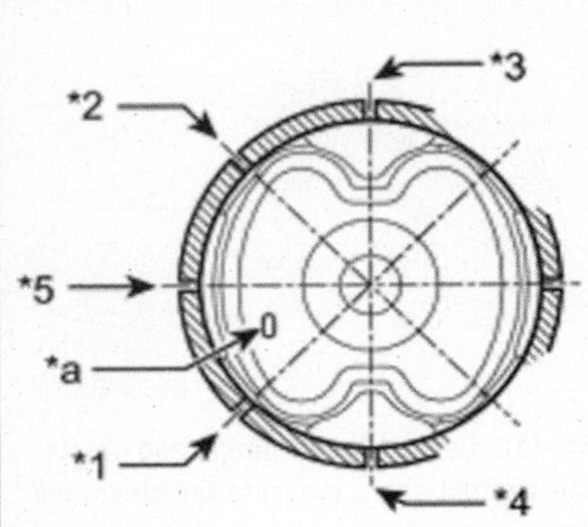

Fig. 112 Piston ring positioning—4.0L (1GR-FE) engine

TIMING CHAIN FRONT COVER

REMOVAL & INSTALLATION

See Figures 113 through 119.

1. Before servicing the vehicle, refer to the precautions.
2. Remove battery.
3. Drain engine coolant.
4. Drain engine oil.
5. Remove power steering rack assembly.
6. On 4WD vehicles, remove front differential carrier assembly.
7. Remove fan.
8. Remove alternator assembly.
9. Remove the air conditioning compressor and set it aside.

➡Do not disconnect the air conditioning lines.

10. Remove the 5 bolts, then remove the V-ribbed belt tensioner.
11. Remove the oil level gauge.
12. Remove the power steering pump assembly and set it aside.

➡Do not disconnect the power steering lines.

13. Remove crankshaft pulley.
14. Remove No. 2 oil pan.
    a. Remove the 10 bolts and 2 nuts.
    b. Insert the blade of SST: 09032-00100 between the oil pan and No. 2 oil pan, cut off applied sealer and remove the No. 2 oil pan.

⁂ WARNING

Be careful not to damage the contact surfaces of the oil pan and No. 2 oil pan. Be careful not to damage the No. 2 oil pan flange.

15. Remove the 2 nuts, then remove the oil strainer and gasket.
16. Remove oil pan.
    a. Remove the 4 housing bolts.
    b. Remove the flywheel housing under cover.
    c. Remove the 17 bolts and 2 nuts.
    d. Using a screwdriver, remove the oil pan by prying between the oil pan and cylinder block.

⁂ WARNING

Be careful not to damage the contact surfaces of the cylinder block and oil pan.

    e. Remove the O-ring from the oil pump.
17. Remove air cleaner assembly.
18. Remove throttle body bracket.
19. Remove oil baffle plate.
20. Remove surge tank bracket.
21. Remove intake air surge tank.
22. Remove ignition coil assembly.
23. Remove camshaft timing oil control valve assembly.
    a. Disconnect the 2 connectors.
    b. Remove the 2 bolts, then remove the 2 camshaft timing oil control valves.
24. Remove VVT sensor.
    a. For the right cylinder head sensor, disconnect the VVT sensor connector and remove the sensor bolt. Then remove the sensor.
    b. For the left cylinder head sensor, disconnect the No. 4 water by-pass hose and No. 5 water by-pass hose. Disconnect the VVT sensor connector and remove the sensor bolt. Then remove the sensor.

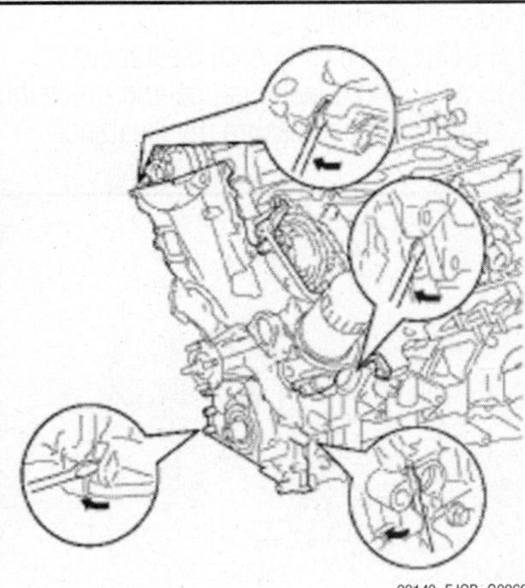

Fig. 113 Remove the timing chain cover by prying between the timing chain cover, cylinder head or cylinder block with a screwdriver

25. Remove the water inlet.
26. Remove the valve covers and gaskets.
27. Remove timing chain cover.
    a. Remove the 24 bolts and 2 nuts.
    b. Remove the timing chain cover by prying between the timing chain cover, cylinder head or cylinder block with a screwdriver.

⁂ WARNING

Be careful not to damage the contact surfaces of the timing chain cover, cylinder block and cylinder head.

    c. Remove the O-ring from the left cylinder head.
28. Pry out the seal from the cover with a flat-bladed tool.
29. It is a good idea to remove the oil pump from the timing cover and replace the O-ring.

To install:

30. Clean and inspect the timing cover area.
31. Apply multi-purpose grease to the new oil seal lip.
32. Tap the seal into place with SST: 09226-10010 or equivalent, and a hammer. Do this until the seal surface is flush with the cover edge.
33. Install timing chain cover.
    a. Remove any old sealant.

⁂ WARNING

Do not drop any oil on the contact surfaces of the timing chain cover, cylinder head and cylinder block.

    b. Install a new O-ring onto the left cylinder head as shown in the illustration.
    c. Apply a continuous bead of Three Bond 1207B or the equivalent sealant 0.12–0.16 in. (3–4 mm) diameter to the 4 locations shown in the illustration.
    d. Keep the seal surface between the cylinder block and the cylinder head shown in the illustration free of oil before installing the chain cover.
    e. Apply a continuous bead of Three Bond or the equivalent sealant 0.12–0.16 in. (3–4 mm) diameter to the timing

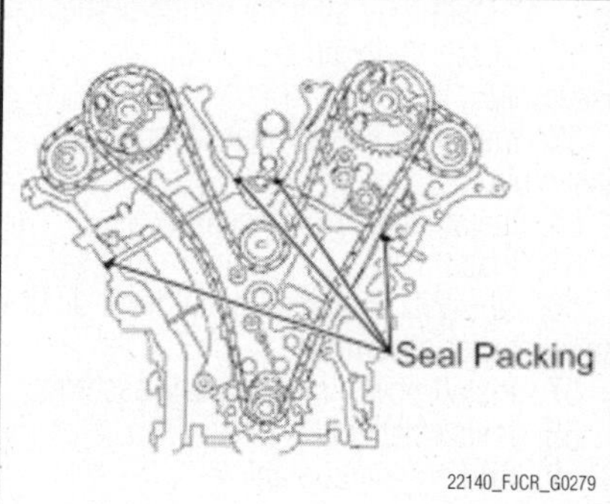

Fig. 114 Apply a continuous bead of Three Bond 1207B or the equivalent sealant 0.12–0.16 in. (3–4 mm) diameter to the 4 locations shown

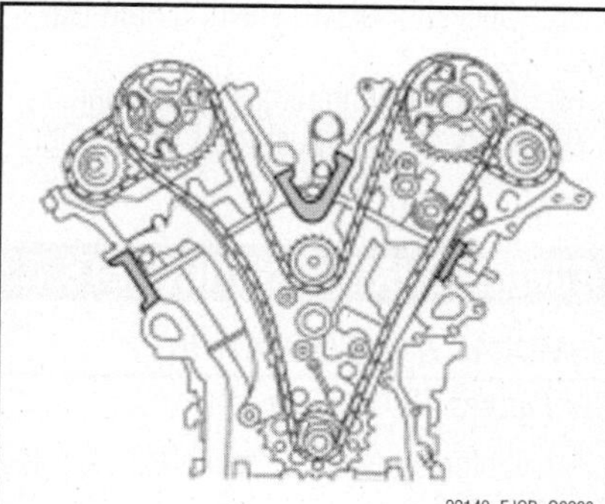

Fig. 115 Keep the seal surface between the cylinder block and the cylinder head shown in the illustration free of oil before installing the chain cover

chain cover as shown in the illustration. Use Three Bond 1282B for the water pump part and Three Bond 1207B for the remainder.

⁂ WARNING

Install the timing chain cover within 3 minutes of applying the sealant. The timing chain cover bolts and nuts must be tightened within 15 minutes of the installation. Otherwise the sealant must be removed and reapplied. Do not apply sealant to portion A shown in the illustration.

f. Align the key way of the oil pump drive rotor with the rectangular portion of the crankshaft timing gear, and slide the timing chain cover into place.

g. Install the timing chain cover with the 24 bolts and 2 nuts. Tighten the bolts and nuts to 17 ft. lbs. (23 Nm) uniformly in several steps.

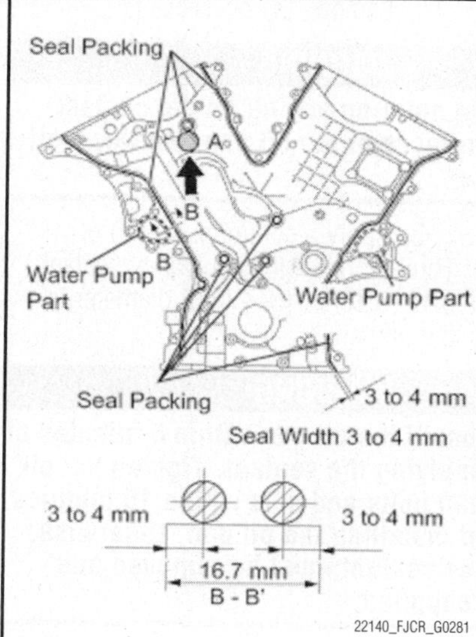

Fig. 116 Apply a continuous bead of Three Bond or the equivalent sealant 0.12–0.16 in. (3–4 mm) diameter to the timing chain cover as shown

➡Pay attention not to wrap the chain and slipper over the timing chain cover seal line.

⁂ WARNING

Place bolts in their proper positions. Bolt A length is 0.98 in. (25 mm) and bolt B length is 2.17 in. (55 mm).

34. Install cylinder head covers.
35. Install water inlet.
36. Install VVT sensor. Tighten to 71 inch lbs. (8 Nm).
37. Install camshaft timing oil control valve assembly. Tighten to 80 inch lbs. (9 Nm).
38. Install ignition coil assembly.
39. Install intake air surge tank.
40. Install surge tank brackets.
41. Install oil baffle plate.
42. Install throttle body bracket.
43. Install air cleaner assembly.
44. Install oil pan.
    a. Remove any old sealant.

⁂ WARNING

Do not drop any oil on the contact surfaces of the cylinder block, rear oil seal and oil pan.

b. Install a new O-ring onto the oil pump.

c. Apply a continuous bead of Three Bond or the equivalent sealant 0.12–0.16 in. (3–4 mm) diameter to the oil pan as shown in the illustration.

⁂ WARNING

Install the oil pan within 3 minutes of applying the sealant. Tighten the oil pan bolts and nuts within 15 minutes of installing the oil pan. Otherwise, the sealant must be removed and reapplied.

d. Install the oil pan with the 17 bolts and 2 nuts, and tighten the bolts and

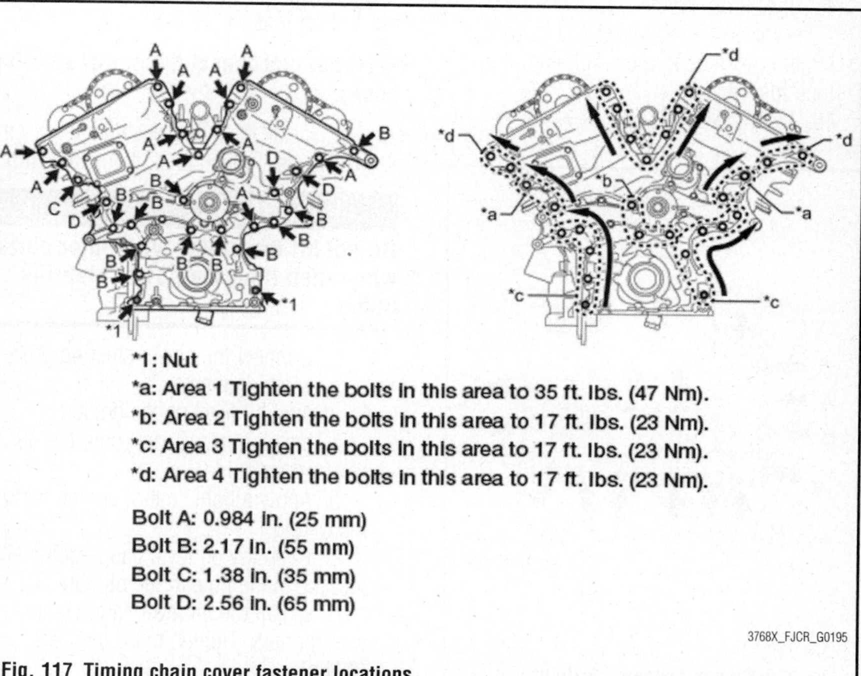

Fig. 117 Timing chain cover fastener locations

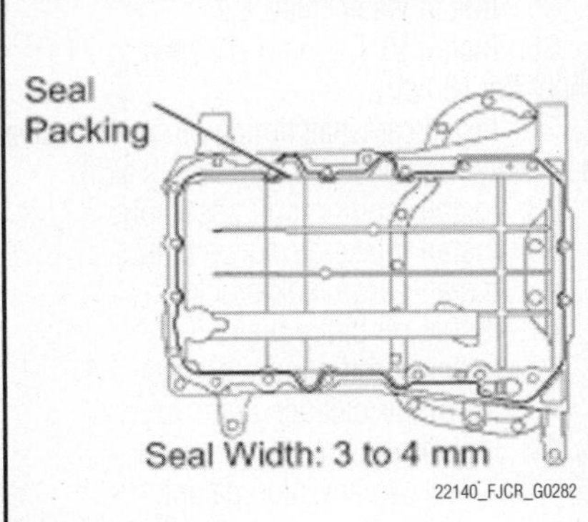

Fig. 118 Apply a continuous bead of Three Bond or the equivalent sealant 0.12–0.16 in. (3–4 mm) diameter to the oil pan as shown

nuts uniformly in several steps. Tighten the 0.39 in. (10 mm) head oil pan bolt to 7 ft. lbs. (10 Nm), the 0.47 in. (12 mm) head oil pan bolt to 16 ft. lbs. (21 Nm) and the nuts to 16 ft. lbs. (21 Nm).

**** WARNING**

Install bolts in their proper locations. Bolt lengths are as follows: A is 0.98 in. (25 mm) B 1.77 in. (45 mm) and C 0.55 in. (14 mm).

e. Install the 4 housing bolts. Tighten to 27 ft. lbs. (37 Nm).

f. Install the flywheel housing under cover.

45. Install oil strainer.

a. Install a new gasket and the oil strainer with the 2 nuts. Tighten to 80 inch lbs. (9 Nm).

46. Install No. 2 oil pan.

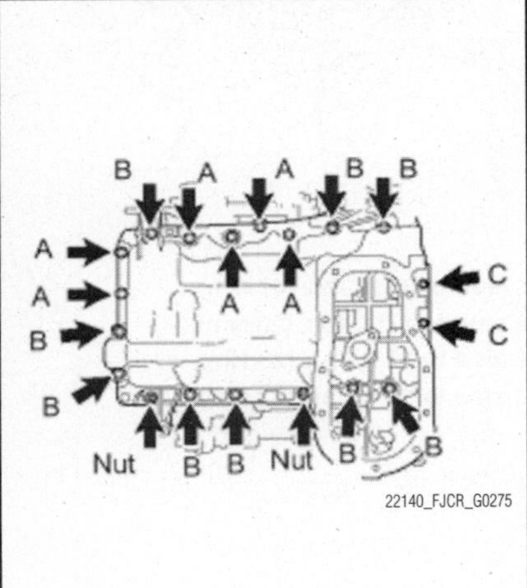

Fig. 119 Oil pan fastener locations

a. Remove any old sealant.

**** WARNING**

Do not drop any oil on the contact surfaces of the oil pan and No. 2 oil pan.

b. Apply a continuous bead of Three Bond or the equivalent sealant 0.12–0.16 in. (3–4 mm) diameter to the oil pan.

**** WARNING**

Install the oil pan within 3 minutes of applying the sealant. Tighten the oil pan bolts and nuts within 15 minutes of installing the oil pan. Otherwise, the sealant must be removed and reapplied.

c. Install the oil pan with the 17 bolts and 2 nuts, and tighten the bolts and nuts uniformly in several steps. Tighten the bolts to 80 inch lbs. (9 Nm) and the nuts to 7 ft. lbs. (10 Nm).

47. Install crankshaft pulley.

48. Install No. 1idler pulley.

a. Install the idler pulley with the bolt. Tighten to 29 ft. lbs. (39 Nm).

➡DOUBLE is marked on the No. 1 idler pulley to distinguish it from the No. 2 idler pulley.

49. Install No. 2idler pulley.

a. Install the 2 No. 2 idler pulleys with the 2 bolts. Tighten to 40 ft. lbs. (54 Nm).

50. Install power steering pump assembly and set it aside.

➡Do not disconnect the power steering hoses.

a. Install the power steering pump with the 2 bolts. Tighten to 32 ft. lbs. (43 Nm).

**** WARNING**

Do not hit the pulley with other parts when installing the power steering pump.

b. Connect the power steering pressure switch connector.

51. Install oil level gauge guide.

a. Install a new O-ring onto the oil level gauge guide.

b. Apply a light coat of engine oil to the O-ring.

c. Push the oil level gauge guide end into the guide hole in the oil pan.

d. Install the oil level gauge guide with the bolt. Tighten to 80 inch lbs. (9 Nm).

e. Install the oil level gauge guide.

52. Install V-ribbed belt tensioner assembly.

53. Install air conditioner compressor assembly.

54. Install alternator assembly.

55. Install fan.

56. On 4WD vehicles, install front differential carrier assembly.

57. Install power steering link assembly.

58. Install battery.

59. Add engine coolant.

60. Add engine oil.

61. Add power steering fluid and bleed the system, as necessary.

62. On 4WD vehicles, inspect and adjust the differential oil.

63. Check for engine coolant leakage.

64. Check for engine oil leakage.

65. Check for power steering fluid leakage.

66. Check for differential oil leakage.

67. Inspect and adjust front wheel alignment.

TIMING CHAIN & SPROCKETS

REMOVAL & INSTALLATION

See Figures 120 through 124.

1. Before servicing the vehicle, refer to the precautions.

2. Remove battery.

3. Drain engine coolant.

4. Drain engine oil.

5. Remove power steering rack assembly

6. On 4WD vehicles, remove front differential carrier assembly.

7. Remove fan.

8. Remove alternator assembly.

9. Remove the air conditioning compressor and set it aside.

➡Do not disconnect the air conditioning lines.

10. Remove the 5 bolts, then remove the V-ribbed belt tensioner.

11. Remove the oil level gauge.

12. Remove the power steering pump assembly and set it aside.

➡Do not disconnect the power steering lines.

13. Remove idler pulleys.

14. Remove crankshaft pulley.

15. Remove No. 2 oil pan.

a. Remove the 10 bolts and 2 nuts.

b. Insert the blade of SST: 09032-00100 between the oil pan and No. 2 oil pan, cut off applied sealer and remove the No. 2 oil pan.

WARNING

Be careful not to damage the contact surfaces of the oil pan and No. 2 oil pan. Be careful not to damage the No. 2 oil pan flange.

16. Remove the 2 nuts, then remove the oil strainer and gasket.
17. Remove oil pan.
   a. Remove the 4 housing bolts.
   b. Remove the flywheel housing under cover.
   c. Remove the 17 bolts and 2 nuts.
   d. Using a screwdriver, remove the oil pan by prying between the oil pan and cylinder block.

WARNING

Be careful not to damage the contact surfaces of the cylinder block and oil pan.

   e. Remove the O-ring from the oil pump.
18. Remove air cleaner assembly.
19. Remove throttle body bracket.
20. Remove oil baffle plate.
21. Remove surge tank bracket.
22. Remove intake air surge tank.
23. Remove ignition coil assembly.
24. Remove camshaft timing oil control valve assembly.
   a. Disconnect the 2 connectors.
   b. Remove the 2 bolts, then remove the 2 camshaft timing oil control valves.
25. Remove VVT sensor.
   a. For the right cylinder head sensor, disconnect the VVT sensor connector

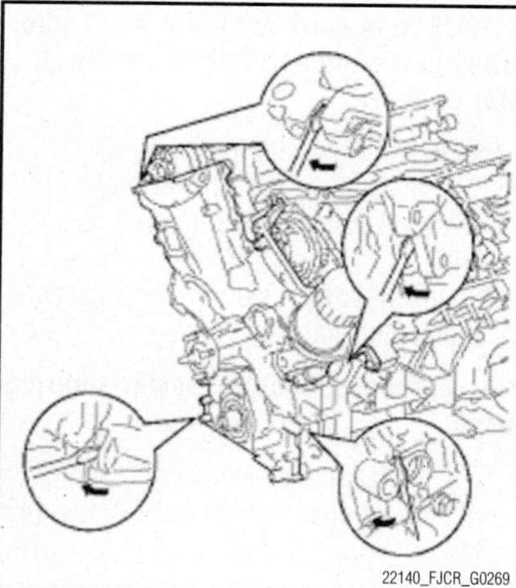

Fig. 120 Remove the timing chain cover by prying between the timing chain cover, cylinder head or cylinder block with a screwdriver

Fig. 121 Timing chain cover plate and gasket

and remove the sensor bolt. Then remove the sensor.
   b. For the right cylinder head sensor, disconnect the No. 4 water by-pass hose and No. 5 water by-pass hose. Disconnect the VVT sensor connector and remove the sensor bolt. Then remove the sensor.
26. Remove the water inlet.
27. Remove the 10 bolts, 3 seal washers, 2 nuts, cylinder head cover and gasket.
28. Remove timing chain cover.
   a. Remove the 24 bolts and 2 nuts.
   b. Remove the timing chain cover by prying between the timing chain cover, cylinder head or cylinder block with a screwdriver.

WARNING

Be careful not to damage the contact surfaces of the timing chain cover, cylinder block and cylinder head.

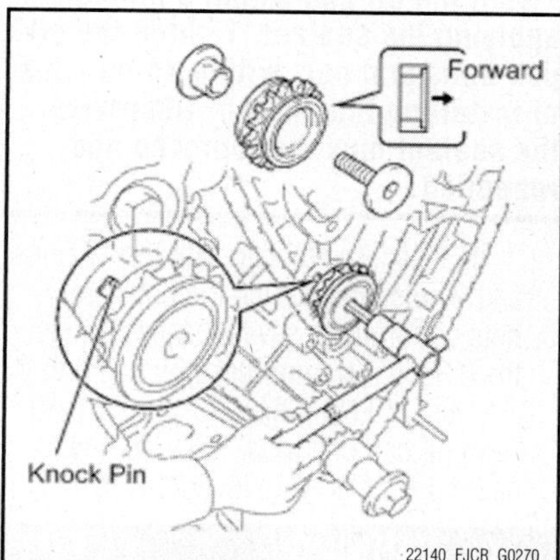

Fig. 122 Using a 10 mm hexagon wrench, remove the No. 2 idle gear shaft, No. 1 idle gear and No. 1 idle gear shaft

   c. Remove the O-ring from the left cylinder head.
29. Remove timing gear case or timing chain case oil seal.
30. Set No. 1 cylinder to TDC on the compression stroke.
   a. Turn the crankshaft pulley until its groove and the "0" timing mark of the timing chain cover are aligned.
   b. Check that the timing marks of the camshaft timing gears are aligned with the timing marks of the bearing caps as shown in the illustration.
31. If not, turn the crankshaft 1 complete revolution (360°) and align the timing marks above.
   a. Place paint marks on the No. 1 chain links corresponding to the timing marks of the camshaft timing gears.
32. Remove No. 1 chain tensioner assembly.

WARNING

Never rotate the crankshaft with the chain tensioner removed. When rotating the camshaft with the timing chain removed, rotate the crankshaft counterclockwise 40° from the TDC first.

   a. Remove the 4 bolts, then remove the timing chain cover plate and gasket.
   b. While turning the stopper plate of the tensioner upward, push in the plunger of the chain tensioner.
   c. While turning the stopper plate of the tensioner downward, insert a pin of 0.138 in. (3.5 mm) diameter into the holes in the stopper plate and tensioner to fix the stopper plate.
   d. Remove the 2 bolts, then remove the chain tensioner.

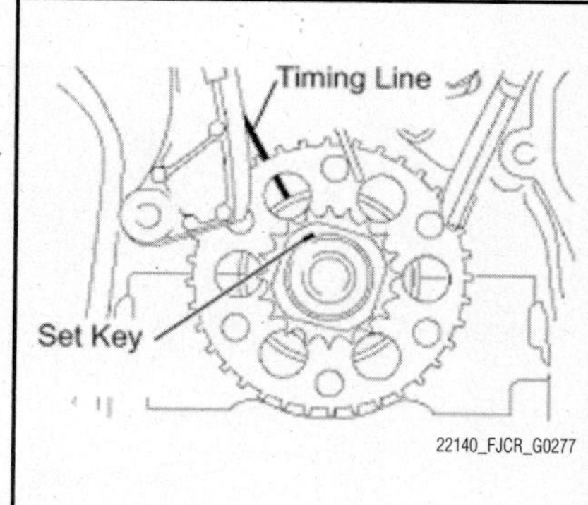

Fig. 123 Using the crankshaft pulley set bolt, turn the crankshaft to align the crankshaft set key with the timing line of the cylinder block

33. Remove chain tensioner slipper.
34. Remove No. 1 idle gear.
    a. Using a 10 mm hexagon wrench, remove the No. 2 idle gear shaft, No. 1 idle gear and No. 1 idle gear shaft.
35. Remove the 2 No. 2 chain vibration dampers.
36. Remove chain.

To install:

37. Install chain tensioner slipper.
38. Install No. 1 chain tensioner assembly.
    a. While turning the stopper plate of the No. 1 chain tensioner clockwise, push in the plunger of the No. 1 chain tensioner.
    b. While turning the stopper plate of the tensioner counterclockwise, insert a pin of 0.138 in. (3.5 mm) diameter into the holes in the stopper plate and No. 1 chain tensioner to fix the stopper plate.
    c. Install the No. 1 chain tensioner with the 2 bolts. Tighten to 7 ft. lbs. (10 Nm).
39. Install the timing chain.
    a. Set the No. 1 cylinder to TDC on the compression stroke.
    b. Align the timing marks of the camshaft timing gears and bearing caps.
    c. Using the crankshaft pulley set bolt, turn the crankshaft to align the crankshaft set key with the timing line of the cylinder block.
    d. Align the yellow mark link with the timing mark of the crankshaft timing link.
    e. Align the orange mark links with the timing marks of the camshaft timing gears, and install the chain.
40. Install the 2 No. 2 chain vibration dampers.
41. Install the idle gear.
    a. Apply a light coat of engine oil to rotating surface of the No. 1 idle gear shaft.
    b. Temporarily install the No. 1 idle gear shaft together with the No. 2 idle gear shaft with the knock pin of the No. 1 idle gear shaft and the knock pin groove of the cylinder block are aligned.

➡Orient the idle gear shafts correctly.

    c. Using a 10 mm hexagon wrench, tighten the No. 2 idle gear shaft to 44 ft. lbs. (60 Nm).
    d. Remove the bar from the chain tensioner.
42. Install the timing gear case oil seal.
43. Install timing chain cover.
44. Install cylinder head covers.
45. Install water inlet.
46. Install VVT sensor. Tighten to 71 inch lbs. (8 Nm).
47. Install camshaft timing oil control valve assembly. Tighten to 80 inch lbs. (9 Nm).
48. Install ignition coil assembly.
49. Install intake air surge tank.
50. Install surge tank brackets.
51. Install oil baffle plate.
52. Install throttle body bracket.
53. Install air cleaner assembly.
54. Install oil pan.
    a. Remove any old sealant.

✻✻ WARNING

Do not drop any oil on the contact surfaces of the cylinder block, rear oil seal and oil pan.

    b. Install a new O-ring onto the oil pump.
    c. Apply a continuous bead of Three Bond or the equivalent sealant 0.12–0.16 in. (3–4 mm) diameter to the oil pan.

✻✻ WARNING

Install the oil pan within 3 minutes of applying the sealant. Tighten the oil pan bolts and nuts within 15 minutes of installing the oil pan. Otherwise, the sealant must be removed and reapplied.

    d. Install the oil pan with the 17 bolts and 2 nuts, and tighten the bolts and nuts uniformly in several steps. Tighten the 0.39 in. (10 mm) head oil pan bolt to 7 ft. lbs. (10 Nm), the 0.47 in. (12 mm) head oil pan bolt to 16 ft. lbs. (21 Nm) and the nuts to 16 ft. lbs. (21 Nm).

✻✻ WARNING

Install bolts in their proper locations. Bolt lengths are as follows: A is 0.98 in. (25 mm) B 1.77 in. (45 mm) and C 0.55 in. (14 mm). Refer to the Oil Pan procedure for bolt locations.

    e. Install the 4 housing bolts. Tighten to 27 ft. lbs. (37 Nm).
    f. Install the flywheel housing under cover.
55. Install oil strainer.
    a. Install a new gasket and the oil strainer with the 2 nuts. Tighten to 80 inch lbs. (9 Nm).
56. Install No. 2 oil pan.
    a. Remove any old sealant.

✻✻ WARNING

Do not drop any oil on the contact surfaces of the oil pan and No. 2 oil pan.

    b. Apply a continuous bead of Three Bond or the equivalent sealant 0.12–0.16 in. (3–4 mm) diameter to the oil pan.

✻✻ WARNING

Install the oil pan within 3 minutes of applying the sealant. Tighten the oil pan bolts and nuts within 15 minutes of installing the oil pan. Otherwise, the sealant must be removed and reapplied.

    c. Install the oil pan with the 17 bolts and 2 nuts, and tighten the bolts and nuts uniformly in several steps. Tighten the bolts to 80 inch lbs. (9 Nm) and the nuts to 7 ft. lbs. (10 Nm).
57. Install crankshaft pulley.
58. Install No. 1idler pulley.
    a. Install the idler pulley with the bolt. Tighten to 29 ft. lbs. (39 Nm).

➡DOUBLE is marked on the No. 1 idler pulley to distinguish it from the No. 2 idler pulley.

59. Install No. 2idler pulley.
    a. Install the 2 No. 2 idler pulleys with the 2 bolts. Tighten to 40 ft. lbs. (54 Nm).
60. Install power steering pump assembly and set it aside.

➡Do not disconnect the power steering hoses.

    a. Install the power steering pump with the 2 bolts. Tighten to 32 ft. lbs. (43 Nm).

✻✻ WARNING

Do not hit the pulley with other parts when installing the power steering pump.

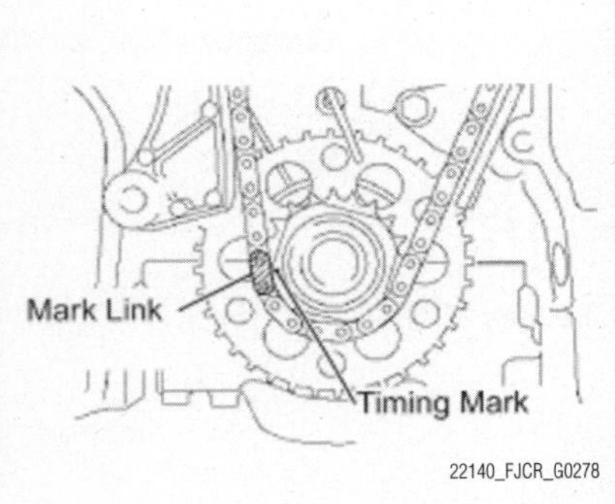

Fig. 124 Align the yellow mark link with the timing mark of the crankshaft timing link

b. Connect the power steering pressure switch connector.

61. Install oil level gauge guide.
 a. Install a new O-ring onto the oil level gauge guide.
 b. Apply a light coat of engine oil to the O-ring.
 c. Push the oil level gauge guide end into the guide hole in the oil pan.
 d. Install the oil level gauge guide with the bolt. Tighten to 80 inch lbs. (9 Nm).
 e. Install the oil level gauge guide.
62. Install V-ribbed belt tensioner assembly.
63. Install air conditioner compressor assembly.
64. Install alternator assembly.
65. Install fan.
66. On 4WD vehicles, install front differential carrier assembly.
67. Install power steering link assembly.
68. Install battery.
69. Add engine coolant.
70. Add engine oil.
71. Add power steering fluid and bleed the system, as necessary.
72. On 4WD vehicles, inspect and adjust the differential oil.
73. Check for engine coolant leakage.
74. Check for engine oil leakage.
75. Check for power steering fluid leakage.
76. Check for differential oil leakage.
77. Inspect and adjust front wheel alignment.

VALVE LASH

ADJUSTMENT

See Figures 125 and 127.

1. Before servicing the vehicle, refer to the precautions.
2. Remove the cylinder head covers.
3. Set No. 1 cylinder to TDC on the compression stroke.
 a. Turn the crankshaft pulley until its groove and the "0" timing mark of the timing chain cover are aligned.
 b. Check that the timing marks of the camshaft timing gears are aligned with the timing marks of the bearing caps as shown in the illustration.
4. If not, turn the crankshaft 1 complete revolution (360°) and align the timing marks above.
5. Inspect valve clearance.
 a. Check the valves indicated in the illustration.
 b. Using a feeler gauge, measure the clearance between the valve lifter and camshaft.

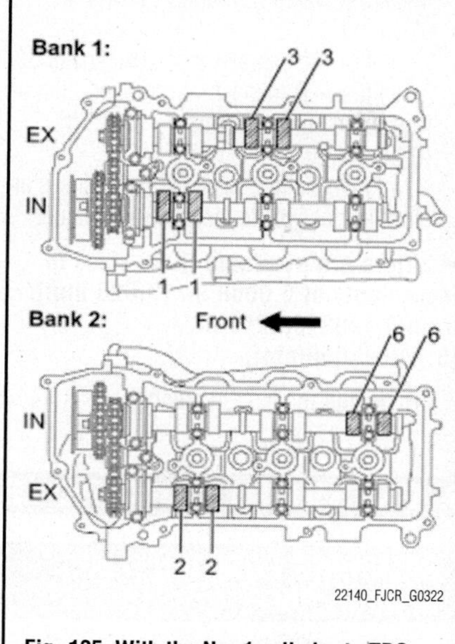

Fig. 125 With the No. 1 cylinder to TDC on the compression stroke check the valves indicated

 c. Record any out-of-specification valve clearance measurements. They will be used later to determine the required replacement valve lifter.
 d. Turn the crankshaft 240° clockwise (first time), and check the valves indicated in the illustration.
 e. Using a feeler gauge, measure the clearance between the valve lifter and camshaft.
 f. Record any out-of-specification valve clearance measurements. They will be used later to determine the required replacement valve lifter.

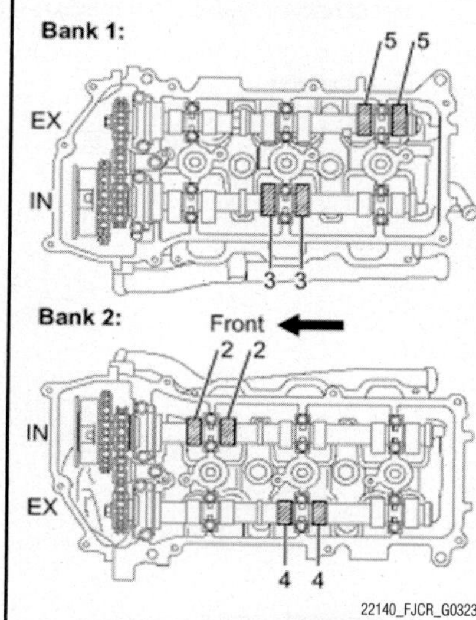

Fig. 126 After rotating the crankshaft 240° clockwise (first time) check the valves indicated

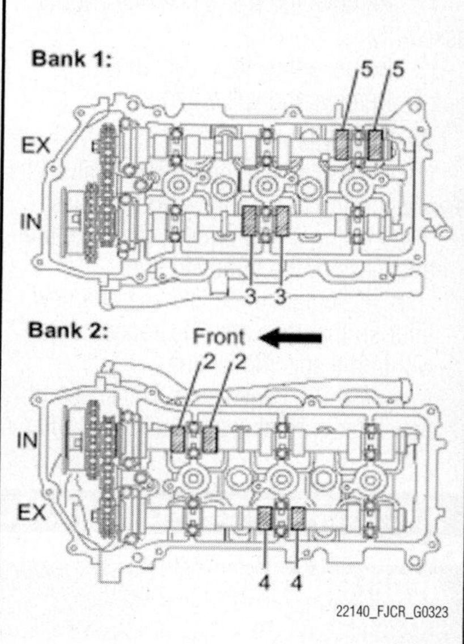

Fig. 127 After rotating the crankshaft 240° clockwise (second time) check the valves indicated

 g. Turn the crankshaft 240° clockwise, and check the valves indicated in the illustration.
 h. Using a feeler gauge, measure the clearance between the valve lifter and camshaft.
 i. Record any out-of-specification valve clearance measurements. They will be used later to determine the required replacement valve lifter.
6. Adjust valve clearance.
 a. Set the No. 1 cylinder to TDC/compression.
 b. Turn the crankshaft pulley until its groove and the "0" timing mark of the timing chain cover are aligned.
 c. Check that the timing marks of the camshaft timing gears are aligned with the timing marks of the bearing caps as shown in the illustration.
 d. If not, turn the crankshaft 1 complete revolution (360°) and align the timing marks as above.
 e. Place paint marks on the No. 1 chain links corresponding to the timing marks of the camshaft timing gears.
7. Remove the No. 1 chain tensioner assembly.
8. Remove the No. 2 camshaft.
9. Remove the No. 2 chain tensioner assembly.
10. Remove the camshaft.
11. Remove the No. 4 camshaft.

12. Remove the No. 3 chain tensioner assembly.
13. Remove the No. 3 camshaft.
14. Remove the valve lifters.

To install:

15. Determine the replacement valve lifter size according to the following formulas:
    a. Using a micrometer, measure the thickness of the removed lifter.
    b. Calculate the thickness of a new lifter so that the valve clearance comes within the specified value.
    c. For the intake side: Thickness of new lifter = Thickness of removed lifter + Measured valve clearance - 0.008 in. (0.20 mm).
    d. For the exhaust side: Thickness of new lifter = Thickness of removed lifter + Measured valve clearance - 0.013 in. (0.34 mm).
    e. Select a new lifter with a thickness as close as possible to the calculated value.

➡Lifters are available in 35 sizes in increments of 0.0008 in. (0.020 mm), from 0.1992–0.2260 in. (5.740–5.060mm).

16. Install the No. 3 camshaft.
17. Install the No. 3 chain tensioner assembly.
18. Install the No. 4 camshaft.
19. Install the camshaft.
20. Install the No. 2 chain tensioner assembly.
21. Install the No. 2 camshaft.
22. Install the No. 1 chain tensioner assembly.
23. Check that the timing marks of the camshaft timing gears are aligned with the timing marks of the bearing cap as shown in the illustration.
24. Install the cylinder head covers.

ENGINE PERFORMANCE & EMISSION CONTROLS

ACCELERATOR PEDAL POSITION (APP) SENSOR

LOCATION

See Figure 128.

Refer to the accompanying illustration for sensor location.

REMOVAL & INSTALLATION

1. Before servicing the vehicle, refer to the precautions section.
2. Disconnect the negative battery cable.
3. Disconnect the sensor connector.
4. Remove the sensor attaching bolts, then remove the accelerator pedal.

To install:

5. Install the accelerator pedal. Tighten the pedal attaching bolts to 44 inch lbs. (5 Nm).
6. Connect the sensor connector.
7. Connect the negative battery cable.

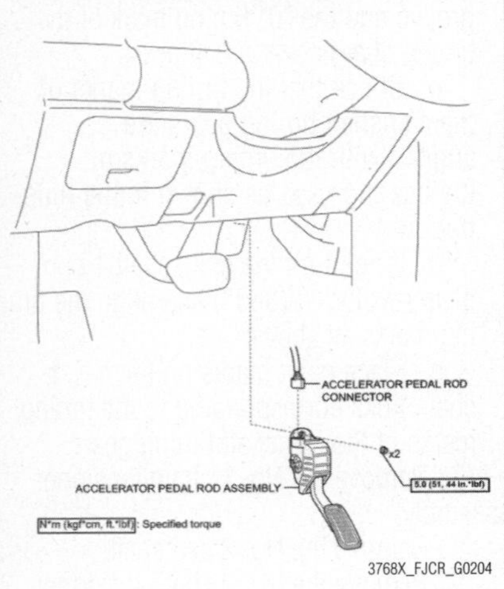

Fig. 128 Accelerator Pedal Position (APP) Sensor component location

CAMSHAFT POSITION (CMP) SENSOR (VVT)

LOCATION

See Figure 129.

Refer to the accompanying illustrations for sensor location.

REMOVAL & INSTALLATION

1. Remove the v-bank cover.

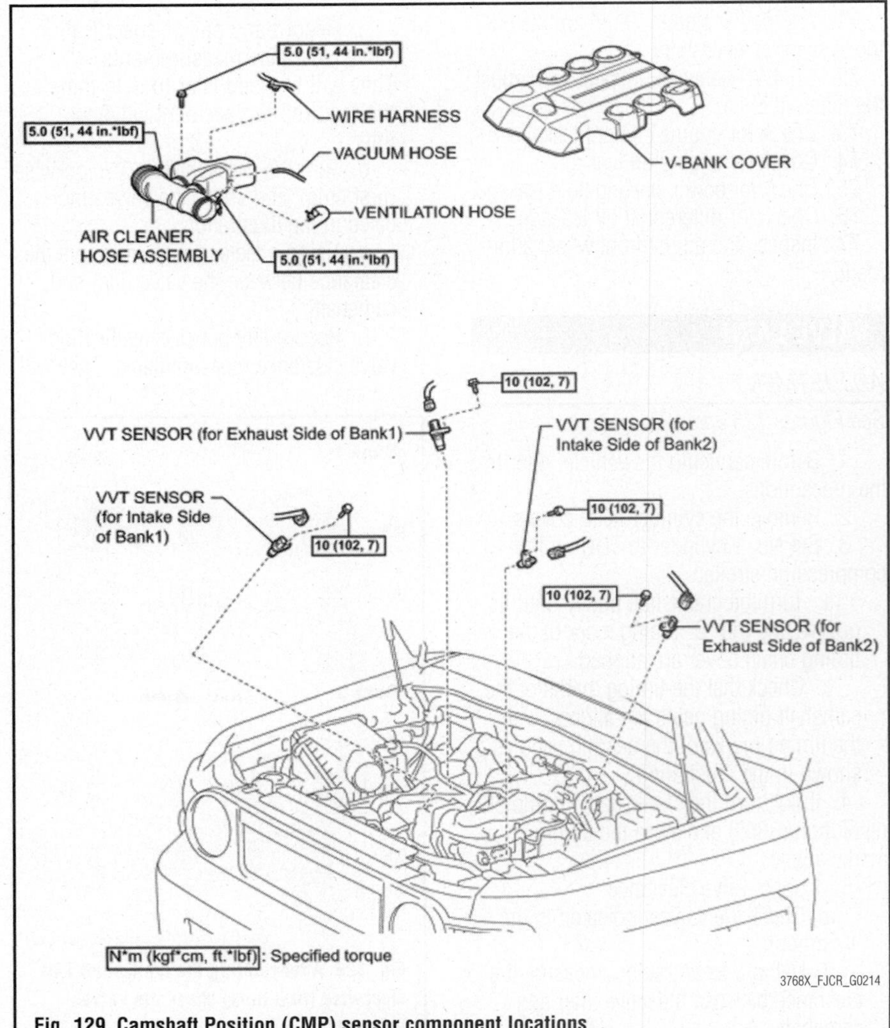

Fig. 129 Camshaft Position (CMP) sensor component locations

2. Remove air cleaner hose assembly.
3. Remove CMP (VVT) sensor:
   a. Disconnect the sensor connector.
   b. Remove the bolt and sensor.

To install:

➡When reusing the sensor, inspect the O-ring. If the O-ring has scratches or cuts, replace the sensor.

4. Perform the following for all 4 sensors:
   a. Apply a light coat of engine oil to O-ring.
   b. Install the sensor with the bolt and tighten to 7 ft. lbs. (10 Nm).
   c. Connect the camshaft timing oil control valve connector.
5. To complete installation, reverse remaining removal procedure.

CRANKSHAFT POSITION (CKP) SENSOR

LOCATION

See Figure 130.

Refer to the accompanying illustrations for sensor location.

REMOVAL & INSTALLATION

1. Before servicing the vehicle, refer to the precautions section.
2. Remove the air conditioning compressor and set it aside.

➡Do not disconnect the air conditioning hoses from the compressor.

3. Disconnect the sensor electrical connector.
4. Remove the attaching bolt and sensor.

To install:

5. Install the sensor into cylinder head. Tighten the sensor to 80 inch lbs. (9 Nm).
6. Connect the sensor electrical connector.
7. Install the air conditioning compressor.
8. Install the alternator.
9. Install the fan.

ELECTRONIC CONTROL MODULE (ECM)

LOCATION

See Figures 131.

Refer to the accompanying illustrations for ECM location.

REMOVAL & INSTALLATION

1. Disconnect the negative battery cable.
2. Remove glove compartment door assembly.
3. On 4WD automatic transmission vehicles, remove four wheel drive control ECU.
   a. Disconnect the connector.
   b. Remove the bolt and four wheel drive control ECU.
4. Remove ECM.
   a. Disconnect the 5 connectors.
   b. Remove the 3 nuts, then remove the ECM.

To install:

5. Install the ECM with the 3 nuts. Tighten to 49 inch lbs. (5.5 Nm).
   a. Connect the 6 connectors.
6. On 4WD automatic transmission vehicles, install four wheel drive control ECU.
   a. Install the four wheel drive control ECU with the bolt. Tighten to 71 inch lbs. (8 Nm).
   b. Connect the connector.
7. Install glove compartment door assembly.
8. Connect the negative battery cable.
9. Connect a scan tool and perform the reset memory automatic transmission initialization.
10. Connect a scan tool and perform the registration (VIN registration) when replacing the ECM.
11. Using a scan tool, set up the func-

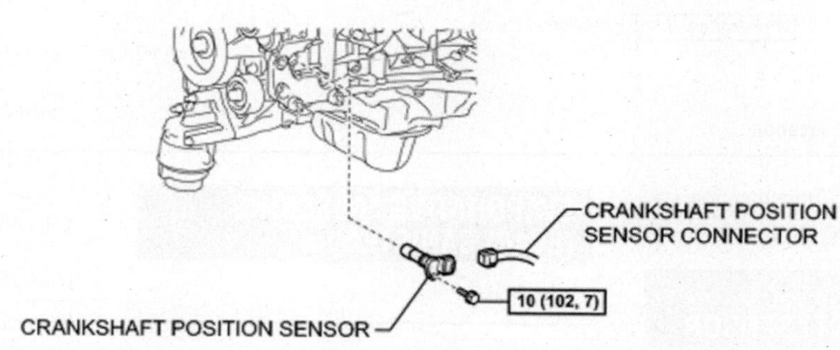

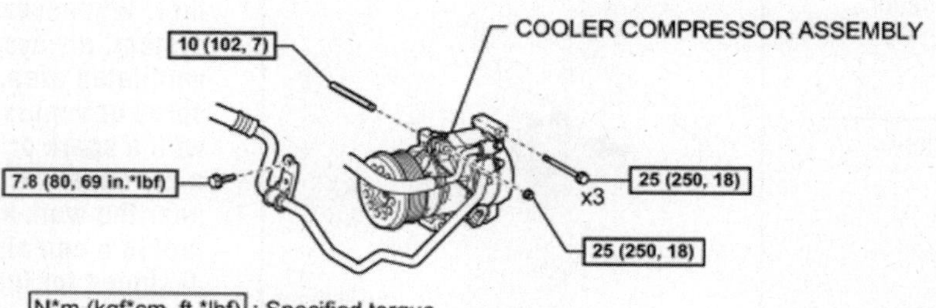

3768X_FJCR_G0220

Fig. 130 Crankshaft Position (CKP) Sensor component locations

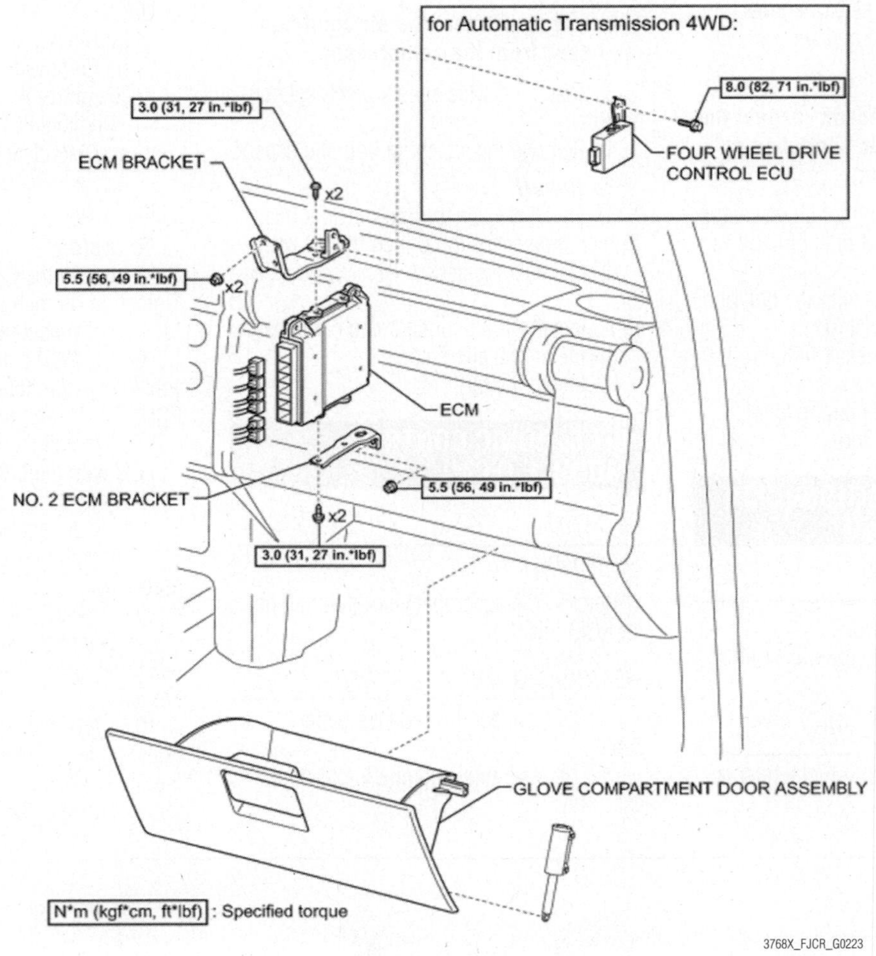

for Automatic Transmission 4WD:
8.0 (82, 71 in.*lbf)
FOUR WHEEL DRIVE CONTROL ECU
3.0 (31, 27 in.*lbf)
ECM BRACKET
x2
5.5 (56, 49 in.*lbf)
x2
ECM
NO. 2 ECM BRACKET
5.5 (56, 49 in.*lbf)
x2
3.0 (31, 27 in.*lbf)
GLOVE COMPARTMENT DOOR ASSEMBLY
N*m (kgf*cm, ft*lbf) : Specified torque
3768X_FJCR_G0223

Fig. 131 ECM component location

tion of the ATF (Automatic Transmission Fluid) temperature warning lamp.

ENGINE COOLANT TEMPERATURE (ECT) SENSOR

LOCATION

See Figure 132.

Refer to the accompanying illustrations for sensor location.

REMOVAL & INSTALLATION

1. Disconnect the negative battery cable.
2. Remove the intake manifold. Refer to Intake Manifold in Engine Mechanical.
3. Disconnect the sensor connector.
4. Using a 19 mm deep socket wrench, remove the sensor.
5. Remove the gasket from the sensor.

EVAPORATIVE EMISSIONS (EVAP) CANISTER

LOCATION

See Figure 133.

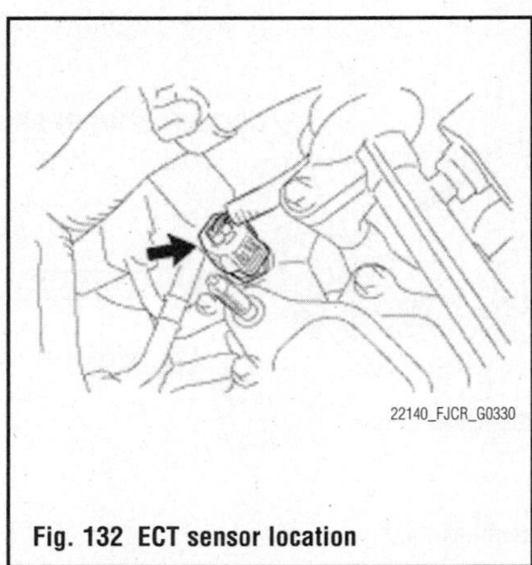

Fig. 132 ECT sensor location

Refer to the accompanying illustrations for EVAP canister location.

REMOVAL & INSTALLATION

See Figures 134 through 136.

**** CAUTION**

Observe all applicable safety precautions when working around fuel. Whenever servicing the fuel system, always work in a well-ventilated area. Do not allow fuel spray or vapors to come in contact with a spark or open flame. Keep a dry chemical fire extinguisher near the work area. Always keep fuel in a container specifically designed for fuel storage; also, always properly seal fuel containers to avoid the possibility of fire or explosion.

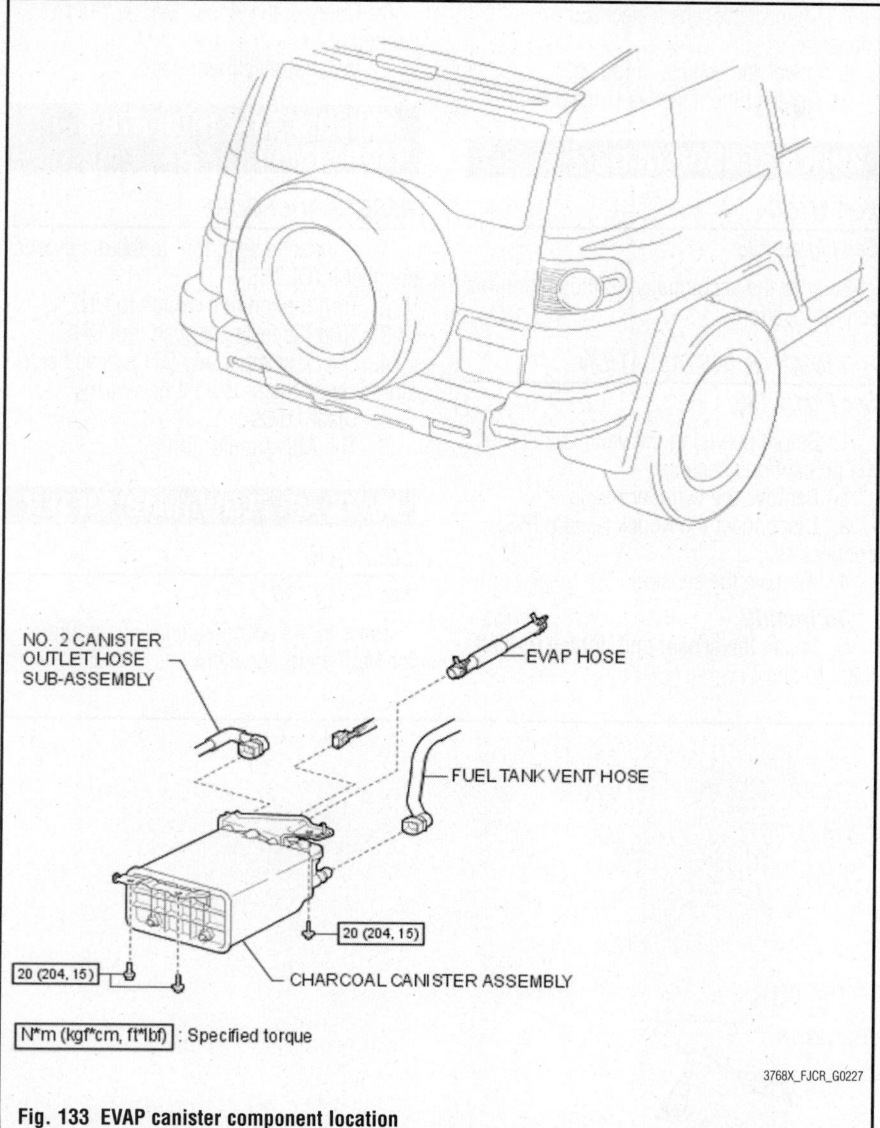

Fig. 133 EVAP canister component location

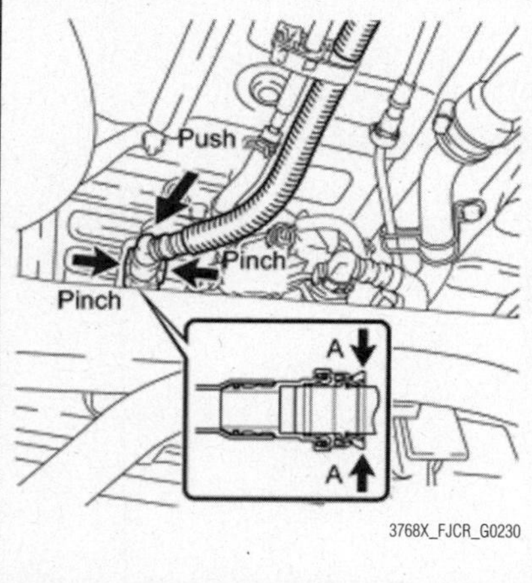

Fig. 134 Disconnect the fuel tank breather tube—Push the connector deep inside, pinch portion A as shown, pull out the connector

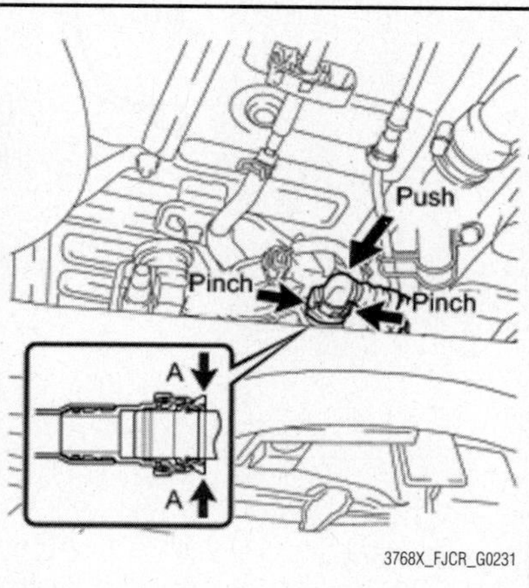

Fig. 135 Disconnect the No. 2 canister outlet hose—Push the connector deep inside, pinch portion A as shown, pull out the connector

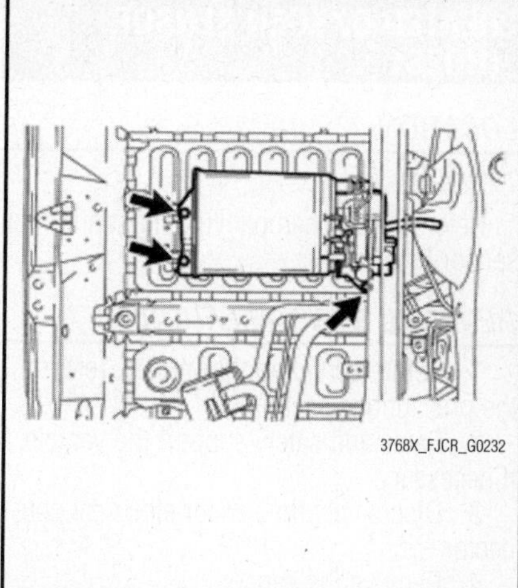

Fig. 136 Remove the 3 bolts, then remove the charcoal canister assembly

➡Note the following:

- Remove any dirt and foreign objects from the fuel tank vent hose connector before performing the work.
- Do not allow any scratches or foreign objects on the parts when disconnecting, as the fuel tank vent hose connector has the O-ring that seals the pipe.
- Perform the work by hand. Do not use any tools.
- Do not forcibly bend, twist or turn the nylon tube.
- Protect the disconnected part by covering it with a vinyl bag after disconnecting the fuel tank vent hose.
- If the fuel vent connector and pipe are stuck, push and pull them to release.

1. Properly relieve the fuel system pressure.
2. Remove charcoal canister assembly:
   a. Disconnect the fuel tank breather tube from the fuel tank.
   b. Disconnect the No. 2 canister outlet hose sub-assembly.
3. Disconnect the connector.
4. Disconnect the EVAP hose.
5. Remove the 3 bolts, then remove the charcoal canister assembly.

To install:

6. Install the canister with the 3 bolts. Tighten to 15 ft. lbs. (20 Nm).
7. To complete installation, reverse remaining removal procedure.
8. When connecting the hose connectors, make sure you hear a "click" sound.
9. Inspect for fuel leaks.

HEATED OXYGEN SENSOR (HO2S)

LOCATION

See Figure 137.

Refer to the accompanying illustration for sensor location.

REMOVAL & INSTALLATION

1. Before servicing the vehicle, refer to the precautions section.
2. Raise and safely support the vehicle, if necessary.
3. Disconnect the sensor electrical connector.
4. Remove the sensor.

To install:

5. Coat the threads of the heated oxygen sensor with the anti-seize compound, if necessary. Some new sensors will come pre-coated.
6. Install the heated oxygen sensor tighten to 35 inch. lbs. (4 Nm).
7. Connect the sensor electrical connector.
8. Lower the vehicle, if raised.
9. Connect the negative battery cable.

KNOCK SENSOR (KS)

LOCATION

See Figure 138.

Refer to the accompanying illustration for sensor location.

REMOVAL & INSTALLATION

See Figure 139.

1. Before servicing the vehicle, refer to the precautions section.
2. Remove the cylinder heads.
3. Disconnect the Knock Sensor (KS) connectors.
4. Remove the sensors.

To install:

5. Install the sensor and tighten to 15 ft. lbs. (20 Nm).
6. Connect the Knock Sensor (KS) connector.
7. Install the cylinder heads.

MALFUNCTION INDICATOR LIGHT (MIL)

RESET PROCEDURE

1. Connect a scan tool to the diagnostic connector (DLC3).
2. Turn the ignition switch to ON.
3. Turn the tester or scan tool ON
4. Check whether any DTCs have been stored. Note them down if necessary.
5. Clear DTCs.
6. The MIL should turn off.

MASS AIR FLOW (MAF) METER

LOCATION

See Figure 140.

Refer to the accompanying illustrations for MAF meter location.

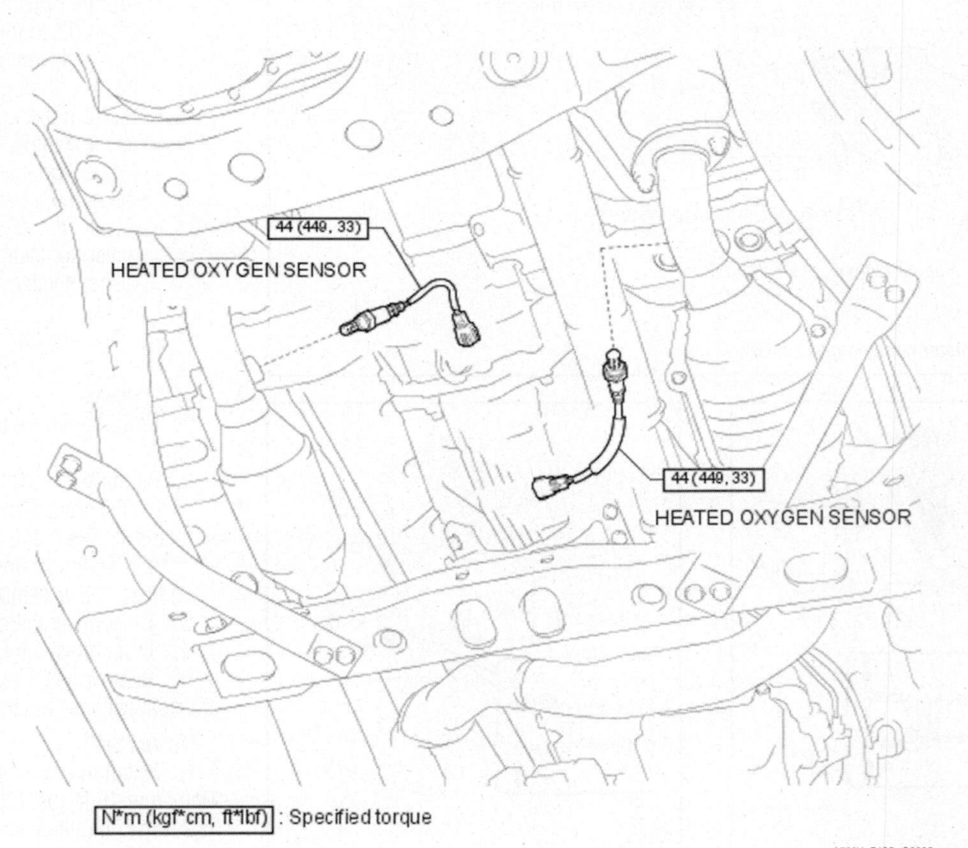

Fig. 137 Heated oxygen sensor component location

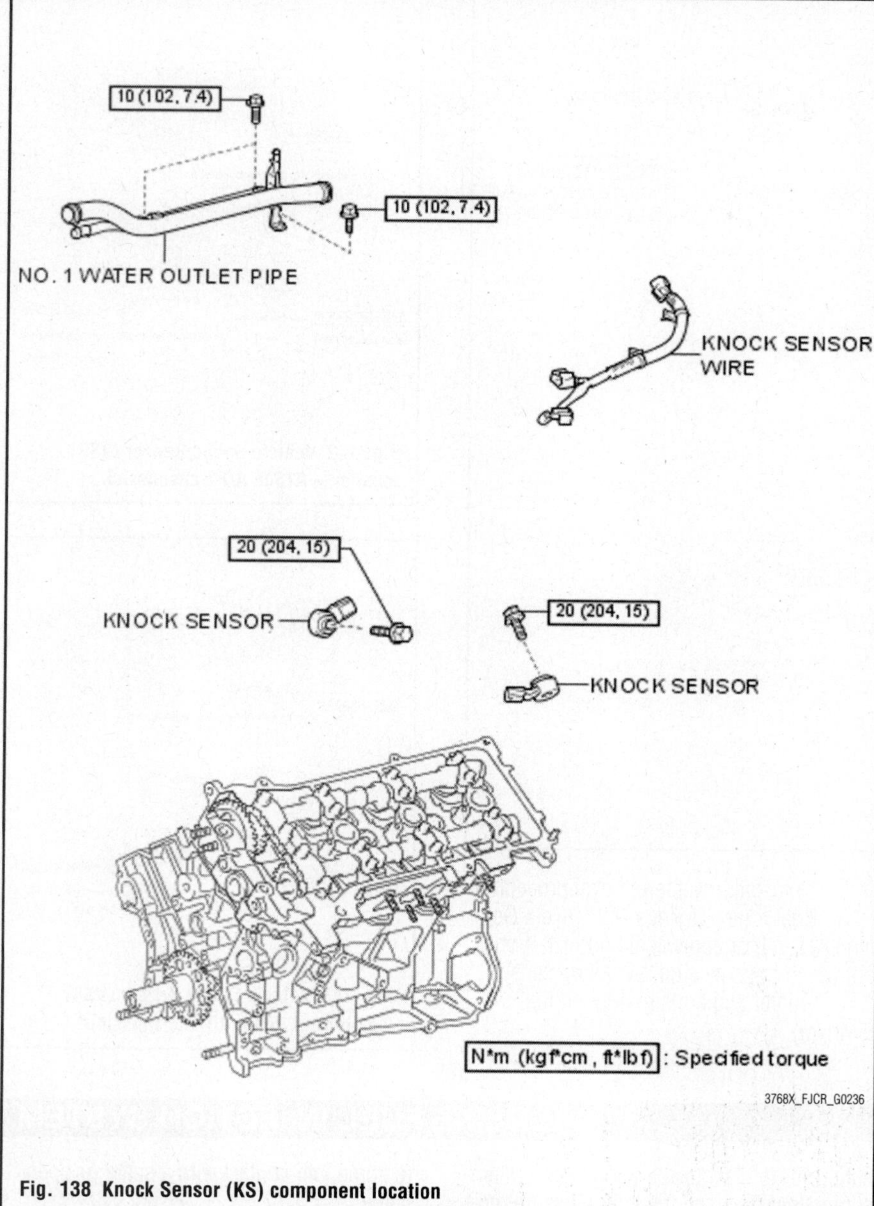

Fig. 138 Knock Sensor (KS) component location

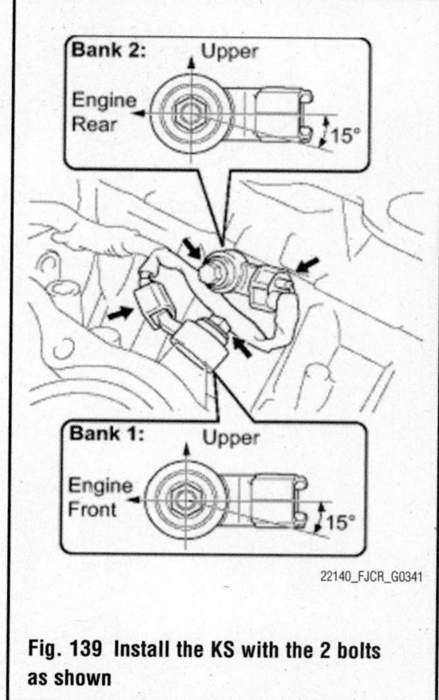

Fig. 139 Install the KS with the 2 bolts as shown

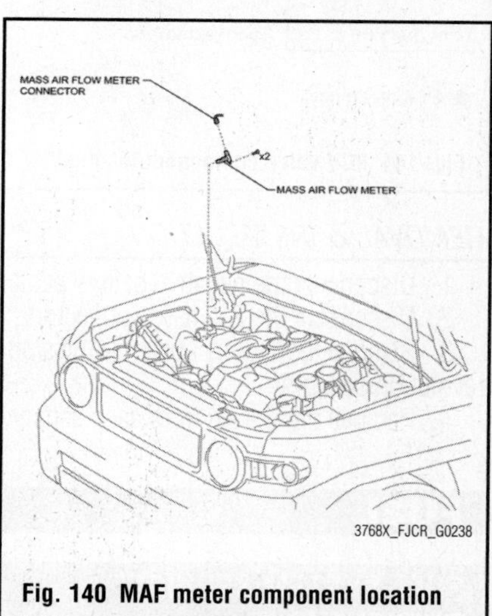

Fig. 140 MAF meter component location

REMOVAL & INSTALLATION

1. Before servicing the vehicle, refer to the precautions section.
2. Disconnect the negative battery cable.
3. Remove V-bank cover.
4. Remove Mass Air Flow (MAF) meter.
   a. Disconnect the connector.
   b. Remove the 2 screws, then remove the MAF meter.

To install:

5. Install MAF meter.
6. Connect the connector.
7. Install V-bank cover.
8. Connect the negative battery cable.

POSITIVE CRANKCASE VENTILATION (PCV) VALVE

LOCATION

See Figure 141.

Refer to the accompanying illustrations for PCV valve location.

REMOVAL & INSTALLATION

a. Disconnect the ventilation hose.
b. Remove the ventilation valve.
c. Disconnect the heater hose clamp.
d. Disconnect the ventilation hose from the ventilation valve.
e. Remove the ventilation valve.

To install:

1. To install, reverse removal procedure.
2. Apply adhesive to 2 or 3 threads of the ventilation valve. Use Toyota Genuine Adhesive 1324, Three Bond® 1324 or equivalent
3. Tighten the valve to 20 ft. lbs. (27 Nm).

VEHICLE SPEED SENSOR (VSS)

LOCATION

See Figures 142 and 143.

Refer to the accompanying illustrations for sensor location.

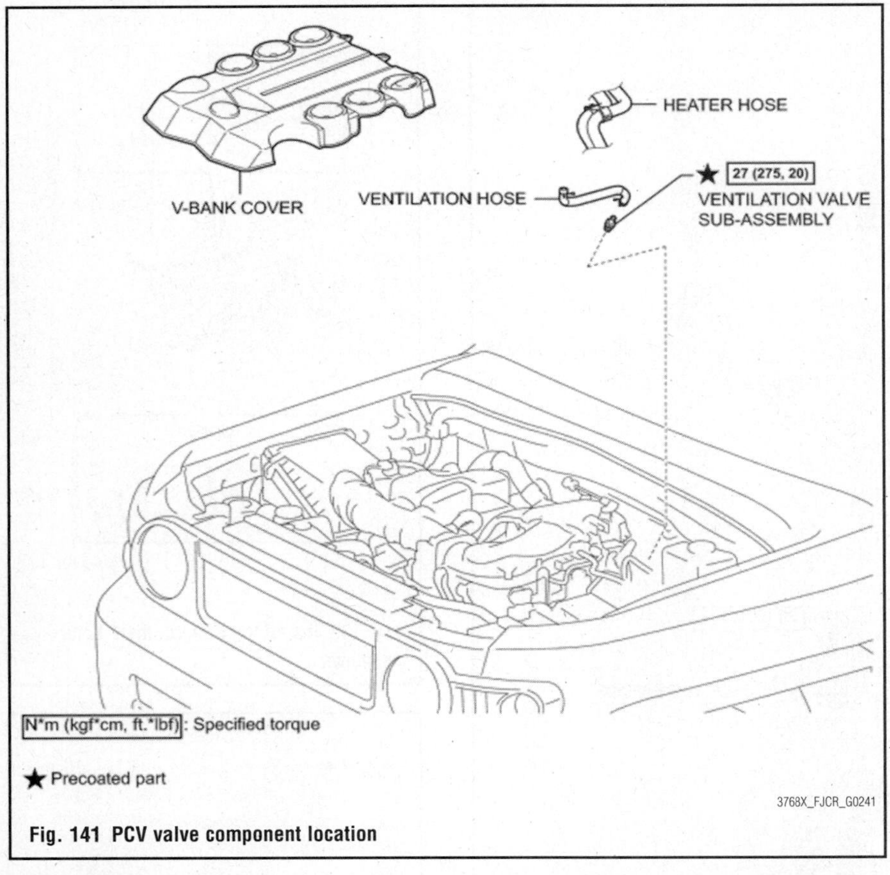

Fig. 141 PCV valve component location

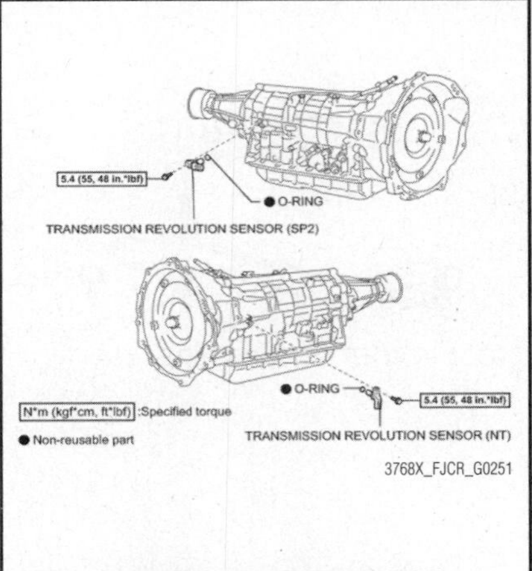

Fig. 142 Vehicle Speed Sensor (VSS) location—A750E A/T transmission

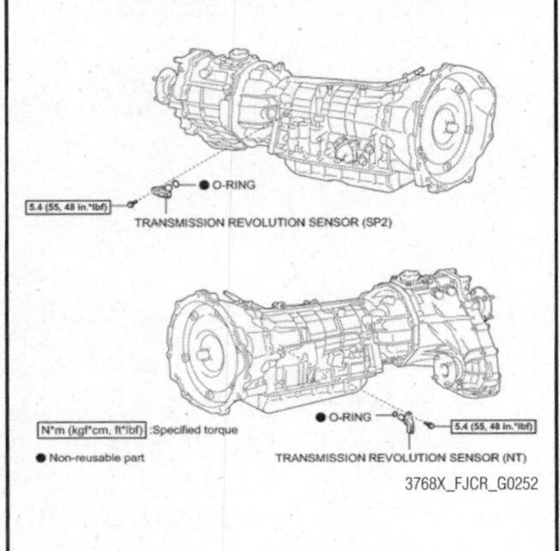

Fig. 143 Vehicle Speed Sensor (VSS) location—A750EF A/T transmission

REMOVAL & INSTALLATION

1. Disconnect the negative battery cable.
2. Disconnect the 2 sensor connectors.
3. Remove the 2 bolts and transmission revolution sensors.
4. Remove the O-ring from each sensor.
5. To install, reverse removal procedure.
6. Coat 2 new O-rings with Toyota Genuine ATF WS or equivalent and install one onto each transmission revolution sensor.
7. Tighten the bolts to 48 inch lbs. (5.4 Nm).

FUEL

GASOLINE FUEL INJECTION SYSTEM

FUEL SYSTEM SERVICE PRECAUTIONS

Safety is the most important factor when performing not only fuel system maintenance but any type of maintenance. Failure to conduct maintenance and repairs in a safe manner may result in serious personal injury or death. Maintenance and testing of the vehicle's fuel system components can be accomplished safely and effectively by adhering to the following rules and guidelines.

- To avoid the possibility of fire and personal injury, always disconnect the negative battery cable unless the repair or test procedure requires that battery voltage be applied.
- Always relieve the fuel system pressure prior to disconnecting any fuel system component (injector, fuel rail, pressure regulator, etc.), fitting or fuel line connection. Exercise extreme caution whenever relieving fuel system pressure to avoid exposing skin, face and eyes to fuel spray. Please be advised that fuel under pressure may penetrate the skin or any part of the body that it contacts.
- Always place a shop towel or cloth around the fitting or connection prior to loosening to absorb any excess fuel due to spillage. Ensure that all fuel spillage (should it occur) is quickly removed from engine surfaces. Ensure that all fuel soaked cloths or towels are deposited into a suitable waste container.
- Always keep a dry chemical (Class B) fire extinguisher near the work area.
- Do not allow fuel spray or fuel vapors to come into contact with a spark or open flame.
- Always use a back-up wrench when loosening and tightening fuel line connection fittings. This will prevent unnecessary stress and torsion to fuel line piping.
- Always replace worn fuel fitting O-rings with new Do not substitute fuel hose or equivalent where fuel pipe is installed.

Before servicing the vehicle, make sure to also refer to the precautions in the beginning of this section as well.

RELIEVING FUEL SYSTEM PRESSURE

See Figure 144.

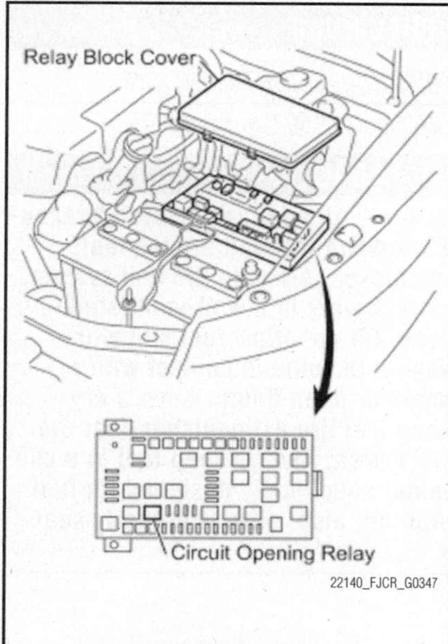

Fig. 144 Circuit opening relay location

⁂ CAUTION

Observe all applicable safety precautions when working around fuel. Whenever servicing the fuel system, always work in a well-ventilated area. Do not allow fuel spray or vapors to come in contact with a spark or open flame. Keep a dry chemical fire extinguisher near the work area. Always keep fuel in a container specifically designed for fuel storage; also, always properly seal fuel containers to avoid the possibility of fire or explosion.

⁂ WARNING

Discharge fuel system pressure procedures must be performed before disconnecting any part of the fuel system. As some pressure remains in the fuel line even after taking precautions to prevent gasoline spillage, use a shop rag or piece of cloth to prevent gasoline splashes when disconnecting the fuel line.

1. Disconnect the negative battery cable.
2. Remove the circuit opening relay.
3. Connect the negative battery cable.
4. Start the engine.
5. Turn the ignition switch to **ON** after the engine stops.

➡DTC P0171 (system too lean (Bank 1)) or DTC P0174 (system too learn (Bank 2)) may be set.

6. Crank the engine again and check that the engine stops.
7. Remove the fuel tank cap and discharge the pressure in the fuel tank completely.
8. Install the circuit opening relay.

FUEL FILTER

REMOVAL & INSTALLATION

See Figures 145 through 148.

⁂ CAUTION

Observe all applicable safety precautions when working around fuel. Whenever servicing the fuel system, always work in a well-ventilated area. Do not allow fuel spray or vapors to come in contact with a spark or open flame. Keep a dry chemical fire extinguisher near the work area. Always keep fuel in a container specifically designed for fuel storage; also, always properly seal fuel containers to avoid the possibility of fire or explosion.

➡The fuel filter is an integral part of the fuel pump unit and is not normally serviced.

1. Before servicing the vehicle, refer to the precautions.
2. Discharge fuel system pressure.
3. Disconnect the negative battery cable.
4. Remove the fuel tank.
5. Remove fuel suction with pump and gauge tube assembly.

⁂ WARNING

Protect the connector and tube joint with masking tape or the equivalent to prevent any foreign matter from sticking to them. Clean any dirt and foreign matter from the fuel suction tube assembly before removing.

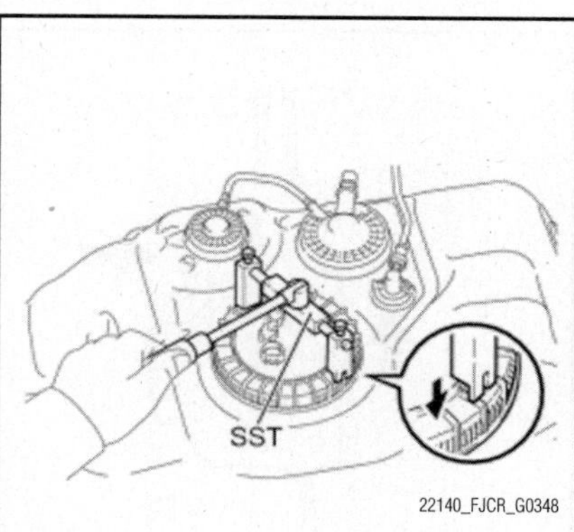

Fig. 145 Using a SST: 09808-14020, loosen the retainer

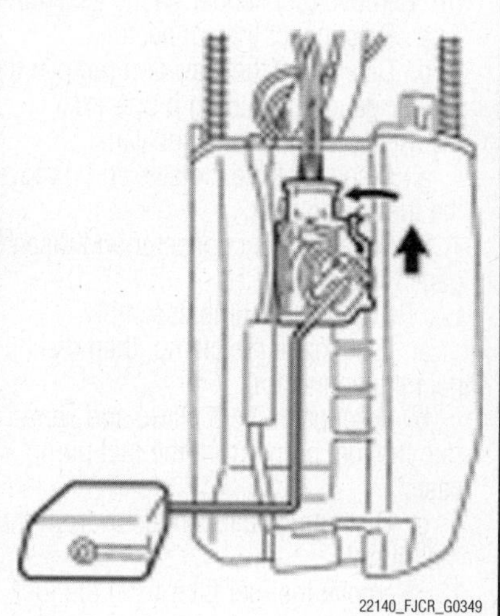

Fig. 146 Disengage the claw and remove the sender gauge by sliding it in the direction shown

6. Using a SST: 09808-14020, loosen the retainer. Align the tips of the SST with the ribs on the retainer.
7. Remove the retainer.
8. Pull the fuel pump assembly out of the fuel tank.

⁂ WARNING

Do not bend the arm of the sender gauge.

9. Remove the gasket from the fuel tank.

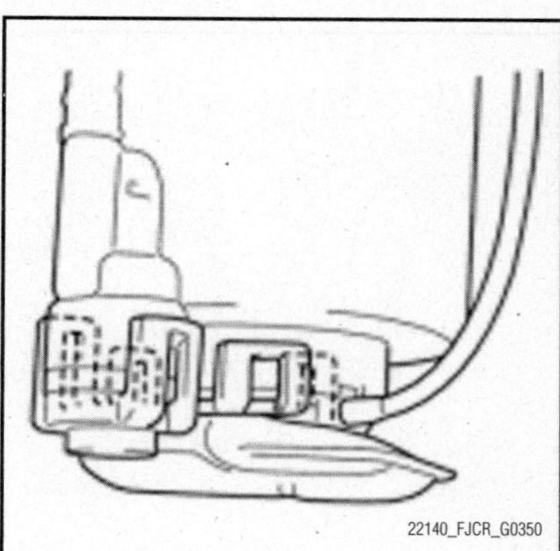

Fig. 147 Disengage the 5 claws and separate the fuel pump from the fuel pump case

10. Remove fuel sender gauge assembly.
   a. Disconnect the connector.
   b. Disengage the claw and remove the sender gauge by sliding it upward.
11. Remove No. 1 fuel sub-tank.
   a. Disengage the 5 claws and remove the fuel pump tank.
   b. Separate the connector and disengage the clamp.
12. Remove fuel pump assembly.
   a. Disengage the clamp, then disconnect the connector.
   b. Disengage the 5 claws and separate the fuel pump from the fuel pump case.
   c. Disconnect the connector from the fuel pump.
13. Remove the fuel filter from the fuel pump.

To install:

14. Install the fuel filter onto the fuel pump.
15. Install fuel pump assembly.
   a. Connect the fuel filter to the fuel pump.
   b. Engage the 5 claw fittings and install the fuel pump onto the fuel pump case.
   c. Engage the clamp, then connect the connector.
16. Install No. 1 fuel sub-tank.
   a. Install the connector and engage the clamp.
   b. Engage the 5 claws and install the fuel pump tank.
17. Install fuel sender gauge assembly.
   a. Engage the claw and install the sender gauge by sliding it in the direction as shown in the illustration.
   b. Connect the connector.
18. Install the fuel tank.
19. Connect the negative battery cable.

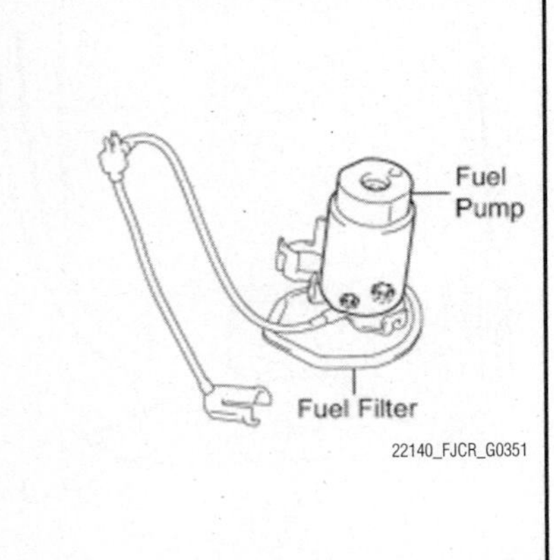

Fig. 148 Fuel filter location

FUEL LEVEL SENDING UNIT

REMOVAL & INSTALLATION

See Figure 149.

** CAUTION

Observe all applicable safety precautions when working around fuel. Whenever servicing the fuel system, always work in a well-ventilated area. Do not allow fuel spray or vapors to come in contact with a spark or open flame. Keep a dry chemical fire extinguisher near the work area. Always keep fuel in a container specifically designed for fuel storage; also, always properly seal fuel containers to avoid the possibility of fire or explosion.

1. Before servicing the vehicle, refer to the precautions.
2. Discharge fuel system pressure.
3. Disconnect the negative battery cable.
4. Remove the fuel tank.
5. Disconnect the fuel sender gauge connector.
6. Unlock the fuel sender gauge, and slide and remove it.

To install:

7. Install the fuel sender gauge onto the fuel suction with pump and gauge tube.
8. Connect the fuel sender gauge connector.
9. Install the fuel tank.
10. Connect the negative battery cable.

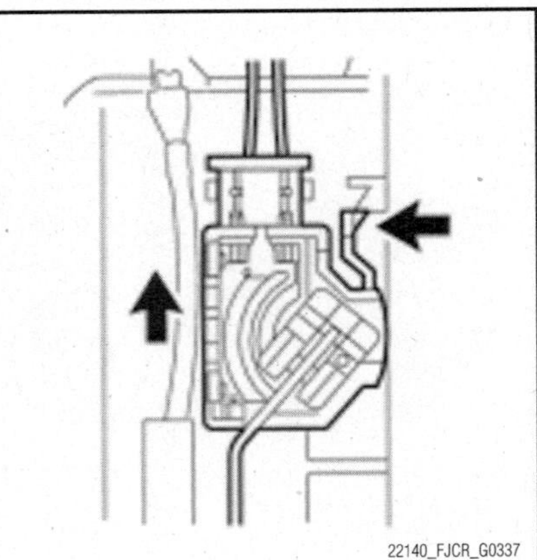

Fig. 149 Unlock the fuel sender gauge, and slide and remove it

FUEL PUMP

REMOVAL & INSTALLATION

See Figures 150 through 152.

** CAUTION

Observe all applicable safety precautions when working around fuel. Whenever servicing the fuel system, always work in a well-ventilated area. Do not allow fuel spray or vapors to come in contact with a spark or open flame. Keep a dry chemical fire extinguisher near the work area. Always keep fuel in a container specifically designed for fuel storage; also, always properly seal

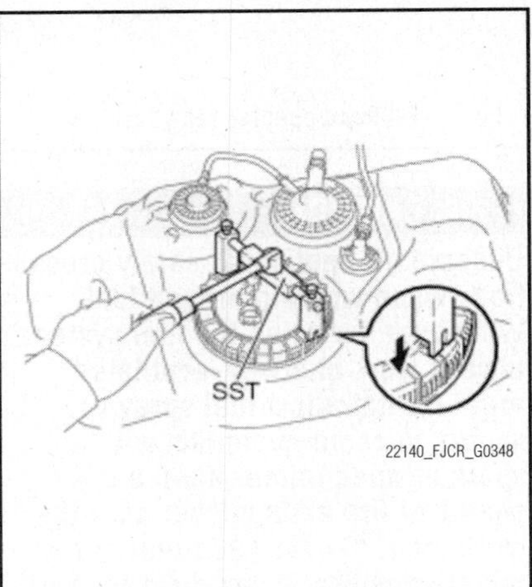

Fig. 150 Using a SST: 09808-14020, loosen the retainer

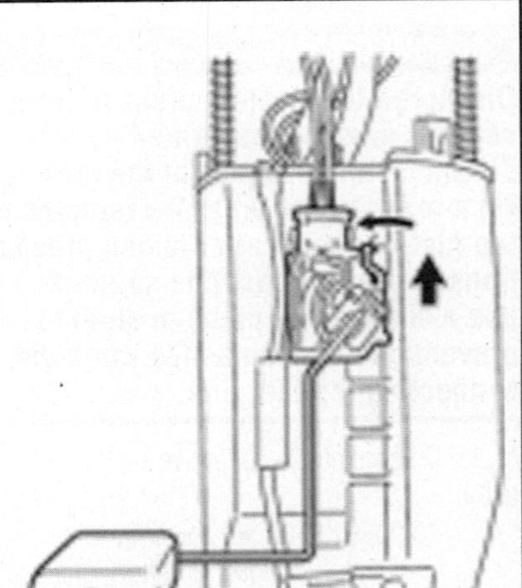

Fig. 151 Disengage the claw and remove the sender gauge by sliding it in the direction shown

fuel containers to avoid the possibility of fire or explosion.

1. Before servicing the vehicle, refer to the precautions.
2. Discharge fuel system pressure.
3. Disconnect the negative battery cable.
4. Remove the fuel tank.
5. Remove fuel suction with pump and gauge tube assembly.

⁂ WARNING

Protect the connector and tube joint with masking tape or the equivalent to prevent any foreign matter from sticking to them. Clean any dirt and foreign matter from the fuel suction tube assembly before removing.

6. Using a SST: 09808-14020, loosen the retainer. Align the tips of the SST with the ribs on the retainer.
7. Remove the retainer.
8. Pull the fuel pump assembly out of the fuel tank.

⁂ WARNING

Do not bend the arm of the sender gauge.

9. Remove the gasket from the fuel tank.
10. Remove fuel sender gauge assembly.
    a. Disconnect the connector.
    b. Disengage the claw and remove the sender gauge by sliding it upward.
11. Remove No. 1 fuel sub-tank.
    a. Disengage the 5 claws and remove the fuel pump tank.
    b. Separate the connector and disengage the clamp.
12. Remove fuel pump assembly.

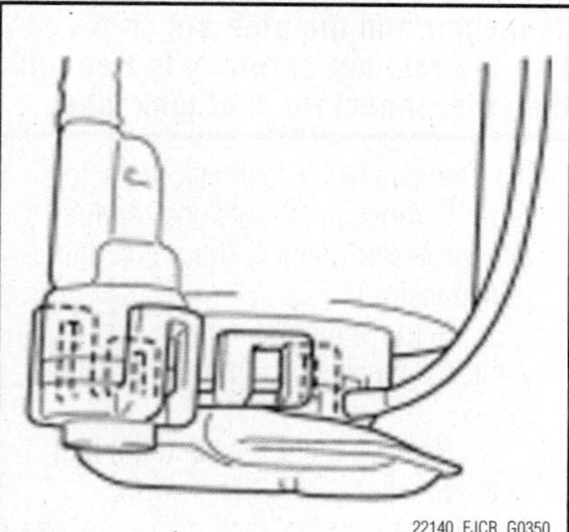

Fig. 152 Disengage the 5 claws and separate the fuel pump from the fuel pump case

    a. Disengage the clamp, then disconnect the connector.
    b. Disengage the 5 claws and separate the fuel pump from the fuel pump case.
    c. Disconnect the connector from the fuel pump.

To install:

13. Install fuel pump assembly.
    a. Connect the fuel filter to the fuel pump.
    b. Engage the 5 claw fittings and install the fuel pump onto the fuel pump case.
    c. Engage the clamp, then connect the connector.
14. Install No. 1 fuel sub-tank.
    a. Install the connector and engage the clamp.
    b. Engage the 5 claws and install the fuel pump tank.
15. Install fuel sender gauge assembly.
    a. Engage the claw and install the sender gauge by sliding it in the direction as shown in the illustration.
    b. Connect the connector.
16. Install the fuel tank.
17. Connect the negative battery cable.

FUEL PRESSURE REGULATOR

REMOVAL & INSTALLATION

See Figure 153.

⁂ CAUTION

Observe all applicable safety precautions when working around fuel. Whenever servicing the fuel system, always work in a well-ventilated area. Do not allow fuel spray or vapors to come in contact with a spark or open flame. Keep a dry chemical fire extinguisher near the work area. Always keep fuel in a container specifically designed for fuel storage; also, always properly seal fuel containers to avoid the possibility of fire or explosion.

1. Before servicing the vehicle, refer to the precautions.
2. Discharge fuel system pressure.
3. Disconnect the negative battery cable.
4. Remove v-bank cover.
5. Remove the intake air surge tank.
6. Disconnect no. 2 fuel pipe sub-assembly.
7. Remove fuel pressure regulator assembly.

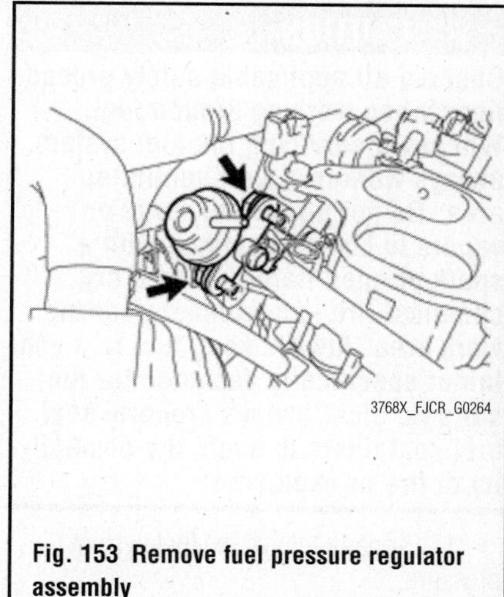

Fig. 153 Remove fuel pressure regulator assembly

    a. Remove the vacuum hose.
    b. Remove the 2 bolts, then remove the fuel pressure regulator.
    c. If necessary, remove the O-ring from the fuel pressure regulator.

To install:

8. Apply a light coat of spindle oil or gasoline to a new O-ring and install it onto the fuel pressure regulator.
9. To complete installation, reverse remaining removal procedure.
10. Check for fluid leaks.

FUEL RAIL AND INJECTOR

REMOVAL & INSTALLATION

See Figures 154 through 156.

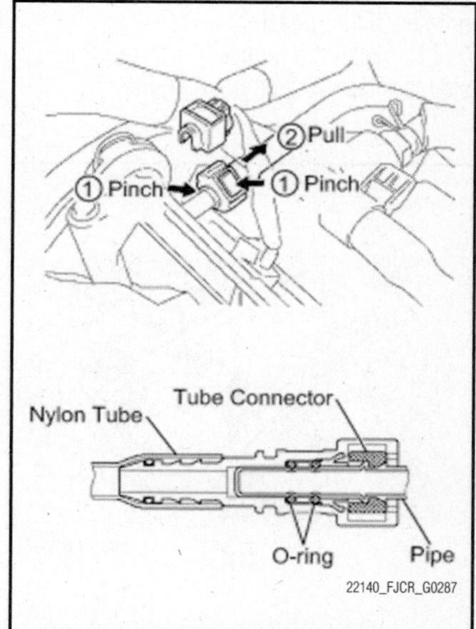

Fig. 154 Pinch the tube connector, and then pull the fuel pipe out of the delivery pipe as shown

※※ CAUTION

Observe all applicable safety precautions when working around fuel. Whenever servicing the fuel system, always work in a well-ventilated area. Do not allow fuel spray or vapors to come in contact with a spark or open flame. Keep a dry chemical fire extinguisher near the work area. Always keep fuel in a container specifically designed for fuel storage; also, always properly seal fuel containers to avoid the possibility of fire or explosion.

1. Properly relieve the fuel system pressure.
2. Remove the intake air surge tank.
3. Remove the 3 bolts and rear cylinder head cover.
4. Disconnect fuel pipes.
    a. Remove the fuel pipe clamp.
    b. Pinch the tube connector, and then pull the fuel pipe out of the delivery pipe as shown in the illustration.
5. Disconnect the 6 fuel injector connectors.
6. Remove the 4 bolts and fuel delivery pipe together with the 6 fuel injectors.
7. Pull the 6 fuel injectors from the fuel delivery pipe.
8. Remove the O-ring and injector vibration insulator from each fuel injector.

To install:

9. Install a new insulator onto each fuel injector.
10. Apply a light coat of spindle oil or gasoline to each new O-ring and install one onto each fuel injector.
11. Install the 6 injectors by twisting side to side and then pushing in.

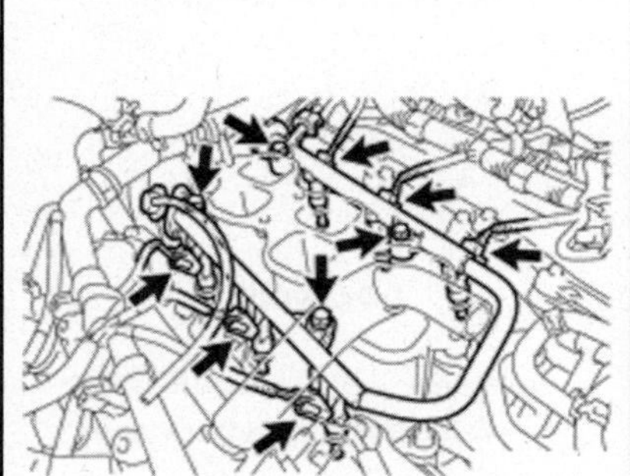

Fig. 155 Identifying the 4 bolts and fuel delivery pipe together with the 6 fuel injectors

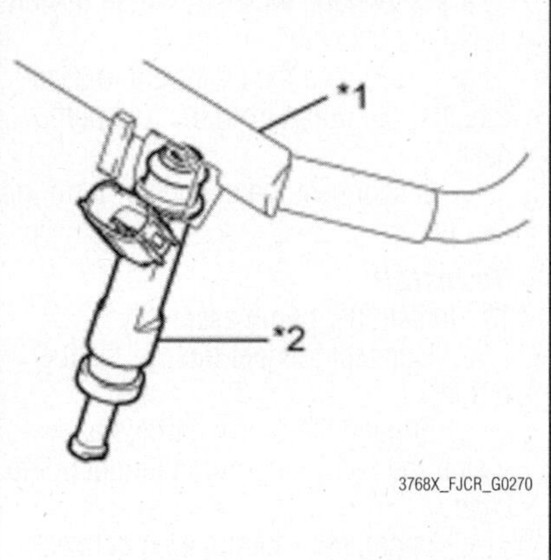

Fig. 156 Pulling the fuel injector from the fuel delivery pipe

12. While turning each fuel injector left and right, install it onto the fuel delivery pipe.
13. Position the fuel injector connectors facing outward.
14. Place the fuel delivery pipe together with the 6 fuel injectors on the intake manifold.
15. Temporarily install the 6 bolts, which are used to hold the fuel delivery pipe, to the intake manifold.
16. Check that the fuel injectors rotate smoothly.

➡If the fuel injectors do not rotate smoothly, replace the O-ring of any injector that does not rotate smoothly.

17. Position the fuel injector connectors facing outward.
18. Tighten the 6 bolts, which are used to hold the fuel delivery pipe, to the intake manifold. Tighten to 15 ft. lbs. (21 Nm).
19. Connect the 6 fuel injector connectors.
20. To complete the installation, reverse remaining removal procedure.
21. Inspect for fuel leaks.

FUEL TANK

REMOVAL & INSTALLATION

See Figures 157 and 158.

※※ CAUTION

Observe all applicable safety precautions when working around fuel. Whenever servicing the fuel system, always work in a well-ventilated area. Do not allow fuel spray or vapors to come in contact with a spark or open flame. Keep a dry chemical fire extinguisher near the work area. Always keep fuel in a container specifically designed for fuel storage; also, always properly seal fuel containers to avoid the possibility of fire or explosion.

1. Before servicing the vehicle, refer to the precautions.
2. Discharge fuel system pressure.
3. Disconnect the negative battery cable.
4. Remove left front door scuff plate.
5. Remove left rear door scuff plate.
6. Remove left rear seat cushion assembly.
7. Remove rear floor service hole cover.
    a. Pull back the floor carpet.
    b. Remove the left center floor silencer pad.
    c. Remove the 3 screws from the service hole cover.
    d. Disconnect the connector.
8. Remove fuel tank to filler pipe hose.
    a. Loosen the clamp bolt, and disconnect the fuel tank to filler pipe hose.
9. Remove fuel tank breather tube.
    a. Pinch the retainer to disengage the lock claws and pull out the No.1 fuel tank breather tube.

※※ WARNING

Check that there is no dirt or mud around the quick connector before performing this work, because the quick connector has an O-ring which seals the pipe and the connector. Clean the connector if necessary.

※※ WARNING

Do not use any tools in this work. Do not bend or twist the nylon tube. To protect the tube, cover it with a vinyl bag after disconnecting it. When the connector and the pipe are stuck, turn the retainer carefully to free and then disconnect the fuel tank tube.

10. Remove No. 1 fuel tank protector.
    a. Remove the 2 nuts and the fuel tank protector bracket. (for automatic transmission)
    b. Remove the nut and the fuel tank protector bracket. (for manual transmission)
    c. Remove the 4 nuts and the fuel tank protector.
11. Disconnect fuel tank main tube and fuel tank return tube.
    a. Pinch the retainer to disengage the lock claws and pull out the 2 fuel tank tubes.

12. Disconnect fuel tank vent hose.
   a. Disconnect the fuel tank breather tube from the fuel tank.
   b. Push the connector deep inside.
   c. Pinch portion A, as shown in the illustration.
   d. Pull out the connector.
13. Remove fuel tank assembly.
   a. Hold the fuel tank using a mission jack.
   b. Remove the 2 fuel tank bands.
   c. Remove the 2 bolts.
   d. Remove the 2 clips and 2 pins, then remove the 2 fuel tank bands.
14. Remove No. 3 fuel tank protector.
   a. Remove the 2 bolts and disengage the 3 claws, then remove the No. 3 fuel tank protector.
15. Remove fuel tank cushion.
16. Remove fuel tank main tube and fuel tank return tube.
   a. Remove the 2 joint clips, and pull out the 2 fuel tank tubes.

➡After disconnecting, cover the fuel tube joint with a vinyl bag.

⁂ WARNING

When the fuel tube joint and fuel suction plate are stuck, turn the fuel tank main tube carefully to free and then disconnect it. Likewise, disconnect the fuel tank return tube.

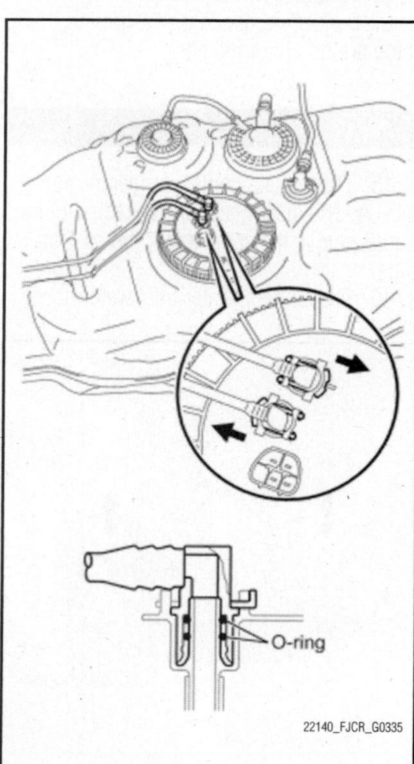

Fig. 157 Remove fuel tank main tube and fuel tank return tube connector at the fuel sender

17. Remove fuel suction with pump and gauge tube assembly.
18. Drain fuel.
19. Remove fuel tank breather tube.
   a. Pinch the retainer to disengage the lock claws and pull out the fuel tank breather tube.
   b. Loosen the clamp bolt, and disconnect the fuel tank to filler pipe hose.

To install:

20. Install fuel tank to filler pipe hose.
   a. Connect a new fuel tank to filler pipe hose, as shown in the illustration, with the clamp. install fuel tank breather tube.
   b. Push the tube connector into the pipe until the connector makes a "click" sound and install the retainer.

➡Check if there is any damage or foreign matter on the connected part of the pipe. After connecting, check if the pipe and the connector are securely connected by pulling them.

21. Install fuel suction with pump and gauge tube assembly.
22. Install fuel tank main tube and fuel tank return tube.
   a. Install the fuel tank main tube and return tube with the 2 joint clips.

➡Check that there are no scratches or foreign matter around the connected part of the fuel tube joint and plug before performing this work. Check that the fuel tube joint is securely inserted into the end. Check that the tube joint clips are on the collar of the fuel tube joint. After installing the tube joint clip, check that the fuel tank main tube and return tube can be pulled out.

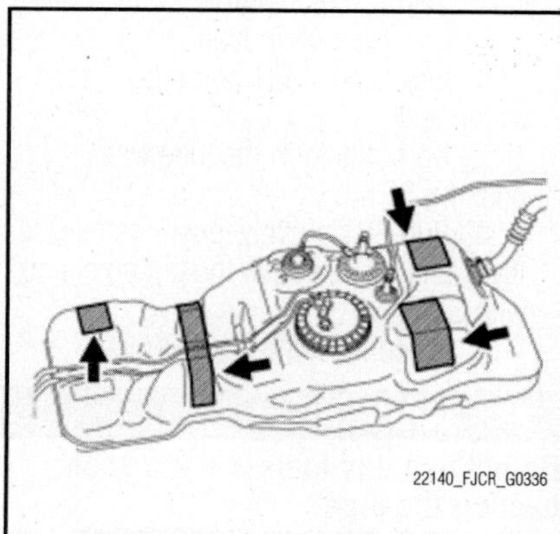

Fig. 158 Fuel tank cushion locations

23. Install 4 new fuel tank cushions the fuel tank.
24. Install No. 3 fuel tank protector.
   a. Install the No. 3 fuel tank protector with the 2 bolts and 3 claws. Tighten to 44 inch lbs. (5.0 Nm).
25. Install fuel tank assembly.
   a. Set a transmission jack to the fuel tank.
   b. Install the fuel tank and 2 fuel tank bands with the 2 clips, 2 pins and the 2 bolts. Tighten to 30 ft. lbs. (40 Nm).
26. Install fuel tank vent hose.
   a. Connect the fuel tank vent hose.
   b. Align the fuel tank vent hose connector with the pipe, then push in the fuel tank vent connector until the retainer makes a "click" sound to connect the fuel tank vent hose to the charcoal canister.
27. Connect fuel tank main tube and fuel tank return tube.
   a. Push the tube connector into the pipe until the connector makes "click" sound.
   b. Push the tube connector into the pipe until the connector makes "click" sound.
   c. Connect the fuel tank to filler pipe hose, as shown in the illustration, with the clamp.
28. Install No. 1 fuel tank protector.
   a. Install the fuel tank protector with the 4 nuts. Tighten to 15 ft. lbs. (20 Nm).
   b. Install the fuel tank protector bracket with the 2 nuts. Tighten to 49 inch lbs. (6 Nm).
29. Install rear floor service hole cover.
   a. Connect the connector.
   b. Install the 3 screws from the center floor.
   c. Install the left center floor silencer pad.
   d. Install the floor carpet.
30. Install left rear door scuff plate.
   a. Engage the 6 claws and install the rear door scuff plate.
31. Install left front door scuff plate.
32. Install rear seat cushion assembly. Tighten to 27 ft. lbs. (37 Nm).
33. Connect the negative battery cable.
34. Check for fuel leaks.

IDLE SPEED

ADJUSTMENT

The idle speed is controlled by the Electronic Control Module (ECM). No adjustment is possible.

THROTTLE BODY

REMOVAL & INSTALLATION

See Figure 159.

1. Disconnect the negative battery cable.
2. Drain engine coolant.
3. Remove the 2 nuts, then remove the V-bank cover.
4. Remove air cleaner assembly.
5. Remove throttle with motor body assembly.
    a. Disconnect the No. 5 water by-pass hose.
    b. Disconnect the No. 4 water by-pass hose.
    c. Disconnect the throttle motor connector.
    d. Remove the 4 bolts, then remove the throttle w/ motor body and gasket.

To install:

6. Install throttle with motor body assembly.
    a. Install a new gasket and the throttle with motor body with the 4 bolts. Tighten to 9 ft. lbs. (11 Nm).
    b. Connect the throttle motor connector.
    c. Connect the No. 4 water by-pass hose.
    d. Connect the No. 5 water by-pass hose.

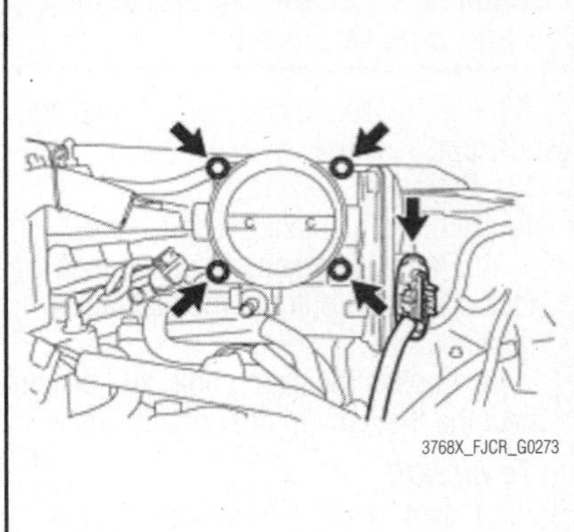

Fig. 159 Locating the throttle body fasteners and electrical connector

7. Install air cleaner assembly.
8. Connect the negative battery cable.
9. Add engine coolant.
10. Check for engine coolant leaks.
11. Install v-bank cover.

INITIALIZATION

➡Be sure to perform this procedure after reassembling the throttle body or removing and reinstalling any throttle body component. Perform the following procedure after replacing the ECM, throttle body or any throttle body components. The following procedure should also be performed if the throttle body is cleaned. Be sure to perform this procedure after replacing the ECM and reconnecting the battery cable.

1. Disconnect the cable from the negative (-) battery terminal. Wait at least 60 seconds and reconnect the cable.
2. Turn the ignition switch to ON without operating the accelerator pedal.

➡If the accelerator pedal is operated, perform the above steps again.

3. Connect the Techstream to the DLC3 and clear the DTC's.
4. Start the engine and check that the MIL is not illuminated and that the idle speed is within the specified range when the A/C is switched off after the engine is warmed up.
    a. Standard:
        - Condition: A/C switched off
        - Engine idle speed: 580 to 680 rpm

➡Be sure to perform this step with all accessories off. Make sure that the shift lever is in neutral.

5. Enter the following menus: Powertrain / Engine and ECT / Data List / All Data / Throttle Sensor Position. Fully depress the accelerator pedal and check that the value is 60% or more.
6. Perform a road test and confirm that there are no abnormalities.

HEATING & AIR CONDITIONING SYSTEM

AIR CONDITIONING UNIT

REMOVAL & INSTALLATION

See Figures 160 through 163.

1. Before servicing the vehicle, refer to the precautions.
2. Disconnect the negative battery cable.

➡Wait for at least 90 seconds after disconnecting the cable to prevent the airbag from working.

3. Discharge refrigerant from refrigeration system.
4. Drain engine coolant.
5. Remove roof antenna pole.
6. Remove windshield wiper arm cover.
7. Remove front wiper arm and blade assembly.
8. Remove left front fender side panel upper.
9. Remove antenna ornament.
10. Remove front fender side panel upper.
11. Remove cowl top ventilator louver assembly.
12. Remove cowl top ventilator louver.
13. Disconnect cooler refrigerant suction pipe **A** using SST: 09870-00015.
    a. Check the directions of the piping clamp and SST by referring to the illustration on the caution label.
    b. Push down SST and release the clamp lock.
    c. Do not deform the tube when pushing SST.
    d. Pull SST slightly, push the release lever, and then remove the piping clamp with SST.
    e. Disconnect the suction pipe.

⁂ WARNING

Do not use any tools when disconnecting the pipe.

14. Seal the openings of the disconnected parts using vinyl tape to prevent moisture and foreign matter from entering.
15. Disconnect cooler refrigerant liquid pipe **A**. The disconnection procedure of the liquid pipe is the same as for the suction pipe.
16. Disconnect heater inlet water hose.

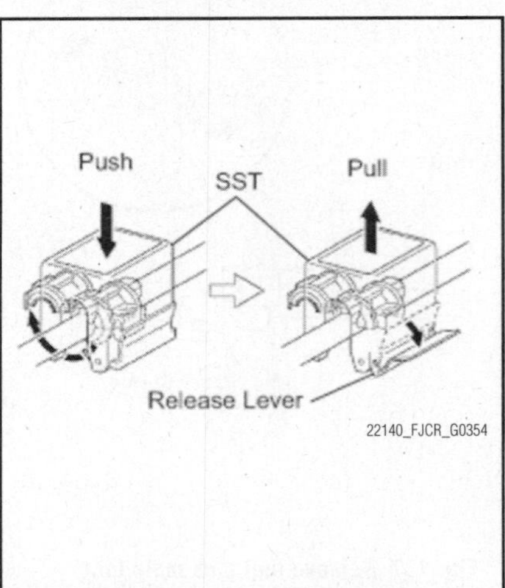

Fig. 160 Disconnect cooler refrigerant suction pipe

a. Using pliers, grip the claw of the clip, slide the clip and disconnect the heater water inlet hose from the heater unit.

17. Using pliers, grip the claw of the clip, slide the clip and disconnect the heater water outlet hose from the heater unit.

18. Position front wheels facing straight ahead.

19. Remove lower steering wheel covers.

20. Remove steering pad.

21. Remove steering wheel assembly.

22. Remove lower steering column cover.

23. Remove upper steering column cover.

24. Remove combination switch assembly.

25. Remove front door scuff plates.

26. Remove front floor footrest.

27. Remove footrest clip.

28. Remove cowl side trim boards.

29. Remove front door opening trim weatherstrips.

30. Remove assist grip plug.

31. Remove assist grip assembly.

32. Remove front pillar garnishes.

33. Remove instrument panel garnishes.

34. Remove integration control and panel assembly.

35. Remove radio receiver assembly.

36. Remove parking brake hole cover.

37. On manual transmission vehicles, remove shift lever knob.

38. On 4WD vehicles, remove shift lever knob.

39. Remove console upper rear panel.

40. Remove box bottom mat.

41. Remove front console box.

42. Remove console upper panel no. 1 garnish.

43. Remove instrument lower cover.

44. Remove no 1 instrument panel register assembly.

45. Separate hood lock control lever.

46. Remove instrument panel finish plate.

47. Remove lower instrument panel finish panel.

48. Remove instrument cluster finish panel.

49. Remove combination meter assembly.

50. Remove glove compartment door assembly.

51. Remove instrument panel lower finish panel.

52. Remove no. 2 instrument panel register assembly.

53. Remove instrument panel speaker panel.

54. Remove front no 2 speaker assembly.

55. Remove assist grip retainers.

56. Disconnect passenger airbag connector.

57. Remove instrument panel.

58. Remove left instrument panel finish panel end.

59. Remove no. 1 heater to register duct.

60. Remove no. 2 heater to register duct.

61. Remove rear no. 1 air duct.

62. Remove rear no. 2 air duct.

63. Remove no. 1 air duct.

64. Remove no. 2 air duct.

65. Remove no. 1 instrument panel brace mounting bracket.

a. On the left side remove the bolt and nut and remove the instrument panel brace mounting bracket.

b. On the right side remove the bolt and nut and remove the instrument panel brace mounting bracket.

66. Remove ECM.

67. Remove steering column hole cover.

68. Remove the bolt and separate the steering shaft thrust stopper from the steering intermediate shaft assembly.

a. Mark matchmarks on the steering column assembly and steering intermediate shaft.

b. Pull the intermediate shaft assembly and steering shaft thrust stopper out of the steering column assembly.

69. Remove steering column assembly.

70. Remove the bolt and remove the instrument panel side bracket.

71. Separate main body ECU.

a. Remove the 2 nuts and remove the driver side junction block.

72. Remove the cooler unit drain hose.

73. Remove instrument panel reinforcement.

a. Remove the 3 bolts and 5 nuts and disconnect the wire harness.

b. Disconnect the connectors.

c. Disengage the clamps.

d. Remove the 5 bolts and the 2 nuts of the air conditioning unit.

e. Remove the 2 caps and the 7 bolts of the reinforcement.

f. Disengage the reinforcement hook of the air conditioning unit, and remove the reinforcement.

g. Remove the air conditioning unit.

74. Remove air conditioning unit assembly.

a. Remove the 2 screws.

b. Remove the air conditioning unit as shown in the illustration.

75. Remove heater to register duct assembly.

a. Disengage the 4 claws and remove the heater to register duct.

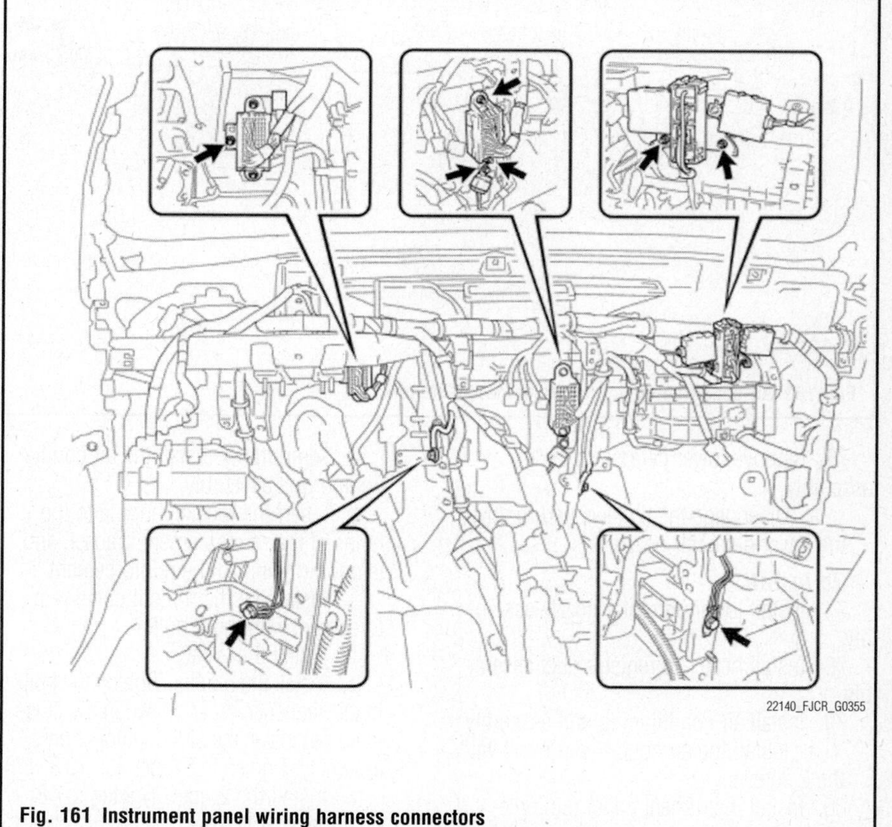

Fig. 161 Instrument panel wiring harness connectors

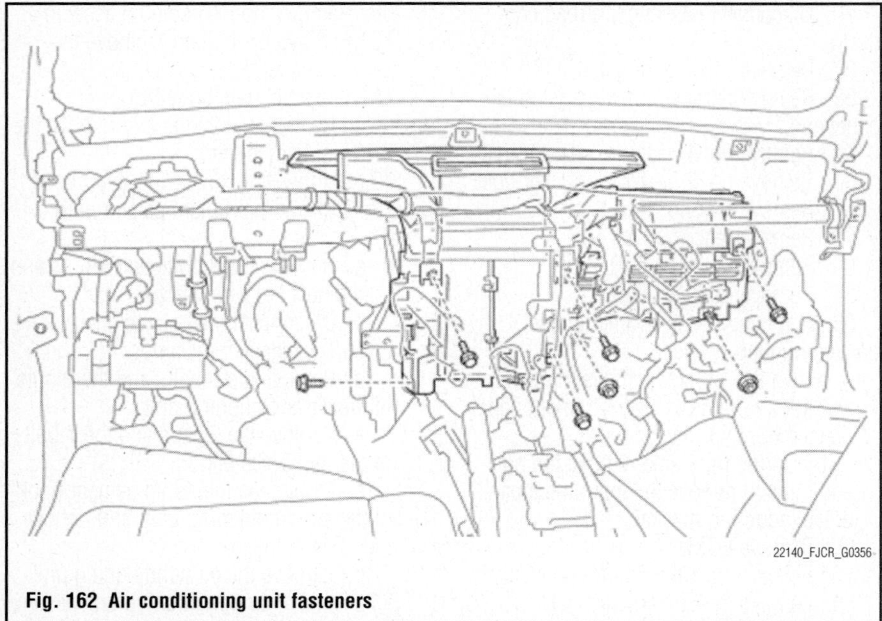

Fig. 162 Air conditioning unit fasteners

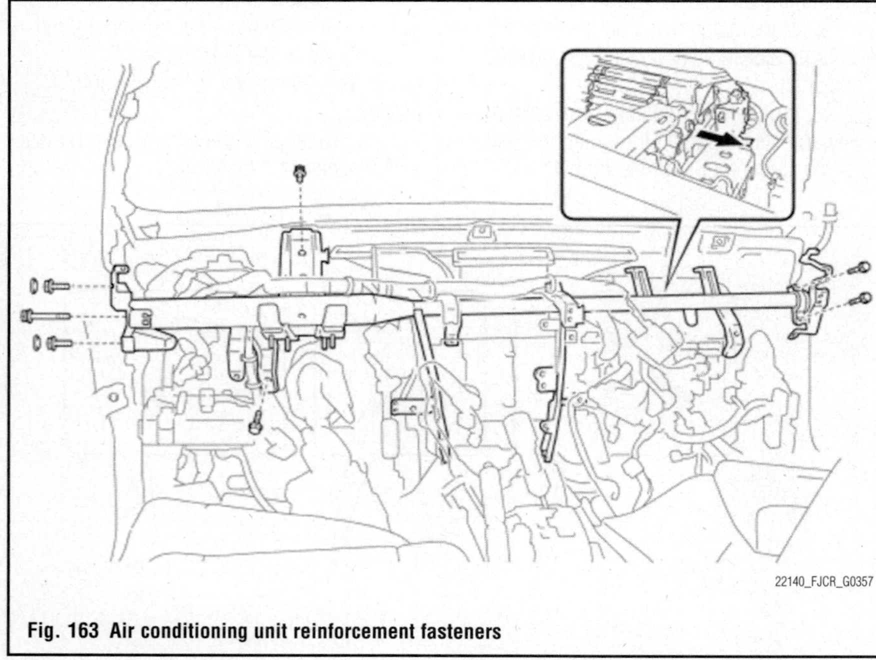

Fig. 163 Air conditioning unit reinforcement fasteners

76. Remove lower defroster nozzle assembly.
   a. Disengage the 4 claws and remove the lower defroster nozzle.

To install:

77. Install lower defroster nozzle assembly.
78. Install heater to register duct assembly.
79. Install air conditioning unit assembly.
   a. Install the air conditioning unit with the 2 screws.
80. Install instrument panel reinforcement.
   a. Temporarily install the air conditioning unit assembly.
   b. Insert the bracket hook into the holes of the reinforcement bracket, and temporarily install the reinforcement.
   c. Install the instrument panel reinforcement with the 7 bolts.
   d. Install the 2 caps.
   e. Install the 5 bolts. Tighten the bolts in the sequence order shown in the illustration to install the air conditioner unit assembly. Tighten to 87 inch lbs. (9.8 Nm).
   f. Install the 2 nuts. Tighten to 48 inch lbs. (5.4 Nm).
   g. Engage the clamps.
   h. Connect the connectors.
   i. Connect the wire harness with the 3 bolts and 5 nuts. Tighten to 65 inch lbs. (7.3 Nm).
81. Install cooler unit drain hose.
   a. Install the cooler unit drain hose.
82. Install the main body ECU with the 2 nuts. Tighten to 74 inch lbs. (8.4 Nm).
83. Install the instrument panel side bracket with the bolt. Tighten to 74 inch lbs. 8.4 Nm).
84. Install steering column assembly.
85. Install steering intermediate shaft assembly.
   a. Align the matchmarks on the steering column assembly and the steering intermediate shaft assembly.
   b. Install the steering intermediate shaft assembly and thrust stopper onto the steering column assembly with the bolt. Tighten to 27 ft. lbs. (36 Nm).
86. Install steering column hole cover.
87. Install ECM.
88. Install no. 1 instrument panel brace mounting bracket.
   a. On the left side, install the instrument panel brace mounting bracket with the bolt and nut.
   b. On the right side, install the instrument panel brace mounting bracket with the bolt and nut.
89. Install no. 1 air duct.
90. Install no. 2 air duct.
91. Install rear no. 1 air duct.
92. Install rear no. 2 air duct.
93. Install no. 1 heater to register duct.
94. Install no. 2 heater to register duct.
95. Install instrument panel.
96. Connect passenger airbag connector.
97. Install left instrument panel finish panel end.
98. Install assist grip retainers.
99. Install front no. 2 speaker assembly.
100. Install no. 2 instrument panel speaker panels.
101. Install no. 2 instrument panel register assembly.
102. Install right instrument panel lower finish panel.
103. Install glove compartment door assembly.
104. Install combination meter assembly.
105. Install instrument cluster finish panel.
106. Install lower instrument panel.
107. Install left lower instrument panel finish panel.
108. Install instrument panel finish plate.
109. Connect hood lock control lever.
110. Install no 1 instrument panel register assembly.

111. Install instrument lower cover.
112. Install console upper panel no 1 garnish.
113. Install front console box.
114. Install box bottom mat.
115. Install console upper rear panel.
116. On manual transmission vehicles, install shift lever knob.
117. On 4WD vehicles, install shift lever knob.
118. Install parking brake hole cover.
119. Install radio receiver assembly.
120. Install integration control and panel assembly.
121. Install instrument panel garnishes.
122. Install front pillar garnishes.
123. Install assist grip assembly.
124. Install assist grip plug.
125. Install front door opening trim weatherstrips.
126. Install cowl side trim boards.
127. Install footrest clip.
128. Install front floor footrest.
129. Install front door scuff plates.
130. Install combination switch assembly.
131. Install upper steering column cover.
132. Install lower steering column cover.
133. Install steering wheel assembly.
134. Install steering pad.
135. Install lower no 2 steering wheel cover.
136. Install lower no 3 steering wheel cover.
137. Connect heater water outlet hose to heater unit.
138. Perform the installation with the hose clip and mark at the correct angle as shown in the illustration.
139. Connect heater inlet water hose.
140. Perform the installation with the hose clip and mark at the correct angle as shown in the illustration.
141. Install cooler refrigerant liquid pipe a.

a. Remove the vinyl tape from liquid tube **A** and the connecting portion of the unit.

b. Apply sufficient ND-OIL8 compressor oil to a new O-ring and the connecting part of the liquid pipe.

c. Install the O-ring onto the liquid pipe **A**.

d. Install the liquid and piping clamp.

142. After connection, check the claw fitting of the piping clamp.
143. Install cooler refrigerant suction pipe **A**.
144. Connection procedure of the suction pipe is the same as for the liquid pipe.
145. Install cowl top ventilator louvers.
146. Install cowl top ventilator louver.
147. Install front fender side upper panels.
148. Install antenna ornament.
149. Install roof antenna pole.
150. Install front wiper arm and blade assemblies.
151. Install windshield wiper arm cover.
152. Add engine coolant.
153. Connect cable to negative battery terminal.
154. Check SRS warning light.
155. Charge refrigerant.
156. Warm up engine.
157. Check for engine coolant leaks.
158. Check for refrigerant leaks.
159. Position front wheels facing straight ahead.

BLOWER MOTOR

REMOVAL & INSTALLATION

See Figure 164.

1. Before servicing the vehicle, refer to the precautions.
2. Disconnect the negative battery cable.
3. Disconnect the connector and the clamp.
4. Remove the 3 screws and the blower motor.
5. To install, reverse removal procedure.

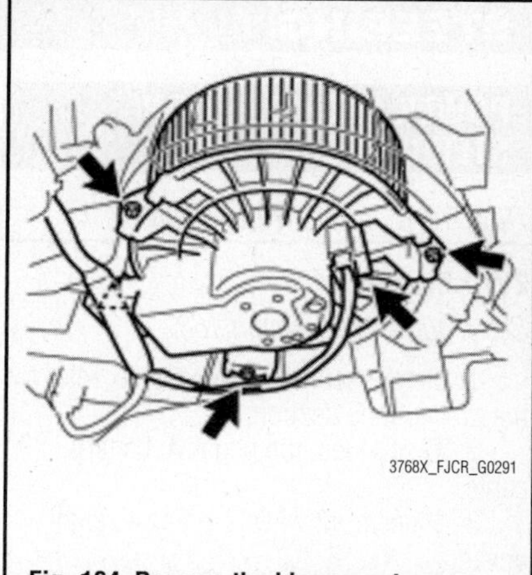

Fig. 164 Remove the blower motor as shown

HEATER CORE

REMOVAL & INSTALLATION

See Figure 165.

1. Before servicing the vehicle, refer to the precautions.
2. Remove the air conditioning unit.
3. Remove the screw and clamp.
4. Remove the heater radiator unit from the heater case.
5. Installation is the reverse of the removal procedure.

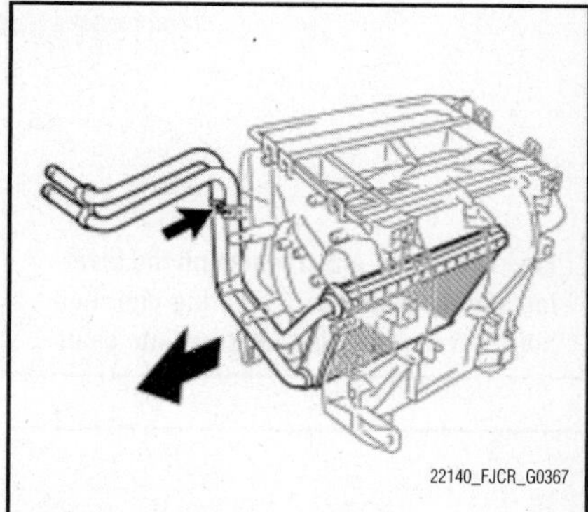

Fig. 165 Removing the heater core

STEERING

POWER RACK & PINION STEERING GEAR

REMOVAL & INSTALLATION

2WD Models

See Figures 166 through 168.

1. Before servicing the vehicle, refer to the precautions section.
2. Disconnect the negative battery cable.
3. Place front wheels facing straight ahead.
4. Remove no.1 engine under cover.
5. Remove fan and generator V-belt.
6. Remove front wheels.
7. Remove no.2 steering intermediate shaft.
   a. Fix the steering wheel with the seat belt in order to prevent it from rotating.

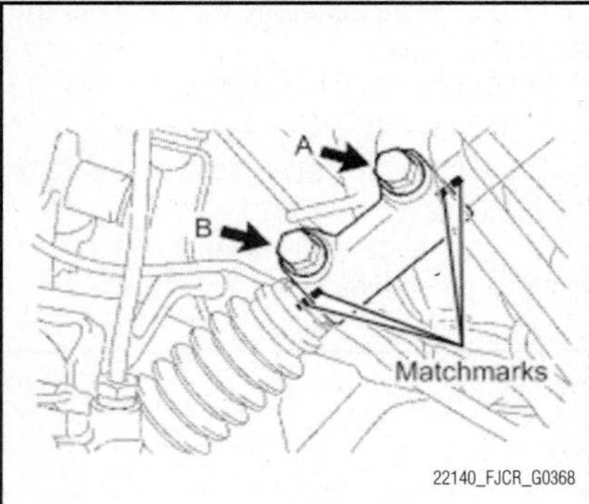

Fig. 166 Place matchmarks on the steering sliding yoke, No. 2 steering intermediate shaft and steering intermediate shaft

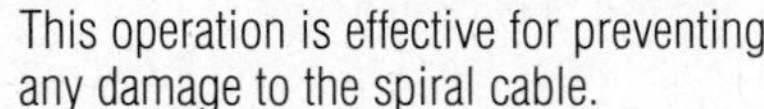

This operation is effective for preventing any damage to the spiral cable.

   b. Place matchmarks on the steering sliding yoke, No. 2 steering intermediate shaft and steering intermediate shaft.
   c. Remove bolts from the steering sliding yoke.
   d. Slide the steering sliding yoke up and separate it from the No. 2 steering intermediate shaft.
   e. Pull down the steering sliding yoke from the steering intermediate shaft to remove it.
   f. Place matchmarks on the No. 2 steering intermediate shaft and power steering link.
   g. Remove bolt C from the No. 2 steering intermediate shaft.
   h. Slide the No. 2 steering intermediate shaft up and remove it from the power steering link.
8. Separate tie rod end.
   a. Remove the cotter pin and nut.
   b. Using SST: 09628-62011, separate the tie rod end from the steering knuckle arm.
9. Separate pressure feed tube assembly.
   a. Remove the 2 bolts and separate the tube support brackets.
   b. Loosen the flare nut and separate the pressure feed tube.
   c. Disengage the clip and disconnect the return hose.
10. Remove the 4 bolts and remove the air conditioning compressor and magnetic clutch.
11. Remove power steering link.
   a. Remove the 2 bolts and 2 nuts.

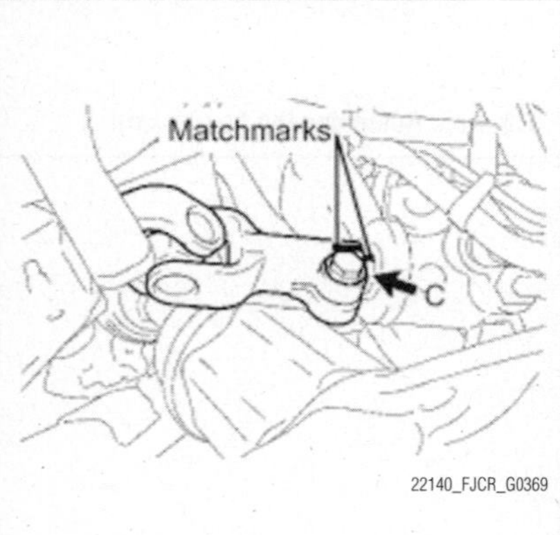

Fig. 167 Place matchmarks on the No. 2 steering intermediate shaft and power steering link

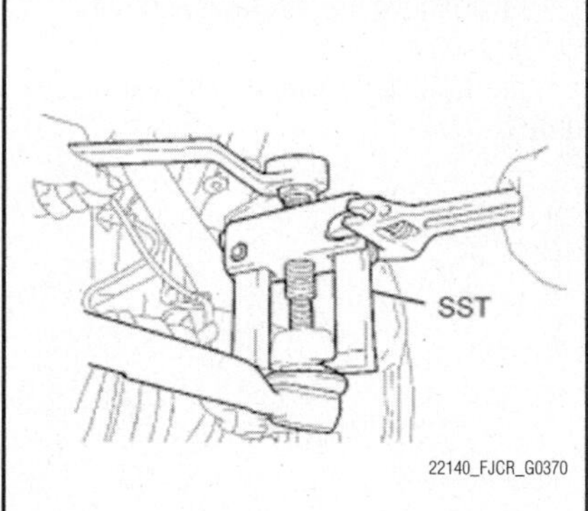

Fig. 168 Using SST: 09628-62011, separate the tie rod end from the steering knuckle arm

✲✲ WARNING

The nut has a detent, so never turn the nut. Be sure to turn only the bolt.

   b. Tilt the transmission and remove the power steering link.

To install:

12. Install the power steering link with the 2 bolts and 2 nuts. Tighten to 74 ft. lbs. (100 Nm).

✲✲ WARNING

Never turn the nut since it has a detent. Be sure to turn only the bolt.

13. Install pressure feed tube assembly.
   a. Connect the return hose with the clip.
   b. Tighten the flare nut to 33 ft. lbs. (44 Nm) and connect the pressure feed tube.
   c. Install the tube support brackets with the 2 bolts. Tighten to 21 ft. lbs. (28 Nm).
14. Install the air conditioning compressor and magnetic clutch with the 4 bolts.
15. Install tie rod end.
   a. Install the tie rod end onto the steering knuckle arm.
   b. Install the nut. Tighten to 67 ft. lbs. (91 Nm).
   c. Install a new cotter pin.
16. Install No. 2 steering intermediate shaft.
   a. Align the matchmarks on the No. 2 steering intermediate shaft and power steering link.
   b. Install the No. 2 steering intermediate shaft onto the power steering link. Tighten to 27 ft. lbs. (36 Nm).
   c. Align the matchmarks on the steering intermediate shaft and steering sliding yoke.
   d. Install the steering sliding yoke onto the steering intermediate shaft and slide it upward.
   e. Align the matchmarks on the steering sliding yoke and No. 2 steering intermediate shaft.
   f. Install the steering sliding yoke. Tighten to 27 ft. lbs. (36 Nm).
17. Install fan and generator V-belt.
18. Install front wheels.
19. Place front wheels facing straight ahead.
20. Inspect steering wheel center point.
21. Connect the negative battery cable.
22. Add power steering fluid.
23. Bleed power steering fluid.

24. Check for power steering fluid leakage.
25. Inspect and adjust front wheel alignment.
26. Install No. 1 engine under cover

4WD Models

See Figures 166 through 168

1. Before servicing the vehicle, refer to the precautions section.
2. Disconnect the negative battery cable.
3. Place front wheels facing straight ahead.
4. Remove no.1 engine under cover.
5. Remove fan and generator V-belt.
6. Remove front wheels.
7. Remove no. 1 engine under cover.
8. Remove rear engine under cover assembly.
9. Remove no. 2 exhaust front pipe assembly.
10. Remove front exhaust pipe assembly.
11. Remove driveshaft heat insulator.
12. Remove front halfshaft assembly.
13. Remove driveshaft assembly.
14. Remove front suspension member brackets.
15. Remove no. 3 frame crossmember.
16. Remove No. 2 steering intermediate shaft.
    a. Fix the steering wheel with the seat belt in order to prevent it from rotating. This operation is effective for preventing any damage to the spiral cable.
    b. Place matchmarks on the steering sliding yoke, No. 2 steering intermediate shaft and steering intermediate shaft.
    c. Remove bolts from the steering sliding yoke.
    d. Slide the steering sliding yoke up and separate it from the No. 2 steering intermediate shaft.
    e. Pull down the steering sliding yoke from the steering intermediate shaft to remove it.
    f. Place matchmarks on the No. 2 steering intermediate shaft and power steering link.
    g. Remove bolt C from the No. 2 steering intermediate shaft.
    h. Slide the No. 2 steering intermediate shaft up and remove it from the power steering link.
17. Separate tie rod end.
    a. Remove the cotter pin and nut.
    b. Using SST: 09628-62011, separate the tie rod end from the steering knuckle arm.
18. Separate pressure feed tube assembly.
    a. Remove the 2 bolts and separate the tube support brackets.
    b. Loosen the flare nut and separate the pressure feed tube.
    c. Disengage the clip and disconnect the return hose.
19. Remove the 4 bolts and remove the air conditioning compressor and magnetic clutch.
20. Remove power steering link.
    a. Remove the 2 bolts and 2 nuts.

⁂ WARNING

The nut has a detent, so never turn the nut. Be sure to turn only the bolt.

    b. Tilt the transmission and remove the power steering link.

To install:

21. Install the power steering link with the 2 bolts and 2 nuts. Tighten to 74 ft. lbs. (100 Nm).

⁂ WARNING

Never turn the nut since it has a detent. Be sure to turn only the bolt.

22. Install pressure feed tube assembly.
    a. Connect the return hose with the clip.
    b. Tighten the flare nut to 33 ft. lbs. (44 Nm) and connect the pressure feed tube.
    c. Install the tube support brackets with the 2 bolts. Tighten to 21 ft. lbs. (28 Nm).
23. Install the air conditioning compressor and magnetic clutch with the 4 bolts.
24. Install tie rod end.
    a. Install the tie rod end onto the steering knuckle arm.
    b. Install the nut. Tighten to 67 ft. lbs. (91 Nm).
    c. Install a new cotter pin.
25. Install No. 2 steering intermediate shaft.
    a. Align the matchmarks on the No. 2 steering intermediate shaft and power steering link.
    b. Install the No. 2 steering intermediate shaft onto the power steering link. Tighten to 27 ft. lbs. (36 Nm).
    c. Align the matchmarks on the steering intermediate shaft and steering sliding yoke.
    d. Install the steering sliding yoke onto the steering intermediate shaft and slide it upward.
    e. Align the matchmarks on the steering sliding yoke and No. 2 steering intermediate shaft.
    f. Install the steering sliding yoke. Tighten to 27 ft. lbs. (36 Nm).
26. Install no. 3 frame crossmember.
27. Install front suspension member brackets.
28. Install driveshaft assembly.
29. Install front halfshaft assembly.
30. Install driveshaft heat insulator.
31. Install front exhaust pipe assembly.
32. Install no. 2 exhaust front pipe assembly.
33. Install rear engine under cover assembly.
34. Install no. 1 engine under cover.
35. Install fan and generator V-belt.
36. Install front wheels.
37. Place front wheels facing straight ahead.
38. Inspect steering wheel center point.
39. Connect the negative battery cable.
40. Add power steering fluid.
41. Bleed power steering fluid.
42. Check for power steering fluid leakage.
43. Inspect and adjust front wheel alignment.
44. Install No. 1 engine under cover.

POWER STEERING PUMP

REMOVAL & INSTALLATION

See Figures 169 through 171.

1. Remove the RH front wheel.
2. Remove RH front fender apron seal.
3. Remove fan and alternator v belt.
4. Drain power steering fluid.
5. Separate pressure feed tube assembly:
    a. Disconnect the oil pressure switch connector.
    b. Remove the bolt and the wire harness clamp bracket.

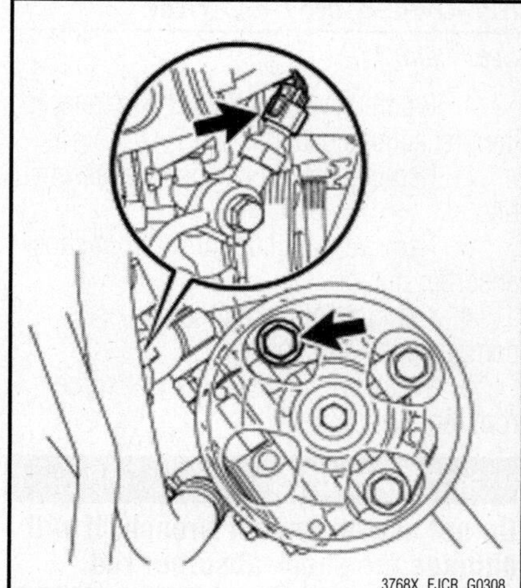

Fig. 169 Disconnect the oil pressure switch connector and remove the bolt and the wire harness clamp bracket

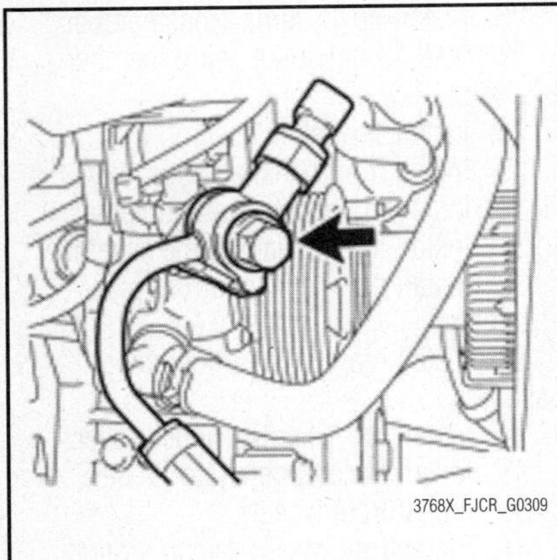

Fig. 170 Remove the union bolt, then separate the pressure feed tube

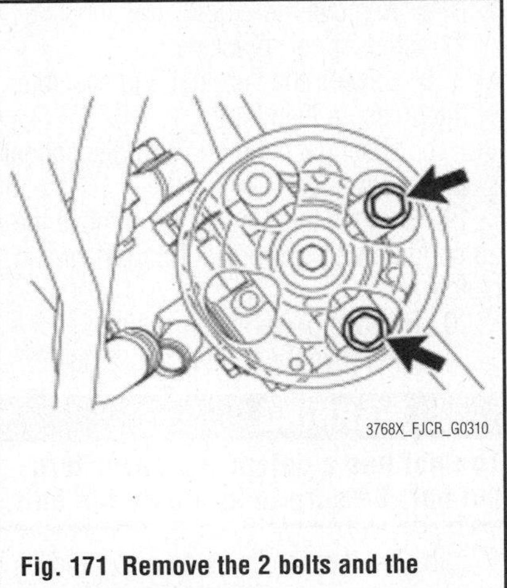

Fig. 171 Remove the 2 bolts and the power steering pump

c. Remove the union bolt, then separate the pressure feed tube.

d. Remove the gasket from the pressure feed tube.

6. Disengage the clip and separate the No. 1 oil reservoir to pump hose.

7. Remove the 2 bolts and the power steering pump.

To install:

8. Install the power steering pump assembly with the 2 bolts. Tighten to 32 ft. lbs (43 Nm).

9. Connect the No. 1 oil reservoir to pump hose with the clip.

10. Install pressure feed tube assembly:

a. Install a new gasket onto the pressure feed tube.

b. Install the pressure feed tube with the union bolt. Tighten to 37 ft. lbs. (50 Nm).

c. Install the wire harness clamp bracket with the bolt. Tighten to 32 ft. lbs. (43 Nm).

➡ Make sure that no oil adheres to the connector.

d. Connect the oil pressure switch connector.

11. To complete installation, reverse remaining removal procedure.

12. Add power steering fluid.

13. Bleed power steering fluid.

14. Check for power steering fluid leakage.

BLEEDING

1. Check the fluid level.
2. Raise and safely support the vehicle securely on jackstands.
3. Turn the steering wheel.
4. With the engine stopped, turn the wheel slowly from lock to lock several times.
5. Lower the vehicle.
6. Start the engine.
7. Run the engine at idle for a few minutes.
8. Turn the steering wheel.
9. With the engine idling, turn the wheel to the left or right full lock position and keep it there for 2 to 3 seconds. Then turn the wheel to the opposite full lock position and keep it there for 2 to 3 seconds.
10. Repeat this step several times.
11. Stop the engine.
12. Check for foaming or emulsification.

⁂ WARNING

If the system has to be bled twice because of foaming or emulsification, check for fluid leakage in the system.

13. Check the fluid level.

SUSPENSION

FRONT SUSPENSION

COIL SPRING

REMOVAL & INSTALLATION

See Figure 172.

1. Before servicing the vehicle, refer to the precautions section.
2. Remove the shock absorber assembly.
3. Remove front support to front shock absorber nut
4. Using a coil spring compressor, compress the coil spring.
5. While holding the shock absorber rod, remove the nut.

⁂ WARNING

Do not use an impact wrench. It will damage the shock absorber rod.

6. Remove front shock absorber cushion retainer.
7. Remove front shock absorber no. 1 cushion.
8. Remove front suspension support.
9. Remove front shock absorber cushion retainer.
10. Remove front coil spring.

To install:

11. Using a coil spring compressor, compress the coil spring.
12. Install the coil spring onto the shock absorber.

➡Fit the lower end of the coil spring into the gap of the spring lower seat.

13. Install front coil spring.
14. Install front shock absorber cushion retainer.
15. Install front suspension support.
16. Install front shock absorber no. 1 cushion.
17. Install front shock absorber cushion retainer.
18. Install front support to front shock absorber nut.
19. Align the suspension support and the absorber bushing as shown in the illustration.
20. Fit and tighten a new lock nut. Tighten to 18 ft. lbs. (25 Nm).

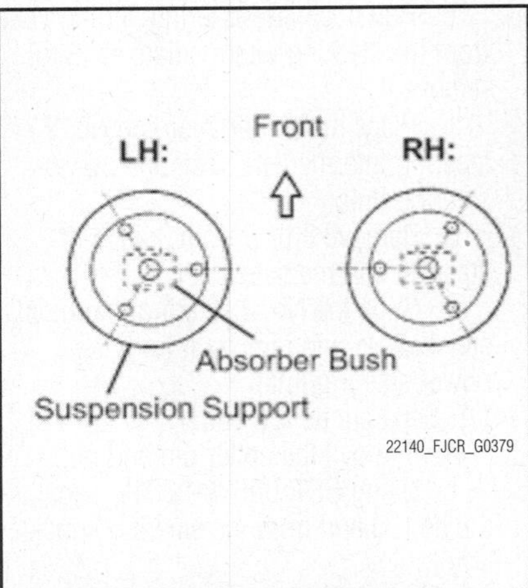

Fig. 172 Align the suspension support and the absorber bushing as shown

21. Release the coil spring while checking the position of the suspension support.

LOWER BALL JOINT

REMOVAL & INSTALLATION

The lower ball joint is an integral part of the lower control arm and is not serviced separately. See Front Suspension, Lower Control Arm, Removal and Installation.

LOWER CONTROL ARM

REMOVAL & INSTALLATION

See Figures 173 through 175.

1. Before servicing the vehicle, refer to the precautions section.
2. Remove front wheel.
3. Inspect front lower suspension arm.
   a. Install the hub nuts onto the disc.
   b. Using a dial indicator, check the lower ball joint for excessive play when you push the hub nuts up and down with a force of 66 lbs. (294 N). Maximum play is 0.020 in. (0.5 mm).
   c. If it is not within the specification, replace the lower arm.
4. Separate front shock absorber with coil spring.
   a. Remove the bolt, nut and washer.
   b. Separate the front shock absorber with coil spring from the lower arm.
5. Remove front lower suspension arm.
   a. Remove the 2 bolts and separate the lower ball joint attachment from the steering knuckle.
   b. Place matchmarks on the No. 2 camber adjust cam and toe adjust cam.
   c. Remove the nut, No. 2 camber adjust cam, No. 1 camber adjust cam, bolt, toe adjust cam, toe adjust plate and lower arm.
   d. Remove the cotter pin and the nut.
   e. Using SST: 09628-00011, remove the lower ball joint attachment.

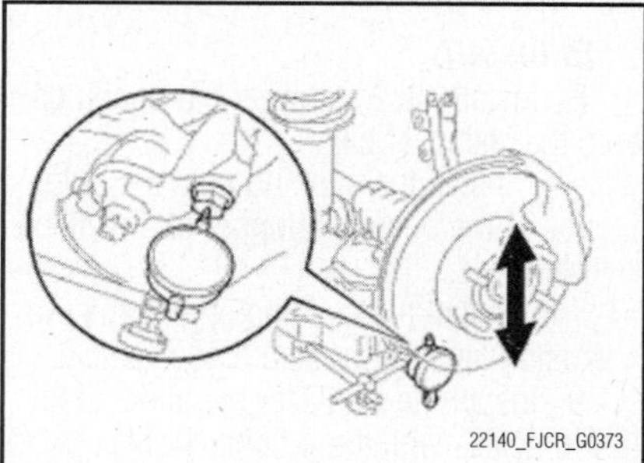

Fig. 173 Using a dial indicator, check the lower ball joint for excessive play when you push the hub nuts up and down with a force of 66 lbs. (294 N). Maximum play is 0.020 in. (0.5 mm)

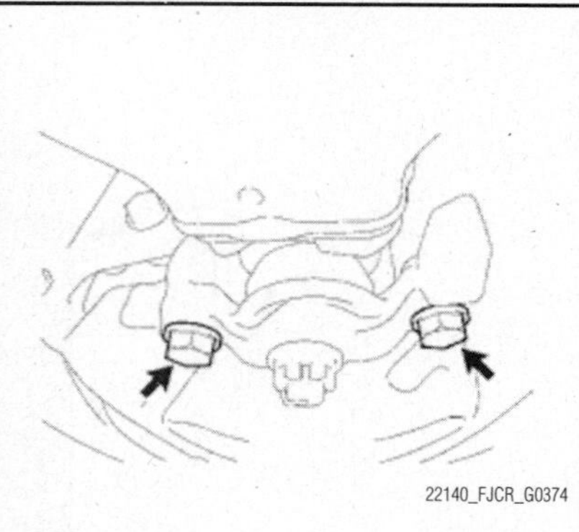

Fig. 174 Lower ball joint attachment to steering knuckle mounting bolts

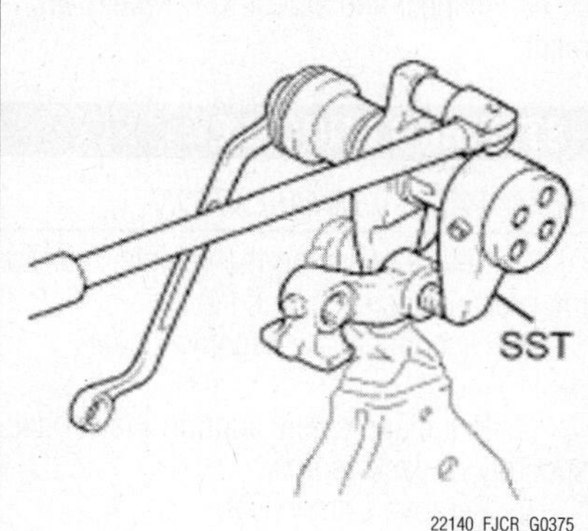

Fig. 175 Using SST: 09628-00011, remove the lower ball joint attachment

To install:

6. Temporarily tighten front lower suspension arm.
   a. Align the matchmarks on the No. 2 camber adjust cam and toe adjust cam.
   b. Temporarily tighten the bolt and the nut.
   c. Install the lower ball joint attachment, a new nut and a new cotter pin. Tighten to 103 ft. lbs. (140 Nm).
   d. Install the lower ball joint attachment with the 2 bolts. Tighten to 118 ft. lbs. (160 Nm).
7. Temporarily tighten front shock absorber with coil spring.
   a. Install the front shock absorber with coil spring, bolt and washer, and temporarily tighten the nut.
8. Install front wheel and lower the vehicle.
9. Bounce the vehicle up and down several times to stabilize the suspension.
10. Fully tighten front lower suspension arm to 100 ft. lbs. (135 Nm).
11. Fully tighten front shock absorber with coil spring to 100 ft. lbs. (135 Nm).
12. Inspect and adjust front wheel alignment.

SHOCK ABSORBERS

REMOVAL & INSTALLATION

See Figures 176 through 178.

1. Before servicing the vehicle, refer to the precautions section.
2. Raise and safely support the vehicle securely on jackstands.
3. Remove front wheels.
4. Remove engine under cover sub-assembly.

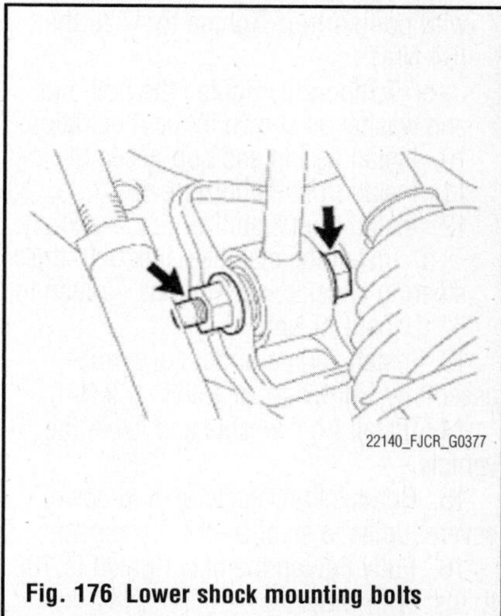

Fig. 176 Lower shock mounting bolts

Fig. 177 Upper shock mounting nuts

5. Remove the nut and separate the stabilizer link from the steering knuckle.

➡If the ball joint turns together with the nut, use a 6 mm hexagon wrench to hold the stud.

6. Remove front stabilizer bar.
7. Separate tie rod end sub-assembly.
8. Remove front shock absorber with coil spring.
   a. Remove the bolt, nut and washer.
   b. Remove the 3 nuts on the upper side of the front shock absorber with coil spring.
   c. Remove the front shock absorber with coil spring.

To install:

9. Temporarily tighten front shock absorber with coil spring.
   a. Install the coil spring onto the body with the lower end of the coil spring facing the rear side of the vehicle.
   b. Install the 3 nuts onto the upper side of the front shock absorber with coil spring. Tighten to 47 ft. lbs. (64 Nm).
   c. Temporarily tighten the bolt, nut and washer as shown in the illustration.
10. Install tie rod end sub-assembly.
11. Install front stabilizer bar.
12. Install front stabilizer link assembly.
   a. Install the stabilizer link onto the steering knuckle with the nut. Tighten to 52 ft. lbs. (70 Nm).
13. Install engine under cover sub-assembly. Tighten to 21 ft. lbs. (29 Nm).
14. Install front wheels and lower the vehicle.
15. Bounce the vehicle up and down several times to stabilize the suspension.
16. Fully tighten the nut. Tighten to 100 ft. lbs. (135 Nm).
17. Inspect and adjust front wheel alignment.

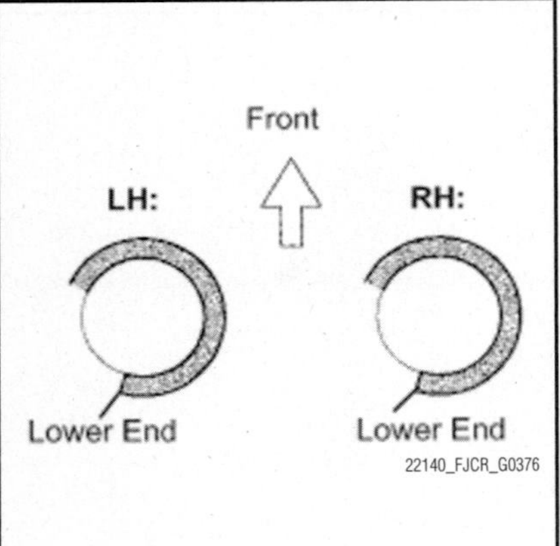

Fig. 178 Install the coil spring onto the body with the lower end of the coil spring facing the rear side of the vehicle

STEERING KNUCKLE

REMOVAL & INSTALLATION

1. Before servicing the vehicle, refer to the precautions section.
2. Disconnect the negative battery cable.
3. Raise and safely support the vehicle securely on jackstands.
4. Remove front wheel.
5. Remove front disc brake caliper assembly and suspend it out of the way.
6. Remove front rotor.
7. Separate front speed sensor.
8. On 4WD vehicles, remove front axle hub grease cap.
9. On 4WD vehicles, remove front axle hub nut.
10. Separate front stabilizer link assembly.
11. Separate tie rod end sub-assembly.
12. Separate front lower ball joint attachment.
13. Separate front upper suspension arm.
   a. Support the lower arm with a jack.
   b. Remove the clip and nut.
   c. Using SST: 09628-62011, separate the upper ball joint from the steering knuckle.
14. Remove front steering knuckle.
   a. On 4WD vehicles, use a plastic hammer, separate the front axle hub from the front drive shaft.
   b. Remove the front steering knuckle.

To install:

15. Install front steering knuckle.
16. Install front upper suspension arm.
   a. Install a new nut and a new clip. Tighten to 81 ft. lbs. (110 Nm).
17. Install front lower ball joint attachment.
18. Install tie rod end sub-assembly.
19. Install front stabilizer link assembly.
20. On 4WD vehicles, install front axle hub nut.
21. Inspect front axle hub bearing.
22. On 4WD vehicles, install front axle hub grease cap.
23. Install front speed sensor.
24. Install front disc.
25. Install front disc brake caliper assembly.
   a. Install the front disc brake caliper assembly with the 2 bolts. Tighten to 91 ft. lbs. (123 Nm).
26. Connect the negative battery cable.
27. Check fluid level in reservoir.
28. Check VSC sensor signal.
29. Inspect and adjust front wheel alignment.

STABILIZER BAR & LINKS

REMOVAL & INSTALLATION

See Figures 179 through 181.

1. Before servicing the vehicle, refer to the precautions section.
2. Raise and safely support the vehicle securely on jackstands.
3. Remove front wheels.
4. Remove engine under cover sub-assembly.
5. Remove the 2 nuts and stabilizer link.

➡If the ball joint turns together with the nut, use a 6 mm hexagon wrench to hold the stud.

6. Remove the 4 bolts and 2 stabilizer brackets and remove the stabilizer bar.
7. Remove the 2 stabilizer bar bushings from the stabilizer bar.

To install:

8. Install the 2 stabilizer bar bushings onto the stabilizer bar.
   a. Install the bushing onto the outer side of the bushing stopper on the stabilizer bar.
   b. Install the bushing so that the protrusion faces inner side of the vehicle.
9. Install the stabilizer bar and 2 stabilizer brackets with the 4 bolts. Tighten to 30 ft. lbs. (40 Nm).
10. Install the stabilizer link with the 2 nuts. Tighten to 52 ft. lbs. (70 Nm).
11. Install engine under cover sub-assembly. Tighten to 21 ft. lbs. (29 Nm).
12. Install front wheels.
13. Lower the vehicle.

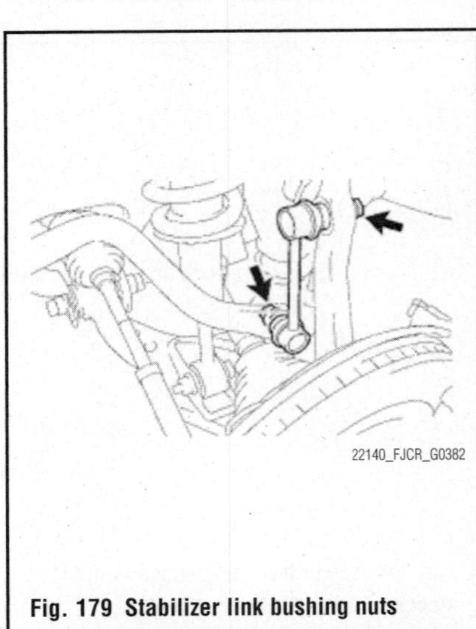

Fig. 179 Stabilizer link bushing nuts

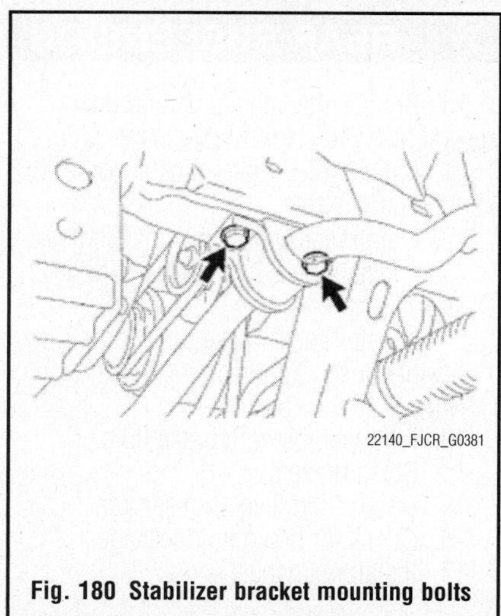

Fig. 180 Stabilizer bracket mounting bolts

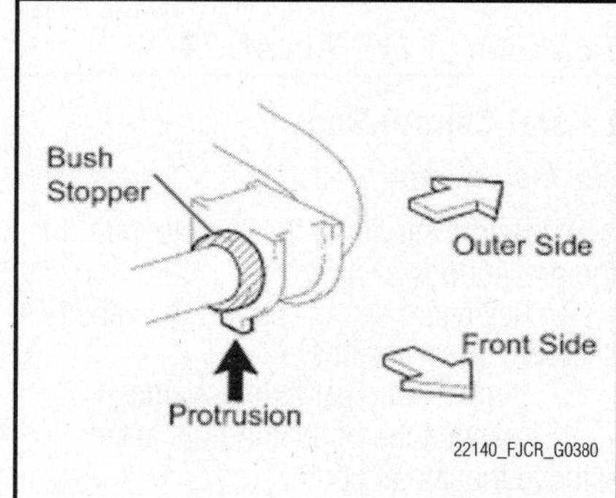

Fig. 181 Install the bushing onto the outer side of the bushing stopper and so the protrusion faces inner side of the vehicle

UPPER BALL JOINT

REMOVAL & INSTALLATION

The upper ball joint is an integral part of the lower control arm and is not serviced separately. See Front Suspension, Upper Control Arm.

UPPER CONTROL ARM

REMOVAL & INSTALLATION

See Figure 182, 183.

1. Remove front wheel.
2. Check that there is no slack on the ball joint by shaking the upper arm up and down by hand.

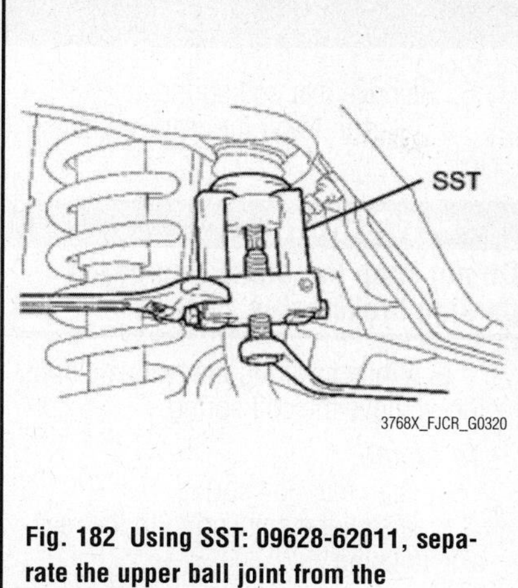

Fig. 182 Using SST: 09628-62011, separate the upper ball joint from the steering knuckle

3. Remove the 2 bolts and separate the skid control sensor wire.
4. Remove front upper suspension arm:
   a. Support the lower arm with a jack.
   b. Remove the clip and nut.
   c. Using SST: 09628-62011, separate the upper ball joint from the steering knuckle.
   d. Remove the bolt and separate the bracket.
   e. Remove the bolt, 2 washers and nut.
   f. Remove the upper arm.

To install:

5. Install the upper arm and temporarily tighten the bolt, 2 washers and nut.
6. Install the bracket with the bolt.
7. Install a new nut and a new clip. Tighten the nut to 81 ft. lbs. (110 Nm).
8. Install the skid control sensor wire with the 2 bolts.
9. Install front wheel.
10. Stabilize the suspension:
    a. Jack down the vehicle.
    b. Bounce the vehicle up and down several times to stabilize the suspension.
11. Fully tighten front upper suspension arm nut to 85 ft. lbs. (115 Nm).
12. Inspect and adjust front wheel alignment.

WHEEL HUB & BEARING

REMOVAL & INSTALLATION

See Figure 184.

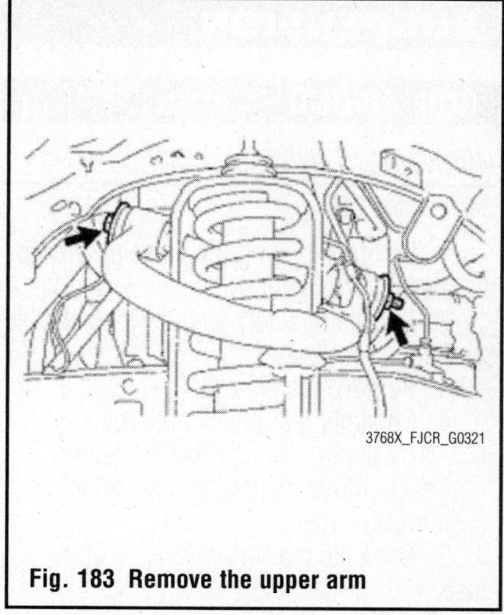

Fig. 183 Remove the upper arm

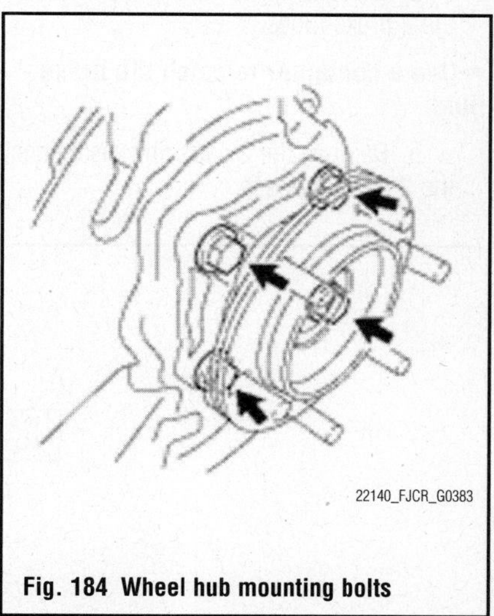

Fig. 184 Wheel hub mounting bolts

1. Before servicing the vehicle, refer to the precautions section.
2. Remove the front steering knuckle.
3. Remove the 4 bolts, wheel hub and dust cover from the steering knuckle.
4. Remove the O-ring from the wheel hub.

To install:

5. Apply MP grease to a new O-ring.
6. Install the new O-ring onto the axle hub.
7. Install the dust cover and axle hub onto the steering knuckle with the 4 bolts. Tighten to 59 ft. lbs. (80 Nm).
8. Install the front steering knuckle.

COIL SPRING

REMOVAL & INSTALLATION

See Figure 185.

1. Before servicing the vehicle, refer to the precautions section.
2. Raise and safely support the vehicle securely on jackstands.
3. Remove rear wheel.
4. Separate rear shock absorber.
   a. Support the rear axle housing.
   b. Remove the bolt and separate the shock absorber.
5. Separate rear brake tube flexible hose.
   a. Using a union nut wrench, separate the 2 brake tubes.

➡Use a container to catch the brake fluid.

   b. Remove the 2 clips and disconnect the 2 flexible hoses.

6. Remove rear coil spring.
   a. Start to lower the rear axle housing.

⁂ WARNING

Do not snap the brake line or the parking brake cable.

   b. While lowering the rear axle housing, remove the coil spring.

To install:

7. Install rear coil spring.
   a. Install the coil spring to the rear axle housing.

➡Fit the lower end of the coil spring into the gap of the spring lower seat.

8. Temporarily tighten rear shock absorber.
   a. Install the shock absorber and temporarily tighten the bolt.
9. Lower the vehicle.
10. Bounce the vehicle up and down several times to stabilize the suspension.
11. Fully tighten rear shock absorber to 72 ft. lbs. (98 Nm).
12. Install rear brake tube flexible hose.
   a. Install the 2 flexible hoses with 2 new clips.
   b. Install the 2 brake tubes onto the flexible hose. Tighten to 11 ft. lbs. (15 Nm).
13. Fill reservoir with brake fluid.
14. Bleed brake line.
15. Check fluid level in reservoir.
16. Check for brake fluid leakage.
17. Install rear wheel.

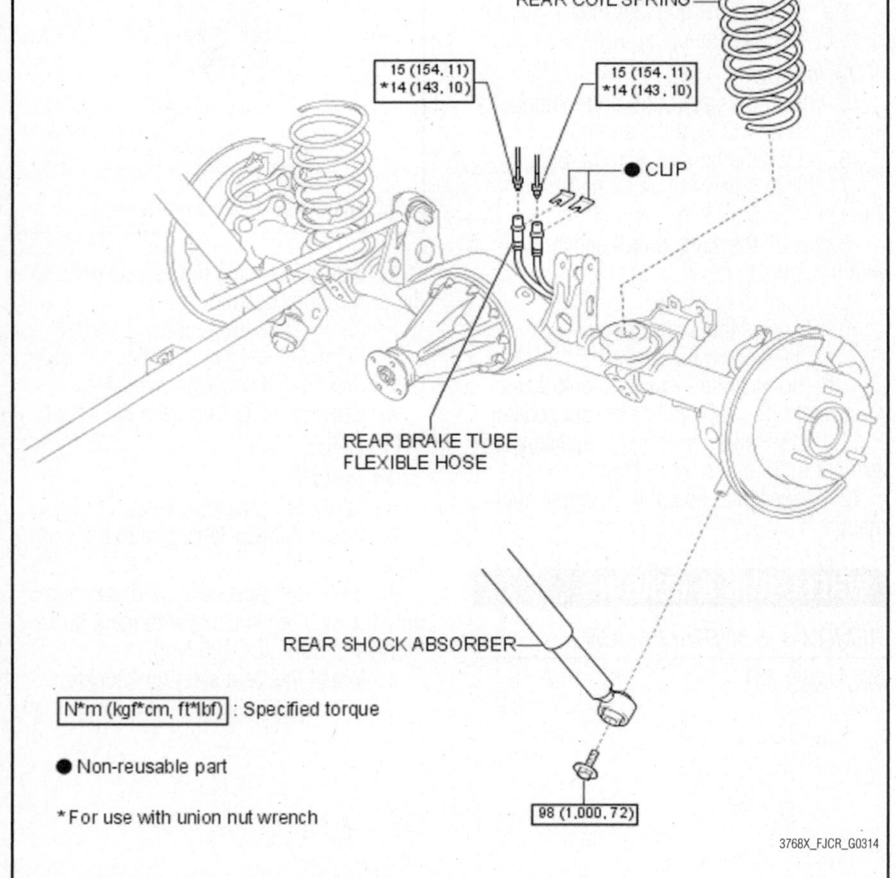

Fig. 185 Rear coil spring component locations

CONTROL ARMS/LINKS

REMOVAL & INSTALLATION

Lateral Control Rod

See Figure 186.

1. Before servicing the vehicle, refer to the precautions section.
2. Raise and safely support the vehicle securely on jackstands.
3. Support the rear axle housing.
4. Remove the bolts and nuts and remove the lateral control rod.

➡While holding the nut, turn and remove the bolt.

To install:

5. Install the lateral control rod and temporarily tighten the bolt.

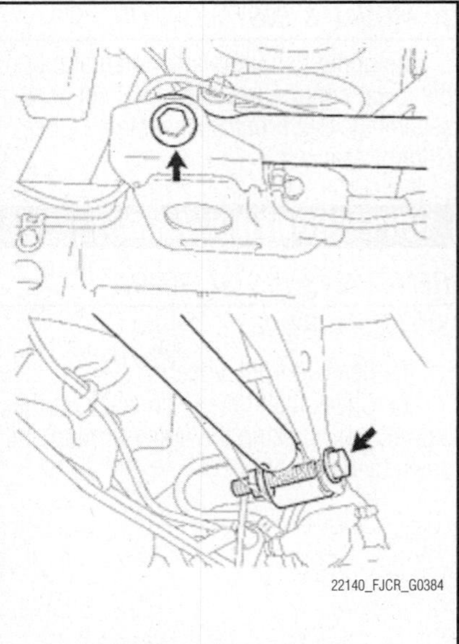

Fig. 186 Lateral control rod mounting bolts

6. Temporarily tighten the bolt and nut.
7. Lower the vehicle.
8. Bounce the vehicle up and down several times to stabilize the suspension.
9. While holding the nut fully tighten the 2 bolts to 96 ft. lbs. (130 Nm).

Lower Control Arm

See Figure 187.

1. Before servicing the vehicle, refer to the precautions section.
2. Raise and safely support the vehicle securely on jackstands.
3. Remove the bolt and separate the parking brake cable.
4. Support the rear axle housing.
5. Remove the bolts and nuts and remove the lateral control rod.

➡While holding the nut, turn and remove the bolt.

To install:

6. Install the lateral control rod and temporarily tighten the bolt.
7. Install the parking brake cable onto the lower control arm. Tighten to 9 ft. lbs. (13 Nm).
8. Lower the vehicle.
9. Bounce the vehicle up and down several times to stabilize the suspension.
10. While holding the nut fully tighten the 2 bolts to 96 ft. lbs. (130 Nm).

SHOCK ABSORBER

REMOVAL & INSTALLATION

See Figures 188 and 189.

1. Before servicing the vehicle, refer to the precautions section.
2. Raise and safely support the vehicle securely on jackstands.

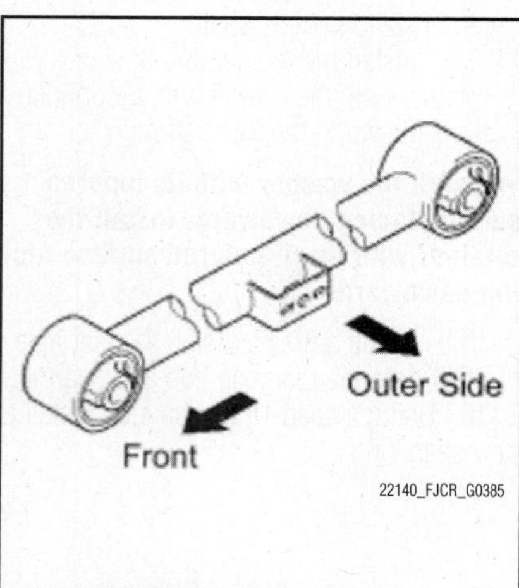

Fig. 187 Lateral control rod mounting bolts

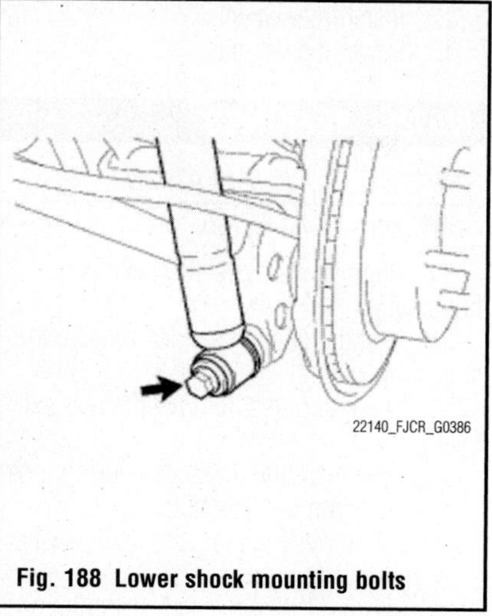

Fig. 188 Lower shock mounting bolts

Fig. 189 Upper shock mounting nuts

3. Remove rear wheels.
4. Support the rear axle housing.
5. Remove the bolt and separate the bottom of the shock absorber.
6. Remove the nut while keeping the piston rod from rotating.
7. Remove the 3 cushion retainers, the No. 1 cushion, the No. 2 cushion.
8. Remove the rear shock absorber.

To install:

9. Install the rear shock absorber.
10. Install the 3 cushion retainers, the No. 1 cushion, the No. 2 cushion.
11. Temporarily tighten rear shock absorber with coil spring.
12. While holding the piston rod, fully tighten a new nut to 18 ft. lbs. (25 Nm).
13. Install rear wheels and lower the vehicle.
14. Bounce the vehicle up and down several times to stabilize the suspension.
15. Fully tighten the nut. Tighten to 72 ft. lbs. (98 Nm).

STABILIZER BAR & LINKS

REMOVAL & INSTALLATION

See Figures 190 through 193.

1. Before servicing the vehicle, refer to the precautions section.
2. Raise and safely support the vehicle securely on jackstands.
3. Remove rear wheels.
4. Remove the nut and separate the stabilizer link from the stabilizer bar.

➡If the ball joint turns together with the nut, use a 6 mm hexagon wrench to hold the stud.

5. While holding the stabilizer link with a wrench, remove the nut.
6. Remove the 2 No. 1 retainers, 2 cushions, No. 2 retainer and stabilizer link from the chassis.
7. Remove the 4 bolts and stabilizer brackets and remove the stabilizer bar.

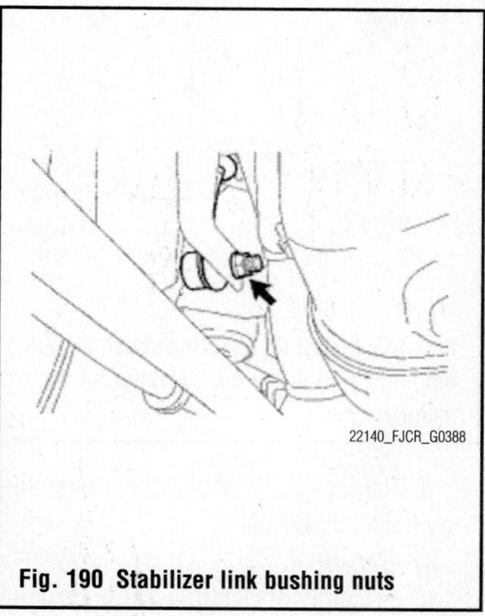

Fig. 190 Stabilizer link bushing nuts

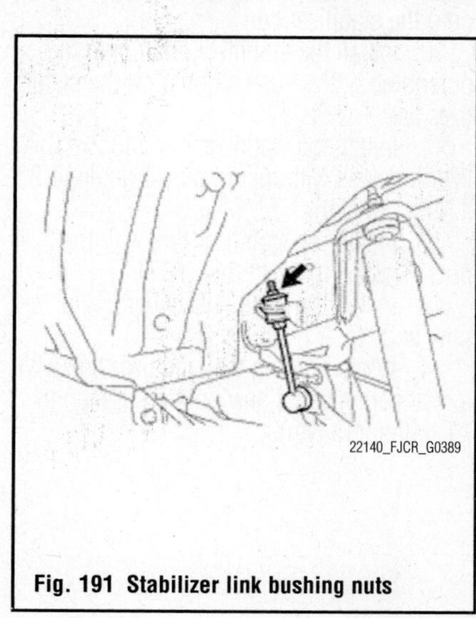

Fig. 191 Stabilizer link bushing nuts

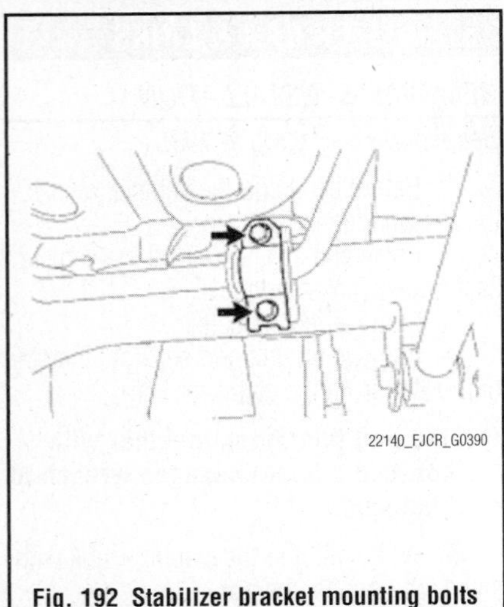

Fig. 192 Stabilizer bracket mounting bolts

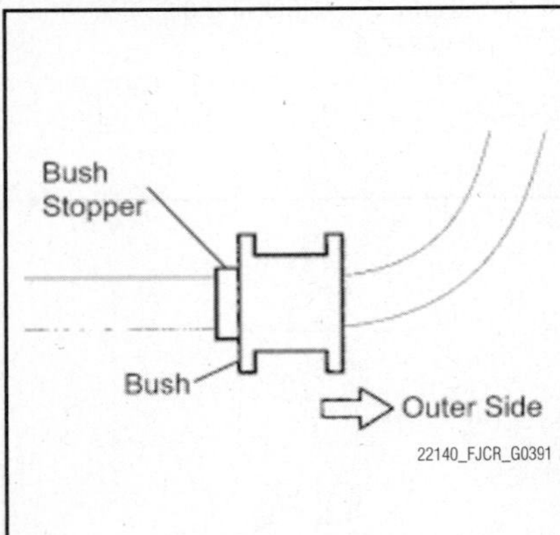

Fig. 193 Install the stabilizer bush onto the outer side of the bush stopper on the stabilizer bar

8. Remove the 2 stabilizer bar bushings from the stabilizer bar.

To install:

9. Install the 2 stabilizer bar bushings onto the stabilizer bar.
10. Install the stabilizer bush onto the outer side of the bush stopper on the stabilizer bar.
11. Install the stabilizer bar and 2 stabilizer brackets with the 4 bolts. Tighten to 22 ft. lbs. (30 Nm).
12. Install the stabilizer link with the nuts. Tighten to 52 ft. lbs. (70 Nm).
13. Install the 2 No. 1 retainers, 2 cushions and No. 2 retainer.
14. While holding the stabilizer link with a spanner, install a new nut and tighten to 11 ft. lbs. (15 Nm).
15. Install rear wheels.
16. Lower the vehicle.

WHEEL BEARINGS

REMOVAL & INSTALLATION

See Figures 194 and 195.

1. Remove rear axle shaft from the housing.
2. Using a snap expander, remove the snap-ring.
3. Using a press, remove the rear axle shaft.
    a. Remove the rear axle bearing inner retainer from the axle hub.
    b. Remove the rear axle shaft washer from the axle hub.
    c. Grind the rear axle bearing inner race surface using a grinder, then remove it with a chisel.
    d. Remove the rear axle shaft oil seal from the rear axle shaft.
4. Remove rear axle hub and bearing assembly.
    a. Attach the 4 nuts to the housing bolts.
    b. Using a hammer, remove the 4 housing bolts and rear axle hub and bearing assembly.

WARNING

Do not reuse the nuts previously removed from the vehicle.

5. Remove brake drum oil deflector.

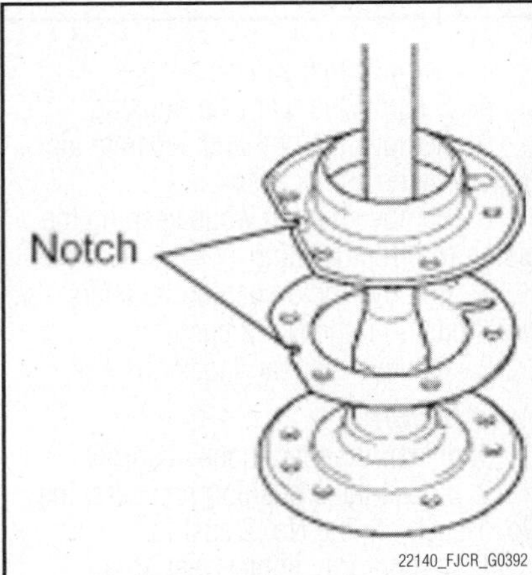

Fig. 194 Install a new deflector gasket and deflector onto the rear axle shaft and align the 2 notches

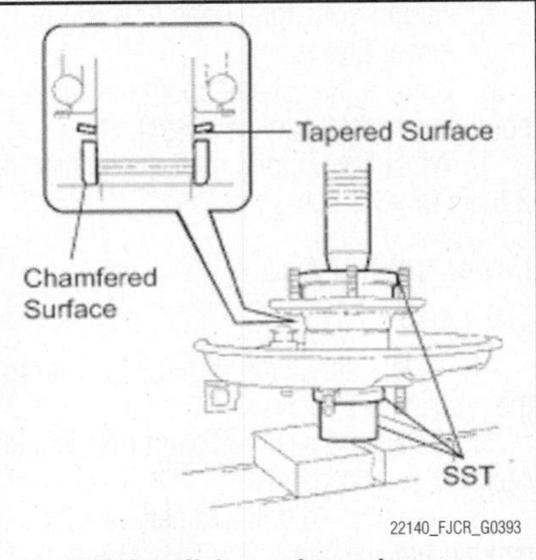

Fig. 195 Install the washer and a new retainer onto the axle hub in the orientations shown

6. Remove the deflector and deflector gasket to the rear axle shaft.

To install:

7. Install a new deflector gasket and deflector onto the rear axle shaft.
    a. Align the 2 notches.
    b. Install the washer and nut onto a new hub bolt, as shown in the illustration.
    c. Install the hub bolt by tightening the nut.
8. Install rear axle hub and bearing assembly.
    a. Position the parking brake plate on a new rear axle hub and bearing assembly and install the 4 housing bolts using 2 socket wrenches and a press.

WARNING

The left and right side bearing assemblies have different part numbers and are not interchangeable side to side.

9. Install rear axle shaft.
    a. Install the washer and a new retainer onto the axle hub in the orientations shown in the illustration.

➡Install the washer with its tapered surface facing downward. Install the retainer with its chamfered surface facing downward.

    b. Using a press, install the rear axle shaft onto the rear axle hub and bearing.
10. Using a snap-ring expander, install a new snap-ring.

TOYOTA

8

Highlander • Highlander Hybrid

SPECIFICATIONS AND MAINTENANCE CHARTS

ENGINE AND VEHICLE IDENTIFICATION

	Engine					
Code ①	Liters (cc)	Cu. In.	Cyl.	Fuel Sys.	Engine Type	Eng. Mfg.
1AR-FE	2.7 (2672)	163	4	SFI	DOHC	Toyota
3MZ-FE	3.3 (3311)	202.1	6	SFI	DOHC	Toyota
2GR-FE	3.5 (3456)	210	6	SFI	DOHC	Toyota

Model Year	
Code ②	Year
B	2011
C	2012

SFI: Sequential Fuel Injection

DOHC: Double Overhead Camshaft

① 10th digit of the VIN

② 10th digit of the Vehicle Identification Number (VIN)

71099_HIGH_C0001

GENERAL ENGINE SPECIFICATIONS

Year	Model	Engine Displacement Liters	Engine Series ID	Net Horsepower @ rpm	Net Torque @ rpm (ft. lbs.)	Bore x Stroke (in.)	Com-pression Ratio	Oil Pressure @ rpm
2011	Highlander	2.7	1AR-FE	187@5800	186@4100	3.54x4.13	10:01	38@4000
		3.5	2GR-FE	270@6200	248@4700	3.70x3.27	10.8:1	55@3000
	Highlander Hybrid	3.3	3MZ-FE	268@5600	212@4400	3.62x3.27	10.8:1	36-78@3000
2012	Highlander	2.7	1AR-FE	187@5800	186@4100	3.54x4.13	10:01	38@4000
		3.5	2GR-FE	270@6200	248@4700	3.70x3.27	10.8:1	55@3000
	Highlander Hybrid	3.3	3MZ-FE	268@5600	212@4400	3.62x3.27	10.8:1	36-78@3000

71099_HIGH_C0002

ENGINE TUNE-UP SPECIFICATIONS

Year	Engine Displacement Liters	Engine ID	Spark Plug Gap (in.)	Ignition Timing (deg.) ①	Fuel Pump (psi)	Idle Speed (rpm)	Valve Clearance	
							Intake	Exhaust
2011	2.7	1AR-FE	0.043	8-12B	44-50	600-700	NA	NA
	3.3	3MZ-FE	0.039-0.043	8-12B	44-50	850-950	0.006-0.010	0.010-0.014
	3.5	2GR-FE	0.043	8-12B	44-50	600-700	NA	NA
2012	2.7	1AR-FE	0.043	8-12B	44-50	600-700	NA	NA
	3.3	3MZ-FE	0.039-0.043	8-12B	44-50	850-950	0.006-0.010	0.010-0.014
	3.5	2GR-FE	0.043	8-12B	44-50	600-700	NA	NA

NA: Not Available

NOTE: The Vehicle Emission Control Information label often reflects specification changes made during production. The label figures must be used if they differ from those in this chart.

B: Before top dead center

① With terminals TC and CG connected to DLC3 (ODB II connector)

71099_HIGH_C0003

CAPACITIES

Year	Model	Engine Displacement Liters	Engine ID	Engine Oil with Filter (qts.)	Auto Transaxle (pts.) *	Rear Transaxle (pts.)	Transfer Case (pts.)	Rear Drive Axle (pts.)	Fuel Tank (gal.)	Cooling System (qts.)
2011	Highlander	2.7	1AR-FE	4.6	3.7	NE	2.0	2.0	19.1	①
		3.5	2GR-FE	6.4	3.7	NE	2.0	2.0	19.1	①
	Highlander Hybrid	3.3	3MZ-FE	5.0	②	4.2	NE	NA	17.2	③
2012	Highlander	2.7	1AR-FE	4.6	3.7	NE	2.0	2.0	19.1	①
		3.5	2GR-FE	6.4	3.7	NE	2.0	2.0	19.1	①
	Highlander Hybrid	3.3	3MZ-FE	5.0	②	4.2	NE	NA	17.2	③

NA: Not Available

NE: Not Equipped

*After draining, add the following amounts, then, fill to the cold full line.

① Non-towing pkg. 7.3 qt. w/o rear air, 9.6 qts. w/ rear air

Towing pkg. 8 qt. w/o rear air, 10.4 qts. w/ rear air

② With towing package: 8.8 pts.

Without towing package: 8.2 pts.

③ With Rear Heater: 12.8 qts.

Without Rear Heater: 10.6 qts.

71099_HIGH_C0004

FLUID SPECIFICATIONS

Year	Model	Engine Displ. Liters	Engine Oil	Auto. Trans.	Drive Axle Rear ①	Transfer Case	Power Steering Fluid ②	Brake Master Cylinder	Cooling System
2011	Highlander	2.7	5W-20	ATF WS	80W-90	80W-90	NE	DOT 3	SLLC ③
		3.5	5W-30	ATF WS	80W-90	80W-90	NE	DOT 3	SLLC ③
	Highlander Hybrid	3.3	5W-30	ATF-WS	75W-90	NA	NE	DOT 3	SLLC ③
2012	Highlander	2.7	5W-20	ATF WS	80W-90	80W-90	NE	DOT 3	SLLC ③
		3.5	5W-30	ATF WS	80W-90	80W-90	NE	DOT 3	SLLC ③
	Highlander Hybrid	3.3	5W-30	ATF-WS	75W-90	NA	NE	DOT 3	SLLC ③

NE: Not Equipped

NA: Not Available

DOT: Department Of Transpotation

① Oil grade: Hypoid gear oil API GL-5

② The Highlander is not equipped with a power steering pump. It utilizes a power steering motor.

③ Toyota Super Long Life Coolant

71099_HIGH_C0005

VALVE SPECIFICATIONS

Year	Engine Displacement Liters	Engine ID	Seat Angle (deg.)	Face Angle (deg.)	Spring Test Pressure (lbs. @ in.)	Spring Installed Height (in.)	Stem-to-Guide Clearance (in.)		Stem Diameter (in.)	
							Intake	Exhaust	Intake	Exhaust
2011	2.7	1AR-FE	45	44.5	NA	NA	0.0010-0.0024	0.0012-0.0026	0.2154-0.2159	0.2152-0.2157
	3.3	3MZ-FE	45	40.5	41.9-46.3@ 1.331	1.331	0.0010-0.0024	0.0012-0.0026	0.2154-0.2159	0.2152 0.2157
	3.5	2GR-FE	45	44.5	NA	NA	0.0010-0.0024	0.0012-0.0026	0.2154-0.2159	0.2151-0.2157
2012	2.7	1AR-FE	45	44.5	NA	NA	0.0010-0.0024	0.0012-0.0026	0.2154-0.2159	0.2152-0.2157
	3.3	3MZ-FE	45	40.5	41.9-46.3@ 1.331	1.331	0.0010-0.0024	0.0012-0.0026	0.2154-0.2159	0.2152 0.2157
	3.5	2GR-FE	45	44.5	NA	NA	0.0010-0.0024	0.0012-0.0026	0.2154-0.2159	0.2151-0.2157

NA: Information not available

71099_HIGH_C0006

CAMSHAFT AND BEARING SPECIFICATIONS

All measurements are given in inches.

Year	Engine Displacement Liters	Engine VIN	Journal Diameter	Brg. Oil Clearance	Shaft End-play	Runout	Journal Bore	Lobe Lift	
								Intake	Exhaust
2011	2.7	1AR-FE	①	②	NA	0.0012	NA	1.73870-1.74429	1.73795-1.74354
	3.3	3MZ-FE	1.0614-1.0620	③	0.0016-0.0035	0.0024	NA	1.6981-1.7020	1.6933-1.6972
	3.5	2GR-FE	④	⑤	NA	0.0016	NA	1.7447-1.7487	1.7426-1.7465
2012	2.7	1AR-FE	①	②	NA	0.0012	NA	1.73870-1.74429	1.73795-1.74354
	3.3	3MZ-FE	1.0614-1.0620	③	0.0016-0.0035	0.0024	NA	1.6981-1.7020	1.6933-1.6972
	3.5	2GR-FE	④	⑤	NA	0.0016	NA	1.7447-1.7487	1.7426-1.7465

NA: Not Available

① No. 1 journal: 1.35626 to 1.53689 inches
Other journals: 0.90390 to 090453 inches

② Intake No. 1: 0.00137 to 0.00283 inches
Exhaust No. 1 journal: 0.00193 to 0.00339 inches
Other journals: 0.000984 to 0.00244 inches

③ Intake Journals 4 and 5: 0.0010 - 0.0022 in.
All Others: 0.0010 - 0.0024 in.

④ No. 1 journal: 1.4152 to 1.4157 inches
Other journals: 1.0220 to 1.0226 inches

⑤ No. 1 journal: 0.0016 to 0.0031 inches
Other journals: 0.00098 to 0.0024 inches

71099_HIGH_C0007

CRANKSHAFT AND CONNECTING ROD SPECIFICATIONS

All measurements are given in inches.

Year	Engine Displacement Liters	Engine ID	Crankshaft Main Brg. Journal Dia.	Main Brg. Oil Clearance	Shaft End-play	Thrust on No.	Connecting Rod Journal Diameter	Oil Clearance	Side Clearance
2011	2.7	1AR-FE	2.16531-2.16535	0.0006-0.0015	0.0016-0.0094	2	1.8894-1.8898	0.0011-0.0024	NA
	3.3	3MZ-FE	2.4011-2.4016	①	0.0016-0.0094	2	2.0863-2.0866	0.0015-0.0026	0.0059-0.0118
	3.5	2GR-FE	2.4011-2.4016	0.0010-0.0019	0.0016-0.0095	2	2.0863-2.0866	0.0018-0.0026	0.0059-0.0157
2012	2.7	1AR-FE	2.16531-2.16535	0.0006-0.0015	0.0016-0.0094	2	1.8894-1.8898	0.0011-0.0024	NA
	3.3	3MZ-FE	2.4011-2.4016	①	0.0016-0.0094	2	2.0863-2.0866	0.0015-0.0026	0.0059-0.0118
	3.5	2GR-FE	2.4011-2.4016	0.0010-0.0019	0.0016-0.0095	2	2.0863-2.0866	0.0018-0.0026	0.0059-0.0157

NA: Not Available

71099_HIGH_C0008

PISTON AND RING SPECIFICATIONS

All measurements are given in inches.

Year	Engine Displ. Liters	Engine ID	Piston Clearance	Ring Gap Top Comp.	Ring Gap Bottom Comp.	Ring Gap Oil Control	Ring Side Clearance Top Comp.	Ring Side Clearance Bottom Comp.	Ring Side Clearance Oil Control
2011	2.7	1AR-FE	0.0003-0.0013	0.0087-0.0106	0.0146-0.0165	0.0039-0.0079	0.0008-0.0028	0.0008-0.0024	0.000787-0.00276
	3.3	3MZ-FE	0.0013-0.0023	0.0118-0.0157	0.0197-0.0236	0.0059-0.0157	0.0012-0.0031	0.0008-0.0024	0.0012-0.0043
	3.5	2GR-FE	0.0018-0.0020	0.0098-0.0138	0.0197-0.0236	0.0039-0.0157	0.0008-0.0028	0.0008-0.0024	0.0028-0.0059
2012	2.7	1AR-FE	0.0003-0.0013	0.0087-0.0106	0.0146-0.0165	0.0039-0.0079	0.0008-0.0028	0.0008-0.0024	0.000787-0.00276
	3.3	3MZ-FE	0.0013-0.0023	0.0118-0.0157	0.0197-0.0236	0.0059-0.0157	0.0012-0.0031	0.0008-0.0024	0.0012-0.0043
	3.5	2GR-FE	0.0018-0.0020	0.0098-0.0138	0.0197-0.0236	0.0039-0.0157	0.0008-0.0028	0.0008-0.0024	0.0028-0.0059

71099_HIGH_C0009

TORQUE SPECIFICATIONS

All readings in ft. lbs.

Year	Engine Displacement Liters	Engine ID	Cylinder Head Bolts	Main Bearing Bolts	Rod Bearing Bolts	Crankshaft Damper Bolts	Flywheel Bolts	Manifold Intake	Manifold Exhaust	Spark Plugs	Oil Pan Drain Plug
2011	2.7	1AR-FE	①	29	②	192	72	15	26	18	30
	3.3	3MZ-FE	③	④	⑤	162	61	11	36	18	33
	3.5	2GR-FE	⑥	⑦	⑧	184	61	15	15	18	33
2012	2.7	1AR-FE	①	29	②	192	72	15	26	18	30
	3.3	3MZ-FE	③	④	⑤	162	61	11	36	18	33
	3.5	2GR-FE	⑥	⑦	⑧	184	61	15	15	18	33

① Step 1: 27 ft. lbs.
Step 2: 27 ft. lbs.
Step 3: Plus 90 degrees
Step 4: Plus 90 degrees

② Step 1: 15 ft. lbs.
Step 2: 30 ft. lbs.
Step 3: Plus 90 degrees

③ Step 1: 12 point bolts to 40 ft. lbs.
Step 2: 12 point bolts plus 90 degrees
Step 3: Hex head recessed bolt to 14 ft. lbs.

④ Step 1: 12 point cap bolts to 16 ft. lbs.
Step 2: 12 point cap bolts plus 90 degrees
Step 3: Hex head side bolts to 20 ft. lbs.

⑤ Step 1: 18 ft. lbs.
Step 2: Plus 90 degrees

⑥ Step 1: 10mm bolts to 27 ft. lbs.
Step 2: 10mm point cap bolts plus 90 degrees
Step 3: 10mm point cap bolts plus 90 degrees
Step 4: Front bolts to 22 ft. lbs.

⑦ Step 1: 16 cap bolts to 45 ft. lbs.
Step 2: 16 cap bolts plus 90 degrees
Step 3: 8 side bolts to 38 ft. lbs.

⑧ Step 1: 18 ft. lbs.
Step 2: Plus 90 degrees

71099_HIGH_C0010

WHEEL ALIGNMENT

Year	Model		Caster Range (+/-Deg.)	Caster Preferred Setting (Deg.)	Camber Range (+/-Deg.)	Camber Preferred Setting (Deg.)	Toe-in (in.)	Steering Axis Inclination (Deg.)
2011	Highlander	Front	0.75	+2.62	0.75	-0.63	0+/-0.08	11.02+/-0.75
		2WD R	NA	NA	0.75	-1.00	0.12+/-0.08	NA
		4WD R	NA	NA	0.75	-0.60	0.12+/-0.08	NA
	Highlander Hybrid	2WD F	0.75	+2.75	0.75	-0.58	0+/-0.08	NA
		4WD F	0.75	+2.50	0.75	-0.58	0+/-0.08	NA
		2WD R	NA	NA	0.75	-1.17	0.12+/-0.08	NA
		4WD R	NA	NA	0.75	-0.67	0.12+/-0.08	NA
2012	Highlander	Front	0.75	+2.62	0.75	-0.63	0+/-0.08	11.02+/-0.75
		2WD R	NA	NA	0.75	-1.00	0.12+/-0.08	NA
		4WD R	NA	NA	0.75	-0.60	0.12+/-0.08	NA
	Highlander Hybrid	2WD F	0.75	+2.75	0.75	-0.58	0+/-0.08	NA
		4WD F	0.75	+2.50	0.75	-0.58	0+/-0.08	NA
		2WD R	NA	NA	0.75	-1.17	0.12+/-0.08	NA
		4WD R	NA	NA	0.75	-0.67	0.12+/-0.08	NA

NA: Not Available

71099_HIGH_C0011

TIRE, WHEEL AND BALL JOINT SPECIFICATIONS

Year	Model	OEM Tires		Tire Pressures (psi)		Wheel	Ball Joint	Lug Nut Torque
		Standard	Optional	Front	Rear	Size	Inspection	(ft. lbs.)
2011	Highlander	P245/65R17	P245/55R19	30	30	7.5-J	①	76
	Highlander Hybrid	P225/65R17	NA	32	32	6.5-J	①	76
2012	Highlander	P245/65R17	P245/55R19	30	30	7.5-J	①	76
	Highlander Hybrid	P225/65R17	NA	32	32	6.5-J	①	76

OEM: Original Equipment Manufacturer

PSI: Pounds Per Square Inch

① Replace if any measurable movement is found.

71099_HIGH_C0012

BRAKE SPECIFICATIONS

All measurements in inches unless noted

Year	Model		Brake Disc Original Thickness	Brake Disc Minimum Thickness	Brake Disc Maximum Runout	Minimum Lining Thickness	Brake Caliper Bracket Bolts (ft. lbs.)	Brake Caliper Mounting Bolts (ft. lbs.)
2011	Highlander	Front	1.102	1.024	0.0020	0.039	76	25
		Rear	0.394	0.335	0.0059	0.039	57	25
	Highlander Hybrid	Front	1.102	1.024	0.0020	0.039	78	25
		Rear	0.394	0.335	0.0059	0.039	56	32
2012	Highlander	Front	1.102	1.024	0.0020	0.039	76	25
		Rear	0.394	0.335	0.0059	0.039	56	25
	Highlander Hybrid	Front	1.102	1.024	0.0020	0.039	78	25
		Rear	0.394	0.335	0.0059	0.039	56	32

71099_HIGH_C0013

SCHEDULED MAINTENANCE INTERVALS
TOYOTA HIGHLANDER & HIGHLANDER HYBRID

TO BE SERVICED	TYPE OF SERVICE	VEHICLE MILEAGE INTERVAL (x1000)												
		7.5	15	22.5	30	37.5	45	52.5	60	67.5	75	82.5	90	97.5
Engine oil & filter	R	✓	✓	✓	✓	✓	✓	✓	✓	✓	✓	✓	✓	✓
Automatic transmission fluid	S/I		✓		✓		✓		✓		✓		✓	
Ball joints & dust covers	S/I		✓		✓		✓		✓		✓		✓	
Bolts & nuts on chassis & body	S/I		✓		✓		✓		✓		✓		✓	
Brake linings & drums	S/I		✓		✓		✓		✓		✓		✓	
Brake line pipes & hoses	S/I		✓		✓		✓		✓		✓		✓	
Brake pads & discs (front & rear)	S/I		✓		✓		✓		✓		✓		✓	
Propeller shaft grease	S/I		✓		✓		✓		✓		✓		✓	
Steering knuckle & chassis grease	S/I		✓		✓		✓		✓		✓		✓	
Steering linkage	S/I		✓		✓		✓		✓		✓		✓	
Air cleaner filter	R				✓				✓				✓	
Spark plugs ①	R				✓				✓				✓	
Drive belts	S/I				✓				✓				✓	
Exhaust pipes & mountings	S/I				✓				✓				✓	
Fuel lines & connections	S/I				✓				✓				✓	
Engine coolant	R						✓				✓			
Charcoal canister	R								✓					
Fuel tank cap gasket	R								✓					
Heated oxygen sensors (except Calif.) ②	R													

R: Replace S/I: Service or Inspect

① Platinum plugs are replaced at 100,000 mile intervals

② Heated oxygen sensors (except Calif.): replace every 80,000 miles.

FREQUENT OPERATION MAINTENANCE (SEVERE SERVICE)

If a vehicle is operated under any of the following conditions it is considered severe service:

- Extremely dusty areas.

- 50% or more of the vehicle operation is in 32°C (90°F) or higher temperatures, or constant temperatures below 0°C (32°F).

- Prolonged idling (vehicle operation in stop and go traffic).

- Frequent short running periods (engine does not warm to normal operating temperatures).

- Police, taxi, delivery usage or trailer towing usage.

Air cleaner filter: service or inspect every 3750 miles

Engine oil & filter: replace every 3750 miles.

Ball joints & dust covers: service or inspect every 7500 miles.

Bolts & nuts on chassis & body: service or inspect every 7500 miles.

Brake pads & discs (front & rear): service or inspect every 7500 miles.

Steering knuckle & chassis grease: service or inspect every 7500 miles.

Steering linkage: service or inspect every 7500 miles.

Exhaust pipes & mountings: service or inspect every 15,000 miles.

71099_HIGH_C0014

PRECAUTIONS

Before servicing any vehicle, please be sure to read all of the following precautions, which deal with personal safety, prevention of component damage, and important points to take into consideration when servicing a motor vehicle:

- Never open, service or drain the radiator or cooling system when the engine is hot; serious burns can occur from the steam and hot coolant.
- Observe all applicable safety precautions when working around fuel. Whenever servicing the fuel system, always work in a well-ventilated area. Do not allow fuel spray or vapors to come in contact with a spark, open flame, or excessive heat (a hot drop light, for example). Keep a dry chemical fire extinguisher near the work area. Always keep fuel in a container specifically designed for fuel storage; also, always properly seal fuel containers to avoid the possibility of fire or explosion. Refer to the additional fuel system precautions later in this section.
- Fuel injection systems often remain pressurized, even after the engine has been turned **OFF**. The fuel system pressure must be relieved before disconnecting any fuel lines. Failure to do so may result in fire and/or personal injury.
- Brake fluid often contains polyglycol ethers and polyglycols. Avoid contact with the eyes and wash your hands thoroughly after handling brake fluid. If you do get brake fluid in your eyes, flush your eyes with clean, running water for 15 minutes. If eye irritation persists, or if you have taken brake fluid internally, IMMEDIATELY seek medical assistance.
- The EPA warns that prolonged contact with used engine oil may cause a number of skin disorders, including cancer. You should make every effort to minimize your exposure to used engine oil. Protective gloves should be worn when changing oil. Wash your hands and any other exposed skin areas as soon as possible after exposure to used engine oil. Soap and water, or waterless hand cleaner should be used.
- All new vehicles are now equipped with an air bag system, often referred to as a Supplemental Restraint System (SRS) or Supplemental Inflatable Restraint (SIR) system. The system must be disabled before performing service on or around system components, steering column, instrument panel components, wiring and sensors. Failure to follow safety and disabling procedures could result in accidental air bag deployment, possible personal injury and unnecessary system repairs.
- Always wear safety goggles when working with, or around, the air bag system. When carrying a non-deployed air bag, be sure the bag and trim cover are pointed away from your body. When placing a non-deployed air bag on a work surface, always face the bag and trim cover upward, away from the surface. This will reduce the motion of the module if it is accidentally deployed. Refer to the additional air bag system precautions later in this section.
- Clean, high quality brake fluid from a sealed container is essential to the safe and proper operation of the brake system. You should always buy the correct type of brake fluid for your vehicle. If the brake fluid becomes contaminated, completely flush the system with new fluid. Never reuse any brake fluid. Any brake fluid that is removed from the system should be discarded. Also, do not allow any brake fluid to come in contact with a painted surface; it will damage the paint.
- Never operate the engine without the proper amount and type of engine oil; doing so WILL result in severe engine damage.
- Timing belt maintenance is extremely important. Many models utilize an interference-type, non-freewheeling engine. If the timing belt breaks, the valves in the cylinder head may strike the pistons, causing potentially serious (also time-consuming and expensive) engine damage. Refer to the maintenance interval charts for the recommended replacement interval for the timing belt, and to the timing belt section for belt replacement and inspection.
- Disconnecting the negative battery cable on some vehicles may interfere with the functions of the on-board computer system(s) and may require the computer to undergo a relearning process once the negative battery cable is reconnected.
- When servicing drum brakes, only disassemble and assemble one side at a time, leaving the remaining side intact for reference.
- Only an MVAC-trained, EPA-certified automotive technician should service the air conditioning system or its components.

BRAKES — ANTI-LOCK BRAKE SYSTEM (ABS)

GENERAL INFORMATION

PRECAUTIONS

- Certain components within the ABS system are not intended to be serviced or repaired individually.
- Do not use rubber hoses or other parts not specifically specified for and ABS system. When using repair kits, replace all parts included in the kit. Partial or incorrect repair may lead to functional problems and require the replacement of components.
- Lubricate rubber parts with clean, fresh brake fluid to ease assembly. Do not use shop air to clean parts; damage to rubber components may result.
- Use only DOT 3 brake fluid from an unopened container.
- If any hydraulic component or line is removed or replaced, it may be necessary to bleed the entire system.
- A clean repair area is essential. Always clean the reservoir and cap thoroughly before removing the cap. The slightest amount of dirt in the fluid may plug an orifice and impair the system function. Perform repairs after components have been thoroughly cleaned; use only denatured alcohol to clean components. Do not allow ABS components to come into contact with any substance containing mineral oil; this includes used shop rags.
- The Anti-Lock control unit is a microprocessor similar to other computer units in the vehicle. Ensure that the ignition switch is **OFF** before removing or installing controller harnesses. Avoid static electricity discharge at or near the controller.
- If any arc welding is to be done on the vehicle, the control unit should be unplugged before welding operations begin.

SPEED SENSORS

REMOVAL & INSTALLATION

Front

See Figures 1 through 3.

➡If the sensor rotor needs to be replaced, replace it together with the front drive outboard joint shaft assembly.

1. Disconnect the negative battery cable.
2. Raise and safely support the vehicle.
3. Remove the front wheel and tire assembly.

4. Remove No. 1 engine under cover. It is secured with 6 bolts and 2 clips.

5. Remove the front fender molding sub-assembly, as follows:

a. Remove the clip.

b. Using a 4 mm hexagon wrench, remove the screw.

c. Peel off the front fender side protector and disengage the 3 clips, and then remove the front fender molding sub-assembly.

d. Remove the pad from the front fender molding sub-assembly.

e. Remove the 2 No. 4 clips from the front fender molding sub-assembly.

f. Remove the front fender side protector from the front fender molding sub-assembly.

6. Remove the front wheel opening extension pad.

7. Remove the front fender liner, as follows:

a. Remove the screw.

b. Using a screwdriver, turn the pin 90 degrees and remove the pin hold clip.

c. Using a 4 mm hexagon wrench, remove the 2 screws.

d. Remove the 2 grommets.

➡The grommets need to be replaced with new ones because they will break when they are removed.

e. Remove the 5 clips.

f. For left hand side, remove 8 screws and front fender liner.

g. For right hand side, remove 7 screws and front fender liner.

8. Disconnect the front speed sensor connector and remove the 2 clamps.

9. Remove the bolt and No. 2 sensor clamp from the body.

10. Remove the bolt, No. 1 sensor clamp, and flexible hose together from the shock absorber assembly.

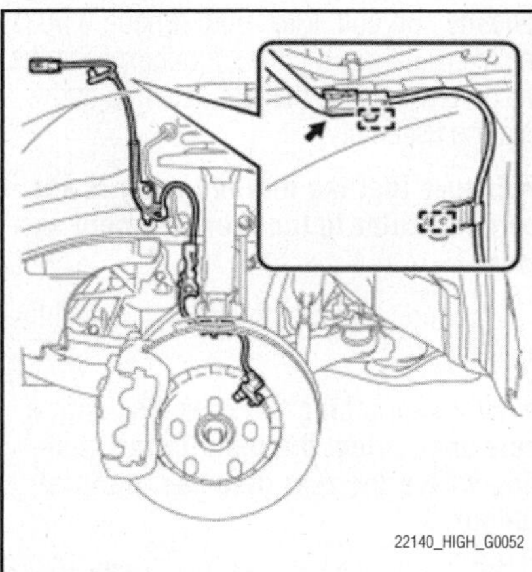

Fig. 1 Front speed sensor connector and clamps

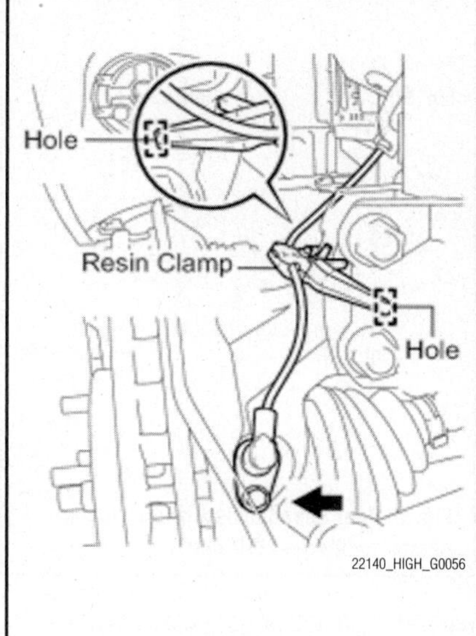

Fig. 2 Front speed sensor, bolt, and resin clamp removal

11. Remove the bolt, resin clamp, and the front speed sensor.

⁂ WARNING

Do not allow any foreign matter to come in contact with the tip of the sensor.

➡Clean the installation hole and the contact surface for the front speed sensor every time it is removed.

To install:

12. Install the resin clamp and the front speed sensor with bolt. Tighten to 71 inch lbs. (8 Nm).

⁂ WARNING

Do not allow any foreign matter to come in contact with the tip of the sensor.

➡Firmly insert the front speed sensor body into the knuckle before tightening the bolt.

➡After installing the front speed sensor to the knuckle, make sure that there is no clearance between the front speed sensor stay and knuckle. Also make sure that no foreign matter is stuck between the parts.

➡Before installing the clamp, firmly insert the points of the clamp into the installation holes.

13. Temporarily install the No. 1 sensor clamp.

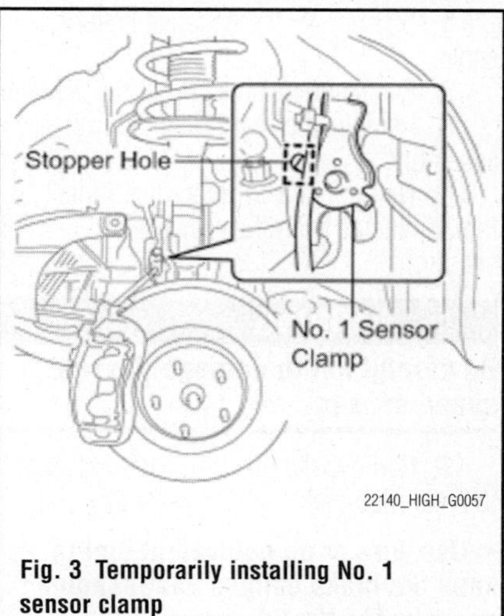

Fig. 3 Temporarily installing No. 1 sensor clamp

➡Be sure to insert the No. 1 sensor clamp claw into the stopper hole while installing the No. 1 sensor clamp.

14. Install the front flexible hose and the No. 1 sensor clamp together to the shock absorber with the bolt. Tighten to 14 ft. lbs. (19 Nm).

➡Do not twist the wire harness for the front speed sensor when installing it.

➡A bolt tightens the brake flexible hose and front speed sensor together. Make sure that the flexible hose is positioned over the front speed sensor.

15. Install the No. 2 sensor clamp to the body with the bolt. Tighten to 44 inch lbs. (5 Nm).

16. Install the 2 clamps and connect the front speed sensor connector.

17. Install the front fender liner.

18. Install the front wheel opening extension pad.

19. Install the front fender molding sub-assembly.

20. Install the No. 1 engine under cover.

21. Install the front wheel and tire assembly and tighten the lug nuts finger-tight.

22. Lower the vehicle, then final tighten the lug nuts to 76 ft. lbs. (103 Nm).

23. Connect the negative battery cable.

24. Check for speed sensor signal.

Rear–2WD Models

See Figures 4 through 6.

➡Use the same procedures for the Left side and Right side. The following procedure is for the Left side.

➡If the sensor rotor needs to be replaced, replace it together with the rear axle hub and bearing assembly with rear speed sensor.

1. Disconnect the negative battery cable.
2. Raise and safely support the vehicle.
3. Remove the rear tire and wheel assembly.
4. Using a screwdriver, disconnect the connector from the rear speed sensor.

⁂ WARNING

Be careful not to damage the rear speed sensor.

5. Remove the two bolts and separate the rear disc brake caliper assembly.

➡Use wire or an equivalent tool to keep the brake caliper from hanging down by the flexible hose.

6. Put matchmarks on the rear disc and the axle hub.
7. Release the parking brake and remove the rear disc.

➡If the disc cannot be removed easily, turn and press firmly the shoe adjuster until the wheel comes free.

8. Remove the 4 bolts and the rear axle hub and bearing assembly, along with the wheel speed sensor.

➡Use wire or an equivalent tool to keep the parking brake assembly from hanging down by the parking brake cable assembly.

9. Install the hub nuts and mount the rear axle hub and bearing assembly in a vise using aluminum plates.

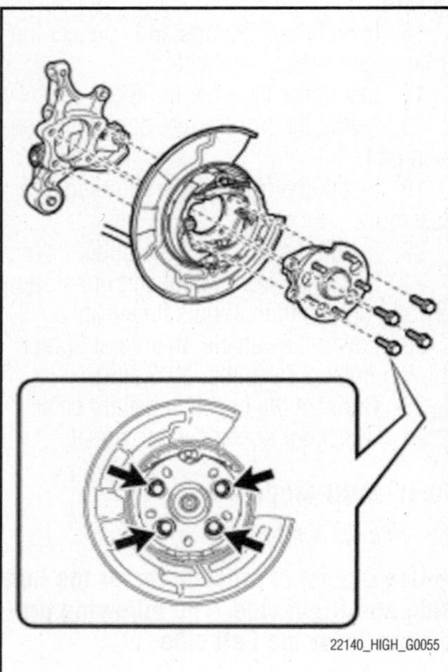

Fig. 4 Rear axle hub and bearing assembly—2WD models

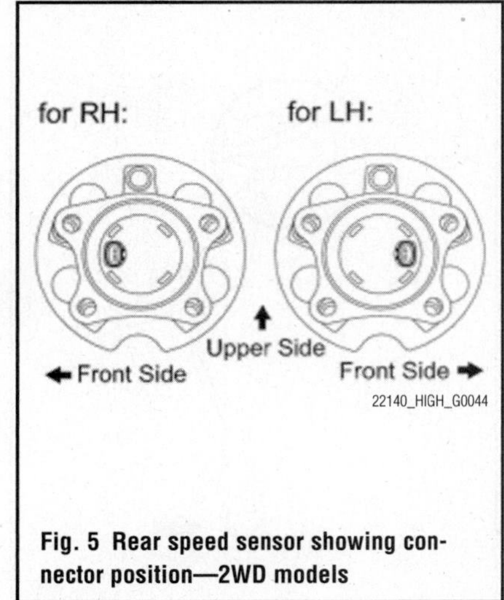

Fig. 5 Rear speed sensor showing connector position—2WD models

⁂ WARNING

Replace the rear axle hub and bearing assembly if it is dropped or receives a strong shock.

10. Using appropriate tool, remove the rear speed sensor from the rear axle hub and bearing assembly.

To install:

11. Clean the contact surface between the rear axle hub and bearing assembly and a new rear speed sensor.

➡Do not allow foreign matter to attach to the sensor rotor.

12. Place the rear speed sensor on the axle hub so that the connector is positioned as shown in the illustration.
13. Using appropriate tool, steel plates, V-blocks and press, install a new rear speed sensor to the rear axle hub and bearing assembly.

⁂ WARNING

Keep the rear speed sensor away from magnets.

⁂ WARNING

Do not use a hammer to install the rear speed sensor.

⁂ WARNING

Check that there is no foreign matter such as iron chips on the detecting portion of the rear speed sensor.

➡Slowly press the rear speed sensor in straight.

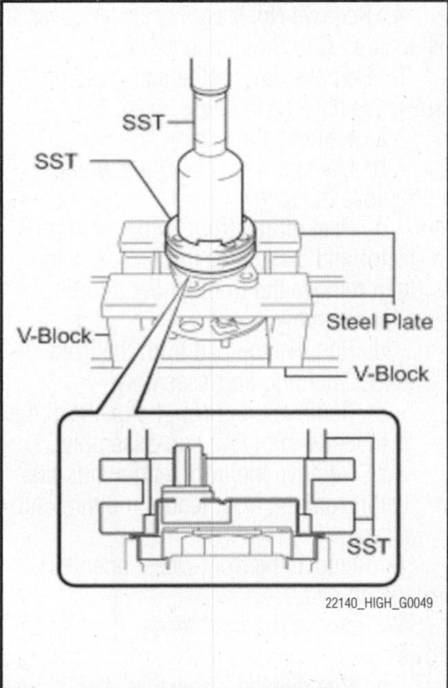

Fig. 6 Installing new rear speed sensor—2WD models

14. Install the parking brake assembly and the rear axle hub and bearing assembly with the 4 bolts, and tighten to 55 ft. lbs. (75 Nm).

➡Do not twist the No. 3 parking brake cable assembly when installing it.

15. Using a dial indicator, check for looseness near the center of the axle hub. Maximum looseness: 0 inches (0 mm). If the looseness exceeds the maximum, replace the rear axle hub and bearing assembly.

➡Ensure that the dial indicator is set perpendicular to the measurement surface.

16. Using a dial indicator, check for runout on the surface of the axle hub outside the hub bolt. Maximum runout: 0.0031 inches (0.08 mm). If the runout exceeds the maximum, replace the rear axle hub and bearing assembly.

➡Ensure that the dial indicator is set perpendicular to the measurement surface.

17. Align the matchmarks and install the rear disc.

➡When replacing the rear disc with a new one, select the installation position where the rear disc has minimal runout.

18. Install the rear disc brake caliper assembly with the 2 bolts, and tighten to 57 ft. lbs. (78 Nm).

19. Connect the connector to the rear speed sensor.
20. Install the rear wheel and tire assembly and tighten the lug nuts finger-tight.
21. Lower the vehicle, then final tighten the lug nuts to 76 ft. lbs. (103 Nm).
22. Connect the negative battery cable.
23. Inspect and adjust rear wheel alignment.
24. Check for speed sensor signal.

Rear–4WD Models

See Figures 7 through 9.

➡Use the same procedures for the Left side and Right side. The following procedure is for the Left side.

➡If the sensor rotor needs to be replaced, replace it together with the front drive outboard joint shaft assembly.

1. Disconnect the negative battery cable.
2. Remove the left rear door scuff plate.
3. Remove the left rear door opening trim weatherstrip.
4. Remove deck board assembly.
5. Remove the No. 3 deck board sub-assembly.
6. Remove the No. 2 deck board sub-assembly.
7. Remove the tonneau cover assembly, as applicable.
8. Remove the rear No. 1 floor board, as applicable.
9. Remove both rear seat side covers, as applicable.
10. Remove both deck side trim boxes.
11. Remove jack carrier support.
12. Remove jack carrier cushion.
13. Remove jack assembly.
14. Remove jack carrier assembly.
15. Remove deck floor board assembly, as applicable.
16. Remove rear mat.
17. Remove rear deck floor box, as applicable.
18. Remove the rear No. 2 seat inner belt assembly, as applicable.
19. Disconnect both rear seat lap type belt assemblies, as applicable.
20. Remove the rear No. 2 seat assembly, as applicable.
21. Remove rear floor finish plate.
22. Remove left deck side trim cover and trim.
23. Remove left side trim cover or power outlet socket bezel, as applicable.
24. Remove left rear combination light service cover.
25. Remove rear power point socket assembly.
26. Remove rear power outlet socket cover.
27. Remove rear deck trim cover or left reclining remote control bezel, as applicable.
28. Remove left rope hook assembly.
29. Remove No. 2 deck side trim hook.
30. Remove left front deck side trim cover.
31. Remove left rear No. 1 seat outer belt assembly.
32. Remove left deck trim side panel assembly.

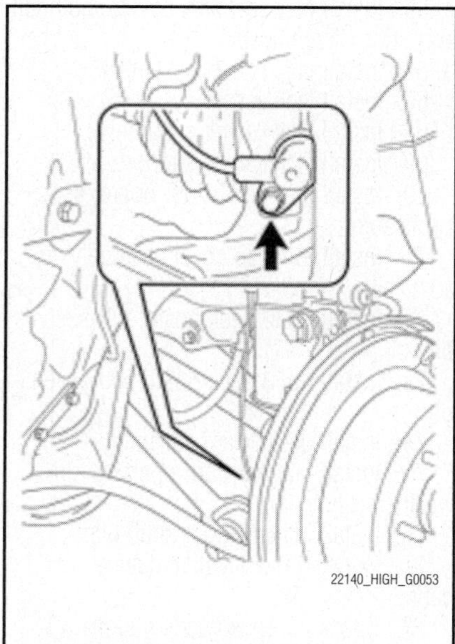

Fig. 7 Rear speed sensor—Non-Hybrid Models

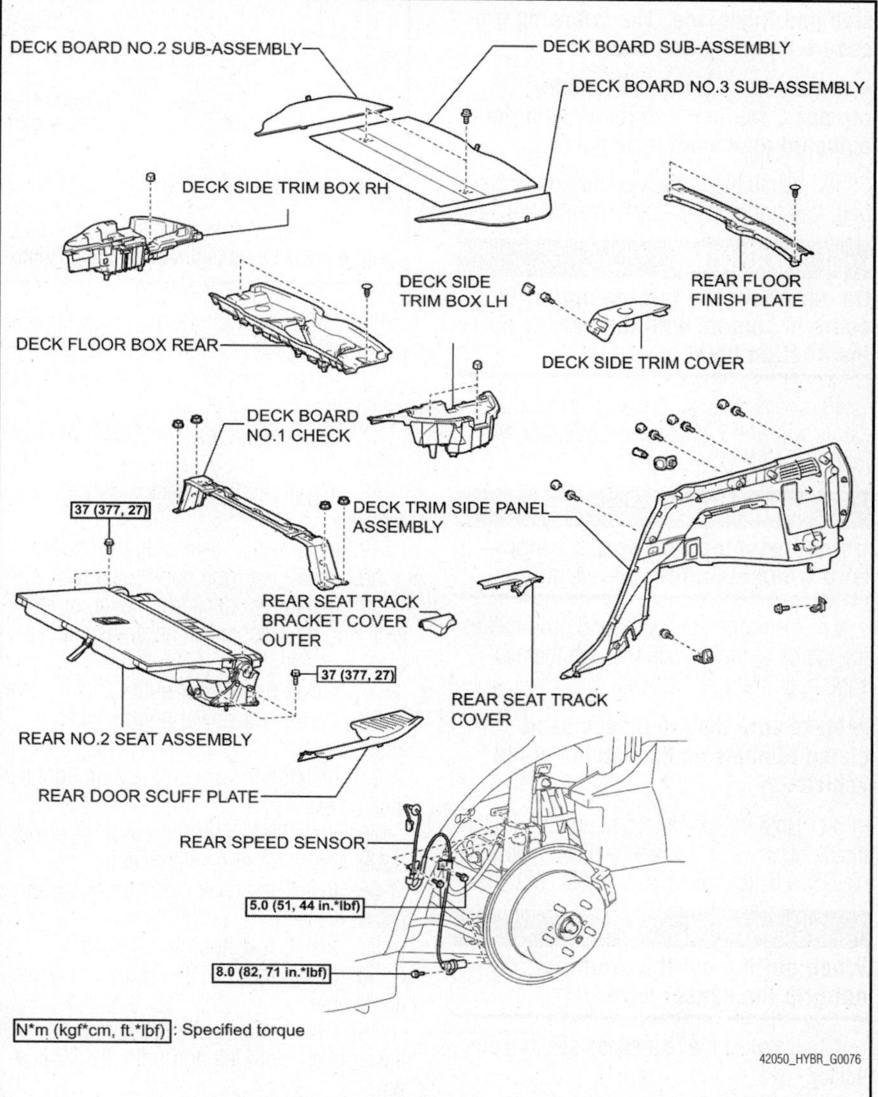

Fig. 8 Exploded view of the rear speed sensor and related components—Hybrid model with no. 2 rear seat shown

33. Raise and safely support the vehicle.
34. Remove rear wheel and tire assembly.
35. Disconnect the rear speed sensor connector.
36. Disconnect the grommet of the rear speed sensor wire from the hole of the wheel house.
37. Remove the 2 bolts, No. 1 clamp and No. 2 clamp from the body and absorber.
38. Remove the bolt and rear speed sensor body from the carrier.

WARNING

Do not allow any foreign matter to come in contact with the sensor tip or installation hole.

To install:

➡Use the same procedures for the Left side and Right side. The following procedure is for the Left side.

➡If the sensor rotor needs to be replaced, replace it together with the outboard joint shaft assembly.

39. Install the rear speed sensor with the bolt, and tighten to 71 inch lbs. (8 Nm).

WARNING

Do not allow any foreign matter to come in contact with the sensor tip or installation hole.

40. Install the No. 1 clamp and No. 2 clamp with the 2 bolts, and tighten to 44 inch lbs. (5 Nm).

WARNING

Do not twist the rear speed sensor wire when installing the clamp.

41. Insert the connector and grommet to the inside of the vehicle through the passage hole in the wheel well.

➡Make sure the grommet's band clamp remains on the outside of the vehicle.

42. Hold the grommet and pull it from the inside of the vehicle to the outside of the vehicle. Then fix it in place so that it is not tilted.

WARNING

When pulling out the grommet, do not grip the sensor wire.

43. Connect the rear speed sensor connector.
44. Install the rear wheel and tire assembly, and tighten the lug nuts finger-tight.

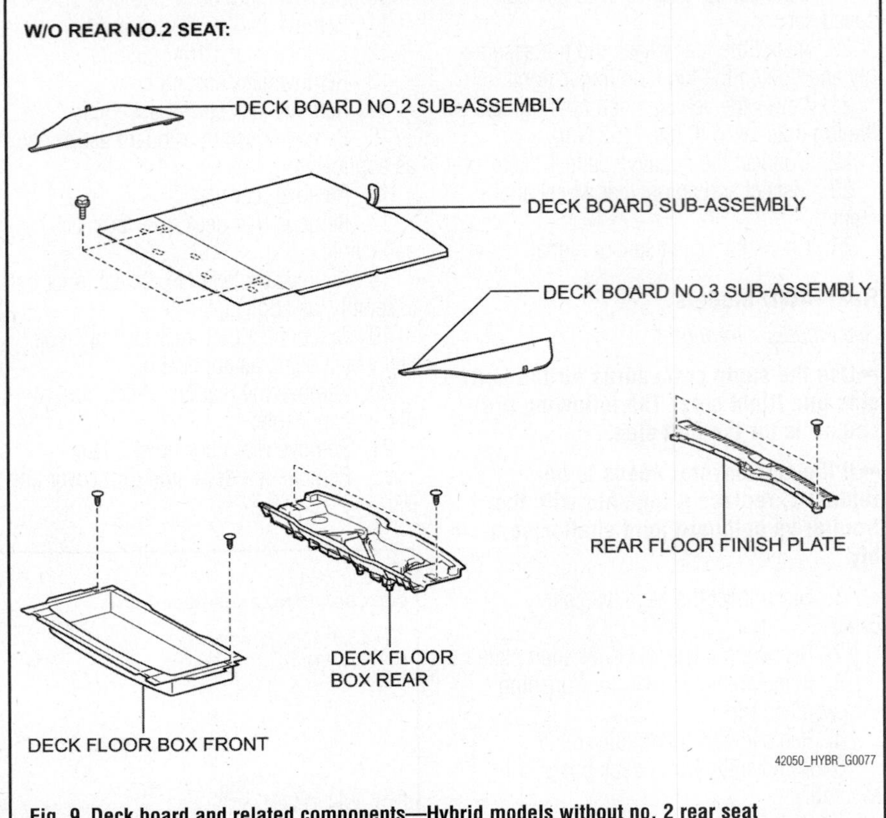

Fig. 9 Deck board and related components—Hybrid models without no. 2 rear seat

45. Lower the vehicle, then final tighten the lug nuts to 76 ft. lbs. (103 Nm).
46. Install left deck trim side panel assembly.
47. Connect left rear no. 1 seat outer belt assembly.
48. Install left front deck side trim cover.
49. Install no. 2 deck side trim hook.
50. Install left rope hook assembly.
51. Install rear deck trim cover, or left reclining remote control bezel, as applicable.
52. Install rear power outlet socket cover.
53. Install rear power point socket assembly.
54. Install left rear combination light service cover.
55. Install left side trim cover, or power outlet socket bezel, as applicable.
56. Install left deck side trim and trim cover.
57. Install rear floor finish plate.
58. Install rear no. 2 seat assembly, as applicable.
59. Connect rear seat lap type belt assemblies, both left and right, as applicable.
60. Install rear no. 2 seat inner belt assembly, as applicable.
61. Install rear deck floor box, as applicable.
62. Install deck floor board assembly, as applicable.
63. Install rear mat.
64. Install deck side trim boxes, both left and right.
65. Install rear seat side covers, both left and right, as applicable.
66. Install jack carrier assembly.
67. Install jack assembly.
68. Install jack carrier cushion.
69. Install jack carrier support.
70. Install rear no. 1 floor board, as applicable.
71. Install tonneau cover assembly, as applicable.
72. Install no. 2 deck board sub-assembly.
73. Install no. 3 deck board sub-assembly.
74. Install deck board assembly.
75. Install left rear door opening trim weatherstrip.
76. Install left rear door scuff plate.
77. Connect the negative battery cable.
78. Check for speed sensor signal.
79. If necessary, perform the initialization procedure.

BLEEDING PROCEDURE

BLEEDING PROCEDURE

Non-Hybrid

See Figure 10.

➡After bleeding the air from the brake system, if the height or feel of the brake pedal cannot be obtained, perform air bleeding in the brake actuator assembly with a hand-held tester by following the procedures below.

1. Depress the brake pedal more than 20 times with the engine off.
2. Connect the hand-held tester to the DLC3, then turn the ignition switch to the **ON** position, but do NOT start the engine.
3. Select "AIR BLEEDING" on the hand-held tester.

➡Refer to the hand-held tester operator's manual for more details.

4. Bleed the air out of the regular brake line when "Step 1: Increase" appears on the hand-held tester display, as follows:

➡Bleed the air by following the steps displayed on the hand-held tester. Make sure that the brake fluid in the master cylinder reservoir tank does not become empty.

a. Connect the vinyl tube to either one of the bleeder plugs.
b. Have an assistant depress the brake pedal several times, then loosen the bleeder plug connected to the vinyl tube with the pedal depressed.
c. When fluid stops coming out, tighten the bleeder plug and release the brake pedal.
d. Repeat the previous 2 steps until all the air in the fluid is completely bled out.
e. Tighten the bleeder plug completely to 73 inch lbs. (8.3 Nm).
f. Repeat the above procedures for each wheel to bleed the air out of the brake line.

5. Bleed the air out of the suction line when "Step 2: Inhalation" appears on the hand-held tester display, as follows:

➡Bleed the air by following the steps displayed on the hand-held tester. Make sure that the brake fluid in the master cylinder reservoir tank does not become empty.

a. Connect the vinyl tube to the bleeder plug at the right front wheel or the right rear wheel and loosen the bleeder plug.
b. Operate the brake actuator assembly to bleed the air using the hand-held tester.

➡This operation stops automatically after 4 seconds. At this time, be sure to release the brake pedal.

c. Check if the operation has stopped by referring to the hand-held tester display.
d. Repeat the previous 2 steps until all air in the fluid is completely bled out.
e. Tighten the bleeder plug completely to 73 inch lbs. (8.3 Nm).
f. Repeat the above procedures to bleed the air out of the brake line for each wheel.

6. Bleed the air out of the pressure reduction line when "Step 3: Decrease" appears on the hand-held tester display, as follows:

➡Bleed the air by following the steps displayed on the hand-held tester. Make sure that the brake fluid in the master cylinder reservoir tank does not become empty.

a. Connect a vinyl tube to either one of the bleeder plugs.
b. Loosen the bleeder plug.
c. Using the hand-held tester, operate the brake actuator assembly, completely depress the brake pedal and keep it depressed.

➡The operation stops automatically after 4 seconds. When performing this procedure continuously, set an interval of at least 20 seconds. When the operation is complete, the brake pedal goes down slightly. This is a normal phenomenon caused when the solenoid opens. During this procedure, the pedal will feel heavy, but completely depress it so that the brake fluid comes out from the bleeder plug. Be sure to keep depressing the brake pedal. Do not depress and release the pedal repeatedly.

d. Tighten the bleeder plug, then release the brake pedal.
e. Repeat the previous 3 steps until all the air in the fluid is completely bled out.
f. Tighten the bleeder plug completely to 73 inch lbs. (8.3 Nm).
g. Repeat the above procedures for each wheel to bleed the air out of the brake line.

7. Bleed the air out of the regular brake line again when "Step 4: Increase" appears on the hand-held tester display, as follows:

➡Bleed the air by following the steps displayed on the hand-held tester. Make sure that the brake fluid in the master cylinder reservoir tank does not become empty.

a. Connect the vinyl tube to either one of the bleeder plugs.
b. Depress the brake pedal several times, then loosen the bleeder plug connected to the vinyl tube with the pedal depressed.

8. When fluid stops coming out, tighten the bleeder plug, then release the brake pedal.

a. Repeat the previous 2 steps until all the air in the fluid is completely bled out.
b. Tighten the bleeder plug completely to 73 inch lbs. (8.3 Nm).
c. Repeat the above procedures for each wheel to bleed the air out of the brake line.

9. Check the fluid level and add fluid if necessary. Use SAE J1703 or FMVSS No. 116 DOT3 Brake fluid.

Hybrid

See Figures 11 through 13.

➡This procedure requires specialized tools. Please read through the procedure and make sure you have access to the proper equipment before beginning the bleeding procedure.

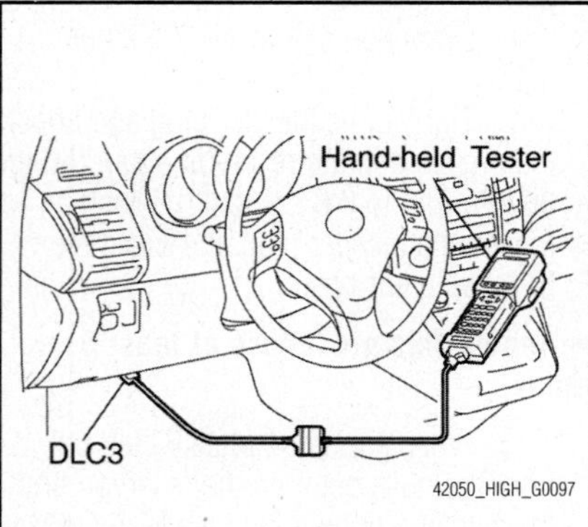

Fig. 10 Connect a suitable hand-held tester to the DLC3 to bleed the actuator assembly

※※ CAUTION

Never bleed air from the brake hydraulic system without using the intelligent tester. failure to use the intelligent tester could cause serious injury or an accident.

Note the following before bleeding the brake system:

- Move the shift lever to the P position and apply the parking brake before bleeding.
- Add brake fluid carefully and check that the reservoir level remains between the min and max lines while bleeding the brakes.
- Do not stand the fluid can on the reservoir inlet when bleeding the brake actuator. doing so will cause brake fluid to overflow.
- The actuator pump motor and solenoid can be operated by the driver even if the ignition switch is off.
- If the pump motor operates while air still remains inside the brake actuator hose, air will enter the actuator, making it more difficult to bleed the brakes. If there is concern about air remaining in the actuator hose, remove the two motor relays (skid control relay no.2) until instructed to reinstall them.
    - Although a buzzer may sound due to a decline in the accumulator pressure while bleeding, it is not necessary to stop bleeding.
    - DTCs indicating a malfunction in the motor relays (skid control relay no.2) or the pressure sensor are stored after bleeding. clear the DTCs when instructed during or after bleeding.

1. Add SAE J1703 or FMVSS no. 116 DOT 3 brake fluid to the max line in the reservoir.

※※ CAUTION

Add brake fluid carefully and check that the reservoir level remains between the min and max lines while bleeding the brakes. Do not stand the fluid can on the reservoir inlet when bleeding the brake actuator. Doing so will cause brake fluid to overflow.

2. Disable the brake control (ECB). When using the intelligent tester:

➡When using the intelligent tester , refer to the intelligent tester operator's manual for further details. Bleed the air by following the steps displayed on the intelligent tester.

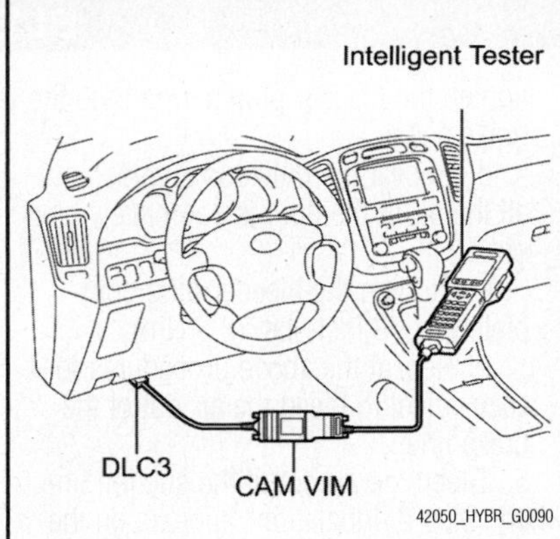

Fig. 11 Connect the intelligent tester to the DLC3 with the ignition switch OFF as shown in the illustration

a. Move the shift lever to the P position and apply the parking brake.

b. Connect the intelligent tester to the DLC3 with the ignition switch **OFF** as shown in the illustration.

c. Turn the ignition switch to the ON position and turn on the intelligent tester.

➡Do not start the engine.

d. Enter the following menus: DIAGNOSIS / OBD/ MOBD / ABS/TRAC/VSC / ECB UTILITY / ECB INVALID.

※※ WARNING

If the pump motor operates while air remains inside the brake actuator hose, air will enter the actuator, and this will make bleeding the brakes more difficult.

e. When removing the ABS motor relay: Remove the 2 ABS motor relays with the ignition switch off in order to disable brake control.

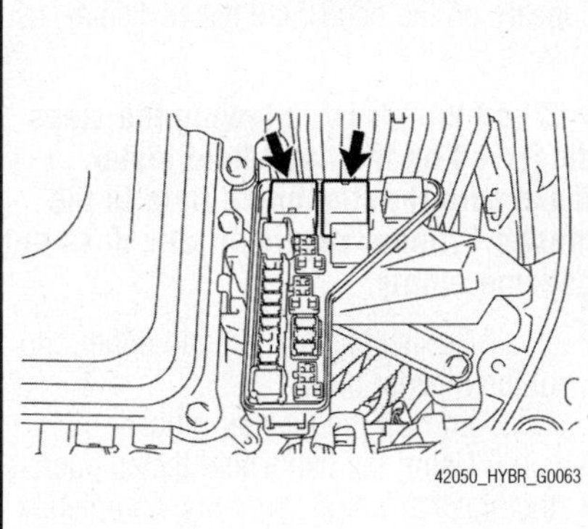

Fig. 12 Location of the 2 ABS motor relays

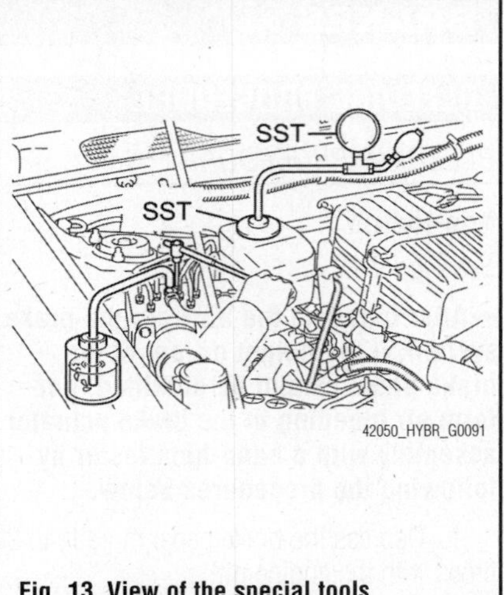

Fig. 13 View of the special tools

※※ WARNING

If the pump motor operates while air remains inside the brake actuator hose, air will enter the actuator, and this will make bleeding the brakes more difficult.

➡After the brake actuator assembly has been replaced, remove the ABS motor relay before bleeding the brakes.

3. **Bleed the brake actuator hose, as follows:**

a. Connect Special Tool 09992-00242, 0992-00350 or equivalent, to the reservoir with the brake reservoir pressure adapter.

b. Using Special Tool 09023-00101, loosen the bleeder plug of the actuator.

c. Connect a vinyl tube to the bleeder plug of the actuator.

d. Use the SST to boost pressure in the reservoir. Standard pressure is 50 to 80 kPa (0.5 to 0.8 kgf/cm2, 7.3 to 11.6 psi)

e. Drain approximately 100 cc of fluid.

f. Tighten the bleeder plug and boost the pressure in the reservoir again (50 to 80 kPa (0.5 to 0.8 kgf/ cm2)). Then, loosen the bleeder plug and bleed the brake actuator hose.

➡Repeat this procedure at least 5 times.

g. When air is completely bled out from the hose between the reservoir and the actuator, tighten the bleeder plug to 74 inch lbs. (8.3 Nm).

4. **Bleed the master cylinder, as follows:**

➡If the master cylinder has been dis-

assembled or if the reservoir becomes empty, bleed the air from the master cylinder.

a. Enter the following menus: DIAGNOSIS / OBD/MOBD / ABS/TRAC/VSC / AIR BLEEDING.

b. Select "USUAL" if the front/rear brakes are removed, installed or disassembled.

c. Select "ACTUATOR" if the actuator is removed, installed or replaced.

d. Select "MASTER CYLINDER" if the brake master cylinder or the brake stroke simulator is removed, installed or replaced.

e. Disconnect the brake lines from the master cylinder.

f. Slowly depress and hold the brake pedal (Procedure A).

g. Cover the outer holes with fingers, and release the brake pedal (Procedure B).

h. Repeat procedure A and B 3 or 4 times.

i. Connect the brake lines to the master cylinder and tighten to 11 ft. lbs. (15 Nm).

5. **Bleed the front brake system, as follows:**

➡Air can be easily bled from the front brake system if air has been bled from the master cylinder when replacing the brake master cylinder assembly.

✲✲ WARNING

If brake fluid leaks onto any painted surface of the vehicle, wash or otherwise remove it completely.

➡Bleed the air by following the steps displayed on the intelligent tester. (a) Depress the brake pedal several times and bleed the front brake system from the bleeder plugs on the front brake cylinder RH and LH.

➡Repeat the procedure until air is completely bled from the front brake system.

6. Tighten the bleeder plugs to 74 inch lbs. (8.3 Nm) after bleeding.

7. Cancel brake control (ECB) disable

a. Install the 2 motor relays (skid control relay No.2) if they have been removed.

b. Complete brake control prevention following the prompts on the tester screen. (If brake control has been prevented using the intelligent tester.)

8. Clear the DTC(s).

9. Bleed the rear brake system, as follows:

✲✲ WARNING

Never bleed air from the brake hydraulic system without using the intelligent tester. Failure to use the intelligent tester could cause serious injury or an accident.

➡Bleed the air by following the steps displayed on the intelligent tester.

a. Connect the intelligent tester to the DLC3 with the ignition switch off.

b. Check that the parking brake is applied and turn the ignition switch to the **ON** position.

c. Enter the following menus: DIAGNOSIS / OBD/ MOBD / ABS/TRAC/VSC / ECB UTILITY / ECB INVALID.

d. With the brake pedal depressed, bleed the rear brake system from the bleeder plug on the rear disc brake cylinder LH while the pump motor and solenoid are operating.

✲✲ WARNING

Keep the fluid inside the reservoir above the LOW level by replenishing.

➡Depress and hold the brake pedal. After the solenoid operates for approximately 30 seconds, release the brake pedal to stop the solenoid. Repeat the procedures until air is completely bled from the rear brake system. The ECB warning light comes on and the buzzer sounds while bleeding, but they do not indicate a malfunction.

e. Tighten the bleeder plug to 74 inch lbs. (8.3 Nm) after bleeding.

f. Enter the following menus: DIAGNOSIS / OBD/ MOBD / ABS/TRAC/VSC / ECB UTILITY / ECB INVALID.

g. With the brake pedal depressed, bleed the rear brake system from the bleeder plug on the rear disc brake cylinder RH while the pump motor and solenoid are operating.

✲✲ WARNING

Keep the fluid inside the reservoir above the LOW level by replenishing.

➡Depress and hold the brake pedal. After the solenoid operates for approximately 30 seconds, release the brake pedal to stop the solenoid. Repeat the procedures until air is completely bled from the rear brake system. The ECB warning light comes on and the buzzer sounds while bleeding, but they do not indicate a malfunction.

h. Tighten the bleeder plug to 74 inch lbs. (8.3 Nm) after bleeding.

10. Perform the accumulator zero down:

✲✲ WARNING

Never bleed air from the brake hydraulic system without using the intelligent tester. Failure to use the intelligent tester could cause serious injury or an accident. Be sure to perform this procedure before replacement, removal, or installation of the actuator.

➡Perform accumulator zero down by following the steps displayed on the intelligent tester.

a. Connect the intelligent tester to the DLC3 with the ignition switch **OFF**.

b. Depressurize the accumulator:

- Check that the parking brake is applied and turn the ignition switch to the **ON** position.
- Enter the following menus: DIAGNOSIS / OBD/ MOBD / ABS/TRAC/VSC / ECB UTILITY / ZERO DOWN.
- When the buzzer sounds, turn the ignition switch **OFF**.

c. Circulate the fluid in the accumulator.

d. Depressurize the accumulator 5 times.

➡Accumulator pressure is released and accumulated repeatedly, which circulates the fluid inside the accumulator, when repeating accumulator zero down. The pump motor rotates and the accumulator is pressurized every time the ignition switch is turned from off to on.

11. Check the brake fluid level:

a. After performing accumulator zero down (accumulator depressurizing), return the fluid in the accumulator back to the reservoir and then adjust the fluid level in the master cylinder reservoir to the MAX level.

➡After performing accumulator zero down (accumulator depressurizing), fluid is built up in the accumulator by turning the ignition switch to the ON position and the fluid level of the reservoir lowers. If the fluid level is adjusted without performing accumulator zero down (accumulator depressurizing), fluid is sent from the accumulator to the reservoir. The fluid level may exceed the MAX level, but it is normal.

12. Clear the DTC(s).

13. When the brake actuator assembly is replaced, perform linear valve offset learning after bleeding is completed.

MASTER CYLINDER BLEEDING

Non-Hybrid

See Figure 14.

If the master cylinder has been disassembled or if the reservoir becomes empty, bleed the air from the master cylinder.

1. Remove the air cleaner assembly with hose.
2. Disconnect the brake lines from the master cylinder, using 12mm union nut wrench, or a suitable brake line wrench.
3. Have an assistant slowly depress the brake pedal and hold it.
4. Cover the outer holes with your fingers, and have your assistant release the brake pedal.
5. Repeat steps 3 and 4 several times.
6. Connect the brake lines. Tighten as follows:
   a. Use a torque wrench with a fulcrum length of 250 mm (9.84 inches) and tighten to 14 ft. lbs. (19 Nm). This torque value is effective when the union nut wrench is parallel to the torque wrench.
7. Install the air cleaner assembly with hose.

BRAKE LINE BLEEDING

Non-Hybrid

See Figure 15.

1. Raise and safely support the vehicle.
2. Connect a piece of vinyl tubing to the brake caliper.
3. Have an assistant depress the brake pedal several times, then loosen the bleeder plug while the pedal is depressed.

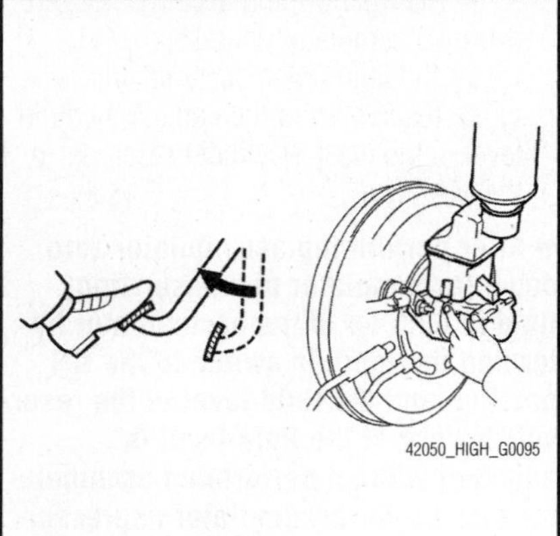

Fig. 14 Use your fingers to cover the outer holes, then release the brake pedal

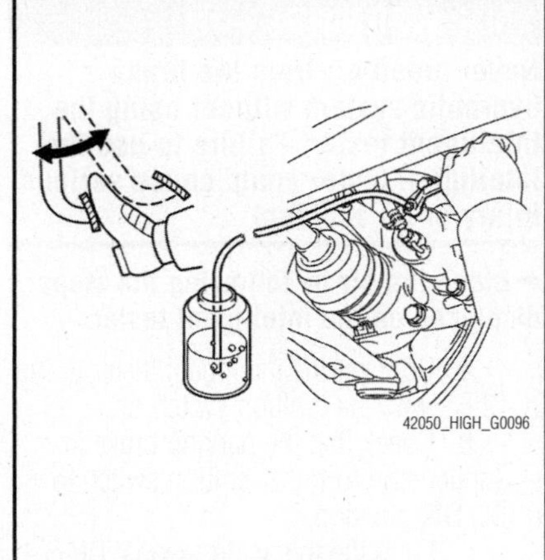

Fig. 15 Bleeding the brake lines at each wheel

4. When fluid stops coming out, tighten the bleeder plug, then release the brake pedal.
5. Repeat steps 2 and 3 until all the air in the fluid has been bled out.
6. Tighten the brake bleeder plug to 73 inch lbs. (8.3 Nm).
7. Repeat the above steps to bleed the air out of the brake line for each wheel.

BLEEDING THE ABS SYSTEM

Hybrid

See Figures 16 through 18.

➡This procedure requires specialized tools. Please read through the procedure and make sure you have access to the proper equipment before beginning the bleeding procedure.

⁂ CAUTION

Never bleed air from the brake hydraulic system without using the intelligent tester. failure to use the intelligent tester could cause serious injury or an accident.

Note the following before bleeding the brake system:

- Move the shift lever to the P position and apply the parking brake before bleeding.
- Add brake fluid carefully and check that the reservoir level remains between the min and max lines while bleeding the brakes.
- Do not stand the fluid can on the reservoir inlet when bleeding the brake actuator. doing so will cause brake fluid to overflow.
- The actuator pump motor and solenoid can be operated by the driver even if the ignition switch is off.
- If the pump motor operates while air still remains inside the brake actuator hose, air will enter the actuator, making it more difficult to bleed the brakes. If there is concern about air remaining in the actuator hose, remove the two motor relays (skid control relay no. 2) until instructed to reinstall them.
  - Although a buzzer may sound due to a decline in the accumulator pressure while bleeding, it is not necessary to stop.
  - DTCs indicating a malfunction in the motor relays (skid control relay no.2) or the pressure sensor are stored after bleeding. clear the DTCs when instructed during or after bleeding.

1. Add SAE J1703 or FMVSS no. 116 DOT 3 brake fluid to the max line in the reservoir.

⁂ CAUTION

Add brake fluid carefully and check that the reservoir level remains between the min and max lines while bleeding the brakes. Do not stand the fluid can on the reservoir inlet when bleeding the brake actuator. Doing this will cause brake fluid to overflow.

2. Disable the brake control (ECB). When using the intelligent tester:

➡When using the intelligent tester, refer to the intelligent tester operator's manual for further details Bleed the air by following the steps displayed on the intelligent tester.

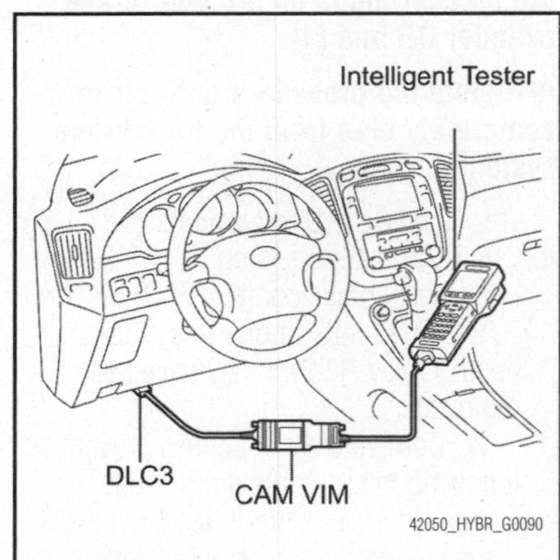

Fig. 16 Connect the intelligent tester to the DLC3 with the ignition switch OFF as shown in the illustration

a. Move the shift lever to the P position and apply the parking brake.

b. Connect the intelligent tester to the DLC3 with the ignition switch **OFF** as shown in the illustration.

c. Turn the ignition switch to the ON position and turn on the intelligent tester.

➡Do not start the engine.

d. Enter the following menus: DIAGNOSIS / OBD/ MOBD / ABS/TRAC/VSC / ECB UTILITY / ECB INVALID.

WARNING

If the pump motor operates while air remains inside the brake actuator hose, air will enter the actuator, and this will make bleeding the brakes more difficult.

e. When removing the ABS motor relay: Remove the 2 ABS motor relays with the ignition switch off in order to disable brake control.

WARNING

If the pump motor operates while air remains inside the brake actuator hose, air will enter the actuator, and this will make bleeding the brakes more difficult.

➡After the brake actuator assembly has been replaced, remove the ABS motor relay before bleeding the brakes.

3. **Bleed the brake actuator hose, as follows:**

a. Connect Special Tool 09992-00242, 0992-00350 or equivalent, to the reservoir with the brake reservoir pressure adapter.

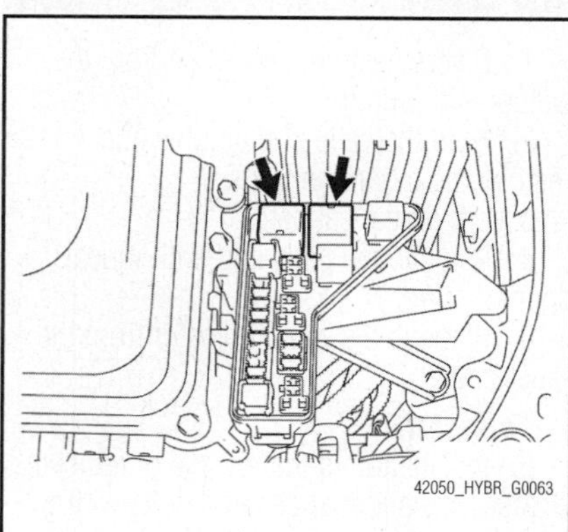

Fig. 17 Location of the 2 ABS motor relays

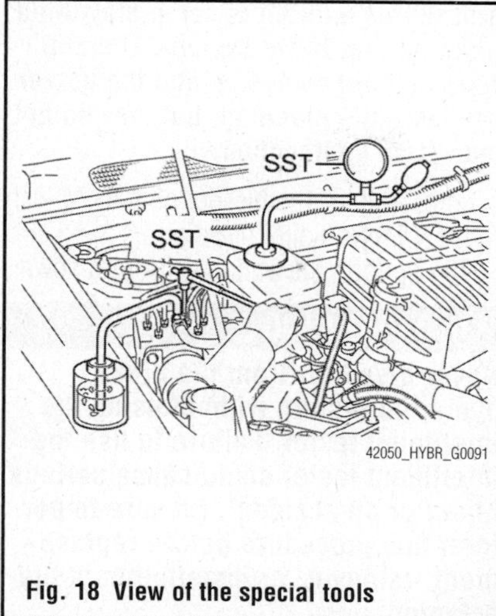

Fig. 18 View of the special tools

b. Using Special Tool 09023-00101, loosen the bleeder plug of the actuator.

c. Connect a vinyl tube to the bleeder plug of the actuator.

d. Use the SST to boost pressure in the reservoir. Standard pressure is 50 to 80 kPa (0.5 to 0.8 kgf/cm2, 7.3 to 11.6 psi)

e. Drain approximately 100 cc of fluid.

f. Tighten the bleeder plug and boost the pressure in the reservoir again (50 to 80 kPa (0.5 to 0.8 kgf/ cm2)). Then, loosen the bleeder plug and bleed the brake actuator hose.

➡Repeat this procedure at least 5 times.

g. When air is completely bled out from the hose between the reservoir and the actuator, tighten the bleeder plug to 74 inch lbs. (8.3 Nm).

4. **Bleed the master cylinder, as follows:**

➡If the master cylinder has been disassembled or if the reservoir becomes empty, bleed the air from the master cylinder.

a. Enter the following menus: DIAGNOSIS / OBD/MOBD / ABS/TRAC/VSC / AIR BLEEDING.

b. Select "USUAL" if the front/rear brakes are removed, installed or disassembled.

c. Select "ACTUATOR" if the actuator is removed, installed or replaced.

d. Select "MASTER CYLINDER" if the brake master cylinder or the brake stroke simulator is removed, installed or replaced.

e. Disconnect the brake lines from the master cylinder.

f. Slowly depress and hold the brake pedal (Procedure A).

g. Cover the outer holes with fingers, and release the brake pedal (Procedure B).

h. Repeat procedure A and B 3 or 4 times.

i. Connect the brake lines to the master cylinder and tighten to 11 ft. lbs. (15 Nm).

5. **Bleed the front brake system, as follows:**

➡Air can be easily bled from the front brake system if air has been bled from the master cylinder when replacing the brake master cylinder assembly.

WARNING

If brake fluid leaks onto any painted surface of the vehicle, wash or otherwise remove it completely.

➡Bleed the air by following the steps displayed on the intelligent tester. (a) Depress the brake pedal several times and bleed the front brake system from the bleeder plugs on the front brake cylinder RH and LH.

➡Repeat the procedure until air is completely bled from the front brake system.

6. Tighten the bleeder plugs to 74 inch lbs. (8.3 Nm) after bleeding.

7. Cancel brake control (ECB) disable

a. Install the 2 motor relays (skid control relay No.2) if they have been removed.

b. Complete brake control prevention following the prompts on the tester screen. (If brake control has been prevented using the intelligent tester.)

8. Clear the DTC(s).

9. Bleed the rear brake system, as follows:

WARNING

Never bleed air from the brake hydraulic system without using the intelligent tester. Failure to use the intelligent tester could cause serious injury or an accident.

➡Bleed the air by following the steps displayed on the intelligent tester.

a. Connect the intelligent tester to the DLC3 with the ignition switch off.

b. Check that the parking brake is applied and turn the ignition switch to the **ON** position.

c. Enter the following menus: DIAG-

NOSIS / OBD/ MOBD / ABS/TRAC/VSC / ECB UTILITY / ECB INVALID.

d. With the brake pedal depressed, bleed the rear brake system from the bleeder plug on the rear disc brake cylinder LH while the pump motor and solenoid are operating.

⁂ WARNING

Keep the fluid inside the reservoir above the LOW level by replenishing.

➡Depress and hold the brake pedal. After the solenoid operates for approximately 30 seconds, release the brake pedal to stop the solenoid. Repeat the procedures until air is completely bled from the rear brake system. The ECB warning light comes on and the buzzer sounds while bleeding, but they do not indicate a malfunction.

e. Tighten the bleeder plug to 74 inch lbs. (8.3 Nm) after bleeding.

f. Enter the following menus: DIAGNOSIS / OBD/ MOBD / ABS/TRAC/VSC / ECB UTILITY / ECB INVALID.

g. With the brake pedal depressed, bleed the rear brake system from the bleeder plug on the rear disc brake cylinder RH while the pump motor and solenoid are operating.

⁂ WARNING

Keep the fluid inside the reservoir above the LOW level by replenishing.

➡Depress and hold the brake pedal. After the solenoid operates for approximately 30 seconds, release the brake pedal to stop the solenoid. Repeat the procedures until air is completely bled from the rear brake system. The ECB warning light comes on and the buzzer sounds while bleeding, but they do not indicate a malfunction.

h. Tighten the bleeder plug to 74 inch lbs. (8.3 Nm) after bleeding.

10. Perform the accumulator zero down:

⁂ WARNING

Never bleed air from the brake hydraulic system without using the intelligent tester. Failure to use the intelligent tester could cause serious injury or an accident. Be sure to perform this procedure before replacement, removal, or installation of the actuator.

➡Perform accumulator zero down by following the steps displayed on the intelligent tester.

a. Connect the intelligent tester to the DLC3 with the ignition switch **OFF**.

b. Depressurize the accumulator:

- Check that the parking brake is applied and turn the ignition switch to the **ON** position.
- Enter the following menus: DIAGNOSIS / OBD/ MOBD / ABS/TRAC/VSC / ECB UTILITY / ZERO DOWN.
- When the buzzer sounds, turn the ignition switch**OFF**.

c. Circulate the fluid in the accumulator.

d. Depressurize the accumulator 5 times.

➡Accumulator pressure is released and accumulated repeatedly, which circulates the fluid inside the accumulator, when repeating accumulator zero down. The pump motor rotates and the accumulator is pressurized every time the ignition switch is turned from off to on.

11. Check the brake fluid level:

a. After performing accumulator zero down (accumulator depressurizing), return the fluid in the accumulator back to the reservoir and then adjust the fluid level in the master cylinder reservoir to the MAX level.

➡After performing accumulator zero down (accumulator depressurizing), fluid is built up in the accumulator by turning the ignition switch to the ON position and the fluid level of the reservoir lowers. If the fluid level is adjusted without performing accumulator zero down (accumulator depressurizing), fluid is sent from the accumulator to the reservoir. The fluid level may exceed the MAX level, but it is normal.

12. Clear the DTC(s).

13. When the brake actuator assembly is replaced, perform linear valve offset learning after bleeding is completed.

FLUID FILL PROCEDURE

1. Add brake fluid carefully and check that the reservoir level remains between the MIN and MAX lines.

⁂ WARNING

Do not stand the fluid can on the reservoir inlet. Doing so will cause brake fluid to overflow.

BRAKES — FRONT DISC BRAKES

⁂ CAUTION

Dust and dirt accumulating on brake parts during normal use may contain asbestos fibers from production or aftermarket brake linings. Breathing excessive concentrations of asbestos fibers can cause serious bodily harm. Exercise care when servicing brake parts. Do not sand or grind brake lining unless equipment used is designed to contain the dust residue. Do not clean brake parts with compressed air or by dry brushing. Cleaning should be done by dampening the brake components with a fine mist of water, then wiping the brake components clean with a dampened cloth. Dispose of cloth and all residue containing asbestos fibers in an impermeable container with the appropriate label. Follow practices prescribed by the Occupational Safety and Health Administration (OSHA) and the Environmental Protection Agency (EPA) for the handling, processing, and disposing of dust or debris that may contain asbestos fibers.

BRAKE CALIPER

REMOVAL & INSTALLATION

Non-Hybrid Models

See Figure 19.

1. Disconnect the brake line from the caliper and plug it.
2. Hold the caliper slide pins and remove the mounting bolts.
3. Lift off the caliper.
4. Remove the pads and anti-squeal shims.
5. Remove the wear indicator from the inner pad.

To install:

6. Installation is the reverse of removal. Grease the caliper slides and bolts with lithium grease or equivalent. Apply disc brake grease to the anti-squeal shims. Torque the caliper bolts to 25 ft. lbs. (34 Nm); the brake line union bolt to 21 ft. lbs. (29 Nm).

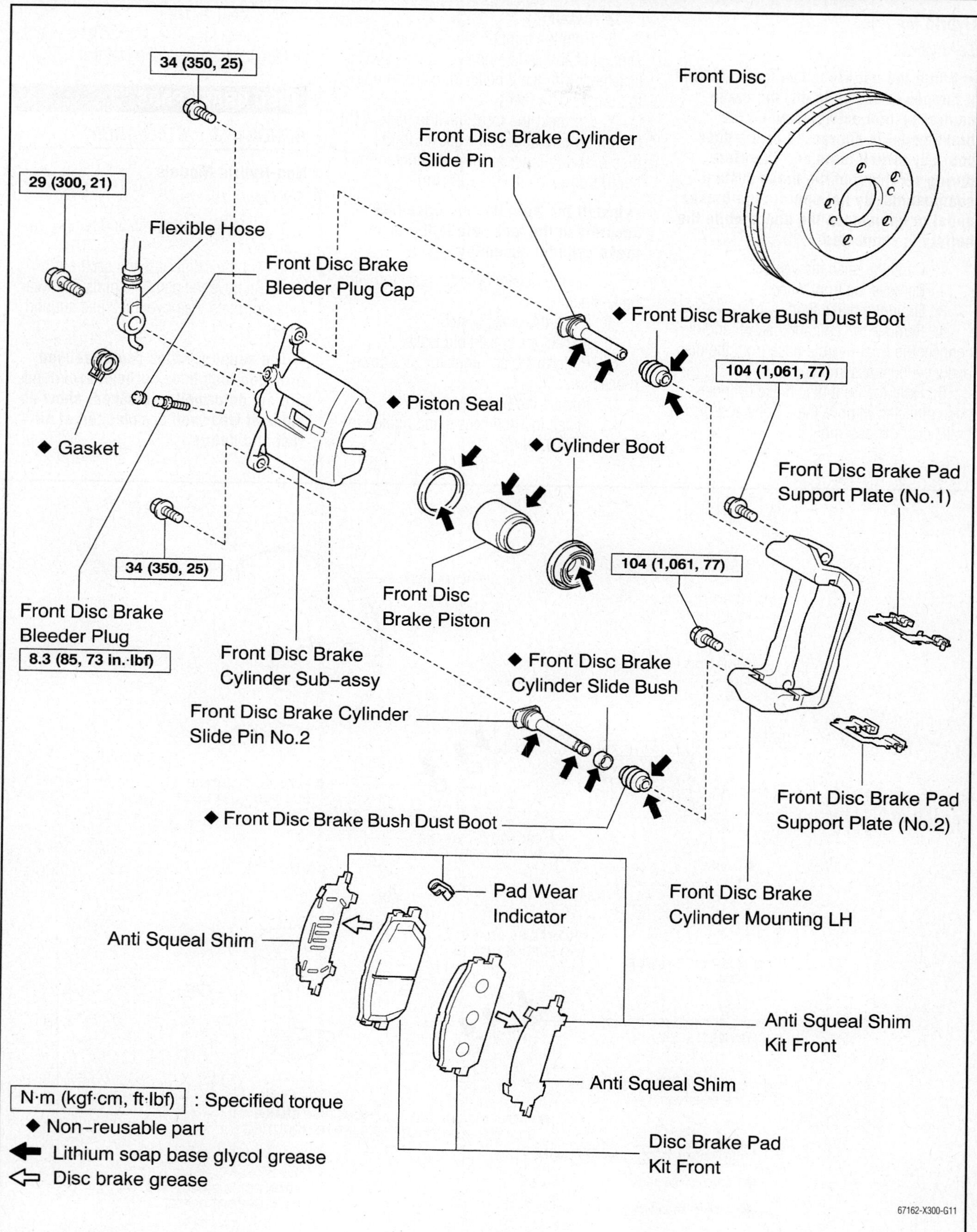
34 (350, 25)
Front Disc
Front Disc Brake Cylinder
Slide Pin
29 (300, 21)
Flexible Hose
Front Disc Brake
Bleeder Plug Cap
◆ Front Disc Brake Bush Dust Boot
104 (1,061, 77)
◆ Piston Seal
◆ Gasket
◆ Cylinder Boot
Front Disc Brake Pad
Support Plate (No.1)
34 (350, 25)
104 (1,061, 77)
Front Disc
Brake Piston
Front Disc Brake
Bleeder Plug
8.3 (85, 73 in.·lbf)
Front Disc Brake
Cylinder Sub-assy
◆ Front Disc Brake
Cylinder Slide Bush
Front Disc Brake Cylinder
Slide Pin No.2
Front Disc Brake Pad
Support Plate (No.2)
◆ Front Disc Brake Bush Dust Boot
Pad Wear
Indicator
Front Disc Brake
Cylinder Mounting LH
Anti Squeal Shim
Anti Squeal Shim
Kit Front
Anti Squeal Shim
N·m (kgf·cm, ft·lbf) : Specified torque
◆ Non-reusable part
Lithium soap base glycol grease
Disc brake grease
Disc Brake Pad
Kit Front
67162-X300-G11

Fig. 19 Front disc brake components

Hybrid Models

See Figure 20.

➡While the battery is connected, even if the power switch is off, the brake control system activates when the brake pedal is depressed or the door courtesy switch turns on. Therefore, during servicing of the brake system components, do not operate the brake pedal or open/close the doors while the battery is connected.

1. Carefully raise the vehicle.
2. Remove the front wheel.
3. Drain the brake fluid.
4. Remove the bolt and gasket, and disconnect the front flexible hose from the disc brake cylinder assembly.
5. Hold the front disc brake cylinder slide pins and remove the 2 bolts and disc brake cylinder assembly.

To install:

6. Hold the front disc brake cylinder slide pins and install the disc brake cylinder assembly with the 2 bolts. Tighten the bolts to 25 ft. lbs. 34 (Nm).
7. Connect the front flexible hose with the union bolt and a new gasket to the disc brake cylinder assembly. Tighten the banjo bolt to 20 ft. lbs. (30 Nm).

➡Install the front flexible hose lock securely in the lock hole in the disc brake cylinder assembly.

8. Fill the master cylinder reservoir with brake fluid.
9. Bleed the brake system.
10. Inspect for brake fluid leaks.
11. Perform the accumulator zero down procedure.
12. Inspect the brake fluid level.
13. Install the front wheel and tighten to 76 ft. lbs. (103 Nm).
14. Clear the DTCs
15. Check for DTCs. If any DTC is output, perform the troubleshooting for that DTC.

DISC BRAKE PADS

REMOVAL & INSTALLATION

Non-Hybrid Models

See Figure 19.

1. Hold the sliding pin and remove the lower bolt.
2. Lift the caliper up and secure it.
3. Remove the pads, 4 shims and wear indicator plate. Remove the 2 pad support plates.

➡The support plates can be reused, provided they have sufficient rebound, are not deformed or cracked, show no signs of wear and are cleaned of all rust and debris.

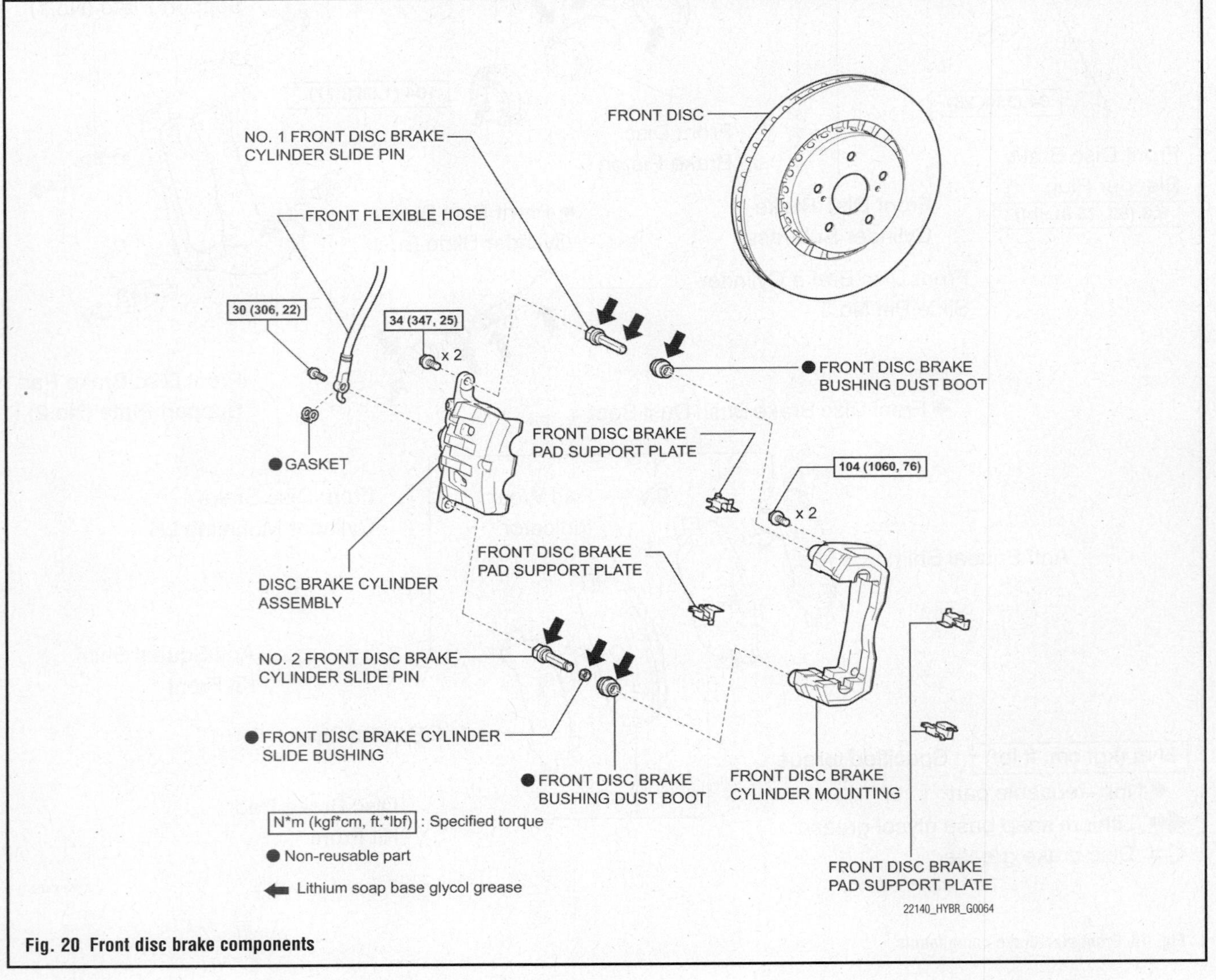

Fig. 20 Front disc brake components

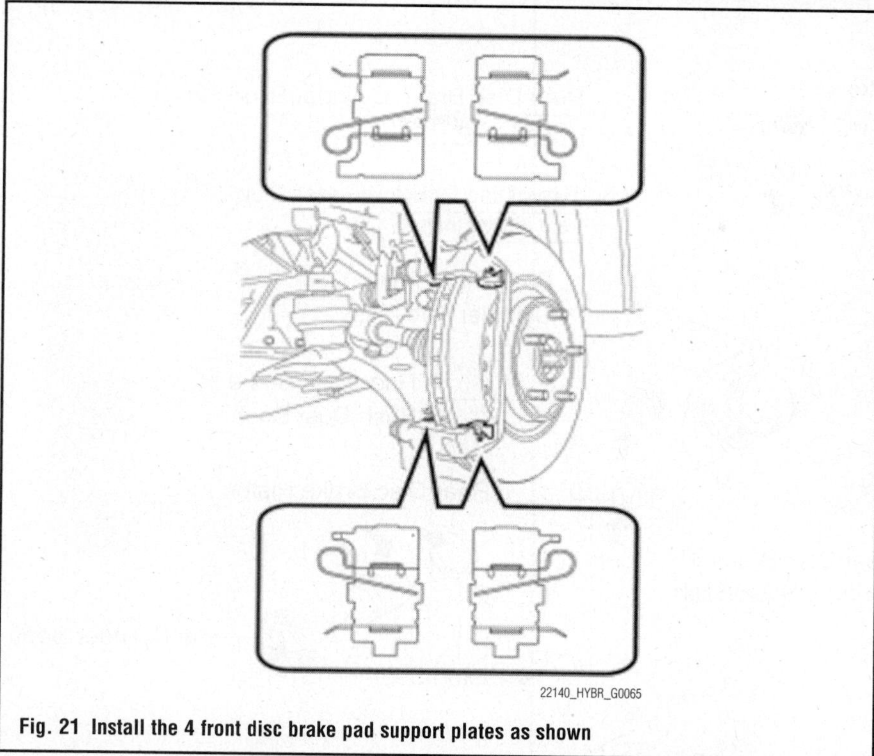

Fig. 21 Install the 4 front disc brake pad support plates as shown

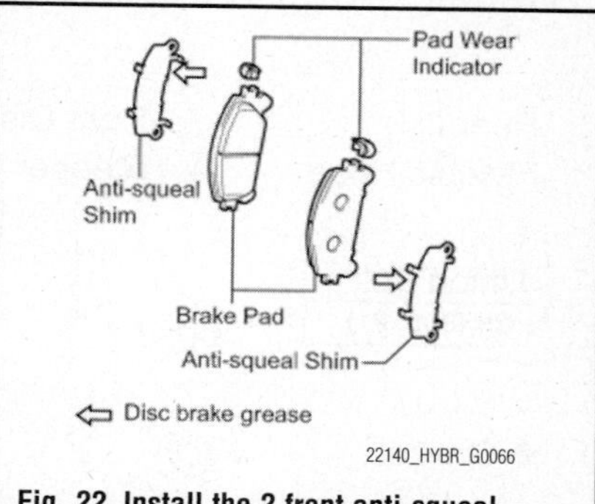

Fig. 22 Install the 2 front anti-squeal shims and the 2 pad wear indicators

To install:

➡Always use new shims and wear indicators, even when re-installing the original pads.

4. Install a wear indicator plate on the inner pad.
5. Apply disc brake grease to both sides of the inner anti-squeal shims and install the shims.
6. Install the inner pad with the wear indicator plate facing upwards.
7. Install the outer pad.
8. Install the caliper. Torque the bolt to 25 ft. lbs. (34 Nm).
9. Install the wheel and tire assembly and carefully lower the vehicle.

Hybrid Models

See Figures 21 and 22.

1. Carefully raise the vehicle.
2. Remove the front wheel.
3. Hold the front disc brake cylinder slide pins and remove the 2 bolts and disc brake cylinder assembly.
4. Remove the 2 anti-squeal springs.
5. Remove the 2 brake pads from the front disc brake cylinder mounting.
6. Remove the 2 anti-squeal shims and the 2 pad wear indicators from the pads.
7. Remove the 4 front disc brake pad support plates from the front disc brake cylinder mounting.

To install:

8. Install the 4 front disc brake pad support plates to the front disc brake cylinder mounting as shown in the illustration.
9. Apply disc brake grease to the inside of the 2 front anti-squeal shims.
10. Install the 2 front anti-squeal shims and the 2 pad wear indicators to the pads.
11. Install the 2 brake pads with the front anti-squeal shims to the front disc brake cylinder mounting.
12. Install the 2 anti-squeal springs to the front disc brake pads.
13. Hold the front disc brake cylinder slide pins and install the disc brake cylinder assembly with the 2 bolts. Tighten the bolts to 25 ft. lbs. (34 Nm).
14. Tighten the wheel to 76 ft. lbs. (103 Nm).

BRAKES

REAR DISC BRAKES

⁂ CAUTION

Dust and dirt accumulating on brake parts during normal use may contain asbestos fibers from production or aftermarket brake linings. Breathing excessive concentrations of asbestos fibers can cause serious bodily harm. Exercise care when servicing brake parts. Do not sand or grind brake lining unless equipment used is designed to contain the dust residue. Do not clean brake parts with compressed air or by dry brushing. Cleaning should be done by dampening the brake components with a fine mist of water, then wiping the brake components clean with a dampened cloth. Dispose of cloth and all residue containing asbestos fibers in an impermeable container with the appropriate label. Follow practices prescribed by the Occupational Safety and Health Administration (OSHA) and the Environmental Protection Agency (EPA) for the handling, processing, and disposing of dust or debris that may contain asbestos fibers.

BRAKE CALIPER

REMOVAL & INSTALLATION

Non-Hybrid Models

See Figure 23.

1. Disconnect the brake line from the caliper and plug it.
2. Remove the caliper mounting bolts.
3. Lift off the caliper.
4. Remove the pads and anti-squeal shims.
5. Remove the wear indicators from each pad.

To install:

6. Installation is the reverse of removal. Grease the caliper slides and bolts with lithium grease or equivalent. Apply disc brake grease to the anti-squeal shims. Torque the caliper bolts to 25 ft. lbs. (34 Nm); the brake line union bolt to 24 ft. lbs. (33 Nm).

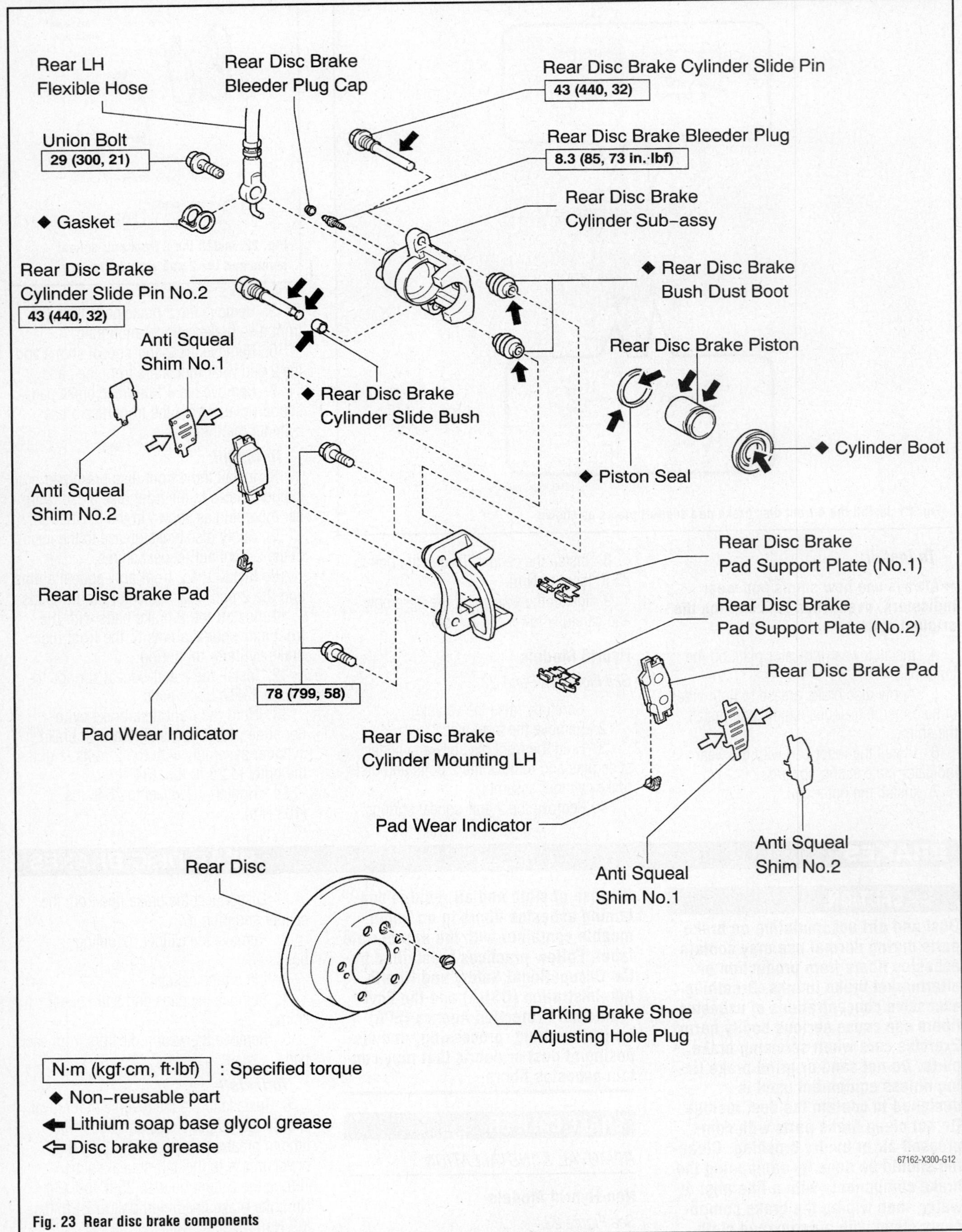
Rear LH
Flexible Hose
Rear Disc Brake
Bleeder Plug Cap
Rear Disc Brake Cylinder Slide Pin
43 (440, 32)
Union Bolt
29 (300, 21)
Rear Disc Brake Bleeder Plug
8.3 (85, 73 in.·lbf)
◆ Gasket
Rear Disc Brake
Cylinder Sub-assy
Rear Disc Brake
Cylinder Slide Pin No.2
43 (440, 32)
◆ Rear Disc Brake
Bush Dust Boot
Anti Squeal
Shim No.1
Rear Disc Brake Piston
◆ Rear Disc Brake
Cylinder Slide Bush
◆ Cylinder Boot
◆ Piston Seal
Anti Squeal
Shim No.2
Rear Disc Brake
Pad Support Plate (No.1)
Rear Disc Brake Pad
Rear Disc Brake
Pad Support Plate (No.2)
Rear Disc Brake Pad
78 (799, 58)
Pad Wear Indicator
Rear Disc Brake
Cylinder Mounting LH
Pad Wear Indicator
Anti Squeal
Shim No.2
Rear Disc
Anti Squeal
Shim No.1
Parking Brake Shoe
Adjusting Hole Plug
N·m (kgf·cm, ft·lbf) : Specified torque
◆ Non-reusable part
Lithium soap base glycol grease
Disc brake grease
67162-X300-G12

Fig. 23 Rear disc brake components

Hybrid Models

See Figure 24.

1. Carefully raise the vehicle.
2. Remove the front wheel and tire assembly.
3. Disconnect the brake line from the caliper and plug it.
4. Remove the caliper mounting bolts.
5. Lift off the caliper.
6. Remove the pads and anti-squeal shims.
7. Remove the wear indicators from each pad.

To install:

8. Install the 2 rear brake pads with rear anti-squeal shims to the rear disc brake cylinder mounting.
9. Hold the rear disc brake cylinder slide pins and install the disc brake cylinder assembly with the 2 bolts. Tighten the mounting bolts to 25 ft. lbs. (34 Nm).
10. Install the brake hose to the caliper, tighten the banjo bolt to 24 ft. lbs. (33 Nm).
11. Tighten the wheel to 76 ft. lbs. (103 Nm).
12. Fill the master cylinder reservoir with brake fluid.
13. Bleed the brake system.
14. Inspect for brake fluid leaks.
15. Perform the accumulator zero down procedure.
16. Inspect the brake fluid level.
17. Install the front wheel and tighten to 76 ft. lbs. (103 Nm).
18. Clear the DTCs.

DISC BRAKE PADS

REMOVAL & INSTALLATION

Non-Hybrid Models

See Figure 23.

1. Disconnect the brake line from the caliper and plug it.
2. Remove the caliper mounting bolts.
3. Lift off the caliper.
4. Remove the pads and anti-squeal shims.
5. Remove the wear indicators from each pad.

To install:

6. Installation is the reverse of removal. Grease the caliper slides and bolts with lithium grease or equivalent. Apply disc brake grease to the anti-squeal shims. Torque the caliper bolts to 25 ft. lbs. (34 Nm); the brake line union bolt to 24 ft. lbs. (33 Nm).

Hybrid Models

See Figure 25.

1. Disconnect the brake line from the caliper and plug it.
2. Remove the caliper mounting bolts.
3. Lift off the caliper.
4. Remove the pads and anti-squeal shims.
5. Remove the wear indicators from each pad.

To install:

6. Installation is the reverse of removal. Grease the caliper slides and bolts with lithium grease or equivalent. Apply disc brake grease to the anti-squeal shims. Tighten the caliper bolts to 32 ft. lbs. (43 Nm); the brake line union bolt to 24 ft. lbs. (33 Nm).

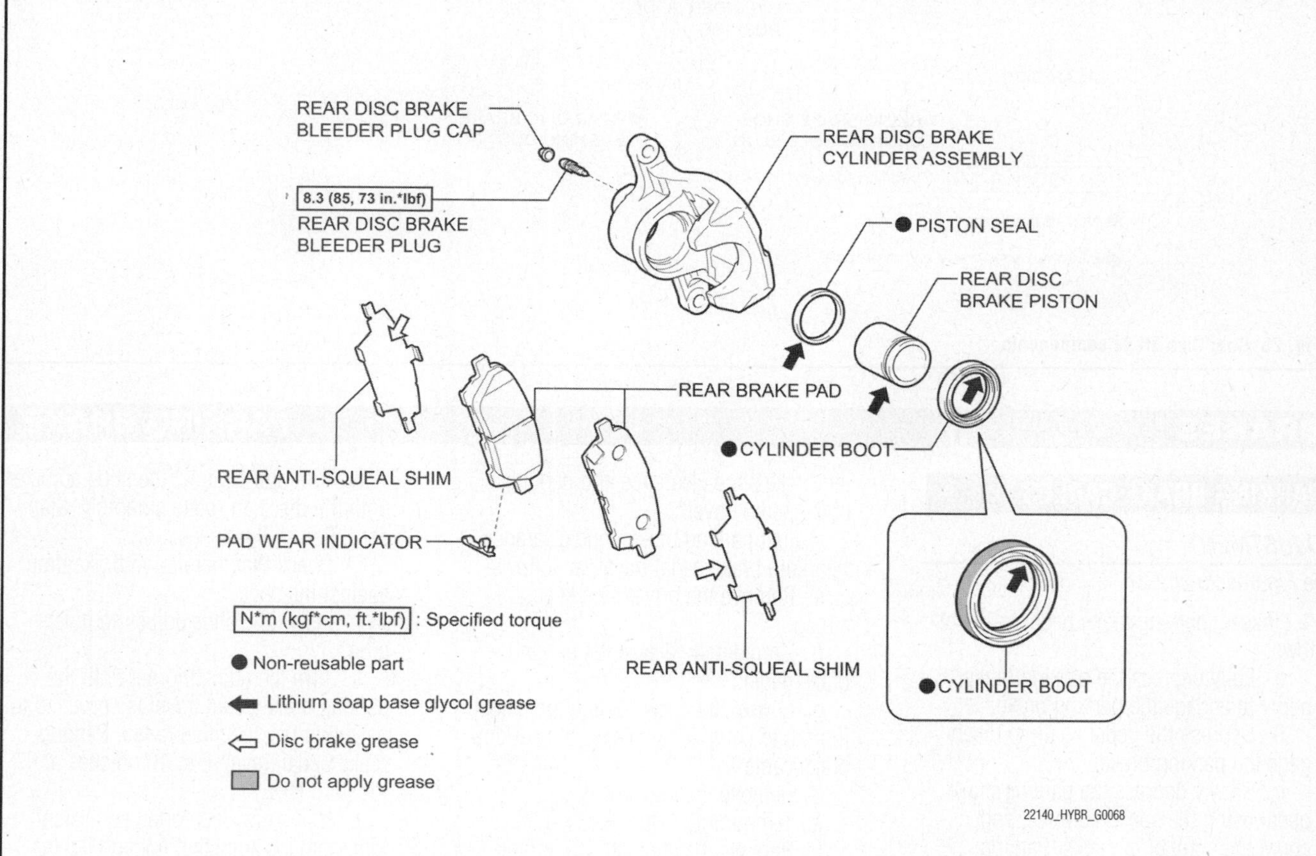

Fig. 24 Rear disc brake components

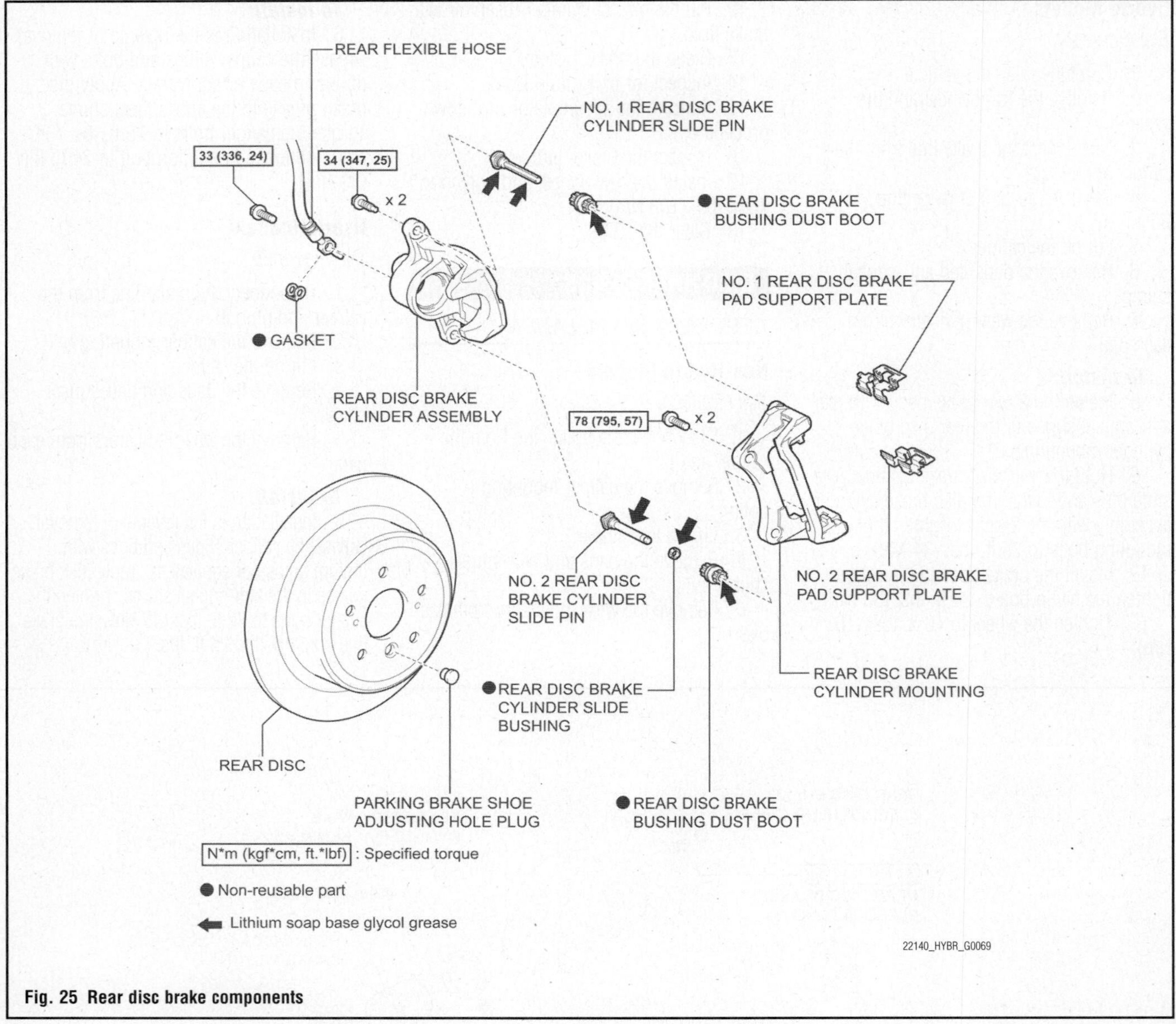

Fig. 25 Rear disc brake components

BRAKES

PARKING BRAKE

PARKING BRAKE CABLES

ADJUSTMENT

See Figures 26 and 27.

1. Inspect parking brake pedal travel, as follows:
   a. Fully depress the parking brake pedal to engage the parking brake.
   b. Depress the pedal again to disengage the parking brake.
   c. Slowly depress the parking brake pedal using the specified force, and count the number of clicks. Parking brake pedal travel: 8 to 10 notches at 67 lbs (300 N). If the parking brake pedal travel is not as specified, adjust the parking brake shoe clearance and parking brake pedal travel.
2. Adjust parking brake shoe clearance and parking brake pedal travel, as follows:
   a. Remove the driver side knee airbag.
   b. Completely release the parking brake pedal.
   c. Loosen the lock nut and the adjusting nut to completely release the parking brake cable.
   d. Remove the rear wheel.
   e. Temporarily install the hub nuts.
   f. Remove the shoe adjusting hole plug.
   g. Turn the shoe adjuster and expand the shoe until the disc locks.
   h. Turn and contract the shoe adjuster until the disc can rotate smoothly. Standard: Return 8 notches.
   i. Check that there is no brake drag against the shoe.
   j. Install the shoe adjusting hole plug.
   k. Turn the adjusting nut until the parking brake pedal travel is corrected to be within the specified range. Parking brake pedal travel: 8 to 10 notches at 67 lbs (300 N).
   l. Using a wrench or an equivalent tool, hold the adjusting nut and tighten the lock nut and tighten to 62 inch lbs. (7 Nm).
   m. Operate the parking brake pedal 3

Fig. 26 Parking brake lock nut and adjusting nut

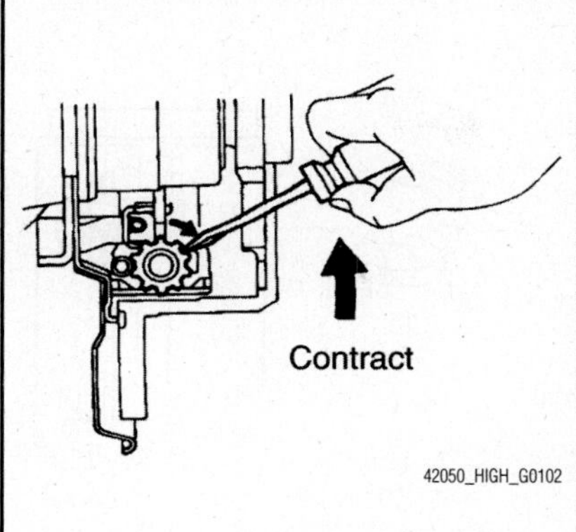

Fig. 28 If you have difficulty removing the rotor, turn the shoe adjuster until the wheel turns freely

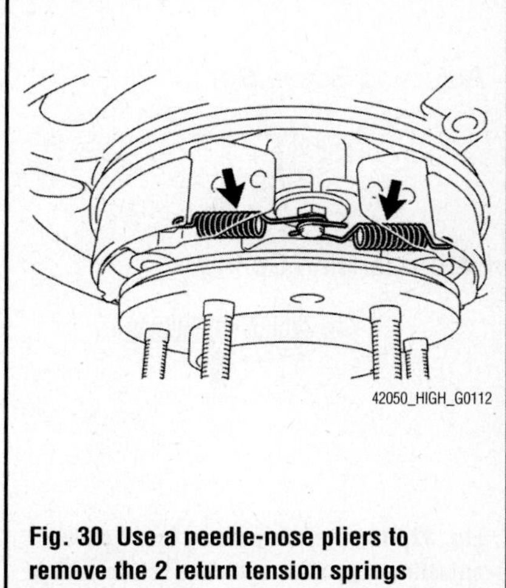

Fig. 30 Use a needle-nose pliers to remove the 2 return tension springs

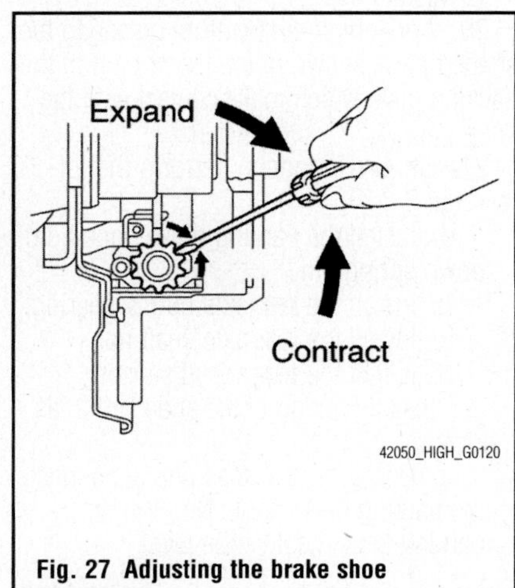

Fig. 27 Adjusting the brake shoe clearance

to 4 times, and check the parking brake pedal travel.

n. Check that there is no brake drag against the shoe.

o. Remove the hub nuts.

p. Install the rear wheel and tighten the lug nuts to 76 ft. lbs. (103 Nm).

q. Install the driver side knee airbag.

3. When operating the parking brake pedal, check that the brake warning light illuminates. Standard: the brake warning light always illuminates at the first click.

PARKING BRAKE SHOES

REMOVAL & INSTALLATION

See Figures 28 through 38.

1. Raise and safely support the vehicle.
2. Remove the rear wheel and tire assemblies.
3. Unbolt and remove the rear caliper, but do not disconnect the fluid line. Suspend the caliper out of the way with a piece of wire.
4. Matchmark the brake disc (rotor) to the axle hub.
5. Make sure the parking brake is fully released, then remove the rear brake disc (rotor).

➡If the rotor cannot be easily removed, turn the shoe adjuster until the wheel turns freely.

6. Inspect the brake disc (rotor) inside diameter, as follows:
   a. Using a brake drum gauge or equivalent, measure the inside diameter of the disc and compare with the following: Standard inside diameter: 190 mm (7.48 in.). Maximum inside diameter: 191 mm (7.52 in.)
   b. If the inside diameter exceeds the maximum, replace the brake disc.
7. Use needle-nose pliers to remove the 2 parking brake shoe return tension springs.
8. Remove the parking brake shoe strut, as follows:
   a. Remove the parking brake shoe strut and the parking brake shoe strut compression spring.
9. Remove parking brake shoe No. 1, as follows:
   a. Remove the parking brake shoe hold down spring cup No. 1, parking brake shoe hold down spring and parking brake shoe hold down spring cup No. 2.
   b. FWD vehicles, remove the parking brake shoe hold down spring pin No. 1.
   c. Disconnect the parking brake shoe return spring No. 2 and remove the parking brake shoe assembly lh No. 1.
10. Remove parking brake shoe adjusting screw set:
   a. Remove the parking brake shoe adjusting screw set.
   b. Remove the parking brake shoe return tension spring No. 2.

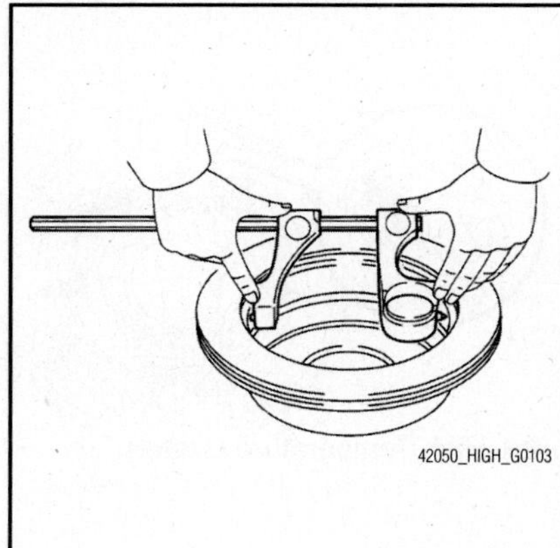

Fig. 29 Use a brake drum gauge to measure the inside of the disc

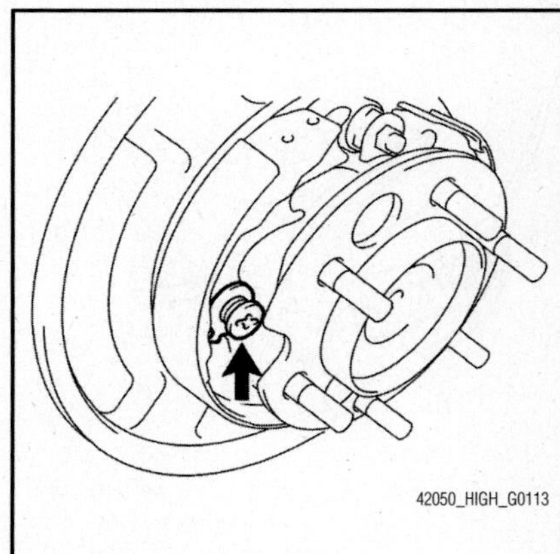

Fig. 31 Remove the parking brake shoe hold-down spring cups

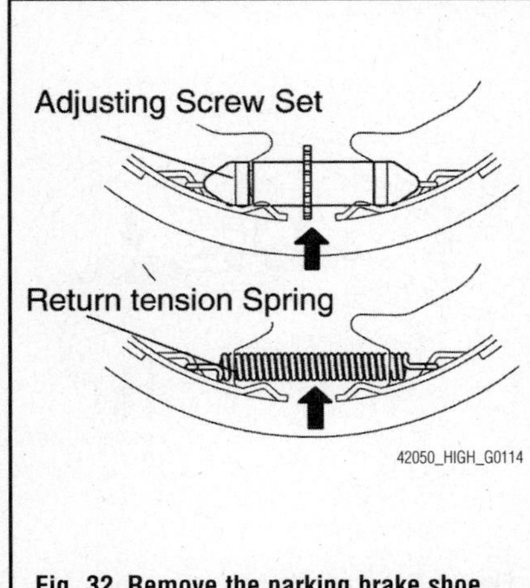

Fig. 32 Remove the parking brake shoe adjusting screw set and shoe return spring

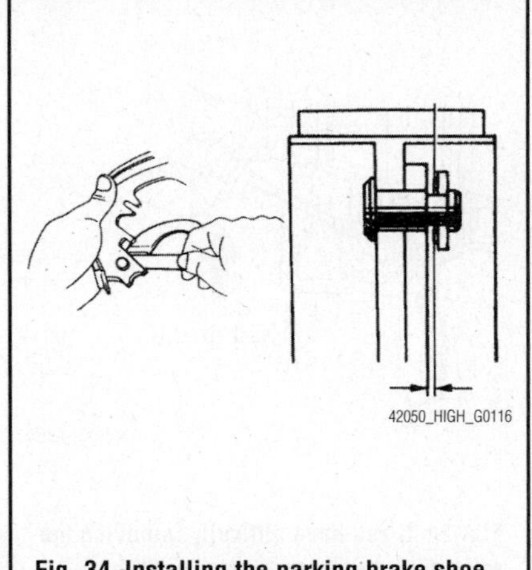

Fig. 34 Installing the parking brake shoe type C-washer

11. Remove parking brake shoe assembly No. 2:
    a. Remove the parking brake shoe hold down spring cup No. 1, parking brake shoe hold down spring, parking brake shoe hold down spring cup No. 2 and parking brake shoe hold down spring pin No. 2.
    b. Remove the parking brake shoe assembly lh No. 2.
    c. Using needle-nose pliers, disconnect the parking brake cable No. 3 from the parking brake cable shoe lever.

WARNING

Be careful not to damage parking brake cable No. 3.

12. On 4WD models, separate the rear speed sensor.
13. On 4WD models, remove the rear axle shaft nut.
14. On 4WD models, remove rear axle hub & bearing assembly
15. On 4WD models, remove parking brake shoe hold down spring pin.
16. Remove parking brake shoe type C-washer, as follows:
    a. Using a screwdriver, remove the c-washer.
    b. Remove the shim and parking brake shoe lever from the parking brake shoe No. 2.
17. Inspect parking brake shoe lining thickness:
    a. Using a ruler, measure the thickness of the shoe lining. Standard thickness is 2.5 mm (0.098 in.) and minimum thickness is 1.0 mm (0.039 in.). If the lining thickness is less than or equal to the minimum, or if there is severe or uneven wear, replace the brake shoe.
18. Inspect brake disc and parking brake shoe lining for proper contact
    a. Apply chalk to the inside surface of the disc, then grind down the brake shoe lining to fit disc.
    b. If the contact between the brake disc and the shoe lining is improper, repair it using a brake shoe grinder or replace the brake shoe assembly.

To install:

19. Install the parking brake shoe type C-washer, as follows
    a. Using a feeler gauge, measure the clearance. Standard clearance: less than 0.35 mm (0.014 in.). If the clearance is not within the specifications, replace the shim with one of the correct size. The shim sizes: 0.3 mm (0.012 in.), 0.9 mm (0.035 in.) or 0.6 mm (0.024 in.).
    b. Using pliers, install the parking brake shoe lever and the shim with a new C-washer.
20. Apply high temperature grease to the shaded parts shown in the illustration of the backing plate which make contact with the shoe.
21. On 4WD models, perform the following:
    a. Install the parking brake shoe hold down spring pin.
    b. Install the rear axle hub & bearing.
    c. Install the rear axle shaft nut.
    d. Install the rear speed sensor.
22. Install parking brake shoe No. 2, as follows:
    a. Using needle-nose pliers, connect the parking brake cable No. 3 to the parking brake cable shoe lever.

➡Be careful not to damage the parking brake cable No. 3.

    b. Install the parking brake shoe No. 2 with the parking brake shoe hold down spring, parking brake shoe hold down spring cup No. 1, parking brake shoe

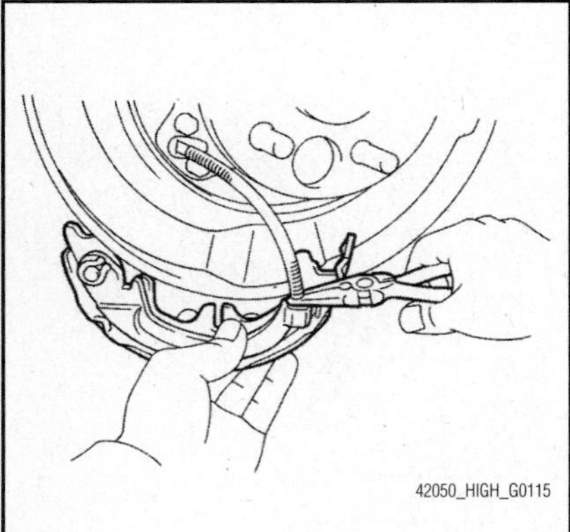

Fig. 33 Using needle-nose pliers, disconnect the parking brake cable No. 3 from the parking brake cable shoe lever

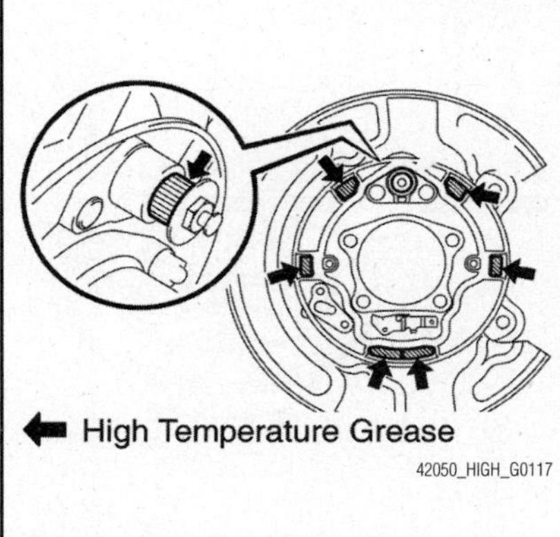

Fig. 35 Apply high temperature grease to the shaded parts of the backing plate which make contact with the shoe

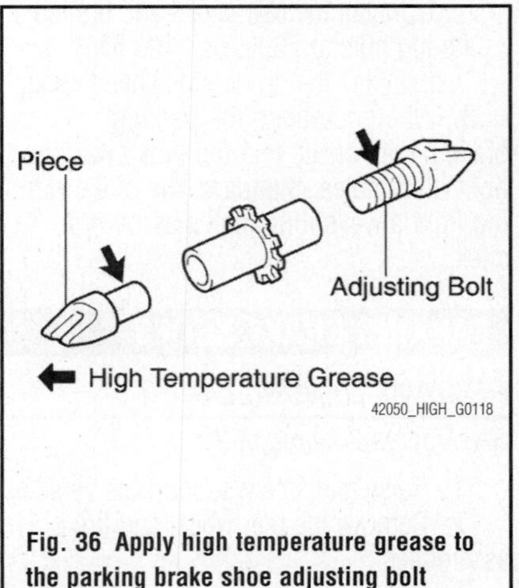

Fig. 36 Apply high temperature grease to the parking brake shoe adjusting bolt and piece

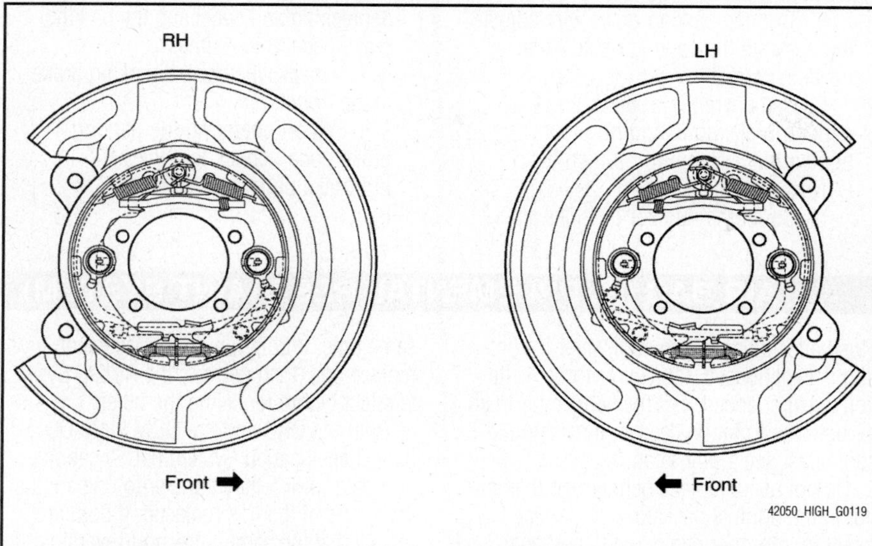

Fig. 37 Check that the parking brake components are properly installed. There should be no oil or grease on the friction surface of the shoe lining and disc

hold down spring cup No. 2 and parking brake shoe hold down spring pin No. 2.

23. Install the parking brake shoe adjusting screw set, as follows:
    a. Apply high temperature grease to the parking brake shoe adjusting bolt and piece.
    b. Attach the parking brake shoe return tension spring No. 2 to the parking brake shoe No. 1 and parking brake shoe assembly No. 2.
    c. Attach the parking brake shoe adjusting screw set to the parking brake shoe No. 1 and parking brake shoe No. 2.
24. Install parking brake shoe No. 1:
    a. For FWD models, install the parking brake shoe hold down spring pin No. 1.
    b. Install the parking brake shoe No. 1 with the parking brake shoe hold down spring, parking brake shoe hold down spring cup No. 2, parking brake shoe hold down spring cup No. 2.
25. Attach the parking brake shoe strut and the parking brake shoe strut compression spring to parking brake shoe No. 2 and parking brake shoe No. 1.
26. Install parking brake shoe return tension spring using needle-nose pliers as shown in the illustration.

➡First install the front side spring then the rear side spring.

27. Check that the parking brake components are properly installed.

**** WARNING**

There should be no oil or grease on the friction surface of the shoe lining and disc.

28. For 4WD models, inspect the bearing backlash and axle hub deviation.
29. Install the rear disc (rotor), aligning the matchmarks made during removal.
30. Adjust parking brake shoe clearance, as follows:
    a. Temporarily install the hub nuts.
    b. Remove the hole plug, turn the adjuster and expand the shoes until the disc locks.
    c. Contract the shoe adjuster until the disc rotates smoothly. Standard : return 8 notches
    d. Check that the shoe has no brake drag.
    e. Install the hole plug.
31. Install the caliper, as outlined earlier in this section.
32. Install the rear wheel and tire assembly and tighten the lug nuts to 76 ft. lbs. (103 Nm).
33. Inspect and adjust the parking brake pedal travel, as outlined in this section.
34. For 4WD models, check the ABS speed sensor signal.

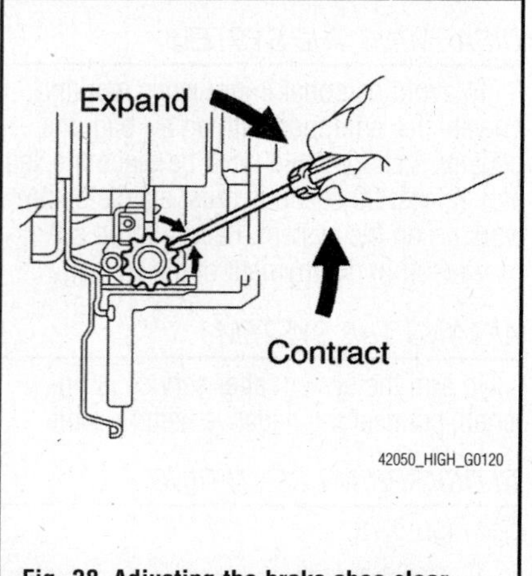

Fig. 38 Adjusting the brake shoe clearance

ADJUSTMENT

See Figures 38 and 39.

1. Raise and safely support the vehicle.
2. Remove the rear wheel and tire assemblies.
3. Adjust parking brake shoe clearance, as follows:
    a. Temporarily install the hub nuts.
    b. Remove the hole plug, turn the adjuster and expand the shoes until the disc locks.
    c. Contract the shoe adjuster until the disc rotates smoothly. Standard : return 8 notches
    d. Check that the shoe has no brake drag.
    e. Install the hole plug.
4. Install the rear wheel and tire assembly and tighten the lug nuts to 76 ft. lbs. (103 Nm).
5. Inspect the parking brake pedal travel, as follows:
    a. Firmly step on the parking brake pedal.
    b. Release the parking brake.
    c. Once more, slowly depress the parking brake pedal all the way, and count the number of clicks The parking brake pedal should travel 8 to 10 clicks at 67 lbs. (300 N).
6. If necessary, adjust parking brake pedal travel, as follows:
    a. Remove the lower instrument panel finish panel sub-assembly.

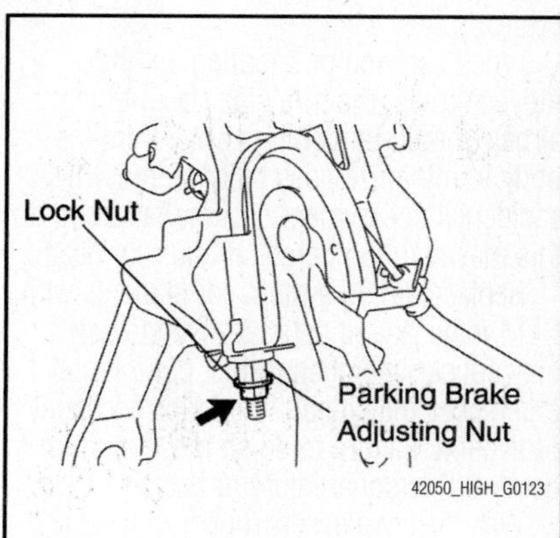

Fig. 39 Use a wrench to hold the parking brake adjusting nut secure while tightening the lock nut

b. Remove the lower instrument panel insert sub- assembly.

c. Depress the parking brake pedal 5 clicks to make room for the procedure, and loosen the lock nut with fixing adjusting nut by wrench.

d. Release the parking brake pedal to the original position.

e. Turn the parking brake wire adjusting nut until the parking brake pedal travel is correct.

f. Use a wrench to hold the parking brake adjusting nut, then tighten the lock nut to 62 inch lbs. (7 Nm).

g. Count the number of clicks after depressing and canceling the parking brake pedal 3 to 4 times.

h. Check whether the parking brake drags or not.

i. When operating the parking brake pedal, check that the parking brake pedal indicator light is lit.

CHASSIS ELECTRICAL

AIR BAG (SUPPLEMENTAL RESTRAINT SYSTEM)

GENERAL INFORMATION

CAUTION

These vehicles are equipped with an air bag system. The system must be disarmed before performing service on, or around, system components, the steering column, instrument panel components, wiring and sensors. Failure to follow the safety precautions and the disarming procedure could result in accidental air bag deployment, possible injury and unnecessary system repairs.

SERVICE PRECAUTIONS

Disconnect and isolate the battery negative cable before beginning any airbag system component diagnosis, testing, removal, or installation procedures. Allow system capacitor to discharge for two minutes before beginning any component service. This will disable the airbag system. Failure to disable the airbag system may result in accidental airbag deployment, personal injury, or death.

Do not place an intact undeployed airbag face down on a solid surface. The airbag will propel into the air if accidentally deployed and may result in personal injury or death.

When carrying or handling an undeployed airbag, the trim side (face) of the airbag should be pointing away from the body to minimize possibility of injury if accidental deployment occurs. Failure to do this may result in personal injury or death.

Replace airbag system components with OEM replacement parts. Substitute parts may appear interchangeable, but internal differences may result in inferior occupant protection. Failure to do so may result in occupant personal injury or death.

Wear safety glasses, rubber gloves, and long sleeved clothing when cleaning powder residue from vehicle after an airbag deployment. Powder residue emitted from a deployed airbag can cause skin irritation. Flush affected area with cool water if irritation is experienced. If nasal or throat irritation is experienced, exit the vehicle for fresh air until the irritation ceases. If irritation continues, see a physician.

Do not use a replacement airbag that is not in the original packaging. This may result in improper deployment, personal injury, or death.

The factory installed fasteners, screws and bolts used to fasten airbag components have a special coating and are specifically designed for the airbag system. Do not use substitute fasteners. Use only original equipment fasteners listed in the parts catalog when fastener replacement is required.

During, and following, any child restraint anchor service, due to impact event or vehicle repair, carefully inspect all mounting hardware, tether straps, and anchors for proper installation, operation, or damage. If a child restraint anchor is found damaged in any way, the anchor must be replaced. Failure to do this may result in personal injury or death.

Deployed and non-deployed airbags may or may not have live pyrotechnic material within the airbag inflator.

Do not dispose of driver/passenger/curtain airbags or seat belt tensioners unless you are sure of complete deployment. Refer to the Hazardous Substance Control System for proper disposal.

Dispose of deployed airbags and tensioners consistent with state, provincial, local, and federal regulations.

After any airbag component testing or service, do not connect the battery negative cable. Personal injury or death may result if the system test is not performed first.

If the vehicle is equipped with the Occupant Classification System (OCS), do not connect the battery negative cable before performing the OCS Verification Test using the scan tool and the appropriate diagnostic information. Personal injury or death may result if the system test is not performed properly.

Never replace both the Occupant Restraint Controller (ORC) and the Occupant Classification Module (OCM) at the same time. If both require replacement, replace one, then perform the Airbag System test before replacing the other.

Both the ORC and the OCM store Occupant Classification System (OCS) calibration data, which they transfer to one another when one of them is replaced. If both are replaced at the same time, an irreversible fault will be set in both modules and the OCS may malfunction and cause personal injury or death.

If equipped with OCS, the Seat Weight Sensor is a sensitive, calibrated unit and must be handled carefully. Do not drop or handle roughly. If dropped or damaged, replace with another sensor. Failure to do so may result in occupant injury or death.

If equipped with OCS, the front passenger seat must be handled carefully as well. When removing the seat, be careful when setting on floor not to drop. If dropped, the sensor may be inoperative, could result in occupant injury, or possibly death.

If equipped with OCS, when the passenger front seat is on the floor, no one should sit in the front passenger seat. This uneven force may damage the sensing ability of the seat weight sensors. If sat on and damaged, the sensor may be inoperative, could result in occupant injury, or possibly death.

DISARMING THE SYSTEM

To avoid personal injury when working on vehicles equipped with an air bag, the negative battery cable must be disconnected and at least 90 seconds must elapse before working on the system. Failure to do so may result in deployment of the air bag.

ARMING THE SYSTEM

To arm the system after service is finished, connect the negative battery cable.

CLOCKSPRING CENTERING

See Figure 40.

1. Check that the front wheels are facing straight ahead.

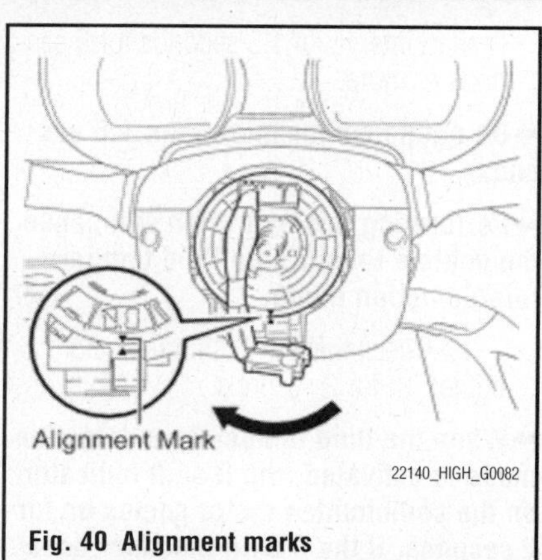

Fig. 40 Alignment marks

2. Set the turn signal switch to the neutral position.

WARNING

If it is not in the neutral position, the turn signal switch pin may be snapped.

3. Check that the battery negative (-) cable is disconnected.

CAUTION

Wait for at least 90 seconds after disconnecting the cable to prevent airbag deployment.

4. Rotate the spiral cable (clockspring) counterclockwise slowly by hand until it stops.

WARNING

Do not turn the spiral cable using the airbag wire harness.

5. Rotate the spiral cable clockwise approximately 2.5 turns to align the marks.

➡The spiral cable will rotate approximately 2.5 turns to both the left and right from the center.

DRIVE TRAIN

AUTOMATIC TRANSAXLE FLUID

DRAIN AND REFILL

U760E Automatic Transaxle

See Figure 42.

1. The U760E automatic transaxle does not have an oil filler tube and oil level gauge. When adding fluid, add fluid through the refill hole on the transaxle case. The fluid level can be adjusted by draining excess fluid (allowing excess fluid to overflow) through the overflow tube of the oil pan.
2. Remove the drain plug and drain the fluid into a suitable container.

➡"Overflow" indicates the condition under which fluid comes out of the overflow plug hole.

➡Before adjusting the fluid level, add the specified amount of fluid when the engine is cold and warm up the engine to circulate the fluid in the transaxle. Ensure that the fluid temperature is as specified and the engine is idling.

➡The U760E automatic transaxle requires Toyota Genuine ATF WS.

➡The adjustment should be performed according to the procedures referenced in the work flow below.

➡Set the vehicle on a lift so that the vehicle is kept level when it is lifted up.

3. Before filling transaxle with fluid:
   a. Lift the vehicle.
   b. Remove the front fender liner LH and front fender apron seal LH.

➡If the transaxle is hot (ATF temperature is high), wait until the fluid tem-

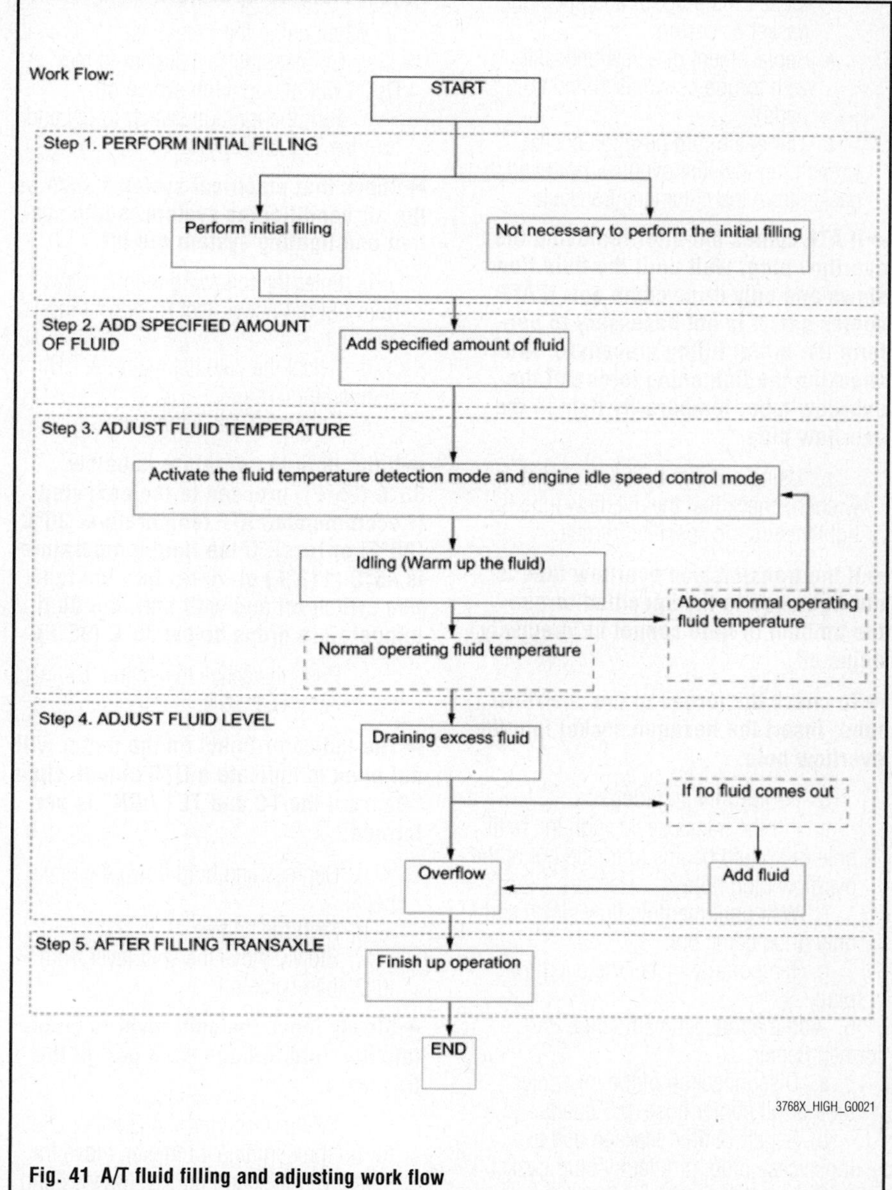

Fig. 41 A/T fluid filling and adjusting work flow

perature becomes the same as the ambient temperature before starting the following procedure. (Recommended ATF temperature: around 68° F (20° C).

4. Perform initial filling:
   a. Remove the refill plug and gasket from the automatic transaxle.

➡After performing any of the following operations, it is not necessary to perform the initial filling procedure. Proceed to the next "ADD SPECIFIED AMOUNT OF FLUID" procedure.

➡Operation that does not require initial filling:

- Disconnection of the oil cooler tube or oil cooler hose
- Repair of fluid leakage due to a loose case plug, or a faulty plug gasket or O-ring
- Replacement of a new transaxle with torque converter (filled fluid parts)

   b. Using a 6 mm hexagon socket wrench, remove the overflow plug and gasket from the automatic transaxle.

➡If ATF comes out after removing the overflow plug, wait until the fluid flow slows and only drips come out. If ATF comes out, it is not necessary to perform the initial filling procedure. After checking the tightening torque of the overflow tube, temporarily tighten the overflow plug.

   c. Using a 6 mm hexagon socket wrench, check that the overflow tube is tightened to the specified torque.

➡If the transmission overflow tube is not tightened to the specified torque, the amount of fluid cannot be precisely adjusted.

➡To check the torque of the overflow tube, insert the hexagon socket into the overflow hole.

   d. Perform initial filling.
   e. Fill the transaxle through the refill hole until fluid begins to trickle out of the overflow plug hole.
   f. Wait until the fluid flow slows and only drips come out.
   g. Temporarily install the overflow plug.

5. Add the following amounts: Performed Repair:
   a. Disconnection of the oil cooler tube or oil cooler hose: 0.5 quarts.
   b. Repair of fluid leakage due to a loose case plug, or a faulty plug gasket or O-ring: 0.5 quarts.
   c. Replacement of a new transaxle with torque converter (filled fluid parts): 0.5 quarts.
   d. Removal & installation of drive shaft: 2.9 quarts.
   e. Removal & installation of transaxle case oil seal: 2.9 quarts.
   f. Removal & installation of oil pan: 2.9 quarts.
   g. Removal & installation of transaxle (remove torque converter): 3.6 quarts.
   h. Replacement of transaxle (torque converter new): 5.4 quarts.
   i. Replacement of transaxle (torque converter reused): 3.7 quarts
6. Temporarily install the refill plug to avoid fluid splash.
7. Lower the vehicle.

Adjust Fluid Temperature

1. When using the Techstream:
   a. Connect the Techstream to the DLC3 with the ignition switch off.
   b. Turn the ignition switch to ON and turn the Techstream on.

➡Check that electrical systems such as the air conditioning system, audio system and lighting system are off.

   c. Enter the following menus: Powertrain / ECT / Active Test / Connect the TC and TE1.
   d. Select the date list menu: A/T Oil Temperature 1.
   e. Check A/T Oil Temperature 1.

➡If the fluid temperature is below 35°C (95°F), proceed to the next step. (Recommended ATF temperature: 30°C [86°F] or less). If the fluid temperature is 45°C (113°F) or more, turn the ignition switch off and wait until the fluid temperature drops below 35°C (95°F).

   f. Push the active test menu: Connect the TC and TE1 / ON.

➡The indicator lights on the meter will not blink to indicate a DTC output when "Connect the TC and TE1 / ON" is performed.

   g. Depress and hold down the brake pedal.
   h. Start the engine.
   i. Slowly move the shift lever from P to D, then back to P.

➡Slowly move the shift lever to circulate the fluid through each part of the transaxle.

   j. While observing the D shift indicator on the combination meter, move the shift lever back and forth between N and D at an interval of 1.5 seconds for 6 seconds or more.

➡Do not pause for more than 1.5 seconds.

➡Performing this operation will cause the vehicle to enter the fluid temperature detection mode.

   k. Check that the D shift indicator comes on for 2 seconds.

➡When the fluid temperature detection mode is activated, the D shift indicator on the combination meter comes on for 2 seconds. If the D shift indicator does not come on for 2 seconds, return to the step where terminal TC is first turned on and perform the procedure again.

   l. Move the shift lever from N to P.
   m. Release the brake pedal.
   n. Push the active test menu: TC and TE1/ OFF.

➡Be sure to turn terminals TC and TE1 off. If the terminals are on, the fluid level cannot be precisely adjusted due to fluctuations in idle speed.

➡Turning terminals TC and TE1 off activates the engine idle speed control mode. In the engine idle speed control mode, engine idle speed control starts when the fluid temperature becomes 35°C (95°F) or more and the engine speed is maintained at approximately 800 rpm. Even after terminals TC and TE1 are turned off, the fluid temperature detection mode is active until the ignition switch is turned off.

   o. Warm up the engine with the engine idling until the fluid temperature reaches the normal operating temperature (35 to 45°C [95 to 113°F]).
   - Below normal operating temperature: 95° F or less
   - Normal operating temperature: 95 to 113° F
   - Above normal operating temperature: 113° F or more

➡If the fluid temperature is within the normal operating temperature range, immediately proceed to the "ADJUST FLUID LEVEL" procedure. If the fluid temperature is 45°C (113°F) or more, stop the engine and wait until the fluid temperature drops to 35°C (95°F) or less. Then perform the "ADJUST FLUID TEMPERATURE" procedure again from the beginning.

➡In the fluid temperature detection mode, the D shift indicator comes on,

goes off, or blinks depending on the fluid temperature.

p. D shift indicator:

- Below normal operating temperature: 95° F or less, the indicator is OFF.
- Normal operating temperature: 95 to 113° F, the indicator is ON.
- Above normal operating temperature: 113° F or more, the indicator is BLINKING.

2. When not using the Techstream:

a. Using SST 09843-18040, connect terminals 13 (TC) and 4 (CG) of the DLC3 with the ignition switch off.

b. Depress and hold down the brake pedal.

c. Start the engine.

➡Check that electrical systems such as the air conditioning system, audio system and lighting system are off.

➡Indicator lights of the meter do the DTC output blinking when terminals TC and CG are connected.

d. Slowly move the shift lever from P to D, then back to P.

➡Slowly move the shift lever to circulate the fluid through each part of the transaxle.

e. While observing the D shift indicator on the combination meter, move the shift lever back and forth between N and D at an interval of 1.5 seconds for 6 seconds or more.

➡Do not pause for more than 1.5 seconds.

➡Performing this operation will cause the vehicle to enter the fluid temperature detection mode.

- Check that the D shift indicator comes on for 2 seconds.

➡When the fluid temperature detection mode is activated, the D shift indicator on the combination meter comes on for 2 seconds. If the D shift indicator does not come on for 2 seconds, return to the step where terminal TC is first turned on and perform the procedure again.

f. Move the shift lever from N to P.

g. Release the brake pedal.

h. Remove SST 09843-18040 from terminals 13 (TC) and 4 (CG).

➡Be sure that terminals TC and CG are not connected. If the terminals are connected, the fluid level cannot be precisely adjusted due to fluctuations in engine speed.

➡When terminals TC and CG are disconnected, the engine idle control mode is activated. In the engine idle speed control mode, engine idle speed control starts when the fluid temperature becomes 35°C (95°F) or more and the engine speed is maintained at approximately 800 rpm. Even after terminals TC and CG are disconnected, the fluid temperature detection mode is active until the ignition switch is turned off.

i. Allow the engine to idle until the D shift indicator comes on again.

➡If the fluid temperature is within the normal operating temperature range, immediately proceed to the "ADJUST FLUID LEVEL" procedure. If the fluid temperature is 45°C (113°F) or more, stop the engine and wait until the fluid temperature drops to 35°C (95°F) or less. Then perform the "ADJUST FLUID TEMPERATURE" procedure again from the beginning.

➡In the fluid temperature detection mode, the D shift indicator comes on, goes off, or blinks depending on the fluid temperature.

j. D shift indicator:

- Below normal operating temperature: 95° F or less, the indicator is OFF.
- Normal operating temperature: 95 to 113° F, the indicator is ON.
- Above normal operating temperature: 113° F or more, the indicator is BLINKING.

Adjust Fluid Level

⁂ CAUTION

Use caution as the engine is running and the radiator fan is operating during the following procedure.

1. Lift the vehicle.
2. Adjust fluid level.

a. Using a 6 mm hexagon socket wrench, remove the overflow plug and gasket.

⁂ CAUTION

Be careful as the fluid coming out of the overflow hole is hot.

b. Check the amount of fluid that comes out of the overflow plug hole.

➡If only a small amount of fluid (approximately 1 cc) comes out of the overflow hole, then only fluid remaining in the overflow tube has come out. This condition is not considered as overflow, so it is necessary to add fluid.

c. If the amount of fluid that comes out of the overflow hole is large, wait until the fluid flow slows and only drips come out.

d. If no fluid comes out of the overflow hole, remove the refill plug and gasket. Then add transaxle fluid through the refill hole until fluid comes out of the overflow hole. Wait until the fluid flow slows and only drips come out.

➡Use Toyota Genuine ATF WS.

e. Check that the fluid flow has slowed and only drips come out.

➡The fluid flow will not completely stop because the fluid expands as its temperature increases.

3. Install the overflow plug with a new gasket. Tighten to 30 ft. lbs. (40 Nm).
4. Install the refill plug with a new gasket. Tighten to 36 ft. lbs. (49 Nm).
5. Lower the vehicle.
6. Turn the ignition switch off.

➡Turning the ignition switch off exits the fluid temperature detection mode.

7. Remove the Techstream from the DLC3 (when using the Techstream).

After Filling

1. Lift the vehicle.
2. Clean each part.
3. Check for fluid leaks.
4. Install the front fender liner LH and front fender apron seal LH.
5. Lower the vehicle.

FILTER REPLACEMENT

U760e Transaxle

See Figures 42 through 44.

1. Drain the ATF fluid into a suitable container.
2. Remove the 19 bolts, automatic transaxle oil pan sub-assembly, and automatic transaxle oil pan gasket from the automatic transaxle assembly.
3. Remove the 2 transmission oil cleaner magnets from the automatic transaxle oil pan sub-assembly. Examine the chips and particles in the pan and on the magnet to determine what type of wear has occurred in the transaxle:

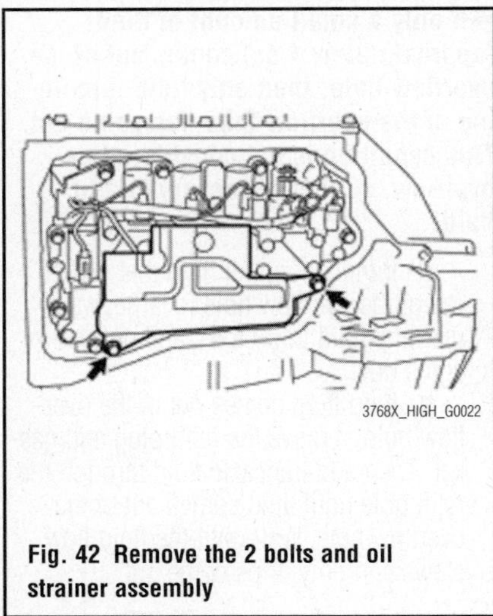

Fig. 42 Remove the 2 bolts and oil strainer assembly

a. Steel (magnetic): Wear of the bearing, gear, and plate.

b. Brass (non-magnetic): Wear of the bushing.

4. Remove the 2 bolts and oil strainer assembly from the valve body assembly.

5. Remove the O-ring from the oil strainer assembly.

To install:

6. Coat a new O-ring with ATF and install it to the oil strainer assembly.

➡Ensure that the O-ring is not twisted or pinched.

7. Install the oil strainer assembly to the valve body assembly with 2 new bolts. Tighten to 85 inch lbs. (9.6 Nm).

8. Install the 2 magnets in the automatic transaxle oil pan sub-assembly.

9. Install the automatic transaxle oil pan sub-assembly and a new gasket to the transaxle case with the 17 bolts. Tighten to 66 inch lbs. (7.5 Nm).

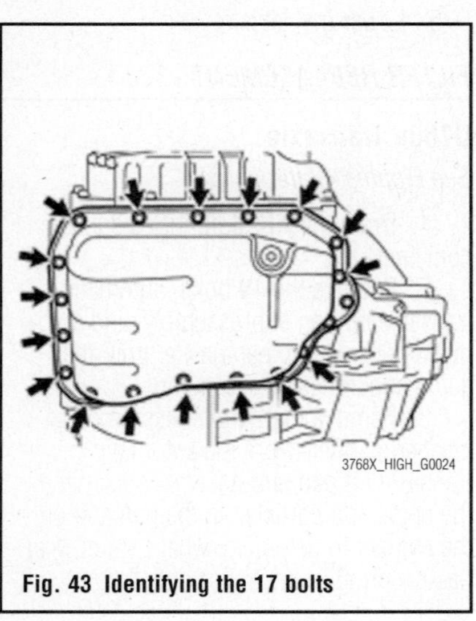

Fig. 43 Identifying the 17 bolts

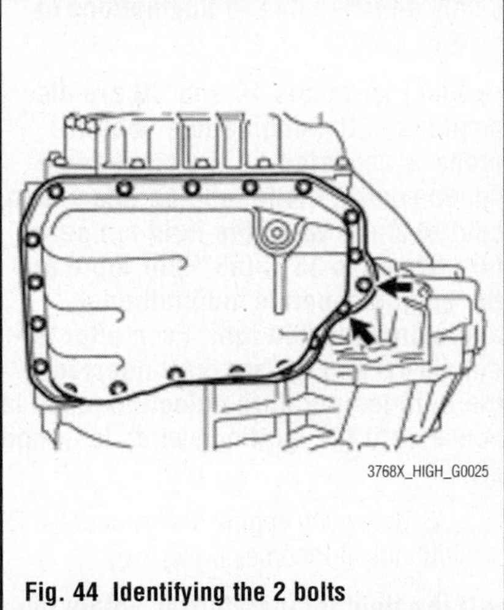

Fig. 44 Identifying the 2 bolts

➡Completely remove any oil or grease from the contact surfaces of the transaxle case and oil pan sub-assembly with the gasket before installation.

10. Apply adhesive to 2 new bolts. Sealant: Toyota Genuine Adhesive 1344, Three Bond 1344 or equivalent.

11. Install the automatic transaxle oil pan sub-assembly with the 2 bolts. Tighten to 62 inch lbs (7 Nm).

➡In order to ensure proper sealing of the transmission pan bolts, apply adhesive to the bolts and install them within 10 minutes of adhesive application.

U151e & U151F Transaxle

See Figure 45.

1. Drain the ATF fluid into a suitable container.

2. Remove the 18 bolts, oil pan, and gasket.

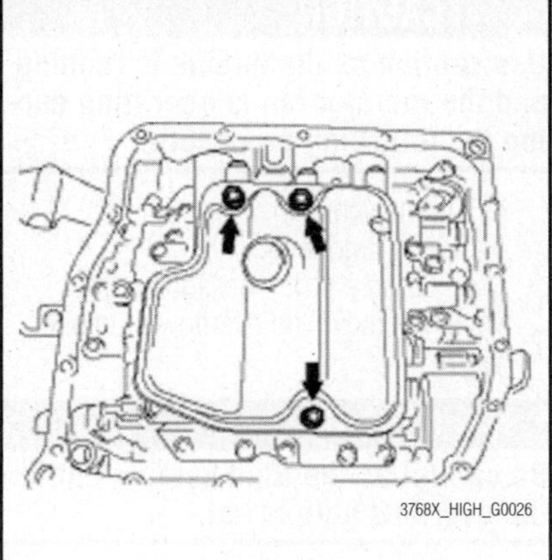

Fig. 45 Removing the filter and 3 bolts

3. Remove the 2 magnets from the oil pan.

4. Remove the magnets and use them to collect any steel chips. Examine the chips and particles in the pan and on the magnet to determine what type of wear has occurred in the transaxle:

a. Steel (magnetic): Wear of the bearing, gear, and plate.

b. Brass (non-magnetic): Wear of the bushing.

5. Remove the 3 bolts and oil strainer.

6. Remove the O-ring from the oil strainer.

To install:

7. Coat a new O-ring with ATF and install it to the oil strainer assembly.

➡Ensure that the O-ring is not twisted or pinched. Apply sufficient ATF to the O-ring before installation.

8. Install the oil strainer assembly to the valve body assembly with 3 new bolts. Apply ATF to the bolts. Tighten to 8 ft. lbs. (11 Nm).

9. Install the 2 magnets to the oil pan.

10. Apply adhesive to 18 new bolts. Adhesive: Toyota Genuine Adhesive 1344, Three Bond 1344 or equivalent.

11. Install the oil pan and new oil pan gasket to the transaxle case with the 18 bolts. Tighten to 69 inch lbs. (7.8 Nm).

➡Apply adhesive to the bolts and tighten them within 10 minutes of application. Completely remove any oil or grease from the contact surface of the transaxle case and the oil pan with the gasket before installing the oil pan to the case.

FRONT DRIVESHAFT

REMOVAL & INSTALLATION

Refer to Driveshaft in Rear Drive Axle in this section.

FRONT HALFSHAFT

REMOVAL & INSTALLATION

2.7L Engine

See Figure 46.

1. Before servicing the vehicle, refer to the precautions section.

2. Remove or disconnect the following:

- Front wheels
- Fender apron seal
- Transaxle fluid
- Transfer case oil (4WD)
- Hub nut

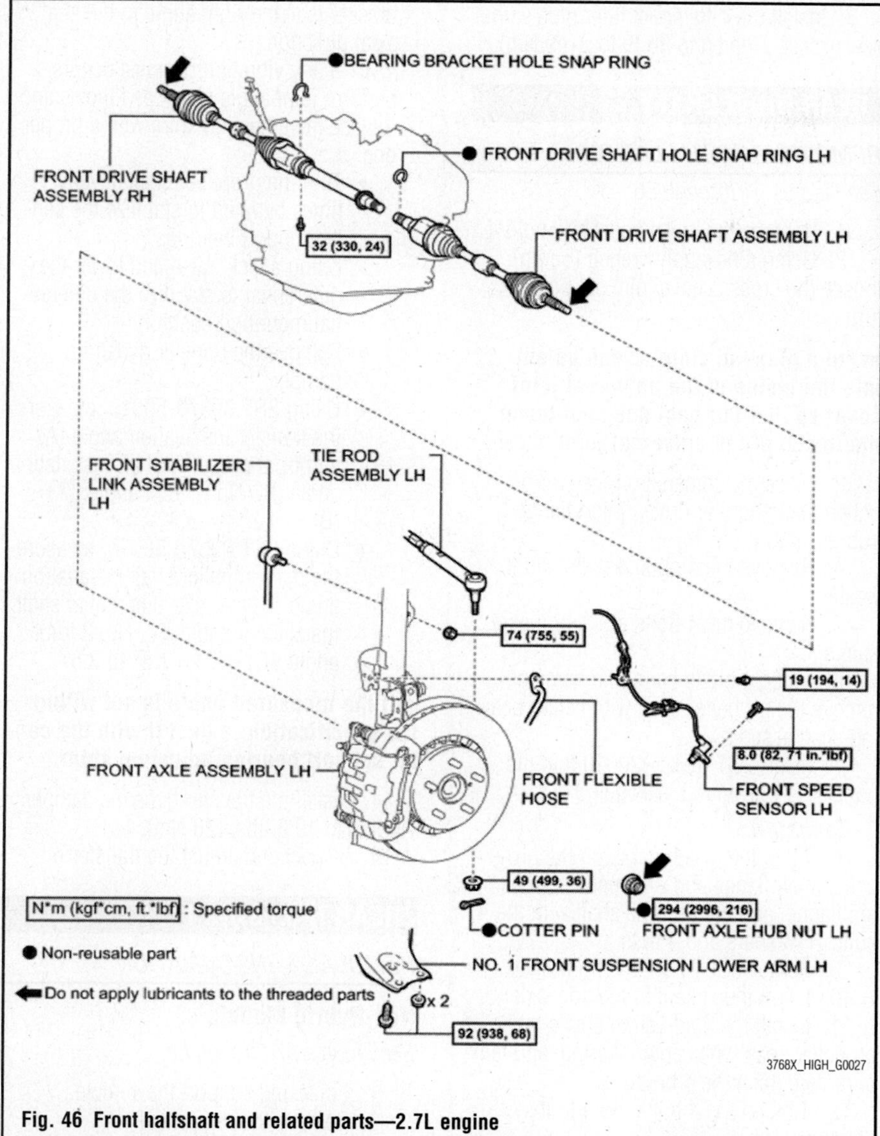

Fig. 46 Front halfshaft and related parts—2.7L engine

- Stabilizer bar link
- Speed sensor
- Tie rod end
- Lower arm from the ball joint

3. Slide the halfshaft from the hub, then, carefully, pry the shaft from the transaxle.

To install:

4. Installation is the reverse of removal. Torque the hub nut to 217 ft. lbs. (294 Nm).

3.3L Hybrid Engine

See Figures 47 through 49.

1. Before servicing the vehicle, refer to the Precautions Section.
2. Remove or disconnect the following:

- Engine under covers
- Front wheels
- Drain HV transaxle fluid
- Cotter pin and hub nut
- Front speed sensor
- Brake caliper
- Brake disc
- Tie rod end, from the steering knuckle
- Steering knuckle, from the lower control arm
- Halfshaft from the axle hub, using a plastic hammer
- Stabilizer link

3. Using Special Tool 095020-01010, remove the halfshaft from the transaxle.
4. For RH axle remove the bearing bracket hole snap ring from the drive shaft bearing bracket.
5. Remove the bolt and front drive shaft assembly RH from the drive shaft bearing bracket.

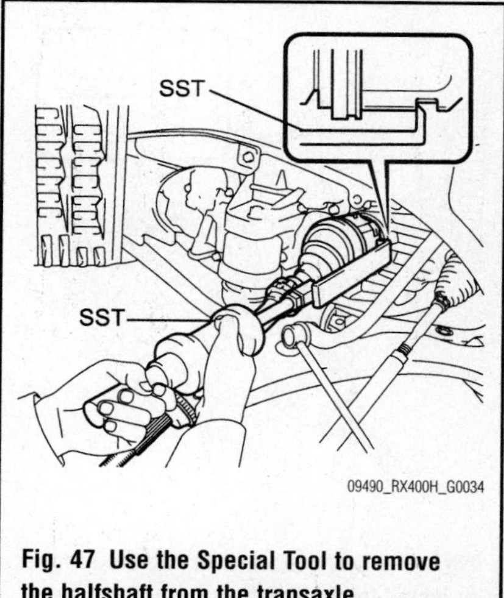

Fig. 47 Use the Special Tool to remove the halfshaft from the transaxle.

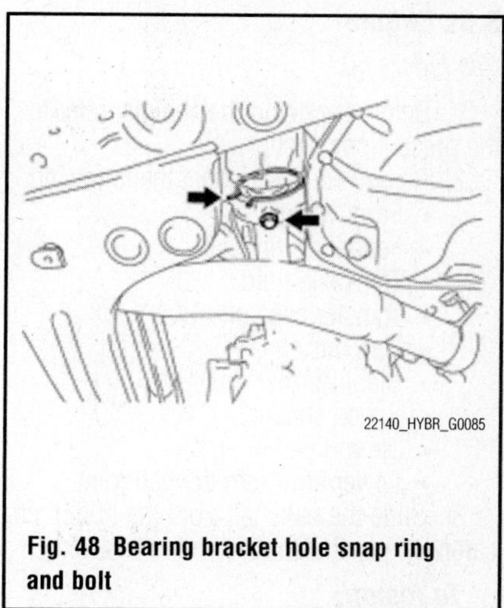

Fig. 48 Bearing bracket hole snap ring and bolt

To install:

6. Install a new halfshaft hole snapring.
7. Coat the splines of the inboard joint shaft assembly with ATF.
8. Align the shaft splines and install the halfshaft assembly with a brass drift and hammer.
9. For RH side axle install the bracket hole snap ring and bolt. Tighten the bolt to 24 ft. lbs. (32 Nm).
10. Tighten the axle shaft nut and tighten to 216 ft. lbs. (294 Nm).
11. The remainder of installation is the reverse order of removal.
12. Fill the HV transaxle with gear oil, install the engine under covers, check front end alignment and test drive.

➡If the cotter pin holes do not align, always correct by tightening the nut until the next hole aligns.

13. Install a new cotter pin.
14. Tighten the front wheels to 76 ft. lbs. (103 Nm).

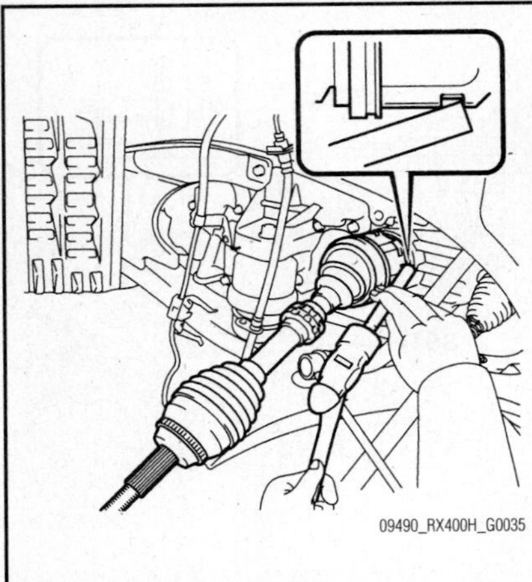

Fig. 49 Insert a brass drift into the groove to install the halfshaft.

3.5L Engine

See Figure 50.

1. Before servicing the vehicle, refer to the precautions section.
2. Remove or disconnect the following:
    - Front wheels
    - Fender apron seal
    - Transaxle fluid
    - Transfer case oil (4WD)
    - Hub nut
    - Stabilizer bar link
    - Speed sensor
    - Tie rod end
    - Lower arm from the ball joint
3. Slide the halfshaft from the hub, then, carefully, pry the shaft from the transaxle.

To install:

4. Installation is the reverse of removal. Torque the hub nut to 217 ft. lbs. (294 Nm).

REAR AXLE FLUID

DRAIN & REFILL

➡After changing the oil seal, drive the vehicle and then check the oil level again. Too much or too little oil will lead to differential problems.

➡If necessary, fill the differential carrier assembly with hypoid gear oil.

1. Stop the vehicle on a level place.
2. Remove the rear differential drain plug and gasket and drain the fluid in to a suitable container.
3. Fill the housing with the recommended fluid.
4. Check that the oil level is between 0 to 0.20 inches. (0 to 5 mm) from the bottom lip of the differential filler plug hole.
5. Check for oil leakage when the oil level is low.
6. Install the differential filler plug with a new gasket. Tighten to 36 ft. lbs. (49 Nm).

REAR DRIVESHAFT

REMOVAL & INSTALLATION

See Figures 51 through 56.

1. Depress the brake pedal and hold it.
2. Using a hexagon wrench (6 mm), loosen the cross groove joint set bolts 1/2 turn.

➡Put a piece of cloth or equivalent into the inside of the universal joint cover so that the boot does not touch the inside of the universal joint cover.

3. Place matchmarks on the rear propeller shaft and rear drive pinion flange sub-assembly.
4. Remove the 4 nuts, 4 bolts and 4 washers.
5. Remove the 4 bolts and 4 adjusting shims.
6. Using a brass bar and a hammer, remove the propeller shaft with center bearing shaft assembly.
7. Insert SST 09325-20010 into the transfer to prevent oil leakage.

To install:

8. Align the matchmarks on the propeller shaft flange and differential companion flange, and connect the shaft with the 4 bolts, 4 washers and 4 nuts.
9. Remove SST from the transaxle.
10. Insert the yoke into the transaxle.
11. Install the 4 adjusting shims and propeller shaft with center bearing, and temporarily tighten the 4 bolts.
12. Tighten the 4 bolts to 54 ft. lbs. (74 Nm).
13. Fully tighten propeller with center bearing shaft assembly:
    a. Remove the piece of cloth from the joint.
    b. Using a hexagon wrench (6 mm), tighten the 6 bolts. Tighten to 19 ft. lbs. (26 Nm).
    c. With the vehicle unloaded, adjust the dimension between the rear side of the cover and shaft as shown in the illustration. (A): 58.0 +/- 0.5 mm (2.283 +/- 0.02 in.)
    d. Under the same condition as above, adjust the front and rear dimensions between the edge surface of the center support bearing and the edge surface of the cushion respectively as shown, and then tighten the bolts. (B): 12.5 +/- 1.0 mm (0.492 +/- 0.039 in.). Tighten to 27 ft. lbs. (37 Nm).
    e. Check that the center line of the bracket is at the right angle in the shaft axial direction.
    f. If any vibration or noise occurs, perform joint angle check as follows and replace the adjusting shim with a proper one.
        - Turn the propeller shaft several times by hand to stabilize the center support bearings.
        - Using a jack, raise and lower the differential to stabilize the differential mounting cushion.
        - Remove the transfer dynamic damper.
        - Using SST 09370-50010, measure the transfer installation angle (A) and front propeller shaft installation angle (B). No. 1 joint angle: (A) - (B) = -1.3° to -3.3°
        - Using SST 09370-50010, measure the rear propeller shaft installation angle (C) and rear differential shaft installation angle (D). No. 2 joint angle: (C) - (D) = 1.8° to 3.5°

➡If the measured angle is not within the specification, adjust it with the center support bearing adjusting shim.

14. Install the transfer dynamic damper. Tighten to 19 ft. lbs. (26 Nm).
15. Inspect and adjust the transfer oil.

REAR HALFSHAFT

REMOVAL & INSTALLATION

Non-Hybrid Models

See Figures 57 through 63.

1. Raise and support the vehicle.
2. Remove the rear wheel.
3. Using a hexagon wrench (10 mm), remove the rear differential filler plug and rear differential filler plug gasket.
4. Using a hexagon wrench (10 mm), remove the rear differential drain plug and rear differential drain plug gasket, and drain the oil.
5. Remove the bolt and separate the rear speed sensor from the rear axle carrier sub-assembly.

➡Keep the sensor tip and rear speed sensor installation hole free from foreign matter.

6. Using Special Service Tool (SST) 09930-00010 or equivalent, along with a hammer, release the staked part of the rear axle shaft nut.

➡Loosen the staked part of the nut completely, otherwise the threads of the drive shaft may be damaged.

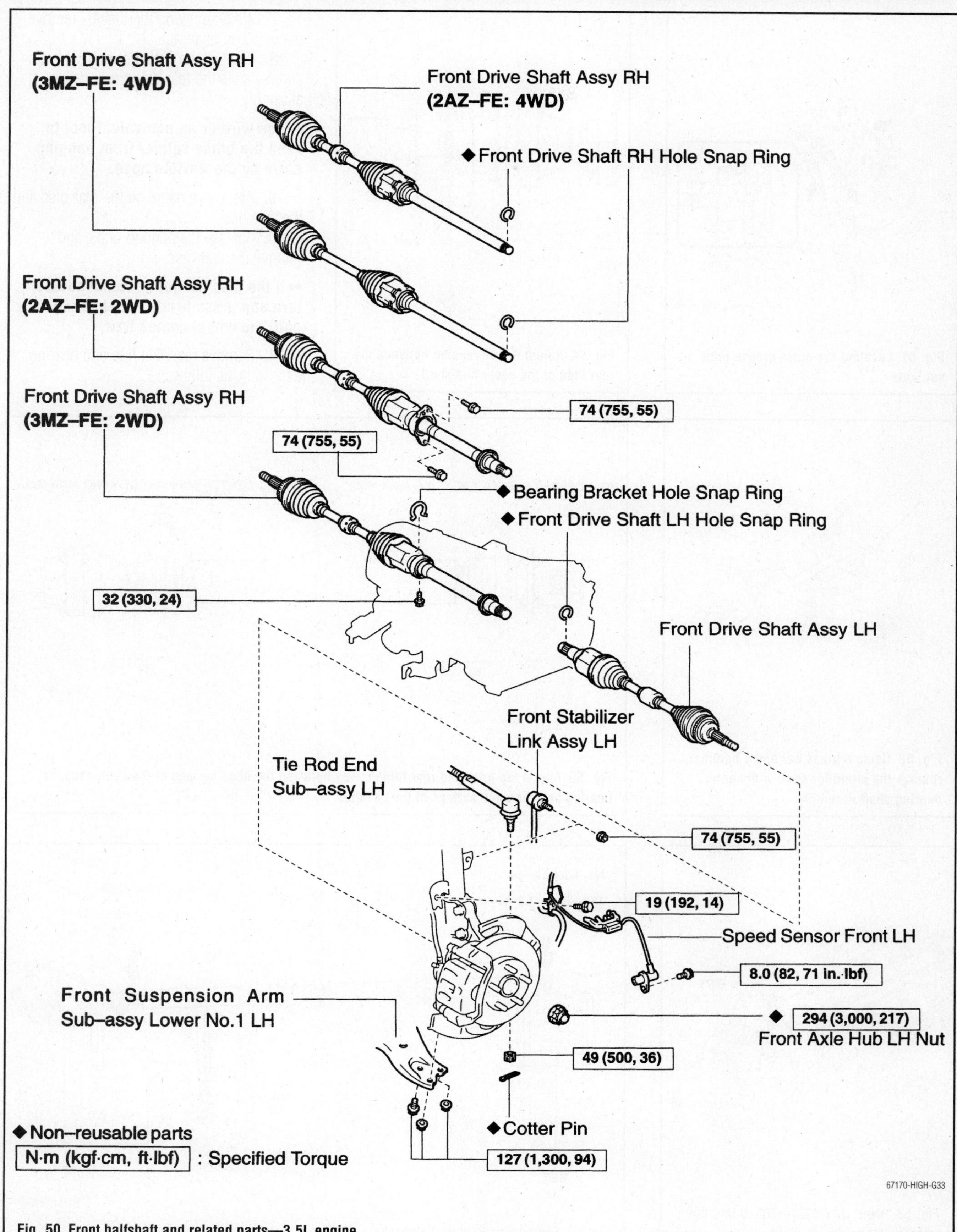
Front Drive Shaft Assy RH
(3MZ–FE: 4WD)
Front Drive Shaft Assy RH
(2AZ–FE: 4WD)
◆Front Drive Shaft RH Hole Snap Ring
Front Drive Shaft Assy RH
(2AZ–FE: 2WD)
Front Drive Shaft Assy RH
(3MZ–FE: 2WD)
74 (755, 55)
74 (755, 55)
◆Bearing Bracket Hole Snap Ring
◆Front Drive Shaft LH Hole Snap Ring
32 (330, 24)
Front Drive Shaft Assy LH
Front Stabilizer
Link Assy LH
Tie Rod End
Sub–assy LH
74 (755, 55)
19 (192, 14)
Speed Sensor Front LH
8.0 (82, 71 in.·lbf)
Front Suspension Arm
Sub–assy Lower No.1 LH
◆ 294 (3,000, 217)
Front Axle Hub LH Nut
49 (500, 36)
◆Cotter Pin
◆Non–reusable parts
N·m (kgf·cm, ft·lbf) : Specified Torque
127 (1,300, 94)
67170-HIGH-G33

Fig. 50 Front halfshaft and related parts—3.5L engine

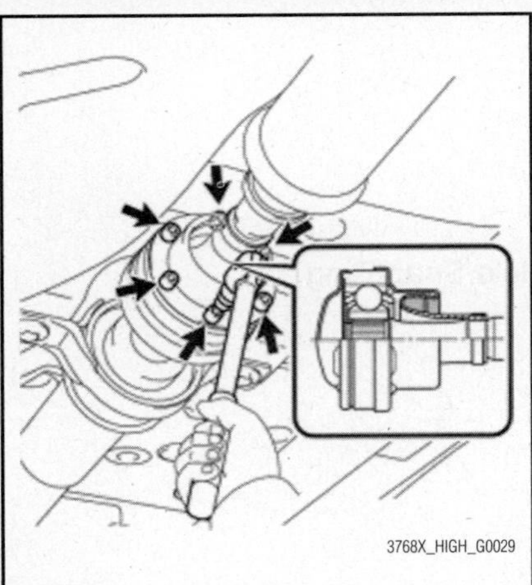
Fig. 51 Locating the cross groove joint set bolts

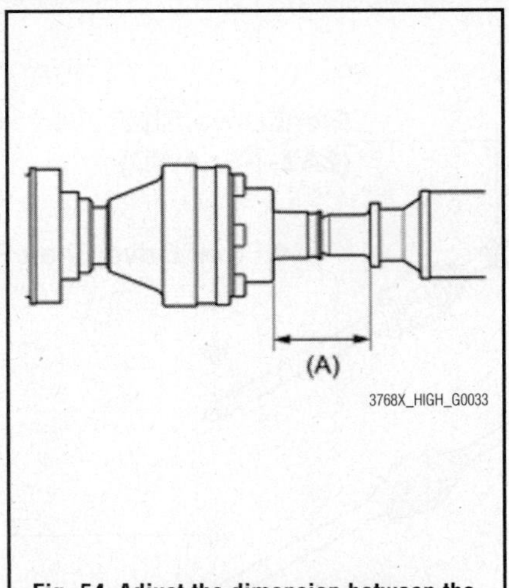

Fig. 54 Adjust the dimension between the rear side of the cover and shaft

7. While applying the brakes, remove the rear axle shaft nut.

8. Remove the 2 caliper bracket bolts and separate the rear disc brake caliper assembly.

➡Use wire or an equivalent tool to keep the brake caliper from hanging down by the flexible hose.

9. Put matchmarks on the rear disc and the axle hub.

10. Release the parking brake and remove the rear disc.

➡If the disc cannot be removed easily, turn and press firmly the shoe adjuster until the wheel comes free.

11. Remove rear axle hub and bearing assembly, as follows:

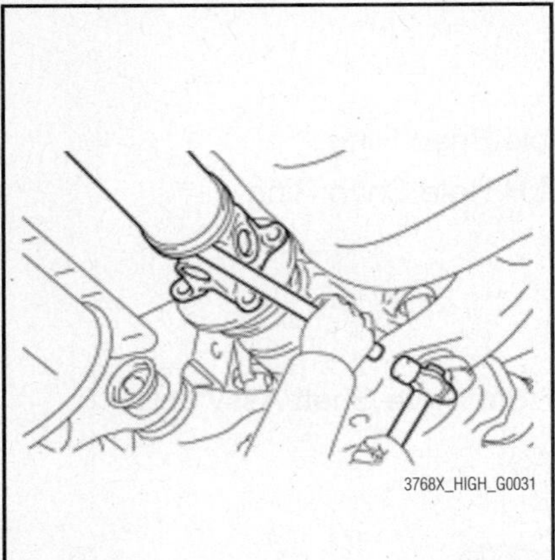
Fig. 52 Using a brass bar and a hammer, remove the propeller shaft with center bearing shaft assembly

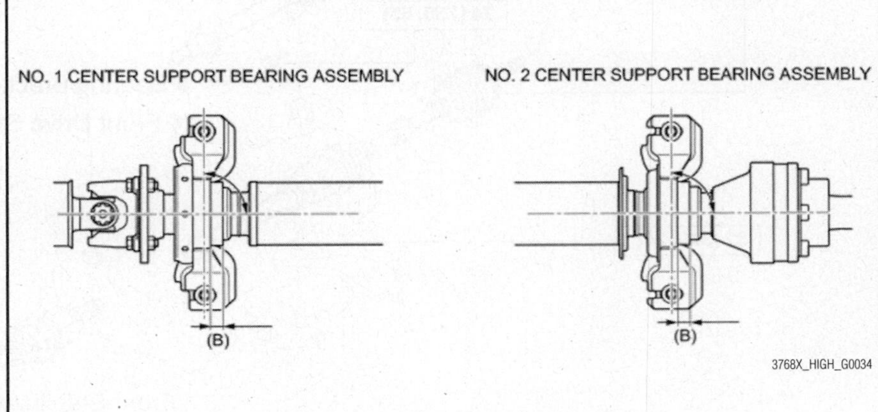

Fig. 55 Adjust the front and rear dimensions between the edge surface of the center support bearing and the edge surface of the cushion

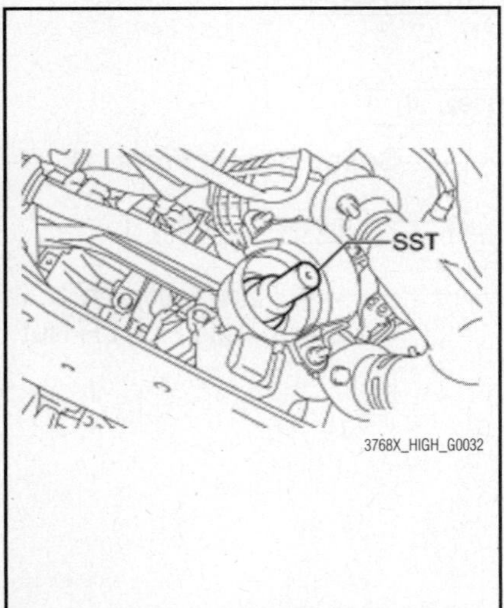

Fig. 53 Insert SST 09325-20010 into the transfer to prevent oil leakage

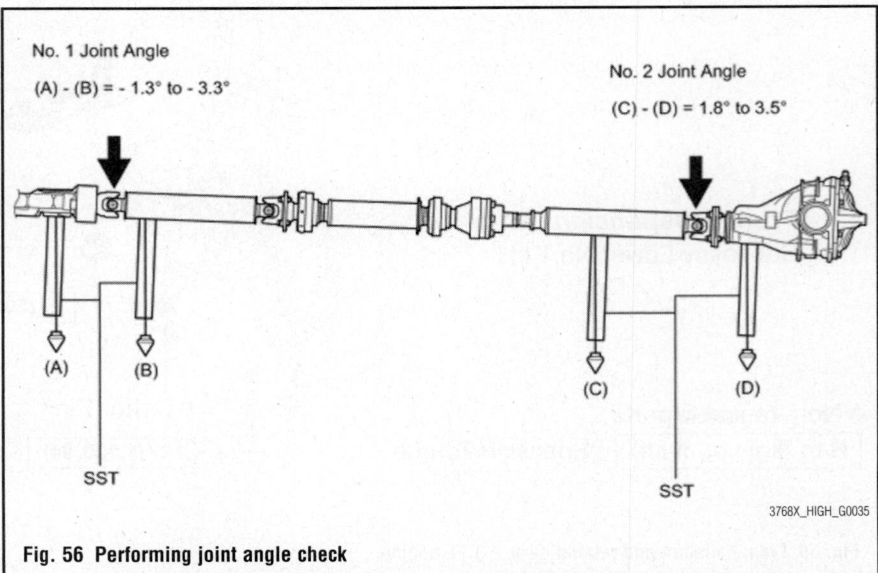

Fig. 56 Performing joint angle check

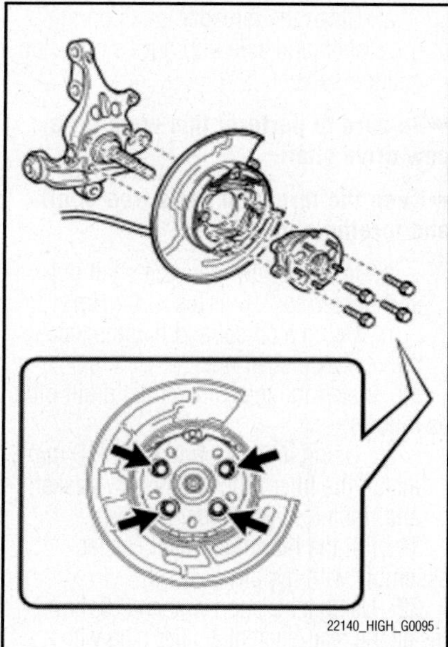

Fig. 57 Rear axle hub and bearing assembly and bolts

a. Put matchmarks on the drive shaft and axle hub.

➡**Do not punch the marks.**

b. Remove the 4 bolts and the rear axle hub and bearing assembly.

➡**Do not rotate the drive shaft with the rear axle hub and bearing assembly removed.**

➡**Use wire or an equivalent tool to keep the parking brake assembly from hanging down by the parking brake cable assembly.**

12. Remove the bolt and the nut, and separate the No. 3 parking brake cable assembly.

13. Remove the 2 bolts, the 2 nuts, and the rear strut rod assembly.

➡**Since stopper nuts are used, loosen the bolts.**

14. Remove the rear axle carrier sub-assembly, as follows:

a. Loosen the 2 bolts.

➡**Since stopper nuts are used, loosen the bolts.**

b. Remove the 2 bolts and 2 nuts, and separate the rear shock absorber with coil spring (lower side) from the rear axle carrier sub-assembly.

➡**Be careful not to damage the outboard joint boot or the speed sensor rotor.**

c. Remove the 2 bolts, the 2 nuts, and the rear axle carrier sub-assembly.

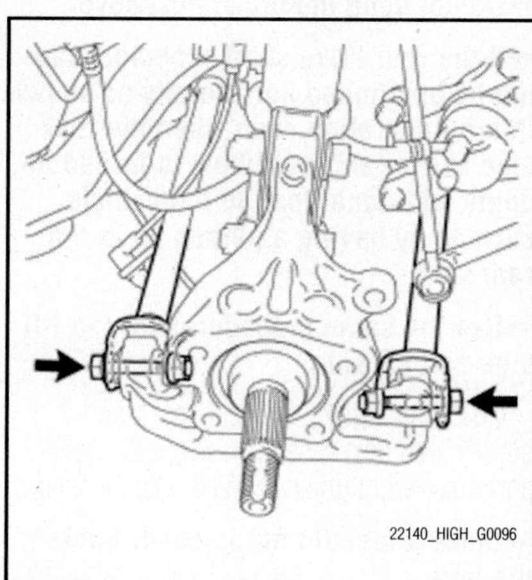

Fig. 59 Rear axle carrier sub-assembly bolts

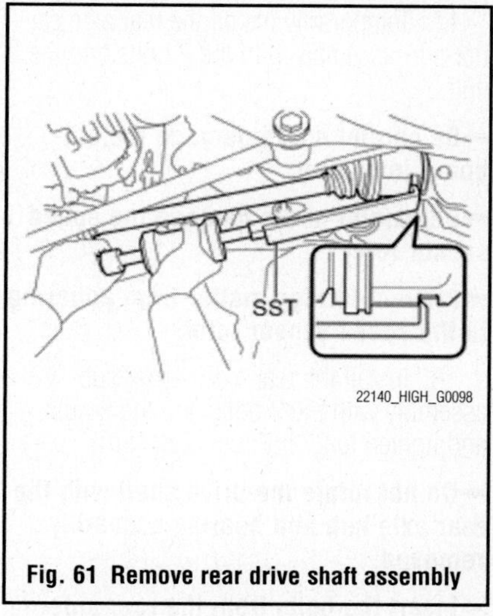

Fig. 61 Remove rear drive shaft assembly

➡**Be careful not to damage the outboard joint boot or the speed sensor rotor.**

15. Using a slide hammer (SST: 09520-01010, SST: 09520-24010 or equivalent), remove the rear drive shaft assembly (halfshaft) as shown in the illustration.

➡**Remove the rear drive shaft assembly while keeping it level.**

To install:

16. Align the shaft splines and install the rear drive shaft assembly (halfshaft) with a brass bar and hammer.

➡**Set the snap ring with the opening facing downward.**

➡**Be careful not to damage the oil seal, boot, or dust cover.**

➡**Install the drive shaft assembly while keeping it level.**

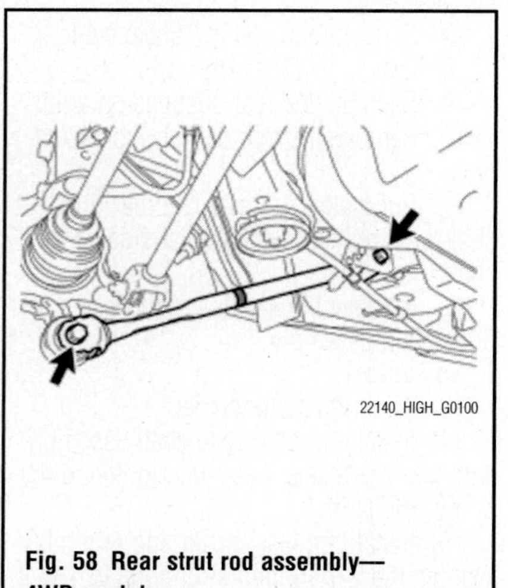

Fig. 58 Rear strut rod assembly—4WD models

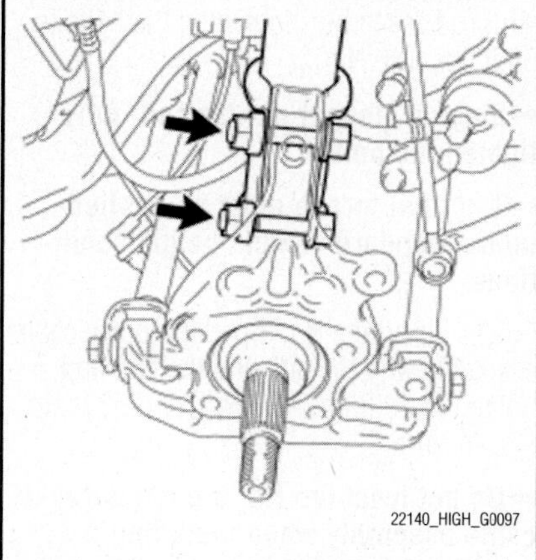

Fig. 60 Rear shock absorber with coil spring bolts

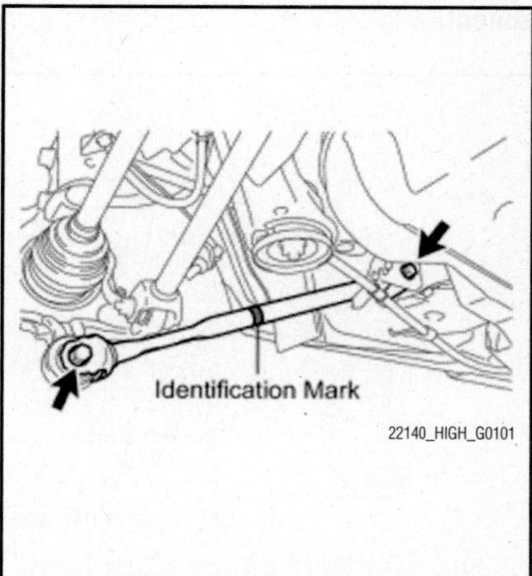

Fig. 62 Rear strut rod assembly identification mark—4WD models

17. Temporarily install the rear axle carrier sub-assembly with the 2 bolts and the 2 nuts.

➡Be careful not to damage the outboard joint boot.

➡Be careful not to damage the speed sensor rotor.

➡Prevent foreign matter from adhering to the speed sensor rotor.

18. Install the rear axle carrier sub-assembly with the 2 bolts and the 2 nuts, and tighten to 213 ft. lbs. (290 Nm).

➡Do not rotate the drive shaft with the rear axle hub and bearing assembly removed.

➡Insert the bolts from the rear side.

19. Check that the identification mark of the rear strut rod assembly is positioned on the inner side of the vehicle.
20. Temporarily install the rear strut rod assembly with the 2 bolts and the 2 nuts.

➡Since stopper nuts are used, temporarily tighten the bolts.

21. Install the rear axle hub and bearing assembly, as follows:
   a. Align the matchmarks on the drive shaft and rear axle hub.

➡Do not rotate the drive shaft.

   b. Install the parking brake assembly and the rear axle hub and bearing assembly with the 4 bolts, and tighten to 55 ft. lbs. (75 Nm).
22. Align the matchmarks and install the rear disc.

➡When replacing the rear disc with a new one, select the installation position where the rear disc has minimal runout.

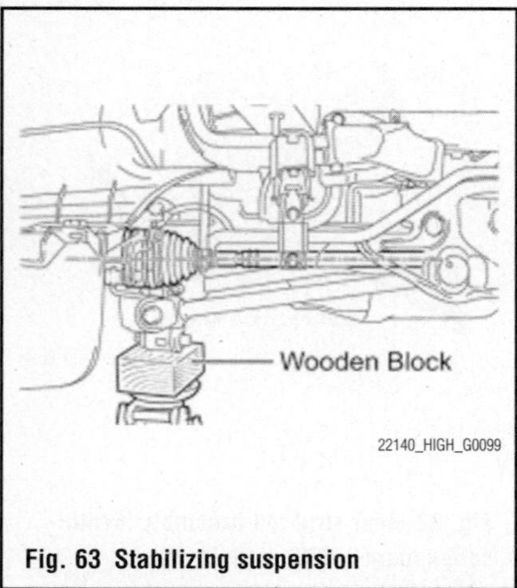

Fig. 63 Stabilizing suspension

23. Install the rear disc brake caliper assembly with the 2 bolts, and tighten to 57 ft. lbs. (78 Nm).
24. Install the rear speed sensor to the rear axle carrier sub-assembly with the bolt, and tighten to 71 inch lbs. (8 Nm).

➡Keep the rear speed sensor tip and sensor installation hole free from foreign matter.

➡Do not twist the rear speed sensor wire when installing.

25. Jack up the rear axle carrier sub-assembly, placing a wooden block underneath to avoid damage. Apply load to the suspension so that the rear drive shaft assembly is positioned horizontally.

CAUTION

Do not jack up the rear axle carrier sub-assembly too high as the vehicle may fall.

➡Do not bend the brake dust cover.

➡If the rear drive shaft assembly cannot be positioned horizontally as shown in the illustration even when the rear axle carrier sub-assembly is jacked up, apply additional load to the vehicle such as by having a person sit in the rear seat.

➡Use the same procedures for the RH side and LH side.

26. Fully tighten the rear No. 1 and 2 suspension arm assemblies with the bolts and nuts, and tighten to 82 ft. lbs. (112 Nm).

➡Since a stopper nut is used, tighten the bolt.

➡The final torque must be applied under standard vehicle height conditions.

27. Complete the installation of the rear strut rod assembly, tightening the bolts to 59 ft. lbs. (80 Nm).

➡Since a stopper nut is used, fully tighten the bolt.

➡The final torque must be applied under standard vehicle height conditions.

28. Install the No. 3 parking brake cable assembly with the bolt and the nut, and tighten to 29 ft. lbs. (39 Nm) and 53 inch lbs. (6 Nm).

➡Do not twist the No. 3 parking brake cable assembly when installing it.

29. Install the rear axle shaft nut, as follows:
   a. Clean the threaded parts on the drive shaft and axle hub nut using a non-residue solvent.

➡Be sure to perform this work for a new drive shaft.

➡Keep the threaded parts free of oil and foreign objects.

   b. Install a new rear axle shaft nut, and tighten to 216 ft. lbs. (294 Nm).
   c. Using a chisel and hammer, stake the rear axle shaft nut.
30. Install the rear differential drain plug, as follows:
   a. Using a hexagon wrench (10 mm), install the filler plug with a new gasket, and tighten to 36 ft. lbs. (49 Nm).
31. Fill the rear differential carrier assembly with hypoid gear oil.
32. Using a hexagon wrench (10 mm), install the rear differential filler plug with a new gasket, and tighten to 36 ft. lbs. (49 Nm).
33. Install the front wheel and tire assembly and tighten the lug nuts finger-tight.
34. Lower the vehicle, then final tighten the lug nuts to 76 ft. lbs. (103 Nm).
35. Inspect and adjust rear wheel alignment.
36. Check ABS speed sensor signal.

Hybrid Models

See Figure 64.

1. Before servicing the vehicle, refer to the Precautions Section.
2. Remove the rear wheel.
3. Disconnect and remove the speed sensor.
4. Remove the brake disc and brake caliper assembly.
5. Remove the axle shaft nut.
6. Separate the rear axle hub and bearing assembly.
7. Separate the parking brake cable.
8. Disconnect the strut rod.
9. Separate the rear suspension arms.
10. Remove the rear carrier sub assembly.
11. Put matchmarks on the rear drive shaft assembly and differential side gear shaft.
12. Remove the 4 nuts, washers and rear drive shaft assembly.

To install

13. Align the matchmarks.
14. Install the rear drive shaft assembly with the 4 nuts and washers. Tighten to 41 ft. lbs. (56 Nm).
15. Install the rear carrier sub assembly. Tighten the 2 mounting strut bolts to 133 ft. lbs. (180 Nm).

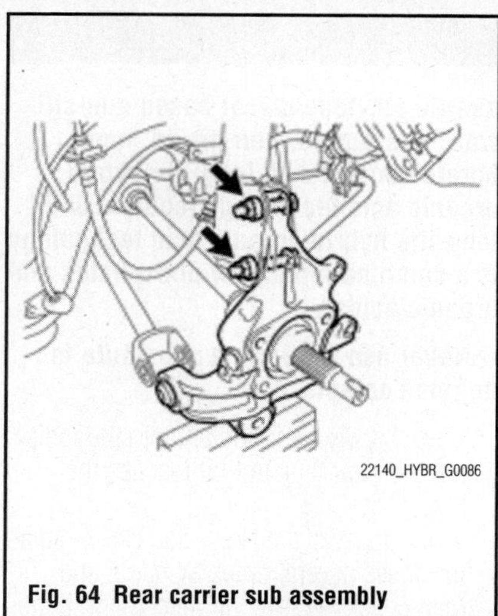

Fig. 64 Rear carrier sub assembly

16. Temporarily tighten the rear suspension arm assembly No.2 with the bolt and nut.
17. Temporarily tighten the rear suspension arm assembly No.1 with the bolt and nut.
18. Temporarily tighten the strut rod assembly rear with the bolt and nut.
19. Install the hub and bearing assembly with the 4 bolts and tighten to 55 ft. lbs. (75 Nm).
20. Install the rear brake disc.
21. Install the rear disc brake caliper assembly with the 2 bolts to the rear axle carrier. Tighten the bolts to 58 ft. lbs. (78 Nm).
22. Install the speed sensor and tighten the mounting bolt to 71 inch lbs. (8 Nm).
23. Using a socket wrench (30mm), install a new rear axle shaft nut. Tighten the axle nut to 217 ft. lbs. (294 Nm).
24. Using a chisel and hammer, stake the rear axle shaft nut.
25. Stabilize the suspension. Then tighten the control arms to 83 ft. lbs. (112 Nm). Tighten the strut rod to 133 ft. lbs. (180 Nm).
26. Install the parking brake cable assembly No.3 with the nut. Tighten the nut to 53 inch lbs. (6 Nm)
27. Install the rear wheel and tighten to 76 ft. lbs. (103 Nm)
28. Check rear wheel alignment.
29. Verify speeds sensor operation.

REAR PINION SEAL

REMOVAL & INSTALLATION

Non-Hybrid Models

See Figure 65.

1. Before servicing the vehicle, refer to the precautions section.
2. Drain the differential oil.
3. Remove or disconnect the following:
   - Exhaust pipe
   - Driveshaft by matchmarking it
   - Companion flange nut, by loosen the staked portion
   - Companion flange, using a screw-type extractor
   - Oil seal, using an extractor
   - Slinger
   - Front bearing
   - Spacer

To install:

4. Install or connect the following
   - New spacer
   - Bearing
   - Slinger
   - New seal

➡Seal installation depth: 2.0mm +/- 0.3mm

   - Companion flange
   - New nut. Coat the threads with clean differential oil. Torque the nut to 80 ft. lbs. (108 Nm).
5. The remainder of installation is the reverse of removal.

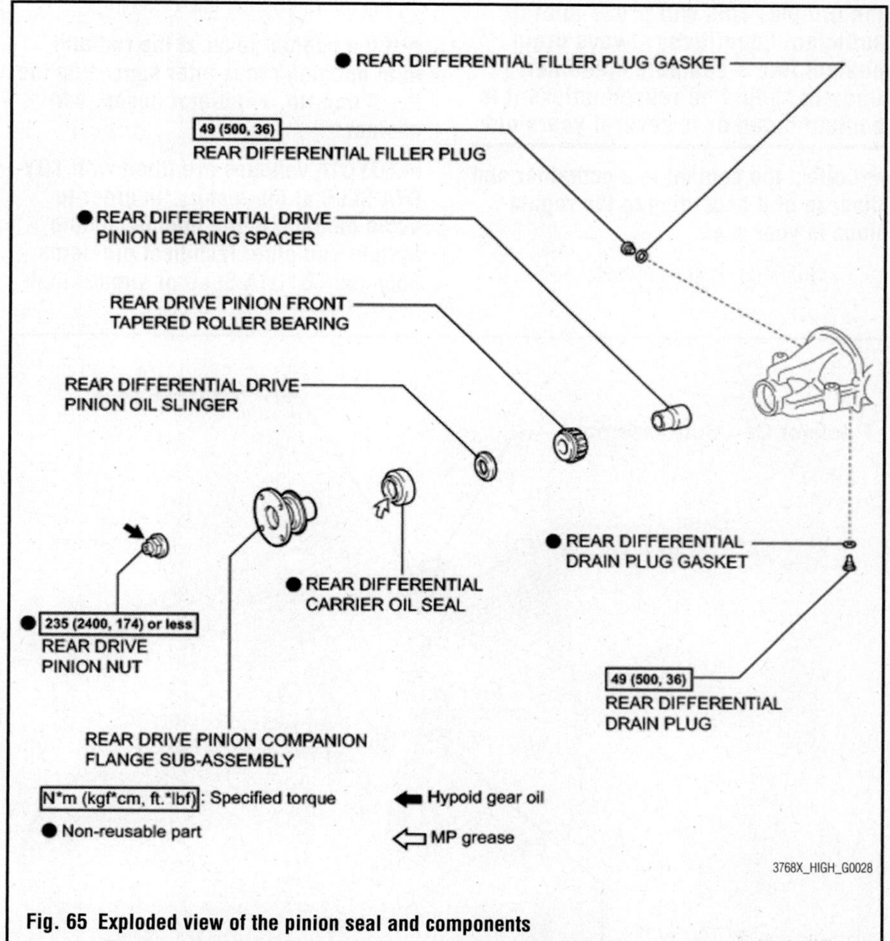

Fig. 65 Exploded view of the pinion seal and components

ENGINE COOLING

ENGINE COOLANT

DRAIN & REFILL PROCEDURE

2.7L Engine

See Figure 66.

CAUTION

Never open, service or drain the radiator or cooling system when hot; serious burns can occur from the steam and hot coolant. Also, when draining engine coolant, keep in mind that cats and dogs are attracted to ethylene glycol antifreeze and could drink any that is left in an uncovered container or in puddles on the ground. This will prove fatal in sufficient quantities. Always drain coolant into a sealable container. Coolant should be reused unless it is contaminated or is several years old.

➡Collect the coolant in a container and dispose of it according to the regulations in your area.

1. Drain the engine coolant.
   a. Remove no. 1 engine under cover.
   b. Loosen the radiator drain cock plug.
   c. Remove the radiator cap sub-assembly from the radiator assembly.
2. Add engine coolant:
   a. Tighten the radiator drain cock plug by hand.
   b. Tighten the cylinder block drain cock plug. Tighten to 9 ft. lbs. (13 Nm).
   c. Loosen the air drain plug at the top of the radiator 3 or 4 turns.
   d. Add TOYOTA Super Long Life Coolant (SLLC) to the radiator inlet opening until coolant overflows from the radiator air drain hole. Then close the air drain plug at the top of the radiator. Tighten to 13 inch lbs. (1.5 Nm).

➡If the coolant level at the radiator inlet opening drops after squeezing the No. 1 and No. 2 radiator hoses, add coolant.

➡TOYOTA vehicles are filled with TOYOTA SLLC at the factory. In order to avoid damage to the engine cooling system and other technical problems, only use TOYOTA SLLC or similar high quality ethylene glycol based non-silicate, non-amine, non-nitrite, non-borate coolant with long-life hybrid organic acid technology (coolant with long-life hybrid organic acid technology is a combination of low phosphates and organic acids).

➡Never use water as a substitute for engine coolant.

   e. Slowly pour coolant into the radiator reservoir tank until it reaches the FULL line.
   f. Squeeze the No. 1 and No. 2 radiator hoses several times by hand, and then check the level of the coolant. If the coolant level is low, add coolant.

➡The thermostat open timing can be confirmed by squeezing the No. 1 radiator hose by hand, and checking when the engine coolant starts to flow inside the hose.

   g. Bleed air from the cooling system:
      - Warm up the engine until the thermostat opens. While the thermostat is open, circulate the coolant for several minutes.
      - Maintain the engine speed at 2500 to 3000 rpm.
      - Squeeze the No. 1 and No. 2 radiator hoses several times by hand to bleed air.

CAUTION

When squeezing the radiator hoses:

- Avoid contact with the cooling fan motor insulator.
- Wear protective gloves.
- Be careful as the radiator hoses are hot.
- Keep your hands away from the radiator fan.

➡Note the following:

- Make sure that the radiator reservoir still has some coolant in it.
- If the coolant temperature gauge indicates an excessive temperature, turn off the engine and let it cool.
- If there is not enough coolant, the engine may overheat or be seriously damaged.
- If the radiator reservoir does not have enough coolant, perform the following: 1) stop the engine, 2) wait until the coolant has cooled down, and 3) add coolant until the reservoir is filled to the FULL line.

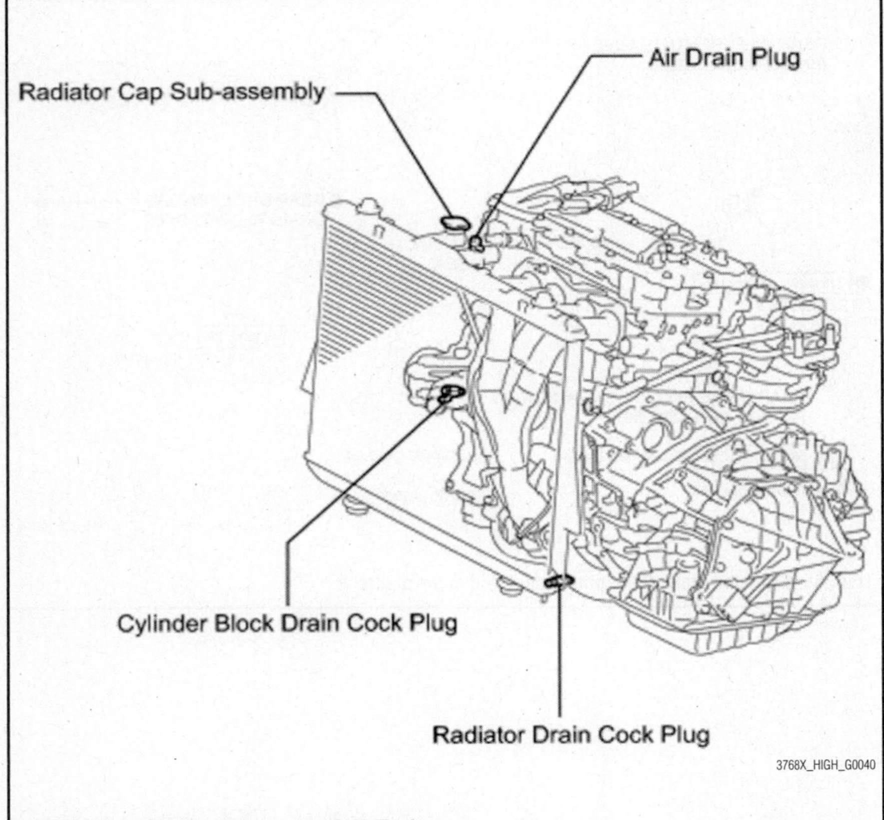

Fig. 66 Locating the drain plugs—2.7L Engine

h. Stop the engine and wait until the engine coolant cools down.

i. Add engine coolant to the FULL line on the radiator reservoir.

3. Inspect for leaks.

3.3L Hybrid Engine

See Figure 67.

⁂ CAUTION

Never open, service or drain the radiator or cooling system when hot; serious burns can occur from the steam and hot coolant. Also, when draining engine coolant, keep in mind that cats and dogs are attracted to ethylene glycol antifreeze and could drink any that is left in an uncovered container or in puddles on the ground. This will prove fatal in sufficient quantities. Always drain coolant into a sealable container. Coolant should be reused unless it is contaminated or is several years old.

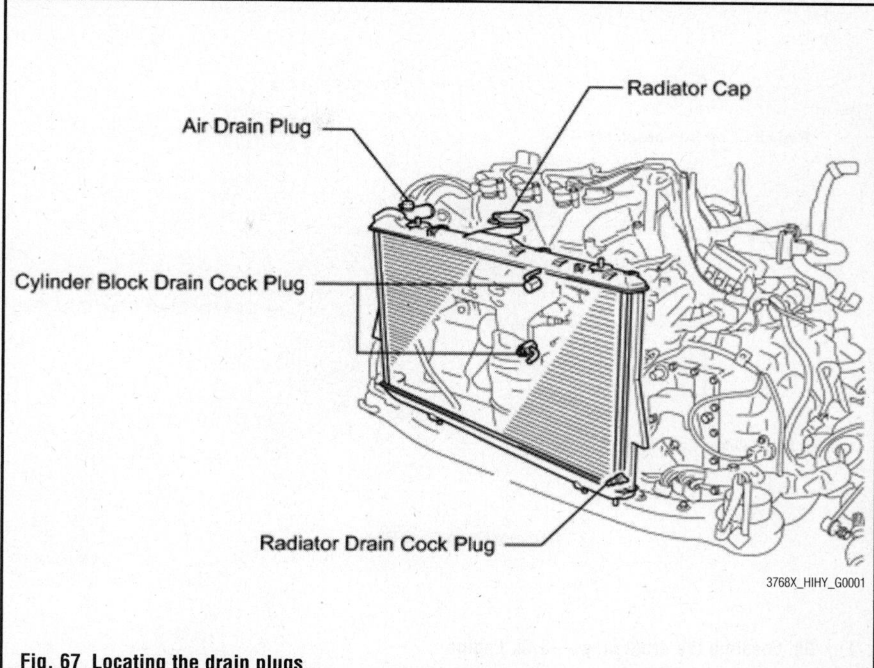

Fig. 67 Locating the drain plugs

⁂ CAUTION

To avoid the danger of being burned, do not remove the radiator cap and drain plugs while the engine and radiator are still hot. Thermal expansion will cause hot engine coolant and steam to blow out from the radiator.

➡Collect the coolant in a container and dispose of it according to the regulations in your area.

1. Loosen the radiator drain cock plug.
2. Remove the radiator cap.
3. Tighten the radiator drain cock plug.
4. Tighten the cylinder block drain cock plugs.
5. Loosen the air drain plug.
6. Install a vinyl tube to the vent tube located on the air drain plug.
7. Fill the radiator with engine coolant until the vinyl tube is filled with the coolant.
8. Tighten the air drain plug.

⁂ CAUTION

Do not use an alcohol type coolant or plain water alone. The coolant must be mixed with plain water (preferably demineralized water).

9. Coolant capacity:
   a. With rear heater: 13.5 quarts.
   b. Without rear heater: 11.2 quarts.

➡Note the following:

- Improper coolants may damage the engine cooling system.
- TOYOTA vehicles are filled with TOYOTA SLLC at the factory. In order to avoid damage to the engine cooling system and other technical problems, only use TOYOTA SLLC or similar high quality ethylene glycol based non-silicate, non-amine, non-nitrite, non-borate coolant with long-life hybrid organic acid technology (coolant with long-life hybrid organic acid technology consists of a combination of low phosphates and organic acids)
- Use of coolant which includes more than 50% (freezing protection down to -35°C (-31°F)) or 60% (freezing protection down to -50°C (-58°F)) of ethylene-glycol is recommended but not more than 70%
- Observe the coolant level inside the radiator by pressing the inlet and outlet radiator hoses several times by hand. If the coolant level goes down, add the coolant

➡Never use water as a substitute for engine coolant.

10. Install the radiator cap securely.
11. Fill the radiator reservoir tank with coolant.
12. Warm up the engine until the thermostat opens. While the thermostat is open, allow the coolant to circulate for several minutes.

➡The thermostat opening timing can be confirmed by squeezing the inlet radiator hose by hand, and sensing vibrations when the engine coolant starts to flow inside the hose.

13. Stop the engine and wait until the coolant cools down.
14. Remove the radiator cap and check the coolant level inside the radiator.
15. If the coolant level is below the full level, perform the steps from (C) through (K) and repeat the operation until the coolant level stays at the full level.
16. Recheck the coolant level inside the radiator reservoir tank. If it is below the full level, add the coolant.
17. Inspect for coolant leaks

3.5L Engine

See Figure 68.

⁂ CAUTION

Never open, service or drain the radiator or cooling system when hot; serious burns can occur from the steam and hot coolant. Also, when draining engine coolant, keep in mind that cats and dogs are attracted to ethylene glycol antifreeze and could drink any that is left in an uncovered container or in puddles on the ground. This will prove fatal in sufficient quantities. Always drain coolant into a sealable container. Coolant should be reused unless it is contaminated or is several years old.

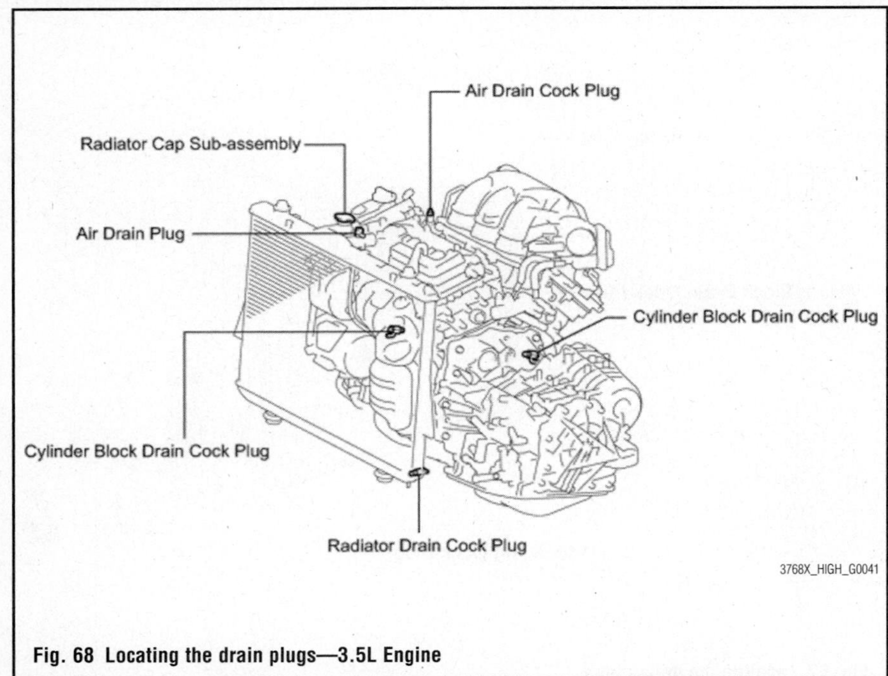

Fig. 68 Locating the drain plugs—3.5L Engine

➡Collect the coolant in a container and dispose of it according to the regulations in your area.

1. Remove v-bank cover sub-assembly.
2. Remove engine under cover assembly.
3. Remove no. 1 engine under cover.
4. Drain engine coolant:
   a. Loosen the radiator drain cock plug.
   b. Remove the radiator cap sub-assembly from the radiator assembly.
   c. Loosen the 2 cylinder block drain cock plugs.
5. Add engine coolant:
   a. Tighten the radiator drain cock plug by hand.
   b. Tighten the cylinder block drain cock plug. Tighten to 9 ft. lbs. (13 Nm).
   c. Loosen the air drain plug at the top of the radiator 3 or 4 turns.
   d. Add TOYOTA Super Long Life Coolant (SLLC) to the radiator inlet opening until coolant overflows from the engine air drain cock hole. Then tighten the air drain cock plug to the water inlet housing. Tighten the air drain cock plug to 9 ft. lbs. (13 Nm).
   e. Continue to add TOYOTA Super Long Life Coolant (SLLC) to the radiator inlet opening until coolant overflows from the radiator air drain hole. Then close the air drain plug at the top of the radiator. Tighten the air drain plug to Tighten to 13 inch lbs. (1.5 Nm).

➡If the coolant level at the radiator inlet opening drops after squeezing the No. 1 and No. 2 radiator hoses, add coolant.

➡TOYOTA vehicles are filled with TOYOTA SLLC at the factory. In order to avoid damage to the engine cooling system and other technical problems, only use TOYOTA SLLC or similar high quality ethylene glycol based non-silicate, non-amine, non-nitrite, non-borate coolant with long-life hybrid organic acid technology (coolant with long-life hybrid organic acid technology is a combination of low phosphates and organic acids).

➡Never use water as a substitute for engine coolant.

   f. Slowly pour coolant into the radiator reservoir tank until it reaches the FULL line.
   g. Squeeze the No. 1 and No. 2 radiator hoses several times by hand, and then check the level of the coolant. If the coolant level is low, add coolant.

➡The thermostat open timing can be confirmed by squeezing the No. 1 radiator hose by hand, and checking when the engine coolant starts to flow inside the hose.

   h. Bleed air from the cooling system:
   - Warm up the engine until the thermostat opens. While the thermostat is open, circulate the coolant for several minutes.
   - Maintain the engine speed at 2500 to 3000 rpm.
   - Squeeze the No. 1 and No. 2 radiator hoses several times by hand to bleed air.

⁂ CAUTION

When squeezing the radiator hoses:

- Avoid contact with the cooling fan motor insulator.
- Wear protective gloves.
- Be careful as the radiator hoses are hot.
- Keep your hands away from the radiator fan.

➡Note the following:

- Make sure that the radiator reservoir still has some coolant in it.
- If the coolant temperature gauge indicates an excessive temperature, turn off the engine and let it cool.
- If there is not enough coolant, the engine may overheat or be seriously damaged.
- If the radiator reservoir does not have enough coolant, perform the following: 1) stop the engine, 2) wait until the coolant has cooled down, and 3) add coolant until the reservoir is filled to the FULL line.

   i. Stop the engine and wait until the engine coolant cools down.
   j. Add engine coolant to the FULL line on the radiator reservoir.
6. Inspect for leaks.

BLEEDING

Bleed air from the cooling system:

- Warm up the engine until the thermostat opens. While the thermostat is open, circulate the coolant for several minutes.
- Maintain the engine speed at 2500 to 3000 rpm.
- Squeeze the No. 1 and No. 2 radiator hoses several times by hand to bleed air.

Observe the following safety precautions:

- Avoid contact with the cooling fan motor insulator.
- Wear protective gloves.
- Be careful as the radiator hoses are hot.
- Keep your hands away from the radiator fan.
- Make sure that the radiator reservoir still has some coolant in it.
- If the coolant temperature gauge indicates an excessive temperature, turn off the engine and let it cool.
- If there is not enough coolant, the engine may overheat or be seriously damaged.

• If the radiator reservoir does not have enough coolant, perform the following: 1) stop the engine, 2) wait until the coolant has cooled down, and 3) add coolant until the reservoir is filled to the FULL line.

LEAK INSPECTION

CAUTION

Do not remove the radiator cap while the engine and radiator are still hot. Pressurized, hot engine coolant and steam may be released and cause serious burns.

➡Before performing each inspection, turn the A/C switch off.

1. Fill the radiator with coolant and attach a radiator cap tester.
2. Warm up the engine.
3. Using the radiator cap tester, increase the pressure inside the radiator to 118 kPa (17 psi), and check that the pressure does not drop.
4. If the pressure drops, check the hoses, radiator and water pump for leaks. If no external leaks are found, check the heater core, cylinder block and cylinder head.
5. Inspect engine coolant level in reservoir:
   a. Check that the engine coolant level is between the LOW and FULL lines when the engine is cold.
   b. If the engine coolant level is low, check for leaks and add "TOYOTA Super Long Life Coolant" or similar high quality ethylene glycol based non-silicate, non-amine, non-nitrite and non-borate coolant with long-life hybrid organic acid technology to the FULL line.

➡Do not substitute plain water for engine coolant.

6. Inspect engine coolant quality:
   a. Remove the radiator cap.

CAUTION

Do not remove the radiator cap while the engine and radiator are still hot. Pressurized, hot engine coolant and steam may be released and cause serious burns.

   b. Check if there are any excessive deposits of rust or scales around the radiator cap and radiator filler hole. Also, the coolant should be free of oil. If excessively dirty, clean the coolant passage and replace the coolant.
   c. Install the radiator cap.

ENGINE FAN

REMOVAL & INSTALLATION

Except 3.3L Hybrid Engine

See Figures 69 through 73.

1. Remove the engine under cover assembly.
2. Remove the No. 1 engine under cover.
3. Drain engine coolant.
4. Remove the V-bank cover sub-assembly.
5. Remove the cool air intake duct seal.
6. Remove the battery and battery tray.
7. Remove the No. 1 and 2 air cleaner inlets.
8. Disconnect the No. 1 and 2 radiator hoses from the radiator.
9. Disconnect the oil cooler hoses from the radiator.
10. Detach the wire harness clamps from the left side of the fan shroud.
11. Detach the wire harness clamps from the right side of the fan shroud and disconnect the cooling fan Electronic Control Unit (ECU) connector.
12. Remove the radiator grill, as follows:
    a. Put protective tape around the radiator grill.
    b. Remove the 2 bolts and 4 clips.
    c. Disengage the 6 claws and 4 guides, and remove the radiator grill.
13. Remove the hood lock assembly.
14. Disconnect the low pitched and high pitched horn connectors.
15. Detach the hood lock control cable clamp and remove the 6 bolts and upper radiator support sub-assembly.
16. Remove the 4 bolts and move the cooler condenser assembly to remove the radiator assembly and fan assembly with motor.

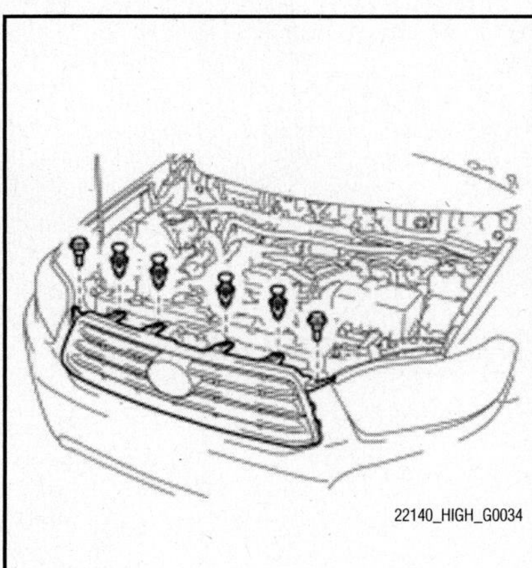

Fig. 69 Radiator grill, bolts, and clips

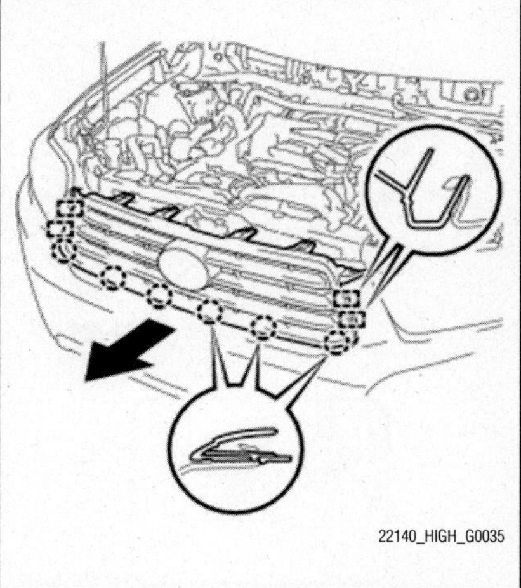

Fig. 70 Radiator grill, claws, and guides

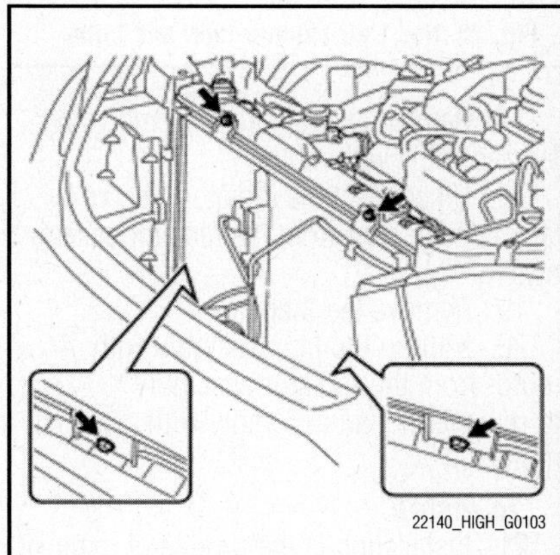

Fig. 71 Cooler condenser assembly attachment points

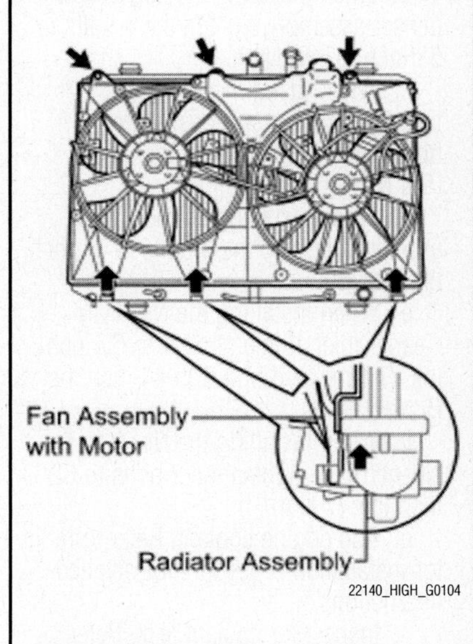

Fig. 72 Remove bolts and fan assembly with motor

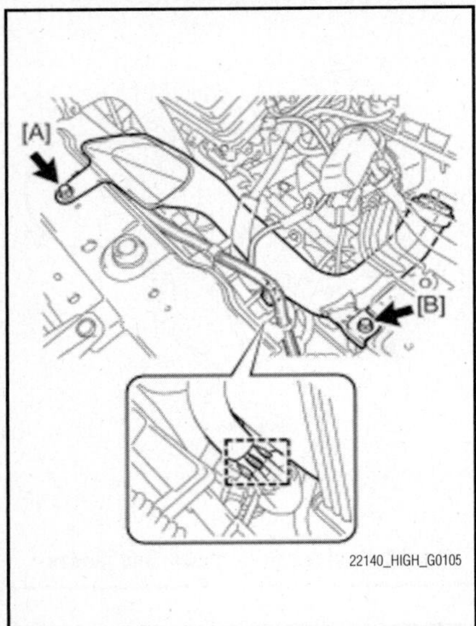

Fig. 73 No. 1 air cleaner inlet and bolts

17. Remove the radiator assembly and fan assembly with motor.
18. Disconnect the radiator reserve tank hose or pipe from the radiator assembly.
19. Remove the 3 bolts.
20. Pull up the fan assembly with motor from the radiator assembly to remove the fan assembly with motor.

To install:

21. Installation is the reverse of removal, noting the following:
   a. When installing the fan assembly with motor, tighten the 3 bolts to 69 inch lbs. (7.8 Nm).
   b. When installing the cooler condenser assembly, tighten the 4 bolts to 53 inch lbs. (6 Nm).
   c. When installing the upper radiator support sub-assembly, tighten the 6 bolts to 87 inch lbs. (9.8 Nm).
   d. When installing the hood lock assembly, tighten the 3 bolts to 71 inch lbs. (8 Nm).
   e. When installing the No. 1 air cleaner inlet, tighten bolt A to 62 inch lbs. (7 Nm), and bolt B to 44 inch lbs. (5 Nm).
   f. When installing the No. 2 air cleaner inlet, tighten the 2 bolts to 62 inch lbs. (7 Nm).
   g. Add engine coolant. Refer to radiator installation procedure for detailed instructions.
   h. Inspect for coolant leak. Refer to radiator installation procedure for detailed instructions.
   i. Check the automatic transaxle fluid level.
   j. Inspect for oil leaks.

3.3L Hybrid Engine

See Figure 74.

1. Before servicing the vehicle, refer to the Precautions Section.
2. Disconnect the radiator reserve tank hose from the radiator assembly.
3. Remove the 3 bolts.
4. Pull up the fan assembly with motor from the radiator assembly to remove the fan assembly with motor.

To install:

5. Put in the fan assembly to the radiator assembly.
6. Install the fan assembly and tighten the 3 bolts to 69 inch lbs. (7.8 Nm).
7. Install the radiator reserve tank hose to the radiator assembly.

RADIATOR

REMOVAL & INSTALLATION

Except Hybrid Models

See Figures 69 through 72 and 75.

1. Remove the engine under cover assembly.
2. Remove the No. 1 engine under cover.
3. Drain engine coolant.
4. Remove the V-bank cover sub-assembly.

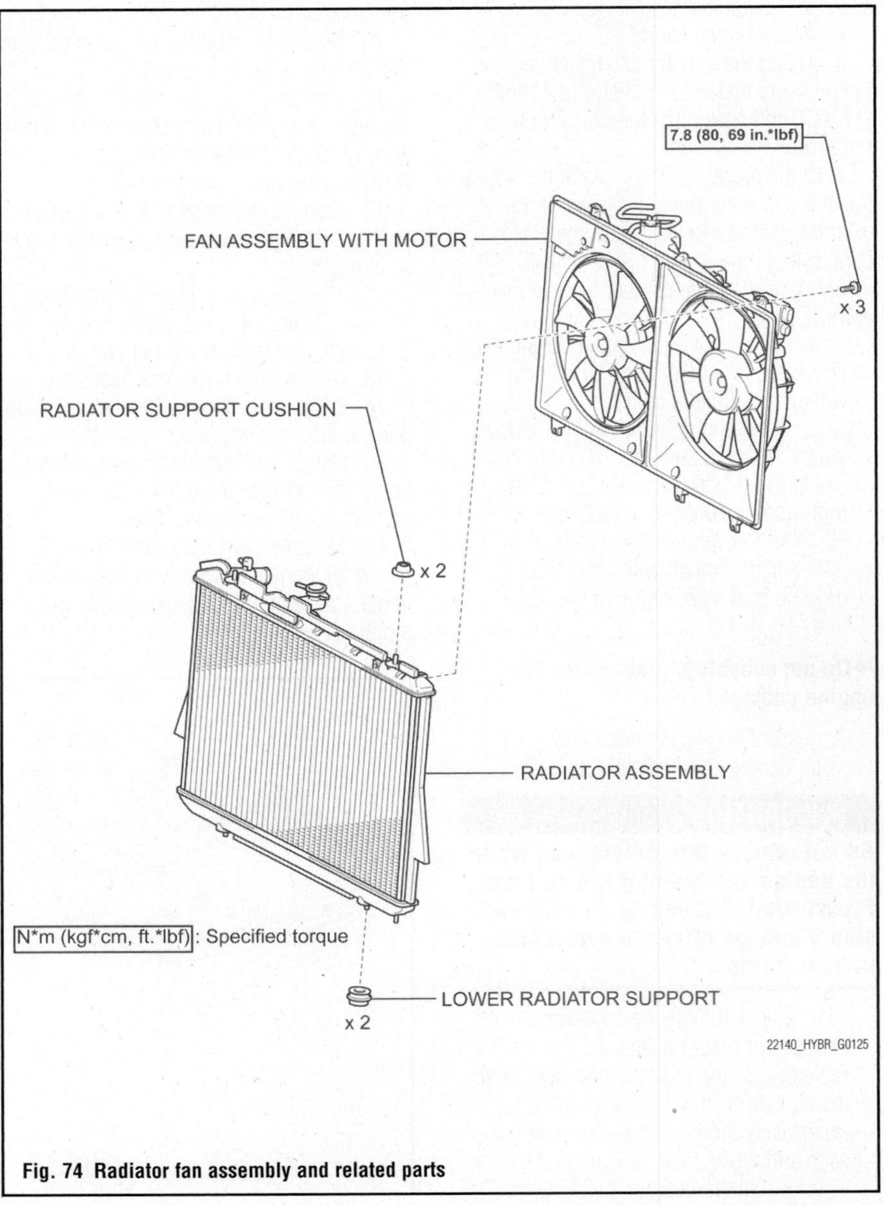

Fig. 74 Radiator fan assembly and related parts

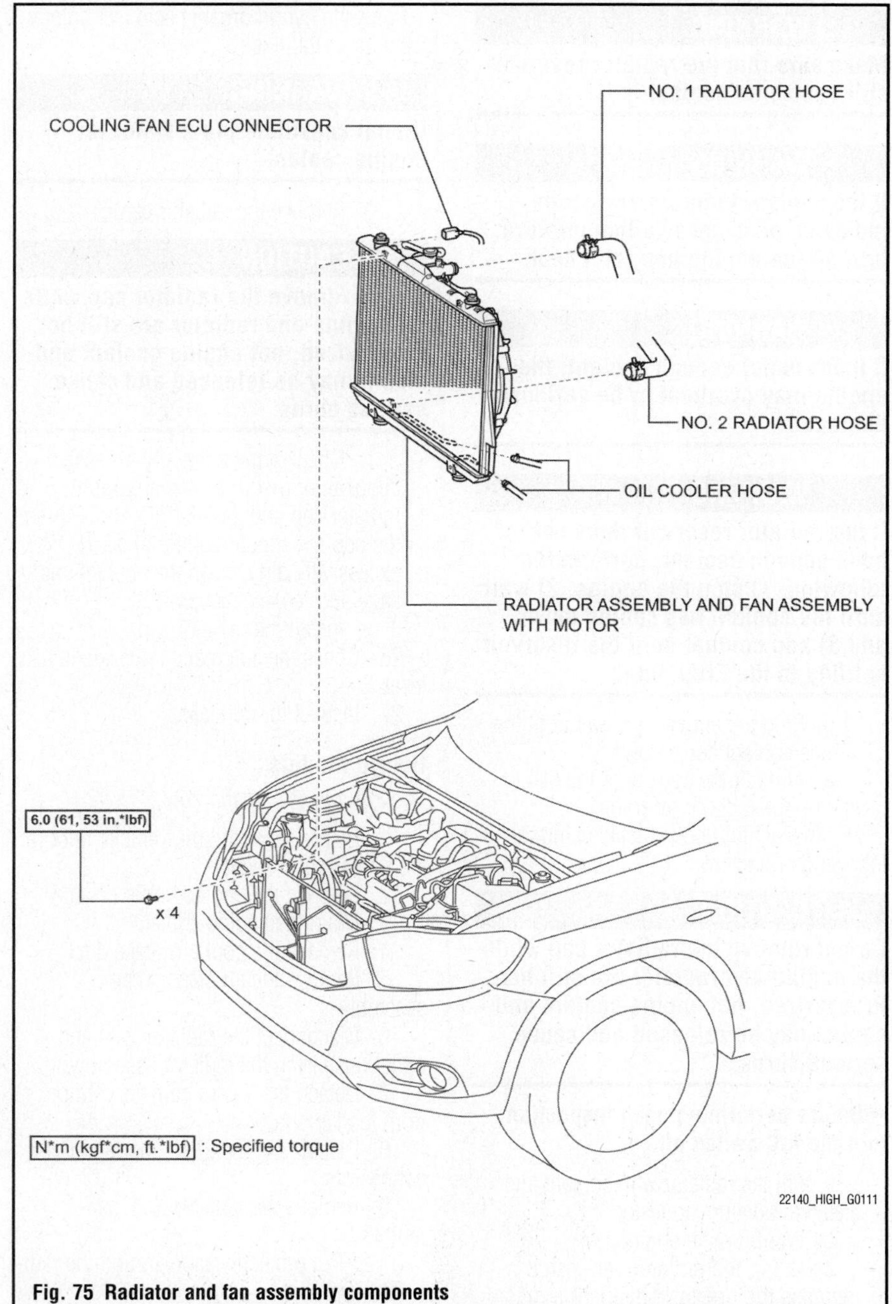

Fig. 75 Radiator and fan assembly components

5. Remove the cool air intake duct seal.

6. Remove the battery and battery tray.

7. Remove the No. 1 and 2 air cleaner inlets.

8. Disconnect the No. 1 and 2 radiator hoses from the radiator.

9. Disconnect the oil cooler hoses from the radiator.

10. Detach the wire harness clamps from the left side of the fan shroud.

11. Detach the wire harness clamps from the right side of the fan shroud and disconnect the cooling fan Electronic Control Unit (ECU) connector.

12. Remove the radiator grill, as follows:

  a. Put protective tape around the radiator grill.

  b. Remove the 2 bolts and 4 clips.

  c. Disengage the 6 claws and 4 guides, and remove the radiator grill.

13. Remove the hood lock assembly.

14. Disconnect the low pitched and high pitched horn connectors.

15. Detach the hood lock control cable clamp and remove the 6 bolts and upper radiator support sub-assembly.

16. Remove the 4 bolts and move the cooler condenser assembly to remove the radiator assembly and fan assembly with motor.

17. Remove the radiator assembly and fan assembly with motor.

18. Disconnect the radiator reserve tank hose or pipe from the radiator assembly.

19. Remove the 3 bolts.

20. Pull up the fan assembly with motor from the radiator assembly to remove the fan assembly with motor.

21. Remove the 2 radiator support cushions.

22. Remove the 2 lower radiator supports.

To install:

23. The installation procedure is the reverse of removal, with fluid levels being checked or fluids added, as noted. Use the following tightening specifications:

  a. When installing the fan assembly with motor, tighten the 3 bolts to 69 inch lbs. (7.8 Nm).

  b. When installing the cooler condenser assembly, tighten the 4 bolts to 53 inch lbs. (6 Nm).

  c. When installing the upper radiator support sub-assembly, tighten the 6 bolts to 87 inch lbs. (9.8 Nm).

  d. When installing the hood lock assembly, tighten the 3 bolts to 71 inch lbs. (8 Nm).

  e. When installing the No. 1 air cleaner inlet, tighten bolt A to 62 inch lbs. (7 Nm), and bolt B to 44 inch lbs. (5 Nm).

  f. When installing the No. 2 air cleaner inlet, tighten the 2 bolts to 62 inch lbs. (7 Nm).

24. Add engine coolant using the following procedure:

  a. Tighten the radiator drain cock plug by hand.

  b. Tighten the 2 cylinder block drain cock plugs to 9 ft. lbs. (13 Nm).

  c. Loosen the air drain cock plug from the water inlet housing.

  d. Loosen the air drain plug at the top of the radiator 3 or 4 turns.

➡TOYOTA vehicles are filled with TOYOTA Super Long Life Coolant (SLLC) at the factory. In order to avoid damage to the engine cooling system and other technical problems, only use TOYOTA SLLC or similar high quality ethylene glycol based non-silicate, non-amine, non-nitrite, non-borate coolant with

long-life hybrid organic acid technology (coolant with long-life hybrid organic acid technology consists of a combination of low phosphates and organic acids). Contact your TOYOTA dealer for further details.

WARNING

Never use water as a substitute for engine coolant.

e. Add TOYOTA SLLC to the radiator inlet opening until coolant overflows from the engine air drain cock hole. Then tighten the air drain cock plug to the water inlet housing. Tighten the plug to 9 ft. lbs. (13 Nm).

f. Continue to add TOYOTA Super Long Life Coolant (SLLC) to the radiator inlet opening until coolant overflows from the radiator air drain hole. Then close the air drain plug at the top of the radiator and tighten to 13 inch lbs. (1.5 Nm).

➡If the coolant level at the radiator inlet opening drops after squeezing the No. 1 and No. 2 radiator hoses, add coolant.

g. Slowly fill the radiator with TOYOTA Super Long Life Coolant (SLLC). For quantity information, refer to the capacities section.

h. Slowly pour coolant into the radiator reservoir tank until it reaches the FULL line.

i. Squeeze the No. 1 and No. 2 radiator hoses several times by hand, and then check the level of the coolant. If the coolant level is low, add coolant.

j. Warm up the engine until the thermostat opens. While the thermostat is open, circulate the coolant for several minutes.

➡The thermostat open timing can be confirmed by squeezing the inlet radiator hose by hand, and checking when the engine coolant starts to flow inside the hose.

k. Maintain the engine speed at 2500 to 3000 RPM.

l. Squeeze the inlet and outlet radiator hoses several times by hand to bleed air.

CAUTION

When squeezing the radiator hoses: Wear protective gloves. Be careful as the radiator hoses are hot. Keep your hands away from the radiator fan.

WARNING

Make sure that the radiator reservoir still has some coolant in it.

WARNING

If the coolant temperature gauge indicates an excessive temperature, turn off the engine and let it cool.

WARNING

If there is not enough coolant, the engine may overheat or be seriously damaged.

WARNING

If the radiator reservoir does not have enough coolant, perform the following: 1) stop the engine, 2) wait until the coolant has cooled down, and 3) add coolant until the reservoir is filled to the FULL line.

m. Stop the engine and wait until the engine coolant cools down.

n. Add engine coolant to the FULL line on the radiator reservoir.

25. Inspect for coolant leak, using the following procedure:

CAUTION

Do not remove the radiator cap while the engine and radiator are still hot. Pressurized, hot engine coolant and steam may be released and cause serious burns.

➡Before performing each inspection, turn the A/C switch off.

a. Fill the radiator with coolant and attach a radiator cap tester.

b. Warm up the engine.

c. Using the radiator cap tester, increase the pressure inside the radiator to 17 psi (118 kPa) and check that the pressure does not drop. If the pressure drops, check the hoses, radiator and water pump for leaks. If no external leaks are found, check the heater core, cylinder block, and cylinder head.

d. Check that the engine coolant level in the reservoir is between the LOW and FULL lines when the engine is cold. If the engine coolant level is low, check for leaks and add "TOYOTA Super Long Life Coolant" or similar high quality ethylene glycol based non-silicate, non-amine, non-nitrite and non-borate coolant with long-life hybrid organic acid technology to the FULL line.

WARNING

Do not substitute plain water for engine coolant.

e. Remove the radiator cap.

CAUTION

Do not remove the radiator cap while the engine and radiator are still hot. Pressurized, hot engine coolant and steam may be released and cause serious burns.

f. Check if there are any excessive deposits of rust or scales around the radiator cap and radiator filler hole. Also, the coolant should be free of oil. If excessively dirty, clean the coolant passage and replace the coolant.

g. Install the radiator cap.

26. Check the automatic transaxle fluid level.

27. Inspect for oil leaks.

Hybrid Models

See Figures 69, 70 and 76.

1. Before servicing the vehicle, refer to the Precautions Section.
2. Remove the engine under cover.
3. Drain the engine coolant.
4. Remove the cool air intake duct.
5. Remove the air cleaner cap assembly.
6. Disconnect the radiator inlet and outlet hose from the radiator assembly.
7. Detach the 7 wire harness clamps from the fan shroud.
8. Disconnect the 2 cooling fan ECU connectors.
9. Remove the radiator grill, as follows:
   a. Put protective tape around the radiator grill.
   b. Remove the 2 bolts and 4 clips.
   c. Disengage the 6 claws and 4 guides, and remove the radiator grill.
10. Remove the hood lock assembly and support.
11. Remove the upper radiator support.
12. Remove the 4 bolts and move the cooler condenser assembly to remove the radiator assembly and fan assembly with motor.
13. Separate the 3 water hose clamps from the fan shroud.
14. Remove the radiator assembly and fan assembly with motor.

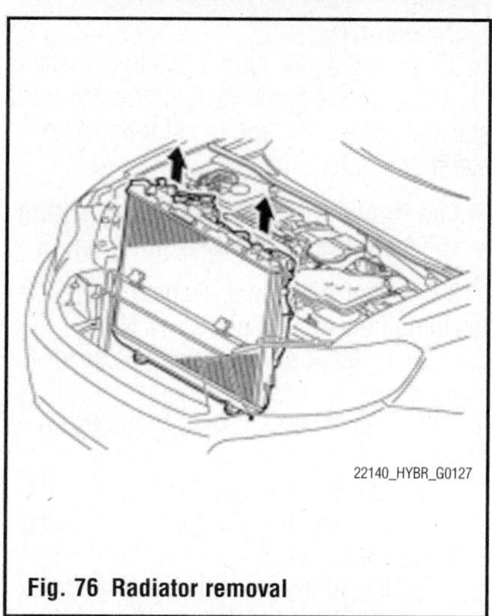

Fig. 76 Radiator removal

15. Disconnect the radiator reserve tank hose from the radiator assembly. Remove the 3 bolts.
16. Pull up the fan assembly with motor from the radiator assembly to remove the fan assembly with motor.
17. Remove the upper and lower radiator support cushions.

To install:

18. Install the upper and lower radiator supports.
19. Install the fan assembly, and tighten bolts to 69 inch lbs. (7.8 Nm).
20. Connect the radiator reserve tank hose to the radiator assembly.
21. Install the radiator assembly and fan assembly.
22. Install the cooler condenser assembly, and tighten the 4 bolts to 53 inch lbs. (6 Nm).
23. Connect the 3 water hose clamps to the fan shroud.
24. Install the upper radiator support sub-assembly. Tighten the bolts to 62 inch lbs. (7 Nm).
25. Install the discharge hose bracket with the bolt and tighten to 87 inch lbs. (9.8 Nm).
26. Install the hood lock support sub-assembly and the bolts to 62 inch lbs. (7 Nm).
27. Install the hood lock assembly and tighten to 71 inch lbs. (8 Nm).
28. Install the radiator grill.
29. Reconnect the fan assembly wiring clamps and connectors.
30. Install and tighten the inlet and outlet radiator hoses.
31. Install the air cleaner cap assembly.
32. Install the intake cool air duct seal.
33. Replace the lost engine coolant. Check for coolant leaks.
34. Install the engine under cover.

THERMOSTAT

REMOVAL & INSTALLATION

2.7L Engine

See Figures 77 and 78.

1. Disconnect the negative battery cable.
2. Raise and safely support the vehicle.
3. Remove the engine under cover assembly.
4. Disconnect the negative battery cable.
5. Remove v-ribbed belt.
6. Remove alternator assembly.
7. Disconnect no. 2 radiator hose.
8. Remove the 2 nuts and water inlet.
9. Remove the thermostat.
10. Remove the gasket from the thermostat.

To install:

11. Install a new gasket to the thermostat. Install the thermostat with the jiggle valve facing upward.

➡The jiggle valve may be set to within 10° on either side of the prescribed position.

12. Install the water inlet with the 2 nuts. Tighten to 7 ft. lbs. (10 Nm).

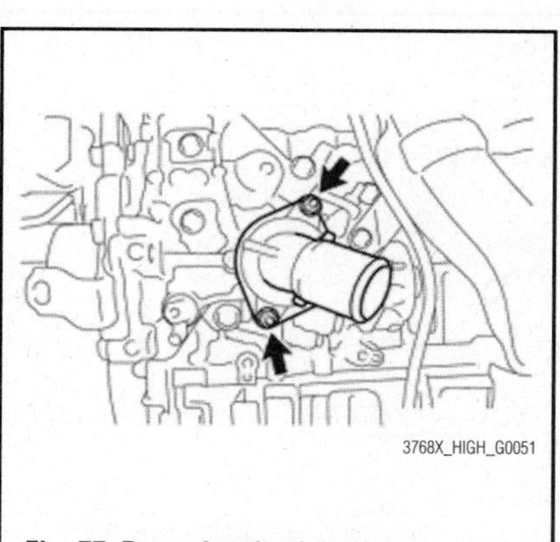

Fig. 77 Removing the thermostat

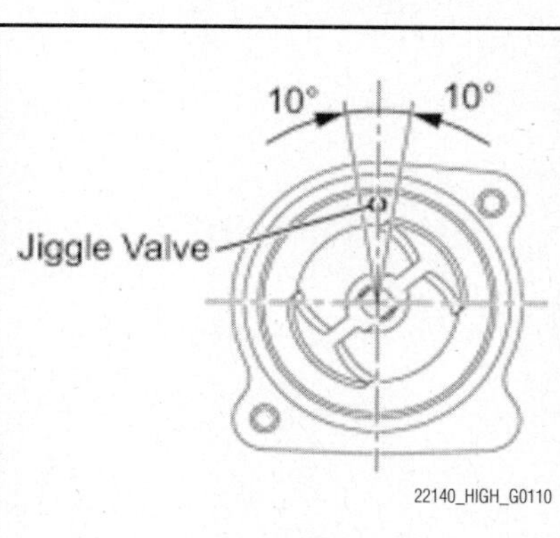

Fig. 78 Thermostat positioning

13. To complete installation, reverse remaining removal procedure.

3.3L Hybrid Engine

See Figures 79 and 80.

1. Before servicing the vehicle, refer to the Precautions Section.

⁂ CAUTION

The HIGHLANDER HV has a hybrid system that operates at voltages up to 650 volts. Be sure to follow the instructions in this manual to handle the system correctly. Failure to do so may result in serious injury or electrocution. Engineer must undergo special training to be able to perform high-voltage system inspection and servicing.

⁂ CAUTION

All high-voltage wire harness connectors are colored orange. The HV battery and other high-voltage components have "High Voltage" caution labels. Do not carelessly touch these wires and components.

2. Before inspecting or servicing the high-voltage system, be sure to follow safety measures, such as wearing insulated gloves and removing the service plug to prevent electrocution. Carry the removed service plug in your pocket to prevent other technicians from reinstalling it while you are servicing the vehicle.
3. After removing the service plug, wait 5 minutes before touching any of the high-voltage connectors and terminals.
4. Disconnect the negative battery cable.
5. Remove the engine room left side cover.
6. Remove the engine room cover.
7. Drain the coolant for the inverter.
8. Remove the wiper arm assembly LH and RH.
9. Remove the cowl top ventilator top louver sub-assembly.
10. Remove the wiper motor and link assembly.
11. Remove the cowl top outer panel sub-assembly.
12. Remove the cool air intake duct seal.
13. Remove the air cleaner cap with inlet.
14. Remove the air cleaner with resonator.
15. Remove the bolt and inverter bracket No.5.

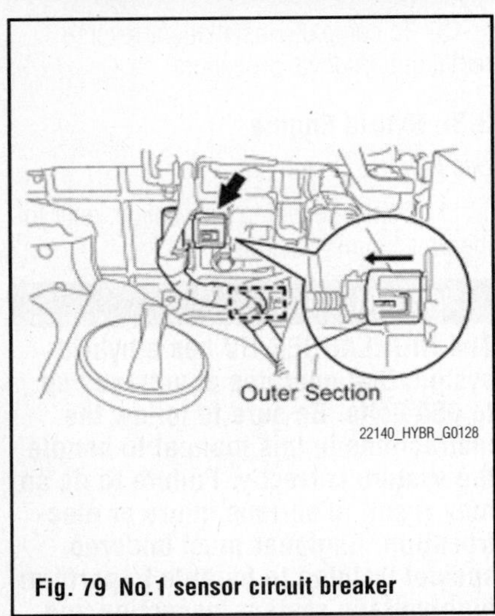

Fig. 79 No.1 sensor circuit breaker

16. Remove bolt and ground cable terminal from power steering ECU assembly.
17. Move the outer section to the wire harness side as illustrated, then disconnect the circuit breaker sensor No.1.
18. Remove the inverter reserve tank subassembly.
19. Disconnect water hose.
20. Disconnect the power steering ECU bracket.
21. Remove the inverter cover.
22. Verify voltage of w/converter inverter assembly is 0 v.
23. Separate engine wire no.4.
24. Disconnect high voltage cable from the front motor.
25. Disconnect the MG ECU connector.
26. Disconnect the no.3 wire frame.
27. Install the inverter cover.
28. Separate engine room relay block assembly
29. Remove the inverter bracket no.4.
30. Remove w/converter inverter assembly. Since the inverter with converter assembly is very heavy, 2 people are needed to remove the inverter with converter assembly. When removing the inverter with converter assembly, do not damage the parts around it.
31. Remove the water inlet pipe, as follows:
32. Remove the bolt and the water inlet pipe.
33. Remove the O-ring from the water inlet pipe.
34. Disconnect the wire harness clamp.
35. Remove the 3 nuts and the water inlet.
36. Remove the thermostat and gasket.

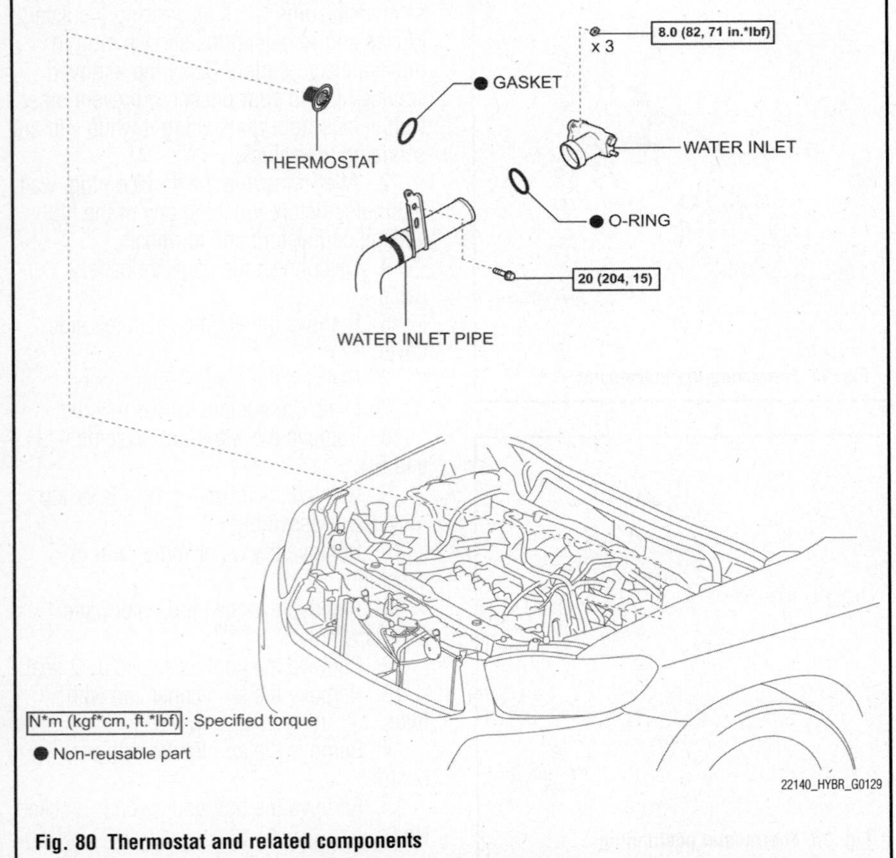

Fig. 80 Thermostat and related components

To install:

37. Install a new gasket and thermostat.
38. Align the jiggle valve of the thermostat and water inlet, and insert the thermostat in the water inlet housing.

➡The jiggle valve should be set within +-15° from the prescribed position.

39. Install water inlet. Tighten the 3 retaining nuts to 71 inch lbs. (8 Nm).
    a. Install the water inlet pipe as follows:
    b. Install a new O-ring to the water inlet pipe.
    c. Apply soapy water to the O-ring.
    d. Connect the water inlet pipe to the water inlet.
    e. Install the bolt which is used to fix the water inlet pipe to the cylinder head with the bolt. Tighten to 15 ft. lbs. (20 Nm).
40. Install w/converter inverter assembly.
41. Install inverter bracket No.4.
42. Install engine compartment relay block assembly.
43. Remove inverter cover.
44. Connect no.3 wire frame.
45. Connect MG ECU connector. Check that each connector and terminal is firmly installed.
46. Connect the high voltage cable of the generator (MG1) with new 5 bolts to the inverter with converter assembly. Tighten the bolts to 7 ft. lbs. (10 Nm).
47. Connect engine wire No.4 and tighten the bolts to 48 inch lbs. (5.4 Nm).
48. Check high voltage cable connection.
49. Install inverter cover. Tighten the cover bolts to 7 ft. lbs. (10 Nm).
50. Install the power steering ECU bracket.
51. Connect the water hose.
52. Install inverter reserve tank subassembly.
53. Connect engine room wire No.2.
54. Connect circuit breaker sensor No.1.
55. Install the inverter bracket No.5.
56. Install air cleaner w/resonator.
57. Install air cleaner cap w/inlet.
58. Install air cleaner w/resonator.
59. Install air cleaner cap w/inlet.
60. Install the cool air intake duct seal.
61. Install the outer cowl top panel.
62. Install wiper motor and link assembly.
63. Install cowl top ventilator louver sub-assembly.
64. Install the LH and RH wiper arm assembly.
65. Install the service plug grip.
66. Connect the negative battery cable.

67. Refill Coolant for inverter and engine.
68. Check for leaks.
69. If necessary, perform the initialization procedure.

3.5L Engine

See Figures 81 through 85.

1. Remove the V-bank cover sub-assembly.
2. Raise and safely support the vehicle.
3. Remove the engine under cover assembly.
4. Remove the No. 1 engine under cover.
5. Remove the right front wheel.
6. Remove right front fender molding sub-assembly.
7. Remove the right front fender liner.
8. Remove the 2 bolts, clip and right front fender apron seal.
9. Drain the engine coolant.
10. Using a Special Service Tool (SST: 09961-00950), release the V-ribbed belt tension by turning the V-ribbed belt tensioner assembly counterclockwise, and remove the V-ribbed belt from the V-ribbed belt tensioner assembly.
11. While turning the V-ribbed belt tensioner assembly counterclockwise, align with its holes, and then insert the 5 mm bi-hexagon wrench into the holes to fix the V-ribbed belt tensioner assembly.
12. Disconnect the No. 2 radiator hose from the engine.
13. Remove the bolt, idler pulley cover plate, and idler pulley sub-assembly.
14. Remove the 2 nuts and the water inlet.
15. Remove the thermostat from the water inlet housing.

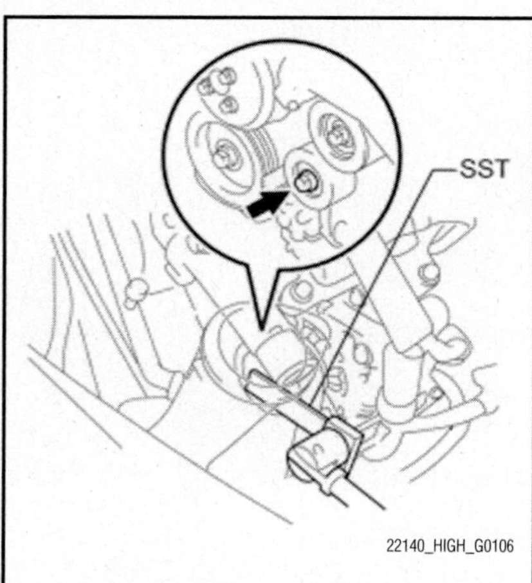

Fig. 81 Releasing tension on V-ribbed belt—3.5L engine

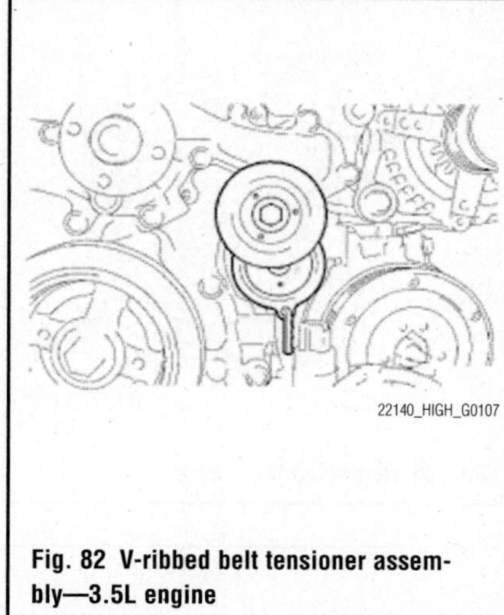

Fig. 82 V-ribbed belt tensioner assembly—3.5L engine

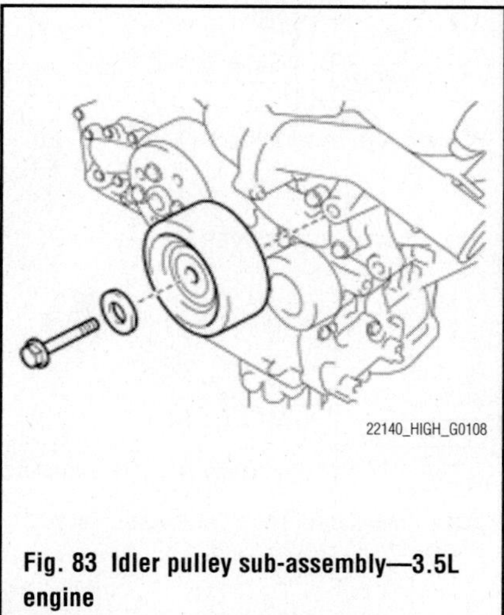

Fig. 83 Idler pulley sub-assembly—3.5L engine

16. Remove the gasket from the thermostat.

To install:

17. Install a new gasket to the thermostat.
18. Install the thermostat with the jiggle valve facing up.

➡The jiggle valve may be set within 10° on either side of the prescribed position.

19. Install the water inlet with the 2 nuts, and tighten to 7 ft. lbs. (10 Nm).
20. Install the idler pulley cover plate and idler pulley sub-assembly with the bolt, and tighten to 32 ft. lbs. (43 Nm).
21. Connect the No. 2 radiator hose to the engine.
22. Using SST: 09961-00950, turn the V-ribbed belt tensioner assembly counterclockwise and remove the bar.
23. If it is difficult to install the V-ribbed belt, perform the following procedure:
    a. Put the V-ribbed belt on every pulley except the tensioner pulley as shown in the illustration.
    b. Release the V-ribbed belt tension by turning the V-ribbed belt tensioner assembly counterclockwise, and put the V-ribbed belt on the V-ribbed tensioner assembly pulley.

➡Put the backside of the V-ribbed belt on the V-ribbed belt tensioner assembly pulley and the No. 2 idler pulley sub-assembly.

➡Check that the V-ribbed belt is properly set to each pulley.

    c. After installing the V-ribbed belt, check that it fits properly in the ribbed grooves. Confirm that the belt has not slipped out of the grooves on the bottom of the crankshaft pulley by hand.

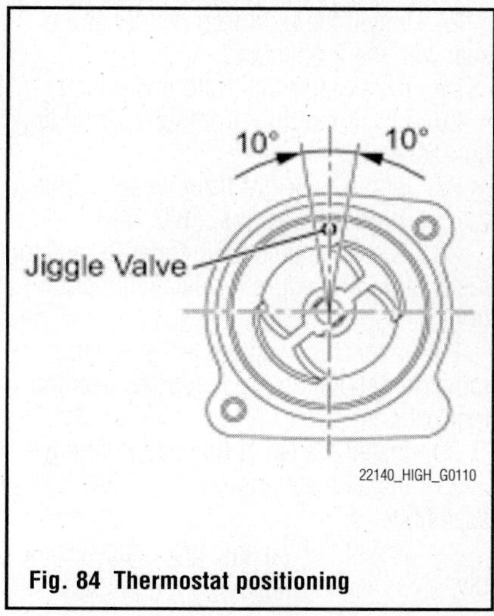

Fig. 84 Thermostat positioning

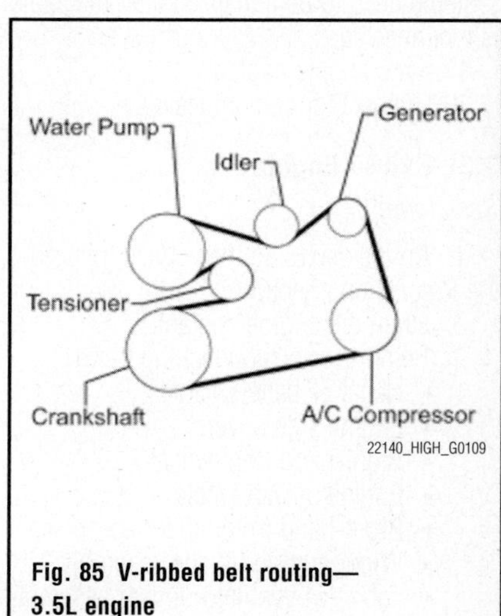

Fig. 85 V-ribbed belt routing—3.5L engine

24. Install the right front fender apron seal with the 2 bolts and clip.
25. Install the right front fender liner.
26. Install the right front fender molding sub-assembly.
27. Install the right front wheel. Tighten the lug nuts to 76 ft. lbs. (103 Nm).
28. Add engine coolant. Refer to radiator installation procedure for detailed instructions.
29. Inspect for coolant leaks. Refer to radiator installation procedure for detailed instructions.
30. Install the No. 1 engine under cover.
31. Install the engine under cover assembly.
32. Install the V-bank cover sub-assembly.

WATER PUMP

REMOVAL & INSTALLATION

2.4L Engine

See Figure 86.

1. Disconnect the negative battery cable.
2. Remove no. 1 engine under cover.
3. Drain and recycle the engine coolant.
4. Remove v-ribbed belt.
5. Remove alternator assembly.
6. Remove v-ribbed belt tensioner assembly.
7. Remove the 7 bolts, water pump and water pump gasket.

To install:

8. Install a new gasket and the water pump with the 7 bolts. Tighten to 15 ft. lbs. (21 Nm).
9. To complete installation, reverse remaining removal procedure.
10. When disconnecting the cable, some systems need to be initialized after the cable is reconnected.
11. Add engine coolant.
12. Inspect for coolant leaks.

3.3L Hybrid Engine

See Figure 87.

1. Before servicing the vehicle, refer to the Precautions Section.
2. Drain the engine coolant.
3. Remove or disconnect the following:
   - Negative battery cable
   - Engine side covers
   - Right-hand front wheel
   - Engine splash shield
   - Right-hand front fender apron seal
   - Wiper arm and blade assembly
   - Top cowl ventilator louver assembly
   - Wiper motor and link assembly
   - Battery and battery tray
   - Air intake assembly
   - Brake master cylinder reservoir
   - Reservoir support bracket
   - Air cleaner support bracket
   - Engine moving control rod
   - Right-hand engine mount
   - Crankshaft pulley
   - Timing belt cover No.1 and 2
   - Timing belt guide
   - Timing belt
   - Timing belt idler sub-assembly No.1

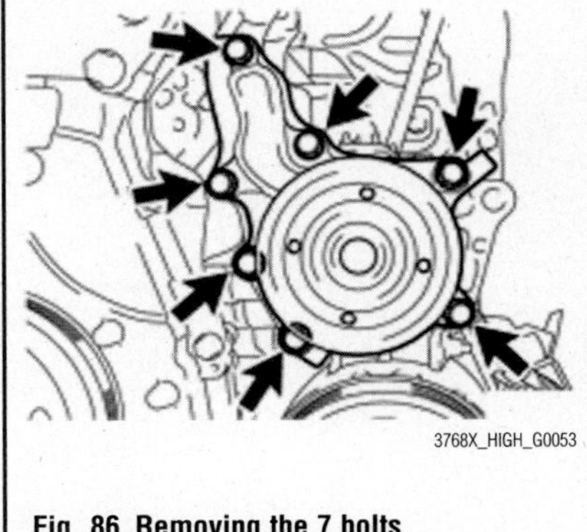

Fig. 86 Removing the 7 bolts

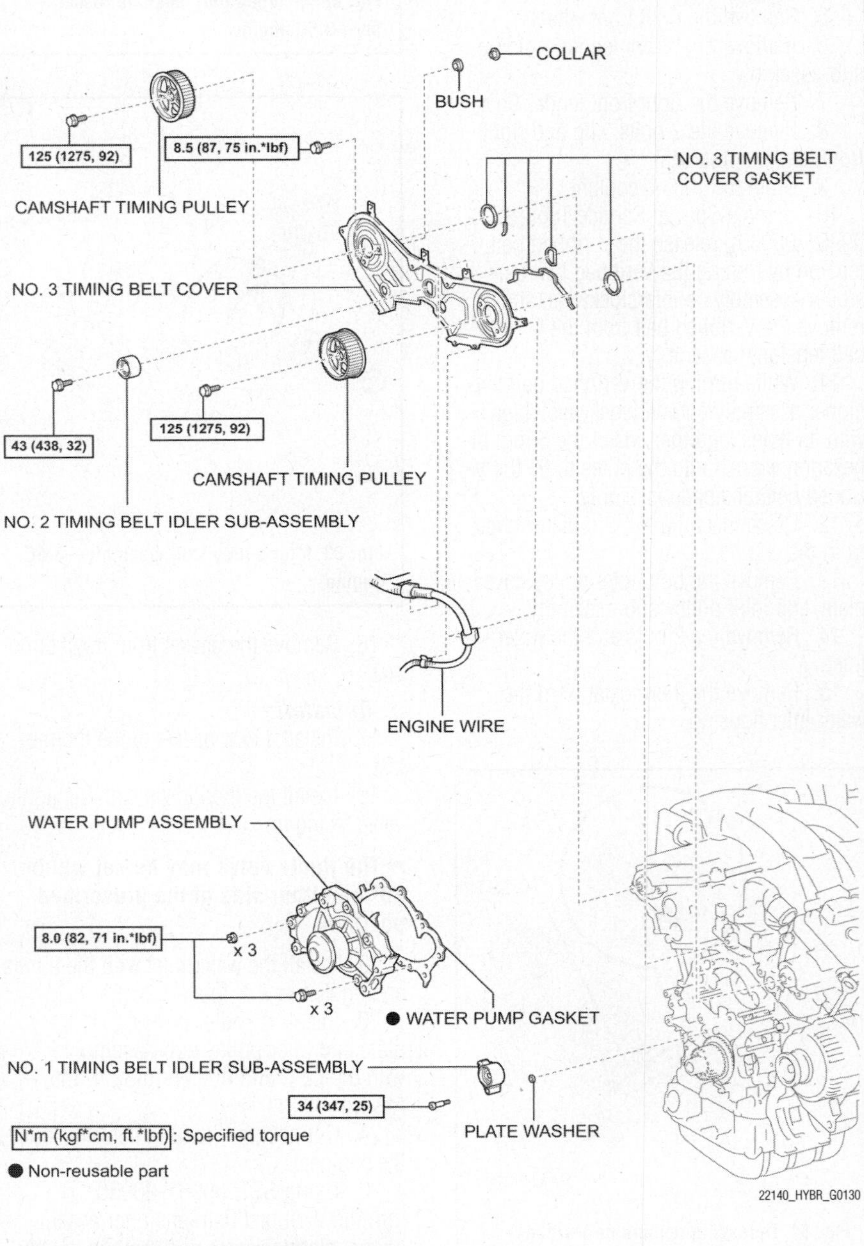

Fig. 87 Water pump and related components

- Camshaft timing pulley
- Timing belt cover No.3
- Timing belt idler sub-assembly No.2
- Water pump

4. Installation is the reverse order of removal. Tighten the water pump with a new gasket to 71 inch lbs. (8 Nm).
5. Refill the coolant to the correct level.
6. Start the engine and check for leaks.
7. If necessary, perform the initialization procedure.

ENGINE ELECTRICAL — BATTERY SYSTEM

BATTERY

REMOVAL & INSTALLATION

Non-Hybrid Models

See Figure 88.

➡For Hybrid models, refer to HV Battery in this section.

1. Loosen the nut, and disconnect the negative battery terminal.
2. Loosen the nut, and disconnect the positive battery terminal.
3. Loosen the nut, and remove the bolt and battery clamp.
4. Remove the battery and battery tray.

To install:

5. To install, reverse removal procedure.

BATTERY RECONNECT/RELEARN PROCEDURE

Non-Hybrid Models

➡If the back door is closed when disconnecting the cable from the battery terminal, it is not necessary to reset it.

1. Necessary Procedure: Reset back door close position.
2. Procedure Detail: Fully close the back door to turn off the courtesy switch.
3. Effect / Inoperative Function When Necessary Procedure is not Performed: Power back door function.

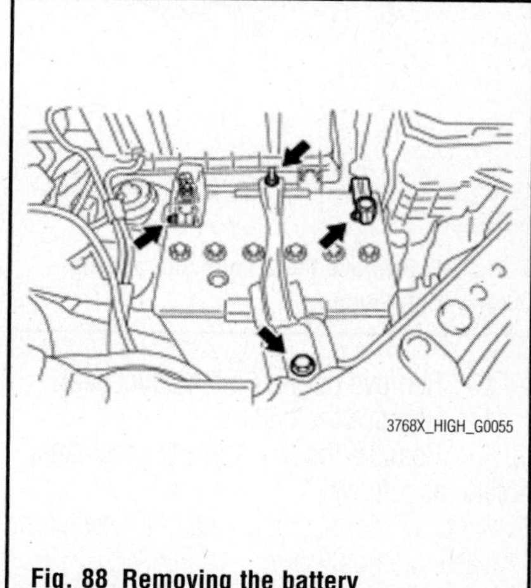

Fig. 88 Removing the battery

ENGINE ELECTRICAL — CHARGING SYSTEM

ALTERNATOR

REMOVAL & INSTALLATION

2.7L Engine

See Figures 89 and 90.

1. Disconnect the negative battery cable.
2. Remove the v-belt.
3. Disconnect the alternator connector.
4. Remove the terminal cap.
5. Remove the nut and disconnect the alternator wire.
6. Remove the bolt and wire harness clamp bracket.
7. Remove the 2 bolts and alternator.

To install:

8. To install, reverse removal procedure.
9. Tighten the alternator bolts to 38 ft. lbs. (52 Nm).

3.5L Engine

See Figures 91 and 92.

1. Raise and safely support the vehicle.
2. Remove the front wheel.
3. Remove the engine under cover assembly.
4. Remove the No. 1 engine under cover.
5. Remove the right front fender molding sub-assembly.
6. Remove the right front fender liner.
7. Remove the right front fender apron seal.
8. Drain engine coolant.
9. Remove the V-bank cover sub-assembly

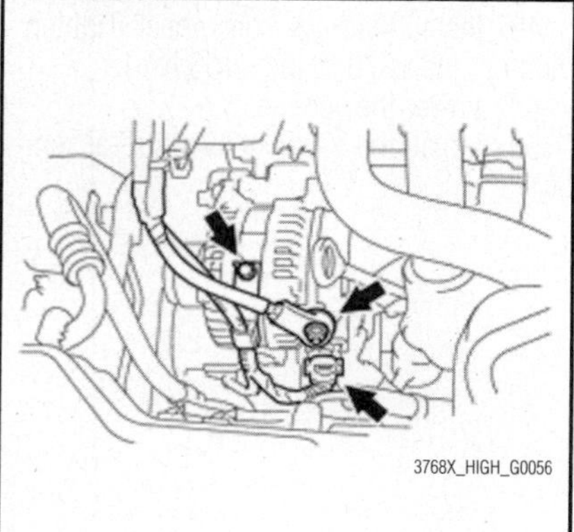

Fig. 89 Disconnect the alternator connector—2.7L Engine

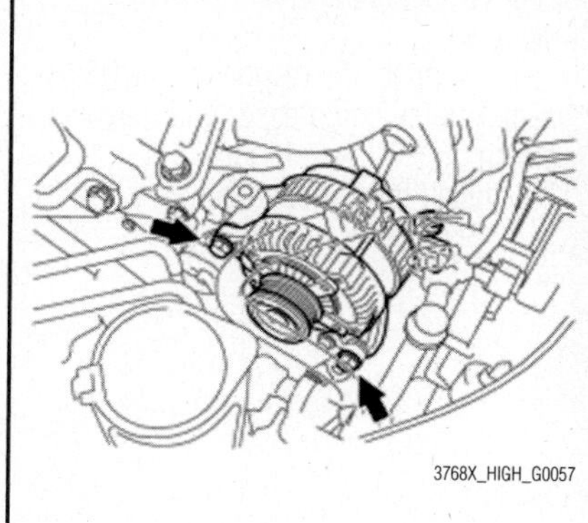

Fig. 90 Remove the 2 bolts and alternator—2.7L Engine

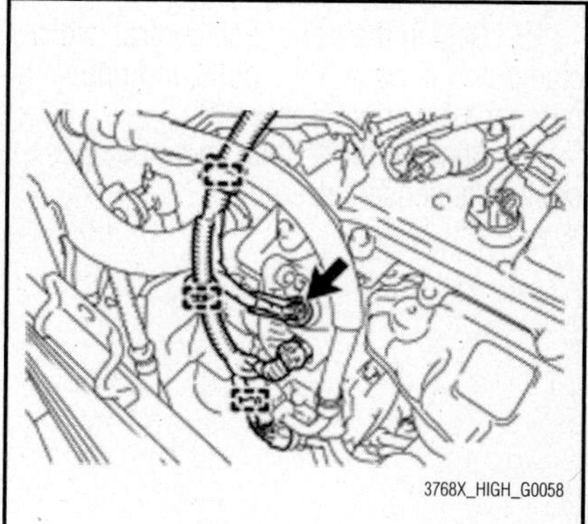

Fig. 91 Disconnecting the electrical connectors—3.5L Engine

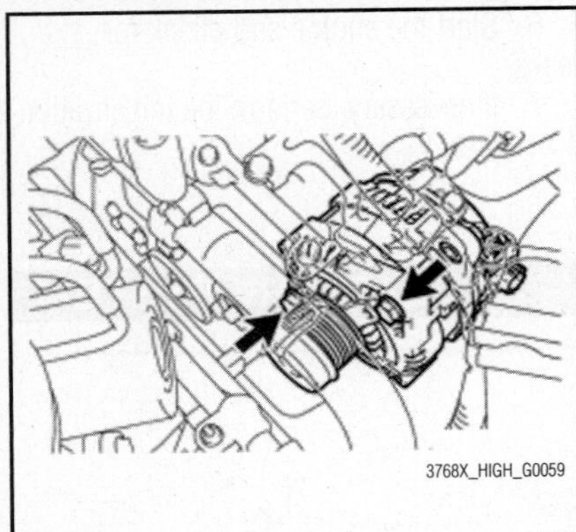

Fig. 92 Remove the 2 bolts and alternator—3.5L Engine

10. Remove cool air intake duct seal.
11. Remove the battery.
12. Remove the No. 1 and 2 air cleaner inlets, as follows:
    a. Disconnect the 2 vacuum switching valve clamps.
    b. Disconnect the 2 vacuum hoses.
    c. Remove the 2 bolts and No. 2 air cleaner inlet.
    d. Disconnect the 2 vacuum hoses, and remove the 2 bolts and No. 1 air cleaner inlet.
13. Disconnect the No. 1 and 2 radiator hoses.
14. Disconnect the oil cooler hoses.
15. Detach the wire harness clamps from both sides of the fan shroud and disconnect the cooling fan ECU connector.
16. Remove the radiator grill. Refer to radiator removal procedure for detailed instructions.
17. Remove the hood lock assembly.
18. Disconnect the low pitched horn and high pitched horn connectors.
19. Detach the hood lock control cable clamp and remove the 6 bolts and upper radiator support sub-assembly.
20. Remove the 4 bolts and move the cooler condenser assembly.
21. Remove the radiator assembly and fan assembly with motor.
22. Remove the bolt and the No. 2 oil level dipstick guide.
23. Remove the V-ribbed belt.
24. Remove the alternator assembly, as follows:
    a. Remove the terminal cap.
    b. Remove the nut and disconnect the wire harness from terminal B.
    c. Disconnect the alternator connector from the alternator assembly.
    d. Disconnect the connector from the compressor and magnetic clutch.
    e. Disconnect the 3 wire harness clamps.
    f. Remove the 2 bolts, and then disconnect the bracket.
    g. Remove the 2 bolts and the alternator assembly.
    h. Disconnect the wire harness clamp, and then remove the alternator bracket.
    i. Remove the bolt and the wire harness clamp stay.

To install:

25. Install the alternator assembly, as follows:
    a. Install the wire harness clamp stay with the bolt, and tighten to 15 ft. lbs. (20 Nm).
    b. Connect the alternator bracket with the wire harness clamp.
    c. Install alternator assembly with the 2 bolts, and tighten to 32 ft. lbs. (43 Nm).
    d. Temporally install the 2 bolts., then fully tighten the 2 bolts to 15 ft. lbs. (20 Nm).
    e. Connect the alternator connector to the alternator assembly.
    f. Install the alternator wire with the nut, and tighten to 87 ft. lbs. (9.8 Nm).
    g. Install the terminal cap.
    h. Connect the 3 wire harness clamps.
    i. Connect the magnetic clutch connector to the compressor and magnetic clutch.
26. Install the V-ribbed belt.
27. Install the No. 2 oil level dipstick guide, as follows:
    a. Install a new O-ring to the No. 2 oil level dipstick guide.
    b. Apply a light coat of engine oil to the O-ring.
    c. Push in the No. 2 oil level dipstick guide end into the No. 1 oil level dipstick guide.
    d. Install the No. 2 oil level dipstick guide with the bolt, and tighten to 15 ft. lbs. (20 Nm).
28. Install the radiator assembly and fan assembly with motor.
29. Install the cooler condenser assembly.
30. Install the upper radiator support sub-assembly.
31. Install the hood lock assembly.
32. Install the radiator grill.
33. Connect the cooling fan ECU connector.
34. Connect the oil cooler hose.
35. Connect the No. 1 and 2 radiator hose.
36. Install the No. 1 and 2 air cleaner inlet.
37. Install the battery.
38. Install the cool air intake duct seal.
39. Install the right front fender apron seal.
40. Install the right front fender liner.
41. Install the right front fender molding sub-assembly.
42. Add engine coolant. Refer to radiator installation procedure for detailed instructions.
43. Inspect for coolant leaks. Refer to radiator installation procedure for detailed instructions.
44. Inspect automatic transaxle fluid.
45. Inspect for oil leaks.
46. Install the No. 1 engine under cover.
47. Install the engine under cover assembly.
48. Install the right front wheel. Tighten the lug nuts to 76 ft. lbs. (103 Nm).
49. Lower the vehicle.
50. Install the V-bank cover sub-assembly.

PRECAUTIONS

Before working on any part of the Hybrid high voltage system, observe the following precautions:

✱✱ CAUTION

The nominal high voltage traction battery voltage is 330 volts DC. The buffer zone must be set up and insulated rubber gloves and a face shield must be worn. Failure to follow these instructions may result in severe injury or death.

✱✱ CAUTION

The high voltage traction battery and charging system contains high voltage components and wiring. High voltage insulated safety gloves and a face shield must be worn when carrying out any diagnostics on this vehicle. Failure to follow these instructions may result in severe personal injury or death.

✱✱ CAUTION

Before carrying out any removal and installation procedures of the high voltage traction battery system, the high voltage traction battery must be Disarmed. Failure to follow these instructions may result in severe personal injury or death.

✱✱ CAUTION

The rubber insulating gloves that are to be worn while working on the high voltage system should be of the appropriate safety and protection rating for use on the high voltage system. They must be inspected before use and must always be worn in conjunction with the leather outer gloves. Any hole in the rubber insulating glove is a potential entry point for high voltage. Failure to follow these instructions may result in severe personal injury or death.

➡The high voltage insulated safety gloves must be re-certified every 6 months to remain within Occupational Safety and Health Administration (OSHA) guidelines:

- Roll the glove up from the open end until the lower portion of the glove begins to balloon from the resulting air pressure. If the glove leaks any air, it must not be used.
- The gloves should not be used if they exhibit any signs of wear and tear.
- The leather gloves must always be worn over the rubber insulating gloves in order to protect them.
- The rubber insulating gloves must be class "00" and meet all of the American Society for Testing and Materials (ASTM) standards

✱✱ CAUTION

High voltage insulated safety gloves and a face shield must be worn when working with high voltage cables. The ignition switch must be OFF for a minimum of 5 minutes before removing high voltage cables. Failure to follow these instructions may result in severe personal injury or death.

✱✱ CAUTION

Establish a buffer zone before servicing the high voltage system. The buffer zone is required only when working with the high voltage system. See the text for buffer zone establishment. Failure to follow these instructions may result in severe personal injury or death. Do not allow any unauthorized personnel into the buffer zone during repairs involving the high voltage system. Only personnel trained for repair on the high voltage system are to be permitted in the buffer zone.

✱✱ CAUTION

Disarm the high voltage traction battery (HVTB) before working on the high voltage system. See the text for the Disarming procedure. Failure to follow these instructions may result in severe personal injury or death.

BUFFER ZONE

1. Before servicing the vehicle, refer to the Precautions Section.

✱✱ CAUTION

Before proceeding, read and observe all of the High Voltage System Precautions.

2. Establish a buffer zone around the vehicle:
    a. Position the vehicle in the repair bay.
    b. Position 4 orange cones at the corners of the vehicle to mark off a 1 m (3 ft.) perimeter around the vehicle.
    c. Do not allow any unauthorized personnel into the buffer zone during repairs involving the high voltage system. Only personnel trained for repair on the high voltage system are to be permitted in the buffer zone.

DISARMING THE HIGH VOLTAGE TRACTION BATTERY

✱✱ WARNING

When disconnecting the cable, some systems need to be initialized after the cable is reconnected.

✱✱ WARNING

After disconnecting the service plug grip, wait for at least 10 minutes before touching any of the high-voltage connectors or terminals.

HIGH VOLTAGE TRACTION BATTERY

REMOVAL AND INSTALLATION

See Figures 93 through 132.

1. Read output DTC.
2. Remove deck board assembly.
3. Remove battery service hole cover assembly.
4. Disconnect cable from negative battery terminal.

➡When disconnecting the cable, some systems need to be initialized after the cable is reconnected.

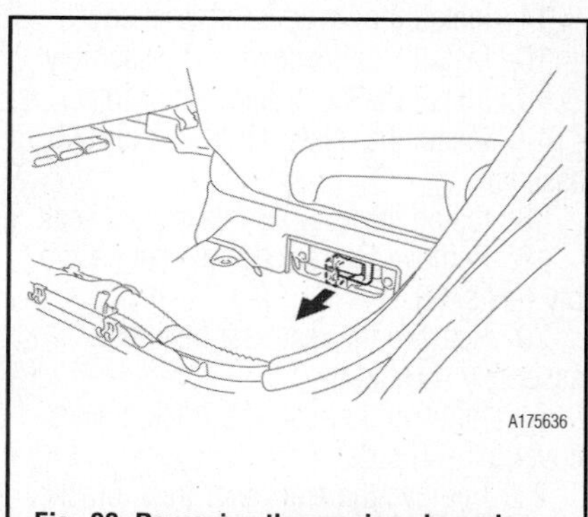

Fig. 93 Removing the service plug grip

5. Remove rear door scuff plate LH.
6. Remove reclining hinge cover LH.
7. Remove service plug grip.

WARNING

Remove the service plug to interrupt a high voltage circuit at the time of the check.

WARNING

Keep the removed service plug in your pocket to prevent other technicians from accidentally reconnecting it while you are servicing the vehicle.

WARNING

All the high voltage wiring connectors are colored in orange.

a. Wear insulated glove and remove the service plug grip, after sliding up the lever of the service plug grip.

WARNING

Keep the removed service plug in your pocket to prevent other technicians from accidentally reconnecting it while you are servicing the vehicle.

WARNING

After removing the service plug grip, do not touch the high voltage connectors and terminals for 10 minutes.

➡After removing the service plug grip, do not operate the ignition switch as it may damage the hybrid vehicle control ECU.

8. Remove the cool air intake duct seal.
9. Remove the intake air resonator sub assembly.
10. Remove the No. 2 air cleaner inlet.
11. Remove V-bank cover sub assembly.
12. Remove the inverter terminal cover.
13. Check the terminal voltage.
14. Install the inverter terminal cover.
15. Install V-bank cover sub assembly.
16. Install the No. 2 air cleaner inlet.
17. Install the intake air resonator sub assembly.
18. Install the cool air intake duct seal.
19. Remove the rear center seat assembly (LH side).
20. Remove the rear seat headrest assembly (LH side).
21. Remove the rear seat track bracket cover LH (LH side).
22. Remove the rear outer track bracket cover LH (LH side).
23. Remove the rear inner track bracket cover LH (LH side).
24. Remove rear seat leg side cover LH (LH side).
25. Disconnect the rear No. 1 seat lock cable assembly LH (LH side).
26. Remove the rear No. 1 seat assembly LH (LH side).
27. Remove the rear seat headrest assembly (RH side).
28. Remove rear seat track bracket cover RH (RH side).
29. Remove the rear outer track bracket cover RH (RH side).
30. Remove the rear inner track bracket cover RH (RH side).
31. Remove the rear seat leg side cover RH (RH side).
32. Disconnect the rear No. 1 seat lock cable assembly RH (RH side).
33. Remove the rear No. 1 seat assembly RH (RH side).
34. Remove the No. 2 deck board sub assembly.
35. Remove the No. 2 deck board.
36. Remove the No. 1 deck board.
37. Remove the jack carrier support.
38. Remove the jack carrier cushion.
39. Remove the jack assembly.
40. Remove the jack carrier assembly.
41. Remove the deck side trim box RH.
42. Remove the rear mat.
43. Remove the rear deck floor box.
44. Disconnect the rear No. 2 seat outer belt assembly RH and LH.
45. Remove the center seat cushion sub assembly.
46. Remove the rear No. 2 seat assembly.
47. Remove the rear door scuff plate RH.
48. Remove the rear door opening trim weatherstrip LH and RH.
49. Remove the rear floor finish plate.
50. Remove the deck side trim cover No. 2.
51. Remove the deck side trim LH.
52. Remove the side trim cover LH (w/o rear seat entertainment system).
53. Remove the power outlet socket bezel (w/rear seat entertainment system).
54. Remove the rear combination light service cover LH.
55. Remove the rear power point socket assembly.
56. Remove the rear power outlet socket cover.
57. Remove reclining remote control lever bezel LH.
58. Remove rope hook assembly (LH side).
59. Remove the No. 2 deck side trim hook.

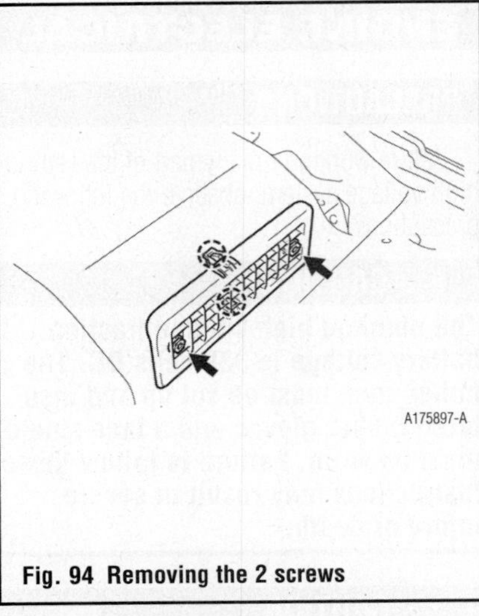

Fig. 94 Removing the 2 screws

60. Remove the front deck side trim cover LH.
61. Disconnect the rear No. 1 seat outer belt assembly LH.
62. Remove the deck trim side panel assembly LH.
63. Remove the rear seat side garnish cap.
64. Remove the deck side trim cover No. 1.
65. Remove deck side trim RH.
66. Remove side trim cover RH (for Manual Air Conditioning System).
67. Remove rear room temperature sensor (for automatic air conditioning system).
68. Remove the rear combination light service cover RH.
69. Remove rope hook assembly (RH side).
70. Remove the No. 1 luggage compartment trim hook.
71. Remove the front deck side trim cover RH.

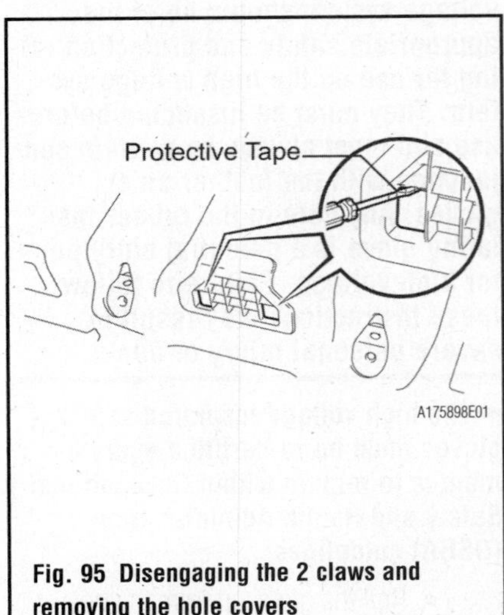

Fig. 95 Disengaging the 2 claws and removing the hole covers

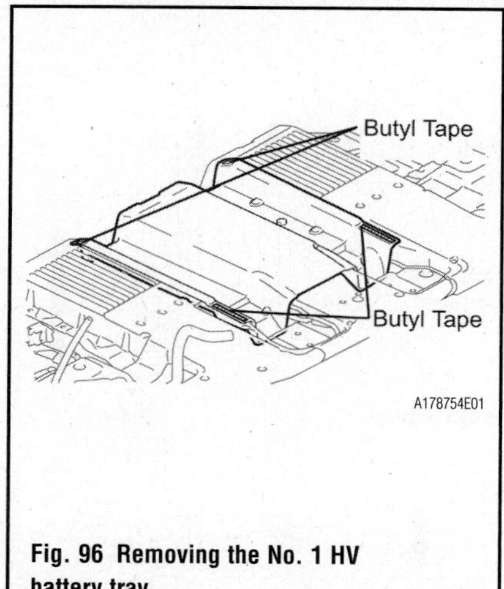

Fig. 96 Removing the No. 1 HV battery tray

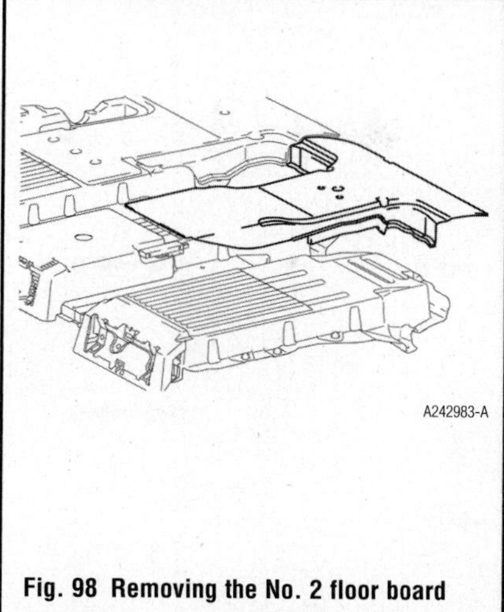

Fig. 98 Removing the No. 2 floor board

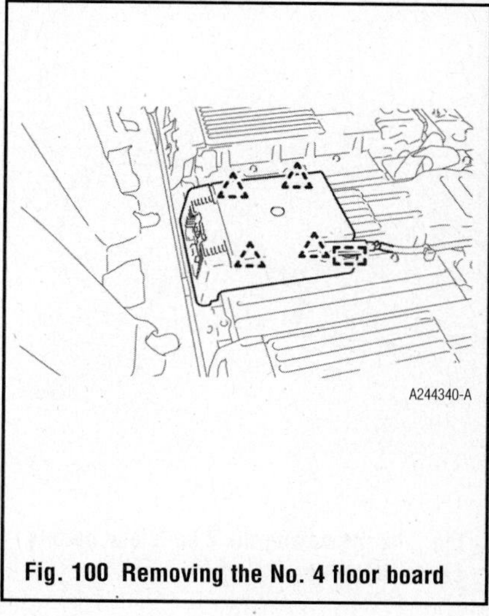

Fig. 100 Removing the No. 4 floor board

72. Disconnect the rear No. 1 seat outer belt assembly RH.
73. Remove the deck trim side panel assembly RH.
74. Remove the rear center seat inner belt assembly.
75. Remove the air intake cover.
   a. Remove the air intake cover LH.
   b. Using a screwdriver, disengage the 2 claws and remove the 2 hole covers.

➡Tape the screwdriver tip before use.

   c. Remove the 2 screws.
   d. Disengage the 2 claws and remove the air intake cover LH.
   e. Remove the air intake cover RH.
   f. Using a screwdriver, disengage the 2 claws and remove the 2 hole covers.

➡Tape the screwdriver tip before use.

   g. Remove the 2 screws.
   h. Disengage the 2 claws and remove the air intake cover RH.

76. Remove the center air intake cover.
   a. Using a screwdriver, disengage the 2 claws and remove the 2 hole covers.

➡Tape the screwdriver tip before use.

   b. Disengage the 2 claws and remove the center air intake cover.

77. Remove the front floor carpet assembly.
   a. Partially remove the front floor carpet assembly.

➡It is not necessary to fully remove the front floor carpet assembly. Partially remove it so that the lower HV battery carrier panel can be removed.

78. Remove the No. 1 HV battery tray.

➡Remove any remaining butyl tape on the No. 1 HV battery tray and HV battery.

79. Remove the No. 1 floor board.
80. Remove the No. 2 floor board.
81. Removing the No. 3 floor board.
82. Remove the No. 4 floor board.
   a. Disengage the clamp.
   b. Disengage the 4 clips and remove the No. 4 floor board.
83. Remove the battery carrier duct.
   a. Remove the 3 bolts and battery carrier duct LH.
   b. Remove the 2 bolts and battery carrier duct RH.
84. Remove the electric vehicle fuse.

⁂ CAUTION

Be sure to wear insulated gloves and protective goggles.

   a. Disengage the 2 claws and remove the fuse block cover.
   b. Remove the 2 bolts and electric vehicle fuse.
85. Remove the HV battery service hole cover.

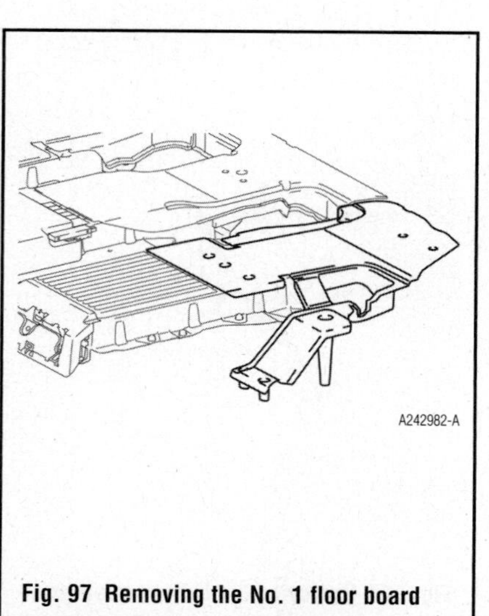

Fig. 97 Removing the No. 1 floor board

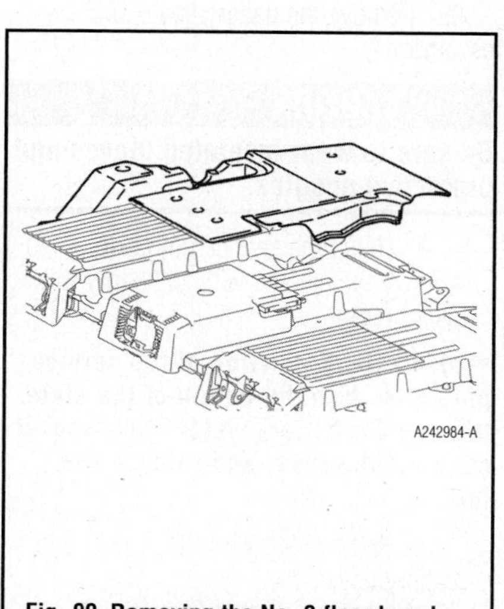

Fig. 99 Removing the No. 3 floor board

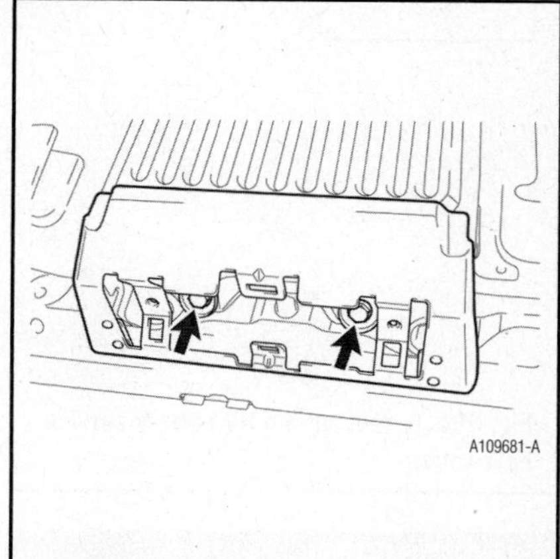

Fig. 101 Removing the 2 bolts and battery carrier duct LH

Fig. 102 Removing the 2 bolts and battery carrier duct RH

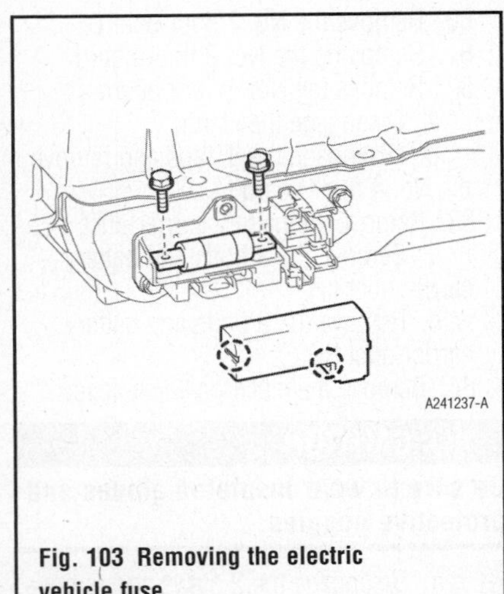

Fig. 103 Removing the electric vehicle fuse

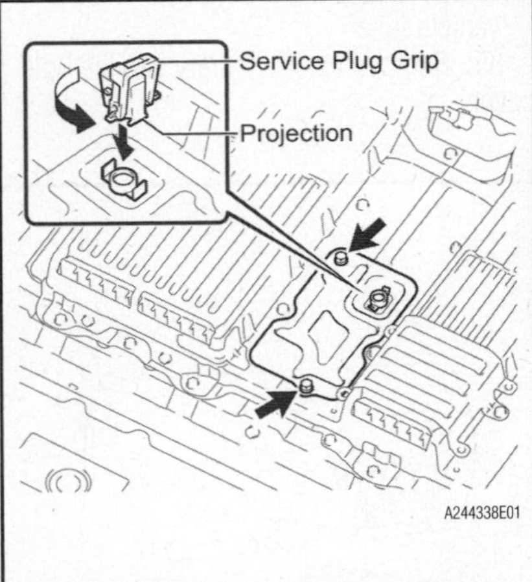

Fig. 104 Removing the HV battery service hole cover

⁂ CAUTION

Be sure to wear insulated gloves and protective goggles.

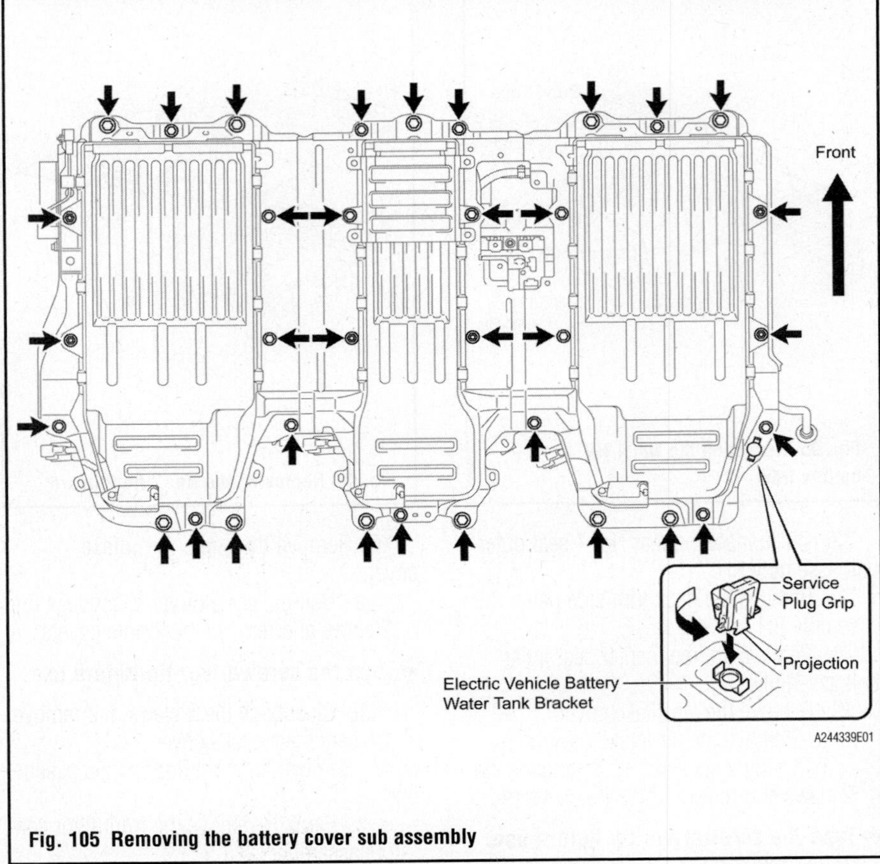

Fig. 105 Removing the battery cover sub assembly

a. Using the service plug grip, remove the electric vehicle battery water tank bracket.

➡Insert the projection of the service plug grip, turn the button of the electric vehicle battery water tank bracket counterclockwise, and release the lock.

b. Remove the 2 bolts and battery service hole cover.

86. Remove the battery cover sub assembly.

⁂ CAUTION

Be sure to wear insulated gloves and protective goggles.

a. Using the service plug grip, remove the electric vehicle battery water tank bracket.

➡Insert the projection of the service plug grip, turn the button of the electric vehicle battery water tank bracket counterclockwise, and release the lock.

b. Remove the 28 bolts, 6 nuts and battery cover.

87. Remove the battery smart unit.

⁂ CAUTION

Be sure to wear insulated gloves and protective goggles.

a. Disconnect the 3 connectors.

b. Remove the 2 nuts and battery smart unit.

88. Remove the hybrid vehicle relay assembly.

89. Remove the HV battery.

Fig. 106 Removing the battery smart unit

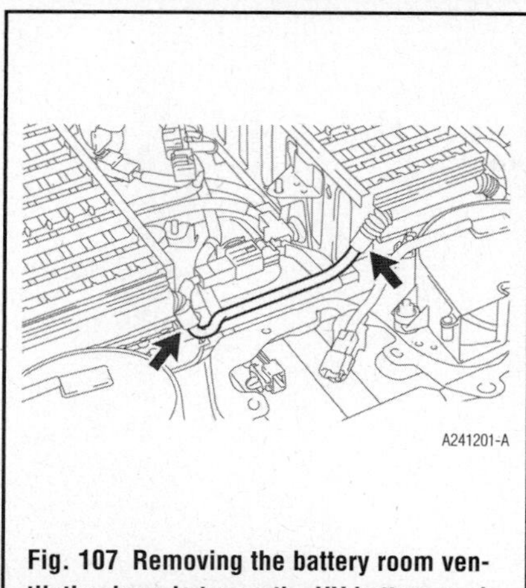

Fig. 107 Removing the battery room ventilation hose between the HV battery module LH and center HV battery module

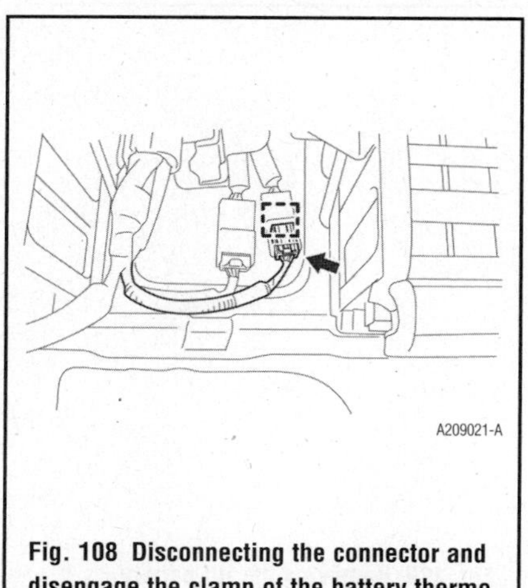

Fig. 108 Disconnecting the connector and disengage the clamp of the battery thermo sensor

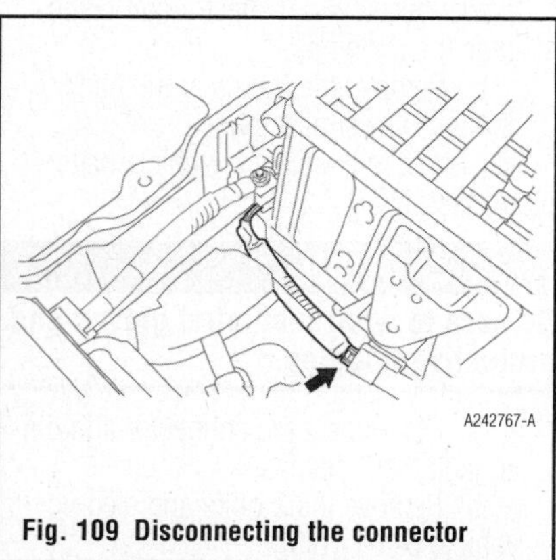

Fig. 109 Disconnecting the connector

⁑ CAUTION

Be sure to wear insulated gloves and protective goggles.

➡Insulate the removed terminals with ventilation tape.

a. Remove the HV battery module LH.

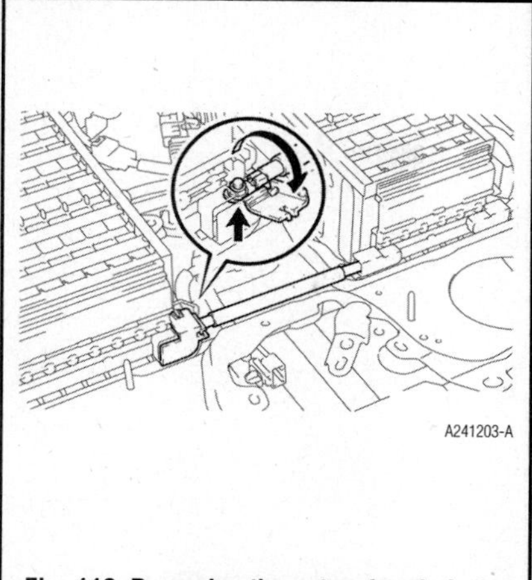

Fig. 110 Removing the nut and main battery cable from the HV battery module LH

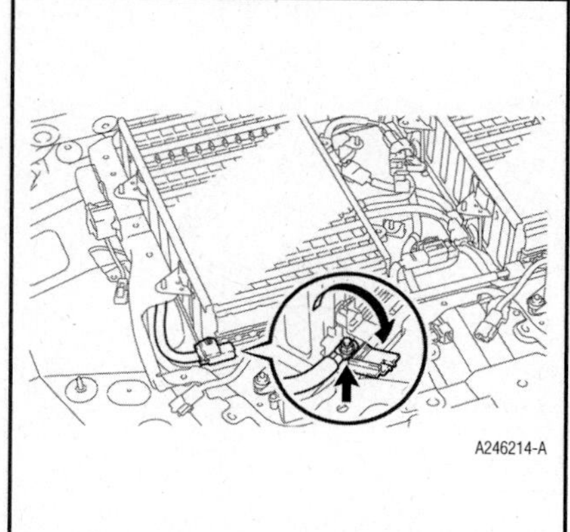

Fig. 111 Removing the nut, and then disconnect the electric vehicle battery plug from the HV battery module LH

b. Remove the battery room ventilation hose between the HV battery module LH and center HV battery module.

c. Disconnect the connector and disengage the clamp of the battery thermo sensor.

d. Disconnect the connector.

e. Remove the nut and main battery cable from the HV battery module LH.

➡Insulate the tool with insulating tape.

➡Do not use a universal joint. Using a universal joint may cause the socket extension to drop and create a short circuit.

f. Remove the nut, and then disconnect the electric vehicle battery plug from the HV battery module LH.

➡Insulate the tool with insulating tape.

➡Do not use a universal joint. Using a universal joint may cause the socket

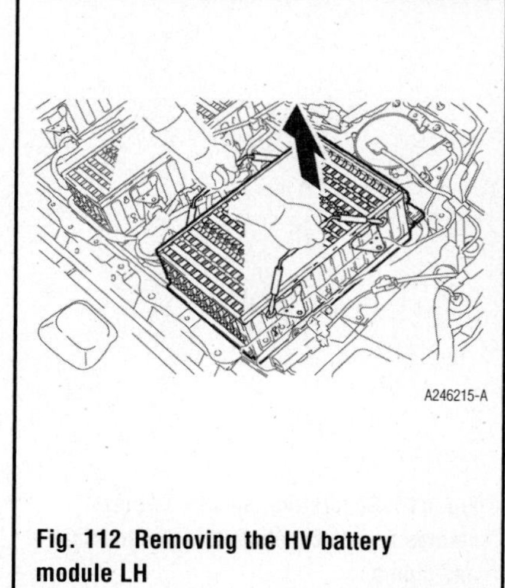

Fig. 112 Removing the HV battery module LH

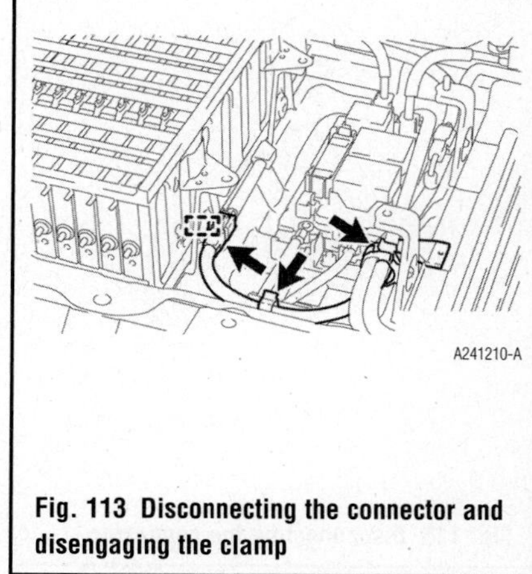

Fig. 113 Disconnecting the connector and disengaging the clamp

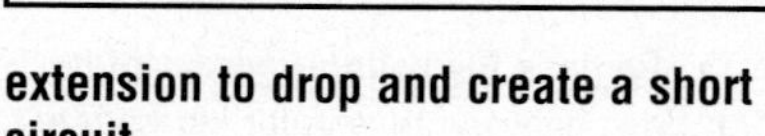
extension to drop and create a short circuit.

g. Remove the carry belts from the center HV battery module and install them to the HV battery module LH.

➡Insert the ends of the carry belts into the installation holes. Pull the carry belts upward to securely install them.

h. Remove the HV battery module LH.

i. Remove the carry belts from the HV battery module LH.

j. Remove the center HV battery module.

k. Disconnect the connector and disengage the clamp.

l. Separate the 2 wire harness clamps.

m. Disconnect the 2 connectors and disengage the clamp.

n. Disconnect the connector.

o. Disconnect the connector and disengage the clamp.

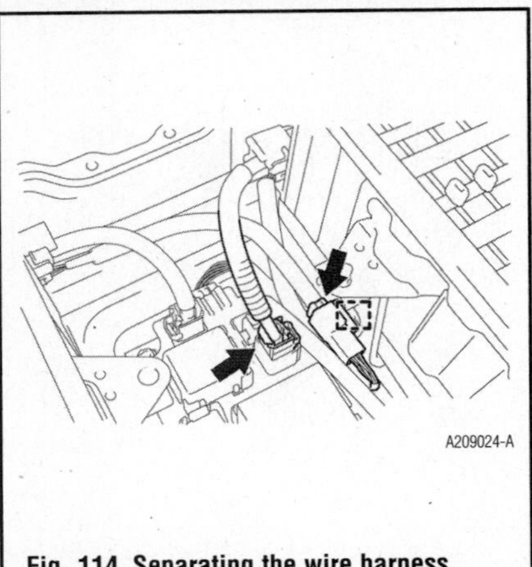

Fig. 114 Separating the wire harness clamps and disconnecting the connectors and clamp

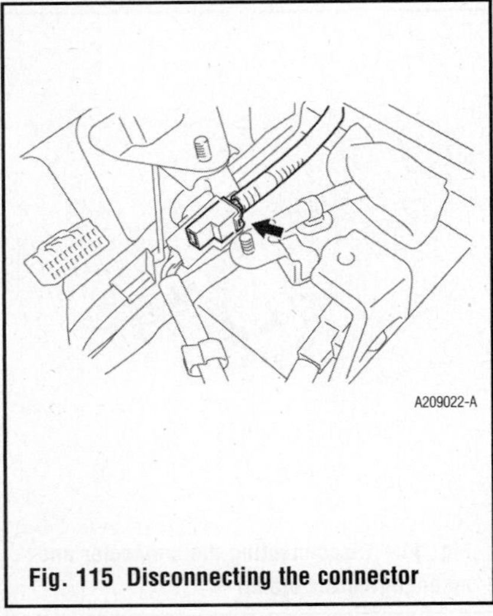

Fig. 115 Disconnecting the connector

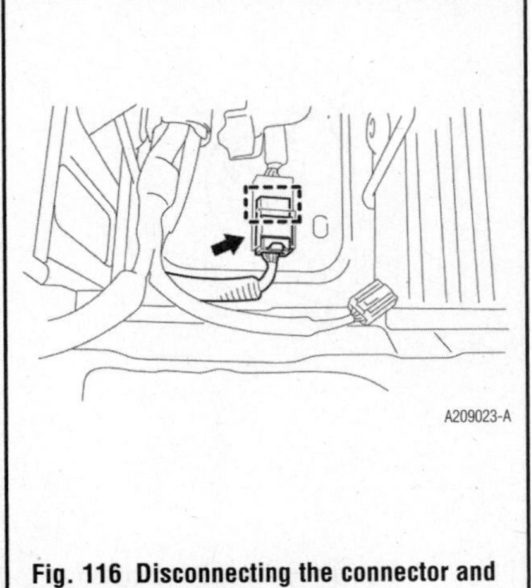

Fig. 116 Disconnecting the connector and disengage the clamp

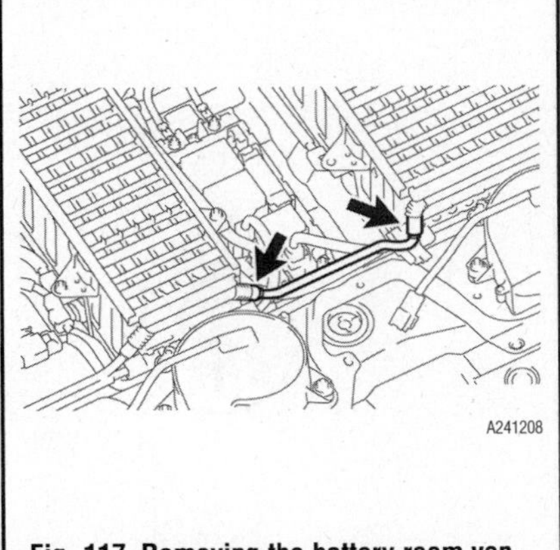

Fig. 117 Removing the battery room ventilation hose between the center HV battery module and HV battery module RH

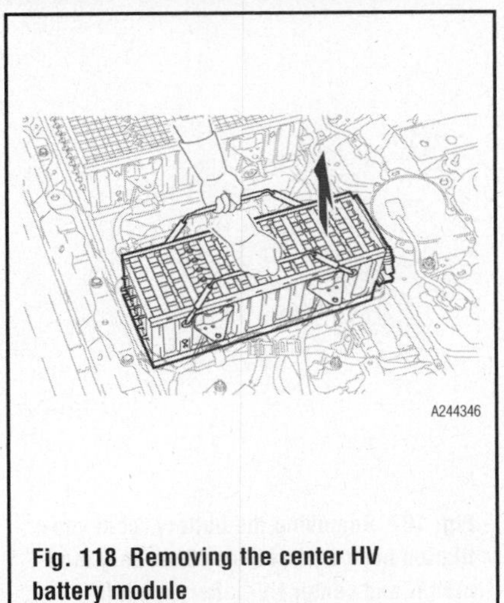

Fig. 118 Removing the center HV battery module

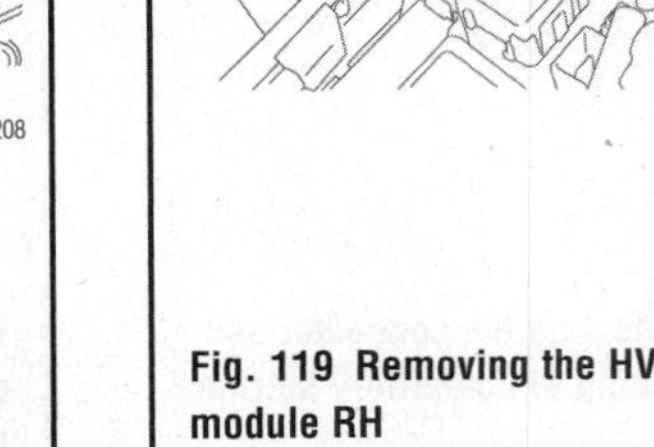

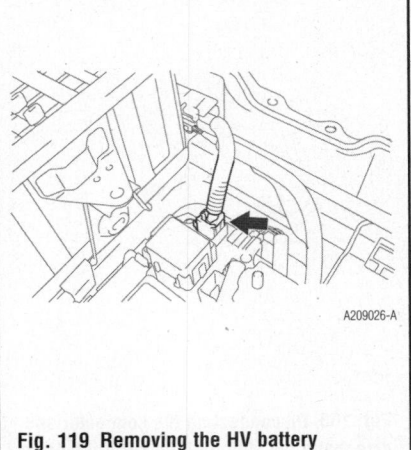

Fig. 119 Removing the HV battery module RH

p. Remove the battery room ventilation hose between the center HV battery module and HV battery module RH.

q. Install the carry belts to the center HV battery module.

➡Insert the ends of the carry belts into the installation holes. Pull the carry belts upward to securely install them.

r. Remove the center HV battery module.

s. Remove the carry belts from the center HV battery module.

t. Remove the HV battery module RH.

u. Disconnect the connector.

v. Disconnect the main battery cable connector.

w. Remove the battery room ventilation hose.

x. Install the carry belts to the HC battery module RH.

➡Insert the ends of the carry belts into the installation holes. Pull the carry belts upward to securely install them.

y. Remover the HV battery module RH.

z. Remove the carry belts from the HV battery module RH.

aa. Remove the nut and No. 2 main battery cable from the HV battery module RH.

bb. Remove the nut and main battery cable from the HV battery module RH.

90. Remove the battery cooling blower assembly.

⁂ CAUTION

Be sure to wear insulated gloves and protective goggles.

➡Replace the lower HV battery carrier panel and the battery cooling blower assemblies as a set when liquids have collected in the lower HV battery carrier panel or there is power adhering to it.

a. Disconnect each battery cooing blower assembly connector and disengage the 3 clamps.

b. Remove the 9 nuts and 3 battery cooling blower assemblies.

91. Remove the electric vehicle battery plug assembly.

⁂ CAUTION

Be sure to wear insulated gloves and protective goggles.

a. Disconnect the connector and disengage the 2 clamps.

b. Remove the 2 bolts and electric vehicle battery plug.

92. Remove the lower HV battery carrier panel.

⁂ CAUTION

Be sure to wear insulated gloves and protective goggles.

a. Disconnect the connector, and then remove the lower HV battery carrier panel.

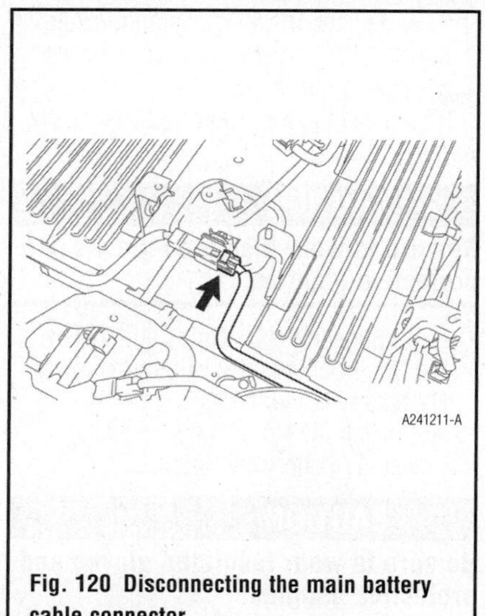

Fig. 120 Disconnecting the main battery cable connector

Fig. 121 Removing the battery room ventilation hose

A244347

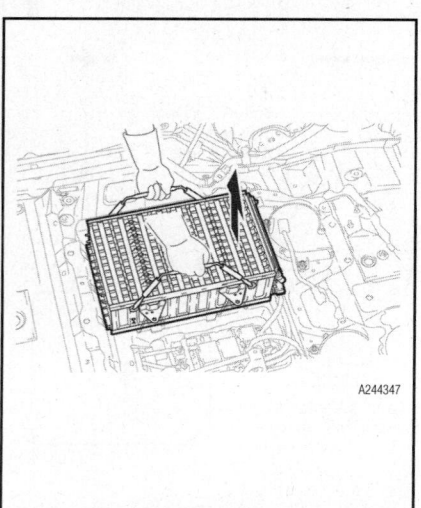

Fig. 122 Removing the HV battery module RH

Fig. 123 Removing the nut and No. 2 main battery cable from the HV battery module RH

Fig. 124 Removing the nut and main battery cable from the HV battery module RH

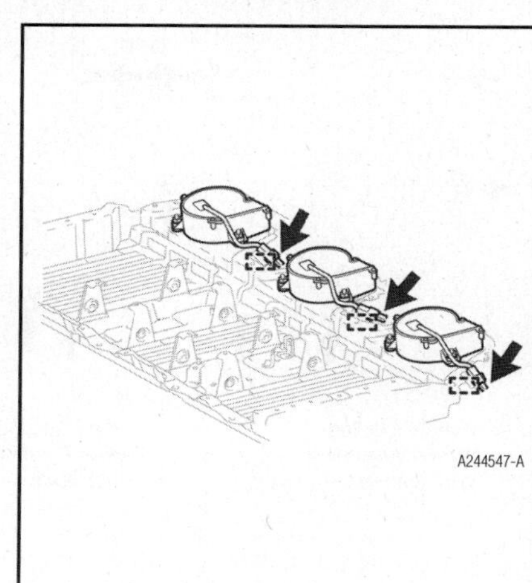

Fig. 125 Disconnecting the battery cooling blower assembly connectors and clamps

A244548-A

Fig. 126 Removing the 9 nuts and 3 battery cooling blower assemblies

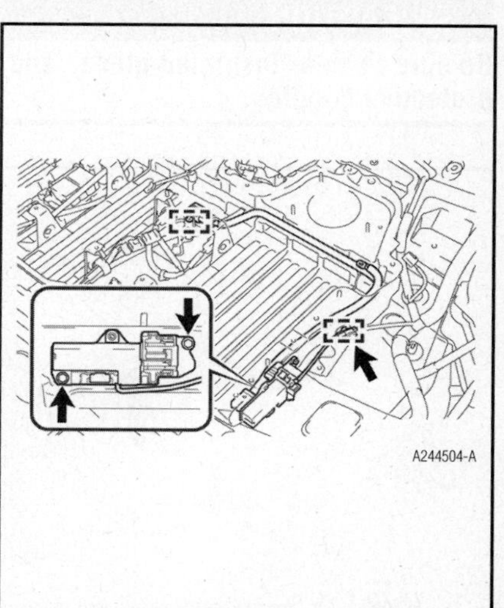

Fig. 127 Removing the electric vehicle battery plug assembly

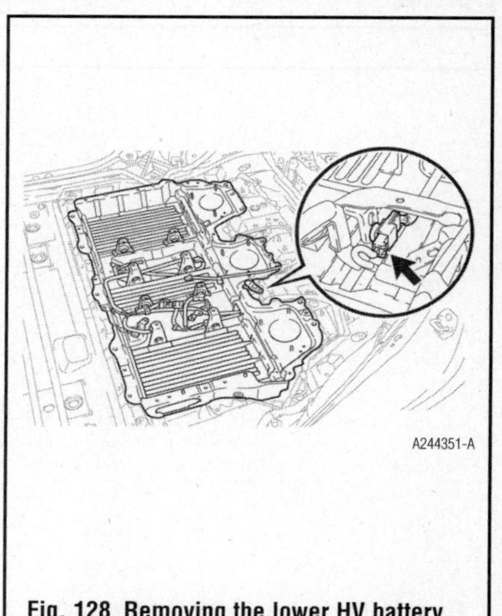

Fig. 128 Removing the lower HV battery carrier panel

93. Remove the No. 2 battery packing.

CAUTION

Be sure to wear insulated gloves and protective goggles.

a. Remove the No. 2 battery packing from the lower HV batter carrier panel.

94. Remove the No. 1 hybrid battery pack wire.

CAUTION

Be sure to wear insulated gloves and protective goggles.

a. Disengage the 3 clamps and remove the No. 1 battery pack wire from the lower HV battery carrier panel.

95. Remove the No. 2 hybrid battery pack wire.

CAUTION

Be sure to wear insulated gloves and protective goggles.

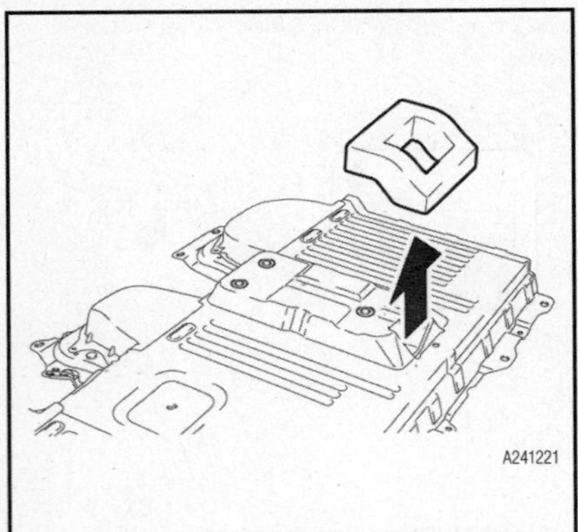

Fig. 129 Removing the No. 2 battery packing

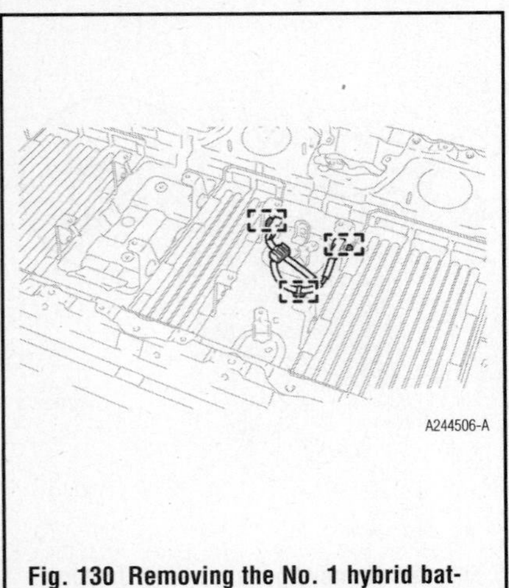

Fig. 130 Removing the No. 1 hybrid battery pack wire

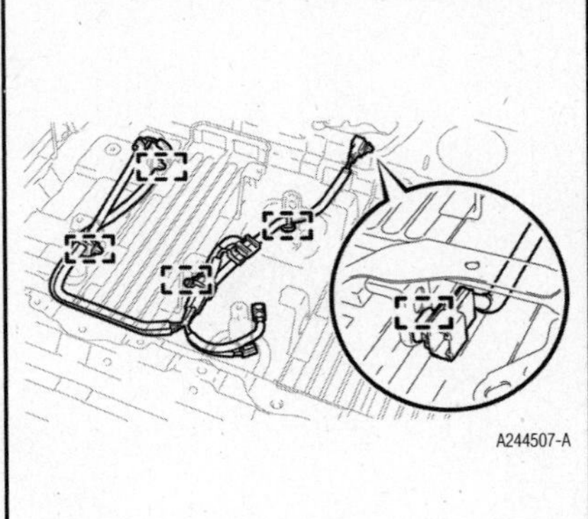

Fig. 131 Removing the No. 2 hybrid battery pack wire

a. Disengage the 5 clamps and remove the No. 2 battery pack wire from the lower HV battery carrier panel.

To install:

96. Install the No. 2 hybrid battery pack wire.

CAUTION

Be sure to wear insulated gloves and protective goggles.

a. Engage the 5 clamps and install the No. 2 battery pack wire to the lower HV battery carrier panel.

97. Install the No. 1 hybrid battery pack wire.

CAUTION

Be sure to wear insulated gloves and protective goggles.

a. Engage the 3 clamps and install the No. 1 battery pack wire to the lower HV battery carrier panel.

98. Install the No. 2 battery packing to the lower HV battery carrier panel.

CAUTION

Be sure to wear insulated gloves and protective goggles.

➡Make sure that there is no clearance between the battery packing and the lower carrier panel.

99. Install the lower HV battery carrier panel.

CAUTION

Be sure to wear insulated gloves and protective goggles.

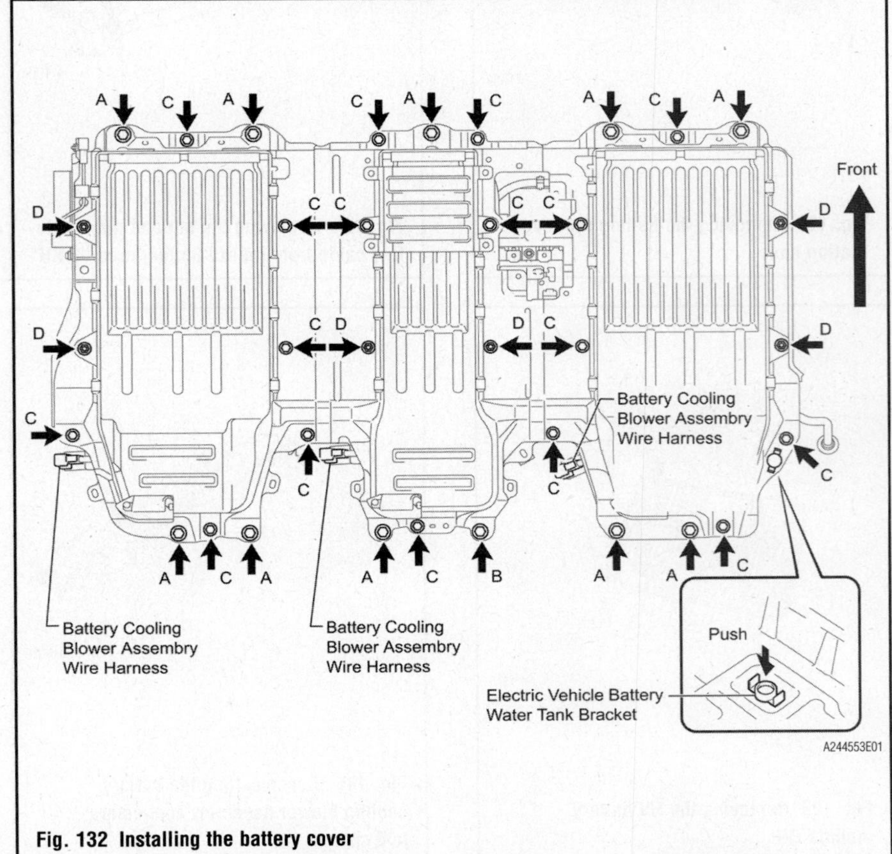

Fig. 132 Installing the battery cover

a. Install the lower HV battery carrier panel on the vehicle.

b. Connect the connector on the lower HV battery carrier panel.

100. Install the electric vehicle battery plug assembly.

CAUTION

Be sure to wear insulated gloves and protective goggles.

a. Install the electric vehicle battery plug with the 2 bolts. Tighten to 48 inch lbs. (5.4 Nm).

b. Connect the connector and engage the 2 clamps.

101. Install the batter cooling blower assembly.

CAUTION

Be sure to wear insulated gloves and protective goggles.

a. Install the 3 battery cooling blower assemblies with the 9 nuts. Tighten to 40 inch lbs. (4.5 Nm).

b. Connect each batter cooling blower assembly connector and engage the 3 clamps.

102. Install the HV battery.

CAUTION

Be sure to wear insulated gloves and protective goggles.

a. Attach the HV battery name label. (When the lower carrier panel is not replaced).

b. Attach a new label to immediate left of the original one.

➡An HV battery name label is supplied with a new battery.

c. Install the HV battery module RH.

d. Install the main battery cable to the HV battery module RH with a new nut. Tighten to 48 inch lbs. (5.4 Nm).

➡Be sure to engage the plastic cover securely.

e. Install the No. 2 main battery cable to the HV battery module RH with a new nut. Tighten to 48 inch lbs. (5.4 Nm).

➡Be sure to engage the plastic cover securely.

f. Install the carry belts to the HV battery module RH.

g. Install the HV battery module RH.

➡After installing the HV battery module RH, remove the carry belts from the HV battery module RH.

h. Install the battery room ventilation hose.

➡Align the projections of the battery room ventilation hose and the rubber tube, and install the hose.

i. Connect the main battery cable connector.

j. Connect the 2 connectors.

k. Install the center HV battery module.

l. Install the carry belts to the center HV battery module.

m. Install the HV battery module.

➡After installing the center HV battery module, remove the carry belts from the center HV battery module.

n. Install the battery room ventilation hose between the center HV battery module and the HV battery module RH.

➡Align the projections of the battery room ventilation hose and the rubber tube, and install the hose.

o. Connect the connector and engage the clamp.

p. Connect the connector.

q. Connect the 2 connectors and engage the clamp.

r. Connect the 2 wire harness clamps.

s. Connect the connector and engage the clamp.

t. Install the HV battery module LH.

u. Install the carry belts to the HV battery module LH.

v. Install the HV battery module LH.

➡After installing the HV battery module LH, remove the carry belts from the HV battery module LH.

w. Install the carry belts to the center HV battery module.

x. Connect the electric vehicle battery plug to the HV battery module LH with a new nut. Tighten to 48 inch lbs. (5.4 Nm).

➡Be sure to engage the plastic cover securely.

➡Insulate the tool with insulating tape.

➡Do not use a universal joint. Using a universal joint may cause the socket extension to drop and create a short circuit.

y. Connect the main battery cable to the HV battery module LH with a new nut. Tighten to 48 inch lbs. (5.4 Nm).

➡Be sure to engage the plastic cover securely.

➡Insulate the tool with insulating tape.

➡Do not use a universal joint. Using a universal joint may cause the socket extension to drop and create a short circuit.

z. Connect the connector.

aa. Connect the connector and engage the clamp of the battery thermo sensor.

bb. Connect the battery room ventilation hose between the HV battery module LH and the center HV battery module.

➡Align the projections of the battery room ventilation hose and the rubber tube, and install the hose.

103. Install the hybrid vehicle relay assembly.

104. Install the battery smart unit.

105. Check high voltage cable connection condition.

CAUTION

Be sure to wear insulated gloves and protective goggles.

a. Check that each wire harness is installed securely.

WARNING

Make sure that the No. 3 wire frames do not cross each other.

WARNING

Be sure to connect the No. 3 frame wires to the correct terminals as shown in the illustration.

➡The connectors should be connected securely.

➡The nuts should be tightened securely.

➡Make sure that the 4 plastic covers are engaged securely.

106. Install the battery cover sub assembly.

CAUTION

Be sure to wear insulated gloves and protective goggles.

a. Install the electric vehicle battery water tank bracket, and then push the button to lock it.

b. Install the battery cover with the 28 bolts and 6 nuts.

➡When installing the battery cover, check that the battery cooling blower assembly wire harness is not pinched

before tightening the bolts and nuts that secure the cover.

➡Use a ground bolt for position B.

c. Tighten A bolts to 19 ft. lbs. (25 Nm).
d. Tighten B bolts to 16 ft. lbs. (22 Nm).
e. Tighten C bolts to 66 inch lbs. (7.5 Nm).
f. Tighten D nuts to 66 inch lbs. (7.5 Nm).

107. Install HV battery service hole cover.

⁂ CAUTION

Be sure to wear insulated gloves and protective goggles.

a. Install the electric vehicle battery water tank bracket, and then push the button to lock it.
b. Install the battery service hole cover with the 2 bolts. Tighten to 66 inch lbs. (7.5 Nm).

108. Install the electric vehicle fuse.

⁂ CAUTION

Be sure to wear insulated gloves and protective goggles.

a. Install the electric vehicle fuse with the 2 bolts. Tighten to 48 inch lbs. (5.4 Nm).
b. Engage the 2 claws and install the fuse block cover.

109. Install the battery carrier RH duct with the 2 bolts. Tighten to 66 inch lbs. (7.5 Nm).
110. Install the battery carrier LH duct with the 2 bolts. Tighten to 66 inch lbs. (7.5 Nm).
111. Install the No. 4 floor board.
a. Engage the 4 clips and install the No. 4 floor board.
b. Engage the clamp.
112. Install the No. 3 floor board.
113. Install the No. 2 floor board.
114. Install the No. 1 floor board.
115. Install the No. 1 HV battery tray.
a. Apply butyl tape to the No. 1 HV battery tray.
b. Install the No. 1 HV battery tray.

➡Temporarily install the No. 1 HV battery tray with butyl tape on the 4 corners, and then install the front floor carpet. If the butyl tape cannot be reused due to deterioration, replace the No. 1 HV battery tray with a new one.

116. Install the front floor carpet assembly.
117. Install the center air intake cover.
a. Engage the 2 claws and install the air intake cover.
b. Install the 2 screws.
c. Install the 2 hole covers.
118. Install the air intake cover.
a. Engage the 2 claws and install the air intake cover LH.
b. Install the 2 screws.
c. Install the 2 hole covers.
d. Engage the 2 claws and install the air intake cover RH.
e. Install the 2 screws.
f. Install the 2 hole covers.
119. Install the rear center seat inner belt assembly.
120. Install the deck trim side panel assembly LH.
121. Connect the rear No. 1 seat outer belt assembly LH.
122. Install the front deck side trim cover LH.
123. Install the No. 2 deck side trim hook.
124. Install the rope hook assembly LH.
125. Install reclining remote control lever bezel LH.
126. Install the rear power outlet socket cover.
127. Install the rear power point socket assembly.
128. Install the rear combination light service cover LH.
129. Install the side trim cover LH (w/o rear seat entertainment system).
130. Install the power outlet socket bezel (w/rear seat entertainment system).
131. Install the deck side trim LH.
132. Install the deck side trim cover No. 2.
133. Install the deck trim side panel assembly RH.
134. Connect the rear No. 1 seat outer belt assembly RH.
135. Install the front deck side trim cover RH.
136. Install the No. 1 luggage compartment trim hook.
137. Install the rope hook assembly RH.
138. Install the rear combination light service cover RH.
139. Install the rear room temperature sensor (for automatic air conditioning system).
140. Install the deck side trim RH.
141. Install the deck side trim cover No. 1.
142. Install the rear seat side garnish cap.
143. Install the rear floor finish plate.
144. Install the rear door opening trim weatherstrip LH and RH.
145. Install the rear door scuff plate RH.
146. Install the rear No. 2 seat assembly.
147. Install the center seat cushion cover sub assembly.
148. Connect the rear No. 2 seat outer belt assemblies LH and RH.
149. Install the rear deck floor box.
150. Install the rear mat.
151. Install the deck side trim box RH.
152. Install the jack carrier assembly.
153. Install the jack assembly.
154. Install the jack carrier cushion.
155. Install the jack carrier support.
156. Install the No. 1 deck board.
157. Install the No. 2 deck board.
158. Install the No. 2 deck board sub assembly.
159. Install the rear No. 1 seat assembly LH.
160. Connect the rear No. 2 seat lock cable assembly LH.
161. Inspect the rear seat leg side cover LH.
162. Install the rear inner track bracket cover LH.
163. Install the rear outer track bracket cover LH
164. Install the rear seat track bracket cover LH.
165. Install the rear seat headrest assembly LH.
166. Install the rear center seat assembly LH.
167. Install the rear No. 1 seat assembly RH.
168. Connect the rear No. 1 seat lock cable assembly RH.
169. Inspect the rear seat slide adjuster lock RH.
170. Install the rear seat leg side cover RH.
171. Install the rear inner track bracket cover RH.
172. Install the rear outer track bracket cover RH.
173. Install the rear seat track bracket cover RH.
174. Install the rear seat headrest assembly RH.
175. Install the service plug grip.
176. Install the reclining hinge cover LH.
177. Install the rear door scuff plate LH.
178. Connect the cable to the negative battery terminal.

➡When disconnecting the cable, some systems need to be initialized after the cable is reconnected.

179. Install the battery service hole cover assembly.
180. Install the deck board assembly.

FIRING ORDER

See Figure 133.

Firing order for 2.7L engine: 1–3–4–2
Firing order for 3.5L engine: 1–2–3–4–5–6

IGNITION COIL & SPARK PLUGS

REMOVAL & INSTALLATION

2.7L Engine

See Figures 134 and 135.

➡Attempting to disengage both front and rear clips at the same time may cause the No. 1 engine cover sub-assembly to break.

1. Disconnect the negative battery cable.
2. Lift the rear of the No. 1 engine cover sub-assembly to detach the cover from the 2 pins, and then lift the front of the No. 1 engine cover sub-assembly to detach the cover from the pin and remove the No. 1 engine cover sub-assembly.
3. Disconnect the 4 ignition coil assembly connectors.
4. Remove the 4 bolts and 4 ignition coil assemblies.
5. Using a spark plug wrench, remove the 4 spark plugs.

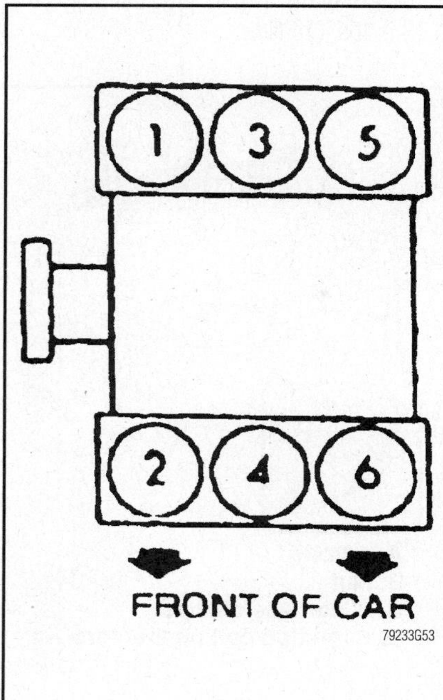

Fig. 133 3.3L (3MZ-FE) Engine firing order

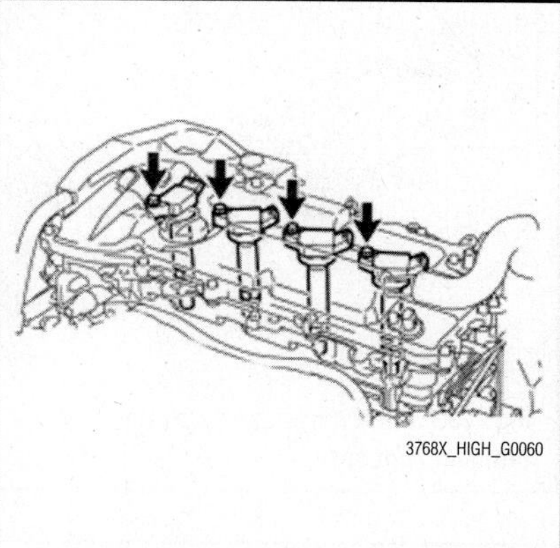

Fig. 134 Remove the 4 bolts and 4 ignition coil assemblies

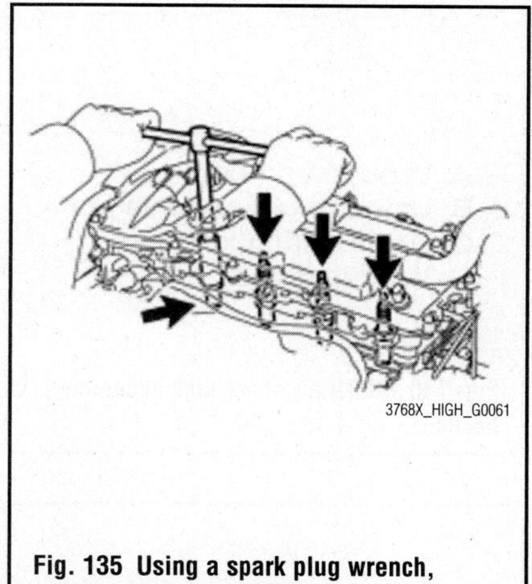

Fig. 135 Using a spark plug wrench, remove the 4 spark plugs

To install:

6. To install, reverse removal procedure.

3.3L Engine

See Figures 136 and 137.

1. Disconnect the negative battery cable.
2. Remove the RH and LH wiper arm assembly.
3. Remove the cowl and seal.
4. Remove the wiper motor and link assembly.
5. Remove the outer cowl top panel.
6. Remove the cool air intake duct seal.
7. Remove the LH engine side cover.
8. Drain the engine coolant.
9. Remove air cleaner cap and case assembly.
10. Remove the air cleaner bracket.
11. Remove the emission control valve set.

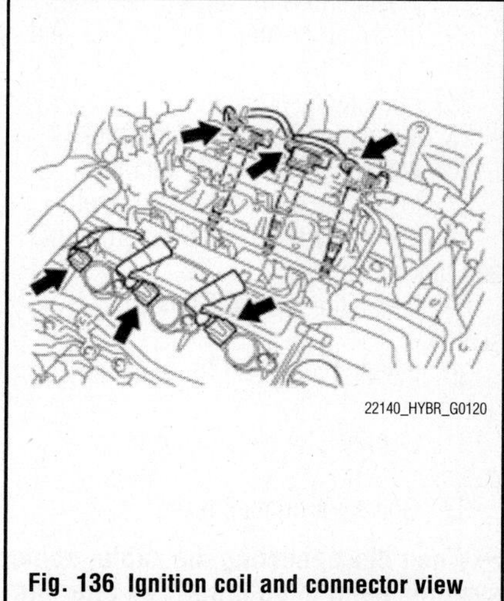

Fig. 136 Ignition coil and connector view

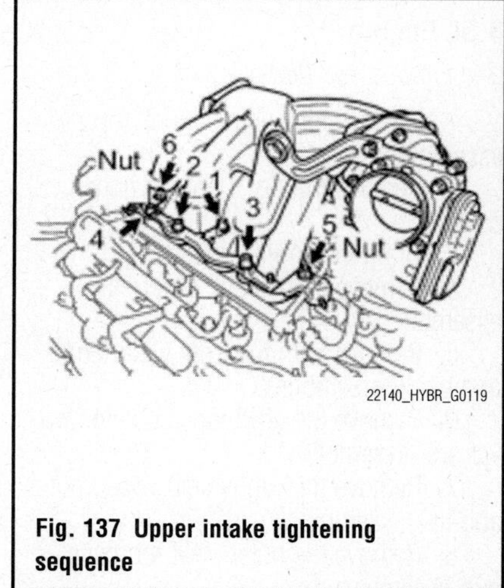

Fig. 137 Upper intake tightening sequence

12. Remove the upper intake plenum.
13. Disconnect the 6 ignition coil connectors.
14. Remove the 6 bolts and 6 ignition coils.
15. Remove the 6 spark plugs.

To install:

16. Install the spark plug and tighten to 13 ft. lbs. (18 Nm).
17. Install the ignition coils and tighten retaining bolts to 71 inch lbs. (8 Nm).
18. Connect the ignition coil connectors.
19. Install intake plenum with a new gasket
20. Using a socket hexagon wrench 8 mm, install the upper intake plenum with the 4 bolts and 2 nuts.
21. Using several steps, tighten the bolts and nuts uniformly to 21 ft. lbs. (28 Nm)

follow the sequence shown in the illustration.

22. Install emission control vale set.
23. Install air cleaner bracket, case and cap.
24. Add engine coolant.
25. Install engine room side cover.
26. Install cool air intake duct seal.
27. Install the outer cowl top panel.
28. Install the windshield wiper motor and link assembly.
29. Install the top cowl and hood seal.
30. Install the LH and RH wiper arm assembly.
31. Connect the negative battery cable.
32. Check for coolant leaks.

➡When disconnecting the cable, some systems need to be initialized after the cable is reconnected.

3.5L Engine

See Figures 138 through 147.

1. Remove the engine under cover assembly.
2. Remove the No. 1 engine under cover.
3. Drain the engine coolant.
4. Remove the V-bank cover sub-assembly.
5. Remove both the front wiper arms and blade assemblies.
6. Remove the cowl top ventilator louver sub-assembly.
7. Remove the windshield wiper motor and link assembly.
8. Remove the outer cowl top panel sub-assembly.
9. Disconnect the engine room main wire.

Fig. 138 Throttle body bracket bolts removal sequence—3.5L engine

Fig. 139 No. 1 surge tank stay bolt removal sequence

Fig. 140 Intake air surge tank assembly hoses

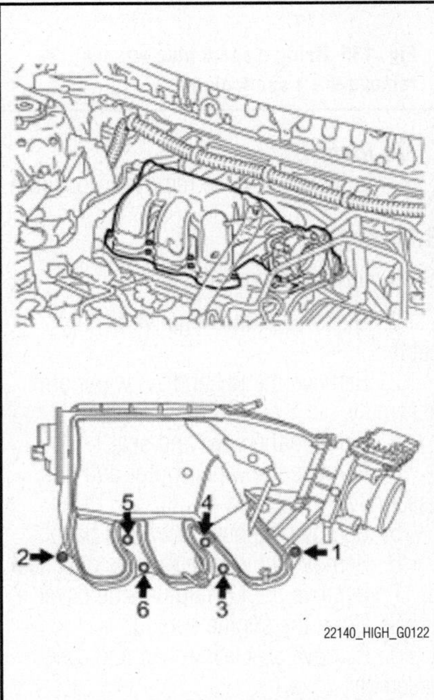

Fig. 141 Intake air surge tank assembly, bolts, and nuts removal sequence

10. Remove the 2 bolts in the order shown in the illustration and remove the throttle body bracket.
11. Remove the 2 bolts in the order shown in the illustration and remove the No. 1 surge tank stay.
12. Remove the air cleaner cap sub-assembly.
13. Remove the intake air surge tank assembly, as follows:
    a. Disconnect the 4 hoses.
    b. Disconnect the throttle body connector and clamp.
    c. Disconnect the connector.
    d. Disconnect the 2 water by-pass hoses from the throttle body.
    e. Remove the 4 bolts and 2 nuts in the order shown in the illustration.

➡Use a 5 mm hexagon socket wrench to remove the 4 bolts.

    f. Remove the gasket from the intake air surge tank.
14. Remove the ignition coil assembly, as follows:
    a. Remove the 4 bolts.
    b. Remove the nut.
    c. Disconnect the 2 harness clamps.
    d. Disconnect the 6 ignition coil connectors.
    e. Remove the 6 bolts and 6 ignition coils
15. Remove the 6 spark plugs.

To install:

16. Install the 6 spark plugs and tighten to 13 ft. lbs. (18 Nm).

Fig. 142 Ignition coil assembly harness clamps

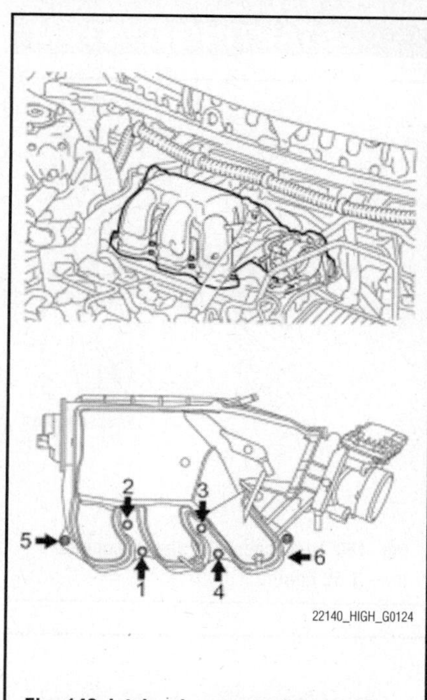

Fig. 143 Intake air surge tank assembly installation sequence

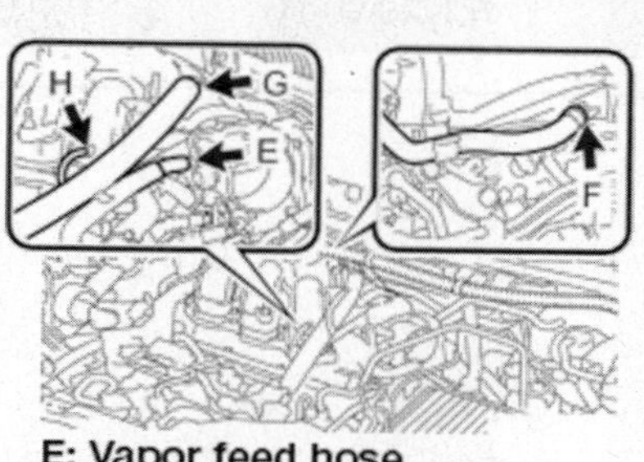

Fig. 144 Intake air surge tank assembly hoses

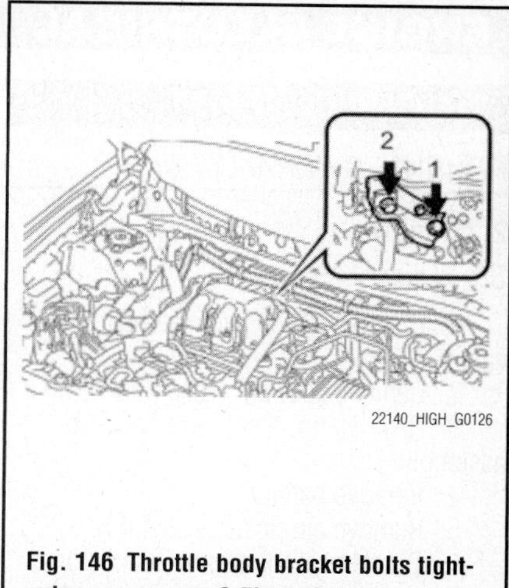

Fig. 146 Throttle body bracket bolts tightening sequence—3.5L engine

Fig. 145 No. 1 surge tank stay bolt tightening sequence

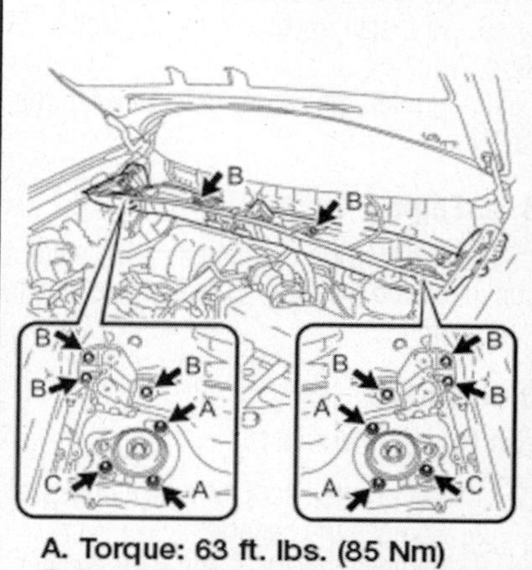

Fig. 147 Outer cowl top panel sub-assembly, bolts, and nuts—3.5L engine

17. Install the ignition coil assembly as follows:
   a. Install the 6 ignition coils with the 6 bolts and tighten to 7 ft. lbs. (10 Nm).
   b. Connect the 6 ignition coil connectors.
   c. Install the 4 bolts and tighten to 73 inch lbs. (8.3 Nm).
   d. Install the nut and tighten to 73 inch lbs. (8.3 Nm).
   e. Install the 2 clamps.
18. Install the intake air surge tank assembly as follows:
   a. Install the surge tank with the 4 bolts and 2 nuts in the order shown in the illustration, and tighten to 12 ft. lbs. (16 Nm) and 13 ft. lbs. (18 Nm), using a 5 mm hexagon socket wrench. **DO NOT** apply oil to the bolts.
   b. Connect the 2 water by-pass hoses to the throttle with motor body assembly.
   c. Connect the connector.
   d. Install the clamp and connect the throttle with motor body assembly connector.
   e. Connect the 4 hoses.
19. Temporarily install the No. 1 surge tank stay as follows:
   a. Temporarily install the intake air surge tank assembly with 3 new gaskets on the intake manifold.

➡Do not allow the gaskets to slip out of place during installation.

   b. Temporarily install the No. 1 surge tank stay with the 2 bolts. **DO NOT** apply oil to the bolts.
20. Temporarily install the throttle body bracket with the 2 bolts. **DO NOT** apply oil to the bolts.
21. Fully tighten the No. 1 surge tank stay, as follows:
   a. Fully tighten the 2 bolts in the order shown in the illustration. Tighten to 15 ft. lbs. (21 Nm). **DO NOT** apply oil to the bolts.
22. Fully tighten the throttle body bracket, as follows:
   a. Fully tighten the 2 bolts in the order shown in the illustration. Tighten to 15 ft. lbs. (21 Nm). **DO NOT** apply oil to the bolts.
23. Connect the engine room main wire.
24. Install the air cleaner cap sub-assembly.
25. Install the outer cowl top panel sub-assembly, as follows:
   a. Install the outer cowl top panel sub-assembly with the 8 bolts and 6 nuts.
   b. Engage the 4 clamps.
26. Install the windshield wiper motor and link assembly.
27. Install the cowl top ventilator louver sub-assembly.
28. Install both front wiper arm and blade assemblies.
29. Install the V-bank cover sub-assembly.
30. Add engine coolant.
31. Inspect for engine coolant leak.
32. Install the No. 1 engine under cover.
33. Install the engine under cover assembly.

IGNITION TIMING

ADJUSTMENT

All engines are equipped with a Distributorless Ignition System (DIS). No timing adjustment is possible.

STARTER

REMOVAL & INSTALLATION

2.7L Engine

See Figures 148 and 149.

1. Disconnect the negative battery cable.
2. Remove cool air intake duct seal.
3. Remove no. 1 engine cover sub-assembly.
4. Remove battery.
5. Remove air cleaner assembly.
6. Disconnect the starter connector.
7. Open the terminal cap, remove the nut and disconnect the starter wire.
8. Remove the 2 bolts and starter.
9. To install, reverse removal procedure.
10. Tighten the starter bolts to 27 ft. lbs. (37 Nm).

3.3L Engine

The 3.3L Hybrid engine has a DC electric converter and therefore does not use a standard starter.

3.5L Engine

See Figures 150 and 151.

1. Remove the cool air intake duct seal.
2. Remove the battery.
3. Remove the No. 1 and 2 air cleaner inlets.
4. Remove the air cleaner cap sub-assembly.
5. Remove the air cleaner case sub-assembly.
6. Remove the 2 bolts and the air cleaner bracket.
7. Disconnect the starter connector.
8. Open the terminal cap, remove the nut and disconnect the starter wire.
9. Remove the 2 bolts and starter.

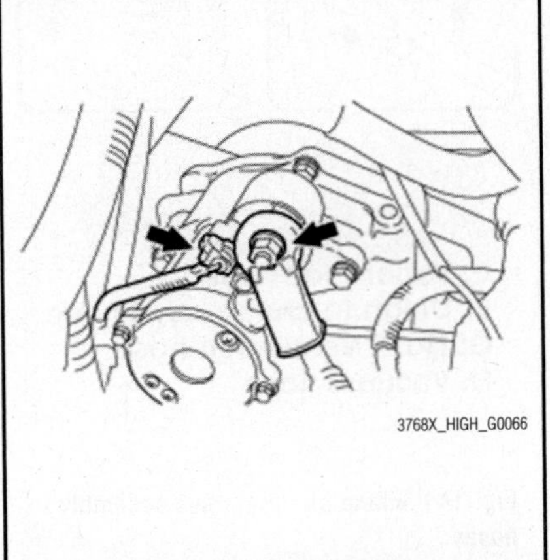

Fig. 148 Disconnect the starter connector—2.7L Engine

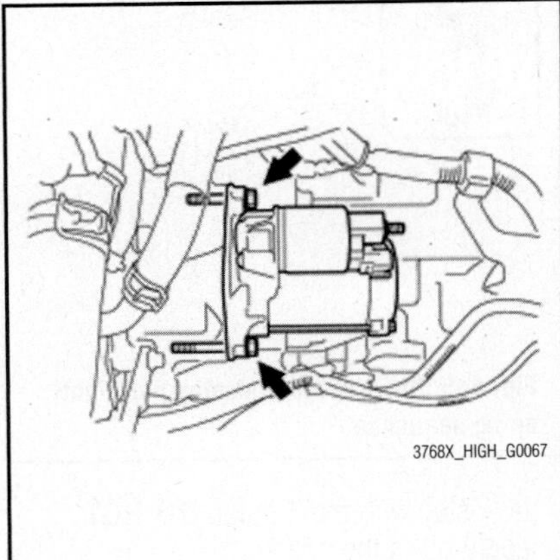

Fig. 149 Remove the 2 bolts and starter—2.7L Engine

To install:

10. Installation is the reverse of removal procedure, noting the following:
    a. After installing the starter assembly with the 2 bolts, tighten to 27 ft. lbs. (37 Nm).

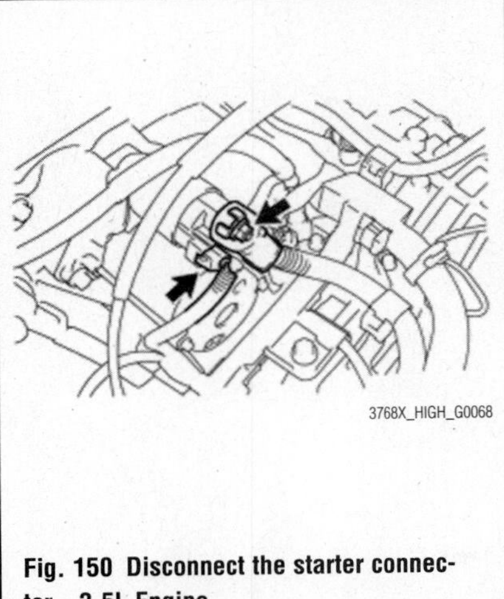

Fig. 150 Disconnect the starter connector—3.5L Engine

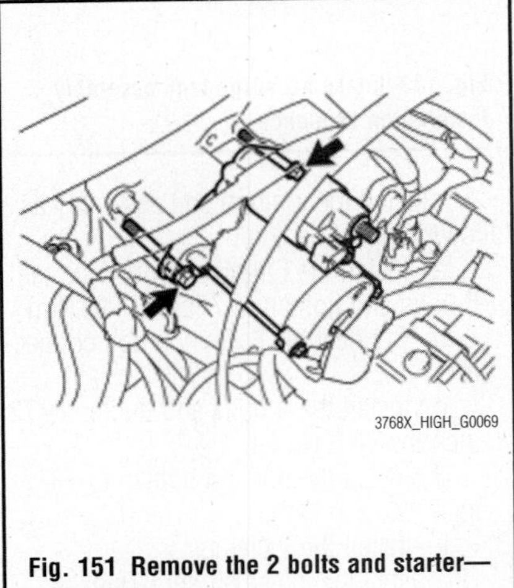

Fig. 151 Remove the 2 bolts and starter—3.5L Engine

    b. After connecting the starter wire with the nut, tighten to 87 ft. lbs. (9.8 Nm).
    c. After installing the bracket with the 2 bolts, tighten to 9 ft. lbs. (12 Nm).

ENGINE MECHANICAL

➡Disconnecting the negative battery cable may interfere with the functions of the on board computer systems and may require the computer to undergo a relearning process, once the negative battery cable is reconnected.

ACCESSORY DRIVE BELTS

ACCESSORY BELT ROUTING

See Figures 152 and 153.

INSPECTION

Inspect the drive belt for signs of glazing or cracking. A glazed belt will be perfectly smooth from slippage, while a good belt will have a slight texture of fabric visible. Cracks will usually start at the inner edge of the belt and run outward. All worn or damaged drive belts should be replaced immediately.

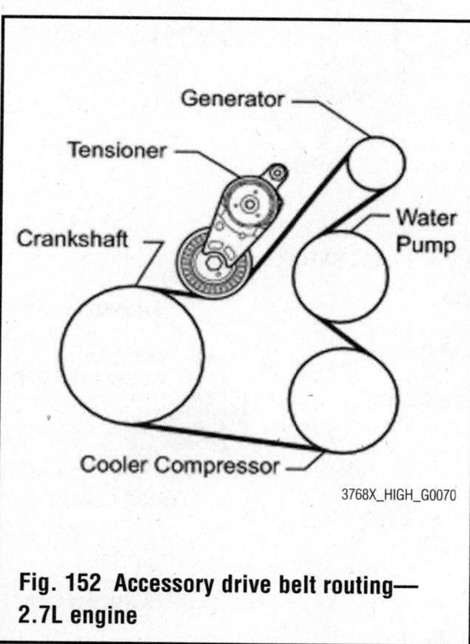

Fig. 152 Accessory drive belt routing—2.7L engine

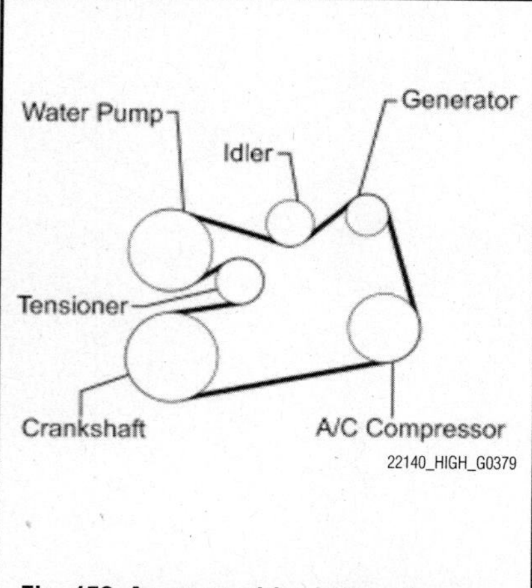

Fig. 153 Accessory drive belt routing—3.5L engine

ADJUSTMENT

Belt tension is maintained by an automatic tensioner. No adjustment is necessary of possible.

REMOVAL & INSTALLATION

2.7L & 3.5L Engines

See Figures 154 and 155.

1. Raise and safely support the vehicle.
2. Remove the right front wheel.
3. Remove the engine under cover assembly.
4. Remove the No. 1 engine under cover.
5. Remove the right front fender molding sub-assembly, as follows:
    a. Remove the clip.
    b. Using a 4 mm hexagon wrench, remove the screw.
    c. Peel off the front fender side protector and disengage the 3 clips, and then remove the right front fender molding sub-assembly.
    d. Remove the pad from the right front fender moulding sub-assembly.
    e. Remove the 2 clips No. 4 from the right front fender molding sub-assembly.
    f. Remove the front fender side protector from the right front fender molding sub-assembly.
6. Remove the right front fender liner.
7. Remove the right front fender apron seal.
8. For 2.7L engines, remove the V-ribbed belt as follows: Attach a wrench to the hexagonal portion of the belt tensioner as shown in the illustration, rotate the belt tensioner clockwise, and remove the V-ribbed belt.
9. For 3.5L engines, remove the V-ribbed belt as follows:
    a. Using SST: 09961-00950 or equivalent, release the V-ribbed belt tension by turning the V-ribbed belt tensioner assembly counterclockwise, and remove the V-ribbed belt from the V-ribbed belt tensioner assembly.
    b. While turning the V-ribbed belt tensioner assembly counterclockwise, align with its holes, and then insert the 5 mm bi-hexagon wrench into the holes to fix the V-ribbed belt tensioner assembly.

To install:

10. Install the V-ribbed belt.
11. Using SST: 09961-00950 or equivalent, turn the V-ribbed belt tensioner assembly counterclockwise and remove the bar.
12. If it is difficult to install the V-ribbed belt, perform the following procedure:
    a. Put the V-ribbed belt on every pulley except the tensioner pulley as shown in the illustration.
    b. Release the V-ribbed belt tension by turning the V-ribbed belt tensioner assembly counterclockwise, and put the

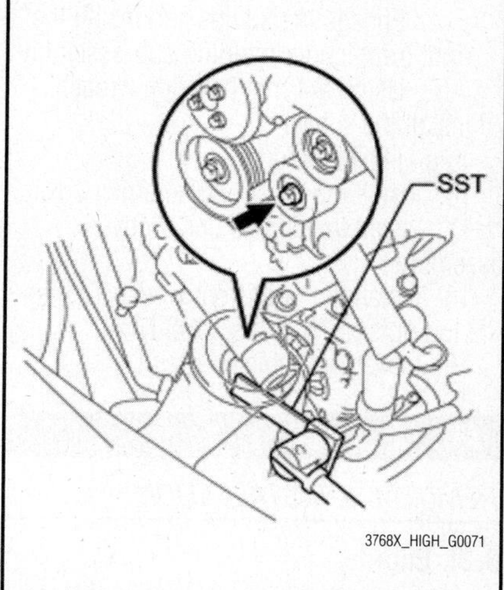

Fig. 155 Using SST: 09961-00950 or equivalent, remove the V-ribbed belt—3.5L Engines

3768X_HIGH_G0072

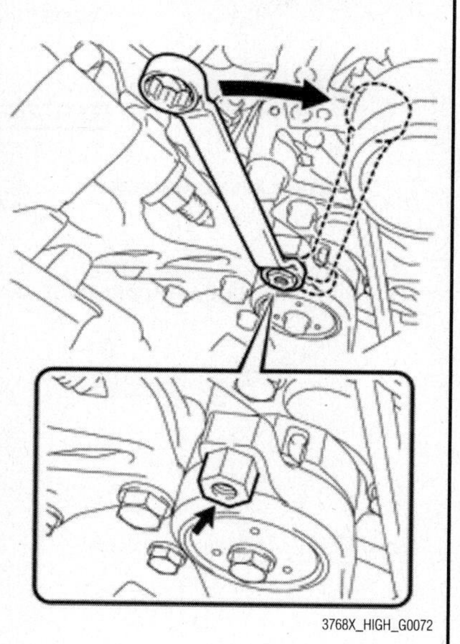

Fig. 154 Attach a wrench to the hexagonal portion of the belt tensioner rotate the belt tensioner clockwise, and remove the V-ribbed belt—2.7L Engines

V-ribbed belt on the V-ribbed tensioner assembly pulley.

➡Put the backside of the V-ribbed belt on the V-ribbed belt tensioner assembly pulley and No. 2 idler pulley sub-assembly.

➡Check that the V-ribbed belt is properly set to each pulley.

c. After installing the V-ribbed belt, check that it fits properly in the ribbed grooves. Confirm that the belt has not slipped out of the grooves on the bottom of the crankshaft pulley by hand.

13. Install the right front fender apron seal.
14. Install the right front fender liner.
15. Install the right front fender molding sub-assembly, as follows:

a. Clean the vehicle body surface by heating the vehicle body surface with a heat light, removing the front fender side protector from the vehicle body, and wiping off any tape adhesive residue with cleaner.

b. If reusing the right front fender molding sub-assembly, heat it with a heat light, remove the front fender side protector, wipe off any tape adhesive residue with cleaner, and install a new front fender side protector to the front fender molding sub-assembly.

c. Using a heat light, heat the vehicle body and the front fender moulding sub-assembly.

d. Remove the release paper from the front fender molding sub-assembly.

➡After removing the release paper, keep the exposed adhesive free from foreign matter.

e. Engage the 3 clips and install the right front fender molding sub-assembly.

f. Using a 4 mm hexagon wrench, install the screw.

g. Install the clip.

16. Install the no. 1 engine under cover.
17. Install the engine under cover assembly.
18. Install the right front wheel. Tighten the lug nuts to 76 ft. lbs. (103 Nm).
19. Lower the vehicle.

AIR CLEANER

REMOVAL & INSTALLATION

3.3L Engine

See Figure 156.

1. Remove front wiper arm and blade assembly.
2. Remove hood to cowl top seal.
3. Remove cowl top ventilator louver sub-assembly.
4. Remove windshield wiper motor and link assembly.
5. Remove outer cowl top panel sub-assembly.
6. Remove cool air intake duct seal.
7. Remove LH no. 2 engine room side cover.
8. Remove the 2 bolts, 4 clamps and air cleaner cap sub-assembly.
9. Remove the air cleaner filter element from the air cleaner case sub-assembly.
10. Remove air cleaner case sub-assembly:

a. Separate the No. 1 ventilation hose.

b. Disconnect the intake air flow meter connector.

c. Disconnect the 2 wire harness clamps from the air cleaner case sub-assembly.

d. Remove the 5 bolts from the air cleaner case sub-assembly.

e. Loosen the hose clamp and separate the No. 1 air cleaner hose.

f. Remove the air cleaner case sub-assembly.

To install:

11. To install, reverse the removal procedure.

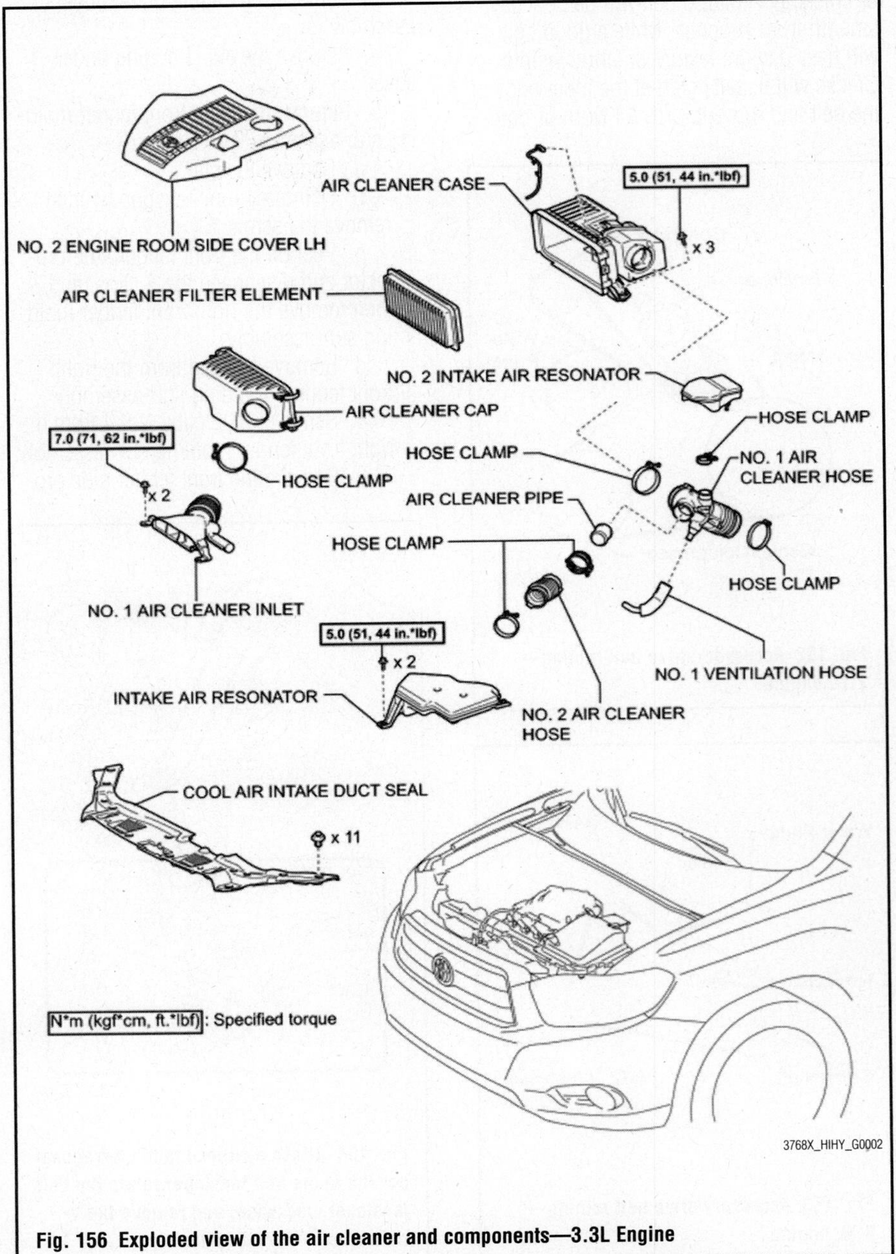

Fig. 156 Exploded view of the air cleaner and components—3.3L Engine

3.5L Engine

See Figure 157.

⁂ CAUTION

Wait at least 90 seconds after disconnecting the cable from the negative (-) battery terminal to prevent airbag and seat belt pretensioner activation.

1. Remove the 11 clips and cool air intake duct seal.
2. Loosen the nut, and disconnect the negative battery terminal.
3. Loosen the nut, and disconnect the positive battery terminal.
4. Loosen the nut, and remove the bolt and battery clamp.
5. Remove the battery and battery tray.
6. Disconnect the 2 vacuum switching valve clamps.
7. Disconnect the 2 vacuum hoses.
8. Remove the 2 bolts and No. 2 air cleaner inlet.
9. Disconnect the 2 vacuum hoses, and remove the 2 bolts and No. 1 air cleaner inlet.
10. Remove air cleaner cap sub-assembly:
    a. Disconnect the 3 vacuum hoses.
    b. Loosen the No. 1 air cleaner hose clamp.
    c. Disconnect the hose clamps and No. 2 ventilation hose.
    d. Disconnect the Mass Air Flow (MAF) meter connector.
    e. Disconnect the Mass Air Flow (MAF) meter connector.
    f. Remove the air cleaner cap sub-assembly and air cleaner filter element.

To install:

11. To install, reverse the removal procedure.

FILTER/ELEMENT REPLACEMENT

3.3L Engine

1. Remove the 2 bolts, 4 clamps and air cleaner cap sub-assembly.
2. Remove the air cleaner filter element from the air cleaner case sub-assembly.
3. To install, reverse removal procedure.

3.5L Engine

Remove the air cleaner cap sub-assembly and air cleaner filter element.

CAMSHAFT AND VALVE LIFTERS

REMOVAL & INSTALLATION

2.7L Engine

See Figures 158 through 173.

1. Before servicing the vehicle, refer to the precautions section.
2. Disconnect the negative battery cable.
3. Drain the engine coolant.
4. Remove or disconnect the following:
    - Right front wheel
    - Right fender splash shield
    - Right fender apron seal
    - No. 1 engine undercover
    - Coil pack
    - Cylinder head cover
5. Remove the timing chain cover and timing chain assembly.
6. Remove camshaft timing gear assembly. Hold the hexagonal portion of the camshaft with a wrench and remove the bolt and camshaft timing gear.

➡Be careful not to damage the cylinder head or spark plug tube with the wrench. Do not disassemble the camshaft timing gear.

7. Remove camshaft timing exhaust gear assembly. Hold the hexagonal portion of the camshaft with a wrench and remove the bolt and camshaft timing gear.

➡Be careful not to damage the cylinder head or spark plug tube with the wrench. Do not disassemble the camshaft timing gear.

8. Remove camshaft housing sub-assembly:
    a. Uniformly loosen and remove the 20 bearing cap bolts in the sequence shown in the illustration.
    b. Remove the camshaft housing by prying between the cylinder head and camshaft housing with a screwdriver.

➡Tape the screwdriver tip before use. Be careful not to damage the contact surfaces of the cylinder head and camshaft housing.

9. Remove camshaft bearing cap:
    a. Remove the 11 bearing cap bolts in the sequence shown in the illustration.
    b. Remove the 5 bearing caps.

➡Arrange the removed parts in the correct order.

10. Remove the oil control valve filter from the No. 1 camshaft bearing cap.
11. Remove the No. 1 camshaft bearing.
12. Remove the No. 1 and No. 2 camshafts.

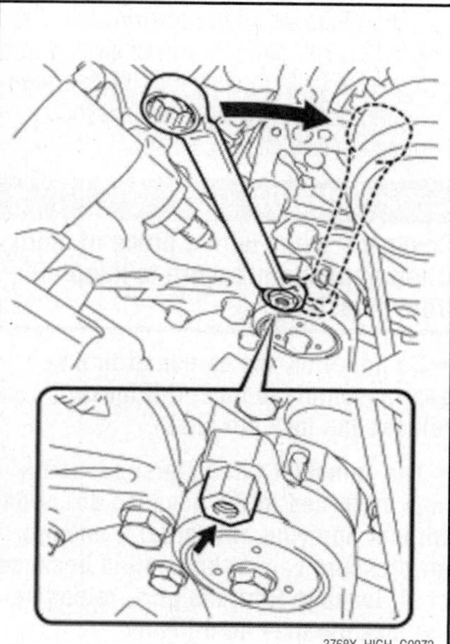

Fig. 157 Exploded view of the air cleaner assembly and related components—3.5L Engines

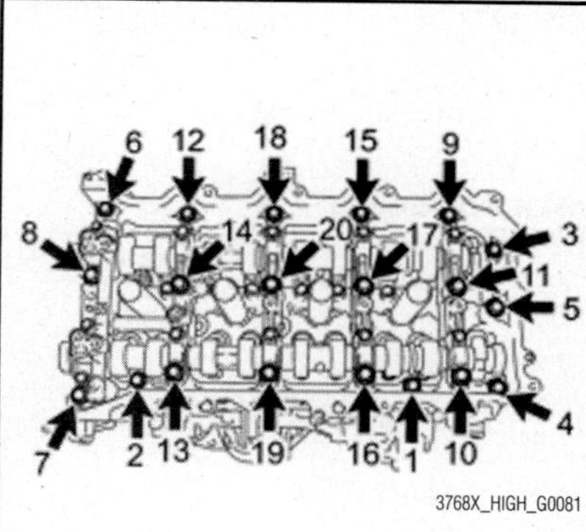

Fig. 158 Uniformly loosen and remove the 20 bearing cap bolts in the sequence shown

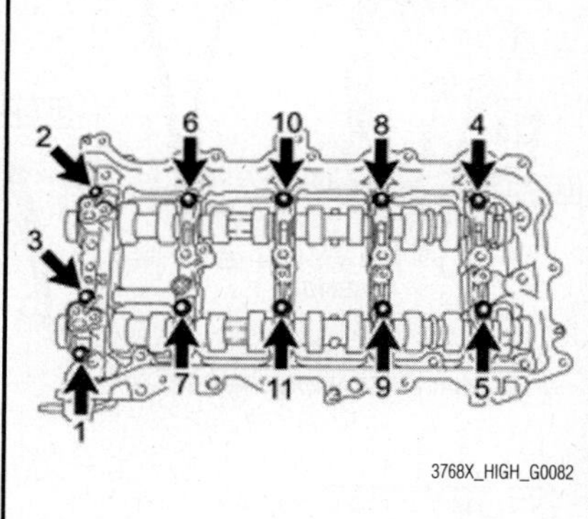

Fig. 159 Remove the 11 bearing cap bolts in the sequence shown

13. Remove the No. 2 camshaft bearing.
14. Remove the 16 valve rocker arms from the cylinder head.

➡Arrange the removed parts in the correct order.

15. Remove the 16 valve lash adjusters from the cylinder head.

➡Arrange the removed parts in the correct order.

To install:

➡When installing the camshaft timing gear, release the lock pin and set the camshaft timing gear to the advanced position before installation.

➡If the camshaft timing gear is set to the advanced position, do not let the camshaft timing gear rotate clockwise during installation. If the camshaft timing gear rotates to the retarded position, release the lock pin and set the camshaft timing gear to the advanced position.

16. Check the camshaft timing gear position.
17. Align and attach the knock pin of the No. 1 camshaft with the pin hole of the camshaft timing gear.
18. Check that there is no clearance between the camshaft timing gear and camshaft flange.
19. Hold the camshaft in place by hand, and then install the installation bolt of the camshaft timing gear by hand.

➡Do not use any tools to install the bolt. If the bolt is installed using a tool, the lock pin will be damaged.

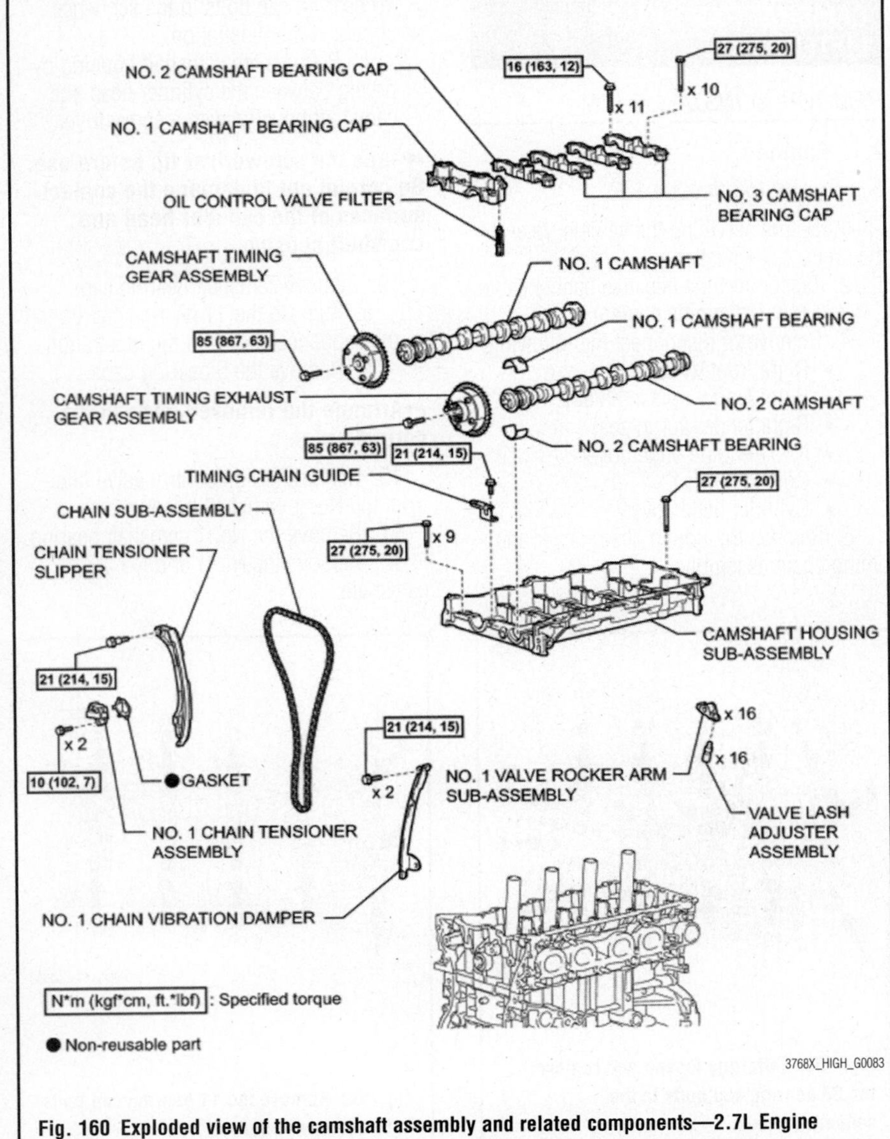

Fig. 160 Exploded view of the camshaft assembly and related components—2.7L Engine

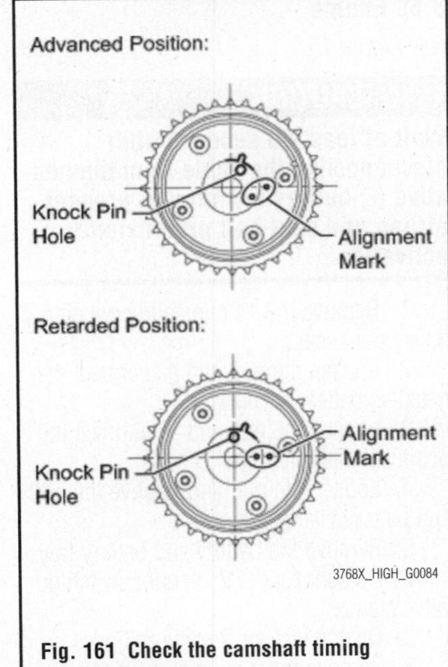

Fig. 161 Check the camshaft timing gear position

20. Release the lock pin:
   a. Clean the camshaft journal with a non-residue solvent.
   b. Cover the 4 oil paths of the cam journal with vinyl tape as shown in the illustration.

➡There are 4 oil paths in the grooves of the camshaft. Plug 3 of the paths with pieces of rubber.

   c. Open a hole at port A shown in the illustration.
   d. While applying approximately 200 kPa (2.0 kgf/ cm2, 29 psi) of air pressure to the oil path, forcibly turn the camshaft timing gear assembly in the advance direction (counterclockwise).

CAUTION

Cover the paths with a piece of cloth when applying pressure to keep oil from splashing.

➡Do not allow the camshaft timing gear assembly to lock. If it locks, release the lock pin again.

➡The camshaft timing gear assembly may be turned in the advance direction without applying any force. If enough air pressure cannot be applied because of air leakage from the port, releasing the lock pin may be difficult.

   e. Remove the vinyl tape and rubber pieces from the camshaft.

21. Remove the bolt and camshaft timing gear.

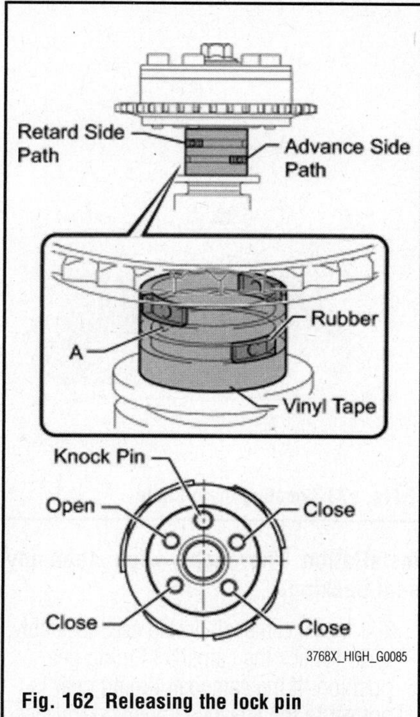

Fig. 162 Releasing the lock pin

➡**Do not allow the camshaft timing gear assembly to lock. If it locks, release the lock pin again.**

22. Install valve lash adjuster assembly:
    a. Inspect the valve lash adjuster before installing it.
    b. Install the 16 lash adjusters to the cylinder head.

➡**Install the lash adjuster to the same place it was removed from.**

23. Install no. 1 valve rocker arm sub-assembly:
    a. Apply engine oil to the lash adjuster tips and valve stem caps.

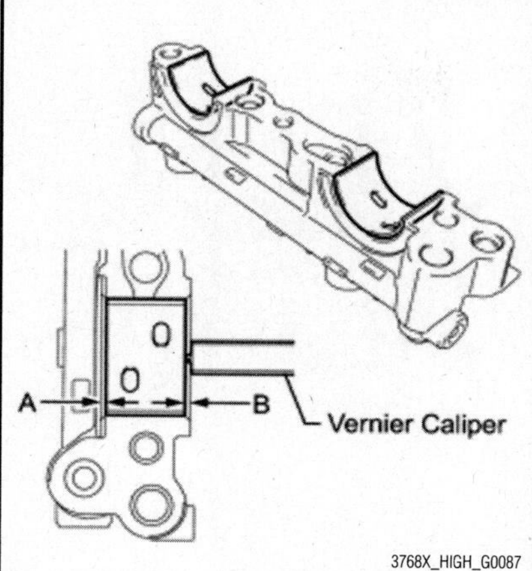

Fig. 164 Measure the distance between the No. 1 camshaft bearing cap edge and the camshaft bearing edge

    b. Install the 16 valve rocker arms as shown in the illustration.

24. Remove no. 2 camshaft bearing:
    a. Clean the No. 2 camshaft bearing.
    b. Install the camshaft bearing to the camshaft housing.
    c. Using a vernier caliper, measure the distance between the camshaft housing edge and the camshaft bearing edge. Standard distance: 1.15 to 1.85 mm (0.0453 to 0.0728 in.)
25. Install no. 1 camshaft bearing:
    a. Clean the No. 1 camshaft bearing.
    b. Install the camshaft bearing to the No. 1 camshaft bearing cap.
    c. Using a vernier caliper, measure the distance between the camshaft bearing cap edge and the camshaft bearing edge. Dimension A - B or B - A: 0 to 0.7 mm (0 to 0.0276 in.).

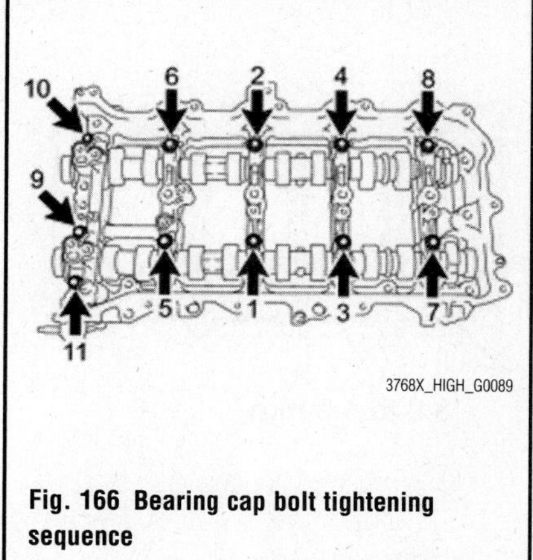

Fig. 166 Bearing cap bolt tightening sequence

26. Install the oil control valve filter to the No. 1 camshaft bearing cap.
27. Install camshaft:
    a. Clean the camshaft journals, camshaft housing and bearing caps.
    b. Apply a light coat of engine oil to the camshaft journal, camshaft housing and bearing caps.
    c. Install the No. 1 and No. 2 camshafts to the camshaft housing.
28. Install camshaft bearing cap:
    a. Confirm the marks and numbers on the camshaft bearing caps and place them in their proper positions and directions.
    b. Install the 11 bolts in the order shown in the illustration. Tighten to 12 ft. lbs. (16 Nm).

➡**Make sure that the camshaft rotates smoothly after installing the bearing caps.**

29. Install camshaft housing sub-assembly:
    a. Check that the valve rocker arms are installed as shown in the illustration.

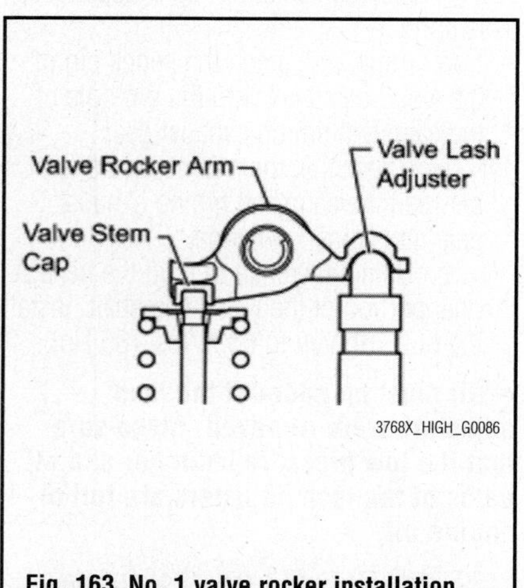

Fig. 163 No. 1 valve rocker installation orientation

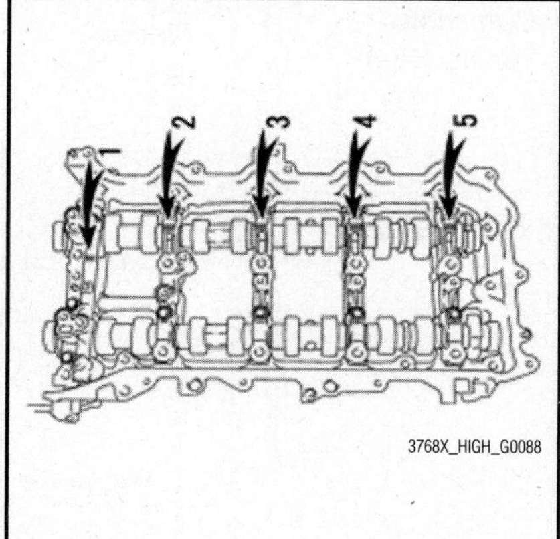

Fig. 165 Identifying correct camshaft bearing cap position and direction

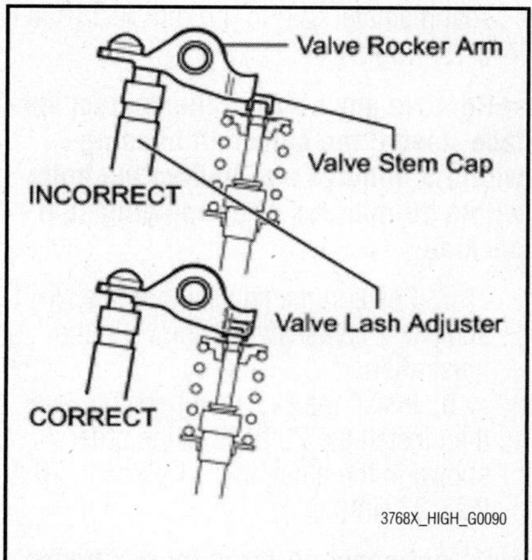

Fig. 167 Valve rocker arm installation orientation

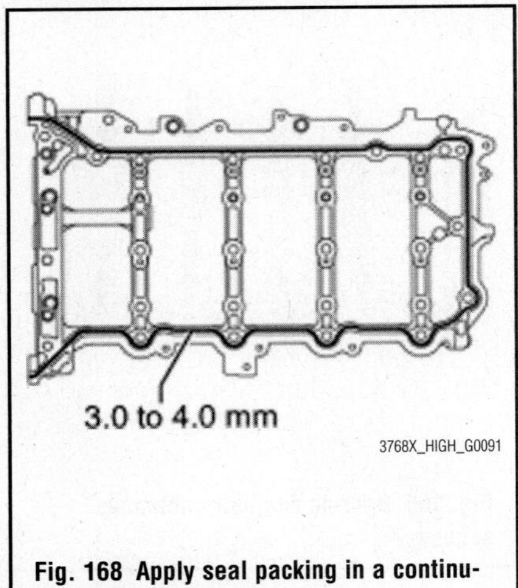

Fig. 168 Apply seal packing in a continuous line as shown

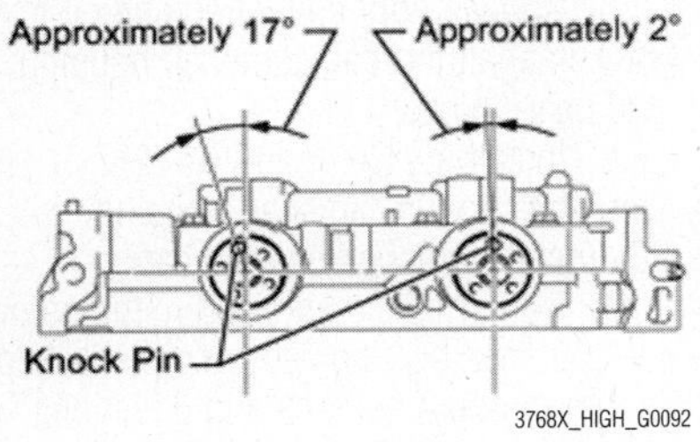

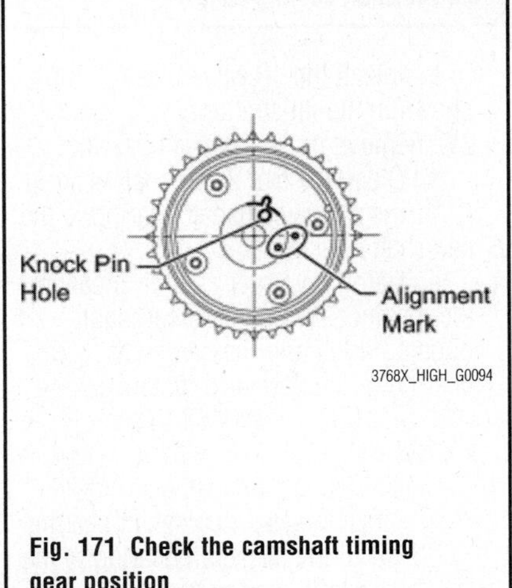

Fig. 169 Position the knock pin of the No. 1 and No. 2 camshafts as shown

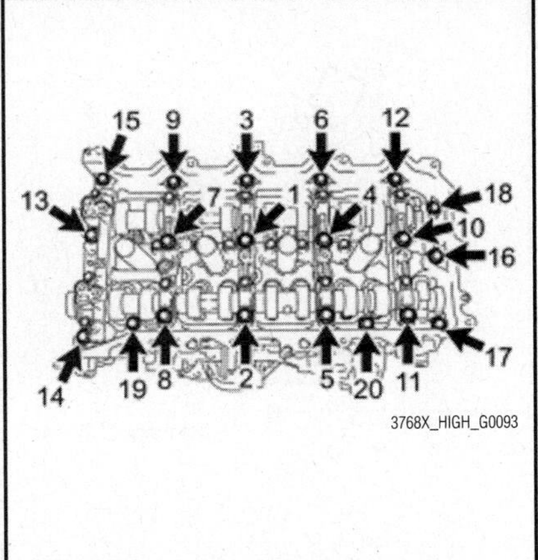

Fig. 170 Camshaft housing tightening sequence

Fig. 171 Check the camshaft timing gear position

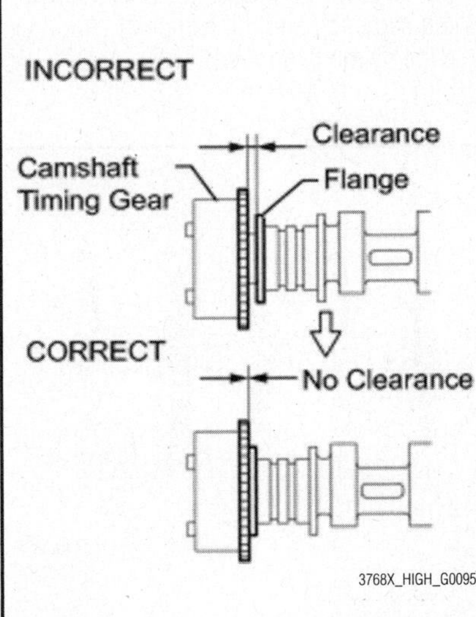

Fig. 172 Check that there is no clearance between the camshaft timing gear and camshaft flange

3768X_HIGH_G0096

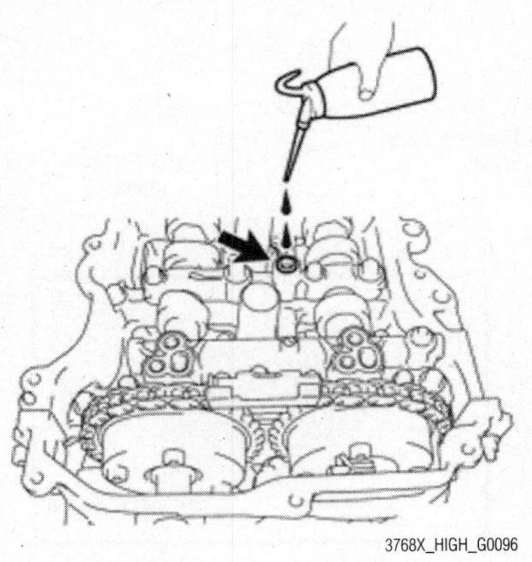

Fig. 173 Locating the oil hole

b. Apply seal packing in a continuous line as shown in the illustration. Seal packing: Toyota Genuine Seal Packing Black, Three Bond 1207B or equivalent. Standard seal diameter: 3.0 to 4.0 mm (0.118 to 0.157 in.)

➡Remove any oil from the contact surface. Install the camshaft housing within 3 minutes and tighten the bolts within 10 minutes after applying seal packing.

c. Position the knock pin of the No. 1 and No. 2 camshafts as shown in the illustration.

d. Install the camshaft housing, and then install the 20 bolts in the order shown in the illustration. Tighten to 20 ft. lbs. (27 Nm).

➡Do not apply oil for at least 4 hours after the installation. Do not start the engine for at least 4 hours after the installation. Thoroughly wipe clean any seal packing.

30. Install camshaft timing gear assembly:

a. Check the camshaft timing gear position. If the camshaft timing gear is not set to the advanced position, release the lock pin and reset the camshaft timing gear (Refer to the "SET CAMSHAFT TIMING GEAR ASSEMBLY" procedures").

b. Align and attach the knock pin of the No. 1 camshaft with the pin hole of the camshaft timing gear.

c. Check that there is no clearance between the camshaft timing gear and camshaft flange.

d. Using a wrench to hold the hexagonal portion of the No. 1 camshaft, install the bolt. Tighten to 63 ft. lbs. (85 Nm).

➡Be careful not to damage the cylinder head or spark plug tube with the wrench. Do not disassemble the camshaft timing gear.

31. Install camshaft timing exhaust gear assembly:

a. Align and attach the knock pin of the No. 2 camshaft with the pin hole of the camshaft timing exhaust gear.

b. Check that there is no clearance between the camshaft timing exhaust gear and camshaft flange.

c. Using a wrench to hold the hexagonal portion of the No. 2 camshaft, install the bolt. Tighten to 63 ft. lbs. (85 Nm).

➡Oil must be added if the lash adjusters were removed. Make sure that the low pressure chamber and oil paths of the lash adjusters are full of engine oil.

32. Add 50 cc (3.1 cu. in) of engine oil into the oil hole shown in the illustration.

33. Set no. 1 cylinder to TDC/compression.
34. Install the timing chain and cover assembly.

3.3L Engine

See Figures 174 through 189.

WARNING

The thrust clearance on both the intake and exhaust camshafts is very small; the camshafts must be kept level during removal. If the camshafts are removed without being kept level, the camshaft may be caught in the cylinder head, causing the head to break or the camshaft to seize.

1. Before servicing the vehicle, refer to the Precautions Section.
2. Relieve the fuel system pressure.
3. Drain the engine oil.
4. Drain the coolant from the engine radiator and hybrid transaxle.
5. Remove or disconnect the following:
   - Negative battery cable
   - Engine cover
   - Right-hand front wheel
   - Engine splash shields
   - Right-hand front fender apron seal
   - Wiper and blade assembly
   - Top cowl ventilator louver assembly
   - Wiper motor and link assembly
   - Battery and battery tray
   - Air intake assembly
   - Emission control valve hoses
   - Air intake surge tank
   - Radiator intake hose
   - Brake master cylinder reservoir and bracket
   - Air cleaner bracket
   - Engine moving control rod
   - Right-hand engine No.2 mounting stay
   - Ignition coil
   - Valve covers
   - Crankshaft pulley
   - Timing belt No.1 and No.2 covers
   - Right-hand engine mounting bracket
   - No.2 timing belt guide
   - Timing belt
   - Timing belt idler
6. Using Special Tool 09960-10010, remove the camshaft timing pulleys.

➡**Keep all valvetrain components in order for reassembly.**

7. Disconnect the engine wiring harness clamps from the No.3 timing belt cover and remove the cover.
8. Remove the left camshafts as follows:
   a. Align the timing marks (2-dot mark) of the camshaft drive and the driven gears by turning the camshaft with a wrench.
   b. Secure the exhaust camshaft sub-gear to the main gear with a service bolt. A bolt 0.63–0.79 in. (16–20mm) long with a 6mm thread diameter and a 1mm pitch is recommended. Tighten bolt to 48 inch lbs. (5.4 Nm).

➡**When removing the camshaft, make certain that the torsional spring force of the sub-gear has been eliminated by installing the service bolt.**

   c. Using several steps, loosen and remove the 10 bearing cap bolts uniformly in the sequence shown.
   d. Remove the 5 bearing caps and the exhaust camshaft.
   e. Using several steps, loosen and remove the 10 bearing cap bolts uniformly in the sequence shown.
   f. Remove the 5 bearing caps and the intake camshaft.

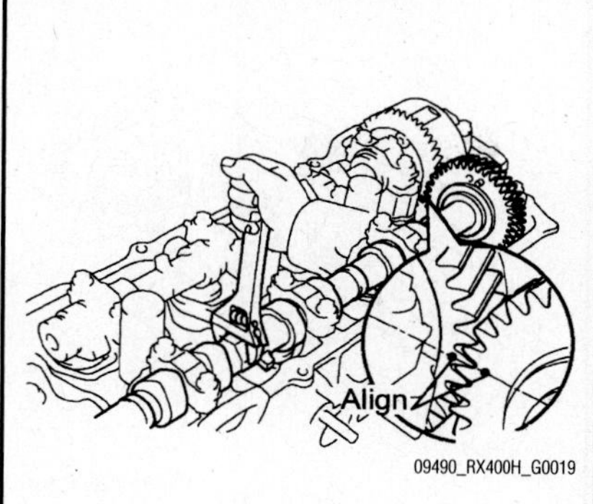

Fig. 175 Align the timing marks of the camshaft gears.

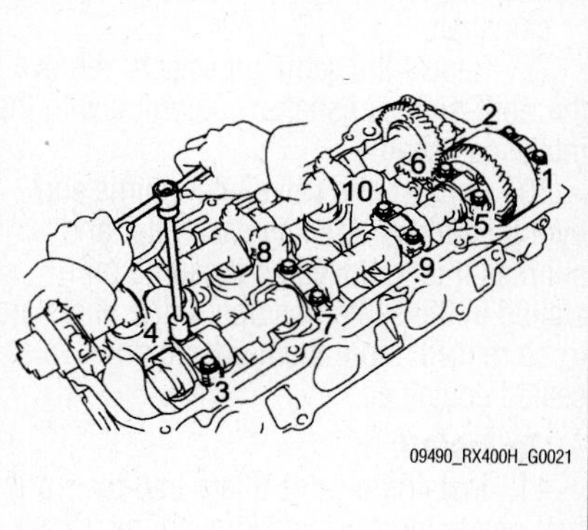

Fig. 177 Intake camshaft bearing cap loosening sequence

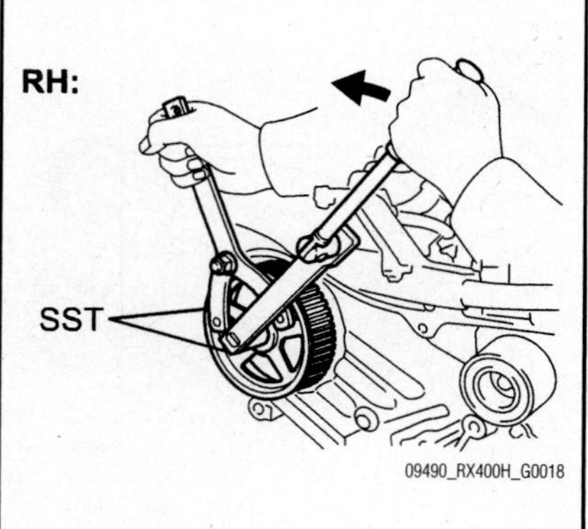

Fig. 174 Removing the right-hand camshaft timing pulley, left-hand similar.

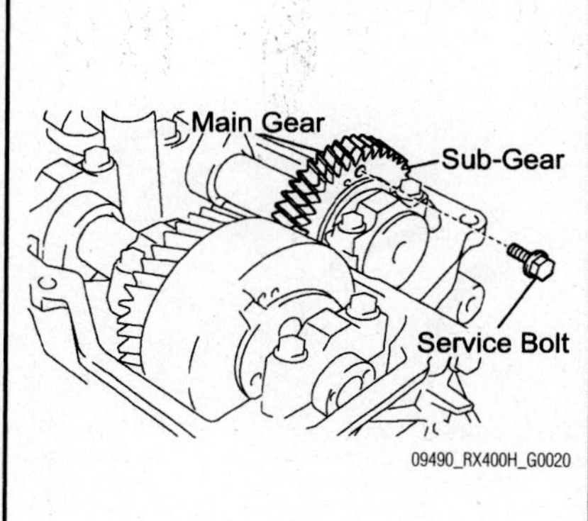

Fig. 176 Install a service bolt to secure the camshaft gears.

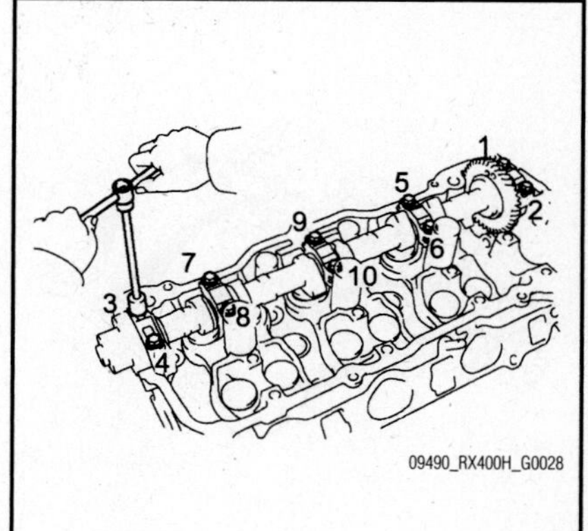

Fig. 178 Exhaust camshaft bearing cap loosening sequence

g. Remove the oil seal from the intake camshaft.

9. Repeat the same process to remove the right-side camshafts, beginning with the intake camshaft.

10. Remove the valve lifter shims and hydraulic lifters. Identify each lifter and shim as it is removed so it can be reinstalled in the same position. If the lifters are to be reused, store them upside down in a sealed container.

To install:

11. Install the valve lifters into their original positions and install the shims. Check valve clearance and replace the shims as necessary.

12. When reinstalling, remember that the camshafts must be handled carefully and kept straight and level to avoid damage.

13. Install the right camshafts, as follows:

a. Apply new engine oil to the thrust portion and journal of the camshaft.

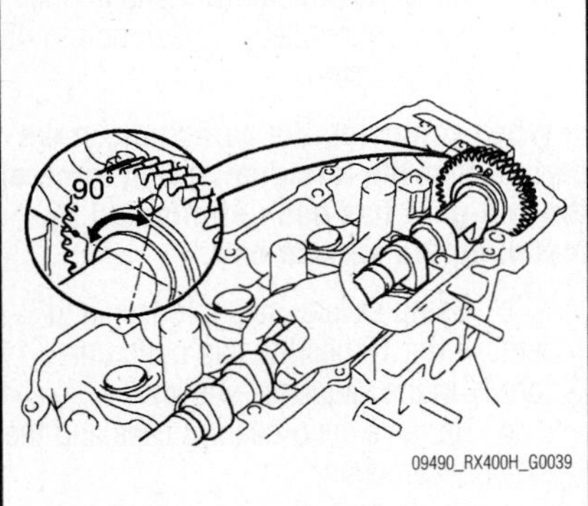

Fig. 179 Install the right exhaust camshaft with the alignment marks in the correct position.

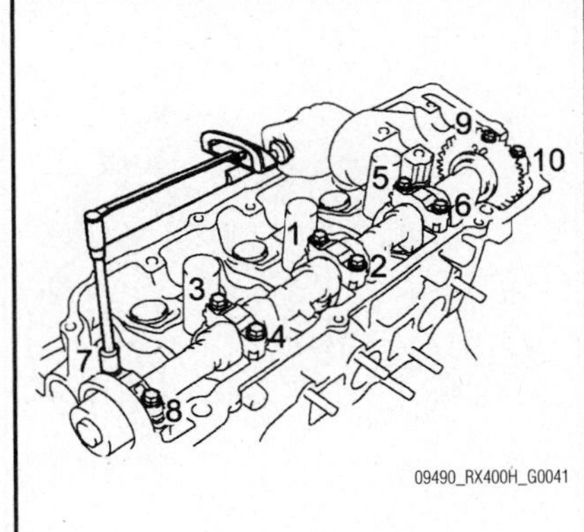

Fig. 181 Right exhaust camshaft bearing cap torque sequence

b. Position the exhaust camshaft on the head so that the alignment marks are at a 90 degrees angle from vertical.

c. Apply multi-purpose grease to the lip of a new oil seal.

d. Install the oil seal to the camshaft.

e. Apply sealant to the No. 1 bearing cap.

f. Apply a light coat of clean engine oil to the bolt threads and under the bolt head. Install the bearing caps to their proper position. Tighten the bolts evenly and in several passes to 12 ft. lbs. (16 Nm) in the proper sequence.

g. Position the intake camshaft on the head so that the alignment marks are at a 90 degrees angle from vertical. The mark should be at the "9 o'clock" position and must align with the marks on the other gear.

h. Apply a light coat of clean engine oil to the bolt threads and under the bolt head. Install the bearing caps to their proper position. Tighten the bolts evenly and in several passes to 12 ft. lbs. (16 Nm) in the proper sequence.

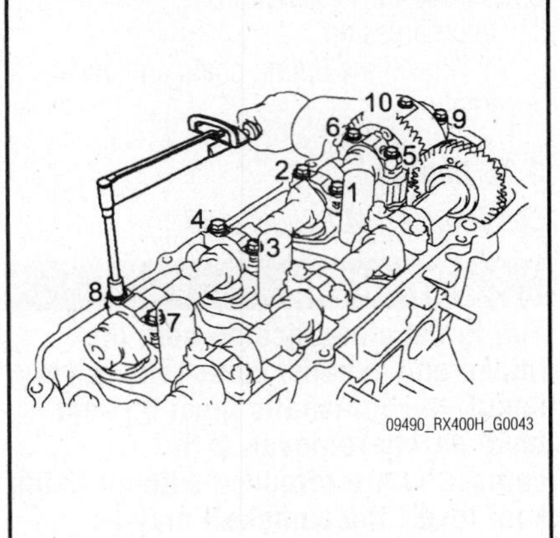

Fig. 183 Right intake camshaft bearing cap bolt tightening sequence

i. Remove the service bolt.

14. Install the left camshafts, as follows:

a. Apply new engine oil to the thrust portion and journal of the camshaft

b. Position the exhaust camshaft on the head so that the alignment mark is at a 90 degrees angle from vertical. The mark should be at the "9 o'clock" position.

c. Apply multi-purpose grease to the oil seal lip and install the new oil seal to the camshaft.

d. Apply sealant to the No. 1 bearing cap.

e. Apply a light coat of clean engine oil to the bolt threads and under the bolt head. Install the bearing caps to their proper position. Tighten the bolts evenly and in several passes to 12 ft. lbs. (16 Nm) in the proper sequence.

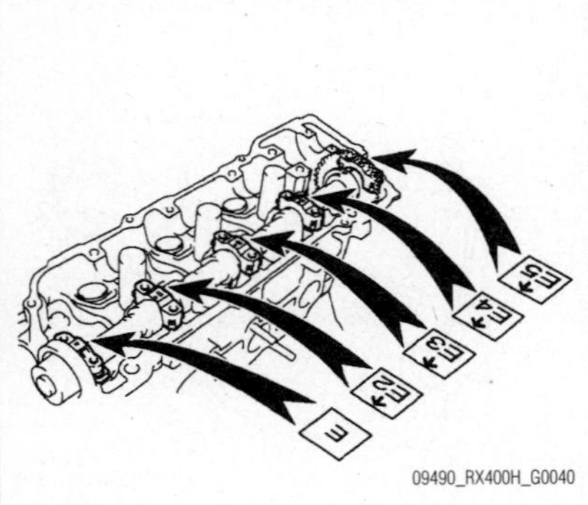

Fig. 180 Right exhaust camshaft bearing caps must be placed in their proper locations.

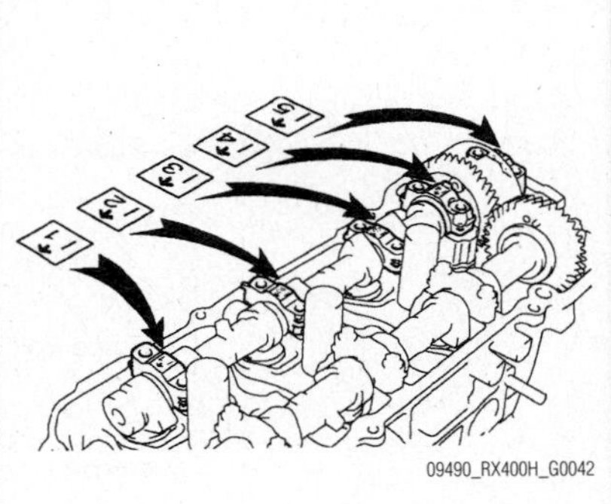

Fig. 182 Right intake camshaft bearing caps must be placed in their proper locations

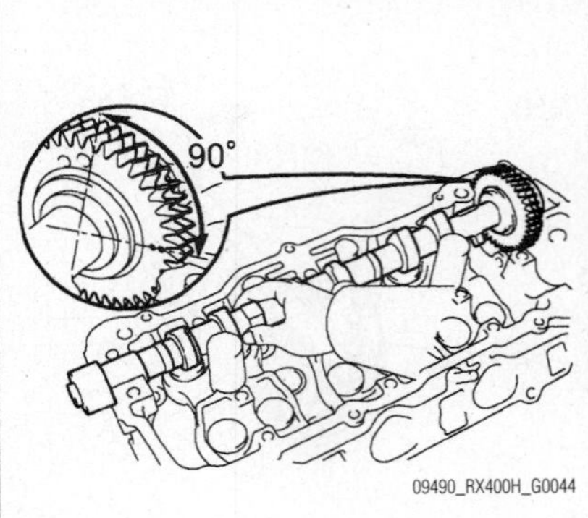

Fig. 184 Install the left exhaust camshaft with the alignment mark in the correct position.

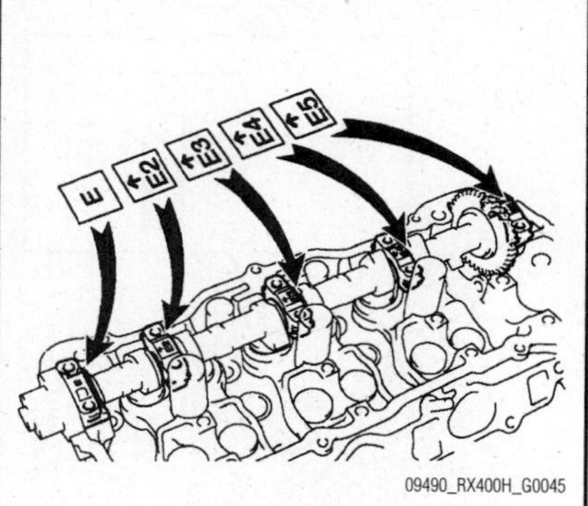

Fig. 185 Left exhaust camshaft bearing caps must be placed in their proper locations

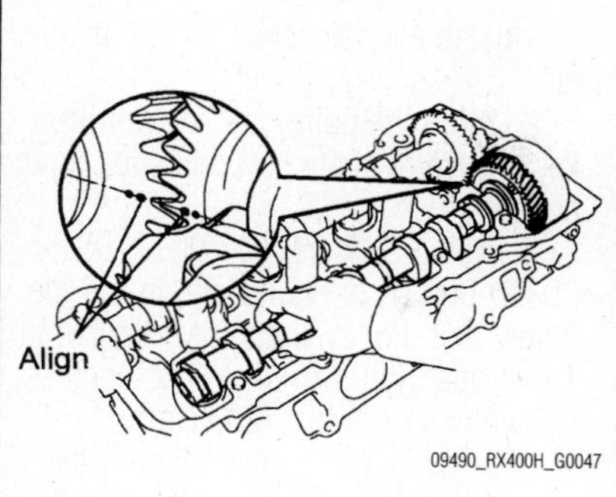

Fig. 187 Install the left intake camshaft with the alignment mark in the correct position.

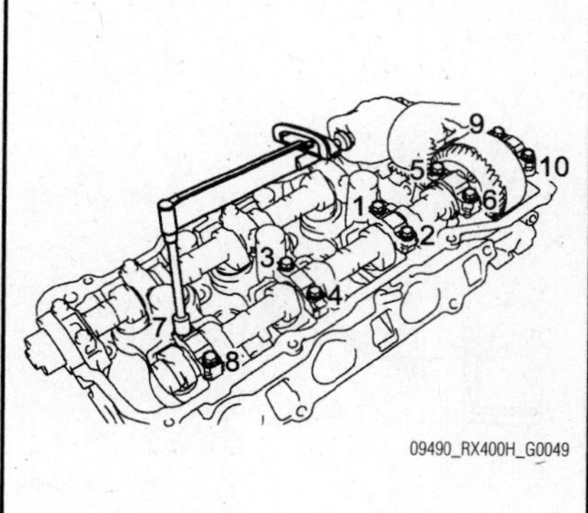

Fig. 189 Left exhaust bearing cap torque sequence

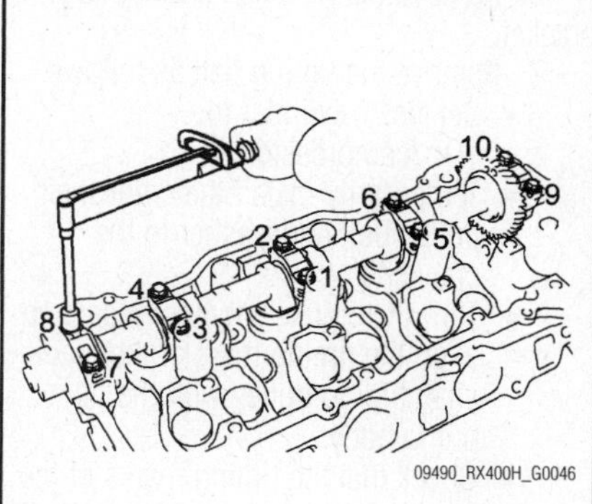

Fig. 186 Left exhaust camshaft bearing cap torque sequence

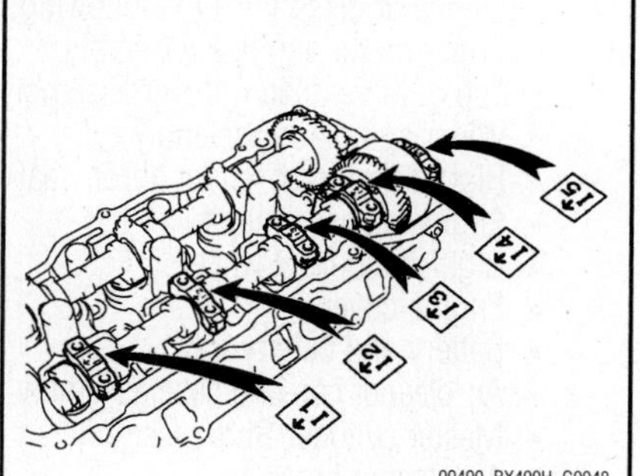

Fig. 188 Left exhaust camshaft bearing caps must be placed in their proper locations.

f. Position the intake camshaft on the head so that the alignment marks are at a 90 degrees angle from vertical. The mark should be at the "3 o'clock" position and must align with the marks on the exhaust camshaft gear.

g. Apply a light coat of clean engine oil to the bolt threads and under the bolt head. Install the bearing caps to their proper position. Tighten the bolts evenly and in several passes to 12 ft. lbs. (16 Nm) in the proper sequence.

h. Remove the service bolt.

15. Install or connect the following:

16. Install the timing belt cover No. 3. Tighten to 76 inch lbs. (8.5 Nm)

17. Using Special Tool 09960-10010, install the camshaft timing pulleys. Tighten to 92 ft. lbs. (125 Nm).

18. Install the timing belt idler and tighten to 32 ft. lbs. (43 Nm).

19. Install or connect the following:

- Timing belt
- No. 2 Timing belt guide
- Right-hand engine mounting bracket
- Timing belt covers Nos. 1 and 2
- Crankshaft pulley. Tighten to 162 ft. lbs. (220 Nm).
- Right-hand engine mounting stay No. 2
- Engine moving control rod
- Air cleaner bracket
- Brake master cylinder reservoir and bracket
- Valve covers
- Ignition coil
- Radiator intake hose
- Air intake surge tank
- Emission control valve hoses
- Air intake assembly
- Battery and battery tray
- Wiper motor and link assembly
- Top cowl ventilator louver assembly
- Wiper and blade assembly
- Right-hand front fender apron seal
- Engine splash shields
- Right-hand front wheel
- Engine covers
- Negative battery cable

20. Refill the cooling system to the correct level.

21. Refill the engine oil to the correct level.

22. Start the engine and check for leaks.

23. If necessary, perform the initialization procedure.

CATALYTIC CONVERTER

REMOVAL & INSTALLATION

The catalytic converter is integrated with the exhaust manifold. Refer to Exhaust Manifold ion this section.

CRANKSHAFT FRONT SEAL

REMOVAL & INSTALLATION

2.7L & 3.5L Engines

See Figures 190 and 191.

1. Raise and support the vehicle.
2. Remove the right front wheel.
3. Remove the engine under cover assembly.
4. Remove the No. 1 engine under cover.
5. Remove the right front fender molding sub-assembly.
6. Remove the right front fender liner.
7. Remove the right front fender apron seal.
8. Remove the V-ribbed belt.
9. Using a special service tool (SST: 09213-70011, SST: 09330-00021 or equivalent), loosen the crankshaft pulley bolt.
10. Using SST: 09950-50013 or equiva-

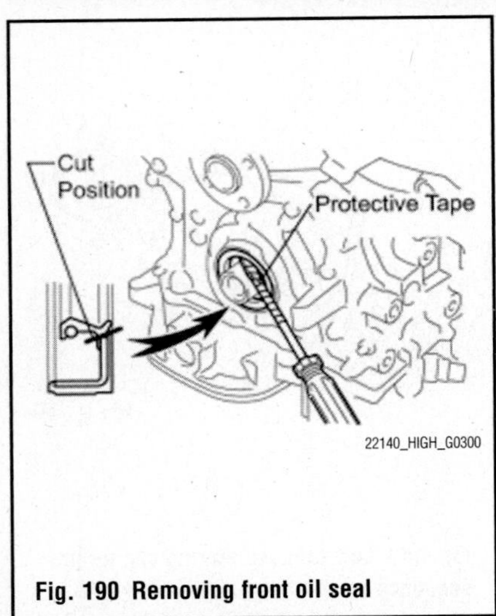

Fig. 190 Removing front oil seal

lent, remove the crankshaft pulley bolt and crankshaft pulley.

11. Using a screwdriver, pry out the timing chain case oil seal.

➡Tape the screwdriver tip before use.

➡After the removal, check the crankshaft for damage. If it is damaged, smooth the surface with 400-grit sandpaper.

To install:

12. Install timing chain case oil seal, as follows:
   a. Apply MP grease to a new oil seal lip.
   b. Using a Special Service Tool (SST: 09223-22010, SST: 09506-35010 or equivalent) and a hammer, tap in the oil seal until its surface is flush with the timing chain cover edge.

➡Keep the lip free of foreign matter.

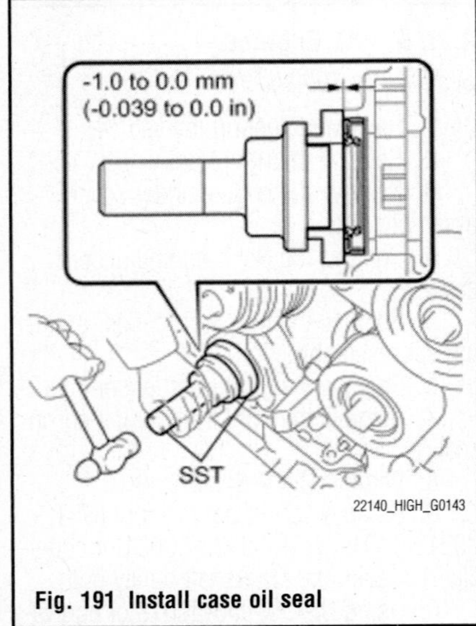

Fig. 191 Install case oil seal

➡Do not tap the oil seal at an angle.

13. Install the crankshaft pulley, as follows:
   a. Align the pulley set key with the key groove of the pulley, and slide on the pulley.
   b. Using SST: 09213-70011, SST: 09330-00021, or equivalent, install the pulley bolt. For 2.7L engines, tighten to 192 ft. lbs. (260 Nm). For 3.5L engines, tighten to 184 ft. lbs. (250 Nm).
14. The remainder of installation is the reverse of removal. When installing the wheel, tighten the lug nuts to 76 ft. lbs. (103 Nm).

3.3L Engine

See Figures 192 and 193.

1. Before servicing the vehicle, refer to the precautions section.
2. Remove or disconnect the following:
   - Wiper motor and link assembly
   - Top cowl ventilator louver assembly
   - Wiper and blade assembly
   - Right-hand front fender apron seal
   - Engine splash shields
   - Right-hand front wheel
   - Engine covers
   - Battery and battery tray
   - Air cleaner cap and case assembly
   - Master cylinder Sub-assembly
   - Air cleaner bracket
   - Engine moving control rod
   - RH engine mounting stay
3. Use Special Tool 09213-54015 to hold the crankshaft pulley in order to loosen the pulley bolt.
4. Use Special Tool 09950-50013 to remove the crankshaft pulley.
5. Remove the timing belt covers No.1 and No.2.

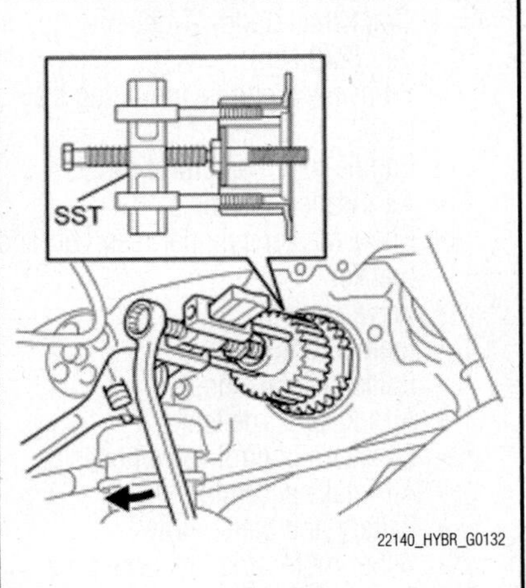

Fig. 192 Remove the timing crankshaft pulley

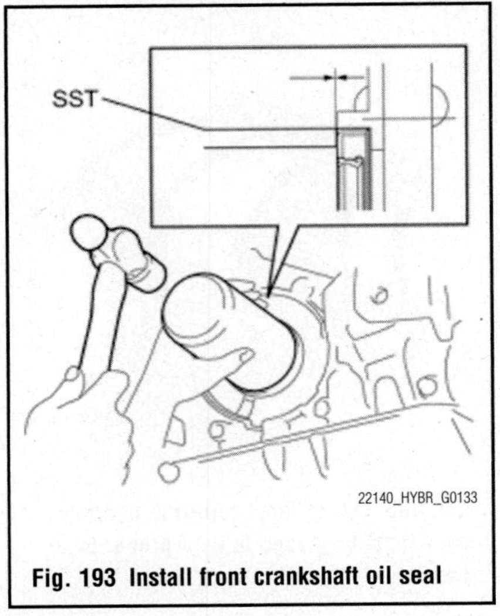

Fig. 193 Install front crankshaft oil seal

6. Remove the RH engine mounting bracket.
7. Remove the timing belt as follows:
   - Set No. 1 cylinder to TDC/compression.
   - Temporarily install the crankshaft pulley bolt and washer to the crankshaft.
   - Turn the crankshaft clockwise, and align the timing mark of the crankshaft timing pulley with the oil pump body.
   - Check that the timing marks of the camshaft timing pulleys and No. 3 timing belt cover are aligned.
   - If not, turn the crankshaft 1 revolution (360°).
8. Remove the bolt and the timing belt plate.
9. Install the pulley bolt to the crankshaft.
10. Using SST, remove the crankshaft timing pulley.
11. Using a knife, cut off the oil seal lip.
12. Using a screwdriver with its tip taped, pry out the oil seal.

To install:

13. Apply MP grease to a new oil seal lip.
14. Using SST and a hammer, tap in a new oil seal until its surface is flush with the oil pump edge.
15. Reverse the removal procedure for installation and note the following:
   - Install new cover gaskets
   - Tighten timing belt covers to 74 inch lbs. (8 Nm).
   - Crankshaft pulley. Tighten the bolt to 162 ft. lbs. (219 Nm).
16. Start the vehicle and check for any leaks.
17. Recheck the ignition timing.

18. If necessary, perform the initialization procedure.

CYLINDER HEAD

REMOVAL & INSTALLATION

3.3L Engine

See Figures 194 through 198.

1. Before servicing the vehicle, refer to the Precautions Section.

CAUTION

The HIGHLANDER HV has a hybrid system that operates at voltages up to 650 volts. Be sure to follow the instructions in this manual to handle the system correctly. Failure to do so may result in serious injury or electrocution. Engineer must undergo special training to be able to perform high-voltage system inspection and servicing.

CAUTION

All high-voltage wire harness connectors are colored orange. The HV battery and other high-voltage components have "High Voltage" caution labels. Do not carelessly touch these wires and components.

2. Before inspecting or servicing the high-voltage system, be sure to follow safety measures, such as wearing insulated gloves and removing the service plug to prevent electrocution. Carry the removed service plug in your pocket to prevent other technicians from reinstalling it while you are servicing the vehicle.
3. After removing the service plug, wait 5 minutes before touching any of the high-voltage connectors and terminals.

➡Wear insulating gloves and protective glasses.

4. Relieve the fuel system pressure.
5. Drain the engine oil.
6. Drain the coolant from the engine radiator and hybrid transaxle.
7. Remove the service plug grip, found underneath the Battery Service cover on the rear seat. Wait 5 minutes to discharge the high voltage capacitor.
8. Remove or disconnect the following:
    - Negative battery cable
    - Engine cover
    - Right-hand front wheel
    - Engine splash shields
    - Right-hand front fender apron seal
    - Wiper and blade assembly
    - Top cowl ventilator louver assembly
    - Wiper motor and link assembly
    - Battery and battery tray
    - Air intake assembly
    - Converter with Inverter assembly
    - Emission control valve set
    - Intake air surge tank
    - Fuel supply hose
    - Heater inlet hose
    - Intake manifold
    - Radiator hoses
    - Water outlet from the cylinder heads
    - Brake master cylinder reservoir
    - Air cleaner bracket
    - Engine moving control rod
    - Right-hand No.2 engine mounting stay
    - Crankshaft pulley
    - Timing belt No.1 and No.2 covers
    - Right-hand engine mounting bracket
    - Timing belt guide No.2
    - Timing belt
    - Timing belt No.2 idler
    - Camshaft timing pulley
    - Timing belt No.3 cover
    - Front exhaust pipe assembly
    - Exhaust manifold heat insulator
    - Exhaust manifold stay
    - Right-hand exhaust manifold and gasket
    - Ignition coil
    - Right-hand cylinder head cover
    - Camshaft
    - VVT sensor connector
    - Camshaft timing oil control valve connecter
9. Loosen the right-hand cylinder head bolts in several steps in the sequence shown.

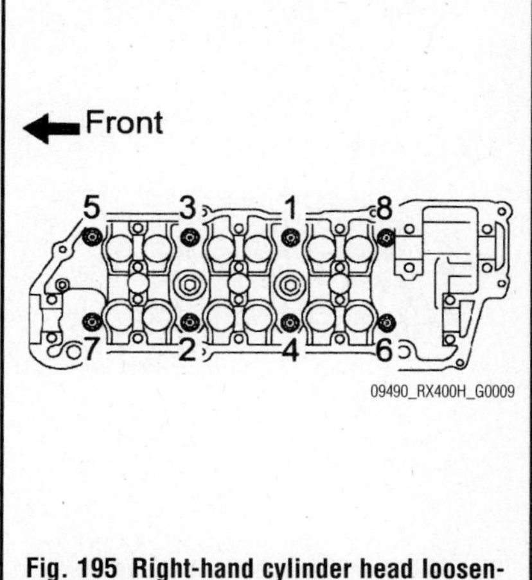

Fig. 195 Right-hand cylinder head loosening sequence

10. Remove the cylinder head bolts and plate washers.
11. Remove the right-hand cylinder head and gasket
12. Remove the manifold converter No.3 insulator.
13. Remove the exhaust manifold No.2 heat insulator.
14. Separate the cooling fan ECU and hang securely with mechanic's wire.
15. Remove or disconnect the following:
    - Left-hand exhaust manifold
    - Oil level gauge guide
    - Water inlet pipe
    - Left-hand cylinder head cover
    - Camshaft
    - Loosen the Right-hand cylinder head bolts in several steps in the sequence shown.
16. Remove the cylinder head bolts and plate washers.
17. Remove the Left-hand cylinder head and gasket.

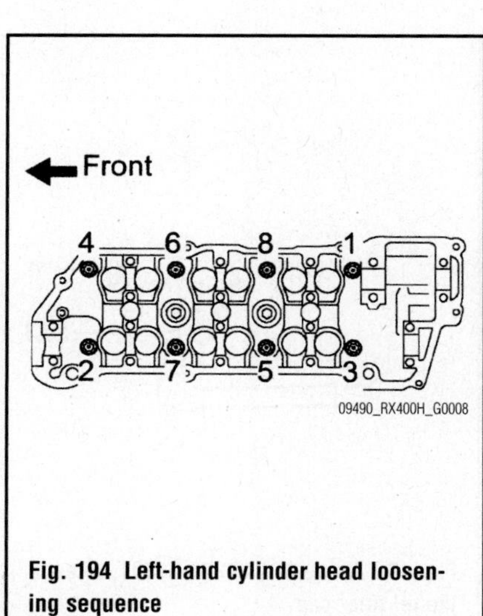

Fig. 194 Left-hand cylinder head loosening sequence

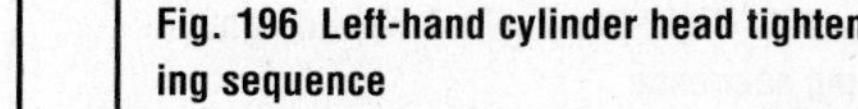

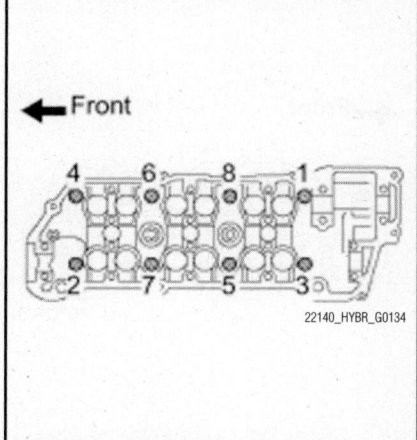

Fig. 196 Left-hand cylinder head tightening sequence

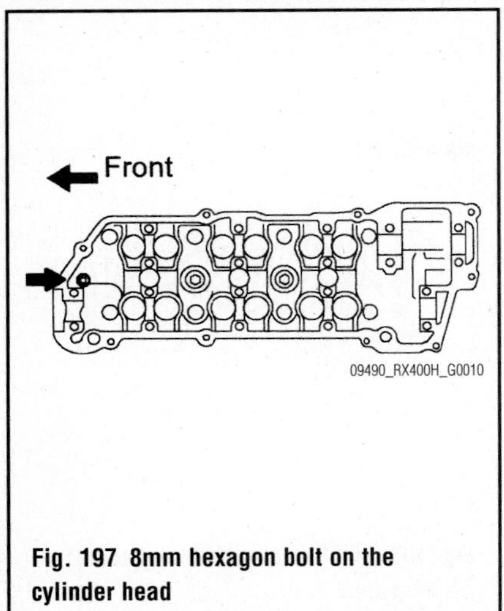

Fig. 197 8mm hexagon bolt on the cylinder head

To install:

18. Install the Left-hand cylinder head with a new gasket. Tighten the cylinder head bolts as follows:
   - Step 1: Tighten the 8 cylinder head bolts to 40 ft. lbs. (54 Nm)
   - Step 2: Tighten each bolt 90°
   - Step 3: Tighten each bolt an additional 90°
   - Step 4: Tighten the single 8mm hexagon bolt to 14 ft. lbs. (19 Nm)
19. Install the or connect the following:
   - Camshaft
   - Left-hand cylinder head cover. Tighten to 71 inch lbs. (8 Nm).
   - Water inlet pipe
   - Oil level gauge guide. Tighten to 71 inch lbs. (8 Nm).
   - Cooling fan ECU
   - Exhaust manifold No.2 heat insulator
   - Manifold converter No.2 insulator
20. Install the Right-hand cylinder head with a new gasket. Tighten the cylinder head bolts as follows:
   - Step 1: Tighten the 8 cylinder head bolts to 40 ft. lbs. (54 Nm)
   - Step 2: Tighten each bolt 90°
   - Step 3: Tighten each bolt an additional 90°
   - Step 4: Tighten the single 8mm hexagon bolt to 14 ft. lbs. (19 Nm).
21. The remainder of installation is the reverse order of removal.
22. Refill the engine oil to the correct level.
23. Refill the coolant to the engine radiator and hybrid transaxle to the correct level.
24. Replace the service plug grip.
25. Start the engine and check for leaks.
26. If necessary, perform the initialization procedure.

ENGINE OIL & FILTER

REPLACEMENT

2.7L & 3.5L Engines

See Figure 199.

✲✲ CAUTION

Prolonged and repeated contact with engine oil will result in the removal of natural oils from the skin, leading to dryness, irritation and dermatitis. In addition, used engine oil contains potentially harmful contaminants which may cause skin cancer.

✲✲ CAUTION

Precautions should be taken when replacing engine oil to minimize the risk of your skin making contact with used engine oil. Protective clothing and gloves that cannot be penetrated by oil should be worn. The skin should be washed with soap and water, or use waterless hand cleaner, to remove any used engine oil thoroughly. Do not use gasoline, thinners, or solvents.

✲✲ WARNING

In order to protect the environment, used oil and used oil filters must be disposed of at designated disposal sites

1. Drain engine oil:
   a. Remove the oil filler cap.
   b. Remove the oil pan drain plug and drain the engine oil into a container.
   c. Install a new gasket and the oil pan drain plug. Tighten to 30 ft. lbs. (40 Nm).
2. Remove oil filter element:
   a. Connect a hose with an inside diameter of 15 mm (0.591 in.) to the pipe.
   b. Remove the oil filter drain plug.
   c. Install the pipe to the oil filter cap.

➡If the O-ring is removed with the drain plug, install the O-ring together with the pipe.

➡Use a container to catch the draining oil.

   d. Check that oil is drained from the oil filter. Then disconnect the pipe and remove the O-ring.
   e. Using SST 09228-06501, remove the oil filter cap.

➡Do not remove the oil filter bracket clip.

   f. Remove the oil filter element and O-ring from the oil filter cap.

➡Be sure to remove the O-ring (for the cap) by hand, without using any tools, to prevent damage to the groove for the O-ring on the cap.

3. Install oil filter element:
   a. Clean the inside of the oil filter cap, the threads and O-ring groove.
   b. Apply a small amount of engine oil to a new O-ring and install it to the oil filter cap.
   c. Set a new oil filter element to the oil filter cap.
   d. Remove any dirt or foreign matter from the installation surface of the engine.
   e. Apply a small amount of engine oil to the O-ring again and temporarily install the oil filter cap.

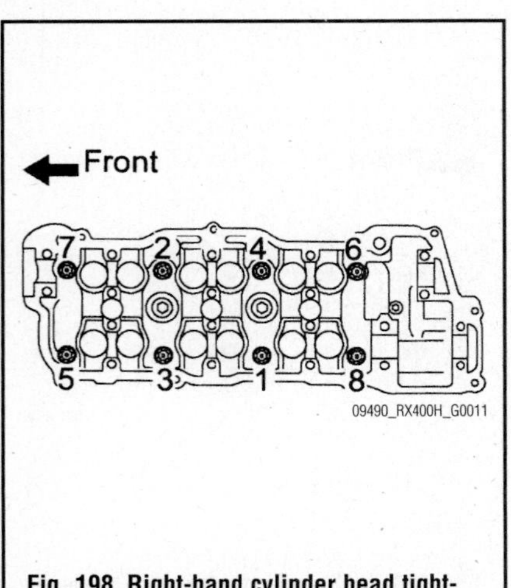

Fig. 198 Right-hand cylinder head tightening sequence

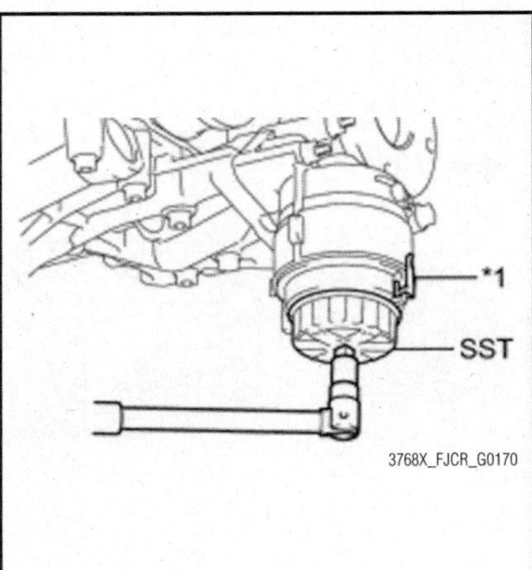

Fig. 199 Using SST 09228-06501, remove the oil filter cap

f. Using SST 09228-06501, tighten the oil filter cap. Tighten to 18 ft. lbs. (25 Nm).

➡When tightening the oil filter cap, do not remove the oil filter bracket clip. Make sure that the oil filter is installed securely. Be careful that the O-ring does not get caught between the parts. Apply a small amount of engine oil to a new drain plug O-ring, and install it to the oil filter cap.

➡Before installing the O-ring, remove any dirt or foreign matter from the installation surface of the oil filter cap.

g. Install the oil filter drain plug. Tighten to 9 ft. lbs. (13 Nm).

➡Be careful that the O-ring does not get caught between the parts.

4. Add 6.4 quarts of fresh oil.
5. Check engine oil level.
a. Start the engine. Make sure that there are no oil leaks from the area that was worked on.

3.3L Engines

See Figure 202.

⁂ CAUTION

Prolonged and repeated contact with engine oil will result in the removal of natural oils from the skin, leading to dryness, irritation and dermatitis. In addition, used engine oil contains potentially harmful contaminants which may cause skin cancer.

⁂ CAUTION

Precautions should be taken when replacing engine oil to minimize the risk of your skin making contact with used engine oil. Protective clothing and gloves that cannot be penetrated by oil should be worn. The skin should be washed with soap and water, or use waterless hand cleaner, to remove any used engine oil thoroughly. Do not use gasoline, thinners, or solvents.

⁂ WARNING

In order to protect the environment, used oil and used oil filters must be disposed of at designated disposal sites

1. Remove cool air intake duct seal.
2. Remove air cleaner cap sub-assembly.
3. Remove the oil filler cap.
4. Remove the oil drain plug and drain the oil into a container.
5. Clean and install the oil pan drain plug with a new gasket. Tighten to ft. lbs. 33 ft. lbs. (45 Nm).
6. Using SST 09228-07501, remove the oil filter.
7. Install oil filter:
a. Check and clean the oil filter installation surface.
b. Apply clean engine oil to the gasket of a new oil filter.
c. Lightly screw the oil filter into place, and tighten it until the gasket contacts the seat.
d. Using SST, tighten the oil filter.
e. Depending on the work space available, choose from the following:
- If enough space is available, use a torque wrench to tighten the oil filter. Tighten to 13 ft. lbs. (18 Nm).
- If enough space is not available to use a torque wrench, tighten the oil filter a 3/4 turn by hand or use a common wrench.

8. Add 5 quarts of ILSAC multi-grade engine 5W-20 oil.
9. Inspect for oil leak.
10. To complete installation, reverse the remaining removal procedure.

EXHAUST MANIFOLD & CATALYTIC CONVERTER

REMOVAL & INSTALLATION

2.7L Engine

See Figure 201.

1. Before servicing the vehicle, refer to the precautions section.

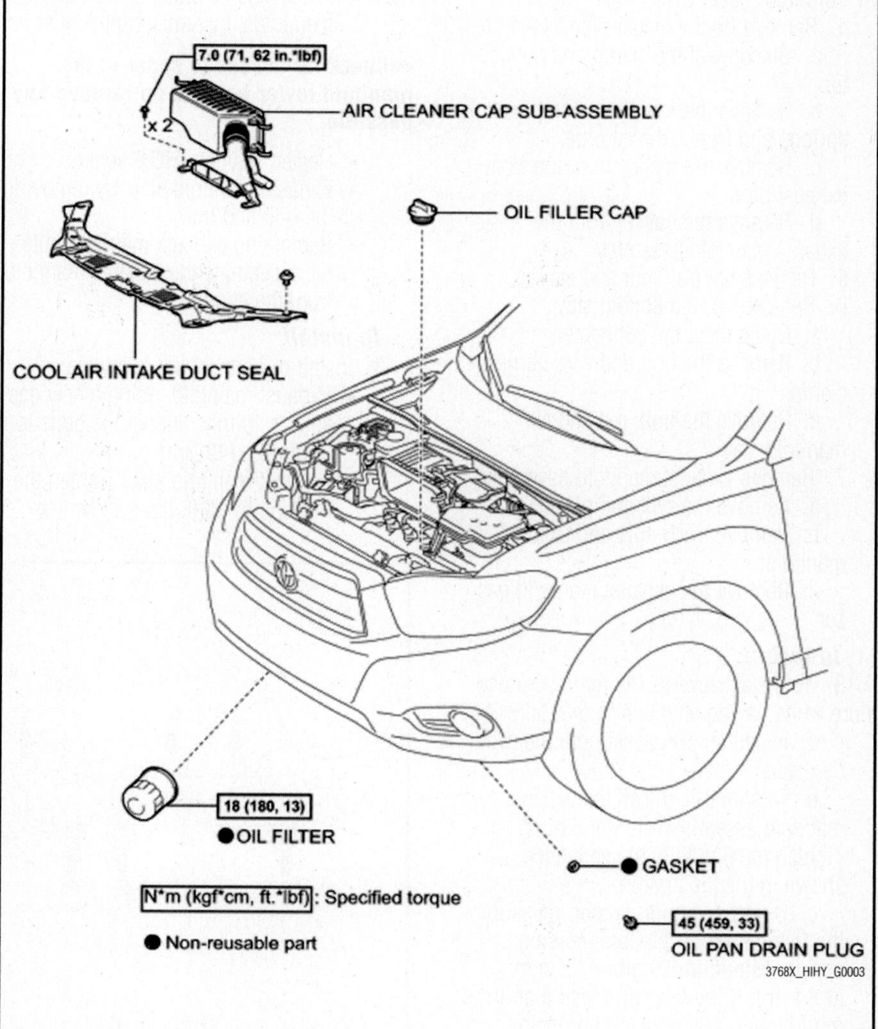

Fig. 200 Locating the oil filter and components

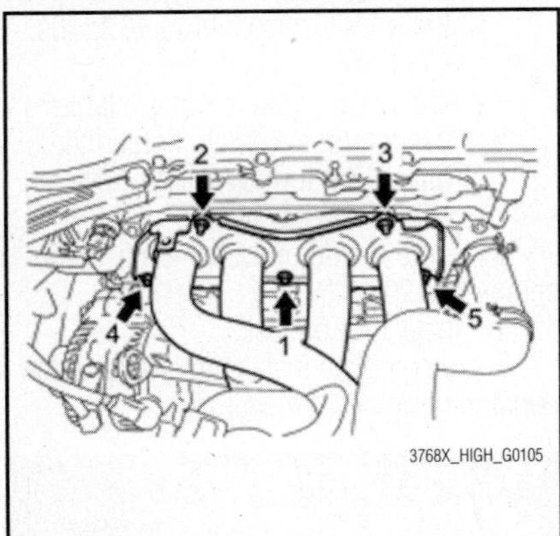

Fig. 201 Exhaust manifold tightening sequence

2. Remove under no. 1 engine cover.
3. Remove front floor cover LH:
   a. Disconnect the 3 clips.
   b. Remove the 2 screws, 4 bolts and front floor cover LH.
4. Remove front exhaust pipe assembly:
   a. Disconnect the clamp and connector.
   b. Remove the 4 bolts, 2 compression springs and front exhaust pipe.
   c. Remove the gasket from the front exhaust pipe.
   d. Remove the gasket from the exhaust manifold assembly.
5. Remove the bolt, nut and stay.
6. Remove no. 2 manifold stay:
   a. Disconnect the connector.
   b. Remove the bolt and wire harness clamp.
   c. Remove the bolt, nut and No. 2 manifold stay.
7. Remove exhaust manifold assembly:
   a. Remove the bolt and plate.
   b. Remove the 5 nuts and exhaust manifold.
   c. Remove the exhaust manifold gasket.

To install:

8. To install, reverse the removal procedure while paying attention to the following:
   a. Install a new gasket onto the cylinder head.
   b. Temporarily install the exhaust manifold assembly with the 5 nuts. Tighten the 5 nuts in the sequence shown in the illustration.
   c. Using a vernier caliper, measure the free length of the compression spring. Minimum length: 41.5 mm (1.64 in.). If the length is less than the minimum, replace the compression spring.
   d. Using a plastic hammer and wooden block, tap in the new gasket until its surface is flush with the exhaust manifold assembly.

➡Be sure to install the gasket in the correct direction. Do not reuse the gasket. Do not damage the gasket. Do not push in the gasket by using the exhaust pipe when connecting it.

3.3L Engine

Front

See Figure 202.

➡Removing the oil filter helps gain access to a lower bolt in the front exhaust manifold.

1. Before servicing the vehicle, refer to the Precautions Section.
2. Remove or disconnect the following:
   - Negative battery cable
   - Engine under covers
   - Front exhaust pipe from the exhaust manifolds, by removing the nuts

➡Check for access to some of the manifold lower bolts, if so remove any possible.

   - Heated Oxygen (HO_2) sensor
   - Exhaust manifold stay, by removing the bolt and nut
   - Remaining exhaust manifold nuts; then, separate the exhaust manifold from the engine

To install:

3. Install or connect the following:
   - Exhaust manifold, using a new gasket. Uniformly, tighten the bolts to 36 ft. lbs. (49 Nm).
   - Exhaust manifold stay. Tighten the nut/bolt to 15 ft. lbs. (20 Nm).
   - Heated Oxygen (HO_2) sensor to the exhaust manifold
   - Front exhaust pipe to the exhaust manifold, using a new gasket. Tighten both nuts to 41 ft. lbs. (56 Nm).
   - Engine under covers
   - Negative battery cable
4. If necessary, perform the initialization procedure.

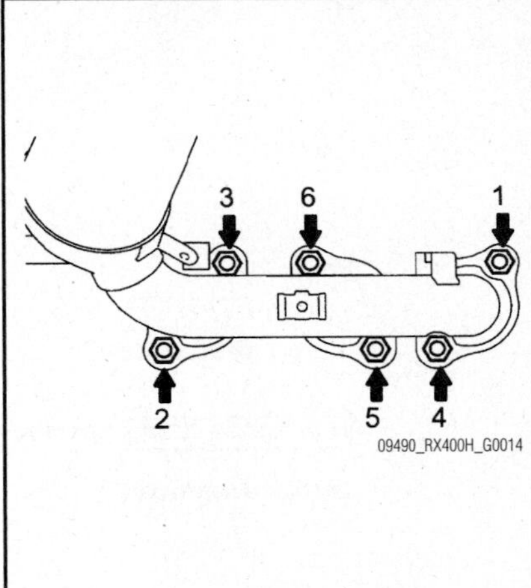

Fig. 202 Front manifold nut locations

Rear

See Figure 203.

1. Before servicing the vehicle, refer to the Precautions Section.
2. Remove or disconnect the following:
   - Negative battery cable
   - Engine under covers
   - Front exhaust pipe from both exhaust manifolds, from below the engine
   - Exhaust Gas Recirculation (EGR) pipe from the rear exhaust manifold, by removing the 4 nuts
   - Heated Oxygen (HO_2) sensor wiring, from the right exhaust manifold
   - Exhaust manifold stay
   - 6 exhaust manifold nuts and the exhaust manifold

To install:

3. Install or connect the following:
   - Exhaust manifold to the engine, using a new gasket. Tighten the 6 nuts to 36 ft. lbs. (49 Nm).
   - Exhaust manifold stay. Tighten the nut/bolt to 25 ft. lbs. (34 Nm).

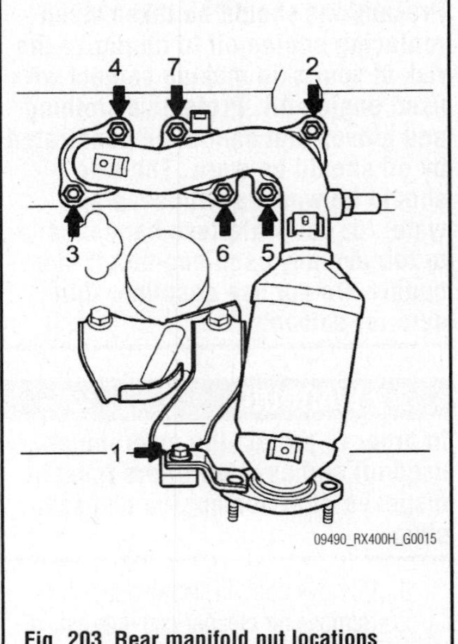

Fig. 203 Rear manifold nut locations

- HO_2sensor wiring to the exhaust manifold
- EGR pipe to the exhaust manifold and the engine, using new gaskets. Tighten the 4 nuts to 108 inch lbs. (12 Nm).
- Front exhaust pipe to the exhaust manifold, use a new gasket. Tighten both nuts to 41 ft. lbs. (56 Nm).
- Engine under covers
- Negative battery cable

4. If necessary, perform the initialization procedure.

3.5L Engine

See Figures 204 through 207.

1. Before servicing the vehicle, refer to the precautions section.
2. Remove the right front wheel.
3. Remove the V-bank cover sub-assembly.
4. Remove the engine under cover assembly.
5. Remove the No. 1 and 2 engine under covers.
6. Drain engine coolant.
7. Disconnect the No. 1 radiator hose.
8. Remove the radiator reserve tank assembly, as follows:
   a. Disconnect the hose.
   b. Remove the 2 bolts and the radiator reserve tank assembly.
9. Remove the No. 2 oil level dipstick guide.
10. Remove the air fuel ratio sensor (for Bank 2 Sensor 1).
11. Remove the 3 bolts and No. 2 exhaust manifold heat insulator.
12. For 4WD vehicles, remove the propeller with center bearing shaft assembly.
13. Remove the tail exhaust pipe assembly.
14. Remove the center exhaust pipe assembly.
15. Remove the front No. 3 exhaust pipe sub-assembly.
16. Remove the front exhaust pipe assembly.
17. Remove the bolt, nut and No. 2 manifold stay.
18. Remove the 6 nuts and left hand exhaust manifold sub-assembly.
19. Remove the gasket.
20. Remove the bolt, nut and manifold stay.
21. Remove the right hand exhaust manifold sub-assembly, as follows:
   a. Disconnect the air fuel ratio sensor (for bank 1 sensor 1) connector and remove the clamp.

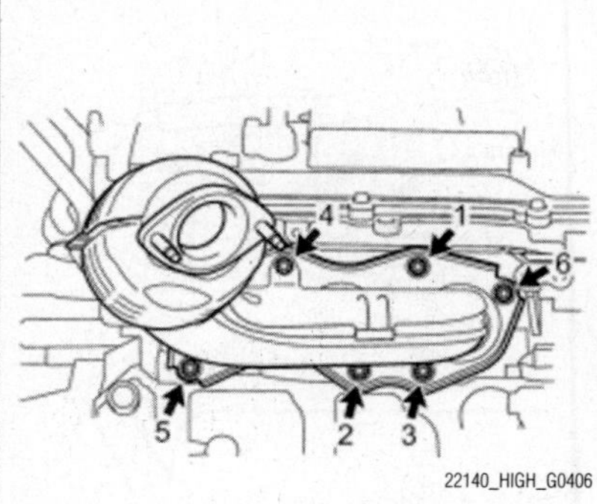

Fig. 204 Right exhaust manifold sub-assembly tightening sequence—3.5L engine

   b. Remove the 6 nuts and the right hand exhaust manifold sub-assembly.
   c. Remove the gasket.
22. Remove the air fuel ratio sensor (for Bank 1 Sensor 1).

To install:

23. Install the air fuel ratio sensor (for Bank 1 Sensor 1).
24. Install the right hand exhaust manifold sub-assembly, as follows:
   a. Install a new gasket.
   b. Install the right hand exhaust manifold sub-assembly by tightening the 6 nuts in the order shown to 15 ft. lbs (21 Nm).
   c. Connect the air fuel ratio sensor (for Bank 1 Sensor 1) connector and install the clamp.
25. Install the manifold stay with the bolt and nut and tighten to 25 ft. lbs (34 Nm), 26 ft. lbs (35 Nm).
26. Install the left hand exhaust manifold sub-assembly, as follows:

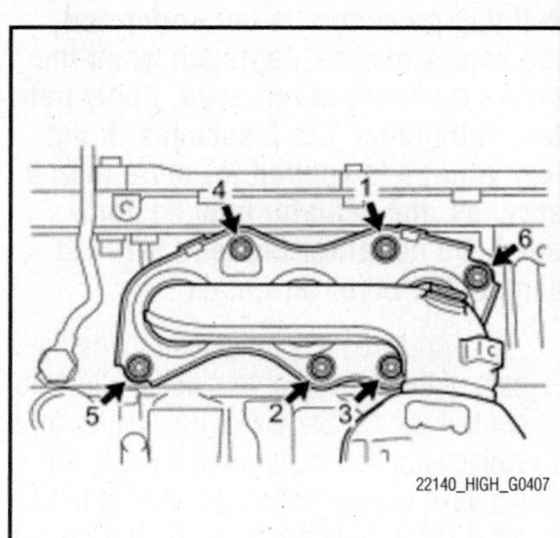

Fig. 205 Left exhaust manifold sub-assembly tightening sequence—3.5L engine

Fig. 206 No. 2 manifold stay tightening sequence—3.5L engine

   a. Install a new gasket.
   b. Install the left hand exhaust manifold sub-assembly by tightening the 6 nuts in the order shown to 15 ft. lbs (21 Nm).
27. Install the No. 2 manifold stay by tightening the bolt and nut in the order shown to 25 ft. lbs (34 Nm).
28. Install the front exhaust pipe assembly.
29. Install the front No. 3 exhaust pipe sub-assembly.
30. Install the center exhaust pipe assembly.
31. Install the tail exhaust pipe assembly.
32. For 4WD vehicles, temporarily tighten the propeller with center bearing shaft assembly.
33. For 4WD vehicles, fully tighten the propeller with center bearing shaft assembly.
34. Install the No. 2 exhaust manifold

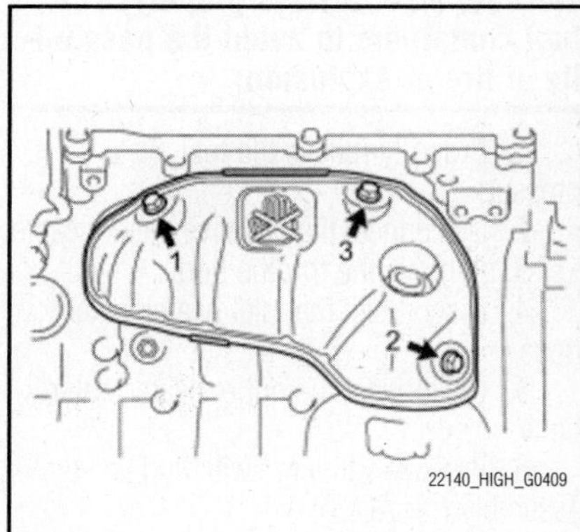

Fig. 207 No. 2 exhaust manifold heat insulator tightening sequence—3.5L engine

heat insulator by tightening the 3 bolts in the order shown to 75 inch lbs (8.5 Nm).

35. Install the air fuel ratio sensor (for Bank 2 Sensor 1).
36. Install the No. 2 oil level dipstick guide.
37. Install the radiator reserve tank assembly with the 2 bolts and tighten to 48 inch lbs (5.4 Nm).
38. Connect the hose.
39. Connect the No. 1 radiator hose.
40. Add engine coolant.
41. Inspect for coolant leaks.
42. Inspect for gas leaks, and repair as necessary.
43. For 4WD vehicles, inspect and adjust transfer oil.
44. Install the No. 1 and 2 engine under covers.
45. Install the engine under cover assembly.
46. Install the V-bank cover sub-assembly.
47. Install the right front wheel.

INTAKE MANIFOLD

REMOVAL & INSTALLATION

2.7L Engine

See Figures 208 through 210.

CAUTION

Observe all applicable safety precautions when working around fuel. Whenever servicing the fuel system, always work in a well-ventilated area. Do not allow fuel spray or vapors to come in contact with a spark or open flame. Keep a dry chemical fire extinguisher near the work area. Always keep fuel in a container specifically designed for fuel storage; also, always properly seal fuel containers to avoid the possibility of fire or explosion.

1. Properly relieve the fuel system pressure.
2. Disconnect the negative battery cable.
3. Remove the throttle body.
4. Disconnect the inlet heater water hose.
5. Disconnect the outlet heater water hose.
6. Remove vacuum switching valve assembly (for ACIS):
    a. Disconnect the 2 vacuum hoses, 2 union to connector tube hoses, clamp and connector.
    b. Remove the bolt and vacuum switching valve assembly (for ACIS).

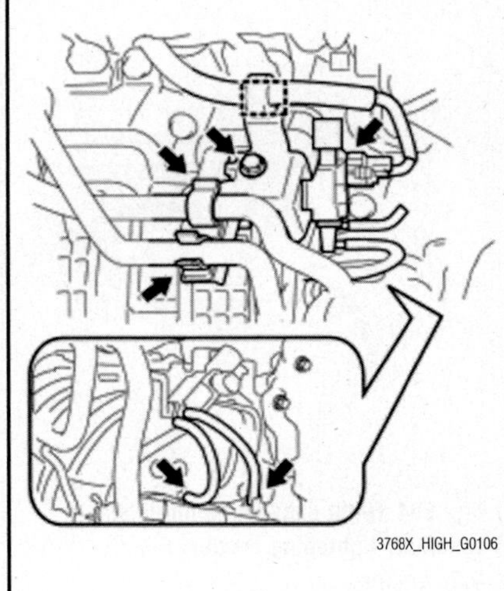

Fig. 208 Remove vacuum switching valve assembly (for ACIS)

7. Disconnect the No. 2 ventilation hose from the intake manifold.
8. Disconnect the union to connector tube hose from the intake manifold.
9. Remove no. 1 vacuum switching valve assembly:
    a. Slide the 2 clips and disconnect the 2 vacuum hoses and connector.
    b. Remove the bolt and No. 1 vacuum switching valve assembly.
10. Remove the fuel rail and injectors.
11. Remove intake manifold:
    a. Disconnect the 4 clamps and wire harness.
    b. Remove the 2 bolts and 2 wire harness brackets.
    c. Disconnect the fuel vapor feed hose, vacuum hose, clamp and connector.
    d. Remove the bolt and wire harness bracket.

➡If this procedure is not performed, the valves may be damaged when the intake manifold is removed. Apply battery voltage for 1 to 3 seconds. If battery voltage is applied for more than 3 seconds, the actuator may be damaged. Do not allow the lead wires to contact the other terminals.

    e. Apply battery voltage to the terminals of the connector to close the tumble control valves. Standard: Positive battery voltage applied to terminal 8 (M-), and negative battery voltage applied to terminal 4 (M+). Specified Condition: Open → Closed.
    f. Remove the bolt.
    g. Detach the 2 clamps from the intake manifold and bracket.

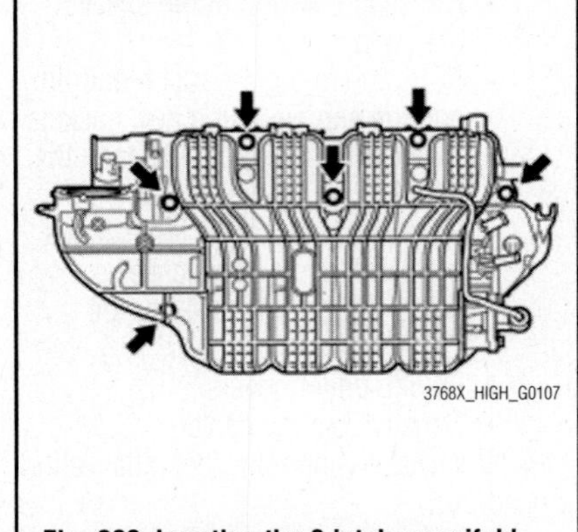

Fig. 209 Locating the 6 intake manifold bolts

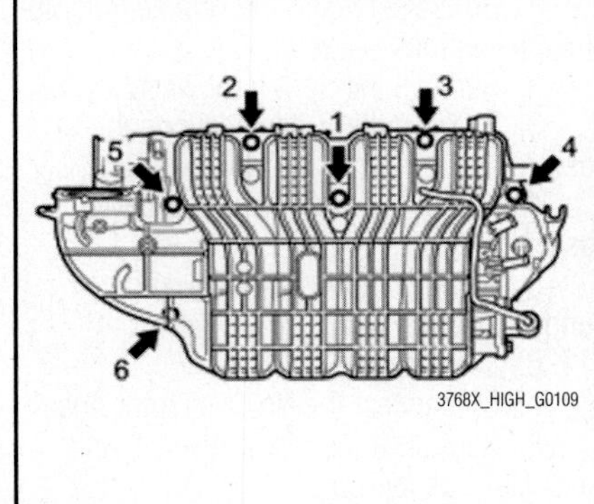

Fig. 210 Intake manifold tightening sequence—2.7L Engines

    h. Disconnect the intake air control valve actuator connector.

➡The valves may be damaged if they are not closed before removing the intake manifold.

    i. Remove the 6 bolts and intake manifold.
    j. Remove the intake manifold gasket from the intake manifold.

To install:

12. To install, reverse the removal procedure while taking note of the following:
    a. Close the tumble control valves. The valves may be damaged if they are not closed before installing the intake manifold. Connect the battery to the terminals of the actuator and operate the motor to close the valves.
    b. Install a new intake manifold gasket. Tighten the intake manifold bolts in sequence to 15 ft. lbs. (21 Nm).

3.3L Engine

See Figures 211 and 212.

1. Before servicing the vehicle, refer to the Precautions Section.

2. Relieve the fuel system pressure.
3. Drain the engine oil.
4. Drain the coolant from the engine radiator and hybrid transaxle.
5. Remove the service plug grip, found underneath the Battery Service cover on the rear seat. Wait 5 minutes to discharge the high voltage capacitor.
6. Remove or disconnect the following:

- Negative battery cable
- Engine cover
- Right-hand front wheel
- Engine splash shields
- Right-hand front fender apron seal
- Wiper and blade assembly
- Top cowl ventilator louver assembly
- Wiper motor and link assembly
- Battery and battery tray
- Air intake assembly
- Converter with Inverter assembly
- Emission control valve set
- Intake air surge tank
- Fuel supply hose
- Heater inlet hose
- Intake manifold ground cable
- Fuel injector connectors

7. Loosen the intake manifold mounting bolts in several steps, in sequence as shown.
8. Remove the intake manifold and gaskets.

To install:

9. Install the intake manifold and gaskets. Tighten the bolts in sequence to 11 ft. lbs. (15 Nm).
10. Install or connect the following:

- Fuel injector connectors

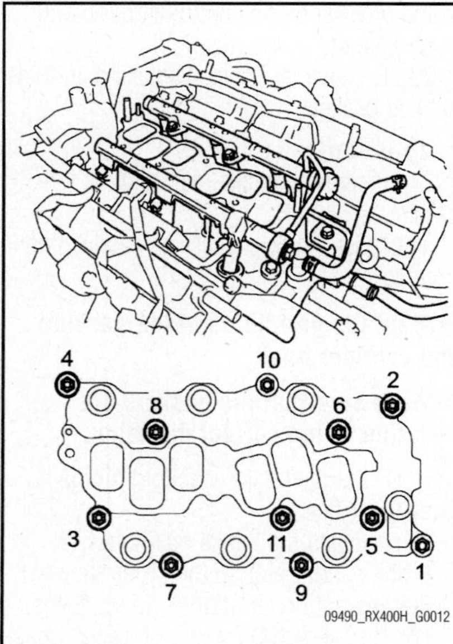

Fig. 211 Intake manifold removal sequence

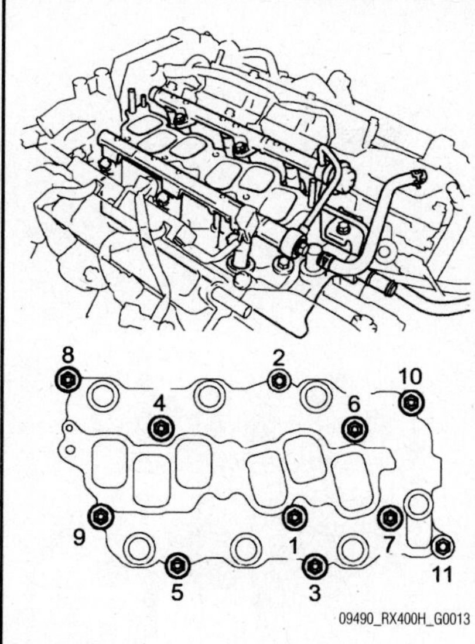

Fig. 212 Intake manifold installation sequence

- Intake manifold ground cable. Tighten to 11 ft. lbs. (15 Nm).
- Heater inlet hose. Tighten to 74 inch lbs. (8.4 Nm).
- Fuel supply hose
- Intake air surge tank
- Emission control valve set
- Converter with Inverter assembly
- Air intake assembly
- Battery and battery tray
- Wiper motor and link assembly
- Top cowl ventilator louver assembly
- Wiper and blade assembly
- Right-hand front fender apron seal
- Engine splash shields
- Right-hand front wheel
- Engine cover
- Negative battery cable

11. Refill the engine oil to the correct level.
12. Refill the coolant to the engine radiator and hybrid transaxle to the correct level.
13. Replace the service plug grip.
14. If necessary, perform the initialization procedure.
15. Start the engine and check for leaks.

3.5L Engine

See Figures 213 through 223.

1. Discharge fuel system pressure.
2. Remove the engine under cover assembly.
3. Remove the No. 1 engine under cover.
4. Drain engine coolant.
5. Remove the V-bank cover sub-assembly.

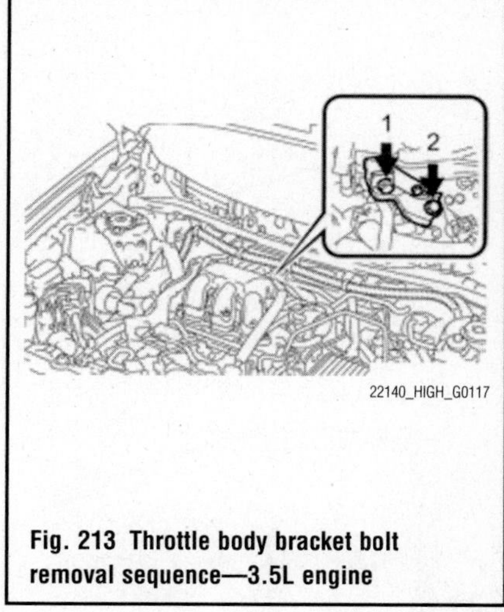

Fig. 213 Throttle body bracket bolt removal sequence—3.5L engine

6. Remove both front wiper arm and blade assemblies.
7. Remove the cowl top ventilator louver sub-assembly.
8. Remove the windshield wiper motor and link assembly.
9. Remove the outer cowl top panel sub-assembly.
10. Remove the air cleaner cap sub-assembly.
11. Disconnect the engine room main wire, as follows:
    a. Disconnect the 5 harness clamps.
12. Remove throttle body bracket:
    a. Remove the 2 bolts in the order shown in the illustration and remove the throttle body bracket.
13. Remove the 2 bolts in the order shown in the illustration and remove the No. 1 surge tank stay.
14. Remove the intake air surge tank assembly, as follows:
    a. Disconnect the 4 hoses.

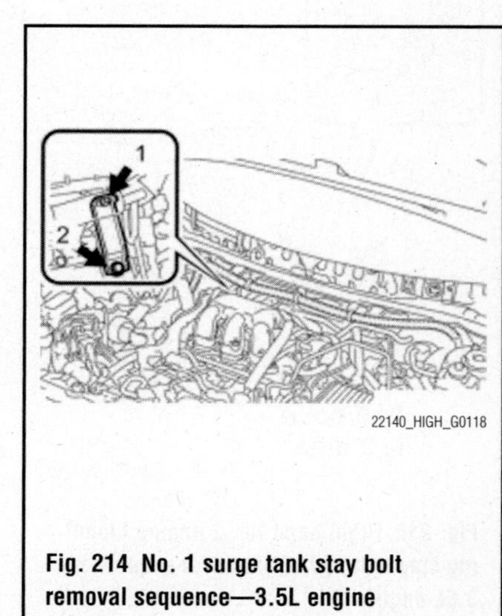

Fig. 214 No. 1 surge tank stay bolt removal sequence—3.5L engine

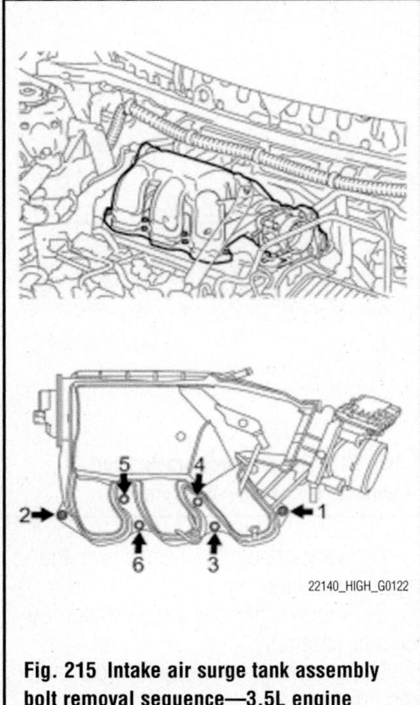

Fig. 215 Intake air surge tank assembly bolt removal sequence—3.5L engine

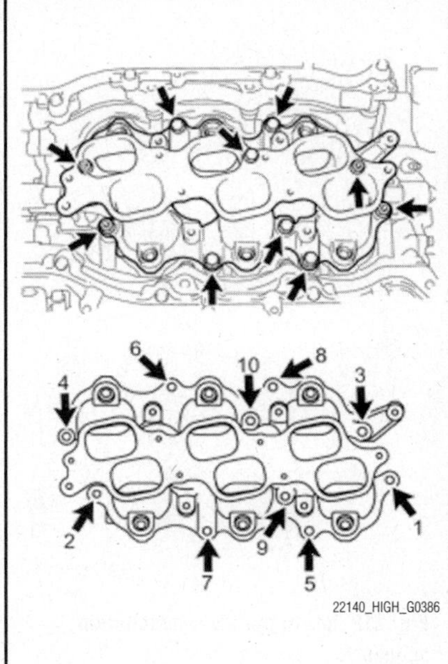

Fig. 217 Intake manifold bolt removal sequence—3.5L engine

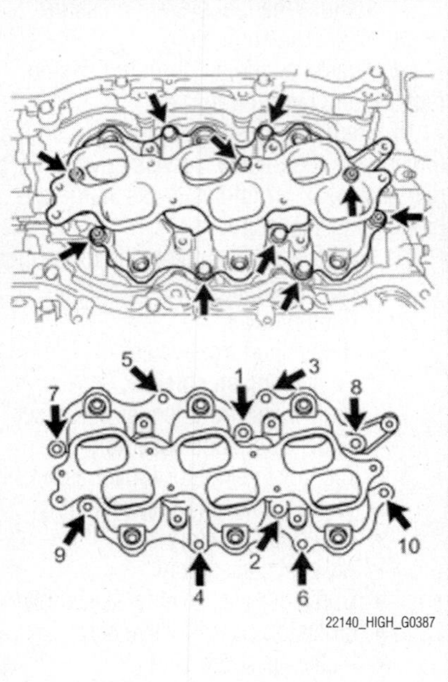

Fig. 218 Intake manifold bolt installation sequence—3.5L engine

b. Disconnect the throttle body connector and clamp.

c. Disconnect the connector.

d. Disconnect the 2 water by-pass hoses from the throttle body.

e. Remove the 4 bolts and 2 nuts in the order shown in the illustration.

➡Use a 5 mm socket hexagon wrench to remove the 4 bolts

f. Remove the gasket from the intake air surge tank.

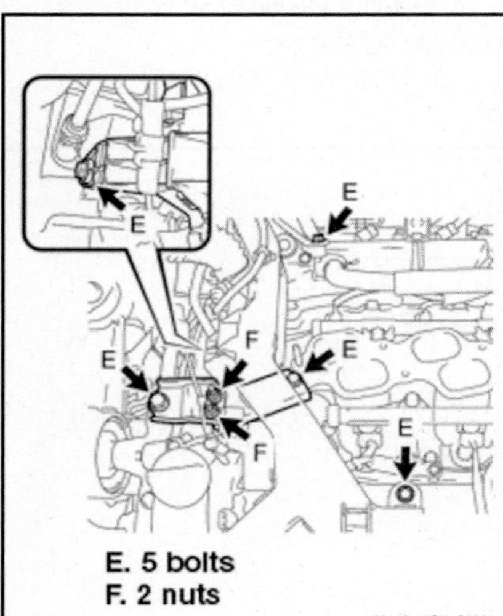

Fig. 216 Right hand No. 2 engine mounting stay bolt and nut removal sequence—3.5L engine

15. Remove the right hand No. 2 engine mounting stay, as follows:

a. Remove the 5 bolts (E).

b. Remove the 2 nuts (F).

c. Remove the right hand No. 2 engine mounting stay.

16. Disconnect the fuel main tube, as follows:

a. Remove the No. 2 fuel pipe clamp.

b. Pinch the tube connector and pull out the fuel pipe.

➡Check that there is no dirt or other foreign objects around the connector before disconnecting it. Clean the connector as necessary.

➡It is necessary to prevent dirt or foreign objects from entering the quick connector. If dirt or foreign objects enter the connector, the O-rings may not seal properly.

➡Only disconnect the quick connector by hand.

➡Do not bend, kink or twist the nylon tubes. Protect the connector by covering it with a plastic bag.

➡If the pipe and the connector are stuck, carefully try wiggling or pushing and pulling on the connector to release it. Pull the connector off the pipe carefully.

17. Remove the fuel delivery pipe sub-assembly, as follows:

a. Disconnect the 6 fuel injector connectors.

18. Remove the 5 bolts and fuel delivery pipe sub-assembly together with the 6 fuel injectors.

➡Be careful not to drop the fuel injectors when removing the fuel delivery pipe sub-assembly.

a. Remove the 6 injector vibration insulators from the intake manifold.

19. Remove the intake manifold, as follows:

a. Remove the 6 bolts and 4 nuts in the order shown in the illustration and remove the intake manifold.

20. Remove the 2 No. 1 intake manifold to head gaskets.

To install:

21. Install the intake manifold, as follows:

a. Set 2 new gaskets on each cylinder head.

➡Align the port holes of the gaskets and cylinder head.

➡Make sure that the gaskets are installed in the correct direction.

b. Set the intake manifold on the cylinder heads.

c. Install the intake manifold with the 6 bolts and 4 nuts in the order shown in the illustration and tighten to 15 ft. lbs. (21 Nm). **DO NOT** apply oil to the bolts.

22. Install the right hand No. 2 engine mounting stay.

23. Install the bolt (A), and tighten to 15

ft. lbs. (21 Nm). **DO NOT** apply oil to the bolt.

24. Install the 2 nuts (B), and tighten to 17 ft. lbs. (23 Nm).

25. Install the bolt (C), and tighten to 28 ft. lbs. (38 Nm). **DO NOT** apply oil to the bolt.

26. Install the 3 bolts (D), and tighten to 73 inch lbs. (8.3 Nm). **DO NOT** apply oil to the bolts.

27. Install the fuel delivery pipe sub-assembly, as follows:

a. Install 6 new insulators to the intake manifold.

b. Place the fuel delivery pipe which has the 6 fuel injectors installed to it in position on the intake manifold.

➡Be careful not to drop the fuel injectors when installing the fuel delivery pipe.

c. Temporarily install the 5 bolts which are used to hold the fuel delivery pipe to the intake manifold. **DO NOT** apply oil to the bolts.

➡After installing the fuel injectors, check that they turn smoothly. If not, reinstall the injectors with new O-rings.

d. Tighten the 5 bolts which are used to hold the fuel delivery pipe to the intake manifold to 15 ft. lbs. (21 Nm). **DO NOT** apply oil to the bolts.

e. Connect the 6 fuel injector connectors.

28. Connect the fuel main tube, as follows:

a. Push in the tube connector onto the pipe until the tube connector clicks.

➡Before connecting the tube, make sure that it is not damaged. Make sure that there is no dirt present on the connecting surfaces.

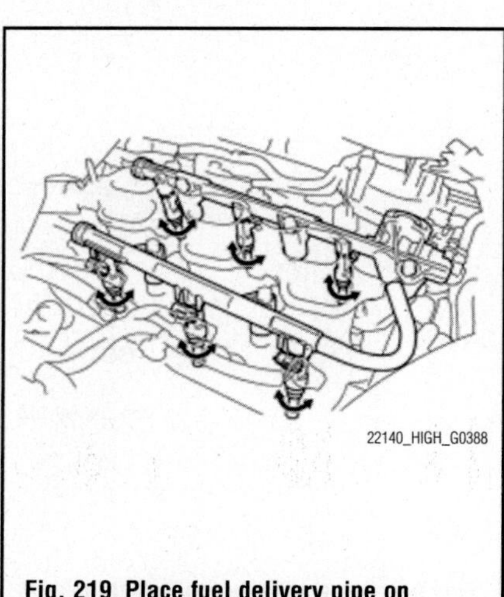

Fig. 219 Place fuel delivery pipe on intake manifold—3.5L engine

➡After connecting, check that the fuel tube connector and the pipe are securely connected by pulling on them.

b. Install the No. 2 fuel pipe clamp.

29. Temporarily install the No. 1 surge tank stay, as follows:

a. Temporarily install the intake air surge tank assembly with 3 new gaskets on the intake manifold.

➡Do not allow the gaskets to slip out of place during installation.

b. Temporarily install the No. 1 surge tank stay with the 2 bolts. **DO NOT** apply oil to the bolts.

30. Temporarily install the throttle body bracket with the 2 bolts. **DO NOT** apply oil to the bolts.

31. Install the intake air surge tank assembly, as follows:

a. Install the surge tank with the 4 bolts and 2 nuts in the order shown in the illustration and tighten to 12 ft. lbs. (16 Nm) and 13 ft. lbs. (18 Nm). **DO NOT** apply oil to the bolts.

➡Use a 5 mm hexagon socket wrench to tighten the 4 bolts.

b. Connect the 2 water by-pass hoses to the throttle with motor body assembly.

c. Connect the connector.

d. Install the clamp and connect the throttle with motor body assembly connector.

e. Connect the 4 hoses.

32. Fully tighten the No. 1 surge tank stay, as follows:

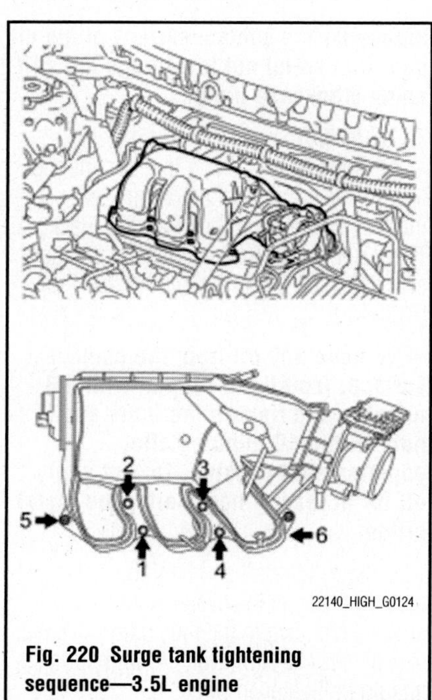

Fig. 220 Surge tank tightening sequence—3.5L engine

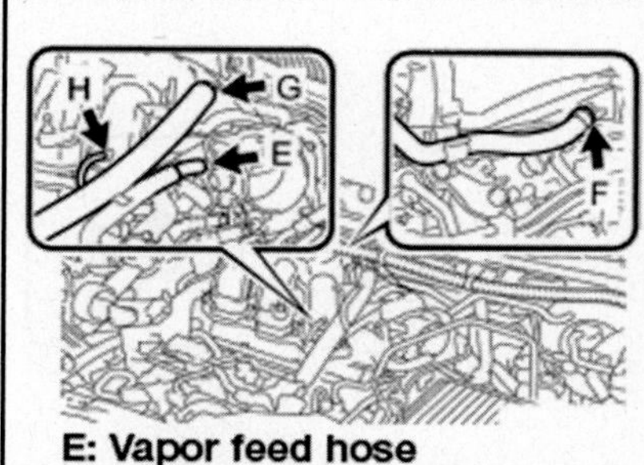

Fig. 221 Intake air surge tank assembly hoses—3.5L engine

a. Fully tighten the 2 bolts in the order shown in the illustration to 15 ft. lbs. (21 Nm). **DO NOT** apply oil to the bolts.

33. Fully tighten the throttle body bracket, as follows:

a. Fully tighten the 2 bolts in the order shown in the illustration to 15 ft. lbs. (21 Nm). **DO NOT** apply oil to the bolts.

34. Connect the 5 engine room main wire harness clamps.

35. Install the air cleaner cap sub-assembly.

36. Install the outer cowl top panel sub-assembly.

37. Install the windshield wiper motor and link assembly.

38. Install the cowl top ventilator louver sub-assembly.

39. Install the both front wiper arm and blade assemblies.

40. Add engine coolant.

Fig. 222 No. 1 surge tank stay tightening sequence—3.5L engine

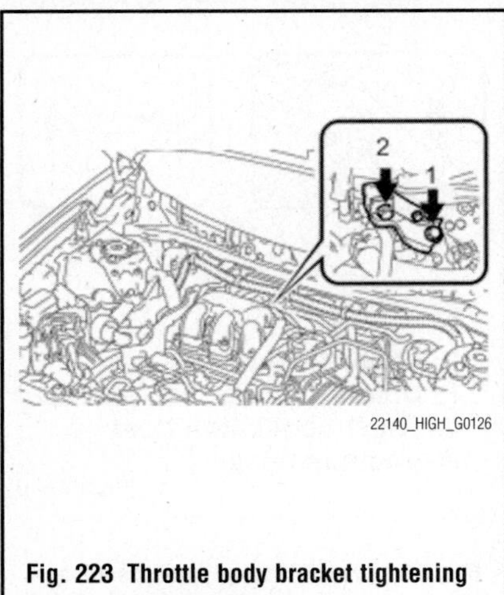

Fig. 223 Throttle body bracket tightening sequence—3.5L engine

41. Inspect for engine coolant leak.
42. Inspect for fuel leak.
43. Install the V-bank cover sub-assembly.
44. Install the No. 1 engine under cover.
45. Install the engine under cover assembly.

OIL PAN

REMOVAL & INSTALLATION

2.7L Engine

See Figures 224 through 226.

1. Drain the oil in to a suitable container.
2. Remove the 11 bolts and 2 nuts.
3. Insert the blade of an oil pan seal cutter between the oil pan and stiffening crankcase, cut off the applied sealer and remove the oil pan.

➡Be careful not to damage the stiffening crankcase contact surface of the oil pan. Be careful not to damage the stiffening crankcase flange.

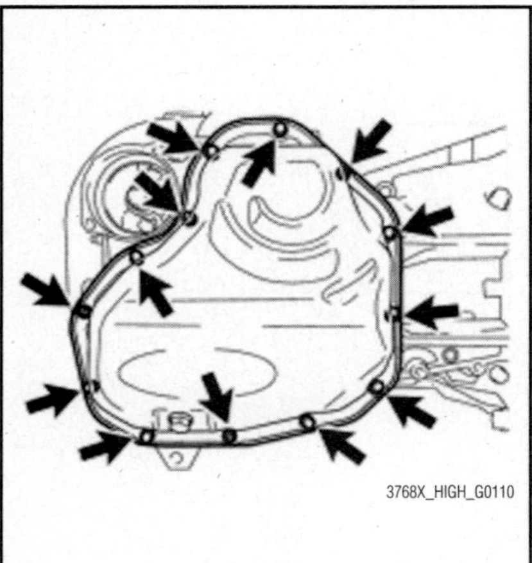

Fig. 224 Remove the 11 bolts and 2 nuts

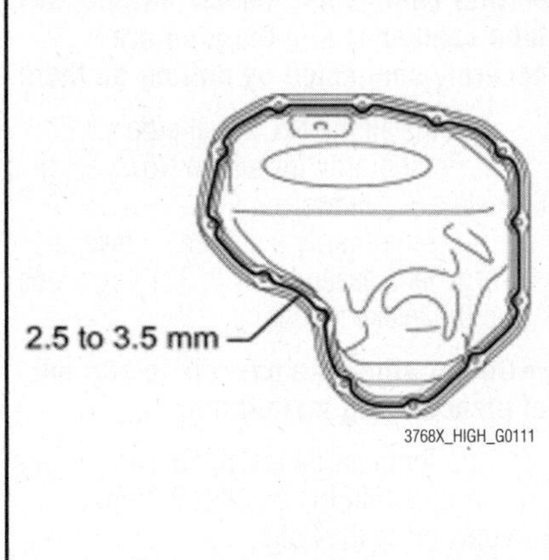

Fig. 225 Apply seal packing in a continuous line as shown

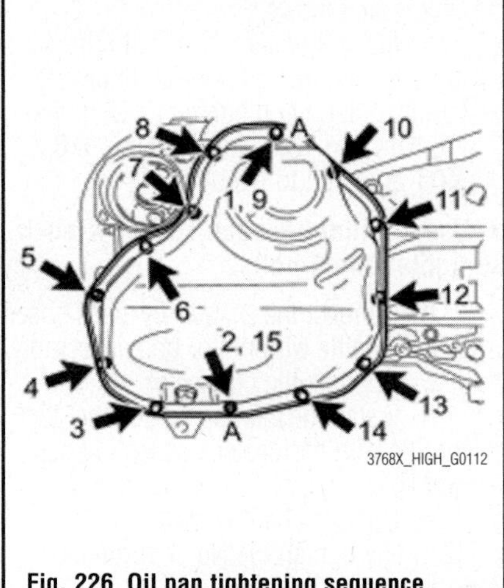

Fig. 226 Oil pan tightening sequence

To install:

4. Apply seal packing in a continuous line as shown in the illustration. Seal packing: Toyota Genuine Seal Packing Black, Three Bond 1207B or equivalent. Standard seal diameter: 2.5 to 3.5 mm (0.0984 to 0.138 in.)

➡Remove any oil from the contact surface. Install the oil pan within 3 minutes and tighten the bolts and nuts within 10 minutes after applying seal packing. Do not apply oil for at least 4 hours after the installation.

5. Install the oil pan by tightening the 11 bolts and 2 nuts in several steps, in the sequence shown in the illustration. Tighten to 7 ft. lbs. (10 Nm). Bolt "A" and nut "A" should be tightened twice.

3.3L Engine

See Figures 227 through 230.

1. Before servicing the vehicle, refer to the Precautions Section.
2. Remove or disconnect the following:
    - Engine/transaxle assembly from the vehicle
    - Right-hand exhaust manifold
    - Transaxle mass damper
    - Front frame assembly
    - Halfshafts
    - Flywheel housing undercover
    - Front engine mounting bracket
    - Transaxle assembly from the engine
    - Transmission input damper assembly
    - Flywheel
3. Install the engine to a suitable engine stand.
4. Remove or disconnect the following:
    - Remaining exhaust manifold heat shields
    - Right-hand engine mounting bracket
    - Compressor mounting bracket
    - Crankshaft pulley
    - Timing belt
    - Timing belt idler
    - Crankshaft timing pulley
    - Oil level gauge assembly
5. Remove the lower oil pan as follows:
    a. Remove the mounting bolts and nuts
    b. Using Special Tool 09032-00100 or suitable seal cutter, cut the sealant between the upper and lower oil pans.
    c. Remove the lower oil pan.
6. Remove the oil strainer and gasket.
7. Remove the upper oil pan as follows:
    a. Uniformly loosen and remove the mounting bolts.
    b. Using a suitable pry tool, pry the upper oil pan from the cylinder block.

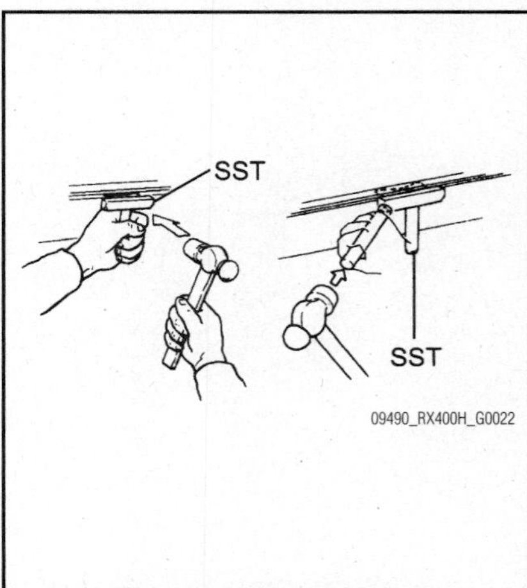

Fig. 227 Use a suitable tool to cut the sealant between the oil pans.

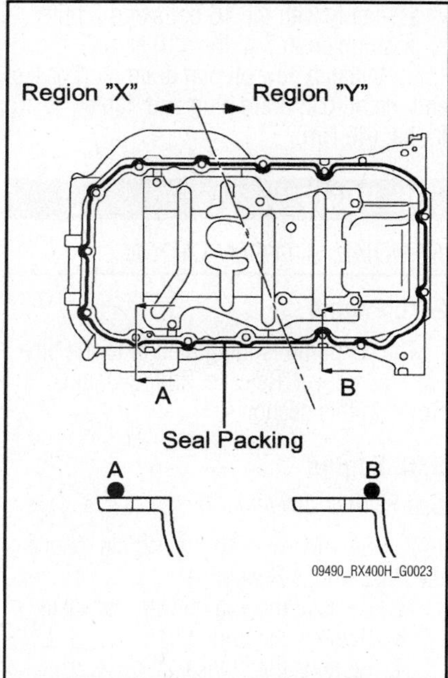

Fig. 228 Apply the sealant to the upper oil pan as shown.

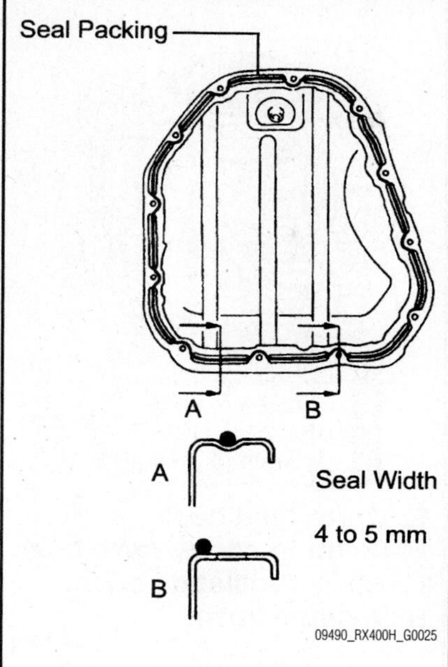

Fig. 230 Apply sealant to the lower oil pan as shown.

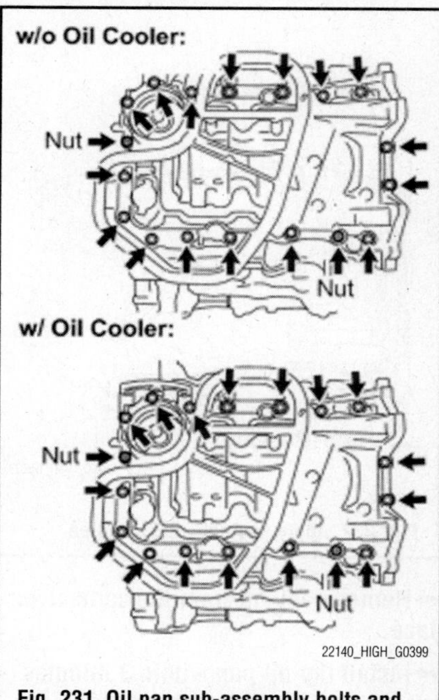

Fig. 231 Oil pan sub-assembly bolts and nuts—3.5L engine

To install:

8. Remove any old sealant from the mating surface of the oil pans.

9. Install the upper oil pan as follows:

a. Apply a 0.12–0.16 inch (3–4 mm) wide continuous bead of sealant to the mating surface as shown in the illustration.

b. Install the upper oil pan mounting bolts and tighten in several steps. Tighten bolts 'A' to 71 inch lbs. (8 Nm) and bolts 'B' to 14 ft. lbs. (20 Nm).

10. Install the oil strainer with a new gasket. Tighten to 71 inch lbs. (8 Nm).

11. Install the lower oil pan as follows:

a. Apply a 0.16–0.20 inch (4–5 mm) wide continuous bead of sealant to the mating surface as shown in the illustration.

b. Install the lower oil pan mounting bolts and nuts. Tighten to 71 inch lbs. (8 Nm).

12. The remainder of the installation is the reverse order of removal.

13. Refill the engine with oil to the correct level.

14. Start the engine and check for leaks.

15. If necessary, perform the initialization procedure.

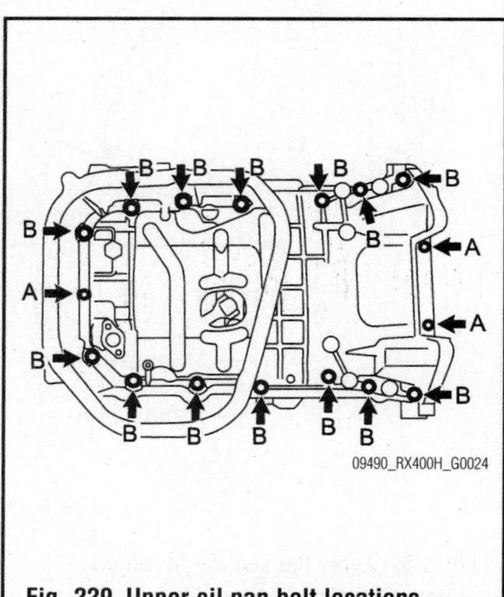

Fig. 229 Upper oil pan bolt locations

3.5L Engine

See Figures 231 through 235.

1. Before servicing the vehicle, refer to the precautions section.

2. Drain the engine oil.

3. Remove the No. 2 oil pan sub-assembly, as follows:

a. Remove the 16 bolts and 2 nuts.

b. Insert the blade of oil pan seal cutter between the oil pans. Cut through the applied sealer and remove the No. 2 oil pan sub-assembly.

➡Be careful not to damage the contact surfaces of the oil pans.

c. Using a Torx® socket wrench E6, remove the 2 stud bolts.

4. Remove the oil strainer sub-assembly, as follows:

a. Remove the bolt, 2 nuts, oil strainer sub-assembly and gasket.

b. Using a Torx® socket wrench E6, remove the 2 stud bolts.

5. Remove the oil pan sub-assembly, as follows:

a. Remove the 16 bolts and 2 nuts.

➡Be sure to clean the bolts and stud bolts and check the threads for cracks or other damage.

b. Remove the oil pan sub-assembly by prying between the oil pan sub-assembly and cylinder block sub-assembly with a screwdriver.

➡Be careful not to damage the contact surfaces of the cylinder block and oil pans.

➡Tape the screwdriver tip before use.

c. Remove the 2 O-rings.

d. Using a Torx® socket wrench E8, remove the 2 stud bolts.

To install:

6. Install the oil pan sub-assembly, as follows:

a. When replacing a stud bolt, install it by using an E8 Torx® socket wrench. Tighten to 7 ft. lbs. (10 Nm).

b. Install 2 new O-rings.

c. Apply seal packing in a continuous line as shown in the illustration. Seal packing: Toyota Genuine Seal Packing Black, Three Bond 1207B or equivalent. Seal diameter: 3.0 to 4.0 mm (0.118 to .0156 inches).

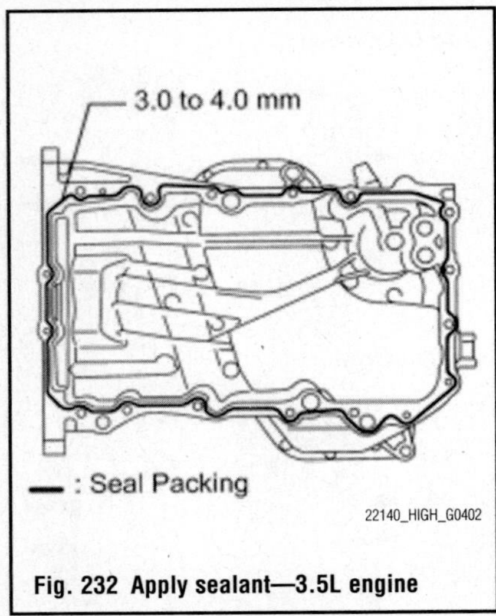

Fig. 232 Apply sealant—3.5L engine

➡Remove any oil from the contact surface.

➡Install the oil pan within 3 minutes after applying seal packing.

➡Do not start the engine for at least 2 hours after installing.

d. Install the oil pan with the 16 bolts and 2 nuts and tighten to 7 ft. lbs. (10 Nm), 15 ft. lbs. (21 Nm).

7. Install the oil strainer sub-assembly, as follows:

a. Using an E6 Torx® socket, install the stud bolts as shown in the illustration and tighten to 35 inch lbs. (4 Nm).

b. Install a new gasket and the oil strainer sub-assembly with the bolt and 2 nuts and tighten to 7 ft. lbs. (10 Nm).

8. Install the No. 2 oil pan sub-assembly, as follows:

a. Using an E6 Torx® socket, install the stud bolts as shown in the illustration and tighten to 35 inch lbs. (4 Nm).

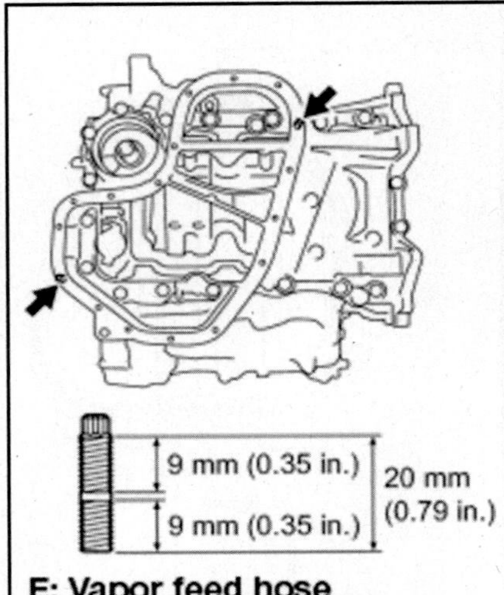

Fig. 234 No. 2 oil pan sub-assembly—3.5L engine

b. Apply seal packing in a continuous line as shown in the illustration. Seal packing: Toyota Genuine Seal Packing Black, Three Bond 1207B or equivalent. Seal diameter: 3.0 to 4.0 mm (0.118 to .0156 inches).

➡Remove any oil from the contact surface.

➡Install the No. 2 oil pan sub-assembly within 3 minutes after applying seal packing.

➡Do not start the engine for at least 2 hours after installing.

c. Install the No. 2 oil pan sub-assembly with the 16 bolts and 2 nuts and tighten to 7 ft. lbs. (10 Nm).

9. Install a new oil pan drain plug gasket and the oil pan drain plug and tighten to 30 ft. lbs. (40 Nm).

OIL PUMP

REMOVAL & INSTALLATION

2.7L Engine

The oil pump is integrated in to the timing chain cover. Refer to Timing Chain Cover in this section.

3.3L Engine

See Figures 236 and 237.

1. Before servicing the vehicle, refer to the Precautions Section.
2. Remove the engine with transaxle.
3. Remove the timing belt.
4. Remove the crankshaft gear.
5. Remove or disconnect the following:
   - Upper and lower oil pans
   - Crankshaft Position (CKP) sensor
   - 9 oil pump bolts

➡Make a note of the position of the each bolt. When replacing the bolts into the oil pump body, place each bolt in the position from which it was removed.

- Oil pump body, by prying between the oil pump and main bearing cap
- O-ring from the cylinder block
- Plug, gasket, spring and relief valve from the oil pump body
- 9 screws, pump body cover, drive and driven rotors

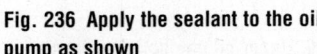

Timing Chain Cover:

16 mm (0.63 in.)

27 mm (1.06 in.)

9 mm (0.35 in.)

22140_HIGH_G0392

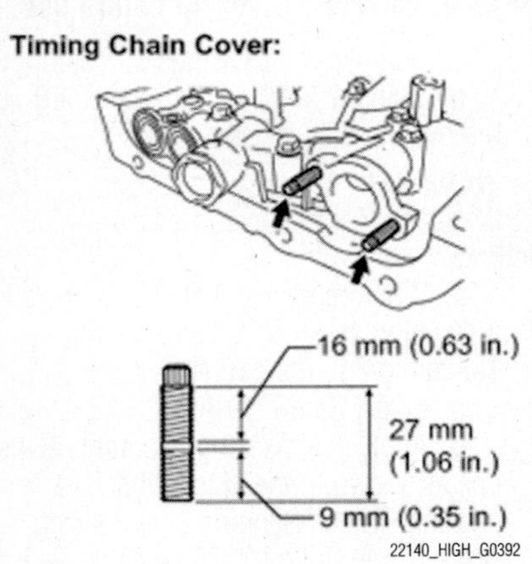

Fig. 233 Oil strainer sub-assembly bolts—3.5L engine

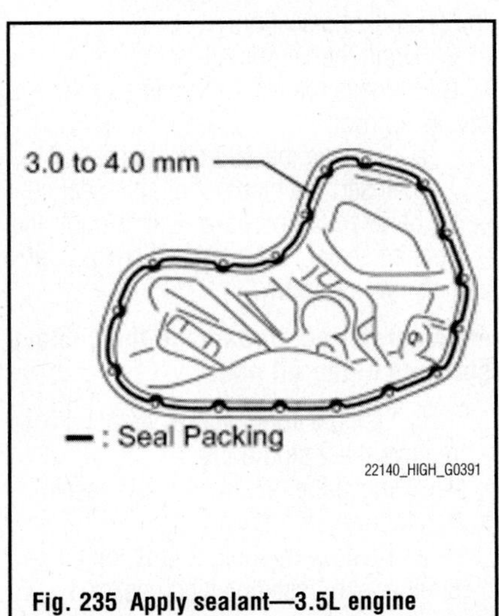

Fig. 235 Apply sealant—3.5L engine

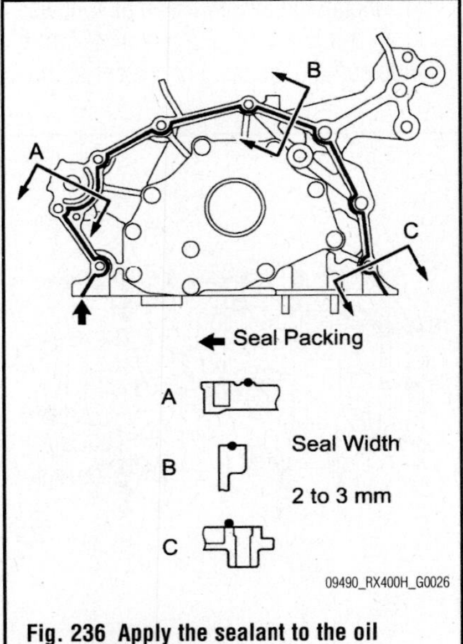

Fig. 236 Apply the sealant to the oil pump as shown

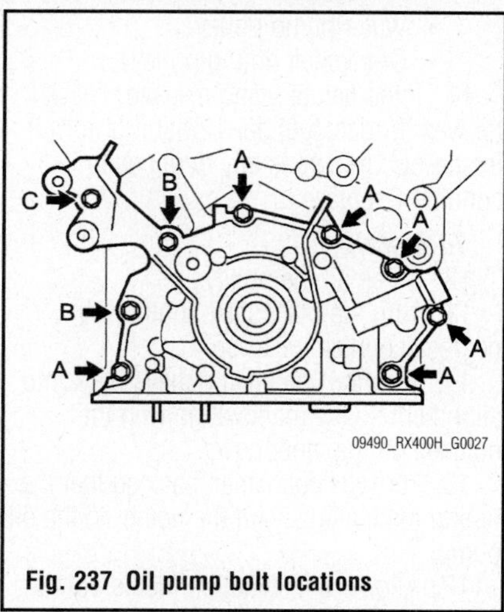

Fig. 237 Oil pump bolt locations

To install:

6. Install or connect the following:
   - Driven rotors, drive, pump body cover, using the 9 screws
   - Oil pump relief valve, spring, gasket and the plug to the oil pump body
   - New O-ring on the cylinder block
7. Using a non-residue solvent, clean both sealing surfaces to the oil pump.
8. Apply liquid sealant to the oil pump and engine block.
9. Install or connect the following:
   - Oil pump

➡Be sure to engage the splined teeth of the oil pump drive gear with the large teeth of the crankshaft.

   - 9 oil pump bolts. Tighten the bolts in several passes to 71 inch lbs. (8 Nm) for bolt 'A'; 14 ft. lbs. (20 Nm), for bolts 'B'; 32 ft. lbs. (43 Nm) for bolt 'C'
   - CKP sensor. Tighten the bolt to 71 inch lbs. (8 Nm).
   - Upper and lower oil pans
   - crankshaft gear
   - Timing belt, and covers
   - Engine with the transaxle.
10. Refill the engine with oil to the correct level.
11. Start the engine and inspect for leaks.
12. Recheck the engine oil level.
13. If necessary, perform the initialization procedure.

PISTON AND RING

POSITIONING

See Figures 238 through 242.

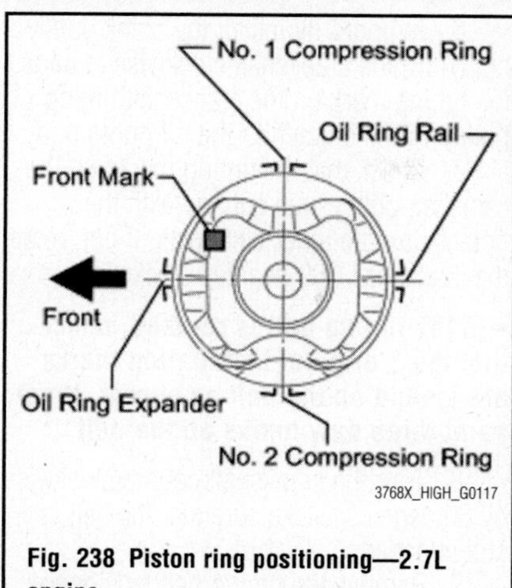

Fig. 238 Piston ring positioning—2.7L engine

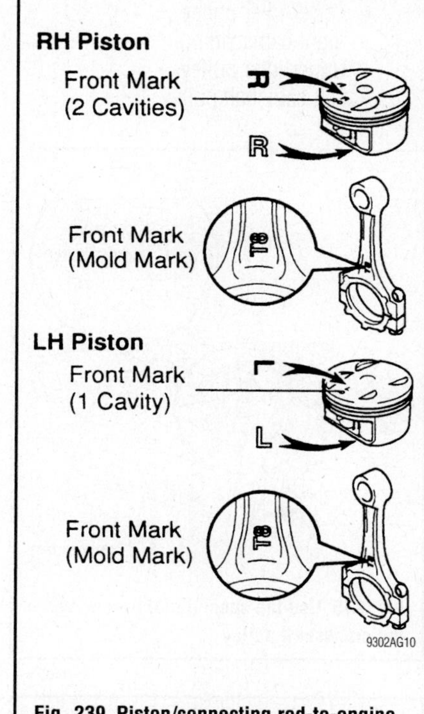

Fig. 239 Piston/connecting rod-to-engine positioning—3.3L engine

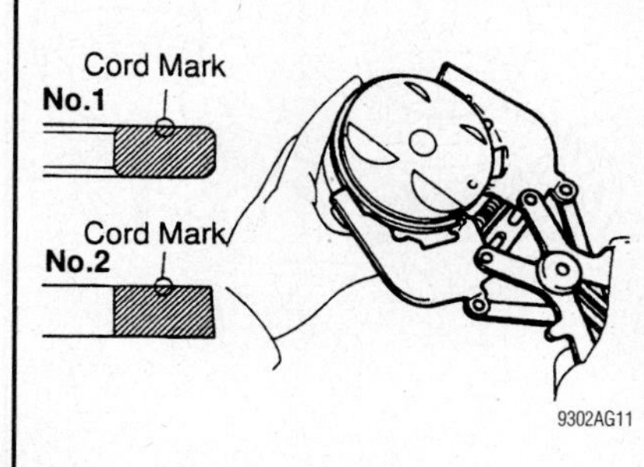

Fig. 240 Piston ring positioning—3.3L engine

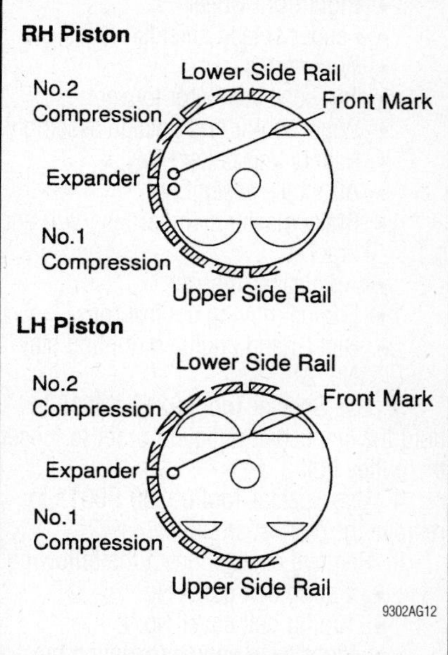

Fig. 241 Piston ring identification—3.3L engine

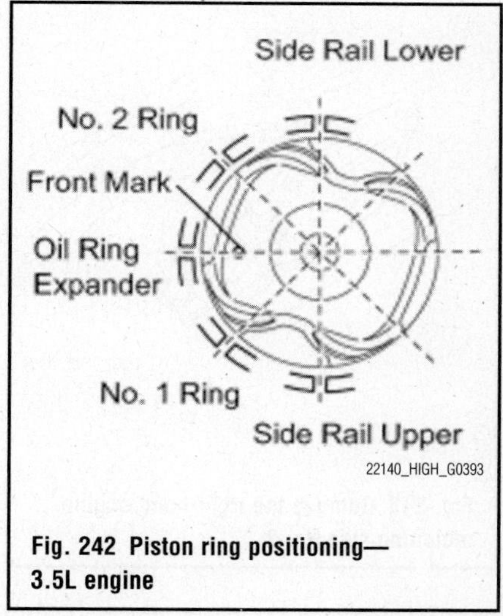

Fig. 242 Piston ring positioning—3.5L engine

TIMING BELT FRONT COVER

REMOVAL & INSTALLATION

Refer to Timing Belt & Sprockets.

TIMING BELT & SPROCKETS

REMOVAL & INSTALLATION

3.3L Engine

See Figures 243 through 255.

1. Before servicing the vehicle, refer to the Precautions Section.
2. Remove or disconnect the following:
   - Negative battery cable
   - Engine covers

- Right front wheel
- Fender splash shields
- Wiper arms
- Top cowl ventilator louver
- Wiper motor and linkage assembly
- Battery and battery tray
- Air intake assembly
- Brake master cylinder reservoir and bracket
- Air cleaner bracket
- Engine moving control rod
- Right-hand engine mounting stay No. 2

3. Use Special Tool 09213-54015 to hold the crankshaft pulley in order to loosen the pulley bolt.
4. Use Special Tool 09950-50013 to remove the crankshaft pulley.
5. Remove or disconnect the following:
   - Timing belt cover No. 1
   - Timing belt cover No. 2
   - Right-hand engine mounting bracket
   - Timing belt guide No. 2
6. Temporarily install the crank pulley bolt. Turn the crankshaft clockwise to align the timing mark on the crankshaft timing pulley with the notch in the oil pump body.
7. Check that the timing marks on the camshaft pulleys are aligned with the notches on the inner belt cover. If not, rotate the crankshaft 360 degrees clockwise.

➡If the timing belt is re-used, check that the 3 original installation marks are visible on the belt as shown. If not, paint three new marks on the belt.

8. Turn the crankshaft counterclockwise by 60 degrees. Make sure that the belt is still engaged.
9. Remove the timing belt tensioner.
10. Remove the belt from the pulleys in this order:
    - Lower idler pulley
    - Right camshaft pulley
    - Upper idler pulley
    - Left camshaft pulley
    - Water pump pulley
    - Crankshaft timing pulley
11. If the belt is being re-used, check it for wear or damage; don't twist it or turn it inside-out. If there is any doubt as to its condition, replace it.

To install:

12. Clean all the pulleys.
13. Turn the crankshaft another 60 degrees counterclockwise.
14. Turn the camshaft pulleys back into alignment so the marks align with the notches on the inner cover.
15. Turn the crankshaft back so that the timing mark aligns with the notch on the oil pump.
16. Align the installation marks on the belt with the timing marks on the pulleys.
17. Install the belt in this order:
    - Crankshaft
    - Water pump
    - Left camshaft

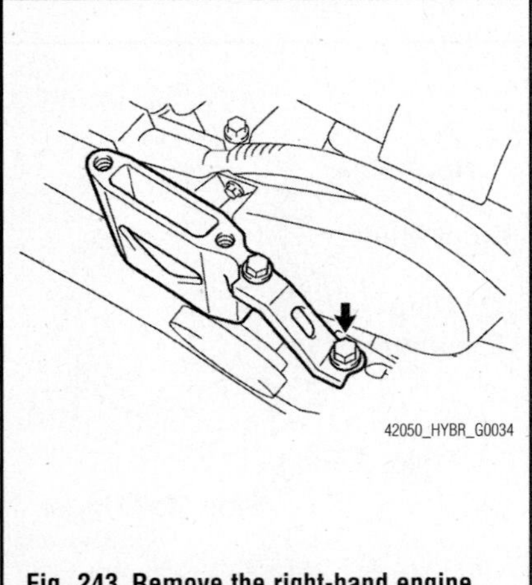

Fig. 243 Remove the right-hand engine mounting stay No. 2

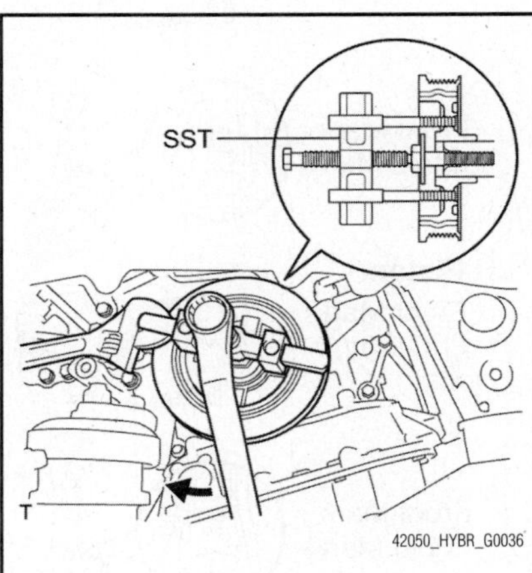

Fig. 245 Use the special tool to remove the crankshaft pulley

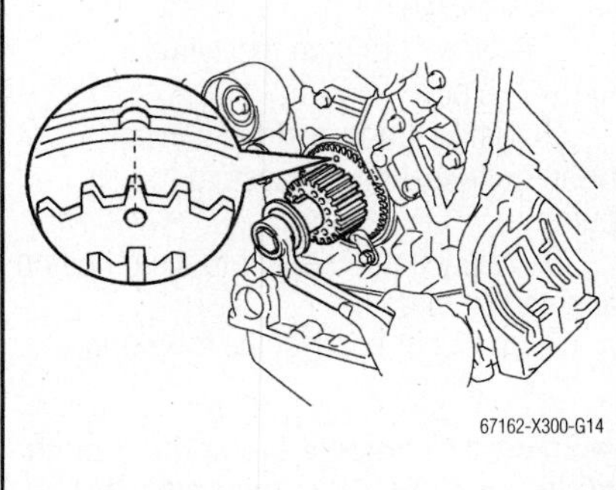

Fig. 247 Turn the crankshaft clockwise to align the timing mark on the crankshaft timing pulley with the notch in the oil pump body

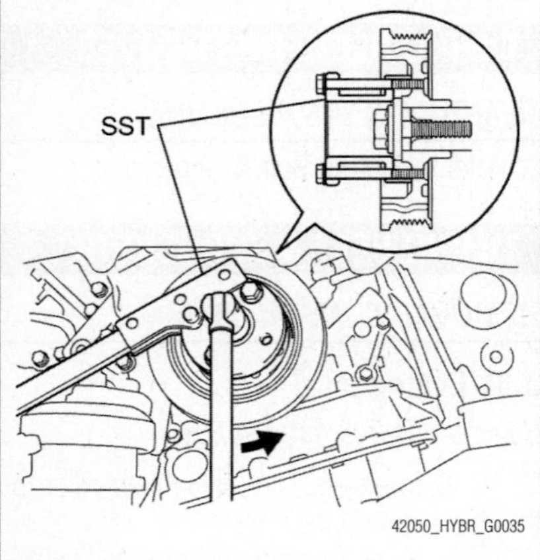

Fig. 244 Use the special tool to hold the pulley in order to loosen the pulley bolt

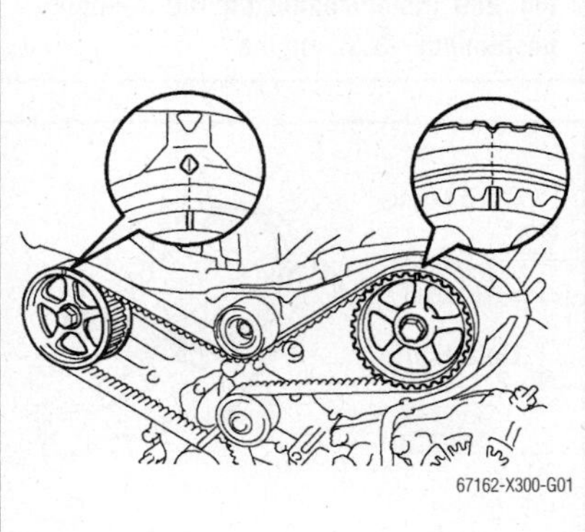

Fig. 246 Check that the timing marks on the camshaft pulleys are aligned with the notches on the inner belt cover

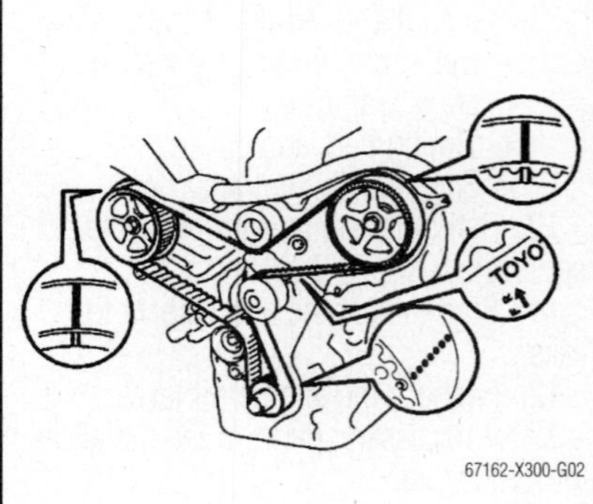

Fig. 248 If the timing belt is re-used, check that the 3 original installation marks are visible on the belt as shown

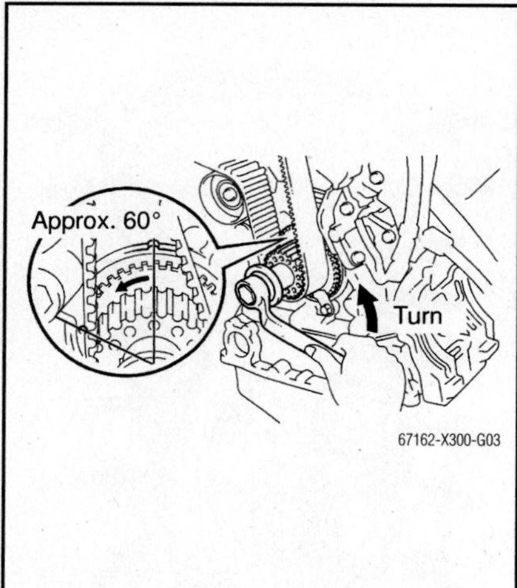

Fig. 249 Turn the crankshaft counterclockwise by 60 degrees

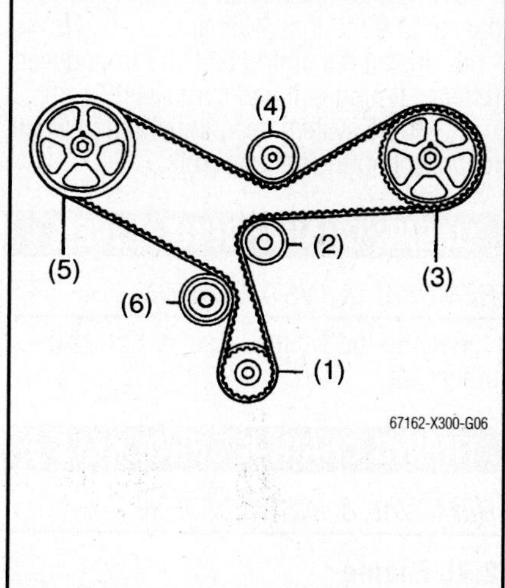

Fig. 252 Install the belt in this order

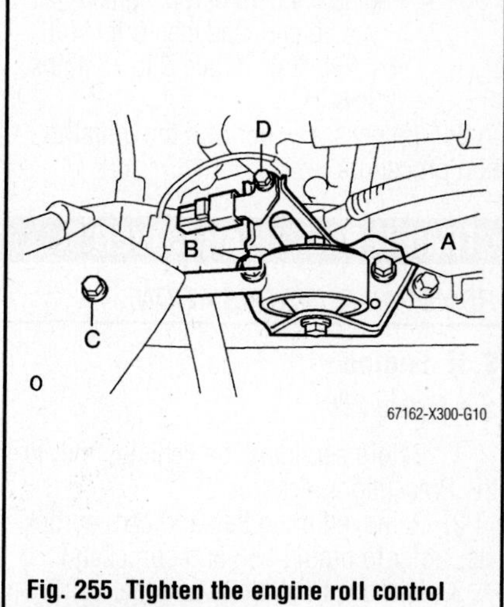

Fig. 255 Tighten the engine roll control rod bolts in this order

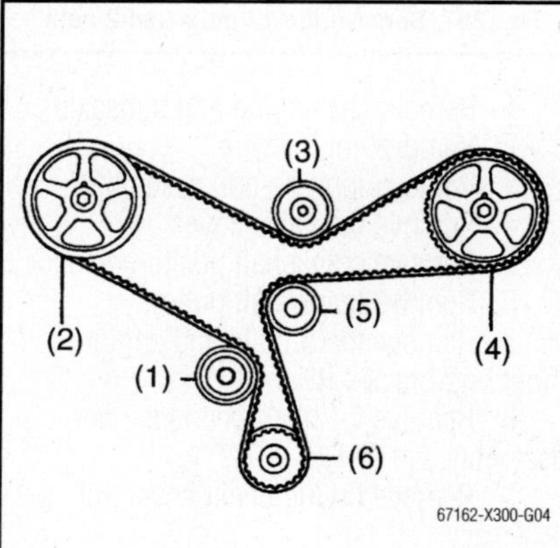

Fig. 250 Remove the belt from the pulleys in this order

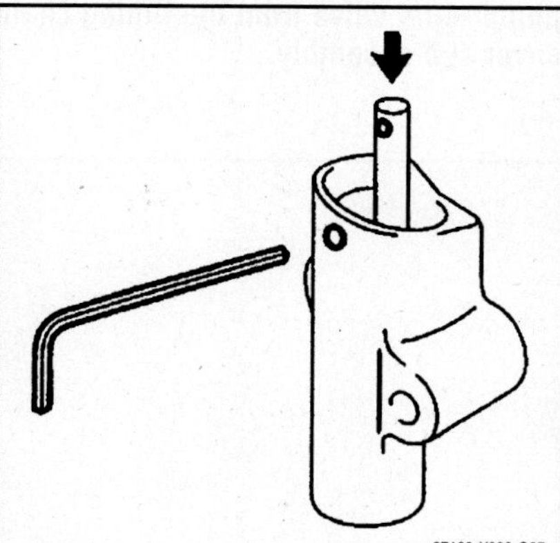

Fig. 253 Set the tensioner in a press and collapse the plunger. Do not apply more than 2,205 lbs (9.8 kn) of force. Insert a suitable metal rod through the holes to hold the plunger in position

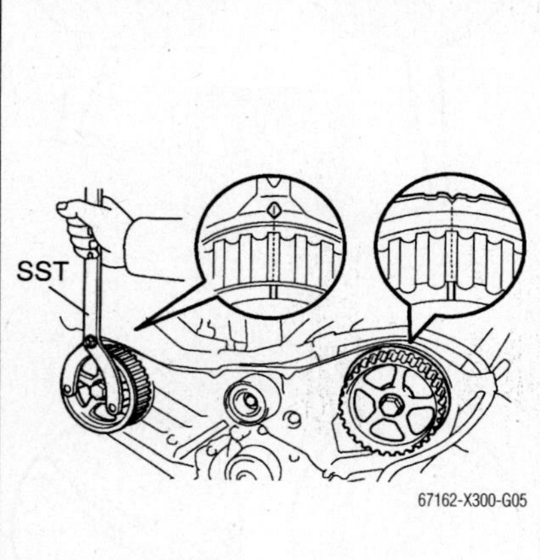

Fig. 251 Turn the camshaft pulleys back into alignment so the marks align with the notches on the inner cover

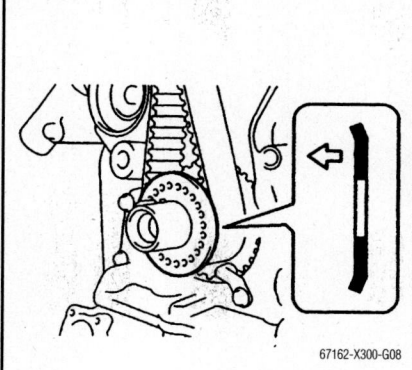

67162-X300-G08

Fig. 254 Install the timing belt guide with the cupped side facing front

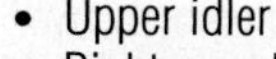

- Upper idler
- Right camshaft
- Lower idler

18. Set the tensioner in a press and collapse the plunger. Do not apply more than 2,205 lbs (9.8 kn) of force. Insert a suitable metal rod through the holes to hold the plunger in position.

19. Install the tensioner and torque the 2 bolts alternately to 20 ft. lbs. (27 Nm).

WARNING

Be sure to tighten to bolts alternately and evenly so the tensioner seats flat.

20. Remove the metal rod from the tensioner.

21. Turn the crankshaft 2 full revolutions clockwise (720 degrees), and align the timing mark on the crank pulley with the notch on the oil pump.

22. Check the timing marks on the camshaft pulleys for alignment with the notches on the inner cover. If they do not align, remove the belt and align the mismatched mark(s).

23. The remainder of installation is the reverse of removal. Observe the following torques:

- Right engine mount bracket: 21 ft. lbs. (28 Nm)
- Right engine mount insulator: 70 ft. lbs. (95 Nm)
- Timing belt covers: 75 inch lbs. (8.5 Nm)
- Crankshaft pulley: 162 ft. lbs. (220 Nm)
- Alternator bracket: 21 ft. lbs. (28 Nm)
- Right engine mount stay: 47 ft. lbs. (64 Nm)

- Engine roll control rod: tighten first A, then B, and then C to 47 ft. lbs. (64 Nm). Torque D to 17 ft. lbs. (23 Nm)

24. If necessary, perform the initialization procedure.

TIMING BELT REAR COVER

REMOVAL & INSTALLATION

3.3L Engine

See Figure 256.

1. Before servicing the vehicle, refer to the Precautions Section.
2. Remove timing belt and cam sprockets. (refer to timing belt and sprockets)
3. Remove the 6 bolts and the rear timing belt cover.

To install:

4. Install the rear timing belt cover No.3 with the 6 bolts. Tighten the bolts to 76 inch lbs. (8.5 Nm).
5. Install the camshaft sprockets and tighten to 92 ft. lbs. (125 Nm).
6. Install the timing belt and front cover. (refer to timing belt and sprockets)
7. Some system need initialization when reconnecting the battery cable.

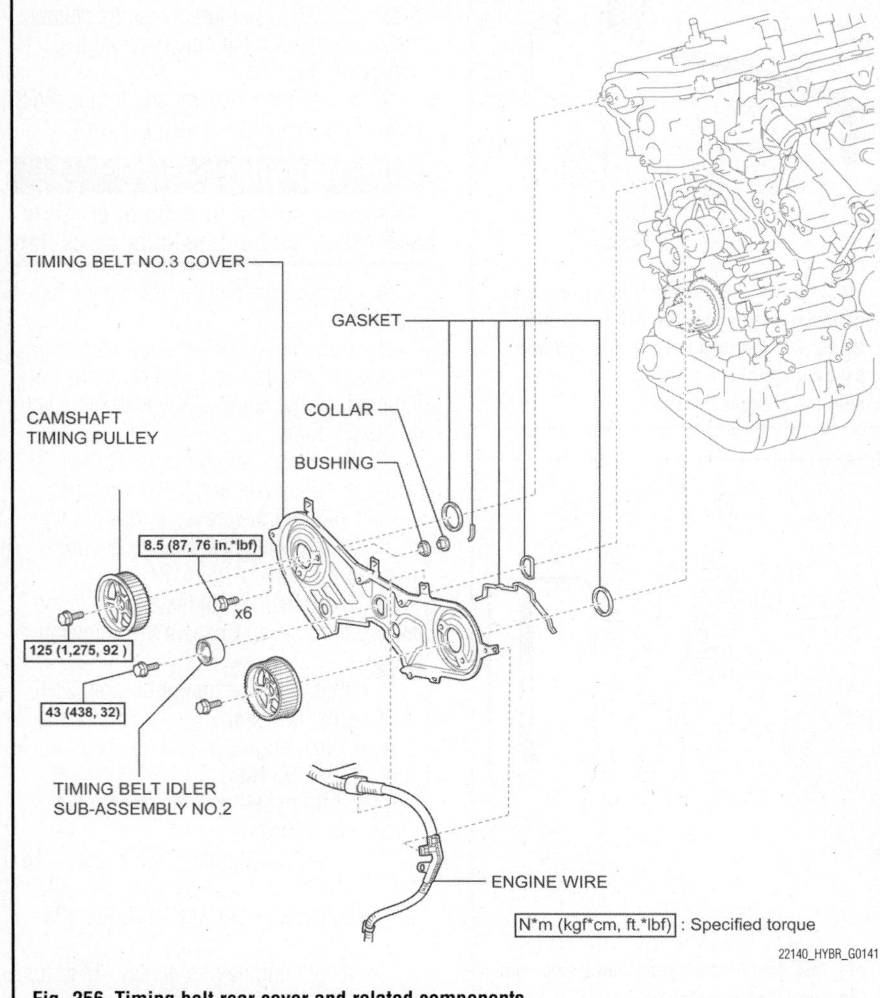

Fig. 256 Timing belt rear cover and related components

TIMING CHAIN FRONT COVER

REMOVAL & INSTALLATION

Refer to the Timing Chain & Sprocket procedure.

TIMING CHAIN & SPROCKETS

REMOVAL & INSTALLATION

2.7L Engine

See Figures 257 through 272.

➡Do not remove the oil pump or oil pump relief valve from the timing chain cover sub-assembly.

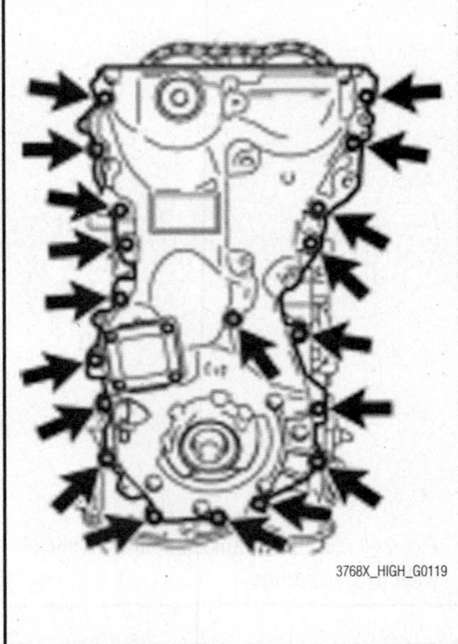

Fig. 257 Remove the 17 bolts and 2 nuts

1. Remove the engine and transaxle.
2. Remove engine wire.
3. Remove ignition coil assembly.
4. Remove the valve cover.
5. Remove crankshaft position sensor.
6. Remove crankshaft pulley.
7. Remove the 5 bolts and engine mounting bracket RH.
8. Remove v-ribbed belt tensioner assembly.
9. Remove timing chain cover sub-assembly:

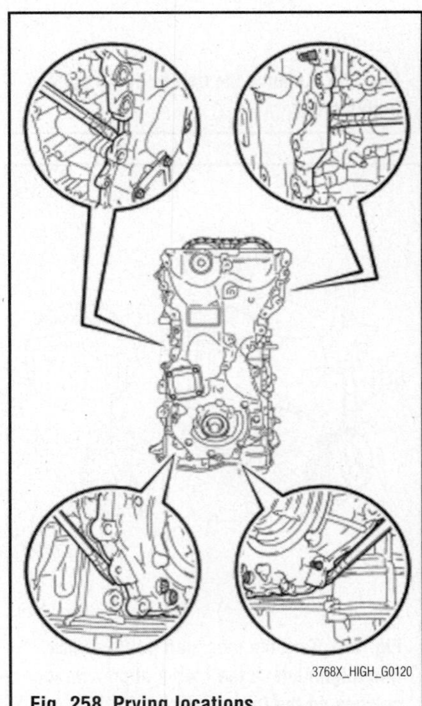

Fig. 258 Prying locations

a. Remove the 17 bolts and 2 nuts.

b. Remove the timing chain cover by prying between the timing chain cover and cylinder head, camshaft housing, cylinder block and stiffening crankcase with a screwdriver as shown in the illustration.

c. Remove the 3 gaskets from the stiffening crankcase.

10. Remove the 4 bolts, timing chain cover plate and gasket.

11. Set no. 1 cylinder to TDC/compression:

a. Temporarily install the crankshaft pulley bolt.

b. Rotate the crankshaft clockwise so that the timing marks on the crankshaft timing gear and camshaft timing gears are as shown in the illustration.

c. If the timing marks do not align, rotate the crankshaft clockwise again and align the timing marks.

d. Remove the crankshaft pulley bolt.

12. Remove the bolt and timing chain guide.

13. Remove no. 1 chain tensioner assembly:

a. Allow the plunger to extend slightly, and then rotate the stopper plate counterclockwise to release the lock. Once the lock is released, push the plunger into the tensioner.

b. Move the stopper plate clockwise to set the lock, and insert a pin into the stopper plate hole.

c. Remove the 2 bolts, chain tensioner and gasket.

14. Remove the bolt and chain tensioner slipper.

15. Remove chain sub-assembly.

16. Remove the 2 bolts and chain vibration damper.

17. Remove the crankshaft timing sprocket from the crankshaft.

To install:

18. Install the crankshaft timing sprocket to the crankshaft.

19. Set no. 1 cylinder to TDC/compression. Refer to Camshaft and Valve Lifters.

➡Make sure the mark plate of the chain faces away from the engine. It is not necessary to install the chain to the teeth of the gears and sprocket.

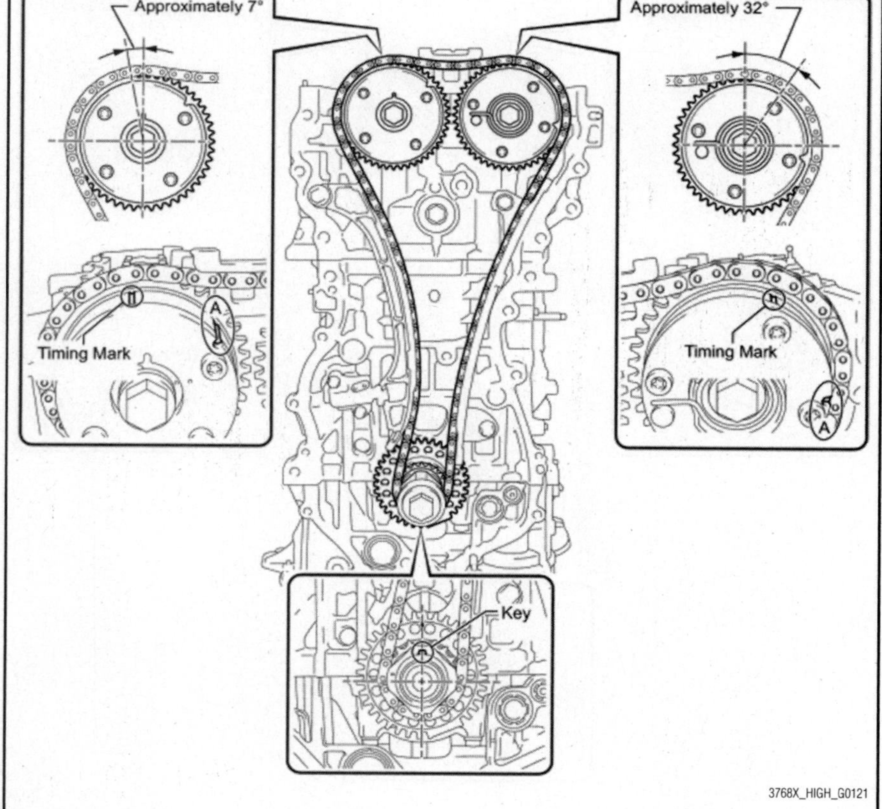

Fig. 259 Set no. 1 cylinder to TDC/compression—"A" is not a timing mark.

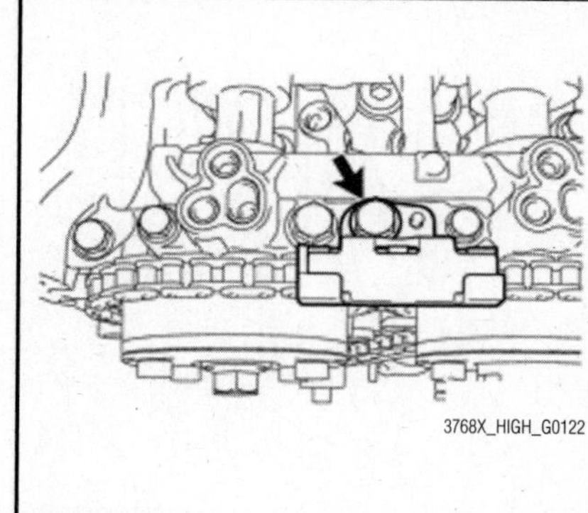

Fig. 260 Remove the bolt and timing chain guide

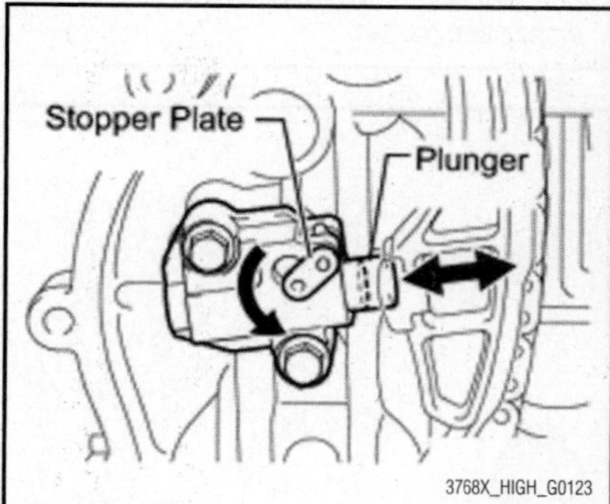

Fig. 261 Allow the plunger to extend slightly, and then rotate the stopper plate counterclockwise to release the lock

20. Install the chain vibration damper with the 2 bolts. Tighten to 15 ft. lbs.

21. Install chain sub-assembly:

a. Place the chain onto the camshaft timing gears and crankshaft timing sprocket.

b. Align the mark plate (yellow or gold) of the chain with the timing mark of the camshaft timing exhaust gear and install the chain to the camshaft timing exhaust gear.

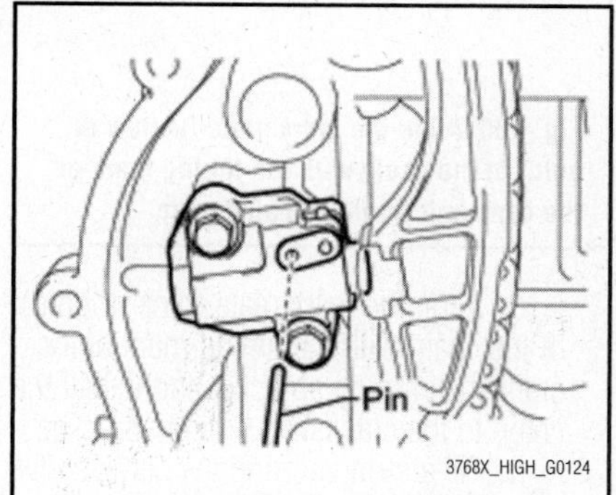

Fig. 262 Move the stopper plate clockwise to set the lock, and insert a pin into the stopper plate hole

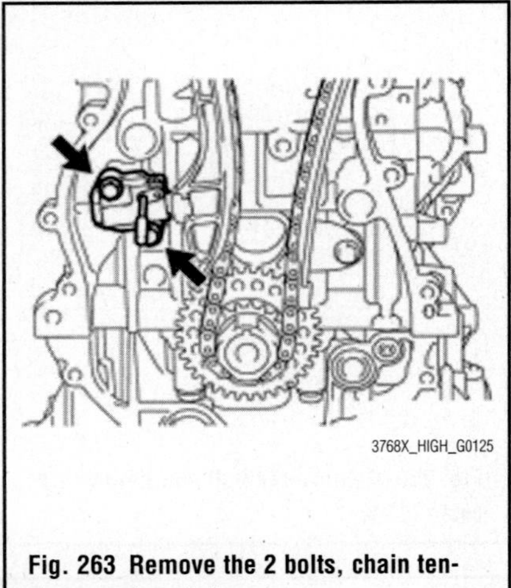
Fig. 263 Remove the 2 bolts, chain tensioner and gasket

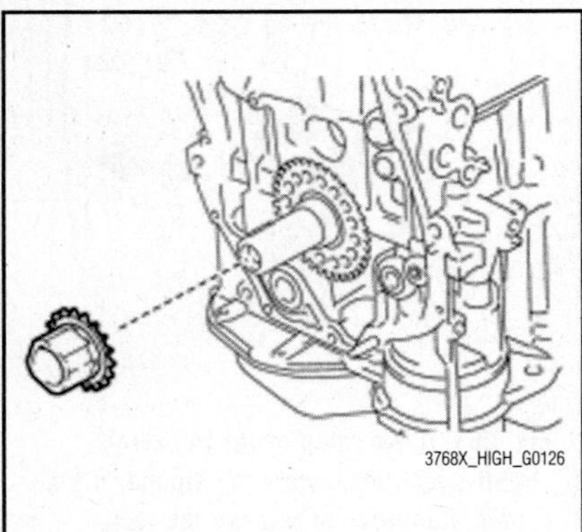
Fig. 264 Remove the crankshaft timing sprocket from the crankshaft

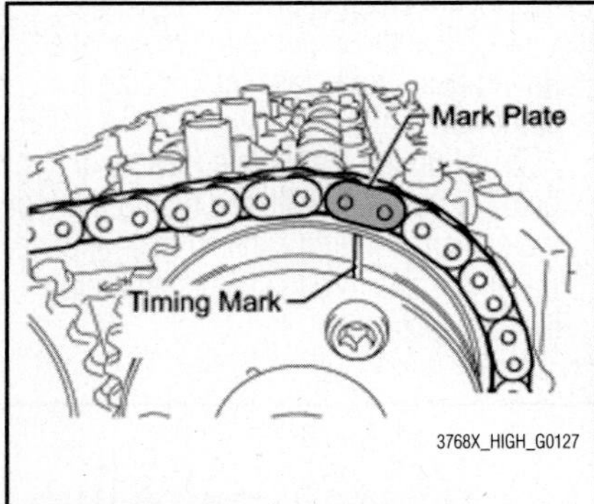

Fig. 265 Align the mark plate (yellow or gold) of the chain with the timing mark of the camshaft timing exhaust gear

c. Align the mark plate (pink or gold) of the chain with the timing mark of the crankshaft timing sprocket and install the chain to the crankshaft timing sprocket.

d. Tie a string around the chain above the crankshaft timing sprocket so that the chain is secure.

e. Using the hexagonal portion of the intake camshaft, rotate the intake

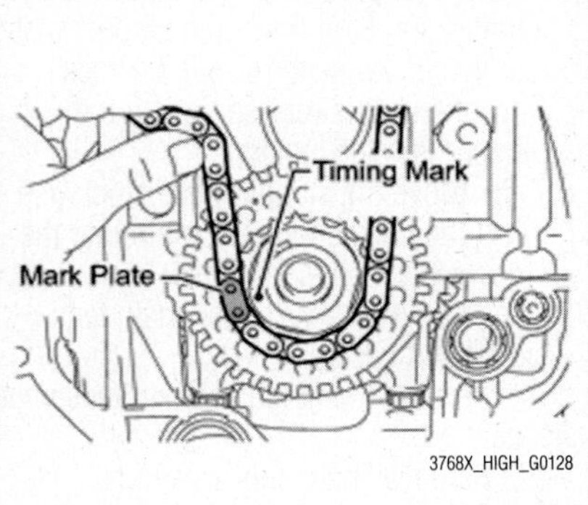

Fig. 266 Align the mark plate (pink or gold) of the chain with the timing mark of the crankshaft timing sprocket

camshaft counterclockwise with a wrench, align the timing mark of the camshaft timing gear with the mark plate (yellow or gold) of the chain and install the chain to the camshaft timing gear.

➡**Make sure the chain is secure.**

f. Remove the string above the crankshaft timing sprocket, rotate the crank-

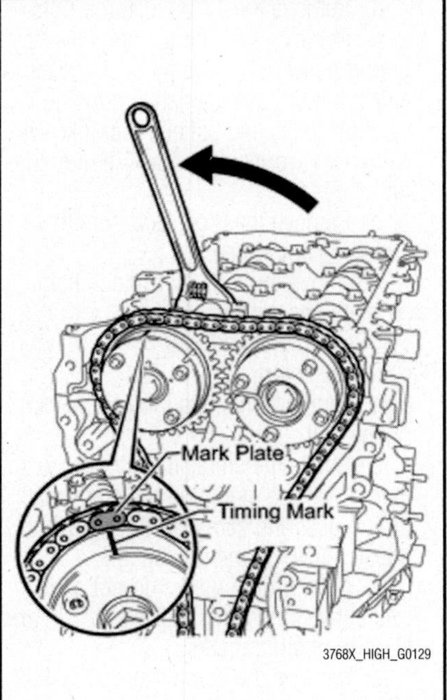

Fig. 267 Rotate the intake camshaft counterclockwise with a wrench

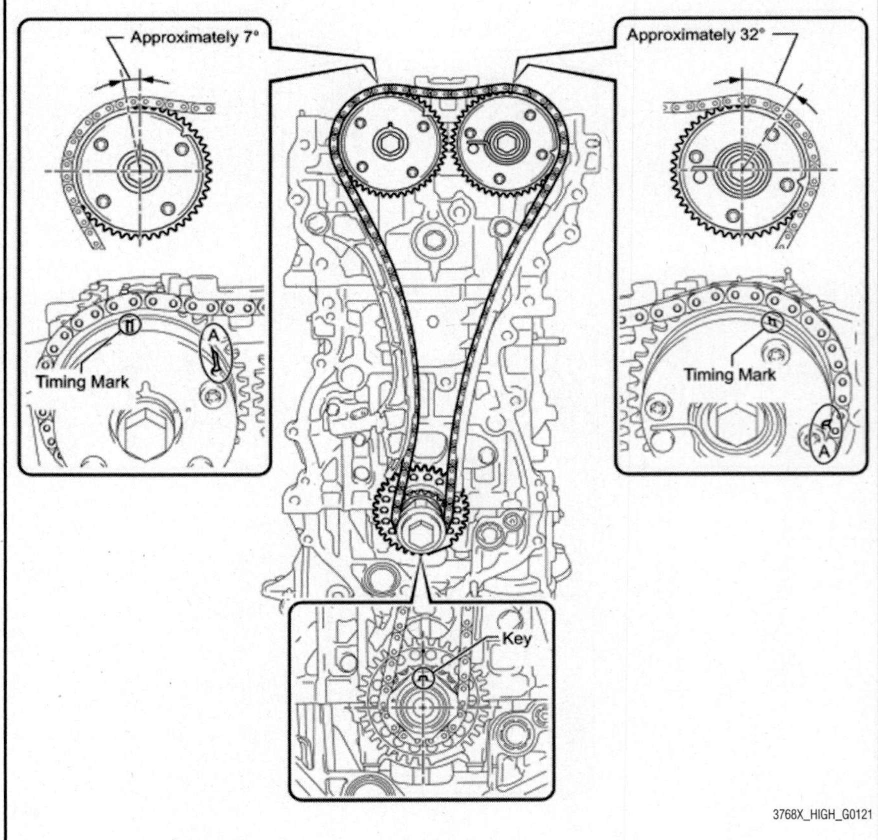

Fig. 268 Set no. 1 cylinder to TDC/compression—"A" is not a timing mark.

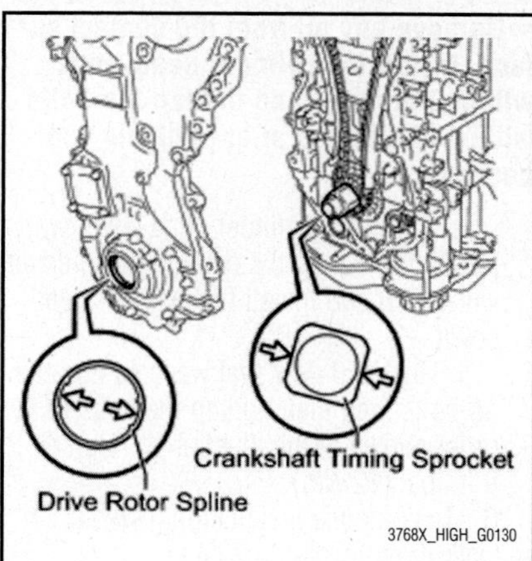

Fig. 269 Align the drive rotor spline and the crankshaft timing sprocket

shaft clockwise, and loosen the chain so that the chain tensioner slipper can be installed.

22. Install the chain tensioner slipper with the bolt. Tighten to 15 ft. lbs. (21 Nm).

23. Install a new gasket and the chain tensioner with the 2 bolts. Tighten to 7 ft. lbs. (10 Nm). Remove the pin from the stopper plate.

24. Install the timing chain guide with the bolt Tighten to 15 ft. lbs. (21 Nm).

25. Check no. 1 cylinder TDC/compression:

a. Temporarily install the crankshaft pulley bolt.

b. Rotate the crankshaft clockwise, and check that the timing marks on the crankshaft timing sprocket and camshaft timing gears are as shown in the illustration.

c. Remove the crankshaft pulley bolt.

26. Install timing chain cover sub-assembly:

a. Apply a light coat of engine oil to 3 new gaskets.

b. Install the 3 gaskets to the stiffening crankcase.

c. Align the drive rotor spline and the crankshaft timing sprocket as shown in the illustration.

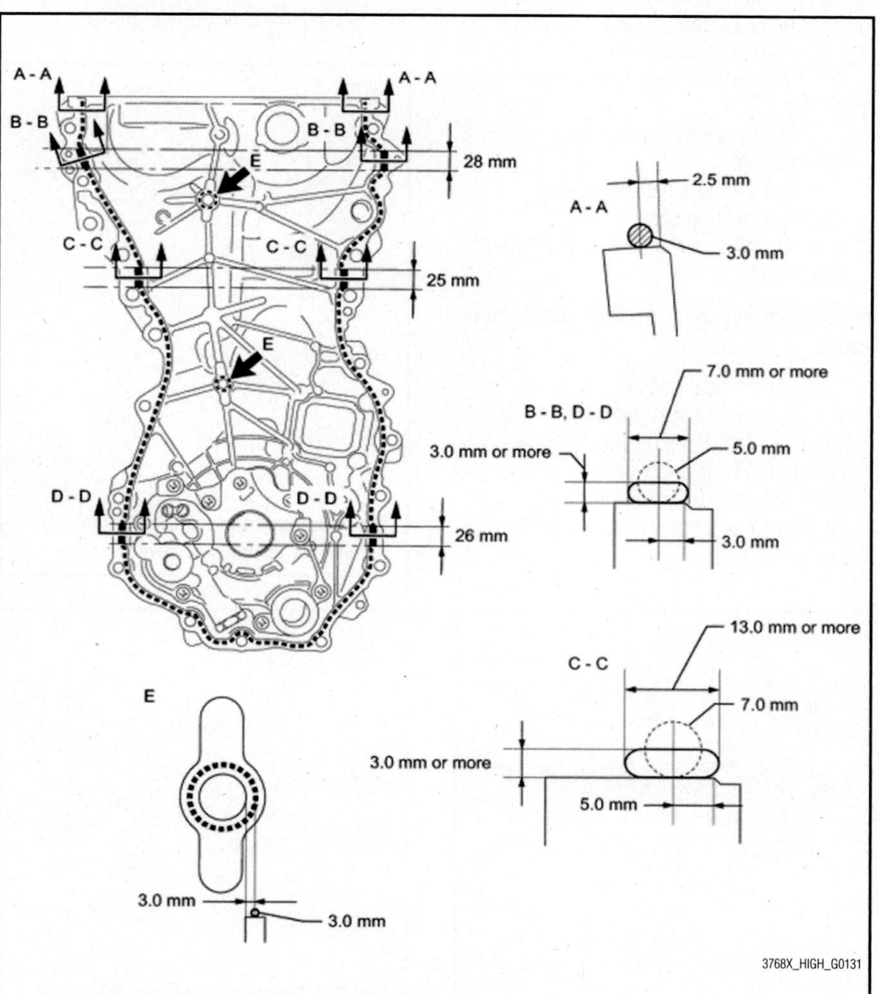

Fig. 270 Apply seal packing in a line to the timing chain cover as shown

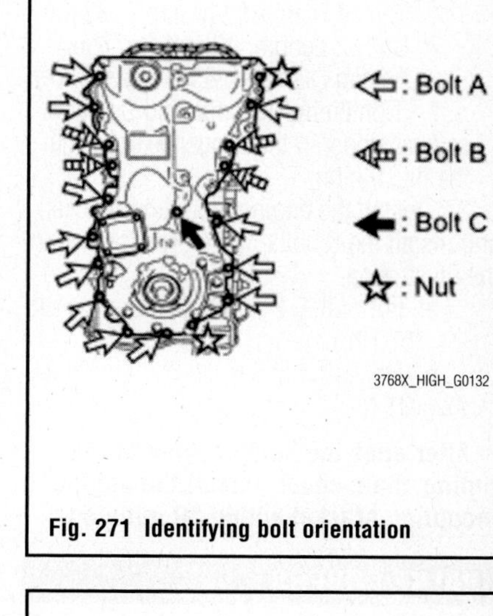

Fig. 271 Identifying bolt orientation

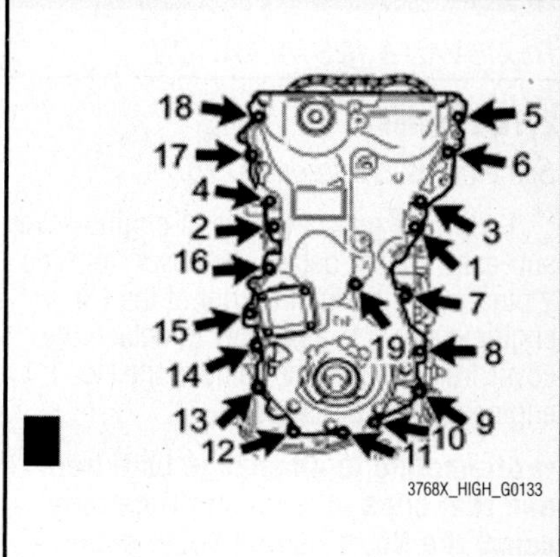

Fig. 272 Timing cover bolt and nut tightening sequence

d. Apply seal packing in a line to the timing chain cover as shown in the following illustration. Seal packing: Toyota Genuine Seal Packing Black, Three Bond 1207B or equivalent.

➡When the contact surfaces are wet, clean the surfaces with non-residue solvent before applying seal packing. Install the timing chain cover within 3 minutes and tighten the bolts within 10 minutes after applying seal packing. After applying seal packing to the timing chain cover, install the engine mounting bracket within 10 minutes. Do not apply oil for at least 4 hours after the installation. Do not start the engine for at least 4 hours after the installation.

e. Temporarily install the timing chain cover with the 17 bolts and 2 nuts. Bolt lengths:

- Bolt A: Length: 30 mm (1.18 in.). Thread Diameter: 8 mm (0.315 in.)

- Bolt B: Length: 35 mm (1.38 in.). Thread Diameter: 10 mm (0.394 in.)
- Bolt C: Length: 45 mm (1.77 in.). Thread Diameter: 8 mm (0.315 in.)

f. Tighten the 17 bolts and 2 nuts in several steps, in the sequence shown in the illustration.

27. Install the engine mounting bracket, and install the 5 bolts in the order shown in the illustration:

a. For bolt 1, 2 and 3, tighten to 41 ft. lbs. (55 Nm).

b. For bolt 4 and 5, , tighten to 15 ft. lbs. (21 Nm).

➡After applying seal packing to the timing chain cover, install the engine mounting bracket within 10 minutes.

VALVE COVERS

REMOVAL & INSTALLATION

2.7L Engine

See Figures 273 through 275.

1. Lift the rear of the No. 1 engine cover sub-assembly to detach the cover from the 2 pins, and then lift the front of the No. 1 engine cover sub-assembly to detach the cover from the pin and remove the No. 1 engine cover sub-assembly.

➡Attempting to disengage both front and rear clips at the same time may cause the No. 1 engine cover sub-assembly to break.

2. Remove the ignition coil and spark plugs.
3. Remove the 16 bolts, 3 seal washers, cylinder head cover and gasket.
4. Remove the 3 gaskets from the camshaft bearing caps.

To install:

5. Visually check the spark plug tube gasket. If there are scratches or deformations on the upper surface, outer lip and inner lip replace the spark plug tube gasket.
6. Install the 4 plug tube gaskets to the cylinder head cover. After pressing in the spark plug tube gasket, make sure the gasket protrudes 1.0 mm (0.0394 in.) or less from the cylinder head cover.
7. Install cylinder head cover sub-assembly:

a. Apply a light coat of engine oil to 3 new gaskets.

b. Install the 3 gaskets to the camshaft bearing caps.

c. Install a new gasket to the cylinder head cover.

➡Remove any oil from the contact surface.

d. Apply seal packing as shown in the illustration. Seal packing: Toyota Genuine Seal Packing Black, Three Bond 1207B or equivalent. Standard seal diameter: 3.0 to 6.0 mm (0.118 to 0.236 in.). Application width A: 5.0 mm (0.197 in.)

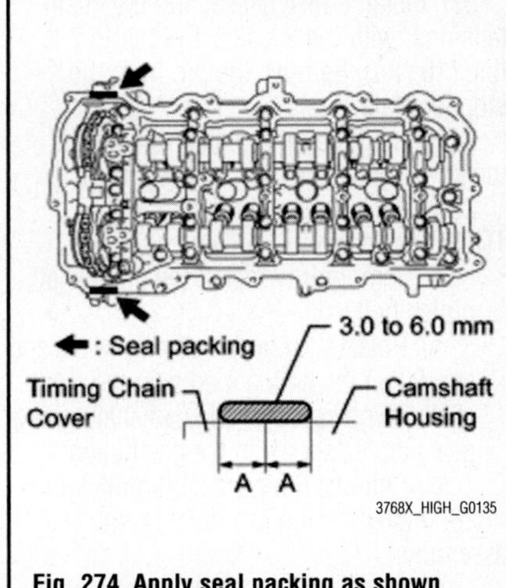

Fig. 274 Apply seal packing as shown

➡Remove any oil from the contact surface. Install the cylinder head cover within 3 minutes and tighten the bolts within 15 minutes after applying seal packing.

e. Align the cylinder head cover with pin A. Then align the cylinder head cover with pin B and install the cylinder head cover.

f. Install 3 new seal washers and the 16 bolts, and then tighten the bolts in the order shown in the illustration. Tighten to 9 ft. lbs. (12 Nm).

8. To complete installation, reverse remaining removal procedure.

3.3L Engine

See Figures 276 through 284.

1. Disconnect the negative battery cable.
2. Using a clip remover, remove the engine side cover.
3. Drain the coolant and engine oil.
4. Remove front upper center suspension brace subassembly, as follows:

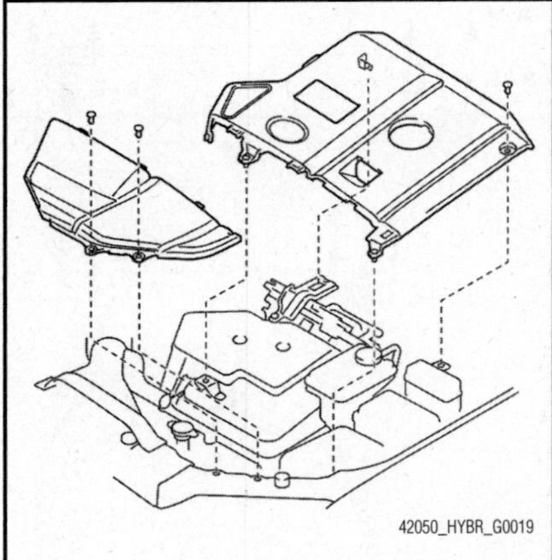

Fig. 276 Use a clip removal tool to remove the engine side cover

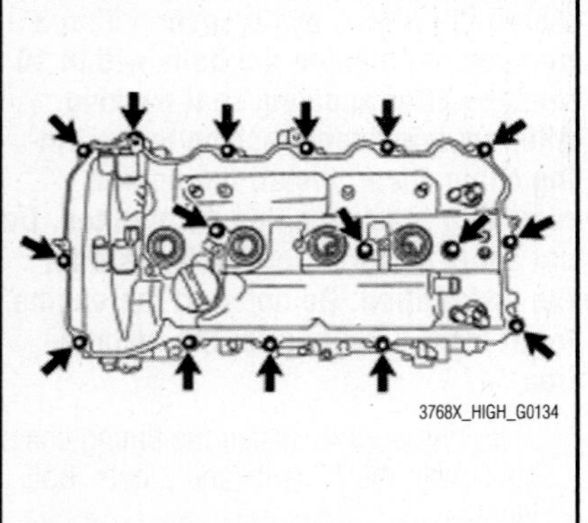

Fig. 273 Remove the 16 bolts

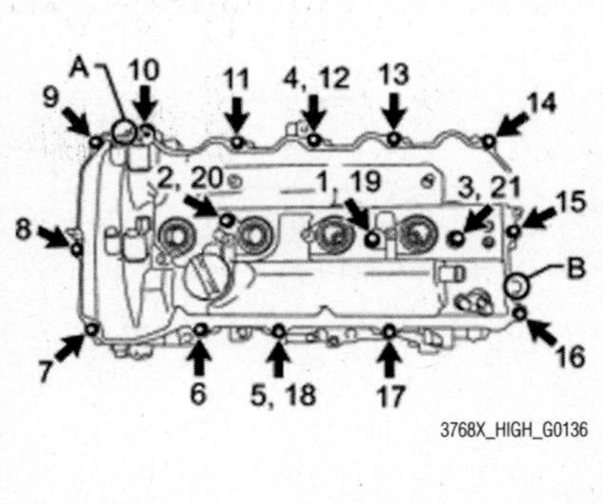

Fig. 275 Valve cover tightening sequence

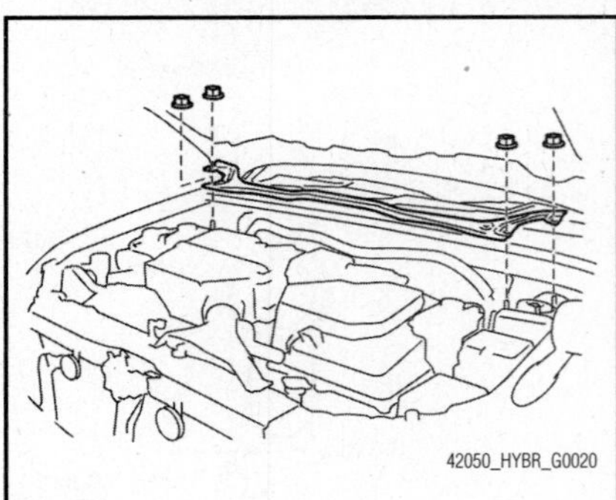

Fig. 277 Remove front upper center suspension brace subassembly

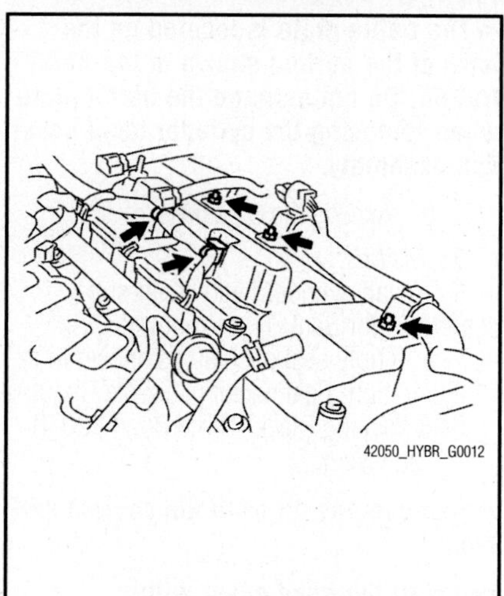

Fig. 278 Remove the 2 clamps and 3 nuts

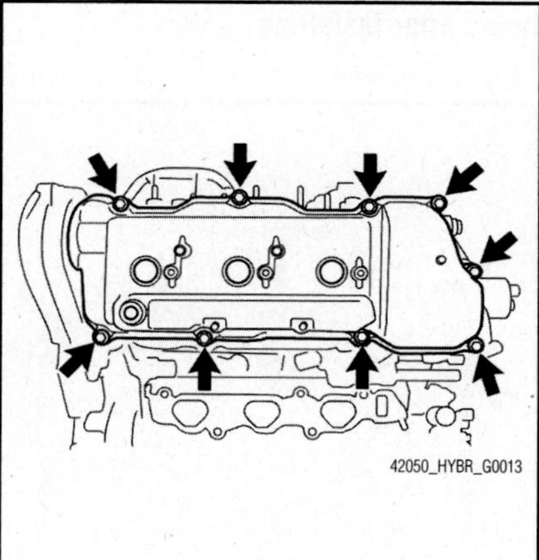

Fig. 279 Remove the 9 bolts and the cylinder head cover

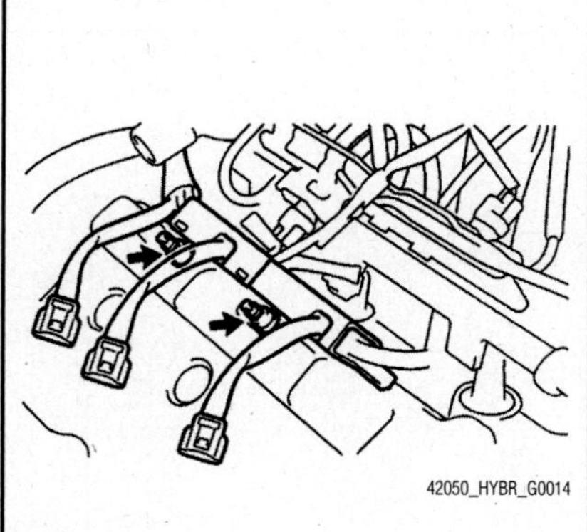

Fig. 280 Use a Torx® wrench to remove the 2 bolts, then disconnect the engine wire harness protector

a. Remove the 4 nuts and front suspension brace subassembly.

b. Install the 4 shock absorber nuts and tighten to 59 ft .lbs. (80 Nm).

5. Remove the air cleaner cap w/ inlet, as follows:

a. Remove the 2 bolts, 4 clamps and air cleaner cap w/ inlet.

b. Remove the air cleaner filter element from the air cleaner case.

6. Remove the air cleaner case w/ resonator.

7. Remove the emission control valve set.

8. Remove the intake air surge tank.

9. Disconnect the inlet radiator hose.

10. Remove the ignition coil assembly

11. Remove the cylinder head cover subassembly, as follows:

a. Remove the 2 engine wire harness clamps.

b. Remove the 3 nuts and disconnect the engine wire harness.

c. Remove the 9 bolts and the cylinder head cover.

12. Remove left cylinder head cover subassembly:

a. Using an E6 Torx® socket wrench, remove the 2 bolts and disconnect the engine wire harness protector.

b. Put on insulating gloves, then remove the 2 engine wire harness clamps.

c. Remove the 2 bolts and 2 brackets.

d. Remove the 9 bolts and the cylinder head cover.

13. Thoroughly clean the gasket mating surfaces.

To install:

14. Install the cylinder head cover subassembly, as follows:

a. Apply seal packing (part no. 08826-00080, or equivalent) to the cylinder head as shown in the illustration. Make sure all oil residue is removed from the contact surfaces prior to installing the seal packing.

WARNING

Install the cylinder head cover within 3 minutes after applying seal packing. Do NOT start the engine within 2 hours after installing.

b. Install the cylinder head cover with the 9 bolts. Tighten the bolts uniformly in several steps to a final torque of 71 inch lbs. (8 Nm).

c. Install the engine wire harness with the 3 nuts and 2 clamps. Tighten to 74 inch lbs. (8.4 Nm).

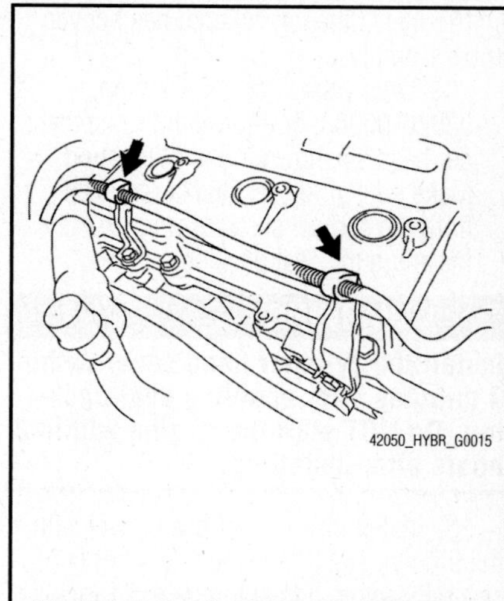

Fig. 281 Put on insulating gloves, then remove the 2 engine wire harness clamps

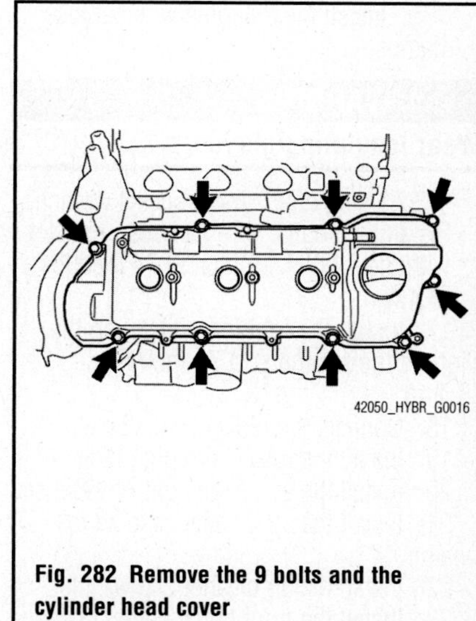

Fig. 282 Remove the 9 bolts and the cylinder head cover

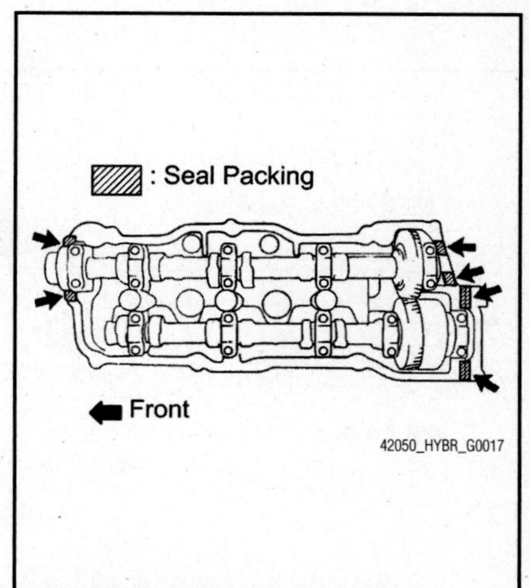

Fig. 283 Apply seal packing (part no. 08826-00080, or equivalent) to the cylinder head as shown

15. Install the left cylinder head cover sub-assembly:
    a. Apply seal packing (part no. 08826-00080, or equivalent) to the cylinder head as shown in the illustration. Make sure all oil residue is removed from the contact surfaces prior to installing the seal packing.

WARNING

Install the cylinder head cover within 3 minutes after applying seal packing. Do NOT start the engine within 2 hours after installing.

16. Install the cylinder head cover with the 9 bolts. Tighten the bolts uniformly in several steps to a final torque of 71 inch lbs. (8 Nm).
    a. Install the 2 brackets with the 2 bolts.
    b. Install the 2 engine wire harness clamps.

WARNING

Wear insulating gloves.

    c. Using an E6 Torx® socket wrench, install the engine wire harness protector with the 2 bolts. Tighten to 74 inch lbs. (8.4 Nm).
17. Install the ignition coil assembly. Tighten the retainers to 71 inch lbs. (8 Nm).
18. Connect the radiator inlet hose.
19. Install the intake air surge tank.
20. Install the emission control valve set.
21. Install the air cleaner case w/ resonator.
22. Install the air cleaner cap w/ inlet.
23. Install the front upper center suspension brace subassembly.
24. Add the proper and amount of engine oil and coolant to the engine.
25. Check for engine coolant and/or oil leaks.
26. Install the engine side cover:
27. If necessary, perform the initialization procedure.

3.5L Engine

See Figures 285 through 292.

1. Remove cylinder head cover sub-assembly (for Bank 1), as follows:
    a. Remove the 12 bolts, seal washer, cylinder head cover sub-assembly and cylinder head cover gasket.
    b. Remove the 3 gaskets.
2. Remove cylinder head cover sub-assembly (for Bank 2), as follows:
    a. Remove the 12 bolts, seal washer, cylinder head cover sub-assembly and cylinder head cover gasket.

➡The baffle plate is located on the back of the portion shown in the illustration. Do not damage the baffle plate when removing the cylinder head cover sub-assembly.

    b. Remove the 3 gaskets.

To install:

3. Install cylinder head cover sub-assembly (for Bank 1), as follows:
    a. Apply seal packing as shown in the illustration. Seal packing: Toyota Genuine Seal Packing Black, Three Bond 1207B or equivalent.

➡Remove any oil from the contact surface.

➡Install the head cover within 3 minutes after applying seal packing.

➡Do not start the engine for at least 2 hours after installing.

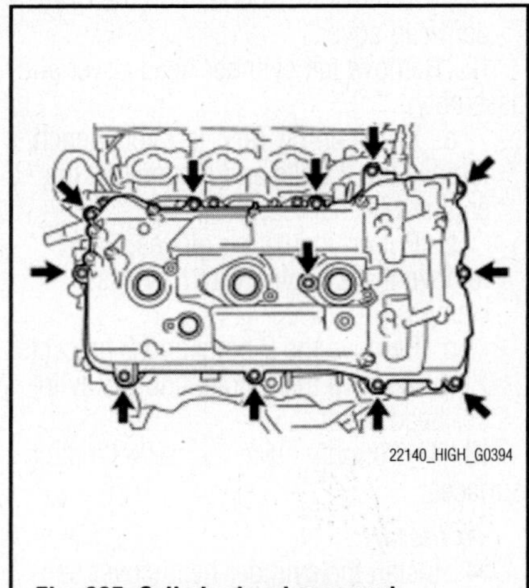

Fig. 285 Cylinder head cover sub-assembly Bank 1—3.5L engine

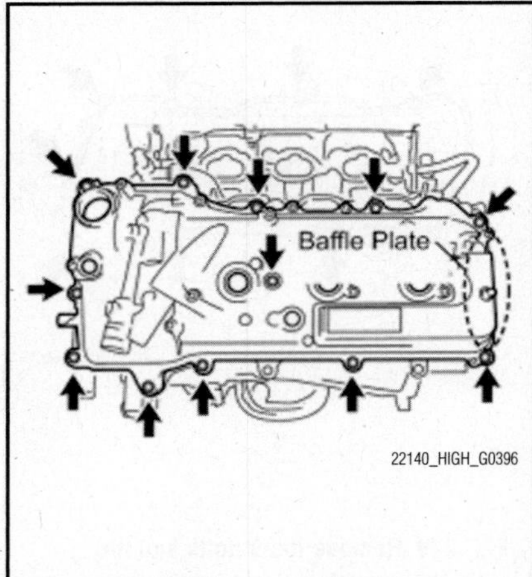

Fig. 287 Cylinder head cover sub-assembly Bank 2—3.5L engine

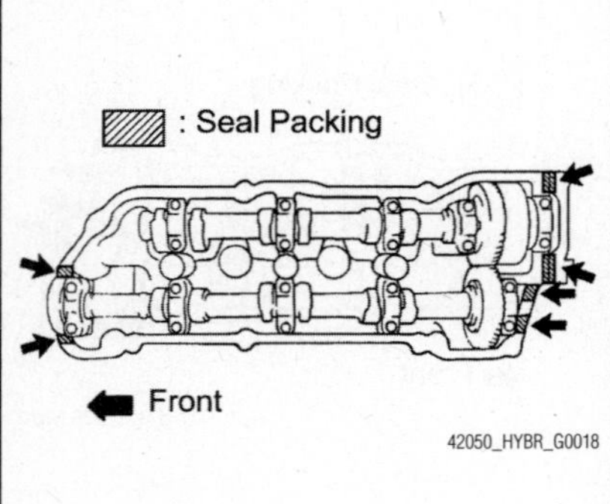

Fig. 284 Apply seal packing (part no. 08826-00080, or equivalent) to the cylinder head as shown

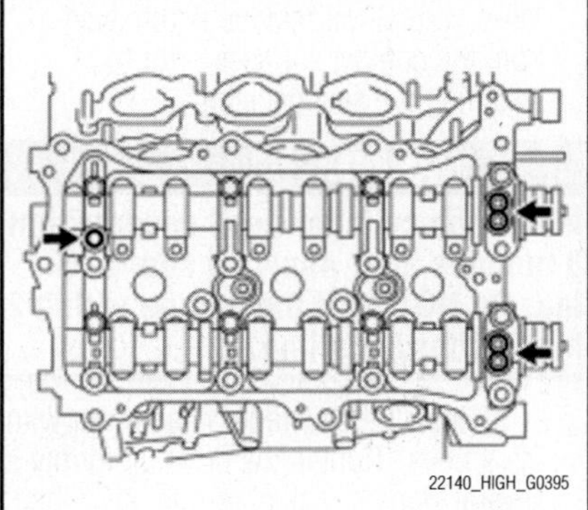

Fig. 286 Cylinder head cover sub-assembly gaskets Bank 1—3.5L engine

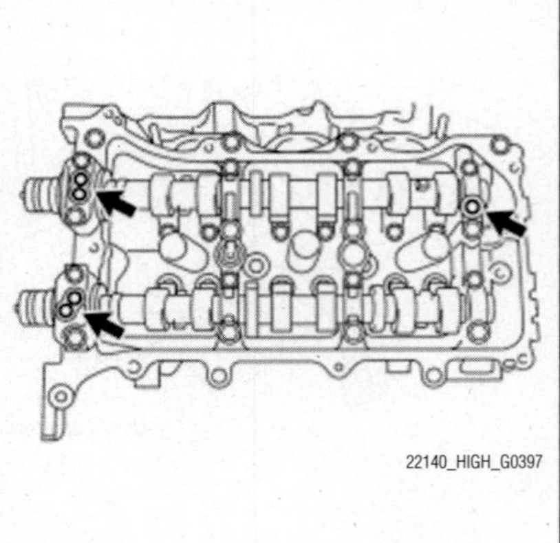

Fig. 288 Cylinder head cover sub-assembly gaskets Bank 2—3.5L engine

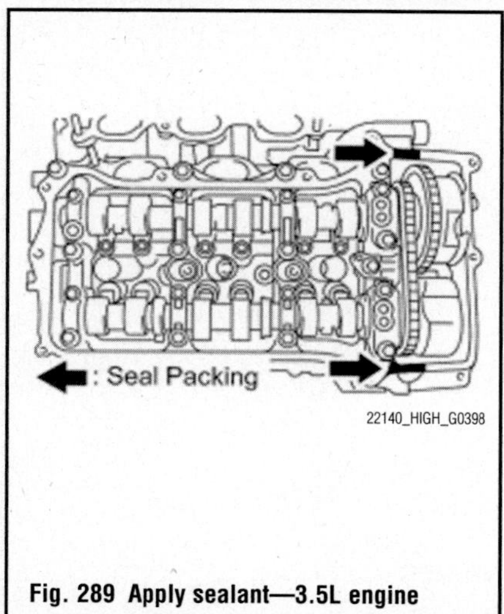

Fig. 289 Apply sealant—3.5L engine

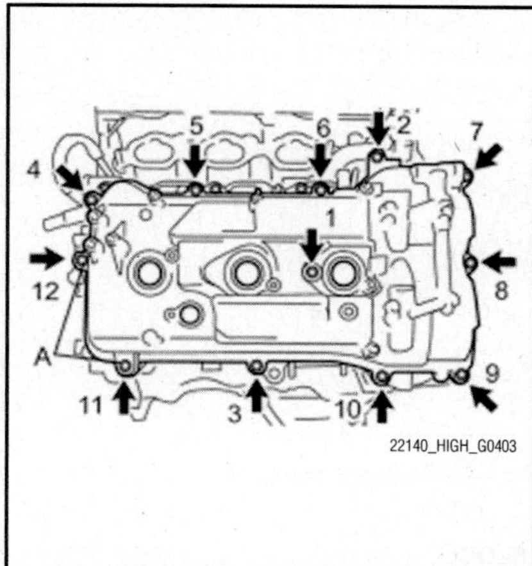

Fig. 290 Cylinder head cover tightening sequence Bank 1—3.5L engine

b. Install 3 new gaskets as previously removed.

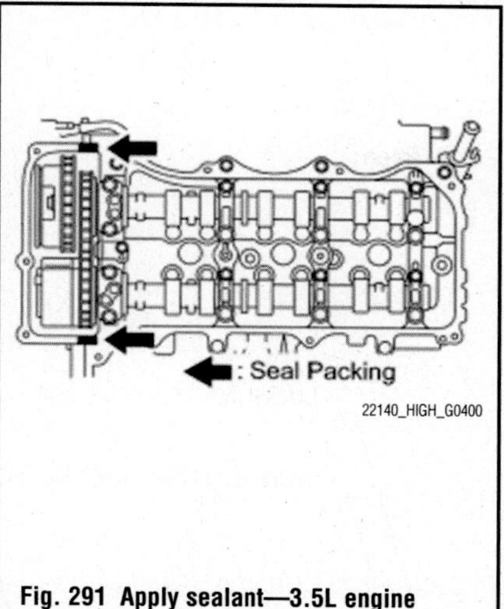

Fig. 291 Apply sealant—3.5L engine

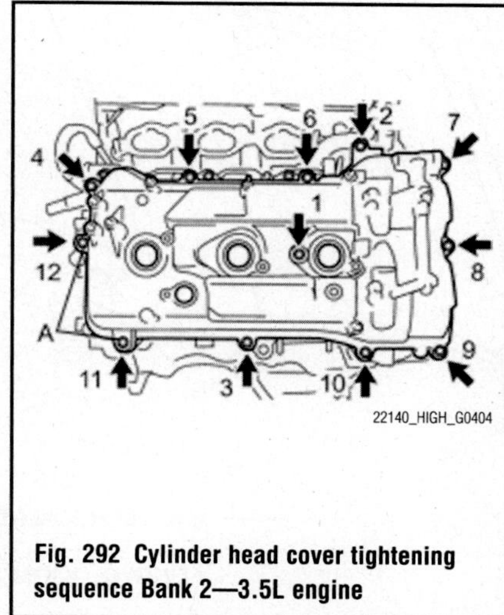

Fig. 292 Cylinder head cover tightening sequence Bank 2—3.5L engine

c. Install a new gasket to the head cover.

d. Install a head cover with the 12 bolts and a new seal washer and tighten to 15 ft. lbs. (21 Nm), 7 ft. lbs. (10 Nm).

➡After tightening all bolts, check the tightening torque of 1 and 11. Retighten the bolt if necessary.

4. Install cylinder head cover sub-assembly (for Bank 2), as follows:

a. Apply seal packing as shown in the illustration. Seal packing: Toyota Genuine Seal Packing Black, Three Bond 1207B or equivalent.

➡Remove any oil from the contact surface.

➡Install the head cover within 3 minutes after applying seal packing.

➡Do not start the engine for at least 2 hours after installing.

b. Install 3 new gaskets as previously removed.

c. Install a new cylinder head cover gasket to the cylinder head cover sub-assembly.

d. Install the cylinder head cover sub-assembly with the 12 bolts and a new seal washer and tighten to 15 ft. lbs. (21 Nm), 7 ft. lbs. (10 Nm).

➡After tightening all bolts, check the tightening torque of 1 and 10. Retighten the bolt if necessary.

VALVE LASH

ADJUSTMENT

No adjustment is necessary on these engines.

ENGINE PERFORMANCE & EMISSION CONTROLS

COMPONENT LOCATIONS

See Figures 293 through 301.

CAMSHAFT POSITION (CMP) SENSOR

LOCATION

See Figures 302 and 303.

REMOVAL & INSTALLATION

2.7L Engine

See Figures 304 and 305.

1. Disconnect the negative battery cable.
2. Remove no. 1 engine cover sub-assembly.
3. Remove Camshaft Position (CMP) sensor (for exhaust side):
   a. Disconnect the sensor connector.
   b. Remove the bolt and sensor.
4. Remove Camshaft Position (CMP) sensor (for intake side):
   a. Disconnect the sensor connector.
   b. Remove the bolt and sensor.

To install:

➡Make sure that the O-ring is not cracked or does not jump out of position during installation.

5. Apply a light coat of engine oil to the O-ring of the sensor.
6. Apply adhesive to 2 or 3 threads of the bolt. Use Toyota Genuine Adhesive 1344, Three Bond 1344 or equivalent.
7. Install the sensor with the bolt. Tighten to 7 ft. lbs. (10 Nm).
8. Connect the sensor connector.
9. Inspect for oil leak.
10. Install no. 1 engine cover sub-assembly.

CRANKSHAFT POSITION (CKP) SENSOR

LOCATION

See Figures 306 and 307.

REMOVAL & INSTALLATION

2.7L Engine

See Figure 308.

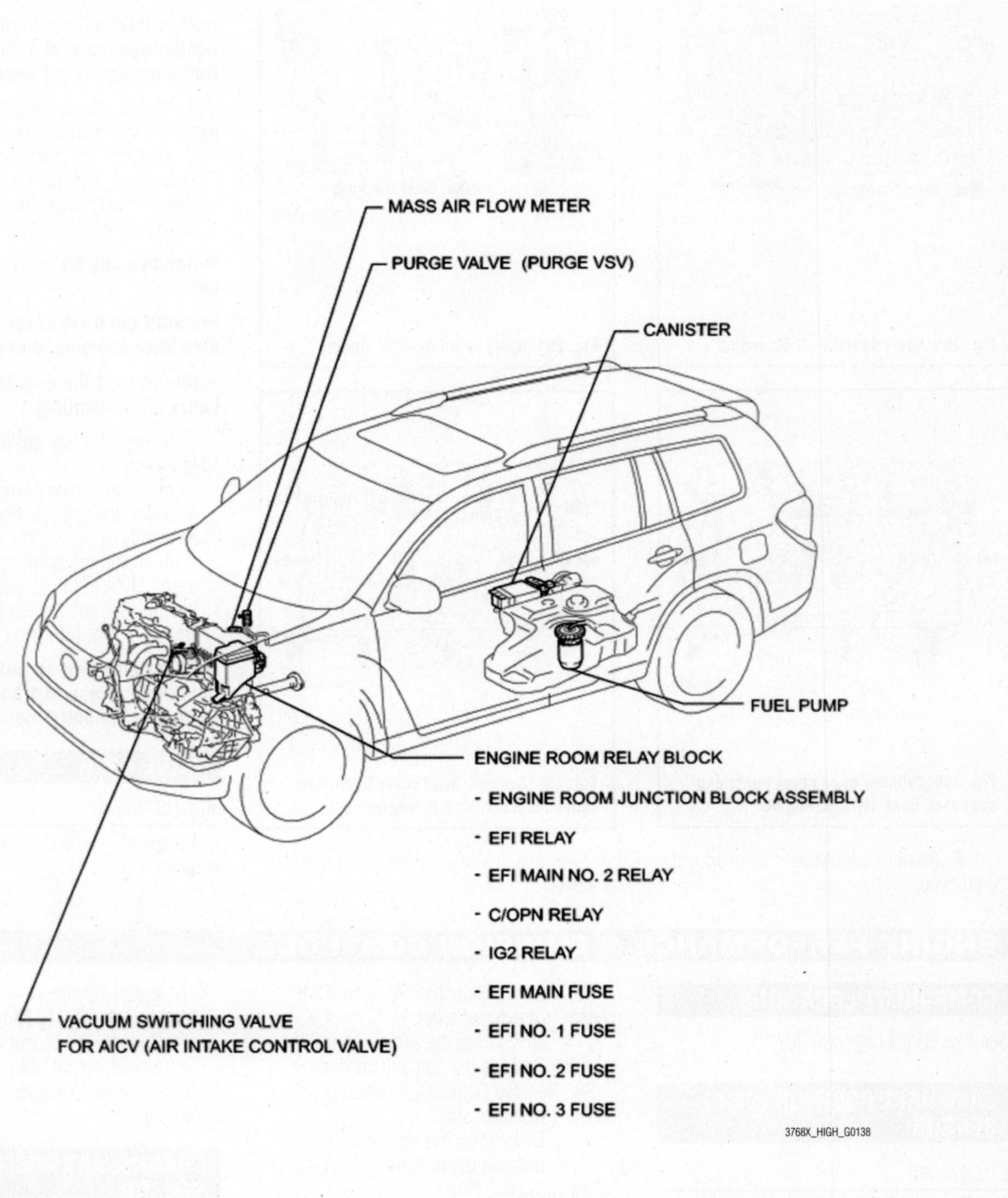

Fig. 293 Highlander component locations—2.7L Engine

IGNITION COIL ASSEMBLY
CAMSHAFT POSITION SENSOR
(FOR EXHAUST CAMSHAFT)
CAMSHAFT POSITION SENSOR
(FOR INTAKE CAMSHAFT)
ENGINE COOLANT
TEMPERATURE SENSOR
INTAKE AIR CONTROL
VALVE ACTUATOR
(FOR TUMBLE CONTROL
VALVE)
AIR FUEL RATIO SENSOR
(BANK 1 SENSOR 1)
TCM
PARK / NEUTRAL POSITION
SWITCH
3768X_HIGH_G0139

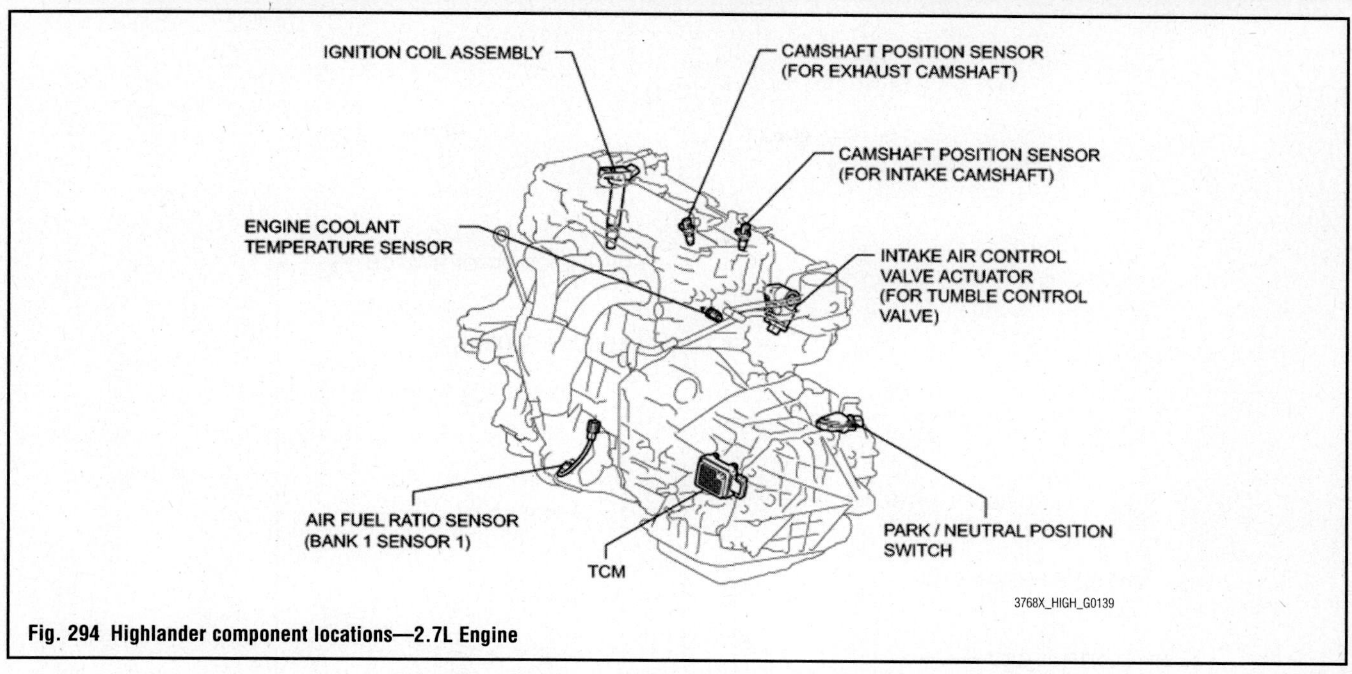

Fig. 294 Highlander component locations—2.7L Engine

CAMSHAFT TIMING OIL
CONTROL VALVE ASSEMBLY
(FOR INTAKE CAMSHAFT)
VACUUM SWITCHING
VALVE
(ACOUSTIC CONTROL
INDUCTION SYSTEM)
CAMSHAFT TIMING OIL
CONTROL VALVE ASSEMBLY
(FOR EXHAUST CAMSHAFT)
FUEL INJECTOR
THROTTLE BODY
CRANKSHAFT POSITION
SENSOR
KNOCK SENSOR
HEATED OXYGEN SENSOR
(BANK 1 SENSOR 2)
3768X_HIGH_G0140

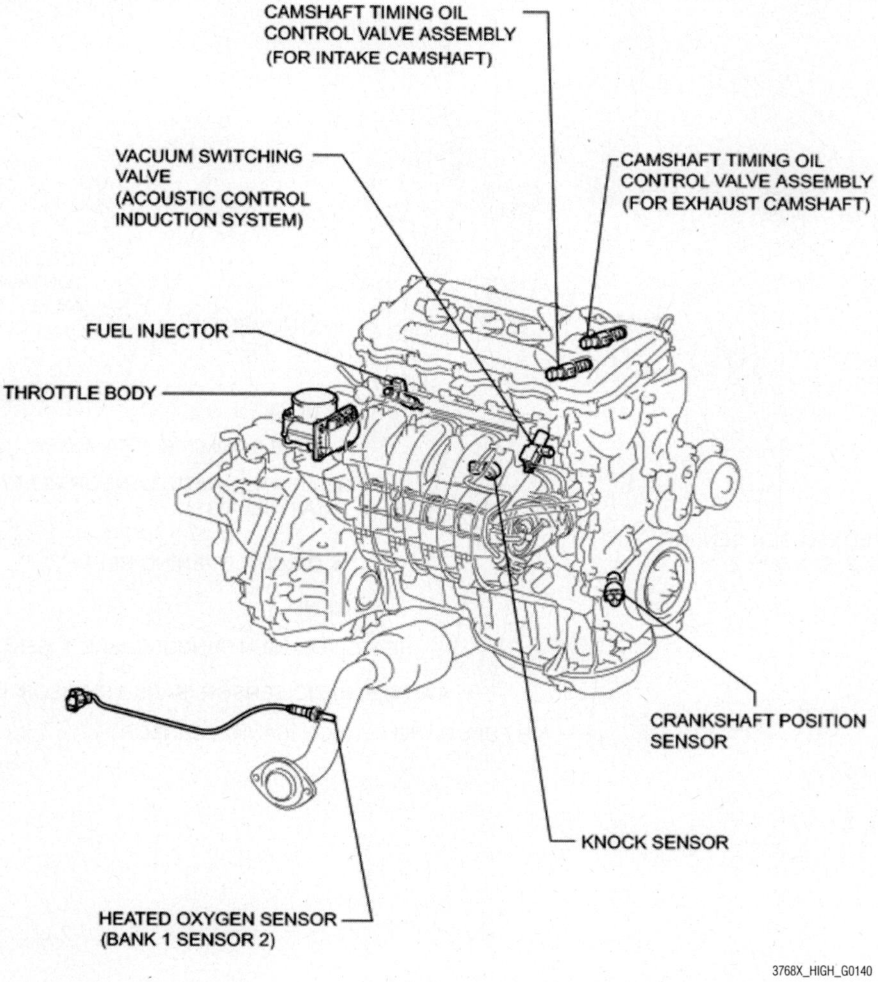

Fig. 295 Highlander component locations—2.7L Engine

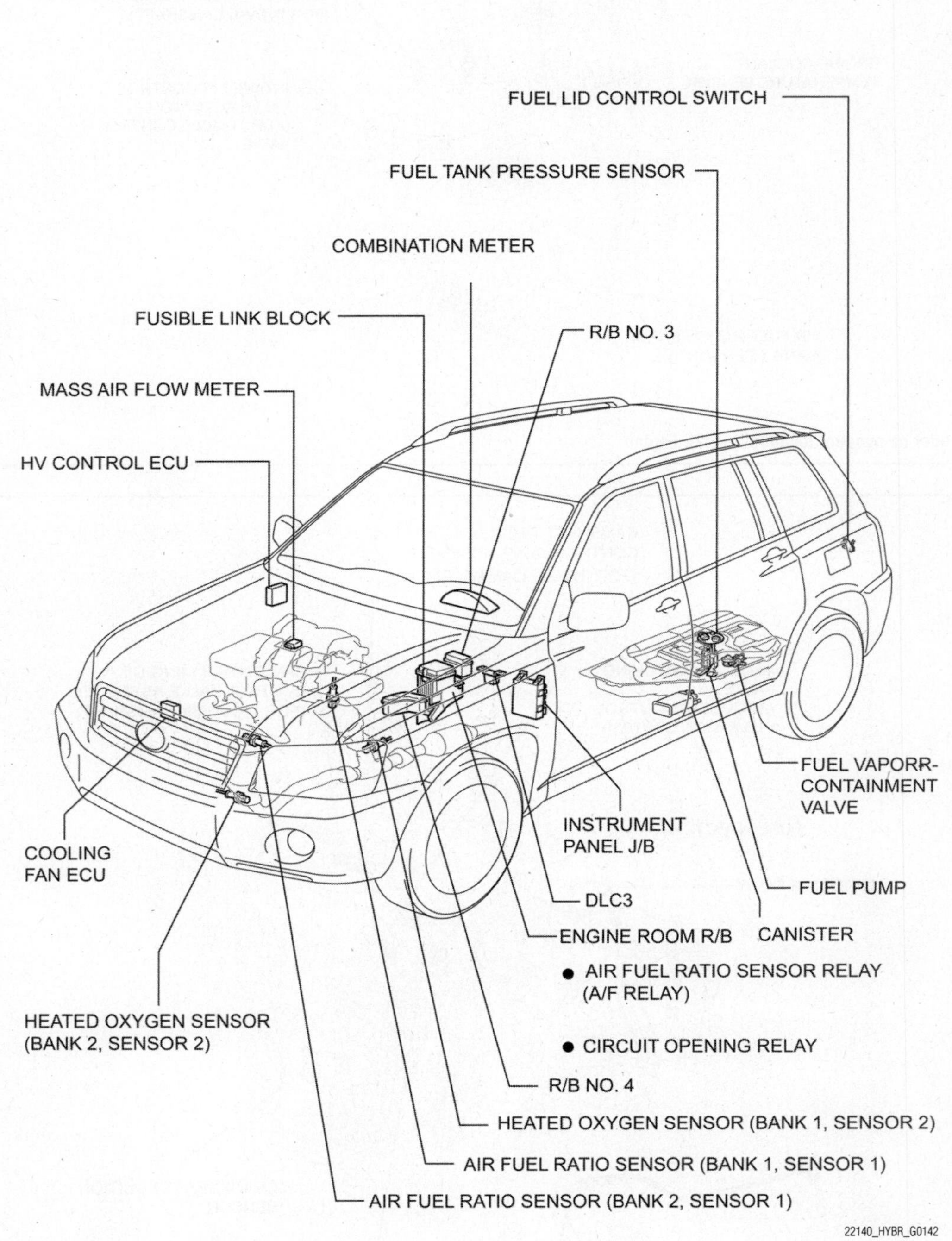

Fig. 296 Highlander Hybrid component locations—3.3L Engine

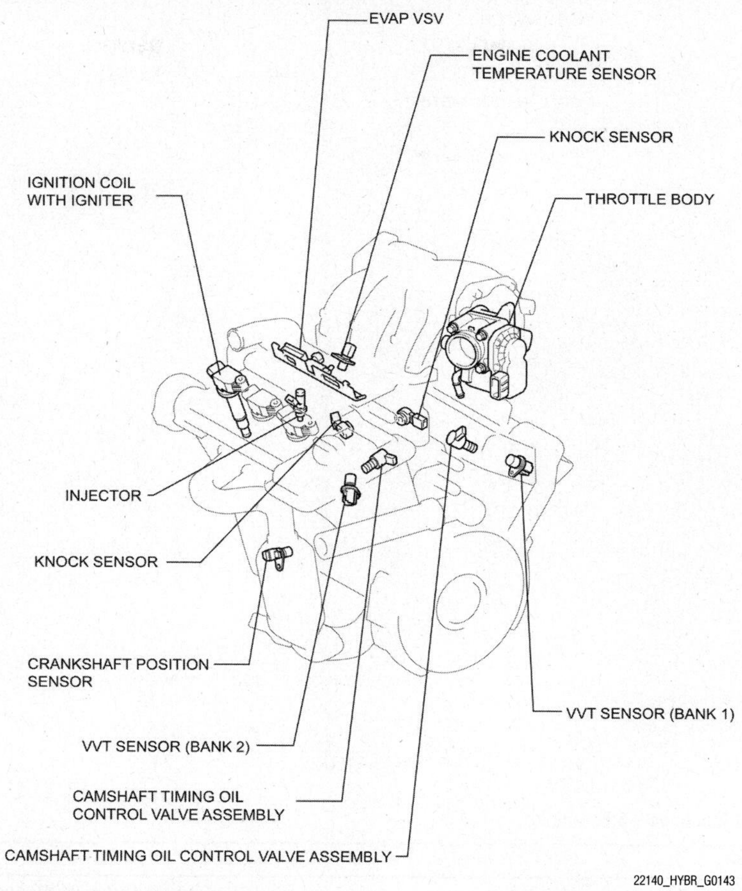

EVAP VSV
ENGINE COOLANT TEMPERATURE SENSOR
KNOCK SENSOR
IGNITION COIL WITH IGNITER
THROTTLE BODY
INJECTOR
KNOCK SENSOR
CRANKSHAFT POSITION SENSOR
VVT SENSOR (BANK 1)
VVT SENSOR (BANK 2)
CAMSHAFT TIMING OIL CONTROL VALVE ASSEMBLY
CAMSHAFT TIMING OIL CONTROL VALVE ASSEMBLY
22140_HYBR_G0143

Fig. 297 Highlander Hybrid component locations—3.3L Engine

CANISTER
HEATED OXYGEN SENSOR (BANK 2 SENSOR 2)
FUEL PUMP
HEATED OXYGEN SENSOR (BANK 1 SENSOR 2)
ENGINE ROOM RELAY BLOCK
- INTEGRATION RELAY
- FUEL PUMP RELAY
22140_HIGH_G0203

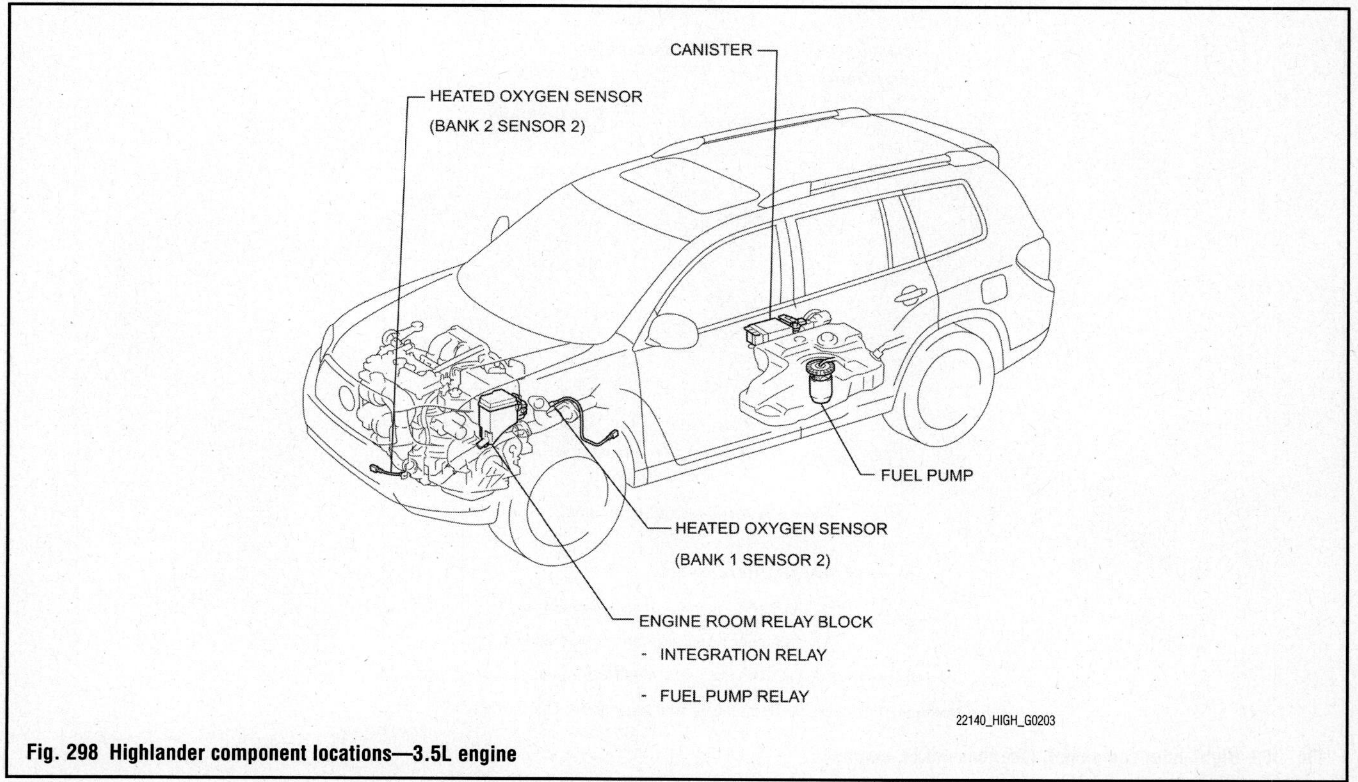

Fig. 298 Highlander component locations—3.5L engine

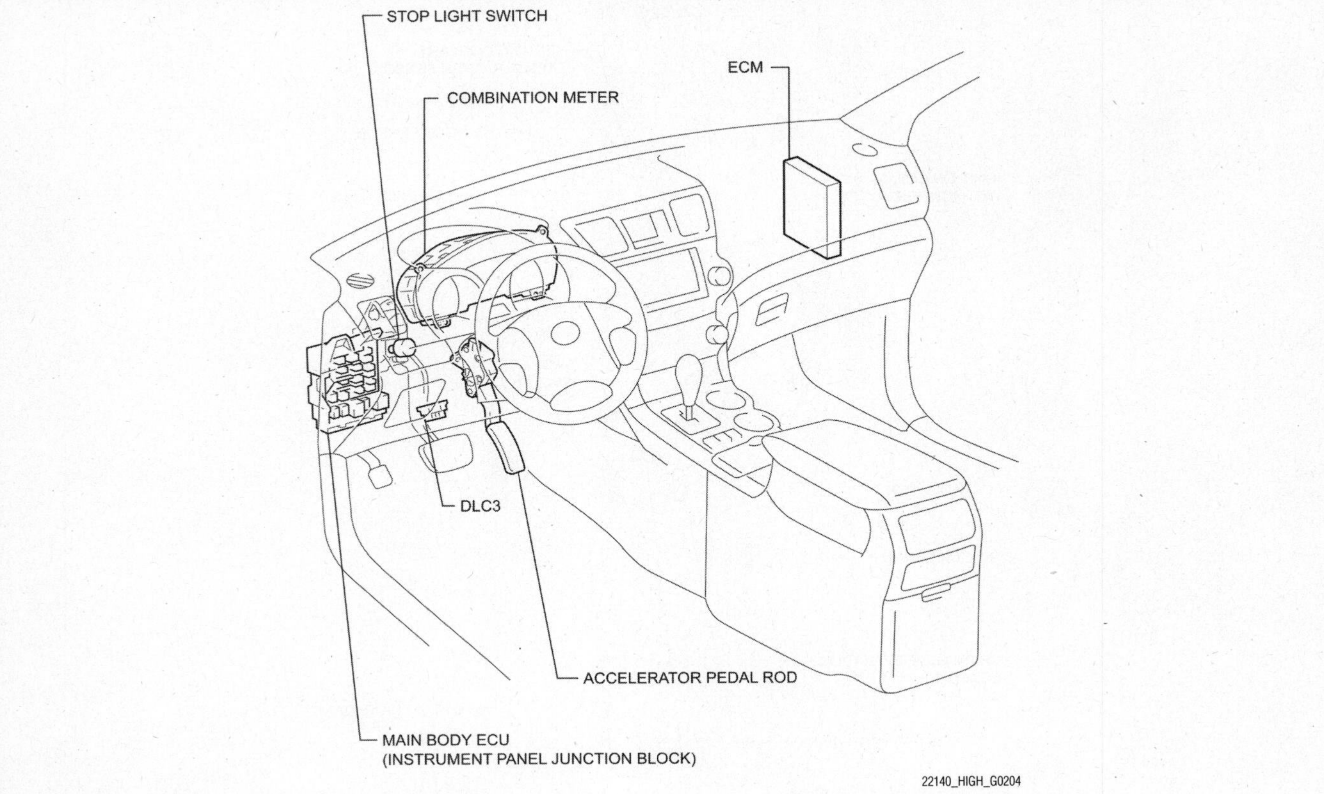

Fig. 299 Highlander component locations—3.5L engine

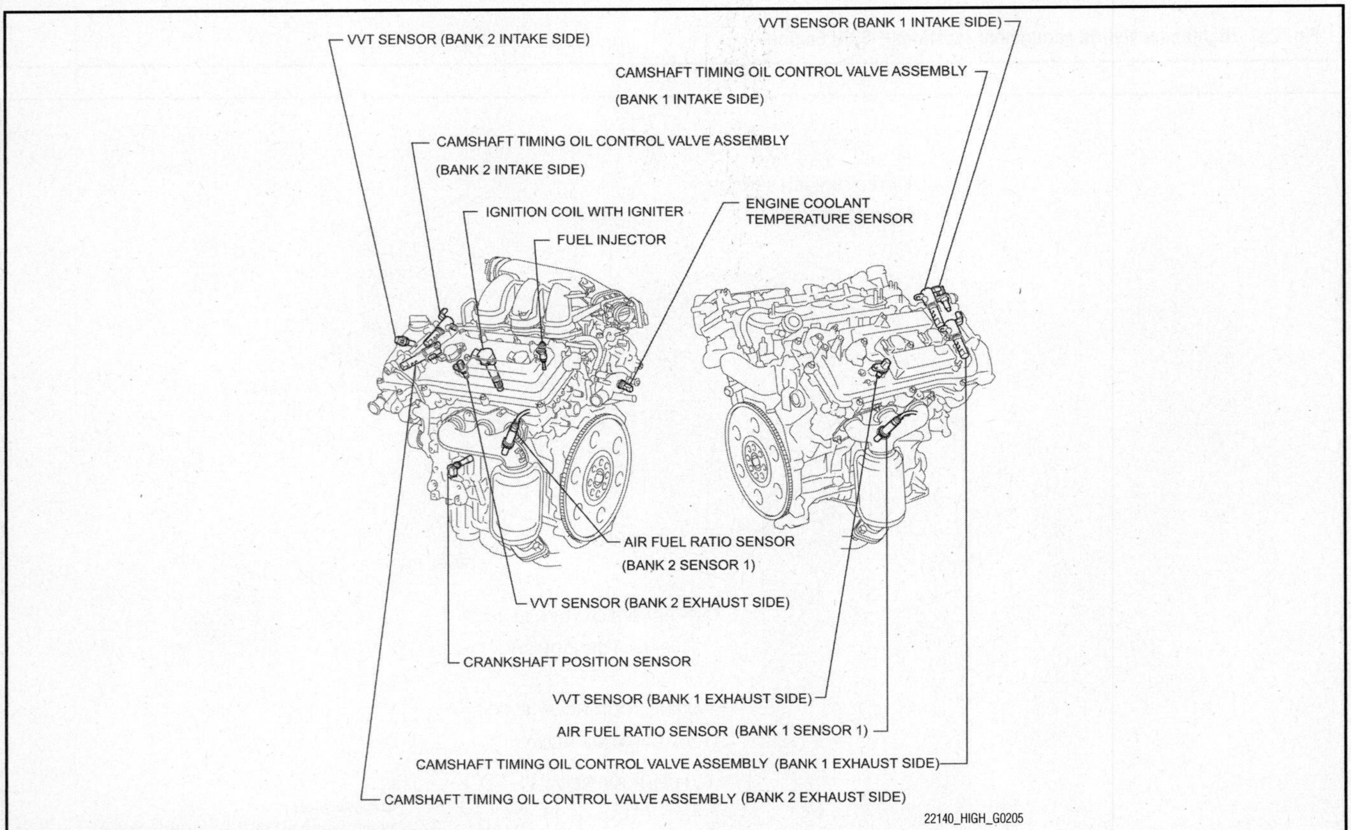

Fig. 300 Highlander component locations—3.5L engine

THROTTLE BODY
(THROTTLE POSITION SENSOR)
INTAKE AIR CONTROL VALVE
KNOCK SENSOR (BANK 1)
KNOCK SENSOR (BANK 2)
22140_HIGH_G0206

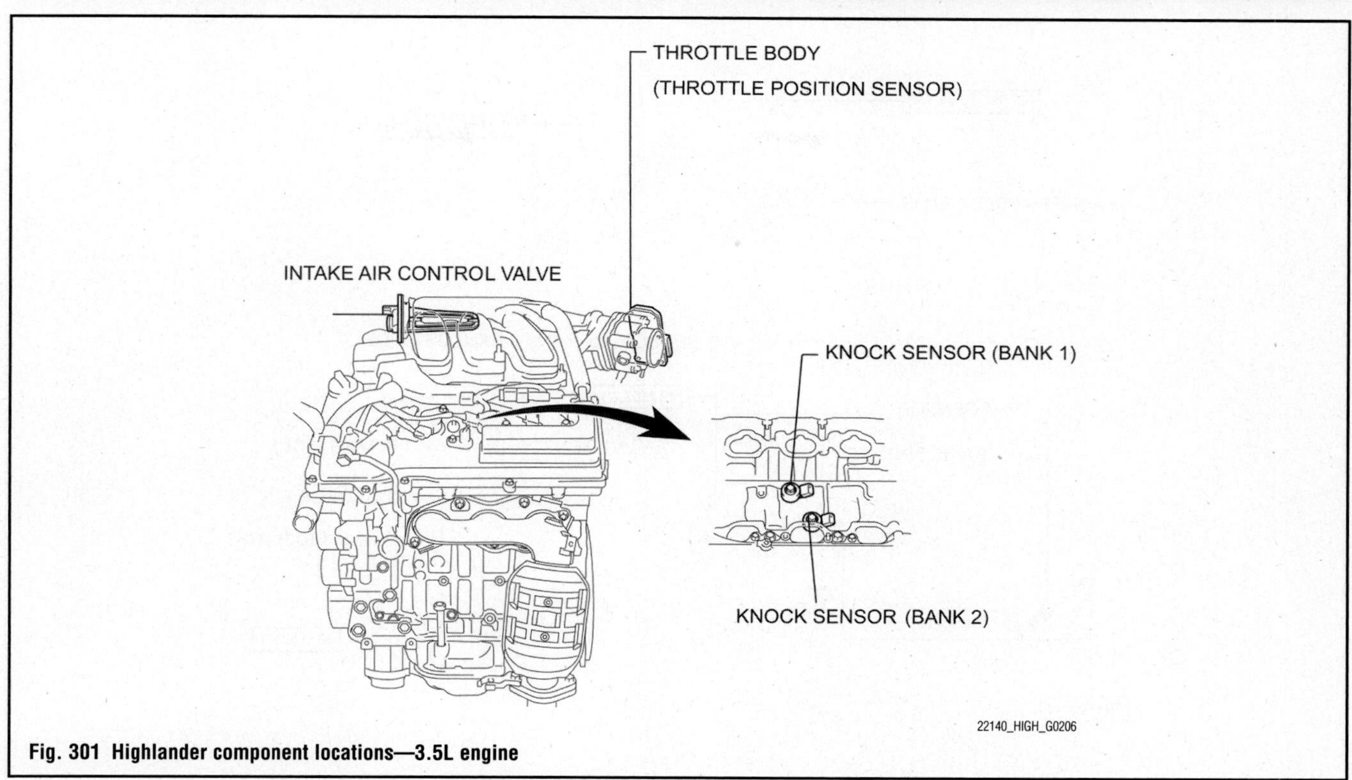

Fig. 301 Highlander component locations—3.5L engine

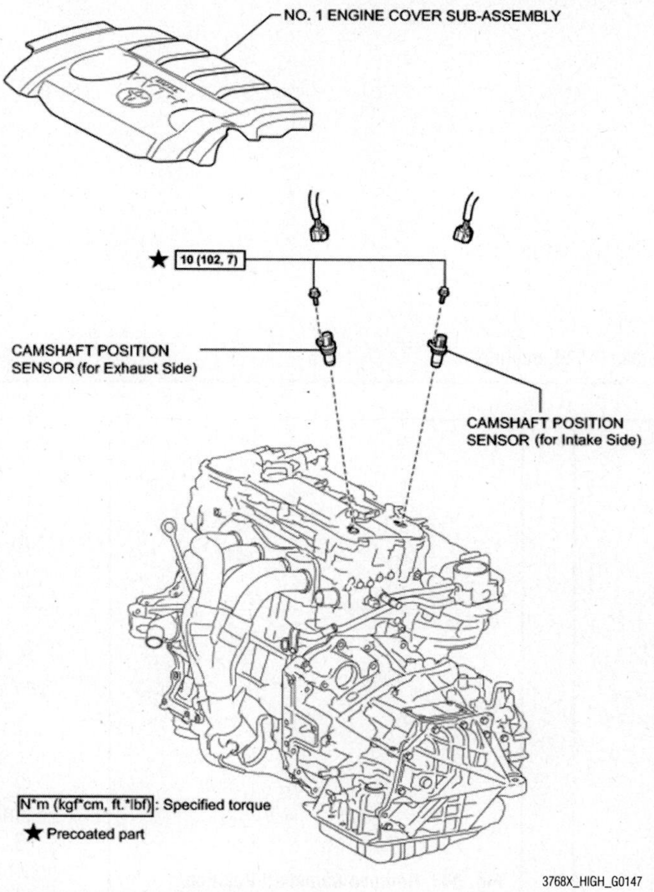

NO. 1 ENGINE COVER SUB-ASSEMBLY
★ 10 (102, 7)
CAMSHAFT POSITION
SENSOR (for Exhaust Side)
CAMSHAFT POSITION
SENSOR (for Intake Side)
N*m (kgf*cm, ft.*lbf): Specified torque
★ Precoated part
3768X_HIGH_G0147

Fig. 302 Camshaft Position (CMP) sensor location—2.7L engine

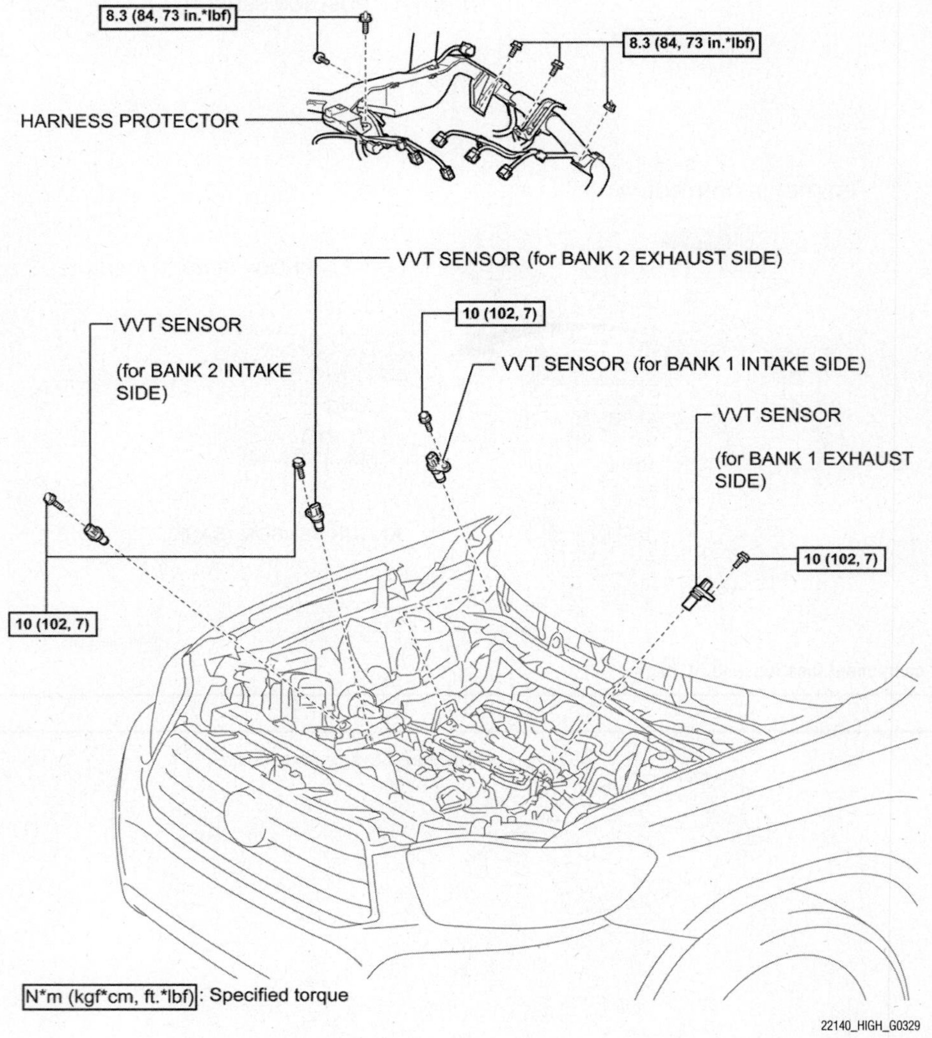

8.3 (84, 73 in.*lbf)
8.3 (84, 73 in.*lbf)
HARNESS PROTECTOR
VVT SENSOR (for BANK 2 EXHAUST SIDE)
10 (102, 7)
VVT SENSOR
(for BANK 2 INTAKE SIDE)
VVT SENSOR (for BANK 1 INTAKE SIDE)
VVT SENSOR
(for BANK 1 EXHAUST SIDE)
10 (102, 7)
10 (102, 7)
N*m (kgf*cm, ft.*lbf): Specified torque
22140_HIGH_G0329

Fig. 303 VVT (Camshaft position) sensor—3.5L engine

3768X_HIGH_G0148

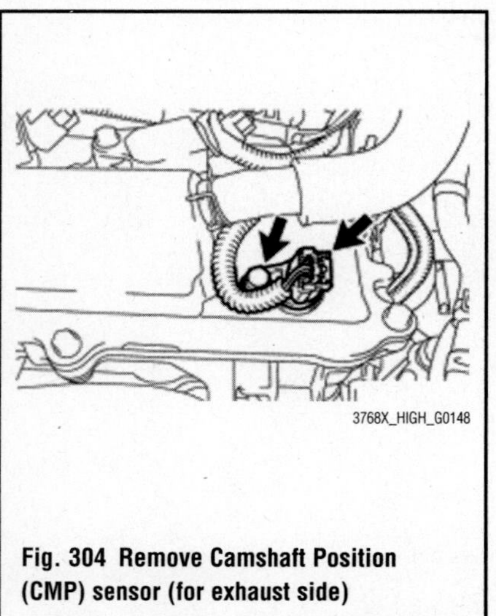

Fig. 304 Remove Camshaft Position (CMP) sensor (for exhaust side)

3768X_HIGH_G0149

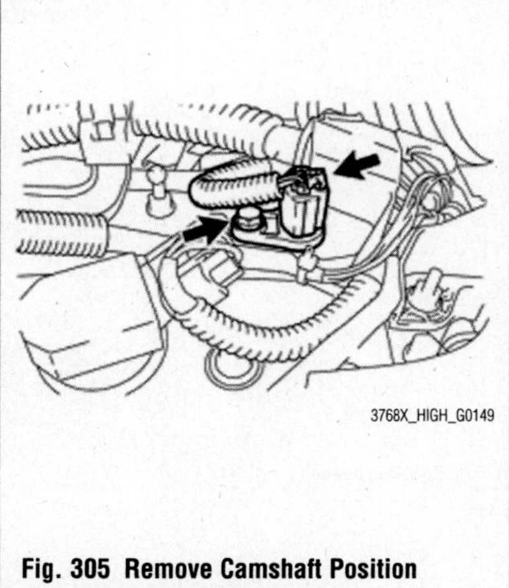

Fig. 305 Remove Camshaft Position (CMP) sensor (for intake side)

1. Disconnect the negative battery cable.
2. Remove front fender apron seal RH.
3. Disconnect the sensor connector.
4. Remove the bolt and sensor.

To install:

5. To install, reverse the removal procedure.

3.3L & 3.5L Engines

See Figures 309 and 310.

1. Recover the refrigerant from refrigeration system.
2. Remove the V-bank cover sub-assembly.
3. Remove the engine under cover assembly.
4. Remove the No. 1 engine under cover.

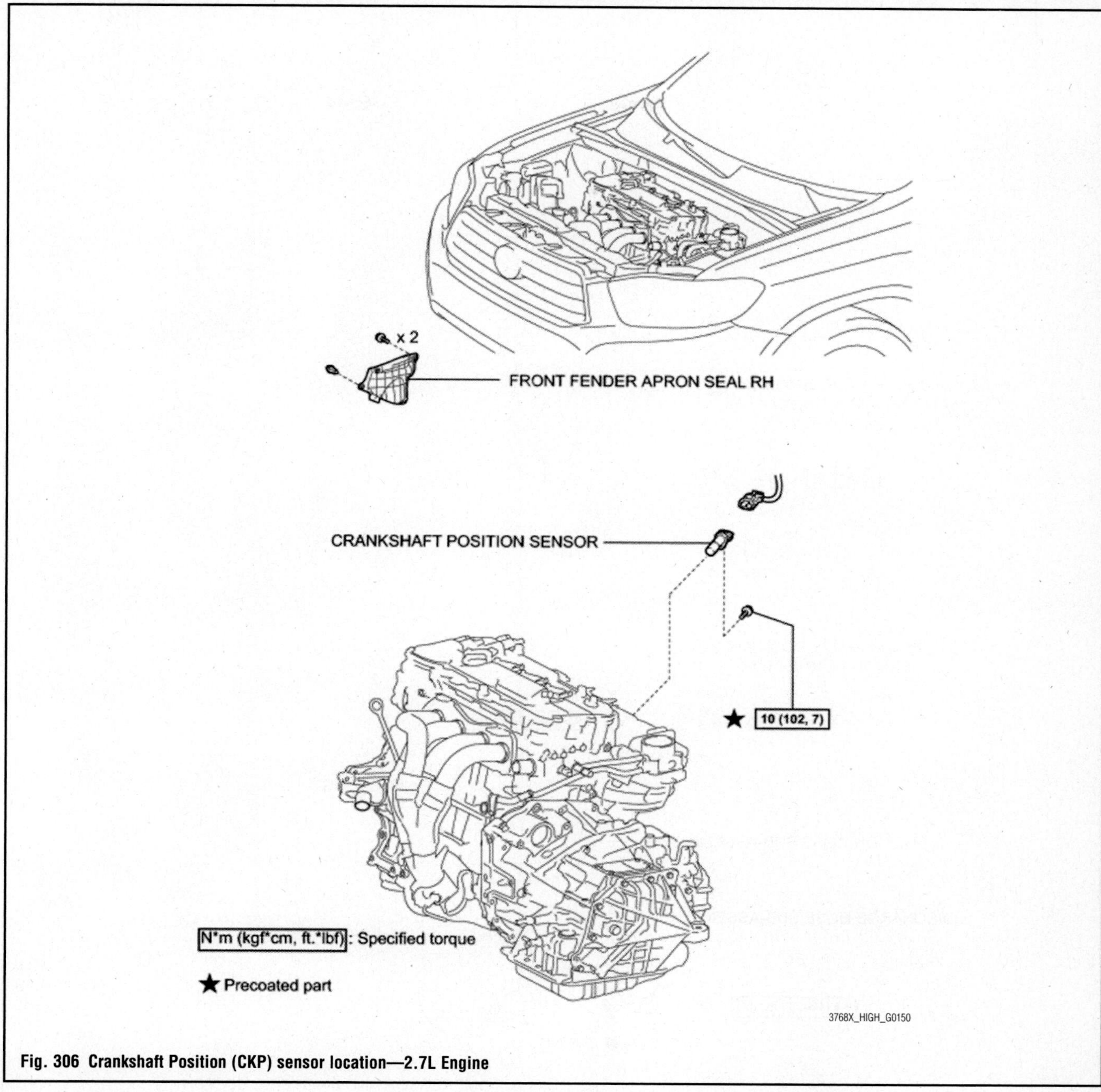

Fig. 306 Crankshaft Position (CKP) sensor location—2.7L Engine

5. Drain engine coolant.
6. Remove the cool air intake duct seal.
7. Remove the battery.
8. Remove the No. 1 and 2 air cleaner inlets.
9. Disconnect the No. 1 and 2 radiator hoses.
10. Disconnect the oil cooler hose.
11. Disconnect the cooling fan ECU connector.
12. Remove the radiator grill.
13. Remove the hood lock assembly.
14. Remove the upper radiator support sub-assembly.
15. Separate the cooler condenser assembly.
16. Remove the radiator assembly and fan assembly with motor.
17. Disconnect the discharge hose sub-assembly.

➡Seal the openings of the disconnected parts using vinyl tape to prevent entry of moisture and foreign matter.

18. Disconnect the suction hose sub-assembly.
19. Remove the compressor and magnetic clutch as follows:
20. Disconnect the Crankshaft Position (CKP) sensor connector.
21. Remove the bolt and crankshaft position sensor, as follows:
    a. Disconnect the Crankshaft Position (CKP) sensor connector.
    b. Remove the bolt and crankshaft position sensor.

To install:

22. Apply a light coat of engine oil to the O-ring on the crankshaft position sensor.
23. Install the Crankshaft Position (CKP) sensor with the bolt and tighten to 7 ft. lbs. (10 Nm).

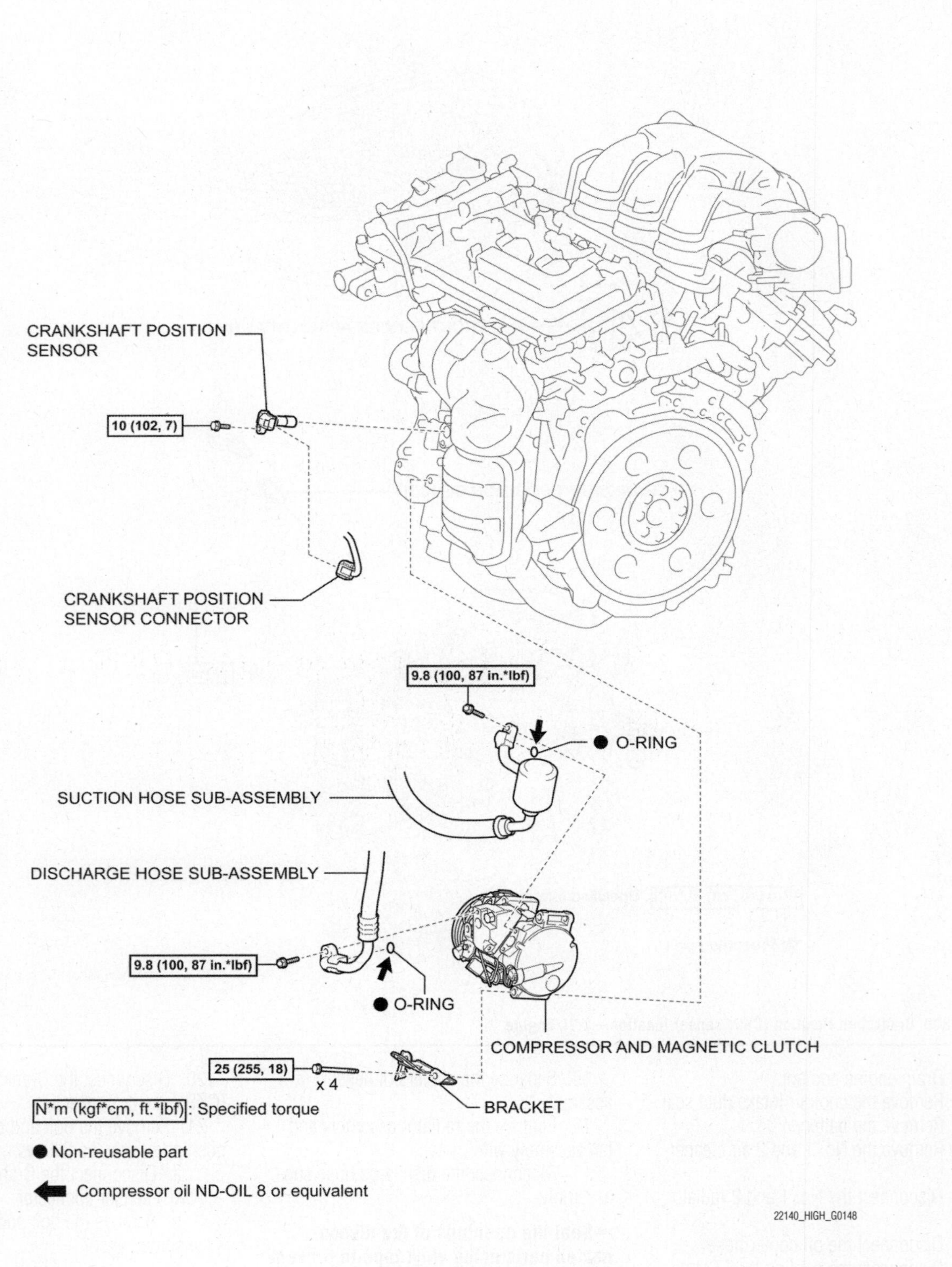

Fig. 307 Crankshaft Position (CKP) sensor location—3.3L & 3.5L Engines

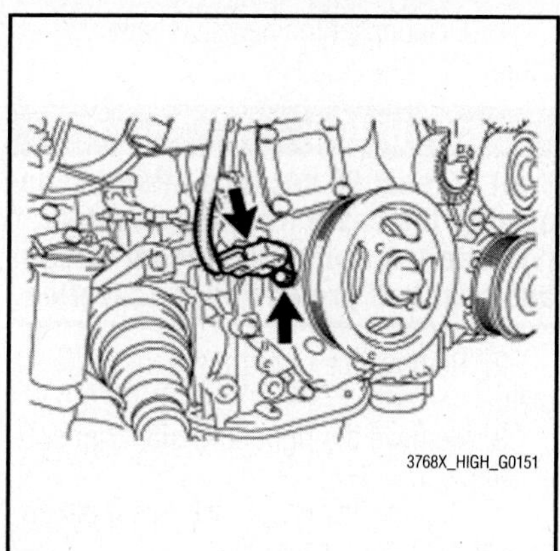

Fig. 308 Remove the bolt and sensor—2.7L Engine

24. Connect the Crankshaft Position (CKP) sensor connector.
25. Temporarily tighten compressor and magnetic clutch with the bolts, in the order shown.
26. Install the compressor and magnetic clutch with the 4 bolts. Tighten the bolts in the order shown above, and torque to 18 ft. lbs. (25 Nm).
27. Connect the suction hose sub-assembly.
28. Connect the discharge hose sub-assembly.
29. Install the radiator assembly and fan assembly with motor.
30. Install the cooler condenser assembly.
31. Install the upper radiator support sub-assembly.

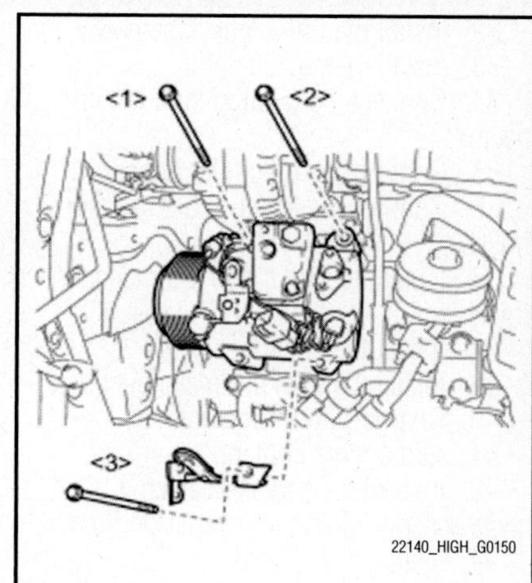

Fig. 310 Compressor and magnetic clutch

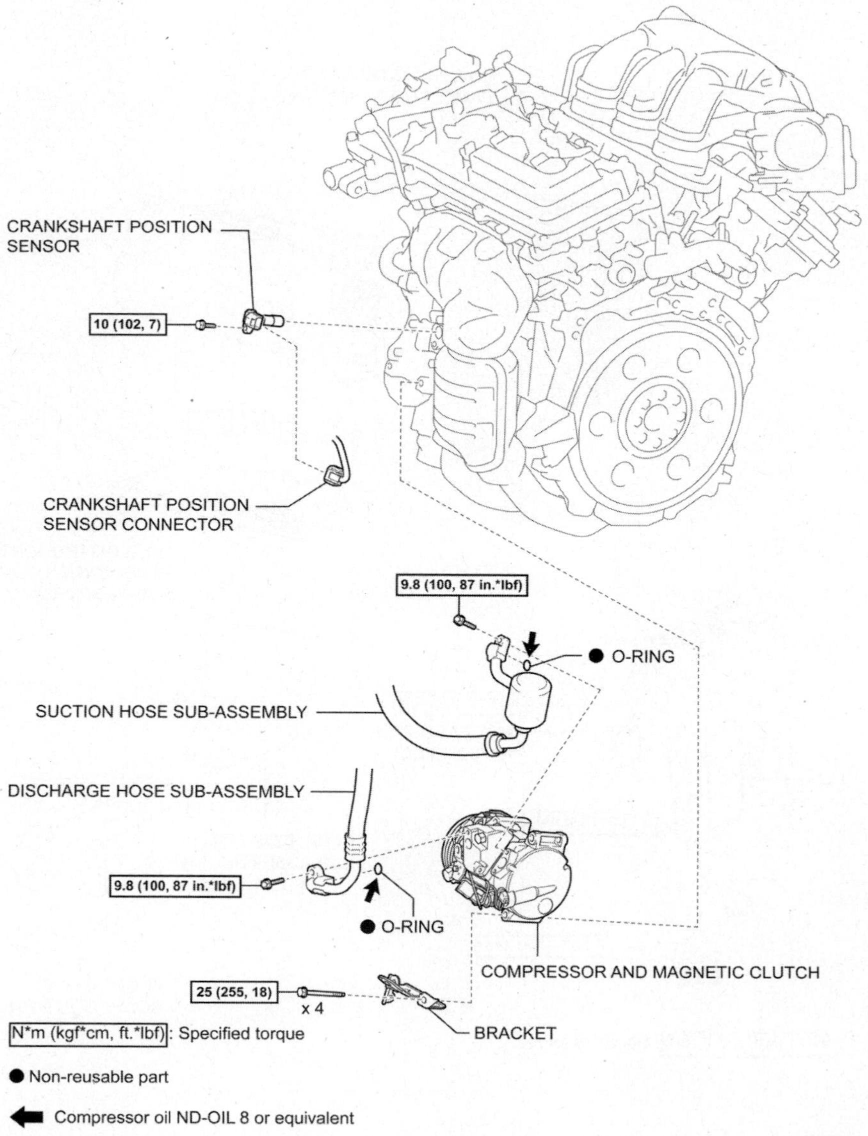

Fig. 309 Crankshaft Position (CKP) sensor and related components

32. Install the hood lock assembly.
33. Install the radiator grill.
34. Connect the cooling fan ECU connector.
35. Connect the oil cooler hose.
36. Connect the No. 1 and 2 radiator hoses.
37. Install the No. 1 and 2 air cleaner inlets.
38. Install the battery.
39. Install the cool air intake duct seal.
40. Charge with refrigerant.
41. Add engine coolant.
42. Inspect for coolant leaks.
43. Inspect automatic transaxle fluid.
44. Inspect for oil leak.
45. Warm up the engine.
46. Inspect for refrigerant leak.
47. Install the No. 1 engine under cover.
48. Install the engine under cover assembly.
49. Install the V-bank cover sub-assembly.

ELECTRONIC CONTROL MODULE (ECM)

LOCATION

See Figures 311 through 313

REMOVAL & INSTALLATION

2.7L & 3.5L Engines

See Figures 314 through 316.

1. Disconnect the negative battery cable.

CAUTION

Wait at least 90 seconds after disconnecting the cable from the negative (-) battery terminal to prevent airbag and seat belt pretensioner activation.

2. Remove the right front door scuff plate.
3. Remove the right cowl side trim sub-assembly.
4. Remove the No. 2 instrument panel under cover sub-assembly.
5. Remove the lower instrument panel sub-assembly.

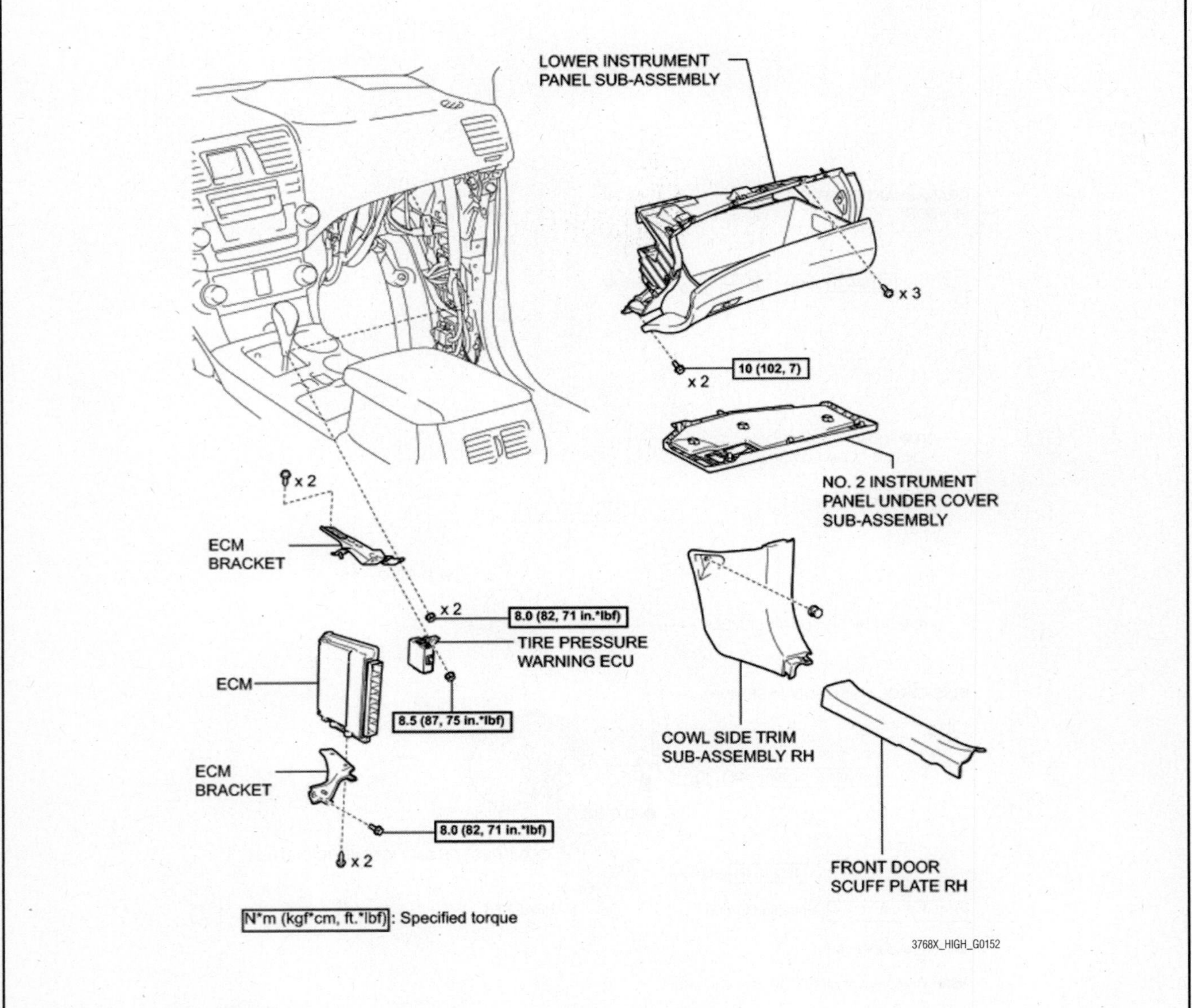

Fig. 311 ECM location—2.7L Engines

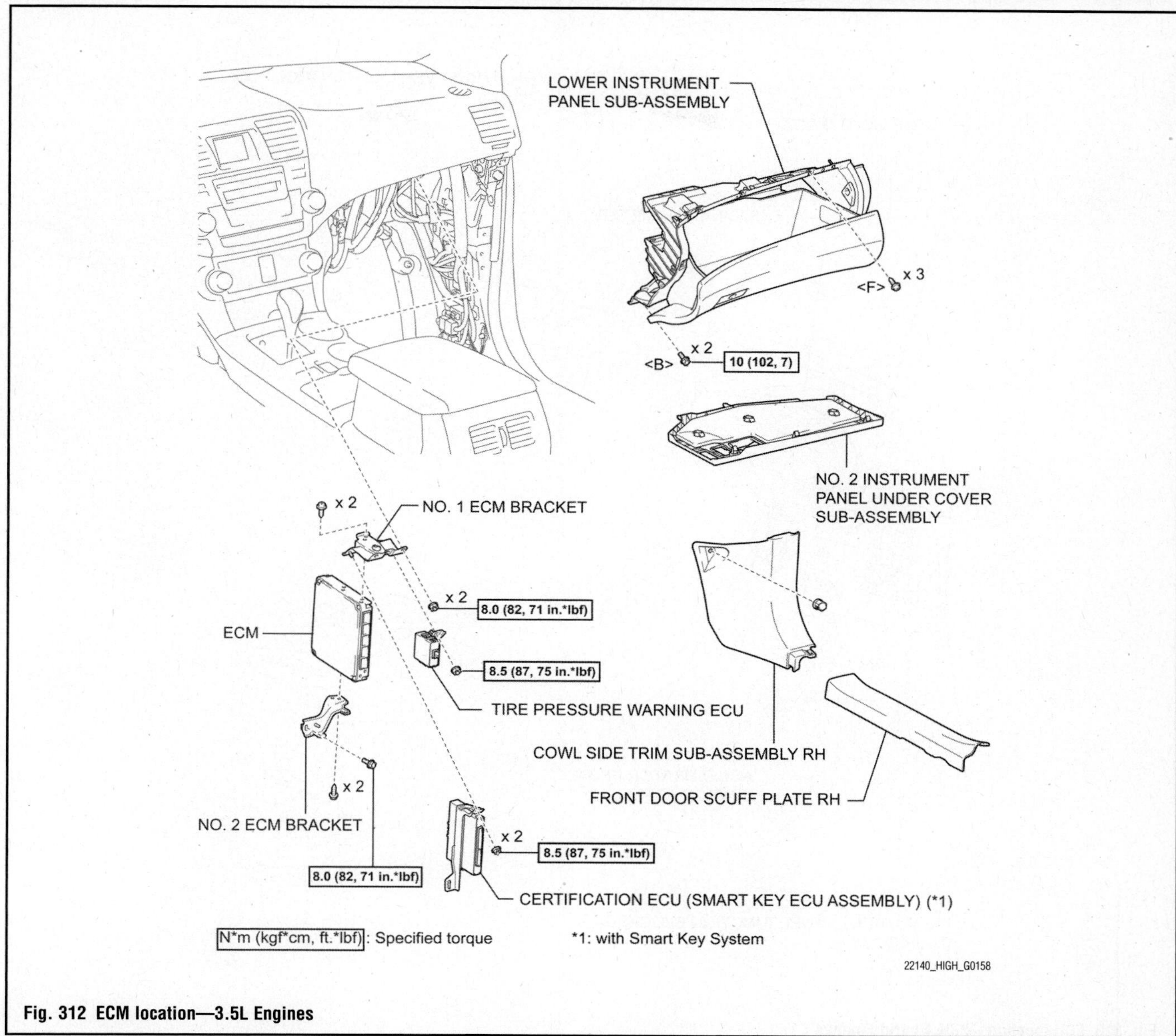

Fig. 312 ECM location—3.5L Engines

6. Remove the certification ECU (Smart Key ECU assembly), if applicable.

WARNING

Before removing the tire pressure warning ECU, read the registered transmitter IDs of all wheels and write them down to use for re-registration of transmitter IDs.

7. Remove the tire pressure warning ECU.
8. Remove the ECM, as follows:
   a. Separate the harness connector.
   b. Disconnect the 5 or 6 ECM connectors.
   c. Remove the wire harness clamp.
   d. Remove the 2 nuts, bolt and ECM.
9. Remove the 2 screws and the No. 1 ECM bracket.
10. Remove the 2 screws and the No. 2 ECM bracket.

To install:

11. Installation is the reverse of the removal procedure. After connecting the negative battery cable, perform the following procedures:
   a. Register the transmitter ID.
   b. Inspect the tire pressure warning system.
   c. Initialize tire pressure warning system.

➡Be sure to register the transmitter IDs of all tires in the ECU before initialization.

➡Be sure to inflate all tires to the proper inflation pressure before initialization.

   d. Register immobilizer communication ID.

➡If the ECM is replaced, register the ECU communication ID for the immobilizer system.

   e. Perform initialization.

➡If the ECM is replaced, perform RESET MEMORY (at initialization).

3.3L Engine

1. Disconnect the negative battery cable.
2. Remove the instrument under cover.
3. Remove the right door scuff plate.
4. Remove the cowl side trim.

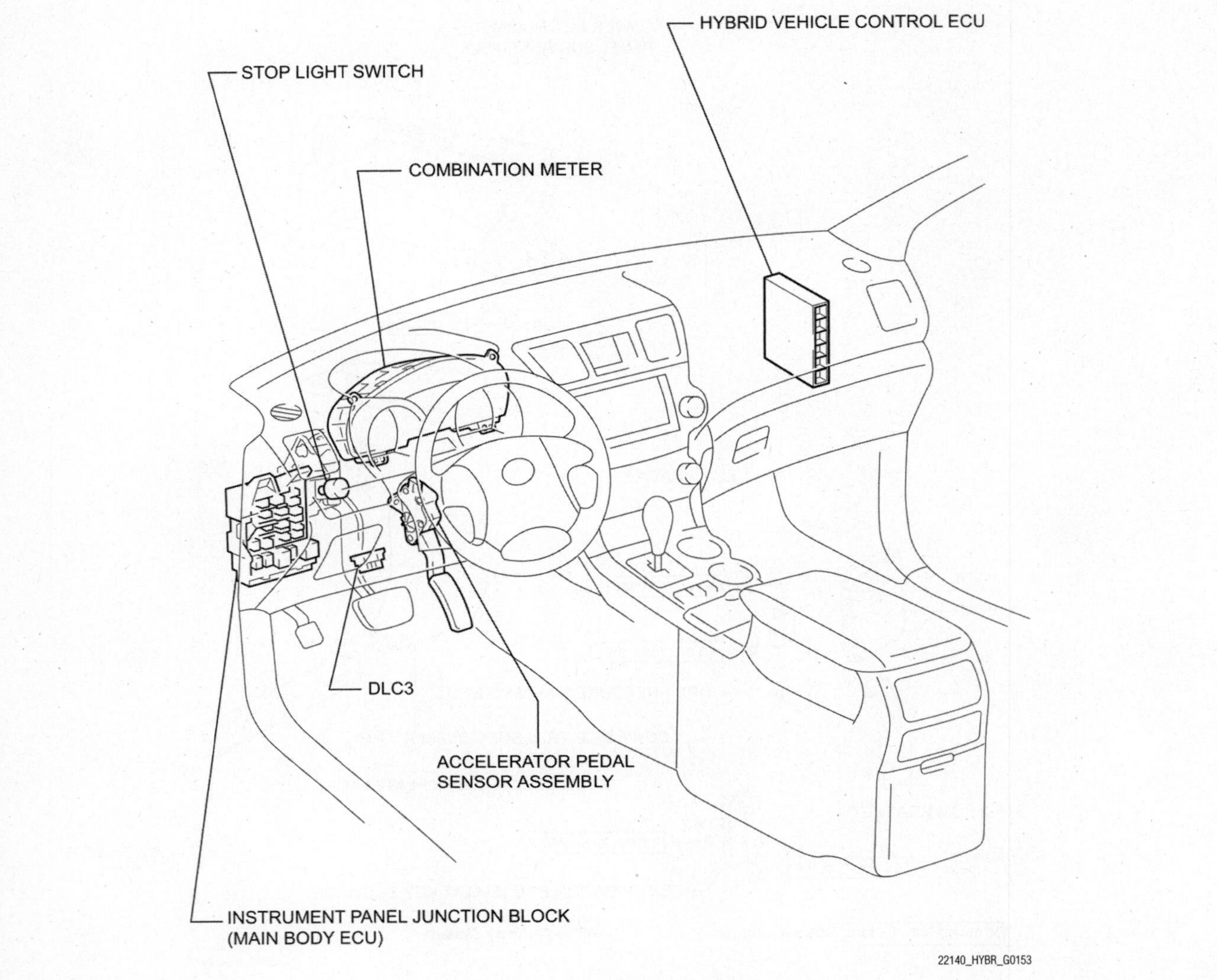

Fig. 313 ECU location—3.3L Hybrid Engines

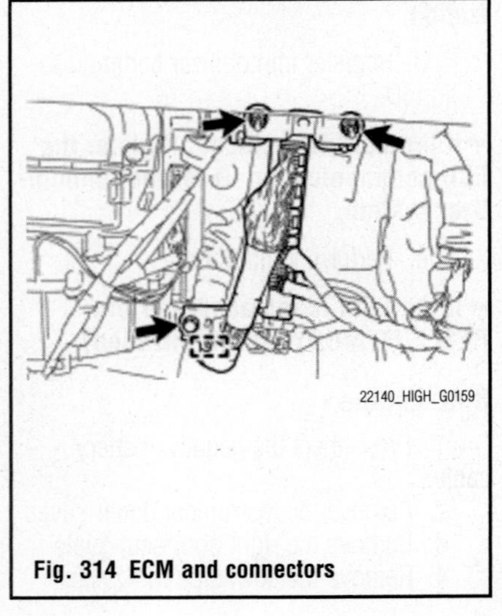

Fig. 314 ECM and connectors

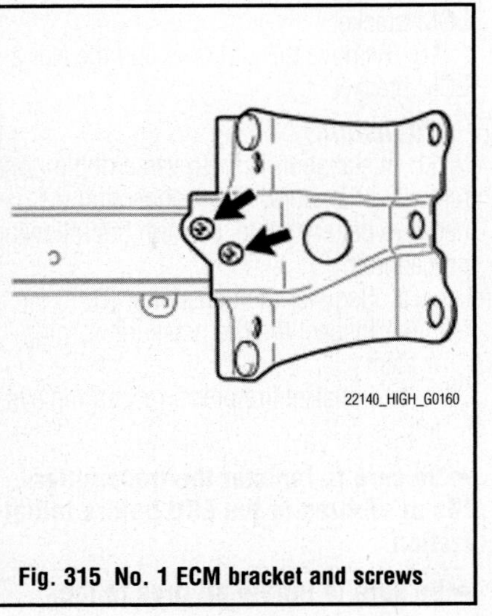

Fig. 315 No. 1 ECM bracket and screws

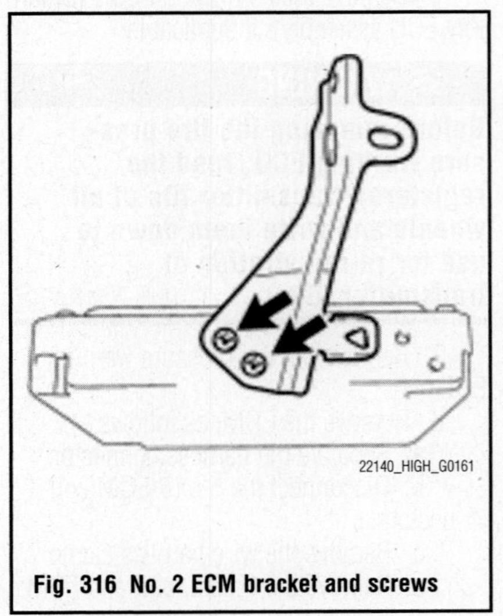

Fig. 316 No. 2 ECM bracket and screws

5. Remove the glove compartment door assembly.
6. Disconnect the 3 wire harness clamps and 6 connectors from the hybrid vehicle control ECU
7. Remove the 2 nuts and hybrid vehicle control ECU.
8. Remove the ECU.
9. If ECU is to be changed remove the brackets.

To install:

10. Install brackets to the ECU if previously removed.
11. Install the wire harness clamp bracket with the 2 screws.
12. Install the hybrid vehicle control ECU with the 2 nuts. Tighten the mounting nuts to 49 inch lbs. (5.5 Nm).
13. Connect the 6 connectors to the hybrid vehicle control ECU.
14. Connect the 3 wire harness clamps.
15. Install the glove compartment door assembly.
16. Install the undercover.
17. Install the cowl side trim.
18. Install the scuff plate.
19. Connect the negative battery cable.

➡After replacing the hybrid vehicle control ECU on vehicles with a dynamic laser cruise control system, it is necessary to initialize the hybrid vehicle control ECU so that the ECU can recognize the dynamic laser cruise control system.

Be sure to perform the following procedures after replacing the ECU.

- Turn the ignition switch to the on position.
- Turn the cruise main switch on.
- With the brake pedal depressed, push the cruise control main switch to RES/ACC 3 times within 3 seconds. Check that the buzzer sounds at this time.

➡Do not turn the headlight dimmer switch on at this time because the optical axis automatic adjustment mode has already started, which may lead to an incorrect optical axis setting. If the headlight dimmer switch is turned on by mistake, readjust the optical axis.

20. Some systems need initialization when disconnecting the cable from the negative battery terminal.

ENGINE COOLANT TEMPERATURE (ECT) SENSOR

LOCATION

See Figures 317 through 319.

REMOVAL & INSTALLATION

2.7L Engine

See Figure 320.

1. Remove windshield wiper motor and link.
2. Separate ejector tube.
3. Remove outer cowl top panel sub-assembly.
4. Remove no. 1 engine under cover.
5. Drain engine coolant.
6. Remove no. 1 engine cover sub-assembly.
7. Remove no. 1 vacuum switching valve.
8. Remove air cleaner cap sub-assembly.
9. Remove air cleaner case.
10. Disconnect the Engine Coolant Temperature (ECT) sensor connector.

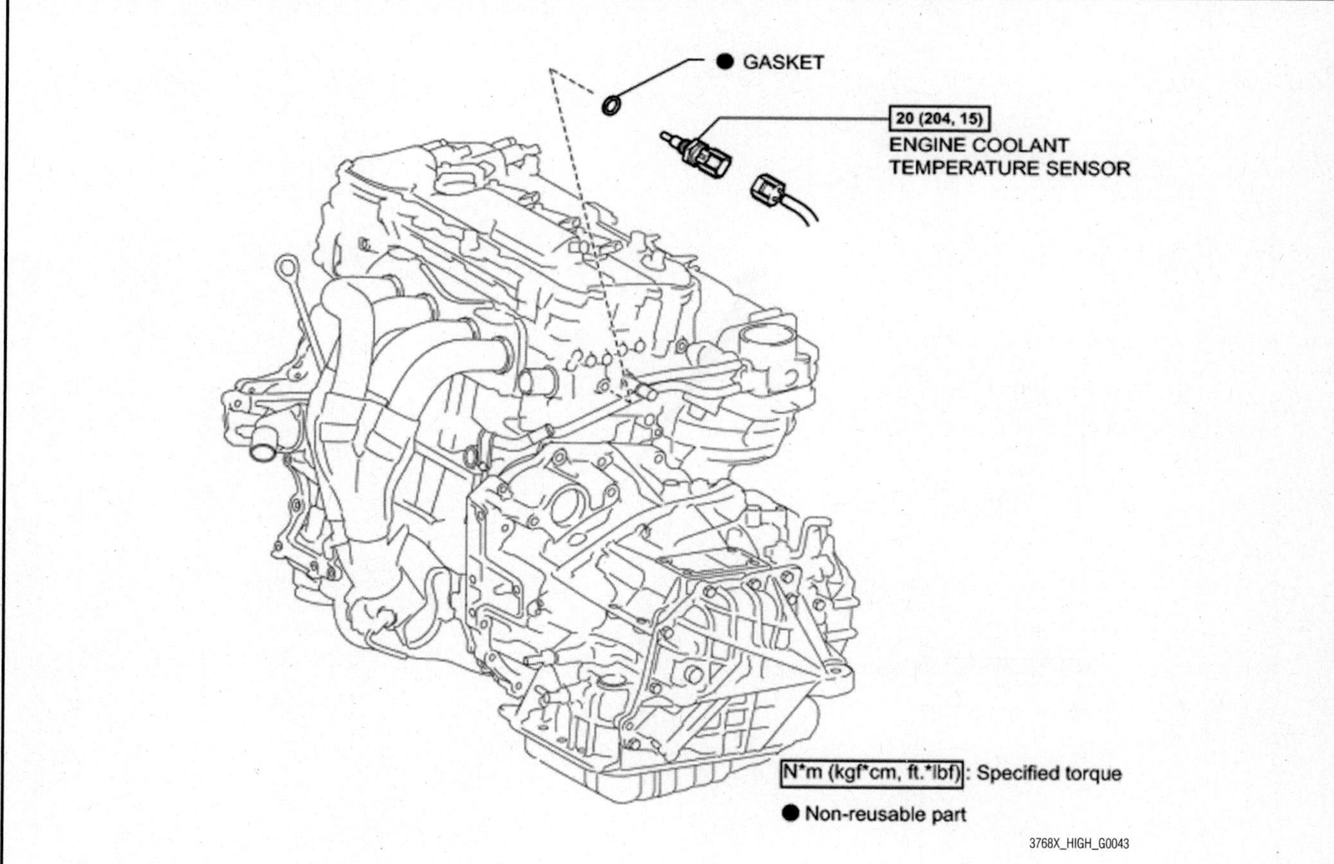

Fig. 317 Engine Coolant Temperature (ECT) sensor location—2.7L engine

11. Remove the Engine Coolant Temperature (ECT) sensor and gasket.

To install:

12. Install a new gasket onto the Engine Coolant Temperature (ECT) sensor.
13. Install the Engine Coolant Temperature (ECT) sensor and tighten to 15 ft. lbs. (20 Nm).
14. The remainder of installation is the reverse of the removal procedure.

3.3L Engine

1. Remove the engine room covers.
2. Remove the RH and LH wiper arm assembly.
3. Remove the cowl and seal.
4. Remove the wiper motor and link assembly.
5. Remove the cool air intake duct seal.
6. Remove the air cleaner cap with inlet.
7. Remove the air cleaner case with resonator.
8. Air cleaner bracket.
9. Disconnect the ECT sensor connector.
10. Using a deep socket, remove the sensor from the top of the engine.
11. Remove the gasket from the sensor.

To install:

12. Installation is the reverse of the removal procedure.

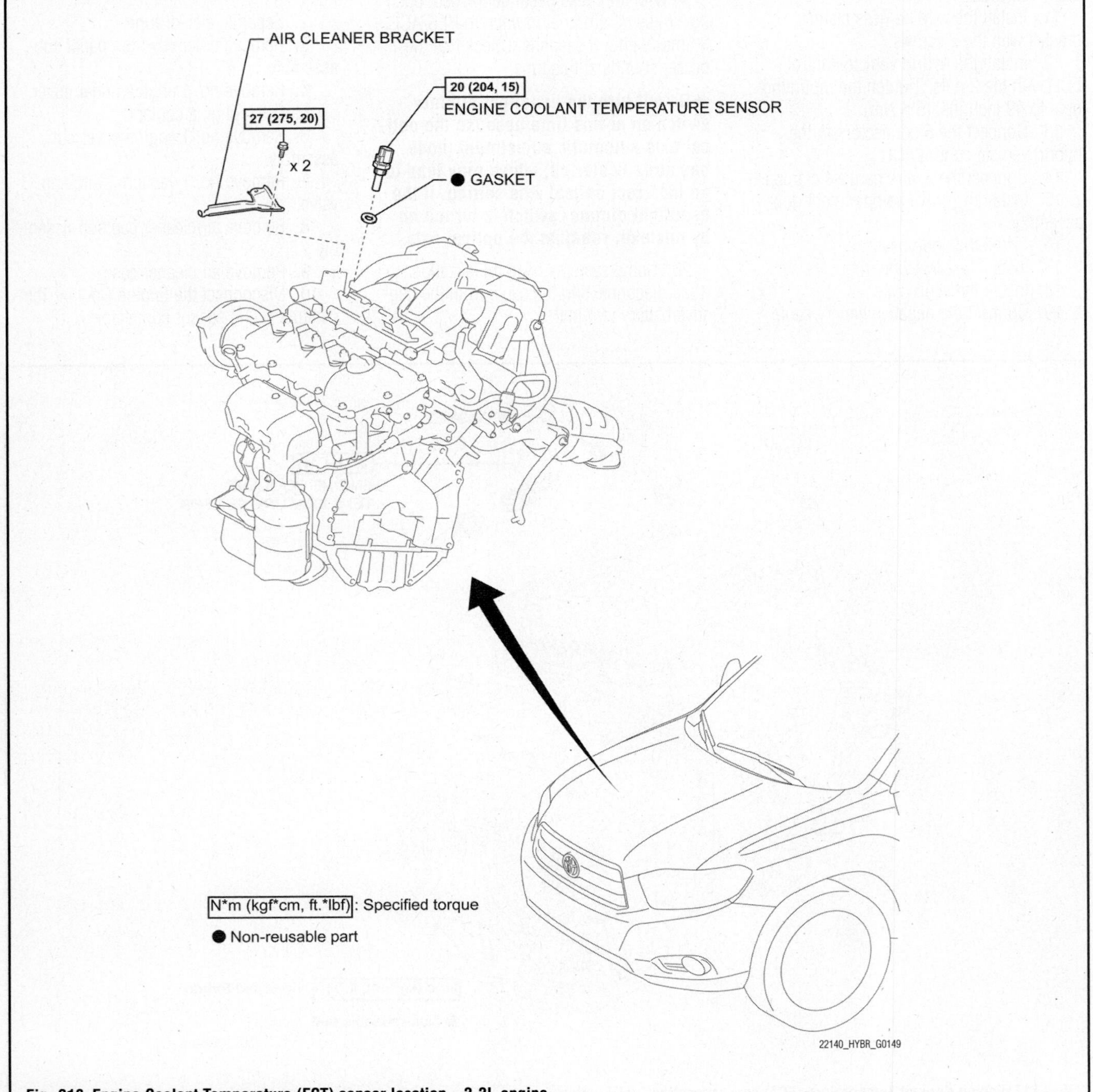

Fig. 318 Engine Coolant Temperature (ECT) sensor location—3.3L engine

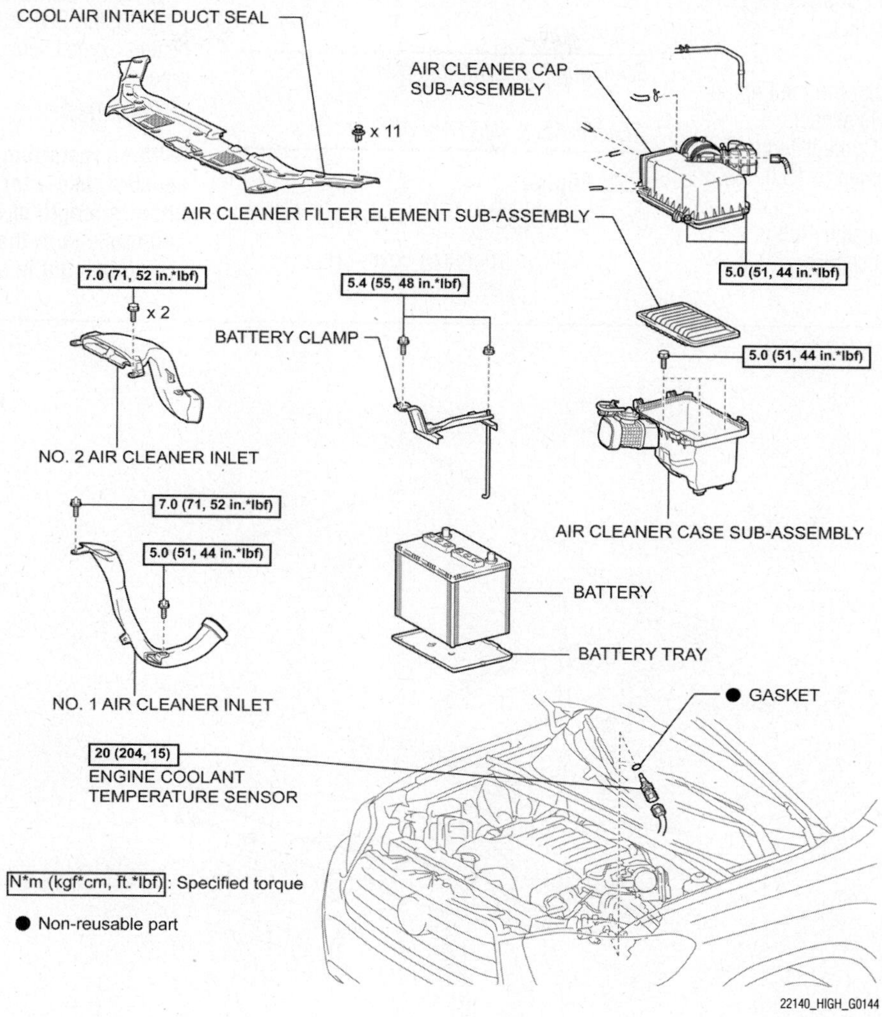

COOL AIR INTAKE DUCT SEAL
x 11
AIR CLEANER CAP SUB-ASSEMBLY
AIR CLEANER FILTER ELEMENT SUB-ASSEMBLY
7.0 (71, 52 in.*lbf)
x 2
5.4 (55, 48 in.*lbf)
5.0 (51, 44 in.*lbf)
BATTERY CLAMP
5.0 (51, 44 in.*lbf)
NO. 2 AIR CLEANER INLET
7.0 (71, 52 in.*lbf)
5.0 (51, 44 in.*lbf)
AIR CLEANER CASE SUB-ASSEMBLY
BATTERY
BATTERY TRAY
NO. 1 AIR CLEANER INLET
● GASKET
20 (204, 15)
ENGINE COOLANT TEMPERATURE SENSOR
N*m (kgf*cm, ft.*lbf): Specified torque
● Non-reusable part
22140_HIGH_G0144

Fig. 319 Engine Coolant Temperature (ECT) sensor location—3.5L engine

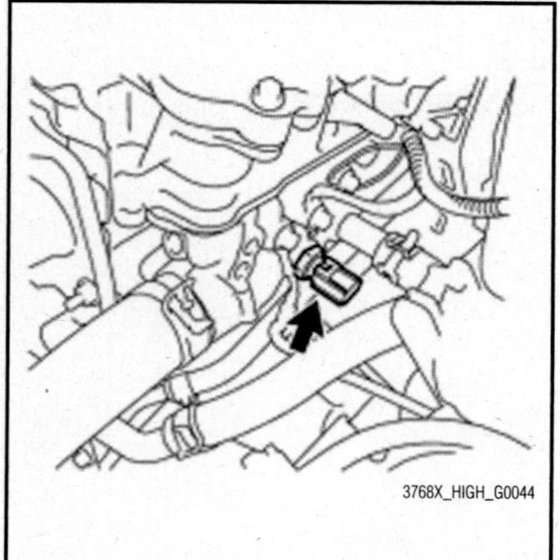

Fig. 320 Removing the Engine Coolant Temperature (ECT) sensor and gasket

13. Tighten the ECT sensor to 15 ft. lbs. (20 Nm).

3.5L Engine

See Figure 321.

1. Remove the V-bank cover sub-assembly.
2. Remove the engine under cover assembly.
3. Remove the No. 1 engine under cover.
4. Drain engine coolant.
5. Remove the cool air intake duct seal.
6. Remove the battery.
7. Remove the No. 1 and 2 air cleaner inlets.
8. Remove the air cleaner cap and case sub-assemblies.

Fig. 321 Engine Coolant Temperature (ECT) sensor, connector, and gasket—3.5L engine

9. Disconnect the Engine Coolant Temperature (ECT) sensor connector.
10. Remove the Engine Coolant Temperature (ECT) sensor and gasket.

To install:

11. Install a new gasket onto the Engine Coolant Temperature (ECT) sensor.
12. Install the Engine Coolant Temperature (ECT) sensor and tighten to 15 ft. lbs. (20 Nm).
13. The remainder of installation is the reverse of the removal procedure.

HEATED OXYGEN (HO2S) SENSOR

LOCATION

See Figures 322 through 324

REMOVAL & INSTALLATION

2.7L Engine

See Figure 325.

1. Remove front floor cover LH.
2. Disconnect the heated oxygen sensor connector.
3. Separate the 2 wire harness clamps.
4. Using SST: 09224-00010, remove the heated oxygen sensor from the front exhaust pipe.

To install:

➡When installing the heated oxygen sensors, use a torque wrench with a fulcrum length of 30 cm (11.81 inches), and make sure that SST and the wrench are connected in a straight line.

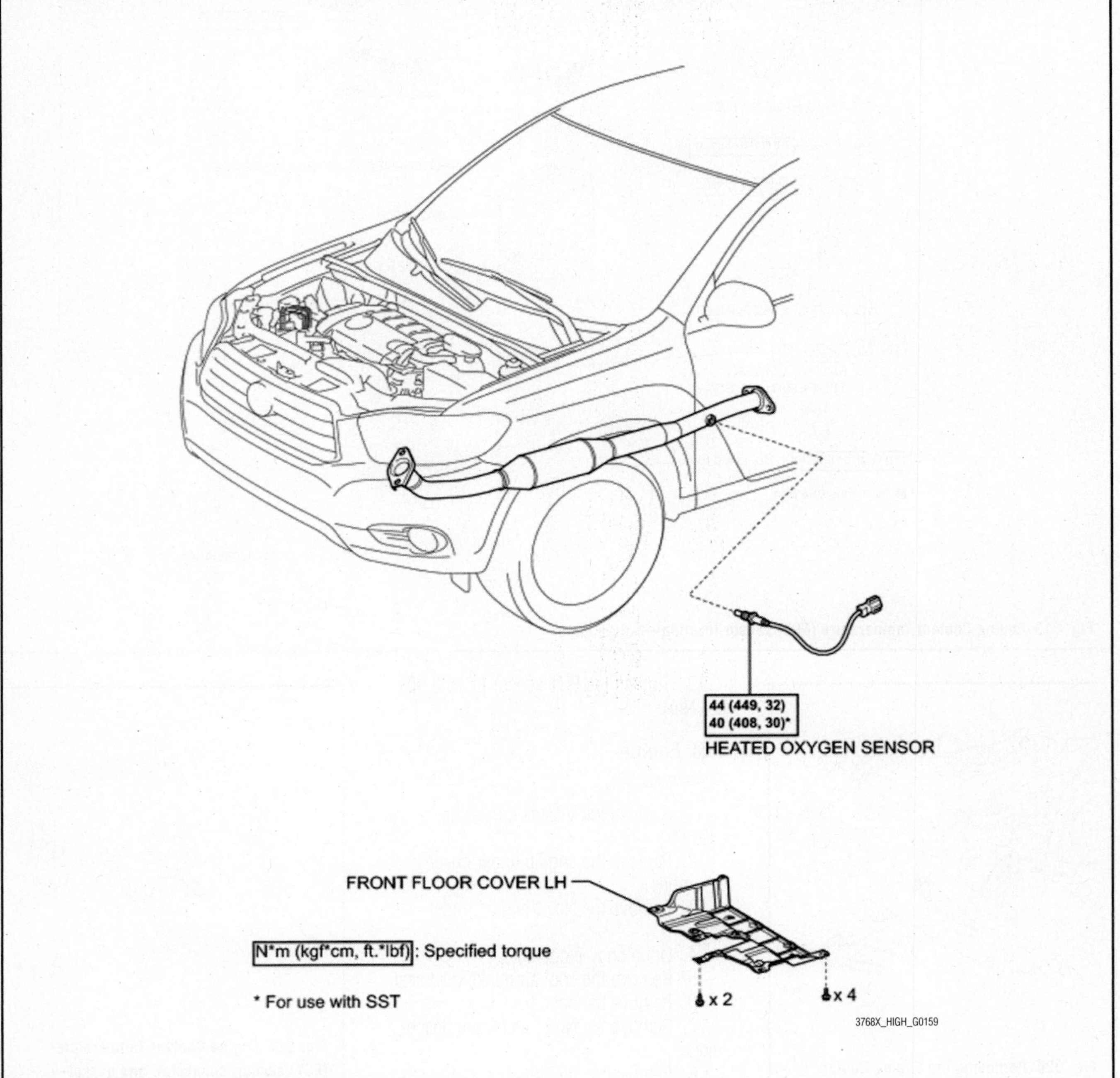

Fig. 322 Heated oxygen sensor—2.7L engine

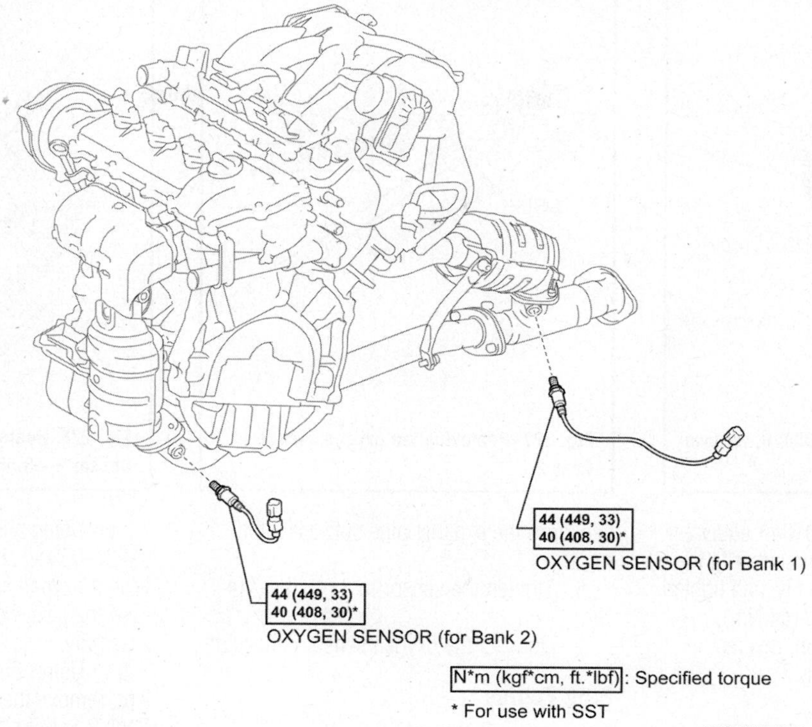

44 (449, 33)
40 (408, 30)*
OXYGEN SENSOR (for Bank 1)
44 (449, 33)
40 (408, 30)*
OXYGEN SENSOR (for Bank 2)
N*m (kgf*cm, ft.*lbf): Specified torque
* For use with SST
22140_HYBR_G0151

Fig. 323 Heated oxygen sensors are located below the converter—3.3L engine

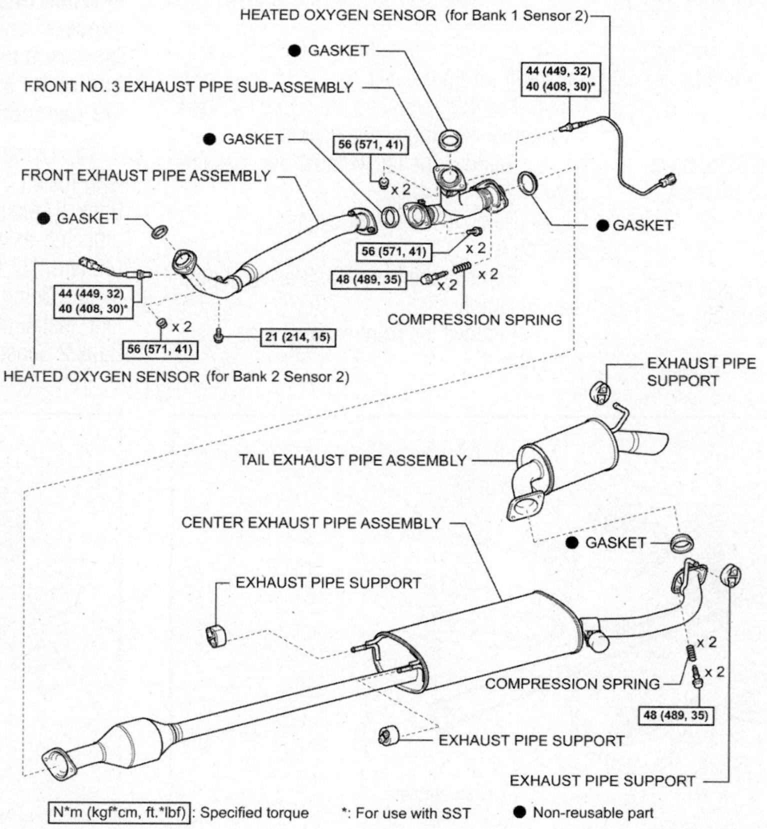

HEATED OXYGEN SENSOR (for Bank 1 Sensor 2)
● GASKET
FRONT NO. 3 EXHAUST PIPE SUB-ASSEMBLY
44 (449, 32)
40 (408, 30)*
● GASKET
56 (571, 41)
x 2
FRONT EXHAUST PIPE ASSEMBLY
● GASKET
● GASKET
56 (571, 41)
x 2
48 (489, 35)
x 2
x 2
44 (449, 32)
40 (408, 30)*
x 2
21 (214, 15)
56 (571, 41)
COMPRESSION SPRING
HEATED OXYGEN SENSOR (for Bank 2 Sensor 2)
EXHAUST PIPE SUPPORT
TAIL EXHAUST PIPE ASSEMBLY
CENTER EXHAUST PIPE ASSEMBLY
● GASKET
EXHAUST PIPE SUPPORT
x 2
x 2
COMPRESSION SPRING
48 (489, 35)
EXHAUST PIPE SUPPORT
EXHAUST PIPE SUPPORT
N*m (kgf*cm, ft.*lbf): Specified torque *: For use with SST ● Non-reusable part
22140_HIGH_G0165

Fig. 324 Heated oxygen sensor—3.5L engine

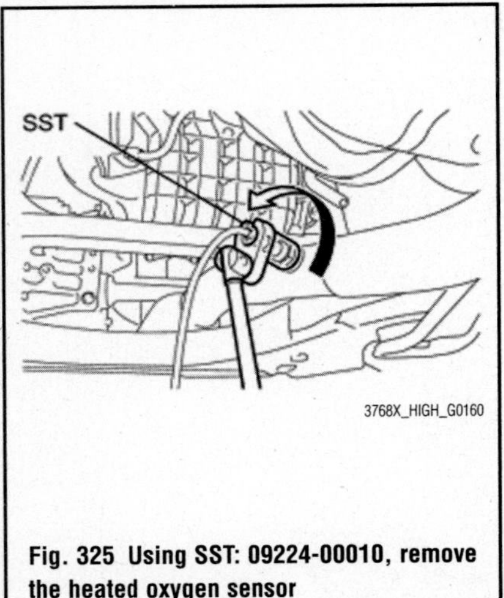

Fig. 325 Using SST: 09224-00010, remove the heated oxygen sensor

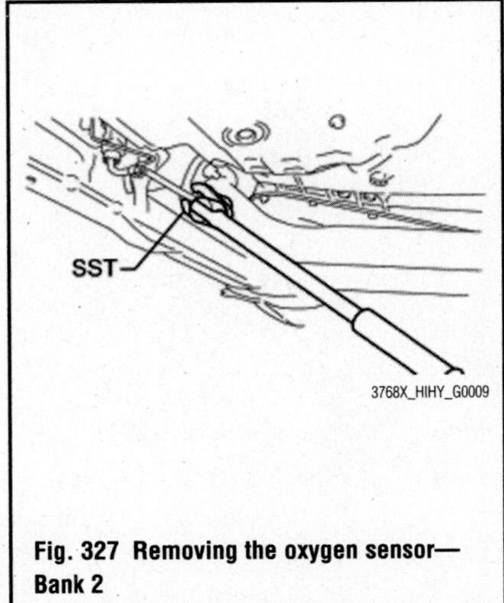

Fig. 327 Removing the oxygen sensor—Bank 2

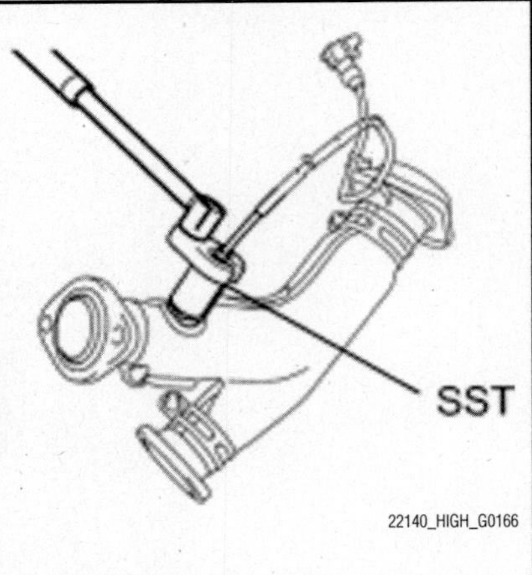

Fig. 228 Heated oxygen sensor Bank 1 Sensor 2—3.5L engine

5. Using SST: 09224-00010 or equivalent, install the heated oxygen sensor to the front exhaust pipe sub-assembly, and tighten 30 ft. lbs. (40 Nm), 32 ft. lbs. (44 Nm).

6. To complete installation, reverse remaining removal procedure.

3.3L Engine

See Figures 326 and 327.

1. For Bank 1:
   a. Disconnect the oxygen sensor connector.
   b. Using SST 09224-00010, remove the oxygen sensor from the front No. 3 exhaust pipe sub-assembly.
2. For Bank 2:
   a. Using SST 09224-00010, remove the oxygen sensor from the front exhaust pipe sub-assembly.

To install:

3. Temporarily tighten the oxygen sensor to the exhaust pipe sub-assembly front.
4. Using SST, fully tighten the oxygen sensor to the exhaust pipe sub-assembly front.
5. Tighten the sensor to 33 ft. lbs. (44 Nm),
6. Connect the oxygen sensor connector.

3.5L Engine

See Figures 228 and 229.

1. Remove the engine under cover assembly.
2. Remove the No. 1 and 2 engine under covers.
3. Remove the 3 clamps, 6 bolts and the left front floor cover.
4. For 4WD, remove the propeller shaft with center bearing shaft assembly.
5. Remove the tail exhaust pipe assembly.
6. Remove the center exhaust pipe assembly.
7. Remove the front No. 3 exhaust pipe sub-assembly.
8. Remove the front exhaust pipe assembly.
9. Using a Special Service Tool (SST: 09224-00010 or equivalent), remove the heated oxygen sensor (for Bank 1 Sensor 2) from the front No. 3 exhaust pipe sub-assembly.
10. Using SST: 09224-00010 or equivalent, remove the heated oxygen sensor (for Bank 2 Sensor 2) from the front exhaust pipe assembly.

To install:

➡When installing the heated oxygen sensors, use a torque wrench with a fulcrum length of 30 cm (11.81 inches), and make sure that SST and the wrench are connected in a straight line.

11. Using SST: 09224-00010 or equivalent, install the heated oxygen sensor (for Bank 1 Sensor 2) to the front No. 3 exhaust pipe sub-assembly, and tighten 30 ft. lbs. (40 Nm), 32 ft. lbs. (44 Nm).
12. Using SST: 09224-00010 or equivalent, install the heated oxygen sensor (for Bank 2 Sensor 2) to the front exhaust pipe

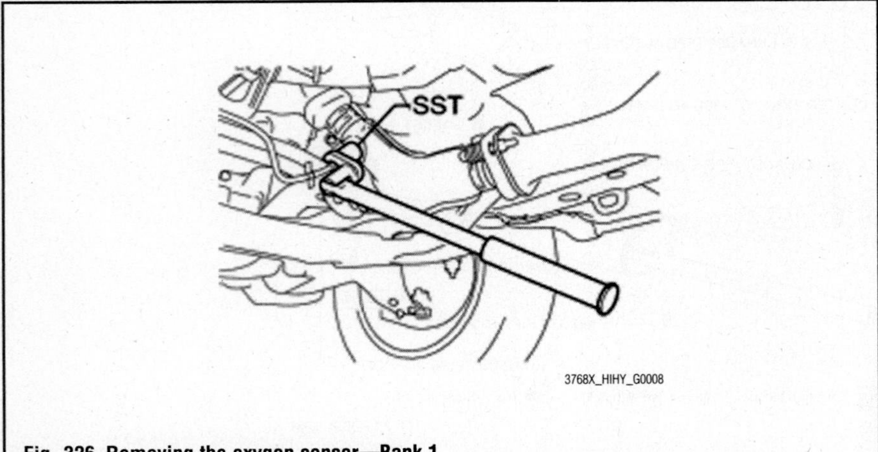

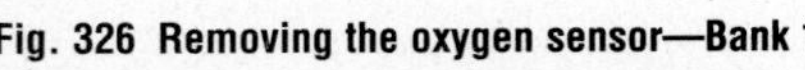

Fig. 326 Removing the oxygen sensor—Bank 1

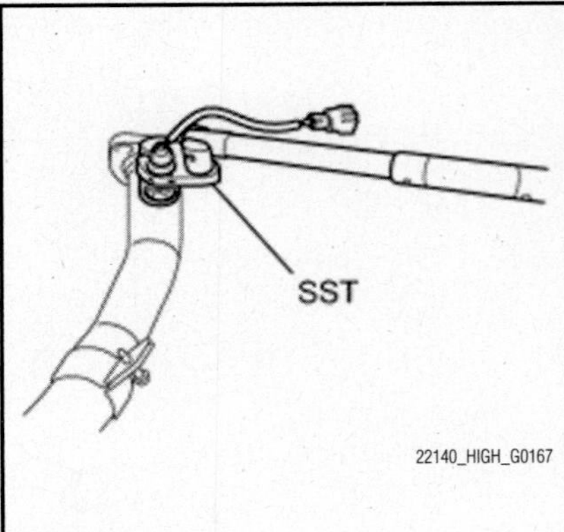

Fig. 229 Heated oxygen sensor Bank 2 Sensor 2—3.5L engine

assembly, and tighten 30 ft. lbs. (40 Nm), 32 ft. lbs. (44 Nm).

13. Install the front exhaust pipe assembly.
14. Install the front No. 3 exhaust pipe sub-assembly.
15. Install the center exhaust pipe assembly.
16. Install the tail exhaust pipe assembly.
17. For 4WD, temporarily tighten the propeller with center bearing shaft assembly.
18. For 4WD, fully tighten the propeller with center bearing shaft assembly.
19. Inspect for exhaust gas leak.
20. For 4WD, inspect and adjust transfer oil.
21. Install the 3 clamps, 6 bolts and the left front floor cover.
22. Install the No. 1 and 2 engine under covers.
23. Install the engine under cover assembly.

KNOCK SENSOR (KS)

LOCATION

See Figures 330 through 332.

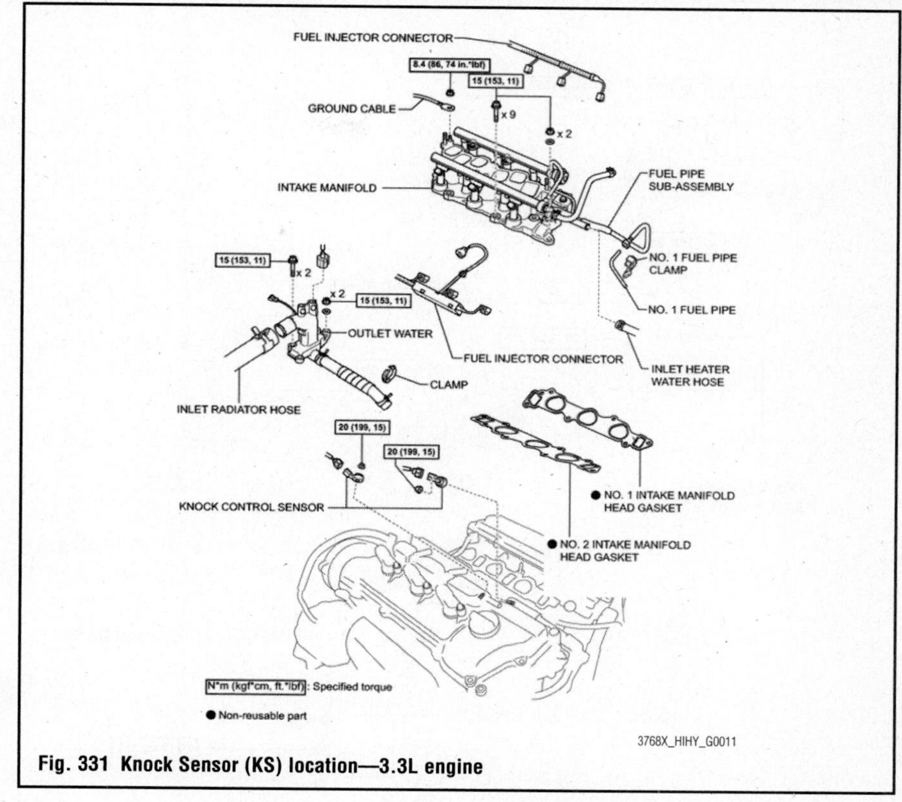

Fig. 331 Knock Sensor (KS) location—3.3L engine

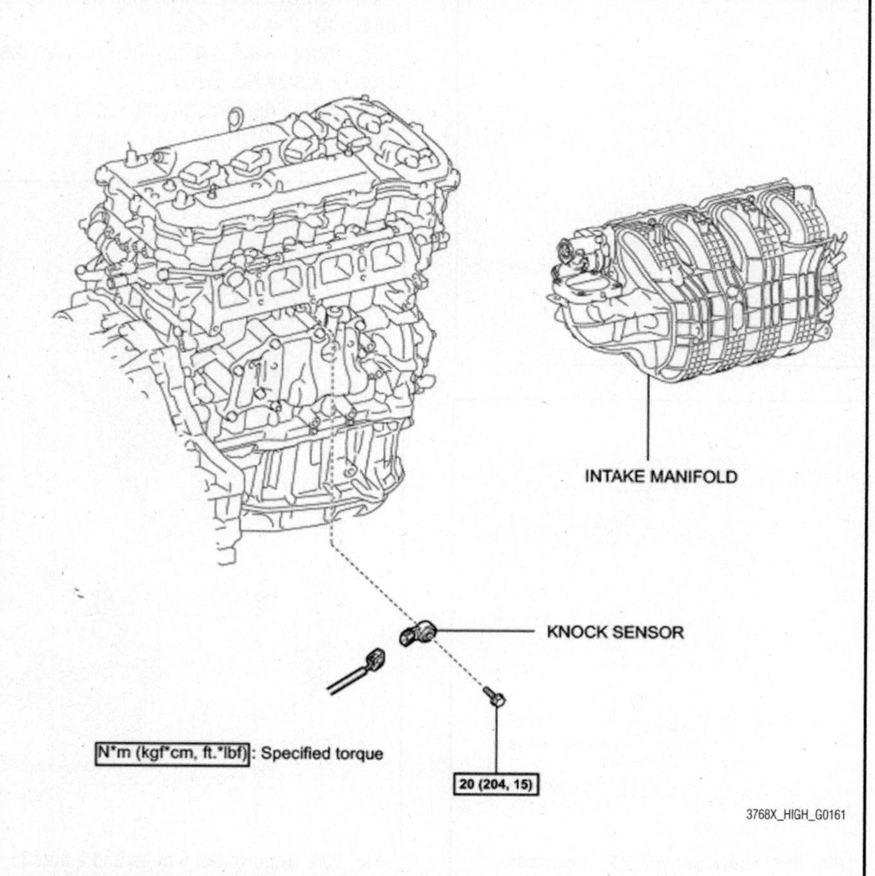

Fig. 330 Knock Sensor (KS) location—2.7L engine

REMOVAL & INSTALLATION

2.7L Engine

See Figures 333 and 334.

1. Disconnect the negative battery cable.
2. Remove the intake manifold.
3. Disconnect the sensor connector.
4. Remove the bolt and sensor.

To install:

➡The acceptable installation angle of the sensor is between 7° upward and 10° downward from the horizontal position.

5. Install the sensor with the bolt so that the sensor is angled as shown in the illustration. Tighten to 15 ft. lbs. (20 Nm).
6. Connect the sensor connector.
7. Install the intake manifold.

3.3L Engine

See Figures 335 through 338.

1. Discharge the fuel system.
2. Remove the engine room covers.
3. Disconnect the negative battery cable.
4. Drain the engine coolant.
5. Remove LH and RH wiper arm assembly.
6. Remove the cowl top ventilator louver.

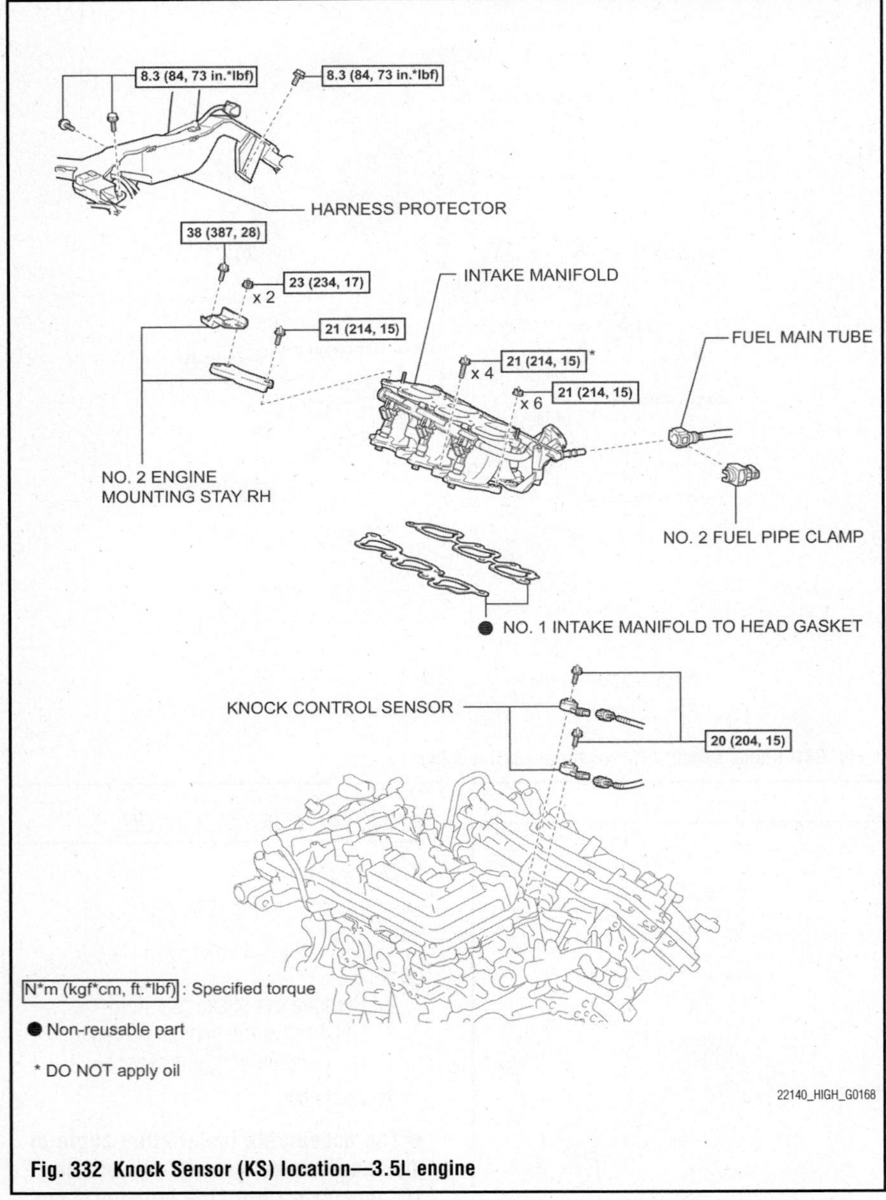

Fig. 332 Knock Sensor (KS) location—3.5L engine

7. Remove the wiper motor and link assembly.
8. Remove the cowl top panel.
9. Remove the cool air intake duct seal.
10. Remove the air cleaner assembly.
11. Remove the brake master cylinder reservoir.
12. Remove the air cleaner bracket.
13. Remove the air filter bracket.
14. Remove engine moving control rod.
15. Disconnect the VSV connector.
16. Remove the wire harness clamp.
17. Disconnect the fuel vapor feed hose No. 1.
18. Disconnect the fuel vapor feed hose No. 2.
19. Remove the 2 nuts (E), then remove the emission control valve set.
20. Disconnect the throttle motor connector.
21. Separate the water by-pass hose No. 2.
22. Separate the water by-pass hose No. 3.
23. Disconnect the ventilation hose.
24. Remove the 2 bolts, then remove the engine hanger No.1
25. Remove the 2 bolts, then remove the engine hanger No.1
26. Remove the 2 bolts, then remove the surge tank stay No. 1 (B).
27. Remove the 2 bolts, then remove the surge tank stay No. 2 (C).
28. Disconnect the ground cable connector.

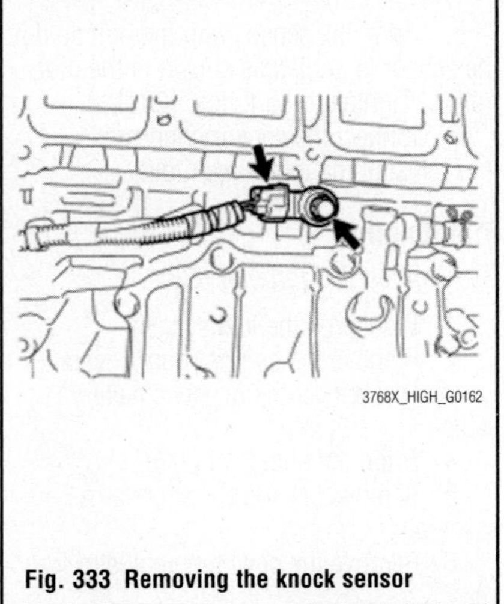

Fig. 333 Removing the knock sensor

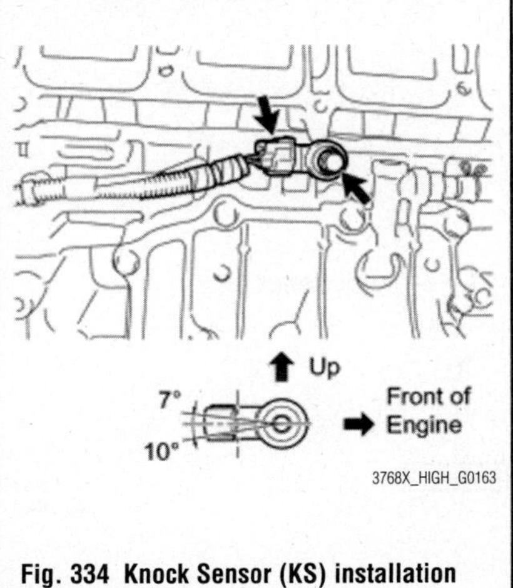

Fig. 334 Knock Sensor (KS) installation orientation

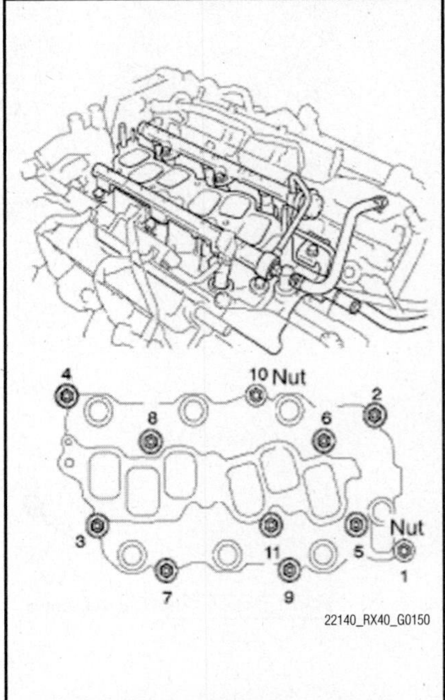

Fig. 335 Remove the 9 bolts and 2 nuts in the sequence shown

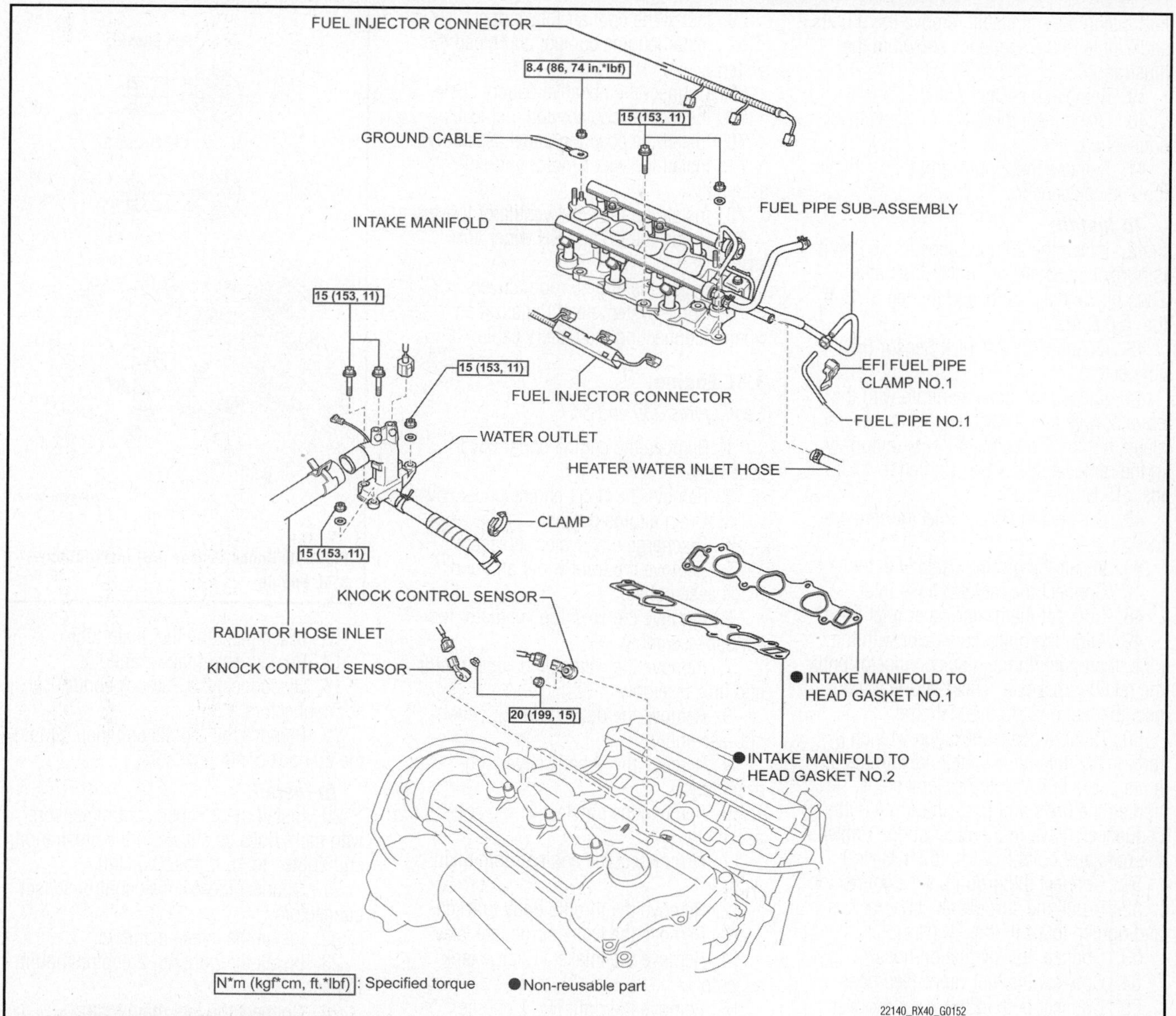

Fig. 336 Knock Sensor (KS) and related components

29. Using a socket hexagon wrench 8 mm, remove the 4 bolts.

30. Remove the 2 nuts, then remove the emission control valve bracket and the intake air surge tank.

31. Remove the gasket from the intake air surge tank.

32. Remove the EFI fuel pipe clamp No.1.

33. Pinch the quick connector and then pull out the fuel pipe No. 1.

34. Disconnect the heater water inlet hose.

35. Disconnect the radiator hose inlet.

36. Remove the nut and ground cable.

37. Disconnect the 6 fuel injector connectors.

38. In order to remove the intake mani-

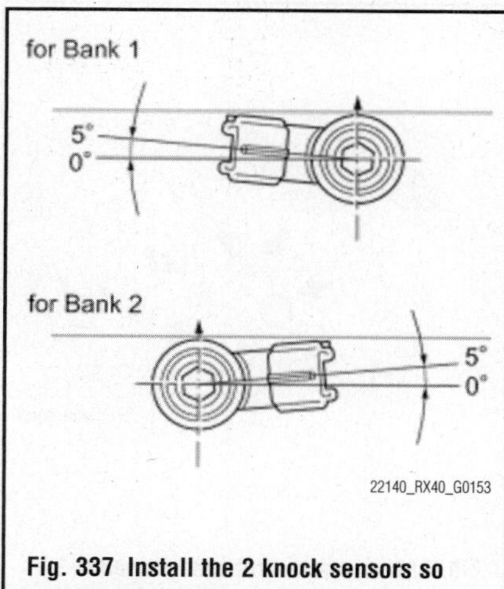

Fig. 337 Install the 2 knock sensors so that it is horizontal

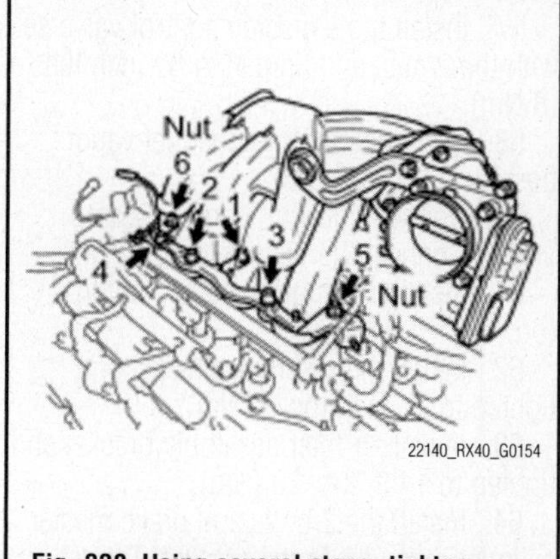

Fig. 338 Using several steps, tighten the bolts and nuts uniformly in the sequence shown

fold, using several steps, remove the 9 bolts and 2 nuts in the sequence shown in the illustration.

39. Remove the water outlet.
40. Disconnect the 2 Knock Sensor (KS) connectors.
41. Remove the 2 nuts, and then remove the 2 knock sensors.

To install:

42. Install the 2 knock sensors so that it is horizontal as shown in the illustration. Then install the 2 bolts and tighten to 15 ft. lbs. (20 Nm).
43. Connect the 2 Knock Sensor (KS) connectors.
44. Install the intake manifold with the 9 bolts, 2 nuts and 2 washers. Using several steps, tighten the bolts and nuts uniformly in the removal sequence. Tighten to 11 ft. lbs. (15 Nm).
45. Connect the 6 fuel injector connectors.
46. Install the ground cable with the nut.
47. Connect the radiator hose inlet.
48. Connect the heater water inlet hose.
49. Align the quick connector with the pipe, then push in the quick connector until the retainer makes a "click" sound to connect the fuel hose to the fuel pipe.
50. Using a socket hexagon wrench 8 mm, install the intake manifold with the 4 bolts and 2 nuts . Using several steps, tighten the bolts and nuts uniformly in the sequence shown in the illustration. Tighten the nuts and bolts to 21 ft. lbs. (28 Nm).
51. Connect the ground cable connector.
52. Install the surge tank stay brackets and tighten to 14 ft. lbs. 20 (Nm).
53. Connect the ventilation hose.
54. Connect the fuel vapor feed hose.
55. Connect both of the water by-pass hoses.
56. Reconnect the throttle motor connector.
57. Install the emission control valve set with the 2 nuts and tighten to 80 inch lbs. (8 Nm).
58. Reconnect both of the fuel vapor hose.
59. Connect the wire harness clamp.
60. Connect the VSV connector.
61. Install the engine moving control rod.
62. Install the reservoir bracket and tighten to 71 inch lbs. (9 Nm).
63. Install air filter assembly bracket and tighten to 14 ft. lbs. 20 (Nm).
64. Install the 2 bolts and brake master cylinder reservoir to the bracket. Tighten to 80 inch lbs. (9 Nm).
65. Install the air cleaner assembly.
66. Install the cool air intake duct.
67. Install engine coolant and bleed the system.
68. Connect the negative battery cable.
69. Inspect for coolant and fuel leaks.
70. Install the cowl top panel assembly.
71. Install the wiper motor and link assembly.
72. Install the top cowl ventilator louver.
73. Install the LH and RH wiper arm assembly.
74. Install the engine room covers.
75. Some system need initialization when reconnecting the battery cable.

3.5L Engine

See Figures 339 and 340.

1. Remove the engine under cover assembly.
2. Remove the No. 1 engine under cover.
3. Drain engine coolant.
4. Discharge fuel system pressure.
5. Remove the front wiper arm and blade assemblies.
6. Remove the cowl top ventilator louver sub-assembly.
7. Remove the windshield wiper motor and link assembly.
8. Remove the outer cowl top panel sub-assembly.
9. Remove the V-bank cover sub-assembly.
10. Remove the air cleaner cap sub-assembly.
11. Disconnect the engine room main wire.
12. Remove the throttle body bracket.
13. Remove the No. 1 surge tank stay.
14. Remove the intake air surge tank assembly.
15. Remove the right No. 2 engine mounting stay.

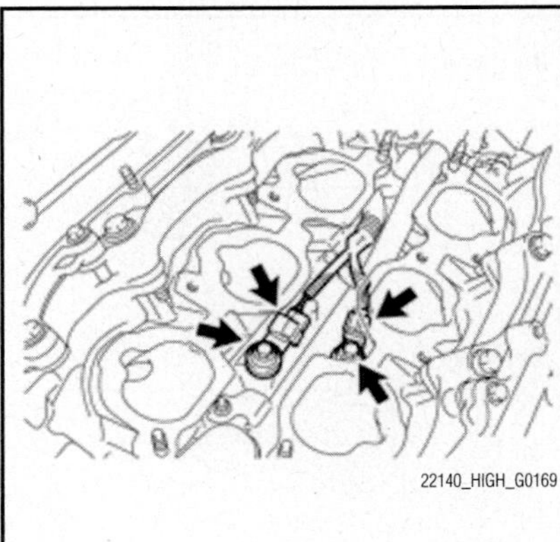

Fig. 339 Knock control sensor connectors—3.5L engine

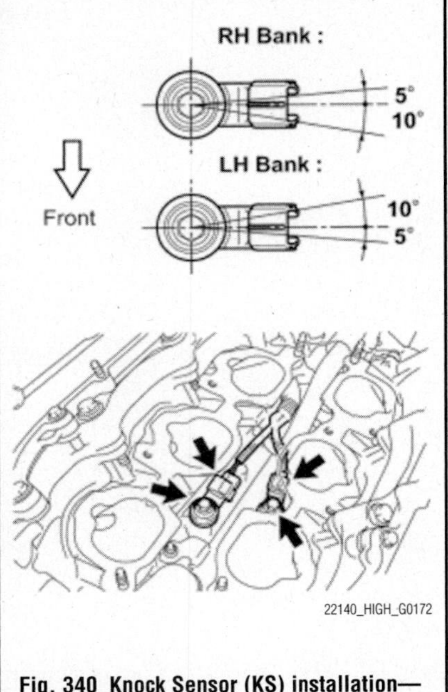

Fig. 340 Knock Sensor (KS) installation—3.5L engine

16. Disconnect the fuel main tube.
17. Remove the intake manifold.
18. Disconnect the 2 knock control sensor connectors.
19. Remove the 2 bolts and then remove the 2 knock control sensors.

To install:

20. Install the 2 knock control sensors with the 2 bolts as shown in the illustration, and tighten to 15 ft. lbs. (20 Nm).
21. Connect the 2 knock control sensor connectors.
22. Install the intake manifold.
23. Install the right no. 2 engine mounting stay.
24. Connect the fuel main tube.
25. Temporarily install the No. 1 surge tank stay.
26. Temporarily install the throttle body bracket.
27. Install the intake air surge tank assembly.
28. Fully tighten the No. 1 surge tank stay.
29. Fully tighten the throttle body bracket.
30. Connect the engine room wire.
31. Install the air cleaner cap sub-assembly.
32. Add engine coolant.
33. Inspect for coolant leaks.
34. Inspect for fuel leak.
35. Install the outer cowl top panel sub-assembly.
36. Install the windshield wiper motor and link assembly.

37. Install the cowl top ventilator louver sub-assembly.
38. Install the front wiper arm and blade assemblies.
39. Install the No. 1 engine under cover.
40. Install the engine under cover assembly.
41. Install the V-bank cover sub-assembly.

MASS AIR FLOW (MAF) SENSOR

LOCATION

See Figures 341 and 342.

REMOVAL & INSTALLATION

2.7L & 3.5L Engines

See Figure 343.

1. Disconnect the Mass Air Flow (MAF) meter connector.
2. Remove the 2 screws and Mass Air Flow (MAF) meter.

To install:

3. Installation is the reverse of removal.

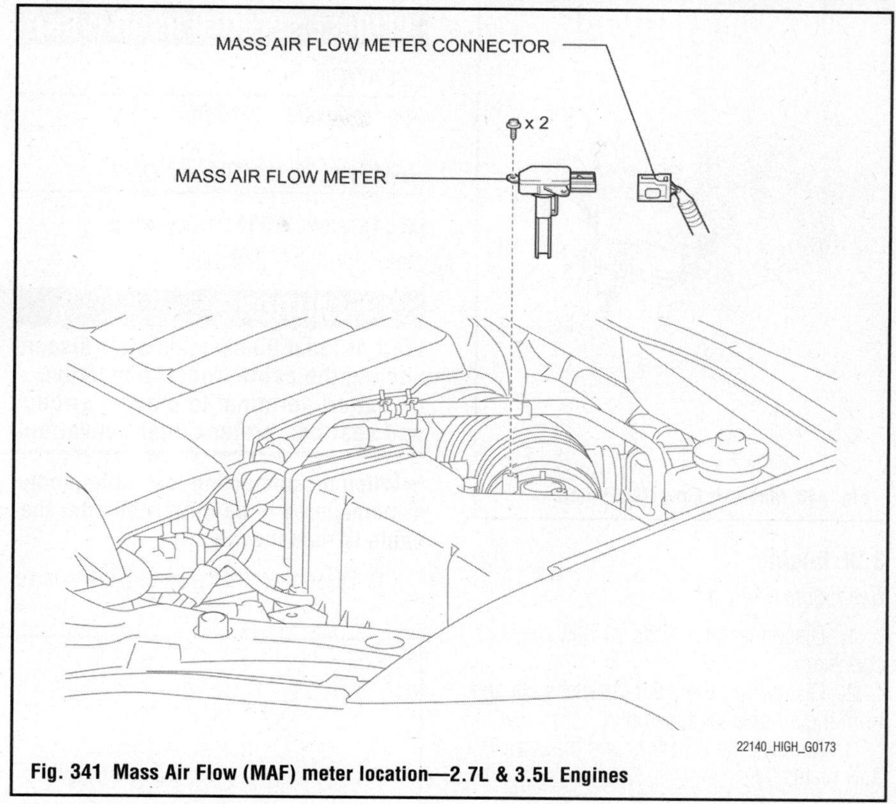

Fig. 341 Mass Air Flow (MAF) meter location—2.7L & 3.5L Engines

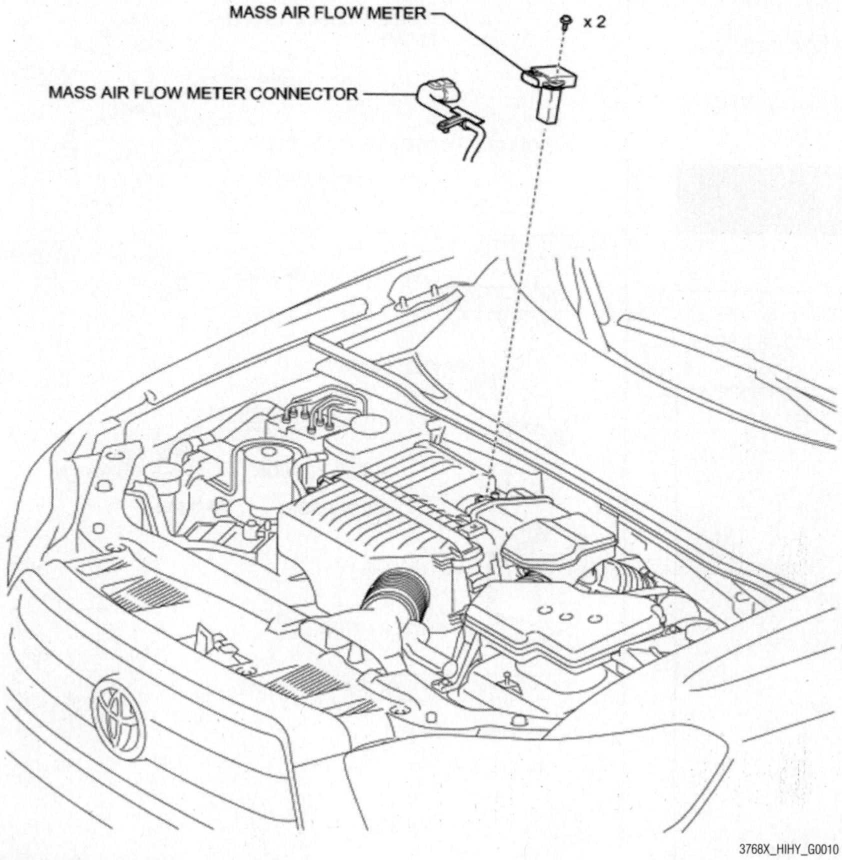

MASS AIR FLOW METER
x 2
MASS AIR FLOW METER CONNECTOR
3768X_HIHY_G0010

Fig. 342 Mass Air Flow (MAF) Sensor location—3.3L engine

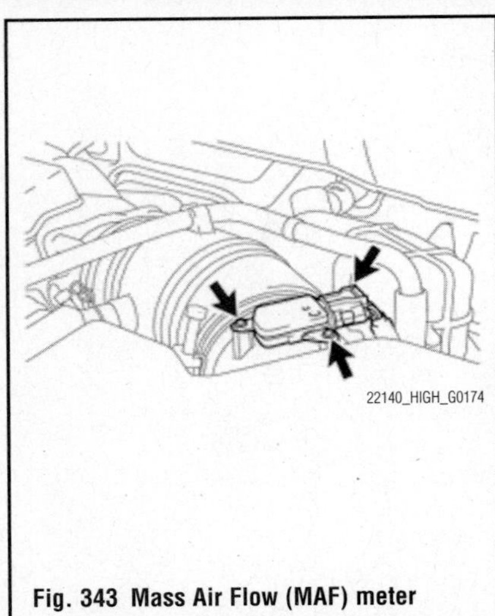

Fig. 343 Mass Air Flow (MAF) meter

3.3L Engine

See Figure 344.

1. Disconnect the mass air flow meter connector.
2. Disconnect the 2 wire harness clamps from the air cleaner assembly.
3. Remove the 2 screws and mass air flow meter.

To install:

4. Install the mass air flow meter with the 2 screws.
5. Connect the mass air flow meter connector.
6. Connect the 2 wire harness clamps to the air cleaner assembly.

THROTTLE POSITION SENSOR (TPS)

LOCATION

The Throttle Position Sensor (TPS) is integral to the electric throttle body assembly.

Fig. 344 Mass Air Flow (MAF) sensor location.

VEHICLE SPEED SENSOR (VSS)

LOCATION

See Figures 344 and 346.

REMOVAL & INSTALLATION

U151e and U151F Transaxles

See Figures 347 and 348.

CAUTION

Wait at least 90 seconds after disconnecting the cable from the negative (-) battery terminal to prevent airbag and seat belt pretensioner activation.

➡When disconnecting the cable, some systems need to be initialized after the cable is reconnected.

1. Disconnect the negative battery cable.
2. Remove battery.
3. Remove cool air intake duct seal.
4. Remove no. 2 air cleaner inlet.
5. Remove no. 1 air cleaner inlet.
6. Remove air cleaner cap sub-assembly.
7. Remove air cleaner filter element sub-assembly.
8. Remove air cleaner case sub-assembly.
9. Remove air cleaner bracket.
10. Remove speed sensors:
    a. Disconnect the speed sensor connector.
    b. Remove the bolt and speed sensor.

To install:

11. Coat a new O-ring with ATF.
12. Install the speed sensor with the bolt. Tighten to 8 ft. lbs. (11 Nm).

➡Make sure to install the same manufacturer's sensor.

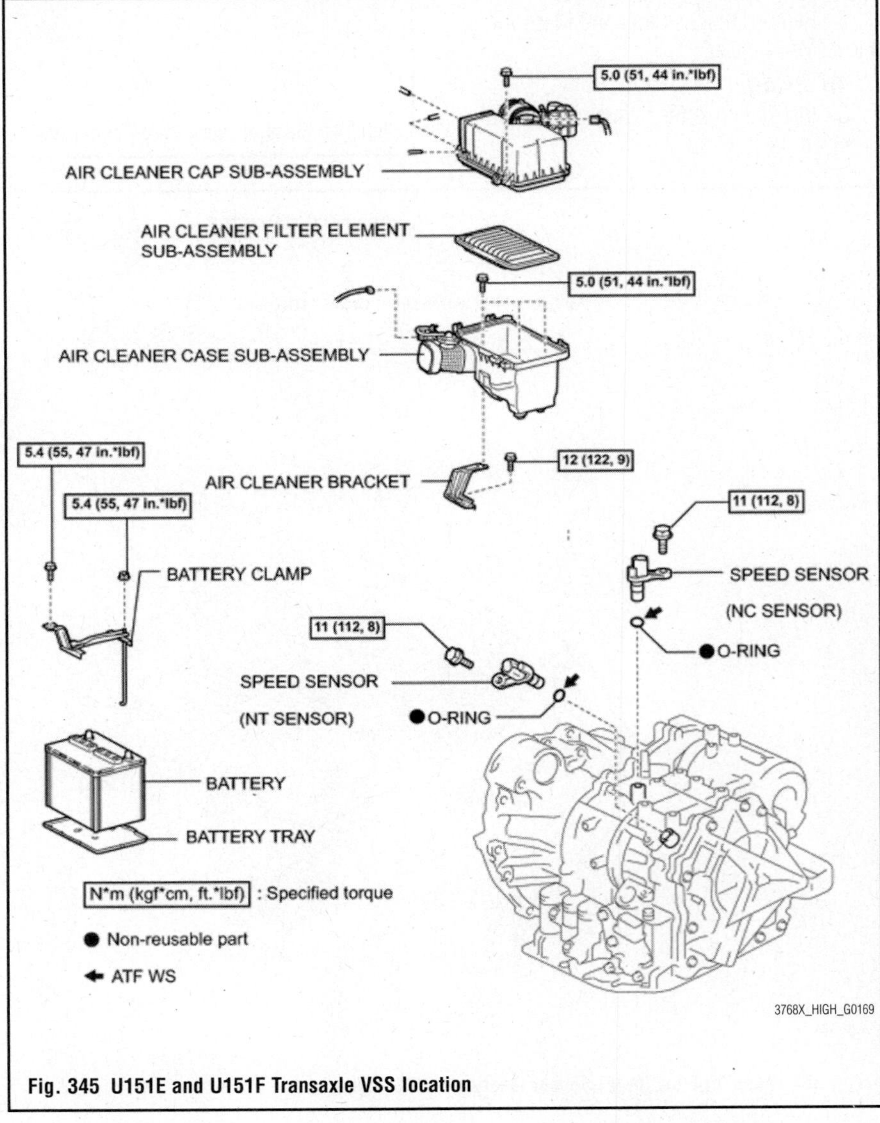

Fig. 345 U151E and U151F Transaxle VSS location

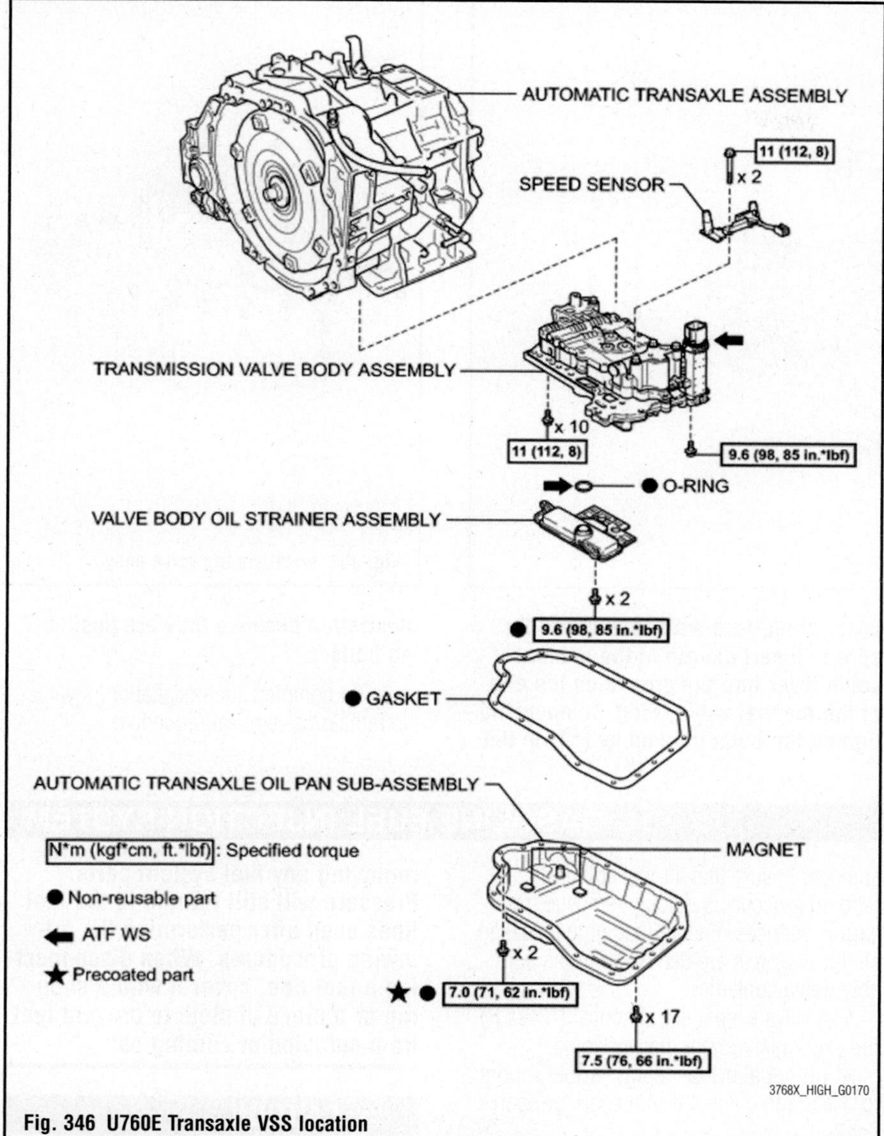

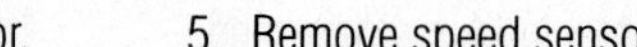

Fig. 346 U760E Transaxle VSS location

Fig. 347 Remove the bolt and speed sensor—NT sensor

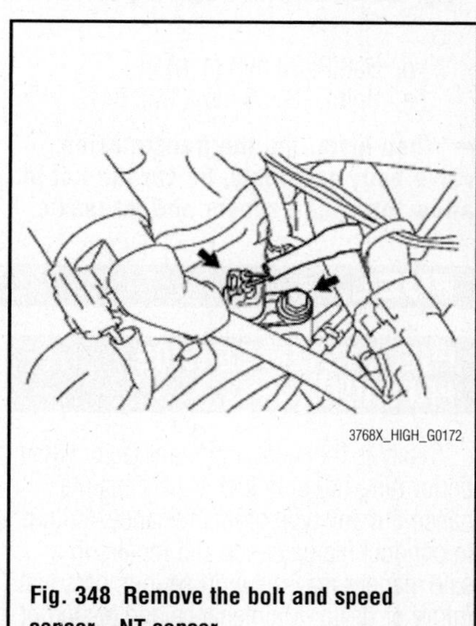

Fig. 348 Remove the bolt and speed sensor—NT sensor

13. Connect the speed sensor connector.
14. To complete the installation, reverse the remaining removal procedure.

U760e Transaxle

See Figures 349 and 351

1. Remove automatic transaxle assembly.
2. Remove automatic transaxle oil pan sub-assembly
3. Remove valve body oil strainer assembly.

➡When removing the transmission valve body assembly, be careful not to allow the speed sensor and the transaxle case to interfere with each other.

4. Remove the 11 bolts and valve body from the transaxle.
5. Remove speed sensor:
   a. Disconnect the connector.
   b. Remove the 2 bolts and speed sensor from the valve body.

To install:

6. Install the speed sensor to the valve body with the 2 bolts. Tighten to 8 ft. lbs. (11 Nm). Connect the connector.
7. Install transmission valve body assembly:
   a. Coat the O-ring of the transmission wire with ATF.
   b. Confirm that the manual valve lever is positioned as shown in the illustration and install the valve body assembly to the transaxle case with the 11 bolts.
      - Bolt A, B, C: Tighten to 8 ft. lbs. (11 Nm).
      - Bolt D: 85 inch lbs. (7 Nm)
   c. Bolt A: 25 mm (0.984 in.)

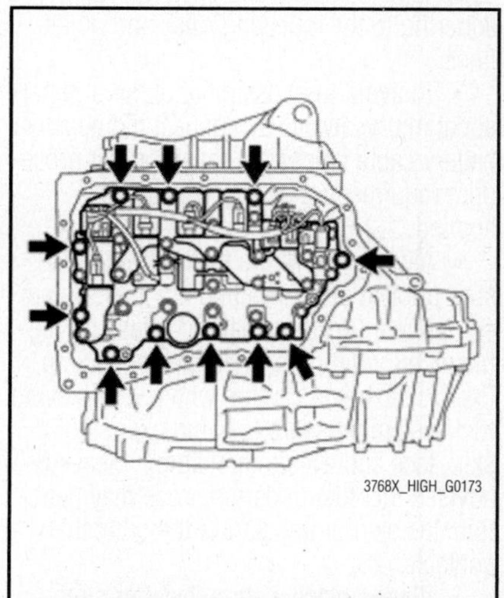

Fig. 349 Remove the 11 bolts and valve body

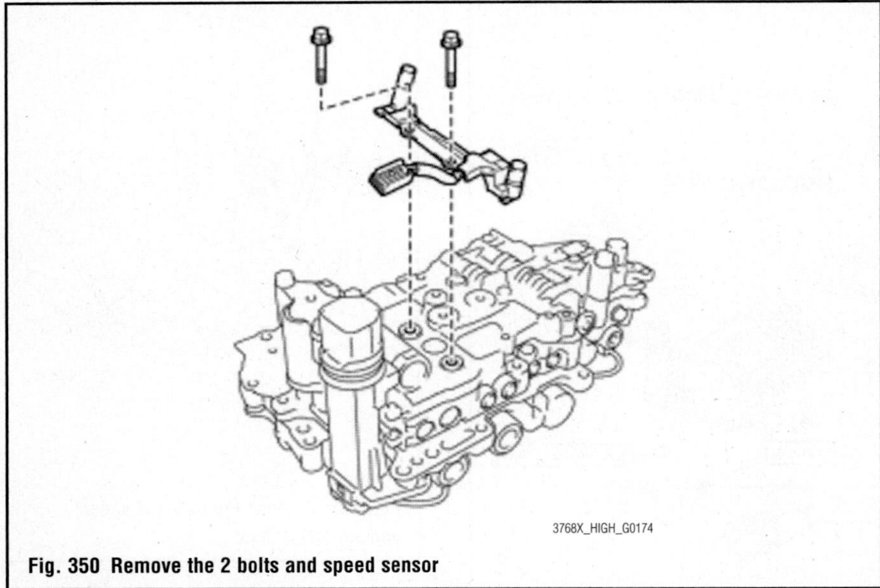

Fig. 350 Remove the 2 bolts and speed sensor

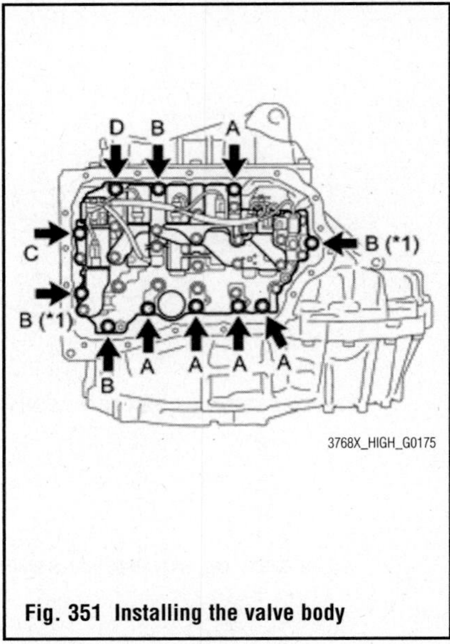

Fig. 351 Installing the valve body

d. Bolt B: 30 mm (1.18 in.)
e. Bolt C, D: 35 mm (1.38 in.)

➡**When installing the transmission valve body assembly, be careful not to allow the speed sensor and transaxle case to interfere with each other. Be sure to insert the pin of the manual valve lever into the groove on the end of the manual valve. First, temporarily tighten the bolts marked by (*1) in the illustration because they are positioning bolts.**

8. To complete the installation, reverse the remaining removal procedure.

FUEL — GASOLINE FUEL INJECTION SYSTEM

FUEL SYSTEM SERVICE PRECAUTIONS

Safety is the most important factor when performing not only fuel system maintenance but any type of maintenance. Failure to conduct maintenance and repairs in a safe manner may result in serious personal injury or death. Maintenance and testing of the vehicle's fuel system components can be accomplished safely and effectively by adhering to the following rules and guidelines.

- To avoid the possibility of fire and personal injury, always disconnect the negative battery cable unless the repair or test procedure requires that battery voltage be applied.
- Always relieve the fuel system pressure prior to disconnecting any fuel system component (injector, fuel rail, pressure regulator, etc.), fitting or fuel line connection. Exercise extreme caution whenever relieving fuel system pressure to avoid exposing skin, face and eyes to fuel spray. Please be advised that fuel under pressure may penetrate the skin or any part of the body that it contacts.
- Always place a shop towel or cloth around the fitting or connection prior to loosening to absorb any excess fuel due to spillage. Ensure that all fuel spillage (should it occur) is quickly removed from engine surfaces. Ensure that all fuel soaked cloths or towels are deposited into a suitable waste container.
- Always keep a dry chemical (Class B) fire extinguisher near the work area.
- Do not allow fuel spray or fuel vapors to come into contact with a spark or open flame.
- Always use a back-up wrench when loosening and tightening fuel line connection fittings. This will prevent unnecessary stress and torsion to fuel line piping.
- Always replace worn fuel fitting O-rings with new Do not substitute fuel hose or equivalent where fuel pipe is installed.

Before servicing the vehicle, make sure to also refer to the precautions in the beginning of this section as well.

RELIEVING FUEL SYSTEM PRESSURE

2.7L Engine

See Figure 352.

⁂ CAUTION

Perform the following procedures to prevent fuel from spilling out before removing any fuel system parts. Pressure will still remain in the fuel lines even after performing the following procedures. When disconnecting a fuel line, cover it with a shop rag or a piece of cloth to prevent fuel from spraying or coming out.

⁂ CAUTION

Observe all applicable safety precautions when working around fuel. Whenever servicing the fuel system, always work in a well-ventilated area. Do not allow fuel spray or

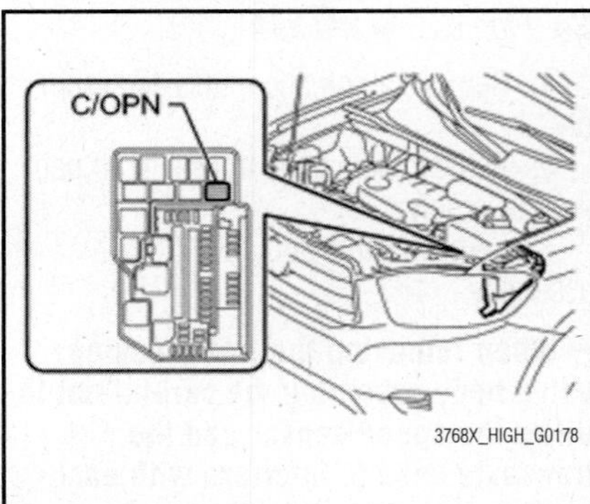

Fig. 352 Locating the C/OPN relay—2.7L Engines

vapors to come in contact with a spark or open flame. Keep a dry chemical fire extinguisher near the work area. Always keep fuel in a container specifically designed for fuel storage; also, always properly seal fuel containers to avoid the possibility of fire or explosion.

1. Separate the brake master cylinder reservoir assembly.
2. Remove the reservoir bracket.
3. Remove the No. 1 relay block cover.
4. Remove the C/OPN relay.
5. Start the engine.
6. After the engine stops, turn the ignition switch off.

➡DTC P0171/25 (fuel problem) may be detected.

7. Crank the engine again. Check that the engine does not start.
8. Remove the fuel tank cap to discharge pressure from the fuel tank.
9. Disconnect the cable from the negative (-) battery terminal.
10. Install the C/OPN relay.
11. Install the No. 1 relay block cover.
12. Install the reservoir bracket.
13. Install the brake master cylinder reservoir assembly.

3.3L Engine

See Figure 353.

1. Before servicing the vehicle, refer to the Precautions Section.
2. Disconnect the No. 3 relay block.
3. Remove the No. 2 junction block cover.
4. Remove the C/OPN RLY.
5. Put the vehicle in Inspection Mode and start the engine.
6. Turn the ignition switch to**OFF**immediately after the engine comes to 'rough idle state'.

➡The hybrid system has a complicated process from an 'out of gas' to 'engine stall' condition. Therefore, 'rough idle' is regarded as 'stop'.

7. Disconnect the negative battery cable.
8. Reinstall the C/OPN RLY.

3.5L Engine

See Figure 354.

⁂ CAUTION

Perform the following procedures to prevent fuel from spilling out before removing any fuel system parts. Pressure will still remain in the fuel lines even after performing the following procedures. When disconnecting a fuel line, cover it with a shop rag or a piece of cloth to prevent fuel from spraying or coming out.

⁂ CAUTION

Observe all applicable safety precautions when working around fuel. Whenever servicing the fuel system, always work in a well-ventilated area. Do not allow fuel spray or vapors to come in contact with a spark or open flame. Keep a dry chemical fire extinguisher near the work area. Always keep fuel in a container specifically designed for fuel storage; also, always properly seal fuel containers to avoid the possibility of fire or explosion.

1. Remove the relay block cover.
2. Remove the FUEL PUMP relay.
3. Start the engine.
4. After the engine stops, turn the ignition switch off.

➡DTC P0171/25 (fuel problem) may be detected.

5. Crank the engine again. Check that the engine does not start.
6. Remove the fuel tank cap to discharge pressure from the fuel tank.
7. Disconnect the cable from the negative (-) battery terminal.
8. Install the FUEL PUMP relay.

FUEL FILTER

REMOVAL & INSTALLATION

The fuel filter is part of the fuel suction tube/fuel pump assembly and is located in the fuel tank. It is not a normally serviced item.

FUEL INJECTORS

REMOVAL & INSTALLATION

2.7L Engine

See Figures 355 through 358.

⁂ CAUTION

Observe all applicable safety precautions when working around fuel. Whenever servicing the fuel system, always work in a well-ventilated area. Do not allow fuel spray or vapors to come in contact with a spark or open flame. Keep a dry chemical fire extinguisher near the work area. Always keep fuel in a container specifically designed for fuel storage; also, always properly seal fuel containers to avoid the possibility of fire or explosion.

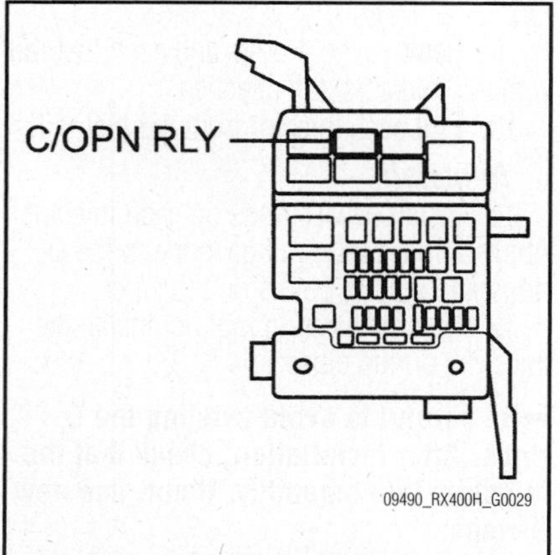

Fig. 353 Location of the C/OPN relay in the junction box.

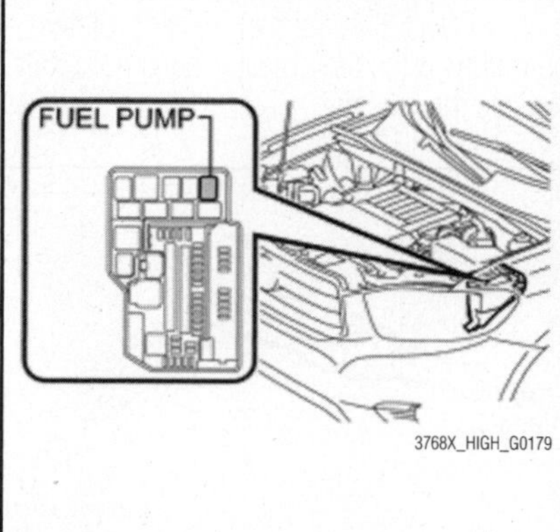

Fig. 354 Locating the fuel pump relay—3.5L Engines

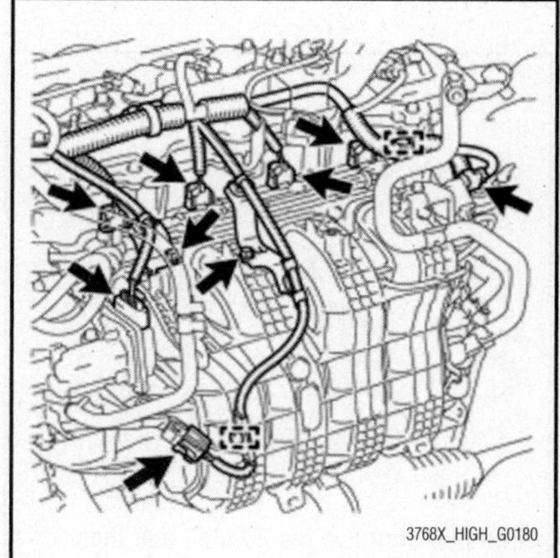

Fig. 355 Disconnect the wire harness

1. Properly relieve the fuel system pressure.
2. Disconnect the negative battery cable.
3. Remove no. 1 engine cover sub-assembly.
4. Remove air cleaner cap sub-assembly.
5. Disconnect fuel tube sub-assembly:
   a. Remove the No. 1 fuel pipe clamp.
   b. Pinch the tube connector, and then pull the tube connector off of the pipe.

➡Note the following:

- Check for foreign matter in the fuel tube around the fuel tube connector. Clean it if necessary. Foreign matter can affect the ability of the O-ring to seal the connector and fuel pipe.
- Do not use any tools to separate the connector and pipe.
- Do not forcefully bend, kink or twist the hose.
- Keep the connector and pipe free from foreign matter.
- If the connector and pipe are stuck together, pinch the connector and turn it carefully to disconnect it.
- Put the connector in a plastic bag to prevent damage and contamination.

   c. Remove the fuel tube sub-assembly from the fuel hose clamp.
6. Disconnect wire harness:
   a. Disconnect the 4 fuel injector connectors.
   b. Disconnect the 3 connectors.
   c. Remove the 2 bolts and 2 wire harness brackets.
   d. Detach the 2 clamps to disconnect the wire harness.
7. Remove vacuum switching valve assembly (for ACIS).
8. Remove fuel delivery pipe sub-assembly:

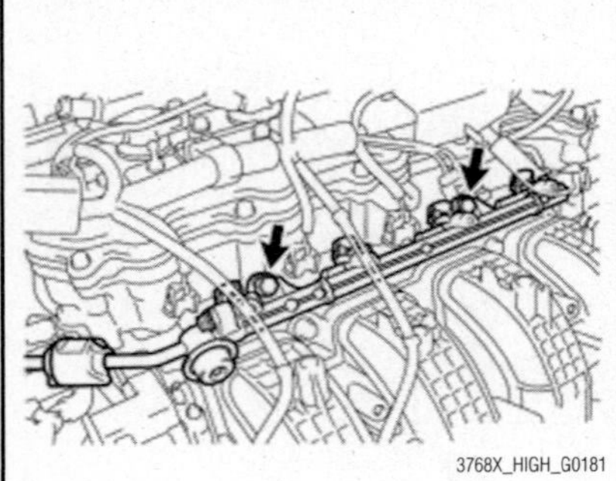

Fig. 356 Remove the 2 bolts, and then remove the fuel delivery pipe together with the 4 fuel injectors

Fig. 357 Remove the 2 fuel delivery spacers from the cylinder head

   a. Remove the 2 bolts, and then remove the fuel delivery pipe together with the 4 fuel injectors.

➡Be careful not to drop the fuel injectors when removing the fuel delivery pipe.

   b. Remove the 2 fuel delivery spacers from the cylinder head.
   c. Remove the 4 injector vibration insulators from the cylinder head.
9. Pull the 4 fuel injectors out of the fuel delivery pipe.

To install:

10. Apply a light coat of gasoline or spindle oil to new O-rings, and then install one onto each fuel injector.
11. Apply a light coat of gasoline or spindle oil to the part of the fuel delivery pipe which comes into contact with the O-ring of the fuel injector.
12. Apply a light coat of gasoline or spindle oil to the O-ring again, and then install the fuel injectors onto the fuel delivery pipe.

➡Make sure that the O-ring is not cracked or jammed when installing the injector.

13. Check that the fuel injector rotates smoothly. If the fuel injector does not rotate, replace the O-ring.

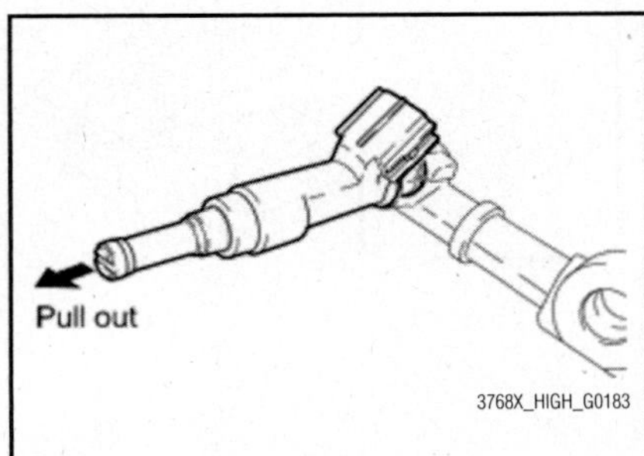

Fig. 358 Pull the 4 fuel injectors out of the fuel delivery pipe

14. Install fuel delivery pipe sub-assembly:
   a. Install 4 new injector vibration insulators to the cylinder head.
   b. Install the 2 fuel delivery spacers onto the cylinder head.

➡Install the fuel delivery spacer so that the longer protrusion is on the cylinder head side.

   c. Install the fuel delivery pipe together with the 4 fuel injectors to the cylinder head, and then temporarily install the 2 bolts.
   d. Check that the fuel injector rotates smoothly. If the fuel injector does not rotate, replace the O-ring.
   e. Tighten the 2 bolts to 15 ft. lbs. (21 Nm).
15. The remainder of installation is the reverse of the removal procedure.

3.3L Engine

See Figure 359.

1. Before servicing the vehicle, refer to the Precautions Section.
2. Relieve the fuel system pressure.
3. Drain the cooling system.
4. Remove or disconnect the following:
   - Negative battery cable
   - Engine cover
   - Wiper arm and blade assembly
   - Top cowl ventilator louver assembly
   - Wiper motor and linkage assembly
   - Air intake assembly
   - Emission control valve hoses
   - Air intake surge tank
5. Remove the mounting bolt and separate fuel hose No. 1
6. Remove the fuel pressure pulsation damper and gaskets.
7. Remove the fuel hose No. 2 union bolt and gaskets.
8. Disconnect the wiring at the injectors.
9. Remove the 4 bolts and each fuel rail with the injectors still attached.
10. Pull each injector from the fuel rail..

To install:

11. Install new O-rings on each injector. Apply a light coating of gasoline to the O-rings and mating points on the pipes.
12. Using a twisting motion, install the injectors on the pipes.

➡Be careful to avoid twisting the O-rings. After installation, check that the injectors turn smoothly. If not, use new O-rings.

13. Install the pipes and injectors.
14. Loosely install the bolts and make

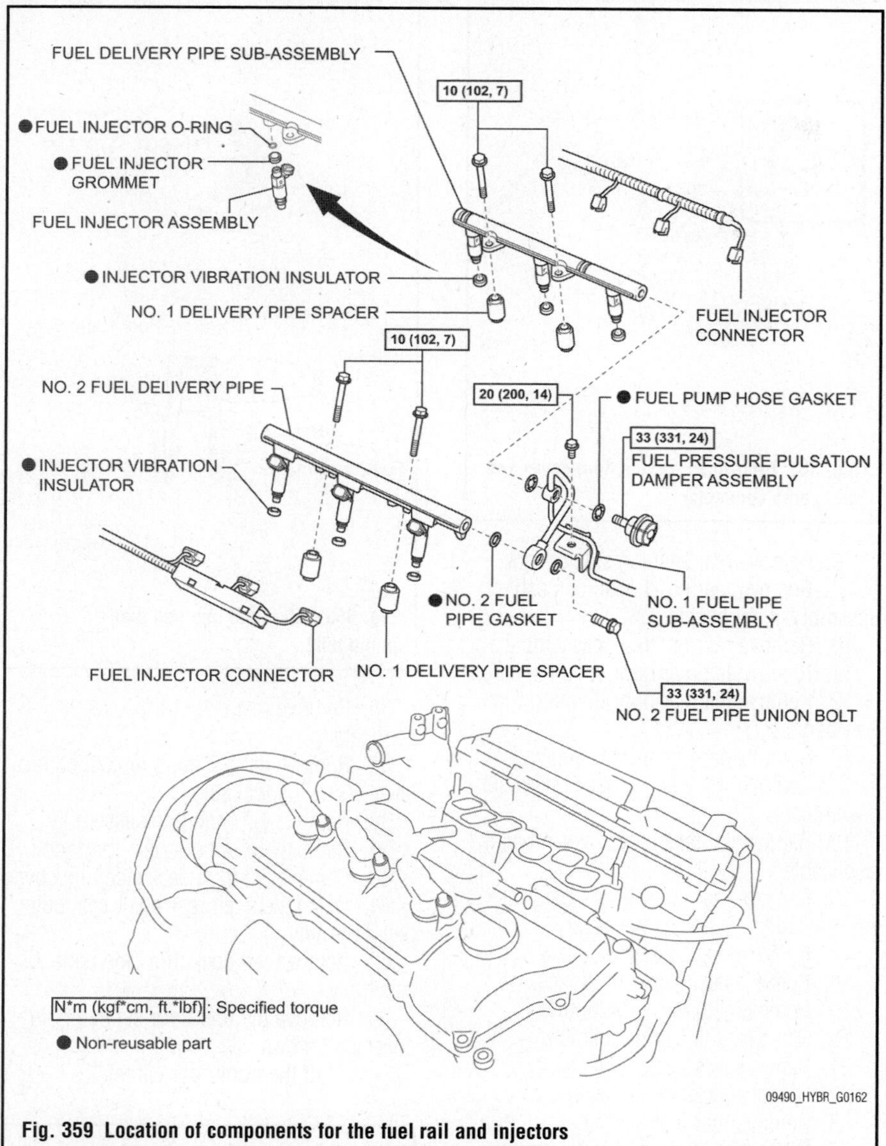

Fig. 359 Location of components for the fuel rail and injectors

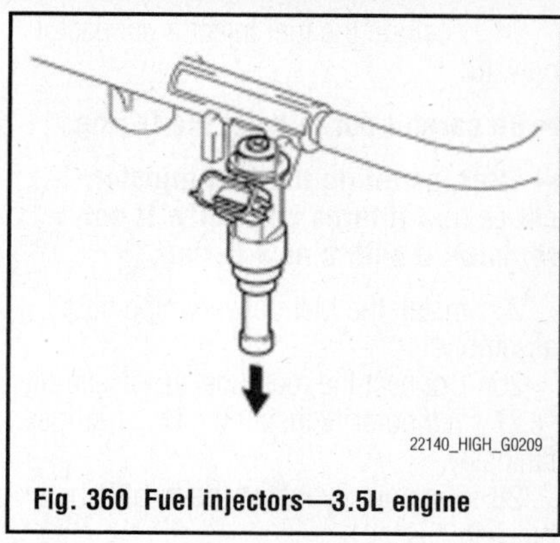

Fig. 360 Fuel injectors—3.5L engine

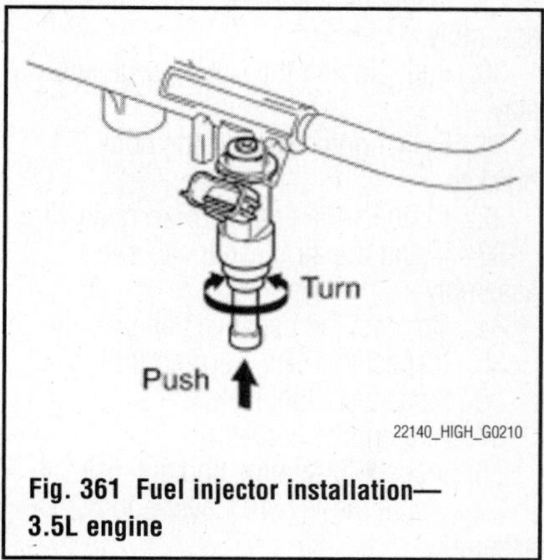

Fig. 361 Fuel injector installation—3.5L engine

sure that the injectors still turn freely. If not, replace the O-rings.

15. Torque the bolts to 84 inch lbs. (10 Nm).

16. The remainder of installation is the reverse of removal. Observe the following torques:

- Fuel hose No. 2 union bolt: 24 ft. lbs. (33 Nm)
- Pulsation damper: 24 ft. lbs. (33 Nm)
- Fuel hose No.1: 14 ft. lbs. (20 Nm)

17. If necessary, perform the initialization procedure.

3.5L Engine

See Figures 360 and 361.

1. Before servicing the vehicle, refer to the precautions section.
2. Relieve the fuel system pressure.
3. Disconnect the negative battery cable.
4. Remove the engine under cover assembly.
5. Remove the No. 1 engine under cover.
6. Drain the coolant.
7. Remove the front wiper arm and blade assemblies.
8. Remove the cowl top ventilator louver sub-assembly.
9. Remove the wiper motor and link assembly.
10. Remove the outer cowl top panel sub-assembly.
11. Remove the V-bank cover sub-assembly.
12. Remove the air cleaner cap sub-assembly.
13. Disconnect the engine room main wire.
14. Remove the throttle body bracket.
15. Remove the No. 1 surge tank stay.
16. Remove the intake air surge tank assembly.
17. Disconnect the fuel tube sub-assembly.
18. Remove the fuel delivery pipe sub-assembly.
19. Pull out the fuel injectors from the fuel delivery pipe.

➡If the injectors are to be reused, reinstall them to the same cylinder they came from.

20. Remove the 6 O-rings from the injectors.

To install:

21. Apply a light coat of spindle oil or gasoline to new O-rings, and install them to each injector.

➡The wound or the foreign body must not adhere in the ditch of O-ring.

22. Apply a light coat of spindle oil or gasoline where the fuel delivery pipe contacts the O-ring.
23. Push the fuel injector while turning it to install the injector in the fuel delivery pipe.

24. Position the fuel injector connector outward.

➡**Be careful not to twist the O-ring.**

➡**After installing the fuel injector, check that it turns smoothly. If not, reinstall it with a new O-ring.**

25. Install the fuel delivery pipe sub-assembly.
26. Connect the fuel tube sub-assembly.
27. Temporarily install the No. 1 surge tank stay.
28. Temporarily install the throttle body bracket.
29. Install the intake air surge tank assembly.
30. Fully tighten the No. 1 surge tank stay.
31. Fully tighten the throttle body bracket.
32. Connect the engine room main wire.
33. Install the air cleaner cap sub-assembly.
34. Connect the negative battery cable.
35. Inspect the SRS warning light.
36. Inspect for fuel leak.
37. Add engine coolant.
38. Inspect for engine coolant leak.
39. Install the V-bank cover sub-assembly.
40. Install the outer cowl top panel sub-assembly.
41. Install the windshield wiper motor and link assembly.
42. Install the cowl top ventilator louver sub-assembly.
43. Install both front wiper arm and blade assemblies.
44. Install the No. 1 engine under cover.
45. Install the engine under cover assembly.

FUEL PUMP

REMOVAL & INSTALLATION

2.7L & 3.5L Engines

See Figures 362 through 370.

1. Before servicing the vehicle, refer to the precautions section.
2. Discharge fuel system pressure.
3. Disconnect cable from negative battery terminal.
4. Remove the rear center seat assembly.
5. Remove the rear seat headrest assemblies.
6. Remove the seat track bracket covers.
7. Remove rear inner and outer track bracket covers.

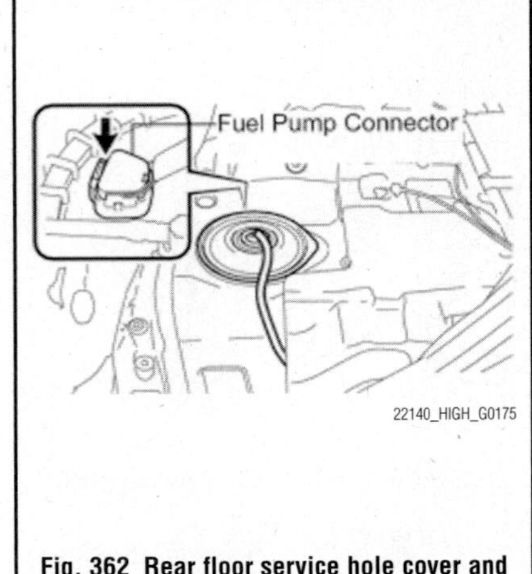

Fig. 362 Rear floor service hole cover and fuel pump connector

8. Remove rear seat leg side covers.
9. Remove rear no. 1 seat lock cable assembly, if applicable.
10. Remove rear no. 1 seat assemblies.
11. Remove left rear door scuff plate.
12. Remove left rear door opening trim weatherstrip.
13. Remove deck board assembly.
14. Remove no. 2 and 3 deck board sub-assemblies.
15. Remove tonneau cover assembly, if applicable.
16. Remove rear no. 1 floor board, or rear seat side covers, as necessary.
17. Remove rear seat side covers.
18. Remove deck side trim boxes.
19. Remove jack carrier support.
20. Remove jack carrier cushion.
21. Remove jack assembly.
22. Remove jack carrier assembly.
23. Remove rear mat.
24. Remove deck floor board assembly, if applicable.
25. Remove rear deck floor box, if applicable.
26. Remove rear no. 2 seat inner belt assembly, if applicable.
27. Disconnect rear seat lap type belt assemblies.
28. Remove rear no. 2 seat assembly, if applicable.
29. Remove rear floor finish plate.
30. Remove deck side trim covers.
31. Remove left side trim cover, if applicable.
32. Remove power outlet socket bezel, if applicable.
33. Remove rear combination light service cover.
34. Remove rear power point socket assembly.
35. Remove rear power outlet socket cover

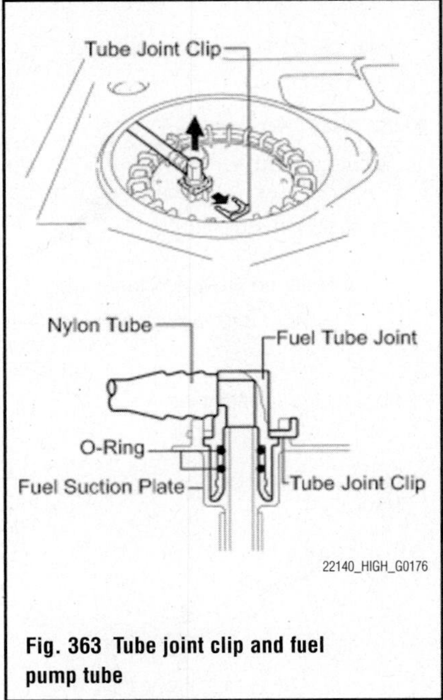

Fig. 363 Tube joint clip and fuel pump tube

36. Remove rear deck trim cover, if applicable.
37. Remove left reclining remote control lever bezel, if applicable.
38. Remove left rope hook assembly.
39. Remove no. 2 deck side trim hook.
40. Remove left front deck side trim cover.
41. Disconnect left rear no. 1 seat outer belt assembly.
42. Remove left deck trim side panel assembly.
43. Remove the rear floor service hole cover, as follows:
    a. Lift the front floor carpet.

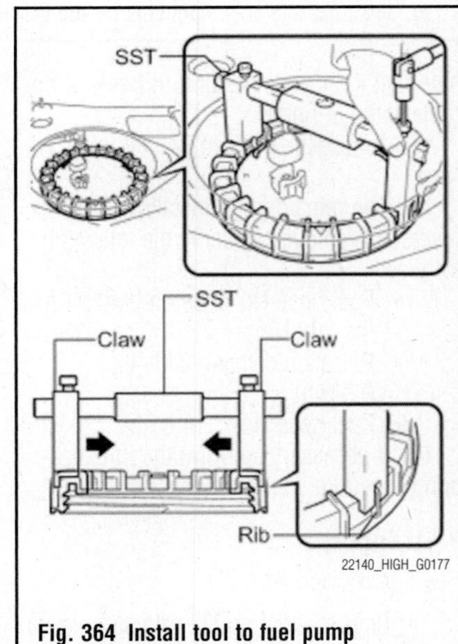

Fig. 364 Install tool to fuel pump gauge retainer

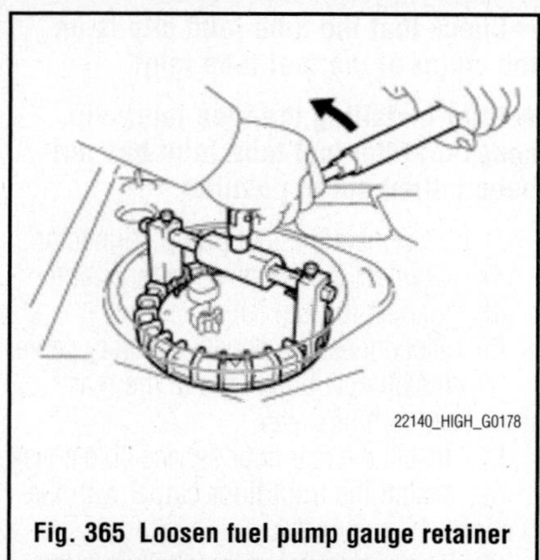

Fig. 365 Loosen fuel pump gauge retainer

b. Using the appropriate tool, remove the 3 clips and lift up the front floor carpet.

c. Remove the rear floor service hole cover.

d. Disconnect the fuel pump connector.

44. Remove the fuel suction tube assembly with pump and gauge, as follows:

a. Remove the tube joint clip, and pull out the fuel pump tube.

➡Check that there is no dirt or other foreign objects around the connector before disconnecting it. Clean the connector if necessary.

➡It is necessary to prevent mud or dirt from entering the quick connector. If mud or dirt gets in the connector, the O-rings may not seal properly.

➡Disconnect the quick connector by hand. Do not use any tools.

➡Do not bend, kink or twist the nylon tubes. Protect the connector by covering it with a plastic bag.

➡If the pipe and connector are stuck, carefully try wiggling or pushing and pulling on the connector to release it. Pull the connector off carefully.

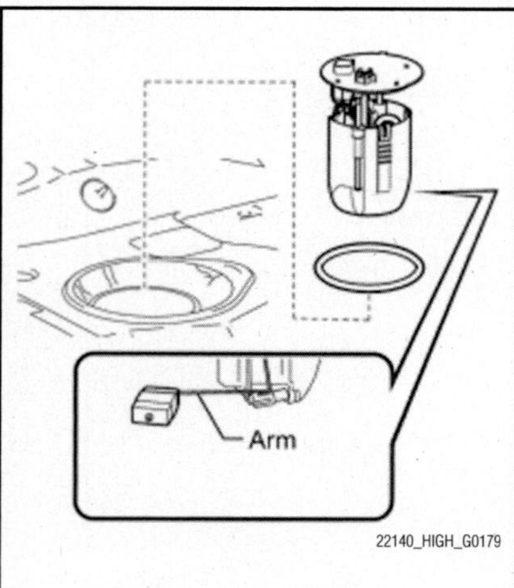

Fig. 366 Remove fuel pump gauge retainer

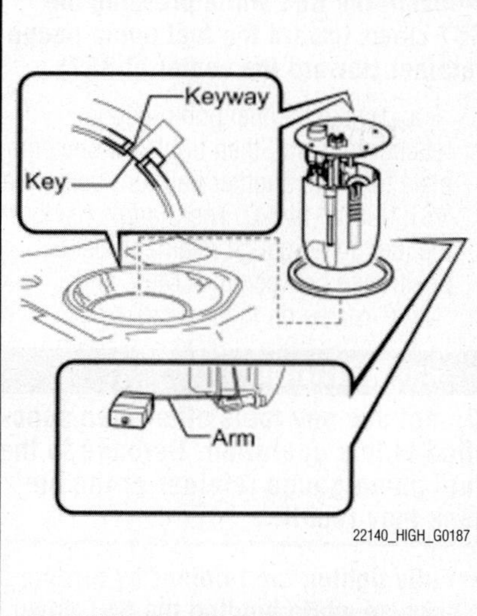

Fig. 367 Fuel suction tube assembly with pump and gauge

b. Using a 6 mm socket hexagon wrench, install SST: 09808-14020 to the fuel pump gauge retainer.

➡Engage the SST claws securely with the fuel pump gauge retainer ribs to secure the SST.

➡Install the SST while pressing the SST claws toward the fuel pump gauge retainer (toward the center of SST).

c. Using SST: 09808-14020, loosen the fuel pump gauge retainer.

d. Remove the fuel pump gauge retainer.

WARNING

Do not use any tools other than specified in this operation. Damage to the fuel pump gauge retainer or the fuel tank may result.

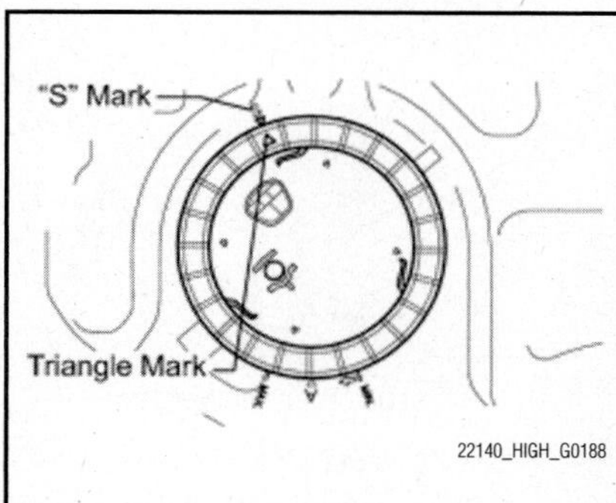

Fig. 368 Align and attach new fuel pump gauge retainer

➡Loosen the retainer by turning it counterclockwise while holding SST down. Do not allow the claw of the tank suction tube support to slip out of its groove on the fuel tank.

➡The ribs on the fuel pump gauge retainer can be fitted into the tips of SST.

e. Remove the fuel suction tube with pump and gauge.

➡Be careful not to bend the arm of the fuel sender gauge.

f. Remove the gasket from the fuel tank.

To install:

45. Install the fuel suction tube assembly with pump and gauge, as follows:

a. Install a new gasket to the fuel tank.

b. Attach the fuel suction tube with pump and gauge to the fuel tank.

➡Be careful not to bend the arm of the fuel sender gauge.

c. Align the keyway of the fuel suction tube support with the key of the fuel suction tube with pump and gauge.

d. Align the triangle mark on a new fuel pump gauge retainer with the "S" mark on the fuel tank while pushing down the fuel suction tube with pump and gauge, and attach the fuel pump gauge retainer.

46. Using a 6 mm socket hexagon wrench, install SST: 09808-14020 to the fuel pump gauge retainer.

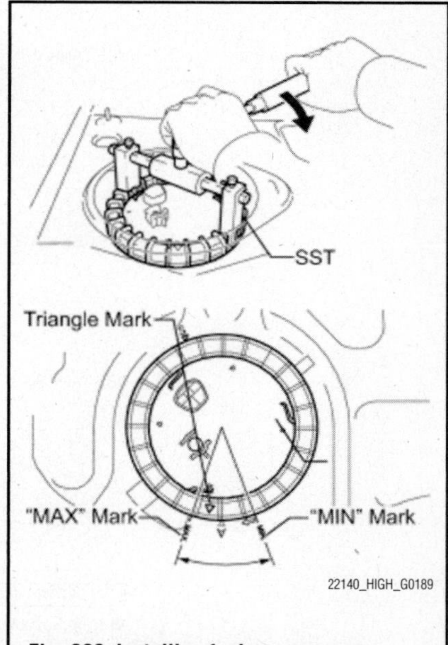

Fig. 369 Installing fuel pump gauge retainer

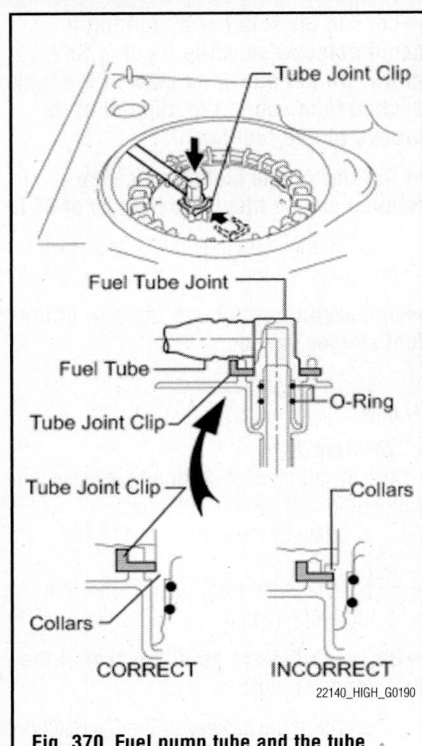

Fig. 370 Fuel pump tube and the tube joint clip

➡Engage the SST claws securely with the fuel pump gauge retainer ribs to secure the SST.

➡Install the SST while pressing the SST claws toward the fuel pump gauge retainer (toward the center of SST).

a. Rotate the fuel pump gauge retainer by hand, then tighten it one complete turn and another half turn using the SST: 09808-14020. The triangle mark on the fuel pump gauge retainer must be positioned between the "MIN" and "MAX" marks on the fuel tank.

⁂ WARNING

Do not use any tools other than specified in this operation. Damage to the fuel pump gauge retainer or the fuel tank may result.

➡Fully tighten the retainer by turning it clockwise while holding the SST down. Do not allow the claw of the tank suction tube support to slip out of its groove on the fuel tank.

➡The ribs on the fuel pump gauge retainer can be fitted into the tips of SST.

b. Install the fuel pump tube and the tube joint clip.

➡Check that there are no scratches or foreign objects on the connecting part.

➡Check that the fuel tube joint is inserted securely.

➡Check that the tube joint clip is on the collar of the fuel tube joint.

➡After installing the tube joint clip, check that the fuel tube joint has not been pulled out of position.

c. Connect the fuel pump connector.

47. Connect the negative battery cable.
48. Inspect for fuel leak.
49. Disconnect the negative battery cable.
50. Install new butyl tape to the rear floor service hole cover.
51. Install the rear floor service hole cover.
52. Install the front floor carpet with the 3 clips.
53. The remainder of installation is the reverse of the removal procedure.
54. After installing all components and connecting the negative battery cable, inspect the SRS warning light.

3.3L Engine

See Figure 371.

1. Before servicing the vehicle, refer to the Precautions Section.
2. Relieve the fuel system pressure, as outlined in this section.
3. Disconnect the negative battery cable.
4. Remove the front floor heat insulator no.3.
5. Remove the fuel tank protector sub-assembly no.1.

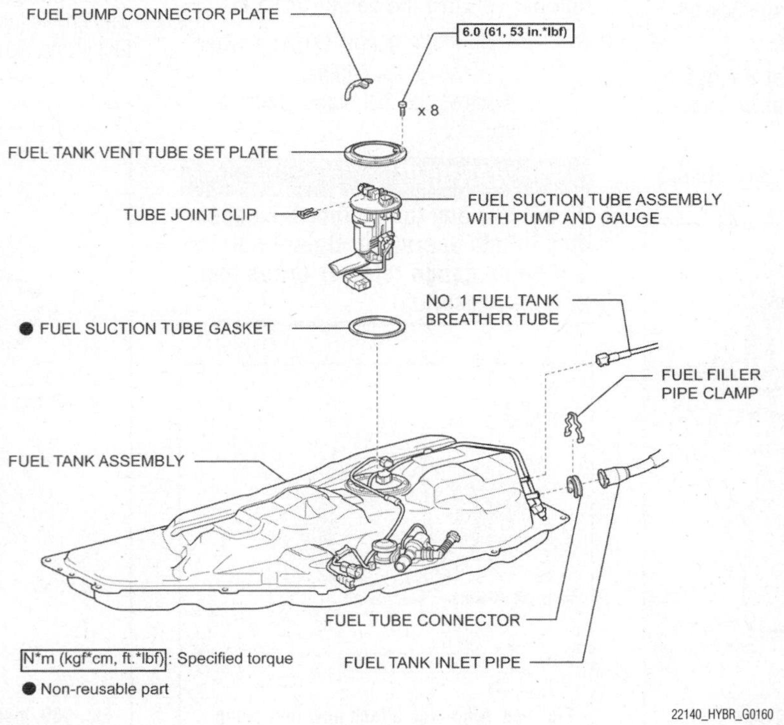

FUEL PUMP CONNECTOR PLATE
6.0 (61, 53 in.*lbf)
x 8
FUEL TANK VENT TUBE SET PLATE
TUBE JOINT CLIP
FUEL SUCTION TUBE ASSEMBLY WITH PUMP AND GAUGE
● FUEL SUCTION TUBE GASKET
NO. 1 FUEL TANK BREATHER TUBE
FUEL FILLER PIPE CLAMP
FUEL TANK ASSEMBLY
FUEL TUBE CONNECTOR
FUEL TANK INLET PIPE
N*m (kgf*cm, ft.*lbf): Specified torque
● Non-reusable part
22140_HYBR_G0160

Fig. 371 Fuel pump and components

6. Separate parking brake cable assembly no.3.
7. Disconnect the fuel tank wire.
8. Remove the charcoal canister protector.
9. Disconnect the charcoal canister fuel hose.
10. Disconnect the fuel tank main tube subassembly.
11. Disconnect the breather lower tube.
12. Remove the fuel tank assembly.
13. Remove the fuel pump and gauge with the suction tube, as follows:
    a. Disconnect the fuel pump connector.
    b. Disconnect the clamp and vapor pressure sensor connector.
14. Remove the tube joint clip and clamp, and pull out the fuel pump tube.

⁂ WARNING

Check that there is no dirt or foreign matter around the fuel tube joint and after cleaning off any excess dirt, proceed with the work. Be careful not to allow dirt or foreign matter to scratch or come into contact with the fuel tube connector and fuel suction plate that are sealed by O-rings. Do not use any tools for this work. Do not bend or twist the nylon tube by force. Cover the disconnected fuel tube joint with a plastic bag. When the fuel tube joint and fuel suction plate are stuck, pinch the fuel tank tube with the fingers, and turn it carefully to release. Disconnect the fuel tank tube.

a. Remove the 8 bolts, fuel tank vent tube set plate and fuel pump connector plate.
b. Pull out the fuel pump assembly from the fuel tank.

To install:

c. Install the fuel pump assembly.
d. Install the fuel tank vent tube set plate by aligning it with the cutout on the fuel pump assembly.
e. Install the fuel tank vent tube set plate and fuel pump connector plate with the 8 bolts. Tighten to 53 inch lbs. (6.0 Nm).
f. Connect the fuel tank main tube with the tube joint clip and clamp.

➡Check that there are no scratches or foreign objects on the connecting part. Check that the fuel tube joint is inserted securely. Check that the tube joint clip is on the collar of the fuel tube joint. after installing the tube joint clip, check that the fuel tube joint is not pulled off.

g. Connect the fuel tank wire connector.
h. Connect the vapor pressure sensor connector and clamp.

15. Install the fuel tank assembly.
16. Connect the breather lower tube,
17. Connect the fuel tank main tube subassembly.
18. Connect charcoal canister fuel hose.
19. Install charcoal canister protector.
20. Connect the fuel tank wire.
21. Install the parking brake cable assembly no.3.
22. Install the fuel tank protector subassembly no.1.
23. Install front floor heat insulator no.3.
24. Install exhaust pipe assembly center.
25. Connect the negative battery cable.
26. If necessary, perform the initialization procedure.
27. Check for fuel leaks.

FUEL TANK

REMOVAL & INSTALLATION

2.7L Engine

See Figures 372 through 384.

⁂ CAUTION

Observe all applicable safety precautions when working around fuel. Whenever servicing the fuel system, always work in a well-ventilated area. Do not allow fuel spray or vapors to come in contact with a spark or open flame. Keep a dry chemical fire extinguisher near the work area. Always keep fuel in a container specifically designed for fuel storage; also, always properly seal fuel containers to avoid the possibility of fire or explosion.

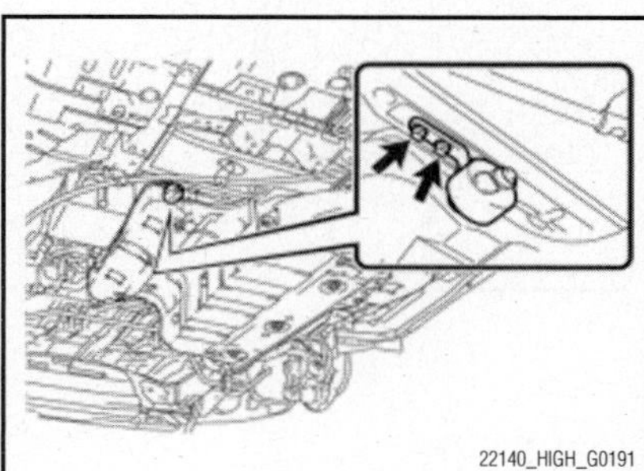

Fig. 372 No. 4 exhaust pipe support bracket

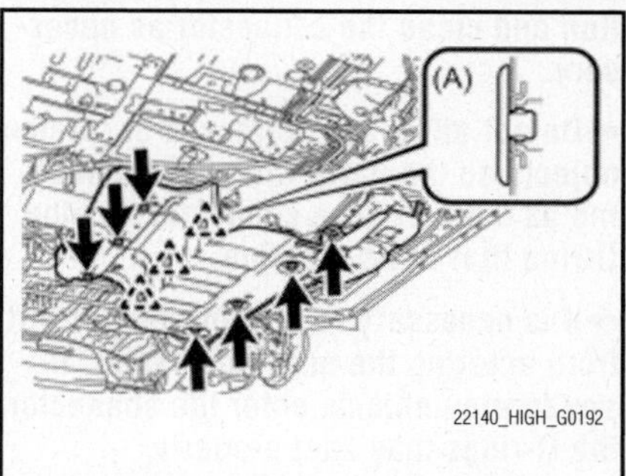

Fig. 373 No. 1 fuel tank protector subassembly, clips (A), and nuts

1. Remove the fuel suction tube assembly with pump and gauge.
2. Drain fuel.
3. Remove center front floor cover.
4. Remove tail exhaust pipe assembly
5. Remove center exhaust pipe assembly.
6. Remove the No. 4 exhaust pipe support bracket, as follows:
    a. Remove the 2 bolts, and then remove the No. 4 exhaust pipe support bracket.
7. Remove the No. 1 fuel tank protector sub-assembly, as follows:
    a. Remove the 3 clips (A) and 7 nuts, and then remove the No. 1 fuel tank protector sub-assembly.
8. Remove the fuel tank assembly, as follows:

➡Check if there is any dirt or mud around the connector before this opera-

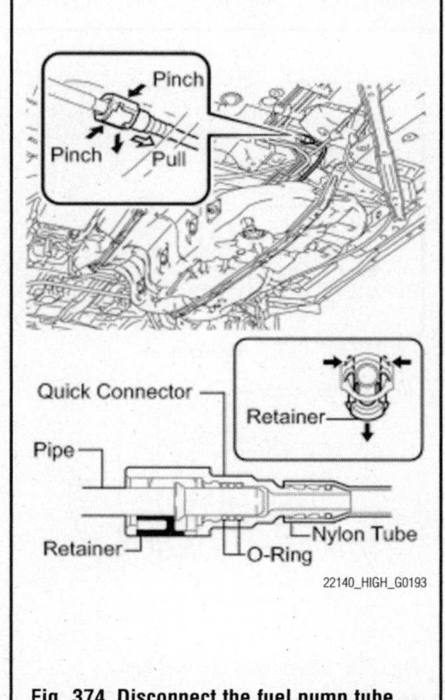

Fig. 374 Disconnect the fuel pump tube

tion and clean the connector as necessary.

➡Do not allow any scratches or foreign objects on the parts when disconnecting as the fuel hose connector has the O-ring that seals the pipe.

➡It is necessary to prevent mud or dirt from entering the quick connector. If any foreign objects enter the connector, the O-rings may seal properly.

➡Perform this work by hand. Do not use any tools.

➡Do not forcibly bend, twist or turn the nylon tube.

➡Protect the connected part by covering it with a plastic bag after disconnecting the fuel tank vent hose.

➡If the connectors or pipe are stuck, push and pull them to release, and pull the connector out carefully.

a. Disconnect the fuel pump tube by pinching the tab of the retainer to disengage the lock claws and pull it down as shown.

b. Pull out the fuel pump tube.

c. Pinch the retainer and pull out the quick connector while pushing the quick connector against the pipe to disconnect the fuel tank vent hose from the charcoal canister assembly.

d. Pinch the tube connector and then pull out the No. 3 fuel tank breather tube.

e. Loosen the hose clamp bolt and disconnect the fuel tank to filler pipe hose.

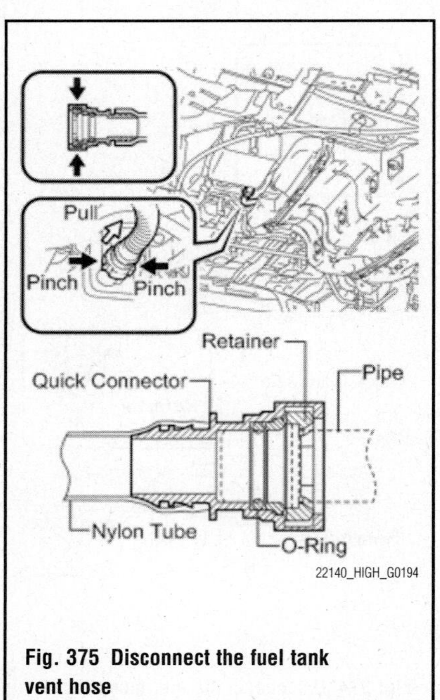

Fig. 375 Disconnect the fuel tank vent hose

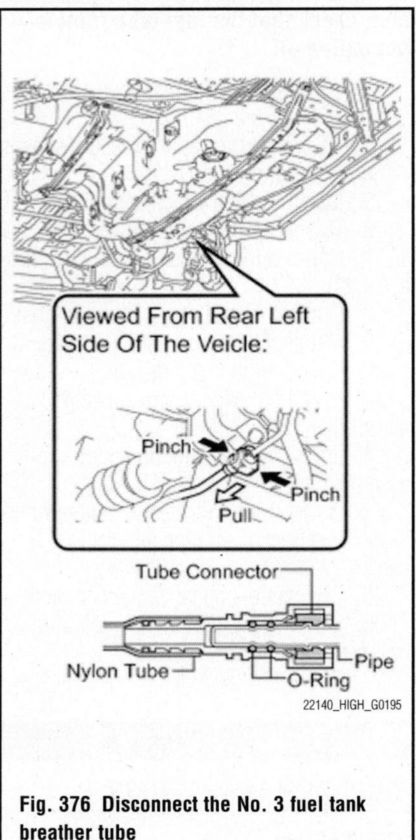

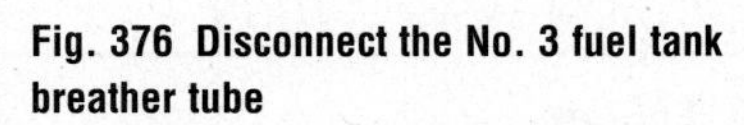

Fig. 376 Disconnect the No. 3 fuel tank breather tube

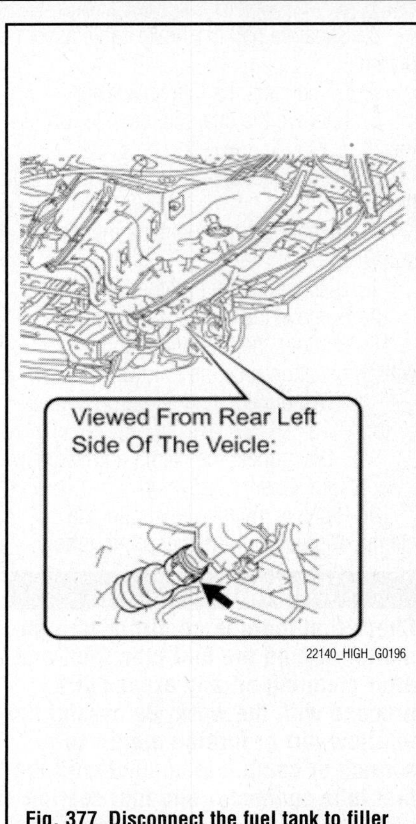

Fig. 377 Disconnect the fuel tank to filler pipe hose

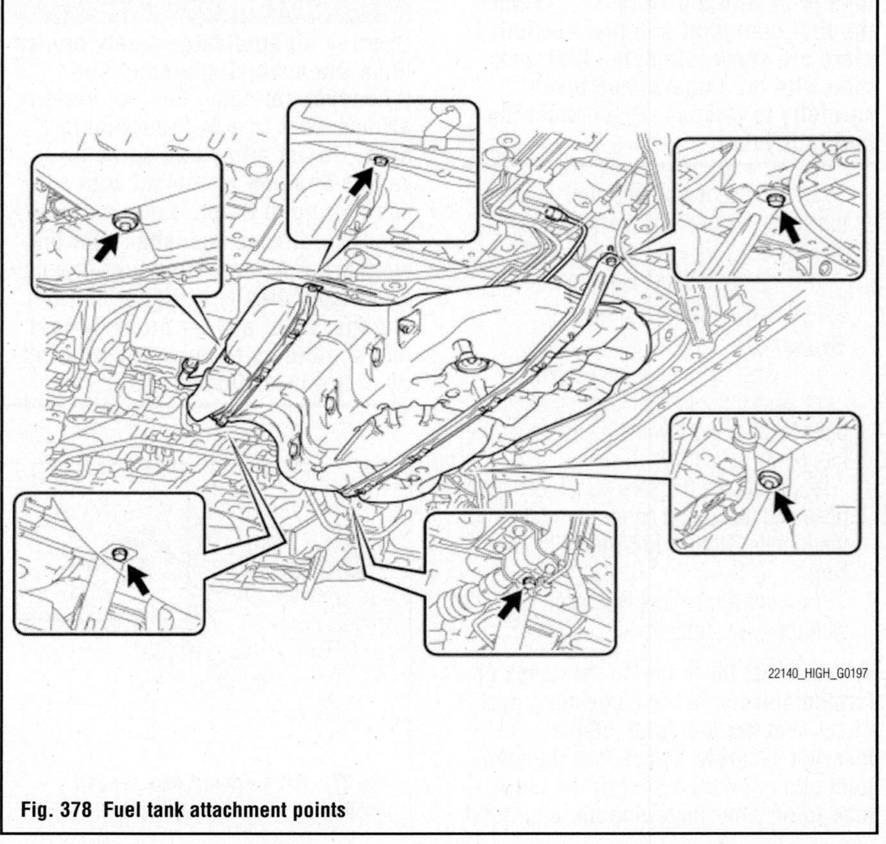

Fig. 378 Fuel tank attachment points

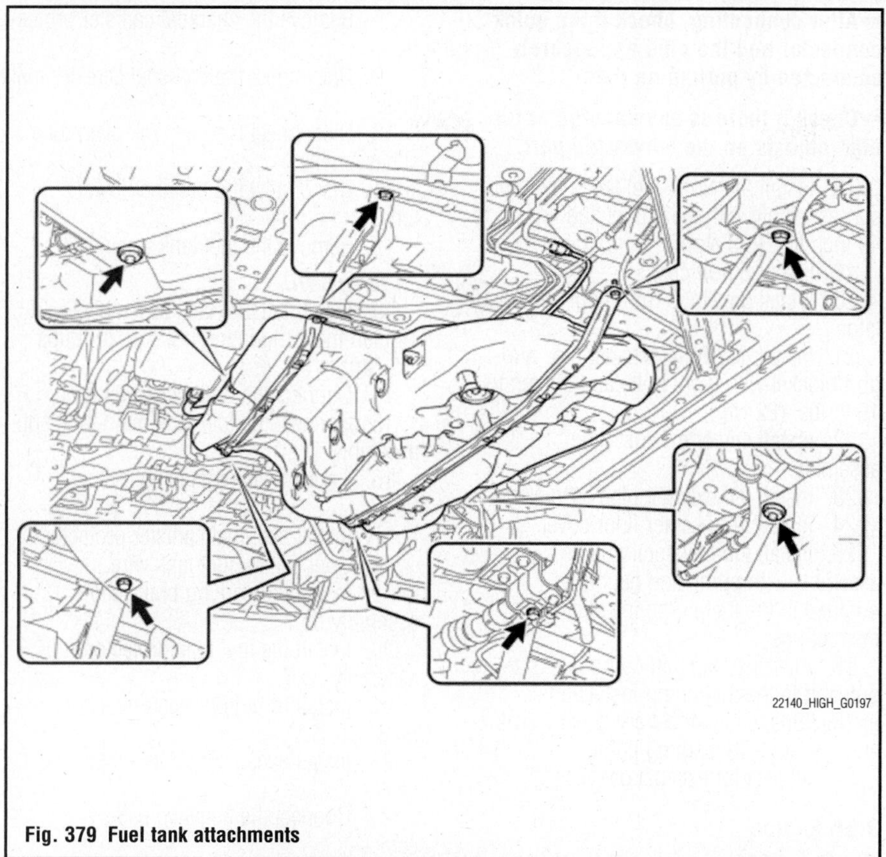

Fig. 379 Fuel tank attachments

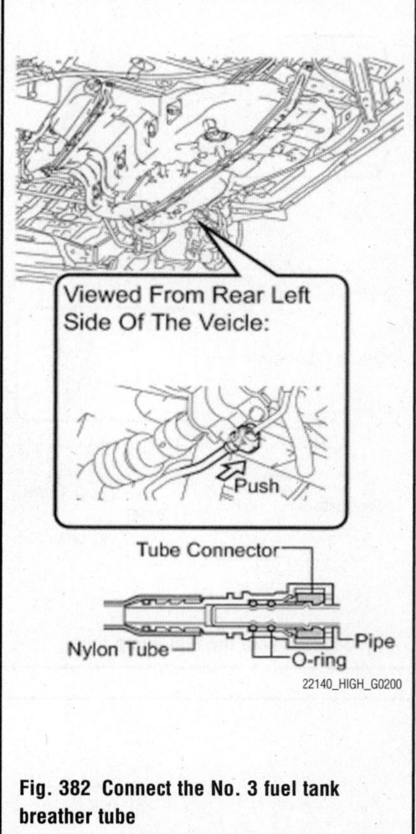

Fig. 382 Connect the No. 3 fuel tank breather tube

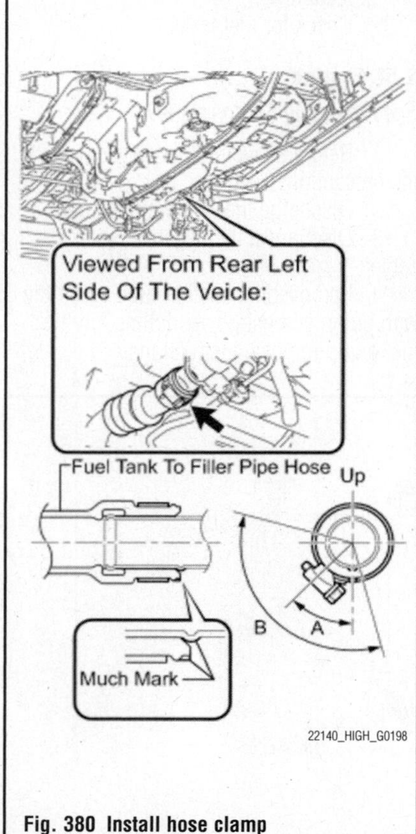

Fig. 380 Install hose clamp

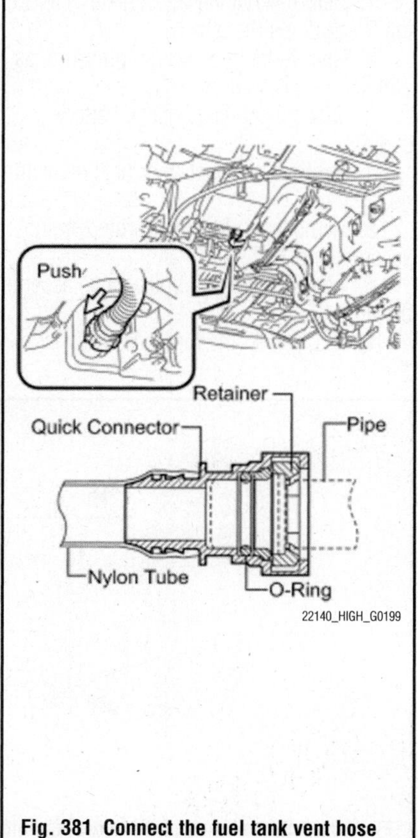

Fig. 381 Connect the fuel tank vent hose

f. Set a transmission jack under the fuel tank.

g. Remove the 4 bolts, and then remove the 2 fuel tank bands.

h. Remove the 2 nuts.

i. Operate the transmission jack to remove the fuel tank.

To install:

9. Set the fuel tank onto a transmission jack.

10. Operating the transmission jack, install the fuel tank.

11. Tighten the 2 nuts to 14 ft. lbs. (20 Nm).

12. Install the 2 fuel tank bands with the 4 bolts and tighten to 29 ft. lbs. (39 Nm).

13. Align the matchmarks and install the fuel tank to filler pipe hose to the fuel tank.

14. Install the hose clamp within the range shown in the illustration.

15. Connect the fuel tank vent hose:

a. Align the quick connector with the pipe, then push in the quick connector until the retainer makes a click sound to connect the charcoal canister fuel hose to the charcoal canister assembly.

➡After connecting the charcoal canister fuel hose, check if the quick connector and the pipe are securely connected by pulling on them.

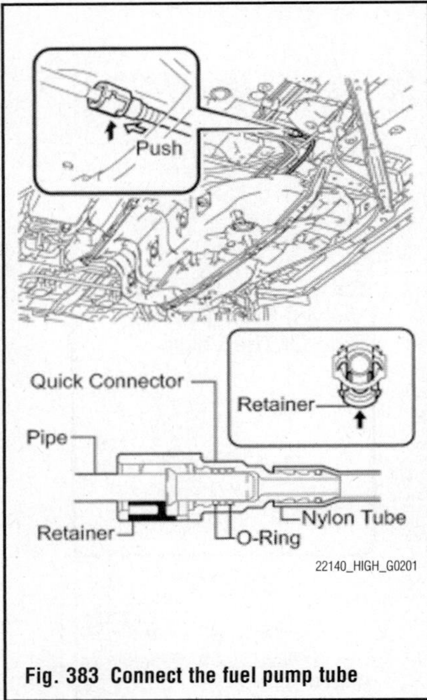

Fig. 383 Connect the fuel pump tube

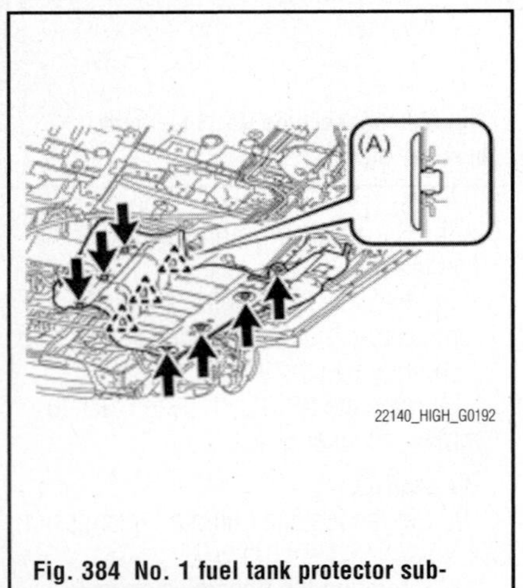

Fig. 384 No. 1 fuel tank protector sub-assembly, clips (A), and nuts

➡Check that there are no scratches or foreign objects around the connected part of the quick connector and pipe before this work.

16. Connect the No. 3 fuel tank breather tube:
   a. Push the quick connector to the pipe until it makes a click sound.

➡After connecting, check if the quick connector and the pipe are securely connected by pulling on them.

➡Check if there is any damage or foreign objects on the connected part.

17. Connect the fuel pump tube:
   a. Push the quick connector and push in the retainer to lock the claws.

➡After connecting, check if the quick connector and the pipe are securely connected by pulling on them.

➡Check if there is any damage or foreign objects on the connected part.

18. Install the No. 1 fuel tank protector sub-assembly with the 7 nuts and tighten to 49 inch lbs. (5.5 Nm).
19. Install 3 new clips.
20. Install the No. 3 front floor heat insulator.
21. Install the No. 4 exhaust pipe support bracket with the 2 bolts and tighten to 16 ft. lbs. (22 Nm).
22. Install center exhaust pipe assembly.
23. Install tail exhaust pipe assembly.
24. Install front center floor cover.
25. Install the fuel suction tube assembly with pump and gauge, as outlined in the fuel pump installation instructions.
26. After installing all components as outlined in the fuel pump installation instructions, connect battery negative cable and inspect SRS warning light.
27. Inspect for exhaust gas leak.

3.3L Engine

See Figure 385.

1. Before servicing the vehicle, refer to the Precautions Section.
2. Relieve the fuel system pressure, as outlined in this section.
3. Disconnect the negative battery cable.
4. Remove the front floor heat insulator no.3.
5. Remove the fuel tank protector sub-assembly no.1.
6. Separate parking brake cable assembly no.3.
7. Disconnect the fuel tank wire.
8. Remove the charcoal canister protector.
9. Disconnect the charcoal canister fuel hose.
10. Disconnect the fuel tank main tube subassembly.
11. Disconnect the breather lower tube.
12. Remove the fuel tank assembly.

To install:

13. Install the fuel tank assembly. Tighten the 4 mounting bolts to 29 ft. lbs. (39 Nm).
14. Connect the breather lower tube,
15. Connect the fuel tank main tube sub-assembly.
16. Connect charcoal canister fuel hose.
17. Install charcoal canister protector.
18. Connect the fuel tank wire.
19. Install the parking brake cable assembly no.3.
20. Install the fuel tank protector sub-assembly no.1.
21. Install front floor heat insulator no.3.
22. Install exhaust pipe assembly center.
23. Connect the negative battery cable.
24. If necessary, perform the initialization procedure.
25. Check for fuel leaks.

3.5L Engine

See Figures 386 through 397.

1. Before servicing the vehicle, refer to the precautions section.
2. Discharge fuel system pressure.
3. Disconnect the negative battery cable.
4. Remove fuel suction tube assembly with pump and gauge, as outlined in the fuel pump removal instructions.

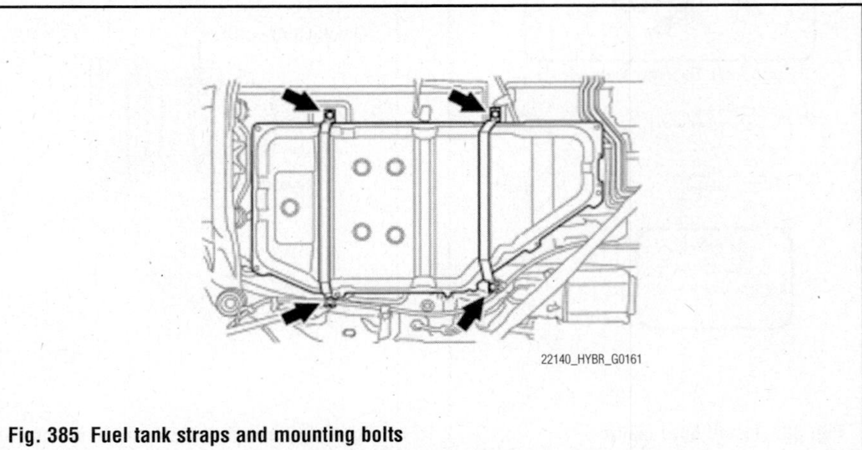

Fig. 385 Fuel tank straps and mounting bolts

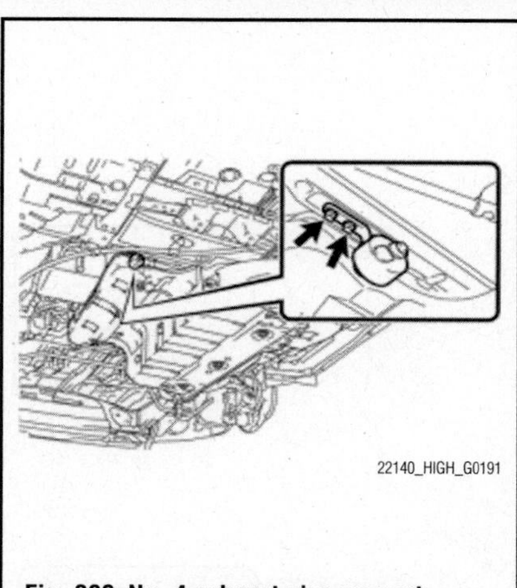

Fig. 386 No. 4 exhaust pipe support bracket

5. Drain fuel.
6. Remove the front center floor cover.
7. Remove the tail exhaust pipe assembly.
8. Remove the center exhaust pipe assembly.
9. Remove the No. 4 exhaust pipe support bracket, as follows:
   a. Remove the 2 bolts, and then remove the No. 4 exhaust pipe support bracket.
10. Remove the No. 3 front floor heat insulator.
11. For 4WD vehicles, remove the following:
   a. Remove the propeller with center bearing shaft assembly.
   b. Remove both rear wheels.
   c. Separate both rear speed sensors.
   d. Remove the rear axle shaft nuts.
   e. Remove both rear disc brake caliper assemblies.
   f. Remove both rear discs.
   g. Remove both rear axle hub and bearing assemblies.
   h. Remove the No. 2 and 3 parking brake cable assemblies.
   i. Remove both rear strut rod assemblies.
   j. Remove both rear axle carrier sub-assemblies.
   k. Remove both rear No. 2 suspension arm assemblies.
   l. Remove the right rear No. 1 suspension arm assembly.
   m. Remove the rear differential filler plug.
   n. Remove the rear differential drain plug.
   o. Remove the rear drive shaft assemblies and snap rings, left and right.
   p. Remove the rear suspension member.
12. Remove the No. 1 fuel tank protector sub-assembly, as follows:
   a. Remove the 3 clips (A) and 7 nuts, and then remove the No. 1 fuel tank protector sub-assembly.
13. Remove the fuel tank assembly, as follows:

➡Check if there is any dirt or mud around the connector before this operation and clean the connector as necessary.

➡Do not allow any scratches or foreign objects on the parts when disconnecting as the fuel hose connector has the O-ring that seals the pipe.

➡It is necessary to prevent mud or dirt from entering the quick connector. If any foreign objects enter the connector, the O-rings may seal properly.

➡Perform this work by hand. Do not use any tools.

➡Do not forcibly bend, twist or turn the nylon tube.

➡Protect the connected part by covering it with a plastic bag after disconnecting the fuel tank vent hose.

➡If the connectors or pipe are stuck, push and pull them to release, and pull the connector out carefully.

   a. Disconnect the fuel pump tube by pinching the tab of the retainer to disengage the lock claws and pull it down as shown.

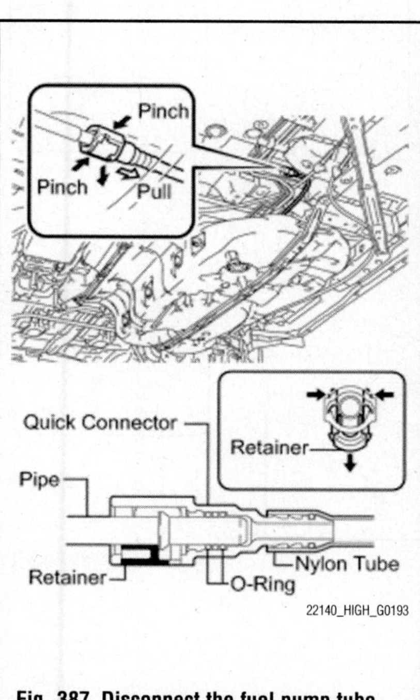

Fig. 387 Disconnect the fuel pump tube

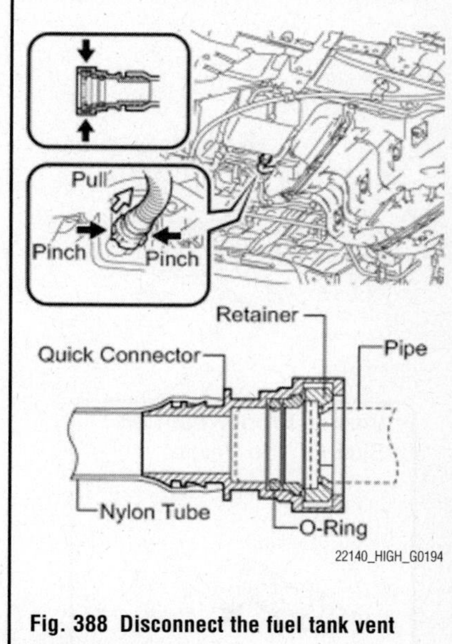

Fig. 388 Disconnect the fuel tank vent hose

   b. Pull out the fuel pump tube.
   c. Pinch the retainer and pull out the quick connector while pushing the quick connector against the pipe to disconnect the fuel tank vent hose from the charcoal canister assembly.
   d. Pinch the tube connector and then pull out the No. 3 fuel tank breather tube.

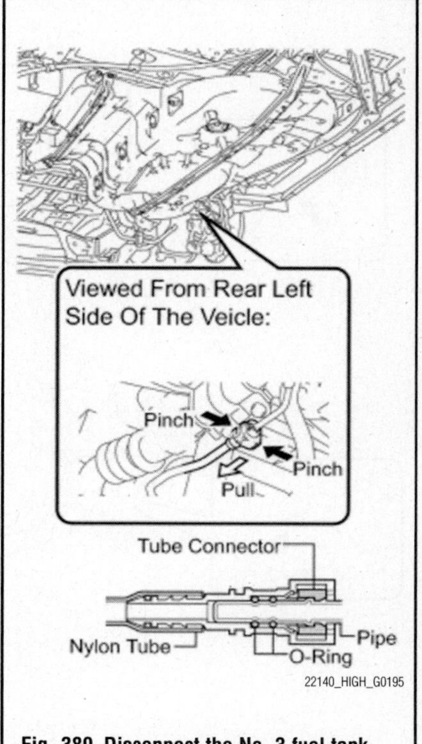

Fig. 389 Disconnect the No. 3 fuel tank breather tube

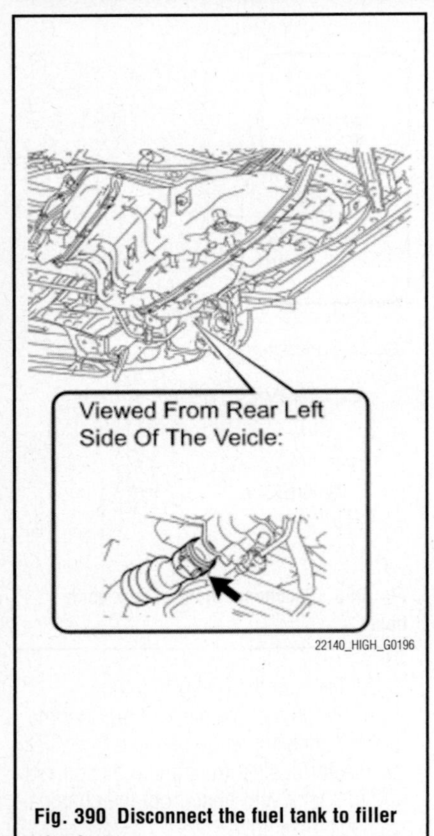

Fig. 390 Disconnect the fuel tank to filler pipe hose

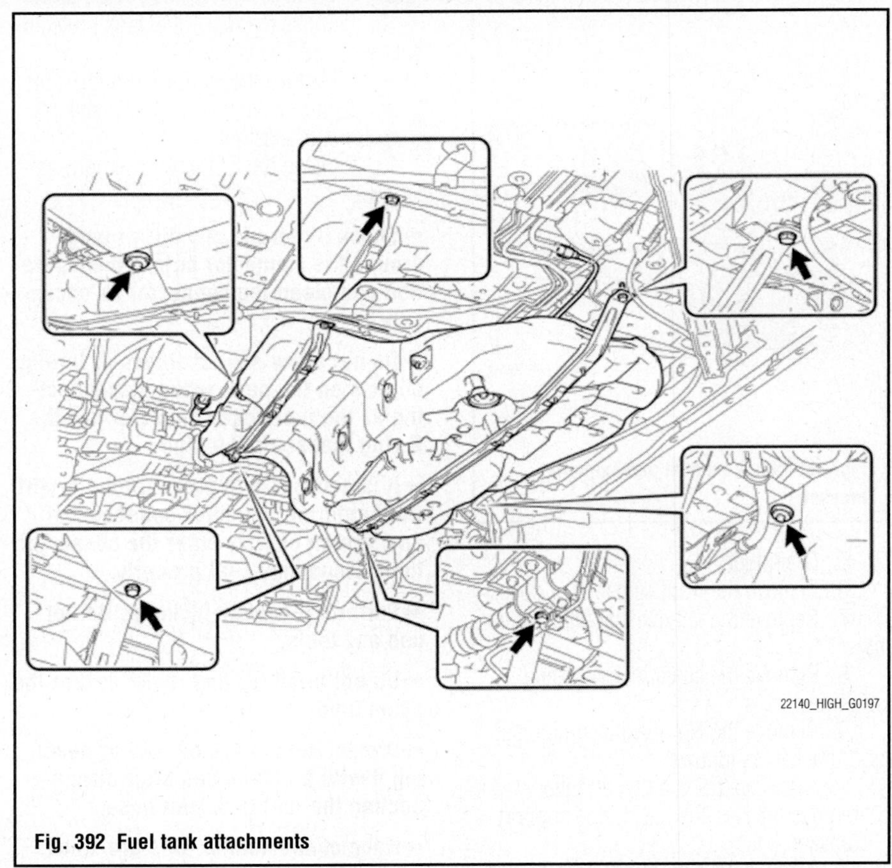

Fig. 392 Fuel tank attachments

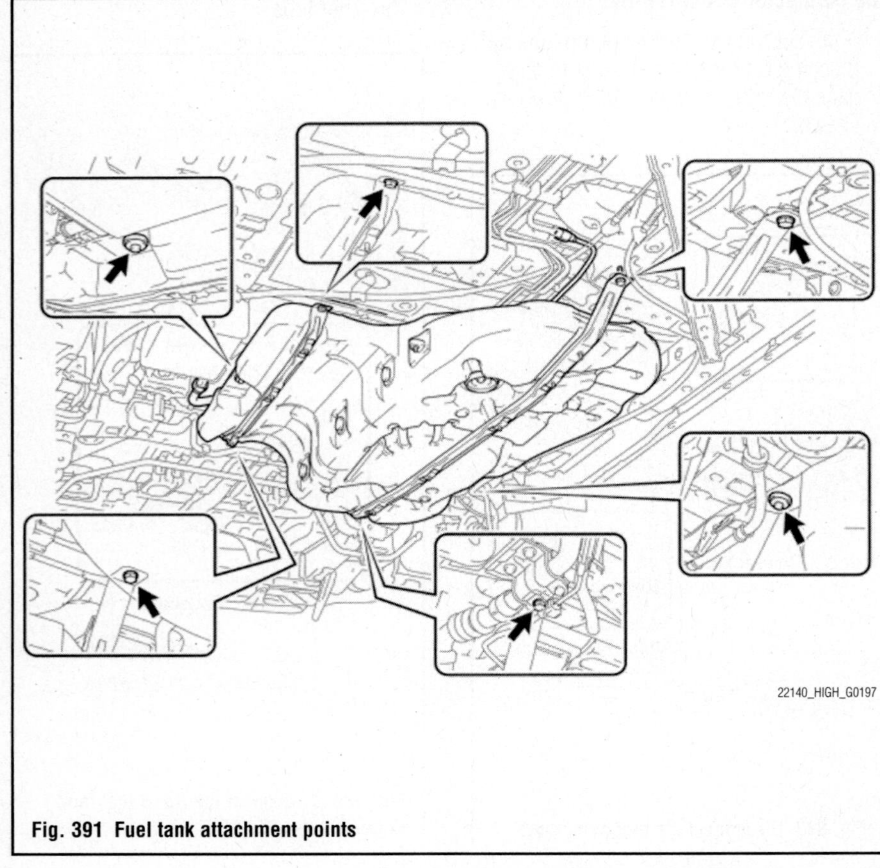

Fig. 391 Fuel tank attachment points

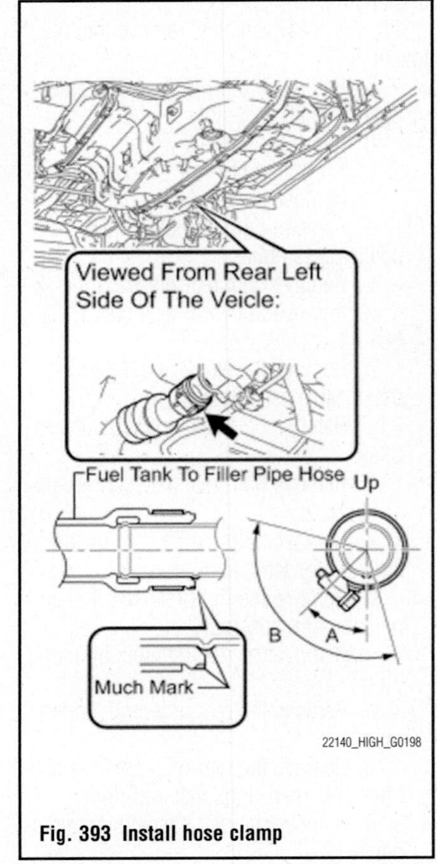

Fig. 393 Install hose clamp

e. Loosen the hose clamp bolt and disconnect the fuel tank to filler pipe hose.

f. Set a transmission jack under the fuel tank.

g. Remove the 4 bolts, and then remove the 2 fuel tank bands.

h. Remove the 2 nuts.

i. Operate the transmission jack to remove the fuel tank.

To install:

14. Set the fuel tank onto a transmission jack.
15. Operating the transmission jack, install the fuel tank.
16. Tighten the 2 nuts to 14 ft. lbs. (20 Nm).
17. Install the 2 fuel tank bands with the 4 bolts and tighten to 29 ft. lbs. (39 Nm).
18. Align the matchmarks and install the fuel tank to filler pipe hose to the fuel tank.
19. Install the hose clamp within the range shown in the illustration.
20. Connect the fuel tank vent hose:

a. Align the quick connector with the pipe, then push in the quick connector until the retainer makes a click sound to connect the charcoal canister fuel hose to the charcoal canister assembly.

➡After connecting the charcoal canister fuel hose, check if the quick connector and the pipe are securely connected by pulling on them.

➡Check that there are no scratches or foreign objects around the connected part of the quick connector and pipe before this work.

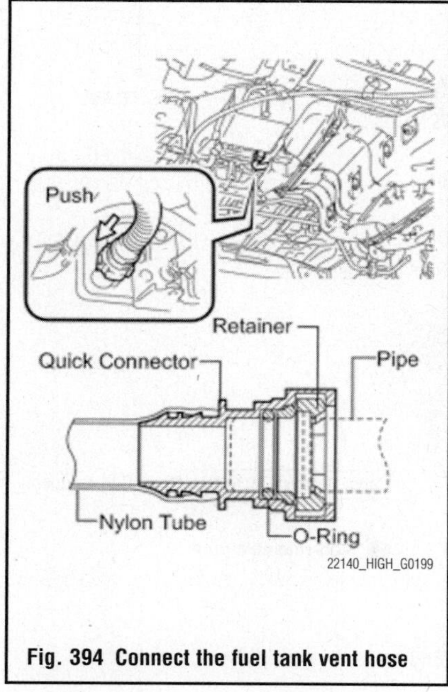

Fig. 394 Connect the fuel tank vent hose

21. Connect the No. 3 fuel tank breather tube:

a. Push the quick connector to the pipe until it makes a click sound.

➡After connecting, check if the quick connector and the pipe are securely connected by pulling on them.

➡Check if there is any damage or foreign objects on the connected part.

22. Connect the fuel pump tube:

a. Push the quick connector and push in the retainer to lock the claws.

➡After connecting, check if the quick connector and the pipe are securely connected by pulling on them.

➡Check if there is any damage or foreign objects on the connected part.

23. Install the No. 1 fuel tank protector sub-assembly with the 7 nuts and tighten to 49 inch lbs. (5.5 Nm).
24. Install 3 new clips.
25. For 4WD vehicles, install the following:

a. Temporarily install the rear suspension member.

b. Fully tighten the rear suspension member.

c. Install rear drive shaft assemblies and snap rings on left and right.

d. Install the rear differential drain plug.

Fig. 395 Connect the No. 3 fuel tank breather tube

e. Add differential oil.

f. Inspect differential oil.

g. Install the rear differential filler plug.

h. Temporarily install the right rear No. 1 suspension arm assembly.

i. Temporarily install both rear No. 2 suspension arm assemblies.

j. Install both rear axle carrier sub-assemblies.

k. Temporarily install both rear strut rod assemblies.

l. Install both rear axle hub and bearing assemblies.

m. Install both rear discs.

n. Install rear disc brake caliper assemblies.

o. Temporarily install both rear axle shaft nuts.

p. Install rear speed sensors.

q. Stabilize suspension.

r. Fully tighten both rear no. 1 suspension arm assemblies.

s. Fully tighten both rear no. 2 suspension arm assemblies.

t. Fully tighten rear strut rod assemblies.

u. Install No. 2 and 3 parking brake cable assemblies.

v. Separate both rear disc brake caliper assemblies.

w. Remove both rear discs.

x. Inspect rear axle hub bearing looseness, left and right.

y. Inspect rear axle hub runout, left and right.

z. Install both rear discs.

26. Install rear disc brake caliper assemblies.

a. Install rear axle shaft nuts.

b. Install rear wheels.

27. Install the No. 3 front floor heat insulator.
28. Install the No. 4 exhaust pipe support bracket with the 2 bolts and tighten to 16 ft. lbs. (22 Nm).
29. Install center exhaust pipe assembly.
30. Install tail exhaust pipe assembly.
31. For 4WD, temporarily tighten propeller with center bearing shaft assembly.
32. For 4WD, fully tighten propeller with center bearing shaft assembly.
33. Install front center floor cover.
34. Install the fuel suction tube assembly with pump and gauge, as outlined in the fuel pump installation instructions.
35. After installing all components as outlined in the fuel pump installation instructions, connect battery negative cable and inspect SRS warning light.
36. Inspect for exhaust gas leak.
37. For 4WD, inspect and adjust transfer oil.

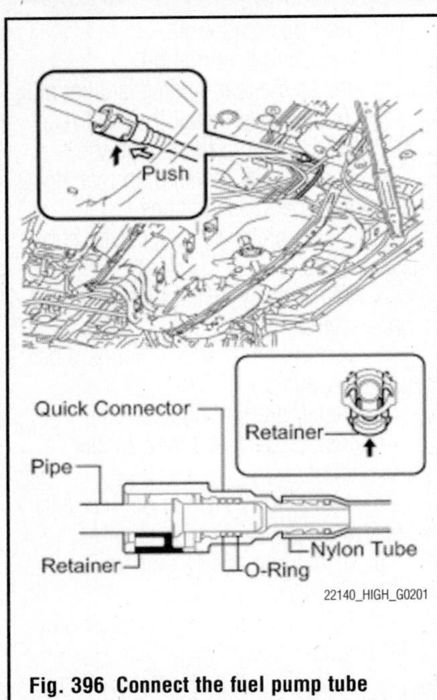

Fig. 396 Connect the fuel pump tube

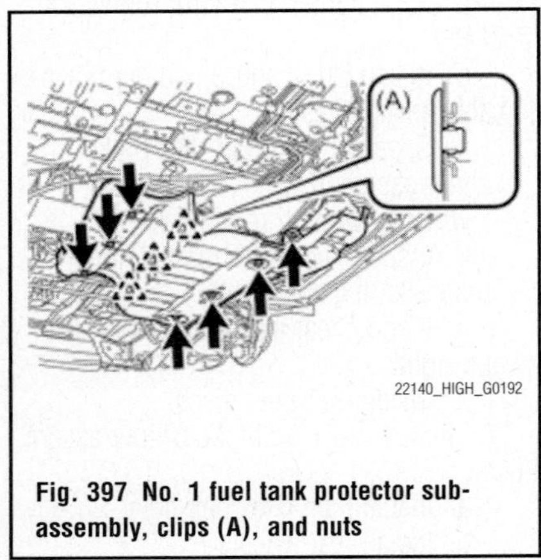

Fig. 397 No. 1 fuel tank protector sub-assembly, clips (A), and nuts

38. For 4WD, inspect rear wheel alignment.
39. For 4WD, check for speed sensor signal.

IDLE SPEED

ADJUSTMENT

Idle speed is maintained by the Powertrain Control Module (PCM). No adjustment is necessary or possible.

THROTTLE BODY

REMOVAL & INSTALLATION

2.7L Engine

See Figure 398.

1. Remove windshield wiper motor and link.
2. Separate ejector tube.
3. Remove outer cowl top panel sub-assembly.
4. Remove cool air intake duct seal.
5. Remove no. 1 engine under cover.
6. Drain engine coolant.
7. Remove no. 1 engine cover sub-assembly.
8. Remove no. 1 vacuum switching valve.
9. Remove air cleaner cap sub-assembly.
10. Remove throttle body assembly:
    a. Disconnect the throttle body assembly connector.
    b. Disconnect the fuel tube from the clamp.
    c. Disconnect the 2 water by-pass hoses from the throttle body assembly.
    d. Remove the 4 bolts and the throttle body assembly with the fuel tube bracket.
    e. Remove the bolt and fuel tube bracket.
    f. Remove the gasket from the intake manifold.

To install:

11. Install a new gasket to the intake manifold.
12. Install the fuel tube bracket with the bolt.
13. Install the throttle body assembly with the 4 bolts. Tighten to 7 ft. lbs. (10 Nm).
14. Connect the 2 water by-pass hoses to the throttle body.
15. Connect the throttle body assembly connector.
16. Connect the fuel tube to the clamp.
17. The remainder of installation is the reverse of the removal procedure.

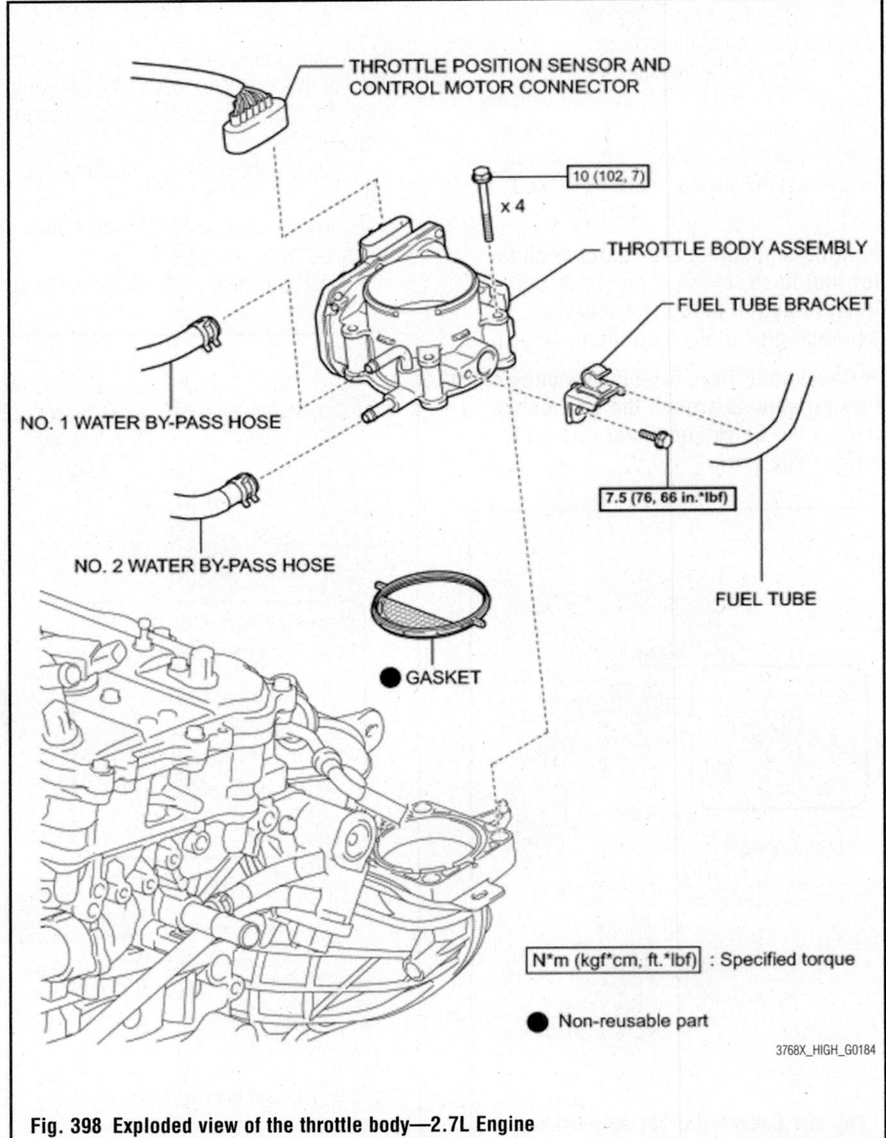

Fig. 398 Exploded view of the throttle body—2.7L Engine

3.3L Engine

See Figure 399.

1. Before servicing the vehicle, refer to the Precautions Section.
2. Relieve the fuel system pressure.
3. Drain the cooling system.
4. Remove or disconnect the following:
   - Negative battery cable
   - Engine cover
   - Wiper arm and blade assembly
   - Top cowl ventilator louver assembly
   - Wiper motor and linkage assembly
   - Cool air intake duct seal
   - Air cleaner cap and case assembly
5. Remove the throttle body as follows:
   a. Disconnect the throttle motor connector (A).
   b. Separate the No. 2 water by-pass hose (B).
   c. Separate the No. 3 water by-pass hose (C).
   d. Separate the fuel vapor feed hose (D).
   e. Remove the 4 bolts and the throttle body assembly.
   f. Remove the throttle body gasket from the intake air surge tank.

To install:

6. Install the throttle body assembly as follows:
7. Install a new throttle body gasket to the intake air surge tank.
   a. Tighten the 4 mounting bolts to 8 ft. lbs. (11 Nm).
   b. Connect the fuel vapor feed hose (A).
   c. Connect the No. 3 water by-pass hose (B).
   d. Connect the No. 2 water by-pass hose (C).
   e. Connect the throttle motor connector (D).
8. Install air cleaner case and cap assembly.
9. Add the engine coolant.

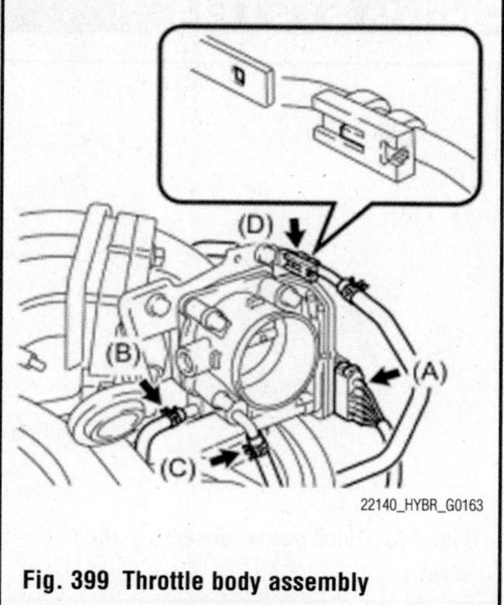

Fig. 399 Throttle body assembly

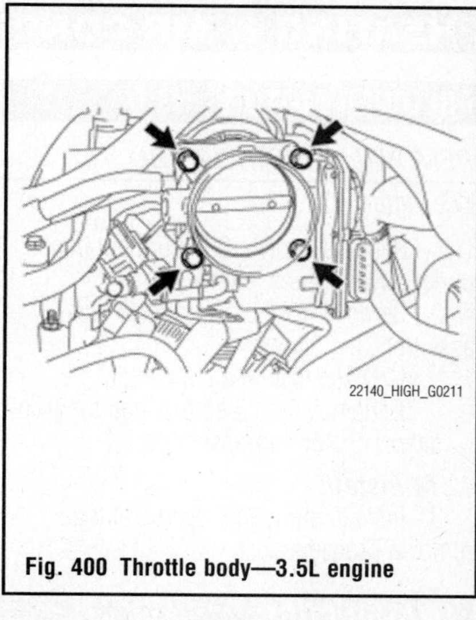

Fig. 400 Throttle body—3.5L engine

10. Inspect for any coolant leaks.
11. Connect the negative battery cable.
12. Install or connect the following:
    - Air cleaner cap and case assembly
    - Cool air intake duct seal
    - Wiper motor and linkage assembly
    - Top cowl ventilator louver assembly
    - Wiper arm and blade assembly
    - Engine cover
    - Negative battery cable

3.5L Engine

See Figure 400.

1. Remove the engine under cover assembly.
2. Remove the No. 1 engine under cover.
3. Drain engine coolant.
4. Remove both front wiper arm and blade assemblies.
5. Remove the cowl top ventilator louver sub-assembly.
6. Remove the windshield wiper motor and link assembly.
7. Remove the outer cowl top panel sub-assembly.
8. Remove the V-bank cover sub-assembly.
9. Remove the air cleaner cap sub-assembly.
10. Disconnect the throttle body connector and clamp.
11. Disconnect the 2 water by-pass hoses from the throttle body.
12. Remove the 4 bolts and throttle body.
13. Remove the throttle body gasket from the intake air surge tank.

To install:

14. Install a new throttle body gasket to the intake air surge tank.
15. Install the throttle body with the 4 bolts and tighten to 7 ft. lbs. (10 Nm).
16. Connect the 2 water by-pass hoses.
17. Connect the throttle body connector and clamp.
18. Install the air cleaner cap sub-assembly.
19. Add engine coolant.
20. Inspect for coolant leaks.
21. The remainder of installation is the reverse of the removal procedure.

HEATING & AIR CONDITIONING SYSTEM

BLOWER MOTOR

REMOVAL & INSTALLATION

See Figure 401.

1. Remove the No. 2 instrument panel under cover sub-assembly.
2. Remove front blower motor sub-assembly, as follows:
   a. Disconnect the connector.
   b. Remove the 3 screws and the front blower motor sub-assembly.

To install:

3. Installation is the reverse of the removal procedure.

HEATER CORE

REMOVAL & INSTALLATION

Non-Hybrid Models

See Figures 402 and 403.

1. Remove the center instrument panel register assembly.
2. Remove the center instrument cluster finish panel assembly.

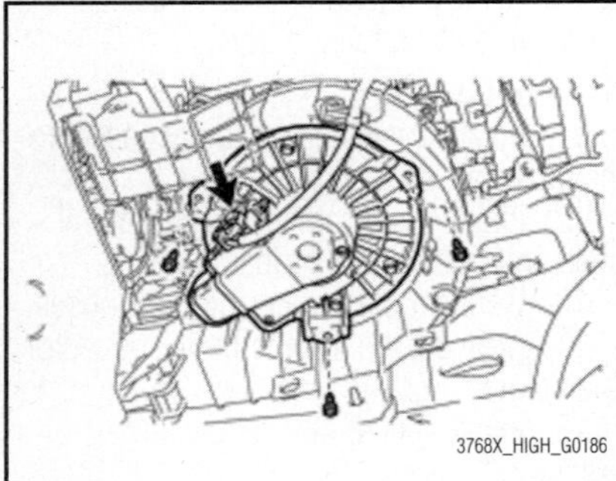

Fig. 401 Remove the 3 screws and the front blower motor sub-assembly

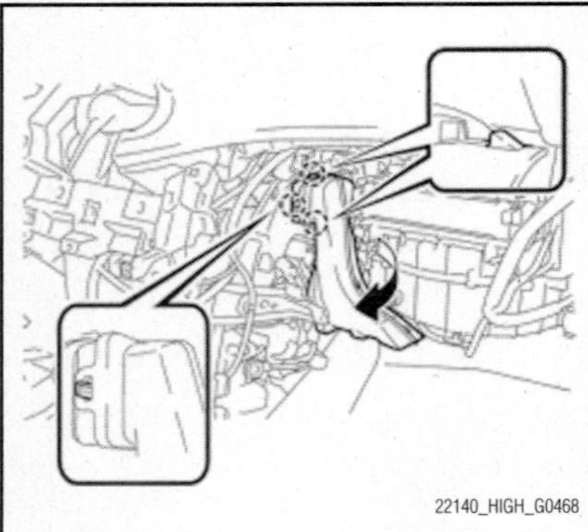

Fig. 402 No. 3 air duct sub-assembly

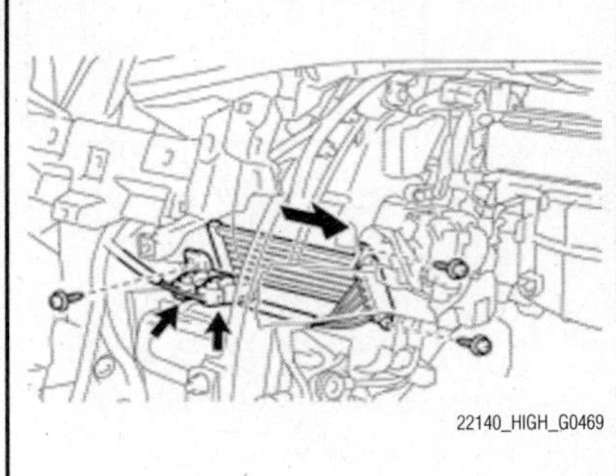

Fig. 403 Quick heater assembly and screws

3. Remove the heater control and accessory assembly (for manual air conditioning system).
4. Remove the air conditioning control assembly (for automatic air conditioning system).
5. Remove the radio receiver assembly with bracket (w/o navigation system).
6. Remove the navigation receiver assembly with bracket (w/ navigation system).
7. Remove the right front door scuff plate.
8. Remove the right cowl side trim sub-assembly.
9. Remove the No. 2 instrument panel under cover sub-assembly.
10. Remove the lower instrument panel sub-assembly.
11. Remove the upper console panel sub-assembly.
12. Remove the lower rear console box.
13. Remove the console box assembly.
14. Remove the front No. 2 console box insert.
15. Disengage the 3 claws and remove the No. 3 air duct sub-assembly as shown in the illustration.
16. Remove quick heater assembly, as follows:
    a. Disconnect the 2 connectors.
    b. Remove the 3 screws and the quick heater assembly as shown in the illustration.

To install:

17. Installation is the reverse of the removal procedure.

Hybrid Models

See Figures 404 through 403.

1. Before servicing the vehicle, refer to the Precautions Section.
2. Recover refrigerant from refrigeration system.
3. Position front wheels straight ahead.
4. Remove both front wheels.

**** CAUTION**

Wait for 90 seconds after disconnecting the cable to prevent the airbag from deploying.

5. Remove both front wiper arm and blade assemblies.
6. Remove the cowl top ventilator louver sub-assembly.
7. Remove the windshield wiper motor and link assembly.
8. Remove the cowl top outer panel sub-assembly.
9. Disconnect the heater inlet and outlet water hoses.
10. Disconnect the cooler refrigerant liquid pipe.
11. Disconnect the No. 1 cooler refrigerant suction pipe.
12. Remove the lower No. 2 and 3 steering wheel covers.
13. Remove the steering pad.
14. Remove the steering wheel assembly.
15. Remove the steering column cover.
16. Remove the turn signal switch assembly with spiral cable sub-assembly.
17. Remove the instrument cluster finish panel assembly.
18. Remove the combination meter assembly.
19. Remove the center instrument panel register assembly.
20. Remove the center instrument cluster finish panel assembly.
21. Remove the air conditioning control assembly.

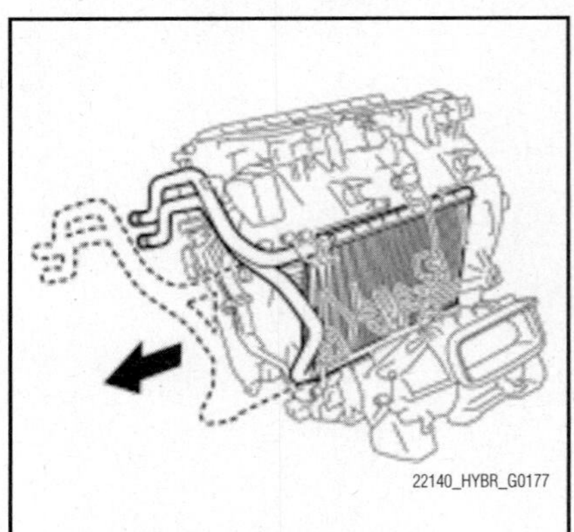

Fig. 404 Remove the heater core

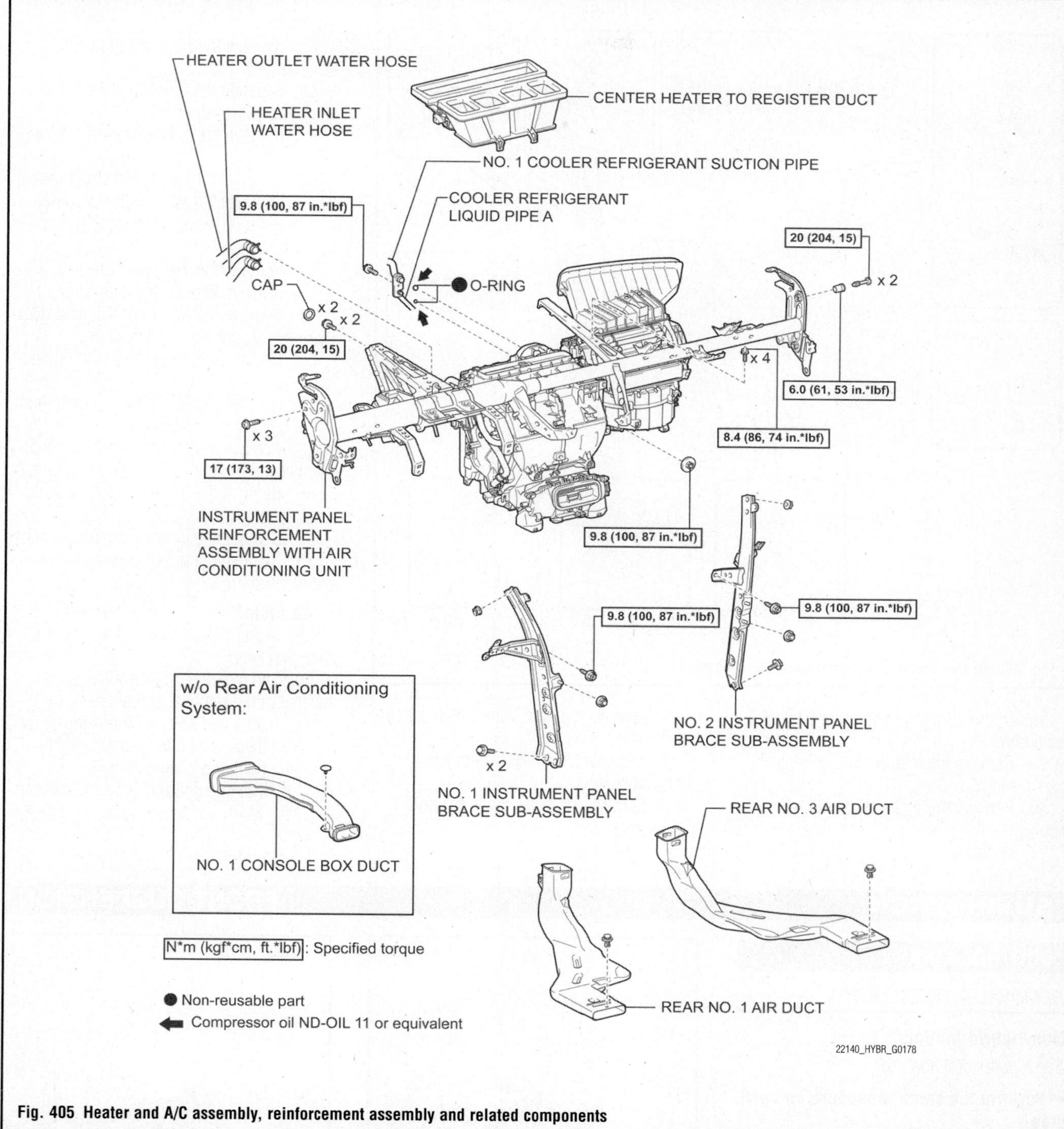

Fig. 405 Heater and A/C assembly, reinforcement assembly and related components

22. Remove the radio receiver assembly with bracket.
23. Remove both front door scuff plates.
24. Remove both cowl side trim sub-assemblies.
25. Remove the lower instrument panel finish panel sub-assembly.
26. Remove the No. 2 instrument panel under cover sub-assembly.
27. Remove the lower instrument panel sub-assembly.
28. Remove the upper console panel sub-assembly.
29. Remove the No. 2 console box duct (w/o rear air conditioning system).
30. Remove the lower rear console box.
31. Remove the console box assembly.
32. Remove the front No. 1 No. 2 console box insert.
33. Remove the engine switch (W/ Smart Key System).
34. Remove the front pillar garnish.
35. Disconnect the front door opening trim weather strip.
36. Remove the No. 1 instrument panel speaker panel sub-assembly.
37. Remove the front No. 2 speaker assembly.

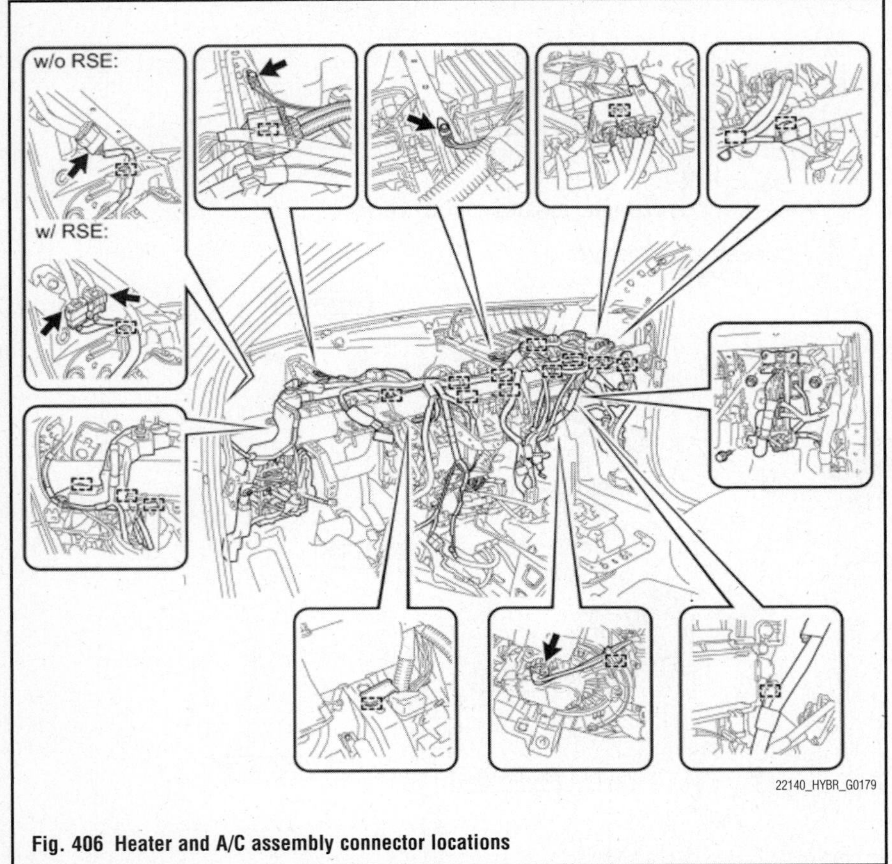

Fig. 406 Heater and A/C assembly connector locations

38. Disconnect the instrument panel wire assembly.
39. Remove the instrument panel safety pad assembly.
40. Remove the brake pedal return spring.
41. Remove the stop light switch assembly.
42. Separate the brake master cylinder push rod clevis.
43. Remove the brake pedal support sub-assembly.
44. Remove the driver side knee airbag assembly.
45. Remove the No. 1 air duct sub-assembly.
46. Separate the steering intermediate shaft sub-assembly.
47. Remove the steering column assembly.
48. Remove the certification ECU (smart key ECU assembly) (W/ smart key system).
49. Remove the air conditioning amplifier assembly.
50. Remove the rear No. 1 air duct.
51. Remove the rear No. 3 air duct.
52. Remove the No. 1 console box duct (w/o rear air conditioning system).
53. Remove the center heater to register duct.
54. Remove the No. 1 and 2 instrument panel brace sub-assemblies.
55. Remove instrument panel reinforcement assembly with air conditioning unit.
56. Remove the 2 caps and the 2 bolts from the engine compartment side.
57. Remove the heater core tube clamps.
58. Remove the heater core as shown in the illustration

To install:

59. Install heater core and tighten the tube brackets.
60. The remainder of installation is reverse of the removal procedure.
61. Check and note the following:
    a. Inspect the SRS warning light.
    b. Refill the engine coolant.
    c. Vacuum and recharge A/C system.
    d. Warm up the engine and check for leaks.

SUSPENSION

FRONT SUSPENSION

CONTROL LINKS

REMOVAL & INSTALLATION

Non-Hybrid Models

See Figures 407 and 408.

➡Perform the same procedure on each side.

1. Remove wheel and tire assembly.
2. For 2.7L engines, remove the two nuts and separate the front stabilizer link assembly.
3. For 3.5L engines, remove the nut and separate the front stabilizer link assembly.

➡If the ball joint turns together with the nut, use a hexagon wrench (6 mm) to hold the stud bolt.

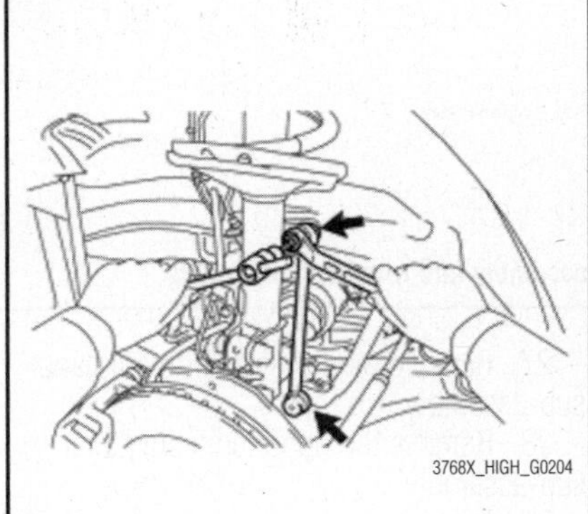

Fig. 407 Remove the two nuts and separate the front stabilizer link assembly—2.7L Engines

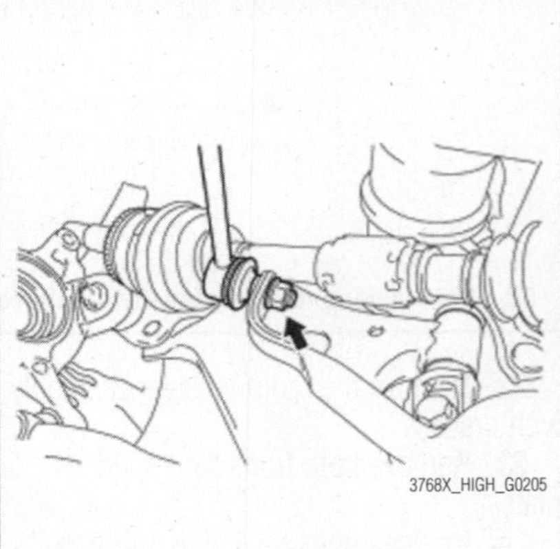

Fig. 408 Remove the nut and separate the front stabilizer link assembly—3.5L Engines

To install:

4. Install the front stabilizer link assembly with the nut and tighten to 55 ft. lbs. (74 Nm).
5. Install the wheel and tire assembly.

Hybrid Models

See Figure 409.

➡Perform the same procedure on each side.

1. Remove wheel and tire assembly.
2. Remove the nut and separate the front stabilizer link assembly.

➡If the ball joint turns together with the nut, use a hexagon wrench (6 mm) to hold the stud bolt.

To install:

3. Install the front stabilizer link assembly with the nut and tighten to 55 ft. lbs. (74 Nm).
4. Install the wheel and tire assembly.

LOWER BALL JOINT

REMOVAL & INSTALLATION

See Figures 410 through 411.

1. Remove the front wheel.
2. Remove the front axle hub nut.
3. Remove the bolt and resin clamp, and separate the front speed sensor.

➡Be sure to completely separate the front speed sensor from the front shock absorber with coil spring.

➡Clean the installation hole and the surface for the speed sensor every time the speed sensor is removed.

➡Be careful not to damage the front speed sensor.

4. Put matchmarks on the front drive shaft assembly and the front axle hub sub-assembly.

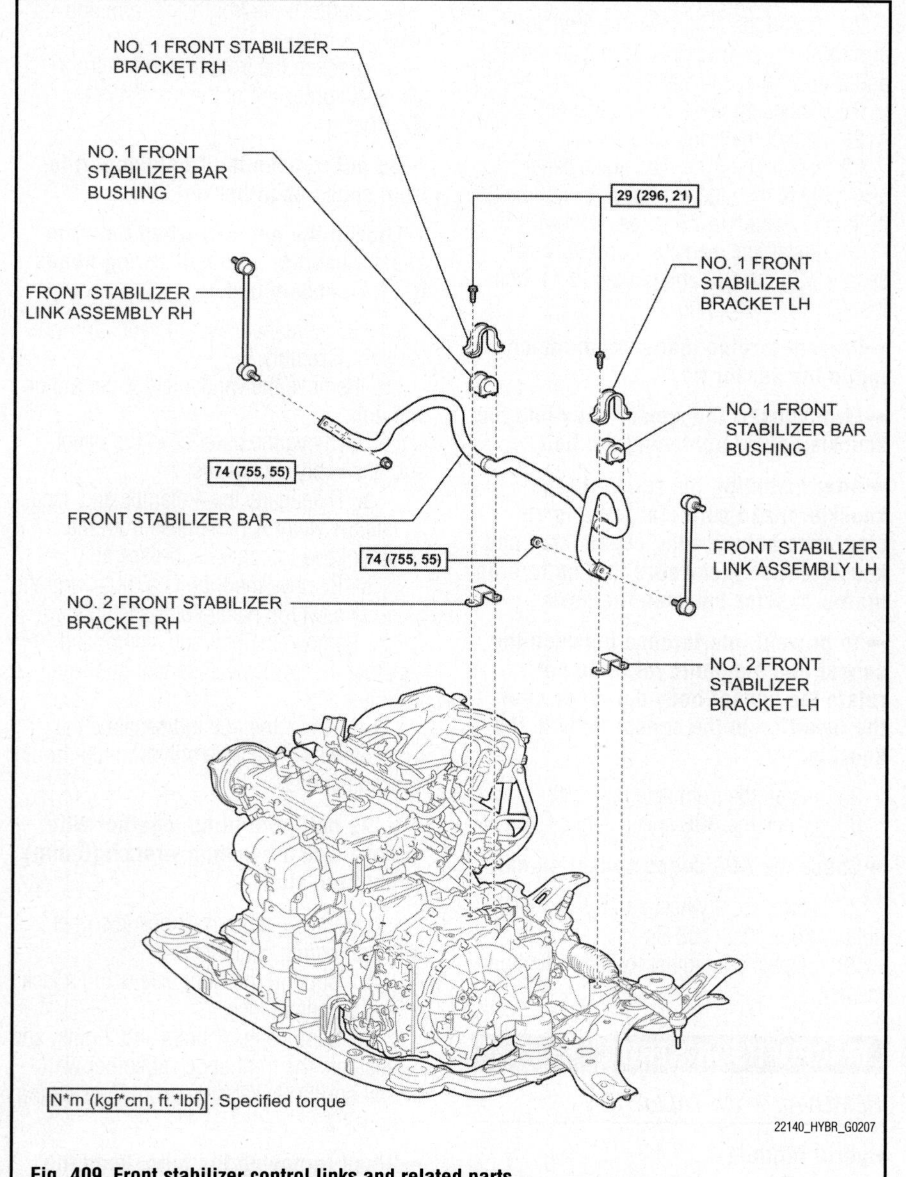

Fig. 409 Front stabilizer control links and related parts.

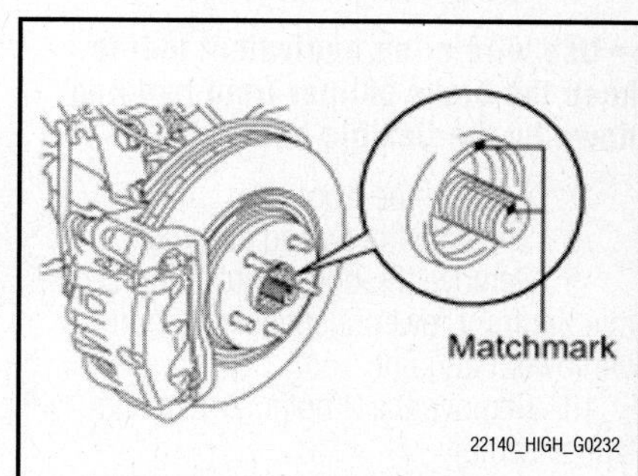

Fig. 410 Matchmark the front drive shaft assembly and the front axle hub sub-assembly

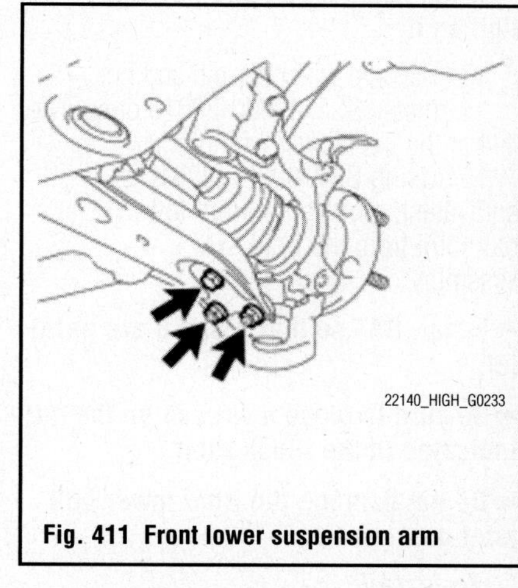

Fig. 411 Front lower suspension arm

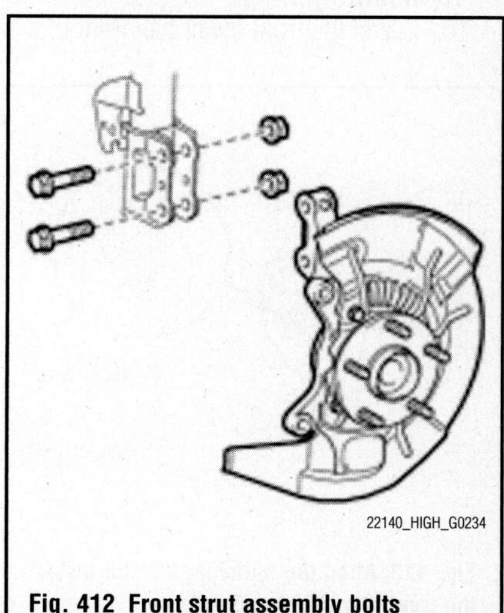

Fig. 412 Front strut assembly bolts

5. Using a plastic hammer, separate the front drive shaft assembly from the front axle assembly.

➡Loosen the staked part of the front axle hub nut completely, otherwise the threads of the drive shaft may be damaged.

6. Remove the 2 bolts and separate the front disc brake caliper assembly.

➡Use wire or an equivalent tool to keep the brake caliper from hanging down by the flexible hose.

7. Remove the front disc.
8. Separate the tie rod assembly.
9. Remove the bolt, 2 nuts, and separate the front lower suspension arm from the lower ball joint.
10. Remove the 2 bolts, 2 nuts and front axle assembly.
11. Remove the front lower ball joint, as follows:
12. Secure the front axle assembly in a vise using aluminum plates.

➡When using a vise, do not overtighten it.

13. Remove the cotter pin and nut.
14. Install SST: 09960-20010 or equivalent to the front lower ball joint.
15. Using SST: 09960-20010 or equivalent, remove the front lower ball joint from the front axle assembly.

➡Install SST so that A and B are parallel.

➡Be sure to place a wrench on the part indicated in the illustration.

➡Do not damage the front lower ball joint dust cover.

To install:

16. Install the front lower ball joint to the steering knuckle with the nut and tighten to 91 ft. lbs. (123 Nm).

➡Prevent oil from adhering to the screw and tapered parts.

17. Install a new cotter pin.

➡If the holes for the cotter pin are not aligned, tighten the nut further up to 60°.

18. Install the front axle assembly to the front shock absorber with the 2 bolts and 2 nuts and tighten to 213 ft. lbs. (290 Nm).

➡Only when reusing the bolts and nuts, apply a small amount of engine oil to the threads of the nuts.

19. Align the matchmarks and install the front drive shaft assembly to the front axle hub sub-assembly.
20. Install the front lower suspension arm to the front lower ball joint with the bolt and 2 nuts 68 ft. lbs. (92 Nm).
21. Connect the tie rod assembly.
22. Install the front disc.
23. Install the front disc brake caliper assembly to the steering knuckle with the 2 bolts and tighten to 77 ft. lbs. (104 Nm).
24. Install the clamp and front speed sensor with the bolt and tighten to 71 inch lbs. (8 Nm).

➡Prevent foreign matter from attaching to the sensor tip.

➡Firmly insert the sensor body into the knuckle before tightening the bolt.

➡After installing the sensor to the knuckle, make sure that there is no clearance between the sensor stay and knuckle. Also make sure that no foreign matter is stuck between the parts.

➡To prevent interference between the sensor and magnetic rotor, do not rotate the sensor body during or after the insertion of the sensor body to the knuckle.

25. Install the front axle hub nut.
26. Check the ABS speed sensor signal

➡Check the ABS speed sensor signal.

27. Install front wheel and tighten lug nuts to 76 ft. lbs. (103 Nm).
28. Inspect and adjust front wheel alignment.

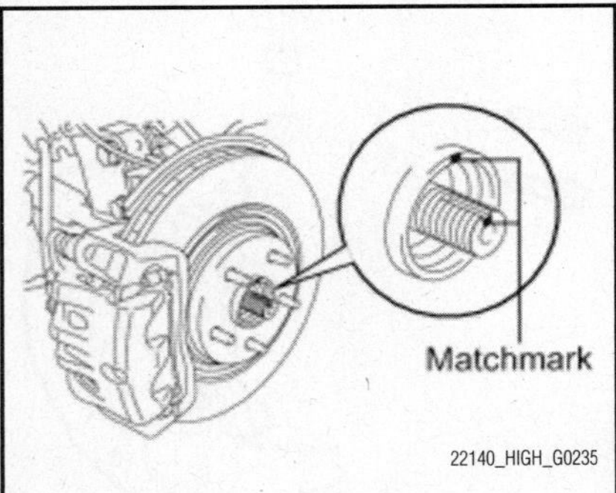

Fig. 413 Align the matchmarks and install the front drive shaft assembly

MACPHERSON STRUT

REMOVAL & INSTALLATION

Hybrid Models

See Figures 414 through 421.

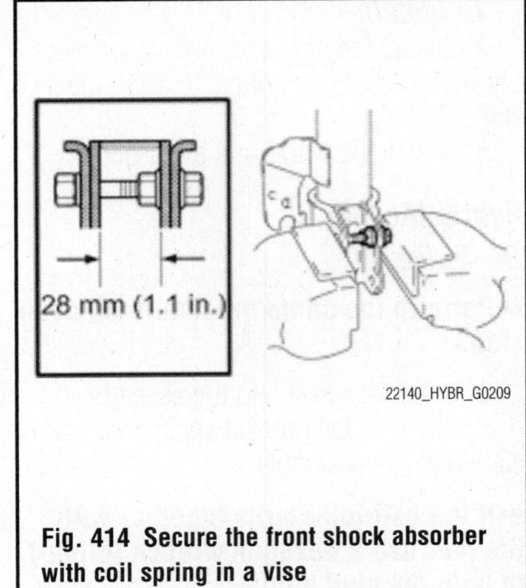

Fig. 414 Secure the front shock absorber with coil spring in a vise

1. Remove the front wheel.
2. Remove the front wiper arm and blade assemblies.
3. Loosen the front support to front shock absorber nut of the front shock absorber.

➡Do not remove the front support to front shock absorber nut.

➡Loosen the nut only when the front shock absorber with coil spring needs to be disassembled.

4. Remove the cowl top ventilator louver sub-assembly.
5. Remove the windshield wiper motor and link.
6. Remove the outer cowl top panel sub-assembly, as follows:
   a. Disengage the 4 clamps and separate the wiper wire harness from the outer cowl top panel sub-assembly.
   b. Remove the 8 bolts, 6 nuts, and the outer cowl top panel sub-assembly.
7. Remove the bolt and clamp, and separate the front speed sensor and front flexible hose.
8. Remove the nut and separate the front stabilizer link assembly from the front shock absorber.

➡If the ball joint turns together with the nut, use a hexagon wrench (6 mm) to hold the stud.

9. Remove the front shock absorber with coil spring, as follows:
   a. Support the front axle using a jack and wooden block.
   b. Remove the 2 bolts and 2 nuts, and separate the front shock absorber with coil spring (lower side) from the steering knuckle.'

➡When removing the nuts, keep the bolts from rotating.

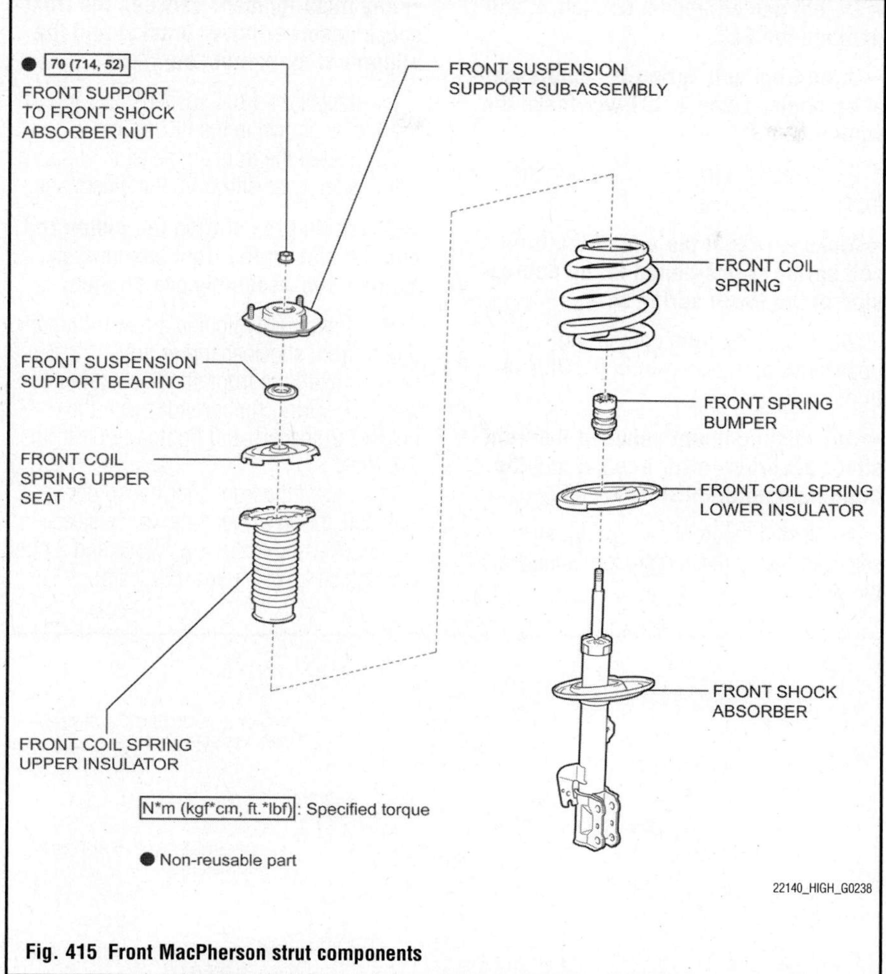

Fig. 415 Front MacPherson strut components

c. Remove the nut and 2 spacers on the upper side of the front shock absorber with coil spring.

➡Make sure that the front speed sensor is completely separated from the front shock absorber with coil spring.

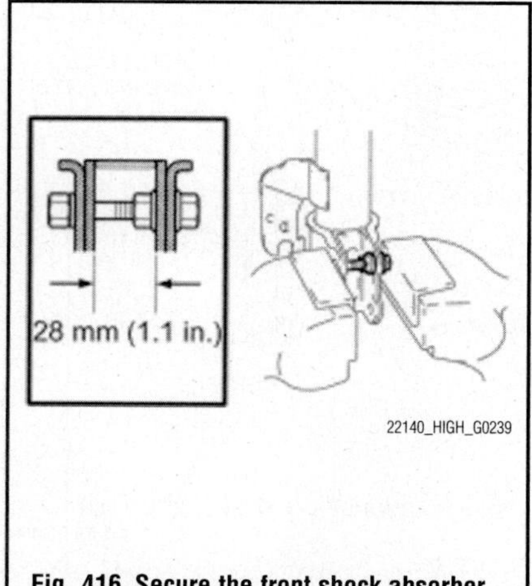

Fig. 416 Secure the front shock absorber with coil spring in a vise

10. Secure the front shock absorber with coil spring, as follows:

a. As shown in the illustration, secure the front shock absorber with coil spring in a vise using aluminum plates by clamping onto a double nut bolt affixed to the bracket at the bottom of the absorber.

11. Remove the front support to front shock absorber nut, as follows:

a. Using A Special Service Tool (SST: 09727-30021, SST: 09727-30021), compress the front coil spring.

➡Do not use an impact wrench. It will damage the SST.

➡If the front coil spring is compressed at an angle, using 2 SST will make the work easier.

b. Check that the front coil spring is fully compressed.

c. Remove the front support to front shock absorber nut.

12. Remove the front suspension support sub-assembly.

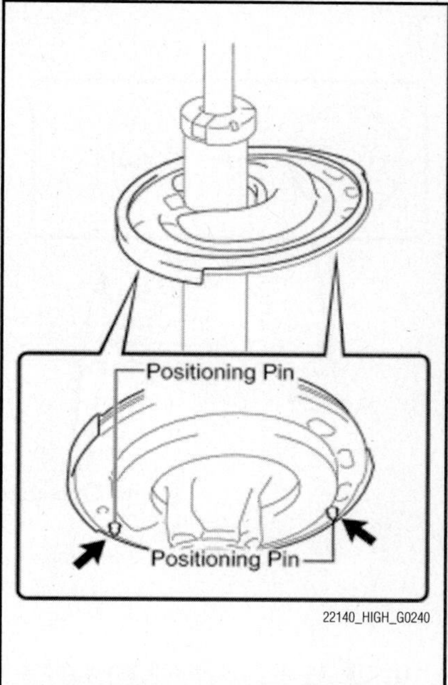

Fig. 417 Front coil spring lower insulator positioning pin

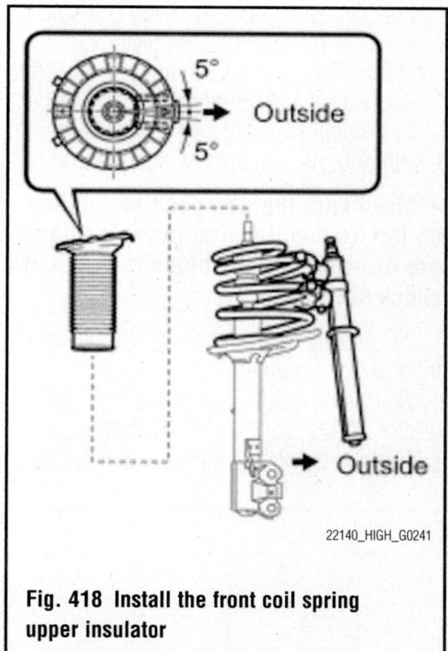

Fig. 418 Install the front coil spring upper insulator

13. Remove the front suspension support bearing.

14. Remove the front coil spring upper seat.

15. Remove the front coil spring upper insulator.

16. Remove the front coil spring.

17. Remove the front spring bumper.

18. Remove the front coil spring lower insulator.

To install:

19. Secure the front shock absorber assembly, as follows:

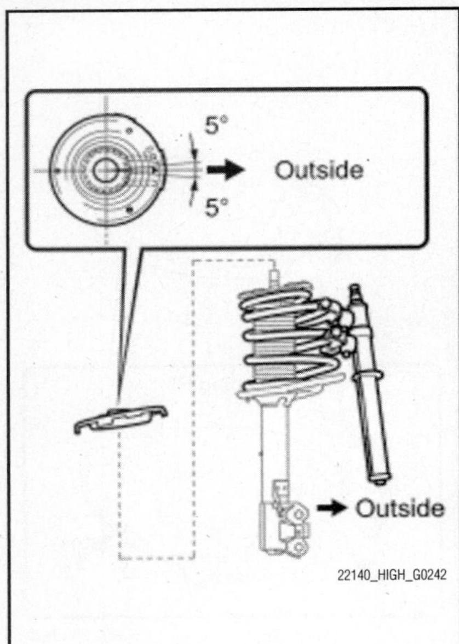

Fig. 419 Install the front coil spring upper seat

a. As shown in the illustration, secure the front shock absorber with coil spring in a vise using aluminum plates by clamping onto a double nut bolt affixed to the bracket at the bottom of the absorber.

20. Install the front coil spring lower insulator to the front shock absorber.

➡Make sure that the positioning pins on the front coil spring lower insulator are inserted into the holes in the front shock absorber.

21. Install the front spring bumper to the front shock absorber.

22. Using a Special Service Tool (SST: 09727-30021, SST: 09727-00050) or equivalent, compress the front coil spring.

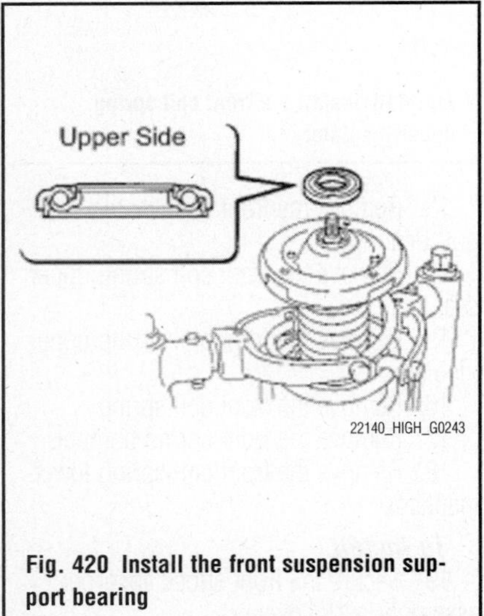

Fig. 420 Install the front suspension support bearing

➡Do not use an impact wrench. It will damage the SST.

➡If the front coil spring is compressed at an angle, using 2 SST will make the work easier.

23. Install the front coil spring to the front shock absorber.

➡Make sure that the end of the front coil spring is positioned in the depression of the lower spring seat.

24. Install the front coil spring upper insulator as shown in the illustration.

➡Any misalignment between the front shock absorber lower bracket and the alignment mark must be +/- 5° .

25. Install the front coil spring upper seat with the mark facing to the outside of the vehicle.

➡Any misalignment between the front shock absorber lower bracket and the alignment mark must be +/- 5° .

26. Install the front suspension support bearing as shown in the illustration.

27. Install the front suspension support sub-assembly as shown in the illustration.

➡Check that the slot on the piston rod and the slot on the front suspension support sub-assembly are aligned.

28. Temporarily tighten a new front support to front shock absorber nut.

29. Install the front shock absorber with coil spring (upper side) with the nut and 2 spacers and tighten to 63 ft. lbs. (85 Nm).

30. Install the front shock absorber with coil spring (lower side) to the steering knuckle and insert the 2 bolts and 2 nuts and tighten to 214 ft. lbs. (290 Nm).

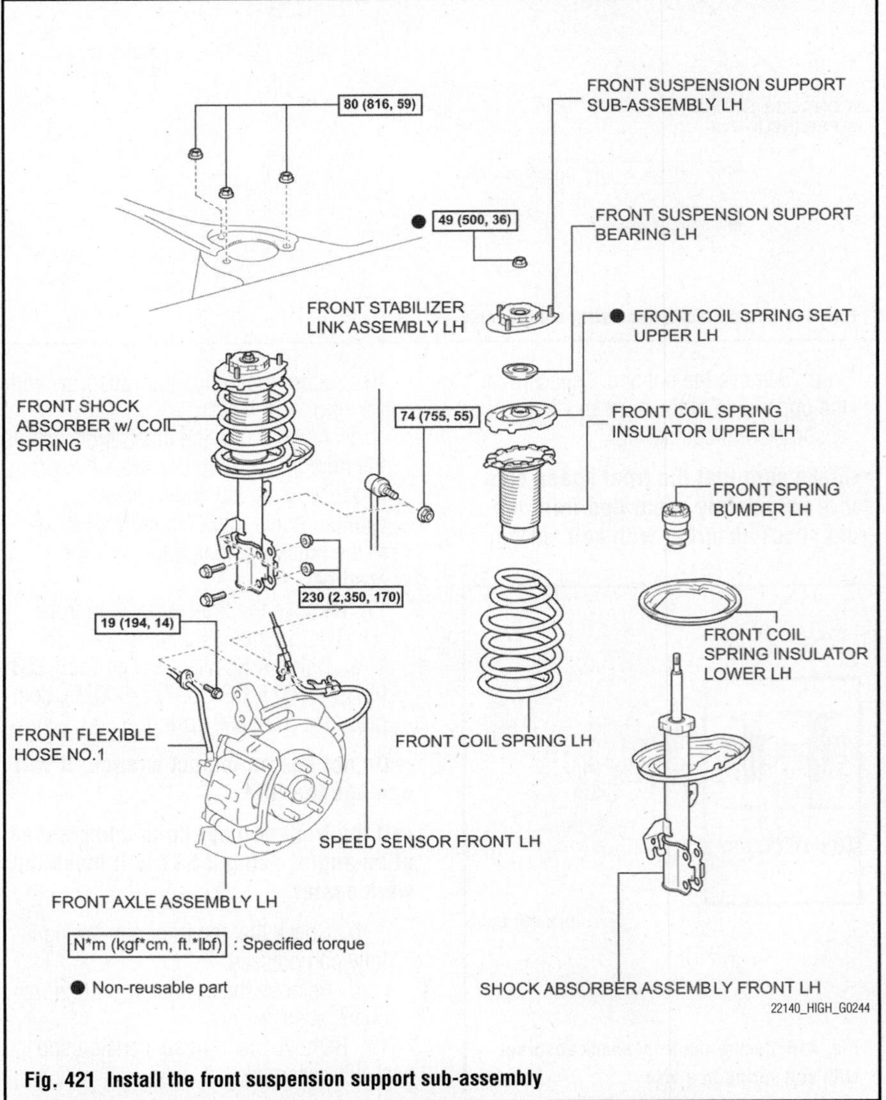

Fig. 421 Install the front suspension support sub-assembly

➡**When installing the nuts, keep the bolts from rotating.**

31. Install the front stabilizer link assembly to the front shock absorber with the nut and tighten to 55 ft. lbs. (74 Nm).

➡**If the ball joint turns together with the nut, use a hexagon wrench (6 mm) to hold the stud bolt.**

32. Install the front speed sensor and front flexible hose with the bolt and tighten to 14 ft. lbs. (19 Nm).

➡**Do not twist the front speed sensor when installing it.**

33. Install the clamp.
34. Install the outer cowl top panel sub-assembly with the 8 bolts and 6 nuts and tighten to 63 ft. lbs. (85 Nm), 78 inch lbs. (8.8 Nm), 78 inch lbs. (8.8 Nm).
35. Engage the 4 clamps.
36. Fully tighten the front support to front shock absorber nut and tighten to 52 ft. lbs. (70 Nm).
37. Install the windshield wiper motor and link.
38. Install the cowl top ventilator louver sub-assembly.
39. Install the front wiper arm and blade assemblies.
40. Install the front wheel and tighten lug nuts to 76 ft. lbs.

STABILIZER BAR

REMOVAL & INSTALLATION

Non-Hybrid Models

See Figure 422.

➡**This procedure requires engine removal.**

1. Remove the engine assembly with transaxle.
2. Remove the nuts and separate both front stabilizer link assemblies.

➡**If the ball joint turns together with the nut, use a hexagon wrench (6 mm) to hold the stud.**

3. Remove the 2 bolts and both No. 1 front stabilizer brackets from the front frame assembly.
4. Remove the front stabilizer bar.
5. Remove both No. 2 front stabilizer brackets from the front stabilizer bar bushing.
6. Remove the 2 No. 1 front stabilizer bar bushings from the front stabilizer bar.

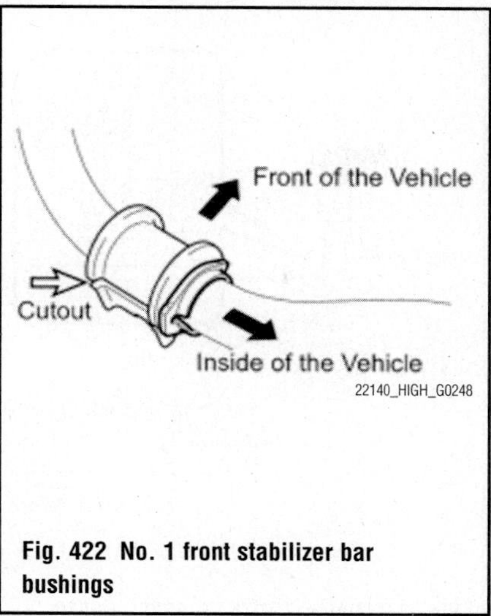

Fig. 422 No. 1 front stabilizer bar bushings

To install:

7. Install the 2 No. 1 front stabilizer bar bushings to the front stabilizer bar as shown in the illustration.

➡**Install the No. 1 front stabilizer bar bushings so that the cutout faces the rear of the vehicle.**

8. Install both No. 2 front stabilizer brackets to the No. 1 front stabilizer bar bushing.
9. Install the front stabilizer bar by inserting it from the right side of the vehicle.
10. Install both No. 1 front stabilizer brackets to the front frame assembly with the 2 bolts and tighten to 21 ft. lbs. (29 Nm).
11. Install both front stabilizer link assemblies with the nuts and tighten to 55 ft. lbs. (74 Nm).

➡**If the ball joint turns together with the nut, use a hexagon wrench (6 mm) to hold the stud bolt.**

12. Install the engine assembly with transaxle.
13. Inspect and adjust the front wheel alignment.

Hybrid Models

See Figure 423.

1. Make sure the vehicle's front wheels are in the straight-ahead position.
2. Disconnect the negative battery cable.
3. Raise and safely support the vehicle.
4. Remove the remove front wheel and tire assemblies.
5. Remove the engine under cover.
6. Separate the steering intermediate shaft subassembly.

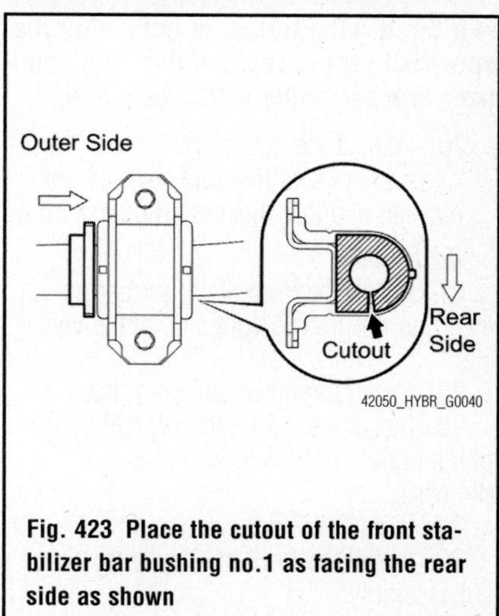

Fig. 423 Place the cutout of the front stabilizer bar bushing no.1 as facing the rear side as shown

7. Separate the tie rod.
8. Remove the 2 nuts, then remove the front left stabilizer link assembly:

➡**If the ball joint turns together with the nut, use a hexagon wrench (6 mm) to hold the stud.**

9. Remove the 2 nuts, then remove the front right stabilizer link assembly:

➡**If the ball joint turns together with the nut, use a hexagon wrench (6 mm) to hold the stud.**

10. Remove the 2 bolts, then remove the left stabilizer bracket no. 1 from the front frame assembly.
11. Remove the 2 bolts, then remove the right stabilizer bracket no. 1 from the front frame assembly.
12. Remove the left and right no. 2 stabilizer brackets from the bushings.
13. Remove the 2 stabilizer bar bushings no.1 from the stabilizer bar.
14. Remove the front no.3 exhaust pipe subassembly
15. Remove the front exhaust pipe/
16. Remove the bolt, nut and the manifold stay.
17. Remove the power steering link.
18. Remove the front stabilizer bar from the right side of the vehicle.
19. Inspect the turning of the stabilizer link ball joint:
    a. secure the front stabilizer link assembly in a vise using aluminum plates.
    b. Install the nut to the front stabilizer link assembly stud.
    c. If using a torque wrench, turn the nut continuously at a rate of 3 to 5 seconds per turn and take the torque reading on the 5th turn. The turning torque should be 18 inch lbs. (2.0 Nm) or less.

➡If the turning torque is not within the specified range, replace the front stabilizer link assembly with a new one.

20. Inspect the dust cover:
    a. Check that the dust cover is not cracked and that there is no grease on it.

To install:

21. Install the front stabilizer bar by inserting it from the right side of the vehicle.
22. Install the power steering link.
23. Install the manifold stay with the bolt and nut. Tighten to 25 ft. lbs. (34 Nm).
24. Install the front exhaust pipe.
25. Install the front no.3 exhaust pipe subassembly.
26. Install the 2 front stabilizer bar bushings no.1 to the outer side of the bushing stopper on the front stabilizer bar.

➡Place the cutout of the front stabilizer bar bushing no.1 as facing the rear side as shown in the illustration.

27. Install the right and left no.2 stabilizer brackets to the stabilizer bushings.
28. Install the right and left no.1 stabilizer bracket to the front frame assembly with the bolts. Tighten to 12 ft. lbs. (16 Nm).
29. Install the right and left stabilizer link assemblies with the nuts and tighten to 55 ft. lbs. (75 Nm).

➡If the ball joint turns together with the nut, use a hexagon (6 mm) wrench to hold the stud.

30. Connect the tie rod.
31. Connect steering intermediate shaft subassembly.
32. Make sure the front wheels are facing straight-ahead.
33. Install the engine under cover.
34. Install the front wheels.
35. Connect the negative battery cable.
36. If necessary, perform the initialization procedure.
37. Initialize the rotation angle sensor and calibrate torque sensor zero point.
38. Inspect and adjust the front wheel alignment.

STEERING KNUCKLE

REMOVAL & INSTALLATION

Non-Hybrid Models

See Figure 424.

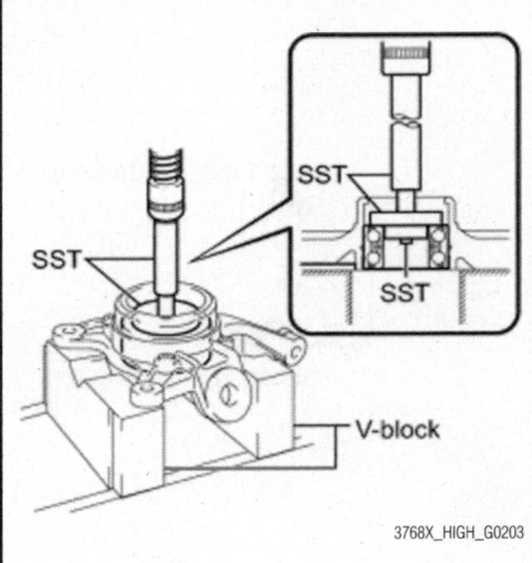

Fig. 424 Using SST (SST: 09950-60010, SST: 09950-70010, SST: 09950-60020, or equivalent),V-blocks and a press, remove the front axle hub bearing from the steering knuckle

1. Remove the front wheel.
2. Remove the front axle hub nut.
3. Separate the front speed sensor.
4. Separate the front disc brake caliper assembly.
5. Remove the front disc.
6. Separate tie rod assembly.
7. Separate the front lower suspension arm.
8. Separate the front drive shaft assembly.
9. Remove the front axle assembly.
10. Remove the front lower ball joint.
11. Remove the No. 1 front wheel bearing dust deflector.
12. Remove the front axle hub hole snap ring.
13. Remove the front axle hub sub-assembly.
14. Remove the front disc brake dust cover.
15. Remove the steering knuckle, as follows:
    a. Place the bearing inner race (outside) on the front axle hub bearing.
    b. Using SST (SST: 09950-60010, SST: 09950-70010, SST: 09950-60020, or equivalent),V-blocks and a press, remove the front axle hub bearing from the steering knuckle.

➡Keep the steering knuckle level.

To install:

16. Using SST (SST: 09950-70010, SST: 09950-60020, or equivalent), install a new front axle hub bearing to the steering knuckle.
17. Install the front disc brake dust cover.
18. Install the front axle hub sub-assembly.
19. Install the front axle hub hole snap ring.
20. Install the No. 1 front wheel bearing dust deflector.
21. Install the front lower ball joint.
22. Install the front axle assembly.
23. Install the front drive shaft assembly.
24. Install the front lower suspension arm.
25. Connect tie rod assembly.
26. Install the front disc.
27. Install the front disc brake caliper assembly.
28. Install the front axle hub nut.
29. Separate the front disc brake caliper assembly.
30. Remove the front disc.
31. Inspect the front axle bearing looseness.
32. Inspect front axle hub runout.
33. Install the front disc.
34. Install the front disc brake caliper assembly.
35. Install the front speed sensor.
36. Stake front axle hub nut.
37. Install the front wheel.
38. Inspect and adjust front wheel alignment.
39. Check for speed sensor signal.

Hybrid Models

See Figures 425 and 426.

1. Raise and support the vehicle.
2. Remove the tire and wheel assembly.
3. Remove the drive axle retaining nut.
4. Remove the speed sensor.
5. Remove the brake caliper assembly.
6. Remove the front brake rotor.
7. Disconnect outer tie rod end.
8. Remove the 2 bolts and 2 nuts, and separate the front shock absorber with coil

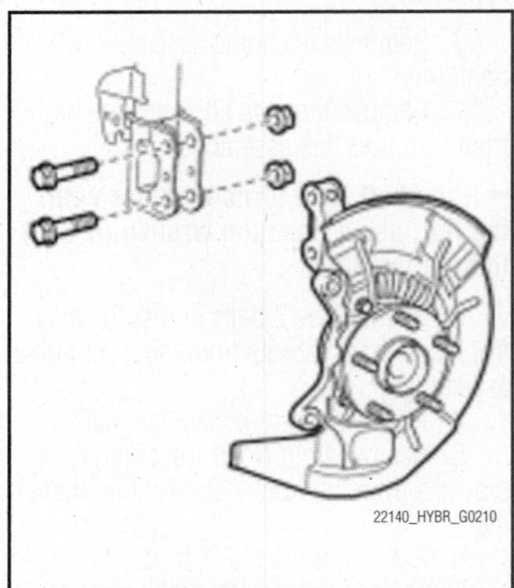

Fig. 425 Remove the 2 bolts and 2 nuts

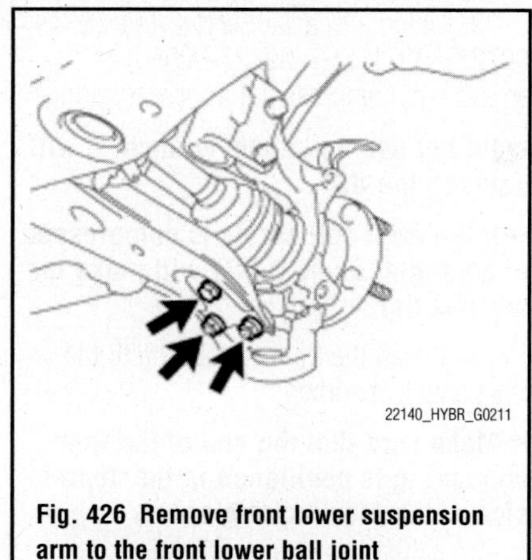

Fig. 426 Remove front lower suspension arm to the front lower ball joint

spring (lower side) from the steering knuckle.

9. Remove the front lower ball joint from the steering knuckle.
10. Move the drive axle to the side.
11. Remove the lower arm to knuckle nuts and bolts.
12. Remove the steering knuckle.

To install:

13. Install steering knuckle and drive axle.
14. Install ball joint retaining nut and tighten to 36 ft. lbs. (49 Nm). Install cotter pin.
15. Install the lower arm to knuckle nuts and bolts. Tighten to 68 ft. lbs. (92 Nm).
16. Install the 2 bolts and 2 nuts, and tighten the front strut with coil spring (lower side) to the steering knuckle. Tighten to 213 ft. lbs. (290 Nm).
17. Connect the outer tie rod end and tighten the castle nut to 36 ft. lbs. (49 Nm). Install a new cotter pin.
18. Install the front brake rotor.
19. Install front brake caliper assembly and tighten the mounting bolts to 77 ft. lbs. (104 Nm).
20. Install the speed sensor and tighten mounting bolt to 71 inch lbs. (8 Nm).
21. Install the new drive axle retaining nut and tighten to 216 ft. lbs. (294 Nm). Using a chisel and hammer, stake the front axle hub nut.
22. Install and tighten the front wheel assembly to 76 ft. lbs. (103 Nm).
23. Check and adjust wheel alignment.
24. Check speed sensor operation.

STRUT & SPRING ASSEMBLY

REMOVAL & INSTALLATION

Non-Hybrid Models

See Figures 427 through 433.

1. Remove the front wheel.

● 70 (714, 52)
FRONT SUPPORT TO FRONT SHOCK ABSORBER NUT
FRONT SUSPENSION SUPPORT SUB-ASSEMBLY
FRONT COIL SPRING
FRONT SUSPENSION SUPPORT BEARING
FRONT SPRING BUMPER
FRONT COIL SPRING UPPER SEAT
FRONT COIL SPRING LOWER INSULATOR
FRONT SHOCK ABSORBER
FRONT COIL SPRING UPPER INSULATOR
N*m (kgf*cm, ft.*lbf): Specified torque
● Non-reusable part
22140_HIGH_G0238

Fig. 427 Front MacPherson strut components

2. Remove the front wiper arm and blade assemblies.
3. Loosen the front support to front shock absorber nut of the front shock absorber.

➡Do not remove the front support to front shock absorber nut.

➡Loosen the nut only when the front shock absorber with coil spring needs to be disassembled.

4. Remove the cowl top ventilator louver sub-assembly.
5. Remove the windshield wiper motor and link.
6. Remove the outer cowl top panel sub-assembly, as follows:
   a. Disengage the 4 clamps and separate the wiper wire harness from the outer cowl top panel sub-assembly.
   b. Remove the 8 bolts, 6 nuts, and the outer cowl top panel sub-assembly.
7. Remove the bolt and clamp, and separate the front speed sensor and front flexible hose.
8. Remove the nut and separate the front stabilizer link assembly from the front shock absorber.

➡If the ball joint turns together with the nut, use a hexagon wrench (6 mm) to hold the stud.

9. Remove the front shock absorber with coil spring, as follows:
   a. Support the front axle using a jack and wooden block.
   b. Remove the 2 bolts and 2 nuts, and separate the front shock absorber with coil spring (lower side) from the steering knuckle.'

➡When removing the nuts, keep the bolts from rotating.

   c. Remove the nut and 2 spacers on the upper side of the front shock absorber with coil spring.

➡Make sure that the front speed sensor is completely separated from the front shock absorber with coil spring.

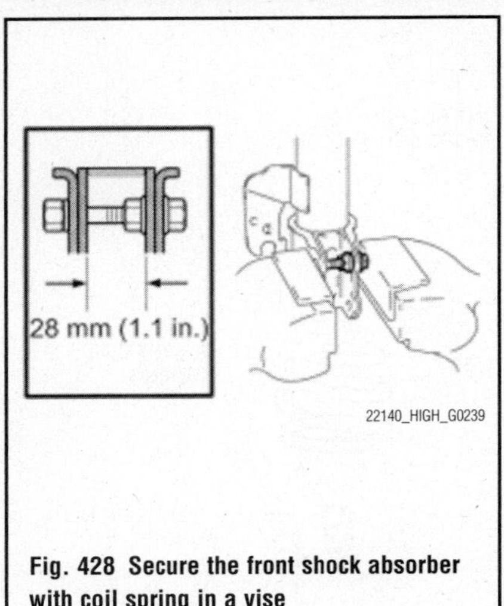

Fig. 428 Secure the front shock absorber with coil spring in a vise

10. Secure the front shock absorber with coil spring, as follows:
    a. As shown in the illustration, secure the front shock absorber with coil spring in a vise using aluminum plates by clamping onto a double nutted bolt affixed to the bracket at the bottom of the absorber.
11. Remove the front support to front shock absorber nut, as follows:
    a. Using A Special Service Tool (SST: 09727-30021, SST: 09727-30021), compress the front coil spring.

➡Do not use an impact wrench. It will damage the SST.

➡If the front coil spring is compressed at an angle, using 2 SST will make the work easier.

    b. Check that the front coil spring is fully compressed.
    c. Remove the front support to front shock absorber nut.
12. Remove the front suspension support sub-assembly.
13. Remove the front suspension support bearing.
14. Remove the front coil spring upper seat.
15. Remove the front coil spring upper insulator.
16. Remove the front coil spring.
17. Remove the front spring bumper.
18. Remove the front coil spring lower insulator.

To install:

19. Secure the front shock absorber assembly, as follows:
    a. As shown in the illustration, secure the front shock absorber with coil spring in a vise using aluminum plates by clamping onto a double nutted bolt affixed to the bracket at the bottom of the absorber.
20. Install the front coil spring lower insulator to the front shock absorber.

➡Make sure that the positioning pins on the front coil spring lower insulator are inserted into the holes in the front shock absorber.

21. Install the front spring bumper to the front shock absorber.
22. Using a Special Service Tool (SST: 09727-30021, SST: 09727-00050) or equivalent, compress the front coil spring.

➡Do not use an impact wrench. It will damage the SST.

➡If the front coil spring is compressed at an angle, using 2 SST will make the work easier.

23. Install the front coil spring to the front shock absorber.

➡Make sure that the end of the front coil spring is positioned in the depression of the lower spring seat.

24. Install the front coil spring upper insulator as shown in the illustration.

➡Any misalignment between the front shock absorber lower bracket and the alignment mark must be +/- 5° .

25. Install the front coil spring upper seat with the mark facing to the outside of the vehicle.

➡Any misalignment between the front shock absorber lower bracket and the alignment mark must be +/- 5° .

26. Install the front suspension support bearing as shown in the illustration.
27. Install the front suspension support sub-assembly as shown in the illustration.

➡Check that the slot on the piston rod and the slot on the front suspension support sub-assembly are aligned.

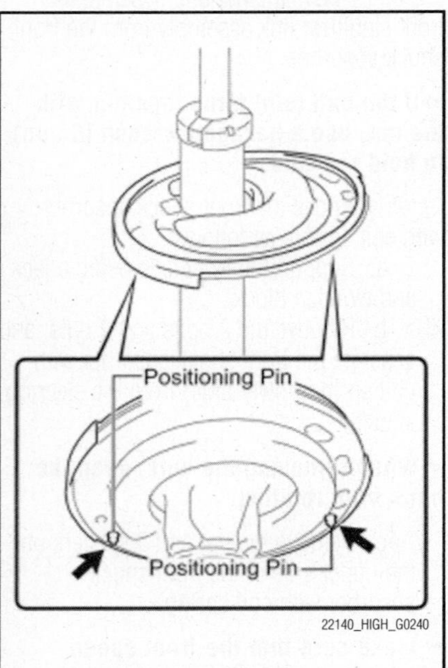

Fig. 429 Front coil spring lower insulator positioning pin

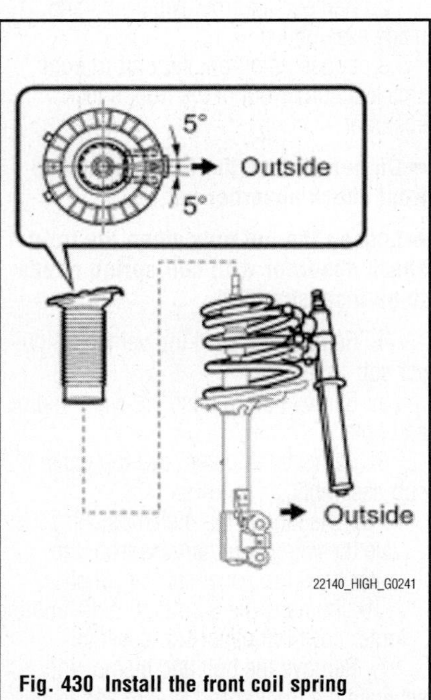

Fig. 430 Install the front coil spring upper insulator

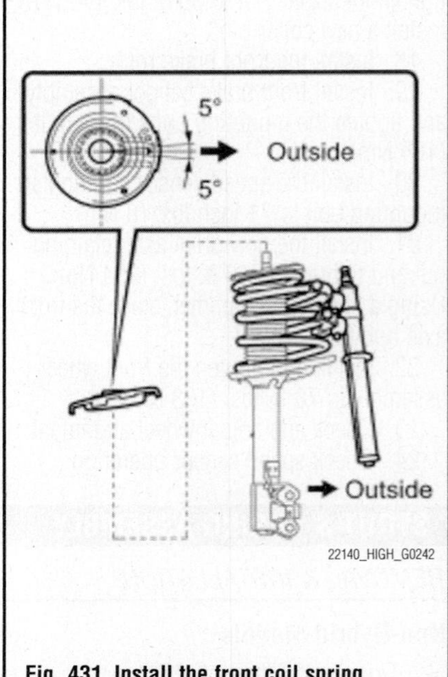

Fig. 431 Install the front coil spring upper seat

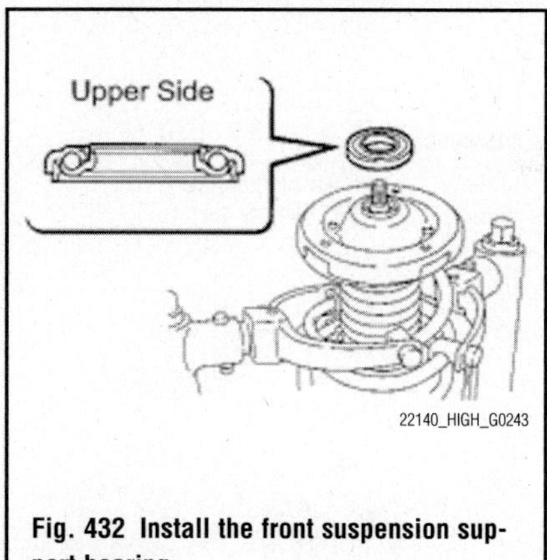

Fig. 432 Install the front suspension support bearing

28. Temporarily tighten a new front support to front shock absorber nut.

29. Install the front shock absorber with coil spring (upper side) with the nut and 2 spacers and tighten to 63 ft. lbs. (85 Nm).

30. Install the front shock absorber with coil spring (lower side) to the steering knuckle and insert the 2 bolts and 2 nuts and tighten to 214 ft. lbs. (290 Nm).

➡When installing the nuts, keep the bolts from rotating.

31. Install the front stabilizer link assembly to the front shock absorber with the nut and tighten to 55 ft. lbs. (74 Nm).

➡If the ball joint turns together with the nut, use a hexagon wrench (6 mm) to hold the stud bolt.

32. Install the front speed sensor and front flexible hose with the bolt and tighten to 14 ft. lbs. (19 Nm).

➡Do not twist the front speed sensor when installing it.

33. Install the clamp.

34. Install the outer cowl top panel sub-assembly with the 8 bolts and 6 nuts and tighten to 63 ft. lbs. (85 Nm), 78 inch lbs. (8.8 Nm), 78 inch lbs. (8.8 Nm).

35. Engage the 4 clamps.

36. Fully tighten the front support to front shock absorber nut and tighten to 52 ft. lbs. (70 Nm).

37. Install the windshield wiper motor and link.

38. Install the cowl top ventilator louver sub-assembly.

39. Install the front wiper arm and blade assemblies.

40. Install the front wheel and tighten lug nuts to 76 ft. lbs. (103 Nm).

41. Inspect and adjust front wheel alignment.

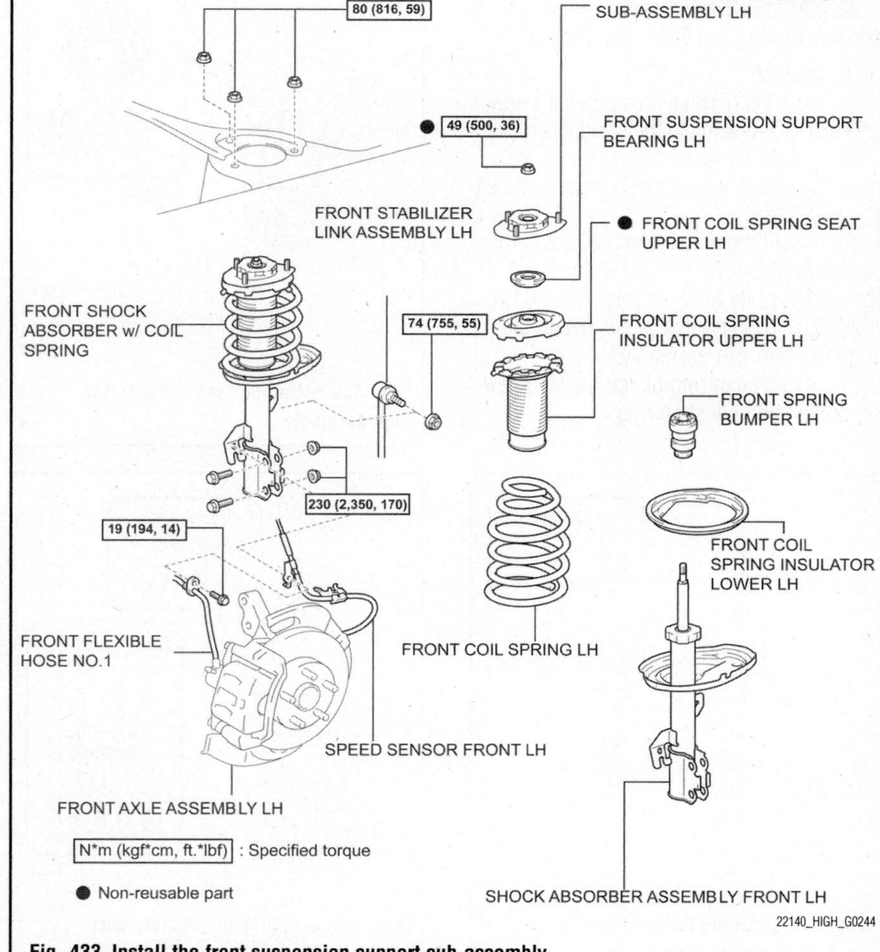

Fig. 433 Install the front suspension support sub-assembly

WHEEL BEARINGS

REMOVAL & INSTALLATION

Non-Hybrid Models

See Steering Knuckle procedure.

Hybrid Models

See Figures 434 through 440.

1. Before servicing the vehicle, refer to the Precautions Section.
2. Remove the front wheel.
3. Remove the front axle hub nut.
4. Separate the front speed sensor.
5. Separate the front disc brake caliper assembly.
6. Remove the front brake rotor.
7. Separate and remove the tie rod end.
8. Remove the bolt and 2 nuts, and separate the front suspension arm sub-assembly lower No.1 from the lower ball joint.
9. Using a plastic hammer, separate the drive shaft from the axle hub.
10. Remove the 2 bolts, nuts and steering knuckle.

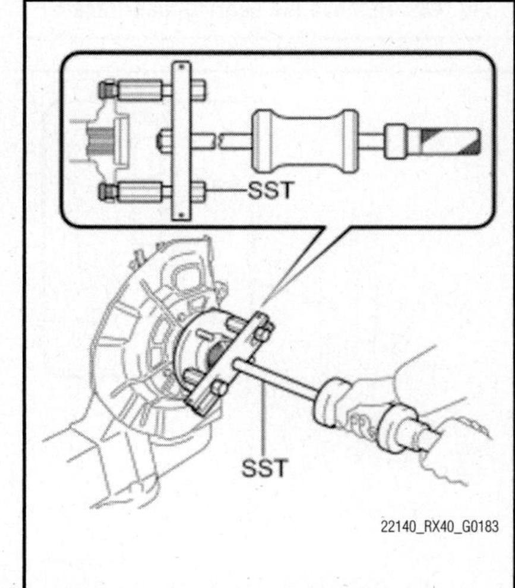

Fig. 434 Front axle hub sub-assembly removal

⁂ WARNING

Be careful not to damage the boot and speed sensor rotor.

11. Using a screwdriver with its tip wrapped with vinyl tape, remove the bearing dust deflector No.1.
12. Using snap ring pliers, remove the front axle hub hole snap ring.
13. Hold the front axle assembly between aluminum plates in a vise.
14. Using SST 09520-00031, remove the front axle hub sub-assembly.
15. Using SST and a press, remove the bearing inner race (outside) from the front axle hub sub-assembly.
16. Remove the 4 bolts and disc brake dust cover from steering knuckle.
17. Remove the lower ball joint from the hub assembly.
18. Remove the front axle hub bearing as follows:
19. Place the bearing inner race (outside) on the front axle hub bearing.
20. Using SST and a press, press the front axle hub bearing until it contacts the SST.
21. Using SST to make the steering knuckle horizontal, fix it to the V-block, as shown in the illustration.
22. Using SST and a press, remove a front axle hub bearing to the steering knuckle.

To install:

23. Using SST and a press, install a new front axle hub bearing to the steering knuckle.
24. Install the lower ball joint into the steering knuckle and tighten castle nut to 91 ft. lbs. (93 Nm).
25. Install the disc brake dust cover to the steering knuckle with the 4 bolts. Tighten the bolts to 73 inch lbs. (8.3 Nm).
26. Using SST and a press, install the front axle hub sub-assembly.
27. Using snap ring pliers, install a new front axle hub hole snap ring.

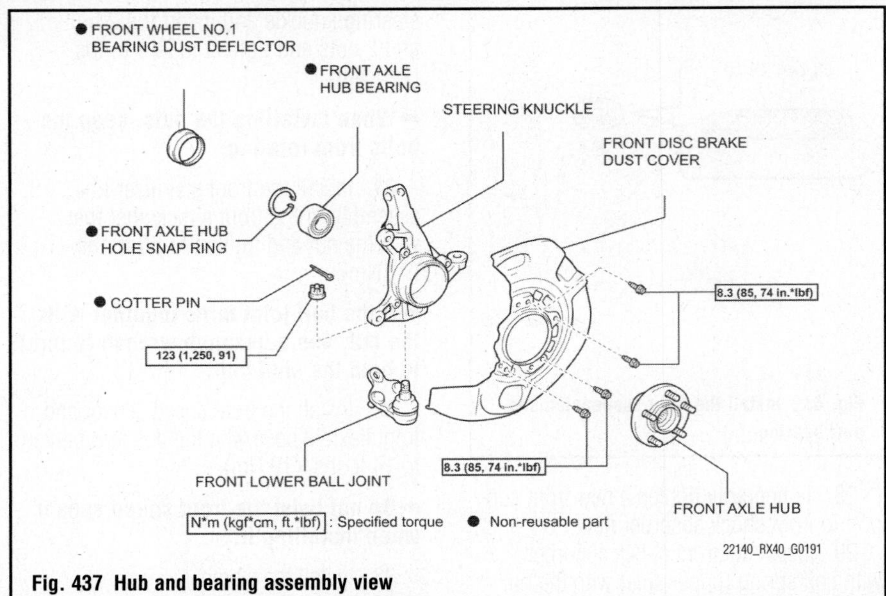

Fig. 437 Hub and bearing assembly view

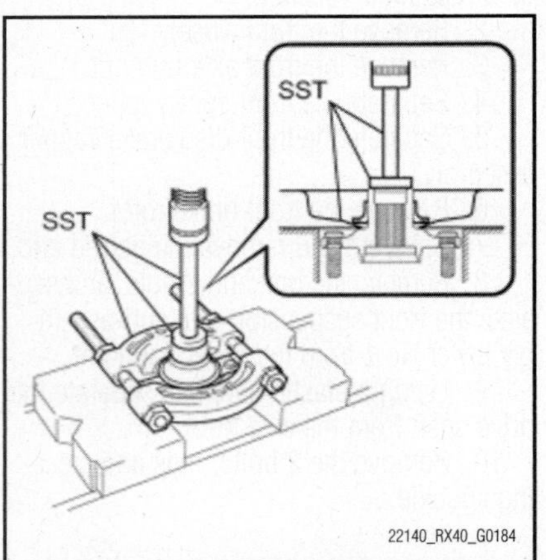

Fig. 435 Remove the bearing inner race

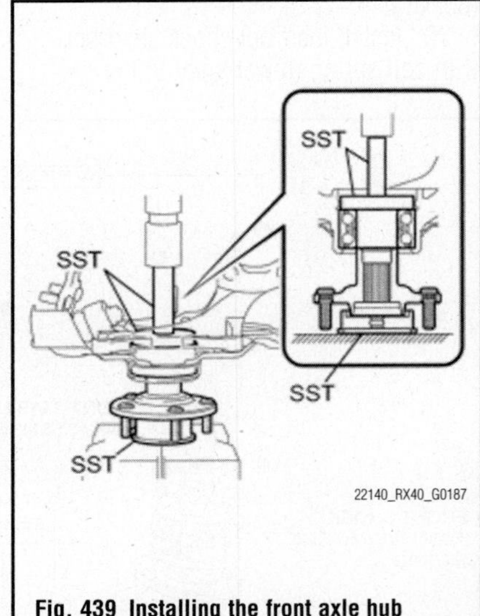

Fig. 439 Installing the front axle hub sub-assembly

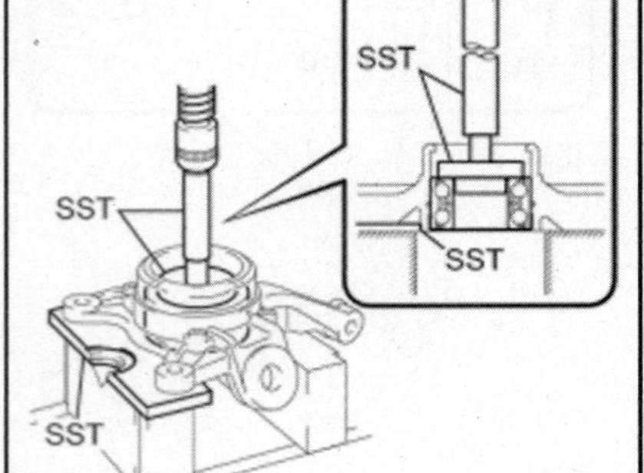

Fig. 436 Fix it to the V-block, as shown

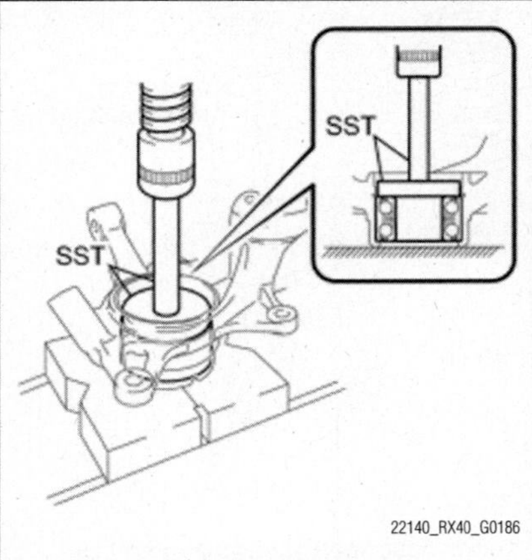

Fig. 438 Installing the front axle hub bearing to the steering knuckle

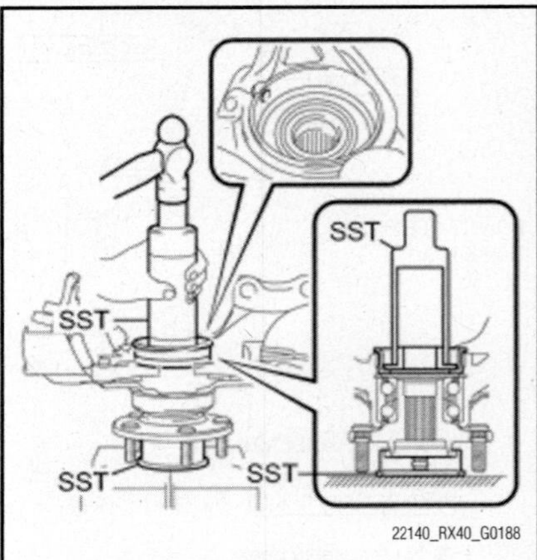

Fig. 440 Installing the bearing dust deflector No.1.

28. Using SST and a hammer, install the bearing dust deflector No.1.

➡Align the hole for the speed sensor in the bearing dust deflector No.1 with the steering knuckle.

29. Install the front axle assembly to the front drive shaft assembly.
30. Install the front axle assembly to the front shock absorber assembly with the 2 bolts and nuts. Tighten the bolts and nuts to 217 ft. lbs. (290 Nm).
31. Install the lower ball joint to the front suspension arm sub-assembly lower with the bolt and 2 nuts. Tighten the nuts and bolts to 94 ft. lbs. (127 Nm).
32. Install the tie rod end to the steering knuckle with the nut. Tighten the nut to 36 ft. lbs. (49 Nm).
33. Install the front brake rotor.
34. Install the front brake caliper assembly.
35. Install the front axle hub nut and tighten to 217 ft. lbs. (294 Nm). Using a chisel and hammer, stake the axle hub nut.
36. Install the speed sensor to the steering knuckle with the bolt and tighten to 71 inch lbs. (8 Nm).
37. Install the front wheel and tighten to 76 ft. lbs. (103 Nm).
38. Adjust the front wheel alignment.
39. Check the speed sensor operation.

SUSPENSION

REAR SUSPENSION

LOWER CONTROL ARM

REMOVAL & INSTALLATION

Non-Hybrid Models

2WD Vehicles

See Figures 441 and 442.

➡Use the same procedures for the RH side and LH side.

1. Remove the deck board assembly.
2. Remove the No. 2 and 3 deck board sub-assemblies, as applicable.
3. Remove the tonneau cover assembly, as applicable.
4. Remove the rear mat.
5. Remove the deck trim service hole cover.
6. Remove the lower spare wheel carrier hinge cover, as applicable.
7. Remove the spare tire.
8. Remove the spare wheel carrier lock cover, as applicable.
9. Remove rear wheel.
10. Remove the nuts and separate the rear stabilizer link assemblies from the rear stabilizer bar.

➡If the ball joint turns together with the nut, use a hexagon wrench (5 mm) to hold the stud bolt.

11. Remove the 4 bolts and rear stabilizer bar.
12. Remove the rear No. 2 suspension arm assembly, as follows:
    a. Remove the bolt and the nut, and separate the rear No. 2 suspension arm assembly from the rear axle carrier sub-assembly.

➡Since a stopper nut is used, loosen the bolt.

    b. Remove the bolt and the rear No. 2 suspension arm assembly.
13. Remove the rear No. 1 suspension arm assembly, as follows:

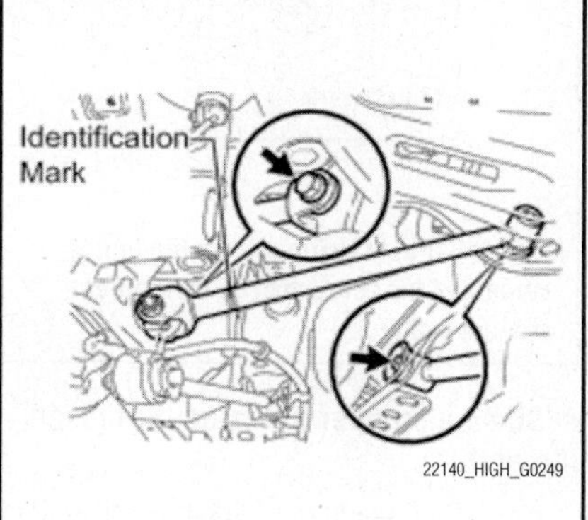

Fig. 441 Temporarily install the rear No. 1 and 2 suspension arm assemblies

    a. Remove the bolt and the nut, and separate the rear No. 1 suspension arm assembly from the rear axle carrier sub-assembly.

➡Since a stopper nut is used, loosen the bolt.

    b. Remove the bolt and the rear No. 1 suspension arm assembly.

To install:

14. Temporarily install the rear No. 1 and 2 suspension arm assemblies to the rear suspension member with the bolts.

➡Ensure that the identification marks face the rear side of the vehicle.

15. Temporarily install the rear No. 1 and 2 suspension arm assemblies to the rear axle carrier sub-assembly with the bolts and the nuts.

➡Since a stopper nut is used, temporarily tighten the bolts.

16. Jack up the rear axle carrier, placing a wooden block underneath to avoid damage. Apply load to the suspension so that the installed bolt of the rear No. 1 suspension arm (inner side) is horizontally aligned with the center of the rear axle hub.

⁂ CAUTION

Do not jack up the rear axle carrier sub-assembly too high as the vehicle may fall.

➡Do not bend the brake dust cover.

➡If the rear drive shaft assembly cannot be positioned horizontally as shown in the illustration even when the rear axle carrier sub-assembly is jacked up, apply additional load to the vehicle such as by having a person sit in the rear seat.

➡Use the same procedures for the RH side and LH side.

17. Fully tighten the rear No. 1 suspension arm assembly, as follows:
    a. Using SST: 09961-00950 or equivalent and a socket wrench (19 mm), fully tighten the bolt to 88 ft. lbs. (120 Nm), 65 ft. lbs. (89 Nm).

➡Use a torque wrench with a fulcrum length of 425 mm (16.73 in.).

➡This torque value is effective when SST is parallel to the torque wrench.

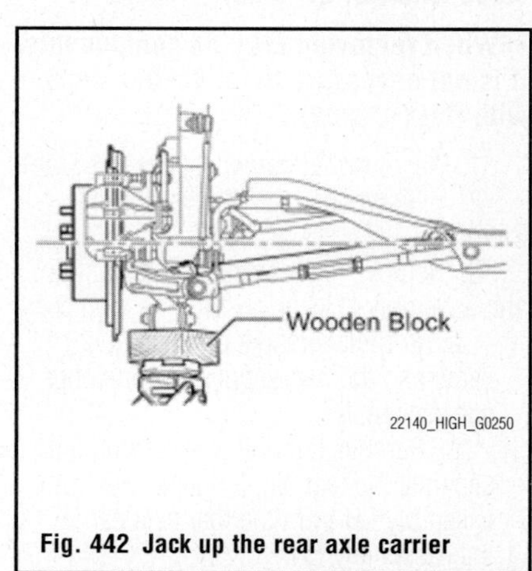

Fig. 442 Jack up the rear axle carrier

➡The final torque must be applied under standard vehicle height conditions.

b. Fully tighten the bolt to 82 ft. lbs. (112 Nm).

➡Since a stopper nut is used, fully tighten the bolt.

➡The final torque must be applied under standard vehicle height conditions.

18. Fully tighten the rear No. 2 suspension arm assembly, as follows:

a. Fully tighten the bolts to 88 ft. lbs. (120 Nm), 82 ft. lbs. (112 Nm).

➡Since a stopper nut is used, fully tighten the bolt.

➡The final torque must be applied under standard vehicle height conditions.

19. Install the rear stabilizer bar with the 4 bolts and tighten to 14 ft. lbs. (19 Nm).

20. Install both rear stabilizer link assemblies to the rear stabilizer bar with the nuts and tighten to 29 ft. lbs. (39 Nm).

➡If the ball joint turns together with the nut, use a hexagon wrench (6 mm) to hold the stud bolt.

21. Install the wheels and tighten lug nuts to 76 ft. lbs. (103 Nm).

22. Inspect and adjust rear wheel alignment.

23. The remainder of installation is the reverse of the removal procedure.

4WD Vehicles

See Figures 443 through 449.

➡The removal procedures for the LH and RH sides are different.

➡When removing RH side components, it is not necessary to follow the steps with (for LH Side).

➡When removing LH side components, it is not necessary to follow the steps with (for RH side).

1. Remove the rear wheel (for RH Side).

2. Remove the rear wheels (for LH Side).

3. Remove the rear no. 2 suspension arm assembly LH (for LH Side), as follows:

a. Put matchmarks on the adjust cams and the rear suspension member sub-assembly.

b. Remove the bolt and the nut, and separate the rear No. 2 suspension arm assembly LH from the rear axle carrier sub-assembly LH.

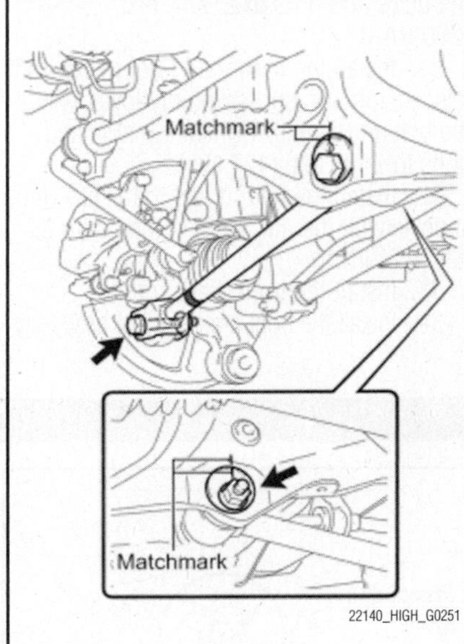

Fig. 443 Put matchmarks on the adjust cams and the rear suspension member sub-assembly

➡Since a stopper nut is used, loosen the bolt.

c. Remove the nut, the No. 2 camber adjust cam, the rear suspension toe adjust cam sub-assembly, and the rear No. 2 suspension arm assembly LH.

➡When removing the nut, keep the rear suspension toe adjust cam sub-assembly from rotating.

4. Remove the tail exhaust pipe assembly.

5. Remove the center exhaust pipe assembly.

6. Remove the rear no. 2 suspension arm assembly RH (for RH Side).

➡Perform the same procedure as the rear No. 2 suspension arm assembly LH.

7. Remove the propeller with center bearing shaft assembly (for LH Side).

8. Remove the rear axle shaft nut LH (for LH Side).

9. Remove the rear axle shaft nut RH (for LH Side).

10. Separate the No. 3 parking brake cable assembly (for LH Side).

11. Separate the No. 2 parking brake cable assembly.

12. Remove the rear strut rod assembly LH (for LH Side).

13. Remove the rear strut rod assembly RH.

14. Remove the rear No. 1 suspension arm assembly RH (for RH Side). As follows:

a. Remove the bolt and the nut, and separate the rear No. 1 suspension arm assembly RH from the rear axle carrier sub-assembly RH.

b. Remove the bolt, the nut, and the rear No. 1 suspension arm assembly RH from the rear suspension member sub-assembly

➡Since stopper nuts are used, loosen the bolts.

15. Remove the rear speed sensor LH (for LH Side).

16. Remove the rear speed sensor RH (for LH Side).

17. Separate the rear disc brake caliper assembly LH (for LH Side).

18. Separate the rear disc brake caliper assembly RH (for LH Side).

19. Remove the rear disc (for LH Side), as follows:

a. Remove the rear disc from the rear axle hub and bearing assembly LH.

b. Remove the rear disc from the rear axle hub and bearing assembly RH.

➡Perform the same procedure as the LH side.

20. Remove the rear axle hub and bearing assembly LH (for LH Side).

21. Remove the rear axle hub and bearing assembly RH (for LH Side).

22. Remove the rear axle carrier sub-assembly LH (for LH Side), as follows:

a. Loosen the bolt.

➡Since a stopper nut is used, loosen the bolt.

b. Remove the 2 bolts and 2 nuts, and separate the rear shock absorber with coil spring from the rear axle carrier sub-assembly LH.

c. Remove the bolt, the nut, and the rear axle carrier sub-assembly LH.

➡Be careful not to damage the outboard joint boot.

➡Be careful not to damage the speed sensor rotor.

➡Use a rope or equivalent to hang the rear drive shaft assembly. Remove rear axle carrier sub-assembly RH (for LH Side), as follows:

d. Loosen the 2 bolts.

e. Remove the 2 bolts and 2 nuts, and separate the rear shock absorber with coil spring from the rear axle carrier sub-assembly RH.

f. Remove the 2 bolts, the 2 nuts, and the rear axle carrier sub-assembly RH.

➡Since stopper nuts are used, loosen the bolts.

➡Be careful not to damage the outboard joint boot.

➡Be careful not to damage the speed sensor rotor.

➡Use a rope or equivalent to hang the rear drive shaft assembly.

23. Remove the rear differential filler plug (for LH Side).
24. Remove the rear differential drain plug (for LH Side).
25. Remove the rear drive shaft assembly LH (for LH Side).
26. Remove the rear drive shaft snap ring LH (for LH Side).
27. Remove the rear drive shaft assembly RH (for LH Side).
28. Remove the rear drive shaft snap ring RH (for LH Side).
29. Remove the rear suspension member (for LH Side).
30. Remove the rear No. 1 suspension arm assembly LH (for LH Side), as follows:
    a. Remove the bolt, the nut, and the rear No. 1 suspension arm assembly LH from the rear suspension member sub-assembly.

➡Since a stopper nut is used, loosen the bolt.

To install:

31. Temporarily install rear No. 1 suspension arm assembly LH (for LH Side), as follows:

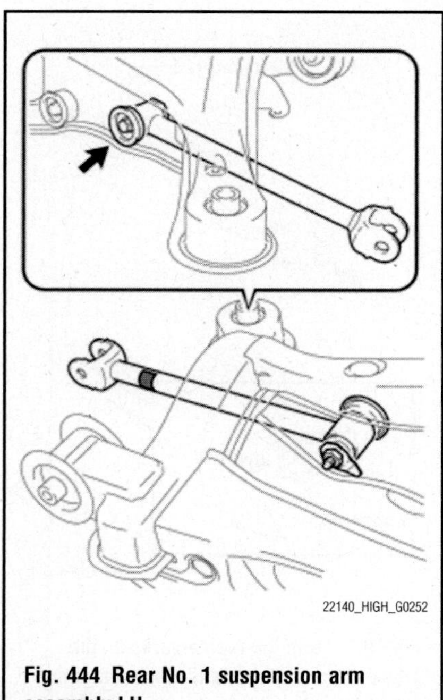

Fig. 444 Rear No. 1 suspension arm assembly LH

    a. Temporarily install the rear No. 1 suspension arm assembly LH to the rear suspension member sub-assembly with the bolt and the nut.

➡Ensure that the identification mark faces the rear side of the vehicle.

    b. Set the rear No. 1 suspension arm LH in the position shown. Length A: 20 mm (0.787 in.)
    c. Fully tighten the bolt to 59 ft. lbs. (80 Nm).

➡Since stopper nuts are used, temporarily tighten the bolts.

32. Install the rear suspension member (for LH Side), as follows:
    a. Support the rear suspension member with a jack using a wooden block.

➡Use a properly sized wooden block to keep the jack and suspension member level.

➡Support the suspension member until retightening of the suspension member is complete.

    b. Raise the rear suspension member with a jack.
    c. Temporarily install the rear suspension member, 2 rear upper suspension member stoppers, and 2 rear lower suspension member stopper retainers with the 4 nuts and 2 bolts.
    d. Fully tighten the 2 nuts 85 ft. lbs. (115 Nm).
    e. Using SST: 09961-00950 or equivalent and a socket wrench (19 mm), fully tighten the nut (LH side) to 133 ft. lbs. (181 Nm), 98 ft. lbs. (134 Nm).
    f. Using the same tools, fully tighten the nut (RH side) to the same specifications.

➡Use a torque wrench with a fulcrum length of 425 mm (16.73 in.).

➡These torque values are effective when SST is parallel to the torque wrench .

33. Install the rear drive shaft snap ring LH (for LH Side).

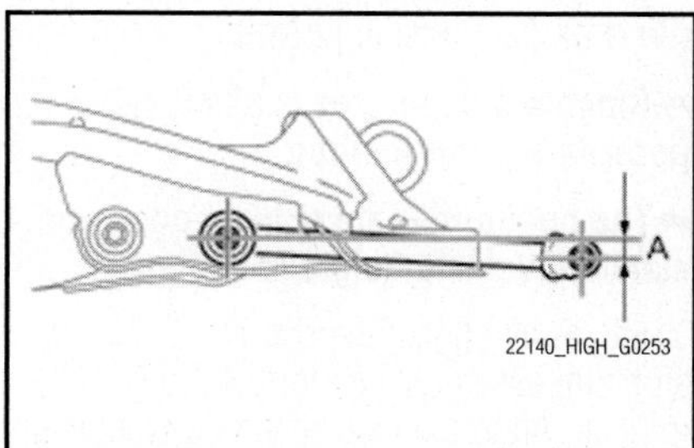

Fig. 445 Rear No. 1 suspension arm LH

34. Install the rear drive shaft assembly LH (for LH Side).
35. Install the rear drive shaft snap ring RH (for LH Side).
36. Install the rear drive shaft assembly RH (for LH Side).
37. Install the rear differential drain plug (for LH Side).
38. Add differential oil (for LH Side).
39. Inspect differential oil (for LH Side).
40. Install the rear differential filler plug (for LH Side).
41. Install the rear axle carrier sub-assembly LH (for LH Side), as follows:
    a. Temporarily install the rear axle carrier sub-assembly LH with the bolt and the nut.
    b. Install the rear axle carrier sub-assembly LH with the 2 bolts and the 2 nuts and tighten to 213 ft. lbs. (290 Nm).

➡Be careful not to damage the outboard joint boot.

➡Be careful not to damage the speed sensor rotor.

➡Prevent foreign matter from adhering to the speed sensor rotor.

➡Do not rotate the rear drive shaft assembly without the rear axle hub and bearing assembly installed.

➡Insert the bolts from the rear side.

42. Install the rear axle carrier sub-assembly RH (for LH Side), as follows:
    a. Temporarily install the rear axle carrier sub-assembly LH with the 2 bolts and the 2 nuts.
    b. Install the rear axle carrier sub-assembly LH with the 2 bolts and the 2 nuts and tighten to 213 ft. lbs. (290 Nm).

➡Be careful not to damage the outboard joint boot.

➡Be careful not to damage the speed sensor rotor.

➡Prevent foreign matter from adhering to the speed sensor rotor.

➡Do not rotate the rear drive shaft assembly without the rear axle hub and bearing assembly installed.

➡Insert the bolts from the rear side.

43. Temporarily install the rear No. 1 suspension arm assembly RH to the rear suspension member sub-assembly with the bolt and the nut.

➡Ensure that the identification mark faces the rear side of the vehicle.

44. Temporarily install the rear No. 1

suspension arm assembly RH to the rear axle carrier sub-assembly RH with the bolt and the nut.

➡Since stopper nuts are used, temporarily tighten the bolts.

45. Temporarily install the rear strut rod assembly LH (for LH Side).
46. Temporarily install the rear strut rod assembly RH.
47. Temporarily install the rear No. 2 suspension arm assembly LH (for LH Side).
   a. Temporarily install the rear No. 2 suspension arm assembly LH to the rear suspension member sub-assembly with the rear suspension toe adjust cam sub-assembly, the No. 2 camber adjust cam and the nut.

➡Ensure that the identification mark faces the rear side of the vehicle.

➡When temporarily tightening the nut, keep the rear suspension toe adjust cam sub-assembly from rotating.

   b. Temporarily install the rear No. 2 suspension arm assembly LH to the rear axle carrier sub-assembly LH with the bolt and the nut.

➡Since a stopper nut is used, temporarily tighten the bolt.

48. Temporarily install the rear No. 2 suspension arm assembly RH (for LH Side).

➡Perform the same procedure as the rear No. 2 suspension arm assembly LH.

49. Install the rear axle hub and bearing assembly LH (for LH Side).
50. Install the rear axle hub and bearing assembly RH (for LH Side).
51. Install the rear disc to the rear axle hub and bearing assembly LH.
52. Install the rear disc to the rear axle hub and bearing assembly RH.

➡Perform the same procedure as the LH side.

53. Install the rear disc brake caliper assembly LH (for LH Side).
54. Install the rear disc brake caliper assembly RH (for LH Side).
55. Temporarily the install rear axle shaft nut LH (for LH Side).
56. Temporarily the install rear axle shaft nut RH (for LH Side).
57. Install the rear speed sensor LH (for LH Side).
58. Install the rear speed sensor RH (for LH Side).
59. Jack up the rear axle carrier sub-assembly, placing a wooden block underneath to avoid damage. Apply load to the suspension so that the rear drive shaft assembly is positioned horizontally.

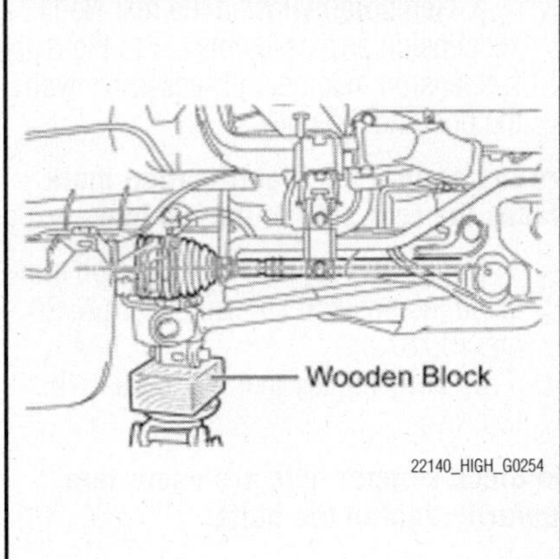

Fig. 446 Jack up the rear axle carrier sub-assembly

⁂ CAUTION

Do not jack up the rear axle carrier sub-assembly too high as the vehicle may fall.

➡Do not bend the brake dust cover.

➡If the rear drive shaft assembly cannot be positioned horizontally as shown in the illustration even when the rear axle carrier sub-assembly is jacked up, apply additional load to the vehicle such as by having a person sit in the rear seat.

➡Use the same procedures for the RH side and LH side.

60. Fully tighten the rear No. 1 suspension arm assembly LH (for LH Side).
61. Fully tighten the rear No. 1 suspension arm assembly RH (for LH Side).

➡Perform the same procedure as the rear No. 1 suspension arm assembly LH.

62. Fully tighten the rear No. 1 suspension arm assembly RH (for RH Side).
63. Fully tighten the 2 bolts to 59 ft. lbs. (80 Nm), 82 ft. lbs. (112 Nm).

➡Since a stopper nut is used, temporarily tighten the bolt.

➡The final torque must be applied under standard vehicle height conditions.

64. Fully tighten the rear No. 2 suspension arm assembly LH (for LH Side).
   a. Align the matchmarks on the adjust cams and rear suspension member sub-assembly.
   b. Fully tighten the nut to 74 ft. lbs. (100 Nm).

➡The final torque must be applied under standard vehicle height conditions.

➡When fully tightening the nut, keep the rear suspension toe adjust cam sub-assembly from rotating.

   c. Fully tighten the bolt to 82 ft. lbs. (112 Nm).

➡Since a stopper nut is used, temporarily tighten the bolt.

➡The final torque must be applied under standard vehicle height conditions.

65. Fully tighten the rear No. 2 suspension arm assembly RH (for RH Side).

➡Perform the same procedure as the rear No. 2 suspension arm assembly LH.

66. Fully tighten rear strut rod assembly LH (for LH Side).
67. Fully tighten rear strut rod assembly RH.
68. Install the No. 3 parking brake cable assembly (for LH Side).
69. Install the No. 2 parking brake cable assembly.
70. Separate the rear disc brake caliper assembly LH (for LH Side).
71. Separate the rear disc brake caliper assembly RH (for LH Side).

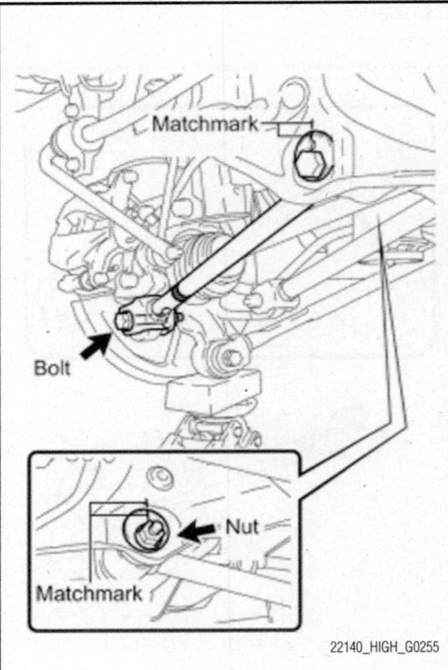

Fig. 447 Align the matchmarks on the adjust cams and rear suspension member sub-assembly

72. Remove the rear disc from the rear axle hub and bearing assembly LH.

73. Remove the rear disc from the rear axle hub and bearing assembly RH.

➡Perform the same procedure as the LH side.

74. Inspect the rear axle hub bearing looseness.

75. Inspect the rear axle hub bearing runout.

➡Use the same procedures for the RH side and LH side.

76. Install the rear disc to the rear axle hub and bearing assembly LH.

77. Install the rear disc to the rear axle hub and bearing assembly RH.

➡Perform the same procedure as the LH side.

78. Install the rear disc brake caliper assembly LH (for LH Side).

79. Install the rear disc brake caliper assembly RH (for LH Side).

80. Install the rear axle shaft nut LH (for LH Side).

81. Install the rear axle shaft nut RH (for LH Side).

82. Temporarily tighten the propeller with center bearing shaft assembly (for LH Side).

83. Fully tighten the propeller with center bearing shaft assembly (for LH Side).

84. Inspect and adjust transfer oil (for LH Side).

85. Install the center exhaust pipe assembly.

86. Install the tail exhaust pipe assembly.

87. Inspect for exhaust gas leak

88. Install the wheels and tighten lug nuts to 76 ft. lbs. (103 Nm).

89. Inspect and adjust rear wheel alignment.

90. Check for rear speed sensor signal (for LH Side).

Hybrid Models

Rear Lower Arm

See Figure 448.

1. Remove the rear wheel.
2. Remove the rear exhaust.
3. Separate the No.3 parking brake cable.
4. Remove the 2 bolts, the 2 nuts, and the rear strut rod assembly.
5. Put matchmarks on the adjust cams and the rear suspension member sub-assembly.
6. Remove the bolt and nut, and separate the rear No. 2 suspension arm assembly from the rear axle carrier sub-assembly.
7. Remove the nut, the No. 2 camber adjust cam, the rear suspension toe adjust cam sub-assembly, and the rear No. 2 suspension arm assembly.
8. Remove the 2 bolts, the 2 nuts, and the rear No. 1 suspension arm assembly.

To install:

9. Temporarily install the rear No. 1 suspension arm assembly with the 2 bolts and the 2 nuts.
10. Ensure that the identification mark faces the rear side of the vehicle.
11. Temporarily install the rear No. 2 suspension arm assembly to the rear suspension member sub-assembly with the rear suspension toe adjust cam sub-assembly, the No. 2 camber adjust cam and the nut.
12. Temporarily install the rear No. 2 suspension arm assembly to the rear axle carrier sub-assembly with the bolt and the nut.
13. Temporarily install the rear strut rod assembly with the 2 bolts and the 2 nuts.
14. Stabilize the suspension.
15. Fully tighten the rear strut rod assembly to 59 ft. lbs. (80 Nm).
16. Install the No.3 parking brake cable.
17. Install the rear wheel and tighten to 76 ft. lbs. (103 Nm).

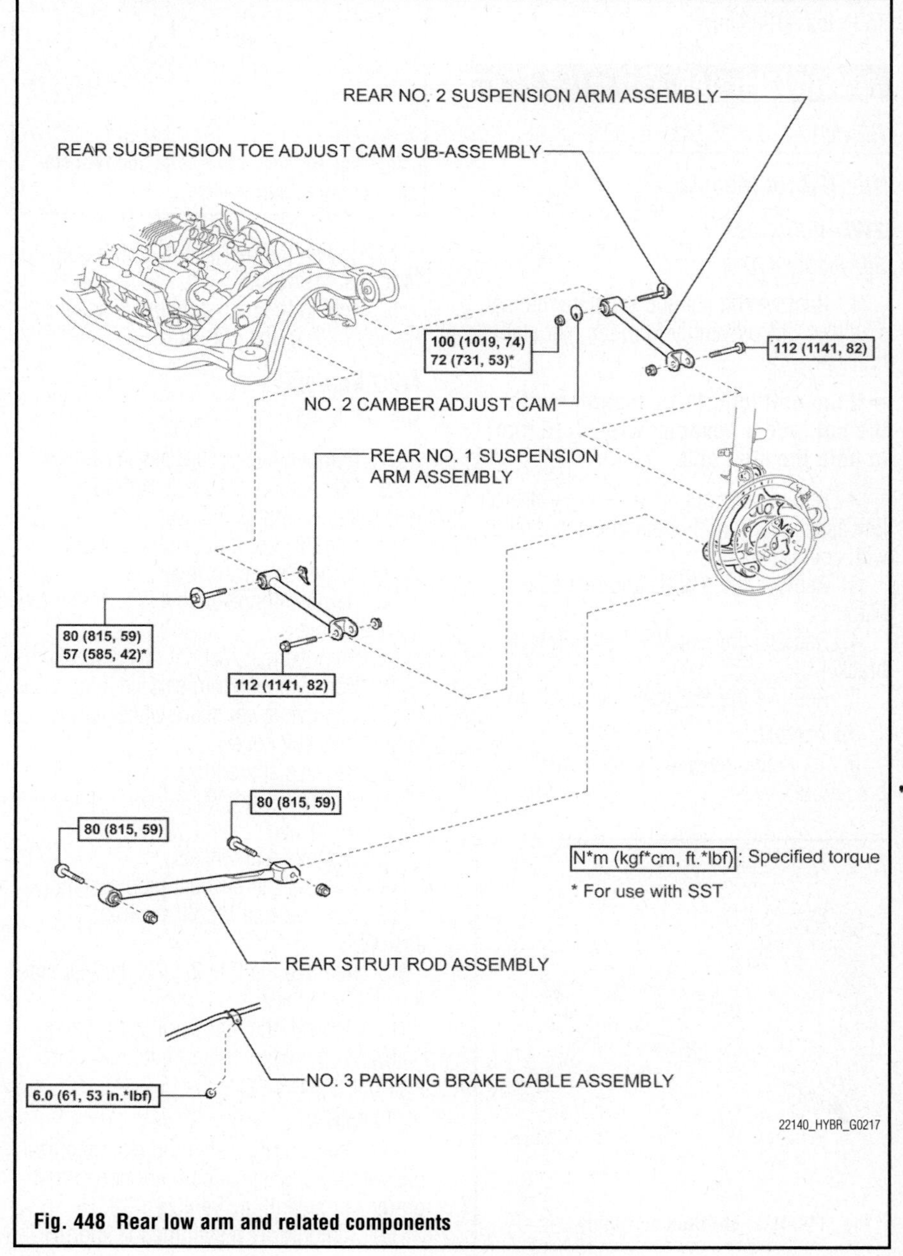

Fig. 448 Rear low arm and related components

Rear Strut Rod Assembly

See Figure 449.

1. Remove the rear wheel.
2. Separate the No.3 parking brake cable.
3. Remove the 2 bolts, the 2 nuts, and the rear strut rod assembly.

To install:

4. Check that the identification mark of the rear strut rod assembly is positioned on the inner side of the vehicle.
5. Temporarily install the rear strut rod assembly with the 2 bolts and the 2 nuts.
6. Stabilize the suspension.
7. Fully tighten the rear strut rod assembly to 59 ft. lbs. (80 Nm).
8. Install the No.3 parking brake cable.
9. Install the rear wheel and tighten to 76 ft. lbs. (103 Nm).

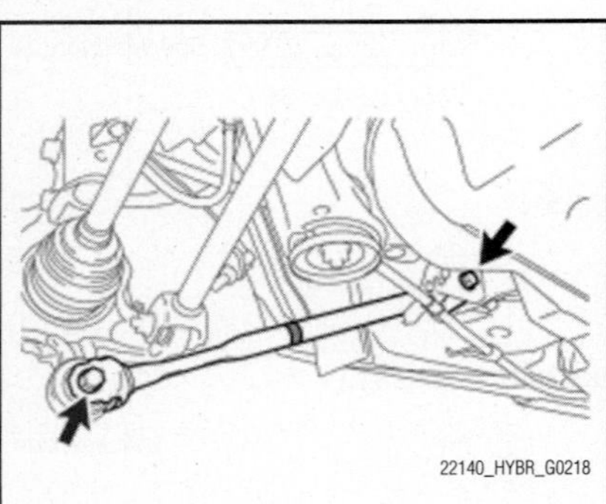

Fig. 449 Rear strut rod assembly

STABILIZER BAR

REMOVAL & INSTALLATION

Non-Hybrid Models

2WD Vehicles

See Figure 450.

1. Remove the nut and separate the rear stabilizer link assembly from the rear stabilizer bar.

➡If the ball joint turns together with the nut, use a hexagon wrench (5 mm) to hold the stud bolt.

2. Remove the nut and the rear stabilizer link assembly from the rear shock absorber with coil spring.
3. Remove the 4 bolts and rear stabilizer bar.
4. Remove the rear No. 1 stabilizer bar bracket.
5. Remove the rear stabilizer bushing.

To install:

6. To install, reverse the removal procedure.

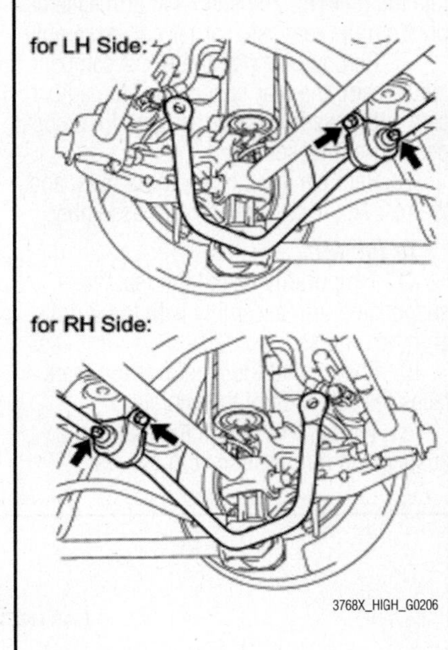

Fig. 450 Remove the 4 bolts and rear stabilizer bar—2WD vehicles

7. Tighten the rear stabilizer bar bolts to 14 ft. lbs. (19 Nm).
8. Tighten the rear stabilizer link assembly nut to 29 ft. lbs. (39 Nm).

4WD Vehicles

See Figure 451.

1. Remove deck board assembly.
2. Remove no. 3 deck board sub-assembly (w/ Tonneau Cover).
3. Remove no. 2 deck board sub-assembly (w/ Tonneau Cover).
4. Remove tonneau cover assembly (w/ Tonneau Cover).
5. Remove rear mat.
6. Remove deck trim service hole cover.
7. Remove lower spare wheel carrier hinge cover (w/ cover).
8. Remove spare tire.
9. Remove spare wheel carrier lock cover (w/ cover).
10. Remove rear wheels.
11. Remove tail exhaust pipe assembly.
12. Remove rear stabilizer link assembly.
13. Remove the 2 bolts and the rear stabilizer bar bracket.
14. Remove the rear stabilizer bushing.
15. Remove the bolt and the rear stabilizer bar bracket.

To install:

16. Temporarily install the rear stabilizer bar with the identification mark positioned on the left side of the vehicle.
17. Temporarily install the LH and RH rear stabilizer bar bracket (front side) with the bolt.

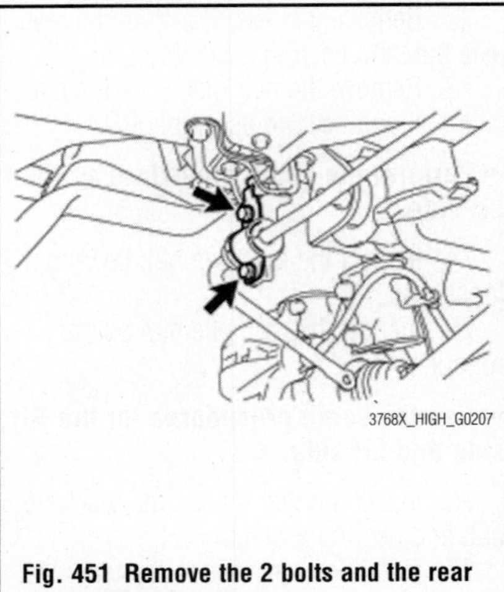

Fig. 451 Remove the 2 bolts and the rear stabilizer bar bracket

18. Loosely tighten the bolt so that the bracket can be moved by hand.
19. Install the rear stabilizer bushings to the rear stabilizer bar.

➡Make sure that the cutout of the rear stabilizer bushing is positioned towards the rear of the vehicle.

20. Install the rear stabilizer bar bracket LH and RH (rear side) with the 2 bolts. Tighten to 14 ft. lbs. (19 Nm).
21. Fully tighten the LH rear stabilizer bar brackets. Tighten to 40 ft. lbs. (54 Nm).
22. Fully tighten the RH rear stabilizer bar brackets. Tighten to 14 ft. lbs. (19 Nm).
23. Install the rear stabilizer link assembly RH and LH to the rear shock absorber with coil spring RH and LH with the nut. Tighten to 29 ft. lbs. (39 Nm).
24. To complete installation, reverse remaining removal procedure.

Hybrid Models

Rear Lower Arm

See Figure 452.

1. Remove the rear wheel.
2. Remove the rear exhaust.
3. Separate the No.3 parking brake cable.
4. Remove the 2 bolts, the 2 nuts, and the rear strut rod assembly.
5. Put matchmarks on the adjust cams and the rear suspension member sub-assembly.
6. Remove the bolt and nut, and separate the rear No. 2 suspension arm assembly from the rear axle carrier sub-assembly.
7. Remove the nut, the No. 2 camber adjust cam, the rear suspension toe adjust

cam sub-assembly, and the rear No. 2 suspension arm assembly.

8. Remove the 2 bolts, the 2 nuts, and the rear No. 1 suspension arm assembly.

To install:

9. Temporarily install the rear No. 1 suspension arm assembly with the 2 bolts and the 2 nuts.

10. Ensure that the identification mark faces the rear side of the vehicle.

11. Temporarily install the rear No. 2 suspension arm assembly to the rear suspension member sub-assembly with the rear suspension toe adjust cam sub-assembly, the No. 2 camber adjust cam and the nut.

12. Temporarily install the rear No. 2 suspension arm assembly to the rear axle carrier sub-assembly with the bolt and the nut.

13. Temporarily install the rear strut rod assembly with the 2 bolts and the 2 nuts.

14. Stabilize the suspension.

15. Fully tighten the rear strut rod assembly to 59 ft. lbs. (80 Nm).

16. Install the No.3 parking brake cable.

17. Install the rear wheel and tighten to 76 ft. lbs. (103 Nm).

Rear Strut Rod Assembly

See Figure 449.

1. Remove the rear wheel.

2. Separate the No.3 parking brake cable.

3. Remove the 2 bolts, the 2 nuts, and the rear strut rod assembly.

To install:

4. Check that the identification mark of the rear strut rod assembly is positioned on the inner side of the vehicle.

5. Temporarily install the rear strut rod assembly with the 2 bolts and the 2 nuts.

6. Stabilize the suspension.

7. Fully tighten the rear strut rod assembly to 59 ft. lbs. (80 Nm).

8. Install the No.3 parking brake cable.

9. Install the rear wheel and tighten to 76 ft. lbs. (103 Nm).

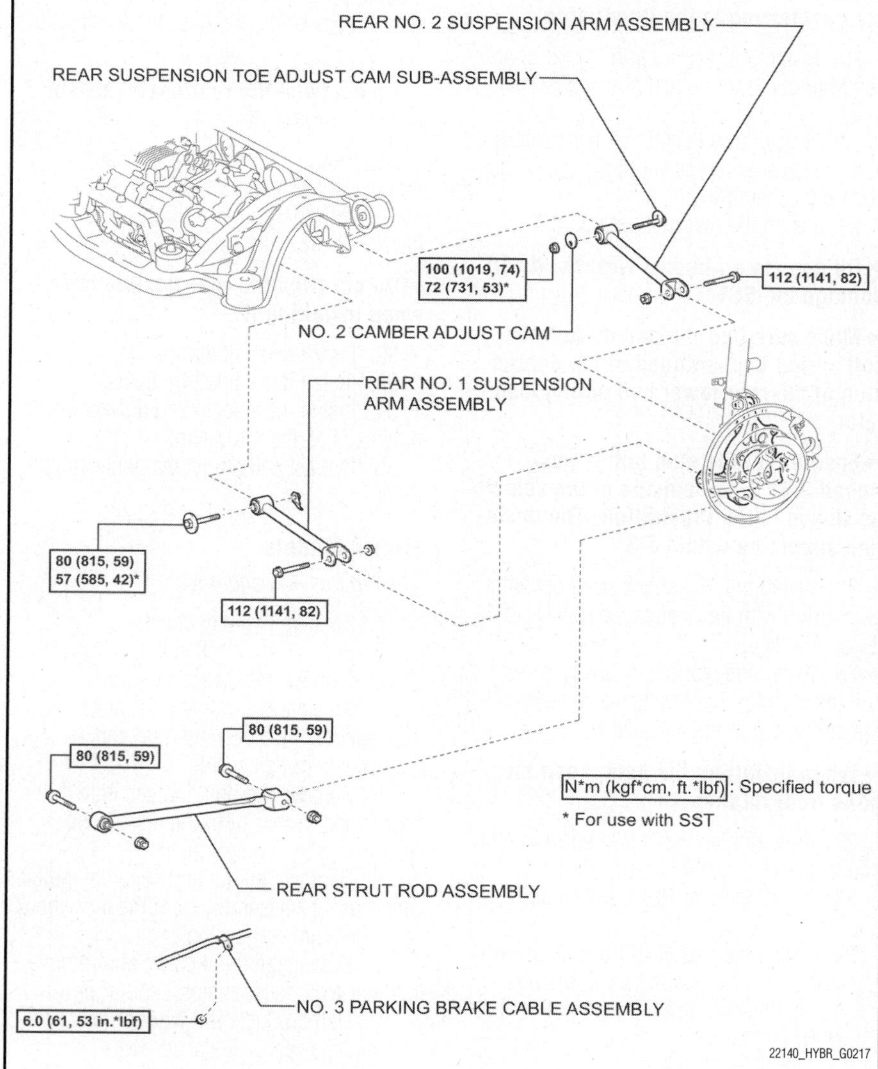

Fig. 452 Rear low arm and related components

STRUT & SPRING ASSEMBLY

REMOVAL & INSTALLATION

Non-Hybrid Models

See Figures 453 and 454.

1. Remove the wheel.

2. Remove the deck side trim cover.

3. Remove the deck side trim.

4. For 4WD vehicles, remove the bolt and separate the rear flexible hose from the rear shock absorber with coil spring.

5. Remove the bolt and separate the rear speed sensor from the rear shock absorber with coil spring.

6. Remove the nut and separate the rear stabilizer link assembly from the rear shock absorber with coil spring.

➡If the ball joint turns together with the nut, use a hexagon wrench (5 mm) to hold the stud bolt.

7. Disengage the 4 claws and remove the rear No. 1 suspension support cover.

8. Using a jack and wooden block, support the rear axle carrier sub-assembly.

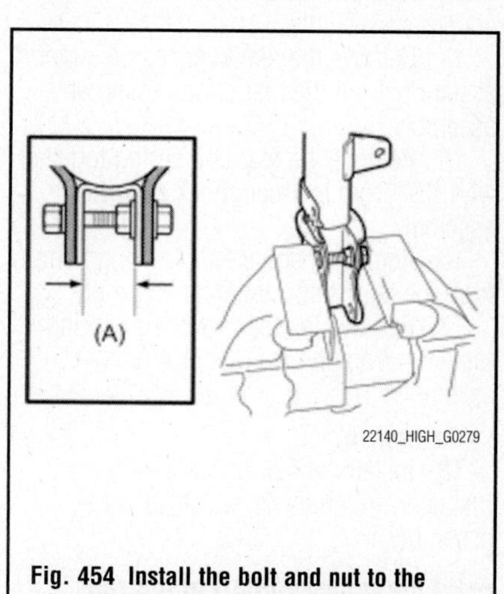

Fig. 454 Install the bolt and nut to the rear shock absorber with coil spring 2WD

➡**Do not deform the dust cover.**

➡**Support the rear axle carrier sub-assembly until reinstallation of the rear shock absorber with coil spring is complete.**

9. Loosen the rear support to rear shock absorber nut.

➡**Do not remove the rear support to rear shock absorber nut.**

➡**Loosen the nut only when the rear shock absorber with coil spring needs to be disassembled.**

10. Remove the 2 bolts and 2 nuts, and separate the rear shock absorber with coil spring from the rear axle carrier sub-assembly,

➡**When removing the nuts, keep the bolts from rotating.**

11. Remove the 3 nuts and rear shock absorber with coil spring.

➡**Make sure that the rear speed sensor and rear flexible hose are disconnected from the rear shock absorber with coil spring.**

12. Remove the rear support to rear shock absorber nut, as follows:

a. Using SST: 09727-30021, compress the rear coil spring.

➡**Do not use an impact wrench. It will damage SST.**

b. Check that the front coil spring is fully compressed.

c. Hold the rear suspension support assembly and remove the rear support to rear shock absorber nut from the rear shock absorber assembly.

13. Remove the rear support to rear shock absorber collar from the rear shock absorber assembly.

14. Remove the rear suspension support assembly from the rear shock absorber assembly.

15. Remove the rear coil spring together with SST from the rear shock absorber assembly.

16. Remove the rear No. 1 spring bumper from the rear shock absorber assembly.

17. Remove the rear lower coil spring insulator from the rear shock absorber assembly.

To install:

18. Install the rear lower coil spring insulator onto the rear shock absorber assembly.

➡**Fit the recessed part of the rear lower coil spring insulator into the recession on the shock absorber assembly.**

19. Install the rear No. 1 spring bumper to the rear shock absorber assembly.

20. Temporarily install rear coil spring, as follows:

a. Using SST: 09727-30021 or equivalent, compress the rear coil spring.

➡**Do not use an impact wrench. It will damage the SST.**

b. Temporarily install the rear coil spring together with SST to the rear shock absorber assembly.

21. Install the rear suspension support assembly to the rear shock absorber assembly.

➡**Align the cutout on the rear shock absorber assembly with the protrusion on the rear suspension support assembly by referring to the illustration.**

22. Install the rear support to rear shock absorber collar to the rear shock absorber assembly.

23. Temporarily install the rear support to rear shock absorber nut to the rear shock absorber assembly.

24. Install the rear coil spring.

➡**Do not use an impact wrench. It will damage the SST.**

➡**Make sure that the end of the rear coil spring is positioned in the depression of the rear lower coil spring insulator.**

➡**Ensure that the stud bolt is positioned 3.5° to the outside of the vehicle as shown in the illustration. The deviation should be within +-5° .**

25. Install the rear shock absorber with coil spring with the 3 nuts and tighten to 43 ft. lbs. (58 Nm).

26. Install the rear shock absorber with coil spring with the 2 bolts and 2 nuts and tighten to 213 ft. lbs. (290 Nm).

➡**When installing the nuts, keep the bolts from rotating.**

27. Fully tighten the rear support to rear shock absorber nut 40 ft. lbs. (55 Nm).

28. Install the rear No. 1 suspension support cover.

29. Install the rear stabilizer link assembly to the rear shock absorber with coil spring with the nut and tighten to 29 ft. lbs. (39 Nm).

➡**If the ball joint turns together with the nut, use a hexagon wrench (5 mm) to hold the stud bolt.**

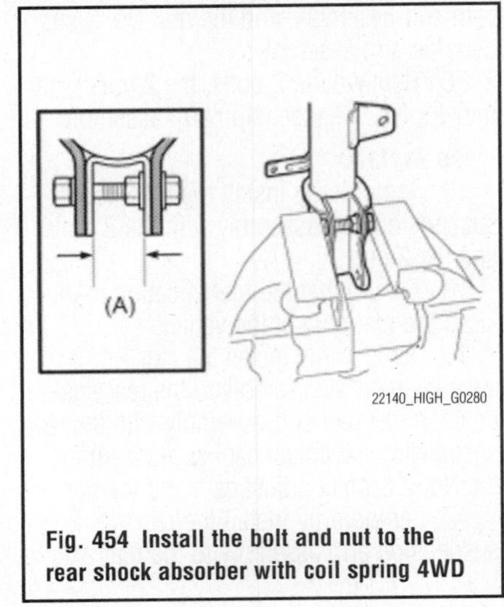

Fig. 454 Install the bolt and nut to the rear shock absorber with coil spring 4WD

30. Install the rear speed sensor wire to the rear shock absorber with coil spring with the bolt and tighten to 44 inch lbs. (5 Nm).

➡**Do not twist the rear speed sensor wire when installing it.**

31. Install the rear flexible hose to the rear shock absorber with coil spring with the bolt and tighten to 14 ft. lbs. (19 Nm).

➡**Do not twist the rear flexible hose when installing it.**

32. Install deck side trim.

33. Install deck side trim cover.

34. Install the wheels and tighten lug nuts to 76 ft. lbs. (103 Nm).

35. Inspect and adjust rear wheel alignment.

Hybrid Models

See Figures 455 and 456.

1. Remove the rear wheel.
2. Remove the deck side trim cover.
3. Remove the deck side trim.
4. Remove the bolt and separate the rear flexible hose from the rear shock absorber with coil spring.
5. Remove the bolt and separate the rear speed sensor from the rear shock absorber with coil spring.
6. Remove the nut and separate the rear stabilizer link assembly from the rear shock absorber with coil spring.
7. Disengage the 4 claws and remove the rear No. 1 suspension support cover.
8. Using a jack and wooden block, support the rear axle carrier sub-assembly.
9. Loosen the rear support to rear shock absorber nut.

CAUTION

Do not remove the rear support to rear shock absorber nut.

10. Remove the 2 bolts and 2 nuts, and separate the rear shock absorber with coil spring from the rear axle carrier sub-assembly.
11. Remove the 3 nuts and rear shock absorber with coil spring.
12. Install the bolt and nut to the rear shock absorber with coil spring as shown in the illustration and secure the rear shock absorber with coil spring in a vise.
13. Using a spring compressor, compress the rear coil spring.
14. Check that the front coil spring is fully compressed.
15. Remove the rear support to rear shock absorber collar from the rear shock absorber assembly.
16. Remove the rear suspension support assembly from the rear shock absorber assembly.
17. Remove the rear coil spring together with SST from the rear shock absorber assembly.
18. Remove the rear No. 1 spring bumper from the rear shock absorber assembly.
19. Remove the rear lower coil spring insulator from the rear shock absorber assembly.

To install:

20. Install the rear lower coil spring insulator onto the rear shock absorber assembly. Fit the recessed part of the rear lower coil spring insulator into the recession on the shock absorber assembly.
21. Install the rear No. 1 spring bumper to the rear shock absorber assembly.
22. Using spring, compress the rear coil spring.
23. Temporarily install the rear coil spring together with SST to the rear shock absorber assembly.
24. Install the rear suspension support assembly to the rear shock

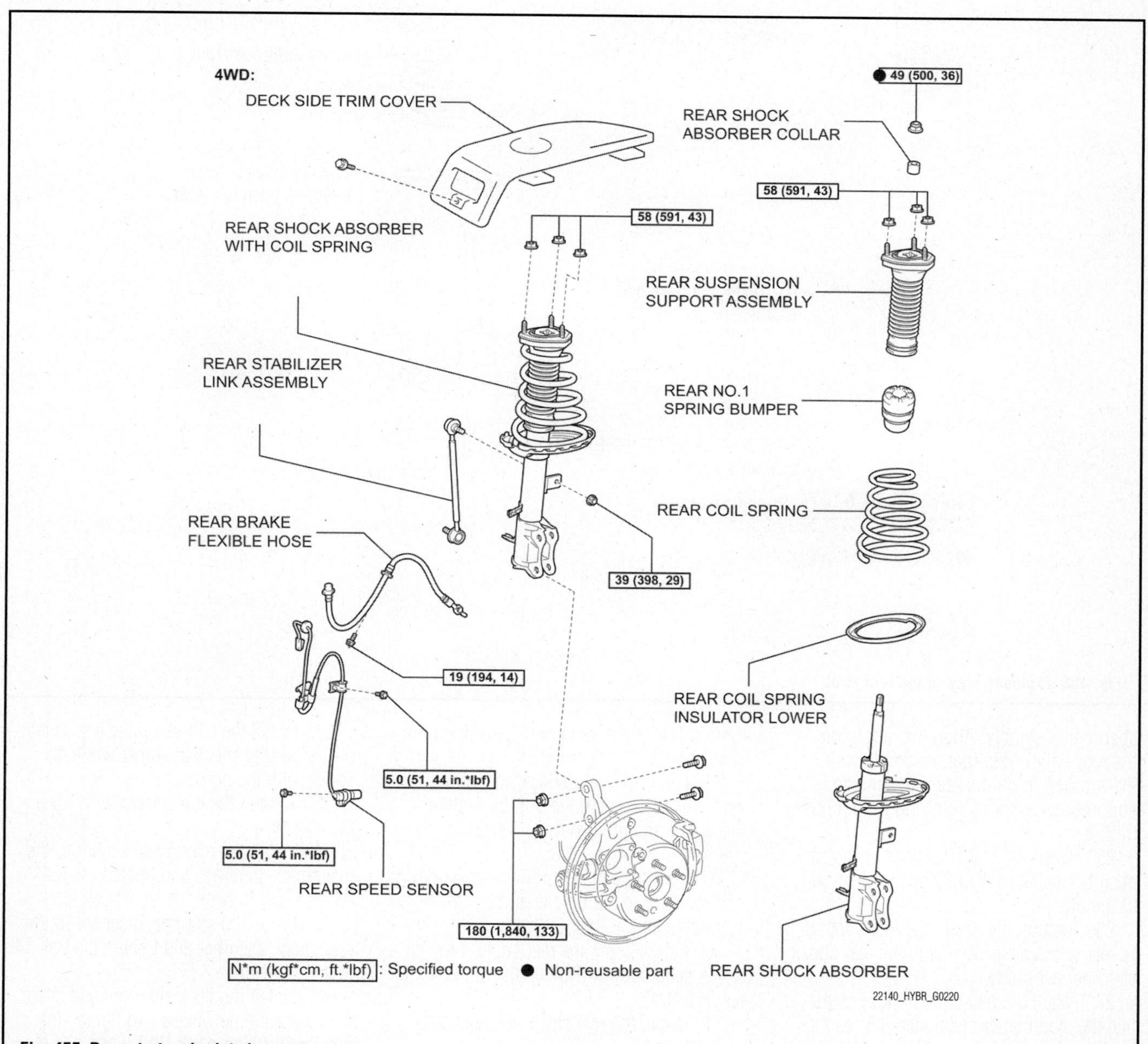

Fig. 455 Rear strut and related components

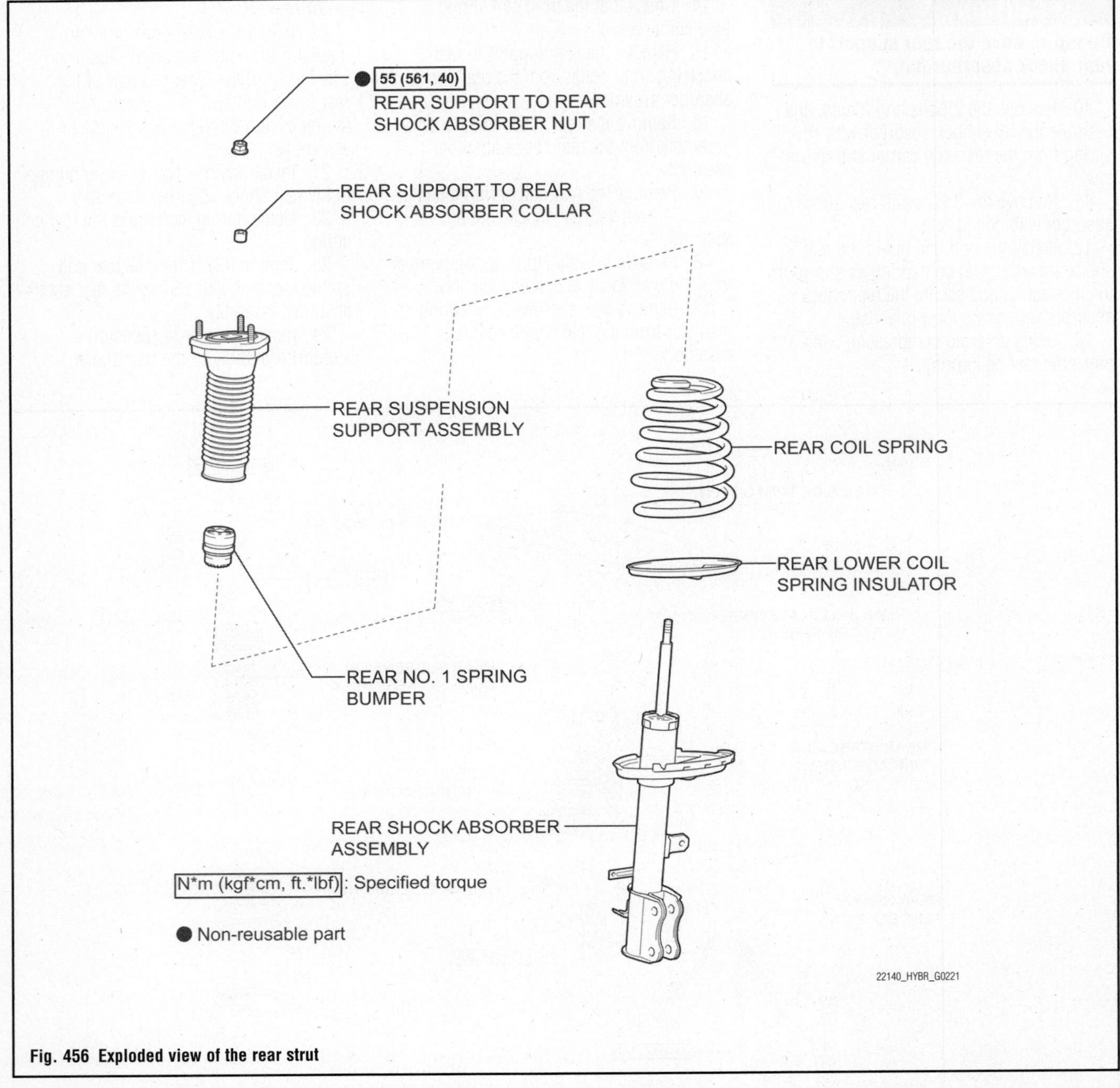

Fig. 456 Exploded view of the rear strut

absorber assembly. Align the cutout on the rear shock absorber assembly with the protrusion on the rear suspension support assembly by referring to the illustration.

25. Install the rear support to rear shock absorber collar to the rear shock absorber assembly.

26. Temporarily install the rear support to rear shock absorber nut to the rear shock absorber assembly.

27. Install the rear coil spring. Ensure that the stud bolt is positioned 3.5° to the outside of the vehicle as shown in the illustration. The deviation should be within plus or minus 5° .

28. Install the rear shock absorber with coil spring with the 3 nuts. Tighten the upper mounting nuts to 43 ft. lbs. (58 Nm).

29. Install the rear shock absorber with coil spring with the 2 bolts and 2 nuts. Tighten to 213 ft. lbs. (290 Nm).

30. Fully tighten the rear support to rear shock absorber nut. Tighten the nut to 40 ft. lbs. (55 Nm).

31. Install the rear No. 1 suspension support cover.

32. Install the rear stabilizer link assembly to the rear shock absorber with coil spring with the nut.

33. Tighten the link assembly to 29 ft. lbs. (39 Nm).

34. Install the rear speed sensor to the rear shock absorber, and tighten the bolt to 54 inch lbs. (5 Nm).

35. Install the rear speed sensor to the rear shock absorber and tighten the bolt 14 ft. lbs. (19 Nm).

36. Install the deck side trim and cover.

37. Install the wheels and tighten lug nuts to 76 ft. lbs. (103 Nm).

38. Inspect and adjust rear wheel alignment.
39. Check the ABS speed sensor signal.

WHEEL BEARINGS

REMOVAL & INSTALLATION

Non-Hybrid Models

See Figure 457.

1. Disconnect the negative battery cable.
2. Remove the wheel.
3. Separate the rear flexible hose.
4. Remove the 2 bolts and separate the rear disc brake caliper assembly.
5. Remove the rear disc.
6. For 4WD vehicles: Using a screwdriver, disconnect the connector from the rear speed sensor.
7. Remove the 4 bolts and the rear axle hub & bearing assembly.

To install:

8. Install the rear axle hub and bearing assembly with the 4 bolts and tighten to 55 ft. lbs. (75 Nm).
9. Inspect rear axle hub bearing looseness.
10. Inspect rear axle hub runout.
11. Connect the connector to the rear speed sensor.
12. Install the rear disc.
13. Install the rear disc brake caliper assembly with the 2 bolts and tighten to 57 ft. lbs. (78 Nm).
14. Install the rear flexible hose to the shock absorber with coil spring with the bolt and tighten to 14 ft. lbs. (19 Nm).
15. Install the wheel and tighten lug nuts to 76 ft. lbs. (103 Nm).
16. Connect the negative battery cable.
17. Inspect and adjust rear wheel alignment.
18. Check the ABS speed sensor signal.

Hybrid Models

See Figure 458.

1. Disconnect the negative battery cable.
2. Remove the wheel.
3. Separate the rear flexible hose.
4. Remove the 2 bolts and separate the rear disc brake caliper assembly.
5. Remove the rear disc.
6. Using a screwdriver, disconnect the connector from the rear speed sensor.
7. For 4WD vehicles,
8. Remove the 4 bolts and the rear axle hub & bearing assembly.

To install:

9. Install the rear axle hub and bearing assembly with the 4 bolts and tighten to 55 ft. lbs. (75 Nm).
10. Inspect rear axle hub bearing looseness.
11. Inspect rear axle hub runout.
12. Connect the connector to the rear speed sensor.

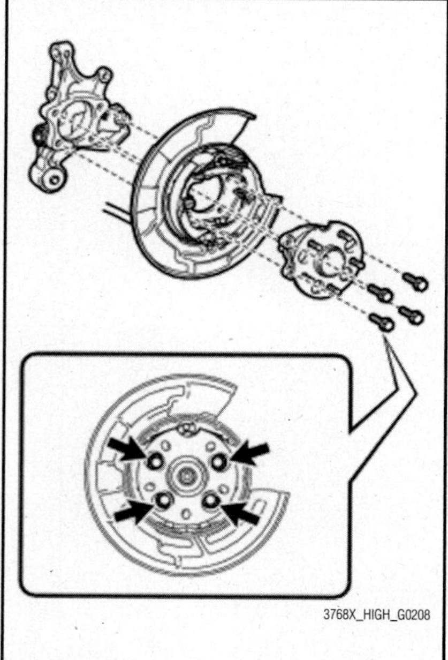

Fig. 457 Remove the 4 bolts and the rear axle hub & bearing assembly

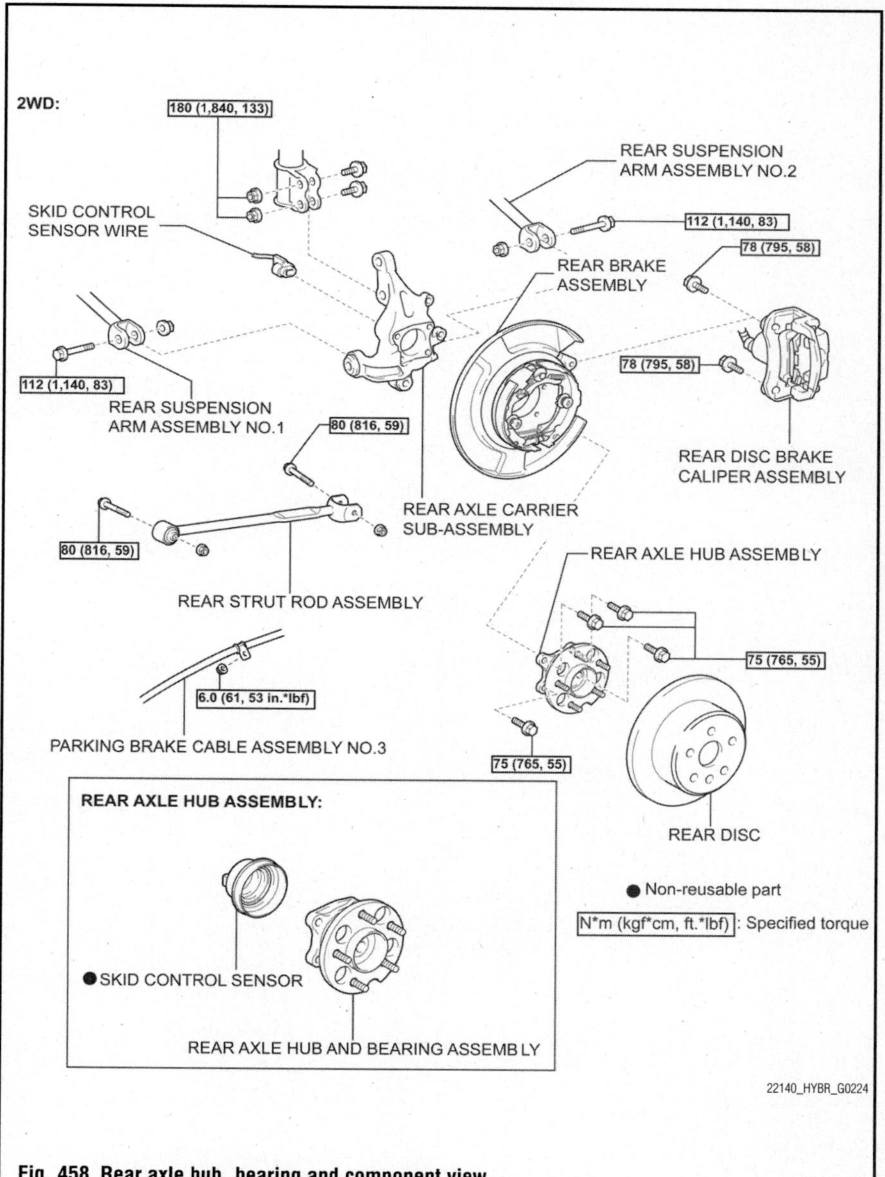

Fig. 458 Rear axle hub, bearing and component view

13. Install the rear disc.
14. Install the rear disc brake caliper assembly with the 2 bolts and tighten to 57 ft. lbs. (78 Nm).
15. Install the rear flexible hose to the shock absorber with coil spring with the bolt and tighten to 14 ft. lbs. (19 Nm).
16. Install the wheel and tighten lug nuts to 76 ft. lbs. (103 Nm).
17. Connect the negative battery cable.
18. Inspect and adjust rear wheel alignment.
19. Check the ABS speed sensor signal.

TOYOTA

Land Cruiser

9

SPECIFICATIONS AND MAINTENANCE CHARTS

ENGINE AND VEHICLE IDENTIFICATION

Engine						
Code ①	Liters (cc)	Cu. In.	Cyl.	Fuel Sys.	Engine Type	Eng. Mfg.
3UR-FE	5.7 (5700)	345.6	8	SFI	DOHC	Toyota

Model Year	
Code ②	Year
B	2011
C	2012

SFI: Sequential Fuel Injection

DOHC: Double Overhead Camshaft

① Stamped on the left side of the engine block

71099_LAND_C0001

GENERAL ENGINE SPECIFICATIONS

Year	Model	Engine Displacement Liters	Engine Series ID	Net Horsepower @ rpm	Net Torque @ rpm (ft. lbs.)	Bore x Stroke (in.)	Com-pression Ratio	Oil Pressure @ rpm
2011	Land Cruiser	5.7	3UR-FE	381@5600	401@3600	3.70x4.02	10.2:1	32 @ 2500
2012	Land Cruiser	5.7	3UR-FE	381@5600	401@3600	3.70x4.02	10.2:1	32 @ 2500

71099_LAND_C0002

ENGINE TUNE-UP SPECIFICATIONS

Year	Engine Displacement Liters	Engine ID	Spark Plug Gap (in.)	Ignition Timing (deg.)*	Fuel Pump (psi)	Idle Speed (rpm)		Valve Clearance	
						MT	AT	Intake	Exhaust
2011	5.7	3UR-FE	0.043	8-12 B at idle	41-42	—	650-750	NA	NA
2012	5.7	3UR-FE	0.043	8-12 B at idle	41-42	—	650-750	NA	NA

NOTE: The Vehicle Emission Control Information label often reflects specification changes made during production.

The label figures must be used if they differ from those in this chart.

B: Before top dead center

* With terminals TC and CG connected to DLC3 (Transmission in Neurtal and A/C switch OFF)

71099_LAND_C0003

CAPACITIES

Year	Model	Engine Disp. Liters	Engine ID	Engine Oil with Filter (qts.)	Transmission (pts.) 5-Spd	Transmission (pts.) Auto.*	Transfer Case (pts.)	Drive Axle Front (pts.)	Drive Axle Rear (pts.)	Fuel Tank (gal.)	Cooling System (qts.)
2011	Land Cruiser	5.7	3UR-FE	7.4	—	10.6	3.1	3.9-5.2	8.8-8.9	24.6	16.3
2012	Land Cruiser	5.7	3UR-FE	7.4	—	10.6	3.1	3.9-5.2	8.8-8.9	24.6	16.3

*After draining, add the following amounts, then fill to the cold full line

① Without rear heater: 15.6

With rear heater: 16.2

71099_LAND_C0004

FLUID SPECIFICATIONS

Year	Model	Engine Displacement Liters	Engine ID/VIN	Engine Oil	Auto. Trans.	Power Steering Fluid	Brake Master Cylinder
2011	Land Cruiser	5.7	3UR-FE	5W-20	NA	ATF Dexron® II or III	DOT 3
2012	Land Cruiser	5.7	3UR-FE	5W-20	NA	ATF Dexron® II or III	DOT 3

DOT: Department Of Transpotation

NA: Not Available

71099_LAND_C0013

VALVE SPECIFICATIONS

Year	Engine Displacement Liters	Engine ID	Seat Angle (deg.)	Face Angle (deg.)	Spring Test Pressure (lbs. @ in.)	Spring Installed Height (in.)	Stem-to-Guide Clearance (in.) Intake	Stem-to-Guide Clearance (in.) Exhaust	Stem Diameter (in.) Intake	Stem Diameter (in.) Exhaust
2011	5.7	3UR-FE	45	44.5	45.9-50.7@ 1.378	1.380	0.0010-0.0024	0.0012-0.0026	0.2154-0.2159	0.2152-0.2157
2012	5.7	3UR-FE	45	44.5	45.9-50.7@ 1.378	1.380	0.0010-0.0024	0.0012-0.0026	0.2154-0.2159	0.2152-0.2157

71099_LAND_C0005

CAMSHAFT AND BEARING SPECIFICATIONS CHART

All measurements are given in inches.

Year	Engine Displ. Liters	Engine ID/VIN	Journal Dia.	Brg. Oil Clearance	Shaft End-play	Runout	Journal Bore	Lobe Height Intake	Lobe Height Exhaust
2011	5.7	3UR-FE	①	②	NA	0.00157	NA	1.744-1.7500	1.740-1.7460
2012	5.7	3UR-FE	①	②	NA	0.00157	NA	1.744-1.7500	1.740-1.7460

NA: Not Available

① No. 1 journal: 1.179-1.180

Other Journals: 1.022-1.023 in.

② No. 1 journal: 0.00118-0.00256 in.

Other journal: 0.000984-0.00244 in.

Other Maximum Journals: 0.00276 in.

71099_LAND_C0014

CRANKSHAFT AND CONNECTING ROD SPECIFICATIONS

All measurements are given in inches.

Year	Engine Displacement Liters	Engine ID	Crankshaft Main Brg. Journal Dia.	Crankshaft Main Brg. Clearance	Crankshaft Shaft End-play	Crankshaft Thrust on No.	Connecting Rod Journal Diameter	Connecting Rod Oil Clearance	Connecting Rod Side Clearance
2011	5.7	3UR-FE	2.6373-2.6378	①	0.0008-0.0087	3	②	0.0010-0.0020	0.0059-0.0217
2012	5.7	3UR-FE	2.6373-2.6378	①	0.0008-0.0087	3	②	0.0010-0.0020	0.0059-0.0217

① No. 1 and No. 5: 0.0007-0.0012

All others: 0.0009-0.0015

② Mark 1: 2.32283-2.32307

Mark 2: 2.32307-2.32330

Mark 3: 2.32330-2.32354

Mark 2: 2.32354-2.32377

71099_LAND_C0006

PISTON AND RING SPECIFICATIONS

All measurements are given in inches.

Year	Engine Displacement Liters	Engine ID	Piston Clearance	Ring Gap Top Comp.	Ring Gap Bottom Comp.	Ring Gap Oil Control	Ring Side Clearance Top Comp.	Ring Side Clearance Bottom Comp.	Ring Side Clearance Oil Control
2011	5.7	3UR-FE	0.0007-0.0028	0.0091-0.0130	0.0157-0.0197	0.0039-0.0157	0.00078-0.0028	0.00078-0.0024	0.0027-0.0057
2012	5.7	3UR-FE	0.0007-0.0028	0.0091-0.0130	0.0157-0.0197	0.0039-0.0157	0.00078-0.0028	0.00078-0.0024	0.0027-0.0057

71099_LAND_C0007

TORQUE SPECIFICATIONS

All readings in ft. lbs.

Year	Engine Displacement Liters	Engine ID	Cylinder Head Bolts	Main Bearing Bolts	Rod Bearing Bolts	Crankshaft Damper Bolts	Flywheel Bolts	Manifold Intake	Manifold Exhaust	Spark Plugs	Oil Pan Drain Plug
2011	5.7	3UR-FE	①	②	③	NA	15	15	15	13	29
2012	5.7	3UR-FE	①	②	③	NA	15	15	15	13	29

① Step 1: 27 ft. lbs.
Step 2: Plus 90 degrees
Step 3: Plus 90 degrees
For 12 mm head: 15 ft. lbs. (21 Nm)

② For inside position: 45 ft. lbs. (61 Nm)
For outside position: 20 ft. lbs.
Step 2: Plus 90 degrees
For cylinder block side position: 33 ft. lbs.

③ Step 1: 30 ft. lbs.
Step 2: Plus 90 degrees

71099_LAND_C0008

WHEEL ALIGNMENT

Year	Model	Caster Range (+/-Deg.)	Caster Preferred Setting (Deg.)	Camber Range (+/-Deg.)	Camber Preferred Setting (Deg.)	Toe-in (in.)	Steering Axis Inclination (Deg.)
2011	Land Cruiser	0.75	+2.90	0.75	+0.13	0.12+/-0.08	12.85+/-0.75
2012	Land Cruiser	0.75	+2.90	0.75	+0.13	0.12+/-0.08	12.85+/-0.75

Note: All algnment specifications are based on nominal ride height and standard tires

71099_LAND_C0009

TIRE, WHEEL AND BALL JOINT SPECIFICATIONS

		OEM Tires		Tire Pressures (psi)				Lug Nut
Year	Model	Standard	Optional	Front	Rear	Wheel Size	Ball Joint Inspection	Torque (ft. lbs.)
2011	Land Cruiser	P285/60R18	None	①	①	18x8J	②	97
2012	Land Cruiser	P285/60R18	None	①	①	18x8J	②	97

OEM: Original Equipment Manufacturer

PSI: Pounds Per Square Inch

STD: Standard

OPT: Optional

NA: Not Available

① See placard on vehicle

② Upper arm ball joint turning torque: 9-39 inch lbs.
Lower arm ball joint turning torque: 27-89 inch lbs.
Stabilizer link ball joint turning torque: 4.4-30 inch lbs.

71099_LAND_C0010

BRAKE SPECIFICATIONS

All measurements in inches unless noted

			Brake Disc				Brake Caliper	
Year	Model		Original Thickness	Minimum Thickness	Maximum Runout	Minimum Lining Thickness	Bracket Bolts (ft. lbs.)	Mounting Bolts (ft. lbs.)
2011	Land Cruiser	Front	1.260	1.150	0.0020	0.039	—	73
		Rear	0.709	0.630	0.0059	0.039	65	70
2012	Land Cruiser	Front	1.260	1.150	0.0020	0.039	—	73
		Rear	0.709	0.630	0.0059	0.039	65	70

71099_LAND_C0011

SCHEDULED MAINTENANCE INTERVALS

Toyota - Land Cruiser

		VEHICLE MILEAGE INTERVAL (x1000)														
TO BE SERVICED	TYPE OF SERVICE	5	10	15	20	25	30	35	40	45	50	55	60	65	70	75
Engine oil & filter	R	✓	✓	✓	✓	✓	✓	✓	✓	✓	✓	✓	✓	✓	✓	✓
Rotate tires	S/I	✓	✓	✓	✓	✓	✓	✓	✓	✓	✓	✓	✓	✓	✓	✓
Automatic transmission fluid & filter	S/I			✓			✓			✓			✓			✓
Ball joints & dust covers	S/I			✓			✓			✓			✓			✓
Bolts & nuts on chassis & body	S/I			✓			✓			✓			✓			✓
Brake line pipes & hoses	S/I			✓			✓			✓			✓			✓
Brake pads & discs	S/I	✓	✓	✓	✓	✓	✓	✓	✓	✓	✓	✓	✓	✓	✓	✓
Cabin air filter	S/I			✓			✓			✓			✓			✓
Propeller shaft grease	S/I			✓			✓			✓			✓			✓
Re-torque propeller shaft bolt	S/I			✓			✓			✓			✓			✓
Steering knuckle & chassis grease	S/I			✓			✓			✓			✓			✓
Steering linkage	S/I			✓			✓			✓			✓			✓
Transfer and differential oil	S/I			✓			✓			✓			✓			✓
Air cleaner filter	R						✓						✓			
Spark plugs ①	R						✓						✓			
Drive belts	S/I						✓						✓			
Exhaust pipes & mountings	S/I						✓						✓			
Fuel lines & connections	S/I						✓						✓			
Engine coolant	R															✓
Charcoal canister	R						✓						✓			
Fuel tank cap gasket	R						✓						✓			

R: Replace S/I: Service or Inspect

① Platinum plugs, replace every 100,000 miles

② Heated oxygen sensors (except Calif.): replace every 80,000 miles.

FREQUENT OPERATION MAINTENANCE (SEVERE SERVICE)

If a vehicle is operated under any of the following conditions it is considered severe service:

- **Extremely dusty areas.**
- **50% or more of the vehicle operation is in 32°C (90°F) or higher temperatures, or constant operation in temperatures below 0°C (32°F).**
- **Prolonged idling (vehicle operation in stop and go traffic).**
- **Frequent short running periods (engine does not warm to normal operating temperatures).**
- **Police, taxi, delivery usage or trailer towing usage.**

Air cleaner filter: service or inspect every 3750 miles

Engine oil & filter: replace every 3750 miles.

Ball joints & dust covers: service or inspect every 7500 miles.

Bolts & nuts on chassis & body: service or inspect every 7500 miles.

Brake pads & discs (front & rear): service or inspect every 7500 miles.

Steering knuckle & chassis grease: service or inspect every 7500 miles.

Steering linkage: service or inspect every 7500 miles.

Propeller shaft grease: service or inspect every 7500 miles.

Exhaust pipes & mountings: service or inspect every 15,000 miles.

71099_LAND_C0012

PRECAUTIONS

Before servicing any vehicle, please be sure to read all of the following precautions, which deal with personal safety, prevention of component damage, and important points to take into consideration when servicing a motor vehicle:

- Never open, service or drain the radiator or cooling system when the engine is hot; serious burns can occur from the steam and hot coolant.
- Observe all applicable safety precautions when working around fuel. Whenever servicing the fuel system, always work in a well-ventilated area. Do not allow fuel spray or vapors to come in contact with a spark, open flame, or excessive heat (a hot drop light, for example). Keep a dry chemical fire extinguisher near the work area. Always keep fuel in a container specifically designed for fuel storage; also, always properly seal fuel containers to avoid the possibility of fire or explosion. Refer to the additional fuel system precautions later in this section.
- Fuel injection systems often remain pressurized, even after the engine has been turned **OFF**. The fuel system pressure must be relieved before disconnecting any fuel lines. Failure to do so may result in fire and/or personal injury.
- Brake fluid often contains polyglycol ethers and polyglycols. Avoid contact with the eyes and wash your hands thoroughly after handling brake fluid. If you do get brake fluid in your eyes, flush your eyes with clean, running water for 15 minutes. If eye irritation persists, or if you have taken brake fluid internally, IMMEDIATELY seek medical assistance.
- The EPA warns that prolonged contact with used engine oil may cause a number of skin disorders, including cancer. You should make every effort to minimize your exposure to used engine oil. Protective gloves should be worn when changing oil. Wash your hands and any other exposed skin areas as soon as possible after exposure to used engine oil. Soap and water, or waterless hand cleaner should be used.
- All new vehicles are now equipped with an air bag system, often referred to as a Supplemental Restraint System (SRS) or Supplemental Inflatable Restraint (SIR) system. The system must be disabled before performing service on or around system components, steering column, instrument panel components, wiring and sensors. Failure to follow safety and disabling procedures could result in accidental air bag deployment, possible personal injury and unnecessary system repairs.
- Always wear safety goggles when working with, or around, the air bag system. When carrying a non-deployed air bag, be sure the bag and trim cover are pointed away from your body. When placing a non-deployed air bag on a work surface, always face the bag and trim cover upward, away from the surface. This will reduce the motion of the module if it is accidentally deployed. Refer to the additional air bag system precautions later in this section.
- Clean, high quality brake fluid from a sealed container is essential to the safe and proper operation of the brake system. You should always buy the correct type of brake fluid for your vehicle. If the brake fluid becomes contaminated, completely flush the system with new fluid. Never reuse any brake fluid. Any brake fluid that is removed from the system should be discarded. Also, do not allow any brake fluid to come in contact with a painted surface; it will damage the paint.
- Never operate the engine without the proper amount and type of engine oil; doing so WILL result in severe engine damage.
- Timing belt maintenance is extremely important. Many models utilize an interference-type, non-freewheeling engine. If the timing belt breaks, the valves in the cylinder head may strike the pistons, causing potentially serious (also time-consuming and expensive) engine damage. Refer to the maintenance interval charts for the recommended replacement interval for the timing belt, and to the timing belt section for belt replacement and inspection.
- Disconnecting the negative battery cable on some vehicles may interfere with the functions of the on-board computer system(s) and may require the computer to undergo a relearning process once the negative battery cable is reconnected.
- When servicing drum brakes, only disassemble and assemble one side at a time, leaving the remaining side intact for reference.
- Only an MVAC-trained, EPA-certified automotive technician should service the air conditioning system or its components.

BRAKES — ANTI-LOCK BRAKE SYSTEM (ABS)

GENERAL INFORMATION

PRECAUTIONS

- Certain components within the ABS system are not intended to be serviced or repaired individually.
- Do not use rubber hoses or other parts not specifically specified for and ABS system. When using repair kits, replace all parts included in the kit. Partial or incorrect repair may lead to functional problems and require the replacement of components.
- Lubricate rubber parts with clean, fresh brake fluid to ease assembly. Do not use shop air to clean parts; damage to rubber components may result.
- Use only DOT 3 brake fluid from an unopened container.
- If any hydraulic component or line is removed or replaced, it may be necessary to bleed the entire system.
- A clean repair area is essential. Always clean the reservoir and cap thoroughly before removing the cap. The slightest amount of dirt in the fluid may plug an orifice and impair the system function. Perform repairs after components have been thoroughly cleaned; use only denatured alcohol to clean components. Do not allow ABS components to come into contact with any substance containing mineral oil; this includes used shop rags.
- The Anti-Lock control unit is a microprocessor similar to other computer units in the vehicle. Ensure that the ignition switch is **OFF** before removing or installing controller harnesses. Avoid static electricity discharge at or near the controller.
- If any arc welding is to be done on the vehicle, the control unit should be unplugged before welding operations begin.

SPEED SENSORS

REMOVAL & INSTALLATION

Front

See Figures 1 through 7.

1. Disconnect the cable from the negative battery terminal.

CAUTION

Wait at least 90 seconds after disconnecting the cable from the negative (-) battery terminal to disable the SRS system.

2. Remove the front wheel.
3. Remove the front skid control sensor wire LH.
    a. Disconnect the connector from the front speed sensor.
    b. Remove the bolt and harness clamp.
    c. Remove the nut and harness clamp.
    d. Remove the 2 bolts and 2 harness clamps.
    e. Disconnect the connector and then remove the bolt, harness clamp and sensor wire.
4. Remove the front speed sensor LH.
    a. Using a 5 mm hexagon wrench, remove the bolt and speed sensor from the knuckles.

➡Pull out the sensor while trying as much as possible not to rotate it.

5. Remove the front skid control sensor wire RH.
    a. Disconnect the connector from the front speed sensor.
    b. Remove the bolt and harness clamp.
    c. Remove the nut and harness clamp.
    d. Disconnect the connector.
    e. Detach the connector from the skid control sensor clamp.
    f. Remove the bolt and skid control sensor clamp.
    g. Remove the 2 bolts, 2 harness clamps and sensor wire.
6. Remove the front speed sensor RH.
    a. Using a 5 mm hexagon wrench, remove the bolt and speed sensor from the knuckle.

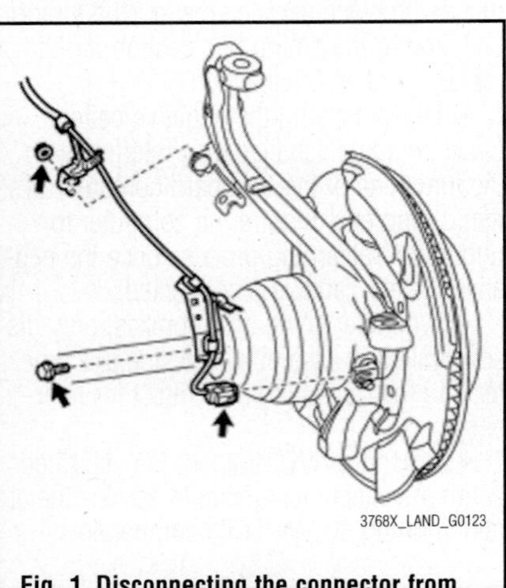

Fig. 1 Disconnecting the connector from the front speed sensor

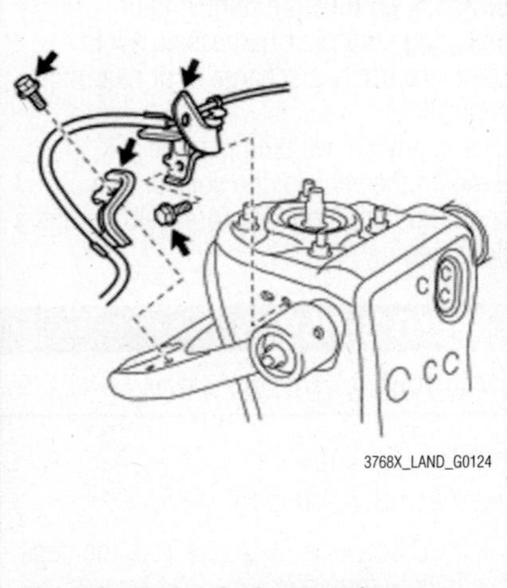

Fig. 2 Removing the harness clamps

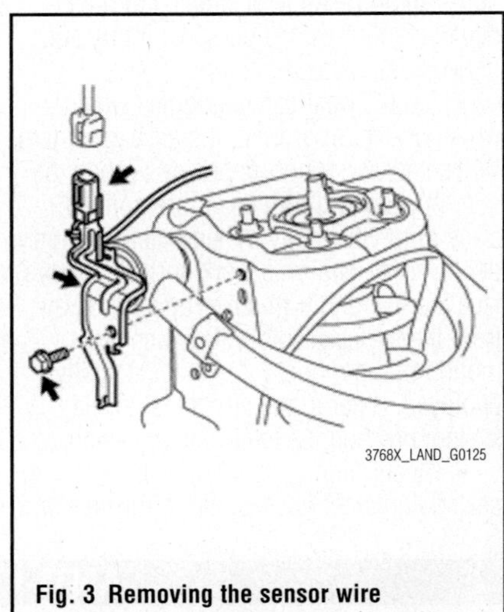

Fig. 3 Removing the sensor wire

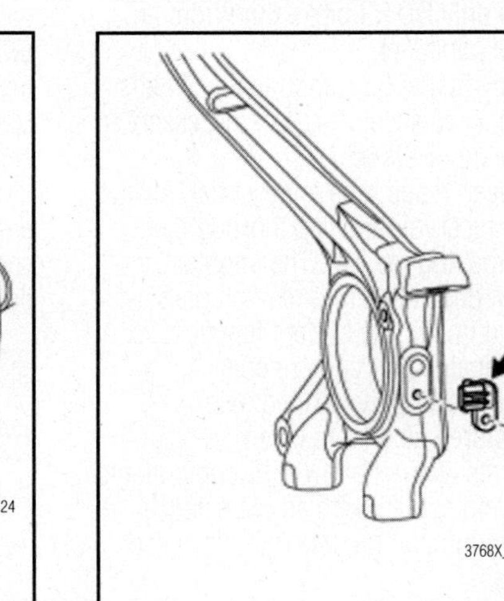

Fig. 4 Removing the LH speed sensor

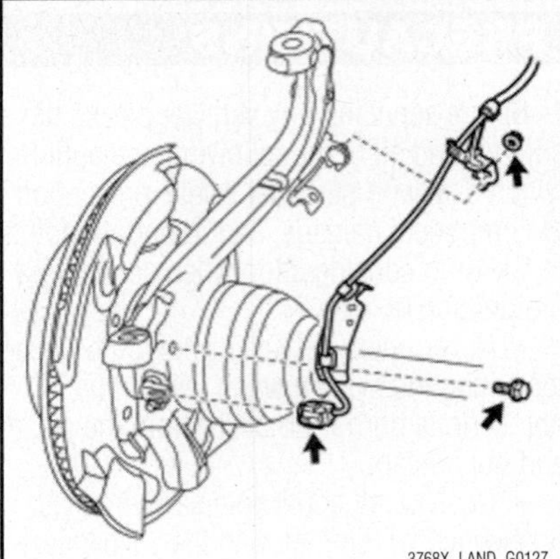

Fig. 5 Disconnecting the connector from the front speed sensor

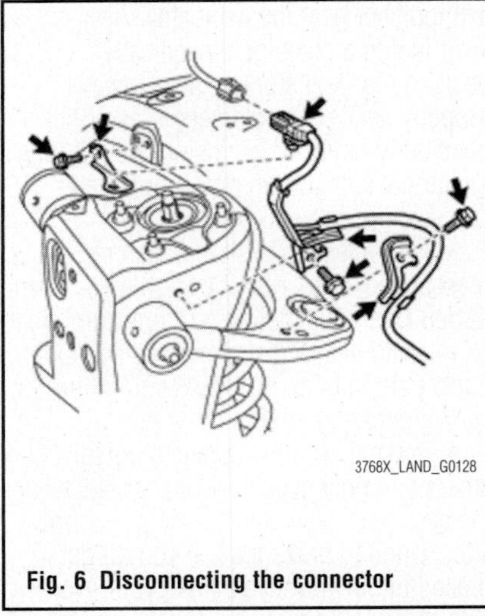

Fig. 6 Disconnecting the connector

➡Pull out the sensor while trying as much as possible not to rotate it.

To install:

7. Install the front speed sensor LH.
    a. Using a 5 mm hexagon wrench, install the speed sensor with the bolt. Tighten to 73 inch lbs. (8.3 Nm).

➡Make sure there are no pieces of iron or other foreign matter attached to the sensor tip.

➡While inserting the speed sensor into the knuckle hole, do not strike or damage the sensor tip

➡After installing the speed sensor, make sure there is no clearance or foreign matter between the sensor stay part and the knuckle.

➡Make sure there is no foreign matter attached to the speed sensor rotor.

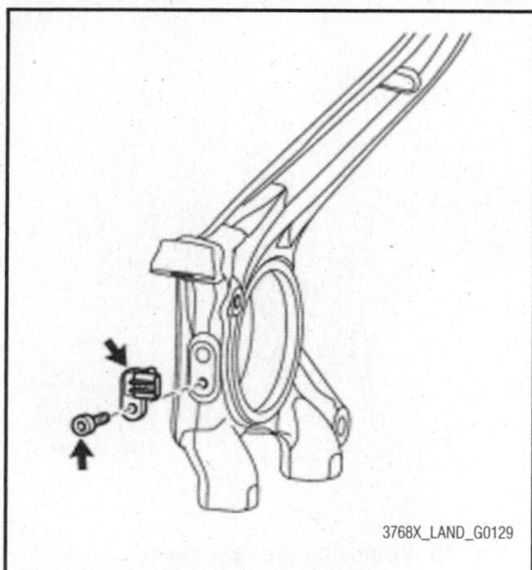
Fig. 7 Removing the front RH speed sensor

8. Install the front skid control sensor wire LH.
   a. Connect the connector.
   b. Install the harness clamp with the bolt. Tighten to 9 ft. lbs. (13 Nm).

➡When installing the clamp, do not twist the wire harness.

➡Make sure the clamp rotation stopper touches the installation position.

   c. Install the 2 harness clamps with the 2 bolts. Tighten to 9 ft. lbs. (13 Nm).

➡When installing the clamps, do not twist the wire harness.

➡Make sure the clamp rotation stopper touches the installation position.

   d. Install the harness clamp with the nut. Tighten to 9 ft. lbs. (13 Nm).

➡When installing the clamps, do not twist the wire harness.

➡Make sure the clamp rotation stopper touches the installation position.

   e. Install the harness clamp with the bolt. Tighten to 9 ft. lbs. (13 Nm).

➡Install the bracket so that the rotation stopper touches the knuckle.

   f. Connect the speed sensor connector.

➡Securely connect the connector.

9. Install the front RH speed sensor.
   a. Using a 5 mm hexagon wrench, install the speed sensor with the bolt. Tighten to 73 inch lbs. (8.3 Nm).

➡Make sure there are no pieces of iron or other foreign matter attached to the sensor tip.

➡While inserting the speed sensor into the knuckle hole, do not strike or damage the sensor tip

➡After installing the speed sensor, make sure there is no clearance or foreign matter between the sensor stay part and the knuckle.

➡Make sure there is no foreign matter attached to the speed sensor rotor.

10. Install the front skid control sensor wire RH.
   a. Install the skid control sensor clamp with the bolt (labeled A). Tighten to 44 inch lbs. (5 Nm).
   b. Attach the connector to the skid control sensor clamp and then connect the connector.
   c. Install the 2 harness clamps with the 2 bolts. Tighten to 9 ft. lbs. (13 Nm).

➡When installing the clamps, do not twist the wire harness.

➡Make sure the clamp rotation stopper touches the installation position.

   d. Install the harness clamp with the nut. Tighten to 9 ft. lbs. (13 Nm).

➡When installing the clamps, do not twist the wire harness.

➡Make sure the clamp rotation stopper touches the installation position.

   e. Install the harness clamp with the bolt. Tighten to 9 ft. lbs. (13 Nm).

➡Install the bracket so that the rotation stopper touches the knuckle.

   f. Connect the speed sensor connector.

➡securely connect the connector.

11. Install the front wheel. Tighten to 97 ft. lbs. (131 Nm).
12. Connect the cable to the negative battery terminal.
13. Check the speed sensor signal.

Rear

See Figures 8 through 18.

1. Disconnect the cable from the negative battery terminal.

CAUTION

Wait at least 90 seconds after disconnecting the cable from the negative (-) battery terminal to disable the SRS system.

2. Remove the rear wheel.
3. Remove the skid control sensor wire.
   a. Disconnect the speed sensor LH connector.
   b. Detach the 2 clamps and remove the bolt and sensor clamp.
   c. Detach the 5 clamps.
4. Detach the 3 clamps.
   a. Disconnect the speed sensor RH connector.
   b. Remove the bolt.
   c. Disconnect the connector, and remove the bolt and skid control sensor wire.
5. Remove the rear speed sensor LH.
   a. Remove the nut and speed sensor.

➡Pull out the sensor while trying as much as possible not to rotate it.

6. Remove the rear speed sensor RH.
   a. Remove the nut and speed sensor.

➡Pull out the sensor while trying as much as possible not to rotate it.

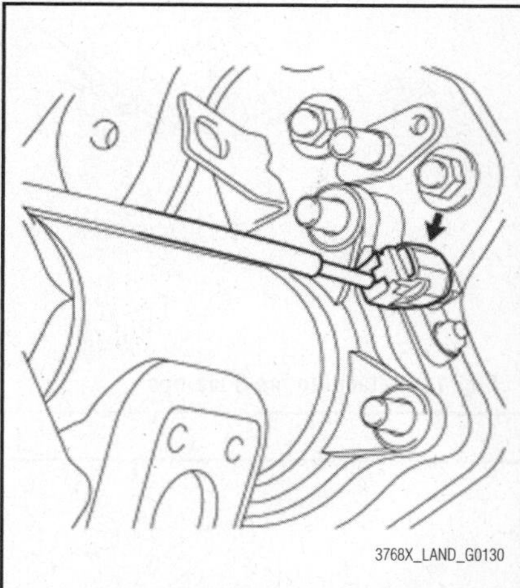
Fig. 8 Disconnecting the speed sensor LH connector

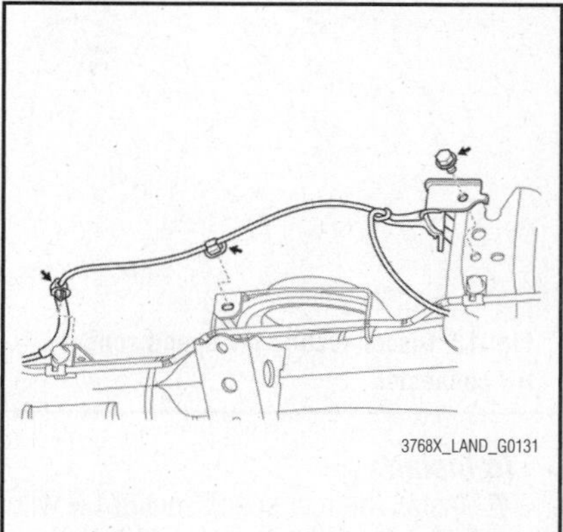
Fig. 9 Detaching the 2 clamps and removing the bolt and sensor clamp

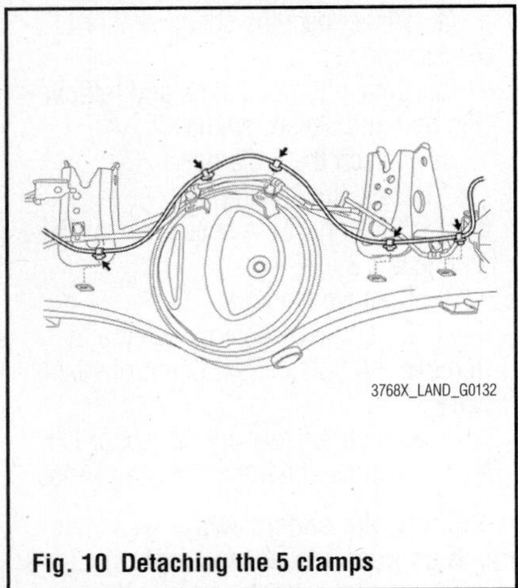

Fig. 10 Detaching the 5 clamps

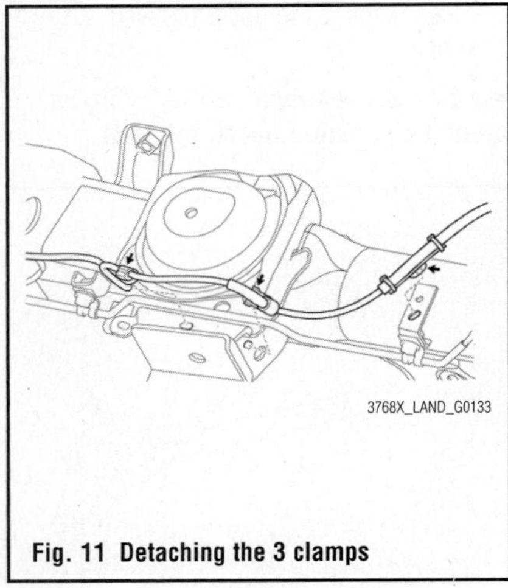

Fig. 11 Detaching the 3 clamps

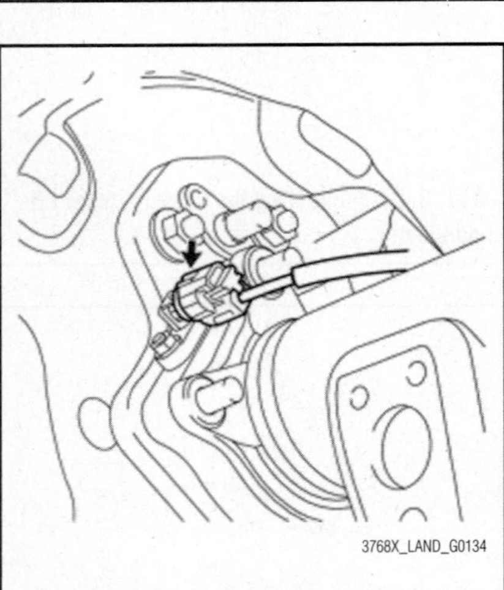

Fig. 12 Disconnecting the speed sensor RH connector

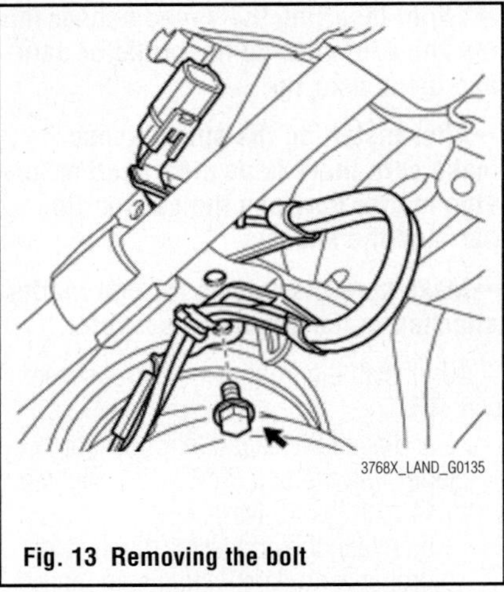

Fig. 13 Removing the bolt

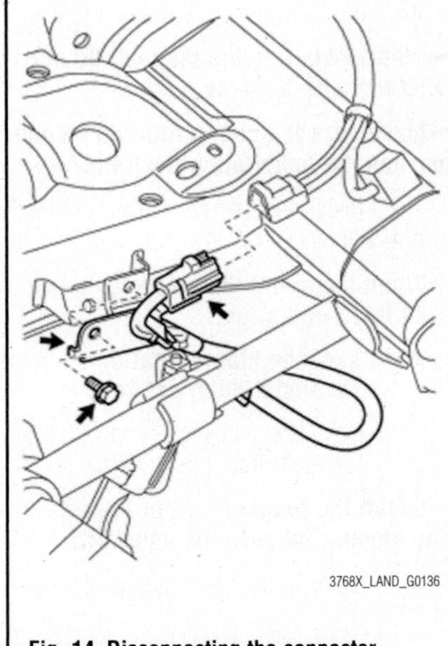

Fig. 14 Disconnecting the connector, and remove the bolt and skid control sensor wire

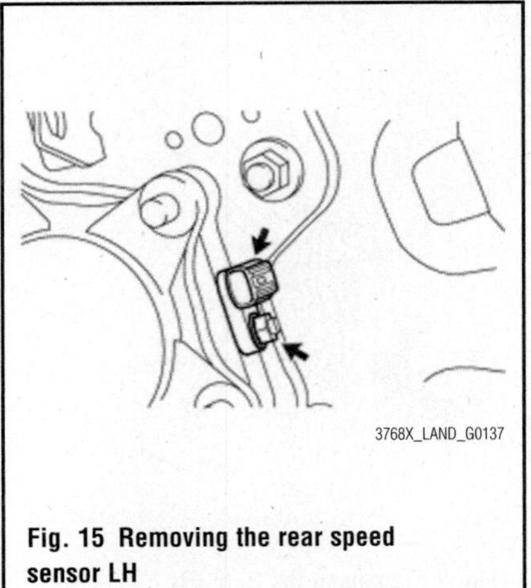

Fig. 15 Removing the rear speed sensor LH

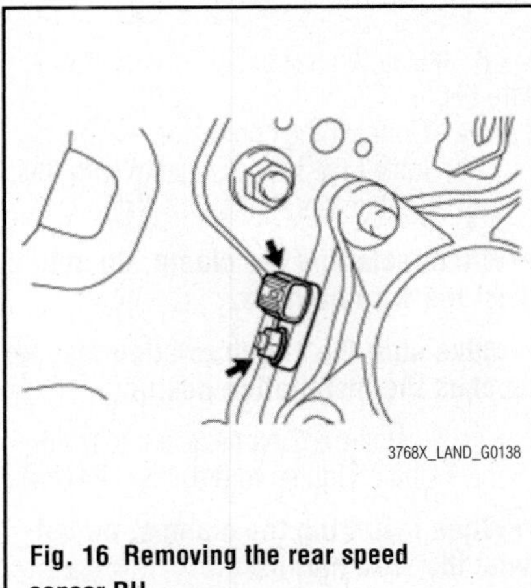

Fig. 16 Removing the rear speed sensor RH

➡While inserting the speed sensor into the knuckle hole, do not strike or damage the sensor tip.

➡After installing the speed sensor, make sure there is no clearance or foreign matter between the sensor stay part and the knuckle.

➡Make sure there is no foreign matter attached to the speed sensor rotor.

To install:

7. Install the rear speed sensor LH with the nut. Tighten to 73 inch lbs. (8.3 Nm).

➡Make sure there are no pieces of iron or other foreign matter attached to the sensor tip.

➡While inserting the speed sensor into the knuckle hole, do not strike or damage the sensor tip.

➡After installing the speed sensor, make sure there is no clearance or foreign matter between the sensor stay part and the knuckle.

➡Make sure there is no foreign matter attached to the speed sensor rotor.

8. Install the rear speed sensor RH with the nut. Tighten to 73 inch lbs. (8.3 Nm).

➡Make sure there are no pieces of iron or other foreign matter attached to the sensor tip.

9. Install the skid control sensor wire.
10. Install the sensor clamp with the bolt. Tighten to 9 ft. lbs. (13 Nm).

➡Make sure the clamp rotation stopper touches the installation position.

a. Connect the connector.

➡Securely insert the connector.

➡When connecting the connector, do not twist the wire harness.

11. Install the sensor clamp with the bolt. Tighten to 9 ft. lbs. (13 Nm).

➡Make sure the clamp rotation stopper touches the installation position.

a. Connect the connector.

➡Securely insert the connector.

➡When connecting the connector, do not twist the wire harness.

b. Connect the speed sensor RH connector.

➡Securely insert the connector.

➡When connecting the connector, do not twist the wire harness.

c. Attach the 3 clamps.

➡When attaching the clamps, do not twist the wire harness.

➡When attaching the clamps, securely insert them as shown in the illustration.

d. Attach the 5 clamps.

➡When attaching the clamps, do not twist the wire harness.

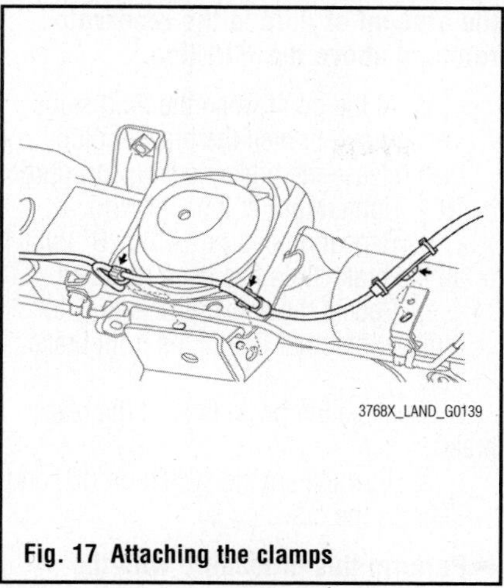

Fig. 17 Attaching the clamps

➡When attaching the clamps, securely insert them as shown in the illustration.

e. Attach the 2 clamps, and install the sensor clamp with the bolt. Tighten to 9 ft. lbs. (13 Nm).

f. Connect the speed sensor LH connector.

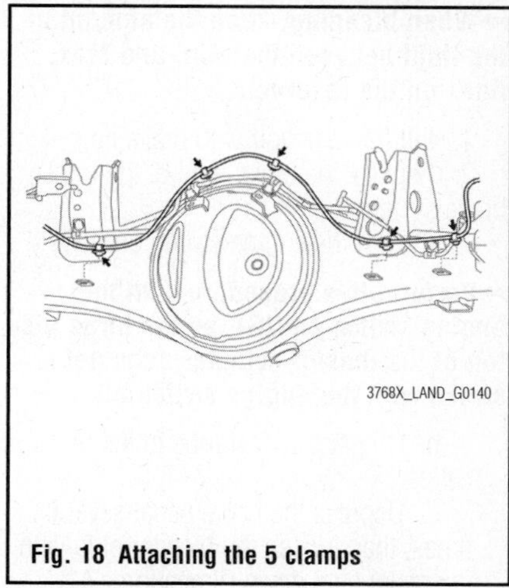

Fig. 18 Attaching the 5 clamps

➡Securely insert the connector.

➡When connecting the connector, do not twist the wire harness.

12. Install the rear wheel. Tighten to 97 ft. lbs. (131 Nm).
13. Connect the cable to the negative battery terminal.
14. Check the speed sensor signal.

BRAKES

BLEEDING THE BRAKE SYSTEM

BLEEDING PROCEDURE

BLEEDING PROCEDURE

➡If any work is done on the brake system or if air is suspected in the brake lines, bleed the air from the system. Bleeding Hydraulic Brake Booster is only possible with a Toyota proprietary scan system.

➡When bleeding, keep the amount of the fluid within the line of reservoir between Min. and Max. Do not let brake fluid remain on a painted surface. Wash it off immediately.

1. Before servicing the vehicle, refer to the precautions in the beginning of this section.
2. Check the fluid level in the reservoir after bleeding each wheel. Add DOT3 fluid, if necessary.
3. If the hydraulic brake booster was disassembled or if the reservoir becomes empty, bleeding Hydraulic Brake Booster is only possible with a Toyota proprietary scan system.
4. Bleeding the air from the hydraulic brake lines can be performed as follows:

a. Turn the ignition switch OFF; depress the brake pedal 40 times or more.

b. Turn the ignition switch to the ON position and start the brake booster pump. The pump stops after approximately 30 to 40 seconds.

➡If the pump does not operate as specified, repeat the above and recheck the operating time.

c. Holding the brake pedal depressed, bleed the right and left rear brake cylinders.

d. Turn the ignition switch ON, depress the brake pedal 20 times or more.

5. Bleeding Front Brake Lines:

a. Turn the ignition switch to the ON position and wait until the pump motor has stopped.

b. Connect the vinyl tube to the brake caliper.

c. Depress the brake pedal several times, then loosen the bleeder plug with the pedal held down.

d. At the point when the fluid stops coming out, tighten the bleeder plug, 8 ft. lbs. (11 Nm) then release the brake pedal.

e. Repeat procedure until all the air in the fluid has been bled out.

f. Repeat the above procedures to bleed the other brake line.

6. Bleeding Rear Brake Lines:

a. Turn the ignition switch to the ON position and depress the brake pedal.

b. Connect the vinyl tube to the brake caliper.

c. Loosen the bleeder plug and release air.

➡Brake fluid is sent through the pump, so keep the brake pedal depressed until the air is completely bled out.

d. When the air is completely bled out of the brake fluid through the bleeder plug, tighten the bleeder plug to 8 ft. lbs. (11 Nm) then release.

e. Repeat the above procedures to bleed the other brake line.

7. Bleeding Hydraulic Brake Booster is only possible with a Toyota proprietary scan system.

FLUID FILL PROCEDURE

➡Wash off brake fluid immediately if it comes into contact with a painted surface.

➡If the brake pedal is depressed with the reservoir cap removed, then brake fluid will spatter.

➡If any work is done on the brake system or if air in the brake lines is suspected, bleed the air from the system.

➡When bleeding, keep the amount of the fluid between the Min. and Max. lines on the reservoir.

1. Fill the reservoir with brake fluid.
2. Replace the brake fluid of the front brake system.
   a. Turn the engine switch on (IG).

➡Perform this procedure with the engine switch on (IG), as the large piston of the master cylinder does not stroke with the engine switch off.

   b. Connect a vinyl tube to the brake caliper.
   c. Depress the brake pedal several times, then loosen the bleeder plug with the pedal held down (procedure "A").

➡While performing the brake fluid replacing procedure, make sure that the amount of fluid in the reservoir remains above the MIN line.

   d. At the point when the fluid stops coming out, tighten the bleeder plug, then release the brake pedal (procedure "B"). Tighten to 8 ft. lbs. (11 Nm).
   e. Repeat procedure "A" to "B" until all the brake fluid has been replaced.
   f. Repeat the above procedures to replace the other line of the front brake fluid.
3. Replace the brake fluid of the rear brake system.
   a. Turn the engine switch on (IG) and depress the brake pedal.

➡Perform this procedure with the engine switch on (IG), as the large piston of the master cylinder does not stroke with the engine switch off.

   b. Connect a vinyl tube to the brake caliper.
   c. Loosen the bleeder plug and replace the brake fluid.

➡While performing the brake fluid replacing procedure, make sure that the amount of fluid in the reservoir remains above the MIN line.

➡Brake fluid is sent through the pump, so keep the brake pedal depressed until all the brake fluid has been replaced.

   d. When all the brake fluid has been replaced, tighten the bleeder plug. Tighten to 8 ft. lbs. (11 Nm).
   e. Repeat the above procedures to replace the other line of the rear brake fluid.

CAUTION

Dust and dirt accumulating on brake parts during normal use may contain asbestos fibers from production or aftermarket brake linings. Breathing excessive concentrations of asbestos fibers can cause serious bodily harm. Exercise care when servicing brake parts. Do not sand or grind brake lining unless equipment used is designed to contain the dust residue. Do not clean brake parts with compressed air or by dry brushing. Cleaning should be done by dampening the brake components with a fine mist of water, then wiping the brake components clean with a dampened cloth. Dispose of cloth and all residue containing asbestos fibers in an impermeable container with the appropriate label. Follow practices prescribed by the Occupational Safety and Health Administration (OSHA) and the Environmental Protection Agency (EPA) for the handling, processing, and disposing of dust or debris that may contain asbestos fibers.

BRAKE CALIPER

REMOVAL & INSTALLATION

See Figures 19 and 20.

➡Use the same procedures for the LH side and RH side.

➡The procedures listed below are for the LH side.

1. Remove the front wheel.
2. Drain the brake fluid.

➡Wash brake fluid off immediately if it is spilled on any painted surface.

3. Remove the brake pads.
4. Disconnect the front flexible hose.
   a. Remove the union bolt and gasket, and then disconnect the flexible hose.
5. Remove the 2 bolts and disc brake caliper from the knuckle.

To install:

6. Install the disc brake caliper assembly.
   a. Install the disc brake caliper with 2 new bolts. Tighten to 73 ft. lbs. (99 Nm).
7. Connect the front flexible hose.

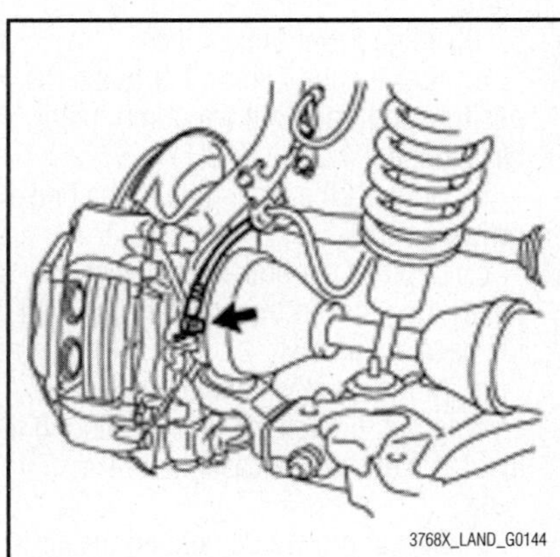

Fig. 19 Disconnecting the flexible hose

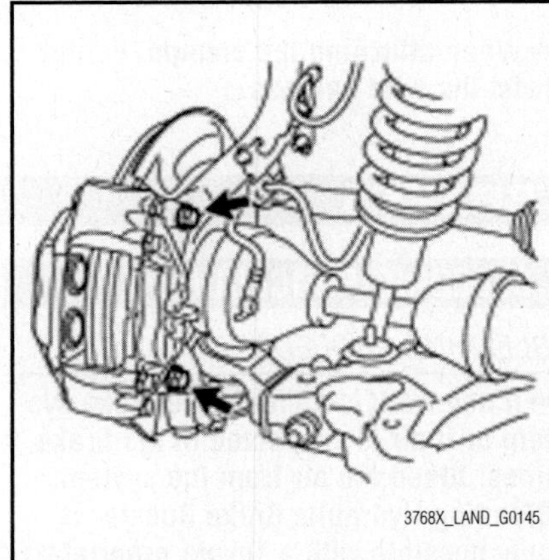

Fig. 20 Removing the disc brake caliper from the knuckle

   a. Install a new gasket and connect the flexible hose with the union bolt. Tighten to 22 ft. lbs. (30 Nm).
8. Fill the reservoir with brake fluid.
9. Bleed the brake line.
10. Check the brake fluid level in the reservoir.
11. Inspect for brake fluid leaks.
12. Install the front wheel. Tighten to 97 ft. lbs. (131 Nm).

DISC BRAKE PADS

REMOVAL & INSTALLATION

See Figures 21 through 23.

➡Use the same procedures for the LH side and RH side.

➡The procedures listed below are for the LH side.

1. Remove the front wheel.
2. Drain the brake fluid.

➡Wash brake fluid off immediately if it is spilled on any painted surface.

3. Remove the pin hold clip.

➡The pin hold clip can be reused if it has sufficient rebound; no deformation or wear; and has had all rust, dirt and foreign matter cleaned off.

4. Remove the 2 hold pins.
   a. Remove the front disc brake anti-rattle spring.

➡The anti-rattle spring can be reused if it has sufficient rebound; no deformation, cracks or wear; and has had all rust, dirt and foreign matter cleaned off.

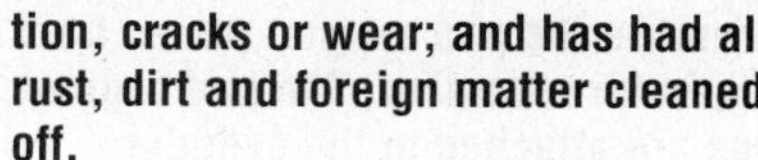

5. Remove the 2 pads from the disc brake caliper.
6. Remove the No. 1 anti-squeal shims from the pads.

To install:

7. Install the No. 1 anti-squeal shims to the pads.
8. Install the 2 pads to the disc brake caliper.
9. Install the front disc brake anti-rattle spring between the 2 pads.

➡The anti-rattle spring can be reused if it has sufficient rebound; no deformation, cracks or wear; and has had all rust, dirt and foreign matter cleaned off.

10. Install the 2 hole pins.
11. Install the pin hold clip.

➡The pin hold clip can be reused if it has sufficient rebound; no deformation or wear; and has had all rust, dirt and foreign matter cleaned off.

12. Fill the reservoir with brake fluid.
13. Bleed the brake line.
14. Check the brake fluid level in the reservoir.
15. Inspect for brake fluid leaks.
16. Install the front wheel. Tighten to 97 ft. lbs. (131 Nm).

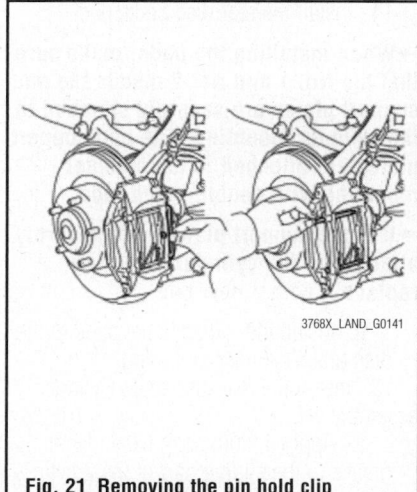

3768X_LAND_G0141

Fig. 21 Removing the pin hold clip

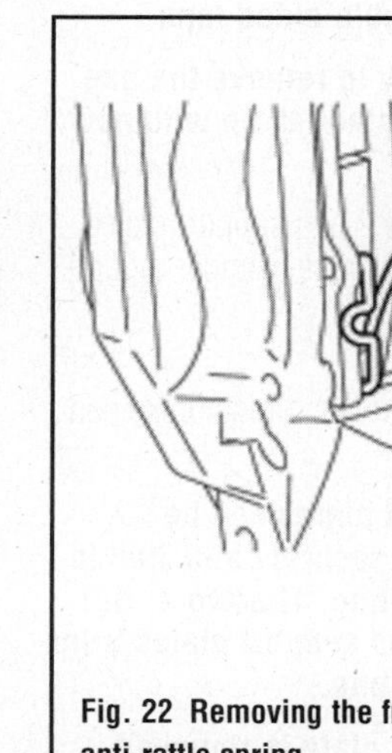

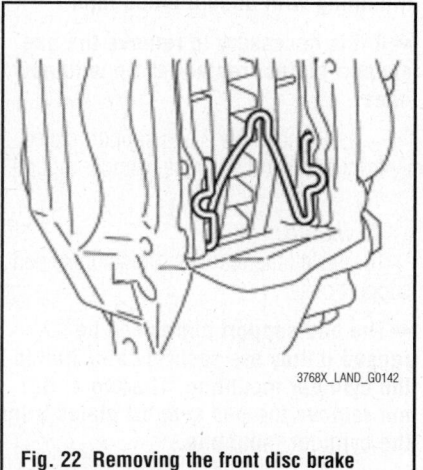

Fig. 22 Removing the front disc brake anti-rattle spring

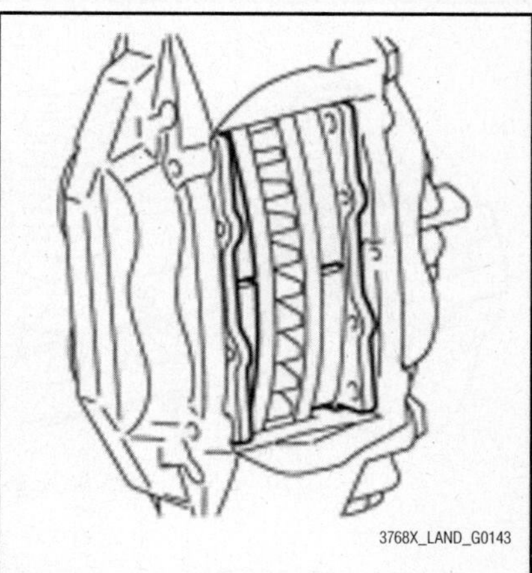

Fig. 23 Removing the 2 pads from the disc brake caliper

BRAKES

REAR DISC BRAKES

⁂ CAUTION

Dust and dirt accumulating on brake parts during normal use may contain asbestos fibers from production or aftermarket brake linings. Breathing excessive concentrations of asbestos fibers can cause serious bodily harm. Exercise care when servicing brake parts. Do not sand or grind brake lining unless equipment used is designed to contain the dust residue. Do not clean brake parts with compressed air or by dry brushing. Cleaning should be done by dampening the brake components with a fine mist of water, then wiping the brake components clean with a dampened cloth. Dispose of cloth and all residue containing asbestos fibers in an impermeable container with the appropriate label. Follow practices prescribed by the Occupational Safety and Health Administration (OSHA) and the Environmental Protection Agency (EPA) for the handling, processing, and disposing of dust or debris that may contain asbestos fibers.

BRAKE CALIPER

REMOVAL & INSTALLATION

See Figures 24 and 25.

➡Use the same procedures for the LH side and RH side.

➡The procedures listed below are for the LH side.

1. Remove the rear wheel.
2. Drain the brake fluid.

➡Wash brake fluid off immediately if it is spilled on any painted surface.

3. Disconnect the rear flexible hose LH.
   a. Remove the union bolt and gasket, and then disconnect the flexible hose from the rear disc brake cylinder. Use a container to catch brake fluid as it drains out.
4. Remove the rear disc brake cylinder assembly LH.
   a. Remove the 2 slide pins.
   b. Remove the cylinder from the rear disc brake cylinder mounting.
5. Remove the rear disc brake pad.
   a. Remove the 2 brake pads from the rear disc brake cylinder mounting.

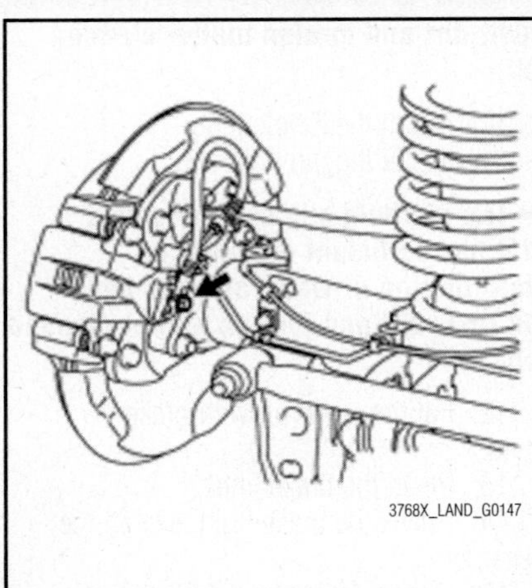

Fig. 24 Disconnecting the rear flexible hose LH

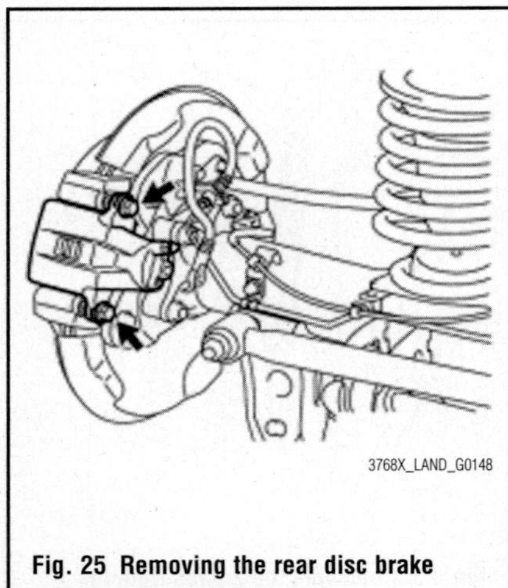

Fig. 25 Removing the rear disc brake cylinder assembly LH

➡When removing the pads, make sure that the No. 1 and No. 2 disc brake pad support plates remain securely attached to the cylinder mounting. The pad support plates are attached to the cylinder mounting with double-sided tape.

➡If a pad support plate is not securely attached to the cylinder mounting, replace it with a new one.

6. Remove the rear disc brake anti-squeal shims from the disc brake pads.
7. Remove the rear disc brake pad wear indicator plate from the inner disc brake pad.
8. Remove the rear No. 1 disc brake pad support plate.

➡The pad support plates can be reused if they are securely attached to the cylinder mounting. Therefore, do not remove the pad support plates from the cylinder mounting. The pad support plates are attached to the cylinder mounting with double-sided tape.

➡If it is necessary to remove the pad support plates, replace them with new ones.

a. Remove the 2 pad support plates from the rear disc brake cylinder mounting.

9. Remove the No. 2 disc brake pad support plate.

➡The pad support plates can be reused if they are securely attached to the cylinder mounting. Therefore, do not remove the pad support plates from the cylinder mounting. The pad support plates are attached to the cylinder mounting with double-sided tape.

➡If it is necessary to remove the pad support plates, replace them with new ones.

a. Remove the 2 pad support plates from the rear disc brake cylinder mounting.

To install:

10. Install the rear No. 1 disc brake pad support plate.

➡The pad support plates can be reused if they are securely attached to the cylinder mounting. Therefore, do not remove the pad support plates from the cylinder mounting.

➡If a pad support plate is removed, replace it with a new one that is supplied with double-sided tape already attached.

➡Make sure to clean the cylinder mounting contact surface before attaching the pad support plates.

a. Clean the cylinder mounting surface where the pad support plates will be attached.

b. Remove the peeling paper from the double-sided tape, and install 2 new pad support plates to the cylinder mounting.

11. Install the rear No. 2 disc brake pad support plate.

➡The pad support plates can be reused if they are securely attached to the cylinder mounting. Therefore, do not remove the pad support plates from the cylinder mounting.

➡If a pad support plate is removed, replace it with a new one that is supplied with double-sided tape already attached.

➡Make sure to clean the cylinder mounting contact surface before attaching the pad support plates.

a. Clean the cylinder mounting surface where the pad support plates will be attached.

b. Remove the peeling paper from the double-sided tape, and install 2 new pad support plates to the cylinder mounting.

12. Install the pad wear indicator plate to the inner disc brake pad.
13. Install the rear disc brake anti-squeal shims to the brake pads.

➡There should be no oil or grease on the friction surfaces of the brake pads and rear disc.

14. Install the rear disc brake pad.

➡When installing the pads, make sure that the No. 1 and No. 2 disc brake pad support plates are securely attached to the cylinder mounting. The pad support plates are attached to the cylinder mounting with double-sided tape.

➡If a pad support plate is not securely attached to the cylinder mounting, replace it with a new one.

a. Install the 2 disc brake pads to the disc brake cylinder mounting.

15. Install the rear disc brake cylinder assembly LH.

a. Apply lithium soap base glycol grease to the sliding part of the 2 cylinder slide pins.

b. Install the cylinder with the 2 cylinder slide pins. Tighten to 65 ft. lbs. (88 Nm).

16. Connect the rear flexible hose LH.

a. Install a new gasket and connect the flexible hose with the union bolt. Tighten to 22 ft. lbs. (30 Nm).

17. Fill the reservoir with brake fluid.
18. Bleed the brake line.
19. Check the brake fluid level in the reservoir.
20. Inspect for brake fluid leaks.
21. Adjust the parking brake lever travel.
22. Install the rear wheel. Tighten to 97 ft. lbs. (131 Nm).

DISC BRAKE PADS

REMOVAL & INSTALLATION

Refer to BRAKE CALIPER for removal procedure.

PARKING BRAKE SHOES

REMOVAL & INSTALLATION

See Figures 26 through 30.

1. Remove the rear brake disc.
2. Remove the parking brake shoe return spring.
   a. Using the special tool (09703-30011), remove the return spring.
3. Remove the No. 1 parking brake shoe assembly LH.
   a. Using special tool (09718-00011), remove the shoe hold down spring cup, compression spring and shoe hold down spring pin.
   b. Disconnect the tension spring.
   c. Remove the No. 1 parking brake shoe and shoe adjuster screw set.

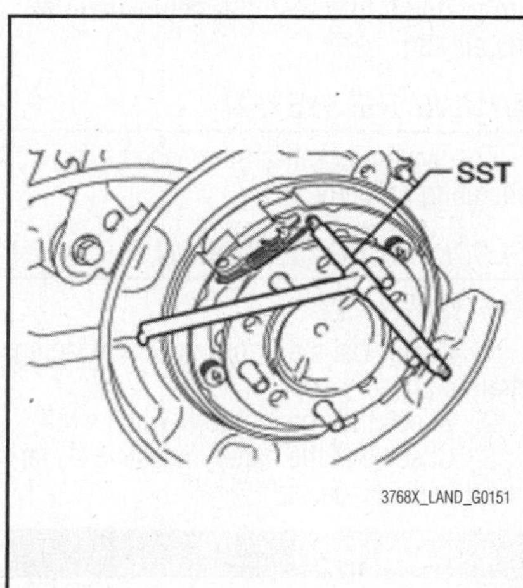

Fig. 26 Removing the parking brake shoe return spring

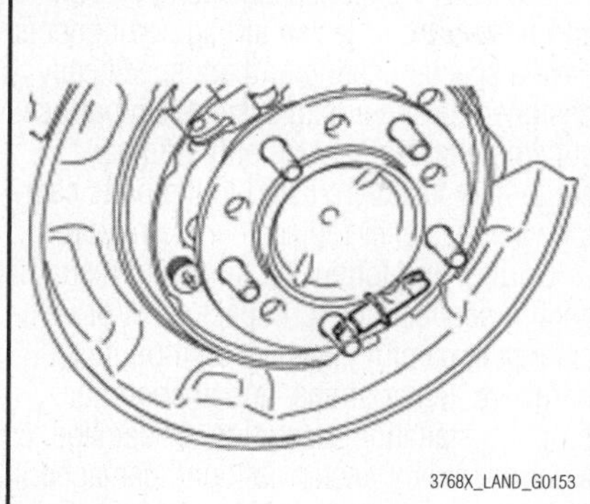

Fig. 28 Removing the No. 1 parking brake shoe and shoe adjuster screw set

4. Remove the No. 2 parking brake shoe assembly LH.
   a. Using the special tool (09718-00011), remove the shoe hold down spring cup, compression spring and shoe hold down spring pin.
   b. Remove the No. 2 parking brake shoe.

To install:

5. Install the No. 2 parking brake shoe assembly LH.
   a. Apply high temperature grease to the areas of the backing plate that contact the shoe.
   b. Using the special tool (09718-00011), install the No. 2 parking brake shoe with the shoe hold down spring pin, compression spring and shoe hold down spring cup.
6. Install the No. 1 parking brake show assembly LH.
   a. Apply high temperature grease to the thread and all joining areas of the parking brake shoe adjuster screw set.
   b. Set the No. 1 parking brake shoe and shoe adjuster screw set in place.
   c. Connect the tension spring.
   d. Using the special tool (09718-00011), install the No. 1 parking brake shoe with the shoe hold down spring pin, compression spring and shoe hold down spring cup.
7. Install the parking brake shoe return spring.
   a. Using the special tool (09703-30011), install the shoe return spring.
8. Check the parking brake installation.
   a. Check that each part is installed properly.
9. Install the rear disc.
10. Check the parking brake lever travel.
11. Adjust the parking brake lever travel.
12. Install the rear wheel LH. Tighten to 97 ft. lbs. (131 Nm).
13. Drive the vehicle for approximately 0.25 miles (400 m) under the following conditions to settle the parking brake shoe and disc.
    a. The vehicle speed is approximately 31 mph (50 km/h) and the vehicle is on a safe, level and dry road.
    b. The parking brake lever is being pulled with a force of 33.7 lbs. (150 N).
14. Repeat the procedure above 2 or 3 times.

➡Set a 5 minute interval between each procedure to prevent the brake assembly from overheating.

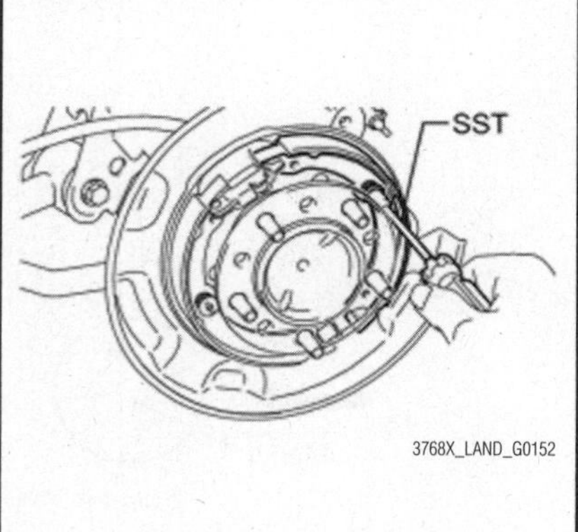

Fig. 27 Removing the shoe hold down spring cup, compression spring and shoe hold down spring pin

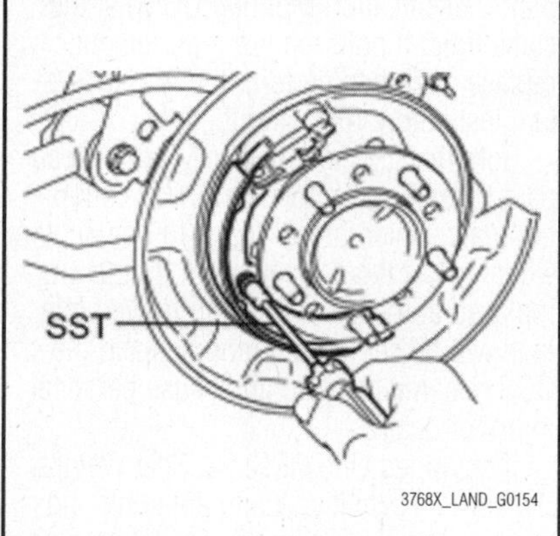

Fig. 29 Removing the No. 2 parking brake shoe assembly LH

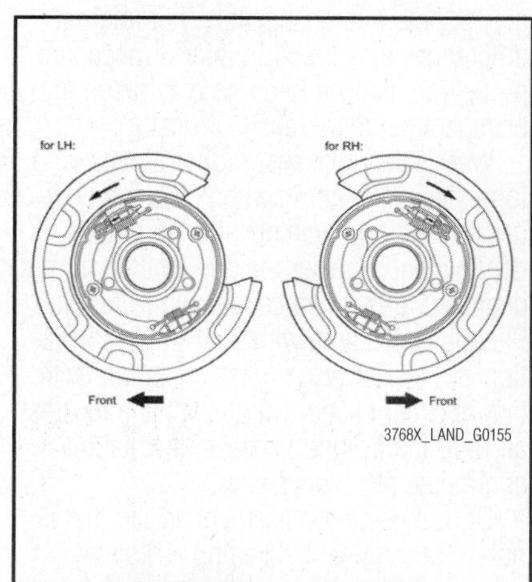

Fig. 30 Checking the parking brake assembly installation

GENERAL INFORMATION

CAUTION

These vehicles are equipped with an air bag system. The system must be disarmed before performing service on, or around, system components, the steering column, instrument panel components, wiring and sensors. Failure to follow the safety precautions and the disarming procedure could result in accidental air bag deployment, possible injury and unnecessary system repairs.

SERVICE PRECAUTIONS

Disconnect and isolate the battery negative cable before beginning any airbag system component diagnosis, testing, removal, or installation procedures. Allow system capacitor to discharge for two minutes before beginning any component service. This will disable the airbag system. Failure to disable the airbag system may result in accidental airbag deployment, personal injury, or death.

Do not place an intact undeployed airbag face down on a solid surface. The airbag will propel into the air if accidentally deployed and may result in personal injury or death.

When carrying or handling an undeployed airbag, the trim side (face) of the airbag should be pointing away from the body to minimize possibility of injury if accidental deployment occurs. Failure to do this may result in personal injury or death.

Replace airbag system components with OEM replacement parts. Substitute parts may appear interchangeable, but internal differences may result in inferior occupant protection. Failure to do so may result in occupant personal injury or death.

Wear safety glasses, rubber gloves, and long sleeved clothing when cleaning powder residue from vehicle after an airbag deployment. Powder residue emitted from a deployed airbag can cause skin irritation. Flush affected area with cool water if irritation is experienced. If nasal or throat irritation is experienced, exit the vehicle for fresh air until the irritation ceases. If irritation continues, see a physician.

Do not use a replacement airbag that is not in the original packaging. This may result in improper deployment, personal injury, or death.

The factory installed fasteners, screws and bolts used to fasten airbag components have a special coating and are specifically designed for the airbag system. Do not use substitute fasteners. Use only original equipment fasteners listed in the parts catalog when fastener replacement is required.

During, and following, any child restraint anchor service, due to impact event or vehicle repair, carefully inspect all mounting hardware, tether straps, and anchors for proper installation, operation, or damage. If a child restraint anchor is found damaged in any way, the anchor must be replaced. Failure to do this may result in personal injury or death.

Deployed and non-deployed airbags may or may not have live pyrotechnic material within the airbag inflator.

Do not dispose of driver/passenger/curtain airbags or seat belt tensioners unless you are sure of complete deployment. Refer to the Hazardous Substance Control System for proper disposal.

Dispose of deployed airbags and tensioners consistent with state, provincial, local, and federal regulations.

After any airbag component testing or service, do not connect the battery negative cable. Personal injury or death may result if the system test is not performed first.

If the vehicle is equipped with the Occupant Classification System (OCS), do not connect the battery negative cable before performing the OCS Verification Test using the scan tool and the appropriate diagnostic information. Personal injury or death may result if the system test is not performed properly.

Never replace both the Occupant Restraint Controller (ORC) and the Occupant Classification Module (OCM) at the same time. If both require replacement, replace one, then perform the Airbag System test before replacing the other.

Both the ORC and the OCM store Occupant Classification System (OCS) calibration data, which they transfer to one another when one of them is replaced. If both are replaced at the same time, an irreversible fault will be set in both modules and the OCS may malfunction and cause personal injury or death.

If equipped with OCS, the Seat Weight Sensor is a sensitive, calibrated unit and must be handled carefully. Do not drop or handle roughly. If dropped or damaged, replace with another sensor. Failure to do so may result in occupant injury or death.

If equipped with OCS, the front passenger seat must be handled carefully as well. When removing the seat, be careful when setting on floor not to drop. If dropped, the sensor may be inoperative, could result in occupant injury, or possibly death.

If equipped with OCS, when the passenger front seat is on the floor, no one should sit in the front passenger seat. This uneven force may damage the sensing ability of the seat weight sensors. If sat on and damaged, the sensor may be inoperative, could result in occupant injury, or possibly death.

DISARMING THE SYSTEM

To avoid personal injury when working on vehicles equipped with an air bag, the negative battery cable must be disconnected and at least 90 seconds must elapse before working on the system. Failure to do so may result in deployment of the air bag.

ARMING THE SYSTEM

The system is self-arming when it is operating properly.

CLOCKSPRING CENTERING

See Figure 31.

1. Check that the front wheels are facing straight ahead.
2. Check that the ignition switch is off.
3. Check that the battery negative (-) terminal is disconnected

CAUTION

After removing the terminal, wait for at least 90 seconds before starting the operation.

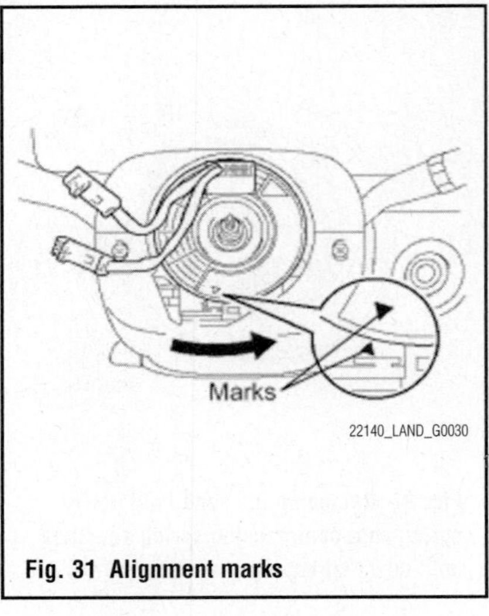

Fig. 31 Alignment marks

4. Rotate the spiral cable clockwise slowly by hand until it feels firm.

➡**Do not turn the spiral cable by the airbag wire harness**

5. Rotate the spiral cable counterclockwise approximately 2.5 turns to align the marks

➡**The spiral cable will rotate approximately 2.5 turns to both the left and right from the center**

DRIVE TRAIN

AUTOMATIC TRANSMISSION FLUID

DRAIN AND REFILL

See Figures 32 through 35.

➡**Maintain the vehicle in a horizontal position while adjusting the fluid level.**

1. Remove the drain plug and gasket, and drain the ATF.
2. Install a new gasket and the drain plug. Tighten to 15 ft. lbs. (20 Nm).
3. Using compressed air or the equivalent, blow dust off the thermostat cap to clean it.
4. Using a screwdriver, push in the shaft of the thermostat 0.217-0.276 inch (5.5-7.0 mm). Push in the shaft until the screwdriver contacts the step inside the cap.
5. With the shaft of the thermostat pressed, insert a pin into a hole on the side of the thermostat cap. Insert the pin until it passes through the hole on the other side of the thermostat cap to fix the shaft in place. The pin diameter is 0.0394-0.0709 inch (1.0-1.8 mm).
6. Fill the transmission with ATF.
7. Remove the refill plug and overflow plug.

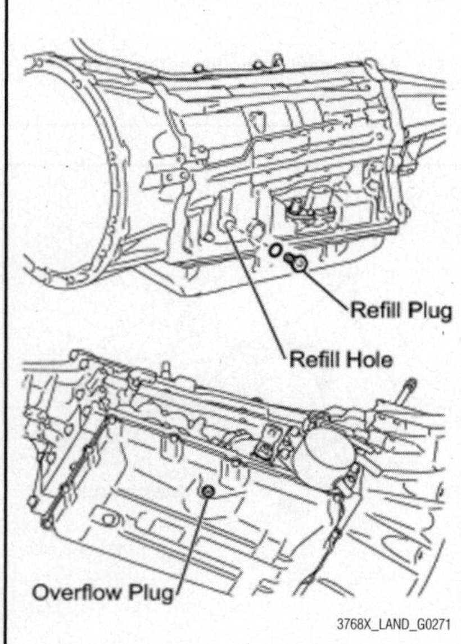

Fig. 32 Locating the refill and overflow plugs

8. Fill the transmission through the refill hole until fluid begins to trickle out of the overflow tube.
9. Reinstall the overflow plug.
10. Fill the transmission with the amount of fluid listed in the table below.
11. Install the refill plug.

➡**If you cannot fill the listed amount of fluid, do the following:**

a. Start the engine and idle it.

➡**Check that electrical systems such as the air conditioning system, audio system and lighting system are off.**

b. Move the shift lever through the entire gear range to circulate the fluid.
c. Wait for 30 seconds with the engine idling.
d. Stop the engine.
e. Remove the refill plug.
f. Fill the transmission with the remaining fluid until the amount in the table has been reached.
g. Install the refill plug.

12. Check the fluid temperature.

➡**The ATF temperature can be confirmed by using the Techstream.**

a. Turn the ignition switch off.
b. Connect the Techstream to the DLC3.
c. Turn the ignition switch to ON.
d. Enter the following menus: Powertrain / Engine and ECT / Data List / A/T Oil Temperature 1.
e. Check the ATF temperature.

➡**If the ATF temperature is above 118° F (48° C), turn the ignition switch off and wait until the fluid temperature drops to between 109 and 118° F (43 and 48° C).**

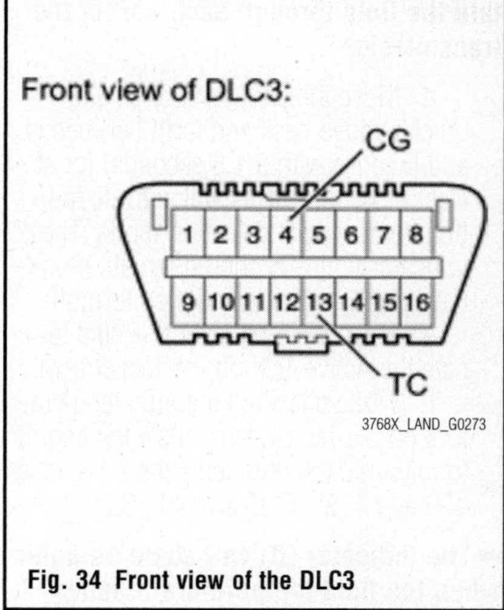

Fig. 34 Front view of the DLC3

13. Check the fluid level.

➡**It is necessary to change to temperature detection mode in order to idle the vehicle appropriately.**

➡**The ATF temperature must be between 109 and 118° F (43 and 48° C) to accurately check the fluid level.**

a. Enter the following menus: Powertrain / Engine and ECT / Active Test / Connect the TC and TE1.
If the condition is correct the indicator lights on the combination meter blink.

➡**When not using the Techstream, using SST, connect terminals 13 (TC) and 4 (CG) of the DLC3.**

b. Start the engine and idle it.

Repair	Fill Amount
Transmission pan and drain plug removal	2.2 qts. (2.1 L)
Transmission valve body removal	5.0 qts. (4.7 L)
Torque converter removal	6.4 qts. (6.1 L)

3768X_LAND_G0272

Fig. 33 Transmission fluid reference capacities

➡**Check that electrical systems such as the air conditioning system, audio system and lighting system are off.**

c. Slowly move the shift lever from P to S, then change the gear from 1st to 6th. Then return the shift lever to P.

➡**Slowly move the shift lever to circulate the fluid through each part of the transmission.**

d. Move the shift lever to D, and quickly move back and forth between N and D (once within 1.5 seconds) for at least 6 seconds. This will activate the fluid temperature detection mode. The indicator light (D) should remain illuminated for 2 seconds and then turn off.

e. Move the shift lever to P and terminate the Active Test on the Techstream.

f. While checking the fluid temperature on the Techstream, allow the engine to idle until the fluid temperature reaches 109 and 118° F (43 and 48° C).

➡**The indicator (D) will come on again when the fluid temperature reaches 109° F (43° C) and will blink if it exceeds 118° F (48° C).**

g. Remove the overflow plug with the engine idling.

h. Wait until the overflow slows to a trickle.

➡**If the fluid does not overflow, perform the following: Remove the refill plug. Refill the transmission through the refill hole until fluid begins to trickle out of the overflow tube. Wait until the overflow slows to a trickle.**

i. Install a new gasket and overflow plug. Tighten to 15 ft. lbs. (20 Nm).

j. Install a new gasket and refill plug. Tighten to 29 ft. lbs. (39 Nm).

14. Turn off the ignition switch.
15. Disconnect the Techstream from the DLC3.
16. Push the shaft with a screwdriver and remove the pin from the thermostat.

➡**Make sure the shaft of the thermostat is protruding from the hole on the cap.**

17. Inspect for automatic transmission fluid leaks.

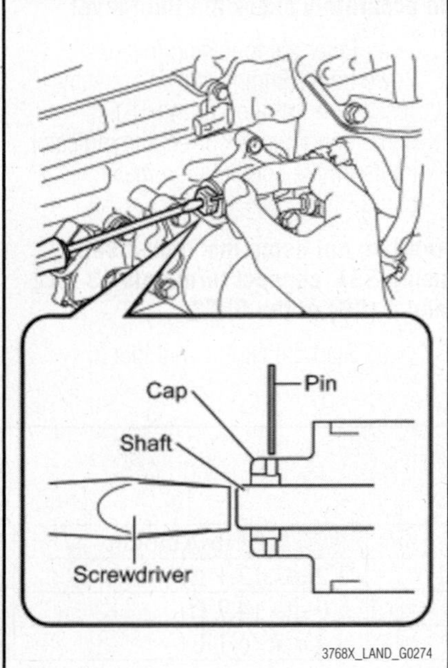

Fig. 35 Removing the pin from the thermostat

TRANSFER CASE ASSEMBLY

REMOVAL & INSTALLATION

See Figure 36.

1. Drain the transfer oil.
2. Remove the automatic transmission with the transfer assembly.
3. Remove the transfer assembly.

a. Remove the 8 transfer adaptor rear mounting bolts.

b. Pull the transfer straight up and remove it from the transmission.

➡**Take care not to damage the adaptor oil seal with the transfer input shaft spline.**

To install:

4. To install, reverse the removal procedure. Tighten the transfer bolts to 30 ft. lbs. (40 Nm). Check for transfer oil leaks.

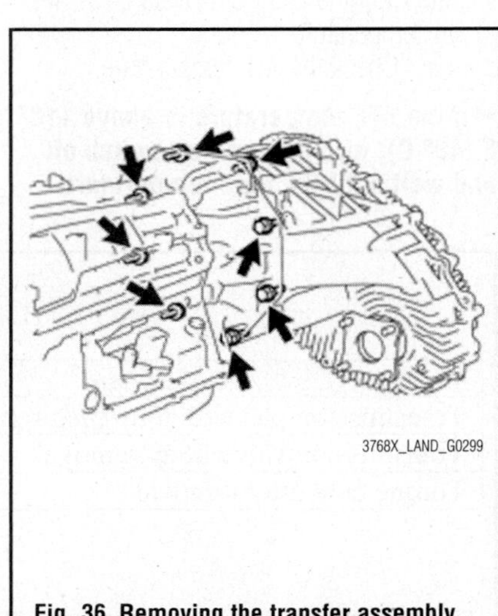

Fig. 36 Removing the transfer assembly

FRONT DRIVESHAFT

REMOVAL & INSTALLATION

See Figures 37 through 39.

1. Remove the front fender splash shield sub assembly LH.
2. Remove the front fender splash shield sub assembly RH.
3. Remove the No. 1 engine under cover sub assembly.
4. Remove the No. 2 engine under cover.
5. Remove the propeller shaft heat insulator.

a. Remove the 2 bolts and insulator.

6. Remove the front propeller shaft assembly.

a. For the transfer side, place matchmarks on the propeller shaft flange and transfer flange. Remove the 4 nuts and 4 washers.

b. For the differential side, place matchmarks on the propeller shaft flange

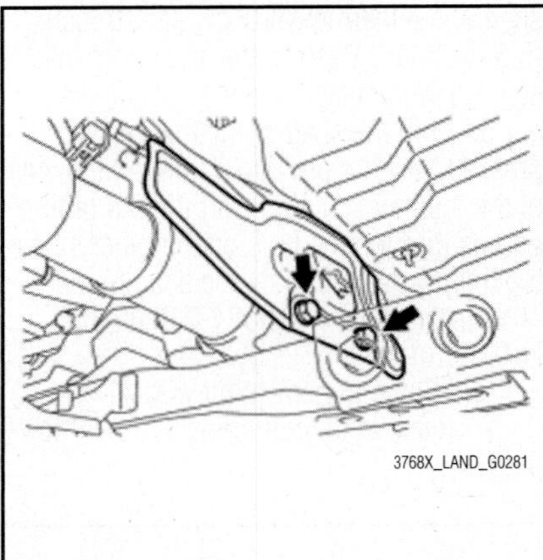

Fig. 37 Removing the propeller shaft heat insulator

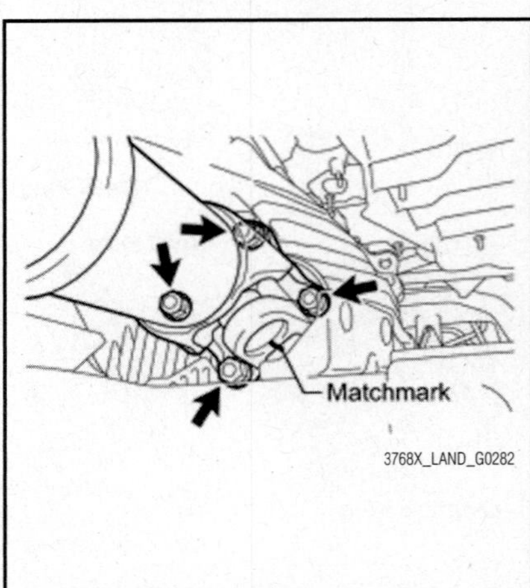

Fig. 38 Marking the propeller shaft flange and transfer flange (transfer side)

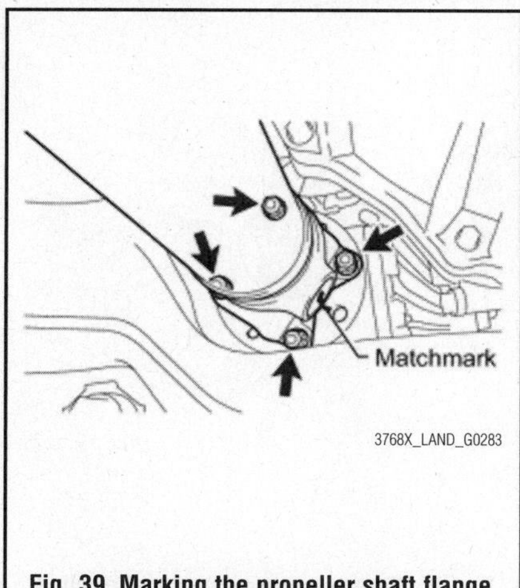

Fig. 39 Marking the propeller shaft flange and differential (differential side)

and differential. Remove the 4 nuts and 4 washers.

c. Remove the front propeller shaft.

To install:

7. Install the front propeller shaft assembly.
   a. For the differential side, align the matchmarks on the yoke and differential flange. Connect the propeller shaft with the 4 nuts. Tighten to 65 ft. lbs. (88 Nm).
   b. For the transfer side, align the matchmarks on the yoke and transfer flange. Install the propeller shaft with the 4 nuts. Tighten to 65 ft. lbs. (88 Nm).
8. Install the propeller shaft heat insulator with the 2 bolts. Tighten to 12 ft. lbs. (16 Nm).
9. Install the No. 2 engine under cover.
10. Install the No. 1 engine under cover sub assembly.
11. Install the front fender splash shield sub assembly LH.
12. Install the front fender splash shield sub assembly RH.

FRONT HALFSHAFT

REMOVAL & INSTALLATION

See Figure 40.

➡Use the same procedures for the RH and LH side.

➡The procedures listed below are for the LH side.

1. Remove the front fender splash shield sub assembly LH and RH.
2. Remove the No. 1 engine under cover sub assembly.
3. Remove the No. 2 engine under cover.
4. Drain the differential oil.
5. Remove the front axle assembly LH.

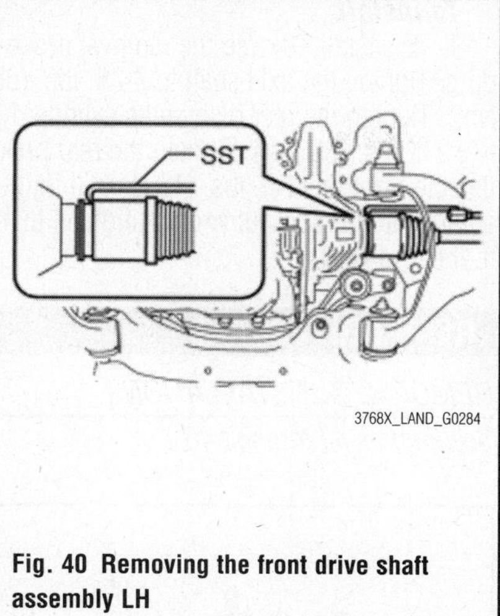

Fig. 40 Removing the front drive shaft assembly LH

6. Remove the front drive shaft assembly LH.
   a. Using the special tool (09520-01010, 09520-024010 or 09520-32040), remove the drive shaft.

➡Be careful not to damage the dust cover and oil seal.

➡Move the drive shaft while keeping it level.

To install:

7. Install the front drive shaft assembly LH.
   a. Coat the lip of the differential oil seal with MP grease.

➡Inspect the oil seal for wear or damage.

   b. Coat the spline of the inboard joint shaft assembly with hypoid gear oil.
   c. Align the shaft splines and install the drive shaft with a brass bar and hammer.

➡Set the snap ring with the opening side facing downward.

➡Be careful not to damage the oil seal, boot and dust cover.

➡If the outboard joint shaft is new, remove the anti-rust oil on the screw end with non-residue solvent.

➡Whether the inboard joint shaft is in contact with the pinion shaft or not can be confirmed from the sound or feeling when driving it.

8. Install the front axle assembly LH.
9. Add differential oil.
10. Check the speed sensor signal.
11. Inspect and adjust the front wheel alignment.
12. Adjust the headlight assembly.
13. Install the No. 2 engine under cover.
14. Install the No. 1 engine under cover sub assembly.
15. Install the front fender splash shield sub assembly LH.
16. Install the front fender splash shield sub assembly RH.

REAR AXLE SHAFT, BEARING & SEAL

REMOVAL & INSTALLATION

See Figures 41 through 43.

➡Use the same procedures for the LH and RH side.

➡The procedures listed below are for the LH side.

1. Remove the stabilizer control valve protector.
2. Open the stabilizer control with the accumulator housing shutter valve.
3. Disconnect the cable from the negative battery terminal.
4. Remove the rear wheel LH.
5. Drain the brake fluid.
6. Disconnect the rear brake flexible hose.
   a. Disconnect the brake tube from the flexible hose with the special tool holding the flexible hose with a wrench.

➡Do not bend or damage the brake tube.

➡Do not allow any foreign matter such as dirt and dust to enter the brake tube from the connecting point.

   b. Remove the clip.
7. Disconnect the rear disc brake cylinder assembly LH.

Fig. 41 Disconnecting the rear brake flexible hose

a. Remove the 2 bolts and disconnect the rear disc brake cylinder.

➡**Do not twist or bend the flexible hose.**

➡**Do not disconnect the flexible hose from the disc brake cylinder.**

b. Remove the O-ring.

8. Remove the rear axle shaft oil seal LH.

a. Using the special tool (09308-00010), tap out the oil seal.

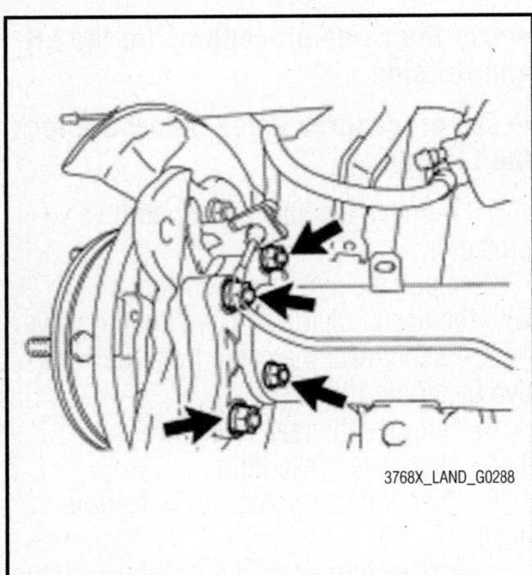

Fig. 42 Disconnecting the rear disc brake cylinder assembly LH

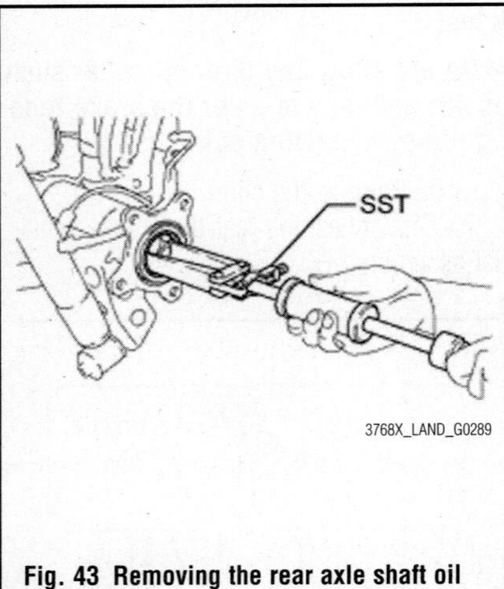

Fig. 43 Removing the rear axle shaft oil seal LH

To install:

9. To install, reverse the removal procedure. Tighten the axle shaft to 44 ft. lbs. (60 Nm). Tighten the rear disc brake cylinder LH to 70 ft. lbs. (95 Nm). Tighten the rear brake flexible hose to 11 ft. lbs. (15 Nm). If tightening with a union nut wrench, tighten to 10 ft. lbs. (14 Nm).

REAR DRIVESHAFT

REMOVAL & INSTALLATION

See Figures 44 through 46.

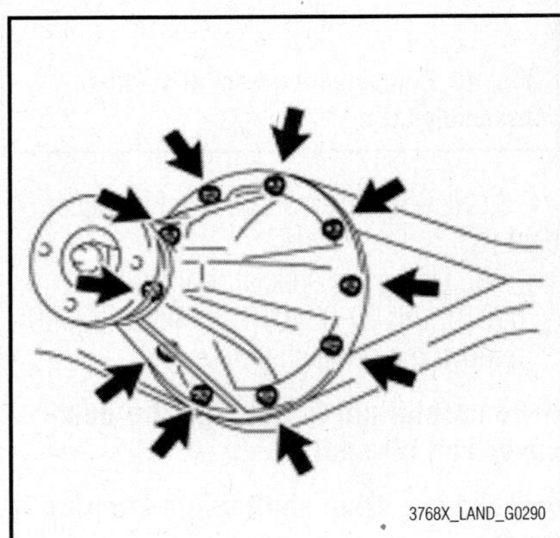

Fig. 44 Removing the transfer heat insulator

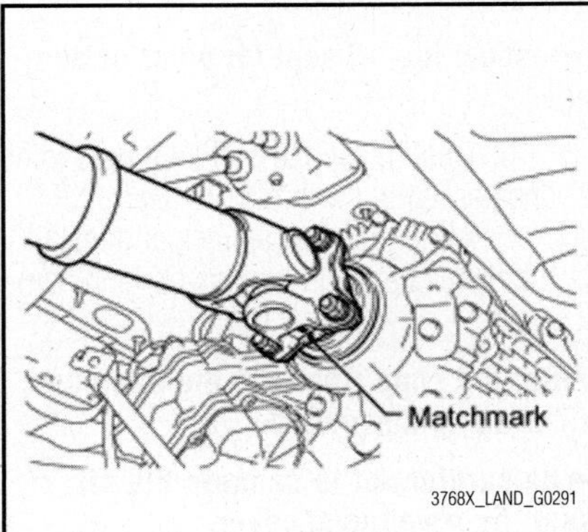

Fig. 45 Removing the rear propeller shaft assembly (transfer side)

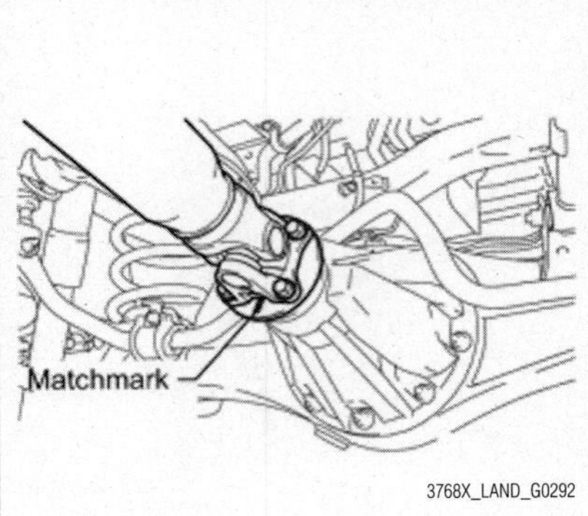

Fig. 46 Removing the rear propeller shaft assembly (differential side)

1. Remove the transfer heat insulator.

a. Remove the 4 bolts and insulator.

2. Remove the rear propeller shaft assembly.

a. For the transfer side, place matchmarks on the propeller shaft flange and transfer flange.

b. Remove the 4 nuts and 4 washers.

c. For the differential side, place matchmarks on the propeller shaft flange and rear differential flange.

d. Remove the 4 nuts, 4 bolts and 4 washers.

e. Remove the rear propeller shaft.

To install:

3. Install the rear propeller shaft.

a. Align the matchmarks on the propeller shaft flange and transfer flange.

b. Connect the propeller shaft with the 4 washers and 4 nuts. Tighten to 65 ft. lbs. (88 Nm).

c. Align the matchmarks on the propeller shaft flange and differential flange.

d. Install the propeller shaft with the 4 bolts, 4 washers and 4 nuts. Tighten to 65 ft. lbs. (88 Nm).

4. Install the transfer heat insulator.

a. Install the insulator with the 4 bolts. Tighten to 22 ft. lbs. (30 Nm).

ENGINE COOLING

ENGINE COOLANT

DRAIN & REFILL PROCEDURE

See Figures 47 and 48.

1. Remove the front fender splash shield sub assembly LH.
2. Remove the front fender splash shield sub assembly RH.
3. Remove the No. 1 engine under cover sub assembly.
4. Drain the engine coolant.

➡Do not remove the radiator cap while the engine and radiator are still hot. Pressurized, hot engine coolant and steam may be released and cause serious burns.

a. Loosen the radiator drain cock plug.

➡Collect the coolant in a container and dispose of it according to the regulations in your area.

b. Remove the radiator cap. Then drain the coolant from the radiator.
c. Loosen the 2 cylinder block drain cock plugs. Then drain the coolant from the engine.
d. Tighten the 2 cylinder block drain cock plugs to 10 ft. lbs. (13 Nm).
e. Tighten the radiator drain cock plug by hand.

5. Add engine coolant. The standard capacity is 17.6 qts. (16.7 L).

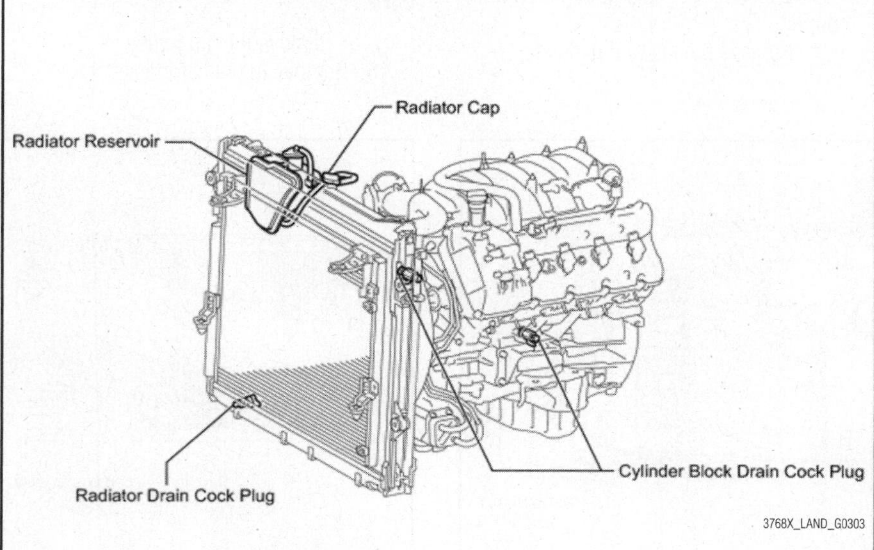

Fig. 47 Draining the coolant

➡Do not substitute plain water for engine coolant.

➡TOYOTA vehicles are filled with TOYOTA SLLC at the factory. In order to avoid damage to the engine cooling system and other technical problems, only use TOYOTA SLLC or similar high quality ethylene glycol based non-silicate, non-amine, non-nitrite, non-borate coolant with long-life hybrid organic acid technology (coolant with long-life hybrid organic acid technology consists of a combination of low phosphates and organic acids).

a. Slowly pour coolant into the radiator reservoir until it reaches the F line.
b. Install the reservoir cap.
c. Press the No. 1 and No. 2 radiator hoses several times by hand, and then check the coolant level. If the coolant level is low, add coolant.
d. Install the radiator cap.
e. Start the engine and warm it up until the thermostat opens.

➡The thermostat opening timing can be confirmed by pressing the radiator inlet hose by hand, and checking when the engine coolant starts to flow inside the hose.

f. Maintain the engine speed at 2000 to 2500 rpm.

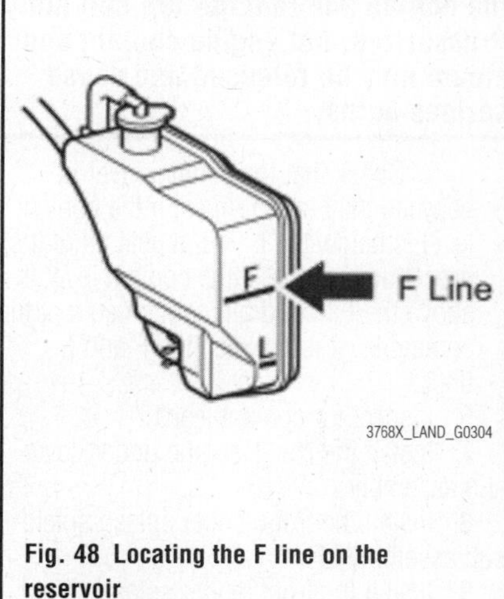

Fig. 48 Locating the F line on the reservoir

➡Make sure that the radiator reservoir still has some coolant in it.

➡Pay attention to the needle of the water temperature meter. Make sure that the needle does not show an abnormally high temperature.

➡If there is not enough coolant, the engine may burn out or overheat.

➡Immediately after starting the engine, if the radiator reservoir does not have any coolant, perform the following: 1) stop the engine, 2) wait until the coolant has cooled down, and 3) add coolant until the coolant is filled to the F line.

➡Run the engine at 2000 rpm until the coolant level has stabilized.

g. Press the No. 1 and No. 2 radiator hoses several times by hand to bleed air.

✻✻ CAUTION

Wear protective gloves.

✻✻ CAUTION

Be careful as the radiator hoses are hot.

✻✻ CAUTION

Keep your hands away from the fan.

h. Stop the engine, and wait until the engine coolant cools down to ambient temperature.

⁂ CAUTION

Do not remove the radiator cap while the engine and radiator are still hot. Pressurized, hot engine coolant and steam may be released and cause serious burns.

i. Check that the coolant level is between the F and L lines. If the coolant level is below the L line, repeat all of the procedures above. If the coolant level is above the F line, drain coolant so that the coolant level is between the F and L lines.

6. Inspect for coolant leaks.
7. Install the No. 1 engine under cover sub assembly.
8. Install the front fender splash shield sub assembly LH.
9. Install the front fender splash shield sub assembly RH.

RADIATOR

REMOVAL & INSTALLATION

See Figures 49 through 55.

1. Remove the upper radiator support seal.
2. Remove the radiator grille.
3. Remove the front bumper cover.
4. Remove the transmission oil cooler air duct (with air cooled transmission oil cooler).
    a. Remove the 4 bolts and oil cooler air duct.
5. Disconnect the radiator side deflector RH (without air cooled transmission oil cooler).
    a. Using a clip remover, remove the 4 clips and disconnect the side deflector.

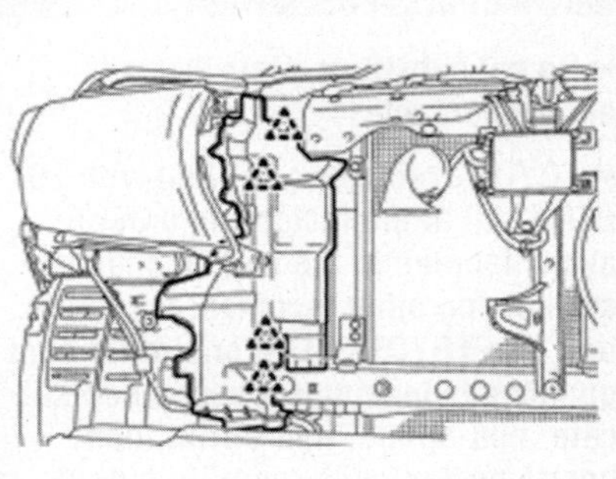

Fig. 50 Disconnecting the radiator side deflector RH

6. Disconnect the radiator side deflector LH.
    a. Using a clip remover, remove the 4 clips and disconnect the side deflector.
7. Remove the front fender splash shield sub assembly LH.
8. Remove the front fender splash shield sub assembly RH.
9. Remove the No. 1 engine under cover sub assembly.
10. Drain the engine coolant.
11. Remove the v-bank cover sub assembly.
12. Remove the air cleaner hose assembly.
13. Remove the No. 1 radiator hose.
14. Remove the No. 2 radiator hose.
15. Remove the fan shroud.
    a. Loosen the 4 nuts holding the fluid coupling fan.
    b. Remove the fan and generator v-belt.

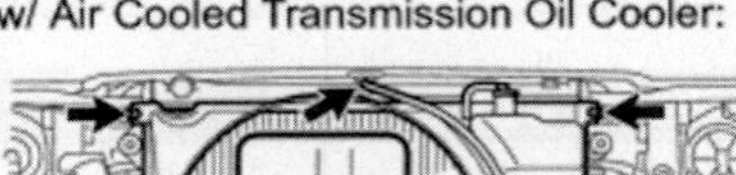

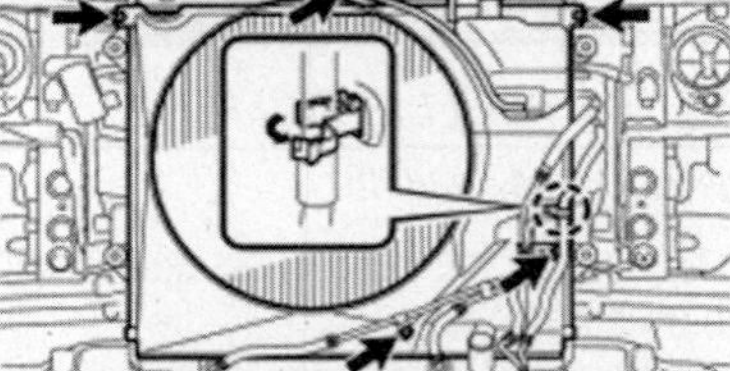

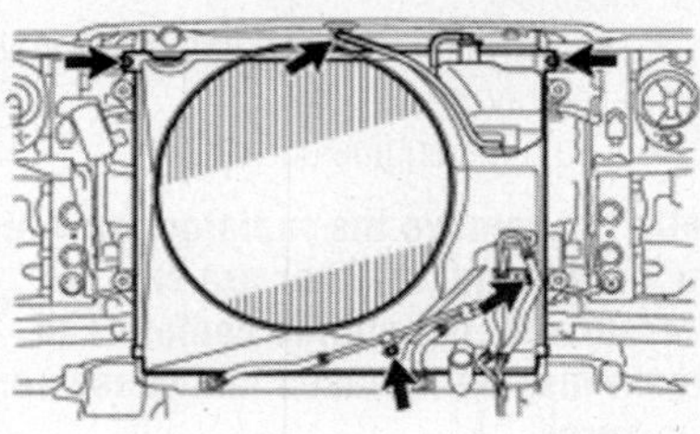

Fig. 52 Disconnecting the reservoir hose from the upper radiator tank

    c. Disconnect the reservoir hose from the upper radiator tank.
    d. For vehicles with air cooled transmission oil coolers, detach the claw to open the flexible hose clamp.
    e. Remove the 2 bolts and disconnect the oil cooler tube from the fan shroud.
    f. Remove the 2 bolts holding the fan shroud.
    g. Remove the 4 nuts of the fluid coupling fan, and then remove the shroud together with the coupling fan.

➡Be careful not to damage the radiator core.

    h. Remove the fan pulley.
16. Remove the radiator assembly.

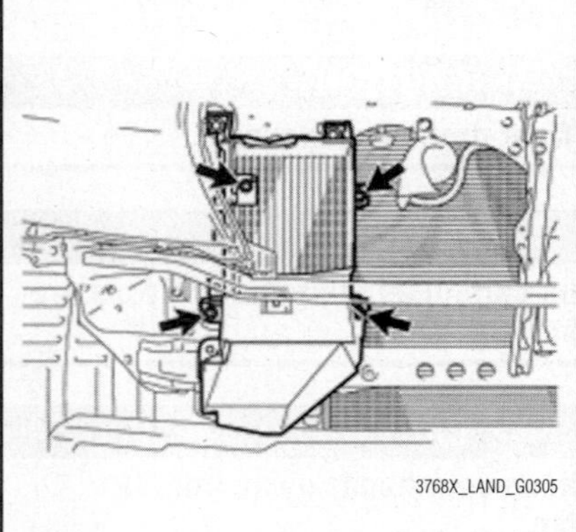

Fig. 49 Removing the transmission oil cooler air duct

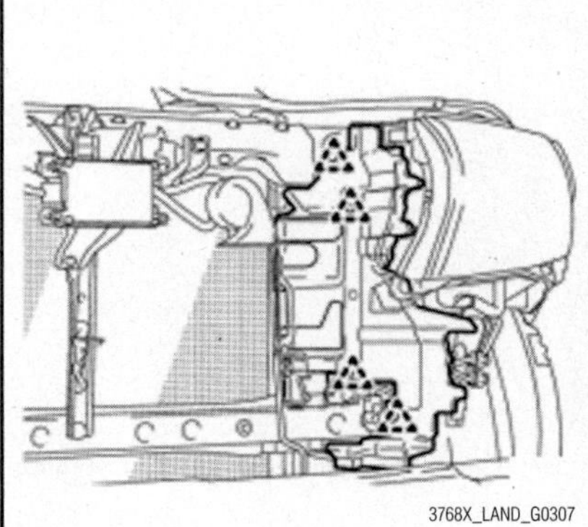

Fig. 51 Disconnecting the radiator side deflector LH

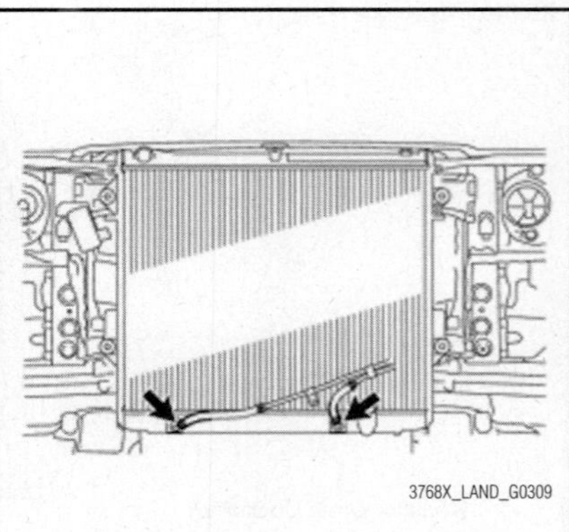

Fig. 53 Disconnecting the oil cooler hoses

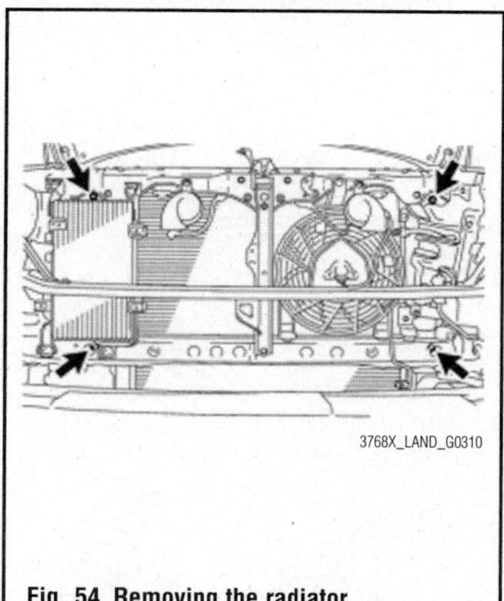
Fig. 54 Removing the radiator

a. Disconnect the 2 oil cooler hoses.
b. Remove the 4 bolts and the radiator.

To install:

17. Install the radiator assembly.
a. Set the radiator bracket hooks to the radiator support holes.
b. Install the radiator with the 4 bolts. Tighten to 13 ft. lbs. (18 Nm).
c. Connect the 2 oil cooler hoses.
18. Install the fan shroud.
a. Install the fan pulley.
b. Place the shroud together with the coupling fan between the radiator and engine.

➡Be careful not to damage the radiator core.

c. Temporarily install the fluid coupling fan to the fluid coupling bracket with the 4 nuts. Tighten the nuts as much as possible by hand.
d. Attach the claws of the shroud to the radiator.
e. Install the shroud with the 2 bolts. Tighten to 71 inch lbs. (8 Nm).
f. Connect the oil cooler tube to the fan shroud with the 2 bolts. Tighten to 44 inch lbs. (5 Nm).
g. Vehicles with air cooled transmission oil cooler, pass the hose through the flexible hose clamp and close the clamp.
h. Connect the reservoir hose to the upper radiator tank.
i. Install the fan and generator v-belt.
j. Tighten the 4 nuts of the fluid coupling fan. Tighten to 15 ft. lbs. (21 Nm).
19. Install the No. 2 radiator hose.
20. Install the No. 1 radiator hose.
21. Install the air cleaner hose assembly.
22. Install the v-bank cover sub assembly.

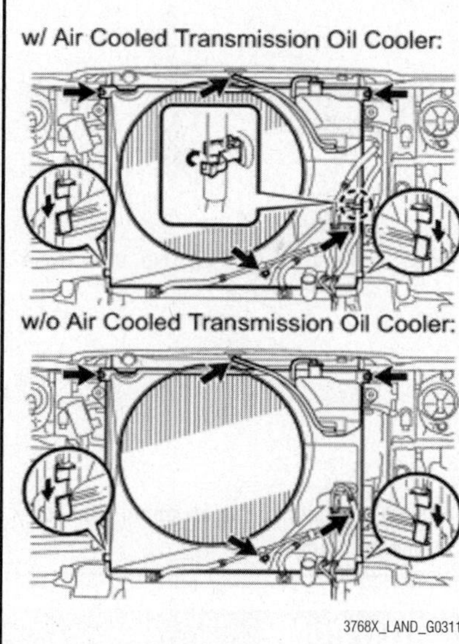

Fig. 55 Attaching the claws of the shroud to the radiator

23. Add engine coolant.
24. Inspect for coolant leaks.
25. Install the No. 1 engine under cover sub assembly.
26. Install the front fender splash shield sub assembly LH.
27. Install the front fender splash shield sub assembly RH.
28. Connect the radiator side deflector LH.
a. Connect the deflector with the 4 clips.
29. Connect the radiator side deflector RH (without air cooled transmission oil cooler).
a. Connect the deflector with the 4 clips.
30. Install the transmission oil cooler air duct (with air cooled transmission oil cooler) with the 4 bolts. Tighten to 43 inch lbs. (4.9 Nm).
31. Install the front bumper cover.
32. Install the radiator grille.
33. Install the upper radiator support seal.

THERMOSTAT

REMOVAL & INSTALLATION

See Figures 56 through 59.

➡If the thermostat is not installed, cooling efficiency decreases. Even if the engine tends to overheat, do not remove the thermostat.

1. Remove the front fender splash shield sub assembly LH.
2. Remove the front fender splash shield sub assembly RH.
3. Remove the No. 1 engine under cover sub assembly.
4. Drain the engine coolant.
5. Remove the v-bank cover sub assembly.
6. Remove the air cleaner hose assembly.
7. Remove the air cleaner assembly.
8. Remove the No. 1 radiator hose.
9. Remove the No. 2 radiator hose.
10. Remove the fan shroud.
11. Remove the water inlet sub assembly with the thermostat.
a. Disconnect the No. 2 and No. 3 air hoses.
b. Disconnect the air pump connector.
c. Disconnect the air pump connector clamp holder.
d. Using a clip remover, detach the wire harness clamp.
e. Disconnect the No. 5 water by pass hose.
f. Remove the air tube bracket bolt.
g. Remove the 3 nuts, water inlet with the thermostat and gasket.

To install:

12. Install the water inlet sub assembly with the thermostat.
a. Install a new gasket and the water inlet with the thermostat with the 3 nuts. Tighten to 7 ft. lbs. (10 Nm).
b. Install the air tube bracket bolt. Tighten to 7 ft. lbs. (10 Nm).
c. Connect the No. 5 water by pass hose.

➡Install the hose so that the direction of the hose clamp is as indicated.

d. Connect the air pump connector clamp holder.

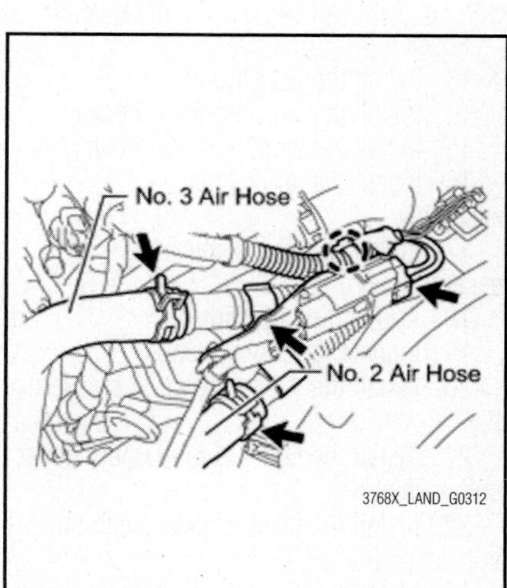

Fig. 56 Disconnecting the air hoses

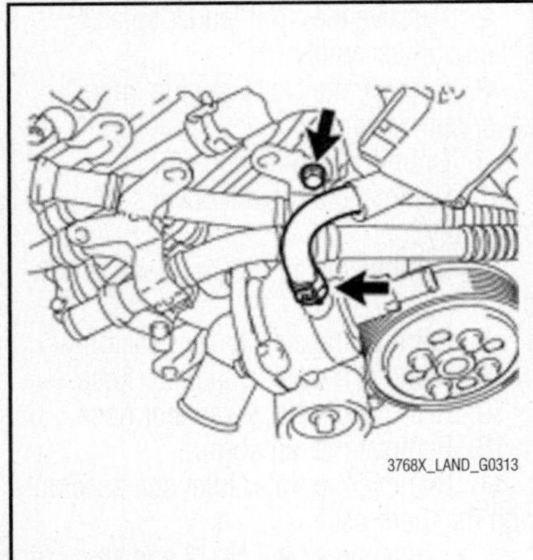

Fig. 57 Disconnecting the No. 5 water by pass hose

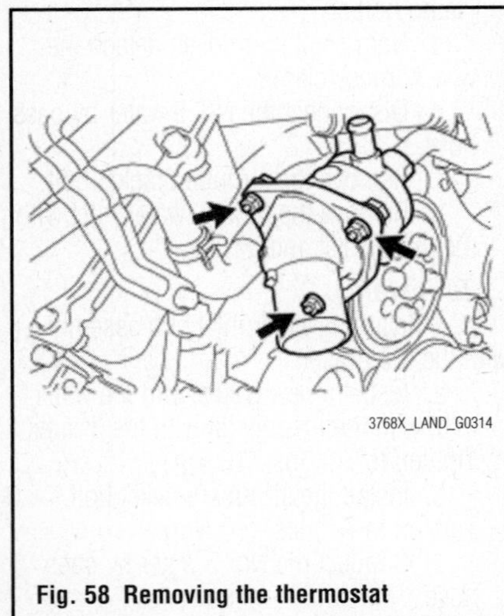

Fig. 58 Removing the thermostat

e. Attach the wire harness clamp.
f. Connect the air pump connector.
g. Connect the No. 2 and No. 3 air hoses.
13. Install the fan shroud.
14. Install the No. 2 radiator hose.
15. Install the No. 1 radiator hose.
16. Install the air cleaner assembly.
17. Install the air cleaner hose assembly.
18. Add engine coolant.
19. Inspect for coolant leaks.
20. Install the v-bank cover sub assembly.
21. Install the No. 1 engine under cover sub assembly.
22. Install the front fender splash shield sub assembly LH.
23. Install the front fender splash shield sub assembly RH.

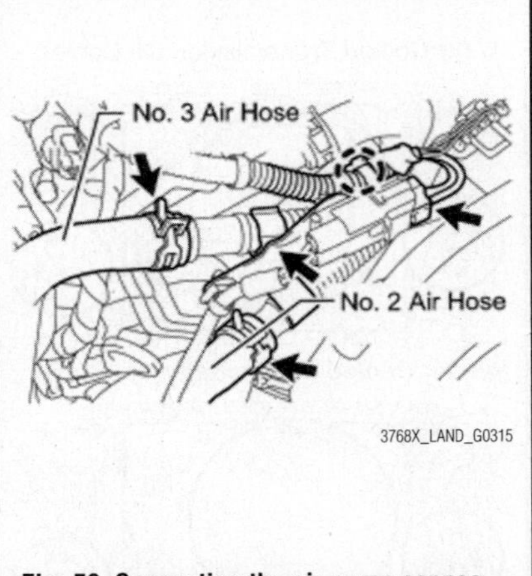

Fig. 59 Connecting the air pump connector clamp holder

WATER PUMP

REMOVAL & INSTALLATION

See Figures 60 through 68.

1. Remove the front fender splash shield sub assembly LH.
2. Remove the front fender splash shield sub assembly RH.
3. Remove the No. 1 engine under cover sub assembly.
4. Drain the engine coolant.
5. Remove the v-bank cover sub assembly.
6. Remove the air cleaner hose assembly.
7. Remove the air cleaner assembly.
8. Remove the No. 1 radiator hose.
9. Remove the No. 2 radiator hose.
10. Remove the fan shroud.
11. Remove the No. 1 water by pass hose.
12. Disconnect the No. 2 water by pass pipe sub assembly (with oil cooler).
a. Disconnect the No. 6 water by-pass hose.
b. Remove the 3 bolts.
c. Disconnect the water by-pass pipe with water hose.
13. Remove the water inlet housing.
a. Disconnect the No. 2 and No. 3 air hoses.
b. Disconnect the air pump connector.
c. Disconnect the air pump connector clamp holder.
d. Using a clip remover, detach the wire harness clamp.
e. Disconnect the No. 5 water by pass hose.
f. Remove the air tube bracket bolt.
g. Disconnect the No. 3 water by pass hose and remove the 3 bolts and water inlet housing.
h. Remove the gasket from the water pump.
14. Remove the water pump pulley.
a. Using the special tool (09960-10010, 09962-01000 or 09963-01000), hold the water pump pulley.
b. Remove the 4 bolts and water pump pulley.
15. Remove the water pump assembly.
a. Remove the 8 bolts, water pump and gasket.

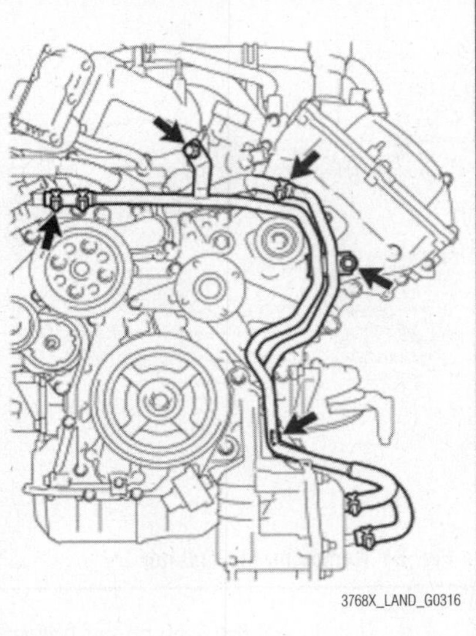

Fig. 60 Disconnecting the No. 2 water by pass pipe sub assembly (with oil cooler)

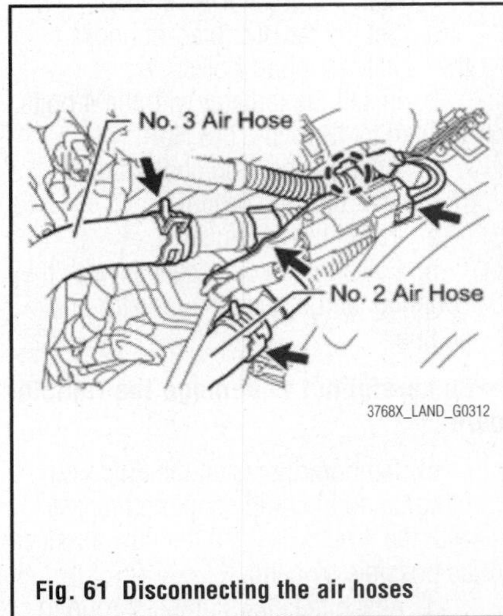

Fig. 61 Disconnecting the air hoses

To install:

16. Install the water pump assembly.
a. Install a new gasket and the water pump with the 8 bolts. Tighten bolt A to

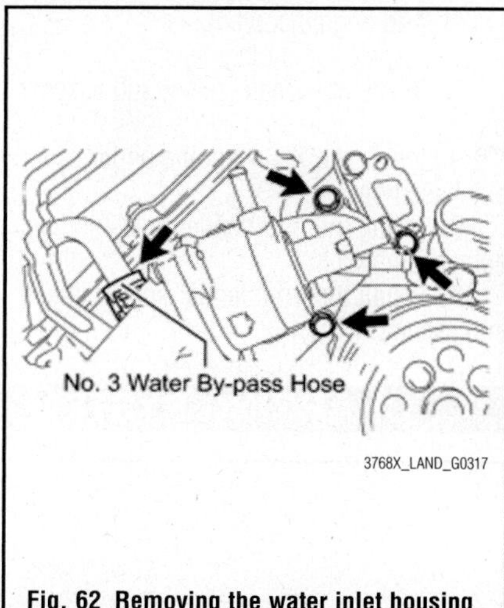

Fig. 62 Removing the water inlet housing

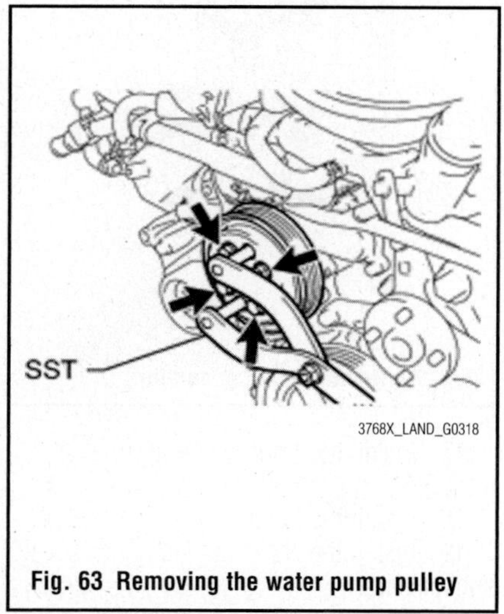

Fig. 63 Removing the water pump pulley

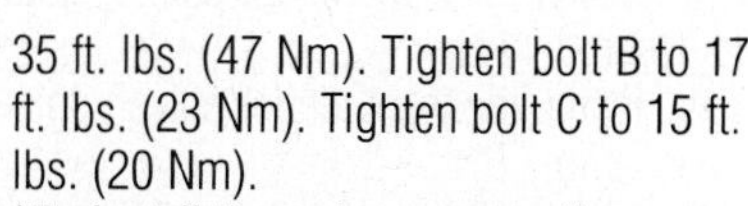

35 ft. lbs. (47 Nm). Tighten bolt B to 17 ft. lbs. (23 Nm). Tighten bolt C to 15 ft. lbs. (20 Nm).

17. Install the water pump pulley.
    a. Temporarily install the pulley with the 4 bolts.
    b. Using the special tool, hold the pulley and tighten the 4 bolts to 15 ft. lbs. (21 Nm).
18. Install the water inlet housing.
    a. Install a new gasket to the water pump.
    b. Install the water inlet housing with the 3 bolts and connect the No. 3 water by pass hose. Tighten to 15 ft. lbs. (21 Nm).

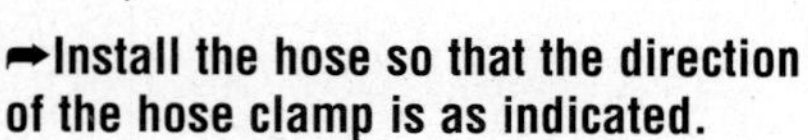

➡Install the hose so that the direction of the hose clamp is as indicated.

c. Install the air tube bracket bolt and tighten to 7 ft. lbs. (10 Nm).
d. Connect the No. 5 water by pass hose.

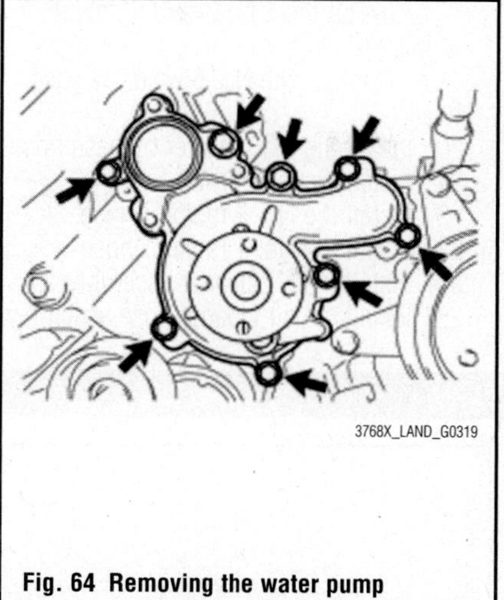
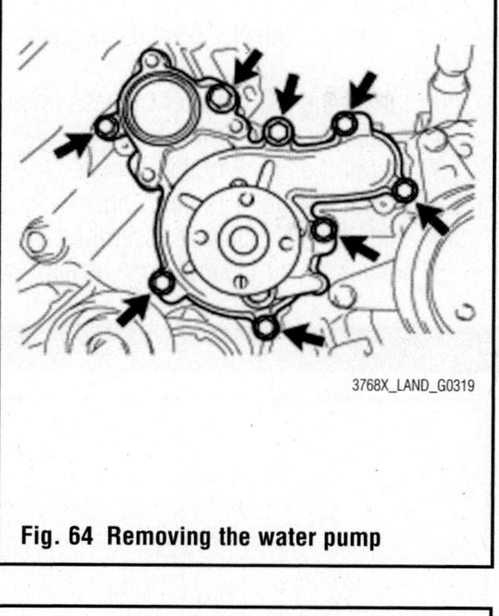
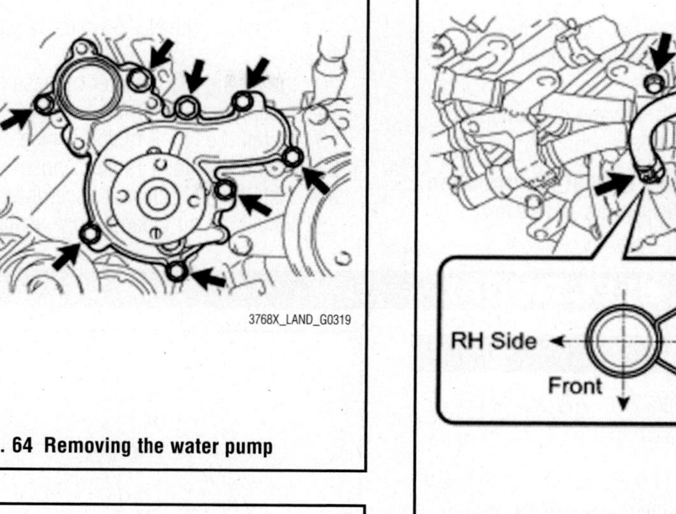

Fig. 64 Removing the water pump

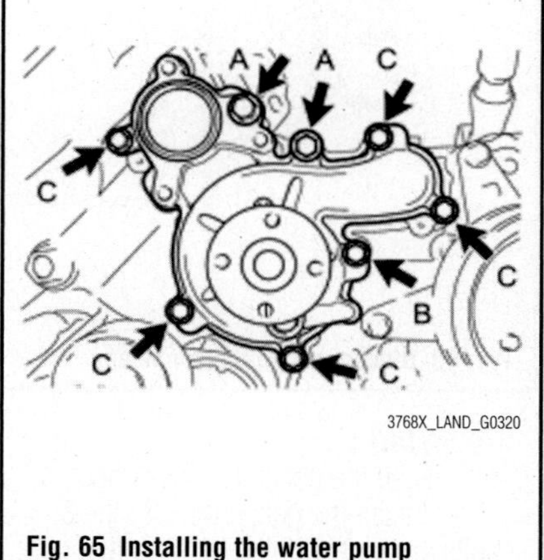

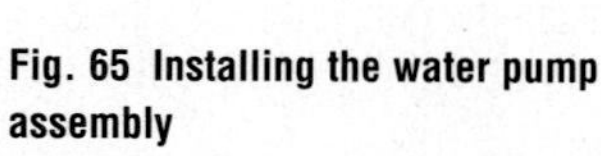

Fig. 65 Installing the water pump assembly

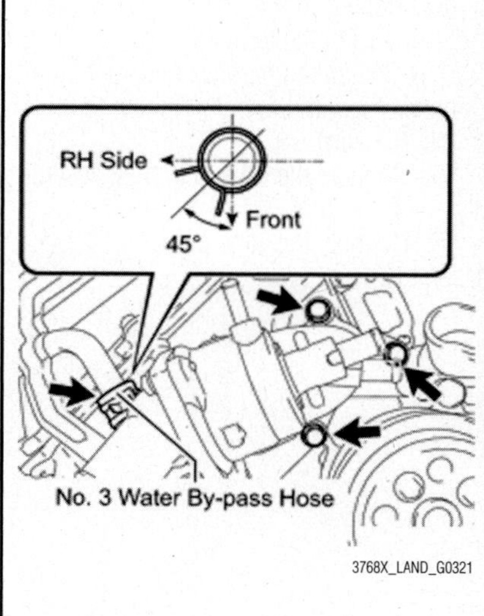

Fig. 66 Installing the water inlet housing

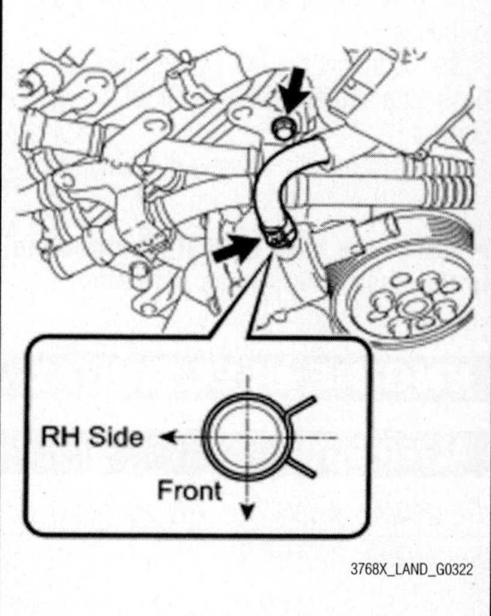

Fig. 67 Installing the air tube bracket bolt and connecting the No. 5 water by pass hose

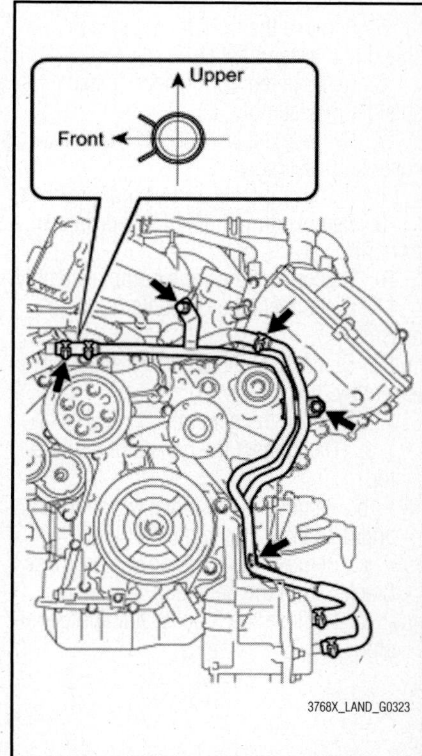

Fig. 68 Connecting the No. 2 water by pass pipe sub assembly (with oil cooler)

➡Install the hose so that the direction of the hose clamp is as indicated.

e. Connect the air pump connector clamp holder.
f. Attach the wire harness clamp.
g. Connect the air pump connector.

h. Connect the No. 2 and No. 3 air hoses.

19. Connect the No. 2 water by pass pipe sub assembly (with oil cooler).

a. Connect the water by pass pipe with the water hose.

➡**Install the hose so that the direction of the hose clamp is as indicated.**

b. Install the 3 bolts and tighten to 7 ft. lbs. (10 Nm).

c. Connect the No. 6 water by pass hose.

20. Install the NO. 1 water by pass hose.
21. Install the fan shroud.
22. Install the No. 2 radiator hose.
23. Install the No. 1 radiator hose.
24. Install the air cleaner assembly.
25. Install the air cleaner hose assembly.
26. Add engine coolant.
27. Inspect for coolant leaks.
28. Install the v-bank cover sub assembly.
29. Install the No. 1 engine under cover sub assembly.
30. Install the front fender splash shield sub assembly LH.
31. Install the front fender splash shield sub assembly RH.

ENGINE ELECTRICAL

CHARGING SYSTEM

ALTERNATOR

REMOVAL & INSTALLATION

See Figures 69 and 70.

1. Disconnect the cable from the negative battery terminal.
2. Remove the v-bank cover sub assembly.
3. Remove the air cleaner hose assembly.
4. Remove the air cleaner assembly.
5. Remove the front fender splash shield sub assembly LH.
6. Remove the front fender splash shield sub assembly RH.
7. Remove the No. 1 engine under cover sub assembly.
8. Remove the fan and generator v-belt.
9. Remove the front fender apron seal front RH.
10. Disconnect the vane pump assembly.
11. Disconnect the oil cooler pipe assembly.
12. Disconnect the oil cooler pipe assembly.
13. Remove the generator.

a. Disconnect the generator connector.

b. Remove the terminal cap and nut, and disconnect the generator wire.

c. Remove the bolt and wire harness bracket from the generator.

d. Remove the 3 bolts, nut and generator.

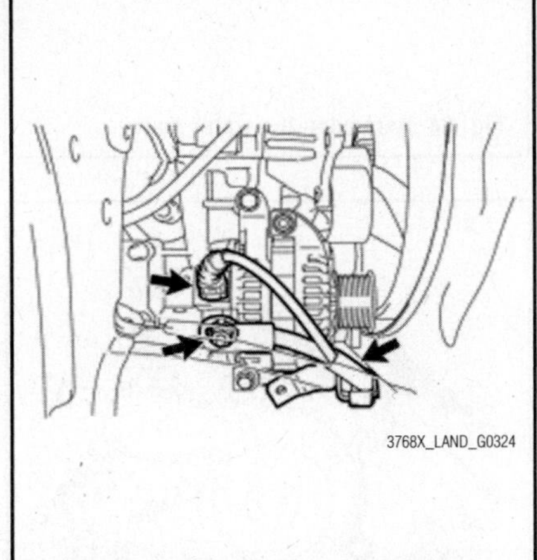

Fig. 69 Disconnecting the connector

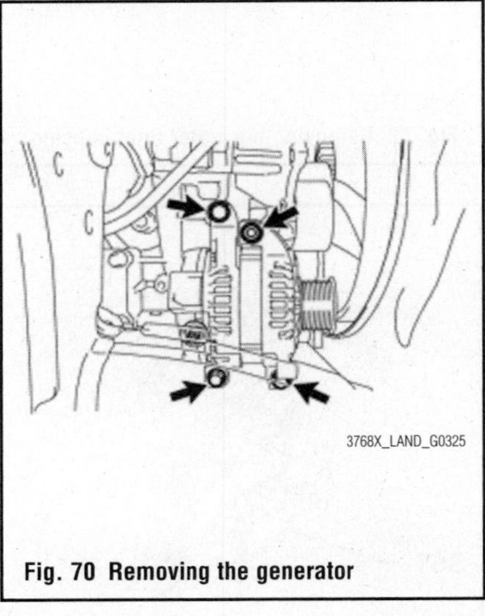

Fig. 70 Removing the generator

To install:

14. Install the generator assembly.

a. Install the generator with the 3 bolts and nut. Tighten to 32 ft. lbs. (43 Nm).

b. Connect the generator connector.

c. Connect the generator wire with the nut. Tighten to 9 ft. lbs. (12 Nm).

d. Install the terminal cap.

e. Install the harness bracket to the generator with the bolt. Tighten to 23 ft. lbs. (31 Nm).

15. Connect the oil cooler pipe assembly.
16. Connect the vane pump assembly.
17. Install the front fender apron seal front RH.
18. Install the fan and generator v-belt.
19. Install the No. 1 engine under cover sub assembly.
20. Install the front fender splash shield sub assembly LH.
21. Install the front fender splash shield sub assembly RH.
22. Install the air cleaner assembly.
23. Install the air cleaner hose assembly.
24. Install the v-bank cover sub assembly.
25. Connect the cable to the negative battery cable.

FIRING ORDER

See Figure 71.

IGNITION COIL

REMOVAL & INSTALLATION

See Figures 72 through 74.

1. Remove the No. 2 engine under cover.
2. Remove the v-bank cover sub assembly.
3. Remove the air cleaner hose assembly.
4. Remove the air cleaner assembly.
5. Disconnect the water by pass pipe assembly.
    a. Remove the 2 bolts and disconnect the water by pass pipe from the cylinder head cover and timing chain cover.
    b. Remove the 3 bolts and disconnect the water by pass pipe from the cylinder head cover.
6. Remove the ignition coil assembly.
    a. Disconnect the 8 ignition coil connectors.
    b. Remove the 8 bolts and 8 ignition coils.
7. Remove the 8 spark plugs.

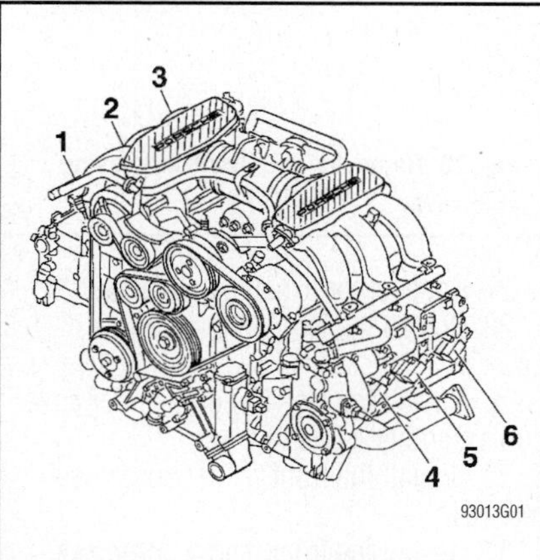

Fig. 71 4.7L Engine Firing order: 1–8–4–3–6–5–7–2 Distributor less ignition system

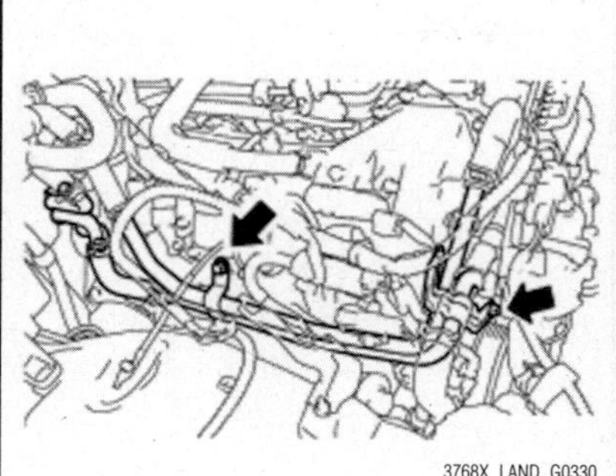

Fig. 72 Disconnecting the water by pass pipe from the cylinder head cover and timing chain cover

To install:

8. Install the spark plugs and tighten to 15 ft. lbs. (21 Nm).
9. Install the ignition coil assembly with the 8 bolts. Tighten to 7 ft. lbs. (10 Nm).
    a. Connect the 8 ignition coil connectors.
10. Connect the water by pass pipe assembly.
    a. Connect the water by pass pipe to the cylinder head cover and automatic transmission with the 3 bolts. Tighten to 13 ft. lbs. (18 Nm).
    b. Connect the water by pass pipe to the cylinder head cover and timing chain cover with the 2 bolts. Tighten to 7 ft. lbs. (10 Nm).
11. Install the air cleaner assembly.
12. Install the air cleaner hose assembly.
13. Install the v-bank cover sub assembly.
14. Install the No. 2 engine under cover.

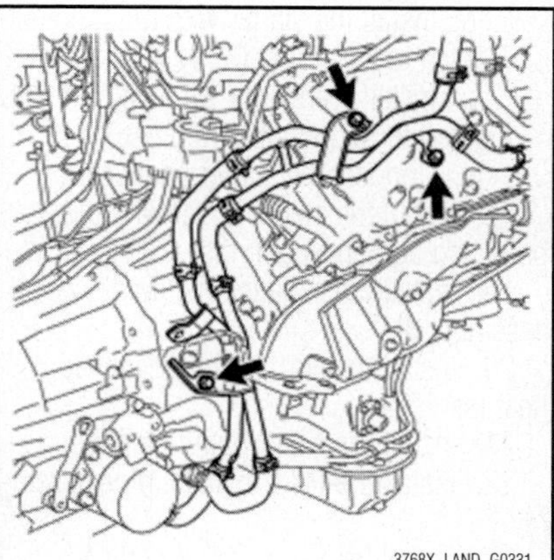

Fig. 73 Disconnecting the water by pass pipe from the cylinder head cover

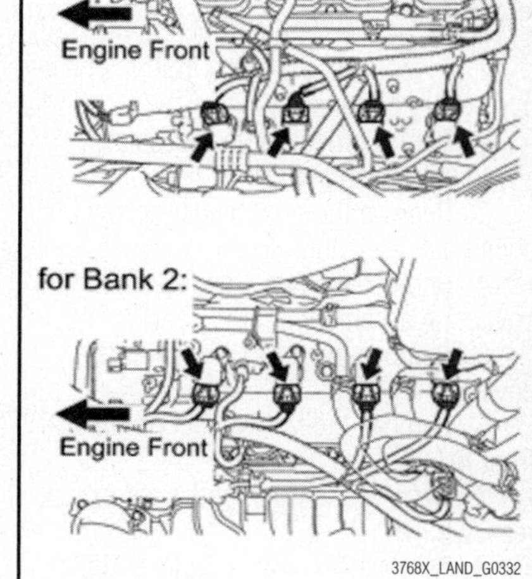

Fig. 74 Removing the ignition coil assembly

IGNITION TIMING

ADJUSTMENT

The ignition timing is controlled by the Powertrain Control Module (PCM). No adjustment is necessary or possible.

SPARK PLUGS

REMOVAL & INSTALLATION

Refer to IGNITION COIL.

STARTER

REMOVAL & INSTALLATION

See Figures 75 through 78.

1. Disconnect the cable from the negative battery terminal.
2. Remove the front fender splash shield sub assembly LH.
3. Remove the front fender splash shield sub assembly RH.
4. Remove the No. 1 engine under cover sub assembly.
5. Remove the No. 2 engine under cover.
6. Remove the front fender apron seal front RH.
7. Remove the front fender apron seal rear RH.
8. Remove the engine oil level dipstick guide.
9. Remove the tailpipe assembly.
10. Remove the center exhaust pipe assembly.
11. Remove the front exhaust pipe assembly.
12. Remove the No. 1 manifold stay.
13. Remove the No. 1 exhaust manifold heat insulator.
14. Remove the exhaust manifold sub assembly RH.
15. Remove the 3 bolts and the starter cover.

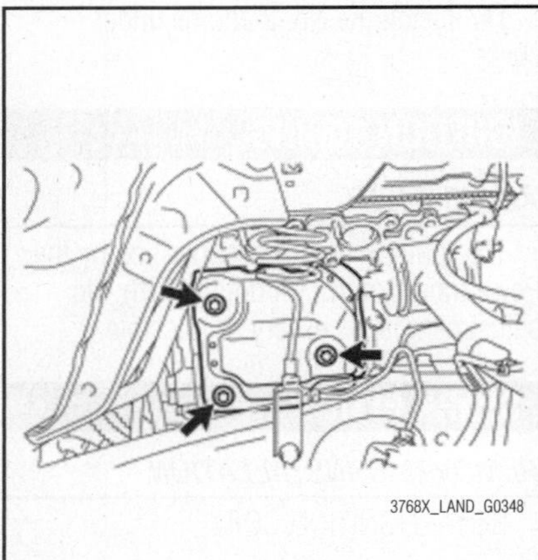

Fig. 75 Removing the starter cover

3768X_LAND_G0349

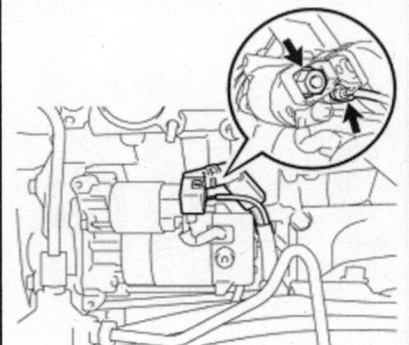

Fig. 76 Disconnecting the starter connector

16. Remove the starter assembly.
    a. Disconnect the starter connector.
    b. Remove the nut and disconnect the starter wire.
    c. Remove the 2 bolts and starter.
    d. Remove the flywheel housing side cover.

To install:

17. Install the starter assembly.
    a. Install the flywheel housing side cover.
    b. Install the starter with the 2 bolts. Tighten to 27 ft. lbs. (37 Nm).
    c. Install the starter wire with the nut. Tighten to 87 inch lbs. (9.8 Nm).
    d. Connect the starter connector.
18. Install the starter cover with the 3 bolts and tighten to 9 ft. lbs. (12 Nm).
19. Install the exhaust manifold sub assembly RH.
20. Install the No. 1 exhaust manifold heat insulator.
21. Install the No. 1 manifold stay.
22. Install the front exhaust pipe assembly.
23. Install the center exhaust pipe assembly.
24. Install the tailpipe assembly.
25. Install the engine oil level dipstick guide.
26. Connect the cable to the negative battery terminal.

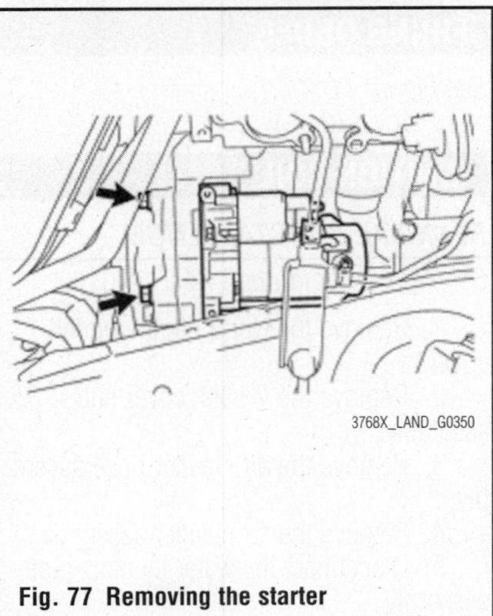

Fig. 77 Removing the starter

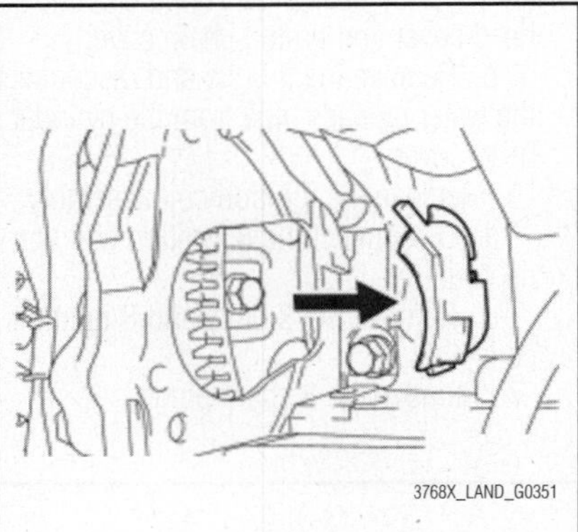

Fig. 78 Removing the flywheel housing side cover

27. Inspect for exhaust gas leaks.
28. Install the No. 2 engine under cover.
29. Install the No. 1 engine under cover sub assembly.
30. Install the front fender apron seal rear RH.
31. Install the front fender apron seal front RH.
32. Install the front fender splash shield sub assembly RH.
33. Install the front fender splash shield sub assembly LH.

ENGINE MECHANICAL

➡Disconnecting the negative battery cable may interfere with the functions of the on board computer systems and may require the computer to undergo a relearning process, once the negative battery cable is reconnected.

ACCESSORY DRIVE BELTS

ACCESSORY BELT ROUTING

See Figure 79.

INSPECTION

See Figure 80.

1. Check the belt for wear, cracks or other signs of damage.
2. If any of the following defects is found, replace the fan and generator V belt.
   - Belt is cracked
   - Belt is worn out to the extent that the cords are exposed
   - Belt has chunks missing from the ribs
3. Check that the belt fits properly in the ribbed grooves.

➡Check with your hand to confirm that the belt has not slipped out of the grooves on the bottom of the pulley. If it has slipped out, replace the fan and generator V belt. Install a new fan and generator V belt correctly.

4. Check that nothing gets caught in the tensioner by turning it clockwise and counterclockwise. If a malfunction exists, replace the tensioner.

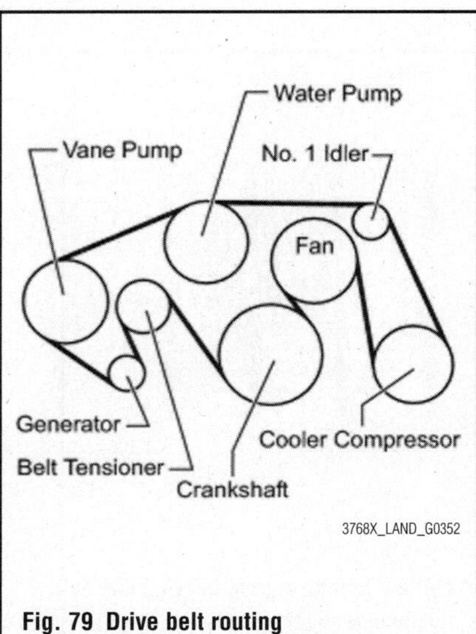

Fig. 79 Drive belt routing

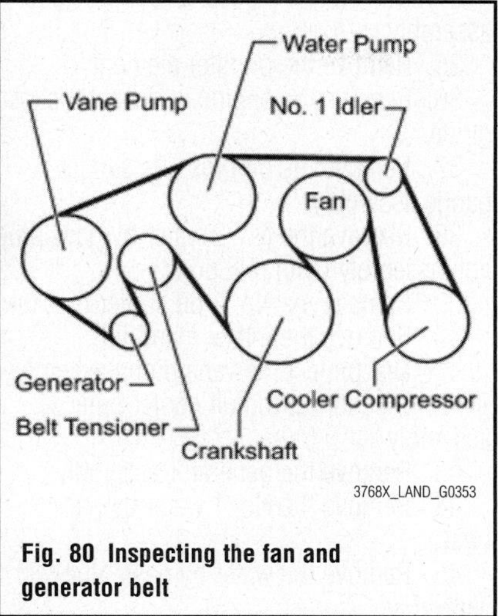

Fig. 80 Inspecting the fan and generator belt

REMOVAL & INSTALLATION

See Figure 81.

1. Remove the front fender splash shield sub assembly LH and RH.
2. Remove the No. 1 engine under cover sub assembly.
3. Remove the v-bank cover sub assembly.
4. Remove the fan and generator v belt.
   a. While turning the belt tensioner counterclockwise, align the service hole for the belt tensioner and the belt tensioner fixing position, and then insert a bar of 5 mm (0.197 in.) into the service hole to fix the belt tensioner in place.

➡The pulley bolt for the belt tensioner has a left hand thread.

   b. Remove the v-belt.

To install:

5. Install the fan and generator v-belt.
   a. Set the v-belt onto every part.
   b. While turning the belt tensioner counterclockwise, remove the bar.

➡Make sure that the v-belt is properly set to each pulley.

   c. After installing the belt, check that it fits properly in the ribbed grooves.

➡Make sure to check by hand that the belt has not slipped out of the grooves on the bottom of the pulley.

6. Install the v-bank cover sub assembly.
7. Install the No. 1 engine under cover sub assembly.
8. Install the front fender splash shield sub assembly LH and RH.

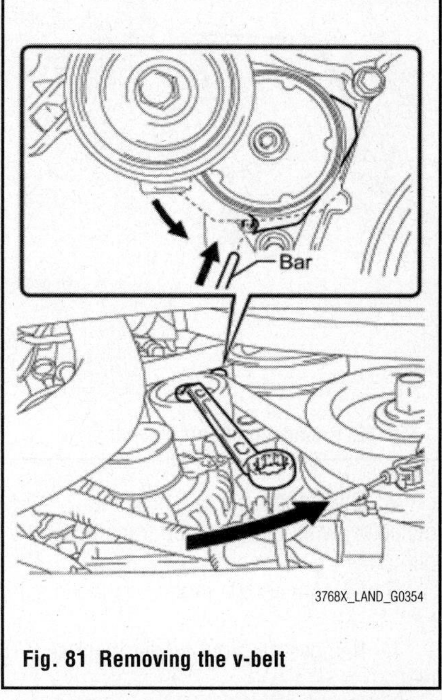

Fig. 81 Removing the v-belt

AIR CLEANER

REMOVAL & INSTALLATION

See Figure 82.

1. Remove the 3 bolts and the air cleaner.

To install:

2. To install, reverse the removal procedure.

CAMSHAFT AND VALVE LIFTERS

REMOVAL & INSTALLATION

See Figures 83 through 102.

1. Discharge the fuel system pressure.
2. Disconnect the cable from the negative battery terminal.

CAUTION

Wait at least 90 seconds after disconnecting the cable from the negative (-) battery terminal to disable to SRS system.

3. Remove the upper radiator support seal.
4. Remove the radiator grille.
5. Remove the front bumper cover.
6. Remove the transmission oil cooler air duct (with air cooled transmission oil cooler).

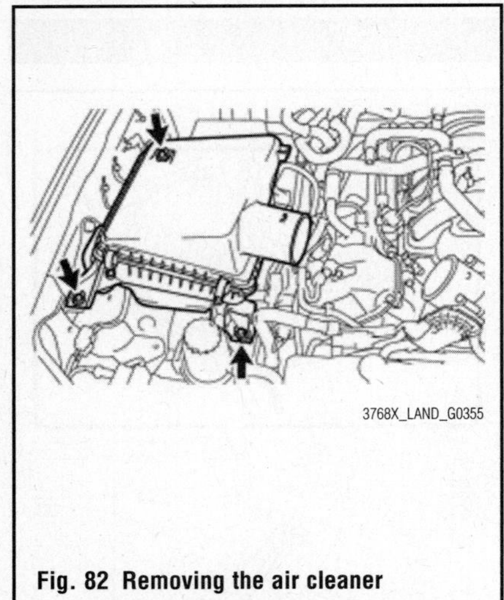
3768X_LAND_G0355

Fig. 82 Removing the air cleaner

7. Disconnect the radiator side deflector (without air cooled transmission oil cooler).
8. Disconnect the radiator side deflector LH.
9. Remove the front fender main seal LH.
10. Remove the front fender main seal RH.
11. Remove the front wiper arm LH.
12. Remove the front wiper arm RH.
13. Remove the hood to cowl top seal.
14. Remove the cowl top ventilator louver sub assembly.
15. Remove the front fender splash shield sub assembly LH.
16. Remove the front fender splash shield sub assembly RH.
17. Remove the No. 1 engine under cover sub assembly.
18. Remove the No. 2 engine under cover.
19. Remove the front fender apron seal LH.
20. Remove the front fender apron seal front RH.
21. Drain the engine oil.
22. Drain the engine coolant.
23. Remove the v-bank cover sub assembly.
24. Remove the air cleaner hose assembly.
25. Remove the air cleaner assembly.
26. Remove the No. 1 radiator hose.
27. Remove the No. 2 radiator hose.
28. Remove the fan shroud.
29. Remove the radiator assembly.
30. Disconnect the engine wire.
31. Disconnect the air pump hose and wire harness.
32. Disconnect the No. 2 water by pass pipe.
33. Disconnect the cooler compressor assembly.
34. Disconnect the No. 2 fuel tube sub assembly.
35. Remove the oil filter element.
36. Remove the engine oil level dipstick guide.
37. Remove the oil pressure sender gauge assembly.
38. Remove the No. 2 water by pass pipe sub assembly (with oil cooler).
39. Remove the No. 1 oil cooler bracket.
40. Remove the intake manifold.
41. Disconnect the vane pump assembly.
42. Disconnect the oil cooler pipe assembly.
43. Remove the generator assembly.
44. Remove the No. 1 water by pass hose.
45. Remove the water by pass pipe sub assembly.
46. Remove the front water by pass joint.
47. Remove the No. 2 engine cover.
48. Remove the No. 1 engine cover.
49. Remove the air tube sub assembly.
50. Remove the water inlet housing.
51. Remove the water pump pulley.
52. Remove the No. 1 idler pulley sub assembly.
53. Remove the fluid coupling bracket.
54. Remove the v-ribbed belt tensioner assembly.
55. Remove the ignition coil assembly.
56. Remove the cylinder head cover sub assembly LH and RH.
57. Remove the spark plug tube gasket.
58. Remove the crankshaft pulley.
59. Disconnect the wire harness clamp bracket.
60. Remove the timing chain cover sub assembly.
61. Remove the water inlet pipe.
62. Set the No. 1 cylinder to TDC/compression.
63. Remove the No. 1 chain tensioner assembly LH.
64. Remove the No. 1 chain tensioner slipper LH.
65. Remove the No. 1 chain vibration damper LH.
66. Remove the No. 1 chain sub assembly LH.
67. Remove the No. 3 chain tensioner assembly.
68. Remove the No. 1 chain tensioner assembly RH.
69. Remove the No. 1 chain tensioner slipper RH.
70. Remove the No. 1 chain vibration damper RH.
71. Remove the No. 1 chain sub assembly RH.
72. Remove the No. 2 chain tensioner assembly.
73. Remove the crankshaft timing gear key.
74. Remove the camshaft bearing cap LH.
   a. Make sure that the knock pin of the camshaft is positioned as shown.
   b. Uniformly loosen and remove the 10 bearing cap bolts in the sequence shown in the illustration.
   c. Uniformly loosen and remove the 18 bearing cap bolts in the sequence shown.

➡Uniformly loosen the bolts while keeping the camshaft level.

   d. Remove the 6 bearing caps.

➡Arrange the removed parts in the correct order.

   e. Remove the No. 3 and No. 4 camshafts.

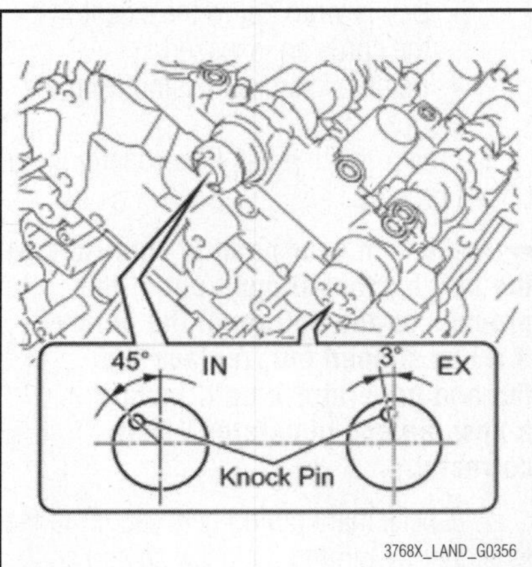

3768X_LAND_G0356

Fig. 83 Positioning the knock pin of the camshaft

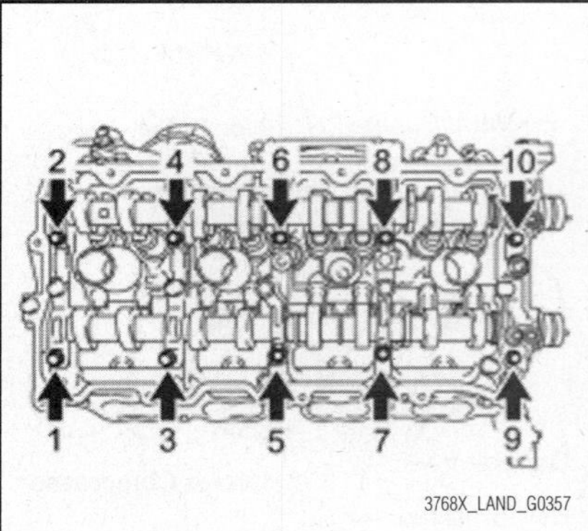

3768X_LAND_G0357

Fig. 84 Identifying the bearing cap bolt removal sequence (1 of 2)

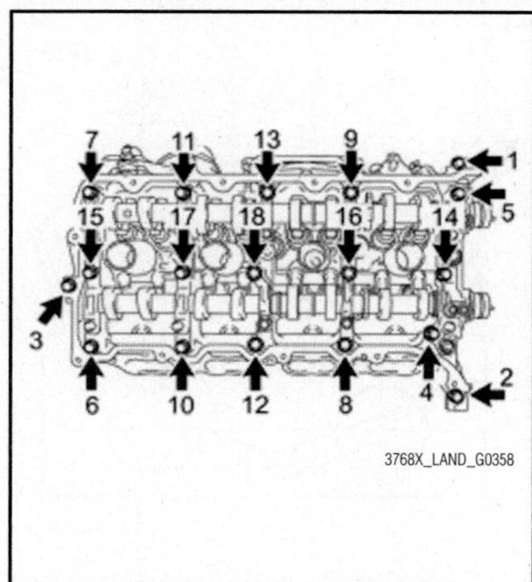

Fig. 85 Identifying the bearing cap bolt removal sequence (2 of 2)

75. Remove the camshaft housing sub assembly LH.

a. Remove the camshaft housing by prying between the cylinder head and camshaft housing with a screwdriver.

➡Be careful not to damage the contact surfaces of the cylinder head and camshaft housing.

➡Tape the screwdriver tip before use.

76. Remove the camshaft bearing cap RH.

a. Make sure that the knock pin of the camshaft is positioned as shown.

b. Uniformly loosen and remove the 10 bearing cap bolts in the sequence shown.

c. Uniformly loosen and remove the 18 bearing cap bolts in the sequence shown.

➡Uniformly loosen the bolts while keeping the camshaft level.

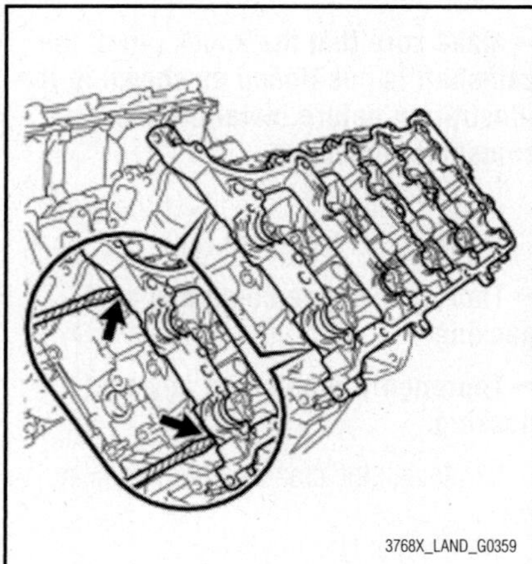

Fig. 86 Removing the camshaft housing sub assembly LH

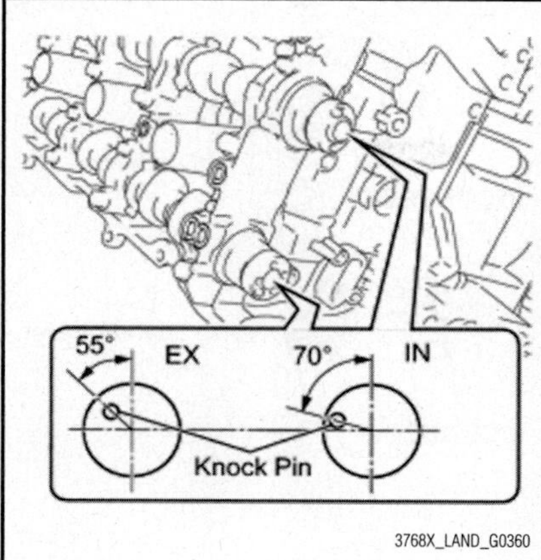

Fig. 87 Positioning the knock pin of the camshaft RH

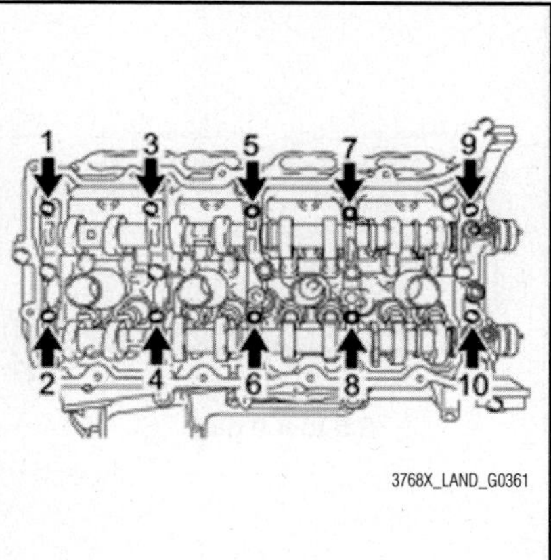

Fig. 88 Removing the 10 bearing cap bolts in sequence

d. Remove the 6 bearing caps.

➡Arrange the removed parts in the correct order.

e. Remove the No. 1 and No. 2 camshafts.

77. Remove the camshaft housing sub assembly RH.

a. Remove the camshaft housing by prying between the cylinder head and camshaft housing with a screwdriver.

➡Be careful not to damage the contact surfaces of the cylinder head and camshaft housing.

➡Tape the screwdriver tip before use.

To install:

78. Install the camshaft bearing cap RH.

a. Apply a light coat of engine oil to the camshaft journals, camshaft housings and bearing caps.

b. Install the No. 1 and No. 2 camshafts to the camshaft housing.

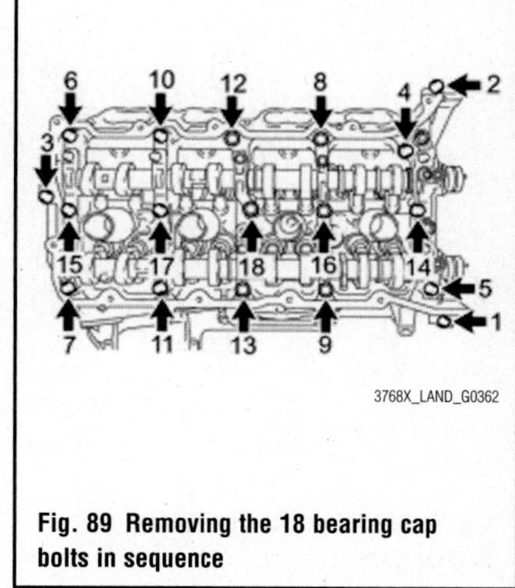

Fig. 89 Removing the 18 bearing cap bolts in sequence

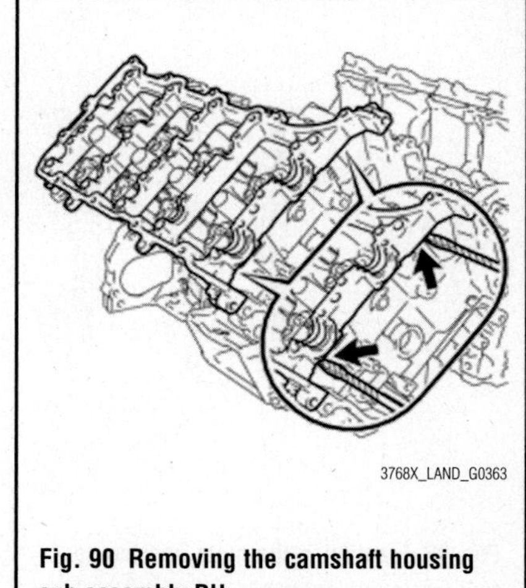

Fig. 90 Removing the camshaft housing sub assembly RH

c. Confirm the marks and numbers on the camshaft bearing caps and place them in their proper positions and directions.

d. Temporarily install the 10 bolts in the order shown.

79. Install the camshaft housing sub assembly RH.

a. Make sure that the valve rocker arms are installed as shown.

b. Apply seal packing in a continuous line as shown with a seal diameter or 0.138-0.157 inch (3.5-4.0 mm).

➡Remove any oil from the contact surface.

➡Install the camshaft housing within 3 minutes and tighten the bolts within 15 minutes after applying seal packing.

c. Install the camshaft housing, and install the 18 bolts in the order shown. Tighten bolt A to 7 ft. lbs. (10 Nm).

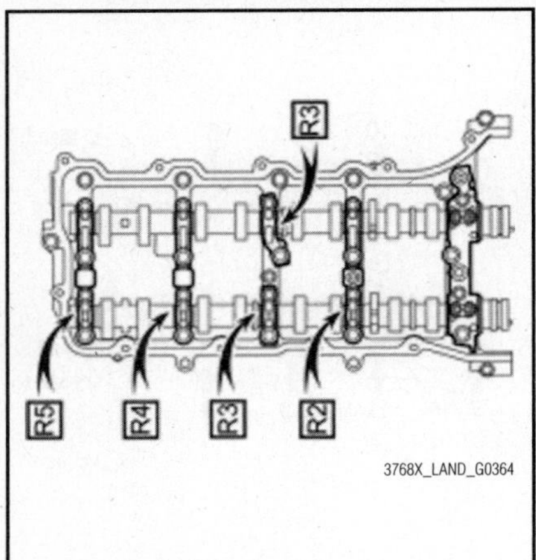

Fig. 91 Positioning the camshaft bearing caps

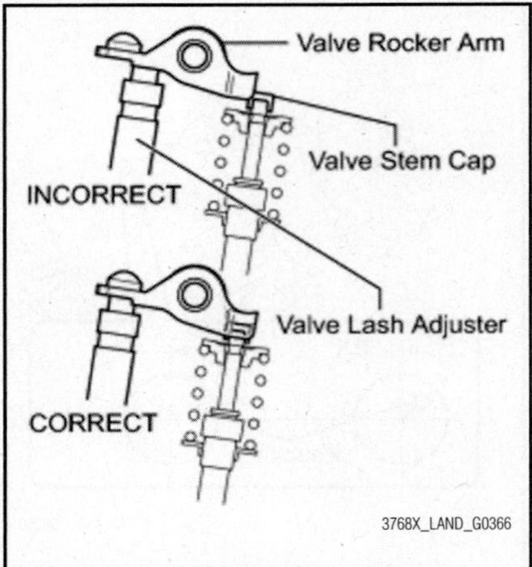

Fig. 93 Checking the valve rocker arm installation

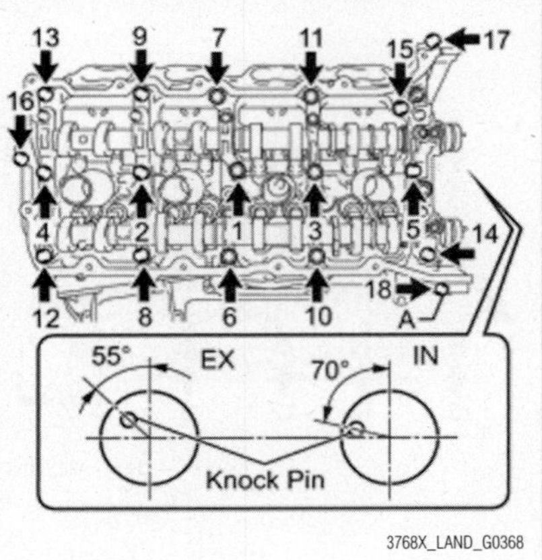

Fig. 95 Identifying the camshaft housing bolt tightening sequence RH

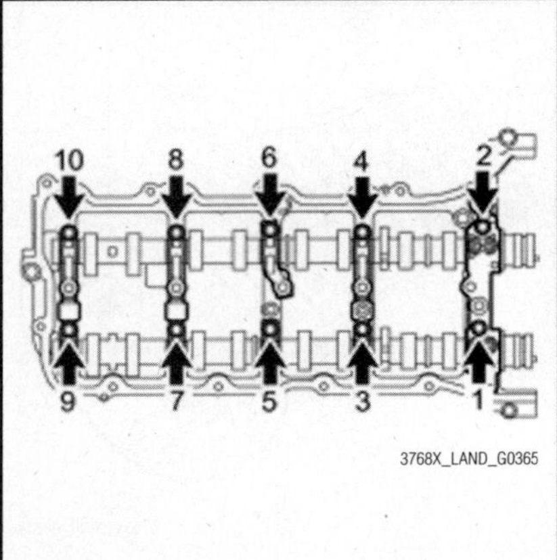

Fig. 92 Identifying the main bearing cap bolt tightening sequence RH

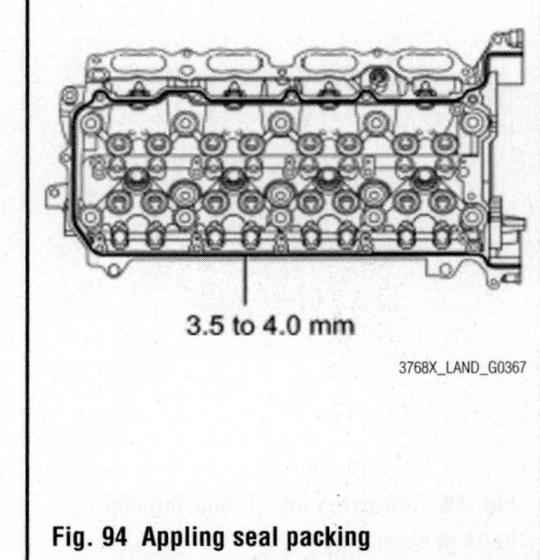

Fig. 94 Appling seal packing

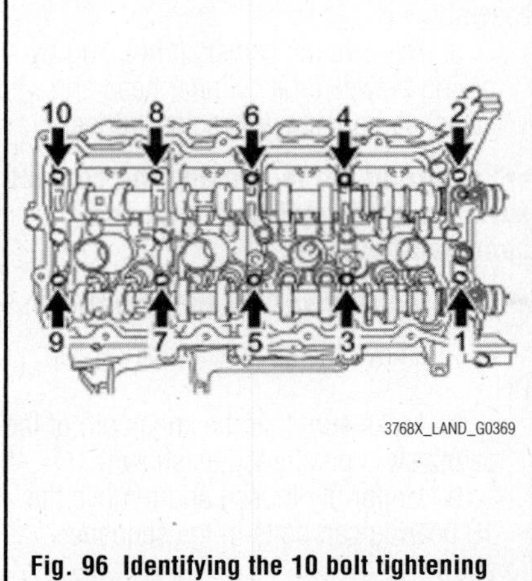

Fig. 96 Identifying the 10 bolt tightening sequence RH

Tighten all bolts except A to 22 ft. lbs. (30 Nm).

➡Do not start the engine for at least 2 hours after the installation.

➡Make sure that the knock pin of the camshaft is positioned as shown in the illustration before installing the camshaft housing.

d. Tighten the 10 bolts in the order shown. Tighten to 12 ft. lbs. (16 Nm).

➡Thoroughly wipe clean any seal packing.

80. Install the camshaft bearing cap LH.

➡Apply a light coat of engine oil to the camshaft journals, camshaft housings and bearing caps.

➡Install the No. 3 and No. 4 camshafts to the camshaft housing.

a. Confirm the marks and numbers on the camshaft bearing caps and place them in their proper positions and directions.

b. Temporarily install the 10 bolts in the order shown.

81. Install the camshaft housing sub assembly LH.

a. Make sure that the valve rocker arms are installed as shown.

b. Apply seal packing in a continuous line with a diameter of 0.138-0.157 inch (3.5-4.0 mm).

➡Remove any oil from the contact surface.

➡Install the camshaft housing within 3 minutes and tighten the bolts within 15 minutes after applying seal packing.

c. Install the camshaft housing, and install the 18 bolts in the order shown. Tighten bolt A to 7 ft. lbs. (10 Nm).

Tighten all other bolts to 22 ft. lbs. (30 Nm).

➡Do not start the engine for at least 2 hours after the installation.

➡Make sure that the knock pin of the camshaft is positioned as shown in the illustration before installing the camshaft housing.

d. Tighten the 10 bolts in the order shown to 12 ft. lbs. (16 Nm).

➡Thoroughly wipe clean any seal packing.

➡Thoroughly wipe clean any seal packing.

82. Install the crankshaft timing gear key.

83. Set the No. 1 cylinder to TDC/compression.

84. Install the No. 2 chain tensioner assembly.

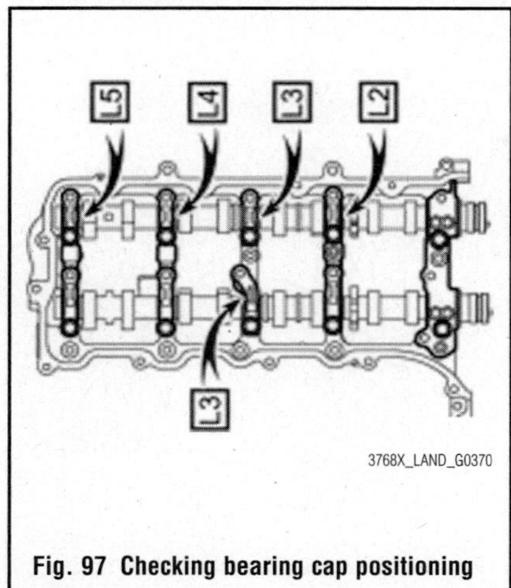

Fig. 97 Checking bearing cap positioning RH

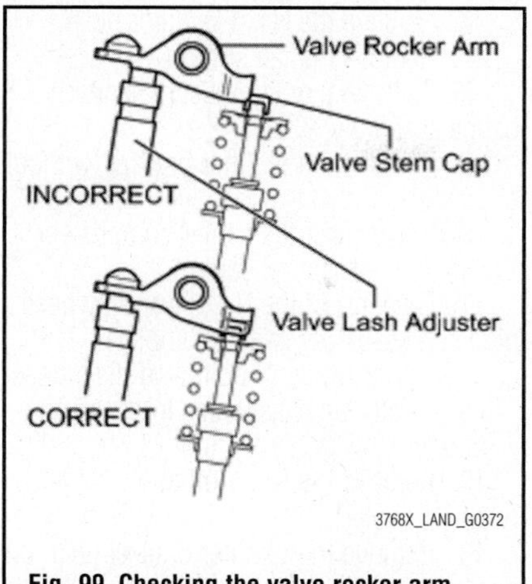

Fig. 99 Checking the valve rocker arm installation LH

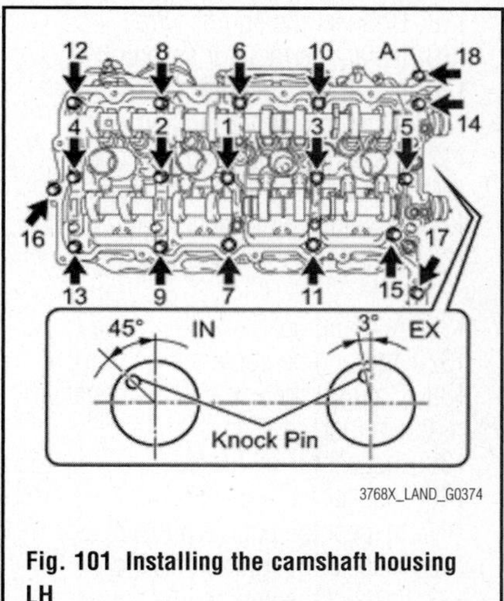

Fig. 101 Installing the camshaft housing LH

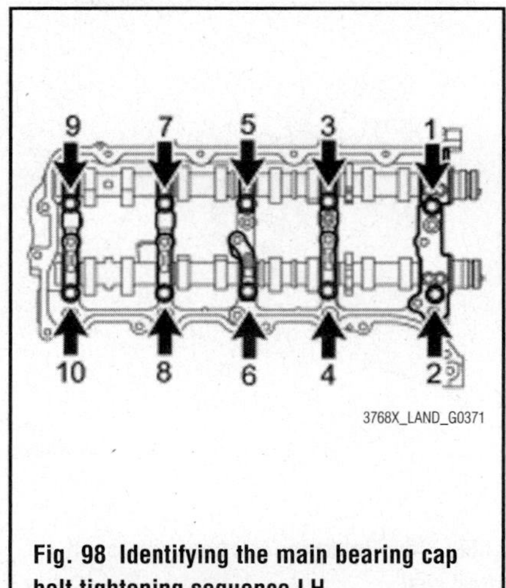

Fig. 98 Identifying the main bearing cap bolt tightening sequence LH

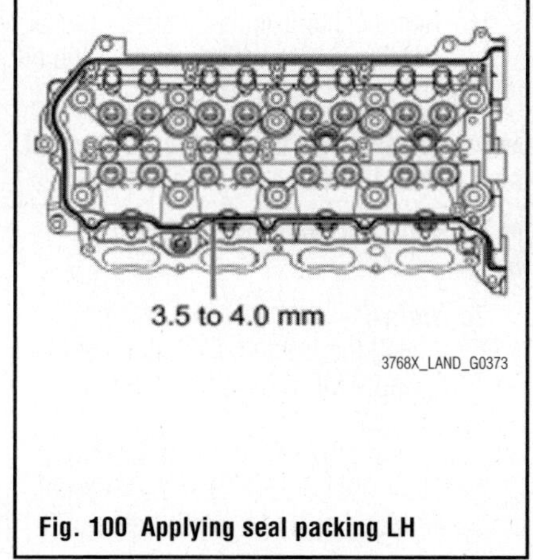
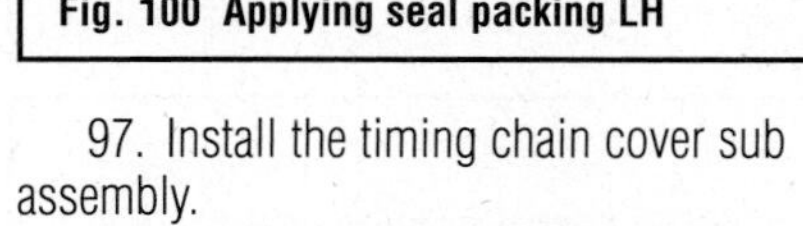

Fig. 100 Applying seal packing LH

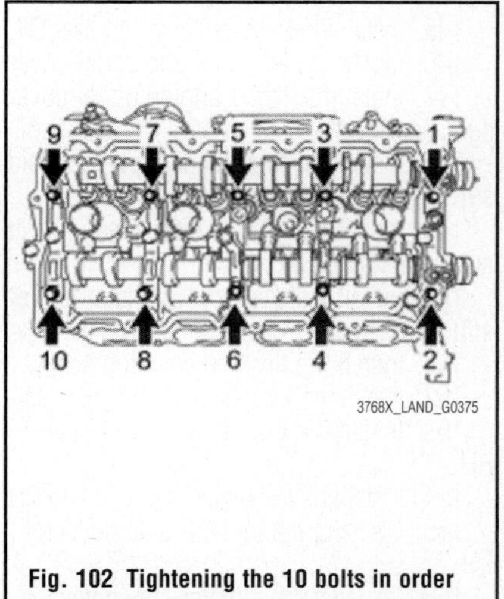

Fig. 102 Tightening the 10 bolts in order

85. Install the No. 1 chain sub assembly RH.
86. Install the No. 1 chain vibration damper RH.
87. Install the No. 1 chain tensioner slipper RH.
88. Install the No. 1 chain tensioner assembly RH.
89. Install the No. 3 chain tensioner assembly.
90. Install the No. 1 chain sub assembly LH.
91. Install the No. 1 chain tensioner slipper LH.
92. Install the No. 1 chain tensioner assembly LH.
93. Install the No. 1 chain vibration damper LH.
94. Tighten the camshaft timing gear assembly.
95. Check the No. 1 cylinder to TCD/compression.
96. Install the water inlet pipe.
97. Install the timing chain cover sub assembly.
98. Install the spark plug tube gasket.
99. Install the cylinder head cover sub assembly LH.
100. Install the cylinder head cover sub assembly RH.
101. Install the ignition coil assembly.
102. Install the crankshaft timing gear key.
103. Install the crankshaft pulley.
104. Connect the wire harness clamp bracket.
105. Install the No. 1 idler pulley sub assembly.
106. Install the water pump pulley.
107. Install the water inlet housing.
108. Install the air tube sub assembly.
109. Install the No. 1 engine cover.
110. Install the No. 2 engine cover.
111. Install the front water by pass joint.
112. Install the water by pass pipe sub assembly.
113. Install the No. 1 water by pass hose.
114. Install the generator assembly.
115. Connect the oil cooler pipe assembly.
116. Connect the vane pump assembly.
117. Install the intake manifold.
118. Install the oil filter bracket.
119. Install the No. 1 oil cooler bracket (with oil cooler).
120. Install the No. 2 water by pass pipe sub assembly (with oil cooler).
121. Install the oil pressure sender gauge assembly.
122. Install the engine oil level dipstick guide.
123. Install the oil filter element.
124. Connect the No. 2 fuel tube sub assembly.
125. Connect the cooler compressor assembly.
126. Connect the No. 2 water by pass pipe.
127. Connect the air pump hose and wire harness.

128. Connect the engine wire.
129. Install the radiator assembly.
130. Install the fan shroud.
131. Install the No. 2 radiator hose.
132. Install the No. 1 radiator hose.
133. Install the air cleaner assembly.
134. Install the air cleaner hose assembly.
135. Install the v-bank cover sub assembly.
136. Add engine oil.
137. Add engine coolant.
138. Connect the cable to the negative battery terminal.
139. Inspect for oil leaks.
140. Inspect for coolant leaks.
141. Inspect the engine oil level.
142. Inspect the ignition timing.
143. Inspect the engine idle speed.
144. Install the front fender apron seal front RH.
145. Install the front fender apron seal LH.
146. Install the No. 2 engine under cover.
147. Install the No. 1 engine under cover sub assembly.
148. Install the front fender splash shield sub assembly LH.
149. Install the front fender splash shield sub assembly RH.
150. Install the cowl top ventilator louver sub assembly.
151. Install the hood to cowl top seal.
152. Install the front fender main seal LH.
153. Install the front fender main seal RH.
154. Install the front wiper arm LH and RH.
155. Connect the radiator side deflector LH.
156. Connect the radiator side deflector RH (without air cooled transmission oil cooler).
157. Install the transmission oil cooler air duct (with air cooled transmission oil cooler).
158. Install the front bumper cover.
159. Install the radiator grille.
160. Install the upper radiator support seal.

CRANKSHAFT FRONT SEAL

REMOVAL & INSTALLATION

See Figures 103 through 105.

1. Remove the front fender splash shield sub assembly LH.
2. Remove the front fender splash shield sub assembly RH.
3. Remove the No. 1 engine under cover sub assembly.
4. Drain the engine coolant.
5. Remove the v-bank cover sub assembly.
6. Remove the No. 1 radiator hose.
7. Remove the fan shroud.
8. Remove the oil pressure sender gauge assembly.
9. Remove the oil filter bracket (without oil cooler).
10. Disconnect the cooler compressor assembly (with the oil cooler).
11. Disconnect the No. 2 water by pass pipe sub assembly (with oil cooler).
    a. Remove the 3 bolts and disconnect the 2 water by pass hoses from the oil cooler.
12. Remove the No. 1 oil cooler bracket (with oil cooler).
13. Remove the oil filter bracket (with oil cooler).
14. Remove the crankshaft pulley.
15. Remove the crankshaft timing gear key.
    a. Remove the crankshaft timing gear key from the crankshaft.
16. Remove the front crankshaft oil seal.
    a. Using a screwdriver, pry out the oil seal.

➡Do not damage the surface of the oil seal press fit hole and crankshaft.

➡Tape the screwdriver tip before use.

To install:

17. Install the front crankshaft oil seal.
    a. Apply MP grease to the lip of a new oil seal.
    b. Using the special tools (09223-22010 or 09506-35010) and a hammer, tap in the oil seal to a depth between 0 and 0.0394 inch (0 and 1 mm) from the timing chain cover edge.

➡Keep the lip free from foreign matter.

➡Do not tap the oil seal at an angle.

18. Install the crankshaft timing gear key.
19. Install the crankshaft pulley.
20. Install the oil filter bracket (with oil cooler).
21. Install the No. 1 oil cooler bracket (with oil cooler).Install the No. 1 oil cooler bracket (with oil cooler).
22. Connect the No. 2 water by pass pipe sub assembly (with oil cooler).Connect the No. 2 water by pass pipe sub assembly (with oil cooler).
    a. Connect the 2 water by pass hoses to the oil cooler.
    b. Install the 3 by pass pipe bolts and tighten to 7 ft. lbs. (10 Nm).
23. Connect the cooler compressor assembly (with oil cooler).

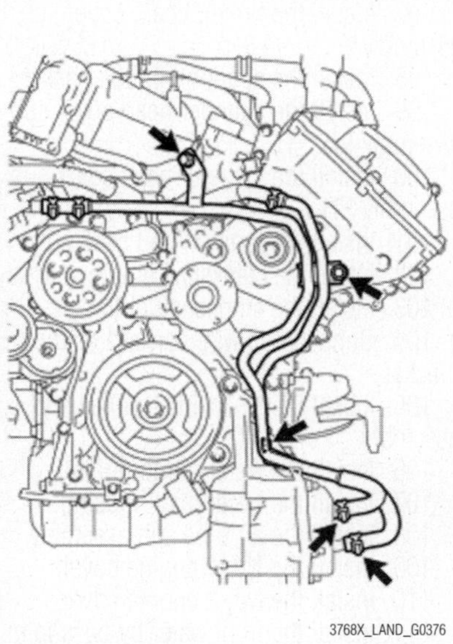

Fig. 103 Disconnecting the 2 water by pass hoses from the oil cooler

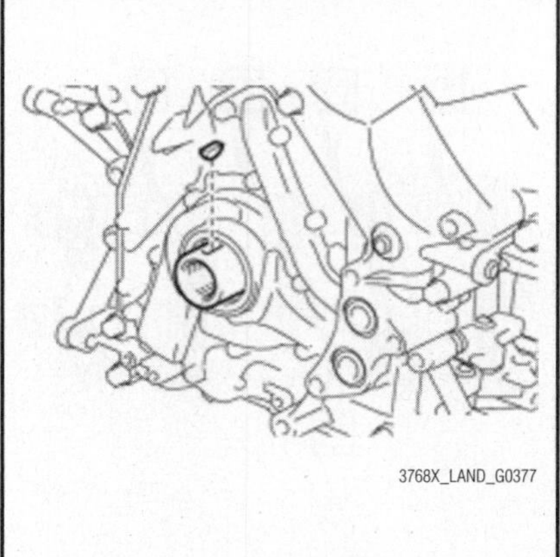

Fig. 104 Removing the crankshaft timing gear key

Fig. 105 Removing the front crankshaft oil seal

24. Install the oil filter bracket (without oil cooler).
25. Install the oil pressure sender gauge assembly.
26. Install the fan shroud.
27. Install the No. 1 radiator hose.
28. Add engine coolant.
29. Inspect for coolant leaks.
30. Inspect for oil leaks.
31. Inspect the engine oil level.
32. Install the v-bank cover sub assembly.
33. Install the No. 1 engine under cover sub assembly.
34. Install the front fender splash shield sub assembly LH and RH.

CYLINDER HEAD

REMOVAL & INSTALLATION

See Figures 106 through 116.

1. Discharge the fuel system pressure.
2. Disconnect the cable from the negative battery terminal.

CAUTION

Wait at least 90 seconds after disconnecting the cable from the negative (-) battery terminal to disable the SRS system.

3. Remove the exhaust manifold sub assembly.
4. Remove the camshaft.
5. Remove the No. 1 valve rocker arm sub assembly.
6. Remove the valve lash adjuster assembly.
7. Remove the valve stem cap.
8. Remove the cylinder head sub assembly LH.
   a. Uniformly loosen and remove the 2 bolts in the sequence shown.
   b. Using a 10 mm bi-hexagon wrench, uniformly loosen the 10 cylinder head bolts in the sequence shown. Remove the 10linder head bolts and plate washers.

➡Be careful not to drop washers into the cylinder head.

➡Head warpage or cracking could result from removing bolts in an incorrect order.

➡Be sure to arrange the removed parts for each installation position separately.

   c. Remove the cylinder head and gasket.

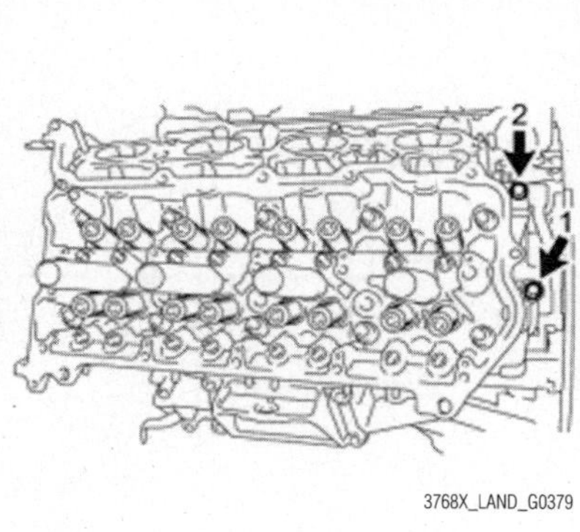

Fig. 106 Loosening the 2 cylinder head sub assembly bolts

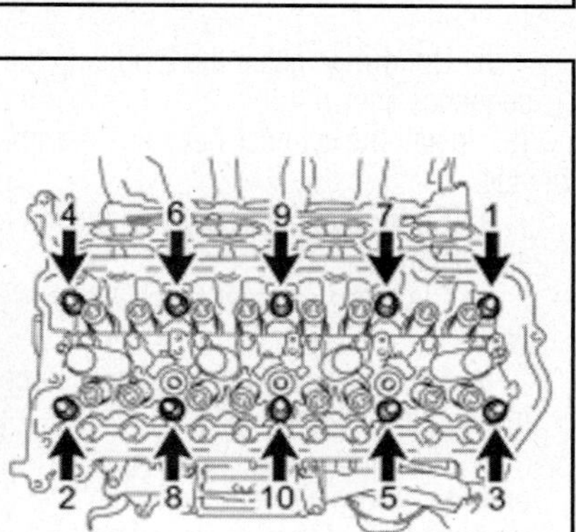

Fig. 107 Identifying the cylinder head bolt loosening sequence LH

9. Remove the cylinder head sub assembly RH.
   a. Uniformly loosen and remove the 2 bolts in the sequence shown.
   b. Using a 10 mm bi-hexagon wrench, uniformly loosen the 10 cylinder head bolts in the sequence shown in the illustration. Remove the 10 cylinder head bolts and plate washers.

➡Be careful not to drop washers into the cylinder head.

➡Head warpage or cracking could result from removing bolts in an incorrect order.

➡Be sure to arrange the removed parts for each installation position separately.

   c. Removing the cylinder head and gasket.

To install:

10. Inspect the cylinder head set bolt.

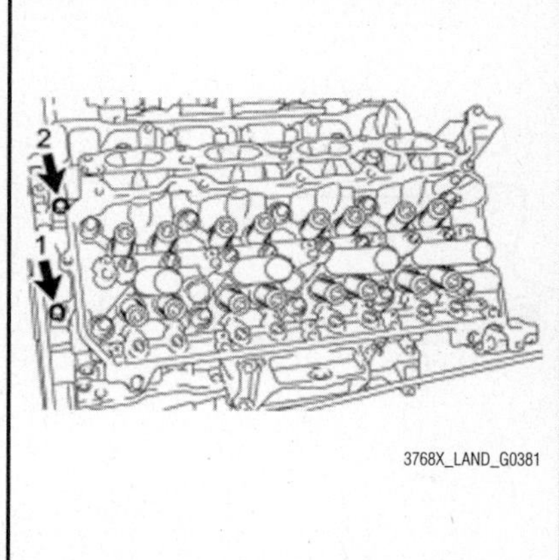

Fig. 108 Loosening the 2 cylinder head sub assembly bolts RH

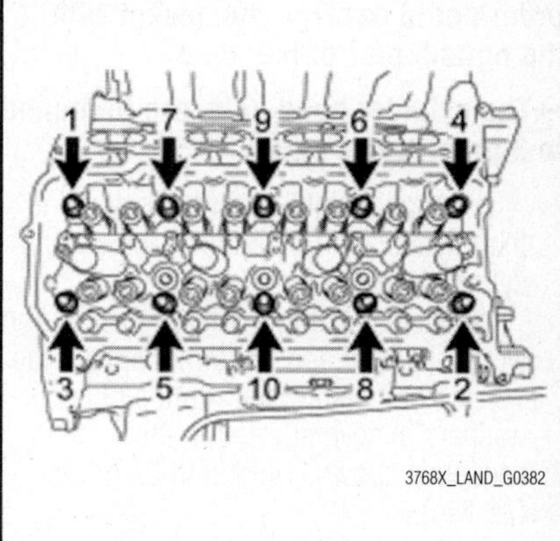

Fig. 109 Identifying the cylinder head bolt loosening sequence RH

11. Inspect the cylinder head sub assembly.
12. Install the cylinder head sub assembly RH.
   a. Check the piston protrusions for each cylinder. Clean the cylinder block with solvent. Set the piston of the cylinder to be measured to slightly ATDS.
   b. Place the cylinder head gasket on the cylinder block surface with the front face of the Lot No. stamp upward.

➡Be careful of the installation direction.

➡Make sure that no oil is on the front end (indicated by the arrows) of the cylinder head gasket.

   c. Place the cylinder head on the cylinder block.

➡Ensure that no oil is on the mounting surface of the cylinder head.

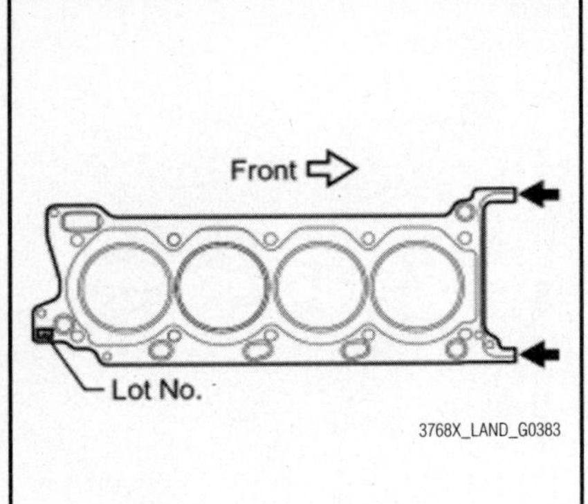

Fig. 110 Placing the cylinder head gasket on the cylinder block surface

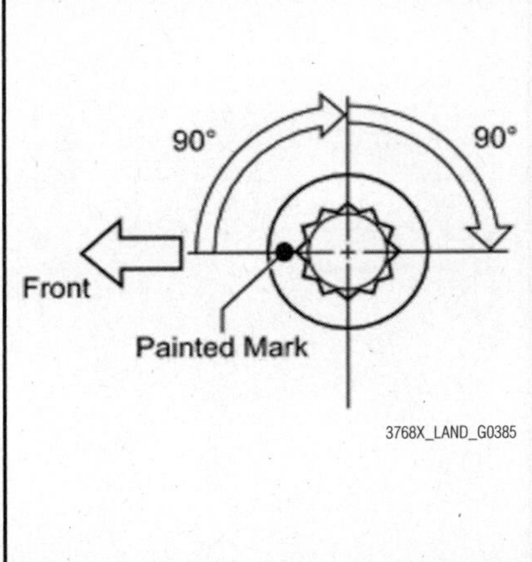

Fig. 112 Marking the cylinder head bolt with paint

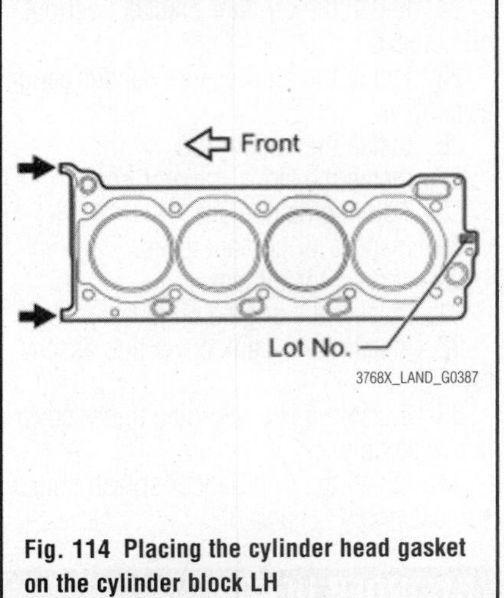

Fig. 114 Placing the cylinder head gasket on the cylinder block LH

➡Gently place the cylinder head in order not to damage the gasket with the bottom part of the head.

➡The cylinder head bolts are tightened in 3 progressive steps.

d. Apply a light coat of engine oil to the threads and under the heads of the cylinder head bolts.

e. Step 1: Using a 10 mm bi-hexagon wrench, install and uniformly tighten the 10 cylinder head bolts with the plate washers in several steps in the sequence shown. Tighten to 27 ft. lbs. (36 Nm).

f. Step 2: Mark each cylinder head bolt with paint. Tighten the cylinder head bolts 90° in the same tightening sequence as the previous step.

g. Step 3: Tighten the cylinder head bolts another 90° in the sequence shown in step 1. Check that the paining marks are facing rearward.

h. Uniformly install the 2 bolts in the sequence shown.

13. Install the cylinder head sub assembly LH.

a. Check the piston protrusions for each cylinder.

b. Clean the cylinder block with solvent.

c. Set the piston of the cylinder to be measured to slightly ATDC.

d. Place the cylinder head gasket on the cylinder block surface with the front face of the Lot No. stamp upward.

➡Be careful of the installation direction.

➡Make sure that no oil is on the front end (indicated by the arrows) of the cylinder head gasket.

e. Place the cylinder head on the cylinder block.

➡Ensure that no oil is on the mounting surface of the cylinder head.

➡Gently place the cylinder head in order not to damage the gasket with the bottom part of the head.

➡The cylinder head bolts are tightened in 3 progressive steps.

f. Apply a light coat of engine oil to the threads and under the heads of the cylinder head bolts.

g. Step 1: Using a 10 mm bi-hexagon wrench, install and uniformly tighten the 10 cylinder head bolts with the plate washers in several steps in the sequence shown. Tighten to 27 ft. lbs. (36 Nm).

h. Step 2: Mark each cylinder head bolt with paint. Tighten the cylinder head bolts 90° in the same tightening sequence as the previous step.

i. Step 3: Tighten the cylinder head bolts another 90° in the sequence shown

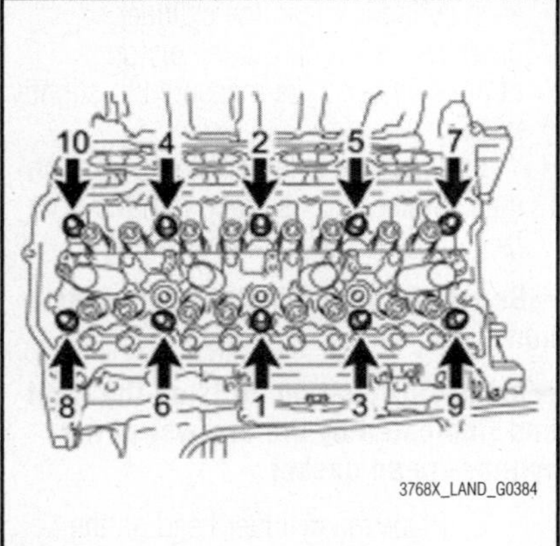

Fig. 111 Installing the RH cylinder head bolts in sequence

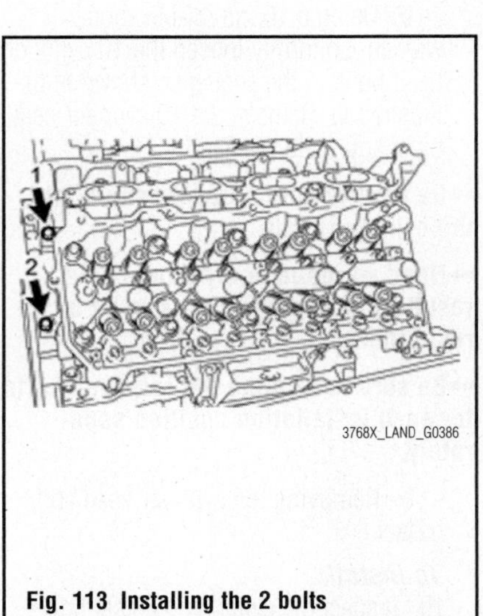

Fig. 113 Installing the 2 bolts

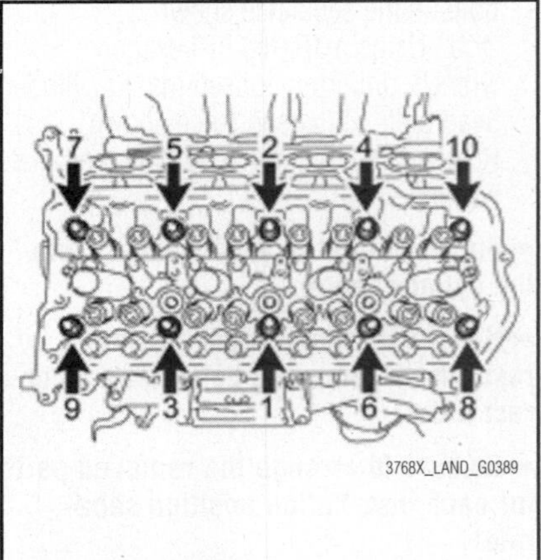

Fig. 115 Installing the LH cylinder head bolts in sequence

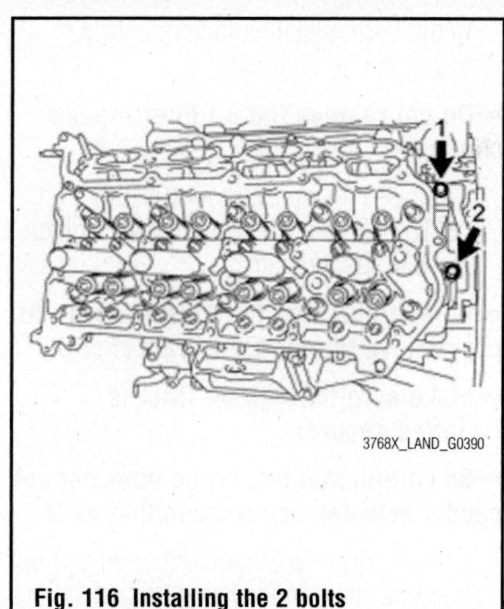

Fig. 116 Installing the 2 bolts

in step 1. Check that the paining marks are facing rearward.

j. Uniformly install the 2 bolts in the sequence shown.

14. Install the valve stem cap.
15. Install the valve lash adjuster assembly.
16. Install the No. 1 valve rocker arm sub assembly.
17. Install the camshaft.
18. Install the exhaust manifold sub assembly.
19. Connect the cable to the negative battery terminal.
20. Inspect for exhaust gas leaks.
21. Inspect the ignition timing.
22. Inspect the engine idle speed.

ENGINE OIL & FILTER

REPLACEMENT

See Figures 117 through 127.

➡Prolonged and repeated contact with engine oil will result in the removal of natural oils from the skin, leading to dryness, irritation and dermatitis. In addition, used engine oil contains potentially harmful contaminants which may cause skin cancer.

➡Precautions should be taken when replacing engine oil to minimize the risk of your skin making contact with used engine oil. Protective clothing and gloves that cannot be penetrated by oil should be worn. The skin should be washed with soap and water, or use waterless hand cleaner, to remove any used engine oil thoroughly. Do not use gasoline, thinners, or solvents.

➡In order to protect the environment, used oil and used oil filters must be disposed of at designated disposal sites.

1. Drain the engine oil.
   a. Remove the oil filler cap.
   b. Remove the 2 bolts and No. 2 engine under cover seal.
   c. Remove the oil pan drain plug and gasket, and drain the engine oil into a container.
   d. Install a new gasket and the oil pan drain plug. Tighten to 30 ft. lbs. (40 Nm).
2. Remove the front fender splash shield sub assembly LH.
3. Remove the No. 1 engine under cover sub assembly.
   a. Remove the 5 bolts and No. 1 engine under cover sub assembly LH.

➡If the No. 1 cooler packing is not divided, remove both sides of the No. 1 engine under cover and cut the packing at the perforation indicated in the illustration.

Fig. 117 Removing the 2 bolts and No. 2 engine under cover seal

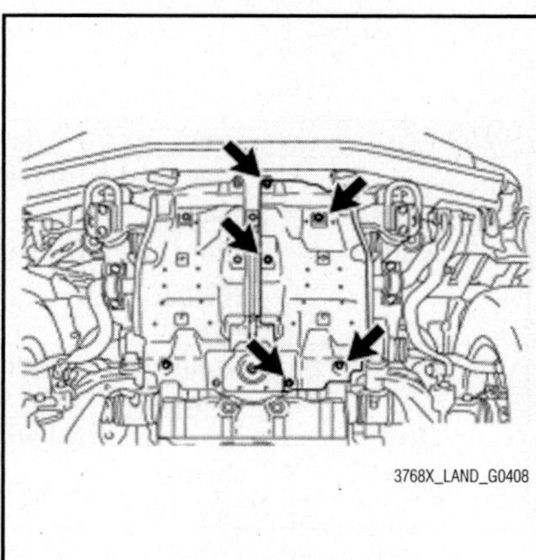

Fig. 118 Removing the No. 1 engine under cover sub assembly

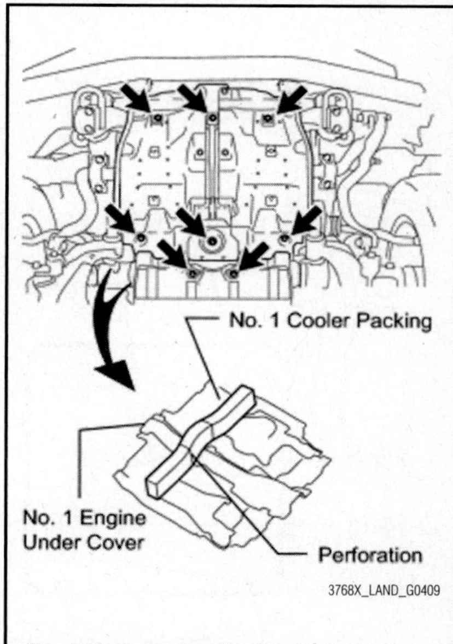

Fig. 119 Removing the No. 1 cooler packing

4. Remove the oil filter element.
   a. Connect a hose with an inside diameter of 0.591 inch (15 mm) to the pipe.
   b. Remove the oil filter drain plug.
   c. Install the pipe to the oil filter cap.

➡If the O-ring is removed with the drain plug, install the O-ring together with the pipe.

➡Use a container to catch the draining oil.

   d. Check that oil is drained from the oil filter. Then disconnect the pipe and remove the O-ring as shown.
   e. Use the special tool (09228-06501), remove the oil filter cap.

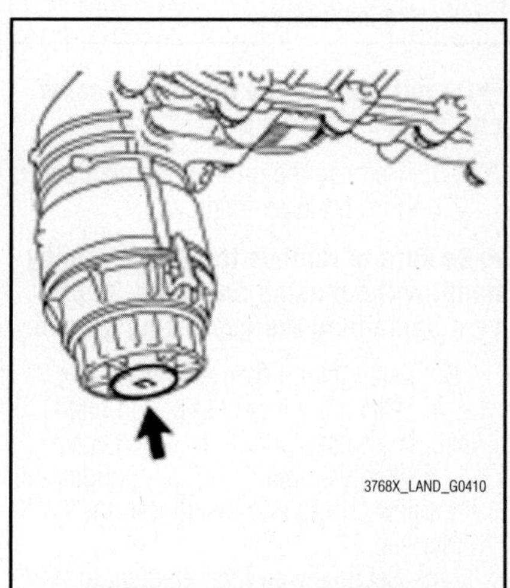

Fig. 120 Removing the oil filter drain plug

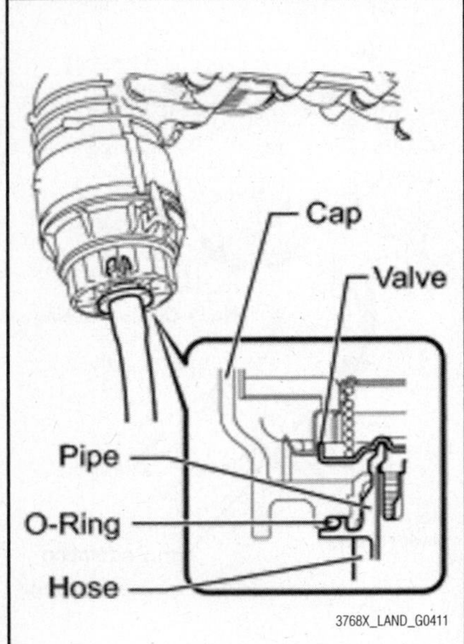

Fig. 121 Installing the pipe to the oil filter cap

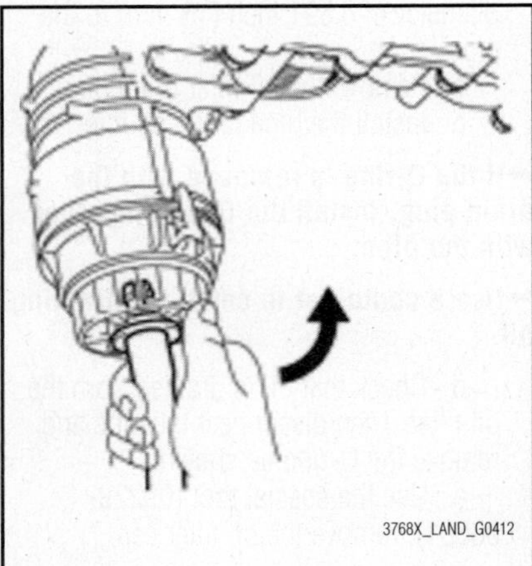

Fig. 122 Disconnecting the pipe and removing the O-ring

➡Do not remove the oil filter bracket clip.

f. Remove the oil filter element and O-ring from the oil filter cap.

➡Be sure to remove the cap O-ring by hand, without using any tools, to prevent damage to the cap O-ring groove.

5. Install the oil filter element.

a. Clean the inside of the oil filter cap, its threads and its O-ring groove.

b. Apply a small amount of engine oil to a new O-ring and install it to the oil filter cap.

c. Set a new oil filter element to the oil filter cap.

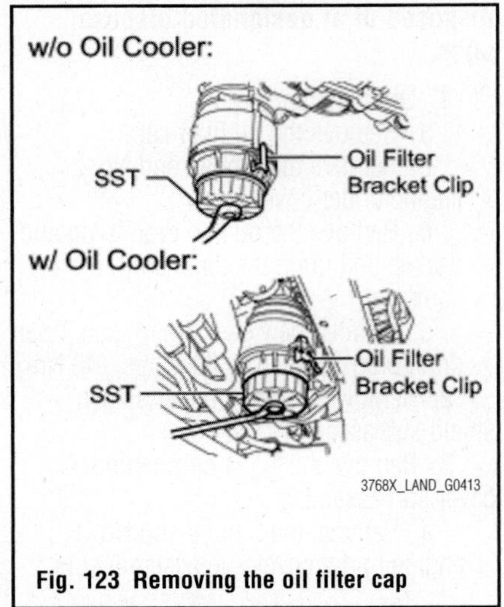

Fig. 123 Removing the oil filter cap

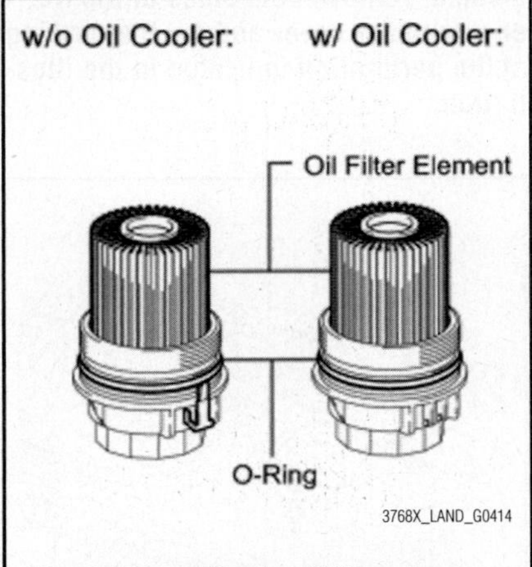

Fig. 124 Removing the oil filter element and O-ring from the oil filter cap

d. Remove any dirt or foreign matter from the installation surface of the engine.

e. Apply a small amount of engine oil to the O-ring again and temporarily install the oil filter cap.

➡Do not remove the oil filter bracket clip.

f. Use the special tool (09228-06501), tighten the oil filter cap. Tighten to 18 ft. lbs. (25 Nm).

➡When tightening the oil filter cap, do not remove the oil filter bracket clip.

➡Make sure that the oil filter is installed securely.

➡Be careful that the O-ring does not get caught between any surrounding parts.

g. Apply a small amount of engine oil to a new drain plug O-ring, and install it to the oil filter cap.

➡Before installing the O-ring, remove any dirt or foreign matter from the installation surface of the oil filter cap.

h. Install the oil filter drain plug. Tighten to 9 ft. lbs. (13 Nm).

➡Be careful that the O-ring does not get caught between any surrounding parts.

i. Install the No. 2 engine under cover seal with the 2 bolts. Tighten to 7 ft. lbs. (10 Nm).

6. Add engine oil and install the oil filler cap.

7. Inspect for oil leaks.

a. Start the engine. Make sure that there are no oil leaks from the area that was worked on.

8. Inspect the engine oil level.

9. Install the No. 1 engine under cover sub assembly.

a. Install the No. 1 engine under cover sub assembly with the 5 bolts.

Paint Mark	Item	Specified Condition
Blue	Drain and refill with oil filter change	7.4 qts (7.0 L)
	Drain and refill without oil filter change	7.0 qts (6.6 L)
	Dry fill	8.9 qts (8.4 L)
Green	Drain and refill with oil filter change	7.9 qts (7.5 L)
	Drain and refill without oil filter change	7.5 qts (7.1 L)
	Dry fill	9.8 qts (9.3 L)

3768X_LAND_G0415

Fig. 125 Standard oil capacity

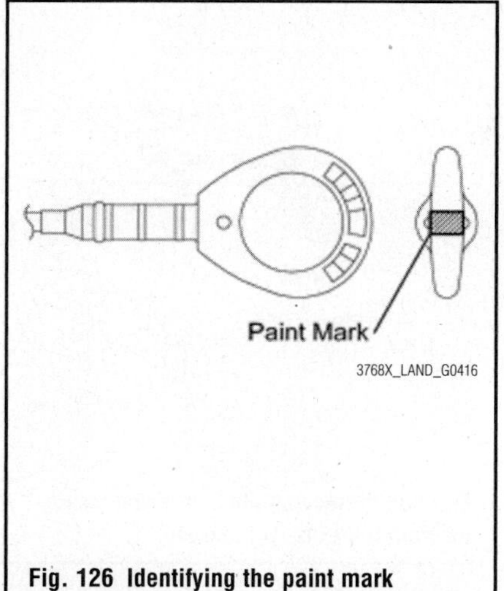

Fig. 126 Identifying the paint mark

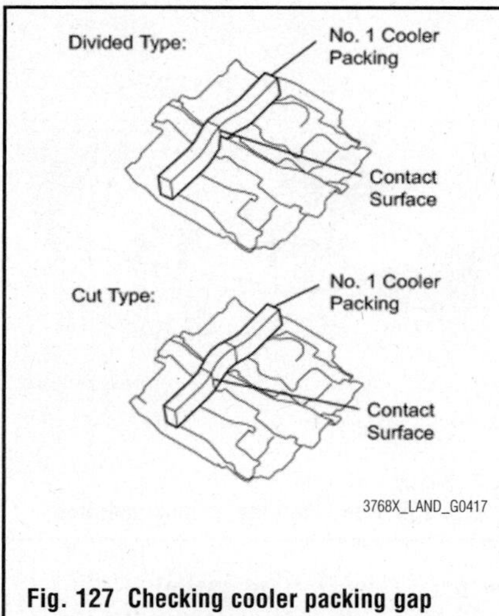

Fig. 127 Checking cooler packing gap

➡**Make sure there is no gap between the ends of the No. 1 cooler packing.**

10. Install the front fender splash shield sub assembly LH.

EXHAUST MANIFOLD

REMOVAL & INSTALLATION

See Figures 128 through 144.

1. Remove the front fender splash shield sub assembly LH.
2. Remove the front fender splash shield sub assembly RH.
3. Remove the No. 1 engine under cover sub assembly.
4. Remove the No. 2 engine under cover.
5. Remove the front fender apron seal front RH.
   a. Using a clip remover, remove the 3 clips and fender apron seal.

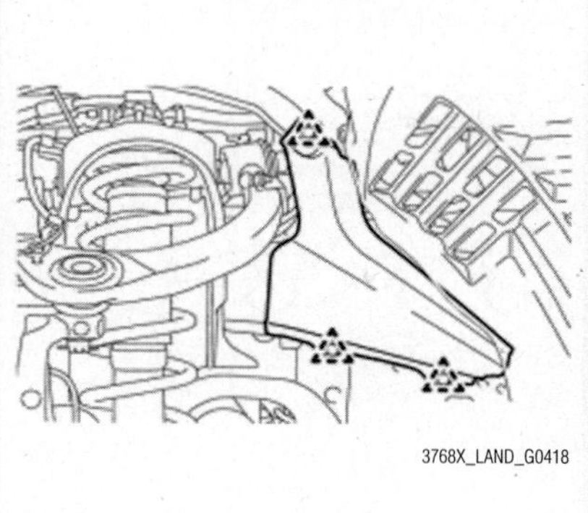

Fig. 128 Removing the front fender apron seal RH

Fig. 130 Removing the front fender apron seal LH

6. Remove the front fender apron seal rear RH.
   a. Using a clip remover, remove the 4 clips and the fender apron seal.
7. Remove the front fender apron seal LH.
   a. Using a clip remover, remove the 3 clips and fender apron seal.
8. Remove the front fender apron seal rear LH.
   a. Using a clip remover, remove the 4 clips and fender apron seal.
9. Remove the engine oil level dipstick guide.
10. Remove the tailpipe assembly.
11. Remove the center exhaust pipe assembly.
12. Remove the front No. 2 exhaust pipe assembly.
13. Remove the front exhaust pipe assembly.
14. Remove the propeller shaft heat insulator.
    a. Remove the 2 bolts and heat insulator.
15. Remove the No. 2 manifold stay.
    a. Remove the 2 bolts and manifold stay.
16. Remove the No. 2 exhaust manifold heat insulator.
    a. Remove the 3 bolts and heat insulator.
17. Remove the exhaust manifold sub assembly LH.
    a. Remove the 10 nuts, exhaust manifold and 2 gaskets.
18. Remove the No. 1 manifold stay.
    a. Remove the 2 bolts and manifold stay.
19. Remove the No. 1 exhaust manifold heat insulator.
    a. Remove the 3 bolts and heat insulator.
20. Remove the exhaust manifold sub assembly RH.

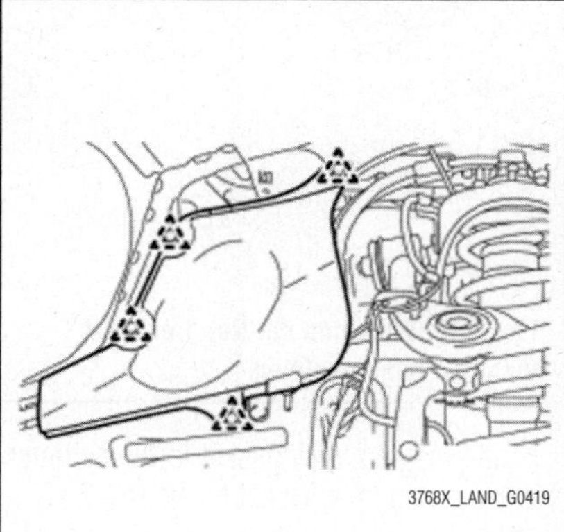

Fig. 129 Removing the front fender apron seal rear RH

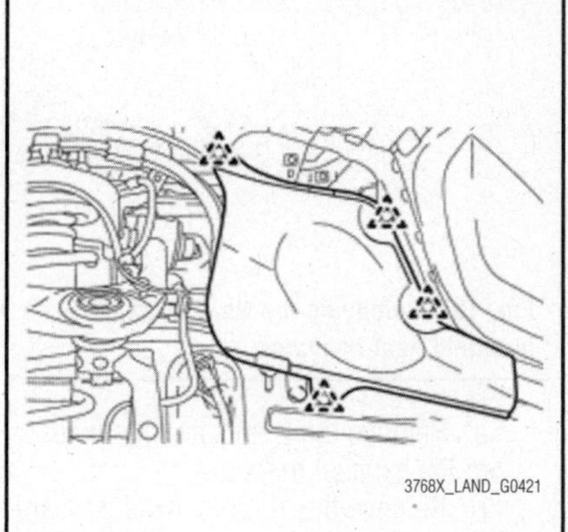

Fig. 131 Removing the front fender apron seal rear LH

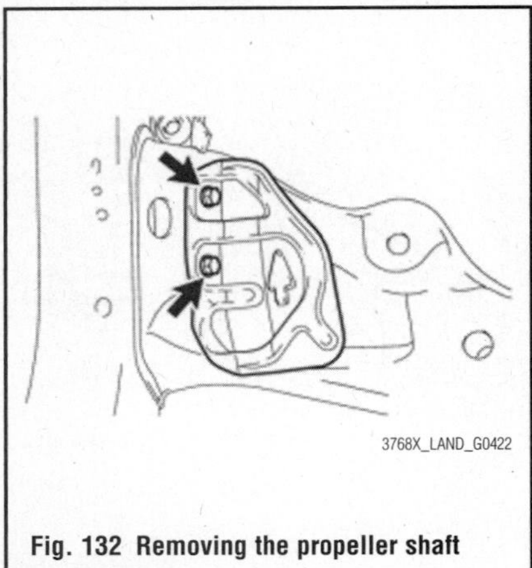

Fig. 132 Removing the propeller shaft heat insulator

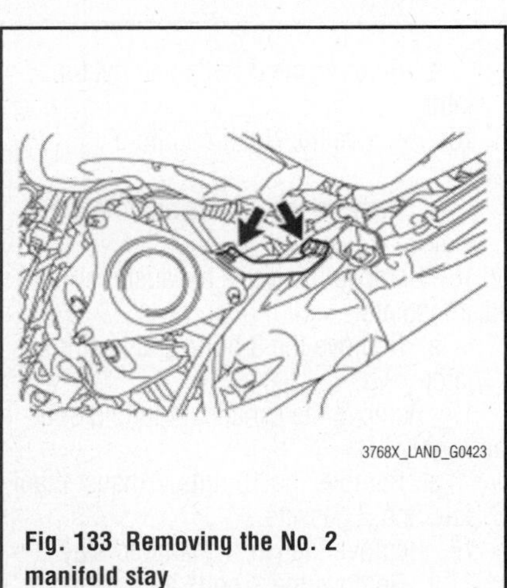

Fig. 133 Removing the No. 2 manifold stay

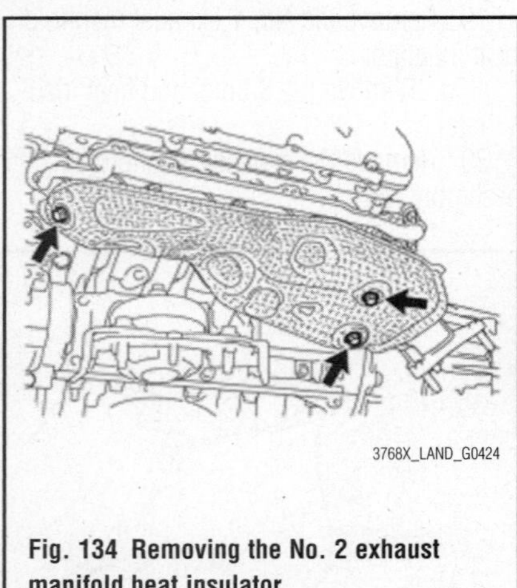

Fig. 134 Removing the No. 2 exhaust manifold heat insulator

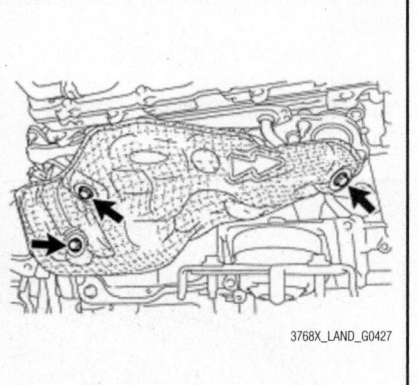

a. Remove the 2 nuts and insulator from the exhaust manifold.

b. Remove the 10 nuts, exhaust manifold and 2 gaskets.

To install:

21. Install the exhaust manifold sub assembly RH.

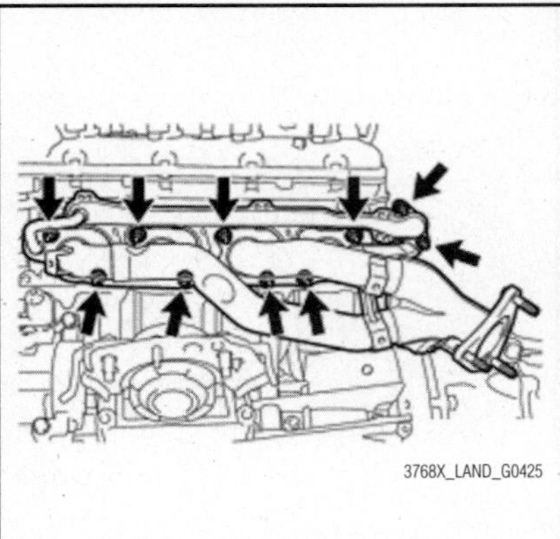

Fig. 135 Removing the exhaust manifold sub assembly LH

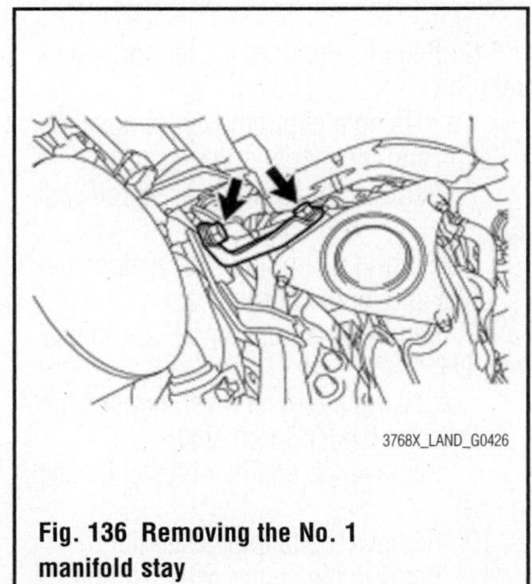

Fig. 136 Removing the No. 1 manifold stay

3768X_LAND_G0427

Fig. 137 Removing the No. 1 exhaust manifold heat insulator

a. Install a new gasket to the cylinder head and a new gasket to the No. 2 air tube.

➡Install the exhaust manifold gasket with the gasket tab facing toward the front of the engine.

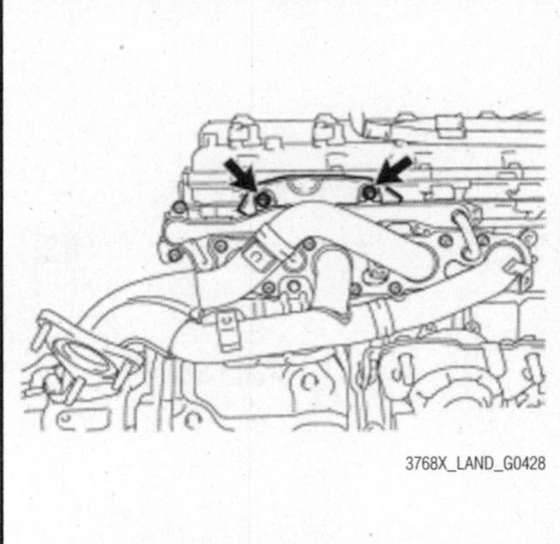

Fig. 138 Removing the 2 nuts and insulator from the exhaust manifold

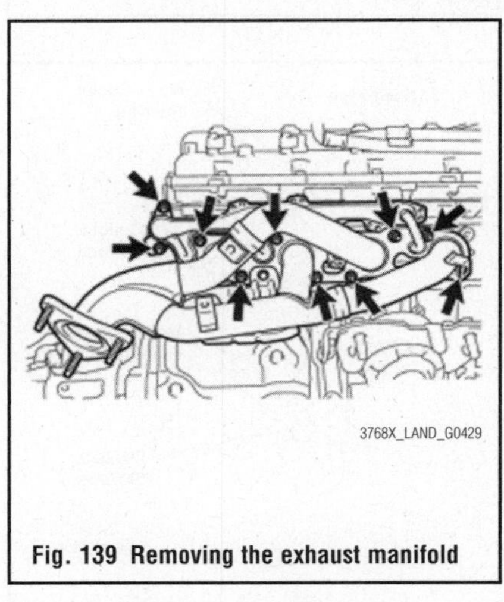

Fig. 139 Removing the exhaust manifold

➡Install the air tube gasket with the gasket claws facing the tube side.

b. Temporarily install the exhaust manifold with the 2 nuts labeled A and 8 new nuts.

c. Uniformly tighten the nuts that are not labeled A, and then tighten the 2 nuts labeled A. Tighten nut A to 7 ft. lbs. (10 Nm). Tighten all the other nuts to 15 ft. lbs. (21 Nm).

d. Install the insulator to the exhaust manifold with the 2 nuts. Tighten to 7 ft. lbs. (10 Nm).

22. Install the No. 1 exhaust manifold heat insulator.

a. Install the heat insulator with the 3 bolts and tighten to 7 ft. lbs. (10 Nm).

23. Install the No. 1 manifold stay.

a. Temporarily install the manifold stay with the 2 bolts.

b. Tighten the 2 bolts in the order shown to 30 ft. lbs. (40 Nm).

24. Install the exhaust manifold sub assembly LH.

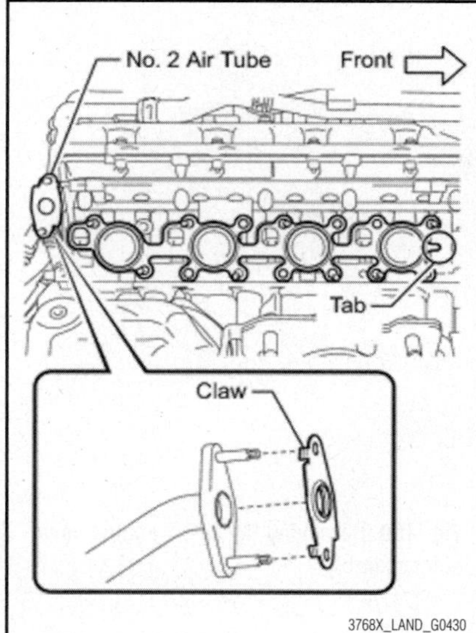

Fig. 140 Installing a new gasket to the cylinder head and No. 2 air tube

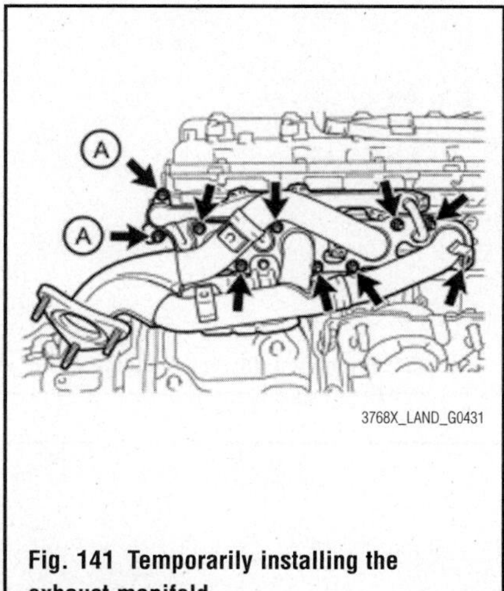

Fig. 141 Temporarily installing the exhaust manifold

a. Install a new gasket to the cylinder head and a new gasket to the No. 3 air tube.

➡Install the exhaust manifold gasket with the gasket tab facing toward the rear of the engine.

➡Install the air tube gasket with the gasket claws facing the tube side.

b. Temporarily install the exhaust manifold with the 2 nuts labeled A and 8 new nuts.

c. Uniformly tighten the nuts that are not labeled A, and then tighten the 2 nuts labeled A. Tighten the A nuts to 7 ft. lbs. (10 Nm). Tighten all other nuts to 15 ft. lbs. (21 Nm).

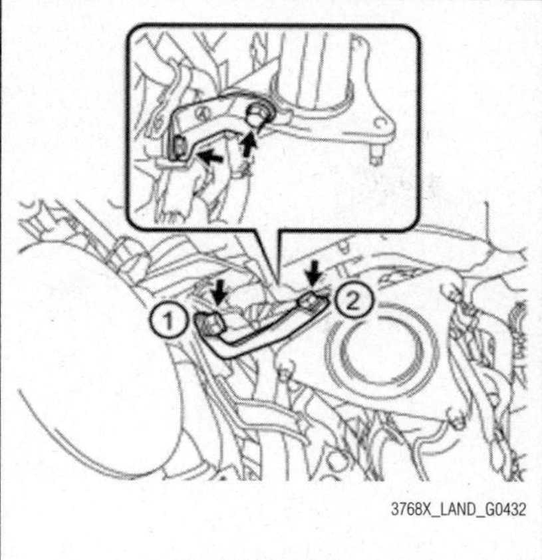

Fig. 142 Installing the No. 1 manifold stay

25. Install the No. 2 exhaust manifold heat insulator.
    a. Install the heat insulator with the 3 bolts to 7 ft. lbs. (10 Nm).
26. Install the No. 2 manifold stay.
    a. Temporarily install the manifold stay with the 3 bolts.
    b. Tighten the bolts in the order shown. Tighten to 30 ft. lbs. (40 Nm).
27. Install the propeller shaft heat insulator with the 2 bolts. Tighten to 12 ft. lbs. (16 Nm).
28. Install the front exhaust pipe assembly.
29. Install the front No. 2 exhaust pipe assembly.
30. Install the center exhaust pipe assembly.
31. Install the tailpipe assembly.
32. Install the engine oil level dipstick guide.
33. Inspect for exhaust gas leaks.

➡If gas is leaking, tighten the areas necessary to stop the leak. Replace any damaged parts as necessary.

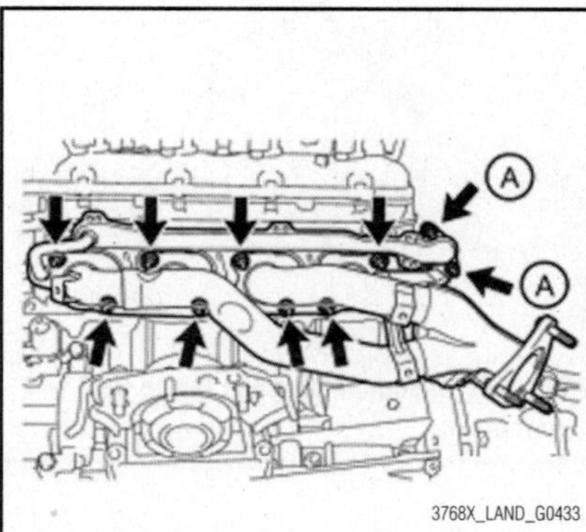

Fig. 143 Temporarily installing the exhaust manifold

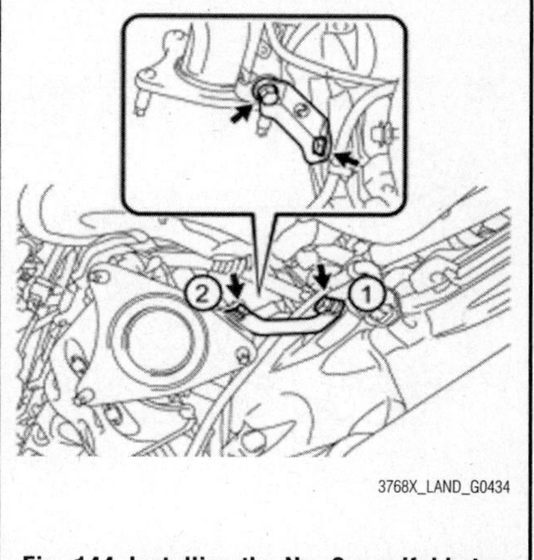

Fig. 144 Installing the No. 2 manifold stay

34. Install the front fender apron seal LH.
    a. Install the fender apron seal with the 3 clips.
35. Install the front fender apron seal rear LH with the 4 clips.
36. Install the front fender apron seal front RH with the 3 clips.
37. Install the front fender apron seal rear RH with the 4 clips.
38. Install the No. 1 engine under cover sub assembly.
39. Install the front fender splash shield sub assembly LH.
40. Install the front fender splash shield sub assembly RH.
41. Install the No. 2 engine under cover.

INTAKE MANIFOLD

REMOVAL & INSTALLATION

See Figures 145 through 161.

1. Disconnect the cable from the negative battery terminal.
2. Remove the cowl top ventilator louver sub assembly.
3. Remove the front fender splash shield sub assembly LH.
4. Remove the front fender splash shield sub assembly RH.
5. Remove the No. 1 engine under cover sub assembly.
6. Drain the engine coolant.
7. Remove the v-bank cover sub assembly.
8. Remove the air cleaner hose assembly.
    a. Disconnect the vacuum hose and No. 2 ventilation hose.
    b. Loosen the 2 hose clamps.
    c. Remove the air cleaner hose.
9. Remove the intake manifold.

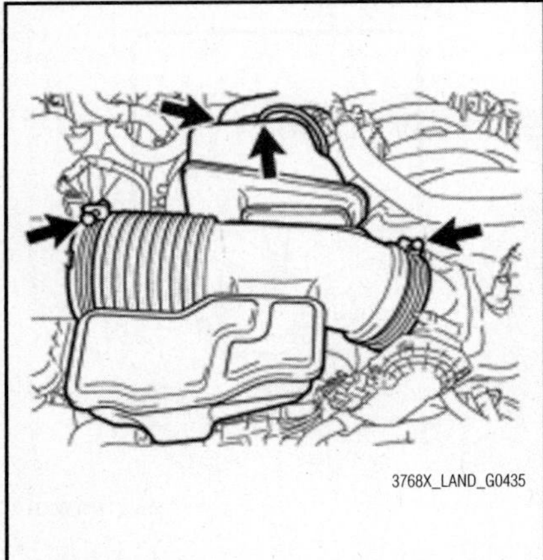

Fig. 145 Removing the air cleaner hose assembly

a. Disconnect the ventilation hose from the ventilation pipe of the cylinder head cover LH and RH.

b. Disconnect the 2 water by pass hoses.

c. Disconnect the throttle body connector.

d. Disconnect the No. 1 ventilation hose.

e. Disconnect the purge VSV connector.

f. Disconnect the purge line hose from the purge VSV.

g. Disconnect the vacuum switching valve connector (for ACIS).

h. Remove the No. 1 engine cover sub assembly.

i. Remove the No. 3 engine cover.

j. Disconnect the 3 wire clamps from the 3 wire brackets.

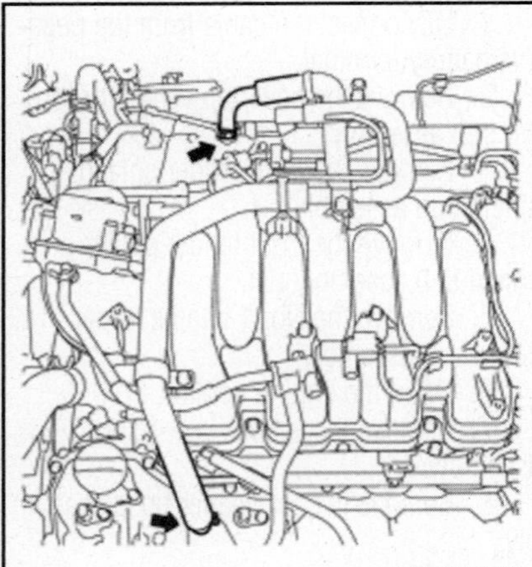

Fig. 146 Disconnecting the ventilation hose from the ventilation pipe of the cylinder head cover LH and RH

Fig. 147 Disconnecting the 2 water by pass hoses

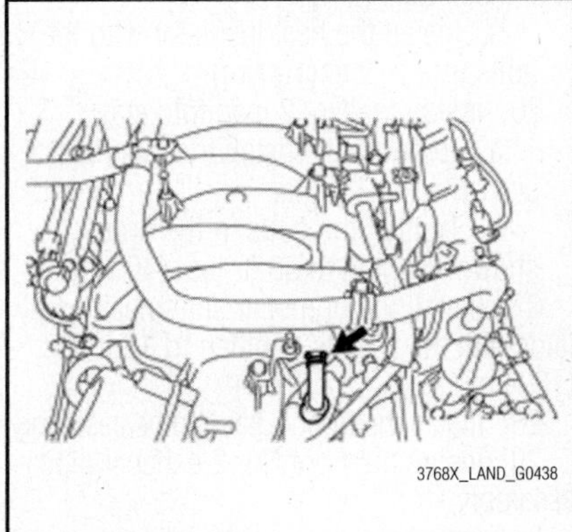

Fig. 148 Disconnecting the No. 1 ventilation hose

Fig. 149 Disconnecting the purge VSV connector

k. Remove the bolt and wire bracket from the intake manifold.

l. Remove the 2 nuts, 8 bolts, intake manifold and 2 gaskets.

10. Remove the ventilation hose assembly.

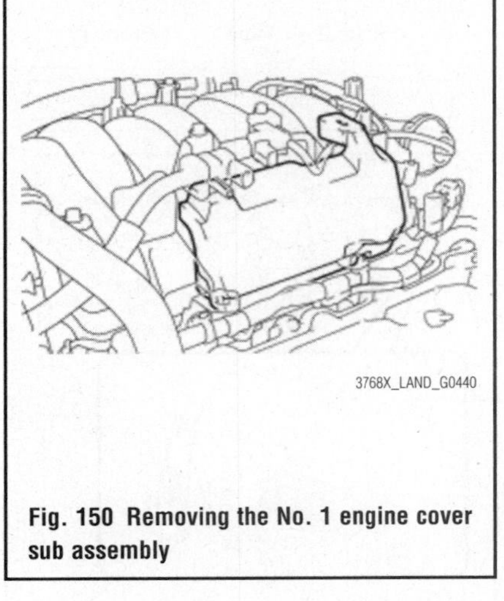

Fig. 150 Removing the No. 1 engine cover sub assembly

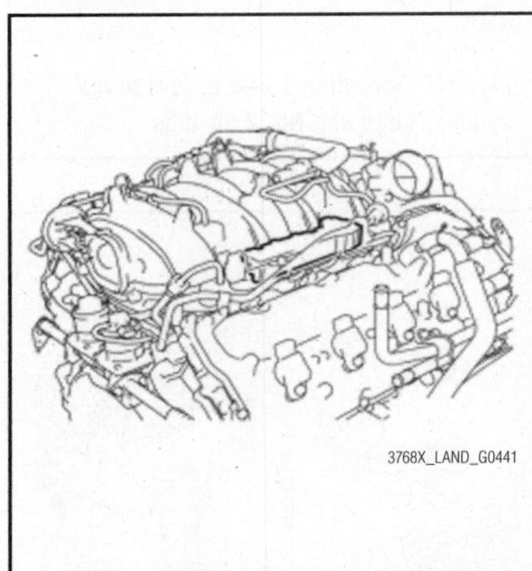

Fig. 151 Removing the No. 3 engine cover

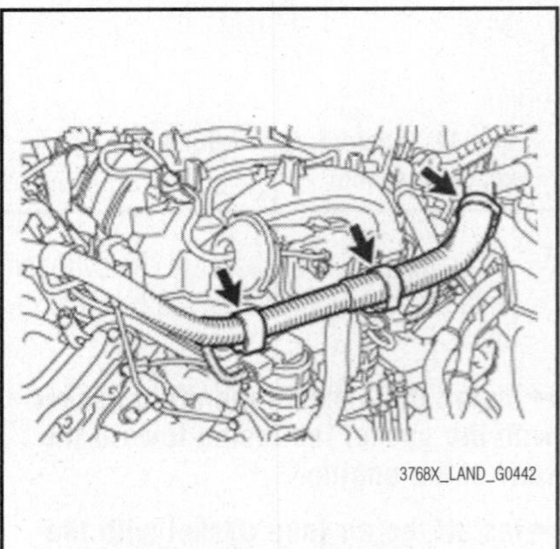

Fig. 152 Disconnecting the 3 wire clamps from the 3 wire brackets

a. Remove the 2 bolts and ventilation hose from the intake manifold.

11. Remove the No. 1 v-bank cover bracket.

a. Remove the 2 bolts and the bracket.

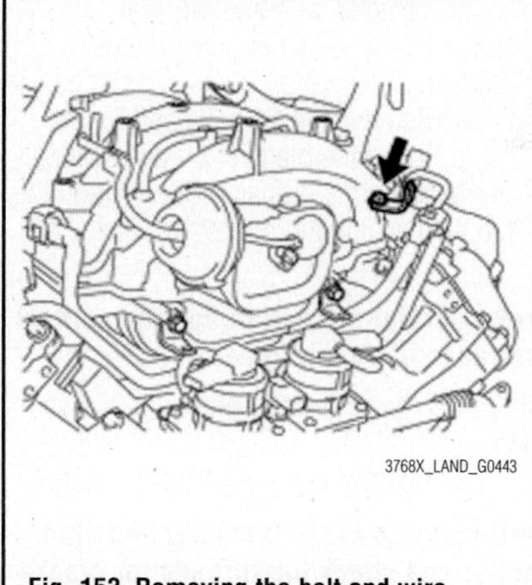

Fig. 153 Removing the bolt and wire bracket from the intake manifold

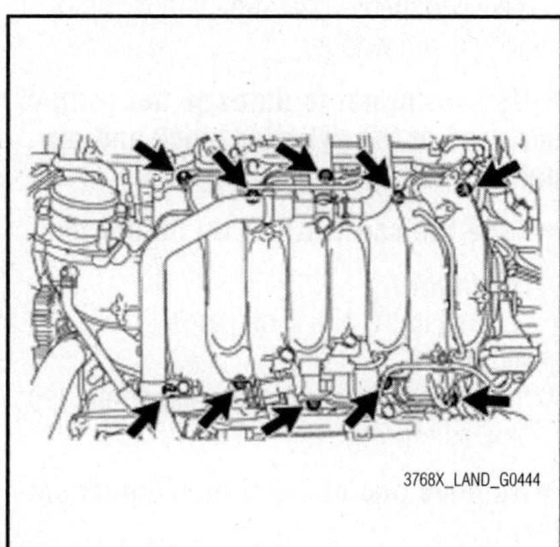

Fig. 154 Removing the 2 nuts, 8 bolts, intake manifold and 2 gaskets

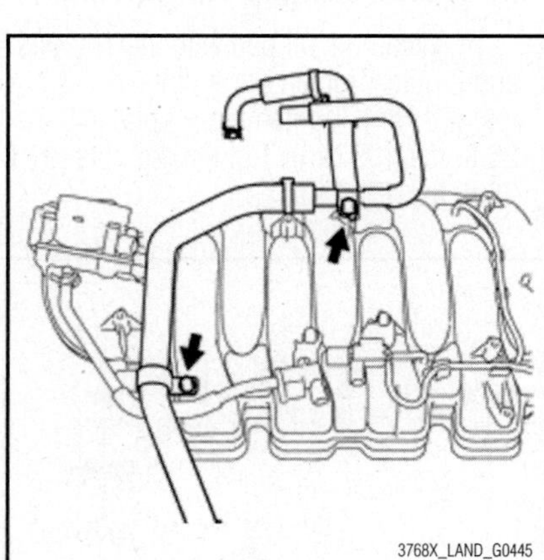

Fig. 155 Removing the 2 bolts and ventilation hose from the intake manifold

12. Remove the v-bank cover bolt from the intake manifold.
13. Remove the No. 2 v-bank cover bracket sub assembly.
    a. Remove the bolt and bracket.
14. Remove the throttle body assembly.

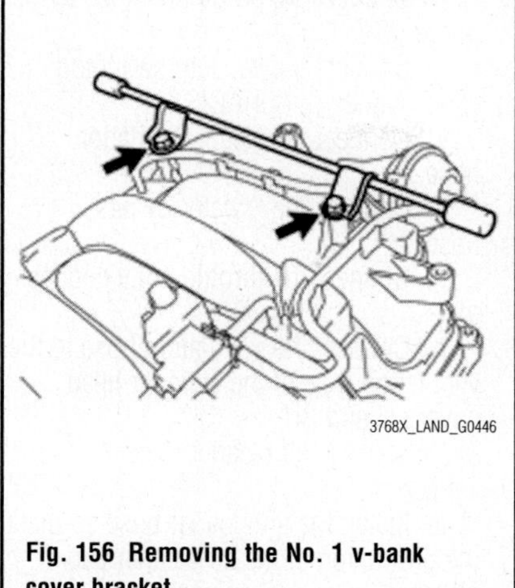

Fig. 156 Removing the No. 1 v-bank cover bracket

Fig. 157 Removing the v-bank cover bolt from the intake manifold

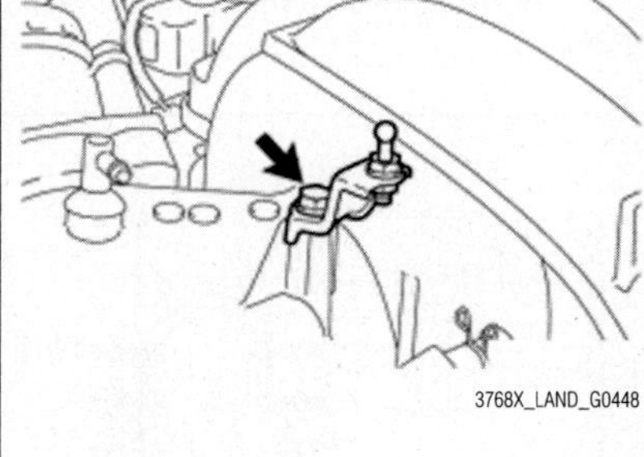

Fig. 158 Removing the No. 2 v-bank cover bracket sub assembly

    a. Remove the 4 bolts, throttle body and gasket.
15. Remove the vacuum switching valve assembly (for ACIS).
    a. Disconnect the 2 vacuum hoses from the vacuum switching valve.

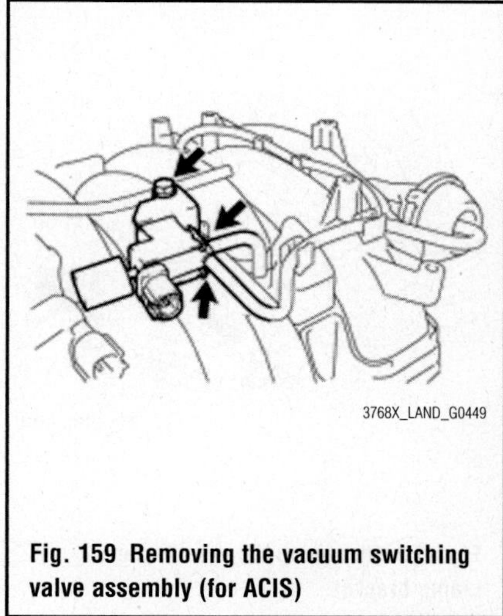

Fig. 159 Removing the vacuum switching valve assembly (for ACIS)

    b. Remove the bolt and vacuum switching valve.
16. Remove the purge VSV.
    a. Disconnect the purge line hose from the intake manifold.
    b. Remove the bolt and purge VSV.
17. Remove the wire harness clamp bracket.
    a. Remove the 2 bolts and 2 wire harness clamp brackets.

To install:

18. Install the wire harness clamp bracket with the 2 bolts. Tighten to 71 inch lbs. (8 Nm).
19. Install the purge VSV.
    a. Connect the purge line hose to the intake manifold.
    b. Install the purge VSV to the intake manifold with the bolt. Tighten to 15 ft. lbs. (21 Nm).
20. Install the vacuum switching valve assembly (for ACIS).

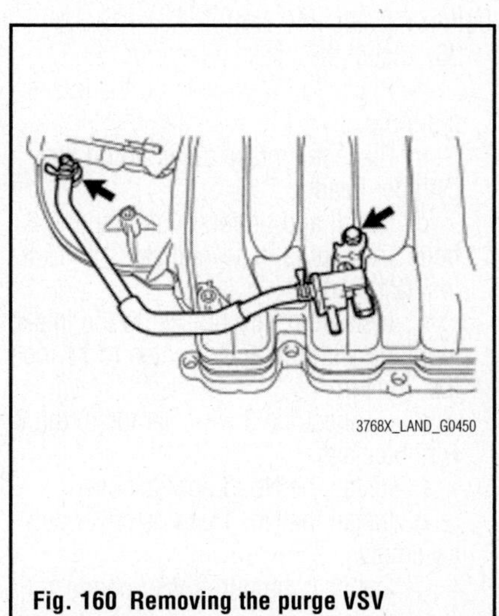

Fig. 160 Removing the purge VSV

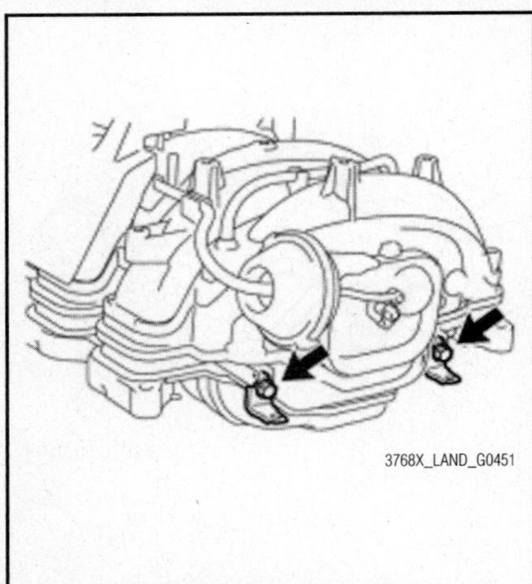

Fig. 161 **Removing the wire harness clamp bracket**

a. Install the vacuum switching valve to the intake manifold with the bolt. Tighten to 80 inch lbs. (9 Nm).

b. Connect the 2 vacuum hoses to the vacuum switching valve.

21. Install the throttle body assembly.

a. Align the protrusion of a new gasket with the groove of the intake manifold, and install the gasket.

b. Install the throttle body with the 4 nuts. Tighten to 7 ft. lbs. (10 Nm).

22. Install the No. 2 v-bank cover bracket sub assembly with the bolt. Tighten to 7 ft. lbs. (10 Nm).

23. Install the v-bank cover bolt to the intake manifold. Tighten to 7 ft. lbs. (10 Nm).

24. Install the No. 1 v-bank cover bracket with the 2 bolts. Tighten to 7 ft. lbs. (10 Nm).

25. Install the ventilation hose to the intake manifold with the 2 bolts. Tighten the A bolt to 15 ft. lbs. (21 Nm). Tighten the B bolt to 7 ft. lbs. (10 Nm).

26. Install the intake manifold.

a. Place 2 new gaskets on the intake manifold.

b. Place the intake manifold on the cylinder head.

c. Install and uniformly tighten the 8 bolts and 2 nuts in several steps to 15 ft. lbs. (21 Nm).

d. Install the wire bracket to the intake manifold with the bolt. Tighten to 71 inch lbs. (8 Nm).

e. Connect the 3 wire clamps to the 3 wire brackets.

f. Install the No. 3 engine cover.

g. Install the No. 1 engine cover sub assembly.

h. Connect the purge VSV connector.

i. Connect the purge line hose to the purge VSV.

j. Connect the vacuum switching valve connector (for ACIS).

k. Connect the No. 1 ventilation hose.

l. Connect the 2 water by pass hoses.

m. Connect the throttle body connector.

n. Connect the ventilation hose to the ventilation pipe of the cylinder head cover LH and RH.

27. Install the air cleaner hose assembly.

a. Install the air cleaner hose so that the protrusion of the air cleaner cap aligns with the groove of the hose.

b. Tighten the 2 clamps to 44 inch lbs. (5 Nm).

c. Connect the vacuum hose.

d. Connect the No. 2 ventilation hose.

28. Add engine coolant.

29. Inspect for coolant leaks.

30. Install the v-bank cover sub assembly.

31. Install the No. 1 engine under cover sub assembly LH and RH.

32. Install the cowl top ventilator louver sub assembly.

33. Connect the cable to the negative battery terminal.

34. Check the throttle body assembly.

OIL PAN

REMOVAL & INSTALLATION

See Figures 162 through 168.

1. Remove the No. 2 oil pan sub assembly.

a. Vehicles with and oil level sensor, remove the engine oil level sensor.

b. Remove the 14 bolts and 2 nuts.

c. Insert the blade of oil pan seal cutter between the oil pans. Cut through the applied sealer and remove the No. 2 oil pan.

➡**Be careful not to damage the contact surfaces of the oil pans**

2. Remove the No. 1 oil pan sub assembly.

a. Remove the 14 bolts and 2 nuts.

➡**Be sure to clean the bolts and stud bolts, and check the threads for cracks or other damage.**

b. Remove the oil pan by prying between the oil pan and cylinder block with a screwdriver.

➡**Be careful not to damage the contact surfaces of the cylinder block and oil pan.**

➡**Tape the screwdriver tip before use.**

To install:

3. Install the No. 1 oil pan sub assembly.

a. Apply seal packing in a continuous line as shown.

➡**Remove any oil from the contact surface.**

➡**Install the oil pan within 3 minutes and tighten the bolts and nuts within 15 minutes after applying seal packing.**

b. Install the oil pan with the 14 bolts and 2 nuts. Tighten the A bolts to 7 ft. lbs. (10 Nm). Tighten the B bolts to 26 ft. lbs. (35 Nm). Tighten the nuts to 26 ft. lbs. (35 Nm).

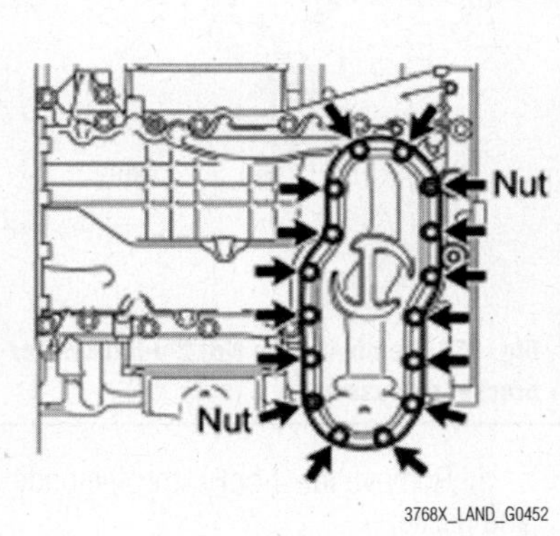

Fig. 162 Removing the No. 2 oil pan sub assembly 14 bolts and 2 nuts

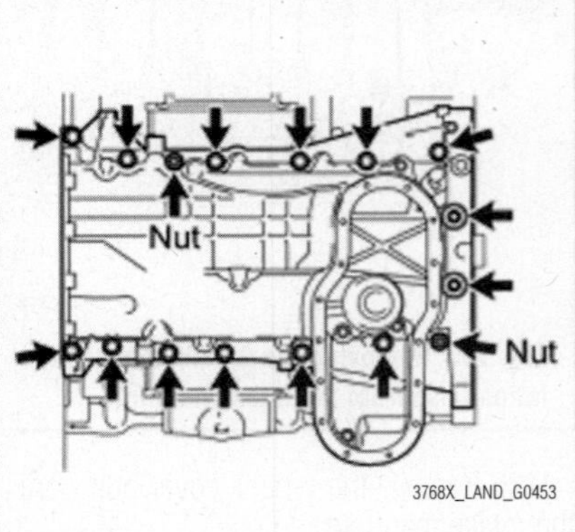

Fig. 163 Removing the No. 1 oil pan sub assembly bolts and nuts

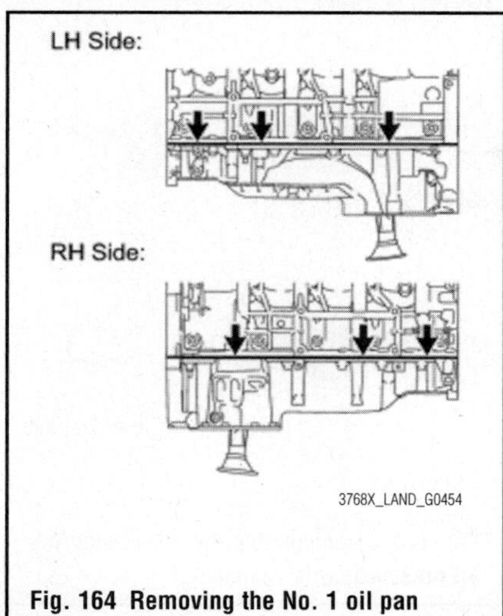

Fig. 164 Removing the No. 1 oil pan

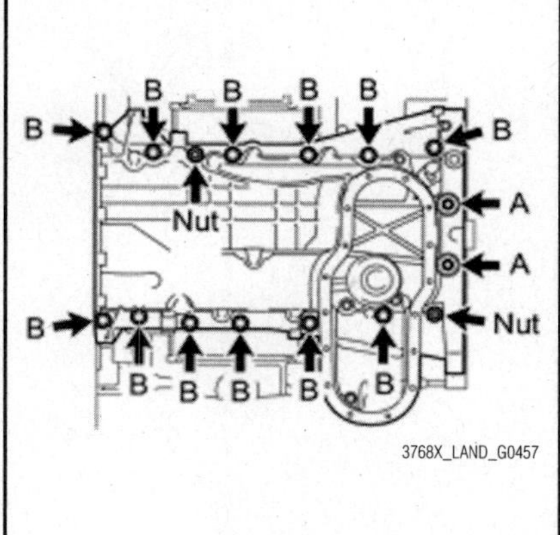

Fig. 166 Installing the No. 1 oil pan sub assembly

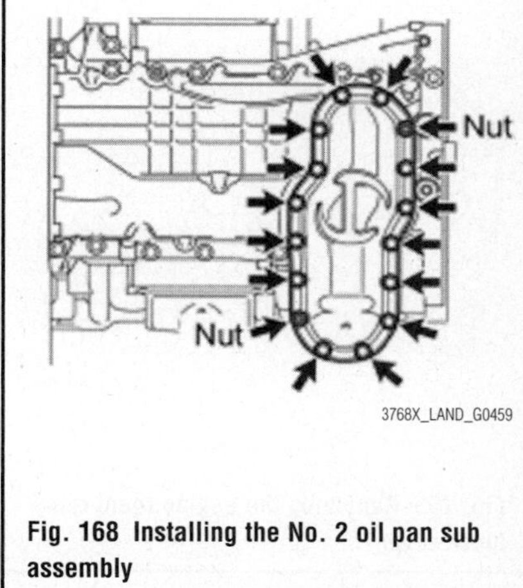

Fig. 168 Installing the No. 2 oil pan sub assembly

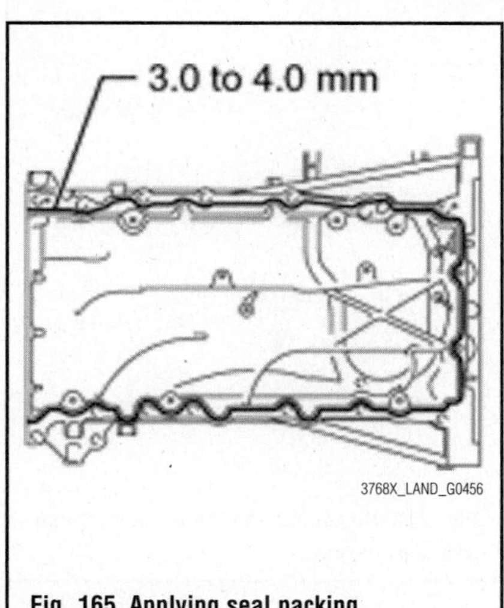

Fig. 165 Applying seal packing

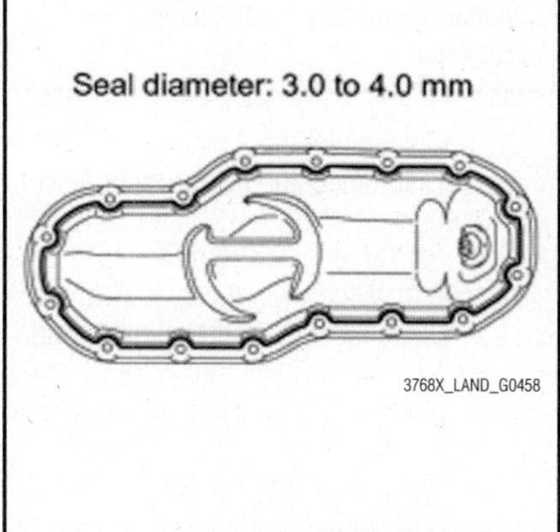

Fig. 167 Applying packing seal to the No. 2 oil pan sub assembly

➡Do not start the engine for at least 2 hours after installing.

4. Install the No. 2 oil pan sub assembly.
   a. Apply seal packing in a continuous line as shown.

➡Remove any oil from the contact surface.

➡Install the oil pan within 3 minutes and tighten the bolts and nuts within 15 minutes after applying seal packing.

   b. Install the oil pan with the 14 bolts and 2 nuts. Tighten to 7 ft. lbs. (10 Nm).

➡Do not start the engine for at least 2 hours after installing.

OIL PUMP

REMOVAL & INSTALLATION

See Figures 169 through 215.

1. Discharge the fuel system pressure.
2. Disconnect the cable from the negative battery terminal.

CAUTION

Wait at least 90 seconds after disconnecting the cable from the negative (-) battery terminal to disable the SRS system.

3. Remove the upper radiator support seal.
4. Remove the radiator grille.
5. Remove the front bumper cover.
6. Remove the transmission oil cooler air duct (with air cooled transmission oil cooler).
7. Disconnect the radiator side deflector RH (without air cooled transmission oil cooler).
8. Disconnect the radiator side deflector LH.
9. Remove the cowl top ventilator louver sub assembly.
10. Remove the front fender apron seal LH.
11. Remove the front fender apron seal front RH.
12. Remove the front fender splash shield sub assembly LH.
13. Remove the front fender splash shield sub assembly RH.
14. Remove the No. 1 engine under cover sub assembly.
15. Remove the No. 2 engine under cover.
16. Drain engine oil.
17. Drain the engine coolant.
18. Remove the v-bank cover sub assembly.
19. Remove the air cleaner hose assembly.
20. Remove the air cleaner assembly.
21. Remove the radiator assembly.
22. Disconnect the engine wire.
    a. For the engine room LH side, remove the engine room relay block cover.
    b. Disconnect the 2 connectors and 2 clips from the engine room junction block.
    c. Disconnect the 4 air injection control driver connectors and wire harness clamp.
    d. Disconnect the injector connector.
    e. Disconnect the 4 ignition coil connectors.
    f. Disconnect the 2 VVT sensor connectors.
    g. Disconnect the 4 clamps.
    h. Remove the 2 bolts and ground wire.
    i. Disconnect the noise filter connector.

Fig. 169 Removing the engine room relay block cover

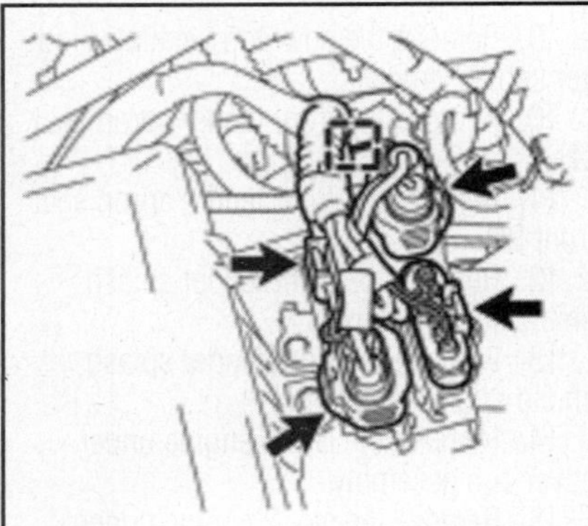

Fig. 170 Disconnecting the air injection control driver connectors and wire harness clamp

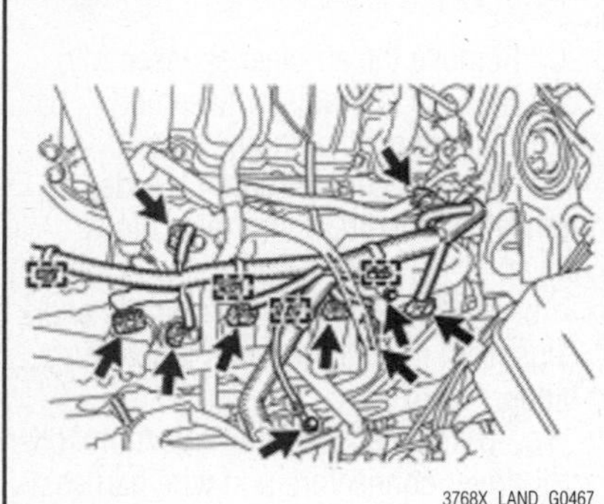

Fig. 171 Disconnecting the injector connector

j. Disconnect the Engine Coolant Temperature (ECT) sensor connector.

k. Disconnect the 2 camshaft timing oil control valve connectors.

l. Disconnect the camshaft position sensor connector.

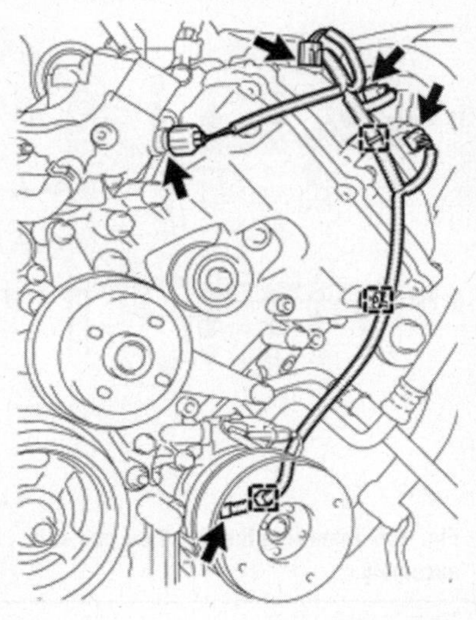

Fig. 172 Disconnecting the Engine Coolant Temperature (ECT) sensor connector

m. Disconnect the 3 clamps.

n. Disconnect the cooler compressor connector.

o. For the engine room RH side, remove or disconnect the following:

- Disconnect the 2 camshaft timing oil control valve connectors
- Disconnect the 4 ignition coil connectors
- Disconnect the injector connector
- Disconnect the 2 VVT sensor connectors
- Disconnect the noise filter connector
- Remove the 2 bolts and ground wire
- Disconnect the 2 air pump connectors
- Disconnect the throttle position sensor and throttle control motor connector
- Disconnect the 5 clamps

p. Disconnect the 2 clamps and power steering oil pressure switch connector.

23. Disconnect air pump hose and wire harness.

a. Disconnect the No. 2 and No. 3 air hoses.

b. Disconnect the 2 clamps.

24. Disconnect the No. 2 water by-pass pipe.

a. Disconnect the 3 hoses.

b. Remove the 3 bolt and disconnect the No. 2 water by pass pipe from the cylinder head cover.

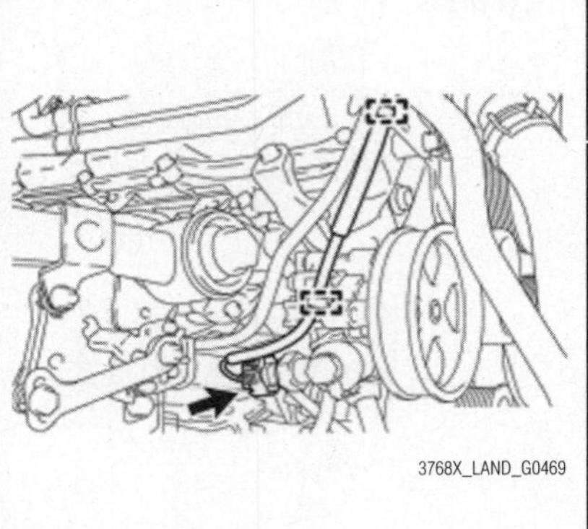

Fig. 173 Disconnecting the power steering oil pressure switch connector

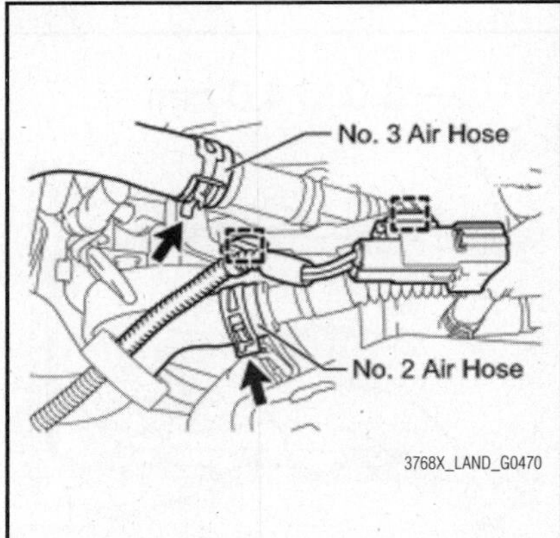

Fig. 174 Disconnecting the air pump hose and wire harness

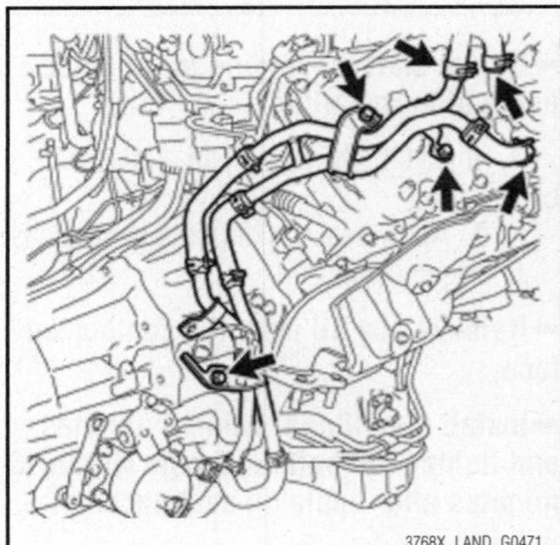

Fig. 175 Disconnect No. 2 water by pass pipe

25. Disconnect cooler compressor assembly.

a. Remove the 3 bolts, nut and stud bolt, and disconnect the cooler compressor.

➡It is not necessary to completely remove the compressor. With the hoses connected to the compressor, hang the compressor on the vehicle body with a rope.

26. Disconnect the No. 2 fuel tube sub assembly.
    a. Remove the 2 bolts and disconnect the fuel tube.
27. Remove the oil filter element.
28. Remove the engine oil level dipstick guide.
29. Remove the oil pressure sender gauge assembly.
30. Remove the No. 2 water by pass pipe sub assembly (with oil cooler).
    a. Remove the 3 bolts.
    b. Disconnect the 4 hoses and remove the water by pass pipe.
31. Remove the No. 1 oil cooler bracket (with oil cooler).
32. Remove the oil filter bracket.
33. Vehicles without oil cooler perform the following.
    a. Remove the 2 bolts, 2 nuts and oil filter bracket.
    b. Remove the 2 O-rings.
34. Remove the intake manifold.
35. Disconnect the vane pump assembly.
    a. Remove the 2 bolts and disconnect the vane pump.
36. Disconnect the oil cooler pipe assembly.
37. Remove the generator assembly.
38. Remove the No. 1 water by pass hose.
    a. Remove the No. 1 water by pass hose by disconnecting the hose from the water inlet housing and front water by pass joint.
39. Remove the water by pass pipe sub assembly.

Fig. 176 Disconnecting the cooler compressor assembly

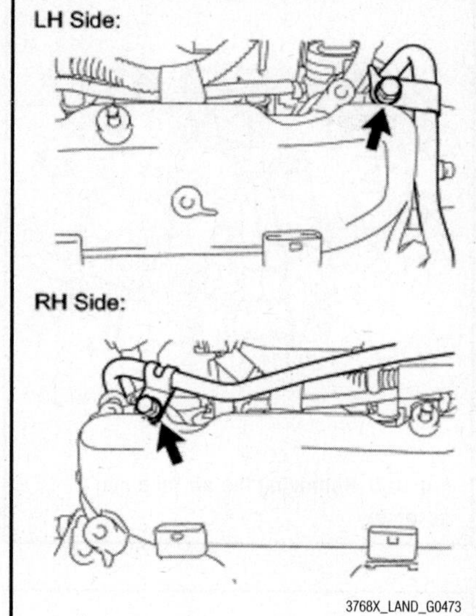

Fig. 177 Disconnecting the No. 2 fuel tube sub assembly

    a. Remove the bolt and disconnect the air tube.
    b. Disconnect the 2 hoses.
    c. Remove the bolts and water by pass pipe.
40. Remove the front water by pass joint.
    a. Disconnect the No. 2 water by pass hose from the water by pass joint.
    b. Remove the 4 nuts, water by pass joint and 2 gaskets.
41. Remove the No. 2 engine cover.
42. Remove the No. 1 engine cover.

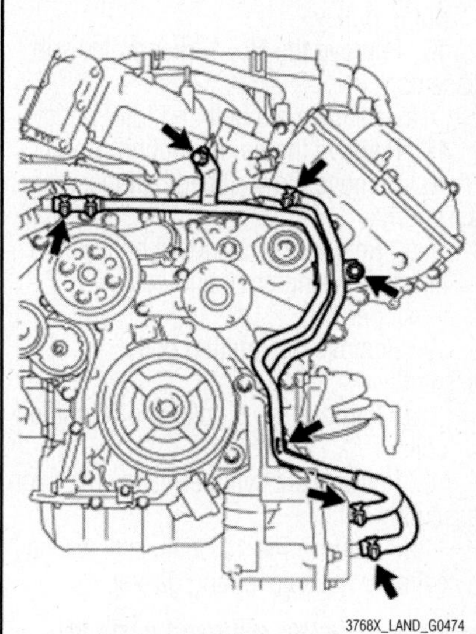

Fig. 178 Remove the No. 2 water by pass pipe sub assembly

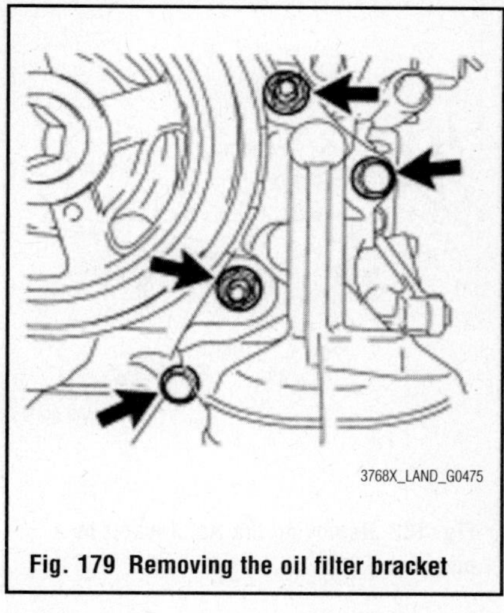

Fig. 179 Removing the oil filter bracket

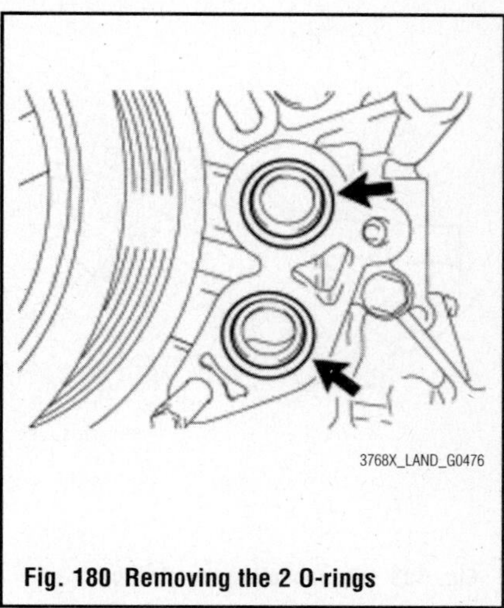

Fig. 180 Removing the 2 O-rings

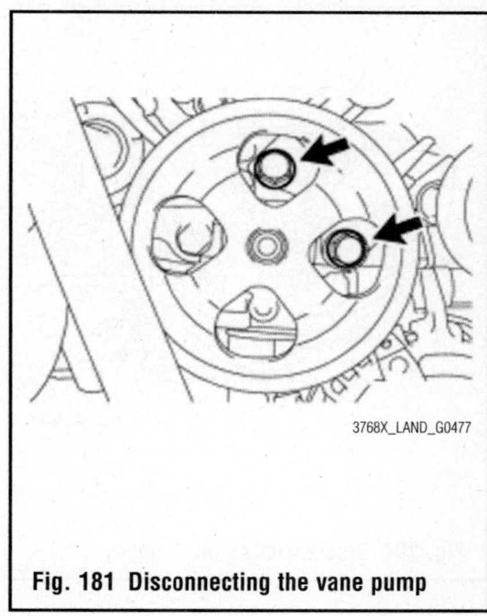

Fig. 181 Disconnecting the vane pump

43. Remove the air tube sub assembly.
    a. Remove the bolt, disconnect the 2 hoses, and remove the air tube.
44. Remove the water inlet housing.

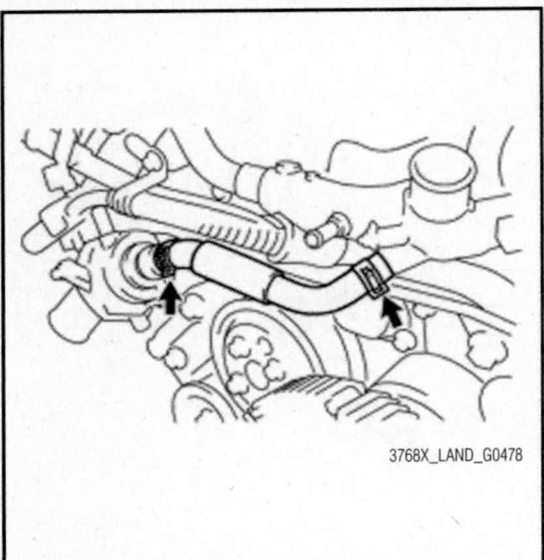

Fig. 182 Removing the No. 1 water by pass hose

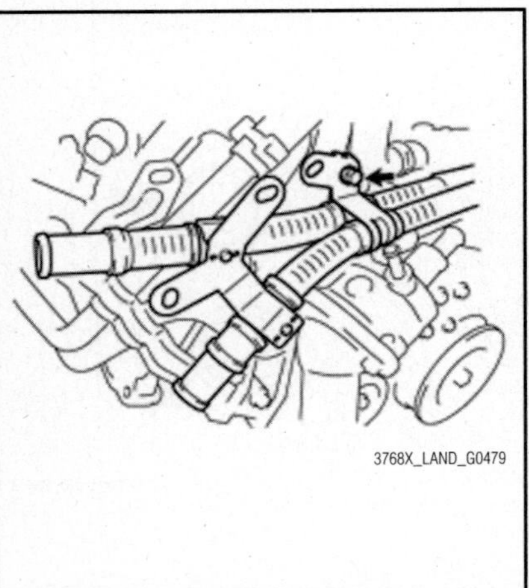

Fig. 183 Disconnecting the air tube

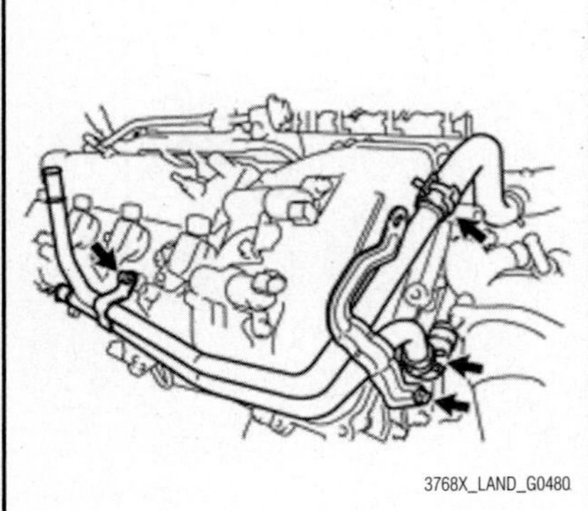

Fig. 184 Disconnecting the 2 hoses

a. Remove the 3 bolts, water inlet housing and gasket.

45. Remove the water pump pulley.

a. Using the special tool (09960-10010, 09962-01000 or 09963-01000), hold the water pump pulley.

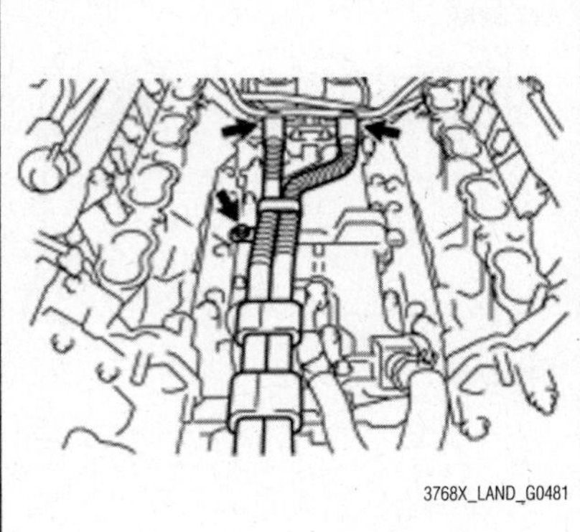

Fig. 185 Removing the air tube sub assembly

Fig. 186 Removing the water inlet housing

b. Remove the 4 bolts and water pump pulley.

46. Remove the No. 1 idler pulley sub assembly.

a. Remove the bolt and idler pulley.

47. Remove the fluid coupling bracket.

48. Remove the v-ribbed belt tensioner assembly.

a. Remove the standard bolt, 6 mm hexagon wrench bolt and belt tensioner.

49. Remove the ignition coil assembly.

a. Remove the 8 bolts and 8 ignition coils.

50. Remove the cylinder head cover sub assembly LH.

a. Remove the 14 bolts, seal washer, cylinder head cover and gasket.

➡Make sure the removed parts are returned to the same places they were removed from.

b. Remove the 5 gaskets from the camshaft bearing caps (No. 2, No. 3).

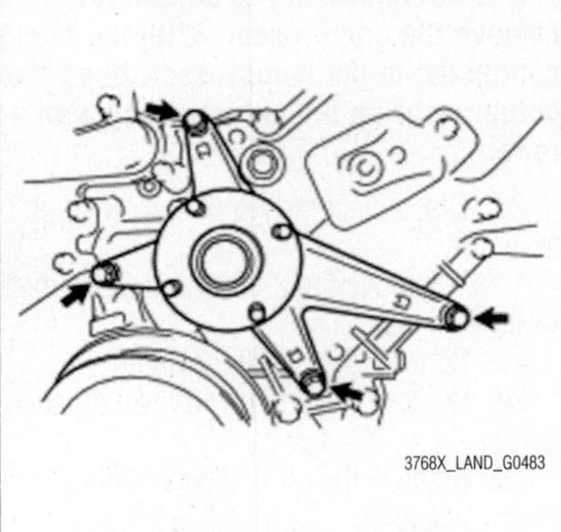

Fig. 187 Removing the fluid coupling bracket

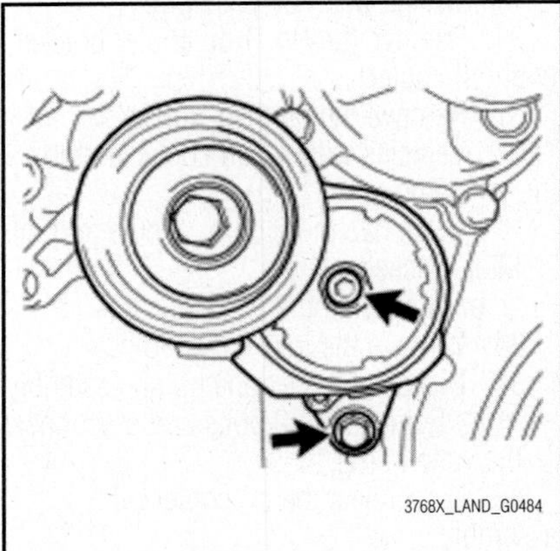

Fig. 188 Removing the v-ribbed belt tensioner assembly

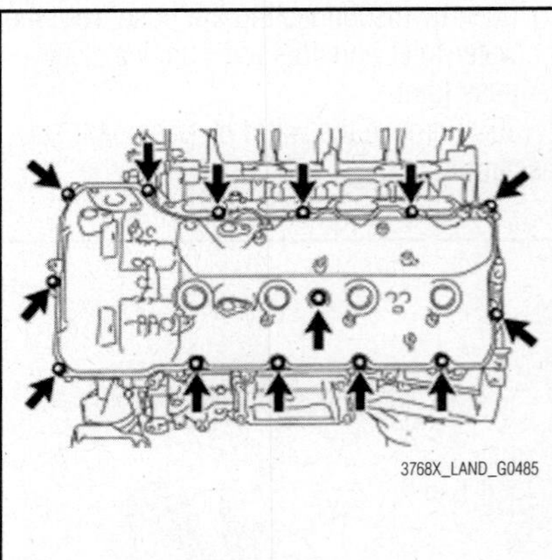

Fig. 189 Removing the bolts, seal washer, cylinder head cover and gasket

51. Remove the cylinder head cover sub assembly RH.

a. Remove the bolt and noise filter.

b. Remove the 14 bolts, seal washer, cylinder head cover and gasket.

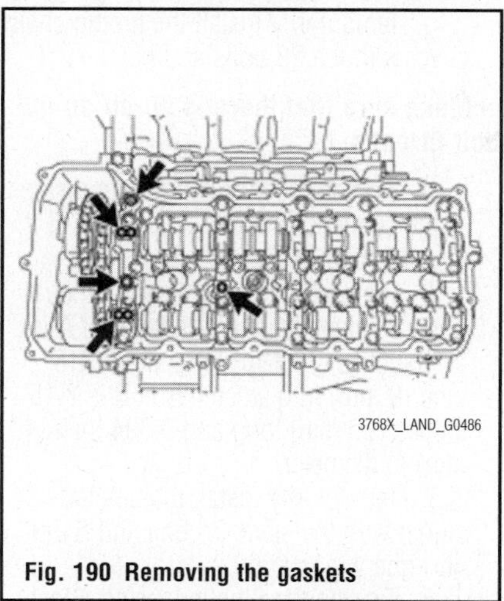

Fig. 190 Removing the gaskets

Fig. 191 Removing the noise filter

➡Make sure the removed parts are returned to the same places they were removed from.

c. Remove the 5 gaskets from the camshaft bearing caps (No. 1, No. 3).

52. Remove the spark plug tube gasket.

53. Remove the crankshaft pulley.

a. Using the special tool (09213-70011 or 09330-00021), loosen the crankshaft pulley set bolt until 2 or 3 threads are engaged.

b. Using the pulley set bolt and special tool (09950-50013 or 09951-05010), remove the crankshaft pulley.

54. Disconnect the wire harness clamp bracket.

a. Remove the bolt and disconnect the bracket.

55. Remove the crankshaft timing gear key.

a. Using a screwdriver, remove the timing gear key from the crankshaft.

56. Remove the timing chain cover sub assembly.

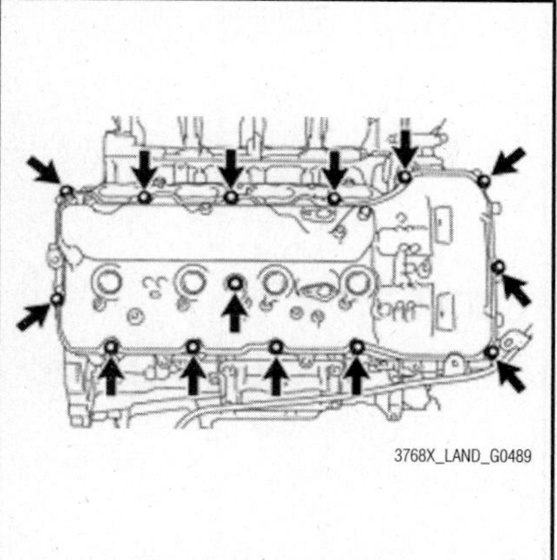

Fig. 192 Removing the seal washer, cylinder head cover and gasket

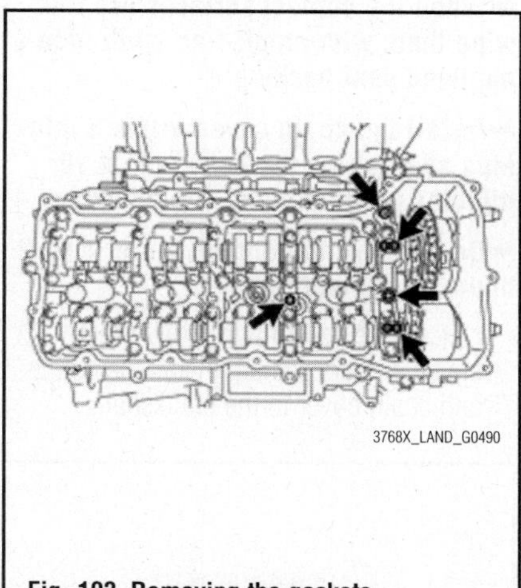

Fig. 193 Removing the gaskets

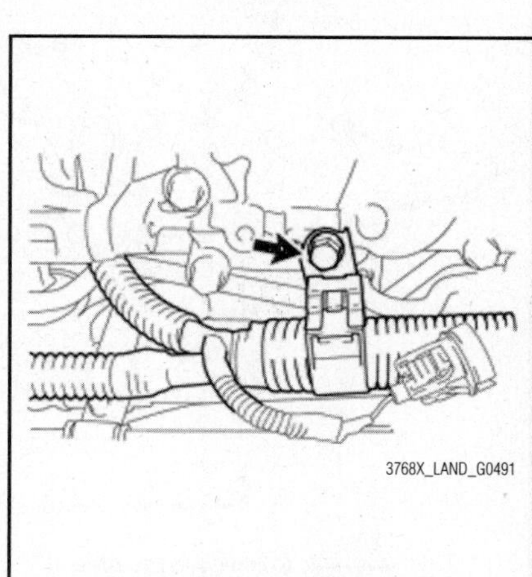

Fig. 194 Removing the wire harness clamp bracket

a. Remove the 28 bolts and nut shown.

b. Remove the timing chain cover by prying between the timing chain cover and cylinder head or cylinder block with a screwdriver as shown in the illustration.

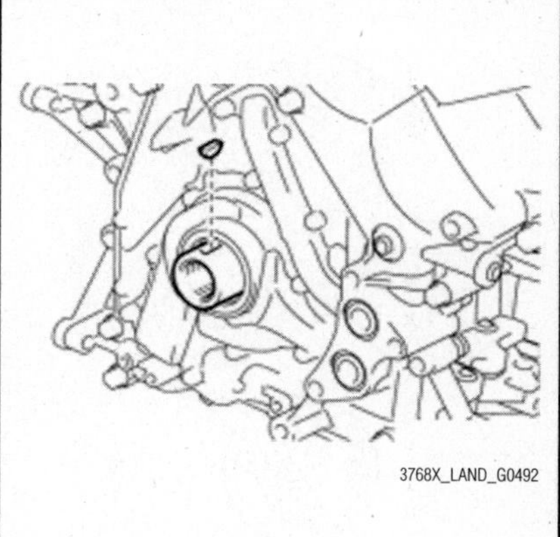

Fig. 195 Removing the crankshaft timing gear key

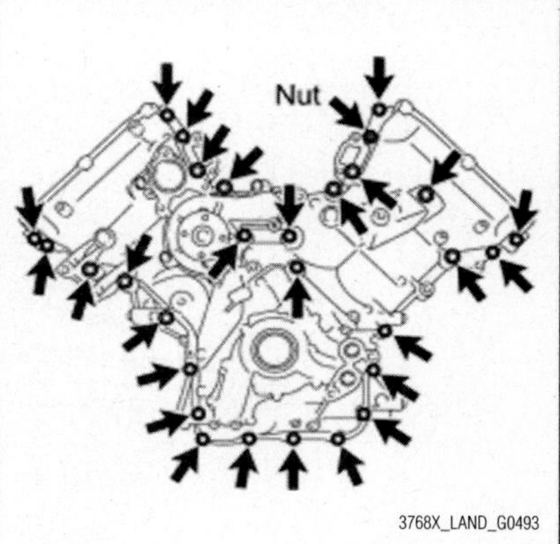

Fig. 196 Removing the timing chain cover sub assembly bolts and nut

➡Be careful not to damage the contact surfaces of the cylinder head, cylinder block and chain cover.

➡Tape the screwdriver tip before use.

c. Remove the oil pump gasket from the cylinder block.

d. Remove the O-ring from the oil pan.

57. Remove the water inlet pipe.

a. Remove the 2 O-rings from the water inlet pipe.

To install:

58. Install the water inlet pipe.

a. Apply soapy water to 2 new O-rings and install them to the inlet pipe.

b. Install the inlet pipe to the No. 1 heat exchanger cover.

59. Install the timing chain cover sub assembly.

a. Apply a light coat of engine oil to a new oil pump gasket.

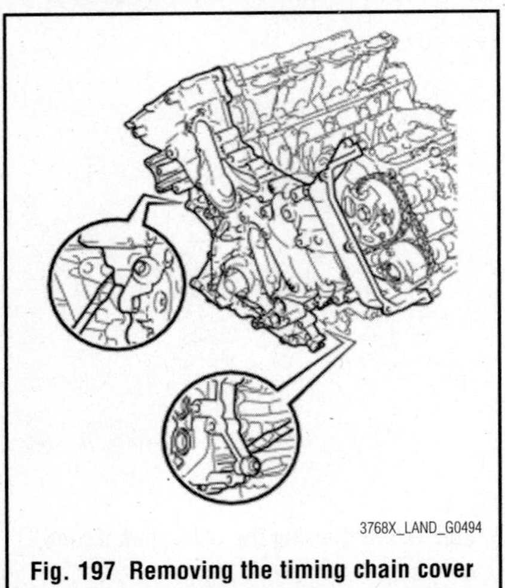

Fig. 197 Removing the timing chain cover

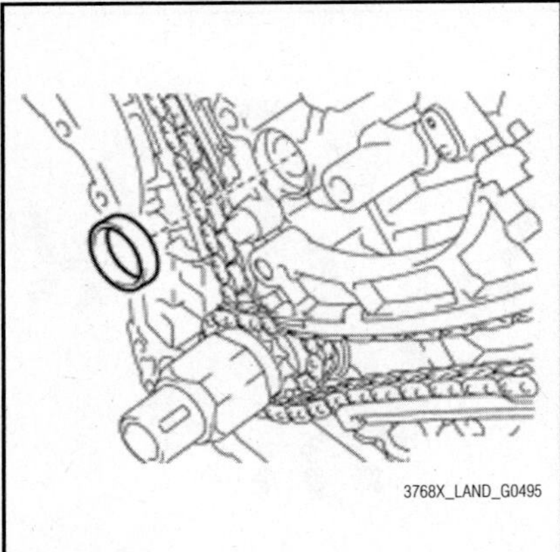

Fig. 198 Removing the oil pump gasket from the cylinder block

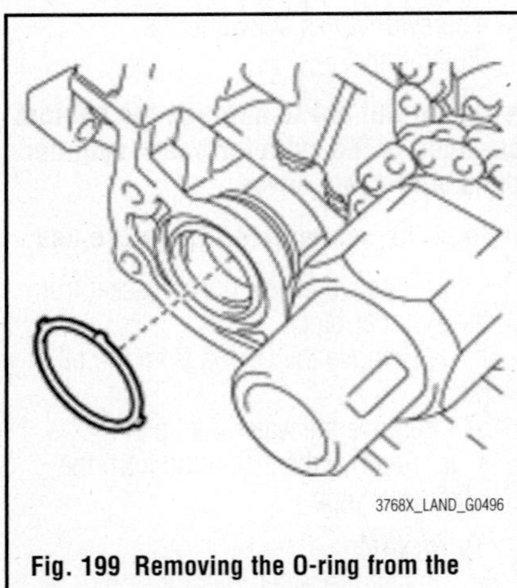

Fig. 199 Removing the O-ring from the oil pan

b. Install the oil pump gasket.

c. Apply a light coat of engine oil to a new O-ring.

d. Install the O-ring.

e. Apply seal packing in a continuous line to the timing chain cover as shown.

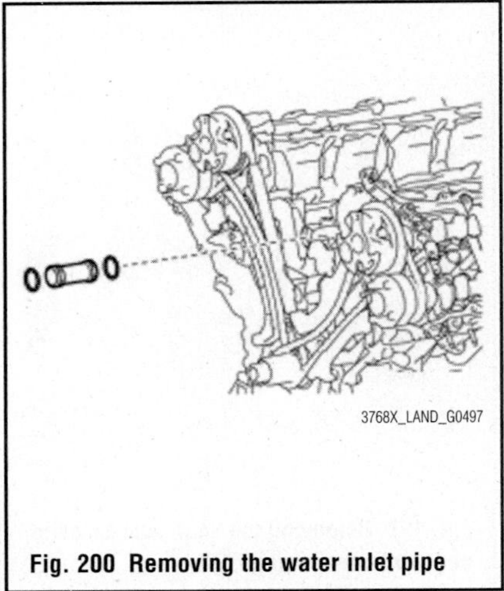

Fig. 200 Removing the water inlet pipe

➡When the contact surfaces are wet, wipe them with an oil-free cloth before applying seal packing.

➡Install the chain cover within 3 minutes and tighten the bolts within 10 minutes after applying seal packing.

➡Do not start the engine for at least 2 hours after installing.

f. Align the oil pump drive rotor spline and crankshaft. Install the spline and chain cover to the crankshaft.

g. Temporarily install the timing chain cover with the 28 bolts and nut.

➡Make sure that there is no oil on the bolt threads.

h. Tighten the 3 bolts in several steps in the sequence shown. Tighten to 35 ft. lbs. (47 Nm).

i. Temporarily install the fluid coupling bracket with the 4 bolts. Bolt A is 2.76 inches (70 mm) long and 0.315 inch (8 mm) in diameter. Bolt B is 3.15 inches (80 mm) long and 0.394 inch (10 mm) in diameter.

j. Temporarily install the belt tensioner with the standard bolt and 6 mm hexagon wrench bolt.

k. Tighten the 8 bolts labeled 4 to 11 in several steps in the sequence shown in the illustration. Tighten to 35 ft. lbs. (47 Nm).

l. Tighten the 23 bolts and nut labeled 12 to 35 in several steps in the sequence shown in the illustration. Tighten to 17 ft. lbs. (23 Nm).

➡After the installation, if the seal packing has seeped out at the areas labeled A shown in the illustration, wipe it off.

60. Install the spark plug tube gasket.

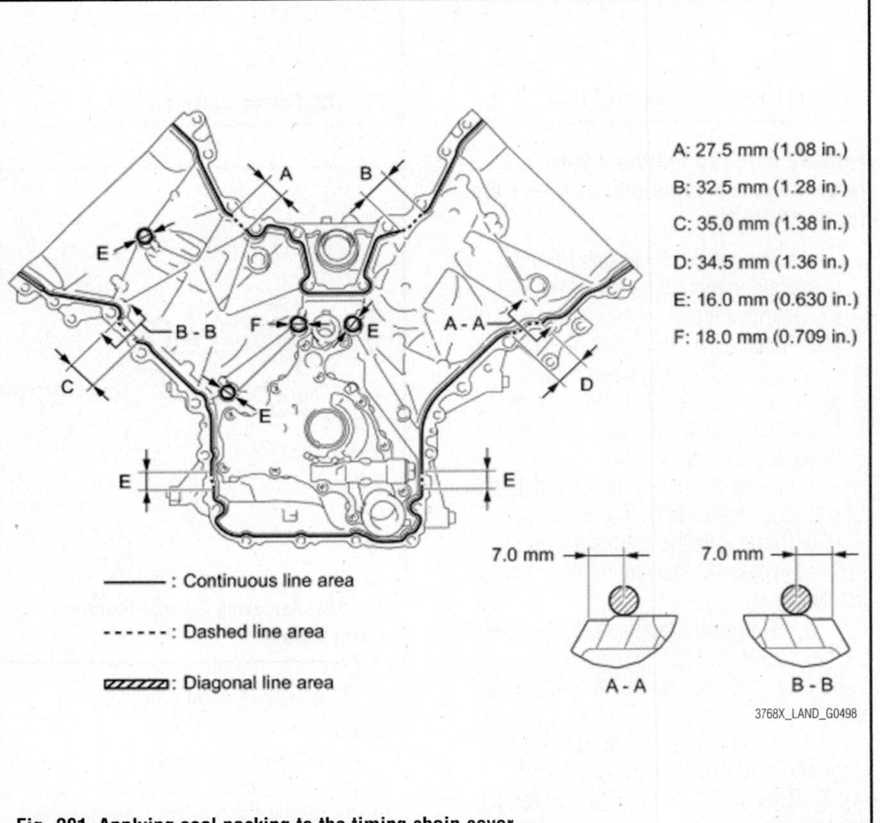

Fig. 201 Applying seal packing to the timing chain cover

Area	Seal Packing Diameter	Application Position From Inside Edge Of Cover
Continuous Line Area	0.118-0.157 inch (3-4 mm)	0.0984 inch (2.5 mm)
Dashed Line Area	0.252 inch (6.4 mm) or more, or within OK area shown in the illustration	0.276 inch (7 mm)
Diagonal Line Area	0.118-0.157 inch (3-4 mm)	0.217 inch (5.5 mm)

3768X_LAND_G0499

Fig. 202 Seal packing specifications

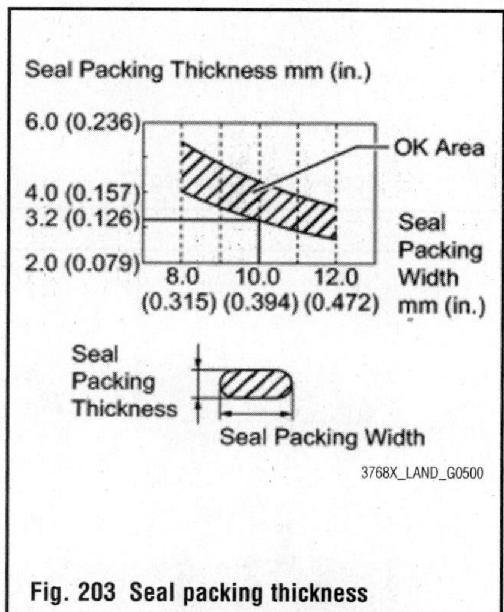

Fig. 203 Seal packing thickness

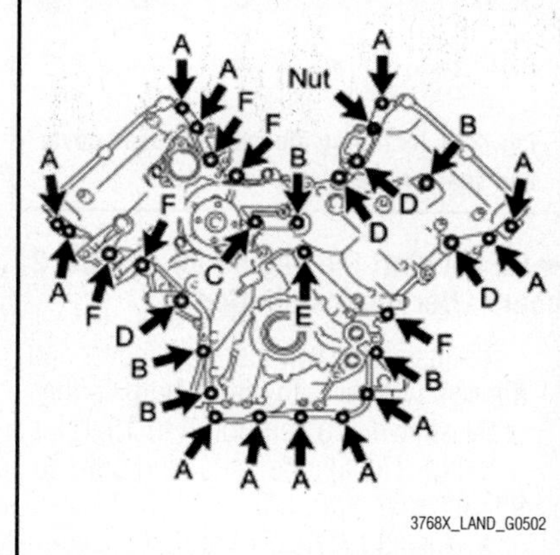

Fig. 205 Temporarily installing the timing chain cover

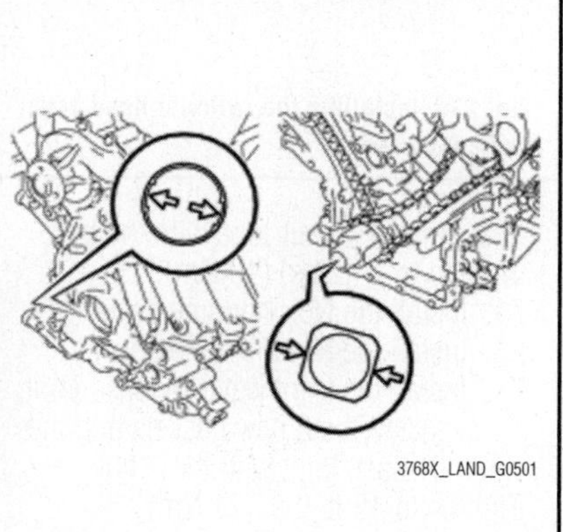

Fig. 204 Aligning the oil pump drive rotor spline and crankshaft

61. Install the cylinder head cover sub assembly LH.

a. Install the 5 new gaskets to the camshaft bearing caps (No. 2, No. 3).

b. Install the gasket to the cylinder head cover.

➡**Remove any oil from the contact surface.**

c. Apply seal packing as shown.

➡**Remove any oil from the contact surface.**

➡**Install the cylinder head cover within 3 minutes and tighten the bolts within 15 minutes after applying seal packing.**

➡**Do not start the engine for at least 2 hours after the installation.**

d. Install the cylinder head cover and the seal washer with the 14 bolts in the order shown. Tighten Bolt A to 15 ft. lbs. (21 Nm). Tighten the remaining bolts to 9 ft. lbs. (12 Nm).

62. Install the cylinder head cover sub assembly RH.

Item	Length	Thread Diameter
Bolt A	0.984 inch (25 mm)	0.315 inch (8 mm)
Bolt B	2.17 inch (55 mm)	0.315 inch (8 mm)
Bolt C	2.76 inch (70 mm)	0.315 inch (8 mm)
Bolt D	1.38 inch (35 mm)	0.394 inch (10 mm)
Bolt E	2.17 inch (55 mm)	0.394 inch (10 mm)
Bolt F	3.15 inch (80 mm)	0.394 inch (10 mm)

3768X_LAND_G0503

Fig. 206 Standard bolts

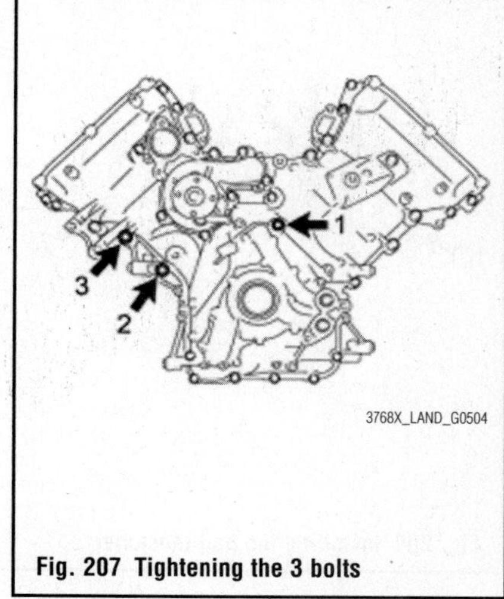

Fig. 207 Tightening the 3 bolts

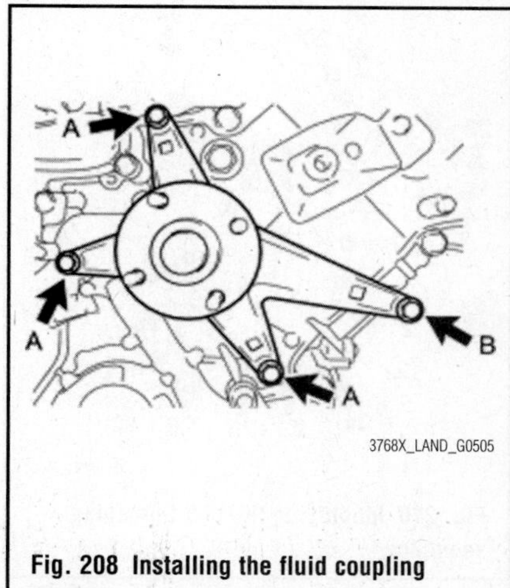

Fig. 208 Installing the fluid coupling bracket

a. Install 5 new gaskets to the camshaft bearing caps (No. 1, No. 3).

b. Install the gasket to the cylinder head cover.

➡**Remove any oil from the contact surface.**

c. Apply seal packing.

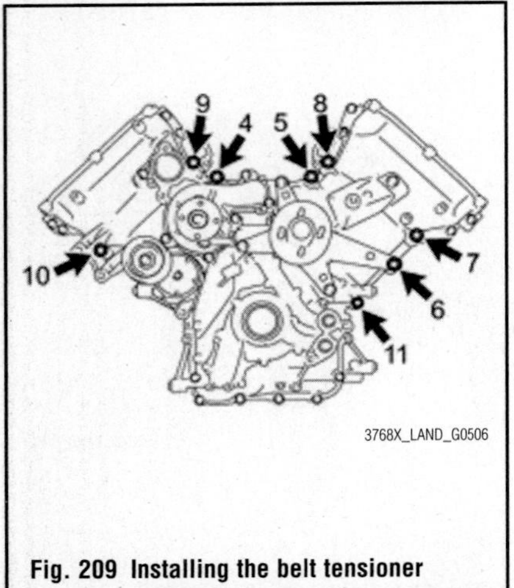

Fig. 209 Installing the belt tensioner

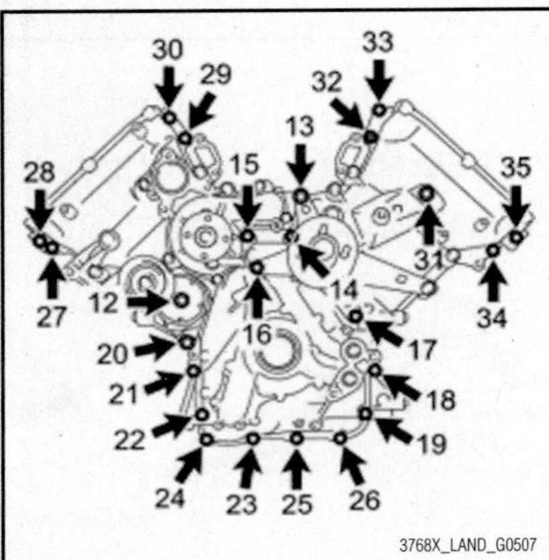

Fig. 210 Identifying the bolt tightening sequence

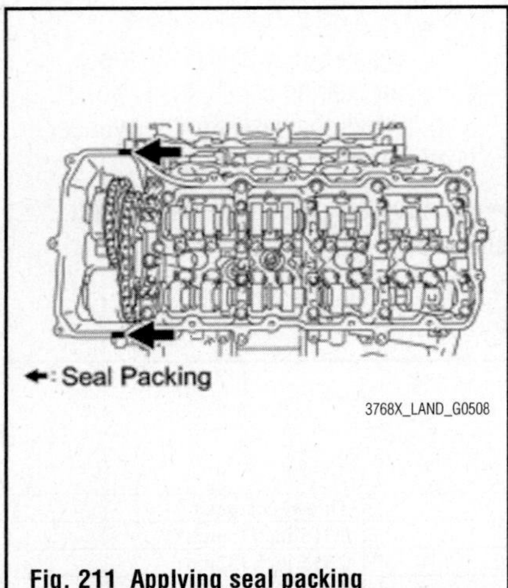

Fig. 211 Applying seal packing

➡Remove any oil from the contact surface.

➡Install the cylinder head cover within 3 minutes and tighten the bolts within 15 minutes after applying seal packing.

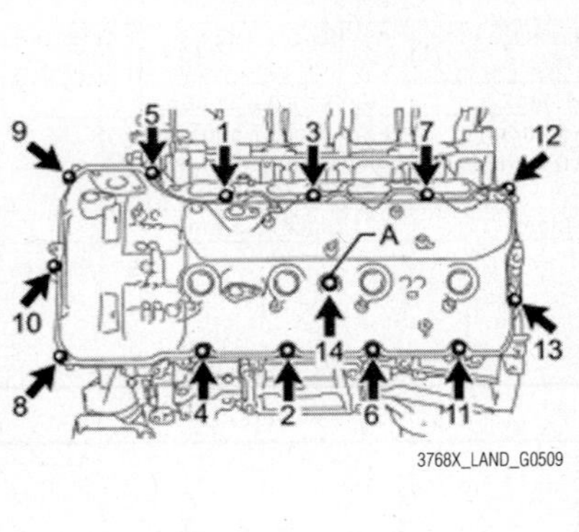

Fig. 212 Installing the cylinder head cover and seal washer

➡Do not start the engine for at least 2 hours after the installation.

d. Install the cylinder head cover and the seal washer with the 14 bolts in the order shown. Tighten Bolt A to 15 ft. lbs. (21 Nm). Tighten the remaining bolts to 9 ft. lbs. (12 Nm).

e. Install the noise filter to the cylinder head cover with the bolt. Tighten to 62 inch lbs. (7 Nm).

63. Install the ignition coil assembly with the 8 bolts. Tighten to 7 ft. lbs. (10 Nm).

64. Install the crankshaft timing gear key to the crankshaft.

65. Connect the wire harness clamp bracket to the timing chain cover with the bolt. Tighten to 71 inch lbs. (8 Nm).

66. Install the crankshaft pulley.

a. Align the pulley set key with the groove of the pulley, and slide on the pulley.

b. Using the special tool (09213-70011 or 09330-00021), install the pulley set bolt. Tighten to 221 ft. lbs. (300 Nm).

67. Install the No. 1 idler pulley sub assembly with the bolt. Tighten to 32 ft. lbs. (43 Nm).

68. Install the water pump pulley.

a. Temporarily install the pulley with the 4 bolts.

b. Using the special tool (09960-10010, 09962-01000 or 09963-01000), hold the pulley and tighten the 4 bolts. Tighten to 15 ft. lbs. (21 Nm).

69. Install the water inlet housing.

a. Install a new gasket to the timing chain cover.

b. Install the water inlet with the 3 bolts. Tighten to 15 ft. lbs. (21 Nm).

70. Install the air tube sub assembly.

a. Connect the 2 hoses.

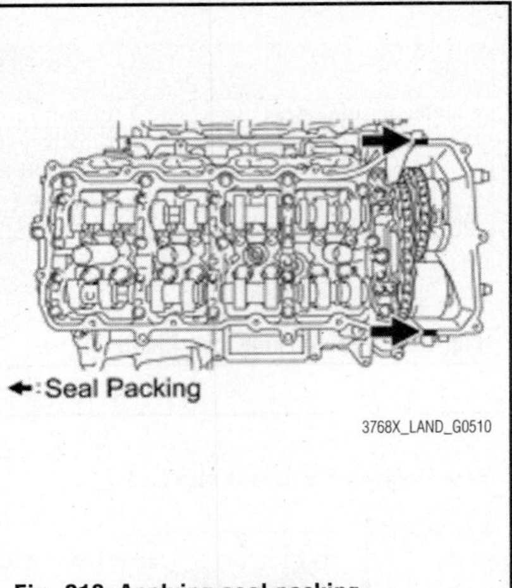

Fig. 213 Applying seal packing

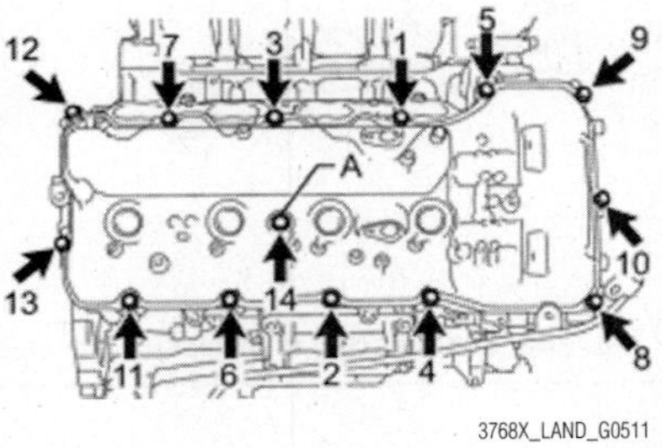

Fig. 214 Installing the cylinder head cover and seal washer

b. Install the air tube with the bolt. Tighten to 7 ft. lbs. (10 Nm).

71. Install the No. 1 engine cover.

72. Install the No. 2 engine cover.

73. Install the front water by pass joint.

a. Install the 2 new gaskets and the water by pass joint with the 4 nuts. Tighten to 15 ft. lbs. (21 Nm).

b. Connect the No. 2 water by pass hose to the water by pass joint.

74. Install the water by pass pipe sub assembly.

a. Connect the 2 hoses.

b. Install the water by pass pipe with the 2 bolts. Tighten to 7 ft. lbs. (10 Nm).

c. Connect the air tube with the bolt. Tighten to 7 ft. lbs. (10 Nm).

75. Install the No. 1 water by pass hose.

a. Install the No. 1 water by pass hose by connecting the hose to the water inlet housing and the front water by pass joint.

76. Install the generator assembly.

77. Connect the oil cooler pipe assembly.

78. Connect the vane pump assembly.

➡**Before performing the following procedures, move the spacer until the vane pump can be installed.**

a. Connect the vane pump to the timing chain cover with the 2 bolts. Tighten to 15 ft. lbs. (21 Nm).
79. Install the intake manifold.
80. Install the oil filter bracket (without oil cooler).
a. Apply a light coat of engine oil to 2 new O-rings.
b. Install the 2 O-rings to the timing chain cover.
c. Install the oil filter bracket with the 2 bolts and 2 nuts. Tighten to 26 ft. lbs. (35 Nm).
81. Install the oil filter bracket (with oil cooler).
82. Install the No. 1 oil cooler bracket (with oil cooler).
83. Install the No. 2 water by pass pipe sub assembly (with oil cooler).
a. Connect the 4 hoses.
b. Install the water by-pass pipe with the 3 bolts. Tighten to 7 ft. lbs. (10 Nm).
84. Install the oil pressure sender gauge assembly.
85. Install the engine oil level dipstick guide.
86. Install the oil filter element.
87. Connect the No. 2 fuel tube sub assembly.
a. Connect the fuel tube with the 2 bolts. Tighten to 7 ft. lbs. (10 Nm).
88. Connect the cooler compressor assembly.
a. Install the cooler compressor with the stud bolt. Tighten to 7 ft. lbs. (10 Nm).
b. Install the 3 bolts and nut. Tighten to 18 ft. lbs. (25 Nm).

➡**Tighten the bolts and nut in the order shown.**

89. Connect the No. 2 water by pass pipe.
a. Connect the No. 2 water by pass pipe to the cylinder head cover with the 3 bolts. Tighten to 13 ft. lbs. (18 Nm).
b. Connect the 3 hoses.
90. Connect the air pump hose and wire harness.
a. Connect the No. 2 and No. 3 air hoses.
b. Connect the 2 clamps and air pump wire harness.
91. Connect the engine wire.
a. For the engine room RH side, connect or install the following:

Fig. 215 Installing the cooler compressor assembly

- 5 clamps
- Throttle position sensor and throttle control motor connector
- 2 air pump connectors
- Ground wire with the 2 bolts (tighten to 71 inch lbs. (8 Nm))
- Noise filter connector
- 2 VVT sensor connectors
- Injector connector
- 4 ignition coil connectors
- 2 camshaft timing oil control valve connectors
- 2 clamps and power steering oil pressure switch connector.

92. For the engine room LH side, connect or install the following:
- Cooler compressor connector
- 3 clamps
- Camshaft position sensor connector
- 2 camshaft timing oil control valve connectors
- Engine Coolant Temperature (ECT) sensor connector
- 4 clamps
- Ground wire with the 2 bolts (tighten to 71 inch lbs. (8 Nm))
- Noise filter connector
- 2 VVT sensor connectors
- 4 ignition coil connectors
- Injector connector
- Wire harness clamp and 4 air injection control driver connectors
- 2 connectors and 2 clips to the engine room junction block
- Engine room relay block cover

93. Install the radiator assembly.
94. Install the air cleaner assembly.
95. Install the air cleaner hose assembly.
96. Install the v-bank cover sub assembly.
97. Add engine oil.
98. Add engine coolant.
99. Connect the cable to the negative battery terminal.
100. Inspect for oil leaks.
101. Inspect for coolant leaks.
102. Inspect the engine oil level.
103. Install the No. 2 engine under cover.
104. Install the No. 1 engine under cover sub assembly.
105. Install the front fender splash shield sub assembly LH.
106. Install the front fender splash shield sub assembly RH.
107. Install the front fender apron seal front RH.
108. Install the front fender apron seal LH.
109. Install the cowl top ventilator louver sub assembly.
110. Connect the radiator side deflector LH.
111. Connect the radiator side deflector RH (without air cooled transmission oil cooler).
112. Install the transmission oil cooler air duct (with air cooled transmission oil cooler).
113. Install the front bumper cover.
114. Install the radiator grille.
115. Install the upper radiator support seal.

REAR MAIN SEAL

REMOVAL & INSTALLATION

See Figure 216.

1. Disconnect the cable from the negative battery terminal.

CAUTION

Wait at least 90 seconds after disconnecting the cable from the negative (-) battery terminal to disable the SRS system.

2. Remove the automatic transmission assembly.
3. Remove the drive plate and ring gear sub assembly.
4. Remove the rear crankshaft oil seal.
a. Using a knife, cut off the lip of the oil seal.
b. Using a screwdriver, pry out the oil seal.

➡**Tape the screwdriver tip before use.**

➡**Do not damage the surface of the oil seal press fit hose and crankshaft.**

To install:

5. To install, reverse the removal procedure.

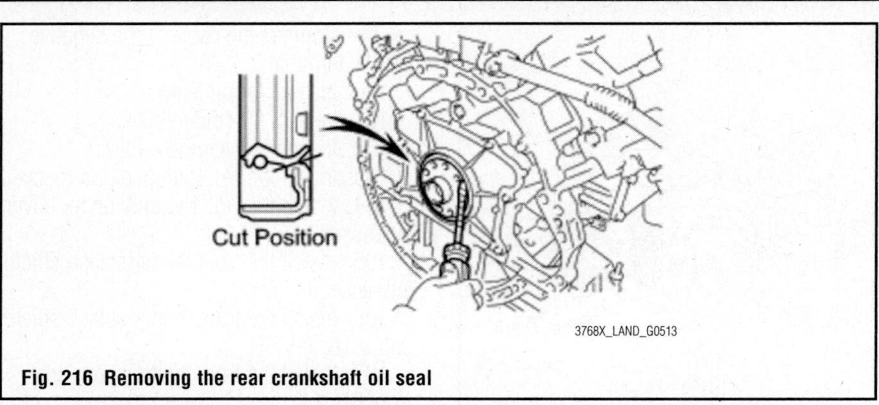

Fig. 216 Removing the rear crankshaft oil seal

TIMING CHAIN FRONT COVER

REMOVAL & INSTALLATION

Refer to OIL PUMP for removal and installation procedures.

TIMING CHAIN & SPROCKETS

REMOVAL & INSTALLATION

Refer to OIL PUMP for removal and installation procedures.

ENGINE PERFORMANCE & EMISSION CONTROLS

CAMSHAFT POSITION (CMP) SENSOR

LOCATION

See Figure 217.

Refer to the accompanying illustration for sensor location.

REMOVAL & INSTALLATION

See Figures 218 and 219.

1. Remove the v-bank cover sub assembly.
2. Remove the air cleaner hose assembly.
3. Remove the camshaft position sensor.
   a. Disconnect the sensor connector.
   b. Remove the bolt and sensor.
4. Remove the VVT sensor.
   a. Disconnect the 4 sensor connectors.
   b. Remove the 4 bolts and 4 sensors.

To install:

5. To install, reverse the removal procedure. Tighten the sensors to 7 ft. lbs. (10 Nm).

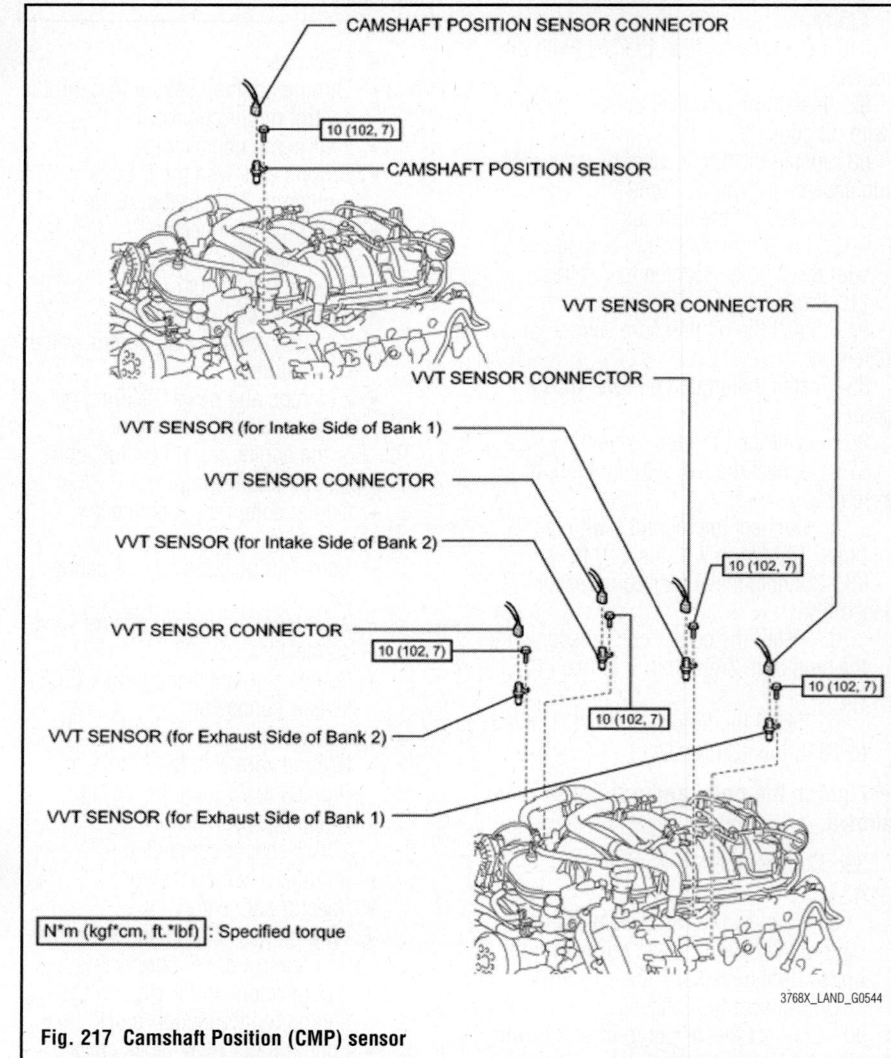

Fig. 217 Camshaft Position (CMP) sensor

CRANKSHAFT POSITION (CKP) SENSOR

LOCATION

See Figure 220.

Refer to the accompanying illustration for sensor location.

REMOVAL & INSTALLATION

See Figures 221 and 222.

1. Remove the No. 2 engine under cover.
2. Remove the crankshaft position sensor protector.
   a. Remove the 2 bolts and sensor protector.
3. Remove the crankshaft position sensor.
   a. Disconnect the sensor connector.
   b. Remove the bolt and sensor.

To install:

4. To install, reverse the removal procedure. Tighten the sensor and sensor protector to 7 ft. lbs. (10 Nm).

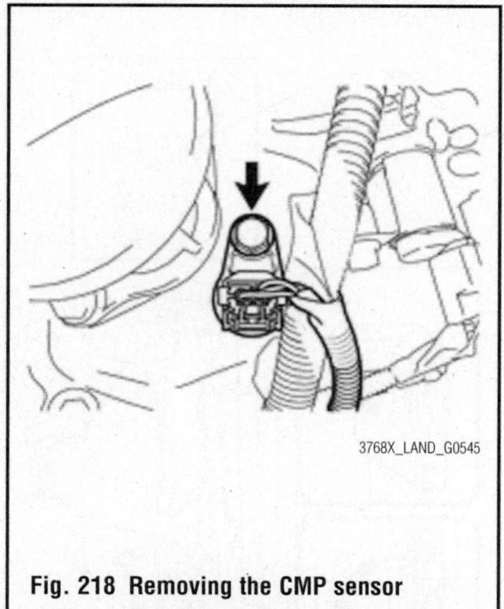

Fig. 218 Removing the CMP sensor

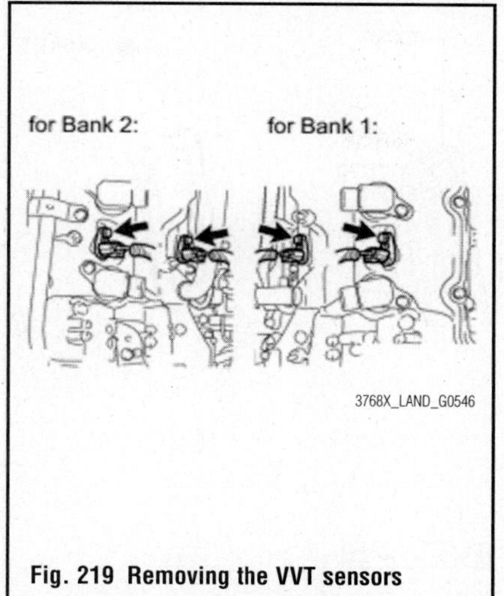

Fig. 219 Removing the VVT sensors

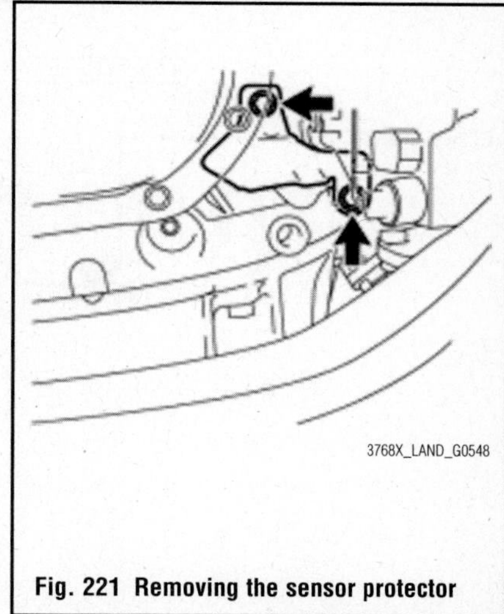

Fig. 221 Removing the sensor protector

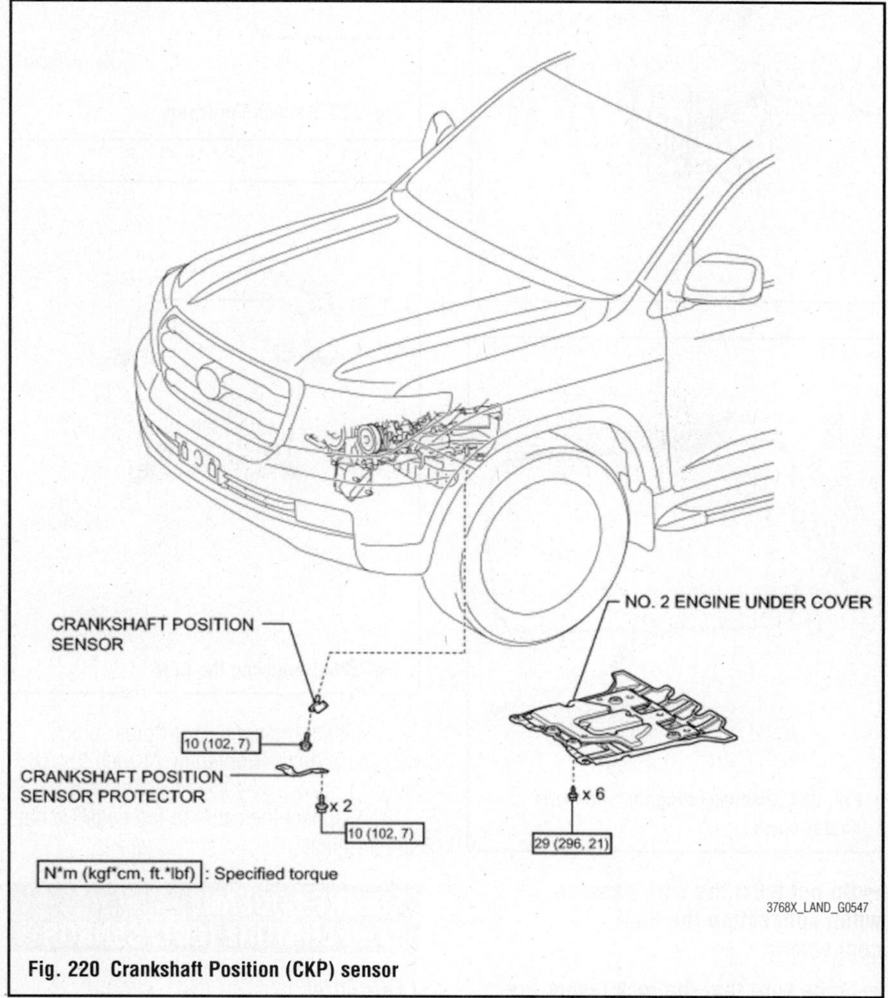

Fig. 220 Crankshaft Position (CKP) sensor

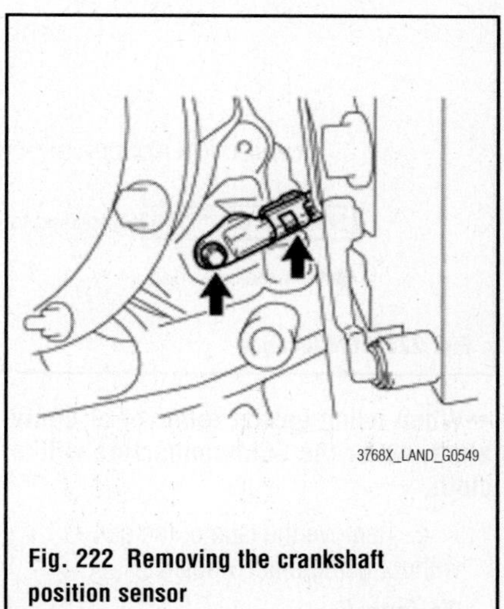

Fig. 222 Removing the crankshaft position sensor

1. Disconnect the cable from the negative battery terminal.
2. Disconnect the connector holder block.
   a. Remove the 3 bolts and move the connector holder block so that the ECM can be removed in the next step.
3. Remove the ECM.
   a. Raise the 2 levers while pushing the locks on the 2 levers.

➡Make sure that the lock levers are raised 90° as shown in the illustration before disconnecting the connectors. Failure to do this may cause the connectors to break.

   b. Disconnect the 2 connectors.
   c. Remove the 4 bolts and ECM.
4. Remove the gasket.
   a. Peel off the gasket.
   b. Spray gasket remover or equivalent on the remaining tape of the gasket.

ELECTRONIC CONTROL MODULE (ECM)

LOCATION

See Figure 223.

Refer to the accompanying illustration for location.

REMOVAL & INSTALLATION

See Figures 224 through 226.

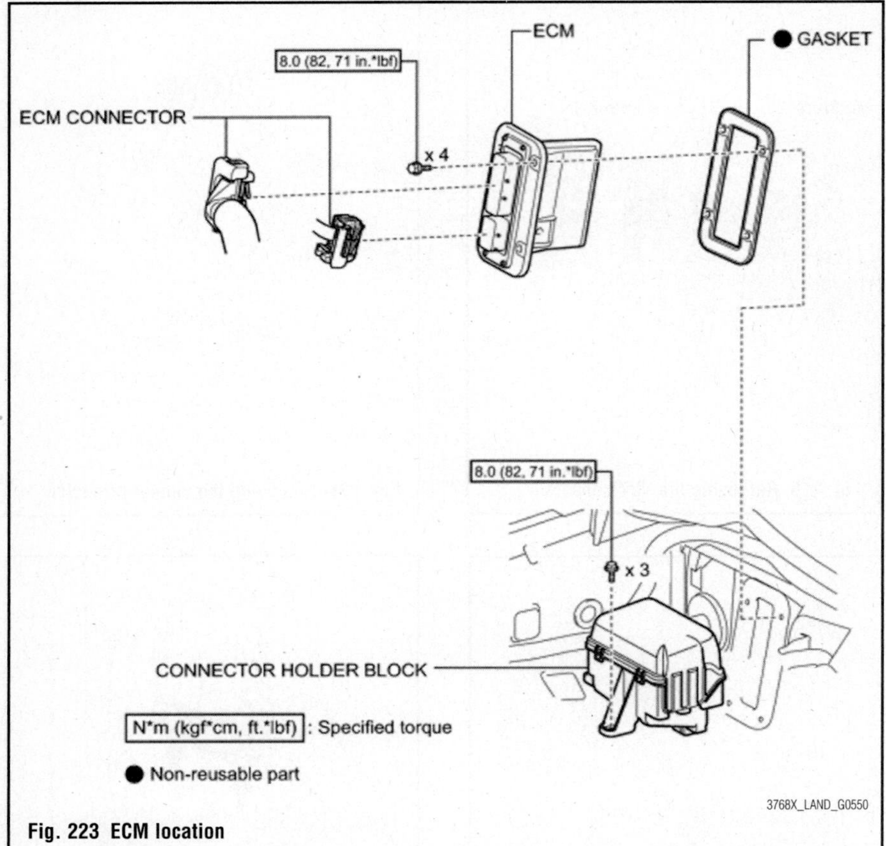

Fig. 223 ECM location

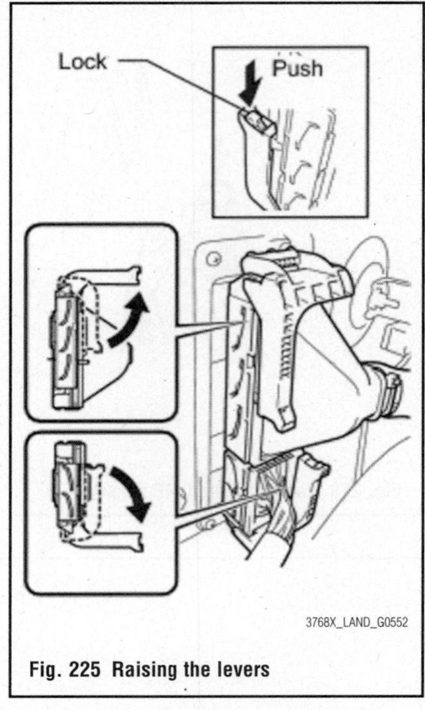

Fig. 225 Raising the levers

➡When using gasket remover or equivalent, cover the ECM connectors with a cloth.

c. Remove the tape of the gasket without using bladed objects.

To install:

5. Install the gasket.
   a. Clean the ECM seal surface with non-residue solvent.
   b. Attach a new gasket to the ECM.
6. Install the ECM with the 4 bolts. Tighten to 71 inch lbs. (8 Nm).

➡Make sure the gasket is not folded or caught on any area of the surrounding part.

➡Make sure there is no foreign matter caught between the gasket and surrounding part.

➡Do not damage the gasket.

a. Connect the 2 ECM connectors and push each lock lever down to lock the ECM connectors.

➡Be careful not to allow dirt, water, or other foreign matter to adhere to the connecting parts when connecting the connectors.

Fig. 224 Disconnecting the connector holder block

➡Do not twist the wire harness when connecting the ECM connectors.

➡Make sure that the lock levers are raised 90° before connecting the ECM connectors. Failure to do this may cause the lock lever and terminals to break.

➡Make sure that the levers are securely locked.

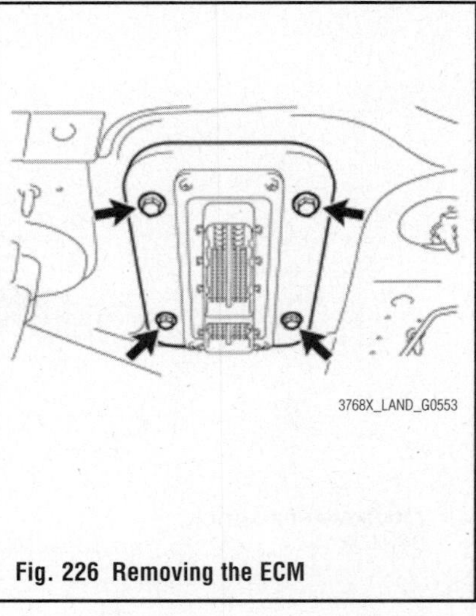

Fig. 226 Removing the ECM

7. Install the connector holder block with the 3 bolts. Tighten to 71 inch lbs. (8 Nm).
8. Connect the cable to the negative battery terminal.

ENGINE COOLANT TEMPERATURE (ECT) SENSOR

LOCATION

See Figure 227.

Refer to the accompanying illustration for sensor location.

REMOVAL & INSTALLATION

See Figure 228.

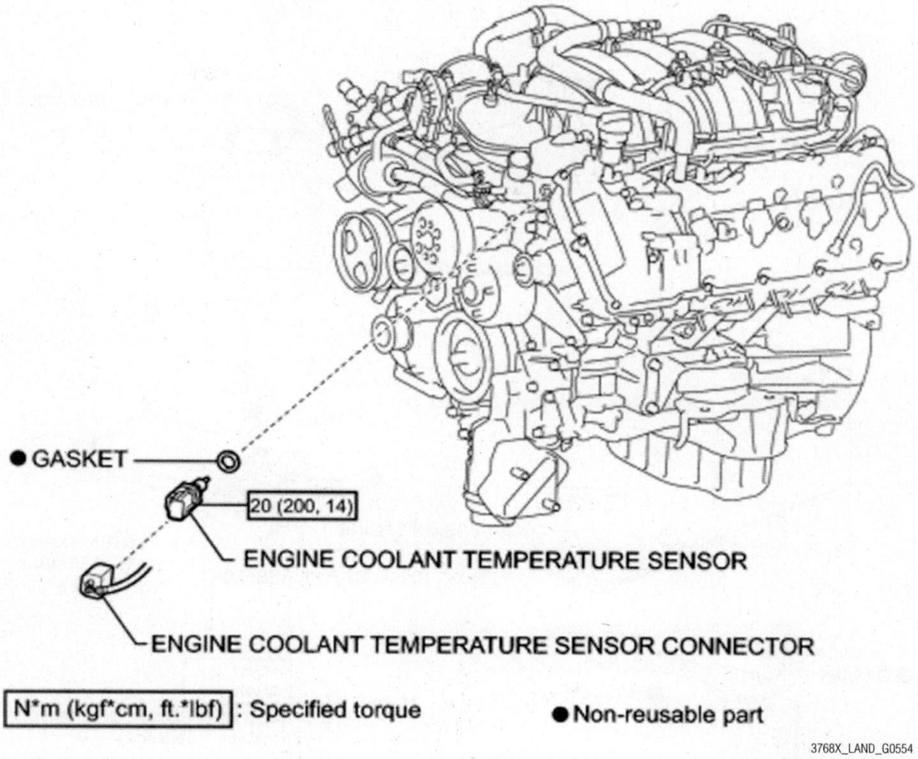

Fig. 227 Engine Coolant Temperature (ECT) sensor

1. Remove the front fender splash shield sub assembly RH.
2. Remove the front fender splash shield sub assembly LH.
3. Remove the No. 1 engine under cover sub assembly.
4. Drain the engine coolant.
5. Remove the v-bank cover sub assembly.
6. Remove the Engine Coolant Temperature (ECT) sensor.

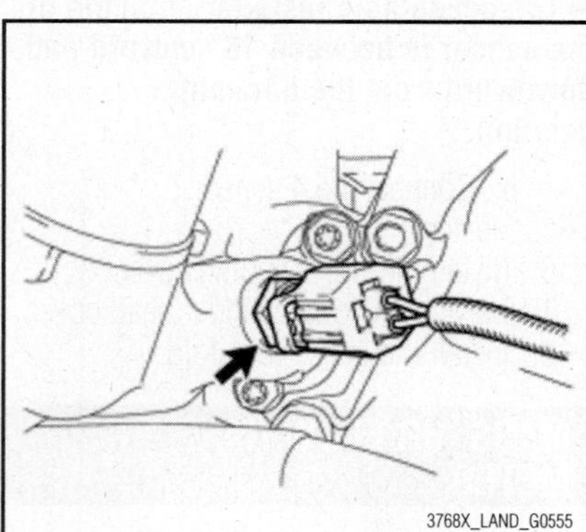

Fig. 228 Removing the Engine Coolant Temperature (ECT) sensor

a. Disconnect the sensor connector.
b. Using a 19 mm deep socket wrench, remove the sensor.
c. Remove the gasket from the sensor.

To install:

7. To install, reverse the removal procedure. Tighten the sensor to 14 ft. lbs. (20 Nm).

HEATED OXYGEN SENSOR (HO2S)

LOCATION

See Figure 229.

Refer to the accompanying illustration for sensor location.

REMOVAL & INSTALLATION

See Figure 230.

CAUTION

The procedures should be performed by at least 2 people.

CAUTION

Wear protective gloves when removing the exhaust pipe.

CAUTION

The exhaust pipe is extremely hot immediately after the engine has stopped.

CAUTION

Confirm that the exhaust pipe has cooled down before removing it.

1. Remove the No. 2 engine under cover.
2. Remove the tailpipe assembly.
3. Remove the center exhaust pipe assembly.
4. Remove the front No. 2 exhaust pipe assembly.
5. Remove the heated oxygen sensor.
   a. Using the special tool (09224-00010), remove the sensor.

To install:

6. To install, reverse the removal procedure. Tighten the sensor to 32 ft. lbs. (44 Nm). If tightening with the special tool, tighten to 30 ft. lbs. (40 Nm).

KNOCK SENSOR (KS)

LOCATION

See Figure 231.

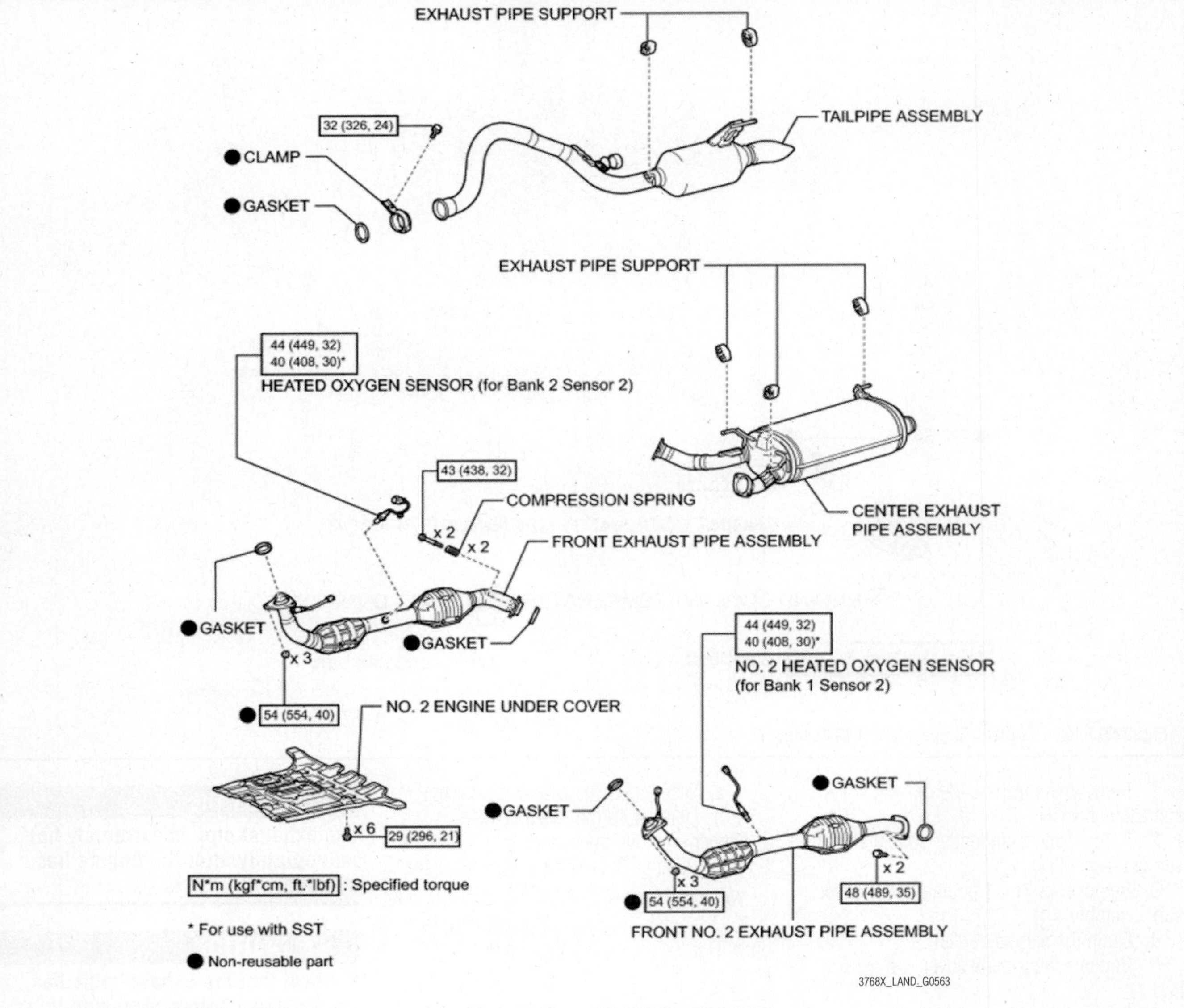

Fig. 229 Heated Oxygen (HO2S) sensor

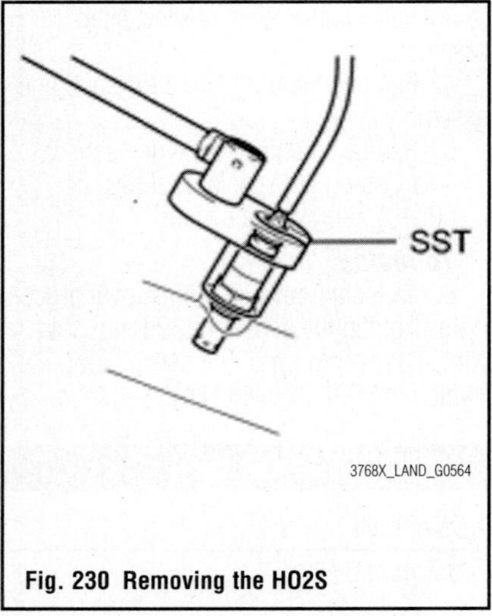

Fig. 230 Removing the HO2S

Refer to the accompanying illustration for sensor location.

REMOVAL & INSTALLATION

See Figures 232 and 233.

1. Remove the intake manifold.
2. Remove the No. 2 cylinder head cover.
3. Remove the No. 1 engine cover.
4. Remove the Knock Sensor (KS):
    a. Disconnect the 4 KS connectors.
    b. Remove the 4 bolts and 4 KS.

To install:

5. Install the KS.
    a. Install the 4 sensors with the 4 bolts so that the sensors are angled as shown. Tighten to 15 ft. lbs. (20 Nm).

➡The acceptable installation angle of the sensor is between 10° upward and downward from the horizontal position.

    b. Connect the 4 sensor connectors.
6. Install the No. 1 engine cover.
7. Install the No. 2 cylinder head cover.
8. Install the intake manifold.

MASS AIR FLOW (MAF) METER

LOCATION

See Figure 234.

Refer to the accompanying illustration for location.

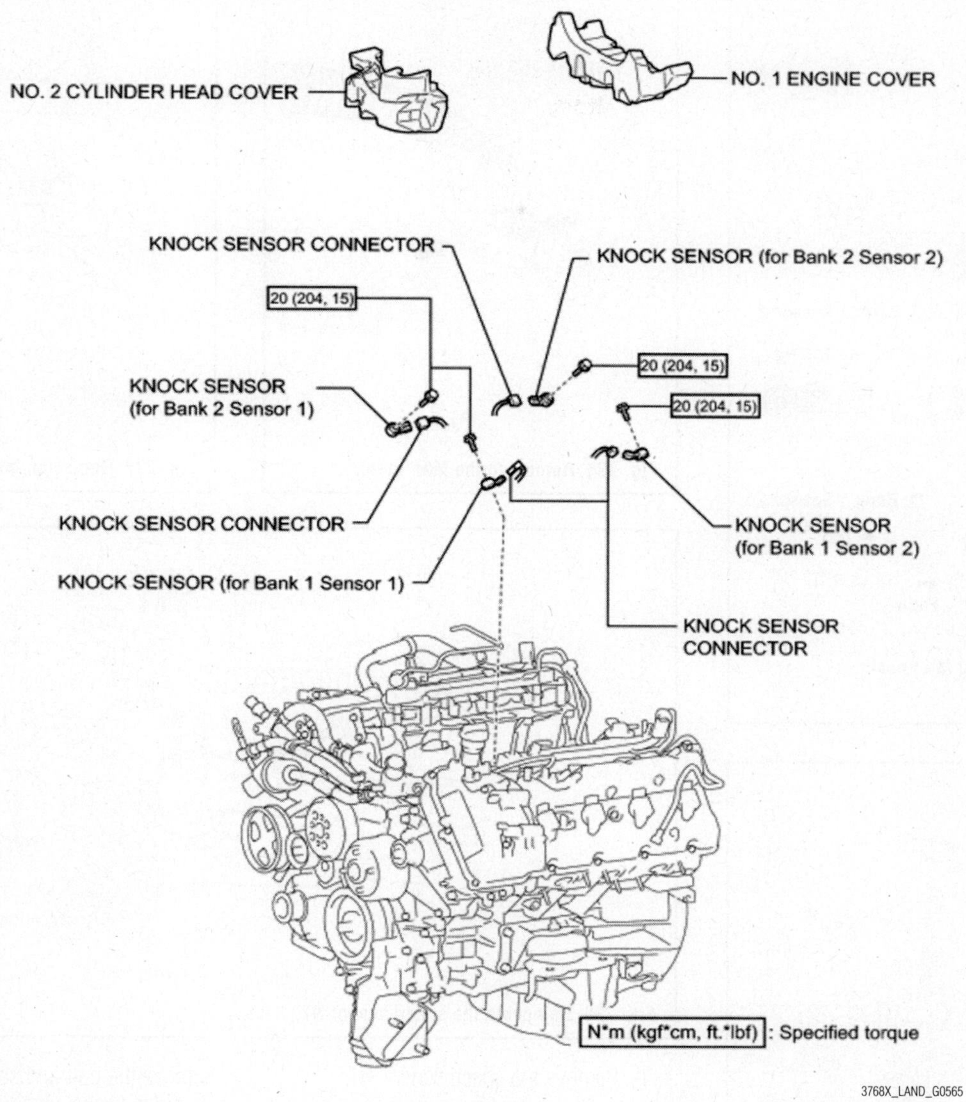

NO. 2 CYLINDER HEAD COVER
NO. 1 ENGINE COVER
KNOCK SENSOR CONNECTOR
KNOCK SENSOR (for Bank 2 Sensor 2)
20 (204, 15)
KNOCK SENSOR (for Bank 2 Sensor 1)
20 (204, 15)
20 (204, 15)
KNOCK SENSOR CONNECTOR
KNOCK SENSOR (for Bank 1 Sensor 2)
KNOCK SENSOR (for Bank 1 Sensor 1)
KNOCK SENSOR CONNECTOR
N*m (kgf*cm, ft.*lbf) : Specified torque
3768X_LAND_G0565

Fig. 231 Knock Sensor (KS)

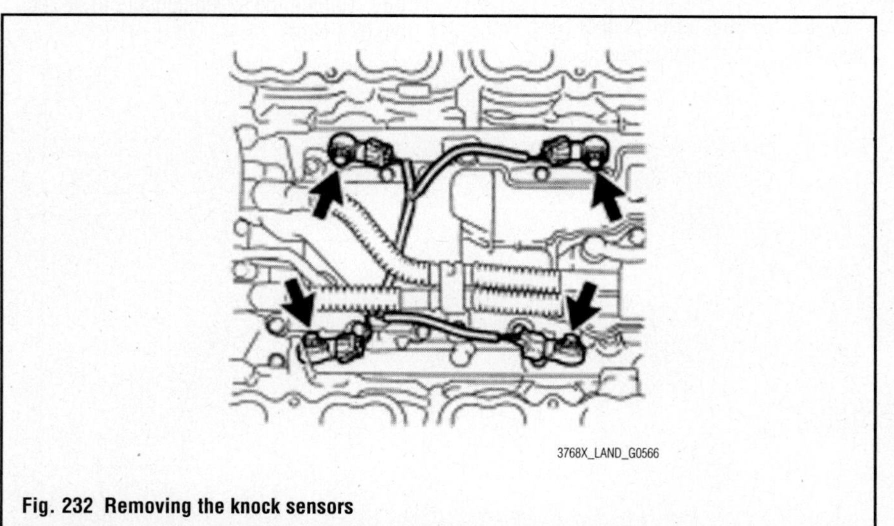

Fig. 232 Removing the knock sensors

REMOVAL & INSTALLATION

See Figure 235.

1. Remove the Mass Air Flow (MAF) meter.
    a. Disconnect the MAF connector.
    b. Remove the 2 screws and the MAF meter.

To install:

2. Install the MAF meter with the 2 screws. Tighten to 15 inch lbs. (1.7 Nm). Connect the MAF connector.

VEHICLE SPEED SENSOR (VSS)

REMOVAL & INSTALLATION

See Figures 236 and 237.

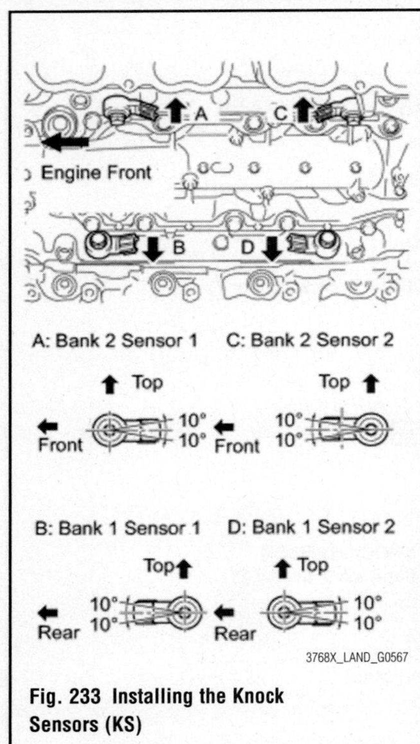

Fig. 233 Installing the Knock Sensors (KS)

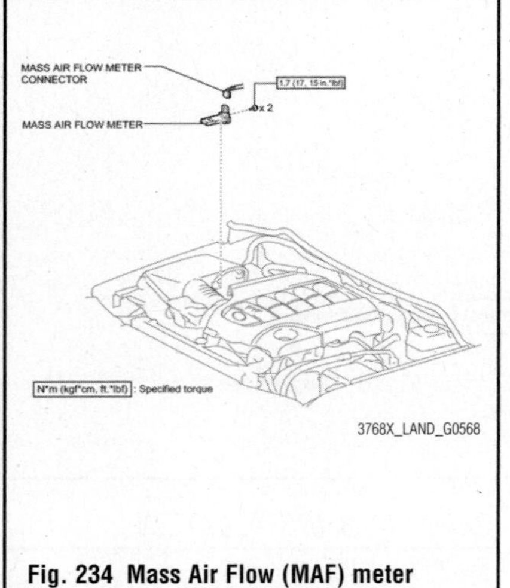

Fig. 234 Mass Air Flow (MAF) meter

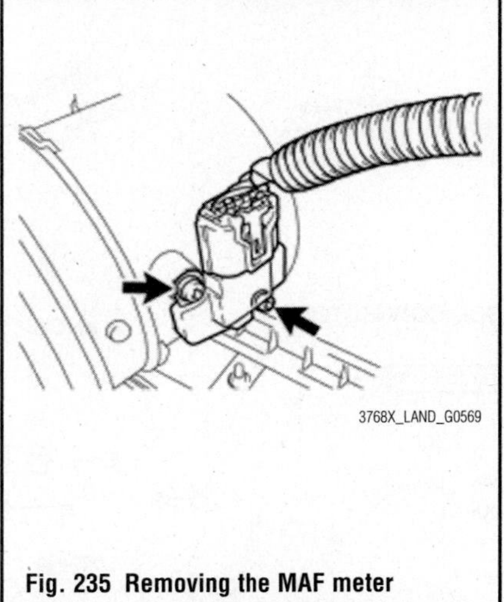

Fig. 235 Removing the MAF meter

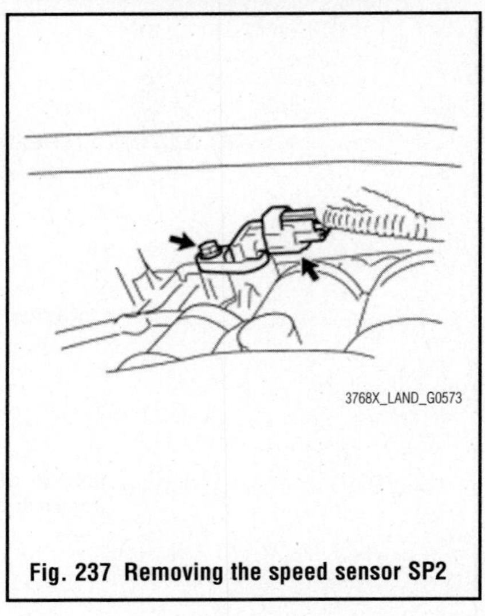

Fig. 237 Removing the speed sensor SP2

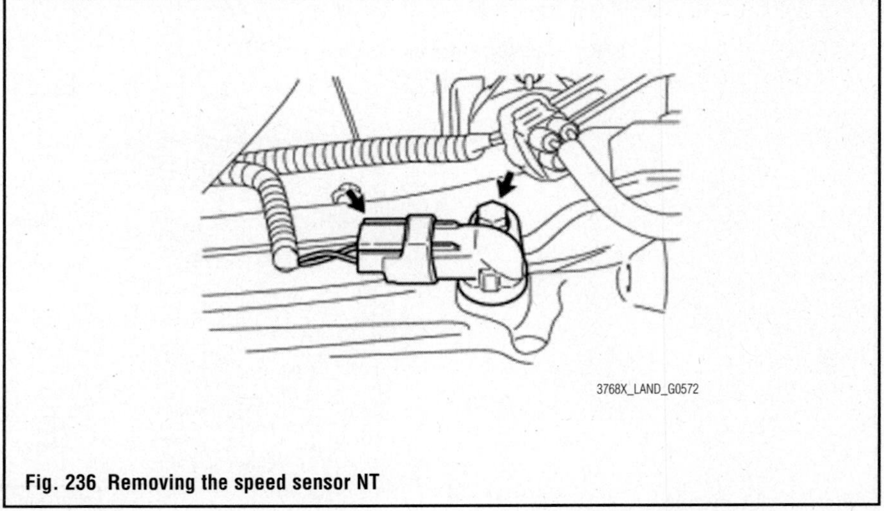

Fig. 236 Removing the speed sensor NT

1. Remove the speed sensor NT.
   a. Disconnect the sensor connector.
   b. Remove the bolt and sensor.
   c. Remove the O-ring from the sensor.
2. Remove the speed sensor SP2. (a) Disconnect the sensor connector. (b) Remove the bolt and sensor. (c) Remove the O-ring from the sensor.

To install:

3. To install, reverse the removal procedure. Tighten the speed sensors to 48 inch lbs. (5.4 Nm).

FUEL — GASOLINE FUEL INJECTION SYSTEM

FUEL SYSTEM SERVICE PRECAUTIONS

Safety is the most important factor when performing not only fuel system maintenance but any type of maintenance. Failure to conduct maintenance and repairs in a safe manner may result in serious personal injury or death. Maintenance and testing of the vehicle's fuel system components can be accomplished safely and effectively by adhering to the following rules and guidelines.

- To avoid the possibility of fire and personal injury, always disconnect the negative battery cable unless the repair or test procedure requires that battery voltage be applied.
- Always relieve the fuel system pressure prior to disconnecting any fuel system component (injector, fuel rail, pressure regulator, etc.), fitting or fuel line connection. Exercise extreme caution whenever relieving fuel system pressure to avoid exposing skin, face and eyes to fuel spray. Please be advised that fuel under pressure may penetrate the skin or any part of the body that it contacts.
- Always place a shop towel or cloth around the fitting or connection prior to loosening to absorb any excess fuel due to spillage. Ensure that all fuel spillage (should it occur) is quickly removed from engine surfaces. Ensure that all fuel soaked cloths or towels are deposited into a suitable waste container.
- Always keep a dry chemical (Class B) fire extinguisher near the work area.
- Do not allow fuel spray or fuel vapors to come into contact with a spark or open flame.
- Always use a back-up wrench when loosening and tightening fuel line connection fittings. This will prevent unnecessary stress and torsion to fuel line piping.
- Always replace worn fuel fitting O-rings with new Do not substitute fuel hose or equivalent where fuel pipe is installed.

Before servicing the vehicle, make sure to also refer to the precautions in the beginning of this section as well.

RELIEVING FUEL SYSTEM PRESSURE

See Figure 238.

✲✲ CAUTION

Do not disconnect any part of the fuel system until you have discharged the fuel system pressure.

✲✲ CAUTION

After discharging the fuel pressure, place a cloth or equivalent over fittings as you separate them to reduce the risk of fuel spray on yourself or in the engine compartment.

1. Remove the engine room relay block cover.
2. Remove the circuit opening relay (C/OPN).
3. Start the engine. After the engine has stopped on its own, turn the engine switch off.

➡If DTC P0230 is output, clear the DTC.

4. Crank the engine, then check that the engine does not start.
5. Loosen the fuel tank cap, then discharge the pressure in the fuel tank completely.
6. Disconnect the cable from the negative (-) battery terminal.
7. Install the circuit opening relay.
8. Install the engine room relay block cover.

FUEL FILTER

REMOVAL & INSTALLATION

See Figures 239 and 240.

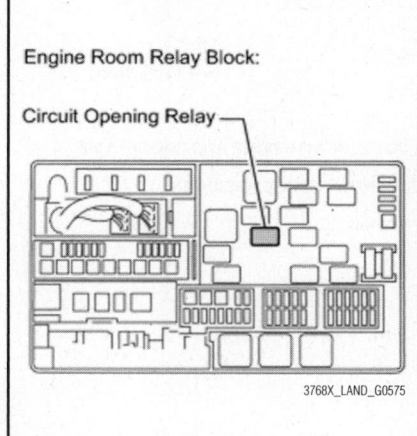

Engine Room Relay Block:

Circuit Opening Relay

3768X_LAND_G0575

Fig. 238 Locating the circuit opening relay

1. Remove the fuel suction with the pump and gauge tube assembly.
2. Remove the fuel sender gauge assembly.
3. Remove the No. 1 fuel sub tank.
4. Remove the fuel pump.
5. Remove the fuel main valve assembly.
   a. Remove the fuel main valve from the fuel filter.
   b. Remove the 2 O-rings from the fuel main valve.
6. Remove the No. 1 fuel suction support.
   a. Detach the 2 claws from the claw holes.

To install:

7. Install the No. 1 fuel suction support.
   a. Attach the 2 claws to the claw holes.
8. Install the fuel main valve assembly.
   a. Apply a light coat of gasoline to 2 new O-rings, and install them onto the fuel main valve.
   b. Install the fuel main valve to the fuel filter.

➡Make sure the O-rings are not cut or pinched during the installation.

9. Install the fuel pump.
10. Install the No. 1 fuel sub tank.
11. Install the fuel sender gauge assembly.
12. Install the fuel suction with the pump and gauge tube assembly.

FUEL PUMP

REMOVAL & INSTALLATION

See Figure 241.

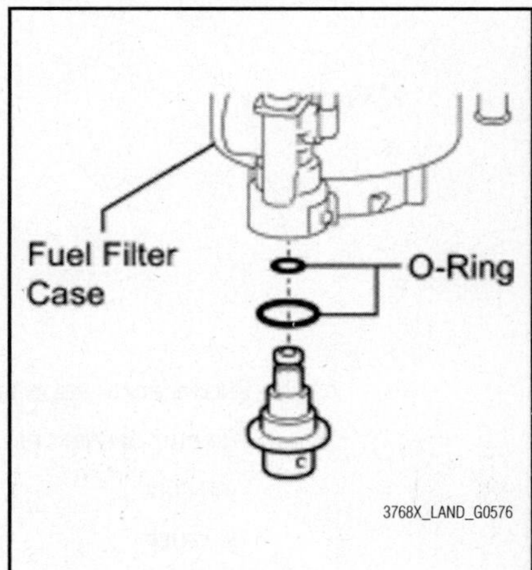

Fig. 239 Removing the fuel main valve and O-rings from the fuel filter

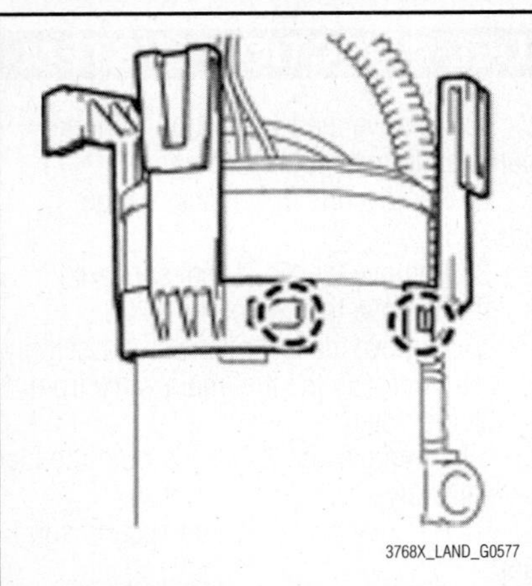

Fig. 240 Removing the No. 1 fuel suction support

1. Remove the fuel suction with pump and gauge tube assembly.

To install:

2. To install, reverse the removal procedure.

FUEL TANK

DRAINING

See Figure 242.

1. Connect the cable to the negative (-) battery terminal.
2. Connect the Techstream to the DLC3.
3. Turn the engine switch on (IG).

➡Do not start the engine.

4. Turn the Techstream main switch on.
5. Enter the following menus: Powertrain/Engine and ECT/Active Test/Control the Fuel Pump/Speed.

CAUTION

Do not smoke or be near an open flame when working on the fuel system.

CAUTION

Secure good ventilation.

CAUTION

Keep gasoline away from rubber or leather parts.

➡If the fuel pump does not operate, remove the fuel tube joint clip and No. 1 fuel tube joint, and drain fuel from the port shown in the illustration.

REMOVAL & INSTALLATION

See Figures 243 through 265.

1. Discharge the fuel system pressure.
2. Remove the fuel tank cap assembly.
3. Remove the No. 1 fuel tank protector sub assembly.
   a. Remove the 5 bolts and fuel tank protector.
4. Disconnect the fuel tank main tube sub assembly.
   a. Detach the fuel tube clamp.
   b. Disconnect the fuel tank main tube.

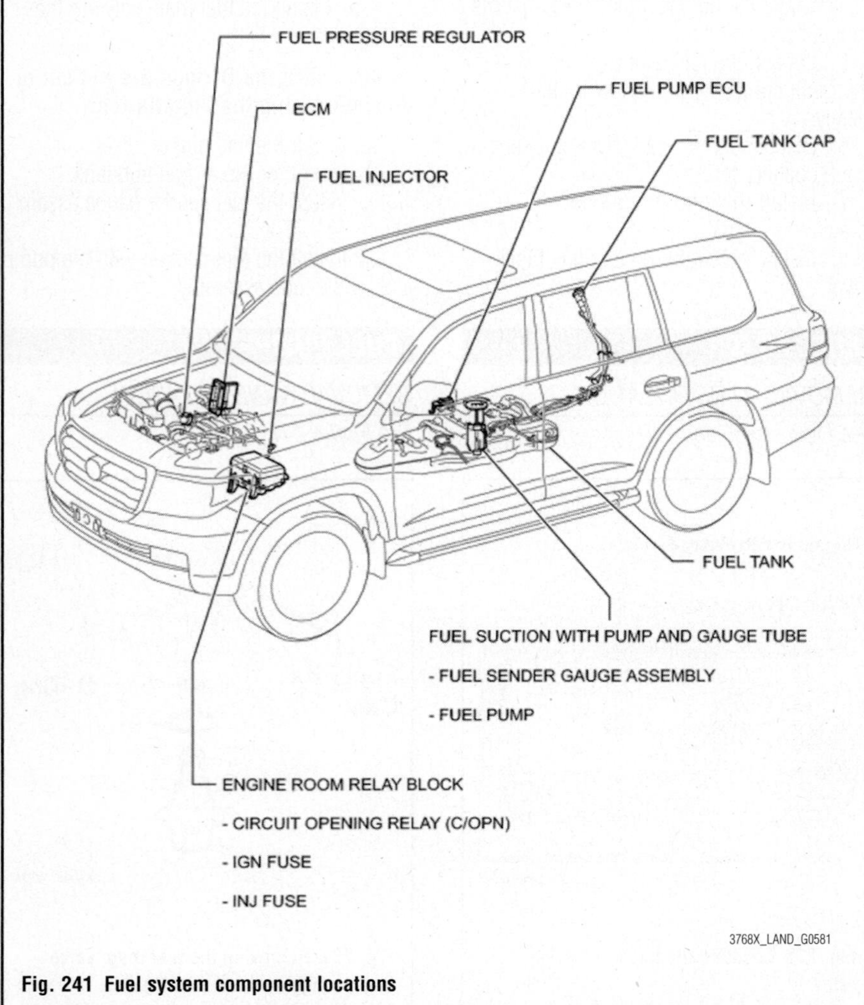

Fig. 241 Fuel system component locations

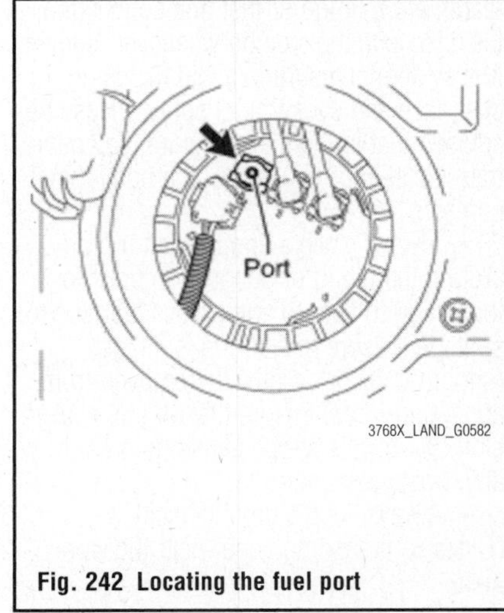

Fig. 242 Locating the fuel port

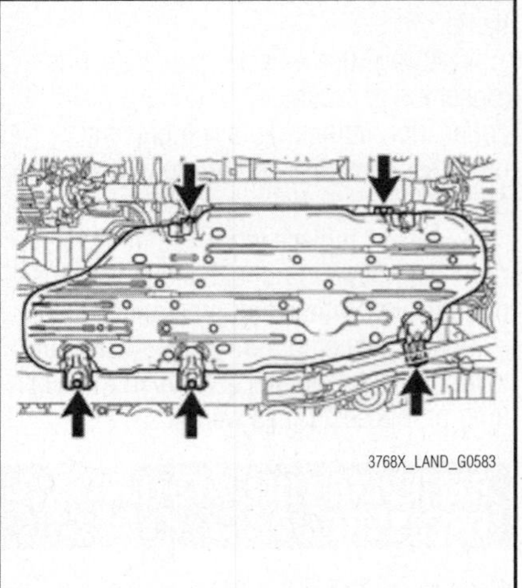

Fig. 243 Removing the fuel tank protector

➡Check for foreign matter in the pipe and around the connector. Clean if necessary. Foreign matter may damage the O-ring or cause leaks in the seal between the pipe and connector.

➡Do not use any tools to separate the pipe and connector.

➡Do not forcefully bend or twist the nylon tube.

➡Check for foreign matter on the pipe seal surface. Clean if necessary.

➡Put the pipe and connector ends in plastic bags to prevent damage and foreign matter contamination.

➡If the pipe and connector are stuck together, pinch the connector between your fingers and turn it carefully to disconnect it.

5. Drain the fuel.
6. Disconnect the cable from the negative battery terminal.

CAUTION

Wait at least 90 seconds after disconnecting the cable from the negative (-) battery terminal to disable the SRS system.

7. Remove the rear No. 1 seat assembly LH.
8. Remove the rear No. 1 seat assembly RH.
9. Remove the rear No. 1 seat protector.
10. Remove the rear No. 2 seat protector.

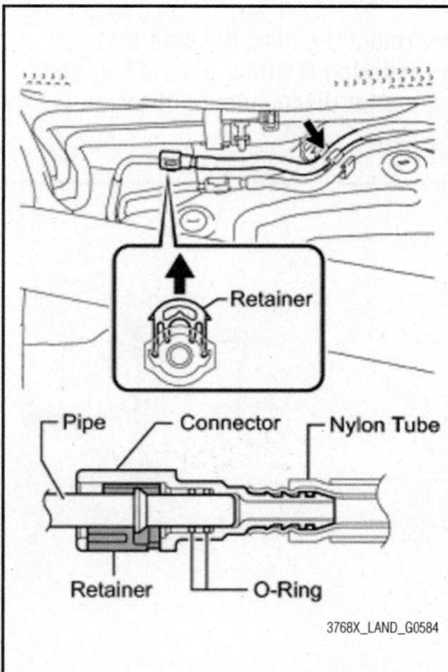

Fig. 244 Disconnecting the fuel tank main tube sub assembly

11. Remove the rear step cover.
12. Remove the rear door scuff plate LH.
13. Remove the rear door scuff plate RH.
14. Remove the rear No. 2 seat assembly.
15. Remove the rear floor mat rear support plate.
16. Remove the luggage compartment No. 1 trim hook.
17. Remove the front quarter trim panel assembly LH.
18. Remove the front quarter trim panel assembly RH.
19. Remove the air duct plug.
20. Remove the rear air duct guide.
21. Remove the front floor carpet assembly.
    a. Fold back the carpet.

➡Fold back the carpet until it is possible to remove the air duct.

22. Remove the rear floor No. 2 service whole cover.
    a. Remove the 2 screws and air duct.
    b. Remove the service hole cover.
    c. Disconnect the fuel pump and fuel sender gauge connector.
23. Disconnect the fuel tank return tube.
    a. Detach the fuel tube clamp.
    b. Disconnect the fuel tank return tube.

➡Check for foreign matter in the pipe and around the connector. Clean if necessary. Foreign matter may damage the O-ring or cause leaks in the seal between the pipe and connector.

➡Do not use any tools to separate the pipe and connector.

➡Do not forcefully bend or twist the nylon tube.

➡Check for foreign matter on the pipe seal surface. Clean if necessary.

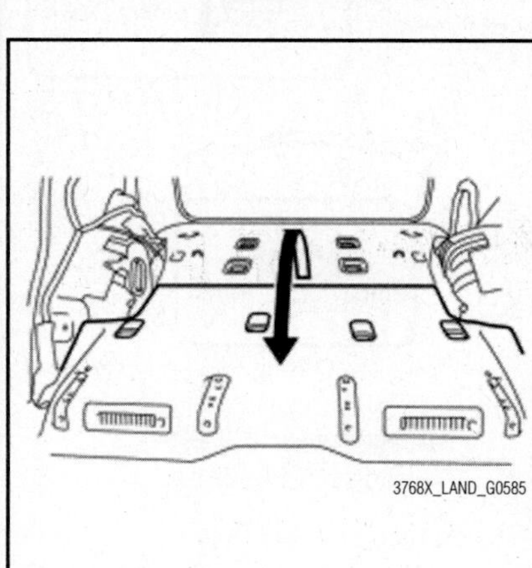

Fig. 245 Removing the front floor carpet

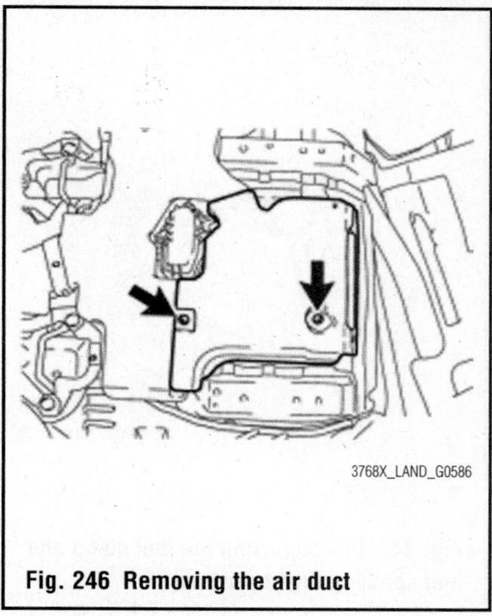

Fig. 246 Removing the air duct

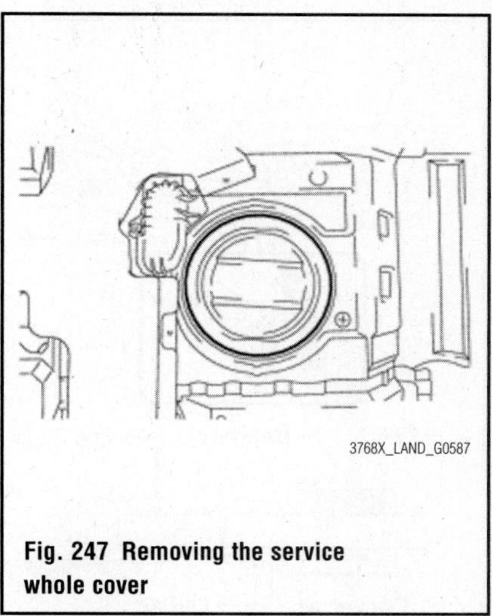

Fig. 247 Removing the service whole cover

➡Put the pipe and connector ends in plastic bags to prevent damage and foreign matter contamination.

➡If the pipe and connector are stuck together, pinch the connector between your fingers and turn it carefully to disconnect it.

24. Disconnect the No. 2 fuel tank breather tube sub assembly.
    a. Detach the 2 fuel tube clamps.
    b. Disconnect the No. 2 fuel tank breather tube.

➡Do not use any tools in this procedure.

➡Check for any dirt and foreign matter contamination in the pipe and around the connector. Clean if necessary. Foreign matter may damage the O-rings or cause leaks in the seal between the pipe and connector.

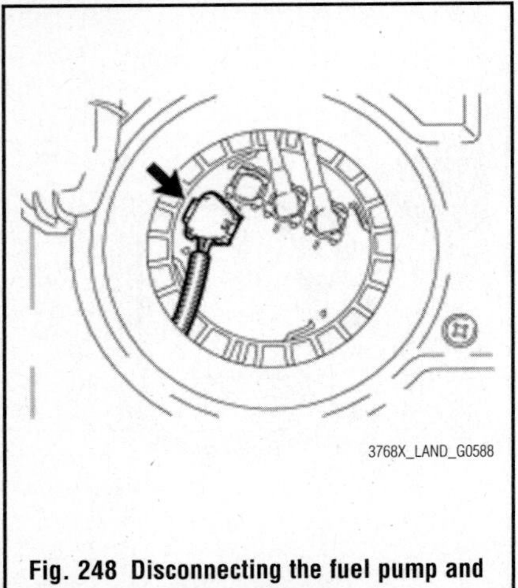

Fig. 248 Disconnecting the fuel pump and fuel sender gauge connector

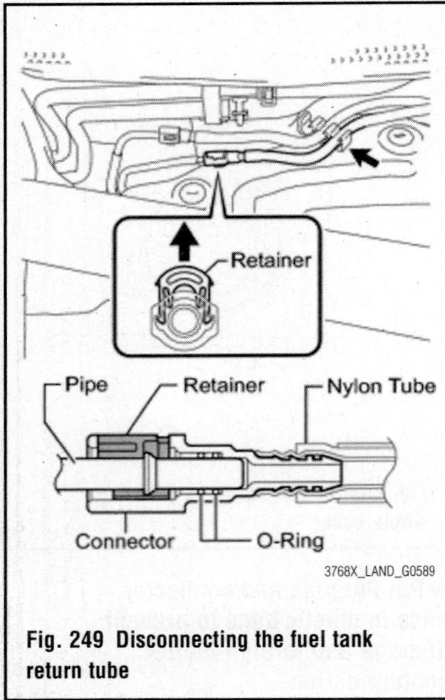

Fig. 249 Disconnecting the fuel tank return tube

25. Disconnect the fuel tank breather tube sub assembly.
    a. Detach the fuel tube clamp.
    b. Disconnect the fuel tank breather tube.

➡Check for foreign matter in the pipe and around the connector. Clean if necessary. Foreign matter may damage the O-ring or cause leaks in the seal between the pipe and connector.

➡Do not use any tools to separate the pipe and connector.

➡Do not forcefully bend or twist the nylon tube.

➡Check for foreign matter on the pipe seal surface. Clean if necessary.

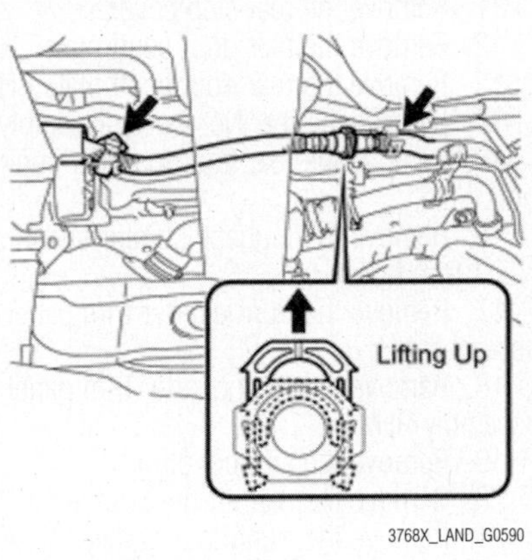

Fig. 250 Disconnecting the No. 2 fuel tank breather tube sub assembly

➡Put the pipe and connector ends in plastic bags to prevent damage and foreign matter contamination.

➡If the pipe and connector are stuck together, pinch the connector between your fingers and turn it carefully to disconnect it.

26. Disconnect the fuel tank to filler pipe hose from the fuel tank filler pipe.
27. Remove the fuel tank sub assembly.
    a. Place a transmission jack under the fuel tank.
    b. Remove the 2 bolts, 2 clips, 2 pins and 2 fuel tank bands.
    c. Slowly lower the transmission jack slightly.

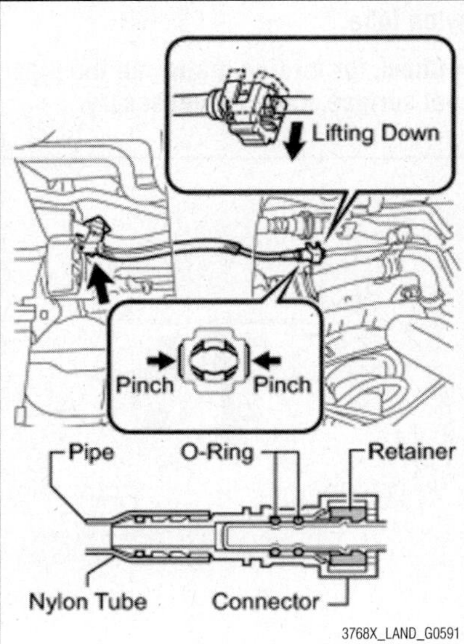

Fig. 251 Disconnecting the fuel tank breather tube sub assembly

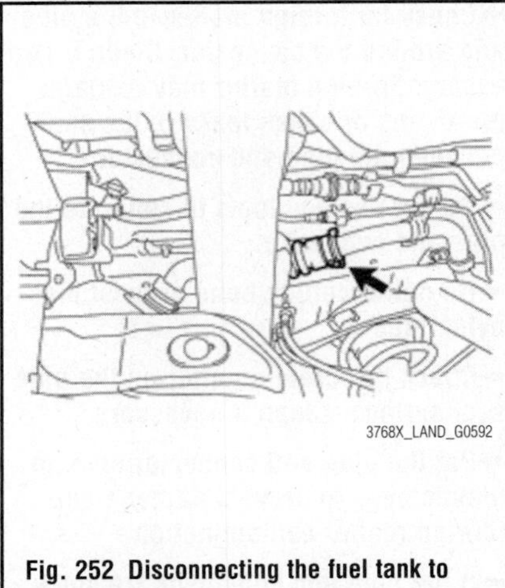

Fig. 252 Disconnecting the fuel tank to filler pipe hose

28. Remove the fuel tank main tube sub assembly and fuel tank return tube sub assembly.
    a. Remove the 3 fuel tube joint clips and pull out the 2 fuel tubes and No. 1 fuel tube joint.

➡Remove any dirt and foreign matter on the fuel tube joint before performing this step.

➡Do not allow any scratches or foreign matter on the parts when disconnecting them, as the fuel tube joint contains the O-rings that seal the plug.

➡Perform this step by hand. Do not use any tools.

➡Do not forcibly bend, twist or turn the nylon tube.

➡Protect the disconnected part by covering it with a plastic bag and tape after disconnecting the fuel tubes.

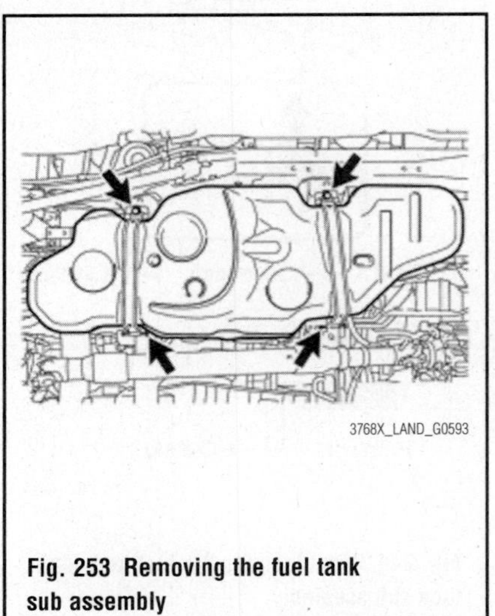

Fig. 253 Removing the fuel tank sub assembly

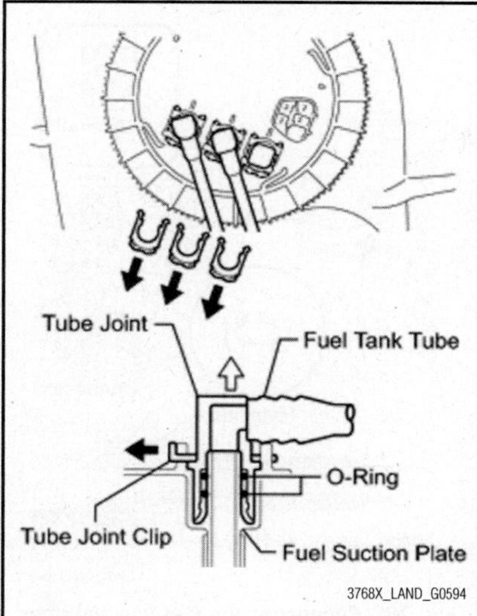

Fig. 254 Removing the 3 fuel tube joint clips and pulling out the 2 fuel tubes and No. 1 fuel tube joint

b. Remove the fuel main tube and fuel return tube from the fuel tank.

29. Remove the fuel suction with the pump and gauge tube assembly.

a. Using the special tool (09808-14020), loosen the retainer.

➡Fit the tips of special tool onto the ribs of the retainer.

➡When the retainer is loosened, be careful as the pump and gauge tube will spring upward from the force of the spring.

➡Remove foreign matter around the fuel suction with pump and gauge before this operation.

b. Remove the retainer.

c. Remove the fuel suction with pump and gauge tube assembly from the fuel tank.

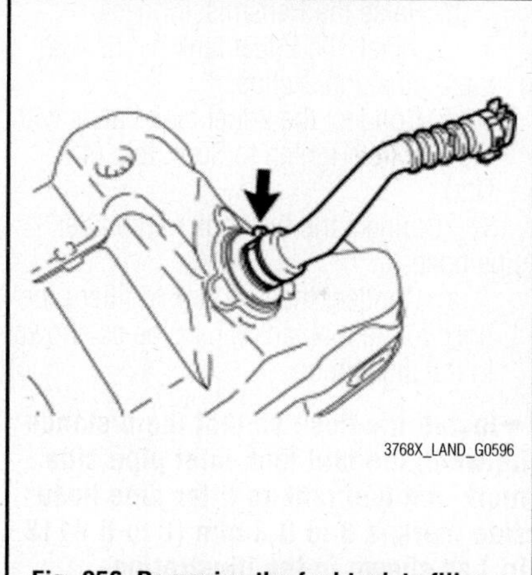

Fig. 256 Removing the fuel tank to filler pipe hose

➡Be careful not to bend the arm of the fuel sender gauge.

d. Remove the gasket from the fuel tank.

30. Remove the fuel tank to filler pipe hose.

a. Remove the fuel tank to filler pipe hose from the fuel tank.

31. Remove the No. 1 fuel tank heat insulator.

a. Remove the fuel tube clamp from the No. 1 fuel tank heat insulator.

b. Using needle nose pliers, remove the 4 clips and then remove the No. 1 fuel tank heat insulator.

To install:

32. Install the No. 1 fuel tank heat insulator with the 4 clips.

a. Install the fuel tube clamp to the No. 1 fuel tank heat insulator.

33. Install the fuel tank to the filler pipe hose.

a. Install the fuel tank to filler pipe hose to the fuel tank as shown in the illustration.

➡Align the fuel tank side mark with the hose side mark when installing the hose.

➡Tighten the hose clamp until the end of the hose clamp contacts the stopper as shown in the illustration.

34. Install the fuel suction with pump and gauge tube assembly.

a. Apply a light coat of gasoline or grease to a new gasket, and install it to the fuel tank.

b. Install the fuel suction with pump and gauge tube into the fuel tank.

➡Be careful not to bend the arm of the fuel sender gauge.

➡Align the protrusion of the fuel suction with pump and gauge tube with the groove of the fuel tank.

c. Put the retainer on the fuel tank. While holding the fuel suction with pump and gauge tube, tighten the retainer 1 complete turn by hand.

d. Using the special tool (09808-14020), tighten the retainer until the mark on the retainer is within range A on the fuel tank, as shown in the illustration.

➡Fit the tips of SST onto the ribs of the retainer.

35. Install the fuel tank main tube sub assembly and the fuel tank return tube sub assembly.

a. Install the 2 fuel tank tubes and No. 1 fuel tube joint with the 3 tube joint clips.

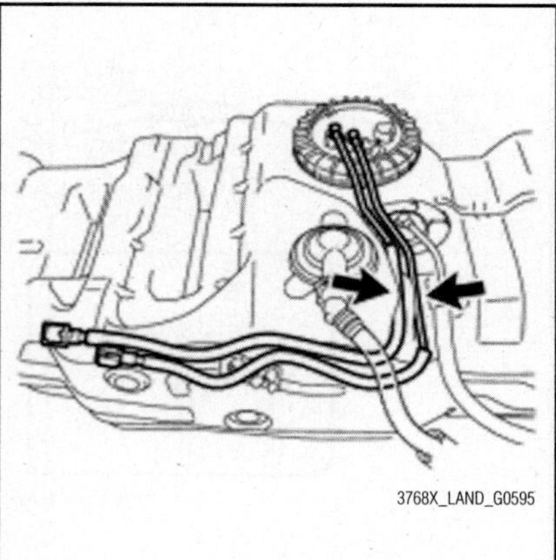

Fig. 255 Removing the fuel main tube and fuel return tube from the fuel tank

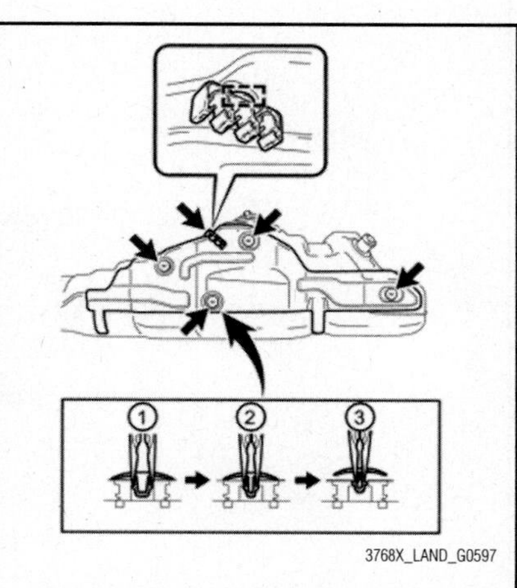

Fig. 257 Removing the No. 1 fuel tank heat insulator

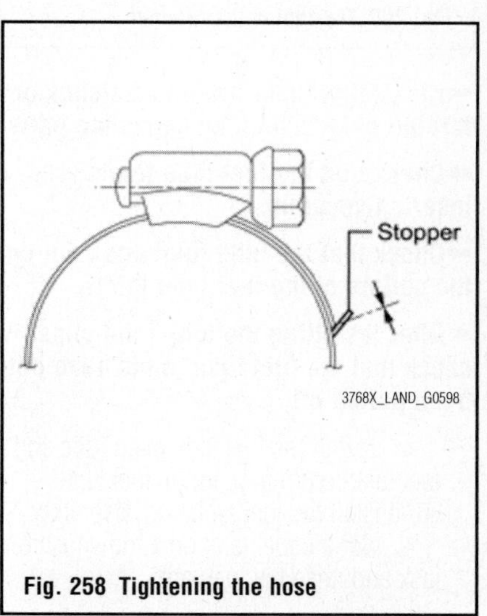

Fig. 258 Tightening the hose

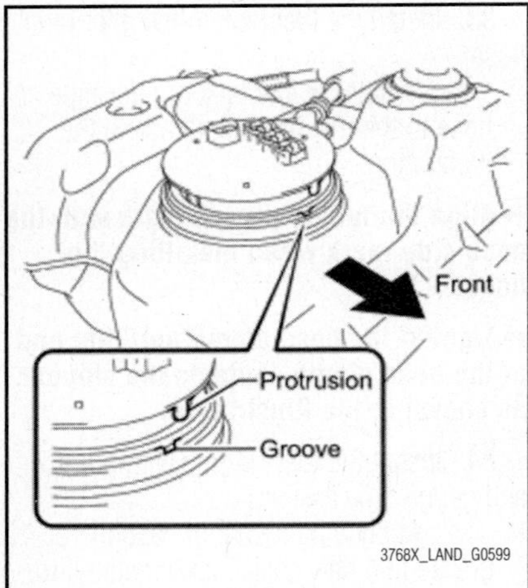

Fig. 259 Aligning the protrusion of the fuel suction with pump and gauge tube with the groove of the fuel tank

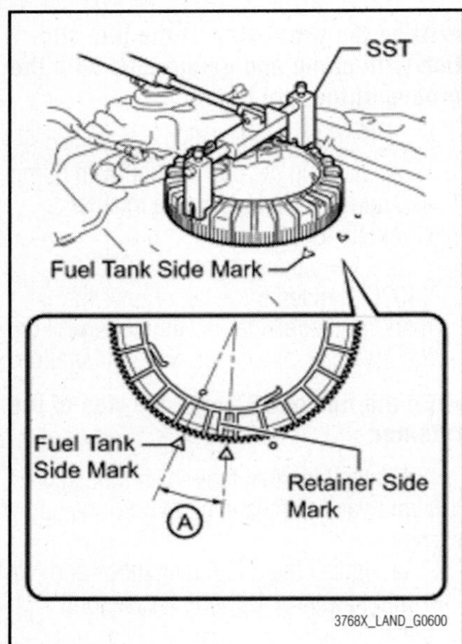

Fig. 260 Tightening the retainer

➡Check that there are no scratches or foreign objects on the connecting parts.

➡Check that the fuel tube joints are inserted securely.

➡Check that the tube joint clips are on the collars of the fuel tube joints.

➡After installing the tube joint clips, check that the fuel tube joints have not been pulled off.

b. Install the fuel tank main tube and fuel tank return tube to the fuel tank.

36. Install the fuel tank sub assembly.

a. Set the fuel tank on a transmission jack and raise the fuel tank.

b. Raise the transmission jack.

c. Install the 2 fuel tank bands with the 2 pins and 2 clips.

d. Connect the 2 fuel tank bands with the 2 bolts. Tighten to 30 ft. lbs. (40 Nm).

37. Connect the fuel tank to the filler pipe hose.

a. Connect the fuel tank to filler pipe hose to the fuel tank filler pipe as shown in the illustration.

➡Install the hose so that the distance between the fuel tank inlet pipe side mark and fuel tank to filler pipe hose side mark is 0 to 0.3 mm (0 to 0.0118 in.) as shown in the illustration.

➡Tighten the hose clamp until the end of the hose clamp contacts the stopper as shown in the illustration.

38. Connect the fuel tank breather tube sub assembly to the fuel tank filler pipe.

➡Push the parts together firmly until a "click" sound is heard.

➡Before installing the tube connector to the pipe, check if there is any damage or foreign matter in the connector.

➡After the connection, check if the connectors and pipes are securely connected by trying to pull them apart.

a. Attach the fuel tube clamp.

39. Connect the No. 2 fuel tank breather tube sub assembly.

➡Before installing the hose, make sure that it is not damaged. Make sure that there is no foreign matter present on the connecting surfaces.

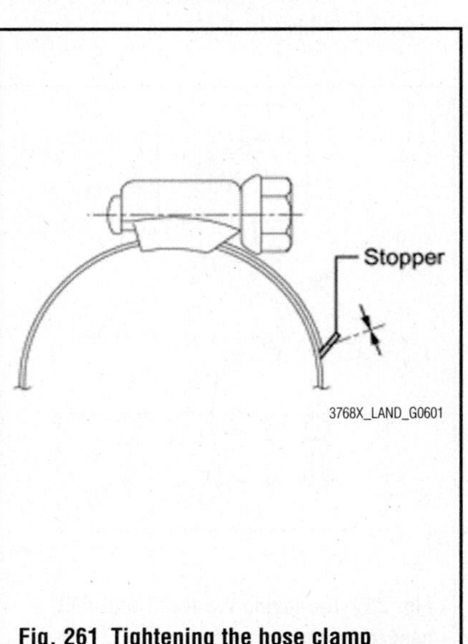

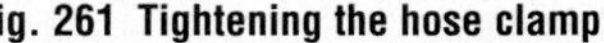

Fig. 261 Tightening the hose clamp

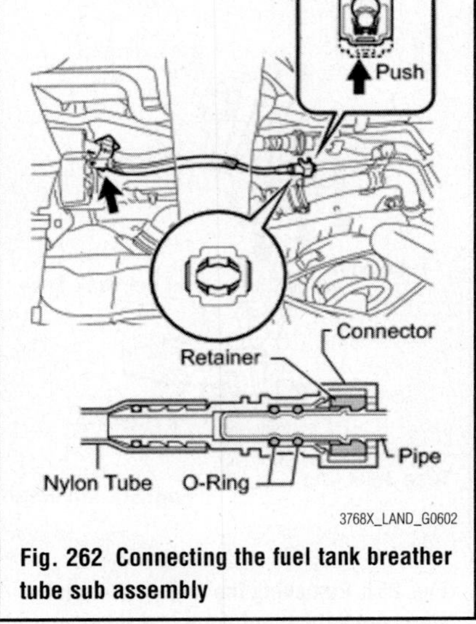

Fig. 262 Connecting the fuel tank breather tube sub assembly

➡After connecting, check if the hose and the connector are securely connected by pulling on them.

a. Attach the 2 fuel tube clamps.

40. Connect the fuel tank return tube.

➡Push the parts together firmly until a "click" sound is heard.

➡Before installing the tube connector to the pipe, check if there is any damage or foreign matter in the connector.

➡After the connection, check if the connectors and pipes are securely connected by trying to pull them apart.

a. Attach the fuel tube clamp.

41. Connect the fuel tank main tube sub assembly.

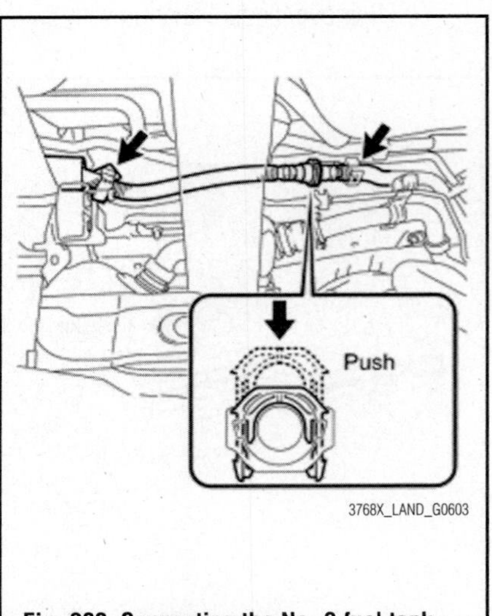

Fig. 263 Connecting the No. 2 fuel tank breather tube sub assembly

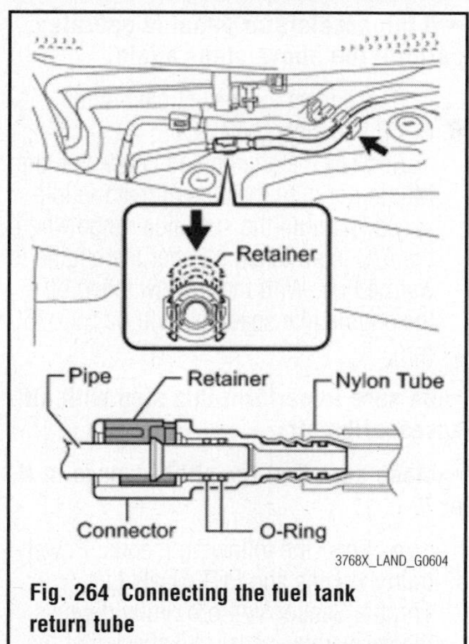

Fig. 264 Connecting the fuel tank return tube

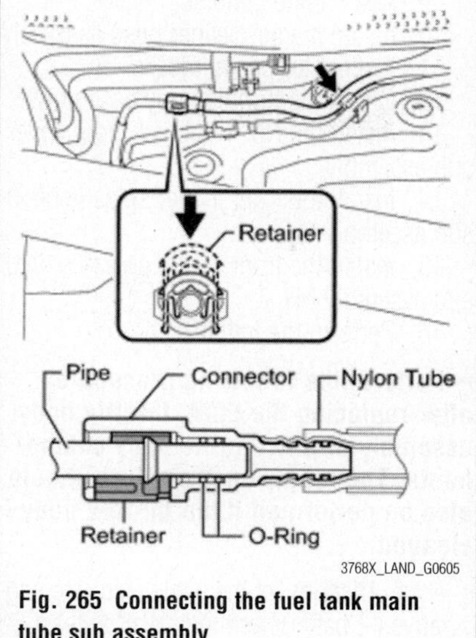

Fig. 265 Connecting the fuel tank main tube sub assembly

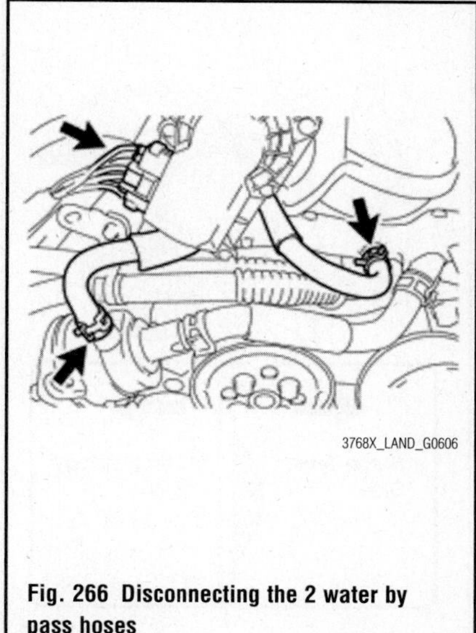

Fig. 266 Disconnecting the 2 water by pass hoses

➡Push the parts together firmly until a "click" sound is heard.

➡Before installing the tube connector to the pipe, check if there is any damage or foreign matter in the connector.

➡After the connection, check if the connectors and pipes are securely connected by trying to pull them apart.

a. Attach the fuel tube clamp.

42. Install the No. 1 fuel tank protector sub assembly with the 5 bolts. Tighten to 15 ft. lbs. (20 Nm).
43. Install the fuel tank cap assembly.
44. Install the rear floor No. 2 service hole cover.
    a. Connect the fuel pump and fuel sender gauge connector.
    b. Install the service hole cover with new butyl tape.
    c. Install the air duct with the 2 screws.
45. Install the front floor carpet assembly.
46. Install the rear air duct guide.
47. Install the air duct plug.
48. Install the front quarter trim panel assembly RH.
49. Install the front quarter trim panel assembly LH.
50. Install the luggage compartment No. 1 trim hook.
51. Install the rear floor mat rear support plate.
52. Install the rear No. 2 seat assembly.
53. Install the rear door scuff plate RH.
54. Install the rear door scuff plate LH.
55. Install the rear step cover.
56. Install the rear No. 2 seat protector.
57. Install the rear No. 1 seat protector.
58. Install the rear No. 1 seat assembly RH.
59. Install the rear No. 1 seat assembly LH.
60. Connect the cable to the negative battery terminal.
61. Check the SRS warning light.
62. Add fuel.
63. Inspect for fuel leaks.
64. Inspect the rear No. 1 seat assembly RH and LH.

IDLE SPEED

ADJUSTMENT

Idle speed is maintained by the ECM. No adjustment is necessary or possible.

THROTTLE BODY

REMOVAL & INSTALLATION

See Figures 266 through 268.

1. Remove the front fender splash shield sub assembly LH.
2. Remove the front fender splash shield sub assembly RH.
3. Remove the No. 1 engine under cover sub assembly.
4. Drain the engine coolant.
5. Remove the v-bank cover sub assembly.
6. Remove the air cleaner hose assembly.
7. Remove the throttle body assembly.
    a. Disconnect the 2 water by pass hoses.
    b. Disconnect the throttle position sensor and control motor connector.
    c. Remove the 4 bolts, throttle body and gasket.
    d. Remove the 2 water by pass hoses from the throttle body.

To install:

8. Install the throttle body assembly.
    a. Install the 2 water by pass hoses to the throttle body.
    b. Install a new gasket to the intake air surge tank.

➡Align the protrusion of the gasket with the groove of the intake air surge tank.

c. Install the throttle body with the 4 bolts. Tighten to 7 ft. lbs. (10 Nm).
d. Connect the throttle position sensor and control motor connector.
e. Connect the 2 water by pass hoses.

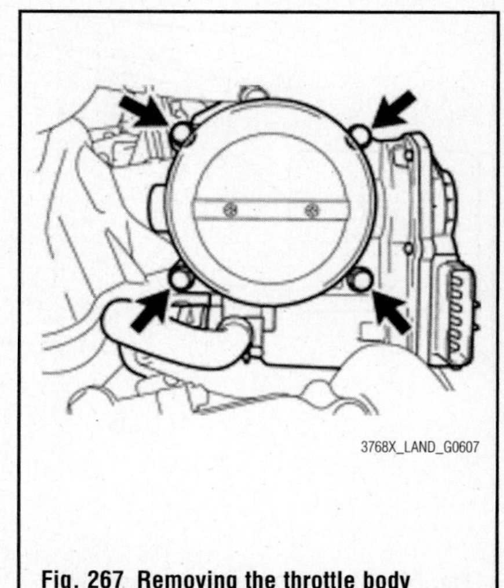

Fig. 267 Removing the throttle body

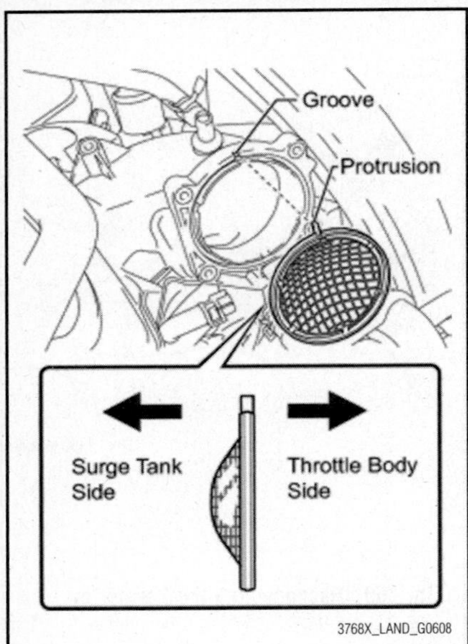

Fig. 268 Installing a new gasket to the intake air surge tank

9. Add engine coolant.
10. Install the air cleaner hose assembly.
11. Inspect for coolant leaks.
12. Check the throttle body assembly.
13. Install the No. 1 engine under cover sub assembly.
14. Install the front fender splash shield sub assembly LH.
15. Install the front fender splash shield sub assembly RH.
16. Perform the initialization.

➡Perform the following procedure after replacing the ECM, throttle body assembly or any throttle body components. The following procedure should also be performed if the throttle body is cleaned.

a. Disconnect the cable from the negative (-) battery terminal. Wait at least 60 seconds and reconnect the cable.
b. Turn the ignition switch on (IG) without operating the accelerator pedal.

➡If the accelerator pedal is operated, perform the above steps again.

c. Connect the Techstream to the DLC3 and clear the DTCs.
d. Start the engine and check that the MIL is not illuminated and that the idle speed is within the specified range when the A/C is switched off after the engine is warmed up. With the A/C switched off the engine idle speed should be 650-750 rpm.

➡Be sure to perform this step with all accessories off.

➡Make sure that the shift lever is in N or P.

e. Enter the following menus: Powertrain / Engine and ECT / Data List / Throttle Sensor Volt o/o. Fully depress the accelerator pedal and check that the value is 60% or more.
f. Perform a road test and confirm that there are no abnormalities.

17. Install the v-bank cover assembly.

HEATING & AIR CONDITIONING SYSTEM

BLOWER MOTOR

REMOVAL & INSTALLATION

Front

See Figure 269.

1. Remove the air conditioning unit.
2. Remove the blower assembly.
   a. Remove the screw.
   b. Disconnect the connector and clamp.
   c. Detach the claw and remove the blower unit.

To install:

3. To install, reverse the removal procedure.

Rear

See Figure 271.

1. Remove the rear cooling unit assembly.
2. Remove the blower with the fan (rear) motor sub assembly.
   a. Disconnect the connector.
   b. Remove the 3 screws and the motor.

To install:

3. To install, reverse the removal procedure.

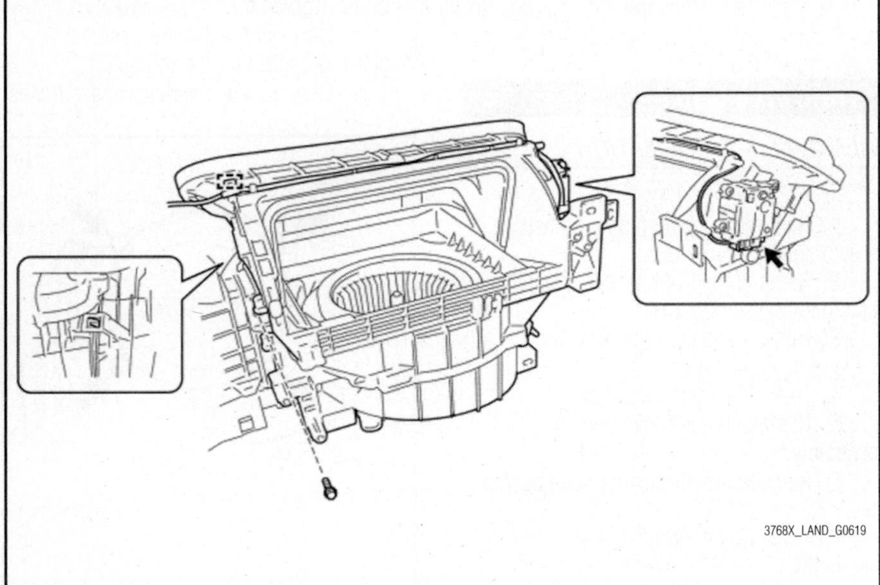

Fig. 269 Removing the blower assembly

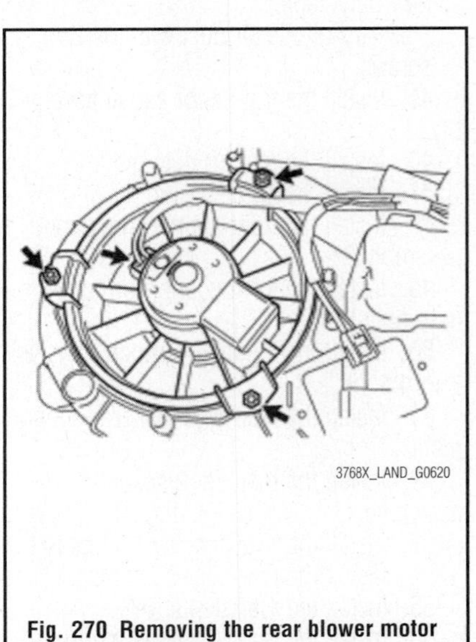

Fig. 270 Removing the rear blower motor

STEERING

POWER STEERING PUMP

REMOVAL & INSTALLATION

See Figures 271 through 273.

1. Disconnect the cable from the negative battery terminal.

CAUTION

Wait at least 90 seconds after disconnecting the cable from the negative (-) battery terminal to disable the SRS system.

➡When disconnecting the cable, some systems need to be initialized after the cable is reconnected.

2. Remove the front wheel RH.
3. Remove the front fender apron seal front RH.
4. Remove the v-bank cover sub assembly.
5. Remove the air cleaner hose assembly.
6. Remove the air cleaner assembly.
7. Remove the fan and generator v-belt.
8. Drain the power steering fluid.
9. Disconnect the suction hose.
   a. Slide the clip and disconnect the suction hose from the vane pump.
10. Disconnect the power steering oil pressure sensor connector.

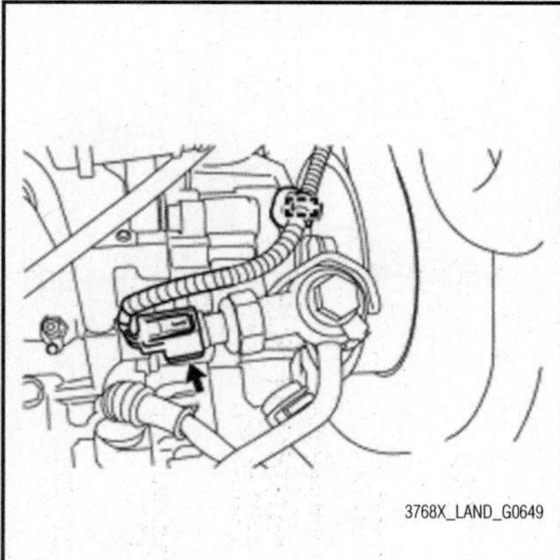

Fig. 271 Disconnecting the power steering oil pressure sensor connector

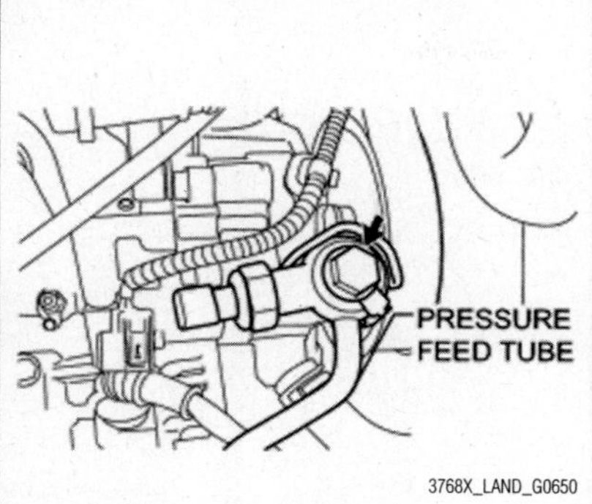

Fig. 272 Disconnecting the pressure feed tube

   a. Detach the wire harness clamp and disconnect the connector.
11. Disconnect the pressure feed tube.
   a. Remove the gasket.
12. Remove the vane pump assembly.
   a. Remove the 2 bolts and vane pump.

To install:

13. Install the vane pump assembly. Tighten the 2 bolts to 15 ft. lbs. (21 Nm).

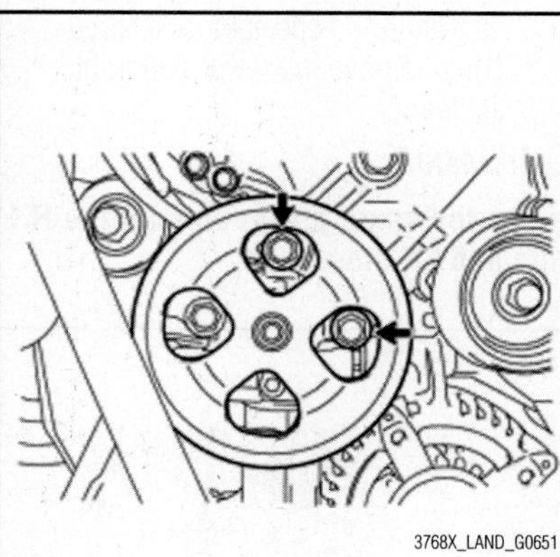

Fig. 273 Removing the vane pump assembly

14. Connect the pressure feed tube. Tighten the union bolt to 37 ft. lbs. (50 Nm).
15. Connect the power steering oil pressure sensor connector. Attach the wire harness clamp.
16. Connect the suction hose to the vane pump assembly with the clip.
17. Install the fan and generator v-belt.
18. Install the air cleaner assembly.
19. Install the air cleaner hose assembly.
20. Install the v-bank cover sub assembly.
21. Install the front fender apron seal front RH.
22. Install the front wheel RH.
23. Add power steering fluid.
24. Bleed the power steering fluid.
25. Inspect for power steering fluid leaks.
26. Connect the cable to the negative battery terminal.

BLEEDING

1. Check the fluid level.
2. Jack up the front of the vehicle and support it with stands.
3. Turn the steering wheel.
   a. With the engine stopped, turn the wheel slowly from lock to lock several times.
4. Lower the vehicle.
5. Start the engine.
6. Idle the engine for a few minutes.
7. Turn the steering wheel.
   a. With the engine idling, turn the wheel left or right to the full lock position and keep it there for 2 to 3 seconds, then turn the wheel to the opposite full lock position and keep it there for 2 to 3 seconds.
   b. Repeat the previous step several times.
8. Stop the engine.
9. Check for foaming or emulsification. If the system has to be bled twice because of foaming or emulsification, check for fluid leaks in the system.
10. Check the fluid level.

LOWER CONTROL ARM

REMOVAL & INSTALLATION

See Figures 274 through 278.

➡**Use the same procedures for the RH side and LH side.**

➡**The procedures listed below are for the LH side.**

1. Remove the stabilizer control valve protector.
2. Open stabilizer control with accumulator housing shutter valve.
3. Remove the front wheel.
4. Remove the front fender splash shield sub assembly LH.
5. Remove the front fender splash shield sub assembly RH.
6. Remove the No. 1 engine under cover sub assembly.
7. Loosen the front No. 1 stabilizer bracket LH.
8. Loosen the front No. 1 stabilizer bracket RH.
9. Remove the front stabilizer link assembly LH.
10. Remove the front stabilizer link assembly RH.
11. Disconnect the front shock absorber with the coil spring LH.
    a. Remove the nut and bolt, and disconnect the shock absorber from the lower side.
12. Disconnect the front lower ball joint attachment LH.
    a. Remove the 2 bolts and disconnect the attachment from the steering knuckle.
13. Remove the front No. 1 suspension arm lower sub assembly LH.

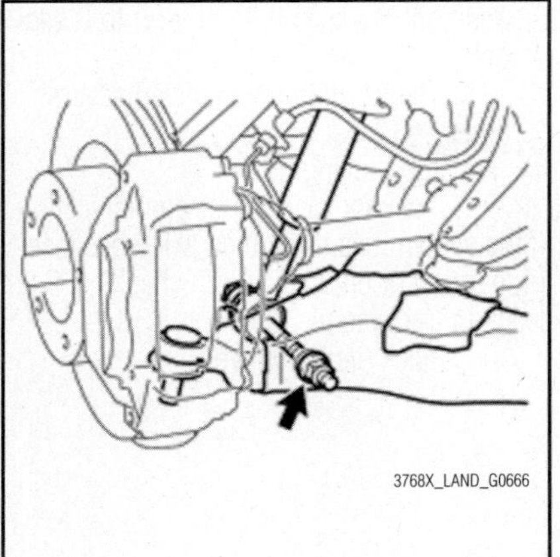

Fig. 274 Disconnecting the front shock absorber with the coil spring

Fig. 275 Disconnecting the front lower ball joint attachment

a. Support the front suspension LH with a jack.

b. Place matchmarks on the No. 2 camber adjusting cam and No. 2 suspension toe adjusting plate.

c. Remove the nut, washer, No. 2 camber adjusting cam, camber adjusting cam assembly, bolt, toe adjusting cam, No. 2 suspension toe adjusting plate and front No. 1 suspension arm lower LH.

14. Remove the front lower ball joint attachment LH.
    a. Remove the cotter pin and nut.
    b. Using the special tool (09950-40011), remove the lower ball joint attachment.

To install:

➡**Use the same procedures for the RH side and LH side.**

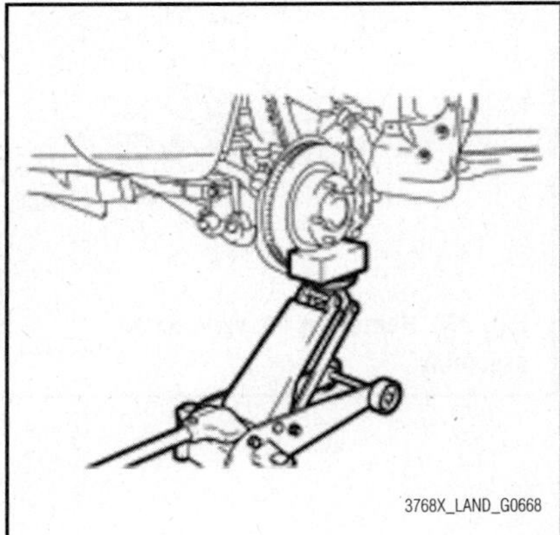

Fig. 276 Removing the front No. 1 suspension arm lower sub assembly

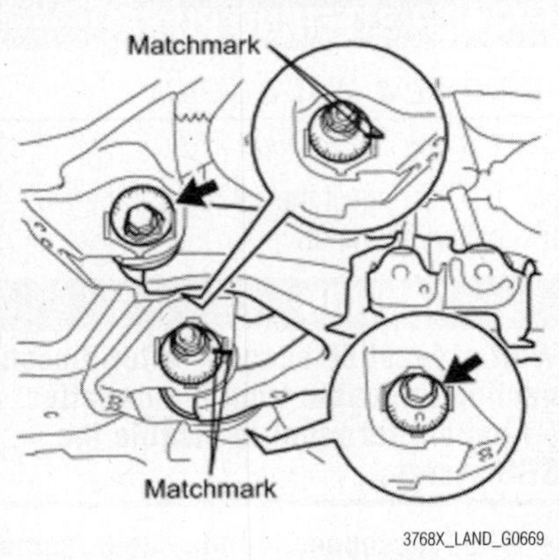

Fig. 277 Placing matchmarks on the No. 2 camber adjusting cam and No. 2 suspension toe adjusting plate

➡**The procedures listed below are for the LH side.**

15. Temporarily install the front No. 1 suspension arm lower sub assembly LH.
    a. Temporarily install the lower suspension arm, camber adjusting cam, No. 2 camber adjusting cam, No. 2 suspen-

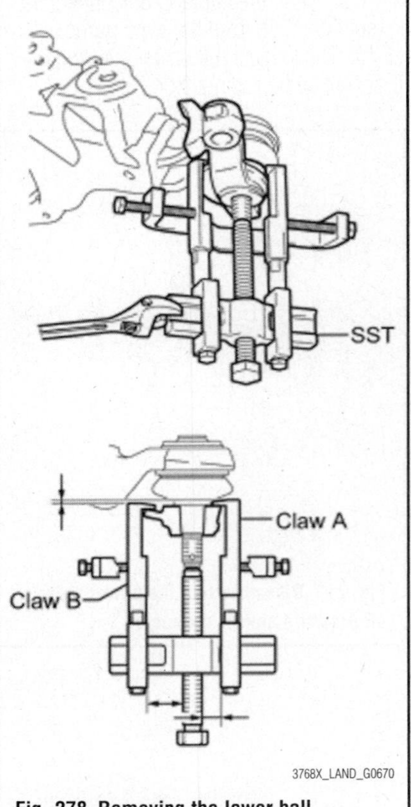

Fig. 278 Removing the lower ball joint attachment

sion toe adjusting plate, toe adjusting cam and washer with the bolt and nut.

b. Align the matchmarks on the No. 2 camber adjusting cam and No. 2 suspension toe adjusting plate with the matchmarks on the vehicle body. Tighten the bolt and nut.

➡The bolt and nut will be tightened to the torque specification in the "Tighten Front No. 1 Suspension Arm Lower Sub-assembly LH" procedure.

16. Connect the front lower ball joint attachment LH.
    a. Connect the attachment to the steering knuckle with the 2 bolts. Tighten to 221 ft. lbs. (300 Nm).
17. Connect the front shock absorber with the coil spring LH.
    a. Connect the shock absorber with the bolt and nut.
18. Temporarily install the front stabilizer link assembly RH.
19. Temporarily install the front stabilizer link assembly LH.
20. Tighten the front No. 1 stabilizer bracket LH.
21. Tighten the front No. 1 stabilizer bracket RH.
22. Install the No. 1 engine under cover sub assembly.
23. Install the front fender splash shield sub assembly LH.
24. Install the front fender splash shield sub assembly RH.
25. Stabilize the suspension.
26. Tighten the front No. 1 suspension arm lower sub assembly LH. Tighten the nut to 207 ft. lbs. (280 Nm).

➡Perform this procedure with all 4 wheels on the ground.

27. Tighten the front shock absorber with the coil spring. Tighten the nut to 133 ft. lbs. (180 Nm).

➡Perform this procedure with all 4 wheels on the ground.

28. Tighten the front stabilizer link assembly LH.
29. Tighten the front stabilizer link assembly RH.
30. Measure the vehicle height.
31. Close the stabilizer control with and accumulator housing shutter valve.
32. Install the stabilizer control valve protector.
33. Inspect and adjust the front wheel alignment.
34. Adjust the headlight assembly.

STABILIZER BAR

REMOVAL & INSTALLATION

See Figures 279 through 294.

1. Remove the stabilize control valve protector.
2. Open the stabilizer control with the accumulator housing shutter valve.
3. Remove the front wheel.
4. Remove the tailpipe assembly.
5. Remove the exhaust center pipe assembly.
6. Discharge the suspension fluid pressure.
7. Remove the front fender splash shield sub assembly LH.
8. Remove the front fender splash shield sub assembly RH.
9. Remove the No. 1 engine under cover sub assembly.
10. Remove the front fender apron seal LH.
11. Remove the front fender apron seal rear LH.
12. Remove the stabilizer control tube protector.
    a. Remove the 2 bolts and tube protector.
13. Disconnect the front No. 1 stabilizer control tube assembly.
    a. Remove the 2 union bolts and 2 gaskets, and disconnect the stabilizer control tube.
14. Loosen the front No. 1 stabilizer bracket LH.
    a. Loosen the 2 bolts of the front stabilizer brackets.
15. Loosen the front No. 1 stabilizer bracket RH.
    a. Loosen the 2 bolts of the front stabilizer brackets.

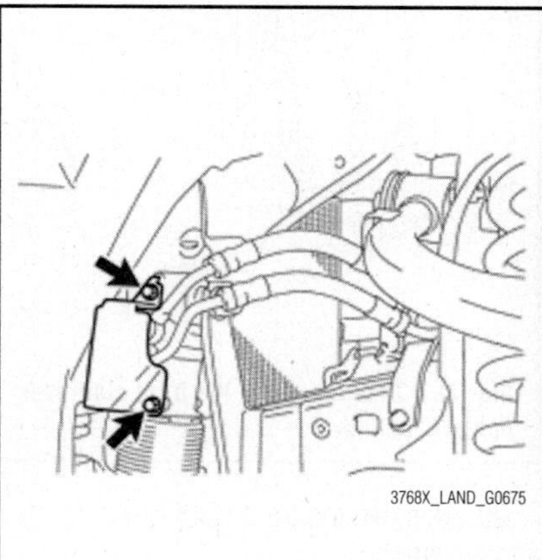

Fig. 279 Removing the stabilizer control tube protector

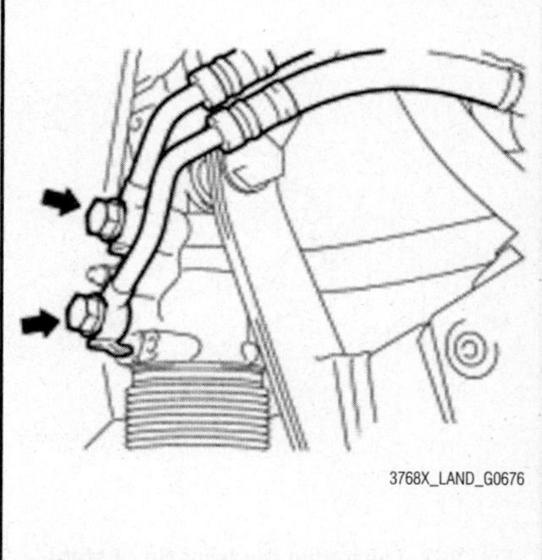

Fig. 280 Disconnecting the front No. 1 stabilizer control tube assembly

Fig. 281 Loosening the front No. 1 stabilizer bracket LH

16. Remove the front stabilizer link assembly LH.
    a. Remove the 2 bolts, nut and stabilizer link.
17. Remove the front stabilizer link assembly RH.
    a. Remove the bolt, nut and stabilizer link.

➡If the ball joint turns together with the nut, use a 6 mm hexagon wrench to hold the stud.

18. Remove the front No. 1 stabilizer bracket LH and RH.
    a. Remove the 2 bolts and stabilizer bracket from the front frame assembly.
19. Disconnect the front stabilizer control cylinder.
    a. Remove the bolt, washer, nut, and stabilizer control cylinder from the frame assembly.
20. Remove the front stabilizer bar.

3768X_LAND_G0678

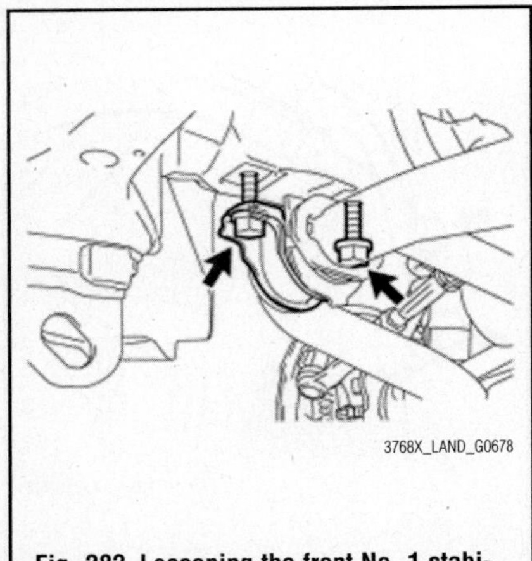

Fig. 282 Loosening the front No. 1 stabilizer bracket RH

3768X_LAND_G0679

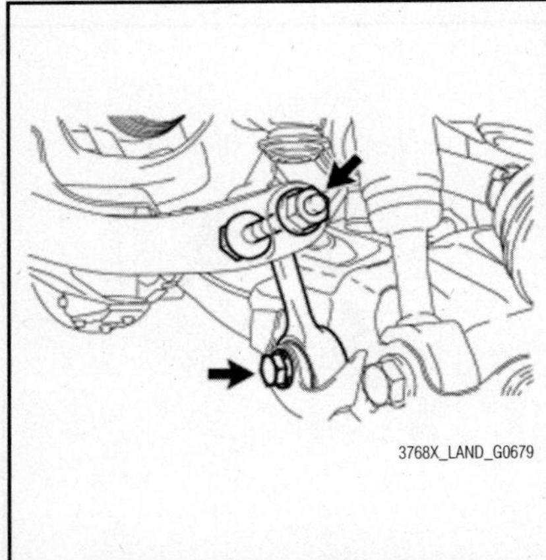

Fig. 283 Removing the front stabilizer link assembly LH

3768X_LAND_G0680

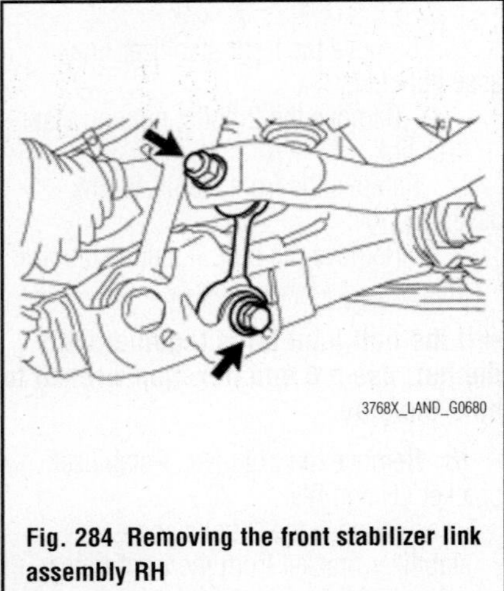

Fig. 284 Removing the front stabilizer link assembly RH

a. Remove the bolt, nut and front stabilizer bar from the stabilizer control arm.

21. Remove the front No. 1 stabilizer bar bush.

a. Remove the 2 stabilizer bar bushes from the front stabilizer bar.

3768X_LAND_G0681

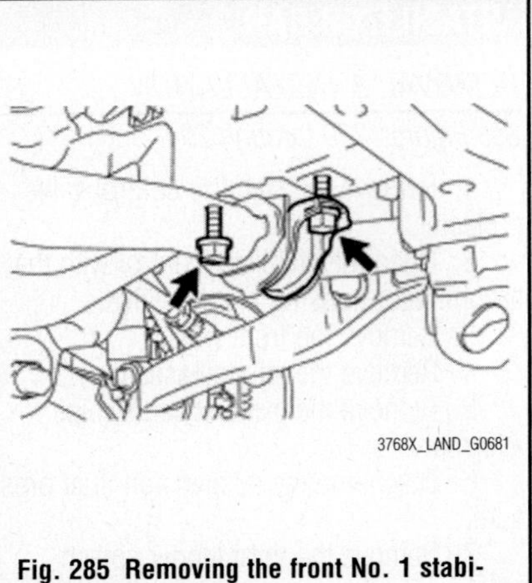

Fig. 285 Removing the front No. 1 stabilizer bracket LH

3768X_LAND_G0682

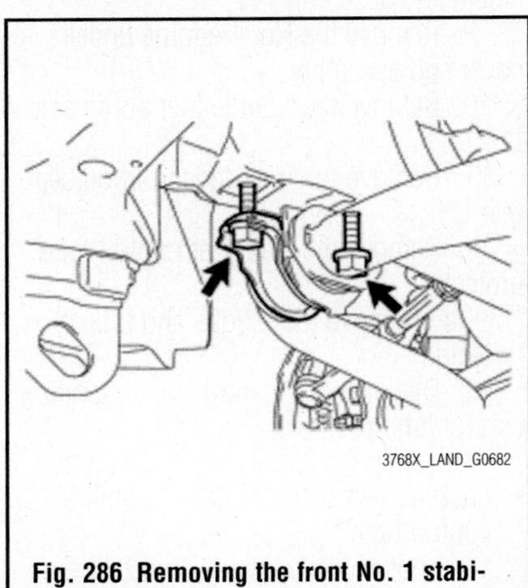

Fig. 286 Removing the front No. 1 stabilizer bracket RH

3768X_LAND_G0683

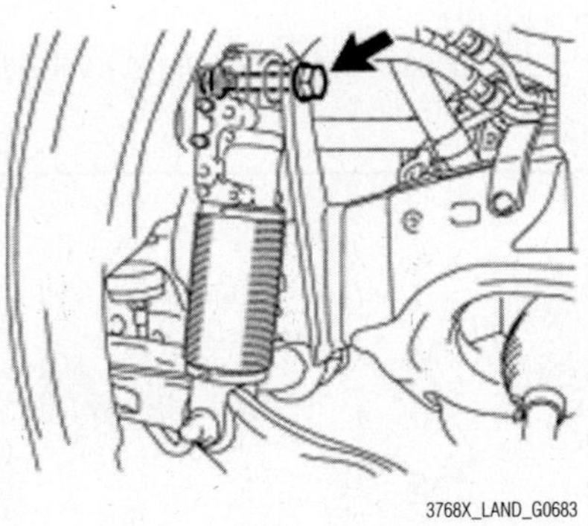

Fig. 287 Disconnecting the front stabilizer control cylinder

22. Remove the front stabilizer control arm assembly.

a. Remove the bolt, nut and front stabilizer control arm from the front stabilizer control cylinder.

3768X_LAND_G0684

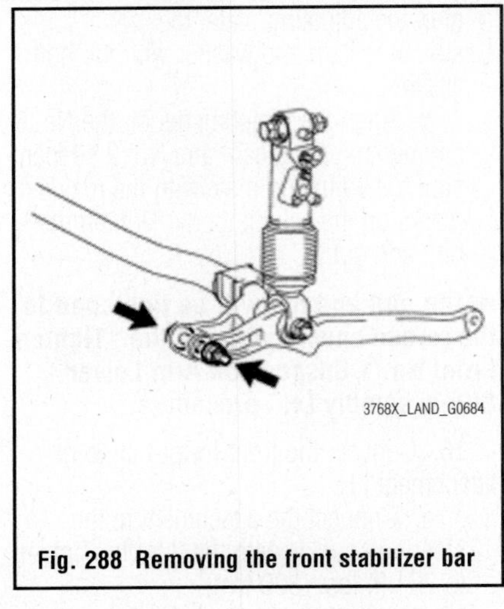

Fig. 288 Removing the front stabilizer bar

3768X_LAND_G0685

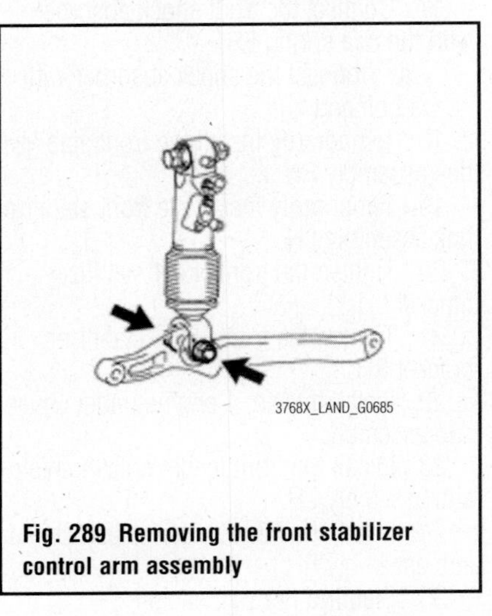

Fig. 289 Removing the front stabilizer control arm assembly

23. Remove the front No. 1 suspension control bleeder plug.

a. Remove the 2 front suspension control bleeder plugs from the front stabilizer control cylinder.

To install:

24. Install the front No. 1 suspension control bleeder plug.

a. Install the 2 suspension control bleeder plugs to the front stabilizer control cylinder. Tighten to 7 ft. lbs. (10 Nm).

25. Install the front stabilizer control cylinder.

a. Install the front stabilizer control cylinder to the frame with the bolt, washer and nut. Tighten to 203 ft. lbs. (275 Nm).

26. Connect the front No. 1 stabilizer control tube assembly.

a. Install 2 new gaskets and the front stabilizer control cylinder to the front stabilizer control cylinder with the 2 union bolts. Tighten to 51 ft. lbs. (69 Nm).

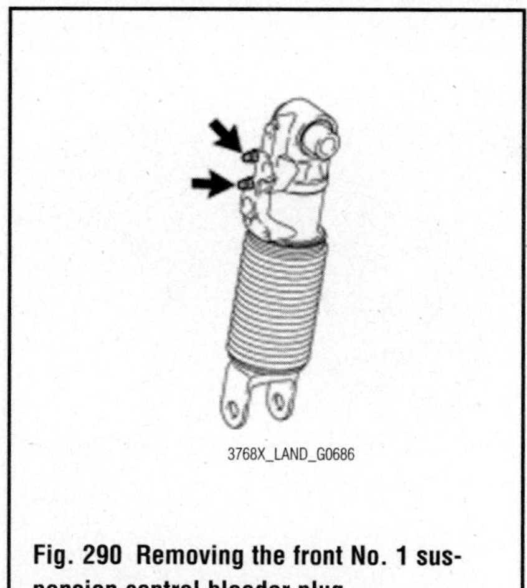

Fig. 290 Removing the front No. 1 suspension control bleeder plug

27. Inspect for suspension fluid leaks.
28. Install the exhaust center pipe assembly.
29. Install the tailpipe assembly.
30. Install the stabilizer control tube protector with the 2 bolts. Tighten to 11 ft. lbs. (15 Nm).
31. Temporarily install the front stabilizer control arm assembly with the nut and bolt.
32. Bleed the air from the suspension fluid.
33. Install the front No. 1 stabilizer bar bush.
    a. Install the 2 bushes to the outer side of the bush stoppers on the front stabilizer bar as shown in the illustration.

➡Be sure to install the front No. 1 stabilizer bar bushes so that the bump of each bush faces the inside of the vehicle.

34. Temporarily install the front stabilizer bar.

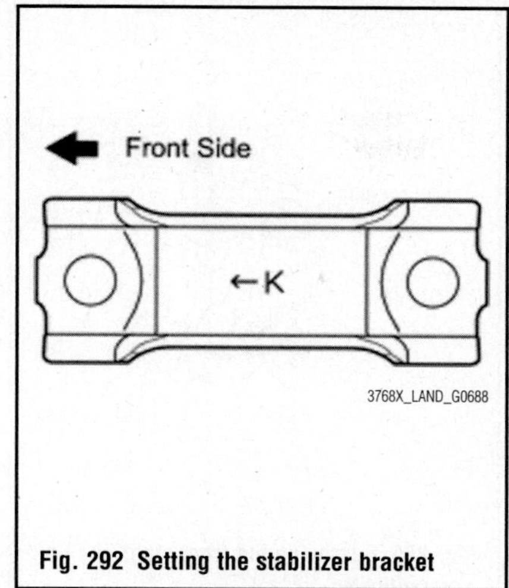

Fig. 292 Setting the stabilizer bracket

    a. Set the stabilizer bracket so that the arrow mark is facing the front side of the vehicle.
    b. Temporarily install the 2 stabilizer brackets and front stabilizer bar with the 4 bolts.
    c. Temporarily install the front stabilizer bar to the stabilizer control arm with the bolt and nut.

35. Temporarily install the front stabilizer link assembly RH with the bolt and nut. Tighten to 94 ft. lbs. (128 Nm).

➡If the ball joint turns together with the nut, use a 6 mm hexagon wrench to hold the stud bolt.

36. Temporarily install the front stabilizer link assembly LH.
    a. Temporarily install the stabilizer link to the lower arm with the bolt.
    b. Temporarily install the stabilizer link to the stabilizer control arm with the bolt and nut.

➡If the front stabilizer control cylinder extends and it is difficult to temporarily install the stabilizer link to the stabilizer control arm, raise the stabilizer control arm with a jack and temporarily install the stabilizer link.

37. Tighten the front No. 1 stabilizer bracket LH with the 2 bolts. Tighten to 64 ft. lbs. (87 Nm).

➡Tighten the bolts in 3 steps, in the order shown in the illustration.

38. Install the front No. 1 stabilizer bracket RH with the 2 bolts. Tighten to 64 ft. lbs. (87 Nm).

➡Tighten the bolts in 3 steps, in the order shown in the illustration.

39. Install the No. 1 engine under cover sub assembly.
40. Install the front fender splash shield sub assembly LH.
41. Install the front fender splash shield sub assembly RH.
42. Stabilize the suspension.
43. Tighten the front stabilizer link assembly LH.

Perform this procedure with all 4 wheels on the ground.

    a. Tighten the bolt to 100 ft. lbs. (135 Nm).
    b. Tighten the nut to 103 ft. lbs. (140 Nm).

44. Tighten the front stabilizer link assembly RH.
    a. Tighten the bolt to 100 ft. lbs. (135 Nm).

➡Perform this procedure with all 4 wheels on the ground.

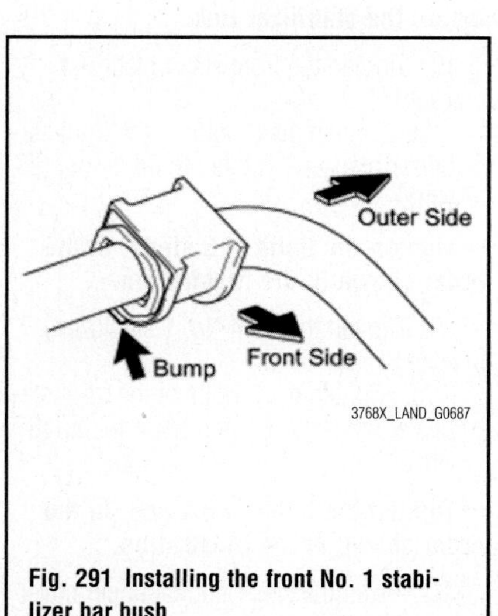

Fig. 291 Installing the front No. 1 stabilizer bar bush

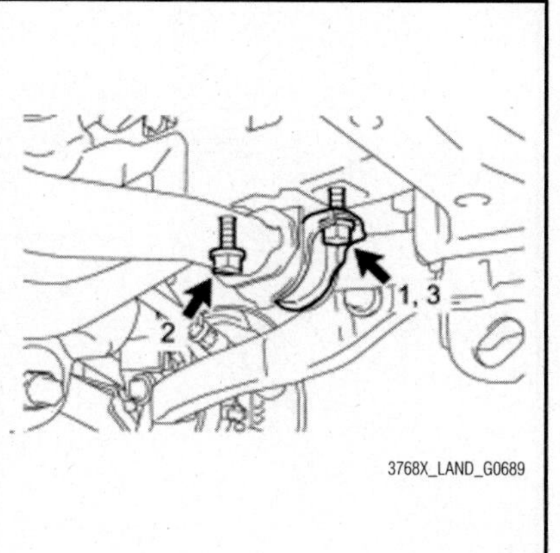

Fig. 293 Tightening the front No. 1 stabilizer bracket LH

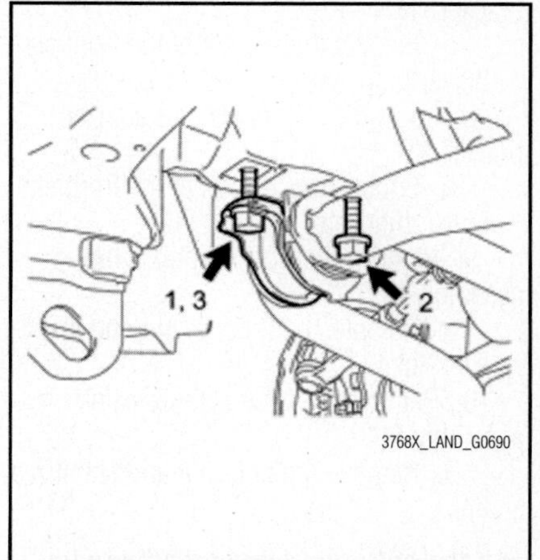

Fig. 294 Tightening the front No. 1 stabilizer bracket RH

45. Tighten the front stabilizer control arm assembly with the nut. Tighten to 140 ft. lbs. (190 Nm).

➡**Perform this procedure with all 4 wheels on the ground.**

46. Tighten the front stabilizer bar nut to 170 ft. lbs. (230 Nm).

➡**Perform this procedure with all 4 wheels on the ground.**

47. Install the front fender apron seal rear LH.
48. Install the front fender apron seal LH.
49. Inspect for suspension fluid leaks.
50. Measure the vehicle height.
51. Close the stabilizer control with the accumulator housing shutter valve.
52. Install the stabilizer control valve protector.
53. Inspect for exhaust gas leaks.

STRUT & SPRING ASSEMBLY

REMOVAL & INSTALLATION

See Figures 295 through 299.

➡**Use the same procedures for the RH side and LH side.**

➡**The procedures listed below are for the LH side.**

1. Remove the stabilizer control valve protector.
2. Open the stabilizer control with the accumulator housing shutter valve.
3. Remove the front wheel.
4. Remove the front fender splash shield sub assembly LH.
5. Remove the front fender splash shield sub assembly RH.
6. Remove the No. 1 engine under cover sub assembly.
7. Loosen the front No. 1 stabilizer bracket LH.
    a. Loosen the 2 bolts of the front stabilizer brackets.
8. Loosen the front No. 1 stabilizer bracket RH.
    a. Loosen the 2 bolts of the front stabilizer brackets.
9. Remove the front stabilizer link assembly LH.
    a. Remove the 2 bolts, nut and stabilizer link.
10. Remove the front stabilizer link assembly RH.
    a. Remove the bolt, nut and stabilizer link.

➡**If the ball joint turns together with the nut, use a 6 mm hexagon wrench to hold the stud.**

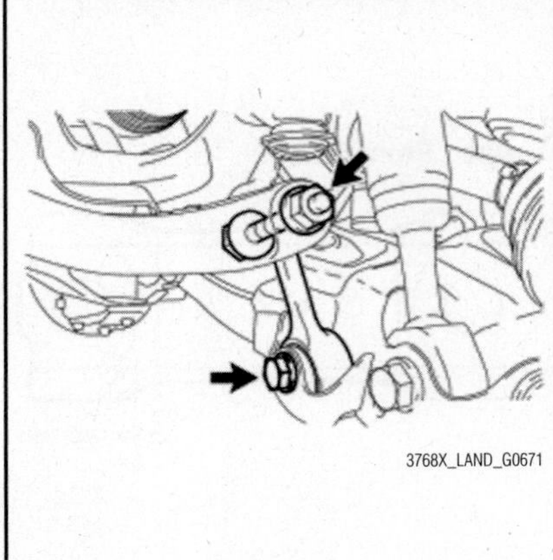

Fig. 295 Removing the front stabilizer link assembly LH

11. Disconnect the skid control sensor wire.
12. Disconnect the steering knuckle LH.
13. Remove the front shock absorber with the coil spring LH.
    a. Remove the nut from the shock absorber lower side.

➡**To prevent the shock absorber with the coil spring from falling, leave the bolt inserted.**

    b. Remove the 4 nuts.
    c. Remove the bolt (lower side) and shock absorber with the coil spring.

To install:

14. Temporarily install the front shock absorber with the coil spring.
    a. Temporarily install the upper side of the shock absorber to the chassis frame with the 4 nuts.
    b. Temporarily install the lower side of the shock absorber to the lower suspension arm with the bolt and nut.

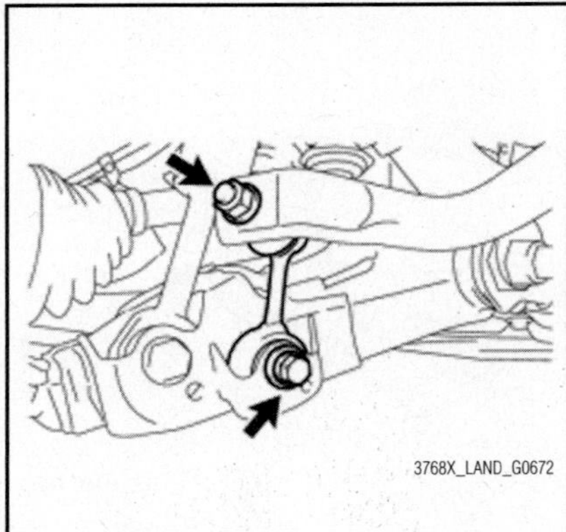

Fig. 296 Removing the front stabilizer link assembly RH

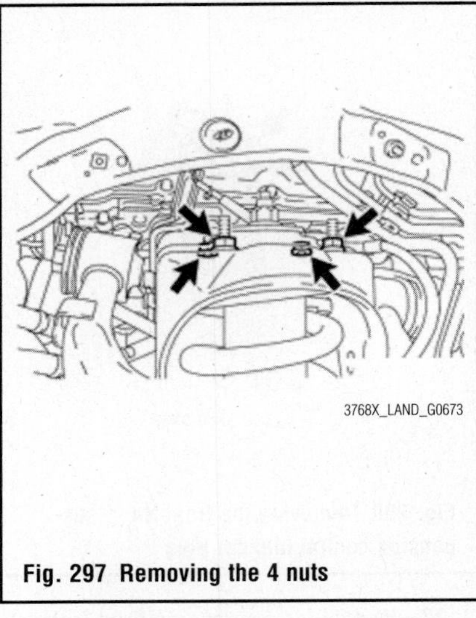

Fig. 297 Removing the 4 nuts

15. Connect the steering knuckle LH.
16. Connect the skid control sensor wire.
17. Temporarily install the front stabilizer link assembly RH.
    a. Temporarily install the stabilizer link with the bolt.
    b. Temporarily install the stabilizer link with the nut.
    c. Tighten the nut. Tighten to 94 ft. lbs. (128 Nm).
18. Temporarily install the front stabilizer link assembly LH.
    a. Temporarily install the stabilizer link to the lower arm with the bolt.
    b. Temporarily install the stabilizer link to the stabilizer control arm with the bolt and nut.

➡**If the front stabilizer control cylinder extends and it is difficult to temporarily install the stabilizer link to the stabilizer control arm, raise the stabilizer control arm with a jack and temporarily install the stabilizer link.**

19. Tighten the front No. 1 stabilizer bracket LH.
    a. Tighten the 2 bolts of the front stabilizer brackets. Tighten to 64 ft. lbs. (87 Nm).

➡**Tighten the bolts in 3 steps, in the order shown in the illustration.**

20. Tighten the front No. 1 stabilizer bracket RH.
    a. Tighten the 2 bolts of the front stabilizer brackets. Tighten to 64 ft. lbs. (87 Nm).

➡**Tighten the bolts in 3 steps, in the order shown in the illustration.**

21. Install the No. 1 engine under cover sub assembly.

Fig. 298 Tightening the front No. 1 stabilizer bracket LH

22. Install the front fender splash shield sub assembly LH and RH.
23. Stabilize the suspension.
   a. Install the front wheels. Tighten to 97 ft. lbs. (131 Nm).
   b. Lower the vehicle.
   c. Press down on the vehicle several times to stabilize the suspension.
24. Tighten the front shock absorber with the coil spring.
   a. Tighten the nut to 133 ft. lbs. (180 Nm).

➡Perform this procedure with all 4 wheel on the ground.

   b. Tighten the 4 upper nuts in diametrically opposite pairs. Tighten to 33 ft. lbs. (45 Nm).
   c. Check that the first nut that was tightened is at the torque specification.

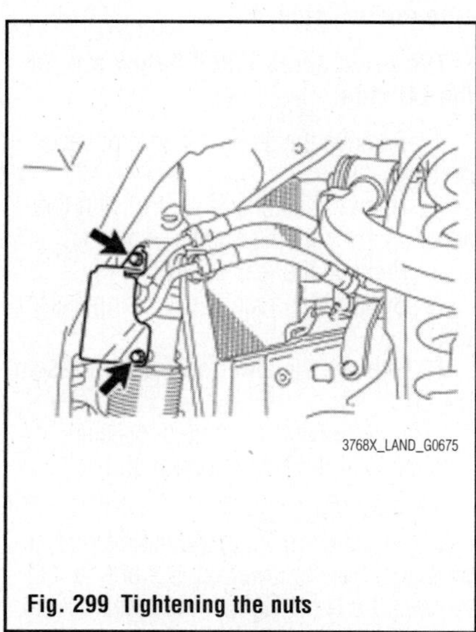

Fig. 299 Tightening the nuts

25. Tighten the front stabilizer link assembly LH.
   a. Tighten the bolt to 100 ft. lbs. (135 Nm).
   b. Tighten the nut to 103 ft. lbs. (140 Nm).

➡Perform this procedure with all 4 wheels on the ground.

26. Tighten the front stabilizer link assembly RH.
   a. Tighten the bolt to 100 ft. lbs. (135 Nm).

➡Perform this procedure with all 4 wheels on the ground.

27. Measure the vehicle height.
28. Close the stabilizer control with an accumulator housing shutter valve.
29. Install the stabilizer control valve protector.

UPPER CONTROL ARM

REMOVAL & INSTALLATION

See Figure 300.

➡Use the same procedures for the RH side and LH side.

➡The procedures listed below are for the LH side.

1. Remove the stabilizer control valve protector.
2. Open the stabilizer control with accumulator housing shutter valve.
3. Remove the front wheel.
4. Disconnect the skid control sensor wire.
   a. Remove the bolt and nut, and disconnect the sensor wire from the steering knuckle and suspension upper arm.
5. Disconnect the steering knuckle LH.
   a. Support the front suspension lower arm LH with a jack.
   b. Remove the clip and the nut.
   c. Using the special tool (09628-62011), disconnect the upper ball joint from the steering knuckle.

➡Do not damage the ball joint dust cover.

6. Remove the front fender apron seal LH.
   a. Using the clip remover, remove the 3 clips and apron seal.
7. Remove the front fender apron seal rear LH.
8. Remove the front suspension upper arm assembly

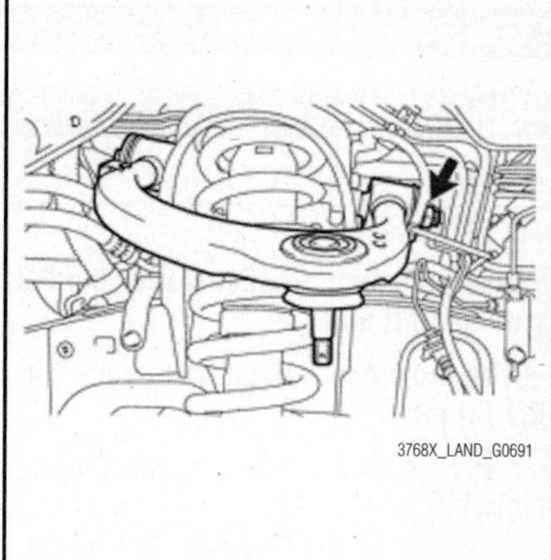

Fig. 300 Removing the front suspension upper arm assembly

   a. Remove the nut, bolt, 2 washers and suspension upper arm.

To install:

9. Temporarily install the front suspension upper arm assembly LH.
   a. Temporarily install the suspension upper arm with the 2 washers, bolt and nut.

➡ After stabilizing the suspension, tighten the nut.

10. Connect the steering knuckle LH.
   a. Connect the steering knuckle to the suspension upper arm.
   b. Install the nut and a new cotter pin. Tighten to 81 ft. lbs. (110 Nm).

➡ If the holes for the cotter pin are not aligned, tighten the nut further up to 60°.

11. Connect the skid control sensor wire to the steering knuckle and upper arm with the bolt and nut. Tighten to 9.6 ft. lbs. (13 Nm).
12. Stabilize the suspension.
13. Tighten the front suspension upper arm assembly LH. Tighten to 136 ft. lbs. (185 Nm).
14. Install the front fender apron seal rear LH with the 4 clips.
15. Install the front fender apron seal LH with the 3 clips.
16. Measure the vehicle height.
17. Close the stabilizer control with the accumulator housing shutter valve.
18. Install the stabilizer control valve protector.
19. Inspect and adjust the front wheel alignment.
20. Adjust the headlight assembly.

COIL SPRING

REMOVAL & INSTALLATION

See Figure 301.

➡Use the same procedures for the RH side and LH side.

➡The procedures listed below are for the LH side.

1. Remove the stabilizer control valve protector.
2. Open the stabilizer control with the accumulator housing shutter valve.
3. Remove the rear wheel.
4. Drain the brake fluid.
5. Disconnect the No. 3 parking brake cable assembly.
6. Disconnect the No. 2 parking brake cable assembly.
7. Disconnect the rear axle breather hose sub assembly.
8. Support rear axle housing assembly.
9. Remove the rear stabilizer end bracket.
10. Disconnect the rear stabilizer link assembly.
    a. Remove the nut and disconnect the stabilizer link assembly from the stabilizer bar.

➡Do not lose the ball joint dust cover protector.

➡If the ball joint turns together with the nut, use a 6 mm hexagon wrench to hold the stud.

11. Disconnect the rear lateral control rod assembly.
12. Disconnect the rear flexible hose.
13. Disconnect the rear shock absorber assembly LH.
    a. Remove the bolt on the lower side of the shock absorber.
    b. Disconnect the shock absorber from the axle housing.
14. Disconnect the rear shock absorber assembly RH.
    a. Remove the bolt on the lower side of the shock absorber.
    b. Disconnect the shock absorber from the axle housing.
15. Remove the rear coil spring LH.
    a. While lowering the axle housing, remove the coil spring.

➡Be careful not to snap the brake line.

16. Remove the hollow spring sub assembly.

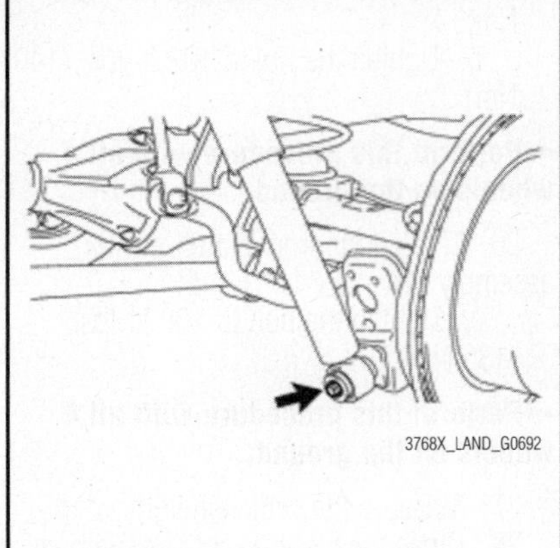

Fig. 301 Disconnecting the rear shock absorber assembly LH

17. Remove the hollow spring from the frame.

To install:

18. Install the hollow spring sub assembly.
19. Install the rear coil spring LH.

➡Before installing the coil spring, check that the coil spring end is in the correct position. If not, reinstall it.

20. Temporarily install the rear shock absorber assembly LH.
    a. Temporarily install the lower side of the shock absorber with the bolt.
21. Temporarily install the rear shock absorber assembly RH.
    a. Temporarily install the lower side of the shock absorber with the bolt.
22. Temporarily install the rear lateral control rod assembly.
23. Connect the rear flexible hose.
24. Stabilize the suspension.
25. Tighten the rear shock absorber assembly LH. Tighten the bolt to 72 ft. lbs. (98 Nm).
26. Tighten the rear shock absorber assembly RH. Tighten the bolt to 72 ft. lbs. (98 Nm).
27. Tighten the rear lateral control rod assembly.
28. Install the rear stabilizer end bracket.
    a. Install the stabilizer end bracket to the vehicle with the 2 bolts. Tighten to 22 ft. lbs. (30 Nm).
29. Connect the rear stabilizer link assembly.
    a. Install the stabilizer link assembly to the ball joint dust cover protector.
    b. Connect the stabilizer link assembly to the stabilizer bar with the nut. Tighten to 74 ft. lbs. (100 Nm).

➡If the ball joint turns together with the nut, use a 6 mm hexagon wrench to hold the stud bolt.

➡If the rear stabilizer control cylinder extends and it is difficult to temporarily install the stabilizer link to the stabilizer bar, raise the stabilizer control arm with a jack and temporarily install the stabilizer link.

30. Connect the No. 3 parking brake cable assembly.
31. Connect the No. 2 parking brake cable assembly.
32. Connect the rear axle breather hose sub assembly.
33. Install the rear wheel. Tighten to 97 ft. lbs. (131 Nm).
34. Fill the reservoir with brake fluid.
35. Bleed the brake line/
36. Check the brake fluid level in the reservoir.
37. Inspect for brake fluid leaks.
38. Measure the vehicle height.
39. Close the stabilizer control with the accumulator housing shutter valve.
40. Install the stabilizer control valve protector.

LOWER CONTROL ARM

REMOVAL & INSTALLATION

See Figure 302.

➡Use the same procedures for the RH side and LH side.

➡The procedures listed below are for the LH side.

1. Remove the stabilizer control valve protector.
2. Open the stabilizer control with the accumulator housing shutter valve.
3. Remove the rear wheel.
4. Support the rear axle housing assembly.
5. Remove the lower control arm assembly LH.
    a. Remove the 2 nuts, 2 washers, 2 bolts, and the lower control arm.

To install:

6. To install, reverse the removal procedure. Tighten the lower control arm to 111 ft. lbs. (150 Nm). Tighten the rear wheel to 97 ft. lbs. (131 Nm).

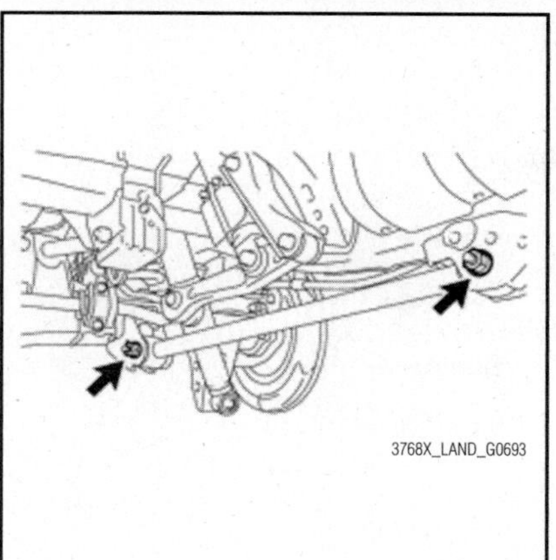

Fig. 302 Removing the lower control arm

STABILIZER BAR

REMOVAL & INSTALLATION

See Figures 303 through 312.

1. Remove stabilizer control valve protector.
2. Open stabilizer control with accumulator housing shutter valve.
3. Remove rear wheel.
4. Remove tailpipe assembly.
5. Remove exhaust center pipe assembly.
6. Discharge suspension fluid pressure.
7. Support rear axle housing assembly.
8. Remove rear stabilizer end bracket.
    a. Remove the 2 bolts and disconnect the stabilizer end bracket.
9. Remove the rear stabilizer bar sub assembly.
    a. Remove the nut and disconnect the stabilizer link assembly from the stabilizer bar.

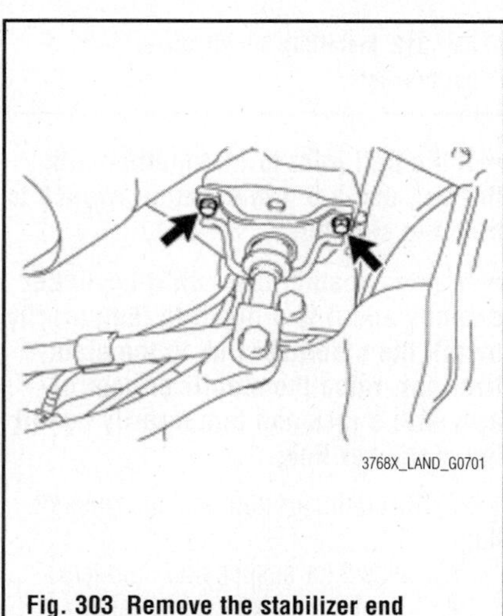

Fig. 303 Remove the stabilizer end bracket

➡Do not miss the ball joint dust cover protector.

➡If the ball joint turns together with the nut, use a 6 mm hexagon wrench to hold the stud.

    b. Remove the 4 bolts and stabilizer bar with the bushes and brackets.
10. Remove the stabilizer link sub assembly.
    a. Remove the nut, 3 retainers and 2 cushions, and disconnect the stabilizer link from the bracket.
    b. Remove the nut, bolt and stabilizer link.
11. Remove the rear stabilizer control tube insulator.
    a. Remove the 2 bolts and rear stabilizer control tube insulator.
12. Remove the rear stabilizer control cylinder.
    a. Remove the 2 union bolts with the 2 gaskets and bolt, and disconnect the stabilizer control tube.
    b. Remove the bolt and stabilizer control cylinder.
13. Remove the rear No. 2 stabilizer link.
    a. Remove the 2 bolts, 2 nuts and No. 2 rear stabilizer link.
14. Remove the rear stabilizer link assembly.
    a. Remove the bolt, nut and rear stabilizer link assembly.

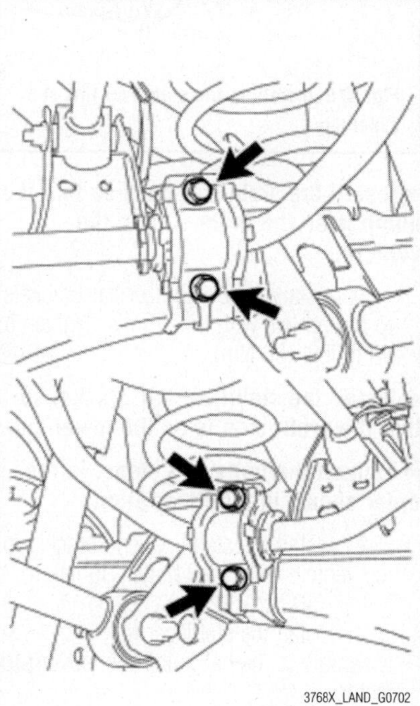

Fig. 304 Removing the rear stabilizer bar sub assembly

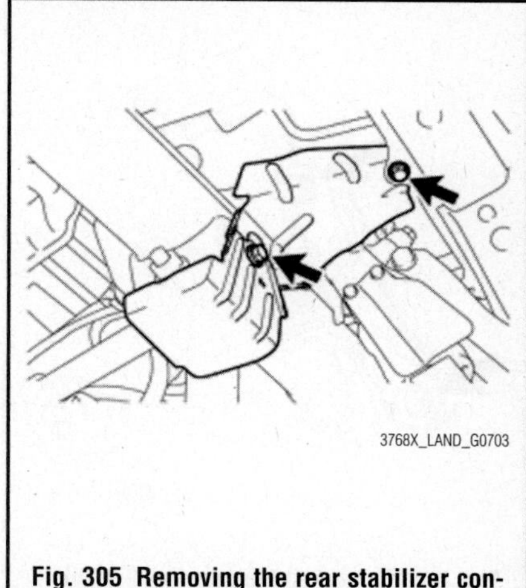

Fig. 305 Removing the rear stabilizer control tube insulator

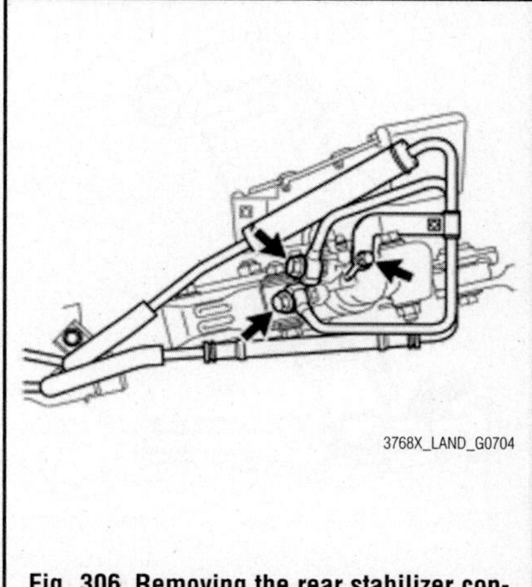

Fig. 306 Removing the rear stabilizer control cylinder

15. Remove the rear suspension control bleeder plug.
    a. Remove the 2 rear suspension control bleeder plugs from the rear stabilizer control cylinder.

To install:

16. Install the rear suspension control bleeder plugs to the rear stabilizer control cylinder. Tighten to 7 ft. lbs. (10 Nm).
17. Install the rear stabilizer link assembly with the bolt to 153 ft. lbs. (208 Nm).

➡Make sure the bolt is facing the correct direction.

18. Install the rear No. 2 stabilizer link with the 2 bolts and 2 nuts. Tighten to 72 ft. lbs. (97 Nm).

➡Make sure the bolt is facing the correct direction.

19. Temporarily install the rear stabilizer control cylinder to the frame with the bolt.

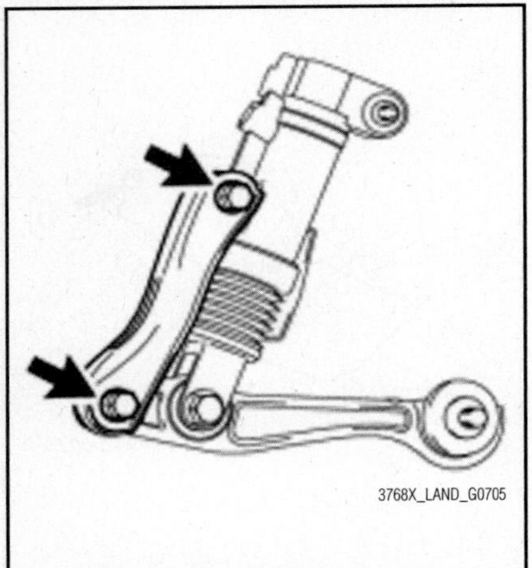

Fig. 307 Removing the rear No. 2 stabilizer link

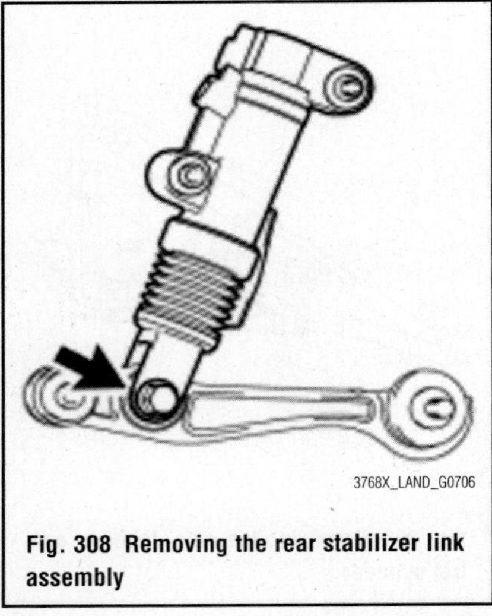

Fig. 308 Removing the rear stabilizer link assembly

a. Install the rear stabilizer control tube and 2 new gaskets to the rear stabilizer control cylinder with the 2 union bolts. Tighten to 51 ft. lbs. (69 Nm).

b. Install the bolt and tighten to 21 ft. lbs. (29 Nm).

20. Install the stabilizer link sub assembly to the stabilizer bar with the bolt and nut. Tighten to 35 ft. lbs. (48 Nm).

➡Make sure the bolt is facing the correct direction.

a. Make sure that all parts are facing as shown in the illustration and install the stabilizer end bracket, retainers and cushions with a new nut. Tighten to 22 ft. lbs. (30 Nm).

21. Install the rear stabilizer bar sub assembly.

a. Install the 2 bushes to the outer side of the bush stoppers on the rear stabilizer bar as shown in the illustration.

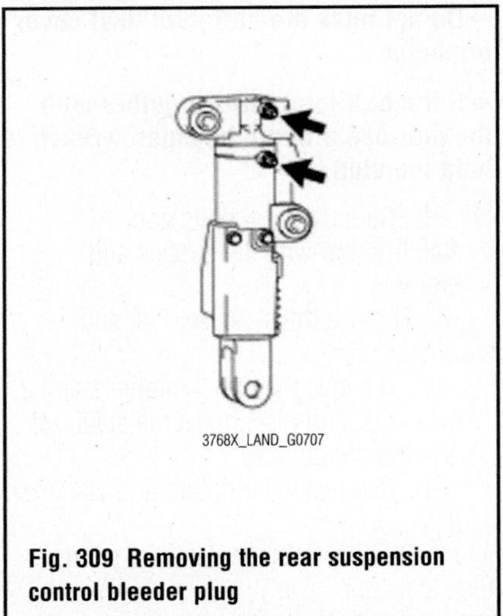

Fig. 309 Removing the rear suspension control bleeder plug

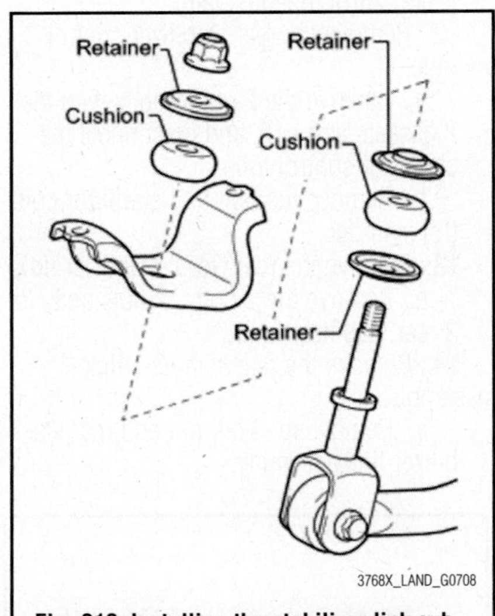

Fig. 310 Installing the stabilizer link sub assembly

➡Install the stabilizer bush so that the bump is on the inner side of the vehicle.

b. Install the 2 stabilizer bar brackets to the vehicle with the 4 bolts. Tighten to 43 ft. lbs. (58 Nm).

➡Install the stabilizer bar bracket so that the protrusion is on the lower side.

➡Tighten the bolts in 4 steps, in the order shown in the illustration.

c. Install the stabilizer end bracket to the vehicle with the 2 bolts. Tighten to 22 ft. lbs. (30 Nm).

d. Install the stabilizer link assembly to the ball joint dust cover protector.

e. Install the stabilizer bar to the stabilizer link assembly with the nut. Tighten to 74 ft. lbs. (100 Nm).

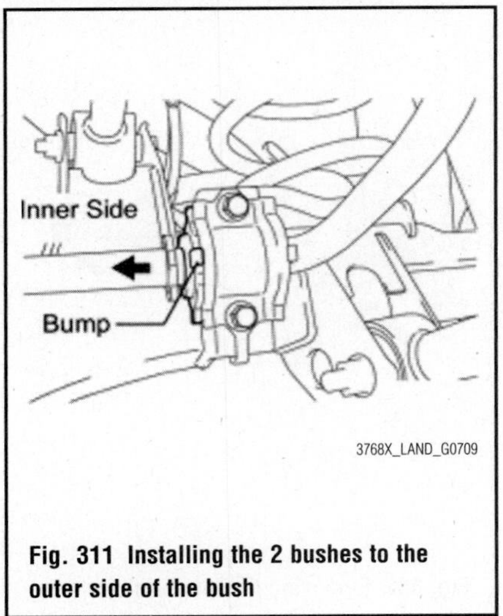

Fig. 311 Installing the 2 bushes to the outer side of the bush

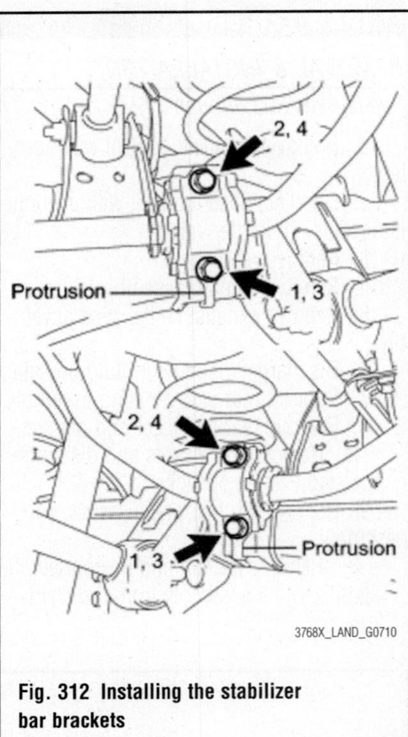

Fig. 312 Installing the stabilizer bar brackets

➡If the ball joint turns together with the nut, use a 6 mm hexagon wrench to hold the stud bolt.

➡If the rear stabilizer control cylinder extends and it is difficult to temporarily install the stabilizer link to the stabilizer bar, raise the stabilizer control arm with a jack and temporarily install the stabilizer link.

22. Bleed the air from the suspension fluid.

23. Inspect for suspension fluid leaks.

24. Install the exhaust center pipe assembly.

25. Install the tailpipe assembly.
26. Stabilize the suspension.
  a. Install the rear wheels and tighten to 97 ft. lbs. (131 Nm).
  b. Lower the vehicle.
  c. Press down on the vehicle several times to stabilize the suspension.
27. Tighten the rear stabilizer control cylinder. Tighten the bolt to 66 ft. lbs. (90 Nm).

➡Perform this procedure with all 4 wheels on the ground.

28. Install the rear stabilizer control tube insulator with the 2 bolts. Tighten to 13 ft. lbs. (18 Nm).
29. Measure the vehicle height.
30. Close the stabilizer control with the accumulator housing shutter valve.
31. Install the stabilizer control valve protector.

STRUT & SPRING ASSEMBLY

REMOVAL & INSTALLATION

See Figures 313 through 317.

➡Use the same procedures for the RH side and LH side.

➡The procedures listed below are for the LH side.

1. Remove the stabilizer control valve protector.
2. Open the stabilizer control with the accumulator housing shutter valve.
3. Remove the rear wheel.
4. Disconnect the No. 3 parking brake cable assembly.
  a. Remove the bolt and disconnect the No. 3 parking brake cable.
5. Disconnect the No. 2 parking brake cable assembly.
  a. Remove the bolt and disconnect the No. 2 parking brake cable.
6. Disconnect the rear axle breather hose sub assembly.
  a. Disconnect the rear axle breather hose from the rear axle housing assembly.
7. Support the rear axle housing assembly.
  a. Support the rear axle housing with a jack using a wooden block to avoid damage.
8. Disconnect the rear lateral control rod assembly.
  a. Remove the bolt and nut, and disconnect the rear lateral control rod from the frame.
9. Disconnect the rear shock absorber assembly RH.

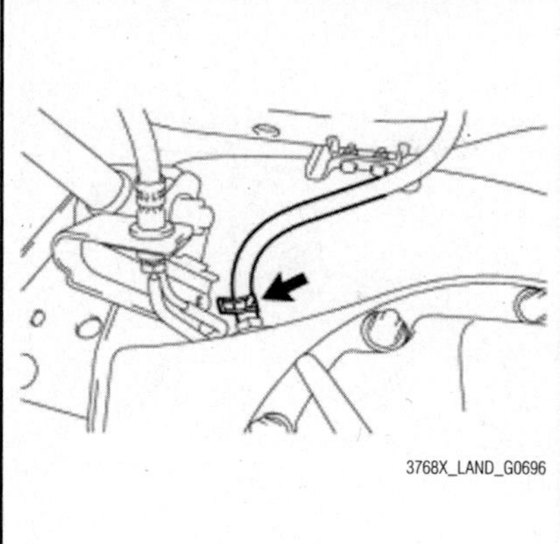

Fig. 313 Disconnect the rear axle breather hose sub assembly

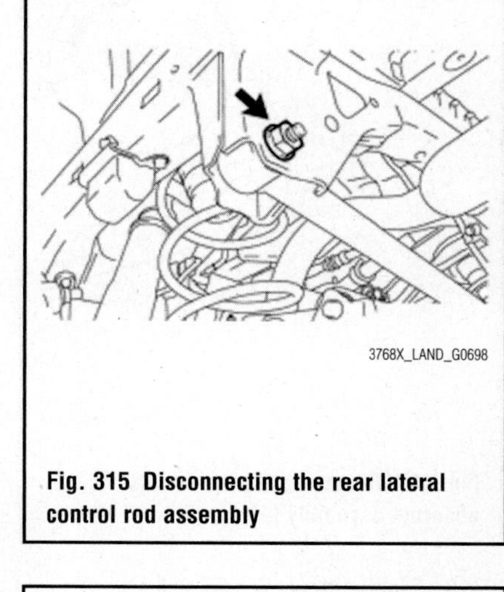

Fig. 315 Disconnecting the rear lateral control rod assembly

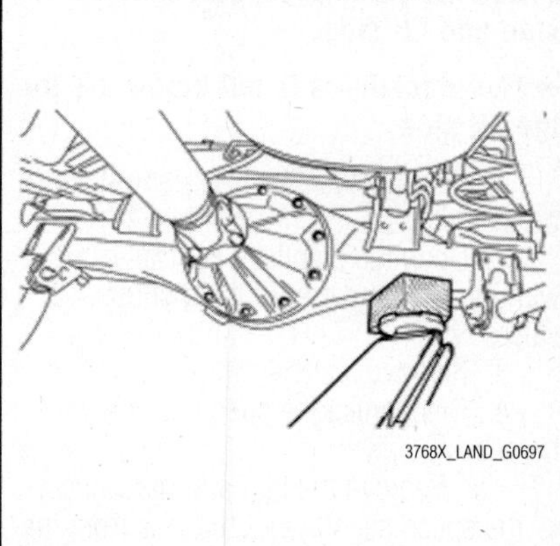

Fig. 314 Support the rear axle housing assembly

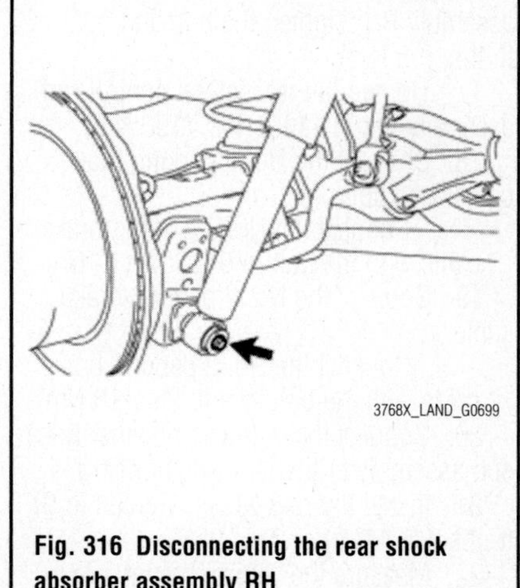

Fig. 316 Disconnecting the rear shock absorber assembly RH

  a. Remove the bolt on the lower side of the shock absorber.
  b. Disconnect the shock absorber from the axle housing.
10. Remove the rear shock absorber assembly LH.
  a. Remove the bolt on the lower side of the shock absorber.
  b. Using the special tool (09922-10010), hold the rear shock absorber in place.
  c. Remove the nut, upper bracket and shock absorber assembly.
  d. Remove the lower bracket from the shock absorber.

To install:

11. Temporarily install the rear shock absorber assembly LH.
  a. Install the lower bracket to the shock absorber.
  b. Using special tool, hold the rear shock absorber in place.
  c. Temporarily install the rear shock absorber and upper bracket with the nut.
  d. Temporarily install the lower side of the shock absorber with the bolt.
12. Temporarily install the rear shock absorber assembly RH.
  a. Temporarily install the lower side of the shock absorber with the bolt.
13. Temporarily install the rear lateral control rod assembly.
  a. Temporarily install the lateral control rod with the nut and bolt.
14. Stabilize the suspension.
  a. Install the rear wheels. Tighten to 97 ft. lbs. (131 Nm).
  b. Lower the vehicle.
  c. Press down on the vehicle several times to stabilize the suspension.
  d. Remove the rear wheels.
15. Tighten the shock absorber assembly LH. Tighten the nut to 39 ft. lbs. (53 Nm). Tighten the bolt to 72 Nm (98 Nm).

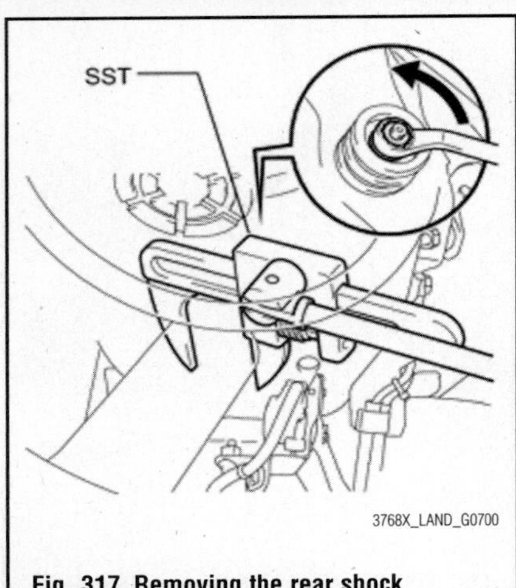

Fig. 317 Removing the rear shock absorber assembly LH

16. Tighten the rear shock absorber assembly RH. Tighten the bolt and nut to 72 ft. lbs. (98 Nm).
17. Tighten the rear lateral control rod assembly nut to 111 ft. lbs. (150 Nm).
18. Connect the No. 3 parking brake cable assembly.
    a. Connect the No. 3 parking brake cable with the bolt to 9 ft. lbs. (13 Nm).
19. Connect the No. 2 parking brake cable.
    a. Connect the No. 2 parking brake cable with the bolt to 9 ft. lbs. (13 Nm).
20. Connect the rear axle breather hose sub assembly to the rear axle housing.
21. Install the rear wheel. Tighten to 97 ft. lbs. (131 Nm).
22. Measure the vehicle height.
23. Close stabilizer control with the accumulator housing shutter valve.
24. Install the stabilizer control valve protector.

UPPER CONTROL ARM

REMOVAL & INSTALLATION

See Figures 318 and 319.

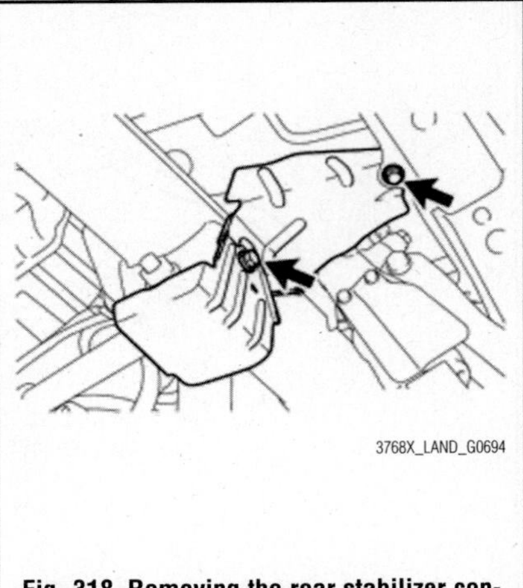

Fig. 318 Removing the rear stabilizer control tube insulator

➡Use the same procedures for the RH side and LH side.

➡The procedures listed below are for the LH side.

1. Remove the stabilizer control valve protector.
2. Open the stabilizer control with the accumulator housing shutter valve.
3. Remove the rear wheel.
4. Disconnect the speed sensor wire harness.
    a. Remove the bolt and disconnect the speed sensor wire harness from the upper control arm.
5. Remove the rear stabilizer control tube insulator.
    a. Remove the 2 bolts and rear stabilizer control tube insulator.
6. Support the rear axle housing assembly.
7. Remove the rear upper control arm assembly LH.
    a. Remove the 2 nuts, 2 washers and 2 bolts, and upper control arm.

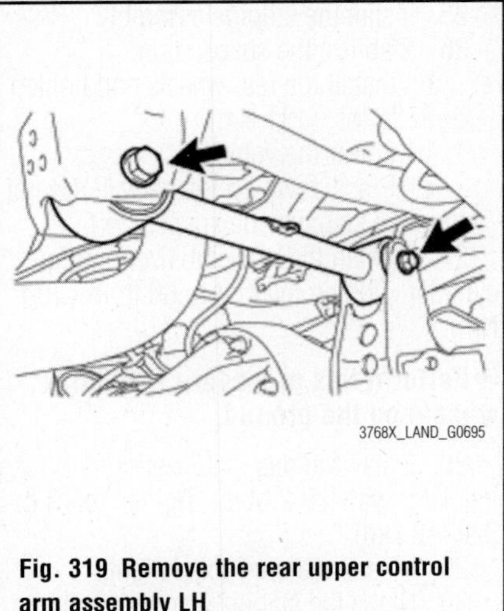

Fig. 319 Remove the rear upper control arm assembly LH

To install:

8. Temporarily install the rear upper control arm assembly LH.
9. Stabilize the suspension.
    a. Install the rear wheels. Tighten to 97 ft. lbs. (131 Nm).
    b. Lower the vehicle.
    c. Press down on the vehicle several times to stabilize the suspension.
    d. Remove the rear wheels.
10. Tighten the rear upper control arm assembly LH.
    a. Tighten the 2 nuts to 111 ft. lbs. (150 Nm).
11. Install the rear stabilizer control tube insulator with the 2 bolts. Tighten to 13 ft. lbs. (18 Nm).
12. Connect the speed sensor wire harness with the bolt. Tighten to 9 ft. lbs. (13 Nm).
13. Install the rear wheel and tighten to 97 ft. lbs. (131 Nm).
14. Measure the vehicle height.
15. Close the stabilizer control with the accumulator housing shutter valve.
16. Install the stabilizer control valve protector.

TOYOTA

10

Matrix

SPECIFICATIONS AND MAINTENANCE CHARTS

ENGINE AND VEHICLE IDENTIFICATION

Engine						
Code ①	Liters (cc)	Cu. In.	Cyl.	Fuel Sys.	Engine Type	Eng. Mfg.
2ZR-FE	1.8 (1794)	109.5	4	EFI	DOHC	Toyota
2AZ-FE	2.4 (2400)	109.5	4	EFI	DOHC	Toyota

Model Year	
Code ②	Year
B	2011
C	2012

EFI: Electronic Fuel Injection

DOHC: Double Overhead Camshaft

① 5th digit of VIN

②10th digit of VIN

71099_MATR_C0001

GENERAL ENGINE SPECIFICATIONS

Year	Model	Engine Displacement Liters (cc)	Engine Series (ID/VIN)	Net Horsepower @ rpm	Net Torque @ rpm (ft. lbs.)	Bore x Stroke (in.)	Com-pression Ratio	Oil Pressure @ rpm
2011	Matrix	1.8 (1794)	2ZR-FE	132@6000	128@4400	3.16x3.48	10.0:1	4.3
		2.4 (2400)	2AZ-FE	158@6000	162@4000	3.48x3.78	9.8	NA
2012	Matrix	1.8 (1794)	2ZR-FE	132@6000	128@4400	3.16x3.48	10.0:1	4.3
		2.4 (2400)	2AZ-FE	158@6000	162@4000	3.48x3.78	9.8	NA

NA: Not Available

EFI: Electronic Fuel Injection

71099_MATR_C0002

ENGINE TUNE-UP SPECIFICATIONS

Year	Engine Displacement Liters	Engine ID/VIN	Spark Plug Gap (in.)	Ignition Timing (deg.)	Fuel Pump (psi)	Idle Speed (rpm)		Valve Clearance	
						MT	AT	In.	Ex.
2011	1.8	2ZR-FE	0.043	①	44-50	600-700	600-700	0.007-0.014	0.015-0.018
	2.4	2AZ-FE	0.043	②	44-50	600-700	600-700	NA	NA
2012	1.8	2ZR-FE	0.043	①	44-50	600-700	600-700	0.007-0.014	0.015-0.018
	2.4	2AZ-FE	0.043	②	44-50	600-700	600-700	NA	NA

Note: The Vehicle Emission Control Information label often reflects specification changes made during production.

NA: Not Available

BTDC: Before Top Dead Center

① With Techstream (ODB-II scanner or equivalent): BTDC 5-15 at idle
Without Techstream: BTDC 8-12 at idle. Connect terminals 13 (TC) and 4 (CG)

② BTDC 8-12 with terminal TC and CG of DLC3 connected

71099_MATR_C0003

CAPACITIES

Year	Model	Engine Displacement Liters	Engine ID/VIN	Engine Oil with Filter	Transmission (pts.) 5-Spd	6-Spd	Auto.	Drive Axle Front (pts.)	Rear (pts.)	Fuel Tank (gal.)	Cooling System (qts.)
2011	Matrix	1.8	2ZR-FE	4.4	4.0	NE	16.0	NA	NE	13.2	6.0
		2.4	2AZ-FE	4.4	4.0	NE	6.0	NA	NE	13.2	6.0
2012	Matrix	1.8	2ZR-FE	4.4	4.0	NE	16.0	NA	NE	13.2	6.0
		2.4	2AZ-FE	4.4	4.0	NE	6.0	NA	NE	13.2	6.0

Note: All capacities are approximate. Add fluid gradually and check to be sure a proper fluid level is obtained. Auto trans is a drain and fill capacity.

NE: Not Equipped

NA: Not Available

71099_MATR_C0004

FLUID SPECIFICATIONS

Year	Model	Engine Displacement Liters	Engine ID/VIN	Engine Oil	Auto. Trans.	Drive Axle	Power Steering Fluid	Brake Master Cylinder
2011	Matrix	1.8	2AZ-FE	5W-20	NA	—	ATF Dexron® II or III	DOT 3
		2.4	2AZ-FE	5W-20	NA	—	ATF Dexron® II or III	DOT 3
2012	Matrix	1.8	2AZ-FE	5W-20	NA	—	ATF Dexron® II or III	DOT 3
		2.4	2AZ-FE	5W-20	NA	—	ATF Dexron® II or III	DOT 3

DOT: Department Of Transpotation

NA: Not Available

71099_MATR_C0005

VALVE SPECIFICATIONS

Year	Engine Displacement Liters	Engine ID/VIN	Seat Angle (deg.)	Face Angle (deg.)	Spring Test Pressure (lbs. @ in.)	Spring Installed Height (in.)	Stem-to-Guide Clearance (in.) Intake	Stem-to-Guide Clearance (in.) Exhaust	Stem Diameter (in.) Intake	Stem Diameter (in.) Exhaust
2011	1.8	2ZR-FE	45	44.5	35.7-39.5@ 1.323	1.323	0.0010-0.0024	0.0012-0.0025	0.2154-0.2159	0.2152-0.2158
	2.4	2AZ-FE	45	44.5	35.7-39.5@ 1.323	1.323	0.0010-0.0024	0.0012-0.0025	0.2154-0.2159	0.2152-0.2158
2012	1.8	2ZR-FE	45	44.5	35.7-39.5@ 1.323	1.323	0.0010-0.0024	0.0012-0.0025	0.2154-0.2159	0.2152-0.2158
	2.4	2AZ-FE	45	44.5	35.7-39.5@ 1.323	1.323	0.0010-0.0024	0.0012-0.0025	0.2154-0.2159	0.2152-0.2158

71099_MATR_C0006

CAMSHAFT AND BEARING SPECIFICATIONS

All measurements are given in inches.

Year	Engine Displacement Liters	Engine VIN	Journal Diameter	Brg. Oil Clearance	Shaft End-play	Runout	Journal Bore	Lobe Lift Intake	Lobe Lift Exhaust
2011	1.8	2ZR-FE	①	0.0012-0.0025	NA	0.0016	NA	1.685-1.6890	1.745-1.7490
	2.4	2AZ-FE	②	③	NA	0.00118	NA	1.865-1.8660	1.813-1.8170
2012	1.8	2ZR-FE	①	0.0012-0.0025	NA	0.0016	NA	1.685-1.6890	1.745-1.7490
	2.4	2AZ-FE	②	③	NA	0.00118	NA	1.865-1.8660	1.813-1.8170

NA: Not Available

① No. 1 journal diameter: 1.356
Other: 0.903 to 0.904

② No. 1 journal diamter: 1.416
Other: 0.903 to 0.904

③ Intake: 0.00276 (Max.)
Exhaust: 0.00394 (Max.)

71099_MATR_C0009

CRANKSHAFT AND CONNECTING ROD SPECIFICATIONS

All measurements are given in inches.

Year	Engine Displacement Liters	Engine ID/VIN	Crankshaft Main Brg. Journal Dia.	Crankshaft Main Brg. Oil Clearance	Crankshaft Shaft End-play	Crankshaft Thrust on No.	Connecting Rod Journal Diameter	Connecting Rod Oil Clearance	Connecting Rod Side Clearance
2011	1.8	2ZR-FE	1.8893-1.8898	0.0006-0.0015	0.0008-0.0087	3	1.7320-1.7323	0.0012-0.0024	0.0063-0.0135
	2.4	2AZ-FE	2.164-2.165	0.00236	NA	3	0.283-0.2870	0.00248	0.0063-0.0143
2012	1.8	2ZR-FE	1.8893-1.8898	0.0006-0.0015	0.0008-0.0087	3	1.7320-1.7323	0.0012-0.0024	0.0063-0.0135
	2.4	2AZ-FE	2.164-2.165	0.00236	NA	3	0.283-0.2870	0.00248	0.0063-0.0143

NA: Not Available

71099_MATR_C0007

PISTON AND RING SPECIFICATIONS

All measurements are given in inches.

Year	Engine Disp. Liters	Engine ID/VIN	Piston Clearance	Ring Gap Top Compression	Ring Gap Bottom Compression	Ring Gap Oil Control	Ring Side Clearance Top Compression	Ring Side Clearance Bottom Compression	Ring Side Clearance Oil Control
2011	1.8	2ZR-FE	0.0011-0.0020	0.0079-0.0118	0.0118-0.0197	0.0039-0.0157	0.0008-0.0028	0.0008-0.0024	0.0008-0.0026
	2.4	2AZ-FE	0.00082-0.0017	0.0094-0.0122	0.0130-0.0169	0.0039-0.0118	0.0007-0.0027	0.0007-0.0023	0.0007-0.0027
2012	1.8	2ZR-FE	0.0011-0.0020	0.0079-0.0118	0.0118-0.0197	0.0039-0.0157	0.0008-0.0028	0.0008-0.0024	0.0008-0.0026
	2.4	2AZ-FE	0.00082-0.0017	0.0094-0.0122	0.0130-0.0169	0.0039-0.0118	0.0007-0.0027	0.0007-0.0023	0.0007-0.0027

71099_MATR_C0008

TORQUE SPECIFICATIONS

All readings in ft. lbs.

Year	Engine Displacement Liters	Engine ID/VIN	Cylinder Head Bolts	Main Bearing Bolts	Rod Bearing Bolts	Crankshaft Damper Bolts	Flywheel Bolts	Manifold Intake	Manifold Exhaust	Spark Plugs	Oil Pan Drain Plug
2011	1.8	2ZR-FE	①	②	③	102	65	21	16	15	27
	2.4	2AZ-FE	④	NA	⑤	NA	⑥	22	27	14	27
2012	1.8	2ZR-FE	①	②	③	102	65	21	16	15	27
	2.4	2AZ-FE	④	NA	⑤	NA	⑥	22	27	14	27

NA: Not Available

① Step 1: 36 ft. lbs.
Step 2: 90 degree turn
Step 3: 45 degree turn

② Inner 12 point bolts:
Step 1: 16 ft. lbs.
Step 2: 32 ft. lbs.
Step 3: 45 degree turn
Step 4: 45 degree turn
Outer cap bolts: 13 ft. lbs.

③ Step 1: 15 ft. lbs.
Step 2: 90 degree turn

④ Step 1: 52 ft. lbs.
Step 2: 90 degree turn

⑤ Step 1: 18 ft. lbs.
Step 2: 90 degree turn

⑥ A/T: 72 ft. lbs.
M/T: 96 ft. lbs.

71099_MATR_C0010

WHEEL ALIGNMENT

Year	Model		Caster Range (+/-Deg.)	Caster Preferred Setting (Deg.)	Camber Range (+/-Deg.)	Camber Preferred Setting (Deg.)	Toe-in (in.)
2011	Matrix - 2WD	F	0.75	①	0.75	-0.57	0+/-0.02
		R	—	—	③ or ⑥	④ or ⑦	⑤ or ⑧
	Matrix - 4WD	F	0.75	②	0.75	-0.58	0+/-0.02
		R	—	—	③ or ⑥	④ or ⑦	⑤ or ⑧
2012	Matrix - 2WD	F	0.75	①	0.75	-0.57	0+/-0.02
		R	—	—	③ or ⑥	④ or ⑦	⑤ or ⑧
	Matrix - 4WD	F	0.75	②	0.75	-0.58	0+/-0.02
		R	—	—	③ or ⑥	④ or ⑦	⑤ or ⑧

① 1.8L engine 205/55R16:+2.93
1.8L engine 215/45R17: +2.92
2.4L engine 205/55R16: +2.90
2.4L engine 215/45R17: +2.88
2.4L engine 215/45R18: +3.00

② 2.4L engine 205/55R16: +3.02
2.4L engine 215/45R17: +3.00

③ Torsion Beam Type Suspension: 0.5

④ Torsion Beam Type Suspension: -1.47

⑤ Torsion Beam Type Suspension: 0.114+/-0.106

⑥ Double Wishbone Type Suspension: 0.75

⑦ Double Wishbone Type Suspension: -1.05

⑧ Double Wishbone Type Suspension: 0.25+/- 0.25

71099_MATR_C0011

TIRE, WHEEL AND BALL JOINT SPECIFICATIONS

Year	Model	OEM Tires		Tire Pressures (psi)		Wheel	Ball Joint	Lug Nuts
		Standard	Optional	Front	Rear	Size	Inspection	(ft. lbs.)
2011	Matrix	205/55R16	—	②	31	6.5-JJ	9-26 in. ①	76
2012	Matrix	205/55R16	—	②	31	6.5-JJ	9-26 in. ①	76

OEM: Original Equipment Manufacturer

PSI: Pounds Per Square Inch

STD: Standard

OPT: Optional

① Torque required in inch lbs. to rotate ball joint when removed from the knuckle

② FWD: 32 PSI , AWD 36 PSI.

71099_MATR_C0012

BRAKE SPECIFICATIONS

All measurements in inches unless noted

Year	Model		Brake Disc			Minimum	Brake Caliper	
			Original Thickness	Minimum Thickness	Maximum Runout	Lining Thickness	Bracket Bolts (ft. lbs.)	Mounting Bolts (ft. lbs.)
2011	Matrix	F	①	②	0.0020	0.039	79	25
		R	③	④	0.0059	0.039	43	20
2012	Matrix	F	①	②	0.0020	0.039	79	25
		R	③	④	0.0059	0.039	43	20

F: Front

R: Rear

① 1.8L engine: 0.866
2.4L engine 1.102

② 1.8L engine:0.748
2.4L engine: 0.984

③ Double Wishbone Type Suspension:0.394
Torsion Beam Type Suspension (1.8L): 0.354
Torsion Beam Type Suspension (2.4L): 0.394

④ Double Wishbone Type Suspension:0.334
Torsion Beam Type Suspension (1.8L): 0.295
Torsion Beam Type Suspension (2.4L): 0.335

71099_MATR_C0013

SCHEDULED MAINTENANCE INTERVALS

TOYOTA MATRIX

TO BE SERVICED	TYPE OF SERVICE	VEHICLE MILEAGE INTERVAL (x1000)													
		5	10	15	20	25	30	35	40	45	50	55	60	90	120
Engine oil & filter	R	✓	✓	✓	✓	✓	✓	✓	✓	✓	✓	✓	✓	✓	✓
Drive belts	S/I						✓						✓	✓	✓
Automatic transaxle fluid & filter	S/I						✓						✓	✓	✓
Brake line pipes & hoses	S/I	✓	✓	✓	✓	✓	✓	✓	✓	✓	✓	✓	✓	✓	✓
Brake linings & drums	S/I	✓	✓	✓	✓	✓	✓	✓	✓	✓	✓	✓	✓	✓	✓
Brake pads & discs (front & rear if equipped)	S/I	✓	✓	✓	✓	✓	✓	✓	✓	✓	✓	✓	✓	✓	✓
Cabin air filter	R				✓				✓				✓		✓
Differential oil	S/I						✓						✓	✓	✓
Drive shaft boots	S/I	✓	✓	✓	✓	✓	✓	✓	✓	✓	✓	✓	✓	✓	✓
Drive shaft bolt (tighten)	S/I	✓	✓	✓	✓	✓	✓	✓	✓	✓	✓	✓	✓	✓	✓
Engine coolant	S/I			✓			✓			✓			✓	✓	✓
Manual transaxle oil	S/I						✓						✓	✓	✓
Steering gear housing oil	S/I	✓	✓	✓	✓	✓	✓	✓	✓	✓	✓	✓	✓	✓	✓
Steering linkage	S/I	✓	✓	✓	✓	✓	✓	✓	✓	✓	✓	✓	✓	✓	✓
Air filter	R						✓						✓	✓	✓
Rotate tires	S/I	✓	✓	✓	✓	✓	✓	✓	✓	✓	✓	✓	✓	✓	✓
Spark plugs	R									✓				✓	
Fuel lines & connections	S/I						✓			✓			✓	✓	✓
Fuel tank cap gasket	R									✓				✓	
Charcoal canister	S/I									✓				✓	

R: Replace S/I: Service or Inspect

FREQUENT OPERATION MAINTENANCE (SEVERE SERVICE)

If a vehicle is operated under any of the following conditions it is considered severe service:

- Extremely dusty areas.

- 50% or more of the vehicle operation is in 32°C (90°F) or higher temperatures, or constant operation in temperatures below 0°C (32°F).

- Prolonged idling (vehicle operation in stop and go traffic).

- Frequent short running periods (engine does not warm to normal operating temperatures).

- Police, taxi, delivery usage or trailer towing usage.

Oil & oil filter: change every 5000 miles.

Bolts & nuts on chassis & body: tighten every 5000 miles.

Ball joints & dust covers: service or inspect every 5,000 miles.

Drive shaft boots & except Supra): service or inspect every 12,000 miles.

Steering linkage: service or inspect every 12,000 miles.

Air filter: service or inspect every 5,000 miles.

Exhaust system: service or inspect every 15,000 miles.

Timing belt: replace every 60,000 miles.

71099_MATR_C0014

PRECAUTIONS

Before servicing any vehicle, please be sure to read all of the following precautions, which deal with personal safety, prevention of component damage, and important points to take into consideration when servicing a motor vehicle:

- Never open, service or drain the radiator or cooling system when the engine is hot; serious burns can occur from the steam and hot coolant.
- Observe all applicable safety precautions when working around fuel. Whenever servicing the fuel system, always work in a well-ventilated area. Do not allow fuel spray or vapors to come in contact with a spark, open flame, or excessive heat (a hot drop light, for example). Keep a dry chemical fire extinguisher near the work area. Always keep fuel in a container specifically designed for fuel storage; also, always properly seal fuel containers to avoid the possibility of fire or explosion. Refer to the additional fuel system precautions later in this section.
- Fuel injection systems often remain pressurized, even after the engine has been turned **OFF**. The fuel system pressure must be relieved before disconnecting any fuel lines. Failure to do so may result in fire and/or personal injury.
- Brake fluid often contains polyglycol ethers and polyglycols. Avoid contact with the eyes and wash your hands thoroughly after handling brake fluid. If you do get brake fluid in your eyes, flush your eyes with clean, running water for 15 minutes. If eye irritation persists, or if you have taken brake fluid internally, IMMEDIATELY seek medical assistance.
- The EPA warns that prolonged contact with used engine oil may cause a number of skin disorders, including cancer. You should make every effort to minimize your exposure to used engine oil. Protective gloves should be worn when changing oil. Wash your hands and any other exposed skin areas as soon as possible after exposure to used engine oil. Soap and water, or waterless hand cleaner should be used.
- All new vehicles are now equipped with an air bag system, often referred to as a Supplemental Restraint System (SRS) or Supplemental Inflatable Restraint (SIR) system. The system must be disabled before performing service on or around system components, steering column, instrument panel components, wiring and sensors. Failure to follow safety and disabling procedures could result in accidental air bag deployment, possible personal injury and unnecessary system repairs.
- Always wear safety goggles when working with, or around, the air bag system. When carrying a non-deployed air bag, be sure the bag and trim cover are pointed away from your body. When placing a non-deployed air bag on a work surface, always face the bag and trim cover upward, away from the surface. This will reduce the motion of the module if it is accidentally deployed. Refer to the additional air bag system precautions later in this section.
- Clean, high quality brake fluid from a sealed container is essential to the safe and proper operation of the brake system. You should always buy the correct type of brake fluid for your vehicle. If the brake fluid becomes contaminated, completely flush the system with new fluid. Never reuse any brake fluid. Any brake fluid that is removed from the system should be discarded. Also, do not allow any brake fluid to come in contact with a painted surface; it will damage the paint.
- Never operate the engine without the proper amount and type of engine oil; doing so WILL result in severe engine damage.
- Timing belt maintenance is extremely important. Many models utilize an interference-type, non-freewheeling engine. If the timing belt breaks, the valves in the cylinder head may strike the pistons, causing potentially serious (also time-consuming and expensive) engine damage. Refer to the maintenance interval charts for the recommended replacement interval for the timing belt, and to the timing belt section for belt replacement and inspection.
- Disconnecting the negative battery cable on some vehicles may interfere with the functions of the on-board computer system(s) and may require the computer to undergo a relearning process once the negative battery cable is reconnected.
- When servicing drum brakes, only disassemble and assemble one side at a time, leaving the remaining side intact for reference.
- Only an MVAC-trained, EPA-certified automotive technician should service the air conditioning system or its components.

BRAKES

ANTI-LOCK BRAKE SYSTEM (ABS)

GENERAL INFORMATION

PRECAUTIONS

- Certain components within the ABS system are not intended to be serviced or repaired individually.
- Do not use rubber hoses or other parts not specifically specified for and ABS system. When using repair kits, replace all parts included in the kit. Partial or incorrect repair may lead to functional problems and require the replacement of components.
- Lubricate rubber parts with clean, fresh brake fluid to ease assembly. Do not use shop air to clean parts; damage to rubber components may result.
- Use only DOT 3 brake fluid from an unopened container.
- If any hydraulic component or line is removed or replaced, it may be necessary to bleed the entire system.
- A clean repair area is essential. Always clean the reservoir and cap thoroughly before removing the cap. The slightest amount of dirt in the fluid may plug an orifice and impair the system function. Perform repairs after components have been thoroughly cleaned; use only denatured alcohol to clean components. Do not allow ABS components to come into contact with any substance containing mineral oil; this includes used shop rags.
- The Anti-Lock control unit is a microprocessor similar to other computer units in the vehicle. Ensure that the ignition switch is **OFF** before removing or installing controller harnesses. Avoid static electricity discharge at or near the controller.
- If any arc welding is to be done on the vehicle, the control unit should be unplugged before welding operations begin.

SPEED SENSORS

REMOVAL & INSTALLATION

Deceleration Sensor (4WD without VSC)

See Figures 1 and 2.

1. Disconnect the cable from the negative battery terminal.

⁂ CAUTION

Wait at least 90 seconds after disconnecting the cable from the negative (-) battery terminal to disable the SRS system.

2. Remove the front seat assembly.
3. Remove the deceleration sensor.
   a. Remove the 2 bolts and yaw rate sensor bracket with deceleration sensor.
   b. Disconnect the connector and clamp from the deceleration sensor.

➡Do not remove the installation nuts.

To install:

4. Install the deceleration sensor.
   a. Connect the connector and clamp to the deceleration sensor.

➡Make sure that the deceleration sensor connector is connected securely.

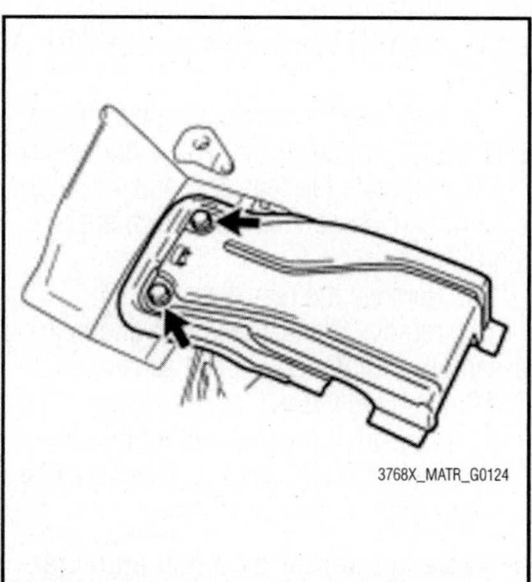

Fig. 1 Removing the yaw rate sensor bracket with the deceleration sensor

   b. Insert the claw on the yaw rate sensor bracket into the deceleration sensor and install the sensor with the 2 bolts. Tighten to 15 ft. lbs. (20 Nm).

➡Keep foreign matter from adhering to the contact surfaces between the deceleration sensor and body.

➡Do not damage the deceleration sensor.

➡Make sure that the deceleration sensor is installed securely.

5. Install the front seat assembly.
6. Connect the cable to the negative battery terminal.
7. Check for the sensor signal.
8. Inspect the SRS warning light.

Front

See Figures 3 through 8.

➡Use the same procedure for the RH side and LH side.

➡The procedure listed below is for the LH side.

➡If the sensor rotor needs to be replaced, replace it together with the front drive shaft assembly.

1. Disconnect the cable from the negative battery terminal.
2. Remove the front wheel.
3. Remove the front fender liner.
4. Remove the front speed sensor.
   a. Disconnect the front speed sensor connector.
   b. Remove the front speed sensor wire harness clamp from the body.
   c. Remove the bolt and No. 2 sensor clamp from the body.
   d. Remove the bolt and separate the brake flexible hose.
   e. Remove the No. 1 sensor clamp from the shock absorber.
   f. Remove the clamp.
   g. Remove the bolt and front speed sensor.

➡Prevent foreign matter from attaching to the sensor tip.

➡Clean the installation hole and the contact surface for the speed sensor every time it is removed.

To install:

➡Use the same procedure for the RH side and LH side.

➡The procedure listed below is for the LH side.

➡If the sensor rotor needs to be replaced, replace it together with the front drive shaft assembly.

5. Install the front speed sensor with the bolt and tighten to 75 inch lbs. (8.5 Nm).

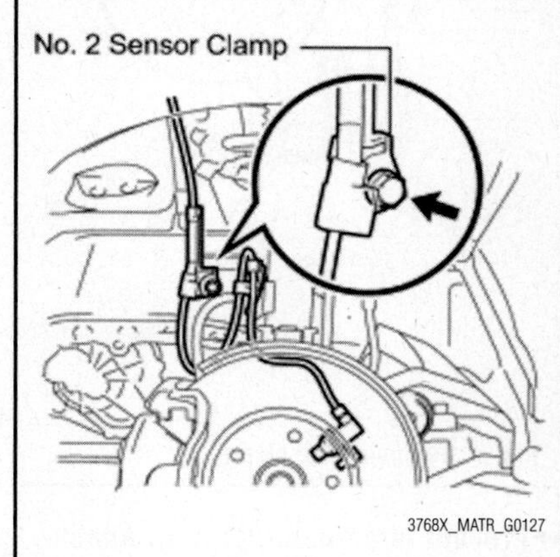

Fig. 4 Removing the bolt and No. 2 sensor clamp from the body

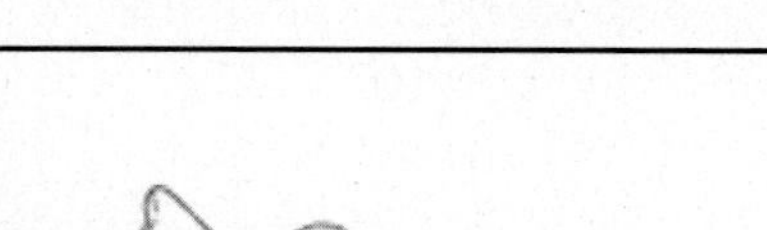
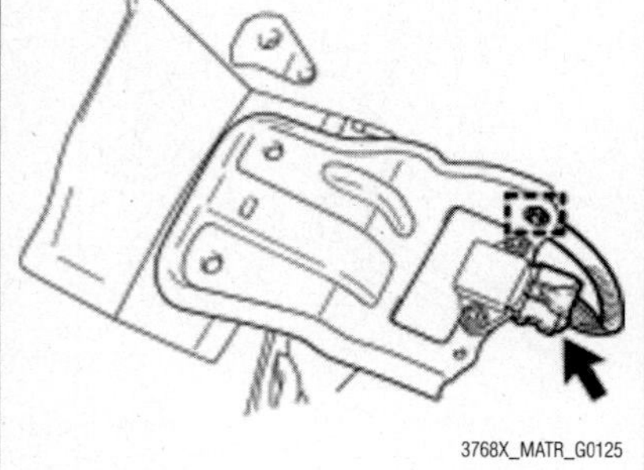

Fig. 2 Disconnecting the connector and clamp from the deceleration sensor

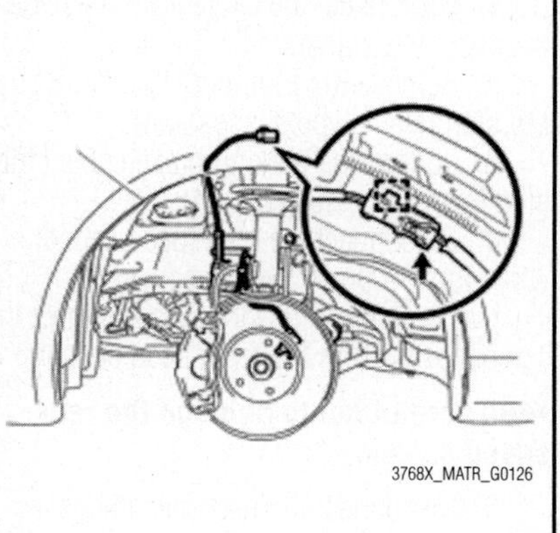

Fig. 3 Disconnecting the front speed sensor connector

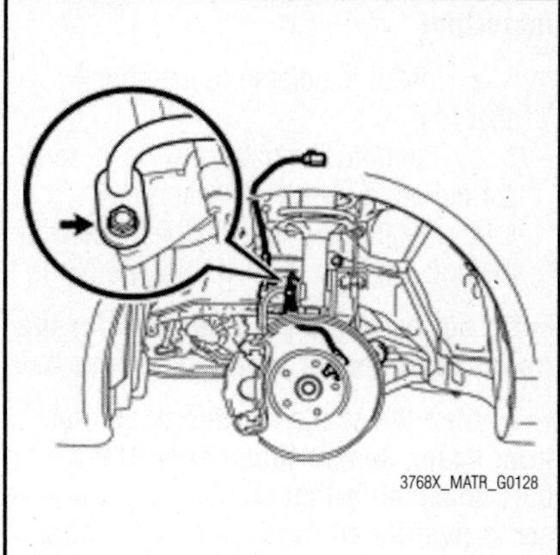

Fig. 5 Removing the bolt and separate the brake flexible hose

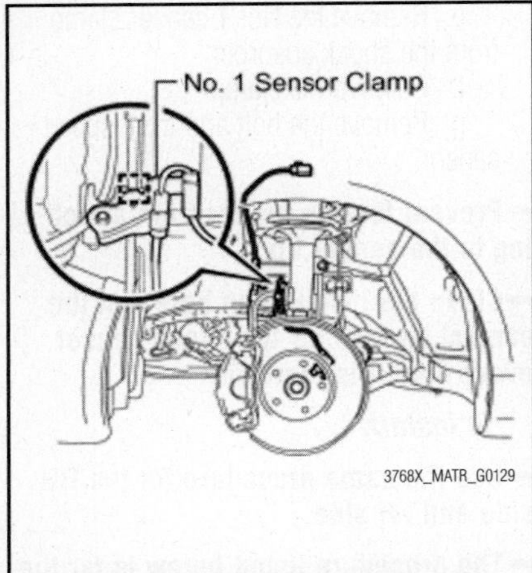

Fig. 6 Removing the No. 1 sensor clamp from the shock absorber

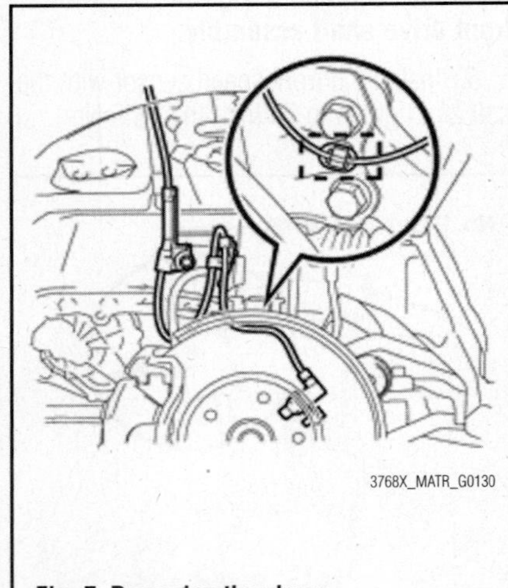

Fig. 7 Removing the clamp

➡**Prevent foreign matter from attaching to the sensor tip.**

➡**Do not file the hole or the contact surface because the gap between the magnet rotor and front speed sensor is important.**

a. Install the clamp to the shock absorber.

b. Temporarily install the No. 1 sensor clamp to the shock absorber.

c. Install the brake flexible hose with the bolt. Tighten to 21 ft. lbs. (29 Nm).

➡**Do not twist the wire harness for the front speed sensor when installing it.**

➡**Tighten the brake flexible hose and front speed sensor together with the bolt. Make sure that the front speed sensor is positioned over the flexible hose.**

d. Install the No. 2 sensor clamp to the body with the bolt. Tighten to 71 inch lbs. (8 Nm).

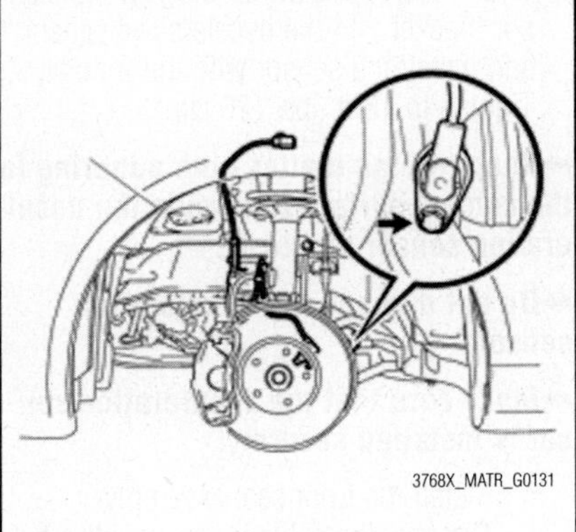

Fig. 8 Removing the bolt and front speed sensor

e. Connect the speed sensor wire harness clamp.

f. Connect the speed sensor connector.

6. Install the front fender liner.
7. Install the front wheel. Tighten to 76 ft. lbs. (103 Nm).
8. Connect the cable to the negative battery terminal.
9. Check for speed sensor signal.

Rear

2WD

See Figures 9 through 12.

➡**Use the same procedure for the RH side and LH side.**

➡**The procedure listed below is for the LH side.**

➡**If the sensor rotor needs to be replaced, replace it together with the rear axle hub and bearing assembly with rear speed sensor.**

1. Disconnect the cable from the negative battery terminal.
2. Remove the front upper console box (Torsion Beam Type suspension).
3. Loosen the parking brake cable (Torsion Beam Type suspension).
4. Disconnect the rear speed sensor wire (Torsion Beam Type suspension).

a. Using a screwdriver, disconnect the connector from the rear speed sensor.

➡**Be careful not to damage the rear speed sensor.**

5. Disconnect the rear speed sensor wire (Double Wishbone Type suspension).

a. Using a screwdriver, disconnect the connector from the rear speed sensor.

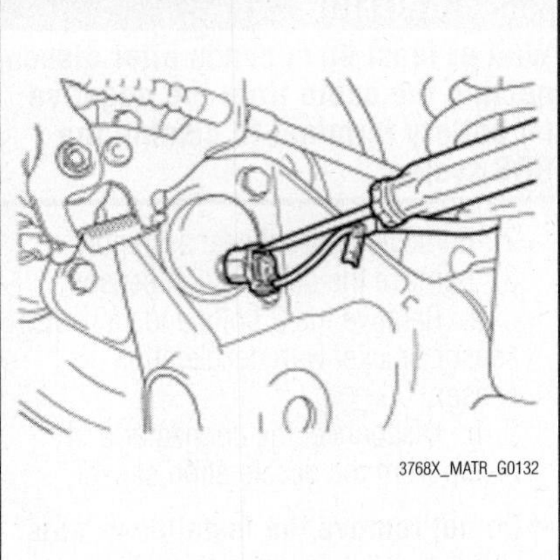

Fig. 9 Disconnecting the rear speed sensor wire (Torsion Beam Type suspension)

➡**Be careful not to damage the rear speed sensor.**

6. Remove the parking brake lever protector (Torsion Beam Type suspension).
7. Separate the No. 3 parking brake cable assembly (Torsion Beam Type suspension).
8. Separate the rear disc brake caliper assembly (Torsion Beam Type suspension).
9. Separate the rear disc brake caliper assembly (Double Wishbone Type suspension).
10. Remove the rear disc.
11. remove the rear axle hub and bearing assembly with the rear speed sensor.
12. Remove the rear speed sensor.

a. Install 3 hub nuts and mount the rear axle hub and bearing assembly in a vise using aluminum plates.

➡**Replace the rear axle hub and bearing assembly if it is dropped or receives a strong shock.**

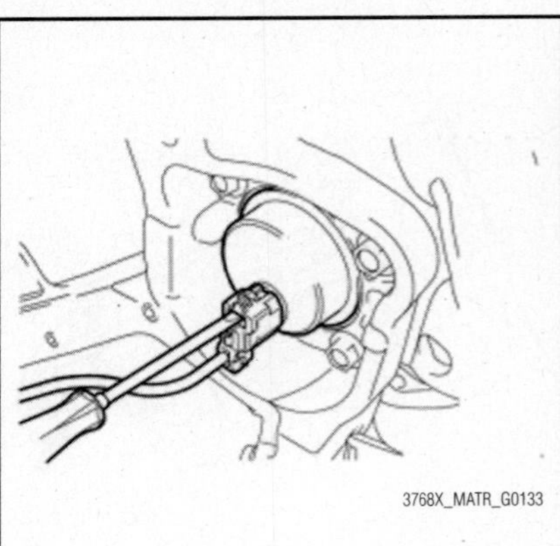

Fig. 10 Disconnecting the rear speed sensor wire (Double Wishbone Type suspension)

b. Using a pin punch and hammer, drive out the 2 pins and remove the 2 attachments from special tool (09520-00031).

c. Using special tool (09520-00031) and 2 bolts (diameter: 12 mm, pitch: 1.5 mm), remove the rear speed sensor from the rear axle hub and bearing assembly.

➡Keep the rear speed sensor away from magnets.

➡Pull the rear speed sensor off straight, taking care not to allow it to contact the rear speed sensor rotor.

➡If the rear speed sensor rotor is damaged or deformed, replace the rear axle hub and bearing assembly.

➡Do not scratch the contact surfaces between the rear axle hub and bearing assembly and the rear speed sensor.

➡Prevent foreign matter from attaching to the speed sensor rotor or tip.

➡If the sensor rotor needs to be replaced, replace it together with the rear axle hub and bearing assembly with rear speed sensor.

To install:

➡Use the same procedure for the RH side and LH side.

➡The procedure listed below is for the LH side.

➡If the sensor rotor needs to be replaced, replace it together with the rear axle hub and bearing assembly with rear speed sensor.

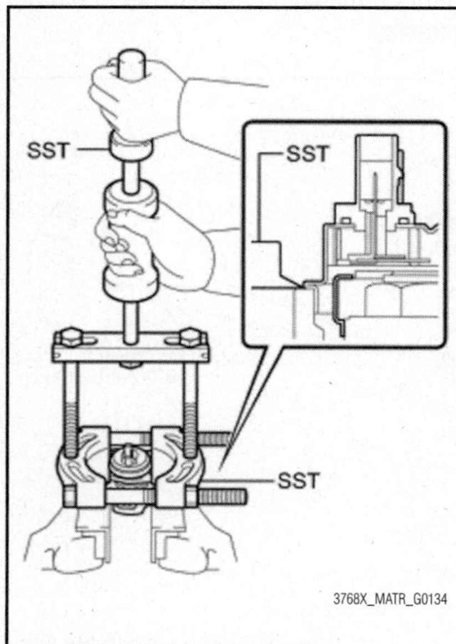

Fig. 11 Removing the rear speed sensor (2WD)

13. Install the rear speed sensor.

a. Clean the contact surface between the rear axle hub and bearing assembly and a new rear speed sensor.

➡Prevent foreign matter from attaching to the sensor rotor.

b. Place the rear speed sensor on the rear axle hub and bearing assembly so that the connector is positioned at the bottom when the sensor is installed on the vehicle.

c. Using the special tool (09214-76011), steel plates and a press, install a new speed sensor to the rear axle hub and bearing assembly.

➡Keep the rear speed sensor away from magnets.

➡Do not use a hammer to install the rear speed sensor.

➡Check that there is no foreign matter such as iron chips on the detecting portion of the rear speed sensor.

➡Slowly press the rear speed sensor in straight.

14. Install the rear axle hub and bearing assembly with the rear speed sensor.
15. Inspect the rear axle hub bearing for looseness.
16. Inspect the rear axle hub runout.
17. Install the rear disc.
18. Install the rear disc brake caliper assembly.
19. Install the parking brake lever protector (Torsion Beam Type suspension).
20. Connect the rear speed sensor wire to the rear speed sensor.

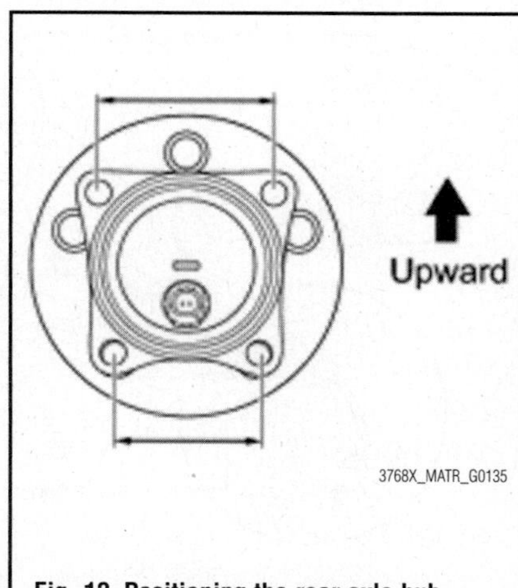

Fig. 12 Positioning the rear axle hub and bearing

21. Inspect the parking brake lever travel.
22. Adjust the parking brake lever travel (Torsion Beam Type suspension).
23. Inspect the rear disc brake cylinder operation lever and stopper clearance (Torsion Beam Type suspension).
24. Adjust the parking brake shoe clearance and parking brake lever travel (Double Wishbone Type suspension).
25. Install the rear wheel and tighten to 76 ft. lbs. (103 Nm).
26. Connect the cable to the negative battery terminal.
27. Install the front upper console box (Torsion Beam Type suspension).
28. Inspect the brake warning light.
29. Check for speed sensor signal.
30. Inspect and adjust the rear wheel alignment (Double Wishbone Type suspension).
31. Inspect the rear wheel alignment.

4WD

See Figures 13 through 16.

➡Use the same procedure for RH side and LH side except for the step to remove the rear sensor clip.

➡If the sensor rotor needs to be replaced, replace it together with the rear drive shaft assembly.

1. Disconnect the cable from the negative battery terminal.
2. Remove the rear seat cushion assembly.
3. Remove the rear seat headrest assembly (LH side).
4. Remove the rear seatback outer hinge cover (LH side).
5. remove the rear seatback inner hinge cover (LH side).
6. Remove the rear seatback assembly LH.
7. Remove the rear seat headrest assembly (RH side).
8. Remove the rear seat center headrest assembly.
9. Disconnect the rear seat center outer belt assembly.
10. Remove the rear seatback outer hinge cover (RH side).
11. Remove the rear seatback inner hinge cover (RH side).
12. Remove the rear seatback assembly RH.
13. Remove the rear door scuff plate (LH side).
14. Remove the rear door opening trim weather-strips (LH side).

15. Remove the rear door scuff plate (RH).
16. Remove the rear door opening trim weather-strip (RH side).
17. Remove the rear seatback hinge sub assembly (LH side).
18. Remove the rear seat side garnish (LH side).
19. Remove the rear seatback hinge sub assembly (RH side).
20. Remove the rear seat side garnish (RH side).
21. Remove the rear wheel.
22. Remove the fuel tank filler pipe protector (LH side).
    a. Remove the 3 bolts and the fuel tank filler pipe protector.
23. Remove the rear speed sensor (LH side).
    a. Disconnect the rear speed sensor connector.
    b. Disconnect the grommet of the rear speed sensor wire from the hole in the wheel well.
    c. Remove bolt A and separate the No. 3 sensor clamp.
    d. Disconnect the sensor clip LH from the breather clamp.
    e. Remove bolt B and separate the No. 2 sensor clamp from the suspension member.
    f. Remove the nut and separate the No. 1 sensor clamp from the upper arm.
    g. Remove bolt C and the rear speed sensor body from the rear axle carrier.

➡Keep the sensor tip and rear speed sensor installation hole free of foreign matter.

24. Remove the rear speed sensor (RH side).

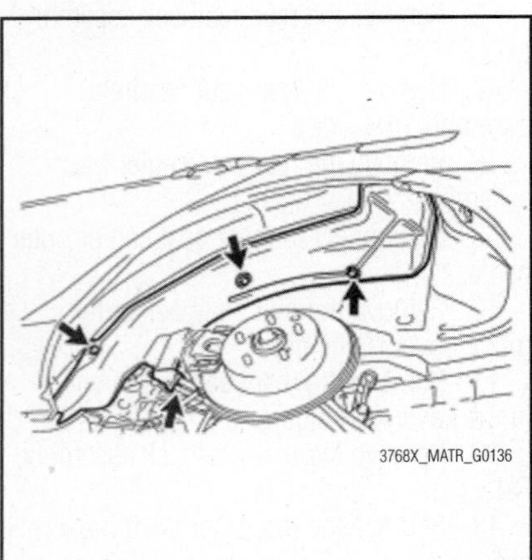

Fig. 13 Removing the fuel tank filler pipe protector (LH side)

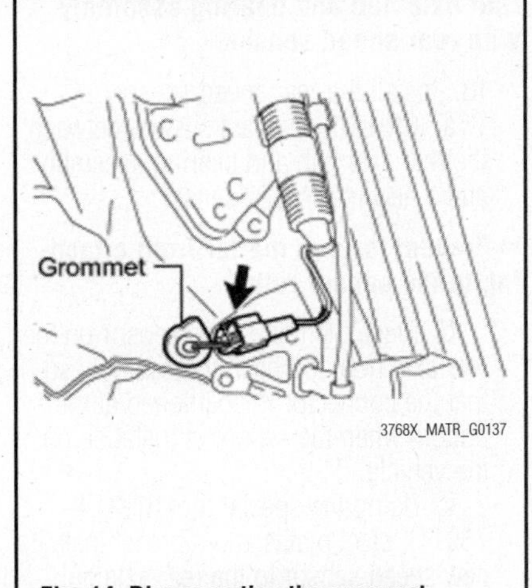

Fig. 14 Disconnecting the grommet

a. Disconnect the rear speed sensor connector.
b. Disconnect the grommet of the rear speed sensor wire from the hole in the wheel well.
c. Remove bolt A and separate the No. 3 sensor clamp.
d. Disconnect the sensor clip from the breather clamp.
e. Remove bolt B and separate the No. 2 sensor clamp from the suspension member.
f. Remove the nut and separate the No. 1 sensor clamp from the upper arm.
g. Remove bolt C and rear speed sensor body from the rear axle carrier.

➡Keep the sensor tip and rear speed sensor installation hole free of foreign matter.

To install:

➡Use the same procedure for RH side and LH side except for the step to install the rear sensor clip.

➡If the sensor rotor needs to be replaced, replace it together with the rear drive shaft assembly.

25. Install the rear speed sensor (LH side).
    a. Install the rear speed sensor with bolt C and tighten to 75 inch lbs. (8.5 Nm).

➡Keep the rear speed sensor tip and sensor installation hole free of foreign matter.

    b. Install the No. 1 sensor clamp with the nut and tighten to 44 inch lbs. (5 Nm).

➡Do not twist the rear speed sensor wire when installing the clamp.

    c. Install the No. 2 sensor clamp with bolt B and tighten to 71 inch lbs. (8 Nm).

➡Do not twist the rear speed sensor wire when installing the clamp.

    d. Install the sensor clip LH.
    e. Install the No. 3 sensor clamp with bolt A and tighten to 71 inch lbs. (8 Nm).

➡Do not twist the rear speed sensor wire when installing the clamp.

    f. Insert the connector and grommet to the inside of the vehicle through the passage hole in the wheel well.

➡Make sure that the grommet's band clamp remains on the outside of the vehicle.

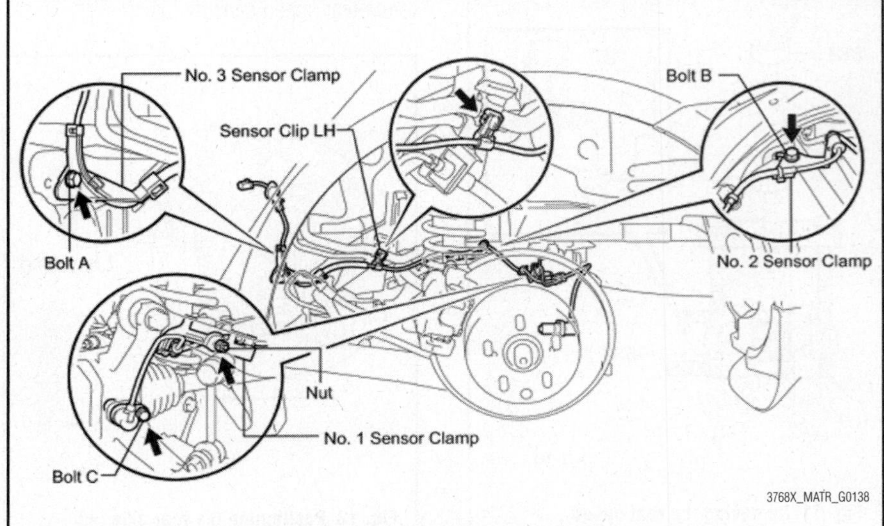

Fig. 15 Removing the rear speed sensor from the rear axle carrier (LH side)

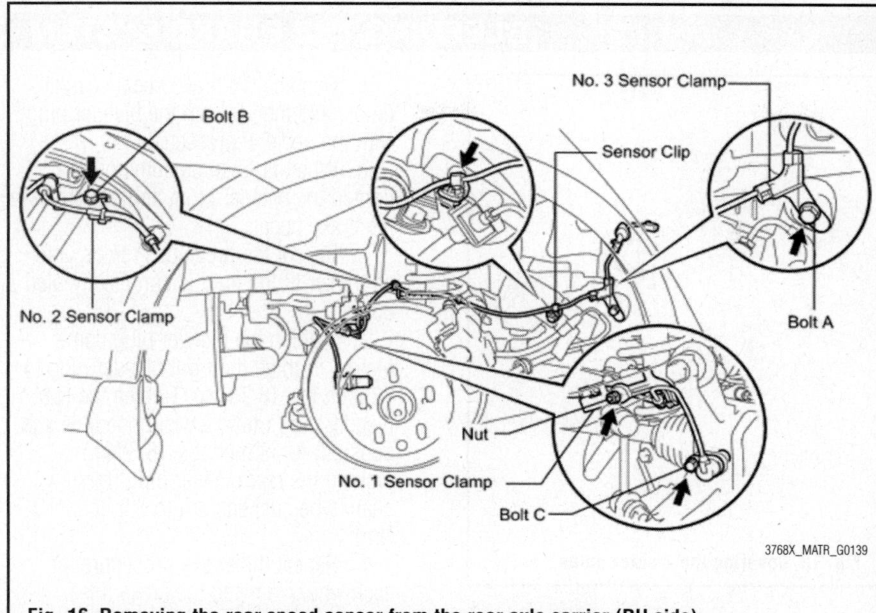

Fig. 16 Removing the rear speed sensor from the rear axle carrier (RH side)

g. Hold the grommet and pull it from the inside of the vehicle to the outside of the vehicle. Then hold in place so that it is not tilted.

➡**When pulling out the grommet, do not pull on the sensor wire.**

h. Connect the rear speed sensor connector.

26. Install the rear speed senor (RH side).

a. Install the rear speed sensor with bolt C and tighten to 75 inch lbs. (8.5 Nm).

➡**Keep the rear speed sensor tip and sensor installation hole free of foreign matter.**

b. Install the No. 1 sensor clamp with the nut and tighten to 44 inch lbs. (5 Nm).

➡**Do not twist the rear speed sensor wire when installing the clamp.**

c. Install the No. 2 sensor clamp with bolt B and tighten to 71 inch lbs. (8 Nm).

➡**Do not twist the rear speed sensor wire when installing the clamp.**

d. Install the sensor clip.

e. Install the No. 3 sensor clamp with bolt A and tighten to 71 inch lbs. (8 Nm).

➡**Do not twist the rear speed sensor wire when installing the clamp.**

f. Insert the connector and grommet to the inside of the vehicle through the passage hole in the wheel well.

➡**Make sure that the grommet's band clamp remains on the outside of the vehicle.**

g. Hold the grommet and pull it from the inside of the vehicle to the outside of the vehicle. Then hold it in place so that it is not tilted.

➡**When pulling out the grommet, do not pull on the sensor wire.**

h. Connect the rear speed sensor connector.

i. Install the fuel tank filler pipe protector (LH side) with the 3 screws.

27. Install the rear wheel and tighten to 76 ft. lbs. (103 Nm).

28. Install the rear seat side garnish LH and RH side.

29. Install the rear seatback hinge sub assembly LH and RH side.

30. Install the rear door opening trim weather-strip RH and LH side.

31. Install the rear door scuff plate RH and LH side.

32. Install the rear seatback assembly RH.

33. Install the rear seatback inner hinge cover (RH side).

34. Install the rear seatback outer hinge cover (RH side).

35. Connect the rear seat center outer belt assembly.

36. Install the rear seat center headrest assembly.

37. Install the rear seat headrest assembly (RH side).

38. Install the rear seatback assembly LH.

39. Install the rear seatback inner hinge cover (LH side).

40. Install the rear seatback outer hinge cover (LH side).

41. Install the rear seat headrest assembly (LH side).

42. Connect the cable to the negative battery terminal.

43. Check for the rear speed sensor signal.

BLEEDING PROCEDURE

BLEEDING PROCEDURE

See Figures 17 through 19.

➡If any work is performed on the brake system or if air in the brake lines is suspected, bleed the brake system.

➡Move the shift lever to P and apply the parking brake before bleeding the brakes.

➡Add brake fluid to keep the level between the MIN and MAX lines of the reservoir while bleeding the brakes.

➡If brake fluid leaks onto any painted surface, immediately wash it off.

1. Remove the center cowl top ventilator louver.
2. Fill the reservoir with brake fluid.
3. Bleed the brake master cylinder.

➡If the master cylinder is reinstalled or runs out of brake fluid, bleed the master cylinder.

➡To prevent brake fluid from damaging painted surface, cover any surrounding parts with a piece of cloth.

a. Using a union nut wrench (10 mm), disconnect the 2 brake lines from the master cylinder.
b. Slowly depress the brake pedal and hold it.
c. Cover the 2 outer holes with fingers, and release the brake pedal.
d. Repeat the previous 2 steps 3 or 4 times.
e. Using a union nut wrench (10 mm), connect the 2 brake lines to the master cylinder. If tightening with a union nut wrench tighten to 11 ft. lbs. (15 Nm). If tightening without a union nut wrench tighten to 10 ft. lbs. (14 Nm).

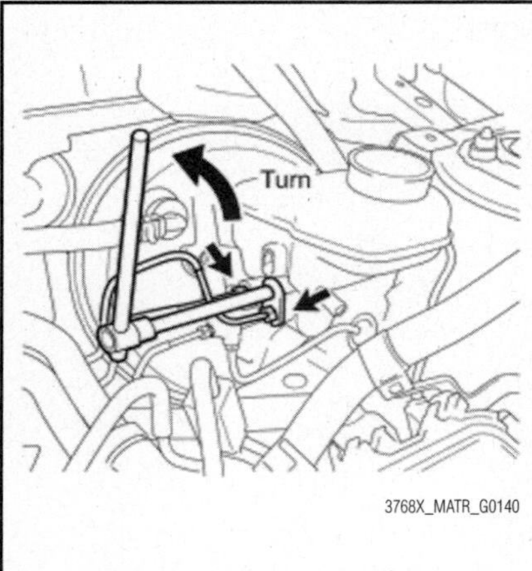

Fig. 17 Disconnecting the 2 brake lines from the master cylinder

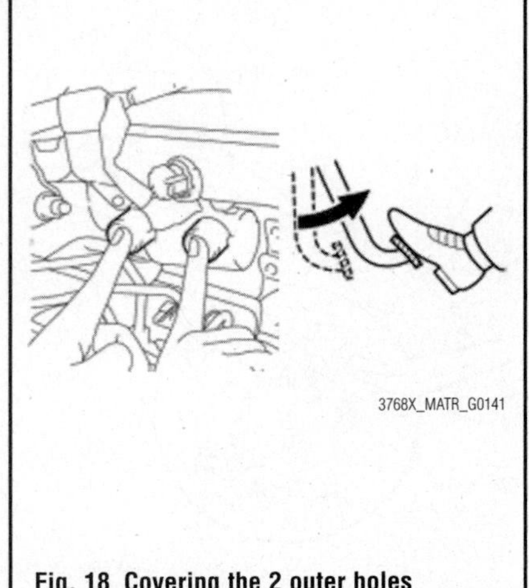

Fig. 18 Covering the 2 outer holes

➡Use a torque wrench with a fulcrum length of 9.84 inch (250 mm).

➡This torque value is effective when the union nut wrench is parallel to the torque wrench.

4. Bleed the brake line.

➡Bleed the brake line of the wheel farthest from the master cylinder first.

➡Add brake fluid to keep the level between the MIN and MAX lines of the reservoir while bleeding the brakes.

a. Connect a vinyl tube to the bleeder plug.

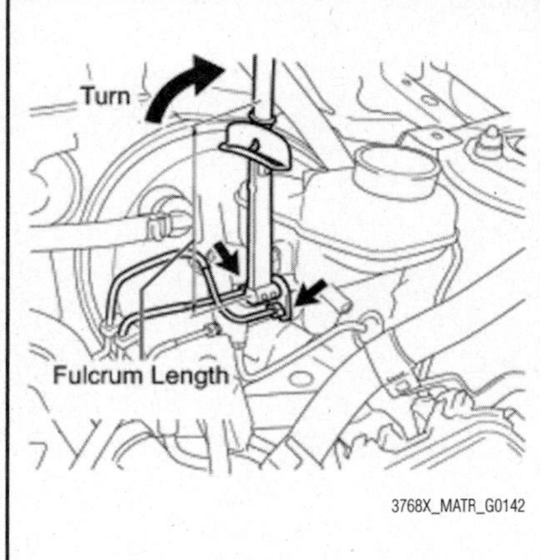

Fig. 19 Connecting the 2 brake lines to the master cylinder

b. Depress the brake pedal several times, and then loosen the bleeder plug with the pedal depressed.
c. When fluid stops coming out, tighten the bleeder plug, and then release the brake pedal.
d. Repeat the previous 2 steps until all the air in the fluid is completely bled out.
e. Tighten the bleeder plug completely. Tighten the front bleeder plug to 73 inch lbs. (8.3 Nm). Tighten the rear bleeder plug (double wishbone type suspension) to 73 inch lbs. (8.3 Nm). Tighten the rear bleeder plug (torsion beam type suspension) to 7 ft. lbs. (10 Nm).
f. Repeat the above procedure for each wheel to bleed the brake line.

5. Bleed the brake actuator (w/VSC).

➡After bleeding the brake system, if the specified height or feel of the brake pedal cannot be obtained, bleed the brake actuator assembly with the Techstream by following the procedure below.

a. Depress the brake pedal more than 20 times with the ignition switch off.
b. Connect the Techstream to the DLC3, and then turn the ignition switch to ON.

➡Do not start the engine.

c. Turn the Techstream on and select "Air Bleeding" on the screen.

➡Refer to the Techstream operator's manual for further details.

➡Bleed air by following the steps displayed on the Techstream.

d. Bleed air according to "Step 1: Increase Line" on the Techstream display.

➡Make sure that the master cylinder reservoir tank does not run out of brake fluid.

➡Add brake fluid to keep the level between the MIN and MAX lines of the reservoir while bleeding the brakes.

e. Connect a vinyl tube to either one of the bleeder plugs.
f. Depress the brake pedal several times, and then loosen the bleeder plug connected to the vinyl tube with the pedal depressed.

g. When fluid stops coming out, tighten the bleeder plug, and then release the brake pedal.

h. Repeat the previous 2 steps until all the air in the fluid is completely bled out.

i. Tighten the bleeder plug completely. Tighten the front bleeder plug to 73 inch lbs. (8.3 Nm). Tighten the rear bleeder plug (double wishbone type suspension) to 73 inch lbs. (8.3 Nm). Tighten the rear bleeder plug (torsion beam type suspension) to 7 ft. lbs. (10 Nm).

j. Repeat the above procedure for each wheel to bleed the brake line.

k. Bleed the suction line according to "Step 2: Inhalation Line" on the Techstream display.

➡Bleed the suction line by following the steps displayed on the Techstream.

➡Add brake fluid to keep the level between the MIN and MAX lines of the reservoir while bleeding the brakes.

l. Connect a vinyl tube to the bleeder plug at the right front wheel or the right rear wheel and loosen the bleeder plug.

m. Operate the brake actuator assembly to bleed air using the Techstream.

➡During this step, be sure to release the brake pedal.

➡The actuator operation stops automatically in 4 seconds.

n. Check that the actuator operation has stopped by referring to the Techstream display, and tighten the bleeder plug.

o. Repeat the previous 2 steps until all the air in the fluid is completely bled out.

p. Tighten the bleeder plug completely. Tighten the front bleeder plug to 73 inch lbs. (8.3 Nm). Tighten the rear bleeder plug (double wishbone type suspension) to 73 inch lbs. (8.3 Nm). Tighten the rear bleeder plug (torsion beam type suspension) to 7 ft. lbs. (10 Nm).

q. For the rest of the wheels, bleed air in the same way as stated in the above procedure.

r. Bleed the pressure reduction line according to "Step 3: Decrease Line" on the Techstream display.

➡Bleed the pressure reduction line by following the steps displayed on the Techstream.

➡Add brake fluid to keep the level between the MIN and MAX lines of the reservoir while bleeding the brakes.

s. Connect a vinyl tube to either one of the bleeder plugs.

t. Loosen the bleeder plug.

u. While keeping the brake pedal fully depressed, operate the brake actuator assembly using the Techstream.

➡The actuator operation stops automatically in 4 seconds. When performing this procedure continuously, an interval of at least 20 seconds is required.

➡After the operation is completed, the brake pedal goes down slightly. This is a normal phenomenon when the solenoid opens.

➡During this procedure, the pedal seems heavy, but completely depress it so that the brake fluid comes out from the bleeder plug.

➡Be sure to keep the brake pedal depressed. Never depress and release the pedal repeatedly.

v. Tighten the bleeder plug, and then release the brake pedal.

w. Repeat previous 3 steps until all the air in the fluid is completely bled out.

x. Tighten the bleeder plug completely. Tighten the front bleeder plug to 73 inch lbs. (8.3 Nm). Tighten the rear bleeder plug (double wishbone type suspension) to 73 inch lbs. (8.3 Nm). Tighten the rear bleeder plug (torsion beam type suspension) to 7 ft. lbs. (10 Nm).

y. Repeat the above procedure for the rest of the brakes to bleed the brake lines.

z. Bleed the brake lines again according to "Step 4: Increase Line" on the Techstream display.

➡Bleed air by following the steps displayed on the Techstream.

➡Add brake fluid to keep the level between the MIN and MAX lines of the reservoir while bleeding the brakes.

aa. Connect a vinyl tube to either one of the bleeder plugs.

bb. Depress the brake pedal several times, and then loosen the bleeder plug connected to the vinyl tube with the pedal depressed.

cc. When fluid stops coming out, tighten the bleeder plug, and then release the brake pedal.

dd. Repeat the previous 2 steps until all the air in the fluid is completely bled out.

ee. Tighten the bleeder plug completely. . Tighten the front bleeder plug to 73 inch lbs. (8.3 Nm). Tighten the rear bleeder plug (double wishbone type suspension) to 73 inch lbs. (8.3 Nm). Tighten the rear bleeder plug (torsion beam type suspension) to 7 ft. lbs. (10 Nm).

ff. Repeat the above procedure for each brake to bleed the brake lines.

gg. Finish "Air Bleeding" on the Techstream, and then turn the Techstream off.

hh. Disconnect the Techstream from the DLC3.

ii. Turn the ignition switch off.

6. Inspect for brake fluid leaks.
7. Inspect the fluid level.
8. Install the center cowl top ventilator louver.

BLEEDING THE ABS SYSTEM

The bleeding procedure for the ABS System is the same as the conventional bleeding procedure. Refer to Bleeding the Brake System.

If after bleeding the air from the brake system using the conventional procedure the height or feel of the brake pedal cannot be obtained, perform air bleeding in the brake actuator assembly with Techstream or equivalent scan tool by following the procedures below.

1. Let the vacuum pressure out from the brake booster assembly.
2. Depress the brake pedal more than 20 times with the engine off.
3. Connect Techstream or equivalent scan tool to the diagnostic connector, and turn the ignition switch to the **ON** position but do not start the engine.
4. Select "Air Bleeding" and bleed the air out of the brake line as usual when "Step 1: Increase" appears on the Techstream or equivalent scan tool display. Bleed the air by following the steps displayed on Techstream or equivalent scan tool.

➡Make sure that the brake fluid in the master cylinder reservoir tank does not become empty.

a. Connect the vinyl tube to either one of the bleeder plugs.

b. Depress the brake pedal several times, then loosen the bleeder plug connected to the vinyl tube with the pedal depressed.

c. When fluid stops coming out, tighten the bleeder plug and release the brake pedal.

d. Repeat the procedure until all air in the fluid is completely bled out.

e. Tighten the bleeder plug completely.

5. Repeat the above procedures for each wheel to bleed the air out of the brake line.

6. Bleed the air out of the suction line when "Step 2: Inhalation" appears on the Techstream or equivalent scan tool display.

a. Connect the vinyl tube to the bleeder plug at the right front wheel or the right rear wheel and loosen the bleeder plug.

b. Operate the brake actuator assembly to bleed the air using Techstream or equivalent scan tool.

➡Make sure to release the brake pedal. This operation stops automatically after 4 seconds.

c. Check if the operation has stopped by referring to the Techstream or equivalent scan tool display.

d. Repeat the procedure until all air in the fluid is completely bled out.

7. Repeat the above procedures for the other wheels to bleed the air out of the brake line.

8. Bleed the air out of the pressure reduction line when "Step 3: Decrease" appears on the Techstream or equivalent scan tool display.

a. Connect a vinyl tube to either one of the bleeder plugs.

b. Loosen the bleeder plug.

c. Using Techstream or equivalent scan tool, operate the brake actuator assembly, completely depress the brake pedal and keep it.

➡During this procedure, the pedal will feel heavy, but completely depress it so that the brake fluid comes out from the bleeder plug. Be sure to keep depressing the brake pedal. Do not depress and release the pedal repeatedly. The operation stops automatically after 4 seconds. When performing this procedure continuously, set an interval of at least 20 seconds.

d. When the operation is complete, the brake pedal goes down slightly. This is a normal phenomenon caused when the solenoid opens.

e. Tighten the bleeder plug, and then release the brake pedal.

f. Repeat the procedure until all the air in the fluid is completely bled out.

9. Repeat the above procedures for the other wheels to bleed the air out of the brake line.

10. Bleed the air out of the brake line as usual again when "Step 4: Increase" appears on the Techstream or equivalent scan tool display.

a. Connect the vinyl tube to either one of the bleeder plugs.

b. Depress the brake pedal several times, and then loosen the bleeder plug connected to the vinyl tube with the pedal depressed.

c. When fluid stops coming out, tighten the bleeder plug, then release the brake pedal.

d. Repeat procedure F until all the air in the fluid is completely bled out.

e. Tighten the bleeder plug.

11. Repeat the above procedures for the other wheels to bleed the air out of the brake line.

12. Make sure that the air bleeding is complete by referring to the Techstream or equivalent scan tool display.

FLUID FILL PROCEDURE

See Figures 20 and 21.

➡Move the shift lever to P and apply the parking brake before bleeding the brakes.

➡Add brake fluid to keep the level between the MIN and MAX lines of the reservoir while bleeding the brakes.

➡If brake fluid leaks onto any painted surface, immediately wash it off.

1. Remove the center cowl top ventilator louver.

a. Slide the hood to cowl top seal and disengage the clip as shown in the illustration.

b. Disengage the 5 claws and remove the center cowl top ventilator louver.

2. Fill reservoir with brake fluid.

➡Add brake fluid to keep the level between the MIN and MAX lines of the reservoir while bleeding the brakes.

3. Bleed the brake line.

4. Inspect for brake fluid leak.

5. Inspect the fluid level.

6. Install the center cowl top ventilator louver.

a. Engage the 5 claws to install the center cowl top ventilator louver.

b. Push and engage the clip, and slide and engage the claw of the hood to cowl top seal.

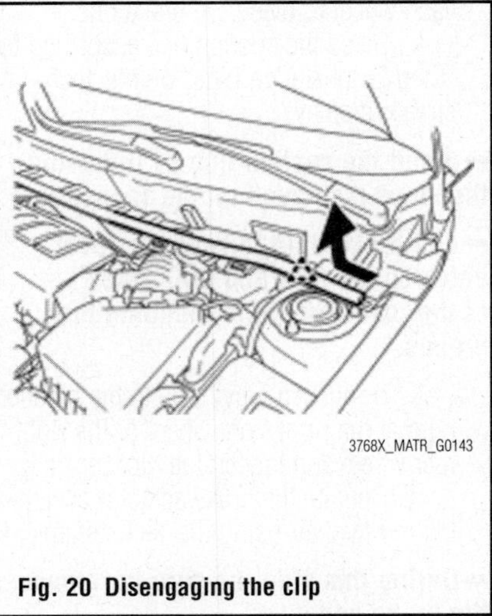

Fig. 20 Disengaging the clip

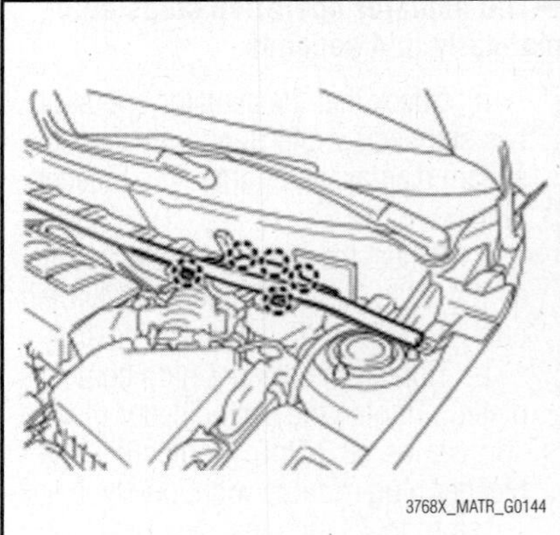

Fig. 21 Disengaging the 5 claws and removing the center cowl top ventilator louver

⁂ CAUTION

Dust and dirt accumulating on brake parts during normal use may contain asbestos fibers from production or aftermarket brake linings. Breathing excessive concentrations of asbestos fibers can cause serious bodily harm. Exercise care when servicing brake parts. Do not sand or grind brake lining unless equipment used is designed to contain the dust residue. Do not clean brake parts with compressed air or by dry brushing. Cleaning should be done by dampening the brake components with a fine mist of water, then wiping the brake components clean with a dampened cloth. Dispose of cloth and all residue containing asbestos fibers in an impermeable container with the appropriate label. Follow practices prescribed by the Occupational Safety and Health Administration (OSHA) and the Environmental Protection Agency (EPA) for the handling, processing, and disposing of dust or debris that may contain asbestos fibers.

BRAKE CALIPER

REMOVAL & INSTALLATION

See Figures 22 through 29.

➡Use the same procedure for the LH side and RH side.

➡The following procedure listed is for the LH side.

1. Remove the front wheel.
2. Drain the brake fluid.

➡If brake fluid leaks onto any painted surface, immediately wash it off.

3. Disconnect the front flexible hose.
   a. Remove the union bolt and gasket, and separate the front flexible hose from the disc brake cylinder assembly.
4. Remove the front disc brake cylinder assembly.
   a. Hold the front disc brake cylinder slide pin, and remove the 2 bolts and front disc brake cylinder assembly.
5. Remove the front disc pad.
   a. Remove the 2 disc brake pads from the disc brake cylinder mounting.
6. Remove the front anti-squeal shim.

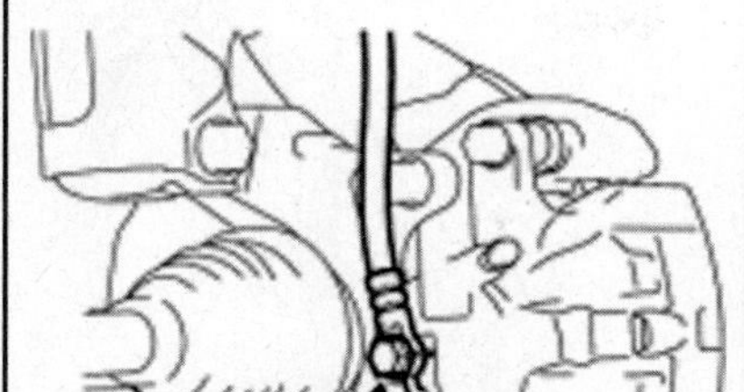

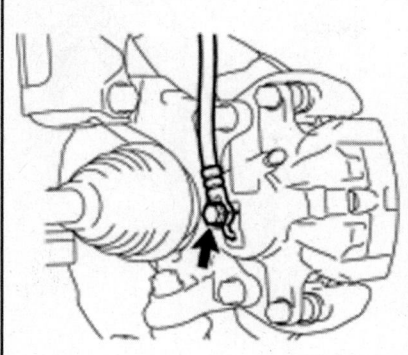

Fig. 22 Disconnecting the front flexible hose

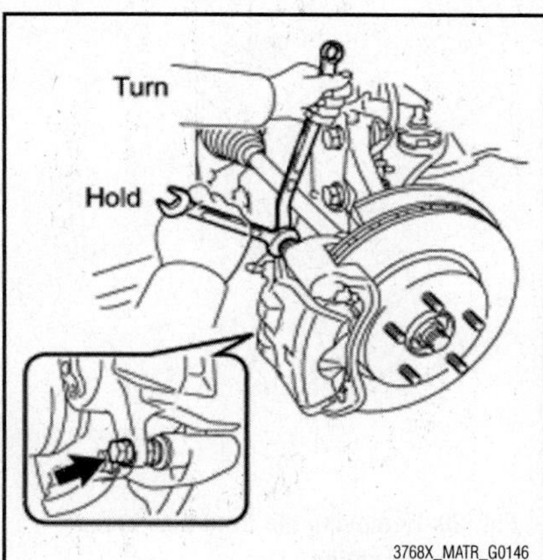

Fig. 23 Removing the front disc brake cylinder assembly

   a. Remove the 2 No. 1 anti-squeal shims and 2 No. 2 anti-squeal shims from the brake pads.
7. Remove the front disc brake pad support plate.
   a. Remove the 2 No. 1 disc brake pad support plates and 2 No. 2 disc brake pad support plates from the front disc brake cylinder mounting.

➡Each front disc brake pad support plate has a different shape. Be sure to put an identification mark on each front disc brake pad support plate so that it can be reinstalled to its original position.

8. Removing the front disc brake cylinder slide pin.
   a. Remove the No. 1 front disc brake cylinder slide pin and No. 2 front disc brake cylinder slide pin from the disc brake cylinder mounting.

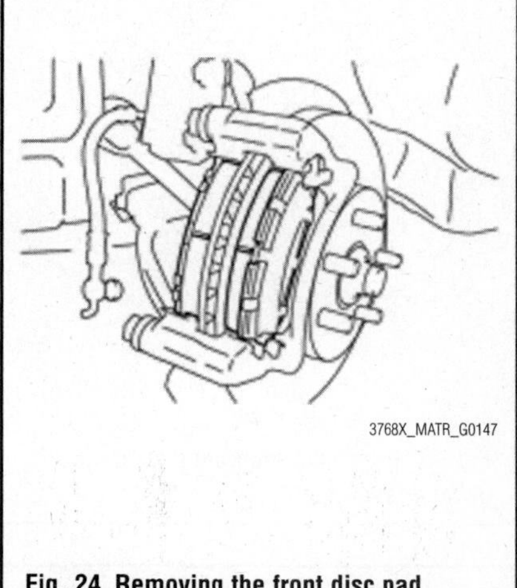

Fig. 24 Removing the front disc pad

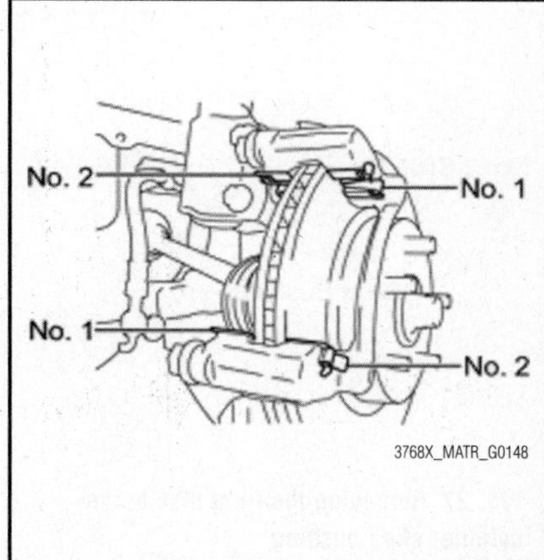

Fig. 25 Removing the front disc brake pad support plate

9. Remove the front disc brake cylinder slide bushing.
   a. Using a screwdriver, remove the front disc brake cylinder slide bushing from the No. 2 front disc brake cylinder slide pin.

➡Do not damage the No. 2 front disc brake cylinder slide pin.

➡Tape the screwdriver tip before use.

10. Remove the front disc brake bushing dust boot.
    a. Remove the 2 front disc brake bushing dust boots from the disc brake cylinder mounting.
11. Remove the front disc brake cylinder mounting.

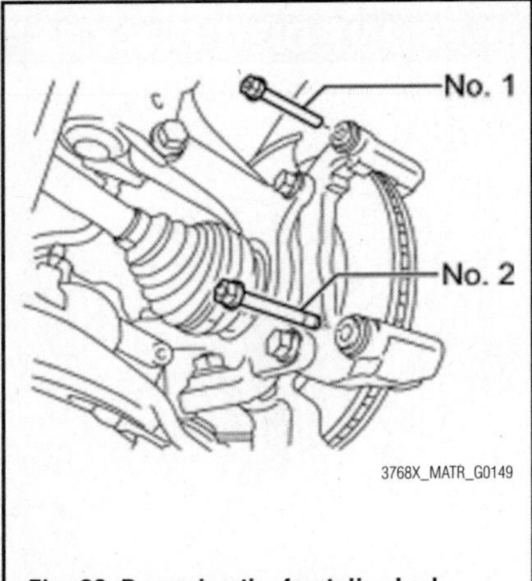

Fig. 26 Removing the front disc brake cylinder slide pin

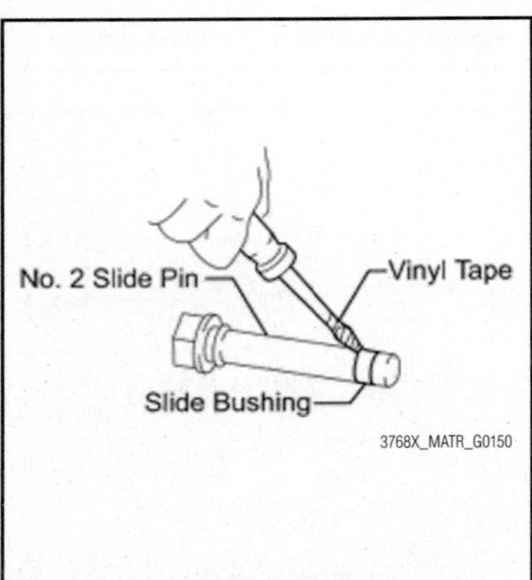

Fig. 27 Removing the front disc brake cylinder slide bushing

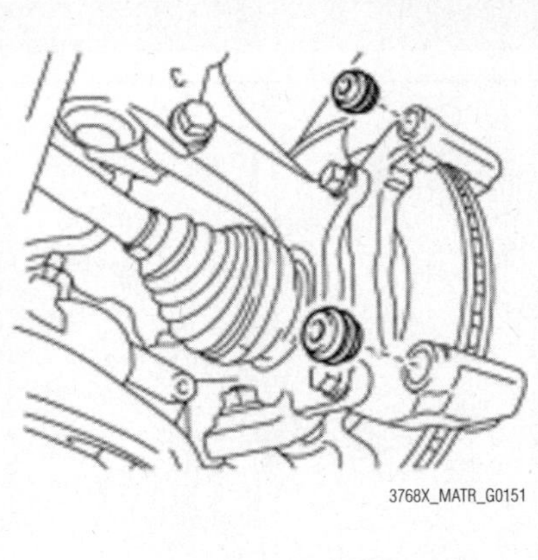

Fig. 28 Removing the front disc brake bushing dust boot

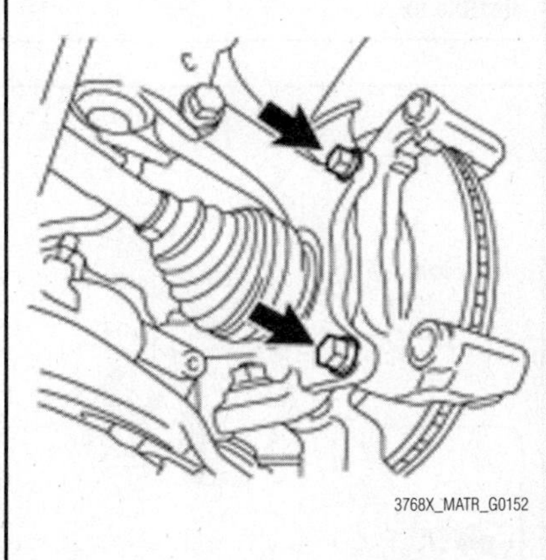

Fig. 29 Removing the front disc brake cylinder mounting

a. Remove the 2 bolts and front disc brake cylinder mounting from the steering knuckle.

To install:

12. Install the front disc.
   a. Align the matchmarks of the disc and axle hub, and install the disc.

➡When replacing the disc with a new one, select the installation position where the front disc has minimal runout.

13. Install the disc brake cylinder mounting to the steering knuckle with the 2 bolts. Tighten to 79 ft. lbs. (107 Nm).
14. Install the front disc brake bushing dust boot.
   a. Apply a light layer of lithium soap base glycol grease to the entire circumference of the 2 new disc brake bushing dust boots.

➡Apply at least 0.01 oz. of lithium soap base glycol grease to each front disc brake bushing dust boot.

   b. Install the 2 front disc brake bushing dust boots to the disc brake cylinder mounting.
15. Install the front disc brake cylinder slide bushing.
   a. Apply a light layer of lithium soap base glycol grease to the contact surface of the No. 2 front disc brake cylinder slide pin.
   b. Install a new front disc brake cylinder slide bushing to the new No. 2 front disc brake cylinder slide pin.
16. Install front disc brake cylinder slide pin.
   a. Apply a light layer of lithium soap base glycol grease to the sliding and sealing surfaces of the No. 1 front disc brake cylinder slide pin and No. 2 front disc brake cylinder slide pin.
   b. Install the No. 1 front disc brake cylinder slide pin and No. 2 front disc brake cylinder slide pin to the cylinder mounting.
17. Install the front disc brake pad support plate.
   a. Install the 2 No. 1 front disc brake pad support plates and 2 No. 2 front disc brake pad support plates to the front disc brake cylinder mounting.

➡Be sure to install each front disc brake pad support plate in the correct position and direction.

18. Install the front anti-squeal shim for (2ZR-FE).
   a. Apply disc brake shim grease to both sides of each No. 1 anti-squeal shim.
   b. Install the No. 1 anti-squeal shim and No. 2 anti-squeal shim to each brake pad.

➡When replacing worn pads, the anti-squeal shims must be replaced together with the pads.

➡Install the shims in the correct positions and directions.

➡Apply disc brake grease to the area that contacts the anti-squeal shims.

➡Disc brake grease can seep out slightly from the area where the anti-squeal shim is installed.

➡Make sure that disc brake grease is not applied onto the lining surface.

19. Install the front anti-squeal shim (2AZ-FE).
   a. Apply disc brake grease to both sides of each No. 2 anti-squeal shim.
   b. Install the pad wear indicator, No. 1 anti-squeal shim and No. 2 anti-squeal shim to each brake pad.

➡When replacing worn pads, the anti-squeal shims must be replaced together with the pads.

➡Install the pad wear indicators and shims in the correct positions and directions.

➡Apply disc brake grease to the area that contacts the anti-squeal shims.

➡Disc brake grease can seep out slightly from the area where the anti-squeal shim is installed.

➡Make sure that disc brake grease is not applied onto the lining surface.

20. Install the front disc brake pad.
21. Install the front disc brake cylinder assembly.
   a. Hold the front disc brake cylinder slide pin, and install the front disc brake cylinder assembly to the front disc brake cylinder mounting with the 2 bolts. Tighten to 25 ft. lbs. (34 Nm).
22. Connect the front flexible hose.
   a. Connect the front flexible hose to the front disc brake cylinder assembly with the union bolt and a new gasket. Tighten to 21 ft. lbs. (29 Nm).

➡Install the flexible hose lock securely into the lock hole in the disc brake cylinder.

23. Fill the reservoir.
24. Bleed the brake line.
25. Inspect for fluid leaks.
26. Inspect the fluid level.
27. Install the front wheel. Tighten to 76 ft. lbs. (103 Nm).

DISC BRAKE PADS

REMOVAL & INSTALLATION

See Figures 30 through 36.

➡Use the same procedure for the LH side and RH side.

➡The following procedure listed is for the LH side.

1. Remove the front wheel.
2. Drain the brake fluid.

➡If brake fluid leaks onto any painted surface, immediately wash it off.

3. Disconnect the front flexible hose.
   a. Remove the union bolt and gasket, and separate the front flexible hose from the disc brake cylinder assembly.
4. Remove the front disc brake cylinder assembly.
   a. Hold the front disc brake cylinder slide pin, and remove the 2 bolts and front disc brake cylinder assembly.
5. Remove the front disc pad.
   a. Remove the 2 disc brake pads from the disc brake cylinder mounting.
6. Remove the front anti-squeal shim.
   a. Remove the 2 No. 1 anti-squeal shims and 2 No. 2 anti-squeal shims from the brake pads.
7. Remove the front disc brake pad support plate.
   a. Remove the 2 No. 1 disc brake pad support plates and 2 No. 2 disc brake pad support plates from the front disc brake cylinder mounting.

➡Each front disc brake pad support plate has a different shape. Be sure to put an identification mark on each front disc brake pad support plate so that it can be reinstalled to its original position.

To install:

8. Install the front disc brake pad support plate.
   a. Install the 2 No. 1 front disc brake pad support plates and 2 No. 2 front disc brake pad support plates to the front disc brake cylinder mounting.

➡Be sure to install each front disc brake pad support plate in the correct position and direction.

9. Install the front anti-squeal shim (1.8L engine).
   a. Apply disc brake shim grease to both sides of each No. 1 anti-squeal shim as shown in the illustration.
   b. Install the No. 1 anti-squeal shim and No. 2 anti-squeal shim to each brake pad.

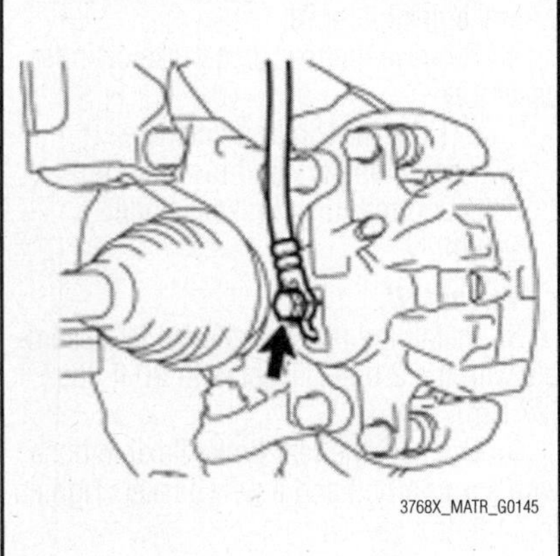

Fig. 30 Disconnecting the front flexible hose

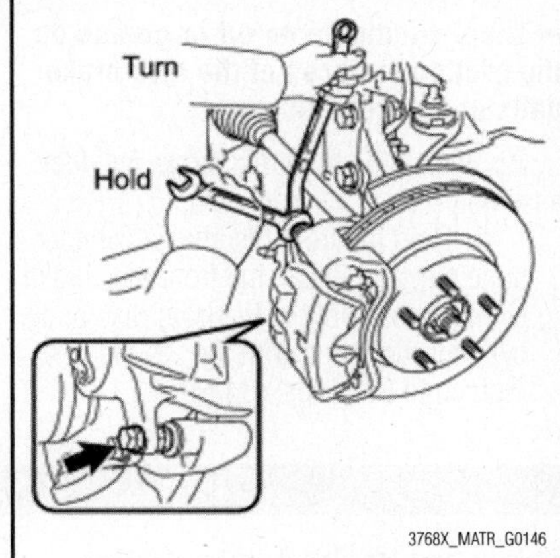

Fig. 31 Removing the front disc brake cylinder assembly

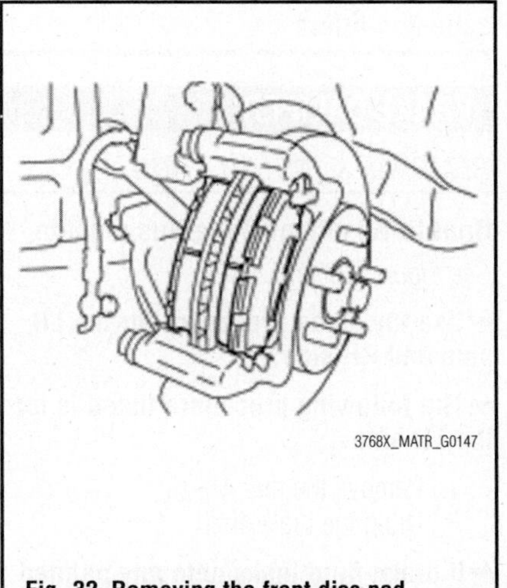

Fig. 32 Removing the front disc pad

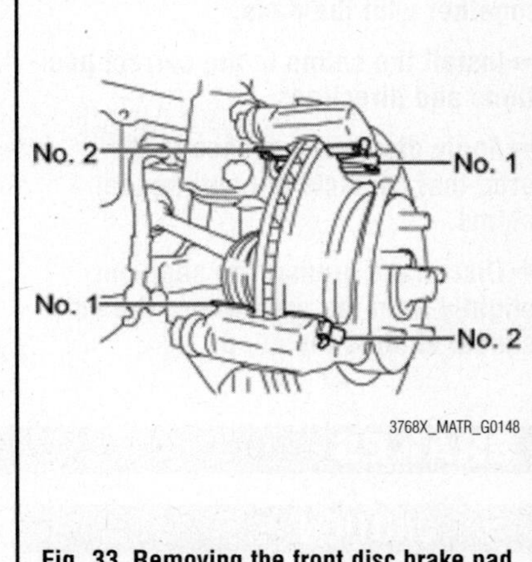

Fig. 33 Removing the front disc brake pad support plate

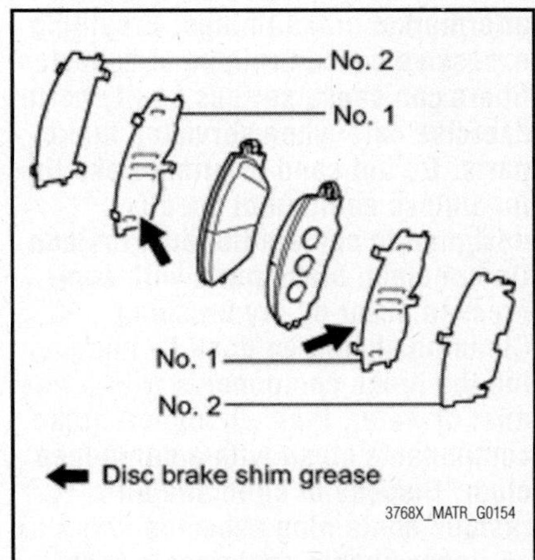

Fig. 34 Applying disc brake shim grease to the anti-squeal shim (1.8L engine)

➡**When replacing worn pads, the anti-squeal shims must be replaced together with the pads.**

➡**Install the shims in the correct positions and directions.**

➡**Apply disc brake grease to the area that contacts the anti-squeal shims.**

➡**Disc brake grease can seep out slightly from the area where the anti-squeal shim is installed.**

➡**Make sure that disc brake grease is not applied onto the lining surface.**

10. Install the front anti-squeal shim (2.4 l engine).
    a. Apply disc brake grease to both sides of each No. 2 anti-squeal shim as shown in the illustration.
    b. Install the pad wear indicator, No. 1 anti-squeal shim and No. 2 anti-squeal shim to each brake pad.

➡**When replacing worn pads, the anti-squeal shims must be replaced together with the pads.**

➡**Install the shims in the correct positions and directions.**

➡**Apply disc brake grease to the area that contacts the anti-squeal shims.**

➡**Disc brake grease can seep out slightly from the area where the anti-squeal shim is installed.**

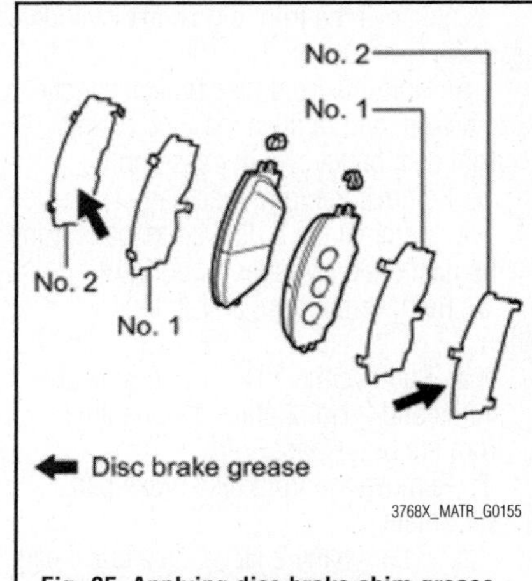

Fig. 35 Applying disc brake shim grease to the anti-squeal shim (2.4L engine)

➡**Make sure that disc brake grease is not applied onto the lining surface.**

11. Install the front disc brake pad.
    a. Install the 2 front disc brake pads to the disc brake cylinder mounting.

➡**There should be no oil or grease on the friction surfaces of the disc brake pads or the front disc.**

12. Install the front disc brake cylinder assembly.
    a. Hold the front disc brake cylinder slide pin, and install the front disc brake cylinder assembly to the front disc brake cylinder mounting with the 2 bolts. Tighten to 25 ft. lbs. (34 Nm).

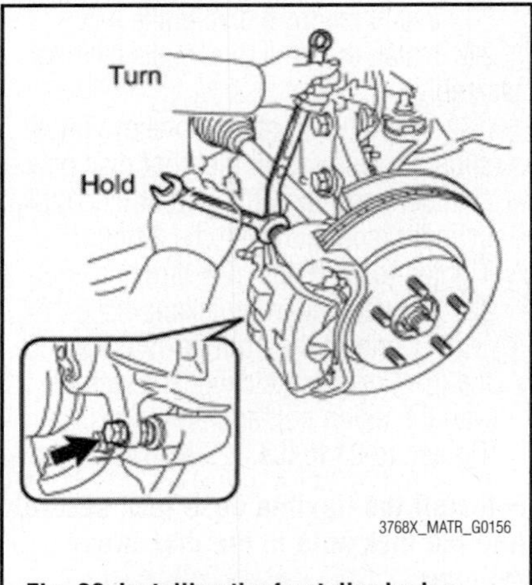

Fig. 36 Installing the front disc brake cylinder assembly

13. Connect the front flexible hose.
    a. Connect the front flexible hose to the front disc brake cylinder assembly with the union bolt and a new gasket. Tighten to 21 ft. lbs. (29 Nm).

➡**Install the flexible hose lock securely into the lock hole in the disc brake cylinder.**

14. Fill the reservoir.
15. Bleed the brake line.
16. Inspect for fluid leaks.
17. Inspect the fluid level.
18. Install the front wheel. Tighten to 76 ft. lbs. (103 Nm).

BRAKES — REAR DISC BRAKES

✲✲ CAUTION

Dust and dirt accumulating on brake parts during normal use may contain asbestos fibers from production or aftermarket brake linings. Breathing excessive concentrations of asbestos fibers can cause serious bodily harm. Exercise care when servicing brake parts. Do not sand or grind brake lining unless equipment used is designed to contain the dust residue. Do not clean brake parts with compressed air or by dry brushing. Cleaning should be done by dampening the brake components with a fine mist of water, then wiping the brake components clean with a dampened cloth. Dispose of cloth and all residue containing asbestos fibers in an impermeable container with the appropriate label. Follow practices prescribed by the Occupational Safety and Health Administration (OSHA) and the Environmental Protection Agency (EPA) for the handling, processing, and disposing of dust or debris that may contain asbestos fibers.

BRAKE CALIPER

REMOVAL & INSTALLATION

Double Wishbone Type Suspension

See Figures 37 and 38.

➡**Use the same procedure for the LH side and RH side.**

➡**The following procedure listed is for the LH side.**

1. Remove the rear wheel.
2. Drain the brake fluid.

➡**If brake fluid leaks onto any painted surface, immediately wash it off.**

3. Disconnect the rear brake flexible hose.
    a. Remove the union bolt and the gasket from the rear disc brake cylinder assembly, and then disconnect the rear brake flexible hose.
4. Remove the rear disc brake cylinder assembly.
    a. Hold the 2 rear disc brake cylinder slide pins and remove the 2 bolts and rear disc brake cylinder assembly.

To install:

5. Install the rear brake cylinder assembly with the 2 bolts. Tighten to 20 ft. lbs. (27 Nm).
6. Connect the rear brake flexible hose with a union bolt and a new gasket. Tighten to 24 ft. lbs. (33 Nm).

Torsion Beam Type Suspension

See Figures 39 through 45.

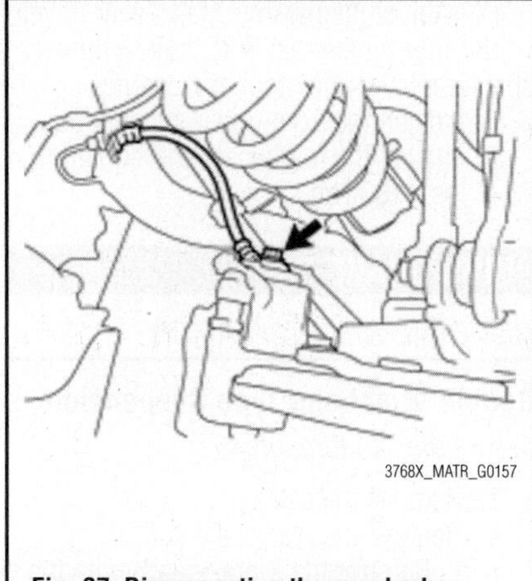

Fig. 37 Disconnecting the rear brake flexible hose

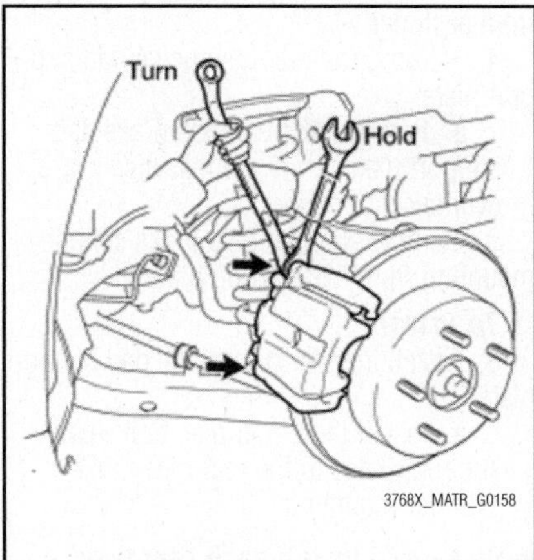

Fig. 38 Removing the rear disc brake cylinder assembly

➡Use the same procedure for the LH side and RH side.

➡The following procedure listed is for the LH side.

1. Remove the rear wheel.
2. Drain the brake fluid.

➡If brake fluid leaks onto any painted surface, immediately wash it off.

3. Remove the front upper console box.
4. Loosen the parking brake cable.
   a. Completely release the parking brake lever.
   b. Loosen the lock nut and the adjusting nut to completely release the parking brake cable.
5. Remove the parking brake lever protector.
6. Disconnect the No. 3 parking brake cable assembly.

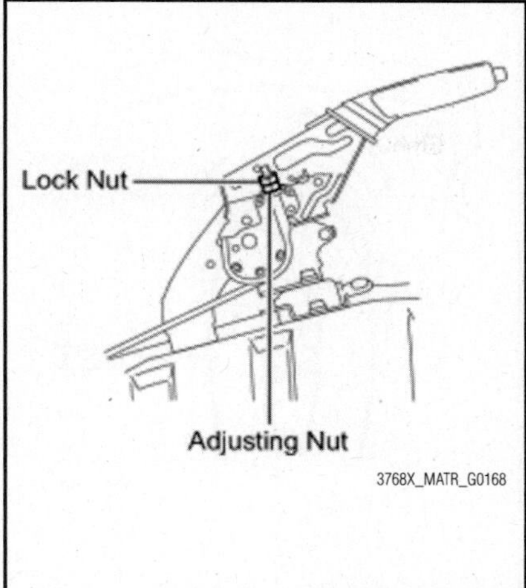

Fig. 39 Loosening the parking brake cable

   a. Disengage the clamp and remove the bolt from the No. 3 parking brake cable assembly.
   b. Separate the tip of the No. 3 parking brake cable assembly from the rear disc brake cylinder assembly.
   c. Separate the No. 3 parking brake cable assembly from the rear disc brake cylinder assembly.

➡Insert an offset wrench (14 mm) at the base of the No. 3 parking brake cable assembly as shown in the illustration to disengage the clip. Pull out the No. 3 parking brake cable assembly from the rear disc brake cylinder assembly.

7. Separate the rear brake flexible hose.

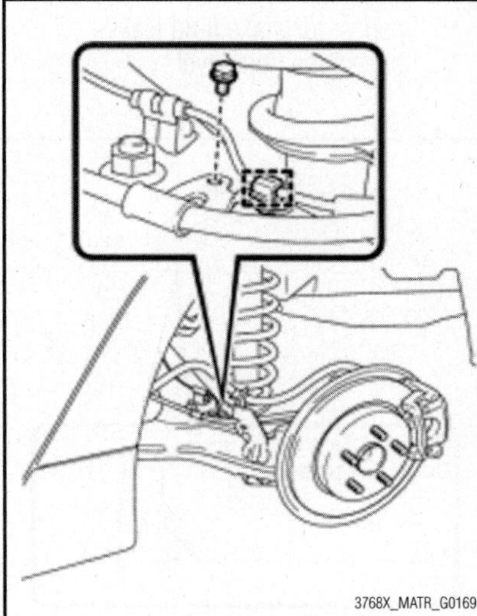

Fig. 40 Disengaging the clamp and removing the bolt from the No. 3 parking brake cable assembly

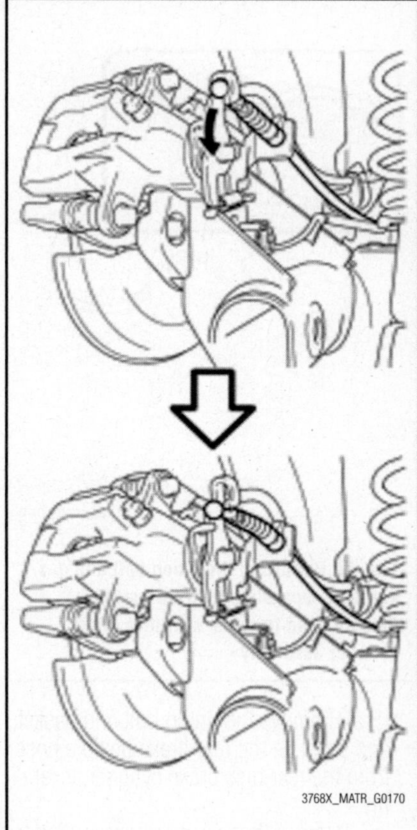

Fig. 41 Separating the tip of the No. 3 parking brake cable assembly from the rear disc brake cylinder assembly

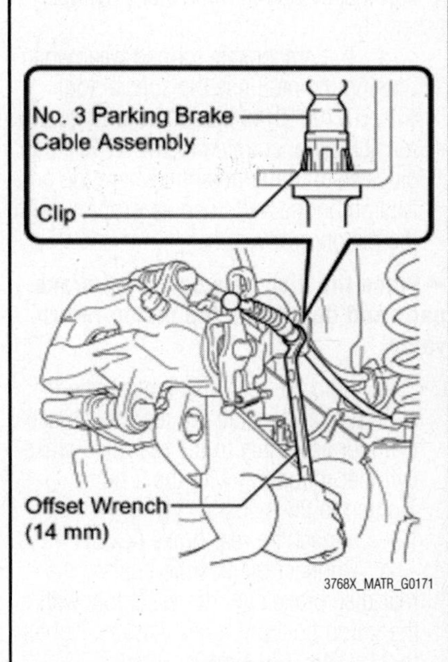

Fig. 42 Separating the No. 3 parking brake cable assembly from the rear disc brake cylinder assembly

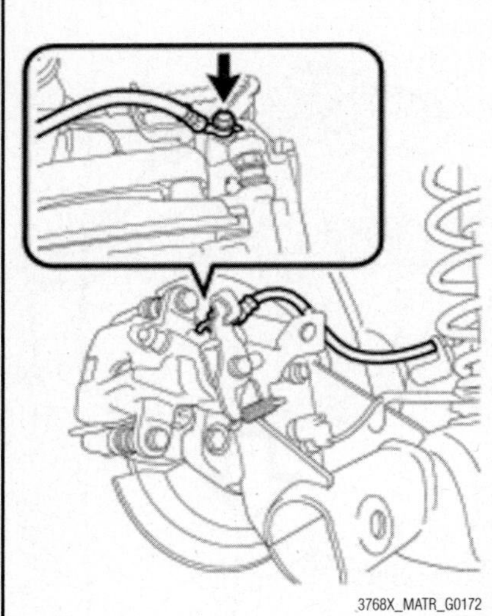

Fig. 43 Removing the union bolt and gasket, and separating the rear brake flexible hose from the rear disc brake cylinder assembly

a. Remove the union bolt and gasket, and separate the rear brake flexible hose from the rear disc brake cylinder assembly.

8. Remove the rear disc brake cylinder assembly.

a. Hold the rear disc brake pad guide pin and remove the 2 bolts and rear disc brake cylinder assembly.

To install:

9. Install the rear disc brake cylinder assembly.

a. To compensate for pad wear when reusing the pad, use the special tool (09960-10010) to push and turn the piston (LH side: counterclockwise, RH side: clockwise) to the position where the protrusion on the pad lines up properly with the piston groove.

➡Place the disc between the 2 brake pads and determine the piston return value.

b. Hold the rear disc brake cylinder slide pin, and install the rear disc brake cylinder assembly to the rear disc brake cylinder mounting with the 2 bolts. Tighten to 26 ft. lbs. (35 Nm).

10. Connect the rear brake flexible hose.

a. Connect the flexible hose to the rear disc brake cylinder assembly with the union bolt and a new gasket. Tighten to 21 ft. lbs. (29 Nm).

➡Install the flexible hose lock securely into the lock hole in the disc brake cylinder.

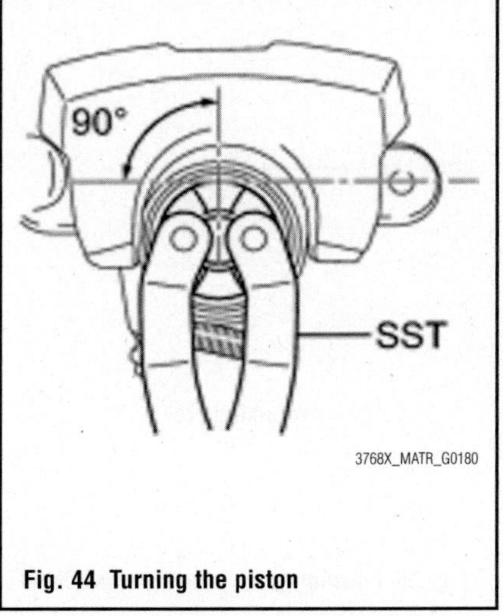

Fig. 44 Turning the piston

11. Connect the No. 3 parking brake cable assembly.

a. Install the No. 3 parking brake cable assembly to the rear disc brake cylinder assembly.

➡Be sure to engage the No. 3 parking brake cable assembly clip onto the rear disc brake cylinder assembly LH as shown in the illustration.

b. Connect the tip of the No. 3 parking brake cable assembly to the rear disc brake cylinder assembly.

c. Install the bolt and tighten to 53 inch lbs. (6 Nm).

d. Engage the clamp to the No. 3 parking brake cable assembly.

12. Install the parking brake lever protector.

13. Fill the reservoir with brake fluid.

14. Bleed the brake line.

15. Inspect for brake fluid leaks.

16. Inspect the fluid level.

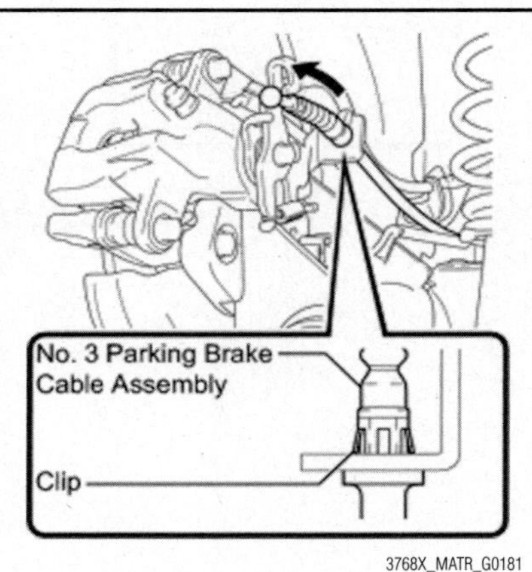

Fig. 45 Connecting the No. 3 parking brake cable assembly

17. Adjust the parking brake lever travel.

18. Inspect the rear disc brake cylinder operation lever and stopper clearance.

19. Install the front upper console box.

20. Install the rear wheel and tighten to 76 ft. lbs. (103 Nm).

DISC BRAKE PADS

REMOVAL & INSTALLATION

Double Wishbone Type Suspension

See Figures 46 through 49.

1. Remove the caliper.

2. Remove the rear brake pad.

a. Remove the 2 brake pads with the rear disc brake anti-squeal shims.

3. Remove the rear disc brake anti-squeal shim and the pad wear indicator from each pad.

4. Remove the rear disc brake pad support plate.

a. Remove the rear disc brake pad support plates from the rear disc brake cylinder mounting.

5. Remove the rear disc brake cylinder mounting from the backing plate.

To install:

6. Install the rear disc brake pad support plate.

a. Install the 2 rear disc brake pad support plates to the rear disc brake cylinder mounting.

➡Be sure to install each rear disc brake pad support plate in the correct position and direction.

7. Install the rear disc brake anti-squeal shim.

a. Apply disc brake grease to the inside of the rear disc brake anti-squeal shims.

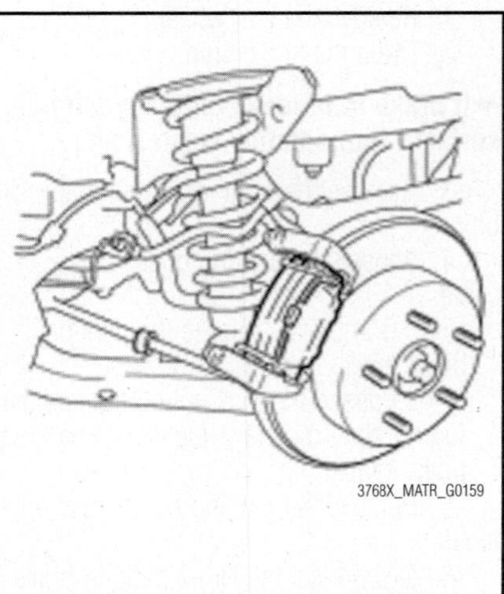

Fig. 46 Removing the 2 brake pads

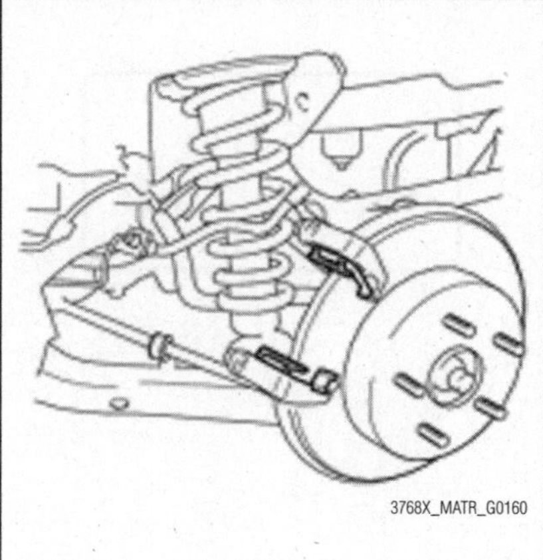

Fig. 47 Removing the rear disc brake pad support plate

➡When replacing worn pads, the rear disc brake anti-squeal shims must be replaced together with the pads.

➡Apply disc brake grease to the area that contacts the rear disc brake anti-squeal shim.

➡Disc brake grease may seep out slightly from the areas where the rear disc brake anti-squeal shims are installed.

➡Make sure that disc brake grease is not applied onto the lining surface.

b. Install the rear disc brake anti-squeal shim and the pad wear indicator to each brake pad.

➡Install the pad wear indicators and rear disc brake anti-squeal shims in the correct positions and directions.

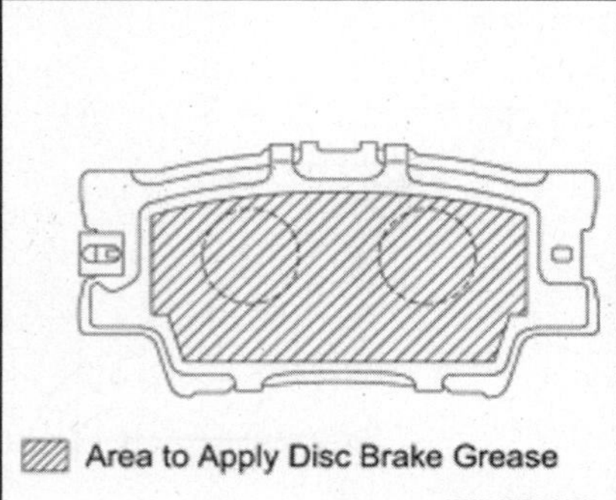

Fig. 48 Applying disc brake grease to the inside of the rear disc brake anti-squeal shims

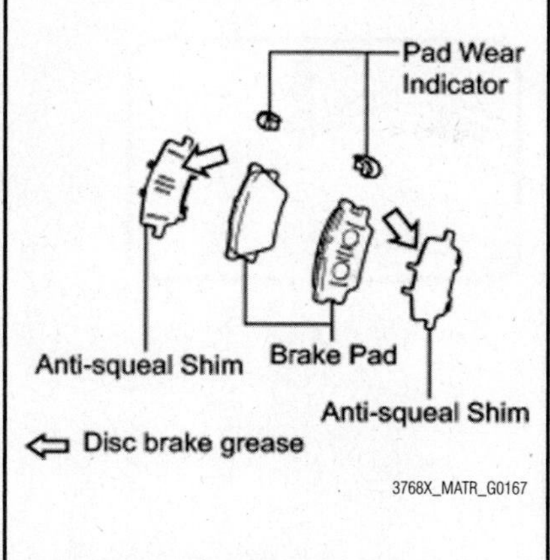

Fig. 49 Installing the rear disc brake anti-squeal shim

8. Install the rear brake pad.
   a. Install the 2 brake pads with rear anti-squeal shims to the rear disc brake cylinder mounting.
9. Install the rear disc brake caliper.

Torsion Beam Type Suspension

See Figures 50 through 59.

➡Use the same procedure for the LH side and RH side.

➡The following procedure listed is for the LH side.

1. Remove the rear wheel.
2. Drain the brake fluid.

➡If brake fluid leaks onto any painted surface, immediately wash it off.

3. Remove the front upper console box.
4. Loosen the parking brake cable.
   a. Completely release the parking brake lever.
   b. Loosen the lock nut and the adjusting nut to completely release the parking brake cable.
5. Remove the parking brake lever protector.
6. Disconnect the No. 3 parking brake cable assembly.
   a. Disengage the clamp and remove the bolt from the No. 3 parking brake cable assembly.
   b. Separate the tip of the No. 3 parking brake cable assembly from the rear disc brake cylinder assembly.
   c. Separate the No. 3 parking brake cable assembly from the rear disc brake cylinder assembly.

➡Insert an offset wrench (14 mm) at the base of the No. 3 parking brake cable assembly as shown in the illustration to disengage the clip. Pull out the No. 3 parking brake cable assembly from the rear disc brake cylinder assembly.

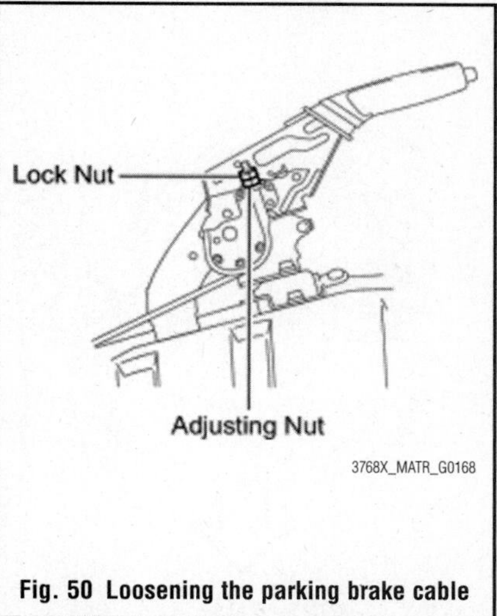

Fig. 50 Loosening the parking brake cable

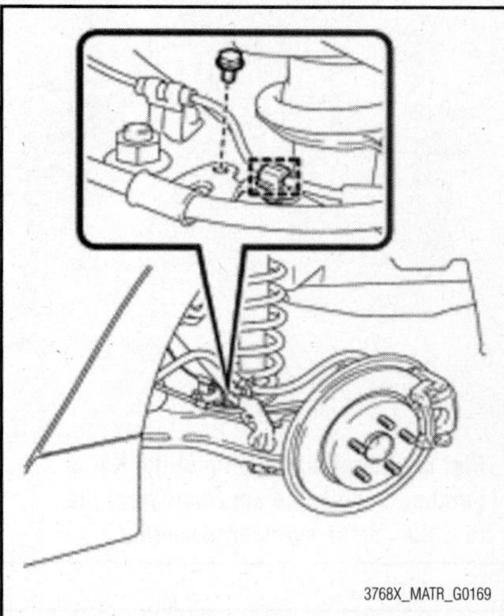

Fig. 51 Disengaging the clamp and removing the bolt from the No. 3 parking brake cable assembly

7. Separate the rear brake flexible hose.
   a. Remove the union bolt and gasket, and separate the rear brake flexible hose from the rear disc brake cylinder assembly.
8. Remove the rear disc brake cylinder assembly.
   a. Hold the rear disc brake pad guide pin and remove the 2 bolts and rear disc brake cylinder assembly.
9. Remove the rear disc brake pad.
   a. Remove the 2 disc brake pads from the rear disc brake cylinder mounting.
10. Remove the rear disc anti-squeal shim.
   a. Remove the 2 anti-squeal shims from each brake pad.

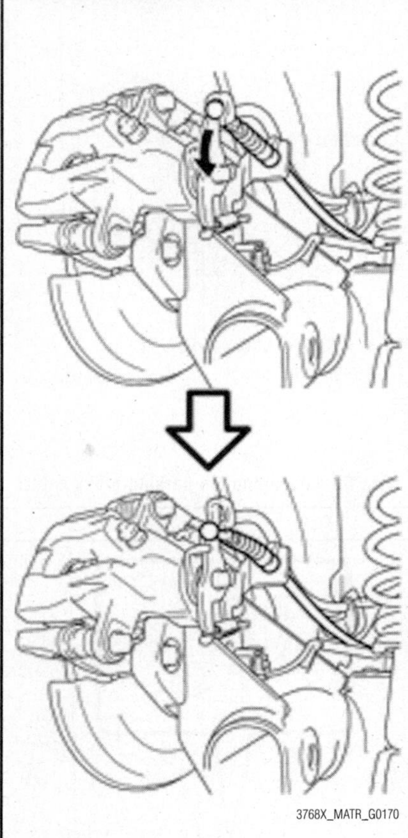

Fig. 52 Separating the tip of the No. 3 parking brake cable assembly from the rear disc brake cylinder assembly

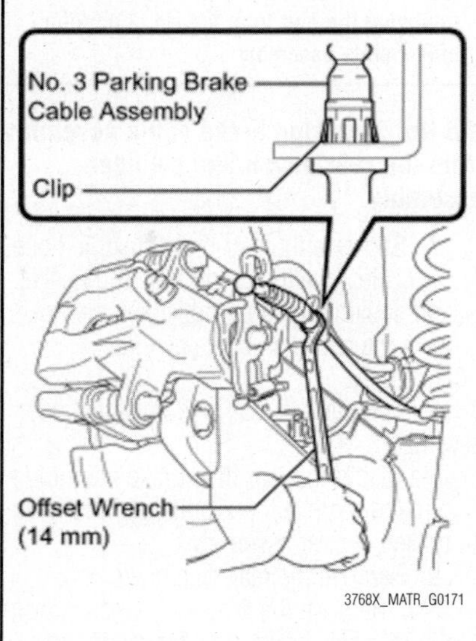

Fig. 53 Separating the No. 3 parking brake cable assembly from the rear disc brake cylinder assembly

Fig. 54 Removing the union bolt and gasket, and separating the rear brake flexible hose from the rear disc brake cylinder assembly

11. Remove the rear disc brake pad support plate.
    a. Remove the rear disc brake pad support plate (upper) and rear disc brake pad support plate (lower) from the disc brake cylinder mounting.

➡Each rear disc brake pad support plate has a different shape. Be sure to put an identification mark on each rear disc brake pad support plate so that it can be installed to its original position.

To install:

12. Install the rear disc brake pad support plate.

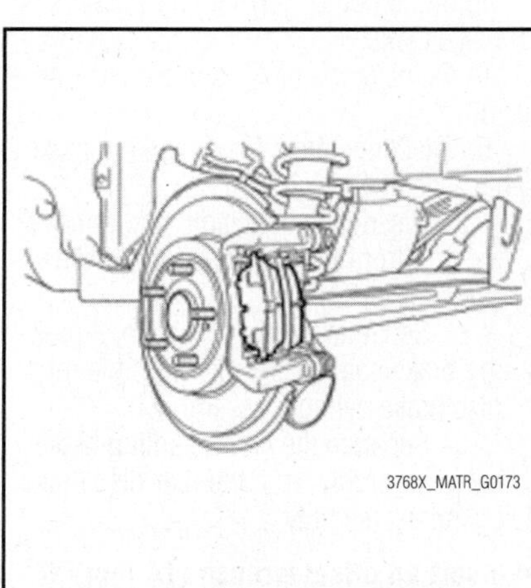

Fig. 55 Removing the 2 disc brake pads from the rear disc brake cylinder mounting

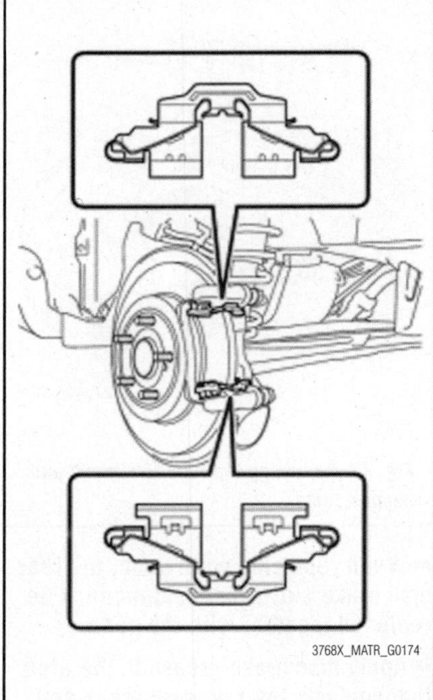

Fig. 56 Removing the rear disc brake pad support plate (upper) and rear disc brake pad support plate (lower) from the disc brake cylinder mounting

    a. Install the 2 rear disc brake pad support plates to the rear disc brake cylinder mounting.

➡Be sure to install each rear disc brake pad support plate in the correct position and direction.

13. Install the rear disc brake anti-squeal shim.
    a. Apply disc brake grease to the 2 No. 1 anti-squeal shims.

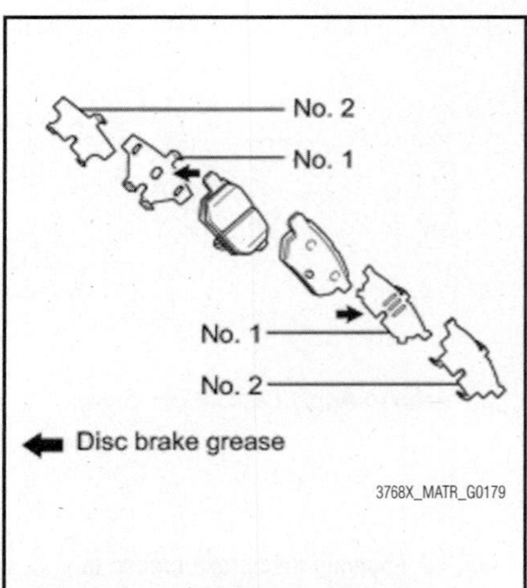

Fig. 57 Apply disc brake grease to the 2 No. 1 anti-squeal shims

b. Install the No. 1 anti-squeal shim and No. 2 anti-squeal shim to each brake pad.

➡When replacing worn pads, the anti-squeal shims must be replaced together with the pads.

➡Apply disc brake grease to the area that contacts the anti-squeal shim.

➡Disc brake grease may seep out slightly from the areas where the anti-squeal shims are installed.

➡Make sure that disc brake grease is not applied onto the lining surface.

14. Install the rear disc brake pad.
   a. Install the 2 rear disc brake pads to the rear disc brake cylinder mounting.

➡There should be no oil or grease on the friction surfaces of the disc brake pads or the rear disc.

15. Install the rear disc brake cylinder assembly.
   a. To compensate for pad wear when reusing the pad, use the special tool (09960-10010) to push and turn the piston (LH side: counterclockwise, RH side: clockwise) to the position where the protrusion on the pad lines up properly with the piston groove.

➡Place the disc between the 2 brake pads and determine the piston return value.

   b. Hold the rear disc brake cylinder slide pin, and install the rear disc brake cylinder assembly to the rear disc brake cylinder mounting with the 2 bolts. Tighten to 26 ft. lbs. (35 Nm).

16. Connect the rear brake flexible hose.
   a. Connect the flexible hose to the rear disc brake cylinder assembly with the union bolt and a new gasket. Tighten to 21 ft. lbs. (29 Nm).

➡Install the flexible hose lock securely into the lock hole in the disc brake cylinder.

17. Connect the No. 3 parking brake cable assembly.
   a. Install the No. 3 parking brake cable assembly to the rear disc brake cylinder assembly.

➡Be sure to engage the No. 3 parking brake cable assembly clip onto the rear disc brake cylinder assembly LH as shown in the illustration.

   b. Connect the tip of the No. 3 parking brake cable assembly to the rear disc brake cylinder assembly.
   c. Install the bolt and tighten to 53 inch lbs. (6 Nm).
   d. Engage the clamp to the No. 3 parking brake cable assembly.

18. Install the parking brake lever protector.
19. Fill the reservoir with brake fluid.
20. Bleed the brake line.
21. Inspect for brake fluid leaks.
22. Inspect the fluid level.
23. Adjust the parking brake lever travel.
24. Inspect the rear disc brake cylinder operation lever and stopper clearance.
25. Install the front upper console box.
26. Install the rear wheel and tighten to 76 ft. lbs. (103 Nm).

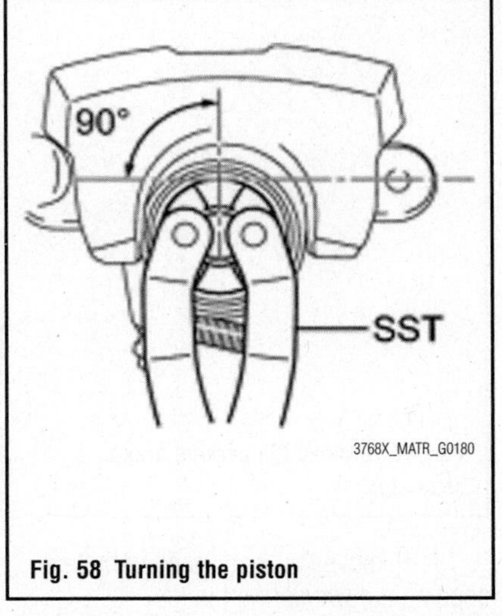

Fig. 58 Turning the piston

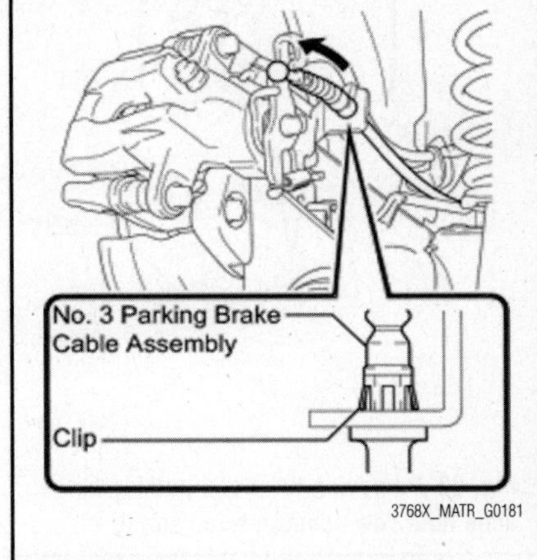

Fig. 59 Connecting the No. 3 parking brake cable assembly

BRAKES — PARKING BRAKE

PARKING BRAKE SHOES

REMOVAL & INSTALLATION

See Figures 60 through 80.

➡Perform the same procedure for the RH side and LH side.

➡The procedures listed below are for the LH side.

1. Remove the rear wheel.
2. Separate the rear disc brake caliper assembly.
3. Remove the rear disc.
4. Remove the No. 3 parking brake shoe return tension spring.
5. Remove the No. 2 parking brake shoe return tension spring.

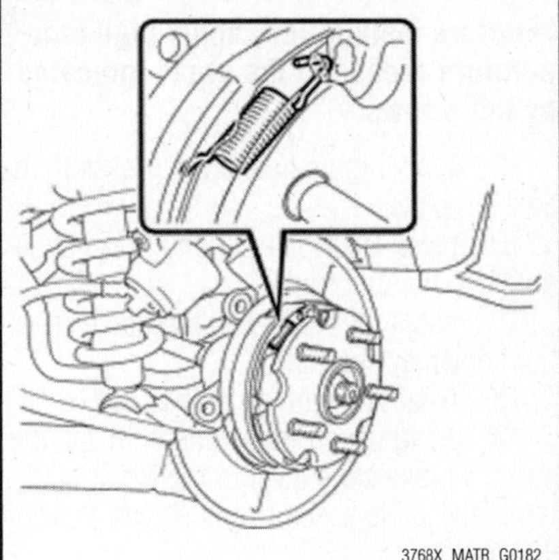

Fig. 60 Removing the No. 3 parking brake shoe return tension spring

6. Remove the No. 1 parking brake shoe hold down compression spring.
   a. Using the special tool (09718-00011), turn the No. 1 parking brake shoe hold down spring pin and remove the No. 1 parking brake shoe hold down compression spring as shown in the illustration.
7. Remove the parking brake shoe strut.
   a. Pull the No. 1 parking brake shoe assembly outward by hand.
   b. Remove the parking brake shoe strut.
8. Separate the No. 2 parking brake shoe hold down compression spring.
   a. Using special tool (09718-00011), turn the No. 2 parking brake shoe hold down spring pin and remove the No. 2

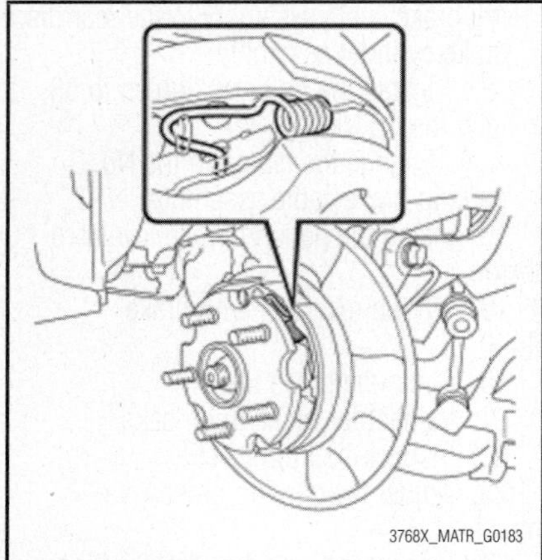
Fig. 61 Removing the No. 2 parking brake shoe return tension spring

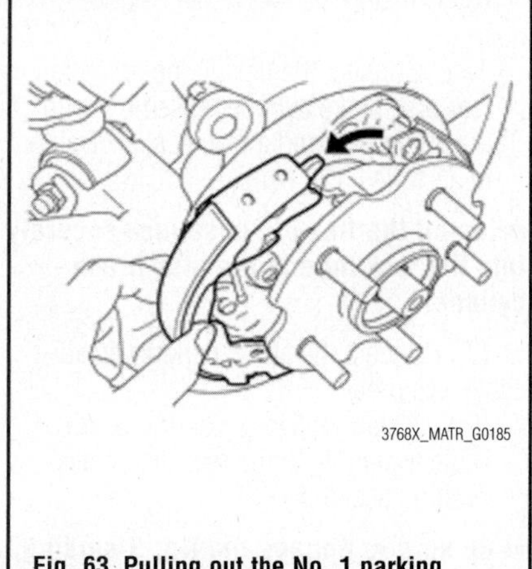
Fig. 63 Pulling out the No. 1 parking brake shoe assembly

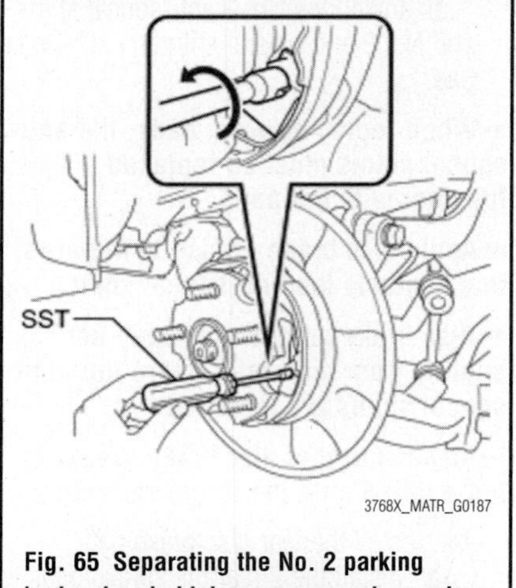

Fig. 65 Separating the No. 2 parking brake shoe hold down compression spring

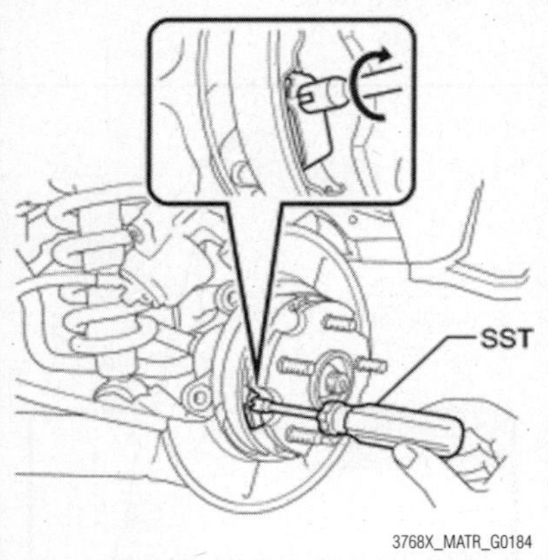

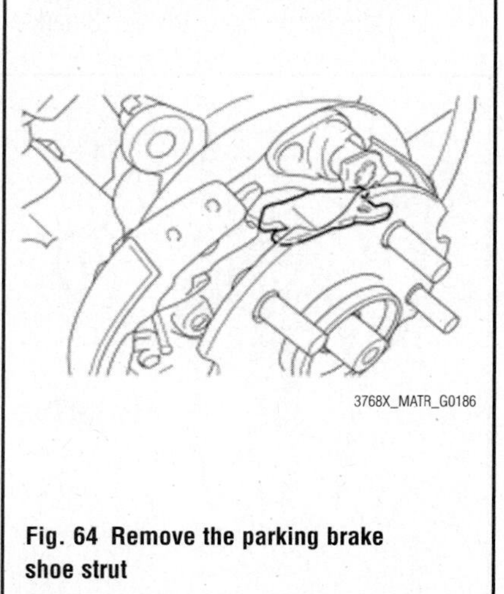

Fig. 62 Removing the No. 1 parking brake shoe hold down compression spring

Fig. 64 Remove the parking brake shoe strut

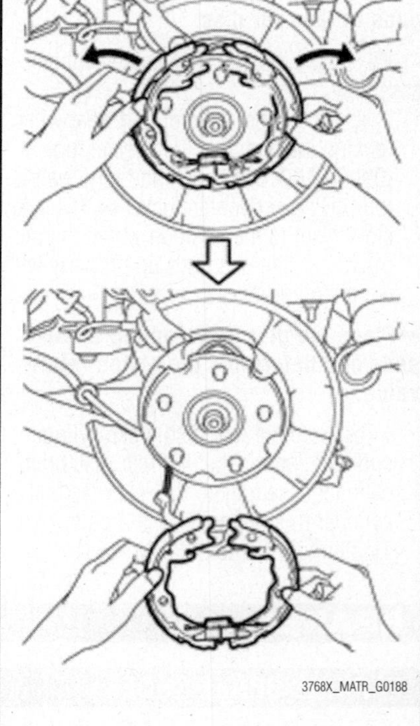
Fig. 66 Separating the parking brake shoe lever

parking brake shoe hold down compression spring as shown in the illustration.

9. Separate the parking brake shoe lever.
   a. Pull the No. 1 and No. 2 parking brake shoe assemblies outward by hand, then pull them out toward the lower of the vehicle.
   b. Separate the parking brake shoe lever from the No. 2 parking brake shoe assembly.
10. Remove the parking brake shoe adjusting screw set.
11. Remove the No. 1 parking brake shoe return tension spring.
12. Remove the parking brake shoe lever.
   a. Using needle-nose pliers, remove the parking brake shoe lever from the No. 3 parking brake cable assembly as shown in the illustration.

➡Be careful not to damage the No. 3 parking brake cable assembly.

13. Remove the No. 1 parking brake shoe hold down spring pin.
14. Remove the No. 2 parking brake shoe hold down spring pin.

To install:

➡Before installation, apply high temperature grease to the areas indicated by the arrows.

15. Apply high temperature grease to the backing plate.
16. Install the No. 1 parking brake shoe hold down spring pin.
17. Install the No. 2 parking brake shoe hold down spring pin.
18. Install the parking brake shoe lever.
19. Using needle-nose pliers, install the parking brake shoe lever to the No. 3 parking brake cable assembly as shown in the illustration.

➡Be careful not to damage the No. 3 parking brake cable assembly.

20. Install the No. 1 parking brake shoe return tension spring.
21. Install the parking brake shoe adjusting screw set.
   a. Apply high temperature grease to the parking brake shoe adjusting screw set as shown in the illustration.
   b. Install the parking brake shoe adjusting screw set.
22. Install the parking brake shoe lever.
   a. Install the parking brake shoe lever to the No. 2 parking brake shoe assembly.

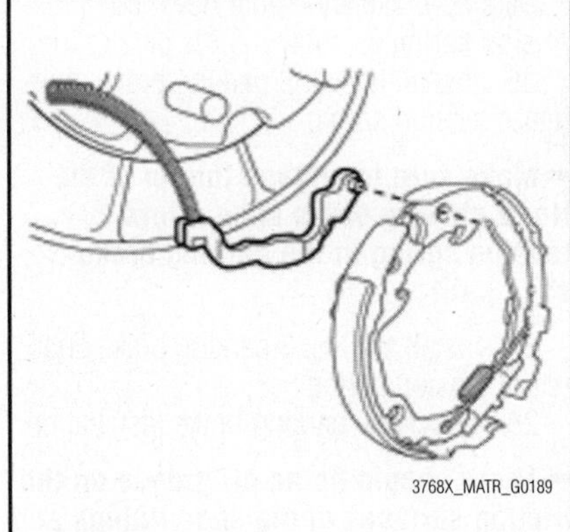

Fig. 67 Separating the parking brake shoe lever from the No. 2 parking brake shoe assembly

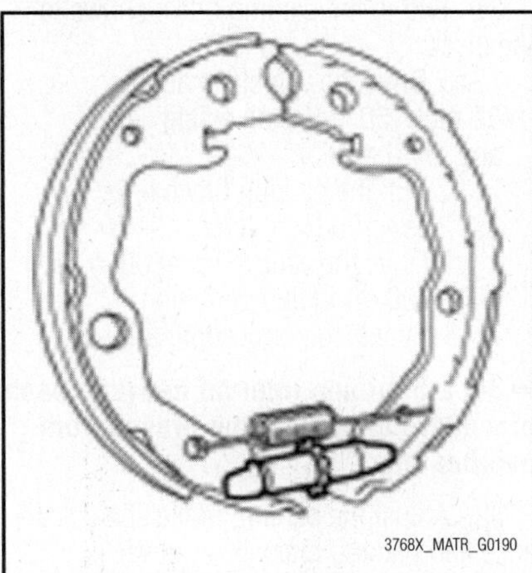

Fig. 68 Removing the parking brake shoe adjusting screw set

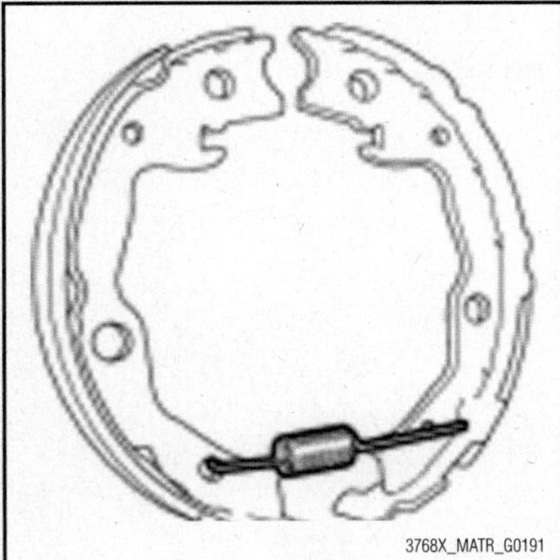

Fig. 69 Removing the No. 1 parking brake shoe return tension spring

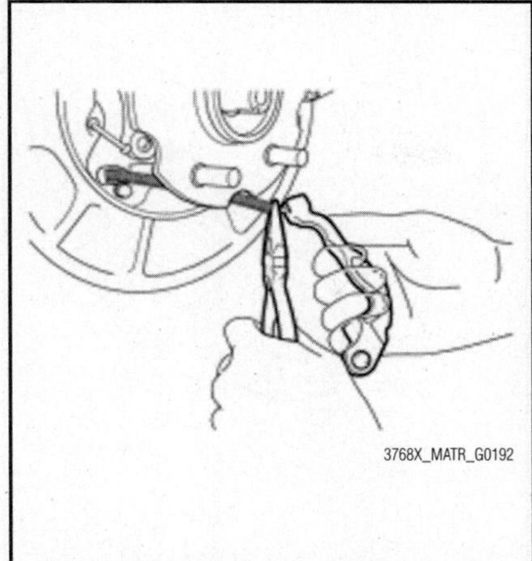

Fig. 70 Removing the parking brake show lever

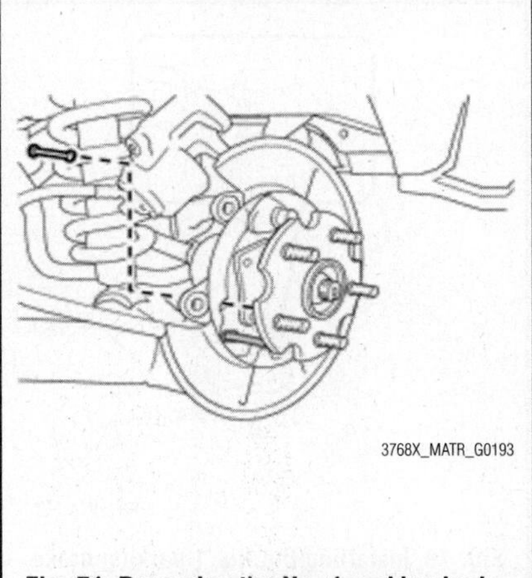

Fig. 71 Removing the No. 1 parking brake shoe hold down spring pin

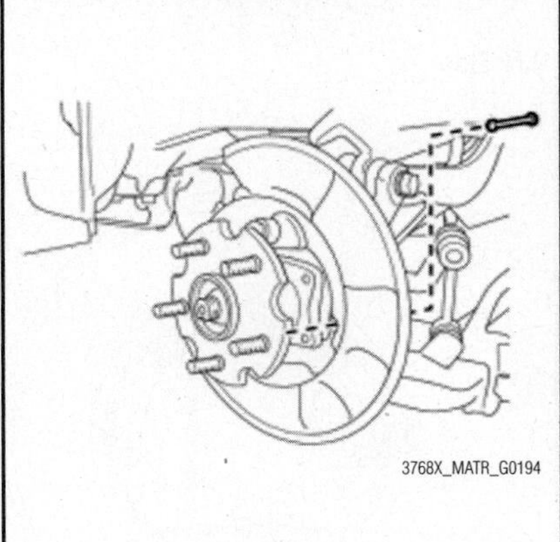

Fig. 72 Removing the No. 2 parking brake shoe hold down spring pin

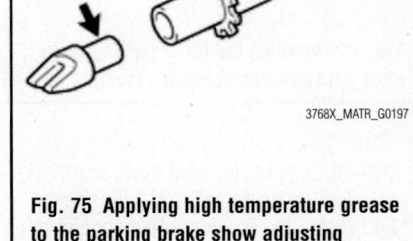

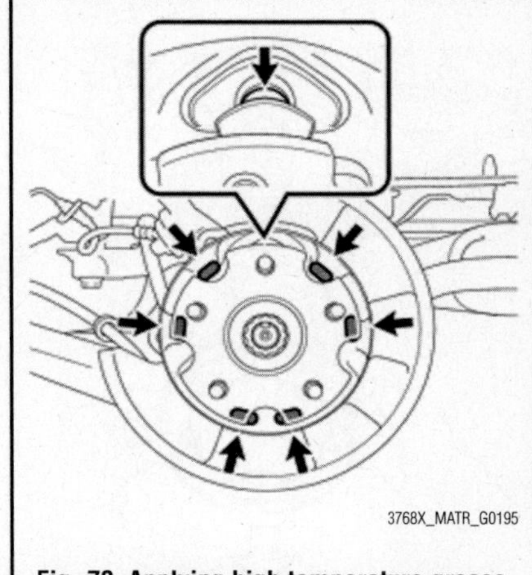

Fig. 73 Applying high temperature grease to the backing plate

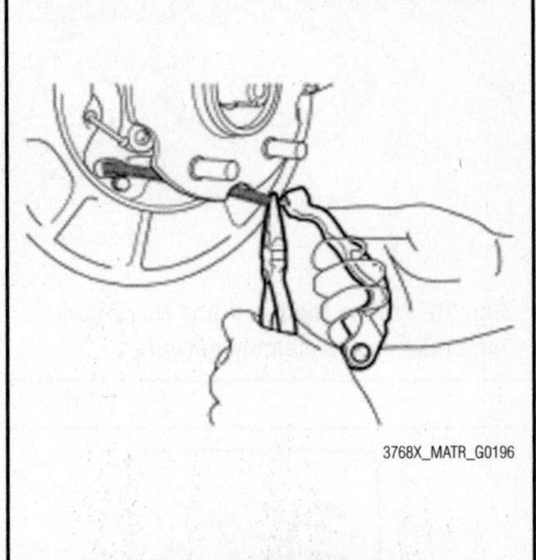

Fig. 74 Installing the parking brake shoe lever

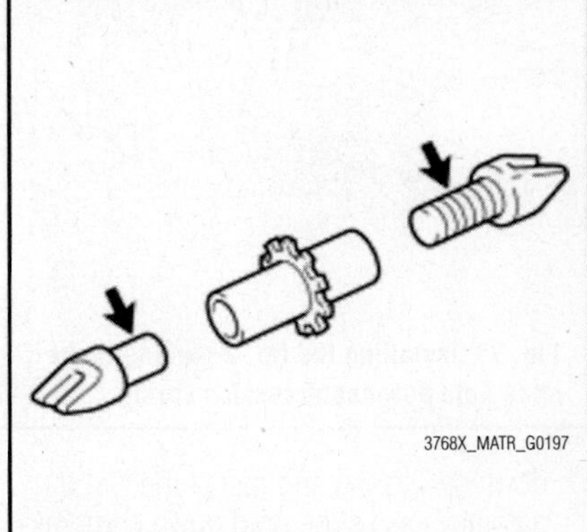

Fig. 75 Applying high temperature grease to the parking brake show adjusting screw set

b. Pull the No. 1 and No. 2 parking brake shoe assemblies outward by hand and align them with the backing plates as shown in the illustration.

➡Do not allow the parking brake shoe lever to come off from the No. 2 parking brake shoe assembly.

23. Install the No. 2 parking brake shoe hold down compression spring.

a. Using special tool (09718-00011), turn the No. 2 parking brake shoe hold

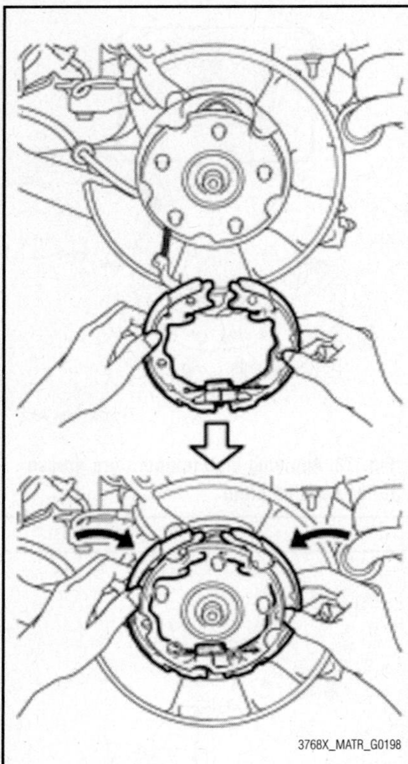

Fig. 76 Pulling the No. 1 and No. 2 parking brake shoe assembly outward

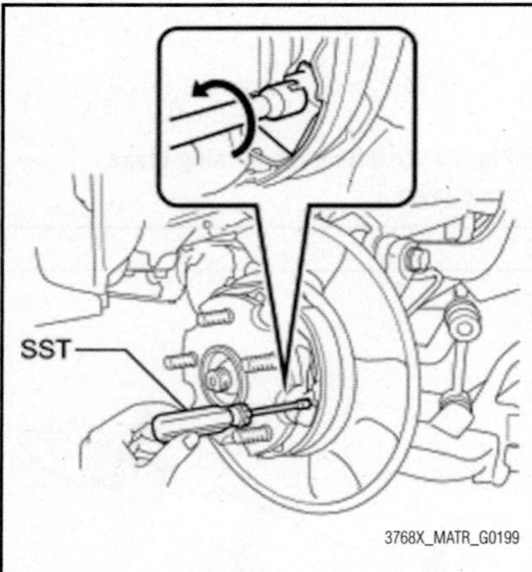

Fig. 77 Installing the No. 2 parking brake shoe hold down compression spring

down spring pin and install the No. 2 parking brake shoe hold down compression spring as shown in the illustration.

24. Install the parking brake shoe strut.
   a. Position the No. 1 parking brake shoe assembly correctly.
25. Install the No. 1 parking brake shoe hold down compression spring.
   a. Using special tool (09718-00011), turn the No. 1 parking brake shoe hold down spring pin and install the No. 1

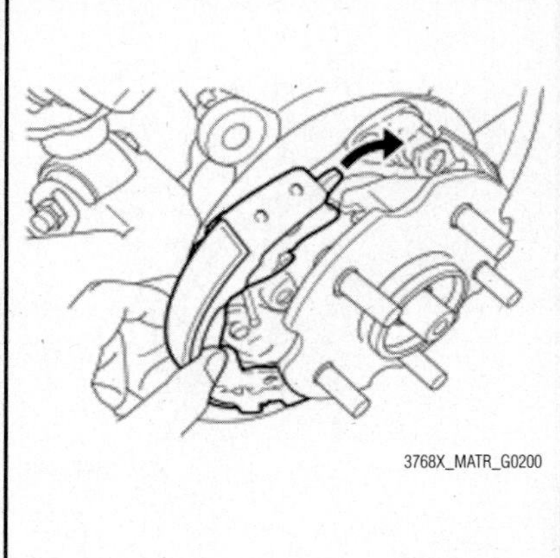

Fig. 78 Positioning the No. 1 parking brake show assembly

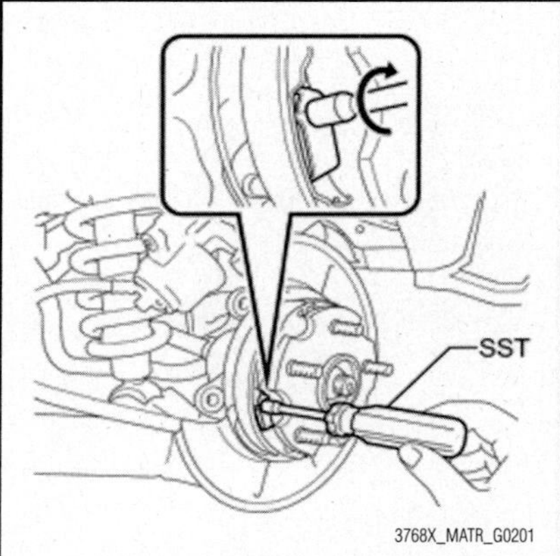

Fig. 79 Installing the No. 1 parking brake shoe hold down compression spring

parking brake shoe hold down compression spring as shown in the illustration.

26. Install the No. 2 parking brake shoe return tension spring.

➡Make sure to engage the tip of the No. 2 parking brake shoe return tension spring in the parking brake shoe strut.

27. Install the No. 3 parking brake shoe return tension spring.
28. Check the parking brake installation.

➡There should be no oil grease on the friction surfaces of the shoe linings and disc.

29. Install the rear disc.
30. Install the rear disc brake caliper assembly.
31. Bed in the parking brake shows to the discs.
   a. Drive the vehicle at about 31 mph (50 km/h) on a safe, level and dry road.
   b. Pull the parking brake lever with 33.7 lbs. (150 N) of force.
   c. Drive the vehicle for about 0.25 mile (400 m) in this condition.
   d. Repeat this procedure 3 times.

➡Set a 5 minute interval between each procedure to prevent the brakes from overheating.

32. Adjust the parking brake shoe clearance and parking brake lever travel.
33. Install the rear wheel and tighten to 76 ft. lbs. (103 Nm).

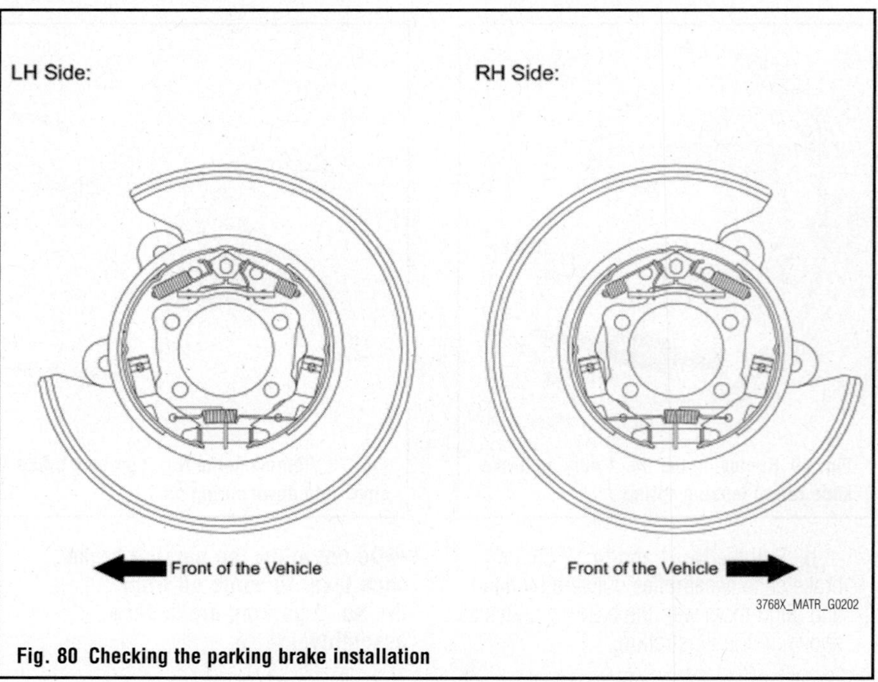

Fig. 80 Checking the parking brake installation

GENERAL INFORMATION

✱✱ CAUTION

These vehicles are equipped with an air bag system. The system must be disarmed before performing service on, or around, system components, the steering column, instrument panel components, wiring and sensors. Failure to follow the safety precautions and the disarming procedure could result in accidental air bag deployment, possible injury and unnecessary system repairs.

SERVICE PRECAUTIONS

Disconnect and isolate the battery negative cable before beginning any airbag system component diagnosis, testing, removal, or installation procedures. Allow system capacitor to discharge for two minutes before beginning any component service. This will disable the airbag system. Failure to disable the airbag system may result in accidental airbag deployment, personal injury, or death.

Do not place an intact undeployed airbag face down on a solid surface. The airbag will propel into the air if accidentally deployed and may result in personal injury or death.

When carrying or handling an undeployed airbag, the trim side (face) of the airbag should be pointing away from the body to minimize possibility of injury if accidental deployment occurs. Failure to do this may result in personal injury or death.

Replace airbag system components with OEMreplacement parts. Substitute parts may appear interchangeable, but internal differences may result in inferior occupant protection. Failure to do so may result in occupant personal injury or death.

Wear safety glasses, rubber gloves, and long sleeved clothing when cleaning powder residue from vehicle after an airbag deployment. Powder residue emitted from a deployed airbag can cause skin irritation. Flush affected area with cool water if irritation is experienced. If nasal or throat irritation is experienced, exit the vehicle for fresh air until the irritation ceases. If irritation continues, see a physician.

Do not use a replacement airbag that is not in the original packaging. This may result in improper deployment, personal injury, or death.

The factory installed fasteners, screws and bolts used to fasten airbag components have a special coating and are specifically designed for the airbag system. Do not use substitute fasteners. Use only original equipment fasteners listed in the parts catalog when fastener replacement is required.

During, and following, any child restraint anchor service, due to impact event or vehicle repair, carefully inspect all mounting hardware, tether straps, and anchors for proper installation, operation, or damage. If a child restraint anchor is found damaged in any way, the anchor must be replaced. Failure to do this may result in personal injury or death.

Deployed and non-deployed airbags may or may not have live pyrotechnic material within the airbag inflator.

Do not dispose of driver/passenger/ curtain airbags or seat belt tensioners unless you are sure of complete deployment. Refer to the Hazardous Substance Control System for proper disposal.

Dispose of deployed airbags and tensioners consistent with state, provincial, local, and federal regulations.

After any airbag component testing or service, do not connect the battery negative cable. Personal injury or death may result if the system test is not performed first.

If the vehicle is equipped with the Occupant Classification System (OCS), do not connect the battery negative cable before performing the OCS Verification Test using the scan tool and the appropriate diagnostic information. Personal injury or death may result if the system test is not performed properly.

Never replace both the Occupant Restraint Controller (ORC) and the Occupant Classification Module (OCM) at the same time. If both require replacement, replace one, then perform the Airbag System test before replacing the other.

Both the ORC and the OCM store Occupant Classification System (OCS) calibration data, which they transfer to one another when one of them is replaced. If both are replaced at the same time, an irreversible fault will be set in both modules and the OCS may malfunction and cause personal injury or death.

If equipped with OCS, the Seat Weight Sensor is a sensitive, calibrated unit and must be handled carefully. Do not drop or handle roughly. If dropped or damaged, replace with another sensor. Failure to do so may result in occupant injury or death.

If equipped with OCS, the front passenger seat must be handled carefully as well. When removing the seat, be careful when setting on floor not to drop. If dropped, the sensor may be inoperative, could result in occupant injury, or possibly death.

If equipped with OCS, when the passenger front seat is on the floor, no one should sit in the front passenger seat. This uneven force may damage the sensing ability of the seat weight sensors. If sat on and damaged, the sensor may be inoperative, could result in occupant injury, or possibly death.

DISARMING THE SYSTEM

To avoid personal injury when working on vehicles equipped with an air bag, the negative battery cable must be disconnected and at least 90 seconds must elapse before working on the system. Failure to do so may result in deployment of the air bag.

ARMING THE SYSTEM

After vehicle service is completed, reattach the battery cables (positive cable first!) to rearm the air bag system.

CLOCKSPRING CENTERING

1. Check that the ignition switch is off.
2. Check that the battery negative (-) terminal is disconnected.
3. Rotate the spiral cable counterclockwise slowly by hand until it feels firm.

➡Do not turn the spiral cable by the airbag wire harness.

4. Rotate the spiral cable clockwise approximately 2.5 turns to align the marks.

➡Do not turn the spiral cable by the airbag wire harness. The spiral cable will rotate approximately 2.5 turns to both the left and right from the center.

DRIVE TRAIN

AUTOMATIC TRANSAXLE FLUID

DRAIN AND REFILL

1. Remove the drain plug and gasket to drain the ATF.
2. Install a new gasket and the drain plug. Tighten to 36 ft. lbs. (49 Nm).
3. Add the correct fluid amount via the dipstick tube.
4. Check the fluid level:
   a. Drive the vehicle so that the engine and transaxle are at normal operating temperature.
   b. Park the vehicle on a level surface and set the parking brake.
   c. With the engine idling and the brake pedal depressed, move the shift lever to all positions from P to S. Then return it to P.
   d. Pull out the dipstick and wipe it clean.
   e. Push the dipstick back fully into the pipe.
   f. Pull the dipstick out again and check that the fluid level is within the HOT range. If the fluid level is below the HOT range, add new fluid and recheck the fluid level. If the fluid level exceeds the HOT range, drain the fluid once, add a proper amount of new fluid and recheck the fluid level.

FILTER REPLACEMENT

See Figures 81 through 85.

1. Remove the LH engine under cover.
2. Drain the ATF fluid.
3. For the U250E transaxles, remove the 18 bolts, oil pan and gasket.
4. For the U341E transaxles, remove the 19 bolts, oil pan and oil pan gasket.
5. Remove the 2 magnets from the oil pan.
6. Examine particles in the pan:
   a. Collect any steel chips with the removed magnets. Look carefully at the chips and particles in the pan and on the magnets to see the type of wear which might be found in the transaxle.
   b. Result:
      - Steel (magnetic): bearing, gear and plate wear
      - Brass (non-magnetic): bushing wear
7. Remove the 3 bolts, oil strainer and O-ring.

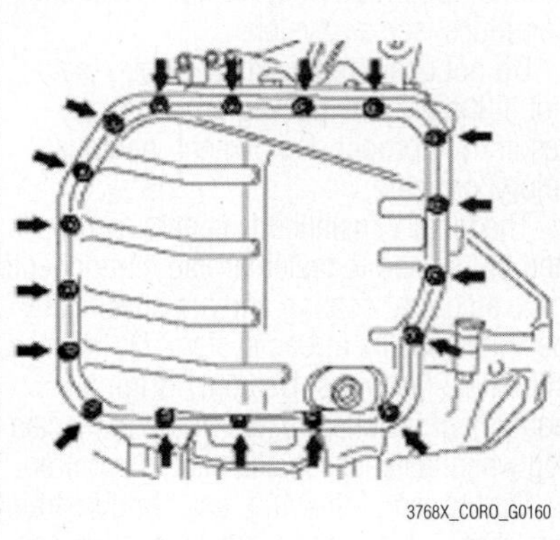

Fig. 81 Removing the 18 bolts—U250E transaxles

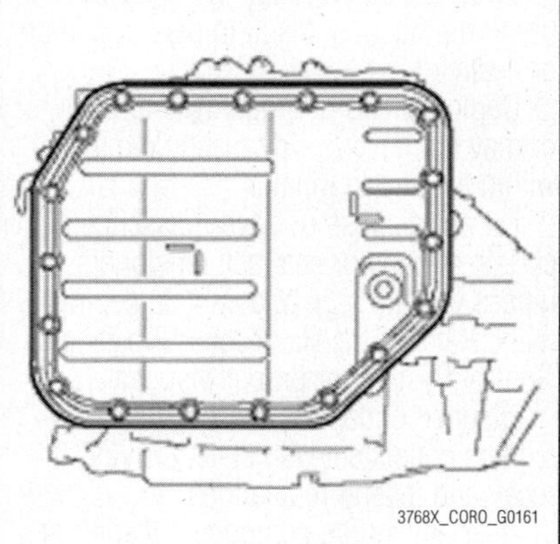

Fig. 82 Removing the 19 bolts—U341E transaxles

To install:

8. Coat a new O-ring with ATF, and install it to the oil strainer.

➡Install the O-ring carefully to avoid twists or pinches. Apply sufficient ATF to the O-ring prior to installation.

➡Apply ATF to the bolts prior to installation.

9. Install the oil strainer to the valve body with the 3 bolts.
10. Tighten the bolts to 8 ft. lbs. (11 Nm).
11. Install the 2 magnets to the oil pan.
12. Install a new oil pan gasket and the oil pan to the transaxle case.
13. Tighten the bolts to 69 inch lbs. (8 Nm).

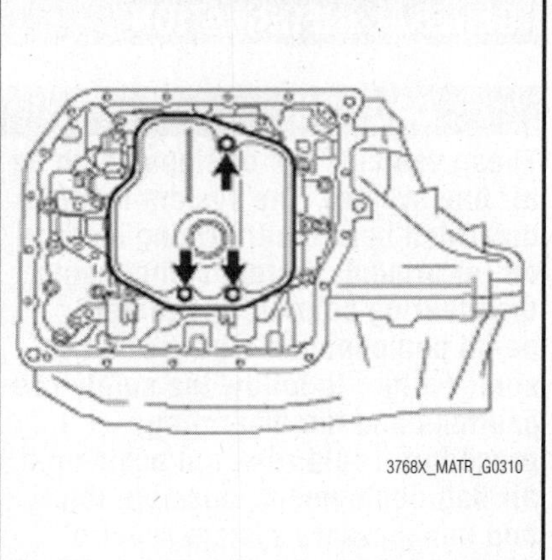

Fig. 83 Removing the 3 bolts and oil strainer—U140F transaxles

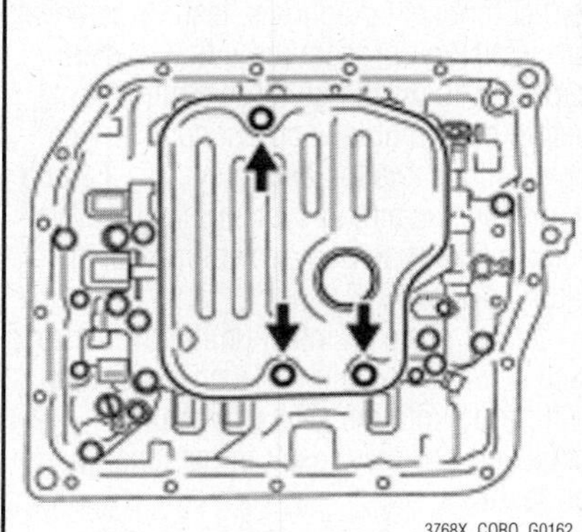

Fig. 84 Removing the 3 bolts and oil strainer—U250E transaxles

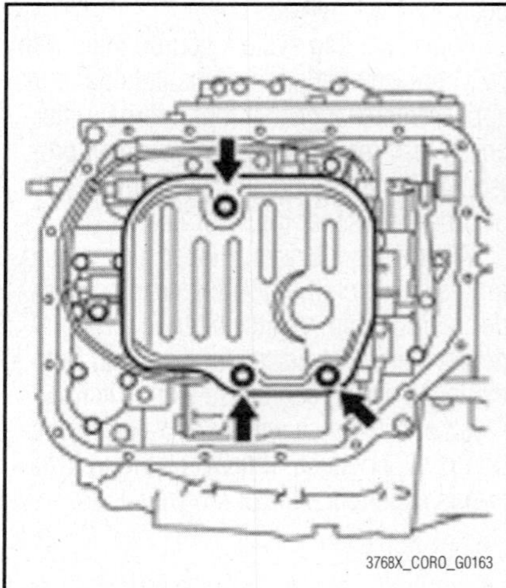

Fig. 85 Removing the 3 bolts and oil strainer—U341E transaxles

MANUAL TRANSAXLE FLUID

DRAIN AND REFILL

1. Remove the filler plug and the gasket.
2. Remove the drain plug and gasket, and then drain the manual transaxle oil in to a suitable container.
3. Install a new gasket to the drain plug. Add manual transaxle oil.
4. Install the transmission filler plug and a new gasket. Tighten to 29 ft. lbs. (39 Nm).

Inspection

See Figure 86.

1. Inspect transaxle oil:
   a. Stop the vehicle in a level place.
   b. Remove the transmission filler plug and the gasket.
   c. Check that the oil surface is within 0.197 inch (5 mm) of the bottom of the transmission filler plug opening.

➡Excessively large or small amounts of oil may cause problems. After replacing the oil, drive the vehicle and check the oil level again.

   d. Check for oil leakage when the oil level is low.

CLUTCH

BLEEDING

➡If any maintenance on the clutch system was performed or the system is suspected of containing air, bleed the system. Use care; brake fluid will remove the paint from any surface. If the brake fluid spills onto any painted surface, wash it off immediately with soap and water.

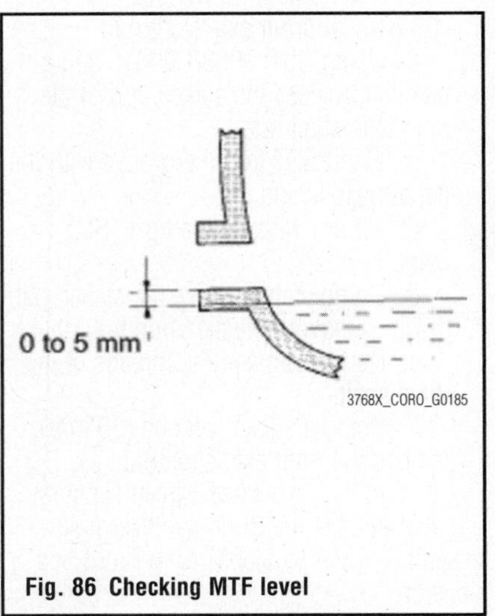

Fig. 86 Checking MTF level

1. Before servicing the vehicle, refer to the Precautions section.
2. Fill the clutch reservoir with brake fluid. Check the reservoir level frequently and add fluid as needed.
3. Connect one end of a vinyl tube to the bleeder plug on the slave cylinder and submerge the other end into a clear container half-filled with brake fluid.
4. Slowly pump the clutch pedal several times.
5. Have an assistant hold the clutch pedal down and loosen the bleeder plug until fluid and/or air starts to run out of the bleeder plug. Close the bleeder plug while the pedal is held to the floor.

➡Do not allow the pedal to raise backup while the bleeder is still open. If this happens, it will allow air to reenter the slave cylinder and cause the clutch system not to work properly.

6. Repeat Steps 2 and 3 until all the air bubbles are removed from the system.
7. Tighten the bleeder plug when all the air is gone.
8. Refill the master cylinder to the proper level as required.
9. Check the system for leaks.

FLUID FILL PROCEDURE

See Figures 87 through 89.

✲✲ WARNING

Clean, high quality brake fluid is essential to the safe and proper operation of the brake system. You should always buy the highest quality brake fluid that is available. If the brake fluid becomes contaminated, drain and flush the system, then refill the master cylinder with new fluid. Never reuse any brake fluid. Any brake fluid that is removed from the system should be discarded. Also, do not allow any brake fluid to come in contact with a painted surface; it will damage the paint.

✲✲ CAUTION

Brake fluid contains polyglycol ethers and polyglycols. Avoid contact with the eyes and wash your hands thoroughly after handling brake fluid. If you do get brake fluid in your eyes, flush your eyes with clean, running water for 15 minutes. If eye irritation persists, or if you have taken brake fluid internally, IMMEDIATELY seek medical assistance.

➡Note the following:

- If any work is performed on the brake system or if air in the brake lines is suspected, bleed the brake system.
- Move the shift lever to P and apply the parking brake before bleeding the brakes.
- Add brake fluid to keep the level between the MIN and MAX lines of the reservoir while bleeding the brakes.
- If brake fluid leaks onto any painted surface, immediately wash it off.

1. Remove center cowl top ventilator louver:
   a. Slide the hood to cowl top seal and disengage the clip as shown in the illustration.
   b. Disengage the 5 claws and remove the center cowl top ventilator louver.
2. Fill the reservoir with brake fluid (SAE DOT 3).

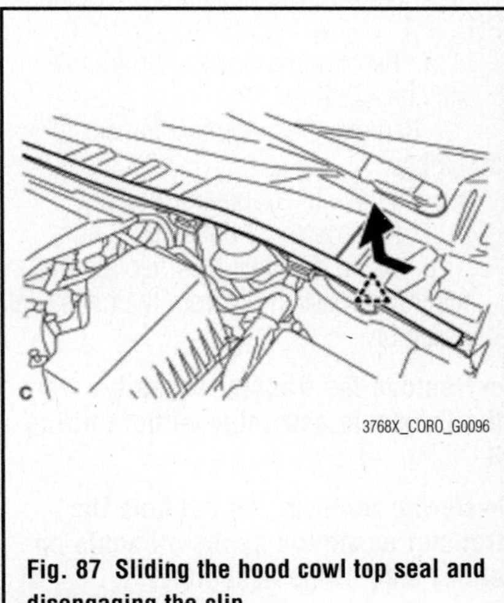

Fig. 87 Sliding the hood cowl top seal and disengaging the clip

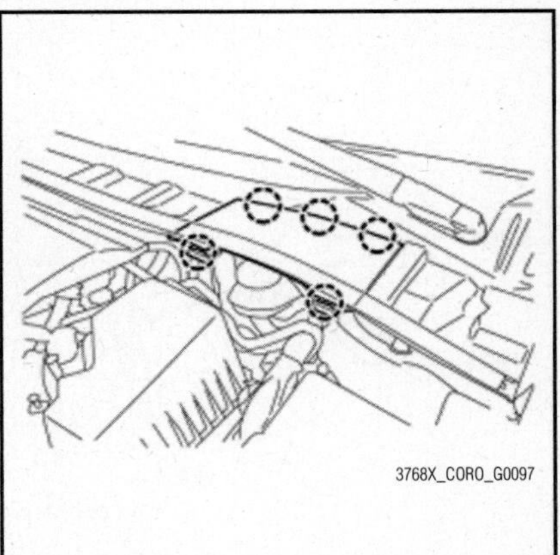

Fig. 88 Disengaging the 5 claws and removing the center cowl top ventilator louver

3. Add brake fluid between the MIN and MAX lines of the reservoir, especially if you are bleeding the brakes.

4. If brake fluid level is lower than the MIN line, check for leaks and inspect the disc brake pads. If necessary, refill the reservoir with brake fluid to the MAX line after repair or replacement.

TRANSFER CASE ASSEMBLY

REMOVAL & INSTALLATION

See Figures 90 through 92.

1. Remove the engine assembly.
2. Remove the rear engine mounting insulator.
3. Install the engine hangers.
4. Remove the starter assembly.
5. Disconnect the wire harness.
6. Remove the rear engine mounting bracket.
7. Remove the rear drive plate and torque converter clutch setting bolt.
8. Remove the transfer stiffener plate RH.
    a. Remove the 4 bolts and transfer stiffener plate RH.
9. Remove the automatic transaxle assembly.
10. Remove the transfer assembly.
    a. Remove the 2 bolts and 6 nuts.
    b. Using a plastic hammer, remove the transfer assembly from the transaxle assembly.

➡Remove the transfer assembly from the transaxle assembly without tilting it.

➡During removal, do not hold the transfer assembly by the oil seals on either side of the assembly.

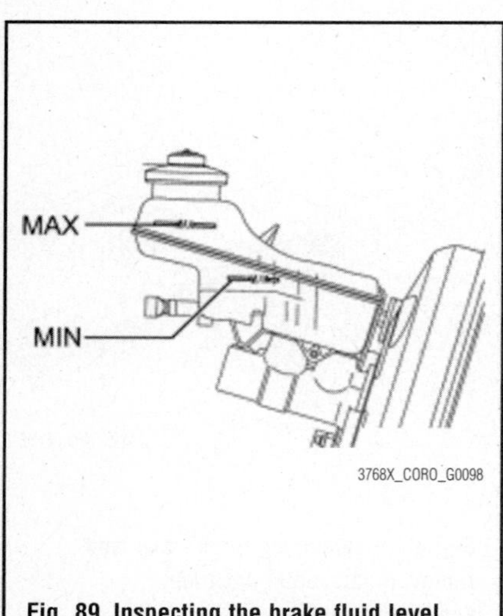

Fig. 89 Inspecting the brake fluid level

3768X_MATR_G0314

Fig. 90 Removing the transfer stiffener plate RH

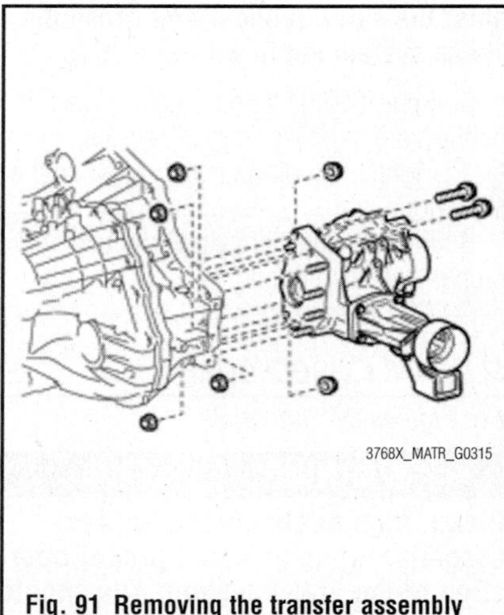

Fig. 91 Removing the transfer assembly

To install:

11. Install the transfer assembly.
    a. Install the transfer assembly to the transaxle assembly with the 2 bolts and 6 nuts. Tighten to 51 ft. lbs. (69 Nm).

➡Install the transfer assembly to the transaxle assembly horizontally.

➡Do not touch the transfer assembly oil seals during installation.

12. Install the automatic transaxle assembly.
13. Install the transfer stiffener plate RH with the 4 bolts. Tighten the bolts to 25 ft. lbs. (34 Nm). The A bolts are 1.7717 inch (45 mm) long. The B bolts are 1.1811 inch (30 mm) long.
14. Install the drive plate and the torque converter clutch setting bolt.
15. Install the rear engine mounting bracket.
16. Connect the wire harness.
17. Install the starter assembly.

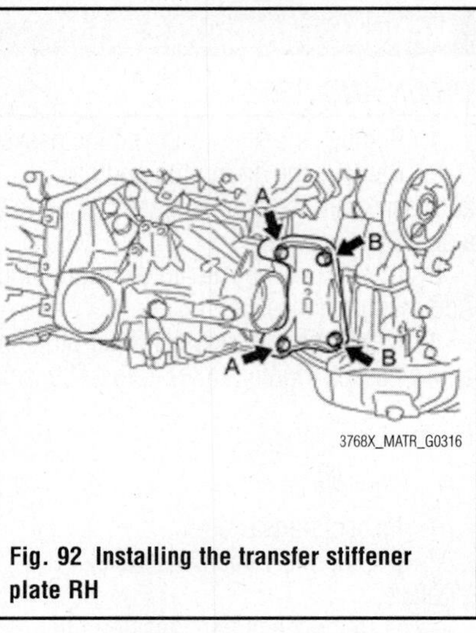

Fig. 92 Installing the transfer stiffener plate RH

18. Temporarily tighten the rear engine mounting insulator.
19. Install the engine assembly with the transaxle.

FRONT HALFSHAFT

REMOVAL & INSTALLATION

See Figures 93 and 94.

➡The hub bearing could be damaged if subjected to the full weight of the vehicle, such as if the vehicle is moved without the halfshafts. If it is absolutely necessary to place the full vehicle weight on the hub bearing, first support the bearing with SST No. 09608–16042.

➡This procedure is for removing both LH and RH halfshafts.

1. Remove the LH and RH engine under covers.
2. Drain transmission fluid.
3. Raise and support the vehicle.
4. Remove the front wheels.
5. Remove front axle shaft nut:
    a. Using SST: 09930-00010 and a hammer, release the staked part of the front axle shaft nut.
    b. Insert SST into the groove with the flat surface facing up.
    c. Do not damage the tip of SST using grinders.
    d. Completely unstake the staked part before removing the axle hub nut.
    e. Do not damage the threads of the drive shaft.
    f. Using a socket wrench (30 mm), remove the front axle shaft nut.
6. Separate the wheel speed sensors.
7. Separate the front stabilizer link assembly. Refer to Stabilizer in Front Suspension.

8. Separate the tie rod end sub-assembly. Refer to Tie Rod in Front Suspension.

9. Separate front lower suspension. Refer to Front Suspension.

10. Separate front axle assembly:

a. Using a plastic hammer, tap the end of the drive shaft and disengage the fitting between the drive shaft and front axle assembly.

b. If it is difficult to disengage the fitting, tap the end of the drive shaft with a brass bar and hammer.

c. Push the front axle assembly out of the vehicle to remove the drive shaft from the front axle assembly.

➡Do not push out the front axle further than necessary. Do not damage the outboard joint boot. Do not damage the speed sensor rotor.

11. If equipped, remove the 2 bolts and the manual transmission case protector.

12. Using a halfshaft puller, remove the LH halfshaft assembly.

13. Remove front RH drive shaft assembly:

a. For 1.8L Engines:

b. Using a brass bar and hammer, remove the front drive shaft assembly.

➡Do not damage the oil seal. Do not damage the inboard joint boot. Do not drop the drive shaft.

c. For 2.4L Engines:

d. Remove the bearing bracket hole snap ring.

e. Remove the bolt and front drive shaft assembly RH from the drive shaft bearing bracket.

➡Do not damage the boot or oil seal.

14. Using a screwdriver, remove the snap ring from the front LH and RH drive inboard joint assembly.

To install:

15. Install a new front drive shaft hole snap ring to the front drive inboard joint assembly.

16. Align the inboard joint splines, and using a brass bar and a hammer, install the front drive shaft assembly.

➡Face the end gap of the front drive inboard joint hole snap ring downward. Do not damage the oil seal. Do not damage the inboard joint boot.

➡Confirm whether the drive shaft is securely driven in by checking the reaction force and sound.

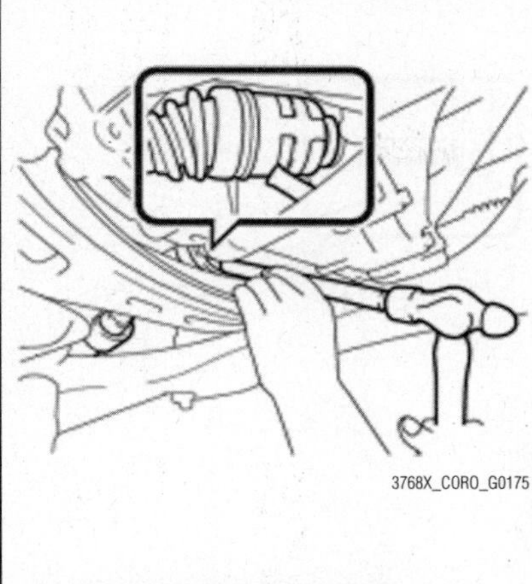

Fig. 93 Removing the halfshaft assembly—1.8L Engine

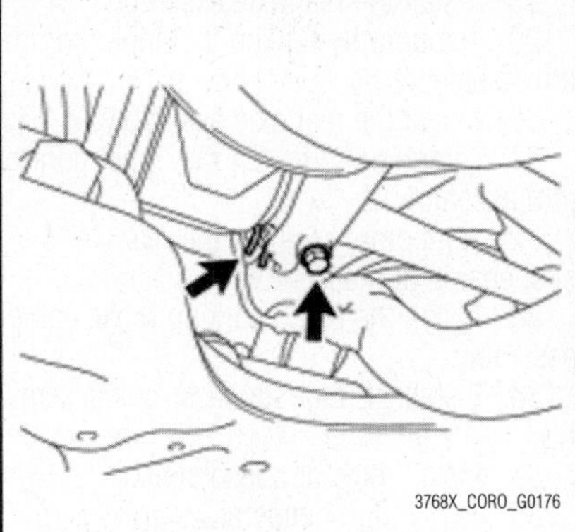

Fig. 94 Removing the bearing bracket hole snap ring and halfshaft assembly—2.4L Engine

17. For 2.4L engines:

a. Using a screwdriver, install a new bearing bracket hole snap ring.

➡Do not damage the boot or oil seal. Move the drive shaft assembly while keeping it level.

b. Install a new bolt. Tighten to 24 ft. lbs. (32 Nm).

18. If equipped, install the manual transmission case protector with the 2 bolts. Tighten to 13 ft. lbs. (18 Nm).

19. To complete installation, reverser remaining removal procedure.

20. Install front axle shaft nut:

a. Clean the threaded parts on the drive shaft and axle shaft nut using a non-residue solvent.

➡Be sure to perform this work for a new drive shaft. Keep the threaded parts free of oil and foreign matter.

b. Using a socket wrench (30 mm), install a new axle shaft nut. Tighten to 160 ft. lbs. (216 Nm).

c. Using a chisel and hammer, caulk the axle shaft nut.

21. Tighten wheel lug nuts to 76 ft. lbs. (103 Nm).

22. Add transmission fluid.

23. Inspect for fluid leaks.

24. Inspect and adjust the wheel alignment.

REAR HALFSHAFT

REMOVAL & INSTALLATION

See Figures 95 through 97.

➡Use the same procedure for the RH side and LH side.

➡The procedure listed below is for the LH side.

1. Remove the rear wheel.

2. Remove the rear axle shaft nut.

a. Using the special tool (09930-00010) and a hammer, release the staked part of the rear axle shaft nut.

➡Loosen the staked part of the nut completely, otherwise the threads of the drive shaft may be damaged.

b. While applying the brakes, remove the rear axle shaft nut.

3. Remove the rear differential carrier cover plug.

4. Remove the rear differential drain plug.

5. Install the rear differential drain plug.

6. Separate the rear disc brake caliper assembly.

7. Remove the rear disc.

8. Remove the rear No. 3 parking brake shoe return tension spring.

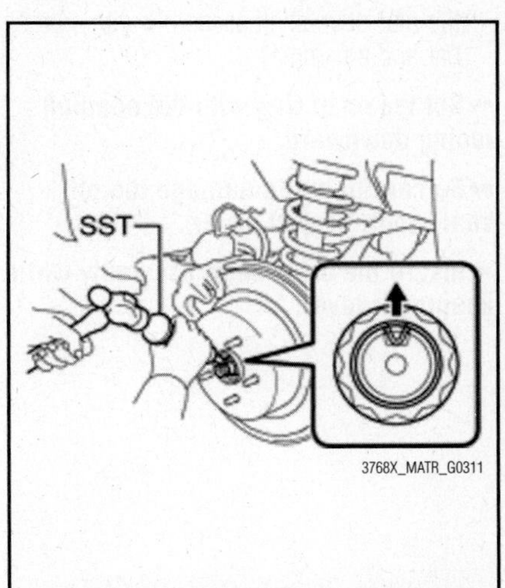

Fig. 95 Removing the rear axle shaft nut

9. Remove the No. 2 parking brake shoe return tension spring.
10. Remove the No. 1 parking brake shoe hold down compression spring.
11. Remove the parking brake shoe strut.
12. Separate the No. 2 parking brake shoe hold down compression spring.
13. Remove the parking brake shoe adjusting screw set.
14. Remove the No. 1 parking brake shoe return tension spring.
15. Remove the parking brake shoe lever.
16. Remove the No. 1 parking brake shoe hold down spring pin.
17. Remove the No. 2 parking brake shoe hold down spring pin.
18. Separate the rear stabilizer link assembly.
19. Separate the rear speed sensor.
20. Separate the No. 3 parking brake cable assembly.
21. Separate the upper control arm assembly.
22. Separate the rear No. 1 suspension arm assembly.
23. Remove the rear axle assembly.
24. Remove the rear drive shaft assembly.
   a. Using a brass bar and a hammer, remove the rear drive shaft assembly.

➡Remove the rear drive shaft assembly while keeping it level.

25. Remove the rear drive shaft snap ring.
   a. Using a screwdriver, remove the rear drive shaft snap ring.

To install:

26. Install the rear drive shaft snap ring.
27. Install the rear drive shaft assembly.
   a. Align the shaft splines and install the rear drive shaft assembly with a brass bar and a hammer.

➡Set the snap ring with the opening facing downward.

➡Be careful not to damage the oil seal, boot, or dust cover.

➡Install the drive shaft assembly while keeping it level.

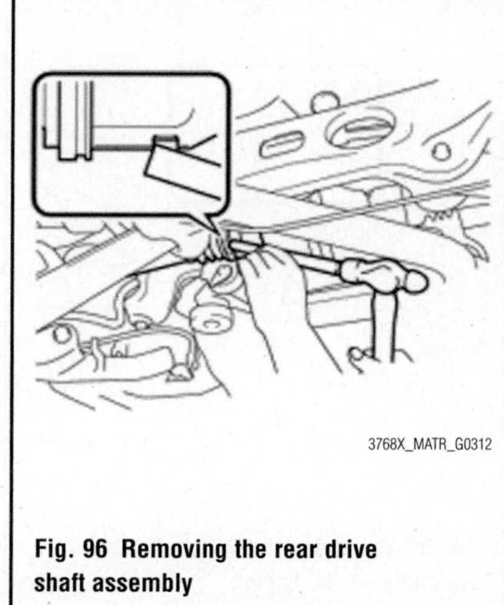

Fig. 96 Removing the rear drive shaft assembly

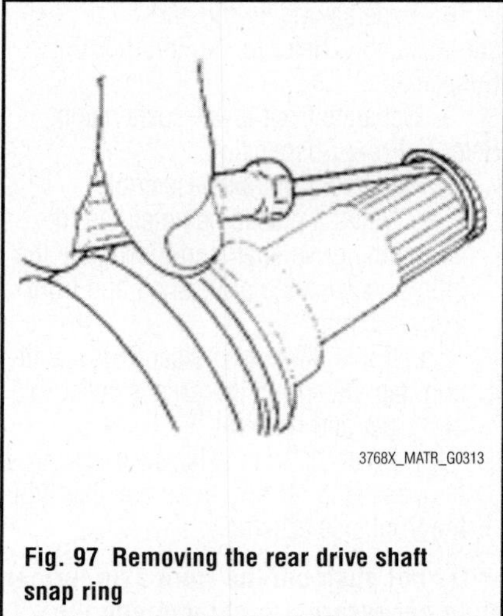

Fig. 97 Removing the rear drive shaft snap ring

28. Install the rear axle assembly.
29. Temporarily tighten the upper control arm assembly.
30. Install the rear axle assembly.
31. Temporarily tighten the upper control arm assembly.
32. Temporarily tighten the rear No. 1 suspension arm assembly.
33. Install the No. 3 parking brake cable assembly.
34. Install the rear stabilizer link assembly.
35. Install the rear speed sensor.
36. Apply high temperature grease.
37. Install the No. 1 parking brake shoe hold down spring pin.
38. Install the No. 2 parking brake shoe hold down spring pin.
39. Install the parking brake shoe lever.
40. Install the No. 1 parking brake shoe return tension spring.
41. Install the parking brake shoe adjusting screw set.
42. Install the parking brake shoe lever.
43. Install the No. 2 parking brake shoe hold down compression spring.
44. Install the parking brake shoe strut.
45. Install the No. 1 parking brake shoe hold down compression spring.
46. Install the No. 2 parking brake shoe return tension spring.
47. Install the No. 3 parking brake shoe return tension spring.
48. Check the parking brake installation.
49. Install the rear disc.
50. Install the rear disc brake caliper assembly.
51. Adjust the parking brake shoe clearance and parking brake pedal travel.
52. Install the rear axle shaft nut.
   a. Clean the threaded parts on the drive shaft and axle shaft nut using a non-residue solvent.

➡Be sure to perform this work for a new drive shaft.

➡Keep the threaded parts free of oil and foreign matter.

   b. Install a new rear axle shaft nut and tighten to 159 ft. lbs. (216 Nm).
   c. Using a chisel and a hammer, stake the rear axle shaft nut LH.
53. Add differential oil.
54. Inspect and adjust the differential oil.
55. Install the rear wheel. Tighten to 76 ft. lbs. (103 Nm).
56. Stabilize suspension.
57. Fully tighten the upper control arm assembly.
58. Fully tighten the rear No. 1 suspension arm assembly.
59. Inspect and adjust the rear wheel alignment.
60. Check the ABS speed sensor signal.

ENGINE COOLING

ENGINE COOLANT

DRAIN & REFILL PROCEDURE

See Figures 98 and 99.

⁂ CAUTION

Do not loosen the lower radiator drain cock plug while the engine and radiator are still hot. Pressurized, hot engine coolant and steam may be released and cause serious burns.

➡Collect the coolant in a container and dispose of it according to the regulations in your area.

1. Loosen the lower radiator drain cock plug.
2. Remove the radiator cap.
3. Loosen the cylinder block drain cock plug.
4. To refill, perform the following:
   a. Tighten the lower radiator drain cock plug.
   b. Tighten the cylinder block drain cock plug. Tighten to 9 ft. lbs. (13 Nm).
   c. Loosen the upper radiator drain cock plug.
   d. Slowly fill the radiator with TOYOTA Super Long Life Coolant (SLLC).
   e. Standard Capacity: 6 quarts (5.7 liters).

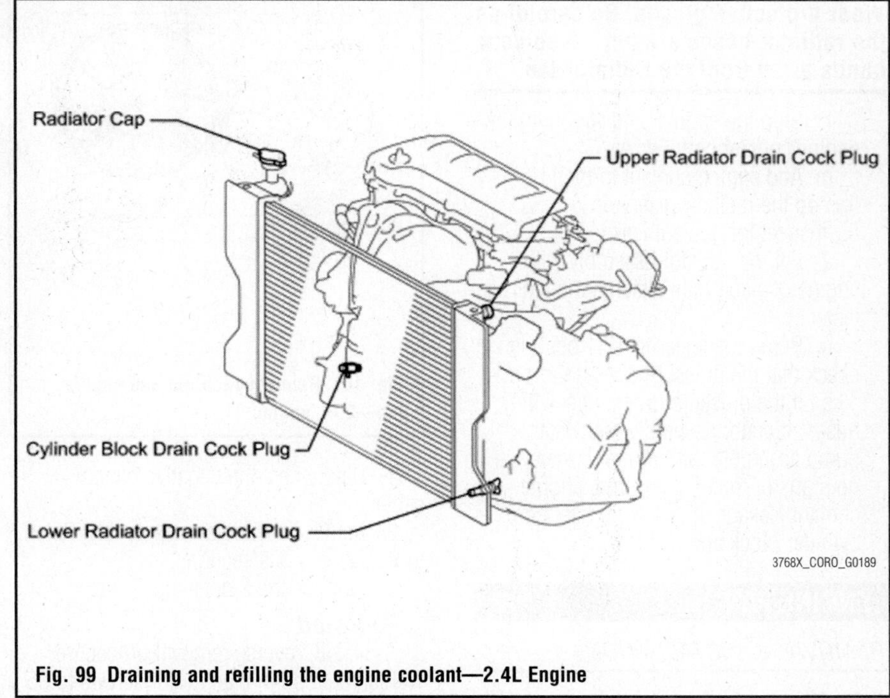

Fig. 99 Draining and refilling the engine coolant—2.4L Engine

➡TOYOTA vehicles are filled with TOYOTA SLLC at the factory. In order to avoid damage to the engine cooling system and other technical problems, only use TOYOTA SLLC or similar high quality ethylene glycol based non-silicate, non-amine, non-nitrite, non-borate coolant with long-life hybrid organic acid technology (coolant with long-life hybrid organic acid technology consists of a combination of low phosphates and organic acids).

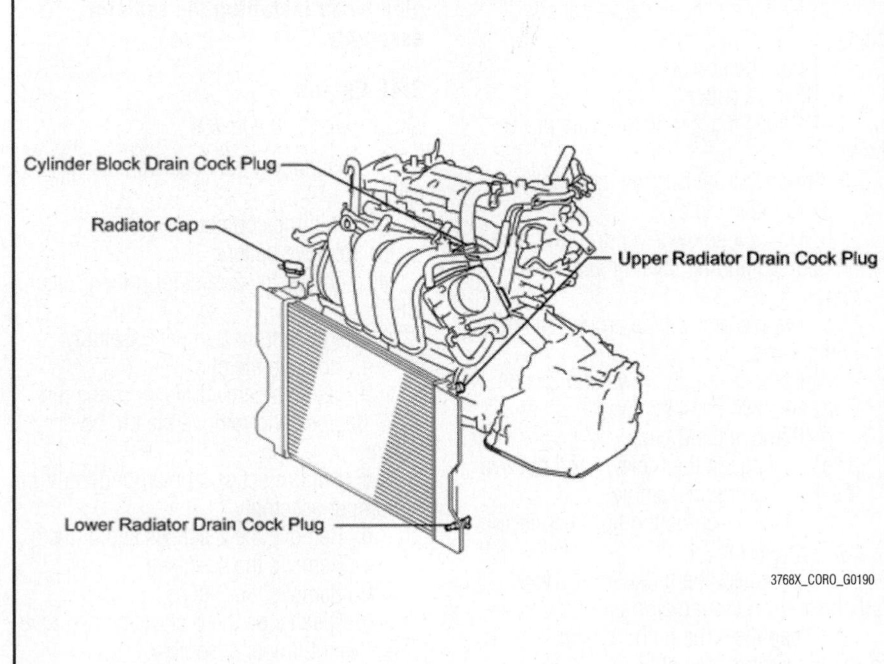

Fig. 98 Draining and refilling the engine coolant—1.8L Engine

➡Never use water as a substitute for engine coolant.

   f. Squeeze the inlet and outlet radiator hoses several times by hand, and then check the level of the coolant. If the coolant level is low, add coolant.
   g. Tighten the upper radiator drain cock plug.
   h. Slowly pour coolant into the radiator reservoir tank until it reaches the FULL line.
   i. Install the radiator cap sub-assembly and reservoir tank cap.

➡Before starting the engine, turn the A/C switch off. Adjust the air conditioning temperature setting to MAX (HOT). Adjust the air conditioning blower setting to LO.

   j. Start the engine and warm it up.
   k. Bleed air from the cooling system.

➡Thermostat opening timing can be determined by squeezing the inlet radiator hose, and sensing vibrations when the engine coolant starts to flow inside the hose.

CAUTION

When squeezing the radiator hoses: Wear protective gloves. Be careful as the radiator hoses are hot. Keep your hands away from the radiator fan.

l. Stop the engine, and wait until the engine coolant cools down.

m. Add engine coolant to the FULL line on the radiator reservoir.

5. Inspect for coolant leaks:

a. Fill the radiator assembly with engine coolant, then attach a radiator cap tester.

b. Pump the tester to 17.1 psi, then check that the pressure does not drop.

c. If the pressure drops, check the hoses, radiator assembly and water pump assembly for leakage. If there are no signs or traces of external engine coolant leakage, check the heater core, cylinder block and head.

ENGINE FAN

REMOVAL & INSTALLATION

1.8L Engine

See Figure 100.

1. Disconnect the negative battery cable.
2. Remove the radiator assembly. Refer to Radiator in this section.
3. Remove the nut, then remove the fan.

To install:

To install, reverse removal procedure. Tighten the cooling fan motor screws to 35 inch lbs. (3.9 Nm).

2.4L Engine

See Figure 101.

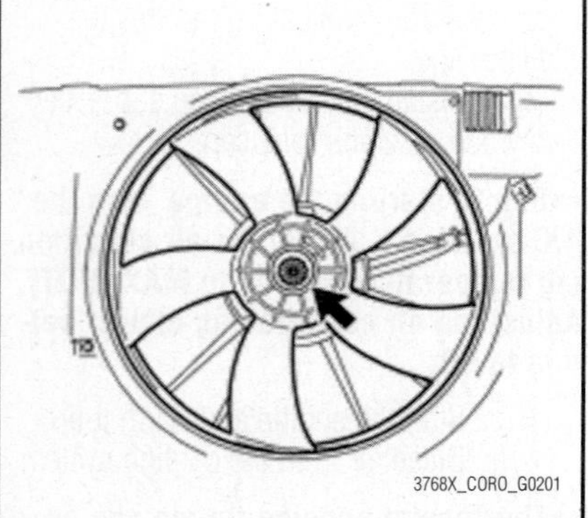
3768X_CORO_G0201

Fig. 100 Removing the nut and engine fan—1.8L engine

3768X_CORO_G0202

Fig. 101 Removing the nut and engine fan—2.4L engine

1. Disconnect the negative battery cable.
2. Remove the radiator assembly. Refer to Radiator in this section.
3. Remove the 2 nuts and 2 fans.

To install:

To install, reverse removal procedure. Tighten the cooling fan motor screws to 35 inch lbs. (3.9 Nm).

RADIATOR

REMOVAL & INSTALLATION

1.8L Engine

See Figures 102 and 103.

1. Disconnect the negative battery cable.
2. Drain the coolant.
3. Remove battery.
4. Remove the 2 radiator grille protectors.
5. Remove front bumper assembly:
   a. Remove the clip.
   b. Using a screwdriver, turn the pin 90 degrees and remove the pin hold clip.
   c. Put protective tape around the front bumper assembly.
   d. Remove the 2 screws and 3 clips.
   e. Remove the 4 screws.
   f. Remove the 2 clips.
   g. Disengage the 6 claws and remove the front bumper assembly.
   h. Disconnect the fog light connector. (w/ Fog Light).
6. Disconnect the radiator reservoir tank hose from the radiator assembly.
7. Disconnect the both radiator hose from the radiator assembly.
8. Remove the 2 bolts and 2 upper radiator supports.
9. Remove the 2 support cushions from the 2 upper radiator supports.
10. Remove the 3 bolts from the hood lock assembly.
11. Disconnect the hood lock control cable and remove the hood lock assembly.
12. Separate the water by-pass hose from the 3 clamps.
13. Disconnect the water by-pass hose from the radiator assembly.
14. Remove the 2 bolts and hood lock support sub-assembly.
15. Disconnect the horn connector.
16. Remove the 4 bolts and upper radiator support sub-assembly.
17. Remove the 2 bolts, disengage the 2 claws, and remove the No. 2 fan shroud from the radiator assembly.
18. Disconnect the cooling fan ECU connector and wire harness clamp.
19. Remove the radiator assembly with the fan shroud.
20. Remove the 2 lower radiator supports.
21. For A/T equipped vehicles:
   a. Disconnect the 2 oil cooler hoses from the radiator.
   b. Remove the 2 bolts and oil cooler hose.
22. Remove the 2 bolts and fan shroud from the radiator assembly.

To install:

To install, reverse removal procedure.

➡Do not apply any excessive force to the cooler condenser assembly or pipe when installing the radiator assembly.

2.4L Engine

See Figures 104 and 105.

1. Disconnect the negative battery cable.
2. Drain the coolant.
3. Remove battery.
4. Remove the 2 radiator grille protectors.
5. Remove front bumper assembly:
   a. Remove the clip.
   b. Using a screwdriver, turn the pin 90 degrees and remove the pin hold clip.
   c. Put protective tape around the front bumper assembly.
   d. Remove the 2 screws and 3 clips.
   e. Remove the 4 screws.
   f. Remove the 2 clips.
   g. Disengage the 6 claws and remove the front bumper assembly.
   h. Disconnect the fog light connector. (w/ Fog Light).

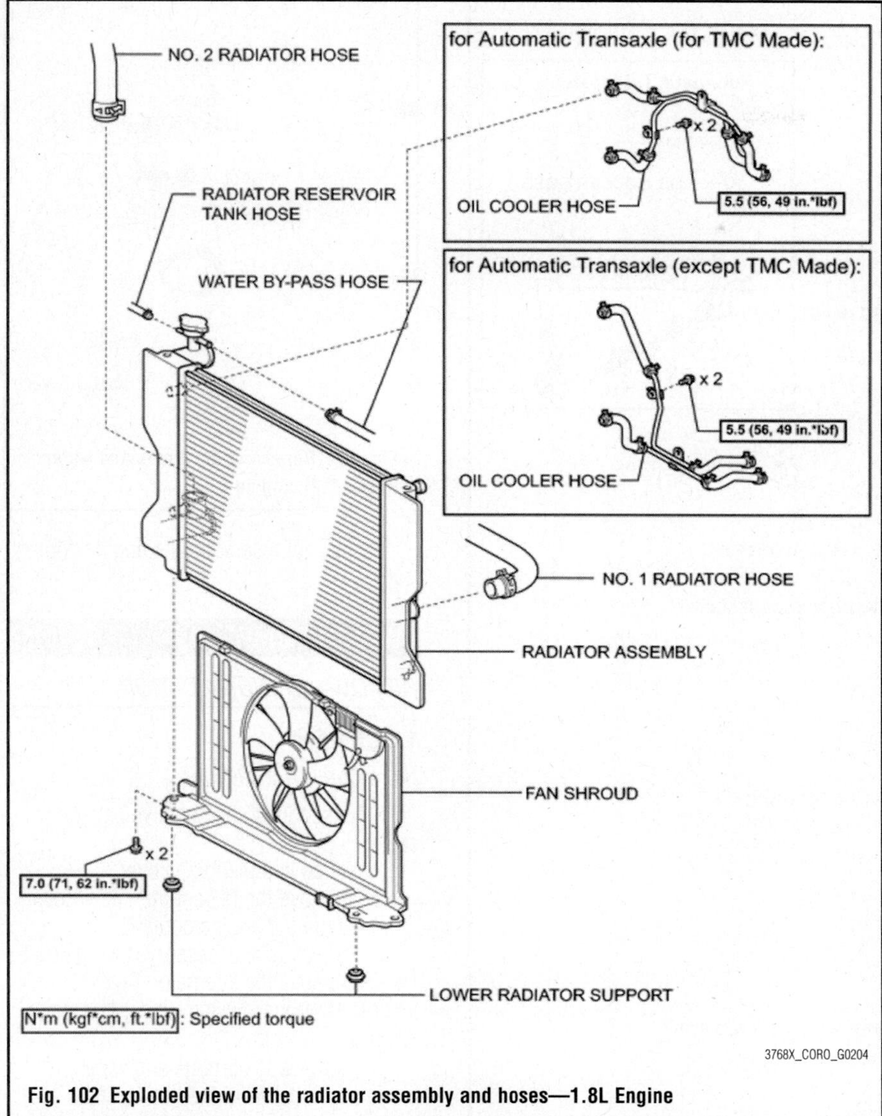

Fig. 102 Exploded view of the radiator assembly and hoses—1.8L Engine

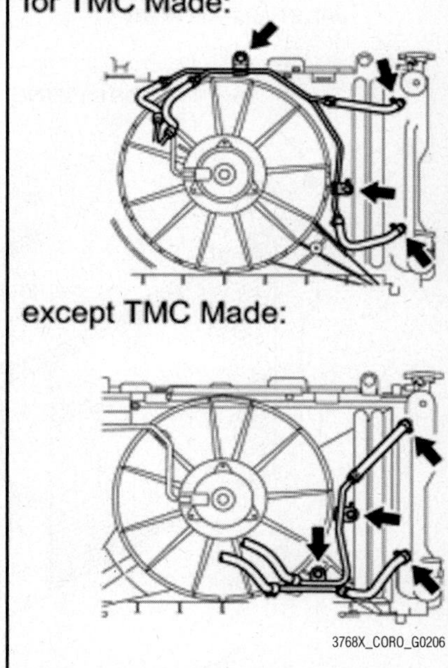

Fig. 103 Disconnecting the 2 oil cooler hoses from the radiator—1.8L Engine

6. Disconnect the radiator reserve tank hose from the radiator assembly.
7. Disconnect both radiator hoses from the radiator assembly.
8. Disconnect oil cooler hose (for automatic transaxle).
9. Remove the 2 bolts and 2 upper radiator supports.
10. Remove the 2 support cushions from the 2 upper radiator supports.
11. Remove the 3 bolts from the hood lock assembly.
12. Disconnect the hood lock control cable and remove the hood lock assembly.
13. Separate the water by-pass hose from the 2 clamps.
14. Disconnect the water by-pass hose from the radiator assembly.
15. Remove the 2 bolts and hood lock support sub-assembly.
16. Disconnect the horn connector.
17. Remove the 4 bolts and upper radiator support sub-assembly.
18. Remove the 2 bolts, disengage the 2 claws, and remove the No. 2 fan shroud from the radiator assembly.
19. Disconnect the 2 cooling fan motor connectors and wire harness clamp.
20. Remove the radiator assembly with the fan shroud.

➡Do not apply any excessive force to the cooler condenser assembly or pipe when removing the radiator assembly.

21. Disconnect the 2 oil cooler hoses from the radiator.
22. Disconnect the clamp from the fan shroud.
23. Remove the 2 bolts and oil cooler hose.
24. Remove the 2 bolts and fan shroud from the radiator assembly.

To install:

To install, reverse removal procedure.

THERMOSTAT

REMOVAL & INSTALLATION

See Figures 106 through 108.

1. Disconnect the negative battery cable.
2. Drain the engine coolant.
3. For 2.4L engines, remove the No. 2 coolant hose.
4. Remove the 2 nuts and water inlet.
5. Remove the thermostat.
6. Remove the gasket from the thermostat.

To install:

7. Install a new gasket on the thermostat.
8. Install the thermostat to the water inlet with the jiggle valve upward.

➡The jiggle valve may be set to within 10°on either side of the indicated position.

9. Install the water inlet with the 2 nuts. Tighten to 7 ft. lbs. (10 Nm).
10. For 2.4L engines, install the No. 2 coolant hose
11. Add coolant and inspect for leaks.

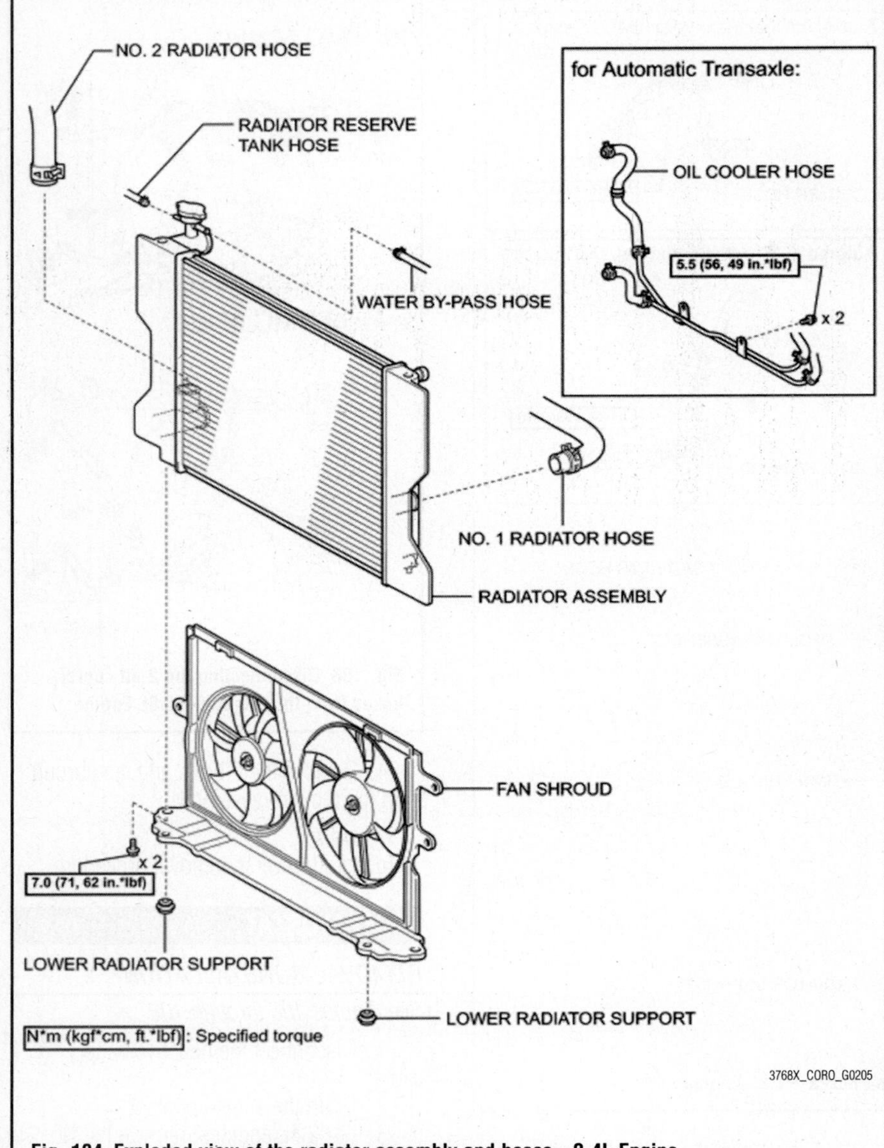
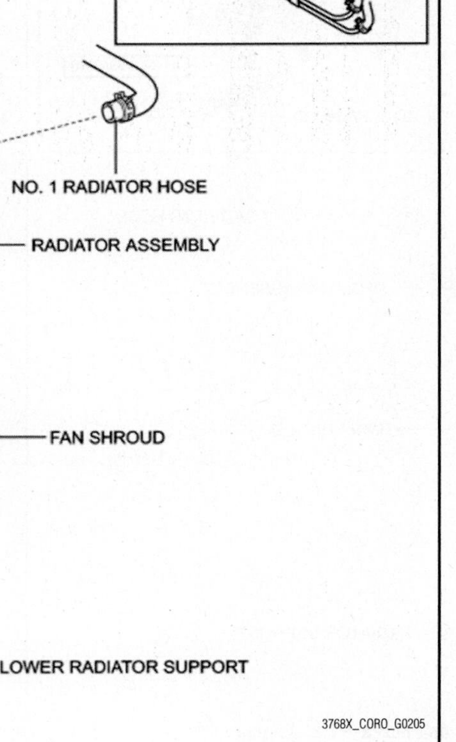

Fig. 104 Exploded view of the radiator assembly and hoses—2.4L Engine

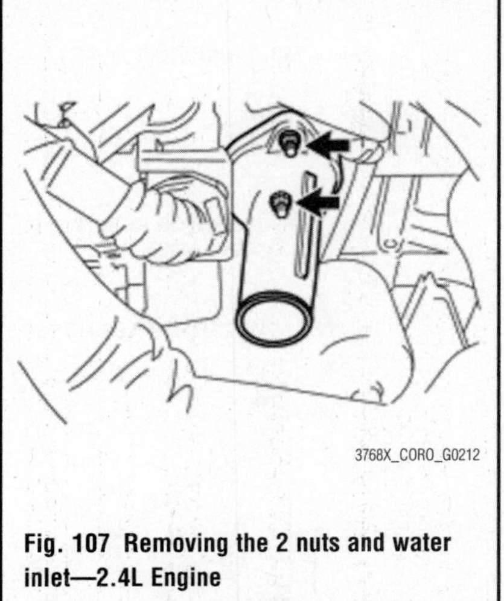

Fig. 107 Removing the 2 nuts and water inlet—2.4L Engine

12. Inspect reservoir tank engine coolant level.

WATER PUMP

REMOVAL & INSTALLATION

1.8L Engine

See Figures 109 and 110.

1. Disconnect the negative battery cable.
2. Remove the engine cover.
3. Remove the RH engine under cover.
4. Drain the engine coolant.
5. Remove the accessory drive belt.
6. Remove the alternator. Refer to Alternator in Charging System under Engine Electrical.
7. Remove the 5 bolts and water pump assembly from the timing chain cover.
8. Remove the water pump gasket from the timing chain cover.

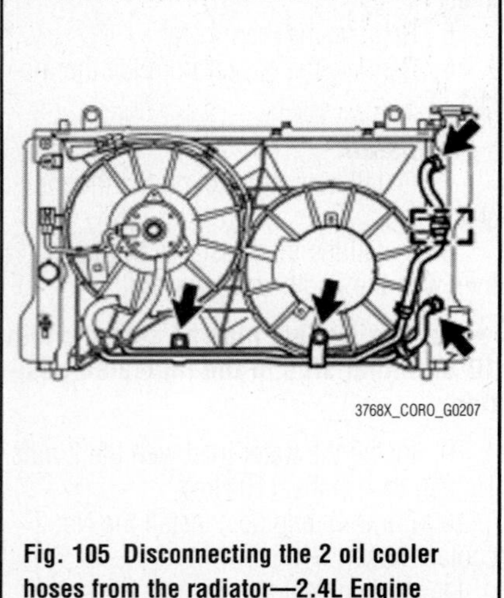

Fig. 105 Disconnecting the 2 oil cooler hoses from the radiator—2.4L Engine

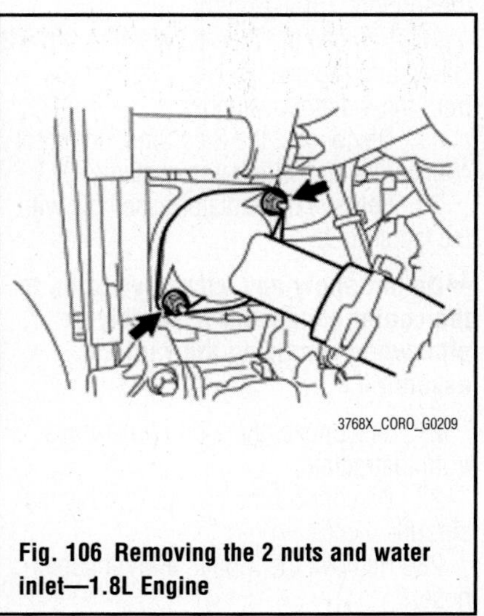

Fig. 106 Removing the 2 nuts and water inlet—1.8L Engine

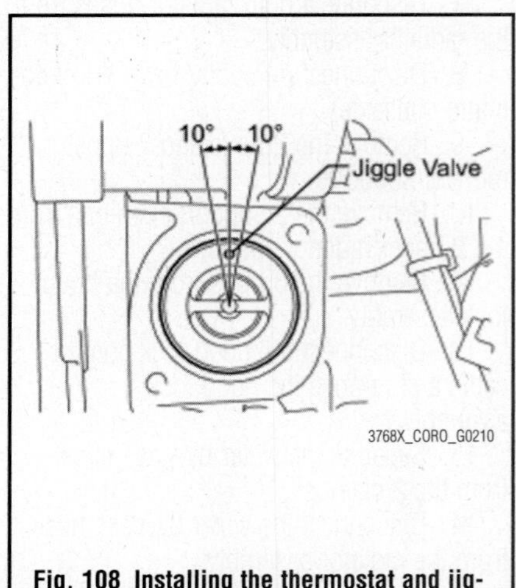

Fig. 108 Installing the thermostat and jiggle valve alignment

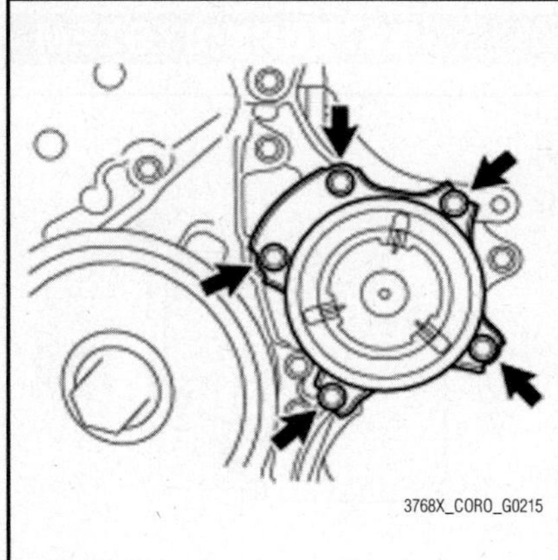

Fig. 109 Location of the 5 water pump bolts—1.8L engine

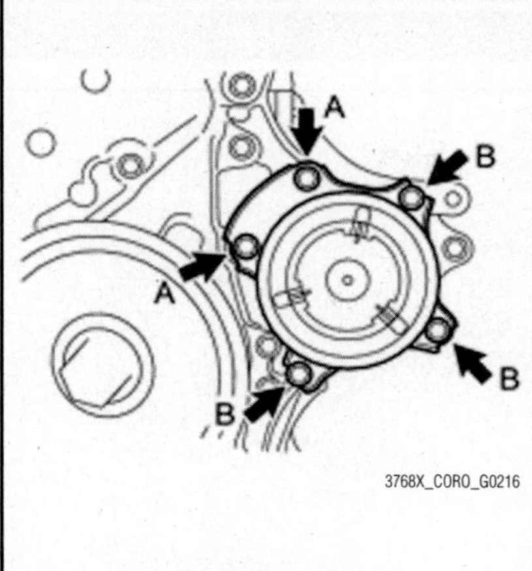

Fig. 110 Identifying the bolts "A" and "B" for installation—1.8L engine

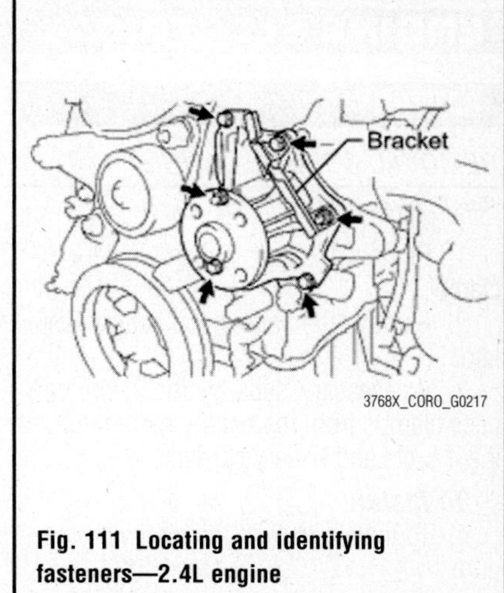

Fig. 111 Locating and identifying fasteners—2.4L engine

To install:

9. Align the protrusion of a new water pump gasket with the cutout in the timing chain cover and install the gasket to the groove of the timing chain cover.

➡Be sure to clean the contact surfaces.

10. Install the water pump assembly to the timing chain cover with the 5 bolts. Tighten to:
    a. Bolt A (1.38 inches): 19 ft. lbs. (26 Nm).
    b. Bolt B: (0.709 inches): 18 ft. lbs. (24 Nm).
11. To complete installation, reverse remaining removal procedure.
12. Inspect for coolant leaks.

2.4L Engine

See Figure 111.

1. Disconnect the negative battery cable.
2. Remove the engine cover.
3. Remove the LH and RH engine under cover.
4. Drain the engine coolant.
5. Remove the accessory drive belt.
6. Remove the alternator. Refer to Alternator in Charging System under Engine Electrical.
7. Remove the RH engine mounting insulator sub-assembly.
8. Using a pulley remover, remove the water pump pulley.
9. Remove the clamp of the Crankshaft Position (CKP) sensor from the water pump.
10. Disconnect the wire of the Crankshaft Position (CKP) sensor from the clamp bracket.
11. Remove the 4 bolts, 2 nuts and clamp bracket.

➡Tape the screwdriver tip before use.

12. Using a screwdriver, pry between the water pump and cylinder block, and then remove the water pump.

➡Be careful not to damage the contact surfaces of the water pump and cylinder block.

➡Remove any oil from the contact surface. The parts must be set within 3 minutes after applying seal packing. Otherwise, the material must be removed and reapplied.

To install:

13. To install, reverse the removal procedure and pay attention to the following:
    a. Apply a continuous line of seal packing. Use Toyota Genuine Seal Packing Black, Three Bond 1207B or equivalent.
    b. Tighten the water pump bolts to 80 inch lbs. (9 Nm).
    c. Tighten the water pump pulley to 19 ft. lbs. (26 Nm).

ENGINE ELECTRICAL — BATTERY SYSTEM

BATTERY

REMOVAL & INSTALLATION

See Figures 112 and 113.

1. Remove the bolt, nut and battery clamp.
2. Remove the battery and battery insulator.
3. If necessary, separate the 2 wire harness clamps from the battery carrier, remove the 4 bolts and battery carrier.

To install:

4. To install, reverse the removal procedure.

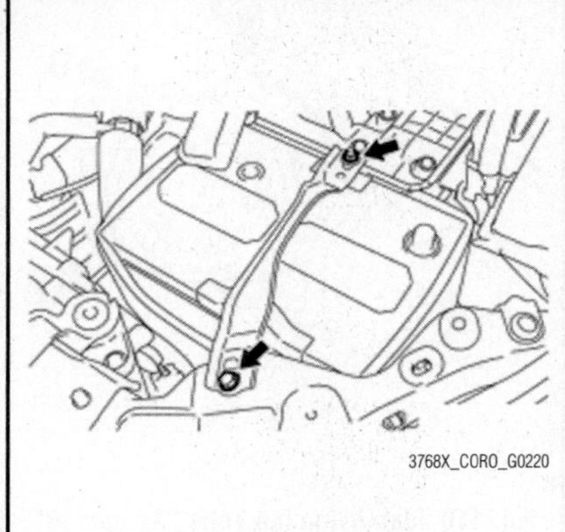

Fig. 112 Removing the bolt, nut and battery clamp

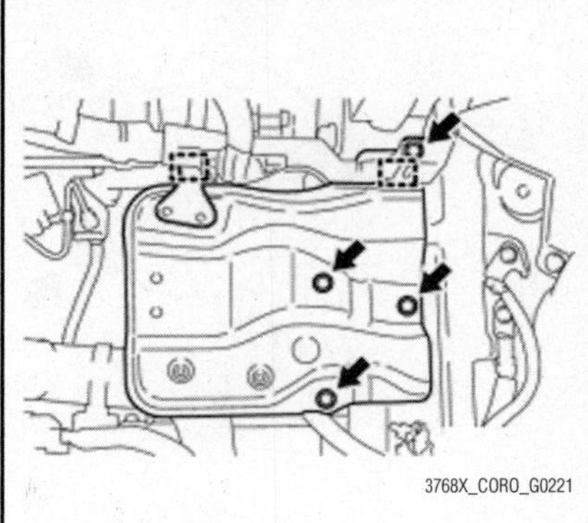

Fig. 113 Separating the 2 wire harness clamps from the battery carrier, removing the 4 bolts and battery carrier

ENGINE ELECTRICAL — CHARGING SYSTEM

ALTERNATOR

REMOVAL & INSTALLATION

1.8L Engine

See Figure 114.

1. Disconnect the negative battery cable.
2. Remove the RH engine under cover.
3. Remove the accessory drive belt.
4. Remove generator assembly:
   a. Remove the terminal cap.

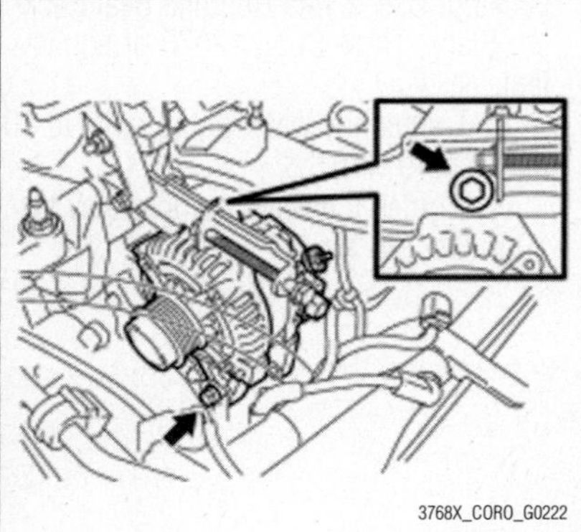

Fig. 114 Removing the alternator assembly—1.8L engine

   b. Remove the nut and disconnect the wire harness from terminal B.
   c. Disconnect the connector and harness clamp.
   d. Remove the 2 bolts and generator assembly.
   e. Remove the bolt and wire harness clamp bracket.

To install:

5. To install, reverse removal procedure. Pay attention to the following:
   a. Alternator upper mounting bolt: Tighten to 32 ft. lbs. (43 Nm).
   b. Alternator lower mounting bolt: Tighten to 14 ft. lbs. (19 Nm).

2.4L Engine

See Figure 115.

1. Disconnect the negative battery cable.
2. Remove the RH engine under cover.
3. Remove the accessory drive belt.
4. Remove generator assembly:
   a. Disconnect the generator connector.
   b. Remove the nut and disconnect the wire harness from terminal B.
   c. Separate the 2 wire harness clamps.
   d. Remove the 2 bolts and generator assembly.

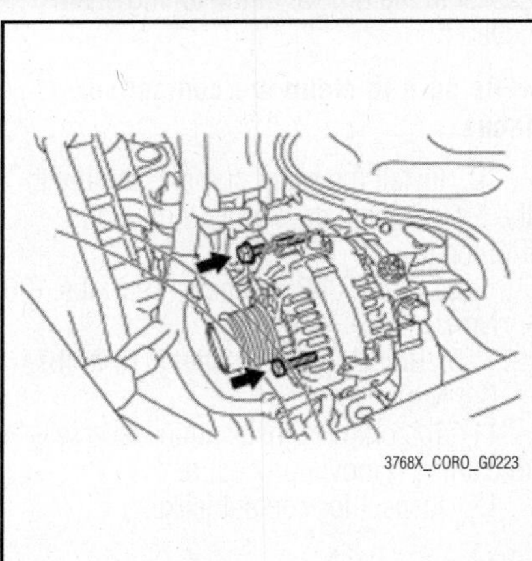

Fig. 115 Removing the alternator assembly—2.4L engine

   e. Remove the bolt and wire harness clamp bracket.

To install:

5. To install, reverse removal procedure. Pay attention to the following:
   a. Alternator upper mounting bolt: Tighten to 16 ft. lbs. (21 Nm).
   b. Alternator lower mounting bolt: Tighten to 38 ft. lbs. (52 Nm).

FIRING ORDER

See Figure 116.

IGNITION COIL

REMOVAL & INSTALLATION

See Figure 117.

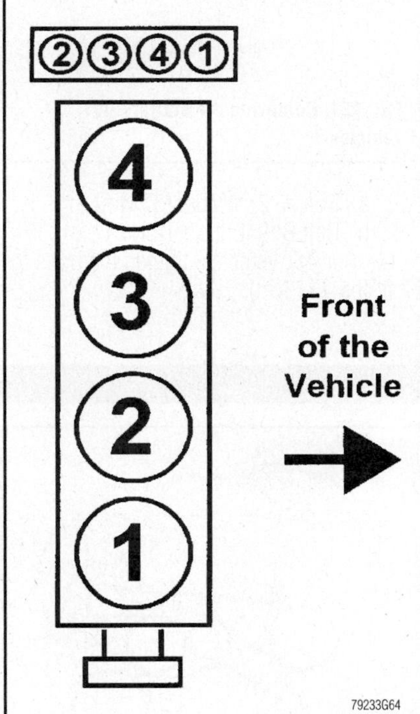

Fig. 116 Firing order: 1–3–4–2 distributorless ignition system

1. Disconnect the battery's negative terminal.
2. Remove the two nuts and clips and remove the cylinder head cover.
3. Disconnect the four ignition coil connectors.
4. Remove the four bolts that retain the coils.

To install:

5. Install the coils and the four bolts.
6. Connect the coil connections.
7. Install the cylinder head cover.

IGNITION TIMING

ADJUSTMENT

The ignition timing is controlled by the Powertrain Control Module (PCM). No adjustment is necessary or possible.

SPARK PLUGS

REMOVAL & INSTALLATION

See Figure 118.

1. Disconnect the negative battery cable.
2. Remove the ignition coil pack. Refer to Ignition Coil Pack under Distributorless Ignition System.
3. Remove the 4 spark plugs.

To install:

4. To install, use a 14 mm spark plug wrench, install the 4 spark plugs and tighten to 15 ft. lbs. (20 Nm).
5. To complete installation, reverse remaining removal procedure.

Fig. 117 Location of the ignition coils

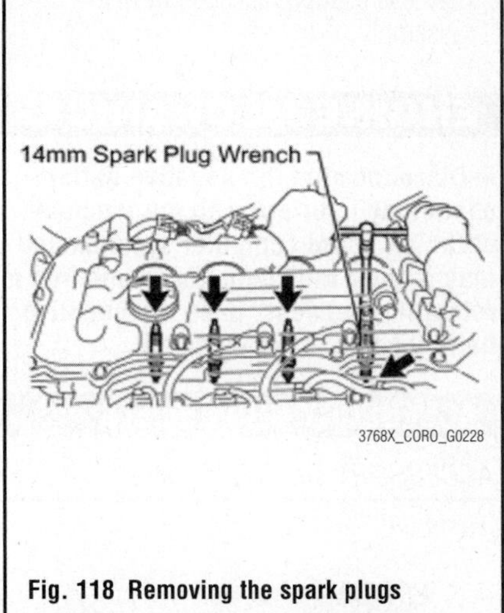

Fig. 118 Removing the spark plugs

STARTER

REMOVAL & INSTALLATION

1.8L Engine

See Figure 119.

1. Disconnect the negative battery cable.
2. Remove transmission oil filler tube sub-assembly (for automatic transaxle).
3. Remove the terminal cap.
4. Remove the nut and disconnect terminal 30.
5. Disconnect the connector.
6. Remove the 2 bolts and starter assembly.

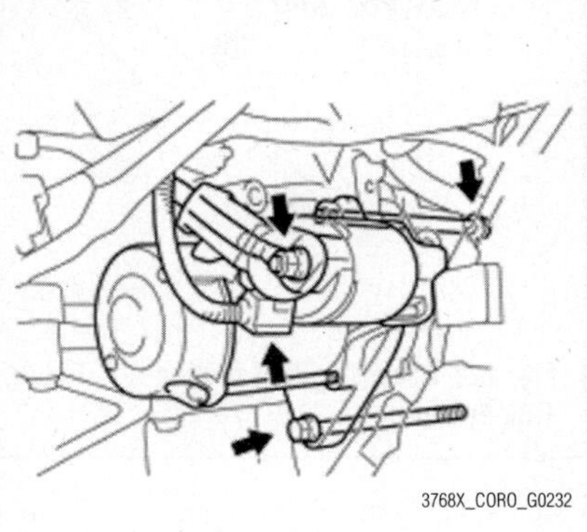

Fig. 119 Removing the starter assembly—1.8L Engine

To install:

7. Install the starter assembly with the 2 bolts and tighten to 27 ft. lbs. (37 Nm).
8. Connect the connector.
9. Connect terminal 30 with the nut. Tighten to 87 inch lbs. (9.8 Nm).
10. Close the terminal cap.
11. To complete the installation, reverse remaining removal procedure.

2.4L Engine

See Figures 120 and 121.

1. Disconnect the negative battery cable.
2. Remove no. 1 engine cover sub-assembly.

3. Remove air cleaner cap sub-assembly with hose and air cleaner case. Refer to Air Cleaner in Engine Mechanical.

4. Remove battery & battery carrier. Refer to Battery in Battery System.

5. Release the claw, and disconnect the wire harness.

6. For M/T vehicles, perform the following:

a. Disconnect the terminal 50 connector from the starter assembly.

b. Remove the nut and disconnect the wire harness from terminal 30.

c. Remove the 3 bolts, clutch accumulator bracket, wire harness clamp bracket and starter assembly.

7. For A/T vehicles:

a. Disconnect the terminal 50 connector from the starter assembly.

b. Remove the nut and disconnect the wire harness from terminal 30.

c. Remove the 2 bolts, wire harness clamp bracket and starter assembly.

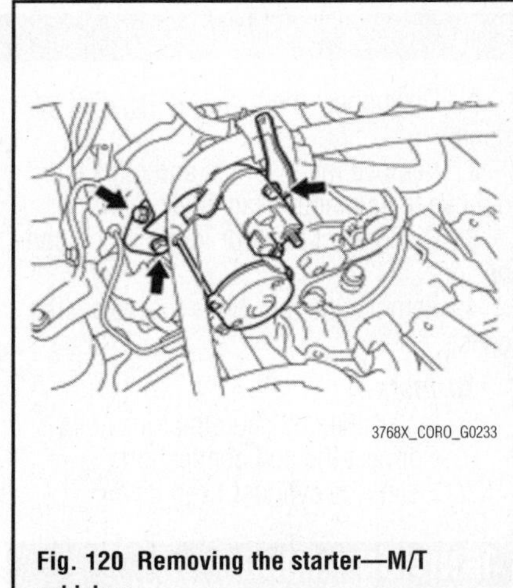

Fig. 120 Removing the starter—M/T vehicles

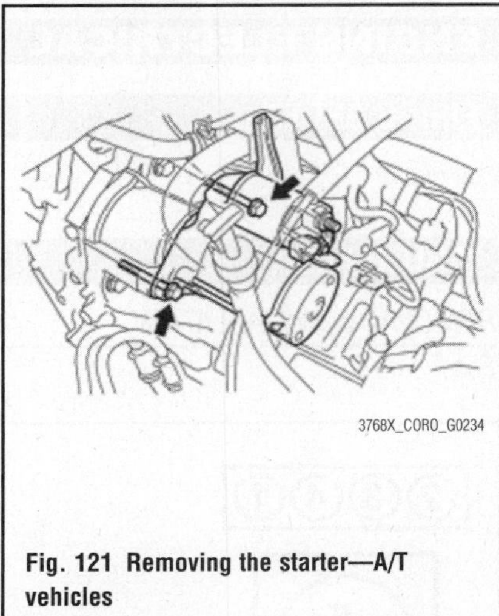

Fig. 121 Removing the starter—A/T vehicles

To install:

8. To install, reverse removal procedure.

9. For M/T vehicles, tighten the starter bolts to:

a. Bolt A: 27 ft. lbs. (37 Nm).

b. Bolt B: 9 ft. lbs. (12 Nm).

10. For A/T vehicles, tighten the bolts to 27 ft. lbs. (37 Nm).

ENGINE MECHANICAL

➡Disconnecting the negative battery cable may interfere with the functions of the on board computer systems and may require the computer to undergo a relearning process, once the negative battery cable is reconnected.

ACCESSORY DRIVE BELTS

ACCESSORY BELT ROUTING

See Figures 122 and 123.

INSPECTION

Inspect the drive belt for signs of glazing or cracking. A glazed belt will be perfectly smooth from slippage, while a good belt will have a slight texture of fabric visible. Cracks will usually start at the inner edge of the belt and run outward. All worn or damaged drive belts should be replaced immediately.

ADJUSTMENT

Only the 1.8L engine belt requires adjustment. Refer to the removal procedure for adjustment procedure.

REMOVAL & INSTALLATION

1.8L Engine

See Figure 124.

➡Refer to the illustration for bolt identification.

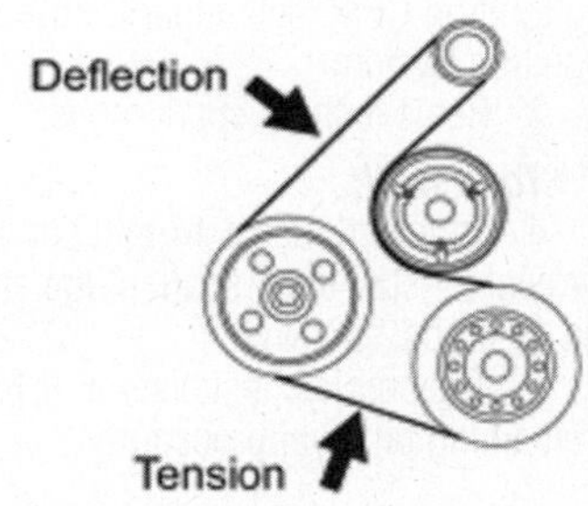

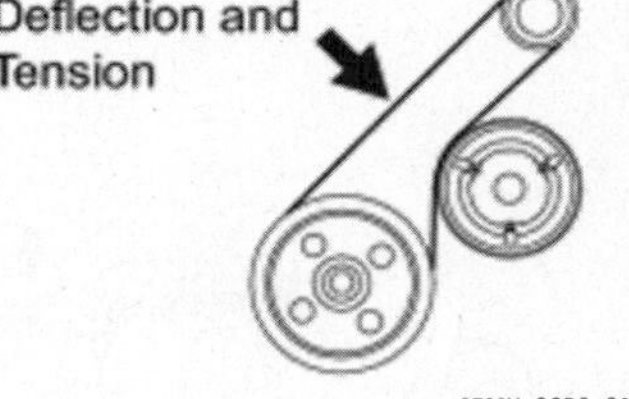

Fig. 122 Accessory belt routing—1.8L Engine

1. Remove the engine cover.

2. Remove the RH engine under cover.

3. Loosen bolts A and B.

4. Loosen bolt C, then remove the V-ribbed belt.

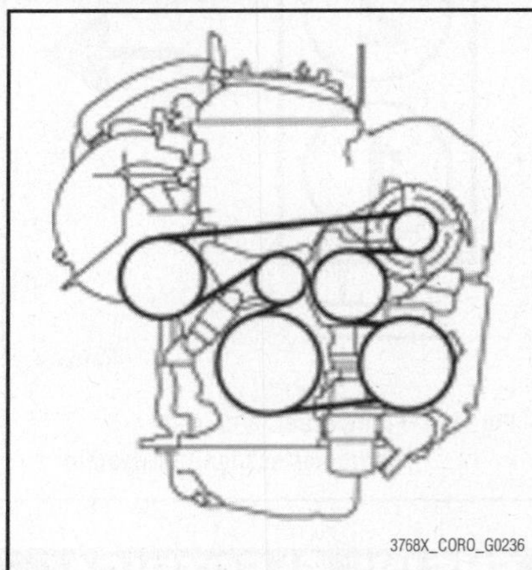

Fig. 123 Accessory belt routing—2.4L Engine

➡Do not loosen bolt D.

To install:

5. Install the belt.

6. Turn bolt C to adjust the tension of the V-ribbed belt.

7. Tighten bolts:

a. A: 14 ft. lbs. (19 Nm).

b. B: 32 ft. lbs. (43 Nm).

➡Confirm that bolt D is not loosened.

8. Inspect the belt.

9. To complete installation, reverse remaining removal procedure.

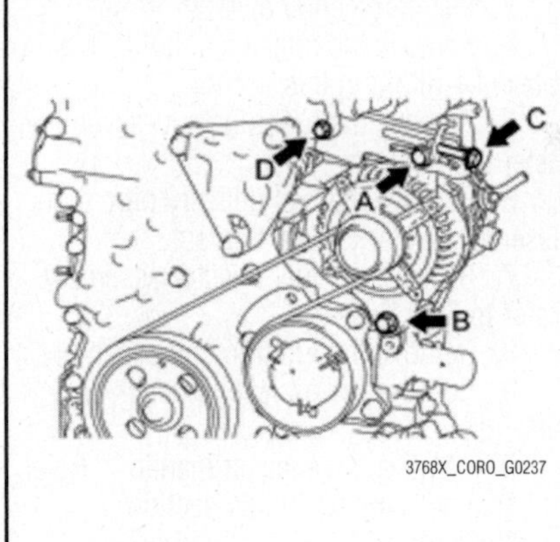

Fig. 124 Removing the accessory belt—1.8L engine

2.4L Engine

See Figure 125.

➡Adjustment is not possible or necessary with the auto-tensioner.

1. Remove the right hand cover under the engine.
2. Turn the drive belt tensioner clockwise to relieve tension on the belt.
3. Remove the fan and generator V belt.
4. Return the tensioner to the unloaded position.

➡When retracting the tensioner, turn it clockwise slowly in 3 sec. or more. Be sure not to apply force rapidly. After the tensioner is retracted all the way, do not apply force any more than necessary.

To install:

5. Turn the drive belt tensioner clockwise.
6. Install the belt.

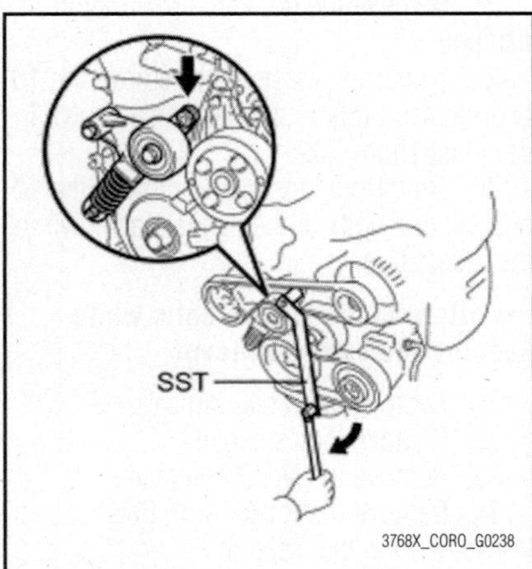

Fig. 125 Removing the serpentine belt—2.4L engine

7. Install the right hand cover under the engine.

AIR CLEANER

REMOVAL & INSTALLATION

1.8L Engine

See Figure 126.

1. Drain and recover the engine coolant.
2. Remove the engine cover.
3. Disconnect the mass air flow meter connector and the 2 wire harness clamps.
4. Disconnect the ventilation hose and loosen the hose clamp.
5. Release the 2 clamps and remove the air cleaner cap sub-assembly with hose.

To install:

6. To install, reverse removal procedure.
7. Check for leaks.

2.4L Engine

See Figure 127.

1. Drain and recover the engine coolant.
2. Remove the engine cover.
3. Disconnect the mass air flow meter connector.
4. Separate the 2 wire harness clamps.
5. Disconnect the No. 1 vacuum switching valve connector and the 2 vacuum hoses.
6. Disconnect the ventilation hose.
7. Loosen the No. 1 air cleaner hose clamp, release the 3 air cleaner assembly clamps and remove the air cleaner cap sub-assembly with No. 1 hose.

To install:

8. To install, reverse removal procedure.
9. Check for leaks.

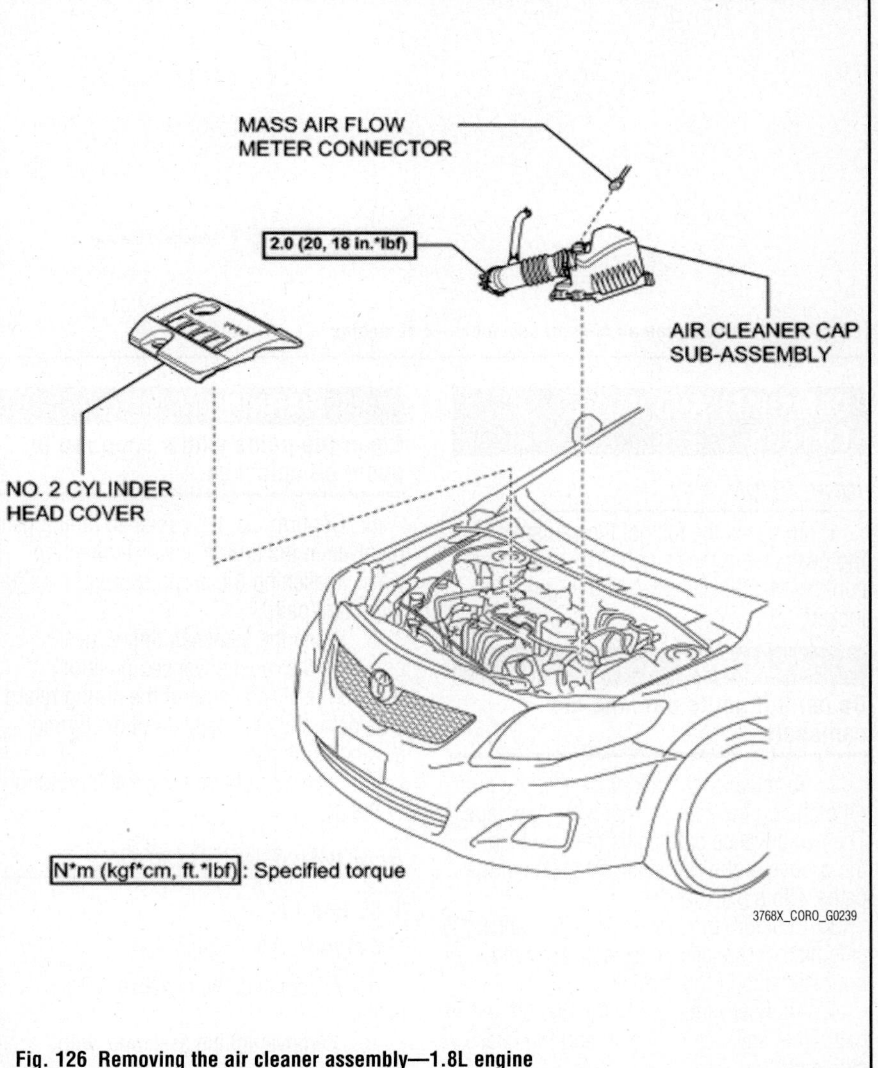

Fig. 126 Removing the air cleaner assembly—1.8L engine

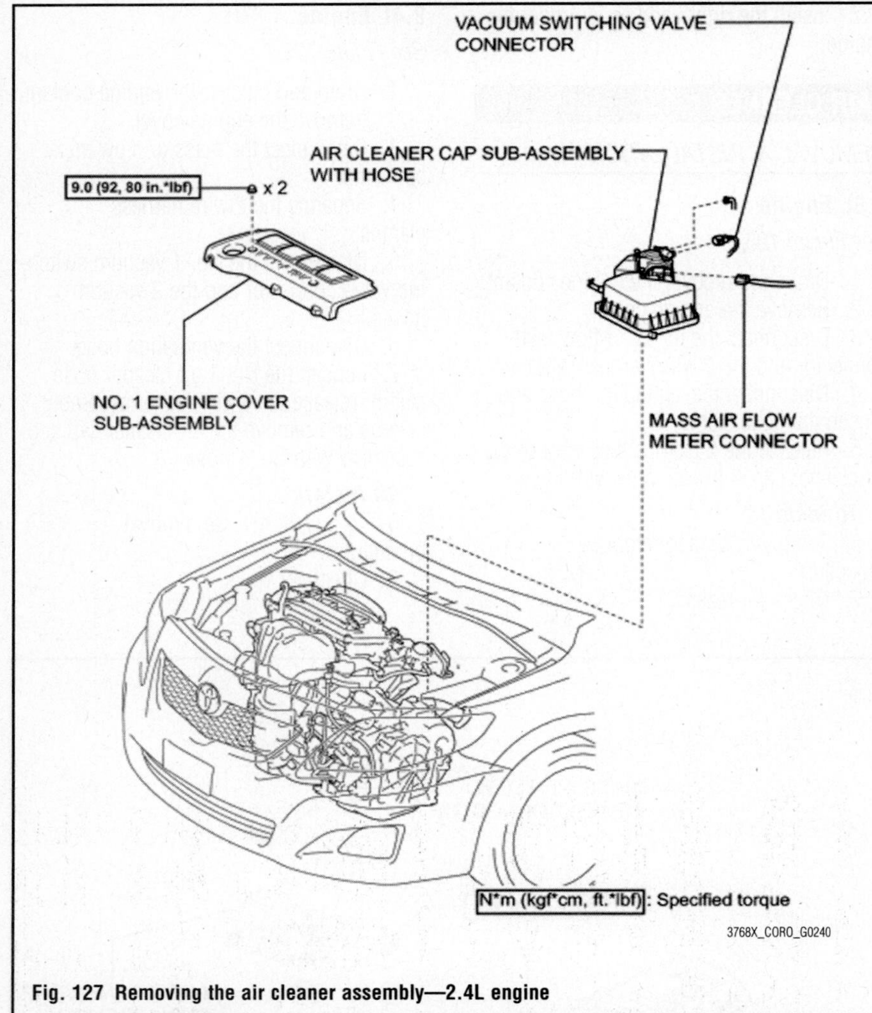

Fig. 127 Removing the air cleaner assembly—2.4L engine

CAMSHAFT AND VALVE LIFTERS

INSPECTION

1. To check the lock of the camshaft timing gear, clamp the camshaft in a vice and confirm that the camshaft timing gear is locked.

**** CAUTION**

Be careful not to damage the camshaft.

2. To release the lock pin, cover the four oil paths on the cam journal with vinyl tape. The two advance side paths are provided in the groove of the camshaft. Plug one of the paths with a piece of rubber.
3. Puncture the tape over the advance side path and retard the side path on the opposite side of the groove.
4. Apply air pressure to the two broken paths (the advance side path and the retard side path).

**** WARNING**

Cover the paths with a shop rag to avoid oil splashes.

5. Confirm that the camshaft timing gear revolves in the timing advance direction when weakening the air pressure of the timing retard path.
6. When the camshaft timing gear comes to the most advanced position, release the air pressure of the timing retard side path, then release that of the timing advance side path.
7. There should be a smooth revolution of the gear.

REMOVAL & INSTALLATION

1.8L Engine

See Figures 128 through 135.

1. Disconnect the negative battery cable.
2. Remove engine assembly with transaxle.
3. Install the engine stand.
4. Remove the intake manifold. Refer to Intake Manifold in this section.
5. Remove the fuel tube sub-assembly. Refer to Fuel System.
6. Remove the fuel delivery pipe sub-assembly. Refer to Fuel System.
7. Remove the fuel injector assembly. Refer to Fuel System.
8. Remove the ignition coil assembly.
9. Remove the oil level dipstick sub-assembly.
10. Remove the exhaust manifold. Refer to Exhaust Manifold in this section.
11. Remove the ventilation hose.
12. Disconnect the no. 3 water by-pass hose.
13. Remove the no. 1 water by-pass pipe.
14. Remove the water by-pass hose.
15. Remove the inlet water hose.
16. Remove the inlet water.
17. Remove the thermostat. Refer to Thermostat in Engine Cooling.
18. Remove the radio setting condenser.
19. Remove the cylinder head cover sub-assembly.
20. Turn the crankshaft so that the No. 1 piston is at TDC on the compression stroke. Check to see that the point marks on the camshaft sprockets are facing each other, if not, rotate the crankshaft 1 full revolution.
21. Remove the crankshaft pulley.
22. Remove the no. 1 chain tensioner assembly.
23. Remove the timing chain cover sub-assembly.
24. Remove the timing chain cover oil seal.
25. Remove the chain tensioner slipper.
26. Remove the no. 1 chain vibration damper.
27. Remove the chain sub-assembly.
28. Remove the no. 2 chain vibration damper.
29. Uniformly loosen and remove the 10 bearing cap bolts in the sequence shown in the illustration.
30. Uniformly loosen and remove the 15 bearing cap bolts in the sequence shown in the illustration.

➡Uniformly loosen the bolts while keeping the camshaft level.

31. Remove the 5 bearing caps.
32. Remove the camshaft.
33. Remove the No. 2 camshaft.
34. Remove the rocker arm. Refer to Rocker Arm in this section.
35. Remove the valve lash adjuster assembly. Refer to Valve Lash Adjuster in this section.

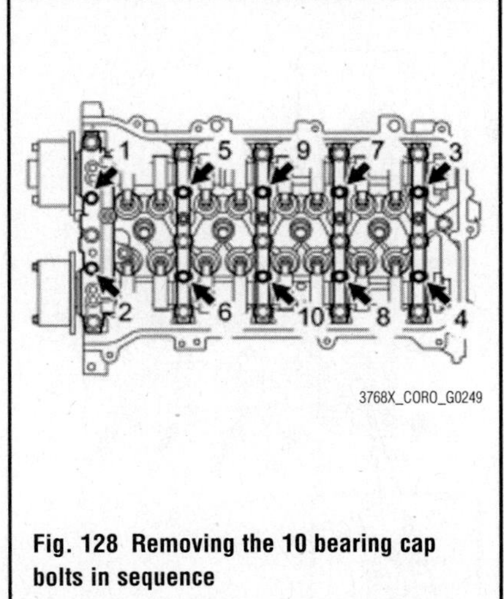

Fig. 128 Removing the 10 bearing cap bolts in sequence

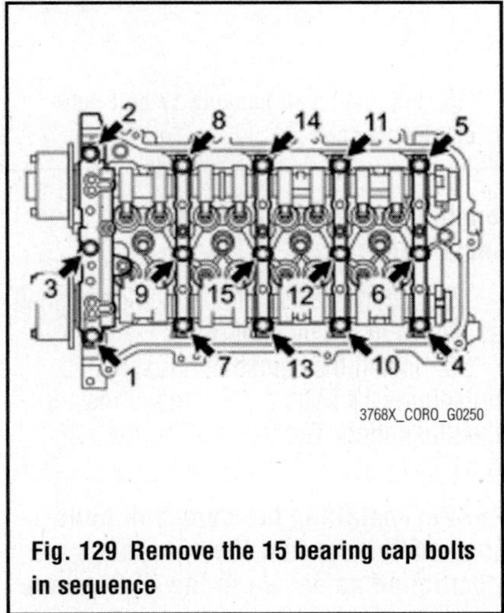

Fig. 129 Remove the 15 bearing cap bolts in sequence

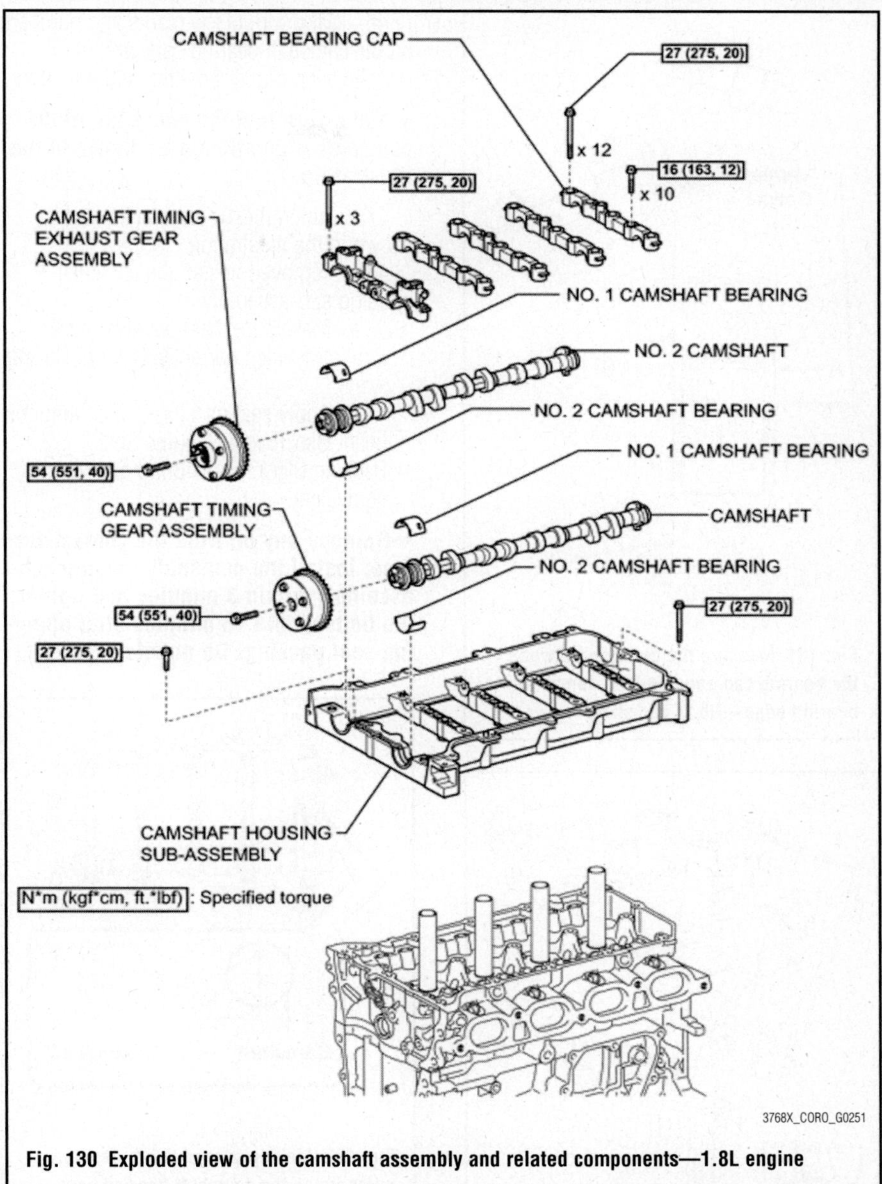

Fig. 130 Exploded view of the camshaft assembly and related components—1.8L engine

36. Remove the 2 No. 1 camshaft bearings.
37. Remove the 2 No. 2 camshaft bearings.
38. If necessary, remove camshaft housing sub-assembly:
   a. Remove the 2 bolts.
   b. Remove the camshaft housing by prying between the cylinder head and camshaft housing with a screwdriver.

➡Be careful not to damage the contact surfaces of the cylinder head and camshaft housing. Tape the screwdriver tip before use.

To install:

39. Remove the valve lash adjuster assembly. Refer to Valve Lash Adjuster in this section.
40. Install the rocker arm. Refer to Rocker Arm in this section.
41. Install the no. 1 camshaft bearing:
   a. Clean both surfaces of the bearings.
   b. Install the 2 No. 1 camshaft bearings.
   c. Using a vernier caliper, measure the distance between the bearing cap edge and the camshaft bearing edge. Dimension A-B: 0.7 mm (0.0276 in.) or less.

➡Position the bearings to the center of the bearing cap by measuring dimensions A.

42. Install the no. 2 camshaft bearing:
   a. Clean both surfaces of the bearings.
   b. Install the 2 No. 2 camshaft bearings.
   c. Using a vernier caliper, measure the distance between the bearing cap edge and the camshaft bearing edge. Dimension: 0.0413 to 0.0689 inch (1.05 to 1.75 mm)

➡Position the bearings to the center of the bearing cap by measuring dimensions A.

43. Install the no. 2 camshaft:
   a. Clean the camshaft journals.
   b. Apply a light coat of engine oil to the camshaft journals, camshaft housings and bearing caps.
   c. Install the No. 2 camshaft to the camshaft housing.
44. Install the camshaft:
   a. Apply a light coat of engine oil to the camshaft journals, camshaft housings and bearing caps.

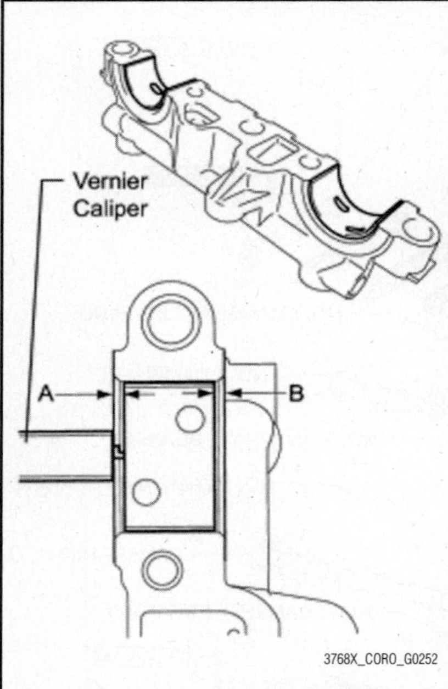

Fig. 131 Measure the distance between the bearing cap edge and the camshaft bearing edge—No. 1 camshaft

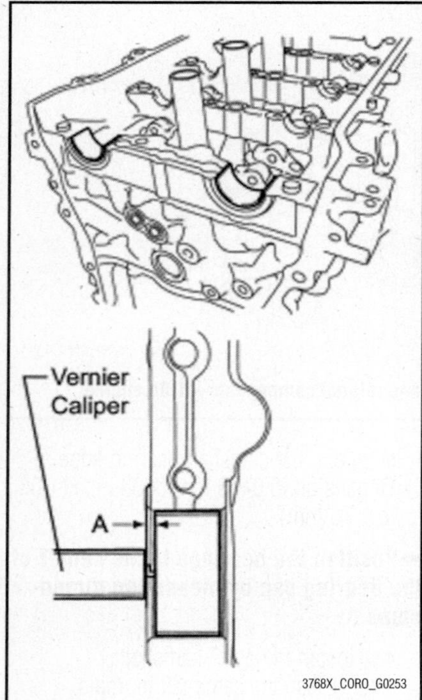

Fig. 132 Measure the distance between the bearing cap edge and the camshaft bearing edge—No. 2 camshaft

b. Install the camshaft to the camshaft housing.

45. Apply engine oil to the camshaft journals, camshaft housings and bearing caps.

46. Make sure of the marks and numbers on the camshaft bearing caps and place them in each proper position and direction.

➡Make sure that the knock pin of the camshaft is positioned as shown in the illustration.

47. Tighten the 10 bolts in the order shown in the illustration.

48. If removed, install the camshaft housing sub-assembly:

a. Make sure that the valve rocker arm is installed correctly. Refer to Rocker Arms.

b. Apply seal packing in a continuous bead. Use Toyota Genuine Seal Packing Black, Three Bond 1207B or equivalent.

➡Remove any oil from the contact surface. Install the camshaft housing sub-assembly within 3 minutes and tighten the bolts within 15 minutes after applying seal packing. Do not start the engine for at least 2 hours after installing.

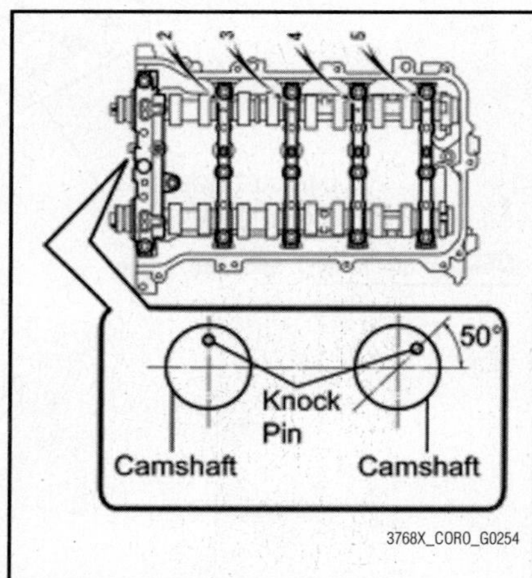

Fig. 133 Making sure of the marks and numbers on the camshaft bearing caps

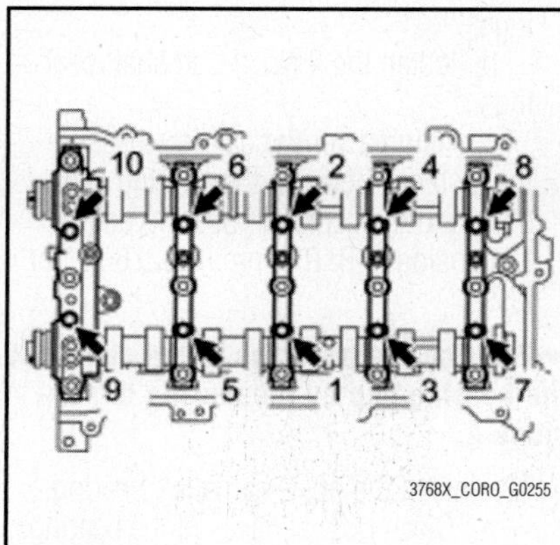

Fig. 134 Bearing cap 10 bolt tightening sequence

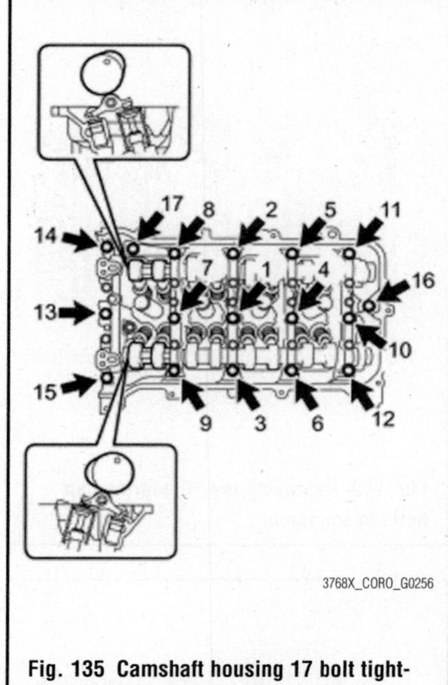

Fig. 135 Camshaft housing 17 bolt tightening sequence

49. Set the camshaft and No. 2 camshaft as shown in the illustration.

50. Install the camshaft housing and tighten the 17 bolts in the order shown in the illustration. Tighten to 20 ft. lbs. (27 Nm).

➡After installing the camshaft housing, make sure that the cam lobes are positioned as shown in the illustration.

➡If any of the bolts are loosened during installation, remove the camshaft housing, clean the installation surfaces, and reapply seal packing.

➡If the camshaft housing is removed because any of the bolts are loosened during installation, make sure that the previously applied seal packing does not enter any oil passages.

➡After installing the camshaft housing, wipe off any seal packing that seeped out from between the housing and the cylinder head.

51. To complete the installation, reverse remaining removal procedure.

2.4L Engine

See Figures 136 through 142.

1. Disconnect the negative battery cable.

2. Remove the RH engine under cover.

3. Remove the No. 1 engine cover sub-assembly. Refer to Valve Cover in this section.

4. Remove the ignition coil assembly. Refer to Ignition Coil in Engine Electrical.

5. Remove the spark plugs. Refer to Spark Plugs in Engine Electrical.

6. Remove the valve cover. Refer to Valve Cover in this section.

7. Set no. 1 cylinder to TDC/compression:

a. Turn the crankshaft pulley until the groove and the timing mark "0" on the timing chain cover are aligned.

b. Check that each timing mark on the camshaft timing gear and sprocket is aligned with each timing mark located on the No. 1 and No. 2 bearing caps.

c. If not, turn the crankshaft pulley 1 revolution (360°) to align the timing marks.

d. Place paint marks on the chain in alignment with the timing marks on the camshaft timing gear and camshaft timing sprocket.

8. While holding the No. 2 camshaft with a wrench, loosen the No. 2 camshaft timing set bolt.

9. Using several steps, uniformly loosen and remove the 10 bearing cap bolts in the sequence shown in the illustration.

10. Remove the 5 bearing caps.

11. While holding the No. 2 camshaft by hand, remove the camshaft timing sprocket set bolt.

12. Remove the camshaft timing sprocket from the No. 2 camshaft with the timing chain wrapped on the sprocket.

13. Remove the camshaft timing sprocket from the timing chain.

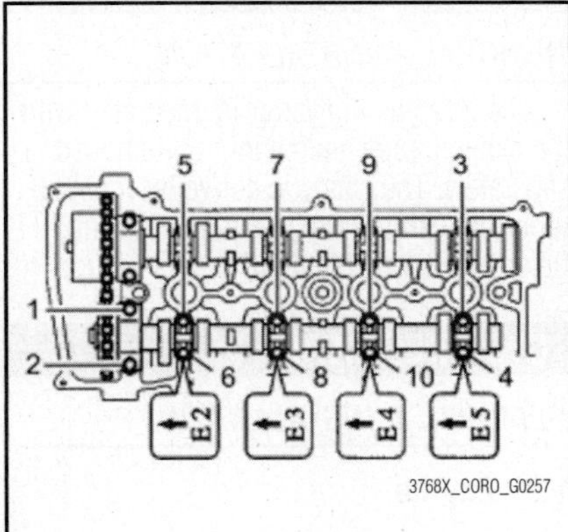

Fig. 136 10 bearing cap bolts removal sequence and identifying the 5 bearing caps

14. In several steps, uniformly loosen and remove the 10 bearing caps bolts in the sequence shown in the illustration.

15. Remove the 5 bearing caps.

16. Remove the camshaft and camshaft timing gear assembly while holding the timing chain by hand.

17. Support the timing chain with a string to prevent it from slipping off the crankshaft sprocket.

➡**Be careful not to drop anything inside the timing chain cover.**

To install:

18. Apply a light coat of engine oil to the journal portion of the camshaft.

19. Install the timing chain onto the camshaft timing gear with the paint mark aligned with the timing mark on the camshaft timing gear.

20. Examine the front marks and numbers, and check that the order is as shown in the illustration. Then install the bearing caps into the cylinder head.

21. Apply a light coat of engine oil on the threads and under the heads of the bearing cap bolts.

22. Using several steps, uniformly tighten the 10 bearing cap bolts in the sequence shown in the illustration.

a. No. 1 bearing: 22 ft. lbs. (30 Nm).

b. No. 3 bearing: 80 inch lbs. (9 Nm).

23. Install the no. 2 camshaft:

a. Apply a light coat of engine oil to the journal portion of the No. 2 camshaft.

b. Put the No. 2 camshaft on the cylinder head with the paint mark on the chain aligned with the timing mark on the camshaft timing sprocket.

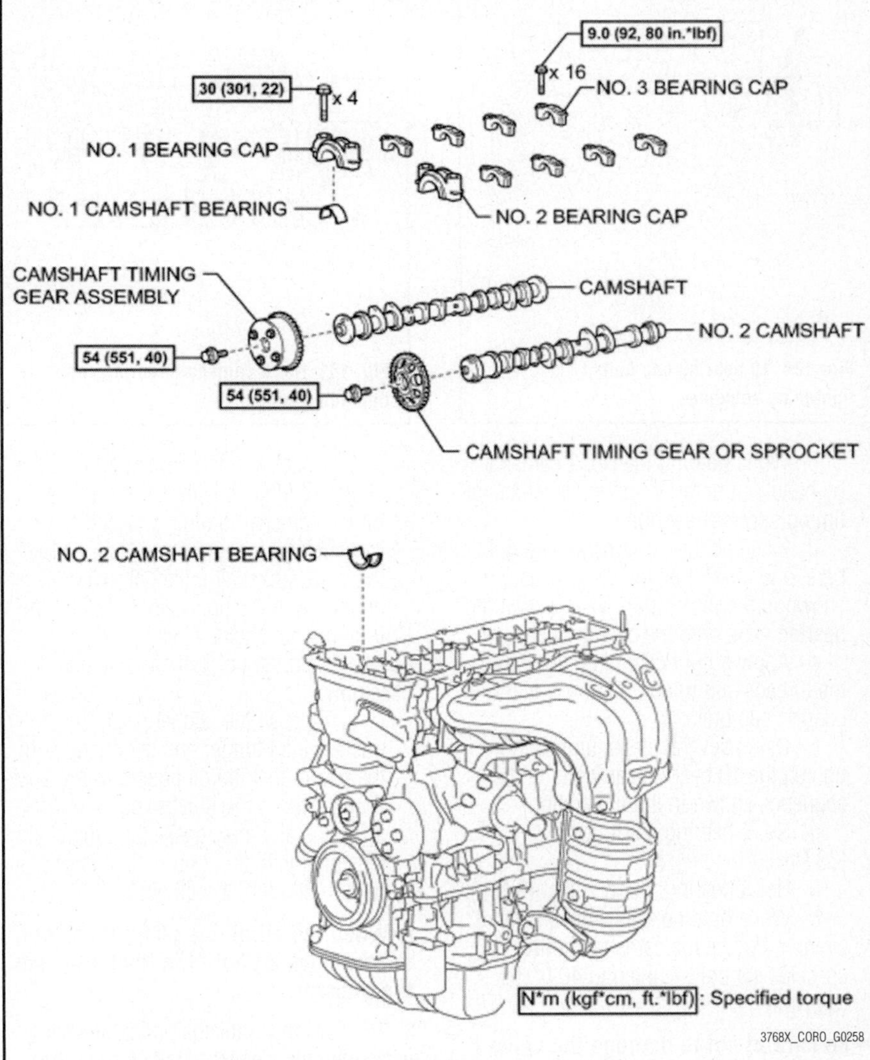

Fig. 137 Exploded view of the camshaft assembly and related components

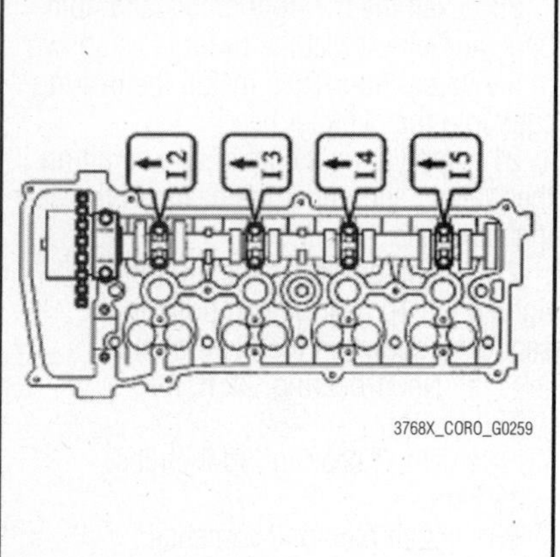

Fig. 138 Checking camshaft No. 1 bearing cap orientation

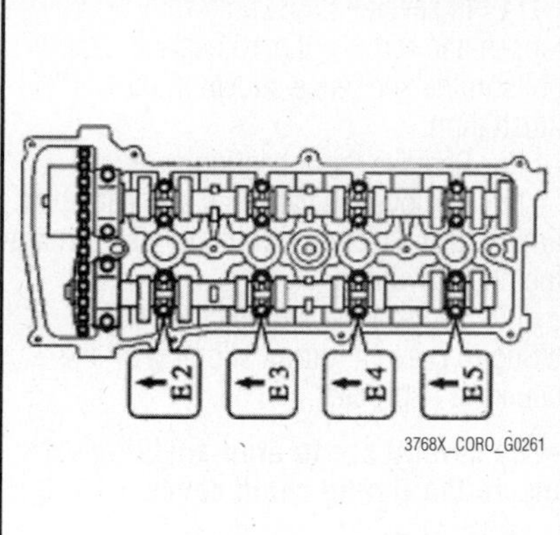

Fig. 140 No. 2 camshaft bearing cap orientation

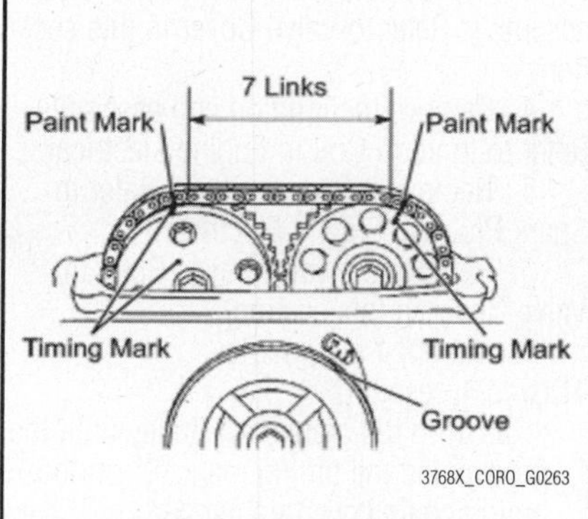

Fig. 142 Check that the paint marks on the chain are aligned with the timing marks on the camshaft timing gear and camshaft timing sprocket

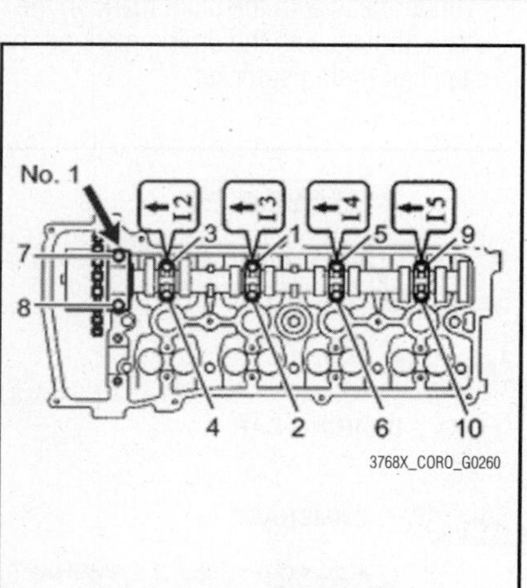

Fig. 139 10 bearing cap bolts tightening sequence

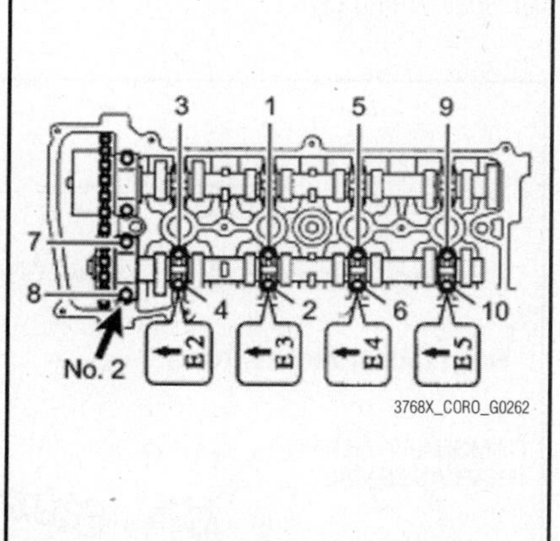

Fig. 141 No. 2 camshaft bearing cap tightening sequence

c. While holding the No. 2 camshaft by hand, temporarily tighten the camshaft timing sprocket set bolt.

d. Examine the front marks and numbers, and check that the order is as shown in the illustration. Then install the bearing caps onto the cylinder head.

e. Apply a light coat of engine oil to the threads and under the heads of the bearing cap bolts.

f. Using several steps, uniformly tighten the 10 bearing cap bolts in the sequence shown in the illustration.

g. No. 2 bearing: 22 ft. lbs. (30 Nm).

h. No. 3 bearing: 80 inch lbs. (9 Nm).

i. While holding the camshaft with a wrench, tighten the camshaft timing sprocket set bolt. Tighten to 40 ft. lbs. (54 Nm.)

➡Be careful not to damage the valve lifter.

j. Check that the paint marks on the chain are aligned with the timing marks on the camshaft timing gear and camshaft timing sprocket. Also, check that the crankshaft pulley groove is aligned with the timing mark "0" on the timing mark chain cover.

24. Install the no. 1 chain tensioner assembly:

a. Release the ratchet pawl, then fully push in the plunger and set the hook to the pin so that the plunger is in the position shown in the illustration.

b. Install a new gasket and the chain tensioner with the 2 nuts. Tighten the nuts to 80 inch. lbs. (9 Nm).

➡When installing the chain tensioner, set the hook again if the hook releases the plunger.

25. Turn the crankshaft counterclockwise, then disconnect the hook from the pin.

26. Turn the crankshaft clockwise, then check that the plunger is extended.

27. Set no. 1 cylinder to TDC/compression:

a. Turn the crankshaft pulley until the groove and the timing mark "0" on the timing chain cover are aligned.

b. Check that each timing mark on the camshaft timing gear and sprocket is aligned with each timing mark located on the No. 1 and No. 2 bearing caps.

c. If not, turn the crankshaft pulley 1 revolution (360°) to align the timing marks.

d. Place paint marks on the chain in alignment with the timing marks on the camshaft timing gear and camshaft timing sprocket.

28. Check valve clearance.

29. Adjust valve clearance.

30. To complete installation, reverse remaining removal procedure.

CATALYTIC CONVERTER

REMOVAL & INSTALLATION

The catalytic converter is integrated with the exhaust pipe and cannot be removed separately. The catalytic converter must be replaced with the exhaust pipe as a unit. Refer to Exhaust Manifold in this section.

CYLINDER HEAD

REMOVAL & INSTALLATION

1.8L Engine

See Figures 143 through 146.

1. Remove the camshaft assembly. Refer to Camshaft in this section.

2. Remove cylinder head sub-assembly:

a. Using several steps, uniformly loosen and remove the 10 cylinder head bolts and 10 plate washers with a 10 mm bi-hexagon wrench in the sequence shown in the illustration.

➡Head warpage or cracking could result from removing the bolts in the wrong order.

b. Using a screwdriver with its tip wrapped with tape, pry between the cylinder head and cylinder block, and remove the cylinder head.

➡Be careful not to damage the contact surfaces of the cylinder head and cylinder block.

3. Remove the cylinder head gasket.
4. Remove the water drain cock plug from the water drain cock sub-assembly.
5. Remove the cylinder block water drain cock sub-assembly from the cylinder block.
6. Remove the ventilation valve.
7. Remove the oil pan drain plug and gasket. Remove the oil pan. Refer to Oil Pan in this section.
8. Remove oil pump assembly. Refer to Oil Pump in this section.
9. Remove rear engine oil seal:
   a. Using a knife, cut off the oil seal lip.
   b. Using a screwdriver with its tip taped, pry out the oil seal.

➡After removing the oil seal, check the crankshaft for damage. If it is damaged, smooth the surface with 400-grit sandpaper.

10. Remove stiffening crankcase assembly:

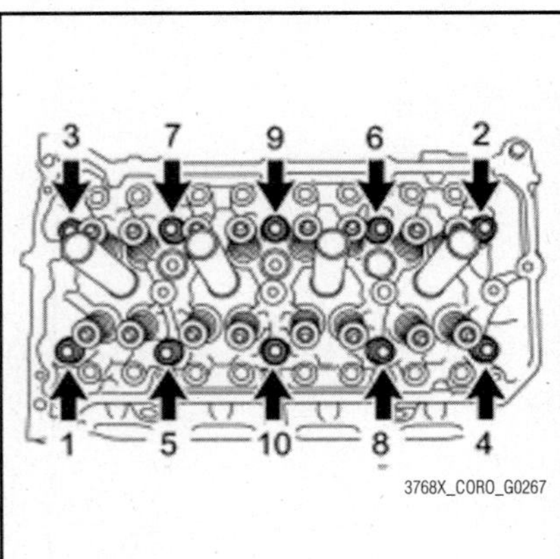

Fig. 143 Cylinder head sub-assembly removal sequence

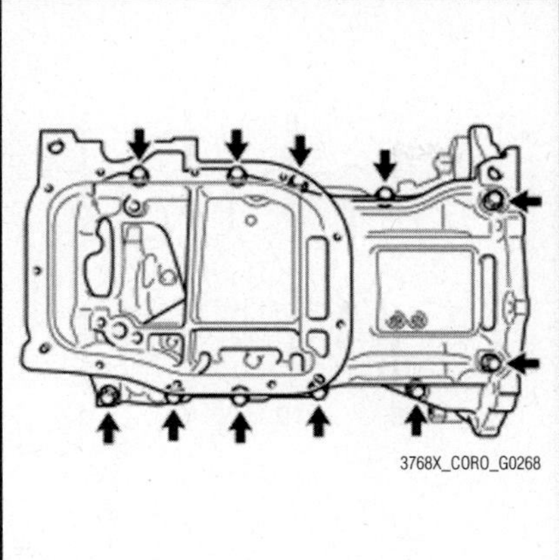

Fig. 144 Location of the 11 stiffening crankcase assembly bolts

a. Uniformly loosen and remove the 11 bolts.

b. Using a screwdriver, remove the crankcase by prying between the crankcase and cylinder block.

➡Be careful not to damage the contact surfaces of the crankcase and cylinder block.

To install:

11. Install the stiffening crankcase with the 11 bolts. Tighten to 16 ft. lbs. (21 Nm).
    a. Bolt A: 5.43 inches.
    b. Bolt B: 1.38 inches.
    c. Bolt C: 2.76 inches.
12. Recheck the torque for bolts 1 and 2. Tighten to 16 ft. lbs. (21 Nm).

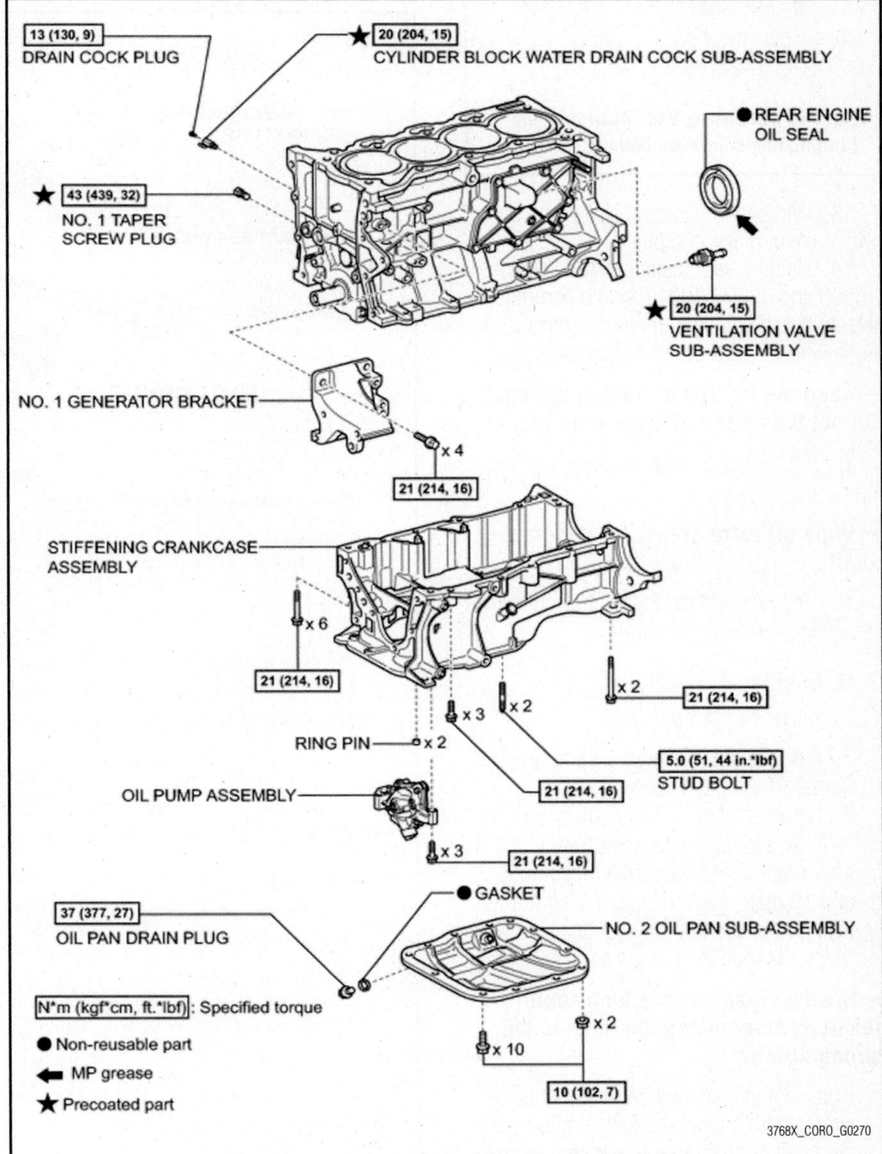

Fig. 145 Exploded view of the cylinder head and related components—1.8L engine

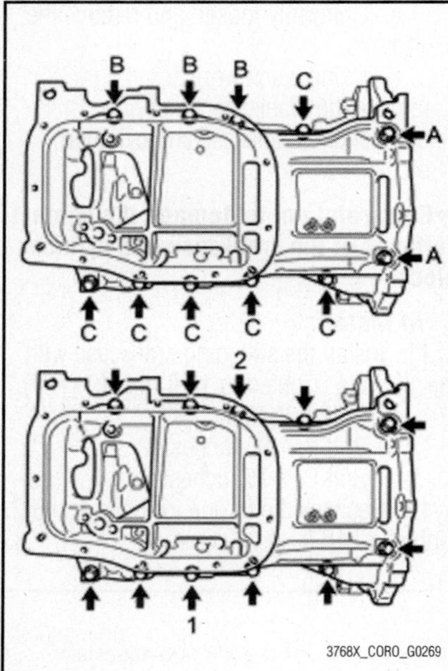

Fig. 146 Identifying and Location of the 11 stiffening crankcase bolts

13. Wipe off any excess seal packing with a clean piece of cloth.

14. Using a seal installer (SST 09223-15030 and 09950-70010) and a hammer, evenly tap the oil seal until its surface is flush with the rear oil seal retainer edge.

➡Keep the lip free of foreign matter. Do not tap on the oil seal at an angle.

15. Apply MP grease to a new oil seal lip.

➡Wipe off extra grease on the crankshaft.

16. To complete the installation, reverse remaining removal procedure.

2.4L Engine

See Figures 147 through 149.

1. Remove the camshaft assembly. Refer to Camshaft in this section.
2. Remove cylinder head sub-assembly:
    a. In several steps, uniformly loosen and remove the 10 cylinder head bolts and 10 plate washers with a 10 mm bi-hexagon wrench in the sequence shown in the illustration.

➡Head warpage or cracking could result from removing the bolts in the wrong order.

b. Using a screwdriver with its tip wrapped with tape, pry between the cylinder head and cylinder block, and remove the cylinder head.

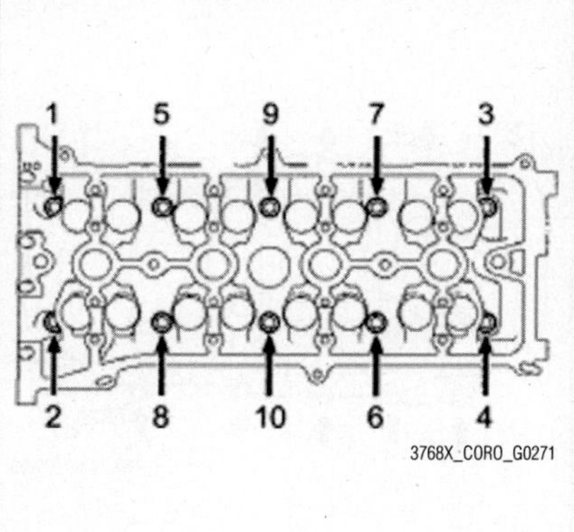

Fig. 147 Cylinder head sub-assembly removal sequence

➡Be careful not to damage the contact surfaces between the cylinder head and cylinder block.

3. Remove cylinder head gasket.

To install:

4. Place a new cylinder head gasket on the cylinder block surface with the Lot No. stamp facing upward.

➡Remove any oil from the contact surface. Be careful of the installation direction.

5. Place the cylinder head on the cylinder head gasket.

➡Place the cylinder head gently in order to avoid damaging the cylinder head gasket.

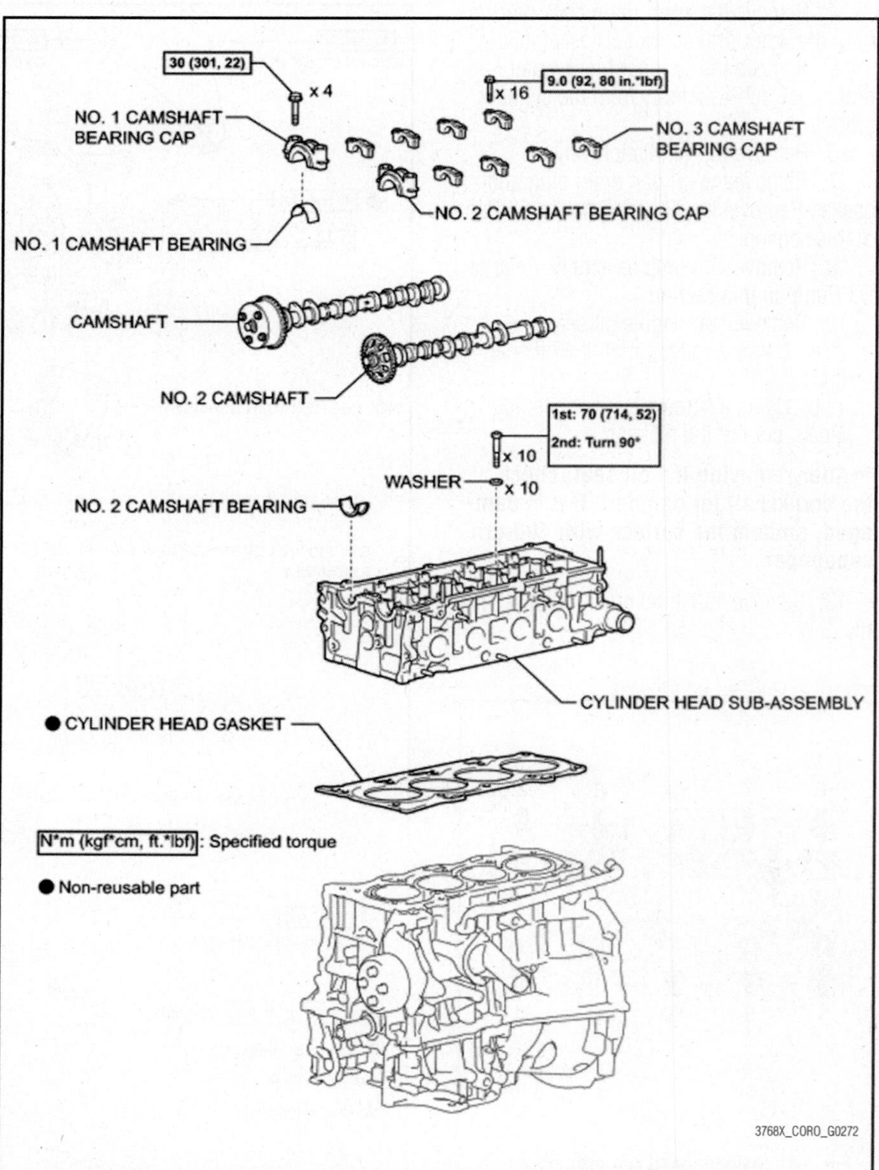

Fig. 148 Exploded view of the cylinder head and related components—2.4L engine

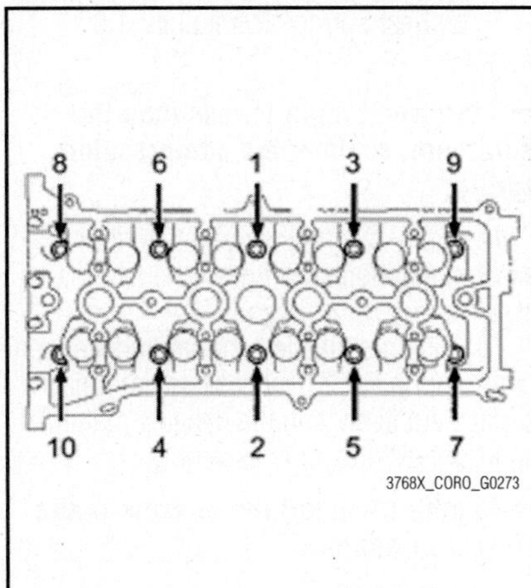

Fig. 149 Cylinder head tightening sequence

➡**The cylinder head bolts are tightened in 2 successive steps.**

6. Install the cylinder head bolts:
   a. Apply a light coat of engine oil to the threads and under the heads of the cylinder head set bolts.
   b. Using several steps, uniformly install and tighten the 10 cylinder head set bolts and plate washers with a 10 mm bi-hexagon wrench in the order shown in the illustration. Tighten to 52 ft. lbs. (70 Nm).
   c. Mark the front of the cylinder head bolts with paint.
   d. Further tighten the cylinder head bolts 90°.
   e. Check that the paint mark is now at a 90° angle to the front.
7. To complete the installation, reverse remaining removal procedure.

ENGINE OIL & FILTER

REPLACEMENT

See Figures 150 through 153.

CAUTION

Prolonged and repeated contact with engine oil will cause removal of natural oils from the skin, leading to dryness, irritation and dermatitis. In addition, used engine oil contains potentially harmful contaminants which may cause skin cancer.

- Precautions should be taken when replacing engine oil to minimize the risk of your skin making contact with used engine oil. Wear protective clothing and gloves. Wash your skin thoroughly with soap and water, or use a waterless hand cleaner to remove any used engine oil. Do not use gasoline, thinners or solvents.
- For environmental protection, used oil and used oil filters must be disposed of at designated disposal sites.

1. Remove the oil filler cap.
2. Remove the oil drain plug and drain the oil into a container.
3. Clean and install the oil drain plug with a new gasket. Tighten to 27 ft. lbs. (37 Nm).
4. Using a wrench or oil filter remover (SST: 09228-06501) loosen the oil filter cap 4 revolutions, align the cap ribs vertically, and drain the remaining engine oil in the oil filter cap.

➡**Set a container below the oil filter cap assembly before loosening the oil filter cap.**

5. Remove the oil filter cap assembly.
6. Remove oil filter element and O-ring from the oil filter cap.

➡**Be sure to remove the O-ring (for the cap) by hand, without using any tools, to prevent damage to the groove for the O-ring on the cap.**

To install:

7. Clean the oil filter cap threads and O-ring groove.
8. Apply a small amount of engine oil to a new O-ring and install it to the oil filter cap.
9. Set a new oil filter element in the oil filter cap.
10. Remove any dirt or foreign matter from the installation surface and inside of the engine.

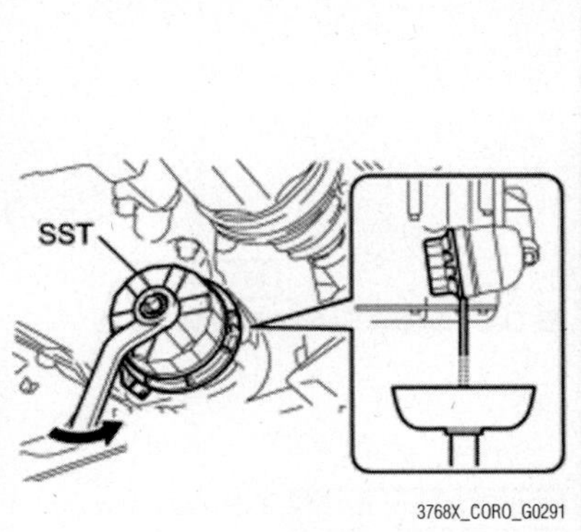

Fig. 150 Removing the oil filter cap assembly—1.8L engine

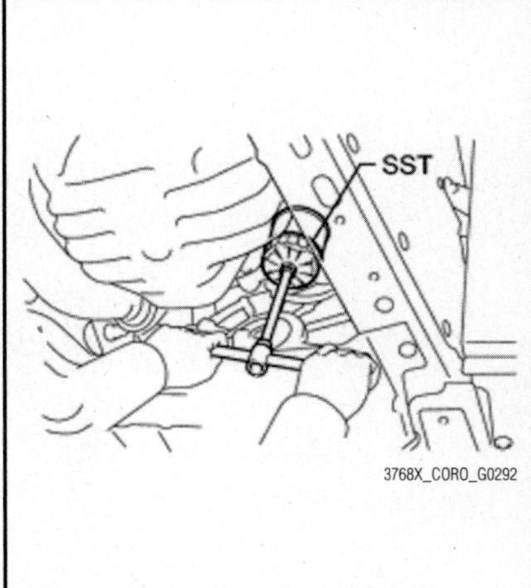

Fig. 151 Removing the oil filter cap assembly—2.4L engine

11. Reapply a small amount of engine oil to the O-ring of the oil filter cap assembly. Align the cutout in the oil filter cap threads 90°to the grooves in the oil filter bracket and temporarily tighten the cap.

➡**Make sure that the O-ring does not get caught between the parts.**

12. Tighten the oil filter cap assembly to 18 ft. lbs. (25 Nm).

➡**After tightening the oil filter cap assembly, check for gaps between the installation surfaces.**

- Do not remove the oil filter bracket clip when installing the oil filter cap assembly.
- Do not cross thread the oil filter cap assembly.

13. Add 5 quarts of fresh engine oil and install the oil filler cap.
14. Start the engine for 30 seconds, turn off and check the fluid level.
15. Inspect for oil leaks.

EXHAUST MANIFOLD

REMOVAL & INSTALLATION

1.8L Engine

See Figures 154 and 155.

1. Disconnect the negative battery cable.
2. Remove front wiper arm motor and assembly.
3. Remove the valve cover.
4. Remove air fuel ratio sensor.
5. Remove the 4 bolts and the exhaust manifold heat insulator.

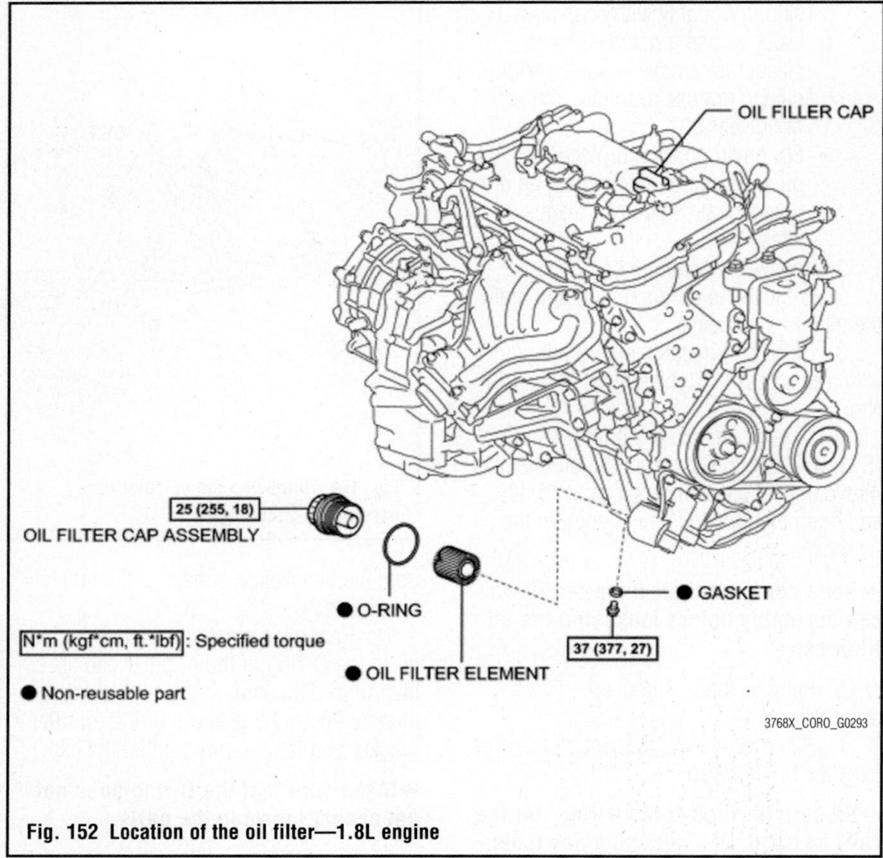

Fig. 152 Location of the oil filter—1.8L engine

6. Disconnect the heated oxygen sensor connector. Refer to Heated Oxygen Sensor in Engine Performance and Emission control.
7. Remove the 4 bolts and 4 compression springs.
8. Remove the front exhaust pipe assembly from the 2 exhaust pipe supports.
9. Remove the 3 bolts and manifold stay.
10. Remove the 5 nuts and exhaust manifold.
11. Remove the exhaust manifold gasket.
12. Remove the 3 bolts and exhaust manifold heat insulator.

To install:

13. Install the No. 2 exhaust manifold heat insulator with the 3 bolts. Tighten to 9 ft. lbs. (12 Nm).
14. Install a new exhaust manifold gasket.
15. Install the exhaust manifold with the 5 nuts. Tighten to 16 ft. lbs. (21 Nm).
16. Install the manifold stay with the 3 bolts. Tighten to 32 ft. lbs. (43 Nm).
17. Using a vernier caliper, measure the free length of the compression springs.
    a. Front springs: Minimum: 1.63 inches.
    b. Rear springs: Minimum: 1.52 inches.

➡If the free length is less than the minimum, replace the compression spring.

18. Fully insert new gaskets to the exhaust manifold and front exhaust pipe assembly.
19. Using a plastic hammer and wooden block, tap in the 2 new gaskets until their surface are flush with the exhaust manifold and front exhaust pipe assembly.

➡Be sure to install the gaskets in the correct direction.

- Do not reuse the gaskets.
- Do not damage the gaskets.
- Do not push in the gaskets by using the exhaust pipe when connecting it.

20. Connect the front exhaust pipe assembly to the 2 exhaust pipe supports.
21. Install the front exhaust pipe assembly with the 4 bolts and 4 compression springs. Tighten to 32 ft. lbs. (43 Nm).
22. Connect the heated oxygen sensor connector.

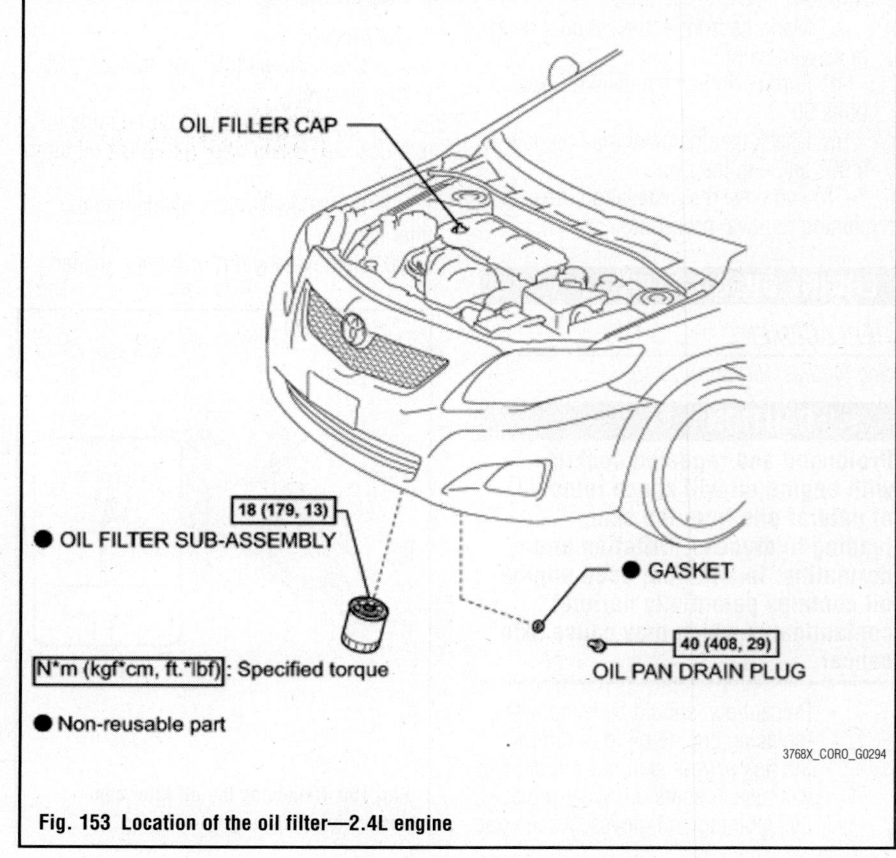

Fig. 153 Location of the oil filter—2.4L engine

Fig. 154 Removing the 4 bolts and 4 compression springs

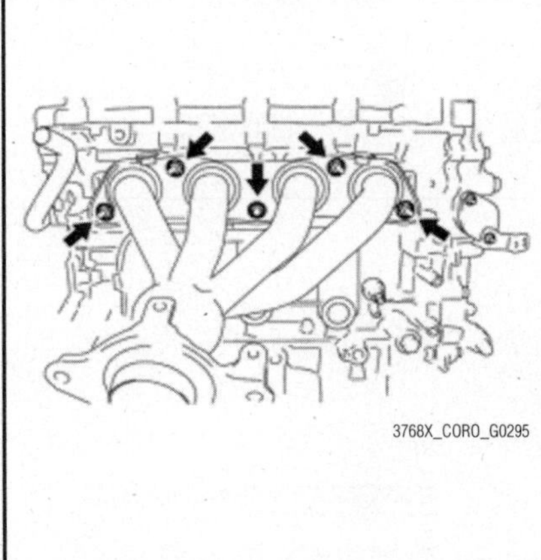

Fig. 155 Removing the 5 nuts and exhaust manifold

23. Install the outer exhaust manifold heat insulator with the 4 bolts. Tighten to 9 ft. lbs. (12 Nm).
24. To complete the installation, reverse remaining removal procedure.
25. Inspect for exhaust gas leak.

2.4L Engine

See Figures 156 through 157.

1. Disconnect the negative battery cable.
2. Remove the drive belt. Refer to Accessory Drive Belts in the section.
3. Remove the alternator. Refer to Alternator in Engine Electrical.
4. Remove the 4 bolts and 2 compression springs.
5. Remove the center exhaust pipe assembly from the 2 exhaust pipe supports.
6. Remove the 2 gaskets.
7. Remove front exhaust pipe assembly:
    a. Disconnect the heated oxygen sensor connector.
    b. Remove the 2 bolts, 2 compression springs and front exhaust pipe assembly.
    c. Remove the gasket from the exhaust manifold.
8. Remove the bolt, nut and LH manifold stay.
9. Remove the bolt, nut and RH manifold stay.
10. Remove the 4 bolts and upper exhaust manifold heat insulator.
11. Remove air fuel ratio sensor.
12. Remove the 5 nuts and exhaust manifold converter sub-assembly.
13. Remove the gasket.
14. Remove the 2 bolts and underside exhaust manifold heat insulator.
15. Remove the 4 bolts and catalytic converter insulator.

To install:

16. Install the 4 bolts and catalytic converter insulator. Tighten to 9 ft. lbs. (12 Nm).
17. Install the underside exhaust manifold heat insulator with the 2 bolts. Tighten to 9 ft. lbs. (12 Nm).
18. Install the exhaust manifold converter sub-assembly:
    a. Install a new gasket.
    b. Install the exhaust manifold converter sub-assembly with the 5 nuts in the order shown in the illustration. Tighten to 27 ft. lbs. (37 Nm).
19. Install the RH and LH manifold stay with the bolt and nut. Tighten to 33 ft. lbs. (44 Nm).
20. Install the air fuel ratio sensor.
21. Install the outer exhaust manifold heat insulator with the 4 bolts. 9 ft. lbs. (12 Nm).
22. Using a vernier caliper, measure the free length of the compression springs.

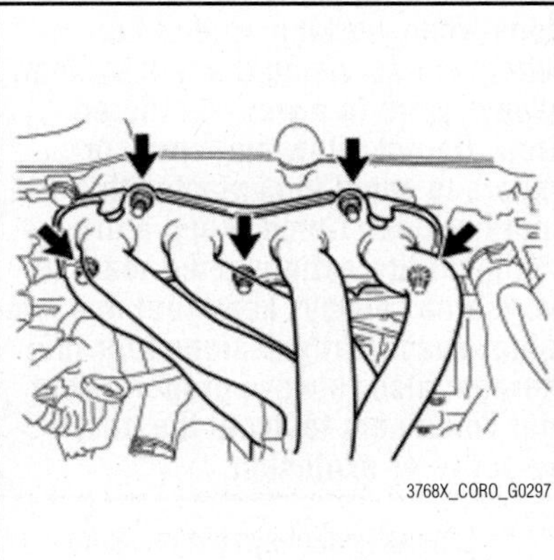

Fig. 156 Remove the 5 nuts and exhaust manifold converter sub-assembly

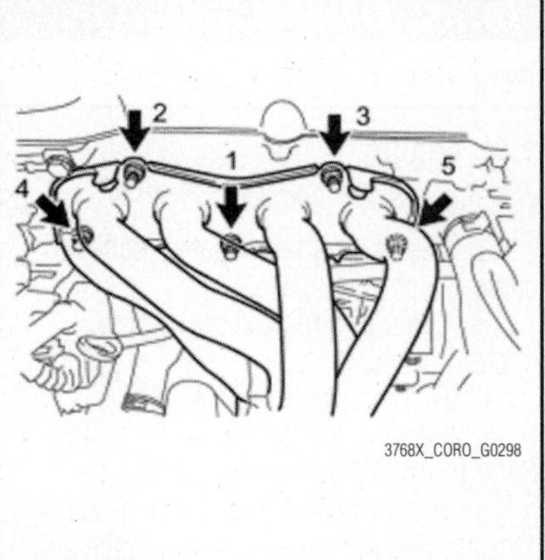

Fig. 157 Exhaust manifold tightening sequence

    a. Front springs: Minimum: 1.63 inches.
    b. Rear springs: Minimum: 1.52 inches.
23. Remove the remains of exhaust manifold converter with wire brash.
24. Fully insert a new gasket to the exhaust manifold.

➡Be sure to install the gaskets in the correct direction.

- Do not reuse the gaskets.
- Do not damage the gaskets.
- Do not push in the gaskets by using the exhaust pipe when connecting it.

25. Install the front exhaust pipe assembly with the 2 bolts and 2 compression springs. Tighten to 32 ft. lbs. (43 Nm).
26. Connect the heated oxygen sensor connector.
27. Fully insert a new gasket to the center exhaust pipe assembly.

➡Be sure to install the gaskets in the correct direction.

- Do not reuse the gaskets.
- Do not damage the gaskets.
- Do not push in the gaskets by using the exhaust pipe when connecting it.

28. Install a new gasket to the front exhaust pipe assembly.
29. Connect the center exhaust pipe assembly to the 2 exhaust pipe supports.
30. Install the center exhaust pipe assembly with the 4 bolts and 2 compression springs. Tighten to 32 ft. lbs. (43 Nm).
31. To complete the installation, reverse remaining removal procedure.
32. Inspect for exhaust gas leak.

INTAKE MANIFOLD

REMOVAL & INSTALLATION

1.8L Engine

See Figures 158 and 159.

1. Disconnect the negative battery cable.
2. Drain and recycle the engine coolant.
3. Remove the engine cover.
4. Remove the air cleaner assembly. Refer to Air Cleaner in this section.
5. Remove the throttle body. Refer to Throttle Body in Engine Performance and Emission Controls.
6. Remove the intake manifold:
   a. Remove the bolt and disconnect the wire harness bracket.
   b. Disconnect the 3 hoses.
   c. Remove the 4 bolts, 2 nuts, intake manifold stay and intake manifold.
   d. Remove the gasket from the intake manifold.
   e. Remove the bolt and bracket.
   f. Using a TORX® socket E6, remove the 2 stud bolts from the intake manifold.

To install:

7. Using a TORX® socket E6, install the 2 stud bolts to the intake manifold. Tighten to 44 inch lbs. (5 Nm).
8. Install the bracket with the bolt. Tighten to 8 ft. lbs. (10 Nm).
9. Install a new gasket into the intake manifold.
10. Install the intake manifold and intake manifold stay with the 4 bolts and 2 nuts. Tighten to 21 ft. lbs. (28 Nm).
11. To complete the installation, reverse remaining removal procedure.

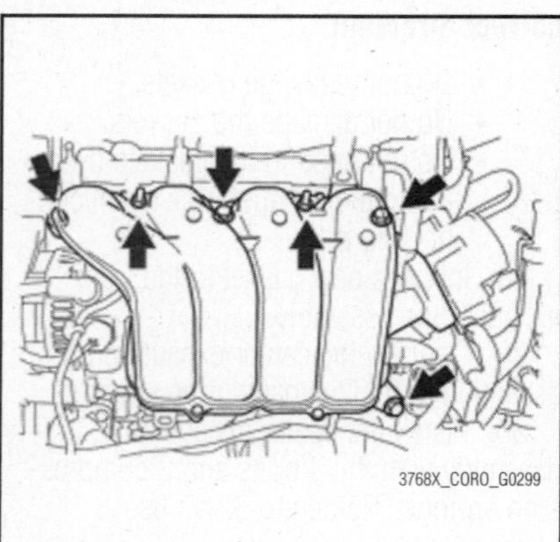

Fig. 158 Removing the 4 bolts, 2 nuts, intake manifold stay and intake manifold

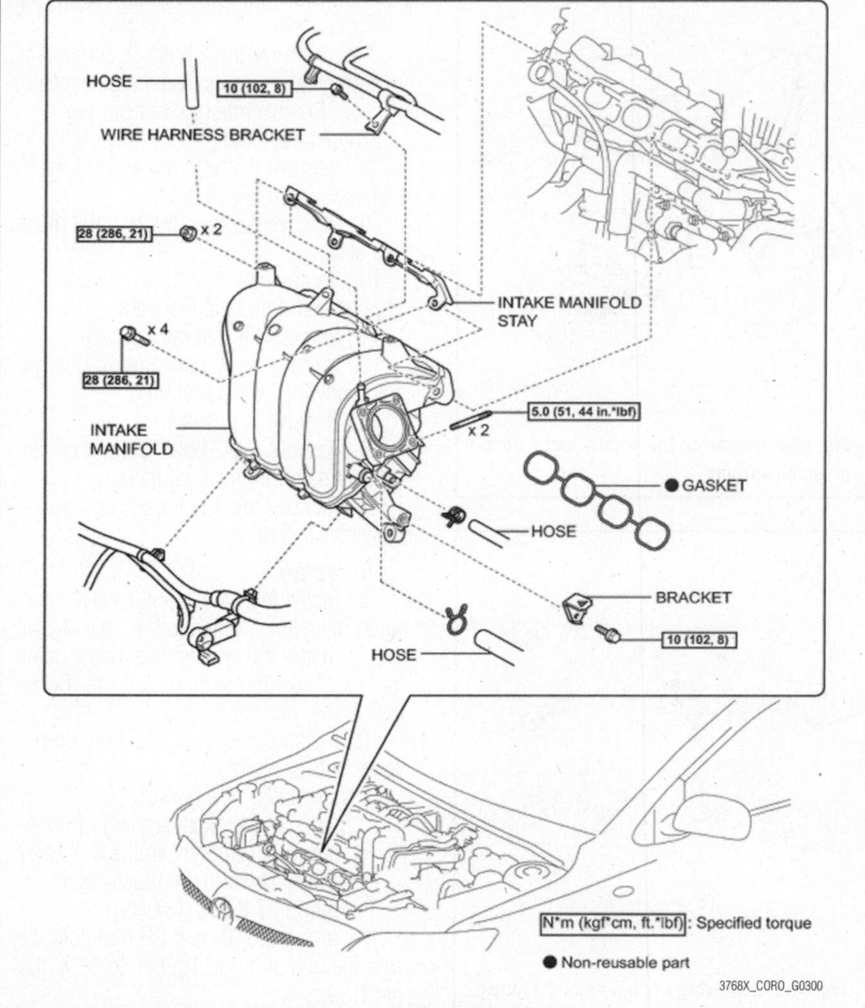

Fig. 159 Exploded view of the intake manifold and related components—1.8L engine

2.4L Engine

See Figures 160 and 161.

CAUTION

Observe all applicable safety precautions when working around fuel. Whenever servicing the fuel system, always work in a well-ventilated area. Do not allow fuel spray or vapors to come in contact with a spark or open flame. Keep a dry chemical fire extinguisher near the work area. Always keep fuel in a container specifically designed for fuel storage; also, always properly seal fuel containers to avoid the possibility of fire or explosion.

1. Discharge fuel system pressure.
2. Disconnect the cable from negative battery terminal.
3. Remove the engine cover.
4. Drain and recycle the engine coolant.
5. Remove the windshield wiper motor and link assembly.
6. Remove suspension tower damper assembly (with front strut bar). Refer to Suspension Tower Damper Assembly in Front Suspension.
7. Remove the air cleaner assembly. Refer to Air Cleaner in this section.
8. Remove the throttle body. Refer to Throttle Body in Engine Performance and Emission Controls.
9. Disconnect the main fuel tube.
10. Disconnect the ventilation hose.
11. Remove fuel delivery pipe with injector. Refer to Fuel Rail and Injectors in Fuel System.
12. Remove intake manifold:
    a. Disconnect the union to connector tube hose from the No. 2 hose to hose tube.
    b. Disconnect the camshaft timing oil control valve connector.

c. Remove the wire harness clamp.
d. Remove the union to connector tube hose from the vacuum hose clamp.
e. Remove the 5 bolts, 2 nuts and intake manifold.
f. Remove the gasket from the intake manifold.

To install:

13. To install, reverse the removal procedure. Refer to exploded view illustration for torque values.

OIL PAN

REMOVAL & INSTALLATION

1.8L Engine

See Figure 162.

1. Remove the oil pan drain plug and gasket and drain the engine oil.
2. Remove the 10 bolts and 2 nuts.
3. Insert the blade of oil pan seal cutter between the crankcase and oil pan. Cut through the sealer and remove the oil pan.

➡Be careful not to damage the contact surfaces of the crankcase and oil pan.

To install:

4. Remove any old packing material and be careful not to drop any oil on the contact surfaces of the cylinder block and oil pan.
5. Apply a continuous bead of seal packing (Diameter 4.0 mm (0.157 in.)). Seal packing: Toyota Genuine Seal Packing Black, Three Bond 1207B or equivalent.

➡Remove any oil from the contact surfaces.

- Install the oil pan within 3 minutes after applying seal packing.
- Do not start the engine for at least 2 hours after installing the oil pan.

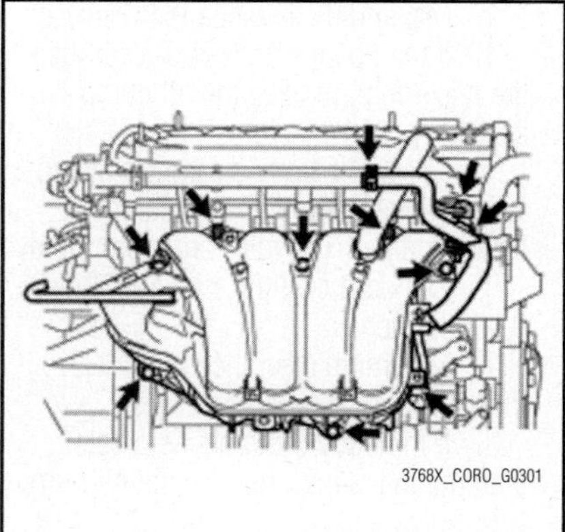

Fig. 160 Removing the 5 bolts, 2 nuts and intake manifold

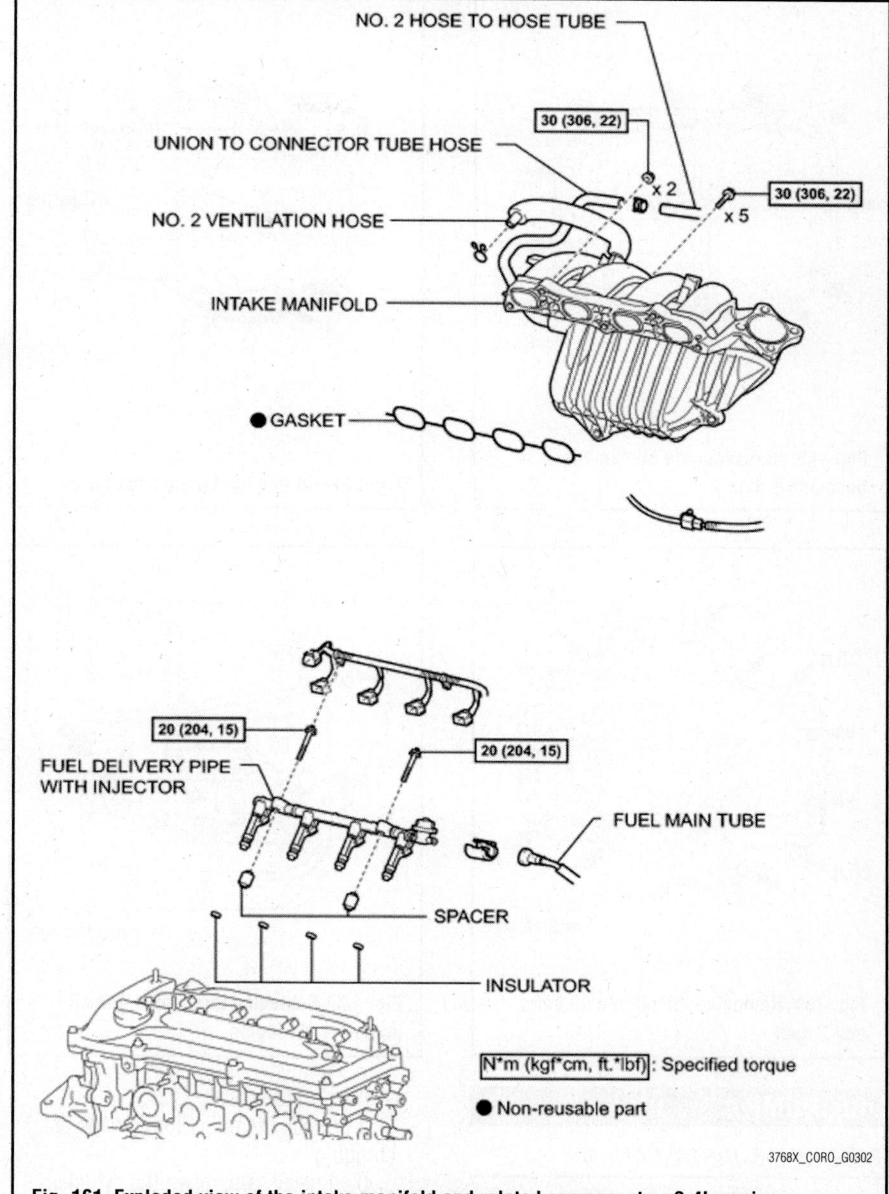

Fig. 161 Exploded view of the intake manifold and related components—2.4L engine

6. Install the oil pan with the 10 bolts and 2 nuts. Tighten to 7 ft. lbs. (10 Nm).

2.4L Engine

See Figures 163 and 164.

1. Remove the oil pan drain plug and gasket and drain the engine oil.
2. Remove the 12 bolts and 2 nuts.
3. Insert the blade of oil pan seal cutter between the crankcase and oil pan. Cut through the sealer and remove the oil pan.

➡Be careful not to damage the contact surfaces of the crankcase and oil pan.

To install:

4. Remove any old packing material and be careful not to drop any oil on the contact surfaces of the cylinder block and oil pan.
5. Apply a continuous bead of seal packing (Diameter 4.0 mm (0.157 in.)). Seal packing: Toyota Genuine Seal Packing Black, Three Bond 1207B or equivalent.

➡Remove any oil from the contact surfaces.

- Install the oil pan within 3 minutes after applying seal packing.
- Do not start the engine for at least 2 hours after installing the oil pan.

6. Uniformly tighten the 12 bolts and 2 nuts in the sequence shown in the illustration. Tighten to 80 inch lbs. (9 Nm).

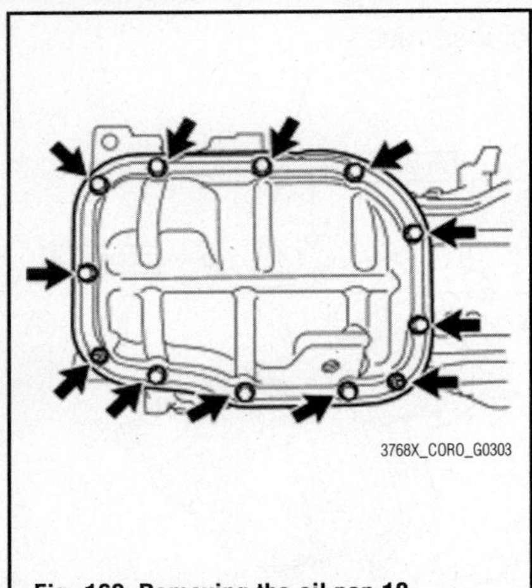

Fig. 162 Removing the oil pan 10 bolts and 2 nuts

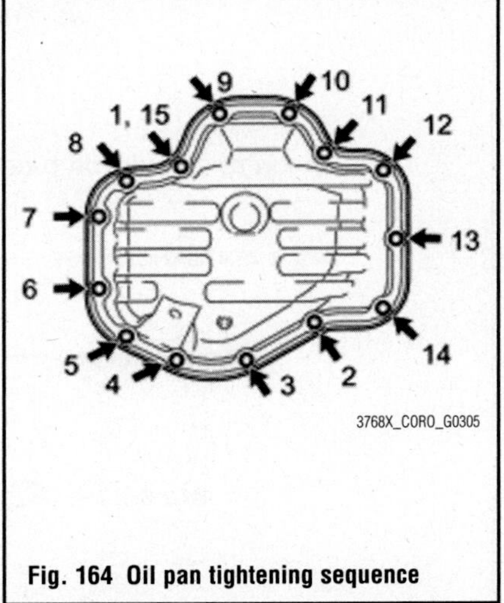

Fig. 164 Oil pan tightening sequence

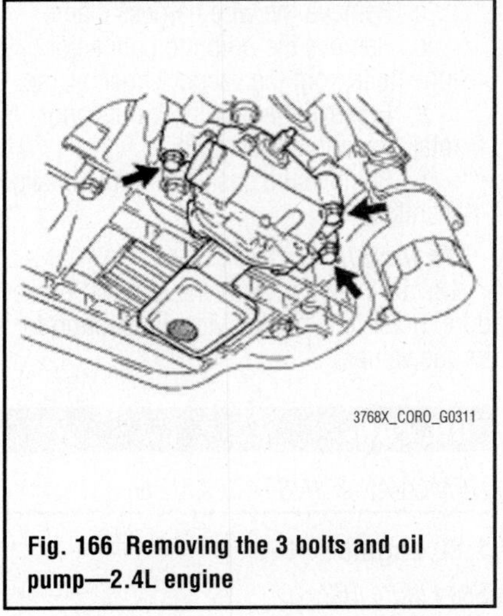

Fig. 166 Removing the 3 bolts and oil pump—2.4L engine

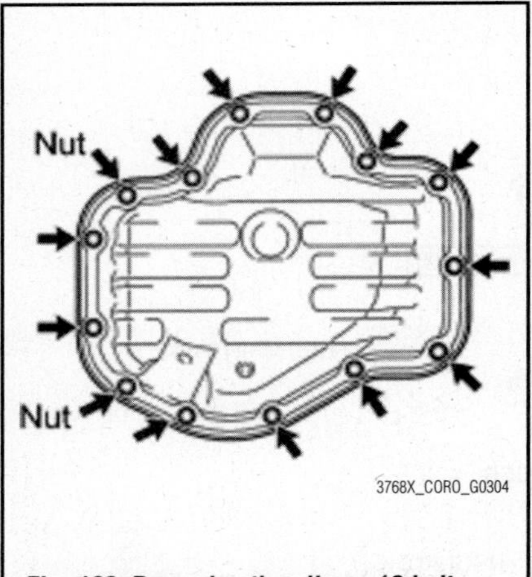

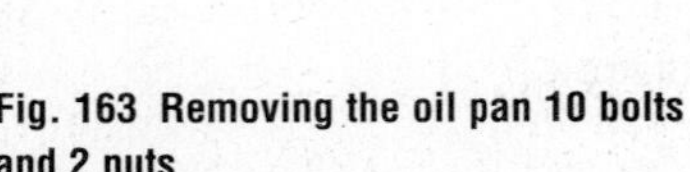

Fig. 163 Removing the oil pan 10 bolts and 2 nuts

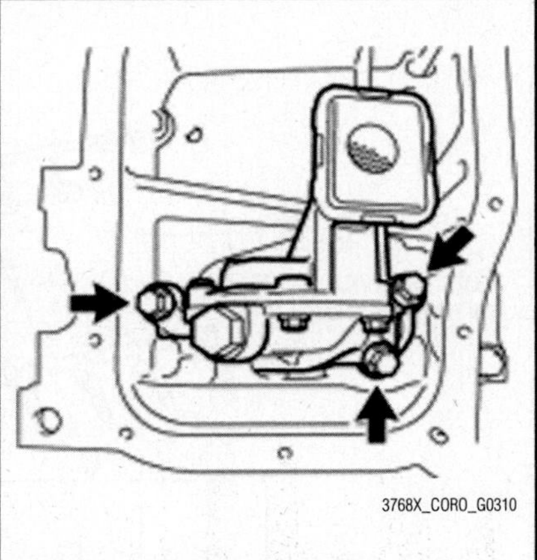

Fig. 165 Removing the 3 bolts and oil pump—1.8L engine

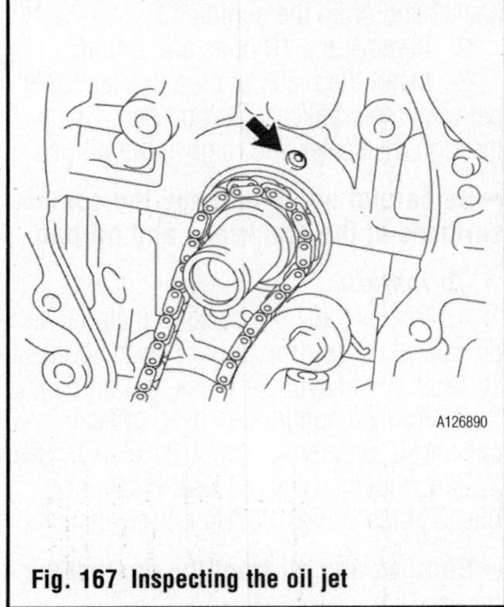

Fig. 167 Inspecting the oil jet

OIL PUMP

REMOVAL & INSTALLATION

See Figures 165 and 166.

1. Remove the engine assembly with the transaxle.
2. Remove the timing chain front cover. Refer to Timing Chain Front Cover in this section.
3. Remove the oil pan. Refer to Oil Pan in this section.
4. Remove the 3 bolts and oil pump.
5. For 2.4L engines, remove the gasket.

To install:

6. To install, reverse the removal procedure. Tighten the oil pump assembly bolts to 16 ft. lbs. (21 Nm).

INSPECTION

See Figures 167 through 172.

1. Inspect the oil jet.
   a. Check the oil jet for damage or clogging.
   b. If necessary, repair the cylinder block.
2. Inspect the oil pump relief valve.
   a. Coat the valve with engine oil, and then check that the valve falls smoothly into the valve hole by its own weight. If it does not, replace the relief valve. If necessary, replace the oil pump assembly.
3. Inspect the oil pump rotor.
   a. Check the side clearance. Using a feeler gauge and precision straightedge, measure the clearance between the rotors and precision straightedge.
   b. Standard measurement is 0.00118-0.00335 inch.
   c. Maximum clearance is 0.00630 inch.
   d. If the side clearance is greater than the maximum, replace the oil pump assembly.
   e. Check the tip clearance. Using a feeler gauge, measure the clearance between the drive and driven rotor tips.
   f. Standard clearance is 0.00315-0.00630 inch.
   g. Maximum clearance is 0.138 inch.
   h. If the tip clearance is greater than the maximum, replace the oil pump assembly.
   i. Check the body clearance. Using a feeler gauge, measure the clearance between the driven rotor and pump body.
   j. Standard clearance is 0.00394-0.00669 inch.
   k. Maximum clearance is 0.0128 inch.
   l. If the body clearance is greater than the maximum, replace the oil pump assembly.
4. Inspect the No. 2 chain sub assembly.
5. Inspect the chain tensioner plate.
   a. Measure the chain tensioner plate wear.

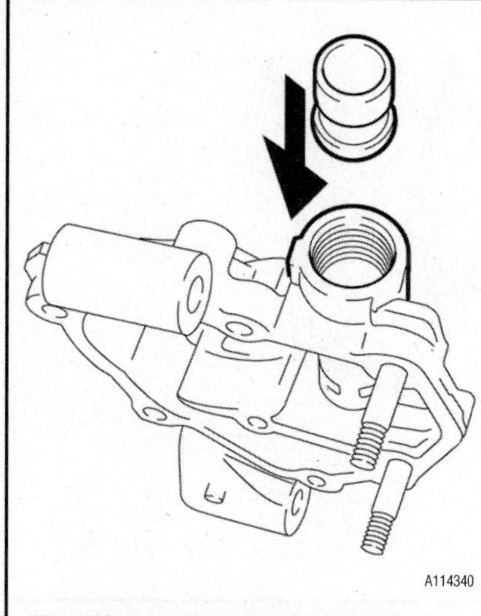
Fig. 168 Inspecting the oil pump relief valve

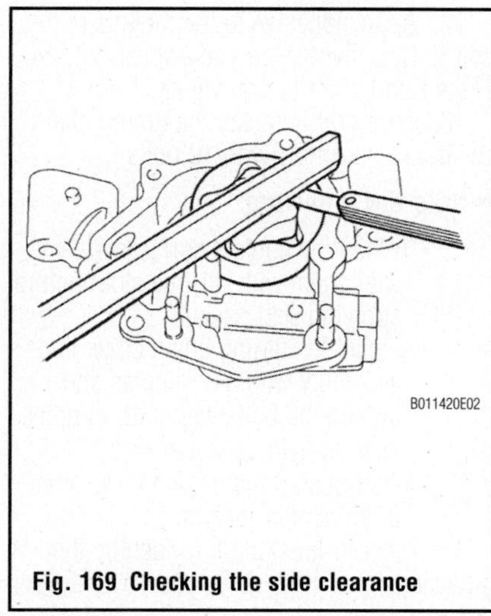
Fig. 169 Checking the side clearance

b. Maximum wear is 0.0197 inch.

c. If the wear is greater than the maximum, replace the chain tensioner plate.

PISTON AND RING

POSITIONING

See Figures 173 through 175.

TIMING CHAIN FRONT COVER

REMOVAL & INSTALLATION

1.8L Engine

See Figures 176 through 181.

1. Remove the 3 bolts and engine mounting bracket.
2. Remove the 4 bolts and oil filter bracket.

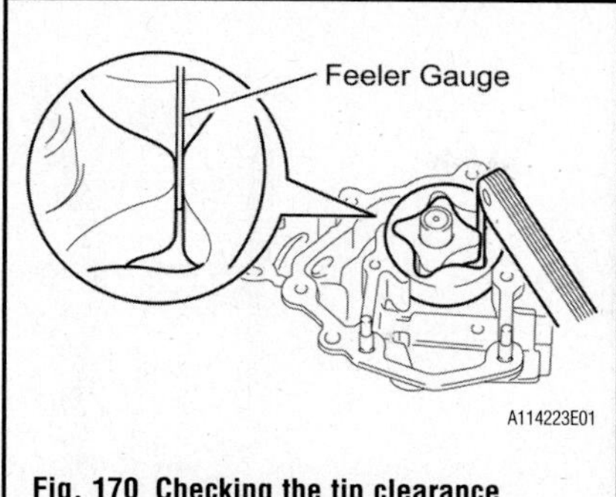

Fig. 170 Checking the tip clearance

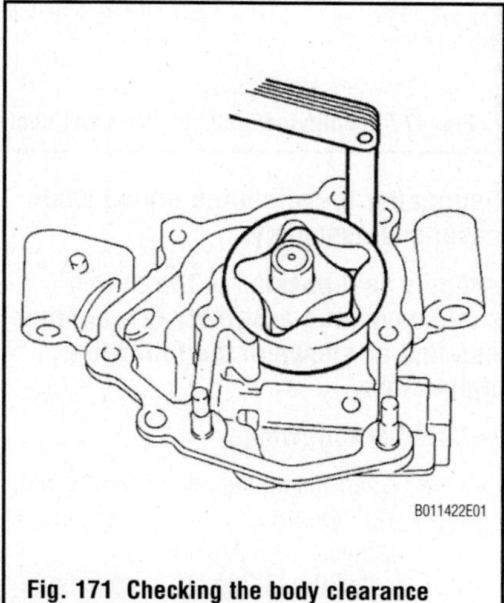
Fig. 171 Checking the body clearance

3. Remove the 2 O-rings.
4. Remove the 19 bolts.
5. Remove the timing chain cover by prying between the timing chain cover and cylinder head or cylinder block with a screwdriver.

➡Be careful not to damage the contact surfaces of the timing chain cover, cylinder block, and cylinder head.

➡Tape the screwdriver tip before use.

6. Place the timing chain cover on wooden blocks.
7. Using a screwdriver and a hammer, pry out the oil seal.

➡Do not damage the surface of the oil seal press fit hole.

8. Remove the 3 O-rings.
9. Remove the 3 bolts and water pump. Remove the gasket.

To install:

10. Using a seal installer (SST 09223-15030) and a hammer, evenly tap the oil seal until its surface is flush with the timing chain cover sub-assembly edge.

➡Keep the lip free from foreign matter. Do not tap the oil seal at an angle.

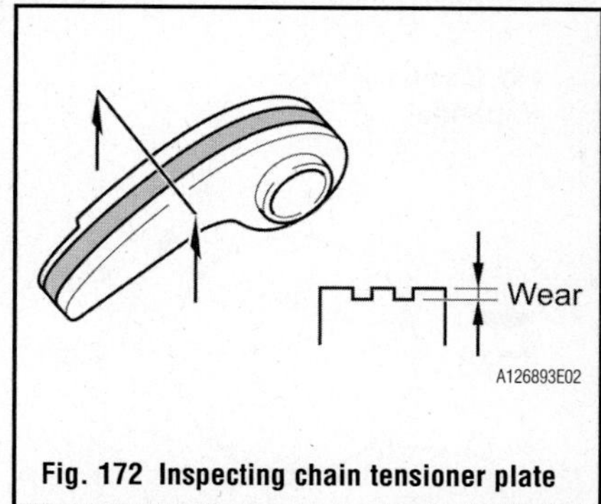

Fig. 172 Inspecting chain tensioner plate

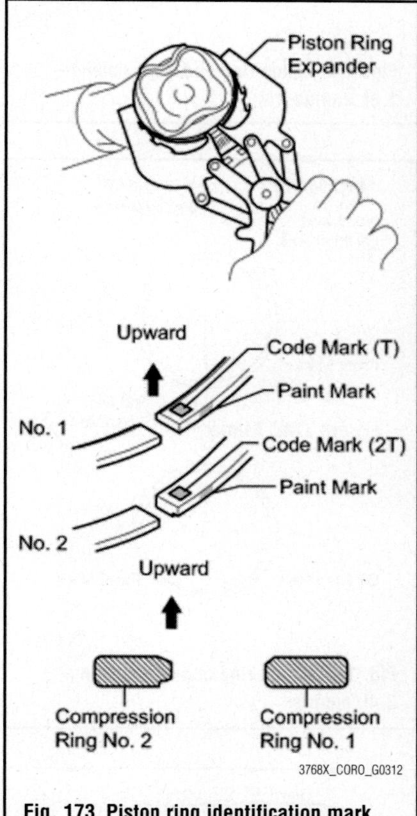

Fig. 173 Piston ring identification mark locations

11. Apply MP grease to the lip of the oil seal.
12. Remove any old packing (FIPG) material and be careful not to drop any oil on the contact surfaces of the timing chain cover sub-assembly, cylinder head, and cylinder block.
13. Install 3 new O-rings.
14. Apply seal packing as shown in the illustration. Seal packing: Toyota Genuine Seal Packing Black, Three Bond 1207B or equivalent.

➡Remove any oil from the contact surfaces. Install the chain cover within 3 minutes after applying seal packing. Do not start the engine for at least 2

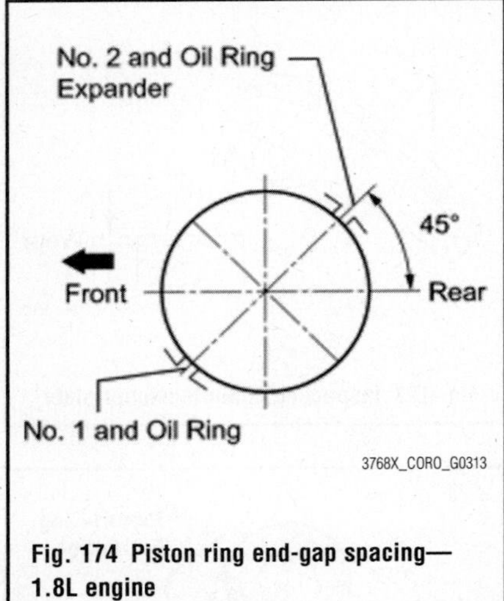

Fig. 174 Piston ring end-gap spacing—1.8L engine

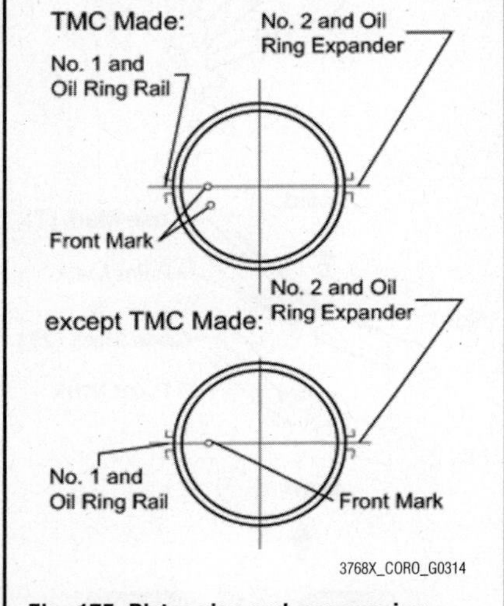

Fig. 175 Piston ring end-gap spacing—2.4L engines

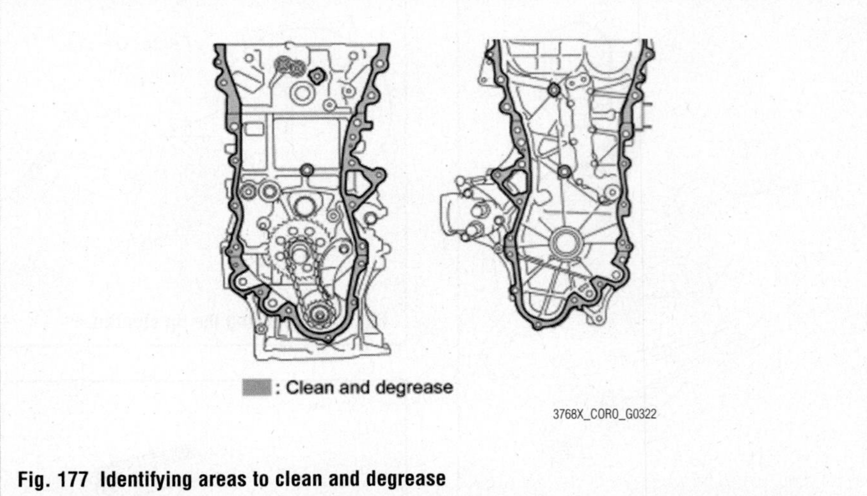

Fig. 177 Identifying areas to clean and degrease

hours after installing the timing chain cover sub-assembly.

➡Apply seal packing to the timing chain cover sub-assembly in a continuous line as shown in the following illustration.

➡Note the following:

- When the contact surfaces are wet, wipe them with oil-free cloth before applying seal packing.
- Install the timing chain cover sub-assembly within 3 minutes and tighten the bolts within 15 minutes after applying seal packing.
- Do not start the engine for at least 2 hours after installing.

15. For the continuous line area, use: Toyota Genuine Seal Packing Black, Three Bond 1207B or equivalent.
16. For the dashed line area, use: Toyota Genuine Seal Packing 1282B, Three Bond 1282B or equivalent.
17. Apply adhesive to the threads of the bolt E. Use Toyota Genuine Adhesive 1324, Three Bond 1324 or equivalent.
18. Temporarily install the timing chain cover sub-assembly with 19 bolts.

➡Note the following:

- When the contact surfaces are wet, wipe them with oil-free cloth before applying seal packing.
- Install the timing chain cover sub-assembly within 3 minutes and tighten the bolts within 15 minutes after applying seal packing.
- Do not start the engine for at least 2 hours after installing.

19. Refer to illustration to identify the following:

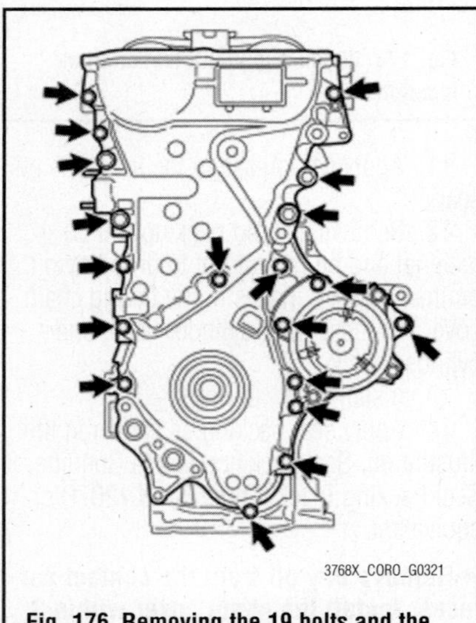

Fig. 176 Removing the 19 bolts and the timing chain front cover

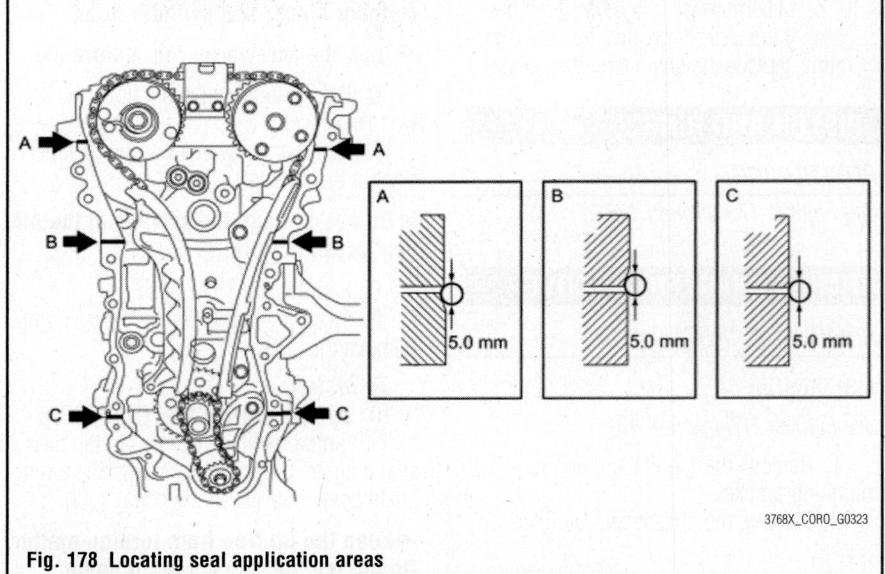

Fig. 178 Locating seal application areas

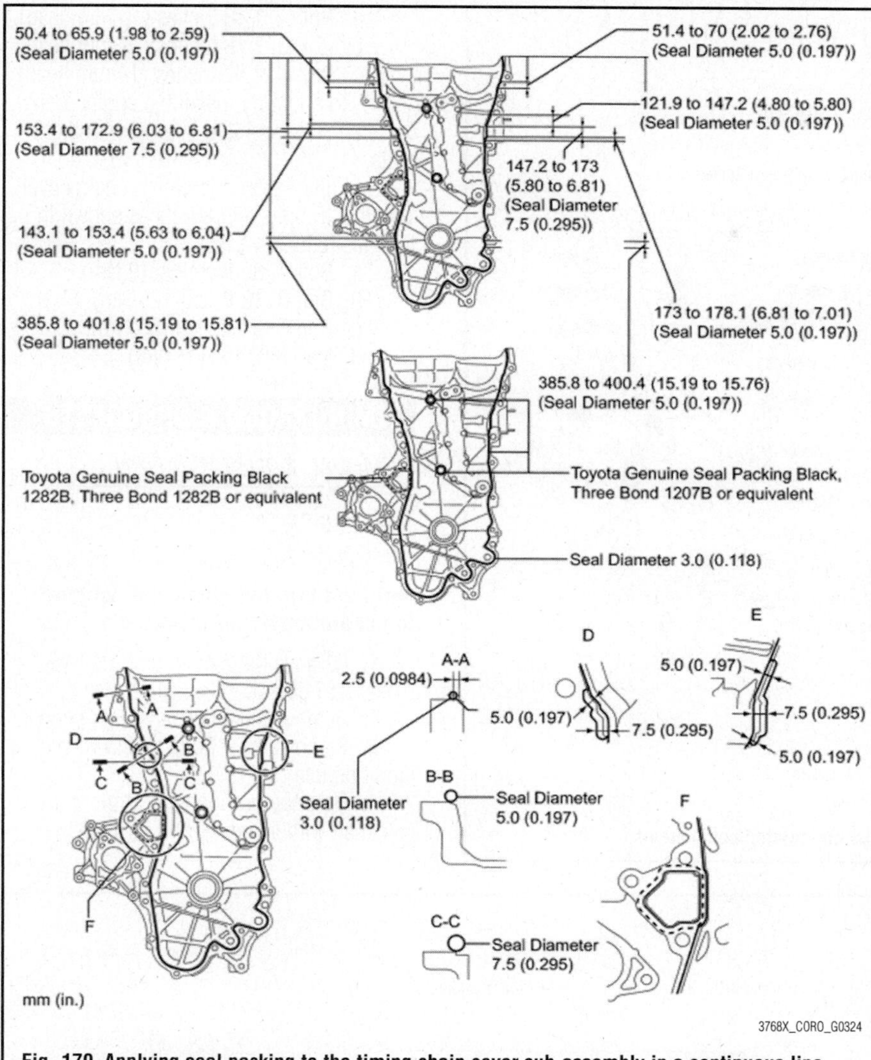

Fig. 179 Applying seal packing to the timing chain cover sub-assembly in a continuous line

a. Bolt A, E: 1.38 inches.
b. Bolt B: 2.16 inches.
c. Bolt C: 3.15 inches.
d. Bolt D: 1.57 inches

20. Install the water pump. Refer to Water Pump in Engine Cooling.
21. Temporarily install the engine mounting bracket with the 3 bolts.
22. Install 2 new O-rings.
23. Temporarily install the oil filter bracket with the 4 bolts.
24. Fully tighten the timing chain cover sub-assembly with the 26 bolts as shown in the illustration.

a. Bolts A, E: 19 ft. lbs. (26 Nm).
b. Bolts B, C: 37 ft. lbs. (51 Nm).
c. Bolt D: 7 ft. lbs. (10 Nm).

➡Note the following:

- When the contact surfaces are wet, wipe them with oil-free cloth before applying seal packing.

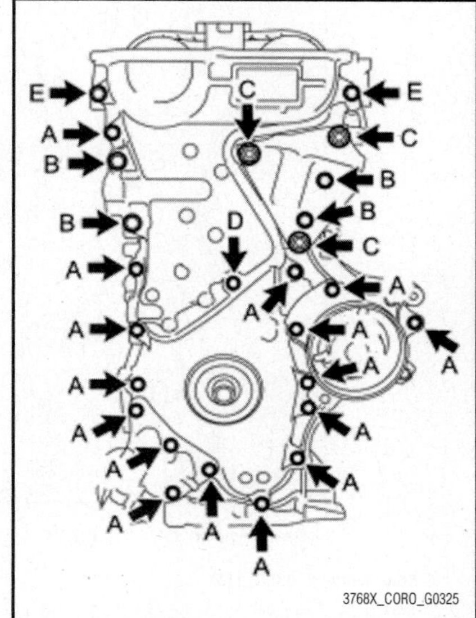

Fig. 180 Identifying the timing chain front cover bolts

- Install the timing chain cover sub-assembly within 3 minutes and tighten the bolts within 15 minutes after applying seal packing.
- Do not start the engine for at least 2 hours after installing.

2.4L Engine

See Figures 182 through 185.

1. Remove the 3 bolts and transverse engine mounting bracket.
2. Using an E10 TORX® socket, remove the stud bolt for the V-ribbed belt tensioner.
3. Remove the 12 bolts and 2 nuts.
4. Remove the timing chain cover by prying the portions between the timing chain cover, cylinder head and cylinder block with a screwdriver.

➡Be careful not to damage the contact surfaces of the timing chain cover, cylinder head and cylinder block.

➡Tape the screwdriver tip before use.

5. Place the timing chain cover on wooden blocks.
6. Using a screwdriver, pry out the oil seal.

➡Do not damage the surface of the oil seal press fit hole.

To install:

7. Using a seal installer (SST 09223-22010) and a hammer, evenly tap the oil seal until its surface is flush with the timing chain cover sub-assembly edge.

➡Keep the lip free from foreign matter. Do not tap the oil seal at an angle.

8. Apply MP grease to the lip of the oil seal.
9. Remove any old packing (FIPG) material and be careful not to drop any oil on the contact surfaces of the timing chain cover sub-assembly, cylinder head, and cylinder block.
10. Apply seal packing (Diameter 4.0 to 4.5 mm (0.157 to 0.177 in.)) as shown in the illustration. Use Toyota Genuine Seal Packing Black, Three Bond 1207B or equivalent.

➡Remove any oil from the contact surfaces. Install the chain cover within 3 minutes of applying seal packing. Do not add engine oil for at least 2 hours after installing the chain cover.

11. Apply a continuous bead of seal packing as shown in the illustration. Use Toyota Genuine Seal Packing Black, Three Bond 1207B or equivalent.

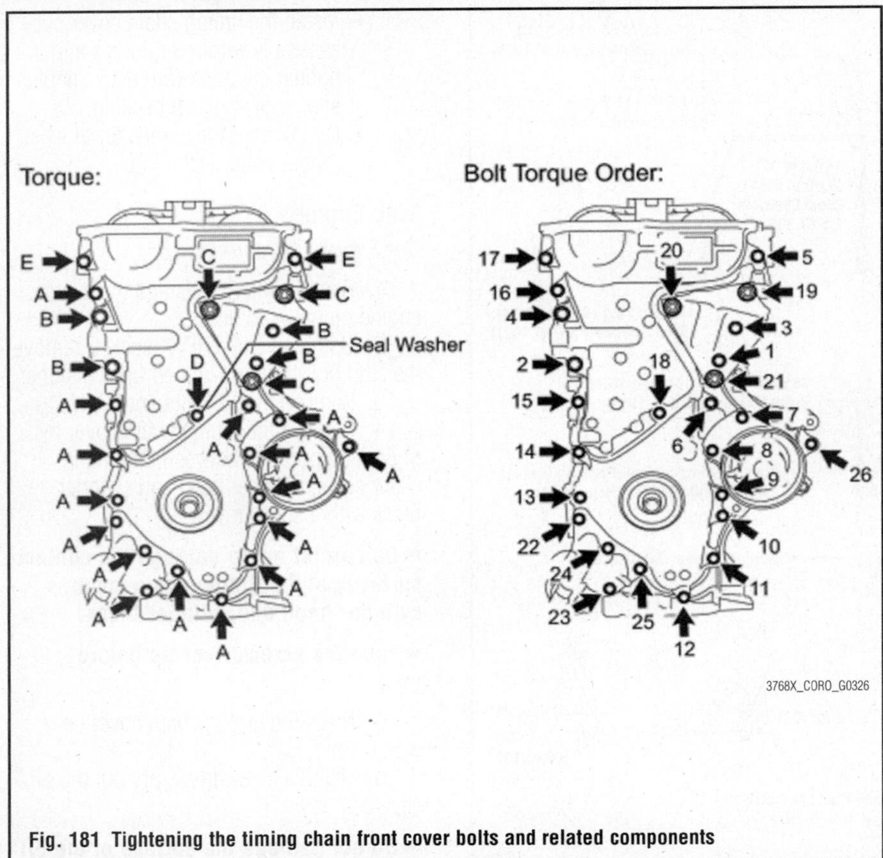

Fig. 181 Tightening the timing chain front cover bolts and related components

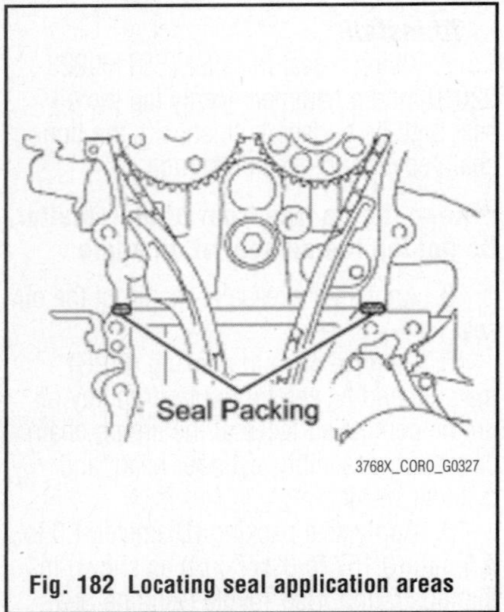

Fig. 182 Locating seal application areas

➡Remove any oil from the contact surfaces. Install the chain cover within 3 minutes of applying seal packing. Do not add engine oil for at least 2 hours after installing the chain cover.

12. Apply adhesive to the threads of the bolt "A". Use Toyota Genuine Adhesive 1324, Three Bond 1324 or equivalent.
13. Temporarily install the timing chain cover with the 12 bolts and 2 nuts.:
    a. Bolt A: 1.18 inches (10 mm head)
    b. Bolt B: 1.18 inches (12 mm head
    c. Bolt C: 1.57 inches (14 mm head)
14. Temporarily install the engine transverse engine mounting bracket with the 3 bolts.
15. Fully tighten the timing chain cover with the 15 bolts and 2 nuts as shown in the illustration.
    a. Bolt A: 80 inch lbs. (9 Nm).
    b. Bolt B: 18 ft. lbs. (25 Nm).
    c. Bolt C: 41 ft. lbs. (55 Nm).
    d. Nut: 8 ft. lbs. (11 Nm).

TIMING CHAIN & SPROCKETS

REMOVAL & INSTALLATION

1.8L Engine

See Figures 186 through 194.

➡Do not turn the crankshaft without the chain tensioner installed.

1. Remove the 2 nuts, bracket, tensioner and gasket.
2. Remove the chain tensioner slipper.
3. Remove the 2 bolts and LH chain vibration damper.
4. Hold the hexagonal portion of the camshaft with a wrench and turn the

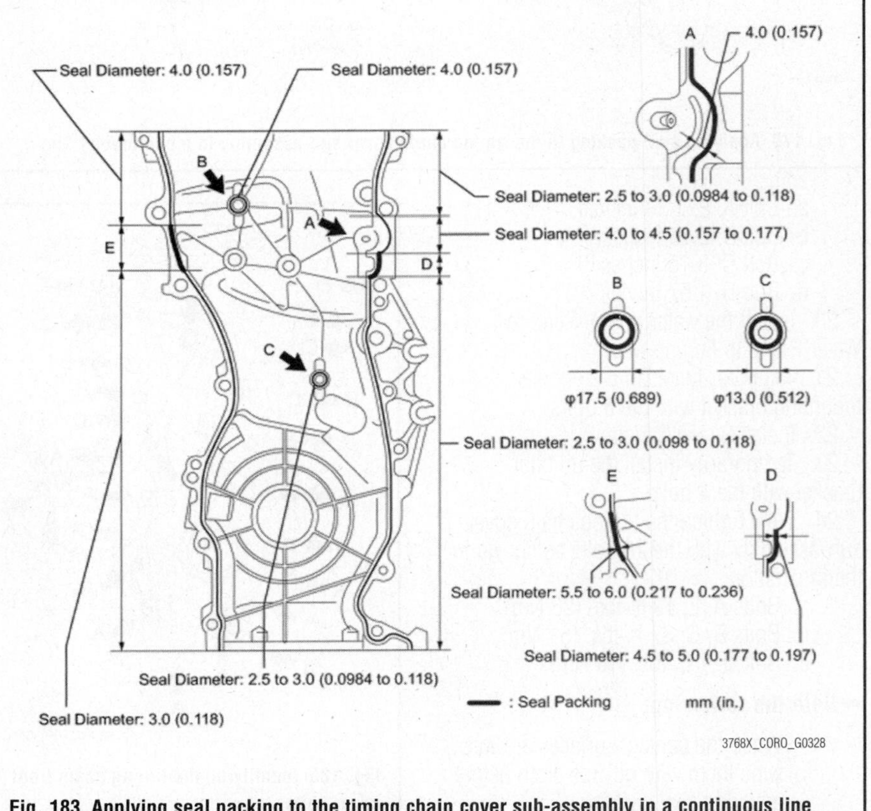

Fig. 183 Applying seal packing to the timing chain cover sub-assembly in a continuous line

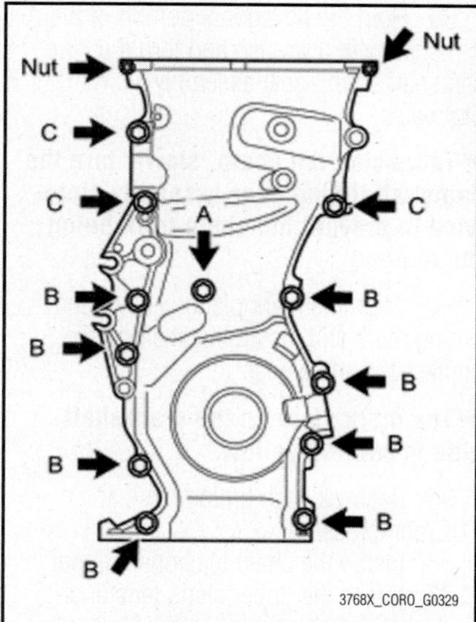

Fig. 184 Identifying the timing chain front cover bolts

camshaft timing gear assembly counter-clockwise to loosen the chain between the camshaft timing gears.

5. With the chain loosened, release the chain from the camshaft timing gear assembly and place it on the camshaft timing gear assembly.

➡**Be sure to release the chain from the sprocket completely.**

6. Turn the camshaft clockwise to return it to the original position and remove the chain.

7. Remove the 2 bolts and No. 2 chain vibration damper.

8. Remove the crankshaft timing sprocket.

⁂ CAUTION

Do not rotate the crankshaft more than 90°. If the crankshaft is rotated too much without the timing chain installed, the valves may hit the pistons and cause damage.

9. Turn the crankshaft 90° clockwise to align the adjusting hole of the oil pump drive shaft sprocket with the groove of the oil pump.

10. Remove the crank pulley bolt.

11. Insert a 3 mm diameter bar into the adjusting hole of the oil pump drive shaft sprocket to lock the gear in position, and then remove the nut.

12. Remove the bolt, chain tensioner plate, and spring.

13. Remove the crankshaft timing sprocket, oil pump drive shaft gear, and No. 2 chain sub-assembly.

To install:

14. Set the crankshaft key as shown in the illustration.

15. Turn the drive shaft so that the cutout faces the right horizontal position.

16. Align the yellow mark links with the timing marks of each gear as shown in the illustration.

17. Install the sprockets onto the crankshaft and oil pump shaft with the chain on the gears.

18. Temporarily tighten the oil pump drive shaft sprocket with the nut.

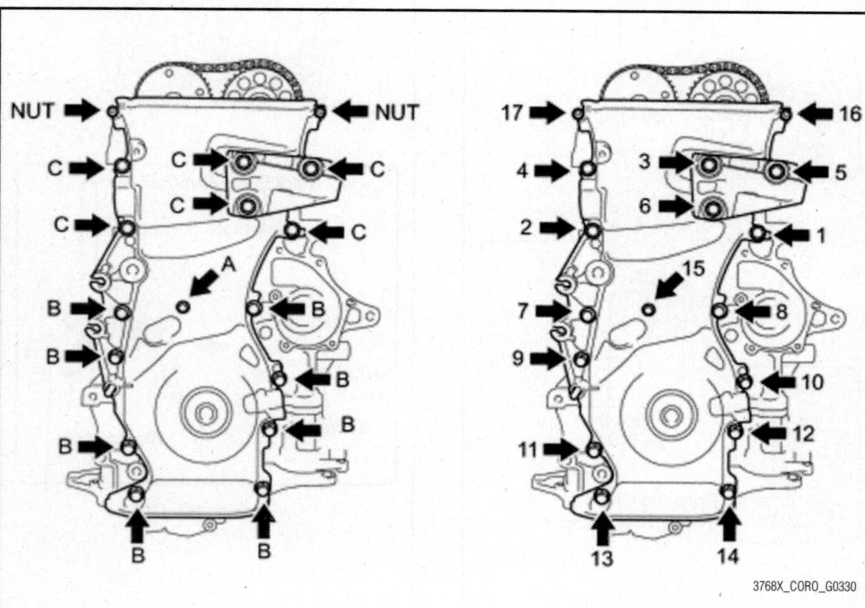

Fig. 185 Tightening the timing chain front cover bolts and related components

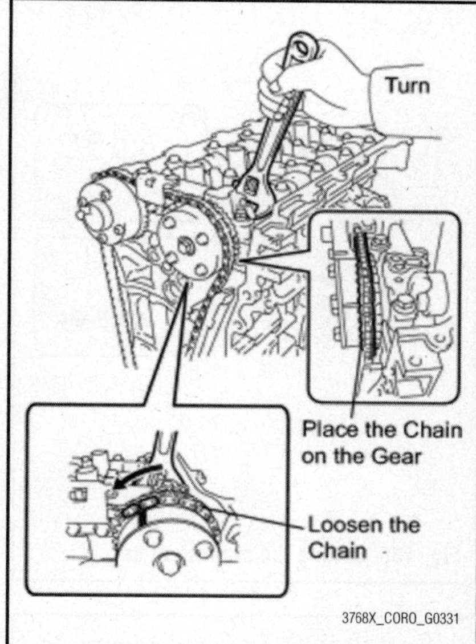

Fig. 186 Removing the upper timing chain

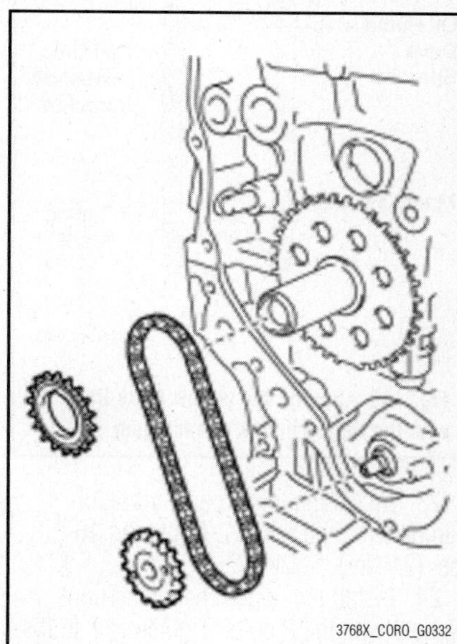

Fig. 187 Removing the crankshaft timing sprocket, oil pump drive shaft gear and the lower timing chain

19. Insert the damper spring into the adjusting hole, and then install the chain tensioner plate with the bolt. Tighten to 7 ft. lbs. (10 Nm).

20. Align the adjusting hole of the oil pump drive shaft sprocket with the groove of the oil pump.

21. Insert a 3 mm diameter bar into the adjusting hole of the oil pump drive shaft gear to lock the gear in position, and then tighten the nut. Tighten to 21 ft. lbs. (28 Nm).

22. Install the crankshaft timing sprocket.

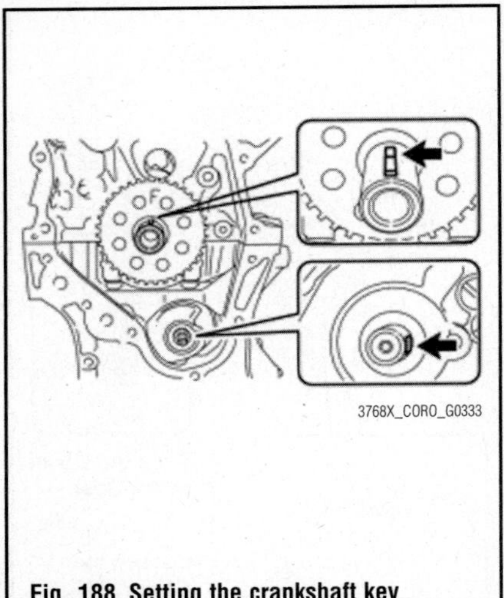

Fig. 188 Setting the crankshaft key

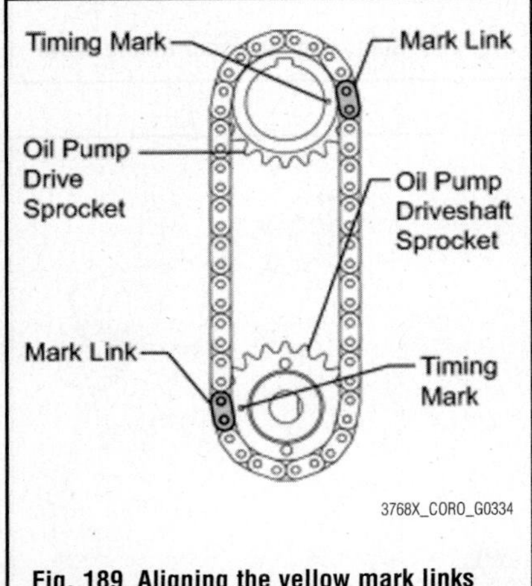

Fig. 189 Aligning the yellow mark links with the timing marks of each gear

23. Install the lower chain vibration damper with the 2 bolts. Tighten to 16 ft. lbs. (21 Nm).
24. Install the upper chain vibration damper with the 2 bolts. Tighten to 7 ft. lbs. (10 Nm).
25. Check the No. 1 cylinder TDC/compression.
26. Temporarily tighten the crankshaft pulley bolt.
27. Turn the crankshaft counterclockwise to position the timing gear key to the top.
28. Remove the crankshaft pulley bolt.
29. Check the timing marks on each camshaft timing gear.
30. Align the mark plate (orange) with the timing mark of the No. 2 camshaft as shown in the illustration and install the chain.

➡Be sure to position the mark plate at the front of the engine. The mark plate on the camshaft side is colored orange.

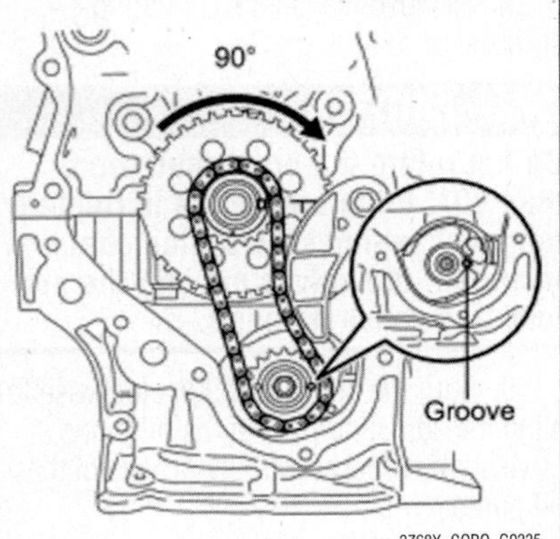

Fig. 190 Aligning the adjusting hole of the oil pump drive shaft sprocket with the groove of the oil pump

Do not pass the chain around the sprocket of the camshaft timing gear assembly. Only place it on the sprocket. Pass the chain through the No. 1 vibration damper.

31. Place the chain on the crankshaft without passing it around the shaft.
32. Hold the hexagonal portion of the camshaft with a wrench and turn the camshaft timing gear assembly counterclockwise to align the mark plate (orange) and timing mark.

➡Be sure to position the mark plate at the front of the engine. The mark plate on the camshaft side is colored orange.

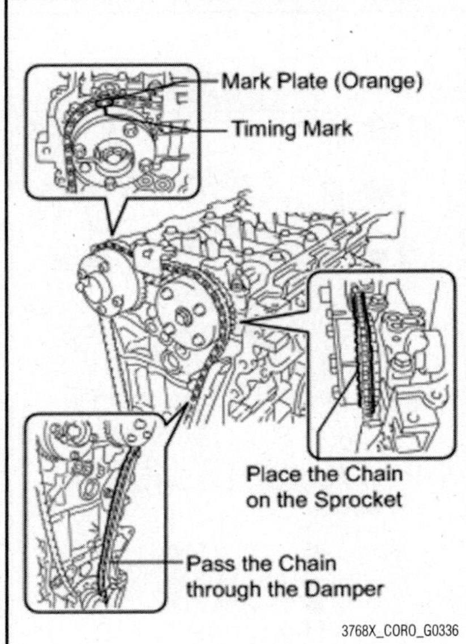

Fig. 191 Align the mark plate (orange) with the timing mark of the No. 2 camshaft

33. Hold the hexagonal portion of the camshaft with a wrench and turn the camshaft timing gear assembly clockwise.

➡To tension the chain, slowly turn the camshaft timing gear assembly clockwise to prevent the chain from being misaligned.

34. Align the mark plate (yellow) and timing mark and install the chain to the crankshaft timing gear.

➡The mark plate on the crankshaft side is colored yellow.

35. Recheck each timing mark at TDC/compression.
36. Install the chain tensioner slipper.
37. Install the upper chain tensioner assembly:
    a. Release the ratchet pawl, then fully push in the plunger and engage the hook to the pin so that the plunger is in the position.

➡Make sure that the cam engages the first tooth of the plunger to allow the hook to pass over the pin.

    b. Install a new gasket, bracket and No. 1 chain tensioner with the 2 nuts.

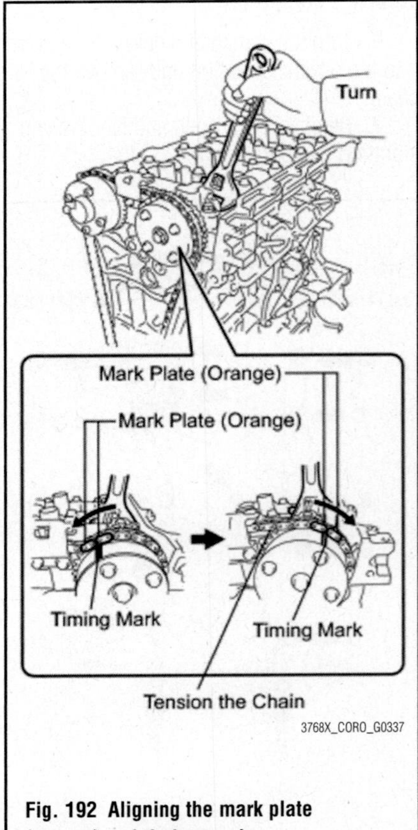

Fig. 192 Aligning the mark plate (orange) and timing mark

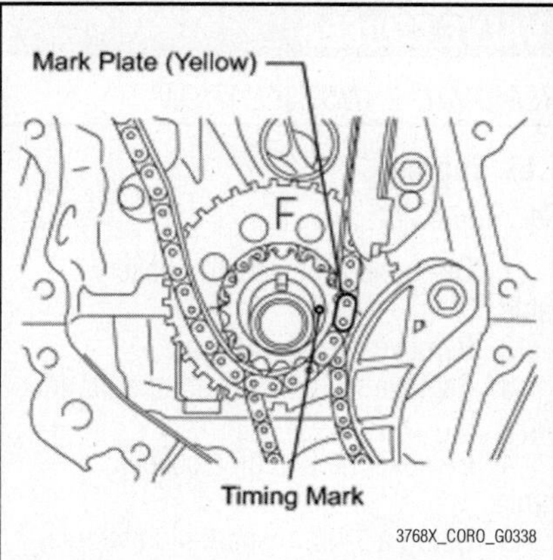

Fig. 193 Aligning the mark plate (yellow) and timing mark on the crankshaft timing gear

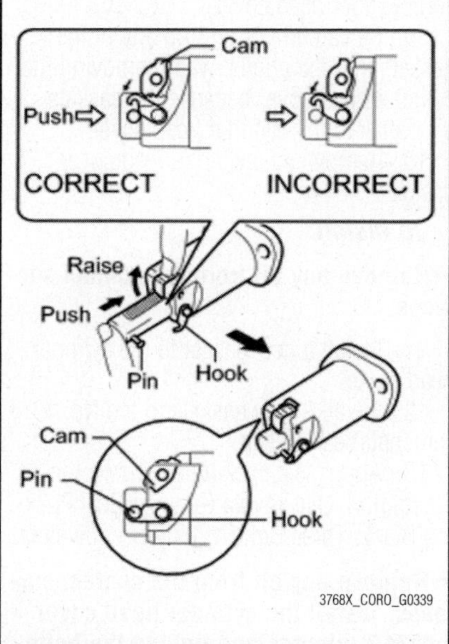

Fig. 194 Correct plunger position and orientation

➡If the hook releases the plunger while the chain tensioner is being installed, engage the hook again.

c. Turn the crankshaft counterclockwise, then disconnect the hook from plunger knock pin.

d. Turn the crankshaft clockwise, then check that the plunger is extended.

38. To complete installation, reverse the remaining removal procedure.

2.4L Engine

See Figures 195 through 201.

1. Turn the crankshaft pulley until the groove and the timing mark "0" on the timing chain cover are aligned.
2. Check that each timing mark on the camshaft timing gear and sprocket is aligned with the timing marks located on the No. 1 and No. 2 bearing caps.
3. Remove the crankshaft pulley.

➡Do not turn the crankshaft without the chain tensioner installed.

➡Do not lift the engine more than necessary.

4. Remove the 2 nuts, bracket, tensioner and gasket.
5. Lift the engine upward.
6. Remove the bolt, nut and V-ribbed belt tensioner assembly.
7. Remove the crankshaft sensor.
8. Remove the timing chain and sprocket cover.
9. Remove the No. 1 Crankshaft Position (CKP) sensor plate.
10. Remove the bolt and timing chain guide.
11. Remove the bolt and chain tensioner slipper.
12. Remove the 2 bolts and No. 1 chain vibration damper.
13. Remove the chain sub-assembly.
14. Remove the crankshaft timing gear or sprocket from the crankshaft.
15. Turn the crankshaft 90°counterclockwise to align the adjusting hole on the oil pump drive shaft sprocket with the groove on the oil pump.
16. Insert a 4 mm diameter bar into the adjusting hole of the oil pump drive shaft sprocket to lock the gear in position, and then remove the nut.
17. Remove the bolt, chain tensioner plate and spring.
18. Remove the oil pump drive sprocket, oil pump drive shaft sprocket and No. 2 chain.

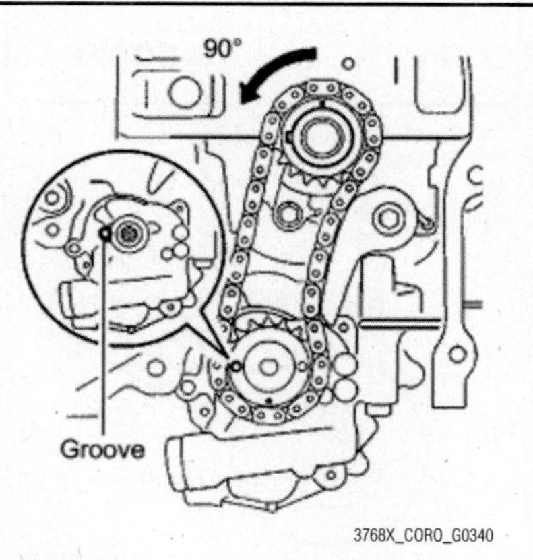

Fig. 195 Aligning the adjusting hole on the oil pump drive shaft sprocket with the groove on the oil pump

To install:

19. Set the crankshaft key in the left horizontal position.
20. Turn the cutout of the drive shaft so that it faces upward.
21. Align the yellow mark links with the timing marks of each gear as shown in the illustration.
22. Install the sprockets onto the crankshaft and oil pump shaft with the chain wrapped on the gears.
23. Temporarily tighten the oil pump drive shaft sprocket with the nut.
24. Insert the damper spring into the adjusting hole, and then install the chain tensioner plate with the bolt. Tighten to 9 ft. lbs. (12 Nm).
25. Align the adjusting hole on the oil pump drive shaft sprocket with the groove on the oil pump.
26. Insert a 4 mm diameter bar into the adjusting hole on the oil pump drive shaft gear to lock the gear in position, and then

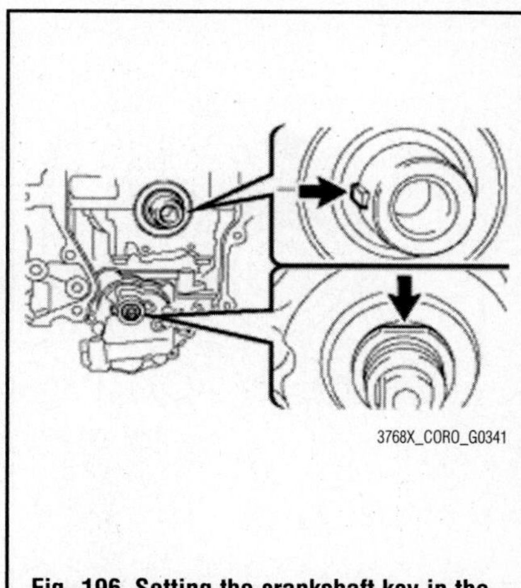

Fig. 196 Setting the crankshaft key in the left horizontal position

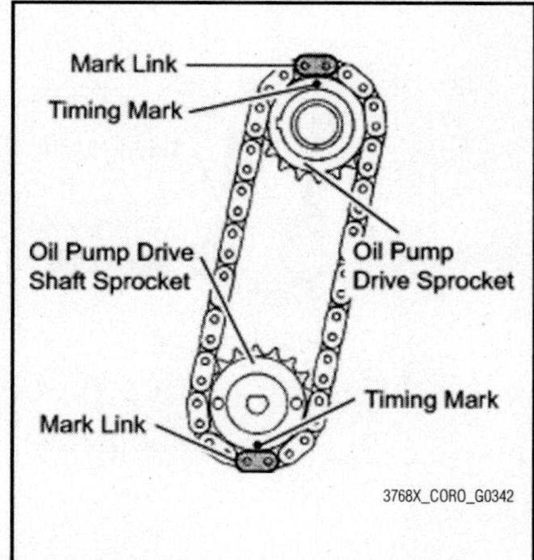

Fig. 197 Aligning the yellow mark links with the timing marks of each gear

tighten the nut. Tighten to 22 ft. lbs. (30 Nm).

27. Install the crankshaft timing gear or sprocket to the crankshaft.

28. Install the No. 1 chain vibration damper with the 2 bolts. Tighten to 80 inch lbs. (9 Nm).

29. Set the No. 1 cylinder to TDC/compression.

30. Turn the camshafts with a wrench (using the hexagonal lobe) to align the timing marks on the camshaft timing gear with the timing marks located on the No. 1 and No. 2 bearing caps

31. Using the crankshaft pulley bolt, turn the crankshaft to position the key on the crankshaft upward.

32. Install the chain onto the crankshaft timing sprocket with the gold or orange mark link aligned with the timing mark on the crankshaft.

33. Using SST 09309-37010 and a hammer, tap in the crankshaft timing sprocket.

Fig. 198 Turning the crankshaft to position the key on the crankshaft upward

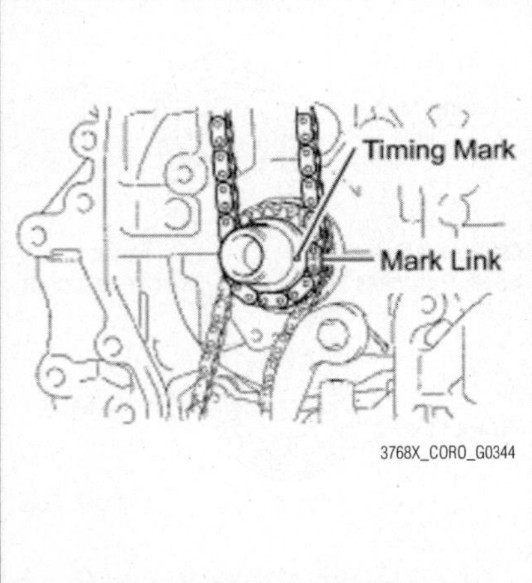

Fig. 199 Install the chain onto the crankshaft timing sprocket

34. Align the gold or yellow links with the timing marks located on the camshaft timing gear and sprocket, then install the chain.

35. Install the chain tensioner slipper with the bolt. Tighten to 14 ft. lbs. (19 Nm).

36. Install the timing chain guide with the bolt. Tighten to 80 inch lbs. (9 Nm).

37. Install the sensor plate with the "F" mark facing forward.

38. Install the timing chain cover.

39. Release the ratchet pawl, then fully push in the plunger and set the hook to the pin so that the plunger is in the position.

40. Install a new gasket and the chain tensioner with the 2 nuts. Tighten to 80 inch lbs. (9 Nm).

41. To complete installation, reverse remaining removal procedure.

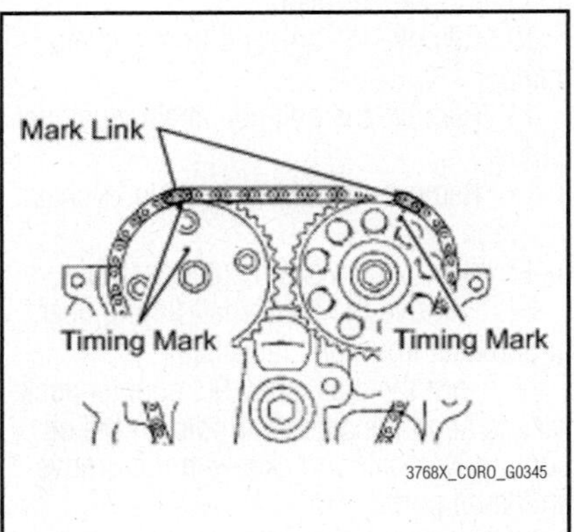

Fig. 200 Aligning the gold or yellow links with the timing marks located on the camshaft timing gear and sprocket

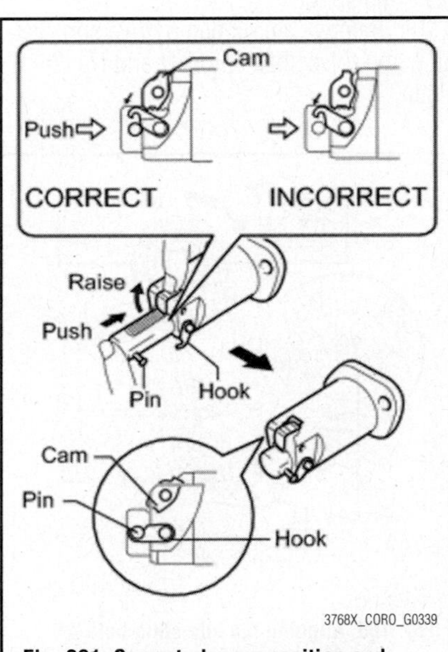

Fig. 201 Correct plunger position and orientation.

VALVE COVERS

REMOVAL & INSTALLATION

1.8L Engine

See Figure 202.

1. Disconnect the negative battery cable.
2. Remove the oil filler cap.
3. Remove the gasket from the oil filler cap.
4. Remove the 2 engine cover joints.
5. Using a 14 mm spark plug wrench, remove the 4 spark plugs.
6. Remove the 2 bolts and 2 sensors.
7. Remove the 2 bolts, O-rings, bracket and 2 oil control valves.
8. Remove the 13 bolts, seal washer and cylinder head cover.
9. Be careful not to drop any of the gaskets into the engine when removing the cylinder head cover because the gaskets may stick to the cylinder head cover.
10. Remove the cylinder head cover gasket.

To install:

➡Remove any oil from the contact surfaces.

11. Install a new gasket to the cylinder head cover.
12. Install 3 new gaskets to the No. 1 camshaft bearing cap.
13. Apply seal packing as shown the illustration. Use Toyota Genuine Seal Packing Black, Three Bond 1207B or equivalent.

➡Remove any oil from the contact surfaces. Install the cylinder head cover within 3 minutes and tighten the bolts within 15 minutes after applying seal

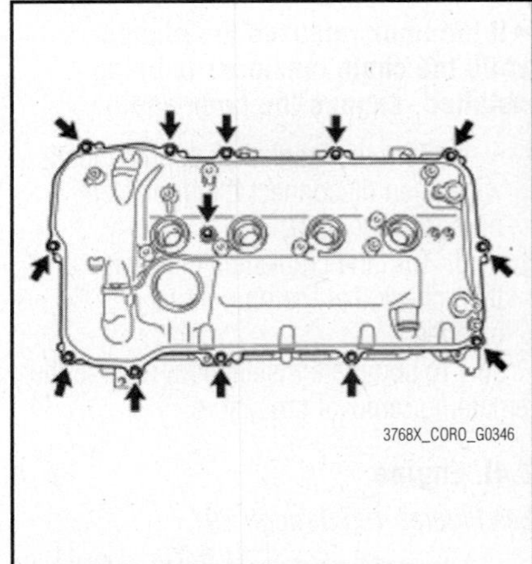

Fig. 202 Removing the valve cover and 13 bolts

packing. Do not start the engine for at least 2 hours after the installation.

14. Install the cylinder head cover with a new seal washer and the 13 bolts. Tighten to 7 ft. lbs. (10 Nm).
15. To complete installation, reverse the remaining removal procedure.

2.4L Engine

See Figures 203 through 205.

1. Disconnect the negative battery cable.
2. Remove the 2 ventilation hoses from the cylinder head cover sub-assembly.
3. Remove the 2 bolts and separate the 2 wire harness brackets.
4. Remove the ignition coil assembly.
5. Remove the oil filler cap from the cylinder head cover.
6. Remove the oil filler cap gasket from the oil filler cap.
7. Using a 22 mm deep socket wrench, remove the ventilation valve from the cylinder head cover.
8. Remove the spark plugs.
9. Remove the 8 bolts and 2 nuts, and remove the cylinder head cover.
10. Remove cylinder head cover gasket from the cylinder head cover.

To install:

11. Install the cylinder head cover gasket onto the cylinder head cover.
12. Remove any old packing material from the contact surface.
13. Apply seal packing to the 2 locations shown in the illustration. Use Toyota Genuine Seal Packing Black, Three Bond 1207B or equivalent.

➡Remove any oil from the contact surfaces. Install the cylinder head cover within 3 minutes of applying seal packing. Do not start the engine for at least 2 hours after installing the cylinder head cover.

14. Install the cylinder head cover with the 8 bolts and 2 nuts.
    a. Bolt A: 8 ft. lbs. (11 Nm).
    b. Bolt B: 10 ft. lbs. (14 Nm).
    c. Bolt C: 8 ft. lbs. (11 Nm).
15. To complete installation, reverse remaining removal procedure.

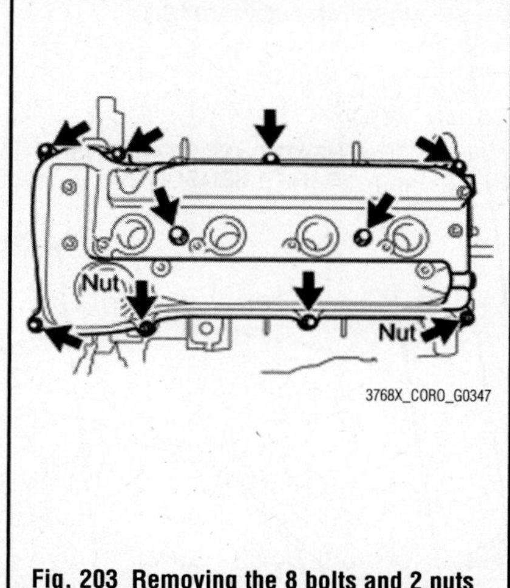

Fig. 203 Removing the 8 bolts and 2 nuts

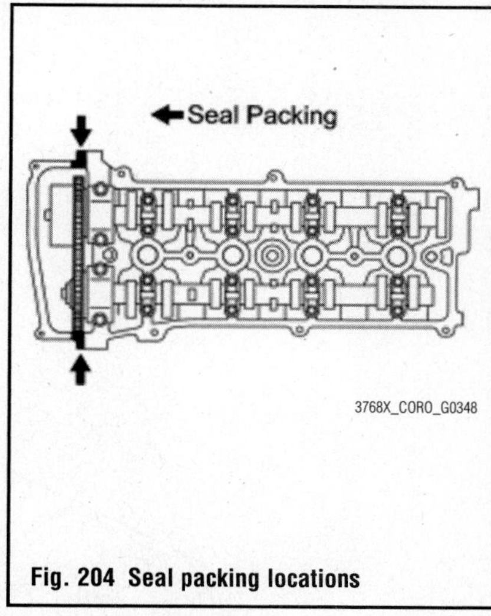

Fig. 204 Seal packing locations

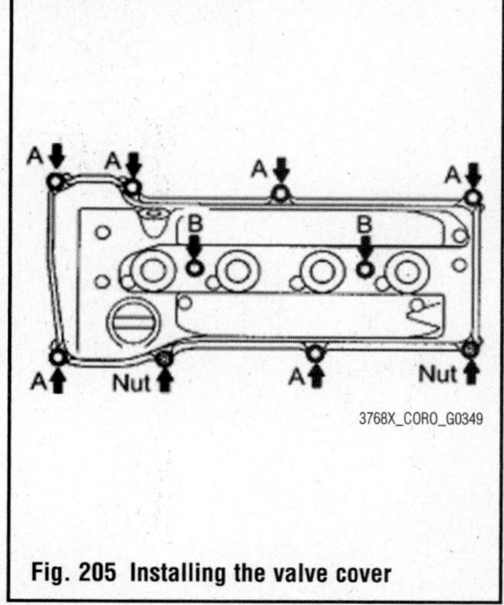

Fig. 205 Installing the valve cover

ENGINE PERFORMANCE & EMISSION CONTROLS

COMPONENT LOCATIONS

See Figures 206 through 208.

CAMSHAFT POSITION (CMP) SENSOR

LOCATION

See Figures 209 and 210.

Refer to the accompanying illustrations for sensor location.

REMOVAL & INSTALLATION

1.8L Engine

See Figures 211 and 212.

➡Although the part name refers to the No. 1 crank position sensors, this procedure is for the camshaft position sensors.

1. Disconnect the negative battery cable.
2. Exhaust camshaft side:
    a. Disconnect the duty vacuum switching valve connector and 3 engine wire harness clamps.
    b. Disconnect the No. 1 crank position sensor connector.
    c. Remove the bolt and No. 1 crank position sensor.
3. Intake camshaft side:
    a. Disconnect the No. 1 crank position sensor connector.
    b. Remove the bolt and No. 1 crank position sensor.

To install:

4. To install, reverse removal procedure.
5. Apply a light coat of engine oil to the O-rings on the No. 1 crank position sensors.
6. Tighten the CMP sensor bolts to 7 ft. lbs. (10 Nm).
7. Inspect for oil leaks.

2.4L Engine

See Figure 213.

1. Disconnect the negative battery cable.
2. Air cleaner cap sub-assembly with hose. Refer to Air Cleaner Box in Engine Mechanical.
3. Disconnect the camshaft position sensor connector.
4. Remove the bolt and the camshaft position sensor.

To install:

5. To install, reverse removal procedure.
6. Apply a light coat of engine oil to the O-ring of the sensor.
7. Tighten the camshaft position sensor with the bolt to 80 inch lbs. (9 Nm).

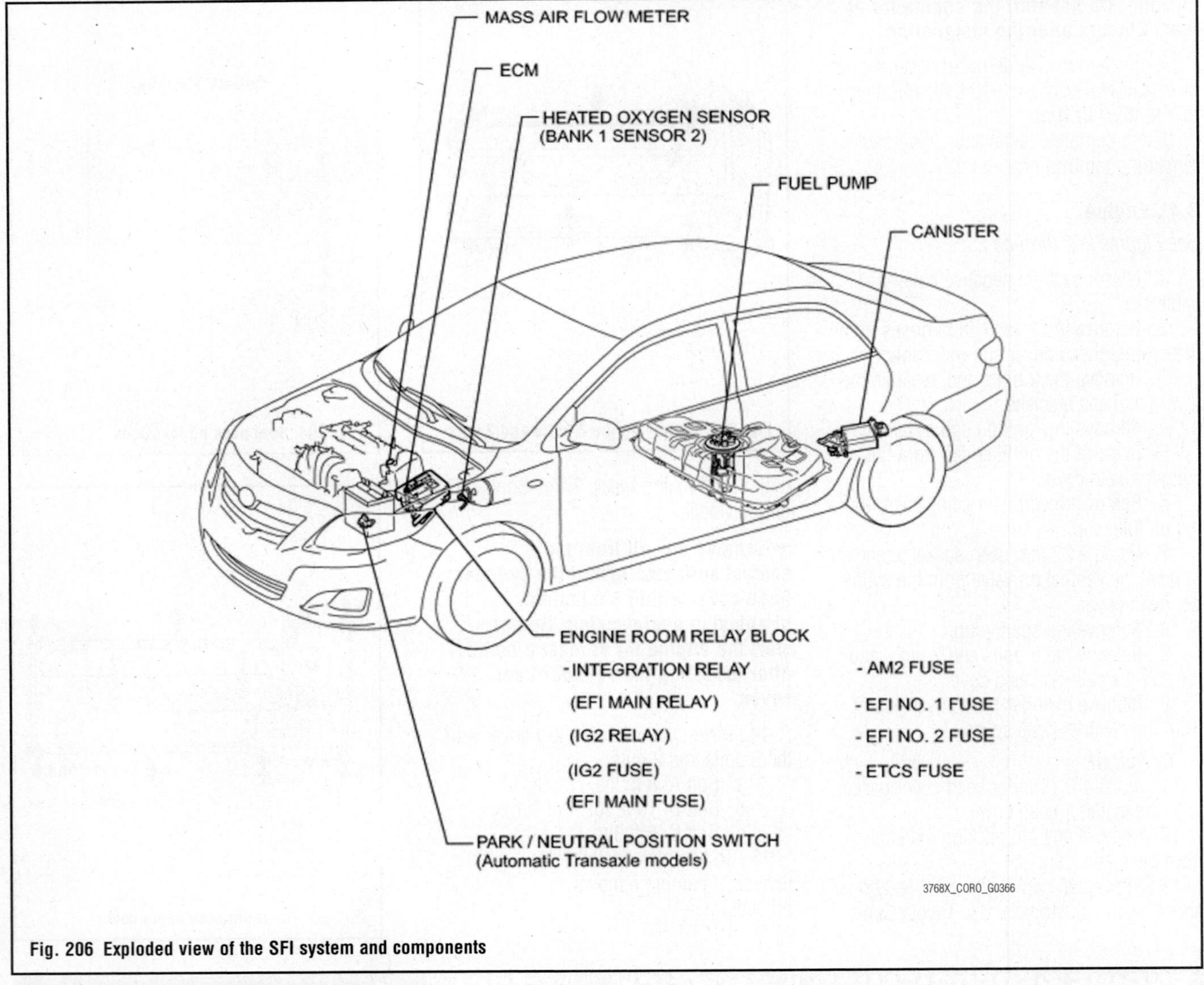

Fig. 206 Exploded view of the SFI system and components

CRANKSHAFT POSITION (CKP) SENSOR

LOCATION

See Figures 214 and 215.

REMOVAL & INSTALLATION

1.8L Engine

See Figure 216.

1. Disconnect the negative battery cable.
2. Remove the RH engine under cover.
3. Disconnect the crank position sensor connector.
4. Remove the bolt and crank position sensor.

To install:

5. To install, apply a light coat of engine oil to the O-ring on the crank position sensor.
6. Install the crank position sensor with the bolt. Tighten to 7 ft. lbs. (10 Nm).
7. Inspect for an oil leak.
8. Install the RH engine under cover.

2.4L Engine

See Figure 217.

1. Disconnect the negative battery cable.
2. Remove the RH engine under cover.
3. Remove the v-ribbed belt. Refer to Accessory Drive Belts in Engine Mechanical.
4. Remove the alternator. Refer to Alternator in Engine Electrical.
5. Disconnect the crank position sensor connector.
6. Separate the crank position sensor connector clamp and wire harness.
7. Remove the bolt and crank position sensor.

To install:

8. To install, apply a light coat of engine oil to the O-ring on the crank position sensor.
9. Install the crank position sensor with the bolt. Tighten to 80 inch lbs. (9 Nm).
10. To complete installation, reverse the remaining removal procedure.
11. Inspect for an oil leak.
12. Install the RH engine under cover.

ELECTRONIC CONTROL MODULE (ECM)

LOCATION

See Figures 218 and 219.

Refer to the accompanying illustrations for location.

REMOVAL & INSTALLATION

See Figure 220.

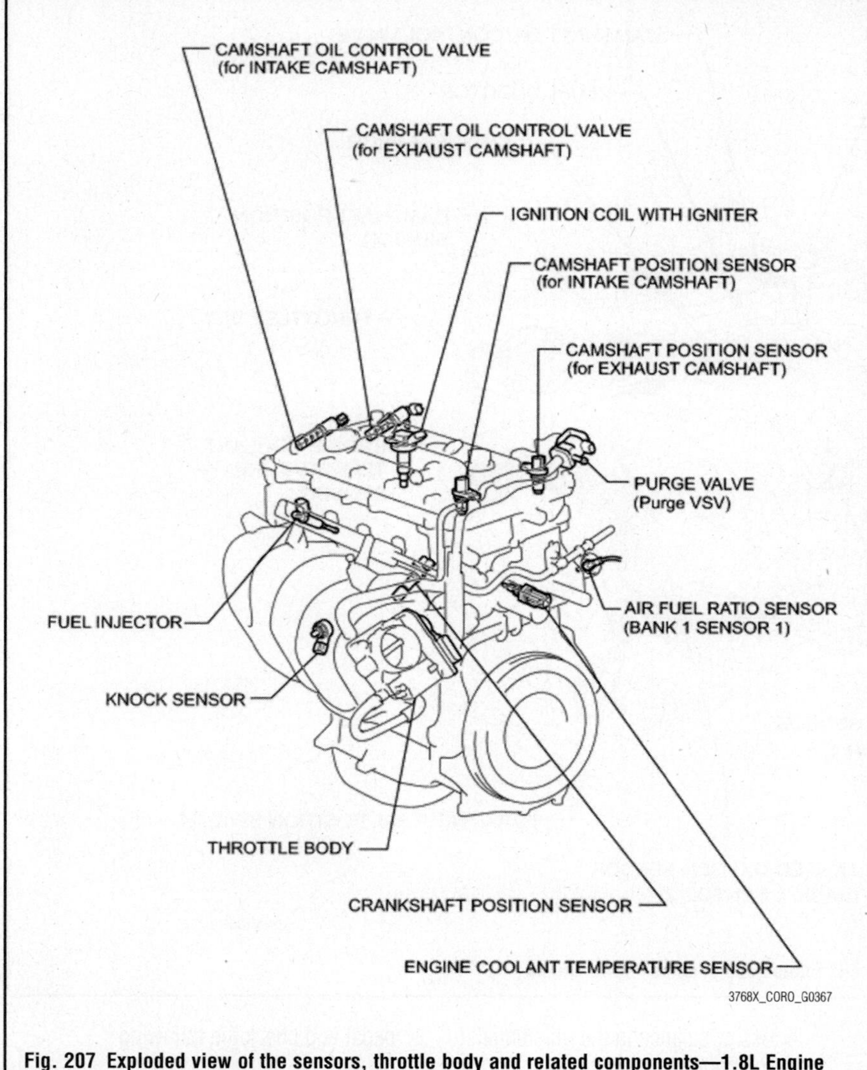

Fig. 207 Exploded view of the sensors, throttle body and related components—1.8L Engine

1. Disconnect the negative battery cable.
2. Remove the air cleaner assembly. Refer to Air Cleaner Assembly in Engine Mechanical.
3. Separate the wire harness clamp.
4. Disconnect the 2 ECM connectors. Push the locks on the 2 levers, then raise the levers, and disconnect the 2 ECM connectors.

➡After disconnecting the connectors, make sure that dirt, water or other foreign matter does not contact the connecting parts of the connectors.

5. Remove the 3 bolts and the ECM.
6. Remove the 4 screws and 2 ECM brackets.

To install:

7. To install, reverse the removal procedure.
8. Perform the VIN registration procedure and/or Automatic Transaxle Initialization.

Automatic Transaxle Initialization

➡The ECM memorizes the condition that the ECT controls the automatic transaxle assembly and engine assembly according to those characteristics. Therefore, when the automatic transaxle assembly, engine assembly, or ECM has been replaced, it is necessary to reset the memory so that the ECM can memorize the new information.

1. Turn the ignition switch off.
2. Connect the Techstream to the DLC3.
3. Turn the ignition switch to ON and the Techstream main switch on.
4. Enter the following menus: Powertrain / Engine and ECT / Utility / Reset Memory. Then, press "Next".

⁂ CAUTION

After performing the reset memory, be sure to perform the road test.

5. The ECM is learned by performing the road test.

VIN Registration

➡The Vehicle Identification Number (VIN) must be input into a replacement ECM.

➡The VIN is a 17-digit alphanumeric vehicle identification number. The Techstream is required to register the VIN.

➡This registration section consists of two parts: Read VIN and Write VIN.

1. Read VIN: This process allows the VIN stored in the ECM to be read in order to confirm that the two VINs, provided with the vehicle and stored in the vehicle's ECM, are the same.
2. Write VIN: This process allows the VIN to be input into the ECM. If the ECM is changed, or the ECM VIN and vehicle VIN do not match, the VIN can be registered, or overwritten in the ECM by following this procedure.
3. Read the VIN:
   a. Confirm the vehicle VIN.
   b. Connect the Techstream to the DLC3.
   c. Turn the ignition switch to ON.
   d. Turn the Techstream on.
   e. Enter the following menus: Powertrain / Engine and ECT / Utility / VIN / VIN Read.
4. Write the VIN:
   a. Confirm the vehicle VIN.
   b. Connect the Techstream to the DLC3.
   c. Turn the ignition switch to ON.
   d. Turn the Techstream on.
   e. Enter the following menus: Powertrain / Engine and ECT / Utility / VIN / VIN Write.

ROAD TEST

1. Perform the test at the ATF (Automatic Transmission Fluid) temperature 122 to 176 °F (50 to 80 °C) in the normal operation.
   a. D position test: Shift into the D position and fully depress the accelerator pedal and check the following points.
      - Check up-shift operation. Check that 1–2, 2–3, 3–4 and 4–5th up-shifts take place, and that the shift points conform to the automatic shift schedule.

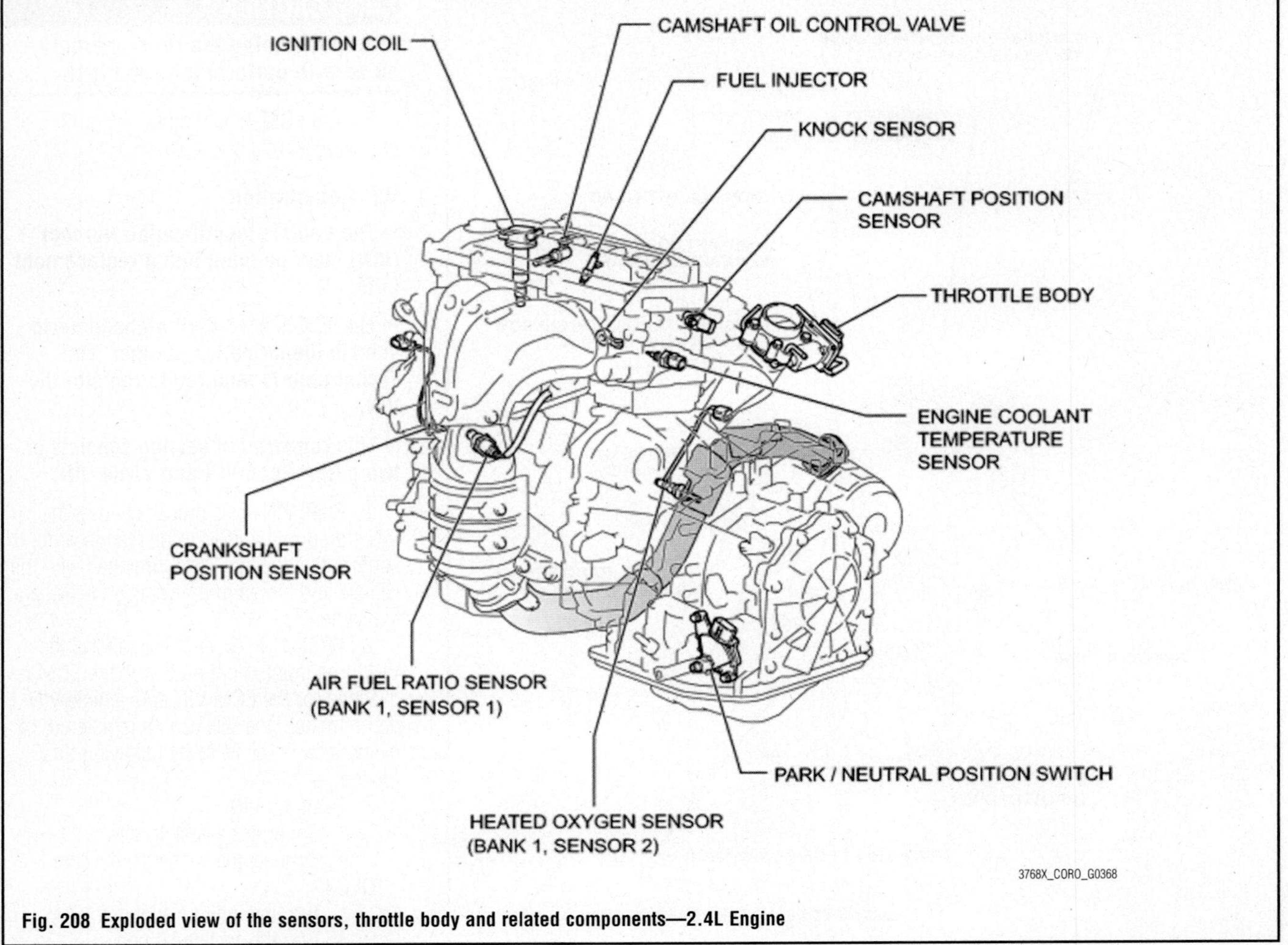

Fig. 208 Exploded view of the sensors, throttle body and related components—2.4L Engine

➡5th Gear Up-shift Prohibition Control: Engine coolant temperature is 131°F (55°C) or less and vehicle speed is at 43 mph (70km/h) or less. ATF temperature is 28°F (-2°C) or less.

➡5th and 4th Gear Lock-up Prohibition Control: Brake pedal is depressed. Accelerator pedal is released. Engine coolant temperature is 140°F (60°C) or less.

- Check for shift shock and slip. Check for shock and slip at the 1–2, 2–3, 3–4, and 4–5th up-shifts.
- Check for abnormal noise and vibration. Check for abnormal noise and vibration when up-shifting from 1–2, 2–3, 3–4, and 4–5 while driving with the shift lever in the D position, and also check while driving in the lock-up condition. The check for the cause of abnormal noise and vibration must be done thoroughly as it could also be due to loss of balance in the differential, torque converter clutch, etc.
- Check kick-down operation.

Check vehicle speeds when the 2nd to 1st, 3rd to 2nd, 4th to 3rd, and 5th to 4th kick-downs take place while driving with the shift lever in the D position. Confirm that each speed is within the applicable vehicle speed range indicated in the automatic shift schedule.

- Check for abnormal shock and slip at kick-down.
- Check the lock-up mechanism: Drive in the D position (5th gear), at a steady speed (lock-up ON). Lightly depress the accelerator pedal and check that the engine speed does not change abruptly.

➡There is no lock-up function in the 1st, 2nd and 3rd gears. If there is a big jump in engine speed, there is no lock-up.

b. S position test: Shift to the S position, depress the accelerator pedal and check the following points:

- Check shift operation.
- While driving in the D position and 5th gear, shift into the S position and back to the D position. Check that the gear change 5–4 down-shift and 4–5 up-shift can be performed.
- With the shift lever in the S position (while the vehicle is stopped), shift into the "+" position to check that the shift position on the combination meter changes as follows: 1–2, 2–3, 3–4, and 4–5.
- While driving in the 4(S) position and 4th gear (at a vehicle speed of approximately 25 to 31mph (40 to 50 km/h)), shift into the "-" position and check if the 3rd gear down-shift occurs and the engine brake performs properly.
- While driving in the 3(S) position and 3rd gear (at a vehicle speed of approximately 19 to 25 mph (30 to 40 km/h)), shift into the "-"

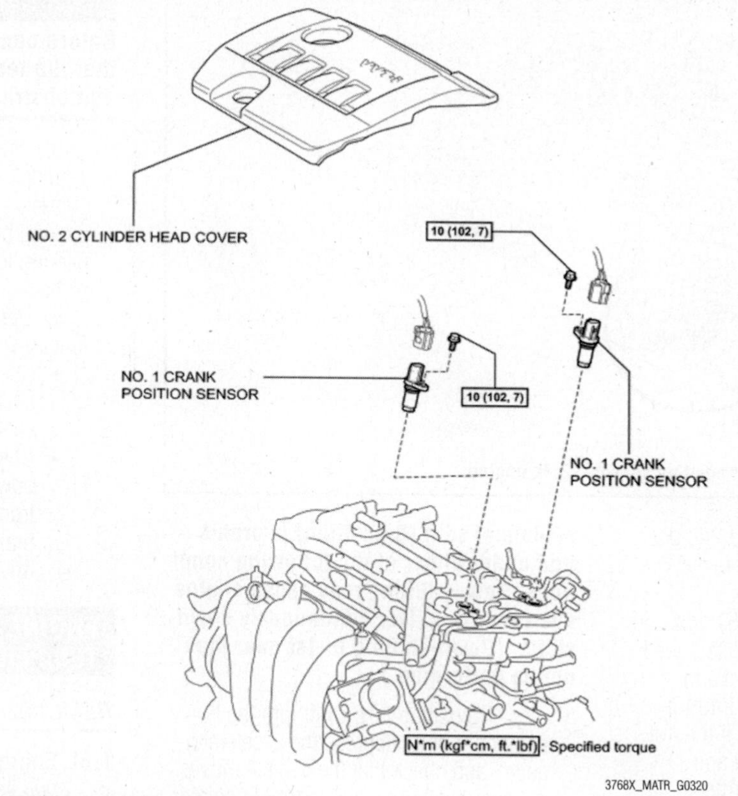

Fig. 209 Camshaft Position (CMP) sensor location—1.8L engine

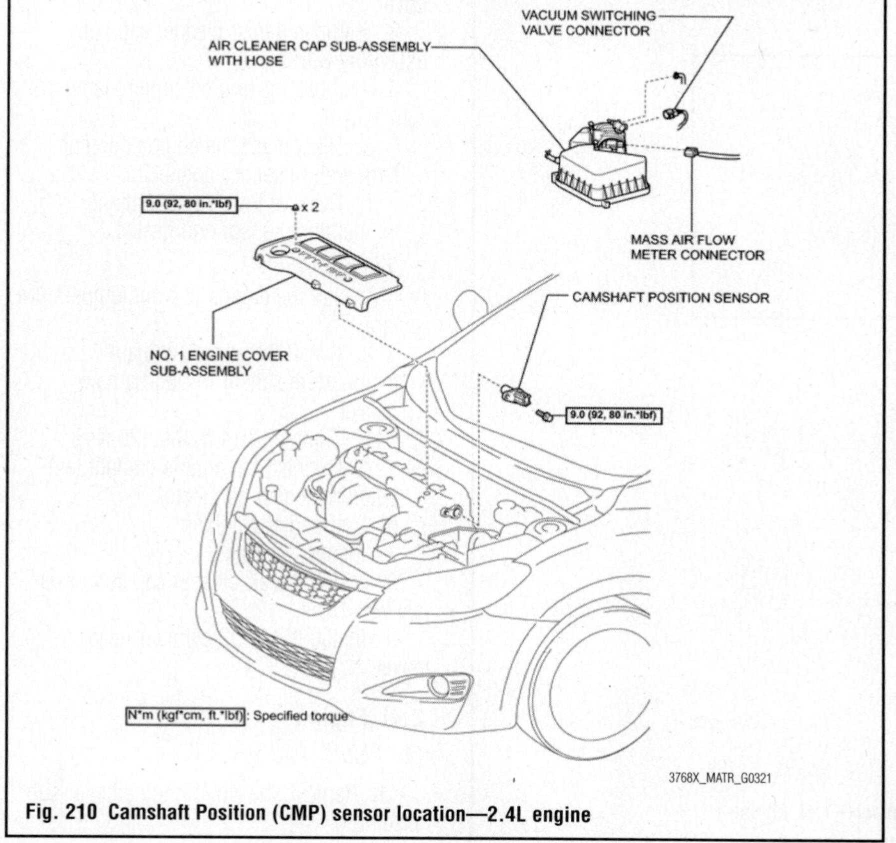

Fig. 210 Camshaft Position (CMP) sensor location—2.4L engine

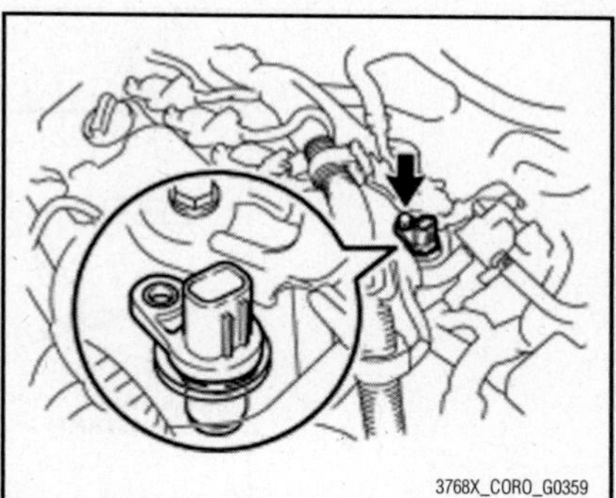

Fig. 211 Removing the exhaust side CMP sensor—1.8L Engine

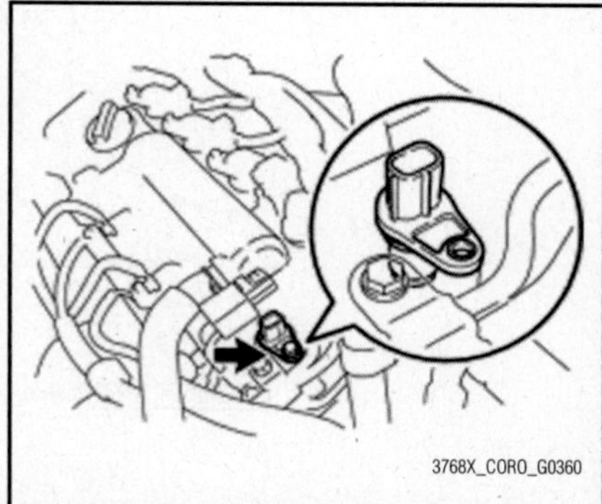

Fig. 212 Removing the intake side CMP sensor—1.8L Engine

Fig. 213 Removing the camshaft position sensor—2.4L engine

position and check if the 2nd gear down-shift occurs and the engine brake performs properly.

- While driving in the 2(S) position and 2nd gear (at a vehicle speed of approximately 12 to 19 mph (20 to 30 km/h)), shift into the "-" position and check if the 1st gear down-shift occurs and the engine brake performs properly.

➡Manual shift (S position) is prohibited under either of the following conditions: Down-shifting may cause engine overrun. The driver continuously down-shifts. (Down-shifting to 1st gear may not be performed.)

c. R position test: Shift into the R position, lightly depress the accelerator pedal, and check that the vehicle moves backward without any abnormal noise or vibration.

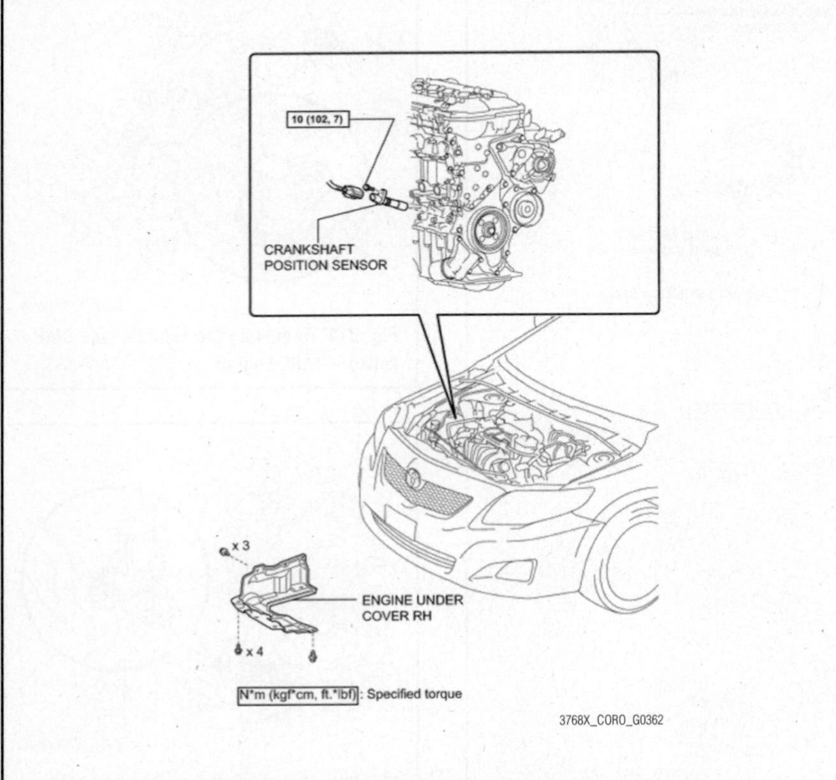

Fig. 214 View of the Crankshaft Position (CKP) sensor—1.8L Engine

⁂ CAUTION

Before conducting this test ensure that the test area is free from people and obstruction.

d. P position test: Stop the vehicle on a grade (more than 5°) and after shifting into the P position, release the parking brake. Then, check that the parking lock pawl holds the vehicle in place.

e. Uphill/downhill control function:

- Check that the gear does not up-shift to the 4th or 5th gear while the vehicle is driving uphill.
- Check that the gear automatically down-shifts from the 5th to 4th or from the 4th to 3rd gear when brake is applied while the vehicle is driving downhill.

ENGINE COOLANT TEMPERATURE (ECT) SENSOR

REMOVAL & INSTALLATION

1.8L Engine

See Figures 221 and 222.

1. Drain the engine coolant.
2. Remove the No. 2 cylinder head cover.
3. Remove the air cleaner cap sub assembly with hose.
4. Remove the engine coolant temperature sensor.
   a. Disconnect the engine coolant temperature sensor connector.
   b. Remove the engine coolant temperature sensor and gasket.

To install:

5. Install the engine coolant temperature sensor.
   a. Install the engine coolant temperature sensor through a new gasket.
   b. Tighten to 14 ft. lbs. (20 Nm).
   c. Connect the engine coolant temperature sensor connector.
6. Add engine coolant.
7. Inspect for leaks.
8. Install the air cleaner cap sub assembly with hose.
9. Install the No. 2 cylinder head cover.

2.4L Engine

See Figures 223 and 224.

1. Remove the No. 1 engine cover sub assembly.

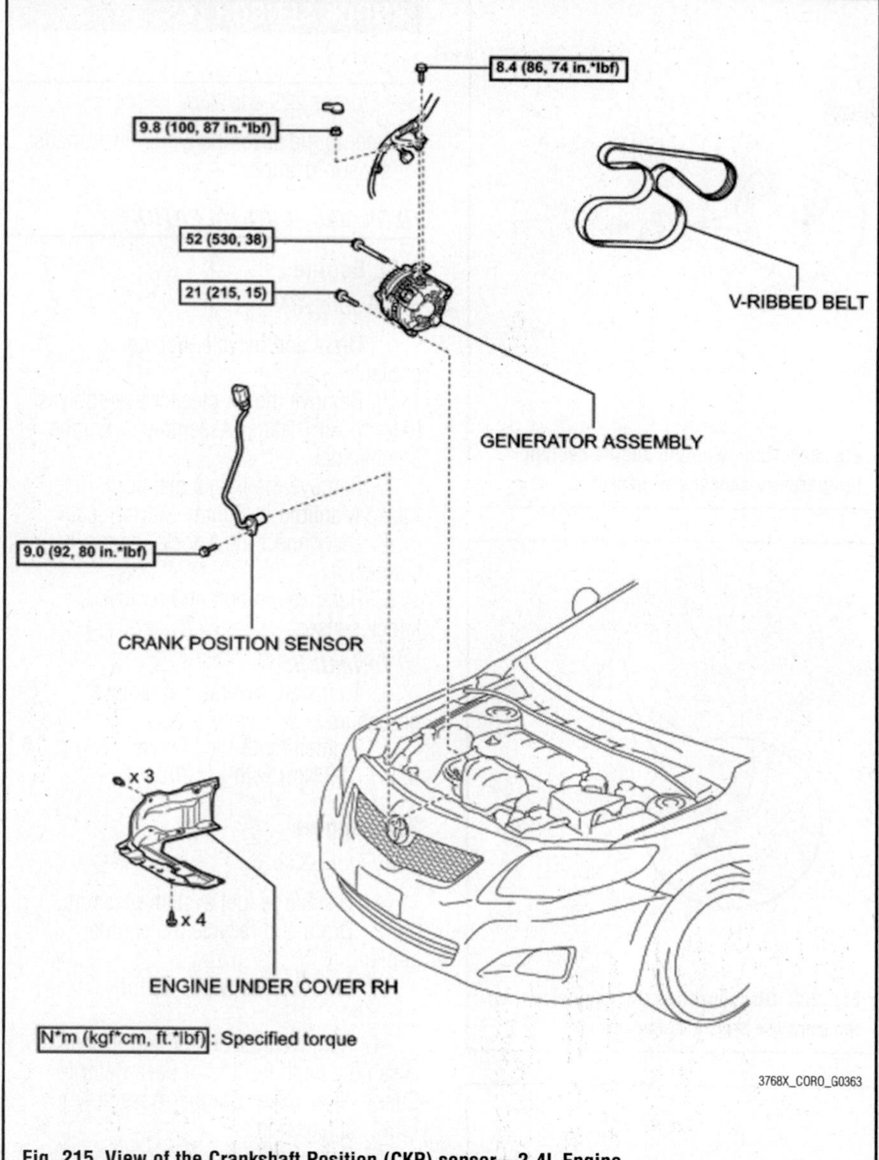

Fig. 215 View of the Crankshaft Position (CKP) sensor—2.4L Engine

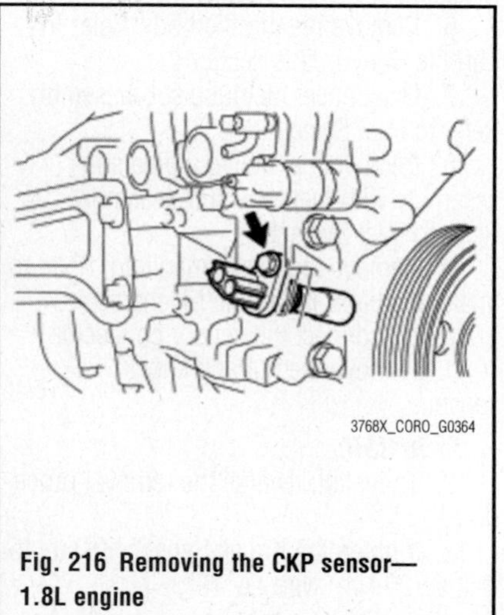

Fig. 216 Removing the CKP sensor—1.8L engine

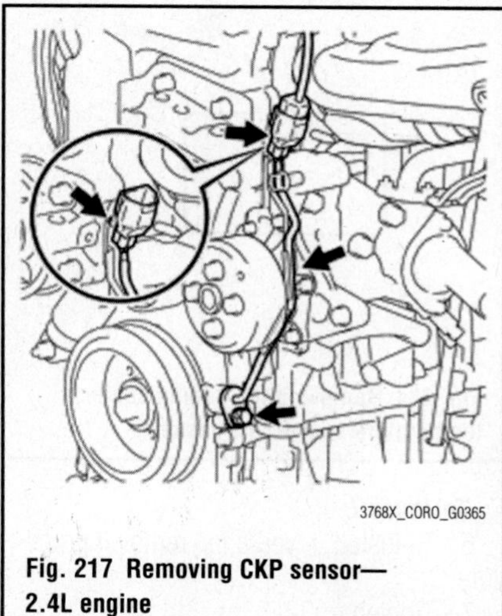

Fig. 217 Removing CKP sensor—2.4L engine

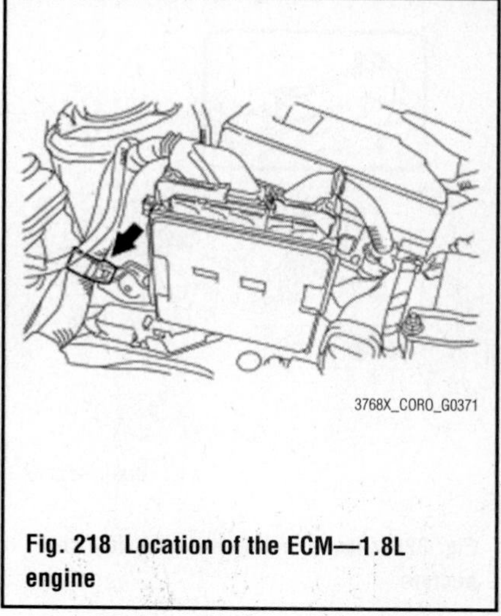

Fig. 218 Location of the ECM—1.8L engine

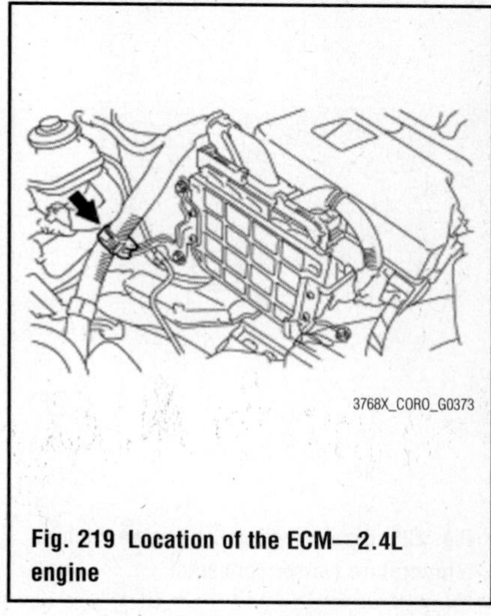

Fig. 219 Location of the ECM—2.4L engine

2. Drain engine coolant.
3. Remove the air cleaner cap sub assembly with hose.
4. Remove air cleaner case.
5. Remove the engine coolant temperature sensor.
   a. Disconnect the engine coolant temperature sensor connector.
   b. Remove the engine coolant temperature sensor and gasket.

To install:

6. Install the engine coolant temperature sensor through a new gasket.
   a. Tighten to 15 ft. lbs. (20 Nm).
   b. Connect the engine coolant temperature sensor connector.
7. Install the air cleaner case.
8. Install the air cleaner cap sub assembly with hose.
9. Add engine coolant.
10. Inspect for coolant leak.
11. Install the No. 2 engine cover sub assembly.

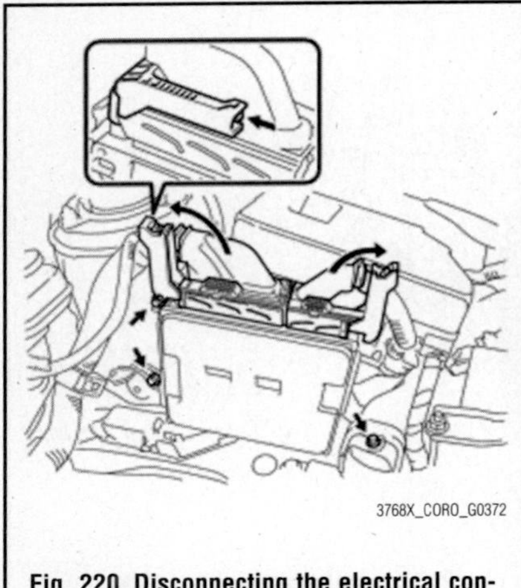

Fig. 220 Disconnecting the electrical connectors

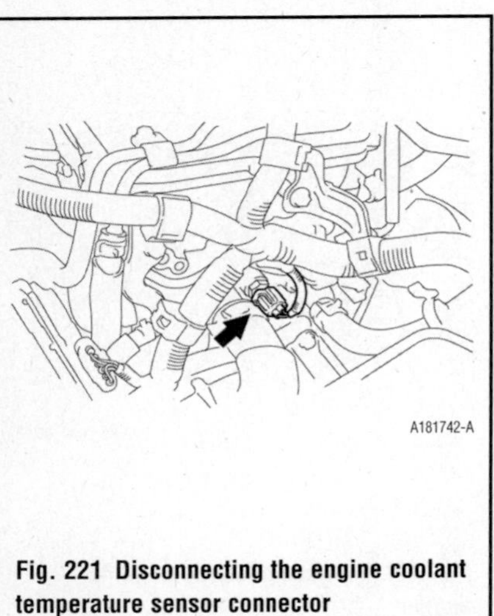

Fig. 221 Disconnecting the engine coolant temperature sensor connector

HEATED OXYGEN SENSOR (HO2S)

LOCATION

See Figures 225 and 226.

Refer to the accompanying illustrations for sensor location.

REMOVAL & INSTALLATION

See Figures 227 and 228.

1. Disconnect the negative battery cable.
2. Disconnect the heated oxygen sensor connector.
3. Remove the grommet and pull the sensor connector out of the cabin through the floor panel.
4. Remove the wire harness clamp bracket and disconnect the wire harness clamp.
5. Using a wrench or SST: 09224-00010, remove the heated oxygen sensor.

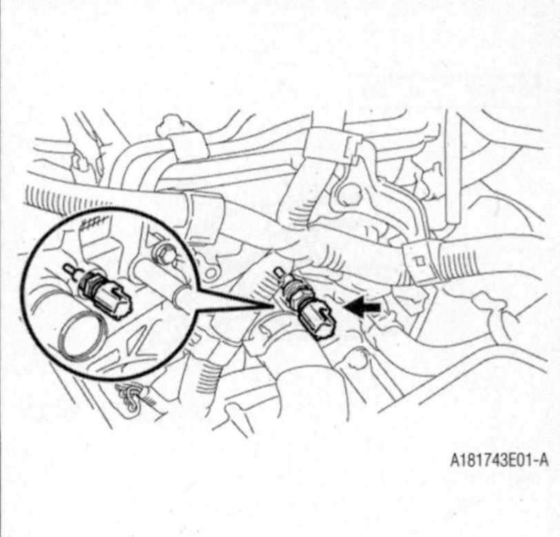

Fig. 222 Removing the engine coolant temperature sensor and gasket

Fig. 223 Disconnecting the engine coolant temperature sensor connector

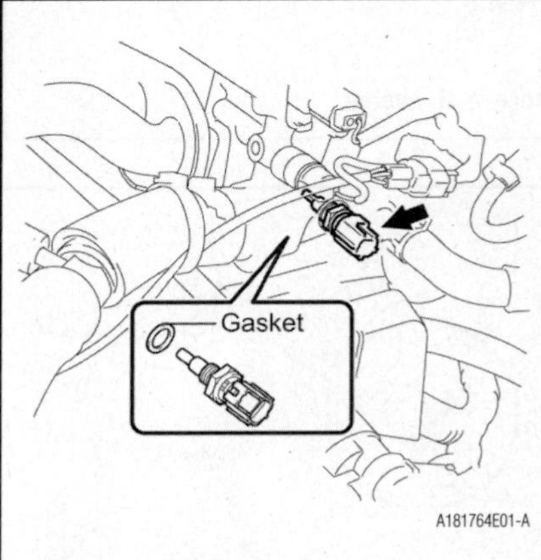

Fig. 224 Removing the engine coolant temperature sensor and gasket

To install:

6. To install, reverse the removal procedure.
7. Tighten to 32 ft. lbs. (44 Nm).

KNOCK SENSOR (KS)

LOCATION

See Figures 229 and 230.

Refer to the accompanying illustrations for sensor location.

REMOVAL & INSTALLATION

1.8L Engine

See Figure 231.

1. Drain and recycle the engine coolant.
2. Remove the air cleaner assembly. Refer to Air Cleaner Assembly in Engine Mechanical.
3. Remove the intake manifold. Refer to Intake Manifold in Engine Mechanical.
4. Disconnect the Knock Sensor (KS) connector.
5. Remove the bolt and remove the knock sensor.

To install:

6. To install, reverse the removal procedure.
7. Tighten the Knock Sensor (KS) to 15 ft. lbs. (20 Nm) with +/- 20°.

2.4L Engine

See Figure 232.

1. Discharge fuel system pressure.
2. Drain and recycle the engine coolant.
3. Remove the wiper motor assembly.
4. Remove suspension tower damper assembly (with front strut bar). Refer to Suspension Tower Damper Assembly in Front Suspension.
5. Remove the air cleaner assembly. Refer to Air Cleaner Assembly in Engine Mechanical.
6. Remove the throttle body. Refer to Throttle Body in this section.
7. Disconnect fuel tube sub-assembly. Refer to Fuel System.
8. Remove fuel delivery pipe sub-assembly with fuel tube sub-assembly. Refer to Fuel System.
9. Remove the intake manifold. Refer to Intake Manifold in Engine Mechanical.
10. Disconnect the sensor connector.
11. Remove the nut and knock sensor.

To install:

12. To install, reverse the removal procedure.
13. Tighten the Knock Sensor (KS) to 15 ft. lbs. (20 Nm) with +/- 10°.

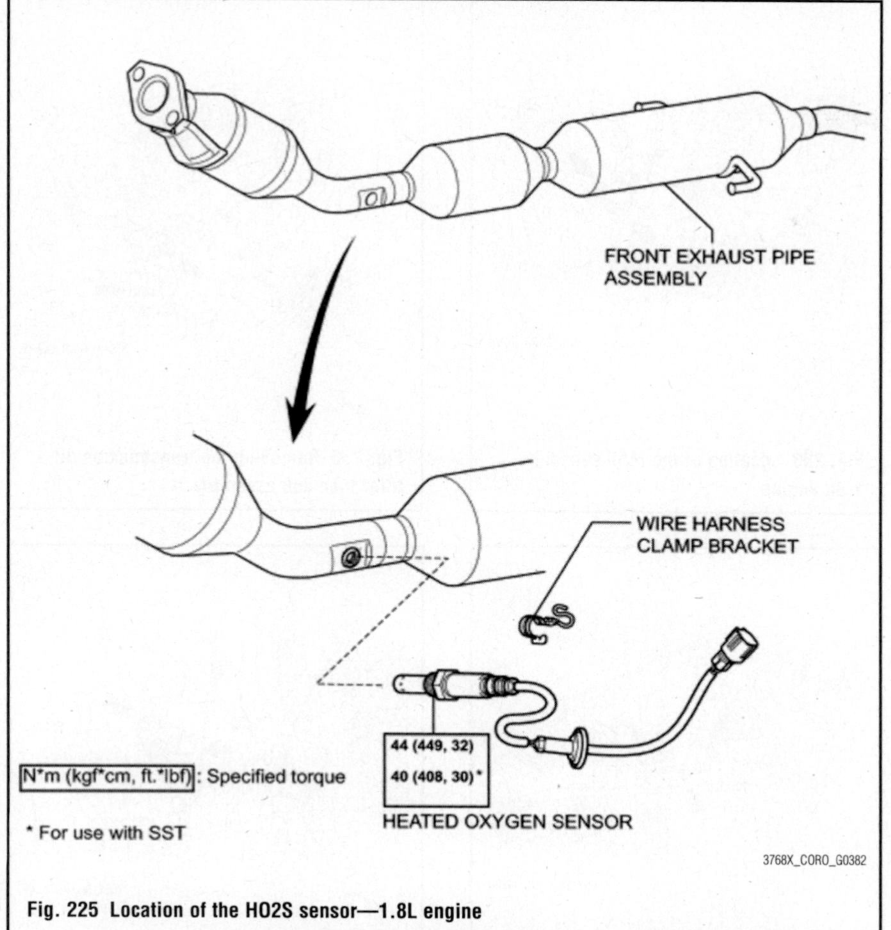

Fig. 225 Location of the HO2S sensor—1.8L engine

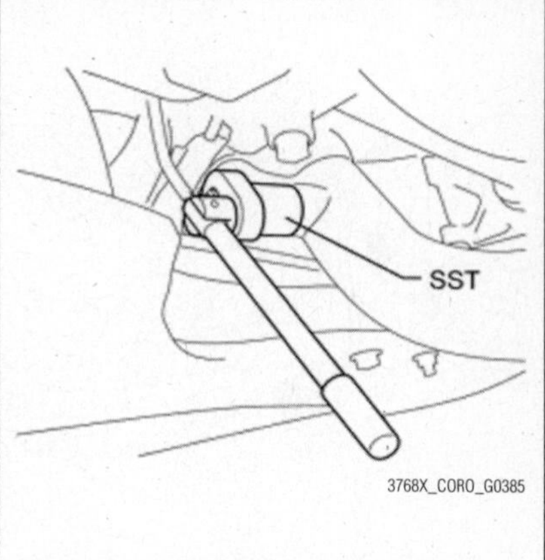

Fig. 228 Removing the heated oxygen sensor—2.4L engine

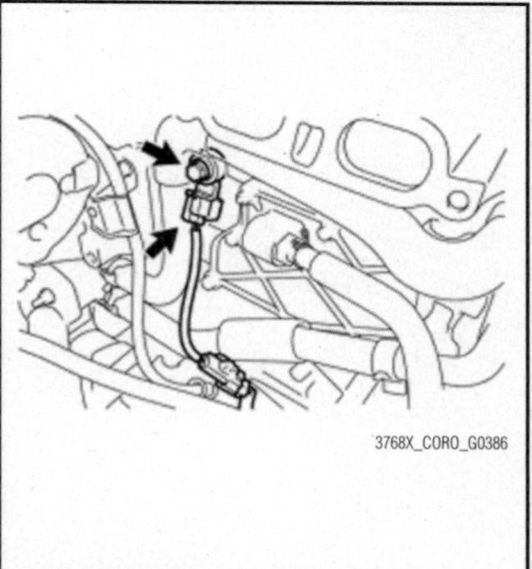

Fig. 229 Location of the Knock Sensor (KS)—1.8L engine

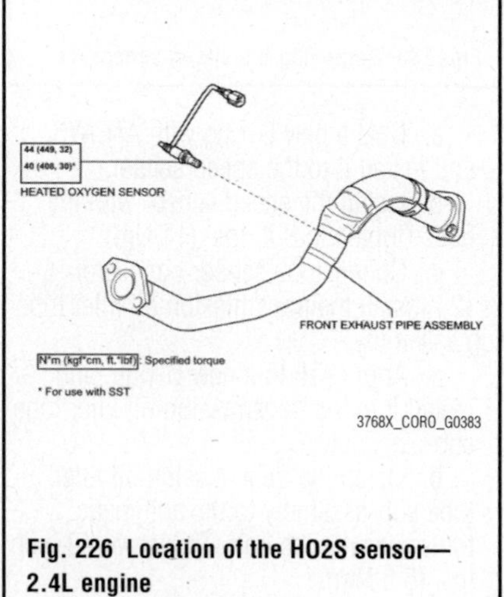

Fig. 226 Location of the HO2S sensor—2.4L engine

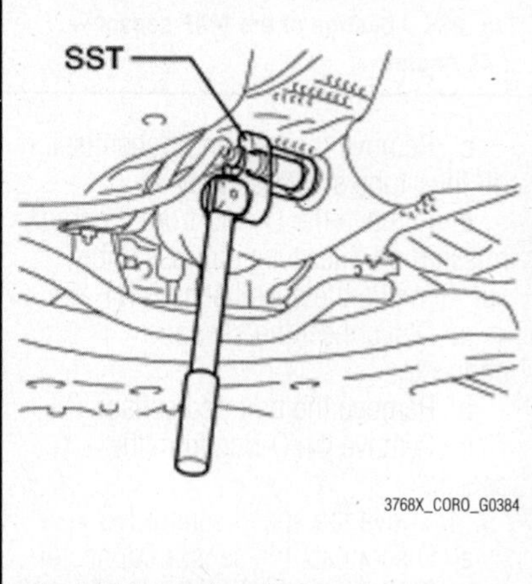

Fig. 227 Removing the heated oxygen sensor—1.8L engine

3768X_CORO_G0387

Fig. 230 Location of the Knock Sensor (KS)—2.4L engine

MASS AIR FLOW (MAF) SENSOR

LOCATION

See Figures 233 and 234.

Refer to the accompanying illustrations for sensor location.

REMOVAL & INSTALLATION

1. Disconnect the negative battery cable.
2. Disconnect the mass air flow meter connector.
3. Remove the 2 screws and the mass air flow meter.

➡Make sure that the O-ring is not cracked or does not jump out of position during installation.

To install:

4. To install, reverse the removal procedure.

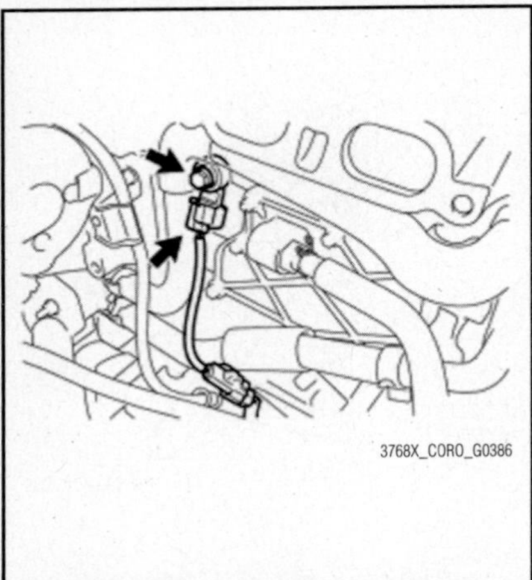

Fig. 231 Location of the knock sensor—1.8L engine

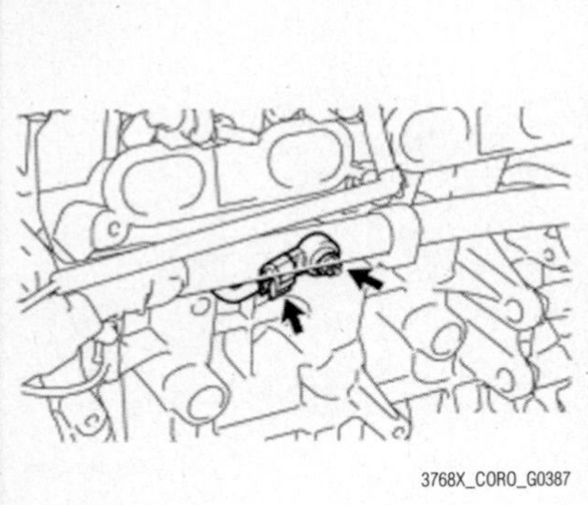

Fig. 232 Location of the knock sensor—2.4L engine

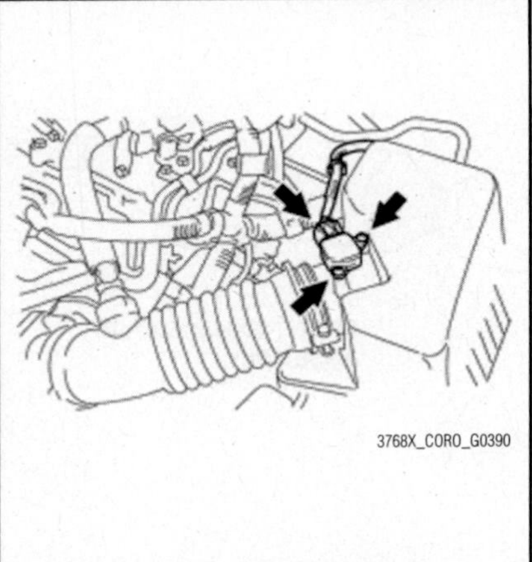

Fig. 233 Location of the MAF sensor—1.8L engine

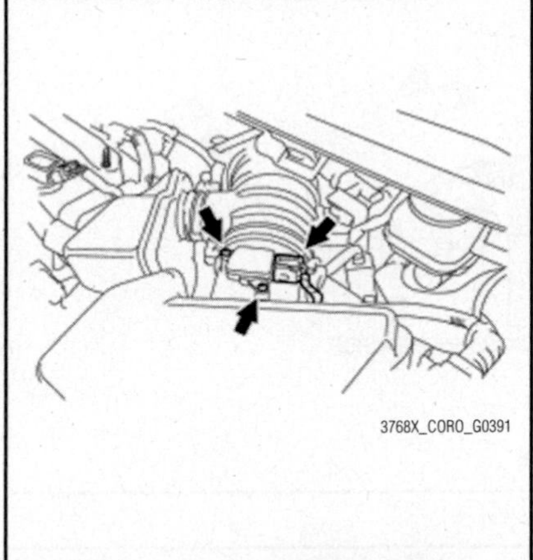

Fig. 234 Location of the MAF sensor—2.4L engine

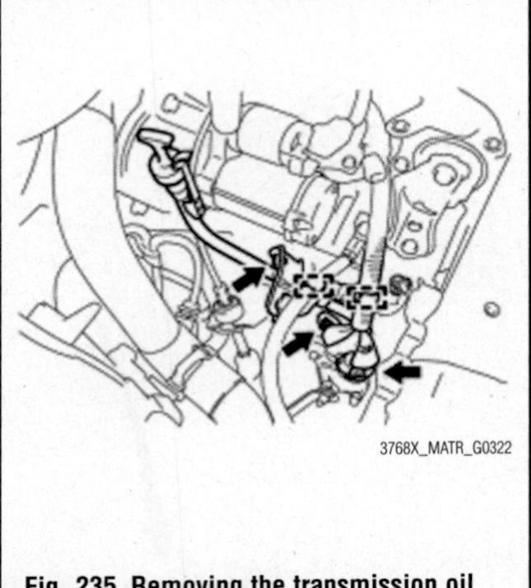

Fig. 235 Removing the transmission oil filler tube sub assembly

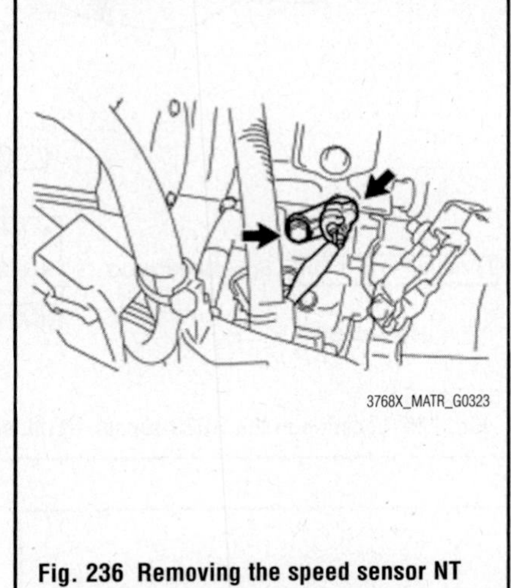

Fig. 236 Removing the speed sensor NT

VEHICLE SPEED SENSOR (VSS)

REMOVAL & INSTALLATION

U140f Automatic Transaxle

See Figures 235 through 237.

1. Disconnect the cable from the negative battery terminal.
2. Remove the No. 1 engine cover sub assembly.
3. Remove the air cleaner cap sub assembly.
4. Remove the air cleaner case.
5. Remove the battery.
6. Remove the battery carrier.
7. Remove the transmission oil filler tube sub assembly.
   a. Remove the transmission oil level gauge sub assembly.
   b. Disconnect the transmission wire connector, park/ neutral position switch connector and 2 wire harness clamps.
   c. Remove the bolt and transmission oil filler tube sub-assembly.
   d. Remove the O-ring from the transmission oil filler tube sub-assembly.
8. Remove the speed sensor NT.
   a. Disconnect the sensor connector.
   b. Remove the bolt and sensor.
   c. Remove the O-ring from the sensor.
9. Remove the speed sensor NC.
   a. Disconnect the sensor connector.
   b. Remove the bolt and sensor.
   c. Remove the O-ring from the sensor.

To install:

10. Install the speed sensor NC.
   a. Coat a new O-ring with ATF WS and install it to the speed sensor.
   b. Install the speed sensor with the bolt. Tighten to 8 ft. lbs. (11 Nm).
   c. Connect the sensor connector.
11. Install the speed sensor NT.
   a. Coat a new O-ring with ATF WS and install it to the speed sensor.
   b. Install the speed sensor with the bolt. Tighten to 8 ft. lbs. (11 Nm).
   c. Connect the sensor connector.
12. Install the transmission oil filler tube sub assembly.
   a. Apply ATF to a new O-ring, and install it to the transmission oil filler tube sub-assembly.
   b. Install the transmission oil filler tube sub-assembly to the automatic transaxle with the bolt. Tighten to 49 inch lbs. (5.5 Nm).
   c. Connect the transmission wire connector, park/ neutral position switch connector and 2 wire harness clamps.
   d. Install the transmission oil level gauge sub-assembly to the transmission oil filler tube sub-assembly.
13. To complete the installation, reverse the remaining removal procedures.

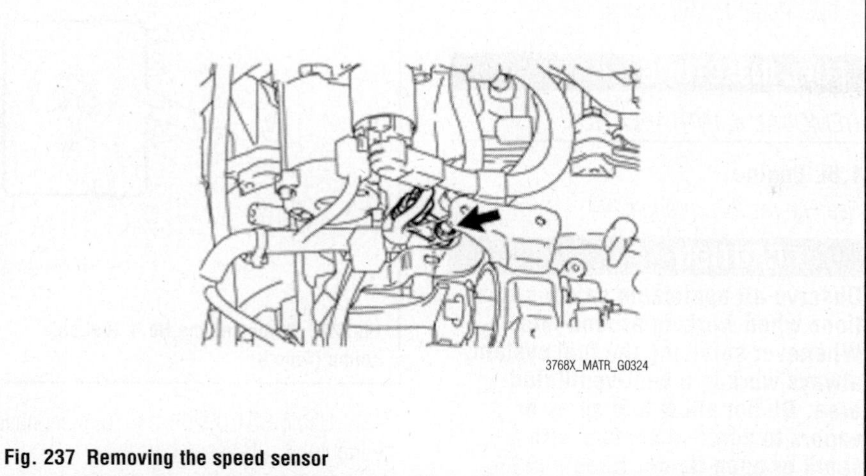

Fig. 237 Removing the speed sensor

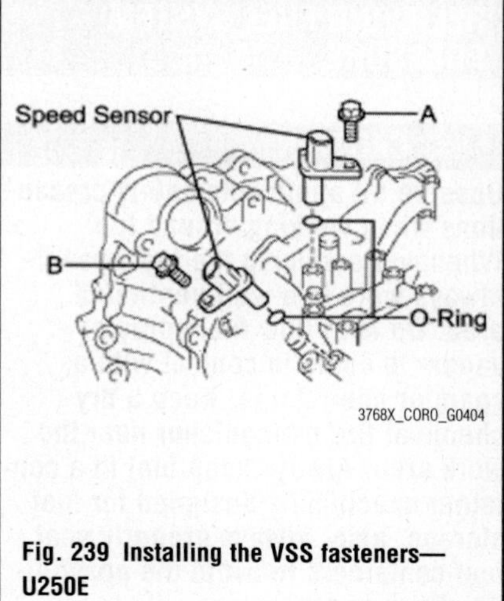

Fig. 239 Installing the VSS fasteners—U250E

U250e Automatic Transaxle

See Figures 238 through 240, and 241

1. Disconnect the negative battery cable.

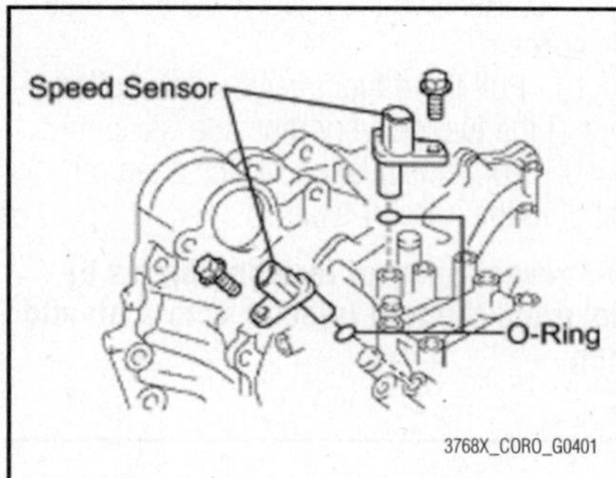

Fig. 238 Locating and removing the 2 speed sensors from the transaxle assembly

2. Remove the 2 bolts and the 2 speed sensors from the transaxle assembly.
3. Remove the 2 O-rings from the 2 speed sensors.

To install:

4. Coat 2 new O-rings with ATF, and install them to the 2 speed sensors.
5. Apply adhesive to the bolts threads. Toyota Genuine Adhesive 1344, Three Bond 1344 or Equivalent.
6. Install the 2 speed sensors to the transaxle case with the 2 bolts:
   a. Bolt A: 78 inch lbs. (9 Nm).
   b. Bolt B: 8 ft. lbs. (11 Nm).

U341E Automatic Transaxle

See Figure 241.

1. Disconnect the negative battery cable.
2. Disconnect the electrical connector
3. Remove the bolt and speed sensor from the transaxle case.

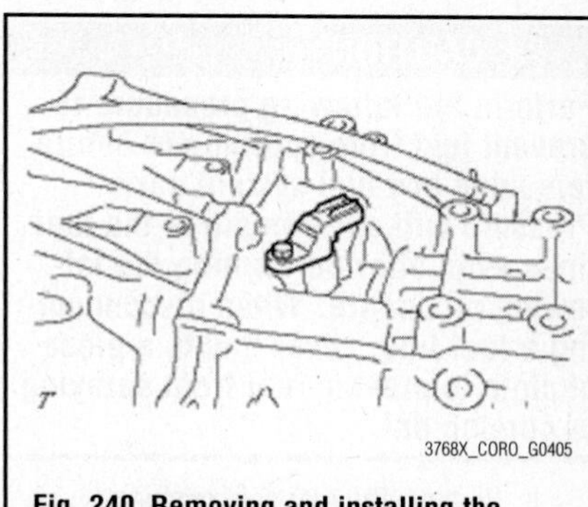

Fig. 240 Removing and installing the VSS—U341E

To install:

4. To install, coat a new O-ring with ATF, and install it to the speed sensor.
   a. Install the speed sensor to the transaxle case with the bolt. Tighten to 48 inch. lbs. (5.4 Nm).

FUEL — GASOLINE FUEL INJECTION SYSTEM

FUEL SYSTEM SERVICE PRECAUTIONS

Safety is the most important factor when performing not only fuel system maintenance but any type of maintenance. Failure to conduct maintenance and repairs in a safe manner may result in serious personal injury or death. Maintenance and testing of the vehicle's fuel system components can be accomplished safely and effectively by adhering to the following rules and guidelines.

- To avoid the possibility of fire and personal injury, always disconnect the negative battery cable unless the repair or test procedure requires that battery voltage be applied.
- Always relieve the fuel system pressure prior to disconnecting any fuel system component (injector, fuel rail, pressure regulator, etc.), fitting or fuel line connection. Exercise extreme caution whenever relieving fuel system pressure to avoid exposing skin, face and eyes to fuel spray. Please be advised that fuel under pressure may penetrate the skin or any part of the body that it contacts.
- Always place a shop towel or cloth around the fitting or connection prior to loosening to absorb any excess fuel due to spillage. Ensure that all fuel spillage (should it occur) is quickly removed from engine surfaces. Ensure that all fuel soaked cloths or towels are deposited into a suitable waste container.
- Always keep a dry chemical (Class B) fire extinguisher near the work area.
- Do not allow fuel spray or fuel vapors to come into contact with a spark or open flame.
- Always use a back-up wrench when loosening and tightening fuel line connection fittings. This will prevent unnecessary stress and torsion to fuel line piping.
- Always replace worn fuel fitting O-rings with new Do not substitute fuel hose or equivalent where fuel pipe is installed.

Before servicing the vehicle, make sure to also refer to the precautions in the beginning of this section as well.

RELIEVING FUEL SYSTEM PRESSURE

✱✱ CAUTION

Observe all applicable safety precautions when working around fuel. Whenever servicing the fuel system, always work in a well-ventilated area. Do not allow fuel spray or vapors to come in contact with a spark or open flame. Keep a dry chemical fire extinguisher near the work area. Always keep fuel in a container specifically designed for fuel storage; also, always properly seal fuel containers to avoid the possibility of fire or explosion.

✱✱ CAUTION

Perform the following procedure to prevent fuel from spilling out before removing any fuel system parts. Pressure will still remain in the fuel lines even after performing the following procedure. When disconnecting a fuel line, cover it with a piece of cloth to prevent fuel from spraying or coming out.

1. Remove the rear seat cushion assembly. Refer to Seats in Body Interior.
2. Remove the rear floor service hole cover.
3. Disconnect the connector from the fuel pump assembly.
4. Start the engine. After the engine stops naturally, turn the ignition switch off.

➡Do not increase engine speed or drive the vehicle while waiting for the engine to stop naturally.

➡DTC P0171/25 (system too lean) may be set.

5. Crank the engine again and make sure that the engine does not start.
6. Remove the fuel tank cap and discharge the pressure from the fuel tank.
7. Disconnect the cable from the negative (-) battery terminal.
8. Connect the connector of the fuel pump assembly.
9. Install the rear floor service hole cover.
10. Install the rear seat cushion assembly.

FUEL FILTER

REMOVAL & INSTALLATION

The fuel filter is integrated in the fuel pump and must be replaced as a unit. Refer to Fuel Pump in this section.

FUEL INJECTORS

REMOVAL & INSTALLATION

1.8L Engine

See Figures 241 through 244.

✱✱ CAUTION

Observe all applicable safety precautions when working around fuel. Whenever servicing the fuel system, always work in a well-ventilated area. Do not allow fuel spray or vapors to come in contact with a spark or open flame. Keep a dry chemical fire extinguisher near the work area. Always keep fuel in a container specifically designed for fuel storage; also, always properly seal fuel containers to avoid the possibility of fire or explosion.

1. Properly relieve the fuel system pressure.
2. Disconnect the negative battery cable.
3. Disconnect the No. 2 ventilation hose.
4. Remove the 2 bolts and disconnect the ground wires.
5. Disconnect the 4 fuel injector assembly connectors.
6. Disconnect the 2 wire harness clamps.
7. Disconnect the 4 wire harness clamps.
8. Remove the 2 bolts and 2 wire harness brackets.
9. Remove the No. 2 fuel pipe clamp (Type A).
10. Remove the No. 2 fuel pipe clamp (Type B).

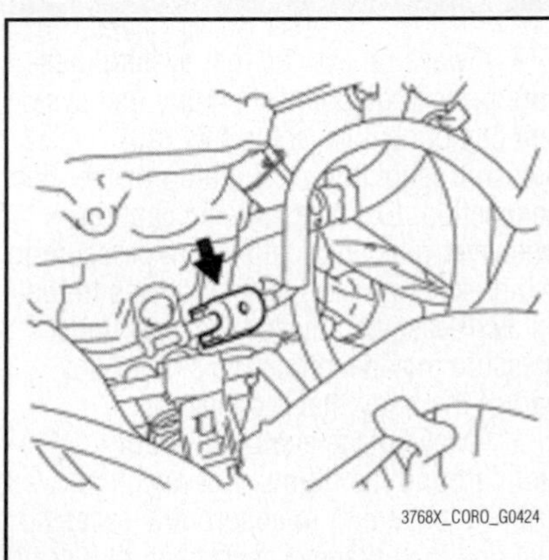

Fig. 241 Removing the No. 2 fuel pipe clamp (Type A)

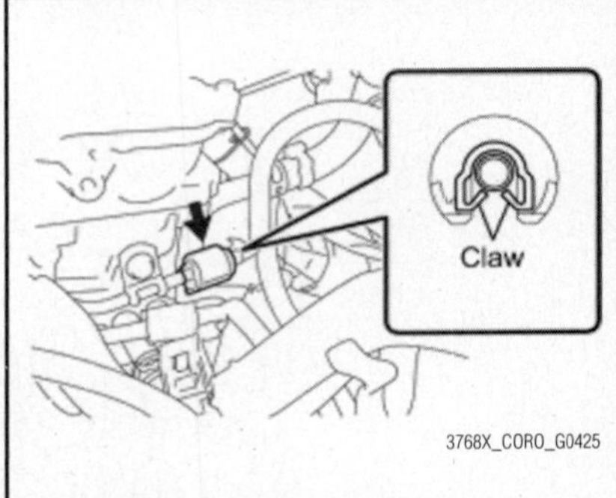

Fig. 242 Removing the No. 2 fuel pipe clamp (Type B)

11. Using SST 09268-21010, disconnect the fuel tube sub-assembly.
12. Remove the fuel delivery pipe sub-assembly:
    a. Remove the bolt and wire harness bracket.
    b. Remove the 2 bolts.
    c. Remove the bolt and the fuel delivery pipe sub-assembly.
    d. Remove the 2 No. 1 delivery pipe spacers.
13. Pull the 4 fuel injector assemblies out of the fuel delivery pipe sub-assembly.
14. For reinstallation, attach a tag or label to the injector shaft.

➡Prevent entry of foreign objects by covering the fuel injector with a plastic bag.

Fig. 243 Using SST 09268-21010, disconnect the fuel tube sub-assembly

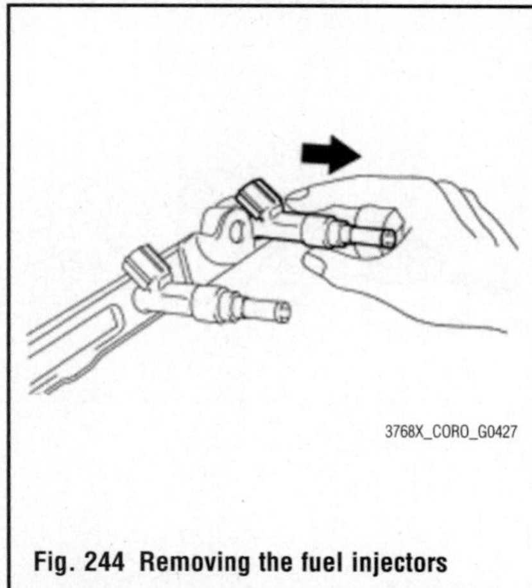

Fig. 244 Removing the fuel injectors

15. Remove the O-rings from the fuel injector assemblies.
16. Remove the 4 injector vibration insulators.

To install:

17. Install 4 new injector vibration insulators to the 4 fuel injector assemblies.
18. Apply a light coat of gasoline or spindle oil to the contact surfaces of the O-rings of the fuel injector assemblies.
19. While turning the fuel injector assembly left and right, install it onto the fuel delivery pipe sub-assembly.

➡Do not twist the O-ring. After installing the fuel injectors, check that they turn smoothly.

20. Install the 2 No. 1 delivery pipe spacers onto the cylinder head.

➡Install the No. 1 delivery pipe spacers in the correct direction.

21. Install the fuel delivery pipe sub-assembly with the 4 fuel injector assemblies, then temporarily install the 2 bolts. Tighten the bolts to 15 ft. lbs. (21 Nm).

➡Do not drop the fuel injectors when installing the fuel delivery pipe sub-assembly. Check that the fuel injector assemblies rotate smoothly after installing the fuel delivery pipe sub-assembly.

22. Install the bolt to secure the fuel delivery pipe sub-assembly. Tighten the bolt to 15 ft. lbs. (21 Nm).
23. Install the wire harness bracket with the bolt. Tighten to 44 inch lbs. (5 Nm).
24. Insert the fuel tube sub-assembly connector into the fuel delivery pipe until a "click" sound can be heard.

➡Check that there are no scratches or foreign matter around the contact surfaces of the fuel tube connector and pipe before performing this work. After connecting the fuel tube, check that the fuel tube connector and pipe are securely connected by pulling on them.

25. Install a new No. 2 fuel pipe clamp (Type B).
26. Install a new No. 2 fuel pump clamp (Type A).
27. Install the 2 wire harness brackets with the 2 bolts. Tighten to 10 ft. lbs. (13 Nm).
28. Connect the 4 wire harness clamps.
29. Connect the 4 fuel injector assembly connectors.
30. Connect the 2 wire harness clamps.
31. Connect the ground wires with the 2 bolts.
32. Install the air cleaner assembly.
33. Connect the No. 2 ventilation hose.
34. Connect cable to negative battery terminal.
35. Inspect for fuel leaks.

2.4L Engine

See Figures 245 through 251.

CAUTION

Observe all applicable safety precautions when working around fuel. Whenever servicing the fuel system, always work in a well-ventilated area. Do not allow fuel spray or vapors to come in contact with a spark or open flame. Keep a dry chemical fire extinguisher near the work area. Always keep fuel in a container specifically designed for fuel storage; also, always properly seal fuel containers to avoid the possibility of fire or explosion.

1. Properly relieve the fuel system pressure.
2. Disconnect the negative battery cable.
3. Remove the air cleaner assembly. Refer to Air Cleaner in Engine Mechanical.

➡Do not forcibly bend, kink or twist the fuel main tube.

4. Remove the fuel tube from the fuel hose clamp.
5. Remove the fuel pipe clamp.
6. Wipe off any dirt on the fuel tube connector.
7. Hold the fuel tube connector, and then install SST 09268-21010.
8. Turn SST to align the retainer inside the fuel tube connector with the chamfered part of SST.

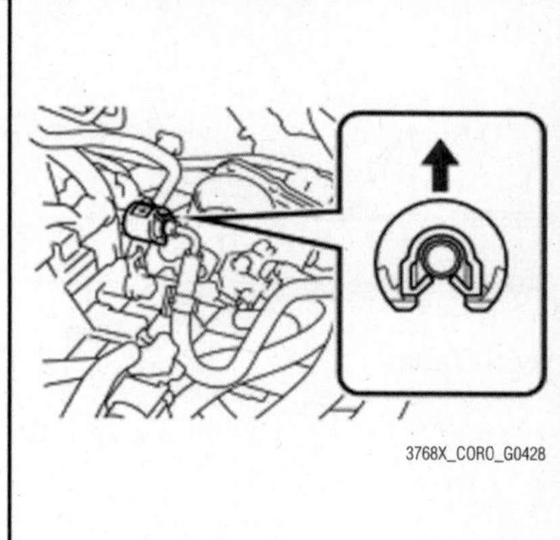

Fig. 245 Removing the fuel tube from the fuel hose clamp

9. Insert SST into the fuel tube and hold it. Then push the fuel tube connector toward SST.
10. Mount the retainer of the fuel tube connector onto the chamfered part of SST.
11. Slide SST and fuel tube connector together towards the fuel tube until they make a "click" sound, and then disconnect the fuel tube.
12. Drain the fuel remaining inside the fuel tube.
13. Cover the fuel tube and fuel pipe with a plastic bag to protect the disconnected part.

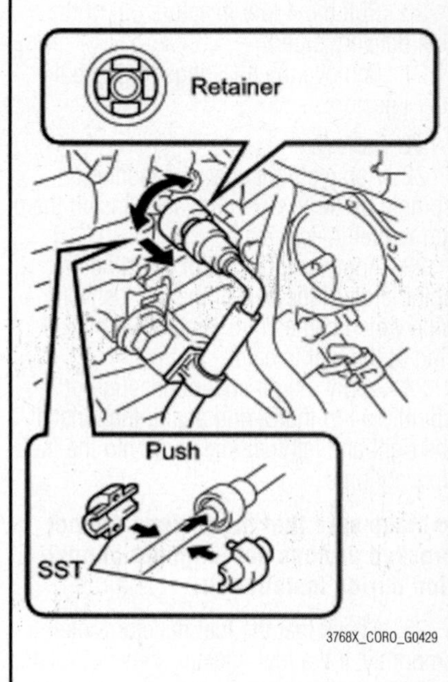

Fig. 246 Installing SST: 09268-21010 to the fuel tube connector

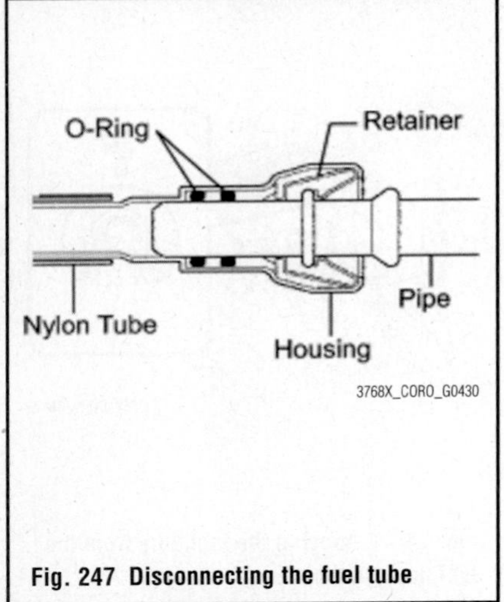

Fig. 247 Disconnecting the fuel tube

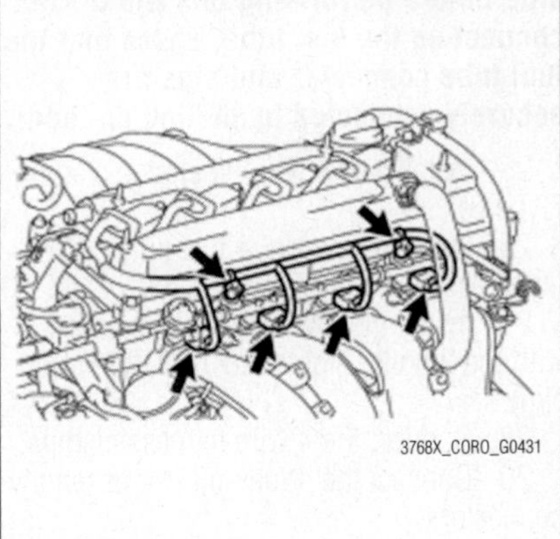

Fig. 248 Disconnecting the 4 fuel injector connectors

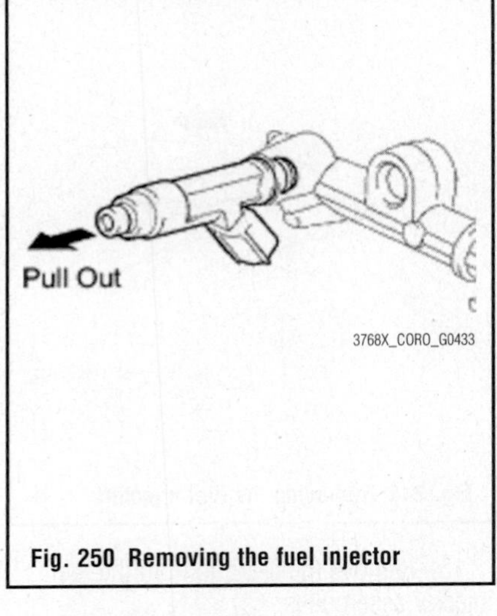

Fig. 250 Removing the fuel injector

14. Disconnect the No. 2 ventilation hose from the ventilation valve.
15. Remove the 2 wire harness clamps.
16. Disconnect the 4 fuel injector connectors.
17. Remove the 2 bolts, then remove the fuel delivery pipe together with the 4 fuel injectors.

➡Be careful not to drop the fuel injectors when removing the fuel delivery pipe.

18. Remove the 2 delivery pipe spacers from the cylinder head.
19. Remove the 4 insulators from the cylinder head.
20. Pull the 4 fuel injectors out of the fuel delivery pipe.
21. Remove the 4 O-rings from the 4 fuel injectors.

To install:

22. Apply a light coat of gasoline or spindle oil to new O-rings, then install them to the fuel injectors.
23. Apply a light coat of gasoline or spindle oil to the part of the fuel delivery pipe which comes into contact with the O-ring of the fuel injector.
24. Apply a light coat of gasoline or spindle oil to the O-ring again, then install the right and left fuel injectors onto the fuel delivery pipe.

➡Make sure that the O-rings are not cracked or does not jump out of position during installation.

25. Check that the fuel injectors rotate smoothly. If the fuel injector does not rotate, replace the O-ring.
26. Install 4 new insulators to the cylinder head.

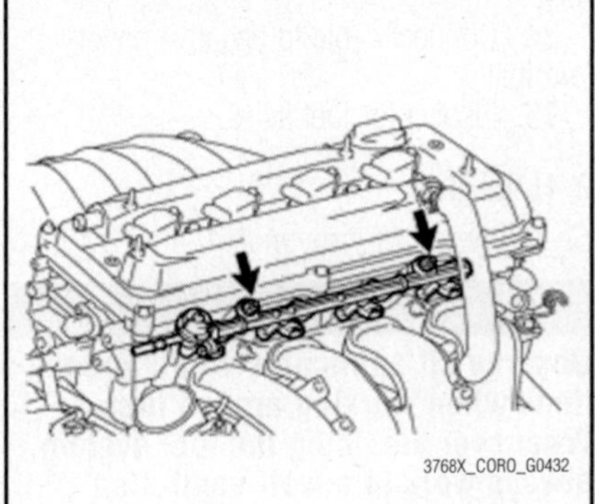

Fig. 249 Removing the 2 bolts, then remove the fuel delivery pipe together with the 4 fuel injectors

27. Install the 2 delivery pipe spacers onto the cylinder head.
28. Install the fuel delivery pipe together with the 4 fuel injectors, then temporarily tighten the 2 bolts.

➡Be careful not to drop the fuel injectors when installing the fuel delivery pipe.

29. Check that the fuel injector rotates smoothly.
30. If the fuel injector does not rotate smoothly, replace the O-ring.
31. Tighten the 2 fuel rail bolts to 15 ft. lbs. (20 Nm).
32. Connect the 4 fuel injector connectors.
33. Install the 2 wire harness clamps.
34. Connect the ventilation hose to the ventilation valve.

➡Make sure that the paint mark and hose clamp are at the correct angle when installing the hose.

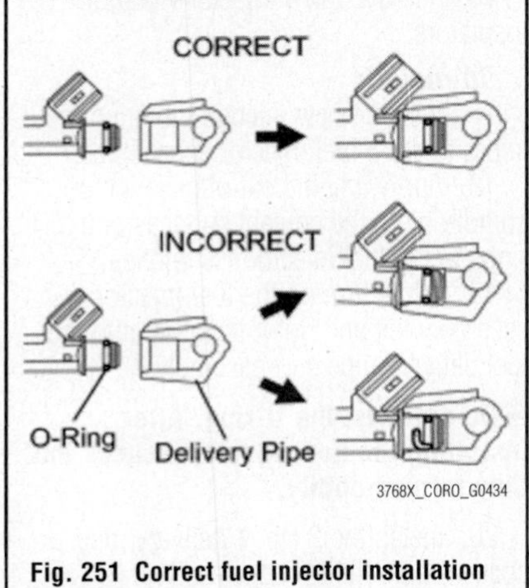

Fig. 251 Correct fuel injector installation position

35. Connect the fuel main tube.
36. Push the fuel tube connector until it makes a “click” sound.
37. Install the fuel pipe clamp.
38. Install the fuel tube to the fuel hose clamp.
39. Install the air cleaner assembly.
40. Connect cable to negative battery terminal.
41. Inspect for fuel leaks.

FUEL PUMP

REMOVAL & INSTALLATION

See Figures 252 and 253.

⁂ CAUTION

Observe all applicable safety precautions when working around fuel. Whenever servicing the fuel system, always work in a well-ventilated area. Do not allow fuel spray or vapors to come in contact with a

spark or open flame. Keep a dry chemical fire extinguisher near the work area. Always keep fuel in a container specifically designed for fuel storage; also, always properly seal fuel containers to avoid the possibility of fire or explosion.

1. Remove the fuel pump module. Refer to Fuel Pump Module in this section.
2. Release the claw and disconnect the fuel pump filter hose.
3. Remove the E-ring and separate the 2 claws, then remove the fuel pump sub-tank.

➡Do not separate the tube indicated in the illustration.

4. Disconnect the fuel pump harness connector. Do not damage the wire harness.
5. Using a screwdriver with its tip wrapped in protective tape, disengage the 2 claws and remove the No. 1 fuel suction support.
6. Using a screwdriver with its tip wrapped in protective tape, disengage the 5 claws, and remove the suction filter and fuel pump from the fuel filter.

➡Do not damage the fuel pump filter or fuel filter. Do not remove the suction filter. Do not use either the fuel pump or the suction filter if the suction filter is removed from the fuel pump.

7. Disconnect the fuel pump connector.
8. Remove the O-ring from the fuel pump.

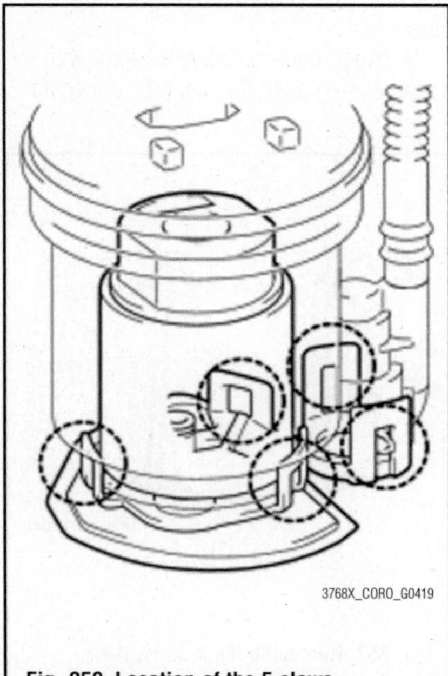

Fig. 252 Location of the 5 claws

➡Do not disassemble the fuel pump and the fuel filter because they are non-reusable parts.

To install:

9. Apply a light coat of gasoline or spindle oil to a new O-ring, then install it into the fuel pump.
10. Connect the fuel pump harness connector.
11. Apply a light coat of gasoline or spindle oil to the O-ring of the fuel pump again.
12. Engage the 5 claws, and install the fuel pump filter onto the fuel pump with fuel filter.

➡Make sure that the O-ring is not cracked or does not jump out of position during installation.

13. Engage the 2 claws of the No. 1 fuel suction support.
14. Connect the fuel pump harness connector.
15. Engage the 2 claws and install a new E-ring and fuel pump sub-tank.
16. Align the groove of the fuel pump filter hose with the cutout of the fuel sub-tank and install the hose.
17. To complete installation reverse remaining removal procedure.

FUEL TANK

REMOVAL & INSTALLATION

See Figures 254 through 257.

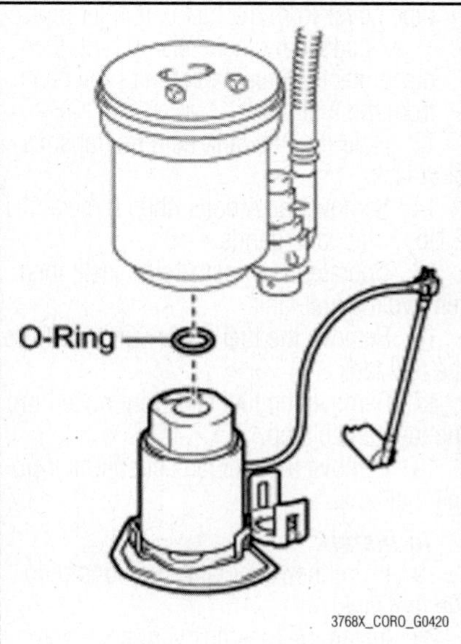

Fig. 253 Removing the fuel pump and O-ring

CAUTION

Observe all applicable safety precautions when working around fuel. Whenever servicing the fuel system, always work in a well-ventilated area. Do not allow fuel spray or vapors to come in contact with a spark or open flame. Keep a dry chemical fire extinguisher near the work area. Always keep fuel in a container specifically designed for fuel storage; also, always properly seal fuel containers to avoid the possibility of fire or explosion.

1. Properly relieve the fuel system pressure.
2. Disconnect the negative battery cable.
3. Remove the fuel pump module. Refer to Fuel Pump Module in this section.
4. Drain fuel.
5. Remove front exhaust pipe assembly. Refer to Exhaust Manifold in Engine Mechanical.
6. Remove the 4 bolts and the No. 1 fuel tank protector.
7. Remove the 4 bolts, and separate the parking brake cables.
8. Disconnect the fuel tank vent hose.
9. Pull the fuel tank vent hose out of the pipe.

➡Check that there is no dirt or other foreign objects around the connector before disconnecting it. Clean the connector if necessary.

➡It is necessary to prevent mud or dirt from entering the connector. If mud or dirt gets in the connector, the O-rings may not seal properly.

➡Only disconnect the quick connector by hand.

➡Do not bend, kink or twist the nylon tubes.

➡Protect the connector by covering it with a plastic bag.

10. Disconnect breather tube fuel hose:
    a. Pinch the retainer of the fuel tube connector, then pull the fuel tube connector out of the pipe.

➡Check that there is no dirt or other foreign objects around the connector before disconnecting it. Clean the connector if necessary.

➡It is necessary to prevent mud or dirt from entering the connector. If mud or

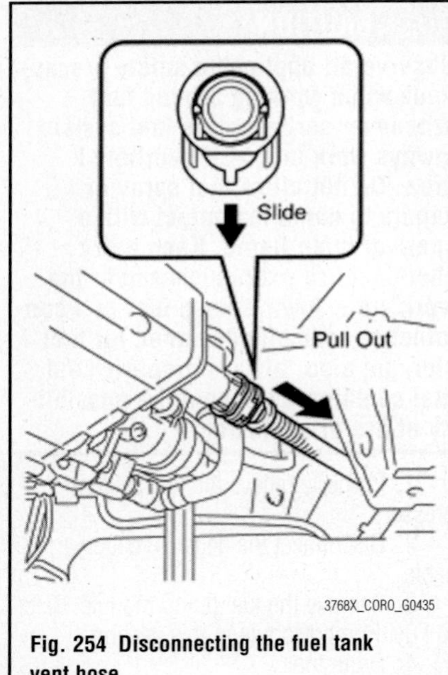

Fig. 254 Disconnecting the fuel tank vent hose

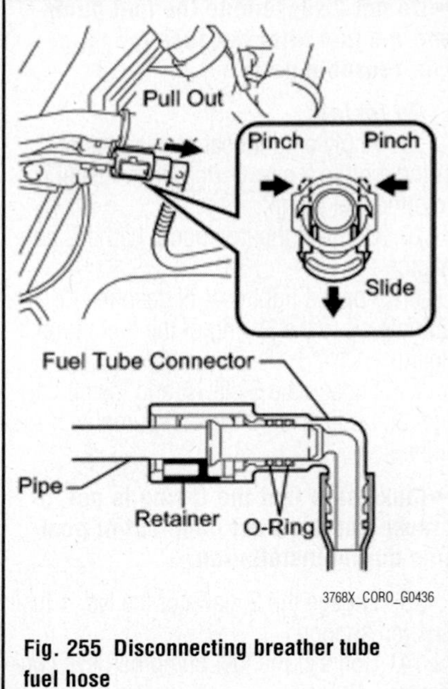

Fig. 255 Disconnecting breather tube fuel hose

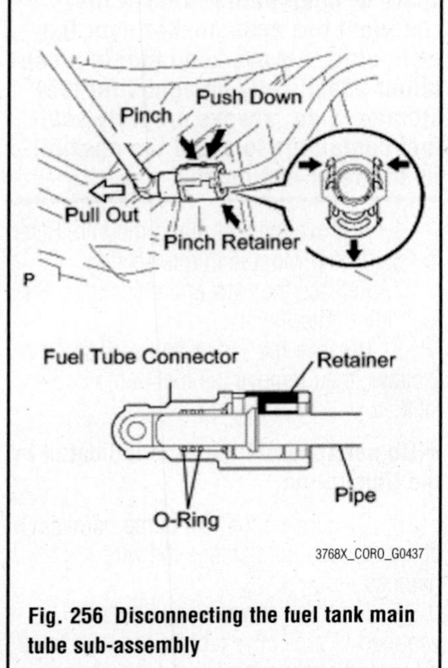

Fig. 256 Disconnecting the fuel tank main tube sub-assembly

dirt gets in the connector, the O-rings may not seal properly.

➡Only disconnect the quick connector by hand.

➡Do not bend, kink or twist the nylon tubes.

➡Protect the connector by covering it with a plastic bag.

➡If the pipe and connector are stuck, carefully try wiggling or pushing and pulling on the connector to release it. Pull the connector off the pipe carefully.

b. Separate the fuel breather tube fuel hose.

11. Disconnect the fuel tank main tube sub-assembly:

a. Pinch the tabs of the retainer of the fuel tube connector to disengage the lock claws and push it down as shown in the illustration.

b. Pull the fuel tank main tube out of the pipe.

➡Check that there is no dirt or other foreign objects around the connector before disconnecting it. Clean the connector if necessary.

➡It is necessary to prevent mud or dirt from entering the connector. If mud or dirt gets in the connector, the O-rings may not seal properly.

➡Only disconnect the quick connector by hand.

➡Do not bend, kink or twist the nylon tubes.

➡Protect the connector by covering it with a plastic bag.

➡If the pipe and connector are stuck, carefully try wiggling or pushing and pulling on the connector to release it. Pull the connector off the pipe carefully.

12. Disconnect fuel tank filler pipe sub-assembly:

a. Using a screwdriver, unfasten the claw. Then remove the fuel tank filler pipe cover from the fuel tank filler pipe.

b. Loosen the hose clamp bolt, then disconnect the fuel tank filler pipe hose from the fuel tank.

13. Hold the fuel tank using a transmission jack.

14. Remove the 4 bolts, then remove the 2 No. 1 fuel tank bands.

15. Operate the transmission jack, then remove the fuel tank.

16. Remove the fuel tank main tube from the fuel tank.

17. Remove the fuel tank vent hose from the fuel tank clamp.

18. Remove the fuel tank cushions from the fuel tank.

To install:

19. Install new fuel tank cushions onto the fuel tank.

20. Install the fuel tank vent hose onto the fuel tank clamp.

21. Install the fuel tank main tube onto the fuel tank.

22. Set the fuel tank on a transmission jack.

23. Operate the transmission jack, then install the fuel tank into the vehicle.

24. Install the 2 No. 1 fuel tank bands with the 4 bolts. Tighten to 29 ft. lbs. (39 Nm).

25. Connect the fuel tank filler pipe to the fuel tank.

➡Make sure that the hose clamp is facing in the correct direction when installing.

26. Engage the claw, then install the fuel tank filler pipe cover onto the fuel tank filler pipe.

27. Align the fuel tube connector with the pipe, then push the fuel tube connector

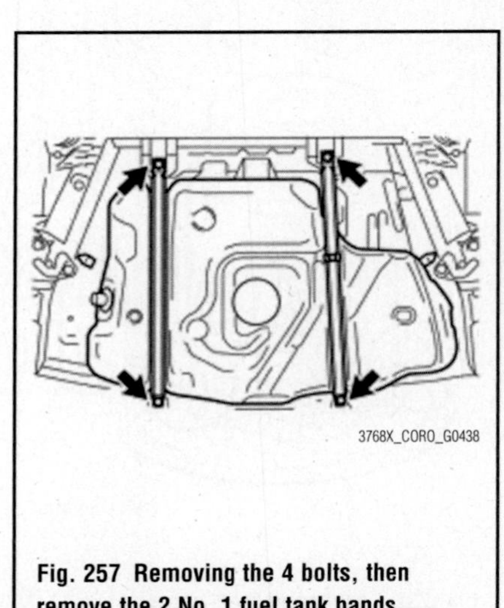

Fig. 257 Removing the 4 bolts, then remove the 2 No. 1 fuel tank bands

in until it comes into contact with the seat to connect the fuel tank main tube to the pipe, then push the retainer up until the claws lock.

➡Check that there are no scratches or foreign objects around the connecting surfaces of the fuel tube connector and pipe before performing this work. After connecting the fuel tank main tube, check that the fuel tank main tube is securely connected by pulling on the fuel tube connector and pipe.

28. Connect the breather tube fuel hose to the clamp.
29. Align the fuel hose connector with the pipe, then push the fuel hose connector in until the retainer makes a "click" sound to connect the fuel tank breather tube fuel hose to the pipe.

➡Check that there are no scratches or foreign objects around the connecting surfaces of the fuel tube connector and pipe before performing this work. After connecting the fuel tank breather tube, check that the fuel pump tube is securely connected by pulling on the fuel tube connector and pipe.

30. Align the fuel tube connector with the pipe, then push the fuel tube connector in until it comes into contact with the seat to connect the fuel tank vent hose to the pipe.
31. Slide the retainer of the fuel tube connector to lock the claws

➡Check that there are no scratches or foreign objects around the connecting surfaces of the fuel tube connector and pipe before performing this work. After connecting the fuel tank breather tube, check that the fuel pump tube is securely connected by pulling on the fuel tube connector and pipe.

32. Install the fuel tank protector sub-assembly with the 4 bolts. Tighten to 49 inch lbs. (5.5 Nm).
33. Install the parking cables with the 4 bolts. Tighten to 49 inch lbs. (5.5 Nm).
34. To complete installation, reverse remaining removal procedure.
35. Inspect for fuel leaks.
36. Inspect for exhaust gas leak.

IDLE SPEED

ADJUSTMENT

Adjustment is not available or necessary.

THROTTLE BODY

REMOVAL & INSTALLATION

1.8L Engine

See Figures 258 and 259.

1. Drain and recycle the engine coolant.
2. Disconnect the negative battery cable.
3. Disconnect the mass air flow meter connector and the 2 wire harness clamps.
4. Remove the air cleaner assembly. Refer to Air Cleaner Assembly in Engine Mechanical.
5. Disconnect the throttle body connector and the 2 water by-pass hoses.
6. Remove the 2 bolts and 2 nuts, and remove the throttle body assembly.
7. Remove the gasket from the intake manifold.

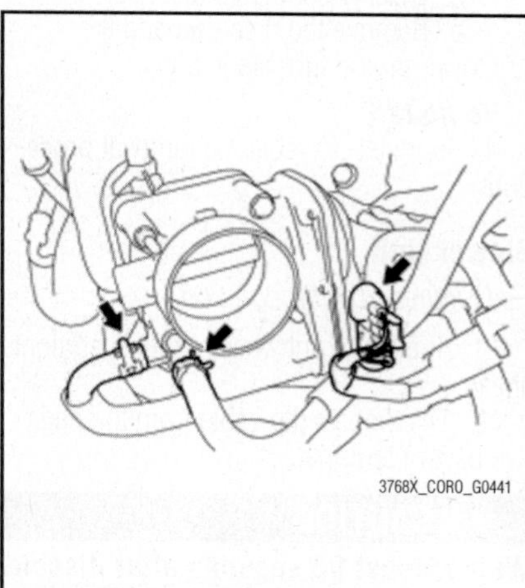

Fig. 258 Disconnecting the throttle body connector and the 2 water by-pass hoses

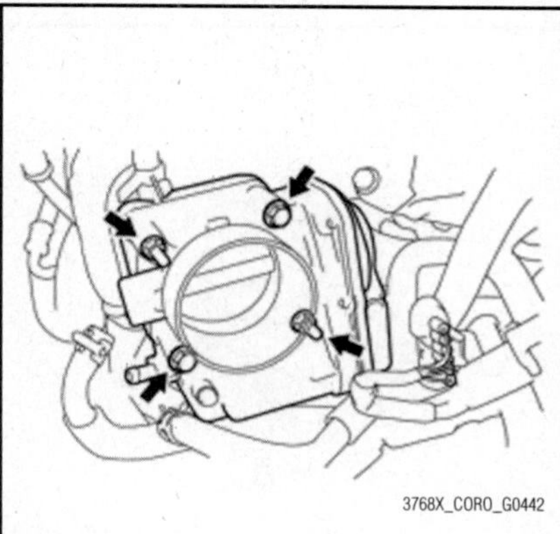

Fig. 259 Removing the 2 bolts and 2 nuts, and removing the throttle body assembly

To install:

8. Install a new gasket onto the intake manifold.
9. Install the throttle body assembly with the 2 bolts and 2 nuts. Tighten to 7 ft. lbs. (10 Nm).
10. Connect the 2 water by-pass hoses and the throttle body connector.
11. To complete installation, reverse remaining removal procedure.
12. Perform the initialization procedure.

2.4L Engine

See Figures 260 and 261.

1. Drain and recycle the engine coolant.
2. Disconnect the negative battery cable.
3. Disconnect the mass air flow meter connector and the 2 wire harness clamps.
4. Remove the air cleaner assembly. Refer to Air Cleaner Assembly in Engine Mechanical.
5. Disconnect the 2 water by-pass hoses.
6. Disconnect the throttle body assembly connector.
7. Disconnect the throttle body hose.
8. Remove the 4 bolts and throttle body assembly.
9. Remove the gasket from the intake manifold.

To install:

10. Install a new gasket onto the intake manifold.
11. Install the throttle body assembly with the 4 bolts. Tighten to 22 ft. lbs. (30 Nm).
12. Connect the 2 water by-pass hoses and the throttle body connector.
13. To complete installation, reverse remaining removal procedure.

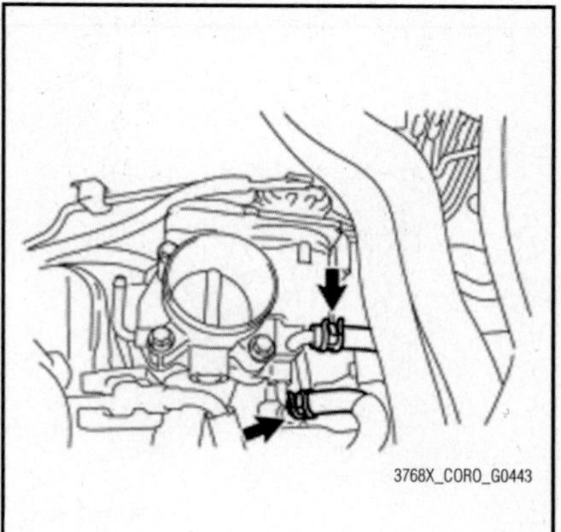

Fig. 260 Disconnecting the 2 water by-pass hoses

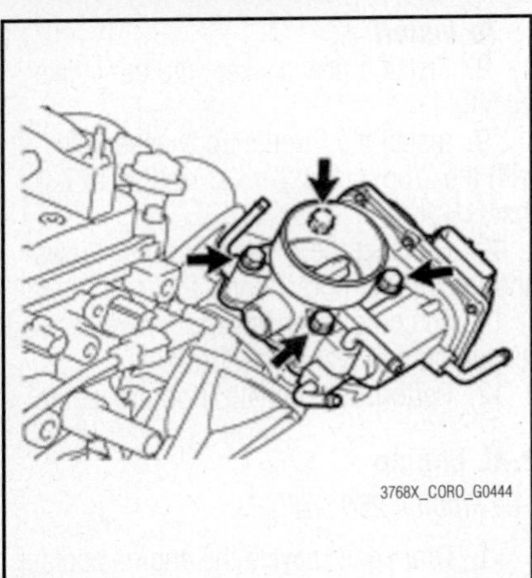

Fig. 261 Removing the 4 bolts and throttle body assembly

14. Perform the initialization procedure.

INITIALIZATION

➡Be sure to perform this procedure after reassembling the throttle body assembly, removing and reinstalling any throttle body component or replacing the ECM.

1. Disconnect the cable from the negative (-) battery terminal. Wait at least 60 seconds and reconnect the cable.
2. Turn the ignition switch to ON without operating the accelerator pedal.

➡If the accelerator pedal is operated, perform the above steps again.

3. Connect the Techstream to the DLC3 and clear the DTC's.
4. Start the engine and check that the MIL is not illuminated and that the idle speed is within the specified range when the air conditioning is switched off after the engine is warmed up.
   a. Standard:
      - Condition: A/C switched off
      - Engine idle speed: 600 to 700 rpm

➡Be sure to perform this step with all accessories off. Make sure that the shift lever is in N or P.

5. Enter the following menus: Powertrain/ Engine and ECT/ Data List/ Throttle Pos. Sensor Output. Fully depress the accelerator pedal and check that the value is 60% or more.
6. Perform a road test and confirm that there are no abnormalities. Refer to Drive Train.

HEATING & AIR CONDITIONING SYSTEM

BLOWER MOTOR

REMOVAL & INSTALLATION

Front

See Figures 262 and 263.

1. Remove the No. 2 instrument panel under cover sub assembly (w/ instrument panel under cover).
2. Remove the blower motor sub assembly (w/o PTC heater).
   a. Disengage the clamp.
   b. Disconnect the connector.
   c. Remove the 3 screws and the blower motor sub assembly.
3. Remove the blower motor sub assembly (w/ PTC heater).
   a. Disengage the clamp.
   b. Remove the quick heater connector screw.
   c. Disconnect the connector.
   d. Remove the 3 screws and the blower motor sub-assembly.

To install:

4. To install, reverse the removal procedure.

Blower Unit

See Figures 264 through 266.

1. Turn the front wheels to face straight ahead.
2. Disconnect the cable from the negative battery terminal.

✲✲ CAUTION

Wait at least 90 seconds after disconnecting the cable from the negative (-) battery terminal to disable the SRS system.

3. Recover the refrigerant from the refrigeration system..
4. Disconnect the suction pipe sub assembly.
5. Disconnect the liquid pipe sub assembly.
6. Disconnect the heater outlet water hose.
7. Disconnect the heater inlet water hose.
8. Remove the lower No. 3 steering wheel cover.
9. Remove the lower No. 2 steering wheel cover.
10. Remove the steering pad.
11. Remove the steering wheel assembly.
12. Remove the center instrument cluster finish panel sub assembly.
13. Remove the instrument cluster finish panel sub assembly.
14. Remove the meter hood sub assembly.
15. Remove the combination meter assembly.
16. Disconnect the front door opening trim weather-strip LH.
17. Remove the front pillar garnish LH.
18. Disconnect the front door opening trim weather strip RH.
19. Remove the front pillar garnish RH.
20. Remove the lower instrument panel finish panel assembly.
21. Remove the shift lever knob sub assembly.
22. Remove the center instrument cluster finish panel assembly.
23. Remove the glove compartment door assembly.

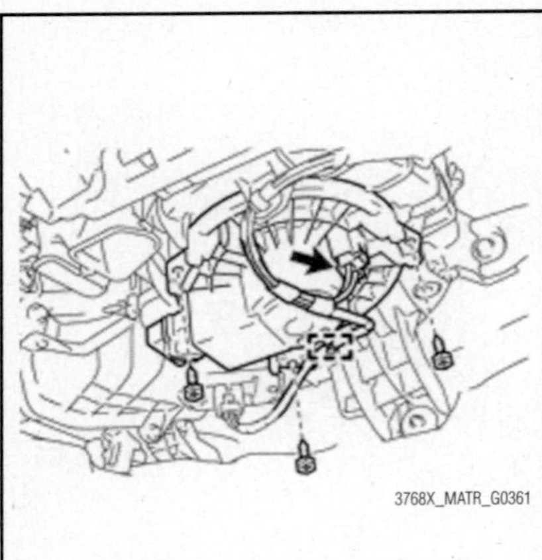

Fig. 262 Removing the blower motor sub assembly (w/o PTC heater)

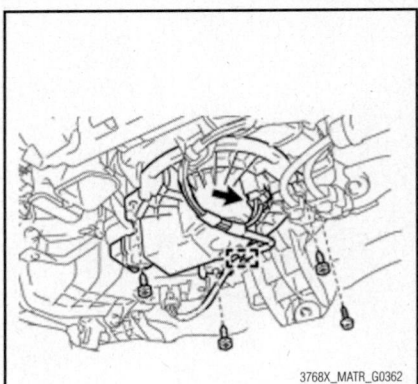

3768X_MATR_G0362

Fig. 263 Removing the blower motor sub assembly (w/ PTC heater)

24. Remove the No. 1 instrument panel box door sub assembly.
25. Disconnect the instrument panel wire assembly.
26. Remove the upper instrument panel sub assembly.
27. Remove the radio receiver assembly with the bracket (w/o navigation system).
28. Remove the navigation receiver assembly with the bracket (w/ navigation system).
29. Remove the navigation antenna cord sub assembly (w/ navigation system).
30. Remove the air conditioning panel assembly.
31. Remove the front upper console box.
32. Remove the No. 2 switch hole base.
33. Remove the console box carpet.
34. Remove the front console box assembly.
35. Remove the front No. 1 console box insert.
36. Remove the front No. 2 console box insert.
37. Remove the lower center instrument panel finish panel.
38. Remove the No. 1 switch hole base.
39. Remove the lower instrument panel finish panel sub assembly.
40. Remove the front door scuff plate LH.
41. Remove the cowl side trim board LH.
42. Remove the No. 2 instrument panel under cover sub assembly (w/ instrument panel under cover).
43. Remove the front door scuff plate RH.
44. Remove the cowl side trim board RH.
45. Disconnect the shift lever assembly.
46. Remove the lower steering column cover.
47. Remove the upper steering column cover.
48. Remove the turn signal switch assembly with the spiral cable sub assembly.
49. Remove the radio wire sub assembly.
50. Remove the lower instrument panel sub assembly.
51. Remove the column hole cover silencer sheet.
52. Separate the No. 2 steering intermediate shaft assembly.
53. Remove the instrument panel sub reinforcement.
54. Remove the No. 2 air duct sub assembly.
55. Remove the transponder key amplifier (w/ engine immobilizer system).
56. Remove the stop light switch assembly.
57. Remove the steering post assembly.
58. Remove the power steering ECU assembly.
59. Remove the air conditioning amplifier assembly.
60. Remove the rear No. 3 air duct (w/ rear air duct).
61. Remove the rear No. 2 air duct (w/ rear air duct).
62. Remove the rear No. 1 air duct (w/ rear air duct).
63. Remove the center instrument panel register connector assembly.
64. Remove the No. 1 air duct sub assembly.
65. Remove the center instrument panel to cowl brace.
66. Remove the theft warning ECU assembly.
67. Remove the instrument panel brace sub assembly.
68. Remove the instrument panel reinforcement assembly.
69. Remove the air conditioning unit.
70. Remove the No. 3 air duct sub assembly.
71. Remove the blower assembly (w/o PTC heater).
    a. Remove the 3 screws and the blower assembly.
72. Remove the blower assembly (w/ PTC heater).
    a. Disengage each clamp.
    b. Remove the screw and disengage the quick heater connector.
    c. Remove the 3 screws and the blower assembly.

To install:

73. To install, reverse the removal procedure.

Fig. 264 Removing the blower assembly (w/o PTC heater)

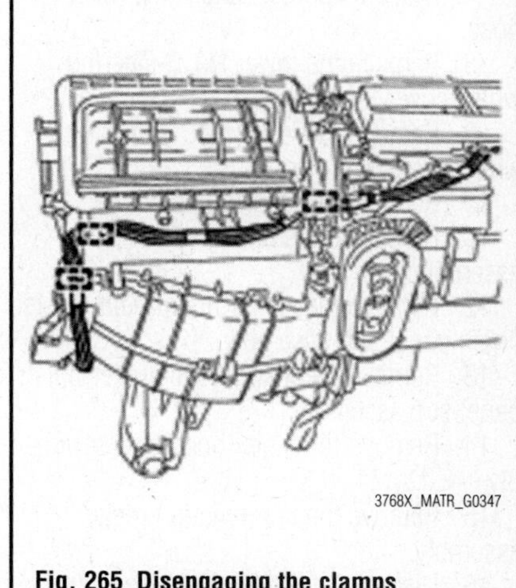

Fig. 265 Disengaging the clamps

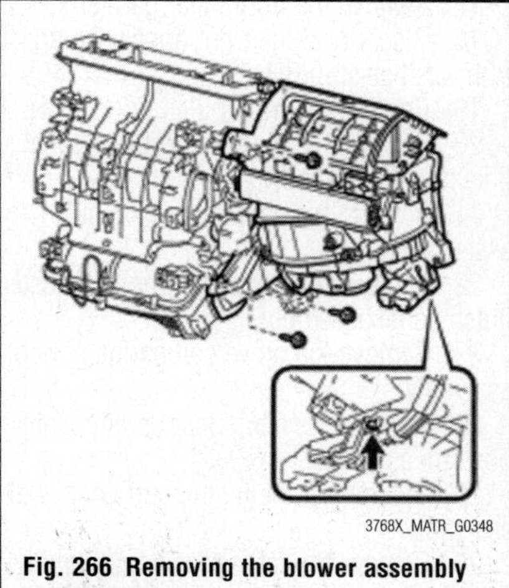

Fig. 266 Removing the blower assembly (w/ PTC heater)

HEATER CORE

REMOVAL & INSTALLATION

See Figures 267 and 268.

1. Turn the front wheels to face straight ahead.
2. Disconnect the cable from the negative battery terminal.

CAUTION

Wait at least 90 seconds after disconnecting the cable from the negative (-) battery terminal to disable the SRS system.

3. Recover the refrigerant from the refrigeration system.
4. Disconnect the suction pipe sub assembly.
5. Disconnect the liquid pipe sub assembly.
6. Disconnect the heater outlet water hose.

7. Disconnect the heater inlet water hose.
8. Remove the lower No. 3 steering wheel cover.
9. Remove the lower No. 2 steering wheel cover.
10. Remove the steering pad.
11. Remove the steering wheel assembly.
12. Remove the center instrument cluster finish panel sub assembly.
13. Remove the instrument cluster finish panel sub assembly.
14. Remove the meter hood sub assembly.
15. Remove the combination meter assembly.
16. Disconnect the front door opening trim weather-strip LH.
17. Remove the front pillar garnish LH.
18. Disconnect the front door opening trim weather-strip RH.
19. Remove the front pillar garnish RH.
20. Remove the lower instrument panel finish panel assembly.
21. Remove the shift lever knob sub assembly.
22. Remove the center instrument cluster finish panel assembly.
23. Remove the glove compartment door assembly.
24. Remove the No. 1 instrument panel box sub assembly.
25. Disconnect the instrument panel wire assembly.
26. Remove the radio receiver assembly with the bracket (w/o navigation system).
27. Remove the navigation receiver assembly with the bracket (w/ navigation system).
28. Remove the navigation antenna cord sub assembly (w/ navigation system).
29. Remove the air conditioning panel assembly.
30. Remove the front upper console box.
31. Remove the No. 2 switch hole base.
32. Remove the console box carpet.
33. Remove the front console box assembly.
34. Remove the front No. 1 console box insert.
35. Remove the front No. 2 console box insert.
36. Remove the lower center instrument panel finish panel.
37. Remove the No. 1 switch hole base.
38. Remove the lower instrument panel finish panel sub assembly.
39. Remove the front door scuff plate LH.
40. Remove the cowl side trim board LH.
41. Remove the No. 2 instrument panel under cover sub assembly (w/ instrument panel under cover).
42. Remove the front door scuff plate RH.
43. Remove the cowl side trim board RH.
44. Disconnect the shift lever assembly.
45. Remove the lower steering column cover.
46. Remove the upper steering column cover.
47. Remove the turn signal switch assembly with the spiral cable sub assembly.
48. Remove the radio wire sub assembly.
49. Remove the lower instrument panel sub assembly.
50. Remove the column hole cover silencer sheet.
51. Separate the No. 2 steering intermediate shaft assembly.
52. Remove the instrument panel sub reinforcement.
53. Remove the No. 2 air duct sub assembly.
54. Remove the transponder key amplifier (w/ engine immobilizer system).
55. Remove the stop light switch assembly.
56. Remove the steering post assembly.
57. Remove the power steering ECU assembly.
58. Remove the air conditioning amplifier assembly.
59. Remove the rear No. 3 air duct (w/ rear air duct).
60. Remove the rear No. 2 air duct (w/ rear air duct).
61. Remove the rear No. 1 air duct (w/ rear air duct).
62. Remove the center instrument panel register connector assembly.
63. Remove the No. 1 air duct sub assembly.
64. Remove the center instrument panel to cowl brace.
65. Remove the theft warning ECU assembly.

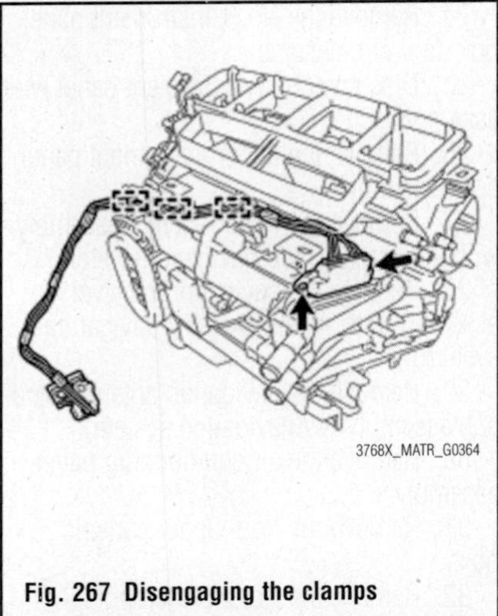

Fig. 267 Disengaging the clamps

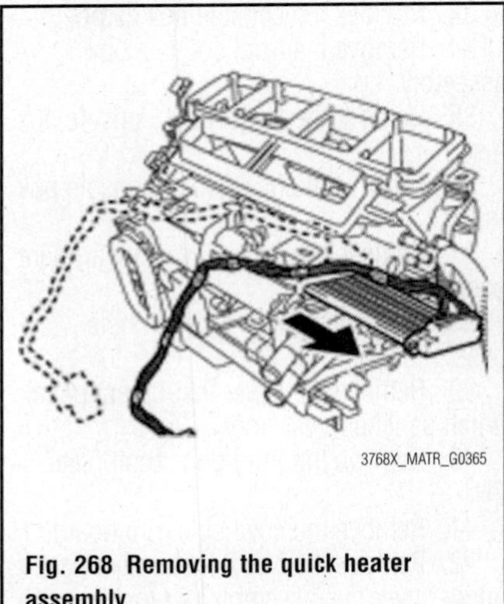

Fig. 268 Removing the quick heater assembly

66. Remove the instrument panel brace sub assembly.
67. Remove the instrument panel reinforcement assembly.
68. Remove the air conditioning unit.
69. Remove the No. 3 air duct sub assembly.
70. Remove the blower assembly.
71. Remove the quick heater assembly.
    a. Disengage each clamp.
    b. Remove the 2 screws.
    c. Remove the quick heater assembly.

To install:

72. To install, reverse the removal procedure.

STEERING

POWER STEERING GEAR

REMOVAL & INSTALLATION

2WD Vehicles

See Figures 269 through 273.

CAUTION

Some models covered by this manual may be equipped with a Supplemental Restraint System (SRS), which uses an air bag. Whenever working near any of the SRS components, such as the impact sensors, the air bag module, steering column and instrument panel, disable the SRS, as described in Section 6.

1. Place front wheels facing straight ahead.
2. Secure the steering wheel with the seat belt in order to prevent rotation.

➡This operation is useful to prevent damage to the spiral cable.

3. Disconnect the negative battery cable.
4. Remove column hole cover silencer sheet.
5. Separate no. 2 steering intermediate shaft assembly.
6. Remove clips A and the No. 1 steering column hole cover sub-assembly and disengage clip B from the body.
7. Remove the front wheels.
8. Remove the engine under covers.
9. Separate the tie rod end sub-assembly.
10. Separate front stabilizer link assembly.
11. Separate front lower suspension arms.
12. Remove front suspension crossmember sub-assembly. Refer to Front Suspension.
13. Remove the No. 1 steering column hole cover sub-assembly from the steering link assembly.
14. Put match marks on the steering intermediate shaft and the steering link assembly.
15. Remove the bolt and the steering intermediate shaft from the steering link assembly.
16. Remove the 4 bolts and steering link assembly from the front suspension crossmember sub-assembly.

➡Tape SST 09612-00012 before use.

17. Using SST 09612-00012, secure the steering link assembly in a vise.
18. Put matchmarks on the tie rod end sub-assembly LH and RH and steering gear assembly.
19. Remove the tie rod end sub-assembly and lock nuts.

To install:

20. Install the lock nut and the tie rod end sub-assembly to the steering gear assembly until the match marks are aligned.

➡After adjusting the toe-in, tighten the lock nut.

21. Install the steering link assembly to the front suspension crossmember sub-assembly with the 4 bolts:
    a. Temporarily tighten the bolts in order of (A), (B), (C), and then (D).
    b. Tighten the bolts in order of (A), (B), (C), and then (D).
    c. Tighten the bolts to 43 ft. lbs. (58 Nm).
22. Align the matchmarks and install the steering intermediate shaft to the steering link assembly.
23. Install the bolt. Tighten to 26 ft. lbs. (35 Nm).
24. To complete installation, reverse remaining removal procedure.

4WD Vehicles

See Figures 274 through 278.

1. Place the front wheels facing straight ahead.
2. Secure the steering wheel.
    a. Secure the steering wheel with the seat belt in order to prevent rotation.

3768X_CORO_G0509

Fig. 271 Remove the 4 bolts and steering link assembly

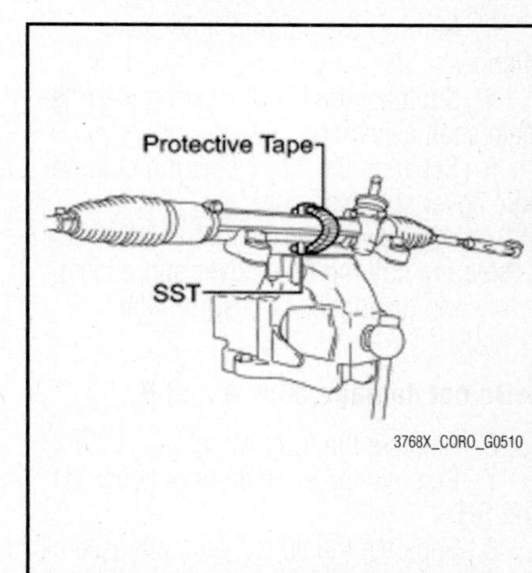

Fig. 272 Installing SST 09612-00012 to the steering link assembly

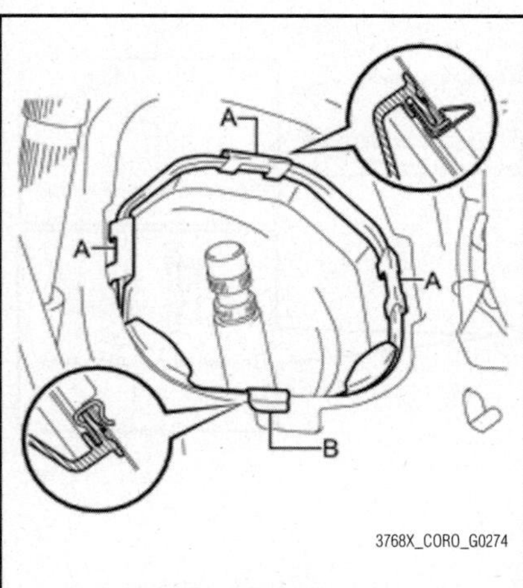

Fig. 269 Remove clips A and the No. 1 steering column hole cover sub-assembly

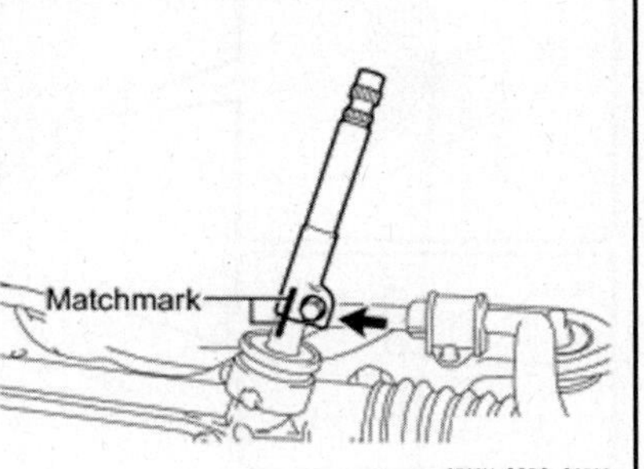

Fig. 270 Putting match marks on the steering intermediate shaft and the steering link assembly

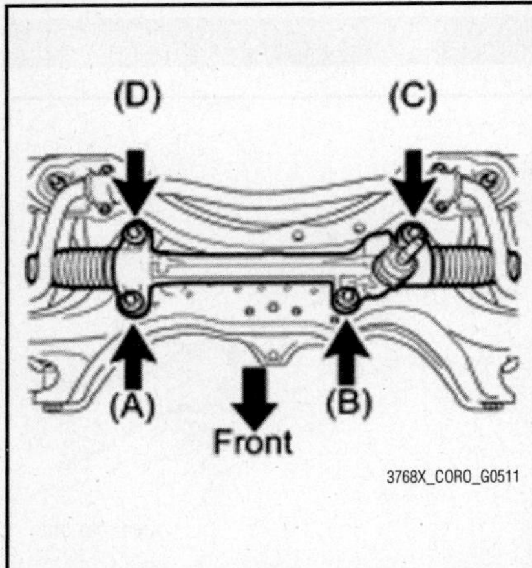

Fig. 273 Steering link assembly bolt tightening sequence

➡This operation is useful to prevent damage to the spiral cable.

3. Remove the column hole cover silencer shaft.
4. Separate the No. 2 steering intermediate shaft assembly.
5. Separate the No. 1 steering column hole cover sub assembly.
   a. Remove clips A and the No. 1 steering column hole cover sub-assembly and disengage clip B from the body.

➡Do not damage clips A and B.

6. Remove the front wheels.
7. Remove the engine under cover LH and RH.
8. Separate the tie rod end sub assembly LH.
   a. Remove the cotter pin and nut.
   b. Install the special tool (09960-20010).

➡Make sure that the upper ends of the tie rod end and SST are aligned.

   c. Using the special tool, separate the tie rod end from the steering knuckle.

⁂ CAUTION

Apply grease to the bolt threads and the tip of the special tool.

➡Be sure to tighten the string firmly to secure SST to the steering knuckle to prevent SST from falling off.

➡Install SST with the center nut so that A and B are parallel. Otherwise, the dust cover may be damaged.

➡Be sure to place the wrench on the part indicated in the illustration.

➡Do not damage the front disc brake dust cover.

➡Do not damage the ball joint dust cover.

➡Do not damage the steering knuckle.

9. Separate the tie rod end sub assembly.
10. Separate the front lower suspension arm LH and RH.
11. Separate the steering link assembly.
   a. Remove the 2 bolts and 2 nuts, and separate the steering link assembly from the front suspension crossmember sub-assembly.

➡Hold the nut when turning the bolt.

   b. Install the stud bolt of each tie rod end to the steering knuckle and temporarily tighten the tie rod ends with the castle nuts.
   c. Secure the steering link assembly to the transfer or stabilizer bar using a string or equivalent.
12. Loosen the front suspension crossmember.
13. Remove the No. 1 front stabilizer bracket LH and RH.
14. Remove the front suspension crossmember.
15. Remove the steering link assembly.
   a. Remove the temporarily installed castle nuts and the string and remove the steering link assembly.
16. Remove the No. 1 steering column hole cover sub assembly.
   a. Remove the No. 1 steering column hole cover sub assembly from the steering link assembly.
17. Secure the steering link assembly.
   a. Using the special tool (09612-00012), secure the steering link assembly in a vise.

➡Tape the special tool before use.

18. Remove the tie rod end sub assembly LH.
   a. Put matchmarks on the tie rod end sub-assembly LH and steering gear assembly.
   b. Remove the tie rod end sub-assembly LH and lock nut.
19. Remove the tie rod end sub assembly RH.
   a. Put matchmarks on the tie rod end sub-assembly RH and steering gear assembly.

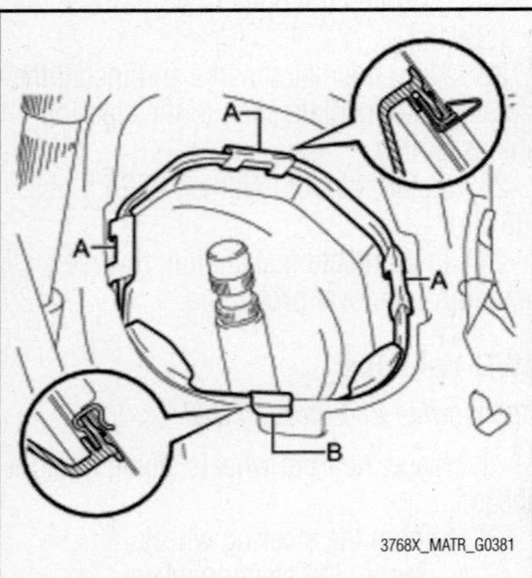

Fig. 274 Separating the No. 1 steering column hole cover sub assembly

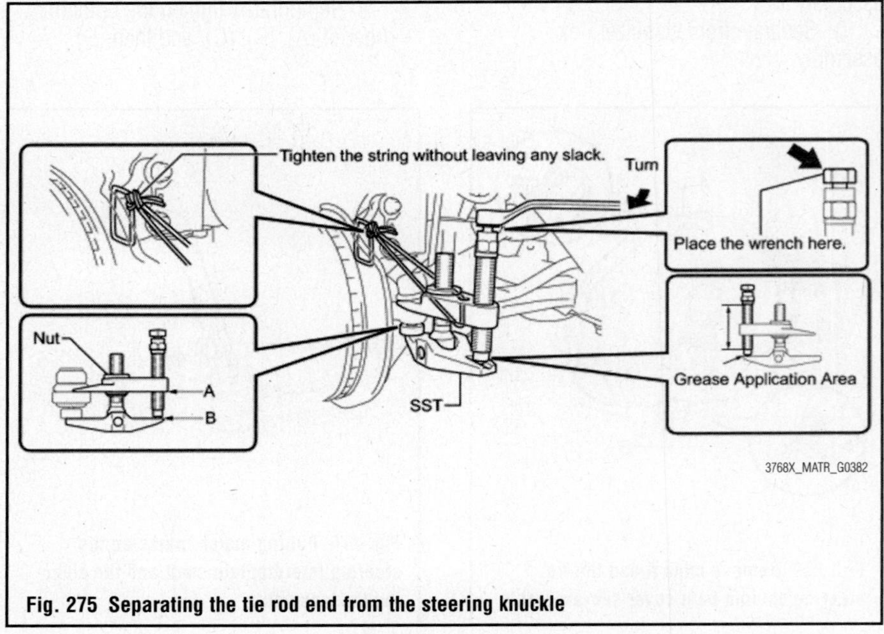

Fig. 275 Separating the tie rod end from the steering knuckle

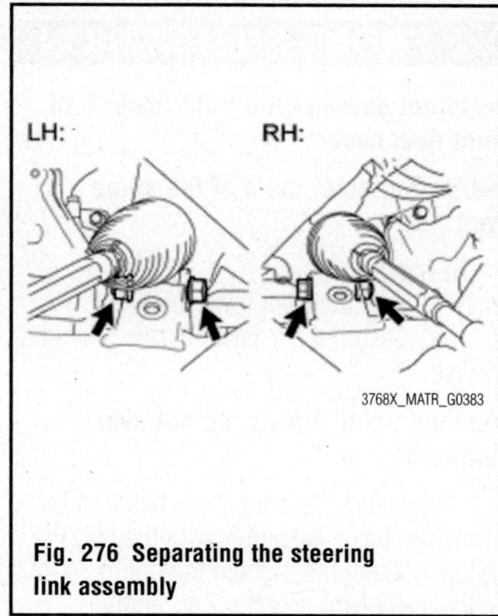

Fig. 276 Separating the steering link assembly

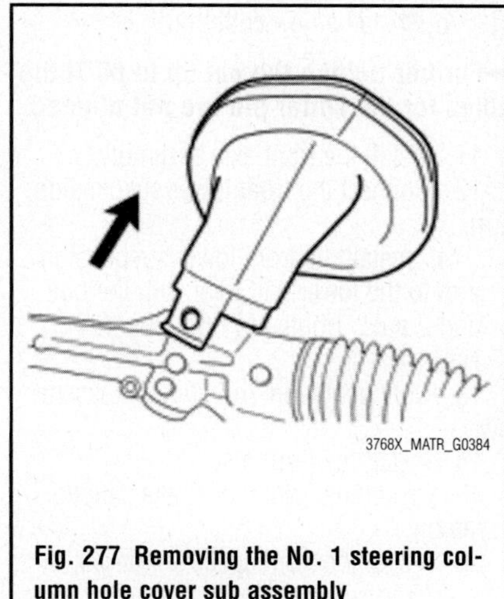

Fig. 277 Removing the No. 1 steering column hole cover sub assembly

b. Remove the tie rod end sub-assembly RH and lock nut.

To install:

20. Install the tie rod end sub assembly LH.

a. Install the lock nut and the tie rod end sub-assembly LH to the steering gear assembly until the matchmarks are aligned.

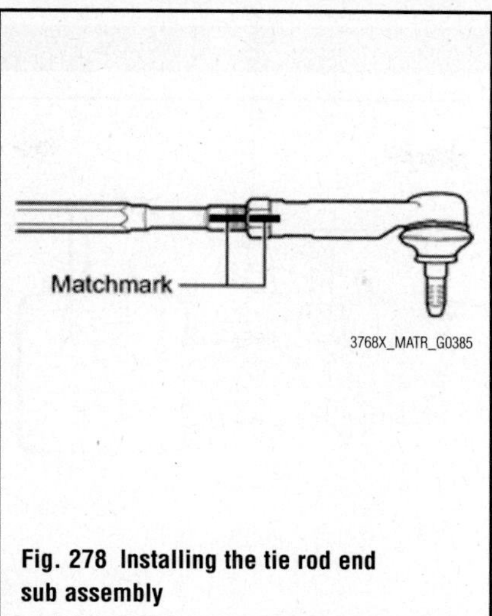

Fig. 278 Installing the tie rod end sub assembly

➡After adjusting the toe-in, tighten the lock nut.

21. Install the tie rod end sub assembly RH.

➡Use the same procedure for the LH side.

22. Install the No. 1 steering column hole cover sub assembly.

a. Align the round hole in the No. 1 steering column hole cover sub-assembly with the protrusion of the steering link assembly to install the cover.

23. Install the steering link assembly.

a. Pass the steering link assembly above the transfer and secure it on top of the transfer using a string.

b. Temporarily install the tie rod to the steering knuckle with the castle nut.

24. Temporarily install the front suspension crossmember.

25. Install the No. 1 front stabilizer bracket LH and RH.

26. Fully tighten the front suspension crossmember.

27. Connect the steering link assembly.

a. Install the steering link assembly to the front suspension crossmember sub-assembly with the 2 bolts and 2 nuts. Tighten to 60 ft. lbs. (82 Nm).

28. Connect the front lower suspension arm LH and RH.

29. Connect the tie rod end sub assembly LH.

a. Connect the tie rod end sub-assembly LH to the steering knuckle with the nut. Tighten to 36 ft. lbs. (49 Nm).

➡Further tighten the nut up to 60°if the holes for the cotter pin are not aligned.

b. Install a new cotter pin.

30. Connect the tie rod end sub assembly RH.

31. Connect the No. 1 steering column hole cover sub assembly.

a. Engage clip B onto the body and install the No. 1 steering column hole cover sub-assembly onto the body with clips A.

➡Make sure that the lip of the No. 1 steering column hole cover sub-assembly is not damaged.

32. Connect the No. 2 steering intermediate shaft assembly.

33. Place the front wheels facing straight ahead.

34. Install the column hole cover silencer sheet.

35. Install the engine under cover LH and RH.

36. Install the front wheels. Tighten to 76 ft. lbs. (103 Nm).

37. Stabilizer the suspension.

38. Adjust the front wheel alignment.

39. Initialize the rotation angle sensor and calibrate the torque sensor zero point.

POWER STEERING PUMP

REMOVAL & INSTALLATION

➡The Matrix does not use a power steering pump.

LOWER BALL JOINT

REMOVAL & INSTALLATION

See Figures 279 through 281.

➡Use the same procedure for the LH side and RH side.

➡The following procedure listed is for the LH side.

1. Remove the front wheel.
2. Remove the front axle shaft nut.
3. Separate the front speed sensor.
4. Separate the front disc caliper assembly.
5. Remove the front disc.
6. Separate the tie rod end sub assembly.
7. Separate the front lower suspension arm.
   a. Remove the bolt and 2 nuts, and separate the front lower suspension arm from the lower ball joint.
8. Remove the front axle assembly.
9. Remove the front lower ball joint.
   a. Secure the front axle assembly in a vise.

➡When using a vise, do not over tighten it.

   b. Remove the cotter pin and nut.
   c. Install the 2 special tools (09960-20010 or 09961-02050) to the front lower ball joint.

➡Check that the clearance measurement is between the special tool and the front axle assembly is 0.0394 inch (1 mm).

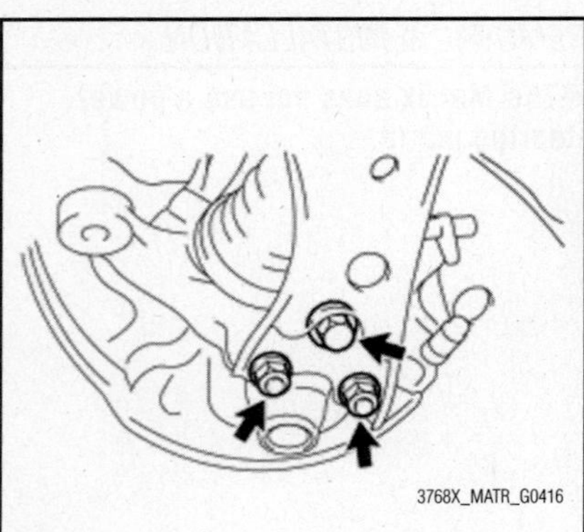

Fig. 279 Separating the front lower suspension arm

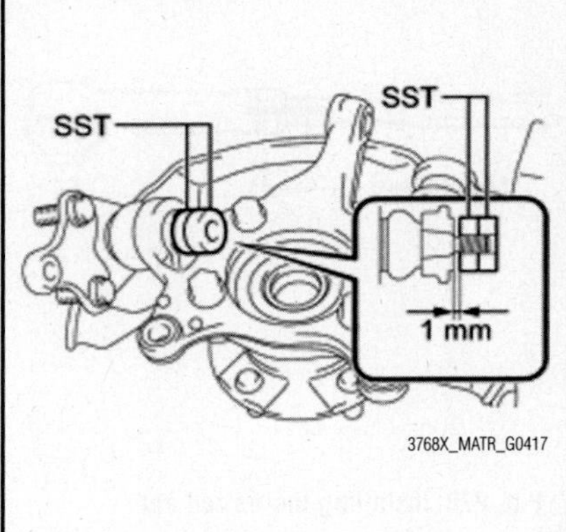

Fig. 280 Installing the special tools to the front lower ball joint

➡Use the 2 special tools of the same type.

   d. Using the special tool, remove the front lower ball joint from the front axle assembly.

⁂ CAUTION

Apply grease to the threads and end of the special tool bolt.

➡Install the special tool so that A and B are parallel.

➡Be sure to place a wrench on the part indicated in the illustration.

➡Do not damage the front lower ball joint dust cover.

➡Use 2 special tools of the same type.

To install:

10. Install the front lower ball joint.
   a. Secure the front axle assembly in a vise.

➡When using a vise, do not over tighten it.

   b. Install the front lower ball join to the front axle assembly with the nut. For the 1.8L engine, tighten the nut to 76 ft. lbs. (103 Nm). For the 2.4L engine, tighten to 91 ft. lbs. (123 Nm).
   c. Install a new cotter pin.

➡Further tighten the nut up to 60°if the holes for the cotter pin are not aligned.

11. Install the front axle assembly.
12. Connect the front lower suspension arm.
   a. Install the front lower suspension arm to the lower ball joint with the bolt and 2 nuts. Tighten to 66 ft. lbs. (89 Nm).
13. Connect the tie rod end sub assembly.
14. Install the front disc.
15. Install the front disc brake caliper assembly.
16. Install the front speed senor.

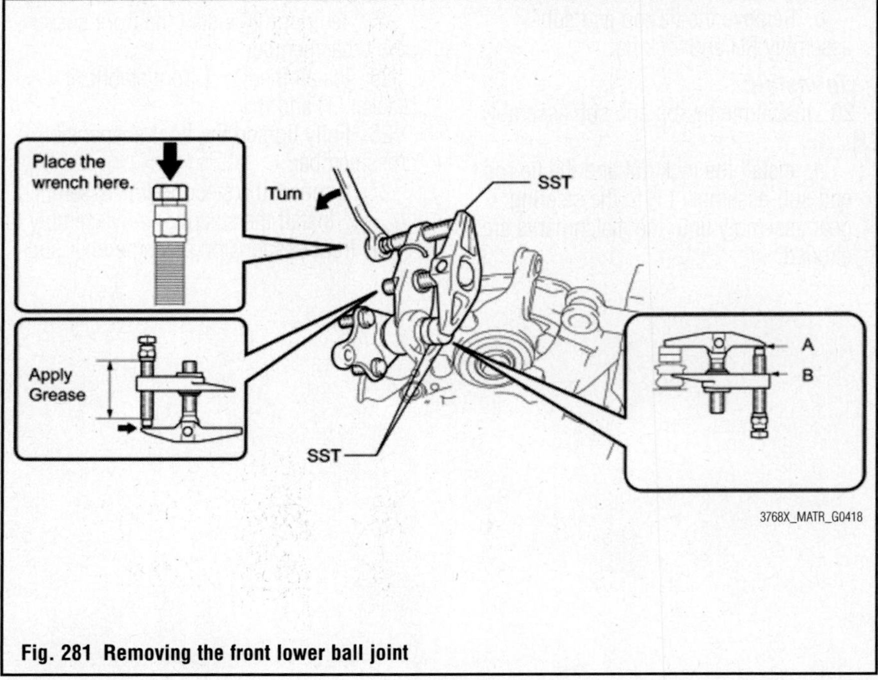

Fig. 281 Removing the front lower ball joint

a. Install the front flexible hose and front speed sensor with the bolt. Tighten the bolt to 21 ft. lbs. (29 Nm).

➡Install the flexible hose and the speed sensor without twisting them.

b. Install the front speed sensor onto the steering knuckle with the bolt. Tighten to 75 inch lbs. (8.5 Nm).

c. Install the front speed sensor with the clamp to the front shock absorber.

17. Install the front axle shaft nut.
18. Install the front wheel. Tighten to 76 ft. lbs. (103 Nm).
19. Inspect and adjust the front wheel alignment.
20. Check for the speed signal.

LOWER CONTROL ARM

REMOVAL & INSTALLATION

See Figures 282 and 283.

➡Use the same procedure for the LH side and RH side.

➡The following procedure listed is for the LH side.

1. Place the front wheels facing straight ahead.
2. Secure the steering wheel.
3. Remove the front wheels.
4. Separate the front lower suspension arm LH (MT).
5. Remove the front lower suspension arm LH (MT).

a. Remove the 2 bolts, nut, and front lower suspension arm LH from the front suspension crossmember.

➡Because the nut has its own stopper, do not turn the nut. Loosen bolt B with the nut secured.

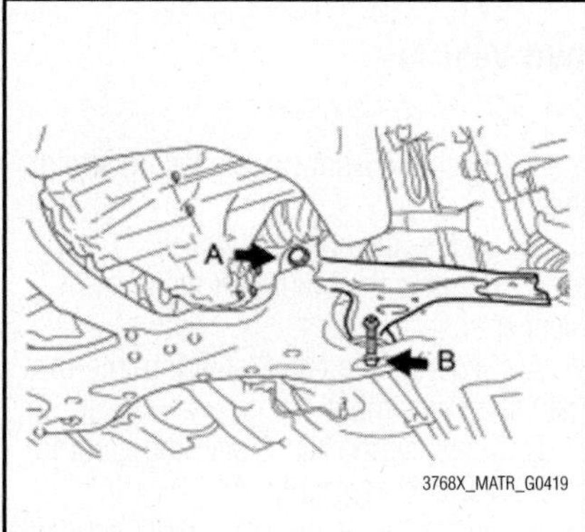

Fig. 282 Removing the front lower suspension arm LH (MT)

6. Remove the column hole cover silencer sheet.
7. Separate the No. 2 steering intermediate shaft assembly.
8. Separating No. 1 steering column hole cover sub assembly.
9. Separate the tie rod end sub assembly LH.
10. Separate the tie rod end sub assembly RH.
11. Separate the front stabilizer link assembly LH and RH.
12. Separate the front lower suspension arm LH and RH.
13. Remove the front suspension crossmember sub assembly.
14. Remove the front lower suspension arm LH.

a. Remove the 2 bolts, nut, and front lower suspension arm LH from the front suspension crossmember.

⁂ CAUTION

Because the nut has its own stopper, do not turn the nut. Loosen the bolt with the nut secured.

To install:

15. Temporarily install the front lower suspension arm LH.

a. Temporarily install the front lower suspension arm LH to the front suspension crossmember with the 2 bolts and nut.

➡Because the nut has its own stopper, do not turn the nut. Tighten the bolt with the nut secured.

➡Fully tighten the 2 bolts after stabilizing the suspension.

16. Temporarily install the front lower suspension arm LH (MT).

a. Temporarily install the front lower suspension arm LH to the front suspension crossmember with the 2 bolts and nut.

➡Because the nut has its own stopper, do not turn the nut. Tighten the bolt with the nut secured.

➡Fully tighten the 2 bolts after stabilizing the suspension.

17. Install the front suspension crossmember sub assembly.
18. Connect the front lower suspension arm LH and RH.
19. Connect the front stabilizer link assembly LH and RH.
20. Connect the tie rod end sub assembly LH and RH.
21. Connect the No. 1 steering column hole cover sub assembly.
22. Connect the No. 2 steering intermediate shaft assembly.
23. Place the front wheels facing straight ahead.
24. Install the column hole cover silencer sheet.
25. Install the front wheels. Tighten to 76 ft. lbs. (103 Nm).
26. Stabilize the suspension.

a. Lower the vehicle and bounce it up and down several times to stabilize the front suspension. Raise the vehicle.

27. Fully tighten the front lower suspension arm LH.

a. Fully tighten bolt A to 101 ft. lbs. (137 Nm).

➡The final torque must be applied under standard vehicle height conditions.

b. Fully tighten bolt B to 101 ft. lbs. (137 Nm).

➡The final torque must be applied under standard vehicle height conditions.

➡Because the nut has its own stopper, do not turn the nut. Tighten bolt B with the nut secured.

28. Inspect and adjust the front wheel alignment.

STABILIZER BAR

REMOVAL & INSTALLATION

2WD Vehicles

See Figures 284 through 286.

1. Place the front wheels facing straight ahead.

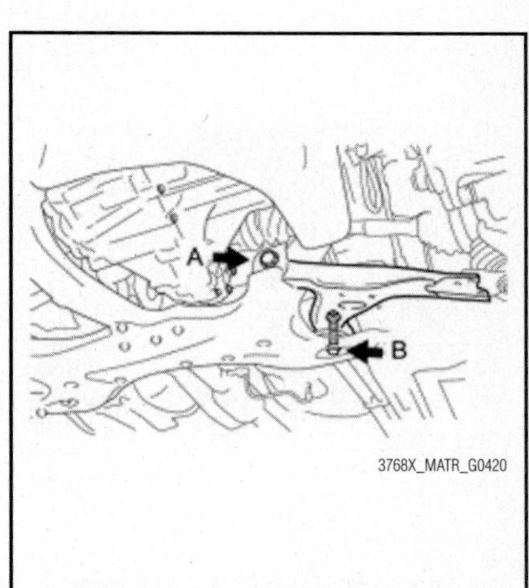

Fig. 283 Tightening the suspension arm bolts

2. Secure the steering wheel.
3. Remove the column hole cover silencer sheet.
4. Separate the No. 2 steering intermediate shaft assembly.
5. Separate the No. 1 steering column hole cover sub assembly.
6. Remove the engine under cover LH and RH.
7. Remove the front wheels.
8. Separate the tie rod end sub assembly LH and RH.
9. Remove the front stabilizer link assembly LH.
   a. Remove the nut and separate the stabilizer link assembly LH from the front shock absorber with coil spring.

➡If the ball joint turns together with the nut, use a hexagon wrench (6 mm) to hold the stud bolt.

   b. Remove the nut and the stabilizer link assembly LH from the front stabilizer bar.

➡If the ball joint turns together with the nut, use a hexagon wrench (6 mm) to hold the stud bolt.

10. Remove the front stabilizer link assembly RH.

➡Perform the same procedure as the LH side.

11. Separate the front lower suspension arm LH and RH.
12. Remove the front suspension crossmember sub assembly.
13. Remove the front stabilizer bar.
   a. Remove the 4 bolts and front stabilizer bar from the front suspension crossmember sub-assembly.

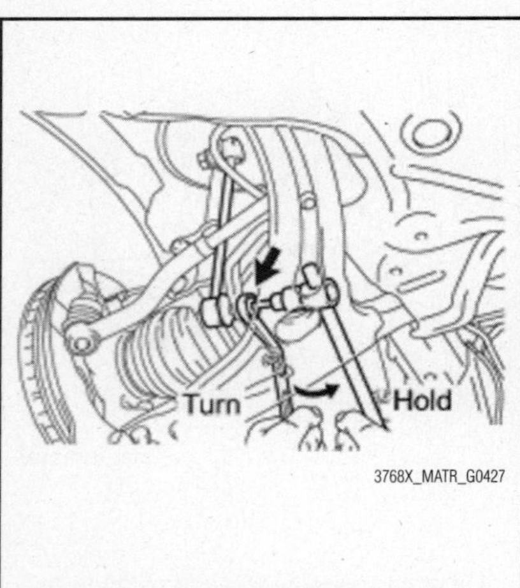

Fig. 284 Removing the stabilizer link assembly from the front stabilizer bar

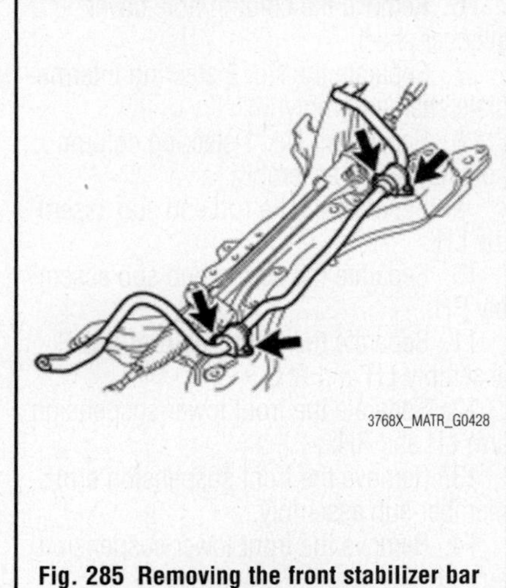

Fig. 285 Removing the front stabilizer bar

14. Remove the No. 1 front stabilizer bar bushing (LH and RH side).
   a. Remove the No. 1 front stabilizer bar bushing from the front stabilizer bar.

To install:

15. Install the No. 1 front stabilizer bar bushing.
   a. Install the No. 1 front stabilizer bar bushing to the front stabilizer bar.

➡Install the No. 1 front stabilizer bar bushing so that the cutout faces the rear of the vehicle.

➡Make sure that the amount of deviation of the front stabilizer bar in the horizontal direction is within +/- 0.197 in. (5 mm).

16. Install the No. 1 front stabilizer bar bushing RH side.
17. Install the front stabilizer bar.
   a. Install the front stabilizer bar and 2 No. 1 front stabilizer brackets to the front suspension crossmember sub-assembly with the 4 bolts. Tighten to 18 ft. lbs. (24 Nm).
18. Install the front suspension crossmember sub assembly.
19. Connect the front lower suspension arm LH and RH.
20. Install the front stabilizer link assembly LH.
   a. Install the front stabilizer link assembly LH to the front shock absorber with coil spring with the nut. Tighten to 55 ft. lbs. (75 Nm).

➡If the ball joint turns together with the nut, use a hexagon wrench (6 mm) to hold the stud bolt.

   b. Install the front stabilizer link assembly LH to the front stabilizer bar with the nut. Tighten to 55 ft. lbs. (75 Nm).

➡If the ball joint turns together with the nut, use a hexagon wrench (6 mm) to hold the stud bolt.

21. Install the front stabilizer link assembly RH.
22. Connect the tie rod end sub assembly LH and RH.
23. Connect the No. 1 steering column hole cover sub assembly.
24. Connect the No. 2 steering intermediate shaft assembly.
25. Install the engine under cover LH and RH.
26. Place the front wheels facing straight ahead.
27. Install the column hole cover silencer sheet.
28. Install the front wheels and tighten to 76 ft. lbs. (103 Nm).
29. Inspect and adjust the front wheel alignment.

4WD Vehicles

See Figures 287 and 288.

1. Place the front wheels facing straight ahead.
2. Secure the steering wheel.
3. Remove the column hole cover silencer sheet.
4. Separate the No. 2 steering intermediate shaft assembly.
5. Separate the No. 1 steering column hole cover sub assembly.
6. Remove the engine under cover LH and RH.
7. Remove the front wheels.
8. Separate the tie rod end sub assembly LH and RH.

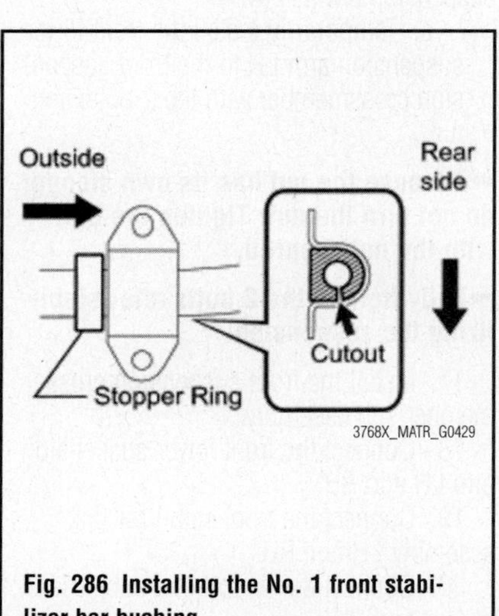

Fig. 286 Installing the No. 1 front stabilizer bar bushing

9. Loosen the front suspension cross-member sub assembly.

10. Remove the No. 1 front stabilizer bracket LH and RH.

11. Separate the steering link assembly.

12. Remove the front suspension cross-member sub assembly.

13. Remove the front stabilizer bar.

a. Remove the 2 nuts and the front stabilizer bar from the front stabilizer link.

➡If the ball joint turn together with the nut, use a hexagon wrench (6 mm) to hole the stud bolt.

14. Remove the front stabilizer link assembly LH.

a. Remove the nut and the stabilizer link assembly LH from the front shock absorber with coil spring.

➡If the ball joint turns together with the nut, use a hexagon wrench (6 mm) to hold the stud bolt.

15. Remove the front stabilizer link assembly RH.

➡Perform the same procedure as the LH side.

16. Remove the No. 1 front stabilizer bar bushing from the front stabilizer bar (LH and RH).

To install:

17. To install, reverse the removal procedure. Tighten the front stabilizer link and bar to 55 ft. lbs. (74 Nm).

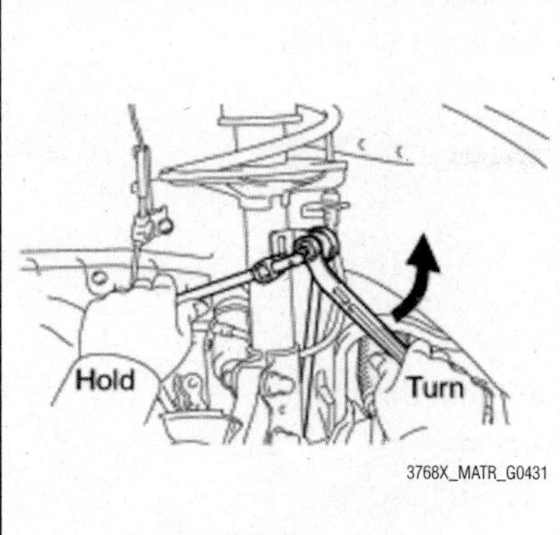

Fig. 288 Removing the front stabilizer link assembly LH

STRUT & SPRING ASSEMBLY

REMOVAL & INSTALLATION

See Figures 289 through 294.

➡Use the same procedure for the LH side and RH side.

➡The following procedure listed is for the LH side.

1. Remove the front wheel.
2. Remove the front wiper arm head cap.
3. Remove the front wiper arm and blade assembly LH and RH.
4. Remove the hood to cowl top seal.
5. Remove the cowl top ventilator louver RH and LH.
6. Remove the windshield wiper motor and link assembly.
7. Remove the outer cowl top panel.
8. Separate the front stabilizer link assembly.
9. Separate the front speed sensor.
10. Remove the front suspension support dust cover.

a. Remove the front suspension support dust cover.

11. Remove the shock absorber with the coil spring.

a. Loosen the front support to front shock absorber nut of the front shock absorber.

➡Do not remove the front support to front shock absorber nut.

➡Loosen the nut only when the front shock absorber with coil spring needs to be disassembled.

b. Support the front axle using a jack and wooden block.

c. Remove the 2 bolts and 2 nuts, and separate the front shock absorber with coil spring (lower side) from the steering knuckle.

d. Remove the 3 nuts and front shock absorber with coil spring (upper side).

➡Make sure that the front speed sensor is completely separated from the front shock absorber with coil spring.

12. Secure the front shock absorber with the coil spring.

a. Install the bolt and nut to the front shock absorber as shown in the illustra-

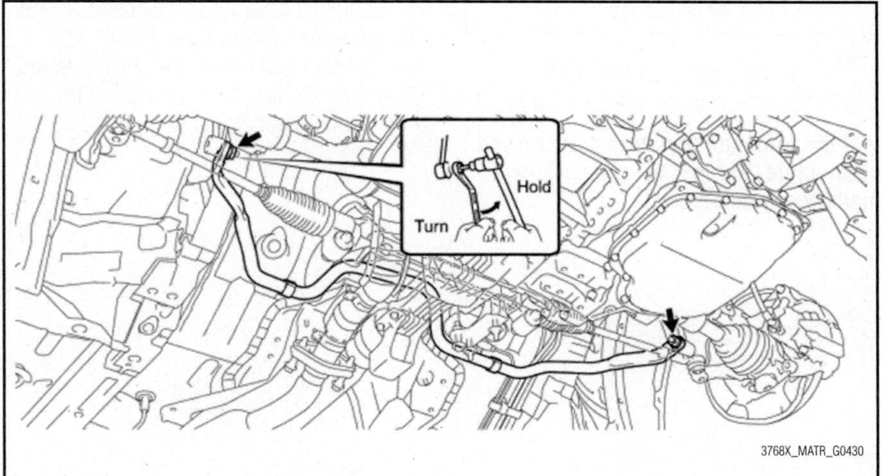

Fig. 287 Removing the front stabilizer bar—4WD

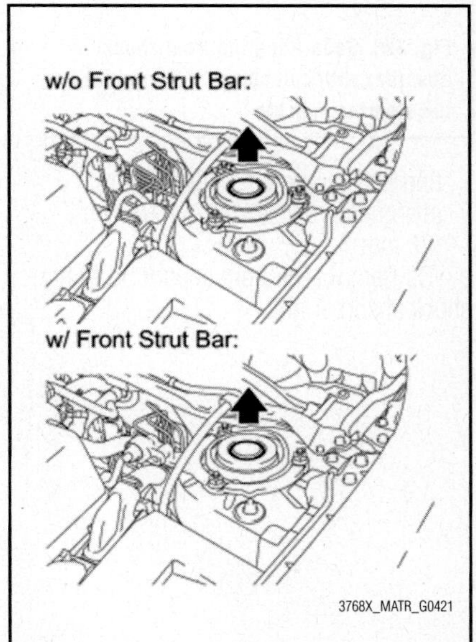

Fig. 289 Removing the front suspension dust cover

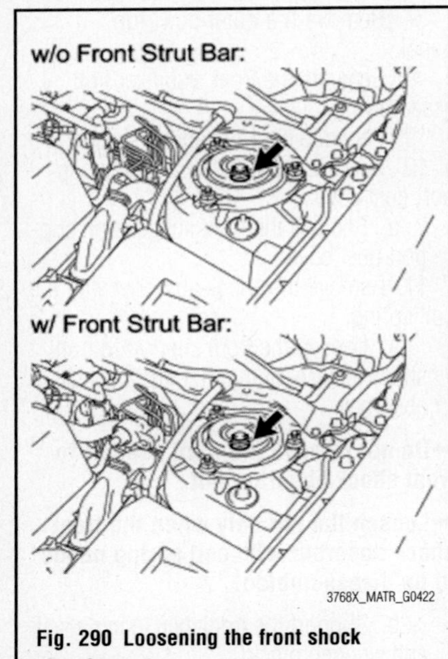

Fig. 290 Loosening the front shock absorber nut

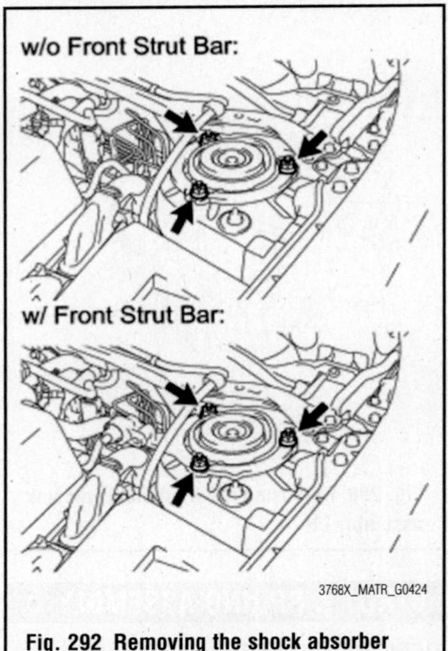

Fig. 292 Removing the shock absorber with the coil spring

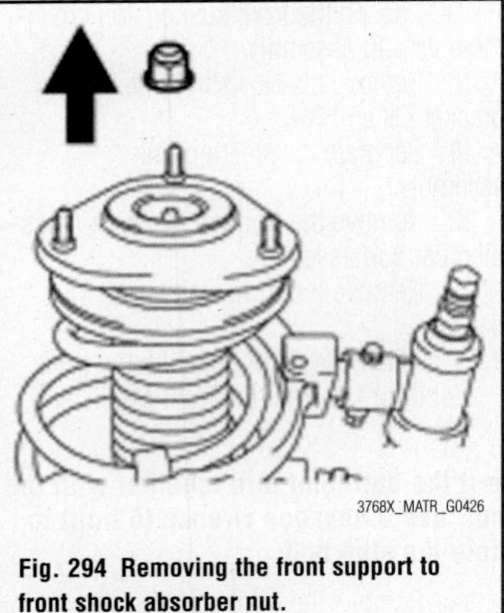

Fig. 294 Removing the front support to front shock absorber nut.

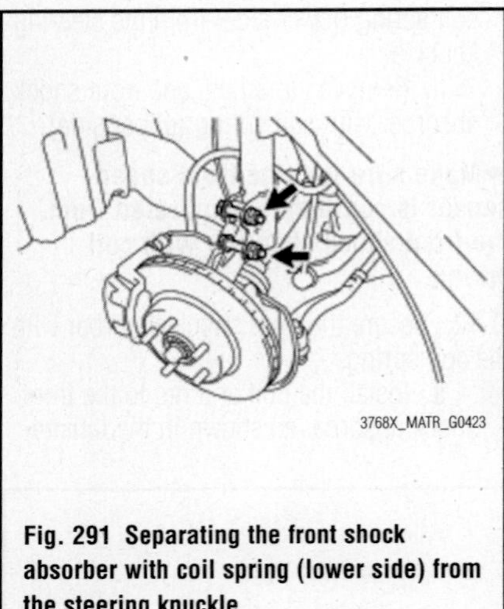

Fig. 291 Separating the front shock absorber with coil spring (lower side) from the steering knuckle

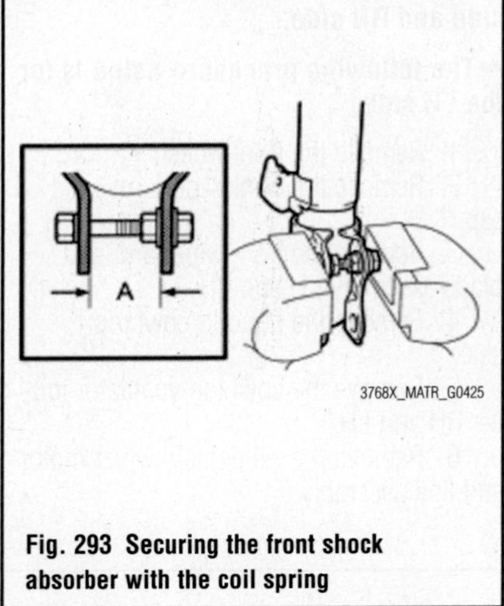

Fig. 293 Securing the front shock absorber with the coil spring

tion and secure the front shock absorber in a vise. "A" = 1.10 inch (28 mm)

13. Remove the front support to front shock absorber nut.

a. Using the special tool (09727-30021), compress the front coil spring.

➡**If the front coil spring is compressed at an angle, using 2 special tools will make the work easier.**

b. Check that the front coil spring is fully compressed.

➡**Do not use an impact wrench. It will damage the special tools.**

c. Remove the front support to front shock absorber nut.

14. Remove the front suspension support sub assembly.

15. Remove the front suspension support dust seal.

16. Remove the front coil spring upper seat.

17. Remove the front coil spring upper insulator.

18. Remove the front coil spring.

19. Remove the front spring bumper.

20. Remove the front coil spring lower insulator.

To install:

21. To install, reverse the removal procedure. Tighten the front shock absorber with the coil spring (upper side) nuts to 37 ft. lbs. (50 Nm) and (lower side) to 177 ft. lbs. (240 Nm). Tighten the front support to front shock absorber nut to 35 ft. lbs. (47 Nm).

LOWER CONTROL ARM

REMOVAL & INSTALLATION

See Figures 295 through 307.

➡Use the same procedure for the RH side and LH side.

➡The procedure listed below is for the LH side.

1. Remove the rear wheel.
2. Remove the rear floor side member brace (LH).
    a. Remove the 2 bolts and rear floor member brace.
3. Separate the No. 3 parking brake cable assembly.
    a. Remove the 2 bolts and separate the No. 3 parking brake cable assembly.
4. Separate the rear stabilizer link assembly.
5. Loosen the rear shock absorber with the coil spring.
    a. Loosen the bolt.

➡Do not remove the bolt and nut.

6. Loosen the rear suspension arm bracket assembly.
    a. Loosen the nut.
7. Remove the rear No. 1 suspension arm assembly.
    a. Remove the bolt and nut.
    b. Support the upper control arm assembly securely.
    c. Put matchmarks on the rear No. 2 suspension toe adjust plate, rear suspension toe adjust cam sub-assembly and rear No. 1 suspension arm assembly.
    d. Remove the nut, the rear No. 2 suspension toe adjust plate and rear suspension toe adjust cam sub-assembly.
    e. Remove the bolt and nut, and disconnect the rear shock absorber with coil spring.
    f. Remove the bolt and nut, and separate the rear part of the rear No. 1 suspension arm assembly.

➡When removing the bolt, stop the nut from rotating and loosen the bolt.

    g. Remove the 3 bolts on the front part, and separate the front part of the rear No. 1 suspension arm assembly.
8. Remove the rear suspension arm bracket assembly.
9. Remove the rear suspension support stopper.
10. Remove the lower control arm bushing.

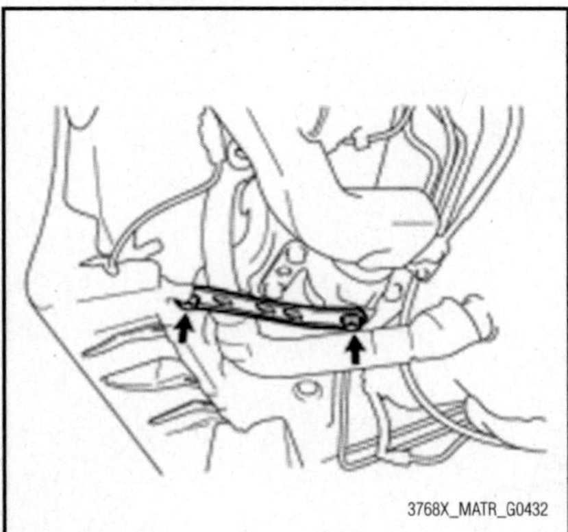

Fig. 295 Removing the rear floor side member brace

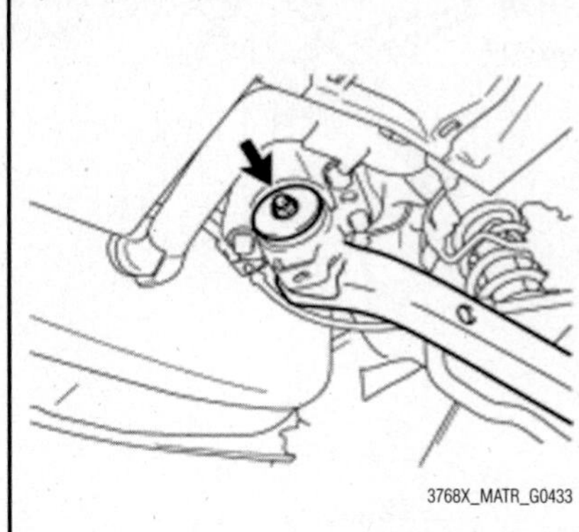

Fig. 296 Loosening the rear suspension arm bracket

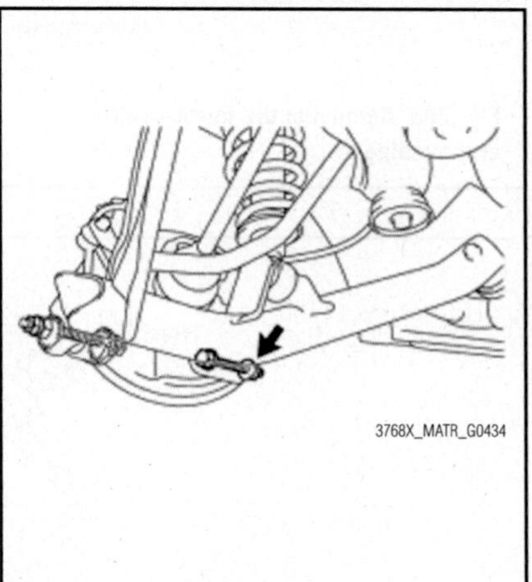

Fig. 297 Removing the bolt and nut

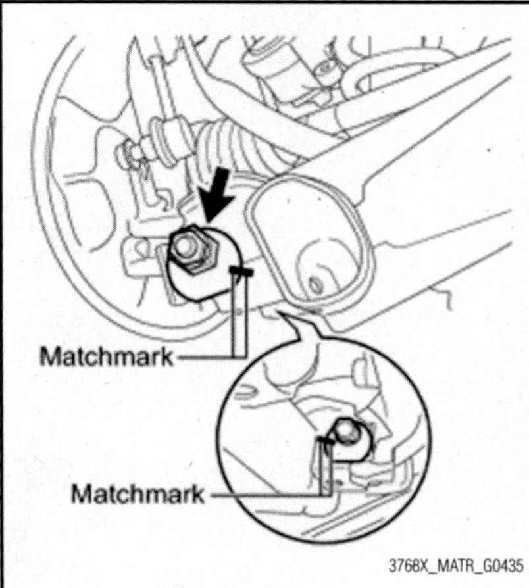

Fig. 298 Putting matchmarks on the rear No. 2 suspension toe adjustment plate, rear suspension toe adjust cam sub assembly and rear No. 1 suspension arm assembly

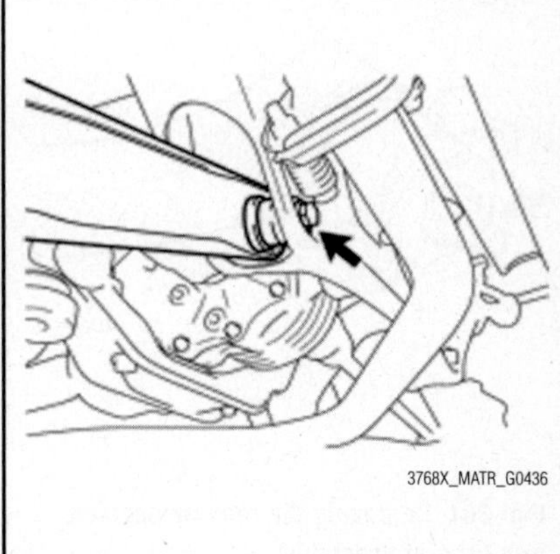

Fig. 299 Separating the rear part of the rear No. 1 suspension arm assembly

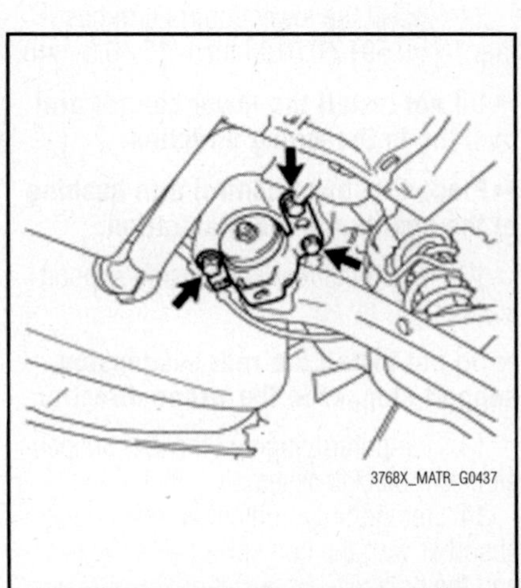

Fig. 300 Separating the front part of the rear No. 1 suspension arm assembly

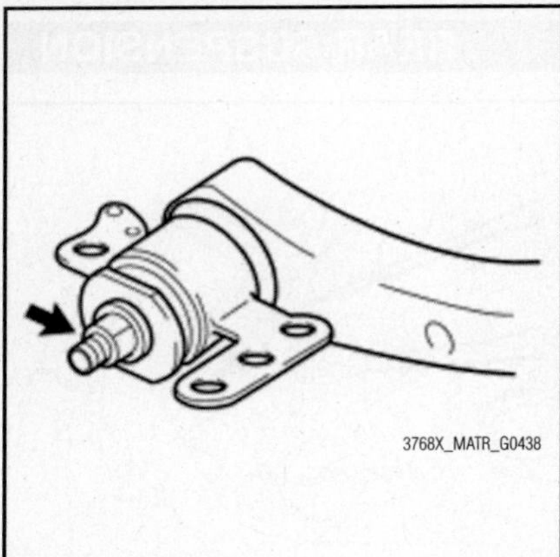

Fig. 301 Removing the rear suspension arm bracket assembly

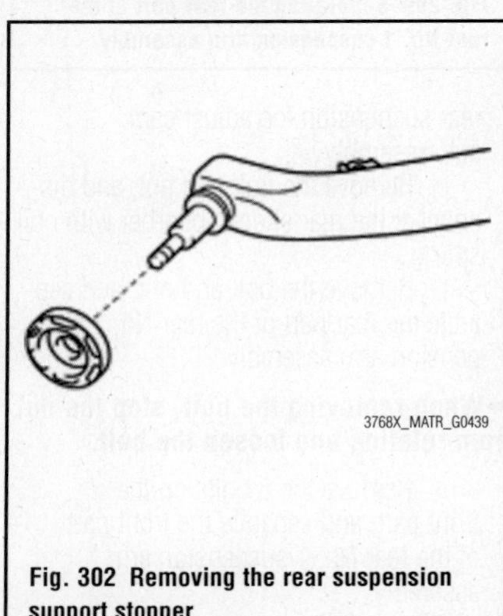

Fig. 302 Removing the rear suspension support stopper

a. Using the special tool (09632-36010), remove the lower control arm bushing.

To install:

11. Install the lower control arm bushing. "A": 0.591/0.0196 inch (15/0.5 mm).

➡Do not install the lower control arm bushing in the wrong direction.

➡Place the lower control arm bushing at the position in the illustration.

12. Install the rear suspension support stopper.

➡Do not install the rear suspension support stopper in the wrong direction.

13. Temporarily tighten the rear suspension arm bracket assembly with the nut.
14. Temporarily tighten the rear shock absorber with the coil spring with the nut and the bolt.
15. Temporarily tighten the rear No. 1 suspension arm assembly with the bolt and nut.

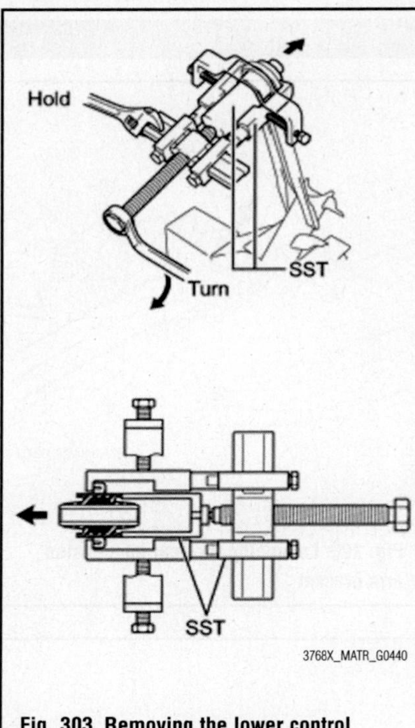

Fig. 303 Removing the lower control arm bushing

➡When installing the bolt, stop the nut from rotating and tighten the bolt.

a. Temporarily install the 3 bolts and fully tighten the 3 bolts in order from 1 to 3. Tighten to 48 ft. lbs. (65 Nm).

b. Temporarily tighten the rear axle assembly to the rear No. 1 suspension arm assembly with the bolt and the nut.

c. Insert the rear suspension toe adjust cam sub-assembly from the front side of the vehicle, and temporarily tighten the nut through the rear No. 2 suspension toe adjust plate.

16. Install the rear stabilizer link assembly.
17. Install the No. 3 parking brake cable assembly with the 2 bolts. Tighten to 53 inch lbs. (6 Nm).
18. Install the rear floor side member brace (LH) with the 2 bolts. Tighten to 22 ft. lbs. (30 Nm).
19. Stabilize the suspension.
20. Fully tighten the rear No. 1 suspension arm assembly.

a. Using the special tool, tighten the bolt. Using the special tool tighten to 64

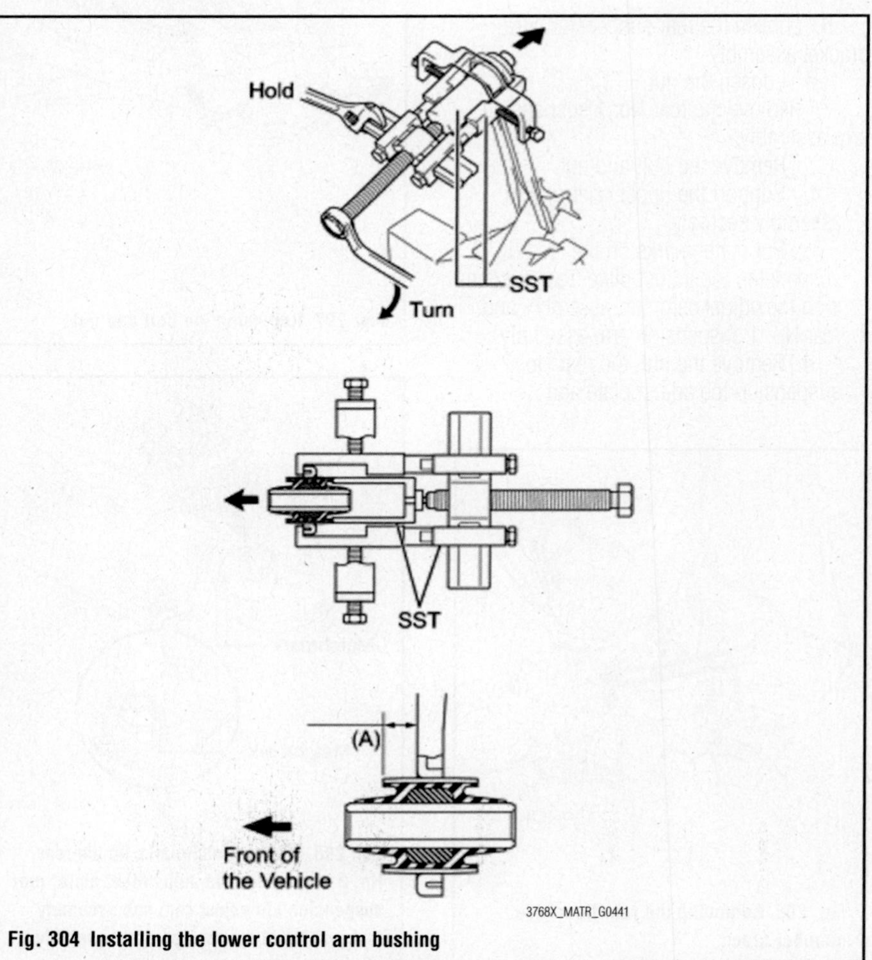

Fig. 304 Installing the lower control arm bushing

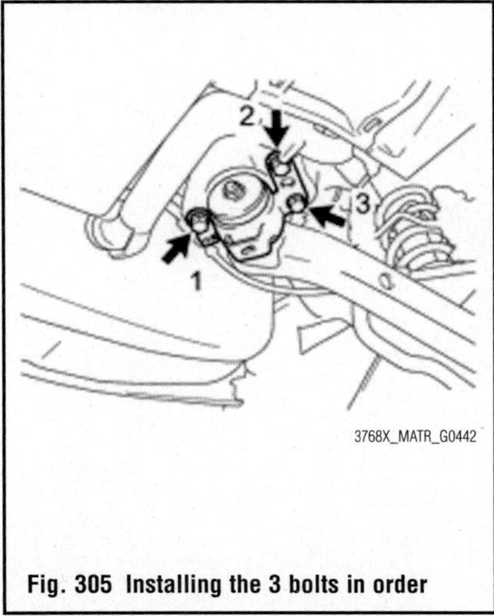

Fig. 305 Installing the 3 bolts in order

ft. lbs. (87 Nm). Without the special tool tighten to 47 ft. lbs. (64 Nm).

➡**Use a torque wrench with a fulcrum length of 425 mm (16.73 in.).**

➡**This torque value is effective when SST is parallel to the torque wrench.**

➡**When installing the bolt, stop the nut from rotating and tighten the bolt.**

➡**The final torque must be applied under standard vehicle height conditions.**

b. Fully tighten the bolt to 103 ft. lbs. (140 Nm).

➡**The final torque must be applied under standard vehicle height conditions.**

c. Align the matchmarks on the rear No. 2 suspension toe adjust plate and fully tighten the nut. Tighten to 55 ft. lbs. (74 Nm).

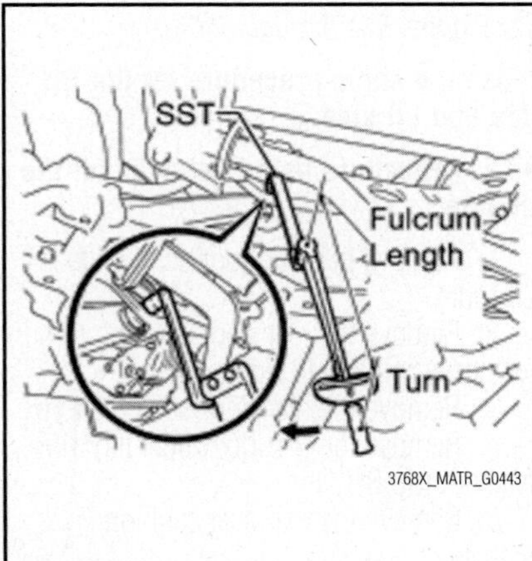

Fig. 306 Tightening the bolt with the special tool

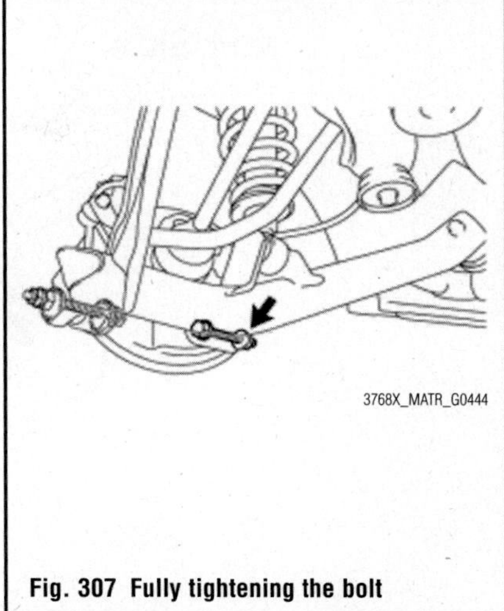

Fig. 307 Fully tightening the bolt

➡**The final torque must be applied under standard vehicle height conditions.**

21. Fully tighten the rear suspension arm bracket assembly. Using the special tool tighten the nut to 81 ft. lbs. (110 Nm). Without the special tool tighten the nut to 60 ft. lbs. (81 Nm).
22. Fully tighten the rear shock absorber with the coil spring.
23. Inspect and adjust the rear wheel alignment.

STABILIZER BAR

REMOVAL & INSTALLATION

Double Wishbone Type Suspension

See Figures 308 and 309.

1. Remove the rear wheels.
2. Remove the center exhaust pipe assembly.
3. Remove the propeller with the center bearing shaft assembly (4WD).
4. Drain the differential oil (4WD).
5. Remove the rear axle shaft nut LH (4WD).
6. Remove the rear axle shaft nut RH (4WD).
7. Remove the charcoal canister assembly (2WD).
8. Separate the rear disc brake caliper assembly LH.
9. Separate the rear disc brake caliper assembly RH.
10. Remove the rear disc LH and RH.
11. Disconnect the rear speed sensor wire LH and RH (2WD).
12. Separate the rear speed sensor wire LH and RH (2WD).
13. Separate the rear speed sensor LH and RH (4WD).
14. Remove the No. 3 parking brake show return tension spring (LH).
15. Remove the No. 2 parking brake shoe return tension spring (LH).
16. Remove the No. 1 parking brake shoe hold down compression spring (LH).
17. Remove the parking brake shoe strut (LH).
18. Remove the No. 2 parking brake shoe hold down compression spring (LH).
19. Separate the parking brake shoe lever (LH).
20. Remove the parking brake shoe adjusting screw set (LH).
21. Remove the No. 1 parking brake shoe return tension spring (LH).
22. Remove the parking brake shoe lever (LH).
23. Remove the No. 1 parking brake shoe hold down spring pin (LH).
24. Remove the No. 2 parking brake shoe hold down spring pin (LH).
25. Remove the No. 3 parking brake shoe return tension spring (RH).
26. Remove the No. 2 parking brake shoe return tension spring (RH).
27. Remove the No. 1 parking brake shoe hold down compression spring (RH).
28. Remove the parking brake shoe strut (RH).
29. Remove the No. 2 parking brake shoe hold down compression spring (RH).
30. Separate the parking brake shoe lever (RH).
31. Remove the parking brake shoe adjusting screw set (RH).
32. Remove the No. 1 parking brake shoe return tension spring (RH).
33. Remove the parking brake shoe lever (RH).
34. Remove the No. 1 parking brake shoe hold down spring pin (RH).
35. Remove the No. 2 parking brake shoe hold down spring pin (RH).
36. Separate the No. 3 parking brake cable assembly.
37. Separate the No. 2 parking brake cable assembly.
38. Remove the rear stabilizer link assembly (LH).
    a. Remove the 2 nuts and rear stabilizer link assembly (LH).

➡**If the ball joint turns together with the nut, use a hexagon wrench (5 mm) to hold the stud.**

39. Remove the rear stabilizer link assembly (RH).

➡**Perform the same procedure as the LH side.**

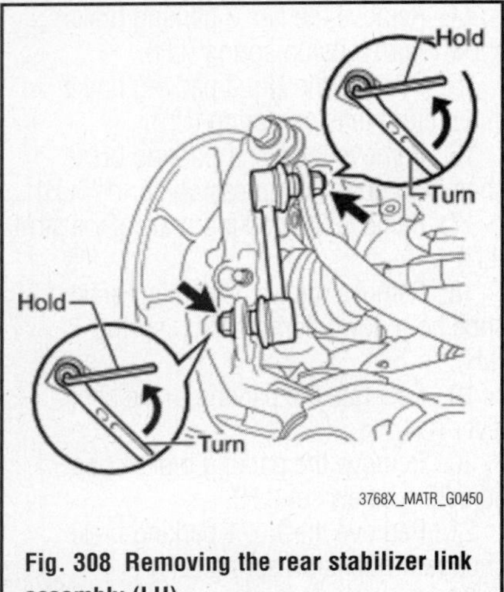

Fig. 308 Removing the rear stabilizer link assembly (LH)

40. Separate the upper control arm assembly LH and RH.
41. Separate the rear No. 1 suspension arm assembly LH.
42. Separate the rear No. 1 suspension arm assembly RH.
43. Remove the rear axle assembly LH and RH.
44. Remove the rear floor side member brace.
45. Loosen the rear shock absorber with the coil spring.
46. Remove the rear No. 1 suspension arm assembly LH and RH.
47. Remove the rear suspension member.
48. Remove the rear No. 1 stabilizer bar bracket.
   a. Remove the 4 bolts and 2 rear No. 1 stabilizer bar brackets from the rear suspension member.
49. Remove the rear stabilizer bushing.
   a. Remove the 2 rear stabilizer bushings from the rear stabilizer bar.
50. Remove the rear stabilizer bar.

To install:

51. To install, reverse the removal procedure. Tighten the rear No. 1 stabilizer bar bracket bolts to 13 ft. lbs. (18 Nm). Tighten the rear stabilizer link assembly nuts to 32 ft. lbs. (44 Nm). Check the speed sensor signal.

Torsion Bean Type Suspension

See Figures 310 through 313.

1. Remove the rear wheels.
2. Remove the rear axle beam damper.
3. Remove the rear stabilizer bar.
   a. Remove the 2 bolts, 2 nuts and rear stabilizer bar.

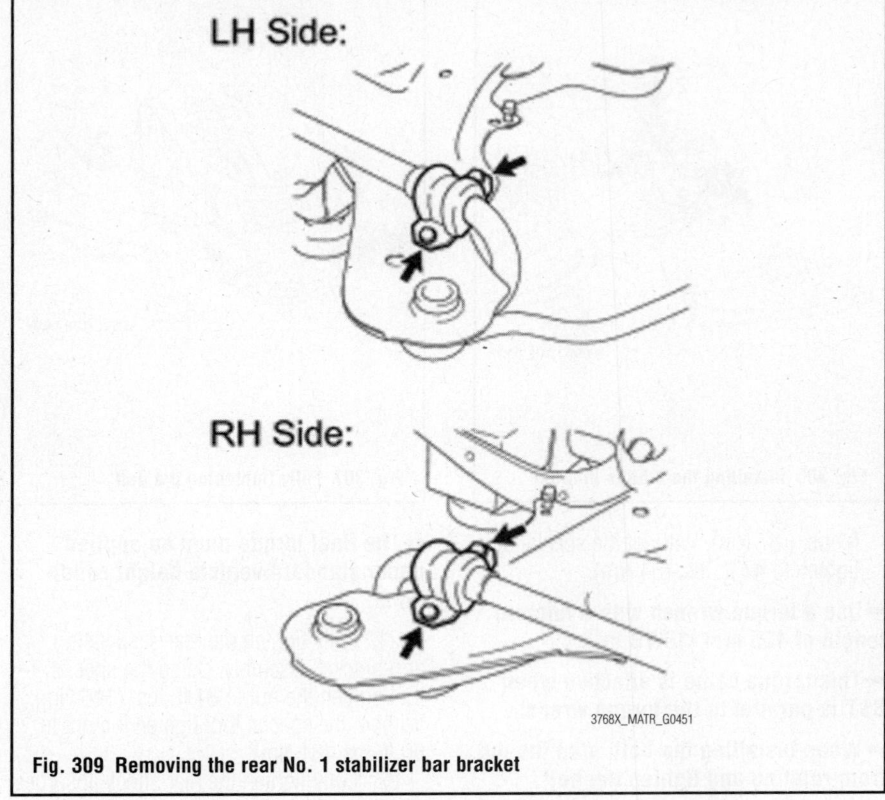

Fig. 309 Removing the rear No. 1 stabilizer bar bracket

➡Be sure to loosen the nuts.

To install:

4. Install the rear stabilizer bar.
   a. Check that the identification mark of the rear stabilizer bar is positioned on the left side of the vehicle.
   b. Install the rear stabilizer bar with the 2 bolts and 2 nuts. Tighten to 184 ft. lbs. (250 Nm).

➡Be sure to tighten the nuts.

➡When reinstalling the bolts, insert them from the upper side of the vehicle.

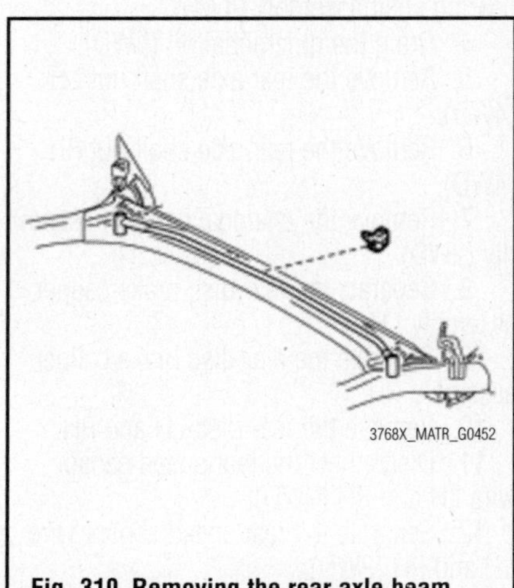

Fig. 310 Removing the rear axle beam damper

5. Install the rear axle beam damper.
   a. Install the rear axle beam damper to the center of the rear stabilizer bar.
6. Install the rear wheels. Tighten to 76 ft. lbs. (103 Nm).

STRUT & SPRING ASSEMBLY

REMOVAL & INSTALLATION

Double Wishbone Type Suspension

See Figures 314 through 316.

➡Use the same procedure for the RH side and LH side.

➡The procedure listed below is for the LH side.

1. Remove the rear door scuff plate LH (LH side).
2. Remove the rear door opening trim weather-strip LH (LH side).
3. Remove the rear door scuff plate RH.
4. Remove the rear door opening trim weather-strip RH.
5. Remove the rear seat cushion assembly.
6. Remove the rear seat headrest assembly (LH side).

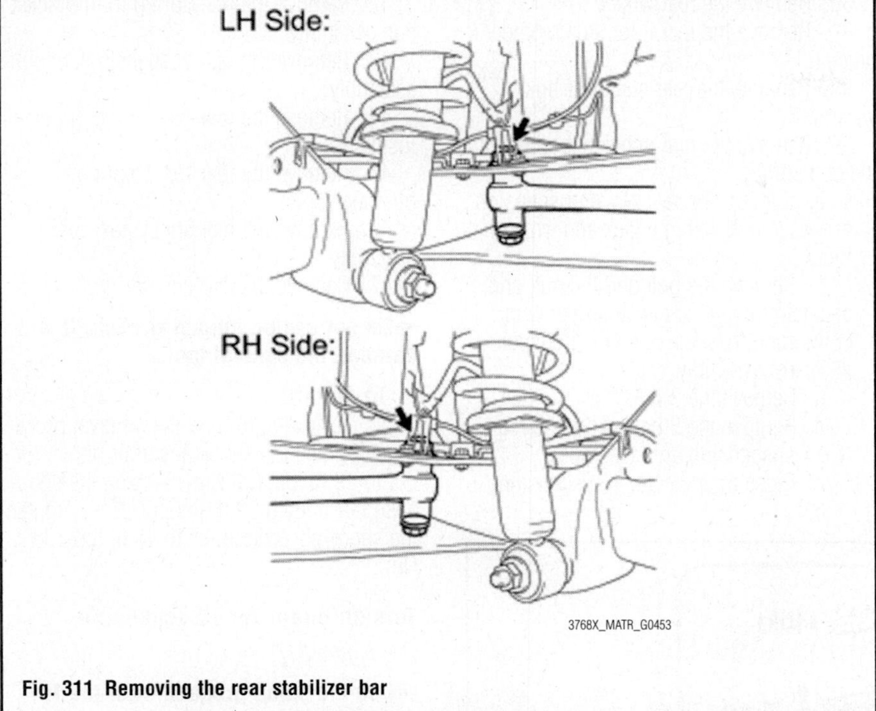

Fig. 311 Removing the rear stabilizer bar

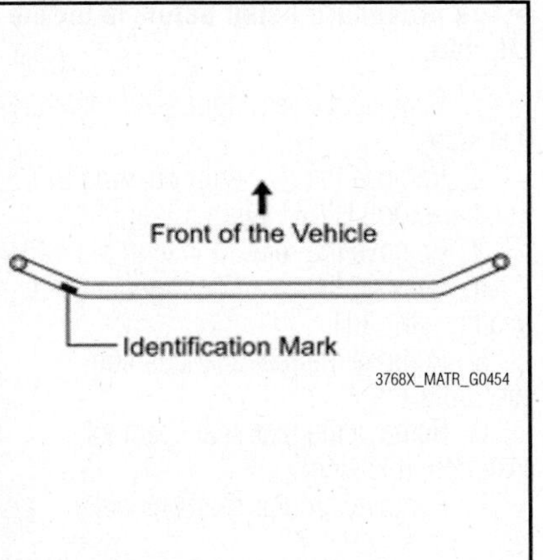

Fig. 312 Checking the identification mark of the rear stabilizer bar

7. Remove the rear seatback outer hinge cover (LH).
8. Remove the rear seatback inner hinge cover (LH).
9. Remove the rear seatback assembly (LH).
10. Remove the rear seat headrest assembly (RH).
11. Remove the rear seat center headrest assembly (RH).
12. Disconnect the rear seat center outer belt assembly (RH).
13. Remove the rear seatback outer hinge cover (RH).
14. Remove the rear seatback inner hinge cover (RH).
15. Remove the tonneau cover assembly (w/ tonneau cover).
16. Remove the deck board assembly.
17. Remove the No. 2 deck board.
18. Remove the jack carrier assembly.
19. Remove the jack assembly.

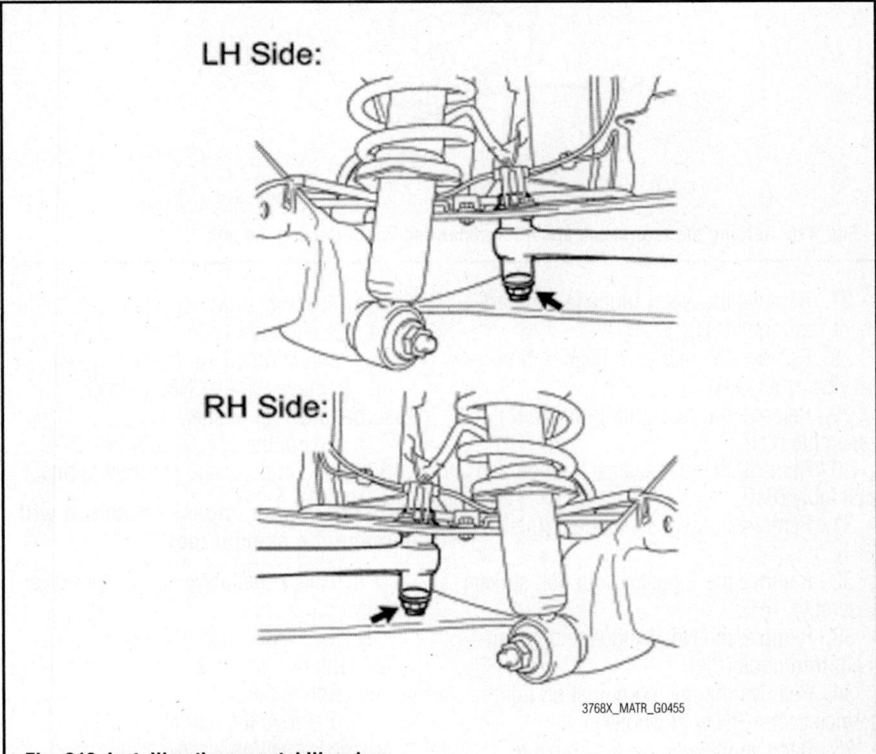

Fig. 313 Installing the rear stabilizer bar

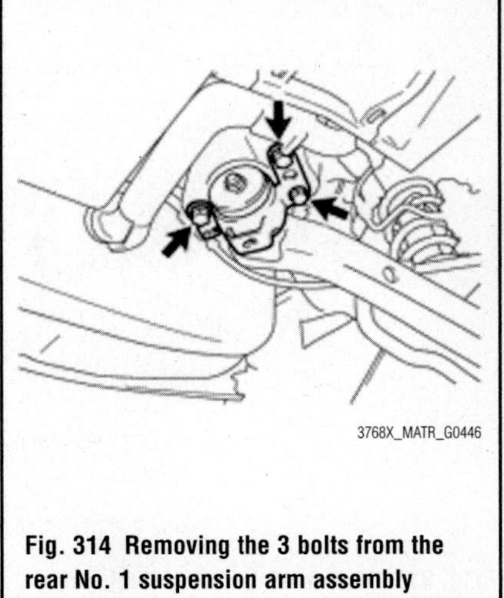

Fig. 314 Removing the 3 bolts from the rear No. 1 suspension arm assembly

20. Remove the deck floor box LH and RH.
21. Remove the child restraint seat tether anchor.
22. Remove the luggage compartment tray.
23. Remove the rear deck trim cover.
24. Remove the seatback hinge sub assembly (LH).
25. Remove the rear seat side garnish (LH).
26. Remove the luggage hold belt striker assembly (LH).

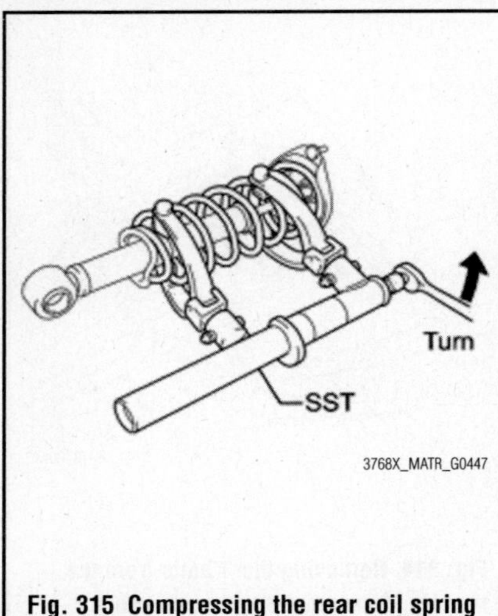

Fig. 315 Compressing the rear coil spring

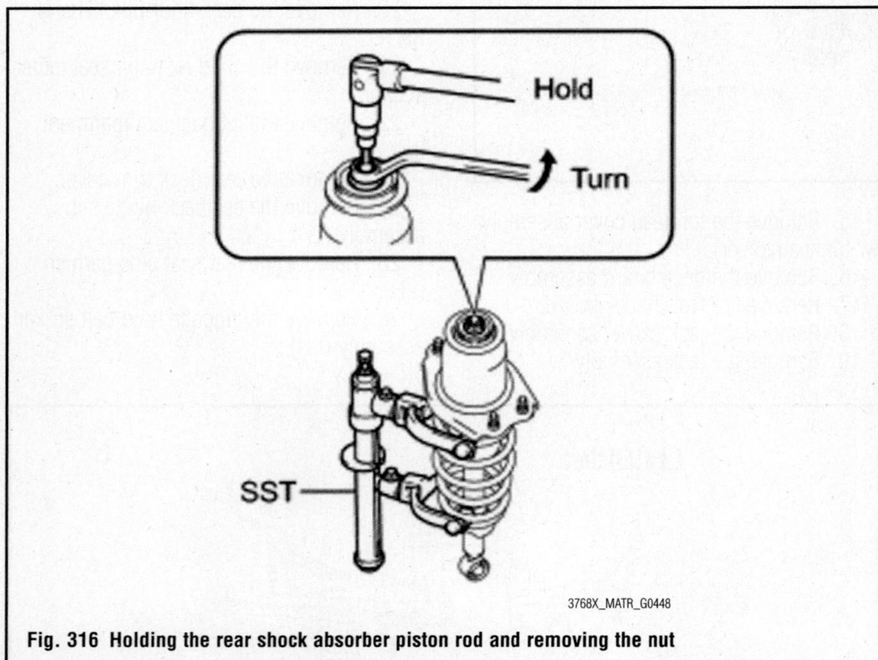

Fig. 316 Holding the rear shock absorber piston rod and removing the nut

27. Remove the No. 1 luggage compartment trim hook (LH).
28. Remove the rear combination light service cover (LH).
29. Remove the rear deck trim panel assembly (LH).
30. Remove the rear seatback hinge sub assembly (RH).
31. Remove the rear seat side garnish (RH).
32. Remove the luggage hold belt striker assembly (RH).
33. Remove the No. 1 luggage compartment trim hook (RH).
34. Remove the rear combination light service cover (RH with woofer).
35. Remove the side deck trim panel assembly (RH).
36. Remove the rear wheel.
37. Remove the rear floor side member brace (LH).
38. Separate the rear stabilizer link assembly.
39. Remove the rear shock absorber with the coil spring.
   a. Support the rear No. 1 suspension arm assembly using a jack and wooden block.
   b. Remove the bolt and the nut, and separate the rear shock absorber with coil spring from the rear No. 1 suspension arm assembly.
   c. Remove the 3 nuts.
   d. Remove the 3 bolts from the rear No. 1 suspension arm assembly.
   e. Press the rear No. 1 suspension arm assembly down to the outside of the vehicle and remove the rear shock absorber with coil spring.
40. Remove the rear No. 1 shock absorber cushion washer.
   a. Using the special tool (09727-30021), compress the rear coil spring.

➡Do not use an impact wrench. It will damage the special tool.

   b. Check that the rear coil spring is fully compressed.
   c. Use a hexagon wrench (6 mm) to hold the rear shock absorber piston rod and remove the nut.
   d. Remove the rear No. 1 shock absorber cushion washer.
41. Remove the rea4r suspension support.
42. Remove the rear spring front bracket sub assembly.
43. Remove the rear suspension support assembly.
44. Remove the rear coil spring upper insulator.
45. Remove the rear No. 1 spring bumper.
46. Remove the rear shock absorber assembly.
47. Remove the rear coil spring.

➡Do not use an impact wrench. It will damage the special tool.

To install:

48. To install, reverse the removal procedure. Tighten the shock absorber piston rod nut to 29 ft. lbs. (39 Nm). Tighten the shock absorber nuts to 59 ft. lbs. (80 Nm). Tighten the shock absorber bolts to 48 ft. lbs. (65 Nm).

Torsion Beam Type Suspension

See Figure 317.

➡Use the same procedure for the RH side and LH side.

➡The procedure listed below is for the LH side.

1. Remove the rear door scuff plate LH (LH side).
2. Remove the rear door opening trim weather-strip LH (LH side).
3. Remove the rear door scuff plate RH.
4. Remove the rear door opening trim weather-strip RH.
5. Remove the rear seat cushion assembly.
6. Remove the rear seat headrest assembly (LH side).
7. Remove the rear seatback outer hinge cover (LH).
8. Remove the rear seatback inner hinge cover (LH).
9. Remove the rear seatback assembly (LH).
10. Remove the rear seat headrest assembly (RH).
11. Remove the rear seat center headrest assembly (RH).
12. Disconnect the rear seat center outer belt assembly (RH).
13. Remove the rear seatback outer hinge cover (RH).
14. Remove the rear seatback inner hinge cover (RH).
15. Remove the tonneau cover assembly (w/ tonneau cover).
16. Remove the deck board assembly.
17. Remove the No. 2 deck board.
18. Remove the jack carrier assembly.

19. Remove the jack assembly.
20. Remove the deck floor box LH and RH.
21. Remove the child restraint seat tether anchor.
22. Remove the luggage compartment tray.
23. Remove the rear deck trim cover.
24. Remove the seatback hinge sub assembly (LH).
25. Remove the rear seat side garnish (LH).
26. Remove the luggage hold belt striker assembly (LH).
27. Remove the No. 1 luggage compartment trim hook (LH).
28. Remove the rear combination light service cover (LH).
29. Remove the rear deck trim panel assembly (LH).
30. Remove the rear seatback hinge sub assembly (RH).
31. Remove the rear seat side garnish (RH).
32. Remove the luggage hold belt striker assembly (RH).
33. Remove the No. 1 luggage compartment trim hook (RH).
34. Remove the rear combination light service cover (RH with woofer).
35. Remove the side deck trim panel assembly (RH).
36. Remove the rear wheel.
37. Remove the rear shock absorber cushion retainer.
   a. Support the rear axle beam assembly using a jack and wooden block.
   b. Remove the nut and rear shock absorber cushion retainer.
38. Remove the rear shock absorber with the coil spring.
   a. Remove the 2 nuts from the rear shock absorber with coil spring (upper side).
   b. Remove the bolt (lower side) from the rear shock absorber with coil spring.
   c. Slowly lower the jack and remove the rear shock absorber with coil spring.
39. Remove the rear No. 1 shock absorber cushion washer.
   a. Using the special tool (09727-30021), compress the rear coil spring.
   b. Check that the rear coil spring is fully compressed.

➡Do not use an impact wrench. It will damage the special tool.

   c. Use a hexagon wrench (6 mm) to hold the rear shock absorber piston rod and remove the nut.
   d. Remove the rear No. 1 shock absorber cushion washer.

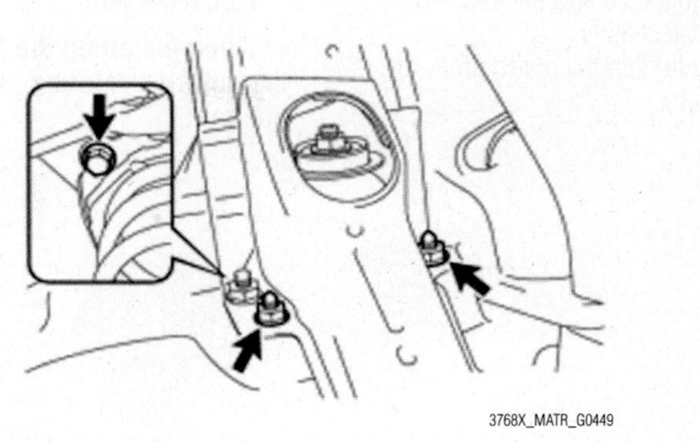

Fig. 317 Removing the 2 nuts from the rear shock absorber with coil spring (upper side)

40. Remove the rear suspension support.
41. Remove the rear spring front bracket sub assembly.
42. Remove the rear suspension support assembly.
43. Remove the rear coil spring upper insulator.
44. Remove the rear No. 1 spring bumper.
45. Remove the rear shock absorber assembly.
46. Remove the rear coil spring.

To install:

47. To install, reverse the removal procedure. Tighten the rear shock absorber piston rod nut to 29 ft. lbs. (39 Nm). Tighten the shock absorber with coil spring bolt (lower side) to 59 ft. lbs. (80 Nm). Tighten the rear shock absorber with coil spring nuts (upper side) to 59 ft. lbs. (80 Nm).

TESTING

1. Compress and extend the rear shock absorber rod, and check that there is no abnormal resistance or unusual sound.
2. If there is any abnormality, replace the rear shock absorber assembly with a new one.

UPPER CONTROL ARM

REMOVAL & INSTALLATION

See Figure 318.

➡Use the same procedure for the RH side and LH side.

➡The procedure listed below is for the LH side.

1. Remove the rear suspension member.
2. Remove the upper control arm assembly LH.
   a. Put matchmarks on the camber adjust cam assembly, No. 2 camber adjust cam and rear suspension member.
   b. Remove the nut, camber adjust cam assembly and No. 2 camber adjust cam.
   c. Remove the upper control arm assembly.

To install:

3. Install the upper control arm assembly LH.
   a. Insert the camber adjust cam assembly from the front side of the vehicle, and temporarily tighten the nut through the No. 2 camber adjust cam.
4. Install the rear suspension member.
5. Fully tighten the upper control arm assembly LH.

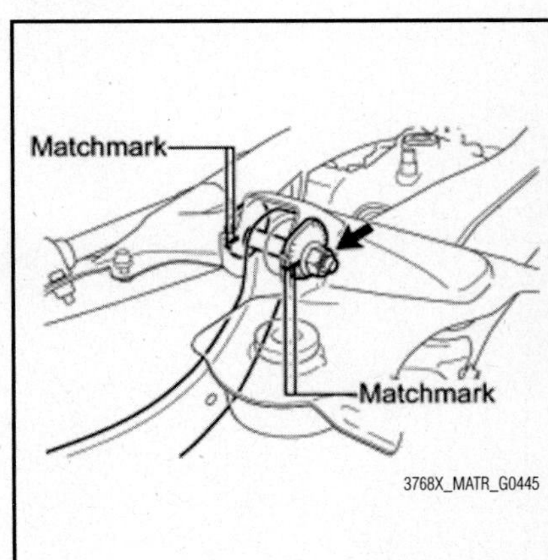

Fig. 318 Removing the upper control arm assembly LH

a. Align the matchmarks on the No. 2 camber adjust cam and the camber adjust cam assembly.

b. Fully tighten the nut to 55 ft. lbs. (74 Nm).

c. Fully tighten the bolt and nut to 55 ft. lbs. (74 Nm).

➡When installing the bolt, stop the nut from rotating and tighten the bolt.

6. Fully tighten the upper control arm assembly RH and LH.

7. Inspect and adjust the rear wheel alignment.

8. Check for the speed sensor signal.

TOYOTA

Prius

11

SPECIFICATIONS AND MAINTENANCE CHARTS

VEHICLE AND ENGINE IDENTIFICATION

Engine						
Code	Liters (cc)	Cu. in.	Cyl.	Fuel Sys.	Engine Type	Eng. Mfg.
1NZ-FXE	1.5 (1497)	91.4	4	SFI	DOHC	Toyota
2ZR-FXE	1.8 (1798)	109.7	4	SFI	DOHC	Toyota

Model Year	
Code	Year
B	2011
C	2012

SFI: Sequential Multiport Fuel Injection

DOHC: Dual Overhead Camshaft

71099_PRIU_C0001

GENERAL ENGINE SPECIFICATIONS

Year	Engine ID/VIN	Engine Displacement Liters (cc)	Fuel System Type	Net Horsepower @ rpm	Net Torque @ rpm (ft. lbs.)	Bore x Stroke (in.)	Com-pression Ratio	Oil Pressure @ rpm
2011	1NZ-FXE	1.5 (1497)	SFI	110@5000	82@4200	2.95x3.33	13.0:1	22-80@2500
	2ZR-FXE	1.8 (1798)	SFI	98@5200	105@4000	3.17x3.48	13.0:1	21@2500
2012	1NZ-FXE	1.5 (1497)	SFI	110@5000	82@4200	2.95x3.33	13.0:1	22-80@2500
	2ZR-FXE	1.8 (1798)	SFI	98@5200	105@4000	3.17x3.48	13.0:1	21@2500

SFI: Sequential Multiport Fuel Injection

71099_PRIU_C0002

GASOLINE ENGINE TUNE-UP SPECIFICATIONS

Year	Engine Displacement Liters	Engine ID/VIN	Spark Plugs Gap (in.)	Ignition Timing (deg.) MT	Ignition Timing (deg.) AT	Fuel Pump (psi)	Idle Speed (rpm) MT	Idle Speed (rpm) AT	Valve Clearance In.	Valve Clearance Ex.
2011	1.5	1NZ-FXE	0.043	—	8-12B	44-50 ①	—	950-1050	0.007-0.009	0.011-0.013
	1.8	2ZR-FXE	0.043	—	8-12B	44-50 ①	—	950-1050	0.007-0.009	0.011-0.013
2012	1.5	1NZ-FXE	0.043	—	8-12B	44-50 ①	—	950-1050	0.007-0.009	0.011-0.013
	1.8	2ZR-FXE	0.043	—	8-12B	44-50 ①	—	950-1050	0.007-0.009	0.011-0.013

Note: The Vehicle Emission Control Information label often reflects specification changes made during production.
The label figures must be used if they differ from those in this chart.

B: Before top dead center

① At idle

71099_PRIU_C0003

CAPACITIES

Year	Model	Engine Displacement Liters	Engine ID/VIN	Engine Oil with Filter (qts.)	Transaxle (pts.) 5-Spd	Transaxle (pts.) Auto.	Fuel Tank (gal.)	Cooling System (qts.)
2011	Prius	1.5	1NZ-FXE	3.9	—	7.6 ①	11.9	②
	Prius	1.8	2ZR-FXE	4.4	—	7.6 ①	11.9	③
2012	Prius	1.5	1NZ-FXE	3.9	—	7.6 ①	11.9	②
	Prius	1.8	2ZR-FXE	4.4	—	7.6 ①	11.9	③

Note: All capacities are approximate. Add fluid gradualy and check to be sure a proper fluid level is obtained.

① Specification for Hybrid transaxle.

② Gasoline engine: 5.2 quarts

Electric motor, inverter and converter: 2.7 quarts

③ 7.7 quarts w/ exhaust heat recirculation system

6.8 quarts w/o exhaust heat recirculation system

71099_PRIU_C0004

FLUID SPECIFICATIONS

Year	Model	Engine Displacement Liters	Engine ID/VIN	Engine Oil	Auto. Trans.	Manual Trans.	Drive Axle	Transfer Case	Power Steering Fluid	Brake Master Cylinder
2011	Prius	1.5	1NZ-FXE	5W-30	ATF WS	—	—	—	—	DOT 3
	Prius	1.8	2ZR-FXE	5W-30	ATF WS	—	—	—	—	DOT 3
2012	Prius	1.5	1NZ-FXE	5W-30	ATF WS	—	—	—	—	DOT 3
	Prius	1.8	2ZR-FXE	5W-30	ATF WS	—	—	—	—	DOT 3

DOT: Department Of Transpotation

71099_PRIU_C0013

VALVE SPECIFICATIONS

Year	Engine ID/VIN	Engine Displacement Liters	Seat Angle (deg.)	Face Angle (deg.)	Spring Test Pressure (lbs. @ in.)	Spring Installed Height (in.)	Stem-to-Guide Clearance (in.) Intake	Stem-to-Guide Clearance (in.) Exhaust	Stem Diameter (in.) Intake	Stem Diameter (in.) Exhaust
2011	1NZ-FXE	1.5	45	45	40.5-44.5@ 0.9880	2.353	0.0001-0.0031	0.0012-0.0039	0.1957-0.1963	0.1955-0.1961
	2ZR-FXE	1.8	NA	NA	NA	NA	NA	NA	0.215-0.2160	0.215-0.2160
2012	1NZ-FXE	1.5	45	45	40.5-44.5@ 0.9880	2.353	0.0001-0.0031	0.0012-0.0039	0.1957-0.1963	0.1955-0.1961
	2ZR-FXE	1.8	NA	NA	NA	NA	NA	NA	0.215-0.2160	0.215-0.2160

71099_PRIU_C0005

CAMSHAFT AND BEARING SPECIFICATIONS CHART

All measurements are given in inches.

Year	Engine Displ. Liters	Engine VIN	Journal Dia.	Brg. Oil Clearance	Shaft End-play	Runout	Journal Bore	Lobe Lift Intake	Lobe Lift Exhaust
2011	1.5	1NZ-FXE	①	0.0001-0.0024	0.0016-0.0037	0.0012	NA	1.6657-1.6697	1.7341-1.7380
	1.8	2ZR-FXE	①	0.0001-0.0024	0.0016-0.0037	0.0012	NA	1.6657-1.6697	1.7341-1.7380
2012	1.5	1NZ-FXE	①	0.0001-0.0024	0.0016-0.0037	0.0012	NA	1.6657-1.6697	1.7341-1.7380
	1.8	2ZR-FXE	①	0.0001-0.0024	0.0016-0.0037	0.0012	NA	1.6657-1.6697	1.7341-1.7380

NA - Not available

① No 1: 1.3563-1.3569

All others: 0.9035-0.9041

71099_PRIU_C0014

CRANKSHAFT AND CONNECTING ROD SPECIFICATIONS

All measurements are given in inches.

Year	Engine ID/VIN	Engine Displacement Liters	Crankshaft Main Brg. Journal Dia.	Crankshaft Main Brg. Oil Clearance	Crankshaft Shaft End-play	Crankshaft Thrust on No.	Connecting Rod Journal Diameter	Connecting Rod Oil Clearance	Connecting Rod Side Clearance
2011	1NZ-FXE	1.5	1.8106-1.8110	0.0004-0.0028	0.0035-0.0120	NA	1.5745-1.5748	0.0006-0.0024	0.0063-0.0142
	2ZR-FXE	1.8	1.7320-1.7323	0.0002-0.0004	0.0035-0.0120	NA	1.5745-1.5748	0.0005-0.0015	0.0063-0.0142
2012	1NZ-FXE	1.5	1.8106-1.8110	0.0004-0.0028	0.0035-0.0120	NA	1.5745-1.5748	0.0006-0.0024	0.0063-0.0142
	2ZR-FXE	1.8	1.7320-1.7323	0.0002-0.0004	0.0035-0.0120	NA	1.5745-1.5748	0.0005-0.0015	0.0063-0.0142

NA: Not Available

71099_PRIU_C0007

PISTON AND RING SPECIFICATIONS

All measurements are given in inches.

Year	Engine ID/VIN	Engine Displacement Liters	Piston Clearance	Ring Gap Top Compression	Ring Gap Bottom Compression	Ring Gap Oil Control	Ring Side Clearance Top Compression	Ring Side Clearance Bottom Compression	Ring Side Clearance Oil Control
2011	1NZ-FXE	1.5	0.0018-0.0032	0.0079-0.0240	0.0118-0.0472	0.0039-0.0453	0.0008-0.0028	0.0008-0.0024	0.0008-0.0024
	2ZR-FXE	1.8	0.0017-0.0032	0.0079-0.0190	0.0118-0.0177	0.0039-0.0138	0.0008-0.0028	0.0008-0.0024	0.0008-0.0024
2012	1NZ-FXE	1.5	0.0018-0.0032	0.0079-0.0240	0.0118-0.0472	0.0039-0.0453	0.0008-0.0028	0.0008-0.0024	0.0008-0.0024
	2ZR-FXE	1.8	0.0017-0.0032	0.0079-0.0190	0.0118-0.0177	0.0039-0.0138	0.0008-0.0028	0.0008-0.0024	0.0008-0.0024

71099_PRIU_C0006

TORQUE SPECIFICATIONS

All readings in ft. lbs.

Year	Engine ID/VIN	Engine Displ. Liters	Cylinder Head Bolts	Main Bearing Bolts	Rod Bearing Bolts	Crankshaft Damper Bolts	Flywheel Bolts	Manifold Intake	Manifold Exhaust	Spark Plugs	Oil Pan Drain Plug
2011	1NZ-FXE	1.5	①	②	③	95	36	15	20	13	13
	2ZR-FXE	1.8	⑤	⑥	③	95	36+90 deg.	21	15	15	27
2012	1NZ-FXE	1.5	①	②	③	95	36	15	20	13	13
	2ZR-FXE	1.8	⑤	⑥	③	95	36+90 deg.	21	15	15	27

① Step 1: 21 ft. lbs.
Step 2: turn head bolts 90 degrees
Step 3: turn head bolts 90 degrees

② Step 1: 16 ft. lbs.
Step 2: turn bearing cap bolts 90 degrees

③ Step 1: 11 ft. lbs.
Step 2: turn bearing cap bolts 90 degrees

④ Step 1: 62 ft. lbs.
Step 2: turn flywheel bolts 90 degrees

⑤ Step 1: 36 ft. lbs.
Step 2: turn head bolts 90 degrees
Step 3: turn head bolts 45 degrees

⑥ Step 1: 30 ft. lbs.
Step 2: turn bearing cap bolts 90 degrees

71099_PRIU_C0008

WHEEL ALIGNMENT

Year	Model		Caster Range (+/-Deg.)	Caster Preferred Setting (Deg.)	Camber Range (+/-Deg.)	Camber Preferred Setting (Deg.)	Toe-in (in.)	Steering Axis Inclination (Deg.)
2011	Prius	F	0.75	+3.17	0.75	-0.58	0+/-0.08	12.58+/-0.75
		R	—	—	0.50	-1.50	0.12+/-0.10	—
	Prius	F	0.75	+5.88	0.75	-0.22	0+/-0.08	12.27
		R	—	—	0.50	-1.48	0.12+/-0.10	—
2012	Prius	F	0.75	+3.17	0.75	-0.58	0+/-0.08	12.58+/-0.75
		R	—	—	0.50	-1.50	0.12+/-0.10	—
	Prius	F	0.75	+5.88	0.75	-0.22	0+/-0.08	12.27
		R	—	—	0.50	-1.48	0.12+/-0.10	—

71099_PRIU_C0009

TIRE, WHEEL AND BALL JOINT SPECIFICATIONS

		OEM Tires		Tire Pressures (psi)		Wheel	Ball Joint	Lug Nut
Year	Model	Standard	Optional	Front	Rear	Size	Inspection	(ft. lbs.)
2011	Prius	P185/65R15	None	35	33	6-JJ	①	76
	Prius	P195/65R15	None	35	32	6-JJ	①	76
2012	Prius	P185/65R15	None	35	33	6-JJ	①	76
	Prius	P195/65R15	None	35	32	6-JJ	①	76

OEM: Original Equipment Manufacturer

PSI: Pounds Per Square Inch

① Replace if any measurable movement is found.

71099_PRIU_C0010

BRAKE SPECIFICATIONS

All measurements in inches unless noted

		Brake Disc			Brake Drum Diameter			Minimum Lining Thickness		Brake Caliper	
Year	Model	Original Thickness	Minimum Thickness	Maximum Runout	Original Inside Diameter	Max. Wear Limit	Max. Machine Diameter	Front	Rear	Bracket bolts (ft. lbs.)	Mounting bolts (ft. lbs.)
2011	Prius	0.866	0.787	0.002	7.874	7.913	7.913	0.039	0.039	81	25
	Prius	F: 0.984	F: 0.866	F: 0.002	NA	NA	NA	0.039	-	81	25
		R: 0.354	R: 0.295	R: 0.006	NA	NA	NA	-	0.039	81	25
2012	Prius	0.866	0.787	0.002	7.874	7.913	7.913	0.039	0.039	81	25
	Prius	F: 0.984	F: 0.866	F: 0.002	NA	NA	NA	0.039	-	81	25
		R: 0.354	R: 0.295	R: 0.006	NA	NA	NA	-	0.039	81	25

71099_PRIU_C0011

SCHEDULED MAINTENANCE INTERVALS
TOYOTA—PRIUS

TO BE SERVICED	TYPE OF SERVICE	VEHICLE MILEAGE INTERVAL (x1000)												
		5	10	15	20	25	30	35	40	45	50	55	60	65
Engine oil & filter	R		✓		✓		✓		✓		✓		✓	
Cabin air filter (solar power ventilation system)	S/I		✓		✓		✓		✓		✓		✓	
Cabin air filter (except solar power ventilation system)	S/I			✓			✓			✓			✓	
Driver's floor mat	S/I	✓	✓	✓	✓	✓	✓	✓	✓	✓	✓	✓	✓	✓
Hybrid transaxle fluid	S/I	✓	✓	✓	✓	✓	✓	✓	✓	✓	✓	✓	✓	✓
Drive axle boots	S/I	✓	✓	✓	✓	✓	✓	✓	✓	✓	✓	✓	✓	✓
Gear shift control operation	S/I	✓	✓	✓	✓	✓	✓	✓	✓	✓	✓	✓	✓	✓
Inspect & rotate tires	S/I	✓	✓	✓	✓	✓	✓	✓	✓	✓	✓	✓	✓	✓
Power steering system	S/I	✓	✓	✓	✓	✓	✓	✓	✓	✓	✓	✓	✓	✓
Suspension system	S/I	✓	✓	✓	✓	✓	✓	✓	✓	✓	✓	✓	✓	✓
Brake discs & pads	S/I	✓	✓	✓	✓	✓	✓	✓	✓	✓	✓	✓	✓	✓
Brake shoes & drums	S/I	✓	✓	✓	✓	✓	✓	✓	✓	✓	✓	✓	✓	✓
Brake hoses & pipes	S/I	✓		✓		✓		✓		✓		✓		✓
Brake fluid	S/I	✓	✓	✓	✓	✓	✓	✓	✓	✓	✓	✓	✓	✓
Brake pedal	S/I		✓		✓		✓		✓		✓		✓	
Cooling system, hoses & connections	S/I		✓		✓		✓		✓		✓		✓	
Fuel tank, cap & lines	S/I		✓		✓		✓		✓		✓		✓	
Air cleaner filter element	R				✓				✓				✓	
Engine coolant	R				✓				✓				✓	
Spark plugs	R				✓				✓				✓	
Drive belt	S/I				✓				✓				✓	
Engine inverter coolant	S/I			✓			✓			✓			✓	
Exhaust system	S/I			✓			✓			✓			✓	

R: Replace S/I: Service or Inspect

① Replace every 60,000 miles.

FREQUENT OPERATION MAINTENANCE (SEVERE SERVICE)

If a vehicle is operated under any of the following conditions it is considered severe service:

- Extremely dusty areas.

- 50% or more of the vehicle operation is in 32°C (90°F) or higher temperatures, or constant operation in temperatures below 0°C (32°F).

- Prolonged idling (vehicle operation in stop and go traffic).

- Frequent short running periods (engine does not warm to normal operating temperatures).

- Police, taxi, delivery usage or trailer towing usage.

Oil & oil filter: change every 3000 miles.

Brake discs & pads: service or inspect initially at 3000 miles, 6000 miles, & every 12,000 miles thereafter.

Brake hoses & pipes: service or inspect initially at 3000 miles, 6000 miles & every 12,000 miles thereafter.

Air cleaner filter element: service or inspect ever 3000 miles & replace every 30,000 miles (if not replaced previously).

Hybrid transaxle fluid: service or inspect every 6000 miles & replace every 15,000 miles (if not replaced previously).

Inspect & rotate tires: service or inspect every 6000 miles.

Power steering system: service or inspect every 6000 miles.

Steering system: service or inspect every 6000 miles.

Suspension system: service or inspect every 6000 miles.

Drive belts: service or inspect every 15,000 miles.

Exhaust system: service or inspect every 15,000 miles.

71099_PRIU_C0012

PRECAUTIONS

Before servicing any vehicle, please be sure to read all of the following precautions, which deal with personal safety, prevention of component damage, and important points to take into consideration when servicing a motor vehicle:

- Never open, service or drain the radiator or cooling system when the engine is hot; serious burns can occur from the steam and hot coolant.
- Observe all applicable safety precautions when working around fuel. Whenever servicing the fuel system, always work in a well-ventilated area. Do not allow fuel spray or vapors to come in contact with a spark, open flame, or excessive heat (a hot drop light, for example). Keep a dry chemical fire extinguisher near the work area. Always keep fuel in a container specifically designed for fuel storage; also, always properly seal fuel containers to avoid the possibility of fire or explosion. Refer to the additional fuel system precautions later in this section.
- Fuel injection systems often remain pressurized, even after the engine has been turned **OFF**. The fuel system pressure must be relieved before disconnecting any fuel lines. Failure to do so may result in fire and/or personal injury.
- Brake fluid often contains polyglycol ethers and polyglycols. Avoid contact with the eyes and wash your hands thoroughly after handling brake fluid. If you do get brake fluid in your eyes, flush your eyes with clean, running water for 15 minutes. If eye irritation persists, or if you have taken brake fluid internally, IMMEDIATELY seek medical assistance.
- The EPA warns that prolonged contact with used engine oil may cause a number of skin disorders, including cancer. You should make every effort to minimize your exposure to used engine oil. Protective gloves should be worn when changing oil. Wash your hands and any other exposed skin areas as soon as possible after exposure to used engine oil. Soap and water, or waterless hand cleaner should be used.
- All new vehicles are now equipped with an air bag system, often referred to as a Supplemental Restraint System (SRS) or Supplemental Inflatable Restraint (SIR) system. The system must be disabled before performing service on or around system components, steering column, instrument panel components, wiring and sensors. Failure to follow safety and disabling procedures could result in accidental air bag deployment, possible personal injury and unnecessary system repairs.
- Always wear safety goggles when working with, or around, the air bag system. When carrying a non-deployed air bag, be sure the bag and trim cover are pointed away from your body. When placing a non-deployed air bag on a work surface, always face the bag and trim cover upward, away from the surface. This will reduce the motion of the module if it is accidentally deployed. Refer to the additional air bag system precautions later in this section.
- Clean, high quality brake fluid from a sealed container is essential to the safe and proper operation of the brake system. You should always buy the correct type of brake fluid for your vehicle. If the brake fluid becomes contaminated, completely flush the system with new fluid. Never reuse any brake fluid. Any brake fluid that is removed from the system should be discarded. Also, do not allow any brake fluid to come in contact with a painted surface; it will damage the paint.
- Never operate the engine without the proper amount and type of engine oil; doing so WILL result in severe engine damage.
- Timing belt maintenance is extremely important. Many models utilize an interference-type, non-freewheeling engine. If the timing belt breaks, the valves in the cylinder head may strike the pistons, causing potentially serious (also time-consuming and expensive) engine damage. Refer to the maintenance interval charts for the recommended replacement interval for the timing belt, and to the timing belt section for belt replacement and inspection.
- Disconnecting the negative battery cable on some vehicles may interfere with the functions of the on-board computer system(s) and may require the computer to undergo a relearning process once the negative battery cable is reconnected.
- When servicing drum brakes, only disassemble and assemble one side at a time, leaving the remaining side intact for reference.
- Only an MVAC-trained, EPA-certified automotive technician should service the air conditioning system or its components.

BRAKES — ANTI-LOCK BRAKE SYSTEM (ABS)

GENERAL INFORMATION

PRECAUTIONS

- Certain components within the ABS system are not intended to be serviced or repaired individually.
- Do not use rubber hoses or other parts not specifically specified for and ABS system. When using repair kits, replace all parts included in the kit. Partial or incorrect repair may lead to functional problems and require the replacement of components.
- Lubricate rubber parts with clean, fresh brake fluid to ease assembly. Do not use shop air to clean parts; damage to rubber components may result.
- Use only DOT 3 brake fluid from an unopened container.
- If any hydraulic component or line is removed or replaced, it may be necessary to bleed the entire system.
- A clean repair area is essential. Always clean the reservoir and cap thoroughly before removing the cap. The slightest amount of dirt in the fluid may plug an orifice and impair the system function. Perform repairs after components have been thoroughly cleaned; use only denatured alcohol to clean components. Do not allow ABS components to come into contact with any substance containing mineral oil; this includes used shop rags.
- The Anti-Lock control unit is a microprocessor similar to other computer units in the vehicle. Ensure that the ignition switch is **OFF** before removing or installing controller harnesses. Avoid static electricity discharge at or near the controller.
- If any arc welding is to be done on the vehicle, the control unit should be unplugged before welding operations begin.

SPEED SENSORS

REMOVAL & INSTALLATION

Front

See Figures 1 and 2.

➡While the battery is connected, even if the power switch is off, the brake control system activates when the brake pedal is depressed or any door courtesy switch is turned on. Therefore, when servicing the brake system components, do not depress the brake pedal or open/ close the doors while the battery is connected.

➡Use the same procedure for the LH side and RH side.

➡The following procedure is for the LH side.

➡If the sensor rotor needs to be replaced, replace it together with the front axle hub and bearing assembly.

1. Remove the rear No. 2 floor board (separate type).
2. Remove the rear deck floor box.
3. Remove the rear No. 3 floor board.
4. Disconnect the cable from the negative battery terminal.

➡When disconnecting the cable, some systems need to be initialized after the cable is reconnected.

5. Remove the front wheel.
6. Remove the front fender liner.
   a. Remove the 12 clips, 4 screw, 3 grommets, and front fender liner.

➡Use the same procedure for the RH side and LH side.

7. Remove the front speed sensor.
   a. Remove the 2 clamps and disconnect the front speed sensor connector.
   b. Remove the bolt and No. 2 sensor clamp (*1) from the body.
   c. Remove the bolt, No. 1 sensor clamp and front brake flexible hose together from the shock absorber assembly.
   d. Remove the clamp from the shock absorber assembly.
   e. Remove the bolt and front speed sensor.

➡Prevent foreign matter from attaching to the front speed sensor tip.

➡Clean the speed sensor installation hole and the contact surfaces every time the front speed sensor is removed.

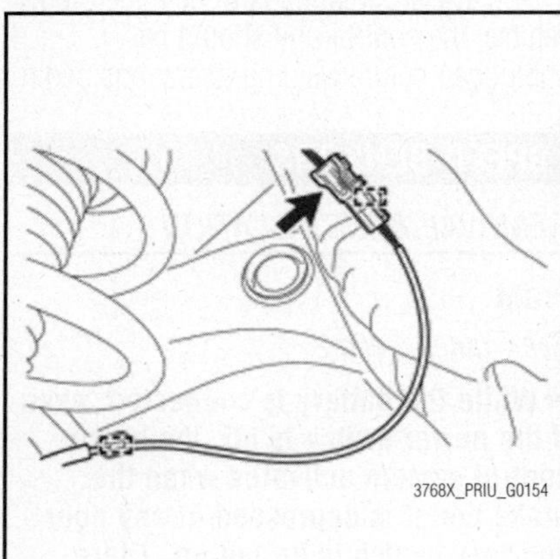

Fig. 1 Disconnecting the front speed sensor connector

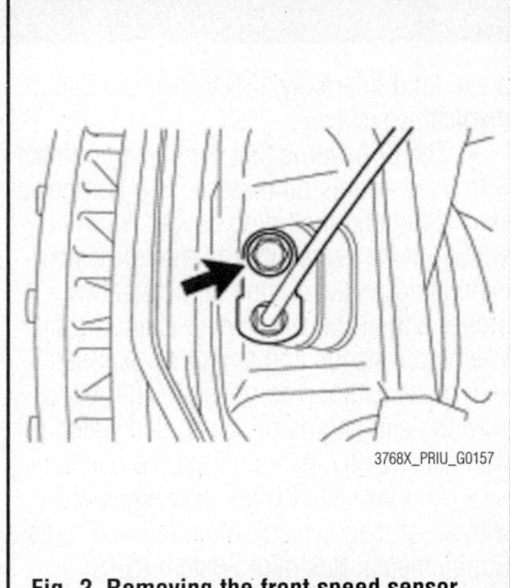

Fig. 2 Removing the front speed sensor

To install:

➡Use the same procedure for the LH side and RH side.

➡The following procedure is for the LH side.

➡If the sensor rotor needs to be replaced, replace it together with the front axle hub and bearing assembly.

8. Install the front speed sensor.
   a. Install the front speed sensor with the bolt. Tighten to 76 inch lbs. (8.5 Nm).

➡Prevent foreign matter from attaching to the front speed sensor tip.

➡Firmly insert the front speed sensor body into the knuckle before tightening the bolt.

➡After installing the front speed sensor to the knuckle, make sure that there is no clearance between the front speed sensor stay and knuckle. Also make sure that no foreign matter is stuck between the parts.

➡Before installing the clamp, firmly insert the points of the clamp into the installation holes.

   b. Temporarily install the front brake flexible hose.
   c. Install the front brake flexible hose and No. 1 sensor clamp together to the shock absorber with the bolt. Tighten to 14 ft. lbs. (19 Nm).

➡Do not twist the wire harness for the front speed sensor when installing the speed sensor.

➡Bolt tightens the brake flexible hose and front speed sensor together. Make sure that the front speed sensor is positioned over the front brake flexible hose.

   d. Install the clamp to the shock absorber assembly.
   e. Install the No. 2 sensor clamp to the body with the bolt. Tighten to 76 inch lbs. (8.5 Nm).
   f. Connect the speed sensor connector.
   g. Connect the 2 speed sensor wire harness clamps to the body.
9. Install the front fender liner with the 12 clips, 3 grommets and 4 screws.
10. Install the front wheel. Tighten to 76 ft. lbs. (103 Nm).
11. Connect the cable to the negative battery terminal.

➡When disconnecting the cable, some systems need to be initialized after the cable is reconnected.

12. Install the rear No. 3 floor board.
13. Install the rear deck floor box.
14. Install the rear No. 2 floor board (separate type).
15. Check for the speed sensor signal.

Rear

➡While the battery is connected, even if the power switch is off, the brake control system activates when the brake pedal is depressed or any door courtesy switch is turned on. Therefore, when servicing the brake system components, do not depress the brake pedal or open/ close the doors while the battery is connected.

➡When the brake pedal is first depressed after replacing the brake pads or pushing back the disc brake piston, DTC C1214 may be output. As there is no malfunction, clear the DTC.

➡Use the same procedure for the RH side and LH side.

➡The procedure listed below is for the LH side.

➡If the sensor rotor needs to be replaced, replace it together with the rear axle hub and bearing assembly.

➡The rear speed sensor is a component of the rear axle hub and bearing assembly. If the sensor malfunctions, replace the rear axle hub and bearing assembly.

1. Remove the rear No. 2 floor board (separate type).
2. Remove the rear deck floor box.
3. Remove the rear No. 3 floor board.
4. Disconnect the cable from the negative battery terminal.

➡**When disconnecting the cable, some systems need to be initialized after the cable is reconnected.**

5. Remove the rear wheel.
6. Remove the front door scuff plate LH.
7. Remove the cowl side trim sub assembly LH.
8. Remove the lower instrument panel finish panel assembly.
9. Loosen the parking brake cable.
10. Disconnect the rear speed sensor wire.
    a. Using a screwdriver, disconnect the connector from the rear speed sensor.

➡**Be careful not to damage the rear speed sensor.**

11. Disconnect the No. 3 parking brake cable assembly.
12. Separate rear disc brake caliper assembly.
13. Remove the rear disc.
14. Remove the rear axle hub and bearing assembly.

➡**The rear speed sensor is a component of the rear axle hub and bearing assembly. If the sensor malfunctions, replace the rear axle hub and bearing assembly.**

To install:

15. Install the rear axle hub and bearing assembly.
16. Inspect the rear axle hub bearing looseness.
17. Inspect the rear axle hub runout.
18. Install the rear disc.
19. Install the rear disc brake caliper assembly.
20. Connect the No. 3 parking brake cable assembly.
21. Connect the rear speed sensor wire.
    a. Connect the rear speed sensor wire connector to the rear speed sensor.
22. Adjust the parking brake lever travel.
23. Inspect the rear disc brake cylinder operation lever and stopper clearance.
24. Install the lower instrument panel finish panel assembly.
25. Install the cowl side trim sub assembly LH.
26. Install the front door scuff plate LH.
27. Install the rear wheel. Tighten to 76 ft. lbs. (103 Nm).
28. Connect the cable to the negative battery terminal.

➡**When disconnecting the cable, some systems need to be initialized after the cable is reconnected.**

29. Install the rear No. 3 floor board.
30. Install the rear deck floor box.
31. Install the rear No. 2 floor board (separate type).
32. Check the speed sensor signal.

BRAKES — BLEEDING THE BRAKE SYSTEM

BLEEDING PROCEDURE

⁂ WARNING

Bleeding the brakes requires the use of a Toyota Techstream, or equivalent scan tool. Follow tool manufacturer's procedures when bleeding the hydraulic brake system.

BLEEDING PROCEDURE

1. Remove the outer cowl top panel sub assembly.
2. Bleed the brake system.
    a. Wait at least 2 minutes with the power switch off, and disconnect the reservoir level switch connector.

➡**Do not depress the brake pedal or open/close the doors until the reservoir level switch connector is disconnected.**

➡**This procedure is not required if the reservoir level switch connector has been disconnected.**

3. Remove the brake master cylinder reservoir filler cap assembly.
4. Add brake fluid into the reservoir between MAX and MIN level on the brake fluid reservoir.

➡**Standard brake fluid is SAE J1703 or FMVSS No. 116 DOT3.**

5. Connect the Techstream to the DLC3 and turn the power switch on (IG).
6. Turn the Techstream on and enter the following menus: Chassis / ABS/VSC/TRAC / Air Bleeding.
7. Select the "ABS actuator has been replaced" on the Techstream display, and bleed air from the brake fluid following the instructions on the Techstream.

➡**Before following the instructions on the Techstream to perform linear valve offset calibration, release the parking brake. When calibration is complete, immediately apply the parking brake.**

8. After air bleeding, tighten each bleeder plug. Tighten the front bleeder plug to 73 inch lbs. (8.3 Nm). Tighten the rear bleeder plug to 8 ft. lbs. (11 Nm). Tighten the stroke simulator bleeder plug to 75 inch lbs. (8.5 Nm).

➡**The stroke simulator bleeder plug is positioned as shown.**

    a. Clear the DTC's.
    b. Turn the Techstream off and turn the power switch off.
9. Install the brake master cylinder reservoir filler cap.
10. Inspect for brake fluid leaks.
11. Install the outer cowl top panel sub assembly.

BRAKE LINE BLEEDING

1. Remove the center cowl top ventilator cover.
    a. Slide the hood to cowl top seal and disengage the claw.
    b. Disengage the 2 claws and 3 guides, and remove the center cowl top ventilator cover.
2. Bleed the brake line.
    a. Remove the brake master cylinder reservoir filler cap assembly.
    b. Add brake fluid into the reservoir between MAX and MIN level on the brake fluid reservoir.

➡**Standard brake fluid is SAE J1703 or FMVSS No. 116 DOT3.**

    c. Connect the Techstream to the DLC3 and turn the power switch on (IG).
    d. Turn the Techstream on and enter the following menus: Chassis / ABS/VSC/TRAC / Air Bleeding.
    e. Select the "Usual air bleeding" on the Techstream display, and bleed air from the brake fluid following the instructions on the Techstream.
    f. After air bleeding, tighten each bleeder plug. Tighten the front bleeder plug to 73 inch lbs. (8.3 Nm). Tighten the rear bleeder plug to 8 ft. lbs. (11 Nm).
    g. Clear the DTC's.
    h. Turn the Techstream off and turn the power switch off.

3. Inspect for brake fluid leaks.
4. Install the brake master cylinder reservoir filler cap.
5. Install the center cowl top ventilator cover.
   a. Engage the 2 claws and 3 guides to install the center cowl top ventilator cover.
   b. Slide the hood to cowl top seal to engage the claw.

FLUID FILL PROCEDURE

➡Performing the following procedure will select ECB (Electronically Controlled Brake system) Invalid Mode without using the Techstream.

➡ECB (Electronically Controlled Brake system) Invalid Mode allows the brake fluid to be replaced without using the Techstream.

➡The brake warning light / yellow will blink to indicate when ECB (Electronically Controlled Brake system) Invalid Mode is selected.

➡Be sure to inspect that the brake warning light / yellow is blinking while replacing the brake fluid.

➡When one of the following conditions is met, ECB (Electronically Controlled Brake system) Invalid Mode is cancelled, and then the DTCs may be stored. So do not cancel the ECB (Electronically Controlled Brake system) Invalid Mode while replacing brake fluid.

c. The shift lever is used to select from P to any other position.
d. Turn the power switch on (READY).
e. Turn the power switch off.
f. The parking brake is released.
g. The vehicle velocity is not 0 mph.

➡Do not rotate the brake disc while ECB (Electronically Controlled Brake system) Invalid Mode is selected.

➡When replacing the brake fluid from the brake line, do not depress the brake pedal to operate the brake booster pump more than 100 seconds. If the brake booster pump is operated more than 100 seconds, ECB (Electronically Controlled Brake system) Invalid Mode is automatically finished and the DTCs may be stored.

➡Add brake fluid carefully and check that the reservoir level remains between the MIN and MAX lines.

➡Do not stand the fluid can on the reservoir inlet. Doing so will cause brake fluid to overflow.

1. Remove the 4 wheels.
2. Select ECB (Electronically Controlled Brake System) Invalid Mode.
3. Perform the following procedures listed below in 1 minute.
   a. Turn the power switch on (IG) with park (P) selected and parking brake applied.
   b. Select N and then depress the brake pedal more than 8 times in 5 seconds.
   c. Push the P position switch and then depress the brake pedal more than 8 times in 5 seconds.
   d. Select N and then depress the brake pedal more than 8 times in 5 seconds.
   e. Push the P position switch.
4. Check that the brake warning light/yellow is blinking.
5. Remove the center cowl top ventilator cover.
   a. Slide the hood to cowl top seal and disengage the claw.
   b. Disengage the 2 claws and 3 guides, and remove the center cowl top ventilator cover.
6. Replace the brake fluid.
   a. Remove the brake master cylinder reservoir filler cap assembly.
   b. Add brake fluid into the reservoir between MAX and MIN level on the brake fluid reservoir.
   c. Connect a vinyl tube to the bleeder plug of the front disc brake cylinder assembly RH.
   d. Depress the brake pedal several times, and then loosen the bleeder plug with the pedal depressed.
   e. When fluid stops coming out, tighten the bleeder plug, and then release the brake pedal.
   f. Repeat the previous 2 steps until all the air in the brake fluid is completely bled out and a new brake fluid comes out.
   g. Tighten the bleeder plug completely. Tighten to 73 inch lbs. (8.3 Nm).
   h. Replace the brake fluid from the front disc brake cylinder assembly LH using the same procedure as for RH.
   i. Connect a vinyl tube to the bleeder plug of the rear disc brake cylinder assembly LH.
   j. Loosen the bleeder plug while depressing and holding the brake pedal, and replace the brake fluid while the brake booster pump assembly and solenoid running.

➡Be sure to keep the brake pedal depressed.

➡Do not depress the brake pedal to operate the brake booster pump more than 100 seconds. When performing this procedure continuously, release the brake pedal to stop the brake booster pump operating and depress the brake pedal again.

   k. Tighten the bleeder plug, then release the brake pedal.
   l. Repeat the previous 2 steps until all the air in the brake fluid is completely bled out and a new brake fluid comes out.
   m. Tighten the bleeder plug completely. Tighten to 8 ft. lbs. (11 Nm).
7. Replace the brake fluid from the rear disc brake cylinder assembly RH using the same procedure as for LH.
   a. Turn the power switch off.
8. Inspect for brake fluid leaks.
9. Adjust the brake fluid level in the reservoir.
10. Install the brake master cylinder reservoir filler cap.
11. Install the center cowl top ventilator cover.
   a. Engage the 2 claws and 3 guides to install the center cowl top ventilator cover.
   b. Slide the hood to cowl top seal to engage the claw.
12. Install the 4 wheels. Tighten to 76 ft. lbs. (103 Nm).

CAUTION

Dust and dirt accumulating on brake parts during normal use may contain asbestos fibers from production or aftermarket brake linings. Breathing excessive concentrations of asbestos fibers can cause serious bodily harm. Exercise care when servicing brake parts. Do not sand or grind brake lining unless equipment used is designed to contain the dust residue. Do not clean brake parts with compressed air or by dry brushing. Cleaning should be done by dampening the brake components with a fine mist of water, then wiping the brake components clean with a dampened cloth. Dispose of cloth and all residue containing asbestos fibers in an impermeable container with the appropriate label. Follow practices prescribed by the Occupational Safety and Health Administration (OSHA) and the Environmental Protection Agency (EPA) for the handling, processing, and disposing of dust or debris that may contain asbestos fibers.

BRAKE CALIPER

REMOVAL & INSTALLATION

See Figure 3.

1. Before servicing the vehicle, refer to the precautions.
2. With the power switch **OFF** to prohibit brake control, remove the No. 1 and No. 2 motor relays.

➡If the pump motor operates while there is air remaining inside the brake actuator hose, the air will enter the actuator, resulting in difficulty in air bleeding.

3. Raise and safely support the front of the vehicle.
4. Remove the front wheel.
5. Remove the bolt and the gasket attaching the brake hose to the caliper.
6. Disconnect the brake hose, and plug the openings in the caliper and the brake hose to prevent fluid loss and contamination.
7. Hold the front disc brake cylinder slide pin using a wrench.
8. Remove the 2 caliper mounting bolts and remove the caliper from the vehicle.
9. Remove the 2 disc brake pads from the front disc brake cylinder mounting.
10. Remove the anti-squeal shims from the disc brake pads.
11. Remove the front disc brake pad support plates.
12. Remove the slide pins and bushing dust boots from the front disc brake cylinder mounting, if necessary.
13. Remove the front disc brake cylinder mounting from the steering knuckle, if necessary.

To install:

14. Install the front disc brake cylinder mounting to the steering knuckle, if removed. Torque the 2 mounting bolts to 81 ft. lbs. (109 Nm).
15. Install new bushing dust boots and slide pins into the front disc brake cylinder mounting, if necessary. Apply lithium soap base glycol grease to the sealing, sliding and fitting areas before installation.
16. Install the front disc brake pad support plates.
17. Install the anti-squeal shims onto the disc brake pads.

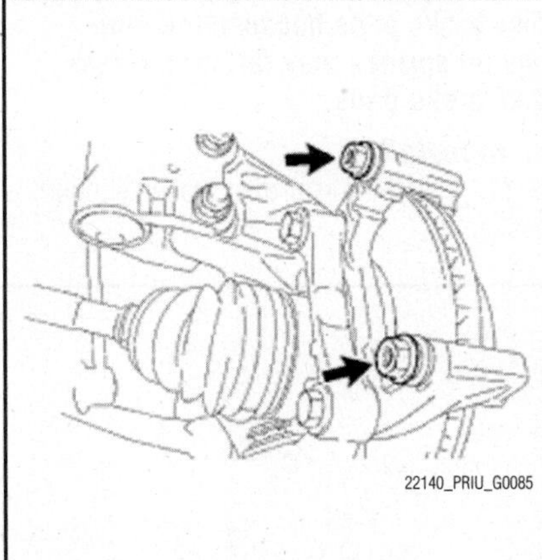

Fig. 3 Remove the front disc brake cylinder slide pins from the mounting

18. Install the disc brake pads onto the front disc brake cylinder mounting.
19. Install the brake caliper onto the mounting along with the 2 caliper mounting bolts. Torque the 2 mounting bolts to 25 ft. lbs. (34 Nm).
20. Connect the brake hose to the caliper with bolt and a new gasket. Tighten the caliper brake hose bolt to 24 ft. lbs. (33 Nm).
21. Fill the master cylinder to the proper level with clean brake fluid.
22. Bleed air from front and rear brake systems.
23. Recheck the fluid level.
24. Install the wheel and lower the vehicle.
25. Repeatedly press the brake pedal to bring the pads in contact with the rotor.
26. Check and clear the DTCs.

DISC BRAKE PADS

REMOVAL & INSTALLATION

1. Before servicing the vehicle, refer to the precautions.
2. Raise and safely support the front of the vehicle.
3. Remove the front wheel.
4. Hold the front disc brake cylinder slide pin using a wrench.

➡Caliper removal is not necessary to service the brake pads.

5. Remove the 2 caliper mounting bolts and separate the caliper from the mounting and support from the vehicle with wire. Do not allow the caliper to hang low enough so to put tension on the brake hose.
6. Remove the brake pads.

To install:

7. Install the anti-squeal shims onto the disc brake pads.
8. Install the disc brake pads.
9. Install the brake caliper onto the mounting along with the 2 caliper mounting bolts. Torque the 2 mounting bolts to 25 ft. lbs. (34 Nm).
10. Install the wheel.
11. Lower the vehicle.
12. Repeatedly press the brake pedal to bring the pads in contact with the rotor.

CAUTION

Dust and dirt accumulating on brake parts during normal use may contain asbestos fibers from production or aftermarket brake linings. Breathing excessive concentrations of asbestos fibers can cause serious bodily harm. Exercise care when servicing brake parts. Do not sand or grind brake lining unless equipment used is designed to contain the dust residue. Do not clean brake parts with compressed air or by dry brushing. Cleaning should be done by dampening the brake components with a fine mist of water, then wiping the brake components clean with a dampened cloth. Dispose of cloth and all residue containing asbestos fibers in an impermeable container with the appropriate label. Follow practices prescribed by the Occupational Safety and Health Administration (OSHA) and the Environmental Protection Agency (EPA) for the handling, processing, and disposing of dust or debris that may contain asbestos fibers.

BRAKE CALIPER

REMOVAL & INSTALLATION

See Figure 4.

1. Before servicing the vehicle, refer to the precautions.

➡When the brake pedal is first depressed after replacing the brake pads or pushing back the disc brake piston, DTC C1214 may be output. As there is no malfunction, clear the DTC.

➡Use the same procedure for the LH side and RH side.

➡The following procedure is for the LH side.

2. Disable the brake control.
3. Remove the rear wheel.

➡If brake fluid leaks onto any painted surface, immediately wash it off.

4. Remove the front door scuff plate LH.
5. Remove the cowl side trim sub assembly LH.
6. Remove the lower instrument panel finish panel assembly.
7. Loosen the parking brake cable.
   a. Completely raise the parking brake pedal.
   b. Loosen the lock nut and adjusting nut to completely release the parking brake cable.
8. Disconnect the No. 3 parking brake cable assembly.
   a. Separate the No. 3 parking brake cable assembly from the rear disc brake cylinder assembly.
   b. Separate the No. 3 parking brake cable assembly from the rear disc brake cylinder assembly.

➡Insert an offset wrench (14 mm) at the base of the No. 3 parking brake cable assembly as shown in the illustration to disengage the clip. Pull out the No. 3 parking brake cable assembly from the rear disc brake cylinder assembly.

9. Separate the rear flexible hose.
   a. Remove the union bolt and gasket, and separate the rear flexible hose from the rear disc brake cylinder assembly.
10. Remove the rear disc brake cylinder assembly.
    a. Hold the rear disc brake pad guide pin, and remove the 2 bolts and rear disc brake cylinder assembly.

➡Remove the rear disc brake cylinder assembly while holding both of the rear disc brake pads because the anti-squeal springs may fall off the rear disc brake pads.

To install:

11. Install the rear disc brake cylinder assembly.

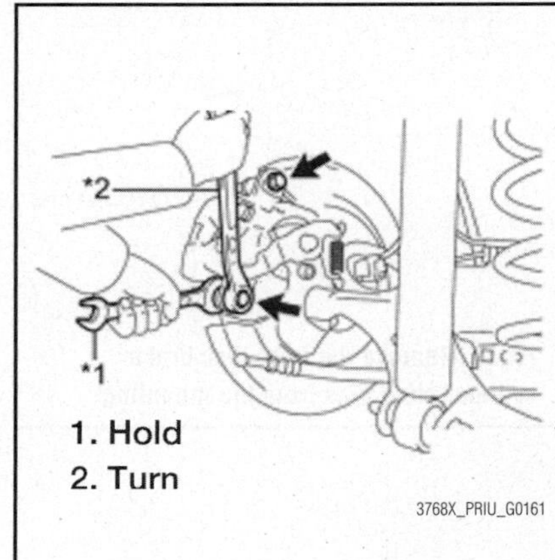

Fig. 4 Removing the rear disc brake cylinder

   a. To compensate for pad wear when reusing the pad, use SST to turn the piston to the position where the protrusion on the pad lines up properly with the piston groove.

➡Place the disc between the 2 brake pads and determine the piston return value.

   b. Hold the rear disc brake cylinder pad guide pin, and install the rear disc brake cylinder assembly to the rear disc brake cylinder mounting with the 2 bolts. Tighten to 25 ft. lbs. (34 Nm).

➡Install the rear disc brake cylinder assembly while holding both of the rear disc brake pads because the anti-squeal springs may fall off the rear disc brake pads.

➡Be sure that the anti-squeal springs are installed to the rear disc brake pads.

12. Connect the rear flexible hose.
    a. Connect the rear flexible hose to the rear disc brake cylinder assembly with a new union bolt and a new gasket. Tighten to 25 ft. lbs. (33 Nm).

➡Install the flexible hose lock securely into the lock hole in the disc brake cylinder.

13. Connect the No. 3 parking brake cable assembly to the rear disc brake cylinder assembly.
    a. Connect the No. 3 parking brake cable assembly to the rear disc brake cylinder assembly.
14. Disconnect the cable from the negative battery terminal.

➡Perform this step only when the Techstream cannot prohibit brake control.

15. Connect the connector.
    a. Connect the connector to the brake booster with the master cylinder assembly.

➡Make sure that the connector can be connected smoothly. Do not allow water, oil or dirt to enter.

➡Make sure that the connector lock is locked securely.

16. Connect the cable to the negative battery terminal.
17. Bleed the brake line.
18. Perform the initialization and calibration of the linear solenoid valve.

19. Adjust the parking brake.
20. Install the lower instrument panel finish panel assembly.
21. Install the cowl side trim sub assembly LH.
22. Install the front door scuff plate LH.
23. Install the rear wheel. Tighten to 76 ft. lbs. (103 Nm).

DISC BRAKE PADS

REMOVAL & INSTALLATION

See Figure 5.

1. Before servicing the vehicle, refer to the precautions.

➡When the brake pedal is first depressed after replacing the brake pads or pushing back the disc brake piston, DTC C1214 may be output. As there is no malfunction, clear the DTC.

➡Use the same procedure for the LH side and RH side.

➡The following procedure is for the LH side.

2. Disable the brake control.
3. Remove the rear wheel.
4. Remove the rear brake caliper.
5. Remove the rear disc brake pad.
   a. Remove the 2 anti-squeal springs from the rear disc brake pads.
   b. Remove the 2 rear disc brake pads from the rear disc brake cylinder mounting.
6. Remove the rear disc brake anti-squeal shim.
   a. Remove the rear No. 1 disc brake anti-squeal shim and rear No. 2 disc brake anti-squeal shim from each rear disc brake pad.
7. Remove the rear disc brake pad support plate.

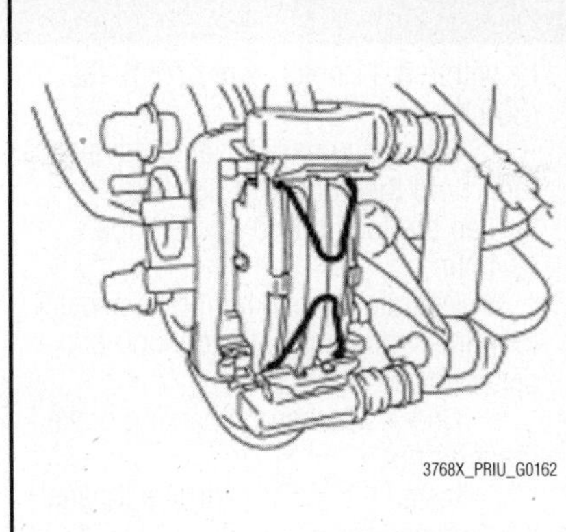

Fig. 5 Removing the rear disc pad

   a. Remove the 2 rear disc brake pad support plates from the disc brake cylinder mounting.

➡Each rear disc brake pad support plate has a different shape. Be sure to put an identification mark on each rear disc brake pad support plate so that it can be installed to its original position.

8. Remove the rear disc brake pad guide pin.
   a. Remove the 2 rear disc brake pad guide pins from the rear disc brake cylinder mounting.

To install:

9. Install the rear disc brake pad guide pin.
   a. Apply a light layer of lithium soap base glycol grease to the sliding and sealing surfaces of the 2 rear disc brake pad guide pins.
   b. Install the 2 rear disc brake pad guide pins to the rear disc brake cylinder mounting.
10. Install the rear disc brake pad support plate.
   a. Install the 2 rear disc brake pad support plates to the rear disc brake cylinder mounting.

➡Be sure to install each rear disc brake pad support plate in the correct position and direction.

11. Install the rear disc brake anti-squeal shim.
   a. Apply disc brake grease to the back plate of the rear disc brake pads.
   b. Install the rear No. 1 disc brake anti-squeal shim and rear No. 2 disc brake anti-squeal shim to each rear disc brake pad.

➡When replacing worn pads, the anti-squeal shims must be replaced together with the pads.

➡Apply disc brake grease to the area that contacts the anti-squeal shim.

➡Disc brake grease may seep out slightly from the areas where the anti-squeal shims are installed.

➡Make sure that disc brake grease is not applied onto the lining surface.

12. Install the rear disc brake pad.
   a. Install the 2 rear disc brake pads to the rear disc brake cylinder mounting.

➡There should be no oil or grease on the friction surface of the disc brake pads or the rear disc.

   b. Install the 2 anti-squeal springs to the rear disc brake pads.

➡When replacing the rear disc brake pads with new ones make sure to replace the anti-squeal springs at the same time.

➡Be sure to install the anti-squeal springs into the rear disc brake pad installation holes as far as they will go.

13. Install the rear disc brake cylinder.

PARKING BRAKE CABLES

ADJUSTMENT

See Figure 6.

1. Remove the lower instrument panel finish panel assembly.
2. Completely release the parking brake pedal.
   a. Loosen the lock nut and the adjusting nut to completely release the parking brake cable.
   b. Turn the adjusting nut until the parking brake pedal travel is corrected to be within 8-11 notches at 67.5 ft. lbs. (300 N).
   c. Using a wrench or an equivalent tool, hold the adjusting nut and tighten the lock nut to 48 inch lbs. (5.4 Nm).
   d. Operate the parking brake pedal 3 to 4 times, and check the parking brake pedal travel.
   e. Check whether the parking brake drags or not.
   f. Install the lower instrument panel finish panel assembly.

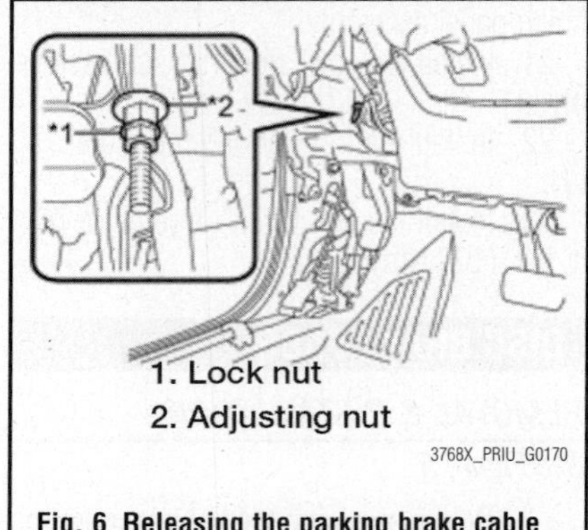

Fig. 6 Releasing the parking brake cable

GENERAL INFORMATION

⁂ CAUTION

These vehicles are equipped with an air bag system. The system must be disarmed before performing service on, or around, system components, the steering column, instrument panel components, wiring and sensors. Failure to follow the safety precautions and the disarming procedure could result in accidental air bag deployment, possible injury and unnecessary system repairs.

SERVICE PRECAUTIONS

Disconnect and isolate the battery negative cable before beginning any airbag system component diagnosis, testing, removal, or installation procedures. Allow system capacitor to discharge for two minutes before beginning any component service. This will disable the airbag system. Failure to disable the airbag system may result in accidental airbag deployment, personal injury, or death.

Do not place an intact undeployed airbag face down on a solid surface. The airbag will propel into the air if accidentally deployed and may result in personal injury or death.

When carrying or handling an undeployed airbag, the trim side (face) of the airbag should be pointing away from the body to minimize possibility of injury if accidental deployment occurs. Failure to do this may result in personal injury or death.

Replace airbag system components with OEM replacement parts. Substitute parts may appear interchangeable, but internal differences may result in inferior occupant protection. Failure to do so may result in occupant personal injury or death.

Wear safety glasses, rubber gloves, and long sleeved clothing when cleaning powder residue from vehicle after an airbag deployment. Powder residue emitted from a deployed airbag can cause skin irritation. Flush affected area with cool water if irritation is experienced. If nasal or throat irritation is experienced, exit the vehicle for fresh air until the irritation ceases. If irritation continues, see a physician.

Do not use a replacement airbag that is not in the original packaging. This may result in improper deployment, personal injury, or death.

The factory installed fasteners, screws and bolts used to fasten airbag components have a special coating and are specifically designed for the airbag system. Do not use substitute fasteners. Use only original equipment fasteners listed in the parts catalog when fastener replacement is required.

During, and following, any child restraint anchor service, due to impact event or vehicle repair, carefully inspect all mounting hardware, tether straps, and anchors for proper installation, operation, or damage. If a child restraint anchor is found damaged in any way, the anchor must be replaced. Failure to do this may result in personal injury or death.

Deployed and non-deployed airbags may or may not have live pyrotechnic material within the airbag inflator.

Do not dispose of driver/passenger/curtain airbags or seat belt tensioners unless you are sure of complete deployment. Refer to the Hazardous Substance Control System for proper disposal.

Dispose of deployed airbags and tensioners consistent with state, provincial, local, and federal regulations.

After any airbag component testing or service, do not connect the battery negative cable. Personal injury or death may result if the system test is not performed first.

If the vehicle is equipped with the Occupant Classification System (OCS), do not connect the battery negative cable before performing the OCS Verification Test using the scan tool and the appropriate diagnostic information. Personal injury or death may result if the system test is not performed properly.

Never replace both the Occupant Restraint Controller (ORC) and the Occupant Classification Module (OCM) at the same time. If both require replacement, replace one, then perform the Airbag System test before replacing the other.

Both the ORC and the OCM store Occupant Classification System (OCS) calibration data, which they transfer to one another when one of them is replaced. If both are replaced at the same time, an irreversible fault will be set in both modules and the OCS may malfunction and cause personal injury or death.

If equipped with OCS, the Seat Weight Sensor is a sensitive, calibrated unit and must be handled carefully. Do not drop or handle roughly. If dropped or damaged, replace with another sensor. Failure to do so may result in occupant injury or death.

If equipped with OCS, the front passenger seat must be handled carefully as well. When removing the seat, be careful when setting on floor not to drop. If dropped, the sensor may be inoperative, could result in occupant injury, or possibly death.

If equipped with OCS, when the passenger front seat is on the floor, no one should sit in the front passenger seat. This uneven force may damage the sensing ability of the seat weight sensors. If sat on and damaged, the sensor may be inoperative, could result in occupant injury, or possibly death.

DISARMING THE SYSTEM

To avoid personal injury when working on vehicles equipped with an air bag, the negative battery cable must be disconnected and at least 90 seconds must elapse before working on the system. Failure to do so may result in deployment of the air bag.

ARMING THE SYSTEM

To rearm the air bag system, reconnect the battery cable (s).

CLOCKSPRING CENTERING

See Figure 7.

1. Before servicing the vehicle, refer to the precautions.

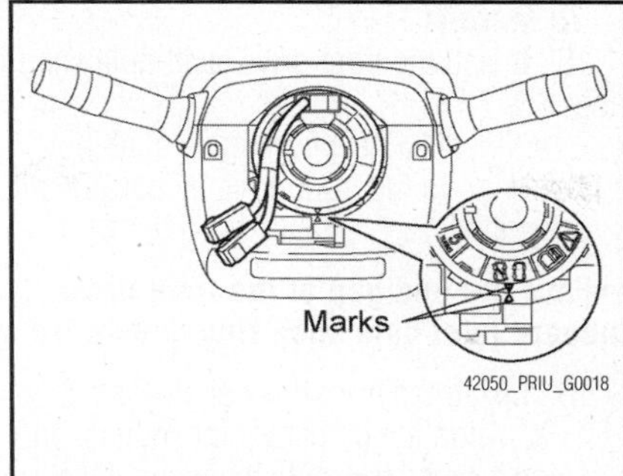

Fig. 7 Spiral cable alignment marks

2. Install the spiral cable.
   a. Check that the front wheels are facing straight ahead.
   b. Set the turn signal switch to the neutral position.

➡Make sure that the turn signal switch is in the neutral position, as the pin of the turn signal switch may be snapped.

   c. Install the spiral cable.

➡When replacing the spiral cable with a new one, remove the lock pin before installing the steering wheel.

   d. Connect the connector to the spiral cable.

⁂ WARNING

When handling the airbag connector, do not damage the airbag wire harness.

3. Slowly rotate the spiral cable counterclockwise by hand until it feels firm.
4. Rotate the spiral cable clockwise approximately 2.5 turns to align the marks.
5. The spiral cable should rotate approximately 2.5 turns to the left and right from the center.

DRIVE TRAIN

AUTOMATIC TRANSAXLE FLUID

DRAIN AND REFILL

See Figure 8.

1. Drain the hybrid transaxle fluid.
   a. Using a 10 mm hexagon socket wrench, remove the filler plug and gasket.
   b. Using a 10 mm hexagon socket wrench, remove the drain plug and gasket.
   c. Using a 10 mm hexagon socket wrench, install the drain plug, and a new gasket. Tighten to 29 ft. lbs. (39 Nm).
2. Add hybrid transaxle fluid.

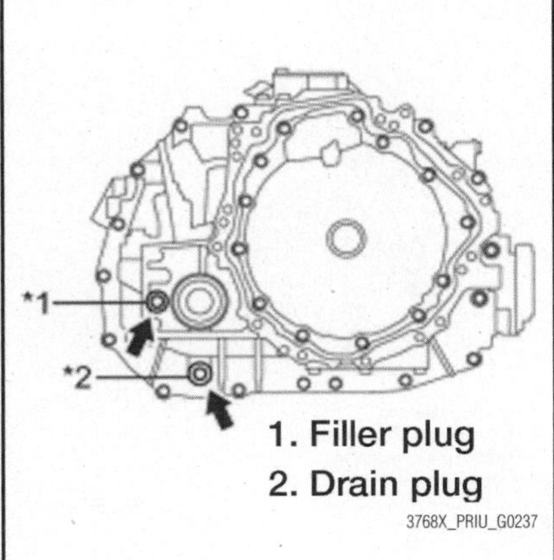

Fig. 8 Locating the filler and drain plug

   a. Add transaxle fluid until the fluid level is between 0 to 10 mm (0 to 0.394 in.) from the bottom lip of the filler plug opening.

➡Stop the vehicle on a flat road.

➡Recheck the transaxle fluid level after driving when exchanging fluid.

➡Insufficient or excessive amounts of transaxle fluid may be the cause of some trouble.

3. Inspect the hybrid transaxle fluid.
   a. Check that the fluid level is between 0-0.394 inch (0 to 10 mm) from the lowest position of the inner surface of the transaxle filler plug opening.

➡Stop the vehicle on a flat road.

➡Recheck the transaxle fluid level after driving when exchanging fluid.

➡Insufficient or excessive amounts of transaxle fluid may be the cause of some trouble.

   b. Check for leaks if the quantity of transaxle fluid is low.
   c. Using a 10 mm hexagon socket wrench, install the filler plug with a new gasket. Tighten to 29 ft. lbs. (39 Nm).

FRONT HALFSHAFT

REMOVAL & INSTALLATION

See Figure 9.

➡Use the same procedure for the RH side and LH side.

➡The procedure listed below is for the LH side.

1. Remove the front wheels.
2. Remove No. 1 engine under cover.
3. Remove the rear engine under cover LH.
4. Remove the rear engine under cover RH.
5. Drain the hybrid transaxle fluid.
6. Remove the front axle shaft nut.
   a. Using the special tool (09930-00010) and a hammer, release the staked part of the front axle shaft nut.

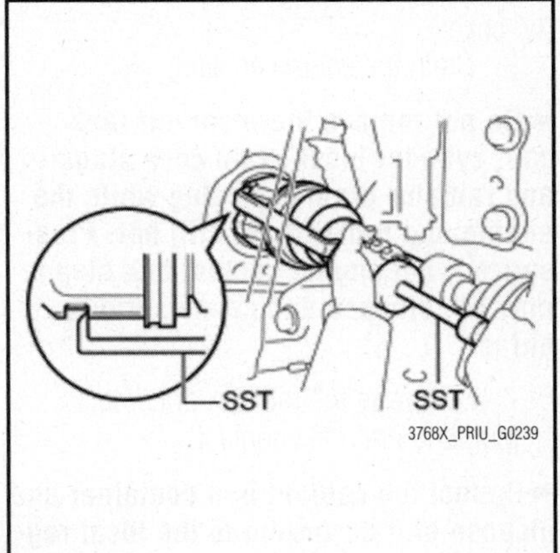

Fig. 9 Removing the front drive shaft assembly

➡Loosen the staked part of the nut completely, otherwise the threads of the drive shaft may be damaged.

b. While applying the brakes, remove the front axle shaft nut.

7. Separate the front speed sensor.
8. Separate the front flexible hose.
9. Separate the tie rod end sub assembly.
10. Separate the front stabilizer link assembly.
11. Separate the front No. 1 lower suspension arm sub assembly.
12. Separate the front drive shaft assembly.
13. Remove the front drive shaft assembly.

a. Using the special tool (09520-00031, 09520-01010, 09521-00020), remove the front drive shaft assembly.

➡Do not damage the transaxle case oil seal.

➡Do not damage the inboard joint boot.

➡Do not drop the front drive shaft assembly.

14. Remove the front drive shaft hole snap ring.

a. Using a screwdriver, remove the front drive shaft hole snap ring.

To install:

15. Install the front drive shaft hole snap ring.

a. Install a new front drive shaft hole snap ring to the front drive inboard joint assembly.

➡Face the end gap of the front drive inboard joint hole snap ring downward.

16. Install the front drive shaft assembly.

a. Align the inboard joint splines, and using a brass bar and a hammer, install the front drive shaft assembly.

➡Face the end gap of the front drive shaft hole snap ring downward.

➡Do not damage the transaxle case oil seal.

➡Do not damage the inboard joint boot.

➡Make sure to center the front drive shaft assembly during installation to prevent damage to the front drive shaft hole snap ring.

➡Confirm whether the drive shaft is securely driven in by checking the reaction force and sound.

b. Align the matchmarks and install the front drive shaft assembly to the front axle hub sub-assembly.

17. Connect the front No. 1 lower suspension arm sub assembly.
18. Connect the front stabilizer link assembly.
19. Connect the tie rod end sub assembly.
20. Install the front flexible hose.
21. Connect the front speed sensor.
22. Install the front axle shaft nut.

a. Clean the threaded parts on the drive shaft and a new axle shaft nut using a non-residue solvent.

➡Be sure to perform this work even when using a new drive shaft.

➡Keep the threaded parts free of oil and foreign matter.

b. Using a socket wrench (30 mm), install the axle shaft nut. Tighten to 159 ft. lbs. (216 Nm).

c. Using a chisel and hammer, stake the front axle shaft nut.

23. Add hybrid transaxle fluid.
24. Inspect hybrid transaxle fluid.
25. Install the front wheels. Tighten to 76 ft. lbs. (103 Nm).
26. Inspect and adjust front wheel alignment.
27. Install the rear engine under cover LH.
28. Install the rear engine under cover RH.
29. Install the No. 1 engine under cover.
30. Inspect the speed signal.

ENGINE COOLING

ENGINE COOLANT

DRAIN & REFILL PROCEDURE

See Figure 10.

1. Remove the front spoiler cover (w/ front cover).
2. Remove the engine under cover (w/ cover).
3. Drain the engine coolant.

➡Do not remove the reservoir tank cap, cylinder block drain cock plugs and radiator drain cock plug while the engine and radiator are still hot. Pressurized, hot engine coolant and steam may be released and cause serious burns.

a. Loosen the radiator drain cock plug and drain the coolant.

➡Collect the coolant in a container and dispose of it according to the local regulations.

b. Loosen the cylinder block drain cock plug.

➡The plug is on the back of the generator on the exhaust manifold side.

4. Add engine coolant.

a. Tighten the radiator drain cock plug.

b. Tighten the cylinder block drain cock plug to 9 ft. lbs. (13 Nm).

c. Remove the reservoir tank cap.

d. Connect the hose to the air release valve.

e. Loosen the air release valve.

f. Add TOYOTA Super Long Life Coolant (SLLC) to the reservoir tank filler opening until it overflows from the air release valve. Then tighten the air release valve. For vehicles with the exhaust heat recirculation system add 7.7 qts (7.3 L). Vehicles without the exhaust heat recirculation system add 6.8 qts (6.5 L).

➡TOYOTA vehicles are filled with TOYOTA SLLC at the factory. In order to avoid damage to the engine cooling system and other technical problems, only use TOYOTA SLLC or similar high quality ethylene glycol based non-silicate, non-amine, non-nitrite, non-borate coolant with long-life hybrid organic acid technology (coolant with long-life hybrid organic acid technology is a combination of low phosphates and organic acids).

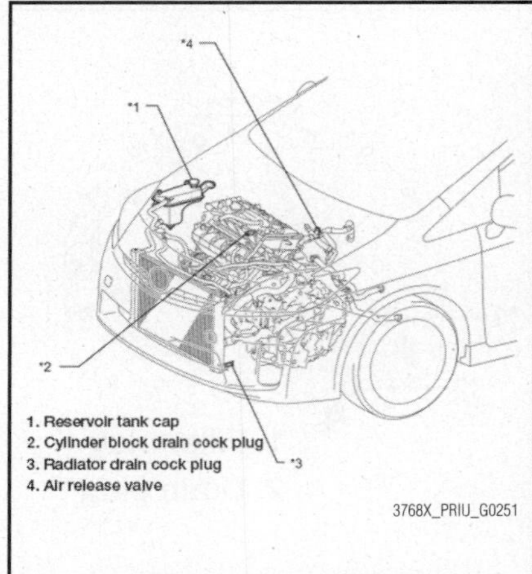

Fig. 10 Adding coolant to the B line of the reservoir tank

➡Contact your TOYOTA dealer for further details.

➡Never use water as a substitute for engine coolant.

g. Disconnect the hose from the air release valve.
h. Add coolant to the B (*1) line of the reservoir tank.
i. Squeeze the inlet and outlet radiator hoses several times by hand, and then check the level of the coolant. If the coolant level is low, add coolant.
j. Put the engine in inspection mode.
k. Install the reservoir tank cap, and warm up the engine sufficiently.
l. Bleed air from the cooling system.

➡Before starting the engine, turn the A/C switch off.

➡Adjust the heater control to the maximum hot setting.

➡Adjust the blower speed to low setting.

m. Warm up the engine until the thermostat opens. While the thermostat is open, allow the coolant to circulate for several minutes.

➡The thermostat opening timing can be confirmed by squeezing the inlet radiator hose by hand, and sensing vibrations when the engine coolant starts to flow inside the hose.

✲✲ WARNING

When squeezing the radiator hose; wear protective gloves.

✲✲ WARNING

When squeezing the radiator hose; be careful as the radiator hoses are hot.

✲✲ WARNING

When squeezing the radiator hose; keep your hands away from the radiator fan.

n. After the engine is warmed up, let it idle for 7 minutes or more.
o. Squeeze the inlet and outlet radiator hoses several times by hand to bleed the air from the system.
p. After the engine has cooled down, check that the coolant level is between full and low. If the coolant level is low, add coolant to the full line on the reservoir.

5. Inspect for coolant leaks.
6. Install the engine under cover.
7. Install the front spoiler.

ENGINE FAN

REMOVAL & INSTALLATION

See Figures 11 and 12.

1. Remove the radiator.
2. Remove the fan.
   a. Remove the nut and fan.
3. Remove the No. 2 fan.

To install:

4. To install, reverse the removal procedure.

RADIATOR

REMOVAL & INSTALLATION

1.5L Engine

See Figures 13 through 17.

✲✲ CAUTION

After turning the ignition switch off, waiting time may be required before disconnecting the cable from the negative (-) auxiliary battery terminal. Therefore, make sure to read the disconnecting the cable from the negative (-) auxiliary battery terminal notice before proceeding with work.

1. Remove the front floor cover RH.
2. Disconnect the cable from the negative auxiliary battery terminal.

➡When disconnecting the cable, some systems need to be initialized after the cable is reconnected.

3. Remove the engine under covers LH and RH.
4. Drain the coolant (inverter and engine).
5. Remove the front bumper assembly.
6. Remove the radiator support opening cover.
7. Remove the air cleaner cap sub assembly.
8. Remove the air cleaner case sub assembly.

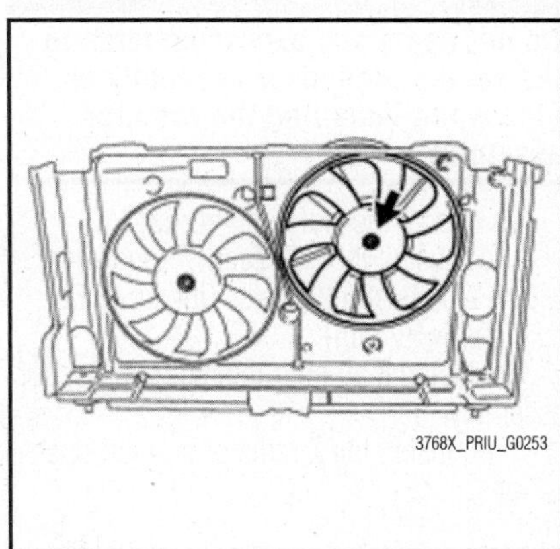

Fig. 11 Removing the fan

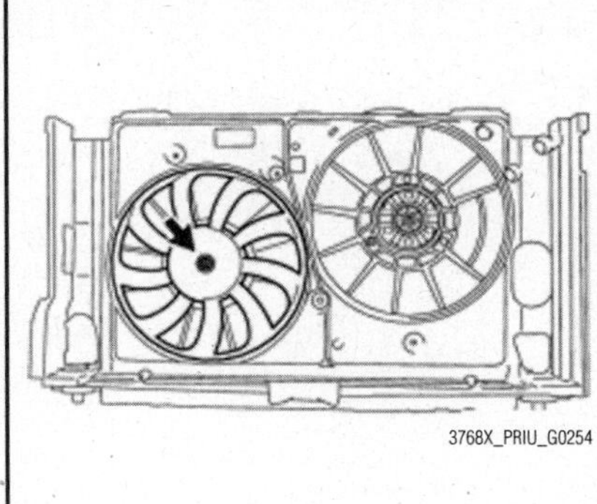

Fig. 12 Removing the No. 2 fan

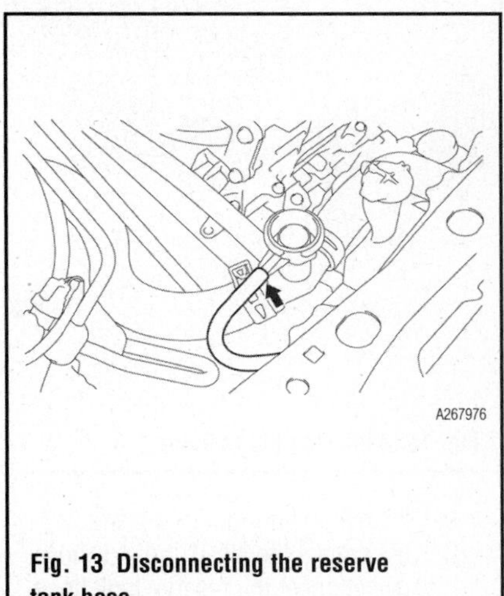

Fig. 13 Disconnecting the reserve tank hose

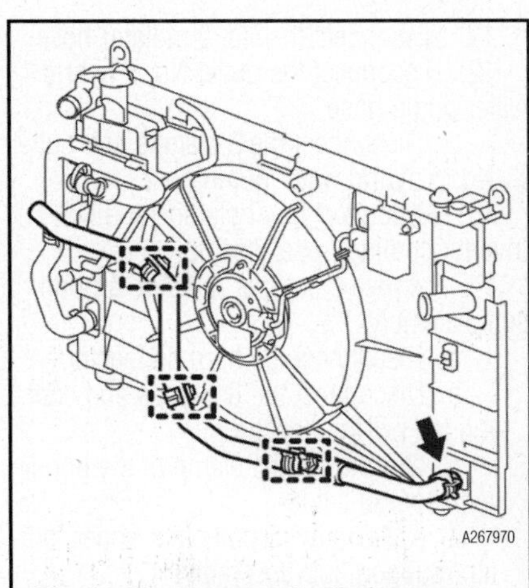

Fig. 14 Disconnecting the 3 clamps and outlet No. 1 hybrid water pump hose

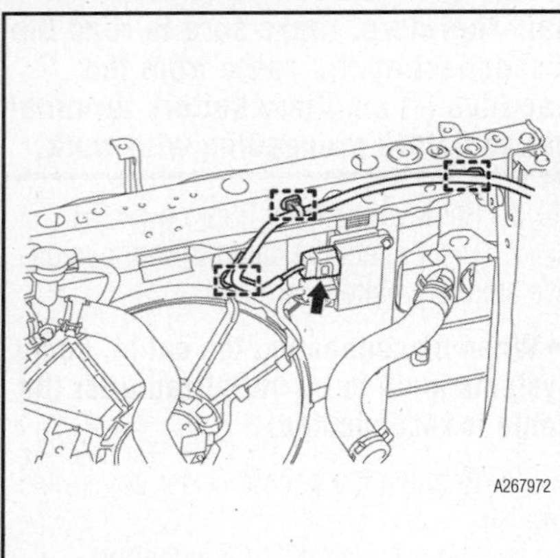

Fig. 15 Disconnecting the 3 clamps and cooling fan ECU connector

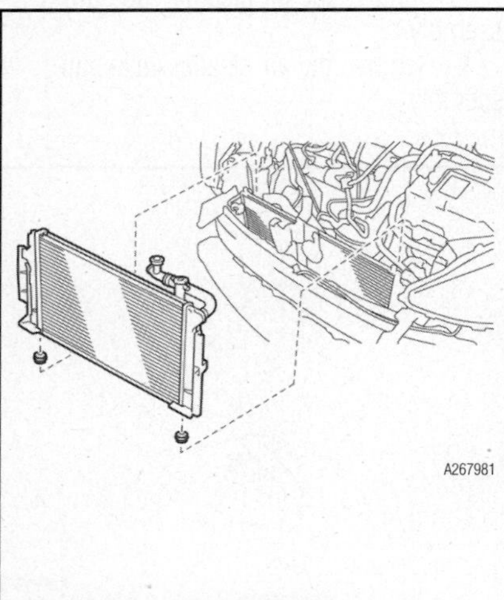

Fig. 16 Removing the radiator

9. Remove the hood lock assembly.
10. Disconnect the No. 1 radiator hose.
    a. Disconnect the reserve tank hose.
    b. Disconnect the No. 1 radiator hose.
11. Disconnect the No. 2 radiator hose.
12. Disconnect the outlet No. 1 hybrid water pump hose.
    a. Disconnect the 3 clamps and outlet No. 1 hybrid water pump hose.
13. Disconnect the 3 clamps and No. 1 inverter cooling hose assembly.
14. Remove the upper radiator support sub assembly.
    a. Disconnect the horn connector.
    b. Disconnect the 3 clamps and cooling fan ECU connector.
    c. Disconnect the clamp of the hood lock control cable.
    d. Remove the 5 bolts and upper radiator support sub-assembly.
    e. Remove the 2 radiator support cushions.
15. Remove the 2 bolts, 2 claws and No. 2 fan shroud.
16. Remove the radiator assembly with the fan shroud.

⁂ WARNING

Do not apply any excessive force to the cooler condenser assembly or pipe when removing the radiator assembly.

    a. Remove the 2 lower radiator supports.
    b. Remove the 2 bolts and No. 3 radiator hose.
    c. Remove the 2 claws and radiator assembly.

To install:

17. Install the radiator assembly.
    a. Install the radiator assembly with the 2 claws.
    b. Install the No. 3 radiator hose with the 2 bolts. Tighten to 66 inch lbs. (7.5 Nm).
    c. Install the 2 lower radiator supports.
    d. Install the radiator assembly with the fan shroud.

⁂ WARNING

Do not apply any excessive force to the cooler condenser assembly or pipe when installing the radiator assembly.

18. Install the No. 2 fan shroud.
    a. Install the No. 2 fan shroud with the 2 bolts and 2 claws. Tighten to 62 inch lbs. (7 Nm).
19. Install the upper radiator support sub assembly.
    a. Install the 2 radiator support cushions.

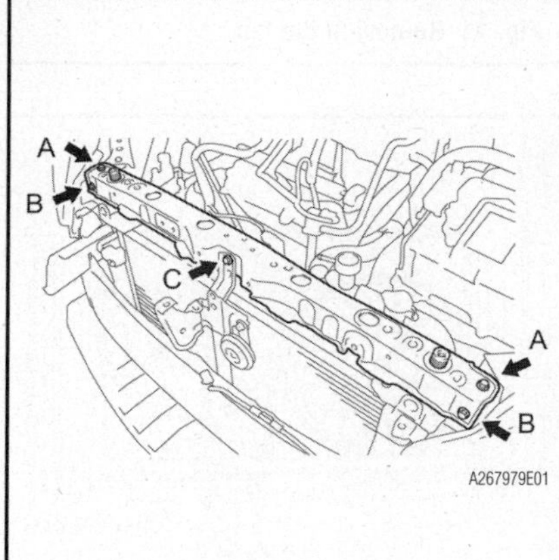

Fig. 17 Installing the upper radiator support sub assembly and identifying bolts A-C

    b. Temporarily install the upper radiator support sub assembly with the 5 bolts.
    c. Fully tighten bolt A to 49 inch lbs. (5.5 Nm).
    d. Fully tighten bolt B to 49 inch lbs. (5.5 Nm).
    e. Fully tighten bolt C to 9 ft. lbs. (13 Nm).
    f. Connect the clamp of the hood lock control cable.
    g. Connect the cooling fan ECU connector and 3 clamps.
    h. Connect the horn connector.
20. Connect the No. 1 inverter cooling hose assembly.
    a. Connect the No. 1 inverter cooing hose asse3mbly and 3 clamps.
21. Connect the outlet No. 1 hybrid water pump hose and 3 clamps.
22. Connect the No. 2 radiator hose.
23. Connect the No. 1 radiator hose and reserve tank hose.
24. Install the hood lock assembly.
25. Install the air cleaner case sub assembly.
26. Install the air cleaner cap sub assembly.
27. Install the radiator support opening cover.
28. Install the front bumper assembly.
29. Connect cable to negative auxiliary battery terminal.

➡When disconnecting the cable, some systems need to be initialized after the cable is reconnected.

30. Install the front floor cover RH.
31. Add coolant (inverter and engine).
32. Inspect for coolant leaks (inverter and engine).
33. Install the engine under covers LH and RH.

1.8L Engine

See Figures 18 through 22.

⁂ CAUTION

After turning the ignition switch off, waiting time may be required before disconnecting the cable from the negative (-) auxiliary battery terminal. Therefore, make sure to read the disconnecting the cable from the negative (-) auxiliary battery terminal notice before proceeding with work.

1. Remove the radiator support opening cover.
2. Remove the rear No. 2 floor board (separate type).
3. Remove the rear deck floor box.

4. Remove the rear No. 3 floor board.

5. Disconnect the cable from the negative battery terminal.

➡When disconnecting the cable, some systems need to be initialized after the cable is reconnected.

6. Remove the No. 1 engine under cover.

7. Drain the engine coolant.

8. Disconnect the No. 1 radiator hose from the radiator assembly.

9. Disconnect the No. 2 radiator hose from the radiator assembly.

10. Remove the front bumper assembly.

11. Remove the millimeter wave radar sensor assembly (w/dynamic radar cruise control system).

12. Remove the millimeter wave radar sensor bracket.

13. Remove the No. 1 inverter bracket.

14. Remove the hood lock support sub assembly.

a. Disconnect the water by pass hose clamp from the radiator support RH.

b. Disconnect the hood lock connector and hood lock control cable wire.

c. Disconnect the 3 wire harness clamps.

d. Disconnect the connector from the No. 2 cooling fan motor.

e. Disconnect the 2 wire harness clamps and connector from the fan shroud and cooling fan motor.

15. Disconnect the 2 horn connectors.

a. Remove the 2 bolts, radiator support RH and radiator support LH with the 2 cushions from the upper radiator support.

b. Remove the 4 bolts and upper radiator support.

16. Remove the air cleaner cap sub assembly.

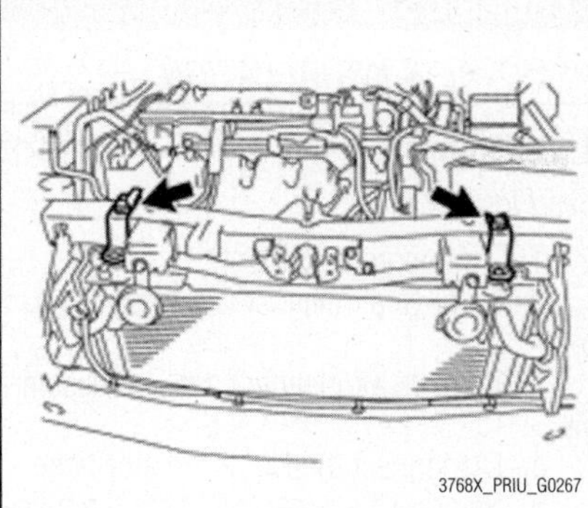

Fig. 19 Removing the radiator supports from the upper radiator support

17. Remove the inlet air cleaner assembly.

18. Remove the air cleaner case.

19. Remove the No. 2 fan shroud.

a. Disconnect the No. 1 water by pass hose from the radiator assembly.

b. Disconnect the water by pass hose from the radiator assembly.

c. Disconnect the 6 water by pass hose clamps from the No. 2 fan shroud.

d. Remove the 2 bolts and No. 2 fan shroud from the radiator assembly.

20. Remove the radiator assembly with the fan shroud.

➡For vehicles with the air conditioning system, do not apply any excessive force to the cooler condenser assembly or pipe when removing the radiator assembly.

a. Remove the 2 lower radiator supports.

b. Remove the 2 bolts.

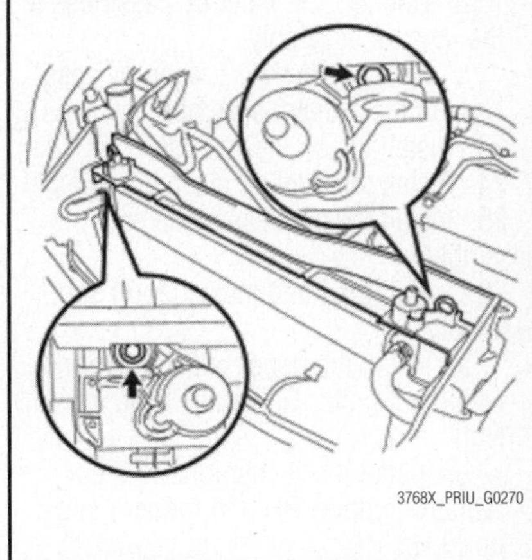

Fig. 21 Removing the 2 bolts and No. 2 fan shroud from the radiator assembly

c. Remove the fan shroud from the radiator assembly.

To install:

21. Install the radiator assembly with the 2 bolts. Tighten to 62 inch lbs. (7 Nm).

a. Install the fan shroud to the radiator assembly.

b. Install the 2 lower radiator supports.

c. Install the radiator assembly with the fan shroud.

➡For vehicles with the air conditioning system, do not apply any excessive force to the cooler condenser assembly or pipe when removing the radiator assembly.

22. Install the No. 2 fan shroud with the 2 bolts. Tighten to 62 inch lbs. (7 Nm).

a. Connect the 6 water by pass hose clamps to the No. 2 fan shroud.

Fig. 18 Disconnecting the horn connectors

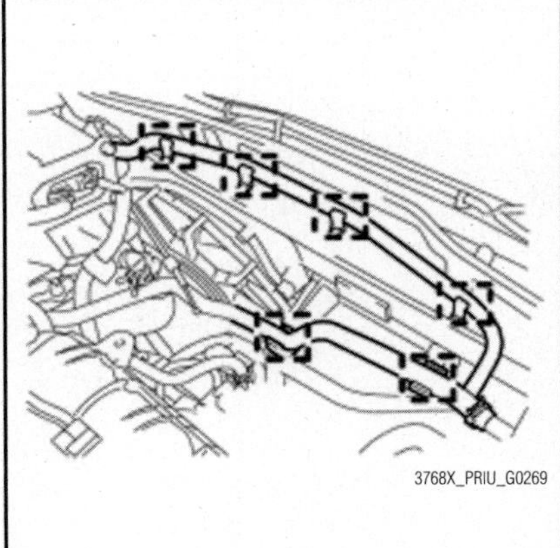

Fig. 20 Disconnecting the water by pass hose clamps from the No. 2 fan shroud

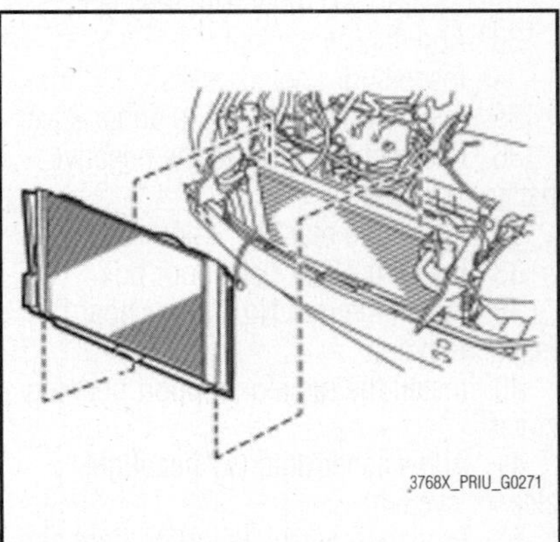

Fig. 22 Removing the radiator with the fan shroud

b. Connect the water by pass hose to the radiator assembly.

c. Connect the No. 1 water by pass hose to the radiator assembly.

23. Install the air cleaner case.
24. Install the inlet air cleaner assembly.
25. Install the air cleaner cap sub assembly.
26. Install the hood lock support sub assembly.

a. Install the upper radiator support with the 4 bolts. Tighten to 10 ft. lbs. (13 Nm).

b. Install the 2 cushions to the radiator support RH and radiator support LH.

c. Install the radiator support RH and radiator support LH with the 2 bolts. Tighten to 14 ft. lbs. (19 Nm).

d. Connect the 2 horn connectors.

e. Connect the 2 wire harness clamps and connector to the fan shroud and cooling fan motor.

f. Connect the connector to the No. 2 cooling fan motor.

g. Connect the hood lock connector and hood lock control cable wire.

h. Connect the 3 wire harness clamps.

i. Connect the water by pass hose clamp to the radiator support RH.

27. Install the No. 1 inverter bracket.
28. Install the millimeter wave radar sensor bracket.
29. Install the millimeter wave radar sensor assembly (w/ dynamic radar cruise control system).
30. Install the front bumper assembly.
31. Connect the No. 2 radiator hose.

a. Connect the No. 2 radiator hose to the radiator assembly with the clamp.

32. Connect the No. 1 radiator hose.

a. Connect the No. 1 radiator hose to the radiator assembly with the clamp.

33. Add engine coolant.
34. Inspect for coolant leaks.
35. Install the No. 1 engine under cover.
36. Connect the cable to the negative battery terminal.
37. Install the rear No. 3 floor board.
38. Install the rear deck floor box.
39. Install the rear No. 2 floor board (separate type).
40. Install the radiator support opening cover.
41. Add washer fluid (w/ headlight cleaner system).
42. Prepare the vehicle for fog light aim adjustment (w/ fog light).
43. Prepare for fog light aiming.
44. Inspect for fog light aiming.
45. Adjust the fog light aiming.

THERMOSTAT

REMOVAL & INSTALLATION

1.5L Engine

See Figure 23.

1. Drain engine coolant.
2. Remove air cleaner cap sub assembly.
3. Remove air cleaner case sub assembly.
4. Disconnect the No. 2 radiator hose.
5. Remove the water inlet with the thermostat sub assembly.

a. Disconnect the water inlet hose and the No. 3 water by-pass hose.

✲✲ WARNING

Do not apply force to the water inlet with thermostat sub-assembly when disconnecting the No. 3 water by-pass hose.

✲✲ WARNING

Do not damage the water inlet with thermostat sub-assembly.

b. Remove the 2 bolts and water inlet with thermostat sub assembly.

c. Remove the gasket from the water inlet with thermostat sub-assembly.

To install:

6. Install the water inlet with thermostat sub assembly.

a. Install a new gasket on the water inlet with thermostat sub assembly.

b. Install the water inlet with thermostat sub assembly with the 2 bolts. Tighten to 7 ft. lbs. (10 Nm).

c. Connect the water inlet hose and No. 3 water by-pass hose.

Fig. 23 Removing water inlet with thermostat sub assembly

7. Connect the No. 2 radiator hose.
8. Install the air cleaner case sub assembly.
9. Install the air cleaner cap sub assembly.
10. Add coolant (engine).
11. Inspect for coolant leaks (engine).

1.8L Engine

See Figure 24.

1. Remove the No. 1 engine under cover.
2. Remove the inlet air cleaner assembly.
3. Remove the engine oil level dipstick guide sub assembly.
4. Drain the engine coolant.
5. Remove the water inlet with the thermostat sub assembly.

a. Disconnect the No. 2 radiator hose and the No. 3 water by pass hose.

➡Do not apply force to the water inlet with thermostat sub-assembly when disconnecting the No. 3 water by-pass hose.

➡Do not damage the water inlet with thermostat sub-assembly.

➡When disconnecting the No. 3 water by-pass hose, pinch the hose clamp, rotate the hose and pull it straight off the pipe.

b. Remove the 2 nuts, bolt and water inlet with thermostat sub assembly.

c. Remove the gasket from the water inlet with thermostat sub assembly.

To install:

6. To install, reverse the removal procedure. Tighten the thermostat sub assembly to 7 ft. lbs. (10 Nm).

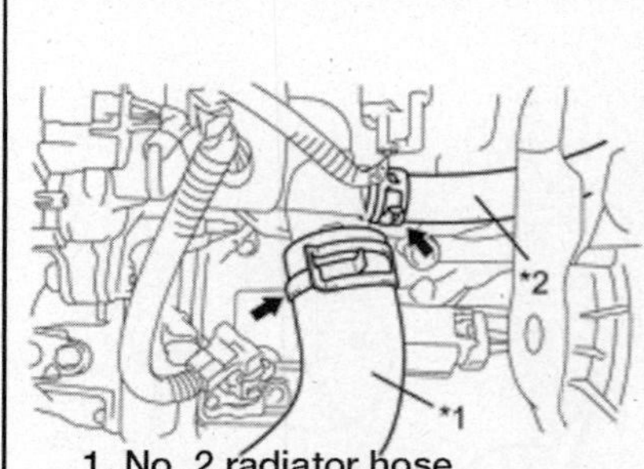

Fig. 24 Disconnecting the No. 2 radiator hose and the No. 3 water by pass hose

WATER PUMP

REMOVAL & INSTALLATION

1.5L Engine

See Figure 25.

1. Drain engine coolant.
2. Remove air cleaner cap sub assembly.
3. Remove air cleaner case sub assembly.
4. Remove the engine water pump assembly.
    a. Disconnect the connector.
    b. Disconnect the 2 clamps.
    c. Remove the 5 bolts and engine water pump assembly.
    d. Remove the gasket from the engine water pump assembly.

To install:

5. Install the engine water pump assembly.
    a. Install a new gasket.

➡Be sure to clean the contact surfaces.

    b. Install the engine water pump assembly with the 5 bolts. Tighten to 15 ft. lbs. (21 Nm).
    c. Connect the 2 clamps.
    d. Connect the connector.

WARNING

Do not start the engine water pump assembly before adding coolant.

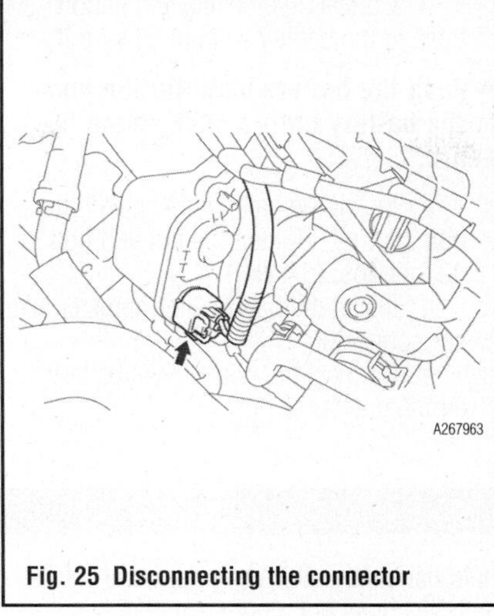

Fig. 25 Disconnecting the connector

6. Install the air cleaner case sub assembly.
7. Install the air cleaner cap sub assembly.
8. Add coolant (engine).
9. Inspect for coolant leaks (engine).

1.8L Engine

See Figure 26.

1. Remove the No. 1 engine under cover.
2. Remove the inlet air cleaner assembly.
3. Drain the engine coolant.
4. Remove the water pump assembly.

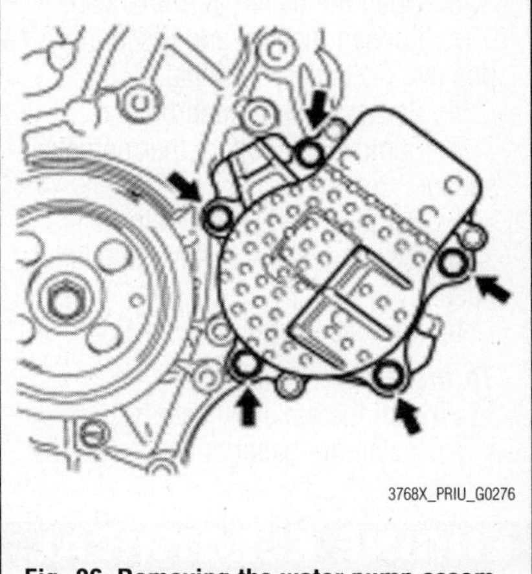

Fig. 26 Removing the water pump assembly from the timing chain cover

    a. Disconnect the water pump connector from the water pump assembly.
    b. Remove the 5 bolts and water pump assembly from the timing chain cover.
    c. Remove the water pump gasket from the water pump.

To install:

5. To install, reverse the removal procedure. Tighten the water pump assembly A bolts to 19 ft. lbs. (26 Nm). Tighten the water pump B bolts to 15 ft. lbs. (21 Nm).

ENGINE ELECTRICAL

BATTERY SYSTEM

BATTERY

REMOVAL & INSTALLATION

1.5L Engine

WARNING

After turning the ignition switch off, waiting time may be required before disconnecting the cable from the negative (-) auxiliary battery terminal. Therefore, make sure to read the disconnecting the cable from the negative (-) auxiliary battery terminal notice before proceeding with work.

1. Remove the front floor cover RH.
2. Disconnect the cable from the auxiliary battery negative terminal.

➡When disconnecting the cable, some systems need to be initialized after the cable is reconnected.

3. Remove the auxiliary battery.
    a. Remove the 2 nuts.
    b. Remove the battery clamp.
    c. Open the battery terminal cap.
    d. Loosen the nut, and disconnect the positive (+) battery terminal.
    e. Disconnect the battery ventilation hose.
    f. Remove the battery.
    g. Remove the battery insulator.

To install:

4. Install the auxiliary battery.
    a. Install the battery insulator.
    b. Install the battery.
    c. Connect the battery ventilation hose to the battery without kinking it.

➡Push in the battery ventilation hose to the battery until a click sound is heard.

    d. Connect the positive (+) battery terminal with the nut. Tighten to 48 inch lbs. (5.4 Nm).
    e. Install the battery terminal cap.
    f. Install the battery clamp with the 2 nuts. Tighten to 48 inch lbs. (5.4 Nm).

5. Connect the cable to the auxiliary battery negative terminal.
    a. Connect the negative battery terminal with the nut. Tighten to 48 inch lbs. (5.4 Nm).

➡When disconnecting the cable, some systems need to be initialized after the cable is reconnected.

6. Install the front floor cover RH.

1.8L Engine

1. Remove the rear No. 2 floor board (separate type).
2. Remove the rear deck floor box.
3. Remove the rear No. 3 floor board.
4. Disconnect the cable from the negative battery terminal.

➡When disconnecting the cable, some systems need to be initialized after the cable is reconnected.

5. Remove the auxiliary battery.
    a. Disengage the 2 claws and remove the rear upper No. 3 floor board plate.

b. Open the battery terminal cap.

c. Loosen the nut, and disconnect the positive (+) battery terminal.

d. Remove the nut and bolt.

e. Remove the battery thermometer sensor connector.

f. Remove the battery clamp.

g. Disconnect the battery ventilation hose.

h. Remove the battery.

To install:

6. Install the auxiliary battery.

a. Install the battery.

b. Connect the battery ventilation hose to the battery without kinking it.

➡Push the battery ve3ntilation hose to the battery until a click sound is heard.

c. Install the battery clamp with the bolt and nut. Tighten the nut and bolt to 48 inch lbs. (5.4 Nm).

d. Install the battery thermometer sensor connector.

e. Connect the positive (+) battery terminal with the nut.

f. Install the battery terminal cap.

g. Engage the 2 claws to install the rear upper No. 3 floor board plate.

7. Connect the cable to the negative battery terminal.

➡When disconnecting the cable, some systems need to be initialized after the cable is reconnected.

8. Install the rear No. 3 floor board.

9. Install the rear deck floor box.

10. Install the rear No. 2 floor board (separate type).

ENGINE ELECTRICAL — CHARGING SYSTEM

ALTERNATOR

REMOVAL & INSTALLATION

The Toyota Prius, being a hybrid vehicle that utilizes both electric and gasoline (internal combustion) engine power for mobility, does not require (or come equipped with) an alternator as a part of its charging system. The Toyota Hybrid system replaces the alternator with a pair of electrical motor-generators, a computerized shunt system to control them, a mechanical power splitter that acts as a second differential, and a battery pack that serves as an energy reservoir.

ENGINE ELECTRICAL — HYBRID SYSTEM

PRECAUTIONS

Before working on any part of the Hybrid high voltage system, observe the following precautions:

⁂ CAUTION

The nominal high voltage traction battery voltage is 330 volts DC. The buffer zone must be set up and insulated rubber gloves and a face shield must be worn. Failure to follow these instructions may result in severe injury or death.

⁂ CAUTION

The high voltage traction battery and charging system contains high voltage components and wiring. High voltage insulated safety gloves and a face shield must be worn when carrying out any diagnostics on this vehicle. Failure to follow these instructions may result in severe personal injury or death.

⁂ CAUTION

Before carrying out any removal and installation procedures of the high voltage traction battery system, the high voltage traction battery must be Disarmed. Failure to follow these instructions may result in severe personal injury or death.

⁂ CAUTION

The rubber insulating gloves that are to be worn while working on the high voltage system should be of the appropriate safety and protection rating for use on the high voltage system. They must be inspected before use and must always be worn in conjunction with the leather outer gloves. Any hole in the rubber insulating glove is a potential entry point for high voltage. Failure to follow these instructions may result in severe personal injury or death.

➡The high voltage insulated safety gloves must be re-certified every 6 months to remain within Occupational Safety and Health Administration (OSHA) guidelines:

- Roll the glove up from the open end until the lower portion of the glove begins to balloon from the resulting air pressure. If the glove leaks any air, it must not be used.
- The gloves should not be used if they exhibit any signs of wear and tear.
- The leather gloves must always be worn over the rubber insulating gloves in order to protect them.
- The rubber insulating gloves must be class "00" and meet all of the American Society for Testing and Materials (ASTM) standards

⁂ CAUTION

High voltage insulated safety gloves and a face shield must be worn when working with high voltage cables. The ignition switch must be OFF for a minimum of 5 minutes before removing high voltage cables. Failure to follow these instructions may result in severe personal injury or death.

⁂ CAUTION

Establish a buffer zone before servicing the high voltage system. The buffer zone is required only when working with the high voltage system. See the text for buffer zone establishment. Failure to follow these instructions may result in severe personal injury or death. Do not allow any unauthorized personnel into the buffer zone during repairs involving the high voltage system. Only personnel trained for repair on the high voltage system are to be permitted in the buffer zone.

⁂ CAUTION

Disarm the high voltage traction battery (HVTB) before working on the high voltage system. See the text for the Disarming procedure. Failure to follow these instructions may result in severe personal injury or death.

- Before inspecting the high-voltage system, take safety precautions to prevent electrical shocks, such as wearing insulated gloves and removing the service plug grip. After removing the service plug grip, put it in your pocket to prevent other technicians from reconnecting it while you are servicing the high-voltage system.
- Turning the power switch **ON (READY)**with the service plug grip removed could cause a malfunction. Therefore, do not turn the power switch **ON (READY)** unless instructed by the repair manual.
- After disconnecting the service plug grip, wait for at least 5 minutes before touching any of high-voltage connectors or terminals No. .
- At least 5 minutes are required to discharge the high-voltage condenser inside the inverter.
- Since liquid leakage may occur, wear protective goggles when checking inside the high voltage battery.
- Wear insulated gloves, turn the power switch **OFF**, and disconnect the negative terminal of the auxiliary battery before touching any of the orange-colored wires of the high-voltage system.
- Turn the power switch **OFF** before performing a resistance check.
- Turn the power switch **OFF** before disconnecting or reconnecting any connector.
- To install the service plug grip, the lever must be flipped and locked downward. Once it is locked in place, it turns the interlock switch **ON**. Make sure to lock it securely because if you leave it unlocked, the system will output a DTC pertaining to the interlock switch system.
- When the warning light is illuminated or the battery has been disconnected and reconnected, pressing the power switch may not start the system on the first attempt. If so, press the power switch again. With the power switches power mode changed to **ON (IG)**, disconnect the battery. If the key is not in the key slot during reconnection, DTC B2799 may be output.

BUFFER ZONE

1. Before servicing the vehicle, refer to the Precautions Section.

CAUTION

Before proceeding, read and observe all of the High Voltage System Precautions.

2. Establish a buffer zone around the vehicle:
   a. Position the vehicle in the repair bay.
   b. Position 4 orange cones at the corners of the vehicle to mark off a 1 m (3 ft.) perimeter around the vehicle.
   c. Do not allow any unauthorized personnel into the buffer zone during repairs involving the high voltage system. Only personnel trained for repair on the high voltage system are to be permitted in the buffer zone.

DISARMING THE HIGH VOLTAGE TRACTION BATTERY

1. Disconnect the negative battery cable.

➡Wait at least 90 seconds after disconnecting the cable from the negative (-) battery terminal to prevent airbag and seat belt pretensioner activation.

DIRECT CURRENT/ALTERNATING CURRENT INVERTER

REMOVAL & INSTALLATION

See Figures 27 through 38.

1. Refer to Precautions before servicing the vehicle.
2. Remove the No. 2 floor board (separate type).
3. Remove the rear deck floor box.
4. Remove the rear No. 3 floor board.
5. Disconnect the cable from the negative battery terminal.

➡When disconnecting the cable, some systems need to be initialized after the cable is reconnected.

6. Remove the service plug grip.
7. Remove the front spoiler cover (w/ front spoiler).
8. Remove the engine under cover (w/ cover).
9. Remove the No. 1 engine under cover.
10. Drain the coolant.
11. Remove the radiator support opening cover.
12. Remove the No. 1 inverter bracket.

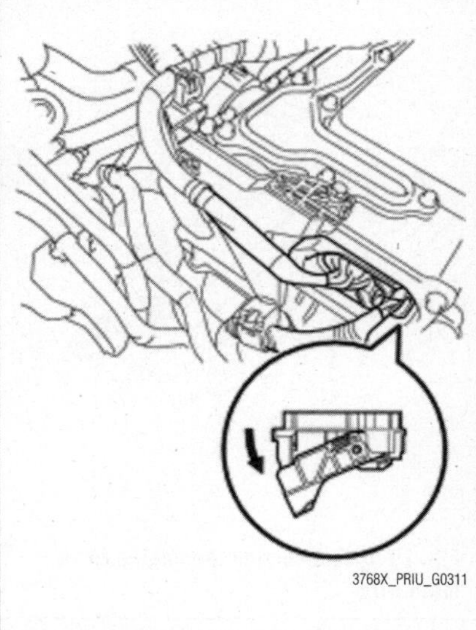

Fig. 27 Disconnecting the inverter with the converter connector

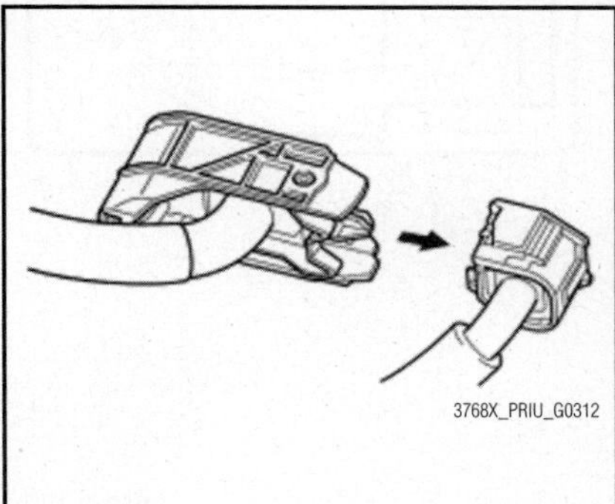

Fig. 28 Disconnecting the engine wire from the engine room main wire

13. Remove the 3 bolts and No. 1 inverter bracket.
14. Disconnect the engine room main wire.
    a. Raise the lock lever and disconnect the inverter with the converter connector.
    b. Disconnect the engine wire from the engine room main wire.
    c. Remove the bolt.
    d. Remove the bolt, clamp and clip, and disconnect the engine room main wire.
15. Remove the inverter terminal cover.

CAUTION

Wear insulating gloves.

a. Remove the 9 bolts and inverter terminal cover.

➡Make sure to pull the inverter terminal cover straight up, as a connector is connected to the bottom of the cover.

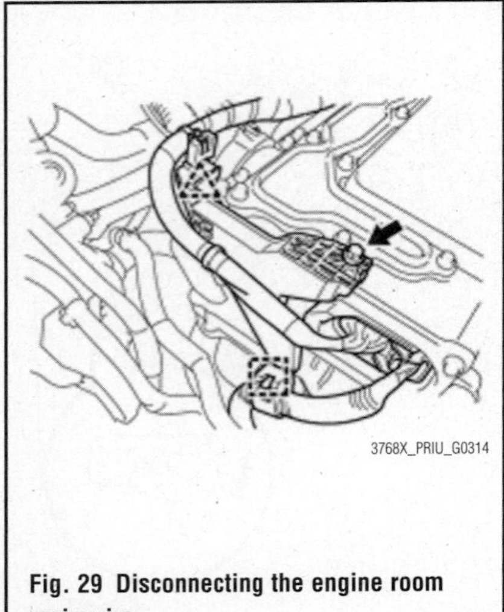

Fig. 29 Disconnecting the engine room main wire

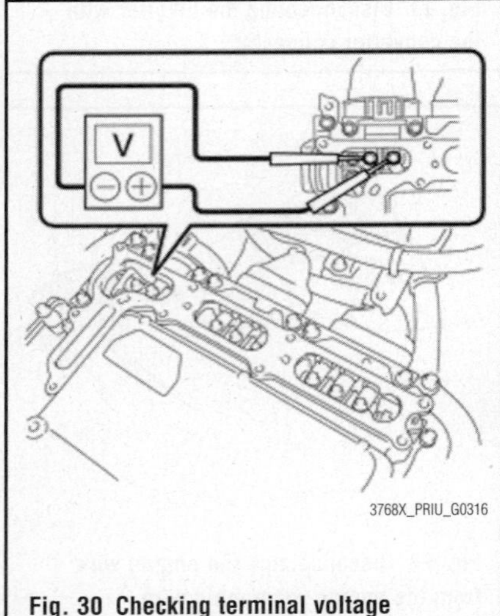

Fig. 30 Checking terminal voltage

16. Check the terminal voltage.

CAUTION

Wear insulating gloves.

➡Do not allow any foreign objects or water to enter the inverter with the converter assembly.

a. Using a voltmeter, measure the voltage between the terminals of the 2 phase connectors. The standard voltage is 0 V.

➡Use measuring range of DC 750 V or more on the voltmeter.

17. Disconnect the frame wire.

CAUTION

Wear insulating gloves.

➡Insulate the removed terminals with insulating tape.

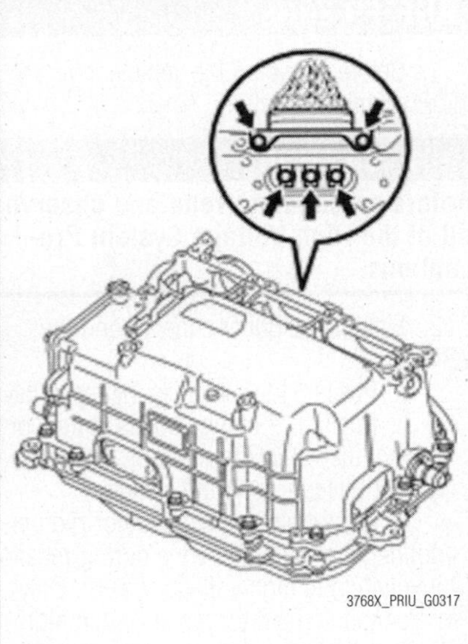

Fig. 31 Disconnecting the high voltage cables from the generator (MG1)

➡Cover the hole where the cable was connected with tape or equivalent (non-residue type) to prevent entry of foreign matter.

18. Disconnect the high voltage cable from the front transaxle.

CAUTION

Wear insulating gloves.

➡Insulate the removed terminals with insulating tape.

➡Cover the hole where the cable was connected with tape or equivalent (non-residue type) to prevent entry of foreign matter.

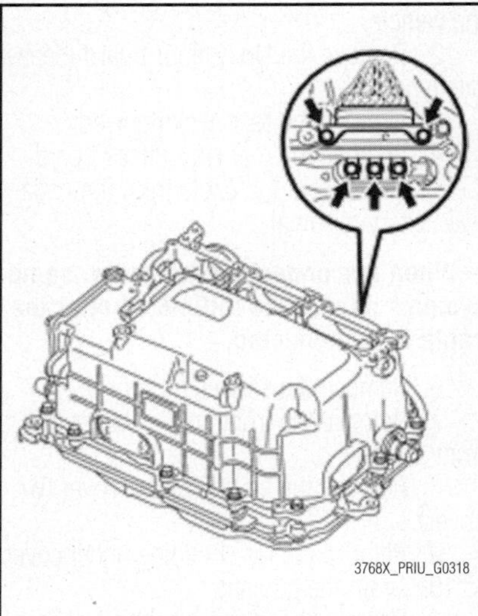

Fig. 32 Disconnecting the high voltage cables of the motor (MG2)

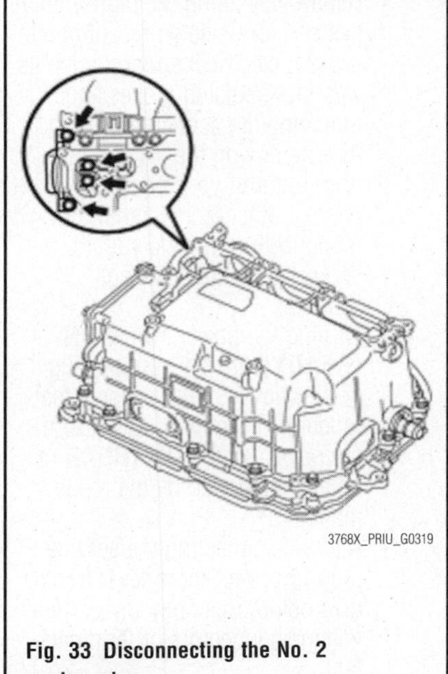

Fig. 33 Disconnecting the No. 2 engine wire

a. Remove the 5 bolts, and disconnect the high voltage cables of the generator (MG1) from the inverter with the converter assembly.

b. Turn back the wire harness cover and release the cable.

c. Remove the 5 bolts, and disconnect the high voltage cables of the motor (MG2) from the inverter with converter assembly.

d. Disconnect the harness clamp.

19. Disconnect the No. 2 engine wire.

CAUTION

Wear insulating gloves.

➡Insulate the removed terminals with insulating tape.

➡Cover the hole where the cable was connected with tape or equivalent (non-residue type) to prevent entry of foreign matter.

a. Remove the 4 bolts, and disconnect the No. 2 engine wire (high voltage cables for the air conditioning compressor) from the inverter with converter assembly.

b. Disconnect the harness clamp.

20. Install the inverter terminal cover.

a. Temporarily install the inverter terminal cover with the 9 bolts to prevent any foreign objects or water from entering the inverter with converter assembly.

21. Disconnect the No. 2 engine room wire.

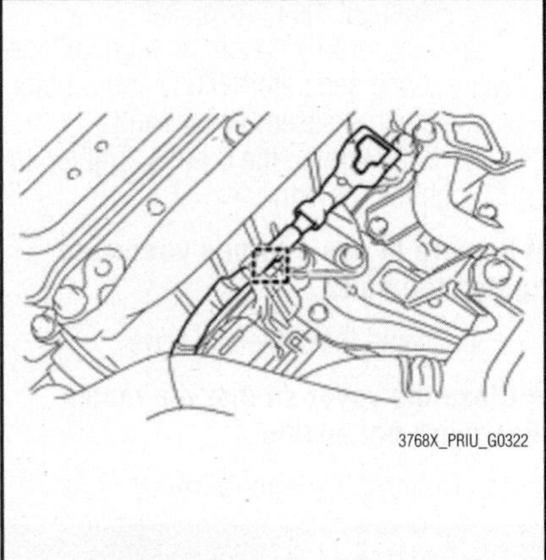

Fig. 34 Connecting the No. 2 engine room wire to the protector

a. Remove the relay block cover.

b. Release the 2 clamps, and remove the No. 1 relay block cover.

c. Remove the bolt from the No. 2 engine room wire.

d. Release the 2 claws, and disconnect the No. 2 engine room wire.

e. Connect the No. 2 engine room wire to the protector.

22. Disconnect the water hose.

a. Release the retainer and disconnect the water hose from the inverter with converter assembly.

b. Release the retainer and disconnect the water hose from the inverter with converter assembly.

c. Disconnect the coolant hose from the inverter with converter assembly. Put a piece of cloth in the pipe and in the disconnected hose or cover the pipe and hose with plastic bags as shown in the illustration, so that foreign matter doesn't stick to the union or the inside of the con-

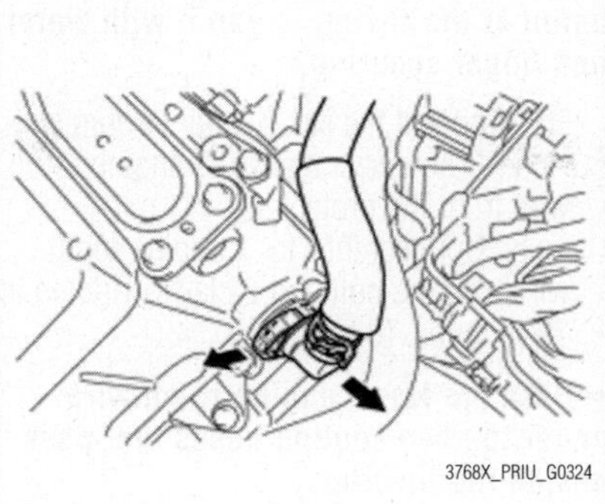

Fig. 36 Releasing the retainer and disconnect the water hose from the inverter with converter assembly

nector and to prevent coolant from spilling near the inverter with converter assembly.

23. Remove the inverter with the converter assembly.

CAUTION

Wear insulating gloves.

a. Remove the 3 bolts and inverter with converter assembly.

➡Since the inverter with converter assembly is very heavy, 2 people are needed to remove the inverter with converter assembly. When removing the inverter with converter assembly, do not damage the parts around it.

➡To prevent damage, do not hold the inverter with converter assembly by the connectors.

➡To prevent damage due to static electricity, do not touch the terminals of the disconnected connectors.

24. Remove the motor cable bracket.

a. Remove the 2 bolts and motor cable bracket.

25. Remove the high voltage fuse.

CAUTION

Wear insulating gloves.

➡Perform this procedure only when replacement of the high voltage fuse is necessary.

a. Remove the 9 bolts and inverter terminal cover.

➡Make sure to pull the inverter terminal cover straight up, as a connector is connected to the bottom of the cover.

b. Remove the 2 bolts and high voltage fuse from the inverter with the converter assembly.

➡Do not allow any foreign objects or water to enter the inverter with converter assembly.

c. Temporarily install the inverter terminal cover with the 9 bolts to prevent any foreign objects or water from entering the inverter with converter assembly.

To install:

26. Install the high voltage fuse.

CAUTION

Wear insulating gloves.

➡Perform this procedure only when replacement of the high voltage fuse is necessary.

a. Remove the 9 bolts and inverter terminal cover.

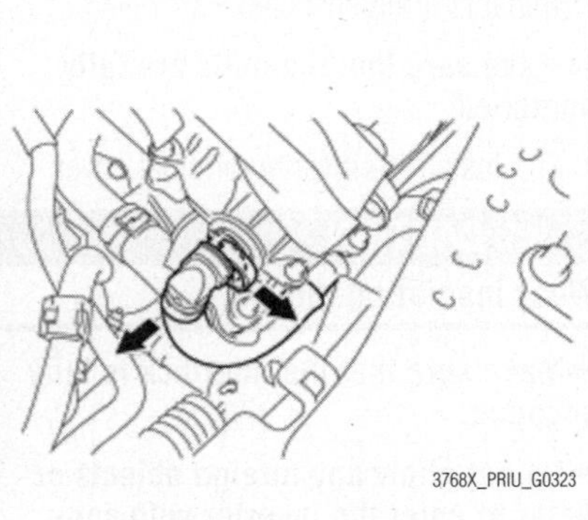

Fig. 35 Releasing the retainer and disconnect the water hose from the inverter with converter assembly

Fig. 37 Covering the hose and pipe

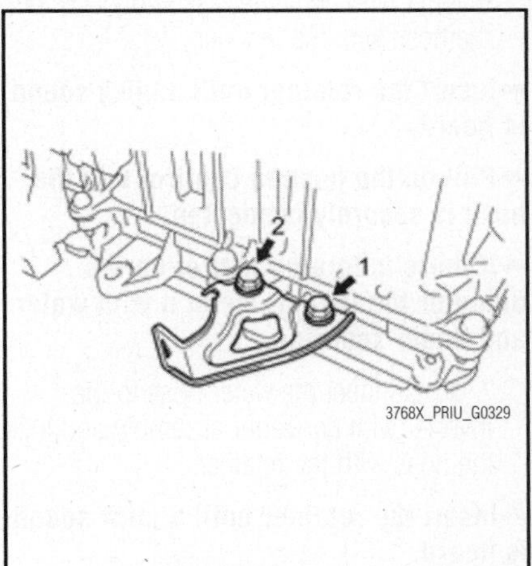

Fig. 38 Identifying the bolt tightening sequence

➡Make sure to pull the inverter terminal cover straight up, as a connector is connected to the bottom of the cover.

b. Install the high voltage fuse with the 2 bolts. Tighten to 35 inch lbs. (4 Nm).

➡Be sure to use a torque wrench to tighten the bolts.

c. Temporarily install the inverter terminal cover with the 9 bolts to prevent any foreign objects or water from entering the inverter with converter assembly.

27. Install the motor cable bracket.

a. Temporarily install the motor cable bracket with the 2 bolts.

b. Tighten the 2 bolts in the order shown in the illustration. Tighten to 71 inch lbs. (8 Nm).

28. Install the inverter with converter assembly.

⁂ CAUTION

Wear insulating gloves.

a. Install the inverter with converter assembly with the 3 bolts. Tighten to 8 ft. lbs. (12 Nm).

➡Since the inverter with converter assembly is very heavy, 2 people are needed to install the inverter with converter assembly. When installing the inverter with converter assembly, do not damage the parts around it.

➡To prevent damage, do not hold the inverter with converter assembly by the connectors.

➡To prevent damage due to static electricity, do not touch the terminals of the disconnected connectors.

29. Connect the water hose.

a. Connect the water hose to the inverter with converter assembly and lock the hose with the retainer.

➡Insert the retainer until a click sound is heard.

➡Pull on the hose to confirm that the hose is securely connected.

➡If there is foreign matter on the union or the O-ring, clean it with water and finger scouring.

b. Connect the water hose to the inverter with converter assembly and lock the hose with the retainer.

➡Insert the retainer until a click sound is heard.

➡Pull on the hose to confirm that the hose is securely connected.

➡If there is foreign matter on the union or the O-ring, clean it with water and finger scouring.

30. Connect the No. 2 engine room wire.

a. Disconnect the No. 2 engine room wire from the protector.

b. Connect the No. 2 engine room wire with the bolt and 2 claws. Tighten to 73 inch lbs. (8.3 Nm).

➡Pass the No. 2 engine room wire under the two cooling hoses that pass beside the inverter.

c. Install the No. 1 relay block cover and 2 clamps.

d. Install the relay block cover.

31. Remove the inverter terminal cover.

⁂ CAUTION

Wear insulating gloves.

a. Remove the 9 bolts and inverter terminal cover.

➡Make sure to pull the inverter terminal cover straight up, as a connector is connected to the bottom of the cover.

32. Connect the No. 2 engine wire.

⁂ CAUTION

Wear insulating gloves.

➡Do not allow any foreign objects or water to enter the inverter with converter assembly.

a. Temporarily install the No. 2 engine wire (high voltage cables of the air conditioning) and 4 bolts to the inverter assembly by hand.

b. Fully tighten the 4 bolts. Tighten to 71 inch lbs. (8 Nm).

➡Be sure to use a torque wrench to tighten the bolts.

c. Connect the harness clamp.

33. Connect the high voltage cable of the front transaxle.

⁂ CAUTION

Wear insulating gloves.

➡Do not allow any foreign objects or water to enter the inverter with converter assembly.

a. Temporarily install the high voltage cable of the motor (MG2) and 5 bolts to the inverter assembly by hand.

b. Fully tighten the 5 bolts. Tighten to 71 inch lbs. (8 Nm).

➡Be sure to use a torque wrench to tighten the bolts.

c. Connect the harness clamp.

d. Temporarily install the high voltage cable of the generator (MG1) and 5 bolts to the inverter assembly by hand.

e. Fully tighten the 5 bolts. Tighten to 71 inch lbs. (8 Nm).

➡Be sure to use a torque wrench to tighten the bolts.

f. Install the cable and cover.

➡Close the cover so that the matchmarks are not visible.

34. Connect the frame wire.

⁂ CAUTION

Wear insulating gloves.

➡Make sure that the interlock is fully engaged.

➡Do not allow any foreign objects or water to enter the inverter with converter assembly.

a. Temporarily install the frame wire (high voltage cables of the hybrid battery) and 4 bolts to the inverter assembly by hand.

b. Fully tighten the 4 bolts. Tighten to 71 inch lbs. (8 Nm).

➡Be sure to use a torque wrench to tighten the bolts.

c. Connect the harness clamp.

35. Check the high voltage cable connection.

⁂ CAUTION

Wear insulating gloves.

➡Do not allow any foreign objects or water to enter the inverter with converter assembly.

a. Check that each connector and terminal is firmly installed.

➡Make sure that the bolts are fully tightened.

36. Install the inverter terminal cover.

⁂ CAUTION

Wear insulating gloves.

➡Make sure that the interlock is fully engaged.

➡Do not allow any foreign objects or water to enter the inverter with converter assembly.

a. Install the inverter terminal cover with the 9 bolts to the inverter with converter assembly. Tighten to 71 inch lbs. (8 Nm).

37. Install the engine room main wire.

➡**Make sure that the interlock is fully engaged.**

➡**Do not allow any foreign objects or water to enter the inverter with converter assembly.**

a. Install the bolt, clamp and clip, and connect the engine room main ware. Tighten to 9 ft. lbs. (12 Nm).
b. Install the bolt. Tighten to 9 ft. lbs. (12 Nm).
c. Connect the engine wire to the engine room main wire.
d. Connect the connector to the inverter with converter assembly and lock the connector with the lock lever.

38. Install the No. 1 inverter bracket.
a. Temporarily install the No. 1 inverter bracket with the 3 bolts.
b. Tighten the 3 bolts in the order shown in the illustration. Tighten to 10 ft. lbs. (14 Nm).

39. Install the service plug grip.
40. Connect the cable to the negative battery terminal.

➡**When disconnecting the cable, some systems need to be initialized after the cable is reconnected.**

41. Install the rear No. 3 floor board.
42. Install the rear deck floor box.
43. Install the rear No. 2 floor board (separate type).
44. Add coolant.
45. Inspect for coolant leaks.
46. Install the No. 1 engine under cover.
47. Install the front spoiler cover (w/ front spoiler).
48. Install the engine under cover (w/ cover).
49. Install the radiator support opening cover.

HIGH VOLTAGE TRACTION BATTERY

REMOVAL & INSTALLATION

1.5L Engine

See Figures 39 through 43.

1. Refer to Precautions before servicing.
2. Read output DTC.

➡**Confirm that P0AA6 (Hybrid Battery Voltage System Isolation Fault) is not output before removing or installing the HV battery. If this DTC is output, perform troubleshooting for this DTC first.**

3. Remove service plug grip.

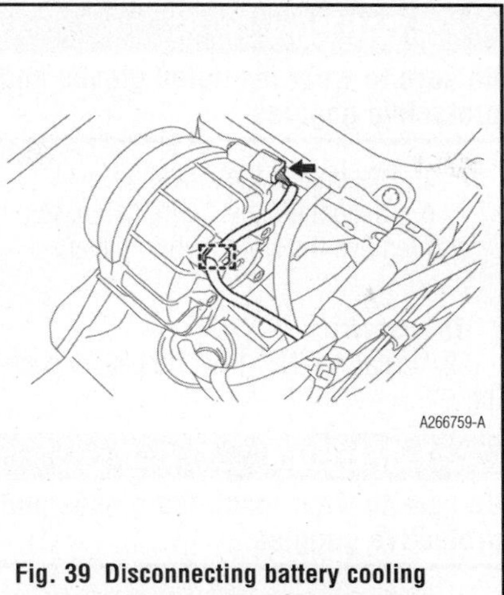

Fig. 39 Disconnecting battery cooling blower assembly connector and clamp

4. Remove rear seat cushion assembly.
5. Remove rear seat cushion leg sub-assembly.
6. Remove inverter terminal cover.
7. Remove inverter cover.
8. Check terminal voltage.
9. Install inverter cover.
10. Install inverter terminal cover.
11. Remove No. 1 hybrid battery exhaust duct.
12. Remove battery cooling blower assembly.

➡**Be sure not to touch the fan part of the battery cooling blower assemblies.**

a. Do not lift the battery cooling blower assemblies using the wire harness.
b. Disconnect the battery cooling blower assembly connector and clamp.
c. Remove the 2 bolts, nut and battery cooling blower assembly.

13. Remove the No. 1 hybrid vehicle battery cover panel LH.
14. Remove the No. 3 indoor electrical key antenna assembly.

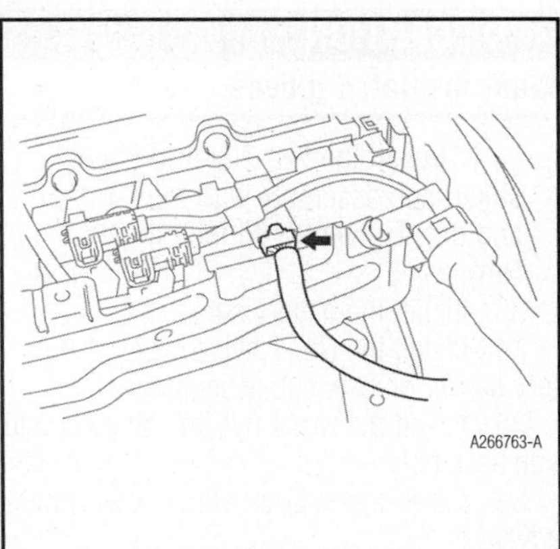

Fig. 40 Disconnecting the connector

15. Remove the wire harness.
a. Disconnect the 2 connectors and 3 clamps.

16. Disconnect the connector.
17. Disconnect the frame wire.
18. Remove the HV battery.

⁂ CAUTION

Wear insulating gloves.

➡**When removing, installing or moving the HV battery, make sure not to tilt it more than 80°.**

a. Remove the 2 clips.
b. Remove the 2 nuts and 2 bolts.
c. Remove the HV battery.

➡**When moving the HV battery, be sure to hold are A.**

➡**Place the HV battery on the attachment.**

19. Remove the hybrid battery service plug cover.
20. Remove the No. 1 hybrid vehicle battery shield sub assembly.
21. Remove the No. 2 hybrid vehicle battery carrier bracket sub assembly.
22. Remove the battery smart unit.
23. Remove the No. 2 hybrid battery shield sub assembly.

⁂ CAUTION

Wear insulating gloves,

a. Remove the 2 bolts, nut and No. 2 hybrid battery shield sub assembly.

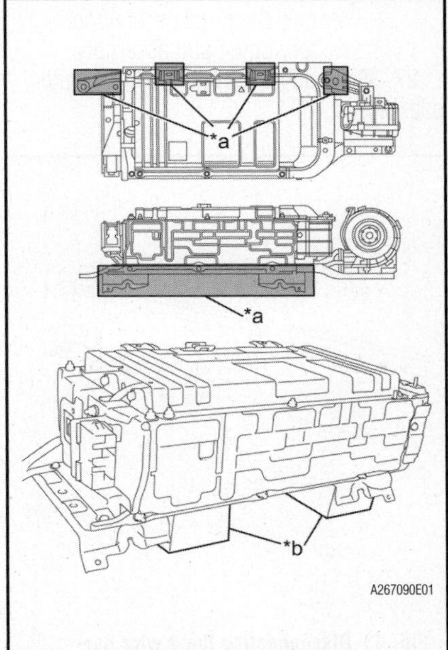

Fig. 41 Identifying area A (a) and attachment (b)

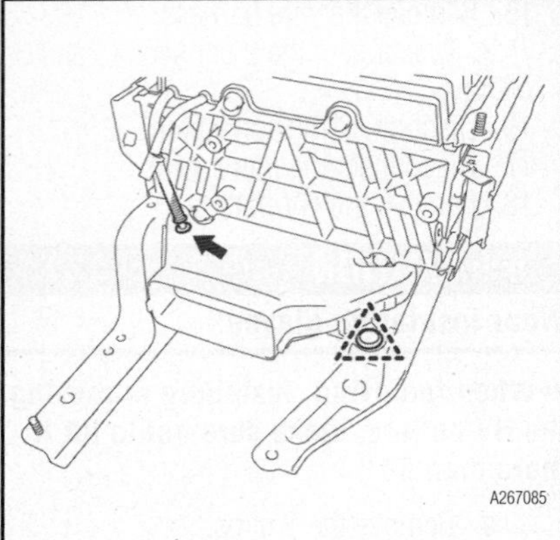

Fig. 42 Disconnecting the HV battery thermistor

24. Remove the hybrid battery junction block assembly.
25. Remove the No. 1 hybrid battery cover intake duct.

✻✻ CAUTION

Wear insulating gloves.

a. Disconnect the HV battery thermistor.
b. Remove the clip and No. 1 hybrid battery cover intake duct.

26. Remove the upper hybrid battery cover sub assembly.

✻✻ CAUTION

Be sure to wear insulated gloves and protective goggles.

a. Remove the 5 nuts and upper hybrid battery cover sub assembly.

27. Disconnect the No. 1 hybrid battery packing.

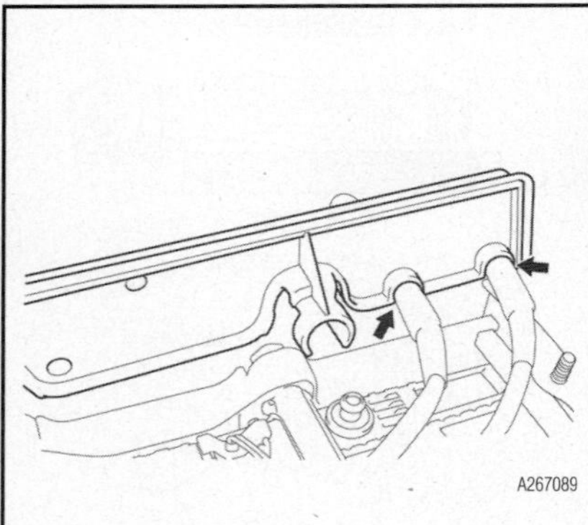

Fig. 43 Disconnecting the 2 wire harnesses and removing the No. 1 hybrid battery packing

✻✻ CAUTION

Be sure to wear insulated gloves and protective goggles.

a. Disconnect the 2 clamps.
b. Disconnect the 2 wire harnesses and remove the No. 1 hybrid battery packing.

To install:

28. Install the No. 1 hybrid battery packing.

✻✻ CAUTION

Be sure to wear insulated gloves and protective goggles.

a. Connect the 2 wire harnesses to the No. 1 hybrid battery packing.
b. Install the No. 1 hybrid battery packing with the 2 clamps.

29. Install the upper hybrid battery cover sub assembly.

✻✻ CAUTION

Be sure to wear insulated gloves and protective goggles.

a. Install the upper hybrid battery cover sub assembly with the 5 nuts. Tighten to 66 inch lbs. (7.5 Nm).

30. Install the No. 1 hybrid battery cover intake duct.

✻✻ CAUTION

Wear insulated gloves.

a. Install the No. 1 hybrid cover intake duct with the clip.
b. Install the HV battery thermistor to the No. 1 hybrid battery cover intake duct.

31. Install the hybrid battery junction block assembly.
32. Install the No. 2 hybrid battery shield sub assembly.

✻✻ CAUTION

Wear insulated gloves.

a. Install the No. 2 hybrid battery shield sub assembly with the 2 bolts and nut. Tighten to 66 inch lbs. (7.5 Nm).

33. Install the battery smart unit.
34. Install the No. 2 hybrid vehicle battery carrier bracket sub assembly.
35. Install the No. 1 hybrid battery shield sub assembly.
36. Install the hybrid battery service plug cover.
37. Install the HV battery.

✻✻ CAUTION

Wear insulated gloves.

➡When removing, installing or moving the HV battery, make sure mot to tilt it more than 80°.

a. Install the HV battery with the 2 bolts and 2 nuts. Tighten to 14 ft. lbs. (19 Nm).
b. Install the 2 clips.

38. Connect the frame wire.
39. Connect the wire harness.

a. Connect the 2 connectors and 3 clamps.

40. install the No. 3 indoor electrical key antenna assembly (w/smart key system).
41. Install the No. 1 hybrid vehicle battery cover panel LH.
42. Install the battery cooling blower assembly.

➡Be sure not to touch the fan part of the battery cooing blower assembly.

➡Do not lift the battery cooling blower assembly using its wire harness.

a. Install the battery cooling blower assembly with the 2 bolts and nut. Tighten to 66 inch lbs. (7.5 Nm).
b. Connect the battery cooling blower assembly connector and engage the clamp to the battery cooling blower assembly.

43. Install the No. 1 hybrid battery exhaust duct with the clip.
44. Install the rear seat cushion leg sub assembly.
45. Install the rear seat cushion assembly.
46. Install the service plug grip.

1.8L Engine

See Figures 43, 44 through 53.

1. Refer to Precautions before servicing.
2. Read output DTC.

➡Check for DTCs and confirm that P0AA6 (Hybrid Battery Voltage System Isolation Fault) is not output before doing removal or installation inside the battery. If this DTC is output, perform troubleshooting first for this DTC.

3. Remove the rear No. 2 floor board (separate type).
4. Remove the rear deck floor box.
5. Remove the rear No. 3 floor board.
6. Disconnect the cable from the negative battery terminal.

➡When disconnecting the cable, some

systems need to be initialized after the cable is reconnected.

7. Remove the service plug grip.
8. Remove the inverter terminal cover.
9. Check the terminal voltage.
10. Install the inverter terminal cover.
11. Remove the tonneau cover assembly (w/ tonneau cover).
12. Remove the rear seat cushion assembly.
13. Remove the rear No. 1 floor board sub assembly.
14. Remove the rear No. 2 floor board sub assembly.
15. Remove the rear No. 1 floor board.
16. Remove the rear door scuff plates (LH and RH.
17. Remove the rear side seat back assemblies (LH and RH).
18. Remove the rear No. 4 floor board.
19. Remove the deck floor box LH.
20. Remove the deck trim service hole cover.
21. Remove the rear deck trim cover.
22. Remove the luggage hold belt striker assembly (LH side).
23. Remove the tonneau cover holder cap (LH side).
24. Remove the deck trim side panel assembly LH.
25. Remove the tonneau cover holder cap (RH side).
26. Remove the luggage hold belt striker assembly (RH side).
27. Remove the deck trim side panel assembly RH.
28. Remove the rear floor board spacer.
    a. Remove the 2 clips and rear floor board spacer.
29. Remove the No. 1 hybrid battery exhaust duct.
    a. Remove the clip and No. 1 hybrid battery exhaust duct.
30. Remove the upper hybrid battery cover sub assembly.

CAUTION

Be sure to wear insulated gloves and protective goggles.

a. Using the service plug grip, remove the battery cover lock striker.

➡Insert the projection part of the service plug grip, and turn the button of the battery cover lock striker counterclockwise, and release the lock.

b. Remove the 4 nuts and upper hybrid battery cover sub assembly.

31. Remove the No. 1 hybrid battery intake duct.

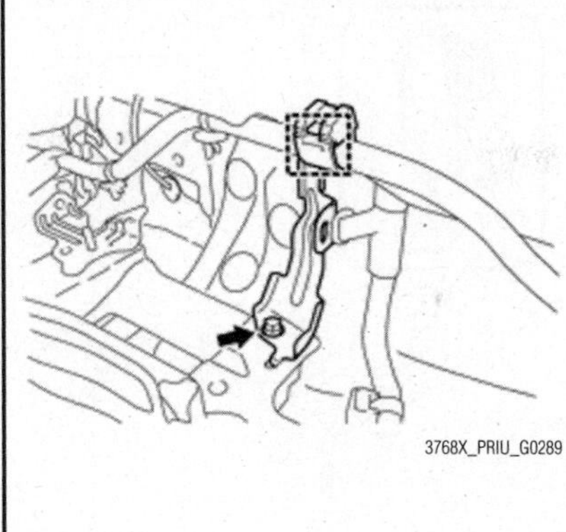

Fig. 44 Removing the No. 7 hybrid vehicle battery upper carrier bracket

a. Remove the 2 clips and No. 1 hybrid battery intake duct.

32. Remove the battery cooling blower assembly.

➡Be sure not to touch the fan part of the battery cooling blower assemblies.

➡Do not lift the battery cooling blower assemblies using the wire harness.

33. Remove the No. 7 hybrid vehicle battery upper carrier bracket.
    a. Disconnect the wire harness clamp.
    b. Remove the bolt and No. 7 hybrid battery upper carrier bracket.
34. Remove the child restraint seat anchor bracket sub assembly LH.
35. Remove the child restraint seat anchor bracket sub assembly RH.
    a. Disconnect the wire harness protector clamp.
    b. Remove the 2 bolts and child restraint seat anchor bracket sub-assembly RH.
36. Remove the frame wire.

CAUTION

Wear insulating gloves.

CAUTION

Insulate the removed terminals with insulating tape.

a. Remove the 2 nuts, then disconnect the frame wire from the hybrid battery junction block assembly.
b. Disconnect the clamp and frame wire.

37. Remove the HV battery assembly.

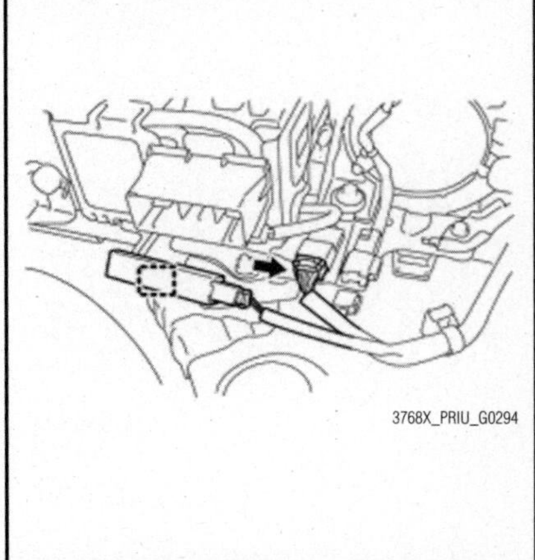

Fig. 45 Disconnecting the connector and electrical key oscillator clamp

CAUTION

Wear insulating gloves.

➡Since the HV battery is very heavy, 2 people are needed to remove the HV battery. When removing the HV battery, do not damage the parts around it.

a. Disconnect the connector and electrical key oscillator clamp.
b. Disconnect the battery room ventilation hose from the floor panel.
c. Remove the 4 bolts shown in the illustration.

38. Remove the hybrid battery junction block.

CAUTION

Wear insulating gloves.

a. Disconnect the 2 connectors from the hybrid battery junction block.

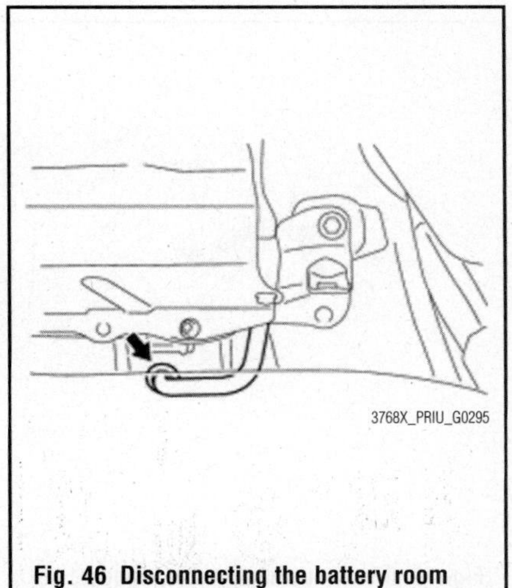

Fig. 46 Disconnecting the battery room ventilation hose from the floor panel

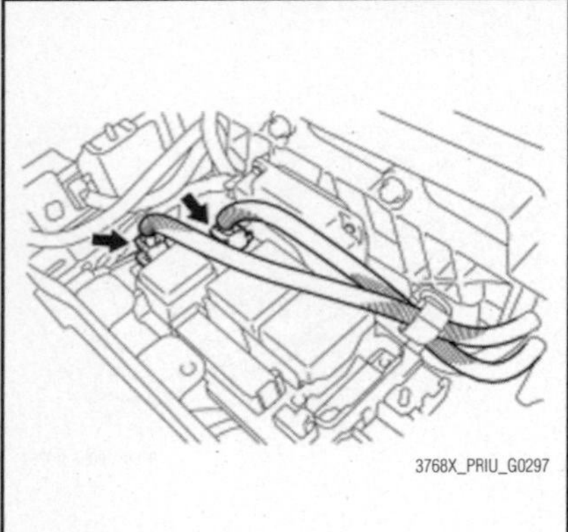

Fig. 47 Disconnecting the 2 connectors from the hybrid battery junction block

b. Disconnect the 2 connectors from the hybrid battery junction block.

c. Remove the 3 nuts and hybrid battery junction block.

39. Remove the battery smart unit.

✻✻ CAUTION

Wear insulating gloves.

40. Remove the No. 1 hybrid vehicle battery carrier bracket assembly.

✻✻ CAUTION

Wear insulating gloves.

a. Disconnect the connector.

b. Remove the bolt and EV battery plug.

c. Using the service plug grip, remove the battery cover lock striker.

➡Insert the projection part of the service plug grip, and turn the button of the battery cover lock striker counterclockwise, and release the lock.

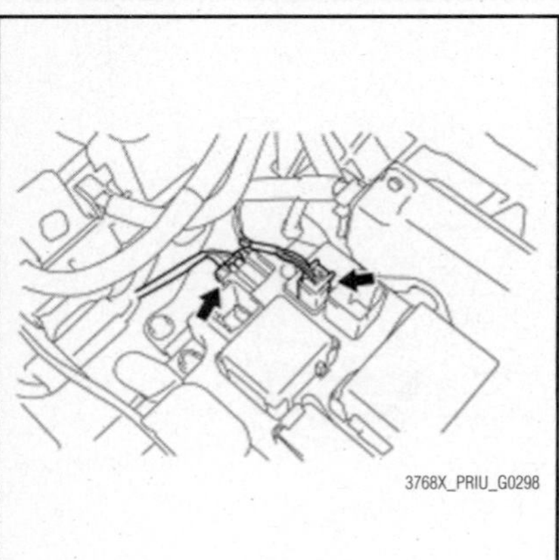

Fig. 48 Disconnecting 2 more connectors from the hybrid battery junction block

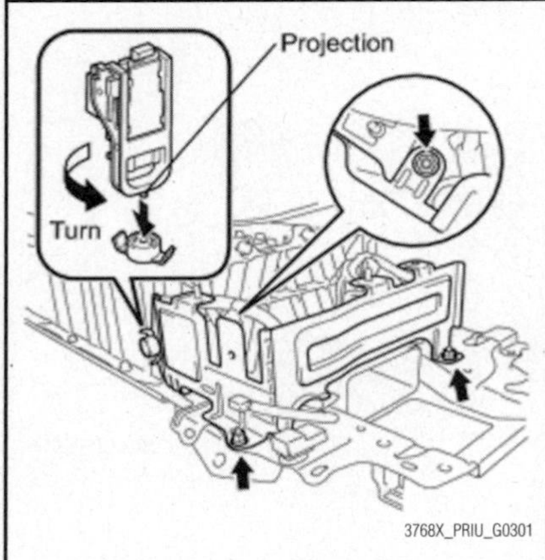

Fig. 49 Removing the No. 1 hybrid vehicle battery carrier bracket sub assembly

d. Remove the 3 nuts and No. 1 hybrid vehicle battery carrier bracket sub assembly.

41. Remove the No. 4 hybrid vehicle batter carrier bracket sub assembly.

✻✻ CAUTION

Wear insulating gloves.

a. Disconnect the 3 wire harness clamps.

b. Remove the 3 bolts and No. 4 hybrid vehicle battery carrier bracket sub assembly.

c. Disengage the 2 claws and remove the HV battery thermistor.

42. Remove the No. 1 hybrid battery cover intake duct.

a. Remove the 2 clips and No. 1 hybrid battery cover intake duct.

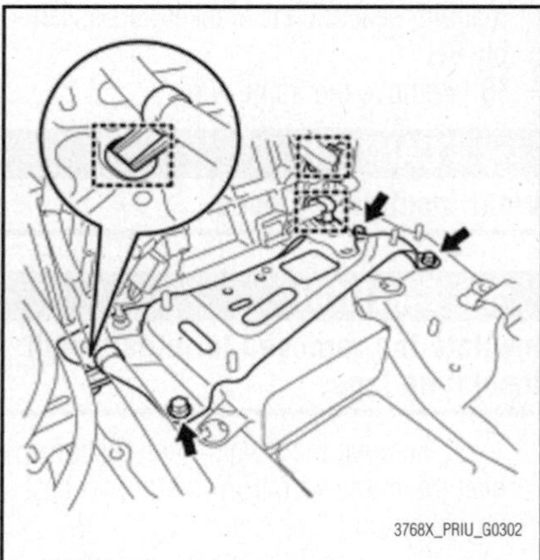

Fig. 50 Disconnecting the 3 wire harness clamps

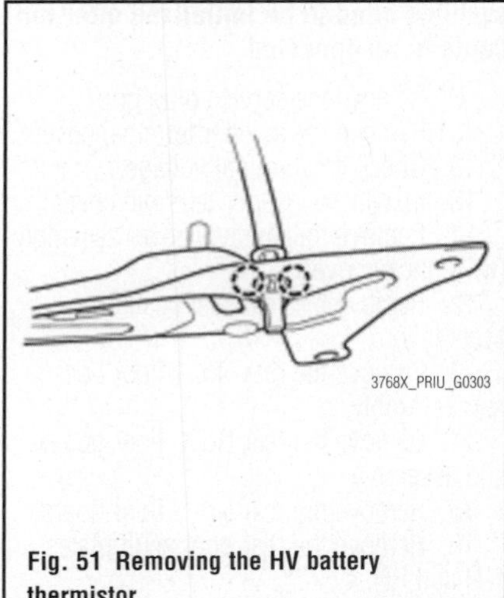

Fig. 51 Removing the HV battery thermistor

43. Remove the upper hybrid battery cover sub assembly.

✻✻ CAUTION

Be sure to wear insulated gloves and protective goggles.

a. Remove the tape (*1).

b. Remove the 5 bolts, 6 nuts and battery cover with the No. 1 hybrid battery shield sub assembly.

44. Remove the No. 1 hybrid battery packing.

✻✻ CAUTION

Be sure to wear insulated gloves and protective goggles.

a. Remove the 2 clamps and No. 1 hybrid battery packing.

45. Remove the hybrid battery hose assembly.

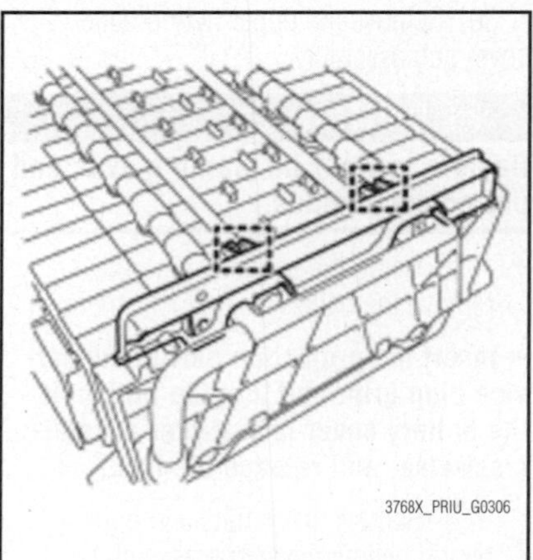

Fig. 52 Removing the No. 1 hybrid battery packing

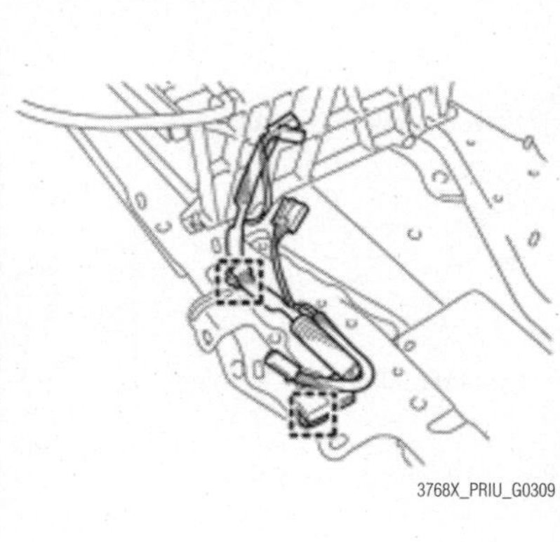

Fig. 53 Removing the No. 2 hybrid battery pack wire

CAUTION

Be sure to wear insulated gloves and protective goggles.

a. Remove the hybrid battery hose assembly from the HV battery.
46. Remove the No. 1 hybrid vehicle battery hose.

CAUTION

Be sure to wear insulated gloves and protective goggles.

a. Remove the 2 No. 1 hybrid vehicle battery hoses from the HV battery.
47. Remove the No. 2 hybrid battery pack wire.
a. Disconnect the 2 clamps, then remove the No. 2 hybrid battery pack wire.

To install:

48. Install the No. 2 hybrid battery pack wire.
a. Connect the 2 clamps and No. 2 hybrid battery pack wire.
49. Install the No. 1hybrid vehicle battery hose.

CAUTION

Be sure to wear insulated gloves and protective goggles.

a. Install the 2 No. 1 hybrid vehicle battery hoses to the HV battery.
50. Install the hybrid battery hose assembly.

CAUTION

Be sure to wear insulated gloves and protective goggles.

a. Install the hybrid battery hose assembly to the HV battery.
51. Install the No. 1 hybrid battery packing.

CAUTION

Be sure to wear insulated gloves and protective goggles.

a. Install the No. 1 hybrid battery packing with the 2 clamps.
52. Install the upper hybrid battery cover sub assembly.

CAUTION

Be sure to wear insulated gloves and protective goggles.

a. Install the battery cover and No. 1 hybrid battery shield sub assembly with the 5 bolts and 6 nuts. Tighten to 66 inch lbs. (7.5 Nm).
b. install the tape.
53. Install the No. 1hybrid battery cover intake duct with the 2 clips.
54. Install the No. 4 hybrid vehicle battery carrier bracket sub assembly.

CAUTION

Wear insulated gloves.

a. Install the HV battery thermistor to the No. 4 hybrid vehicle battery carrier bracket sub assembly.
b. Install the No. 4 hybrid vehicle battery carrier bracket sub assembly with the 3 bolts. Tighten to 66 inch lbs. (7.5 Nm).
c. Connect the 3 wire harness clamps.
55. Install the No. 1 hybrid vehicle battery carrier bracket sub assembly.

CAUTION

Wear insulated gloves.

a. Install the No. 1 hybrid vehicle battery carrier bracket sub-assembly with the 3 nuts. Tighten to 66 inch lbs. (7.5 Nm).
b. Install the battery cover lock striker, then push the button to lock it.
c. Install the electric vehicle battery plug assembly with the bolt. Tighten to 66 inch lbs. (7.5 Nm).
d. Connect the connector.

➡The connector should be connected securely.

56. Install the battery smart unit.

CAUTION

Wear insulated gloves.

a. Install the battery smart unit with the 2 bolts. Tighten to 66 inch lbs. (7.5 Nm).
b. Connect the 3 connectors.

➡The connectors should be connected securely.

57. Install the hybrid battery junction block.
a. Install the hybrid battery junction block with the 3 nuts. Tighten to 66 inch lbs. (7.5 Nm).
b. Connect the 2 connectors to the hybrid battery junction block.

➡The connectors should be connected securely.

c. Connect the 2 connectors to the hybrid battery junction block.

➡The connectors should be connected securely.

58. Install the HV battery assembly.

CAUTION

Wear insulated gloves.

➡Since the HV battery is very heavy, 2 people are needed to install the HV battery. When installing the HV battery, do not damage the parts around it.

a. Install the HV battery to the vehicle with the 4 bolts. Tighten to 14 ft. lbs. (19 Nm).
b. Connect the battery room ventilation hose to the floor panel.

➡Make sure that there is no space or gap between the grommet and the body.

c. Connect the connector and electrical key oscillator clamp.
59. Install the frame wire.

CAUTION

Wear insulated gloves.

a. Install the frame wire on the hybrid battery junction block assembly with the 2 nuts. Tighten to 80 inch lbs. (9 Nm).

➡Make sure that the ends of the frame wire are not crossed over each other.

➡Be sure to connect the frame wires to the correct terminals.

b. Connect the clamp and frame wire.
60. Check the high voltage cable connection condition.

⁂ CAUTION

Wear insulated gloves and protective goggles.

a. Check that each wire harness is being installed securely.

➡Make sure that the end of the frame wire are not crossover each other.

➡Be sure to connect the frame wire to the correct terminals.

➡The connectors should be connected securely.

➡The nuts should be fastened securely.

➡Make sure that the 4 plastic covers are engaged securely.

61. Install the child restraint seat anchor bracket sub assembly LH.
    a. Install the child restraint seat anchor bracket sub assembly LH with the 2 bolts. Tighten to 14 ft. lbs. (20 Nm).
62. Install the child restraint seat anchor bracket sub assembly RH.
    a. Install the child restraint seat anchor bracket sub assembly RH with the 2 bolts. Tighten to 14 ft. lbs. (20 Nm).
    b. Connect the wire harness protector clamp.
63. Install the No. 7 hybrid vehicle battery upper carrier bracket.
    a. Install the No. 7 hybrid battery upper carrier bracket with the bolt. Tighten to 66 inch lbs. (7.5 Nm).
64. Install the battery cooling blower assembly.

➡Be sure not to touch the fan part of the battery cooling blower assemblies.

➡Do not lift the battery cooling blower assemblies using the wire harness.

65. Install the No. 1 hybrid battery intake duct.

➡Ensure that the duct is installed securely.

66. Install the upper hybrid battery cover sub assembly with the 4 nuts. Tighten to 66 inch lbs. (7.5 Nm).
    a. Install the battery cover lock striker, then push the button to lock.
67. Install the No. 1 hybrid battery exhaust duct with the clip.

➡Ensure that the duct is installed securely.

68. Install the rear floor board spacer with the 2 clips.
69. Install the deck trim side panel assembly LH.
70. Install the tonneau cover holder cap (LH side).
71. Install the luggage hold belt striker assembly (LH side).
72. Install the deck trim side panel assembly RH.
73. Install the tonneau cover holder cap (RH side).
74. Install the luggage hold belt striker assembly (RH side).
75. Install the rear deck trim cover.
76. Install the deck trim service hole cover.
77. Install the deck floor box LH.
78. Install the rear No. 4 floor board.
79. Install the rear side seat back assembly LH.
80. Install the rear side seat back assembly RH.
81. Install the rear door scuff plate LH.
82. Install the rear door scuff plate RH.
83. Install the rear No. 1 floor board.
84. Install the rear No. 2 floor board sub assembly.
85. Install the rear No. 1 floor board sub assembly.
86. Install the rear seat cushion assembly.
87. Install the tonneau cover assembly (w/ tonneau cover).
88. Install the service plug grip.
89. Connect the cable to the negative battery terminal.

➡When disconnecting the cable, some systems need to be initialized after the cable is reconnected.

90. Install the rear No. 3 floor board.
91. Install the rear deck floor box.
92. Install the rear No. 2 floor board (separate type).

HIGH VOLTAGE CABLES

REMOVAL & INSTALLATION

See Figure 54.

1. Disconnect the negative battery cable.

➡Wait at least 90 seconds after disconnecting the cable from the negative (-) battery terminal to prevent airbag and seat belt pretensioner activation.

2. Remove service plug grip.
3. Remove the high voltage battery from the vehicle.
4. Remove battery cover.
5. Wearing insulating gloves, remove No. 1 wire harness protector cover.
6. Wearing insulating gloves, remove No. 3 wire harness protector cover.

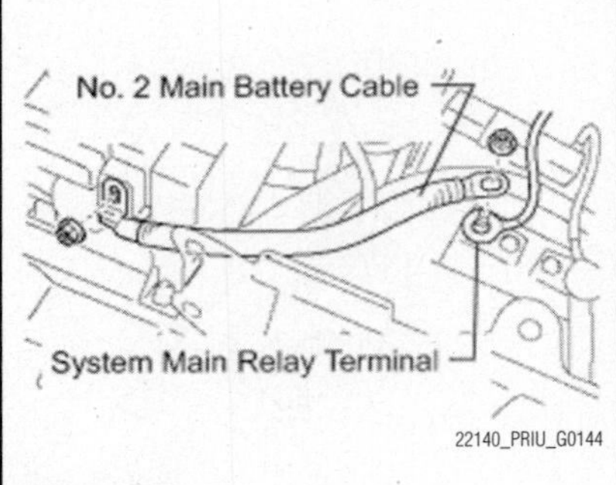

Fig. 54 Disconnect the system main relay terminal and No. 2 main battery cable from the No. 2 system main relay

7. Wearing insulating gloves, peel off the bonded parts, then remove the battery carrier cushion.
8. Wearing insulating gloves, remove main battery cable.
    a. Remove the terminal cover.
    b. Remove the nut, then disconnect the aluminum shield wire.
    c. Remove the nut, then disconnect the main battery cable from the No. 3 system main relay.
    d. Remove the nut, then disconnect the main battery cable from the No. 2 frame wire.
    e. Remove the main battery cable from the high voltage battery.
9. Wearing insulating gloves, remove No. 2 main battery cable.
    a. Remove the terminal cover.
    b. Remove the nut, then disconnect the system main relay terminal and No. 2 main battery cable from the No. 2 system main relay.
    c. Remove the nut, then disconnect the No. 2 main battery cable from the frame wire.
    d. Remove the No. 2 main battery cable from the high voltage battery.

To install:

10. Wearing insulating gloves, install No. 2 main battery cable.
    a. Temporarily install the No. 2 main battery cable to the high voltage battery.
    b. Install the No. 2 main battery cable to the No. 2 frame wire with a new nut. Tighten to 48 inch lbs. (5 Nm).
    c. Temporarily install the main battery cable and system main relay terminal, in that order, to the No. 2 system main relay, then tighten the new nut. Tighten to 50 inch lbs. (6 Nm).
    d. Install the terminal cover.

11. Wearing insulating gloves, install main battery cable.
 a. Temporarily install the main battery cable to the high voltage battery.
 b. Install the main battery cable to the No. 2 frame wire with a new nut. Tighten to 48 inch lbs. (5 Nm).
 c. Install the main battery cable to the No. 3 system main relay with a new nut. Tighten to 50 inch lbs. (6 Nm).
 d. Install the aluminum shield wire with a new nut. Tighten to 29 inch lbs. (3 Nm).
 e. Install the terminal cover.
12. Wearing insulating gloves, install No. 3 battery carrier cushion
 a. Degrease and clean the installation surface of the battery carrier cushion.
 b. Install a new battery carrier cushion.
13. Wearing insulating gloves, install No. 1 wire harness protector cover
14. Wearing insulating gloves, install No. 3 wire harness protector cover
15. Install battery cover
16. Install the high voltage battery to the vehicle.
17. Install service plug grip.
18. Connect the negative battery cable.
19. Perform the power window initialization procedure. Refer to Interior, Power Windows, Initialization Procedure.

MOTOR ELECTRONICS RADIATOR

REMOVAL & INSTALLATION

1. Remove the condenser with receiver assembly.
2. Remove the radiator assembly.
 a. Remove the 4 bolts and radiator assembly from the condenser assembly.

To install:

3. To install, reverse the removal procedure. Tighten the bolts to 80 inch lbs. (9 Nm).

WATER PUMP WITH MOTOR

REMOVAL & INSTALLATION

See Figure 55.

1. Remove the front spoiler cover (w/ front spoiler).
2. Remove the engine under cover (w/ cover).
3. Remove the No. 1 engine under cover.
4. Drain the coolant (inverter).
5. Remove the inverter with converter assembly.
6. Remove the inverter reserve tank assembly.
7. Remove the inverter tray bracket.

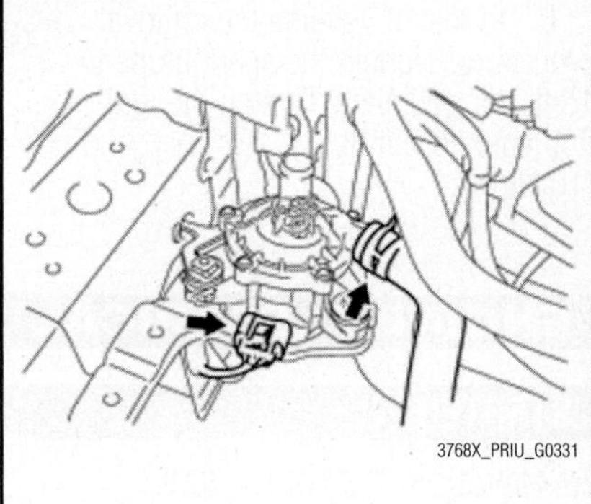

Fig. 55 Disconnecting the water hose and connector

8. Remove the water pump with the motor assembly.
 a. Disconnect the water hose and connector.
 b. Remove the 3 bolts and the water pump with motor assembly with the water pump bracket.

To install:

9. To install, reverse the removal procedure. Tighten the water pump bolts to 54 inch lbs. (6.1 Nm).

ENGINE ELECTRICAL

IGNITION SYSTEM

FIRING ORDER

See Figure 56.

IGNITION COIL

REMOVAL & INSTALLATION

1.5L Engine

1. Remove the ignition coils.

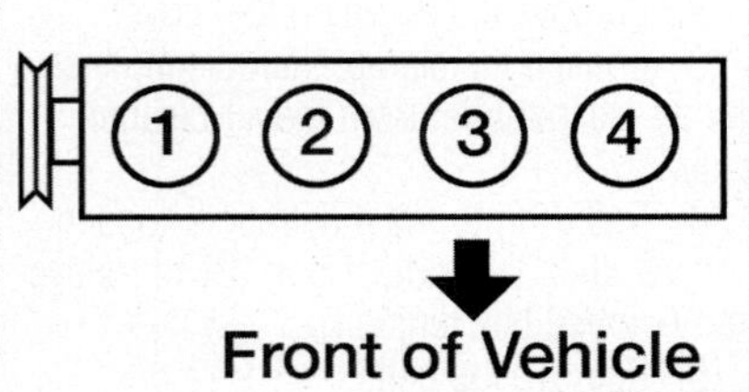

Fig. 56 1.5L I4 Hybrid engine Firing order: 1–3–4–2 Distributorless ignition system

2. Using a 16 mm plug wrench, remove the spark plugs.

To install:

3. Using a 16 mm plug wrench, install the spark plugs and tighten to 17.5 Nm (13 ft. lbs.)
4. Reinstall the ignition coils. Tighten to 80 inch lbs. (9 Nm).

1.8L Engine

See Figure 57.

1. Remove the windshield wiper motor and link.
2. Remove the outer cowl top panel sub assembly.
3. Remove the No. 2 cylinder head cover.
4. Remove the ignition coil assembly.
 a. Disconnect the 4 ignition coil connectors.
 b. Remove the 4 bolts and 4 ignition coils.

➡When removing each ignition coil, do not damage the plug cap on the engine head cover opening or the upper edge of the spark plug tube.

5. Remove the spark plugs.
 a. Using a 14 mm spark plug wrench, remove the 4 spark plugs.

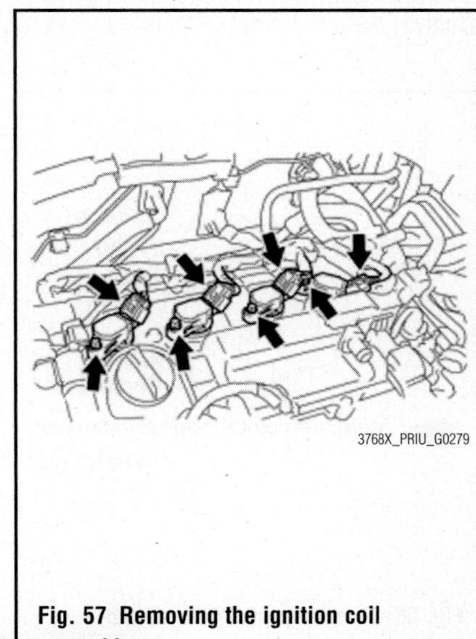

Fig. 57 Removing the ignition coil assembly

To install:

6. To install, reverse the removal procedure. Tighten the spark plugs to 15 ft. lbs. (20 Nm). Tighten the ignition coils to 7 ft. lbs. (10 Nm).

➡When installing each ignition coil, do not damage the plug cap on the engine head cover opening or the upper edge of the spark plug tube.

IGNITION TIMING

ADJUSTMENT

The ignition timing is controlled by the Powertrain Control Module (PCM). No adjustment is possible.

ENGINE ELECTRICAL

STARTING SYSTEM

STARTER

REMOVAL & INSTALLATION

The Toyota Prius, being a hybrid vehicle that utilizes both electric and gasoline (internal combustion) engine power for mobility, does not require (or come equipped with) a starter motor as a part of its starting system. The function of the starter motor is performed by a pair of electrical motor-generators, a computerized shunt system to control them, a mechanical power splitter that acts as a second differential, and a battery pack that serves as an energy reservoir.

ENGINE MECHANICAL

➡Disconnecting the negative battery cable may interfere with the functions of the on board computer systems and may require the computer to undergo a relearning process, once the negative battery cable is reconnected.

ACCESSORY DRIVE BELTS

ACCESSORY BELT ROUTING

See Figures 58 and 59.

INSPECTION

Inspect the drive belt for signs of glazing or cracking. A glazed belt will be perfectly smooth from slippage, while a good belt will have a slight texture of fabric visible. Cracks will usually start at the inner edge of the belt and run outward. All worn or damaged drive belts should be replaced immediately.

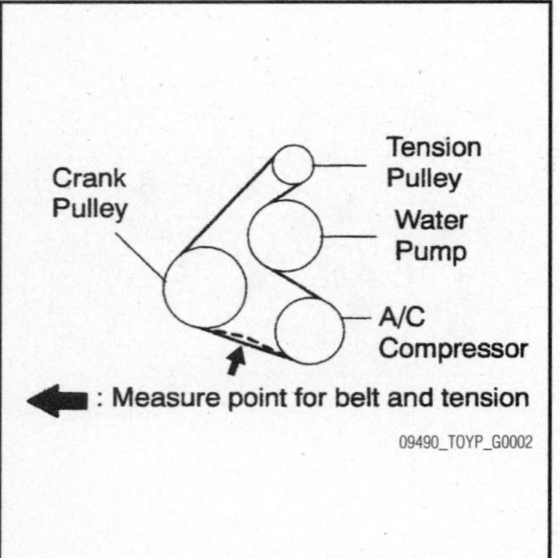

Fig. 58 Accessory drive belt routing—1NZ-FXE engine with A/C

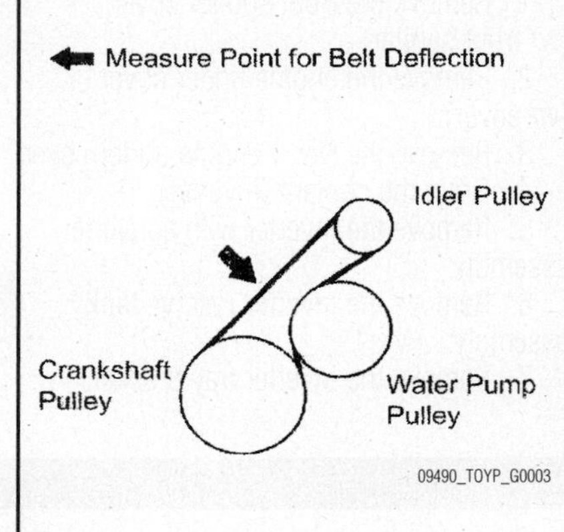

Fig. 59 Accessory drive belt routing—1NZ-FXE engine without A/C

ADJUSTMENT

See Figure 60.

1. Using a belt tension gauge, measure the belt tension.
   a. New belt tension should be 99 to 121 ft. lbs. used belt tension should be 55 to 77 ft. lbs.

➡A "new belt" is a belt which has been used less than 5 minutes on a running engine. A "used belt" is a belt which has been used on a running engine for 5 minutes or more.

   b. If tension is not as specified, adjust the belt by loosening nut "A", then turning adjust bolt "B" to provide the proper belt tension.
2. When not using a belt tension gauge, measure the belt deflection.
   a. Press the belt at the center of the longest span using 22ft. lbs. (10 kg) of force.

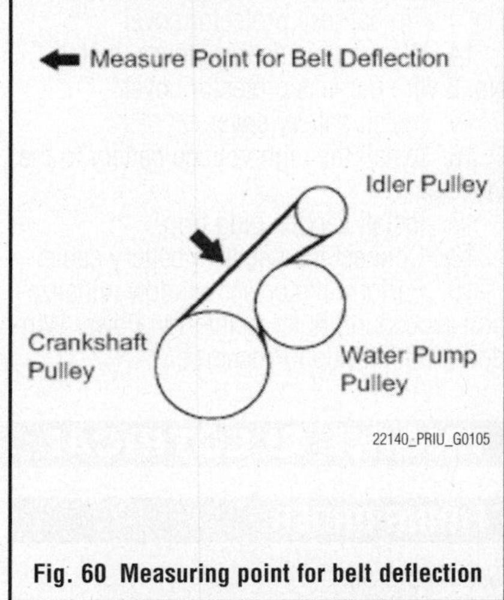

Fig. 60 Measuring point for belt deflection

   b. New belt tension should be 0.35 to 0.47 in. (9.0 to 12.0 mm), used belt tension should be 0.43 to 0.59 in. (11 to 15 mm)

REMOVAL & INSTALLATION

See Figure 61.

1. Before servicing the vehicle, refer to the precautions.
2. Remove the engine under cover.
3. Loosen the clamp, then disconnect the air cleaner inlet from the air cleaner case.
4. Referring to the illustration, loosen nut "A", then turn adjust bolt "B" to relieve the V-ribbed belt tension.
5. Remove the belt.

To install:

6. Install the V-ribbed belt to each pulley.

➡After installing the drive belt, check that it fits properly in the ribbed

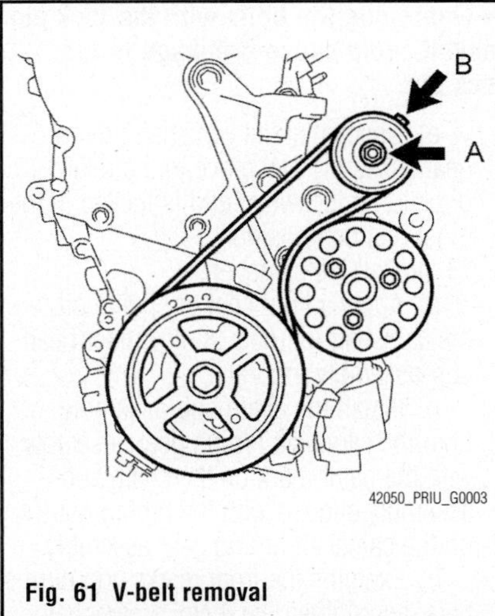

Fig. 61 V-belt removal

grooves. Check with your hands to confirm that the belt has not slipped out of the groove on the bottom of the crankshaft pulley. After installing a new belt, run the engine for approximately 5 minutes and then recheck the tension.

7. Turn adjust bolt "B" to adjust the V-ribbed belt tension.
8. Tighten nut "A" to 40 Nm (30 ft. lbs.).

AIR CLEANER

REMOVAL & INSTALLATION

1.8L Engine

See Figure 62.

1. Remove the air cleaner cap sub assembly.
   a. Disconnect the air flow meter connector.
   b. Disconnect the 2 clamps and hose band, and remove the air cleaner cap sub-assembly.
   c. Remove the air cleaner filter element.
2. Remove the inlet air cleaner assembly.
   a. Separate the water by pass hose from the inlet air cleaner assembly.
   b. Separate the wire harness clamp from the inlet air cleaner assembly.
   c. Remove the 3 bolts and inlet air cleaner assembly.

To install:

3. To install, reverse the removal procedure. Tighten the air cleaner assembly bolts to 62 inch lbs. (7 Nm).

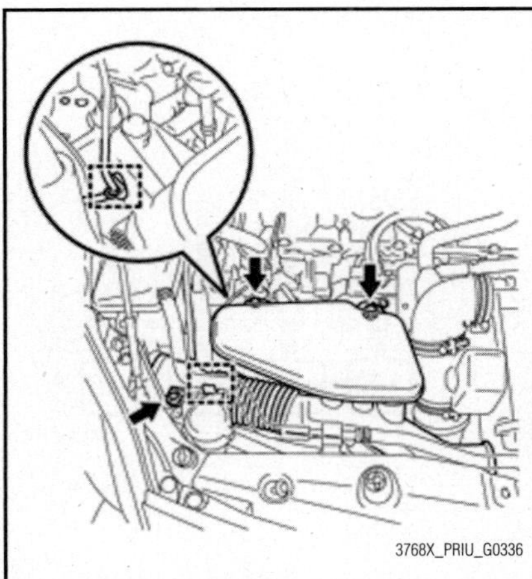

Fig. 62 Removing the inlet air cleaner assembly

CAMSHAFT AND VALVE LIFTERS

REMOVAL & INSTALLATION

1.5L Engine

See Figures 63 through 69.

1. Remove the EGR cooler assembly.
2. Remove the ignition coil assembly.
3. Separate the engine wire.
   a. Disconnect the 4 injector connectors.
   b. Disconnect the wire harness clamp.
   c. Remove the 3 bolts and separate the engine wire.
4. Remove the cylinde3r head cover sub assembly.
   a. Disconnect the No. 1 ventilation hose and No. 2 ventilation hose.
   b. Remove the 9 bolts, 2 seal washers, 2 nuts and cylinder head cover sub assembly.
   c. Remove the gasket from the cylinder head cover sub assembly.
5. Set No. 1 cylinder to TDC/COMPRESSION.
6. Remove the screw plug.
7. Remove the No. 2 camshaft.
   a. Loosen the bolts on the camshaft timing gear or sprocket while holding the hexagonal portion of the No. 2 camshaft.
   b. Insert a screwdriver from the plug hole and turn the stopper plate of the No. 1 chain tensioner assembly clockwise to release the lock, and keep it as it is.

➡The plunger of the No. 1 chain tensioner assembly is locked.

➡If the stopper plate is locked firmly, slightly turn the hexagonal portion of the camshaft to the right and left.

   c. Slightly turn the hexagonal portion of the No. 2 camshaft clockwise so that the plunger of the No. 1 chain tensioner assembly is pushed by the chain sub assembly.
   d. Insert a screwdriver from the plug hole and turn the stopper plate of the No. 1 chain tensioner assembly counterclockwise to the lock, and keep it as it is. Insert the hexagon wrench into the stopper plate hole.

➡Hold the hexagon wrench with tape so that the hexagon wrench does not come off.

   e. Using several steps, uniformly loosen and remove the 11 bolts in sequence. Then remove the No. 1 camshaft bearing cap and 4 No. 2 camshaft bearing caps.

➡Uniformly loosen the bolts while keeping the No. 2 camshaft level.

   f. Remove the bolt with the No. 2 camshaft lifted up.

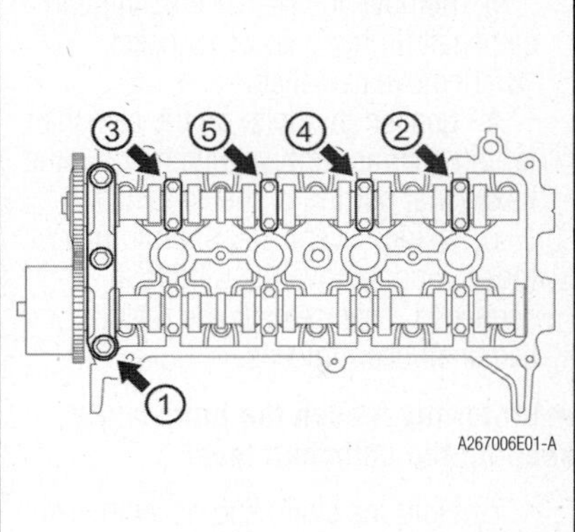

Fig. 63 Identifying bolt loosening sequence

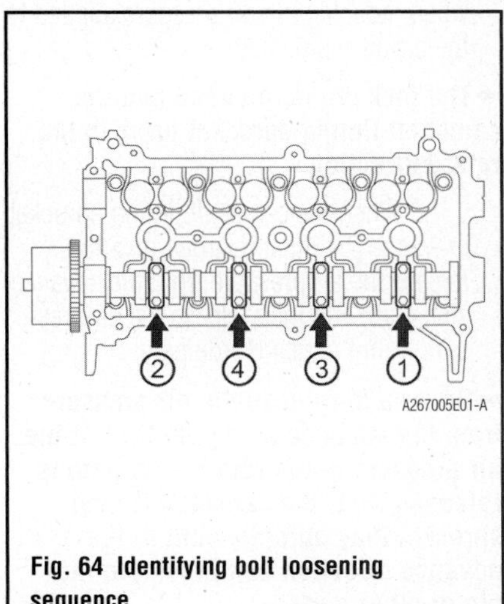

Fig. 64 Identifying bolt loosening sequence

g. Remove the No. 2 camshaft and camshaft timing gear or sprocket.

8. Remove camshaft.

a. Loosen the bolts on the camshaft timing gear assembly while holding the hexagonal portion of the camshaft.

b. Using several steps, uniformly loosen and remove the 8 bolts in sequence. Then remove the 4 No. 2 camshaft bearing caps.

➡Uniformly loosen the bolts while keeping the camshaft level.

c. Hold the chain sub-assembly with one hand, and remove the camshaft.

d. Tie the chain sub-assembly with a string or similar.

➡Be careful not to drop anything inside the timing chain cover.

9. Inspect the camshaft timing gear assembly.

a. Check the camshaft timing gear assembly lock. Clamp the camshaft in a vise, and make sure that the camshaft timing gear assembly does not rotate.

10. Release the lock pin. Cover the 4 oil paths of the cam journal with vinyl tape. Prick a hole in the tape placed on the advance side path. Prick a hole in the tape placed on the retard side path, on the opposite side to that of the advance side path. Apply approximately 22 psi of air pressure to the 2 paths (the advance side path and the retard side path).

➡4 oil paths are provided in the groove. Plug 2 paths with rubber pieces.

➡Cover the paths with a piece of cloth when applying pressure to keep oil from splashing.

a. Make sure that the camshaft timing sprocket turns in the retard direction when reducing the air pressure applied to the advance side path.

➡The lock pin is released and the camshaft timing sprocket turns in the retard direction.

b. When the camshaft timing sprocket moves to the most retarded position, release the air pressure from the advance side path, and then release the air pressure from the retard side path.

➡Be sure to release the air pressure from the advance side path first. If the air pressure of the retard side path is released first, the camshaft timing sprocket may abruptly shift in the advance direction and break the lock pin or other parts.

11. Remove the camshaft timing gear assembly.

a. Remove the bolt and camshaft timing gear assembly.

⁂ WARNING

Be sure not to remove the other 4 bolts.

➡When reusing the camshaft timing gear assembly, release the lock pin first, then install the camshaft timing gear assembly.

To install:

12. Install the camshaft timing gear assembly.

➡Perform inspection after repairs.

a. Put the camshaft timing gear assembly and camshaft together with the straight pin and key groove misaligned.

b. Turn the camshaft timing gear assembly while pushing it gently against the camshaft. Push further at the position where the pin fits into the groove.

➡Be sure not to turn the camshaft timing gear assembly to the retard direction (clockwise).

c. Check that there is no clearance between the camshaft timing gear assembly and camshaft flange.

d. Tighten the bolt with the camshaft timing gear assembly secured in place. Tighten to 47 ft. lbs. (64 Nm).

➡Do not lock the camshaft timing gear assembly when tightening the bolt.

➡Release the lock pin of the camshaft timing gear assembly first, and tighten the bolt when the lock pin is locked in the most retarded position.

➡Tightening the bolts with the lock pin locked could cause breakage of the lock pin.

e. Check that the camshaft timing gear assembly can move to the retard direction (clockwise) and is locked in the most retarded position.

13. Install the camshaft.

a. Apply a light coat of engine oil to the camshaft journals and cylinder head sub-assembly journals.

b. Install the chain sub-assembly onto the camshaft timing gear assembly with the paint mark on the chain sub-assembly aligned with the timing mark on the camshaft timing gear assembly.

c. Examine the front marks and numbers, and tighten the 4 No. 2 camshaft bearing caps with the 8 bolts in sequence. Tighten to 9 ft. lbs. (13 Nm).

➡Tighten each bolt uniformly, keeping the camshaft level.

14. Install the No. 2 camshaft.

a. Apply a light coat of engine oil to the No. 2 camshaft journals and cylinder head sub-assembly journals.

b. Put the camshaft timing gear or sprocket on the cylinder head sub-assembly with the paint mark on the chain sub-assembly aligned with the timing mark on the camshaft timing gear or sprocket.

➡There are 3 marks on the camshaft timing gear or sprocket. Make sure to align the mark plate with the timing mark (rectangle).

c. While holding the No. 2 camshaft by hand, temporarily install the No. 2 camshaft to the camshaft timing gear or sprocket using the bolt.

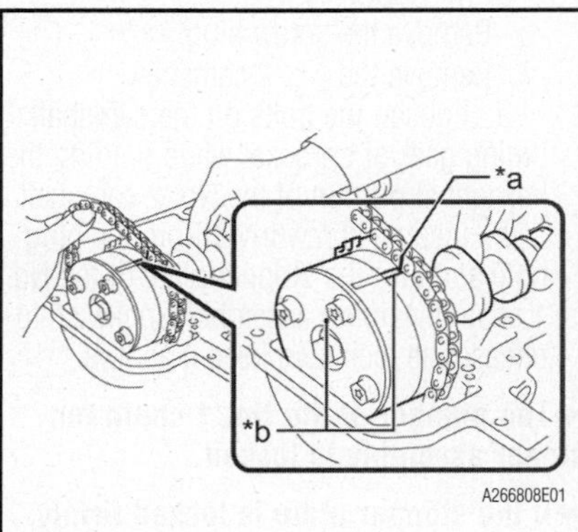

Fig. 65 Installing the chain sub assembly onto the camshaft timing gear assembly (a. paint mark, b. timing mark)

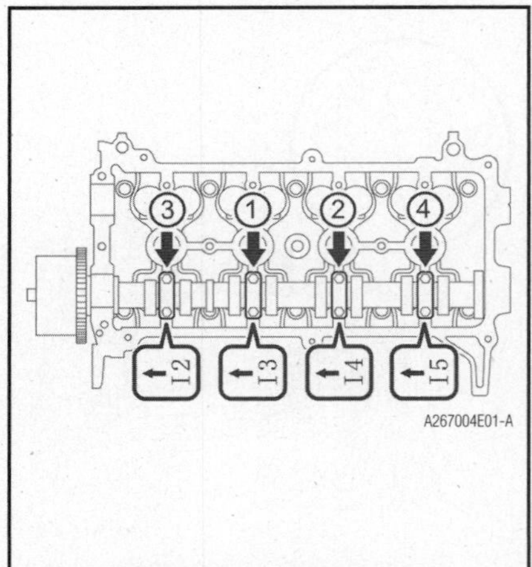

Fig. 66 Identifying bolt tightening sequence

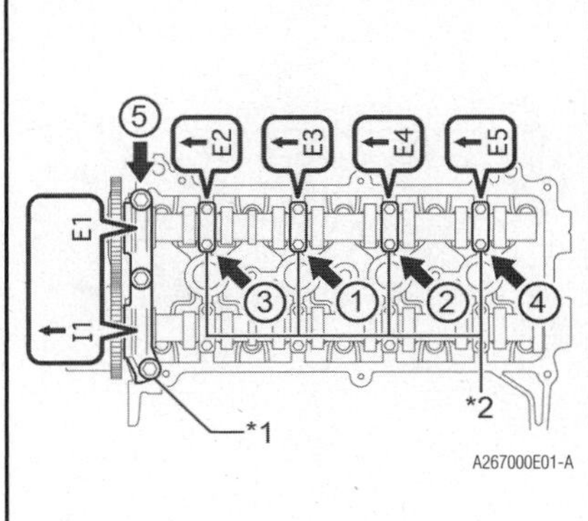

Fig. 67 Identifying bolt tightening sequence and No. 1 camshaft bearing cap (1) and No. 2 camshaft bearing cap (2)

d. Examine the front marks and numbers, and tighten the No. 1 camshaft bearing cap and No. 2 camshaft bearing caps with the 11 bolts in sequence. Tighten the No. 2 camshaft bearing cap to 9 ft. lbs. (13 Nm). Tighten the No. 1 camshaft bearing cap to 17 ft. lbs. (23 Nm).

➡Tighten each bolt uniformly, keeping the No. 2 camshaft level.

e. Using a 14 mm union nut wrench, tighten the bolts onto the camshaft timing gear or sprocket while holding the hexagonal portion of the No. 2 camshaft. Tighten to 47 ft. lbs. (64 Nm).

➡Use the torque value compensation formula to calculate the torque value for use when a torque wrench is combined with a tool such as a union nut wrench.

f. Remove the hexagon wrench from the No. 1 chain tensioner assembly.

g. Turn the crankshaft damper until its notch and timing mark "0" of the timing chain cover are aligned.

➡There are 3 marks on the camshaft timing gear or sprocket. Make sure that the timing mark (rectangle) is at the top.

h. Check that all the pairs of timing marks are aligned.

15. Install the screw plug.

16. Install the cylinder head cover sub assembly.

a. Install a new gasket to the cylinder head cover sub assembly.

b. Apply seal packing to the 2 locations.

➡Remove any oil from the contact surface.

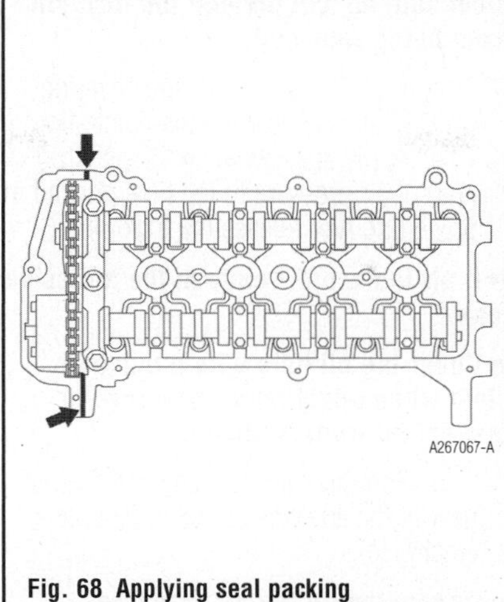

Fig. 68 Applying seal packing

➡Install the cylinder head cover within 3 minutes of applying seal packing.

⁂ WARNING

Do not start the engine for at least 2 hours after installation.

c. Temporarily install the cylinder head cover sub-assembly with the 9 bolts, 2 nuts and 2 new seal washers.

d. Using several steps, uniformly tighten the bolts and nuts in sequence. Tighten to 10 ft. lbs. (13 Nm).

e. Connect the No. 1 ventilation hose and the No. 2 ventilation hose.

17. Connect the engine wire.

a. Install the engine wire with the 3 bolts. Tighten to 74 inch lbs. (8.4 Nm).

b. Connect the wire harness clamp.

c. Connect the 4 injector connectors.

18. Install the ignition coil assembly.

19. Install the EGR cooler assembly.

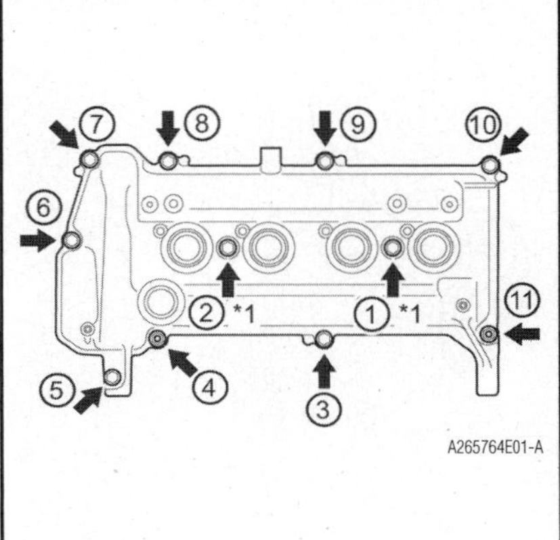

Fig. 69 Identifying bolt and nut tightening sequence

1.8L Engine

See Figures 70 through 82.

1. Remove the engine assembly with the transaxle.
2. Install the engine on an engine stand.
3. Remove the engine hangers.
4. Remove the throttle body assembly.
5. Remove the engine oil level dipstick guide.
6. Remove the EGR pipe assembly.
7. Remove the EGR valve assembly.
8. Remove the EGR with the cooler pipe sub assembly.
9. Remove the intake manifold.
10. Remove the fuel vapor feed pipe.
11. Remove the fuel delivery pipe sub assembly.
12. Remove the No. 1 delivery pipe spacer.
13. Remove the fuel injector assembly.
14. Remove the ignition coil assembly.
15. Remove the cylinder head cover sub assembly.
16. Remove the cylinder head cover gasket.
17. Remove the spark plug tube gasket.
18. Set the No. 1 cylinder to TDC/Compression.

a. Turn the crankshaft pulley until its notch and timing mark "0" of the timing chain cover are aligned.

b. Check that timing marks on both the camshaft timing sprocket and camshaft timing gear are facing upward as shown in the illustration.

c. If not, turn the crankshaft 1 complete revolution (360°) and align the marks.

19. Remove the crankshaft pulley.
20. Remove the No. 1 chain tensioner assembly.

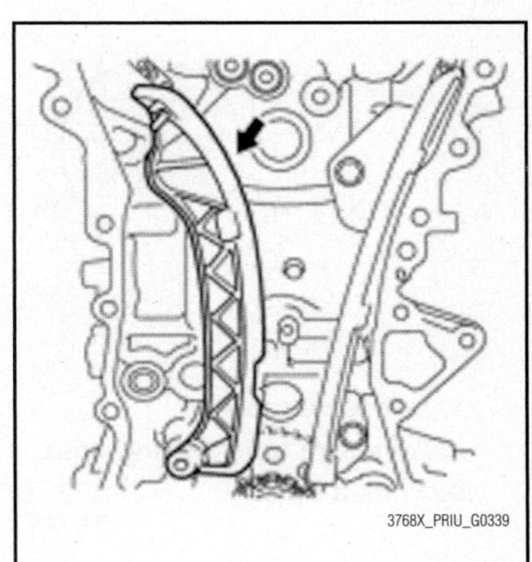

Fig. 70 Removing the chain tensioner slipper

a. Remove the 2 nuts, bracket, chain tensioner and gasket.

➡**Do not turn the crankshaft without the No. 1 chain tensioner installed.**

21. Remove the timing chain cover sub assembly.
22. Remove the timing chain cover oil seal.
23. Remove the chain tensioner slipper.
   a. Remove the chain tensioner slipper from the cylinder block.
24. Remove the 2 bolts and chain vibration damper.
25. Remove the 2 bolts and No. 2 chain vibration damper.
26. Remove the chain sub assembly.
   a. Hold the hexagonal portion of the camshaft with a wrench and turn the camshaft timing gear counterclockwise to loosen the chain between the camshaft timing gears.
   b. With the chain loosened, release the chain from the camshaft timing gear and place it on the camshaft timing gear.

➡**Be sure to release the chain from the sprocket completely.**

   c. Turn the camshaft clockwise to return it to the original position and remove the chain.
27. Inspect the camshaft timing gear assembly.
   a. Inspect the lock of the camshaft timing gear.
   b. After cleaning and degreasing the VVT oil hole on the intake side of the No. 1 camshaft bearing cap, completely seal the oil hole with adhesive tape or equivalent as shown in the illustration to prevent air from leaking.

➡**Be sure to cover the oil hole completely because air leaks due to insufficient sealing will prevent the lock pin from being released.**

   c. Prick a hole in the tape covering the oil hole as shown in the illustration.
   d. Apply approximately 22 psi (150 kPa) of air pressure to the hole pricked in procedure A to release the lock pin.

➡**If air leaks out, reattach the adhesive tape.**

➡**Cover the oil hole with a piece of cloth when applying air pressure to prevent oil from spraying.**

   e. Forcibly turn the camshaft timing gear in the advance direction (counterclockwise).

➡**Depending on the air pressure applied, the camshaft timing gear may turn in the advance direction without assistance.**

   f. Turn the camshaft timing gear within its movable range (26.5 to 28.5°) 2 or 3 times without turning it to the most retarded position. Make sure that the camshaft timing gear turns smoothly.
   g. Remove the adhesive tape from the No. 1 camshaft bearing cap.
28. Remove the camshaft timing gear assembly.
   a. Remove the flange bolt while holding the hexagonal portion of the camshaft with a wrench, and then remove the camshaft timing gear.

➡**Before removing the camshaft timing gear, make sure that the lock pin has been released.**

➡**Be sure not to remove the other 4 bolts.**

➡**Keep the camshaft timing gear horizontal while removing it from the camshaft.**

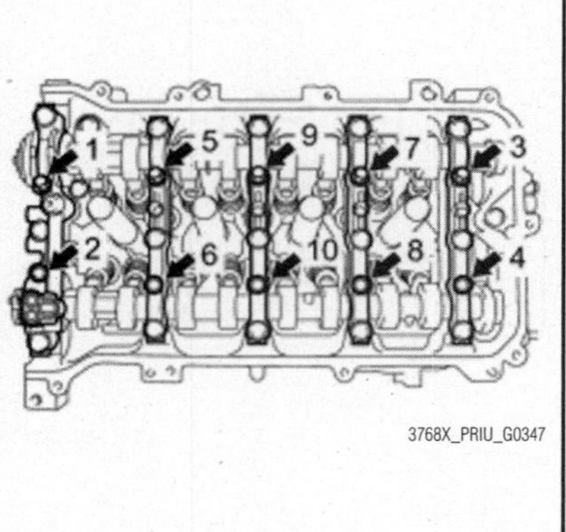

Fig. 73 Identifying the 10 bearing cap removal sequence

29. Remove the camshaft timing sprocket.
   a. Remove the flange bolt while holding the hexagonal portion of the camshaft with a wrench, and then remove the camshaft timing sprocket.
30. Remove the camshaft bearing caps.
   a. Uniformly loosen and remove the 10 bearing cap bolts in the sequence shown in the illustration.
   b. Uniformly loosen and remove the 15 bearing cap bolts in the sequence shown in the illustration.

➡**Uniformly loosen the bearing cap bolts while keeping the camshaft housing level.**

   c. Remove the 5 bearing caps.

➡**Arrange the removed parts in the correct order.**

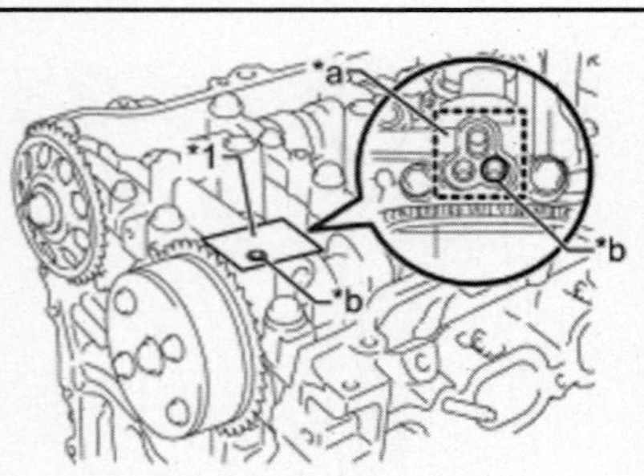

Fig. 71 Sealing the hole with adhesive tape

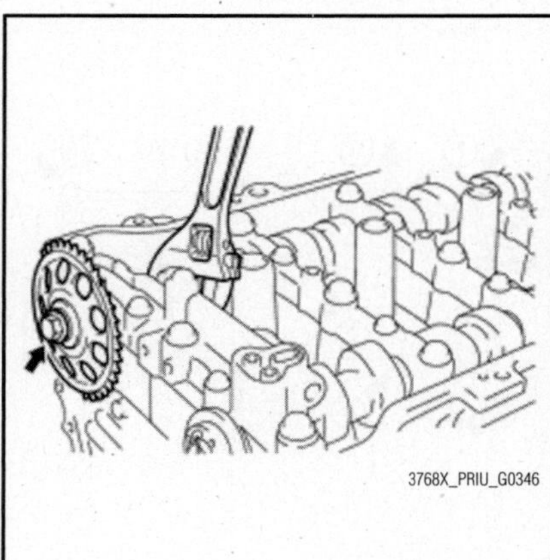

Fig. 72 Removing the camshaft timing sprocket

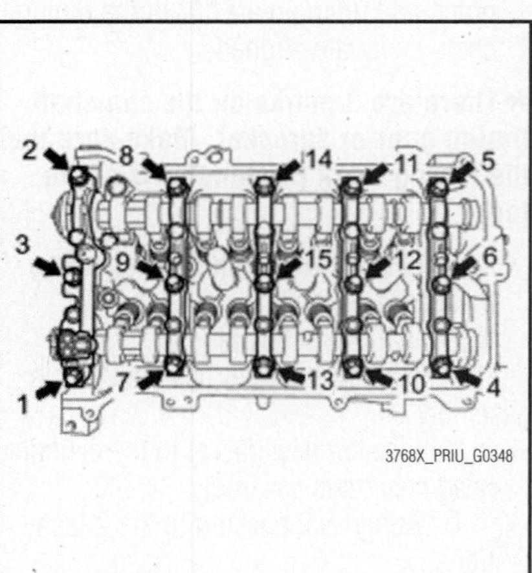

Fig. 74 Identifying the 15 bearing cap removal sequence

31. Remove the camshaft.
32. Remove the No. 2 camshaft.
33. Remove the 2 No. 1 camshaft bearings.
34. Remove the 2 No. 2 camshaft bearings.
35. Remove the camshaft housing sub assembly.
   a. Remove the 2 bolts.
   b. Remove the camshaft housing by prying between the cylinder head and camshaft housing with a screwdriver.

➡Be careful not to damage the contact surfaces of the cylinder head and camshaft housing.

➡Tape (*1) the screwdriver tip before use.

To install:

36. Install the No. 1 camshaft bearing.
   a. Clean both surfaces of the bearings.
   b. Install the 2 No. 1 camshaft bearings.
   c. Using a vernier caliper, measure the distance between the bearing cap edge and the camshaft bearing edge. The standard dimension is (A-B): 0.0276 inch (0.7 mm) or less.

➡Position the bearings to the center of the bearing cap by measuring the dimensions A and B.

37. Install the No. 2 camshaft bearing.
   a. Clean both surfaces of the bearings.
   b. Install the 2 No. 2 camshaft bearings.
   c. Using a vernier caliper, measure the distance between the bearing cap edge and the camshaft bearing edge. The standard dimension for A is 0.0413-0.0689 inch (1.05-1.75 mm).

➡Positing the bearings to the center of the bearing cap by measuring dimension A.

38. Install the No. 2 camshaft.
   a. Clean the camshaft journals.
   b. Apply a light coat of engine oil to the camshaft journals, camshaft housings and bearing caps.
   c. Install the No. 2 camshaft to the camshaft housing.
39. Install the camshaft.
   a. Clean the camshaft journals.
   b. Apply a light coat of engine oil to the camshaft journals, camshaft housings and bearing caps.
   c. Install the camshaft to the camshaft housing.

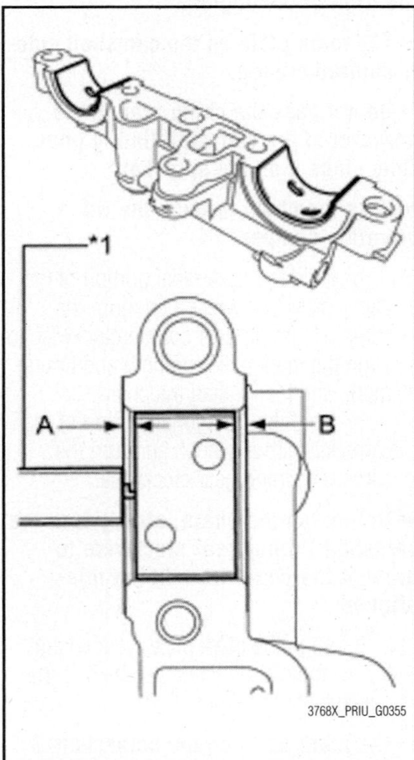

Fig. 75 Measuring the distance between the bearing cap edge and the camshaft bearing edge

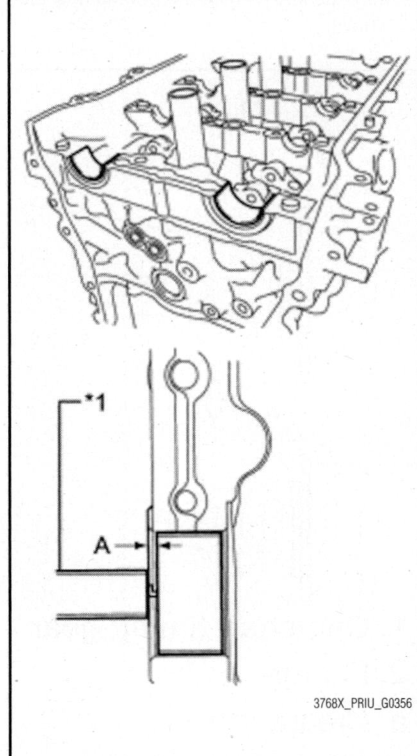

Fig. 76 Installing the No. 2 camshaft bearing

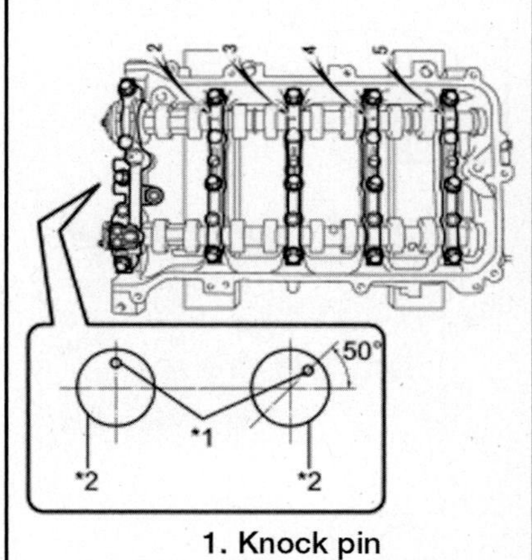

Fig. 77 Positioning the camshaft bearing caps

40. Install the bearing cap.
   a. Apply engine oil to the camshaft journals, camshaft housings and bearing caps.
   b. Make sure of the marks and numbers on the camshaft bearing caps and place them in each proper position and direction.
   c. Tighten the 10 bolts in the order shown. Tighten to 12 ft. lbs. (16 Nm).
41. Install the camshaft housing sub assembly.
   a. Check that the valve rocker arms are installed as shown.
   b. Apply seal packing in a continuous line as shown in the illustration.

➡Remove any oil from the contact surfaces.

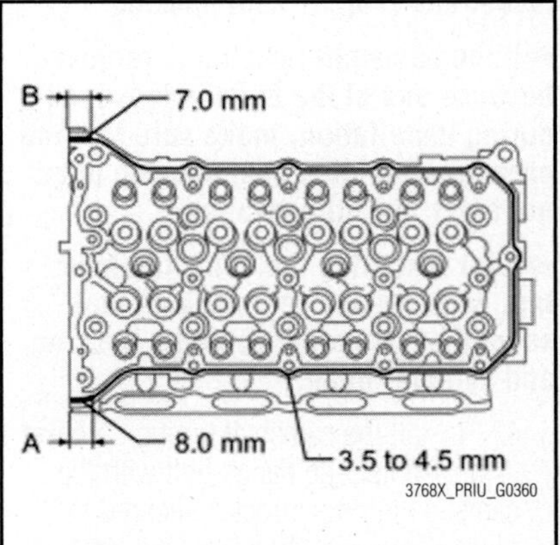

Fig. 78 Applying seal packing in a continuous line

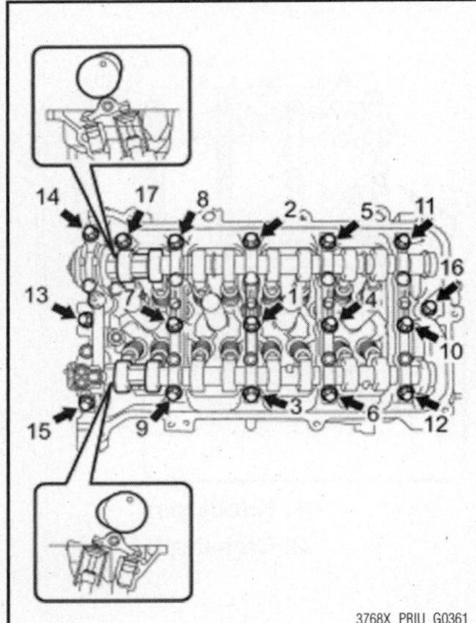

Fig. 79 Setting the camshaft and No. 2 camshaft and identifying the bolt tightening sequence

➡**Install the camshaft housing within 3 minutes and tighten the bolts within 10 minutes of applying seal packing.**

➡**Do not start the engine for at least 2 hours after installation.**

c. Set the camshaft and No. 2 camshaft as shown in the illustration.

d. Install the camshaft housing with the 17 bolts and tighten them in the order shown in the illustration. Tighten to 20 ft. lbs. (27 Nm).

➡**After installing the camshaft housing, make sure that the cam lobes are positioned as shown in the illustration.**

➡**If any of the bolts is loosened during installation, remove the camshaft housing, clean the installation surfaces, and reapply seal packing.**

➡**If the camshaft housing is removed because any of the bolts is loosened during installation, make sure that the previously applied seal packing does not enter any oil passages.**

➡**After installing the camshaft housing, wipe off any seal packing that seeped out from between the housing and cylinder head.**

42. Install the camshaft timing sprocket.

a. Tighten the flange bolt with the camshaft timing sprocket secured in place. Tighten to 40 ft. lbs. (54 Nm).

43. Install the camshaft timing gear assembly.

a. Put the camshaft timing gear and

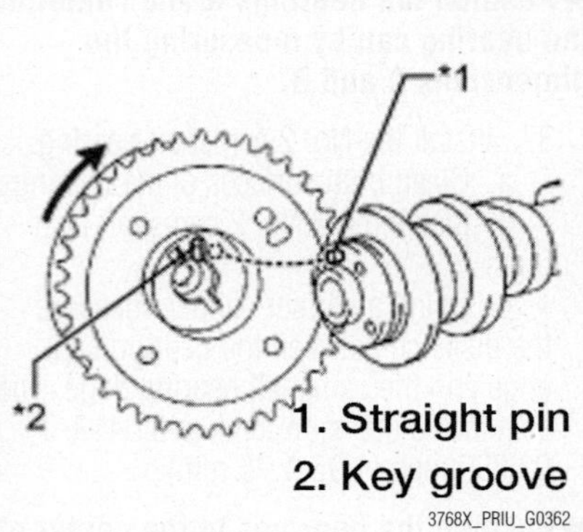

Fig. 80 Putting the camshaft timing gear and camshaft together

camshaft together with the straight pin and key groove misaligned as shown in the illustration.

➡**Do not forcefully push in the camshaft timing gear. This may cause the camshaft straight pin tip to damage the installation surface of the camshaft timing gear.**

b. Turn the camshaft timing gear as shown in the illustration while pushing it gently against the camshaft. Push further at the position where the pin fits into the groove.

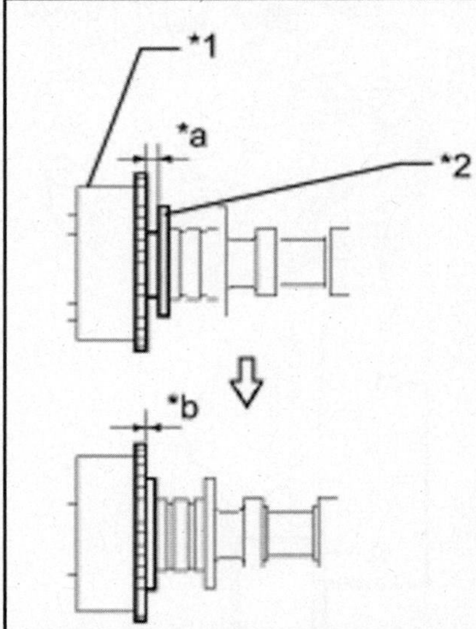

Fig. 81 Checking for clearance

➡**Do not turn the camshaft timing gear in the retard direction (clockwise).**

c. Check that there is no clearance between the camshaft timing gear and camshaft flange.

d. Tighten the flange bolt with the camshaft timing gear secured in place. Tighten to 40 ft. lbs. (54 Nm).

e. Check that the camshaft timing gear can move in the retard direction (clockwise) and is locked in the most retarded position.

44. Install the No. 1 chain vibration damper with the 2 bolts. Tighten to 15 ft. lbs. (21 Nm).

45. Set the No. 1 cylinder to TDC/Compression.

a. Temporarily install the crankshaft pulley bolt.

b. Turn the crankshaft to position the timing gear key to the top.

c. Check that the timing marks (*1) on the camshaft timing gear and camshaft timing sprocket are aligned.

d. Remove the crankshaft pulley bolt.

46. Install the chain sub assembly.

a. Align the mark plate (orange) with the timing mark as shown in the illustration and install the chain.

➡**Be sure to position the mark plate at the front of the engine.**

➡**The mark plate on the camshaft side is colored orange.**

➡**Do not pass the chain around the sprocket of the camshaft timing gear. Only place it on the sprocket.**

➡**Pass the chain through the No. 1 vibration damper.**

b. Hold the hexagonal portion of the camshaft with a wrench and turn the camshaft timing gear counterclockwise to align the mark plate (orange) and timing mark, and then install the chain.

c. Hold the hexagonal portion of the camshaft with a wrench and turn the camshaft timing gear clockwise.

➡**To tension the chain, slowly turn the camshaft timing gear clockwise to prevent the chain from being misaligned.**

d. Align the mark plate (yellow) and timing mark and install the chain to the crankshaft timing gear.

➡**The mark plate on the crankshaft side is colored pink.**

47. Check the No. 1 cylinder to TDC/Compression.

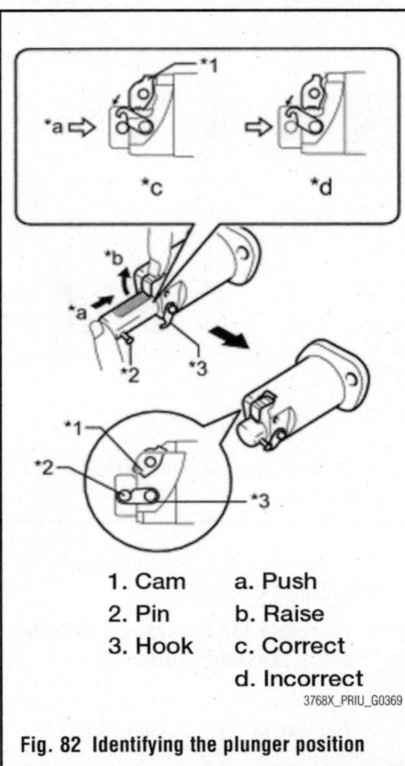

Fig. 82 Identifying the plunger position

a. Check each timing mark at TDC/Compression.

48. Install the No. 2 chain vibration damper with the 2 bolts. Tighten to 7 ft. lbs. (10 Nm).
49. Install the chain tensioner slipper to the cylinder block.
50. Install the timing chain cover oil seal.
51. Install the timing chain cover sub assembly.
52. Install the crankshaft pulley.
53. Install the No. 1 chain tensioner assembly.

a. Release the cam, and then fully push in the plunger and engage the hook to the pin so that the plunger is in the position shown in the illustration.

➡Make sure that the cam engages the first tooth of the plunger to allow the hook to pass over the pin.

b. Install a new gasket, the bracket and chain tensioner with 2 nuts. Tighten to 9 ft. lbs. (12 Nm).

➡If the hook releases the plunger while the chain tensioner is being installed, set the hook again.

c. Rotate the crankshaft counterclockwise slightly and check that the hook becomes released.

d. Turn the crankshaft clockwise and check that the plunger is extended.

54. Install the spark plug tube gasket.
55. Install the cylinder head cover gasket.
56. Install the cylinder head cover sub assembly.
57. Install the ignition coil assembly.
58. Install the fuel injector assembly.
59. Install the No. 1 delivery pipe spacer.
60. Install the fuel delivery pipe sub assembly.
61. Install the fuel vapor feed pipe.
62. Install the intake manifold.
63. Install the EGR with the cooler pipe sub assembly.
64. Install the EGR valve assembly.
65. Install the EGR pipe assembly.
66. Install the engine oil level dipstick guide.
67. Install the throttle body assembly.
68. Install the engine hangers.
69. Remove the engine from the engine stand.
70. Install the engine assembly with the transaxle.

CRANKSHAFT FRONT SEAL

REMOVAL & INSTALLATION

1.5L Engine

See Figure 83.

1. Remove front wheel RH.
2. Remove engine under cover RH.
3. Separate No. 2 engine room relay block and No. 2 junction block assembly.

a. Remove the bolt and 2 nuts, and separate the No. 2 engine room relay block and No. 2 junction block assembly.

4. Remove the engine mounting insulator sub assembly RH.

a. Place a jack underneath the engine, then put a wooden block on the jack.

➡To prevent the No. 2 oil pan from deforming, do not place any attachments onto the No. 2 oil pan.

b. Remove the 4 bolts, 2 nuts and engine mounting insulator sub-assembly RH.

5. Remove crankshaft damper sub assembly.

a. Using SST (09213-58012), hold the crankshaft damper sub-assembly in place and loosen the bolt.

➡When installing SST, be careful that the bolts which holds it does not interfere with the timing chain cover.

b. Using SST (09950-50013 (09951-05010, 09952-05010, 09953-05020, 09954-05021)), remove the crankshaft damper sub assembly and bolt.

6. Remove the oil pump seal.

a. Using a knife, cut off the lip of the oil pump seal.

b. Using a screwdriver with its tip wrapped with tape, pry out the oil seal.

➡Do not damage the surface of the oil seal press fit hole or the crankshaft.

➡Wrap the screwdriver tip with protective tape before use.

➡After removing, check the crankshaft for damage. If damaged, smooth the surface with 400-grit sandpaper.

To install:

7. Install the oil pump seal.

a. Apply MP grease to the lip of a new oil pump seal.

➡Keep the lip free from foreign matter.

b. Using SST (09223-2210) and a hammer, tap in the oil pump seal until its surface is flush with the front oil pump seal retainer edge.

c. Tap in depth: 0-1.0 mm from the edge of the timing chain cover.

➡Do not tap the oil pump seal at an angle.

➡Wipe off extra grease from the crankshaft.

8. Install the crankshaft damper sub assembly.

a. Align the hole of the crankshaft damper sub-assembly with the straight pin, then install the crankshaft damper sub-assembly.

b. Using SST (09213-58012 (91111-50845) 09330-00021), hold the crankshaft damper sub-assembly in place and tighten the bolt. Tighten to 94 ft. lbs. (128 Nm).

➡When installing SST, be careful that the bolt which holds SST does

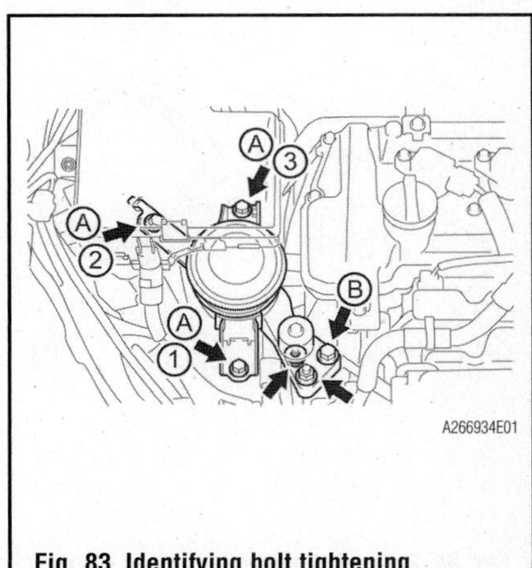

Fig. 83 Identifying bolt tightening sequence

not interfere with the timing chain cover.

9. Install the engine mounting insulator sub assembly RH.
   a. Temporarily install the engine mounting insulator sub assembly RH with the 4 bolts and 2 nuts.
   b. Tighten the 3 bolts (A) in sequence. Tighten to 38 ft. lbs. (52 Nm).
   c. Tighten the bolt (B) and 2 nuts. Tighten to 38 ft. lbs. (52 Nm).
10. Install the No. 2 engine room relay block and No. 2 junction block assembly.
   a. Install the No. 2 engine room relay block and No. 2 junction block with the 2 nuts and bolt. Tighten to 48 inch lbs. (5.4 Nm).
11. Inspect for oil leak.
12. Install the engine under cover RH.
13. Install the front wheel RH. Tighten to 76 ft. lbs. (103 Nm).

1.8L Engine

See Figure 84.

1. Remove the front RH wheel.
2. Remove the rear engine under cover RH.
3. Remove the rear side rail reinforcement sub assembly RH.
4. Remove the crankshaft pulley.
   a. Using the special tool (09213-58014, 91551-80840 or 09330-00021), hold the pulley in place and loosen the pulley bolt.
   b. Using the special tool (09951-05010, 09952-05010, 09953-05020 or 09954-05021), remove the crankshaft pulley and pulley bolt.

➡**If necessary, remove the pulley and pulley bolt using the special tool.**

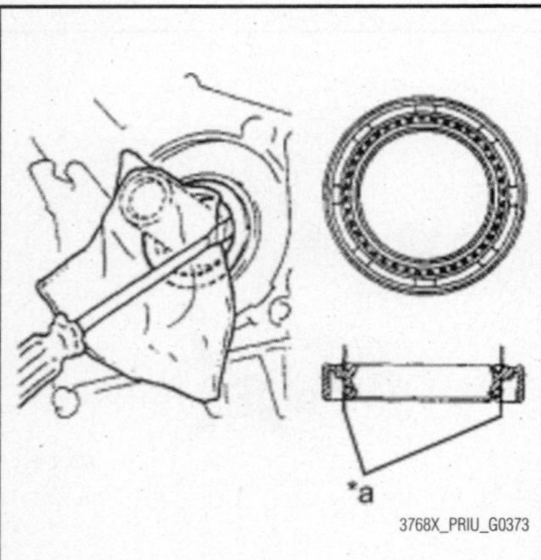

Fig. 84 Removing the timing chain cover oil seal

5. Remove the timing chain cover oil seal.
   a. Using a knife, cut off the lip of the oil seal.
   b. Using a screwdriver with the tip wrapped with tape, pry out the oil seal.

➡**After removing, check the crankshaft for damage. If damage, smooth the surface with 400 grit sandpaper.**

To install:

6. Install the timing chain cover oil seal.
   a. Apply MP grease to the lip of a new oil seal.

➡**Keep the lip free from foreign matter.**

   b. Using the special tool and a hammer, tap in the oil seal retainer edge.

➡**Wipe off extra grease from the crankshaft.**

➡**Do not tap the oil seal at an angle.**

7. Install the crankshaft pulley.
   a. Align the pulley set key with the key groove of the pulley.
   b. Using the special too, hold the pulley in place and tighten the bolt to 140 ft. lbs. (190 Nm).
8. Install the rear side rail reinforcement sub assembly RH.
9. Install the front wheel RH. Tighten to 76 ft. lbs. (103 Nm).
10. Inspect for oil leaks.
11. Install the rear engine under cover RH.

CYLINDER HEAD

REMOVAL & INSTALLATION

1.5L Engine

See Figures 85 through 92.

1. Remove timing chain cover.
2. Remove intake manifold stay.
3. Remove intake manifold.
4. Remove No. 1 exhaust manifold heat insulator.
5. Remove manifold support bracket.
6. Remove exhaust manifold.
7. Remove No. 1 chain tensioner assembly.
   a. Turn the stopper plate of the No. 1 chain tensioner assembly clockwise and push in the plunger with the lock released.
   b. Turn the stopper plate of the No. 1 chain tensioner assembly counterclockwise and insert a hexagon wrench into the hole in the stopper plate to lock with the plunger pushed in.

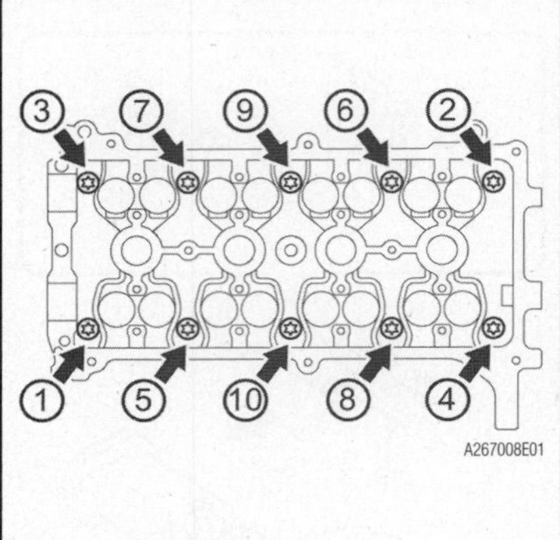

Fig. 85 Removing the cylinder head sub assembly

   c. Remove the 2 bolts and No. 1 chain tensioner assembly.

➡**Do not rotate the crankshaft with the No. 1 chain tensioner removed.**

8. Remove the chain tensioner slipper.
9. Remove the No. 1 chain vibration damper.
   a. Remove the 2 bolts and No. 1 chain vibration damper.
10. Remove the chain sub assembly.

➡**When rotating the camshaft with the chain sub-assembly removed, rotate the crankshaft counterclockwise 40° from TDC and align the oil jet hole with the paint mark to prevent the pistons from coming into contact with the valves.**

11. Remove the camshaft.
12. Remove the cylinder head sub assembly.
   a. Using several steps, uniformly loosen and remove the 10 cylinder head

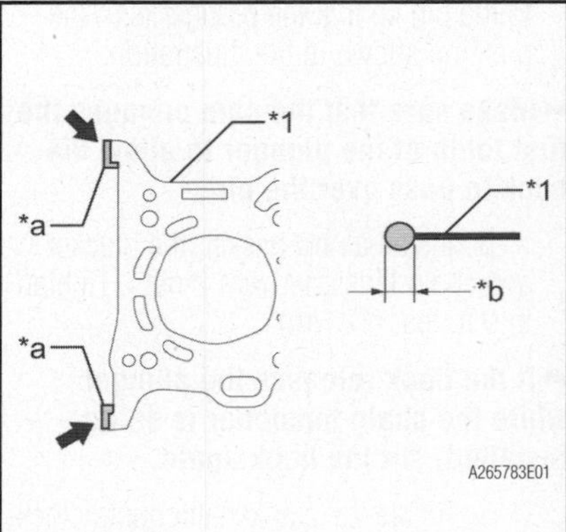

Fig. 86 Applying seal packing (a) to a new cylinder head gasket (1) (b. apply 0.177-0.217 inch)

bolts and 10 washers with an 8 mm bi-hexagon wrench in sequence.

⁂ WARNING

When removing the cylinder head bolt, do not drop the washer into the cylinder head sub-assembly.

⁂ WARNING

Removing the bolts in the wrong order may cause damage to the cylinder head sub-assembly.

13. Remove the cylinder head gasket from the cylinder head.

To install:

➡Perform "Inspection After Repairs" after replacing the cylinder head sub-assembly.

14. Inspect the cylinder head set bolt.
15. Install the cylinder head gasket.
   a. Apply seal packing to a new cylinder head gasket.

➡Use Toyota Genuine Seal Packing Black, Three Bond 1207B or equivalent.

➡Install the cylinder head sub-assembly within 3 minutes and tighten the cylinder head bolts within 15 minutes after applying seal packing.

   b. Place the cylinder head gasket on the cylinder block with the Lot No. facing upward.

⁂ WARNING

Remove any oil from the contact surface.

⁂ WARNING

Be careful of the installation direction.

16. Install cylinder head sub assembly.

➡Perform "Inspection After Repairs" after replacing the cylinder head sub assembly.

➡The cylinder head bolts are tightened in 2 progressive steps.

   a. Place the cylinder head sub assembly on the cylinder block.

⁂ WARNING

Ensure that no oil is on the mounting surface of the cylinder head sub-assembly.

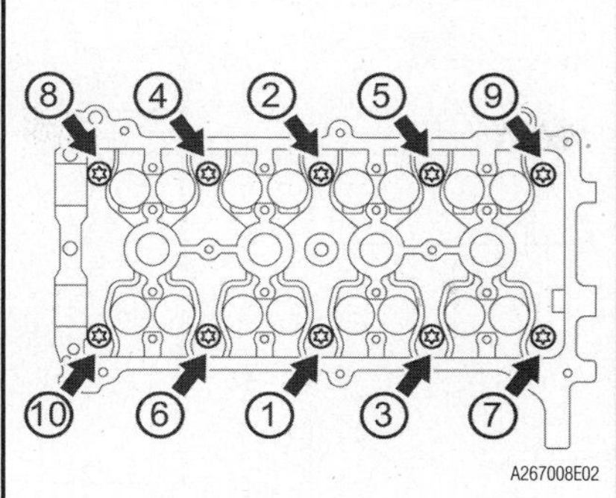

Fig. 87 Identifying the bolt tightening sequence

⁂ WARNING

Place the cylinder head sub-assembly gently in order not to damage the cylinder head gasket.

   b. Apply a light coat of engine oil to the threads and under the heads of the 10 cylinder head bolts.

⁂ WARNING

Ensure that no engine oil drips from any of the cylinder head bolts.

   c. Step 1: Using several steps, uniformly install and tighten the 10 cylinder head bolts and 10 washers with an 8 mm bi-hexagon wrench in order. Tighten to 22 ft. lbs. (29 Nm).
   d. Step 2: Mark each cylinder head bolt head with paint.
   e. Further tighten the cylinder head bolts 90°, and then an additional 90°.
   f. Check that the paint mark is now at a 180° angle to the front.
   g. After tightening the cylinder head bolts, wipe off the seal packing material that seeped out from the contact surface between the cylinder head sub-assembly and cylinder block.

17. Install the camshaft.
   a. Apply a light coat of engine oil to the camshaft and cylinder head sub-assembly journals.
   b. Place the camshaft and No. 2 camshaft on the cylinder head sub-assembly with the timing mark on the camshaft timing gear assembly and camshaft timing sprocket facing upward.

⁂ WARNING

There are 3 marks on the camshaft timing sprocket. Make sure that the timing mark (rectangle) is at the top.

   c. Check the front marks and numbers on the No. 1 camshaft bearing cap and 8 No. 2 camshaft bearing caps, then temporarily install them with the 19 bolts.
   d. Uniformly tighten the 19 bolts in sequence. Tighten No. 2 camshaft bearing cap to 9 ft. lbs. (13 Nm). Tighten No. 1 camshaft bearing cap to 17 ft. lbs. (23 Nm).

➡Uniformly tighten the bolts while keeping the camshaft level.

18. Install the chain sub assembly.
   a. Set the crankshaft in the position shown.

➡The crankshaft will be set to 90° ATDC and the timing mark will be at the location shown.

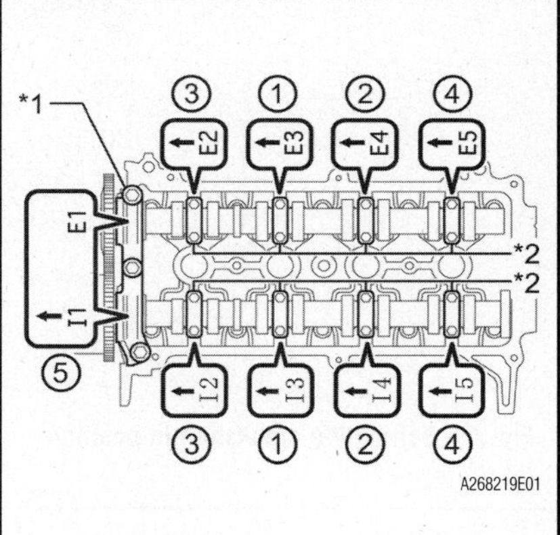

Fig. 88 Identifying the bearing caps and bolts and tightening sequence

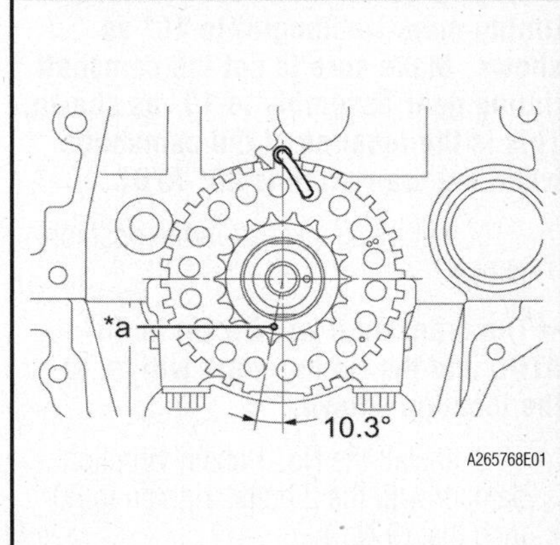

Fig. 89 Setting the crankshaft (a: timing mark)

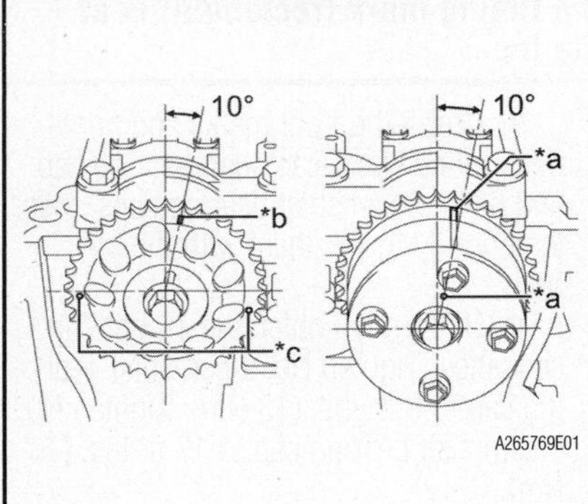

Fig. 90 Setting camshaft timing gear assembly and camshaft timing sprocket (a: timing mark, b: timing mark-rectangle, c: mark-circle)

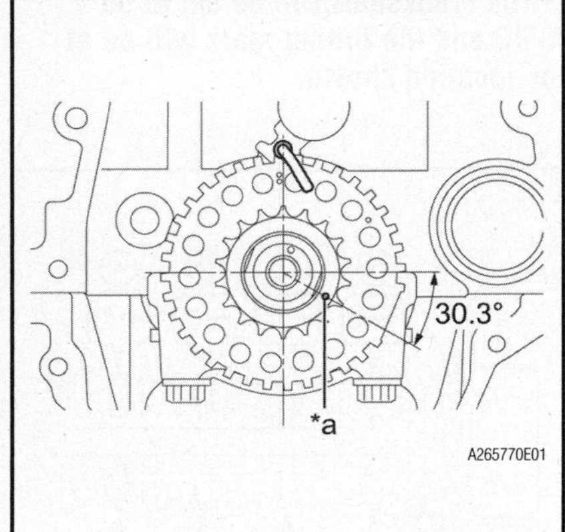

Fig. 91 Setting the crankshaft in position (a: timing mark)

b. Set the camshaft timing gear assembly and camshaft timing sprocket in the positions shown.

➡There are 3 marks on the camshaft timing sprocket. Make sure to set the timing mark (rectangle) to 10° as shown. Make sure to set the camshaft timing gear assembly to 10° as shown. This is the location of the camshafts when the crankshaft is 20° ATDC.

c. Set the crankshaft in the position shown.

➡The crankshaft will be set to 20° ATDC and the timing mark will be at the location shown.

d. Install the No. 1 chain vibration damper with the 2 bolts. Tighten to 80 inch lbs. (9 Nm).

e. Align the timing marks of the camshaft and crankshaft with the mark plates of the timing chain sub-assembly and install the timing chain sub-assembly.

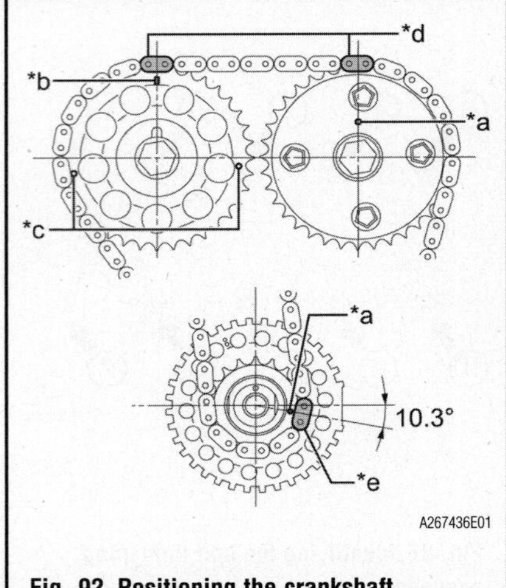

Fig. 92 Positioning the crankshaft

➡To prevent the No. 2 camshaft from springing back, turn it using a wrench and set it at the mark on the chain.

➡There are 3 marks on the camshaft timing sprocket. Make sure to align the mark plate with the timing mark (rectangle).

f. Install the chain tensioner slipper.

g. Install the No. 1 chain tensioner assembly with the 2 bolts. Tighten to 80 inch lbs. (9 Nm).

h. Remove the hexagon wrench from the No. 1 chain tensioner assembly.

i. Turn the crankshaft approximately 20° counterclockwise to set it to TDC. Make sure that the timing marks and mark plates are correctly positioned.

➡There are 3 marks on the camshaft timing sprocket. Make sure that the timing mark (rectangle) is at the top.

19. Install the exhaust manifold.
20. Install the manifold support bracket.
21. Install the No. 1 exhaust manifold heat insulator.
22. Install the intake manifold.
23. Install the intake manifold stay.
24. Install the timing chain cover.

1.8L Engine

See Figure 93.

1. Remove the camshaft housing sub assembly.
2. Remove the No. 1 exhaust manifold heat insulator.
3. Remove the manifold stay.
4. Remove the exhaust manifold.
5. Remove the No. 1 valve rocker arm sub assembly.

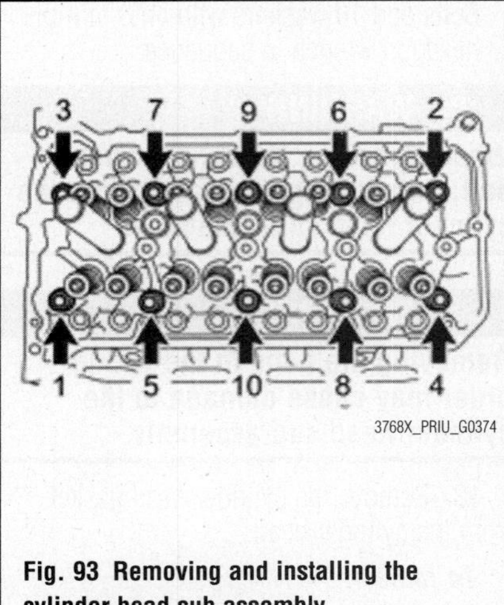

Fig. 93 Removing and installing the cylinder head sub assembly

6. Remove the valve lash adjuster assembly.
7. Remove the valve stem cap.
8. Remove the cylinder head sub assembly.

a. Using a 10 mm bi-hexagon wrench, uniformly loosen and remove the 10 cylinder head bolts and 10 plate washers in several steps in the sequence shown in the illustration.

➡Be careful not to drop washers into the cylinder head.

➡Head warpage or cracking could result from removing the bolts in the wrong order.

b. Using a screwdriver with its tip wrapped with tape, pry between the cylinder head and cylinder block, and remove the cylinder head.

➡Be careful not to damage the contact surfaces of the cylinder head and cylinder block.

To install:

➡The cylinder head bolts are tightened in 3 progressive steps.

c. Place the cylinder head on the cylinder block.

➡Make sure that no oil is on the mounting surface of the cylinder head.

➡Place the cylinder head on the cylinder block gently in order not to damage the gasket with the bottom part of the head.

d. Install the plate washers to the cylinder head bolts.

e. Apply a light coat of engine oil to the threads and under the heads of the cylinder head bolts.

f. Step 1: Using a 10 mm bi-hexagon wrench, install and uniformly tighten the 10 cylinder head bolts in several steps, in the sequence shown in the illustration. (Reverse the sequence shown). Tighten to 36 ft. lbs. (49 Nm).

➡Do not drop the plate washers into the cylinder head.

g. Step 2: Mark each cylinder head bolt head with paint. Tighten the cylinder head bolts 90° in the sequence shown.

h. Step 3: Tighten the cylinder head bolts another 45° in the same sequence as above. Check that the paint mark is now at a 135° angle to the front.

9. Install the valve stem cap.
10. Install the valve lash adjuster assembly.
11. Install the No. 1 valve rocker arm sub assembly.
12. Install the exhaust manifold.
13. Install the manifold stay.
14. Install the No. 2 exhaust manifold heat insulator.
15. Install the camshaft housing sub assembly.

ENGINE OIL & FILTER

REPLACEMENT

1.5L Engine

CAUTION

Prolonged and repeated contact with engine oil will result in the removal of natural oils from the skin, leading to dryness, irritation and dermatitis. In addition, used engine oil contains potentially harmful contaminants which may cause skin cancer.

CAUTION

Wear protective clothing and gloves. Avoid contact with used oil. If contact occurs, wash your skin thoroughly with soap or waterless hand cleaner. Never use gasoline, thinners, or solvents to wash the skin.

➡In order to protect the environment, dispose of used oil and used oil filters at designated disposal sites only.

1. Drain the engine oil.
   a. Remove the oil filler cap.
   b. Remove the oil pan drain plug and gasket, and drain the oil into a container.
   c. Clean and install the oil drain plug with a new gasket. Tighten to 28 ft. lbs. (38 Nm).
2. Remove oil filter sub assembly with SST (09228-06501).
3. Install the oil filter sub assembly.
   a. Check and clean the oil filter installation surface.
   b. Apply clean engine oil to the gasket of a new oil filter.
   c. Lightly screw the oil filter into place by hand. Tighten it until the gasket contacts the seat.
   d. Using SST, tighten the oil filter
   e. Depending on the work space available, choose from the following.
      - If enough space is available, use a torque wrench to tighten the oil filter to 10 ft. lbs. (13 Nm).
      - If enough space is not available to use a torque wrench, tighten the oil filter a ¾ turn by hand or use a common wrench.
4. Add engine oil and install the oil filler cap.
5. Inspect for oil leak.

1.8L Engine

➡Prolonged and repeated contact with engine oil will result in the removal of natural oils from the skin, leading to dryness, irritation and dermatitis. In addition, used engine oil contains potentially harmful contaminants which may cause skin cancer.

➡Wear protective clothing and gloves. Avoid contact with used oil. If contact occurs, wash your skin thoroughly with soap or waterless hand cleaner. Never use gasoline, thinners, or solvents to wash the skin.

➡In order to protect the environment, dispose of used oil and used oil filters at designated disposal sites only.

1. Remove the front spoiler cover (w/ spoiler).
2. Remove the engine under cover (w/ cover).
3. Remove the No. 1 engine under cover.
4. Drain the engine oil.
   a. Remove the oil filler cap.
   b. Remove the oil pan drain plug and gasket, and drain the oil into a container.
   c. Clean and install the oil pan drain plug with a new gasket. Tighten to 27 ft. lbs. (37 Nm).
5. Remove the oil filter cap assembly.
   a. Using the special tool (09228-06501), loosen the oil filter cap 4 revolutions, align the cap ribs vertically, and drain the remaining engine oil in the oil filter cap.

➡Do not remove the oil filter bracket clip when removing the oil filter cap assembly.

➡Set a container below the oil filter cap assembly before loosening the oil filter cap.

   b. Remove the oil filter cap assembly.
   c. Remove the oil filter element and o-ring from the oil filter cap.

➡Be sure to remove the O-ring (for the cap) by hand, without using any tools, to prevent damage to the groove for the O-ring on the cap.

6. Install the oil filter cap assembly.
   a. Clean the oil filter cap threads and O-ring groove.
   b. Apply a small amount of engine oil to a new O-ring and install it to the oil filter cap.
   c. Set a new oil filter element in the oil filter cap.
   d. Remove any dirt or foreign matter from the installation surface and inside of the engine.
   e. Reapply a small amount of engine oil to the O-ring of the oil filter cap assembly. Align the cutout in the oil filter cap threads 90° to the grooves in the oil filter bracket and temporarily tighten the cap.

➡Make sure that the O-ring does not get caught between the parts.

   f. Using the special tool, tighten the oil filter cap assembly to 18 ft. lbs. (25 Nm).

➡After tightening the oil filter cap assembly, check for gaps between the installation surfaces.

➡Do not remove the oil filter bracket clip when installing the oil filter cap assembly.

➡Do not cross thread the oil filter cap assembly.

7. Add engine oil and install the oil filler cap. The standard oil grad is 0W-20.
   a. The standard fill capacity for a drain and refill with an oil filter change is 4.4 qts. (4.2 L). The standard fill capacity for a drain and refill without an oil filter change is 4.1 qts. (3.9 L). The standard capacity for a dry fill is 5 qts. (4.7 L).
8. Inspect for engine oil leaks.
   a. Put the engine in inspection mode.
   b. Start the engine.
   c. Check for engine oil leaks from the connected parts of the oil filter cap and oil filter drain plug.

9. Install the No. 1 engine under cover.
10. Install the engine under cover (w/ cover).
11. Install the front spoiler cover (w/ front spoiler).

EXHAUST MANIFOLD

REMOVAL & INSTALLATION

1.5L Engine

See Figures 94 through 96.

1. Remove the EGR cooler assembly.
2. Remove the front exhaust pipe assembly.

CAUTION

Perform this procedure when the coolant is not hot.

a. Disconnect the 2 heater water hoses.
b. Remove the 4 bolts and 4 compression springs.
c. Remove the front exhaust pipe assembly from the 3 exhaust pipe supports.
d. Remove the 2 gaskets from the front exhaust pipe assembly and exhaust manifold.

3. Remove the air fuel ratio sensor.
4. Remove the No. 1 exhaust manifold heat insulator.
a. Remove the 3 bolts and No. 1 exhaust manifold insulator.
5. Remove the manifold support bracket.
a. Remove the 4 bolts and manifold support bracket.
6. Remove the No. 1 EGR pipe.
a. Remove the gasket.
7. Remove the exhaust manifold.
a. Disconnect the wire harness clamp and oxygen sensor connector.
b. Remove the 3 bolts, 2 nuts and exhaust manifold.
8. Remove the No. 2 exhaust manifold heat insulator.
a. Remove the 3 bolts and No. 2 exhaust manifold heat insulator.
9. Remove the heated oxygen sensor.

To install:

10. Install the heated oxygen sensor.
11. Install the No. 2 exhaust manifold heat insulator with the 3 bolts. Tighten to 71 inch lbs. (8 Nm).
12. Install the exhaust manifold.
a. Set a new gasket to the cylinder head sub-assembly.
b. Temporarily install the exhaust manifold with the 3 bolts and 2 nuts.
c. Temporarily install the manifold support bracket with the 3 bolts.
d. Tighten the exhaust manifold with the 3 bolts and 2 nuts. Tighten to 22 ft. lbs. (30 Nm).
e. Connect the wire harness clamp and heated oxygen sensor connector.
f. Remove the 3 bolts and manifold support bracket.
13. Install the No. 1 EGR pipe.
a. Install a new gasket and No. 1 EGR pipe with the bolt and nut. Tighten to 23 ft. lbs. (31 Nm).
14. Install the manifold support bracket with the 4 bolts. Tighten the A bolts to 32 ft. lbs. (44 Nm). Tighten the B bolts to 7 ft. lbs. (10 Nm).
15. Install the No. 1 exhaust manifold heat insulator with the 3 bolts. Tighten to 71 inch lbs. (8 Nm).
16. Install the air fuel ratio sensor.
17. Install the front exhaust pipe assembly.

WARNING

When installing the water hose, ensure that the exhaust heat recirculation system is filled with coolant. Otherwise, the engine water pump assembly may be damaged.

a. Using a vernier caliper, measure the free length of the compression springs.
- Minimum (front): 1.64 inch (41.5 mm)
- Minimum (rear): 1.52 inch (38.5 mm)

➡If the free length is less than minimum, replace the compression spring.

b. Fully insert 2 new gaskets to the exhaust manifold converter and front exhaust pipe assembly.
c. Using a plastic hammer and wooden block, tap in each new gasket until its surface is flush with the exhaust manifold and front exhaust pipe assembly.

WARNING

Be careful with the installation direction of the gaskets.

WARNING

Do not reuse the gaskets.

WARNING

Do not damage the gaskets.

WARNING

Do not push in the gaskets by using the exhaust pipe when connecting it.

d. Connect the front exhaust pipe assembly to the 3 exhaust pipe supports.
e. Install the front exhaust pipe assembly with the 2 compression springs and 2 bolts. Tighten to 32 ft. lbs. (43 Nm).

➡After installation, check that the gaps between the flanges of the exhaust manifold and front exhaust pipe assembly are consistent front-to-rear and left-to-right.

f. Install the front exhaust pipe assembly with the 2 compression springs and 2 bolts. Tighten to 32 ft. lbs. (43 Nm).

➡After installation, check that the gaps between the flanges of the tail exhaust

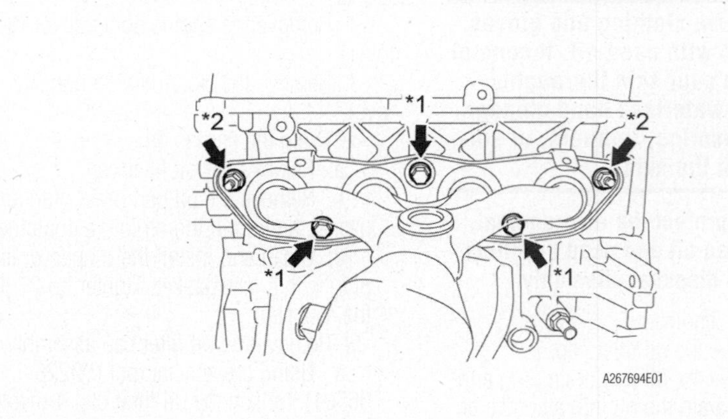

Fig. 95 Tightening the exhaust manifold with the 3 bolts (1) and 2 nuts (2)

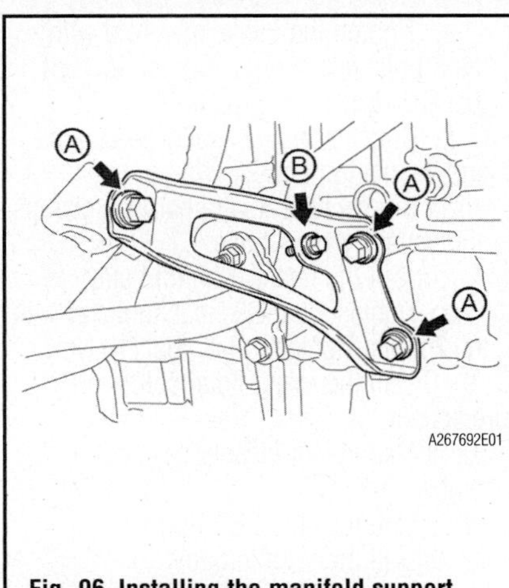

Fig. 96 Installing the manifold support bracket

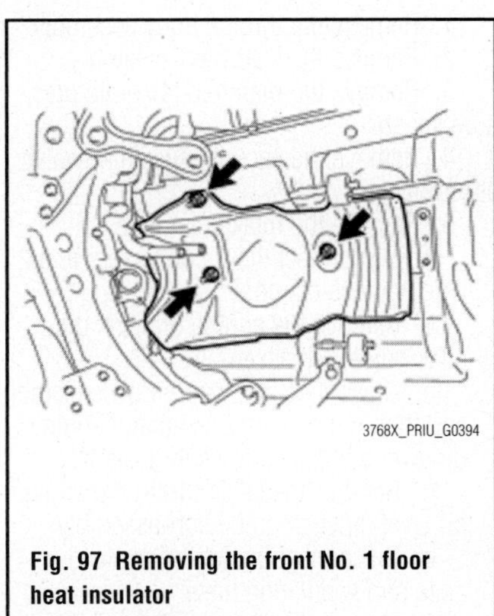

Fig. 97 Removing the front No. 1 floor heat insulator

pipe assembly and front exhaust pipe assembly are consistent front-to-rear and left-to-right.

g. Connect the 2 heater water hoses.
18. Install the EGR cooler assembly.
19. Inspect for exhaust gas leak.

1.8L Engine

See Figures 97 through 99.

1. Remove the No. 1 engine under cover.
2. Remove the No. 2 engine under cover.
3. Remove the front No. 3 engine under cover.
4. Remove the front center floor brace.
5. Drain the coolant (for engine with exhaust heat recirculation system).
6. Remove the front exhaust pipe assembly (w/ exhaust heat recirculation system).
   a. Disconnect the 3 clamps and oxygen sensor connector.
   b. Disconnect the 2 heater water hoses.
   c. Remove the 4 bolts and 4 compression springs.
   d. Remove the front exhaust pipe assembly from the 3 exhaust pipe supports.
   e. Remove the 2 gaskets from the front exhaust pipe assembly and exhaust manifold.
7. Remove the front exhaust pipe assembly (w/o exhaust heat recirculation system).
   a. Remove the 4 bolts and 4 compression springs.
   b. Remove the front exhaust pipe assembly from the 3 exhaust pipe supports.
   c. Remove the 2 gaskets from the front exhaust pipe assembly and exhaust manifold.
8. Remove the front No. 1 floor heat insulator.

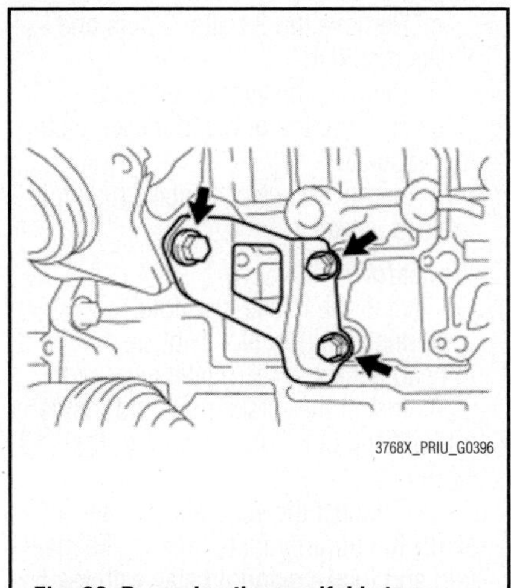

Fig. 98 Removing the manifold stay

Fig. 99 Removing the exhaust manifold

   a. Remove the 3 nuts and No. 1 floor heat insulator.
9. Remove the No. 1 exhaust manifold heat insulator.
   a. Remove the 3 bolts and No. 1 exhaust manifold heat insulator.
10. Remove the air fuel ratio sensor.
11. Remove the manifold stay.
12. Remove the exhaust manifold.
   a. Remove the 7 nuts, exhaust manifold and 2 gaskets.
13. Remove the No. 2 exhaust manifold heat insulator.
   a. Remove the 3 bolts and No. 2 exhaust manifold heat insulator.

To install:

14. Install the No. 2 exhaust manifold heat insulator with the 3 bolts. Tighten to 9 ft. lbs. (12 Nm).
15. Install the exhaust manifold.
   a. Install the 2 new gaskets and the exhaust manifold with the 7 nuts. Tighten to 15 ft. lbs. (21 Nm).
16. Install the manifold stay with the 3 bolts. Tighten to 32 ft. lbs. (43 Nm).
17. Install the air fuel ratio sensor.
18. Install the No. 1 exhaust manifold heat insulator with the 3 bolts. Tighten to 9 ft. lbs. (12 Nm).
19. Install the front No. 1 floor heat insulator with the 3 nuts. Tighten to 49 inch lbs. (5.5 Nm).
20. Install the front exhaust pipe assembly (w/ exhaust heat recirculation system).

➡When installing the water hose, ensure that the exhaust heat recirculation system is filled with coolant. Otherwise, the electric water pump may be damaged.

   a. Using a vernier caliper, measure the free length of the compression rings. Minimum (front) is 1.64 inch (41.5 mm). Minimum (rear) is 1.52 inch (38.5 mm).

➡If the free length is less than the minimum, replace the compression spring.

   b. Fully insert 2 new gaskets to the exhaust manifold and from exhaust pipe assembly.
   c. Using a plastic hammer and wooden block, tap in the new gaskets until its surface is flush with the exhaust manifold and from exhaust pipe assembly.
   d. Connect the front exhaust pipe assembly to the 3 exhaust pipe supports.
   e. Install the front exhaust pipe assembly with the 4 bolts and 4 compression springs. Tighten to 32 ft. lbs. (43 Nm).

f. Connect the 2 heater water hoses.

g. Connect the 3 clamps and oxygen sensor connector.

21. Install the front exhaust pipe assembly (w/o exhaust heat recirculation system).

a. Using a vernier caliper, measure the free length of the compression springs. Minimum (front) is 1.64 inch (41.5 mm). Minimum (rear) is 1.52 inch (38.5 mm).

➡If the free length is less than the minimum, replace the compression spring.

b. Fully insert 2 new gaskets to the exhaust manifold and front exhaust pipe assembly.

c. Using a plastic hammer and wooden block, tap in the new gaskets until its surface is flush with the exhaust manifold and front exhaust pipe assembly.

➡Be careful with the installation direction of the gaskets.

➡Do not reuse the gaskets.

➡Do not damage the gaskets.

➡Do not push in the gasket by using the exhaust pipe when connecting it.

d. Connect the front exhaust pipe assembly to the 3 exhaust pipe supports.

e. Install the front exhaust pipe assembly with the 4 bolts and compression springs. Tighten to 32 ft. lbs. (43 Nm).

22. Add coolant (for engine with exhaust heat recirculation system).

23. Inspect for coolant leak (for engine with exhaust heat recirculation system).

24. Install the front center floor brace.

25. Install the front No. 3 engine under cover.

26. Install the No. 2 engine under cover.

27. Install the No. 1 engine under cover.

28. Inspect for exhaust gas leaks.

INTAKE MANIFOLD

REMOVAL & INSTALLATION

1.5L Engine

See Figures 100 and 101.

1. Remove the throttle body.
2. Remove the No. 2 EGR pipe.
3. Remove the air cleaner case sub assembly/
4. Remove the manifold absolute pressure sensor.
5. Remove the intake manifold stay.

a. Remove the 2 bolts and intake manifold stay.

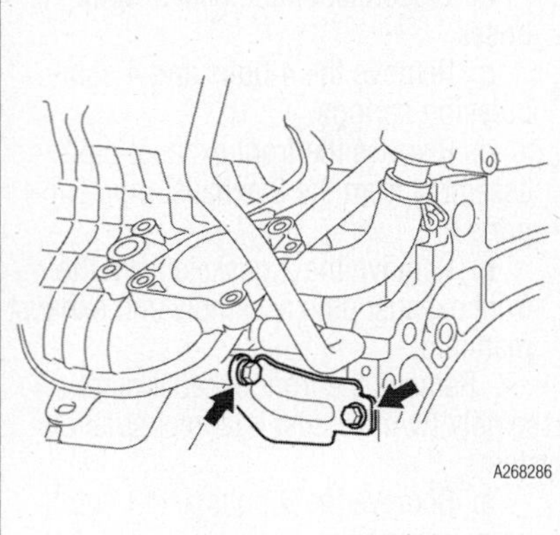

Fig. 100 Removing the intake manifold stay

6. Remove the intake manifold.

a. Disconnect the 4 wire harness clamps and water by-pass hose.

b. Disconnect the fuel vapor feed hose and ventilation hose.

c. Remove the 3 bolts, 2 nuts and intake manifold.

d. Remove the ventilation hose.

e. Remove the bolt and intake manifold insulator.

f. Remove the No. 1 intake manifold to head gasket.

To install:

7. Install the intake manifold.

a. Install a new No. 1 intake manifold to head gasket to the intake manifold.

b. Install the intake manifold insulator with the bolt. Tighten to 7 ft. lbs. (10 Nm).

c. Connect the ventilation hose.

d. Temporarily install the intake manifold and intake manifold stay with the 5 bolts and 2 nuts.

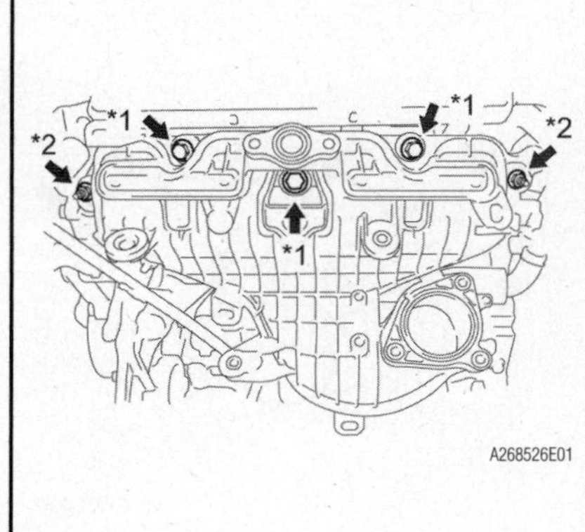

Fig. 101 Identifying the intake manifold bolts (1) and nuts (2)

e. Tighten the intake manifold with the 3 bolts and 2 nuts. Tighten to 15 ft. lbs. (21 Nm)

f. Connect the fuel vapor feed hose and ventilation hose.

g. Connect the 4 wire harness clamps and water by-pass hose.

8. Install the intake manifold stay.

a. Tighten the intake manifold stay with the 2 bolts. Tighten to 7 ft. lbs. (10 Nm).

9. Install the manifold absolute pressure sensor.

10. Install the air cleaner case sub assembly.

11. Install the No. 2 EGR pipe.

12. Install the throttle body.

1.8L Engine

See Figure 102.

1. Remove the throttle body assembly.
2. Remove the EGR pipe assembly.
3. Remove the manifold absolute pressure sensor.
4. Remove the vacuum switching valve assembly.
5. Remove the intake manifold.

a. Disconnect the 3 connectors and 2 wire harness clamps.

b. Remove the engine oil level dipstick guide sub assembly.

c. Disconnect the wire harness clamp, and then remove the 2 bolts and engine oil level dipstick guide sub-assembly.

d. Remove the O-ring from the engine oil level dipstick guide sub-assembly.

e. Disconnect the fuel vapor feed hose and ventilation hose.

f. Remove the 2 bolts, 2 nuts and intake manifold.

g. Remove the No. 1 intake manifold to head gasket.

6. Replace the stud bolt.

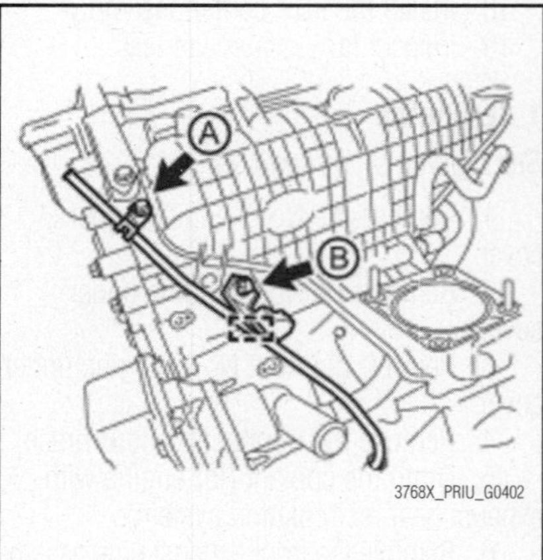

Fig. 102 Installing the engine oil level dipstick guide sub assembly

➡If the stud bolt is deformed or the threads are damaged, replace it.

a. Using an E6 TORX® wrench, replace the 2 stud bolts. Tighten to 44 inch lbs. (5 Nm).

To install:

7. Install the intake manifold.
 a. Install a new No. 1 intake manifold to head gasket to the intake manifold.
 b. Install the intake manifold with the 2 bolts and 2 nuts. Tighten to 21 ft. lbs. (28 Nm).
 c. Connect the fuel vapor feed hose and ventilation hose.
 d. Install a new O-ring to the engine oil level dipstick guide sub-assembly.
 e. Apply a light coat of engine oil to the O-ring.
 f. Install the engine oil level dipstick guide sub-assembly with the 2 bolts and connect the wire harness clamp. Tighten bolt A to 21 ft. lbs. (28 Nm). Tighten bolt B to 15 ft. lbs. (21 Nm).
 g. Install the engine oil level dipstick.
 h. Connect the 2 wire harness clamps and 3 connectors.
8. Install the vacuum switching valve assembly.
9. Install the manifold absolute pressure sensor.
10. Install the EGR pipe assembly.
11. Install the throttle body assembly.

OIL PUMP

REMOVAL & INSTALLATION

1.5L Engine

See Figure 103.

1. Install engine stand.
2. Remove the No. 2 EGR pipe.
3. Remove the EGR valve with cooler assembly.
4. Remove the ignition coil assembly.
5. Separate the engine wire.
6. Remove the cylinder head cover sub assembly.
7. Remove the crankshaft position sensor.
8. Remove the crankshaft damper sub assembly.
9. Remove the engine oil level dipstick.
10. Remove the engine oil level dipstick guide.
 a. Remove the bolt and engine oil level dipstick guide.
 b. Remove the o-ring from the engine oil level dipstick guide.
11. Remove the 4 bolts and engine mounting bracket RH.

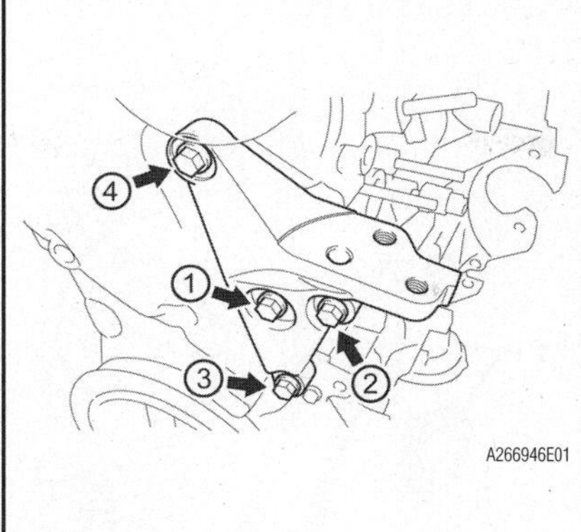

Fig. 103 Identifying the engine mounting bracket bolt tightening sequence

12. Remove the screw plug.
 a. Using an 8 mm hexagon socket wrench, remove the screw plug.
13. Remove the timing chain cover.
14. Remove the oil pump seal.
 a. Using a screwdriver and wooden block, remove the oil pump seal.

⁂ WARNING

Do not damage the surface of the oil seal press fit hole.

➡Tape the screwdriver tip before use.

To install:

15. Install the oil pump seal.
 a. Apply a light coat of MP grease to the timing chain cover oil pump seal lip.

➡Keep the seal lip free of foreign matter.

 b. Using SST and wooden block, tap in a new oil pump seal until its surface is flush with the timing chain cover edge.

➡Tap in 0-1.0 mm from the edge of the timing chain cover.

➡Do not tap the oil pump seal at an angle.

16. Install the timing chain cover.
17. Install the screw plug.
 a. Apply a few drops of adhesive to 2 or 3 threads of the screw plug.
 b. Using a 8 mm hexagon socket wrench, install the screw plug. Tighten to 11 ft. lbs. (15 Nm).
18. Install the engine mounting bracket RH.
 a. Temporarily install the engine mounting bracket RH with the 4 bolts.
 b. Tighten the 4 bolts in sequence. Tighten to 41 ft. lbs. (55 Nm).
19. Install engine oil level dipstick guide.
 a. Apply a light coat of engine oil to a new O-ring and install the O-ring to the engine oil level dipstick guide.
 b. Install the engine oil level dipstick guide with the bolt. Tighten to 7 ft. lbs. (10 Nm).
20. Install the engine oil level dipstick.
21. Install the crankshaft damper sub assembly.
22. Install the crankshaft position sensor.
23. Install the cylinder head cover sub assembly.
24. Connect the engine wire.
25. Install the ignition coil assembly.
26. Install the EGR valve with cooler assembly.
27. Install the No. 2 EGR pipe.
28. Remove the engine stand.

1.8L Engine

See Figures 104 through 111.

1. Remove the engine assembly with the transaxle.
2. Install the engine on an engine stand.
3. Remove the engine hangers.
4. Remove the throttle body assembly.
5. Remove the engine oil level dipstick guide.
6. Remove the EGR pipe assembly.
7. Remove the EGR valve assembly.
8. Remove the EGR cooler sub assembly.

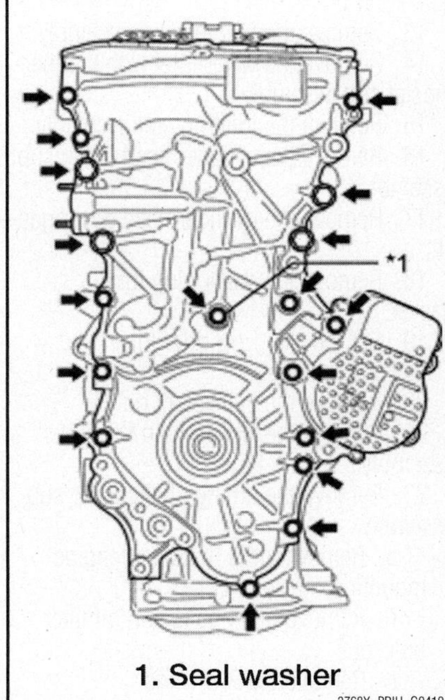

Fig. 104 Removing the 18 bolts and seal washer

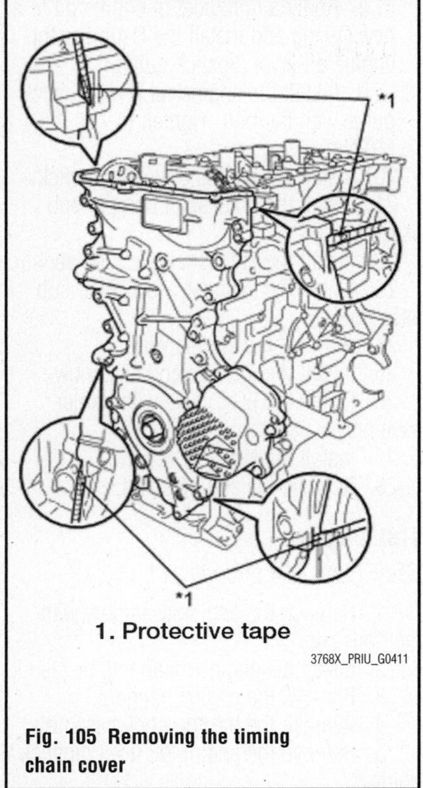

Fig. 105 Removing the timing chain cover

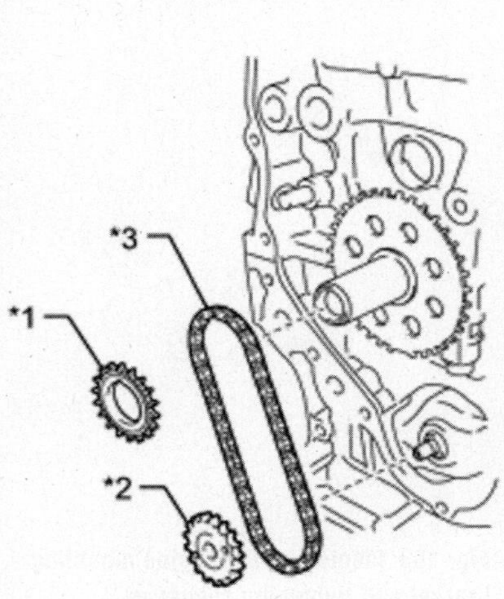

Fig. 106 Removing the oil pump drive gear, oil pump drive shaft gear and No. 2 chain sub assembly

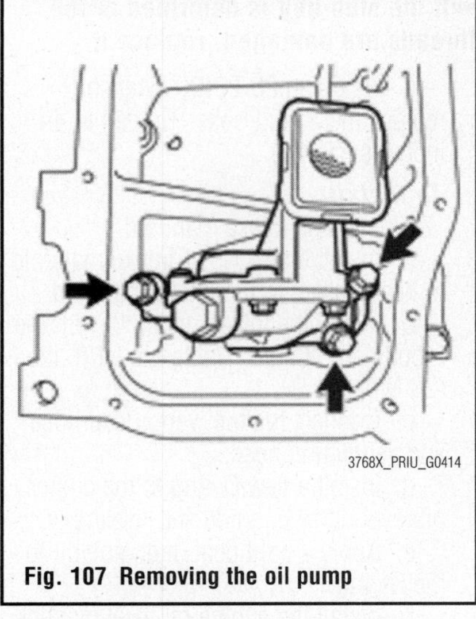

Fig. 107 Removing the oil pump

9. Remove the intake manifold.
10. Remove the fuel vapor feed pipe.
11. Remove the fuel delivery pipe sub assembly.
12. Remove the No. 1 delivery pipe spacer.
13. Remove the fuel injector assembly.
14. Remove the water inlet with the thermostat sub assembly.
15. Remove the ignition coil assembly.
16. Remove the cylinder head cover sub assembly.
17. Remove the cylinder head cover gasket.
18. Remove the spark plug tube gasket.
19. Set the No. 1 cylinder to TDC/Compression.
20. Remove the crankshaft pulley.
21. Remove the No. 1 chain tensioner assembly.
22. Remove the timing chain cover sub assembly.
    a. Remove the 3 bolts and engine mounting bracket RH.
    b. Remove the 4 bolts and oil filter bracket.
    c. Remove the 2 O-rings.
    d. Remove the 18 bolts and seal washer.
    e. Remove the timing chain cover by prying between the timing chain cover and cylinder head or cylinder block with a screwdriver.

➡Be careful not to damage the contact surfaces of the timing chain cover, cylinder block, and cylinder head.

➡Pry the timing chain cover out evenly in order to prevent damaging the knock pins.

➡Tape the screwdriver tip before use.

    f. Remove the 3 O-rings.
23. Remove the timing chain cover oil seal.
    a. Place the timing chain cover on wooden blocks.
    b. Using a screwdriver, pry out the oil seal.

➡Tape the screwdriver tip before use.

➡Do not damage the surface of the oil seal press fit hole.

24. Remove the chain tensioner slipper.
25. Remove the No. 1 chain vibration damper.
26. Remove the No. 2 chain vibration damper.
27. Remove the chain sub assembly.

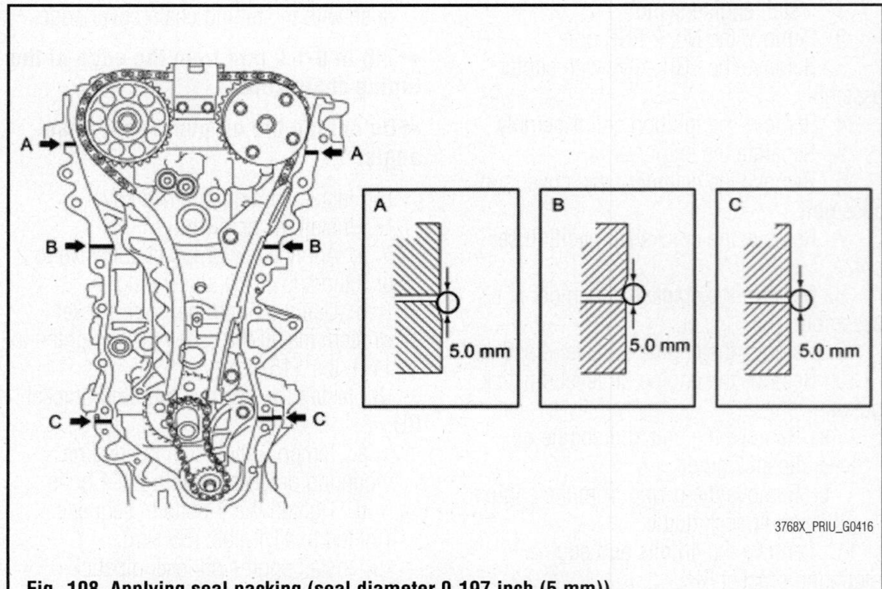

Fig. 108 Applying seal packing (seal diameter 0.197 inch (5 mm))

28. Remove the crankshaft timing sprocket.

29. Remove the No. 2 chain sub assembly.

a. Temporarily tighten the crankshaft pulley and crankshaft pulley bolt.

b. Using the special tool (09213-58014, 91551-80840, or 09330-00021), remove the oil pump drive shaft sprocket nut while holding the crankshaft pulley.

c. Remove the special tool, crankshaft pulley, and crankshaft pulley bolt.

d. Remove the bolt, chain tensioner plate, and spring.

e. Remove the oil pump drive gear, oil pump drive shaft gear, and No. 2 chain sub assembly.

30. Remove the No. 1 Crankshaft Position (CKP) sensor plate.

31. Remove the oil pan drain plug.

32. Remove the No. 2 oil pan sub assembly.

33. Remove the oil pump assembly.

a. Remove the 3 bolts and oil pump.

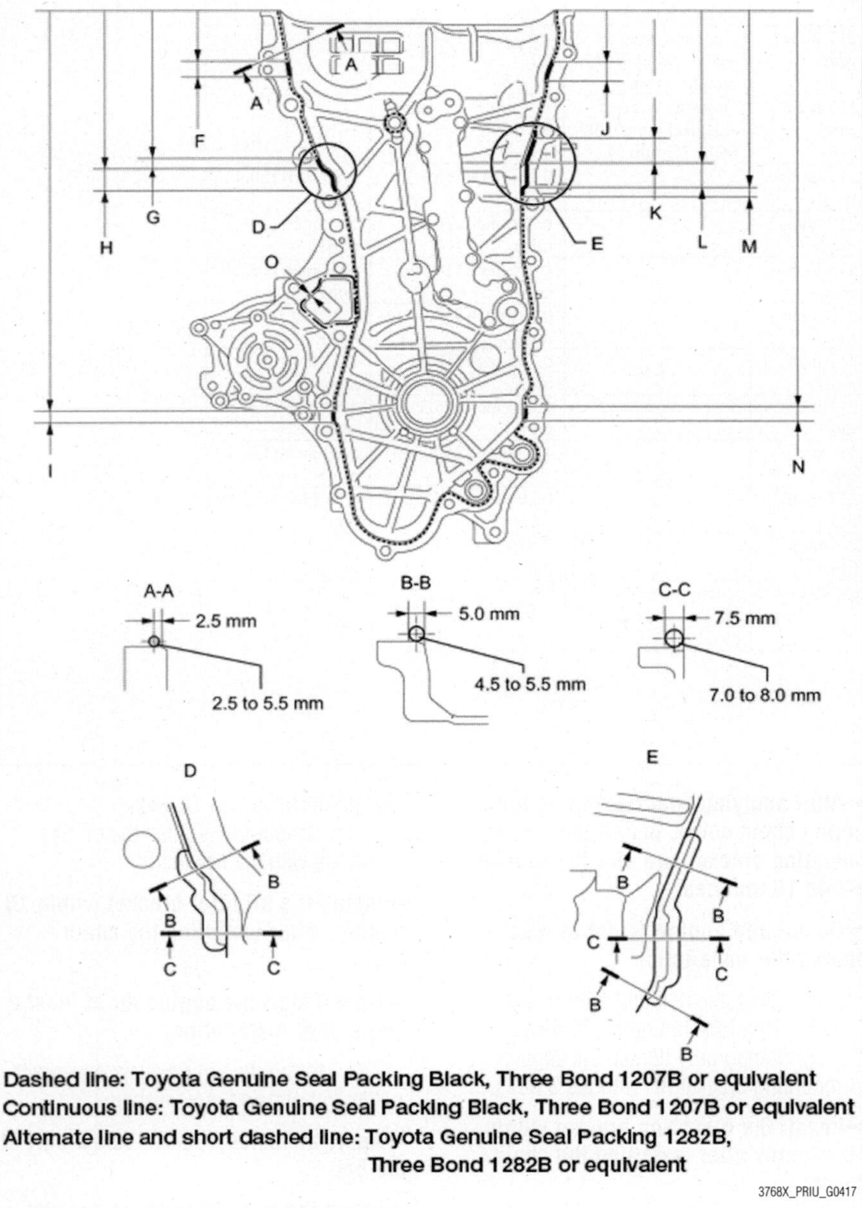
A
F
J
G
H
D
E
K
L
M
O
I
N
A-A
2.5 mm
2.5 to 5.5 mm
B-B
5.0 mm
4.5 to 5.5 mm
C-C
7.5 mm
7.0 to 8.0 mm
D
E
B
C
Dashed line: Toyota Genuine Seal Packing Black, Three Bond 1207B or equivalent
Continuous line: Toyota Genuine Seal Packing Black, Three Bond 1207B or equivalent
Alternate line and short dashed line: Toyota Genuine Seal Packing 1282B, Three Bond 1282B or equivalent
3768X_PRIU_G0417

Fig. 109 Applying seal packing to the timing chain cover

To install:

34. Install the oil pump assembly with the 3 bolts. Tighten to 16 ft. lbs. (21 Nm).

35. Install the No. 2 oil pan sub assembly.

36. Install the oil pan drain plug.

37. Install the crankshaft timing gear key.

38. Install the No. 1 Crankshaft Position (CKP) sensor plate.

39. Install the No. 2 chain sub assembly.

40. Install the crankshaft timing sprocket.

41. Install the No. 1 chain vibration damper.

42. Set the No. 1 cylinder to TDC/Compression.

43. Install the chain sub assembly.

44. Check the No. 1 cylinder to TDC/Compression.

45. Install the chain tensioner slipper.

46. Install the No. 2 chain vibration damper.

47. Install the timing chain cover oil seal.

a. Using the special tool (09223-22010), tap in a new oil seal until its surface is flush with the timing chain cover edge.

➡Keep the lip free from foreign matter.

➡Do not tap on the oil seal at an angle.

➡Make sure that the oil seal edge does not stick out of the timing chain case.

➡Tap in the oil seal so that it is positioned within 1.0 mm from the edge of the timing chain case.

b. Apply MP grease to the lip of the oil seal.

48. Install the timing chain cover sub assembly.

a. Remove any old packing (FIPG) material and be careful not to drop any oil on the contact surfaces of the timing chain cover, cylinder head, and cylinder block.

b. Install 3 new O-rings.

c. Apply seal packing.

➡Remove any oil from the contact surfaces.

➡Install the chain cover within 3 minutes after applying seal packing.

➡Do not start the engine for at least 2 hours after installing the timing chain cover sub-assembly.

d. Apply seal packing to the timing chain cover in a line as shown in the following illustration.

Area	Seal Packing Diameter	Distance From Edge Of Cover To:	Seal Packing Application Length	Distance From Top Of Cover To Top Of Seal Packing
Dashed line	0.0984-0.118 inch (2.5-3.0 mm)	Center of seal packing 0.0984 inch (2.5 mm)	-	-
Continuous line	0.177-0.217 inch (4.5-5.5 mm) or 0.276-0.315 inch (7.0-8.0 mm)	-	-	-
Alternate long and short dashed line	0.157 inch (4.0 mm)	Center of seal packing 0.118 inch (3.0 mm)	-	-
A-A	0.0984-0.118 inch (2.5-3.0 mm)	Center of seal packing 0.0984 inch (2.5 mm)	-	-
B-B	0.177-0.217 inch (4.5-5.5 mm)	Opposite edge of seal packing 0.197 inch (5.0 mm)	-	-
C-C	0.276-0.315 inch (7.0-8.0 mm)	Opposite edge of seal packing (0.295 inch) 7.5 mm	-	-
F	0.177-0.217 inch (4.5-5.5 mm)	-	0.610 inch (15.5 mm)	1.98 inch (50.4 mm)
G	0.177-0.217 inch (4.5-5.5 mm)	-	0.406 inch (10.3 mm)	5.63 inch (143.1 mm)
H	0.276-0.315 inch (7.0-8.0 mm)	-	0.768 inch (19.5 mm)	6.04 inch (153.4 mm)
I	0.177-0.217 inch (4.5-5.5 mm)	-	0.630 inch (16.0 mm)	1.27 ft. (385.8 mm)
J	0.177-0.217 inch (4.5-5.5 mm)	-	0.732 inch (18.6 mm)	2.02 inch (51.4 mm)
K	0.177-0.217 inch (4.5-5.5 mm)	-	0.996 inch (25.3 mm)	4.80 inch (121.9 mm)
L	0.276-0.315 inch (7.0-8.0 mm)	-	1.02 inch (25.8 mm)	5.80 inch (147.2 mm)
M	0.177-0.217 inch (4.5-5.5 mm)	-	0.201 inch (5.1 mm)	6.81 inch (173.0 mm)
N	0.177-0.217 inch (4.5-5.5 mm)	-	0.575 inch (14.6 mm)	1.27 ft. (385.8 mm)
O	0.157 inch (4.0 mm)	Center of seal packing 0.118 inch (3.0 mm	-	-

3768X_PRIU_G0418

Fig. 110 Seal packing application specifications

➡When the contact surfaces are wet, wipe them with oil-free cloth before applying seal packing.

➡Install the chain cover within 3 minutes and tighten the bolts within 15 minutes after applying seal packing.

➡Do not start the engine for at least 2 hours after installation.

➡When the contact surfaces are wet, wipe them with oil-free cloth before applying seal packing.

➡Install the timing chain cover within 3 minutes and tighten the bolts within 10 minutes after applying seal packing.

➡After applying seal packing to the timing chain cover, install the engine mounting bracket and oil filter bracket within 10 minutes.

➡Do not add engine oil for at least 2 hours after installation.

e. Clean the bolt and fitting hole.
f. Install the timing chain cover.
g. Temporarily tighten the engine mounting bracket RH with the 3 bolts.

➡Install the mounting bracket within 10 minutes after installing the chain cover.

➡Do not start the engine for at least 2 hours after installation.

h. Install 2 new O-rings.
i. Temporarily tighten the oil filter bracket with the new bolts.

➡Install the oil filter bracket within 10 minutes after installing the chain cover.

➡Do not start the engine for at least 2 hours after installation.

j. Install the timing chain cover with the 25 bolts and seal washer.

➡Apply adhesive 1324 to screw part of the bolt F.

➡When the contact surfaces are wet, wipe them with oil-free cloth before applying seal packing.

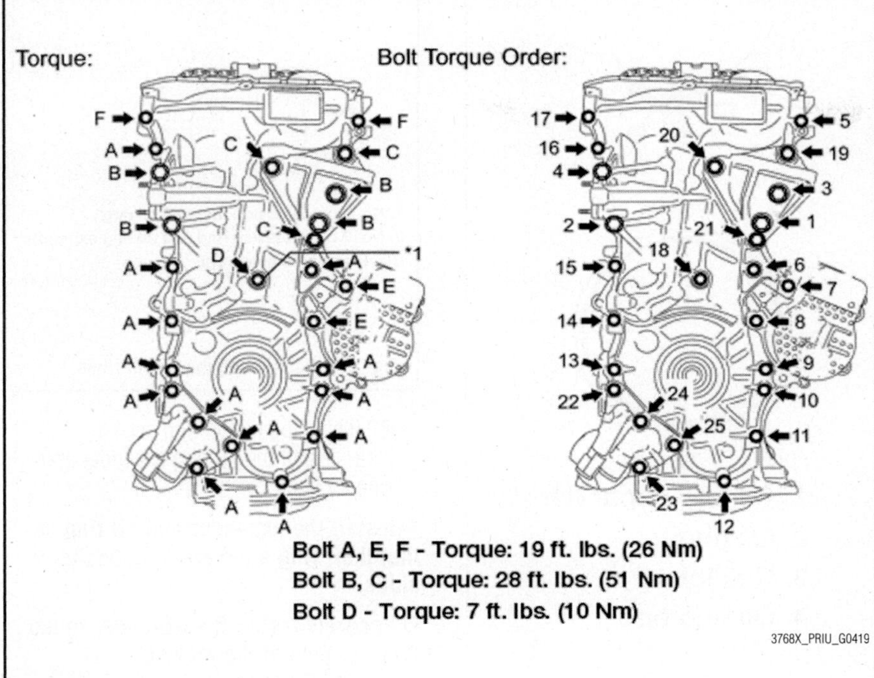

Fig. 111 Installing the timing chain cover

➡Install the chain cover within 3 minutes and tighten the bolts within 15 minutes after applying the seal packing.

➡Do not add engine oil for at least 2 hours after installing the chain cover.

➡Do not start the engine for at least 2 hours after installing the chain cover.

49. Install the crankshaft pulley.
50. Install the No. 1 chain tensioner assembly.
51. Install the spark plug tube gasket.
52. Install the cylinder head cover gasket.
53. Install the cylinder head cover sub assembly.
54. Install the water inlet with the thermostat sub assembly.
55. Install the fuel injector assembly.
56. Install the No. 1 deliver pipe spacer.
57. Install the fuel delivery pipe sub assembly.
58. Install the No. 1 delivery pipe spacer.
59. Install the fuel delivery pipe sub assembly.
60. Install the fuel vapor feed pipe.
61. Install the intake manifold.
62. Install the EGR cooler sub assembly.
63. Install the EGR valve assembly.
64. Install the EGR pipe assembly.
65. Install the engine oil level dipstick guide.
66. Install the throttle body assembly.
67. Install the engine hangers.
68. Remove the engine on engine stand.
69. Install the engine assembly with the transaxle.

INSPECTION

1. Inspect the oil pump relief valve.
   a. Coat the oil pump relief valve with engine oil, then check that it falls smoothly into the valve hole by its own weight. If this does not occur, repair or replace the oil pump assembly.
2. Inspect the oil pump rotor.
   a. Using a feeler gauge, measure the clearance between the drive and driven rotor tips. Standard tip clearance is 0.00315-0.00631 inch (0.08-0.16 mm). The maximum tip clearance is 0.0138 inch (0.35 mm). If the tip clearance is greater than the maximum, replace the oil pump assembly.
   b. Using a feeler gauge and precision straightedge, measure the clearance between the 2 rotors and precision straight edge. The standard side clearance is 0.00335-0.00532 inch (0.085-0.135 mm). The maximum side clearance is 0.00631 inch (0.16 mm). If the side clearance is greater than the maximum, replace the oil pump assembly.
   c. Using a feeler gauge, measure the clearance between the driven rotor and oil pump body. The standard body clearance is 0.00472-0.00750 inch (0.12-0.19 mm). The maximum body clearance is 0.0128 inch (0.325 mm). If the body clearance is greater than the maximum, replace the oil pump assembly.

PISTON AND RING

POSITIONING

1.5L Engine

See Figures 112 through 114.

1.8L Engine

See Figures 115 through 118.

1. Using a screwdriver, install a new snap ring at one end of the piston pin hole.

➡Make sure that the end gap of the snap ring is not aligned with the pin hole cutout portion of the piston.

   a. Gradually heat the piston to approximately 176-194° F (80-90° C).

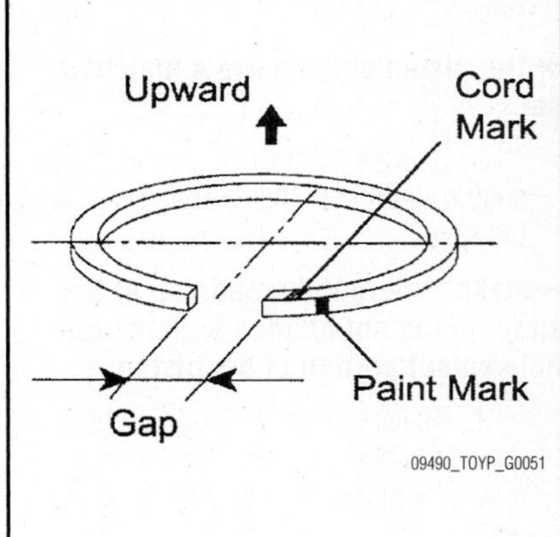

Fig. 112 Piston ring positioning and mark locations—1.5L Hybrid engine

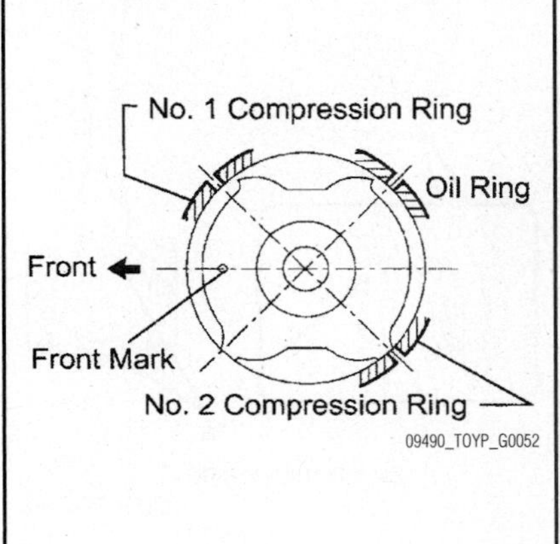

Fig. 113 Piston ring positioning—1.5L Hybrid engine

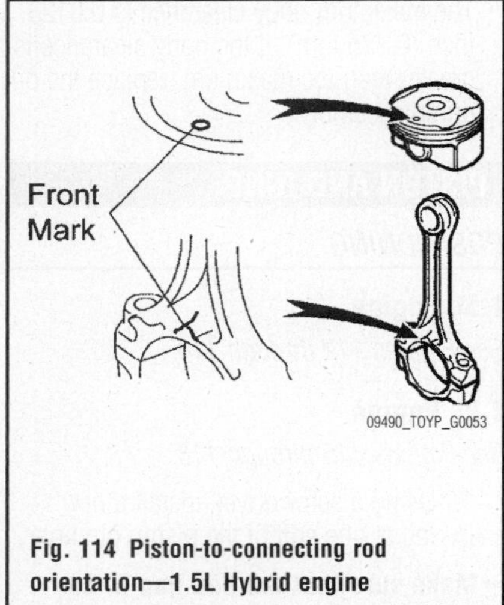

Fig. 114 Piston-to-connecting rod orientation—1.5L Hybrid engine

b. Align the front marks of the piston and connecting rod, insert the connecting rod into the piston, and then push in the piston pin with your thumb until the pin comes into contact with the snap ring.

➡The piston and pin are a matched set.

c. Using a screwdriver, install a new snap ring on the other end of the piston pin hole.

➡Make sure that the end gap of the snap ring is not aligned with the pin hole cutout portion of the piston.

d. Check the fitting condition between the piston and piston pin by trying to move the piston back and forth on the piston pin.

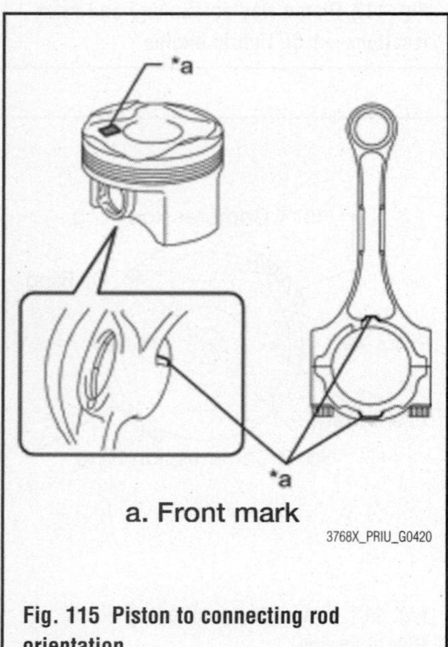

Fig. 115 Piston to connecting rod orientation

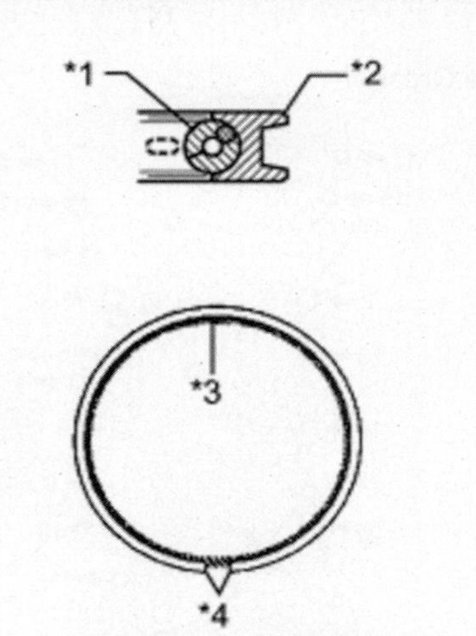

1. Oil ring expander
2. Oil ring
3. Coil joint
4. Oil ring end

3768X_PRIU_G0421

Fig. 116 Installing the piston ring set

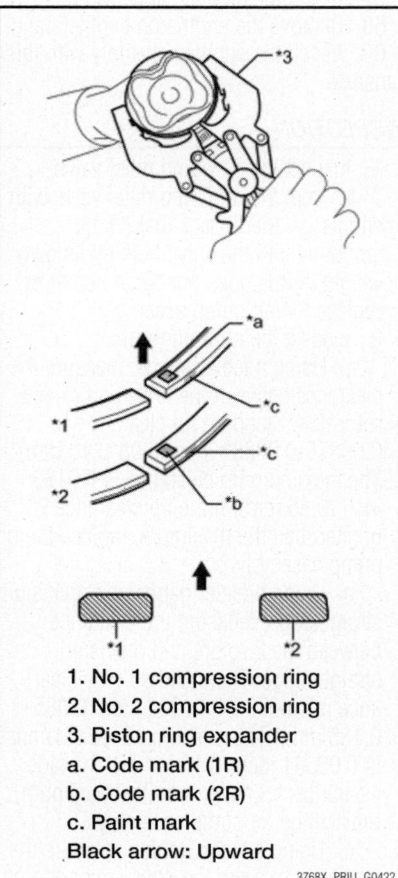

Fig. 117 Installing the compression rings

*2
45°
*1

1. No. 1 compression ring and oil ring
2. No. 2 compression ring and oil ring expander
Black arrow: Engine front

3768X_PRIU_G0423

Fig. 118 Positioning the piston rings

2. Install the piston ring set.

a. Install the oil ring expander and oil ring by hand.

➡Install the expander and oil ring so that their ring ends are at opposite sides.

➡Securely install the expander to the inner groove of the oil ring.

b. Using a piston ring expander, install the 2 compression rings so that the paint marks are positioned as shown in the illustration.

➡Install the No. 1 compression ring with the code mark (1R) facing upward.

➡Install the No. 2 compression ring with the code mark (2R) facing upward.

➡Paint marks can only be checked on new piston rings. When reusing piston rings, check each piston ring profile in order to install them into the correct positions.

c. Position the piston rings so that the ring ends are as shown in the illustration.

TIMING CHAIN FRONT COVER

REMOVAL & INSTALLATION

See Figures 119 through 122.

1. Before servicing the vehicle, refer to the precautions.
2. Remove the crankshaft pulley.
3. Remove the No. 1 chain tensioner assembly.
4. Remove the timing chain cover sub assembly.

a. Remove the 3 bolts and engine mounting bracket RH.

b. Remove the 4 bolts and oil filter bracket.

c. Remove the 2 O-rings.

d. Remove the 18 bolts and seal washer (*1).

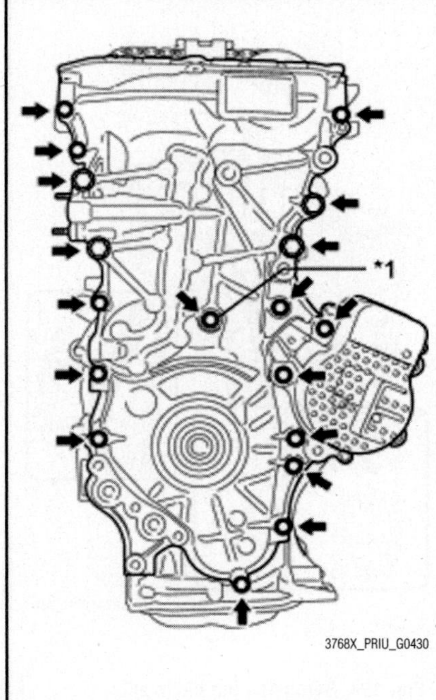

Fig. 119 Removing the 18 bolts and seal washer

e. Remove the timing chain cover by prying between the timing chain cover and cylinder head or cylinder block with a screwdriver.

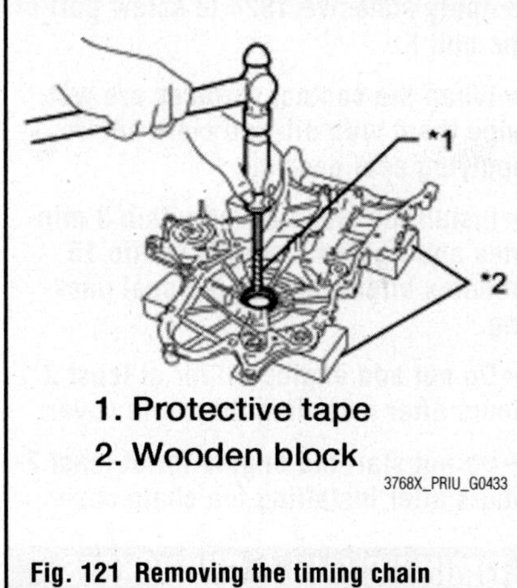

Fig. 121 Removing the timing chain cover oil seal

➡Be careful not to damage the contact surfaces of the timing chain cover, cylinder block, and cylinder head.

➡Pry the timing chain cover out evenly in order to prevent damaging the knock pins.

➡Tape the screwdriver tip before use.

f. Remove the 3 O-rings.

5. Remove the timing chain cover oil seal.

a. Place the timing chain cover on wooden blocks.

b. Using a screwdriver, pry out the oil seal.

➡Tape the screwdriver tip before use.

➡Do not damage the surface of the oil seal press fit hole.

To install:

6. Install the timing chain cover oil seal.

a. Using the special tool (0923-22010), tap in a new oil seal until its surface is flush with the timing chain cover edge.

➡Keep the lip free from foreign matter.

➡Do not tap on the oil seal at an angle.

➡Make sure that the oil seal edge does not stick out of the timing chain case.

➡Tap in the oil seal so that it is positioned within 1.0 mm from the edge of the timing chain case.

b. Apply MP grease to the lip of the oil seal.

7. Install the timing chain cover sub assembly.

a. Remove any old packing (FIPG) material and be careful not to drop any oil on the contact surfaces of the timing chain cover, cylinder head, and cylinder block.

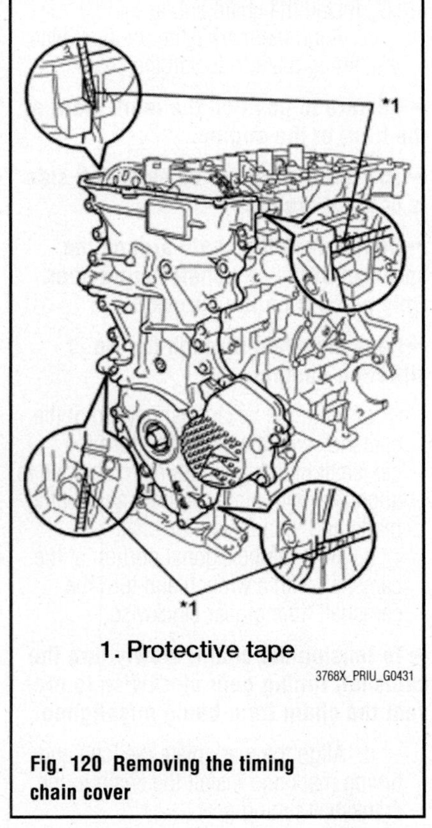

Fig. 120 Removing the timing chain cover

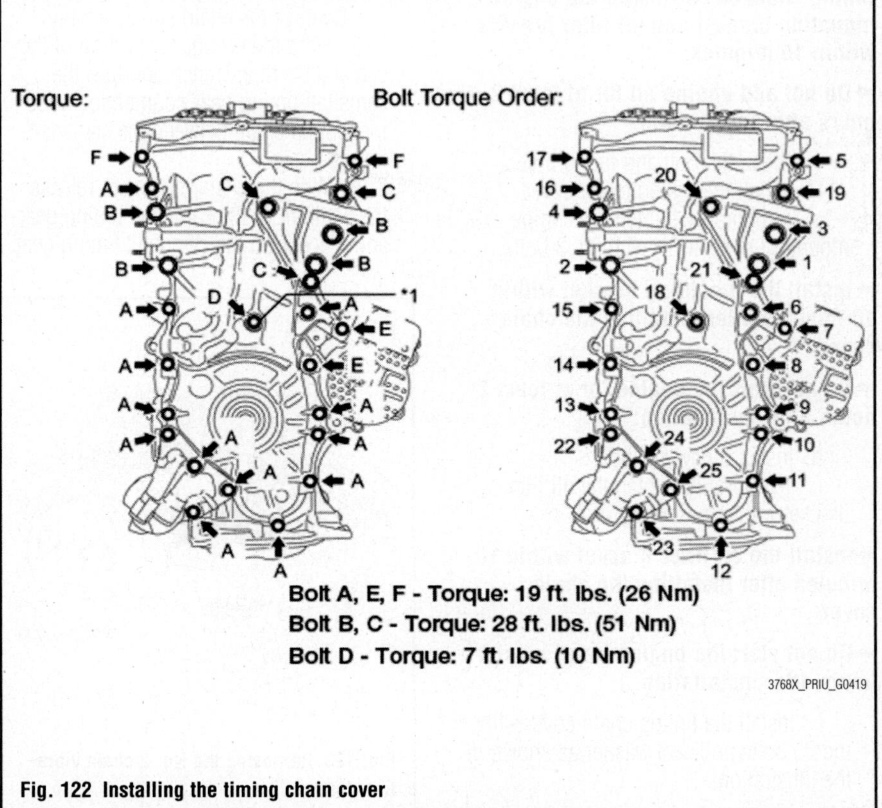

Fig. 122 Installing the timing chain cover

b. Install 3 new O-rings.
c. Apply seal packing.

➡**Remove any oil from the contact surfaces.**

➡**Install the chain cover within 3 minutes after applying seal packing.**

➡**Do not start the engine for at least 2 hours after installing the timing chain cover sub-assembly.**

d. Apply seal packing to the timing chain cover in a line as shown in the following illustration.

➡**When the contact surfaces are wet, wipe them with oil-free cloth before applying seal packing.**

➡**Install the chain cover within 3 minutes and tighten the bolts within 15 minutes after applying seal packing.**

➡**Do not start the engine for at least 2 hours after installation.**

➡**When the contact surfaces are wet, wipe them with oil-free cloth before applying seal packing.**

➡**Install the timing chain cover within 3 minutes and tighten the bolts within 10 minutes after applying seal packing.**

➡**After applying seal packing to the timing chain cover, install the engine mounting bracket and oil filter bracket within 10 minutes.**

➡**Do not add engine oil for at least 2 hours after installation.**

e. Clean the bolt and fitting hole.
f. Install the timing chain cover.
g. Temporarily tighten the engine mounting bracket RH with the 3 bolts.

➡**Install the mounting bracket within 10 minutes after installing the chain cover.**

➡**Do not start the engine for at least 2 hours after installation.**

h. Install 2 new O-rings.
i. Temporarily tighten the oil filter bracket with the 4 bolts.

➡**Install the oil filter bracket within 10 minutes after installing the chain cover.**

➡**Do not start the engine for at least 2 hours after installation.**

j. Install the timing chain cover with the 25 bolts and seal washer as shown in the illustration.

➡**Apply adhesive 1324 to screw part of the bolt F.**

➡**When the contact surfaces are wet, wipe them with oil-free cloth before applying seal packing.**

➡**Install the chain cover within 3 minutes and tighten the bolts within 15 minutes after applying the seal packing.**

➡**Do not add engine oil for at least 2 hours after installing the chain cover.**

➡**Do not start the engine for at least 2 hours after installing the chain cover.**

TIMING CHAIN & SPROCKETS

REMOVAL & INSTALLATION

See Figures 123 through 126.

1. Remove the timing chain cover and seal.
2. Remove the chain tensioner slipper.
  a. Remove the chain tensioner slipper from the cylinder block.
3. Remove the No. 1 chain vibration damper.
  a. Remove the 2 bolts and chain vibration damper.
4. Remove the No. 2 chain vibration damper.
  a. Remove the 2 bolts and No. 2 chain vibration damper.
5. Remove the chain sub assembly.
  a. Hold the hexagonal portion of the camshaft with a wrench and turn the camshaft timing gear counterclockwise to loosen the chain between the camshaft timing gears.
  b. With the chain loosened, release the chain from the camshaft timing gear and place it on the camshaft timing gear.

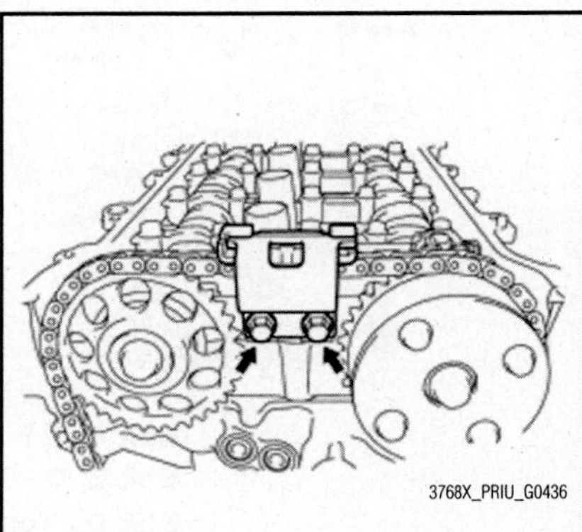

Fig. 123 Removing the No. 2 chain vibration damper

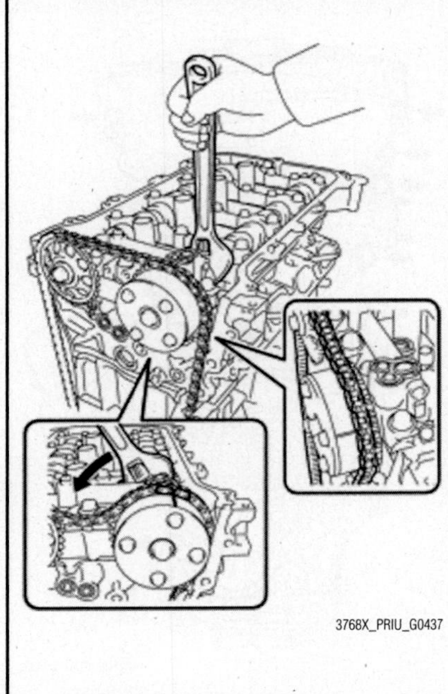

Fig. 124 Removing the chain sub assembly

➡**Be sure to release the chain from the sprocket completely.**

  c. Turn the camshaft clockwise to return it to the original position and remove the chain.

To install:

6. Install the chain sub assembly.
  a. Align the mark plate (orange) with the timing mark to install the chain.

➡**Be sure to position the mark plate at the front of the engine.**

➡**The mark plate on the camshaft side is colored orange.**

➡**Do not pass the chain around the sprocket of the camshaft timing gear. Only place it on the sprocket.**

➡**Pass the chain through the No. 1 vibration damper.**

  b. Hold the hexagonal portion of the camshaft with a wrench and turn the camshaft timing gear counterclockwise to align the mark plate (orange) and timing mark, and then install the chain.
  c. Hold the hexagonal portion of the camshaft with a wrench and turn the camshaft timing gear clockwise.

➡**To tension the chain, slowly turn the camshaft timing gear clockwise to prevent the chain from being misaligned.**

  d. Align the mark plate (yellow) and timing mark and install the chain to the crankshaft timing gear.

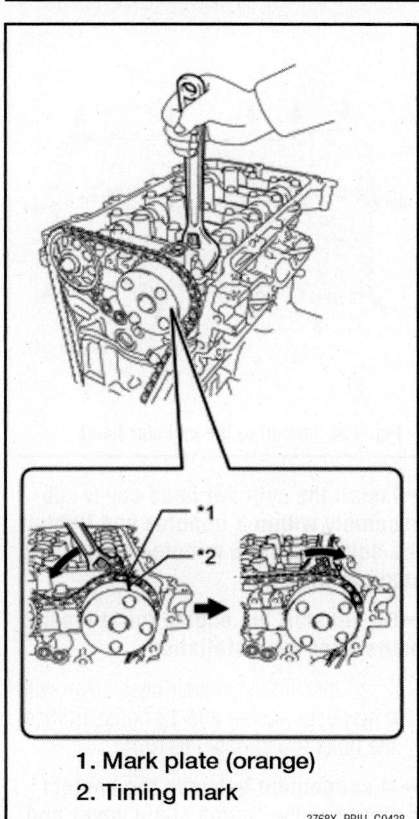

Fig. 125 Aligning the mark plate (orange) with the timing mark

➡**The mark plate on the crankshaft side is colored pink.**

7. Check the No. 1 cylinder TDC/Compression.
   a. Check each timing mark at TDC/Compression.
8. Install the No. 2 chain vibration damper.
   a. Install the No. 2 chain vibration damper with the 2 bolts. Tighten to 7 ft. lbs. (10 Nm).
9. Install the chain tensioner slipper.
   a. Install the chain tensioner slipper to the cylinder block.
10. Install the timing chain cover oil seal.
11. Install the timing chain cover sub assembly.

VALVE COVERS

REMOVAL & INSTALLATION

1.5L Engine

See Figures 127 through 129.

1. Before servicing the vehicle, refer to the precautions.
2. Disconnect the fuel injector connector and ignition coil connector.
3. Remove the 3 bolts and wire harness clamp, then disconnect the wire harness.
4. Remove the brake master cylinder reservoir cover.
5. Disconnect the ventilation hose and ventilation hose No. 2.
6. Remove the bolt and ignition coil.
7. Remove the 9 bolts and 2 nuts, then remove the cylinder head cover.

To install:

8. Apply Three Bond 1207B or equivalent sealant to the 2 locations shown in the illustration, then install the cylinder head cover.

➡**Remove any oil from the contact surface. Install the cylinder head cover within 3 minutes after applying seal packing. Do not start the engine within 2 hours of installing.**

9. Install the cylinder head cover with the 9 bolts and 2 nuts.
10. Using several steps, tighten the bolts and nuts to 7 ft. lbs. (10 Nm) in the sequence shown in the illustration.
11. Install the ignition coil with the bolt. Tighten to 80 inch lbs. (9 Nm).

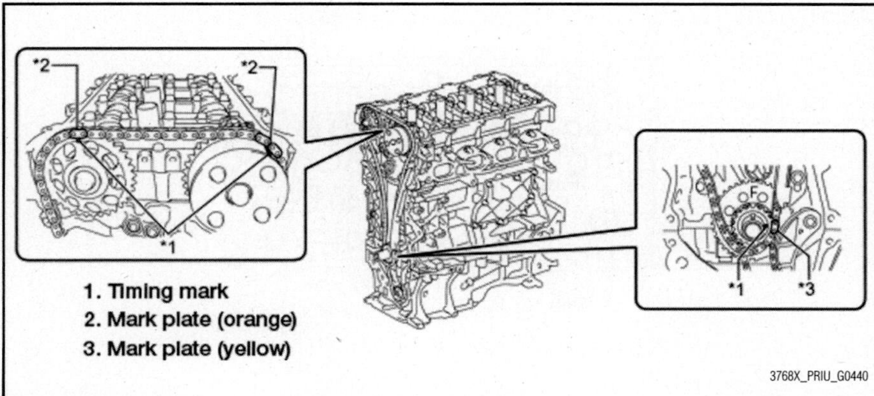

Fig. 126 Checking the No. 1 cylinder TDC/Compression

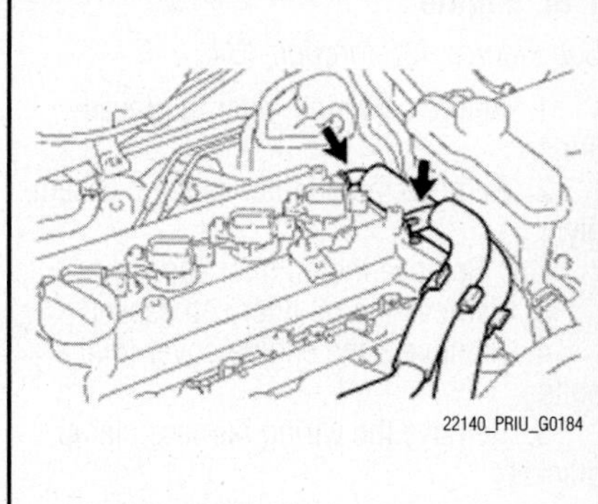

Fig. 127 Disconnect the ventilation hose and No. 2 ventilation hose

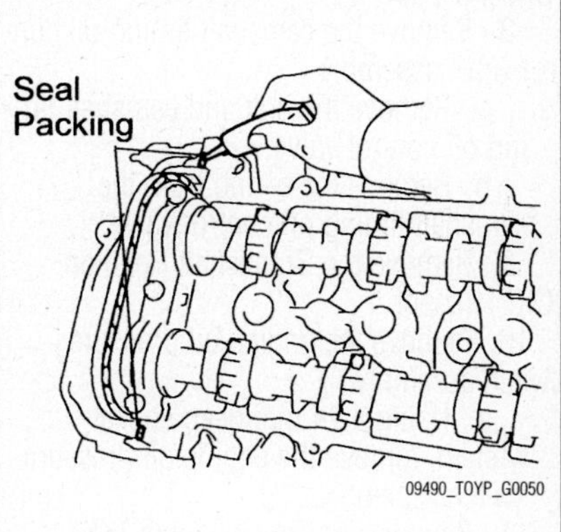

Fig. 128 Prior to installing the cylinder head cover, apply sealant to 2 locations as shown

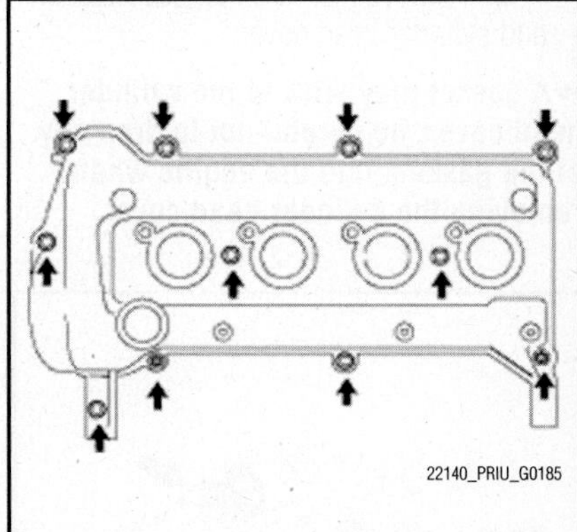

Fig. 129 Cylinder head cover bolt torque sequence

12. Install the brake master cylinder reservoir cover to the cylinder head cover.
13. Install the wire harness and brake master cylinder reservoir cover with the 3 bolts. Tighten to 80 inch lbs. (9 Nm).
14. Connect the fuel injector connector and ignition coil connector.

1.8L Engine

See Figures 130 through 134.

1. Before servicing, refer to precautions.
2. Remove the oil filler cap sub assembly.
   a. Remove the oil filler cap.
3. Remove the oil filler cap gasket.
4. Remove 3 the engine cover joint bolts.
5. Remove the wiring harness clamp bracket.
   a. Remove the bolt and wiring harness clamp bracket.
6. Remove the spark plug.
7. Remove the Camshaft Position (CMP) sensor.
8. Remove the camshaft timing oil control valve assembly.
   a. Remove the bolt and camshaft timing oil control valve.
   b. Remove the o-ring from the camshaft timing oil control valve.
9. Remove the Crankshaft Position (CKP) sensor.
10. Remove the engine oil pressure switch assembly.
    a. Using a 24 mm deep socket wrench, remove the engine oil pressure switch assembly.
11. Remove the Knock Sensor (KS).
12. Remove the Engine Coolant Temperature (ECT) sensor.
13. Remove the cylinder head cover sub assembly.
    a. Remove the 13 bolts, seal washer and cylinder head cover.

➡A gasket may stick to the cylinder head cover, be careful not to drop any of the gaskets into the engine when removing the cylinder head cover.

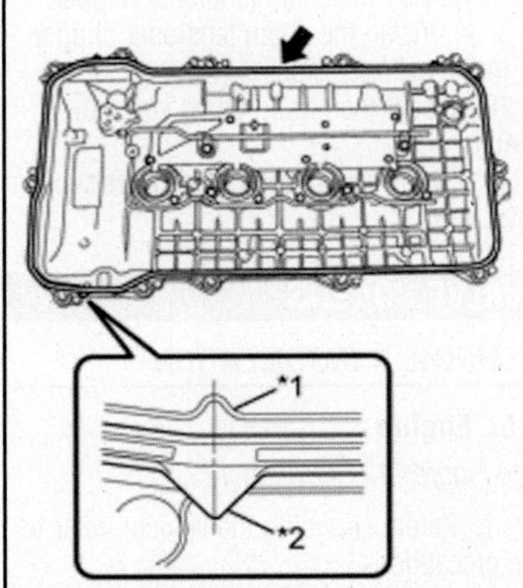

Fig. 132 Installing a new cylinder head cover gasket

   b. Remove the 2 gaskets from the camshaft bearing cap.

To install:

14. Install the cylinder head cover gasket.
15. Remove any oil from the contact surfaces.
16. Misalignment between the center of the cylinder head cover rib and the center of the cylinder head gasket tab is within 0.157 inch (4 mm).
17. Install the cylinder head cover sub assembly.
    a. Install 2 new gaskets to the camshaft bearing cap.
    b. Apply seal packing.

➡Remove any oil from the contact surfaces.

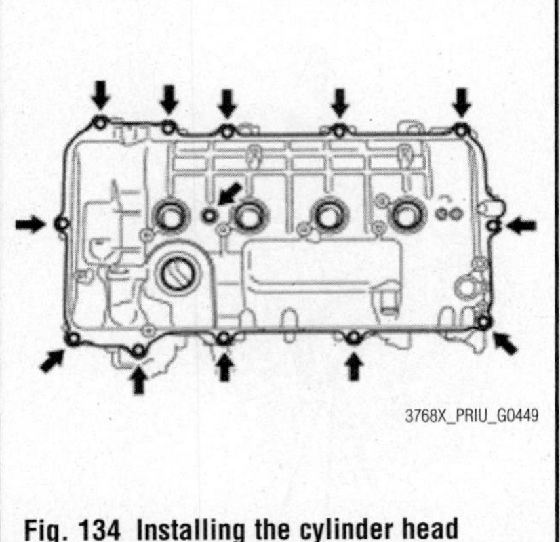

Fig. 134 Installing the cylinder head

➡Install the cylinder head cover sub-assembly within 3 minutes and tighten the bolts within 15 minutes of applying seal packing.

➡Do not start the engine for at least 2 hours after the installation.

   c. Install the cylinder head cover with a new seal washer and 13 bolts. Tighten the bolts to 7 ft. lbs. (10 Nm).

➡Misalignment between the contact surfaces of the timing chain cover and the camshaft housing and the center of the cylinder head gasket tab is within 0.157 inch (4 mm).

18. Install the engine oil pressure switch assembly. Tighten to 11 ft. lbs. (15 Nm).
19. Install the Engine Coolant Temperature (ECT) sensor. Tighten to 15 ft. lbs. (20 Nm).
20. Install the Knock Sensor (KS). Tighten to 15 ft. lbs. (21 Nm).

➡Make sure that the Knock Sensor (KS) is in the correct position.

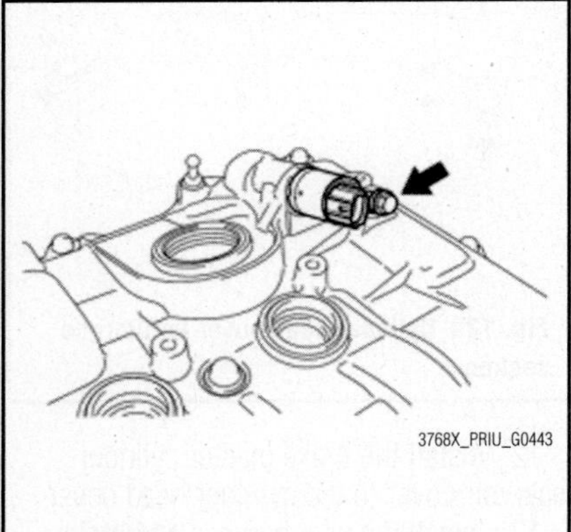

Fig. 130 Removing the camshaft timing oil control valve assembly

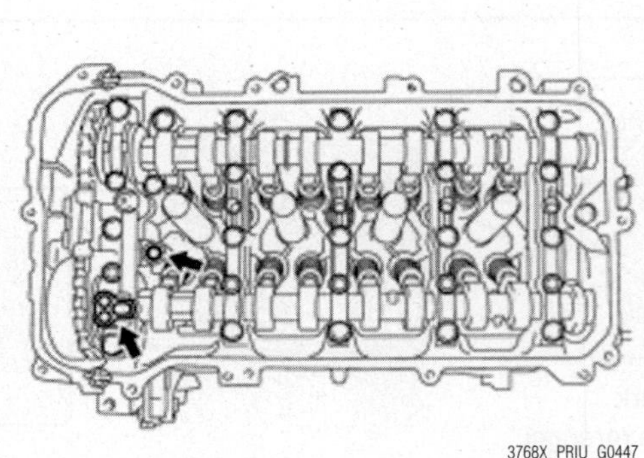

Fig. 133 Installing new gaskets to the camshaft bearing cap

21. Install the Crankshaft Position (CKP) sensor. Tighten to 7 ft. lbs. (10 Nm).

22. Install the camshaft timing oil control valve assembly. Tighten to 7 ft. lbs. (10 Nm).

➡Do not allow foreign matter to contact the oil seal face of the oil control valve (connecting surface with cylinder head cover).

➡Be careful that the O-ring is not cracked or moved out of place when installing the oil control valve.

23. Install the Camshaft Position (CMP) sensor. Tighten to 7 ft. lbs. (10 Nm).

➡Make sure that the o-ring is not cracked or jammed when installing the sensor.

24. Install the spark plug.

25. Install the engine cover joint bolt. Tighten to 7 ft. lbs. (10 Nm).

26. Install the wiring harness clamp bracket. Tighten to 7 ft. lbs. (10 Nm).

27. Install the oil filler cap gasket.

28. Install the oil filler cap sub assembly.

VALVE LASH

ADJUSTMENT

See Figures 135 through 140.

➡Inspect and adjust the valve clearance when the engine is cold.

1. Before servicing the vehicle, refer to the precautions.

2. Set the No. 1 cylinder to TDC/compression.
   - Turn the crankshaft pulley until its timing notch and timing mark 0 of the chain cover are aligned.
   - Check that both timing marks on the camshaft timing sprocket and camshaft timing gear are facing upward as shown in the illustration. If not, turn the crankshaft 1 complete revolution (360 degrees) and align the marks as above.

3. Check the valves indicated in the illustration. Using a feeler gauge, measure the clearance between the valve lifter and camshaft.
   - Standard intake valve clearance (Cold): 0.007–0.009 inch (0.17–0.23mm)
   - Standard exhaust valve clearance (Cold): 0.011–0.013 inch (0.27–0.33mm)

4. Record any out-of-specification valve clearance measurements. They will be used later to determine the required replacement lifter.

5. Turn the crankshaft 1 complete revolution until its timing notch and timing mark 0 of the chain cover are aligned.

6. Check the valves indicated in the illustration. Using a feeler gauge, measure the clearance between the valve lifter and camshaft.
   - Standard intake valve clearance (Cold): 0.007–0.009 inch (0.17–0.23mm)
   - Standard exhaust valve clearance (Cold): 0.011–0.013 inch (0.27–0.33mm)

7. Record any out-of-specification valve clearance measurements. They will be used later to determine the required replacement lifter.

8. Set the No. 1 cylinder to TDC/compression.

9. Turn the crankshaft pulley until its timing notch and timing mark 0 of the chain cover are aligned.

10. Check that both timing marks on the camshaft timing sprocket and valve timing controller assembly are facing upward. If not, turn the crankshaft 1 complete revolution (360 degrees) and align the marks as above.

11. Put paint marks on the timing chain where the timing marks of the camshaft timing sprocket and the camshaft timing gear are located.

12. Using an 8mm hexagon wrench, remove the screw plug.

13. Insert a screwdriver into the service hole of the chain tensioner to hold the stopper plate of the chain tensioner at an upward position. Lifting up the stopper plate of the chain tensioner unlocks the plunger.

14. Keeping the stopper plate of the chain tensioner lifted, slightly rotate the hexagonal lobe of the No. 2 camshaft to the right with an adjustable wrench so the plunger of the chain tensioner is pushed. When the camshaft No. 2 is slightly rotated to the right, the plunger is pushed.

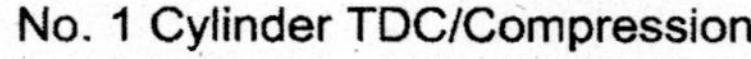

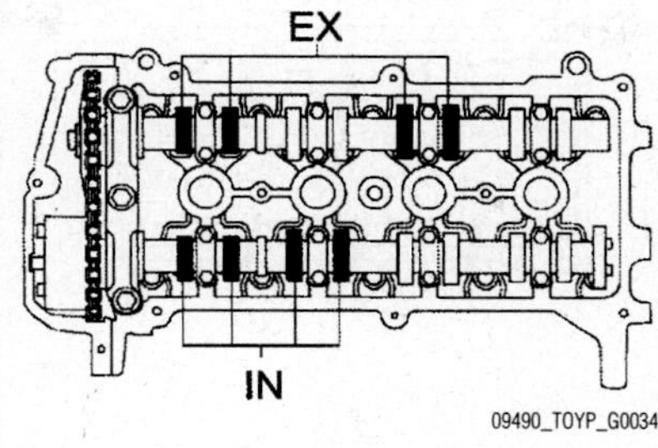

Fig. 136 Check the clearance of the valves indicated (No. 1 Cylinder)

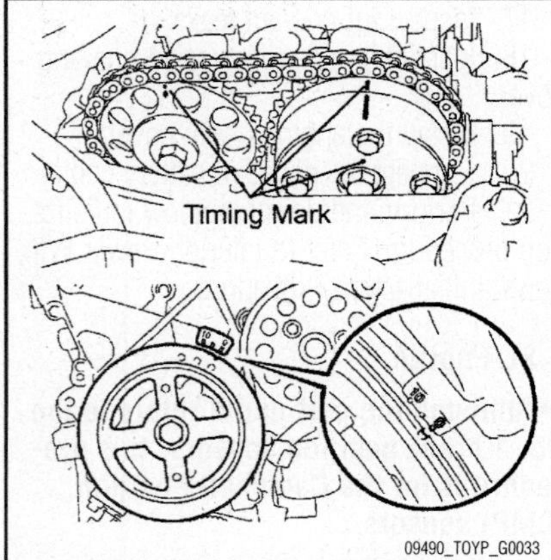

Fig. 135 Check that both timing marks on the camshaft timing sprocket and camshaft timing gear are facing upward

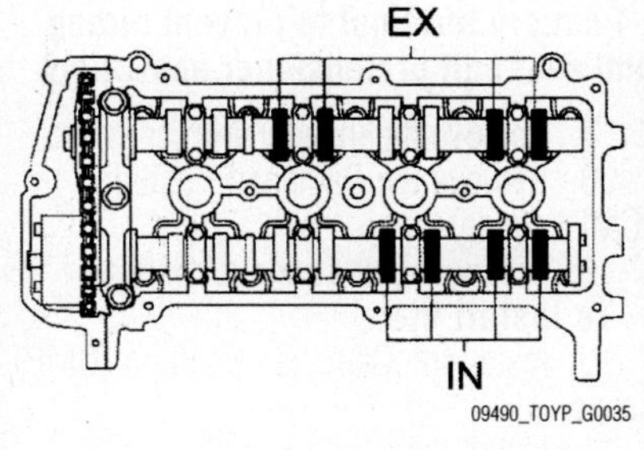

Fig. 137 Check the clearance of the following valves (No. 4 cylinder)

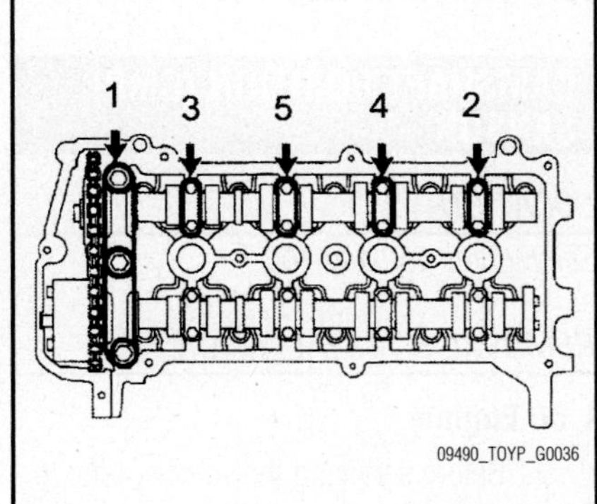

Fig. 138 Remove the 11 bearing cap bolts in the sequence

15. Keeping the adjustable wrench installed, remove the screwdriver with the plunger pushed. Do not move the adjustable wrench.

➡Removing the screwdriver lowers the stopper plate and locks the plunger.

16. Insert a 0.118 inch (3.0mm) diameter bar into the hole of the stopper plate with the stopper plate of the chain tensioner lowered and locked. If the bar cannot be inserted into the hole of the stopper plate, rotate the No. 2 camshaft slightly to the left and right. Then that bar can be inserted easily.
17. Secure the bar with tape.
18. Hold the hexagonal lobe of the camshaft No. 2 with the adjustable wrench.
19. Using SST 09023-38400, loosen the bolt.
20. Using several steps, uniformly loosen and remove the 11 bearing cap bolts in the sequence shown in the illustration. Then remove the 5 bearing caps. Loosen each bolt uniformly, keeping the camshaft level.
21. Remove the flange bolt with the No. 2 camshaft lifted up. Then detach the No. 2 camshaft and the camshaft timing sprocket.
22. Using several steps, uniformly loosen and remove the 8 bearing cap bolts in the sequence shown in the illustration. Then remove the 4 bearing caps. Loosen each bolt uniformly, keeping the camshaft level.
23. Hold the timing chain with one hand, and remove the camshaft and the camshaft timing gear assembly.

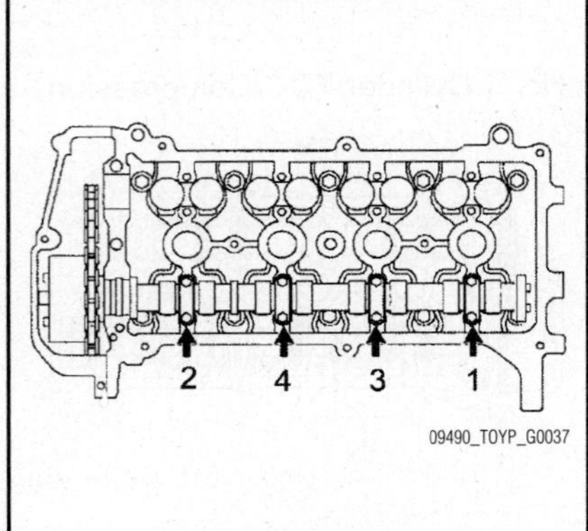

Fig. 139 Remove the 8 bearing cap bolts in the sequence

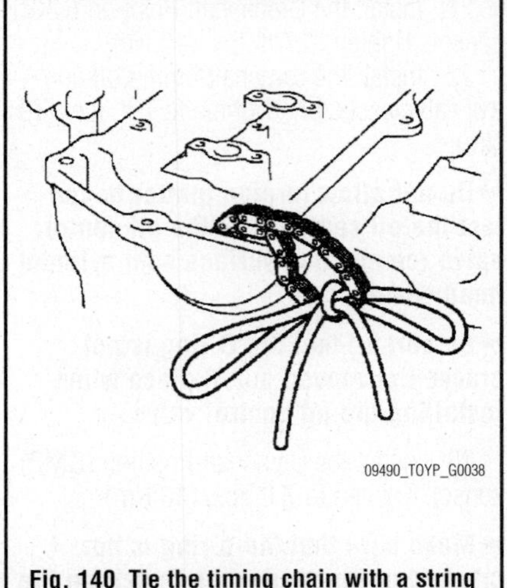

Fig. 140 Tie the timing chain with a string

24. Tie the timing chain with a string as shown in the illustration.

⁂ CAUTION

Be careful not to drop anything inside the timing chain cover.

25. Remove the valve lifters.
26. Using a micrometer, measure the thickness of the removed lifter.
27. Calculate the thickness of a new lifter so that the valve clearance comes within the specified value.
28. Select a new lifter with the thickness as close to the calculated values as possible.

- EXAMPLE: (Intake) Measured valve clearance equals 0.0158 inch (0.40mm)
- 0.0158 inch (0.40mm) minus 0.0079 inch (0.20mm) equals 0.0079 inch (0.20mm) (Measured minus Specification equals Excess clearance)
- Used lifter measurement equals 0.2067 inch (5.25mm)
- 0.0079 inch (0.20mm) plus 0.2067 inch (5.25mm) equals 0.2146 inch (5.45mm) (Excess clearance plus Used lifter equals Ideal new lifter)
- Closest new lifter equals 5.45 mm (0.2146 in.); select lifter (0.2150 inch (5.46mm))

➡Lifters are available in 35 sizes in increments of 0.0008 inch (0.020mm), from 0.1992 inch (5.060mm) to 0.2260 inch (5.740mm).

ENGINE PERFORMANCE & EMISSION CONTROLS

COMPONENT LOCATIONS

See Figures 141 through 158.

CAMSHAFT POSITION (CMP) SENSOR

LOCATION

See Figures 159 and 160.

REMOVAL & INSTALLATION

1.5L Engine

1. Before servicing the vehicle, refer to the precautions.
2. Remove the radiator support opening cover.
3. Remove the engine under cover.
4. Drain engine coolant.
5. Drain high voltage battery coolant.
6. Disconnect the negative battery cable.

➡Wait at least 90 seconds after disconnecting the cable from the negative (-) battery terminal to prevent airbag and seat belt pretensioner activation.

7. Remove the inverter with converter.
8. Remove the Camshaft Position (CMP) sensor.
9. Disconnect the sensor connector.

To Install the:

10. Install the Camshaft Position (CMP) sensor.
11. Install the sensor with the bolt. Tighten to 66 inch lbs. (8 Nm).
12. Connect the sensor connector.
13. Install the inverter with converter.
14. Install the inverter with converter.
15. Add high voltage battery coolant.
16. Add engine coolant.
17. Inspect for coolant leaks.
18. Install the radiator support opening cover.
19. Install the engine under cover.
20. Connect the negative battery cable.
21. Perform the power window initialization procedure. Refer to Interior, Power Windows, Initialization Procedure.

1.8L Engine

➡Although the part name refers to the No. 1 crank position sensors, this procedure is for the Camshaft Position (CMP) sensors.

1. Remove the No. 2 cylinder head cover.
2. Remove the No. 1 crank position sensor.

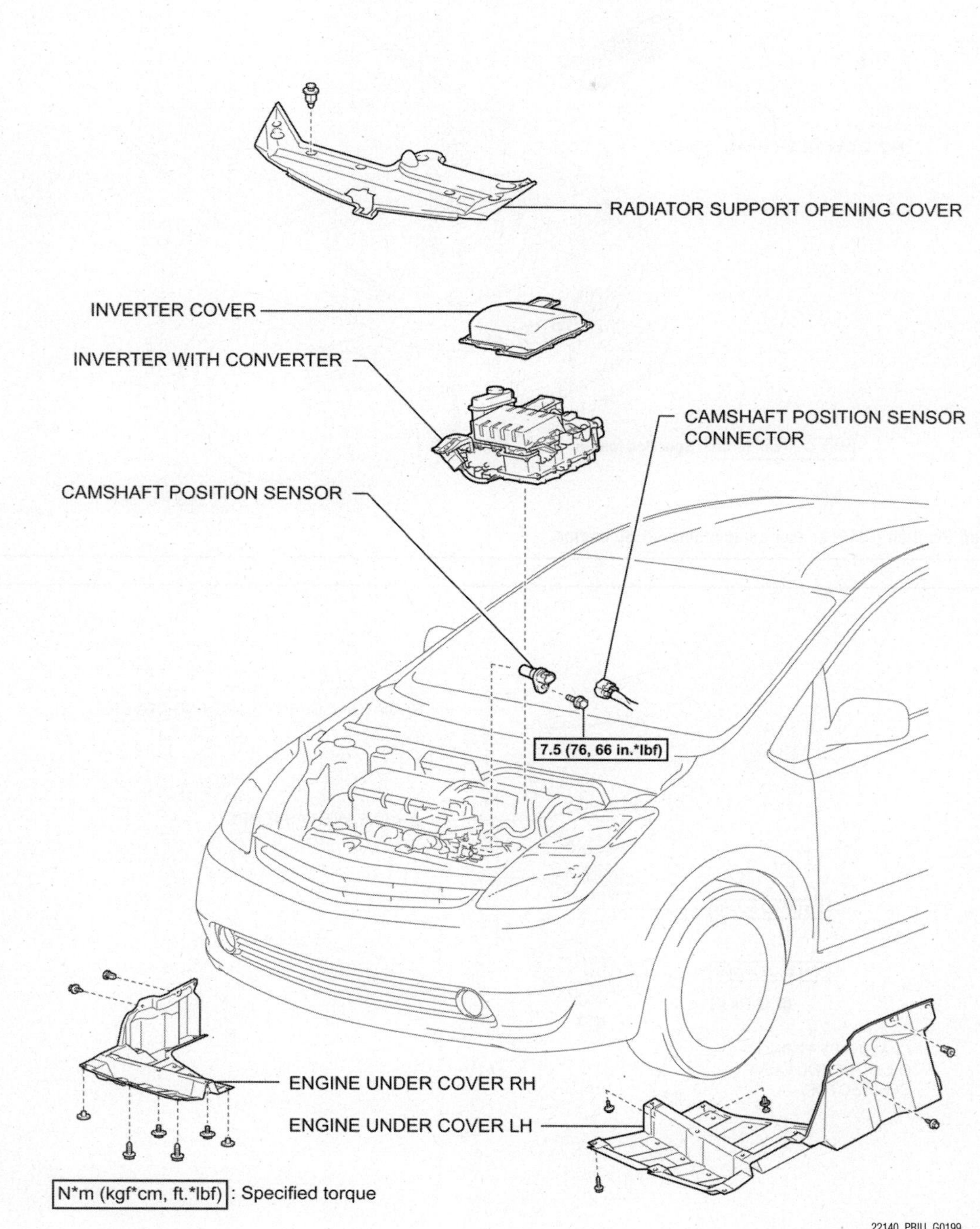

Fig. 141 Camshaft Position (CMP) sensor components—1.5L engine

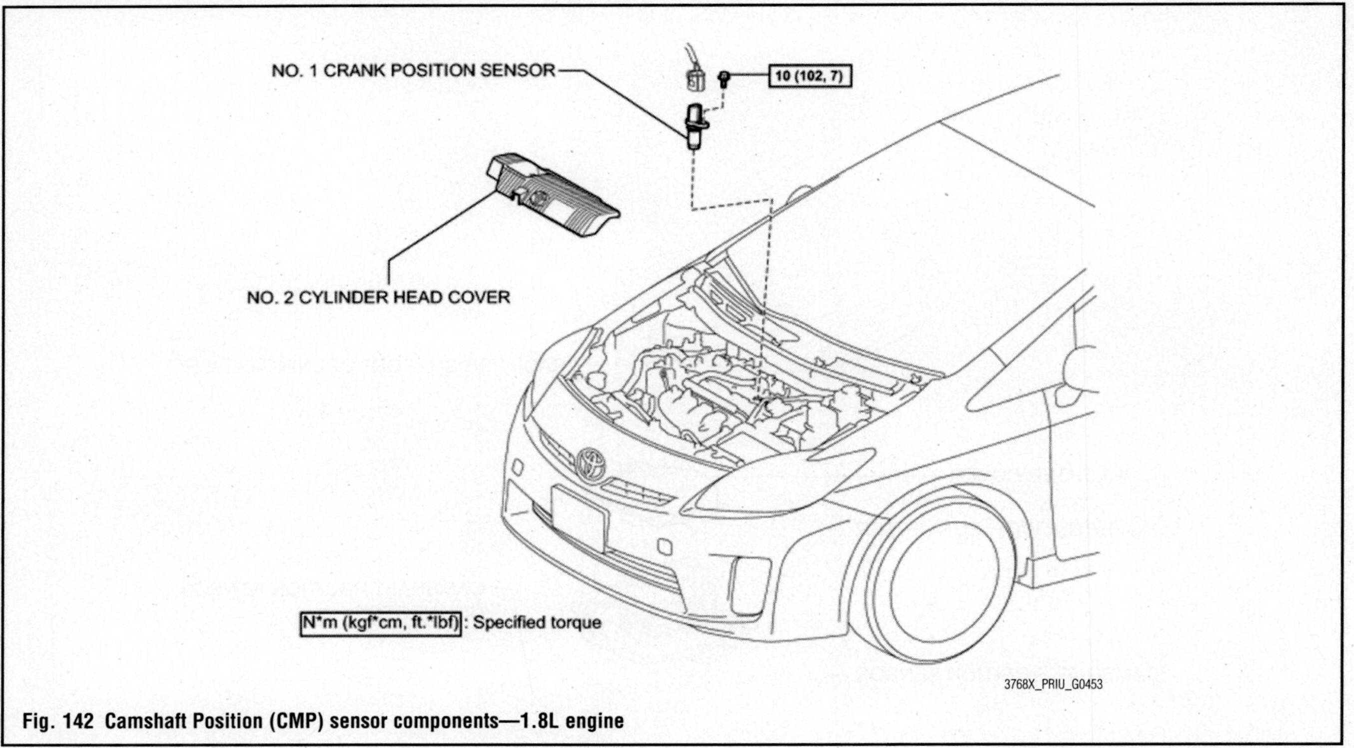

Fig. 142 Camshaft Position (CMP) sensor components—1.8L engine

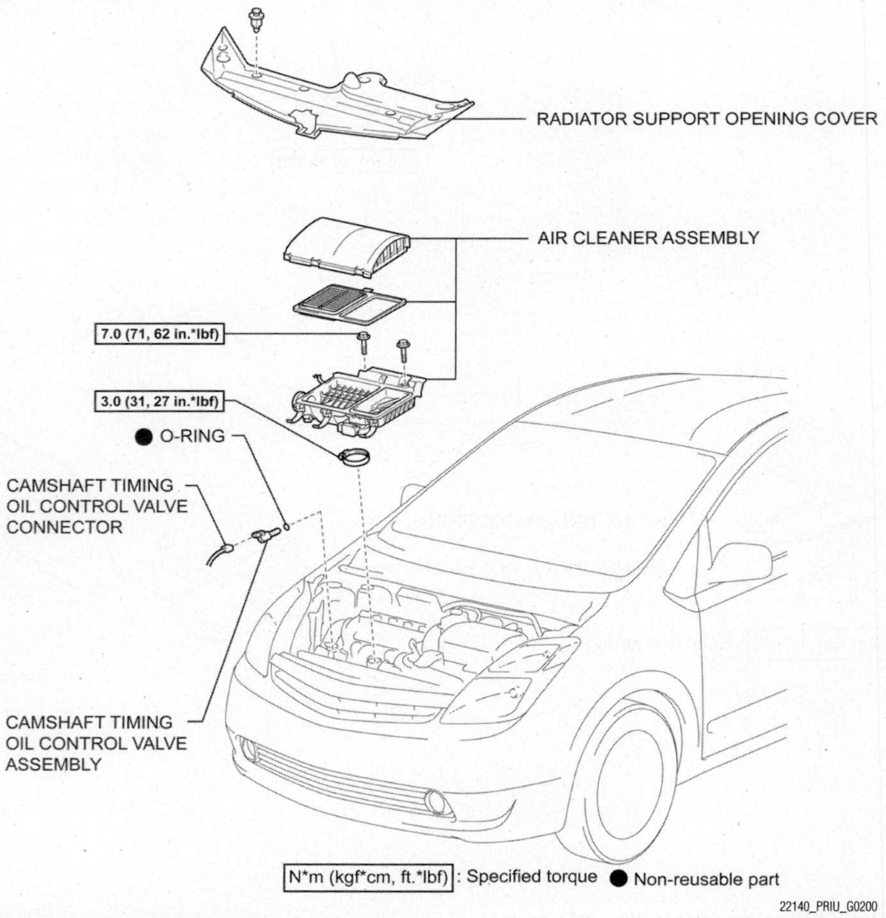

RADIATOR SUPPORT OPENING COVER

AIR CLEANER ASSEMBLY

7.0 (71, 62 in.*lbf)

3.0 (31, 27 in.*lbf)

● O-RING

CAMSHAFT TIMING OIL CONTROL VALVE CONNECTOR

CAMSHAFT TIMING OIL CONTROL VALVE ASSEMBLY

N*m (kgf*cm, ft.*lbf) : Specified torque ● Non-reusable part

22140_PRIU_G0200

Fig. 143 Camshaft timing oil control valve components—1.5L engine

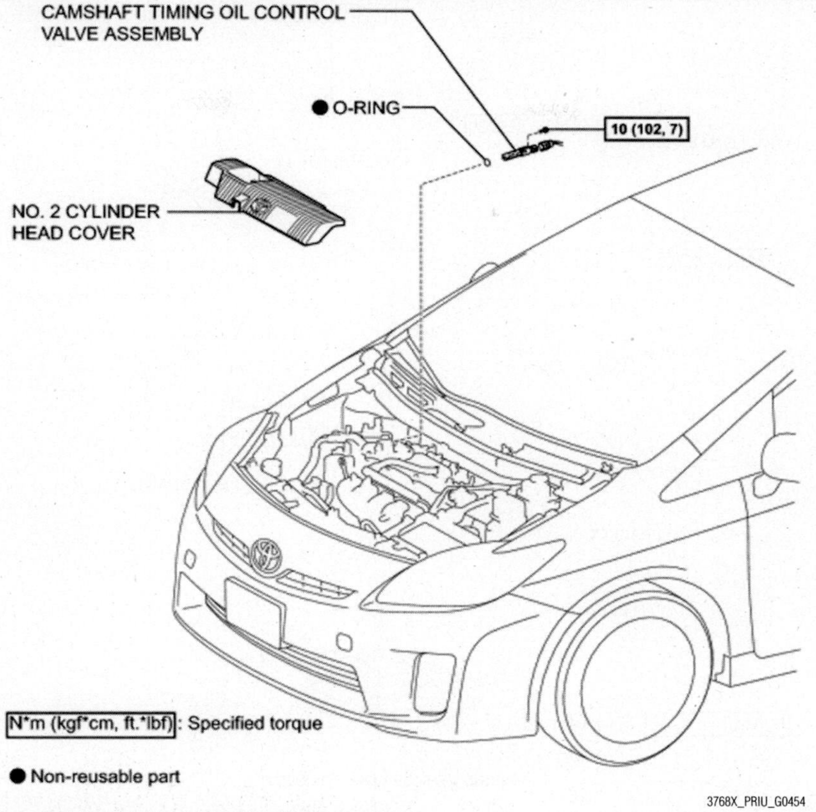

CAMSHAFT TIMING OIL CONTROL VALVE ASSEMBLY
O-RING
10 (102, 7)
NO. 2 CYLINDER HEAD COVER
N*m (kgf*cm, ft.*lbf) : Specified torque
Non-reusable part
3768X_PRIU_G0454

Fig. 144 Camshaft timing oil control valve components—1.8L engine

20 (204, 15)
CRANKSHAFT POSITION SENSOR CONNECTOR
CRANKSHAFT POSITION SENSOR
ENGINE UNDER COVER RH
5.0 (51, 44 in.*lbf)
N*m (kgf*cm, ft.*lbf) : Specified torque
22140_PRIU_G0201

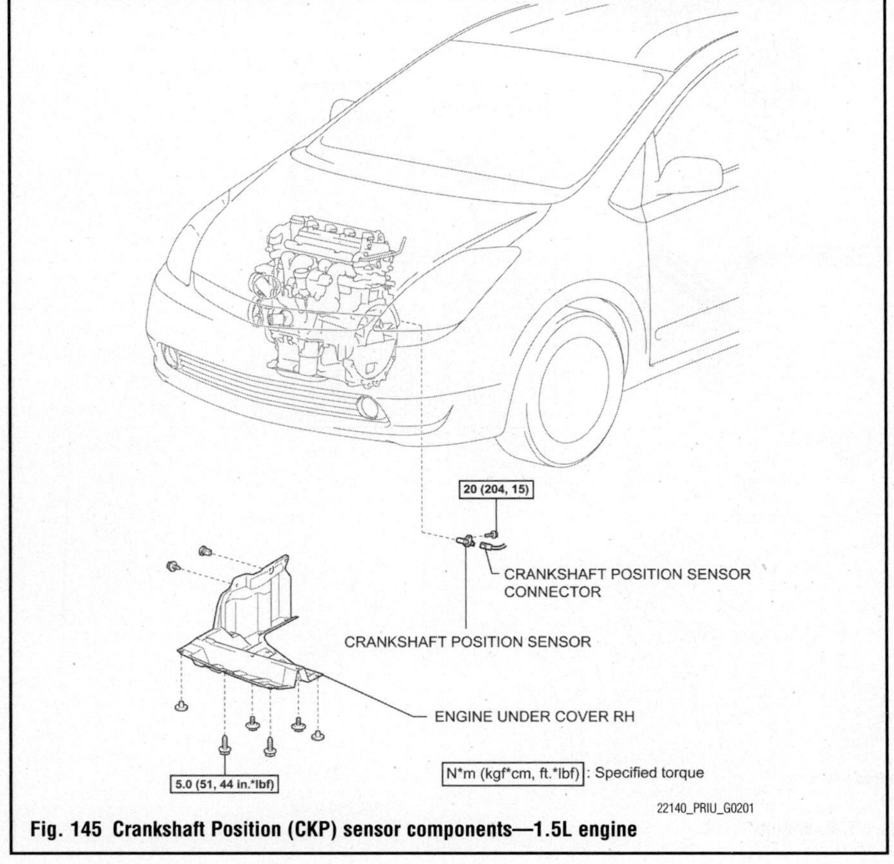

Fig. 145 Crankshaft Position (CKP) sensor components—1.5L engine

10 (102, 7)
CRANK POSITION SENSOR
N*m (kgf*cm, ft.*lbf) : Specified torque
3768X_PRIU_G0455

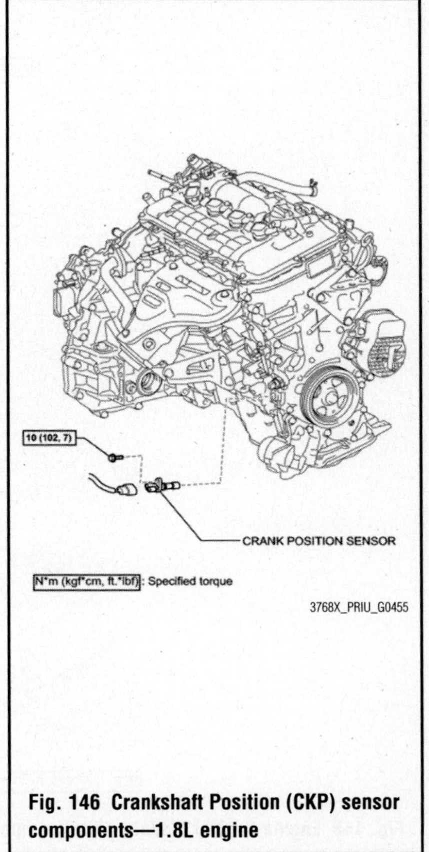

Fig. 146 Crankshaft Position (CKP) sensor components—1.8L engine

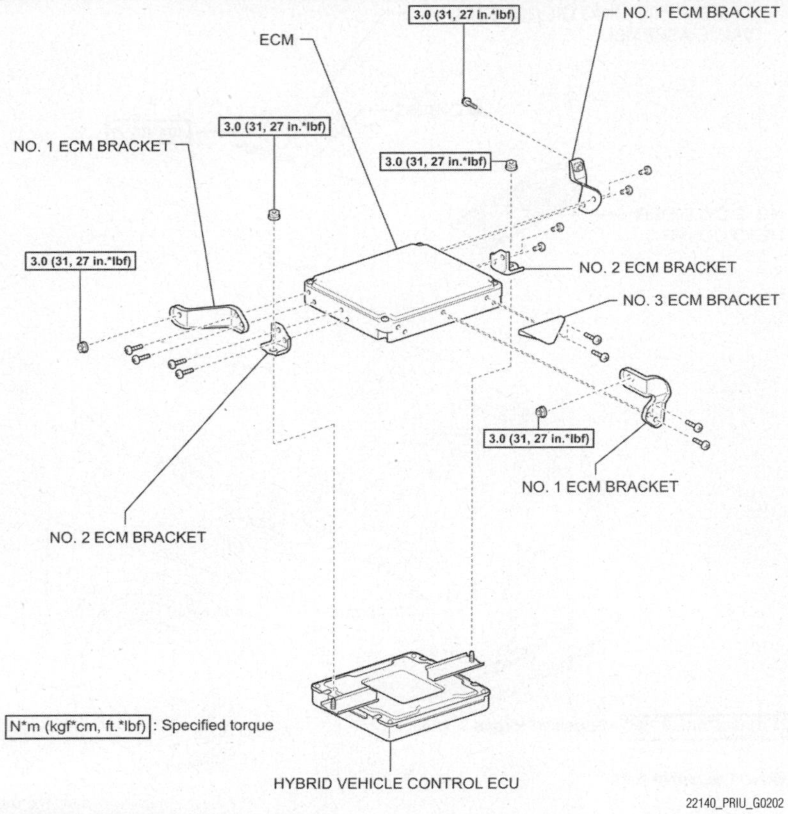
ECM
3.0 (31, 27 in.*lbf)
NO. 1 ECM BRACKET
3.0 (31, 27 in.*lbf)
NO. 1 ECM BRACKET
3.0 (31, 27 in.*lbf)
3.0 (31, 27 in.*lbf)
NO. 2 ECM BRACKET
NO. 3 ECM BRACKET
3.0 (31, 27 in.*lbf)
NO. 1 ECM BRACKET
NO. 2 ECM BRACKET
N*m (kgf*cm, ft.*lbf) : Specified torque
HYBRID VEHICLE CONTROL ECU
22140_PRIU_G0202

Fig. 147 Engine Control Module (ECM) components—1.5L engine

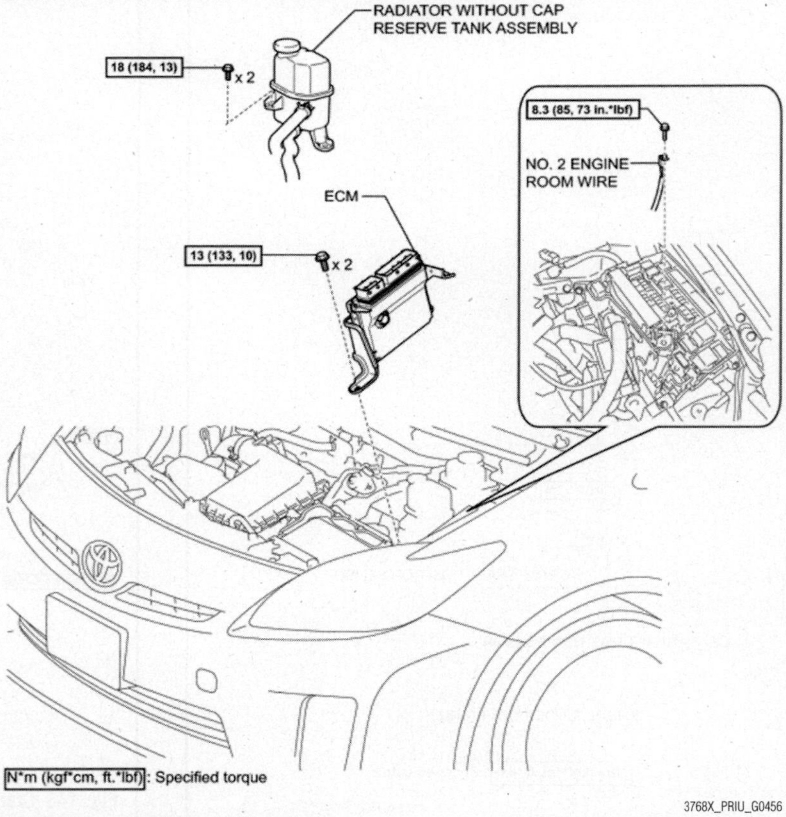
RADIATOR WITHOUT CAP
RESERVE TANK ASSEMBLY
18 (184, 13)
x 2
8.3 (85, 73 in.*lbf)
NO. 2 ENGINE
ROOM WIRE
ECM
13 (133, 10)
x 2
N*m (kgf*cm, ft.*lbf): Specified torque
3768X_PRIU_G0456

Fig. 148 Engine Control Module (ECM) components—1.8L engine

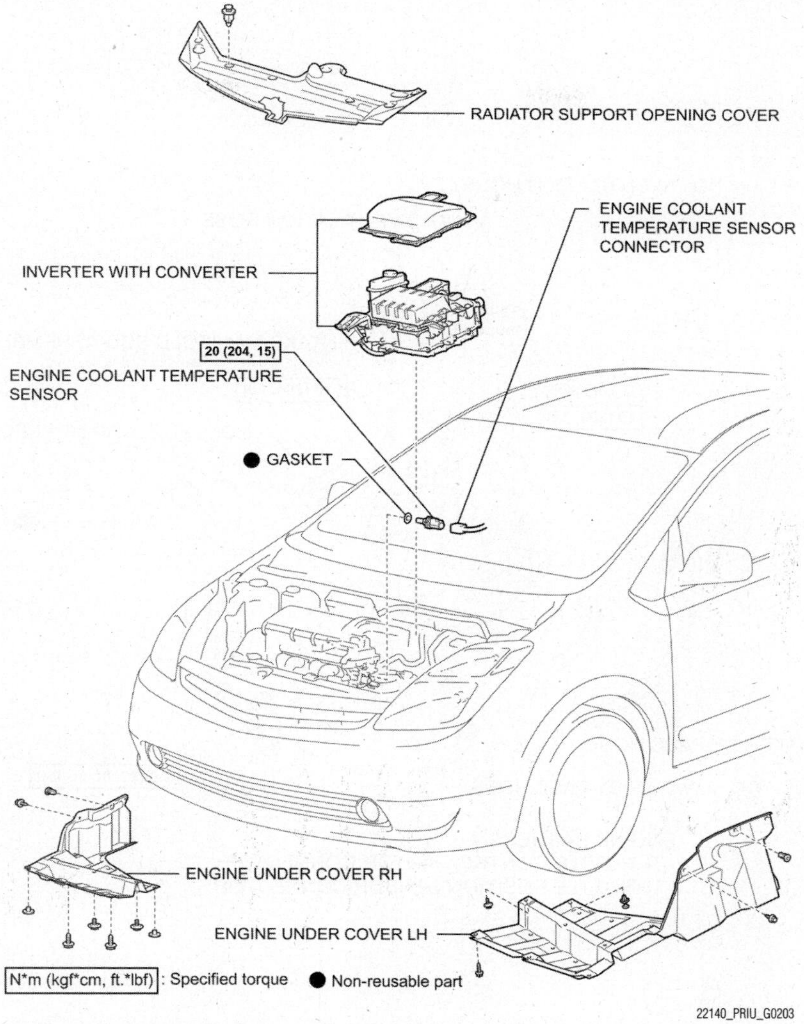

RADIATOR SUPPORT OPENING COVER
ENGINE COOLANT TEMPERATURE SENSOR CONNECTOR
INVERTER WITH CONVERTER
20 (204, 15)
ENGINE COOLANT TEMPERATURE SENSOR
GASKET
ENGINE UNDER COVER RH
ENGINE UNDER COVER LH
N*m (kgf*cm, ft.*lbf) : Specified torque
Non-reusable part
22140_PRIU_G0203

Fig. 149 Engine Coolant Temperature (ECT) sensor components—1.5L engine

GASKET
20 (200, 14)
ENGINE COOLANT TEMPERATURE SENSOR
N*m (kgf*cm, ft.*lbf): Specified torque
Non-reusable part
3768X_PRIU_G0457

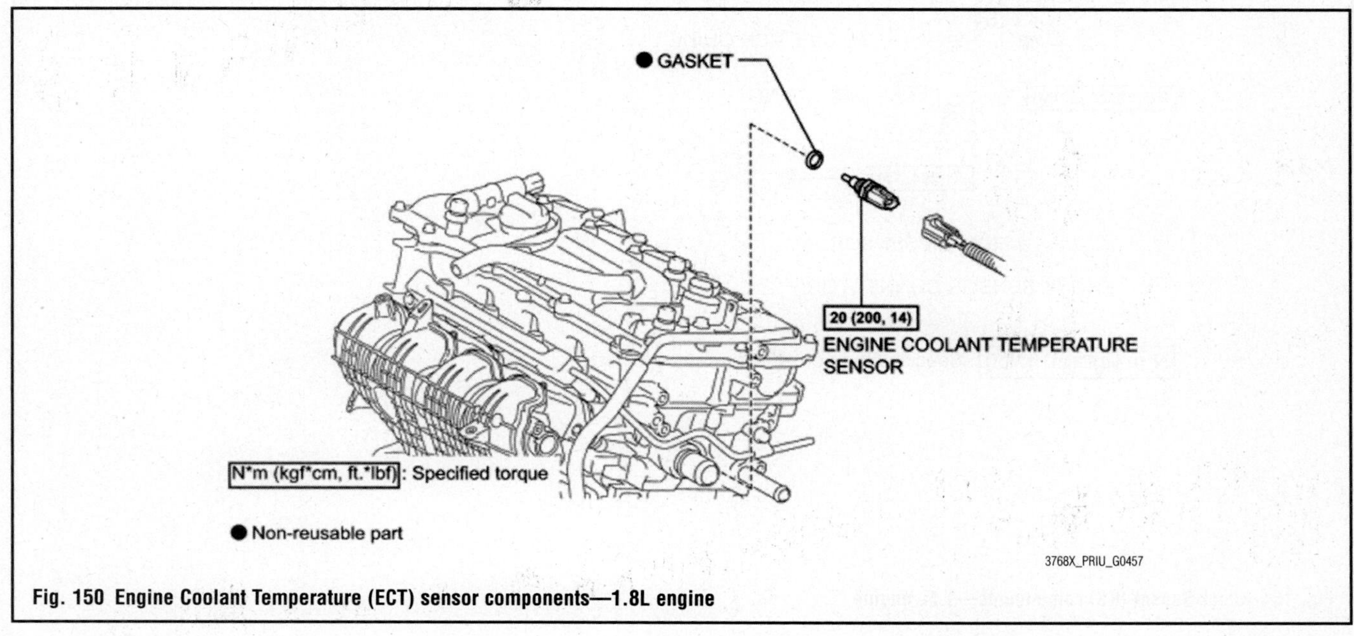

Fig. 150 Engine Coolant Temperature (ECT) sensor components—1.8L engine

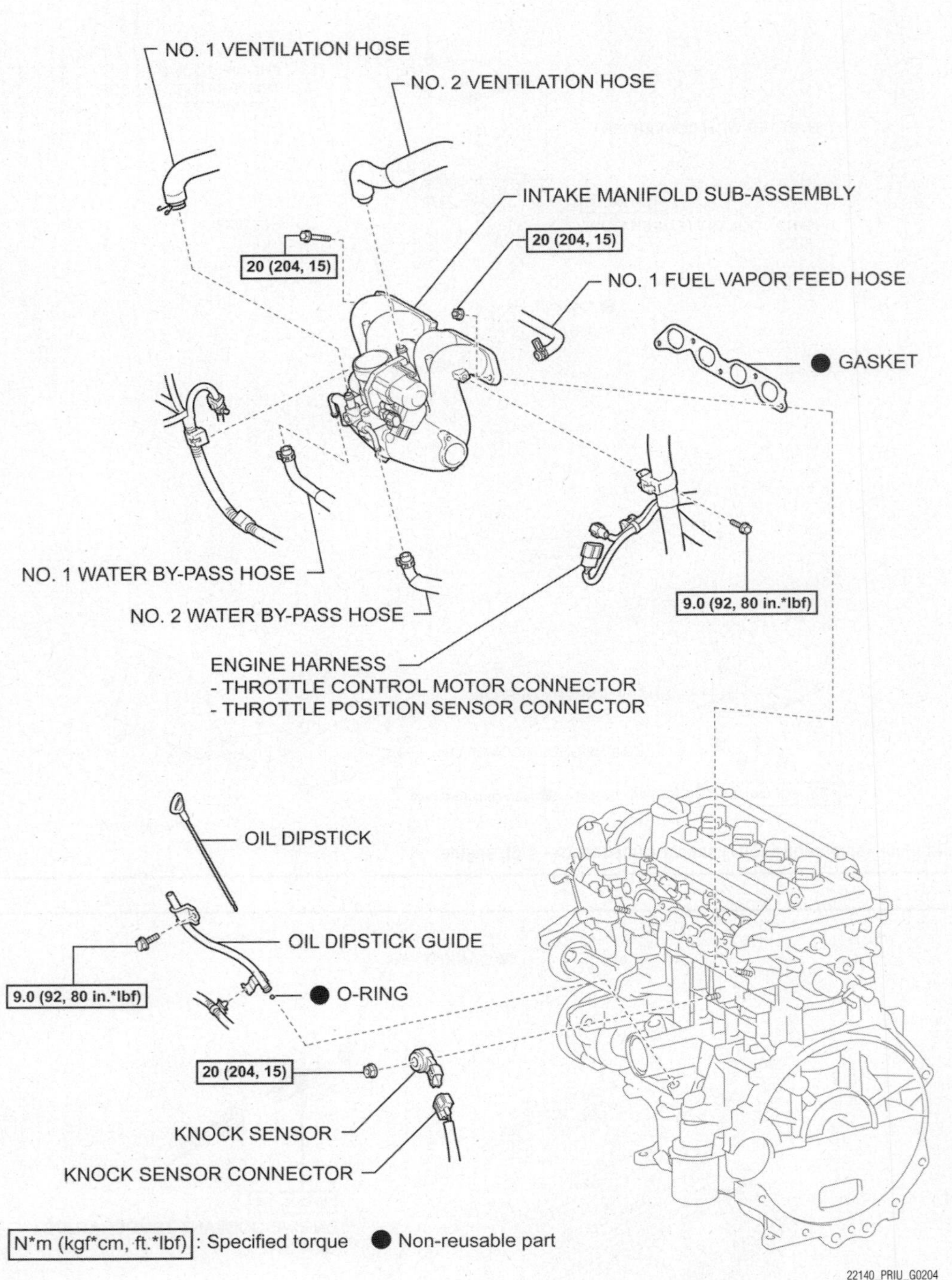

Fig. 151 Knock Sensor (KS) components—1.5L engine

28 (285, 21)
INTAKE MANIFOLD
x 2
28 (285, 21)
x 2
NO. 1 INTAKE MANIFOLD TO HEAD GASKET
KNOCK SENSOR
20 (204, 15)
N*m (kgf*cm, ft.*lbf): Specified torque
Non-reusable part
3768X_PRIU_G0459

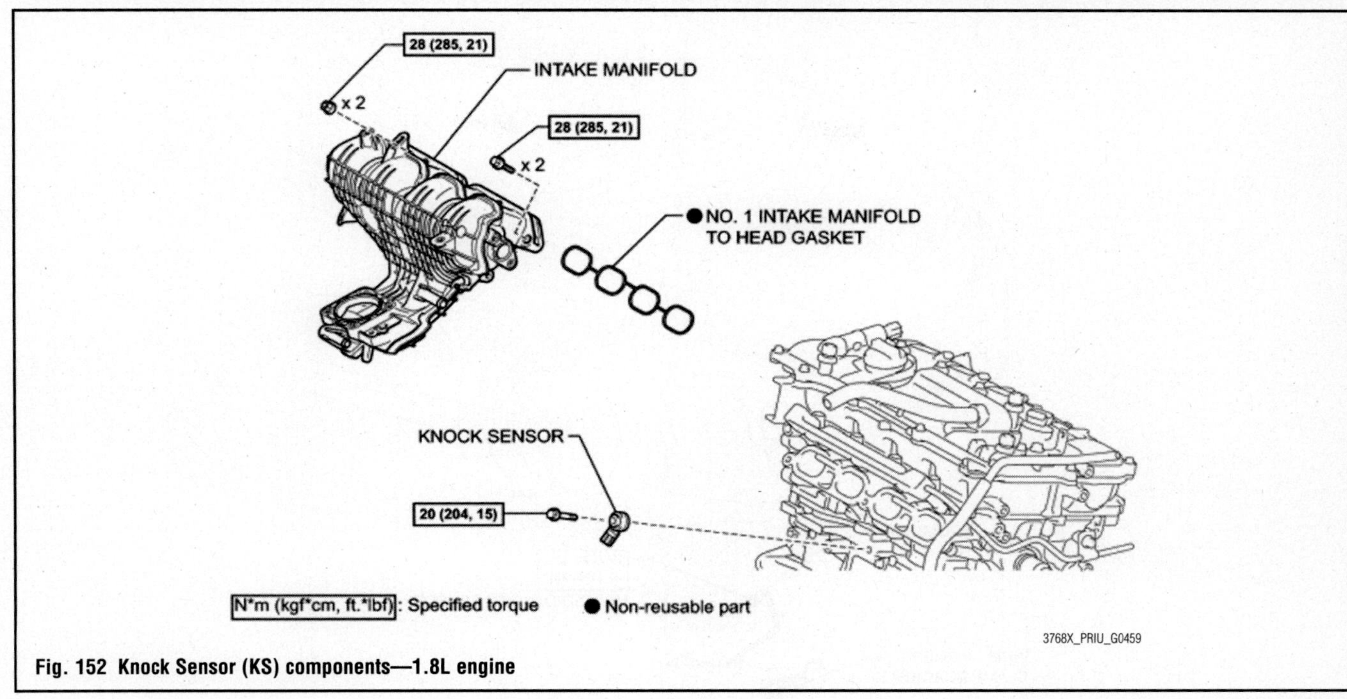

Fig. 152 Knock Sensor (KS) components—1.8L engine

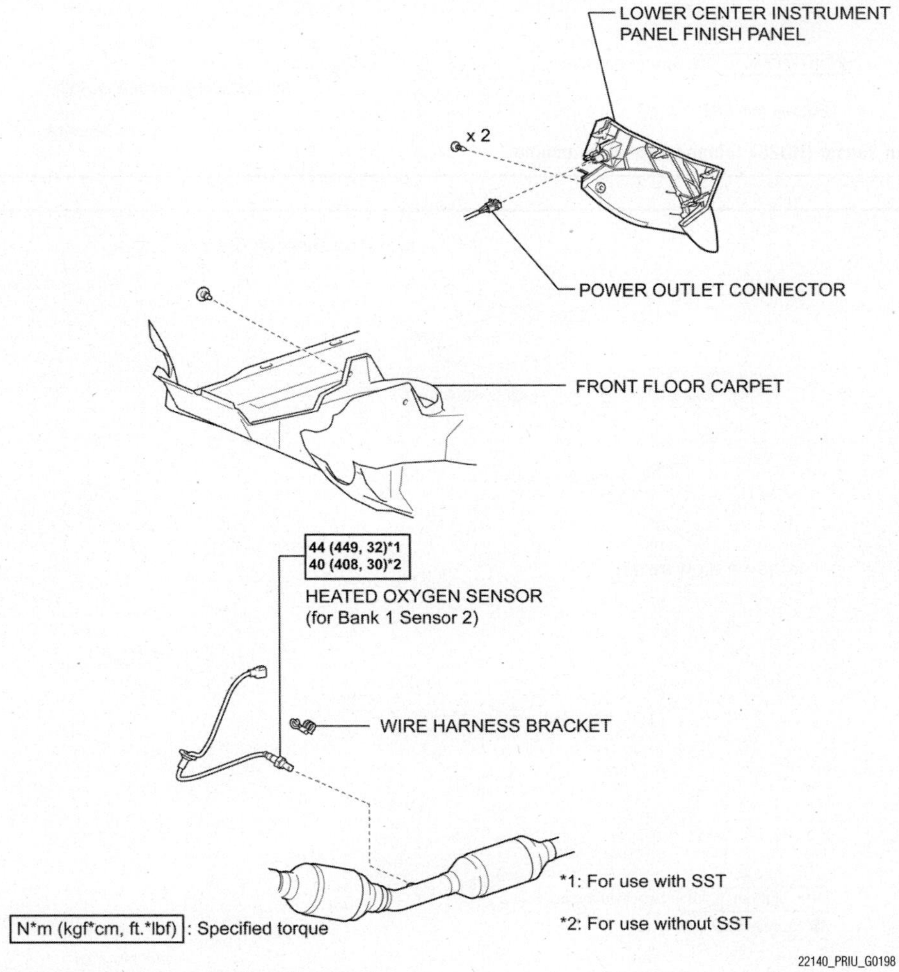

LOWER CENTER INSTRUMENT PANEL FINISH PANEL
x 2
POWER OUTLET CONNECTOR
FRONT FLOOR CARPET
44 (449, 32)*1
40 (408, 30)*2
HEATED OXYGEN SENSOR (for Bank 1 Sensor 2)
WIRE HARNESS BRACKET
*1: For use with SST
*2: For use without SST
N*m (kgf*cm, ft.*lbf): Specified torque
22140_PRIU_G0198

Fig. 153 Heated Oxygen Sensor (HO2S) components—1.5L engine

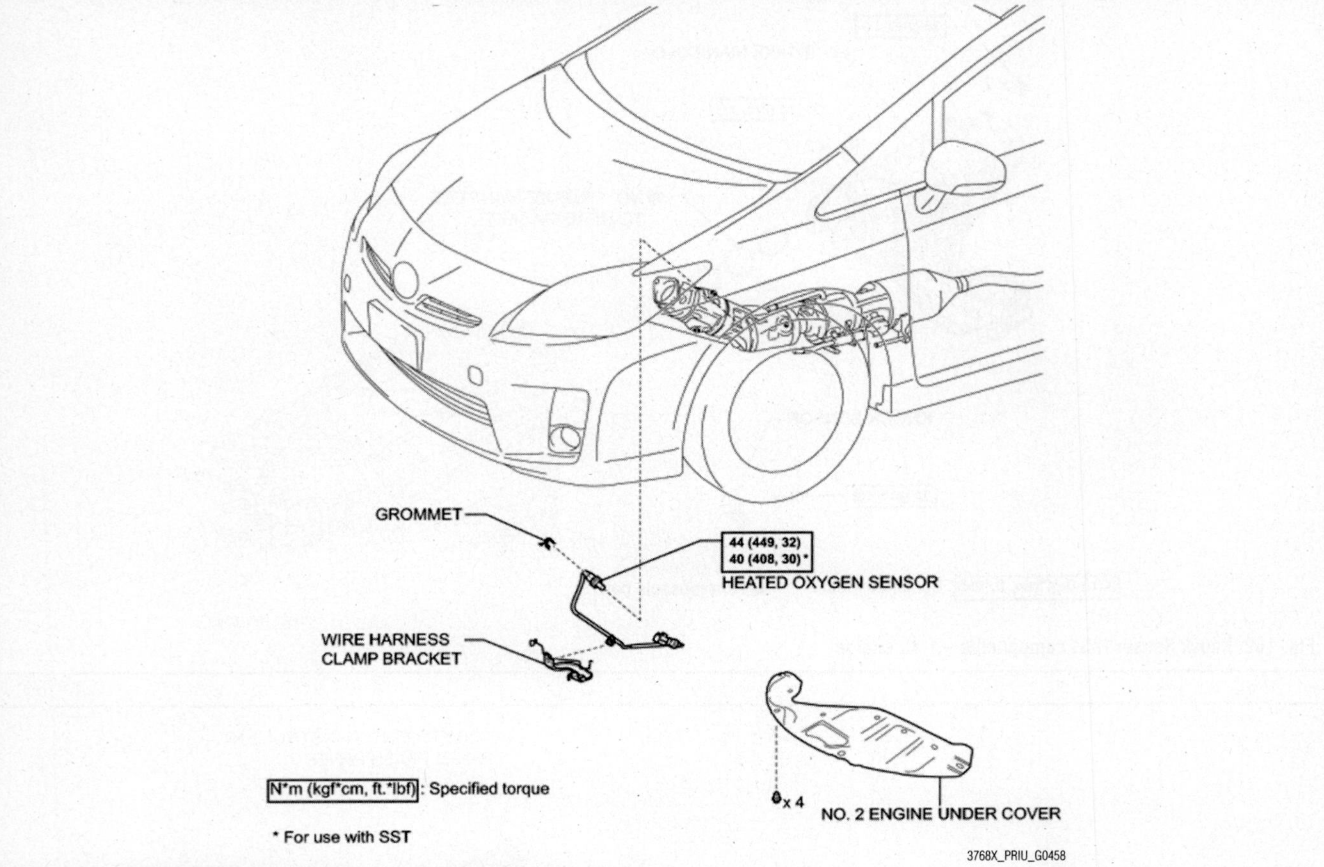

Fig. 154 Heated Oxygen Sensor (HO2S) components—1.8L engine

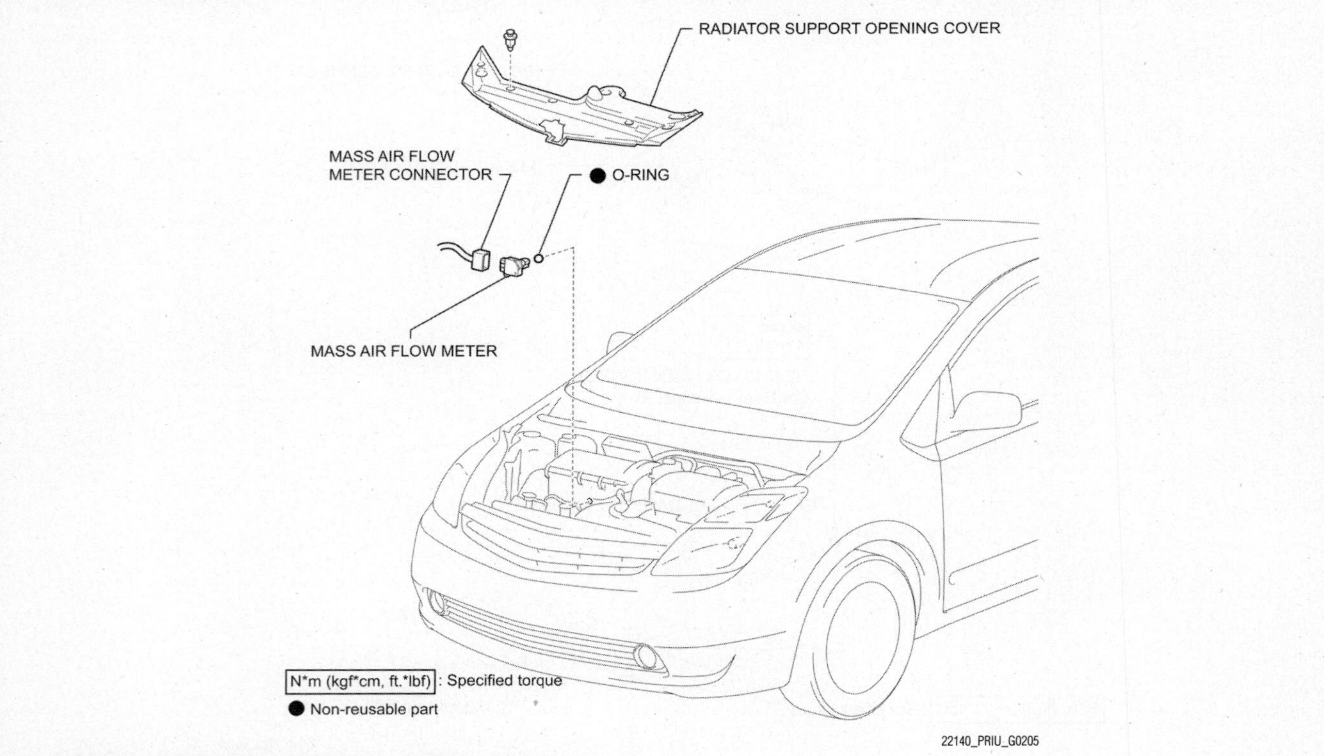

Fig. 155 Mass Air Flow (MAF) sensor components—1.5L engine

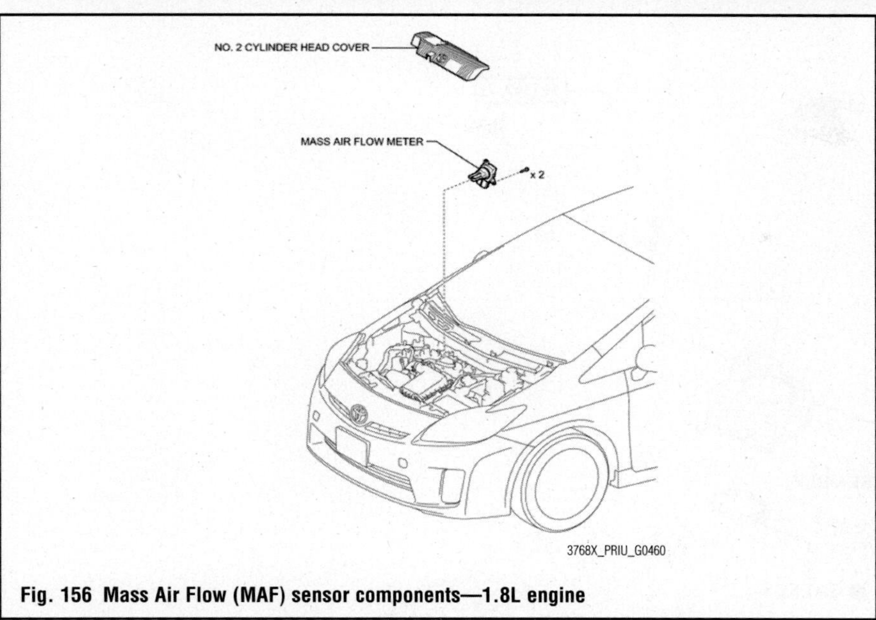

Fig. 156 Mass Air Flow (MAF) sensor components—1.8L engine

a. Disconnect the No.1 crank position sensor connector.

b. Remove the bolt and No. 1 crank position sensor.

To install:

➡Although the part name refers to the No. 1 crank position sensors, this procedure is for the Camshaft Position (CMP) sensors.

3. Install the No. 1 crank position sensor.
   a. Apply a light coat of engine oil to the O-rings on the No. 1 crank position sensor.
   b. Install the No. 1 crank position sensor with the bolt. Tighten to 7 ft. lbs. (10 Nm).
   c. Connect the No. 1 crank position sensor connector.
4. Inspect for oil leaks.
5. Install the No. 2 cylinder head cover.

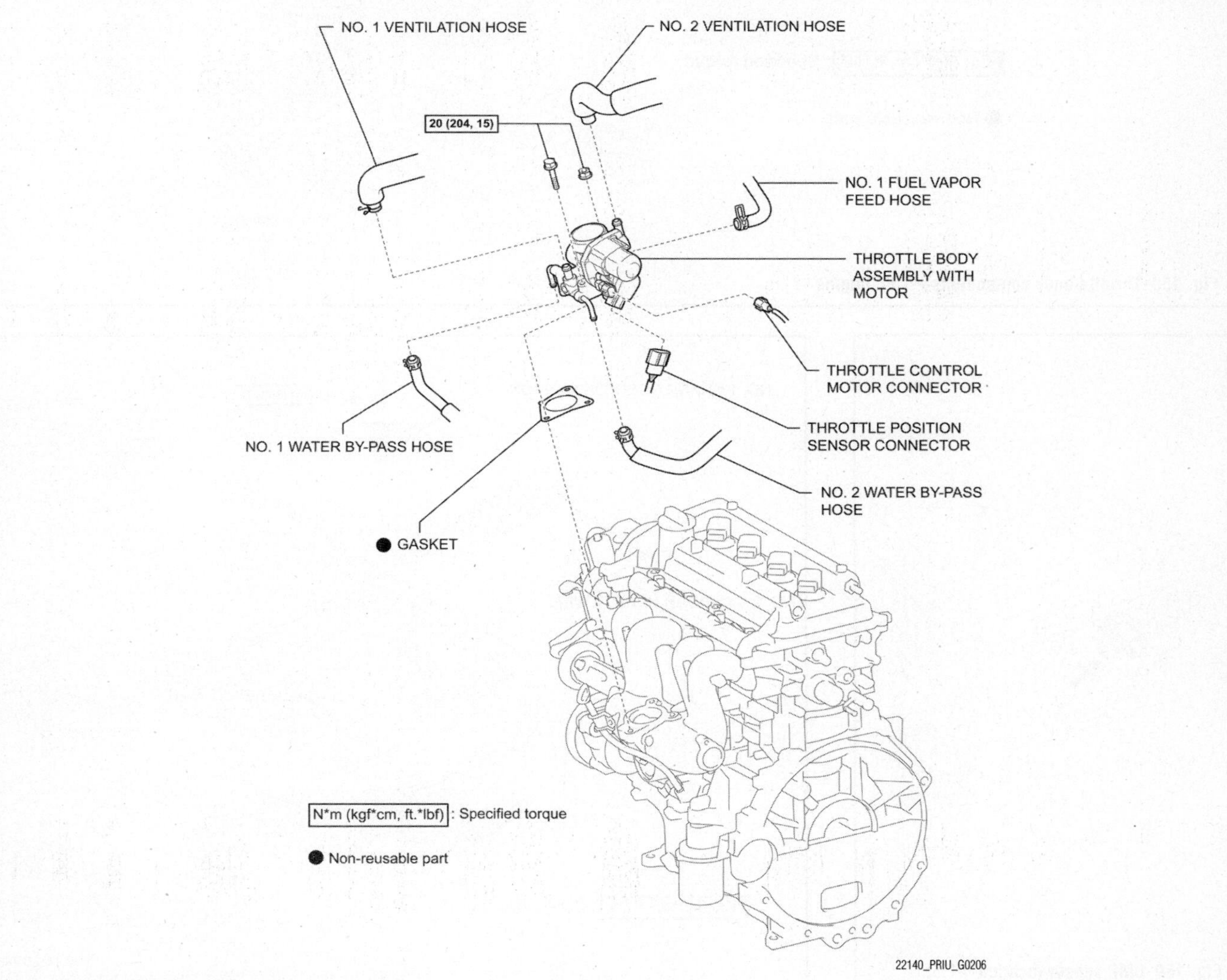

NO. 1 VENTILATION HOSE
NO. 2 VENTILATION HOSE
20 (204, 15)
NO. 1 FUEL VAPOR FEED HOSE
THROTTLE BODY ASSEMBLY WITH MOTOR
THROTTLE CONTROL MOTOR CONNECTOR
THROTTLE POSITION SENSOR CONNECTOR
NO. 1 WATER BY-PASS HOSE
NO. 2 WATER BY-PASS HOSE
● GASKET
N*m (kgf*cm, ft.*lbf) : Specified torque
● Non-reusable part
22140_PRIU_G0206

Fig. 157 Throttle body components—1.5L engine

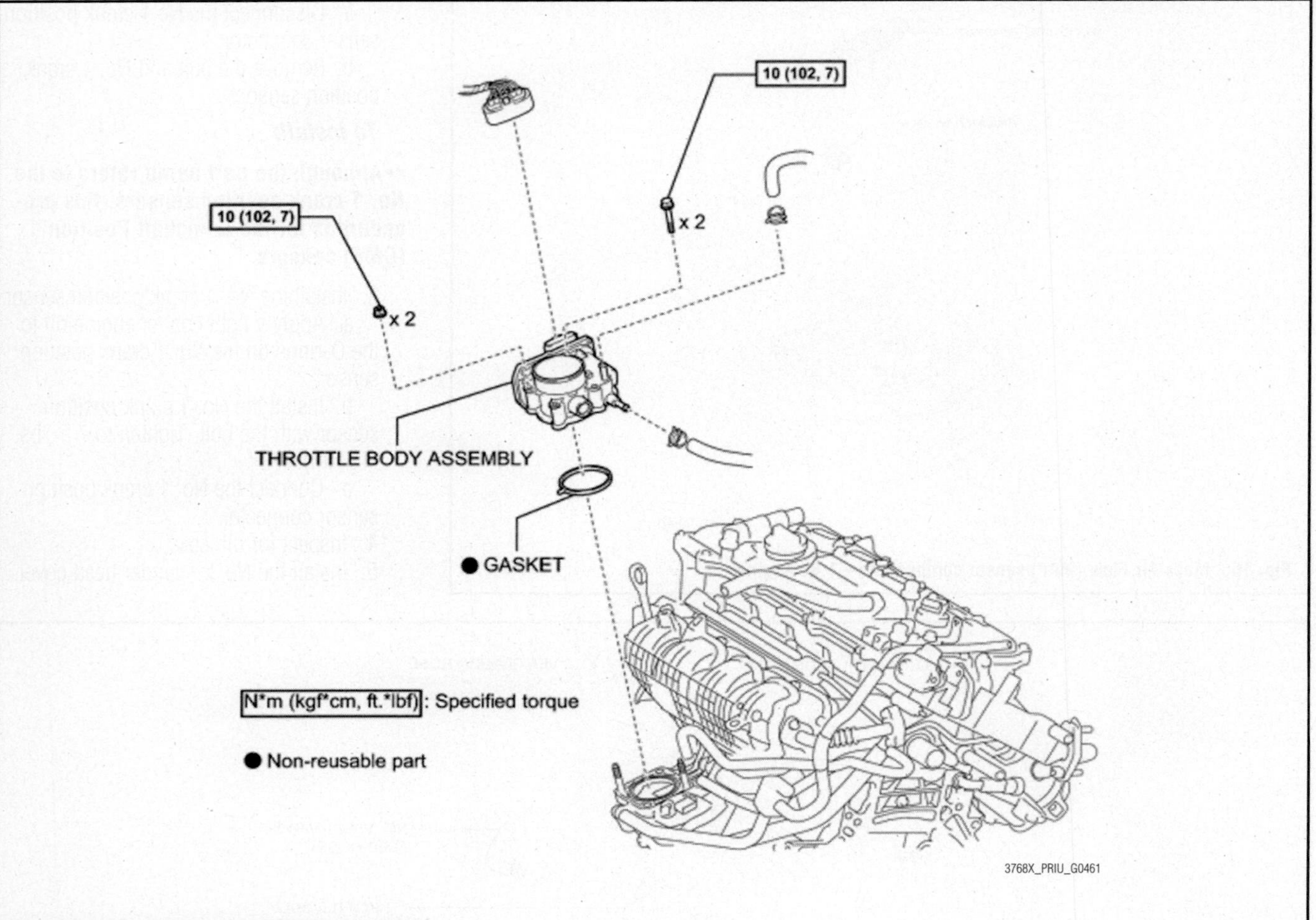

10 (102, 7)
x 2
10 (102, 7)
x 2
THROTTLE BODY ASSEMBLY
● GASKET
N*m (kgf*cm, ft.*lbf): Specified torque
● Non-reusable part
3768X_PRIU_G0461

Fig. 158 Throttle body components—1.8L engine

22140_PRIU_G0187

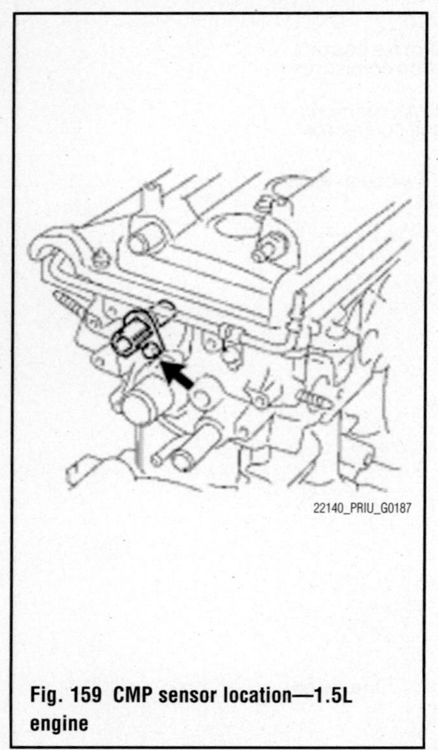

Fig. 159 CMP sensor location—1.5L engine

NO. 1 CRANK POSITION SENSOR
10 (102, 7)
NO. 2 CYLINDER HEAD COVER
N*m (kgf*cm, ft.*lbf): Specified torque
3768X_PRIU_G0453

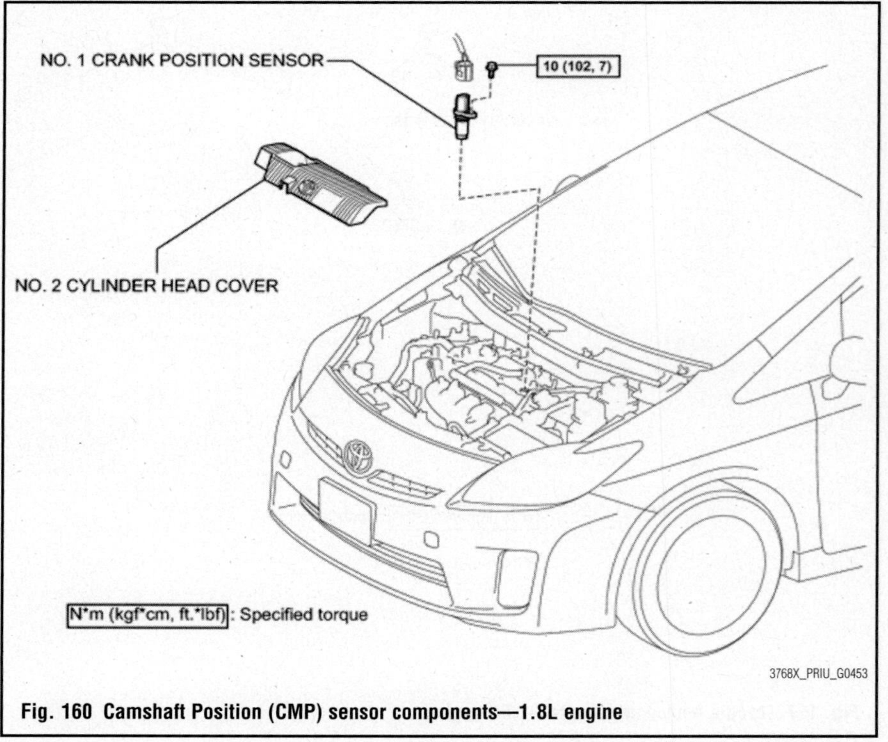

Fig. 160 Camshaft Position (CMP) sensor components—1.8L engine

CRANKSHAFT POSITION (CKP) SENSOR

LOCATION

See Figures 161 and 162.

REMOVAL & INSTALLATION

1.5L Engine

1. Before servicing the vehicle, refer to the precautions.
2. Raise and safely support the vehicle securely on jackstands.
3. Remove engine under cover.
4. Disconnect the sensor connector.
5. Remove the bolt and sensor.

To install:

6. Install the sensor with the bolt. Tighten to 66 inch lbs. (8 Nm).
7. Connect the sensor connector.
8. Install engine under cover.
9. Lower the vehicle.

Fig. 161 CKP sensor location

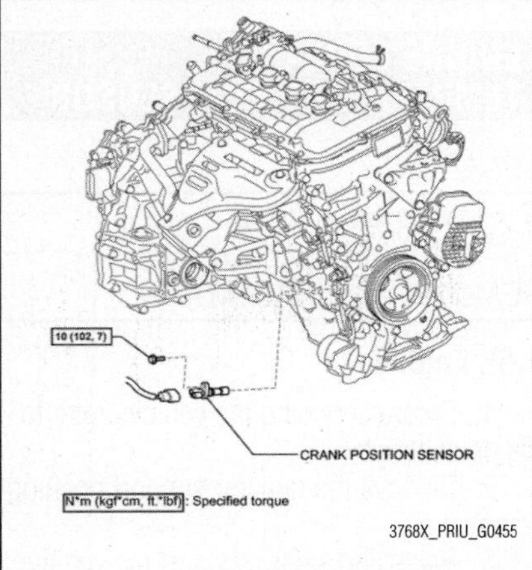

Fig. 162 Crankshaft Position (CKP) sensor components—2010

1.8L Engine

1. Remove the rear engine under cover RH.
2. Remove the No. 1 engine under cover.
3. Remove the crank position sensor.
   a. Disconnect the Crankshaft Position (CKP) sensor connector.
   b. Remove the bolt and Crankshaft Position (CKP) sensor.

To install:

4. Install the Crankshaft Position (CKP) sensor.
   a. Apply a light coat of engine oil to the o-ring on the Crankshaft Position (CKP) sensor.
   b. Install the Crankshaft Position (CKP) sensor with the bolt. Tighten to 7 ft. lbs. (10 Nm).
   c. Connect the Crankshaft Position (CKP) sensor connector.
5. Inspect for oil leaks.
6. Install the No. 1 engine under cover.
7. Install the rear engine under cover RH.

ELECTRONIC CONTROL MODULE (ECM)

LOCATION

See Figures 163 and 164.

REMOVAL & INSTALLATION

1.5L Engine

See Figures 165 and 166.

1. Disconnect the cable from the negative battery terminal.

CAUTION

Wait at least 90 seconds after disconnecting the cable from the negative (-) battery terminal to prevent airbag and seat belt pretensioner activation.

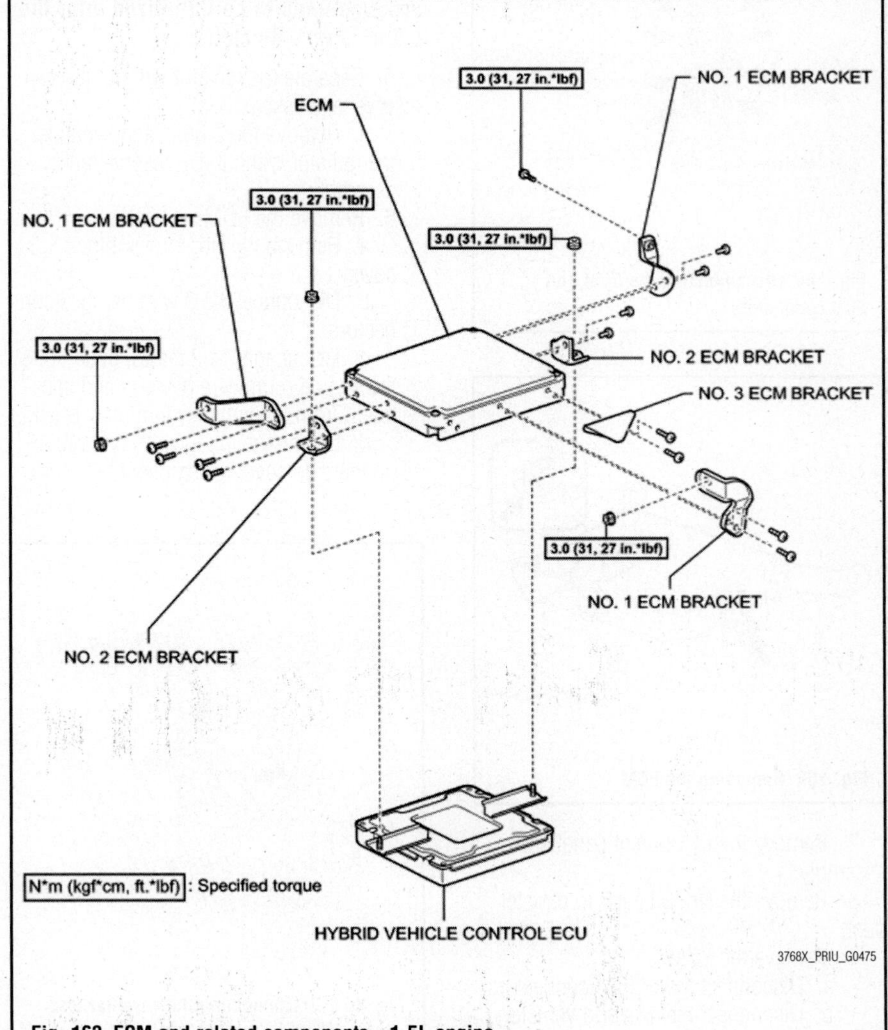

Fig. 163 ECM and related components—1.5L engine

Fig. 164 ECM location—1.8L engine

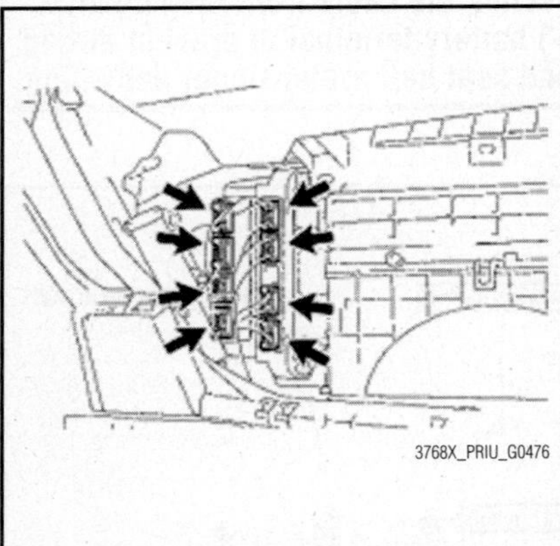

Fig. 165 Disconnecting the ECM and ECU connectors

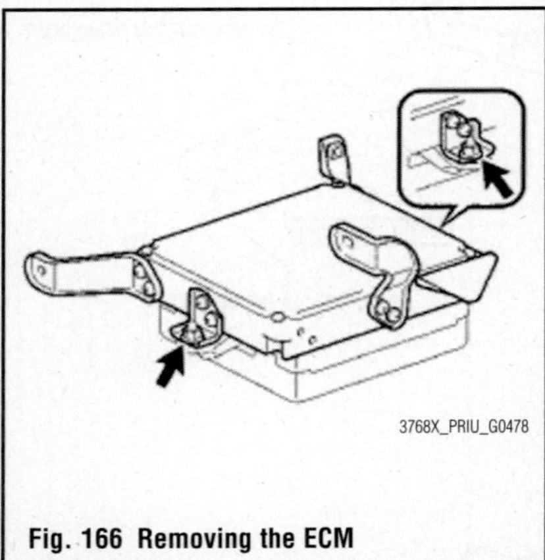

Fig. 166 Removing the ECM

2. Remove the instrument panel sub assembly.
3. Remove the No. 3 heater to register duct.
4. Remove the ECM.
   a. Disconnect the 4 ECM connectors.
   b. Disconnect the 4 hybrid vehicle control ECU connectors.
   c. Remove the 2 nuts and bolt, and ECM with the bracket.
   d. Remove the 2 nuts and ECM.
   e. Remove the 6 screws and 3 No. 1 ECM brackets.
   f. Remove the 4 screws and 2 No. 2 ECM brackets.
   g. Remove the 2 screws and No. 3 ECM bracket.

To install:

5. To install, reverse the removal procedure. Tighten the ECM nuts and bolt to 27 inch lbs. (3 Nm).

1.8L Engine

See Figures 167 and 168.

1. Remove the rear No. 2 floor board (separate type).
2. Remove the rear deck floor box.
3. Remove the rear No. 3 floor board.
4. Disconnect the cable from the negative battery terminal.

➡When disconnecting the cable, some systems need to be initialized after the cable is reconnected.

5. Separate the radiator without the cap reserve tank assembly.
   a. Remove the 2 bolts and separate the radiator without cap reserve tank assembly.
6. Remove the ECM.
   a. Remove the No. 1 relay block cover.
   b. Disconnect the 3 wire harness connectors.
   c. Disengage the 2 claws, push up the engine room wire harness and separate it from the engine room relay block.
   d. Disengage the 2 claws and push up the side cover to remove it.

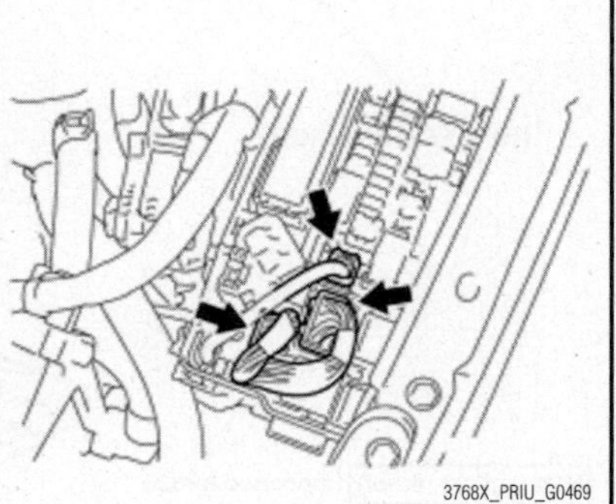

Fig. 167 Disconnecting the wire harness connectors

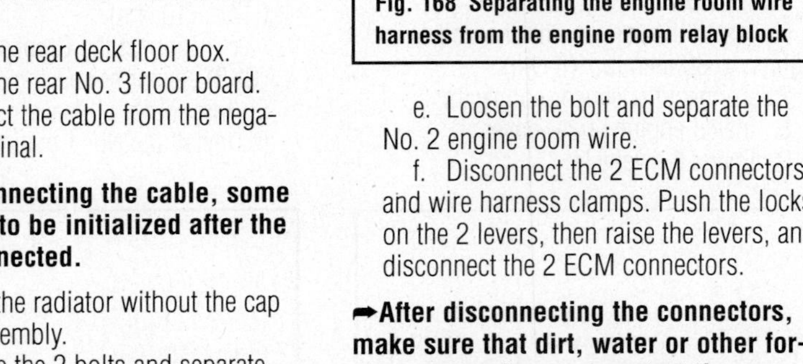

Fig. 168 Separating the engine room wire harness from the engine room relay block

   e. Loosen the bolt and separate the No. 2 engine room wire.
   f. Disconnect the 2 ECM connectors and wire harness clamps. Push the locks on the 2 levers, then raise the levers, and disconnect the 2 ECM connectors.

➡After disconnecting the connectors, make sure that dirt, water or other foreign matter does not contact the connecting parts of the connectors.

   g. Separate the 2 wire harness clamps.
   h. Remove the 2 bolts and ECM with 2 ECM brackets.

To install:

7. To install, reverse the removal procedure. Take note to tighten the following:
   - ECM bolts: 10 ft. lbs. (13 Nm)
   - No. 2 engine room wire bolt: 73 inch lbs. (8.3 Nm)
   - Radiator without cap reserve tank assembly: 13 ft. lbs. (18 Nm)

ENGINE COOLANT TEMPERATURE (ECT) SENSOR

LOCATION

See Figures 169 and 170.

REMOVAL & INSTALLATION

1.5L Engine

1. Before servicing the vehicle, refer to the precautions.
2. Remove the radiator support opening cover.
3. Raise and safely support the vehicle securely on jackstands.
4. Remove the engine under cover.

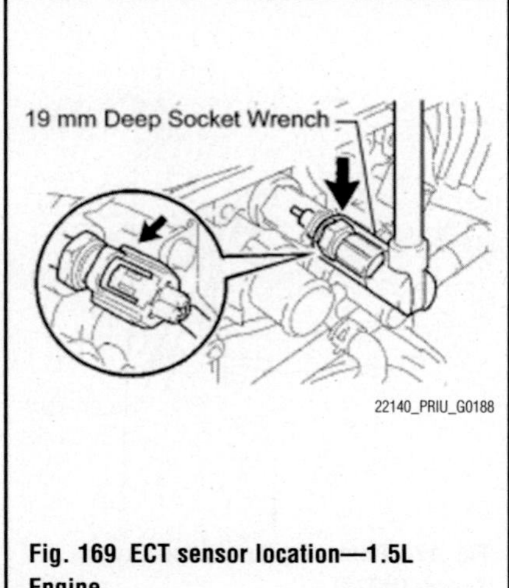

Fig. 169 ECT sensor location—1.5L Engine

5. Drain the engine coolant.
6. Drain the high voltage coolant.
7. Disconnect the negative battery cable.

➡Wait at least 90 seconds after disconnecting the cable from the negative (-) battery terminal to prevent airbag and seat belt pretensioner activation.

8. Remove the inverter with converter.
9. Remove Engine Coolant Temperature (ECT) sensor.
10. Disconnect the sensor connector.
11. Using a 19 mm deep socket wrench, remove the sensor and gasket.

To install:

12. Install Engine Coolant Temperature (ECT) sensor.
13. Using a 19 mm deep socket wrench, install a new gasket and the sensor. Tighten to 15 ft. lbs. (20 Nm).
14. Connect the sensor connector.
15. Install inverter with converter.
16. Install the inverter with converter.
17. Connect the negative battery cable.
18. Perform the power window initialization procedure. Refer to Interior, Power Windows, Initialization Procedure.
19. Add engine coolant.
20. Add high voltage coolant.
21. Inspect for coolant leaks.
22. Install engine under cover.
23. Lower the vehicle.
24. Install radiator support opening cover.

1.8L Engine

1. Remove the EGR with the cooler pipe sub assembly.
2. Remove the radiator without cap reserve tank assembly.
3. Remove the Engine Coolant Temperature (ECT) sensor.
   a. Disconnect the Engine Coolant Temperature (ECT) sensor connector.
   b. Remove the Engine Coolant Temperature (ECT) sensor.

To install:

4. To install, reverse the removal procedure. Tighten the sensor to 14 ft. lbs. (20 Nm).

HEATED OXYGEN (HO2S) SENSOR

LOCATION

See Figures 171 and 172.

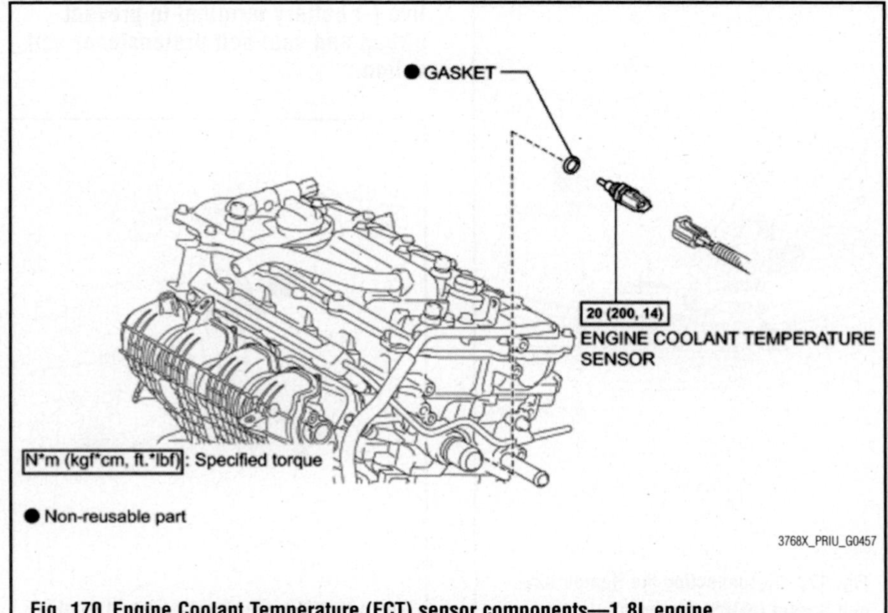

Fig. 170 Engine Coolant Temperature (ECT) sensor components—1.8L engine

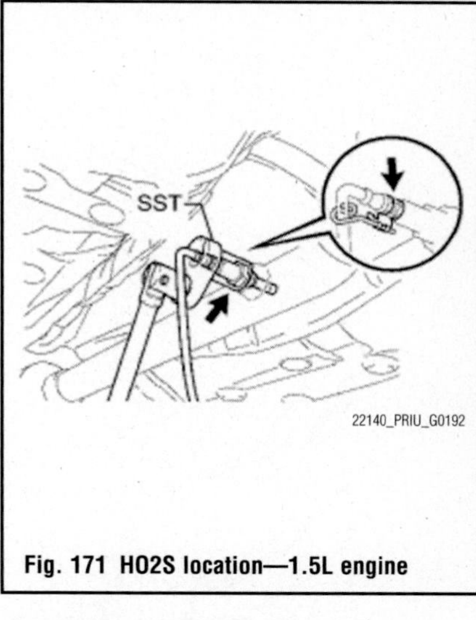

Fig. 171 HO2S location—1.5L engine

REMOVAL & INSTALLATION

1.5L Engine

1. Disconnect the negative battery cable.

➡Wait at least 90 seconds after disconnecting the cable from the negative (-) battery terminal to prevent airbag and seat belt pretensioner activation.

2. Remove lower center instrument panel finish panel.
3. Remove Heated Oxygen Sensor (HO2S).
   a. Using a clip remover, remove the clip.
   b. Fold back the floor carpet front.
   c. Disconnect the sensor connector.
   d. Remove the grommet of the sensor from the vehicle.
   e. Remove the wire harness clamp bracket from the sensor.
   f. Using special tool SST: 09224-00010 remove the sensor.

To install:

4. Install Heated Oxygen Sensor (HO2S).
   a. Using special tool SST: 09224-00010 install the sensor. Tighten to 30ft. lbs. (40 Nm).
   b. Install the wire harness clamp bracket to the sensor.
   c. Install the grommet of the sensor to the vehicle.
   d. Connect the sensor connector.
5. Install the floor carpet front with the clip.
6. Connect cable to negative battery terminal.
7. Inspect for exhaust gas leak.
8. Install lower center instrument panel finish panel.

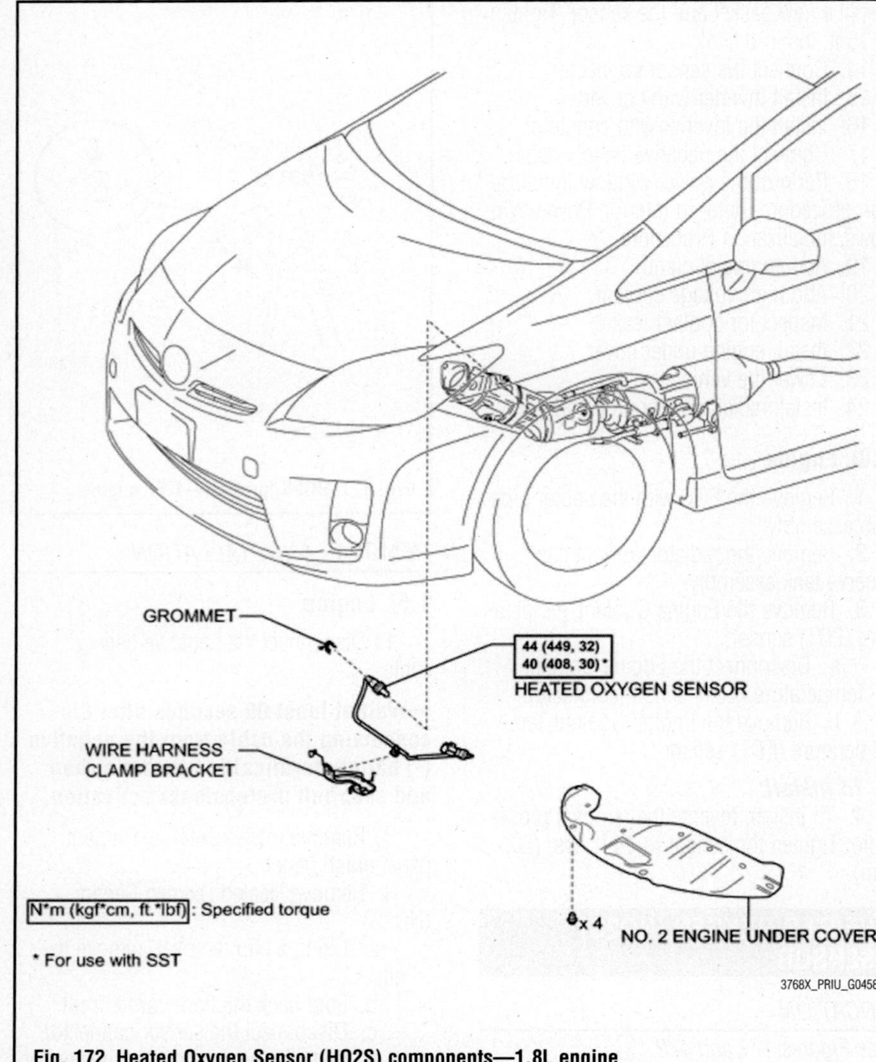

Fig. 172 Heated Oxygen Sensor (HO2S) components—1.8L engine

9. Connect the negative battery cable.
10. Perform the power window initialization procedure. Refer to Interior, Power Windows, Initialization Procedure.

1.8L Engine

See Figures 173 and 174.

1. Remove the No. 2 engine under cover.
   a. Remove the 3 clips and No. 3 engine under cover.
2. Remove the Heated Oxygen Sensor (HO2S).
   a. Disconnect the Heated Oxygen Sensor (HO2S) connector.
   b. Remove the grommet and pull the sensor connector out of the cabin through the floor panel.
   c. Remove the wire harness clamp bracket and disconnect the wire harness clamp.
   d. Using the special tool (09224-

Fig. 173 Disconnecting the Heated Oxygen Sensor (HO2S) connector

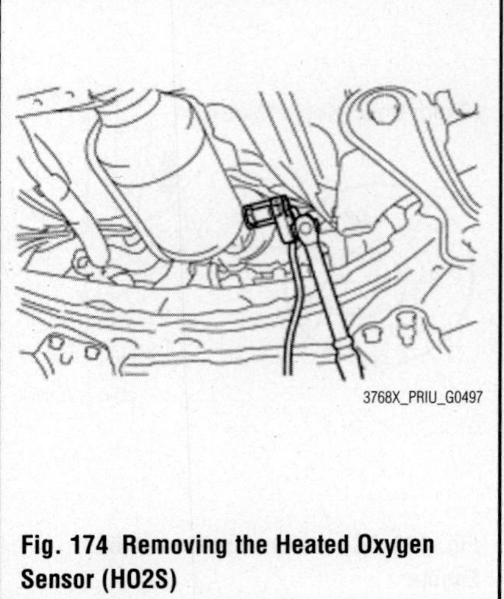

Fig. 174 Removing the Heated Oxygen Sensor (HO2S)

00010), remove the Heated Oxygen Sensor (HO2S).

To install:

3. To install, reverse the removal procedure. Tighten the Heated Oxygen Sensor (HO2S) to 32 ft. lbs. (44 Nm).

KNOCK SENSOR (KS)

LOCATION

See Figures 175 and 176.

REMOVAL & INSTALLATION

1.5L Engine

1. Disconnect the negative battery cable.

➡Wait at least 90 seconds after disconnecting the cable from the negative (-) battery terminal to prevent airbag and seat belt pretensioner activation.

Fig. 175 KS sensor location—1.5L engine

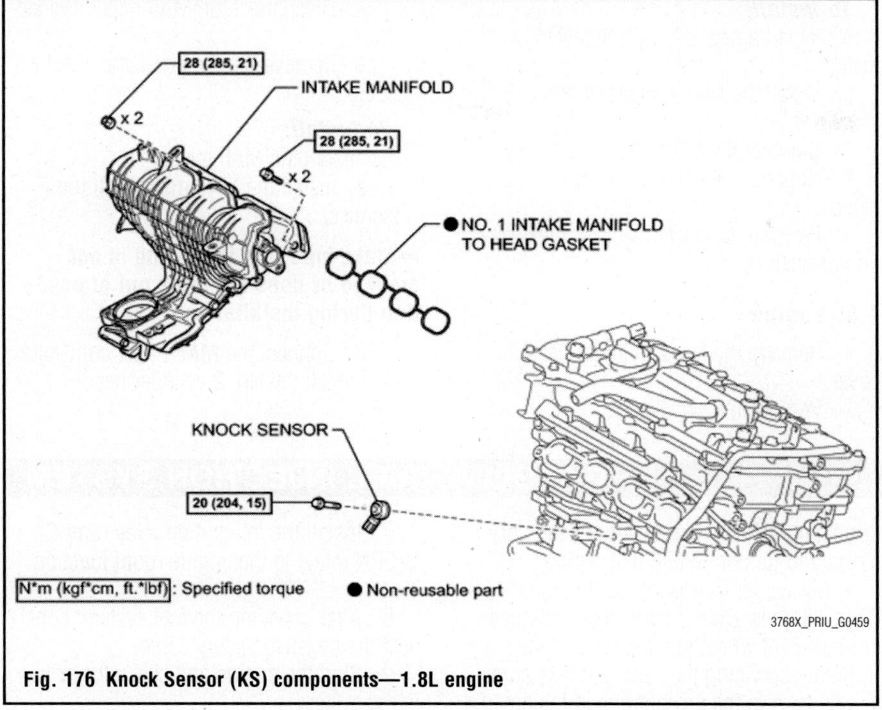

Fig. 176 Knock Sensor (KS) components—1.8L engine

2. Remove radiator support opening cover.
3. Remove engine under cover.
4. Drain engine coolant.
5. Remove air cleaner assembly.
6. Remove oil dipstick guide.
   a. Remove the dipstick.
   b. Disconnect the wire harness clamp.
   c. Remove the bolt and dipstick guide.
7. Remove intake manifold.
8. Disconnect the Knock Sensor (KS) connector.
9. Remove the nut and sensor.

To install:

10. Install Knock Sensor (KS).
11. Install the Knock Sensor (KS) with the nut. Tighten to 15 ft. lbs. (20 Nm).

➡Be careful to install the Knock Sensor (KS) in the correct direction.

12. Connect the Knock Sensor (KS) connector.
13. Install intake manifold.
14. Install oil dipstick guide.
   a. Apply a light coat of engine oil to a new o-ring and install it to the dipstick guide.
   b. Install the dipstick guide with the bolt. Tighten to 80 inch lbs. (9 Nm).

➡Be careful that the o-ring is not cracked or jammed when installing it.

   c. Connect the wire harness clamp.
   d. Install the dipstick.
15. Install air cleaner assembly.
16. Connect the negative battery cable.
17. Perform the power window initialization procedure.
18. Add engine coolant.
19. Inspect for coolant leak.
20. Install engine under cover.
21. Install radiator support opening cover.

1.8L Engine

See Figure 177.

1. Remove the intake manifold.
2. Remove the Knock Sensor (KS).
   a. Disconnect the Knock Sensor (KS) connector.

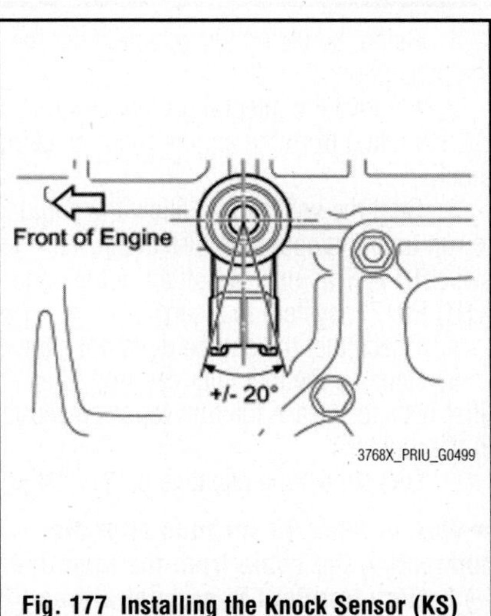

Fig. 177 Installing the Knock Sensor (KS)

   b. Remove the bolt and remove the Knock Sensor (KS).

To install:

3. Install the Knock Sensor (KS).
   a. Install the Knock Sensor (KS) with the bolt. Tighten to 15 ft. lbs. (20 Nm).

➡Make sure that the knock control sensor is in the correct position.

   b. Connect the Knock Sensor (KS) connector.
4. Install the intake manifold.

MASS AIR FLOW (MAF) SENSOR

LOCATION

See Figures 178 and 179.

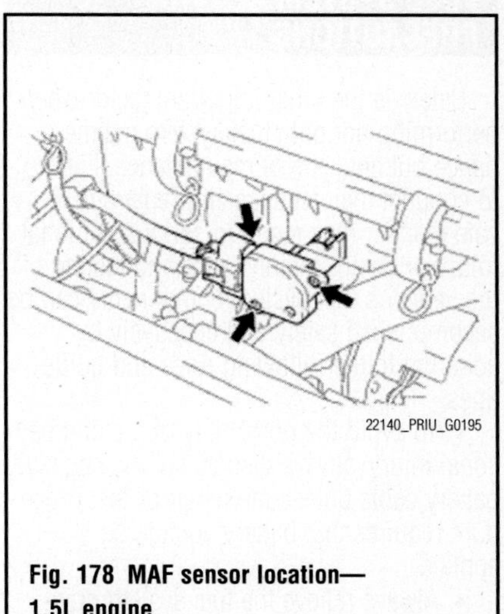

Fig. 178 MAF sensor location—1.5L engine

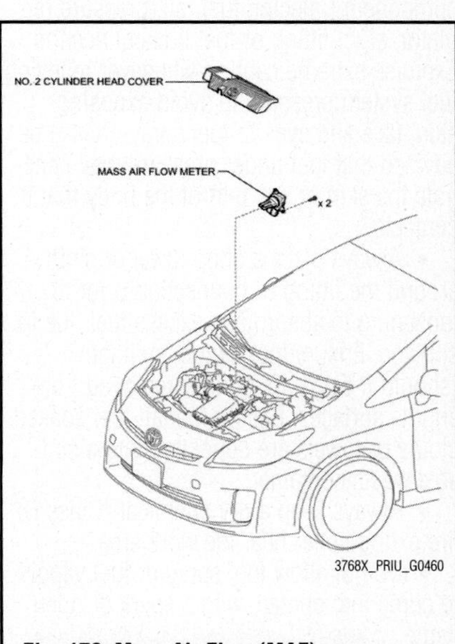

Fig. 179 Mass Air Flow (MAF) sensor components—1.8L engine

REMOVAL & INSTALLATION

1.5L Engine

1. Disconnect the negative battery cable.

➡**Wait at least 90 seconds after disconnecting the cable from the negative (-) battery terminal to prevent airbag and seat belt pretensioner activation.**

2. Remove radiator support opening cover.
3. Disconnect the MAF meter connector.
4. Remove the 2 screws and MAF meter.

To install:

5. Install a new O-ring to the MAF meter.
6. Install the MAF meter with the 2 screws.
7. Connect the MAF meter connector.
8. Connect the negative battery cable.
9. Perform the power window initialization procedure.

1.8L Engine

1. Remove the No. 2 cylinder head cover.
2. Remove the MAF meter.
   a. Disconnect the MAF meter connector.
   b. Remove the 2 screws and MAF meter.

To install:

3. Install the MAF meter.
   a. Install the MAF meter with the 2 screws.

➡**Make sure that the o-ring is not cracked or does not jump out of position during installation.**

   b. Connect the MAF meter connector.
4. Install the No. 2 cylinder head cover.

FUEL

GASOLINE FUEL INJECTION SYSTEM

FUEL SYSTEM SERVICE PRECAUTIONS

Safety is the most important factor when performing not only fuel system maintenance but any type of maintenance. Failure to conduct maintenance and repairs in a safe manner may result in serious personal injury or death. Maintenance and testing of the vehicle's fuel system components can be accomplished safely and effectively by adhering to the following rules and guidelines.

- To avoid the possibility of fire and personal injury, always disconnect the negative battery cable unless the repair or test procedure requires that battery voltage be applied.
- Always relieve the fuel system pressure prior to disconnecting any fuel system component (injector, fuel rail, pressure regulator, etc.), fitting or fuel line connection. Exercise extreme caution whenever relieving fuel system pressure to avoid exposing skin, face and eyes to fuel spray. Please be advised that fuel under pressure may penetrate the skin or any part of the body that it contacts.
- Always place a shop towel or cloth around the fitting or connection prior to loosening to absorb any excess fuel due to spillage. Ensure that all fuel spillage (should it occur) is quickly removed from engine surfaces. Ensure that all fuel soaked cloths or towels are deposited into a suitable waste container.
- Always keep a dry chemical (Class B) fire extinguisher near the work area.
- Do not allow fuel spray or fuel vapors to come into contact with a spark or open flame.
- Always use a back-up wrench when loosening and tightening fuel line connection fittings. This will prevent unnecessary stress and torsion to fuel line piping.
- Always replace worn fuel fitting O-rings with new Do not substitute fuel hose or equivalent where fuel pipe is installed.

Before servicing the vehicle, make sure to also refer to the precautions in the beginning of this section as well.

RELIEVING FUEL SYSTEM PRESSURE

1.5L Engine

⁂ CAUTION

The fuel system pressure relief procedure must be performed before disconnecting any part of the fuel system. After performing this procedure, pressure will remain in the fuel line. When disconnecting the fuel line, place a cloth or equivalent over fittings to reduce the risk of fuel spray.

1. Before servicing the vehicle, refer to the precautions.
2. Remove the integration relay (unit C: C/OPN relay) from the engine room junction block.
3. Start the vehicle and allow the engine to run until it stops, then turn the power switch**OFF**. This may set off a trouble code (DTC P0171: system too lean).
4. Check that the engine does not start.
5. Remove the fuel filler cap from the filler neck to release the fuel vapor pressure in the fuel tank.
6. Disconnect the negative battery cable.

➡**Wait at least 90 seconds after disconnecting the cable from the negative (-) battery terminal to prevent airbag and seat belt pretensioner activation.**

7. Install the integration relay (unit C: C/OPN relay) to the engine room junction block.
8. After servicing the fuel system, connect the negative battery cable.
9. Start the engine and check for leaks in the system.

1.8L Engine

⁂ CAUTION

Perform the following procedure to prevent fuel from spilling out before removing any fuel system parts.

⁂ CAUTION

Pressure will still remain in the fuel lines even after performing the following procedure. When disconnecting a fuel line, cover it with a piece of cloth to prevent fuel from spraying or coming out.

1. Remove the rear seat cushion assembly.
2. Remove the rear floor service hole cover.
3. Disconnect the fuel pump connector.
4. Put the vehicle in the "Inspection Mode".
5. Start the engine.
6. After the engine has stopped on its own, turn the power switch off.

➡**DTCs P0171/25 may be detected.**

7. Crank the engine again and make sure that the engine does not start.
8. Disconnect the cable from the negative (-) battery terminal.

➡**When disconnecting the cable, some systems need to be initialized after the cable is reconnected.**

9. Connect the fuel pump connector.
10. Loosen the fuel tank cap, and then discharge the pressure in the fuel tank completely.

FUEL FILTER

REMOVAL & INSTALLATION

The fuel filter for Prius models is an integral component of the tank-mounted fuel pump assembly and is not regularly serviced.

FUEL INJECTORS

REMOVAL & INSTALLATION

1.5L Engine

See Figure 180.

1. Before servicing the vehicle, refer to the precautions.
2. Relieve the pressure from the fuel system.
3. Disconnect the negative battery cable.
4. Remove the windshield wipers and windshield wiper motor and link assembly, if necessary.
5. Remove the front outer cowl top panel, if necessary.
6. Remove the air cleaner assembly.
7. Remove the brake master cylinder reservoir.
   a. Disconnect the brake fluid level switch connector.
   b. Remove the 2 bolts.
   c. Disconnect the claw fitting, and then remove the brake master cylinder reservoir.
8. Remove the reservoir bracket.
   a. Remove the No. 2 fuel vapor feed hose from the hose clamp.
   b. Remove the connector clamp.
   c. Remove the wire harness clamp, the 3 bolts and reservoir bracket.
9. Remove the cylinder head cover.
10. Remove fuel delivery pipe.
   a. Disconnect the fuel tube from the fuel delivery pipe.
   b. Remove the No. 1 fuel pipe clamp.
   c. Pinch the retainer of the fuel tube connector, and then pull out the fuel tube connector to disconnect the fuel tube from the fuel delivery pipe.

CAUTION

Be careful not to drop the injectors when removing the delivery pipe.

   d. Disconnect the 4 injector connectors from injector.

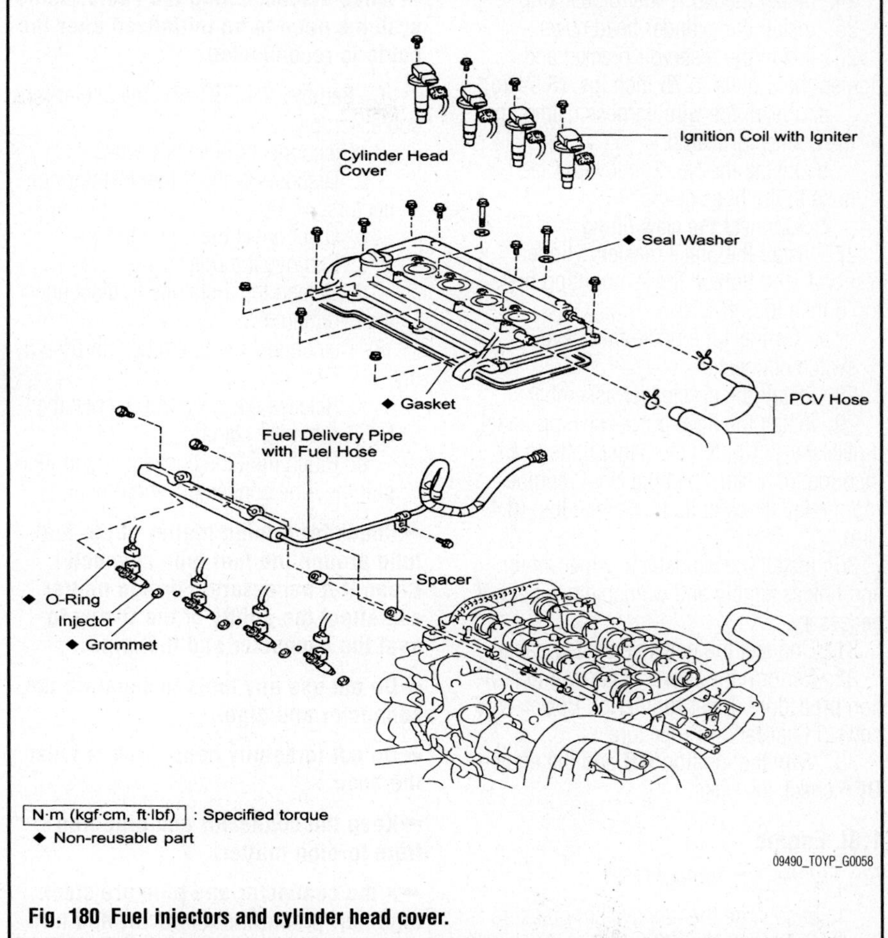

Fig. 180 Fuel injectors and cylinder head cover.

   e. Remove the 3 bolts and delivery pipe together with the 4 injectors and fuel pipe.
   f. Remove the 2 spacers from the cylinder head.
   g. Pull out the 4 injectors from the delivery pipe.
   h. Remove the O-ring and grommet from each injector.

To install:

11. Install the grommet to each injector.
12. Apply a light coat of spindle oil or gasoline to new O-ring and install them to each injector.
13. Apply a light coat of spindle oil or gasoline to the surface of the fuel delivery pipe which contacts the O-ring of the fuel injector.
14. Apply a light coat of spindle oil or gasoline to the O-ring again, and install the fuel injector by turning it right and left while pushing it onto the fuel delivery pipe.

➡Be careful that the O-ring is not cracked or jammed when installing it.

15. Check that the fuel injector rotates smoothly. If the fuel injector does not rotate, replace the O-ring.
16. Install the 4 injectors.
17. Position the injector connector outward.
18. Install 4 new insulators to the cylinder head.
19. Install the 2 delivery pipe spacers to the cylinder head.
20. Place the delivery pipe and fuel pipe together with the 4 injectors in position on the cylinder head and then temporarily tighten the 3 bolts.
21. Check that the injectors rotate smoothly. If the fuel injectors do not rotate, replace the O-ring.
22. Tighten the 2 bolts holding the delivery pipe to the cylinder head and tighten to 14 ft. lbs. (19 Nm). Tighten the bolt holding the fuel pipe to the cylinder head to 80 inch lbs. (9 Nm).
23. Align the fuel tube connector with the pipe, then push in the fuel tube connector until the retainer makes a "click" sound to connect the fuel tube to the fuel delivery pipe.

24. Install the No. 1 fuel pipe clamp.
25. Install the cylinder head cover.
26. Install the reservoir bracket and tighten the 3 bolts to 75 inch lbs. (8.5 Nm).
  a. Install the wire harness clamp and the connector clamp.
  b. Install the No. 2 fuel vapor feed hose to the hose clamp.
  c. Connect the claw fitting.
27. Install the brake master cylinder reservoir and tighten the 2 mounting bolts to 75 inch lbs. (8.5 Nm).
  a. Connect the brake fluid level switch connector.
28. Install the air cleaner assembly.
29. Install the front outer cowl top panel, if necessary. Tighten the panel bolts to 57 inch lbs. (6.4 Nm), and the No. 2 engine room relay block bolts to 74 inch lbs. (8.4 Nm).
30. Install the windshield wiper motor and link assembly and windshield wipers, if necessary.
31. Connect the negative battery cable.
32. Perform the power window initialization procedure. Refer to Interior, Power Windows, Initialization Procedure.
33. With the ignition **ON** and the engine **OFF** check for leaks.

1.8L Engine

See Figures 181 through 183.

1. Discharge the fuel system pressure.
2. Remove the rear No. 2 floor board (separate type).
3. Remove the rear deck floor box.
4. Remove the rear No. 3 floor board.
5. Disconnect the cable from the negative battery terminal.

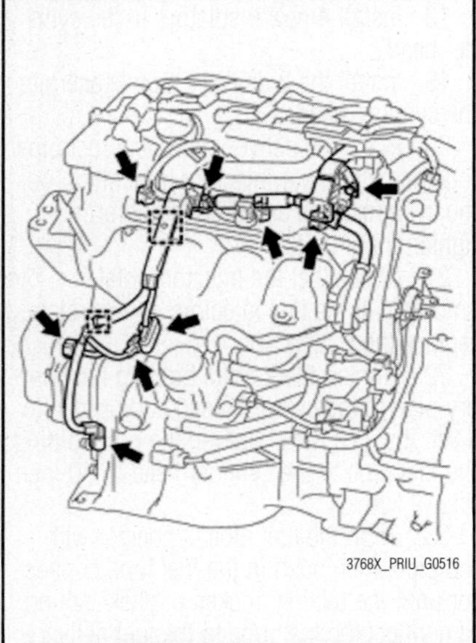

Fig. 181 Disconnecting the engine wire

➡When disconnecting the cable, some systems need to be initialized after the cable is reconnected.

6. Remove the EGR with the cooler pipe sub assembly.
7. Disconnect the engine wire.
  a. Disconnect the 4 fuel injector connectors.
  b. Disconnect the 4 connectors.
  c. Remove the bolt.
  d. Detach the 2 clamps to disconnect the wire harness.
8. Disconnect the fuel tube sub assembly.
  a. Release the claw and remove the No. 1 fuel pipe clamp.
  b. Pinch the tube connector, and then pull the tube connector off of the pipe.

➡Check for foreign matter in the fuel tube around the fuel tube connector. Clean it if necessary. Foreign matter can affect the ability of the O-ring to seal the connector and fuel pipe.

➡Do not use any tools to separate the connector and pipe.

➡Do not forcefully bend, kink or twist the hose.

➡Keep the connector and pipe free from foreign matter.

➡If the connector and pipe are stuck together, pinch the connector and turn it carefully to disconnect it.

➡Put the connector in a plastic bag to prevent damage and contamination.

9. Remove the fuel delivery pipe sub assembly.
  a. Remove the bolt.
  b. Remove the 2 bolts and the fuel delivery pipe sub assembly.

➡Be careful not to drop the fuel injectors when removing the fuel delivery pipe.

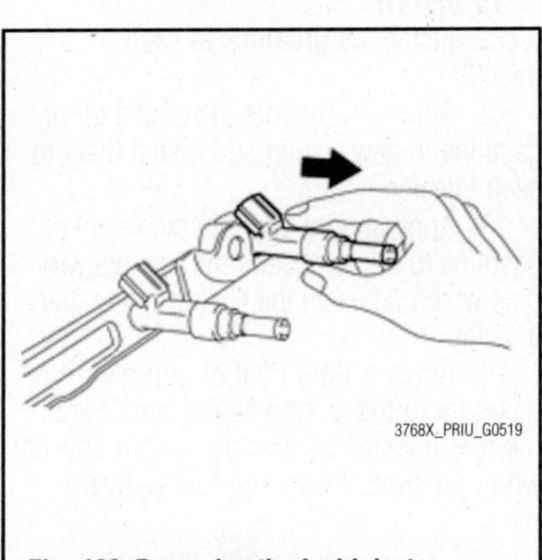

Fig. 182 Removing the fuel injector assemblies

10. Remove the No. 1 delivery pipe spacer.
  a. Remove the 2 delivery pipe spacers from the cylinder head.
11. Remove the fuel injector assembly.
  a. Pull the 4 fuel injector assemblies out of the fuel delivery pipe sub assembly.
  b. Remove the o-ring from each fuel injector.
  c. For reinstallation, attach a tag or label to each injector shaft.

➡Prevent entry of foreign objects by covering the fuel injectors with plastic bags.

  d. Remove the 4 injector vibration insulators.

To install:

12. Install the fuel injector assembly.
  a. Install a new insulator and o-ring to each fuel injector assembly.
  b. Apply a light coat of gasoline or spindle oil to the contact surfaces of the new o-ring on each fuel injector assembly.
  c. While turning the fuel injector assembly left and right, install it onto the fuel delivery pipe sub-assembly.

➡Do not damage the fuel injector assembly or O-ring.

➡Do not twist the O-ring.

➡After installing each fuel injector, check that it turns smoothly. If not, replace the O-ring with a new one.

13. Install the No. 1 delivery pipe spacer.
  a. Install the 2 No. 1 delivery pipe spacers onto the cylinder head.

➡Install the No. 1 delivery pipe spacers in the correct direction.

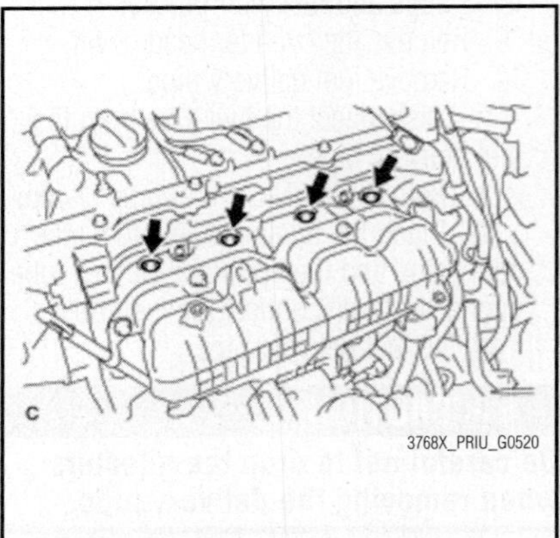

Fig. 183 Removing the injector vibration insulators

14. Install the fuel delivery pipe sub assembly.

a. Install the fuel deliver pipe sub assembly with the 4 fuel injector assemblies and install the 2 bolts. Tighten to 15 ft. lbs. (21 Nm).

➡Do not drop the fuel injectors when installing the fuel delivery pipe sub-assembly.

➡Check that the fuel injector assemblies rotate smoothly after installing the fuel delivery pipe sub-assembly.

b. Install the bolt to secure the fuel delivery pipe sub assembly. Tighten to 15 ft. lbs. (21 Nm).

15. Connect the fuel tube sub assembly.

a. Push the tube connector to the pipe until the tube connector makes a "click" sound.

➡Before connecting the connector and fuel pipe, check that there is no damage or foreign matter on the connecting part of the fuel pipe.

➡After connecting the fuel tube connector and pipe, check that they are securely connected by trying to pull them apart.

b. Engage the lock claw to install the No. 1 fuel pipe clamp.

16. Connect the engine wire.

a. Install the bolt and tighten to 7 ft. lbs. (10 Nm).

b. Connect the 4 fuel injector connectors.

c. Connect the 4 connectors.

d. Attach the 2 clamps to connect the wire harness.

17. Install the EGR with the cooler pipe sub assembly.

18. Install the rear No. 3 floor board.

19. Install the rear deck floor box.

20. Install the rear No. 2 floor board (separate type).Install the rear No. 2 floor board (separate type).

21. Connect the cable to the negative battery terminal.

22. Inspect for fuel leaks.

FUEL PUMP

REMOVAL & INSTALLATION

1.5L Engine

1. Before servicing the vehicle, refer to the precautions.

2. Relieve the pressure from the fuel system.

3. Disconnect the negative battery cable.

➡Wait at least 90 seconds after disconnecting the cable from the negative (-) battery terminal to prevent airbag and seat belt pretensioner activation.

4. Remove the fuel tank.

5. Remove the trap canister with pump module.

a. Disconnect the VSV connector.

b. Remove the clamp from the fuel tank vent hose and canister hose.

c. Remove the fuel tank vent hose from the 2 fuel tube clamps.

d. Remove the 2 bolts and trap canister with pump module and disconnect the ground terminal of the fuel tank wire.

e. Remove the gasket from the fuel tank.

f. Remove the 2 clamps from the trap canister with pump module.

To install:

6. Install the trap canister with pump module.

a. Install a new gasket to the fuel tank.

b. Insert the trap canister with pump module to the fuel tank. Be careful that the gasket does not drop in the fuel tank.

c. Install the 2 clamps to the trap canister with pump module.

d. Install the trap canister with pump module and connect the ground terminal of the fuel tank wire with the 2 bolts. Tighten the 2 bolts to 53 inch lbs. (6 Nm).

e. Install the fuel tank vent hose to the 2 fuel tube clamps.

f. Install the clamp to the fuel tank vent hose and canister hose.

g. Connect the VSV connector.

7. Install the fuel tank.

8. Connect the negative battery cable.

9. Perform the power window initialization procedure. Refer to Interior, Power Windows, Initialization Procedure.

10. Check for fuel leaks and proper fuel pressure.

1.8L Engine

See Figures 184 through 187.

1. Remove the rear seat cushion assembly.

2. Remove the rear floor service hole cover.

a. Disconnect the fuel pump connector.

3. Discharge the fuel system pressure.

4. Remove the No. 2 floor board (separate type).

5. Remove the rear deck floor box.

6. Remove the rear No. 3 floor board.

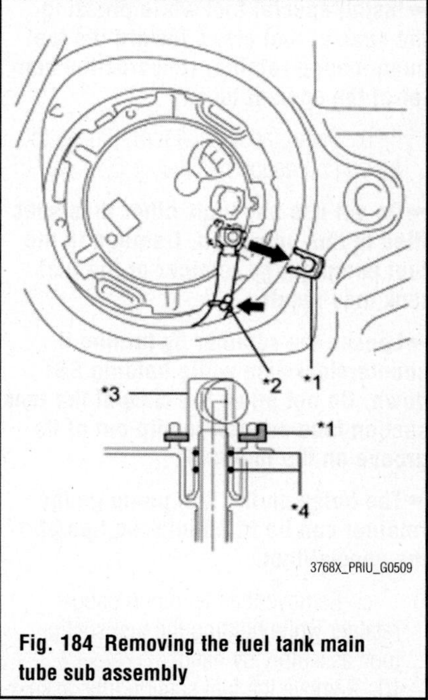

Fig. 184 Removing the fuel tank main tube sub assembly

7. Disconnect the cable from the negative battery terminal.

➡When disconnecting the cable, some systems need to be initialized after the cable is reconnected.

8. Remove the fuel tank main tube sub assembly.

a. Remove the tube joint clip and disengage the fuel tank main tube clamp, then pull the fuel tube joint out of the plug of the fuel suction tube assembly.

➡Check that there is no dirt or other foreign objects around the fuel tube joint before disconnecting it. Clean the joint if necessary.

➡It is necessary to prevent mud or dirt from entering the joint. If mud or dirt gets in the joint, the O-rings may not seal properly.

➡Only disconnect the joint by hand.

➡Do not bend, kink or twist the nylon tubes.

➡Protect the contact surfaces by covering it with a plastic bag.

9. Remove the fuel pump gauge retainer.

a. Using a 6 mm socket hexagon wrench, set the special tool (09808-14020) to the fuel pump gauge retainer.

➡Engage the special tool claws securely with the fuel pump gauge retainer holes to secure special tool.

➡**Install special tool while pressing the special tool claws toward the fuel pump gauge retainer (towards the center of the special tool).**

b. Using the special tool, loosen the fuel pump gauge retainer.

➡**Do not use any tools other than specified in this operation. Damage to the fuel pump gauge retainer or the fuel tank may result.**

➡**Loosen the retainer by turning it counterclockwise while holding SST down. Do not allow the claw of the tank suction tube support to slip out of its groove on the fuel tank.**

➡**The holes on the fuel pump gauge retainer can be fitted into the tips of the special tool.**

c. Remove the fuel pump gauge retainer while holding the fuel suction tube assembly by hand.

10. Remove the fuel suction tube assembly with the pump and gauge.

a. Remove the fuel suction tube assembly with pump and gauge from the fuel tank.

➡**Make sure that the fuel sender gauge arm does not bend.**

b. Remove the gasket from the fuel tank.

To install:

11. Install the fuel suction tube assembly with the pump and gauge.

a. Install a new gasket onto the fuel tank.

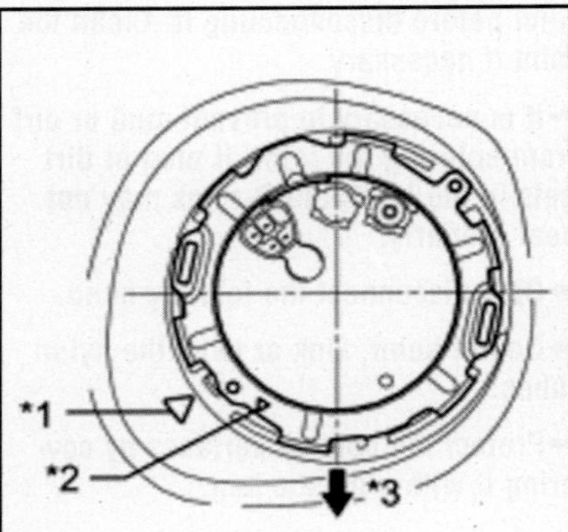

Fig. 185 Positioning the fuel pump gauge retainer

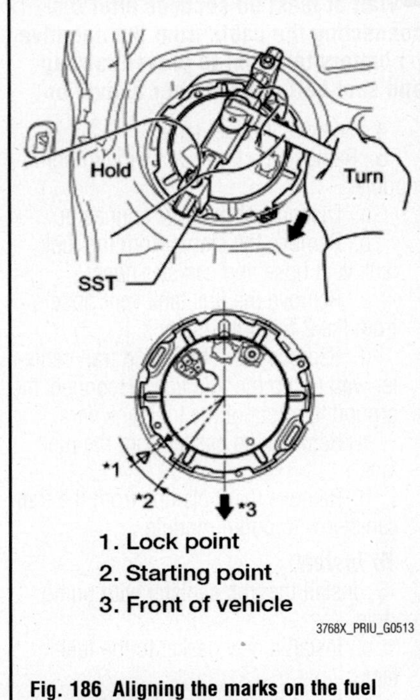

Fig. 186 Aligning the marks on the fuel tank and fuel tank retainer

b. Install the fuel suction tube assembly with pump and gauge to the fuel tank.

➡**Make sure that the fuel sender gauge arm does not bend.**

c. Align the protrusions of the fuel suction tube assembly with pump and gauge with the notches of the fuel tank.

➡**Ensure the fuel suction tube gasket is in the correct position.**

12. Install the fuel pump gauge retainer.

a. While holding the fuel suction tube assembly with pump and gauge by hand, position the fuel pump gauge retainer and tighten it lightly by hand.

➡**Check that the contact surface of the fuel tank retainer is not scratched or damaged, and prevent the entry of foreign objects.**

b. Using a 6 mm socket hexagon wrench, set special tool (09808-14020) to the fuel pump gauge retainer.

➡**Hold the fuel suction tube assembly upright by hand to ensure that the fuel suction tube gasket is not moved out of position.**

➡**Engage the special tool (09808-14020) claws securely with the fuel pump gauge retainer holes to secure special tool (09808-14020).**

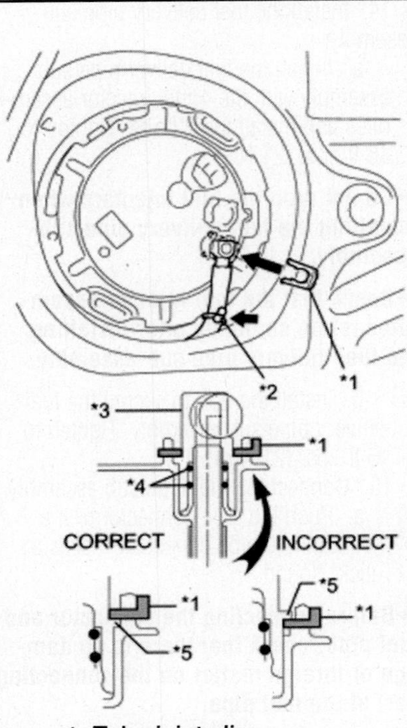

Fig. 187 Connecting the fuel tank main tube sub assembly

➡**Install special tool (09808-14020) while pressing the special tool (09808-14020) claws toward the fuel pump gauge retainer (toward the center of special tool (09808-14020)).**

c. Using the special tool (09808-14020), align the marks on the fuel tank and fuel pump gauge retainer.

➡**Do not use any tools other than specified in this operation. Damage to the fuel pump gauge retainer or the fuel tank may result.**

➡**Tighten the retainer by turning it clockwise while holding SST down.**

13. Connect the fuel tank main tube sub assembly.

a. Push the fuel tube joint in the plug of the fuel suction plate, then install the tube joint clip.

b. Connect the fuel pump connector.

14. Connect the cable to the negative battery terminal.

➡**When disconnecting the cable, some**

systems need to be initialized after the cable is reconnected.

15. Inspect for fuel leaks.
16. Install the rear floor service hole cover with new butyl tape.
17. Install the rear seat cushion assembly.
18. Install the rear No. 3 floor board.
19. Install the rear deck floor board.
20. Install the rear No. 2 floor board (separate type).

FUEL TANK

REMOVAL & INSTALLATION

1.5L Engine

See Figures 188 through 193.

1. Before servicing the vehicle, refer to the precautions.
2. Relieve the pressure from the fuel system.
3. Disconnect the negative battery cable.

➡Wait at least 90 seconds after disconnecting the cable from the negative (-) battery terminal to prevent airbag and seat belt pretensioner activation.

4. Remove instrument panel finish panel lower center.
5. Remove the front floor panel brace and front exhaust pipe assembly.
6. Remove the rear seat cushion.
    a. Detach the seat cushions 2 front hooks from the vehicle body by choosing a hook to detach first. Place your hands near one of the hooks, and then lift the seat cushion to detach the hook.
    b. Repeat for the other hook.
    c. Detach the seat cushions rear hook.
    d. Remove the seat cushion.
7. Remove the butyl tape and rear floor service hole cover.

⁂ CAUTION

Remove dirt or foreign objects on the fuel line connectors before any disconnecting procedures. Do not allow any scratches or foreign objects on the parts when disconnecting them as the fuel connectors have O-rings that seal the pipes. Perform such work by hand. Do not use any tools. Do not forcibly bend, twist or turn the nylon tube. Protect the connecting part by covering it with a plastic bag after disconnecting the tube. If the connector and pipe are stuck, push and pull them to release them.

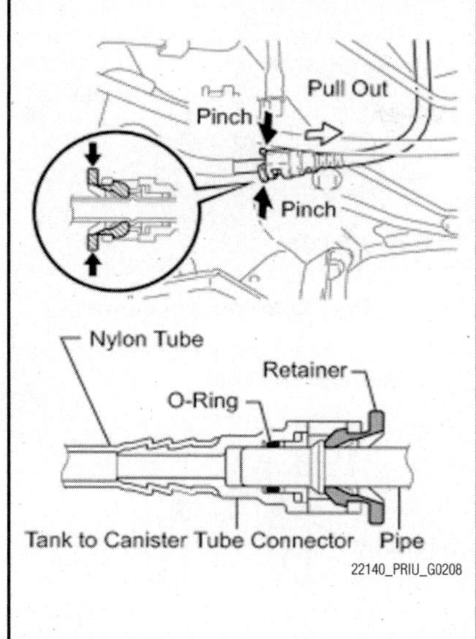

Fig. 188 Disconnect the fuel tank to canister tube from the pipe

8. Disconnect the fuel pump connector.
9. Disconnect the wire-to-wire connector.
10. Pinch the retainer of the fuel tube connector, then pull out the fuel tube connector to disconnect the fuel tank to canister tube from the pipe.
11. Remove the checker of the fuel tube connector from the pipe.
12. Pinch the retainer of the fuel tube connector, and then pull out the fuel tube connector to disconnect the No. 2 fuel tank main tube from the pipe.

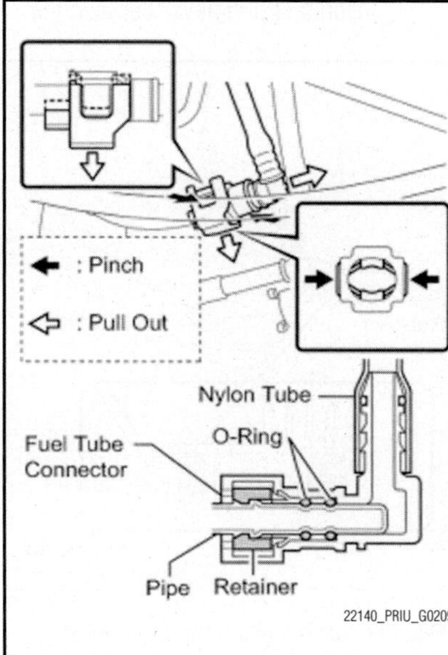

Fig. 189 Disconnect the No. 2 fuel tank main tube from the pipe

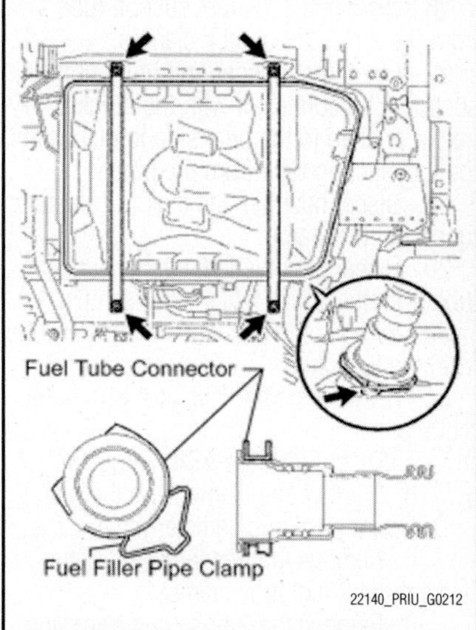

Fig. 190 Remove the fuel filler pipe clamp and fuel tube connector from the fuel tank inlet pipe

13. Pinch the retainer and pull out the fuel tank vent hose connector with the fuel tank vent hose connector pushed to the pipe side to disconnect the fuel tank vent hose from the canister filter.
14. Pinch the retainer and pull out the suction tube connector with the suction tube connector pushed to the pipe side to disconnect the fuel suction tube from the fuel tank to filler pipe.
15. Pinch the retainer and pull out the No. 1 canister tube connector with the No. 1 canister tube connector pushed to the pipe side to disconnect the No. 1 canister tube from the fuel tank to filler pipe.
16. Set a transmission jack to the fuel tank.
17. Remove the fuel filler pipe clamp and fuel tube connector from the fuel tank inlet pipe.
18. Remove the 4 bolts and No. 1 fuel tank band right and left.
19. Operate the transmission jack, and then disconnect the fuel tank inlet pipe.
20. Operate the transmission jack, and then remove the fuel tank.
21. Remove the 3 nuts and rear fuel tank bracket.
22. Disconnect the wire to wire connector from the rear fuel tank bracket.
23. Disconnect the No. 2 fuel tank main tube from the clamp.
24. Remove the checker of the main tube connector from the pipe.
25. Pinch the retainer of the main tube connector, then pull out the fuel tube connector to disconnect the No. 2 fuel tank main tube from the pipe.

26. Disconnect the fuel suction tube from the 2 No. 1 fuel tube clamps.

27. Pinch the retainer and pull out the suction tube connector with the quick connector pushed to the pipe side to disconnect the fuel suction tube from the pipe.

28. Disconnect the fuel tank to canister tube from the clamp.

29. Disconnect the fuel tank to canister tube from the 2 No. 1 fuel tube clamps.

30. Remove the fuel tank to canister tube from the fuel tank.

31. Remove the trap canister with pump module.

a. Disconnect the VSV connector.

b. Remove the clamp from the fuel tank vent hose and canister hose.

c. Remove the fuel tank vent hose from the 2 fuel tube clamps.

d. Remove the 2 bolts and trap canister with pump module and disconnect the ground terminal of the fuel tank wire.

e. Remove the gasket from the fuel tank.

f. Remove the 2 clamps from the trap canister with pump module.

32. Pinch the retainer and pull out the fuel tank vent hose connector with the fuel tank vent hose connector pushed to the fuel tank vent hose side to disconnect the fuel tank vent hose from the trap with outlet valve canister.

33. Remove fuel tank wire.

a. Remove the clamp as shown in the illustration "A".

b. Disconnect the VSV connector as shown in the illustration "B".

c. Disconnect the vapor pressure sensor connector as shown in the illustration "C".

d. Remove the 3 wire harness clamps as shown in the illustration "D".

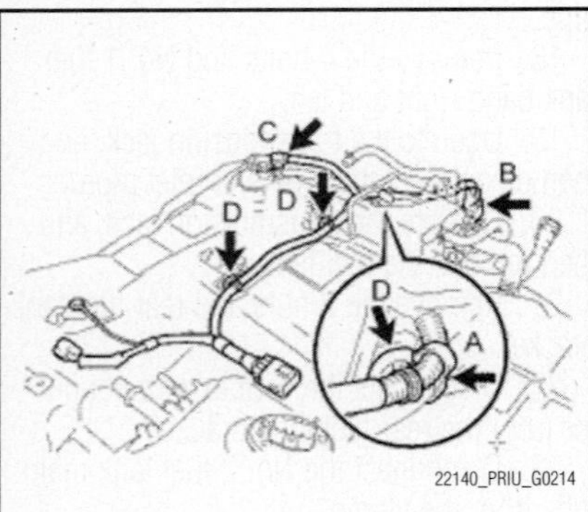

Fig. 191 Remove the clamp (A), disconnect the VSV connector (B), disconnect the vapor pressure sensor connector (C) and remove the 3 wire harness clamps (D).

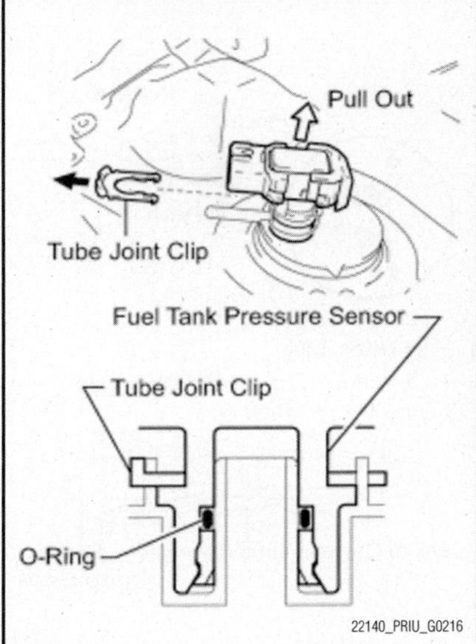

Fig. 192 Remove the tube joint clip, then pull out the fuel tank pressure sensor from the fuel tank retainer

34. Remove canister.

a. Disconnect the canister hose from the fuel tank retainer.

b. Disconnect the No. 1 canister outlet hose from the fuel tank.

c. Remove the bolt, 2 nuts and canister.

d. Remove the nut from the fuel tank.

35. Remove fuel tank pressure sensor.

a. Remove the tube joint clip, then pull out the fuel tank pressure sensor from the fuel tank retainer.

36. Remove fuel tank retainer.

a. Insert a clip remover between the fuel tank retainer and gasket, then remove the fuel tank retainer by lifting it little by little.

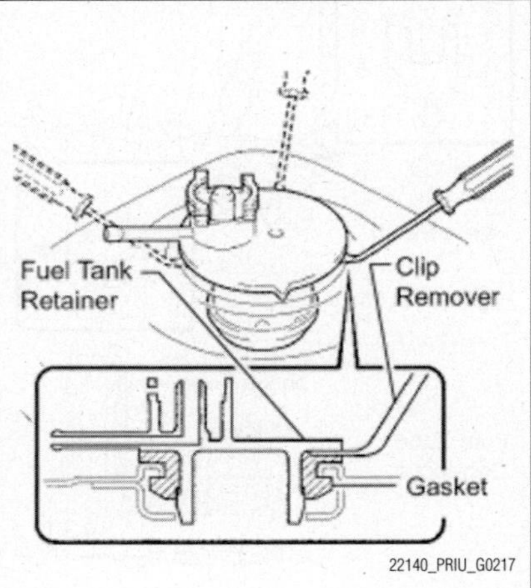

Fig. 193 Insert a clip remover between the fuel tank retainer and gasket, then remove the fuel tank retainer by lifting it little by little

WARNING

The fuel tank retainer is made of resin and easily damaged if removed or installed forcibly. Handle the part correctly to ensure proper sealing. After removing the fuel tank retainer, check that the contact surface of the fuel tank retainer on the fuel tank is not damaged.

b. Remove the gasket from the fuel tank.

37. Remove the 2 clamps from the fuel tank.

38. Remove the 9 cushions from the fuel tank.

To install:

39. Install 9 new cushions to the fuel tank.

40. Install the 2 clamps to the fuel tank.

41. Install fuel tank retainer.

a. Install a new gasket to the fuel tank.

b. While being careful that the gasket does not drop in the fuel tank, insert the fuel tank retainer into the fuel tank so the protrusion of the fuel tank retainer is in the middle of the 2 convex pats of the fuel tank.

42. Push the fuel tank pressure sensor to the plug of the fuel tank retainer, and then install the tube joint clip.

43. Install canister.

a. Install the nut to the fuel tank.

b. Install the canister with the bolt and 2 nuts. Tighten to 53 inch lbs. (6 Nm).

c. Connect the No. 1 canister outlet hose to the fuel tank.

d. Connect the canister hose to the fuel tank retainer.

44. Install fuel tank wire.

a. Install the 3 wire harness clamps.

b. Connect the vapor pressure sensor connector.

c. Connect the VSV connector.

d. Install the clamp.

45. Install fuel tank vent hose.

a. Align the fuel tank vent hose connector with the pipe, then push in the fuel tank vent hose connector until the retainer makes a "click" sound to install the fuel tank vent hose to the trap canister with pump module.

46. Install the trap canister with pump module.

a. Install a new gasket to the fuel tank.

b. Insert the trap canister with pump module to the fuel tank. Be careful that the gasket does not drop in the fuel tank.

c. Install the 2 clamps to the trap canister with pump module.

d. Install the trap canister with pump module and connect the ground terminal of the fuel tank wire with the 2 bolts. Tighten the 2 bolts to 53 inch lbs. (6 Nm).

e. Install the fuel tank vent hose to the 2 fuel tube clamps.

f. Install the clamp to the fuel tank vent hose and canister hose.

g. Connect the VSV connector.

47. Install the fuel tank to canister tube to the canister's hose.

48. Connect the fuel tank to canister tube to the 2 No. 1 fuel tube clamps.

CAUTION

Check that there are no scratches or foreign objects around any connected part of the fuel line connectors and pipes before these procedures. After connecting any fuel line tubes, check that the tube is securely connected by pulling on the connector.

49. Align the suction tube connector with the pipe, and then push in the suction tube connector until the retainer makes a "click" sound to install the fuel suction tube to the pipe.

50. Connect the fuel suction tube to the 2 No. 1 fuel tube clamps.

51. Align the main tube connector with the pipe, and then push in the main tube connector until the retainer makes a "click" sound to install the No. 2 fuel tank main tube to the pipe. Install the checker to the pipe.

52. Connect the No. 2 fuel tank main tube to the clamp.

53. Connect the connector clamp to the rear fuel tank bracket.

54. Install the rear fuel tank bracket with the 3 nuts and tighten to 53 inch lbs. (6 Nm).

55. Set the fuel tank to a transmission jack.

56. Operate the transmission jack, and then install the fuel tank to the vehicle.

57. Operate the transmission jack, and then connect the fuel tank inlet pipe.

58. Install the No. 1 fuel tank band right and left with the 4 bolts. Torque the 4 bolts to 29 ft. lbs. (39 Nm).

59. Install the fuel tube connector and fuel filler pipe clamp to the fuel tank inlet pipe.

60. Align the No. 1 canister tube connector with the pipe, and then push in the No. 1 canister tube connector until the retainer makes a "click" sound to connect the No. 1 canister tube to the fuel tank to filler pipe.

61. Align the suction tube connector with the pipe, and then push in the suction tube connector until the retainer makes a "click" sound to connect the fuel suction tube to the fuel tank to filler pipe.

62. Align the fuel tank vent hose connector with the pipe, and then push in the fuel tank vent hose connector until the retainer makes a "click" sound to connect the fuel tank vent hose to the canister filter.

63. Align the fuel tube connector with the pipe, and then push in the fuel tube connector until the retainer makes a "click" sound to connect the No. 2 fuel tank main tube to the pipe. Install the checker to the pipe.

64. Align the fuel tank to canister tube connector with the pipe, and then push in the fuel tank to canister tube connector until the retainer makes a "click" sound to connect the fuel tank to canister tube to the pipe.

65. Install the front exhaust pipe

66. Connect the negative battery cable.

67. Perform the power window initialization procedure. Refer to Interior, Power Windows, Initialization Procedure.

68. Check for fuel and exhaust leaks.

69. Install the front floor panel brace.

70. Install the instrument panel finish panel lower center.

71. Attach new butyl tape to the rear floor service hole cover.

72. Connect the wire-to-wire connector.

73. Connect the fuel pump connector.

74. Install the rear floor service hole cover while adjusting it to the 3 convex parts of the floor panel.

WARNING

Be careful that the rear floor service hole cover does not overlap the convex parts of the floor panel when installing.

75. Install the rear seat cushion by engaging the 3 seat hooks.

1.8L Engine

See Figures 194 and 195.

1. Remove the fuel suction tube assembly with the pump and gauge.

2. Drain the fuel.

3. Remove the rear floor side member covers LH and RH (w/ floor under cover).

4. Remove the rear suspension brace sub assembly.

5. Remove the rear floor step under cover sub assembly (w/ floor under cover).

6. Remove the rear suspension brace sub assembly.

7. Remove the rear floor step under cover sub assembly (w/ floor under cover).

8. Remove the No. 1 fuel tank protector.

a. Remove the 3 bolts and the No. 1 fuel tank protector.

9. Disconnect the fuel cut-off tube from the charcoal canister assembly.

➡Do not remove the retainer.

➡Remove any dirt or foreign matter on the fuel cut-off tube connector before performing this work.

➡Do not allow any scratches or foreign matter on the parts when disconnecting them as the fuel cut-off tube connector has an O-ring that seals the pipe.

➡Perform this work by hand. Do not use any tools.

➡Do not forcibly bend, twist or turn the fuel cut-off tube.

➡Protect the disconnected part by covering it with a plastic bag after disconnecting the fuel cut-off tube.

➡If the vent hose connector and pipe are stuck, push and pull to release them.

10. Disconnect the fuel tank breather tube.

a. Pinch the tabs of the retainer to remove the lock claws and pull it down as shown in the illustration.

b. Pull out the fuel tank breather tube.

➡Check that there is no dirt or other foreign objects around the connector

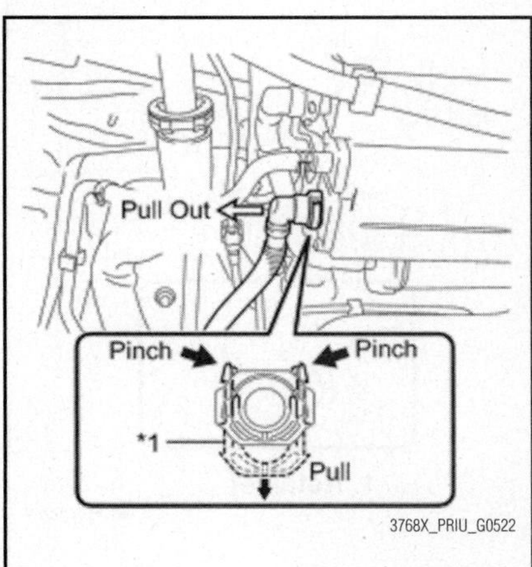

Fig. 194 Disconnecting the fuel cut off tube from the charcoal canister assembly

before this operation and clean the connector as necessary.

➡It is necessary to prevent mud or dirt from entering the connector. If mud or dirt gets in the connector, the O-rings may not seal properly.

➡Do not use any tools in this operation.

➡Do not bend, kink or twist the nylon tube. Protect the connector by covering it with a plastic bag.

➡When the pipe and connector are stuck, push and pull the connector to release and pull the connector out carefully.

11. Disconnect the fuel tank to filler pipe hose.
    a. Pull the tabs of the retainer to disengage the lock claws and pull it down.
    b. Pull out the fuel tank to filler pipe hose.

➡Check that there is no dirt or other foreign objects around the connector before this operation and clean the connector as necessary.

➡It is necessary to prevent mud or dirt from entering the connector. If mud or dirt gets in the connector, the O-rings may not seal properly.

➡Do not use any tools in this operation.

➡Do not bend, kink or twist the nylon tube. Protect the connector by covering it with a plastic bag.

➡When the pipe and connector are stuck, push and pull the connector to release and pull the connector out carefully.

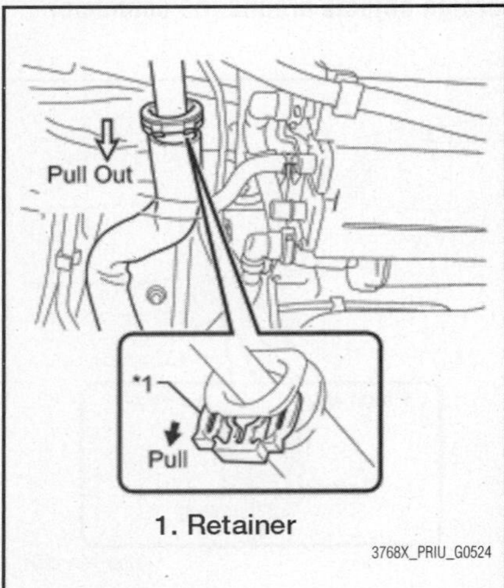

Fig. 195 Disconnecting the fuel tank to filler pipe hose

Fig. 195 Disconnecting the fuel tank to filler pipe hose

3768X_PRIU_G0524

12. Remove the fuel tank main tube sub assembly.
    a. Pinch the tabs of the retainer to remove the lock claws and pull it down as shown in the illustration.
    b. Pull out and remove the fuel tank main tube sub-assembly.

➡Check that there is no dirt or other foreign objects around the connector before this operation and clean the connector as necessary.

➡It is necessary to prevent mud or dirt from entering the connector. If mud or dirt gets in the connector, the O-rings may not seal properly.

➡Do not use any tools in this operation.

➡Do not bend, kink or twist the nylon tube. Protect the connector by covering it with a plastic bag.

➡When the pipe and connector are stuck, push and pull the connector to release and pull the connector out carefully.

13. Remove the fuel tank assembly.
    a. Remove the 2 bolts and disconnect the parking brake cable assembly.
    b. Support the fuel tank using an engine lifter.
    c. Remove the 4 set bolts of the 2 fuel tank bands.
    d. Lower the engine lifter to remove the fuel tank.

➡Slowly operate the engine lifter to lower the fuel tank.

➡Do not drop the fuel tank.

➡When removing the fuel tank, tilt it slightly to prevent it from interfering with the suspension arm or other surrounding parts.

14. Remove the fuel tank cushion.
    a. Remove the 3 No. 1 fuel tank cushions and 2 No. 2 fuel tank cushions.

To install:

15. To install, reverse the removal procedure. Take note to tighten the fuel tank band bolts to 29 ft. lbs. (39 Nm). Tighten the parking brake cable assembly bolts to 53 inch lbs. (6 Nm). Also, tighten the No. 1 fuel tank protector bolts to 49 inch lbs. (5.5 Nm).

IDLE SPEED

ADJUSTMENT

Idle speed is maintained by the Electronic Control Module (ECM). No adjustment is possible.

THROTTLE BODY

REMOVAL & INSTALLATION

1.5L Engine

See Figures 196 and 197.

1. Disconnect the negative battery cable.

➡Wait at least 90 seconds after disconnecting the cable from the negative (-) battery terminal to prevent airbag and seat belt pretensioner activation.

2. Remove engine under cover.
3. Drain engine coolant.
4. Remove the 6 clips and radiator support opening cover.

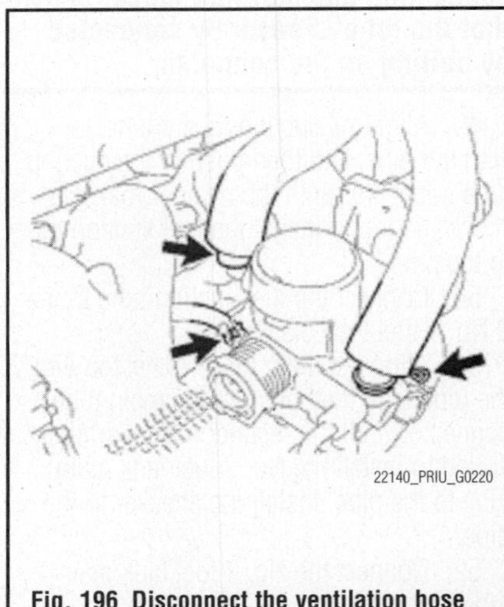

Fig. 196 Disconnect the ventilation hose

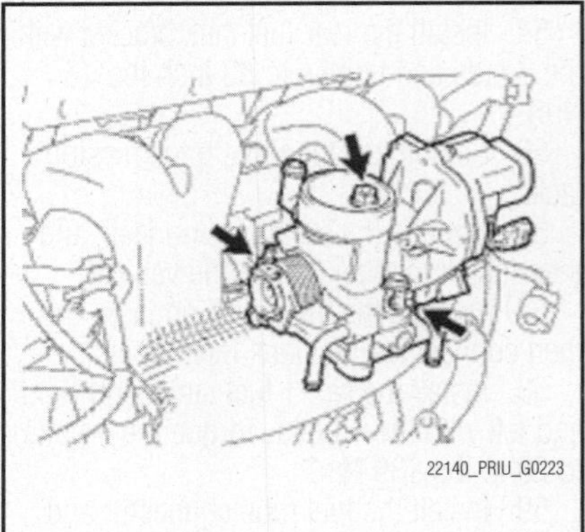

Fig. 197 Throttle body fastener locations

5. Remove air cleaner assembly
   a. Disconnect the MAF sensor connector.
   b. Disconnect the wire harness from the wire harness clamp.
   c. Loosen the hose clamp bolt, and then disconnect the No. 1 air cleaner inlet.
   d. Remove the 2 bolts.
   e. Loosen the hose clamp bolt, and then remove the air cleaner.
6. Remove throttle body assembly with motor.
   a. Disconnect the ventilation hose.
   b. Disconnect the No. 2 ventilation hose.
   c. Disconnect the No. 1 fuel vapor feed hose.
   d. Disconnect the water by-pass hose.
   e. Disconnect the No. 2 water by-pass hose.
   f. Disconnect the throttle control motor connector.
   g. Disconnect the throttle position sensor connector.
   h. Remove the bolt, 2 nuts and throttle with motor body.
   i. Remove the gasket from the intake manifold.

To install:

7. Install throttle body assembly with motor
   a. Install a new gasket to the intake manifold.
   b. Install the throttle with motor body with the bolt and 2 nuts. Tighten to 15 ft. lbs. (20 Nm).
   c. Connect the throttle position sensor connector.
   d. Connect the throttle control motor connector.
   e. Connect the No. 2 water by-pass hose.
   f. Connect the water by-pass hose.
   g. Connect the No. 1 fuel vapor feed hose.
   h. Connect the No. 2 ventilation hose.
   i. Connect the ventilation hose.
8. Install air cleaner assembly
   a. Install the air cleaner with the 2 bolts. Tighten to 62 inch lbs. (7 Nm).
   b. Tighten the hose clamp bolt to 27 inch lbs. (3 Nm).
   c. Connect the No. 1 air cleaner inlet, and tighten the hose clamp bolt to 27 inch lbs. (3 Nm).
   d. Connect the MAF sensor meter connector.
9. Connect the negative battery cable.
10. Perform the power window initialization procedure. Refer to Interior, Power Windows, Initialization Procedure.
11. Add engine coolant
12. Inspect for coolant leaks.
13. Install engine under cover.
14. Install radiator support opening cover.

1.8L Engine

See Figures 198 and 199.

1. Remove the No. 1 engine under cover.
2. Drain the engine coolant (engine).
3. Remove the No. 2 cylinder head cover.
4. Remove the air cleaner cap sub assembly.
5. Remove the inlet air cleaner assembly.
6. Remove the air cleaner case.
7. Remove the air cleaner hose assembly.
8. Remove the throttle body assembly.
   a. Disconnect the throttle body connector and the 2 water by pass hoses.
   b. Remove the 2 bolts, 2 nuts and throttle body assembly.
   c. Remove the gasket from the intake manifold.

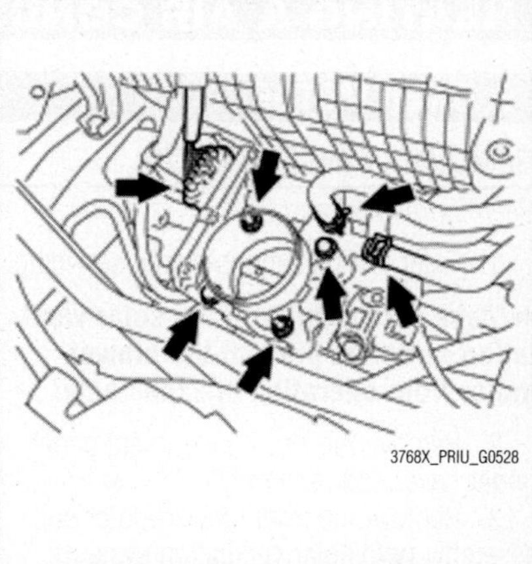

Fig. 198 Disconnecting the throttle body connector and 2 water by pass hoses

Fig. 199 Removing the gasket from the intake manifold

To install:

9. To install, reverse the removal procedure. Tighten the throttle body assembly bolts and nuts to 7 ft. lbs. (10 Nm).

HEATING & AIR CONDITIONING SYSTEM

BLOWER MOTOR

REMOVAL & INSTALLATION

See Figures 200 and 201.

1. Refer to Precautions before servicing.

➡**Make sure to turn off the solar ventilation switch to prevent the blower motor from operating unexpectedly.**

2. Remove the No. 2 instrument panel under cover sub assembly.
3. Remove the front blower motor sub assembly (w/o solar ventilation system).
    a. Remove the quick heater connector screw.
    b. Disengage the clamp and disconnect the connector.
    c. Remove the 3 screws and front blower motor sub-assembly.

➡**Do not remove the front blower motor sub-assembly if it has been damaged or impacted.**

4. Remove the front blower motor sub assembly (w/ solar ventilation system).
    a. Remove the quick heater connector screw.
    b. Disengage the clamp and disconnect the connector.
    c. Remove the 3 screws and front blower motor sub-assembly.

➡**Do not remove the front blower motor sub-assembly if it has been damaged or impacted.**

To install:

5. To install, reverse the removal procedure.

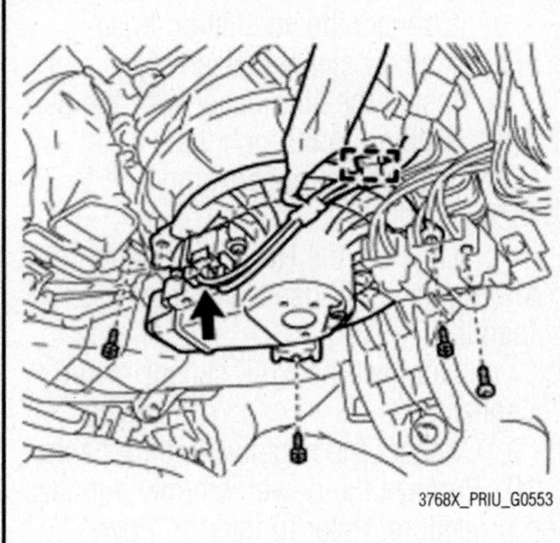

Fig. 200 Removing the front blower motor sub assembly (w/ solar ventilation system)

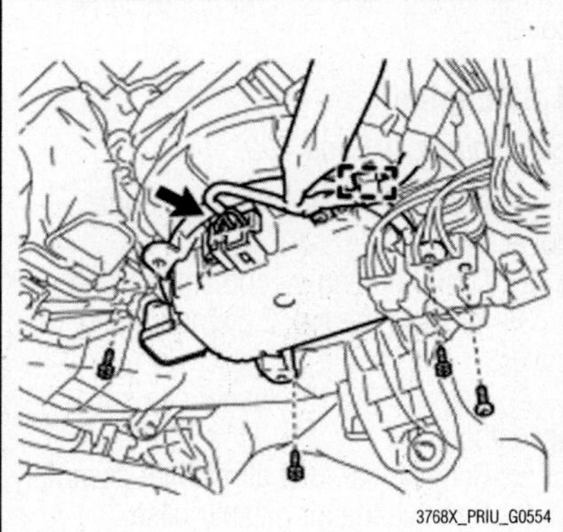

Fig. 201 Removing the front blower motor sub assembly (w/o solar ventilation system)

HEATER CORE

REMOVAL & INSTALLATION

See Figure 202.

1. Before servicing the vehicle, refer to the precautions.
2. Disconnect the negative battery cable.

➡**Wait at least 90 seconds after disconnecting the cable from the negative (-) battery terminal to prevent airbag and seat belt pretensioner activation.**

3. Remove the air conditioning unit.
4. Detach the clamp and disconnect the evaporator temperature sensor connector.
5. Detach the clamp and 2 claws, and remove the heater piping cover.
6. Remove the 4 screws and 4 clamps.
7. Remove the radiator heater unit from the air conditioner radiator.

➡**Prepare a drain pan or cloth for when the cooling water leaks.**

8. Remove the 2 screws and No. 1 air duct.

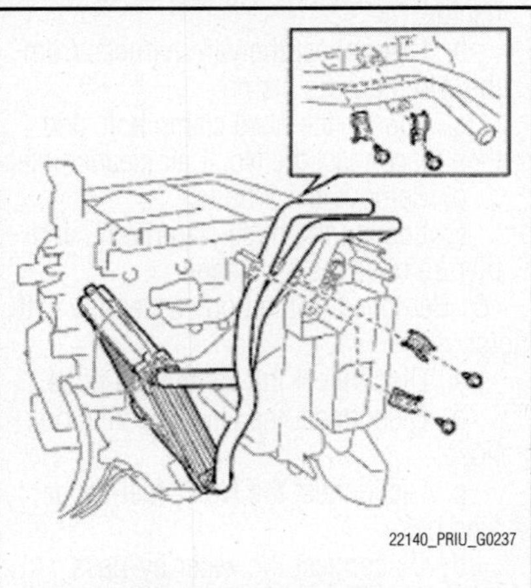

Fig. 202 Remove the radiator heater unit from the air conditioner radiator

9. Remove the 4 screws and quick heater assembly.
10. Remove the 7 screws and No. 1 cooler evaporator from the heater case.
11. Remove the 2 O-rings from the No. 1 cooler evaporator.
12. Detach the 2 claws and remove the evaporator temperature sensor.

To install:

13. Install the air conditioning radiator to the radiator heater unit and install the 4 clamps and 4 screws.
14. Attach the 2 claws and clamp to install the heater piping cover.
15. Attach the clamp and connect the evaporator temperature sensor connector.
16. Install the cooler expansion valve to the No. 1 cooler evaporator.
17. Install the air conditioning unit.
18. Connect the negative battery cable.
19. Perform the power window initialization procedure. Refer to Interior, Power Windows, Initialization Procedure.
20. Operate the engine to normal operating temperatures; then, check the climate control operation and check for coolant and refrigerant leaks.
21. Connect the negative battery cable.
22. Perform the power window initialization procedure. Refer to Interior, Power Windows, Initialization Procedure.
23. Check the SRS warning light.

STEERING

RACK & PINION STEERING GEAR

REMOVAL & INSTALLATION

See Figures 203 and 204.

1. Place the front wheels facing straight ahead.
2. Secure the steering wheel with the seat belt in order to prevent the rotation.

➡This operation is useful to prevent damage to the spiral cable.

3. Remove the column hole cover silencer sheet.
4. Separate the No. 2 steering intermediate shaft assembly.
5. Separate the No. 1 steering column hole cover sub assembly.
   a. Remove clip A, detach clip B from the body and disconnect the No. 1 steering column hole cover sub-assembly.

➡Do not damage the clips.

6. Remove the front wheels.
7. Remove the No. 1 engine under cover.
8. Remove the No. 2 engine under cover.
9. Remove the front No. 3 engine under cover.
10. Remove the rear engine under covers LH and RH.
11. Separate the front stabilizer link assemblies LH and RH.
12. Separate the tie rod end sub assembly LH.

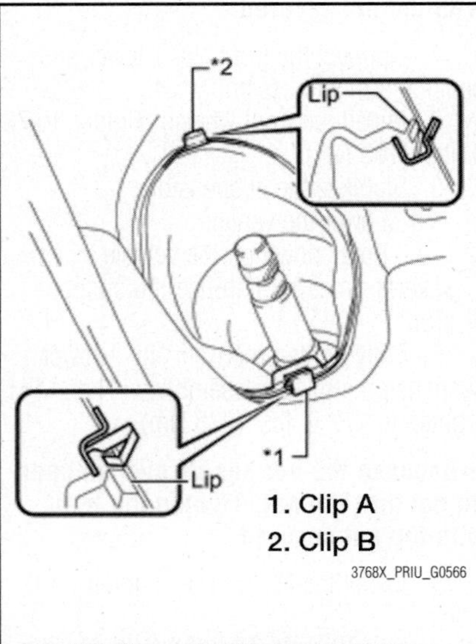

Fig. 203 Separating the No. 1 steering column hole cover sub assembly

13. Separate the tie rod end sub assembly RH.
14. Separate the front No. 1 lower suspension arm sub assemblies LH and RH.
15. Remove the front engine mounting bracket lower reinforcement.
16. Remove the rear side rail reinforcement sub assemblies LH and RH.
17. Remove the front suspension member rear braces LH and RH.
18. Remove the front suspension crossmember sub assembly.
19. Remove the No. 1 steering column hole cover sub assembly.
   a. Remove the No. 1 steering column hole cover sub-assembly from the steering link assembly.
20. Remove the steering intermediate shaft.
   a. Put matchmarks on the steering intermediate shaft and steering link assembly.
   b. Remove the bolt and steering intermediate shaft from the steering link assembly.
21. Remove the steering link assembly.
   a. Remove the 2 bolts, 2 nuts and steering link assembly from the front suspension crossmember sub-assembly.

➡Keep the nut from rotating while turning the bolt because the nut has its own stopper.

22. Secure the steering link assembly.
   a. Using the special tool (09612-00012), secure the steering link assembly in a vise.

➡Tape the special tool before use.

23. Remove the tie rod end sub assembly LH.

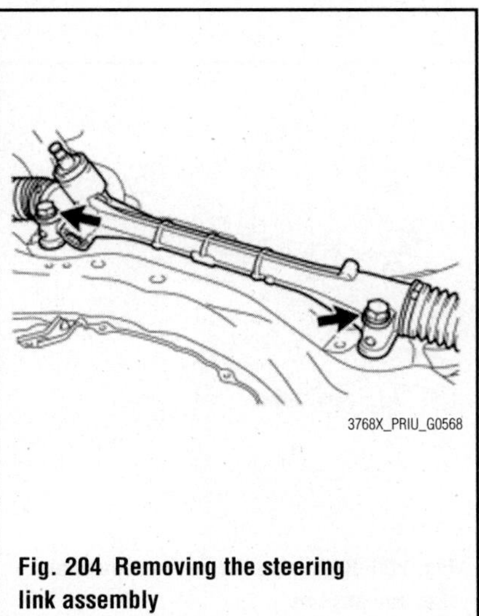

Fig. 204 Removing the steering link assembly

24. Remove the tie rod end sub assembly RH.

To install:

25. Install the tie rod end sub assemblies LH and RH.

➡After adjusting the toe in, tighten the lock nut.

26. Install the steering link assembly.
   a. Install the steering link assembly to the front suspension crossmember sub-assembly with the 2 bolts and 2 nuts. Tighten to 102 ft. lbs. (138 Nm).

➡Keep the nut from rotating while turning the bolt because the nut has its own stopper.

➡Make sure to tighten the bolts starting from the left side of the vehicle.

27. Install the steering intermediate shaft.
   a. Align the matchmarks and install the steering intermediate shaft to the steering link assembly.
   b. Install the bolt and tighten to 26 ft. lbs. (35 Nm).
28. Install the No. 1 steering column hole cover sub assembly.
   a. Align the round hole in the No. 1 steering column hole cover sub-assembly with the protrusion of the steering link assembly to install the cover.
29. Install the front suspension crossmember sub assembly.
30. Install the front suspension member rear brace LH.
31. Install the front suspension member rear brace RH.
32. Install the rear side rail reinforcement sub assemblies LH and RH.
33. Install the front engine mounting bracket lower reinforcement.
34. Connect the front No. 1 lower suspension arm sub assemblies LH and RH.
35. Connect the tie rod end sub assemblies LH and RH. Tighten to 36 ft. lbs. (49 Nm).

➡Further tighten the nut up to 60° if the holes for the clip are not aligned.

36. Connect the front stabilizer link assemblies LH and RH.
37. Connect the No. 1 steering column hole cover sub assembly.

➡Make sure the lips of the No. 1 steering column hole cover sub assembly are not damaged.

38. Connect the No. 2 steering intermediate shaft assembly.
39. Place the front wheels facing straight ahead.
40. Install the column hole cover silencer sheet.
41. Install the rear engine under covers LH and RH.
42. Install the front No. 3 engine under cover.
43. Install the No. 2 engine under cover.
44. Install the No. 1 engine under cover.
45. Install the front wheels. Tighten to 76 ft. lbs. (103 Nm).
46. Stabilize the suspension.
47. Inspect and adjust the front wheel alignment.

SUSPENSION

FRONT SUSPENSION

LOWER BALL JOINT

REMOVAL & INSTALLATION

See Figure 205.

➡Use the same procedure for the LH side and RH side.

➡The procedure listed below is for the LH side.

1. Remove the front axle.
2. Remove the front lower ball joint assembly.
   a. Secure the front axle assembly in a vise.

➡When using a vise, do not over tighten it.

   b. Remove the clip and nut.
   c. Install the special tool (09960-20010) to the front lower ball joint.

➡Check that the clearance measurement between SST and the front axle assembly is 1 mm (0.0394 in.).

   d. Using the special tool (09960-20010), remove the front lower ball joint assembly from the front axle assembly.

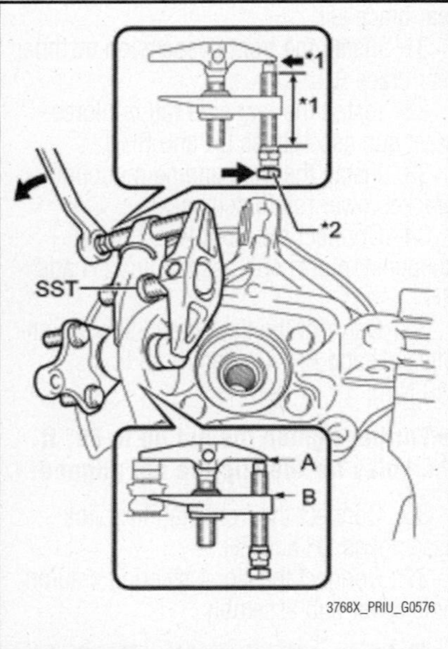

Fig. 205 Removing the front lower ball joint assembly

⁂ CAUTION

Apply grease to the threads and end of the SST bolt.

➡Install the special tool so that A and B are parallel.

➡Be sure to place the wrench on the part indicated in the illustration.

➡Do not damage the front lower ball joint dust cover.

To install:

3. Install the front lower ball joint assembly.
   a. Secure the front axle assembly in a vise.

➡When using a vise, do not over tighten it.

   b. Install the front lower ball joint assembly to the front axle assembly with the nut. Tighten to 52 ft. lbs. (71 Nm).
   c. Install a new clip.

➡Further tighten the nut up to 60° if the holes for the cotter pin are not aligned.

4. Install the front axle assembly.

LOWER CONTROL ARM

REMOVAL AND & INSTALLATION

See Figure 206.

Fig. 206 Removing the front suspension arm sub assembly

1. Remove the front wheels.
2. Remove the No. 1 engine under cover.
3. Separate the front No. 1 lower suspension arm sub assembly LH.
4. Remove the front No. 1 lower suspension arm sub assembly LH.
   a. Remove the 2 bolts, nut, and front No. 1 lower suspension arm sub-assembly LH from the front suspension member.

➡Because the nut has its own stopper, do not turn the nut. Loosen the bolt with the nut secured

5. Remove the front No. 1 lower suspension arm sub assembly RH.

➡Refer to the procedures listed above for the removal of the RH side.

To install:

6. Temporarily tighten the front No. 1 lower suspension arm sub assembly RH and LH.
   a. Temporarily install the front No. 1 lower suspension arm LH and RH to the front suspension crossmember with the 2 bolts and nut.

➡Because the nut has its own stopper, do not turn the nut. Tighten the bolt with the nut secured.

7. Connect the front No. 1 lower suspension arm sub assembly LH.
8. Install the front wheels. Tighten to 76 ft. lbs. (103 Nm).
9. Stabilize the suspension.
   a. Lower the vehicle.
   b. Press down on the vehicle several times to stabilize the suspension.
10. Fully tighten the front No. 1 lower suspension arm sub assemblies LH and RH. Tighten to 172 ft. lbs. (233 Nm).

➡Because the nut has its own stopper, do not turn the nut. Tighten the bolt with the nut secured.

11. Install the No. 1 engine under cover.
12. Inspect and adjust the front wheel alignment.

STABILIZER BAR

REMOVAL & INSTALLATION

See Figures 207 and 208.

1. Remove the front wheels.
2. Remove the No. 1 engine under cover.
3. Remove the No. 2 engine under cover.
4. Remove the front stabilizer link assembly LH.
    a. Remove the 2 nuts and front stabilizer link assemblies LH.

➡If the ball joint turns together with the nut, use a hexagon wrench (6 mm) to hold the stud bolt.

5. Remove the front stabilizer link assembly RH.

➡Perform the same procedure as for the LH side.

6. Separate the front No. 1 lower suspension arm sub assembly LH.
7. Remove the front No. 1 lower suspension arm sub assembly LH.
8. Remove the front suspension member front brace LH.
    a. Remove the 4 bolts and front suspension member front brace LH.
9. Remove the front suspension member front brace RH.

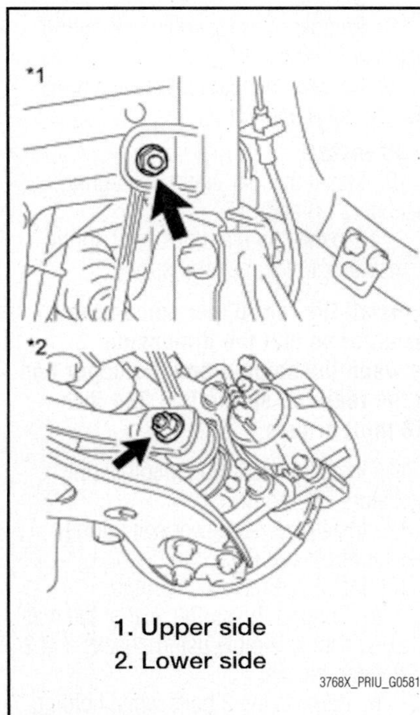

Fig. 207 Removing the front stabilizer link assembly LH

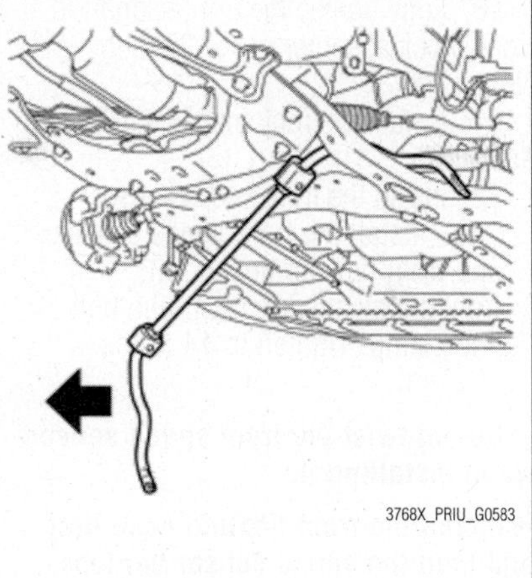

Fig. 208 Removing the front stabilizer bar

➡Perform the same procedure as for the LH side.

10. Remove the front stabilizer bar.
    a. Remove the front stabilizer bar with front stabilizer bar bushing from the front suspension crossmember sub-assembly.
11. Remove the front stabilizer bar bushing LH.
    a. Remove the front stabilizer bar bushing LH from the front stabilizer bar.
12. Remove the front stabilizer bar bushing RH.

➡Perform the same procedure as for the LH side.

To install:

13. To install, reverse the removal procedure. Tighten the front stabilizer link assembly nuts to 55 ft. lbs. (74 Nm). Tighten the front wheels to 76 ft. lbs. (103 Nm).

STRUT & SPRING ASSEMBLY

REMOVAL & INSTALLATION

See Figure 209.

➡Use the same procedure for the LH side and RH side.

➡The procedure listed below is for the LH side.

1. Remove the front wheel.
2. Remove the windshield wiper motor and link assembly.
3. Remove the cowl body mounting reinforcement LH.
4. Remove the outer cowl top panel sub assembly.
5. Separate the front stabilizer link assembly.

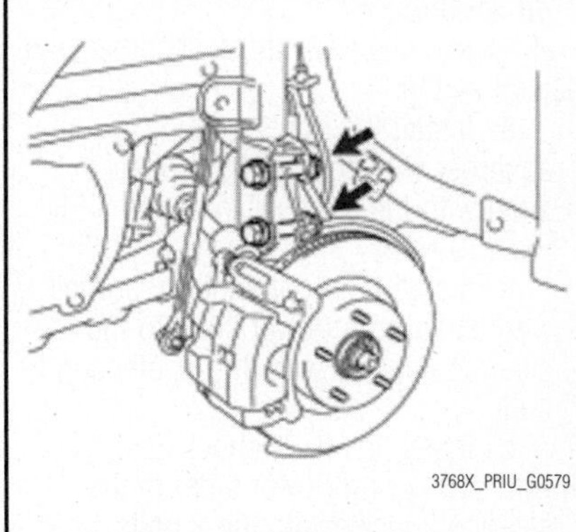

Fig. 209 Removing the 2 bolts and 2 nuts, and separate the front shock absorber with coil spring (lower side) from the steering knuckle

    a. Remove the nut and separate the stabilizer link assembly from the front shock absorber with coil spring.

➡If the ball joint turns together with the nut, use a hexagon wrench (6 mm) to hold the stud bolt.

6. Separate the front speed sensor.
    a. Remove the bolt and clamp, and separate the front speed sensor and front flexible hose from the front shock absorber with coil spring.

➡Be sure to separate the front speed sensor from the front shock absorber with coil spring completely.

7. Remove the front suspension support dust cover.
8. Remove the front shock absorber with the coil spring.
    a. Loosen the front support to front shock absorber nut of the front shock absorber.

➡Do not remove the front support to front shock absorber nut.

➡Loosen the nut only when the front shock absorber with coil spring needs to be disassembled.

    b. Support the front axle using a jack and wooden block.
    c. Remove the 2 bolts and 2 nuts, and separate the front shock absorber with coil spring (lower side) from the steering knuckle.
    d. Remove the 3 nuts and front shock absorber with coil spring.

➡Make sure that the front speed sensor is completely separated from the front shock absorber with coil spring.

To install:

9. Install the front shock absorber with the coil spring.

a. Install the front shock absorber with coil spring (upper side) with the 3 nuts. Tighten to 37 ft. lbs. (50 Nm).

b. Install the front shock absorber with coil spring (lower side) to the steering knuckle with the 2 bolts and 2 nuts.

c. Install the front shock absorber with coil spring (lower side) to the steering knuckle with the 2 bolts and 2 nuts. Tighten to 177 ft. lbs. (240 Nm).

➡**While keeping the bolts from rotating, tighten the nuts.**

10. Fully tighten the front support to front shock absorber nut to 35 ft. lbs. (47 Nm).

11. Install the front suspension support dust cover.

12. Install the front speed sensor.

a. Install the front speed sensor and front flexible hose to the front shock absorber with the bolt and clamp. Tighten to 14 ft. lbs. (19 Nm).

➡**Do not twist the front speed sensor when installing it.**

➡**Install the front flexible hose first and then the speed sensor harness bracket.**

13. Install the front stabilizer link assembly to the front shock absorber with coil spring with the nut. Tighten to 55 ft. lbs. (74 Nm).

➡**If the ball joint turns together with the nut, use a hexagon wrench (6 mm) to hole the stud bolt.**

14. Install the outer cowl top panel sub assembly.

15. Install the cowl body mounting reinforcement LH.

16. Install the windshield wiper motor and link assembly.

17. Install the front wheel and tighten to 76 ft. lbs. (103 Nm).

18. Inspect and adjust the front wheel alignment.

SUSPENSION

REAR SUSPENSION

COIL SPRING

REMOVAL & INSTALLATION

See Figures 210 and 211.

1. Remove the rear wheel.

2. Disconnect the rear speed sensor wire LH and RH.

3. Separate the rear speed sensor wire LH and RH.

a. Remove the nut and separate the 2 clamps and rear speed sensor wire.

4. Separate the rear height control sensor sub assembly RH (w/ height control).

a. Remove the bolt and separate the rear height control sensor sub-assembly RH from the rear axle beam assembly.

b. Using a vinyl tape (*1), secure the rear height control sensor sub-assembly RH as shown in the illustration.

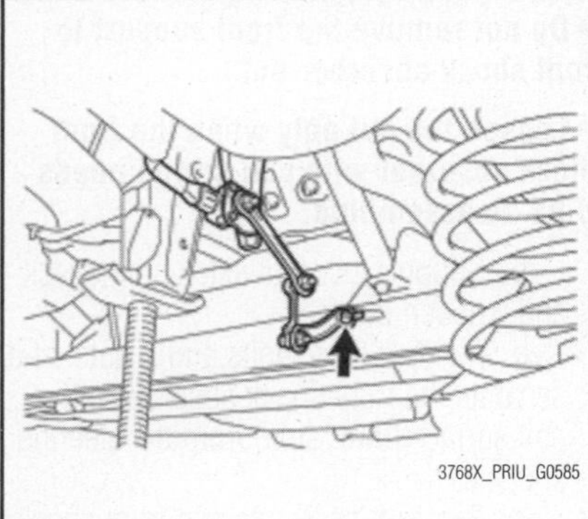

Fig. 210 Removing the bolt and separating the rear height control sensor sub-assembly RH from the rear axle beam assembly

5. Separate the rear wheel house liner LH and RH.

a. Remove the clip and turn back the rear wheel house liner LH to separate the rear wheel house liner.

6. Remove the rear coil spring LH.

a. Loosen the 2 bolts.

➡**Do not remove the bolts.**

b. Support the spring seat of the rear axle beam assembly using 2 jacks and 2 wooden blocks.

CAUTION

Do not jack up the rear axle beam assembly too high as the vehicle may fall.

➡**Support the rear shock absorber at a position where it compresses by approximately 0.787 to 1.18 inch (20 to 30 mm).**

c. Remove the 2 bolts while holding the 2 nuts and separate the rear axle beam assembly from the rear shock absorber assemblies LH and RH.

➡**Since the stopper nuts are used, turn the bolts.**

d. Slowly lower the rear axle beam assembly using 2 jacks and 2 wooden blocks, and remove the rear coil spring LH.

➡**When moving the rear axle beam assembly beyond full rebound, make sure that the rear axle beam assembly is not out of position for more than 10 minutes.**

e. Slowly jack up the rear axle beam assembly using 2 jacks and 2 wooden blocks, and temporarily tighten the rear axle beam assembly to the rear shock absorber assemblies LH and RH with the 2 bolts and 2 nuts.

➡**Since the stopper nuts are used, turn the bolts.**

7. Remove the rear coil spring RH.

➡**Perform the same procedure as the LH side.**

8. Remove the rear upper coil spring insulators LH and RH.

9. Remove the rear lower coil spring insulators LH and RH.

To install:

10. Install the rear upper coil spring insulators LH and RH.

a. Install the rear upper coil spring insulator to the rear coil spring.

➡**Install the rear upper coil spring insulator so that the dimension between the stopper and the upper end of the rear coil spring is 0.394 inch (10 mm) or less.**

11. Install the rear upper coil spring insulators LH and RH.

12. Install the rear lower coil spring insulators LH and RH.

13. Install the rear coil springs.

a. Support the spring seat of the rear axle beam assembly using 2 jacks and 2 wooden blocks.

b. Remove the 2 bolts while holding the 2 nuts and separate the rear axle beam assembly from the rear shock absorber assemblies LH and RH.

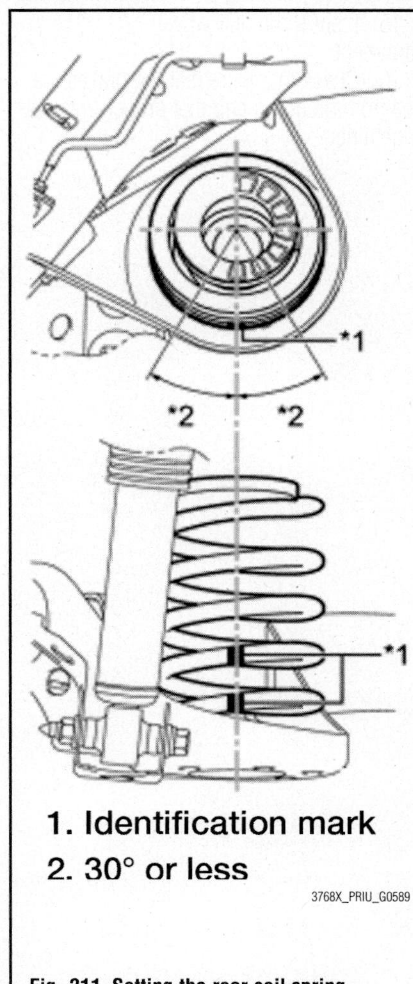

Fig. 211 Setting the rear coil spring

➡Since the stopper nuts are used, turn the bolts.

c. Slowly lower the rear axle beam assembly using 2 jacks and 2 wooden blocks.

d. Set the rear coil spring to the rear axle beam assembly.

➡Set the rear coil spring so that the identification marks are positioned as shown in the illustration.

14. Slowly jack up the rear axle beam assembly using 2 jacks and 2 wooden blocks and temporarily install the rear axle beam assembly and rear coil spring with the 2 bolts and 2 nuts.

➡Since the stopper nuts are used, turn the bolts.

➡Insert the bolt with the threaded end facing the outside of the vehicle.

15. Install the rear height control sensor sub assembly RH (w/ height control sensor) to the rear axle beam assembly with the bolt. Tighten to 71 inch lbs. (8 Nm).

16. Install the rear speed sensor wires LH and RH. Tighten the nuts to 75 inch lbs. (8.5 Nm).

➡Do not twist the rear speed sensor wire when installing it.

17. Connect the rear speed sensor wires LH and RH.

18. Install the rear wheels. Tighten to 76 ft. lbs. (103 Nm).

19. Stabilize the suspension.

20. Fully tighten the rear axle beam assembly bolts to 100 ft. lbs. (135 Nm).

➡The final torque must be applied under the standard vehicle height conditions.

a. Fully tighten the 2 bolts to 66 ft. lbs. (90 Nm).

➡Since the stopper nut are used, turn the bolts.

➡The final torque must be applied under the standard vehicle height conditions.

21. Install the rear wheel house liners LH and RH.

22. Inspect the rear wheel alignment.

23. Place the front wheels facing straight ahead.

24. Perform yaw rate and acceleration calibration.

25. Check for the speed sensor signal.

26. Perform the initialization.

WHEEL BEARINGS

REMOVAL & INSTALLATION

See Figure 212.

1. Before servicing the vehicle, refer to the precautions.
2. Raise and safely support the vehicle.
3. Remove the rear wheel.
4. Remove the rear brake drum.
5. Disconnect the rear ABS speed sensor connector.
6. Remove automatic adjusting lever, shoe adjuster and adjusting lever spring from the brake shoes assembly, if necessary.
7. Remove the hub bolts and hub/bearing assembly.

To install:

8. Install the hub/bearing assembly.

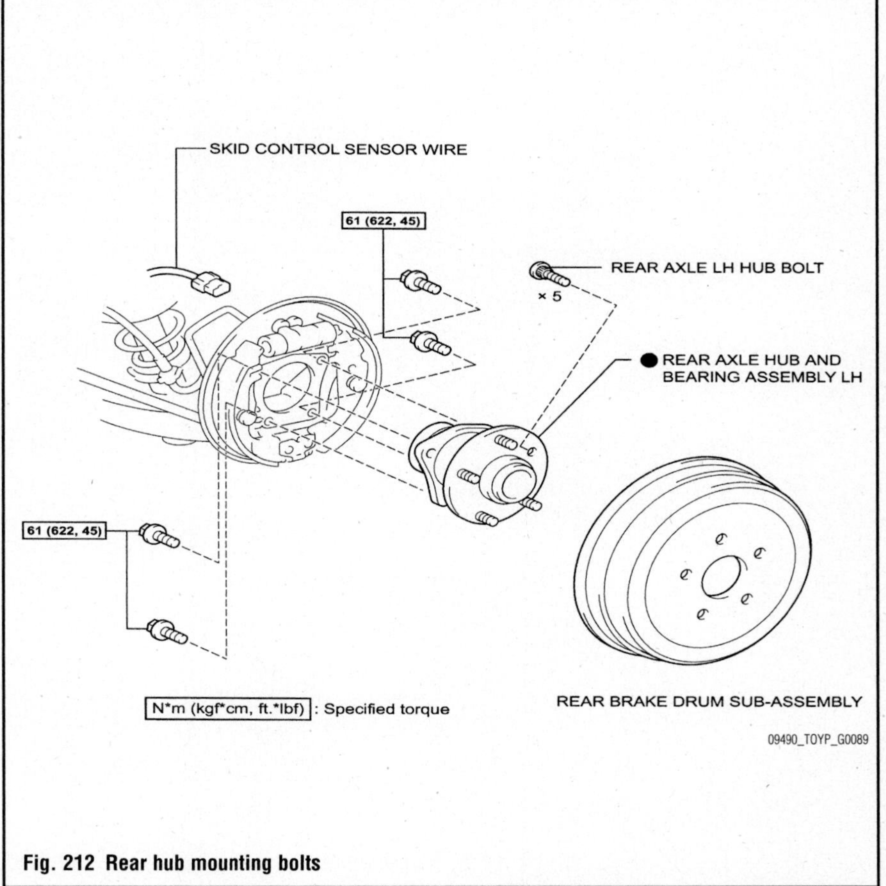

Fig. 212 Rear hub mounting bolts

9. Tighten the hub assembly bolts to 45 ft. lbs. (61 Nm).
10. Install adjusting lever spring, shoe adjuster and automatic adjusting lever from the brake shoes assembly, if removed earlier.
11. Connect the rear ABS speed sensor connector.
12. Install the rear brake drum.
13. Install the rear wheel.
14. Lower the vehicle.
15. Inspect the rear wheel alignment.
16. Check to ensure that the brakes are free from dragging and that proper braking is obtained.

TOYOTA

RAV4

12

SPECIFICATIONS AND MAINTENANCE CHARTS

ENGINE AND VEHICLE IDENTIFICATION

Engine							Model Year	
Code/VIN ①	Liters (cc)	Cu. In.	Cyl.	Fuel Sys.	Engine Type	Eng. Mfg.	Code ②	Year
2AR-FE	2.4 (2362)	144	4	SFI	DOHC	Toyota	B	2011
2GR-FE	3.5 (3498)	213	6	SFI	DOHC	Toyota	C	2012

SFI: Sequential Fuel Injection

DOHC: Double Overhead Camshaft

① 5th digit of the vehicle identification number (VIN)

② 10th digit of the Vehicle Identification Number (VIN)

71099_RAV4_C0001

GENERAL ENGINE SPECIFICATIONS

Year	Model	Engine Displacement Liters	Engine Series Code/VIN	Net Horsepower @ rpm	Net Torque @ rpm (ft. lbs.)	Bore x Stroke (in.)	Com-pression Ratio	Oil Pressure @ rpm
2011	RAV4	2.4	2AR-FE	166@6000	165@4000	3.48x3.78	NA	①
	RAV4	3.5	2GR-FE	269@6200	246@4700	NA	NA	②
2012	RAV4	2.4	2AR-FE	166@6000	165@4000	3.48x3.78	NA	①
	RAV4	3.5	2GR-FE	269@6200	246@4700	NA	NA	②

NA: Not Available

① 4.3 psi or more at idle

② 11.6 psi or more at idle

71099_RAV4_C0002

ENGINE TUNE-UP SPECIFICATIONS

Year	Engine Displacement Liters	Engine Code/VIN	Spark Plug Gap (in.)	Ignition Timing (deg.)	Fuel Pump (psi)	Idle Speed (rpm)		Valve Clearance	
						MT	AT	Intake	Exhaust
2011	2.4	2AR-FE	0.043	①	44-50	—	600-700	0.0075-0.0114	0.0150-0.0189
	3.5	2GR-FE	0.043	②	44-50	—	600-700	NA	NA
2012	2.4	2AR-FE	0.043	①	44-50	—	600-700	0.0075-0.0114	0.0150-0.0189
	3.5	2GR-FE	0.043	②	44-50	—	600-700	NA	NA

NOTE: The Vehicle Emission Control Information label often reflects specification changes made during production.
The label figures must be used if they differ from those in this chart.

NA: Not Available

① 5-15 degrees BTDC when using intelligent tester tool.
8-12 degrees BTDC at idle. Connect SST when not using intelligent tester tool.
5-15 degrees BTDC at idle. Disconnect SST when not using intelligent tester tool.

② 8-12 degrees BTDC when using intelligent tester tool.
8-12 degrees BTDC at idle. Connect SST when not using intelligent tester tool.
5-15 degrees BTDC at idle. Disconnect SST when not using intelligent tester tool.

71099_RAV4_C0004

CAPACITIES

Year	Model	Engine Displacement Liters	Engine Code/VIN	Engine Oil with Filter (qts.)	Transmission (pts.) 5-Spd	Transmission (pts.) Auto.*	Transfer Case (pts.)	Drive Axle Front (pts.)	Drive Axle Rear (pts.)	Fuel Tank (gal.)	Cooling System (qts.)
2011	RAV4	2.4	2AR-FE	4.5	—	7.4	1.0	—	1.0	15.9	7.2
	RAV4	3.5	2GR-FE	6.4	—	NA	1.0	—	1.0	15.9	①
2012	RAV4	2.4	2AR-FE	4.5	—	7.4	1.0	—	1.0	15.9	7.2
	RAV4	3.5	2GR-FE	6.4	—	NA	1.0	—	1.0	15.9	①

*After draining, add the following amounts, then fill to the cold full line.

NA: Not Available

① STD: 9.4
TWG: 9.8

71099_RAV4_C0005

FLUID SPECIFICATIONS

Year	Model	Engine Displacement Liters	Engine Oil	Auto. Trans.	Drive Axle	Power Steering Fluid	Brake Master Cylinder
2011	RAV4	2.4	①	ATF World Standard	②	NA	DOT 3
	RAV4	3.5	5W-30	ATF World Standard	③	NA	DOT 3
2012	RAV4	2.4	①	ATF World Standard	②	NA	DOT 3
	RAV4	3.5	5W-30	ATF World Standard	③	NA	DOT 3

DOT: Department Of Transportation

NA: Not Available

Note: If specification disagrees with specification in owners manual, use specification in owners manaual

① 10W-20 above 40 degrees F. 5W-20 below 40 degrees F

② API GL-5 SAE 90 above 0 degrees F. API GL-5 SAE 80W-90 below 0 degrees

③ API GL-5 SAE 85W-90 above 0 degrees F. API GL-5 SAE 80W-90 below 0 degrees

71099_RAV4_C0003

VALVE SPECIFICATIONS

Year	Engine Displacement Liters	Engine Code/VIN	Seat Angle (deg.)	Face Angle (deg.)	Inner Spring Free lenth (in.)	Spring Installed Height (in.)	Stem-to-Guide Clearance (in.) Intake	Stem-to-Guide Clearance (in.) Exhaust	Stem Diameter (in.) Intake	Stem Diameter (in.) Exhaust
2011	2.4	2AR-FE	45	44.5	NA	1.8670	0.0010-0.0024	0.0012-0.0026	0.2154-0.2159	0.2152-0.2158
	3.5	2GR-FE	45	44.5	NA	1.7898	0.0010-0.0024	0.0012-0.0026	0.2154-0.2159	0.2151-0.2157
2012	2.4	2AR-FE	45	44.5	NA	1.8670	0.0010-0.0024	0.0012-0.0026	0.2154-0.2159	0.2152-0.2158
	3.5	2GR-FE	45	44.5	NA	1.7898	0.0010-0.0024	0.0012-0.0026	0.2154-0.2159	0.2151-0.2157

NA: Not Available

71099_RAV4_C0006

CAMSHAFT SPECIFICATIONS

All measurements in inches unless noted

Year	Engine Displacement Liters	Engine Code/VIN	Journal Dia.	Brg. Oil Clearance	Shaft End-play ①	Circle Runout	Lobe Height Intake	Lobe Height Exhaust
2011	2.4	2AR-FE	②	NA	NA	NA	1.8624 1.8664	1.8104- 1.8143
	3.5	2GR-FE	③	④	0.0031 0.0051	0.0016	1.7447- 1.7487	1.7426- 1.7465
2012	2.4	2AR-FE	②	NA	NA	NA	1.8624 1.8664	1.8104- 1.8143
	3.5	2GR-FE	③	④	0.0031 0.0051	0.0016	1.7447- 1.7487	1.7426- 1.7465

① Thrust clearance

② Mark 1, 2 and 3: 1.4162-1.4167

③ No. 1: 1.4152-1.4157

All others: 1.0220-1.0226

④ No. 1: 0.0016-0.0031

All others: 0.0010-0.0024

71099_RAV4_C0007

CRANKSHAFT AND CONNECTING ROD SPECIFICATIONS

All measurements are given in inches.

			Crankshaft				Connecting Rod		
Year	Engine Displ. Liters	Engine Code/VIN	Main Brg. Journal Dia.	Main Brg. Oil Clearance	Shaft End-play	Thrust on No.	Journal Diameter	Oil Clearance	Side Clearance
2011	2.4	2AR-FE	2.1649- 2.1654	0.0006- 0.0015	0.0016- 0.0095	3	1.8894- 1.8898	0.0012- 0.0025	0.0063- 0.0202
	3.5	2GR-FE	2.4011- 2.4016	0.0010- 0.0019	0.0016- 0.0095	2	2.0863- 2.0866	0.0018- 0.0026	0.0059- 0.0157
2012	2.4	2AR-FE	2.1649- 2.1654	0.0006- 0.0015	0.0016- 0.0095	3	1.8894- 1.8898	0.0012- 0.0025	0.0063- 0.02
	3.5	2GR-FE	2.4011- 2.4016	0.0010- 0.0019	0.0016- 0.0095	2	2.0863- 2.0866	0.0018- 0.0026	0.0059- 0.0157

71099_RAV4_C0008

PISTON AND RING SPECIFICATIONS

All measurements are given in inches.

Year	Engine Displ. Liters	Engine Code/VIN	Piston Clearance	Ring Gap Top Comp.	Ring Gap Bottom Comp.	Ring Gap Oil Control	Ring Side Clearance Top Comp.	Ring Side Clearance Bottom Comp.	Ring Side Clearance Oil Control
2011	2.4	2AR-FE	0.0020-0.0029	0.0094-0.0122	0.0130-0.0169	0.0040-0.0119	0.0008-0.0028	0.0008-0.0024	0.0008-0.0028
	3.5	2GR-FE	0.0018-0.0020	0.0098-0.0138	0.0197-0.0413	0.0039-0.0157	0.0008-0.0028	0.0008-0.0024	0.0028-0.0059
2012	2.4	2AR-FE	0.0020-0.0029	0.0094-0.0122	0.0130-0.0169	0.0040-0.0119	0.0008-0.0028	0.0008-0.0024	0.0008-0.0028
	3.5	2GR-FE	0.0018-0.0020	0.0098-0.0138	0.0197-0.0413	0.0039-0.0157	0.0008-0.0028	0.0008-0.0024	0.0028-0.0059

71099_RAV4_C0009

TORQUE SPECIFICATIONS

All readings in ft. lbs.

Year	Engine Displacement Liters	Engine Code/VIN	Cylinder Head Bolts	Main Bearing Bolts	Rod Bearing Bolts	Crankshaft Damper Bolts	Flywheel Bolts	Manifold Intake	Manifold Exhaust	Spark Plugs	Oil Pan Drain Plug
2011	2.4	2AR-FE	①	②	③	133	72	22	27	18	30
	3.5	2GR-FE	④	⑤	⑥	184	132	15	15	13	30
2012	2.4	2AR-FE	①	②	③	133	72	22	27	18	30
	3.5	2GR-FE	④	⑤	⑥	184	132	15	15	13	30

① Step 1: 52 ft. lbs.
Step 2: plus 90 degrees

② Step 1: 30 ft. lbs.
Step 2: plus 90 degrees

③ Step 1: 30 ft. lbs.
Step 2: plus 90 degrees

④ Step 1: 27 ft. lbs.
Step 2: plus 90 degrees
Step 3: plus 90 degrees
Bolt should be 22 ft. lbs. on Bank 2

⑤ Step 1: 45 ft. lbs.
Step 2: plus 90 degrees
Main bearing cap bolt: 38 ft. lbs.

⑥ Step 1: 45 ft. lbs.
Step 2: plus 90 degrees
Step 3: 8 side bolts to 38 ft. lbs.

71099_RAV4_C0010

WHEEL ALIGNMENT

Year	Model		Caster Range (+/-Deg.)	Caster Preferred Setting (Deg.)	Camber Range (+/-Deg.)	Camber Preferred Setting (Deg.)	Toe-in (in.)	Steering Axis Inclination (Deg.)
2011	RAV4	Front	0.75	+5.72	0.75	-0.13	0+/-0.08	11.27+/-0.50
		Rear	NA	NA	0.75	-0.92	0.04+/-0.09	11.27+/-0.50
2012	RAV4	Front	0.75	+5.72	0.75	-0.13	0+/-0.08	11.27+/-0.50
		Rear	NA	NA	0.75	-0.92	0.04+/-0.09	11.27+/-0.50

71099_RAV4_C0011

TIRE, WHEEL AND BALL JOINT SPECIFICATIONS

Year	Model	OEM Tires Standard	OEM Tires Optional	Tire Pressures (psi) Front	Tire Pressures (psi) Rear	Wheel Size	Ball Joint Inspection	Lug Nut Torque (ft. lbs.)
2011	RAV4	①	①	②	②	NA	NA	76
2012	RAV4	①	①	②	②	NA	NA	76

NA: Not Available

OEM: Original Equipment Manufacturer

PSI: Pounds Per Square Inch

① Base model: P215/70R16, P225/65R17. Optional P235/55R18

Sport model: P235/55R18

Limited model: P225/65R17

② 32 PSI. However if placard on vehicle disagrees with this specification, use the specification on vehicle placard.

71099_RAV4_C0012

BRAKE SPECIFICATIONS

All measurements in inches unless noted

Year	Model		Brake Disc Original Thickness	Brake Disc Minimum Thickness	Brake Disc Maximum Runout	Brake Drum Diameter Original Inside Diameter	Brake Drum Diameter Max. Wear Limit	Brake Drum Diameter Maximum Machine Diameter	Minimum Lining Thickness	Brake Caliper Bracket Bolts (ft. lbs.)	Brake Caliper Mounting Bolts (ft. lbs.)
2011	RAV4	F	①	②	0.0020	—	—	—	0.039	79	25
		R	0.472	0.413	0.0059	—	—	—	0.039	65	20
2012	RAV4	F	①	②	0.0020	—	—	—	0.039	79	25
		R	0.472	0.413	0.0059	—	—	—	0.039	65	20

F: Front

R: Rear

① 275 disc (15" disc): 0.984
296 disc (16" disc) : 1.102

② 275 disc (15" disc): 0.886
296 disc (16" disc): 0.984

71099_RAV4_C0013

SCHEDULED MAINTENANCE INTERVALS

TOYOTA—RAV4

TO BE SERVICED	TYPE OF SERVICE	VEHICLE MILEAGE INTERVAL (x1000)																		
		5	10	15	20	25	30	35	40	45	50	55	60	65	70	75	80	85	90	95
Automatic transmission and differential fluid	S/I			✓			✓			✓			✓			✓			✓	
Ball joints and boots	S/I			✓			✓			✓			✓			✓			✓	
Brake system	S/I	✓	✓	✓	✓	✓	✓	✓	✓	✓	✓	✓	✓	✓	✓	✓	✓	✓	✓	✓
Charcoal canister	S/I												✓							
Drive belts	S/I						✓						✓						✓	
Driveshaft bushing	L						✓						✓						✓	
Engine coolant	R			✓			✓			✓			✓			✓			✓	
Engine oil & filter	R		✓		✓		✓		✓		✓		✓		✓		✓		✓	
Exhaust pipes & mounts	S/I			✓			✓			✓			✓			✓			✓	
Fuel tank cap gasket	S/I						✓						✓						✓	
Halfshaft boots & flange bolts	S/I			✓			✓			✓			✓			✓			✓	
Limited slip differential fluid	R												✓						✓	
Manual transmission and differential fluid	S/I						✓						✓						✓	
Platinum spark plugs	R												✓							
Propeller shaft bolts	S/I			✓			✓			✓			✓			✓			✓	
Rack and pinion assembly	S/I			✓			✓			✓			✓			✓			✓	
Tires (rotate)	S/I	✓	✓	✓	✓	✓	✓	✓	✓	✓	✓	✓	✓	✓	✓	✓	✓	✓	✓	✓
Transfer case and differential fluid	S/I			✓			✓			✓			✓			✓			✓	
Valves	S/I												✓							

R: Replace S/I: Service or Inspect L: Lubricate

FREQUENT OPERATION MAINTENANCE (SEVERE SERVICE)

If a vehicle is operated under any of the following conditions it is considered severe service:

- Towing a trailer or using a camper or car-top carrier.

- Repeated short trips of less than 5 miles in temperatures below freezing.

- Excessive idling or low-speed driving for long distances as in heavy commercial use, such as delivery, taxi or police cars.

- Operating on rough, muddy or salt-covered roads.

- Operating on unpaved or dusty roads.

Oil filter: service or inspect every 5000 miles or 4 months, whichever occurs first.

Brake linings and discs or drums: service or inspect every 5000 miles or 4 months, whichever occurs first.

Steering linkage: service or inspect every 5000 miles or 4 months, whichever occurs first.

Ball joints and boots: service or inspect every 5000 miles or 4 months, whichever occurs first.

Brake discs & pads (front): service or inspect every 6000 miles.

Halfshaft boots: service or inspect every 5000 miles or 4 months. Retighten the flange bolts, whichever occurs first.

Body chassis bolts and nuts: service or inspect every 5000 miles or 4 months, whichever occurs first.

Transmission and differential fluid: replace every 15,000 miles or 12 months, whichever occurs first.

Transfer case and differential fluid: replace every 15,000 miles or 12 months, whichever occurs first.

71099_RAV4_C0014

PRECAUTIONS

Before servicing any vehicle, please be sure to read all of the following precautions, which deal with personal safety, prevention of component damage, and important points to take into consideration when servicing a motor vehicle:

- Never open, service or drain the radiator or cooling system when the engine is hot; serious burns can occur from the steam and hot coolant.
- Observe all applicable safety precautions when working around fuel. Whenever servicing the fuel system, always work in a well-ventilated area. Do not allow fuel spray or vapors to come in contact with a spark, open flame, or excessive heat (a hot drop light, for example). Keep a dry chemical fire extinguisher near the work area. Always keep fuel in a container specifically designed for fuel storage; also, always properly seal fuel containers to avoid the possibility of fire or explosion. Refer to the additional fuel system precautions later in this section.
- Fuel injection systems often remain pressurized, even after the engine has been turned **OFF**. The fuel system pressure must be relieved before disconnecting any fuel lines. Failure to do so may result in fire and/or personal injury.
- Brake fluid often contains polyglycol ethers and polyglycols. Avoid contact with the eyes and wash your hands thoroughly after handling brake fluid. If you do get brake fluid in your eyes, flush your eyes with clean, running water for 15 minutes. If eye irritation persists, or if you have taken brake fluid internally, IMMEDIATELY seek medical assistance.
- The EPA warns that prolonged contact with used engine oil may cause a number of skin disorders, including cancer. You should make every effort to minimize your exposure to used engine oil. Protective gloves should be worn when changing oil. Wash your hands and any other exposed skin areas as soon as possible after exposure to used engine oil. Soap and water, or waterless hand cleaner should be used.
- All new vehicles are now equipped with an air bag system, often referred to as a Supplemental Restraint System (SRS) or Supplemental Inflatable Restraint (SIR) system. The system must be disabled before performing service on or around system components, steering column, instrument panel components, wiring and sensors. Failure to follow safety and disabling procedures could result in accidental air bag deployment, possible personal injury and unnecessary system repairs.
- Always wear safety goggles when working with, or around, the air bag system. When carrying a non-deployed air bag, be sure the bag and trim cover are pointed away from your body. When placing a non-deployed air bag on a work surface, always face the bag and trim cover upward, away from the surface. This will reduce the motion of the module if it is accidentally deployed. Refer to the additional air bag system precautions later in this section.
- Clean, high quality brake fluid from a sealed container is essential to the safe and proper operation of the brake system. You should always buy the correct type of brake fluid for your vehicle. If the brake fluid becomes contaminated, completely flush the system with new fluid. Never reuse any brake fluid. Any brake fluid that is removed from the system should be discarded. Also, do not allow any brake fluid to come in contact with a painted surface; it will damage the paint.
- Never operate the engine without the proper amount and type of engine oil; doing so WILL result in severe engine damage.
- Timing belt maintenance is extremely important. Many models utilize an interference-type, non-freewheeling engine. If the timing belt breaks, the valves in the cylinder head may strike the pistons, causing potentially serious (also time-consuming and expensive) engine damage. Refer to the maintenance interval charts for the recommended replacement interval for the timing belt, and to the timing belt section for belt replacement and inspection.
- Disconnecting the negative battery cable on some vehicles may interfere with the functions of the on-board computer system(s) and may require the computer to undergo a relearning process once the negative battery cable is reconnected.
- When servicing drum brakes, only disassemble and assemble one side at a time, leaving the remaining side intact for reference.
- Only an MVAC-trained, EPA-certified automotive technician should service the air conditioning system or its components.

BRAKES

ANTI-LOCK BRAKE SYSTEM (ABS)

GENERAL INFORMATION

PRECAUTIONS

- Certain components within the ABS system are not intended to be serviced or repaired individually.
- Do not use rubber hoses or other parts not specifically specified for and ABS system. When using repair kits, replace all parts included in the kit. Partial or incorrect repair may lead to functional problems and require the replacement of components.
- Lubricate rubber parts with clean, fresh brake fluid to ease assembly. Do not use shop air to clean parts; damage to rubber components may result.
- Use only DOT 3 brake fluid from an unopened container.
- If any hydraulic component or line is removed or replaced, it may be necessary to bleed the entire system.
- A clean repair area is essential. Always clean the reservoir and cap thoroughly before removing the cap. The slightest amount of dirt in the fluid may plug an orifice and impair the system function. Perform repairs after components have been thoroughly cleaned; use only denatured alcohol to clean components. Do not allow ABS components to come into contact with any substance containing mineral oil; this includes used shop rags.
- The Anti-Lock control unit is a microprocessor similar to other computer units in the vehicle. Ensure that the ignition switch is **OFF** before removing or installing controller harnesses. Avoid static electricity discharge at or near the controller.
- If any arc welding is to be done on the vehicle, the control unit should be unplugged before welding operations begin.

WHEEL SPEED SENSORS

REMOVAL & INSTALLATION

➡Use the same procedure for the RH and LH sides.

➡The procedures listed below are for the LH side.

Front

See Figure 1.

1. Disconnect the cable from the negative battery terminal.

⁂ CAUTION

Wait at least 90 seconds after disconnecting the cable from the negative (-) battery terminal to disable the SRS system.

2. Remove the front wheel.
3. Remove the front fender liner LH.
4. Remove the front speed sensor LH.
   a. Disconnect the sensor connector.
   b. Remove the sensor clip (labeled A), bolt (labeled B) and sensor clamp (labeled C).
   c. Remove the sensor clip (labeled D), bolt (labeled E) and sensor clamp (labeled F).
   d. Remove the bolt and sensor body from the knuckle.

➡Keep the sensor tip and sensor installation hole free from foreign matter.

To install:

5. Install the speed sensor front LH.

➡To prevent interference with other parts, do not twist the painted line areas of the sensor wire when installing it.

a. Set the sensor body into the knuckle, and then install the sensor with the bolt. Tighten to 75 inch lbs. (8.5 Nm).

➡Keep the sensor tip and sensor installation hole free from foreign matter.

➡Firmly insert the sensor body into the knuckle before tightening the bolt.

➡After installing the sensor to the knuckle, make sure that there is no clearance between the sensor stay and knuckle. Also make sure that no foreign matter is stuck between the parts.

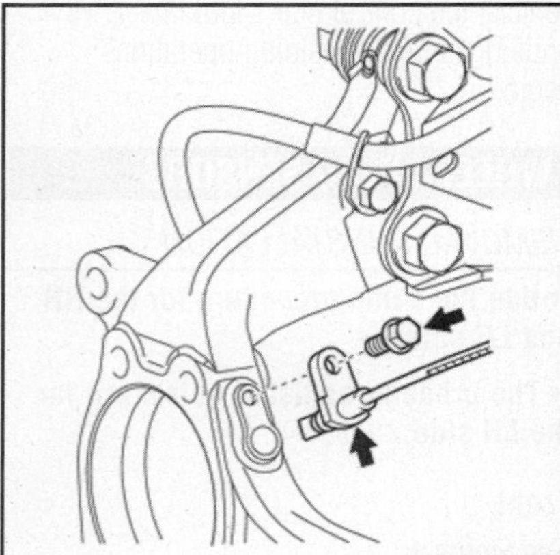

Fig. 1 Removing the speed senor body from the knuckle

➡To prevent interference between the sensor and magnetic rotor, do not rotate the sensor body during or after the insertion of the sensor body to the knuckle.

b. Install the sensor clamp and sensor clip as follows.

c. Simultaneously perform the following: 1) hang the hook part of the sensor clamp (labeled A) on the flexible hose bracket (labeled C); and 2) insert the hook part of the sensor clamp (labeled B) into the flexible hose bracket (labeled D).

➡Do not twist the sensor wire when installing the clamp.

d. Install the sensor clamp to the flexible hose clamp and flexible hose bracket with the bolt (labeled E). Tighten to 14 ft. lbs. (19 Nm).

e. Insert the sensor clip (labeled F) into the hole on the absorber lower bracket.

(c) Install the sensor clamp and sensor clip as follows.

f. Set the sensor clamp (labeled G) on the side member, and then install the bolt (labeled H). Tighten to 75 inch lbs. (8.5 Nm).

➡Do not twist the sensor wire when installing the clamp.

g. Insert the sensor clip (labeled I) into the hole on the apron.

h. Connect the sensor connector.

6. Install the front fender liner LH.

➡Install the fender liner so that the sensor wire harness passes beyond the fender liner installation clip towards the rear side of the vehicle.

7. Install the front wheel. Tighten to 76 ft. lbs. (103 Nm).
8. Connect the cable to the negative battery terminal.
9. Check the ABS speed sensor signal.

Rear

1. Disconnect the cable from the negative battery terminal.

⁂ CAUTION

Wait at least 90 seconds after disconnecting the cable from the negative (-) battery terminal to disable the SRS system.

2. Remove the rear wheel.
3. Remove the deck trim side panel assembly LH.

➡Refer to the procedures from the removal of the rear door scuff plate LH up until the removal of the deck trim side panel assembly LH.

4. Remove the rear speed sensor LH.
   a. Disconnect the speed sensor connector.
   b. Disconnect the grommet of the speed sensor wire from the hole of the wheel house.
   c. Remove the bolt (labeled A) and sensor clamp (labeled B) from the side member.
   d. Remove the 2 nuts (labeled C) and 2 sensor clamps (labeled D) from the upper arm.
   e. Remove the nut (labeled E) and sensor clamp (labeled F) from the trailing arm.
   f. Remove the bolt (labeled G) and sensor body (labeled H) from the carrier.

➡Keep the sensor tip and sensor installation hole free from foreign matter.

To install:

5. Install the rear speed sensor LH.

➡To prevent interference with other parts, do not twist the painted line areas of the sensor wire when installing it.

a. Install the sensor (labeled A) with the bolt (labeled B). Tighten to 75 inch lbs. (8.5 Nm).

➡Keep the sensor tip and sensor installation hole free from foreign matter.

➡To prevent interference with the bearing rotor, do not rotate the sensor body when inserting the sensor body or after inserting the sensor body.

b. Install the sensor clamp (labeled C) with the nut (labeled D). Tighten to 44 inch lbs. (5.0 Nm).

c. Install the 2 sensor clamps (labeled E) with the 2 nuts (labeled F).

➡Do not twist the sensor wire when installing the clamps.

d. Install the sensor clamp (labeled G) with the bolt (labeled H). Tighten to 75 inch lbs. (8.5 Nm).

➡Do not twist the sensor wire when installing the clamps.

e. Insert the connector and grommet into the inside of the vehicle through the passage hole in the wheel house.

➡Make sure the grommet band clamp remains on the outside of the vehicle.

f. Hold the grommet and pull it toward the outside of the vehicle. Then fix the grommet in place so that it is not tilted.

➡When pulling out the grommet, do not grip the sensor wire.

➡Fix the grommet in place within the range shown in the illustration.

g. Connect the speed sensor connector.

6. Install the deck trim side panel assembly LH.

➡Refer to the procedures from the installation of the deck trim side panel LH up until the installation of the rear door scuff plate LH.

7. Install the rear wheel. Tighten to 76 ft. lbs. (103 Nm).

8. Connect the cable to the negative battery terminal.

9. Check the speed sensor signal.

BRAKES — BLEEDING THE BRAKE SYSTEM

BLEEDING PROCEDURE

➡If any work is performed on the brake system or if air in the brake lines is suspected, bleed air from the brake system.

➡Wash off brake fluid immediately if it comes in contact with any painted surface.

1. Remove the cowl top ventilator louver bracket LH.
   a. Detach the 6 claws and remove the bracket.
2. Fill the reservoir with brake fluid.
   a. Set a brake fluid can upside down on the reservoir.

➡Make sure there is sufficient brake fluid in the can.

➡After adding brake fluid, make sure the reservoir is sufficient full.

3. Bleed the air from the brake master cylinder.

➡If the master cylinder has been disassembled or if the reservoir becomes empty, bleed air from the master cylinder.

a. Using a union nut wrench (10 mm), disconnect the 2 brake lines from the master cylinder.

b. Slowly depress and hole the brake pedal.

c. Block the outer holes with your fingers, and release the pedal.

d. Repeat the 2 previous steps 3 or 4 times.

e. Using a union nut wrench (10 mm), connect the 2 brake lines to the master cylinder. Tighten to 11 ft. lbs. (15 Nm).

➡Use a torque wrench with a fulcrum length of 11.8 inch.

➡This torque value is effective when the union nut wrench is parallel to the torque wrench.

4. Bleed the air from the brake line.
   a. Remove the bleeder plug cap.
   b. Connect a vinyl tube to the bleeder plug.
   c. Depress the brake pedal several times, and then loosen the bleeder plug with the pedal depressed.
   d. When fluid stops coming out, immediately tighten the bleeder plug. Then release the pedal.
   e. Repeat the 2 previous steps until all the air in the brake fluid is gone.
   f. Tighten the bleeder plug. Tighten to 73 inch lbs. (8.3 Nm).
   g. Remove the vinyl tube and install the cap.
   h. Bleed air from the brake line for each wheel by repeating the above procedures.
5. Bleed the air from the ABS and traction actuator assembly.

➡After bleeding the air from the brake system, if the height or feel of the brake pedal is not correct, perform air bleeding of the brake actuator with the Techstream by performing the following procedure.

a. Depress the brake pedal more than 20 times with the engine off.

b. Connect the Techstream to the DLC3, and turn the ignition switch to ON.

➡DO NOT start the engine.

c. Select AIR BLEEDING on the Techstream.

➡Refer to the Techstream operator's manual for further details.

Bleed the air from the suction line.

➡Perform the bleeding for the right front wheel and right rear wheel.

➡Bleed the air by following the steps displayed on the Techstream.

a. Connect a vinyl tube to the bleeder plug for the right front wheel.

b. Loosen the bleeder plug.

c. Operate the brake actuator to bleed the air using the Techstream.

➡Do not depress the brake pedal at this time.

➡This operation stops automatically after 4 seconds.

d. Check if the operation has stopped by referring to the Techstream display.

e. Tighten the bleeder plug.

f. Repeat the 4 previous steps until all air in the fluid is completely bled out.

g. Tighten the bleeder plug. Tighten to 73 inch lbs. (8.3 Nm).

h. Repeat all of the above procedures for the right rear wheel to bleed the air out of the suction line.

Bleed the air out of the pressure reduction line.

➡Perform the bleeding for all 4 wheels.

➡Bleed the air by following the steps displayed on the Techstream.

a. Connect a vinyl tube to one of the bleeder plugs.

b. Loosen the bleeder plug.

c. Using the Techstream, operate the brake actuator assembly, completely depress the brake pedal and hold it there.

➡During this procedure, the pedal will feel heavy, but completely depress it so that the brake fluid comes out of the bleeder plug.

➡Depress and hold the brake pedal. Do not depress and release the pedal repeatedly.

➡The operation stops automatically after 4 seconds. When performing this procedure consecutively, wait at least 20 seconds between each operation.

d. Tighten the bleeder plug, then release the brake pedal.

e. Repeat the 3 previous steps until all air in the fluid is completely bled out.

f. Tighten the bleeder plug. Tighten to 73 inch lbs. (8.3 Nm).

g. Repeat all of the above procedures for the other wheels to bleed the air out of the pressure reduction line.

6. Check the brake fluid level in the reservoir.

7. Install the cowl top ventilator louver bracket LH.

a. Attach the 6 claws to install the bracket.

BRAKES — FRONT DISC BRAKES

BRAKE CALIPER

REMOVAL & INSTALLATION

See Figures 2 and 3.

1. Remove the front wheel.
2. Drain the brake fluid.

➡Wash off brake fluid immediately if it comes in contact with any painted surface.

3. Disconnect the front flexible hose.
   a. Remove the union bolt and gasket.
   b. Disconnect the flexible hose from the brake caliper.
4. Remove the disc brake caliper assembly.
   a. Remove the 2 bolts and caliper.

To install:

5. Install the disc brake caliper assembly. Tighten the 2 bolts to 25 ft. lbs. (34 Nm).
6. Connect the front flexible hose.
   a. Install a new gasket and connect the flexible hose with the union bolt. Tighten to 22 ft. lbs. (29 Nm).

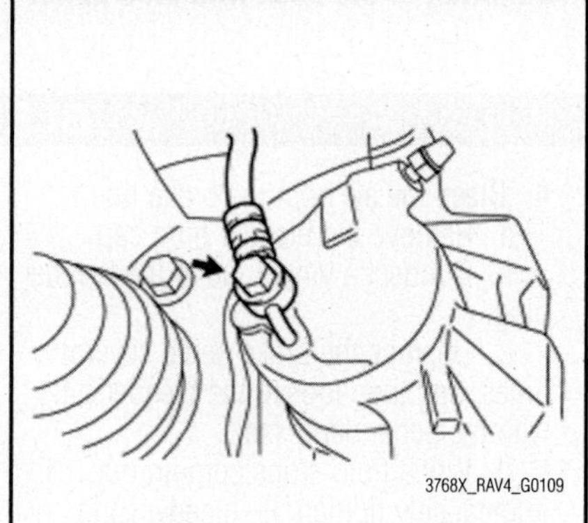

Fig. 2 Disconnecting the flexible hose from the brake cylinder

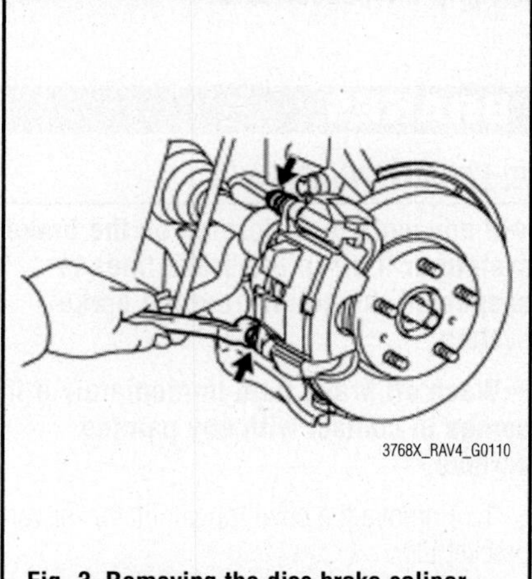

Fig. 3 Removing the disc brake caliper assembly

➡Install the flexible hose lock securely in the lock hole in the cylinder.

7. Fill the reservoir with brake fluid.
8. Bleed air from the brake master cylinder.
9. Bleed air from the brake line.
10. Bleed air from the ABS and traction actuator assembly.
11. Check the brake fluid level in the reservoir.
12. Inspect for fluid leaks.
13. Install the front wheel. Tighten to 76 ft. lbs. (103 Nm).

BRAKES — REAR DISC BRAKES

BRAKE CALIPER

REMOVAL & INSTALLATION

See Figure 4.

1. Remove the rear wheel.
2. Drain brake fluid.

➡Wash off brake fluid immediately if it comes in contact with any painted surface.

3. Disconnect the rear flexible hose LH.
   a. Remove the union bolt and gasket.
   b. Disconnect the flexible hose from the brake cylinder.
4. Remove the rear disc brake caliper assembly LH.
   a. Remove the 2 bolts and caliper.

To install:

5. Install the rear disc brake caliper assembly.

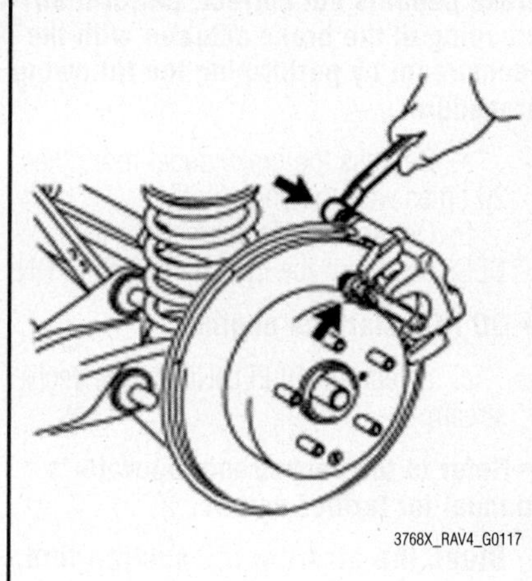

Fig. 4 Removing the rear disc brake caliper assembly LH

   a. Install the cylinder with the 2 bolts and tighten to 20 ft. lbs. (27 Nm).
6. Connect the rear flexible hose LH.
   a. Install a new gasket and connect the flexible hose with the union bolt. Tighten to 22 ft. lbs. (30 Nm).

➡Install the flexible hose lock securely in the lock hole in the cylinder.

7. Fill the reservoir with brake fluid.
8. Bleed air from the brake master cylinder.
9. Bleed air from the brake line.
10. Bleed air from the ABS and traction actuator assembly.
11. Check the brake fluid level in the reservoir.
12. Inspect for brake fluid leaks.
13. Install the rear wheel. Tighten to 76 ft. lbs. (103 Nm).

PARKING BRAKE SHOES

REMOVAL & INSTALLATION

See Figures 5 through 7.

➡Use the same procedures for the LH side and RH side.

➡The procedures listed below are for the LH side.

1. Remove the rear brake discs.
2. Remove the parking brake shoe return tension spring (upper side).
   a. Using needle-nose pliers, remove the 2 shoe return tension springs.
3. Remove the parking brake shoe strut LH.
4. Remove the parking brake shoe return tension spring (lower side).
5. Using needle nose pliers, remove the tension spring.
6. Remove the parking brake shoe adjusting screw set.
7. Remove the No. 1 parking brake shoe assembly LH.
   a. Press the shoe hold down spring to remove the pin, shoe hold down spring and shoe.
8. Remove the No. 2 parking brake shoe assembly LH.
   a. Slide the shoe to remove the shoe hold down spring, pin and shoe.
   b. Disconnect the parking brake cable from the parking brake shoe lever.

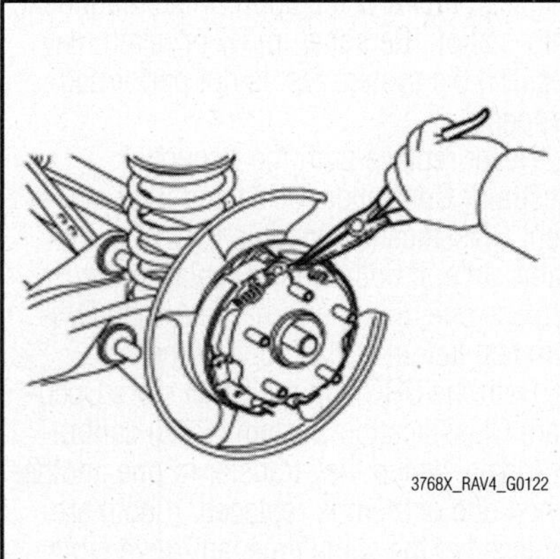

Fig. 5 Removing the parking brake shoe return tension spring (upper side)

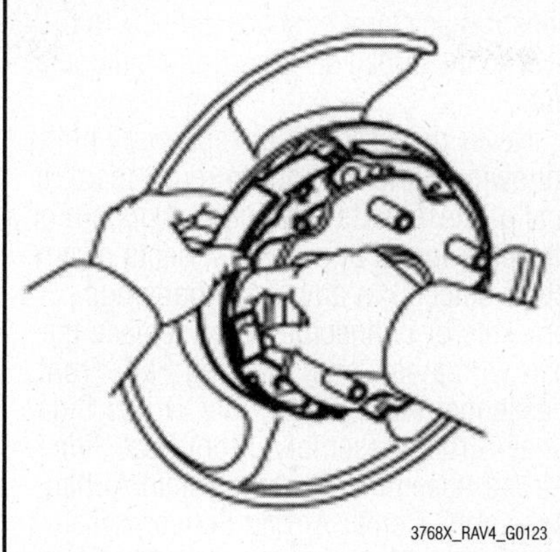

Fig. 6 Removing the No. 1 parking brake shoe assembly LH

9. Remove the parking brake shoe lever LH.

To install:

10. Apply high temperature grease to the areas of the backing plate that contact the shoes.
11. Install the parking brake shoe lever LH.
12. Install the No. 2 parking brake shoe assembly LH.
   a. Connect the parking brake cable to the parking brake shoe lever.
   b. Apply high temperature grease to the following contact surfaces of the shoe:
      - Shoe to anchor pin
      - Shoe to parking brake lever pin
      - Shoe to spring
   c. Install the shoe with the shoe hold down spring and pin.
13. Install the No. 1 parking brake shoe assembly.
   a. Apply high temperature grease to the following contact surfaces of the shoe:
      - Shoe to anchor pin
      - Shoe to spring
   b. Install the shoe with the shoe hold-down spring and pin.
14. Install the parking brake shoe adjusting screw set.
   a. Apply high temperature grease to the areas of the shoe adjusting screw set shown.

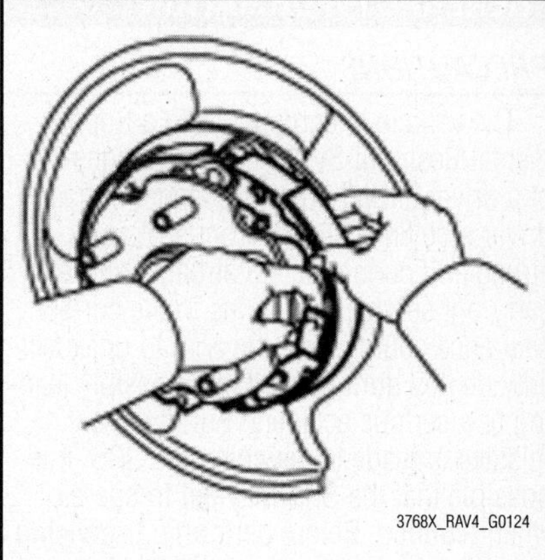

Fig. 7 Removing the No. 2 parking brake shoe assembly LH

   b. Install the screw set.
15. Install the parking brake shoe return tension spring (lower side).
   a. Using needle nose pliers, install the tension spring.
16. Install the parking brake shoe strut.
   a. Apply high temperature grease to the contact surface of the strut and shoe return tension spring.
   b. Install the strut.
17. Install the parking brake shoe return tension spring (upper side).
   a. Using needle nose pliers, install the 2 shoe return tension springs.
18. Check the parking brake installation.
   a. Check that each part is installed properly.

➡There should be no oil or grease adhering to the friction surfaces of the shoe lining and disc.

19. Install the rear brake disc.
20. Adjust the parking brake shoe clearance.

ADJUSTMENT

1. Remove the shoe adjusting hole plug, and then turn the adjuster to expand the shoe adjuster until the disc locks.
2. Turn the shoe adjuster so that it contracts to a point where the disc can rotate smoothly. Standard is 8 notches.
3. Check that the shoe has no brake drag.

GENERAL INFORMATION

PRECAUTIONS

The vehicle is equipped with a Supplemental Restraint System (SRS). It consists of a driver airbag, front passenger airbag, driver side knee airbag, front seat side airbag and curtain shield airbag. Failure to carry out service operations in the correct sequence could cause the SRS to unexpectedly deploy during servicing, possibly leading to a serious accident. Further, if a mistake is made in servicing the SRS, it is possible that the SRS may fail to operate when required. Before performing servicing (including removal or installation of parts, inspection or replacement), be sure to read the following items carefully, then follow the correct procedures indicated in the repair manual.

Disconnect and isolate the battery negative cable before beginning any airbag system component diagnosis, testing, removal, or installation procedures. Wait at least 90 seconds after the ignition switch is turned off and the negative (-) terminal cable is disconnected from the battery before starting the operation. The SRS is equipped with a backup power source, so if work is started within 90 seconds after disconnecting the negative (-) terminal cable from the battery, the SRS may be deployed. Failure to disable the airbag system may result in accidental airbag deployment, personal injury, or death.

Do not expose the steering pad, front passenger airbag assembly, driver side knee airbag assembly, center airbag sensor assembly, front airbag sensor, front seat side airbag assembly, side airbag sensor, curtain shield airbag assembly, rear airbag sensor, seat position airbag sensor or occupant classification ECU directly to hot air or flames.

Malfunction symptoms of the SRS are difficult to confirm, so DTCs are the most important source of information when troubleshooting. When troubleshooting the SRS, always inspect DTCs before disconnecting the battery.

Even in the case of a minor collision when the SRS does not deploy, the steering pad, front passenger airbag assembly, driver side knee airbag assembly, center airbag sensor assembly, front airbag sensor, front seat side airbag assembly, side airbag sensor, curtain shield airbag assembly, rear airbag sensor, seat position airbag sensor and occupant classification ECU should be inspected.

Before repair work, remove the airbag sensor if any kind of shock is likely to o ccur to the airbag sensor during the operation.

Never disassemble or repair any of the following parts in order to reuse them. If any of these parts have been dropped, or a defect is found (e.g. cracks, dents or any other defects) in any of the housings, brackets or connectors, then replace the part with a new one: Steering Pad, Front Passenger Airbag Assembly, Driver Side Knee Airbag Assembly, Front Seat Side Airbag Assembly, Curtain Shield Airbag Assembly, Center Airbag Sensor Assembly, Front Airbag Sensor, Front Seat Inner Belt Assembly, Seat Position Sensor, Occupant Classification ECU, Side Airbag Sensor, Rear Airbag Sensor, Front Seat Outer Belt Assembly.

Use an volt/ohmmeter with high impedance (10 kohm/V minimum) for troubleshooting the electrical circuits.

Information labels are attached near the SRS components. Follow the instructions in the caution.

Do not place an intact undeployed airbag face down on a solid surface. The airbag will propel into the air if accidentally deployed and may result in personal injury or death.

When carrying or handling an undeployed airbag, the trim side (face) of the airbag should be pointing towards the body to minimize possibility of injury if accidental deployment occurs. Failure to do this may result in personal injury or death.

Replace airbag system components with OEM replacement parts. Substitute parts may appear interchangeable, but internal differences may result in inferior occupant protection. Failure to do so may result in occupant personal injury or death.

Wear safety glasses, rubber gloves, and long sleeved clothing when cleaning powder residue from vehicle after an airbag deployment. Powder residue emitted from a deployed airbag can cause skin irritation. Flush affected area with cool water if irritation is experienced. If nasal or throat irritation is experienced, exit the vehicle for fresh air until the irritation ceases. If irritation continues, see a physician.

Do not use a replacement airbag that is not in the original packaging. This may result in improper deployment, personal injury, or death.

The factory installed fasteners, screws and bolts used to fasten airbag components have a special coating and are specifically designed for the airbag system. Do not use substitute fasteners. Use only original equipment fasteners listed in the parts catalog when fastener replacement is required.

During, and following, any child restraint anchor service, due to impact event or vehicle repair, carefully inspect all mounting hardware, tether straps, and anchors for proper installation, operation, or damage. If a child restraint anchor is found damaged in any way, the anchor must be replaced. Failure to do this may result in personal injury or death.

Deployed and non-deployed airbags may or may not have live pyrotechnic material within the airbag inflator.

Do not dispose of driver/passenger/curtain airbags or seat belt tensioners unless you are sure of complete deployment. Refer to the Hazardous Substance Control System for proper disposal.

Dispose of deployed airbags and tensioners consistent with state, provincial, local, and federal regulations.

After any airbag component testing or service, do not connect the battery negative cable. Personal injury or death may result if the system test is not performed first.

If the vehicle is equipped with the Occupant Classification System (OCS), do not connect the battery negative cable before performing the OCS Verification Test using the scan tool and the appropriate diagnostic information. Personal injury or death may result if the system test is not performed properly.

Never replace both the Occupant Restraint Controller (ORC) and the Occupant Classification Module (OCM) at the same time. If both require replacement, replace one, then perform the Airbag System test before replacing the other.

Both the ORC and the OCM store Occupant Classification System (OCS) calibration data, which they transfer to one another when one of them is replaced. If both are replaced at the same time, an irreversible fault will be set in both modules and the OCS may malfunction and cause personal injury or death.

If equipped with OCS, the Seat Weight Sensor is a sensitive, calibrated unit and must be handled carefully. Do not drop or handle roughly. If dropped or damaged, replace with another sensor. Failure to do so may result in occupant injury or death.

If equipped with OCS, the front passenger seat must be handled carefully as well. When removing the seat, be careful when setting on floor not to drop. If dropped, the sensor may be inoperative, could result in occupant injury, or possibly death.

If equipped with OCS, when the passenger front seat is on the floor, no one should sit in the front passenger seat. This uneven force may damage the sensing ability of the seat weight sensors. If sat on and damaged, the sensor may be inoperative, could result in occupant injury, or possibly death.

DISARMING THE SYSTEM

To avoid personal injury when working on vehicles equipped with an air bag, the negative battery cable must be disconnected and at least 90 seconds must elapse before working on the system. Failure to do so may result in deployment of the air bag.

ARMING THE SYSTEM

To arm the system after service is finished, connect the negative battery cable.

CLOCKSPRING CENTERING

1. Rotate the spiral cable counterclockwise by hand until it feels firm

➡Do not use the airbag wire harness to turn the spiral cable.

2. Rotate the spiral cable clockwise approximately 2.5 turns to align the marks.

➡Do not use the airbag wire harness to turn the spiral cable.

➡The spiral cable will rotate approximately 2.5 turns to the left and right from the center.

DRIVE TRAIN

FRONT HALFSHAFT

REMOVAL & INSTALLATION

See Figures 8 and 9.

1. Before servicing the vehicle, refer to the Precautions Section.
2. Disconnect the negative battery cable.

➡Wait at least 90 seconds after disconnecting the negative battery cable before starting any repair work to prevent air bag and seat belt pretensioner activation.

3. Raise and support the vehicle safely.
4. Remove the tire and wheel assembly. Remove the front axle hub nut.
5. Drain the transaxle fluid.
6. Disconnect the left and right speed sensors.
7. Remove the left and right brake calipers.
8. Disconnect the left and right front stabilizer link assemblies.
9. Disconnect the left and right front lower number one arm subassemblies.
10. Matchmark the halfshaft and the axle hub, left and right side.

➡Do not punch the marks.

11. Using a plastic hammer, disconnect the steering knuckle with the axle hub, left and right side.

➡Be careful not to damage the boot and speed sensor rotor. Do not push out excessively the halfshaft from the axle assembly.

12. Disconnect the left and right tie rod subassemblies.
13. On the left side, using tool SST09520-01010, or equivalent remove the front halfshaft.
14. On 2WD, to remove the right halfshaft remove the two bolts and pull out the halfshaft together with the halfshaft bearing case. Remove the halfshaft from the transaxle.

➡Be careful not to damage the boot and speed sensor rotor. Do not push out excessively the halfshaft from the axle assembly.

15. On 4WD, to remove the right halfshaft use a brass bar and hammer to remove the halfshaft.

➡Do not damage the oil seal, boot or allow the halfshaft to fall out.

16. Support the front axle assembly.

➡The hub bearing could be damaged if it is subjected to the vehicle weight. If it is necessary to place weight on the hub bearing, such as moving it when the halfshaft is removed, support it using too SST09608-16042, or equivalent.

To install:

17. Installation is the reverse of the removal procedure.
18. On vehicles manufactured thru 01/2006 tighten the front axle hub nut to 159 ft. lbs. On vehicles manufactured after 01/2006 and equipped with the 2.4L engine, tighten the axle hub nut to 159 ft. lbs. On vehicles manufactured after 01/2006 and equipped with the 3.5L engine, tighten the axle hub nut to 215 ft. lbs.
19. On 2WD, tighten the right side bearing bracket bolts to 47 ft. lbs.
20. On 4WD, tighten the right side bearing bracket bolts to 24 ft. lbs.
21. Check and adjust the wheel alignment, as required.
22. Be sure to fill the transaxle with the proper grade and type transaxle fluid.
23. Start the engine and check for leaks.

REAR AXLE HOUSING

REMOVAL & INSTALLATION

See Figures 10, 11 and 12.

1. Disconnect the negative battery cable.

➡Wait at least 90 seconds after disconnecting the negative battery cable to prevent air bag and seat belt pretensioner activation.

2. Raise and safely support the vehicle.
3. Drain the rear differential fluid. Be sure to properly dispose of used fluid.
4. Remove the tire and wheel assembly.
5. Remove the fuel tank.
6. Remove the tailpipe assembly. Remove the center exhaust pipe assembly.
7. Remove the driveshaft with the center bearing.
8. Remove the two retaining bolts and remove the left side rear suspension member brace.
9. Remove the two retaining bolts and remove the right side rear suspension member brace.
10. Disconnect the harness clamp. Remove the breather tube. Disconnect the connector.
11. Properly support the differential carrier assembly with a transmission jack, or equivalent.
12. Remove the two bolts. See the illustration for location. Do not loosen the nuts, loosen the bolts.
13. Slowly lower the transmission jack and tilt the differential carrier assembly.

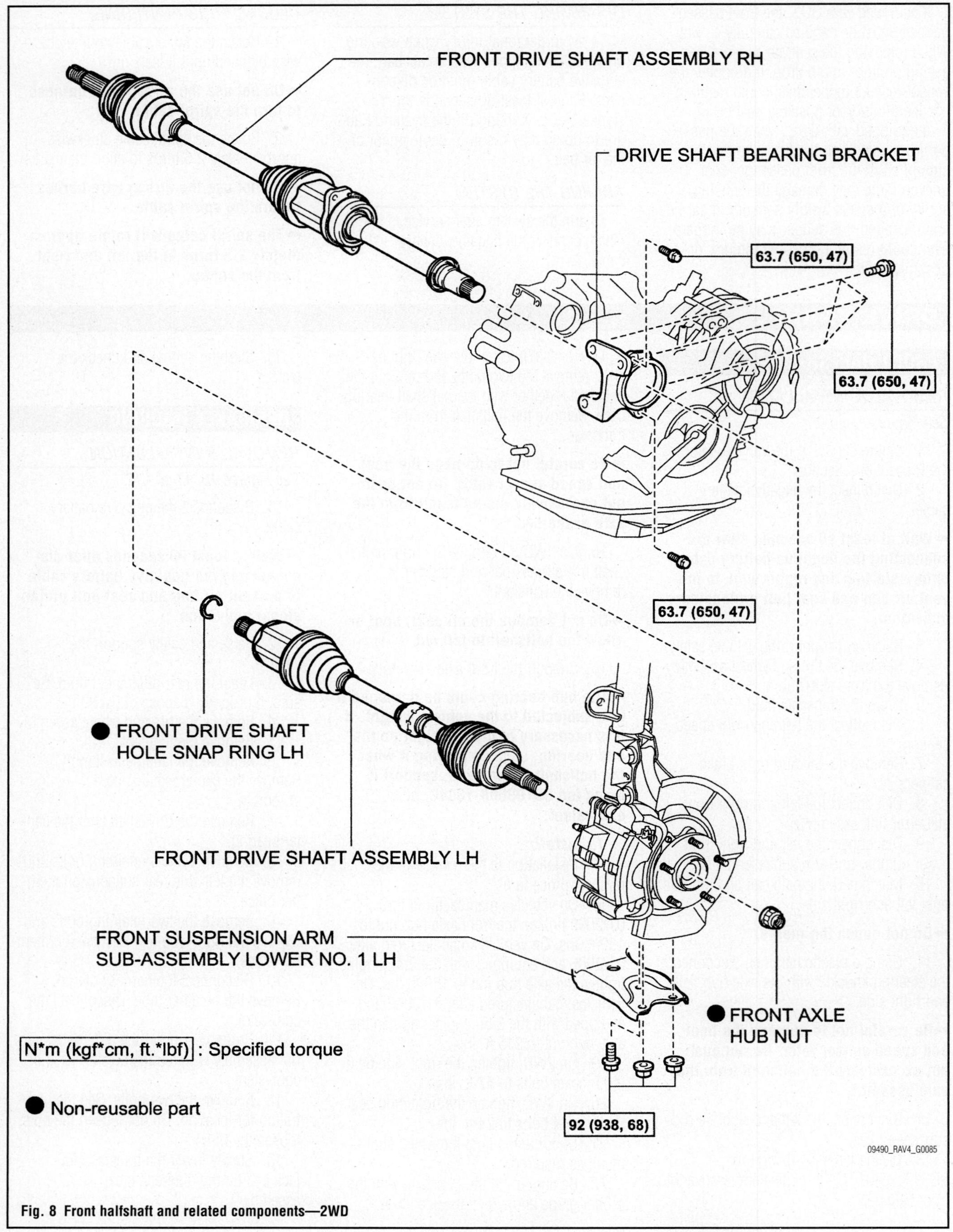

Fig. 8 Front halfshaft and related components—2WD

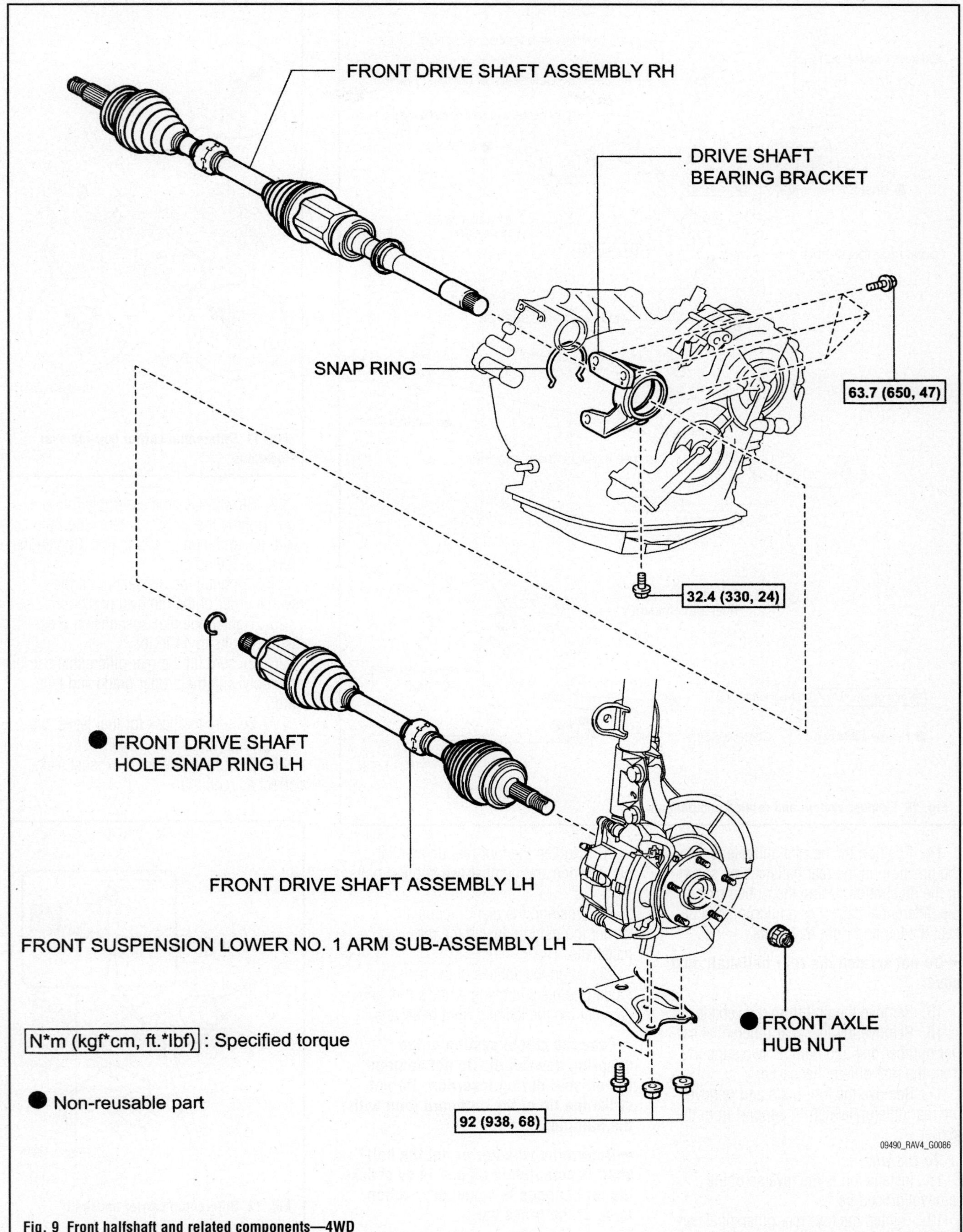

Fig. 9 Front halfshaft and related components—4WD

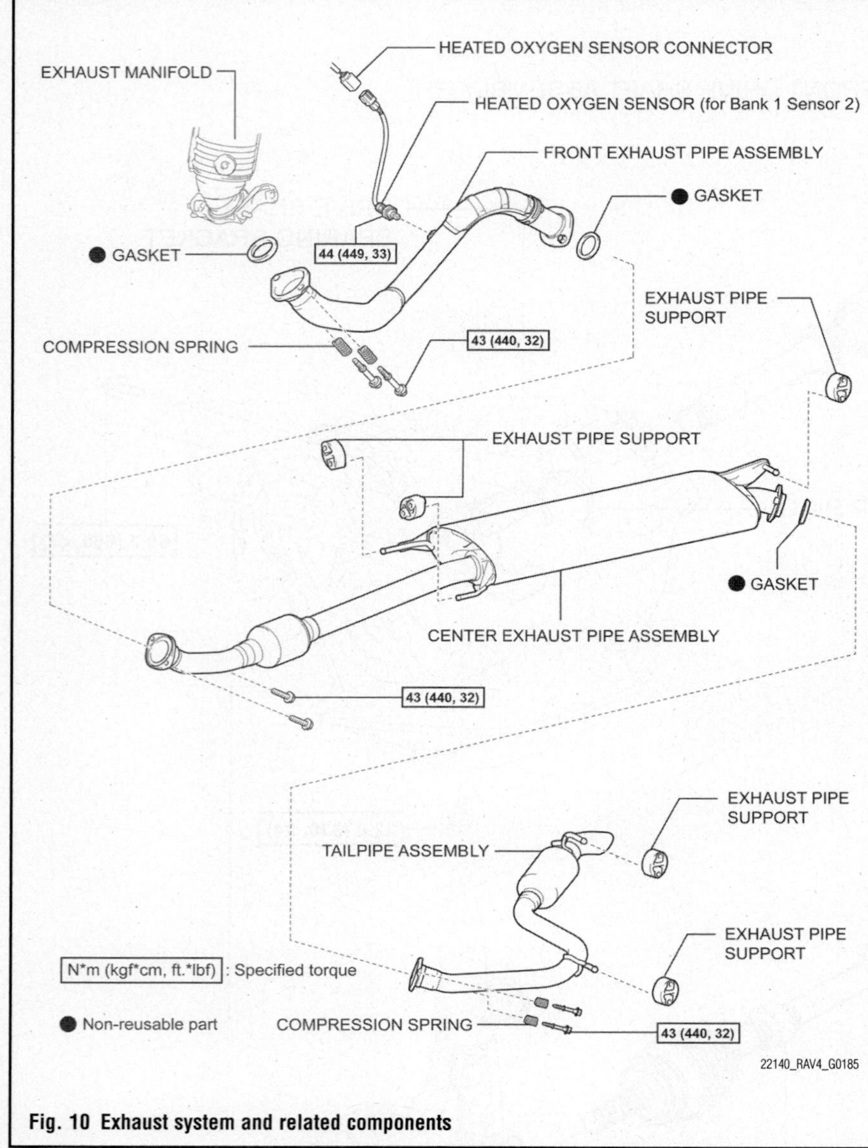

Fig. 10 Exhaust system and related components

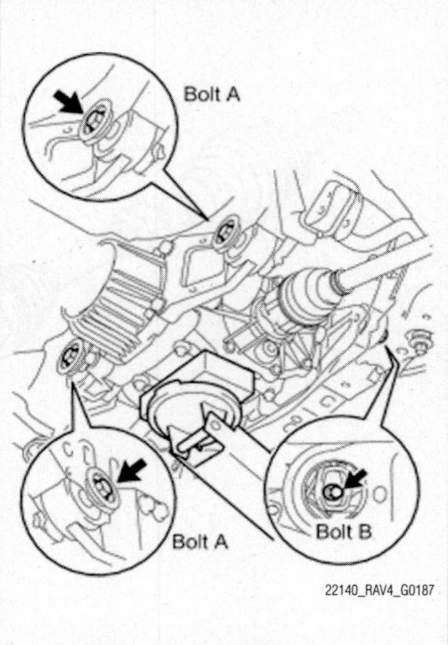

Fig. 11 Differential carrier bolt removal locations

14. Position the tip of a suitable tool to the position on the rear halfshaft, as shown in the illustration. Using the ribbed part of the differential carrier as a fulcrum, disconnect the left and right halfshafts.

➡Do not scratch the rear halfshaft dust cover.

15. Remove the unit from the vehicle.
16. Remove the four rear differential carrier number one and number two supports from the rear differential carrier.
17. Remove the four bolts and remove the rear differential carrier support from the carrier.

To install:

18. Installation is the reverse of the removal procedure.
19. Tighten the four rear differential carrier support bolts to 72 ft. lbs.
20. Tighten the four rear differential number one and number two support bolts 41 ft. lbs.
21. Position the rear differential carrier in position. Install the rear halfshafts.
22. Align the splines of the halfshaft inboard joints and using a brass drift and hammer tap the left and right halfshafts.

➡Face the cutout section of the snapring downward. Do not damage the oil seal during insertion. Do not strike the tip of the outboard joint with the hammer.

➡Determine whether or not the halfshaft is completely tapped in by checking for changes in sound or reaction force of the brass bar.

23. Slowly raise the assembly into position. Tighten the assembly retaining bolts to 63 ft. lbs and then to 103 ft. lbs. Tighten the bolts, not the nuts.
24. Continue the installation in the reverse order of the removal procedure.
25. Tighten the rear suspension crossmember bolts to 44 ft. lbs.
26. Be sure fill the rear differential carrier assembly with the proper grade and type fluid.
27. Be sure to check for fuel leaks, correct as required.
28. Be sure to check for exhaust leaks, correct as required.

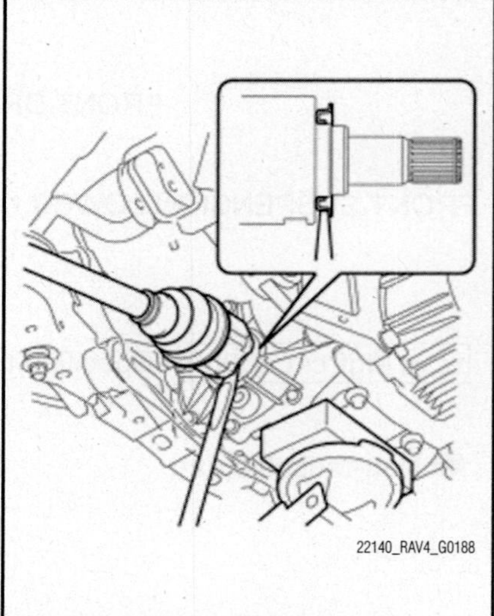

Fig. 12 Differential carrier halfshaft removal point location

REAR HALFSHAFT

REMOVAL & INSTALLATION

See Figures 13 and 14.

1. Before servicing the vehicle, refer to the Precautions Section.
2. Disconnect the negative battery cable.

➡**Wait at least 90 seconds after disconnecting the negative battery cable before starting any repair work to prevent air bag and seat belt pretensioner activation.**

3. Raise and support the vehicle safely.
4. Remove the tire and wheel assembly.
5. Drain the differential oil.
6. Remove the tailpipe assembly.
7. Remove the driveshaft with the center bearing.
8. Remove the rear axle shaft nut.

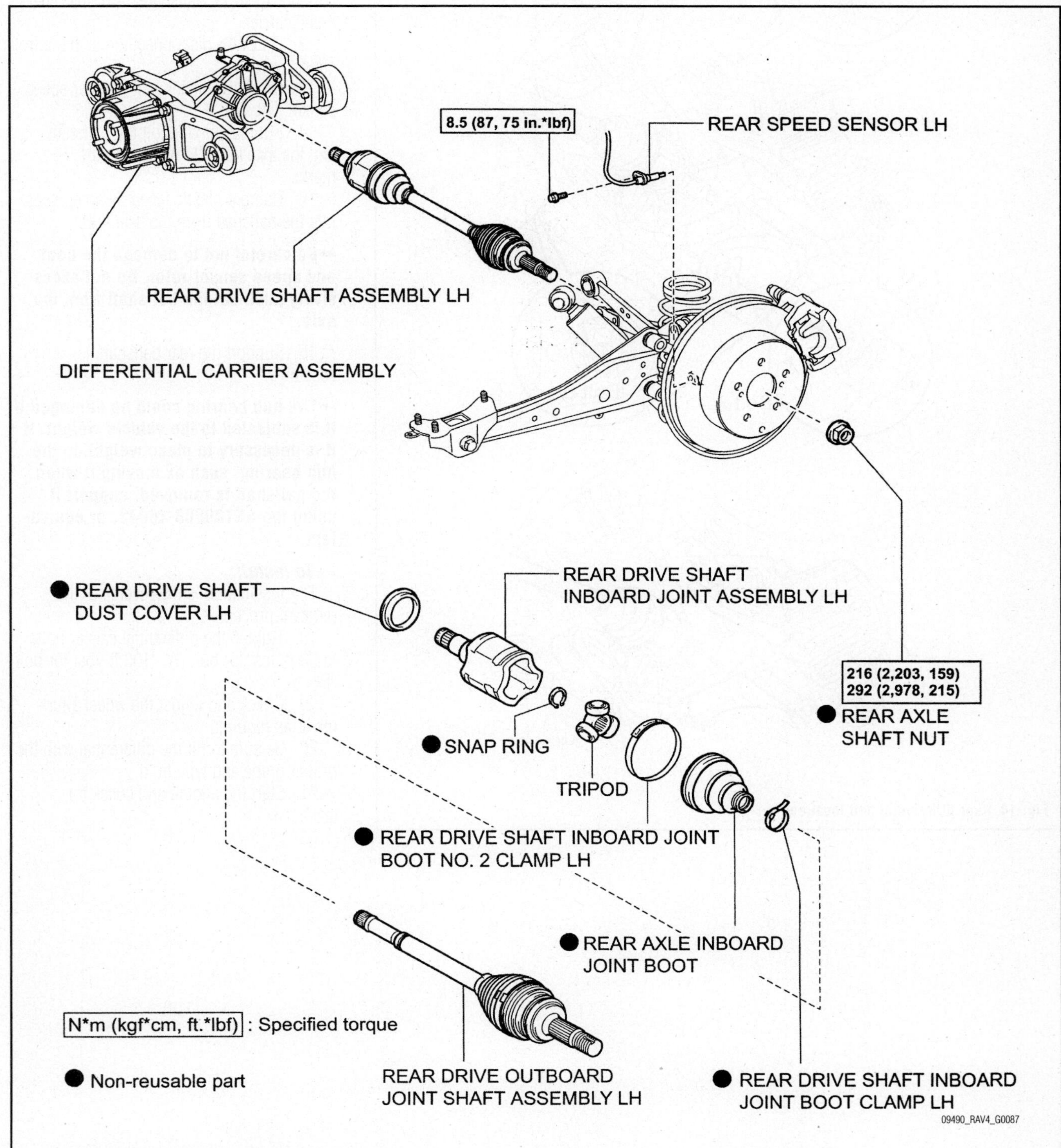

Fig. 13 Rear halfshaft and related components

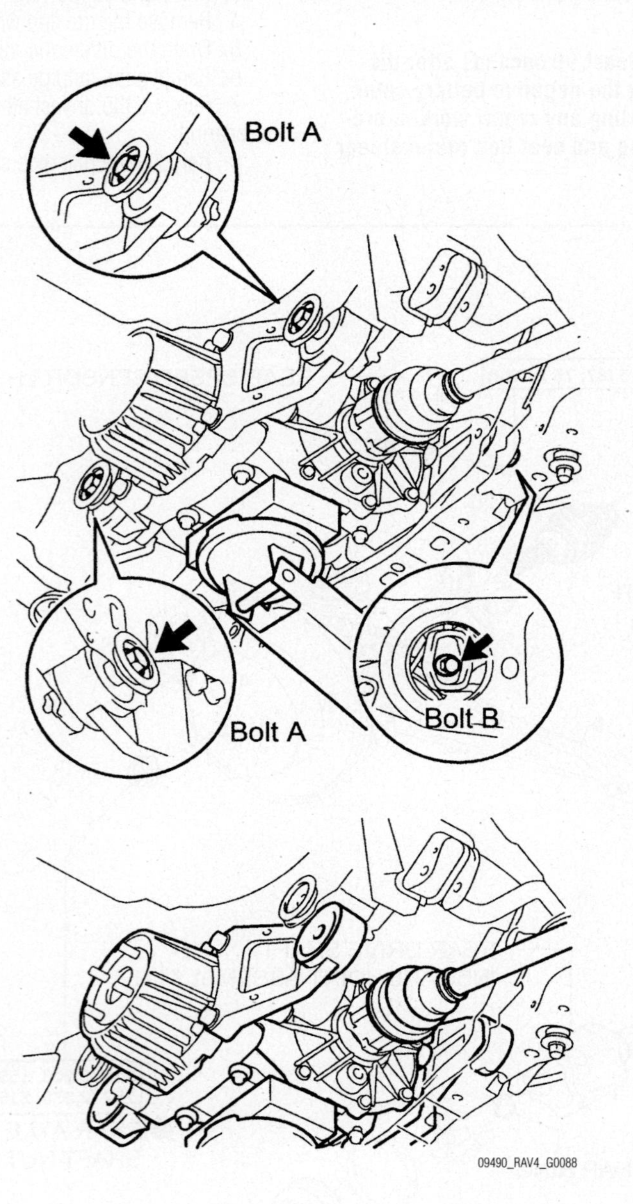

Fig. 14 Rear differential bolt locations

9. Support the rear differential using a suitable jack.

10. Fix the nuts in place and remove bolt "A" "B" and "C". Do not loosen the nuts, loosen the bolts. Slowly lower the jack and tilt the rear differential carrier, as shown in the illustration.

11. Using a suitable tool disconnect the left and right rear halfshafts from the differential carrier.

12. On 2WD, disconnect the skid control sensor wire.

13. On 4WD, disconnect the rear speed sensor, left side.

14. Put matchmarks on the halfshaft and the axle hub. Do not punch the marks.

15. Using a plastic faced hammer, separate the halfshaft from the axle hub.

➡Be careful not to damage the boot and speed sensor rotor. Do not excessively push out the halfshaft from the axle.

16. Support the rear halfshaft assembly.

➡The hub bearing could be damaged if it is subjected to the vehicle weight. If it is necessary to place weight on the hub bearing, such as moving it when the halfshaft is removed, support it using too SST09608-16042, or equivalent.

To install:

17. Installation is the reverse of the removal procedure.

18. Tighten the differential carrier bolts to 63 ft. lbs. for bolt "A", 103 ft. lbs. for bolt "B".

19. Check and adjust the wheel alignment, as required.

20. Be sure to fill the differential with the proper grade and type fluid.

21. Start the engine and check for leaks.

ENGINE COOLING

ENGINE FAN

REMOVAL & INSTALLATION

See Figure 15.

1. Disconnect the negative battery cable.

➡Wait at least 90 seconds after disconnecting the negative battery cable to prevent air bag and seat belt pretensioner activation.

2. Drain the engine coolant.
3. Remove the radiator.
4. Remove the fan retaining nuts.
5. Remove the screws from the fan motor.
6. Remove the fan motor.

To install:

7. Installation is the reverse of the removal procedure.
8. Fill the engine with the proper grade and type coolant.
9. Start the engine and check for leaks, correct as required.

RADIATOR

REMOVAL & INSTALLATION

See Figures 16 through 18.

1. Disconnect the negative battery cable.

➡Wait at least 90 seconds after disconnecting the negative battery cable to prevent air bag and seat belt pretensioner activation.

2. Drain the engine coolant.
3. Remove the radiator support opening cover.
4. Remove the battery clamp. Remove the battery.
5. Remove the battery insulator, if equipped.
6. On 3.5L engine, remove the V-bank cover sub-assembly. Remove the radiator support opening cover.
7. Remove the number one engine cover. Remove the radiator grille sub-assembly (front grille).
8. Disconnect the hood lock switch connector.
9. Disconnect the number one water bypass hose. Disconnect the number five water bypass hose. Remove the number one water bypass pipe.
10. Disconnect the cooling fan electrical connectors from the cooling fans.
11. Detach the five harness clamps from the fan shroud and upper radiator support.
12. Disconnect the horn connector.
13. Remove the upper radiator support bracket retaining bolts. Remove the support bracket.
14. Remove the number two fan shroud retaining bolts. Remove the number two fan shroud.
15. Remove the four bolts and the upper radiator support with the hood lock.
16. Disconnect the radiator hoses.

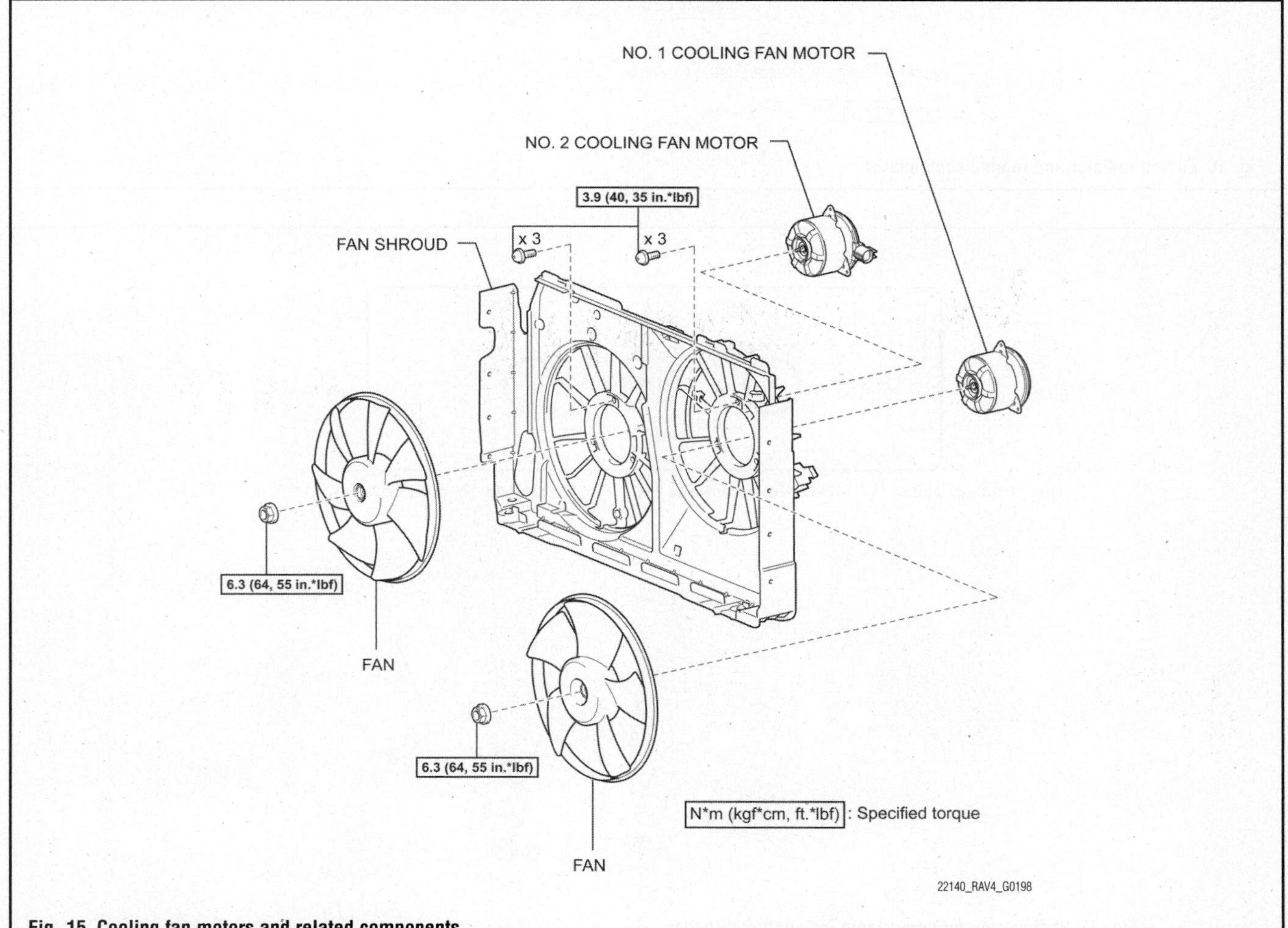

Fig. 15 Cooling fan motors and related components

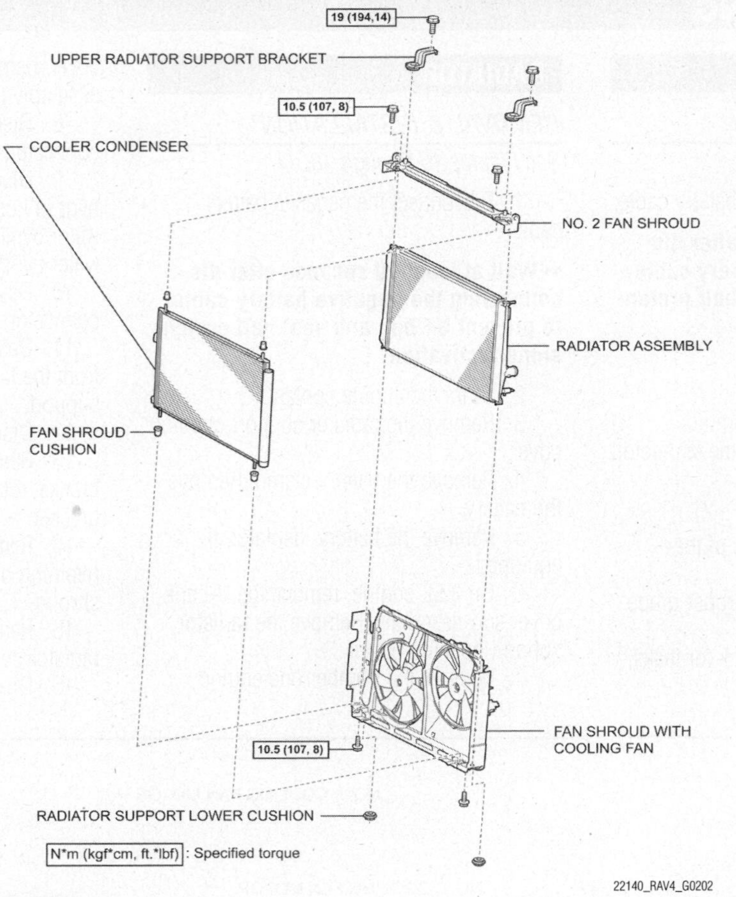

19 (194,14)
UPPER RADIATOR SUPPORT BRACKET
10.5 (107, 8)
COOLER CONDENSER
NO. 2 FAN SHROUD
RADIATOR ASSEMBLY
FAN SHROUD CUSHION
FAN SHROUD WITH COOLING FAN
10.5 (107, 8)
RADIATOR SUPPORT LOWER CUSHION
N*m (kgf*cm, ft.*lbf) : Specified torque
22140_RAV4_G0202

Fig. 16 Engine radiator and related components

Upper
Upper
LH Side
LH Side
A
B
No. 1 Radiator Hose
No. 2 Radiator Hose
No. 2 Water By-pass Hose
A
A
A
A
B
A
A
22140_RAV4_G0204

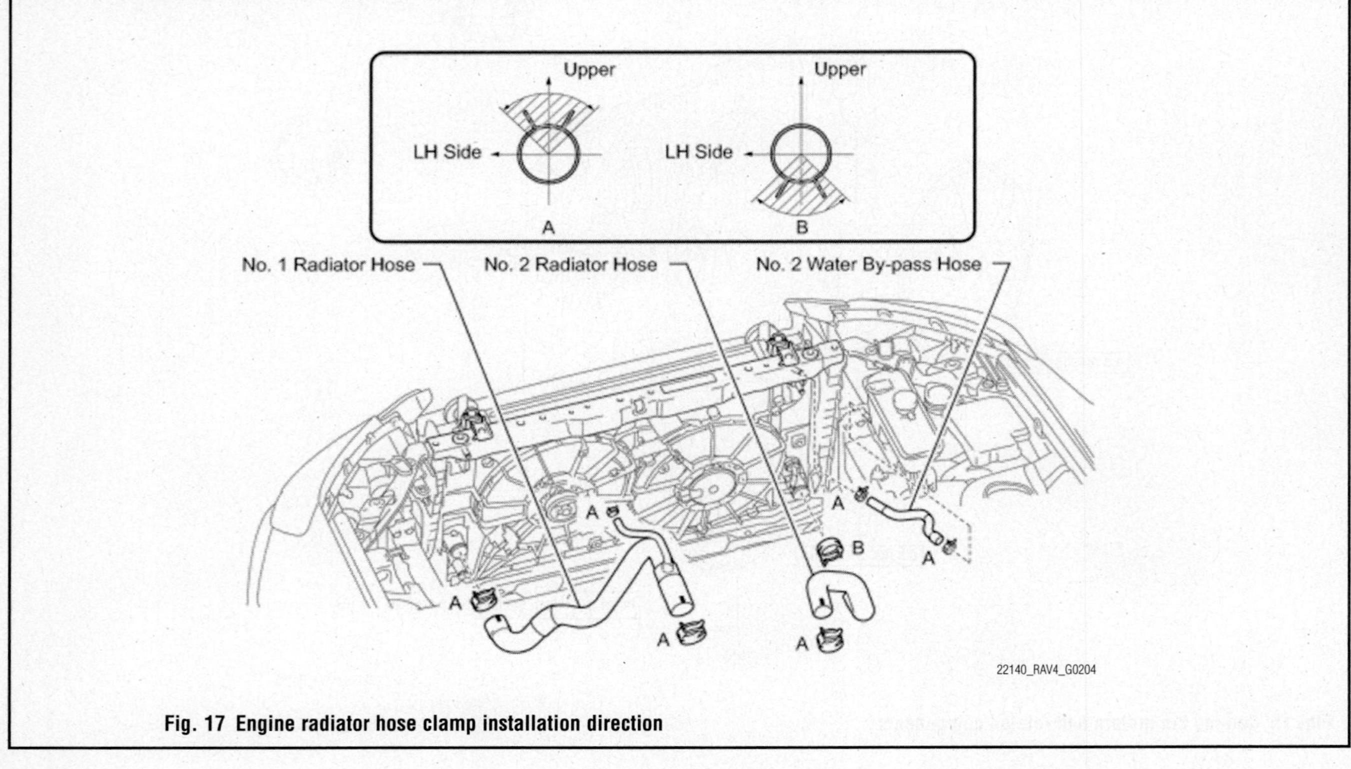

Fig. 17 Engine radiator hose clamp installation direction

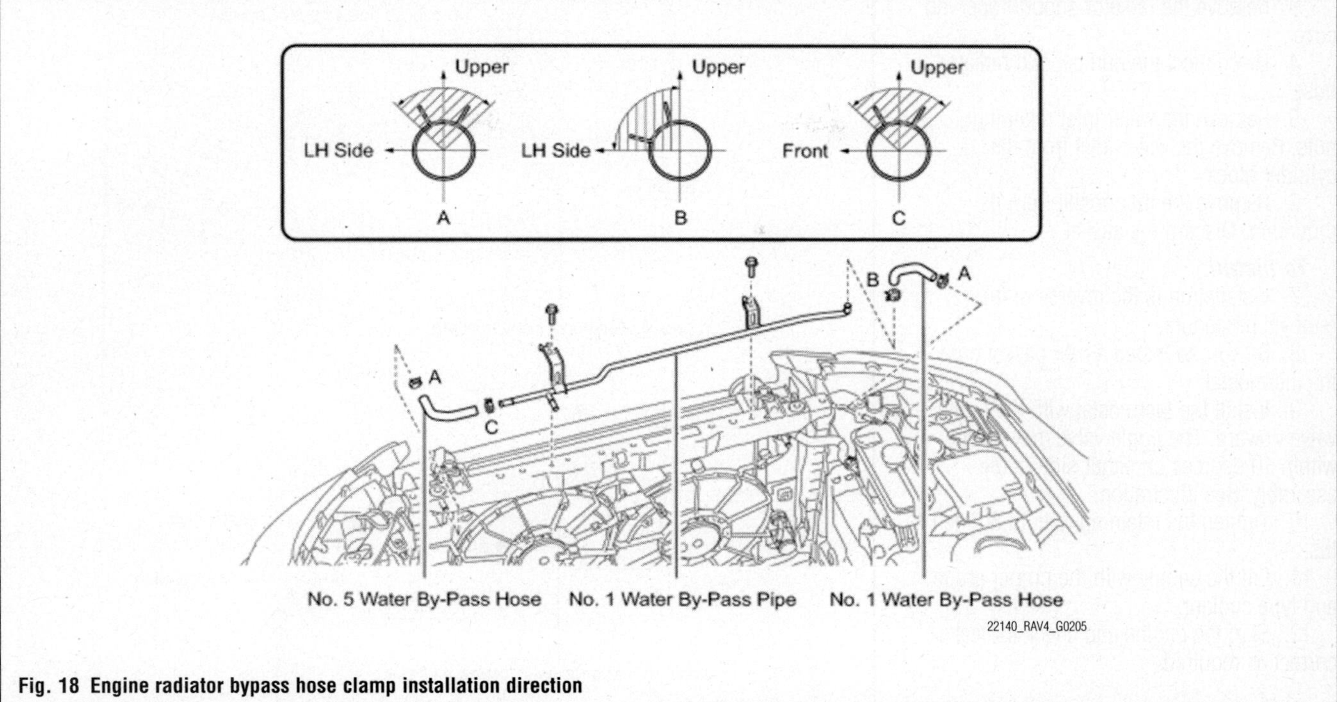

Fig. 18 Engine radiator bypass hose clamp installation direction

Disconnect the number two water bypass hose.

17. Remove the radiator from the vehicle.

18. As required, separate the cooler condenser from the radiator.

19. As required, separate the fan shroud and cooling fans from the radiator assembly.

To install:

20. Installation is the reverse of the removal procedure.

21. Be sure to install hoses clamps as indicated in the illustration.

22. Be sure to install bypass hose clamps as indicated in the illustration.

23. Fill the engine with the proper grade and type coolant.

24. Start the engine and check for leaks, correct as required.

THERMOSTAT

REMOVAL & INSTALLATION

2.4L Engine

See Figure 19.

1. Disconnect the negative battery cable.

➡Wait at least 90 seconds after disconnecting the negative battery cable to prevent air bag and seat belt pretensioner activation.

2. Drain the engine coolant.

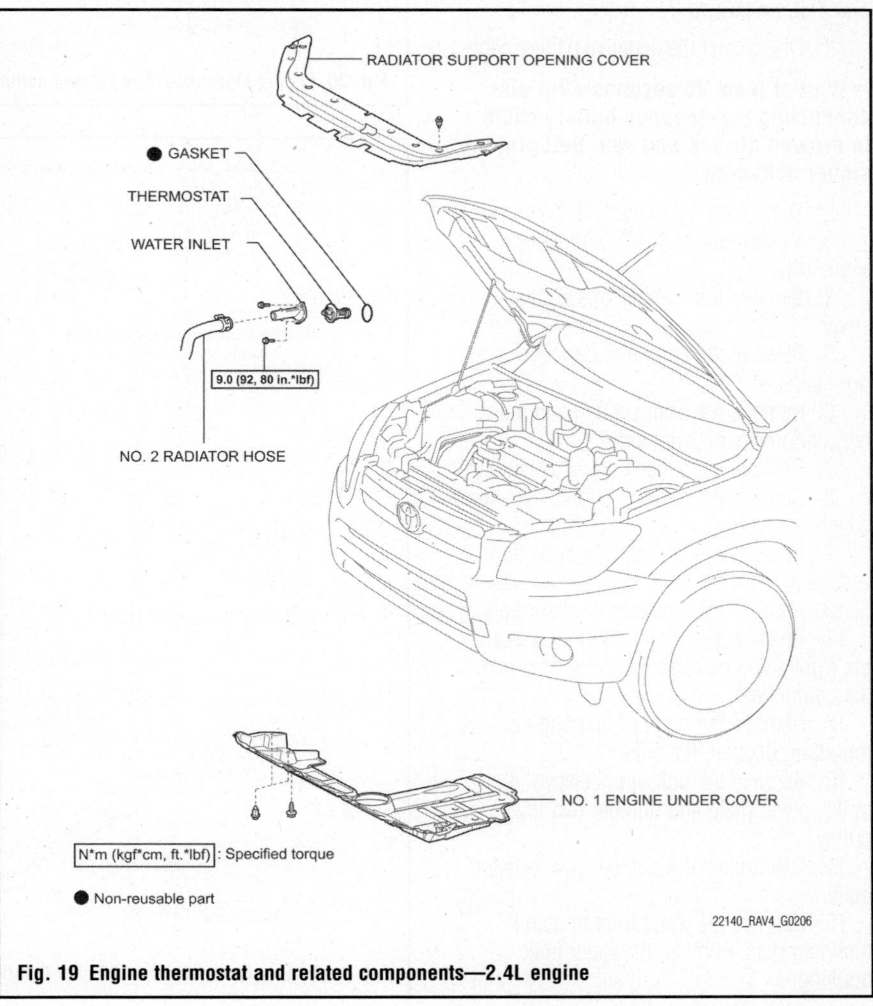

Fig. 19 Engine thermostat and related components—2.4L engine

3. Remove the radiator support opening cover.
4. Disconnect the number two radiator hose.
5. Remove the water inlet retaining nuts. Remove the water inlet from the cylinder block.
6. Remove the thermostat from its mounting. Discard the gasket.

To install:

7. Installation is the reverse of the removal procedure.
8. Be sure to install a new gasket onto the thermostat.
9. Install the thermostat with the jiggle valve upward. The jiggle valve may be set within 10 degrees on either side of the assembly. See illustration.
10. Tighten the retaining nuts to 80 inch lbs.
11. Fill the engine with the proper grade and type coolant.
12. Start the engine and check for leaks, correct as required.

3.5L Engine

See Figures 20 and 21

1. Disconnect the negative battery cable.

➡Wait at least 90 seconds after disconnecting the negative battery cable to prevent air bag and seat belt pretensioner activation.

2. Drain the engine coolant.
3. Remove the right tire and wheel assembly.
4. Remove the number one engine cover.
5. Remove the engine under cover rear, right side.
6. Remove the front suspension member reinforcement, right side.
7. Remove the V-bank cover assembly.
8. Remove the radiator support opening cover.
9. Remove the radiator reservoir tank assembly.
10. Remove the fan and alternator belt.
11. Remove the engine mounting insulator, right side five retaining bolts. Remove the component.
12. Remove the number one engine mounting bracket, left side.
13. Remove the bolt, number two idler pulley cover plate and number two idler pulley.
14. Disconnect the number two radiator hose.
15. Remove the water inlet housing retaining nuts. Remove the water inlet housing.

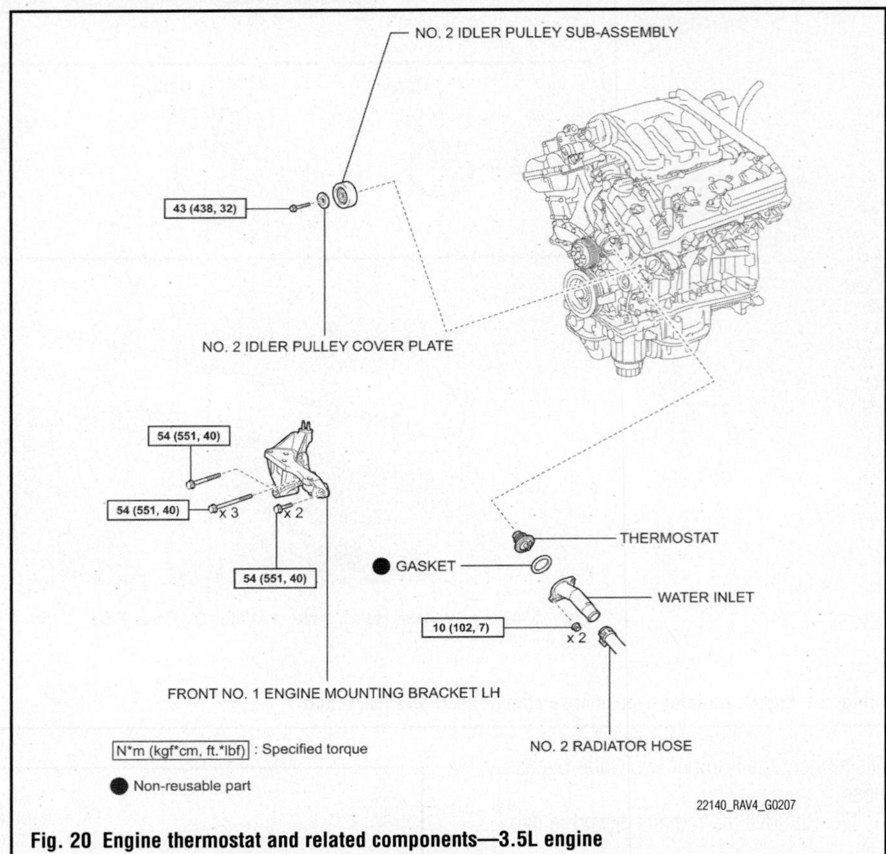

Fig. 20 Engine thermostat and related components—3.5L engine

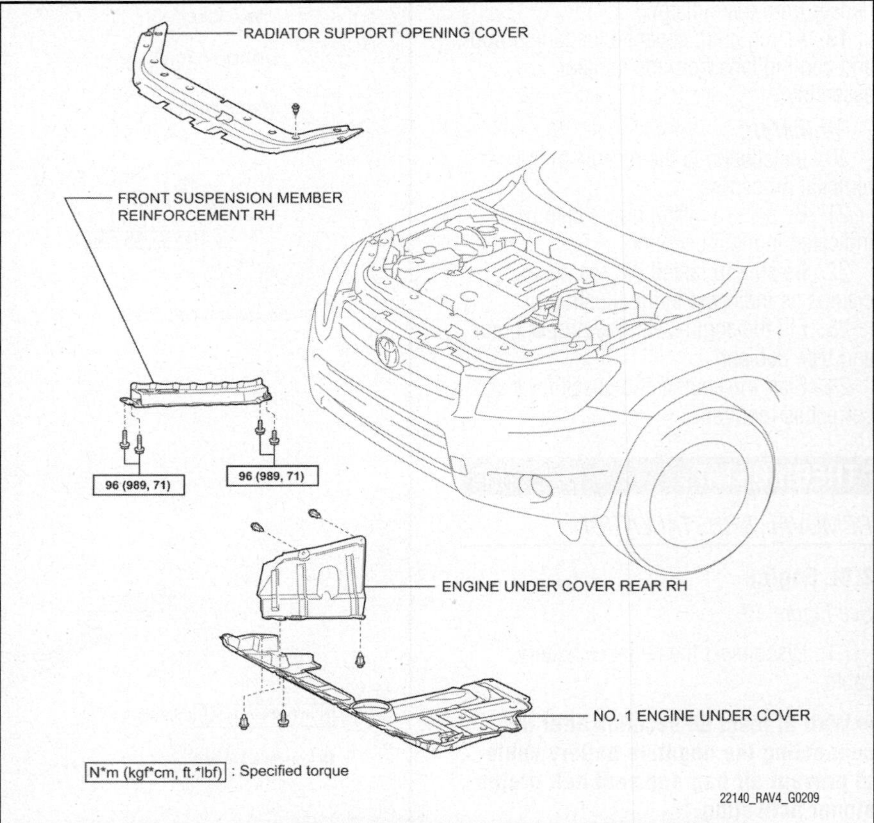

Fig. 21 Engine under cover, right side location and related components—3.5L engine

16. Remove the thermostat from its mounting. Discard the gasket.

To install:

17. Installation is the reverse of the removal procedure.

18. Be sure to install a new gasket onto the thermostat.

19. Install the thermostat with the jiggle valve upward. The jiggle valve may be set within 10 degrees on either side of the assembly.

20. Tighten the retaining nuts to 7 ft. lbs.

21. Tighten the number two idler pulley bolt to 32 ft. lbs.

22. Fill the engine with the proper grade and type coolant.

23. Start the engine and check for leaks, correct as required.

WATER PUMP

REMOVAL & INSTALLATION

2.4L Engine

See Figure 22.

1. Before servicing the vehicle, refer to the Precautions Section.

2. Disconnect the negative battery cable.

➡Wait at least 90 seconds after disconnecting the negative battery cable before starting any repair work to prevent air bag and seat belt pretensioner activation.

3. Remove the number one engine undercover.

4. Remove the front fender apron, right side.

5. Drain the cooling system. Remove the radiator support opening cover.

6. Remove the front suspension member reinforcement, right side.

7. Remove the fan and alternator drive belt. Remove the alternator.

8. Using tool SST09960-10010 remove the four retaining bolts and the water pump pulley.

9. Remove the clamp of the Crankshaft Position (CKP) sensor from the water pump.

10. Disconnect the wire of the sensor from the clamp bracket.

11. Remove the four water pump retaining bolts, two nuts and clamp bracket. Remove the water pump from the engine.

To install:

12. Apply a 2.5mm wide bead of RTV gasket material to the pump sealing surface as shown.

➡Install the pump with 5 minutes of applying the sealer or the sealer will have to be removed and new sealer applied.

13. Install the pump and torque the nuts and bolts to 80 inch lbs. (9 Nm).

14. The remainder of installation is the reverse of removal. Refill the cooling system.

15. Start the vehicle and check for leaks, correct ass required.

3.5L Engine

See Figures 23 and 24.

➡In order to replace the water pump, the manufacturer recommends removing the engine from the vehicle.

1. Before servicing the vehicle, refer to the Precautions Section.

2. Disconnect the negative battery cable.

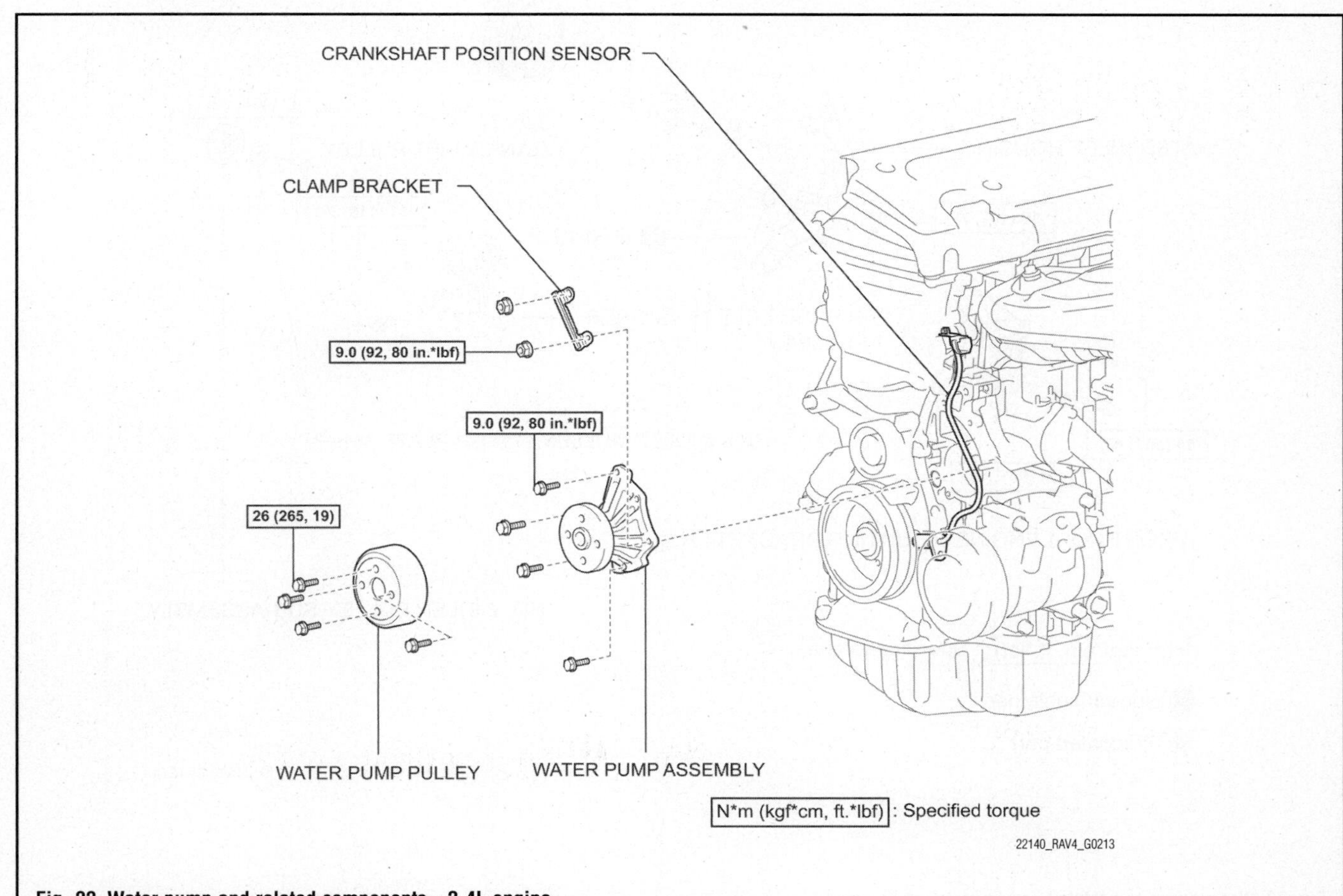

Fig. 22 Water pump and related components—2.4L engine

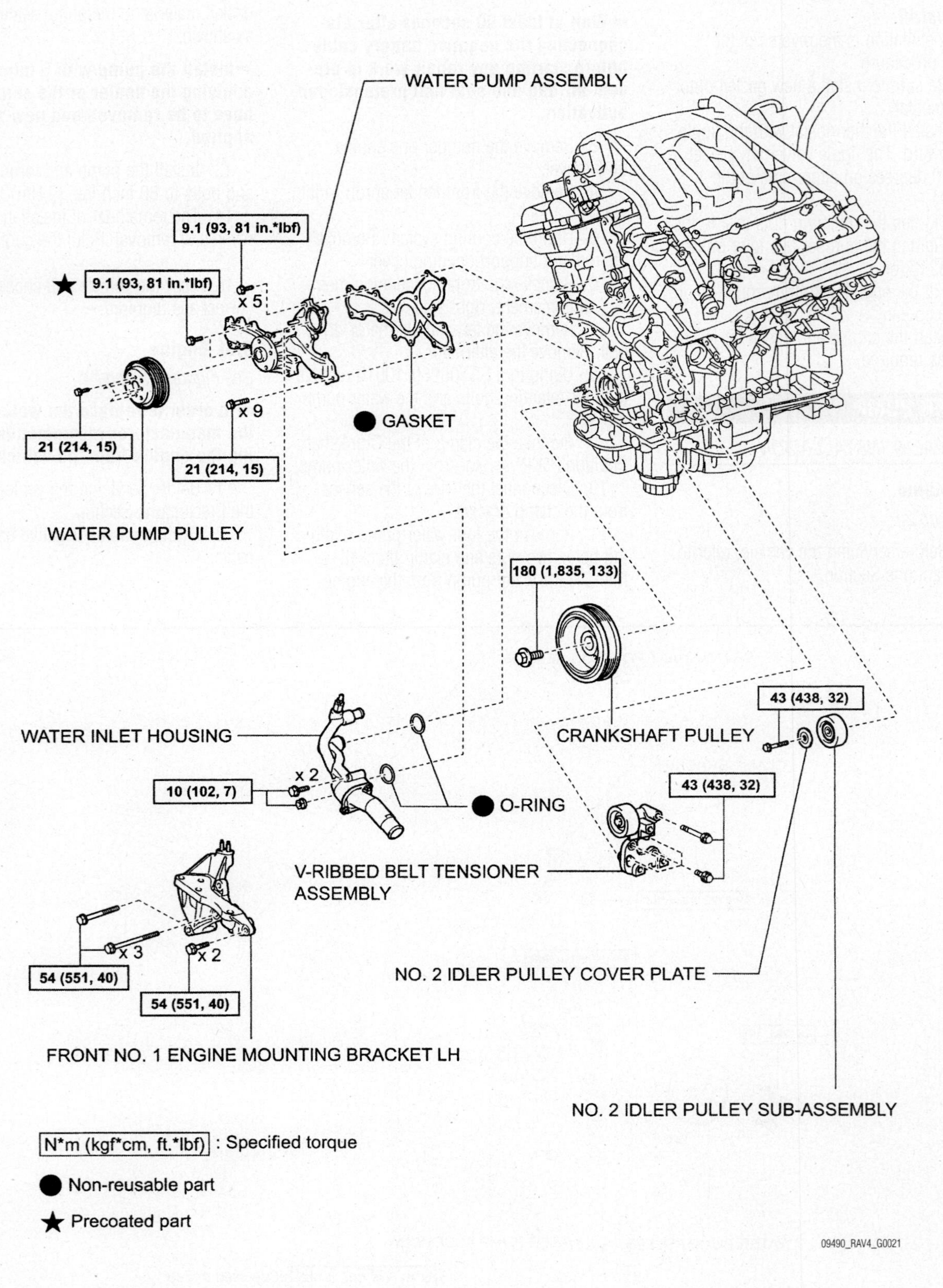

Fig. 23 Water pump and related components—3.5L engine

Fig. 24 Water pump bolt tightening sequence and bolt markings—3.5L engine

➡Wait at least 90 seconds after disconnecting the negative battery cable before starting any repair work to prevent air bag and seat belt pretensioner activation.

3. Remove the engine from the vehicle and position it in a suitable holding fixture.
4. Remove the number one engine mounting bracket, left side.
5. Remove the water inlet housing. Remove the crankshaft pulley.
6. Remove the water pump pulley. Remove the number two idler pulley sub-assembly. Remove the belt tensioner assembly.
7. Remove the 16 water pump retaining bolts. Remove the water pump from its mounting.

To install:

8. Install the water pump to the engine using a new gasket.
9. Tighten the retaining bolts to 15 ft. lbs. for bolts marked "A" and 81 inch lbs for bolts marked "B" and "C" and in the proper sequence.
10. Temporarily install the V-ribbed belt tensioner with the five bolts. Bolt "A" 2.76 inch and bolts "B" 1.30 inch. See illustration for location. Tighten bolts 1 and 2, in the order shown, to 32 ft. lbs. Tighten all other bolts to 32 ft. lbs.
11. Install the number two idler pulley sub-assembly. Tighten the bolt to 32 ft. lbs.
12. Continue the installation in the reverse order of the removal procedure.

ENGINE ELECTRICAL

CHARGING SYSTEM

ALTERNATOR

REMOVAL & INSTALLATION

2.4L Engine

See Figure 25.

1. Disconnect the negative battery cable.
2. Remove the front wheel RH.
3. Remove the engine front cover RH.
4. Remove the front fender apron seal RH.
5. Remove the V-bank cover sub-assembly.
6. Remove the V-ribbed belt.
7. Remove the alternator assembly, as follows:
   a. Disconnect the alternator connector.
   b. Remove the nut and disconnect the wire harness from terminal B.
   c. Remove the bolt and wire harness clamp bracket.
   d. Remove the wire harness clamps.
   e. Remove the 2 bolts and alternator assembly.

To install:

8. Install the alternator assembly, as follows:
   a. Install the bracket with the bolt and tighten to 15 ft. lbs. (20 Nm).
   b. Install the wire harness clamp stay and tighten to 74 inch lbs. (8.4 Nm).
   c. Install the alternator assembly with the 2 bolts and tighten to 32 ft. lbs. (43 Nm).
   d. Install the nut to the cylinder block and tighten to 15 ft. lbs. (20 Nm).

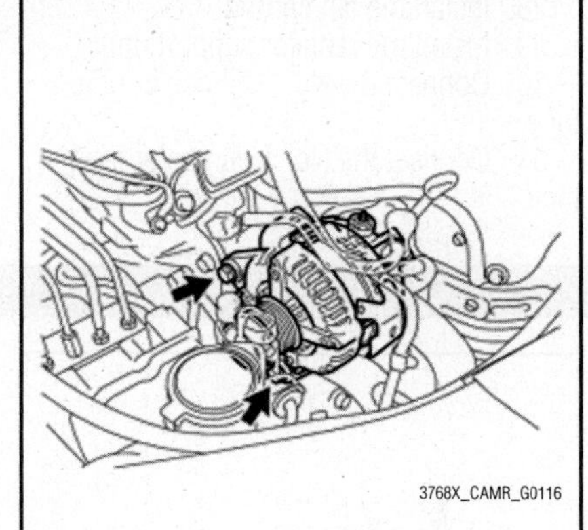

Fig. 25 Removing the alternator

   e. Connect the alternator connector to the alternator assembly.
   f. Install the alternator wire with the nut and tighten to 87 inch lbs. (9.8 Nm).
   g. Install the terminal cap.
   h. Connect the wire harness clamp.
9. Install the V-ribbed belt.
10. Install the V-bank cover sub-assembly.
11. Connect the negative battery cable.
12. Perform initialization.

3.5L Engine

See Figure 26.

1. Disconnect the cable from the negative battery terminal.
2. Remove the front wheel.
3. Remove the front fender apron seal RH.
4. Remove the front wheel opening extension pad RH.
5. Remove the front wheel opening extension pad LH.
6. Remove the engine under cover RH.
7. Remove the engine under cover LH.
8. Drain the engine coolant.
9. Remove the v-bank cover sub assembly.
10. Remove the cool air intake duct seal.
11. Remove the air cleaner inlet assembly.
12. Remove the No. 1 air cleaner inlet.
13. Remove the front bumper assembly.
14. Remove the front bumper energy absorber.
15. Separate the radiator reserve tank hose.
16. Separate the radiator inlet hose.
17. Separate the radiator outlet hose.
18. Separate the No. 1 oil cooler inlet hose.
19. Separate the No. 1 oil cooler outlet hose.
20. Remove the radiator support upper.
21. Remove the fan shroud.
22. Remove the radiator assembly.
23. Remove the v-ribbed belt.
24. Remove the alternator assembly.
   a. Remove the terminal cap.
   b. Remove the nut and disconnect the wire harness from terminal B.
   c. Disconnect the alternator connector from the alternator assembly.
   d. Disconnect the connector from the compressor and magnetic clutch.
   e. Disconnect the 2 wire harness clamps.
   f. Remove the 2 bolts.
   g. Remove the bolt from the cylinder block.

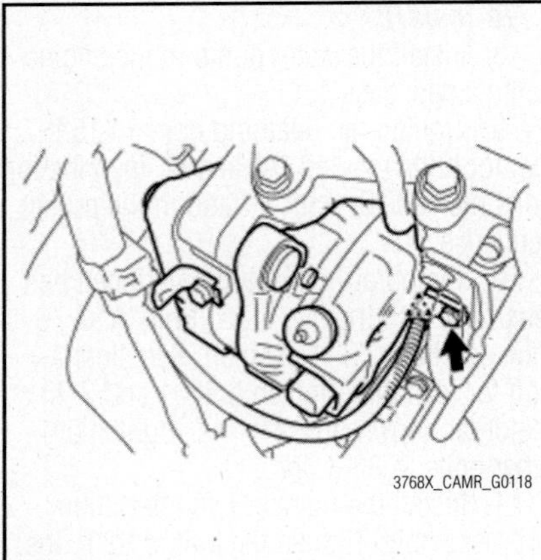

Fig. 26 Removing the bolt from the cylinder block

h. Disconnect the wire harness clamp and remove the alternator assembly.

i. Remove the bolt and wire harness clamp stay.

j. Remove the bolt and bracket.

To install:

25. Install the alternator assembly.

a. Install the bracket with the bolt. Tighten to 15 ft. lbs. (20 Nm).

b. Install the wire harness clamp stay and tighten to 74 inch lbs. (8.4 Nm).

c. Connect the wire harness clamp.

d. Install the alternator assembly to the cylinder block with the bolt. Tighten to 15 ft. lbs. (20 Nm).

e. Install the 2 bolts. Tighten to 32 ft. lbs. (43 Nm).

f. Connect the alternator connector to the alternator assembly.

g. Install the alternator wire with the nut and tighten to 87 inch lbs. (9.8 Nm).

h. Install the terminal cap.

i. Connect the 2 wire harness clamps.

j. Connect the magnetic clutch connector to the compressor and magnetic clutch.

26. Install the v-ribbed belt.
27. Install the radiator assembly.
28. Install the fan shroud.
29. Install the radiator support upper.
30. Connect the No. 1 oil cooler outlet tube.
31. Connect the No. 1 oil cooler inlet tube.
32. Connect the radiator reserve tank hose.
33. Install the front bumper energy absorber.
34. Install the front bumper assembly.
35. Install the No. 1 air cleaner inlet.
36. Install the air cleaner cap sub assembly.
37. Install the air cleaner inlet assembly.
38. Connect the cable to the negative battery terminal.
39. Add engine coolant.
40. Check for engine coolant leaks.
41. Install the v-bank cover sub assembly.
42. Install the cool air intake duct seal.
43. Install the front fender apron seal RH.
44. Install the engine under cover RH.
45. Install the engine under cover LH.
46. Install the front wheel opening extension pad RH.
47. Install the front wheel opening extension pad LH.
48. Install the front wheel and tighten to 76 ft. lbs. (103 Nm).

ENGINE ELECTRICAL

IGNITION SYSTEM

FIRING ORDER

Firing order for the 2.4L engine: 1–3–4–2.

Firing order for 3.5L engine: 1–2–3–4–5–6

IGNITION COIL

REMOVAL & INSTALLATION

2.4L Engine

See Figure 27.

1. Before servicing the vehicle, refer to the Precautions Section.
2. Disconnect the negative battery cable.
3. Remove engine cover(s).
4. Disconnect the 4 ignition coil connectors. Remove the 4 bolts and 4 ignition coils.

To install:

5. To install, reverse removal procedure. Tighten ignition coil bolts to 80 inch lbs. (9 Nm).

3.5L Engine

See Figure 28.

1. Before servicing the vehicle, refer to the Precautions Section.

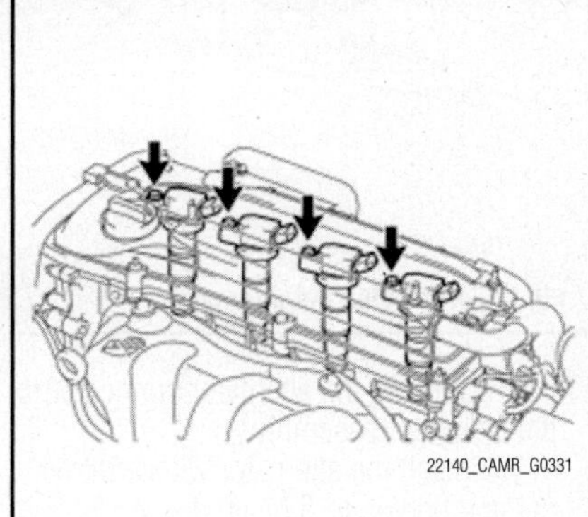

Fig. 27 Removing ignition coils—2.4L engines

2. Disconnect the negative battery cable.
3. Drain and recycle the engine coolant.
4. Remove windshield wiper link assembly.
5. Remove cowl top panel outer sub-assembly.
6. Remove v-bank cover sub-assembly.
7. Remove air cleaner cap sub-assembly.
8. Remove intake air surge tank assembly.
9. Remove No. 1 surge tank stay by performing the following:

a. Remove the bolt and disconnect the harness clamp.

b. Remove the bolt and No. 1 surge tank stay.

10. Disconnect the 6 ignition coil connectors.
11. Remove the 6 bolts and 6 ignition coils.

To install:

12. To install, reverse removal procedure.
13. Tighten the following to specification:

a. 6 ignition coil bolt: 10 ft. lbs. (10 Nm).

b. No. 1 surge tank stay bolt: 15 ft. lbs. (21 Nm).

c. No. 1 surge tank stay bolt and clamp: 62 inch lbs. (7 Nm).

IGNITION TIMING

INSPECTION

3.5L Engine

1. Warm up the engine.
2. Using SST: 09843-18040, connect terminals 13 (TC) and 4 (CG) of the DLC3.

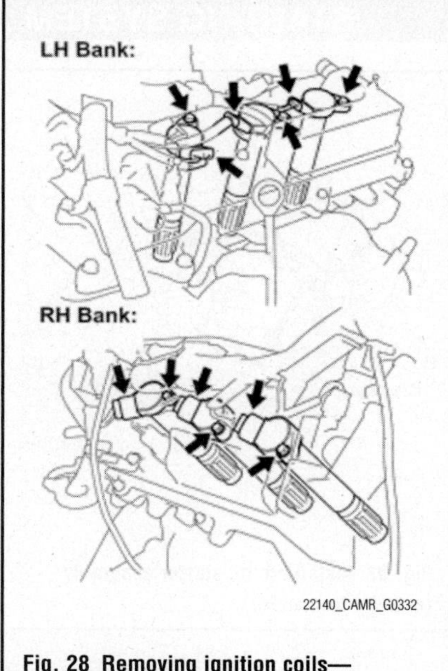

Fig. 28 Removing ignition coils—3.5L engines

WARNING

Confirm the terminal numbers before connecting them. Connecting the wrong terminals can damage the engine.

➡Turn off all electrical systems before connecting the terminals.

➡Perform this inspection after the cooling fan motor is turned off.

3. Remove the v-bank cover.
4. Pull out the red lead wire harness.
5. Connect the tester terminal of the timing light to the red lead wire.

➡Use a timing light which can detect the first signal.

6. Check the ignition timing at idle. Standard ignition timing: 8 to 12°BTDC at idle.

➡When checking the ignition timing, the transmission should be in the neutral position.

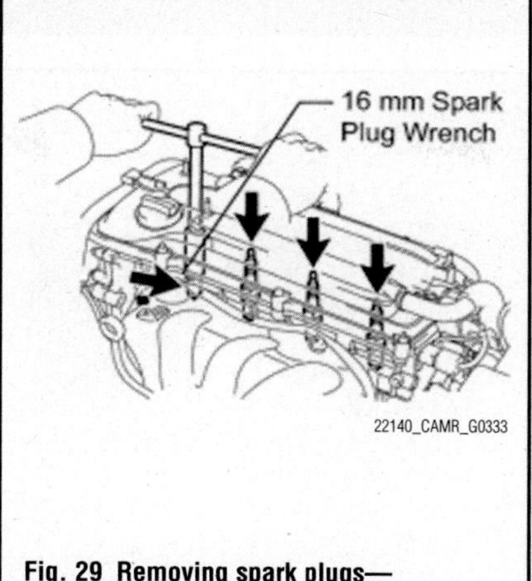

Fig. 29 Removing spark plugs—2.4L engines

➡Run the engine at 1000 to 1300 rpm for 5 seconds, and then check that the engine rpm returns to idle speed.

7. Disconnect terminals 13 (TC) and 4 (CG) of the DLC3.
8. Check the ignition timing at idle. Standard ignition timing: 7 to 24°BTDC at idle.
9. Confirm that the ignition timing advances immediately when the engine rpm is increased.
10. Remove the timing light from the engine.

ADJUSTMENT

All engines are equipped with a Distributorless Ignition System (DIS). No timing adjustment is possible.

SPARK PLUGS

REMOVAL & INSTALLATION

2.4L Engine

See Figure 29.

1. Before servicing the vehicle, refer to the Precautions Section.
2. Remove the plastic engine cover.

Fig. 30 Removing spark plugs—3.5L engines

3. Disconnect the 4 ignition coil connectors and remove the 4 bolts and ignition coils.
4. Using a 16 mm (0.63 in.) spark plug wrench, remove the 4 spark plugs.

To install:

5. Installation is the reverse of removal.
    a. Torque the ignition coils to 66 inch lbs. (7.5 Nm) and the spark plugs to 13 ft. lbs (18 Nm).

3.5L Engine

See Figure 30.

1. Before servicing the vehicle, refer to the Precautions Section.
2. Remove the V-bank cover.
3. Remove the intake air surge tank.
4. Disconnect the 6 ignition coil connectors.
5. Remove the 6 bolts and 6 ignition coils.
6. Using a 16 mm (0.63 in.) plug wrench, remove the spark plugs.

To install:

7. Installation is the reverse of removal, noting the following:
    a. Torque the ignition coils to 66 inch lbs. (7.5 Nm) and the spark plugs to 13 ft. lbs (18 Nm).

STARTER

REMOVAL & INSTALLATION

2.4L Engine

See Figures 31 through 33.

1. Disconnect the cable from the negative battery terminal.
2. Remove the air cleaner inlet assembly.
3. Remove the air cleaner cap sub assembly.
4. Remove the air cleaner case sub assembly.
5. Remove the starter assembly (manual transaxle).
    a. Disconnect the terminal 50 connector from the starter assembly.
    b. Remove the nut and disconnect the wire harness from terminal 30.
    c. Remove the 3 bolts, clutch flexible hose bracket and starter assembly.
6. Remove the starter assembly (automatic transaxle).
    a. Disconnect the terminal 50 connector from the starter assembly.
    b. Remove the nut and disconnect the wire harness from terminal 30.
    c. Remove the 2 bolts and starter assembly.

To install:

7. Install the starter assembly (manual transaxle).
    a. Install the starter assembly and clutch flexible hose bracket with the 3 bolts. Tighten bolt A to 28 ft. lbs. (37 Nm). Tighten bolt B to 9 ft. lbs. (12 Nm).

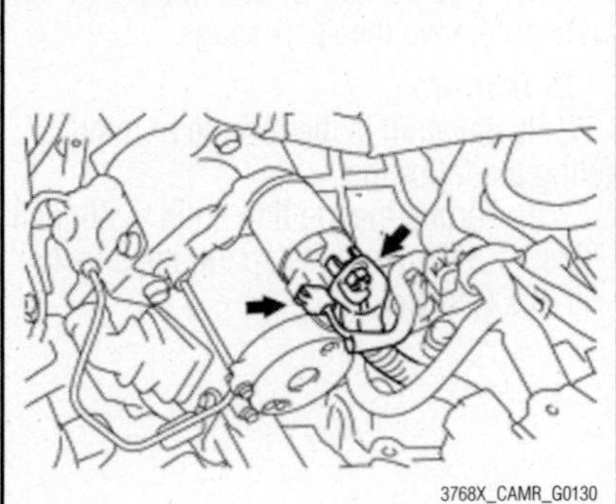

Fig. 31 Disconnecting the terminal 50 connector from the starter assembly (2.4L engine, manual transaxle)

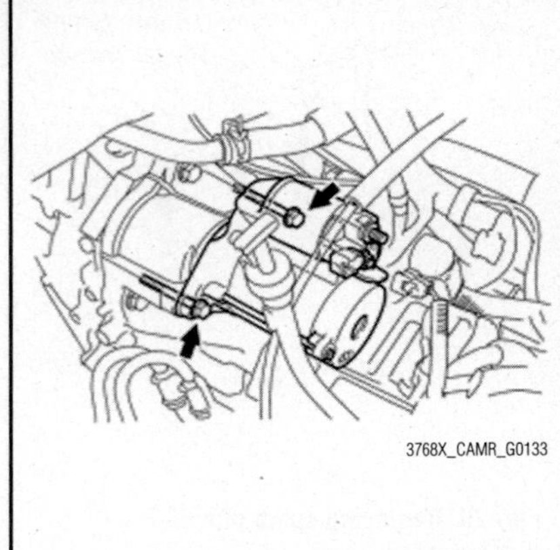

Fig. 32 Removing the starter assembly (2.4L engine, automatic transaxle)

    b. Connect the wire harness to terminal 30 and install the nut. Then, attach the terminal cap. Tighten to 87 inch lbs. (9.8 Nm).
    c. Connect the terminal 50 connector to the starter assembly.
8. Install the starter assembly (automatic transaxle).
    a. Install the starter assembly with the 2 bolts. Tighten to 28 ft. lbs. (37 Nm).
    b. Connect the wire harness to terminal 30 and install the nut. Then, attach the terminal cap. Tighten to 87 inch lbs. (9.8 Nm).
    c. Connect the terminal 50 connector to the starter assembly.
9. Install the air cleaner case sub assembly.
10. Install the air cleaner cap sub assembly.
11. Install the air cleaner inlet assembly.
12. Connect the cable to the negative battery terminal.

3.5L Engine

See Figure 34.

1. Before servicing the vehicle, refer to the Precautions Section.
2. Disconnect the negative battery cable.
3. Remove cool air intake duct seal.
4. Remove v-bank cover sub-assembly.
5. Remove air cleaner inlet assembly.
6. Remove air cleaner cap sub-assembly.
7. Remove air cleaner case sub-assembly.

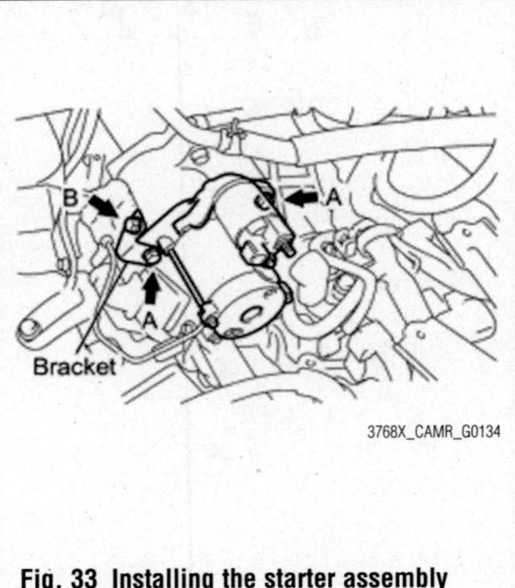

Fig. 33 Installing the starter assembly (manual transaxle)

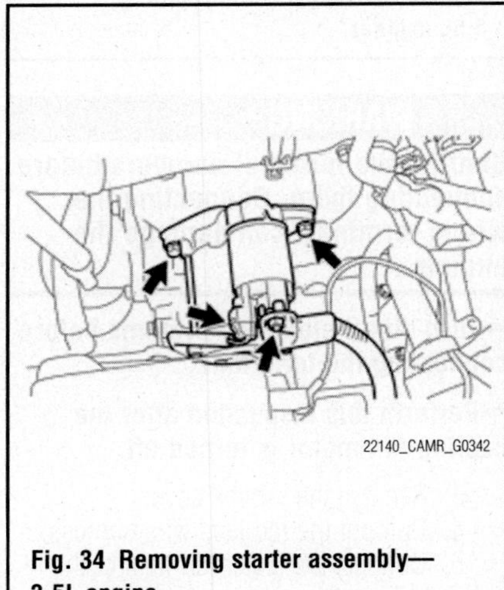

Fig. 34 Removing starter assembly—3.5L engine

8. Remove No. 1 air cleaner inlet.
9. Disconnect the terminal 50 connector from the starter assembly.
10. Remove the nut and disconnect the wire harness from terminal 30.
11. Remove the 2 bolts and starter assembly.

To install:

12. Install the starter assembly with the 2 bolts and tighten to 26 ft. lbs. (37 Nm).
13. Connect the wire harness to terminal 30 and install the nut and tighten to 87 inch lbs. (9.8 Nm).
14. Cover the nut with the cap.
15. Connect terminal 50 to the starter assembly.
16. To complete installation, reverse removal procedure.

ENGINE MECHANICAL

ACCESSORY DRIVE BELTS

ACCESSORY BELT ROUTING

See Figures 35 and 36.

INSPECTION

See Figure 37.

Visually check the V-ribbed belt for excessive wear, frayed cords, etc. If any defect has been found, replace the V-ribbed belt.

- Cracks on the rib side of a belt are considered acceptable. If the belt has chunks missing from the ribs, it should be replaced
- A "new belt" is a belt which has been used for less than 5 minutes with the engine running
- A "used belt" is a belt which has been used for 5 minutes or more with the engine running

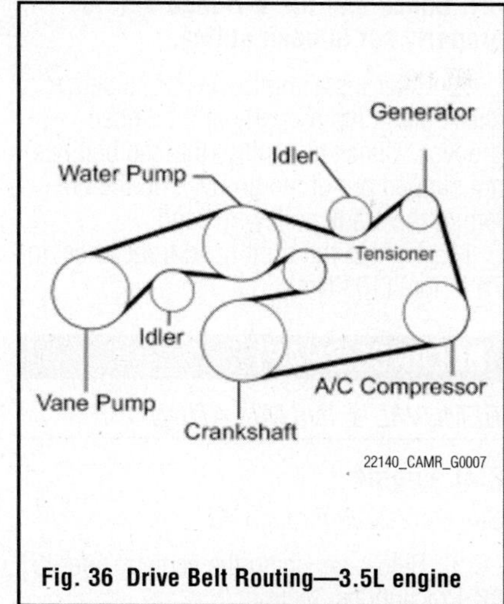
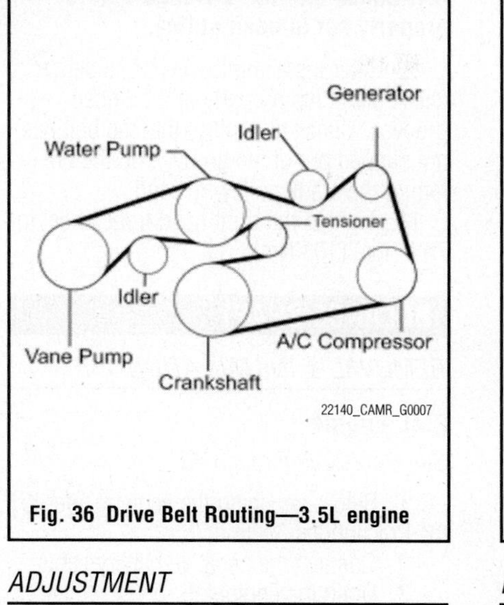

Fig. 36 Drive Belt Routing—3.5L engine

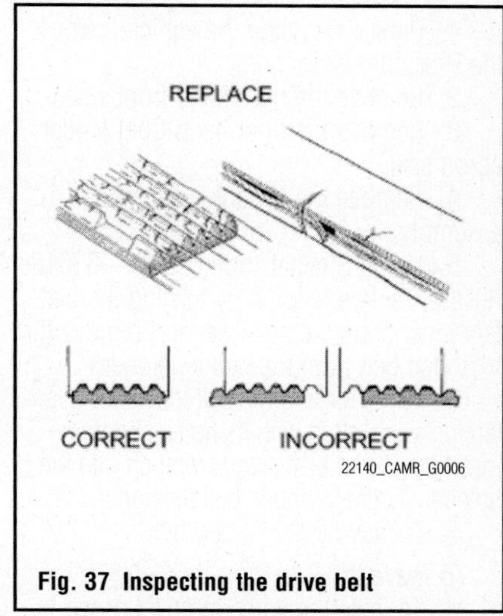

Fig. 37 Inspecting the drive belt

ADJUSTMENT

This vehicle is equipped with an auto-tensioner and cannot be adjusted.

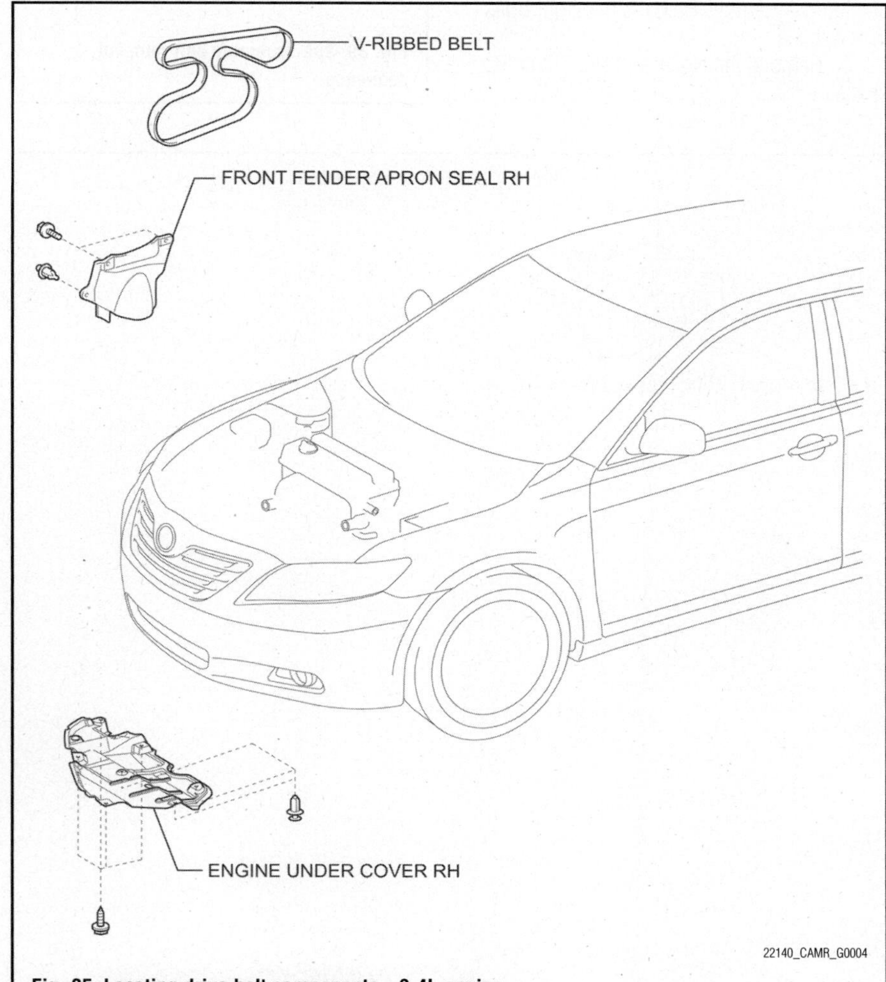

Fig. 35 Locating drive belt components—2.4L engine

REMOVAL & INSTALLATION

2.4L Engine

See Figure 35.

1. Before servicing the vehicle, refer to the Precautions Section.
2. Remove the right hand front wheel.
3. Remove the right hand engine under cover.
4. Remove the right hand front fender apron seal.

➡Before removing, take note of the following:

- Be sure to connect Special Tool: 09216—42010 and the tools so that they are in line during use
- When retracting the tensioner, turn it clockwise slowly for 3 seconds or more. Do not apply force rapidly
- After the tensioner is fully retracted, do not apply force any more than necessary

5. Using the Special Tool and a 19 mm socket wrench, loosen the v-ribbed belt tensioner arm clockwise, then remove the v-ribbed belt.
6. Remove the v-ribbed belt.

To install:

7. To install, reverse removal procedure.
8. After installing the V-ribbed belt, check that it fits properly in the ribbed grooves. Check to confirm that the belt has not slipped out of the grooves on the bottom of the crank pulley by hand.
9. Tighten the right hand front wheel to: 76 ft. lbs. (103 Nm).

3.5L Engine

See Figure 38.

1. Before servicing the vehicle, refer to the Precautions Section.
2. Remove the right hand front wheel.
3. Remove the right hand front fender apron seal.
4. Remove the V-bank cover sub-assembly.
5. Using Special Tool: 09249—63010, release the belt tension by turning the belt tensioner counterclockwise, and remove the V-ribbed belt from the belt tensioner.
6. While turning the belt tensioner counterclockwise, align with its holes and then insert the 5 mm bi-hexagon wrench into the holes to fix the V-ribbed belt tensioner.
7. Remove the v-ribbed belt.

To install:

8. To install, reverse removal procedure.
9. If it is difficult to install the V-ribbed belt, perform the following procedure:
   a. Put the V-ribbed belt on every pulley except the tensioner pulley.
   b. While releasing the belt tension by turning the belt tensioner counterclockwise, put the V-ribbed belt on the tensioner pulley.

➡Put the backside of the V-ribbed belt on the tensioner pulley and idler pulley. Check that the V-ribbed belt is properly set to each pulley.

10. After installing the V-ribbed belt, check that it fits properly in the ribbed grooves. Check to confirm that the belt has not slipped out of the grooves on the bottom of the crank pulley by hand.
11. Tighten the right hand front wheel to: 76 ft. lbs. (103 Nm).

BALANCE SHAFT

REMOVAL & INSTALLATION

2.4L Engine

See Figures 39 through 43.

1. Before servicing the vehicle, refer to the Precautions Section.
2. Connect the negative battery cable.
3. Drain the engine oil.
4. Remove the oil pump. Refer to Oil Pump below for removal procedure.
5. Remove the No. 1 and No. 2 balance shaft sub-assembly. Remove the eight bolts in sequence.
6. Remove the No. 1 and No. 2 balance shafts.
7. Remove the balance shaft bearings if necessary.

To install:

➡Do not apply engine oil to the bearings and the contact surfaces.

8. Install the bearings in the crankcase and balance shaft housing.
9. Apply a light coat of engine oil to the bearings.

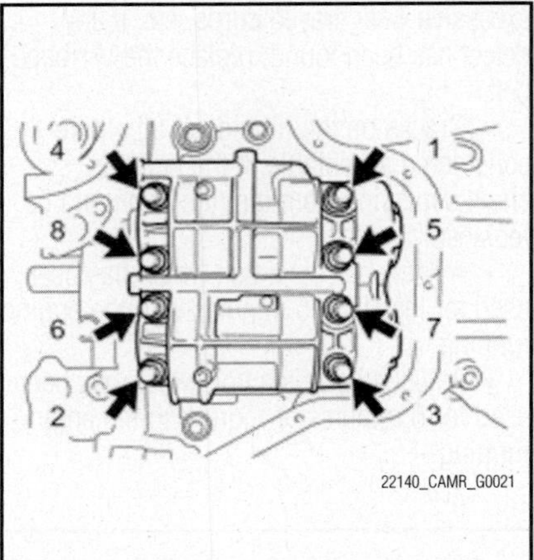

Fig. 39 Sub-assembly bolt removal sequence

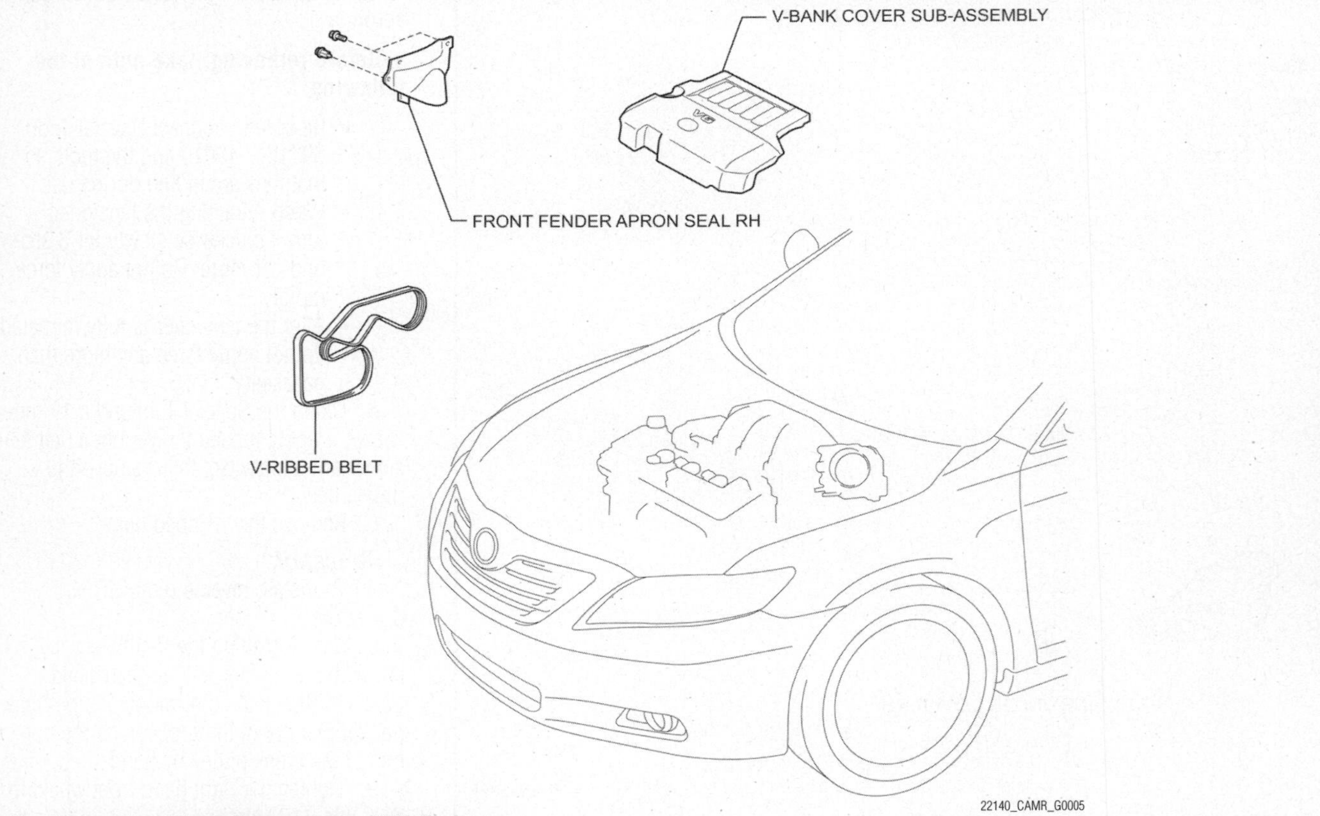

V-BANK COVER SUB-ASSEMBLY

FRONT FENDER APRON SEAL RH

V-RIBBED BELT

22140_CAMR_G0005

Fig. 38 Locating drive belt components—3.5L Engine

➡Confirm that the match marks on driven gears No. 1 and No. 2 are matched.

10. Install No. 1 and No. 2 balance shaft sub-assembly. Rotate the driven gear No. 1 of balance shaft No. 1 in the rotating direction until it hits the stopper.

11. Align the alignment marks of the No. 1 and No. 2 balance shafts as shown.

12. Place the No. 1 and No. 2 balance shafts on the crankcase.

13. Apply a light coat of engine oil under the heads of the balance shaft housing bolts.

14. Install the balance shaft housing bolts. The balance shaft housing bolts should be tightened in 2 progressive steps as follows:

a. Tighten the eight balance shaft housing bolts in sequence to: 16 ft. lbs. (22 Nm).

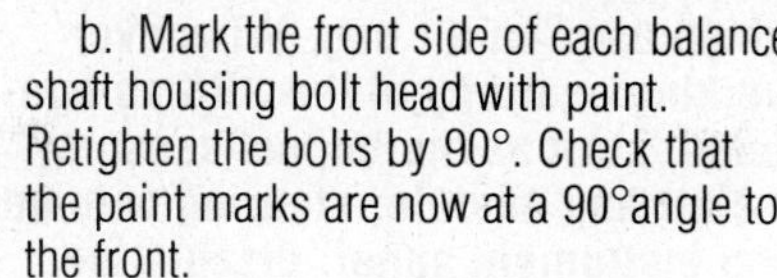

b. Mark the front side of each balance shaft housing bolt head with paint. Retighten the bolts by 90°. Check that the paint marks are now at a 90°angle to the front.

15. To complete installation, reverse remaining removal procedure.

16. Check the engine for leaks.

CAMSHAFT AND TIMING GEAR

INSPECTION

2.4L Engine

➡Be careful not to damage the camshafts.

1. To inspect the No.1 camshaft, perform the following:

a. Clamp the camshaft in a vise, and confirm that the camshaft timing gear is locked.

b. Release the lock pin.

➡The 2 advance side paths are provided in the groove of the camshaft. Plug one of the paths with a rubber piece.

c. Cover the 4 oil paths of the cam journal with vinyl tape.

d. Break through the tape of the advance side path and the retard side path on the opposite side to the hole of the advance side path.

➡Cover the paths with a piece of cloth when applying pressure to keep oil from splashing.

e. Apply approximately 28 psi of air pressure to the two broken paths.

f. Check that the camshaft timing gear revolves in the advance direction when reducing the air pressure of the retard side path.

➡This operation releases the lock pin for the most retarded position.

➡Do not remove the air gun from the advance side path first. The gear may abruptly shift in the retard direction and break the lock pin.

g. When the camshaft timing gear reaches the most advanced position, remove the air gun from the retard side path and advance side path, in that order.

➡Do not use an air gun to check for smooth operation.

h. Rotate the camshaft timing gear within its movable range several times, but do not turn it to the most retarded position. Check that the gear rotates smoothly.

i. Check the lock in the most retarded position. Confirm that the camshaft timing gear is locked at the most retarded position.

j. To inspect the camshaft for runout, place the camshaft on V-blocks.

k. Using a dial indicator, measure the circle runout at the center journal. Maximum circle runout: 0.0012 inches (0.03 mm). If the circle runout is greater than the maximum, replace the camshaft.

l. Inspect the cam lobes by using a micrometer and measure the cam lobe height. Standard cam lobe height: 1.8624 to 1.8664 inches. (47.306 to 47.406 mm). Minimum cam lobe height: 1.8581 inches (47.196 mm).

m. If the cam lobe height is less than the minimum, replace the No.1 camshaft.

n. Inspect the camshaft journals by using a micrometer and measure the journal diameter. If the journal diameter

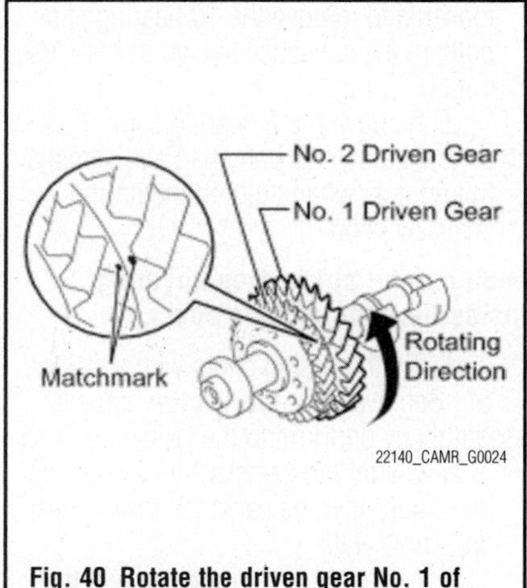

Fig. 40 Rotate the driven gear No. 1 of balance shaft No. 1

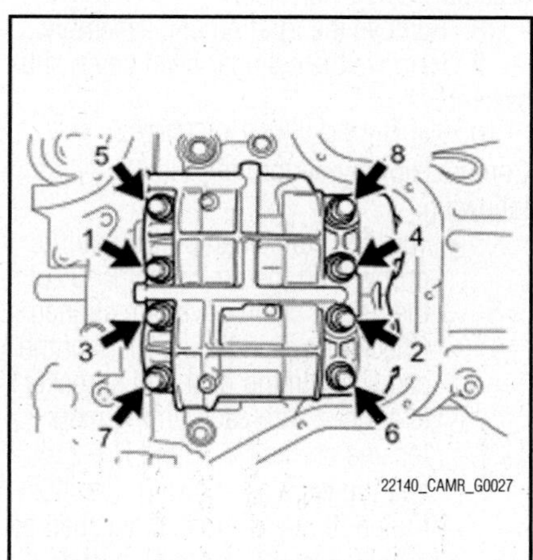

Fig. 42 Balance shaft housing bolt tightening sequence

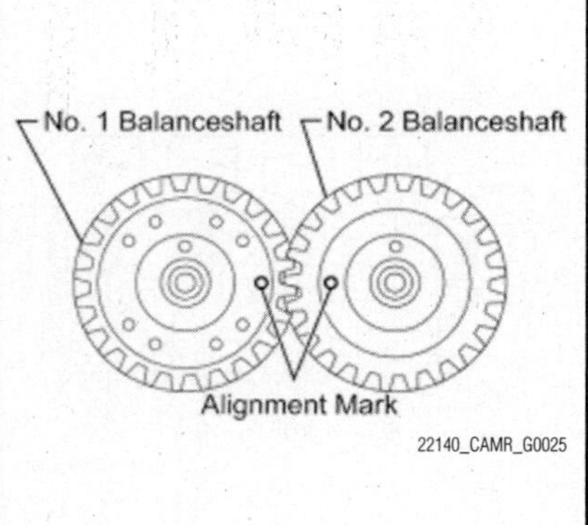

Fig. 41 Aligning marks of the No. 1 and No. 2 balance shafts

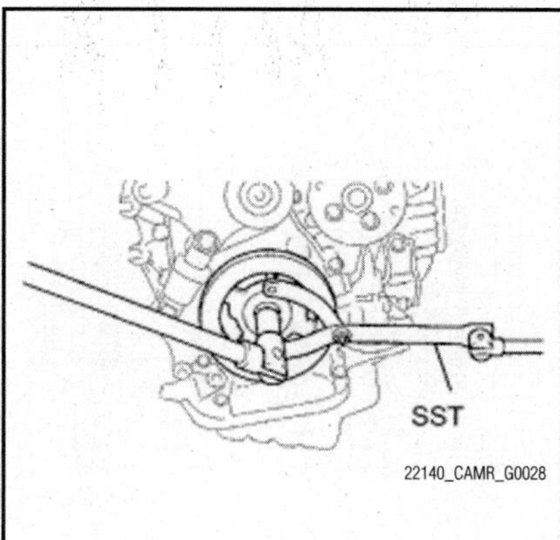

Fig. 43 Marking the bolt head and tightening procedure

is not as specified, check the oil clearance.

2. To inspect the No.2 Camshaft, perform the following:

a. Place the camshaft on V-blocks.

b. Using a dial indicator, measure the circle runout at the center journal. Maximum circle runout: 0.0012 inches. (0.03 mm). If the circle runout is greater than the maximum, replace the No. 2 camshaft.

c. Inspect the cam lobes by using a micrometer and measure the cam lobe height. Standard cam lobe height: 1.8104 to 1.8143 inches. (45.983 to 46.083 mm). Minimum cam lobe height: 1.8060 inches (45.873 mm).

d. If the cam lobe height is less than the minimum, replace the No.2 camshaft.

e. Inspect the camshaft journals by using a micrometer and measure the journal diameter. If the journal diameter is not as specified, check the oil clearance.

3.5L Engine

1. To inspect the camshaft, place the camshaft on V-blocks.

2. Using a dial indicator, measure the circle runout at the center journal.

a. Maximum circle runout: 0.0016 inches (0.04 mm). If the circle runout is greater than the maximum, replace the camshaft.

➡**Check the oil clearance after replacing the camshaft.**

3. Inspect the cam lobes by using a micrometer and measure the cam lobe height.

a. Standard cam lobe height:
- Intake: 1.7447 to 1.7487 inches (44.316 to 44.416 mm)
- Exhaust: 1.7426 to 1.7465 inches. (44.262 to 44.362 mm)

b. Maximum cam lobe height:
- 1.7388 inches (44.166 mm)
- 1.7367 inches (44.112 mm)

4. Inspect the camshaft journals by using a micrometer and measure the journal diameter. If the journal diameter is not as specified, check the oil clearance.

a. No. journal: 1.4152 to 1.4157 inches (35.946 to 35.960 mm).

b. Other journals: 1.0220 to 1.0226 inches (25.959 to 25. 975 mm).

REMOVAL & INSTALLATION

⁂ CAUTION

All models are equipped with a Supplemental Restraint System (SRS), which uses an air bag. Whenever working near any of the SRS components, such as the impact sensors, the air bag module, steering column and instrument panel, disable the SRS.

2.4L Engine

See Figures 44 through 48.

1. Before servicing the vehicle, refer to the Precautions Section.

2. Disconnect the negative battery cable.

3. Loosen the lug nuts on the front right hand wheel.

4. Apply the parking brake, block the rear wheels, then raise and safely support the front of the vehicle securely on jackstands.

5. Remove the front right hand wheel.

6. Remove the left and right hand under cover.

7. Remove the No.1 engine cover sub-assembly and 2 nuts.

8. Remove the ignition coil assembly.

9. Remove the cylinder head cover sub-assembly

10. Set No.1 Cylinder to TDC/ Compression by performing the following:
- Turn the crankshaft pulley until its groove and the timing mark "0" of the timing chain cover are aligned
- Check that each timing mark of the camshaft timing gear and sprocket is aligned with each timing mark located on the No. 1 and No. 2 bearing caps as shown in the illustration. If not, turn the crankshaft by 1 revolution (360°) to align the timing marks

➡**Do not turn the crankshaft without the chain tensioner.**

11. Remove No.1 Chain tensioner assembly. Remove the 2 nuts, tensioner and gasket.

12. Remove the No. 2 camshaft by performing the following:

a. While holding the camshaft with a wrench, loosen the camshaft timing set bolt.

b. Using several steps, uniformly loosen and remove the 10 bearing cap bolts in the sequence shown in the illustration.

c. While holding the No. 2 camshaft by hand, remove the camshaft timing sprocket set bolt.

d. Remove the camshaft timing sprocket from the No. 2 camshaft with the timing chain wrapped on the sprocket.

e. Remove the camshaft timing sprocket from the timing chain.

13. Remove the No. 1 camshaft by performing the following:

a. Using several steps, uniformly loosen and remove the 10 bearing cap bolts in the sequence shown in the illustration.

b. Remove the 5 bearing caps.

c. Remove the camshaft and camshaft timing gear while holding the timing chain by hand.

➡**Be careful not to drop anything inside the timing chain cover.**

d. Tie the timing chain with a string.

14. Remove the camshaft timing gear assembly by performing the following:

a. Clamp the camshaft in a vise, and make sure that the camshaft timing gear does not rotate.

b. Cover all the oil ports except the advance side port shown in the illustration with vinyl tape

Fig. 44 No. 2 camshaft bearing cap bolt removal sequence

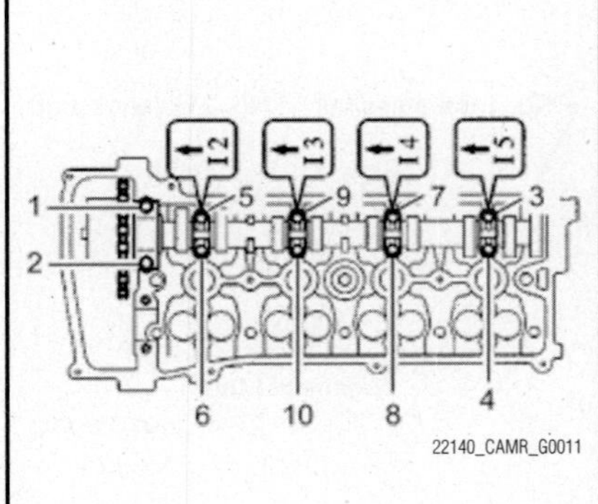

Fig. 45 No.1 camshaft bearing cap bolt removal sequence

➡Cover the paths with a shop rag or piece of cloth to avoid oil splashes.

➡Depending on the air pressure, the camshaft timing gear will turn to the advance angle side without applying force by hand. Also, if the pressure is difficult to apply because of air leakage from the port, the lock may be difficult to release.

c. Apply air pressure of 14 psi to the oil path, then turn the camshaft timing gear in the advance direction (counterclockwise) by hand.

d. Remove the flange bolt of the camshaft timing gear. Be sure not to remove the other four bolts. If planning to reuse the gear, be sure to release the straight pin lock before installing the gear.

To install:

15. Put the camshaft timing gear and camshaft together with the straight pin and key groove misaligned.

➡Be sure not to turn the camshaft timing gear to the retard angle side (the right angle).

16. Turn the camshaft timing gear as shown in the illustration while pushing it gently against the camshaft. Push further at the position where the pin fits into the groove.

17. Check that there is no clearance between the gear and camshaft.

18. Tighten the flange bolt with the camshaft timing gear fixed in place and tighten to 40 ft. lbs. (54 Nm).

19. Check that the camshaft timing gear can move to the retard angle side (the right direction) and is locked in the most retarded position.

20. To install the No.1 camshaft, perform the following:

a. Apply a light coat of engine oil to the journal portion of the camshaft.

b. Install the timing chain onto the camshaft timing gear with the paint mark aligned with the timing mark in the camshaft timing gear.

c. Examine the front marks and numbers, and check that the order is as shown in the illustration below. Then install the bearing caps into the cylinder head.

d. Apply a light coat of engine oil to the threads and under the heads of the bearing cap bolts.

e. Using several steps, uniformly tighten the 10 bearing cap bolts in the sequence shown in the illustration. Tighten the following to specification:

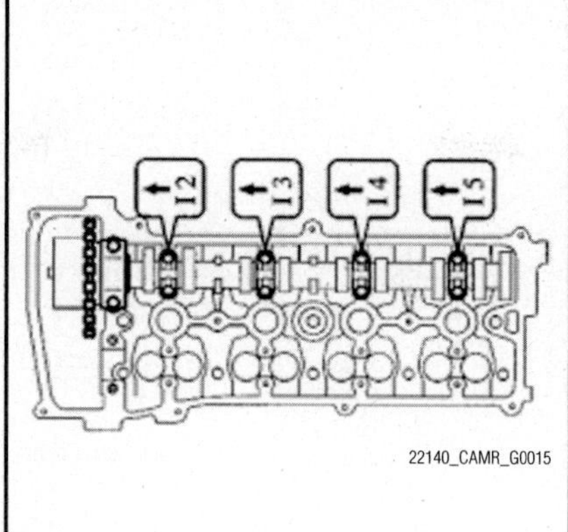

Fig. 46 Checking the No. 1 camshaft bearing cap front marks, numbers and order

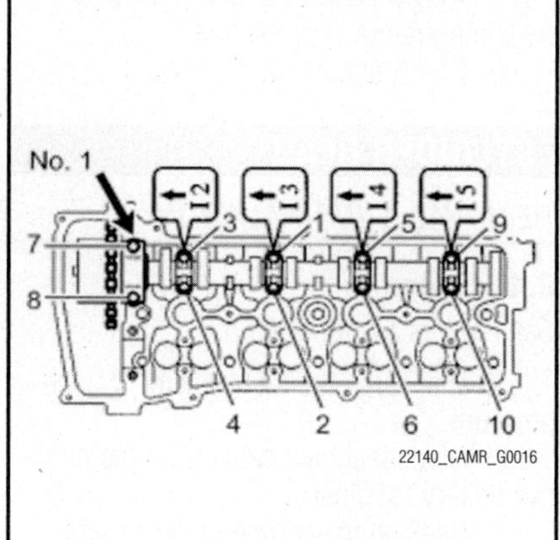

Fig. 47 No. 1 camshaft bearing cap tightening sequence

- No. 1 Bearing cap: 22 ft. lbs. (30 Nm)
- No. 3 Bearing cap: 80 inch lbs. (9 Nm)

21. To install camshaft No. 2, perform the following:

a. Apply a light coat of engine oil to the journal portion of the No. 2 camshaft.

b. Put the No. 2 camshaft on the cylinder head with the paint mark of the chain aligned with the timing mark on the camshaft timing sprocket.

c. While holding the No. 2 camshaft by hand, temporarily tighten the camshaft timing sprocket set bolt.

d. Examine the front marks and numbers, and check that the order is as shown in the illustration. Then install the bearing caps onto the cylinder head.

e. Apply a light coat of engine oil to the threads and under the heads of the bearing cap bolts.

f. Using several steps, uniformly

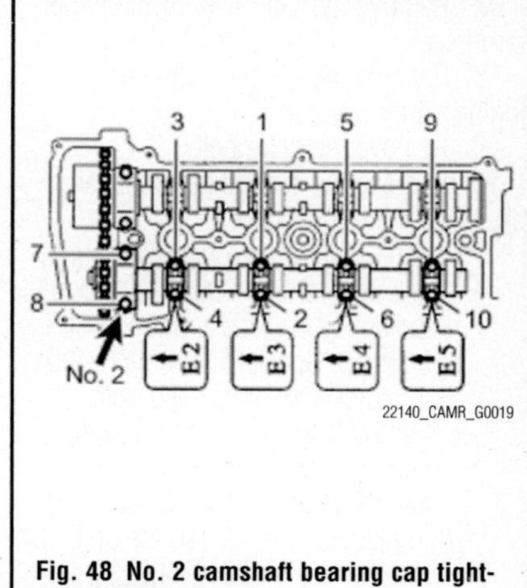

Fig. 48 No. 2 camshaft bearing cap tightening sequence

tighten the 10 bearing cap bolts in the sequence shown in the illustration. Tighten the following to specification:

- No. 1 Bearing cap: 22 ft. lbs. (30 Nm)
- No. 3 Bearing cap: 80 inch lbs. (9 Nm)

g. While holding the camshaft with a wrench, tighten the camshaft timing sprocket set bolt and tighten to 40 ft. lbs. (54 Nm).

h. Check that the paint marks on the chain are aligned with the timing marks on the camshaft timing gear and camshaft timing sprocket. Also, check that the crankshaft pulley groove is aligned with the timing mark "0" of the timing chain cover.

22. To complete installation, reverse remaining removal procedure.

23. Check engine for oil leaks.

24. Connect the negative battery cable.

CRANKSHAFT FRONT SEAL

REMOVAL & INSTALLATION

2.4L Engine

See Figures 49 and 50.

1. Before servicing the vehicle, refer to the Precautions Section.
2. Disconnect the negative battery cable.
3. Drain the engine oil
4. Loosen the lug nuts on the left front wheel.
5. Apply the parking brake, block the rear wheels, then raise and safely support the front of the vehicle securely on jack stands.
6. Remove the left front wheel.

7. Remove the right hand front fender apron seal.

8. Remove the left and right hand engine under cover.

9. Remove the drive belt.

10. Remove the crankshaft pulley and take note of the following:

- For TMMK made crankshaft pulley, use Special Tool 09960-10010 to fix the pulley in place and to loosen the bolt. Use Special Tool 09950-40011 to remove the pulley and bolt.
- For TMC made crankshaft pulley, use Special Tools 09213-54015 and 09330-00021 to fix the pulley in place and to loosen the bolt. Use Special Tools 09950-50013 and 09950-40011 to remove the pulley and bolt.

11. Using a knife, cut off the oil seal lip. Using a screwdriver with the tip taped, pry out the oil seal.

12. After the removal, check the crankshaft for damage. If it is damaged, smooth the surface with 400-grit sandpaper.

To install:

13. Apply MP grease to a new oil seal lip. Keep the lip free from foreign matter.

14. Using Special Tool 09223-22010 and a hammer, tap in the oil seal until its surface is flush with the rear oil seal retainer edge.

15. For TMMK made pulleys, align the pulley set key with the key groove of the pulley. Using Special Tool 09960-10010, keep the pulley in place and tighten the bolt to: 125 ft. lbs. (170 Nm).

16. For TMC made pulleys, align the pulley set key with the key groove of the pulley. Using Special Tools 09213-54015 and 09330-00021, keep the pulley in place and tighten the bolt to 133 ft. lbs. (180 Nm).

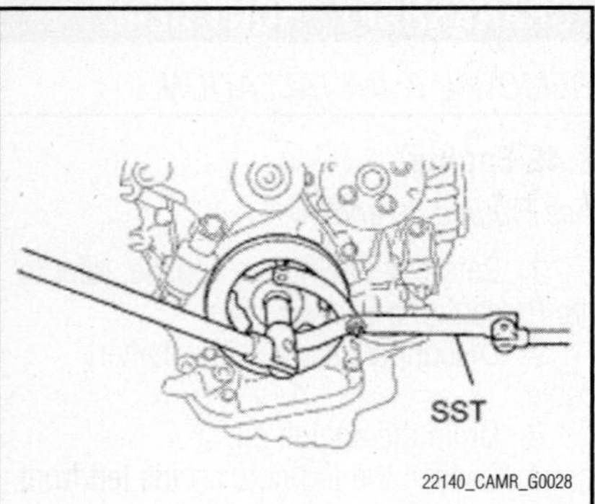

Fig. 49 Using Special Tool 09960-10010, fix the pulley in place and loosen the bolt—TMMK Pulley

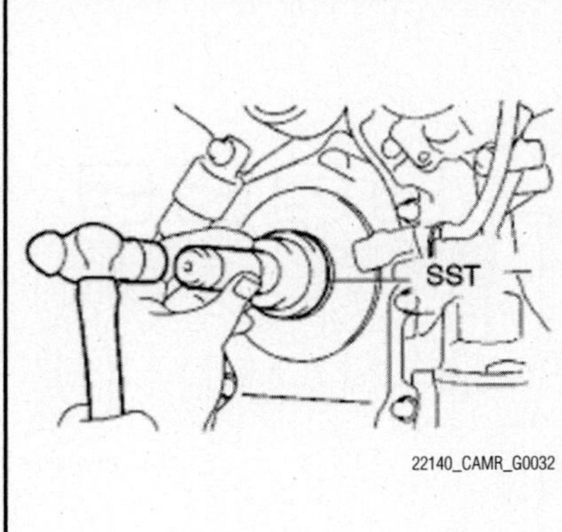

Fig. 50 Using Special Tool 09223-22010 and hammer to install new oil seal

17. To complete installation, reverse remaining removal procedure.

18. Check engine for oil leaks.

CYLINDER HEAD

REMOVAL & INSTALLATION

2.4L Engine

See Figures 51 through 55.

1. Discharge the fuel system pressure.
2. Disconnect the cable from the negative battery terminal.
3. Remove the engine under covers.
4. Remove the front fender apron seal RH.
5. Remove the No. 1 engine cover sub assembly.
6. Drain the engine coolant.
7. Drain the engine oil.
8. Remove the windshield wiper link assembly.
9. Remove the cowl top panel outer sub assembly.
10. Remove the air cleaner inlet assembly.
11. Remove the air cleaner cap sub assembly.
12. Remove the air cleaner case sub assembly.
13. Remove the battery.
14. Remove the throttle body assembly.
15. Disconnect the fuel tube sub assembly.
16. Remove the fuel delivery pipe with the injectors.
17. Remove the intake manifold.
18. Remove the intake air control valve (PZEV).
19. Remove the No. 1 intake manifold insulator.
20. Remove the front exhaust pipe assembly.
21. Remove the No. 2 engine mounting stay RH.
22. Remove the engine moving control rod sub assembly.
23. Remove the No. 2 engine mounting bracket RH.
24. Remove the v-ribbed belt.
25. Remove the generator assembly.
26. Remove the oil level gauge sub assembly.
27. Remove the oil level gauge guide.
28. Remove the manifold stay.
29. Remove the No. 2 manifold stay.
30. Remove the exhaust manifold converter sub assembly.
31. Remove the No. 2 camshaft.
32. Remove the camshaft.
33. Remove the camshaft timing oil control valve assembly.
34. Disconnect the radiator hose inlet.
35. Disconnect the engine wire.
    a. Disconnect the radio setting condenser connector.
    b. Disconnect the engine oil pressure switch connector.
    c. Disconnect the engine coolant temperature sensor connector.
    d. Disconnect the camshaft position sensor connector.
    e. Remove the bolt and ground cable.
36. Remove the No. 2 camshaft bearing.
37. Remove the cylinder head sub assembly.
    a. Using several steps, uniformly loosen and remove the 10 cylinder head bolts and 10 plate washers with a 10 mm bi-hexagon wrench in the sequence shown in the illustration.

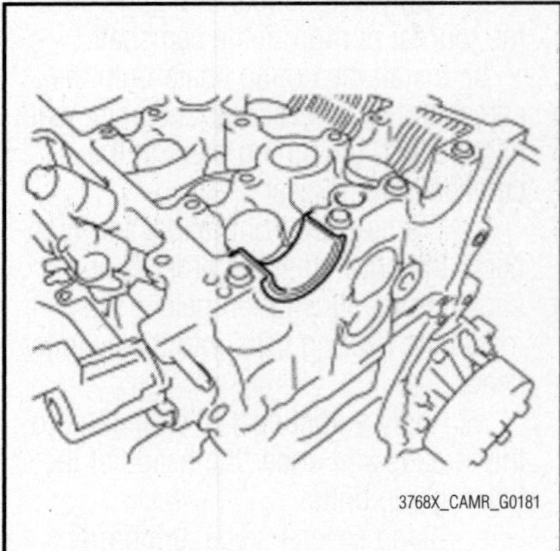

Fig. 51 Removing the No. 2 camshaft bearing

➡Head warpage or cracking could result from removing the bolts in the wrong order.

b. Using a screwdriver with its tip wrapped with tape, pry between the cylinder head and cylinder block, and remove the cylinder head.

➡Be careful not to damage the contact surfaces of the cylinder head and cylinder block.

38. Remove the cylinder head gasket.

To install:

39. Install the cylinder head gasket.

a. Place a new gasket on the cylinder block surface with the Lot No. stamp facing upward.

➡Remove any oil from the contact surfaces.

➡Make sure that the gasket is installed in the correct direction.

40. Install the cylinder head sub assembly.

a. Install the cylinder head on the cylinder block.

➡The cylinder head bolts are tightened in 2 progressive steps.

b. Apply a light coat of engine oil to the bolt threads and the area beneath the bolt heads that come in contact with the washers.

c. Install the bolts and plate washers to the cylinder head.

➡Do not drop the washers into the cylinder head.

d. Using several steps, uniformly install and tighten the 10 cylinder head set bolts and plate washers with a 10 mm bi-hexagon wrench in the order shown in the illustration. Tighten to 52 ft. lbs. (70 Nm).

e. Mark the front side of the cylinder head bolt with paint.

f. Retighten the cylinder head bolts 90°in the same sequence provided before.

g. Check that the paint mark is now at a 90° angle to the front.

41. Connect the engine wire.

a. Connect the ground cable with the bolt and tighten to 74 inch lbs. (8.4 Nm).

b. Connect the camshaft position sensor connector.

c. Connect the engine coolant temperature sensor connector.

d. Connect the engine oil pressure switch connector.

e. Connect the radio setting condenser connector.

42. Connect the radiator hose inlet.

43. Install the No. 2 camshaft bearing.

44. Install the camshaft timing oil control valve assembly.

45. Install the camshafts.

a. Apply a light coat of engine oil to the camshaft journals.

b. Place the 2 camshafts on the cylinder head with the No. 1 cam lobes facing the directions shown in the illustration.

c. Examine the front marks and numbers, and check that the order is as shown in the illustration. Then install the bearing caps onto the cylinder head.

d. Apply a light coat of engine oil to the threads and under the heads of the bearing cap bolts.

e. Using several steps, uniformly tighten the 20 bearing cap bolts in the sequence shown in the illustration. Tighten the No. 1 and No. 2 bearing caps to 22 ft. lbs. (30 Nm). Tighten the No. 3 bearing cap to 80 inch lbs. (9 Nm).

46. To complete installation, reverse the remaining removal procedure.

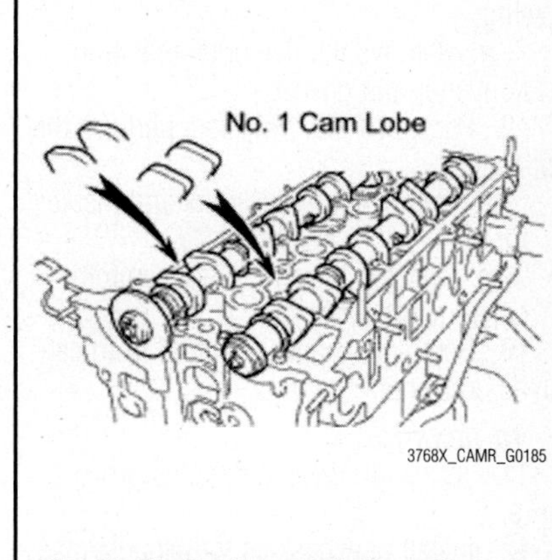

Fig. 54 Placing the camshafts on the cylinder head

EXHAUST MANIFOLD

REMOVAL & INSTALLATION

2.4L Engine

See Figures 56 through 60.

1. Before servicing the vehicle, refer to the Precautions Section.
2. Disconnect the negative battery cable.
3. Remove engine cover.
4. Remove air cleaner assembly.
5. Remove manifold stays.
6. Disconnect the air-fuel ratio sensor connector.
7. Remove or disconnect remaining components from the exhaust manifold.

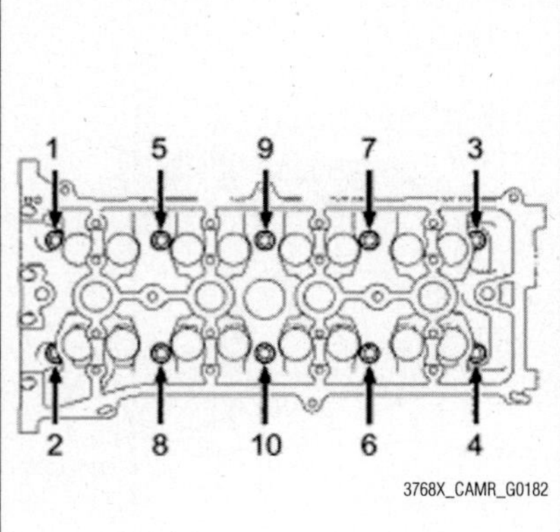

Fig. 52 Identifying the cylinder head bolt and plate washer removal sequence

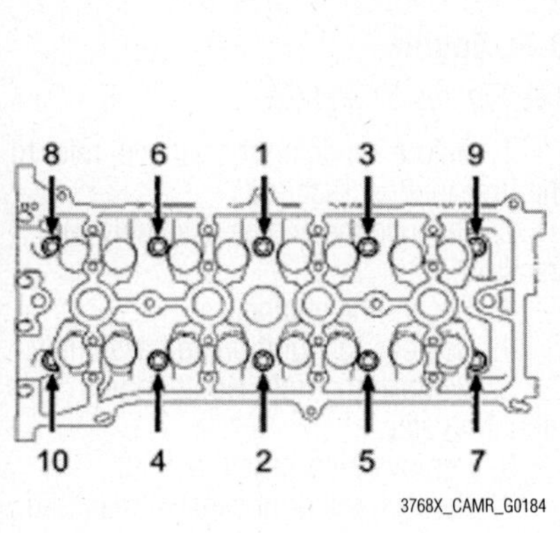

Fig. 53 Identifying the cylinder head set bolts and plate washer installation order

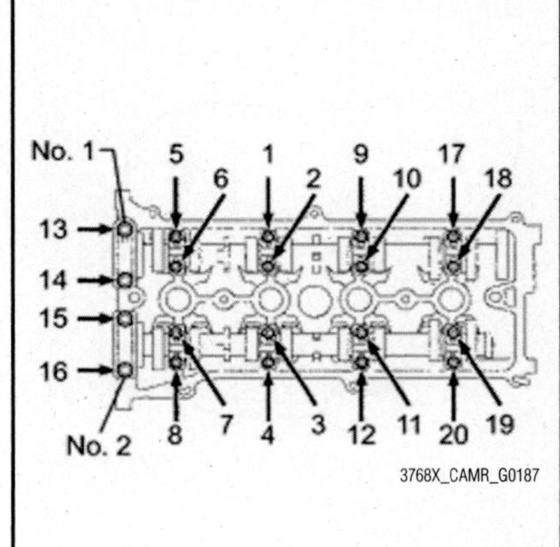

Fig. 55 Bearing cap installation sequence

8. For PZEV vehicles, perform the following:
   a. Remove the five nuts, manifold converter and gasket.
9. For non-PZEV vehicles, perform the following:
   a. Remove the four bolts and insulator.
   b. Remove the five nuts, manifold converter and gasket.
10. Remove exhaust manifold from catalytic converter.

To install:

11. To install, reverse removal procedure.
12. Install new gaskets for exhaust manifolds.
13. For non-PZEV vehicles, tighten exhaust manifold bolts in sequence to: 27 ft. lbs. (37 Nm). Tighten the exhaust manifold heat insulator bolts to: 9 ft. lbs. (12 Nm).

Fig. 56 Removing five nuts, exhaust manifold and gasket—PZEV vehicles

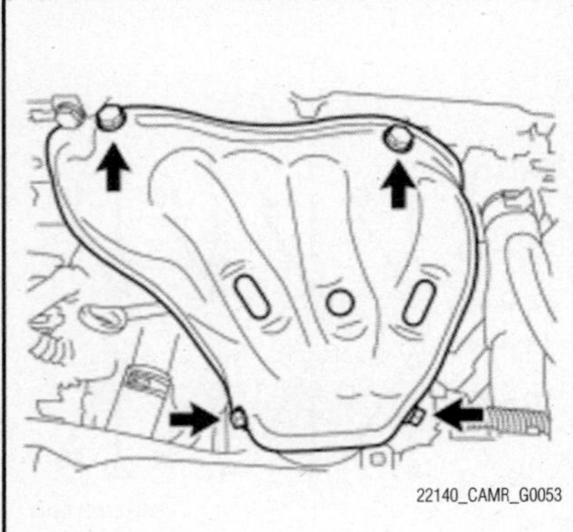

Fig. 57 Removing four bolts and insulator—except PZEV vehicles

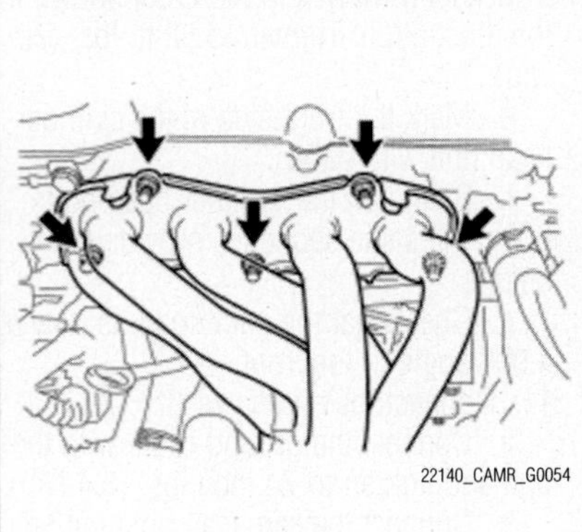

Fig. 58 Removing the five nuts, exhaust manifold and gasket—except PZEV vehicles

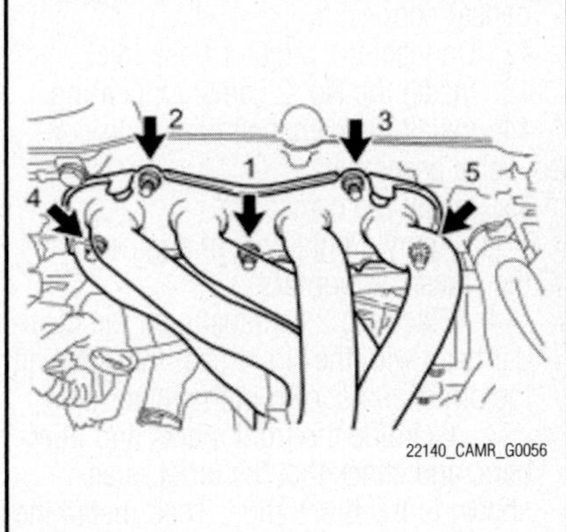

Fig. 59 Exhaust manifold tightening sequence—except PZEV vehicles

14. For PZEV vehicles, tighten exhaust manifold bolts in sequence to: 27 ft. lbs. (37 Nm).
15. Tighten the exhaust manifold stays to 32 ft. lbs. (44 Nm).

3.5L Engine

See Figures 61 and 62.

1. Before servicing the vehicle, refer to the Precautions Section.
2. Remove the engine assembly with transaxle.
3. Secure the engine.
4. Remove the ignition coil assembly.
5. Remove the right No. 2 engine mounting stay.
6. Remove the intake manifold.
7. Remove the right exhaust manifold sub-assembly, as follows:
   a. Uniformly loosen and remove the 6 nuts.
   b. Remove the manifold and gasket.

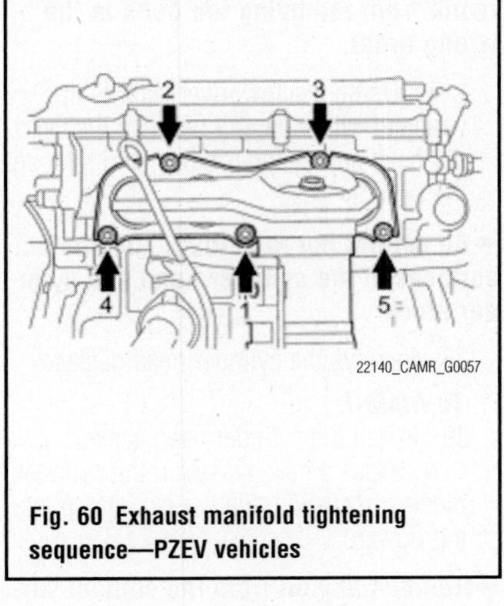

Fig. 60 Exhaust manifold tightening sequence—PZEV vehicles

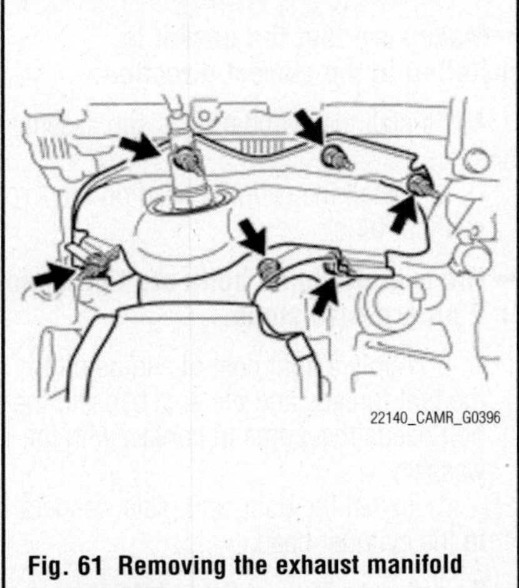

Fig. 61 Removing the exhaust manifold 6 nuts—Right

8. Remove the oil level gauge guide sub-assembly.
9. Remove the bolt, nut and No. 2 manifold stay.

Fig. 62 Removing the exhaust manifold 6 nuts—Left

10. Remove the 3 bolts and No. 2 exhaust manifold heat insulator.

11. Remove the left exhaust manifold sub-assembly, as follows:

a. Uniformly loosen and remove the 6 nuts.

b. Remove the manifold and gasket.

To install:

12. Install the left exhaust manifold sub-assembly, as follows:

a. Install a new gasket.

b. Install the left exhaust manifold sub-assembly with the 6 nuts and tighten to 15 ft. lbs. (21 Nm).

13. Install the No. 2 exhaust manifold heat insulator with the 3 bolts and tighten to 75 inch lbs. (8.5 Nm).

14. Install the No. 2 manifold stay with the bolt and nut and tighten to 25 ft. lbs. (34 Nm).

15. Install the oil level gauge guide sub-assembly.

16. Install the right exhaust manifold sub-assembly, as follows:

a. Install a new gasket.

b. Install the right exhaust manifold sub-assembly with the 6 nuts and tighten to 15 ft. lbs. (21 Nm).

17. Install the intake manifold.

18. Install the right No. 2 engine mounting stay.

19. Install the ignition coil assembly.

20. Install the engine assembly with transaxle.

INTAKE MANIFOLD

REMOVAL & INSTALLATION

3.5L Engine

See Figure 63.

1. Before servicing the vehicle, refer to the Precautions Section.

2. Remove the engine assembly with transaxle.

3. Secure the engine.

4. Remove the ignition coil assembly.

5. Remove the right No. 2 engine mounting stay.

6. Remove the intake manifold, as follows:

a. Uniformly loosen and remove the 6 bolts and 4 nuts.

b. Remove the intake manifold and 2 gaskets.

To install:

7. Install the intake manifold, as follows:

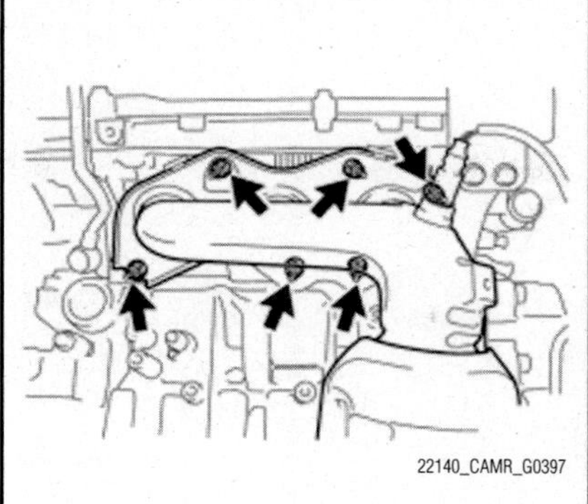

Fig. 63 Locating intake manifold bolts and nuts

WARNING

DO NOTapplies oil to the intake manifold and cylinder head sub-assembly bolts.

a. Set a new gasket on each cylinder head.

➡Align the port holes of the gasket and cylinder head.

➡Make sure that the gasket is installed in the correct direction.

b. Set the intake manifold on the cylinder heads.

c. Install and tighten the 6 bolts and 4 nuts uniformly in several steps to 15 ft. lbs. (21 Nm).

8. Install the right No. 2 engine mounting stay.

9. Install the ignition coil assembly.

10. Install the engine assembly with transaxle.

OIL PAN

REMOVAL & INSTALLATION

2.4L Engine

See Figures 64 and 65.

1. Before servicing the vehicle, refer to the Precautions Section.

2. Disconnect the negative battery cable.

3. Drain the engine oil.

4. Remove the oil pan drain plug and gasket.

5. Remove the oil pan 12 bolts and 2 nuts.

➡Be careful not to damage the contact surfaces of the crankcase, chain cover and oil pan.

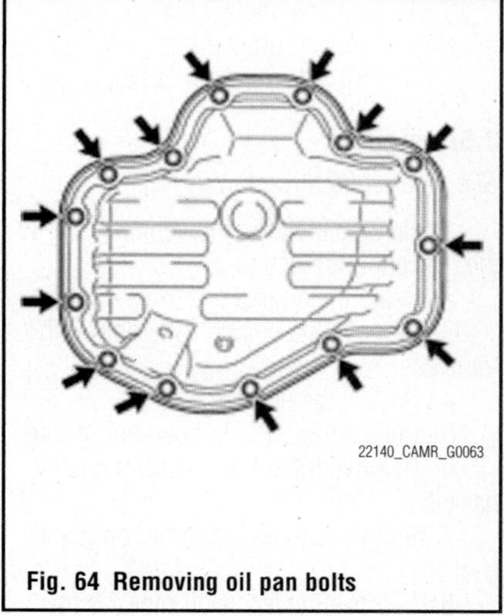

Fig. 64 Removing oil pan bolts

6. Insert the blade of Special Tool: 09032-00100 between the crankcase and oil pan. Cut through the sealer and remove the oil pan.

To install:

7. Remove any old packing material and be careful not to drop any oil on the contact surfaces of the cylinder block and oil pan.

8. Apply a continuous bead of seal packing (Diameter 0.118 to 0.157 inches (3.0 to 4.0 mm)). Use Toyota Genuine Seal Packing Block, Three Bond 1207B or Equivalent.

➡Remove any oil from the contact surfaces. Install the oil pan within 3 minutes after applying seal packing. Do not start the engine for at least 2 hours after installing.

9. Uniformly tighten the 12 bolts and 2 nuts in sequence. Tighten the bolts and nuts to 80 inch lbs. (9 Nm).

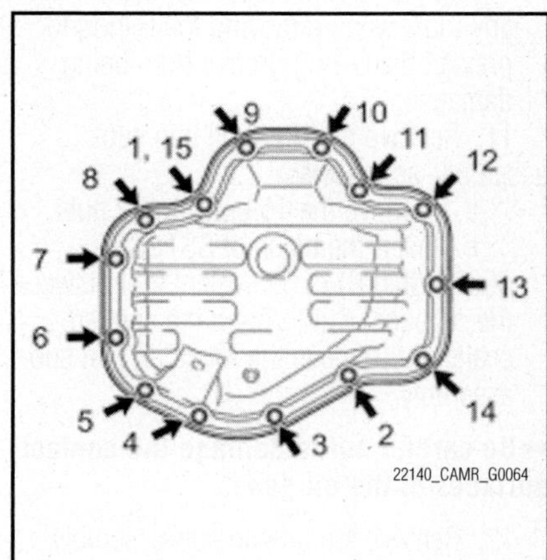

Fig. 65 Oil pan bolt and nut tightening sequence

10. To complete installation, reverse remaining removal procedure.
11. Add oil and check for leaks.

3.5L Engine

See Figures 66 through 70.

1. Before servicing the vehicle, refer to the Precautions Section.
2. Drain the engine oil.
3. Remove the engine assembly with transaxle.
4. Secure the engine.
5. Remove the oil filler cap and gasket.
6. Remove the oil pan drain plug and gasket.
7. Remove the oil pan drain plug and gasket.
8. Remove the No. 1 oil pipe, as follows:
   a. Remove the 2 oil pipe unions and oil pipe.
   b. Remove the left hand oil control valve filter and gaskets.
9. Remove the oil pipe, as follows:
   a. Remove the bolt.
   b. Remove the 2 oil pipe unions and oil pipe.
   c. Remove the right oil control valve filter and gaskets.
10. Remove the oil filter element, as follows:
   a. Remove the drain plug. Do not remove the O-ring.
   b. Connect the hose to the pipe.
   c. Insert the pipe with the hose into the oil filter cap.
   d. Make sure that the oil is completely drained and remove the pipe and O-ring.
   e. Using SST (SST: 09228-06501) or equivalent, remove the oil filter cap.
   f. Remove the oil filter element and O-ring from the oil filter cap. Do not use any tools when removing the O-ring to prevent the O-ring groove from being damaged.
11. Remove the No. 2 oil pan sub-assembly, as follows:
   a. Remove the 16 bolts and 2 nuts.
   b. Insert the blade of SST (SST: 09032-00100) or equivalent tool between the oil pans. Cut through the applied sealer and remove the No. 2 oil pan sub-assembly.

➡Be careful not to damage the contact surfaces of the oil pans.

12. Remove the oil pan sub-assembly, as follows:
   a. Remove the 16 bolts and 2 nuts.

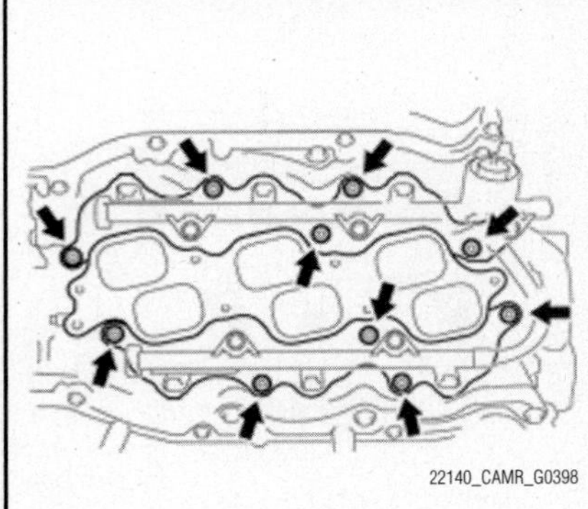

Fig. 66 No. 2 oil pan sub-assembly removal

➡Be sure to clean the bolts and stud bolts and check the threads for cracks or other damage.

   b. Remove the oil pan by prying between the oil pan and cylinder block with a taped screwdriver.

➡Be careful not to damage the contact surfaces of the cylinder block and oil pan.

   c. Remove the 2 O-rings.

To install:

13. Install the oil pan sub-assembly, as follows:
   a. Using an E8 Torx® socket wrench, install the stud bolts as shown in the illustration. Tighten to 7 ft. lbs (10 Nm).
   b. Install 2 new O-rings.

➡Remove any oil from the contact surface.

➡Install the oil pan within 3 minutes after applying seal packing.

➡Do not start the engine for at least 2 hours after installing.

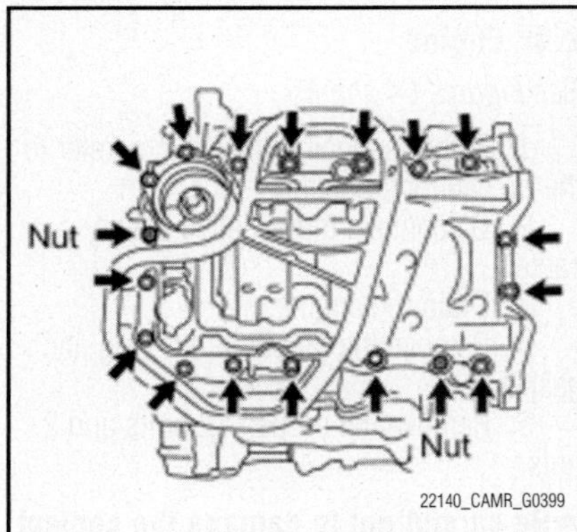

Fig. 67 Oil pan sub-assembly removal

   c. Apply seal packing (Toyota Genuine Seal Packing Black, Three Bond 1207B or equivalent) in a continuous line as shown in the illustration. Seal diameter: 3.0 to 4.0 mm (0.118 to 0.156 in.).
   d. Install the oil pan with the 16 bolts and 2 nuts and tighten to 7 ft. lbs (10 Nm), and 15 ft. lbs (21 Nm).
14. Install the No. 2 oil pan sub-assembly, as follows:

➡Remove any oil from the contact surface.

➡Install the No. 2 oil pan within 3 minutes after applying seal packing.

➡Do not start the engine for at least 2 hours after installing.

   a. Using an E6 Torx® socket wrench, install the stud bolts as shown in the illustration and tighten to 35 inch lbs (4 Nm).
   b. Apply seal packing (Toyota Genuine Seal Packing Black, Three Bond 1207B or equivalent) in a continuous line as shown in the illustration. Seal diameter: 3.0 to 4.0 mm (0.118 to 0.156 in.).
   c. Install the No. 2 oil pan with the 16 bolts and 2 nuts and tighten to 7 ft. lbs (10 Nm).
15. Install the oil pan drain plug and a new gasket. Tighten to 30 ft. lbs (40 Nm).
16. Install the oil filter element, as follows:
   a. Clean the inside of the oil filter cap, the threads and O-ring groove.
   b. Apply a small amount of engine oil to a new O-ring and install it to the oil filter cap.
   c. Set a new oil filter element to the oil filter cap.
   d. Remove dirt or foreign matter from the installation surface and inside of the engine.
   e. Apply a small amount of engine oil to the O-ring again and install the oil filter cap.

➡Be careful that the O-ring does not get caught between the parts. The O-ring must not be twisted on the groove.

   f. Using SST (SST: 09228-06501) or equivalent, install the oil filter cap and tighten to 18 ft. lbs (25 Nm). Make sure that the oil filter is installed securely as shown in the illustration.
17. Install the oil filler cap sub-assembly.
18. Install the engine assembly with transaxle.
19. Check for oil leaks.

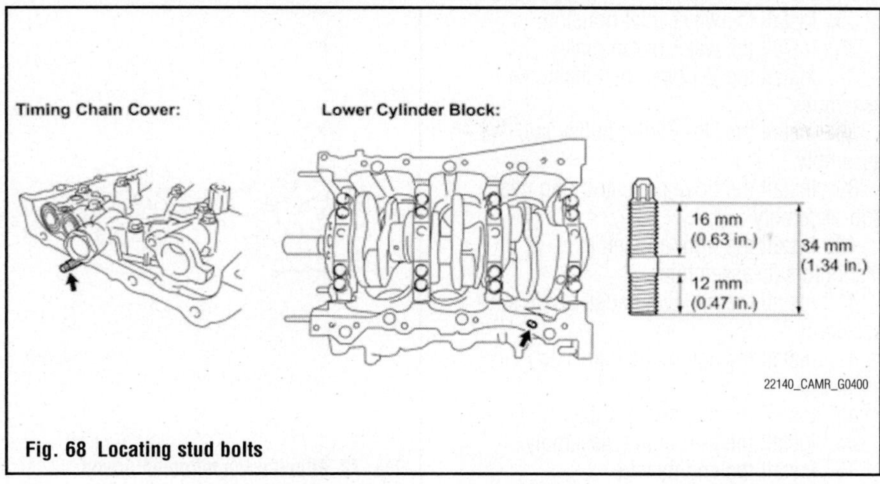

Fig. 68 Locating stud bolts

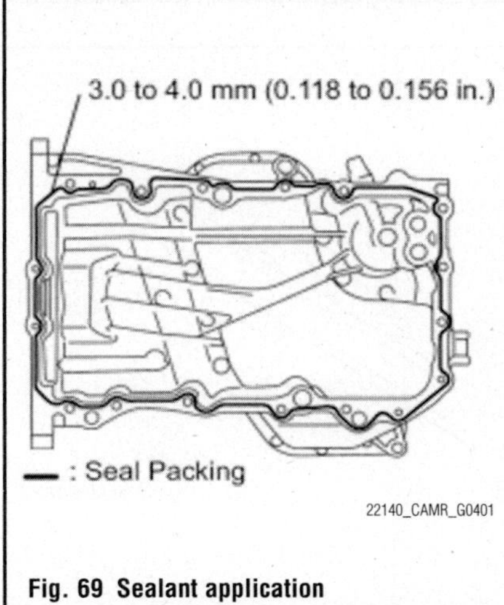

Fig. 69 Sealant application

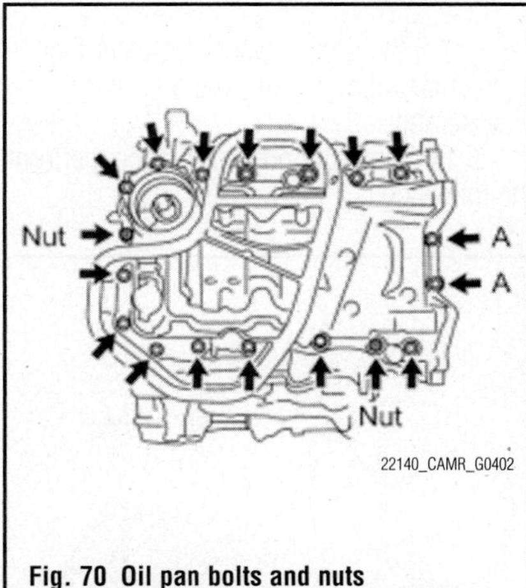

Fig. 70 Oil pan bolts and nuts

OIL PUMP

REMOVAL & INSTALLATION

2.4L Engine

See Figure 71.

1. Before servicing the vehicle, refer to the Precautions Section.
2. Disconnect the negative battery cable.
3. Drain the engine oil.
4. Remove or disconnect the following:
   - Plastic engine cover and two nuts
   - Front right wheel
   - LH and RH engine under cover
   - RH front fender apron seal
   - Front exhaust pipe assembly
   - RH engine mount stay
   - Engine mount
   - RH engine mounting bracket
   - Drive belt
   - Generator assembly
   - Vane pump
   - Ignition coil assembly
   - Ventilation hoses
   - Valve cover
5. Turn the crankshaft pulley until its groove and the timing mark "0" of the timing chain cover are aligned.
6. Check that each timing mark of the camshaft timing gear and sprocket is aligned with each timing mark located on the intake and exhaust bearing caps. If not, turn the crankshaft by 1 revolution (360°) to align the timing marks.
7. Remove or disconnect the following:
   - Crankshaft pulley
   - Crankshaft position sensor
   - Oil pan
   - Chain upper tensioner assembly
8. Install the No. 1 engine hanger (12281-28010) and No. 2 engine hanger (12282-28010) with the bolts (91512-61020) and tighten to: 28 ft. lbs. (38 Nm).
9. Remove or disconnect the following:
   - Drive belt tensioner
   - Engine mount insulator
   - RH engine mount bracket
   - Using an E10 Torx® socket, remove the stud bolt for the drive belt tensioner from the cylinder block
   - Timing chain cover twelve bolts and two nuts

➡Be careful not to damage the contact surfaces of the timing chain cover, cylinder block and cylinder head. Tape the screwdriver tip before use.

   - Timing chain cover by prying between the timing chain cover and cylinder head or cylinder block with a screwdriver

➡Tape the screwdriver tip before use.

   - Using a screwdriver and a hammer, tap out the timing chain case oil seal
   - Crankshaft position sensor plate
   - Chain tensioner slipper
   - Chain vibration damper
   - Timing chain guide
   - Upper chain assembly
   - Crankshaft timing sprocket
   - Lower chain assembly
   - Three bolts, oil pump and gasket

To install:

10. Install a new gasket and the oil pump with the 3 bolts. Tighten the bolts 14 ft. lbs. (19 Nm).
11. To complete installation, reverse removal procedure. Refer to the appropriate sections to install components correctly.

3.5L Engine

1. Before servicing the vehicle, refer to the Precautions Section.
2. Remove the engine assembly with transaxle.
3. Secure engine.
4. Remove the engine wire.
5. Remove the front frame assembly.
6. Remove the starter assembly.

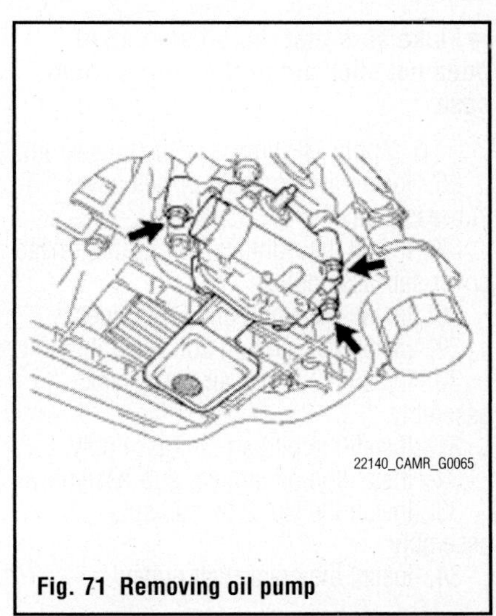

Fig. 71 Removing oil pump

7. Remove the automatic transaxle assembly.
8. Remove the oil level gauge guide sub-assembly.
9. Remove the right and left exhaust manifold sub-assemblies.
10. Remove the drive plate and ring gear sub-assembly.
11. Remove the No. 2 idler pulley sub-assembly.
12. Remove the V-ribbed belt tensioner assembly.
13. Remove the water pump pulley.
14. Remove the water inlet housing.
15. Remove the crankshaft pulley.
16. Remove the No. 2 oil pan sub-assembly.
17. Remove the oil strainer sub-assembly.
18. Remove the oil pan sub-assembly.
19. Remove the intake air surge tank assembly.
20. Remove the ignition coil assembly.
21. Remove the No. 1 and 2 oil pipes.
22. Remove the right and left cylinder head cover sub-assemblies.
23. Remove the timing chain or belt cover sub-assembly.
24. Remove the timing gear case or timing chain case oil seal, as follows:
   a. Using a screwdriver with the tip taped, pry out the oil seal.

To install:

25. Install timing gear case or timing chain case oil seal, as follows:
   a. Using SST (SST: 09316-60011) or equivalent tool, tap in a new oil seal until its surface is flush with the timing chain case edge.

➡Keep the lip free from foreign matter.

➡Do not tap on the oil seal at an angle.

➡Make sure that the oil seal edge does not stick out of the timing chain case.

   b. Apply MP grease to the oil seal lip.
26. Install timing chain or belt cover sub-assembly.
27. Install the right and left cylinder head cover sub-assemblies.
28. Install the No. 1 and 2 oil pipes.
29. Install the ignition coil assembly.
30. Install the intake air surge tank assembly.
31. Install the oil pan sub-assembly.
32. Install the oil strainer sub-assembly.
33. Install the No. 2 oil pan sub-assembly.
34. Install the crankshaft pulley.
35. Install the water inlet housing.
36. Install the water pump pulley.
37. Install the V-ribbed belt tensioner assembly.
38. Install the No. 2 idler pulley sub-assembly.
39. Install the drive plate and ring gear sub-assembly.
40. Install the right and left exhaust manifold sub-assemblies.
41. Install the oil level gauge guide sub-assembly.
42. Install the automatic transaxle assembly.
43. Install the starter assembly.
44. Install the front frame assembly.
45. Install the engine wire.
46. Install the engine assembly with transaxle.

INSPECTION

2.4L Engine

See Figures 72 through 75.

1. Check the oil jet located above the timing chain sprocket for damage or clogging. If necessary, repair the cylinder block.

➡If the valve does not fall smoothly, replace the relief valve. If necessary, replace the oil pump assembly.

2. Coat the relief valve with engine oil, and then check that the valve falls smoothly into the valve hole under its own weight.
3. Install the oil pump rotor. Coat the drive rotor and driven rotor with engine oil.
4. Place the drive and driven rotors into the oil pump with the marks facing the pump covers side.
5. To check the side clearance, perform the following:
   a. Using a feeler gauge and precision straightedge, measure the clearance between the rotors and precision straightedge.
   b. Standard clearance: 0.0012 to 0.0033 inches (0.030 to 0.085 mm).
   c. Maximum clearance: 0.0063 inches (0.16 mm).
   d. If the side clearance is greater than the maximum, replace the oil pump assembly.
6. To check the tip clearance, perform the following:
   a. Using a feeler gauge, measure the clearance between the drive and driven rotor tips.
   b. Standard clearance: 0.0031 to 0.0063 inches (0.080 to 0.160 mm).
   c. Maximum clearance: 0.0138 inches (0.35 mm).

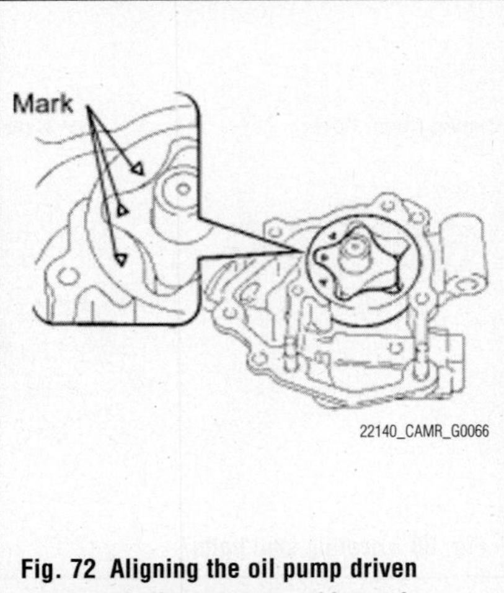

Fig. 72 Aligning the oil pump driven rotors and oil pump cover side marks

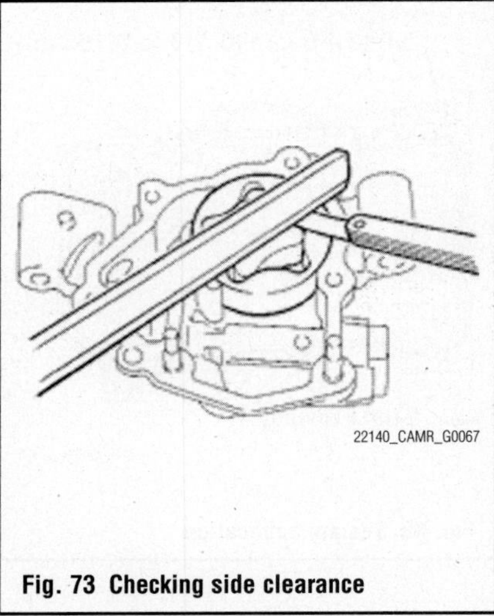

Fig. 73 Checking side clearance

   d. If the tip clearance is greater than the maximum, replace the oil pump assembly.
7. To check the body clearance, perform the following:

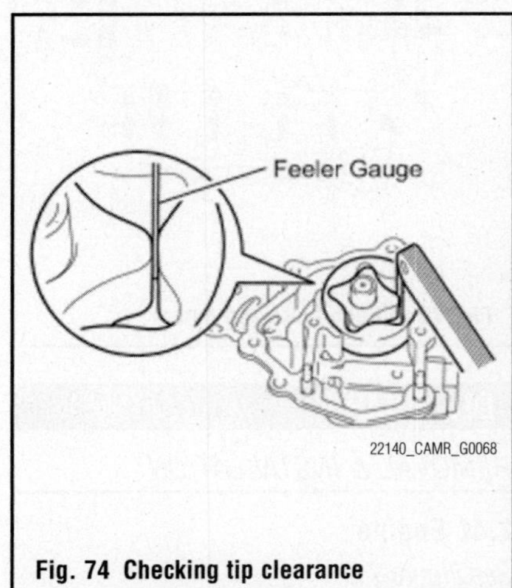

Fig. 74 Checking tip clearance

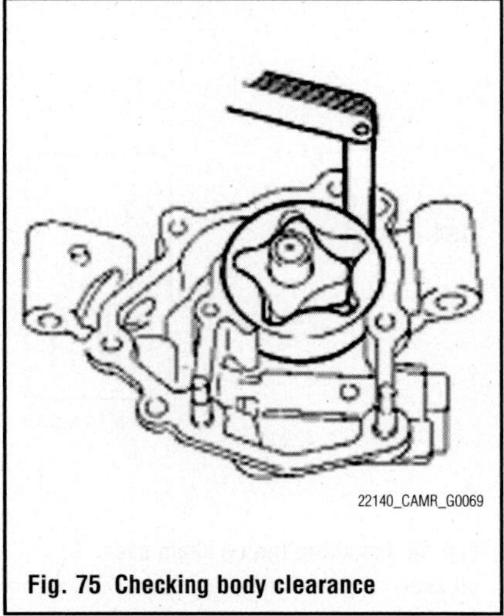

Fig. 75 Checking body clearance

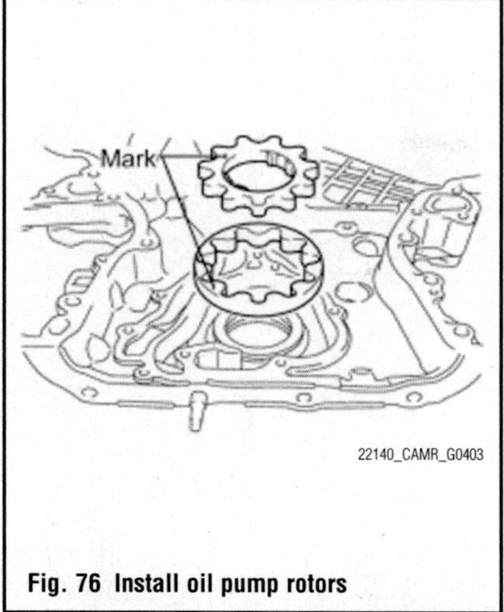

Fig. 76 Install oil pump rotors

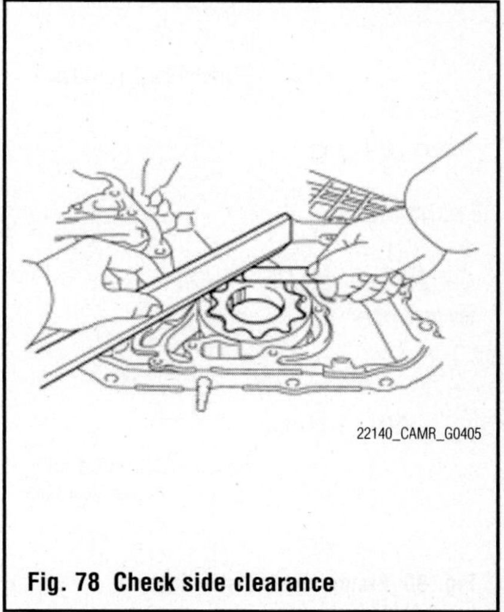

Fig. 78 Check side clearance

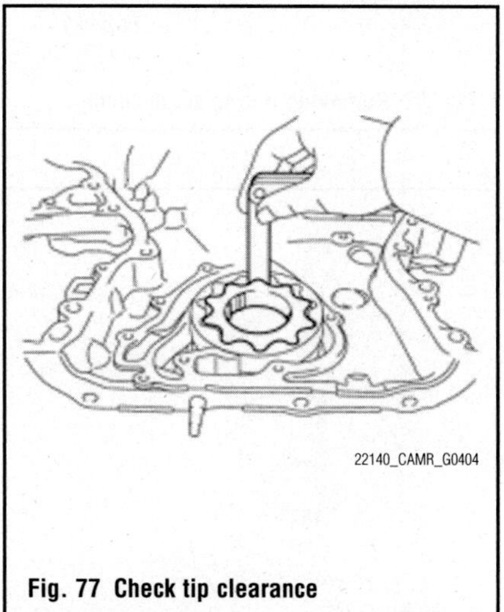

Fig. 77 Check tip clearance

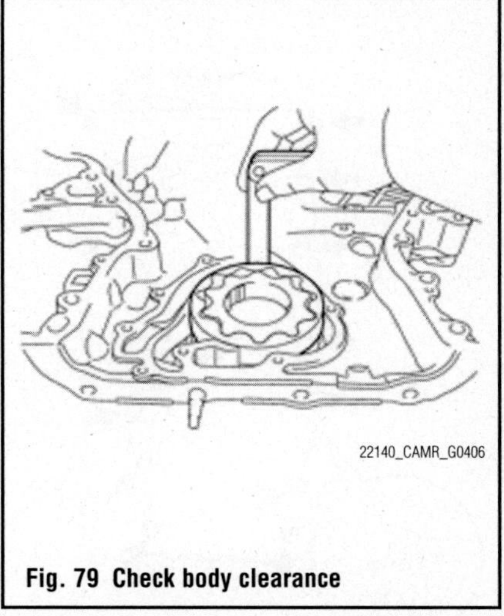

Fig. 79 Check body clearance

a. Using a feeler gauge, measure the clearance between the driven rotor and pump body.

b. Standard clearance: 0.0039 to 0.0067 inches (0.100 to 0.170 mm).

c. Maximum clearance: .0128 inches. (0.325 mm).

d. If the body clearance is greater than the maximum, replace the oil pump assembly.

8. Inspect the lower chain assembly and replace as necessary.

9. Inspect the oil pump drive sprocket and replace as necessary.

10. Inspect the chain tensioner plate and replace as necessary.

3.5L Engine

See Figures 76 through 79.

1. Inspect the oil pump relief valve, as follows:

a. Coat the relief valve with engine oil and check that it falls smoothly into the valve hole by its own weight. If the valve does not fall smoothly, replace the relief valve. If necessary, replace the oil pump assembly.

2. Inspect the oil pump rotor set, as follows:

a. Install the rotors to the timing chain cover with the rotors' marks outward. Check that the rotors rotate smoothly.

b. Check the tip clearance: using a feeler gauge, measure the clearance between the drive and driven rotor tips, as shown in the illustration. If the clearance is greater than the maximum, replace the drive and driven rotors. Standard: 0.0024 to 0.0063 inches (0.060 to 0.160 mm). Maximum: 0.0063 inches (0.16 mm).

c. Check the side clearance: using a

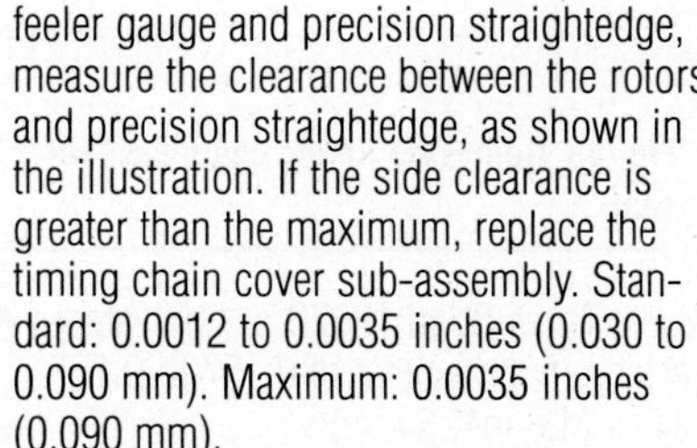

feeler gauge and precision straightedge, measure the clearance between the rotors and precision straightedge, as shown in the illustration. If the side clearance is greater than the maximum, replace the timing chain cover sub-assembly. Standard: 0.0012 to 0.0035 inches (0.030 to 0.090 mm). Maximum: 0.0035 inches (0.090 mm).

d. Check the body clearance: using a feeler gauge, measure the clearance between the timing chain cover and driven rotor, as shown in the illustration. If the body clearance is greater than the maximum, replace the timing chain cover sub-assembly. Standard: 0.0098 to 0.0128 inches (0.250 to 0.325 mm). Maximum: 0.0128 inches (0.325 mm).

PISTON AND RING

POSITIONING

See Figure 80.

TIMING CHAIN FRONT COVER

REMOVAL & INSTALLATION

2.4L Engine

See Figures 81 through 86.

1. Before servicing the vehicle, refer to the Precautions Section.

2. Using an E10 Torx® socket, remove the stud bolt for the drive belt tensioner from the cylinder block.

3. Remove the 12 bolts and 2 nuts.

➡Be careful not to damage the contact surfaces of the timing chain cover, cylinder block and cylinder head.

4. Remove the timing chain cover by prying between the timing chain cover and cylinder head or cylinder block with a screwdriver. Tape the screwdriver tip before use.

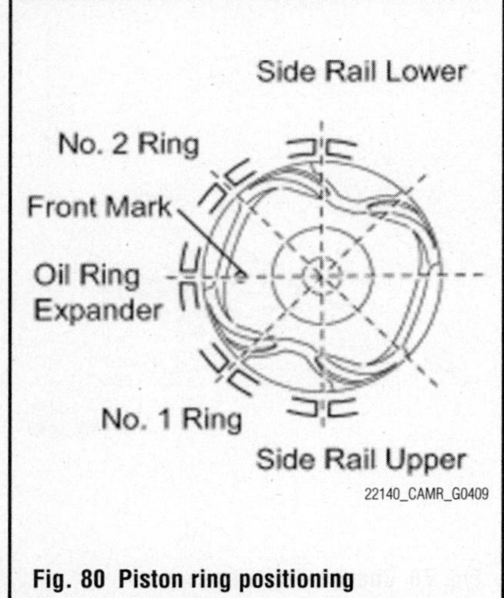

Fig. 80 Piston ring positioning

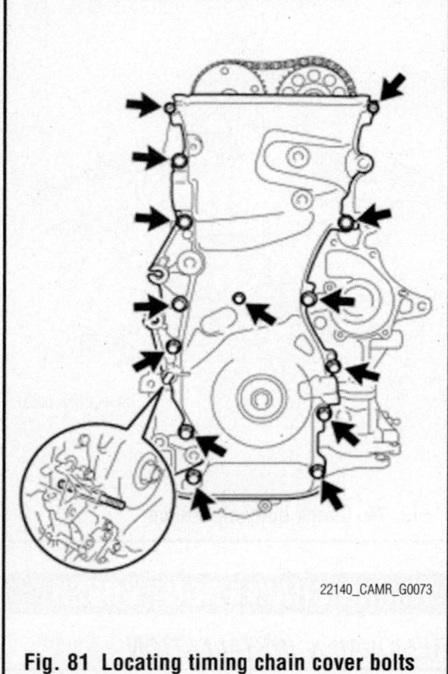

Fig. 81 Locating timing chain cover bolts and nuts

5. Using a screwdriver and a hammer, tap out the oil seal.

To install:

➡Keep the gap between the timing chain cover edge and the oil seal free of foreign matter.

6. Using Special Tool: 09223-22010, tap in a new oil seal until its surface is flush with the timing chain cover edge. Apply a light coat of MP grease to the lip of the oil seal.

7. Remove any old packing (FIPG) material and be careful not to drop any oil on the contact surfaces of the timing chain cover, cylinder head and cylinder block.

8. Apply Toyota Genuine Seal Packing

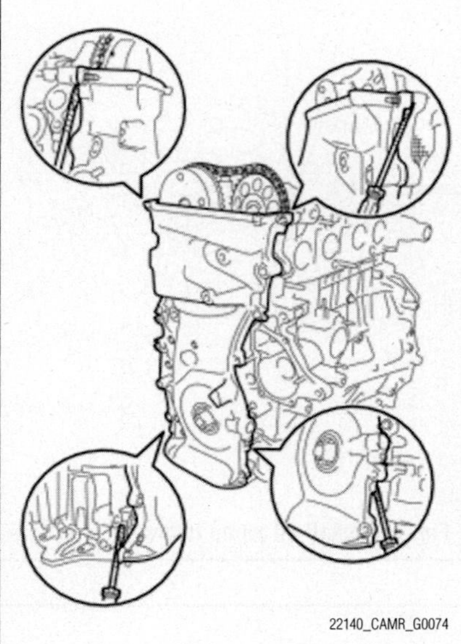

Fig. 82 Removing timing chain cover

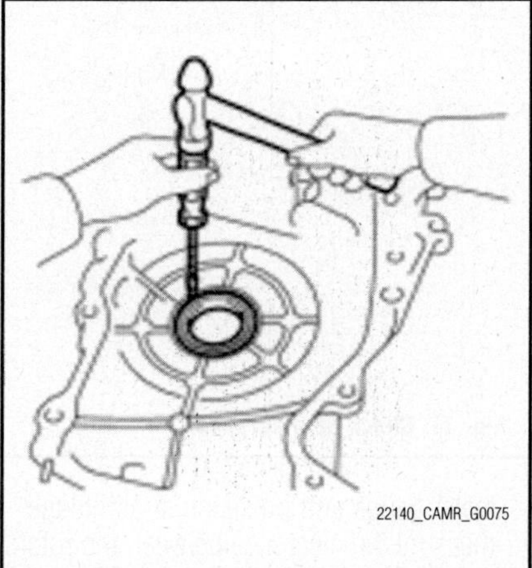

Fig. 83 Removing timing chain case oil seal

Black, Three Bond 1207B or Equivalent seal packing in a diameter of 0.157 to 0.177 inches (4.0 to 4.5 mm).

9. Apply seal packing in a continuous bead as shown in the illustration below.

➡Remove any oil from the contact surface. Install the chain cover within 3 minutes after applying seal packing. Do not start the engine for at least 2 hours after installing.

10. Install the timing chain cover with the twelve bolts and two nuts in sequence to the following torque specification:

- Bolt A length: 1.18 inches (30 mm) for 10 mm head: 80 inch lbs. (9 Nm)

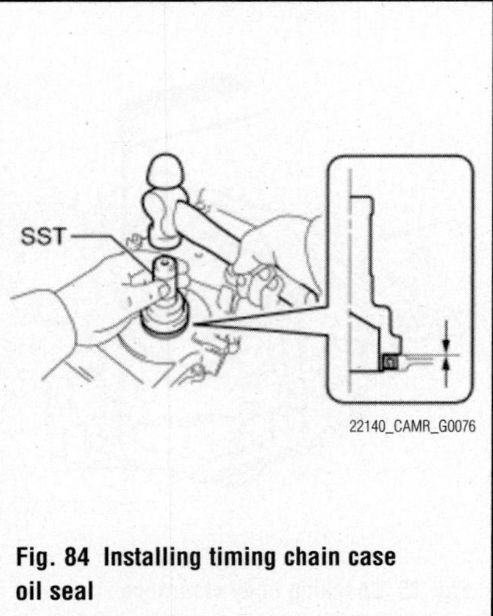

Fig. 84 Installing timing chain case oil seal

- Bolt B length: 1.18 inches (30 mm) for 12 mm head: 18 ft. lbs. (25 Nm)
- Bolt C length: 1.57 inches (40 mm) for 14 mm head: 41 ft. lbs. (55 Nm)
- Nut: 8 ft. lbs. (11 Nm)

11. Using a E10 Torx® socket, install the stud bolt to the drive belt tensioner and tighten to 16 ft. lbs. (22 Nm).

3.5L Engine

See Figures 87 through 92.

1. Before servicing the vehicle, refer to the Precautions Section.
2. Remove the engine assembly with transaxle.
3. Secure engine.
4. Remove the oil filler cap sub-assembly.
5. Remove the spark plugs and ignition coil assembly.
6. Remove the oil pan drain plug and gasket.
7. Remove the ventilation valve sub-assembly.
8. Remove the camshaft position sensor.
9. Remove the camshaft timing oil control valve assembly.
10. Remove crankshaft position sensor.
11. Remove the No. 1 oil pipe.
12. Remove the oil pipe.
13. Remove the cylinder block water drain cock sub-assembly.
14. Remove the oil filter.
15. Remove the crankshaft pulley.
16. Remove the left hand No. 1 front engine mounting bracket.
17. Remove the water inlet housing.
18. Remove the water outlet.
19. Remove the left-hand cylinder head cover sub-assembly and gasket.

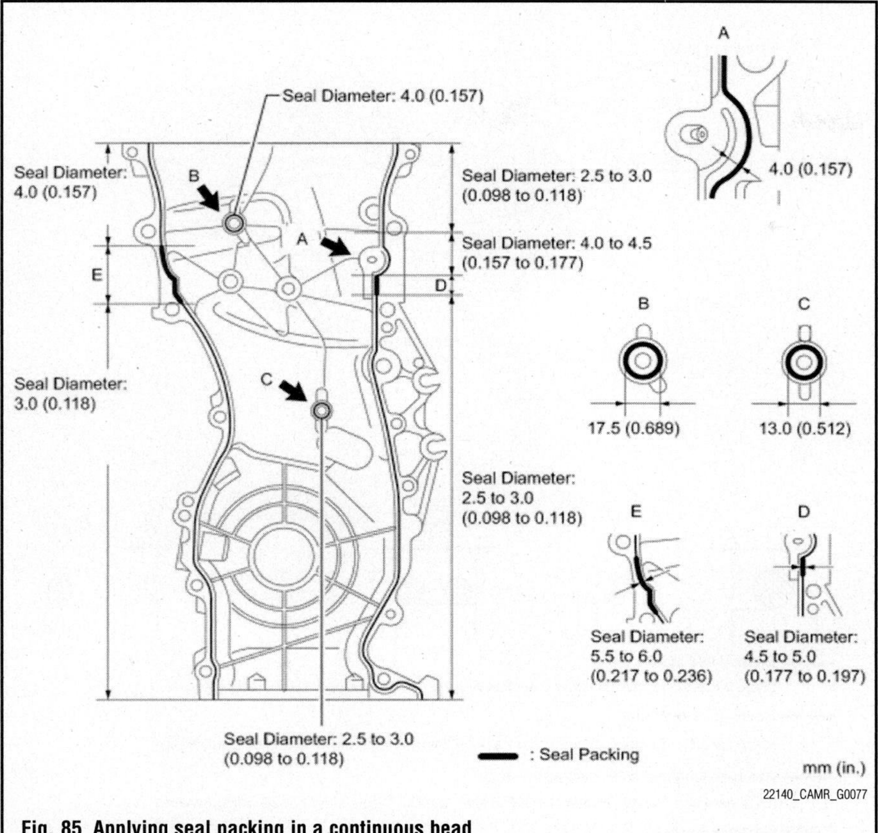

Fig. 85 Applying seal packing in a continuous bead

20. Remove the cylinder head cover sub-assembly and gasket.

21. Remove the No. 2 oil pan sub-assembly.

22. Remove the oil strainer sub-assembly.

23. Remove the oil pan sub-assembly.

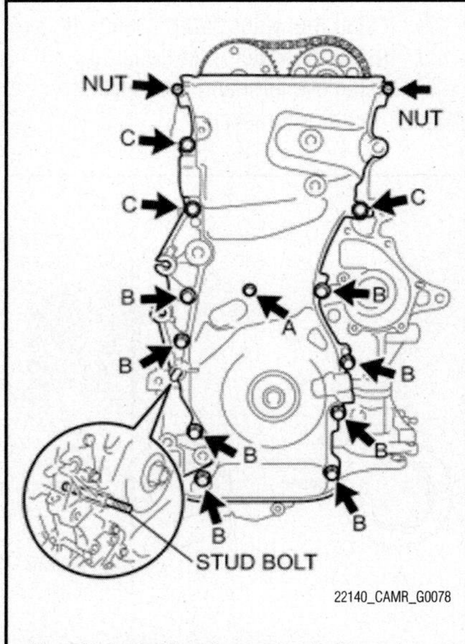

Fig. 86 Installing the twelve bolts and two nuts

24. Remove the water pump assembly.

25. Remove the timing chain cover sub-assembly, as follows:

a. Remove the 15 bolts and 2 nuts as shown in the illustration.

➡Be careful not to damage the contact surfaces of the cylinder head, cylinder block and chain cover.

b. Remove the timing chain cover by prying between the timing chain cover and cylinder head or cylinder block with a screwdriver with the tip taped.

c. Remove the 4 bolts, chain cover plate and gasket.

d. Remove the gasket.

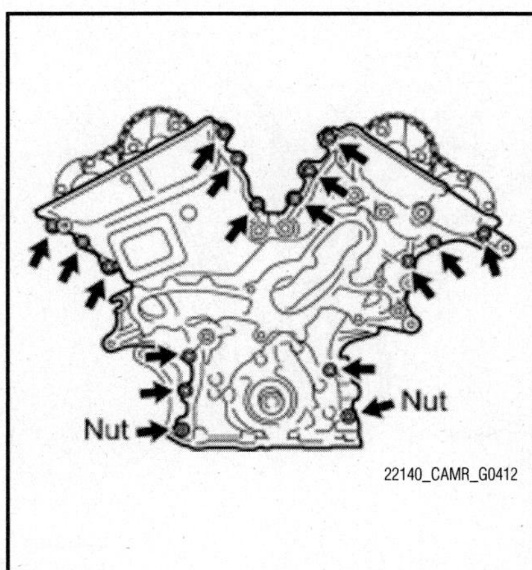

Fig. 87 Locating timing chain cover sub-assembly bolts and nuts

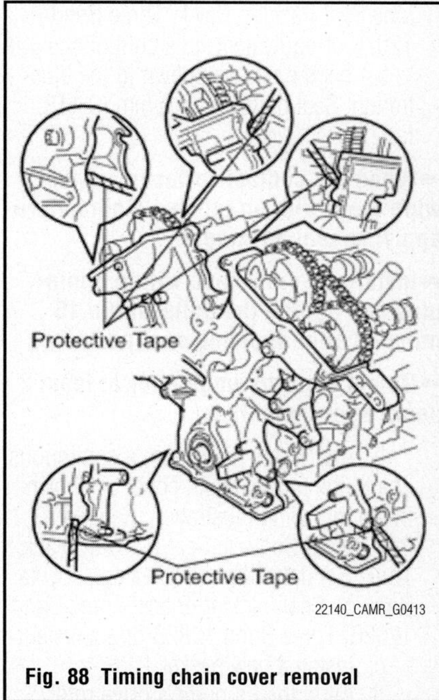

Fig. 88 Timing chain cover removal

To install:

26. Install timing gear case or timing chain cover oil seal, as follows:

➡Keep the lip free from foreign matter.

➡Do not tap on the oil seal at an angle.

➡Make sure that the oil seal edge does not stick out of the timing chain cover.

a. Apply MP grease to a new oil seal lip.

b. Using SST (SST: 09316-60011) and a hammer, tap in the oil seal until its surface is flush with the timing chain cover edge.

➡Be sure to clean and degrease the contact surfaces, especially the surfaces indicated by C in the illustration.

➡When the contact surfaces are wet, wipe them with an oil-free cloth before applying seal packing.

➡Install the chain cover within 3 minutes after applying seal packing.

➡Do not start the engine for at least 2 hours after installing.

27. Install the timing chain cover sub-assembly, as follows:

a. Apply seal packing (Toyota Genuine Seal Packing Black, Three Bond 1207B or equivalent) in a continuous line to the engine unit as shown in the illustration. Seal diameter: 3.0 mm (0.118 in.).

➡When the contact surfaces are wet, wipe them with an oil-free cloth before applying seal packing.

➡Install the crankcase within 3 minutes and tighten the bolts within 15 minutes after applying seal packing.

➡Do not start the engine for at least 2 hours after installing.

b. Apply seal packing in a continuous line to the timing chain cover as shown in the following illustration. Seal packing: Toyota Genuine Seal Packing Black, Three Bond 1207B or equivalent, Toyota Genuine Seal Packing Black, Three Bond 1282B, Three Bond 1282B or equivalent.

c. Install a new gasket.

d. Align the oil pump's drive rotor spline and the crankshaft as shown in the illustration. Install the spline and chain cover to the crankshaft.

e. Loosely install the timing chain cover with the 23 bolts and 2 nuts, but do not tighten the bolts and 2 nuts yet.

CAUTION

Make sure that there is no oil on the bolt and nut threads.

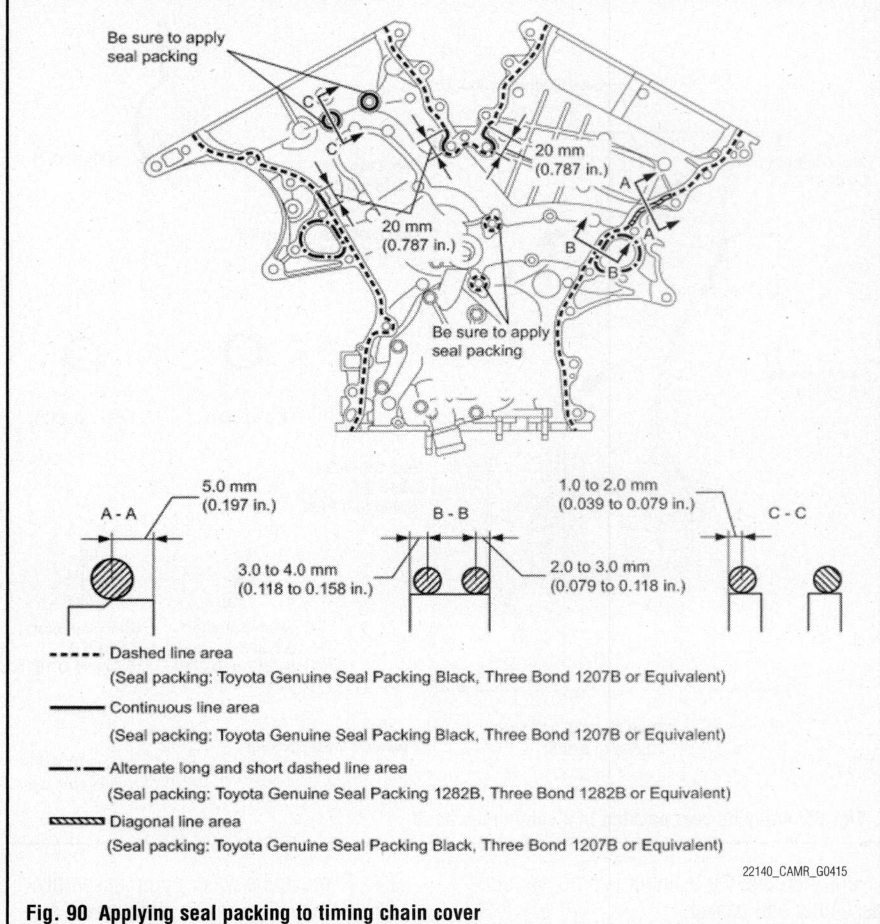

Fig. 90 Applying seal packing to timing chain cover

f. Fully tighten the bolts in this order: Area 1 and Area 2, tighten to 15 ft. lbs. (21 Nm).

g. Fully tighten the bolts in Area 3 to 15 ft. lbs. (21 Nm). Tighten the bolts and nuts in the order of upper to lower as shown in the illustration.

h. Fully tighten the bolts in Area 4 to 32 ft. lbs. (43 Nm), and to 15 ft. lbs. (21 Nm). Tighten the bolts and nuts in the order of lower to upper as shown in the illustration.

- Bolt A: 1.57 inches (40 mm)
- Bolt B: 2.17 inches (55 mm)
- Bolt C: 0.98 inches (25 mm)

28. Install the water pump assembly.
29. Install the water inlet housing.
30. Install the left hand No. 1 front engine mounting bracket.

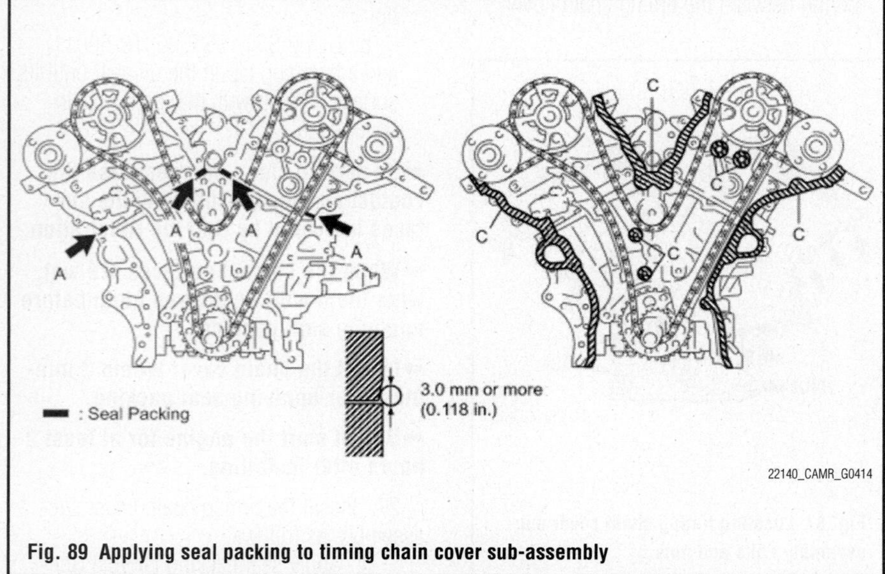

Fig. 89 Applying seal packing to timing chain cover sub-assembly

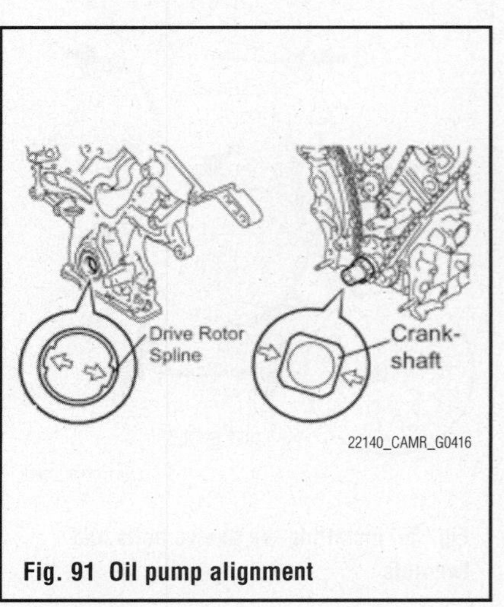

Fig. 91 Oil pump alignment

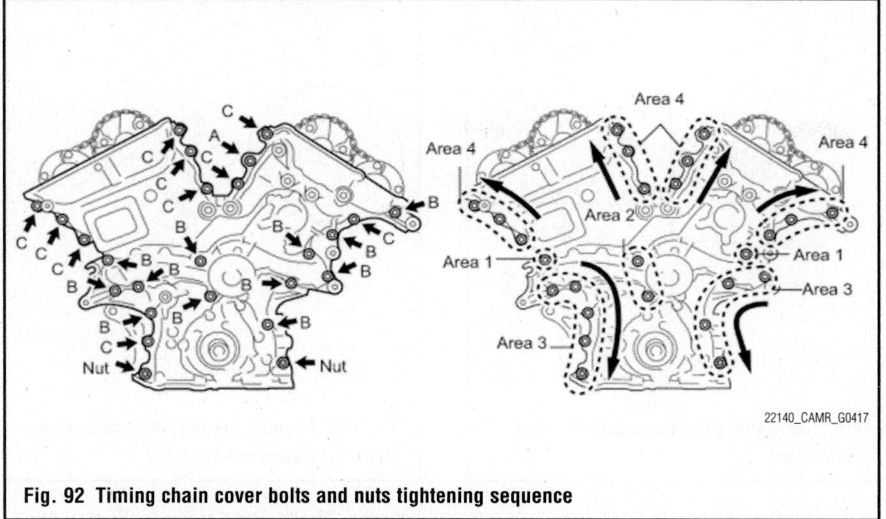

Fig. 92 Timing chain cover bolts and nuts tightening sequence

31. Install the oil pan sub-assembly.
32. Install the oil strainer sub-assembly.
33. Install the No. 2 oil pan sub-assembly.
34. Install the oil pan drain plug and gasket.
35. Install the cylinder head cover sub-assembly.
36. Install the left-hand cylinder head cover sub-assembly.
37. Install the water outlet.
38. Install the crankshaft pulley.
39. Install the oil filter.
40. Install the cylinder block water drain cock sub-assembly.
41. Install the No. 1 oil pipe.
42. Install the oil pipe.
43. Install the crankshaft position sensor.
44. Install the camshaft timing oil control valve assembly.
45. Install the camshaft position sensor.
46. Install the ventilation valve sub-assembly.
47. Install the spark plugs and ignition coil assembly.
48. Install the oil filler cap sub-assembly.
49. Install the water pump assembly.
50. Install the engine assembly with transaxle.

TIMING CHAIN & SPROCKETS

REMOVAL & INSTALLATION

2.4L Engine

See Figures 93 through 112.

1. Before servicing the vehicle, refer to the Precautions Section.
2. Remove the crankshaft position sensor plate.
3. Remove the bolt and chain tensioner slipper.
4. Remove the two bolts and chain vibration damper.
5. Remove the bolt and timing chain guide.
6. Remove the upper chain assembly.
7. Remove the crankshaft timing sprocket.
8. To remove the lower chain assembly. Turn the crankshaft by 90° counterclockwise to align the adjusting hole of the oil pump drive shaft sprocket with the groove of the oil pump.
9. Insert a 4 mm diameter bar into the adjusting hole of the oil pump drive shaft sprocket to lock the gear in position, and then remove the nut.
10. Remove the bolt, chain tensioner plate and spring.
11. Remove the chain tensioner, oil pump driven sprocket and chain.

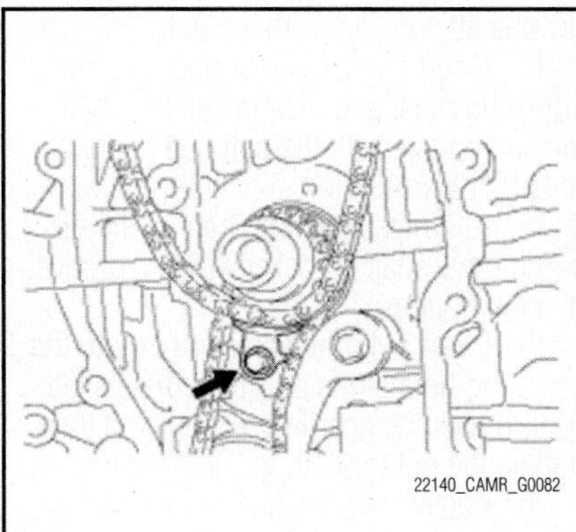

Fig. 93 Removing the bolt and timing chain guide

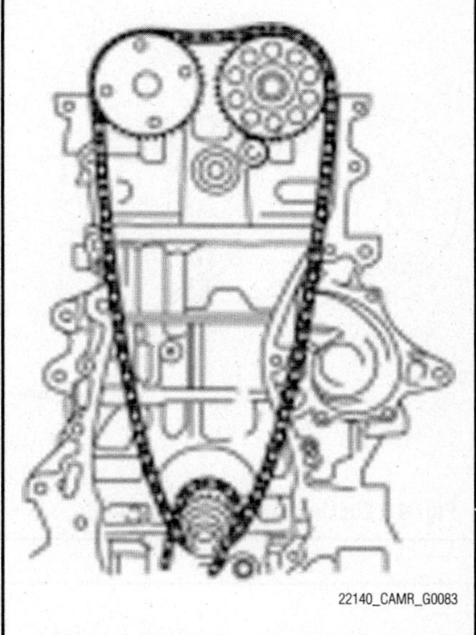

Fig. 94 Removing the upper chain assembly

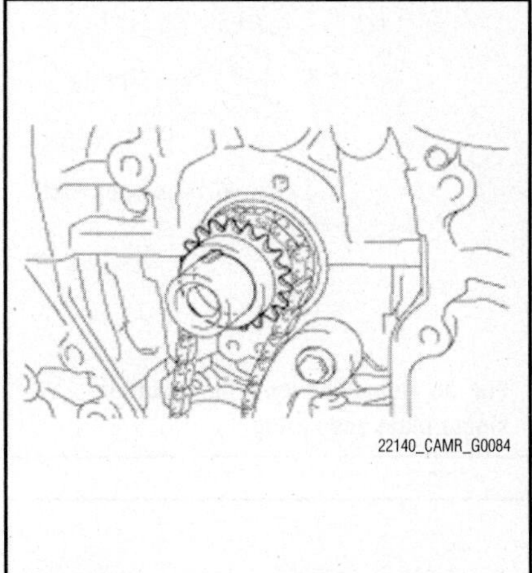

Fig. 95 Removing the crankshaft timing sprocket

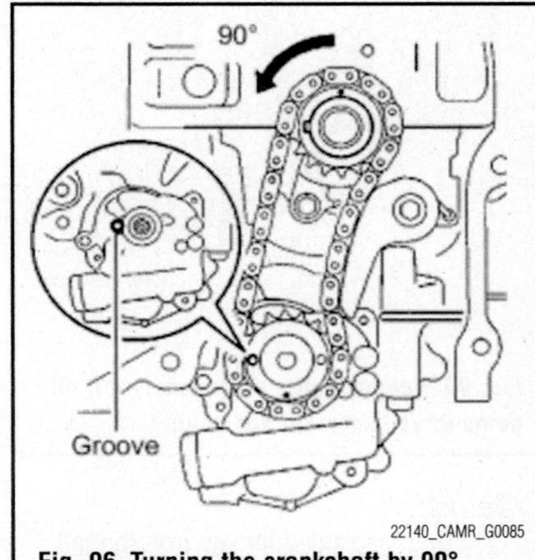

Fig. 96 Turning the crankshaft by 90° counterclockwise to align the adjusting hole

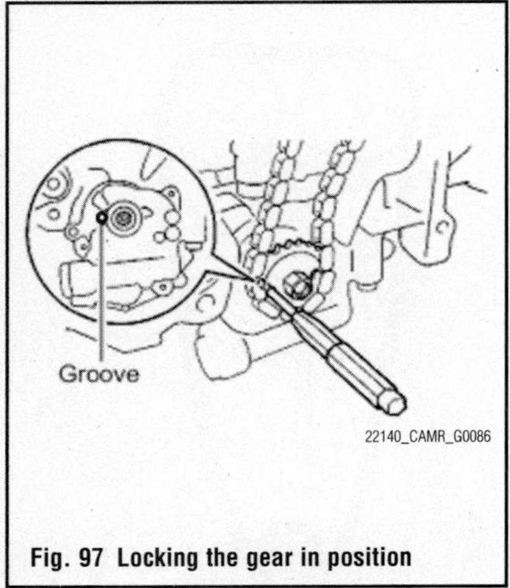

Fig. 97 Locking the gear in position

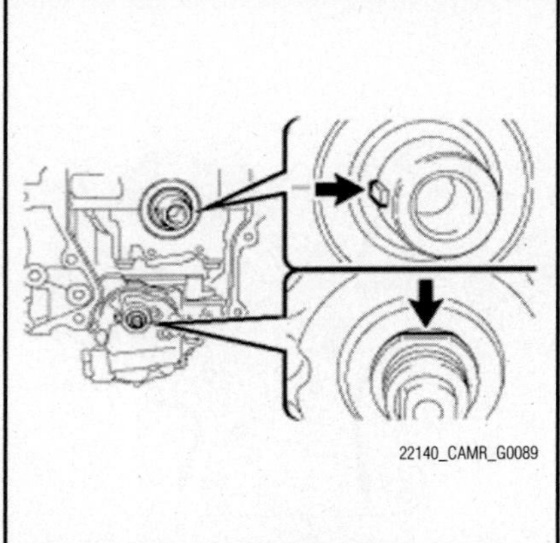

Fig. 100 Setting the crankshaft key and cutout face

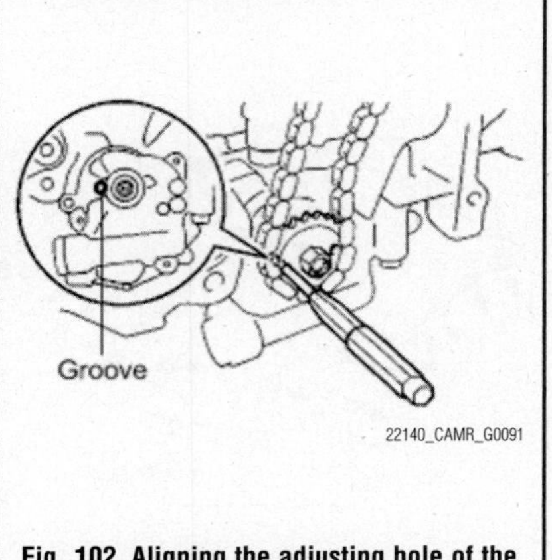

Fig. 102 Aligning the adjusting hole of the oil pump drive shaft sprocket

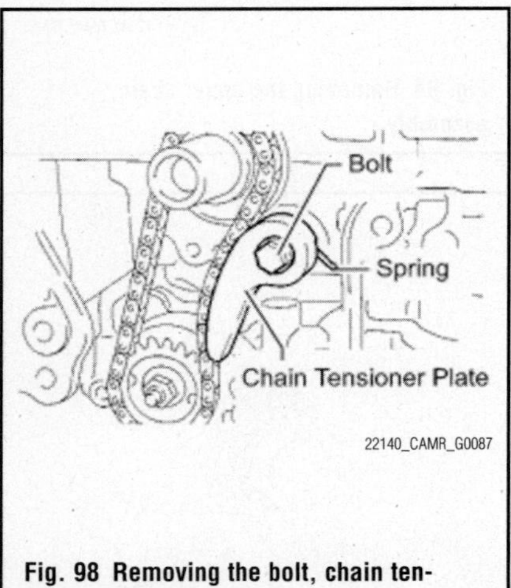

Fig. 98 Removing the bolt, chain tensioner plate and spring

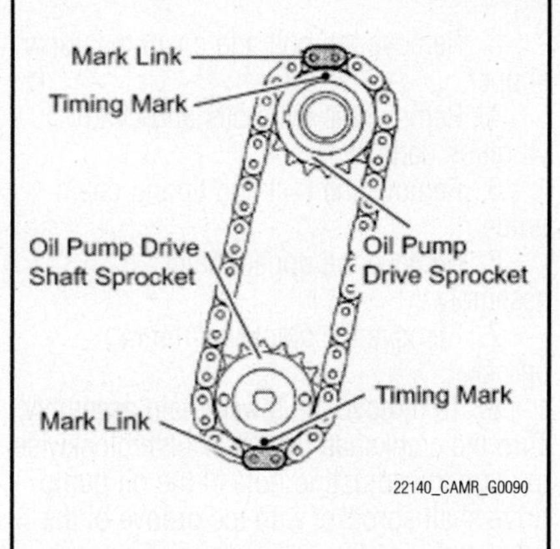

Fig. 101 Aligning the yellow mark links with the timing marks of each gear

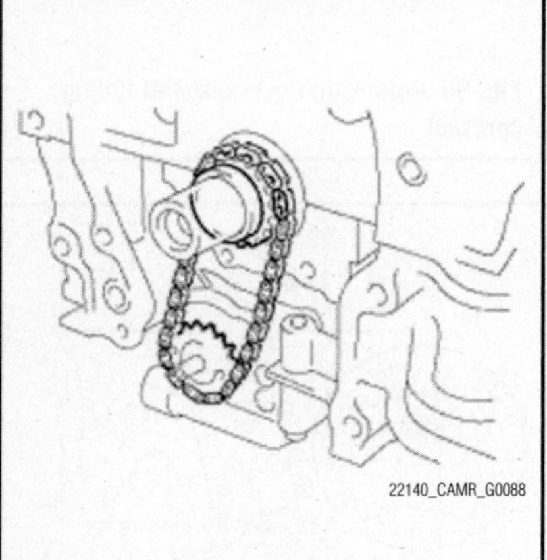

Fig. 99 Removing the chain tensioner, oil pump driven sprocket and chain

To install:

12. Set the crankshaft key into the left horizontal position.
13. Turn the drive shaft so that the cutout faces upward.
14. Align the yellow mark links with the timing marks of each gear.
15. Install the sprockets onto the crankshaft and oil pump shaft with the chain wrapped on the gears
16. Temporarily tighten the oil pump drive shaft sprocket with the nut.
17. Insert the damper spring into the adjusting hole, and then install the chain tensioner plate with the bolt and tighten to 9 ft. lbs. (12 Nm).
18. Align the adjusting hole of the oil pump drive shaft sprocket with the groove of the oil pump.
19. Insert a 4 mm diameter bar into the adjusting hole of the oil pump drive shaft gear to lock the gear in position, and then tighten the nut to 22 ft. lbs. (30 Nm).
20. Rotate the crankshaft clockwise by 90°, and align the crankshaft key to the top.
21. Install the crankshaft timing sprocket.
22. Install the chain vibration damper with the 2 bolts and tighten to 80 inch lbs. (9 Nm).
23. Set the No. 1 cylinder to TDC/compression.
24. Turn the camshafts with a wrench (using the hexagonal lobe) to align the timing marks of the camshaft timing gear with each timing mark located on the No. 1 and No. 2 bearing caps.
25. Using the crankshaft pulley bolt, turn the crankshaft to position with the key on the crankshaft upward.
26. Install the chain onto the crankshaft timing sprocket with the gold or pink mark link aligned with the timing mark on the crankshaft.
27. Using Special Tool: 09309-37010 and a hammer, tap in the crankshaft timing sprocket.
28. Align the gold or yellow link with each timing mark located on the camshaft timing gear and sprocket, then install the chain.
29. Install the chain tensioner slipper

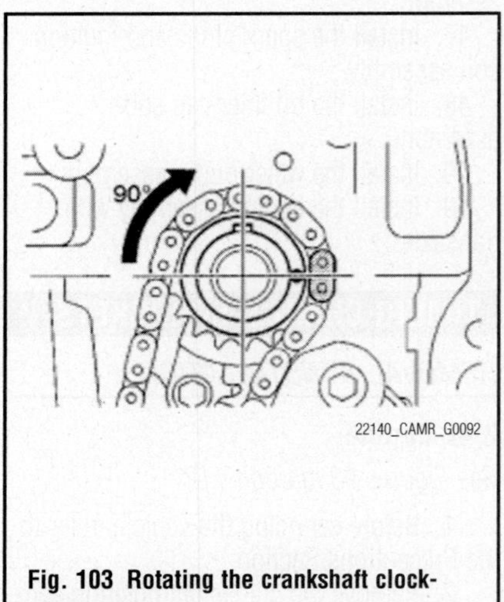

Fig. 103 Rotating the crankshaft clockwise by 90°

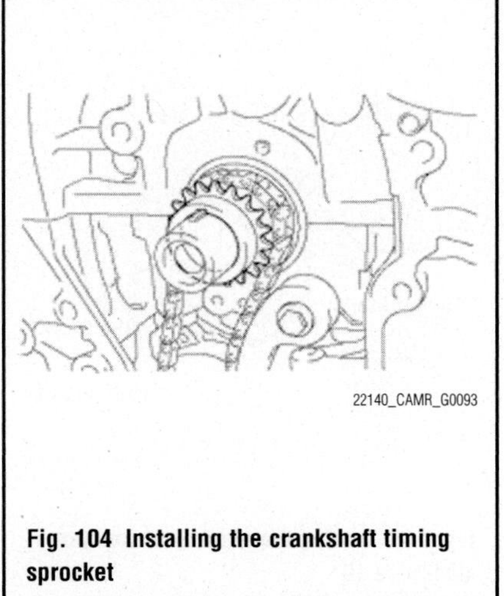

Fig. 104 Installing the crankshaft timing sprocket

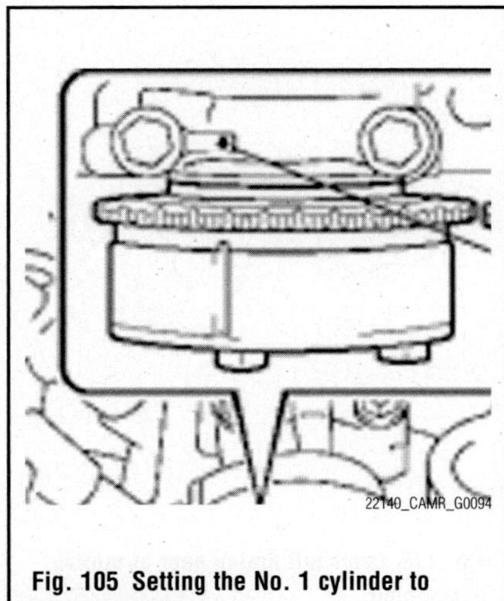

Fig. 105 Setting the No. 1 cylinder to TDC/compression

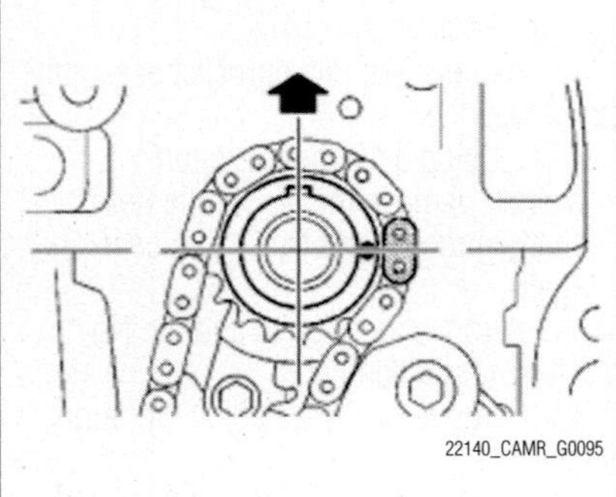

Fig. 106 Turning the crankshaft to position with the key on the crankshaft upward

with the bolt and tighten to: 14 ft. lbs. (19 Nm).

30. Install the timing chain guide with the bolt and tighten to: 80 inch lbs. (9 Nm).

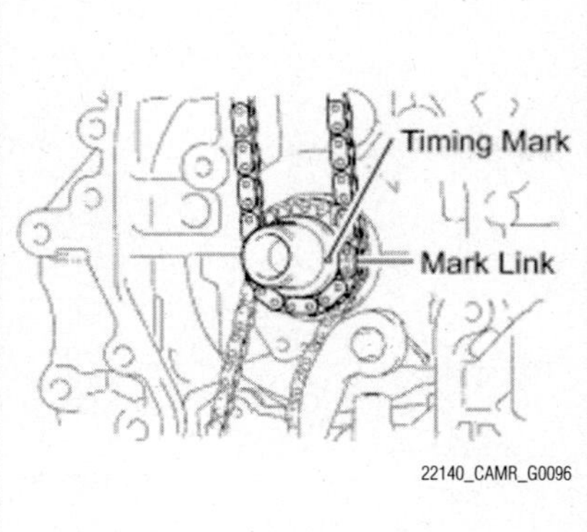

Fig. 107 Installing the chain onto the crankshaft timing sprocket with the gold or pink mark link

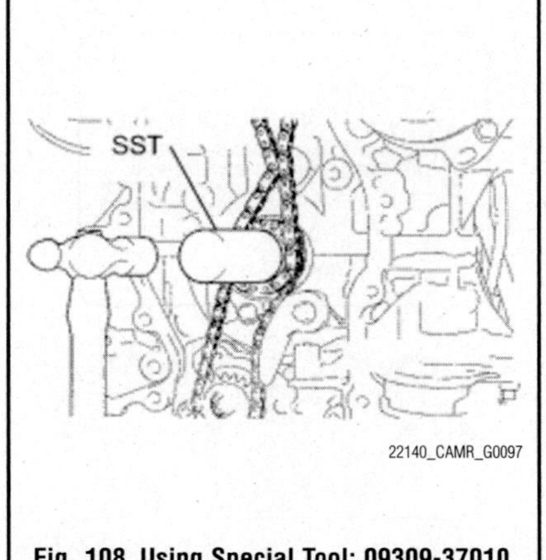

Fig. 108 Using Special Tool: 09309-37010 and a hammer to install the crankshaft timing sprocket

31. Install the crankshaft sensor plate with the "F" mark facing forward.

3.5L Engine

See Figures 113 through 124.

1. Before servicing the vehicle, refer to the Precautions Section.
2. Remove the engine assembly with transaxle.
3. Secure engine.
4. Remove the oil filler cap sub-assembly.
5. Remove the spark plugs and ignition coil assembly.
6. Remove the oil pan drain plug and gasket.
7. Remove the ventilation valve sub-assembly.
8. Remove the camshaft position sensor.
9. Remove the camshaft timing oil control valve assembly.

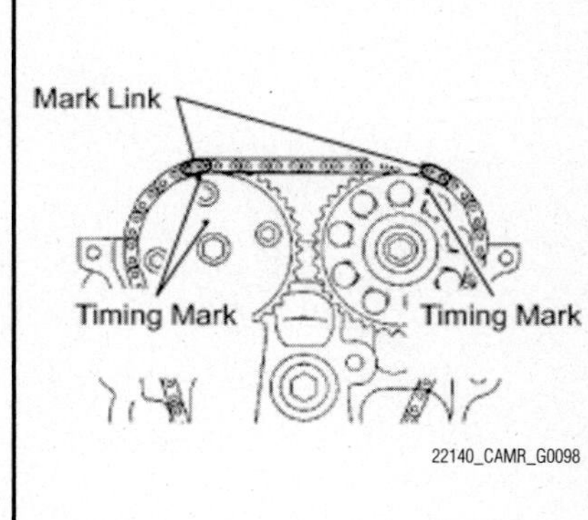

Fig. 109 Aligning the gold or yellow link with each timing mark located on the camshaft timing gear and sprocket

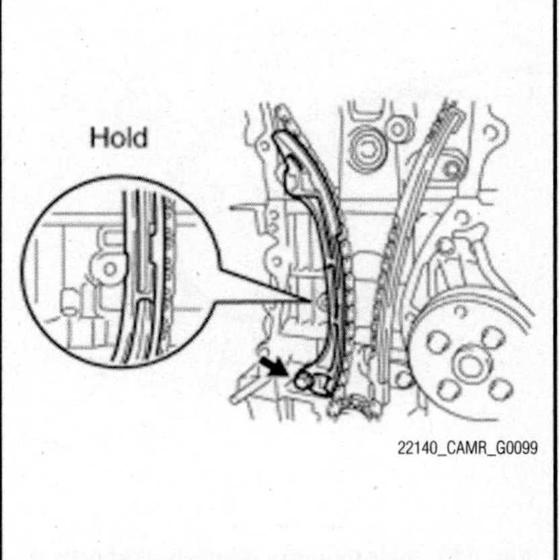

Fig. 110 Installing the chain tensioner slipper and bolt

10. Remove crankshaft position sensor.
11. Remove the No. 1 oil pipe.
12. Remove the oil pipe.
13. Remove the cylinder block water drain cock sub-assembly.
14. Remove the oil filter.
15. Remove the crankshaft pulley.
16. Remove the left hand No. 1 front engine mounting bracket.
17. Remove the water inlet housing.
18. Remove the water outlet.
19. Remove the left-hand cylinder head cover sub-assembly and gasket.
20. Remove the cylinder head cover sub-assembly and gasket.
21. Remove the No. 2 oil pan sub-assembly.
22. Remove the oil strainer sub-assembly.
23. Remove the oil pan sub-assembly.
24. Remove the water pump assembly.
25. Remove the timing chain cover and seal.

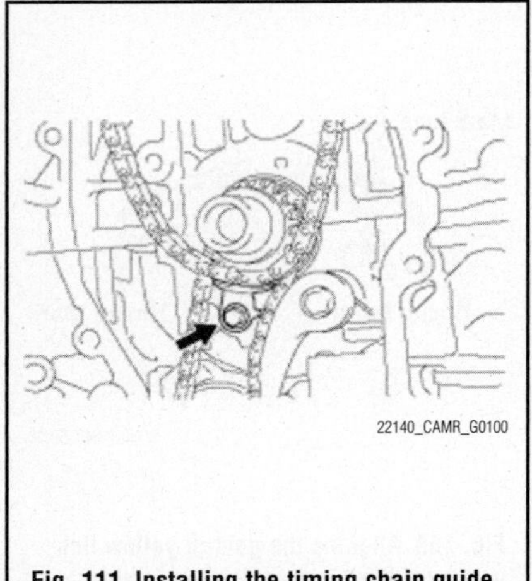

Fig. 111 Installing the timing chain guide and bolt

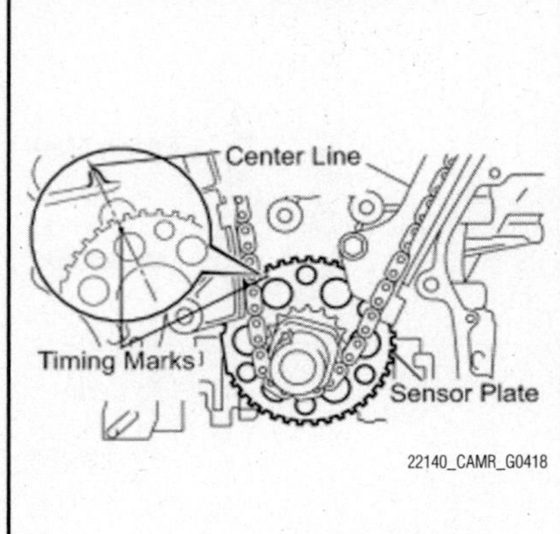

Fig. 113 Set the timing mark on the crank angle sensor plate

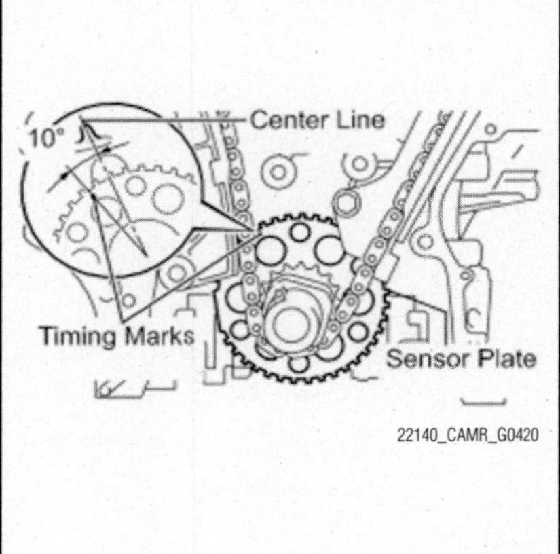

Fig. 115 Turning the crankshaft counter-clockwise 10°

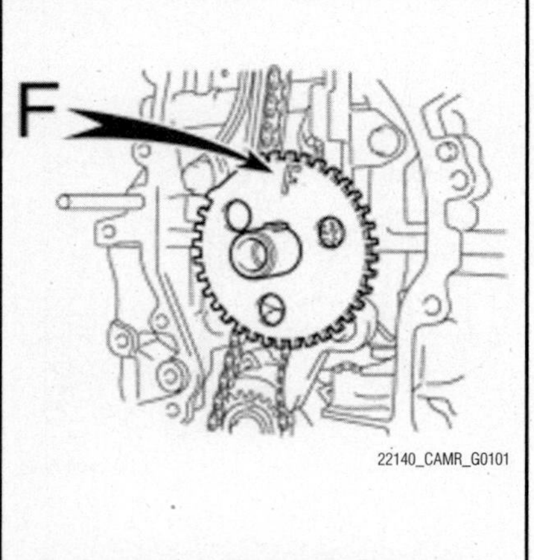

Fig. 112 Installing the crankshaft sensor plate with the "F" mark facing forward

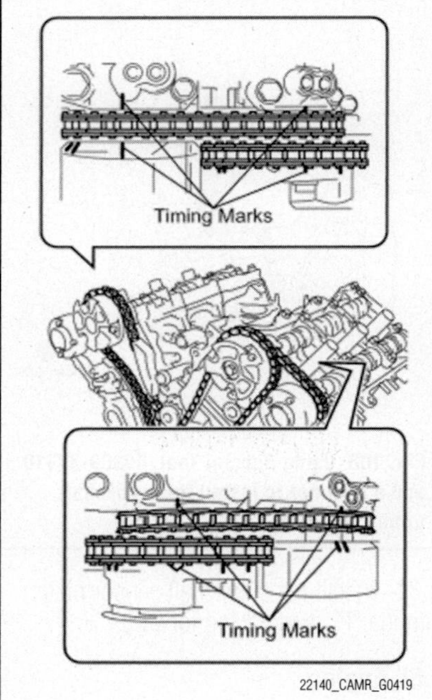

Fig. 114 Check that the timing marks of the camshaft timing gears

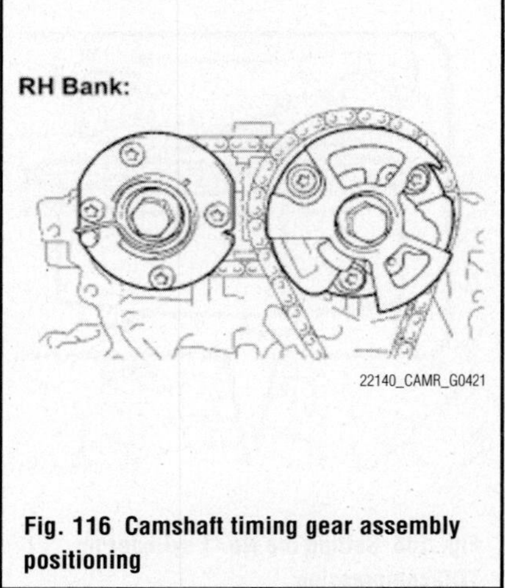

Fig. 116 Camshaft timing gear assembly positioning

26. Set no. 1 cylinder to TDC/compression, as follows:

a. Temporarily tighten the pulley set bolt. Set the timing mark on the crank angle sensor plate to the RH block bore center line (TDC/compression).

b. Check that the timing marks of the camshaft timing gears are aligned with the timing marks of the bearing cap as shown in the illustration. If not, turn the crankshaft 1 revolution (360°) and align the timing marks as above.

27. Remove the No. 1 chain tensioner assembly, as follows:

a. Move the stopper plate upward to release the lock, and push the plunger deep into the tensioner.

b. Move the stopper plate downward to set the lock, and insert a pin of φ1.27 mm (0.05 in.) into the stopper plate's hole.

c. Remove the 2 bolts and chain tensioner.

28. Remove the chain tensioner slipper.

29. Remove the chain sub-assembly, as follows:

a. Turn the crankshaft counterclockwise 10° to loosen the chain of the crankshaft timing sprocket.

b. Remove the pulley set bolt.

c. Remove the chain from the crankshaft timing sprocket and place it on the crankshaft.

d. Turn the camshaft timing gear assembly on the right hand bank clockwise (approximately 60°) and set it as shown in the illustration. Be sure to loosen the chain between the banks.

e. Remove the chain.

30. Remove the idle sprocket assembly, as follows:

a. Using a 10 mm hexagon wrench, remove the No. 2 idle gear shaft, sprocket and No. 1 idle gear shaft.

31. Remove the 2 bolts and the No. 1 chain vibration damper.

32. Remove the No. 2 chain vibration damper.

33. Remove the crankshaft timing sprocket, as follows:

a. Remove the pulley set bolt.

b. Remove the crankshaft timing gear from the crankshaft.

c. Remove the 2 pulley set keys from the crankshaft.

34. Remove the camshaft timing gears and No. 2 chain (for Right-hand Bank), as follows:

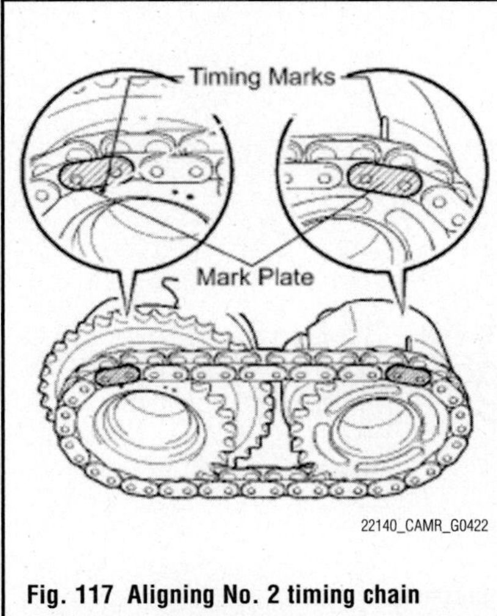

Fig. 117 Aligning No. 2 timing chain

a. While raising up the No. 2 chain tensioner, insert a pin of 1.0 mm (0.039 in.) into the hole to hold it.

b. Hold the hexagonal portion of the camshaft with a wrench, and remove the 2 bolts and 2 camshaft timing gears.

➡Be careful not to damage the cylinder head with the wrench.

➡Do not disassemble the camshaft timing gear assemblies.

c. Remove the No. 2 chain.

35. Remove the bolt and No. 2 chain tensioner.

36. Remove the camshaft timing gears and No. 2 chain (for Left-hand Bank), as follows:

a. While pushing down on the No. 3 chain tensioner, insert a pin of 1.0 mm (0.039 in.) into the hole to hold it.

b. Hold the hexagonal portion of the camshaft with a wrench, and remove the 2 bolts and 2 camshaft timing gears.

➡Be careful not to damage the cylinder head with the wrench.

➡Do not disassemble the camshaft timing gear assemblies.

c. Remove the No. 2 chain.

37. Remove the bolt and the No. 3 chain tensioner.

To install:

38. Install the No. 2 chain tensioner assembly with the bolt and tighten to 15 ft. lbs. (21 Nm).

39. While pushing in the tensioner, insert a pin of 1.0 mm (0.039 in.) into the hole to hold it.

40. Install the camshaft timing gears and No. 2 chain (for Right-hand Bank), as follows:

a. Align the mark plate with the timing marks (1-dot mark) of the camshaft timing gears as shown.

b. Apply a light coat of engine oil to the bolt threads and bolt-seating surface.

c. Align the knock pin of the camshaft with pin hole of the camshaft timing gear. Install the camshaft timing gear and the right camshaft timing exhaust gear with the No. 2 chain installed.

d. Hold the hexagonal portion of the camshaft with the wrench and tighten the two bolts to 74 ft. lbs. (100 Nm).

e. Remove the pin from the No. 2 chain tensioner.

41. Install the No. 3 chain tensioner assembly with the bolt and tighten to 15 ft. lbs. (21 Nm).

42. While pushing in the tensioner, insert a pin of 1.0 mm (0.039 in.) into the hole to hold it.

43. Install the camshaft timing gears and No. 2 chain (for Left-hand Bank), as follows:

a. Align the mark plate (yellow) with the timing marks (2-dot mark) of the camshaft timing gears as shown.

b. Apply a light coat of engine oil to the bolt threads and bolts seating surface.

c. Align the knock pin of the camshaft with pin hole of the camshaft timing gear. Install the camshaft timing gear and the left camshaft timing exhaust gear with the No. 2 chain installed.

d. Hold the hexagonal portion of the camshaft with the wrench and tighten the two bolts to 74 ft. lbs. (100 Nm).

e. Remove the pin from the No 2 chain tensioner.

44. Install the No. 1 and 2 chain vibration dampers.

45. Install the timing gear set keys and crankshaft timing sprocket as shown in the illustration.

46. Install the idle sprocket assembly, as follows:

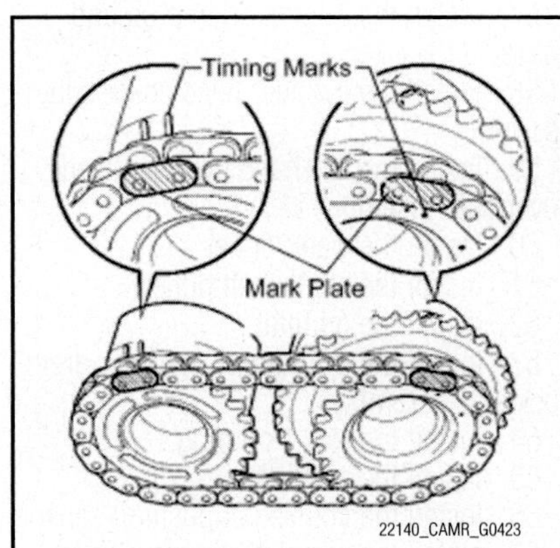

Fig. 118 Aligning No. 2 timing chain (bank 2)

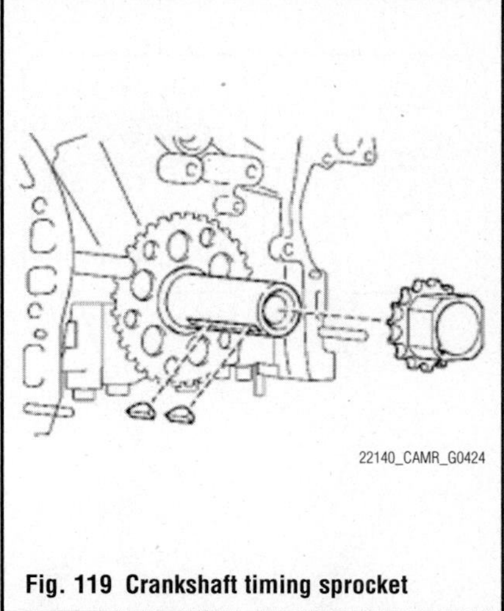

Fig. 119 Crankshaft timing sprocket

a. Apply a light coat of engine oil to the rotating surface of the No. 1 idle gear shaft.

b. Temporarily install the No. 1 idle gear shaft and idle sprocket with the No. 2 idle gear shaft while aligning the knock pin of the No. 1 idle gear with the knock pin groove of the cylinder block. Be careful of the idle gear direction.

c. Using a 10 mm hexagon wrench, tighten the No. 2 idle gear shaft to 44 ft. lbs. (60 Nm).

d. After installing the idle sprocket assembly, check that the idle sprocket turns smoothly.

47. Install the chain sub-assembly, as follows:

a. Align the mark plate and timing marks as shown in the illustration and install the chain. The camshaft mark plate is orange.

b. Do not pass the chain over the crankshaft, just put it on.

c. Turn the camshaft timing gear assembly on the right bank counterclockwise to tighten the chain between the banks.

➡When the idle sprocket assembly is reused, align the timing chain plate with the mark on the sprocket in order to tighten the chain between the banks.

d. Align the mark plate and timing marks as shown in the illustration and install the chain onto the crankshaft timing sprocket. The crankshaft to mark plate is yellow.

e. Temporarily tighten the pulley set bolt.

f. Turn the crankshaft clockwise to set it to the right-hand block bore more centerline. (TDC/compression).

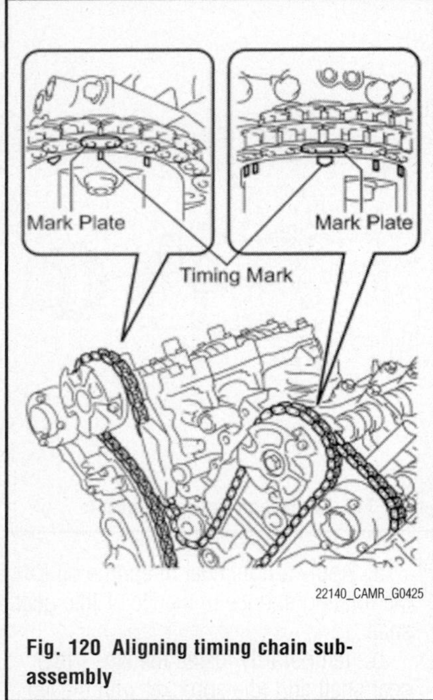

Fig. 120 Aligning timing chain sub-assembly

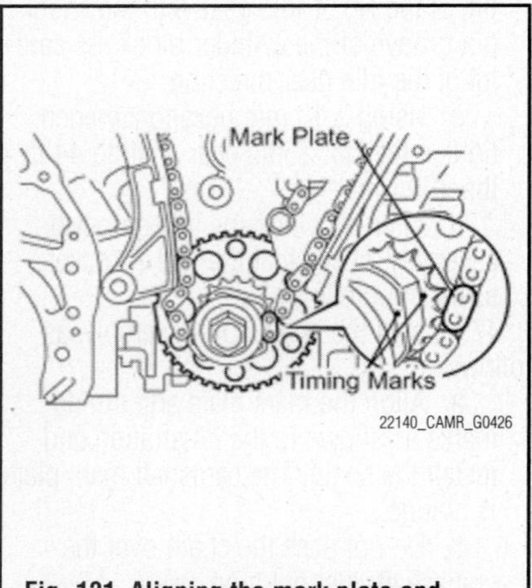

Fig. 121 Aligning the mark plate and timing mark

48. Install the chain tensioner slipper.
49. Install the No. 1 chain tensioner assembly, as follows:
   a. Move the stopper plate upward to release the lock, and push the plunger deep into the tensioner.
   b. Move the stopper plate downward to set the lock, and insert a hexagon wrench into the hole of the stopper plate.
   c. Install the No. 1 chain tensioner with the 2 bolts and tighten to 7 ft. lbs. (10 Nm).
   d. Remove the lock pin of the No. 1 chain tensioner. Check that each timing mark is aligned with the crankshaft at TDC/compression.
   e. Remove the pulley set bolt.

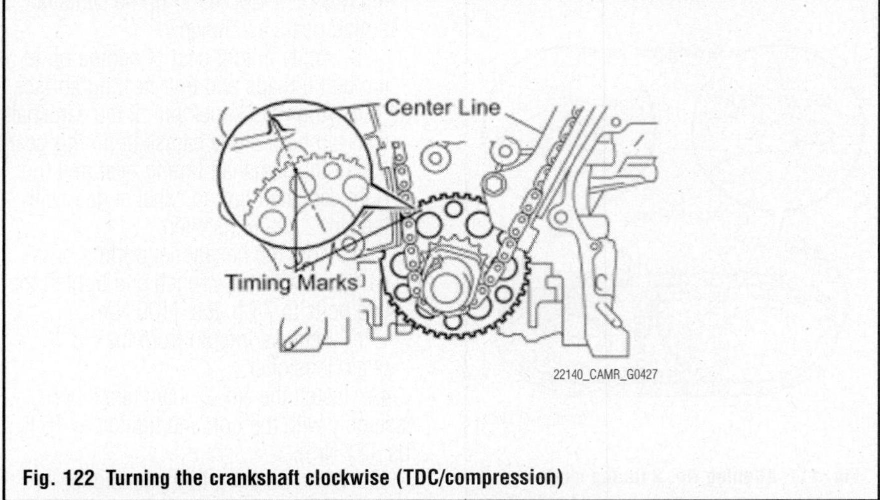

Fig. 122 Turning the crankshaft clockwise (TDC/compression)

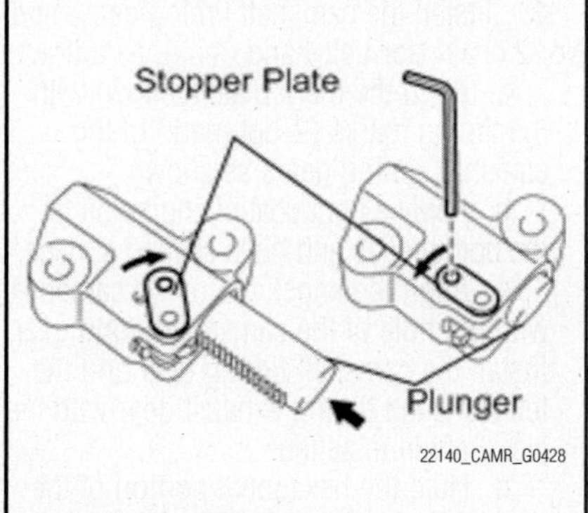

Fig. 123 Set chain tensioner plunger position

50. Install timing chain cover and seal.
51. Install the water pump assembly.
52. Install the water inlet housing.
53. Install the left hand No. 1 front engine mounting bracket.
54. Install the oil pan sub-assembly.
55. Install the oil strainer sub-assembly.
56. Install the No. 2 oil pan sub-assembly.
57. Install the oil pan drain plug and gasket.
58. Install the cylinder head cover sub-assembly.
59. Install the left-hand cylinder head cover sub-assembly.
60. Install the water outlet.
61. Install the crankshaft pulley.
62. Install the oil filter.
63. Install the cylinder block water drain cock sub-assembly.
64. Install the No. 1 oil pipe.
65. Install the oil pipe.
66. Install the crankshaft position sensor.
67. Install the camshaft timing oil control valve assembly.
68. Install the camshaft position sensor.
69. Install the ventilation valve sub-assembly.
70. Install the spark plugs and ignition coil assembly.
71. Install the oil filler cap sub-assembly.
72. Install the engine assembly with transaxle.

VALVE LASH

ADJUSTMENT

➡Keep the lash adjuster free of dirt and foreign objects.

➡Only use clean engine oil.

1. Place the lash adjuster into a container filled with engine oil.
2. Insert the SST's (SST: 09276-75010) tip into the lash adjuster's plunger and use the tip to press down on the check ball inside the plunger.
3. Squeeze the SST and lash adjuster together to move the plunger up and down 5 to 6 times.
4. Check the movement of the plunger and bleed the air. OK: Plunger moves up and down.

➡When bleeding air from the high-pressure chamber, make sure that the tip of the SST is actually pressing the check ball as shown in the illustration. If the check ball is not pressed, air will not bleed.

5. After bleeding the air, remove the SST. Then, try to press the plunger quickly and firmly with a finger. OK: Plunger is very difficult to move. If the result is not as specified, replace the lash adjuster.
6. Install the lash adjusters.

➡Install the lash adjuster to the same place where it was removed from.

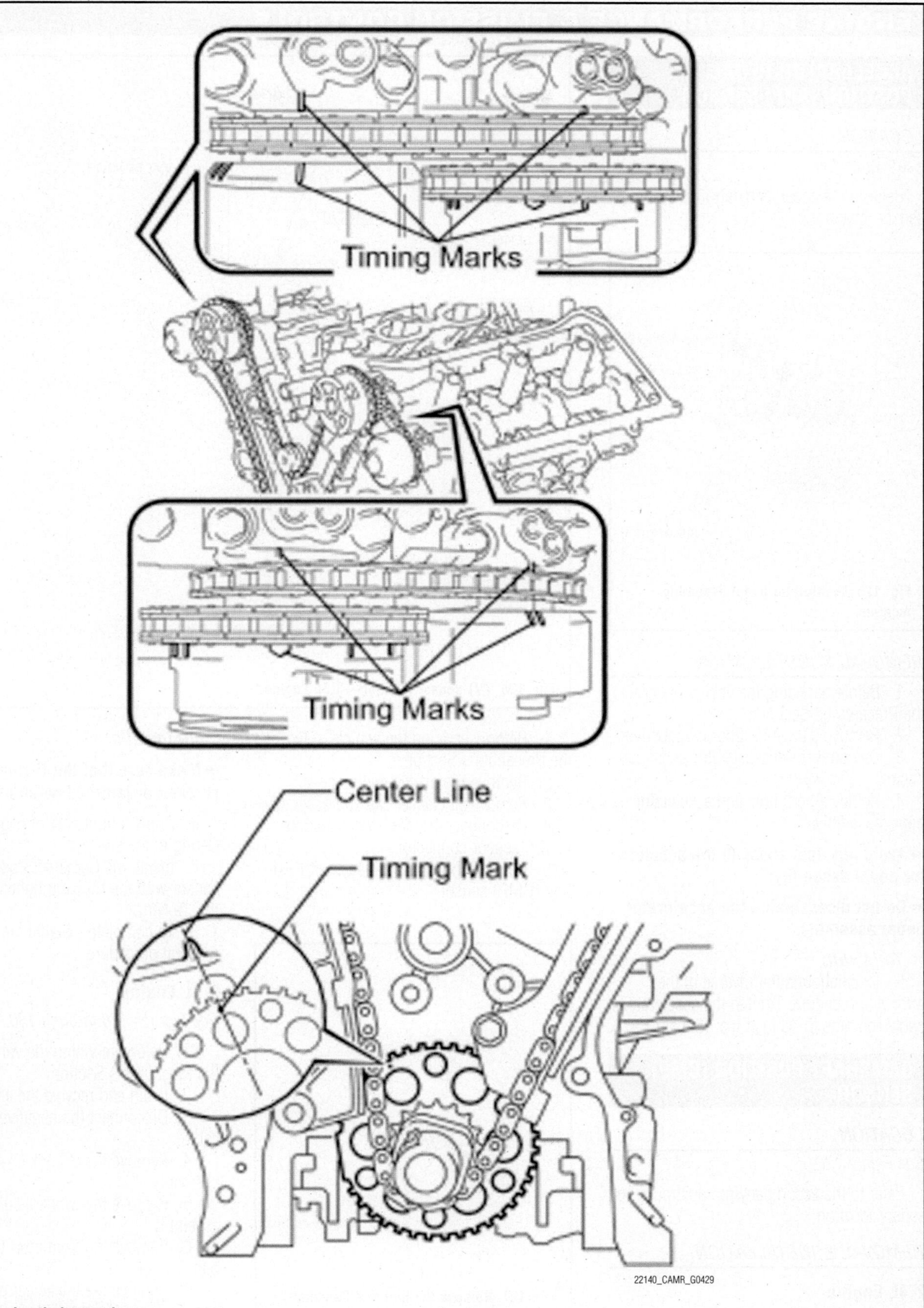

Fig. 124 Aligning timing marks

ENGINE PERFORMANCE & EMISSION CONTROLS

ACCELERATOR PEDAL POSITION (APP) SENSOR

LOCATION

See Figure 125.

Refer to the accompanying illustration for sensor location.

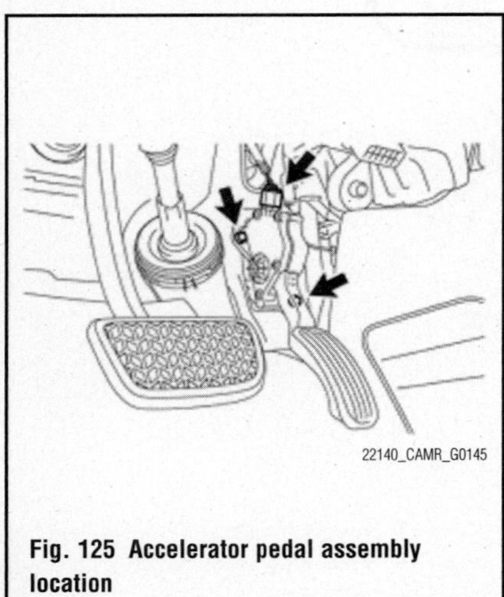

Fig. 125 Accelerator pedal assembly location

REMOVAL & INSTALLATION

1. Before servicing the vehicle, refer to the Precautions Section.
2. Remove left center floor carpet cover.
3. Disconnect the accelerator pedal connector.
4. Remove the 2 nuts and accelerator pedal assembly.

➡Avoid physical shock to the accelerator pedal assembly.

➡Do not disassemble the accelerator pedal assembly.

To install:

5. Installation is the reverse of the removal procedure. Tighten the accelerator pedal rod nuts to 43 inch lbs. (5.4 Nm).

CAMSHAFT POSITION (CMP) SENSOR

LOCATION

See Figure 126.

Refer to the accompanying illustration for sensor location.

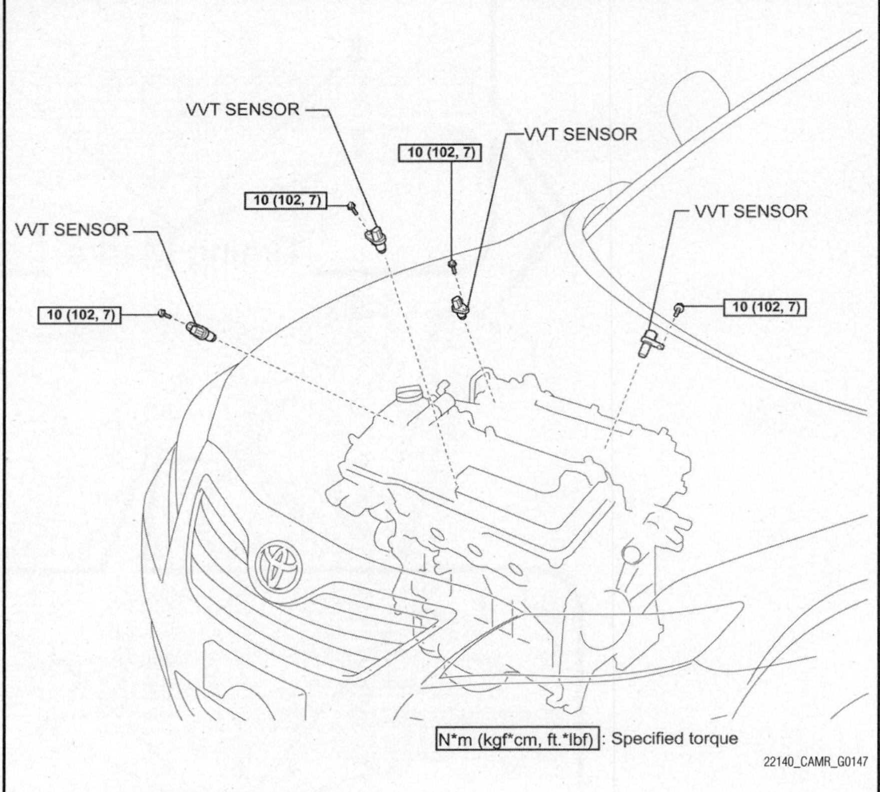

Fig. 126 VVT sensor location—3.5L Engine

REMOVAL & INSTALLATION

2.4L Engine

See Figure 127.

1. Before servicing the vehicle, refer to the Precautions Section.
2. Remove the plastic engine cover.
3. Remove air cleaner cap sub-assembly.
4. Disconnect the Camshaft Position (CMP) sensor connector.
5. Remove the bolt and Camshaft Position (CMP) sensor.

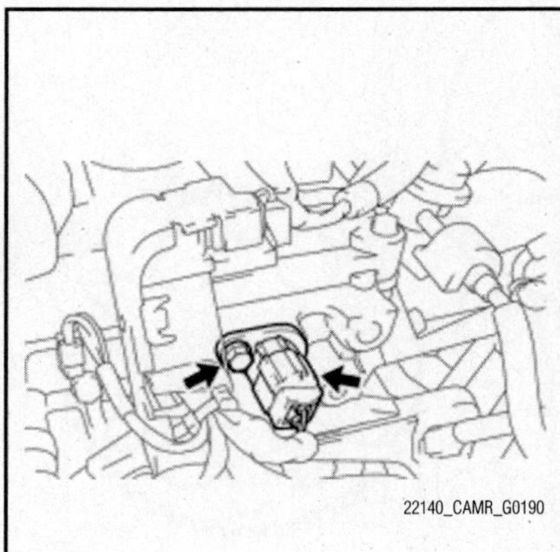

Fig. 127 Remove the bolt and Camshaft Position (CMP) sensor—2.4L Engine

To install:

➡Make sure that the O-ring is not cracked or jammed when installing it.

6. Apply a light coat of engine oil to the O-ring of the sensor.
7. Install the Camshaft Position (CMP) sensor with the bolt and tighten to: 80 inch lbs. (9 Nm).
8. To complete installation, reverse removal procedure.

3.5L Engine

See Figures 128 through 130.

1. Before servicing the vehicle, refer to the Precautions Section.
2. Drain and recycle the engine coolant.
3. Disconnect the negative battery cable.
4. Remove the V-bank cover sub-assembly.
5. Remove the windshield wiper link assembly.
6. Remove the front cowl top outside panel.
7. Remove the Intake camshaft VVT sensor (Bank 1), as follows:

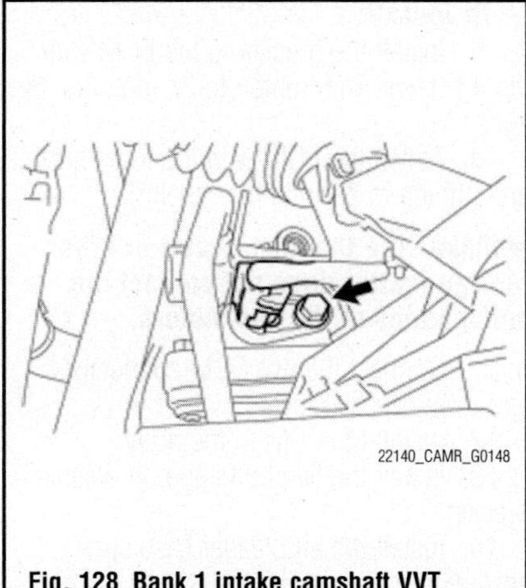

Fig. 128 Bank 1 intake camshaft VVT sensor

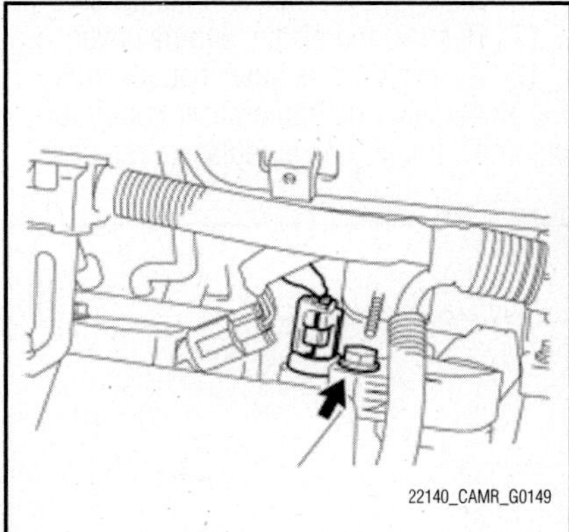

Fig. 129 Bank 1 exhaust camshaft VVT sensor

a. Disconnect the VVT sensor connector.

b. Remove the bolt and VVT sensor.

8. Remove the Exhaust camshaft VVT sensor (Bank 1), as follows:

a. Disconnect the VVT sensor connector.

b. Remove the bolt and VVT sensor.

9. Remove the Exhaust camshaft VVT sensor (Bank 2), as follows:

a. Disconnect the VVT sensor connector.

b. Remove the bolt and VVT sensor.

10. Remove the Intake camshaft VVT sensor (Bank 2), as follows:

a. Disconnect the VVT sensor connector.

b. Remove the bolt and VVT sensor.

To install:

11. Install the VVT sensors and tighten to 7 ft. lbs. (10 Nm).

Fig. 130 Bank 2 intake camshaft VVT sensor

a. Connect the VVT sensor connectors.

12. To complete installation, reverse the remaining removal procedure.

CRANKSHAFT POSITION (CKP) SENSOR

LOCATION

See Figures 131 and 132.

Refer to the accompanying illustrations for sensor location.

REMOVAL & INSTALLATION

2.4L Engine

See Figure 133.

1. Before servicing the vehicle, refer to the Precautions Section.

2. Disconnect the negative battery cable.

3. Remove the front right wheel.

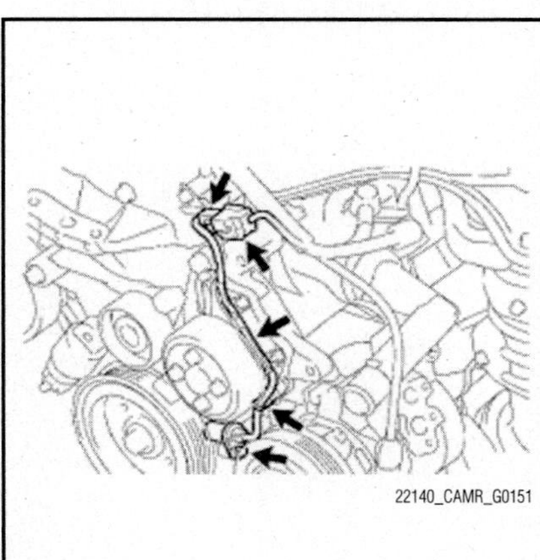

Fig. 131 Crankshaft Position (CKP) sensor location—2.4L engine

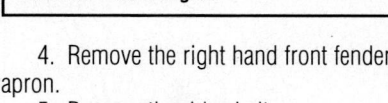

Fig. 132 Crankshaft Position (CKP) sensor location—3.5L engine

4. Remove the right hand front fender apron.

5. Remove the drive belt.

6. Remove the alternator assembly.

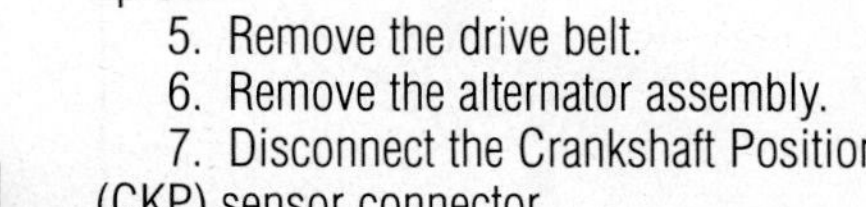

7. Disconnect the Crankshaft Position (CKP) sensor connector.

8. Remove the connector clamp and wire harness clamp.

9. Remove the wire harness clamp bracket from the wire harness.

10. Remove the bolt, then remove the Crankshaft Position (CKP) sensor.

To install:

11. Apply a light coat of engine oil to the O-ring on the Crankshaft Position (CKP) sensor.

12. Install the Crankshaft Position (CKP) sensor with the bolt and tighten to 7ft. lbs. (10 Nm).

13. Connect the Crankshaft Position (CKP) sensor connector.

14. The remainder of installation is the reverse of the removal procedure.

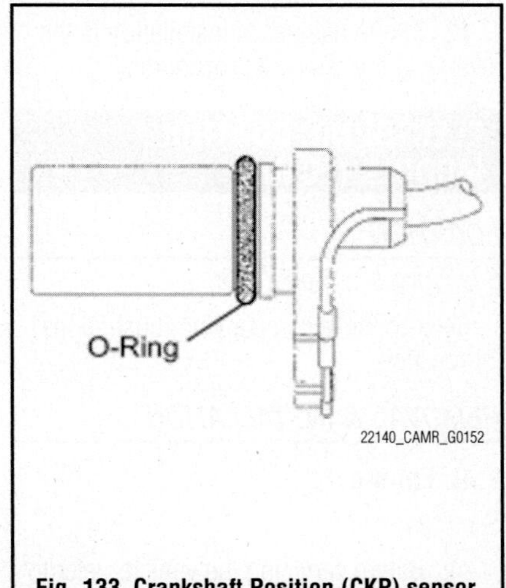

Fig. 133 Crankshaft Position (CKP) sensor

3.5L Engine

See Figure 134.

1. Before servicing the vehicle, refer to the Precautions Section.
2. Disconnect the negative battery cable.
3. Remove alternator assembly.
4. Disconnect the cooler compressor assembly.
5. Remove the Crankshaft Position (CKP) sensor connector.
6. Remove the bolt, and then remove the Crankshaft Position (CKP) sensor.

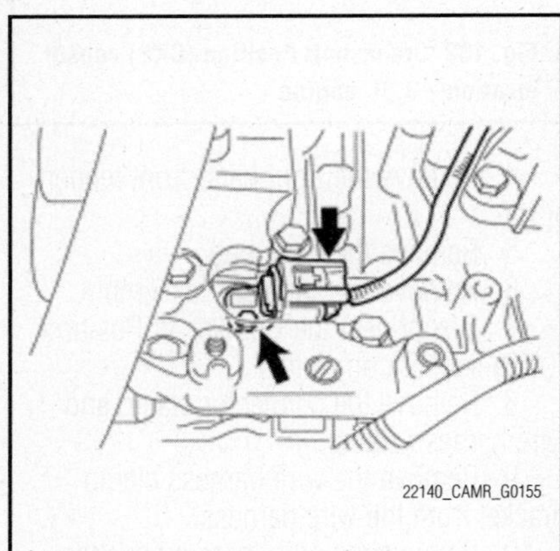

Fig. 134 Installing Crankshaft Position (CKP) sensor—3.5L Engine

To install:

7. Apply a light coat of engine oil to the O-ring on the Crankshaft Position (CKP) sensor.
8. Install the Crankshaft Position (CKP) sensor with the bolt and tighten to 7ft. lbs. (10 Nm).
9. Connect the Crankshaft Position (CKP) sensor connector.
10. The remainder of installation is the reverse of the removal procedure.

ELECTRONIC CONTROL MODULE (ECM)

LOCATION

See Figures 135 and 136.

Refer to the accompanying illustrations for location.

REMOVAL & INSTALLATION

2.4L Engine

See Figure 137.

1. Before servicing the vehicle, refer to the Precautions Section.
2. Disconnect the negative battery cable.
3. Remove the plastic engine cover.
4. Remove the air cleaner inlet assembly.
5. Remove the air cleaner cap sub-assembly.
6. Remove the air cleaner case sub-assembly.
7. Remove the two bolts and air cleaner bracket.
8. Disconnect the two ECM connectors by raising the two levers. While pushing the locks on the two levers, disconnect the two ECM connectors.

➡After disconnecting the connectors, make sure that dirt, water or other foreign matter does not contact the connections of the connectors.

a. Remove the ECM with the bracket and three bolts.
b. Remove the four screws and ECM brackets.

To install:

9. Install the bracket to the ECM with the 4 screws, and tighten to 27 inch lbs. (3 Nm).
10. Attach the ECM with the three nuts and tighten to 71 inch lbs. (8 Nm).

➡Make sure that dirt, water or other foreign matter does not contact the connections of the connectors.

11. Connect the two ECM connectors and lower the two levers.
12. Install the ECM to the body.
13. Install the two bolts and air cleaner bracket.
14. Install the air cleaner case sub-assembly.
15. Install the air cleaner cap sub-assembly.
16. Install the air cleaner inlet assembly.
17. Remove the plastic engine cover.
18. Connect the negative battery cable.
19. Register the immobilizer communication ID. If the ECM is replaced, register

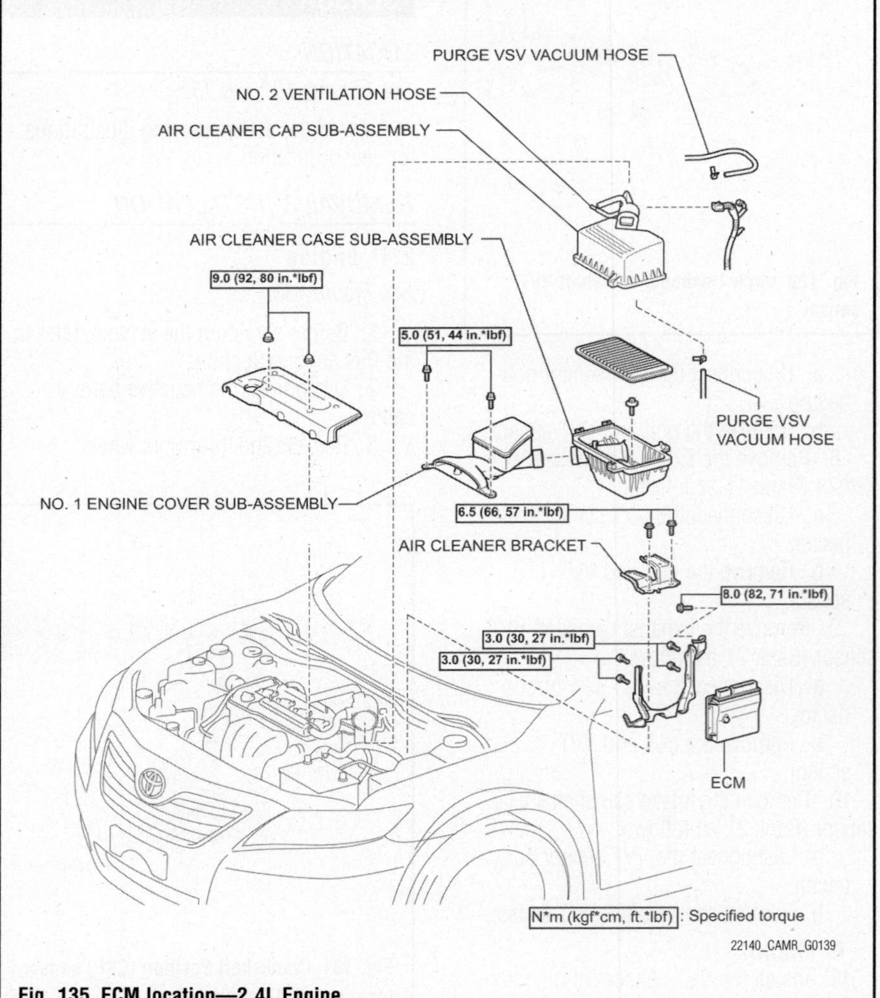

Fig. 135 ECM location—2.4L Engine

the ECM communication ID for the immobilizer system (refer to the Service Bulletin for registration).

20. Perform initialization. After replacing the ECM on vehicles with a dynamic laser cruise control system, it is necessary to initialize the ECM so that the ECM can recognize the dynamic laser cruise control system.

21. Be sure to perform the following procedure after replacing the ECM:

a. Turn the ignition switch on (IG).

b. Turn the cruise control main switch on.

c. With the brake pedal depressed, push the cruise control main switch to RES/ACC 3 times within 3 seconds. Check that the buzzer sounds at this time.

➡Do not turn the headlight dimmer switch on at this time because the optical axis automatic adjustment mode has already started, which may lead to an incorrect optical axis setting. If the headlight dimmer switch is turned on by mistake, readjust the optical axis.

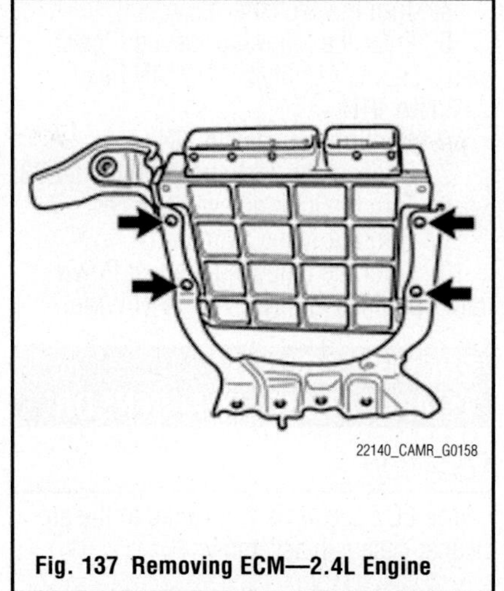

Fig. 137 Removing ECM—2.4L Engine

3.5L Engine

See Figure 138.

WINDSHIELD WIPER ARM AND BLADE ASSEMB LY RH
20 (204, 15)
WINDSHIELD WIPER ARM AND BLADE ASSEMB LY LH
20 (204, 15)
FRONT FENDER TO COWL SIDE SEAL LH
FRONT FENDER TO COWL SIDE SEAL RH
7.5 (77, 66 in.*lbf)
7.5 (77, 66 in.*lbf)
COWL TOP VENTILATOR LOUVER SUB-ASSEMBLY
WINDSHIELD WIPER MOTOR AND LINK
85 (867, 63)
5.0 (51, 44 in.*lbf)
5.0 (51, 44 in.*lbf)
85 (867, 63)
5.0 (51, 44 in.*lbf)
85 (867, 63)
COWL TOP PANEL OUTER SUB-ASSEMBLY
3.0 (31, 27 in.*lbf)
8.0 (82, 71 in.*lbf)
ECM
8.0 (82, 71 in.*lbf)
3.0 (31, 27 in.*lbf)
N*m (kgf*cm, ft.*lbf) : Specified torque
22140_CAMR_G0161

Fig. 136 ECM location—3.5L Engine

1. Before servicing the vehicle, refer to the Precautions Section.

2. Disconnect the negative battery cable.

3. Remove both windshield wiper arm and blade assemblies.

4. Remove the cowl top ventilator louver sub-assembly.

5. Remove the windshield wiper motor and link assembly.

6. Remove the outer cowl top panel.

7. Remove the ECM, as follows:

a. Remove the 3 nuts.

b. Separate the ECM from the body. When separating the ECM, do not apply excessive force to the wire harness.

c. Raise the 2 levers while pushing the locks on the 2 levers, and disconnect the 2 ECM connectors.

➡After disconnecting the connectors, make sure that dirt, water or other foreign matter does not contact the connections of the connectors.

d. Remove the ECM.

e. Remove the 4 screws and 2 ECM brackets.

To install:

8. Install the 2 ECM brackets with the 4 screws, and tighten to 27 inch lbs. (3 Nm).

9. Connect the 2 ECM connectors and lower the 2 levers.

➡Make sure that dirt, water or other foreign matter does not contact the connections of the connectors.

10. Install the ECM to the body.

11. Attach the ECM with the 3 nuts and tighten to 71 inch lbs. (8 Nm).

12. Install the outer cowl top panel.

13. Install the windshield wiper motor and link assembly.

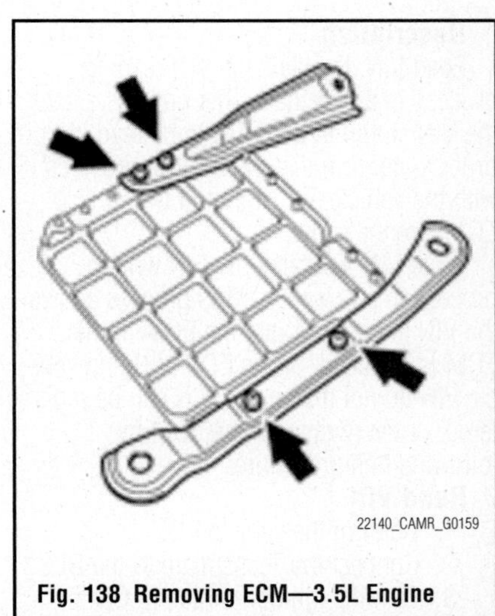

Fig. 138 Removing ECM—3.5L Engine

14. Install the cowl top ventilator louver sub-assembly.
15. Install both windshield wiper arm and blade assemblies.
16. Connect the negative battery cable.
17. Register the immobilizer communication ID. If the ECM is replaced, register the ECM communication ID for the immobilizer system (refer to the Service Bulletin for registration).
18. Perform initialization. After replacing the ECM on vehicles with a dynamic laser cruise control system, it is necessary to initialize the ECM so that the ECM can recognize the dynamic laser cruise control system.
19. Be sure to perform the following procedure after replacing the ECM:
    a. Turn the ignition switch on (IG).
    b. Turn the cruise control main switch on.
    c. With the brake pedal depressed, push the cruise control main switch to RES/ACC 3 times within 3 seconds. Check that the buzzer sounds at this time.

➡Do not turn the headlight dimmer switch on at this time because the optical axis automatic adjustment mode has already started, which may lead to an incorrect optical axis setting. If the headlight dimmer switch is turned on by mistake, readjust the optical axis.

RESET

➡The Vehicle Identification Number (VIN) must be input into the replacement ECM.

➡The VIN is of a 17-digit alphanumeric vehicle identification number. The Techstream is required to register the VIN.

Description

Read VIN: Explains the VIN reading process in a flowchart. This process allows the VIN stored in the ECM to be read in order to confirm that the 2 VINs, provided with the vehicle and stored in the vehicle ECM, are the same.

Write VIN: Explains the VIN writing process in a flowchart. This process allows the VIN to be input into the ECM. If the ECM is replaced, or the ECM VIN and vehicle VIN do not match, the VIN can be registered, or overwritten in the ECM by following this procedure.

Read VIN

1. Confirm the vehicle VIN.
2. Connect the Techstream to the DLC3.
3. Turn the ignition switch to ON.
4. Turn the Techstream on.
5. Enter the following menus: Powertrain / Engine / Utility / VIN / VIN Read.

Write VIN

6. Confirm the vehicle VIN.
7. Connect the Techstream to the DLC3.
8. Turn the ignition switch to ON.
9. Turn the Techs ream on.
10. Enter the following menus: Powertrain / Engine / Utility / VIN / VIN Write.

ENGINE COOLANT TEMPERATURE (ECT) SENSOR

LOCATION

The ECT sensor is connected to the air cleaner case sub assembly.

REMOVAL & INSTALLATION

2.4L Engine

See Figure 139.

1. Drain the engine coolant.
2. Remove the No. 1 engine cover sub assembly.
3. Remove the air cleaner inlet assembly.
4. Remove the air cleaner cap sub assembly.
5. Remove the air cleaner case sub assembly.
6. Remove the Engine Coolant Temperature (ECT) sensor.
    a. Disconnect the ECT sensor connector.
    b. Using the special tool, remove the ECT sensor and gasket.

To install:

To install, reverse the removal procedure. Tighten the ECT sensor to 15 ft. lbs. (20 Nm).

3.5L Engine

See Figure 140.

1. Remove the engine under covers.
2. Drain the engine coolant.
3. Remove the v-bank cover sub assembly.
4. Remove the air cleaner inlet assembly.
5. Remove the air cleaner cap sub assembly.
6. Remove the air cleaner case sub assembly.
7. Remove the No. 1 air cleaner inlet.
8. Remove the Engine Coolant Temperature (ECT) sensor.
    a. Remove the ECT sensor connector.
    b. Remove the engine coolant temperature sensor.

To install:

To install, reverse the removal procedure. Tighten the ECT sensor to 15 ft. lbs. (20 Nm).

EVAPORATIVE EMISSIONS (EVAP) CANISTER

REMOVAL & INSTALLATION

2.4L Engine

See Figure 141.

1. Remove the fuel tank.
2. Remove the charcoal canister assembly.
    a. Disconnect the fuel tank vent hose from the charcoal canister. Push the connector deep inside. Pinch portion A. Pull out the connector.
    b. Disconnect the charcoal canister filter sub assembly from the charcoal canister. Push the connector deep inside. Pinch portion A. Pull out the connector.

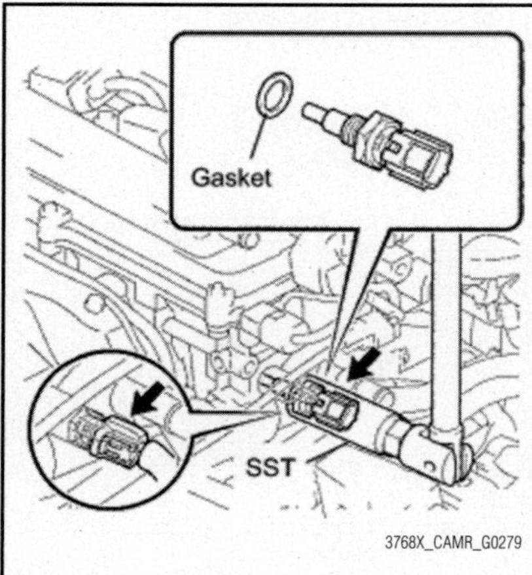

Fig. 139 Removing the ECT sensor—2.4L engine

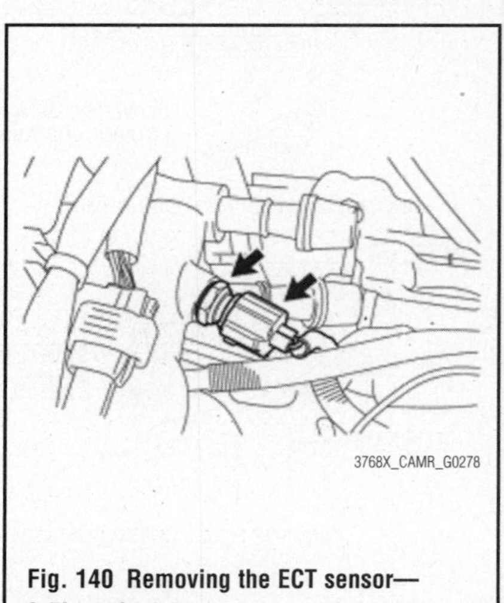

Fig. 140 Removing the ECT sensor—3.5L engine

Fig. 141 Removing the charcoal canister

c. Disconnect the vapor pressure sensor connector.

d. Disconnect the wire harness clamp.

e. Disconnect the purge line hose from the charcoal canister.

f. Remove the 2 bolts, clip and charcoal canister.

To install:

To install, reverse the removal procedure. Tighten the charcoal canister to 29 ft. lbs. (39 Nm).

3.5L Engine

See Figure 142.

1. Remove the fuel tank.
2. Remove the charcoal canister assembly.

a. Disconnect the fuel tank vent hose from the charcoal canister. Push the connector deep inside. Pinch portion A. Pull out the connector.

b. Disconnect the charcoal canister filter sub assembly from the charcoal canister. Push the connector deep inside. Pinch portion A. Pull out the connector.

c. Disconnect the vapor pressure sensor connector.

d. Disconnect the wire harness clamp.

e. Disconnect the purge line hose from the charcoal canister.

f. Remove the 2 bolts, clip and the charcoal canister.

To install:

3. Install the charcoal canister assembly.

a. Install the 2 bolts, clip and charcoal canister. Tighten to 29 ft. lbs. (39 Nm).

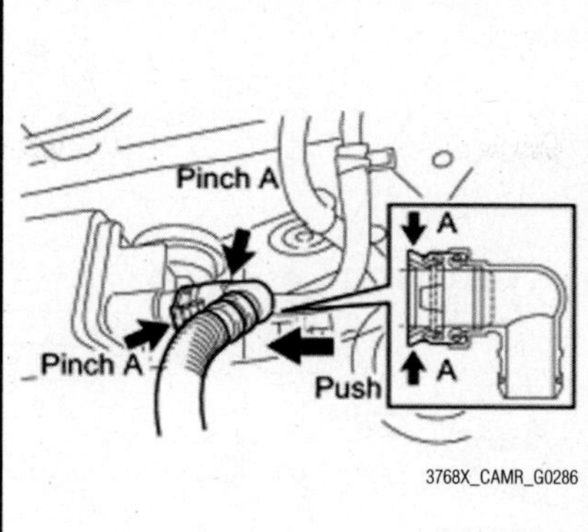

Fig. 142 Disconnecting the fuel tank vent hose from the charcoal canister

b. Connect the purge line hose to the charcoal canister.

c. Connect the wire harness clamp.

d. Connect the vapor pressure sensor connector.

e. Connect the charcoal canister filter sub-assembly to the charcoal canister.

f. Connect the fuel tank vent hose to the charcoal canister.

4. Install the fuel tank.

HEATED OXYGEN SENSOR (HO2S)

LOCATION

See Figures 143 and 144.

Refer to the accompanying illustrations for location.

REMOVAL & INSTALLATION

2.4L Engine

See Figure 145.

1. Before servicing the vehicle, refer to the Precautions Section.
2. Disconnect the oxygen sensor connectors.

➡Do not damage the heated oxygen sensor.

3. Using Special Tool: 09224-00010 or equivalent, remove the two oxygen sensors from the front pipe assembly.

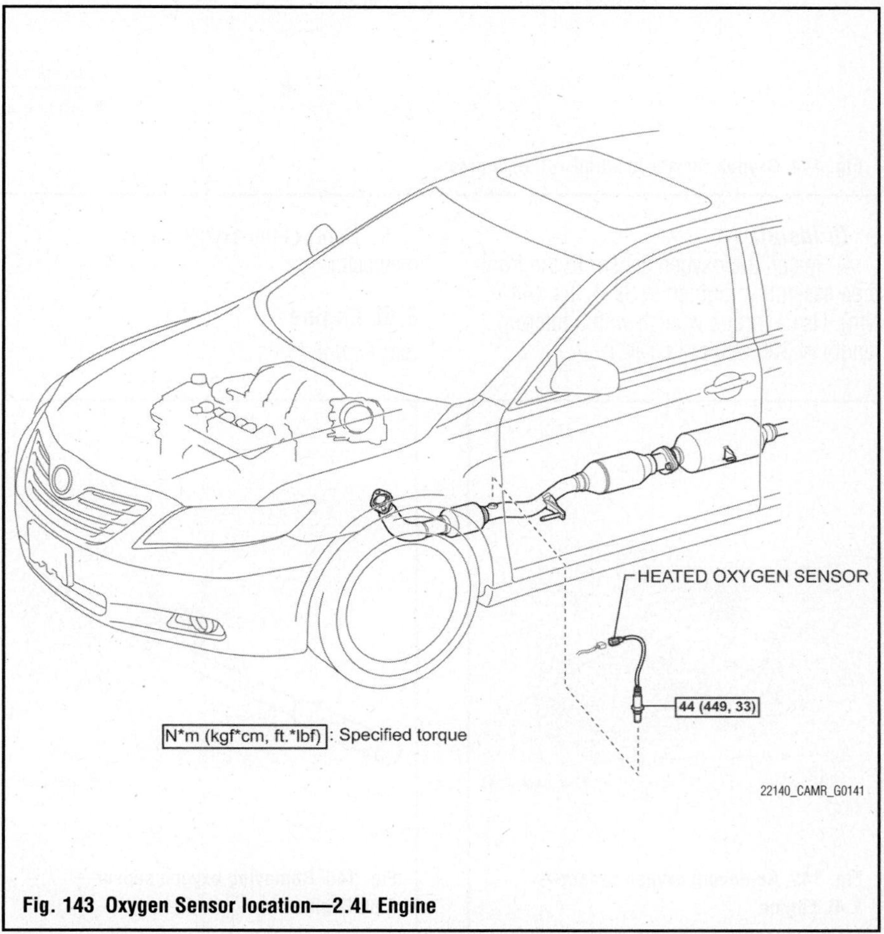

Fig. 143 Oxygen Sensor location—2.4L Engine

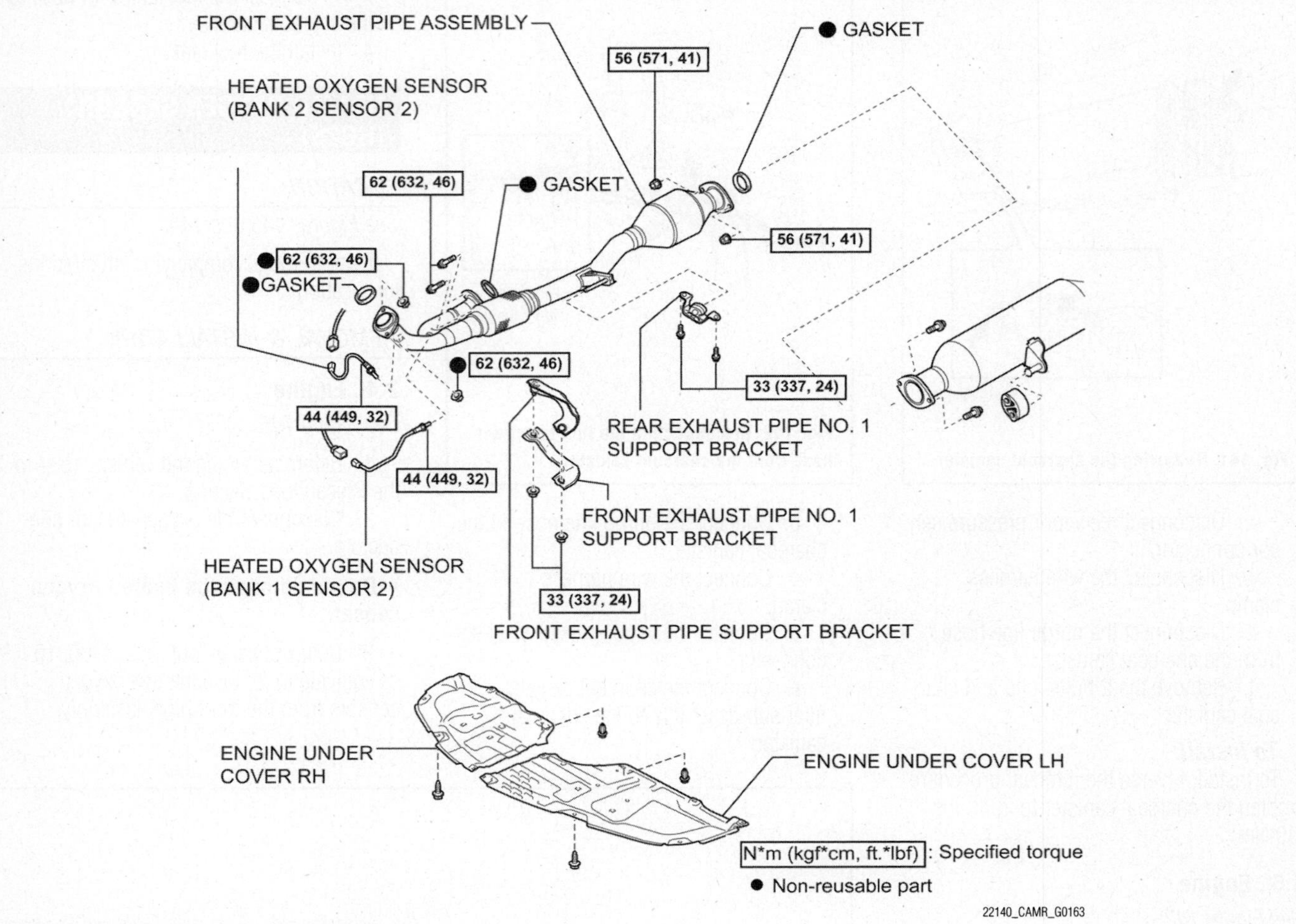

FRONT EXHAUST PIPE ASSEMBLY
● GASKET
56 (571, 41)
HEATED OXYGEN SENSOR
(BANK 2 SENSOR 2)
62 (632, 46)
● GASKET
● 62 (632, 46)
●GASKET
56 (571, 41)
● 62 (632, 46)
33 (337, 24)
44 (449, 32)
REAR EXHAUST PIPE NO. 1
SUPPORT BRACKET
44 (449, 32)
FRONT EXHAUST PIPE NO. 1
SUPPORT BRACKET
HEATED OXYGEN SENSOR
(BANK 1 SENSOR 2)
33 (337, 24)
FRONT EXHAUST PIPE SUPPORT BRACKET
ENGINE UNDER
COVER RH
ENGINE UNDER COVER LH
N*m (kgf*cm, ft.*lbf) : Specified torque
● Non-reusable part
22140_CAMR_G0163

Fig. 144 Oxygen Sensor location—3.5L Engine

To install:

4. Install the oxygen sensor to the front pipe assembly. Tighten to 32 ft. lbs. (44 Nm). Use a torque wrench with a fulcrum length of 300 mm (11.81 in.).

SST
22140_CAMR_G0169

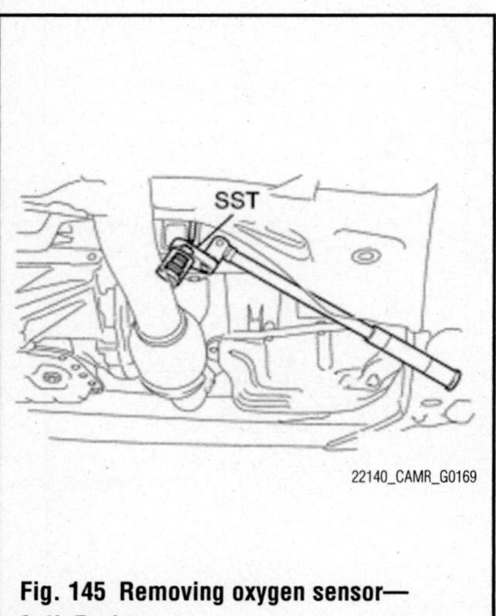

Fig. 145 Removing oxygen sensor—2.4L Engine

5. Connect the oxygen sensor connector.

3.5L Engine

See Figure 146.

SST
22140_CAMR_G0171

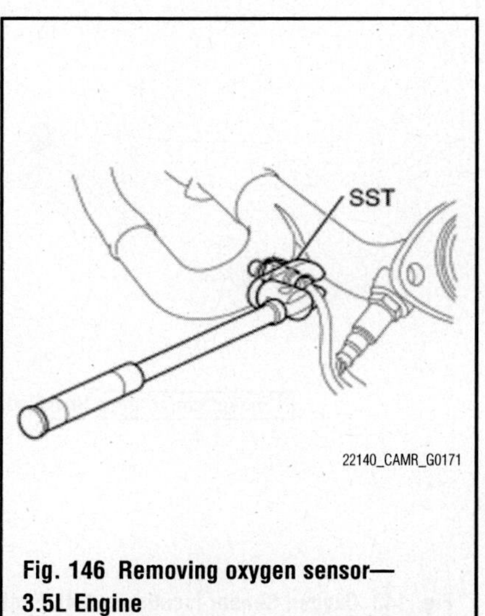

Fig. 146 Removing oxygen sensor—3.5L Engine

1. Before servicing the vehicle, refer to the Precautions Section.
2. Remove the front exhaust pipe assembly.
3. Disconnect the 2 oxygen sensor connectors.
4. Using Special Tool: 09224-00010 or equivalent, remove the 2 oxygen sensors from the front pipe assembly.

To install:

5. Install the 2 oxygen sensors to the front pipe assembly. Tighten to 32 ft. lbs. (44 Nm) and 30 ft. lbs. (40 Nm). Use a torque wrench with a fulcrum length of 300 mm (11.81 in.).
6. Connect the 2 oxygen sensor connectors.
7. Remove the front exhaust pipe assembly.

KNOCK SENSOR (KS)

LOCATION

See Figures 147 and 148.

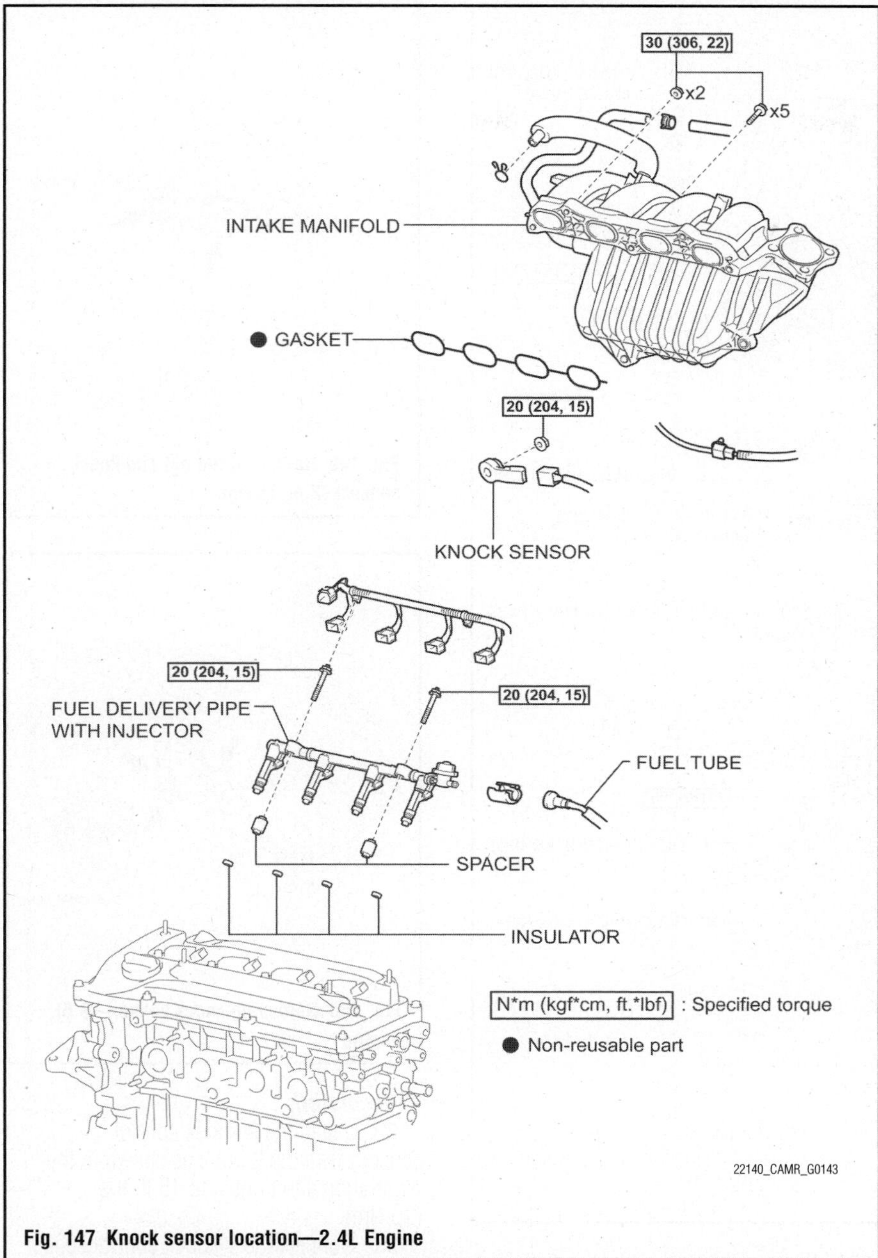

Fig. 147 Knock sensor location—2.4L Engine

Refer to the accompanying illustrations for sensor location.

REMOVAL & INSTALLATION

2.4L Engine

See Figure 149.

CAUTION

Observe all applicable safety precautions when working around fuel. Whenever servicing the fuel system, always work in a well ventilated area. Do not allow fuel spray or vapors to come in contact with a spark or open flame. Keep a dry chemical fire extinguisher near the work area. Always keep fuel in a container specifically designed for fuel storage; also, always properly seal fuel containers to avoid the possibility of fire or explosion.

1. Before servicing the vehicle, refer to the Precautions Section.
2. Properly discharge the fuel system pressure.
3. Disconnect battery negative cable.
4. Remove plastic engine cover.
5. Drain and recycle the engine coolant.
6. Remove windshield wiper arms and blade assemblies.
7. Remove both front fender to cowl side seals.
8. Remove cowl top ventilator louver sub-assembly.
9. Remove windshield wiper motor and link.
10. Remove the four bolts, four nuts and cowl top panel outer sub-assembly.
11. Remove air cleaner cap sub-assembly.
12. Remove air cleaner case sub-assembly.
13. Remove throttle body.
14. Disconnect fuel tube.
15. Remove fuel delivery pipe with injector.
16. Disconnect the union to check valve hose from the brake booster.
17. Disconnect the camshaft timing oil control valve connector.
18. Remove the wire harness clamp.
19. Remove the union to check valve hose from the vacuum hose clamp.
20. Remove the 5 bolts, 2 nuts and intake manifold. Remove the gasket from the intake manifold.
21. Disconnect the knock sensor connector. Remove the nut and knock sensor.

To install:

➡Make sure that the knock sensor is in the correct position.

22. Install the knock control sensor with the nut and tighten to 15 ft. lbs. (20 Nm).
23. Connect the knock control sensor connector.
24. The remainder of installation is the reverse of the removal procedure.
25. Inspect for fuel leak and check the function of throttle body.

3.5L Engine

See Figure 150.

CAUTION

Observe all applicable safety precautions when working around fuel. Whenever servicing the fuel system, always work in a well ventilated area. Do not allow fuel spray or vapors to come in contact with a spark or open flame. Keep a dry chemical fire extinguisher near the work area. Always keep fuel in a container specifically designed for fuel storage; also, always properly seal fuel containers to avoid the possibility of fire or explosion.

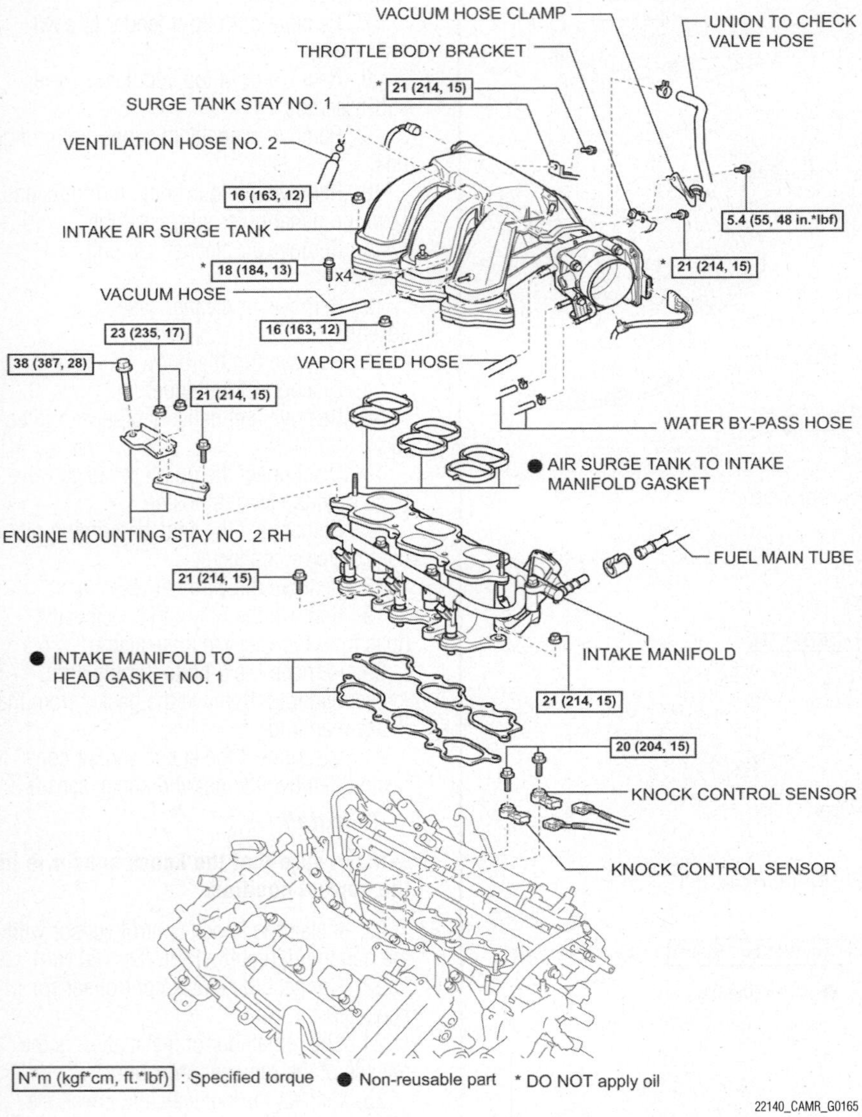

Fig. 148 Knock sensor location—3.5L Engine

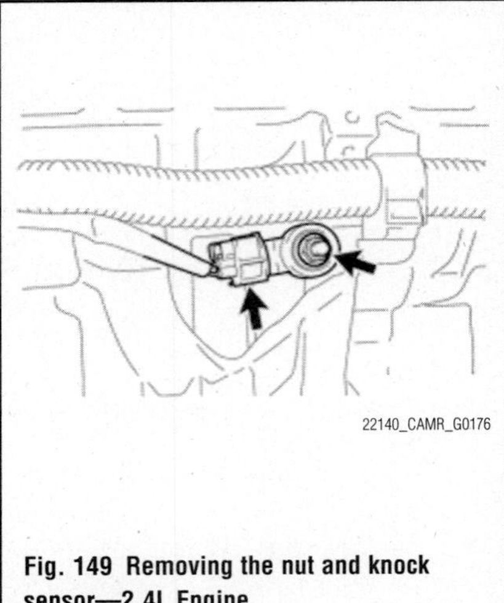

Fig. 149 Removing the nut and knock sensor—2.4L Engine

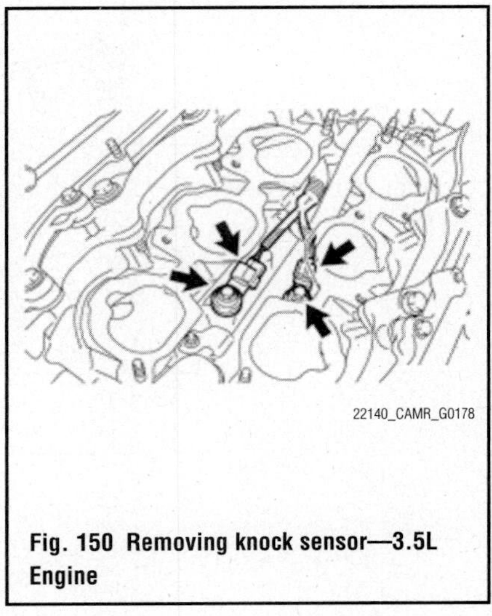

Fig. 150 Removing knock sensor—3.5L Engine

1. Before servicing the vehicle, refer to the Precautions Section.
2. Properly discharge the fuel system pressure.
3. Disconnect battery negative cable.
4. Drain and recycle the engine coolant.
5. Remove both plastic engine under covers.
6. Remove windshield wiper arms and blade assemblies.
7. Remove both of the front fender to cowl side seals.
8. Remove cowl top ventilator louver sub-assembly.
9. Remove the windshield wiper motor and link assembly.
10. Remove the outer cowl top panel.
11. Remove the cool air intake duct seal.
12. Remove the V-bank cover sub-assembly.
13. Remove air cleaner inlet assembly.
14. Remove the air cleaner cap sub-assembly.
15. Remove air cleaner case sub-assembly.
16. Remove the intake air surge tank.
17. Remove no. 1 air cleaner inlet.
18. Separate fuel tube sub-assembly.
19. Remove the intake manifold.
20. Disconnect the 2 knock control sensor connectors.
21. Remove the 2 bolts and 2 knock control sensors.

To install:

22. Install the 2 knock control sensors with the 2 bolts as shown in the illustration and tighten to 15 ft. lbs. (20 Nm).
23. Connect the 2 knock control sensor connectors.
24. The remainder of installation is the reverse of the removal procedure.
25. Inspect for fuel leak and check the function of throttle body.

MASS AIR FLOW (MAF) METER

LOCATION

The MAF meter is between the throttle body and air cleaner housing.

REMOVAL & INSTALLATION

See Figure 151.

1. Before servicing the vehicle, refer to the Precautions Section.

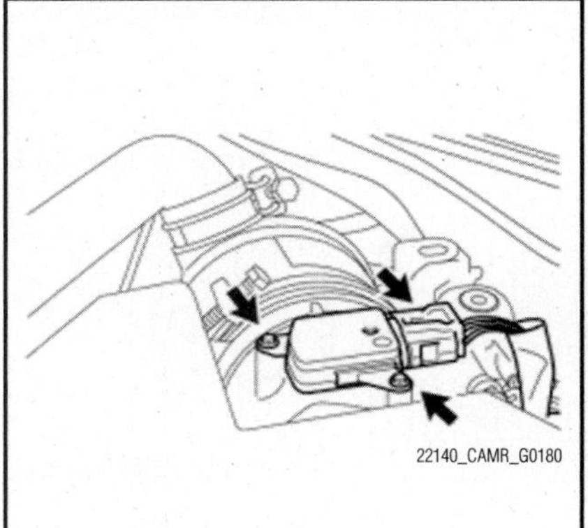

Fig. 151 Mass Air Flow (MAF) meter—2.4L shown, 2.5L and 3.5L are similar

2. Disconnect the Mass Air Flow (MAF) meter connector.
3. Remove the 2 screws and Mass Air Flow (MAF) meter.

To install:

4. Installation is the reverse of the removal procedure.

POSITIVE CRANKCASE VENTILATION (PCV) VALVE

REMOVAL & INSTALLATION

3.5L Engine

See Figure 152.

1. Remove the v-bank cover sub assembly.
2. Disconnect the ventilation hose.
   a. Disconnect the ventilation hose from the ventilation valve.
3. Remove the ventilation valve.

To install:

4. Install the ventilation valve.
   a. Apply adhesive to 2 or 3 threads.

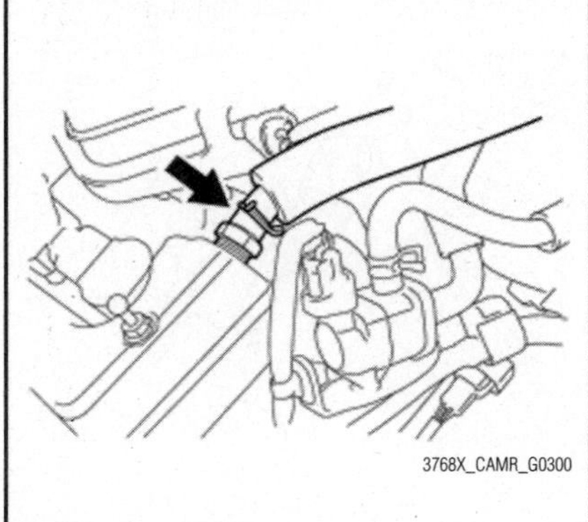

Fig. 152 Disconnecting the ventilation hose from the ventilation valve

   b. Install the ventilation valve. Tighten to 20 ft. lbs. (27 Nm).
5. Connect the ventilation hose to the ventilation hose.
6. Install the v-bank cover sub assembly.

THROTTLE POSITION SENSOR (TPS)

LOCATION

See Figures 153 and 154.

Refer to the accompanying illustrations for sensor location.

REMOVAL & INSTALLATION

Refer to the Throttle Body removal and installation procedures in the Fuel System Section.

VARIABLE CAMSHAFT TIMING OIL CONTROL SOLENOID

REMOVAL & INSTALLATION

2.4L Engine

See Figure 155.

1. Before servicing the vehicle, refer to the Precautions Section.
2. Remove the plastic engine cover.
3. Remove the bolt and disconnect the vacuum hose clamp.
4. Remove the clip nut.
5. Disconnect the camshaft timing oil control valve assembly connector.
6. Remove the bolt and camshaft timing oil control valve assembly.

To install:

7. Apply a light coat of engine oil to an

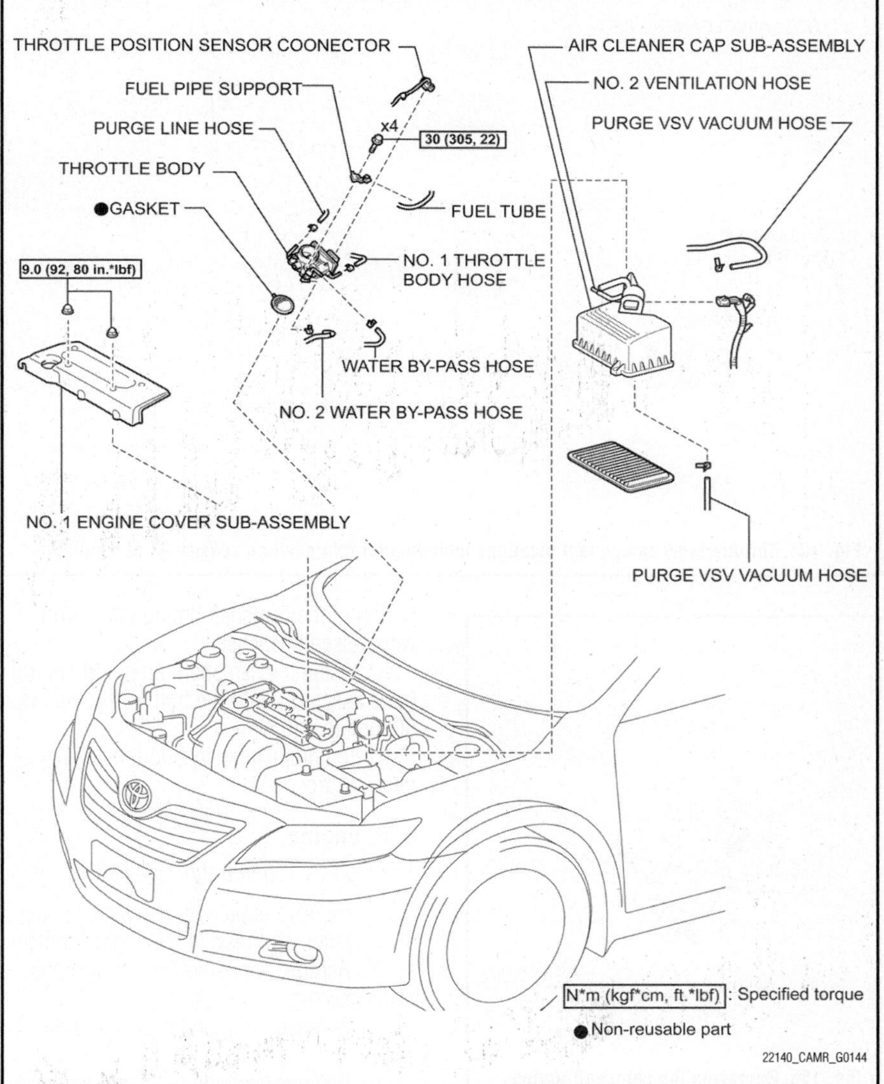

Fig. 153 Throttle body component locations including throttle position sensor—2.4L Engine

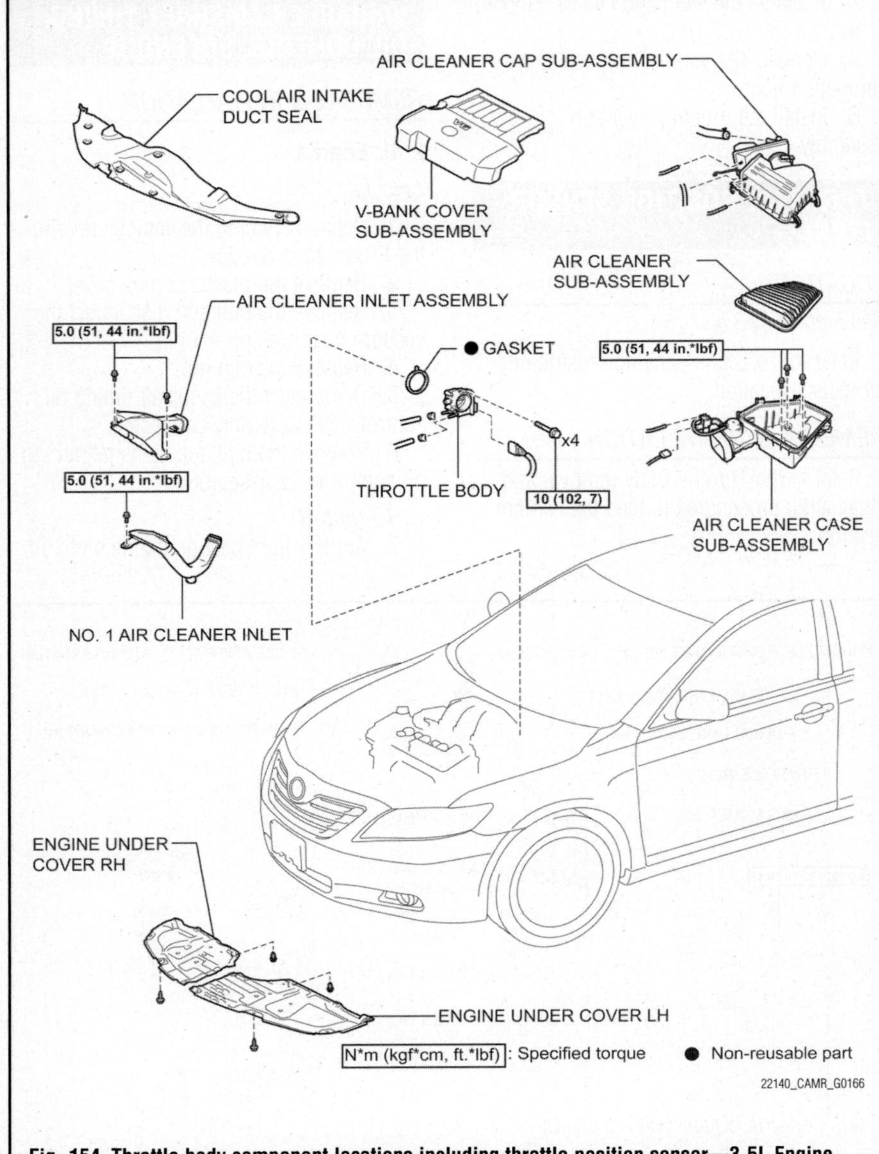

Fig. 154 Throttle body component locations including throttle position sensor—3.5L Engine

Fig. 155 Removing the camshaft timing oil control valve assembly—2.4L Engine

O-ring of the camshaft timing oil control valve assembly sensor.

8. Install the camshaft timing oil control valve assembly with the bolt and tighten to 80 inch lbs. (9 Nm).
9. To complete installation, reverse removal procedure.

3.5L Engine

See Figures 156 through 159.

1. Remove engine under cover.
2. Drain and recycle the engine coolant.
3. Remove windshield wiper arm and blade assembly.
4. Remove front fender to cowl side seal.
5. Remove cowl top ventilator louver sub-assembly.

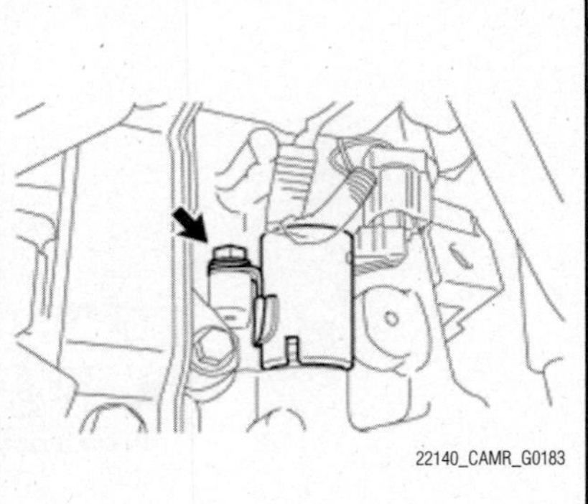

Fig. 156 Removing camshaft timing oil control valve assembly—bank 1, exhaust side—3.5L Engine

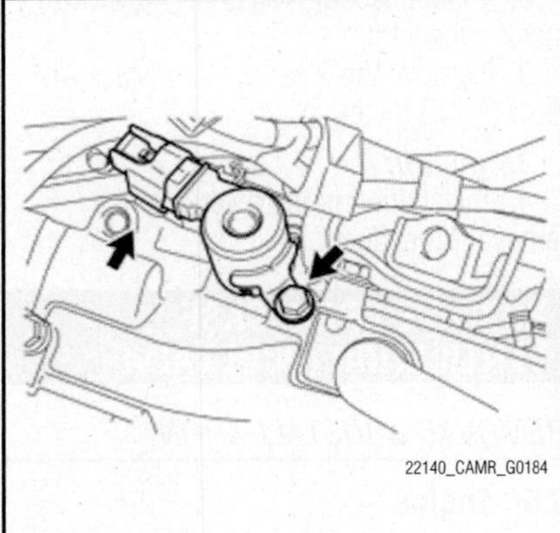

Fig. 157 Removing camshaft timing oil control valve assembly—bank 1, intake side—3.5L Engine

6. Remove windshield wiper motor and link.
7. Remove the 4 bolts, 4 nuts and cowl top panel outer sub-assembly.

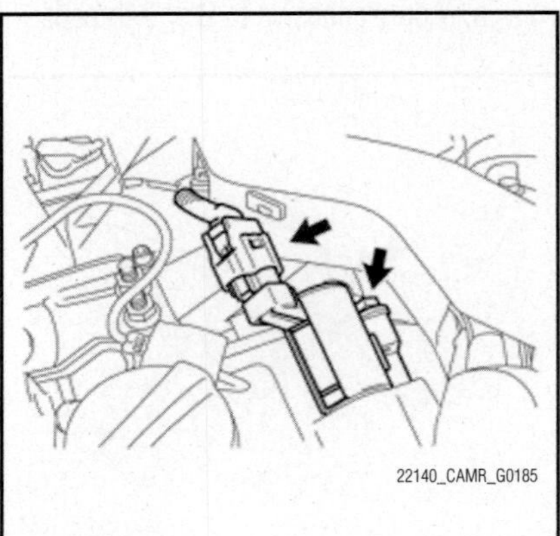

Fig. 158 Removing camshaft timing oil control valve assembly—bank 2, exhaust side—3.5L Engine

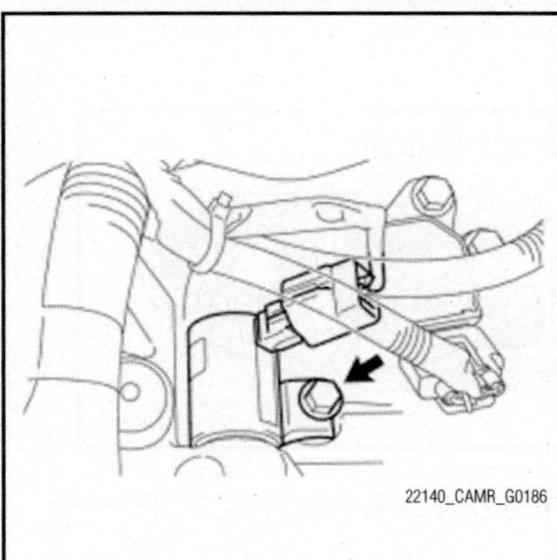

Fig. 159 Removing camshaft timing oil control valve assembly—bank 2, intake side—3.5L Engine

8. Remove cools air intake duct seal.
9. Remove v-bank cover sub-assembly.
10. Remove air cleaner inlet assembly.
11. Remove air cleaner cap sub-assembly.
12. Remove air cleaner case sub-assembly.
13. Remove no. 1 air cleaner inlet.
14. Remove intake air surge tank.
15. Remove the four camshaft timing oil control valves by performing the following:
    a. Disconnect the camshaft timing oil control valve connector.
    b. Remove the bolt and camshaft timing oil control valve.
    c. Remove the O-ring from the camshaft timing oil control valve.

To install:

16. Install the camshaft timing oil control valve assembly, as follows:
    a. Install the 4 oil control valves with the 4 bolts and tighten to 7 ft. lbs. (10 Nm).
17. Install the 4 Camshaft Position (CMP) sensors with the 4 bolts and tighten to 7 ft. lbs. (10 Nm).
18. Apply adhesive (Toyota Genuine Adhesive 1324, Three Bond 1324 or equivalent) around the ventilation valve.
19. Install the ventilation valve sub-assembly and tighten to 20 ft. lbs. (27 Nm).
20. Install the spark plugs and the ignition coil assembly.
21. Install the oil filler cap sub-assembly.
22. Remove the engine stand.
23. Install the engine assembly.

FUEL — GASOLINE FUEL INJECTION SYSTEM

FUEL SYSTEM SERVICE PRECAUTIONS

Safety is the most important factor when performing not only fuel system maintenance, but any type of maintenance. Failure to conduct maintenance and repairs in a safe manner may result in serious personal injury or death. Work on a vehicle's fuel system components can be accomplished safely and effectively by adhering to the following rules and guidelines.

- To avoid the possibility of fire and personal injury, always disconnect the negative battery cable unless the repair or test procedure requires that battery voltage be applied.
- Always relieve the fuel system pressure prior to disconnecting any fuel system component (injector, fuel rail, pressure regulator, etc.) fitting or fuel line connection. Exercise extreme caution whenever relieving fuel system pressure to avoid exposing skin, face and eyes to fuel spray. Please be advised that fuel under pressure may penetrate the skin or any part of the body that it contacts.
- Always place a shop towel or cloth around the fitting or connection prior to loosening to absorb any excess fuel due to spillage. Ensure that all fuel spillage is quickly removed from engine surfaces. Ensure that all fuel-soaked cloths or towels are deposited into a flame-proof waste container with a lid.
- Always keep a dry chemical (Class B) fire extinguisher near the work area.
- Do not allow fuel spray or fuel vapors to come into contact with a spark or open flame.
- Always use a second wrench when loosening or tightening fuel line connection fittings. This will prevent unnecessary stress and torsion on fuel piping. Always follow the proper torque specifications.
- Always replace worn fuel fitting O-rings with new ones. Do not substitute fuel hose where rigid pipe is installed.

FUEL SYSTEM PRESSURE

RELIEVING

2.4L Engine

➡The following procedure must be performed before disconnecting any part of the fuel system. After performing this procedure, pressure will remain in the fuel line. Use care when disconnecting any fuel lines.

1. Before servicing the vehicle, refer to the Precautions Section.
2. Remove the console box.
3. Disconnect the connector.
4. Start the engine. After the engine has stopped, turn the ignition switch to the OFF position.

➡DTC P0171 (system lean) may set.

5. Check that the engine does not start.
6. Disconnect the negative battery cable.
7. Remove the fuel tank cap.
8. Connect the electrical connector. Install the console box.

3.5L Engine

➡The following procedure must be performed before disconnecting any part of the fuel system. After performing this procedure, pressure will remain in the fuel line. Use care when disconnecting any fuel lines.

1. Before servicing the vehicle, refer to the Precautions Section.
2. Remove the console box.
3. Disconnect the connector.
4. Start the engine. After the engine has stopped, turn the ignition switch to the OFF position.

➡DTC P0171/P0172 (system lean) may set.

5. Check that the engine does not start.
6. Disconnect the negative battery cable.
7. Remove the fuel tank cap.
8. Connect the electrical connector. Install the console box.

FUEL FILTER

REMOVAL & INSTALLATION

The filter is part of the fuel pump module and is not normally serviced.

FUEL PUMP

REMOVAL & INSTALLATION

2.4L Engine

See Figure 160.

➡The fuel tank must first be removed from the vehicle. Be sure to check and

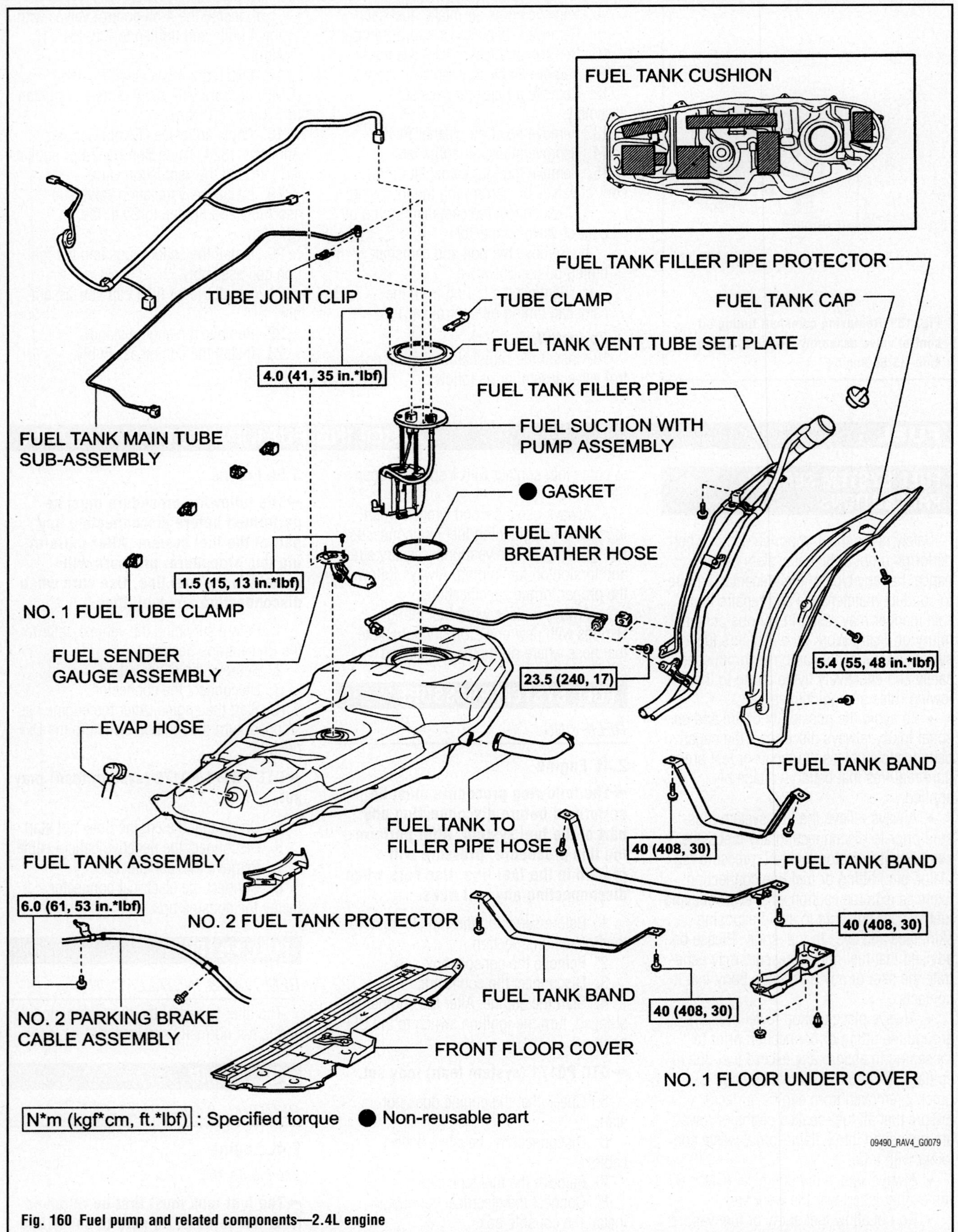

Fig. 160 Fuel pump and related components—2.4L engine

adjust the fuel level as required, before removing the fuel tank. Take all the necessary precautions to avoid safety and fuel disposal problems.

1. Before servicing the vehicle, refer to the Precautions Section.
2. Properly relieve the fuel system pressure.
3. Disconnect the negative battery cable.

➡Wait at least 90 seconds after disconnecting the negative battery cable before starting any repair work to prevent air bag and seat belt pretensioner activation.

4. Remove the front floor carpet.
5. Disconnect the number two parking brake cable assembly.
6. Disconnect the fuel tank main tube assembly.
7. Disconnect the fuel tank to filler pipe hose.
8. Disconnect the fuel tank breather hose.
9. Remove the fuel tank filler pipe.
10. Position a suitable jack under the fuel tank. Remove the six bolts and three fuel tank bands.
11. Slightly lower the suitable jack.

➡Be careful not to cut the wing nuts.

12. Fold back about half of each cushion rubber so that the wire harness can be detached.
13. Disconnect the fuel pump connector and sender gauge connector.
14. Detach the wire harness from the four clamps. Carefully remove the fuel tank from the vehicle.
15. Remove the joint clip and fuel tank main tube.
16. Remove the eight bolts and the fuel tank vent tube set plate.
17. Disconnect the fuel hose and remove the fuel pump assembly from the fuel tank. Discard the gasket.

To install:

18. Installation is the reverse of the removal procedure.
19. Tighten the eight retaining bolts to 35 inch lbs, and in an alternating sequence.
20. Tighten the fuel tank retaining bolts to 30 ft. lbs.
21. Tighten the fuel tank filler pipe bolts to 17 ft. lbs.
22. Start the engine and check for leaks. Correct as required.

3.5L Engine

➡The fuel tank must first be removed from the vehicle. Be sure to check and adjust the fuel level as required, before removing the fuel tank. Take all the necessary precautions to avoid safety and fuel disposal problems.

1. Before servicing the vehicle, refer to the Precautions Section.
2. Properly relieve the fuel system pressure.
3. Disconnect the negative battery cable.

➡Wait at least 90 seconds after disconnecting the negative battery cable before starting any repair work to prevent air bag and seat belt pretensioner activation.

4. Remove the front floor carpet.
5. Disconnect the number two parking brake cable assembly.
6. Disconnect the fuel tank main tube assembly.
7. Disconnect the fuel tank to filler pipe hose.
8. Disconnect the fuel tank breather hose.
9. Remove the fuel tank filler pipe.
10. Position a suitable jack under the fuel tank. Remove the six bolts and three fuel tank bands.
11. Slightly lower the suitable jack.

➡Be careful not to cut the wing nuts.

12. Fold back about half of each cushion rubber so that the wire harness can be detached.
13. Disconnect the fuel pump connector and sender gauge connector.
14. Detach the wire harness from the four clamps. Carefully remove the fuel tank from the vehicle.
15. Remove the joint clip and fuel tank main tube.
16. Remove the eight bolts and the fuel tank vent tube set plate.
17. Disconnect the fuel hose and remove the fuel pump assembly from the fuel tank. Discard the gasket.

To install:

18. Installation is the reverse of the removal procedure.
19. Tighten the eight retaining bolts to 35 inch lbs, and in an alternating sequence.
20. Tighten the fuel tank retaining bolts to 30 ft. lbs.
21. Tighten the fuel tank filler pipe bolts to 17 ft. lbs.
22. Start the engine and check for leaks. Correct as required.

FUEL PRESSURE REGULATOR

REMOVAL & INSTALLATION

See Figures 161 and 162.

The fuel pressure regulator is integrated in the fuel pump assembly. When replacing, be sure to use a new O-ring.

FUEL RAIL & INJECTORS

REMOVAL & INSTALLATION

2.4L Engine

See Figures 163 through 165.

1. Before servicing the vehicle, refer to the Precautions Section.
2. Properly relieve the fuel system pressure.

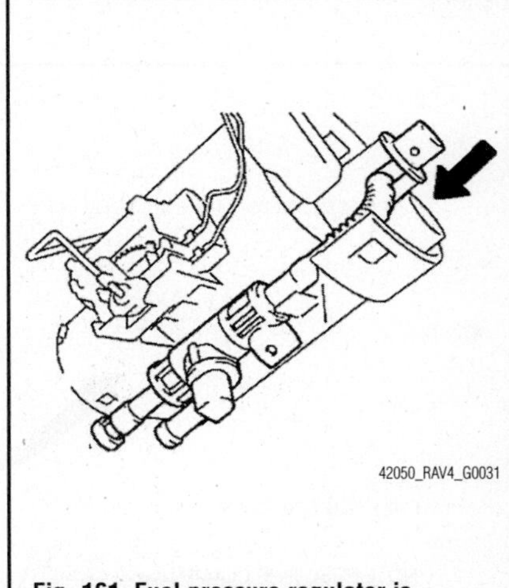

Fig. 161 Fuel pressure regulator is located in the fuel pump assembly (arrow)

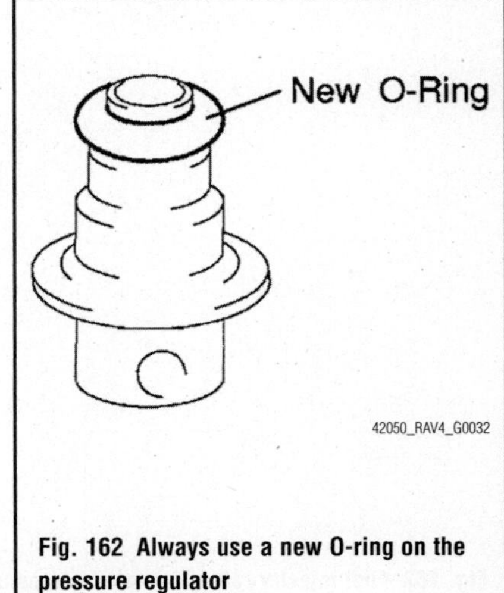

Fig. 162 Always use a new O-ring on the pressure regulator

3. Disconnect the negative battery cable.

➡Wait at least 90 seconds after disconnecting the negative battery cable before starting any repair work to prevent air bag and seat belt pretensioner activation.

4. Remove the air cleaner cap sub-assembly.
5. To disconnect the fuel main tube; remove the fuel tube from the fuel hose clamp. Remove the fuel pipe clamp.

➡Do not forcibly bend, kink or twist the tube.

6. Hold the fuel tube connector and install tool SST 09268-21010, or equivalent. Turn the tool to align the retainer inside the fuel tube connector with the chamfered part of the tool.
7. Insert the tool into the fuel tube and hold it. Push the fuel tube connector toward the tool,
8. Mount the retainer of the fuel tube connector onto the chamfered part of the tool.
9. Slide the tool and the fuel tube connector together toward the fuel tube until they make a "click", then disconnect the fuel tube.
10. Properly drain the remaining fuel inside the fuel tube. Cover the fuel tube with a clean plastic bag, to protect the disconnected part.
11. Disconnect the number two ventilation hose from the ventilation valve.
12. Remove the two wiring harness clamps. Disconnect the four fuel injector electrical connectors.
13. Remove the two retaining bolts. Remove the fuel rail along with the fuel injectors.

➡Be careful not to drop the fuel injectors when removing the injector rail.

14. Remove the two delivery pipe spacers from the cylinder head.
15. Remove the four insulators from the cylinder head.
16. As required, remove the injectors from the fuel rail.

To install:

17. Installation is the reverse of the removal procedure.
18. Installation is the reverse of removal.
19. Apply a light coat gasoline to the new fuel injector O-rings and install them on to the injectors.
20. Apply a light coat of gasoline to the part of the injector rail that contacts the O-ring of the injector.
21. Install the injectors on the fuel rail.

➡Be sure that the O-rings are not damaged or jammed when you are installing them.

22. Check that the injector rotates freely. If it does not, replace the O-ring.
23. Install the four new insulators to the cylinder head. Install the two delivery pipe spacers to the cylinder head.
24. Install the fuel rail together with the fuel injectors, then temporarily tighten the two retaining bolts.
25. Check that the fuel injectors rotate smoothly. If it does not replace the O-ring.
26. Tighten the two retaining bolts to 15 ft. lbs.
27. When installing the number two ventilation, see the illustration for proper positioning.

➡Make sure that the paint mark and hose clamp are at the correct angle when installing the hose.

28. Once the installation is complete be sure to inspect for fuel leaks, using the Techstream tool.
29. Connect the tool to the DLC3.

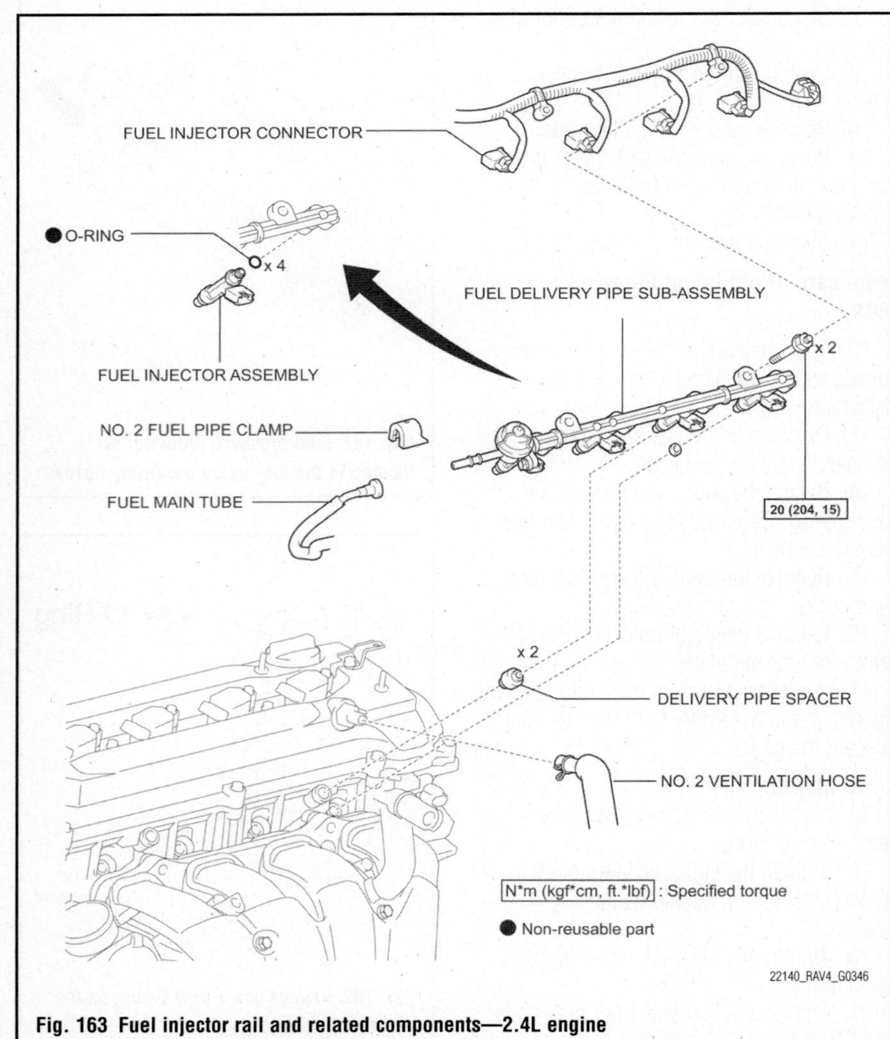

Fig. 163 Fuel injector rail and related components—2.4L engine

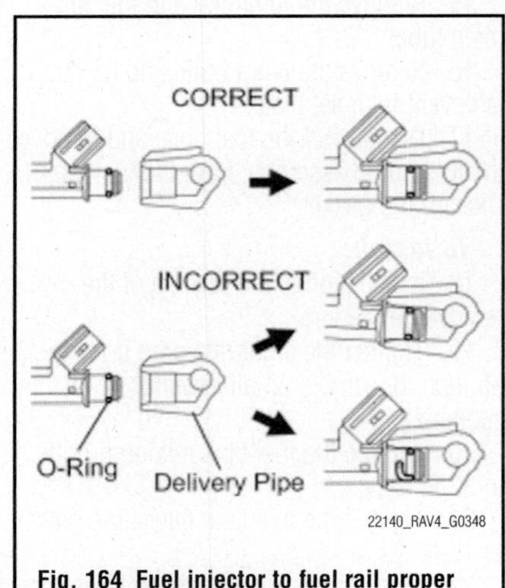

Fig. 164 Fuel injector to fuel rail proper installation—2.4L engine

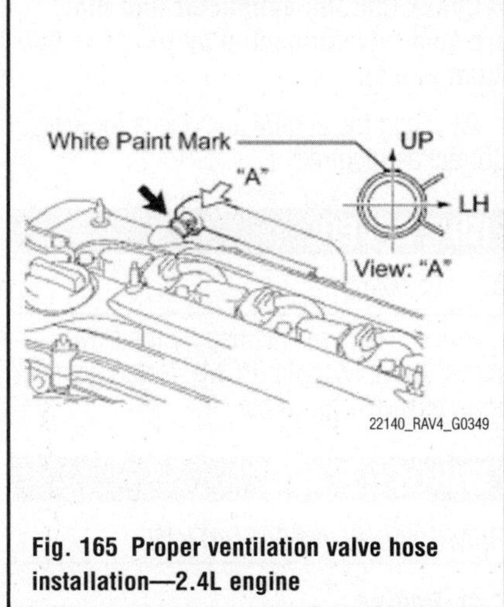

Fig. 165 Proper ventilation valve hose installation—2.4L engine

30. Turn the ignition switch to the ON position. Push the Techstream main switch on.

➡Do not start the engine.

31. Select the active test mode on the tool. Refer to the tools instruction book for additional details.
32. Check that there are no fuel system leaks.
33. Turn the ignition switch OFF.
34. Disconnect the tool.

3.5L Engine

See Figure 166.

➡In order to replace the fuel rail and injectors, the manufacturer recommends removing the engine from the vehicle.

1. Before servicing the vehicle, refer to the Precautions Section.
2. Properly relieve the fuel system pressure.
3. Disconnect the negative battery cable.

➡Wait at least 90 seconds after disconnecting the negative battery cable before starting any repair work to prevent air bag and seat belt pretensioner activation.

4. Remove the engine from the vehicle and position it in a suitable holding fixture.
5. Remove the intake air surge tank assembly.
6. Disconnect the number two main tube. Pinch the tube connector and then pull out the pipe.
7. Disconnect the six injector connectors. Remove the five bolts and the fuel delivery rail together with the injectors.
8. Remove the six insulators from the intake manifold.

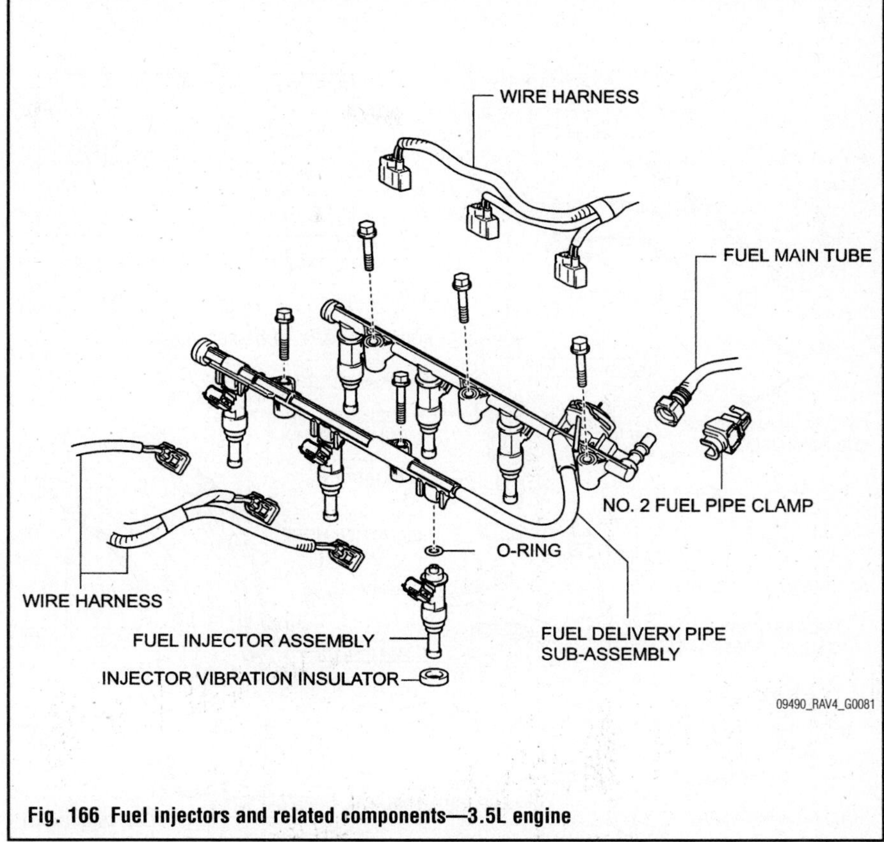

Fig. 166 Fuel injectors and related components—3.5L engine

9. Pull out the injectors from the rail. Remove and discard the O-rings.

To install:

10. Installation is the reverse of removal.
11. Coat the new O-rings with gasoline. Push the injectors onto the pipes and make sure they rotate freely.
12. Position the assembly onto the head and install the bolts finger tight. Make sure that the injectors still rotate freely. If not, replace the O-rings. Torque the bolts to 15 ft. lbs. (20 Nm).
13. Start the engine and check for leaks, correct as required.

FUEL TANK

REMOVAL & INSTALLATION

See Figures 167 and 168.

1. Before servicing the vehicle, refer to the Precautions Section.
2. Properly relieve the fuel system pressure.
3. Disconnect the negative battery cable.

➡Wait at least 90 seconds after disconnecting the negative battery cable before starting any repair work to prevent air bag and seat belt pretensioner activation.

4. Remove the fuel cap.
5. Raise and support the vehicle safely.
6. Remove the nut, bolt and three clips in order to remove the front floor cover.
7. Remove the two bolts and disconnect the parking brake cover.
8. Disconnect the fuel tank main tube sub-assembly.
9. Disconnect the evaporative emission hose from the tank.
10. Disconnect the breather hose from the tank.
11. Remove the three bolts and the filler pipe protector. Remove the two bolts and the filler pipe.
12. Properly position a transmission jack underneath the fuel tank assembly.
13. Remove the six bolts and three fuel tank retaining bands.
14. Carefully lower the transmission jacks.

➡Be careful not to cut any electrical wiring or hoses.

15. Fold back about half of each cushion rubber so that the wire harness can be detached from the fuel tank.

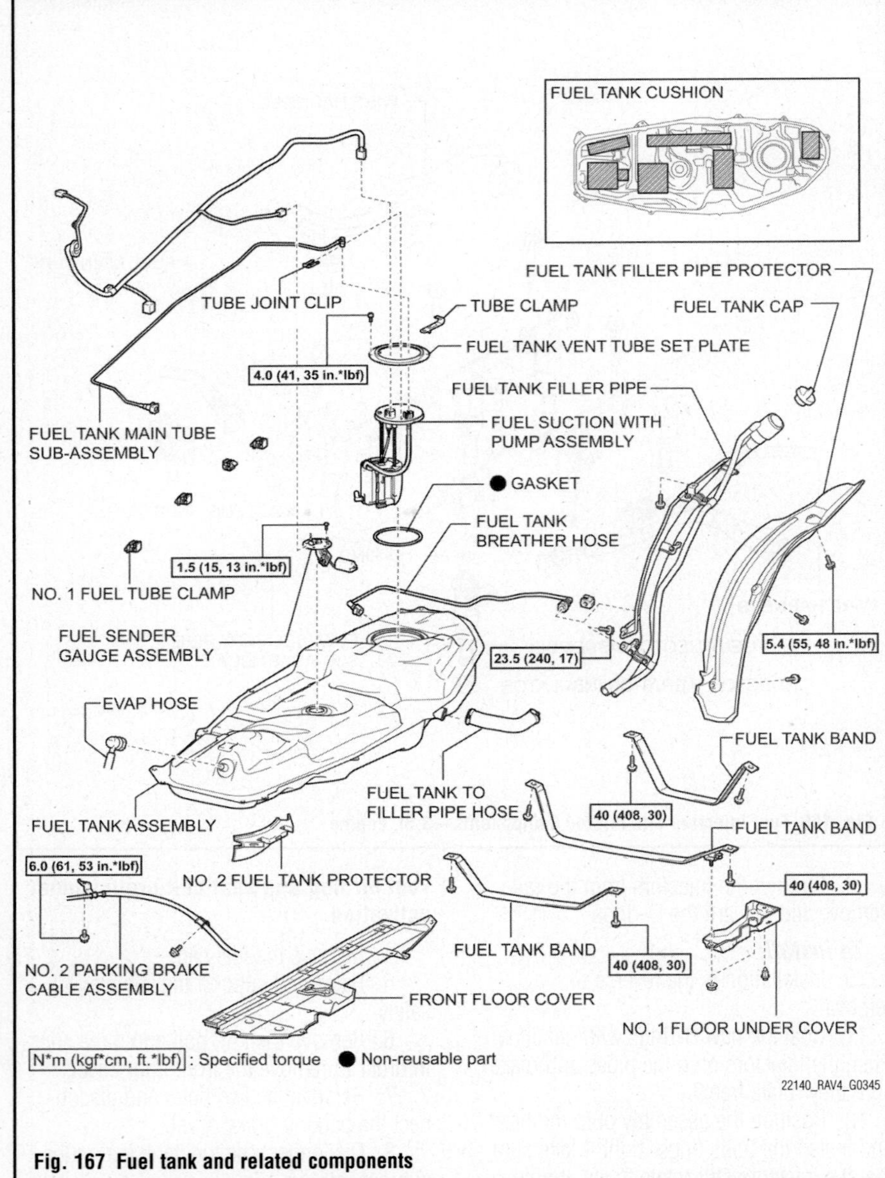

Fig. 167 Fuel tank and related components

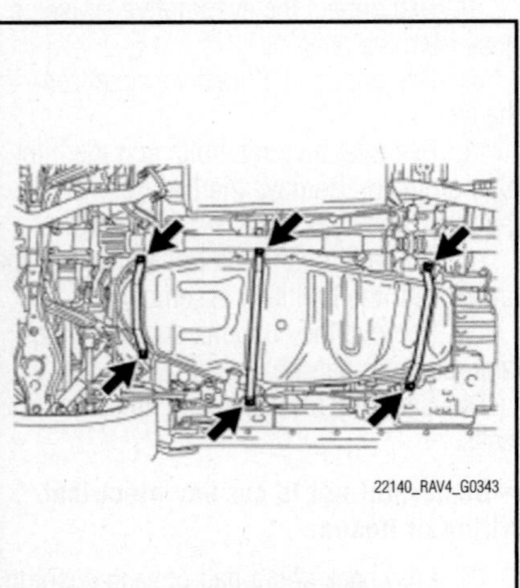

Fig. 168 Fuel tank retaining bolts and bands

16. Disconnect the fuel pump electrical connector and sender gauge connector.

➡**Check the connector for dirt, mud and other contamination, prior to disconnection. Do not use any tools to disconnect the connector.**

17. Detach the wire harness from the four clamps.
18. Remove the fuel tank from the vehicle.

To install:

19. Installation is the reverse of the removal procedure.
20. Tighten the six band retaining bolts to 30 ft. lbs.
21. When connecting tube connectors push firmly until a "click" sound is heard.

➡**Check that the connector and pipe are securely connected by trying to pull them apart.**

22. Start the engine and check for leaks. Correct as required.

IDLE SPEED

ADJUSTMENT

Idle speed is maintained by the Powertrain Control Module (PCM). No adjustment is necessary or possible.

THROTTLE BODY

REMOVAL & INSTALLATION

2.4L Engine

See Figures 169 through 173.

1. Before servicing the vehicle, refer to the Precautions Section.
2. Properly relieve the fuel system pressure.
3. Disconnect the negative battery cable.

➡**Wait at least 90 seconds after disconnecting the negative battery cable before starting any repair work to prevent air bag and seat belt pretensioner activation.**

4. Drain and properly dispose of the engine coolant.
5. Remove the number one engine cover retaining nuts. Remove the cover.
6. Disconnect the Mass Air Flow (MAF) meter connector. Disconnect the purge connector. Disconnect the four wire harness clamps.
7. Disconnect the number two ventilation hose from the air cleaner hose. Disconnect the purge line hose from the clamp.
8. Lock the number one air cleaner hose clamp and then disconnect the number one air cleaner hose from the throttle body.
9. Unfasten the two hook clamps and remove the air cleaner cap.
10. Remove the filter element from the case.
11. Disconnect the purge line hose from the throttle body.
12. Disconnect the water bypass hose from the throttle body. Disconnect the number two water bypass hose from the throttle body.
13. Disconnect the number one water bypass hose from the throttle body.
14. Disconnect the throttle position sensor and the control motor connector. Disconnect the wire harness clamp.
15. Disconnect the fuel tube from the clamp.

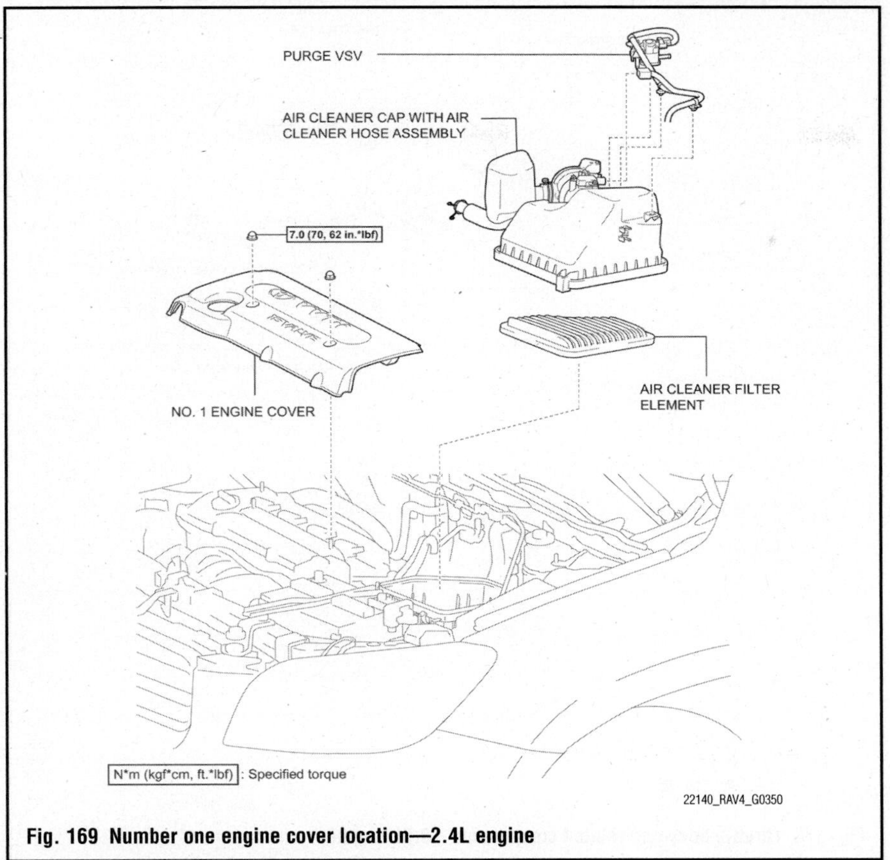

Fig. 169 Number one engine cover location—2.4L engine

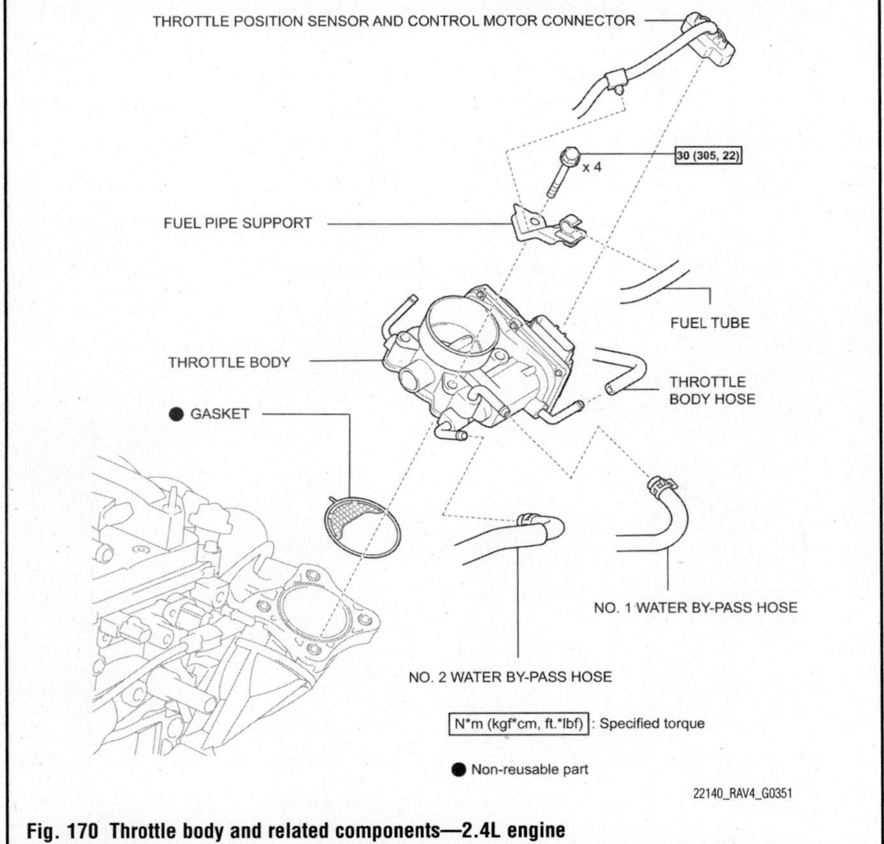

Fig. 170 Throttle body and related components—2.4L engine

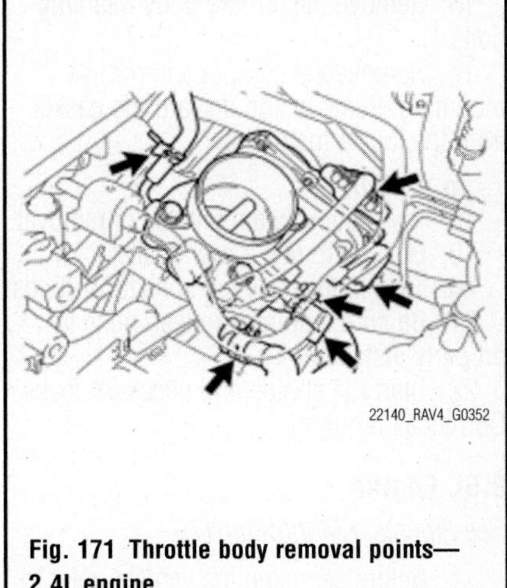

Fig. 171 Throttle body removal points—2.4L engine

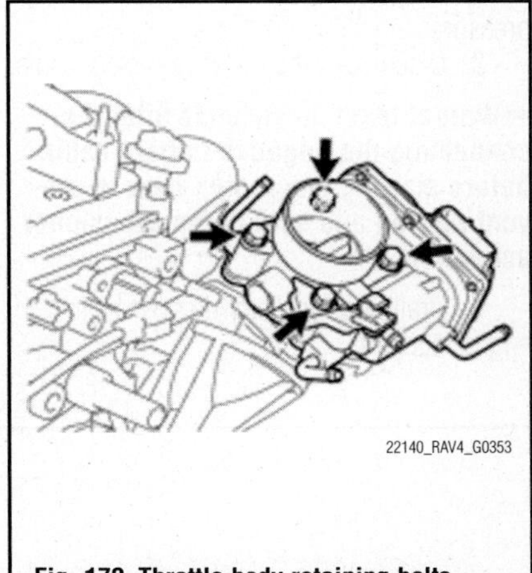

Fig. 172 Throttle body retaining bolts—2.4L engine

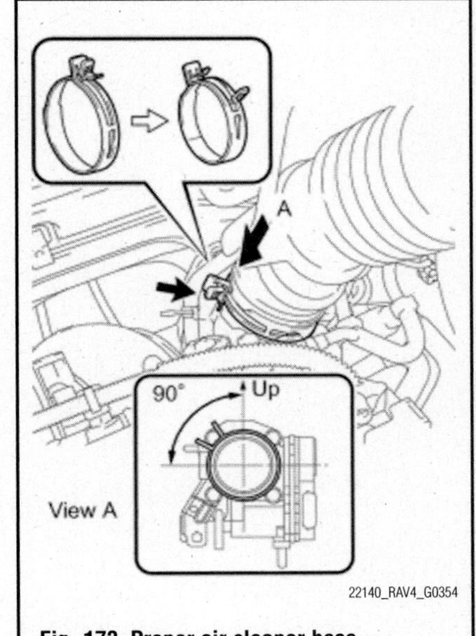

Fig. 173 Proper air cleaner hose installation—2.4L engine

16. Remove the throttle body retaining bolts.

17. Remove the component from its mounting. Remove and discard the gasket from the intake manifold.

To install:

18. Installation is the reverse of removal.

19. Be sure to use a new gasket.

20. Tighten the retaining bolts to 22 ft. lbs.

21. Be sure that the air clean hose is properly installed.

22. Start the engine and check for leaks. Correct as required.

3.5L Engine

See Figures 174 through 176.

1. Before servicing the vehicle, refer to the Precautions Section.

2. Properly relieve the fuel system pressure.

3. Disconnect the negative battery cable.

➡Wait at least 90 seconds after disconnecting the negative battery cable before starting any repair work to prevent air bag and seat belt pretensioner activation.

4. Drain and properly dispose of the engine coolant.

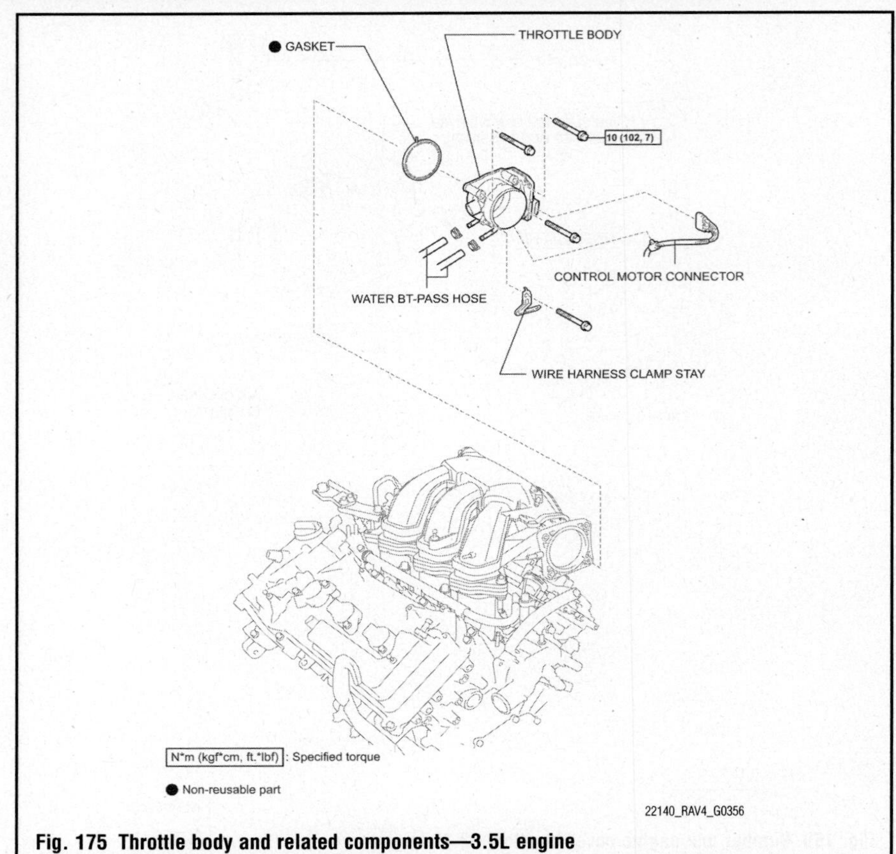

Fig. 175 Throttle body and related components—3.5L engine

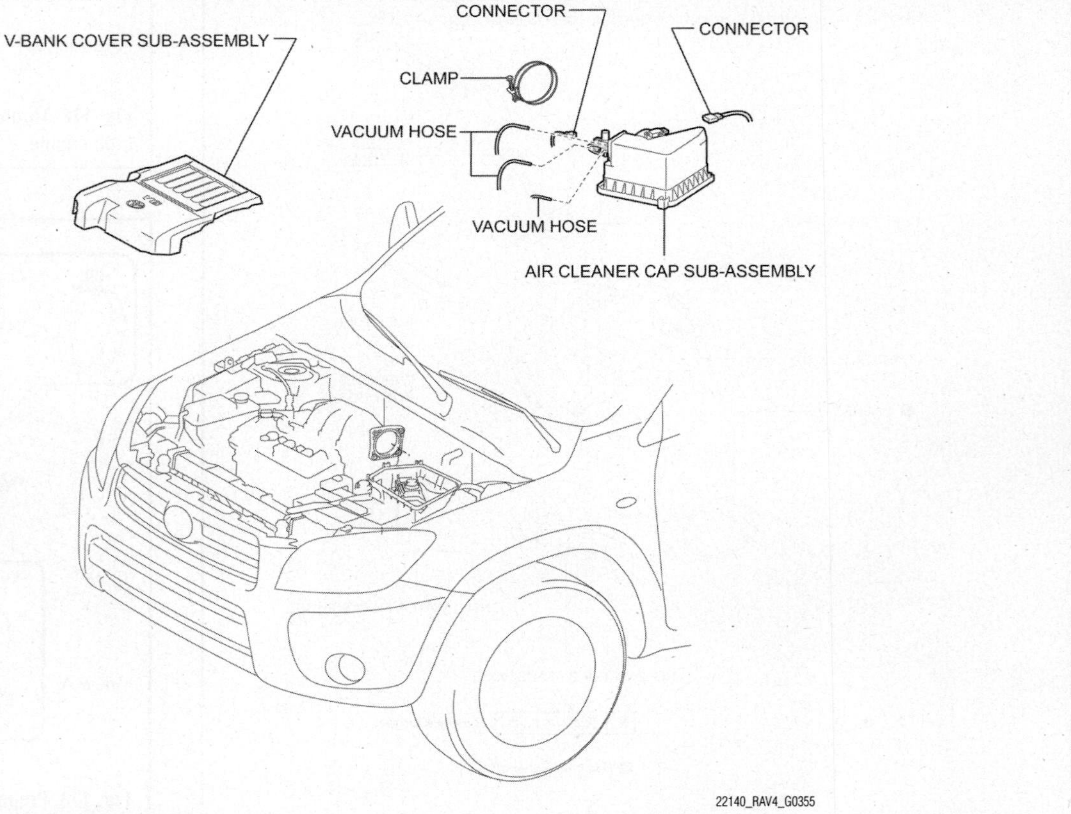

Fig. 174 V-bank cover location—3.5L engine

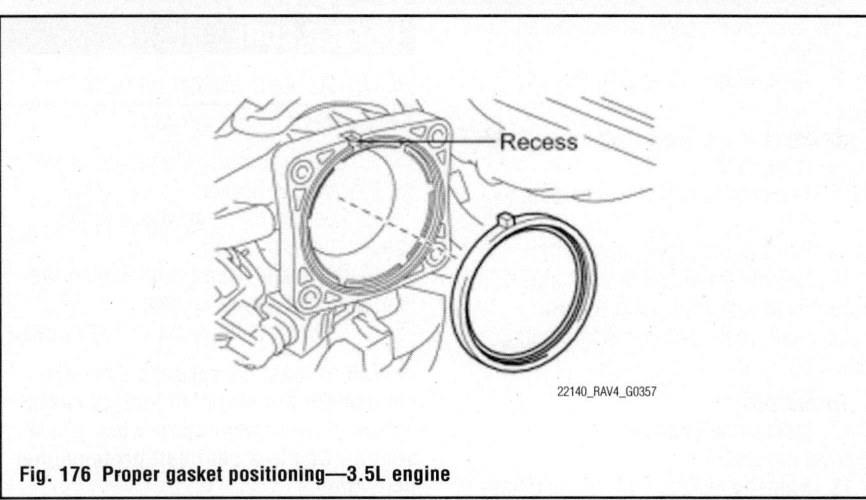

Fig. 176 Proper gasket positioning—3.5L engine

5. Remove the air cleaner sub-assembly.
6. Disconnect the two water bypass hoses from the throttle body.
7. Disconnect the control motor connector.
8. Remove the four throttle body retaining bolts. Remove the wire harness clamp stay.
9. Remove the component from its mounting.
10. Discard the gasket.

To install:

11. Installation is the reverse of removal.
12. Be sure to use a new gasket.
13. Tighten the retaining bolts to 7 ft. lbs.
14. Start the engine and check for leaks. Correct as required.

HEATING & AIR CONDITIONING SYSTEM

BLOWER MOTOR

REMOVAL & INSTALLATION

See Figure 177.

1. Before servicing the vehicle, refer to the Precautions Section.
2. Disconnect the negative battery cable.

➡Wait at least 90 seconds after disconnecting the negative battery cable before starting any repair work to prevent air bag and seat belt pretensioner activation.

3. Properly discharge the air conditioning system. Be sure to plug the refrigerant lines to avoid dirt and moisture from entering the system. Properly recycle the ac refrigerant.
4. Remove the upper instrument panel.
5. Remove the lower instrument panel.
6. Remove the air duct.
7. Remove the heater/air conditioning unit.
8. Remove the blower assembly.
9. Disconnect the electrical connector.
10. Remove the retaining screws.
11. Remove the component from its mounting.
12. Remove the blower motor retaining screws.
13. Remove the blower motor from its mounting.

To install:

14. Installation is the reverse of the removal procedure.
15. Properly recharge the air conditioning system.

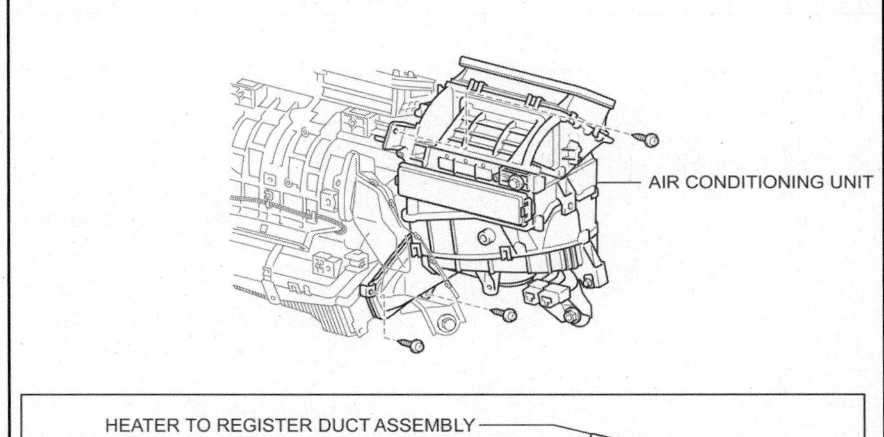

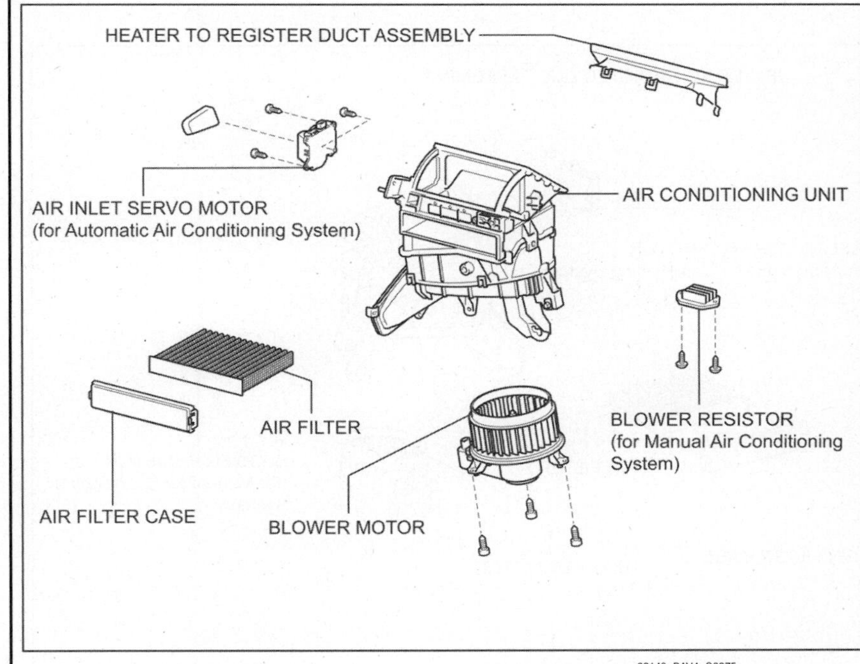

Fig. 177 Blower assembly and related components

16. Check the SRS warning light for proper operation.

BLOWER UNIT

REMOVAL & INSTALLATION

See Figure 178.

1. Before servicing the vehicle, refer to the Precautions Section.
2. Disconnect the negative battery cable.

➡Wait at least 90 seconds after disconnecting the negative battery cable before starting any repair work to prevent air bag and seat belt pretensioner activation.

3. Properly discharge the air conditioning system. Be sure to plug the refrigerant lines to avoid dirt and moisture from entering the system. Properly recycle the ac refrigerant.
4. Remove the upper instrument panel.
5. Remove the lower instrument panel.
6. Remove the air duct.
7. Remove the heater/air conditioning unit.
8. Remove the blower assembly.
9. Disconnect the electrical connector.
10. Remove the retaining screws.
11. Remove the component from its mounting.

To install:

12. Installation is the reverse of the removal procedure.
13. Properly recharge the air conditioning system.
14. Check the SRS warning light for proper operation.

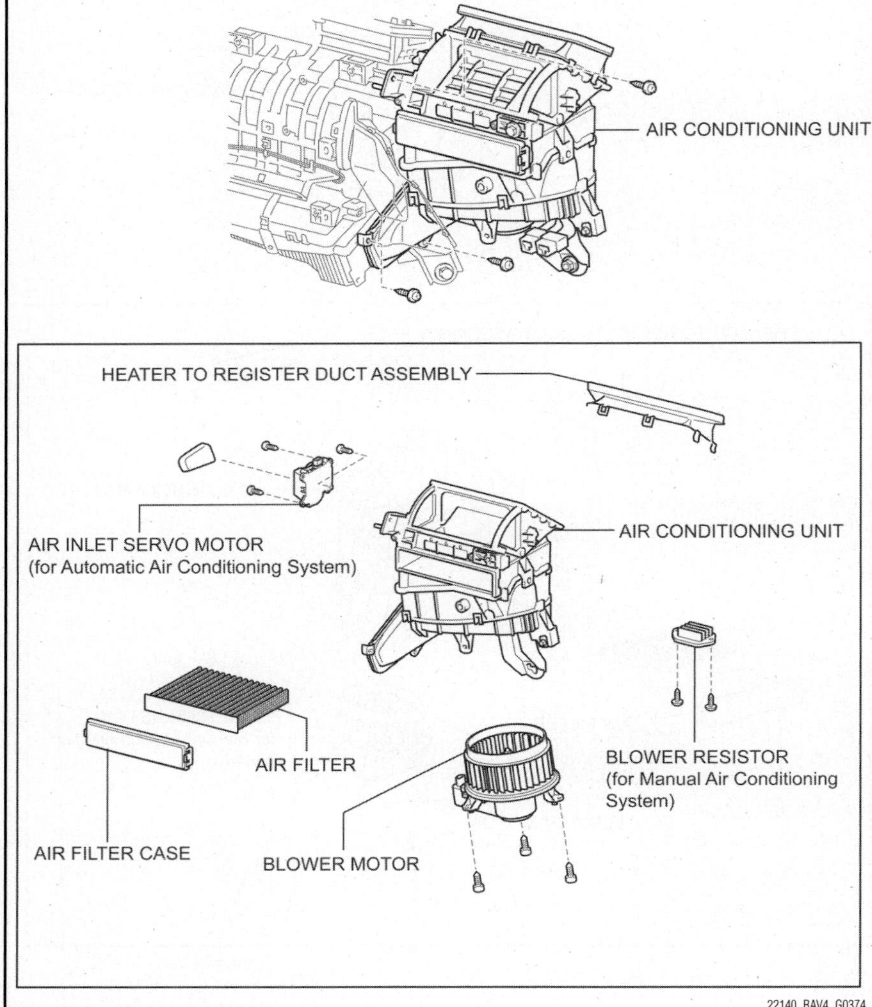

Fig. 178 Blower unit and related components

HEATER CORE

REMOVAL AND INSTALLATION

See Figures 179 through 184.

1. Before servicing the vehicle, refer to the Precautions Section.
2. Position the front wheels in the straight ahead position.
3. Discharge the air conditioning system. Drain the engine coolant.
4. Disconnect the negative battery cable.

➡Wait at least 90 seconds after disconnecting the negative battery cable before starting any repair work to prevent air bag and seat belt pretensioner activation.

5. Disconnect and plug the refrigerant lines at the evaporator. Disconnect the heater hoses at the heater core.
6. Using a Torx® socket, loosen the two Torx® screws at the wheel pad. Pull out the wheel pad from the steering wheel and disconnect the air bag connector. Remove the wheel pad.

➡If the air bag connector is disconnected with the ignition switch "ON", DTC's will be recorded. When storing the wheel pad, keeps the upper surface of the pad facing upward. Never disassemble the wheel pad. When removing the wheel pad, take care not to pull the air bag wire harness.

7. Disconnect the connector. Matchmark the steering wheel. Remove the steering wheel retaining nut. Using the proper removal tool, remove the steering wheel.
8. Detach the four claws, release the tilt lever and remove the lower steering column cover. Detach the claw and remove the upper steering column cover.
9. Disconnect all connectors from the turn signal switch and the spiral cable. Remove the spiral cable.
10. Detach the clamp holding the combination switch in place. Remove the combination switch from the steering column.
11. Remove the instrument panel subassembly.
12. Disconnect the power steering motor wire harness and torque sensor wire harness clamps from the power steering ECU side. Disconnect the two steering column connectors from the power steering ECU.
13. Remove the lower instrument panel finish panel.
14. Turn back the driver's side carpet. Remove the two clips and remove the steering column hole cover silencer sheet.

15. Place matchmarks on the steering intermediate shaft and steering gear. Remove the bolt and detach the steering gear.

16. Place matchmarks on the steering intermediate shaft and the steering column. Remove the bolt and detach the steering intermediate shaft from the steering column.

17. Remove the brake pedal support assembly.

18. Disconnect the connectors and the wire harness clamps from the steering column assembly.

19. Remove the bolts and nuts and remove the steering column from the instrument panel reinforcement.

20. Remove the headlight dimmer switch assembly. Remove the windshield wiper switch assembly.

21. Remove the number one and the number two instrument cluster finish panel center assemblies.

22. Remove the radio.

23. Remove the two screws. Disconnect the connectors and remove the air conditioning control.

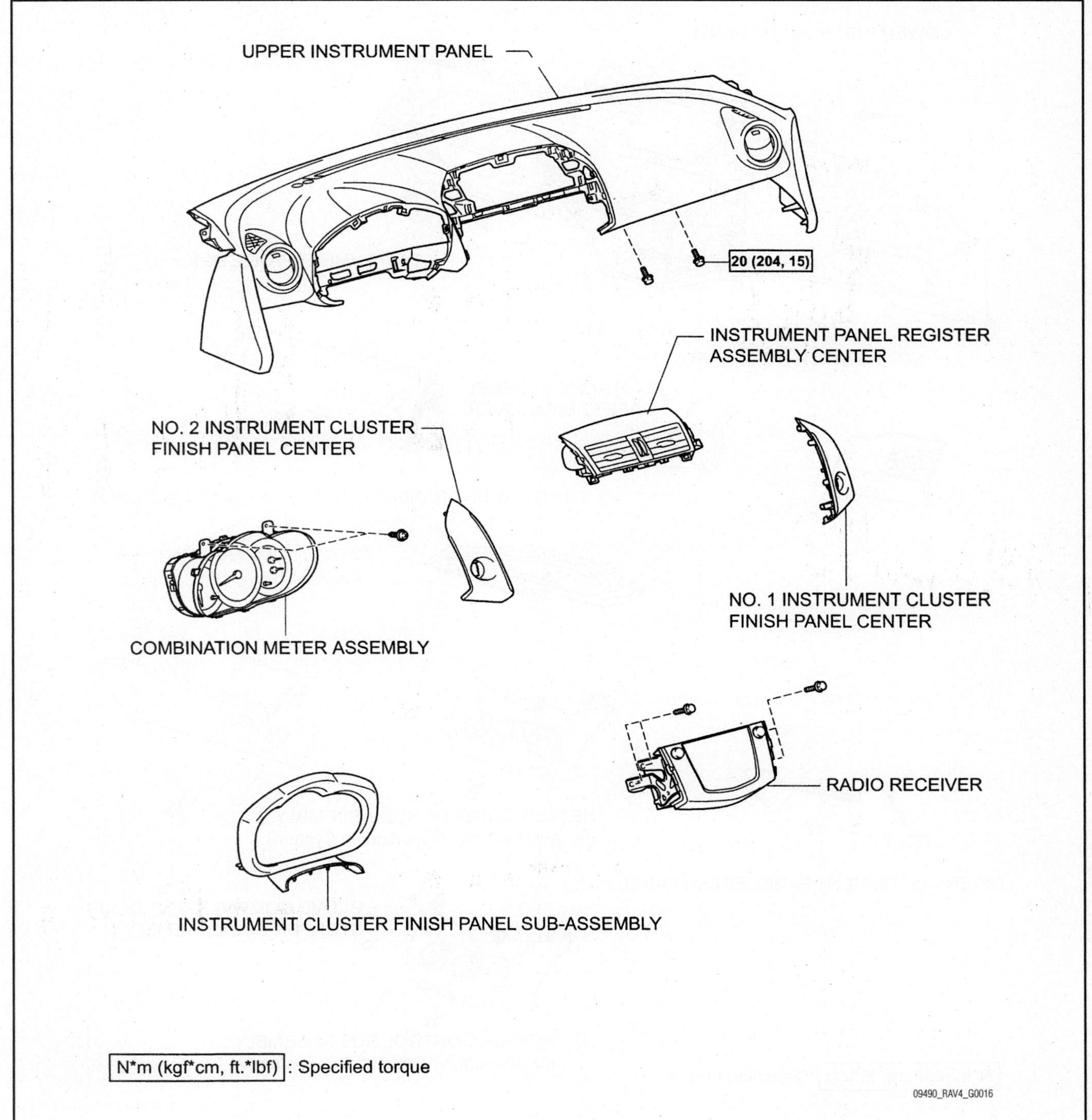

Fig. 179 Upper instrument panel and related components

24. If equipped with automatic climate control, remove the air mix control switch, the blower control switch and the air vent mode control switch.

25. Remove the number one console upper panel garnish. Remove the number two console upper panel garnish.

26. Remove the shift lever knob. Using a pry tool, detach the two clips, four claws and remove the upper console panel. Disconnect the connector.

27. Detach the two clips and two claws. Disconnect the connectors and remove the switch base.

28. Remove the two screws. Detach the two clips and remove the cup holder box.

29. Using a pry tool, detach the four claws and two clips. Disconnect the connector and remove the upper rear console.

30. Using a pry tool, detach the six claws and remove the console rear end panel.

31. Detach the two claws and remove the right instrument panel bracket cover.

32. Detach the two claws and remove the left instrument panel bracket cover.

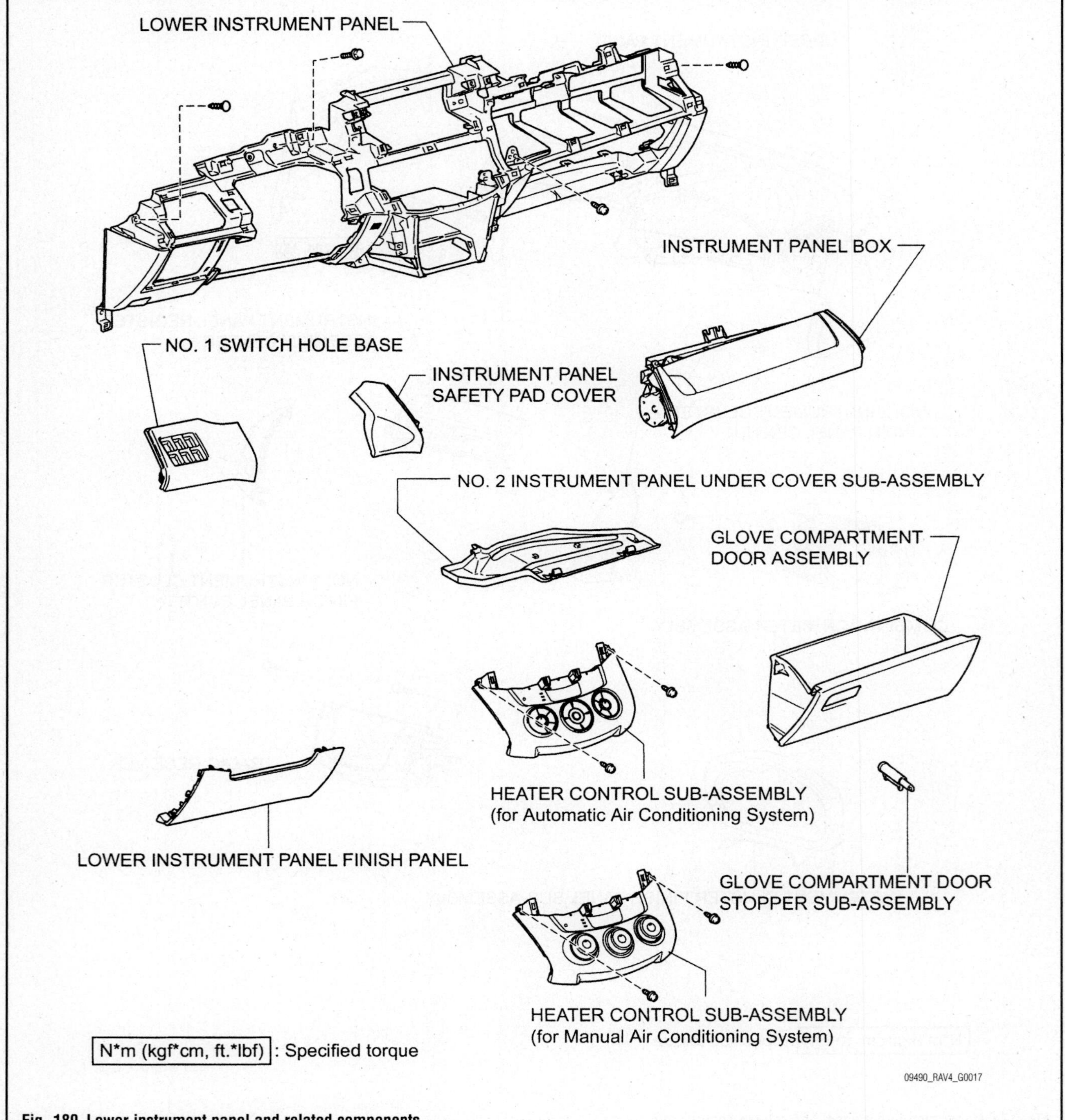

Fig. 180 Lower instrument panel and related components

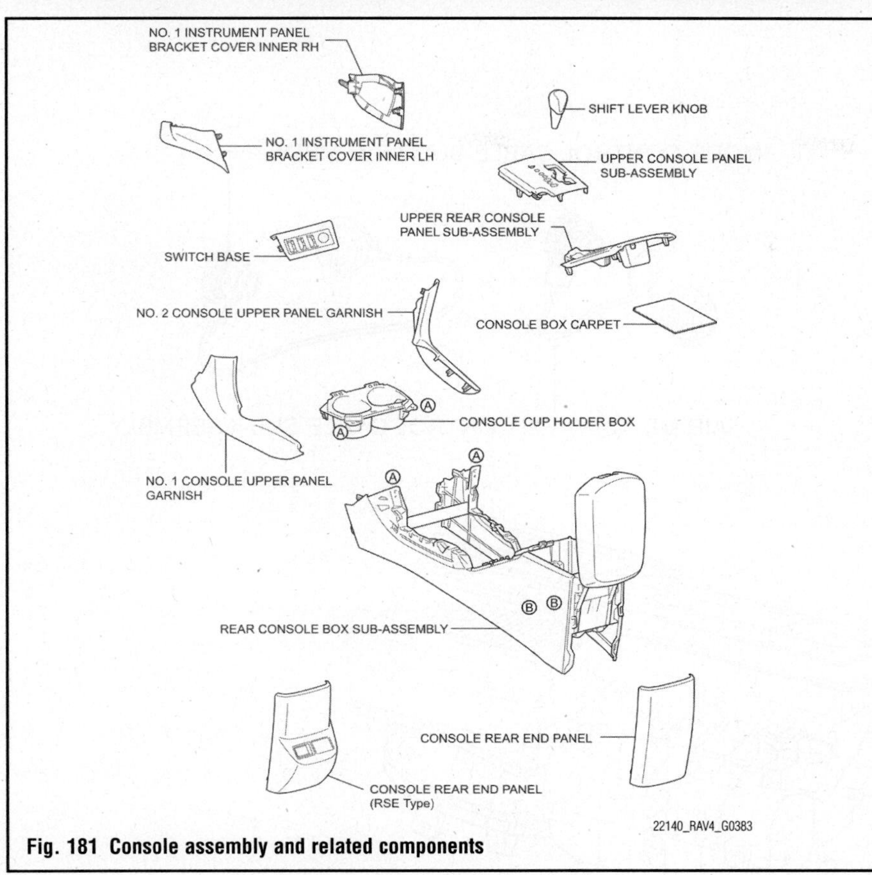

Fig. 181 Console assembly and related components

33. Remove the console box carpet. Remove the two bolts and two screws. Disconnect the connector and remove the rear console box.
34. Detach the two claws and remove the instrument panel under cover.
35. Remove the glove box door assembly.
36. Using a pry tool, detach the four clips. Disconnect the connectors and remove the number one switch hole base.
37. Using a pry tool, detach the four claws and remove the instrument panel safety pad cover.
38. Remove the screw. Using a pry tool, detach the six clips and remove the instrument panel box.
39. Disconnect the connectors. Remove the three nuts and remove the power steering ECU assembly.
40. Remove the two bolts, three screws and two clips. Disconnect the connectors and clamps. Remove the lower instrument panel from the vehicle.
41. Detach the three claws and remove the defroster nozzle assembly.
42. Fold back the carpet. Disconnect the clamp and disconnect the wire harness. Remove the bolt, nut, screw and instrument panel brace.
43. Disconnect the two clamps and disconnect the wire harness. Detach the three claws and remove the rear air duct.
44. Detach the two claws and remove the air duct.
45. Detach the two clamps. Disconnect the connector. Remove the screw and the air conditioning amplifier assembly. Remove the drain hose.
46. Disconnect the twelve retaining clamps. Remove the four bolts and disconnect the ground wire. Remove the six bolts. Remove the instrument panel reinforcement.
47. Remove the bolt, nut and the air conditioner unit assembly.
48. Remove the number three heater register duct. Remove the air duct. Remove the air outlet control servo motor. Remove the air mix control servo motor.
49. If equipped with manual air conditioning remove the mode control cable and the air mix damper control cable.
50. Remove the heater core from its mounting.

To install:

51. Installation is the reverse of the removal procedure.
52. Tighten the steering column retaining bolts to 18 ft. lbs.

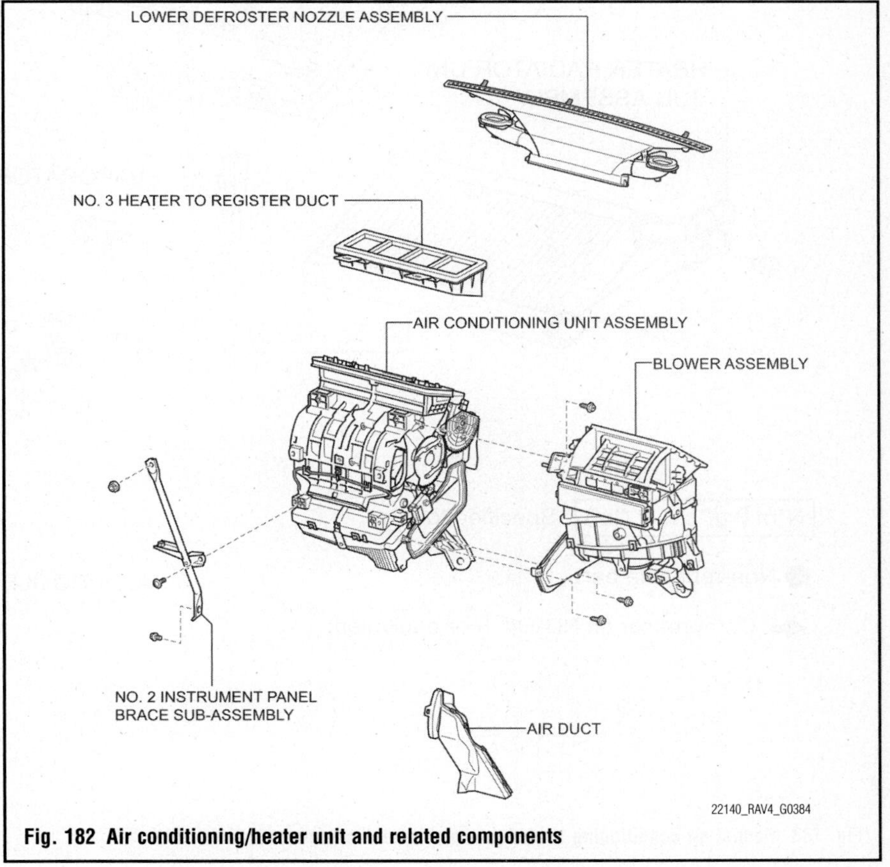

Fig. 182 Air conditioning/heater unit and related components

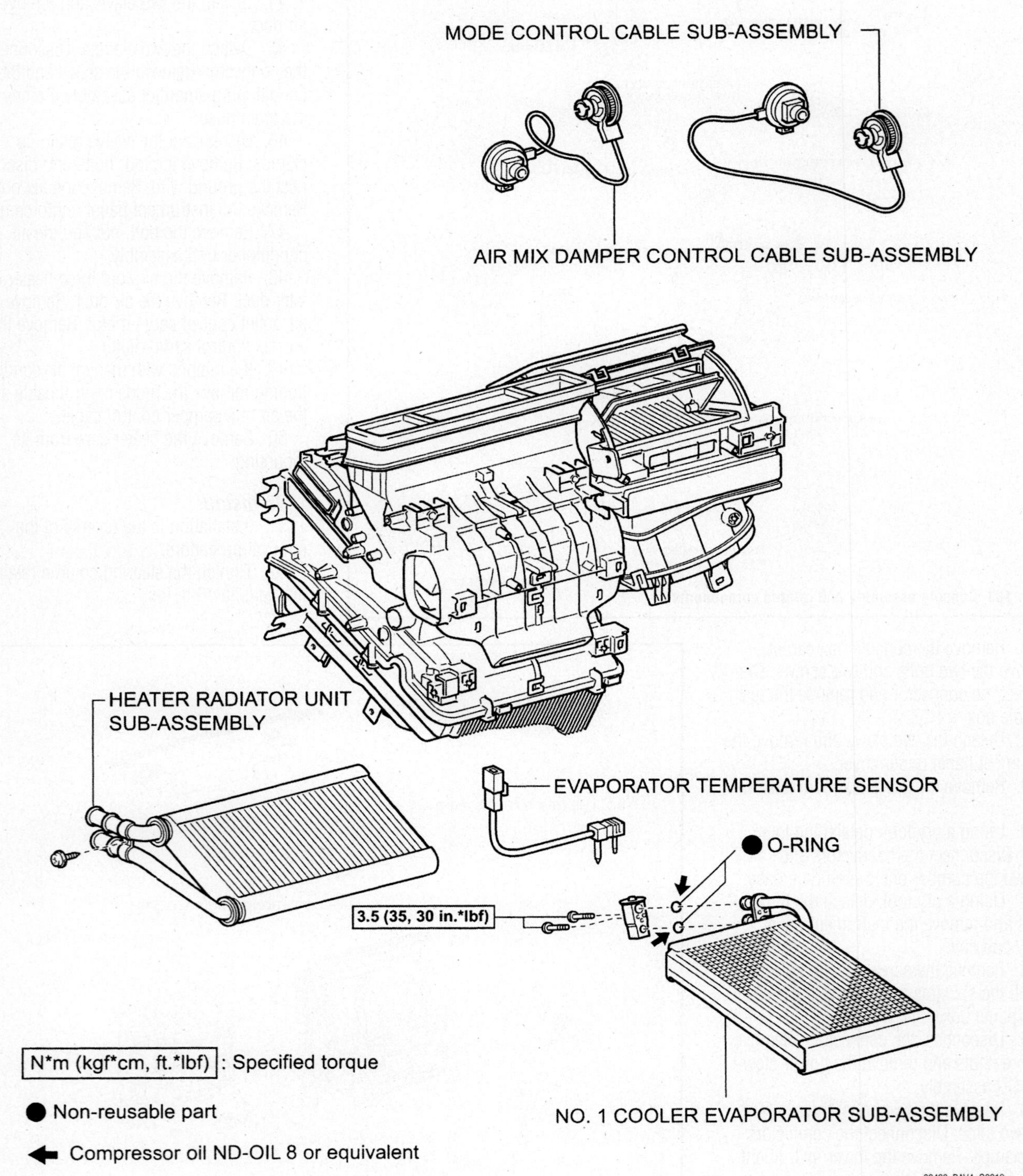

Fig. 183 Manual air conditioning heater core and related components

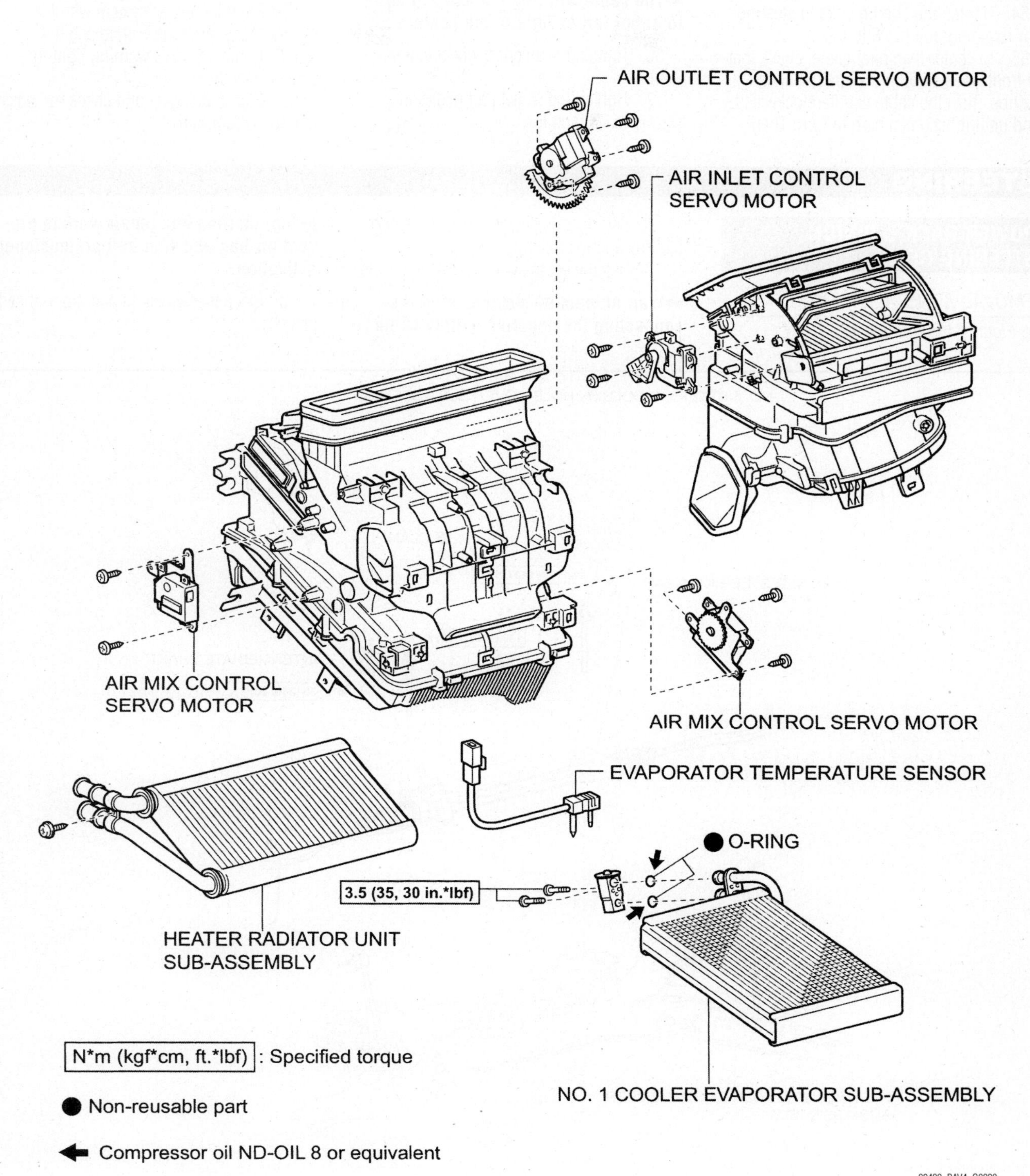

Fig. 184 Automatic air conditioning heater core and related components

53. Tighten the sliding yoke to number intermediate shaft retaining bolt to 26 ft. lbs.

54. Tighten the sliding yoke to steering gear retaining bolt to 26 ft. lbs.

55. To center the spiral cable, check that the front wheels are in the straight ahead position. Turn the cable counterclockwise by hand until it becomes hard to turn. Then rotate the cable clockwise about 2.5 turns to align the marks.

➡The cable will rotate about 2.5 turns to either left or right of the center.

56. Tighten the steering wheel locknut to 37 ft. lbs.

57. Tighten the wheel pad retaining screws to 78 inch lbs.

58. Check the steering wheel center point.

59. Recharge the air conditioning system.

60. Fill the cooling system with the proper grade and type engine coolant.

61. Check the SRS warning light for proper operation.

62. Start the engine and check for leaks, correct as required.

STEERING

POWER RACK & PINION STEERING GEAR

REMOVAL & INSTALLATION

See Figures 185 through 187.

1. Before servicing the vehicle, refer to the Precautions Section.

2. Disconnect the negative battery cable.

➡Wait at least 90 seconds after disconnecting the negative battery cable before starting any repair work to prevent air bag and seat belt pretensioner activation.

3. Place the wheels in a straight-ahead position.

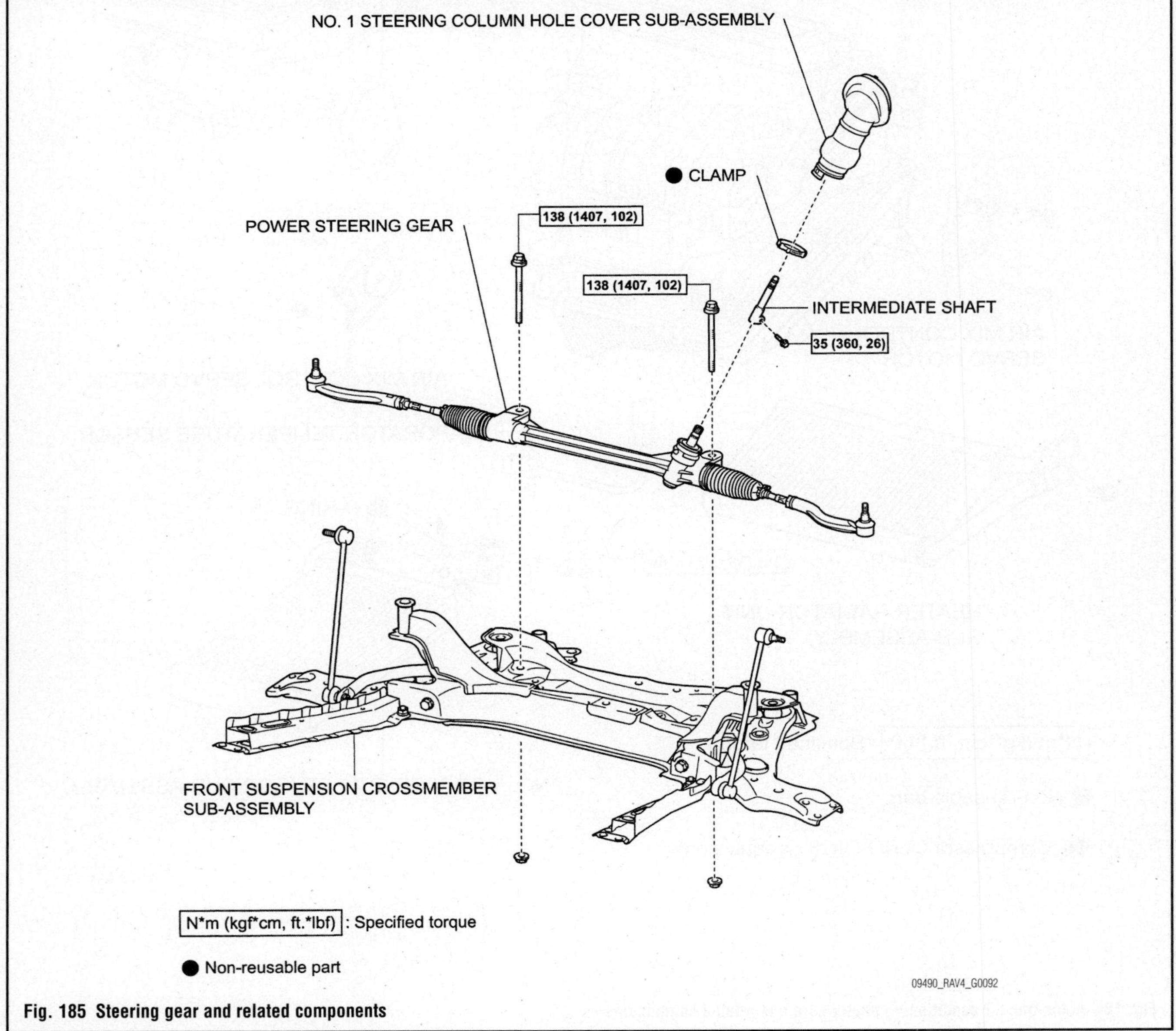

Fig. 185 Steering gear and related components

4. Raise and support the vehicle safely.
5. Remove the tire and wheel assemblies.
6. Disconnect the right and left tie rod ends, using tool SST09628-62011 or equivalent.
7. Remove the floor carpet. Remove the two clips and the column hole silencer cover.
8. Use the seat belt to position the steering wheel in order to avoid breakage of the spiral cable.
9. Matchmark the sliding yoke of the steering intermediate shaft. Remove the bolt and disconnect the sliding yoke.
10. Remove the bottom clip and detach the upper clip from the body and disconnect the number one steering column hole cover. Be careful not to damage the clips.
11. Remove the engine and transaxle assembly from the vehicle and position it in a suitable holding fixture.
12. Remove the clamp and disconnect the number one column hole cover from the steering gear.
13. Matchmark the intermediate shaft of the steering gear. Remove the bolt and disconnect the steering intermediate shaft from the steering gear.

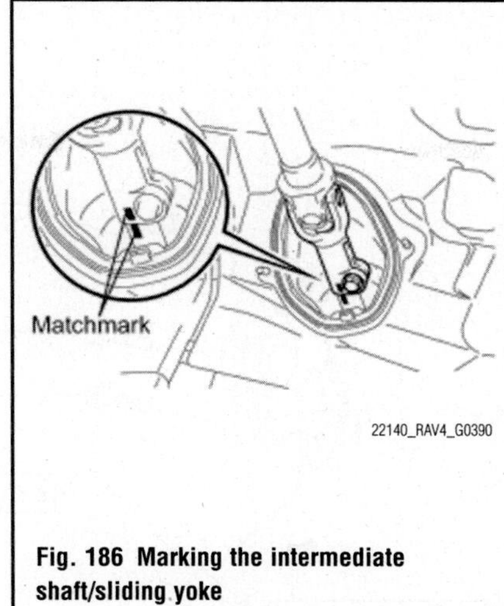

Fig. 186 Marking the intermediate shaft/sliding yoke

14. Remove the two bolts, two nuts and the steering gear from the crossmember. Be sure to keep the nut from rotating while turning the bolt.

To install:

15. Installation is the reverse of the removal procedure.
16. Tighten the steering gear retaining bolts to 102 ft. lbs. Be sure to keep the nut from rotating while turning the bolt.

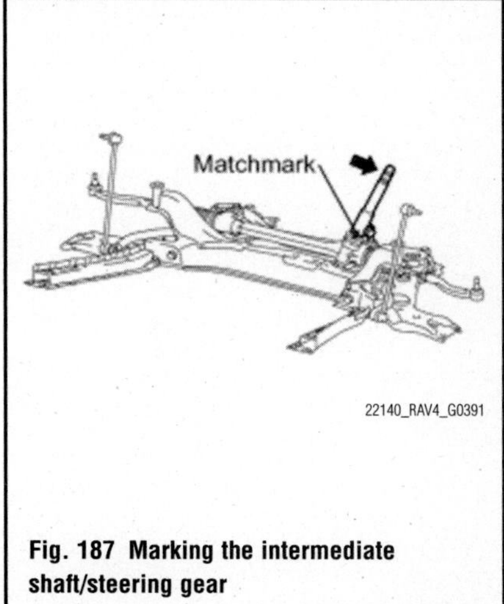

Fig. 187 Marking the intermediate shaft/steering gear

17. Tighten the intermediate shaft retaining bolt to 26 ft. lbs.
18. Tighten the sliding yoke retaining bolt to 26 ft. lbs.
19. Tighten the tie rod end castle nut to 36 ft. lbs. If the holes for the clip are not aligned, tighten the nut an additional 60 degrees. Be sure to use a new cotter pin.
20. Check and adjust the alignment, as required.

SUSPENSION

LOWER BALL JOINT

REMOVAL & INSTALLATION

See Figures 188 and 189.

1. Before servicing the vehicle, refer to the Precautions Section.
2. Raise and support the vehicle safely.
3. Remove the tire and wheel assembly.
4. On 2WD, remove the front speed sensor.
5. Remove the front caliper. Remove the rotor.
6. Remove the front axle hub nut.
7. To disconnect the front suspension lower number one arm, remove the bolt and two nuts. Disconnect the lower arm from the ball joint.
8. Using the proper tools, disconnect the tie rod end.
9. Remove the steering knuckle with the axle hub, using the proper removal tools.
10. Remove the cotter pin and nut. Using tool SST09628-62011, remove the lower ball joint.

To install:

11. Installation is the reverse of the removal procedure.
12. Tighten the two ball joint to the lower arm nuts to 68 ft. lbs. (92 Nm).
13. Tighten the lower ball joint to the steering knuckle to 98 ft. lbs. (133 Nm).
14. Tighten the wheel lug nuts to 76 ft. lbs. (103 Nm).
15. Be sure to check and adjust the alignment, as required.

LOWER CONTROL ARM

REMOVAL & INSTALLATION

See Figures 190 and 191.

1. Before servicing the vehicle, refer to the Precautions Section.
2. Disconnect the negative battery cable.

➡Wait at least 90 seconds after disconnecting the negative battery cable before starting any repair work to prevent air bag and seat belt pretensioner activation.

3. Matchmark and remove the hood.
4. Raise and support the vehicle safely.
5. Remove the tire and wheel assembly.
6. Install engine hanger tools 12281-28010 and 12282-28010. Suspend the engine assembly using the proper engine removal tools.
7. Disconnect the front stabilizer links. Disconnect the front suspension lower number one arm assembly.
8. Remove the two nuts, two bolts and engine mounting rear insulator. Remove the bolt from the suspension member.
9. Support the crossmember, using a suitable jack.
10. Remove the four bolts "A" from the member reinforcement. Remove the six bolts "B" from the member reinforcement.
11. Carefully lower the jack and disconnect the crossmember from the vehicle.
12. Remove the bolt and nut from the suspension member (front). Remove the bolt and nut from the suspension member (rear). Remove the lower control arm.

To install:

13. Temporarily install the front suspension lower arm in its mounting.
14. Connect the front crossmember sub-assembly.
15. Install, but do not fully tighten, the four retaining bolts "A", the six retaining bolts "B", the suspension member with the bolt to the body and the rear mounting insulator bolts.

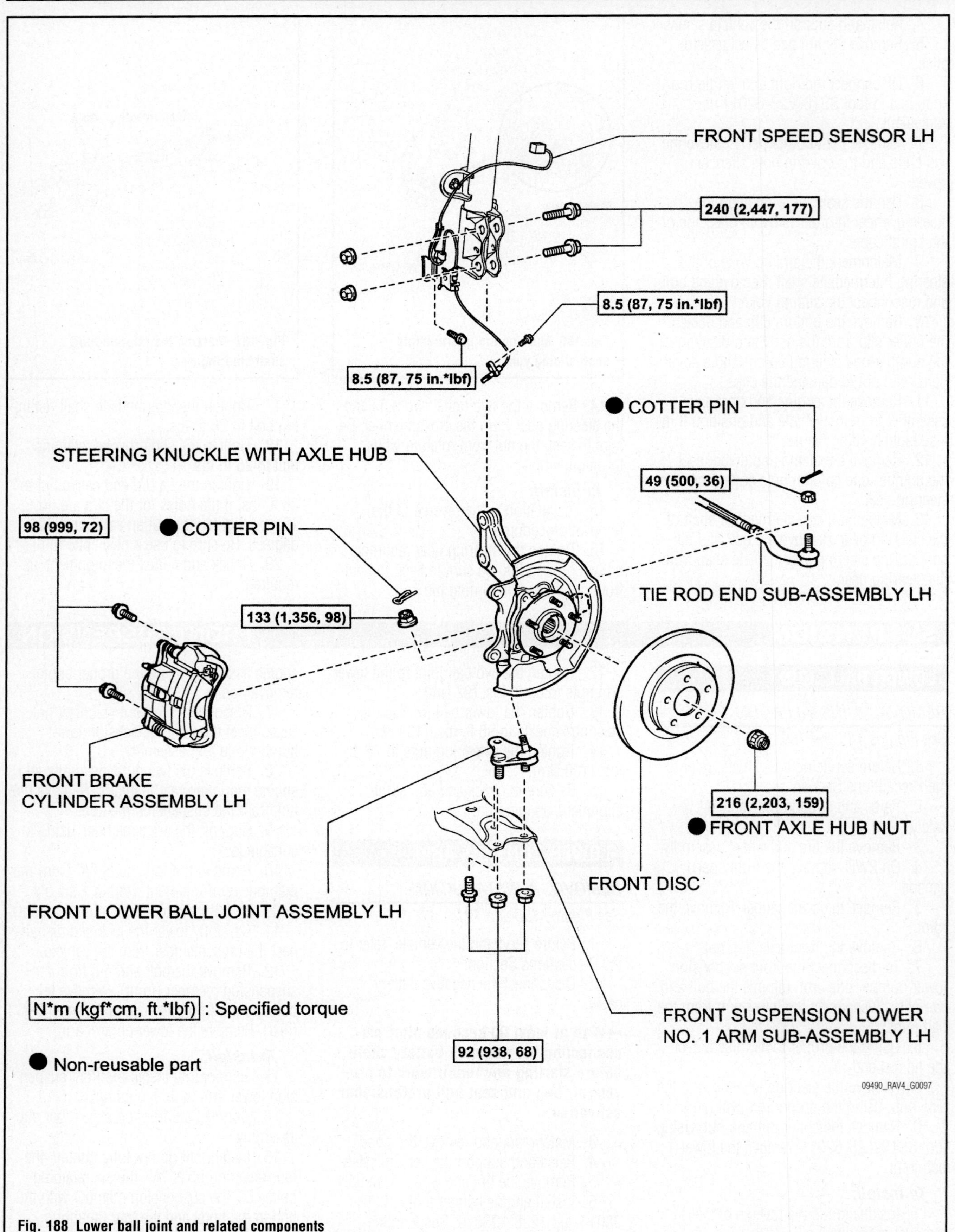

Fig. 188 Lower ball joint and related components

SST
7924ZG81

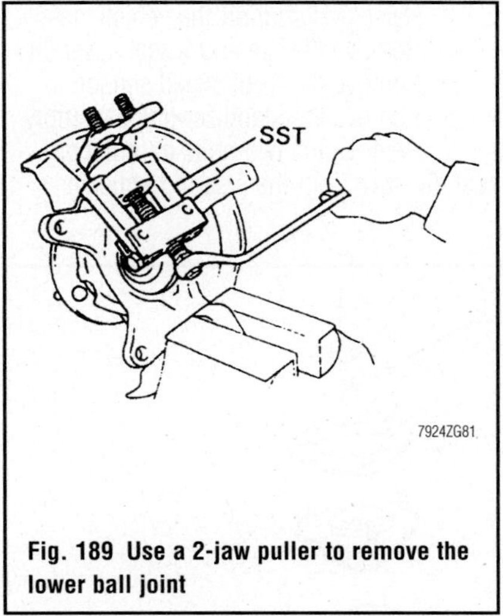

Fig. 189 Use a 2-jaw puller to remove the lower ball joint

A
A
B
B
A
A
B
B
B
B
09490_RAV4_G0099

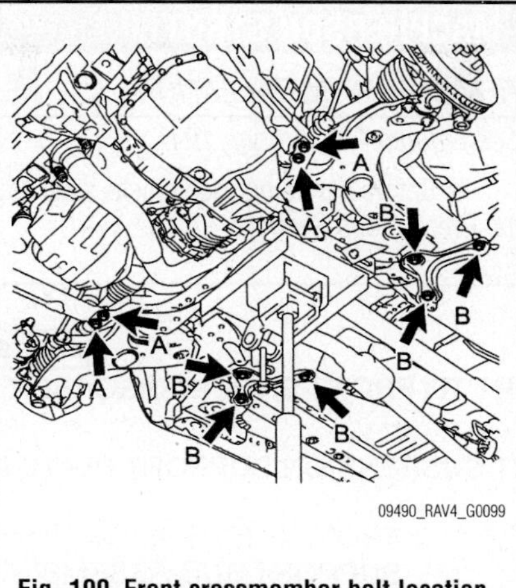

Fig. 190 Front crossmember bolt location and identification

16. Install the front stabilizer link.

17. Connect the front suspension lower number one arm sub-assembly. Tighten the two bolts and nut to 68 ft. lbs.

18. Install the tire and wheel assembly.

19. To stabilize the suspension, lower the vehicle to ground height. Press down on the vehicle several times to stabilize the suspension.

20. Tighten the front crossmember sub-assembly retaining bolts to 64 ft. lbs. for bolt "A", 69 ft. lbs. (93 Nm) for bolt "B", 107 ft. lbs (145 Nm). for the bolt to the body and 70 ft. lbs. (95 Nm) for the rear engine insulator bolts.

21. Tighten the front suspension lower number one arm sub-assembly.

REAR ENGINE MOUNTING INSULATOR
● COTTER PIN
FRONT STABILIZER
LINK ASSEMBLY LH
49 (500, 36)
233 (2,376, 172)
95 (969, 70)
FRONT SUSPENSION
CROSSMEMBER SUB-ASSEMBLY
233 (2,376, 172)
FRONT SUSPENSION
MEMBER BRACE REAR RH
145 (1,479, 107)
93 (948, 69)
FRONT SUSPENSION
MEMBER BRACE REAR LH
92 (938, 68)
145 (1,479, 107)
FRONT SUSPENSION LOWER
NO. 1 ARM SUB-ASSEMBLY LH
N*m (kgf*cm, ft.*lbf) : Specified torque ● Non-reusable part
93 (948, 69)
09490_RAV4_G0098

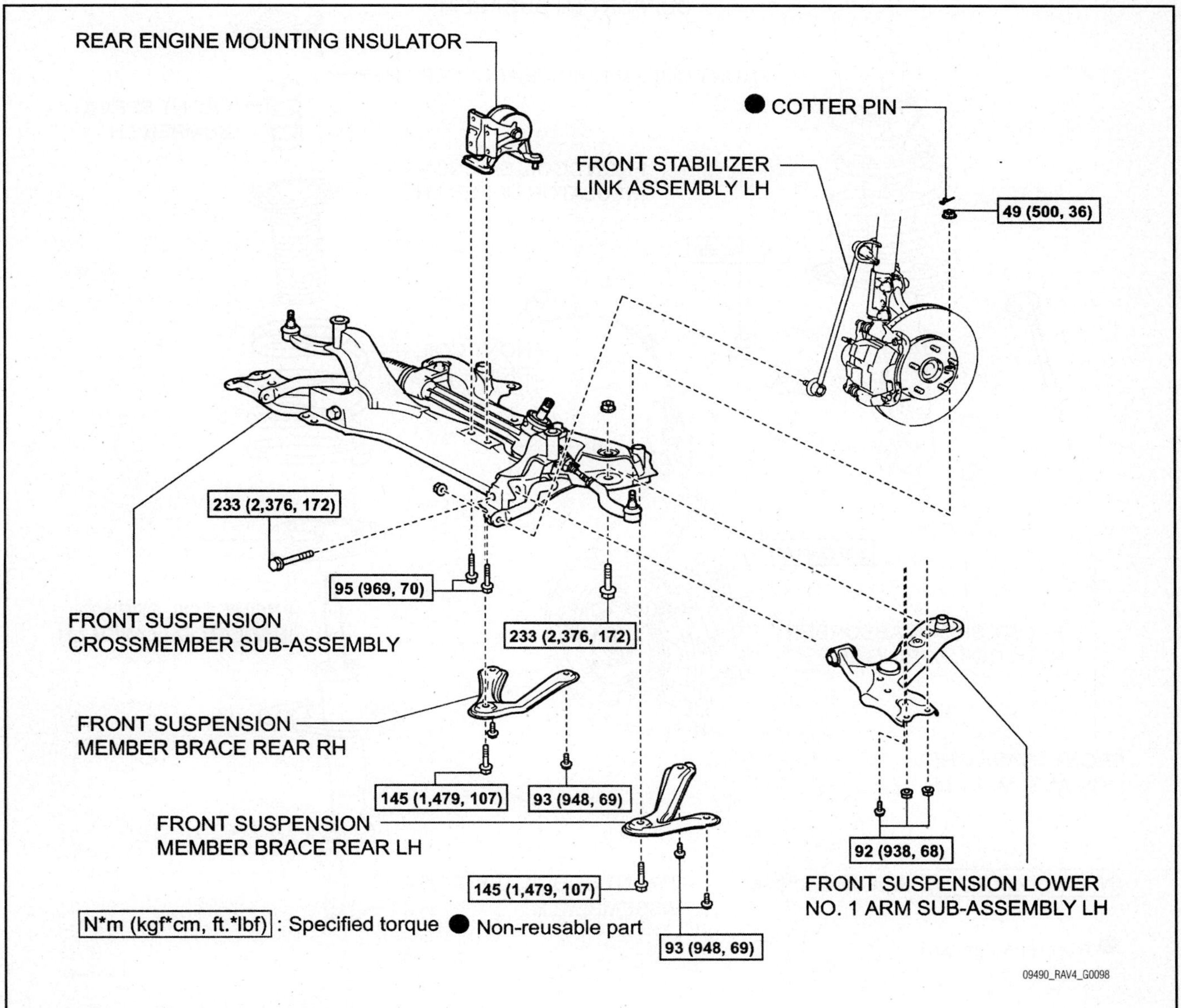

Fig. 191 Lower control arm and related components

22. Continue the installation in the reverse order of the removal procedure.
23. Tighten the wheel lug nuts to 76 ft. lbs. (103 Nm).
24. Be sure to check and adjust the alignment, as required.

MACPHERSON STRUT

REMOVAL & INSTALLATION

See Figures 192 through 194.

1. Before servicing the vehicle, refer to the Precautions Section.
2. Raise and support the vehicle safely.
3. Remove the tire and wheel assembly.
4. Remove the front speed sensor.
5. Remove the stabilizer link assembly.
6. Remove the two bolts and disconnect the strut from the steering knuckle.

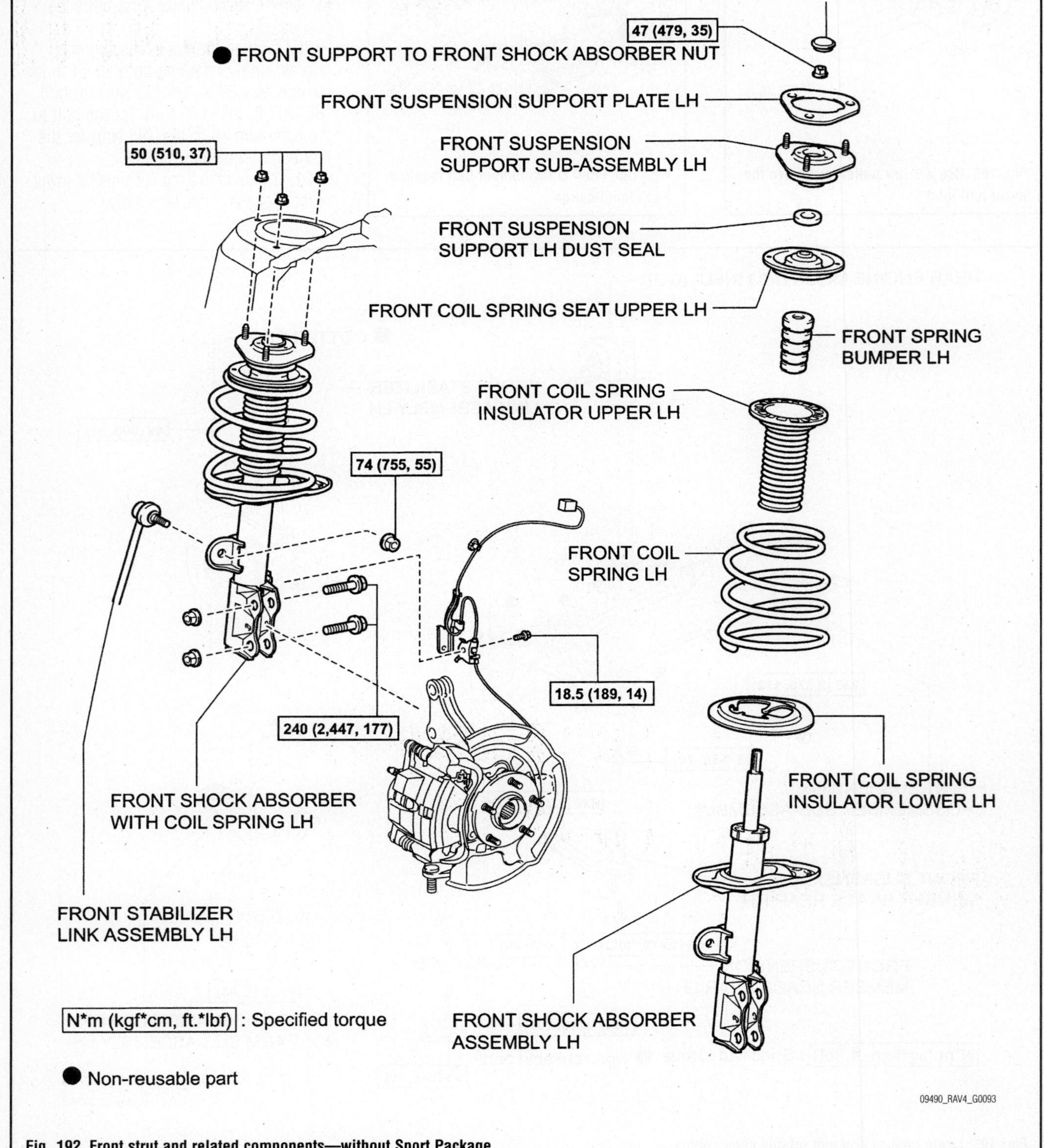

Fig. 192 Front strut and related components—without Sport Package

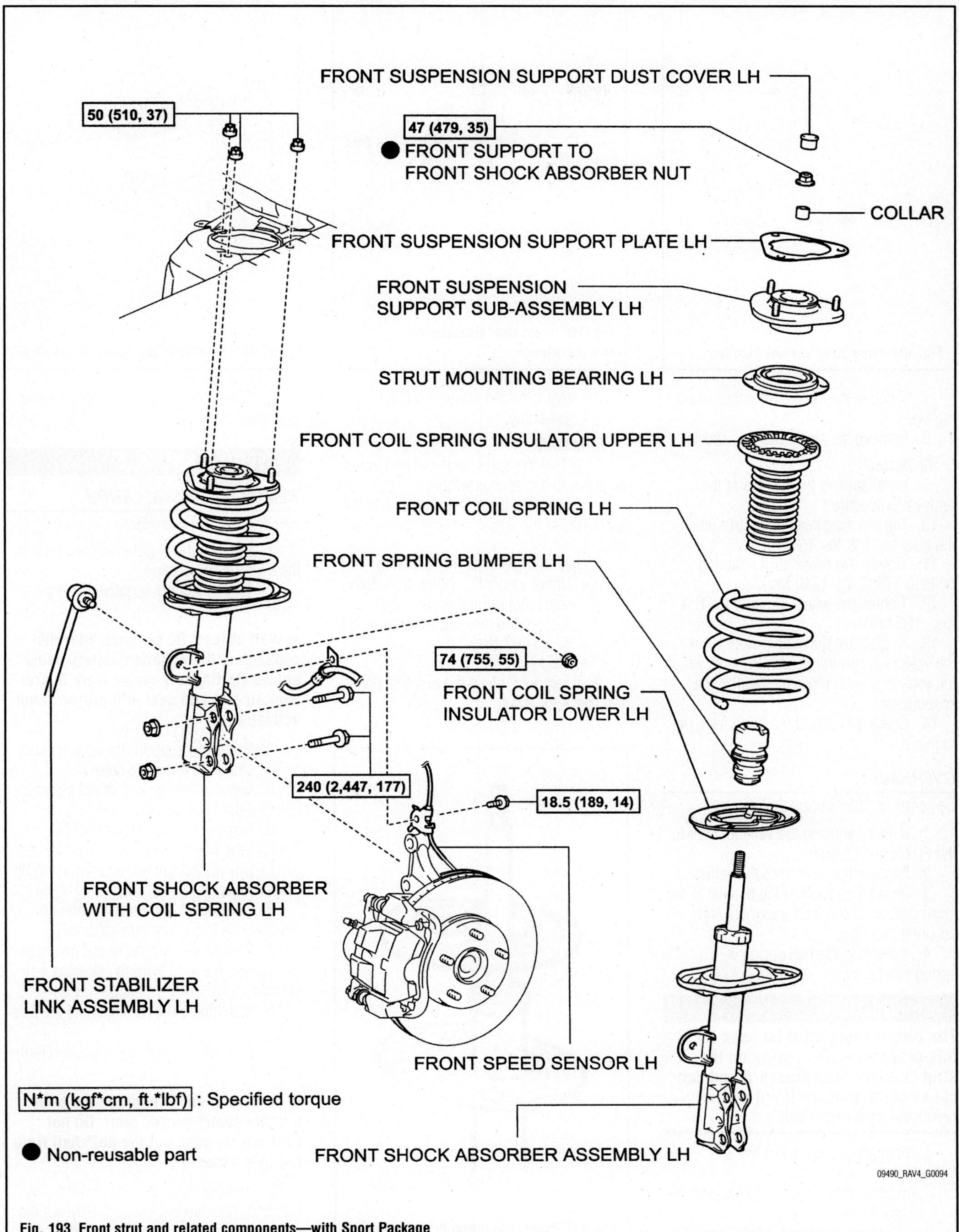

Fig. 193 Front strut and related components—with Sport Package

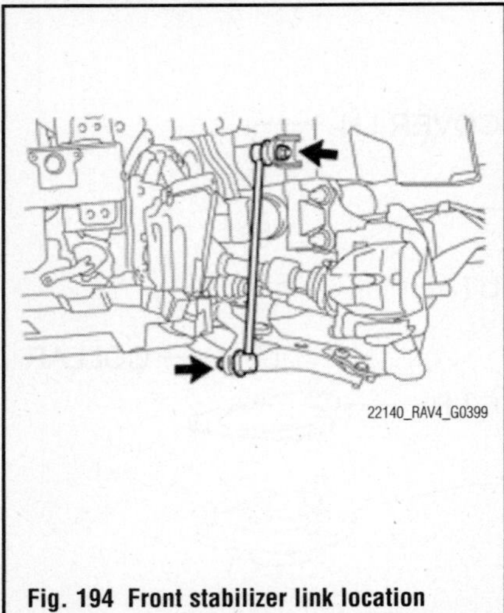

Fig. 194 Front stabilizer link location

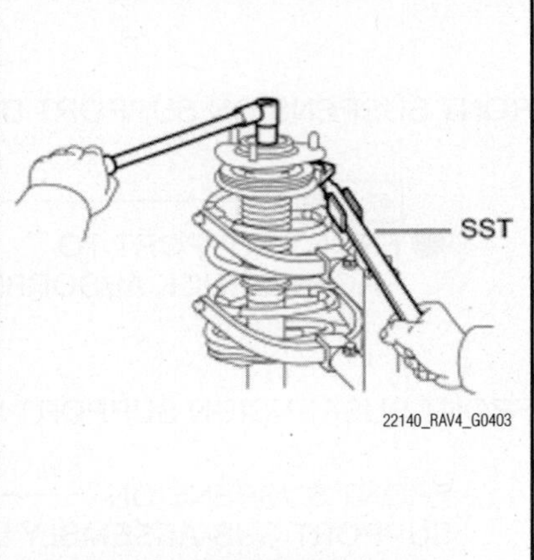

Fig. 195 Front strut disassembly tool installation

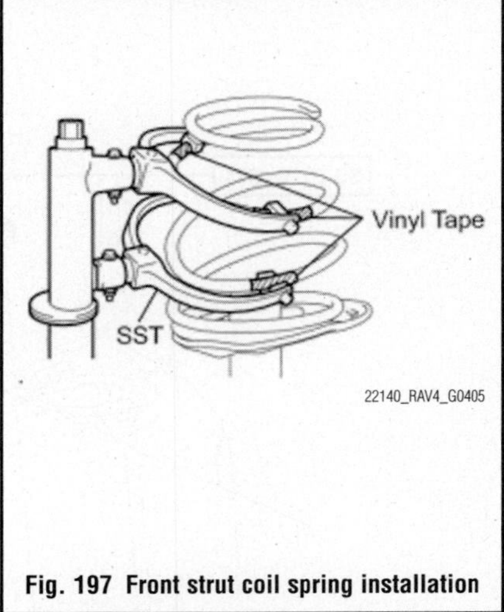

Fig. 197 Front strut coil spring installation

7. Remove the three strut upper retaining bolts.
8. Remove the strut from the vehicle.

To install:

9. Installation is the reverse of the removal procedure.
10. Tighten the three upper strut retaining nuts to 37 ft. lbs. (50 Nm).
11. Tighten the lower strut retaining bolts to 177 ft. lbs. (240 Nm).
12. Tighten the wheel lug nuts to 76 ft. lbs. (103 Nm).
13. To stabilize the suspension, lower the vehicle to ground height. Press down on the vehicle several times to stabilize the suspension.
14. Check and adjust the alignment, as required.

OVERHAUL

See Figures 195 through 197.

1. Before servicing the vehicle, refer to the Precautions Section.
2. Remove the strut from the vehicle.
3. Install a nut/bolt to the bracket at the lower portion of the strut assembly and secure it in a vise.
4. Compress the coil spring with a spring compressor.

CAUTION

The proper tools must be used for this procedure. The spring on the strut is under high pressure and can cause serious injury if not properly removed and installed.

5. Remove or disconnect the following:
   - Center retaining nut, by holding the spring seat
   - Support, dust seal, spring seat, insulator and spring from the strut assembly

To install:

6. Install the spring bumper and lower insulator to the strut assembly.
7. Compress the coil spring and fit the lower end of the spring into the spring seat gap.
8. Install or connect the following:
   - Upper insulator, spring seat, dust seal, support and spring seat. Tighten the new retaining nut to 34 ft. lbs. (47 Nm).
   - Strut
9. If required, bleed the brake system and check for leaks.

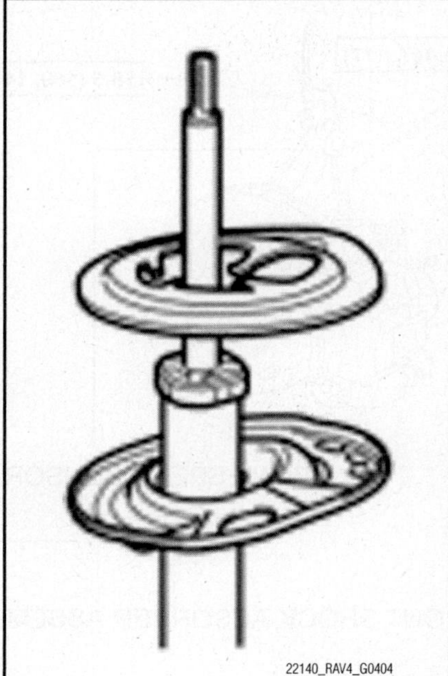

Fig. 197 Front strut insulator installation

10. Check and/or adjust the front wheel alignment.

STEERING KNUCKLE

REMOVAL & INSTALLATION

See Figures 198 and 199.

1. Before servicing the vehicle, refer to the Precautions Section.
2. Disconnect the negative battery cable.

➡Wait at least 90 seconds after disconnecting the negative battery cable before starting any repair work to prevent air bag and seat belt pretensioner activation.

3. Raise and support the vehicle safely.
4. Drain the transaxle fluid.
5. Remove the tire and wheel assembly. Remove the front axle hub nut.
6. Remove the front speed sensor. Remove the brake caliper. Remove the rotor.
7. Disconnect the tie rod end, using the proper tools.
8. Disconnect the front suspension number one lower arm sub-assembly.
9. Remove the two bolts and two nuts. Disconnect the strut from the steering knuckle.
10. Matchmark the halfshaft and the axle hub.
11. Remove the steering knuckle with the axle hub.

➡Be careful not to damage the boot and the speed sensor rotor. Do not excessively push out the halfshaft from the axle assembly.

12. Remove the four bolts and the axle hub from the steering knuckle. Remove the dust cover from the steering knuckle.

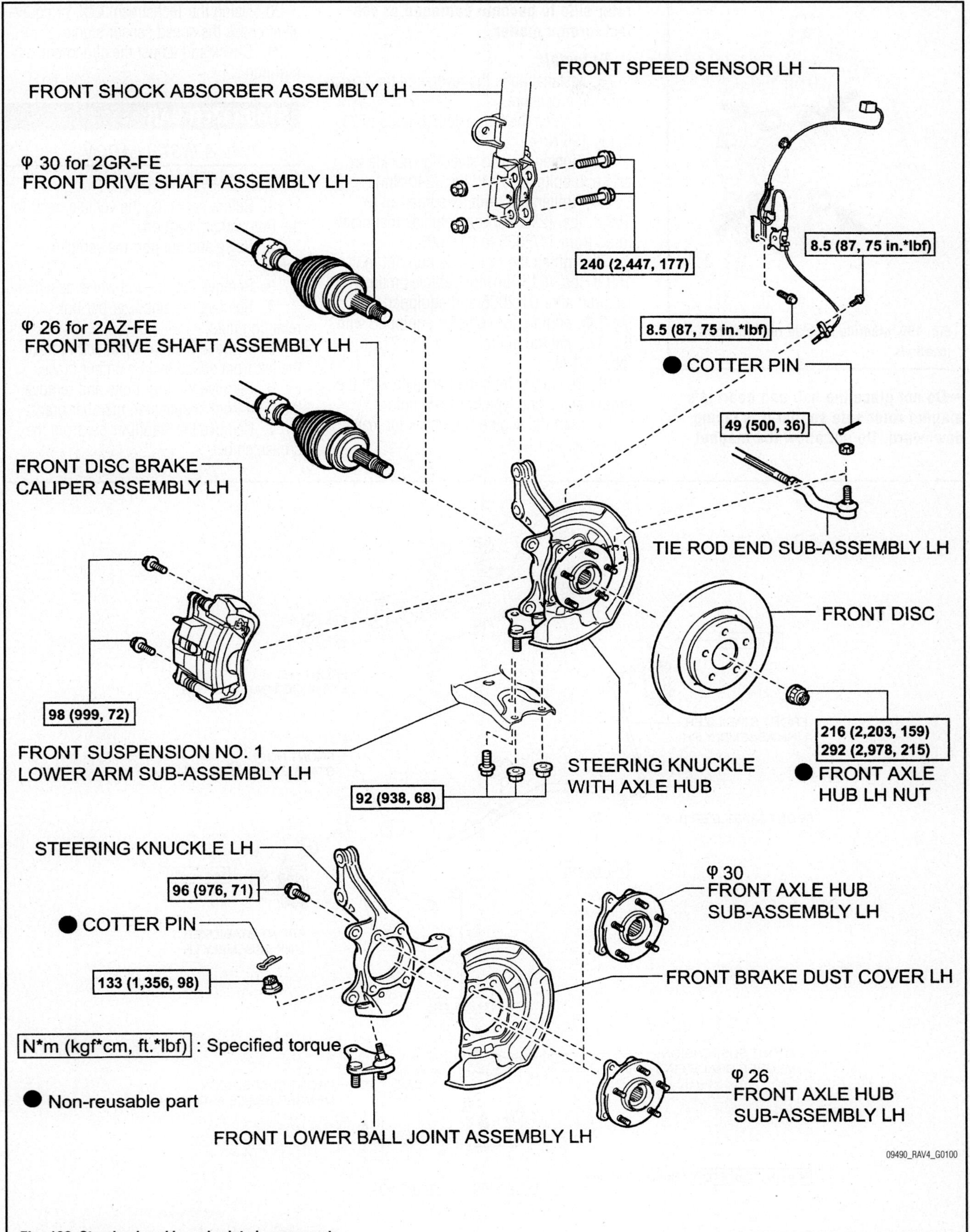

Fig. 198 Steering knuckle and related components

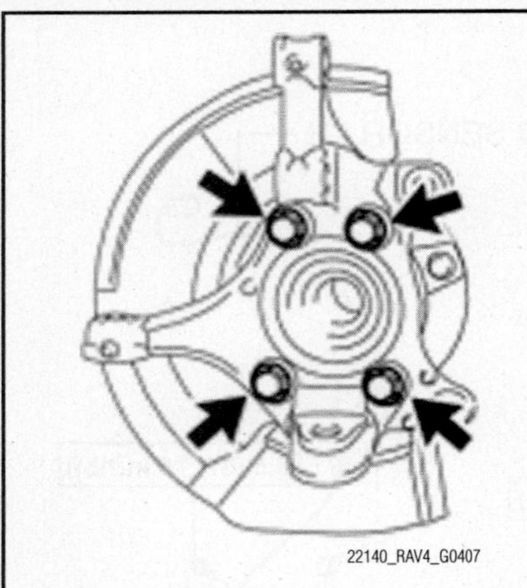

Fig. 199 Steering knuckle retaining bolt locations

➡Do not place the hub and bearing's magnet rotor side so that it is facing downward. Do not allow the magnet rotor side to become damaged or contact foreign matter.

To install:

13. Installation is the reverse of the removal procedure.
14. Tighten the four axle hub bolts to 71 ft. lbs. (96 Nm).
15. Tighten the two steering knuckle to axle hub bolts to 177 ft. lbs. (240 Nm).
16. Tighten the front axle hub nut to 159 ft. lbs. (215 Nm) on vehicles manufactured from 11/2005 to 01/2006.
17. Tighten the front axle hub nut to 159 ft. lbs. (215 Nm) for vehicles manufactured after 01/2006 and equipped with the 2.4L engine, for vehicles equipped with the 3.5L engine tighten the nut to 215 ft. lbs. (291 Nm).
18. Be sure to refill the transaxle with the proper grade and type transaxle fluid.
19. Start the engine and check for leaks, correct as required.
20. Using the Techstream tool, or equivalent check the speed sensor signal.
21. Check and adjust the alignment, as required.

STABILIZER BAR

REMOVAL & INSTALLATION

See Figures 200 and 201.

1. Before servicing the vehicle, refer to the Precautions Section.
2. Raise and support the vehicle safely.
3. Remove the tire and wheel assembly.
4. Remove the stabilizer bar link retaining nuts.
5. Remove the four bolts and remove the left front suspension member brace.
6. Remove the four bolts and remove the right front suspension member brace.
7. Remove the stabilizer bar from the crossmember.

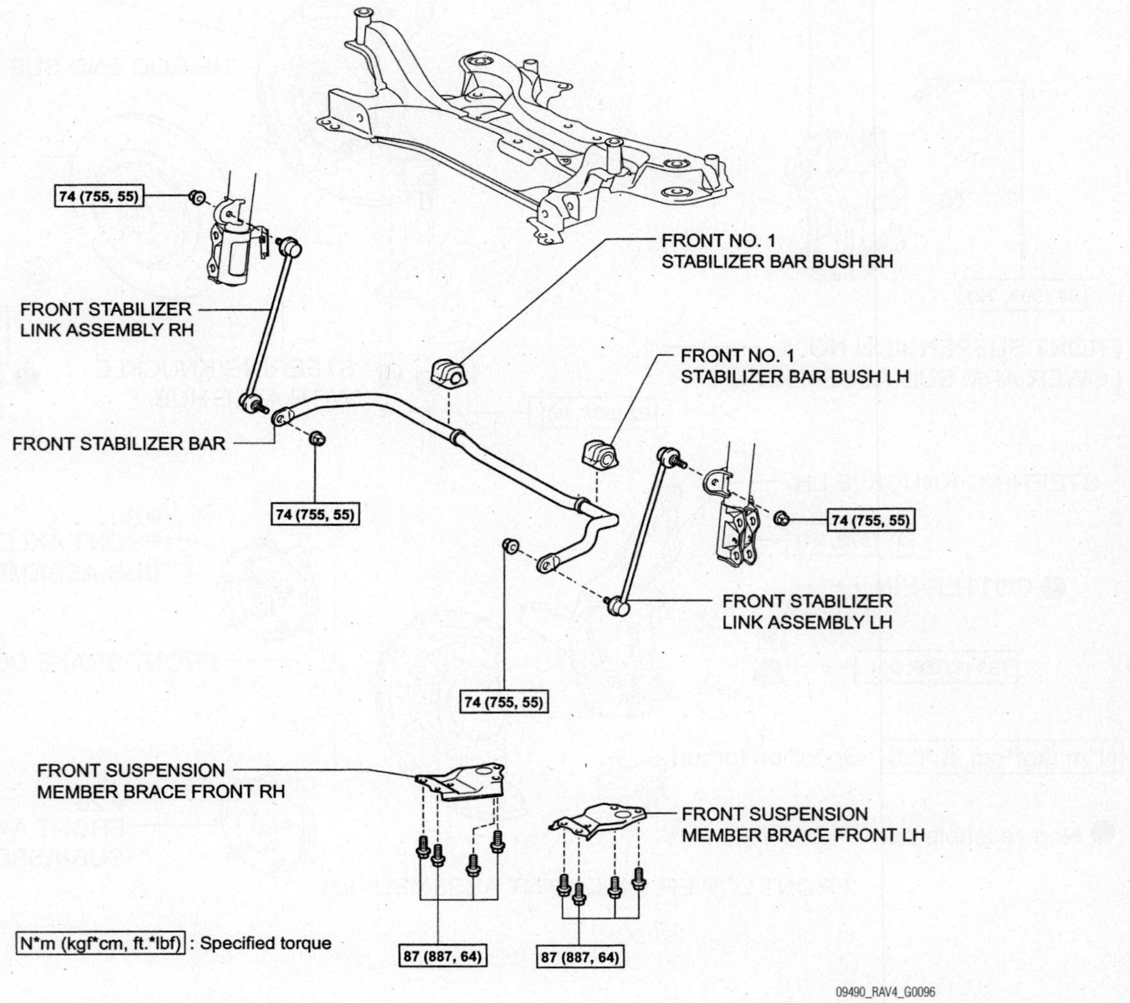

Fig. 200 Front stabilizer bar and related components

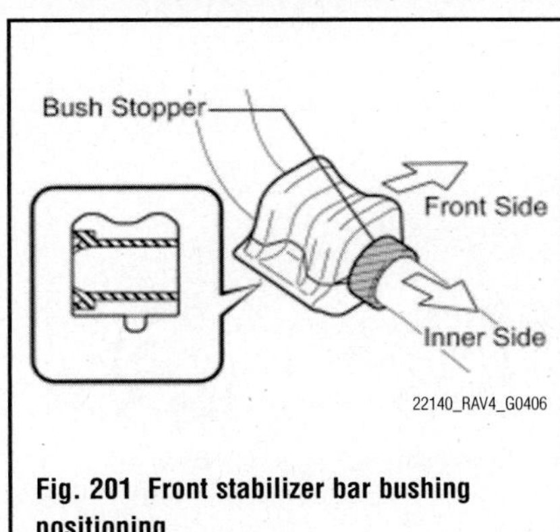

Fig. 201 Front stabilizer bar bushing positioning

8. Remove the bushings from the stabilizer bar.

To install:

9. Installation is the reverse order of the removal procedure.
10. Install the bushings to the inner side of each bushing stopper on the stabilizer bar.
11. Install the bushing with its slit facing the vehicle rear side.
12. Tighten the stabilizer bar link nuts to 55 ft. lbs. (74 Nm).
13. Tighten the front suspension member brace bolts to 64 ft. lbs. (87 Nm).
14. To stabilize the suspension, lower the vehicle to ground height. Press down on the vehicle several times to stabilize the suspension.

WHEEL BEARINGS

REMOVAL & INSTALLATION

See Figure 202.

1. Before servicing the vehicle, refer to the Precautions Section.
2. Disconnect the negative battery cable.

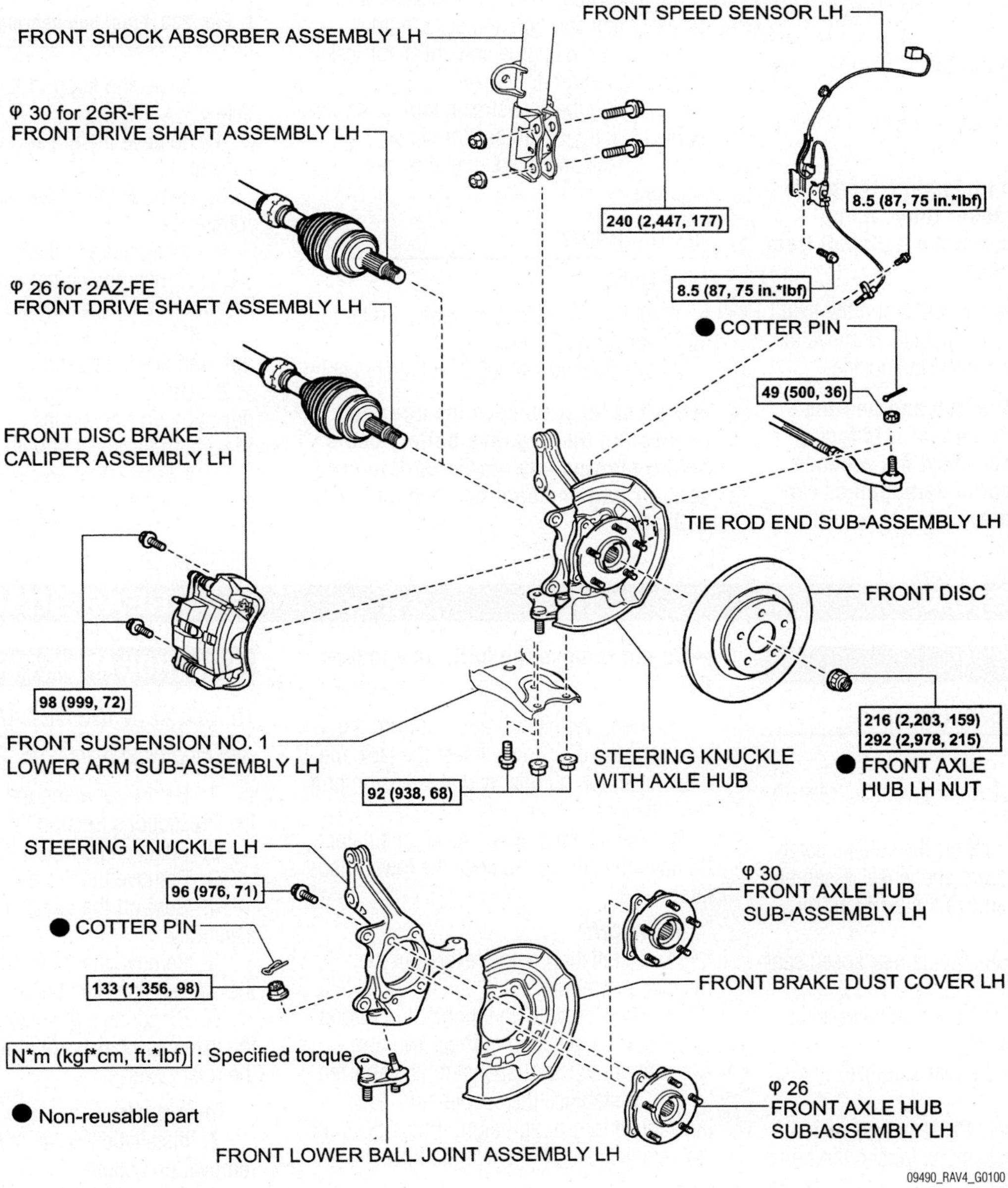

Fig. 202 Front hub and related components

➡Wait at least 90 seconds after disconnecting the negative battery cable before starting any repair work to prevent air bag and seat belt pretensioner activation.

3. Raise and support the vehicle safely.
4. Drain the transaxle fluid.
5. Remove the tire and wheel assembly. Remove the front axle hub nut.
6. Remove the front speed sensor. Remove the brake caliper. Remove the rotor.
7. Disconnect the tie rod end, using the proper tools.
8. Disconnect the front suspension number one lower arm sub-assembly.
9. Remove the two bolts and two nuts. Disconnect the strut from the steering knuckle.
10. Matchmark the halfshaft and the axle hub.
11. Remove the steering knuckle with the axle hub.

➡Be careful not to damage the boot and the speed sensor rotor. Do not excessively push out the halfshaft from the axle assembly.

12. Remove the four bolts and the axle hub from the steering knuckle. Remove the dust cover from the steering knuckle.

➡Do not place the hub and bearing's magnet rotor side so that it is facing downward. Do not allow the magnet rotor side to become damaged or contact foreign matter.

To install:

13. Installation is the reverse of the removal procedure.
14. Tighten the four axle hub bolts to 71 ft. lbs. (96 Nm).
15. Tighten the two steering knuckle to axle hub bolts to 177 ft. lbs. (240 Nm).
16. Tighten the front axle hub nut to 159 ft. lbs. (215 Nm) on vehicles manufactured from 11/2005 to 01/2006.
17. Tighten the front axle hub nut to 159 ft. lbs. (215 Nm) for vehicles manufactured after 01/2006 and equipped with the 2.4L engine, for vehicles equipped with the 3.5L engine tighten the nut to 215 ft. lbs. (291 Nm).
18. Be sure to refill the transaxle with the proper grade and type transaxle fluid.
19. Start the engine and check for leaks, correct as required.
20. Using the Techstream tool, or equivalent check the speed sensor signal.
21. Check and adjust the alignment, as required.

ADJUSTMENT

See Figure 203.

1. Before servicing the vehicle, refer to the Precautions Section.
2. Disconnect the negative battery cable.

➡Wait at least 90 seconds after disconnecting the negative battery cable before starting any repair work to prevent air bag and seat belt pretensioner activation.

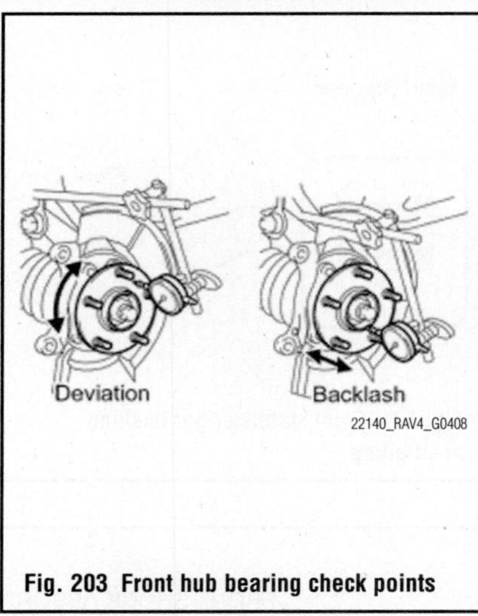

Fig. 203 Front hub bearing check points

3. Raise and support the vehicle safely.
4. Remove the tire and wheel assembly.
5. Remove the brake caliper. Remove the rotor.
6. Using a dial indicator tool, check the backlash near the center of the axle hub. Maximum backlash should be 0.0020 inch.
7. If backlash is greater than specification, replace the bearing.
8. Using a dial indicator tool, check the deviation on the surface of the axle hub. Maximum backlash should be 0.0020 inch.
9. If backlash is greater than specification, replace the bearing.

SUSPENSION

REAR SUSPENSION

COIL SPRING

REMOVAL & INSTALLATION

See Figure 204.

1. Before servicing the vehicle, refer to the Precautions Section.
2. Raise and support the vehicle safely.
3. Remove the tire and wheel assembly.
4. On 2WD, remove the skid control sensor wire.
5. On 4WD, remove the rear speed sensor wire.
6. Disconnect the number two parking brake cable assembly.
7. Disconnect the rear stabilizer link assembly.
8. To disconnect the rear suspension number two arm assembly, loosen the bolt from the suspension member side. Support the number two suspension arm with a suitable jack.

➡Do not remove the bolt, only loosen it.

9. Remove the bolt and nut from the axle carrier side. Slowly lower the jack and disconnect the number two suspension arm from the axle carrier.
10. Remove the upper spring insulator. Remove the spring. Remove the lower insulator.

To install:

11. Installation is the reverse of the removal procedure.
12. Do not apply final tightening torque to the rear suspension number two arm assembly until the suspension is stabilized.
13. To stabilize the suspension, lower the vehicle to ground height. Press down on the vehicle several times to stabilize the suspension.
14. Check and adjust the alignment, as required.

CONTROL ARMS/LINKS

REMOVAL & INSTALLATION

See Figures 205 and 206.

1. Before servicing the vehicle, refer to the Precautions Section.
2. Raise and support the vehicle safely.
3. Remove the tire and wheel assembly.
4. Support the lower control arm assembly.
5. Remove the bolt and two nuts from the suspension member and axle carrier.
6. Disconnect the suspension arm from the axle carrier. Be careful not to damage the dust cover.

To install:

7. Installation is the reverse of the removal procedure.
8. Do not apply final tightening torque to the rear suspension number two arm assembly until the suspension is stabilized.

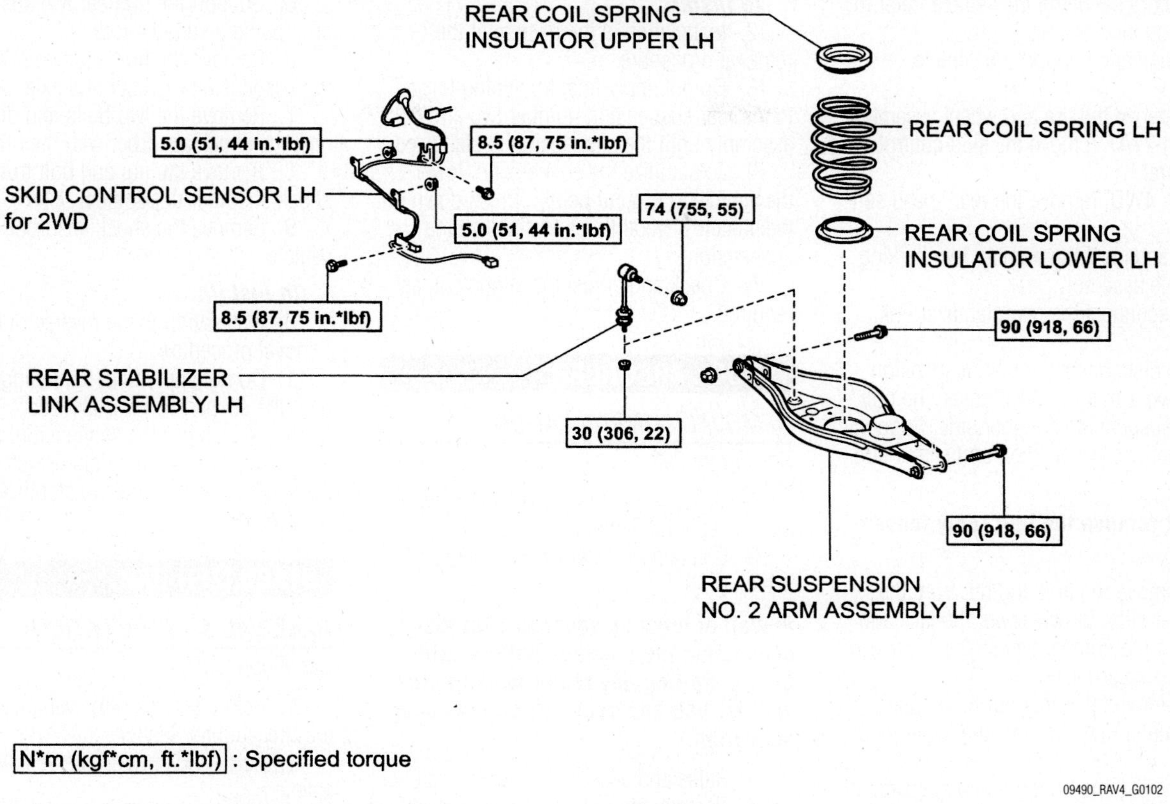

REAR COIL SPRING INSULATOR UPPER LH
5.0 (51, 44 in.*lbf)
8.5 (87, 75 in.*lbf)
REAR COIL SPRING LH
SKID CONTROL SENSOR LH
for 2WD
74 (755, 55)
5.0 (51, 44 in.*lbf)
REAR COIL SPRING
INSULATOR LOWER LH
8.5 (87, 75 in.*lbf)
90 (918, 66)
REAR STABILIZER
LINK ASSEMBLY LH
30 (306, 22)
90 (918, 66)
REAR SUSPENSION
NO. 2 ARM ASSEMBLY LH
N*m (kgf*cm, ft.*lbf) : Specified torque
09490_RAV4_G0102

Fig. 204 Rear coil spring and related components

N*m (kgf*cm, ft.*lbf) : Specified torque
Non-reusable part
90 (918, 66)
56 (571, 41)
100 (1,020, 74)
REAR NO. 1 SUSPENSION ARM ASSEMBLY LH
22140_RAV4_G0410

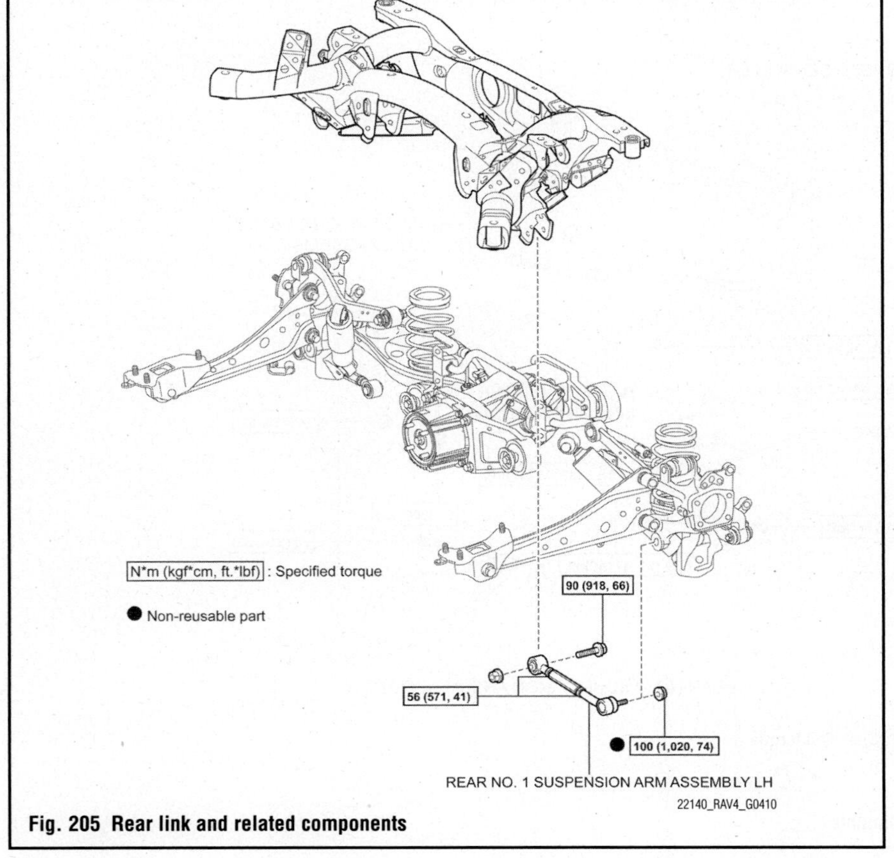

Fig. 205 Rear link and related components

22140_RAV4_G0411

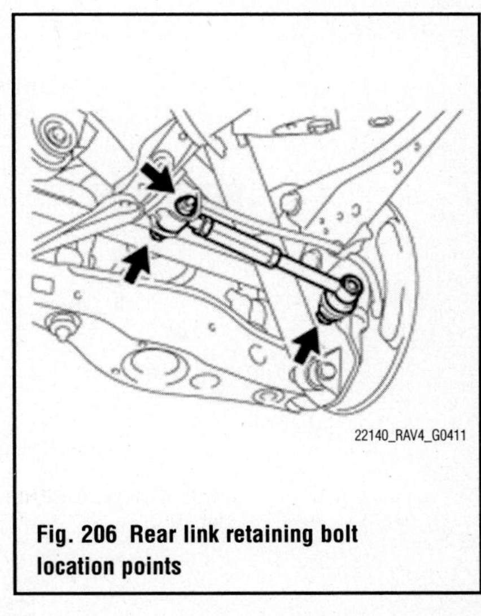

Fig. 206 Rear link retaining bolt location points

9. Tighten the bolts to 66 ft. lbs. (89 Nm). Tighten the nut to 74 ft. lbs. (100 Nm).

10. Tighten the wheel lug nuts to 76 ft. lbs. (103 Nm).

LOWER CONTROL ARM

REMOVAL & INSTALLATION

See Figure 207.

1. Before servicing the vehicle, refer to the Precautions Section.
2. Raise and support the vehicle safely.
3. Remove the tire and wheel assembly.
4. On 2WD, remove the skid control sensor wire.
5. On 4WD, remove the rear speed sensor wire.
6. Disconnect the number two parking brake cable assembly.
7. Disconnect the rear stabilizer link assembly.
8. To disconnect the rear suspension number two arm assembly, loosen the bolt from the suspension member side. Support the number two suspension arm with a suitable jack.

➡Do not remove the bolt, only loosen it.

9. Remove the bolt and nut from the axle carrier side. Slowly lower the jack and disconnect the number two suspension arm from the axle carrier.
10. Remove the upper spring insulator. Remove the spring. Remove the lower insulator.
11. Remove the bolt, nut and suspension arm from the suspension member.

To install:

12. Installation is the reverse of the removal procedure.
13. Do not apply final tightening torque to the rear suspension number two arm assembly until the suspension is stabilized.
14. To stabilize the suspension, lower the vehicle to ground height. Press down on the vehicle several times to stabilize the suspension.
15. Check and adjust the alignment, as required.

SHOCK ABSORBER

REMOVAL & INSTALLATION

See Figure 208.

1. Before servicing the vehicle, refer to the Precautions Section.
2. Disconnect the negative battery cable.

➡Wait at least 90 seconds after disconnecting the negative battery cable before starting any repair work to prevent air bag and seat belt pretensioner activation.

3. Raise and support the vehicle safely.
4. Remove the tire and wheel assembly.
5. Support the number two suspension arm, using a suitable jack.
6. Remove the bolt and two nuts from the suspension member and axle carrier.
7. Remove the two bolts and disconnect the shock absorber with the bracket.
8. Remove the nut and bolt from the shock absorber upper side.
9. Remove the shock absorber from the vehicle.

To install:

10. Installation is the reverse of the removal procedure.
11. Do not apply final tightening torque until the suspension is stabilized.
12. To stabilize the suspension, lower the vehicle to ground height. Press down on the vehicle several times to stabilize the suspension.

STABILIZER BAR

REMOVAL & INSTALLATION

See Figure 209.

1. Before servicing the vehicle, refer to the Precautions Section.
2. Raise and support the vehicle safely.
3. Remove the tire and wheel assembly.

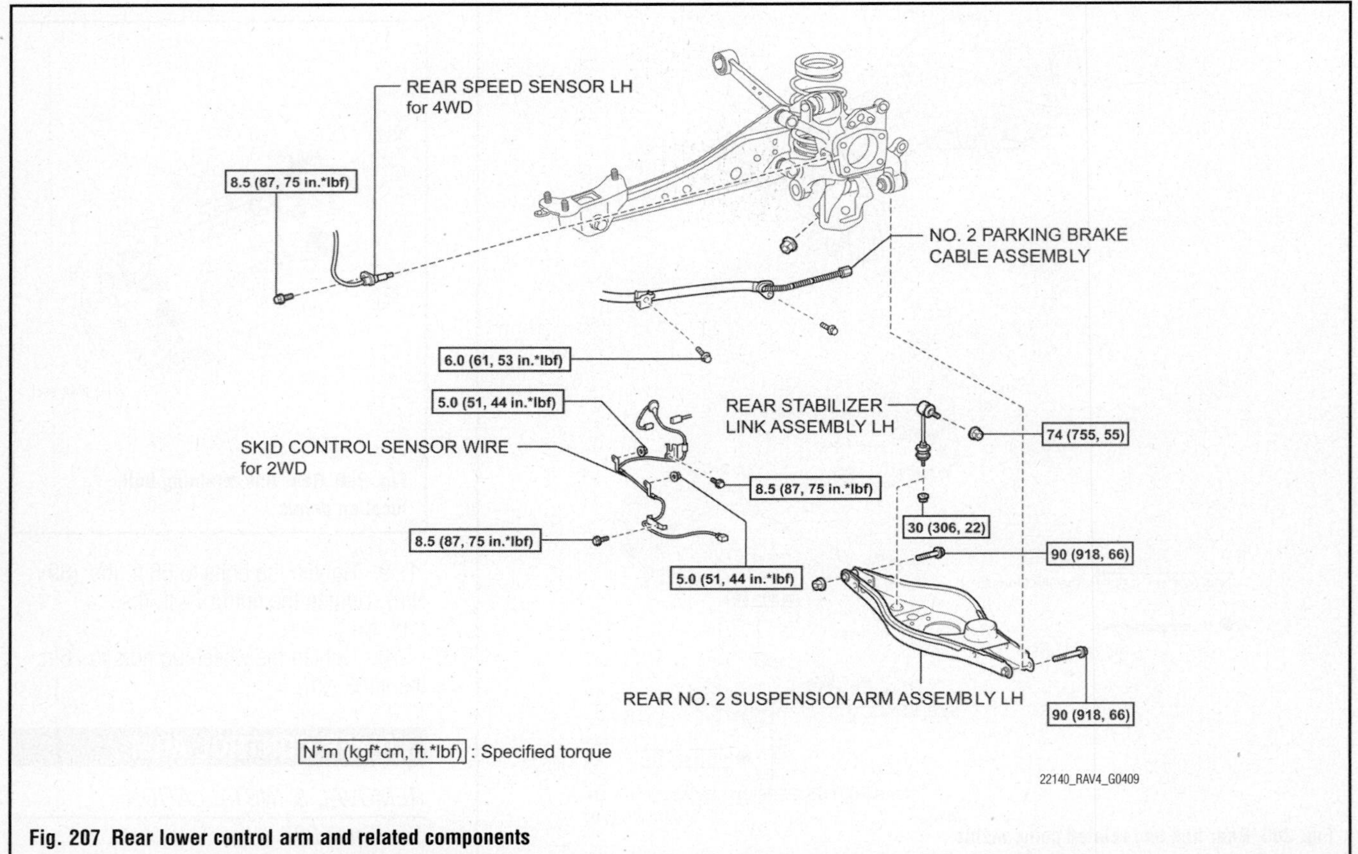

Fig. 207 Rear lower control arm and related components

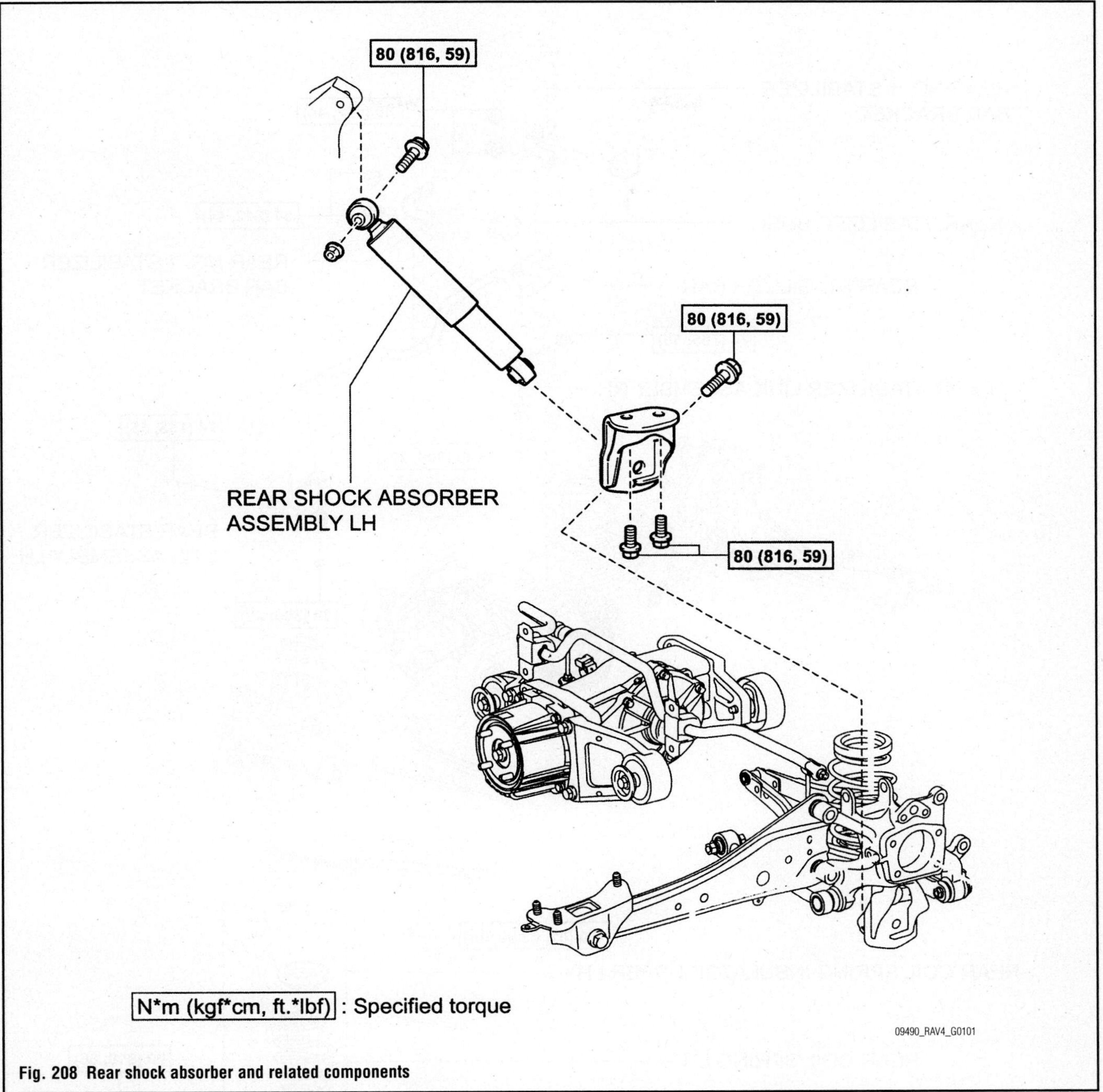

Fig. 208 Rear shock absorber and related components

4. Remove the nut and disconnect the link from the suspension number two arm. Remove the nut and the link from the stabilizer bar.
5. Remove the rear number two suspension arms (lower control arm).
6. Remove the coil springs.
7. Remove the stabilizer bracket retaining bolts. Remove the stabilizer bar from the vehicle. Remove the bushings from the bar.

➡When removing the bar be sure not to damage the sensor wire, brake hose etc.

To install:

8. Installation is the reverse of the removal procedure.
9. Tighten the stabilizer bar links to 55 ft. lbs. (74 Nm).
10. Tighten the stabilizer bracket and bushing bolts to 44 ft. lbs. (60 Nm).
11. Install each bushing to the outer side of the bushing stopper on each stabilizer bar. Install each bushing with its slit facing the vehicle front side.
12. Do not apply final tightening torque until the suspension is stabilized.
13. To stabilize the suspension, lower the vehicle to ground height. Press down on the vehicle several times to stabilize the suspension.
14. Check and adjust the alignment, as required.

UPPER CONTROL ARM

REMOVAL & INSTALLATION

See Figure 210.

1. Before servicing the vehicle, refer to the Precautions Section.

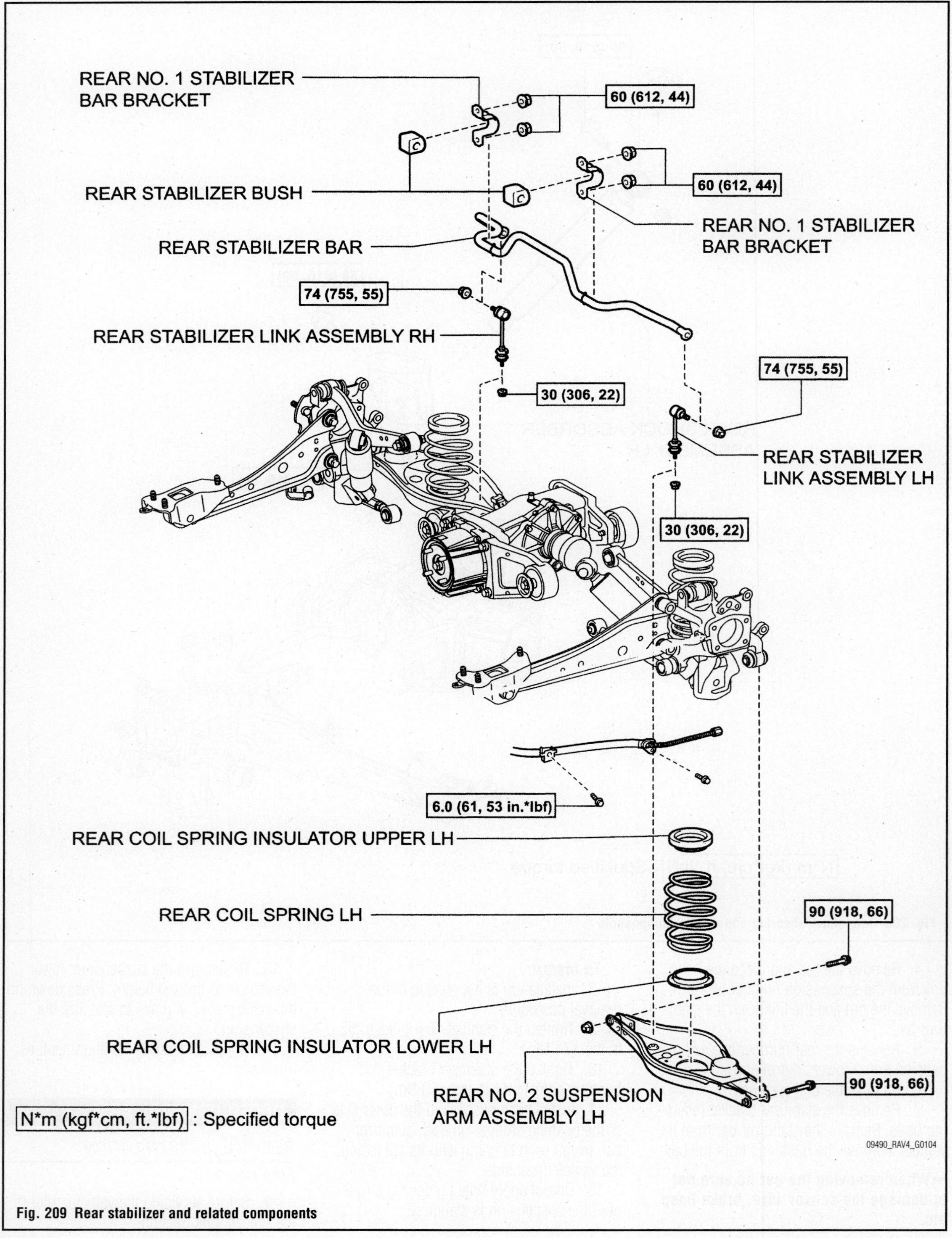

Fig. 209 Rear stabilizer and related components

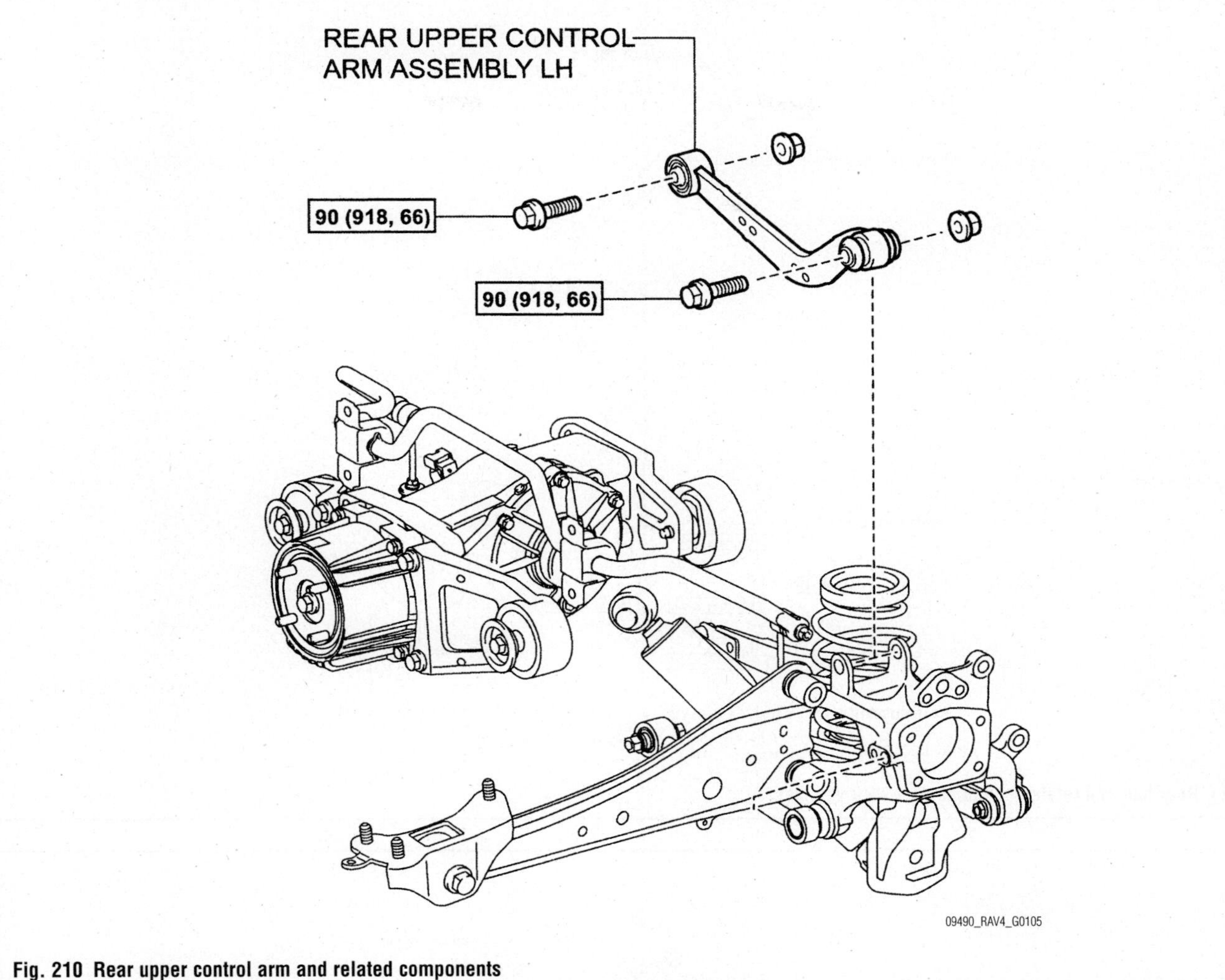

Fig. 210 Rear upper control arm and related components

2. Raise and support the vehicle safely.
3. Remove the tire and wheel assembly.
4. On 2WD, remove the skid control sensor wire.
5. On 4WD, remove the rear speed sensor wire.
6. Remove the upper control arm retaining bolts. Remove the upper control arm from the vehicle.

To install:

7. Installation is the reverse of the removal procedure.
8. Do not apply final tightening torque to the component until the suspension is stabilized.
9. To stabilize the suspension, lower the vehicle to ground height. Press down on the vehicle several times to stabilize the suspension.
10. Check and adjust the alignment, as required.

WHEEL HUB & BEARING

REMOVAL & INSTALLATION

See Figures 211 through 213.

1. Before servicing the vehicle, refer to the Precautions Section.
2. Properly relieve the fuel system pressure.
3. Disconnect the negative battery cable.

➡Wait at least 90 seconds after disconnecting the negative battery cable before starting any repair work to prevent air bag and seat belt pretensioner activation.

4. Raise and support the vehicle safely.
5. Remove the tire and wheel assembly.
6. On 4WD, remove the rear axle shaft nut.
7. Remove the caliper. Remove the rotor.
8. On 2WD remove the skid control sensor wire.
9. On 4WD, remove the rear speed sensor.
10. Remove the rear suspension number one arm.
11. Disconnect the shock absorber from the axle carrier.
12. On 2WD, remove the four bolts, and the axle hub and bearing from the axle carrier.
13. On 4WD, matchmark the halfshaft and the axle hub and bearing. Remove the four bolts, and the axle hub and bearing from the axle carrier.

➡Do not place the hub and bearing's magnet rotor side so that it is facing downward. Do not allow the magnet rotor side to become damaged or contact foreign matter.

To install:

14. Installation is the reverse of the removal procedure.

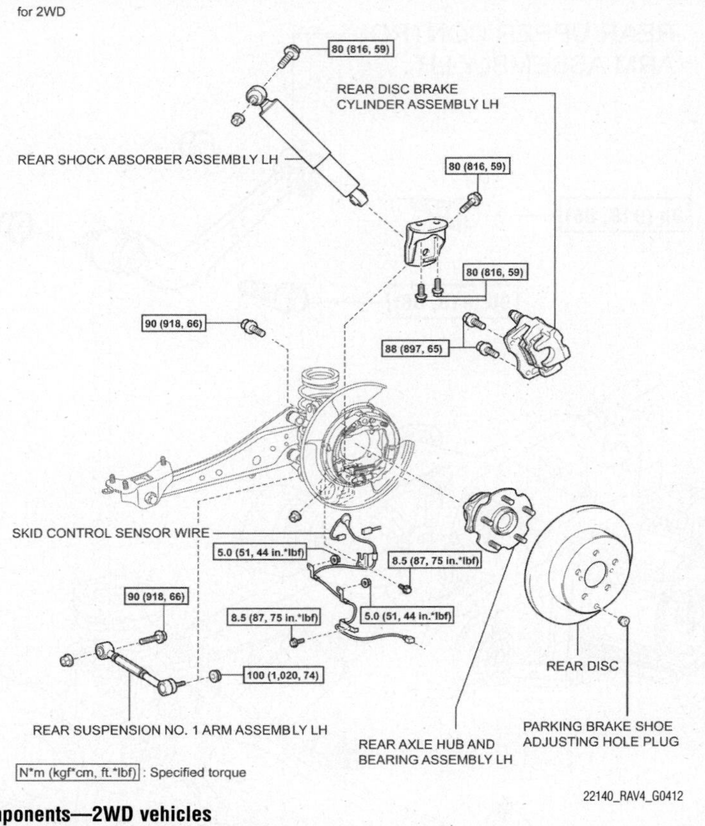

for 2WD

80 (816, 59)
REAR DISC BRAKE CYLINDER ASSEMBLY LH
REAR SHOCK ABSORBER ASSEMBLY LH
80 (816, 59)
80 (816, 59)
90 (918, 66)
88 (897, 65)
SKID CONTROL SENSOR WIRE
5.0 (51, 44 in.*lbf)
8.5 (87, 75 in.*lbf)
90 (918, 66)
8.5 (87, 75 in.*lbf)
5.0 (51, 44 in.*lbf)
100 (1,020, 74)
REAR DISC
REAR SUSPENSION NO. 1 ARM ASSEMBLY LH
REAR AXLE HUB AND BEARING ASSEMBLY LH
PARKING BRAKE SHOE ADJUSTING HOLE PLUG
N*m (kgf*cm, ft.*lbf) : Specified torque
22140_RAV4_G0412

Fig. 211 Rear hub and related components—2WD vehicles

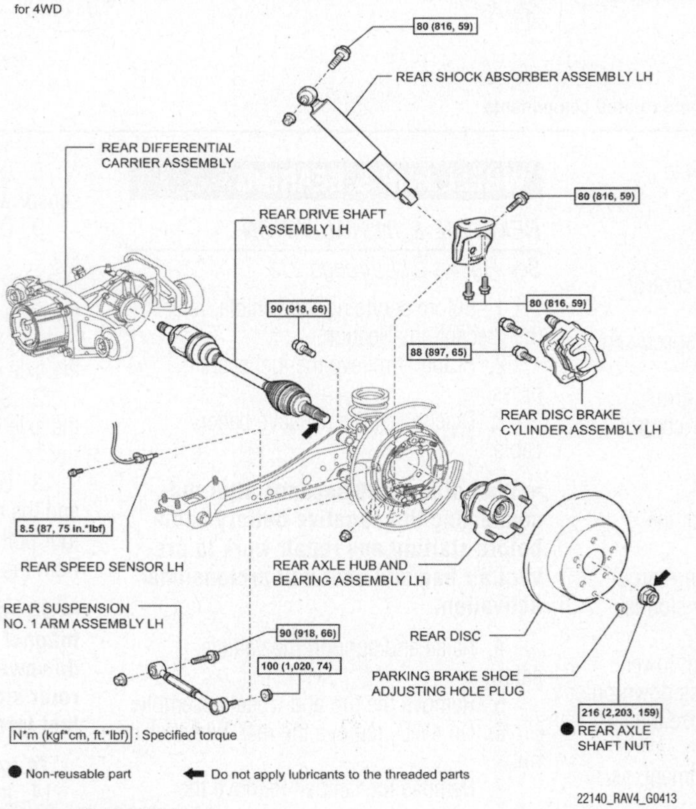

for 4WD

80 (816, 59)
REAR SHOCK ABSORBER ASSEMBLY LH
REAR DIFFERENTIAL CARRIER ASSEMBLY
80 (816, 59)
REAR DRIVE SHAFT ASSEMBLY LH
90 (918, 66)
80 (816, 59)
88 (897, 65)
REAR DISC BRAKE CYLINDER ASSEMBLY LH
8.5 (87, 75 in.*lbf)
REAR SPEED SENSOR LH
REAR AXLE HUB AND BEARING ASSEMBLY LH
REAR SUSPENSION NO. 1 ARM ASSEMBLY LH
90 (918, 66)
REAR DISC
100 (1,020, 74)
PARKING BRAKE SHOE ADJUSTING HOLE PLUG
216 (2,203, 159)
● REAR AXLE SHAFT NUT
N*m (kgf*cm, ft.*lbf) : Specified torque
● Non-reusable part
Do not apply lubricants to the threaded parts
22140_RAV4_G0413

Fig. 212 Rear hub and related components—4WD vehicles

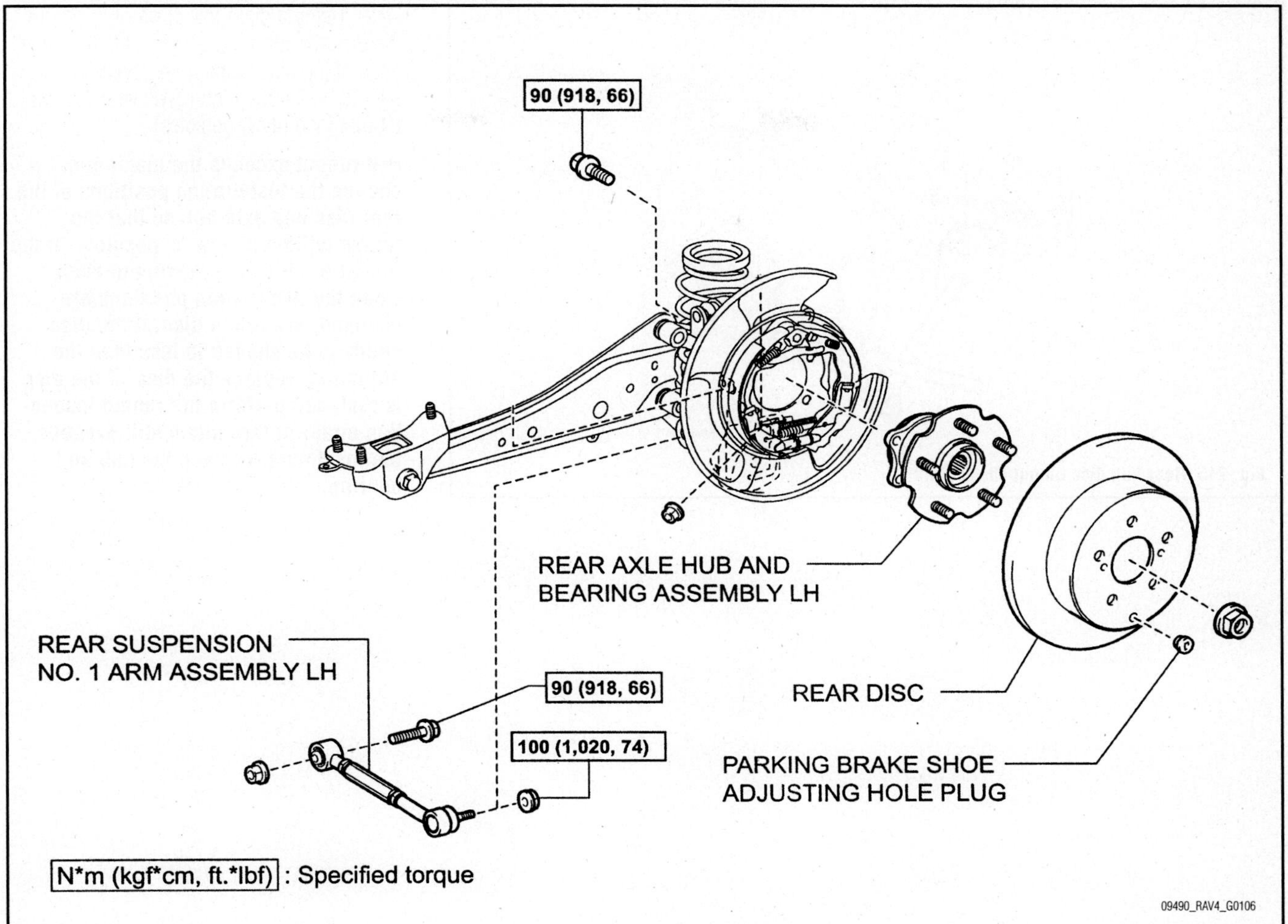

Fig. 213 Rear hub and bearing assembly

15. Tighten the four hub and bearing bolts to 68 ft. lbs. (92 Nm).

16. Do not apply final tightening torque to the component until the suspension is stabilized.

17. To stabilize the suspension, lower the vehicle to ground height. Press down on the vehicle several times to stabilize the suspension.

18. Check and adjust the alignment, as required.

ADJUSTMENT

See Figures 214 and 215.

1. Before servicing the vehicle, refer to the Precautions Section.
2. Disconnect the negative battery cable.

➡Wait at least 90 seconds after disconnecting the negative battery cable before starting any repair work to prevent air bag and seat belt pretensioner activation.

3. Raise and support the vehicle safely.
4. Remove the tire and wheel assembly.
5. On 4WD vehicles, disconnect the halfshaft.
6. Remove the brake caliper. Remove the rotor.
7. Using a dial indicator tool, measure the backlash near the center of the axle hub. Maximum backlash should be 0.0020 inch.
8. If backlash is greater than specification, replace the bearing.

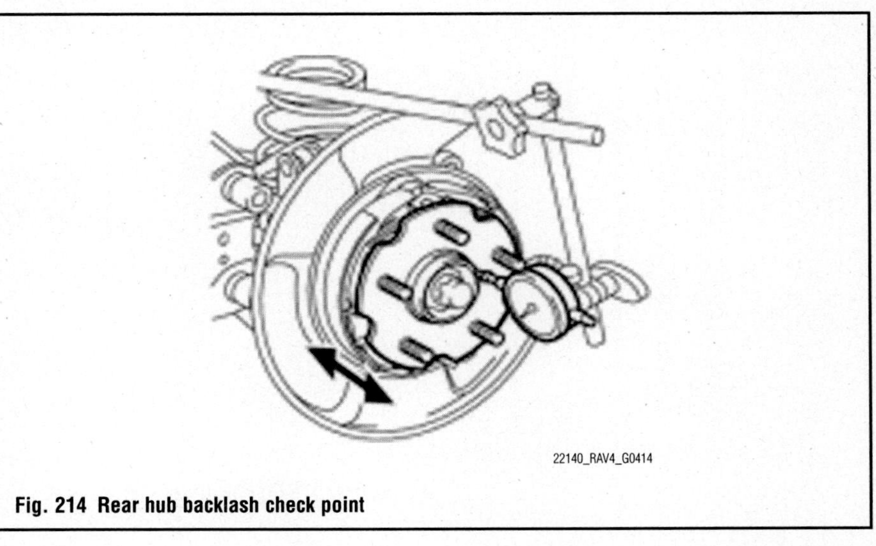

Fig. 214 Rear hub backlash check point

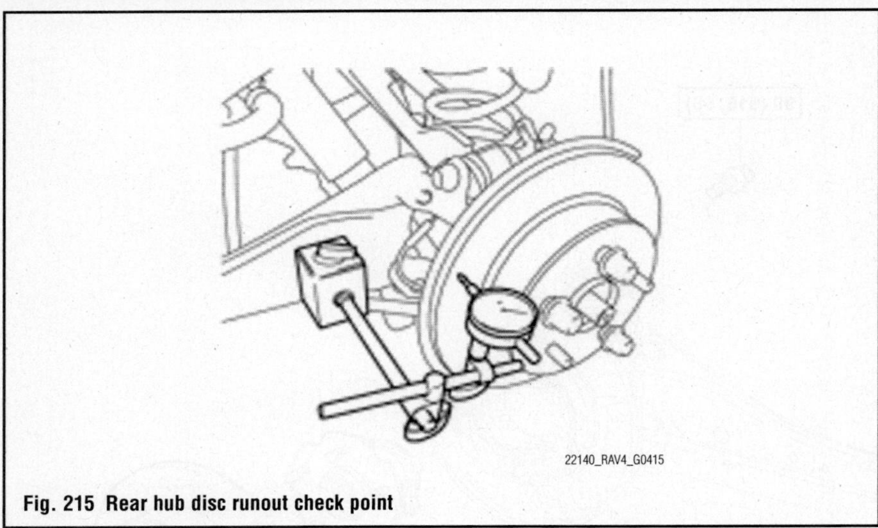

Fig. 215 Rear hub disc runout check point

9. Using a dial indicator tool, measure the disc runout (0.39 inch) inside the outer edge of the disc. Maximum deviation should be 0.0031 inch (2WD vehicles) and 0.0024 inch (4WD vehicles).

➡If runout exceeds the maximum, change the installation positions of the rear disc and axle hub so that the runout will be as low as possible. If the runout exceeds the maximum even when the installation positions are changed, shave the disc. If the disc needs to be shaved to less than the minimum, replace the disc. If the disc is replaced perform the runout inspection again. If the runout still exceeds the maximum, replace the hub and bearing.

TOYOTA

Sequoia

13

SPECIFICATIONS AND MAINTENANCE CHARTS

ENGINE AND VEHICLE IDENTIFICATION

Engine						
Code	Liters (cc)	Cu. In.	Cyl.	Fuel Sys.	Engine Type	Eng. Mfg.
1UR-FE	4.6 (4608)	282	8	SFI	DOHC	Toyota
3UR-FE	5.7 (5663)	346	8	SFI	DOHC	Toyota
3UR-FBE	5.7 (5663)	346	8	SFI	DOHC	Toyota

Model Year	
Code ①	Year
B	2011
C	2012

SFI: Sequential Fuel Injection

DOHC: Double Overhead Camshaft

① 10th digit of the VIN number

71099_SEQU_C0001

GENERAL ENGINE SPECIFICATIONS

Year	Model	Engine Displacement Liters	Engine Series Code/ID	Net Horsepower @ rpm	Net Torque @ rpm (ft. lbs.)	Bore x Stroke (in.)	Com-pression Ratio	Oil Pressure @ rpm
2011	Sequoia	4.6	1UR-FE	310@5600	327@3400	3.70x3.27	10.2:1	32@2500
		5.7	3UR-FE	381@5600	403@3600	3.70x4.02	10.2:1	32@2500
		5.7	3UR-FBE	381@5600	403@3600	3.70x4.02	10.2:1	32@2500
2012	Sequoia	4.6	1UR-FE	310@5600	327@3400	3.70x3.27	10.2:1	32@2500
		5.7	3UR-FE	381@5600	403@3600	3.70x4.02	10.2:1	32@2500
		5.7	3UR-FBE	381@5600	403@3600	3.70x4.02	10.2:1	32@2500

71099_SEQU_C0002

ENGINE TUNE-UP SPECIFICATIONS

Year	Engine Displacement Liters	Engine Code/ID	Spark Plug Gap (in.)	Ignition Timing (deg.)	Fuel Pump (psi)	Idle Speed (rpm)		Valve Clearance	
						MT	AT	Intake	Exhaust
2011	4.6	1UR-FE	0.039-0.043	8-12 ①	41-42	N/A	650-750	N/A	N/A
	5.7	3UR-FE	0.039-0.043	8-12 ①	38-44	N/A	650-750	N/A	N/A
	5.7	3UR-FBE	0.039-0.043	8-12 ①	38-44	N/A	650-750	N/A	N/A
2012	4.6	1UR-FE	0.039-0.043	8-12 ①	41-42	N/A	650-750	N/A	N/A
	5.7	3UR-FE	0.039-0.043	8-12 ①	38-44	N/A	650-750	N/A	N/A
	5.7	3UR-FBE	0.039-0.043	8-12 ①	38-44	N/A	650-750	N/A	N/A

NOTE: The Vehicle Emission Control Information label often reflects specification changes made during production.

The label figures must be used if they differ from those in this chart.

B: Before top dead center

N/A: Not Available

① With terminals TC and CG of the DLC3 connected. Transmission in N and AC off.

71099_SEQU_C0003

CAPACITIES

Year	Model	Engine Displacement Liters	Engine Code/ID	Engine Oil with Filter (qts.)	Transmission (qts.) 5-Spd	Transmission (qts.) Auto.	Transfer Case (qts.)	Drive Axle Front (qts.)	Drive Axle Rear (qts.)	Fuel Tank (gal.)	Cooling System (qts.)
2011	Sequoia	4.6	1UR-FE	7.4	N/A	3.6	1.43	2.17	1.64	26.4	①
		5.7	3UR-FE	7.4	N/A	4.4	1.43	2.17	1.64	26.4	①
		5.7	3UR-FBE	7.4	N/A	4.4	1.43	2.17	1.64	26.4	①
2012	Sequoia	4.6	1UR-FE	7.4	N/A	3.6	1.43	2.17	1.64	26.4	①
		5.7	3UR-FE	7.4	N/A	4.4	1.43	2.17	1.64	26.4	①
		5.7	3UR-FBE	7.4	N/A	4.4	1.43	2.17	1.64	26.4	①

① 14.5 w/o tow package
15.4 with tow package

N/A: Not Available

71099_SEQU_C0004

FLUID SPECIFICATIONS

Year	Model	Engine Displacement Liters	Engine ID/VIN	Engine Oil	Auto. Trans.	Power Steering Fluid	Brake Master Cylinder
2011	Sequoia	4.6	1UR-FE	5W-20	①	②	DOT 3
		5.7	3UR-FE	5W-20	①	②	DOT 3
		5.7	3UR-FBE	0W-20	①	②	DOT 3
2012	Sequoia	4.6	1UR-FE	5W-20	①	②	DOT 3
		5.7	3UR-FE	5W-20	①	②	DOT 3
		5.7	3UR-FBE	0W-20	①	②	DOT 3

DOT: Department Of Transportation

① A750E, A750F: ATF WS

② ATF Dexron II or III

71099_SEQU_C0013

VALVE SPECIFICATIONS

Year	Engine Displ. Liters	Engine Code/ID	Seat Angle (deg.)	Face Angle (deg.)	Spring Free Length (in.)	Spring Maximum Deviation (in.)	Stem-to-Guide Clearance (in.) Intake	Stem-to-Guide Clearance (in.) Exhaust	Stem Diameter (in.) Intake	Stem Diameter (in.) Exhaust
2011	4.6	1UR-FE	N/A	N/A	2.03	0.0394	0.000984-0.00236	0.00118-0.00256	0.2150-0.2160	0.2150-0.2160
	5.7	3UR-FE	N/A	N/A	1.94	0.0394	0.000984-0.00236	0.00118-0.00256	0.2150-0.2160	0.2150-0.2160
	5.7	3UR-FBE	N/A	N/A	1.94	0.0394	0.000984-0.00236	0.00118-0.00256	0.2150-0.2160	0.2150-0.2160
2012	4.6	1UR-FE	N/A	N/A	2.03	0.0394	0.000984-0.00236	0.00118-0.00256	0.2150-0.2160	0.2150-0.2160
	5.7	3UR-FE	N/A	N/A	1.94	0.0394	0.000984-0.00236	0.00118-0.00256	0.2150-0.2160	0.2150-0.2160
	5.7	3UR-FBE	N/A	N/A	1.94	0.0394	0.000984-0.00236	0.00118-0.00256	0.2150-0.2160	0.2150-0.2160

N/A: Not Available

71099_SEQU_C0005

CAMSHAFT SPECIFICATIONS

All measurements in inches unless noted

Year	Engine Displacement Liters	Engine Code/ID	Journal Dia.	Brg. Oil Clearance	Shaft End-play ①	Circle Runout	Lobe Height	
							Intake	Exhaust
2011	4.6	1UR-FE	①	N/A	0.00315-0.00531	0.00157	1.7440-1.7500	1.74000-1.74600
	5.7	3UR-FE	①	N/A	0.00315-0.00531	0.00157	1.7440-1.7500	1.74000-1.74600
	5.7	3UR-FBE	①	N/A	0.00315-0.00531	0.00157	1.7440-1.7500	1.74000-1.74600
2012	4.6	1UR-FE	①	N/A	0.00315-0.00531	0.00157	1.7440-1.7500	1.74000-1.74600
	5.7	3UR-FE	①	N/A	0.00315-0.00531	0.00157	1.7440-1.7500	1.74000-1.74600
	5.7	3UR-FBE	①	N/A	0.00315-0.00531	0.00157	1.7440-1.7500	1.74000-1.74600

N/A: Not Available

① No. 1: 1.4162-1.4167

All others: 0.9039-0.9045

71099_SEQU_C0006

CRANKSHAFT AND CONNECTING ROD SPECIFICATIONS

All measurements are given in inches.

Year	Engine Displ. Liters	Engine Code/ID	Crankshaft				Connecting Rod		
			Main Brg. Journal Dia.	Main Brg. Oil Clearance	Shaft End-play	Thrust on No.	Journal Diameter	Oil Clearance	Side Clearance
2011	4.6	1UR-FE/M	2.6373-2.6378	①	0.0008-0.0087	3	N/A	0.000984-0.00197	0.00591-0.02170
	5.7	3UR-FE/W	2.6373-2.6378	①	0.0008-0.0087	3	2.3230-2.3235	0.000984-0.00197	0.00591-0.02170
	5.7	3UR-FE/V, Y	2.6373-2.6378	①	0.0008-0.0087	3	2.3230-2.3235	0.000984-0.00197	0.00591-0.02170
2012	4.6	1UR-FE/M	2.6373-2.6378	①	0.0008-0.0087	3	N/A	0.000984-0.00197	0.00591-0.02170
	5.7	3UR-FE/W	2.6373-2.6378	①	0.0008-0.0087	3	2.3230-2.3235	0.000984-0.00197	0.00591-0.02170
	5.7	3UR-FE/V, Y	2.6373-2.6378	①	0.0008-0.0087	3	2.3230-2.3235	0.000984-0.00197	0.00591-0.02170

N/A: Not Available

① Nos. 1 and 5: 0.000669-0.00118

All others: 0.000945-0.00146

71099_SEQU_C0007

PISTON AND RING SPECIFICATIONS

All measurements are given in inches.

Year	Engine Displ. Liters	Engine Code/ID	Piston Clearance	Ring Gap Top Compression	Ring Gap Bottom Compression	Ring Gap Oil Control	Ring Side Clearance Top Compression	Ring Side Clearance Bottom Compression	Ring Side Clearance Oil Control
2011	4.6	1UR-FE/M	N/A	0.00906-0.0130	0.0157-0.0197	0.00394-0.0157	0.000787-0.000276	0.000787-0.000236	0.00276-0.00571
	5.7	3UR-FE/W	0.0016-0.0024	0.0098-0.0138	0.0157-0.0197	0.0039-0.0157	0.0007-0.0028	0.0007-0.0027	0.0028-0.0057
	5.7	3UR-FE/V, Y	0.0016-0.0024	0.0098-0.0138	0.0157-0.0197	0.0039-0.0157	0.0007-0.0028	0.0007-0.0027	0.0028-0.0057
2012	4.6	1UR-FE/M	N/A	0.00906-0.0130	0.0157-0.0197	0.00394-0.0157	0.000787-0.000276	0.000787-0.000236	0.00276-0.00571
	5.7	3UR-FE/W	0.0016-0.0024	0.0098-0.0138	0.0157-0.0197	0.0039-0.0157	0.0007-0.0028	0.0007-0.0027	0.0028-0.0057
	5.7	3UR-FE/V, Y	0.0016-0.0024	0.0098-0.0138	0.0157-0.0197	0.0039-0.0157	0.0007-0.0028	0.0007-0.0027	0.0028-0.0057

N/A: Not Available

71099_SEQU_C0008

TORQUE SPECIFICATIONS

All readings in ft. lbs.

Year	Engine Displacement Liters	Engine Code/ID	Cylinder Head Bolts	Main Bearing Bolts	Rod Bearing Bolts	Crankshaft Damper Bolts	Flywheel Bolts	Manifold Intake	Manifold Exhaust	Spark Plugs	Oil Pan Drain Plug
2011	4.6	1UR-FE/M	①	②	③	221	④	15	15	15	30
	5.7	3UR-FE/W	①	②	③	221	④	15	15	15	30
	5.7	3UR-FE/V, Y	①	②	③	221	④	15	15	15	30
2012	4.6	1UR-FE/M	①	②	③	221	④	15	15	15	30
	5.7	3UR-FE/W	①	②	③	221	④	15	15	15	30
	5.7	3UR-FE/V, Y	①	②	③	221	④	15	15	15	30

① Step 1: 27 ft. lbs.
Step 2: Plus 90 degrees
Step 3: Plus 90 degrees
12 mm head 15 ft. lbs.

② Inside position 45 ft. lbs.
Step 1: Outside position 20 ft. lbs.
Step 2: Plus 90 degrees
Step 3: Cylinder block side position 33 ft. lbs.

③ Step 1: 30 ft. lbs.
Step 2: Plus 90 degrees

④ Step 1: 22 ft. lbs.
Step 2: Plus 90 degrees

71099_SEQU_C0009

TIRE, WHEEL AND BALL JOINT SPECIFICATIONS

		OEM Tires		Tire Pressures (psi)		Wheel	Ball Joint	Lug Nut Torque
Year	Model	Standard	Optional	Front	Rear	Size	Inspection	(ft. lbs.)
2011	Sequoia	P275/65R18	P275/55R20	①	①	8J	②	③
2012	Sequoia	P275/65R18	P275/55R20	①	①	8J	②	③

OEM: Original Equipment Manufacturer

PSI: Pounds Per Square Inch

① P275/65R18 tires: 33. P275/55R20 tires: front: 30, rear: 33. Use specification on vehicle placard if different from one given

② Turning torque 89 inch lbs.

③ Steel wheel 154 ft. lbs. aluminum wheels 97 ft. lbs. If specification differs from one given, see owners manual.

71099_SEQU_C0012

BRAKE SPECIFICATIONS

All measurements in inches unless noted

			Brake Disc			Minimum	Brake Caliper	
			Original	Minimum	Maximum	Lining	Bracket Bolts	Mounting Bolts
Year	Model		Thickness	Thickness	Runout	Thickness	(ft. lbs.)	(ft. lbs.)
2011	Sequoia	F	1.260	1.140	0.00197	0.469	—	73
		R	0.709	0.630	0.00590	0.472	70	65
2012	Sequoia	F	1.260	1.140	0.00197	0.469	—	73
		R	0.709	0.630	0.00590	0.472	70	65

F: Front

R: Rear

71099_SEQU_C0010

WHEEL ALIGNMENT

		Caster		Camber			Steering
		Range	Preferred Setting	Range	Preferred Setting	Toe-in	Axis Inclination
Year	Model	(+/-Deg.)	(Deg.)	(+/-Deg.)	(Deg.)	(in.) ①	(Deg.)
2011	Sequoia	3.75+/-0.75	3.45	-1.25 +/- 45'	0.75	0.04+/-0.08	13.14+/-0.75
2012	Sequoia	3.75+/-0.75	3.45	-1.25 +/- 45'	0.75	0.04+/-0.08	13.14+/-0.75

① Rear toe-in 0+/-0.08

71099_SEQU_C0011

SCHEDULED MAINTENANCE INTERVALS

TOYOTA—SEQUOIA

TO BE SERVICED	TYPE OF SERVICE	VEHICLE MILEAGE INTERVAL (x1000)																		
		5	10	15	20	25	30	35	40	45	50	55	60	65	70	75	80	85	90	95
Automatic transmission and differential fluid	S/I			✓			✓			✓			✓			✓			✓	
Ball joints and boots	S/I			✓			✓			✓			✓			✓			✓	
Brake system	S/I			✓			✓			✓			✓			✓			✓	
Cabin filter	S/I			✓			✓			✓			✓			✓			✓	
Charcoal canister	S/I												✓							
Drive belts	S/I						✓						✓						✓	
Driveshaft bushing	L						✓						✓						✓	
Engine coolant	R						✓						✓						✓	
Engine oil & filter	R	✓	✓	✓	✓	✓	✓	✓	✓	✓	✓	✓	✓	✓	✓	✓	✓	✓	✓	✓
Exhaust system	S/I			✓			✓			✓			✓			✓			✓	
Fuel lines	S/I						✓						✓						✓	
Fuel tank cap gasket	S/I						✓						✓						✓	
Halfshaft boots & flange bolts	S/I			✓			✓			✓			✓			✓			✓	
Limited slip differential fluid	R						✓						✓						✓	
Differential fluid	S/I						✓						✓						✓	
Non-platinum spark plugs	R						✓						✓						✓	
Platinum spark plugs	R												✓							
Propeller shaft (4WD)	L			✓			✓			✓			✓			✓			✓	
Propeller shaft bolts	S/I			✓			✓			✓			✓			✓			✓	
Steering gear	S/I			✓			✓			✓			✓			✓			✓	
Steering linkage	S/I			✓			✓			✓			✓			✓			✓	
Tires (rotate)	S/I	✓	✓	✓	✓	✓	✓	✓	✓	✓	✓	✓	✓	✓	✓	✓	✓	✓	✓	✓
Valves	S/I												✓							

R: Replace S/I: Service or Inspect L: Lubricate

FREQUENT OPERATION MAINTENANCE (SEVERE SERVICE)

If a vehicle is operated under any of the following conditions it is considered severe service:

- Towing a trailer or using a camper or car-top carrier.

- Repeated short trips of less than 5 miles in temperatures below freezing.

- Excessive idling or low-speed driving for long distances as in heavy commercial use, such as delivery, taxi or police cars.

- Operating on rough, muddy or salt-covered roads.

- Operating on unpaved or dusty roads.

Oil and filter: service every 2500 miles or 3 months, whichever occurs first.

Brake linings and discs or drums: service or inspect every 5000 miles or 4 months, whichever occurs first.

Steering linkage: service or inspect every 5000 miles or 4 months, whichever occurs first.

Ball joints and boots: service or inspect every 5000 miles or 4 months, whichever occurs first.

Brake discs & pads (front): service or inspect every 6000 miles.

Halfshaft boots: service or inspect every 5000 miles or 4 months. Retighten the flange bolts, whichever occurs first.

Body chassis bolts and nuts: service or inspect every 5000 miles or 4 months, whichever occurs first.

Transmission and differential fluid: replace every 15,000 miles or 12 months, whichever occurs first.

Transfer case and differential fluid: replace every 15,000 miles or 12 months, whichever occurs first.

71099_SEQU_C0014

BRAKES — INFORMATION AND PRECAUTIONS

ANTI-LOCK SYSTEMS

- Certain components within the ABS system are not intended to be serviced or repaired individually.
- Do not use rubber hoses or other parts not specifically specified for and ABS system. When using repair kits, replace all parts included in the kit. Partial or incorrect repair may lead to functional problems and require the replacement of components.
- Lubricate rubber parts with clean, fresh brake fluid to ease assembly. Do not use shop air to clean parts; damage to rubber components may result.
- Use only DOT 3 brake fluid from an unopened container.
- If any hydraulic component or line is removed or replaced, it may be necessary to bleed the entire system.
- A clean repair area is essential. Always clean the reservoir and cap thoroughly before removing the cap. The slightest amount of dirt in the fluid may plug an orifice and impair the system function. Perform repairs after components have been thoroughly cleaned; use only denatured alcohol to clean components. Do not allow ABS components to come into contact with any substance containing mineral oil; this includes used shop rags.
- The Anti-Lock control unit is a microprocessor similar to other computer units in the vehicle. Ensure that the ignition switch is **OFF**before removing or installing controller harnesses. Avoid static electricity discharge at or near the controller.
- If any arc welding is to be done on the vehicle, the control unit should be unplugged before welding operations begin.

DISC AND DRUM SYSTEMS

CAUTION

Dust and dirt accumulating on brake parts during normal use may contain asbestos fibers from production or aftermarket brake linings. Breathing excessive concentrations of asbestos fibers can cause serious bodily harm. Exercise care when servicing brake parts. Do not sand or grind brake lining unless equipment used is designed to contain the dust residue. Do not clean brake parts with compressed air or by dry brushing. Cleaning should be done by dampening the brake components with a fine mist of water, then wiping the brake components clean with a dampened cloth. Dispose of cloth and all residue containing asbestos fibers in an impermeable container with the appropriate label. Follow practices prescribed by the Occupational Safety and Health Administration (OSHA) and the Environmental Protection Agency (EPA) for the handling, processing, and disposing of dust or debris that may contain asbestos fibers.

BRAKES — BLEEDING THE BRAKE SYSTEM

BLEEDING PROCEDURE

Brake System

➡Immediately wash off any brake fluid that comes into contact with any painted surfaces. Depressing the brake pedal with the reservoir cap removed will cause the fluid to spray. When bleeding, maintain the amount of fluid in the reservoir between the MIN. and MAX. lines indicated on the master cylinder.

1. Remove the master cylinder cap.
2. Check and add fluid as required. Install the cap.
3. Disconnect the two brake lines from the master cylinder.
4. Slowly depress and hold the brake pedal.
5. Block the outer holes with your fingers and release the pedal.
6. Repeat the above three or four times.
7. Connect the brake lines to the master cylinder.
8. Tighten to 14 ft. lbs. (20 Nm), without a union nut wrench and 13 ft. lbs. (18 Nm), with a union wrench.
9. Be sure that the brake cylinder is full of fluid.
10. Depress the pedal several times and loosen the bleeder plug. With the brake pedal depressed, bleed fluid from the front calipers, (RH and LH).
11. After bleeding tighten the bleeder plug and release the pedal.
12. Depress the pedal several times and loosen the bleeder plug. With the brake pedal depressed, bleed fluid from the rear calipers, (RH and LH).
13. After bleeding tighten the bleeder plug and release the pedal.

➡After bleeding air from the system, if the height or feel of the pedal cannot be obtained, perform air bleeding of the actuator using the Techstream diagnostic tool, or equivalent. Follow the directions on the tool.

Brake Lines

1. Be sure that the brake cylinder is full of fluid.
2. Depress the pedal several times and loosen the bleeder plug. With the brake pedal depressed, bleed fluid from the front calipers, (RH and LH).
3. After bleeding tighten the bleeder plug and release the pedal.
4. Depress the pedal several times and loosen the bleeder plug. With the brake pedal depressed, bleed fluid from the rear calipers, (RH and LH).
5. After bleeding tighten the bleeder plug and release the pedal.
6. If the following symptoms occur, low or spongy pedal and pedal is depressed but braking is insufficient, bleed the brake system.
7. Bleed the VSC actuator assembly.

➡Depress the pedal more than twenty times with the engine off. Connect the diagnostic tool to the DLC3. Turn the ignition switch ON. Do not start the engine. Select AIR BLEEDING on the tool and follow the directions. Bleed the air out of the suction line.

WHEEL SPEED SENSORS

REMOVAL & INSTALLATION

Front

See Figure 1.

1. Before servicing the vehicle, refer to the Precautions Section.
2. Disconnect the negative battery cable. Tape the cable with insulating tape.
3. Raise and support the vehicle safely.
4. Remove front wheel.
5. Disconnect speed sensor connector.
6. Remove the clips and 3 clamp bolts holding the sensor harness from the frame, upper arm and steering knuckle.
7. Remove the bolt and speed sensor.

To install:

➡Be sure to use new fasteners, as required.

8. Installation is the reverse of the removal procedure.
9. Tighten bolt to 8 ft. lbs (11 Nm).
10. Inspect the sensor signal. Correct as required.

Rear

See Figure 2.

1. Before servicing the vehicle, refer to the Precautions Section.
2. Disconnect the negative battery cable. Tape the cable with insulating tape.
3. Raise and support the vehicle safely.
4. Remove the tire and wheel assembly.
5. Disconnect rear speed sensor connector.
6. Remove the bolt and speed sensor from its mounting.

To install:

➡Be sure to use new fasteners, as required.

7. Installation is the reverse of the removal procedure.
8. Tighten bolt to 73 inch lbs. (8.3 Nm).
9. Inspect the sensor signal. Correct as required.

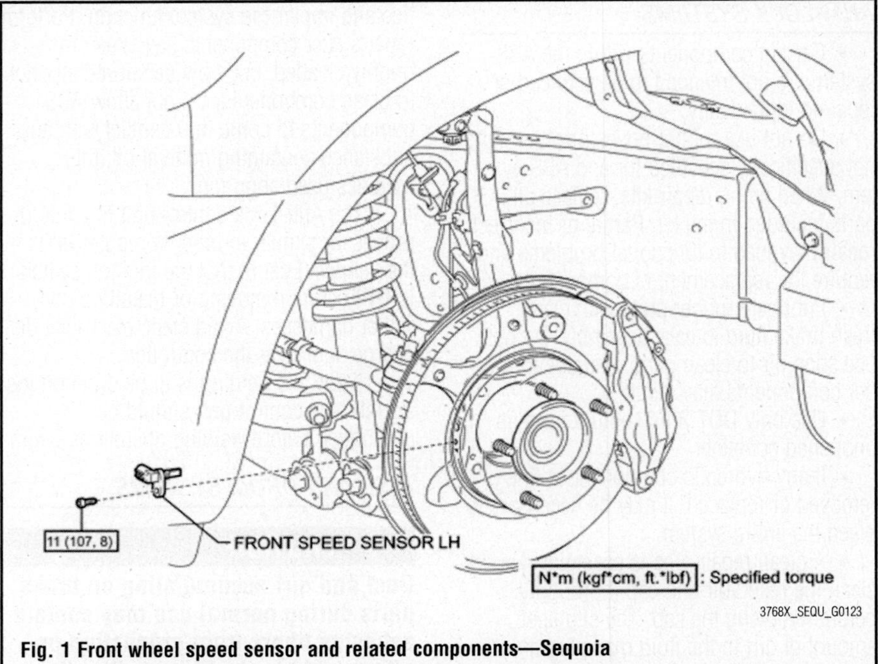

Fig. 1 Front wheel speed sensor and related components—Sequoia

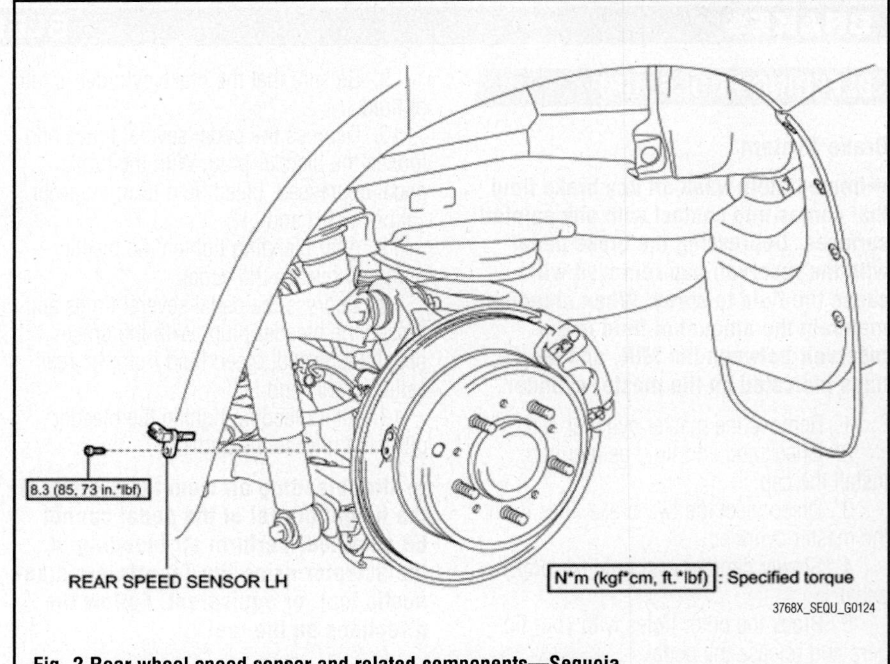

Fig. 2 Rear wheel speed sensor and related components—Sequoia

BRAKE CALIPER

REMOVAL & INSTALLATION

See Figure 3.

1. Before servicing the vehicle, refer to the Precautions Section.
2. Disconnect the negative battery cable. Tape the cable with insulating tape.
3. Raise and support the vehicle safely.
4. Remove the tire and wheel assembly.
5. To remove the front disc brake pad kit, remove the pin holddown clip. Remove the anti rattle spring. Remove the pads from the caliper. Remove the anti squeal shims.

➡The pin holddown clip can be used again if it has sufficient rebound, no deformation or wear and has all rust, dirt and foreign matter removed.

➡The anti rattle spring can be used again if it has sufficient rebound, no deformation or wear and has all rust, dirt and foreign matter removed.

6. Drain the brake fluid to an acceptable level.
7. Remove and plug the brake line hose.
8. Remove the caliper retaining bolts.
9. Remove the caliper from its mounting.

To install:

➡Be sure to use new fasteners, as required.

10. Installation is the reverse of the removal procedure.
11. Bleed the system.

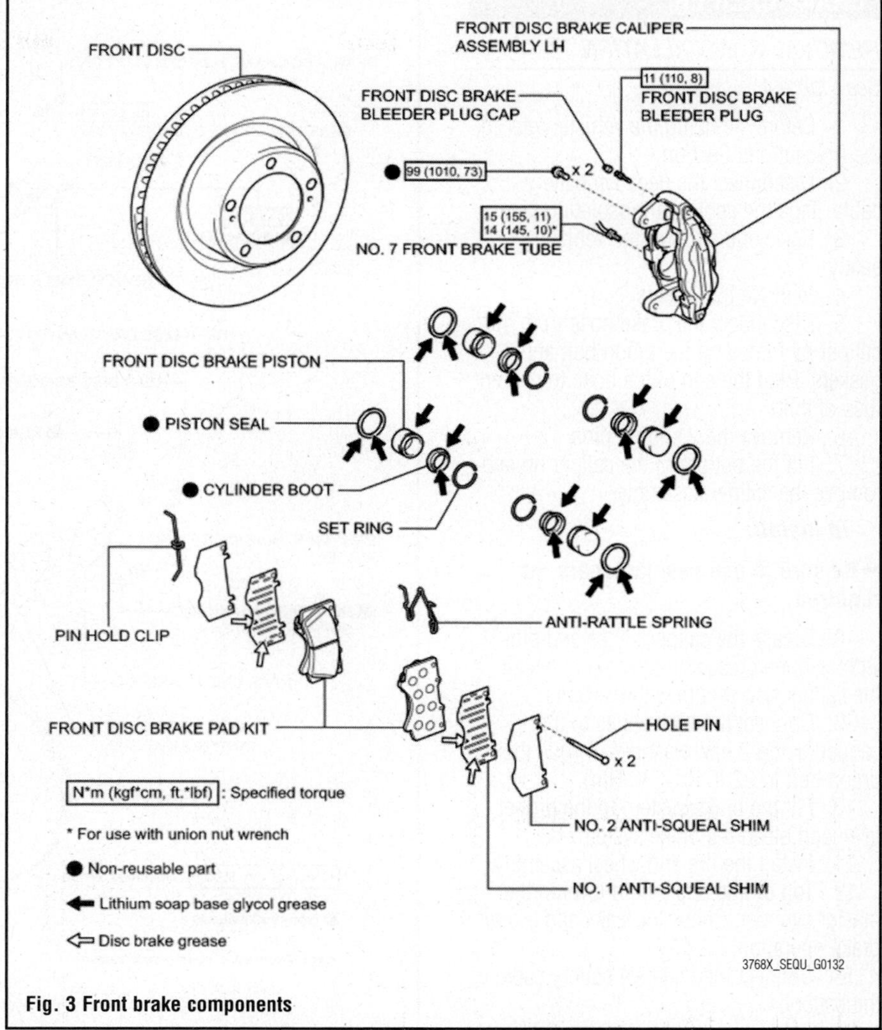

Fig. 3 Front brake components

DISC BRAKE PADS

REMOVAL & INSTALLATION

See Figure 3.

1. Before servicing the vehicle, refer to the Precautions Section.
2. Disconnect the negative battery cable. Tape the cable with insulating tape.
3. Raise and support the vehicle safely.
4. Remove the tire and wheel assembly.
5. To remove the front disc brake pad kit, remove the pin holddown clip. Remove the anti rattle spring. Remove the pads from the caliper. Remove the anti squeal shims.

➡The pin holddown clip can be used again if it has sufficient rebound, no deformation or wear and has all rust, dirt and foreign matter removed.

➡The anti rattle spring can be used again if it has sufficient rebound, no deformation or wear and has all rust, dirt and foreign matter removed.

To install:

➡Be sure to use new fasteners, as required.

6. Installation is the reverse of the removal procedure.

BRAKE CALIPER

REMOVAL & INSTALLATION

See Figure 4.

1. Before servicing the vehicle, refer to the Precautions Section.
2. Disconnect the negative battery cable. Tape the cable with insulating tape.
3. Raise and support the vehicle safely.
4. Remove the wheels.
5. Disconnect the brake hose from the caliper by removing the union bolt and 2 gaskets. Plug the end of the hose to prevent loss of fluid.
6. Remove the 2 sliding pins.
7. Lift the bottom of the caliper up and remove the caliper assembly.

To install:

➡Be sure to use new fasteners, as required.

8. Grease the caliper slides and pins with silicone grease or equivalent. Install the caliper and secure with the bolts.
9. Connect the brake hose to the caliper, using 2 new washers. Torque the union bolt to 22 ft. lbs. (30 Nm).
10. Fill the brake system to the proper level and bleed the brake system.
11. Install the tire and wheel assembly.
12. Top off the brake fluid level in the master cylinder. Check for leaks and proper brake operation.
13. Connect the negative battery cable to the battery.

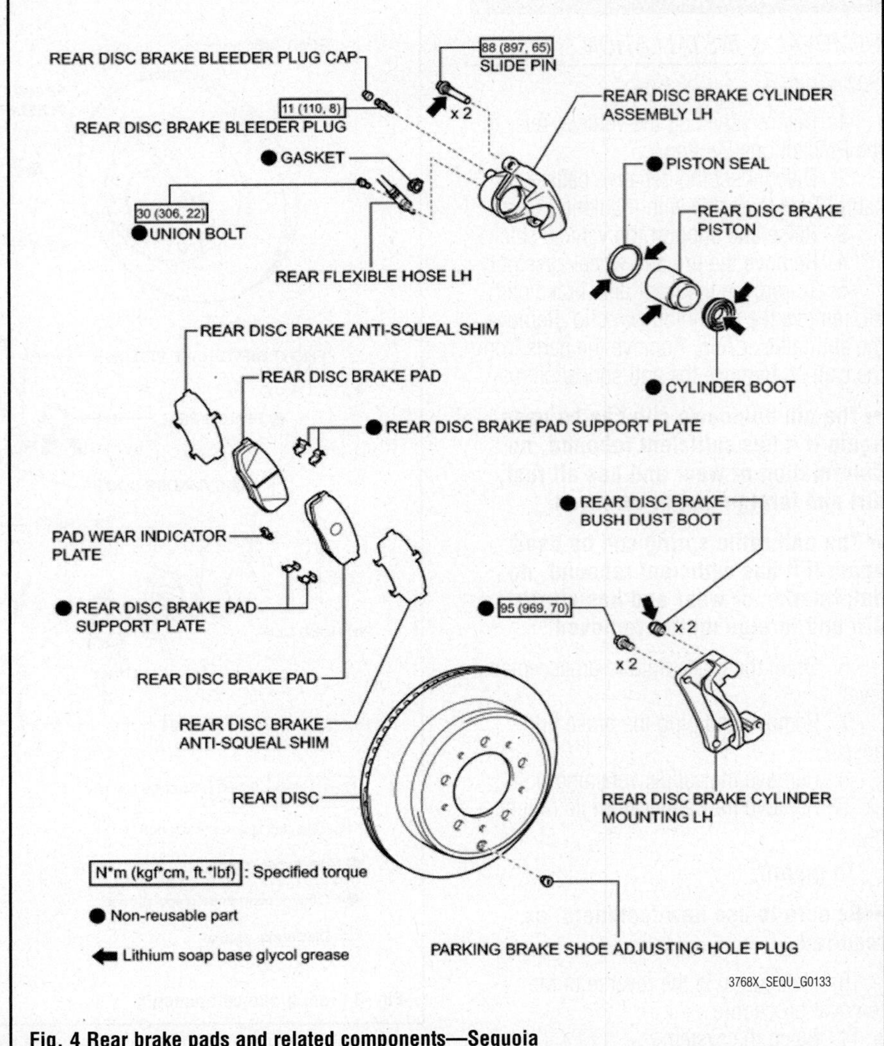

Fig. 4 Rear brake pads and related components—Sequoia

DISC BRAKE PADS

REMOVAL & INSTALLATION

See Figure 4.

1. Before servicing the vehicle, refer to the Precautions Section.
2. Disconnect the negative battery cable. Tape the cable with insulating tape.
3. Raise the vehicle and support it safely.
4. Remove the wheels.
5. Remove the brake caliper and suspend it so the hose is not stretched.
6. Remove the brake pads, anti-squeal shim, pad support plates and wear indicators.

To install:

➡Be sure to use new fasteners, as required.

7. Before installing the new pads, check the disc thickness and disc runout.
8. Install the pad support plates.
9. Install the pad wear indicator plates on each pad.
10. Install the anti-squeal shim to the outer pad. Install the pads.
11. Install the brake caliper.
12. Install the wheels.
13. Apply the brake pedal several times.
14. Road-test the vehicle for proper operation.

PARKING BRAKE CABLES

ADJUSTMENT

➡Before adjusting the parking brake, make sure that the rear brake shoe clearance has been adjusted. For shoe clearance adjustment.

1. Remove the front door scuff plate, cowl side trim board, side panel, lower finish panel and No. 2 heater to register duct.
2. Loosen the lock nut and turn the adjusting nut until the pedal travel is 6 to 9 clicks at 67 ft lbs.
3. Tighten the lock nut to 48 inch lbs. (5.4 Nm).
4. Install the No. 2 heater to register duct, lower finish panel, side panel, cowl side trim and front door scuff plate.

PARKING BRAKE SHOES

REMOVAL & INSTALLATION

See Figure 5.

1. Before servicing the vehicle, refer to the Precautions Section.
2. Disconnect the negative battery cable. Tape the cable with insulating tape.
3. Raise and safely support the vehicle.
4. Remove the rear wheel.
5. Remove the 2 mounting bolts and remove the disc brake assembly.
6. Suspend the disc brake securely and so the hose is not stretched.
7. Release the parking brake lever.
8. Place matchmarks on the rotor and rear axle hub.
9. Remove the rotor.

➡If the disc cannot be removed easily, turn the shoe adjuster until the wheel turns freely.

10. Using the proper tool, remove the 2 shoe return springs.

➡At the time of reassembly, install the strut with the spring facing forward.

11. Slide the front shoe toward outside and remove the shoe adjuster.
12. Using the proper tool, disconnect the anchor spring and tension spring from the front shoe.
13. Using the proper tool, disconnect the anchor spring and tension spring from the rear shoe.

To install:

➡Be sure to use new fasteners, as required.

14. Installation is the reverse of the removal procedure.
15. Be sure to apply high temperature grease to the shoe adjuster.
16. Adjust the parking brake.

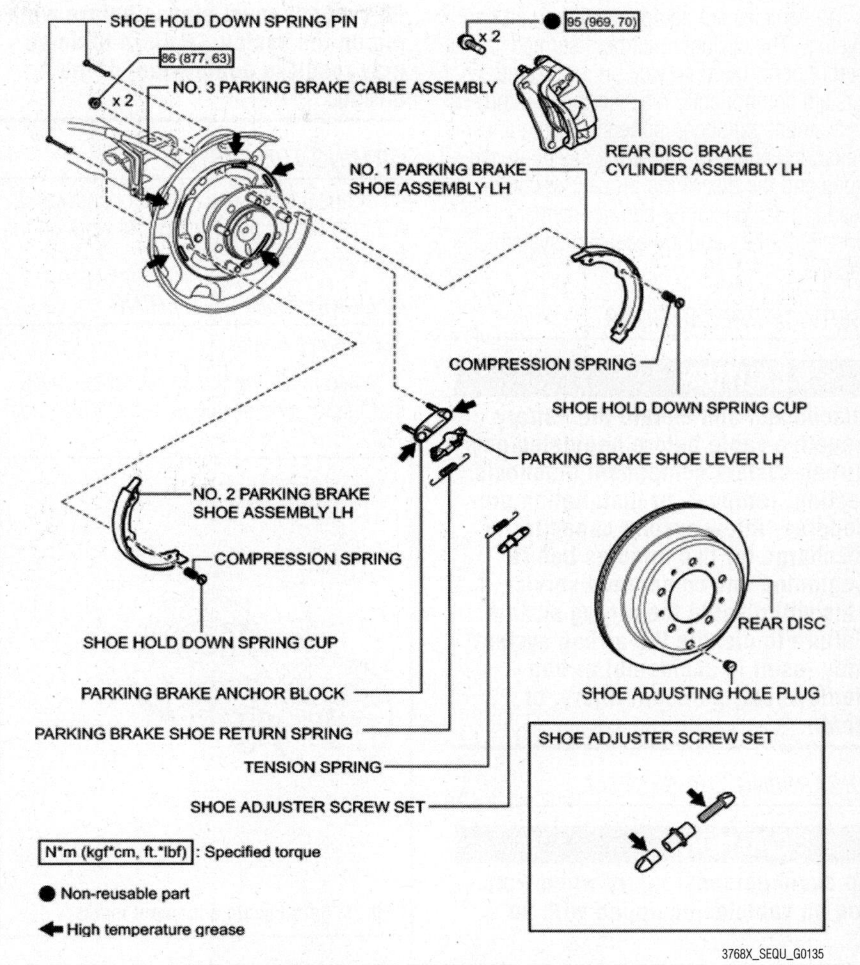

Fig. 5 Rear parking brake assembly and related components

ADJUSTMENT

1. Before servicing the vehicle, refer to the precautions in the beginning of this section.
2. Turn the adjuster and expand the shoes until the disc locks.
3. Return the adjuster 8 notches.
4. Depress the parking brake pedal with 147 N (33 lbs).
5. Drive the vehicle at about 50 km/h (31 mph) on a safe, level and dry road for about 400 meters (0.25 mile) in this condition.
6. Repeat this procedure 2 or 3 times.

CHASSIS ELECTRICAL

AIR BAG (SUPPLEMENTAL RESTRAINT SYSTEM)

GENERAL INFORMATION

All vehicles are equipped with an air bag system. The system must be disarmed before performing service on, or around, system components, the steering column, instrument panel components, wiring and sensors. Failure to follow the safety precautions and the disarming procedure could result in accidental air bag deployment, possible injury and unnecessary system repairs.

SERVICE PRECAUTIONS

⁂ CAUTION

Disconnect and isolate the battery negative cable before beginning any airbag system component diagnosis, testing, removal, or installation procedures. Allow system capacitor to discharge for two minutes before beginning any component service. This will disable the airbag system. Failure to disable the airbag system may result in accidental airbag deployment, personal injury, or death.

DISARMING THE SYSTEM

⁂ CAUTION

To avoid personal injury when working on vehicles equipped with an airbag, the negative battery cable must be disconnected and at least 90 seconds must elapse before working on the system. Failure to do so may result in deployment of the airbag.

ARMING THE SYSTEM

Connect the negative battery and wait 2 minutes before performing and work on the vehicle.

CLOCKSPRING CENTERING

See Figures 6 and 7.

When installing the spiral cable, check that the ignition switch is in the "OFF" position. Turn the cable counterclockwise by hand until it becomes hard to turn. Turn the cable clockwise about 21/2 turns to align the marks.

➡**The cable will rotate about 21/2 turns both left and right from the center.**

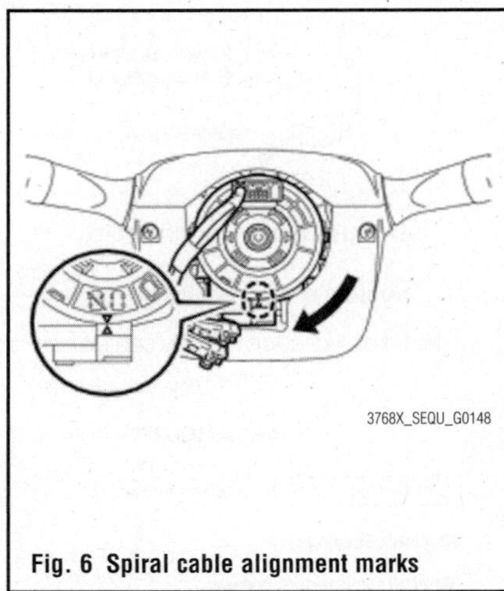

Fig. 6 Spiral cable alignment marks

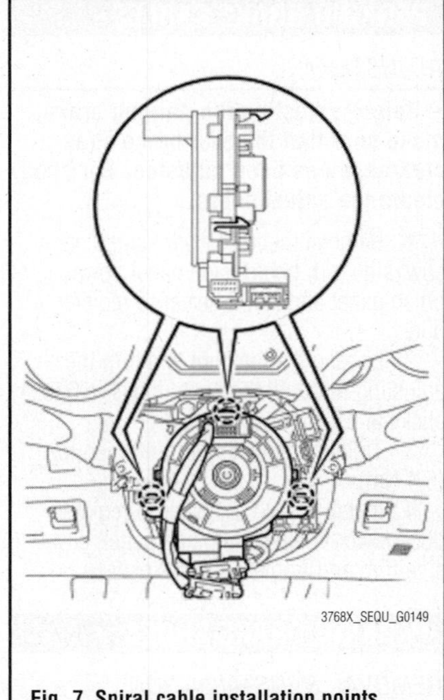

Fig. 7 Spiral cable installation points

DRIVE TRAIN

FRONT DRIVESHAFT

REMOVAL & INSTALLATION

See Figure 8.

1. Before servicing the vehicle, refer to the Precautions Section.
2. Disconnect the negative battery cable. Tape the cable with insulating tape.
3. Raise and safely support the vehicle.
4. Remove the driveshaft heat insulator bracket sub assembly, if equipped.
5. Matchmark and remove the driveshaft retaining bolts.
6. Plug the transmission end to prevent fluid leakage.
7. Remove the component from the vehicle.

To install:

➡**Be sure to use new fasteners, as required.**

8. Installation is the reverse of the removal procedure.

FRONT HALFSHAFTS

REMOVAL & INSTALLATION

See Figure 9.

1. Before servicing the vehicle, refer to the Precautions Section.
2. Disconnect the negative battery cable. Tape the cable with insulating tape.
3. Drain the differential.
4. Remove the steering knuckle assembly.
5. Using tool SST09520-01010, and SST09520-24010 remove the halfshaft. Be careful not to damage the oil seal.

To install:

➡**Be sure to use new fasteners, as required.**

6. Installation is the reverse of the removal procedure.

REAR AXLE HOUSING

REMOVAL & INSTALLATION

See Figures 10 through 12.

1. Before servicing the vehicle, refer to the Precautions Section.
2. Disconnect the negative battery cable. Tape the cable with insulating tape.
3. Remove the driveshaft.
4. Remove the halfshaft assemblies.
5. Drain the differential assembly.
6. Properly support the differential carrier assembly.
7. Remove the carrier retaining bolts.
8. Carefully remove the carrier from its mounting.

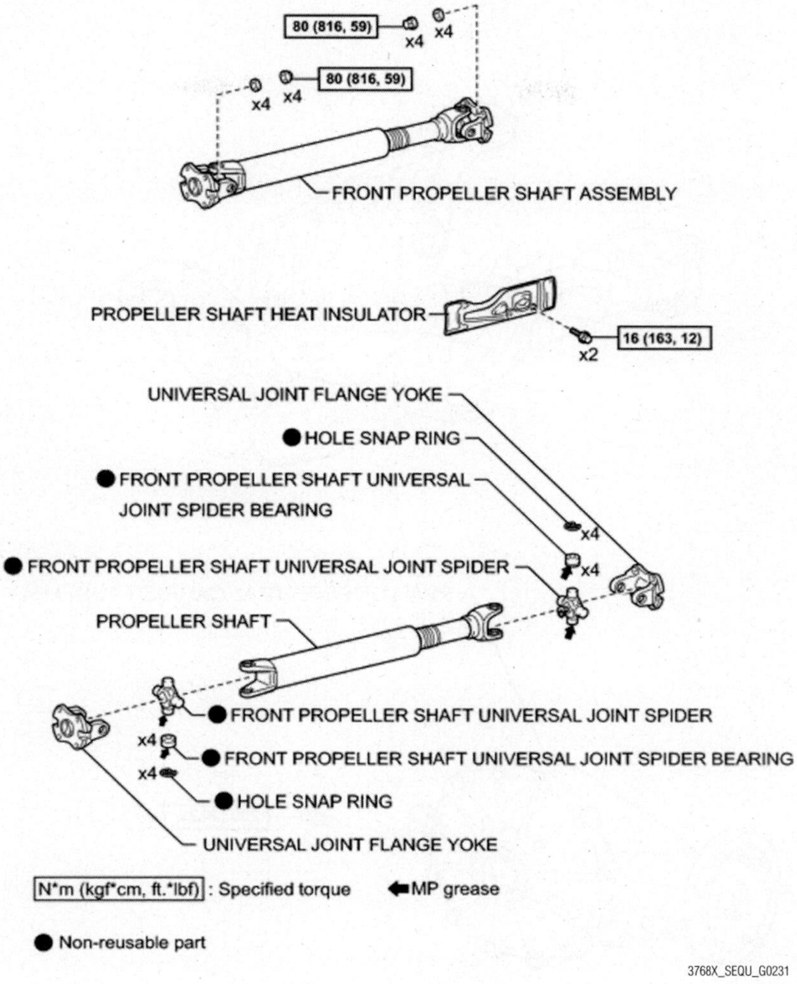

80 (816, 59) x4 x4

80 (816, 59) x4 x4

FRONT PROPELLER SHAFT ASSEMBLY

PROPELLER SHAFT HEAT INSULATOR

16 (163, 12) x2

UNIVERSAL JOINT FLANGE YOKE

● HOLE SNAP RING

● FRONT PROPELLER SHAFT UNIVERSAL JOINT SPIDER BEARING

x4

● FRONT PROPELLER SHAFT UNIVERSAL JOINT SPIDER

x4

PROPELLER SHAFT

● FRONT PROPELLER SHAFT UNIVERSAL JOINT SPIDER

x4 ● FRONT PROPELLER SHAFT UNIVERSAL JOINT SPIDER BEARING

x4 ● HOLE SNAP RING

UNIVERSAL JOINT FLANGE YOKE

N*m (kgf*cm, ft.*lbf) : Specified torque

MP grease

● Non-reusable part

3768X_SEQU_G0231

Fig. 8 Front driveshaft and related components

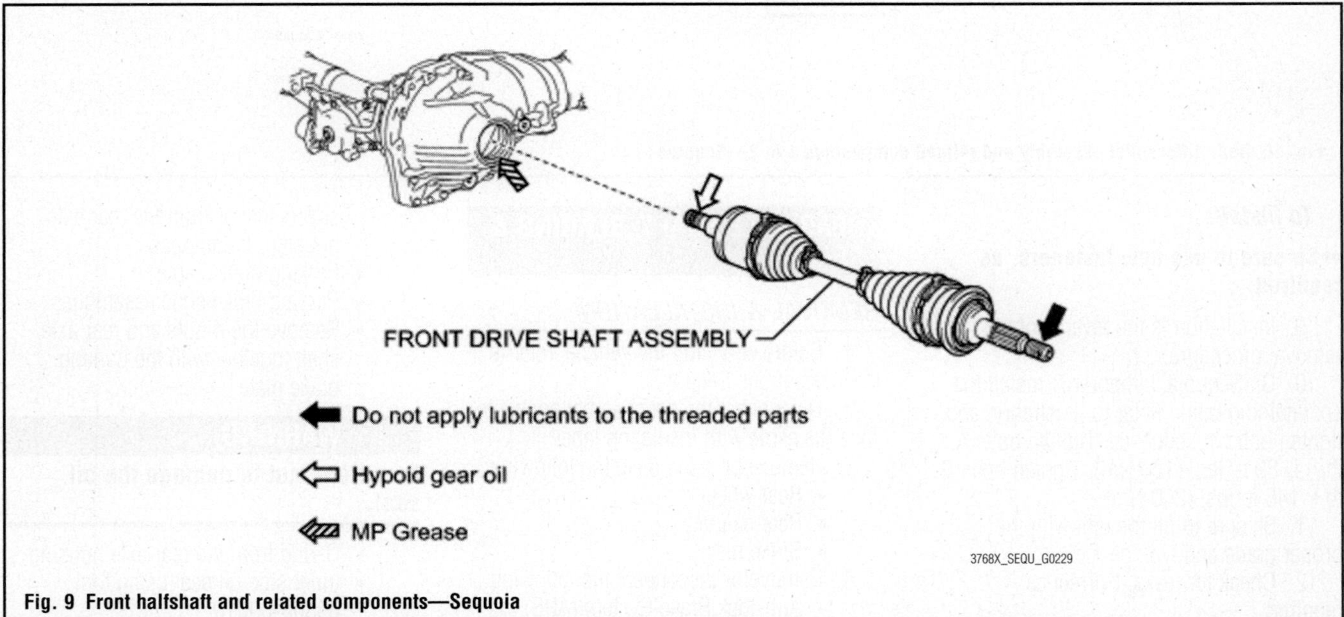

Fig. 9 Front halfshaft and related components—Sequoia

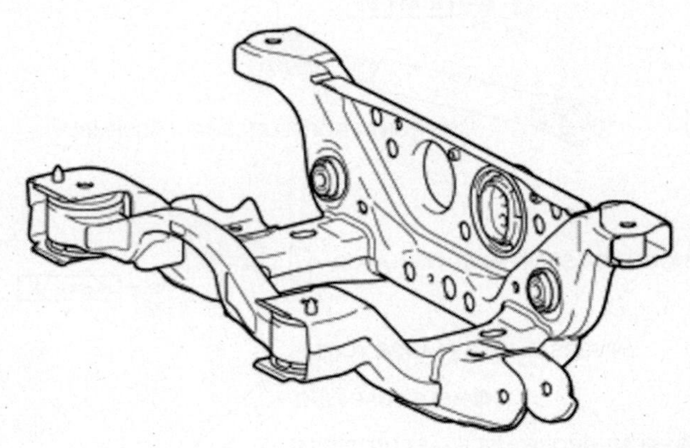

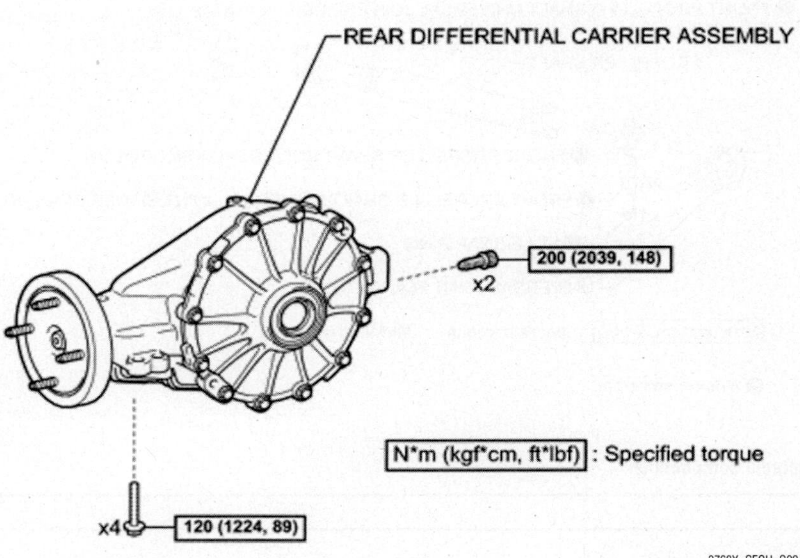

Fig. 10 Rear differential assembly and related components 1 of 2—Sequoia

To install:

➡Be sure to use new fasteners, as required.

9. Installation is the reverse of the removal procedure.
10. On Sequoia, temporarily install the six retaining bolts. Refer to illustration and tighten bolts in sequence. Tighten bolts A thru D 89 ft. lbs. (120 Nm). Tighten bolts E to F 148 ft. lbs. (200 Nm).
11. Be sure to fill the unit with the proper grade and type gear oil.
12. Check for leaks. Correct as required.

REAR AXLE SHAFT, BEARING & SEAL

REMOVAL & INSTALLATION

1. Before servicing the vehicle, refer to the Precautions Section.
2. Disconnect the negative battery cable. Tape the cable with insulating tape.
3. Remove or disconnect the following:
   - Rear wheel
   - Rear caliper
   - Brake disc
4. Remove or disconnect the following:
   - Anti-lock Brake System (ABS) speed sensor from the rear axle housing, if equipped
   - Parking brake cable
   - Parking brake shoe assemblies
   - Remove the 4 nuts and rear axle shaft together with the parking brake plate

WARNING

Be careful not to damage the oil seal.

   - O-ring from the rear axle housing
   - Inner side oil seal using tool 09308-00010

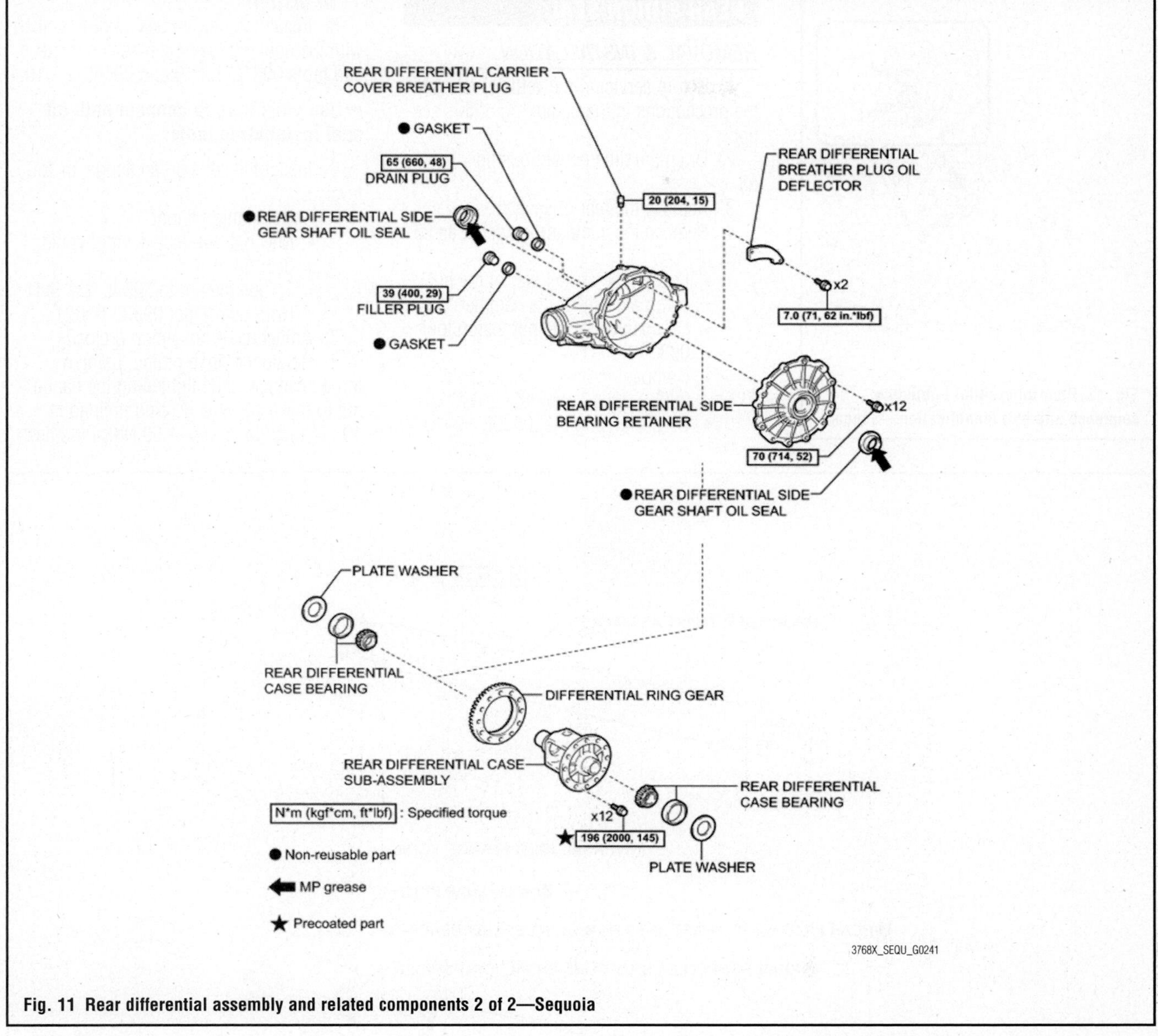

Fig. 11 Rear differential assembly and related components 2 of 2—Sequoia

To install:

➡Be sure to use new fasteners, as required.

5. Install or connect the following:
   - New O-ring to the rear axle
   - Install the rear axle shaft and parking brake plate with the 4 nuts and tighten to 44 ft. lbs. (60 Nm)
   - Anti-lock Brake System (ABS) speed sensor to the rear axle housing, if equipped
   - Parking brake cable assembly
   - Parking brake shoe assembly
   - Install rear disc
   - Adjust parking brake shoe clearance
   - Connect the rear disc brake cylinder and install 2 new bolts and tighten to 70 ft. lbs. (95 Nm)

➡Do not twist the flexible hose. Make sure that the bolts are free from damage and foreign matter. Do not over tighten the bolts.

   - Rear wheel

REAR DRIVESHAFT

REMOVAL & INSTALLATION

See Figures 13 and 14.

1. Before servicing the vehicle, refer to the Precautions Section.
2. Disconnect the negative battery cable. Tape the cable with insulating tape.
3. Raise and safely support the vehicle.
4. Remove the driveshaft heat insulator bracket sub assembly, if equipped.
5. Matchmark and remove the driveshaft retaining bolts.
6. Plug the transmission end to prevent fluid leakage.
7. Remove the component from the vehicle.

To install:

➡Be sure to use new fasteners, as required.

8. Installation is the reverse of the removal procedure.

Fig. 12 Rear differential tightening sequence and bolt identification—Sequoia

REAR PINION SEAL

REMOVAL & INSTALLATION

1. Before servicing the vehicle, refer to the precautions in the beginning of this section.
2. Drain the differential housing oil.
3. Remove the rear driveshaft.
4. Remove the companion flange, as follows:
   - Loosen the staked part of the nut, using a chisel and a hammer
   - Companion flange nut, using tool 09330-00021
   - Companion flange, using tools 09950-30012 and 09955-03050
   - Oil seal, using tool 09308-10010

To install:

5. Install the new oil seal until it is flush with the housing, using a plastic hammer and tools 09316-12010 and 09649-17010

➡Use vinyl tape to connect both oil seal installation tools.

6. Install the companion flange, as follows:
   - Companion flange
   - New nut, lubricated with hypoid gear oil
   - Torque the nut to 325 ft. lbs. (441 Nm), using tool 09330-00021.
7. Adjust the drive pinion preload
8. Rotate the drive pinion, using a torque wrench while tightening the flange nut to make sure the bearing preload is 10–14 inch lbs. (1.04–1.69 Nm) for a new

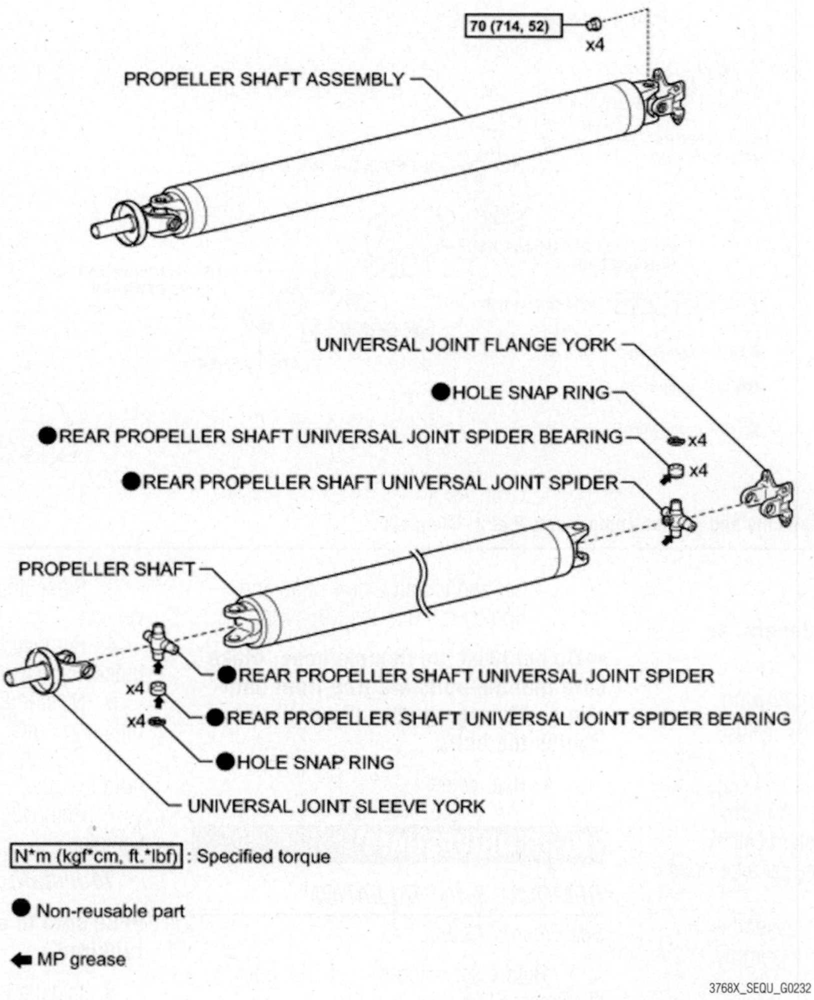

70 (714, 52)
x4
PROPELLER SHAFT ASSEMBLY
UNIVERSAL JOINT FLANGE YORK
● HOLE SNAP RING
● REAR PROPELLER SHAFT UNIVERSAL JOINT SPIDER BEARING
x4
● REAR PROPELLER SHAFT UNIVERSAL JOINT SPIDER
x4
PROPELLER SHAFT
● REAR PROPELLER SHAFT UNIVERSAL JOINT SPIDER
x4
● REAR PROPELLER SHAFT UNIVERSAL JOINT SPIDER BEARING
x4
● HOLE SNAP RING
UNIVERSAL JOINT SLEEVE YORK
N*m (kgf*cm, ft.*lbf) : Specified torque
● Non-reusable part
⬅ MP grease
3768X_SEQU_G0232

Fig. 13 Rear driveshaft and related components 2WD—Sequoia

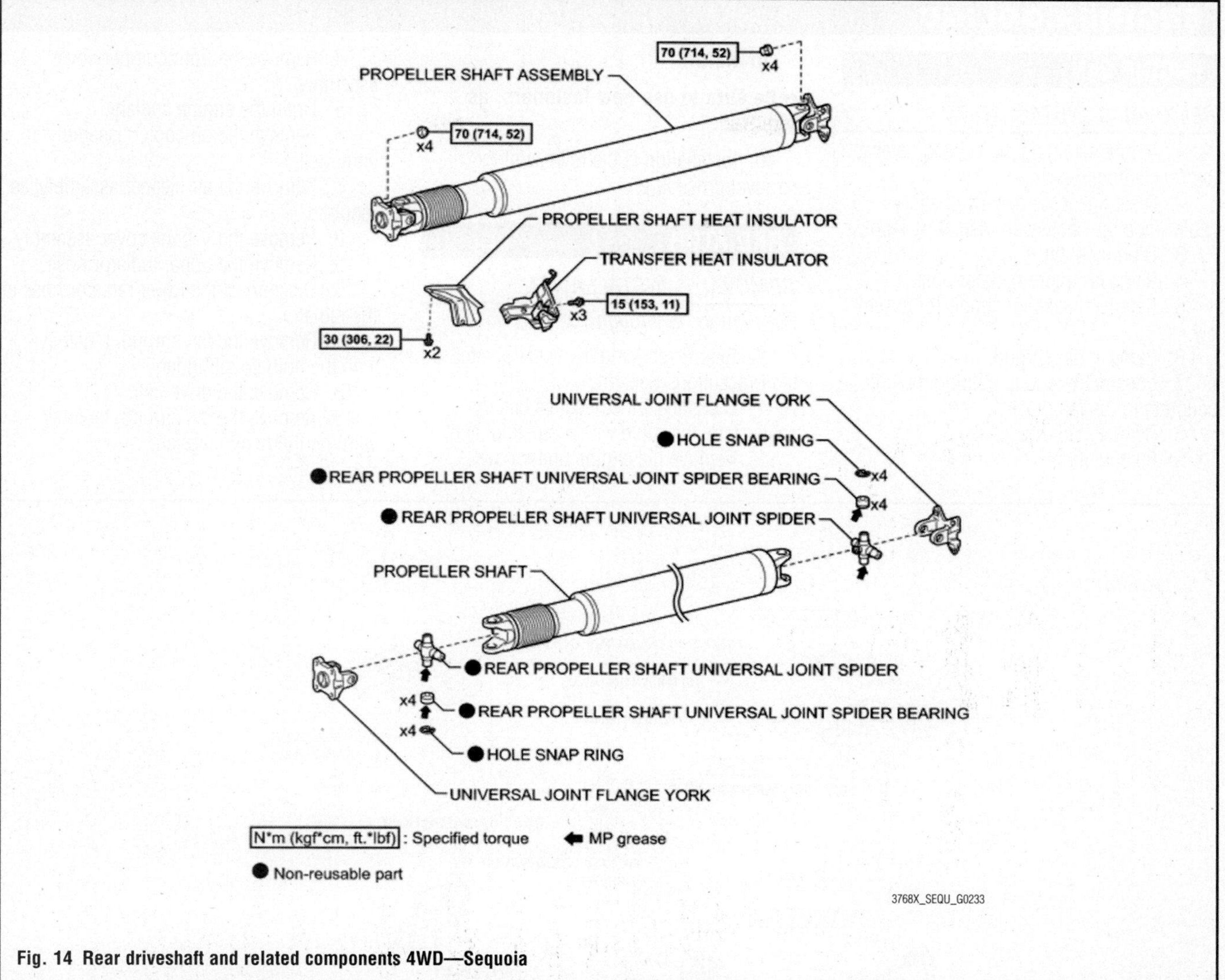

Fig. 14 Rear driveshaft and related components 4WD—Sequoia

bearing or 8–12 inch lbs. (0.85–01.37 Nm) for a used bearing. Tighten the flange nut to achieve the preload torque readings originally recorded.

CAUTION

Never loosen the pinion nut to reduce bearing preload.

9. Install or connect the following:
   - Drive pinion nut, stake it
   - Rear driveshaft.

ENGINE COOLING

ENGINE FAN

REMOVAL & INSTALLATION

1. Before servicing the vehicle, refer to the Precautions Section.
2. Disconnect the negative battery cable. Tape the cable with insulating tape.
3. Drain coolant.
4. Remover upper radiator hose.
5. Remove intake air connector assembly.
6. Remove fan shroud.
7. Loosen the 4 nuts holding the fluid coupling to the fan bracket.
8. Remove the drive belt.
9. Remove fan.

To install:

➡**Be sure to use new fasteners, as required.**

10. Installation is the reverse of the removal procedure.

RADIATOR

REMOVAL & INSTALLATION

See Figures 15 through 18.

1. Before servicing the vehicle, refer to the Precautions Section.
2. Disconnect the negative battery cable. Tape the cable with insulating tape.
3. Remove the engine under cover.
4. Remove the front bumper cover assembly.
5. Drain the engine coolant.
6. Remove the oil cooler assembly, if equipped.
7. Remove the air cleaner assembly, as required.
8. Remove the V-bank cover assembly.
9. Remove the upper radiator hose.
10. Disconnect the lower radiator hose at the radiator.
11. Remove the fan shroud, together with the fluid coupling fan.
12. Remove the drive belt.
13. Remove the fan shroud, together with the fluid coupling fan.

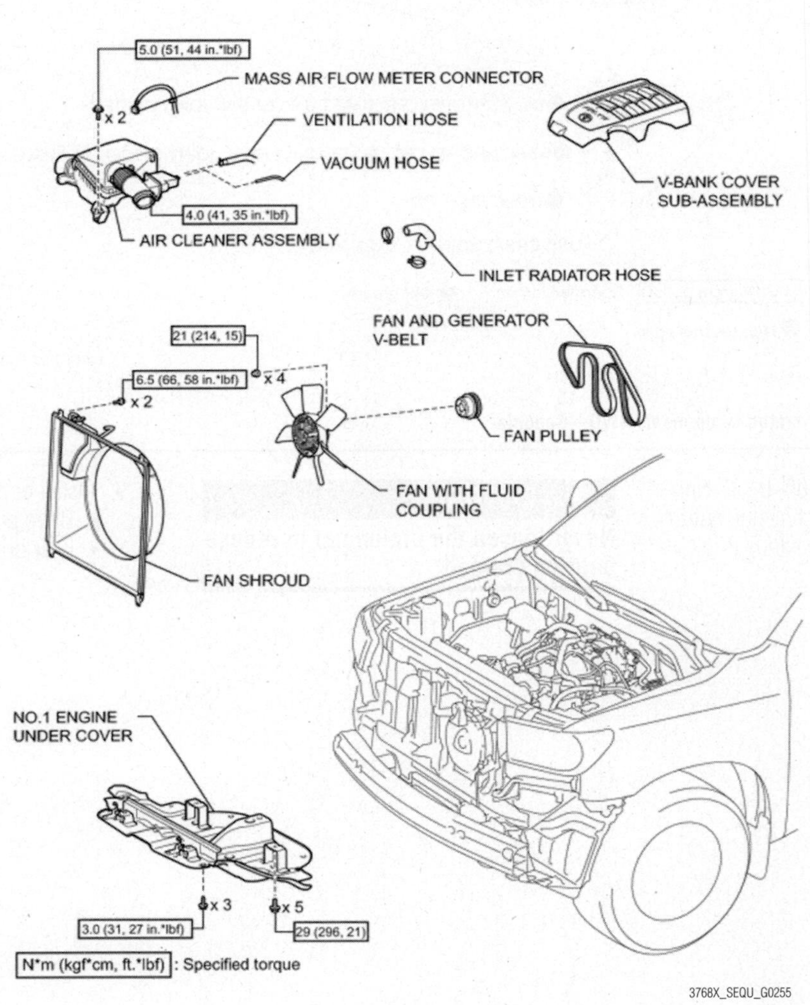

Fig. 15 Radiator and related components 4.6L engine 1 of 2—Sequoia

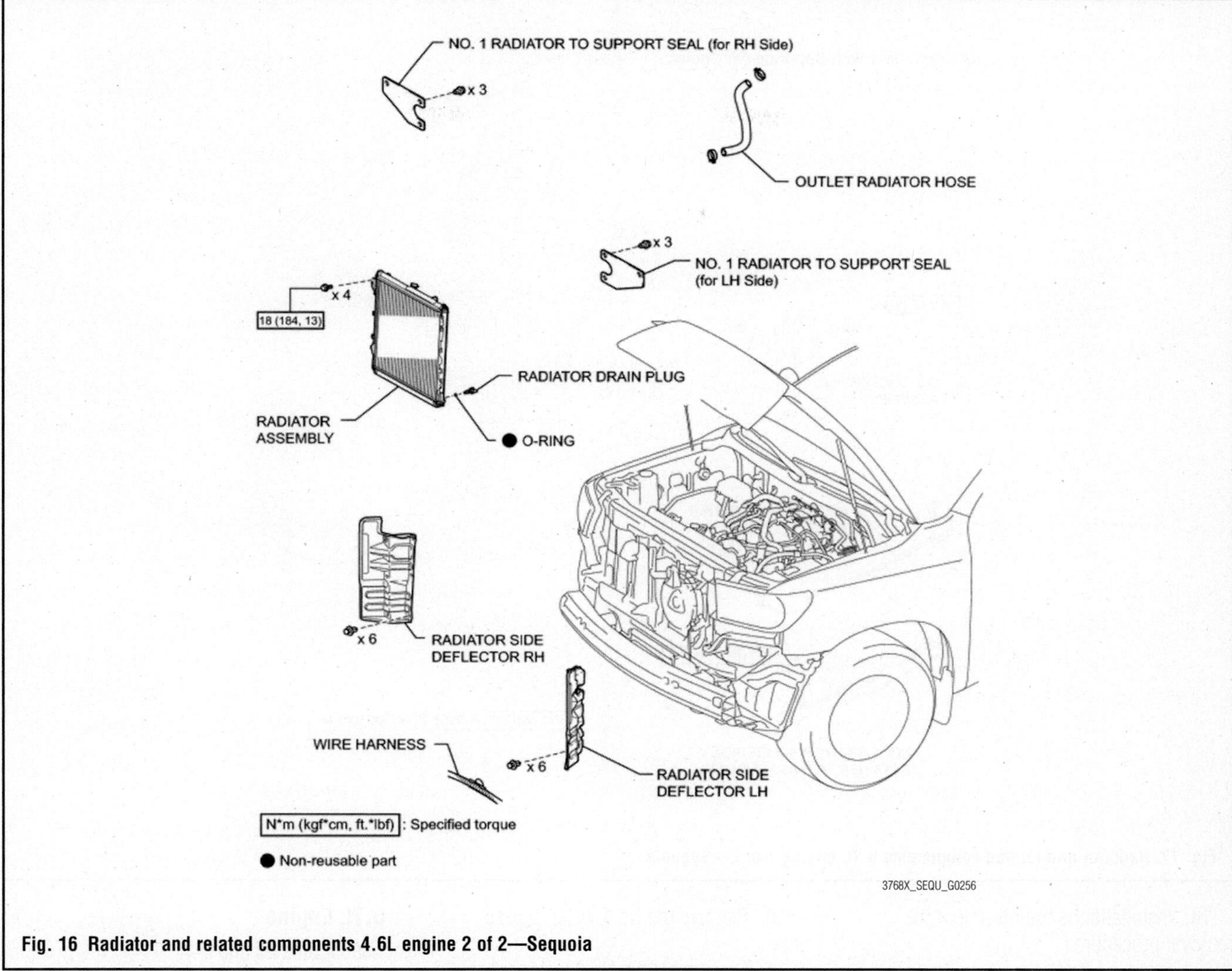

Fig. 16 Radiator and related components 4.6L engine 2 of 2—Sequoia

14. Remove the fan pulley from the fan bracket.
15. Remove the radiator side deflectors.
16. Remove the radiator side support seals.
17. Remove the four radiator retaining bolts and remove the radiator.

To install:

➡Be sure to use new fasteners, as required.

18. Installation is the reverse of the removal procedure.

THERMOSTAT

REMOVAL & INSTALLATION

4.6L Engine

1. Before servicing the vehicle, refer to the Precautions Section.
2. Disconnect the negative battery cable. Tape the cable with insulating tape.
3. Remove the No. 1 engine cover.
4. Drain the coolant. Be sure to properly dispose of used coolant.
5. Remove the V-bank cover sub assembly.
6. Remove the air cleaner assembly.
7. Remove the air tube sub assembly.
8. Disconnect the lower radiator hose.
9. Remove the thermostat housing retaining bolts.
10. Remove the thermostat. Discard the gasket.

To install:

➡Be sure to use new fasteners, as required.

11. Installation is the reverse of the removal procedure.

5.7L Engine

See Figures 19 through 21.

1. Before servicing the vehicle, refer to the Precautions Section.
2. Disconnect the negative battery cable. Tape the cable with insulating tape.
3. Remove No. 1 engine under cover.
4. Drain engine coolant.
5. Remove V-bank cover sub-assembly.
6. Disconnect outlet radiator hose.
7. Disconnect the No. 2 and No. 3 air hoses.
8. Disconnect the air pump's connector.
9. Disconnect the air pump connector clamp's holder.
10. Using a clip remover, detach the wire harness clamp.
11. Disconnect the No. 5 water by-pass hose.
12. Remove the air tube bracket's bolt.
13. Remove the 3 nuts, water inlet with thermostat and gasket.

To install:

➡Be sure to use new fasteners, as required.

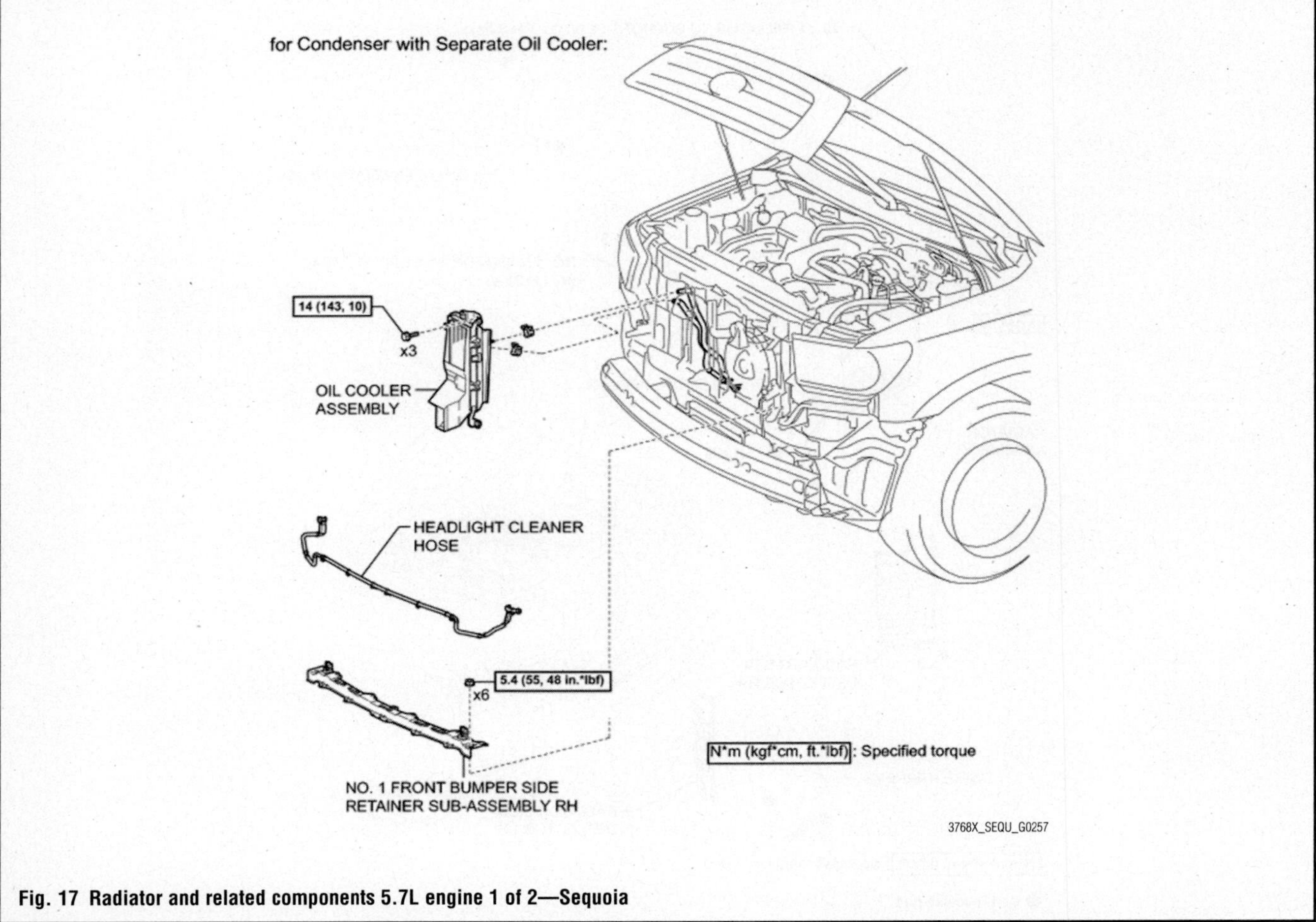

Fig. 17 Radiator and related components 5.7L engine 1 of 2—Sequoia

14. Installation is the reverse of the removal procedure.

WATER PUMP

REMOVAL & INSTALLATION

4.6L Engine

See Figure 22.

1. Before servicing the vehicle, refer to the Precautions Section.
2. Disconnect the negative battery cable. Tape the cable with insulating tape.
3. Drain the cooling system.
4. Remove the water inlet sub assembly.
5. Remove the top radiator hose.
6. Remove the fan shroud.
7. Remove the front fender apron seal, left side.
8. Disconnect the air conditioning compressor assembly. If the compressor has to be removed for clearance, be sure to properly discharge the system.
9. If equipped with oil cooler, remove the water bypass pipe.
10. Remove the No.1 water bypass hose.
11. Disconnect the No.8 water bypass hose.
12. Disconnect the No.5 water bypass hose.
13. Remove the water inlet housing.
14. Remove the water pump pulley, using the proper removal tools.
15. Remove the water pump retaining bolts.
16. Remove the water pump from its mounting. Discard the gasket.

To install:

➡Be sure to use new fasteners, as required.

17. Installation is the reverse of the removal procedure.
18. Be sure to use a new gasket.
19. Tighten the retaining bolts A to 35 ft. lbs. (47 Nm). Tighten the retaining bolts B to 17 ft. lbs. (23 Nm). Tighten the retaining bolts C to 15 ft. lbs. (20 Nm).
20. Be sure to fill the cooling system using the proper grade and type coolant.

5.7L Engine

See Figures 23 and 24.

1. Before servicing the vehicle, refer to the Precautions Section.
2. Disconnect the negative battery cable. Tape the cable with insulating tape.
3. Remove No. 1 engine under cover.
4. Drain engine coolant.
5. Remove V-bank cover sub-assembly.
6. Remove air cleaner assembly with element.
7. Remove air cleaner hose assembly.
8. Disconnect inlet radiator hose.
9. Remove fan shroud.
10. Remove No. 1 water by-pass hose.
11. Disconnect water by-pass pipe.
    a. Disconnect the No. 6 water by-pass hose.
    b. Remove the 3 bolts.
    c. Disconnect the water by-pass pipe with water hose.
12. Disconnect the No. 2 and No. 3 air hoses.
13. Disconnect the air pump's connector.
14. Disconnect the air pump connector clamp's holder.

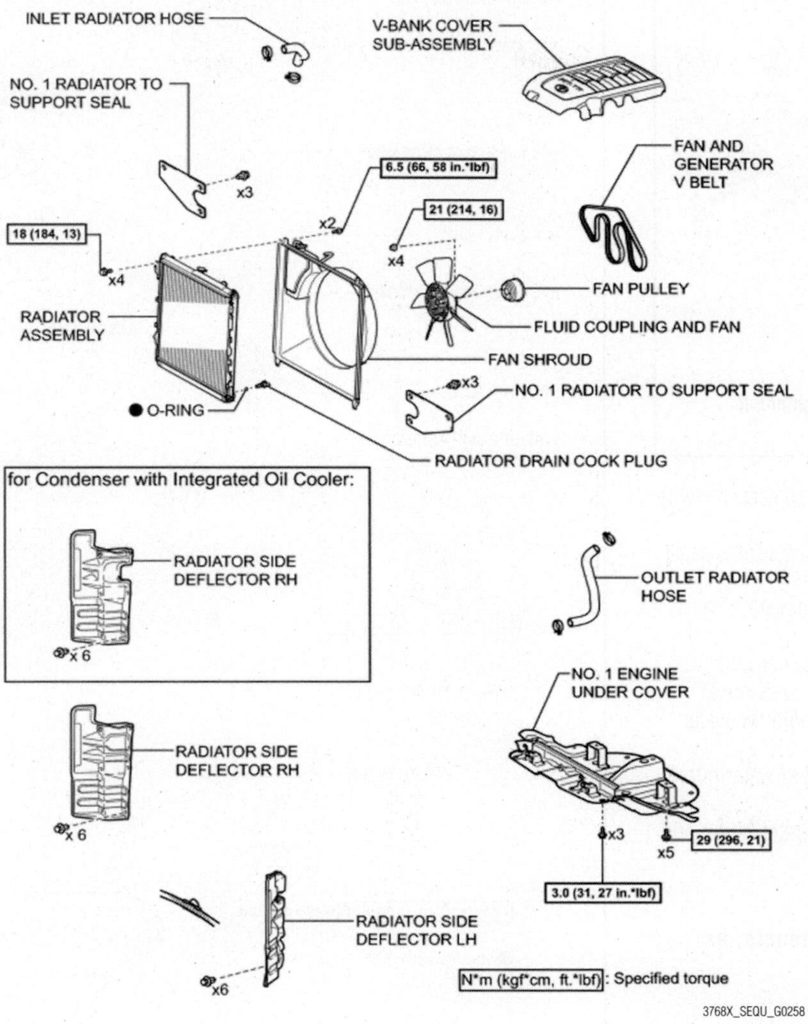

INLET RADIATOR HOSE
V-BANK COVER SUB-ASSEMBLY
NO. 1 RADIATOR TO SUPPORT SEAL
x3
6.5 (66, 58 in.*lbf)
FAN AND GENERATOR V BELT
21 (214, 16)
x2
18 (184, 13)
x4
x4
FAN PULLEY
RADIATOR ASSEMBLY
FLUID COUPLING AND FAN
FAN SHROUD
x3
NO. 1 RADIATOR TO SUPPORT SEAL
O-RING
RADIATOR DRAIN COCK PLUG
for Condenser with Integrated Oil Cooler:
RADIATOR SIDE DEFLECTOR RH
x 6
OUTLET RADIATOR HOSE
NO. 1 ENGINE UNDER COVER
RADIATOR SIDE DEFLECTOR RH
x 6
x3
x5
29 (296, 21)
3.0 (31, 27 in.*lbf)
RADIATOR SIDE DEFLECTOR LH
x6
N*m (kgf*cm, ft.*lbf): Specified torque
3768X_SEQU_G0258

Fig. 18 Radiator and related components 5.7L engine 2 of 2—Sequoia

No. 3 Hose
No. 2 Hose
22140_SEQU_G0138

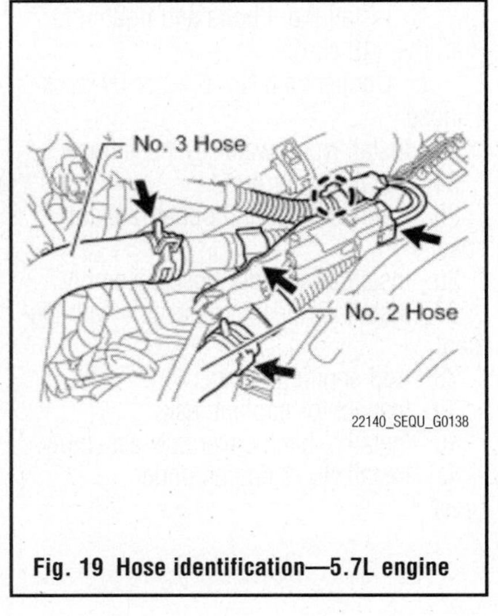

Fig. 19 Hose identification—5.7L engine

22140_SEQU_G0139

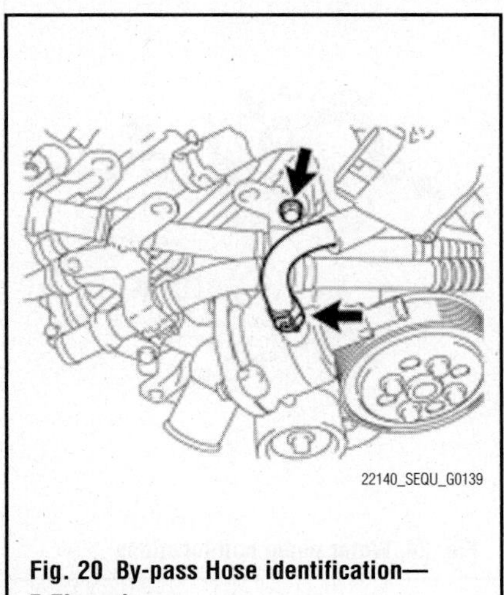

Fig. 20 By-pass Hose identification—5.7L engine

22140_SEQU_G0140

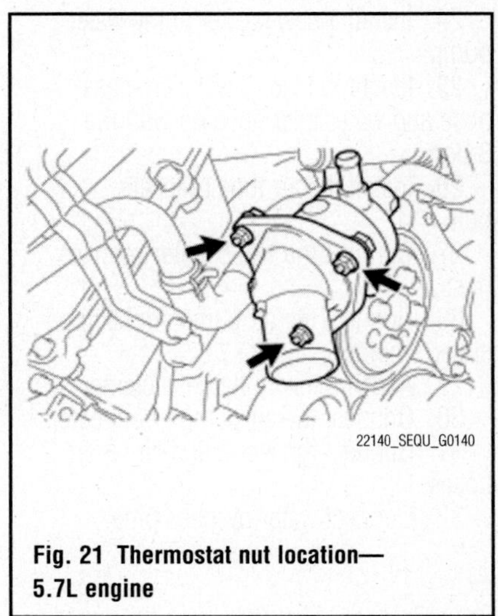

Fig. 21 Thermostat nut location—5.7L engine

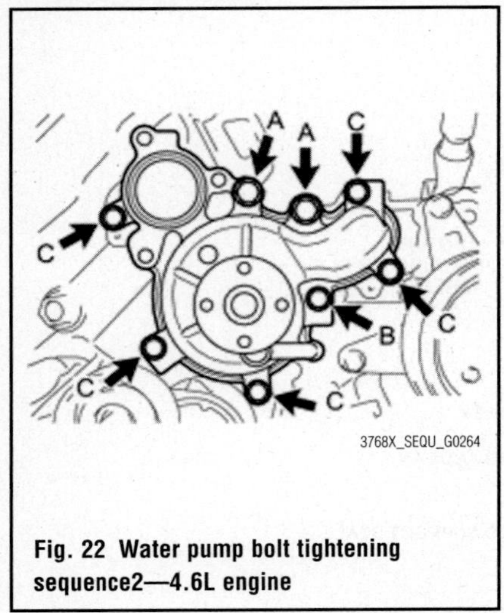

Fig. 22 Water pump bolt tightening sequence2—4.6L engine

15. Using a clip remover, detach the wire harness clamp.
16. Disconnect the No. 5 water by-pass hose.
17. Remove the air tube bracket's bolt.
18. Remove the 3 bolts, water inlet housing and No. 3 water by-pass hose.
19. Using Service Tool, hold the water pump pulley.
20. Remove the 4 bolts and water pump pulley.
21. Remove the 8 bolts, water pump and gasket.

To install:

➡Be sure to use new fasteners, as required.

22. Install a new gasket and the water pump with the 8 bolts and tighten: Bolt A to 35 ft. lbs. (47 Nm), bolt B to 17 ft. lbs. (23 Nm), and bolt C to 15 ft. lbs. (20 Nm).
23. Install the pulley with the 4 bolts and tighten to 15 ft. lbs. (21 Nm).
24. Install a new gasket to the water pump.
25. Install the No. 3 water by-pass hose and water inlet housing with the 3 bolts.
26. Install the air tube bracket's bolt.
27. Connect the No. 5 water by-pass hose.
28. Connect the air pump connector clamp's holder.
29. Attach the wire harness clamp.
30. Connect the air pump's connector.
31. Connect the No. 2 and No. 3 air hoses.
32. Connect water by-pass pipe.

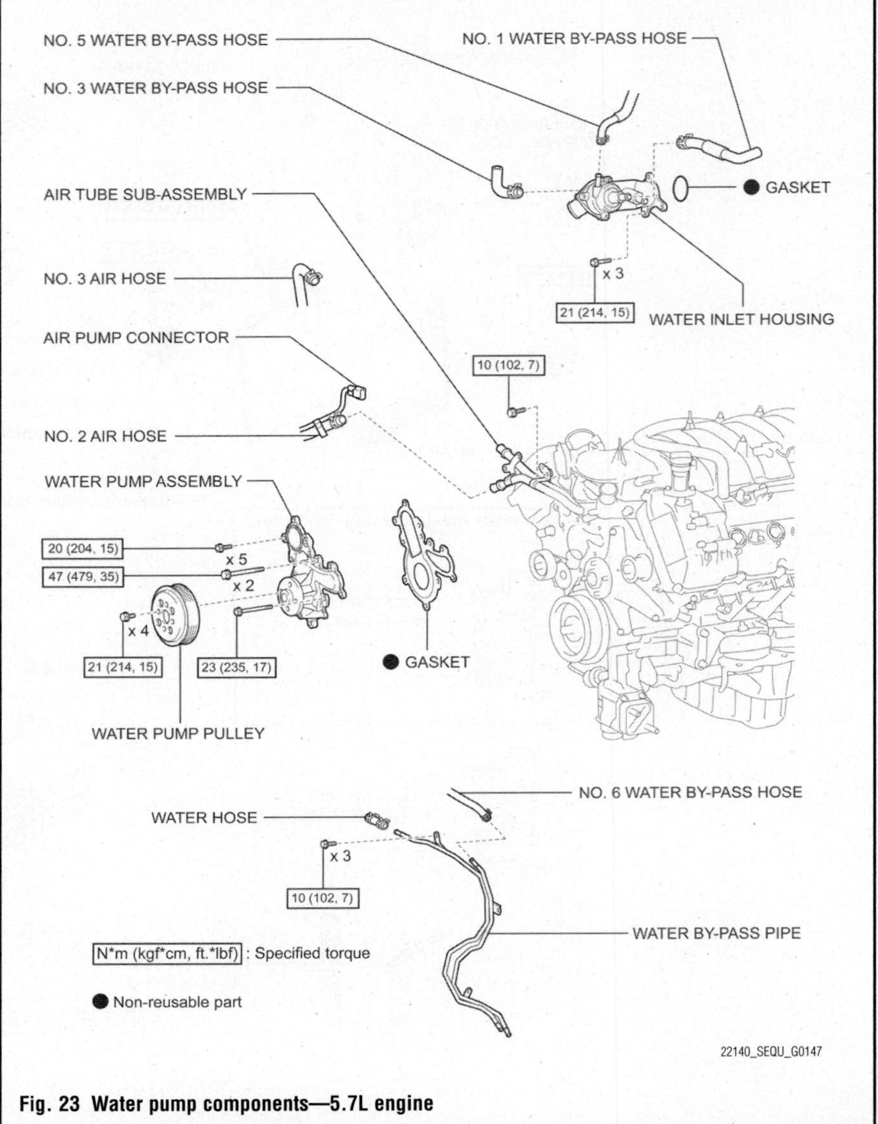

Fig. 23 Water pump components—5.7L engine

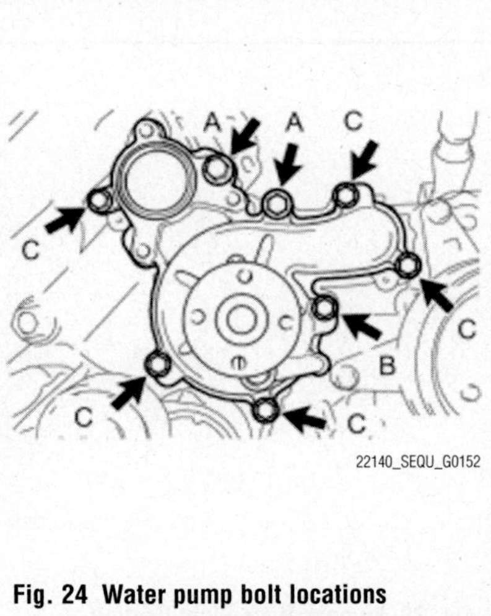

Fig. 24 Water pump bolt locations

a. Connect the water by-pass pipe with water hose.
b. Install the 3 bolts and tighten to 7 ft. lbs. (10 Nm).
c. Connect the No. 6 water by-pass hose.

33. Install No. 1 water by-pass hose.
34. Install fan shroud.
35. Connect inlet and outlet radiator hose.
36. Install air cleaner hose assembly.
37. Install air cleaner assembly with element.
38. Add engine coolant.
39. Inspect for coolant leak.
40. Install V-bank cover sub-assembly.
41. Install No. 1 engine under cover.

ENGINE ELECTRICAL — CHARGING SYSTEM

ALTERNATOR

REMOVAL & INSTALLATION

4.6L Engine

1. Disconnect the negative battery cable. Tape the cable with insulating tape.
2. Remove the fan and alternator V belt
3. Remove the air cleaner assembly.
4. Remove the front fender apron seal.
5. Disconnect the power steering pressure switch connector.
6. Disconnect the solenoid valve connector.
7. Disconnect the oil cooler tube sub assembly, if equipped.
8. Remove the alternator retaining bolts.
9. Disconnect the electrical connectors.
10. Remove the component from its mounting.

To install:

➡Be sure to use new fasteners, as required.

11. Installation is the reverse of the removal procedure.

5.7L Engine

See Figures 25 and 26.

1. Before servicing the vehicle, refer to the Precautions Section.
2. Disconnect the negative battery cable. Tape the cable with insulating tape.
3. Remove air cleaner hose assembly.
4. Remove air cleaner assembly.

Fig. 25 Alternator connector location

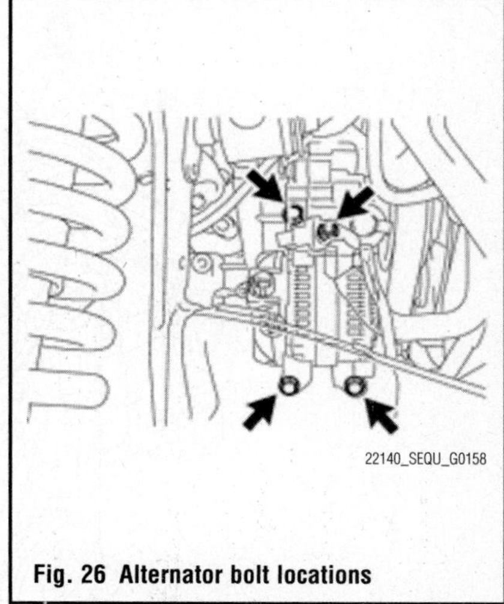

Fig. 26 Alternator bolt locations

5. Remove V-bank cover.
6. Remove No. 1 engine under cover sub-assembly.
7. Remove fan and alternator V belt.
8. Remove the 5 clips and apron seal.
9. Disconnect vane pump assembly.
10. Disconnect the alternator connector.
11. Remove the terminal cap and nut, and disconnect the alternator wire.
12. Remove the bolt and disconnect the wire harness bracket from the alternator.
13. Remove the 3 bolts and alternator.

To install:

➡Be sure to use new fasteners, as required.

14. Install the alternator with the 3 bolts and tighten to 32 ft. lbs. (43 Nm).
15. Connect the alternator wire with the nut and tighten to 87 inch lbs. (9.8 Nm).
16. Install the terminal cap.
17. Connect the alternator connector.
18. Install the wire harness clamp bracket with the nut.
19. Connect vane pump assembly.
20. Install the apron seal with the 5 clips.
21. Install fan and alternator V belt.
22. Install No. 1 engine under cover sub-assembly.
23. Install air cleaner assembly.
24. Install air cleaner hose assembly.
25. Install V-bank cover.
26. Install battery.
27. Connect cable to negative battery terminal.

➡Perform initialization, if necessary.

ENGINE ELECTRICAL — IGNITION SYSTEM

FIRING ORDER

See Figure 27.

IGNITION COIL

REMOVAL & INSTALLATION

1. Before servicing the vehicle, refer to the Precautions Section.
2. Disconnect the negative battery cable. Tape the cable with insulating tape.
3. Remove the V bank cover assembly, as required.
4. Remove the air cleaner assembly, as required.
5. Remove the necessary components in order to gain access to the component.
6. Disconnect ignition coil (with igniter) connectors.
7. Remove the bolt, and pull out the ignition coil (with igniter). Remove the ignition coils (with igniter).

To install:

➡Be sure to use new fasteners, as required.

8. Installation is the reverse of the removal procedure.

IGNITION TIMING

ADJUSTMENT

The ignition timing is controlled by the Powertrain Control Module (PCM). No adjustment is necessary or possible.

SPARK PLUGS

REMOVAL & INSTALLATION

1. Before servicing the vehicle, refer to the Precautions Section.
2. Remove the ignition coils.
3. Using a 16 mm plug wrench, remove the spark plugs.
4. Clean the spark plugs.
5. If the electrode has traces of wet carbon, allow it to dry and then clean with a spark plug cleaner.
6. Check the spark plug for thread damage and insulator damage. If abnormal, replace the spark plug.

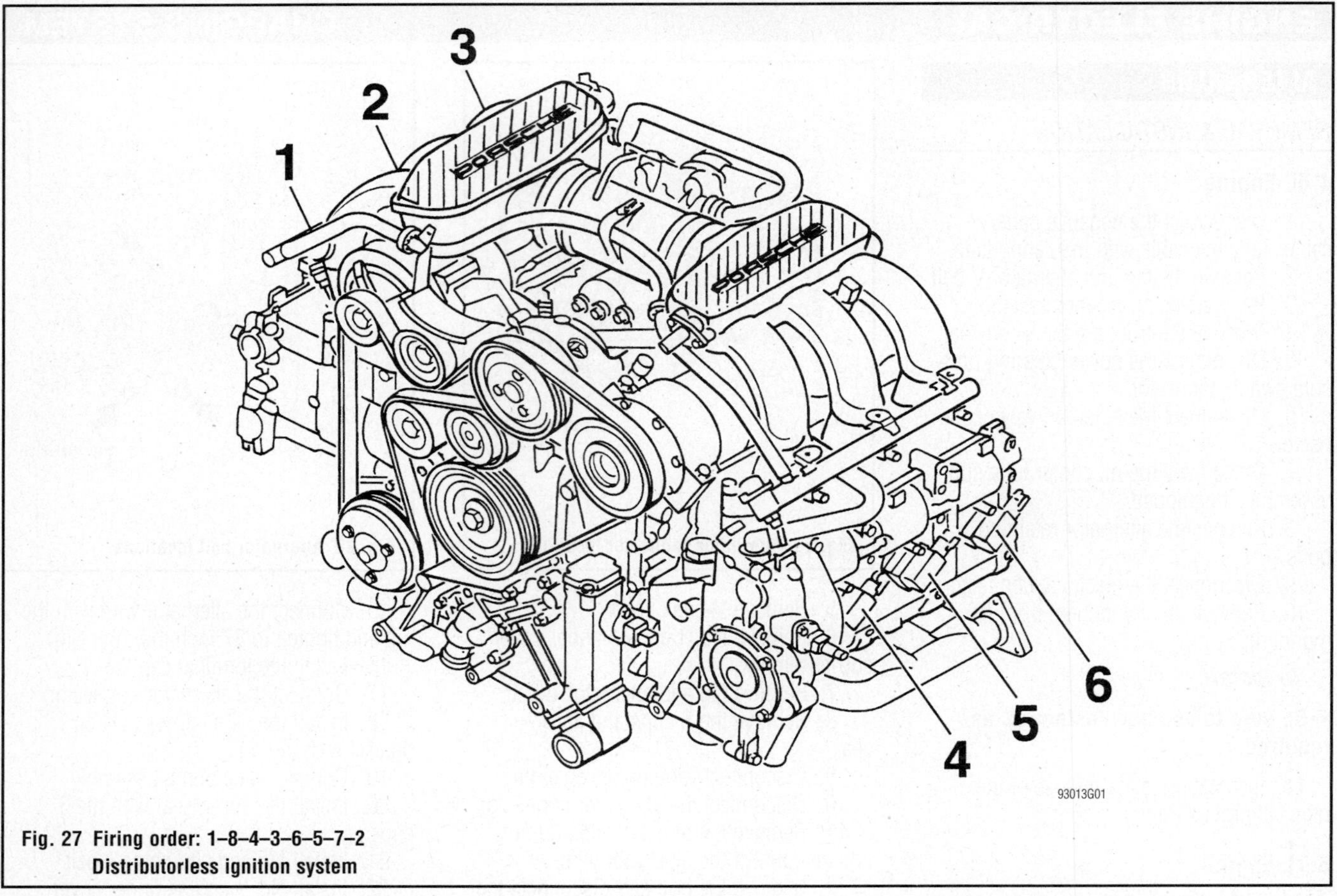

Fig. 27 Firing order: 1-8-4-3-6-5-7-2 Distributorless ignition system

7. Adjust the spark plug electrode gap. Electrode gap for new spark plug is 1.0 to 1.1 mm (0.039 to 0.043 in.).

8. Using a 16 mm plug wrench, install the spark plugs and tighten to specification.

9. Reinstall the ignition coils.

ENGINE ELECTRICAL

STARTER

REMOVAL & INSTALLATION

4.6L Engine

See Figure 28.

1. Before servicing the vehicle, refer to the Precautions Section.
2. Disconnect the negative battery cable. Tape the cable with insulating tape.
3. Remove the exhaust manifold sub assembly.
4. Remove the starter cover.
5. Disconnect the electrical connector.
6. Remove the starter retaining bolts.
7. Remove the starter from its mounting.
8. Remove the flywheel side cover.

To install:

➡Be sure to use new fasteners, as required.

9. Installation is the reverse of the removal procedure.
10. Tighten the retaining bolts to 27 ft. lbs. (37 Nm).

5.7L Engine

See Figure 29.

1. Before servicing the vehicle, refer to the Precautions Section.
2. Disconnect the negative battery cable. Tape the cable with insulating tape.
3. Remove No. 1 engine under cover.
4. Remove V-bank cover.
5. Remove air cleaner hose assembly.
6. Remove air cleaner assembly.
7. Remove fan and alternator V belt.
8. Remove front fender apron seal RH.
9. Disconnect the 2 clamps and power steering oil pressure switch connector.
10. Remove the 2 bolts and disconnect the vane pump
11. Disconnect the alternator connector.
12. Remove the terminal cap and nut, and disconnect the alternator wire.
13. Remove the bolt and disconnect the wire harness bracket from the alternator.
14. Remove the 2 bolts, 2 nuts and alternator.
15. Remove the dipstick.
16. Remove the bolt and dipstick guide.
17. Remove the O-ring from the dipstick guide.
18. Remove front exhaust pipe assembly.
    a. Disconnect the air fuel ratio sensor connector.
    b. Disconnect the heated oxygen sensor connector and 2 clamps.
    c. Remove the 2 bolts.
    d. Remove the 3 nuts and front exhaust pipe.
    e. Remove the 2 gaskets.
19. Remove No. 1 exhaust manifold heat insulator.
20. Remove exhaust manifold sub-assembly RH.

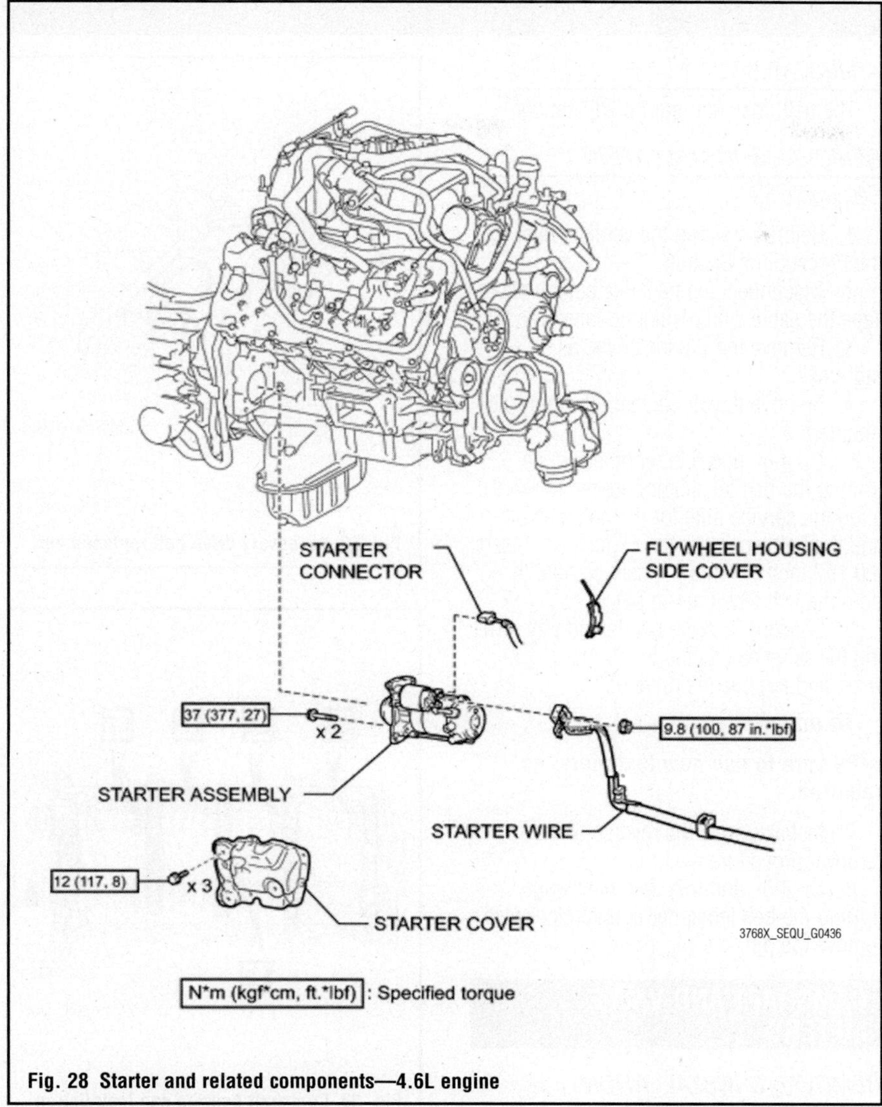

Fig. 28 Starter and related components—4.6L engine

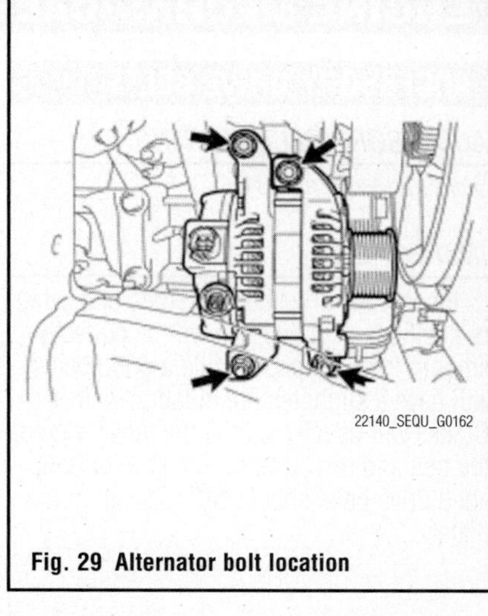

Fig. 29 Alternator bolt location

21. Remove the 3 bolts and starter cover.
22. Disconnect the starter connector.
23. Remove the nut and disconnect the starter wire.
24. Remove the 2 bolts and starter.
25. Remove the flywheel housing side cover.

To install:

➡Be sure to use new fasteners, as required.

26. Install the flywheel housing side cove.
27. Install the starter with the 2 bolts and tighten to 27 ft. lbs. (37 Nm).
28. Install the starter wire with the nut and tighten to 87 inch lbs. (9.8 Nm).
29. Connect the starter connector.
30. Install the starter cover with the 3 bolts and tighten to 9 ft. lbs. (12 Nm).
31. Install exhaust manifold sub-assembly RH.
32. Install No. 1 exhaust manifold heat insulator.
33. Install a new gasket and the exhaust pipe to the exhaust manifold RH with 3 new nuts and tighten to 40 ft. lbs. (54 Nm).
34. Install a new gasket and the exhaust pipe to the center exhaust pipe with the 2 bolts and tighten to 35 ft. lbs. (48 Nm).
35. Connect the air fuel ratio sensor connector.
36. Connect the heated oxygen sensor connector and 2 clamps.
37. Install dipstick.
    a. Apply a light coat of engine oil to a new O-ring.
    b. Install the O-ring to the guide.
    c. Install the dipstick guide with the bolt and tighten to 7 ft. lbs. (10 Nm).
38. Install the alternator with the 2 bolts and 2 nuts and tighten to 32 ft. lbs. (43 Nm).
39. Connect the alternator connector.
40. Connect the alternator wire with the nut and tighten to 87 inch lbs. (9.8 Nm).
41. Install the terminal cap.
42. Install the harness bracket to the alternator with the bolt.
43. Connect vane pump assembly.
    a. Connect the vane pump to the timing chain cover with the 2 bolts and tighten to 21 ft. lbs. (28 Nm).
    b. Connect the 2 clamps and power steering oil pressure switch connector.
44. Install front fender apron seal RH.
45. Install fan and alternator V belt.
46. Install air cleaner assembly.
47. Install air cleaner hose assembly.
48. Install V-bank cover.
49. Install No. 1 engine under cover.
50. Connect cable to negative battery terminal.

➡Some systems need to be initialized after the cable is reconnected.

ENGINE MECHANICAL

ACCESSORY DRIVE BELTS

ACCESSORY BELT ROUTING

See Figures 30 and 31.

INSPECTION

Inspect the drive belt for signs of glazing or cracking. A glazed belt will be perfectly smooth from slippage, while a good belt will have a slight texture of fabric visible. Cracks will usually start at the inner edge of the belt and run outward. All worn or damaged drive belts should be replaced immediately.

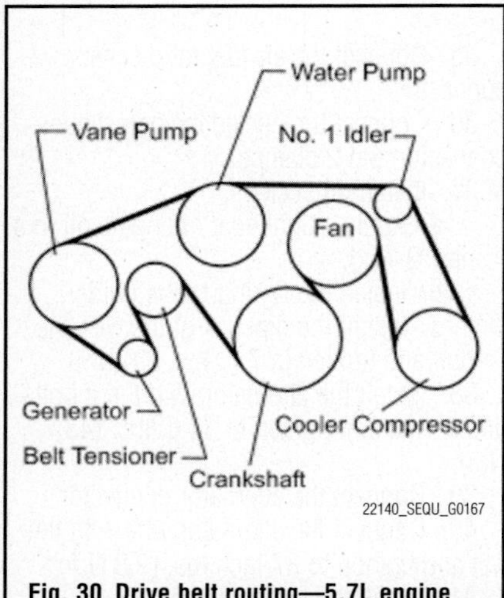

Fig. 30 Drive belt routing—5.7L engine

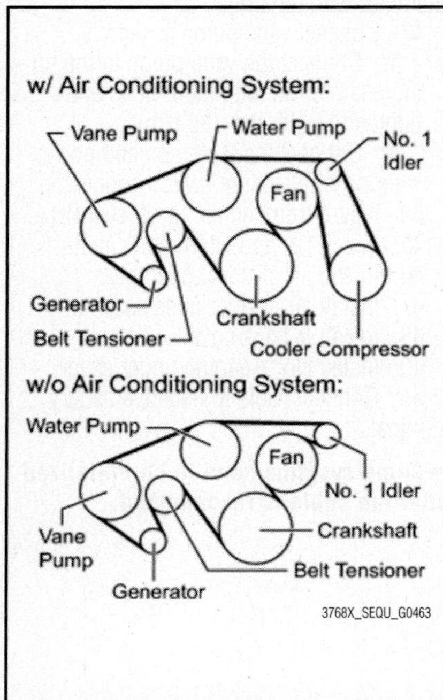

Fig. 31 Drive belt routing—4.6L engine

ADJUSTMENT

The belt does not require adjustment.

REMOVAL & INSTALLATION

See Figure 32.

1. Before servicing the vehicle, refer to the Precautions Section.
2. Disconnect the negative battery cable. Tape the cable with insulating tape.
3. Remove the V bank cover, as required.
4. Remove the air cleaner assembly, as required.
5. On 4.6L and 5.7L engines, while turning the belt tensioner counterclockwise, align the service hole for the belt tensioner and the belt tensioner fixing position. Insert a 0.197 inch bar into the service hole to hold the belt tensioner in place.
6. Loosen the drive belt tension by turning the drive belt tensioner counterclockwise, and remove the drive belt.

To install:

➡Be sure to use new fasteners, as required.

7. Installation is the reverse of the removal procedure.
8. On 4.6L and 5.7L engines, while turning the belt tensioner counterclockwise remove the pin.

CAMSHAFT AND VALVE LIFTERS

REMOVAL & INSTALLATION

4.6L Engine

Bank 1

See Figures 33 through 42.

1. Before servicing the vehicle, refer to the Precautions Section.
2. Disconnect the negative battery cable. Tape the cable with insulating tape.
3. Remove the timing chain cover.
4. Position the No. 1 cylinder at TDC on the compression stroke.
5. Remove the No. 1 chain tensioner assembly.
6. Remove the chain tensioner slipper.
7. Remove the No. 1 chain vibration damper.
8. Remove the No. 1 chain sub assembly.
9. Remove the No. 3 chain tensioner assembly.
10. Remove the camshaft bearing cap.

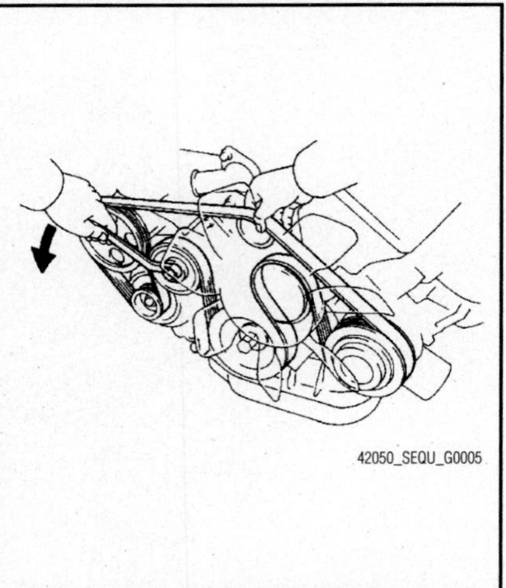

Fig. 32 Accessory drive belt replacement

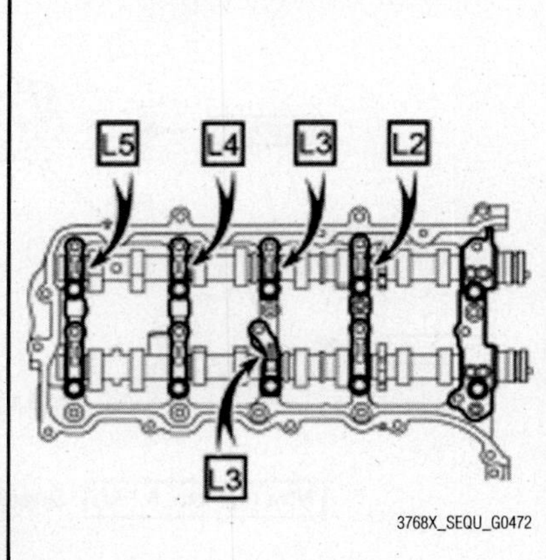

Fig. 33 Camshaft bearing cap installation Bank 1—4.6L engine

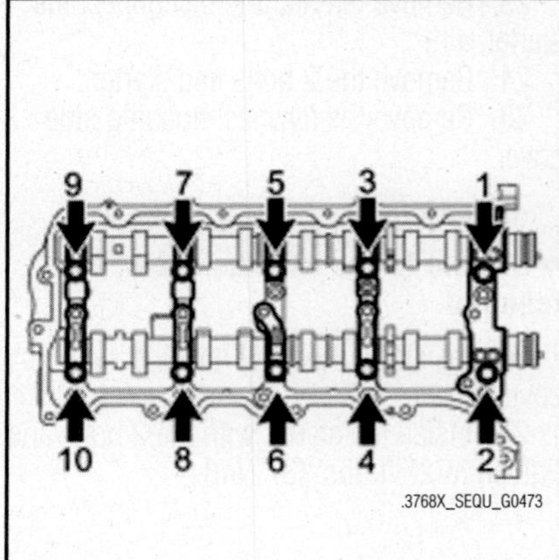

Fig. 34 Camshaft bearing bolt installation Bank 1—4.6L engine

➡ **Be sure that the knock pin is properly positioned. Remove retaining bolts, as shown in illustrations. When removing the bolts loosen them uniformly and keep the camshaft level.**

11. Remove the camshaft housing sub assembly.

To install:

➡**Be sure to use new fasteners, as required.**

12. Install the camshaft bearing cap.

➡**Apply a light coat of clean engine oil to the camshaft journals, housing and caps.**

13. Install the No. 3 and No. 4 camshafts to the housing.
14. Temporarily install the ten bolts as shown in the illustration.
15. To install the camshaft housing sub assembly, install the rocker arms.
16. Apply seal packing in a continuous bead, see illustration.

➡**Install the component within 15 minutes after applying the seal packing.**

17. Install the camshaft housing, and install the eighteen bolts as shown in the illustration. Tighten bolt A to 7 ft. lbs. (10 Nm). Tighten other bolts to 22 ft. lbs. (30 Nm). Be sure that the knock pin of the camshaft is properly positioned.

➡**Do not start the engine for at least two hours after installation.**

18. Tighten the ten bolts to 12 ft. lbs. (16 Nm), see illustration.
19. Install the No. 3 chain tensioner assembly.
20. Install the No. 1 chain sub assembly.
21. Install the No. 1 chain tensioner slipper.
22. Install the No. 1 chain tensioner assembly. Be sure to use a new gasket. Tighten the retaining bolts to 7 ft. lbs. (10 Nm).

➡**Move the stopper plate upward to release the lock and push the plunger deep into the tensioner. Move the stopper plate downward to set the lock and insert a hex wrench into the hole of the stopper plate.**

23. Install the No.1 chain vibration damper.
24. Tighten the camshaft timing gear. Tighten bolt to 74 ft. lbs. (100 Nm).
25. Check that the engine is at TDC, on the compression stroke.
26. Continue the installation in the reverse order of the removal procedure.
27. Use the Techstream diagnostic tool, or equivalent and reprogram the required systems.

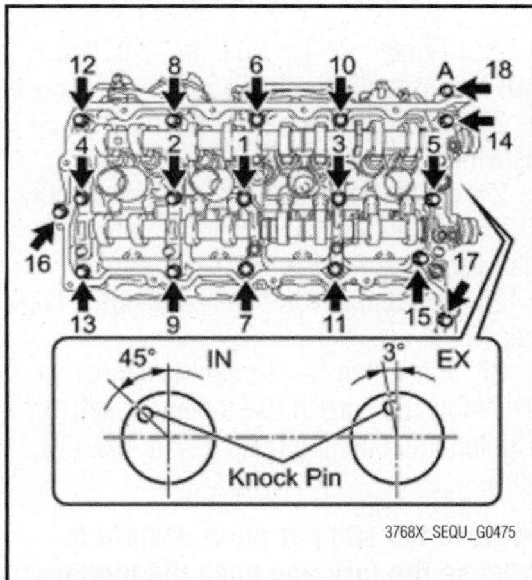

Fig. 35 Camshaft housing bolt (18) installation Bank 1—4.6L engine

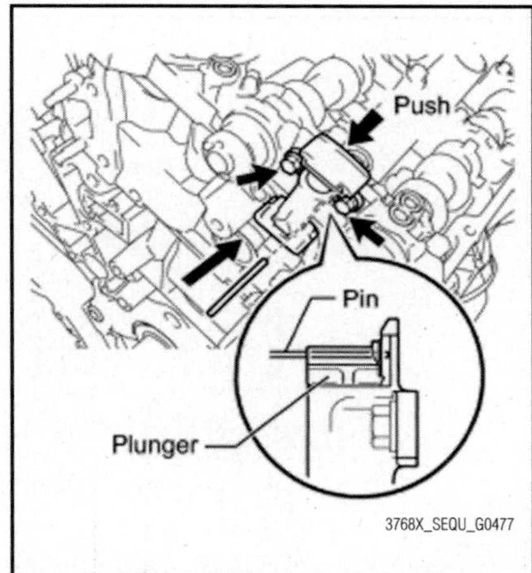

Fig. 37 Camshaft No. 3 chain tensioner installation Bank 1—4.6L engine

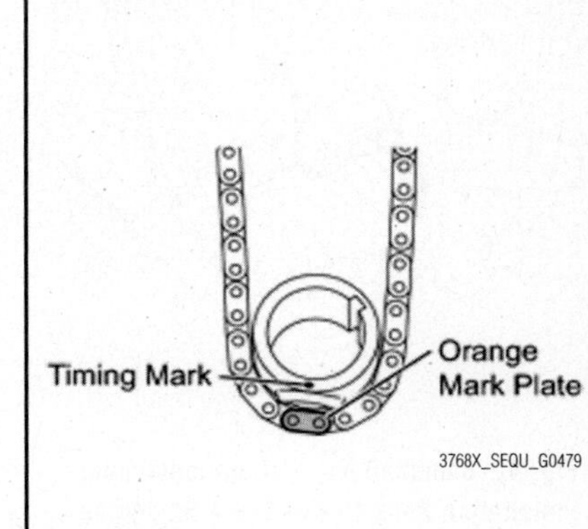

Fig. 39 Camshaft No. 1 chain sub assembly installation Bank 1 (2 of 3)—4.6L engine

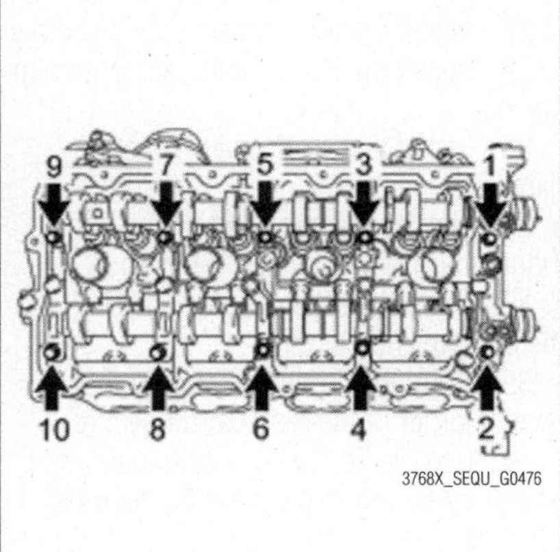

Fig. 36 Camshaft bolt (10) installation Bank 1—4.6L engine

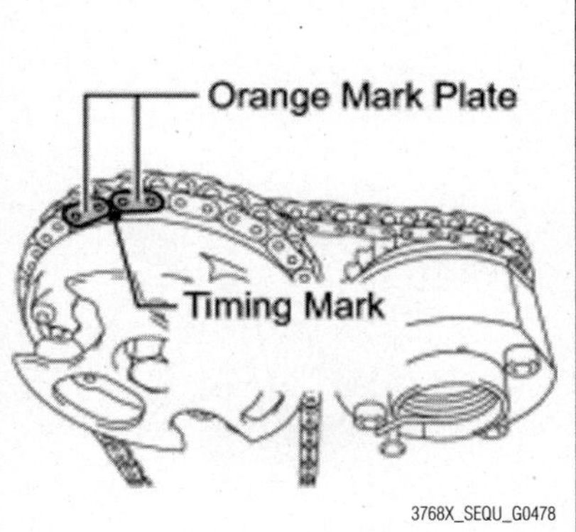

Fig. 38 Camshaft No. 1 chain sub assembly installation Bank 1 (1 of 3)—4.6L engine

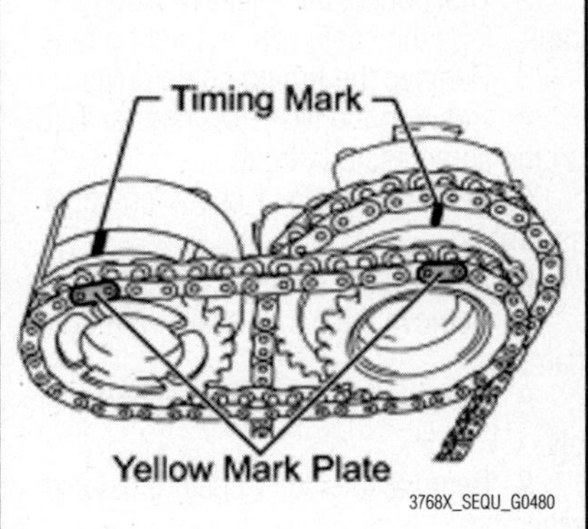

Fig. 40 Camshaft No. 1 chain sub assembly installation Bank 1 (3 of 3)—4.6L engine

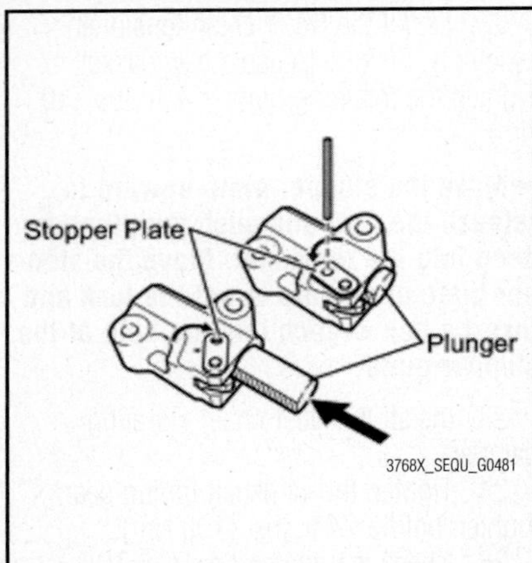

Fig. 41 Camshaft No. 1 chain tensioner installation Bank 1 (1 of 2)—4.6L engine

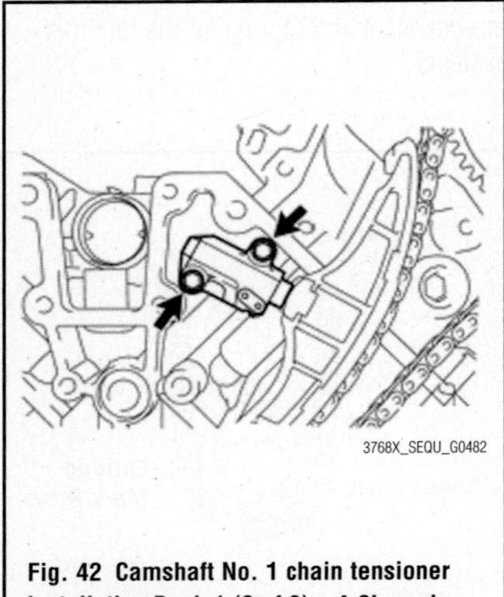

Fig. 42 Camshaft No. 1 chain tensioner installation Bank 1 (2 of 2)—4.6L engine

BANK 2

See Figures 43 through 53.

1. Before servicing the vehicle, refer to the Precautions Section.
2. Disconnect the negative battery cable. Tape the cable with insulating tape.
3. Remove the timing chain cover.
4. Position the No. 1 cylinder at TDC on the compression stroke.
5. Remove the No. 1 chain tensioner assembly.
6. Remove the chain tensioner slipper.
7. Remove the No. 1 chain vibration damper.
8. Remove the No. 1 chain sub assembly.
9. Remove the No. 1 chain tensioner assembly.

➡Move the stopper plate upward to release the lock and push the plunger deep into the tensioner. Move the stopper plate downward to set the lock and insert a hex wrench into the hole of the stopper plate.

10. Remove the No. 1 chain tensioner slipper. Remove the No.1 vibration damper.
11. Remove the No. 1 chain sub assembly.

➡While raising up on the No. 2 chain tensioner, insert a 0.0394 inch pin into the hole to retain it in place.

12. Remove the No.1 and No.2 chains from the gear.
13. Remove the No. 2 chain tensioner assembly.
14. Remove the camshaft bearing cap.
15. Remove the camshaft housing sub assembly.

To install:

➡Be sure to use new fasteners, as required.

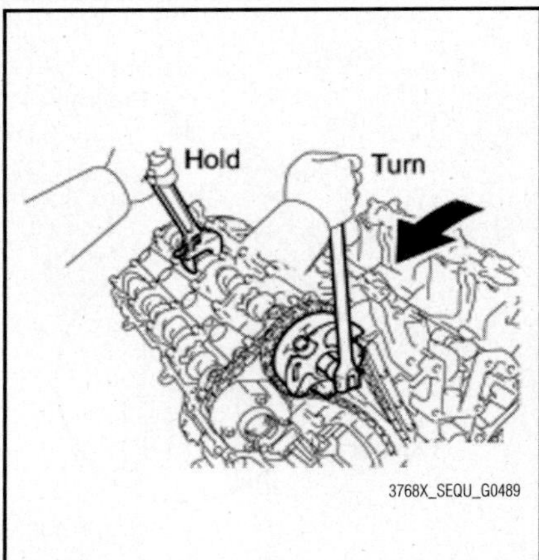

Fig. 43 Camshaft timing chain removal Bank 2 (1 of 2)—4.6L engine

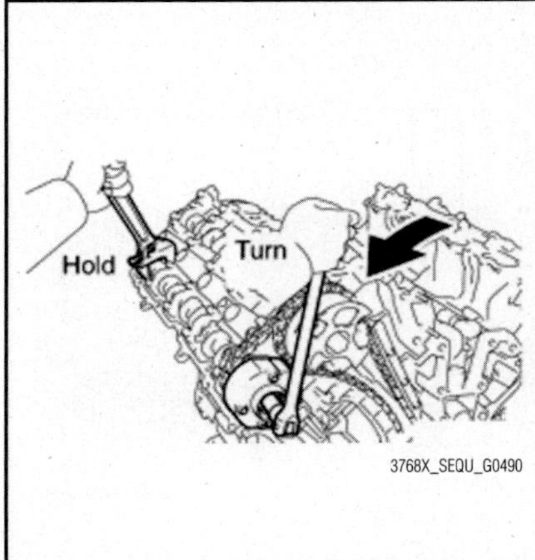

Fig. 44 Camshaft timing chain removal Bank 2 (2 of 2)—4.6L engine

16. Install the camshaft bearing cap.

➡Apply a light coat of clean engine oil to the camshaft journals, housing and caps.

17. Install the No. 1 and No. 2 camshafts to the housing.
18. Temporarily install the ten bolts as shown in the illustration.
19. To install the camshaft housing sub assembly, install the rocker arms.
20. Apply seal packing in a continuous bead, see illustration.

➡Install the component within 15 minutes after applying the seal packing.

21. Install the camshaft housing, and install the eighteen bolts as shown in the illustration. Tighten bolt A to 7 ft. lbs. (10 Nm). Tighten other bolts to 22 ft. lbs. (30 Nm). Be sure that the knock pin of the camshaft is properly positioned.

➡Do not start the engine for at least two hours after installation.

22. Tighten the ten bolts to 12 ft. lbs. (16 Nm), see illustration.
23. Install the No. 2 chain tensioner assembly.
24. Install the No. 1 chain sub assembly.
25. Install the No. 1 chain vibration damper.
26. Install the No. 1 chain tensioner slipper.
27. Install the No. 1 chain tensioner assembly. Be sure to use a new gasket. Tighten the retaining bolts to 7 ft. lbs. (10 Nm).

➡Move the stopper plate upward to release the lock and push the plunger deep into the tensioner. Move the stopper plate downward to set the lock and insert a hex wrench into the hole of the stopper plate.

28. Install the No. 1 chain sub assembly.
29. Install the No. 1 chain tensioner slipper.
30. Install the No. 1 chain vibration damper.
31. Tighten the camshaft timing gear. Tighten bolt to 74 ft. lbs. (100 Nm).
32. Check that the engine is at TDC, on the compression stroke.
33. Continue the installation in the reverse order of the removal procedure.
34. Use the Techstream diagnostic tool, or equivalent and reprogram the required systems.

5.7L Engine

See Figures 54 through 66.

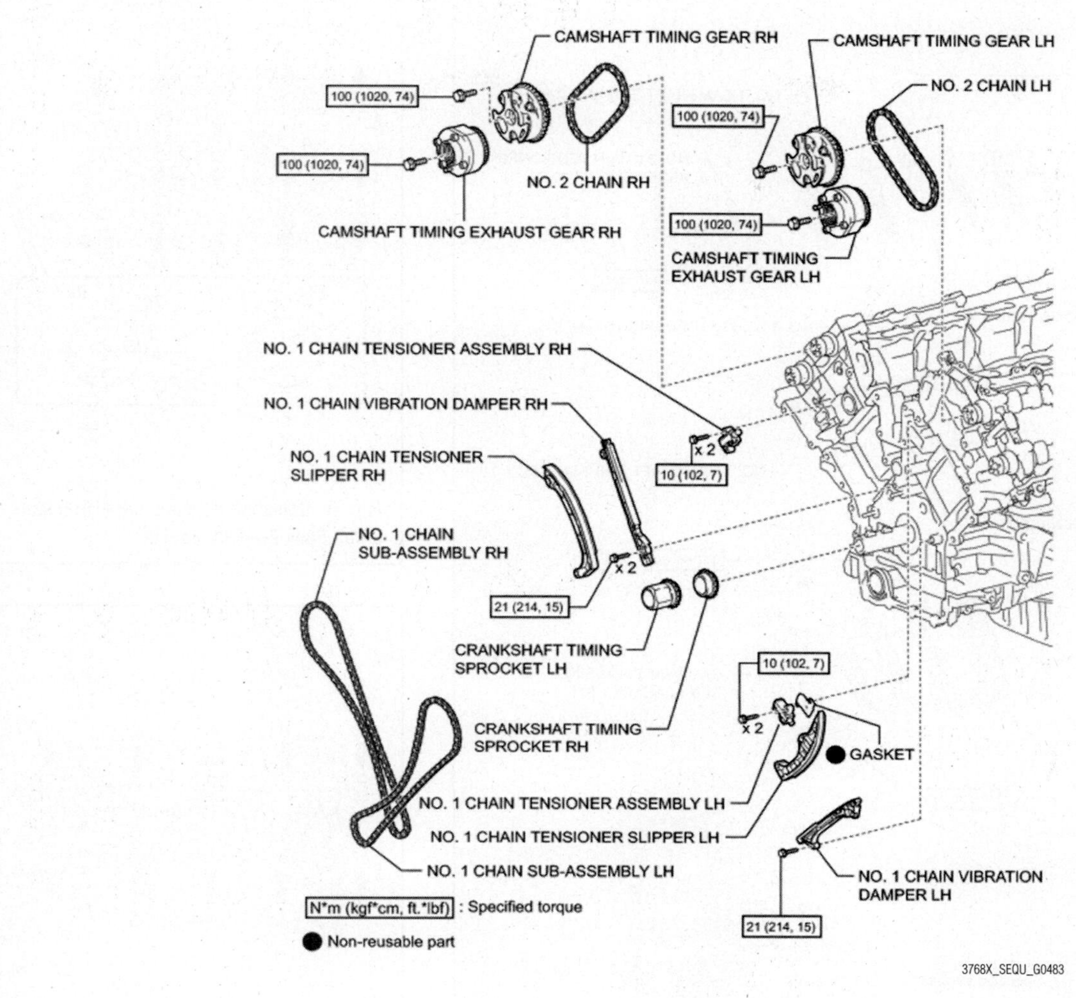

Fig. 45 Camshaft removal points Bank 2—4.6L engine

1. Before servicing the vehicle, refer to the Precautions Section.
2. Disconnect the negative battery cable. Tape the cable with insulating tape.
3. Discharge fuel system pressure.
4. Remove front exhaust pipe assembly.
5. Remove front fender apron seal.
6. Remove front fender apron seal rear.
7. Remove No. 2 steering intermediate shaft sub-assembly.
8. Remove exhaust manifold heat insulator.
9. Remove exhaust manifold sub-assembly.
10. Remove the timing chain cover.
11. Set No. 1 cylinder to TDC / compression.
12. Remove chain tensioner assembly.
13. Remove chain tensioner slipper.
14. Remove chain vibration damper.
15. Remove chain sub-assembly.
16. Remove chain tensioner assembly.
17. Make sure that the knock pin of the camshaft is positioned as shown.
18. Uniformly loosen and remove the 10 bearing cap bolts in the sequence LH.
19. Uniformly loosen and remove the 18 bearing cap bolts in the sequence LH.

➡Uniformly loosen the bolts while keeping the camshaft level.

20. Remove the 6 bearing caps.
21. Remove the No. 3 and No. 4 camshafts.
22. Remove the camshaft housing by prying between the cylinder head and camshaft housing with a screwdriver.

➡Be careful not to damage the contact surfaces of the cylinder head and camshaft housing.

23. Remove the 16 valve rocker arms from the cylinder head.
24. Remove the 16 valve lash adjusters from the cylinder head.
25. Remove the 16 valve stem caps from the cylinder head.

➡Arrange the removed parts in the correct order.

26. Make sure that the knock pin of the camshaft is positioned as shown in the illustration.
27. Uniformly loosen and remove the 10 bearing cap bolts in the sequence RH.
28. Uniformly loosen and remove the 18 bearing cap bolts in the sequence RH.

➡Uniformly loosen the bolts while keeping the camshaft level.

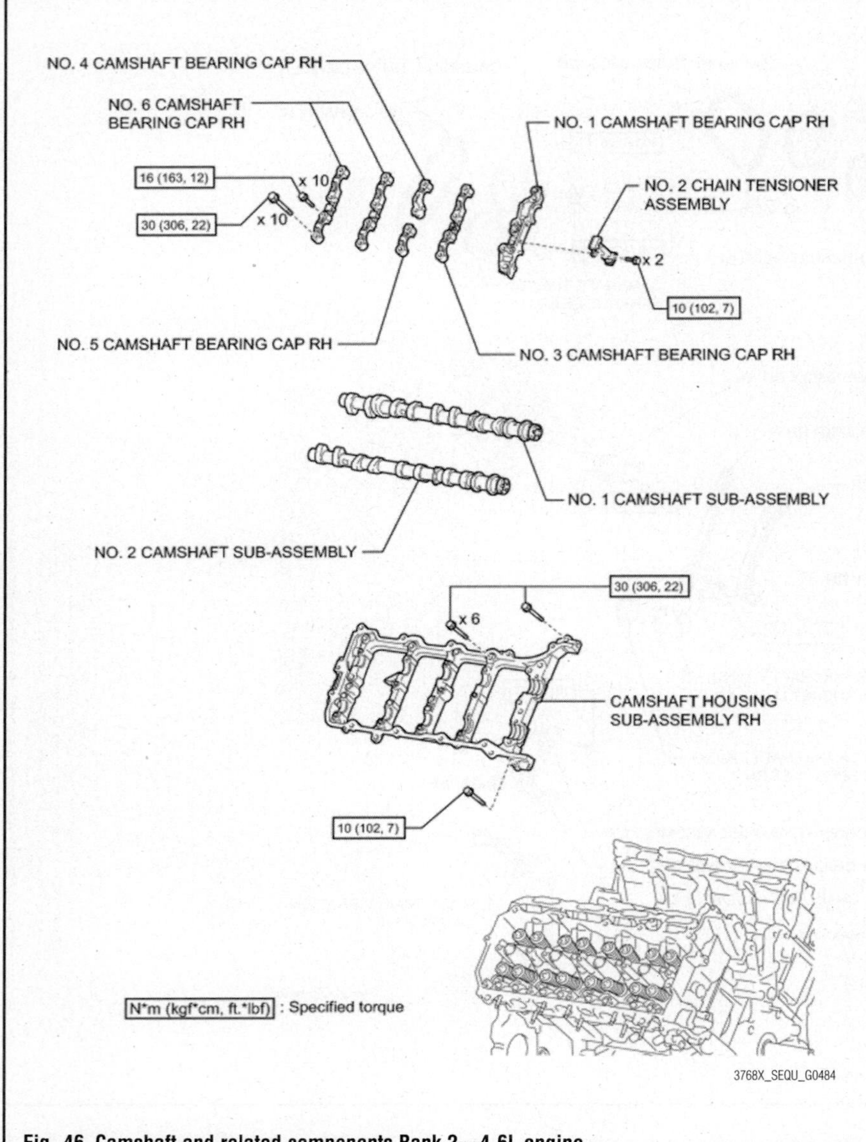

Fig. 46 Camshaft and related components Bank 2—4.6L engine

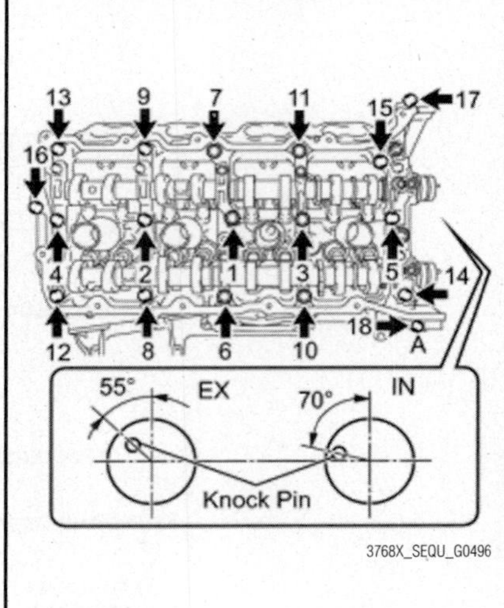

Fig. 49 Camshaft housing bolt (18) installation Bank 2—4.6L engine

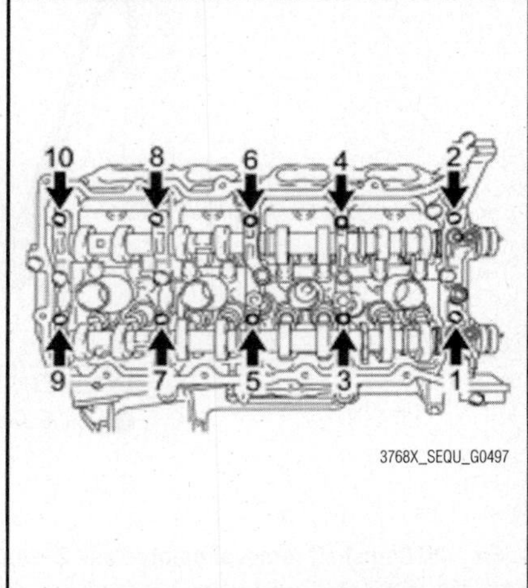

Fig. 50 Camshaft bolt (10) installation Bank 2—4.6L engine

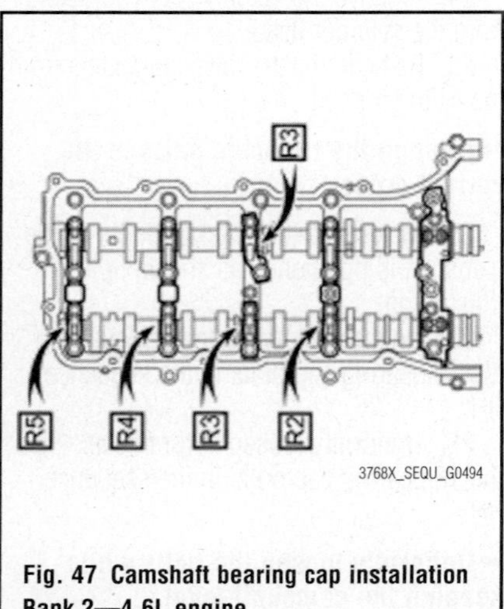

Fig. 47 Camshaft bearing cap installation Bank 2—4.6L engine

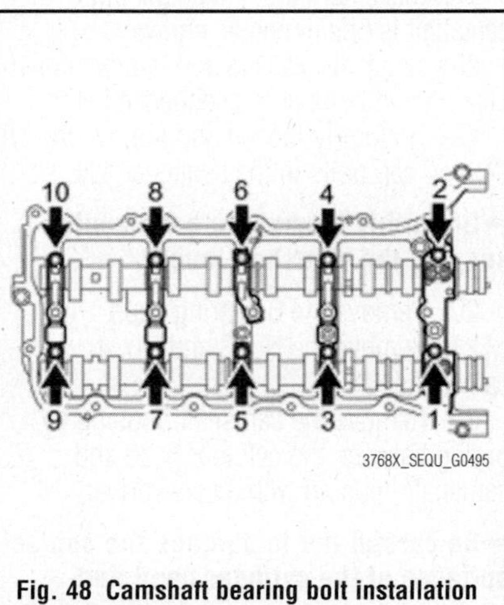

Fig. 48 Camshaft bearing bolt installation Bank 2—4.6L engine

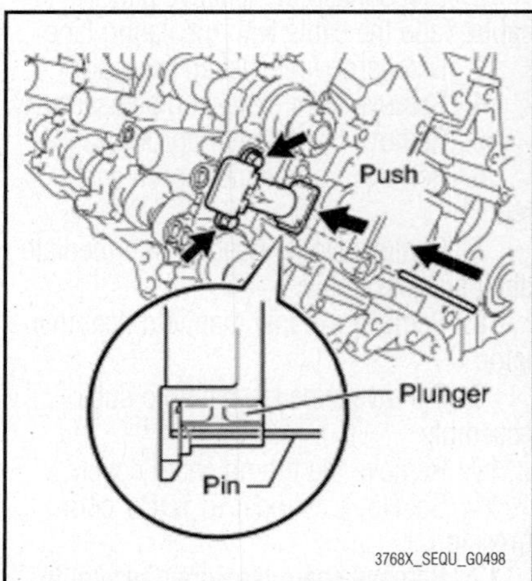

Fig. 51 Camshaft No. 1 chain tensioner installation Bank 2—4.6L engine

29. Remove the 6 bearing caps.
30. Remove the No. 1 and No. 2 camshafts.

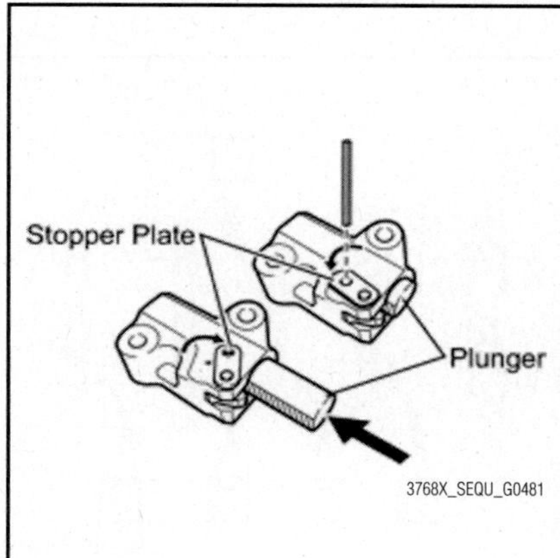

Fig. 52 Camshaft No. 1 chain tensioner installation Bank 2 (1 of 2)—4.6L engine

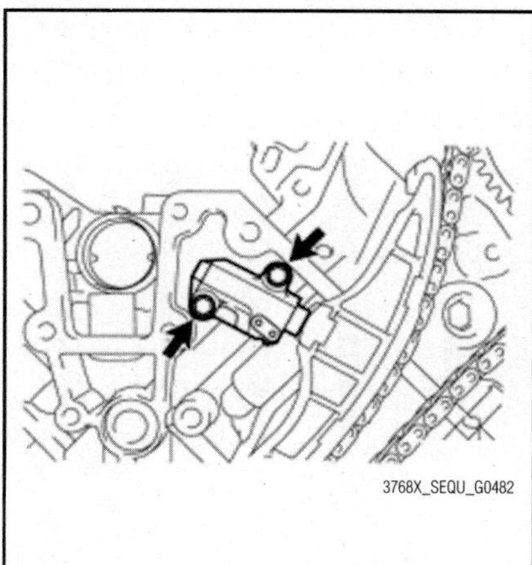

Fig. 53 Camshaft No. 1 chain tensioner installation Bank 2 (2 of 2)—4.6L engine

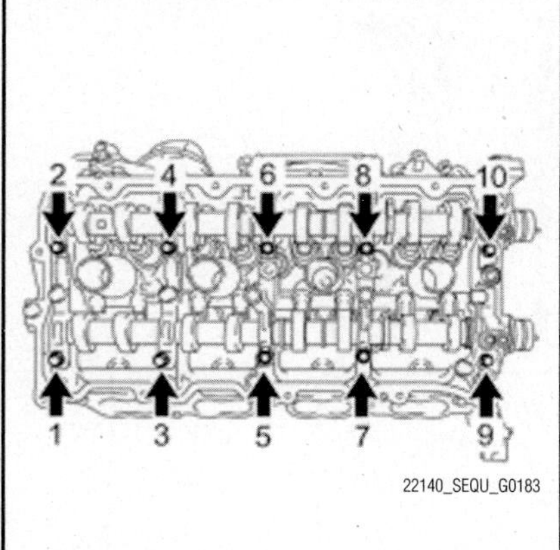

Fig. 54 Camshaft cap removal LH—5.7L engine

31. Remove the camshaft housing by prying between the cylinder head and camshaft housing with a screwdriver.

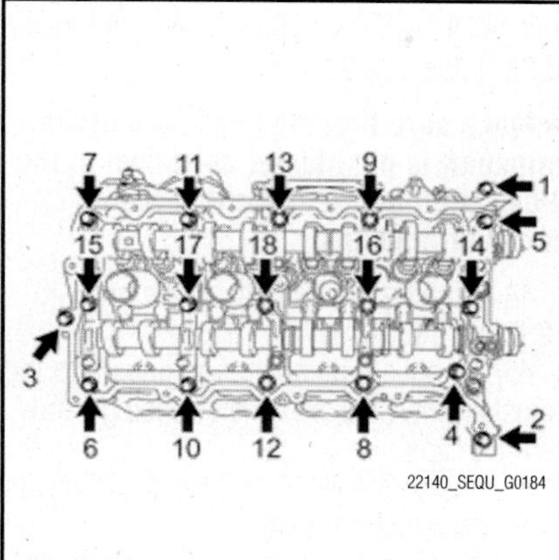

Fig. 55 Camshaft cap removal LH—5.7L engine

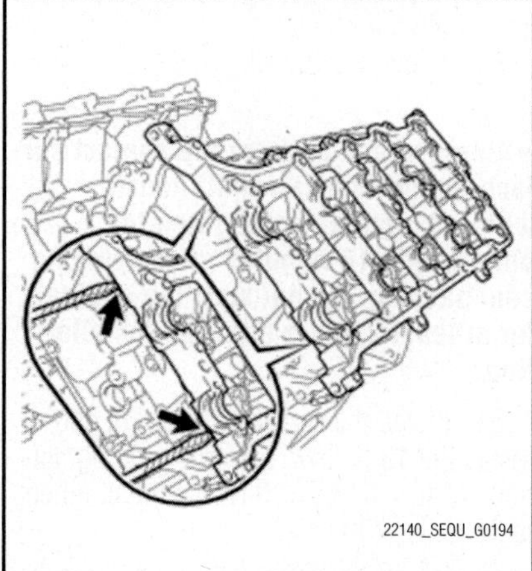

Fig. 56 Camshaft housing removal LH—5.7L engine

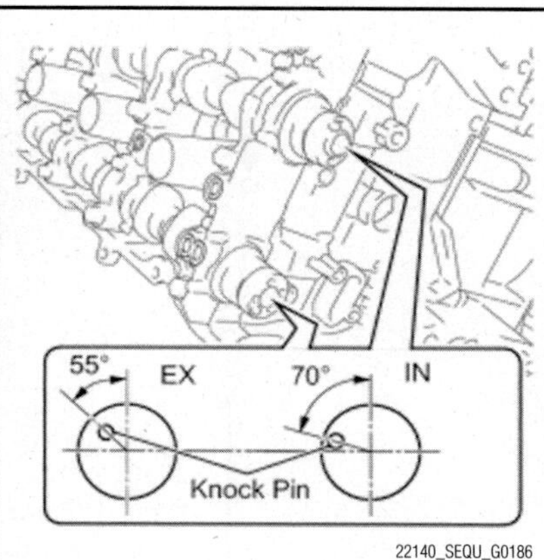

Fig. 57 Camshaft knock pin positioning RH—5.7L engine

➡Be careful not to damage the contact surfaces of the cylinder head and camshaft housing.

32. Remove the 16 valve rocker arms from the cylinder head.
33. Remove the 16 valve lash adjusters from the cylinder head.
34. Remove the 16 valve stem caps from the cylinder head.

➡Arrange the removed parts in the correct order.

To install:

➡Be sure to use new fasteners, as required.

35. Install valve stem cap.
   a. Apply a light coat of engine oil to the valve stem caps.
   b. Install the 32 valve stem caps to the cylinder head.

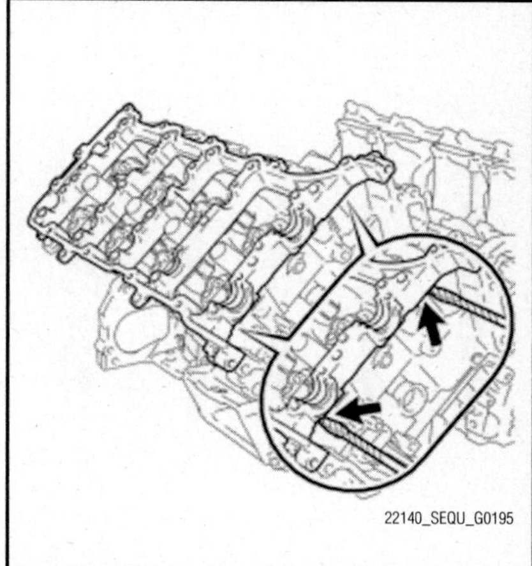

Fig. 58 Camshaft housing removal RH—5.7L engine

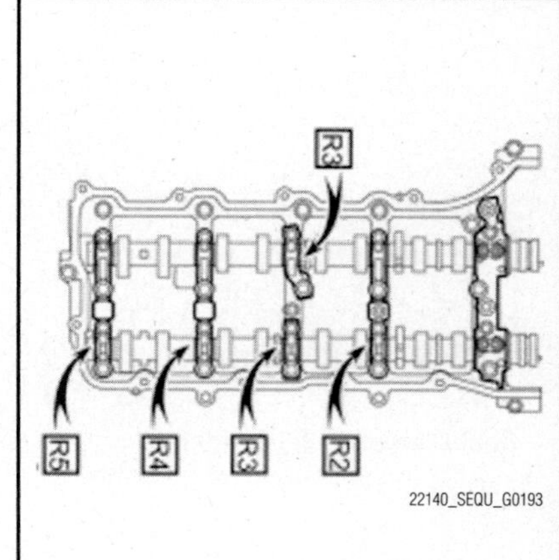

Fig. 59 Camshaft bearing cap position RH—5.7L engine

➡Install the lash adjuster at the same place it was removed from.

36. Install the 32 valve rocker arms.
37. Apply a light coat of engine oil to the camshaft journals, camshaft housings and bearing caps.
38. Install the No. 1 and No. 2 camshafts to the camshaft housing.
39. Confirm the marks and numbers on the camshaft bearing caps and place them in their proper positions and directions.
40. Temporarily install the 10 bolts in the order shown in the illustration.
41. Make sure that the valve rocker arms are installed properly.
42. Apply seal packing in a continuous line.

➡Remove any oil from the contact surface. Install the camshaft housing within 3 minutes and tighten the bolts within 15 minutes after applying seal packing. Do not start the engine for at least 2 hours after the installation.

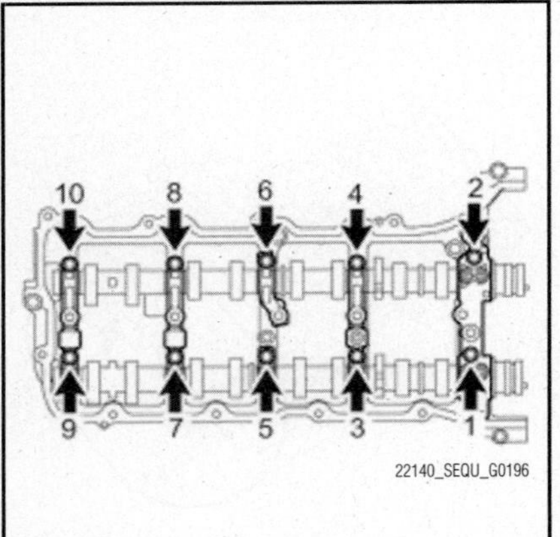

Fig. 60 Camshaft bearing cap tightening sequence RH—5.7L engine

43. Install the camshaft housing, and install the 18 bolts in the order and tighten bolt "A" to 7 ft. lbs. (10 Nm), and all others to 22 ft. lbs. (30 Nm).

➡Make sure that the knock pin of the camshaft is positioned as shown in the illustration before installing the camshaft housing.

44. Tighten the 10 bolts in the order to 12 ft. lbs. (16 Nm).
45. Apply a light coat of engine oil to the camshaft journals, camshaft housings and bearing caps.
46. Install the No. 1 and No. 2 camshafts to the camshaft housing.
47. Confirm the marks and numbers on the camshaft bearing caps and place them in their proper positions and directions.
48. Temporarily install the 10 bolts in the order shown in the illustration.
49. Make sure that the valve rocker arms are installed properly.
50. Apply seal packing in a continuous line.

➡Remove any oil from the contact surface. Install the camshaft housing within 3 minutes and tighten the bolts within 15 minutes after applying seal packing. Do not start the engine for at least 2 hours after the installation.

51. Install the camshaft housing, and install the 18 bolts in the order and tighten bolt "A" to 7 ft. lbs. (10 Nm), and all others to 22 ft. lbs. (30 Nm).

➡Make sure that the knock pin of the camshaft is positioned as shown in the illustration before installing the camshaft housing.

52. Tighten the 10 bolts in the order to 12 ft. lbs. (16 Nm).

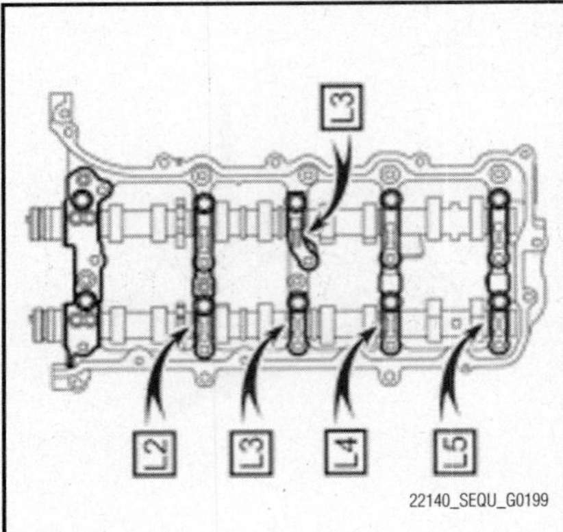

Fig. 63 Camshaft bearing cap position LH—5.7L engine

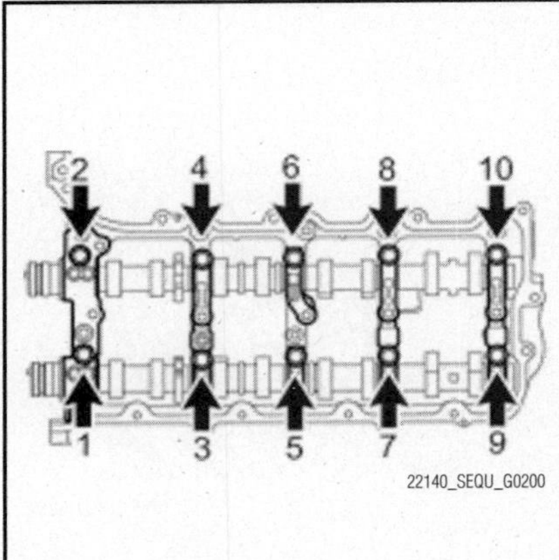

Fig. 64 Camshaft bearing cap tightening sequence LH—5.7L engine

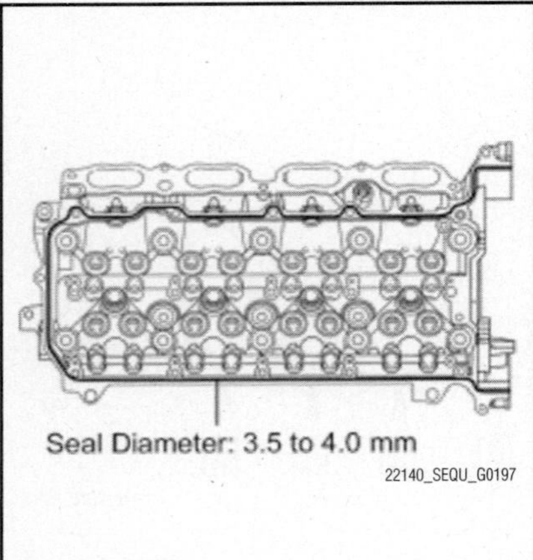

Fig. 61 Camshaft housing sealant position RH—5.7L engine

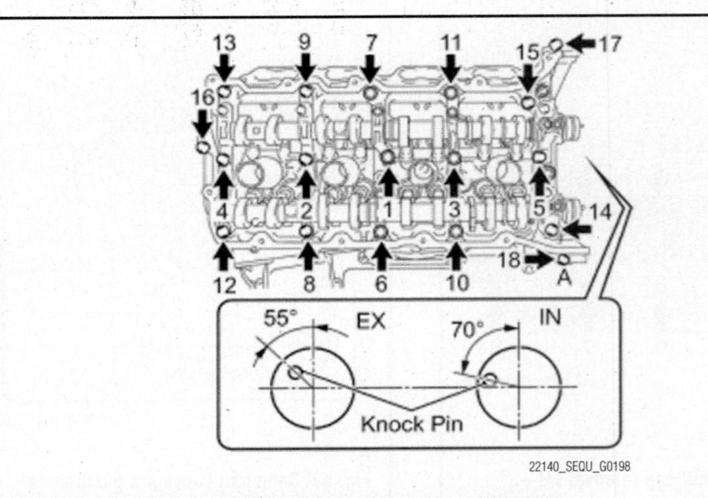

Fig. 62 Camshaft housing tightening sequence RH—5.7L engine

53. Install chain tensioner assembly.
54. Install chain sub-assembly.
55. Install chain tensioner slipper.
56. Install chain tensioner assembly.
57. Install chain vibration damper.
58. Tighten camshaft timing gear.
59. Check No. 1 cylinder to TDC.
60. Install the timing chain cover.
61. Install exhaust manifold sub-assembly.
62. Install exhaust manifold heat insulator.
63. Install No.2 steering intermediate shaft sub-assembly.
64. Install front fender apron seal rear.
65. Install front fender apron seal.
66. Install front exhaust pipe assembly.
67. Connect cable to negative battery terminal.
68. Use the Techstream diagnostic tool, or equivalent and reprogram the required systems.

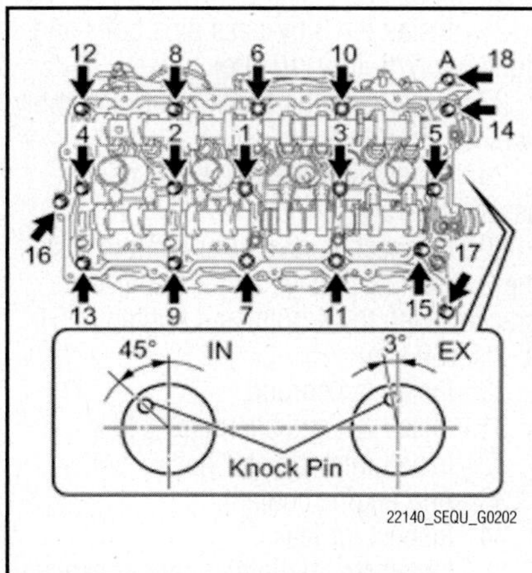

Fig. 65 Camshaft housing tightening sequence LH—5.7L engine

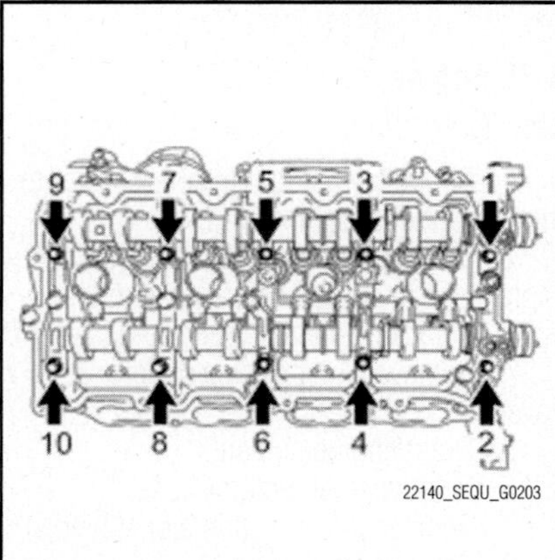

Fig. 66 Camshaft bearing cap tightening sequence LH—5.7L engine

CATALYTIC CONVERTER

REMOVAL & INSTALLATION

See Figures 67 and 68.

At this time the manufacturer does not provide removal and installation procedures for this component. The following procedure is a guideline and may differ from the vehicle you are servicing.

1. Before servicing the vehicle, refer to the Precautions Section.
2. Disconnect the negative battery cable. Tape the cable with insulating tape.
3. Raise and safely support the vehicle.
4. Remove the necessary components in order to gain access to the converter.
5. Disconnect the oxygen sensors, as necessary. Do not drop the sensor, if dropped it must be replaced.
6. Remove the converter mounting bolts.

➡**Be sure to properly support the exhaust system.**

7. Remove the converter from the vehicle.
8. Discard the gaskets.

To install:

➡**Be sure to use new fasteners, as required.**

9. Installation is the reverse of the removal procedure.
10. Be sure to use new gaskets.
11. Be sure to coat the sensor with the proper sealer, if reusing the existing one.

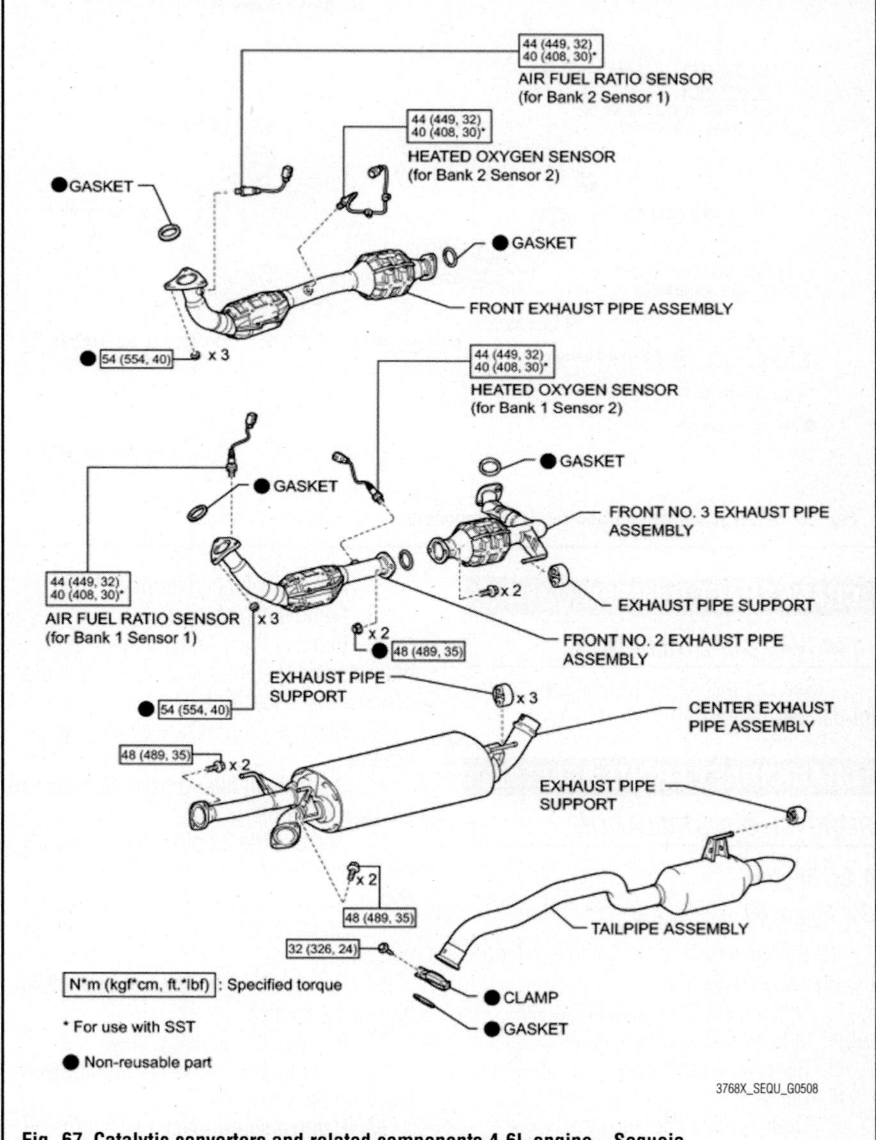

Fig. 67 Catalytic converters and related components 4.6L engine—Sequoia

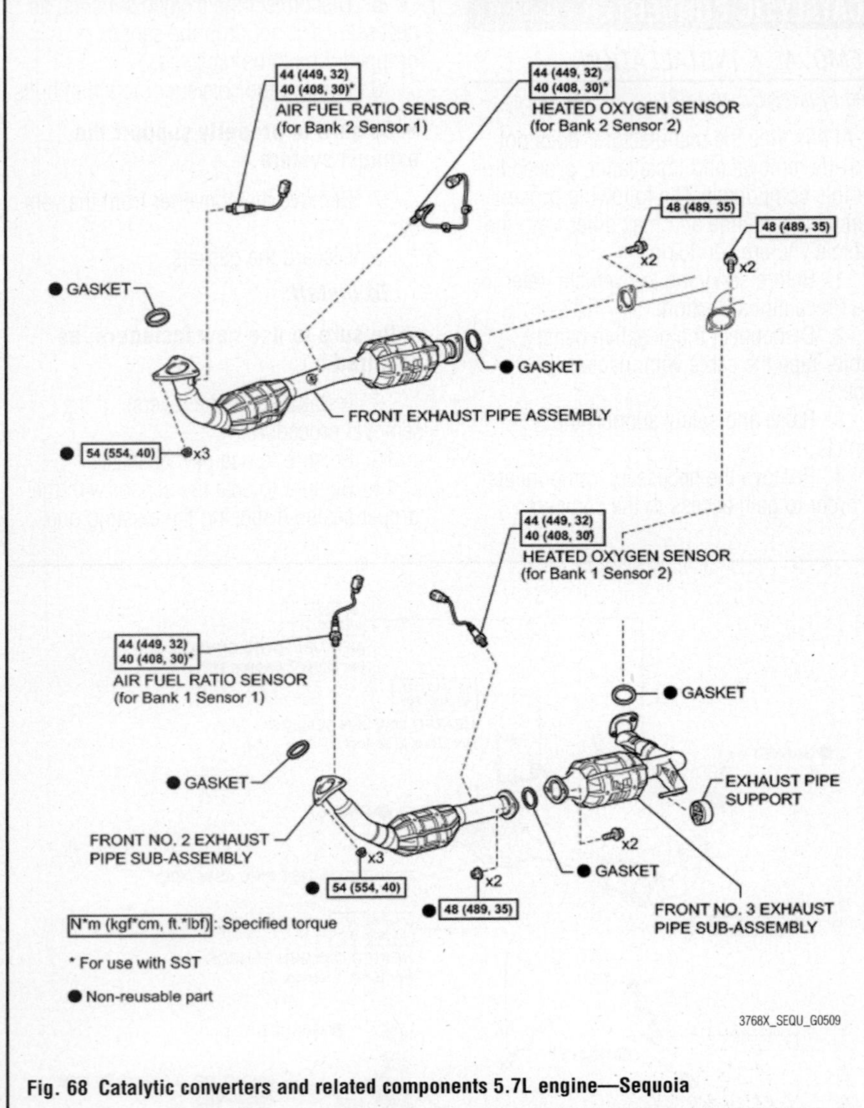

Fig. 68 Catalytic converters and related components 5.7L engine—Sequoia

CRANKSHAFT DAMPER

REMOVAL & INSTALLATION

1. See Crankshaft Front Seal procedure for damper removal.

CRANKSHAFT FRONT SEAL

REMOVAL & INSTALLATION

4.6L Engine

See Figure 69.

1. Before servicing the vehicle, refer to the Precautions Section.
2. Disconnect the negative battery cable. Tape the cable with insulating tape.
3. Remove V-bank cover sub-assembly.
4. Remove No. 1 engine under cover.
5. Drain engine coolant.
6. Remove inlet radiator hose.
7. Remove fan and alternator V belt.
8. Remove fan shroud.
9. Remove the 4 bolts, 2 stabilizer brackets and 2 stabilizer bushes. Then disconnect the stabilizer bar.
10. Remove oil pressure sender gauge assembly.
11. Remove oil filter bracket sub-assembly (w/o oil cooler).
12. Remove the 3 bolts and disconnect the 2 water by-pass hoses from the oil cooler.
13. Remove No. 1 oil cooler bracket (w/ oil cooler).
14. Remove oil filter bracket sub-assembly (w/ oil cooler).
15. Remove crankshaft pulley.
16. Remove the crankshaft timing gear key from the crankshaft.
17. Using a screwdriver, pry out the oil seal.

➡Do not damage the surface of the oil seal press fit hole and crankshaft.

To install:

➡Be sure to use new fasteners, as required.

18. Installation is the reverse of the removal procedure.
19. Apply MP grease to the lip of a new oil seal.
20. Using Service Tool and a hammer, tap in the oil seal to a depth between 0 to 1.0 mm (0 to 0.0394 in.) from the timing chain cover edge.
21. Install the crankshaft timing gear key.
22. Install crankshaft pulley and tighten bolt to 221 ft. lbs. (300 Nm).
23. Install oil filter bracket sub-assembly (w/oil cooler).
24. Install no. 1 oil cooler bracket (w/oil cooler).
25. Connect the 2 water by-pass hoses to the oil cooler.
26. Install the 3 by-pass pipe bolts and tighten to 7 ft. lbs. (10 Nm).
27. Install oil filter bracket sub-assembly (w/o oil cooler).
28. Install oil pressure sender gauge assembly.
29. Connect the stabilizer bar and install the 2 stabilizer bushes and 2 stabilizer brackets with the 4 bolts and tighten to 51 ft. lbs. (69 Nm).
30. Install fan shroud.
31. Install fan and alternator v belt.
32. Install inlet radiator hose.
33. Add engine coolant.
34. Inspect for leaks.
35. Install No. 1 engine under cover.
36. Install V-bank cover sub-assembly.
37. Use the Techstream diagnostic tool, or equivalent and reprogram the required systems.

5.7L Engine

See Figure 70.

1. Before servicing the vehicle, refer to the Precautions Section.
2. Disconnect the negative battery cable. Tape the cable with insulating tape.
3. Remove V-bank cover sub-assembly.
4. Remove No. 1 engine under cover.
5. Drain engine coolant.
6. Remove inlet radiator hose.
7. Remove fan and alternator V belt.
8. Remove fan shroud.
9. Remove the 4 bolts, 2 stabilizer brackets and 2 stabilizer bushes. Then disconnect the stabilizer bar.

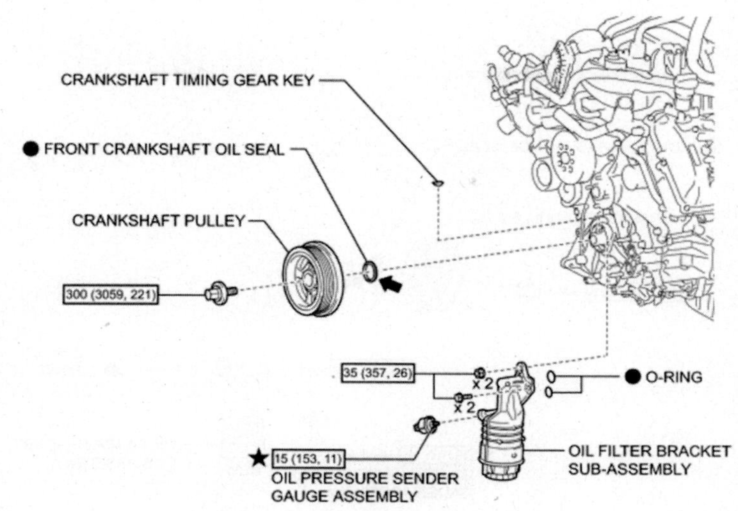

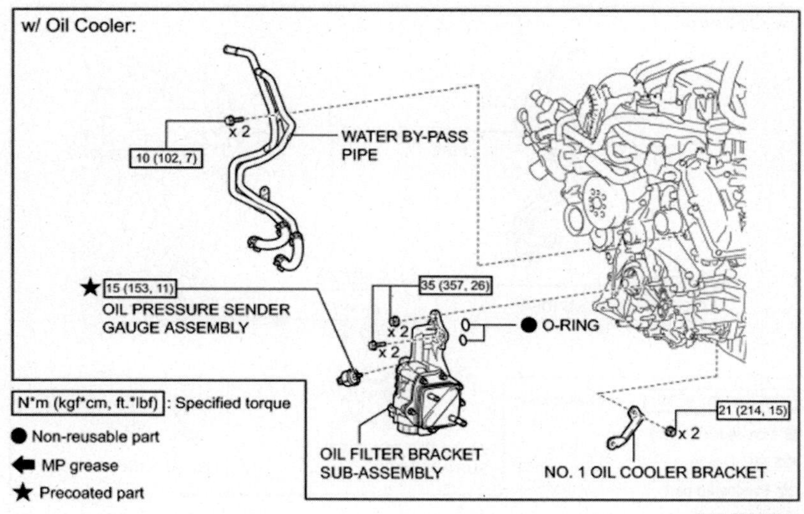

Fig. 69 Front oil seal and related components—4.6L engine

10. Remove oil pressure sender gauge assembly.
11. Remove oil filter bracket sub-assembly (w/o oil cooler).
12. Remove the 3 bolts and disconnect the 2 water by-pass hoses from the oil cooler.
13. Remove No. 1 oil cooler bracket (w/ oil cooler).
14. Remove oil filter bracket sub-assembly (w/ oil cooler).
15. Remove crankshaft pulley.
16. Remove the crankshaft timing gear key from the crankshaft.
17. Using a screwdriver, pry out the oil seal.

➡Do not damage the surface of the oil seal press fit hole and crankshaft.

To install:

➡Be sure to use new fasteners, as required.

18. Apply MP grease to the lip of a new oil seal.
19. Using Service Tool and a hammer, tap in the oil seal to a depth between 0 to 1.0 mm (0 to 0.0394 in.) from the timing chain cover edge.
20. Install the crankshaft timing gear key.
21. Install crankshaft pulley and tighten bolt to 221 ft. lbs. (300 Nm).
22. Install oil filter bracket sub-assembly (w/oil cooler).
23. Install no. 1 oil cooler bracket (w/oil cooler).
24. Connect the 2 water by-pass hoses to the oil cooler.
25. Install the 3 by-pass pipe bolts and tighten to 7 ft. lbs. (10 Nm).
26. Install oil filter bracket sub-assembly (w/o oil cooler).
27. Install oil pressure sender gauge assembly.
28. Connect the stabilizer bar and install the 2 stabilizer bushes and 2 stabilizer brackets with the 4 bolts and tighten to 51 ft. lbs. (69 Nm).
29. Install fan shroud.
30. Install fan and alternator v belt.
31. Install inlet radiator hose.
32. Add engine coolant.
33. Inspect for leaks.

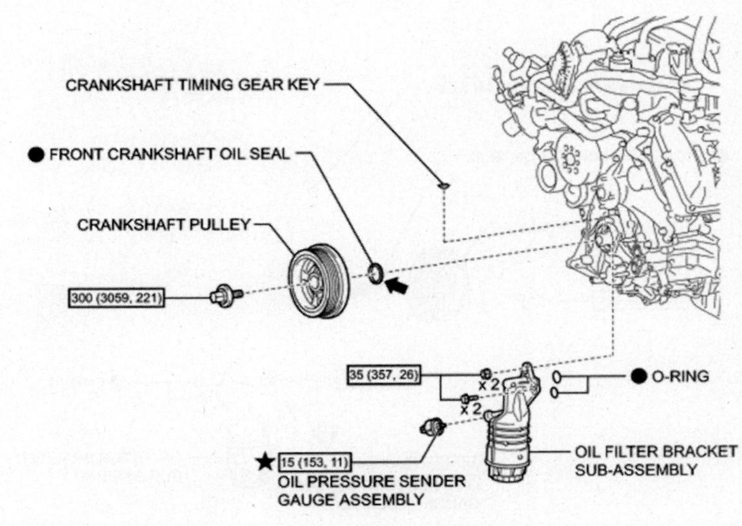

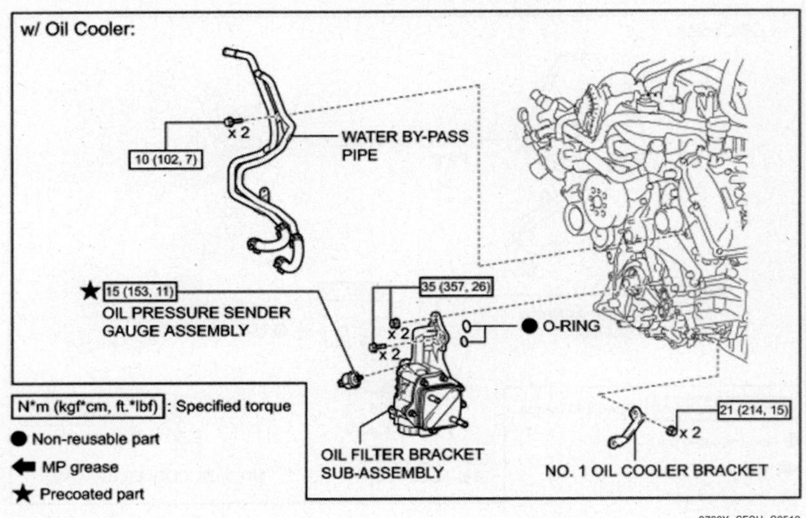

Fig. 70 Front oil seal and related components—5.7L engine

34. Install No. 1 engine under cover.
35. Install V-bank cover sub-assembly.
36. Use the Techstream diagnostic tool, or equivalent and reprogram the required systems.

CYLINDER HEAD

REMOVAL & INSTALLATION

4.6L Engine

BANK 1

See Figures 71 through 74.

1. Before servicing the vehicle, refer to the Precautions Section.
2. Disconnect the negative battery cable. Tape the cable with insulating tape.
3. Remove the exhaust manifold sub assembly.
4. Remove the camshafts.
5. Remove the No. 1 valve rocker arm sub assembly.
6. Remove the 16 valve lash adjusters.
7. Remove the 16 valve stem caps from the cylinder head.
8. Remove bolts 1 and 2 from the head, see illustration. Remove the remaining bolts in the sequence shown in the illustration. Be sure to arrange the removed parts for installation separately.
9. Remove the cylinder head from its mounting.
10. Discard the gasket.

To install:

➡Be sure to use new fasteners, as required.

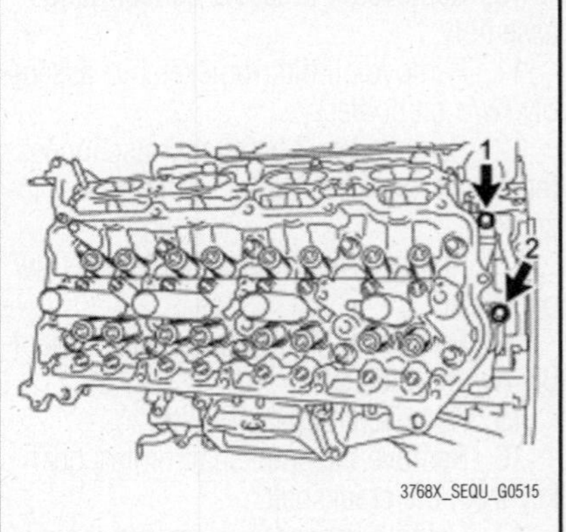

Fig. 71 Cylinder head bolt sequence removal Bank 1 (1 of 2)—4.6L engine

11. Installation is the reverse of the removal procedure.

12. Check for proper gasket positioning. Place the gasket on the head with the lot No. stamp upward. Be sure there is no oil on the gasket.

13. Tighten the bolts to specification and in the proper sequence. Lightly coat the bolt threads with clean engine oil.

14. Once the head bolts are installed the painted mark (on the bolt) should be facing rearward.

15. Install and tighten the remaining two bolts.

16. Use the Techstream diagnostic tool, or equivalent and reprogram the required systems.

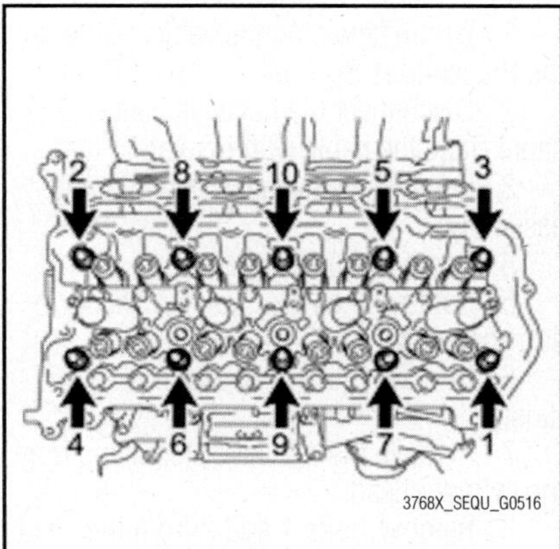

Fig. 72 Cylinder head bolt sequence removal Bank 1 (2 of 2)—4.6L engine

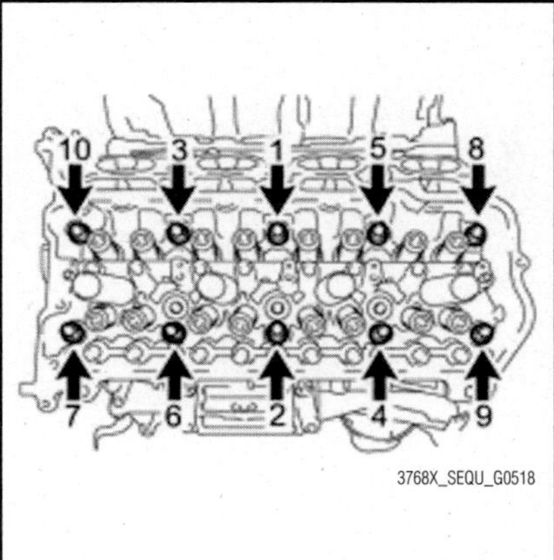

Fig. 73 Cylinder head gasket tightening sequence Bank 1—4.6L engine

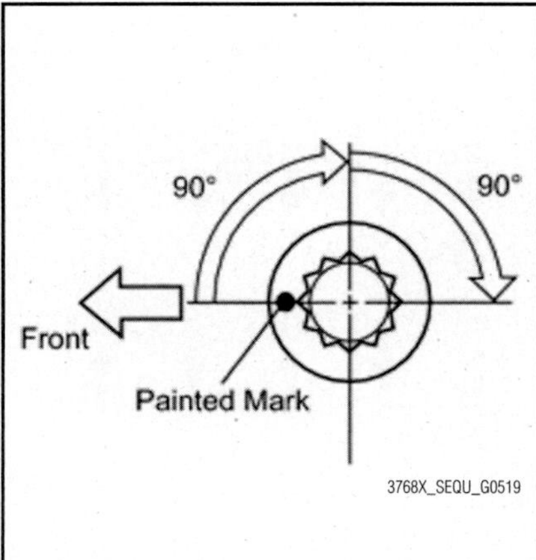

Fig. 74 Cylinder head bolt proper installation Bank 1 and Bank 2—4.6L engine

BANK 2

See Figures 75 through 77.

1. Before servicing the vehicle, refer to the Precautions Section.

2. Disconnect the negative battery cable. Tape the cable with insulating tape.

3. Remove the exhaust manifold sub assembly.

4. Remove the camshafts.

5. Remove the No. 1 valve rocker arm sub assembly.

6. Remove the 16 valve lash adjusters.

7. Remove the 16 valve stem caps from the cylinder head.

8. Remove bolts 1 and 2 from the head, see illustration. Remove the remaining bolts in the sequence shown in the illustration. Be sure to arrange the removed parts for installation separately.

9. Remove the cylinder head from its mounting.

10. Discard the gasket.

To install:

➡Be sure to use new fasteners, as required.

11. Installation is the reverse of the removal procedure.

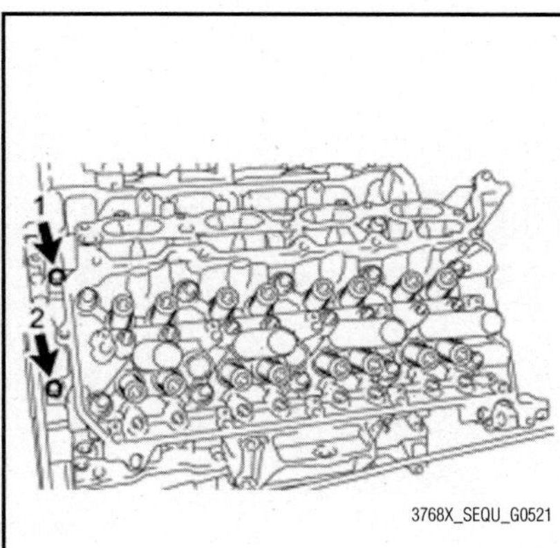

Fig. 75 Cylinder head bolt sequence removal Bank 2 (1 of 2)—4.6L engine

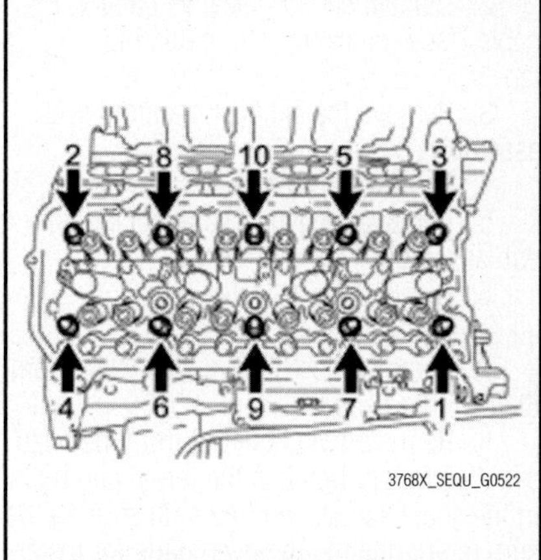

Fig. 76 Cylinder head bolt sequence removal Bank 2 (2 of 2)—4.6L engine

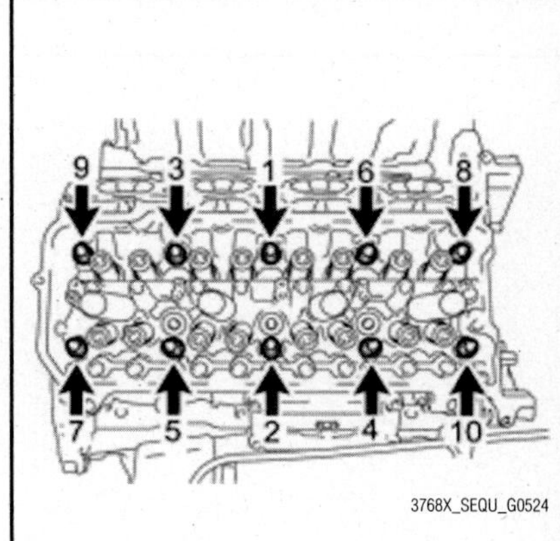

Fig. 77 Cylinder head gasket tightening sequence Bank 2—4.6L engine

12. Check for proper gasket positioning. Place the gasket on the head with the lot No. stamp upward. Be sure there is no oil on the gasket.

13. Tighten the bolts to specification and in the proper sequence. Lightly coat the bolt threads with clean engine oil.

14. Once the head bolts are installed the painted mark (on the bolt) should be facing rearward.

15. Install and tighten the remaining two bolts.

16. Use the Techstream diagnostic tool, or equivalent and reprogram the required systems.

5.7L Engine

BANK 1

See Figures 78 through 81.

1. Before servicing the vehicle, refer to the Precautions Section.

2. Disconnect the negative battery cable. Tape the cable with insulating tape.
3. Remove the exhaust manifold sub assembly.
4. Remove the camshafts.
5. Remove the No. 1 valve rocker arm sub assembly.
6. Remove the 16 valve lash adjusters.
7. Remove the 16 valve stem caps from the cylinder head.
8. Remove bolts 1 and 2 from the head, see illustration. Remove the remaining bolts in the sequence shown in the illustration. Be sure to arrange the removed parts for installation separately.
9. Remove the cylinder head from its mounting.
10. Discard the gasket.

To install:

➡Be sure to use new fasteners, as required.

11. Installation is the reverse of the removal procedure.
12. Check for proper gasket positioning. Place the gasket on the head with the lot No. stamp upward. Be sure there is no oil on the gasket.
13. Tighten the bolts to specification and in the proper sequence. Lightly coat the bolt threads with clean engine oil.
14. Once the head bolts are installed the painted mark (on the bolt) should be facing rearward.
15. Install and tighten the remaining two bolts.
16. Use the Techstream diagnostic tool, or equivalent and reprogram the required systems.

BANK 2

See Figures 82 through 85.

1. Before servicing the vehicle, refer to the Precautions Section.
2. Disconnect the negative battery cable. Tape the cable with insulating tape.
3. Remove the exhaust manifold sub assembly.
4. Remove the camshafts.
5. Remove the No. 1 valve rocker arm sub assembly.
6. Remove the 16 valve lash a djusters.
7. Remove the 16 valve stem caps from the cylinder head.
8. Remove bolts 1 and 2 from the head, see illustration. Remove the remaining bolts

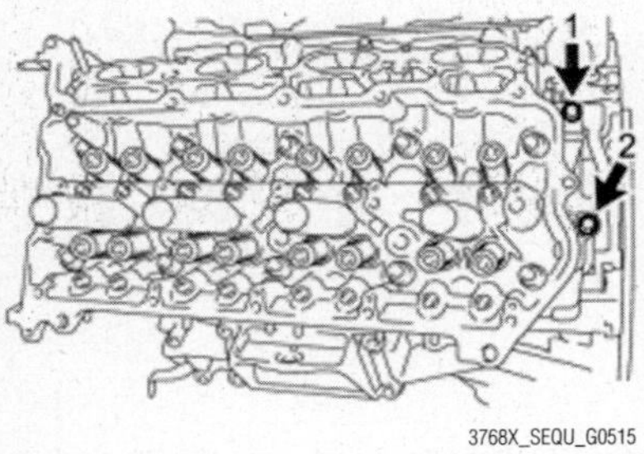

Fig. 78 Cylinder head bolt sequence removal Bank 1 (1 of 2)—5.7L engine

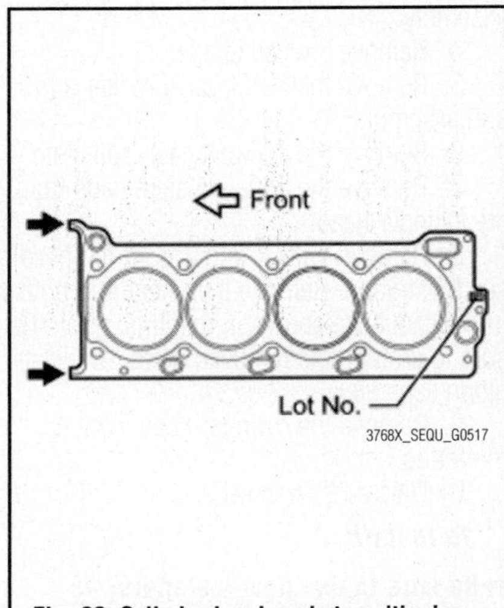

Fig. 80 Cylinder head gasket positioning Bank 1—5.7L engine

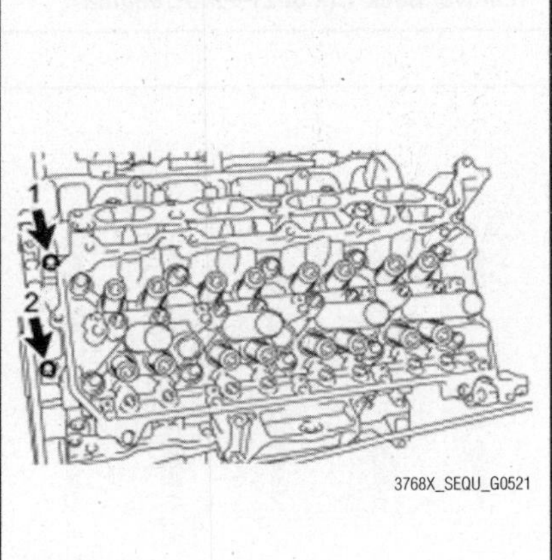

Fig. 82 Cylinder head bolt sequence removal Bank 2 (1 of 2)—5.7L engine

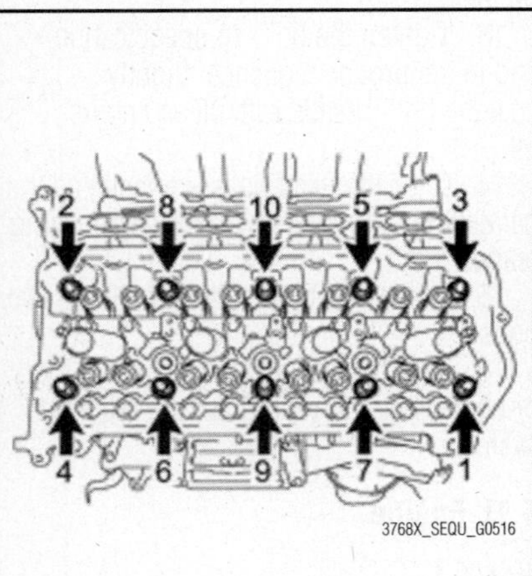

Fig. 79 Cylinder head bolt sequence removal Bank 1 (2 of 2)—5.7L engine

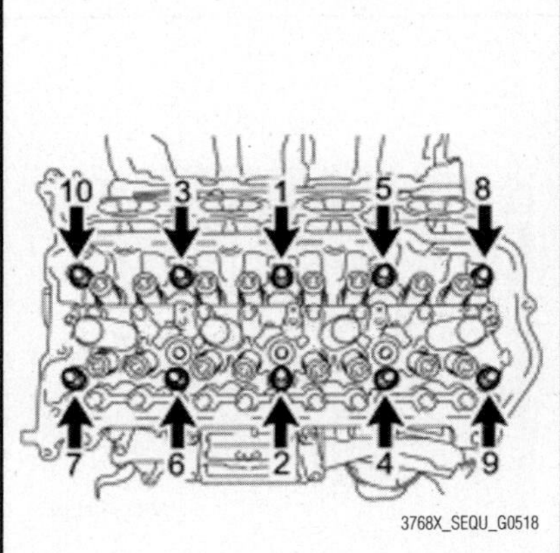

Fig. 81 Cylinder head gasket tightening sequence Bank 1—5.7L engine

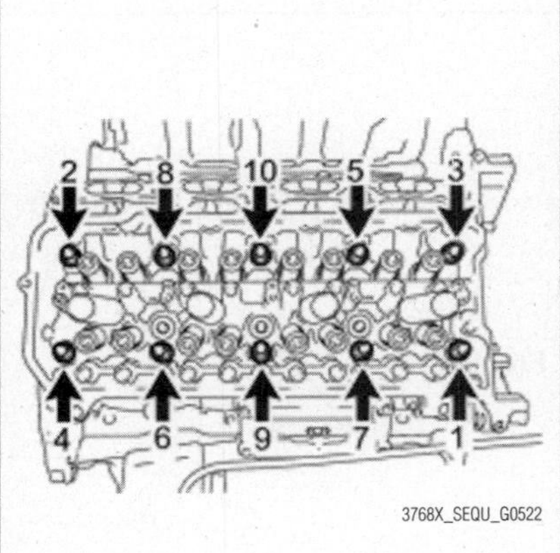

Fig. 83 Cylinder head bolt sequence removal Bank 2 (2 of 2)—5.7L engine

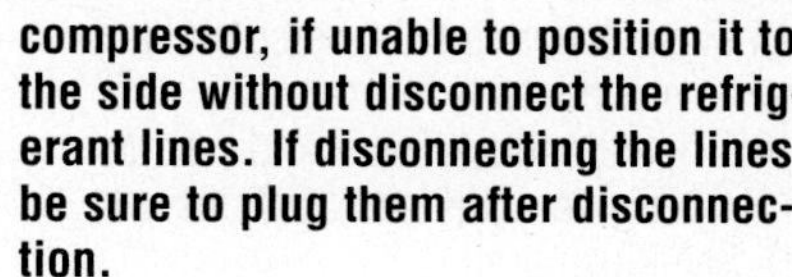

in the sequence shown in the illustration. Be sure to arrange the removed parts for installation separately.

9. Remove the cylinder head from its mounting.

10. Discard the gasket.

To install:

➡Be sure to use new fasteners, as required.

11. Installation is the reverse of the removal procedure.

12. Check for proper gasket positioning. Place the gasket on the head with the lot No. stamp upward. Be sure there is no oil on the gasket.

13. Tighten the bolts to specification and in the proper sequence. Lightly coat the bolt threads with clean engine oil.

14. Once the head bolts are installed the painted mark (on the bolt) should be facing rearward.

15. Install and tighten the remaining two bolts.

16. Use the Techstream diagnostic tool, or equivalent and reprogram the required systems.

EXHAUST MANIFOLD

REMOVAL & INSTALLATION

4.6L Engine

See Figures 86 through 89.

1. Before servicing the vehicle, refer to the Precautions Section.
2. Disconnect the negative battery cable. Tape the cable with insulating tape.
3. Remove the front fender apron seals.
4. Remove the intake manifold assembly.
5. Remove the alternator.
6. Disconnect the compressor.

➡It may be necessary to properly discharge the system and remove the compressor, if unable to position it to the side without disconnect the refrigerant lines. If disconnecting the lines be sure to plug them after disconnection.

7. Remove the engine oil dipstick guide.
8. Remove the exhaust pipe assembly.
9. Remove the front driveshaft, if equipped with 4WD.
10. Remove the No. 3 and No. 4 floor heat insulators.
11. Remove the EGR pipe.
12. Remove the exhaust manifold heat insulators.
13. Remove the manifold retaining nuts. Discard the nuts.
14. Remove the manifold from the engine. Discard the gasket.

To install:

➡Be sure to use new fasteners, as required.

Fig. 84 Cylinder head gasket positioning Bank 2—5.7L engine

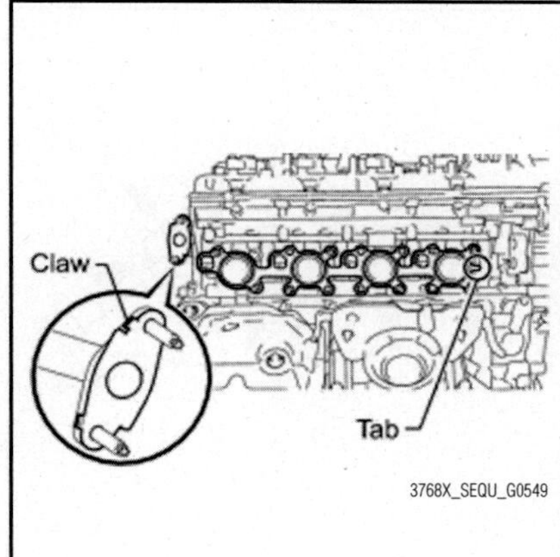

Fig. 86 Exhaust manifold gasket positioning left side—4.6L engine

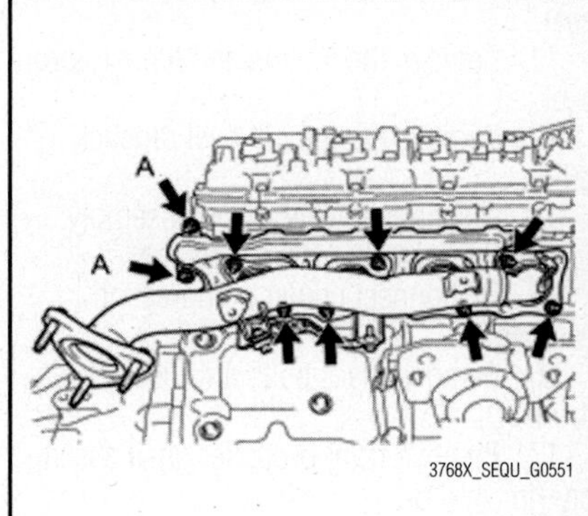

Fig. 88 Exhaust manifold bolt location and tightening sequence left side—4.6L engine

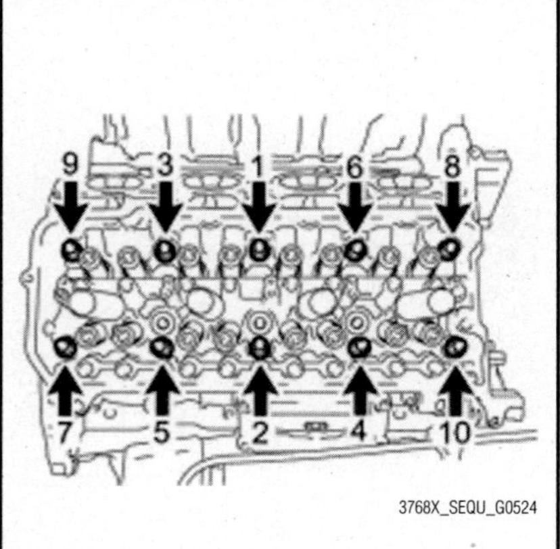

Fig. 85 Cylinder head gasket tightening sequence Bank 2—5.7L engine

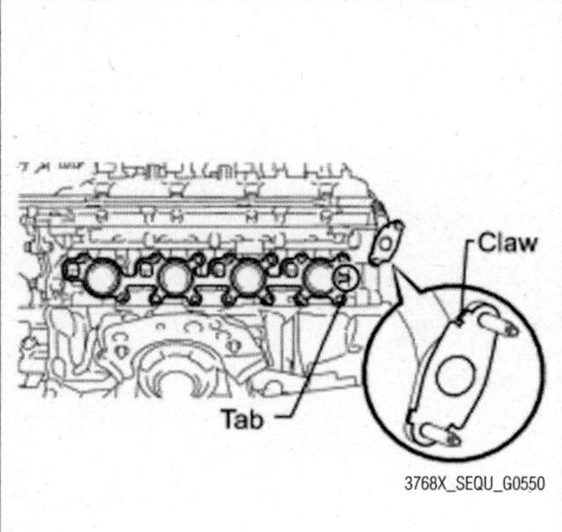

Fig. 87 Exhaust manifold gasket positioning right side—4.6L engine

Fig. 89 Exhaust manifold bolt location and tightening sequence right side—4.6L engine

15. Installation is the reverse of the removal procedure.

16. Tighten nuts labeled A to 7 ft. lbs. (10 Nm). Tighten others to 15 ft. lbs. (21 Nm). See illustration.

17. Use the Techstream diagnostic tool, or equivalent and reprogram the required systems.

5.7L Engine

See Figures 90 through 93.

1. Before servicing the vehicle, refer to the Precautions Section.
2. Disconnect the negative battery cable. Tape the cable with insulating tape.
3. Remove No. 1 engine under cover.
4. Drain engine coolant.
5. Remove V-bank cover sub-assembly.
6. Remove air cleaner hose assembly.
7. Disconnect inlet and outlet radiator hose.
8. Remove fan and alternator v-belt.
9. Remove fan shroud.
10. Remove the 6 clips and fender apron seal.
11. Remove the 5 clips and fender apron seal.
12. Remove engine oil level dipstick guide.
13. Disconnect vane pump assembly.
14. Remove alternator assembly.
15. Disconnect cooler compressor assembly.
16. Remove the 2 bolts and heat insulator 4WD.
17. Remove front propeller shaft assembly (for 4WD).
18. Regular Cab Standard Deck:
    a. Disconnect the air fuel ratio sensor connector.
    b. Disconnect the heated oxygen sensor connector.
    c. Remove the 2 bolts, 5 nuts, No. 2 exhaust pipe and 2 gaskets.
19. Except Regular Cab Standard Deck:
    a. Disconnect the air fuel ratio sensor connector.
    b. Disconnect the heated oxygen sensor connector.
    c. Remove the 2 bolts, 3 nuts, No. 2 exhaust pipe and 2 gaskets.
    d. Disconnect the exhaust support.
20. Remove front exhaust pipe assembly.
    a. Disconnect the air fuel ratio sensor connector.
    b. Disconnect the heated oxygen sensor connector and 2 clamps.
    c. Remove the 2 bolts from the center exhaust pipe.
    d. Remove the 3 nuts, front exhaust pipe and 2 gaskets.
21. Put matchmarks on the No. 2 steering intermediate shaft and steering intermediate shaft.
    a. Remove the bolt and disconnect the No. 2 steering intermediate shaft from the steering intermediate shaft.
    b. Put matchmarks on the No. 2 steering intermediate shaft and the power steering gear.
    c. Remove the bolt and disconnect the No. 2 steering intermediate shaft from the power steering gear.
22. Remove the 3 bolts and heat insulator.
23. Remove the 10 nuts, exhaust manifold and 2 gaskets.
24. Remove the 3 bolts and heat insulator.
25. Remove the 10 nuts, exhaust manifold and 2 gaskets.

To install:

➡Be sure to use new fasteners, as required.

26. Install a new gasket to the cylinder head and a new gasket to the No. 2 air tube.

➡Install the exhaust manifold gasket with the gasket tab facing toward the front of the engine. Install the air tube gasket with the gasket's claws facing the tube side.

27. Temporarily install the exhaust manifold and then uniformly tighten 8 new nuts that are not labeled A.

28. Tighten the new nuts labeled A in the illustration to 7 ft. lbs. (10 Nm) and remaining nuts to 15 ft. lbs. (21 Nm).

29. Install a new gasket to the cylinder head and a new gasket to the No. 3 air tube.

➡Install the exhaust manifold gasket with the gasket tab facing toward the front of the engine. Install the air tube gasket with the gasket's claws facing the tube side.

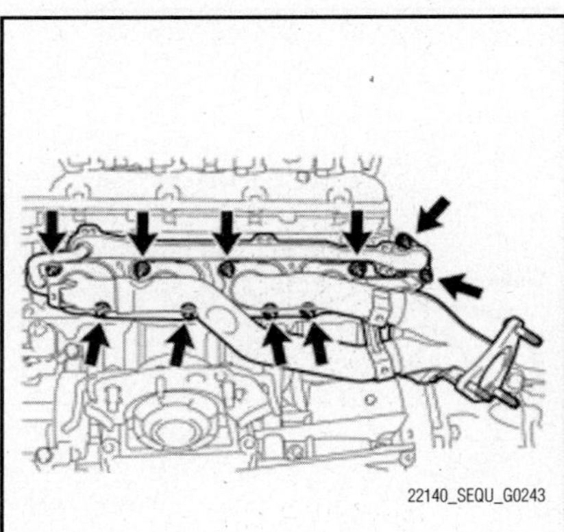

Fig. 90 Exhaust manifold LH—5.7L engine

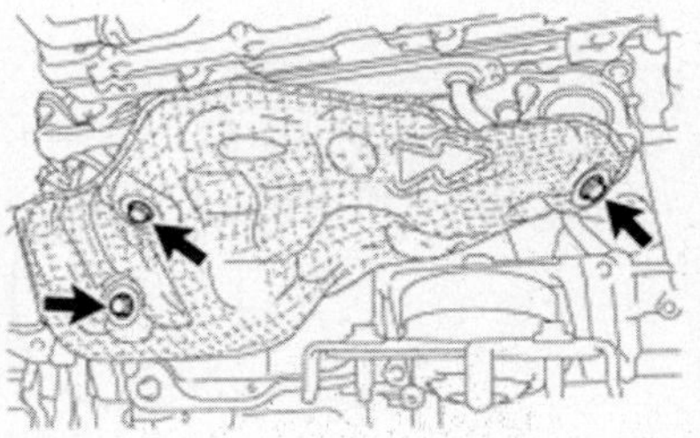

Fig. 92 Exhaust manifold tightening sequence RH—5.7L engine

Fig. 91 Exhaust heat insulator No. 1—5.7L engine

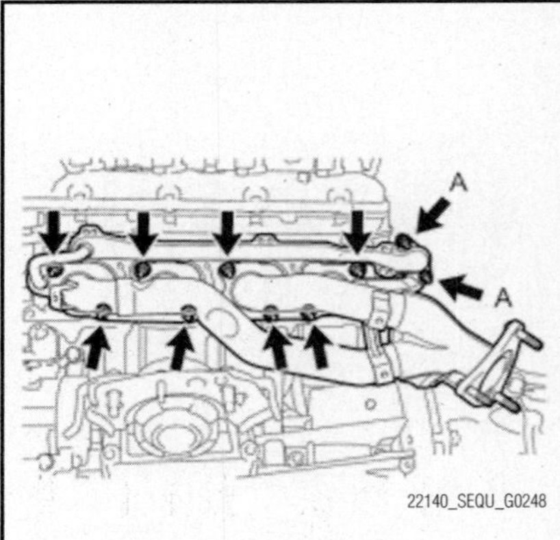

Fig. 93 Exhaust manifold tightening sequence LH—5.7L engine

30. Temporarily install the exhaust manifold and then uniformly tighten 8 new nuts that are not labeled A.

31. Tighten the new nuts labeled A in the illustration to 7 ft. lbs. (10 Nm) and remaining nuts to 15 ft. lbs. (21 Nm).

32. Install the heat insulator with the 3 bolts and tighten to 7 ft. lbs. (10 Nm).

33. Align the matchmarks and insert the No. 2 intermediate shaft into the intermediate shaft.

34. Align the matchmarks and insert the No. 2 intermediate shaft into the power steering gear.

35. Install the 2 bolt and tighten to 26 ft. lbs. (35 Nm).

36. Install front exhaust pipe assembly.

a. Install a new gasket and the front exhaust pipe to the exhaust manifold RH with 3 new nuts and tighten to 40 ft. lbs. (54 Nm).

b. Install a new gasket and the front exhaust pipe to the center exhaust pipe with the 2 bolts and tighten to 35 ft. lbs. (48 Nm).

c. Connect the air fuel ratio sensor connector.

d. Connect the heated oxygen sensor connector and 2 clamps.

37. Except Regular Cab Standard Deck:

a. Connect the front No. 2 exhaust pipe to the exhaust support.

b. Install a new gasket and the front No. 2 exhaust pipe to the exhaust manifold LH with 3 new nuts and tighten to 40 ft. lbs. (54 Nm).

c. Install a new gasket and the front No. 2 exhaust pipe to the center exhaust pipe with the 2 bolts and tighten to 35 ft. lbs. (48 Nm).

d. Connect the air fuel ratio sensor connector.

e. Connect the heated oxygen sensor connector.

38. Regular Cab Standard Deck :

a. Connect the front No. 2 exhaust pipe to the exhaust support.

b. Install a new gasket and the front No. 2 exhaust pipe to the exhaust manifold LH with 3 new nuts and tighten to 40 ft. lbs. (54 Nm).

c. Install a new gasket and the front No. 2 exhaust pipe to the center exhaust pipe with the 2 bolts and tighten to 35 ft. lbs. (48 Nm).

d. Connect the air fuel ratio sensor connector.

e. Connect the heated oxygen sensor connector.

39. Install front propeller shaft assembly (4WD).

40. Install the heat insulator with the 2 bolts (4WD).

41. Connect cooler compressor assembly.

42. Install alternator assembly.

43. Connect vane pump assembly.

44. Install engine oil level dipstick guide.

45. Install fan shroud.

46. Install fan and alternator v belt.

47. Connect outlet and inlet radiator hose.

48. Install air cleaner assembly.

49. Install air cleaner hose assembly.

50. Install front fender apron seal rear LH.

51. Install front fender apron seal LH.

52. Install front fender apron seal rear RH.

53. Install front fender apron seal RH.

54. Install No. 1 engine under cover.

55. Install V-bank cover sub-assembly.

56. Add engine coolant.

57. Inspect for exhaust gas leak.

➡If gas is leaking, tighten the areas necessary to stop the leak. Replace damaged parts as necessary.

58. Use the Techstream diagnostic tool, or equivalent and reprogram the required systems.

INTAKE MANIFOLD

REMOVAL & INSTALLATION

4.6L Engine

See Figures 94 and 95.

1. Before servicing the vehicle, refer to the Precautions Section.

2. Disconnect the negative battery cable. Tape the cable with insulating tape.

3. Properly discharge the fuel system pressure.

4. Remove the EGR valve assembly.

5. Remove the front wiper motor and link assembly.

6. Remove the cowl top outer panel assembly.

7. Remove the ventilation hose.

8. Remove the air tube sub assembly, left side.

9. Disconnect the required electrical connectors.

10. Disconnect the required hoses.

11. Disconnect the fuel lines.

12. Remove the intake manifold retaining bolts.

13. Remove the intake manifold from its mounting.

14. Discard the gasket.

➡Be sure to use new fasteners, as required.

15. Installation is the reverse of the removal procedure.

16. Be sure to use a new gasket.

17. Use the Techstream diagnostic tool, or equivalent and reprogram the required systems.

5.7L Engine

See Figures 96 through 97.

1. Before servicing the vehicle, refer to the Precautions Section.

2. Disconnect the negative battery cable. Tape the cable with insulating tape.

3. Remove the front wiper motor and link.

4. Disconnect the 2 washer hoses.

5. Remove the 7 bolts and outer panel.

6. Remove No. 1 engine under cover.

7. Drain engine coolant.

8. Remove V-bank cover sub-assembly.

9. Remove air cleaner hose assembly.

10. Disconnect the ventilation hose from the ventilation pipe of the cylinder head cover LH and RH.

11. Disconnect the 2 water by-pass hoses.

12. Disconnect the throttle body connector.

13. Disconnect the No. 1 ventilation hose.

14. Remove the No. 1 engine cover sub-assembly.

15. Remove the No. 3 engine cover.

16. Disconnect the purge VSV connector.

17. Disconnect the purge line hose from the purge VSV.

18. Disconnect the vacuum switching valve connector (for ACIS).

19. Remove the 2 bolts.

20. Disconnect the No. 1 tube from the union to connector tube hose, and move the hose aside.

21. Disconnect the 3 wire clamps from the 3 wire brackets.

22. Remove the bolt and wire bracket from the intake manifold.

23. Remove the 2 nuts, 8 bolts, intake manifold and 2 gaskets.

To install:

➡Be sure to use new fasteners, as required.

24. Place 2 new gaskets on the intake manifold.

25. Place the intake manifold on the cylinder head.

26. Install and uniformly tighten the 8 bolts and 2 nuts to 15 ft. lbs. (21 Nm) in several steps.

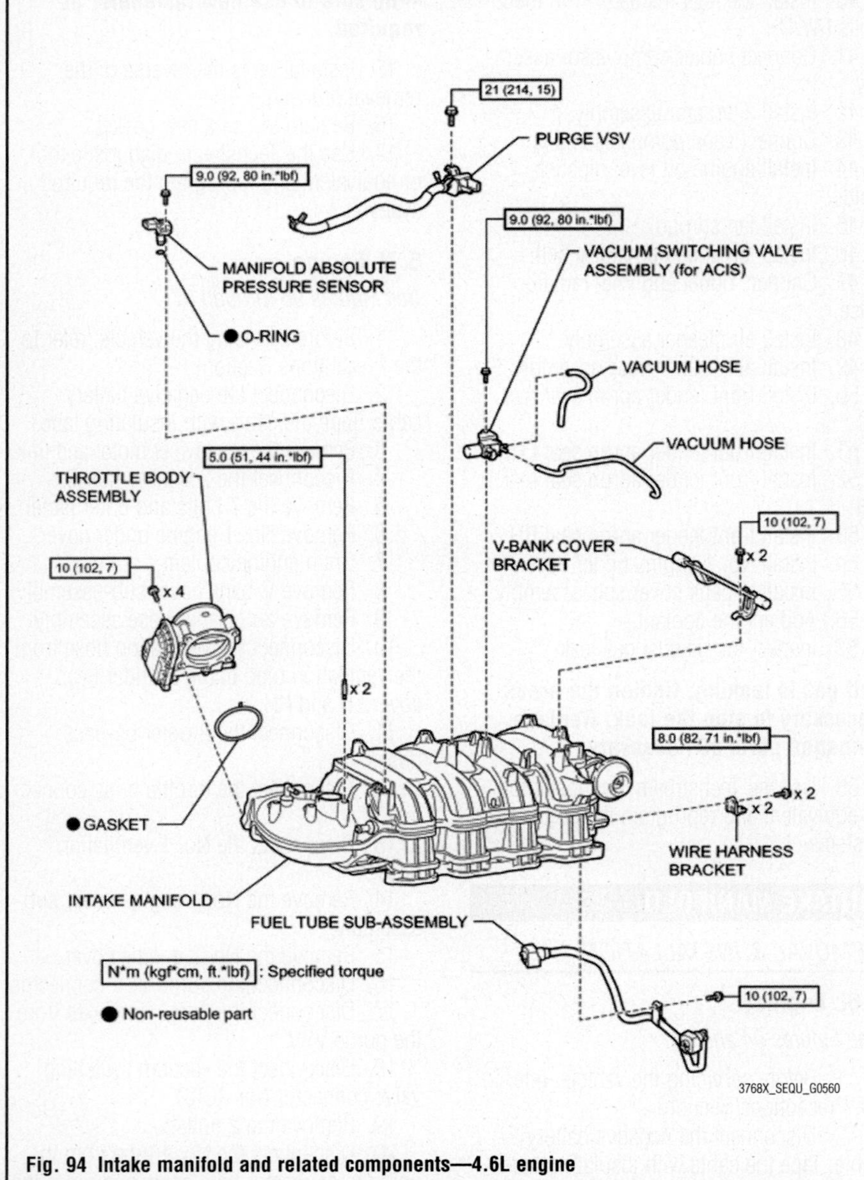

Fig. 94 Intake manifold and related components—4.6L engine

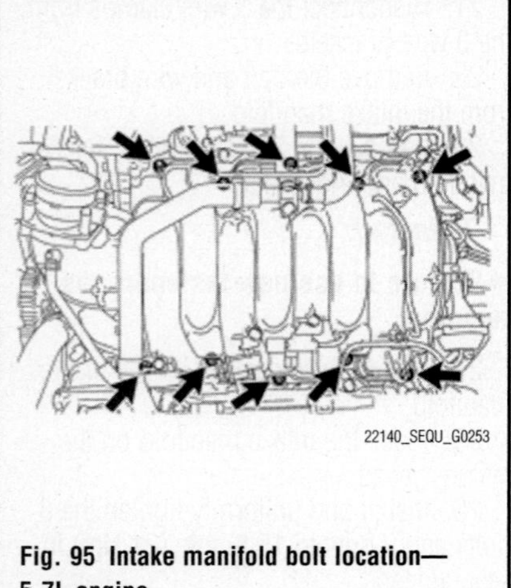

Fig. 95 Intake manifold bolt location—5.7L engine

27. Install the wire bracket to the intake manifold with the bolt.
28. Connect the 3 wire clamps to the 3 wire brackets.
29. Connect the No. 1 tube and install it to the intake manifold with the 2 bolts.
30. Connect the purge VSV connector.
31. Connect the purge line hose to the purge VSV.
32. Connect the vacuum switching valve connector (for ACIS).
33. Install the No. 1 engine cover sub-assembly.
34. Install the No. 3 engine cover.
35. Connect the No. 1 ventilation hose.
36. Connect the 2 water by-pass hoses.
37. Connect the throttle body connector.
38. Connect the ventilation hose to the ventilation pipe of the cylinder head cover LH and RH.
39. Install air cleaner hose assembly.
40. Install the outer panel with the 7 bolts.
41. Connect the 2 washer hoses.
42. Install the front wiper motor and link.
43. Add engine coolant.
44. Install V-bank cover sub-assembly.
45. Install No. 1 engine under cover.
46. Inspect for coolant leak.
47. Use the Techstream diagnostic tool, or equivalent and reprogram the required systems.

OIL PUMP

REMOVAL & INSTALLATION

4.6L Engine

See Figures 98 through 108.

1. Before servicing the vehicle, refer to the Precautions Section.
2. Disconnect the negative battery cable. Tape the cable with insulating tape.
3. Discharge fuel system pressure.
4. Drain engine oil.
5. Drain engine coolant.
6. Remove the radiator.
7. Remove air cleaner hose assembly.
8. Remove air cleaner assembly.
9. Engine Room LH Side:
   a. Remove the engine room relay block cover.
   b. Disconnect the 2 connectors and detach the 2 clamps from the engine room junction block.
   c. Disconnect the 4 air injection control driver connectors.
   d. Disconnect the 2 wire harness clamps.
   e. Disconnect the injector connector.
   f. Disconnect the 4 ignition coil connectors.
   g. Disconnect the 2 VVT sensor connectors.
   h. Disconnect the 4 clamps.
   i. Remove the 2 bolts and ground wire.
   j. Disconnect the noise filter connector.
   k. Disconnect the engine coolant temperature sensor connector.
   l. Disconnect the 2 camshaft timing oil control valve connectors.
   m. Disconnect the camshaft position sensor connector.
   n. Disconnect the 3 clamps.
   o. Disconnect the cooler compressor connector.

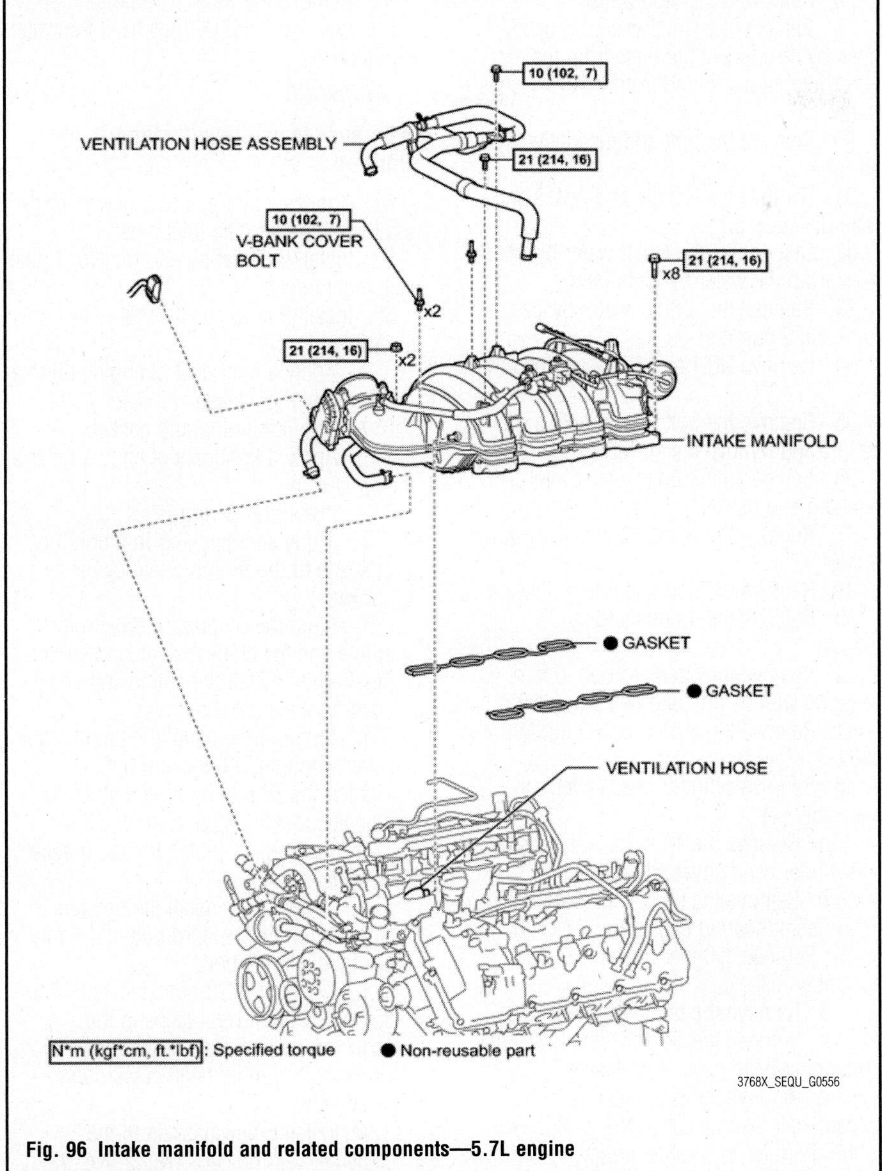

Fig. 96 Intake manifold and related components—5.7L engine

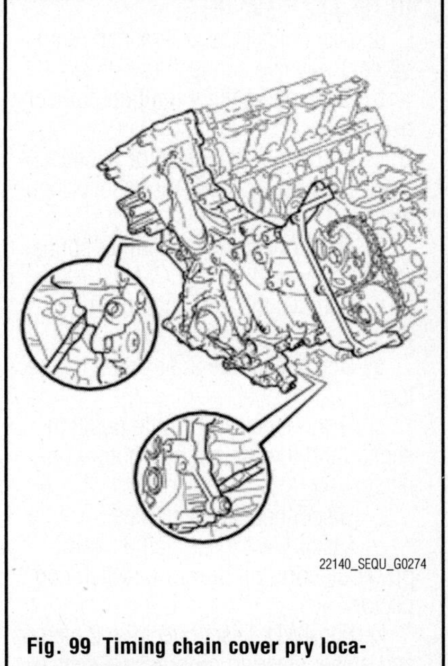

Fig. 99 Timing chain cover pry locations—4.6L engine

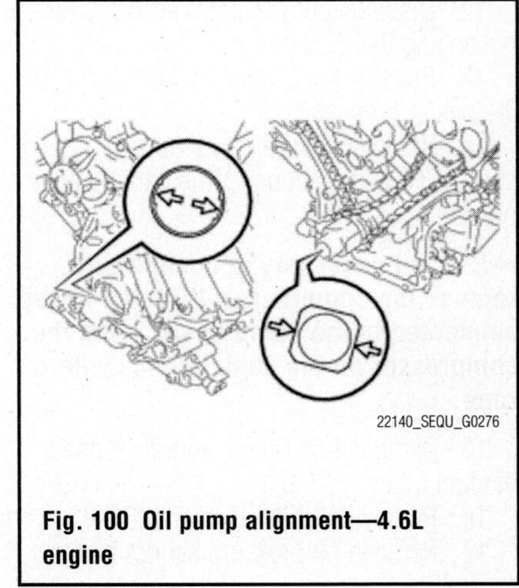

Fig. 100 Oil pump alignment—4.6L engine

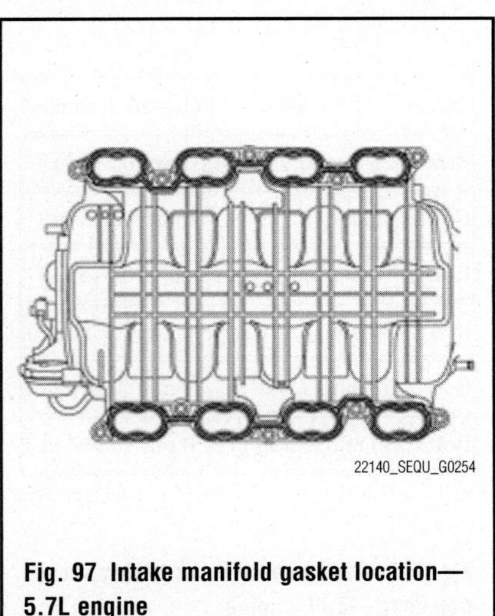

Fig. 97 Intake manifold gasket location—5.7L engine

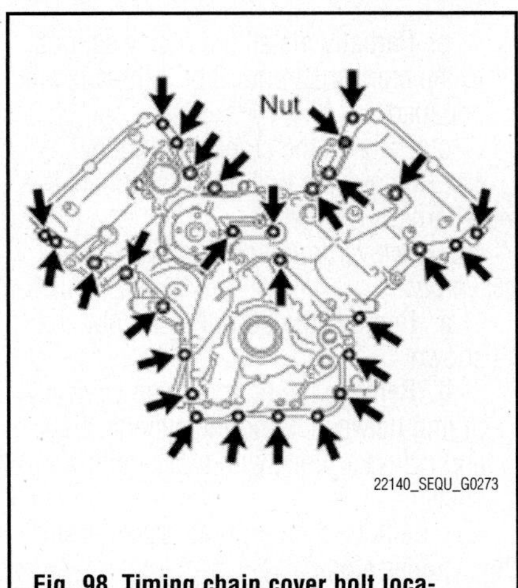

Fig. 98 Timing chain cover bolt locations—4.6L engine

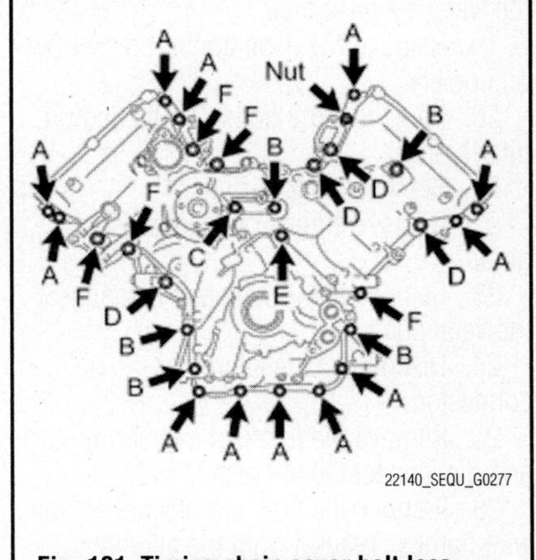

Fig. 101 Timing chain cover bolt locations—4.6L engine

10. Engine Room RH Side.
 a. Disconnect the 2 camshaft timing oil control valve connectors.
 b. Disconnect the 4 ignition coil connectors.
 c. Disconnect the injector connector.
 d. Disconnect the 2 VVT sensor connectors.
 e. Disconnect the noise filter connector.
 f. Remove the 2 bolts and ground wire.
 g. Disconnect the 2 air pump connectors.
 h. Disconnect the throttle position sensor and throttle control motor connector.
 i. Disconnect the 5 clamps.
 j. Disconnect the 2 clamps and power steering oil pressure switch connector.
 k. For 4WD: Disconnect the 2 clamps and power steering oil pressure switch connector.
11. Remove the intake manifold.
12. Disconnect the No. 2 and No. 3 air hoses and the 2 clamps.
13. Remove the 2 bolts and disconnect the heater 3 hoses.
14. Remove the 2 bolts, 2 nuts and 2 stud bolts, and disconnect the cooler compressor.

➡It is not necessary to completely remove the compressor. With the hoses connected to the compressor, hang the compressor on the vehicle body with a rope.

15. Remove the 2 bolts and disconnect the fuel tube.
16. Remove oil filter element.
17. Remove oil pressure sender gauge assembly.
18. Disconnect the 4 hoses and remove the water by-pass pipe.
19. Remove No. 1 oil cooler bracket (w/ oil cooler).
20. Remove the 2 bolts, 2 nuts and oil filter bracket.
21. Remove the 2 O-rings.
22. Remove engine oil level dipstick guide.
23. Remove the 2 bolts and disconnect the vane pump.
24. Disconnect the alternator connector.
25. Remove the terminal cap and nut, and disconnect the alternator wire.
26. Remove the bolt and disconnect the wire harness bracket from the alternator.
27. Remove the 2 bolts, 2 nuts and alternator.
28. Remove the 2 stud bolts.
29. Remove the No. 1 water by-pass hose by disconnect the hose from the water inlet housing and front water by-pass joint.
30. Remove the bolt and disconnect the air tube.
31. Remove the 2 bolts and hoses and water by-pass pipe.
32. Disconnect the No. 2 water by-pass hose from the water by-pass joint.
33. Remove the 4 nuts, water by-pass joint and 2 gaskets.
34. Remove No.1 and No. 2 engine covers.
35. Remove the bolt, disconnect the 2 hoses, and remove the air tube.
36. Remove the 3 bolts, water inlet housing and gasket.
37. Remove the 4 bolts and water pump pulley.
38. Remove the bolt and idler pulley.
39. Remove the 4 bolts and fan bracket.
40. Remove the standard bolt, 6 mm hexagon wrench bolt and belt tensioner.
41. Remove the 8 bolts and 8 ignition coils.
42. Remove cylinder head cover sub-assembly LH.
 a. Remove the 14 bolts, seal washer, cylinder head cover and gasket.
 b. Remove the 5 gaskets from the camshaft bearing caps (No. 2, No. 3).
43. Remove cylinder head cover sub-assembly RH.
 a. Remove the bolt and noise filter.
 b. Remove the 14 bolts, seal washer, cylinder head cover and gasket.
 c. Remove the 5 gaskets from the camshaft bearing caps (No. 1, No. 3).
44. Remove crankshaft pulley.
 a. Loosen the crankshaft pulley set bolt.
 b. Partially install the pulley set bolt to the crankshaft until 2 or 3 threads are engaged.
 c. Remove the crankshaft pulley.
45. Remove the bolt and disconnect the wire harness bracket.
46. Remove timing chain cover sub-assembly.
 a. Remove the 28 bolts and nut shown.
 b. Remove the timing chain cover by prying between it and the cylinder head and cylinder block with a pry tool as shown.
47. Remove the oil pump gasket from the cylinder block.
48. Remove the O-ring from the oil pan.
49. Remove the water inlet pipe.
50. Remove the 2 O-rings from the water inlet pipe.

To install:

➡Be sure to use new fasteners, as required.

51. Apply soapy water to 2 new O-rings and install them to the inlet pipe.
52. Install the inlet pipe to the No. 1 heat exchanger cover.
53. Install timing chain cover sub-assembly.
 a. Apply a light coat of engine oil to a new oil pump gasket.
 b. Install the oil pump gasket.
 c. Apply a light coat of engine oil to a new O-ring.
 d. Install the O-ring.
 e. Apply seal packing in a continuous line to the timing chain cover as shown.
 f. Align the oil pump's drive rotor spline and the crankshaft as shown in the illustration. Install the spline and chain cover to the crankshaft.
 g. Temporarily install the timing chain cover with the 28 bolts and nut.
 Tighten the 5 bolts in several steps in the sequence shown to 35 ft. lbs. (47 Nm).
 h. Temporarily install the fan bracket with the 4 bolts.
 i. Temporarily install the belt tensioner with the standard bolt and 6 mm hexagon wrench bolt.
 j. Tighten the 23 bolts labeled 12 to 35 and nut in several steps in the sequence to 17 ft. lbs. (23 Nm).
54. Install cylinder head cover sub-assembly LH.
 a. Install 5 new gaskets to the camshaft bearing caps (No. 2, No. 3).

Item	Length	Thread diameter
Bolt A	25 mm (0.984 in.)	8 mm (0.315 in.)
Bolt B	55 mm (2.165 in.)	8 mm (0.315 in.)
Bolt C	70 mm (2.756 in.)	8 mm (0.315 in.)
Bolt D	35 mm (1.378 in.)	10 mm (0.394 in.)
Bolt E	55 mm (2.165 in.)	10 mm (0.394 in.)
Bolt F	80 mm (3.150 in.)	10 mm (0.394 in.)

22140_SEQU_G0278

Fig. 102 Timing chain cover bolt application chart—4.6L engine

b. Install the gasket to the cylinder head cover.

c. Apply seal packing as shown.

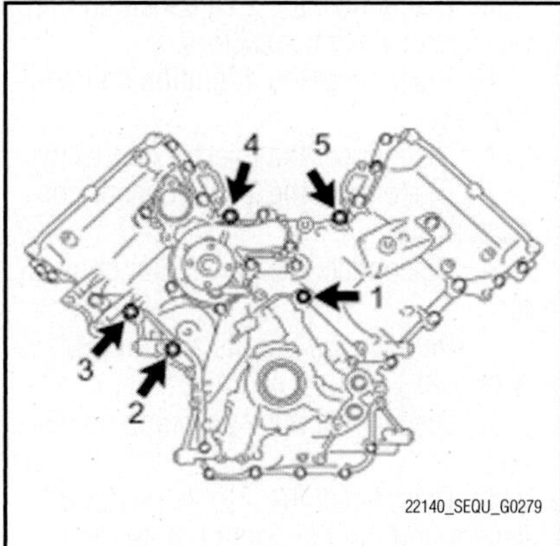

Fig. 103 Timing chain cover bolt tightening sequence No. 1—4.6L engine

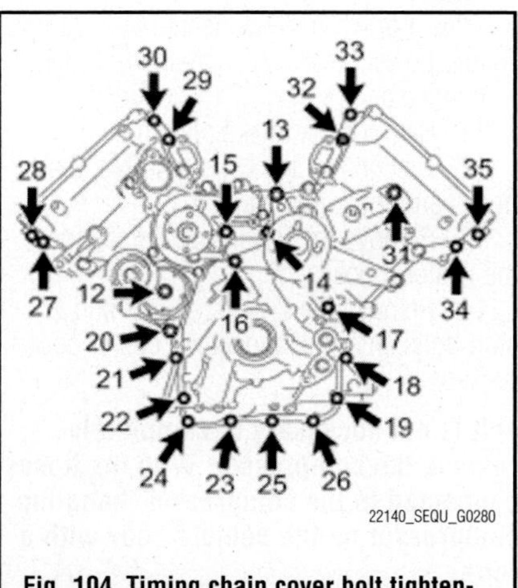

Fig. 104 Timing chain cover bolt tightening sequence No. 2—4.6L engine

d. Install the cylinder head cover washer with a new seal and the 14 bolts and tighten bolt A to 15 ft. lbs. (21 Nm) and remaining bolts to 9 ft. lbs. (12 Nm).

55. Install cylinder head cover sub-assembly RH.

a. Install 5 new gaskets to the camshaft bearing caps (No. 1, No. 3).

b. Install the gasket to the cylinder head cover.

c. Apply seal packing as shown.

d. Install the cylinder head cover washer with a new seal and the 14 bolts and tighten bolt A to 15 ft. lbs. (21 Nm) and remaining bolts to 9 ft. lbs. (12 Nm).

e. Install the noise filter to the cylinder head cover with the bolt and tighten to 62 inch lbs. (7.0 Nm).

56. Install the 8 ignition coils with the 8 bolts and tighten to 7 ft. lbs. (10 Nm).

57. Align the pulley set key with the key groove of the pulley, and slide on the pulley and tighten to 221 ft. Lbs. (300 Nm).

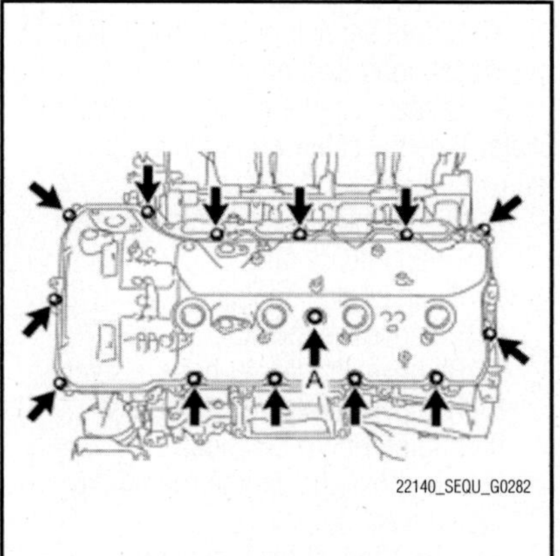

Fig. 106 Cylinder head cover bolt tightening sequence LH—4.6L engine

58. Connect the bracket to the timing chain cover with the bolt.

59. Install the idler pulley with the bolt and tighten to 32 ft. lbs. (43 Nm).

60. Temporarily install the water pump pulley with the 4 bolts and tighten to 15 ft. lbs. (21 Nm).

61. Install a new water cover gasket to the timing chain cover.

62. Install the water inlet with the 3 bolts and tighten to 15 ft. lbs. (21 Nm).

63. Connect the 2 air tube hoses.

64. Connect the 2 hoses.

65. Install No. 1 and No 2 engine covers.

66. Install 2 new gaskets and the water by-pass joint with the 4 nuts and tighten to 15 ft. lbs. (21 Nm).

67. Connect the No. 2 water by-pass hose to the water by-pass joint.

68. Install the water by-pass pipe.

69. Connect the air tube.

70. Install the No. 1 water by-pass hose by connecting the hose to the water inlet housing and front water by-pass joint.

71. Install the 2 stud bolts.

72. Install the alternator with the 2 bolts and 2 nuts and tighten to 32 ft. lbs. (43 Nm).

73. Connect the alternator wire with the nut and tighten to 87 inch lbs. (9.8 Nm).

74. Connect the wire harness bracket to the alternator.

75. Connect the vane pump to the timing chain cover with the 2 bolts and tighten to 21 ft. lbs. (28 Nm).

76. Install engine oil level dipstick guide.

77. Install oil filter bracket (w/o Oil Cooler).

a. Apply a light coat of engine oil to 2 new O-rings.

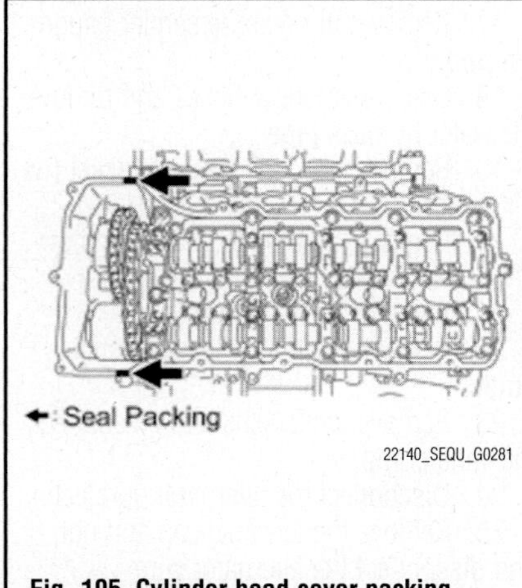

Fig. 105 Cylinder head cover packing location LH—4.6L engine

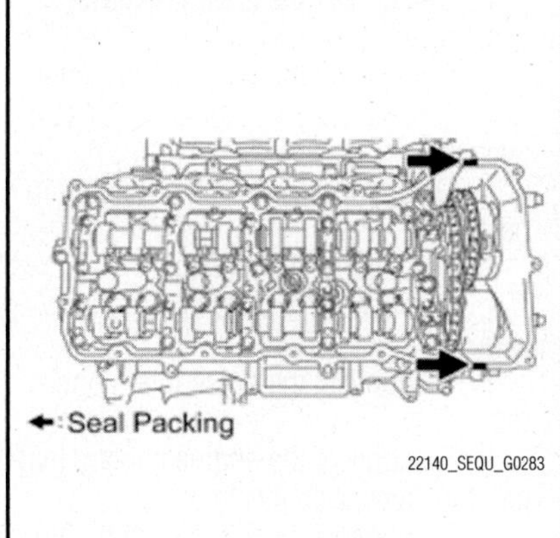

Fig. 107 Cylinder head cover packing location RH—4.6L engine

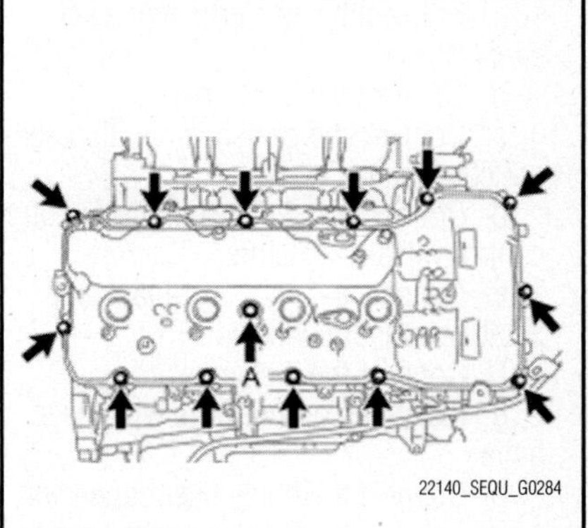

Fig. 108 Cylinder head cover bolt tightening sequence RH—4.6L engine

b. Install the 2 O-rings to the timing chain cover.

c. Install the oil filter bracket with the 2 bolts and 2 nuts and tighten to 26 ft. lbs. (35 Nm).

78. Install No. 1 oil cooler bracket.
79. Install No. 2 water by-pass pipe.
80. Install oil pressure sender gauge assembly.
81. Install oil filter element.
82. Connect No. 2 fuel tube sub-assembly.
83. Install the cooler compressor with the 2 stud bolts and tighten to 7 ft. lbs. (10 Nm).
84. Install the 2 bolts and 2 nuts and tighten to 18 ft. lbs. (235 Nm).
85. Connect the 3 heater hoses.
86. Connect the No. 2 and No. 3 air hoses.
87. Connect the 2 clamps and air pump wire harness.
88. Install the intake manifold.
89. Engine Room RH Side:
    a. Connect the 5 clamps.
    b. Connect the throttle position sensor and throttle control motor connector.
    c. Connect the 2 air pump connectors.
    d. Install the ground wire with the 2 bolts.
    e. Connect the noise filter connector.
    f. Connect the 2 VVT sensor connectors.
    g. Connect the injector connector.
    h. Connect the 4 ignition coil connectors.
    i. Connect the 2 camshaft timing oil control valve connectors.
    j. Connect the 2 clamps and power steering oil pressure switch connector.
    k. For 4WD: Connect the clamp and junction connector.
90. Engine Room LH Side:
    a. Connect the cooler compressor connector.
    b. Connect the 3 clamps.
    c. Connect the camshaft position sensor connector.
    d. Connect the 2 camshaft timing oil control valve connectors.
    e. Connect the engine coolant temperature sensor connector.
    f. Connect the 4 clamps.
    g. Install the ground wire with the 2 bolts.
    h. Connect the noise filter connector.
    i. Connect the 2 VVT sensor connectors.
    j. Connect the 4 ignition coil connectors.
    k. Connect the injector connector.
    l. Connect the 2 wire harness clamps.
    m. Connect the 4 air injection control driver connectors.
    n. Connect the 2 connectors and attach the 2 clamps to the engine room junction block.
    o. Install the engine room relay block cover.
91. Install air cleaner assembly.
92. Install air cleaner hose assembly.
93. Install the radiator.
94. Add engine oil.
95. Add engine coolant.
96. Connect cable to negative battery terminal.
97. Inspect for leaks.
98. Check engine oil level.
99. Use the Techstream diagnostic tool, or equivalent and reprogram the required systems.

5.7L Engine

See Figures 109 through 117.

1. Before servicing the vehicle, refer to the Precautions Section.
2. Disconnect the negative battery cable. Tape the cable with insulating tape.
3. Discharge fuel system pressure.
4. Drain engine oil.
5. Drain engine coolant.
6. Remove the radiator.
7. Remove air cleaner hose assembly.
8. Remove air cleaner assembly.
9. Engine Room LH Side:
    a. Remove the engine room relay block cover.
    b. Disconnect the 2 connectors and detach the 2 clamps from the engine room junction block.
    c. Disconnect the 4 air injection control driver connectors.
    d. Disconnect the 2 wire harness clamps.
    e. Disconnect the injector connector.
    f. Disconnect the 4 ignition coil connectors.
    g. Disconnect the 2 VVT sensor connectors.
    h. Disconnect the 4 clamps.
    i. Remove the 2 bolts and ground wire.
    j. Disconnect the noise filter connector.
    k. Disconnect the engine coolant temperature sensor connector.
    l. Disconnect the 2 camshaft timing oil control valve connectors.
    m. Disconnect the camshaft position sensor connector.
    n. Disconnect the 3 clamps.
    o. Disconnect the cooler compressor connector.
10. Engine Room RH Side.
    a. Disconnect the 2 camshaft timing oil control valve connectors.
    b. Disconnect the 4 ignition coil connectors.
    c. Disconnect the injector connector.
    d. Disconnect the 2 VVT sensor connectors.
    e. Disconnect the noise filter connector.
    f. Remove the 2 bolts and ground wire.
    g. Disconnect the 2 air pump connectors.
    h. Disconnect the throttle position sensor and throttle control motor connector.
    i. Disconnect the 5 clamps.
    j. Disconnect the 2 clamps and power steering oil pressure switch connector.
    k. For 4WD: Disconnect the 2 clamps and power steering oil pressure switch connector.
11. Remove the intake manifold.
12. Disconnect the No. 2 and No. 3 air hoses and the 2 clamps.
13. Remove the 2 bolts and disconnect the heater 3 hoses.
14. Remove the 2 bolts, 2 nuts and 2 stud bolts, and disconnect the cooler compressor.

➡It is not necessary to completely remove the compressor. With the hoses connected to the compressor, hang the compressor on the vehicle body with a rope.

15. Remove the 2 bolts and disconnect the fuel tube.
16. Remove oil filter element.
17. Remove oil pressure sender gauge assembly.
18. Disconnect the 4 hoses and remove the water by-pass pipe.
19. Remove No. 1 oil cooler bracket (w/ oil cooler).
20. Remove the 2 bolts, 2 nuts and oil filter bracket.
21. Remove the 2 O-rings.
22. Remove engine oil level dipstick guide.
23. Remove the 2 bolts and disconnect the vane pump.
24. Disconnect the alternator connector.
25. Remove the terminal cap and nut, and disconnect the alternator wire.
26. Remove the bolt and disconnect the wire harness bracket from the alternator.

27. Remove the 2 bolts, 2 nuts and alternator.
28. Remove the 2 stud bolts.
29. Remove the No. 1 water by-pass hose by disconnect the hose from the water inlet housing and front water by-pass joint.
30. Remove the bolt and disconnect the air tube.
31. Remove the 2 bolts and hoses and water by-pass pipe.
32. Disconnect the No. 2 water by-pass hose from the water by-pass joint.
33. Remove the 4 nuts, water by-pass joint and 2 gaskets.
34. Remove No.1 and No. 2 engine covers.
35. Remove the bolt, disconnect the 2 hoses, and remove the air tube.
36. Remove the 3 bolts, water inlet housing and gasket.
37. Remove the 4 bolts and water pump pulley.
38. Remove the bolt and idler pulley.
39. Remove the 4 bolts and fan bracket.
40. Remove the standard bolt, 6 mm hexagon wrench bolt and belt tensioner.
41. Remove the 8 bolts and 8 ignition coils.
42. Remove cylinder head cover sub-assembly LH.
    a. Remove the 14 bolts, seal washer, cylinder head cover and gasket.
    b. Remove the 5 gaskets from the camshaft bearing caps (No. 2, No. 3).
43. Remove cylinder head cover sub-assembly RH.
    a. Remove the bolt and noise filter.
    b. Remove the 14 bolts, seal washer, cylinder head cover and gasket.
    c. Remove the 5 gaskets from the camshaft bearing caps (No. 1, No. 3).
44. Remove crankshaft pulley.
    a. Loosen the crankshaft pulley set bolt.
    b. Partially install the pulley set bolt to the crankshaft until 2 or 3 threads are engaged.
    c. Remove the crankshaft pulley.
45. Remove the bolt and disconnect the wire harness bracket.
46. Remove timing chain cover sub-assembly.
    a. Remove the 28 bolts and nut shown.
    b. Remove the timing chain cover by prying between it and the cylinder head and cylinder block with a pry tool as shown.
47. Remove the oil pump gasket from the cylinder block.
48. Remove the O-ring from the oil pan.
49. Remove the water inlet pipe.
50. Remove the 2 O-rings from the water inlet pipe.

To install:

➡**Be sure to use new fasteners, as required.**

51. Apply soapy water to 2 new O-rings and install them to the inlet pipe.
52. Install the inlet pipe to the No. 1 heat exchanger cover.
53. Install timing chain cover sub-assembly.
    a. Apply a light coat of engine oil to a new oil pump gasket.
    b. Install the oil pump gasket.

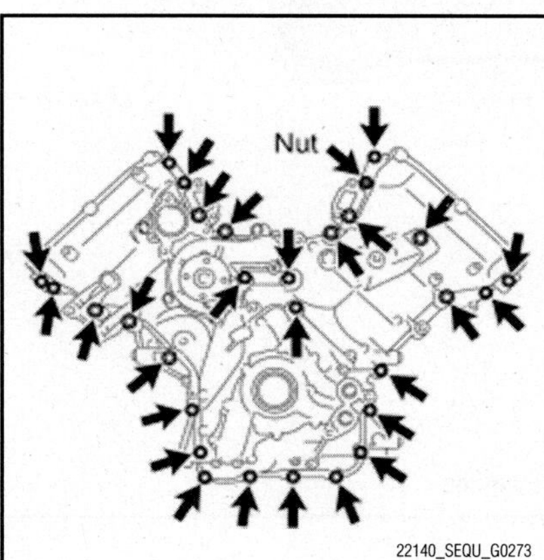

Fig. 109 Timing chain cover bolt locations—5.7L engine

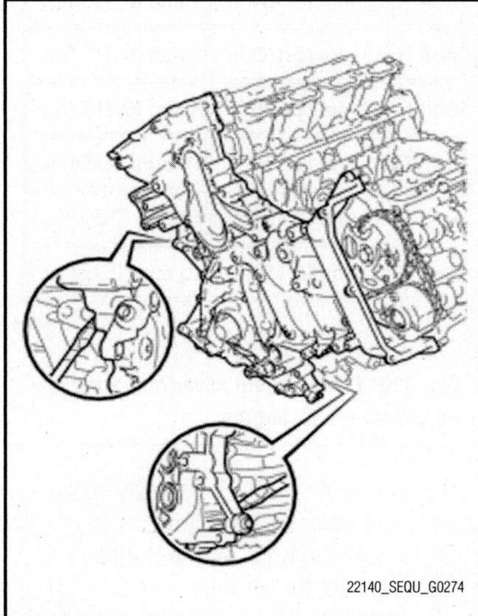

Fig. 110 Timing chain cover pry locations—5.7L engine

    c. Apply a light coat of engine oil to a new O-ring.
    d. Install the O-ring.
    e. Apply seal packing in a continuous line to the timing chain cover as shown.
    f. Align the oil pump's drive rotor spline and the crankshaft as shown in the illustration. Install the spline and chain cover to the crankshaft.
    g. Temporarily install the timing chain cover with the 28 bolts and nut.
Tighten the 5 bolts in several steps in the sequence shown to 35 ft. lbs. (47 Nm).
    h. Temporarily install the fan bracket with the 4 bolts.
    i. Temporarily install the belt tensioner with the standard bolt and 6 mm hexagon wrench bolt.
    j. Tighten the 23 bolts labeled 12 to 35 and nut in several steps in the sequence to 17 ft. lbs. (23 Nm).
54. Install cylinder head cover sub-assembly LH.
    a. Install 5 new gaskets to the camshaft bearing caps (No. 2, No. 3).
    b. Install the gasket to the cylinder head cover.
    c. Apply seal packing.
    d. Install the cylinder head cover washer with a new seal and the 14 bolts and tighten bolt A to 15 ft. lbs. (21 Nm) and remaining bolts to 9 ft. lbs. (12 Nm).
55. Install cylinder head cover sub-assembly RH.
    a. Install 5 new gaskets to the camshaft bearing caps (No. 1, No. 3).
    b. Install the gasket to the cylinder head cover.
    c. Apply seal packing.
    d. Install the cylinder head cover washer with a new seal and the 14 bolts and tighten bolt A to 15 ft. lbs. (21 Nm) and remaining bolts to 9 ft. lbs. (12 Nm).
    e. Install the noise filter to the cylinder head cover with the bolt and tighten to 62 inch lbs. (7.0 Nm).
56. Install the 8 ignition coils with the 8 bolts and tighten to 7 ft. lbs. (10 Nm).
57. Align the pulley set key with the key groove of the pulley, and slide on the pulley and tighten to 221 ft. Lbs. (300 Nm).
58. Connect the bracket to the timing chain cover with the bolt.
59. Install the idler pulley with the bolt and tighten to 32 ft. lbs. (43 Nm).
60. Temporarily install the water pump pulley with the 4 bolts and tighten to 15 ft. lbs. (21 Nm).
61. Install a new water cover gasket to the timing chain cover.
62. Install the water inlet with the 3 bolts and tighten to 15 ft. lbs. (21 Nm).

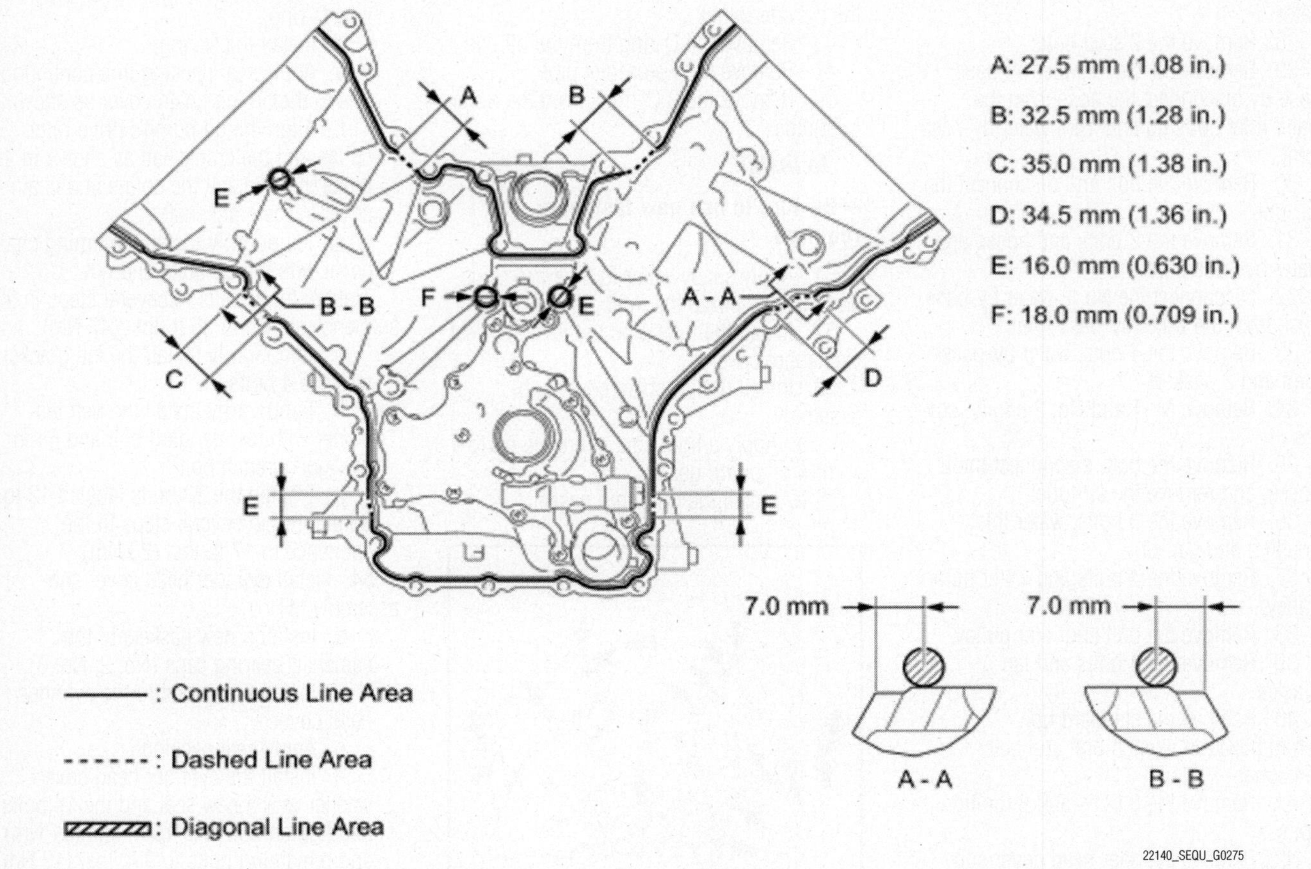

A: 27.5 mm (1.08 in.)
B: 32.5 mm (1.28 in.)
C: 35.0 mm (1.38 in.)
D: 34.5 mm (1.36 in.)
E: 16.0 mm (0.630 in.)
F: 18.0 mm (0.709 in.)
A
B
E
B - B
F
E
A - A
C
D
E
E
7.0 mm
7.0 mm
A - A
B - B
: Continuous Line Area
: Dashed Line Area
: Diagonal Line Area
22140_SEQU_G0275

Fig. 111 Timing chain cover sealant locations—5.7L engine

Nut
A
B
C
D
E
F
22140_SEQU_G0277

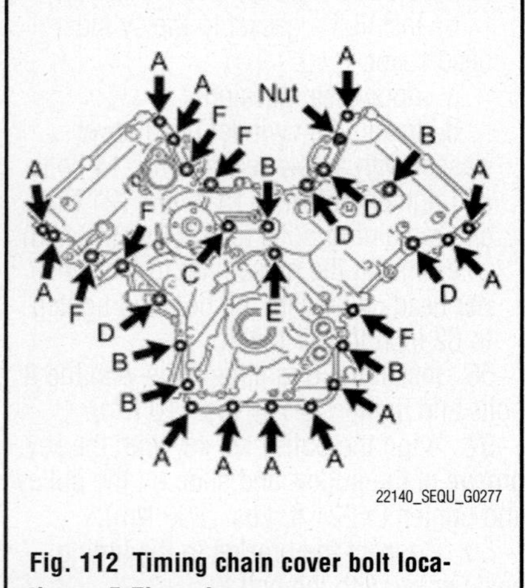

Fig. 112 Timing chain cover bolt locations—5.7L engine

Item	Length	Thread diameter
Bolt A	25 mm (0.984 in.)	8 mm (0.315 in.)
Bolt B	55 mm (2.165 in.)	8 mm (0.315 in.)
Bolt C	70 mm (2.756 in.)	8 mm (0.315 in.)
Bolt D	35 mm (1.378 in.)	10 mm (0.394 in.)
Bolt E	55 mm (2.165 in.)	10 mm (0.394 in.)
Bolt F	80 mm (3.150 in.)	10 mm (0.394 in.)

22140_SEQU_G0278

Fig. 113 Timing chain cover bolt application chart—5.7L engine

1
2
3
4
5
22140_SEQU_G0279

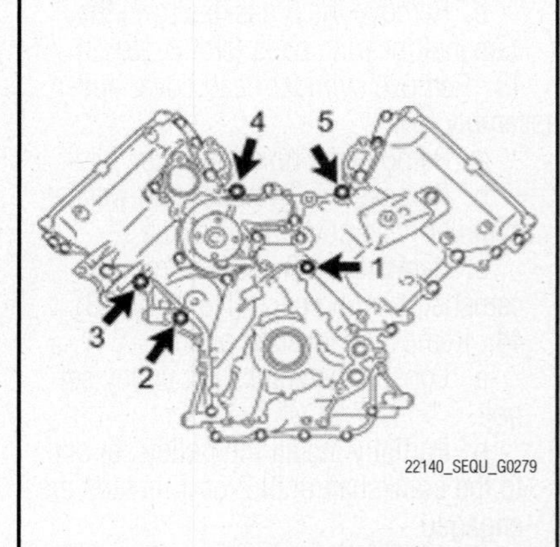

Fig. 114 Timing chain cover bolt tightening sequence No. 1—5.7L engine

63. Connect the 2 air tube hoses.
64. Connect the 2 hoses.
65. Install No. 1 and No 2 engine covers.
66. Install 2 new gaskets and the water by-pass joint with the 4 nuts and tighten to 15 ft. lbs. (21 Nm).
67. Connect the No. 2 water by-pass hose to the water by-pass joint.
68. Install the water by-pass pipe.
69. Connect the air tube.
70. Install the No. 1 water by-pass hose by connecting the hose to the water inlet housing and front water by-pass joint.
71. Install the 2 stud bolts.
72. Install the alternator with the 2 bolts and 2 nuts and tighten to 32 ft. lbs. (43 Nm).
73. Connect the alternator wire with the nut and tighten to 87 inch lbs. (9.8 Nm).
74. Connect the wire harness bracket to the alternator.

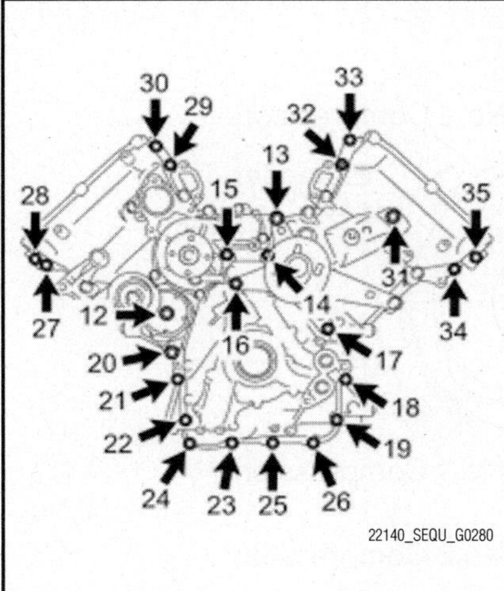

Fig. 115 Timing chain cover bolt tightening sequence No. 2—5.7L engine

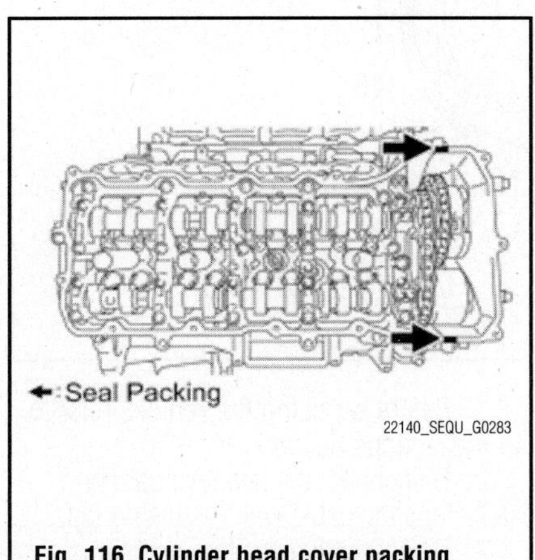

Fig. 116 Cylinder head cover packing location RH—5.7L engine

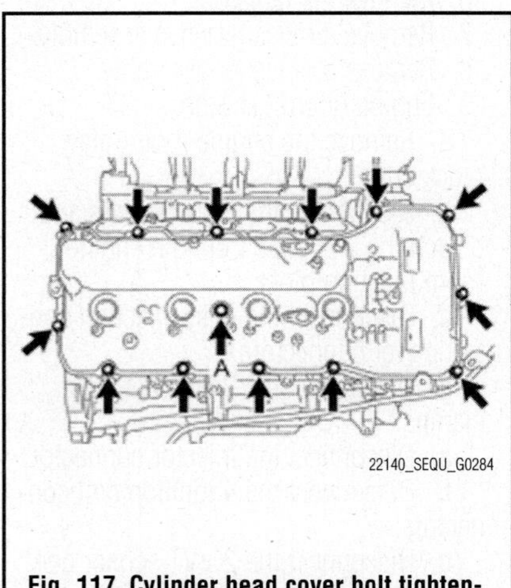

Fig. 117 Cylinder head cover bolt tightening sequence RH—5.7L engine

75. Connect the vane pump to the timing chain cover with the 2 bolts and tighten to 21 ft. lbs. (28 Nm).

76. Install engine oil level dipstick guide.

77. Install oil filter bracket (w/o Oil Cooler).
   a. Apply a light coat of engine oil to 2 new O-rings.
   b. Install the 2 O-rings to the timing chain cover.
   c. Install the oil filter bracket with the 2 bolts and 2 nuts and tighten to 26 ft. lbs. (35 Nm).

78. Install No. 1 oil cooler bracket.

79. Install No. 2 water by-pass pipe.

80. Install oil pressure sender gauge assembly.

81. Install oil filter element.

82. Connect No. 2 fuel tube sub-assembly.

83. Install the cooler compressor with the 2 stud bolts and tighten to 7 ft. lbs. (10 Nm).

84. Install the 2 bolts and 2 nuts and tighten to 18 ft. lbs. (235 Nm).

85. Connect the 3 heater hoses.

86. Connect the No. 2 and No. 3 air hoses.

87. Connect the 2 clamps and air pump wire harness.

88. Install the intake manifold.

89. Engine Room RH Side:
   a. Connect the 5 clamps.
   b. Connect the throttle position sensor and throttle control motor connector.
   c. Connect the 2 air pump connectors.
   d. Install the ground wire with the 2 bolts.
   e. Connect the noise filter connector.
   f. Connect the 2 VVT sensor connectors.
   g. Connect the injector connector.
   h. Connect the 4 ignition coil connectors.
   i. Connect the 2 camshaft timing oil control valve connectors.
   j. Connect the 2 clamps and power steering oil pressure switch connector.
   k. For 4WD: Connect the clamp and junction connector.

90. Engine Room LH Side:
   a. Connect the cooler compressor connector.
   b. Connect the 3 clamps.
   c. Connect the camshaft position sensor connector.
   d. Connect the 2 camshaft timing oil control valve connectors.
   e. Connect the engine coolant temperature sensor connector.
   f. Connect the 4 clamps.
   g. Install the ground wire with the 2 bolts.
   h. Connect the noise filter connector.
   i. Connect the 2 VVT sensor connectors.
   j. Connect the 4 ignition coil connectors.
   k. Connect the injector connector.
   l. Connect the 2 wire harness clamps.
   m. Connect the 4 air injection control driver connectors.
   n. Connect the 2 connectors and attach the 2 clamps to the engine room junction block.
   o. Install the engine room relay block cover.

91. Install air cleaner assembly.

92. Install air cleaner hose assembly.

93. Install the radiator.

94. Add engine oil.

95. Add engine coolant.

96. Connect cable to negative battery terminal.

97. Inspect for leaks.

98. Check engine oil level.

99. Use the Techstream diagnostic tool, or equivalent and reprogram the required systems.

INSPECTION

4.6L Engine

1. Before servicing the vehicle, refer to the Precautions Section.
2. Disconnect the negative battery cable. Tape the cable with insulating tape.
3. Remove the oil pump from the engine and disassemble it.
4. Coat the relief valve with engine oil and check that it falls smoothly into the valve hole by its own weight.
5. If it doesn't, replace the relief valve. |If necessary, replace the oil pump assembly.
6. Place the drive and driven rotors into the oil pump body.
7. Using a feeler gauge, measure the clearance between the drive and driven rotor tips.
8. Standard tip clearance is 0.180 to 0.300 mm (0.00709 to 0.0118 in.).
9. If the tip clearance is greater than maximum, replace the rotors as a set.
10. Using a feeler gauge and precision straight edge, measure the clearance between the rotors and precision straight edge.
11. Standard side clearance is 0.030 to 0.090 mm (0.0012 to 0.0035 in.).
12. If the side clearance is greater than maximum, replace the rotors as a set. If necessary, replace the oil pump assembly.

13. Using a feeler gauge, measure the clearance between the driven rotor and body.
14. Standard body clearance is 0.175 to 0.250 mm (0.00689 to 0.00984 in.).
15. If the body clearance is greater than maximum, replace the rotors as a set. If necessary, replace the oil pump assembly.
16. Use the Techstream diagnostic tool, or equivalent and reprogram the required systems.

5.7L Engine

1. Before servicing the vehicle, refer to the Precautions Section.
2. Disconnect the negative battery cable. Tape the cable with insulating tape.
3. Remove the oil pump from the engine and disassemble it.
4. Coat the relief valve with engine oil and check that it falls smoothly into the valve hole by its own weight.
5. If it doesn't, replace the relief valve. If necessary, replace the oil pump assembly.
6. Place the drive and driven rotors into the oil pump body.
7. Using a feeler gauge, measure the clearance between the drive and driven rotor tips.
8. Standard tip clearance is 0.180 to 0.300 mm (0.00709 to 0.0118 in.).
9. If the tip clearance is greater than maximum, replace the rotors as a set.
10. Using a feeler gauge and precision straight edge, measure the clearance between the rotors and precision straight edge.
11. Standard side clearance is 0.030 to 0.090 mm (0.0012 to 0.0035 in.).
12. If the side clearance is greater than maximum, replace the rotors as a set. If necessary, replace the oil pump assembly.
13. Using a feeler gauge, measure the clearance between the driven rotor and body.
14. Standard body clearance is 0.175 to 0.250 mm (0.00689 to 0.00984 in.).
15. If the body clearance is greater than maximum, replace the rotors as a set. If necessary, replace the oil pump assembly.
16. Use the Techstream diagnostic tool, or equivalent and reprogram the required systems.

PISTON AND RING

POSITIONING

See Figures 118 through 120.

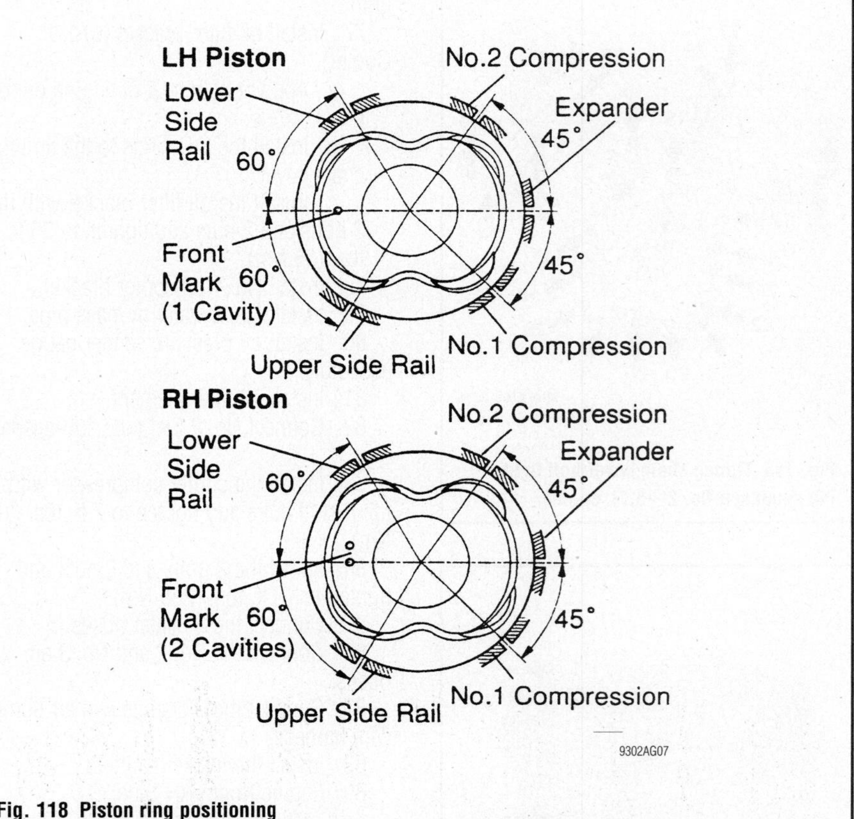

Fig. 118 Piston ring positioning

REAR MAIN SEAL

REMOVAL & INSTALLATION

See Figures 121 and 122.

1. Before servicing the vehicle, refer to the Precautions Section.
2. Disconnect the negative battery cable. Tape the cable with insulating tape.
3. Remove the transmission and flywheel from the vehicle.
4. Cut off the rubber lip portion of the seal with a sharp knife.
5. Pry out the oil seal.

To install:

➡Be sure to use new fasteners, as required.

6. Installation is the reverse of the removal procedure.
7. Use the Techstream diagnostic tool, or equivalent and reprogram the required systems.

TIMING CHAIN & SPROCKETS

REMOVAL & INSTALLATION

4.6L Engine

See Figures 123 through 141.

1. Before servicing the vehicle, refer to the Precautions Section.
2. Disconnect the negative battery cable. Tape the cable with insulating tape.
3. Discharge fuel system pressure.
4. Drain engine oil.
5. Drain engine coolant.
6. Remove the radiator.
7. Remove air cleaner hose assembly.
8. Remove air cleaner assembly.
9. Engine Room LH Side:
   a. Remove the engine room relay block cover.
   b. Disconnect the 2 connectors and detach the 2 clamps from the engine room junction block.
   c. Disconnect the 4 air injection control driver connectors.
   d. Disconnect the 2 wire harness clamps.
   e. Disconnect the injector connector.
   f. Disconnect the 4 ignition coil connectors.
   g. Disconnect the 2 VVT sensor connectors.
   h. Disconnect the 4 clamps.
   i. Remove the 2 bolts and ground wire.
   j. Disconnect the noise filter connector.

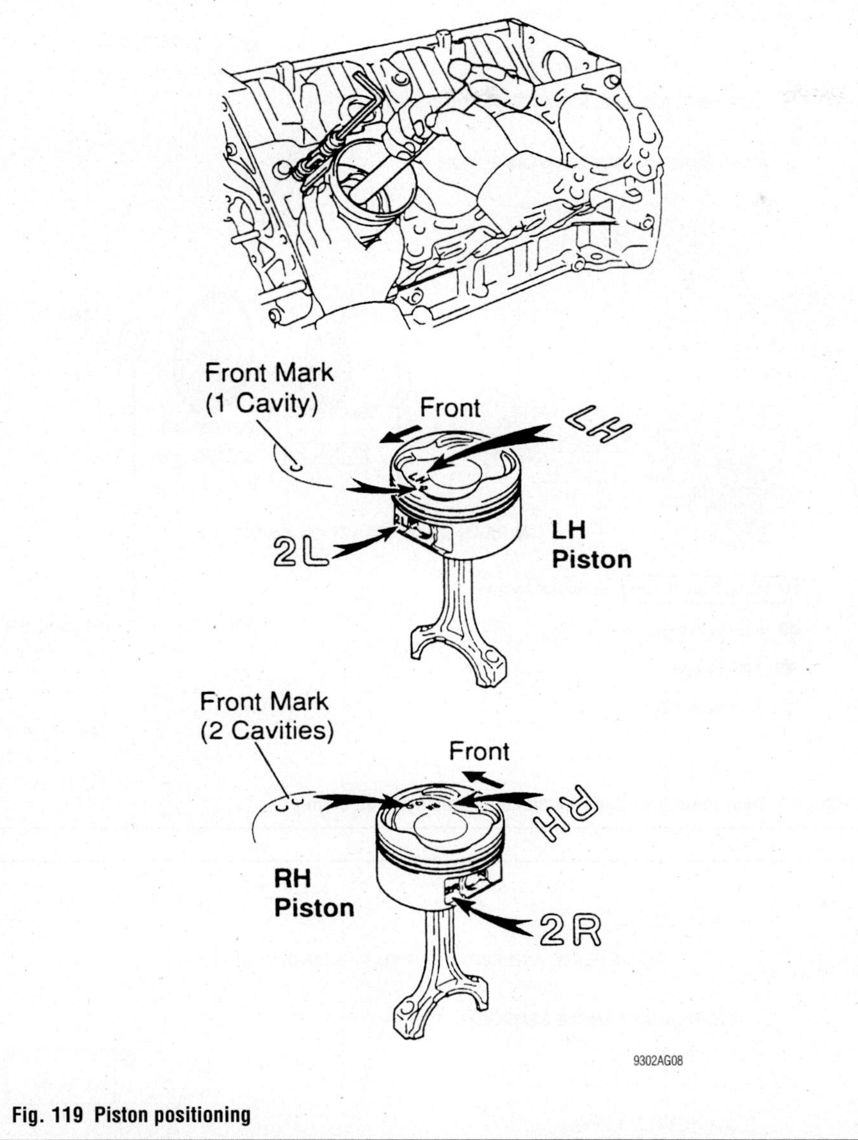

Fig. 119 Piston positioning

k. Disconnect the engine coolant temperature sensor connector.

l. Disconnect the 2 camshaft timing oil control valve connectors.

m. Disconnect the camshaft position sensor connector.

n. Disconnect the 3 clamps.

o. Disconnect the cooler compressor connector.

10. Engine Room RH Side.

a. Disconnect the 2 camshaft timing oil control valve connectors.

b. Disconnect the 4 ignition coil connectors.

c. Disconnect the injector connector.

d. Disconnect the 2 VVT sensor connectors.

e. Disconnect the noise filter connector.

f. Remove the 2 bolts and ground wire.

g. Disconnect the 2 air pump connectors.

h. Disconnect the throttle position sensor and throttle control motor connector.

i. Disconnect the 5 clamps.

j. Disconnect the 2 clamps and power steering oil pressure switch connector.

k. For 4WD: Disconnect the 2 clamps and power steering oil pressure switch connector.

11. Remove the intake manifold.

12. Disconnect the No. 2 and No. 3 air hoses and the 2 clamps.

13. Remove the 2 bolts and disconnect the heater 3 hoses.

14. Remove the 2 bolts, 2 nuts and 2 stud bolts, and disconnect the cooler compressor.

➡It is not necessary to completely remove the compressor. With the hoses connected to the compressor, hang the compressor on the vehicle body with a rope.

15. Remove the 2 bolts and disconnect the fuel tube.

16. Remove oil filter element.

17. Remove oil pressure sender gauge assembly.

18. Disconnect the 4 hoses and remove the water by-pass pipe.

19. Remove No. 1 oil cooler bracket (w/ oil cooler).

20. Remove the 2 bolts, 2 nuts and oil filter bracket.

21. Remove the 2 O-rings.

22. Remove engine oil level dipstick guide.

23. Remove the 2 bolts and disconnect the vane pump.

24. Disconnect the alternator connector.

25. Remove the terminal cap and nut, and disconnect the alternator wire.

26. Remove the bolt and disconnect the wire harness bracket from the alternator.

27. Remove the 2 bolts, 2 nuts and alternator.

28. Remove the 2 stud bolts.

29. Remove the No. 1 water by-pass hose by disconnect the hose from the water inlet housing and front water by-pass joint.

30. Remove the bolt and disconnect the air tube.

31. Remove the 2 bolts and hoses and water by-pass pipe.

32. Disconnect the No. 2 water by-pass hose from the water by-pass joint.

33. Remove the 4 nuts, water by-pass joint and 2 gaskets.

34. Remove No.1 and No. 2 engine covers .

35. Remove the bolt, disconnect the 2 hoses, and remove the air tube.

36. Remove the 3 bolts, water inlet housing and gasket.

37. Remove the 4 bolts and water pump pulley.

38. Remove the bolt and idler pulley.

39. Remove the 4 bolts and fan bracket.

40. Remove the standard bolt, 6 mm hexagon wrench bolt and belt tensioner.

41. Remove the 8 bolts and 8 ignition coils.

42. Remove cylinder head cover sub-assembly LH.

a. Remove the 14 bolts, seal washer, cylinder head cover and gasket.

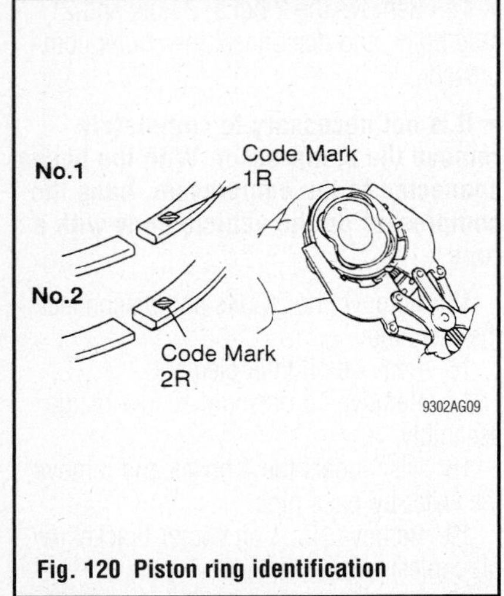

Fig. 120 Piston ring identification

b. Remove the 5 gaskets from the camshaft bearing caps (No. 2, No. 3).

43. Remove cylinder head cover sub-assembly RH.
    a. Remove the bolt and noise filter.
    b. Remove the 14 bolts, seal washer, cylinder head cover and gasket.
    c. Remove the 5 gaskets from the camshaft bearing caps (No. 1, No. 3).
44. Remove crankshaft pulley.
    a. Loosen the crankshaft pulley set bolt.
    b. Partially install the pulley set bolt to the crankshaft until 2 or 3 threads are engaged.
    c. Remove the crankshaft pulley.
45. Remove the bolt and disconnect the wire harness bracket.
46. Remove timing chain cover sub-assembly.
    a. Remove the 28 bolts and nut shown.
    b. Remove the timing chain cover by prying between it and the cylinder head and cylinder block with a screwdriver as shown
47. Remove the oil pump gasket from the cylinder block.
48. Remove the O-ring from the oil pan.
49. Remove the water inlet pipe.
50. Remove the 2 O-rings from the water inlet pipe.
51. Set No. 1 cylinder to TDC / compression.
52. Remove chain tensioner assembly.
53. Remove chain tensioner slipper.
54. Remove chain vibration damper.
55. Remove chain sub-assembly.
56. Remove chain tensioner assembly.

To install:

➡Be sure to use new fasteners, as required.

57. Check No. 1 cylinder to TDC.

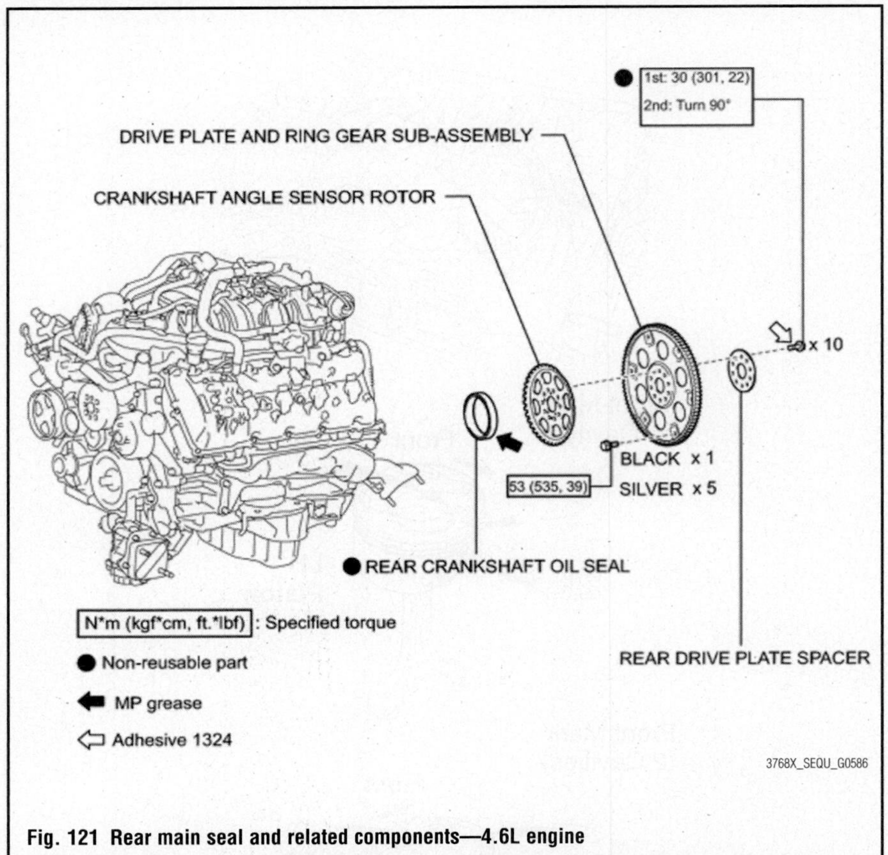

Fig. 121 Rear main seal and related components—4.6L engine

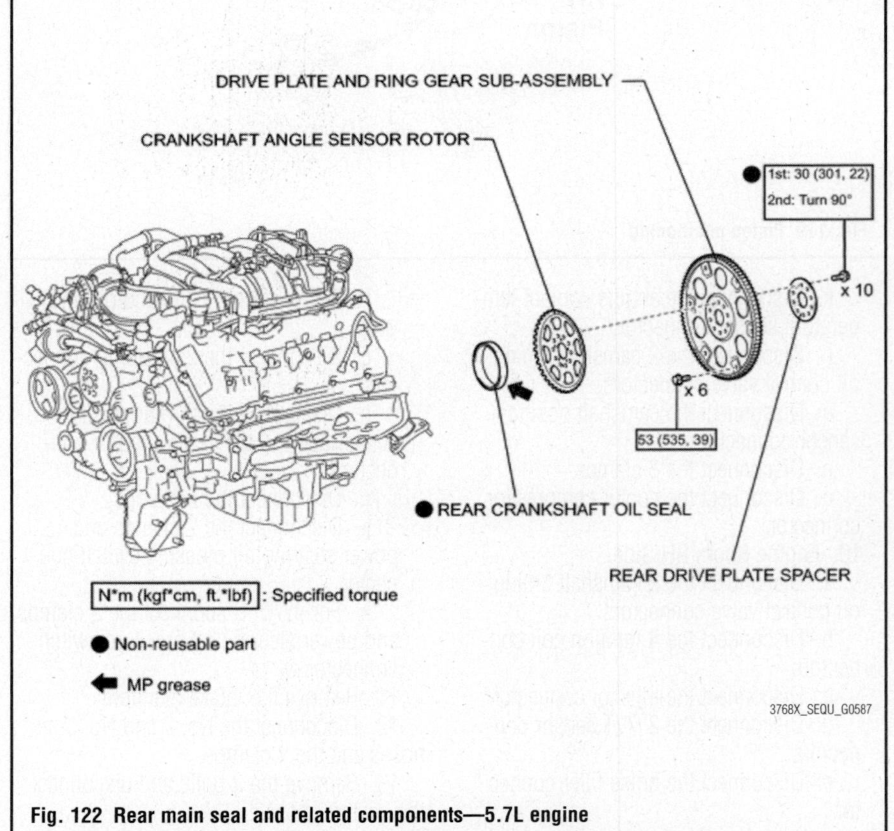

Fig. 122 Rear main seal and related components—5.7L engine

58. Install No.2 chain tensioner assembly. While raising up the No. 2 chain tensioner, insert a pin of _1.0 mm (0.0394 in.) into the hole to fix it in place.

59. Install No. 1 chain sub-assembly RH.

a. Align the No. 1 chain's orange mark plates with the camshaft timing gear's timing mark, and attach the chain to the gear.

b. Align the No. 1 chain's orange mark plate with the crankshaft timing gear's timing mark, and attach the chain to the gear.

c. Align the No. 2 chain's mark plates (yellow) with the timing marks of the camshaft timing gear and camshaft timing exhaust gear, and attach the No. 2 chain to the gears as shown.

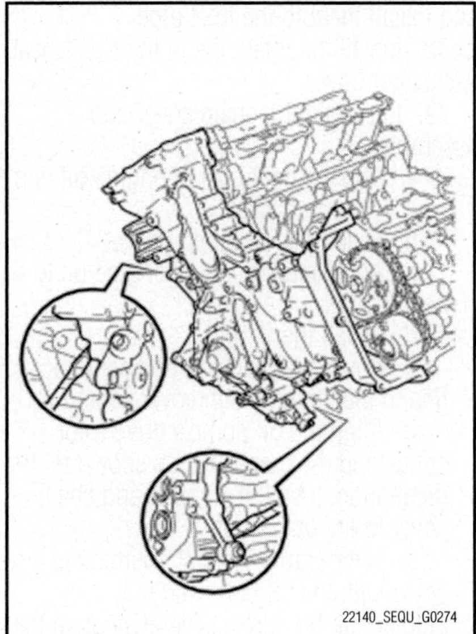

Fig. 123 Timing chain cover pry locations—4.6L engine

➡The crankshaft timing gear and camshaft exhaust gear will be installed with the No. 1 and No. 2 chains connected to the gears.

d. Install the crankshaft timing sprocket to the crankshaft.

e. Align and attach the knock pin of the No. 1 camshaft with the pin hole of the camshaft timing gear.

f. Using the hexagonal portion of the No. 2 camshaft, align and attach the knock pin of the No. 2 camshaft with the pin hole of the camshaft timing exhaust gear.

g. Remove the pin from the No. 2 chain tensioner.

60. Install the vibration damper with the 2 bolts.

61. Install No. 1 chain tensioner slipper.

➡If you cannot install the chain tensioner slipper due to the tension of the chain, use the hexagonal portion of the camshaft to loosen the chain, and then install the chain tensioner slipper

62. Install No. 1 chain tensioner assembly.

a. Move the stopper plate upward to release the lock, and push the plunger deep into the tensioner.

b. Move the stopper plate downward to set the lock, and insert a hexagon wrench into the hole of the stopper plate.

c. Install the chain tensioner with the 2 bolts and tighten to 7 ft. lbs. (10 Nm).

d. Remove the hexagon wrench from the chain tensioner.

63. Install No.3 chain tensioner assembly. While raising up the No. 3 chain tensioner, insert a pin of _1.0 mm (0.0394 in.) into the hole to fix it in place.

64. Install No. 1 chain sub-assembly LH.

a. Align the No. 1 chain's orange mark plates with the camshaft timing gear's timing mark, and attach the chain to the gear as shown.

b. Align the No. 1 chain's orange mark plate with the crankshaft timing gear's timing mark, and attach the chain to the gear as shown

c. Align the No. 2 chain's mark plates (yellow) with the timing marks of the camshaft timing gear and camshaft timing exhaust gear, and attach the No. 2 chain to the gears as shown

➡The crankshaft timing gear and camshaft exhaust gear will be installed with the No. 1 and No. 2 chains connected to the gears.

d. Install the crankshaft timing sprocket to the crankshaft.

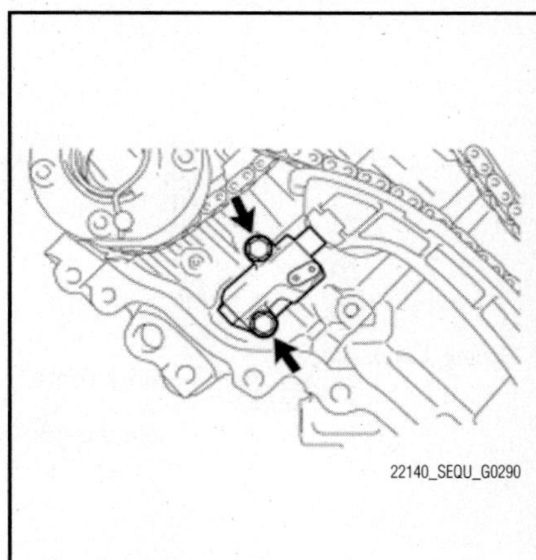

Fig. 126 Timing chain tensioner positioning RH—4.6L engine

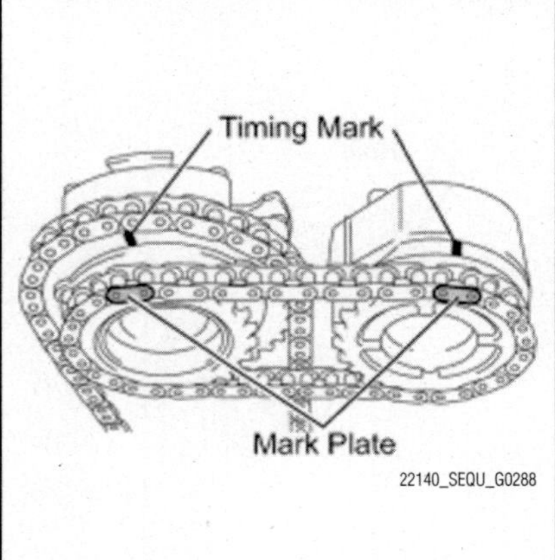

Fig. 124 Timing chain positioning RH—4.6L engine

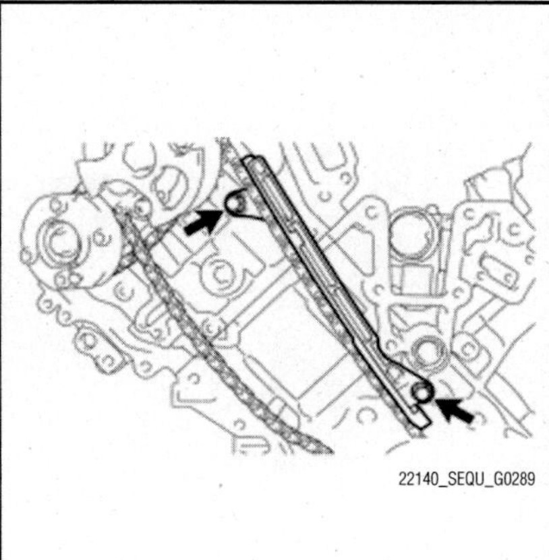

Fig. 125 Timing chain vibrations damper positioning RH—4.6L engine

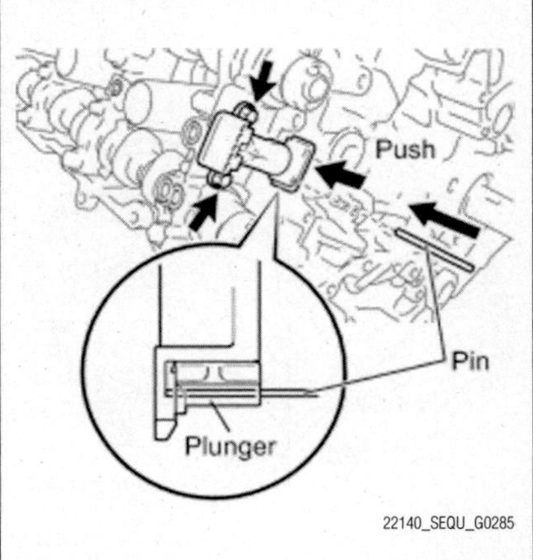

Fig. 127 Timing chain tensioner assembly No. 2—4.6L engine

e. Align and attach the knock pin of the No. 3 camshaft with the pin hole of the camshaft timing gear.

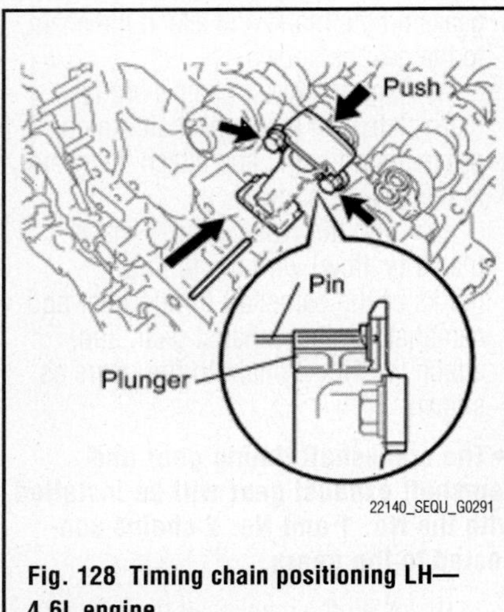

Fig. 128 Timing chain positioning LH—4.6L engine

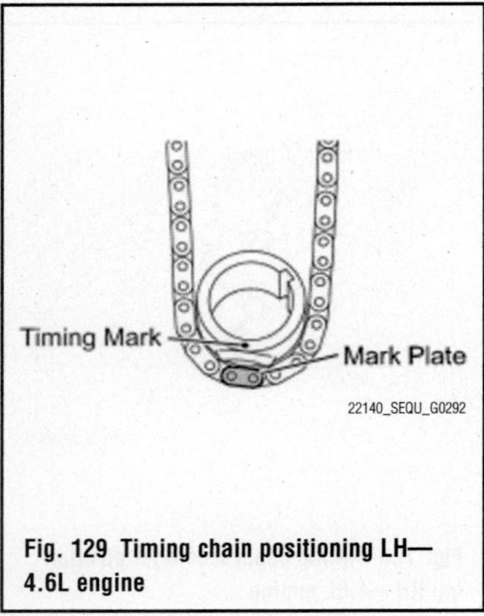

Fig. 129 Timing chain positioning LH—4.6L engine

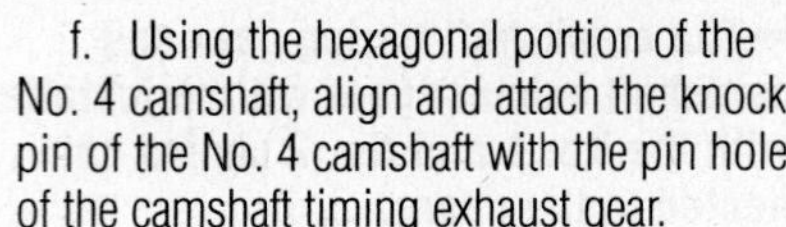

f. Using the hexagonal portion of the No. 4 camshaft, align and attach the knock pin of the No. 4 camshaft with the pin hole of the camshaft timing exhaust gear.

➡**Because the gears' timing mark positions may shift due to looseness of the No. 1 chain, use the hexagonal portion of the camshaft to hold the No. 3 camshaft in place until the No. 1 chain tensioner is installed.**

g. Remove the pin from the No. 2 chain tensioner.

65. Install No. 1 chain tensioner slipper.

➡**If you cannot install the chain tensioner slipper due to the tension of the chain, use the hexagonal portion of the camshaft to loosen the chain, and then install the chain tensioner slipper.**

66. Install No. 1 chain tensioner assembly.

a. Move the stopper plate upward to release the lock, and push the plunger deep into the tensioner.

b. Move the stopper plate downward to set the lock, and insert a hexagon wrench into the hole of the stopper plate.

c. Install the chain tensioner with the 2 bolts and tighten to 7 ft. lbs. (10 Nm).

67. Install the vibration damper with the 2 bolts

a. Remove the hexagon wrench from the chain tensioner.

68. Tighten camshaft timing gears LH.

a. Using a wrench to hold the hexagonal portion of the No. 3 camshaft, tighten the camshaft timing gear with the bolt and tighten to 74 ft. lbs. (100 Nm).

b. Using a wrench to hold the hexagonal portion of the No. 4 camshaft, tighten the camshaft timing gear with the bolt and tighten to 74 ft. lbs. (100 Nm).

69. Tighten camshaft timing gears RH.

a. Using a wrench to hold the hexagonal portion of the No. 1 camshaft, tighten the camshaft timing gear with the bolt and tighten to 74 ft. lbs. (100 Nm).

b. Using a wrench to hold the hexagonal portion of the No. 2 camshaft, tighten the camshaft timing gear with the bolt and tighten to 74 ft. lbs. (100 Nm).

70. Check No. 1 cylinder to TDC / compression.

a. Temporarily install the pulley set bolt.

b. Rotate the crankshaft clockwise, and check that the timing marks on the crankshaft timing gear and camshaft timing gears are as shown.

c. Remove the crankshaft pulley bolt.

71. Apply soapy water to 2 new O-rings and install them to the inlet pipe.

72. Install the inlet pipe to the No. 1 heat exchanger cover.

73. Install timing chain cover sub-assembly.

a. Apply a light coat of engine oil to a new oil pump gasket.

b. Install the oil pump gasket.

c. Apply a light coat of engine oil to a new O-ring.

d. Install the O-ring.

e. Apply seal packing in a continuous line to the timing chain cover as shown.

f. Align the oil pump's drive rotor spline and the crankshaft as shown in the illustration. Install the spline and chain cover to the crankshaft.

g. Temporarily install the timing chain cover with the 28 bolts and nut. Tighten the 5 bolts in several steps in the sequence shown to 35 ft. lbs. (47 Nm).

h. Temporarily install the fan bracket with the 4 bolts.

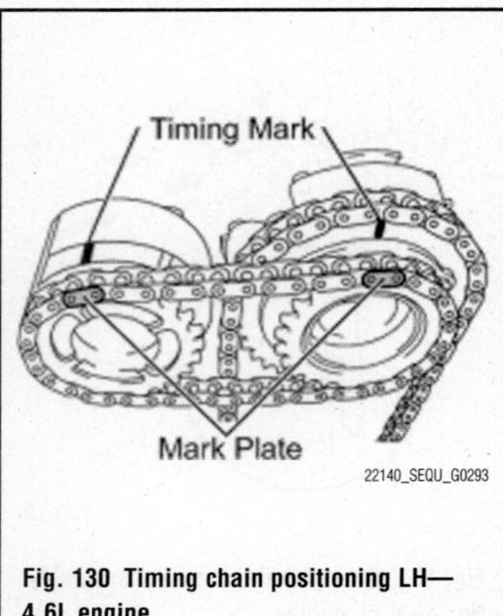

Fig. 130 Timing chain positioning LH—4.6L engine

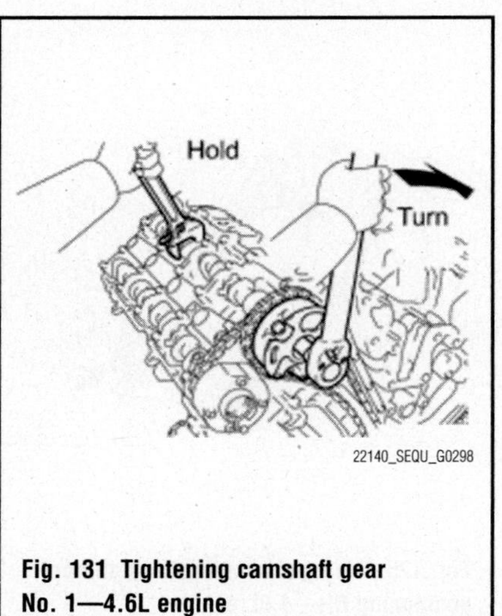

Fig. 131 Tightening camshaft gear No. 1—4.6L engine

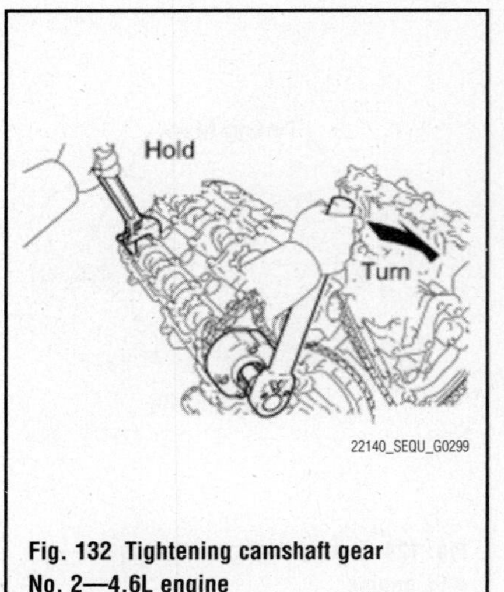

Fig. 132 Tightening camshaft gear No. 2—4.6L engine

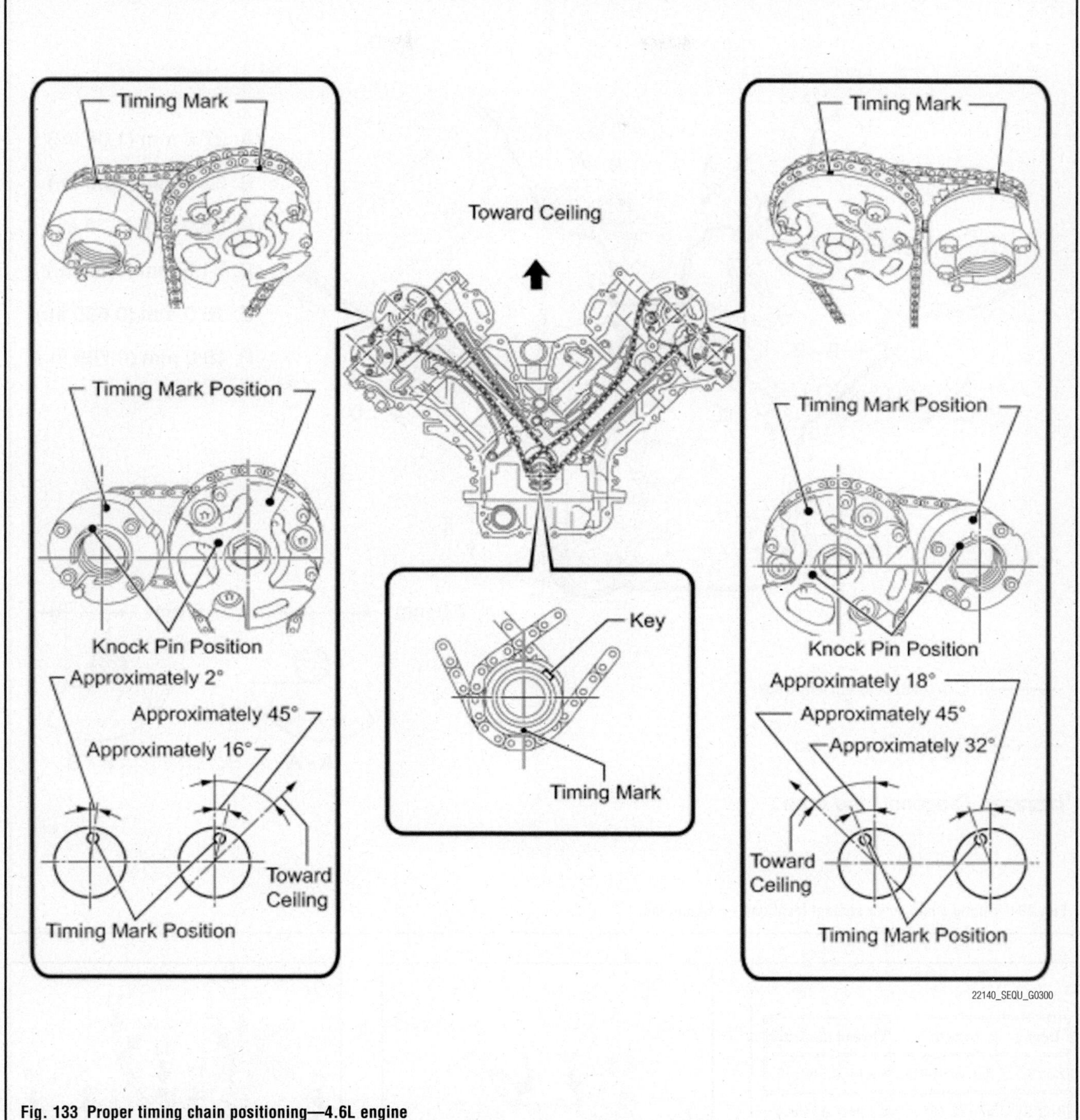

Fig. 133 Proper timing chain positioning—4.6L engine

i. Temporarily install the belt tensioner with the standard bolt and 6 mm hexagon wrench bolt.

j. Tighten the 23 bolts labeled 12 to 35 and nut in several steps in the sequence to 17 ft. lbs. (23 Nm).

74. Install cylinder head cover sub-assembly LH.

a. Install 5 new gaskets to the camshaft bearing caps (No. 2, No. 3).

b. Install the gasket to the cylinder head cover.

c. Apply seal packing as shown.

d. Install the cylinder head cover washer with a new seal and the 14 bolts and tighten bolt A to 15 ft. lbs. (21 Nm) and remaining bolts to 9 ft. lbs. (12 Nm).

75. Install cylinder head cover sub-assembly RH.

a. Install 5 new gaskets to the camshaft bearing caps (No. 1, No. 3).

b. Install the gasket to the cylinder head cover.

c. Apply seal packing as shown.

d. Install the cylinder head cover washer with a new seal and the 14 bolts and tighten bolt A to 15 ft. lbs. (21 Nm) and remaining bolts to 9 ft. lbs. (12 Nm).

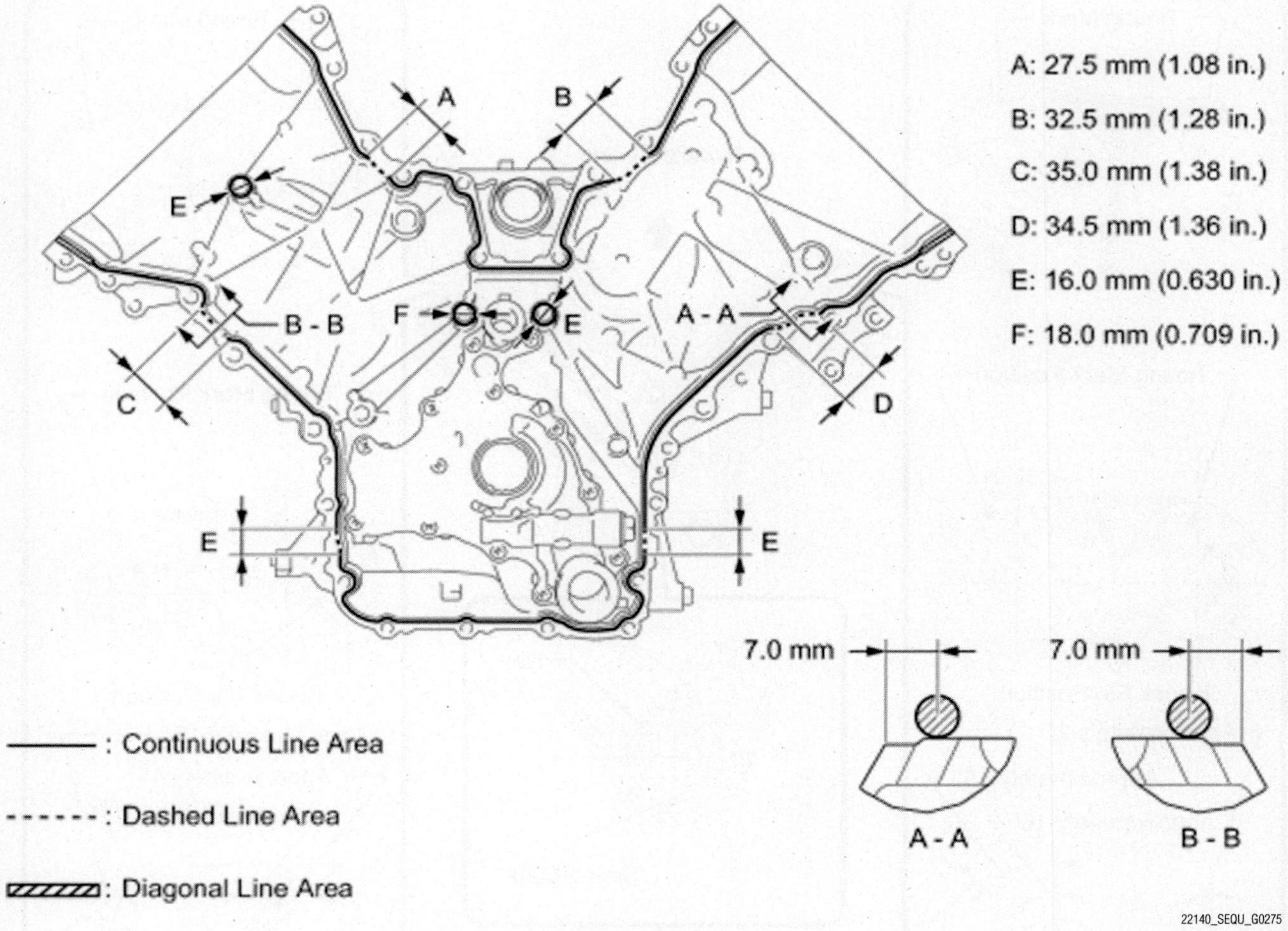

A: 27.5 mm (1.08 in.)
B: 32.5 mm (1.28 in.)
C: 35.0 mm (1.38 in.)
D: 34.5 mm (1.36 in.)
E: 16.0 mm (0.630 in.)
F: 18.0 mm (0.709 in.)

7.0 mm A - A

7.0 mm B - B

——— : Continuous Line Area

- - - - - - : Dashed Line Area

▨▨▨ : Diagonal Line Area

22140_SEQU_G0275

Fig. 134 Timing chain cover sealant locations—4.6L engine

Item	Length	Thread diameter
Bolt A	25 mm (0.984 in.)	8 mm (0.315 in.)
Bolt B	55 mm (2.165 in.)	8 mm (0.315 in.)
Bolt C	70 mm (2.756 in.)	8 mm (0.315 in.)
Bolt D	35 mm (1.378 in.)	10 mm (0.394 in.)
Bolt E	55 mm (2.165 in.)	10 mm (0.394 in.)
Bolt F	80 mm (3.150 in.)	10 mm (0.394 in.)

22140_SEQU_G0278

Fig. 135 Timing chain cover bolt application chart—4.6L engine

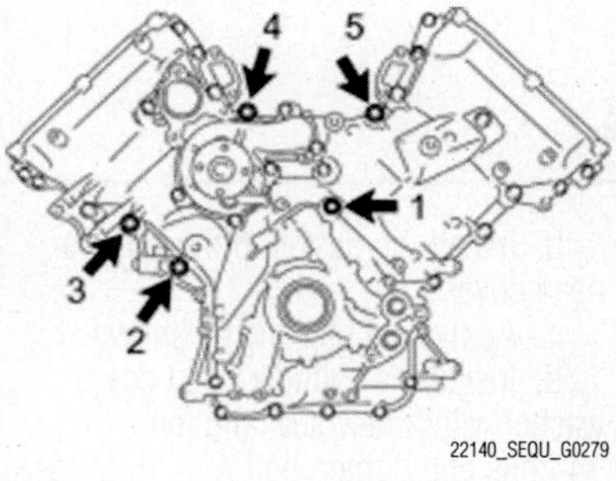

22140_SEQU_G0279

Fig. 136 Timing chain cover bolt tightening sequence No. 1—4.6L engine

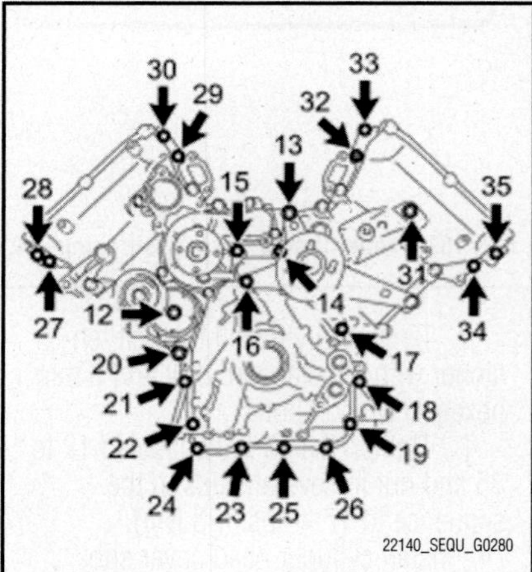

Fig. 137 Timing chain cover bolt tightening sequence No. 2—4.6L engine

e. Install the noise filter to the cylinder head cover with the bolt and tighten to 62 inch lbs. (7.0 Nm).

76. Install the 8 ignition coils with the 8 bolts and tighten to 7 ft. lbs. (10 Nm).

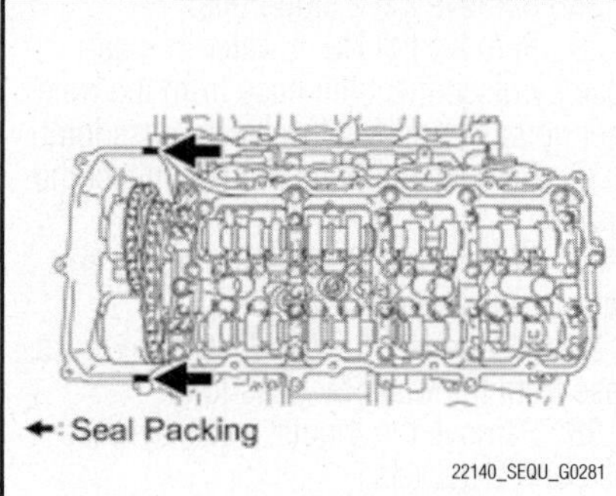

Fig. 138 Cylinder head cover packing location LH—4.6L engine

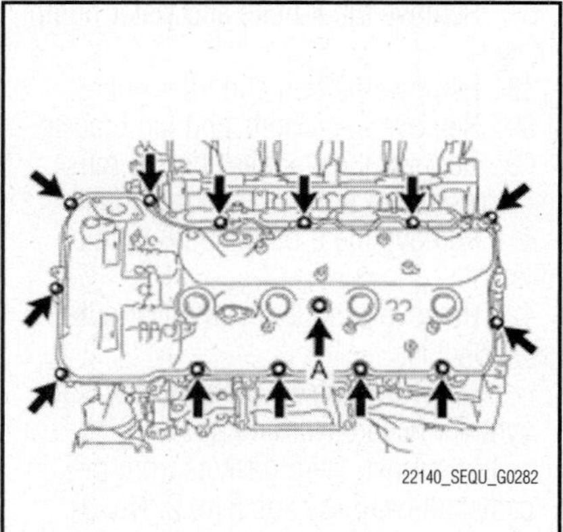

Fig. 139 Cylinder head cover bolt tightening sequence LH—4.6L engine

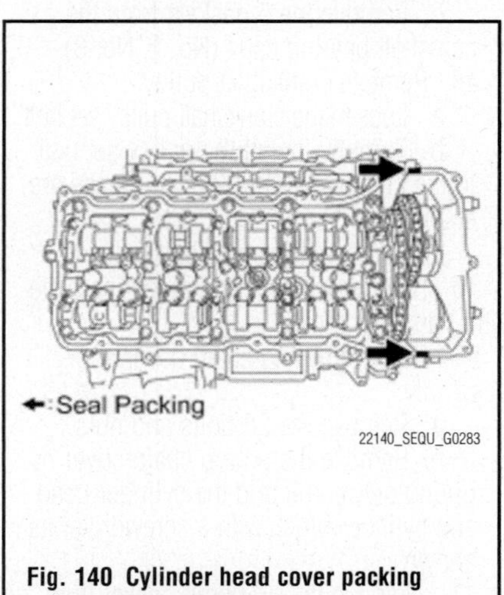

Fig. 140 Cylinder head cover packing location RH—4.6L engine

77. Align the pulley set key with the key groove of the pulley, and slide on the pulley and tighten to 221 ft. Lbs. (300 Nm).

78. Connect the bracket to the timing chain cover with the bolt.

79. Install the idler pulley with the bolt and tighten to 32 ft. lbs. (43 Nm).

80. Temporarily install the water pump pulley with the 4 bolts and tighten to 15 ft. lbs. (21 Nm).

81. Install a new water cover gasket to the timing chain cover.

82. Install the water inlet with the 3 bolts and tighten to 15 ft. lbs. (21 Nm).

83. Connect the 2 air tube hoses.

84. Connect the 2 hoses.

85. Install No. 1 and No 2 engine covers.

86. Install 2 new gaskets and the water by-pass joint with the 4 nuts and tighten to 15 ft. lbs. (21 Nm).

87. Connect the No. 2 water by-pass hose to the water by-pass joint.

88. Install the water by-pass pipe.

89. Connect the air tube.

90. Install the No. 1 water by-pass hose by connecting the hose to the water inlet housing and front water by-pass joint.

91. Install the 2 stud bolts.

92. Install the alternator with the 2 bolts and 2 nuts and tighten to 32 ft. lbs. (43 Nm).

93. Connect the alternator wire with the nut and tighten to 87 inch lbs. (9.8 Nm).

94. Connect the wire harness bracket to the alternator.

95. Connect the vane pump to the timing chain cover with the 2 bolts and tighten to 21 ft. lbs. (28 Nm).

96. Install engine oil level dipstick guide.

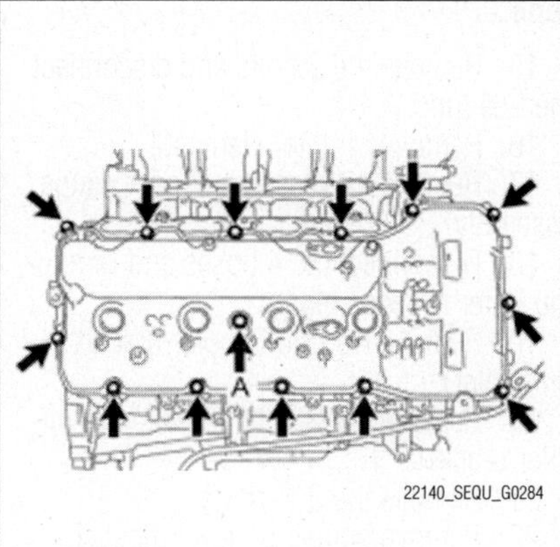

Fig. 141 Cylinder head cover bolt tightening sequence RH—4.6L engine

97. Install oil filter bracket (w/o Oil Cooler).

a. Apply a light coat of engine oil to 2 new O-rings.

b. Install the 2 O-rings to the timing chain cover.

c. Install the oil filter bracket with the 2 bolts and 2 nuts and tighten to 26 ft. lbs. (35 Nm).

98. Install No. 1 oil cooler bracket.

99. Install No. 2 water by-pass pipe.

100. Install oil pressure sender gauge assembly.

101. Install oil filter element.

102. Connect No. 2 fuel tube sub-assembly.

103. Install the cooler compressor with the 2 stud bolts and tighten to 7 ft. lbs. (10 Nm).

104. Install the 2 bolts and 2 nuts and tighten to 18 ft. lbs. (235 Nm).

105. Connect the 3 heater hoses.

106. Connect the No. 2 and No. 3 air hoses.

107. Connect the 2 clamps and air pump wire harness.

108. Install the intake manifold.

109. Engine Room RH Side:

a. Connect the 5 clamps.

b. Connect the throttle position sensor and throttle control motor connector.

c. Connect the 2 air pump connectors.

d. Install the ground wire with the 2 bolts.

e. Connect the noise filter connector.

f. Connect the 2 VVT sensor connectors.

g. Connect the injector connector.

h. Connect the 4 ignition coil connectors.

i. Connect the 2 camshaft timing oil control valve connectors.

j. Connect the 2 clamps and power steering oil pressure switch connector.

k. For 4WD: Connect the clamp and junction connector.

110. Engine Room LH Side:

a. Connect the cooler compressor connector.

b. Connect the 3 clamps.

c. Connect the camshaft position sensor connector.

d. Connect the 2 camshaft timing oil control valve connectors.

e. Connect the engine coolant temperature sensor connector.

f. Connect the 4 clamps.

g. Install the ground wire with the 2 bolts.

h. Connect the noise filter connector.

i. Connect the 2 VVT sensor connectors.

j. Connect the 4 ignition coil connectors.

k. Connect the injector connector.

l. Connect the 2 wire harness clamps.

m. Connect the 4 air injection control driver connectors.

n. Connect the 2 connectors and attach the 2 clamps to the engine room junction block.

o. Install the engine room relay block cover.

111. Install air cleaner assembly.

112. Install air cleaner hose assembly.

113. Install the radiator.

114. Add engine oil.

115. Add engine coolant.

116. Connect cable to negative battery terminal.

117. Inspect for leaks.

118. Check engine oil level.

119. Use the Techstream diagnostic tool, or equivalent and reprogram the required systems.

5.7L Engine

See Figures 144 through 167.

1. Before servicing the vehicle, refer to the Precautions Section.
2. Disconnect the negative battery cable. Tape the cable with insulating tape.
3. Discharge fuel system pressure.
4. Drain engine oil.
5. Drain engine coolant.
6. Remove the radiator.
7. Remove air cleaner hose assembly.
8. Remove air cleaner assembly.
9. Engine Room LH Side:
    a. Remove the engine room relay block cover.
    b. Disconnect the 2 connectors and detach the 2 clamps from the engine room junction block.
    c. Disconnect the 4 air injection control driver connectors.
    d. Disconnect the 2 wire harness clamps.
    e. Disconnect the injector connector.
    f. Disconnect the 4 ignition coil connectors.
    g. Disconnect the 2 VVT sensor connectors.
    h. Disconnect the 4 clamps.
    i. Remove the 2 bolts and ground wire.
    j. Disconnect the noise filter connector.
    k. Disconnect the engine coolant temperature sensor connector.
    l. Disconnect the 2 camshaft timing oil control valve connectors.
    m. Disconnect the camshaft position sensor connector.
    n. Disconnect the 3 clamps.
    o. Disconnect the cooler compressor connector.
10. Engine Room RH Side.
    a. Disconnect the 2 camshaft timing oil control valve connectors.
    b. Disconnect the 4 ignition coil connectors.
    c. Disconnect the injector connector.
    d. Disconnect the 2 VVT sensor connectors.
    e. Disconnect the noise filter connector.
    f. Remove the 2 bolts and ground wire.
    g. Disconnect the 2 air pump connectors.
    h. Disconnect the throttle position sensor and throttle control motor connector.
    i. Disconnect the 5 clamps.
    j. Disconnect the 2 clamps and power steering oil pressure switch connector.
    k. For 4WD: Disconnect the 2 clamps and power steering oil pressure switch connector.
11. Remove the intake manifold.
12. Disconnect the No. 2 and No. 3 air hoses and the 2 clamps.
13. Remove the 2 bolts and disconnect the heater 3 hoses.
14. Remove the 2 bolts, 2 nuts and 2 stud bolts, and disconnect the cooler compressor.

➡It is not necessary to completely remove the compressor. With the hoses connected to the compressor, hang the compressor on the vehicle body with a rope.

15. Remove the 2 bolts and disconnect the fuel tube.
16. Remove oil filter element.
17. Remove oil pressure sender gauge assembly.
18. Disconnect the 4 hoses and remove the water by-pass pipe.
19. Remove No. 1 oil cooler bracket (w/ oil cooler).
20. Remove the 2 bolts, 2 nuts and oil filter bracket.
21. Remove the 2 O-rings.
22. Remove engine oil level dipstick guide.
23. Remove the 2 bolts and disconnect the vane pump.
24. Disconnect the alternator connector.
25. Remove the terminal cap and nut, and disconnect the alternator wire.
26. Remove the bolt and disconnect the wire harness bracket from the alternator.
27. Remove the 2 bolts, 2 nuts and alternator.
28. Remove the 2 stud bolts.
29. Remove the No. 1 water by-pass hose by disconnect the hose from the water inlet housing and front water by-pass joint.
30. Remove the bolt and disconnect the air tube.
31. Remove the 2 bolts and hoses and water by-pass pipe.
32. Disconnect the No. 2 water by-pass hose from the water by-pass joint.
33. Remove the 4 nuts, water by-pass joint and 2 gaskets.
34. Remove No.1 and No. 2 engine covers .
35. Remove the bolt, disconnect the 2 hoses, and remove the air tube.
36. Remove the 3 bolts, water inlet housing and gasket.
37. Remove the 4 bolts and water pump pulley.
38. Remove the bolt and idler pulley.
39. Remove the 4 bolts and fan bracket.
40. Remove the standard bolt, 6 mm hexagon wrench bolt and belt tensioner.
41. Remove the 8 bolts and 8 ignition coils.
42. Remove cylinder head cover sub-assembly LH.
    a. Remove the 14 bolts, seal washer, cylinder head cover and gasket.
    b. Remove the 5 gaskets from the camshaft bearing caps (No. 2, No. 3).
43. Remove cylinder head cover sub-assembly RH.
    a. Remove the bolt and noise filter.
    b. Remove the 14 bolts, seal washer, cylinder head cover and gasket.
    c. Remove the 5 gaskets from the camshaft bearing caps (No. 1, No. 3).
44. Remove crankshaft pulley.
    a. Loosen the crankshaft pulley set bolt.
    b. Partially install the pulley set bolt to the crankshaft until 2 or 3 threads are engaged.
    c. Remove the crankshaft pulley.
45. Remove the bolt and disconnect the wire harness bracket.
46. Remove timing chain cover sub-assembly.
    a. Remove the 28 bolts and nuts.
    b. Remove the timing chain cover by prying between it and the cylinder head and cylinder block with a screwdriver as shown
47. Remove the oil pump gasket from the cylinder block.

48. Remove the O-ring from the oil pan.
49. Remove the water inlet pipe.
50. Remove the 2 O-rings from the water inlet pipe.
51. Set No. 1 cylinder to TDC / compression.
52. Remove chain tensioner assembly.
53. Remove chain tensioner slipper.
54. Remove chain vibration damper.
55. Remove chain sub-assembly.
56. Remove chain tensioner assembly.

To install:

➡Be sure to use new fasteners, as required.

57. Check No. 1 cylinder to TDC.
58. Install No.2 chain tensioner assembly. While raising up the No. 2 chain tensioner, insert a pin of _1.0 mm (0.0394 in.) into the hole to fix it in place.
59. Install No. 1 chain sub-assembly RH.
    a. Align the No. 1 chain's orange mark plates with the camshaft timing gear's timing mark, and attach the chain to the gear as shown.
    b. Align the No. 1 chain's orange mark plate with the crankshaft timing gear's timing mark, and attach the chain to the gear.
    c. Align the No. 2 chain's mark plates (yellow) with the timing marks of the camshaft timing gear and camshaft timing exhaust gear, and attach the No. 2 chain to the gears as shown.

➡The crankshaft timing gear and camshaft exhaust gear will be installed with the No. 1 and No. 2 chains connected to the gears.

    d. Install the crankshaft timing sprocket to the crankshaft.
    e. Align and attach the knock pin of the No. 1 camshaft with the pin hole of the camshaft timing gear.
    f. Using the hexagonal portion of the No. 2 camshaft, align and attach the knock pin of the No. 2 camshaft with the pin hole of the camshaft timing exhaust gear.
    g. Remove the pin from the No. 2 chain tensioner.
60. Install the vibration damper with the 2 bolts.
61. Install No. 1 chain tensioner slipper.

➡If you cannot install the chain tensioner slipper due to the tension of the chain, use the hexagonal portion of the camshaft to loosen the chain, and then install the chain tensioner slipper

62. Install No. 1 chain tensioner assembly.
    a. Move the stopper plate upward to release the lock, and push the plunger deep into the tensioner.
    b. Move the stopper plate downward to set the lock, and insert a hexagon wrench into the hole of the stopper plate.
    c. Install the chain tensioner with the 2 bolts and tighten to 7 ft. lbs. (10 Nm).
    d. Remove the hexagon wrench from the chain tensioner.
63. Install No.3 chain tensioner assembly. While raising up the No. 3 chain tensioner, insert a pin of _1.0 mm (0.0394 in.) into the hole to fix it in place.
64. Install No. 1 chain sub-assembly LH.
    a. Align the No. 1 chain's orange mark plates with the camshaft timing gear's timing mark, and attach the chain to the gear as shown.
    b. Align the No. 1 chain's orange mark plate with the crankshaft timing gear's timing mark, and attach the chain to the gear as shown
    c. Align the No. 2 chain's mark plates (yellow) with the timing marks of the camshaft timing gear and camshaft timing exhaust gear, and attach the No. 2 chain to the gears as shown

➡The crankshaft timing gear and camshaft exhaust gear will be installed with the No. 1 and No. 2 chains connected to the gears.

    d. Install the crankshaft timing sprocket to the crankshaft.
    e. Align and attach the knock pin of the No. 3 camshaft with the pin hole of the camshaft timing gear.
    f. Using the hexagonal portion of the No. 4 camshaft, align and attach the knock pin of the No. 4 camshaft with the pin hole of the camshaft timing exhaust gear.

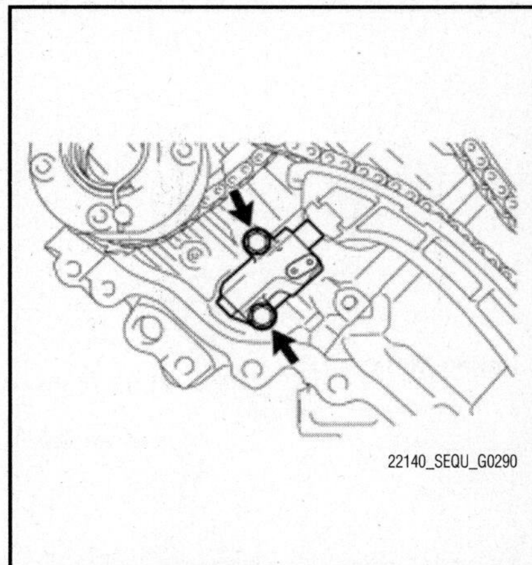

Fig. 144 Timing chain tensioner positioning RH—5.7L engine

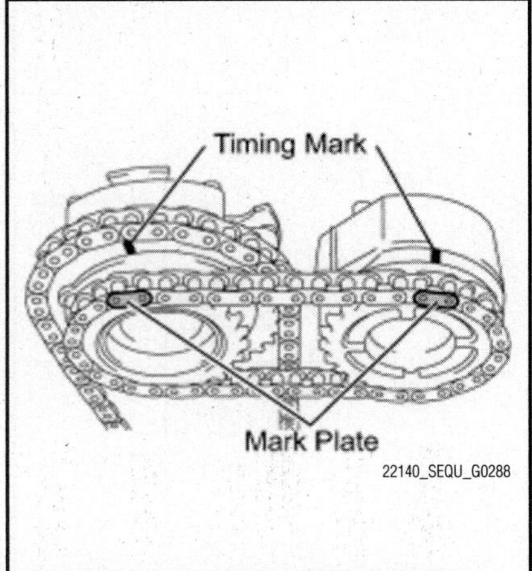

Fig. 142 Timing chain positioning RH—5.7L engine

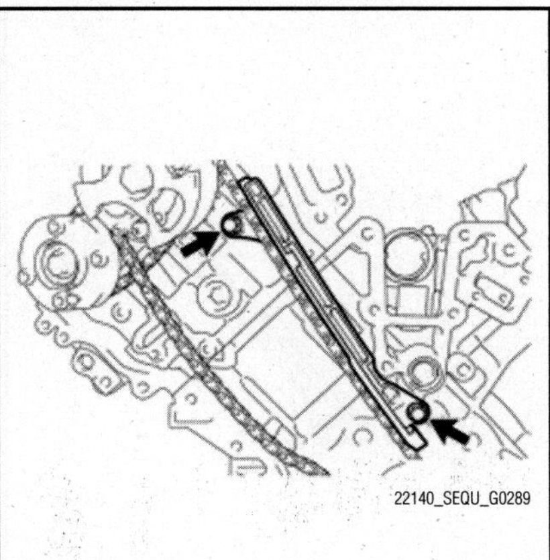

Fig. 143 Timing chain vibrations damper positioning RH—5.7L engine

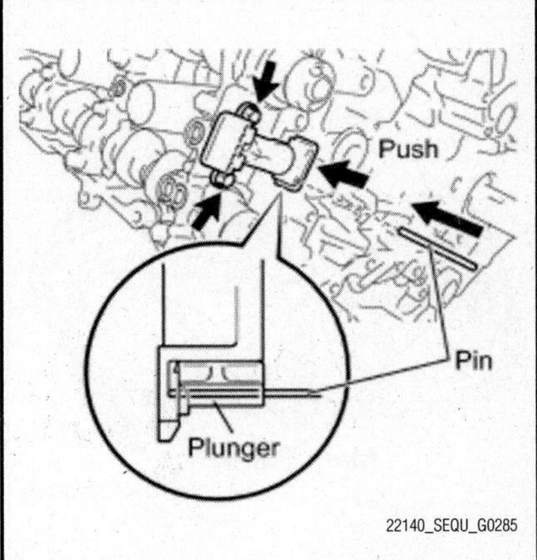

Fig. 145 Timing chain tensioner assembly No. 2—5.7L engine

➡Because the gears' timing mark positions may shift due to looseness of the No. 1 chain, use the hexagonal portion of the camshaft to hold the No. 3 camshaft in place until the No. 1 chain tensioner is installed.

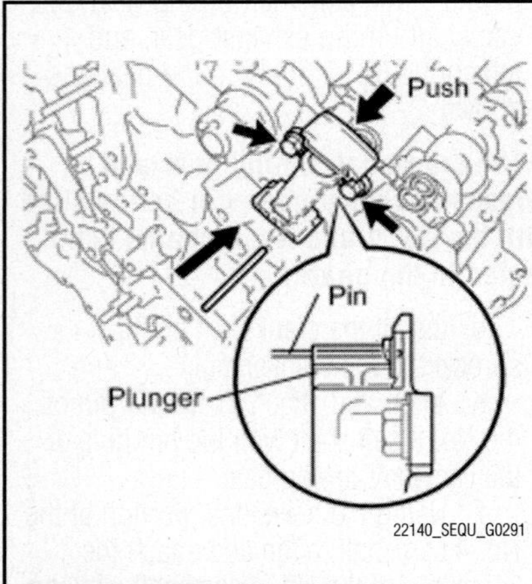

Fig. 146 Timing chain positioning LH—5.7L engine

g. Remove the pin from the No. 2 chain tensioner.

65. Install No. 1 chain tensioner slipper.

➡If you cannot install the chain tensioner slipper due to the tension of the chain, use the hexagonal portion of the camshaft to loosen the chain, and then install the chain tensioner slipper.

66. Install No. 1 chain tensioner assembly.

a. Move the stopper plate upward to release the lock, and push the plunger deep into the tensioner.

b. Move the stopper plate downward to set the lock, and insert a hexagon wrench into the hole of the stopper plate.

c. Install the chain tensioner with the 2 bolts and tighten to 7 ft. lbs. (10 Nm).

67. Install the vibration damper with the 2 bolts

a. Remove the hexagon wrench from the chain tensioner.

68. Tighten camshaft timing gears LH.

a. Using a wrench to hold the hexagonal portion of the No. 3 camshaft, tighten the camshaft timing gear with the bolt and tighten to 74 ft. lbs. (100 Nm).

b. Using a wrench to hold the hexagonal portion of the No. 4 camshaft, tighten the camshaft timing gear with the bolt and tighten to 74 ft. lbs. (100 Nm).

69. Tighten camshaft timing gears RH.

a. Using a wrench to hold the hexagonal portion of the No. 1 camshaft, tighten the camshaft timing gear with the bolt and tighten to 74 ft. lbs. (100 Nm).

b. Using a wrench to hold the hexagonal portion of the No. 2 camshaft, tighten the camshaft timing gear with the bolt and tighten to 74 ft. lbs. (100 Nm).

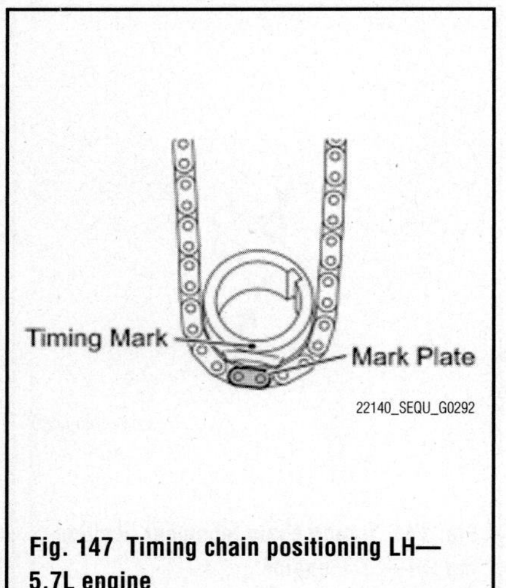

Fig. 147 Timing chain positioning LH—5.7L engine

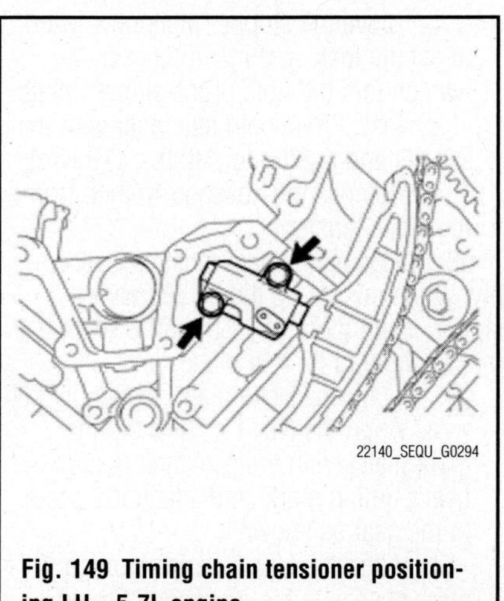

Fig. 149 Timing chain tensioner positioning LH—5.7L engine

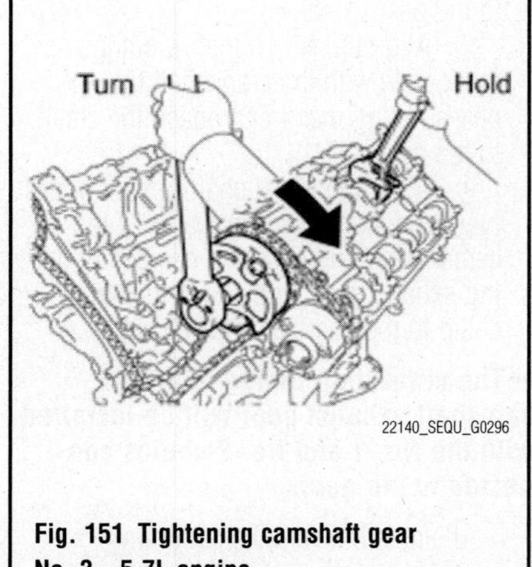

Fig. 151 Tightening camshaft gear No. 3—5.7L engine

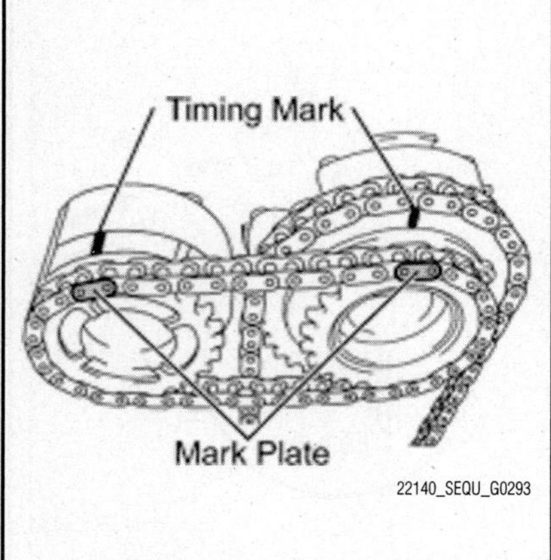

Fig. 148 Timing chain positioning LH—5.7L engine

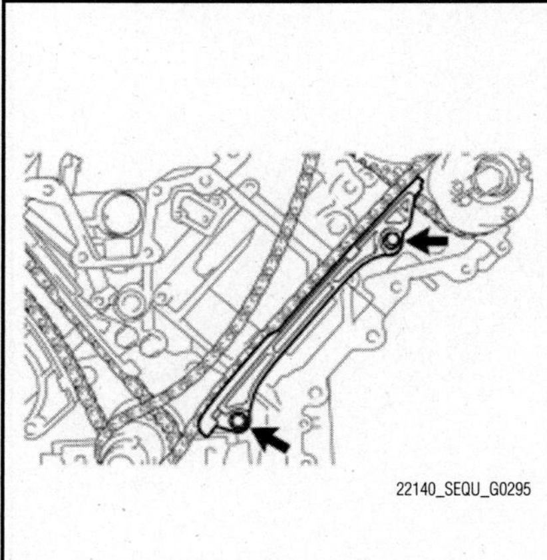

Fig. 150 Timing chain vibrations damper positioning RH—5.7L engine

Fig. 152 Tightening camshaft gear No. 4—5.7L engine

70. Check No. 1 cylinder to TDC / compression.
a. Temporarily install the pulley set bolt.
b. Rotate the crankshaft clockwise, and check that the timing marks on the crankshaft timing gear and camshaft timing gears are as shown.
c. Remove the crankshaft pulley bolt.
71. Apply soapy water to 2 new O-rings and install them to the inlet pipe.
72. Install the inlet pipe to the No. 1 heat exchanger cover.
73. Install timing chain cover sub-assembly.
a. Apply a light coat of engine oil to a new oil pump gasket.
b. Install the oil pump gasket.
c. Apply a light coat of engine oil to a new O-ring.
d. Install the O-ring.
e. Apply seal packing in a continuous line to the timing chain cover as shown.

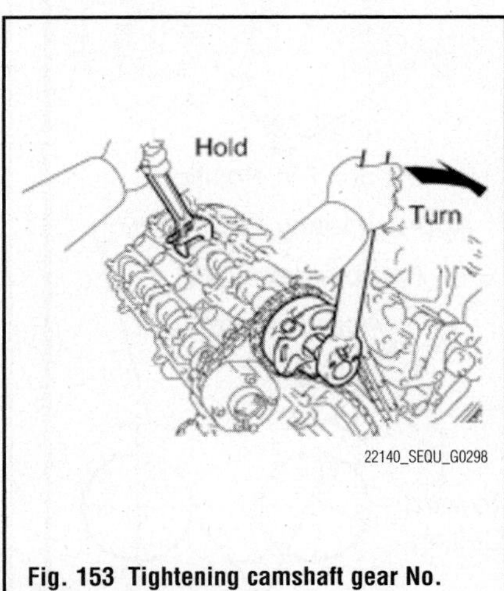

Fig. 153 Tightening camshaft gear No. 1—5.7L engine

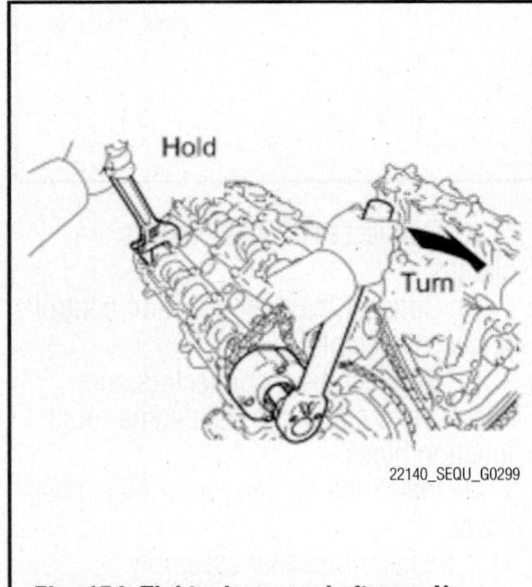

Fig. 154 Tightening camshaft gear No. 2—5.7L engine

f. Align the oil pump's drive rotor spline and the crankshaft as shown in the illustration. Install the spline and chain cover to the crankshaft.
g. Temporarily install the timing chain cover with the 28 bolts and nut.
Tighten the 5 bolts in several steps in the sequence shown to 35 ft. lbs. (47 Nm).
h. Temporarily install the fan bracket with the 4 bolts.
i. Temporarily install the belt tensioner with the standard bolt and 6 mm hexagon wrench bolt.
j. Tighten the 23 bolts labeled 12 to 35 and nut in several steps in the sequence to 17 ft. lbs. (23 Nm).
74. Install cylinder head cover sub-assembly LH.
a. Install 5 new gaskets to the camshaft bearing caps (No. 2, No. 3).
b. Install the gasket to the cylinder head cover.
c. Apply seal packing as shown.
d. Install the cylinder head cover washer with a new seal and the 14 bolts and tighten bolt A to 15 ft. lbs. (21 Nm) and remaining bolts to 9 ft. lbs. (12 Nm).
75. Install cylinder head cover sub-assembly RH.
a. Install 5 new gaskets to the camshaft bearing caps (No. 1, No. 3).
b. Install the gasket to the cylinder head cover.
c. Apply seal packing as shown.
d. Install the cylinder head cover washer with a new seal and the 14 bolts and tighten bolt A to 15 ft. lbs. (21 Nm) and remaining bolts to 9 ft. lbs. (12 Nm).
e. Install the noise filter to the cylinder head cover with the bolt and tighten to 62 inch lbs. (7.0 Nm).
76. Install the 8 ignition coils with the 8 bolts and tighten to 7 ft. lbs. (10 Nm).
77. Align the pulley set key with the key groove of the pulley, and slide on the pulley and tighten to 221 ft. Lbs. (300 Nm).
78. Connect the bracket to the timing chain cover with the bolt.
79. Install the idler pulley with the bolt and tighten to 32 ft. lbs. (43 Nm).
80. Temporarily install the water pump pulley with the 4 bolts and tighten to 15 ft. lbs. (21 Nm).
81. Install a new water cover gasket to the timing chain cover.
82. Install the water inlet with the 3 bolts and tighten to 15 ft. lbs. (21 Nm).
83. Connect the 2 air tube hoses.
84. Connect the 2 hoses.
85. Install No. 1 and No 2 engine covers.
86. Install 2 new gaskets and the water by-pass joint with the 4 nuts and tighten to 15 ft. lbs. (21 Nm).
87. Connect the No. 2 water by-pass hose to the water by-pass joint.
88. Install the water by-pass pipe.
89. Connect the air tube.
90. Install the No. 1 water by-pass hose by connecting the hose to the water inlet housing and front water by-pass joint.
91. Install the 2 stud bolts.
92. Install the alternator with the 2 bolts and 2 nuts and tighten to 32 ft. lbs. (43 Nm).
93. Connect the alternator wire with the nut and tighten to 87 inch lbs. (9.8 Nm).
94. Connect the wire harness bracket to the alternator.
95. Connect the vane pump to the timing chain cover with the 2 bolts and tighten to 21 ft. lbs. (28 Nm).
96. Install engine oil level dipstick guide.
97. Install oil filter bracket (w/o Oil Cooler).
a. Apply a light coat of engine oil to 2 new O-rings.
b. Install the 2 O-rings to the timing chain cover.
c. Install the oil filter bracket with the 2 bolts and 2 nuts and tighten to 26 ft. lbs. (35 Nm).
98. Install No. 1 oil cooler bracket.
99. Install No. 2 water by-pass pipe.
100. Install oil pressure sender gauge assembly.
101. Install oil filter element.
102. Connect No. 2 fuel tube sub-assembly.
103. Install the cooler compressor with the 2 stud bolts and tighten to 7 ft. lbs. (10 Nm).
104. Install the 2 bolts and 2 nuts and tighten to 18 ft. lbs. (235 Nm).
105. Connect the 3 heater hoses.
106. Connect the No. 2 and No. 3 air hoses.
107. Connect the 2 clamps and air pump wire harness.
108. Install the intake manifold.
109. Engine Room RH Side:
a. Connect the 5 clamps.
b. Connect the throttle position sensor and throttle control motor connector.
c. Connect the 2 air pump connectors.
d. Install the ground wire with the 2 bolts.
e. Connect the noise filter connector.
f. Connect the 2 VVT sensor connectors.
g. Connect the injector connector.
h. Connect the 4 ignition coil connectors.
i. Connect the 2 camshaft timing oil control valve connectors.

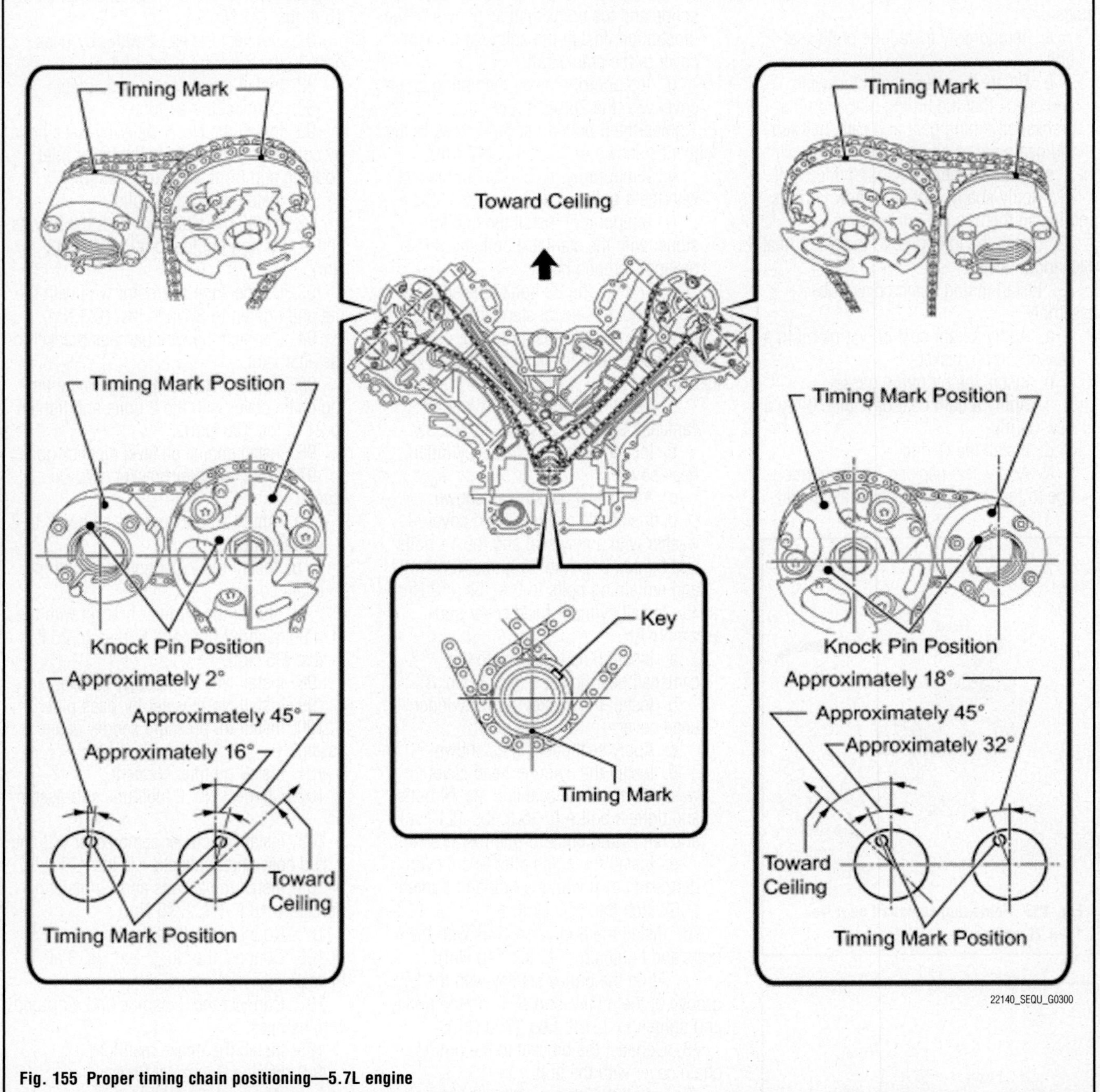

Fig. 155 Proper timing chain positioning—5.7L engine

j. Connect the 2 clamps and power steering oil pressure switch connector.

k. For 4WD: Connect the clamp and junction connector.

110. Engine Room LH Side:

a. Connect the cooler compressor connector.

b. Connect the 3 clamps.

c. Connect the camshaft position sensor connector.

d. Connect the 2 camshaft timing oil control valve connectors.

e. Connect the engine coolant temperature sensor connector.

f. Connect the 4 clamps.

g. Install the ground wire with the 2 bolts.

h. Connect the noise filter connector.

i. Connect the 2 VVT sensor connectors.

j. Connect the 4 ignition coil connectors.

k. Connect the injector connector.

l. Connect the 2 wire harness clamps.

m. Connect the 4 air injection control driver connectors.

n. Connect the 2 connectors and attach the 2 clamps to the engine room junction block.

o. Install the engine room relay block cover.

111. Install air cleaner assembly.

112. Install air cleaner hose assembly.

113. Install the radiator.

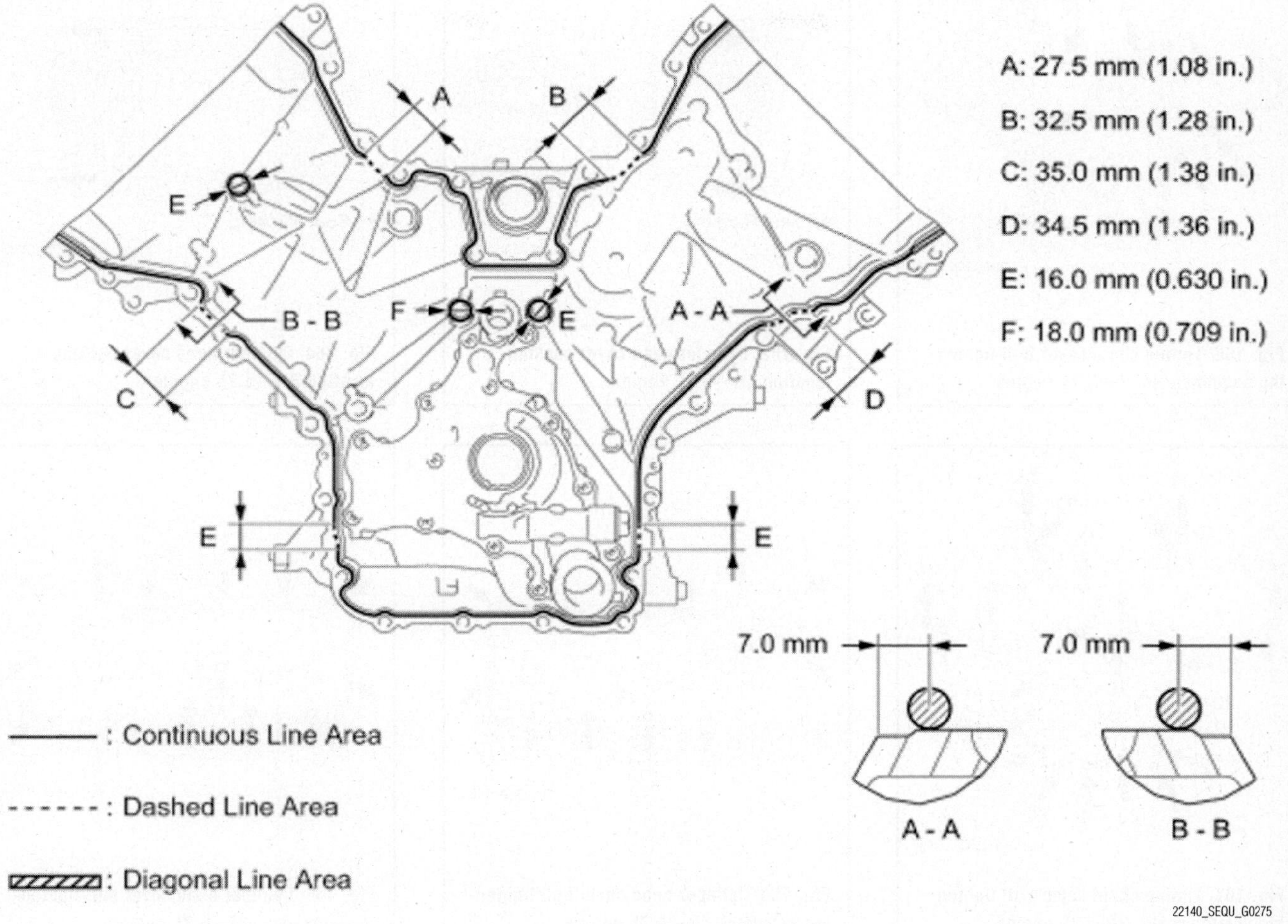

A: 27.5 mm (1.08 in.)
B: 32.5 mm (1.28 in.)
C: 35.0 mm (1.38 in.)
D: 34.5 mm (1.36 in.)
E: 16.0 mm (0.630 in.)
F: 18.0 mm (0.709 in.)
A
B
C
D
E
F
B - B
A - A
7.0 mm
7.0 mm
A - A
B - B
: Continuous Line Area
: Dashed Line Area
: Diagonal Line Area
22140_SEQU_G0275

Fig. 156 Timing chain cover sealant locations—5.7L engine

22140_SEQU_G0276

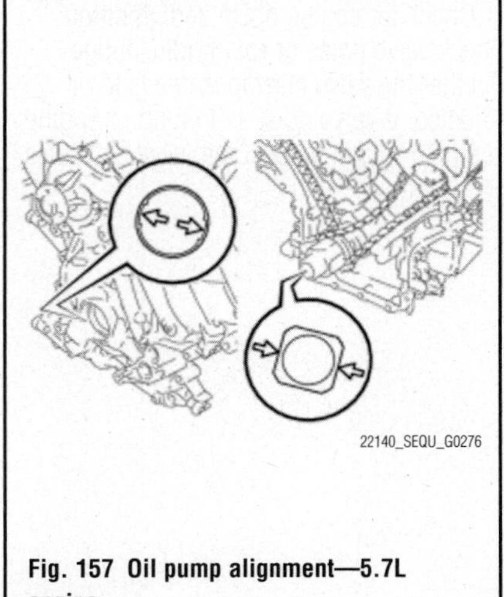

Fig. 157 Oil pump alignment—5.7L engine

Nut
A
B
C
D
E
F
22140_SEQU_G0277

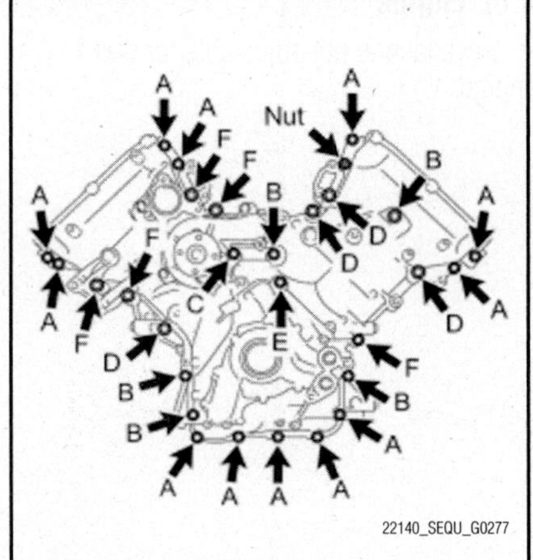

Fig. 158 Timing chain cover bolt locations—5.7L engine

Item	Length	Thread diameter
Bolt A	25 mm (0.984 in.)	8 mm (0.315 in.)
Bolt B	55 mm (2.165 in.)	8 mm (0.315 in.)
Bolt C	70 mm (2.756 in.)	8 mm (0.315 in.)
Bolt D	35 mm (1.378 in.)	10 mm (0.394 in.)
Bolt E	55 mm (2.165 in.)	10 mm (0.394 in.)
Bolt F	80 mm (3.150 in.)	10 mm (0.394 in.)

22140_SEQU_G0278

Fig. 159 Timing chain cover bolt application chart—5.7L engine

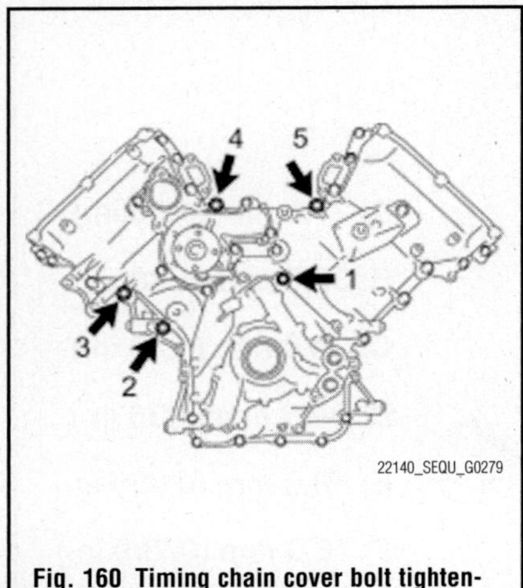

Fig. 160 Timing chain cover bolt tightening sequence No. 1—5.7L engine

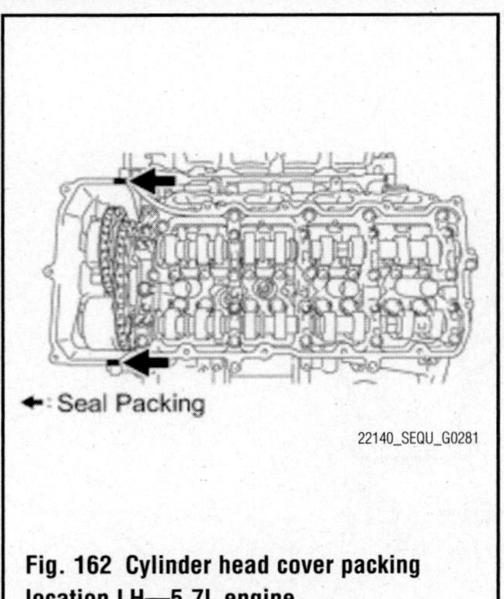

Fig. 162 Cylinder head cover packing location LH—5.7L engine

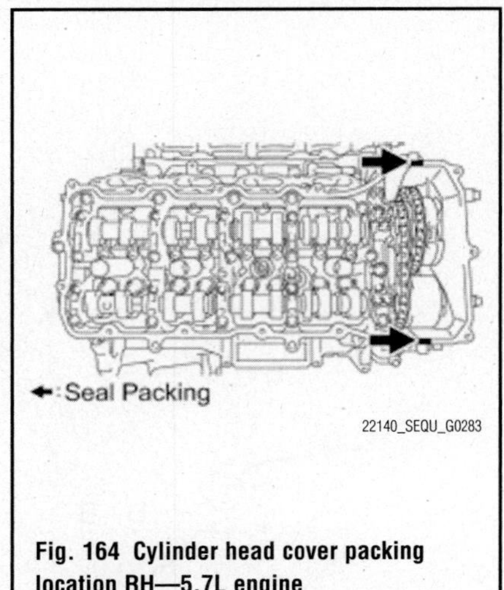

Fig. 164 Cylinder head cover packing location RH—5.7L engine

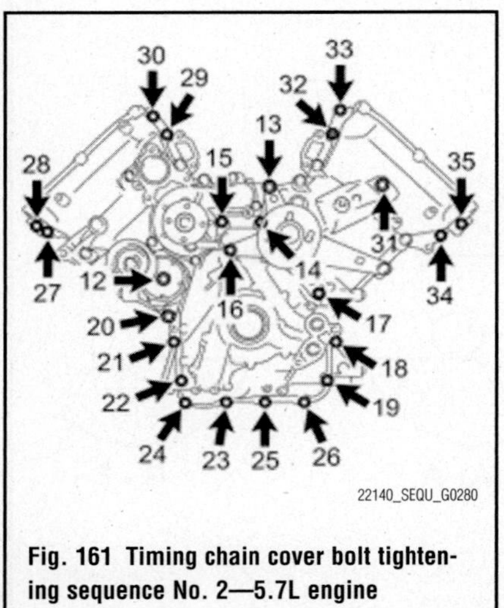

Fig. 161 Timing chain cover bolt tightening sequence No. 2—5.7L engine

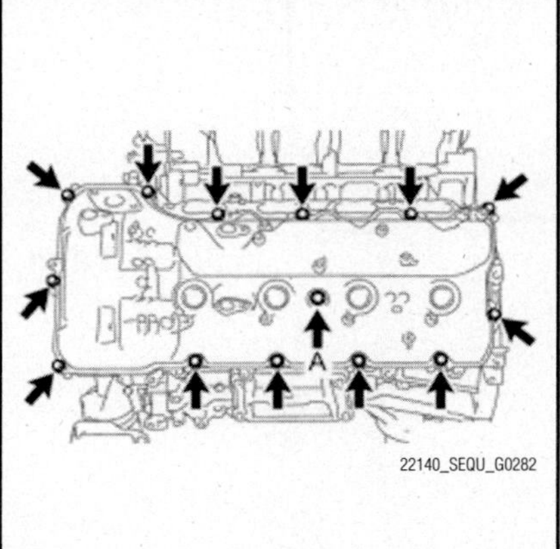

Fig. 163 Cylinder head cover bolt tightening sequence LH—5.7L engine

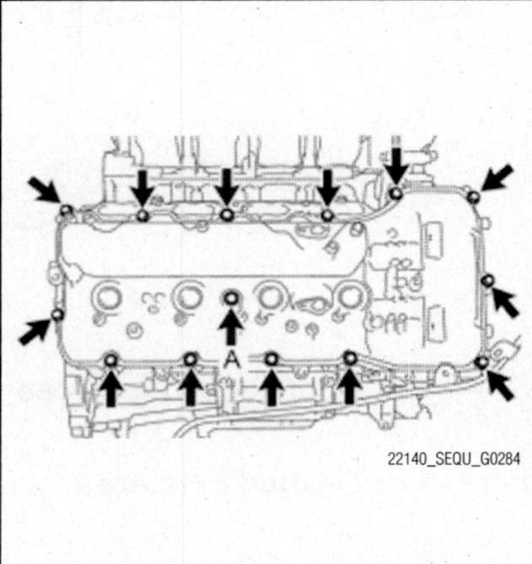

Fig. 165 Cylinder head cover bolt tightening sequence RH—5.7L engine

114. Add engine oil.
115. Add engine coolant.
116. Connect cable to negative battery terminal.
117. Inspect for leaks.
118. Check engine oil level.
119. Use the Techstream diagnostic tool, or equivalent and reprogram the required systems.

VALVE LASH

ADJUSTMENT

4.6L Engine

Valve lash is not adjustable for this engine.

5.7L Engine

Valve lash is not adjustable for this engine.

INSPECTION

Check for engine noise and vibration. Check valve noise or rough idle. Judge whether the valve clearance needs to be adjusted. If valve noise is to loud or engine idle is to rough, inspect the valve clearance.

ENGINE PERFORMANCE & EMISSION CONTROLS

ACCELERATOR PEDAL POSITION (APP) SENSOR

LOCATION

See Figure 166.

REMOVAL & INSTALLATION

1. Before servicing the vehicle, refer to the Precautions Section.
2. Disconnect the negative battery cable. Tape the cable with insulating tape.
3. Disconnect a accelerator pedal connector.
4. Remove the 2 nuts and accelerator pedal assembly.

To install:

➡Be sure to use new fasteners, as required.

➡Be careful not to give a shock to the accelerator pedal assembly. Be careful not to disassemble the accelerator pedal assembly.

5. Install the accelerator pedal assembly with the 2 nuts. Tighten nuts to 44 inch lbs. (5.0 Nm)
6. Connect a accelerator pedal connector.
7. Use the Techstream diagnostic tool, or equivalent and reprogram the required systems.

CAMSHAFT POSITION (CMP) SENSOR

LOCATION

See Figure 167.

REMOVAL & INSTALLATION

4.6L Engine

1. Before servicing the vehicle, refer to the Precautions Section.
2. Disconnect the negative battery cable. Tape the cable with insulating tape.
3. Remove V-bank cover sub-assembly.
4. Disconnect the sensor connector.

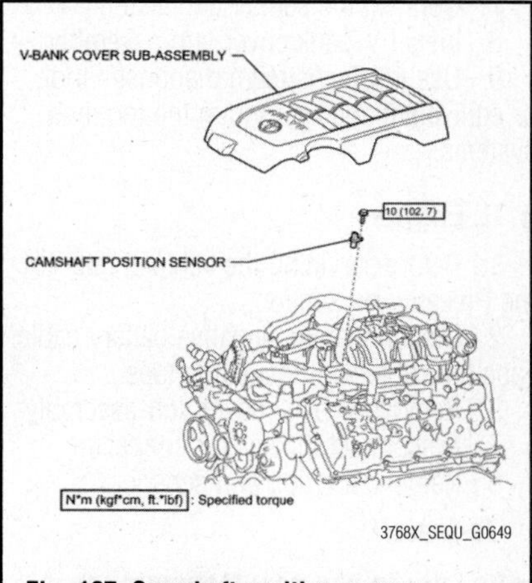

Fig. 167 Camshaft position sensor location

5. Remove the bolt and sensor.

To install:

➡Be sure to use new fasteners, as required.

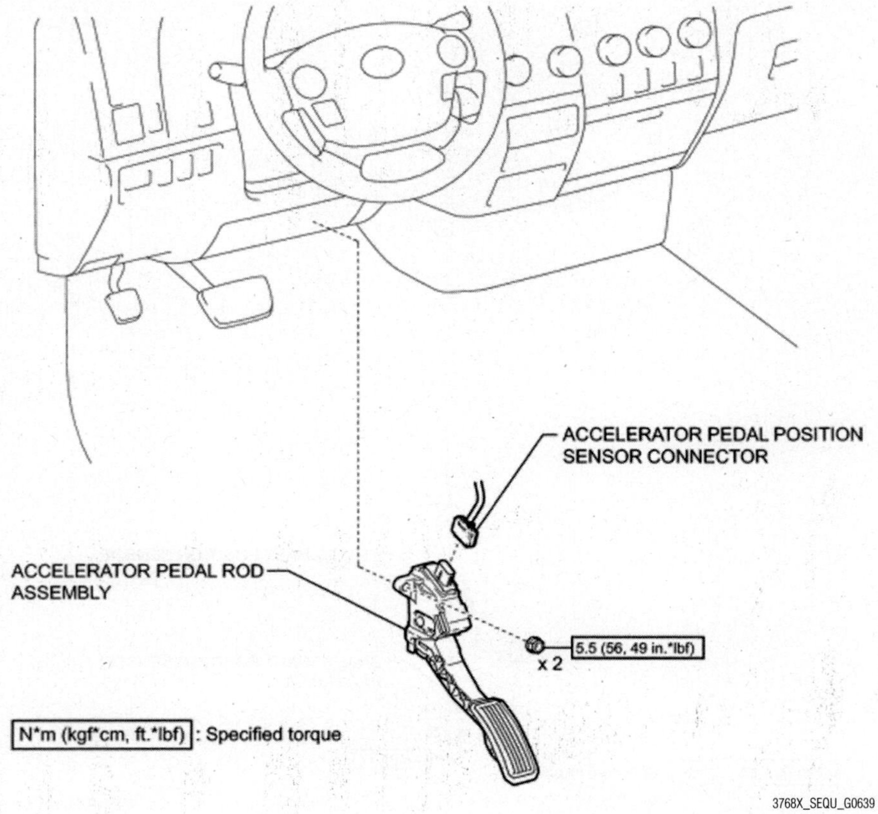

Fig. 166 Accelerator pedal location

6. Install the sensor with the bolt and tighten to 7 ft. lbs. (10 Nm).
7. Connect the sensor connector.
8. Install V-bank cover sub-assembly.
9. Use the Techstream diagnostic tool, or equivalent and reprogram the required systems.

5.7L Engine

1. Before servicing the vehicle, refer to the Precautions Section.
2. Disconnect the negative battery cable. Tape the cable with insulating tape.
3. Remove V-bank cover sub-assembly.
4. Disconnect the sensor connector.
5. Remove the bolt and sensor.

To install:

➡Be sure to use new fasteners, as required.

6. Install the sensor with the bolt and tighten to 7 ft. lbs. (10 Nm).
7. Connect the sensor connector.
8. Install V-bank cover sub-assembly.
9. Use the Techstream diagnostic tool, or equivalent and reprogram the required systems.

CRANKSHAFT POSITION (CKP) SENSOR

LOCATION

See Figures 168 and 169.

REMOVAL & INSTALLATION

4.6L Engine

1. Before servicing the vehicle, refer to the Precautions Section.
2. Disconnect the negative battery cable. Tape the cable with insulating tape.
3. Remove the sensor protector.
4. Disconnect the electrical connector.
5. Remove the retaining bolts.
6. Remove the component from its mounting.

To install:

➡Be sure to use new fasteners, as required.

7. Installation is the reverse of the removal procedure.
8. Install the sensor with the bolt and tighten to 7 ft. lbs. (10 Nm)
9. Use the Techstream diagnostic tool, or equivalent and reprogram the required systems.

5.7L Engine

1. Before servicing the vehicle, refer to the Precautions Section.
2. Disconnect the negative battery cable. Tape the cable with insulating tape.
3. Remove the 2 bolts and sensor protector
4. Disconnect the sensor connector.
5. Remove the bolt and sensor

To install:

➡Be sure to use new fasteners, as required.

6. Install the sensor with the bolt and tighten to 7 ft. lbs. (10 Nm)
7. Connect the sensor connector
8. Install the sensor protector with the 2 bolts and tighten to 7 ft. lbs. (10 Nm)
9. Use the Techstream diagnostic tool, or equivalent and reprogram the required systems.

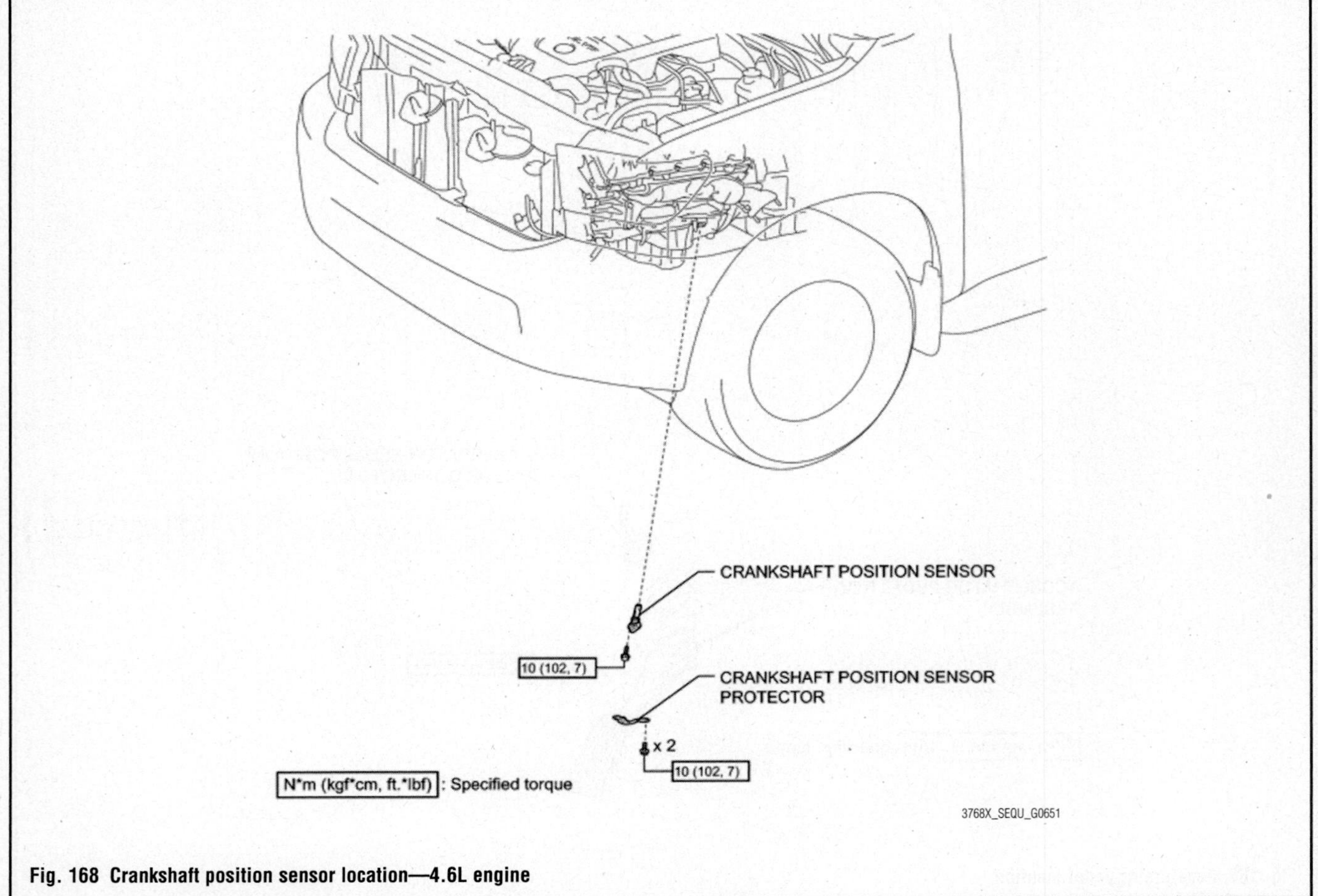

Fig. 168 Crankshaft position sensor location—4.6L engine

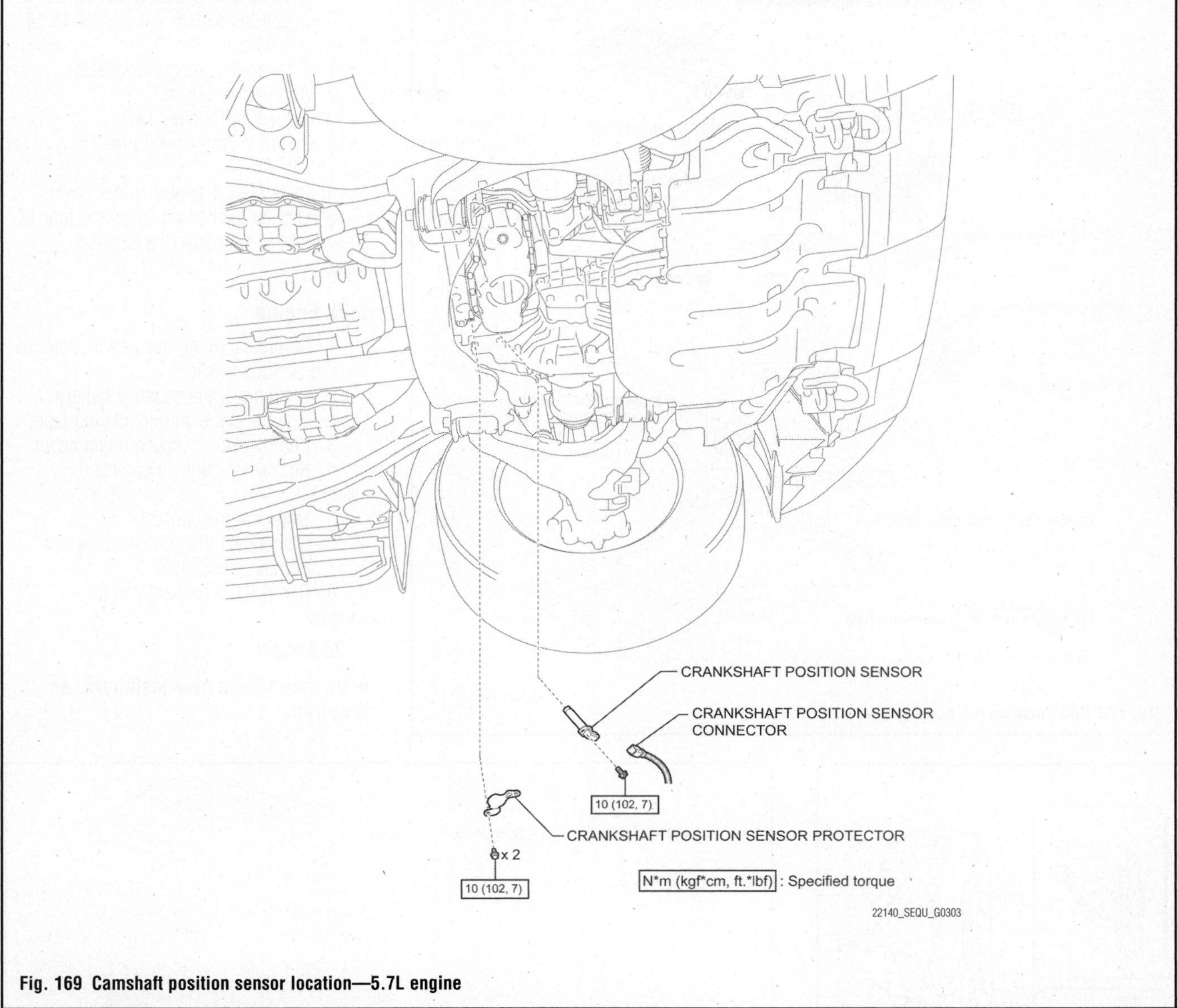

Fig. 169 Camshaft position sensor location—5.7L engine

ELECTRONIC CONTROL MODULE (ECM)

LOCATION

See Figure 170.

REMOVAL & INSTALLATION

See Figure 171.

1. Before servicing the vehicle, refer to the Precautions Section.
2. Disconnect the negative battery cable. Tape the cable with insulating tape.
3. On 4.6L engine remove the V- back cover sub assembly.
4. Remove the air cleaner assembly, as required on Sequoia.
5. Disconnect the connector holder block.
6. Remove the retaining bolts and move the connector block.
7. Raise the retaining levers, as shown in the illustration.

➡Make sure that the lock lever is raised 90 degrees, before disconnecting the connectors. Failure to do so may cause the connectors to break.

8. Remove the retaining bolts. Remove the component from its mounting. Discard the gasket.

To install:

➡Be sure to use new fasteners, as required.

9. Installation is the reverse of the removal procedure.
10. Use the Techstream diagnostic tool, or equivalent and reprogram the required systems.

ENGINE COOLANT TEMPERATURE (ECT) SENSOR

LOCATION

See Figure 172.

REMOVAL & INSTALLATION

4.6L Engine

1. Before servicing the vehicle, refer to the Precautions Section.
2. Disconnect the negative battery cable. Tape the cable with insulating tape.
3. Remove No. 1 engine under cover.

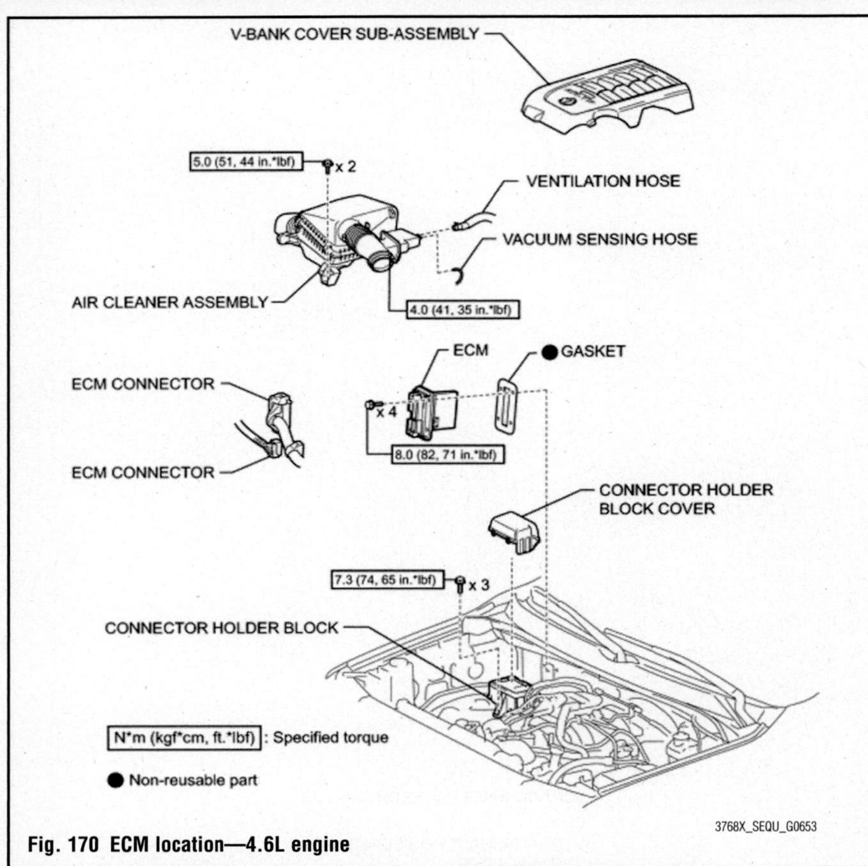

Fig. 170 ECM location—4.6L engine

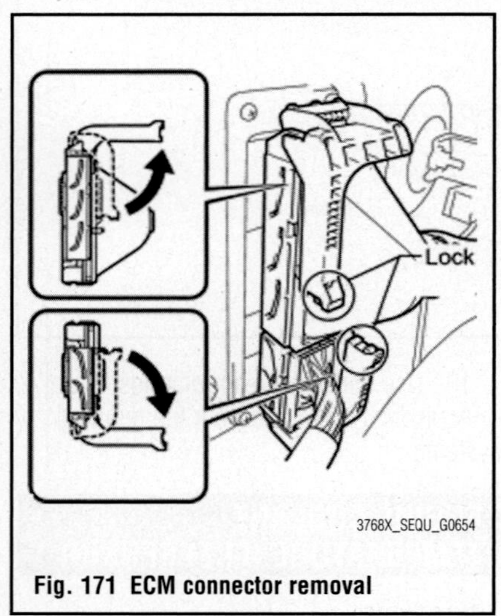

Fig. 171 ECM connector removal

4. Remove V-bank cover sub-assembly.
5. Drain engine coolant.
6. Disconnect the sensor connector.
7. Remove the sensor.
8. Remove the gasket from the sensor.

To install:

➡Be sure to use new fasteners, as required.

9. Install a new gasket to the sensor.
10. Install the sensor and tighten to 14 ft. lbs. (20 Nm).
11. Connect the sensor connector.
12. Add engine coolant.
13. Inspect for coolant leak.
14. Install V-bank cover sub-assembly.
15. Install No. 1 engine under cover.
16. Use the Techstream diagnostic tool, or equivalent and reprogram the required systems.

5.7L Engine

1. Before servicing the vehicle, refer to the Precautions Section.
2. Disconnect the negative battery cable. Tape the cable with insulating tape.
3. Remove No. 1 engine under cover.
4. Remove V-bank cover sub-assembly.
5. Drain engine coolant.
6. Disconnect the sensor connector.
7. Remove the sensor.
8. Remove the gasket from the sensor.

To install:

➡Be sure to use new fasteners, as required.

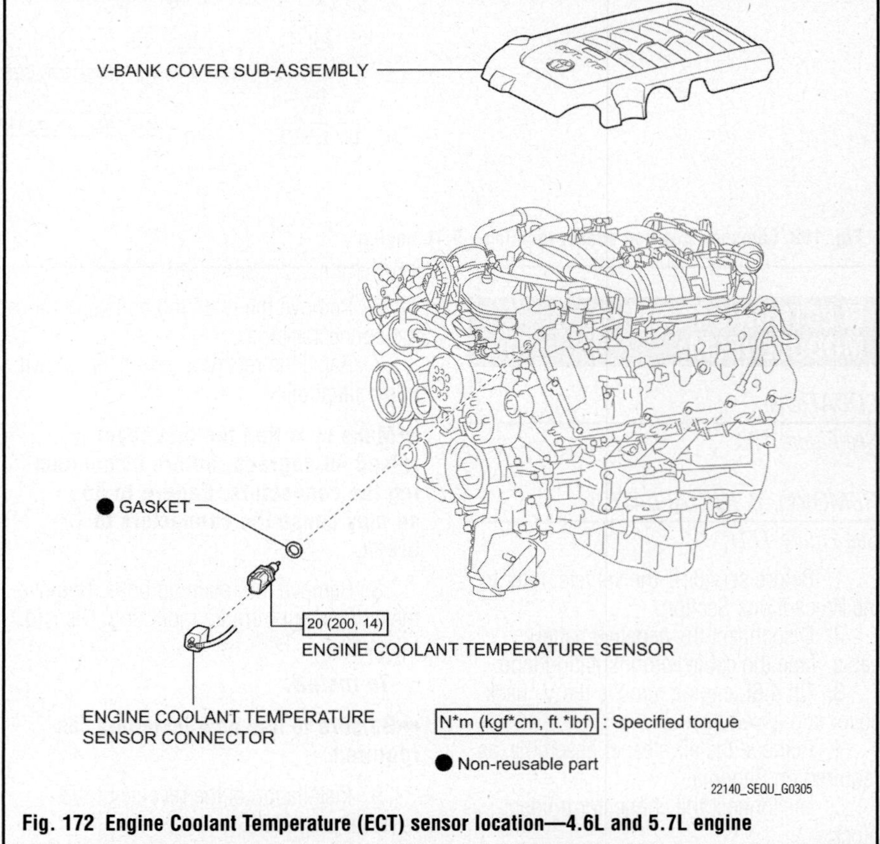

Fig. 172 Engine Coolant Temperature (ECT) sensor location—4.6L and 5.7L engine

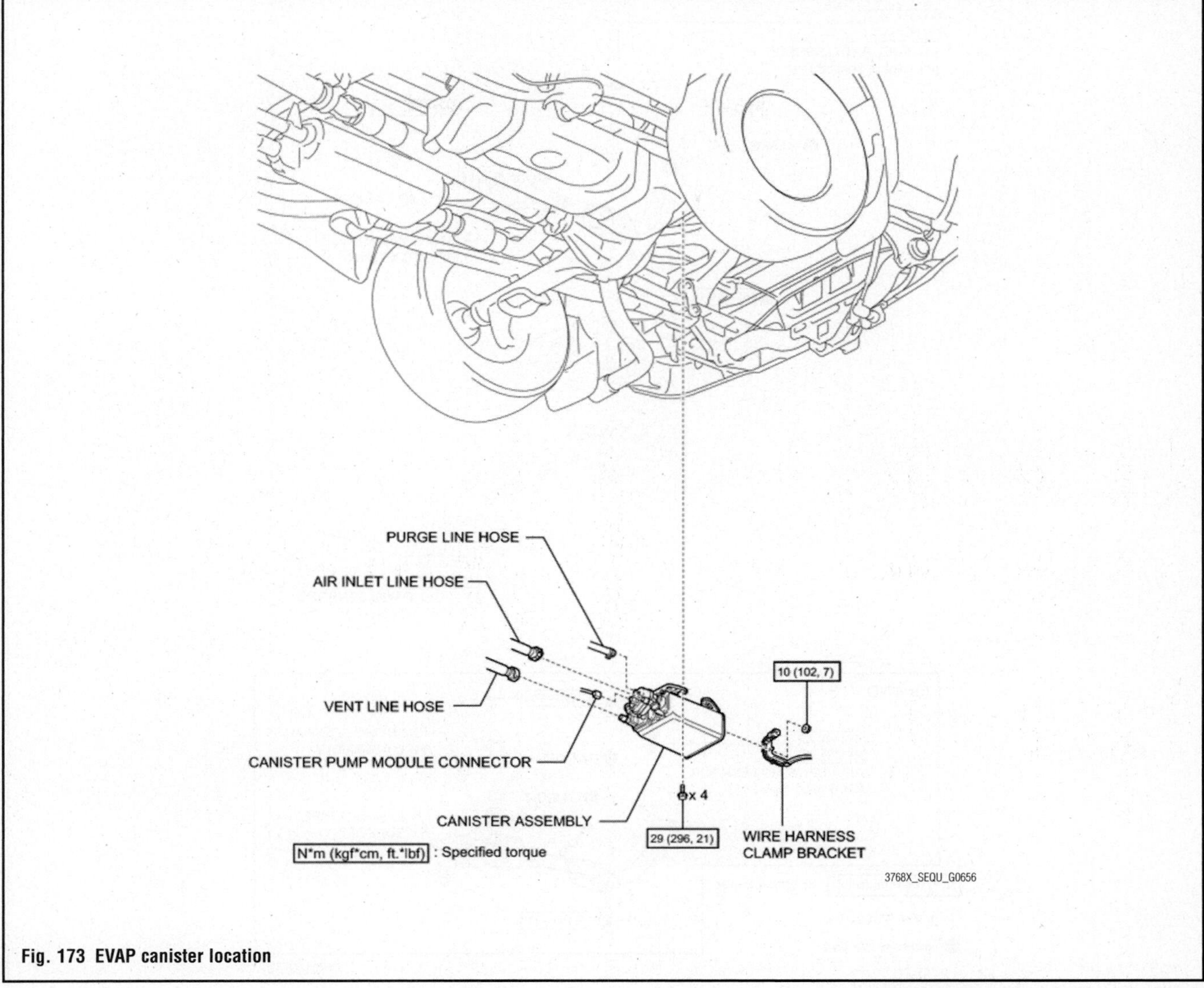

Fig. 173 EVAP canister location

9. Install a new gasket to the sensor.
10. Install the sensor and tighten to 14 ft. lbs. (20 Nm).
11. Connect the sensor connector.
12. Add engine coolant.
13. Inspect for coolant leak.
14. Install V-bank cover sub-assembly.
15. Install No. 1 engine under cover.
16. Use the Techstream diagnostic tool, or equivalent and reprogram the required systems.

EVAPORATIVE EMISSIONS (EVAP) CANISTER

LOCATION

See Figure 173.

REMOVAL & INSTALLATION

1. Before servicing the vehicle, refer to the Precautions Section.
2. Disconnect the negative battery cable. Tape the cable with insulating tape.
3. raise and safely support the vehicle.
4. Remove the necessary components in order to gain access to the EVAP canister.
5. Disconnect the electrical connectors.
6. Disconnect the hoses.
7. Remove the retaining bolts.
8. Remove the component from its mounting.

To install:

➡Be sure to use new fasteners, as required.

9. Installation is the reverse of the removal procedure.
10. Use the Techstream diagnostic tool, or equivalent and reprogram the required systems.

HEATED OXYGEN SENSOR (HO2S)

LOCATION

See Figures 174 and 175.

REMOVAL & INSTALLATION

1. Before servicing the vehicle, refer to the Precautions Section.
2. Disconnect the negative battery cable. Tape the cable with insulating tape.
3. Raise and support the vehicle safely.
4. Remove the engine under cover, as required.
5. Remove the heat insulator, as required.
6. Disconnect the sensor electrical connector.
7. Remove the sensor from its mounting.

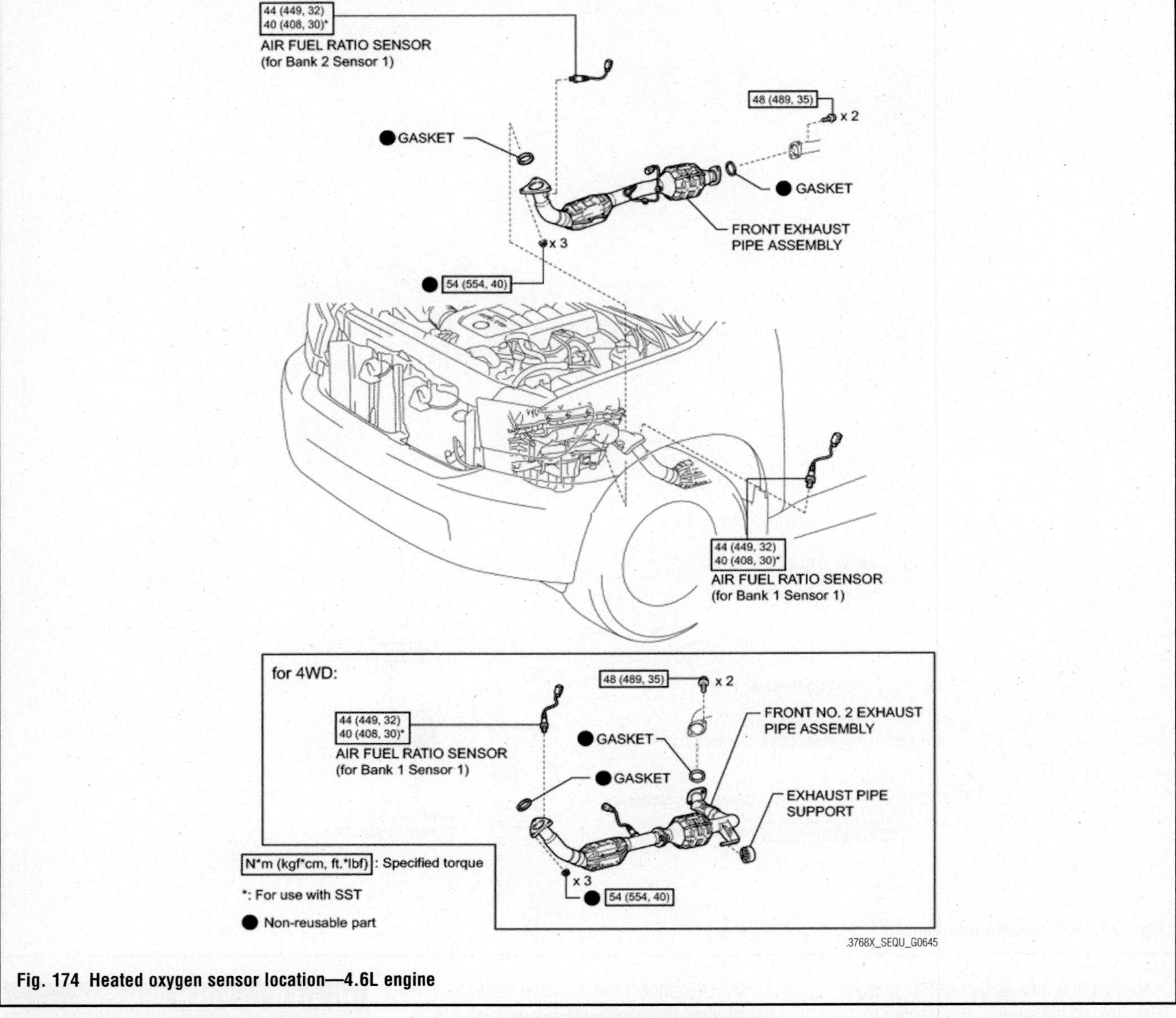

Fig. 174 Heated oxygen sensor location—4.6L engine

To install:

➡**Be sure to use new fasteners, as required.**

8. Installation is the reverse of the removal procedure.
9. Use the Techstream diagnostic tool, or equivalent and reprogram the required systems.

INTAKE AIR TEMPERATURE (IAT) SENSOR

LOCATION

The Intake Air Temperature (IAT) sensor, built into the Mass Air Flow (MAF) meter.

REMOVAL & INSTALLATION

See Mass Air Flow Meter.

KNOCK SENSOR (KS)

LOCATION

See Figure 176.

REMOVAL & INSTALLATION

4.6L Engine

See Figure 177.

1. Before servicing the vehicle, refer to the Precautions Section.
2. Disconnect the negative battery cable. Tape the cable with insulating tape.
3. Remove the intake manifold.
4. Remove No. 2 cylinder head cover.
5. Remove No. 1 engine cover.
6. Remove the 4 bolts and separator case.
7. Disconnect the 4 knock sensor connectors.
8. Remove the 4 bolts and 4 knock sensors.

To install:

➡**Be sure to use new fasteners, as required.**

9. Install the 4 sensors with the 4 bolts so that the sensors are angled as shown in the illustration.
10. Connect the 4 sensor connectors.
11. Install the separator case.
12. Install No. 1 engine cover.
13. Install No. 2 cylinder head cover.
14. Install the intake manifold.
15. Use the Techstream diagnostic tool, or equivalent and reprogram the required systems.

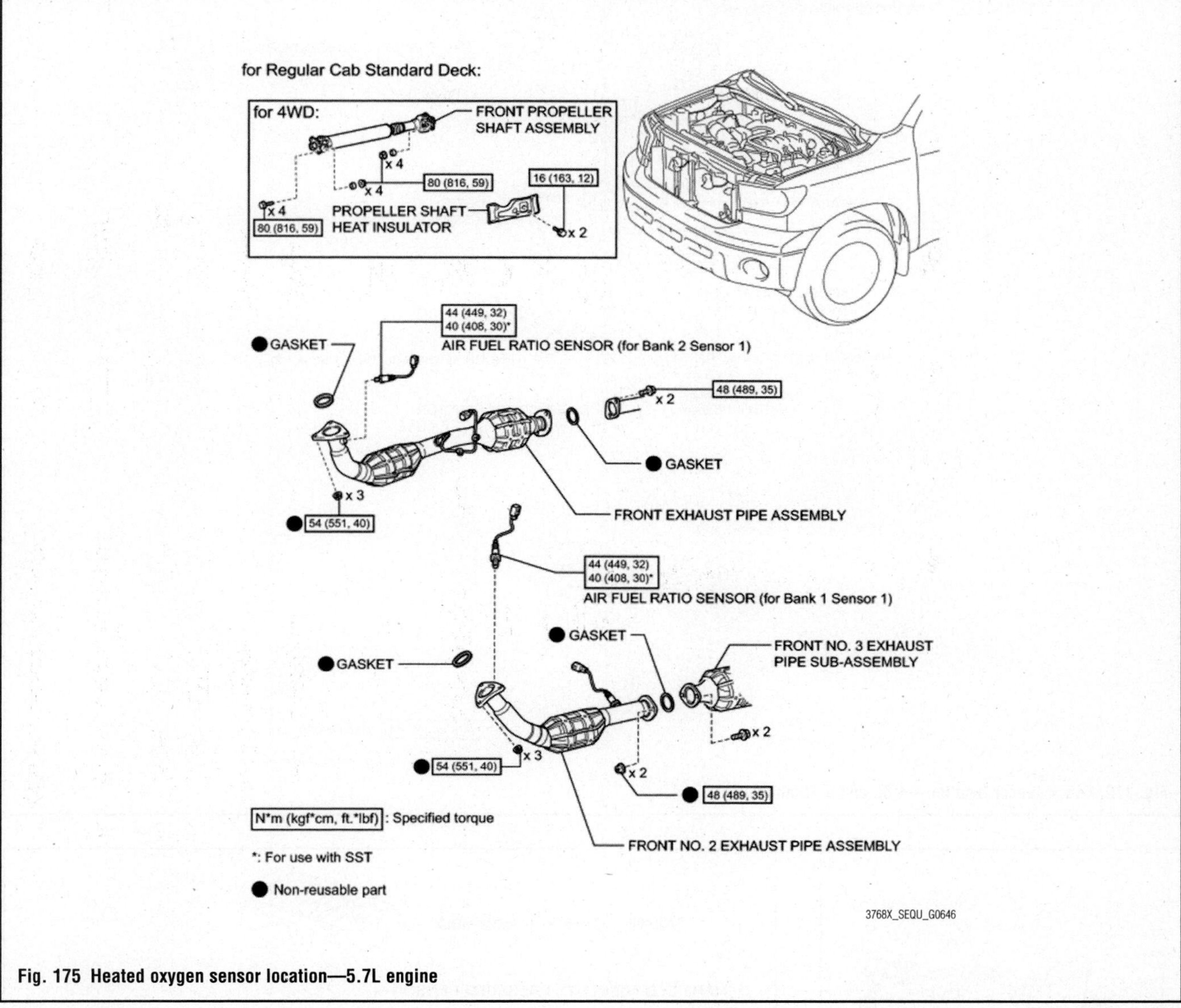

Fig. 175 Heated oxygen sensor location—5.7L engine

5.7 Engine

1. Before servicing the vehicle, refer to the Precautions Section.
2. Disconnect the negative battery cable. Tape the cable with insulating tape.
3. Remove the intake manifold.
4. Remove No. 2 cylinder head cover.
5. Remove No. 1 engine cover.
6. Remove the 4 bolts and separator case.
7. Disconnect the 4 knock sensor connectors.
8. Remove the 4 bolts and 4 knock sensors.

To install:

➡Be sure to use new fasteners, as required.

9. Install the 4 sensors with the 4 bolts so that the sensors are angled as shown in the illustration.
10. Connect the 4 sensor connectors.
11. Install the separator case.
12. Install No. 1 engine cover.
13. Install No. 2 cylinder head cover.
14. Install the intake manifold.
15. Use the Techstream diagnostic tool, or equivalent and reprogram the required systems.

MASS AIR FLOW (MAF) SENSOR

LOCATION

The MAF sensor is located in the air intake snorkel.

REMOVAL & INSTALLATION

1. Before servicing the vehicle, refer to the Precautions Section.
2. Disconnect the negative battery cable. Tape the cable with insulating tape.
3. Disconnect connector.
4. Remove attaching screws an remove MAF.

To install:

➡Be sure to use new fasteners, as required.

5. Installation is the reverse of the removal procedure.
6. Use the Techstream diagnostic tool, or equivalent and reprogram the required systems.

MANIFOLD ABSOLUTE PRESSURE (MAP) SENSOR

LOCATION

See Figure 178.

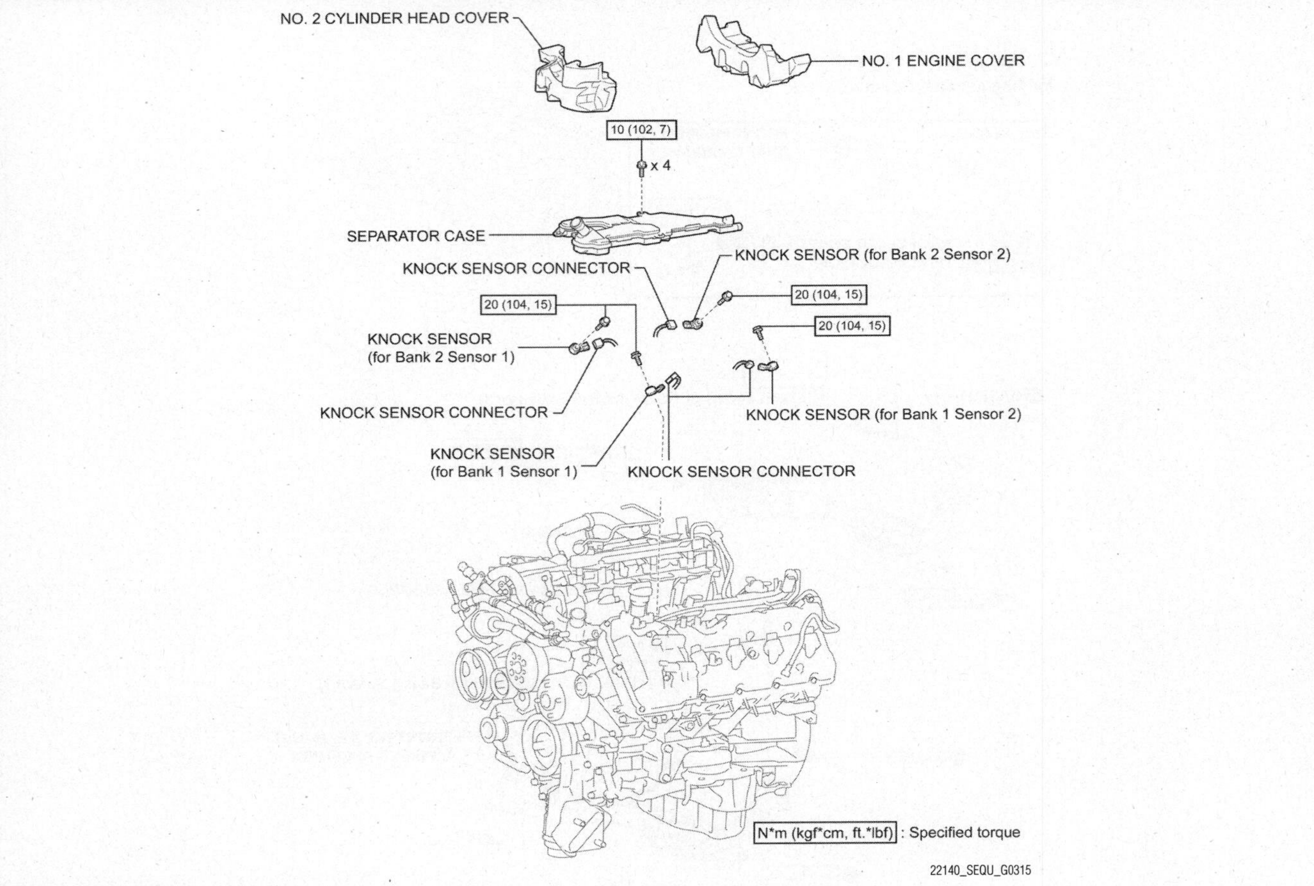

NO. 2 CYLINDER HEAD COVER
NO. 1 ENGINE COVER
10 (102, 7)
x 4
SEPARATOR CASE
KNOCK SENSOR CONNECTOR
KNOCK SENSOR (for Bank 2 Sensor 2)
20 (104, 15)
20 (104, 15)
20 (104, 15)
KNOCK SENSOR (for Bank 2 Sensor 1)
KNOCK SENSOR CONNECTOR
KNOCK SENSOR (for Bank 1 Sensor 2)
KNOCK SENSOR (for Bank 1 Sensor 1)
KNOCK SENSOR CONNECTOR
N*m (kgf*cm, ft.*lbf) : Specified torque
22140_SEQU_G0315

Fig. 176 Knock sensor location—4.6L and 5.7L engines

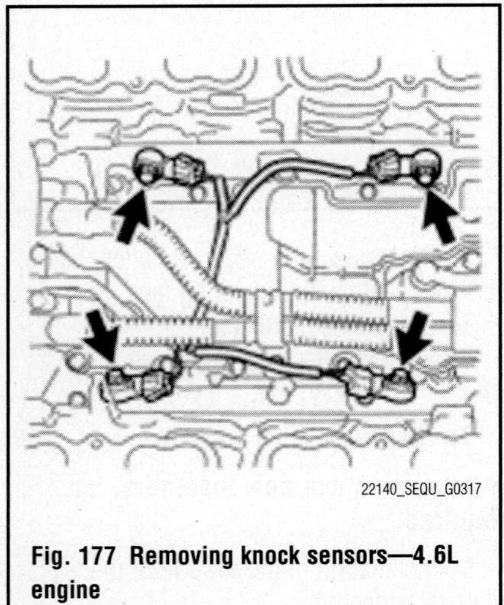

Fig. 177 Removing knock sensors—4.6L engine

REMOVAL & INSTALLATION

1. Before servicing the vehicle, refer to the Precautions Section.
2. Disconnect the negative battery cable. Tape the cable with insulating tape.
3. Remove the V- bank cover assembly.
4. Disconnect the electrical connector.

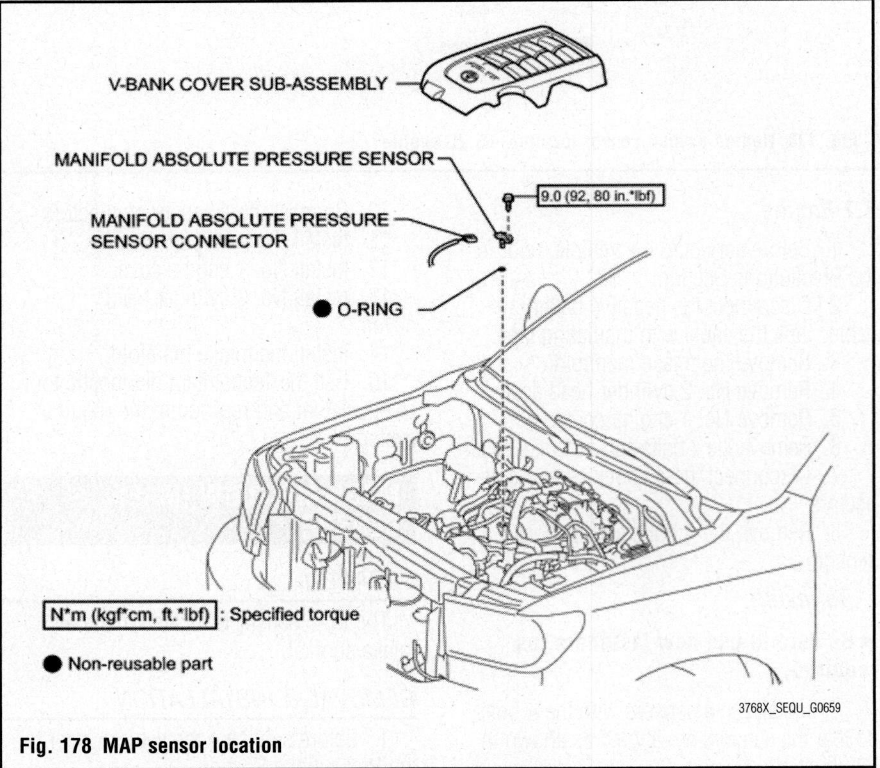

Fig. 178 MAP sensor location

5. Remove the sensor from its mounting.
6. Discard the O-ring.

To install:

➡Be sure to use new fasteners, as required.

7. Installation is the reverse of the removal procedure.
8. Use the Techstream diagnostic tool, or equivalent and reprogram the required systems.

FUEL

GASOLINE FUEL INJECTION SYSTEM

FUEL SYSTEM SERVICE PRECAUTIONS

Safety is the most important factor when performing not only fuel system maintenance, but any type of maintenance. Failure to conduct maintenance and repairs in a safe manner may result in serious personal injury or death. Work on a vehicle's fuel system components can be accomplished safely and effectively by adhering to the following rules and guidelines.

- To avoid the possibility of fire and personal injury, always disconnect the negative battery cable unless the repair or test procedure requires that battery voltage be applied.
- Always relieve the fuel system pressure prior to disconnecting any fuel system component (injector, fuel rail, pressure regulator, etc.) fitting or fuel line connection. Exercise extreme caution whenever relieving fuel system pressure to avoid exposing skin, face and eyes to fuel spray. Please be advised that fuel under pressure may penetrate the skin or any part of the body that it contacts.
- Always place a shop towel or cloth around the fitting or connection prior to loosening to absorb any excess fuel due to spillage. Ensure that all fuel spillage is quickly removed from engine surfaces. Ensure that all fuel-soaked cloths or towels are deposited into a flame-proof waste container with a lid.
- Always keep a dry chemical (Class B) fire extinguisher near the work area.
- Do not allow fuel spray or fuel vapors to come into contact with a spark or open flame.
- Always use a second wrench when loosening or tightening fuel line connection fittings. This will prevent unnecessary stress and torsion on fuel piping. Always follow the proper torque specifications.
- Always replace worn fuel fitting O-rings with new ones. Do not substitute fuel hose where rigid pipe is installed.

FUEL SYSTEM PRESSURE

RELIEVING

1. Before servicing the vehicle, refer to the Precautions Section.
2. Disconnect the negative battery cable. Tape the cable with insulating tape.
3. Disconnect the fuel pump ECU connectors.
4. Connect the battery cable.
5. Start the engine, after the engine stops (on its own) turn the ignition switch off.
6. Crank the engine again, to be sure it does not start.
7. Loosen the fuel tank cap, then discharge the pressure in the fuel tank.
8. Connect the fuel pump ECU connectors.

FUEL LEVEL SENDING UNIT

REMOVAL & INSTALLATION

1. Before servicing the vehicle, refer to the Precautions Section.
2. Disconnect the negative battery cable. Tape the cable with insulating tape.
3. Properly discharge the fuel system.
4. Remove the fuel tank.
5. Remove the joint tube clips and pull out the tank tubes.
6. Using tool SST:09808-14020, or equivalent loosen the retainer.

➡When the retainer is loosened, be careful as the pump and gauge tube will spring upward from the force of the spring.

7. Remove the retainer.
8. Remove the component from the tank. Discard the gasket.

To install:

➡Be sure to use new fasteners, as required.

9. Installation is the reverse of the removal procedure.
10. Be sure to align the pump with the tank groove.
11. When tightening the retainer be sure the alignment is correct (within range A).
12. Be sure to proper install the return tubes.
13. Start the engine and check for leaks.
14. Correct as required.

FUEL PUMP MODULE

REMOVAL & INSTALLATION

1. Before servicing the vehicle, refer to the Precautions Section.
2. Disconnect the negative battery cable. Tape the cable with insulating tape.
3. Properly discharge the fuel system.
4. Remove the fuel tank.
5. Remove the joint tube clips and pull out the tank tubes.
6. Using tool SST:09808-14020, or equivalent loosen the retainer.

➡When the retainer is loosened, be careful as the pump and gauge tube will spring upward from the force of the spring.

7. Remove the retainer.
8. Remove the component from the tank. Discard the gasket.

To install:

➡Be sure to use new fasteners, as required.

9. Installation is the reverse of the removal procedure.
10. Be sure to align the pump with the tank groove.
11. When tightening the retainer be sure the alignment is correct (within range A).
12. Be sure to proper install the return tubes.
13. Start the engine and check for leaks.
14. Correct as required.

FUEL PRESSURE REGULATOR

REMOVAL & INSTALLATION

See Figure 179.

1. Before servicing the vehicle, refer to the Precautions Section.
2. Disconnect the negative battery cable. Tape the cable with insulating tape.
3. Properly relieve the fuel system pressure.
4. Remove the V bank cover.
5. Remove the air cleaner assembly.
6. Disconnect the tube from the regulator.

➡Place a shop towel under the component.

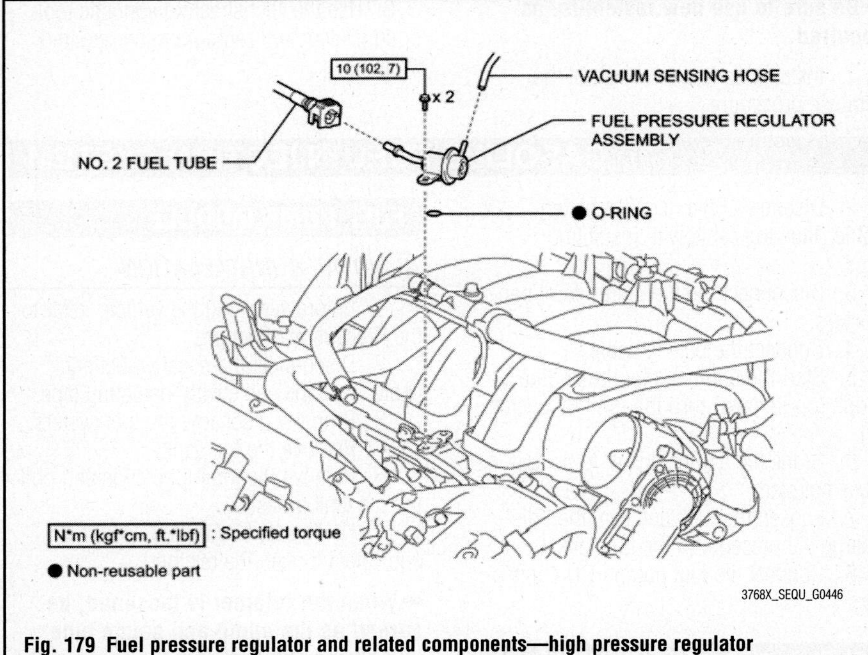

Fig. 179 Fuel pressure regulator and related components—high pressure regulator

7. Disconnect the vacuum hose.
8. Remove the retaining bolts.
9. Remove the component from its mounting.
10. Discard the O-ring.

To install:

➡**Be sure to use new fasteners, as required.**

11. Installation is the reverse of the removal procedure.

FUEL RAIL AND INJECTOR

REMOVAL & INSTALLATION

4.6L Engine

See Figure 180.

1. Properly discharge the fuel system.
2. Disconnect cable from negative battery terminal.
3. Remove the EGR valve bracket.
4. Disconnect the No. 2 fuel tube from the fuel pressure regulator.
5. Disconnect the No. 1 fuel tube from the fuel delivery pipes.
6. Disconnect the fuel tube from the fuel delivery pipe LH.
7. Disconnect the ventilation hose and No. 6 wire harness connector.
8. Remove the 2 bolts and fuel delivery pipe RH.
9. Remove the 2 delivery pipe spacers and 4 insulators from the intake manifold.
10. Disconnect the No. 7 wire harness connector.
11. Remove the No. 3 engine cover.
12. Remove the 2 bolts and fuel delivery pipe LH.
13. Remove the 2 delivery pipe spacers and 4 insulators from the intake manifold.
14. Remove the fuel injector from the fuel delivery pipe, and then disconnect the injector connector.

To install:

15. Attach the 2 clamps to install the No. 6 wire harness to the delivery pipe.
16. Attach the 3 clamps to install the No. 7 wire harness to the delivery pipe LH.
17. Apply gasoline or spindle oil to a new O-ring and install it to the injector.
18. Connect the injector connector.
19. Check that each injector is installed to the delivery pipe facing the direction shown in the illustration.
20. Install the 2 delivery pipe spacers and 4 insulators to the cylinder head LH.
21. Install the delivery pipe (with injectors) to the cylinder head LH.

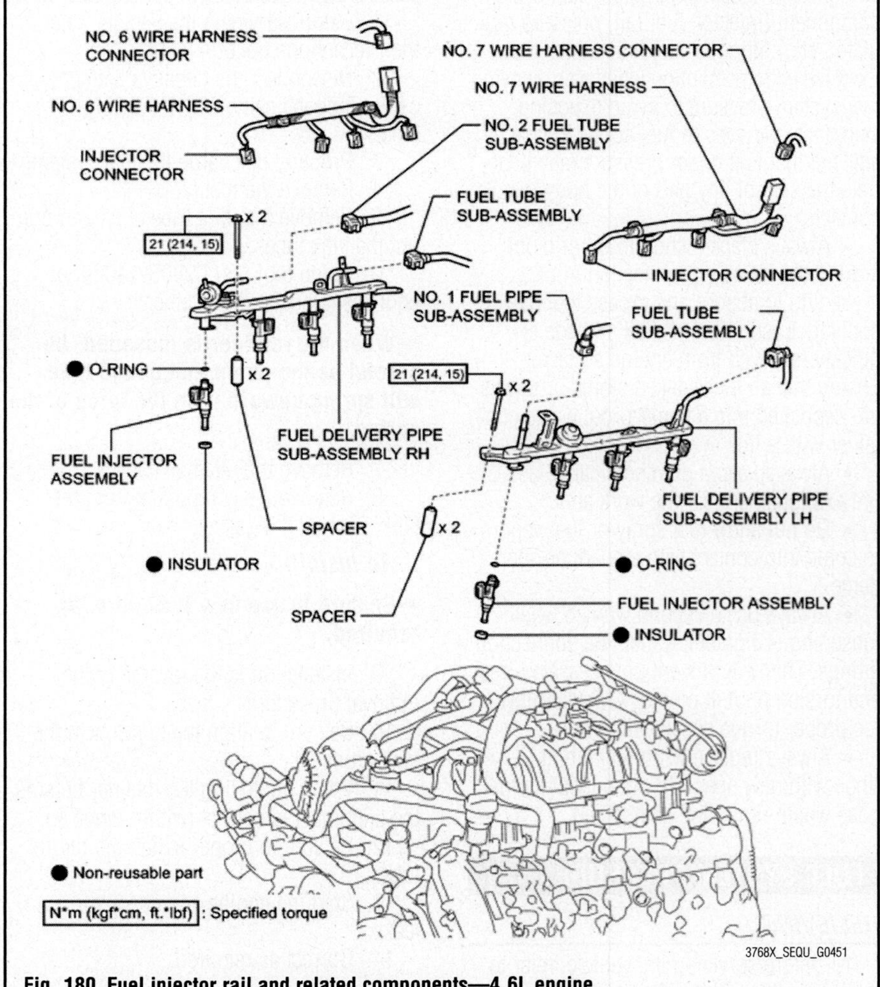

Fig. 180 Fuel injector rail and related components—4.6L engine

22. Install the 2 bolts and tighten to 15 ft. lbs. (21 Nm).

➡Make sure that the part of the injector labeled B is between the parts of the delivery pipe labeled A.

23. Connect the No. 6 wire harness connector.
24. Install the No. 3 engine cover.
25. Install the 2 delivery pipe spacers and 4 insulators to the cylinder head RH.
26. Install the delivery pipe (with injectors) to the cylinder head RH.
27. Install the 2 bolts and tighten to 15 ft. lbs. (21 Nm).
28. Install the No. 1 engine cover.
29. Connect the No. 7 wire harness connector.
30. Connect the ventilation hose.
31. Connect the fuel tube to the fuel delivery pipe LH.
32. Connect the No. 1 fuel tube to the delivery pipes.
33. Connect the No. 2 fuel tube to the fuel pressure regulator.
34. Install air cleaner hose assembly.
35. Install V-bank cover.
36. Connect cable to negative battery terminal.
37. Check for fuel leak.
38. Perform initialization, if necessary.

5.7L Engine

See Figures 181 through 183.

1. Properly discharge the fuel system.
2. Disconnect cable from negative battery terminal.
3. Remove V-bank cover. Remove air cleaner hose assembly.
4. Disconnect the No. 2 fuel tube from the fuel pressure regulator.
5. Disconnect the No. 1 fuel tube from the fuel delivery pipes.
6. Disconnect the fuel tube from the fuel delivery pipe LH.
7. Disconnect the ventilation hose and No. 6 wire harness connector.
8. Remove the 2 bolts and fuel delivery pipe RH.
9. Remove the 2 delivery pipe spacers and 4 insulators from the intake manifold.
10. Disconnect the No. 7 wire harness connector.
11. Remove the No. 3 engine cover.
12. Remove the 2 bolts and fuel delivery pipe LH.
13. Remove the 2 delivery pipe spacers and 4 insulators from the intake manifold.
14. Remove the fuel injector from the fuel delivery pipe, and then disconnect the injector connector.

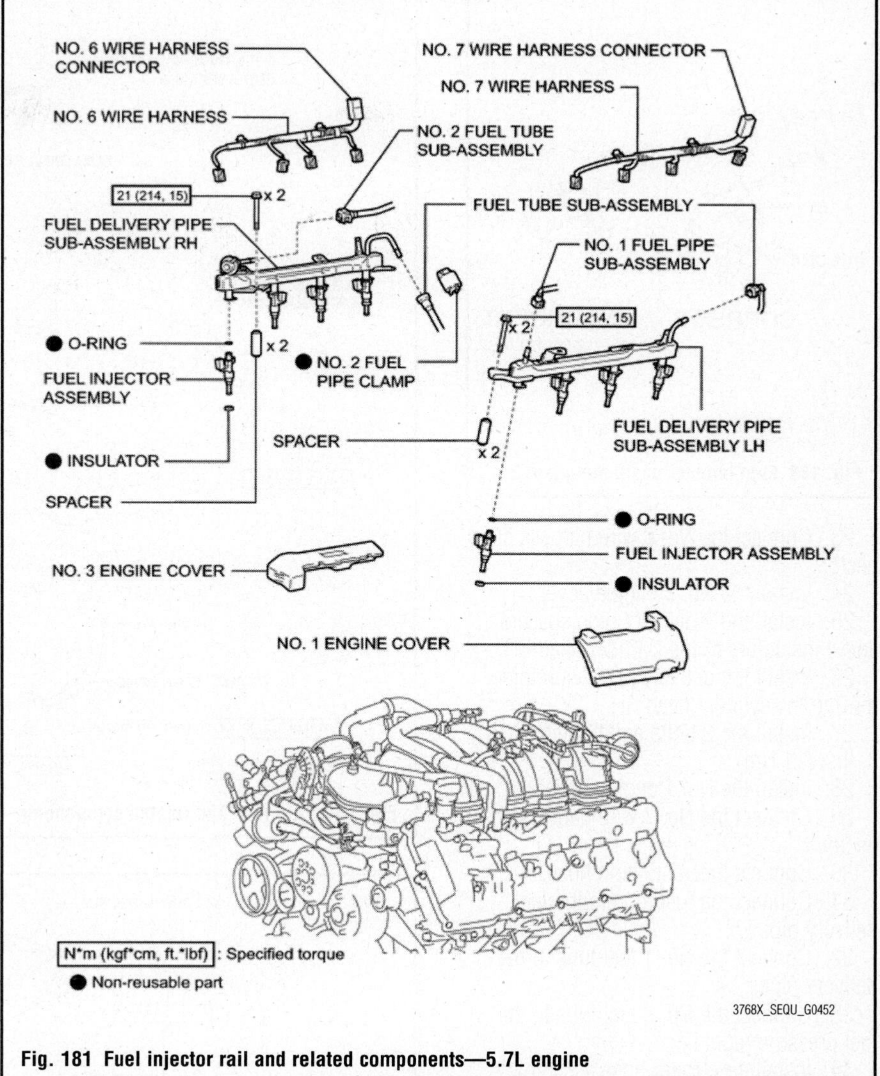

Fig. 181 Fuel injector rail and related components—5.7L engine

To install:

15. Attach the 2 clamps to install the No. 6 wire harness to the delivery pipe.
16. Attach the 3 clamps to install the No. 7 wire harness to the delivery pipe LH.
17. Apply gasoline or spindle oil to a new O-ring and install it to the injector.
18. Connect the injector connector.
19. Check that each injector is installed to the delivery pipe facing the direction shown in the illustration.
20. Install the 2 delivery pipe spacers and 4 insulators to the cylinder head LH.
21. Install the delivery pipe (with injectors) to the cylinder head LH.
22. Install the 2 bolts and tighten to 15 ft. lbs. (21 Nm).

➡Make sure that the part of the injector labeled B is between the parts of the delivery pipe labeled A.

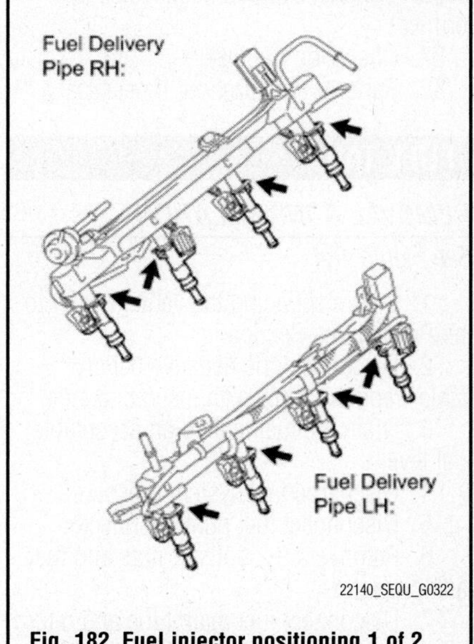

Fig. 182 Fuel injector positioning 1 of 2

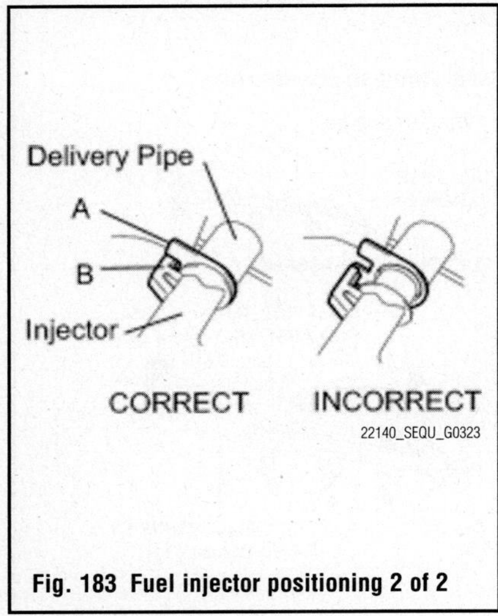

Fig. 183 Fuel injector positioning 2 of 2

23. Connect the No. 6 wire harness connector.
24. Install the No. 3 engine cover.
25. Install the 2 delivery pipe spacers and 4 insulators to the cylinder head RH.
26. Install the delivery pipe (with injectors) to the cylinder head RH.
27. Install the 2 bolts and tighten to 15 ft. lbs. (21 Nm).
28. Install the No. 1 engine cover.
29. Connect the No. 7 wire harness connector.
30. Connect the ventilation hose.
31. Connect the fuel tube to the fuel delivery pipe LH.
32. Connect the No. 1 fuel tube to the delivery pipes.
33. Connect the No. 2 fuel tube to the fuel pressure regulator.
34. Install air cleaner hose assembly.
35. Install V-bank cover.
36. Connect cable to negative battery terminal.
37. Check for fuel leak.
38. Perform initialization, if necessary.

FUEL TANK

REMOVAL & INSTALLATION

See Figure 184.

1. Before servicing the vehicle, refer to the Precautions Section.
2. Disconnect the negative battery cable. Tape the cable with insulating tape.
3. Drain the fuel tank to an acceptable fuel level.
4. Discharge fuel system pressure.
5. Disconnect fuel pump connector.
6. Remove the 2 bolts, 2 nuts and fuel tank protector.
7. Disconnect fuel main tube and return tube.

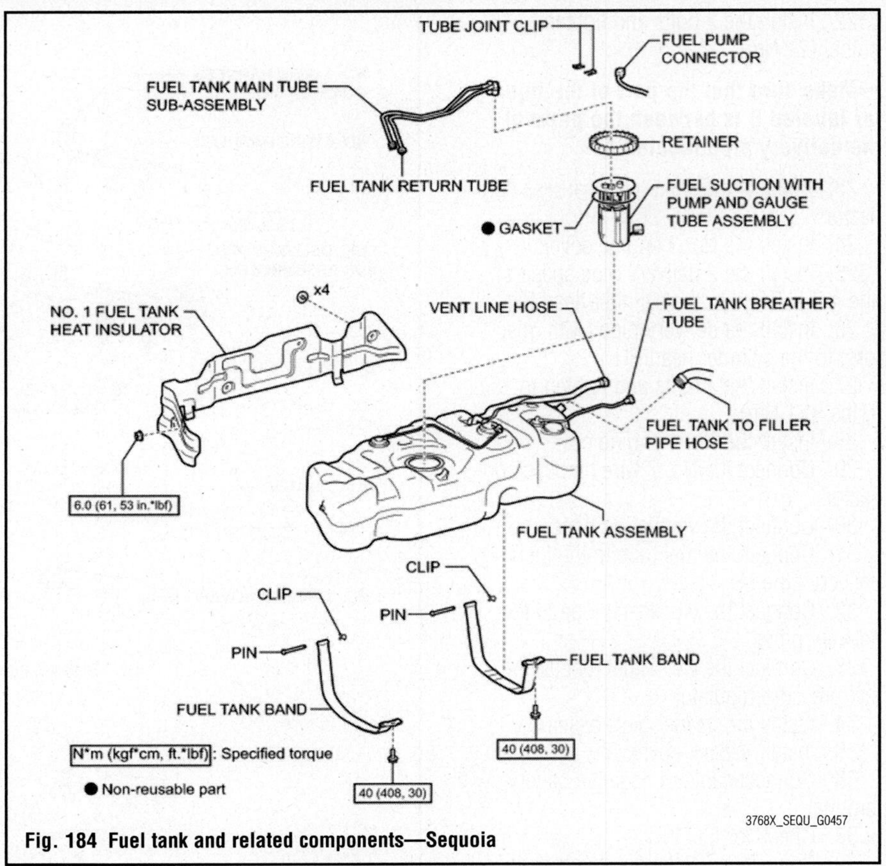

Fig. 184 Fuel tank and related components—Sequoia

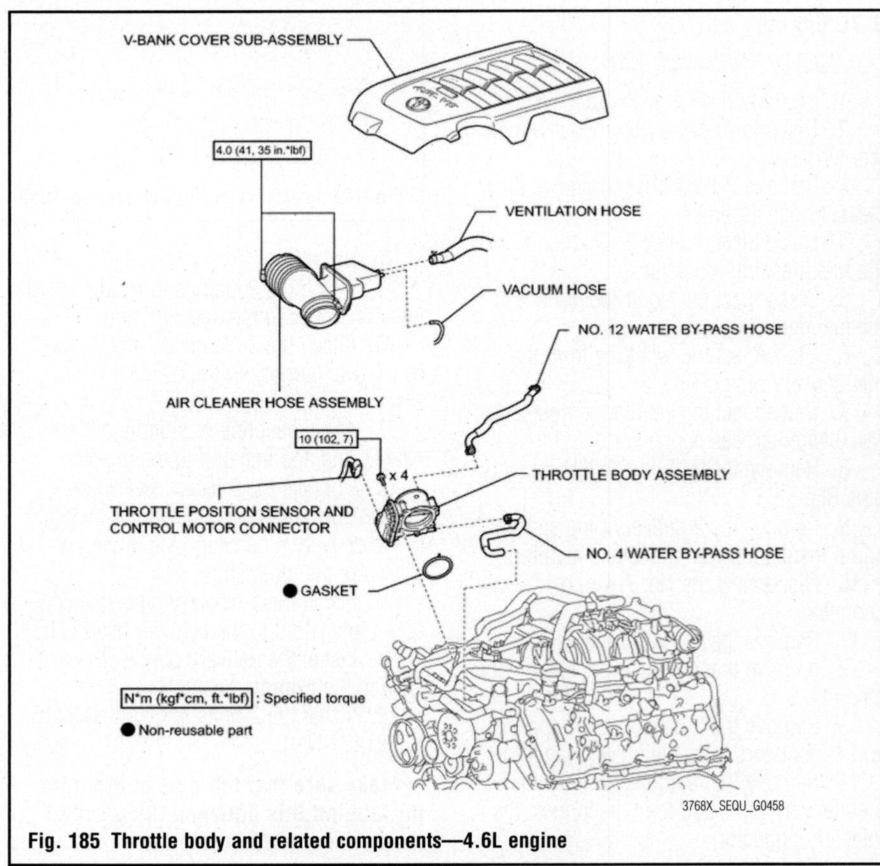

Fig. 185 Throttle body and related components—4.6L engine

8. Push the connector deep into the charcoal canister to release the locking tab.
9. Pinch and pull out the connector.
10. Loosen the hose clamp bolt and disconnect the fuel inlet hose from the fuel filler pipe.
11. Disconnect the breather tube.
12. Properly position a suitable jack under the fuel tank.
13. Remove the 2 bolts and disconnect the 2 fuel tank bands from the fuel tank.
14. Operate the jack and carefully lower the tank, check to see that nothing is interfering with the removal (wires, hoses etc).
15. Lower the jack and remove the fuel tank.

To install:

➡Be sure to use new fasteners, as required.

16. Installation is the reverse of the removal procedure.
17. Set up the fuel tank to the transmission jack.
18. Operate the transmission jack and install the fuel tank.
19. Install the 2 fuel tank bands with the 2 bolts and tighten to 45 ft. lbs. (62 Nm).
20. Connect the fuel inlet hose to the filler pipe and install the clamp.
21. Connect the breather tube.
22. Connect the fuel tank vent hose to the charcoal canister.
23. Connect fuel tank main tube and return tube.
24. Install the fuel tank protector with the 2 bolts and 2 nuts.
25. Connect fuel pump connector.
26. Check for fuel leaks.
27. Install spare tire.

IDLE SPEED

ADJUSTMENT

Idle speed is maintained by the Powertrain Control Module (PCM). No adjustment is necessary or possible.

THROTTLE BODY

REMOVAL & INSTALLATION

See Figures 185 through 187

1. Before servicing the vehicle, refer to the Precautions Section.
2. Disconnect the negative battery cable. Tape the cable with insulating tape.

➡On 5.7L engines the throttle body is removed along with the throttle body motor.

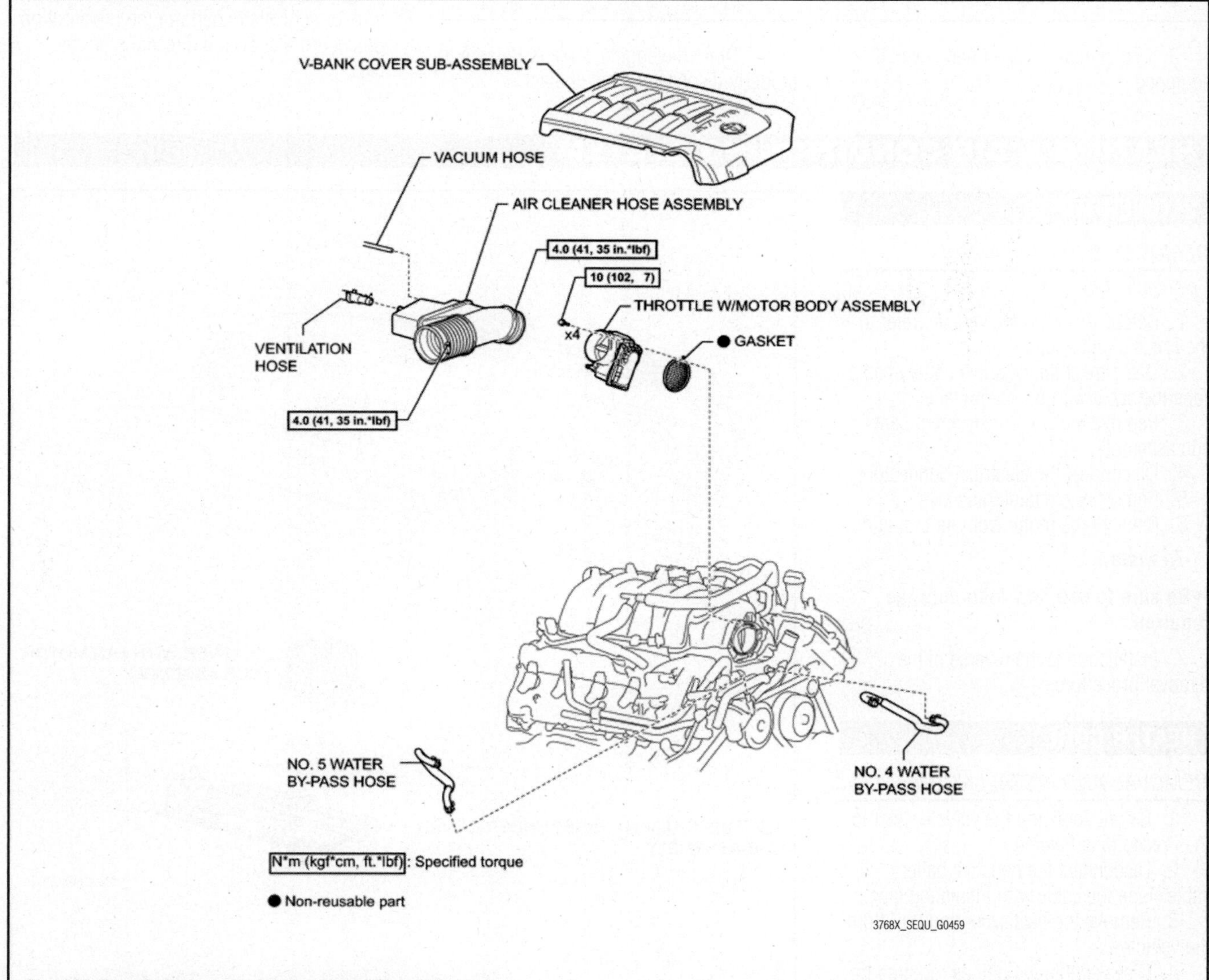

Fig. 186 Throttle body and related components—5.7L engine

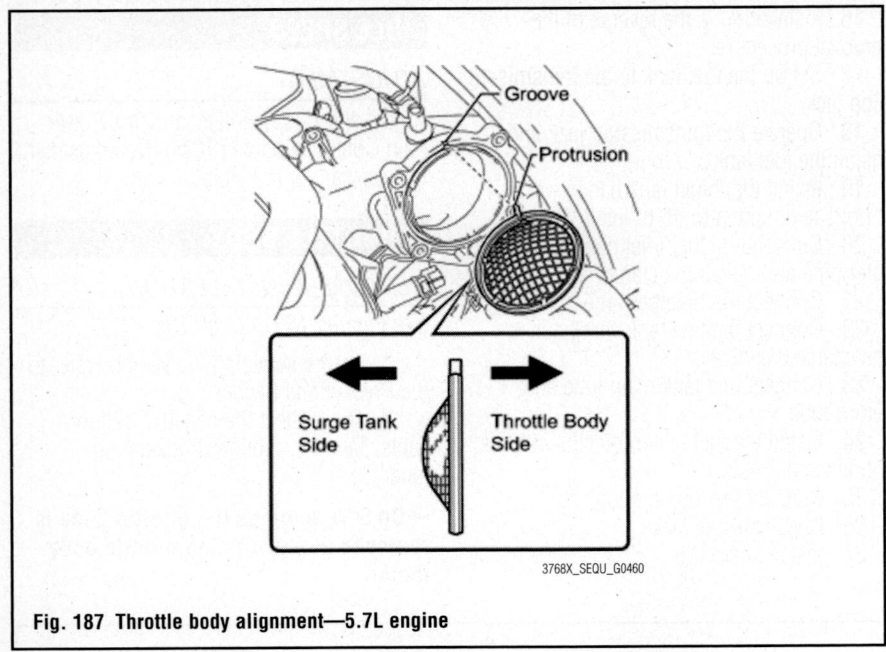

Fig. 187 Throttle body alignment—5.7L engine

3. Remove the engine undercover, if equipped.

4. Drain the engine coolant. Be sure to properly dispose of used coolant.

5. Remove the V- bank cover assembly.
6. Remove the air cleaner assembly.
7. Disconnect the coolant by pass hoses.
8. Disconnect the electrical connector.
9. Remove the component retaining bolts.
10. Remove the component from its mounting. Discard the gasket.

To install:

➡Be sure to use new fasteners, as required.

11. Installation is the reverse of the removal procedure.
12. On 4.6L and 5.7L engines tighten the retaining bolts to 7 ft. lbs. (10 Nm).
13. Be sure to fill the cooling system with the proper grade and type coolant.
14. Start the engine and check for leaks. Correct as required.
15. Using the Techstream diagnostic tool, or equivalent, perform the initialization procedure. Follow the directions on the tool.

HEATING & AIR CONDITIONING SYSTEM

BLOWER MOTOR

REMOVAL & INSTALLATION

See Figure 188.

1. Before servicing the vehicle, refer to the Precautions Section.
2. Disconnect the negative battery cable. Tape the cable with insulating tape.
3. Remove the No. 2 instrument panel sub assembly.
4. Disconnect the electrical connector.
5. Remove the retaining screws.
6. Remove the motor from its mounting.

To install:

➡Be sure to use new fasteners, as required.

7. Installation is the reverse of the removal procedure.

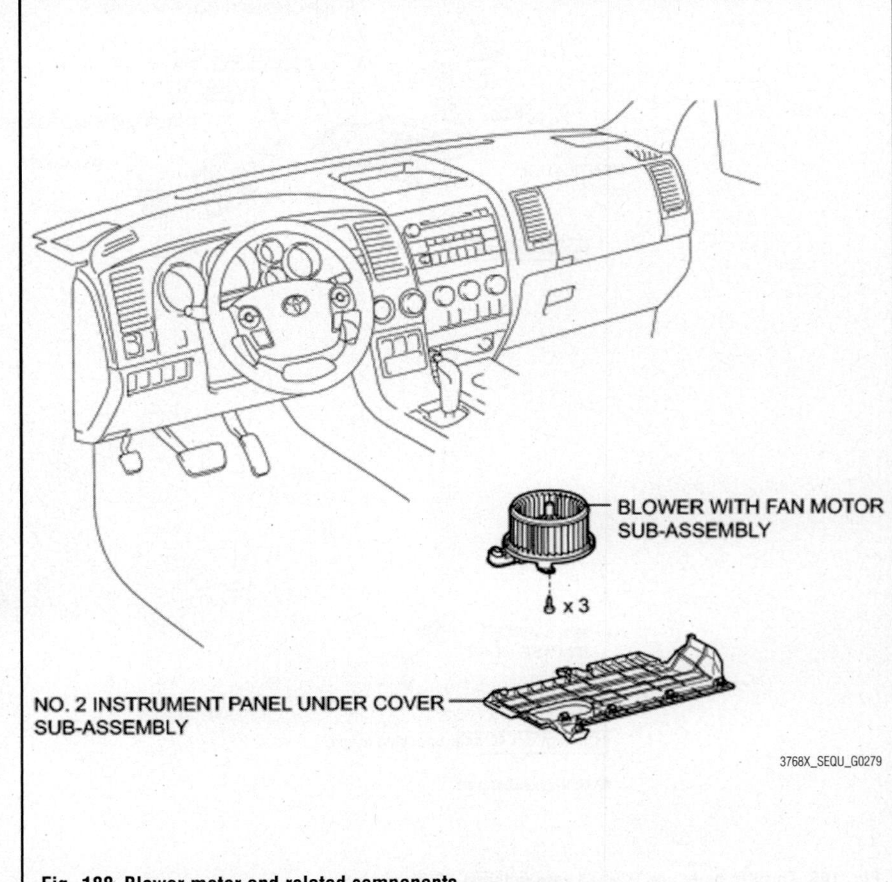

Fig. 188 Blower motor and related components

HEATER CORE

REMOVAL AND INSTALLATION

1. Before servicing the vehicle, refer to the Precautions Section.
2. Disconnect the negative battery cable. Tape the cable with insulating tape.
3. Remove the heater/cooling unit from the vehicle.
4. Disassemble the unit as required to remove the heater core.

To install:

➡Be sure to use new fasteners, as required.

5. Installation is the reverse of the removal procedure.
6. Be sure to refill the cooling system, with the proper grade and type coolant.
7. Be sure to properly recharge the system.
8. Be sure to adjust the spiral cable, as required.
9. Check the SRS system, for proper operation.
10. Using the Techstream diagnostic tool or equivalent, reprogram the required systems.

AUXILIARY HEATING & AIR CONDITIONING SYSTEM

BLOWER MOTOR

REMOVAL & INSTALLATION

1. Before servicing the vehicle, refer to the Precautions Section.
2. Disconnect the negative battery cable. Tape the cable with insulating tape.
3. Remove the No. 2 deck board assembly.
4. Remove the luggage compartment trim box.
5. Remove the No. 1 deck board assembly.
6. Remove the rope hook assembly.
7. Remove the deck board reinforcement.
8. Remove the rear seat assembly.
9. Remove the scuff plates and weatherstrip. Remove the floor mat support plate.
10. Remove the quarter panel trim assembly.
11. Disconnect the electrical connectors.
12. Remove the retaining bolts and screws.
13. Remove the component from the vehicle.

To install:

➡Be sure to use new fasteners, as required.

14. Installation is the reverse of the removal procedure.

HEATER CORE

REMOVAL & INSTALLATION

1. Before servicing the vehicle, refer to the Precautions Section.
2. Disconnect the negative battery cable. Tape the cable with insulating tape.
3. Remove the heater/cooling unit assembly.
4. Disassemble the unit as required to remove the heater core.

To install:

➡Be sure to use new fasteners, as required.

5. Installation is the reverse of the removal procedure.
6. Be sure to refill the cooling system, with the proper grade and type coolant.
7. Be sure to properly recharge the system.
8. Using the Techstream diagnostic tool or equivalent, reprogram the required systems.

STEERING

POWER RACK & PINION STEERING GEAR

REMOVAL & INSTALLATION

See Figure 189.

1. Before servicing the vehicle, refer to the Precautions Section.
2. Disconnect the negative battery cable. Tape the cable with insulating tape.
3. Position the front tires in the straight ahead position.
4. Drain the power steering fluid.
5. Raise and support the vehicle safely.
6. Remove the tire and wheel assemblies.
7. Remove the engine from the vehicle.
8. Matchmark and disconnect the No. 2 steering intermediate shaft sub assembly.
9. Disconnect the tie rod ends, using the proper tools.
10. Disconnect and plug the fluid lines.
11. Remove the bolts holding lines in place.
12. Remove the gear retaining bolts.
13. Remove the gear assembly from its mounting.

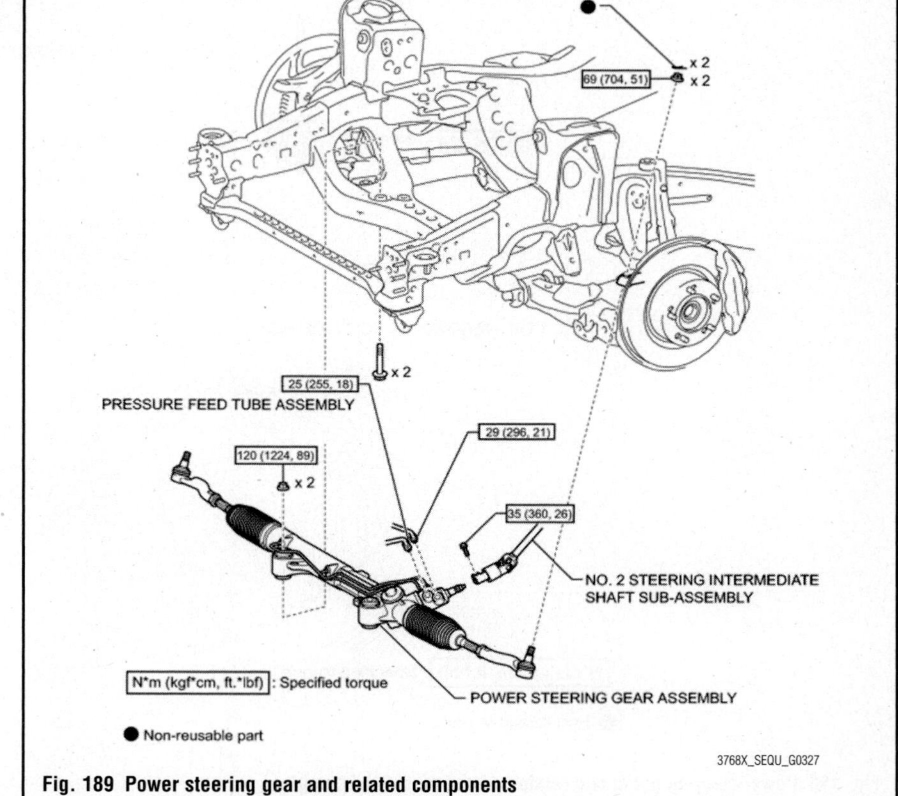

Fig. 189 Power steering gear and related components

To install:

➡Be sure to use new fasteners, as required.

14. Installation is the reverse of the removal procedure.
15. Tighten the gear retaining bolts to 89 ft. lbs. (120 Nm).
16. Use the matchmarks when installing the No.2 steering intermediate shaft to the steering gear. Tighten the retaining bolt to 26 ft. lbs. (35 Nm).
17. Adjust the front end alignment, as required.
18. Be sure to fill the power steering system, with the correct grade and type fluid.

POWER STEERING PUMP

REMOVAL & INSTALLATION

4.6L Engine

See Figures 190 and 191.

1. Before servicing the vehicle, refer to the Precautions Section.
2. Disconnect the negative battery cable. Tape the cable with insulating tape.
3. Remove the drive belt.
4. Drain the cooling system. Remove the radiator.
5. Separate the harness bracket.
6. Drain the power steering system. Be sure to properly dispose of used fluid.
7. Disconnect and plug the fluid lines.
8. Disconnect the pressure sensor.
9. Disconnect the solenoid valve connector.
10. Remove the pump retaining bolts.
11. Remove the pump from its mounting.

To install:

➡Be sure to use new fasteners, as required.

12. Installation is the reverse of the removal procedure.
13. Move the space so that the pump can be installed.
14. Tighten the retaining bolts to 21 ft. lbs. (29 Nm).
15. Be sure to fill the cooling system, with the proper grade and type coolant.
16. Be sure to fill the power steering system, with the proper grade and type fluid.

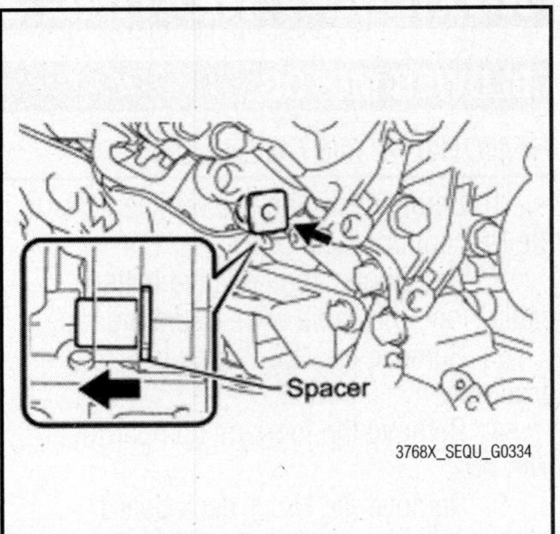

Fig. 191 Power steering pump spacer location—4.6L engine

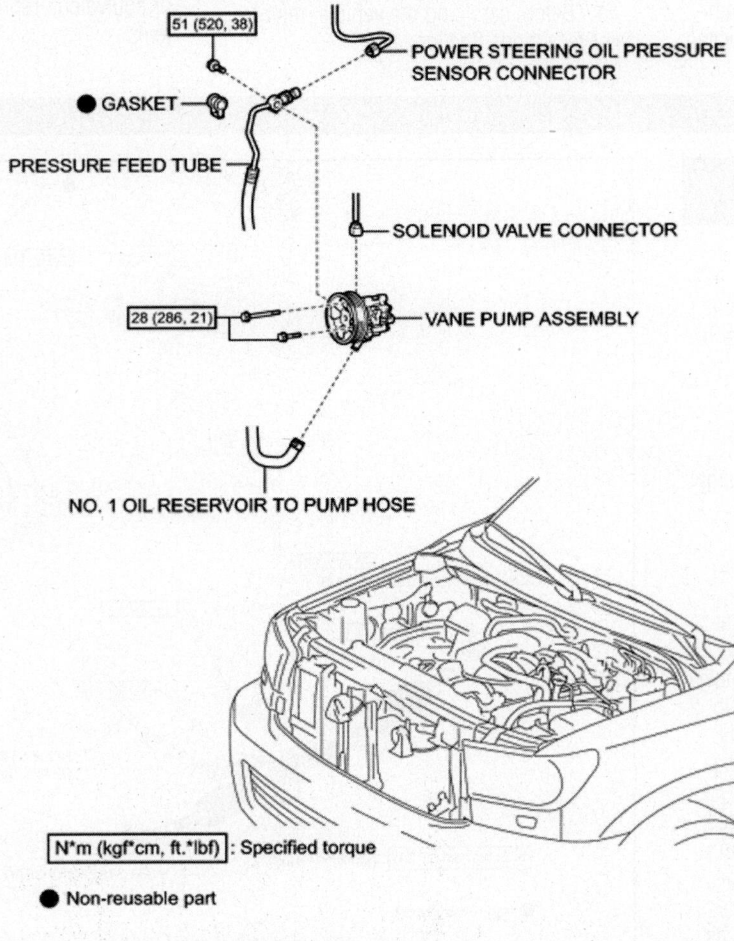

Fig. 190 Power steering pump and related components—4.6L engine

5.7L Engine

See Figure 192.

1. Before servicing the vehicle, refer to the Precautions Section.
2. Disconnect the negative battery cable. Tape the cable with insulating tape.
3. Remove the drive belt.
4. Remove the V- bank sub assembly.
5. Remove the air cleaner assembly.
6. Drain the power steering system. Be sure to properly dispose of used fluid.
7. Disconnect and plug the fluid lines.
8. Disconnect the pressure sensor.
9. Remove the pump retaining bolts.
10. Remove the pump from its mounting.

To install:

➡Be sure to use new fasteners, as required.

11. Installation is the reverse of the removal procedure.
12. Move the space so that the pump can be installed.
13. Tighten the retaining bolts to 21 ft. lbs. (28 Nm).
14. Be sure to fill the cooling system, with the proper grade and type coolant.
15. Be sure to fill the power steering system, with the proper grade and type fluid.

BLEEDING

See Figures 193 through 195.

1. Before servicing the vehicle, refer to the Precautions Section.
2. Disconnect the negative battery cable. Tape the cable with insulating tape.
3. Check the fluid level.
4. Raise and properly support the vehicle.
5. With the engine stopped, turn the wheels slowly from stop to stop several times.

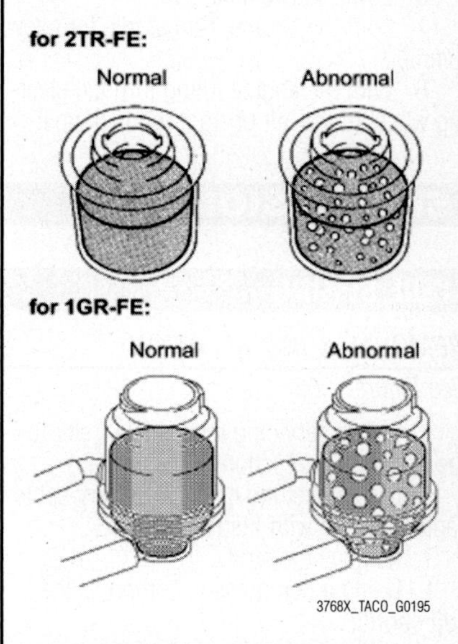

Fig. 193 Power steering fluid foaming identification

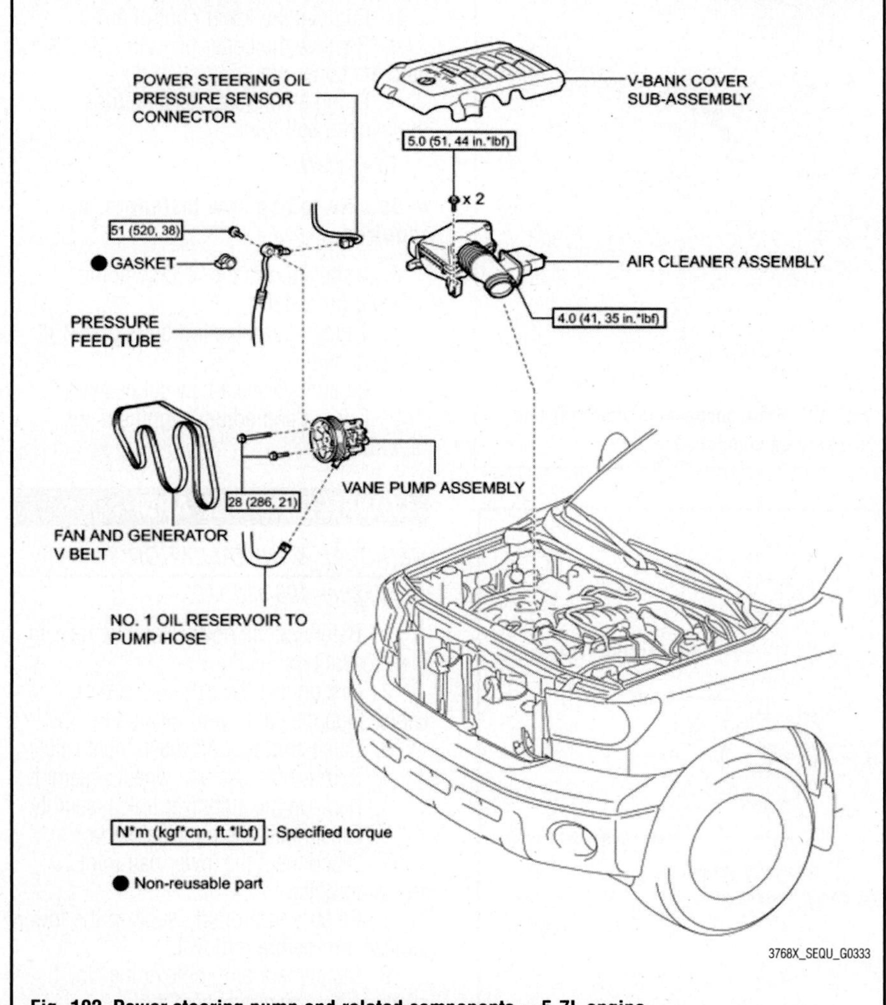

Fig. 192 Power steering pump and related components— 5.7L engine

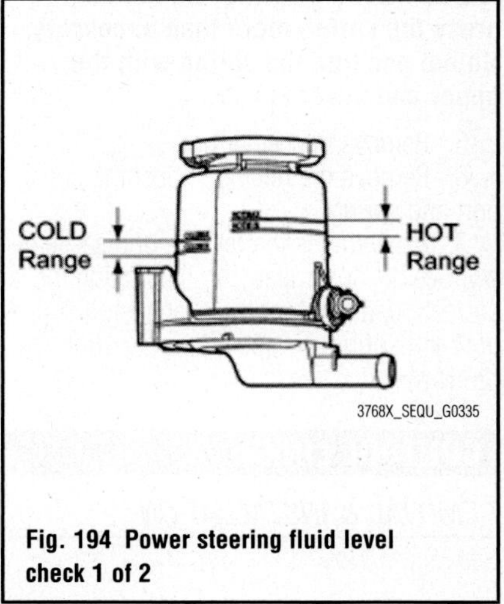

Fig. 194 Power steering fluid level check 1 of 2

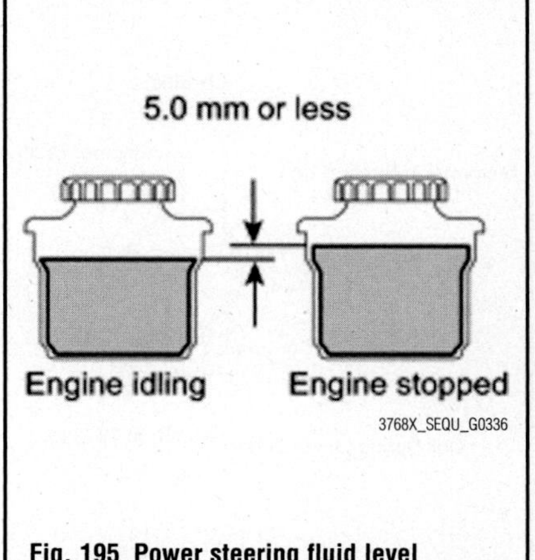

Fig. 195 Power steering fluid level check 2 of 2

6. Lower the vehicle.
7. Start the engine. Run at idle for a few minutes.
8. With the engine idling turn the steering wheel to the left or right full lock position and keep it there for two or three seconds. Repeat for the other side.
9. Repeat the above step several times.
10. Stop the engine. Check for foaming.
11. If foaming exists, repeat the procedure.
12. With the engine stopped check the fluid level.
13. Adjust as necessary. Be sure to use the proper grade and type fluid.

SUSPENSION

FRONT SUSPENSION

COIL SPRING

REMOVAL & INSTALLATION

See Figure 196.

1. Before servicing the vehicle, refer to the Precautions Section.
2. Disconnect the negative battery cable. Tape the cable with insulating tape.
3. Remove the strut.
4. Using a compressor, compress the coil spring.

➡A compressor with a force of 2,860 lbs. (12,740 N) or more must be used. Make sure that the suspension support is free from the spring. Do not compress the spring more than necessary. Do not position the spring with the upper end towards you.

5. Remove the center nut.
6. Remove the retainers, bushing support and spring.
7. Assembly is the reverse of disassembly. See the accompanying illustration for correct positioning of the suspension support and spring. Torque the nut to 18 ft. lbs. (25 Nm).

CONTROL LINKS

REMOVAL & INSTALLATION

See Figure 197.

1. Before servicing the vehicle, refer to the Precautions Section.
2. Disconnect the negative battery cable. Tape the cable with insulating tape.
3. Raise and support the vehicle safely.
4. Remove the tire and wheel assembly.
5. Remove the bolt, nut and stabilizer link from its mounting.

To install:

➡Be sure to use new fasteners, as required.

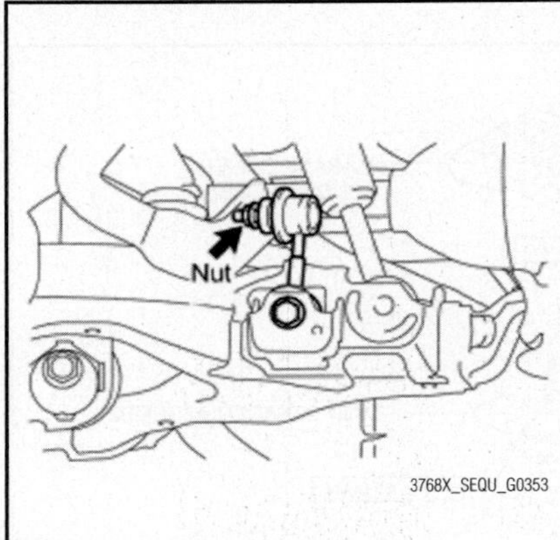

Fig. 197 Front suspension stabilizer link and related components

6. Installation is the reverse of the removal procedure.
7. Temporarily install the component.
8. Stabilize the suspension.
9. Tighten the retaining nut to 111 ft. lbs. (150 Nm).
10. Tighten the retaining bolt to 89 ft. lbs. (120 Nm).

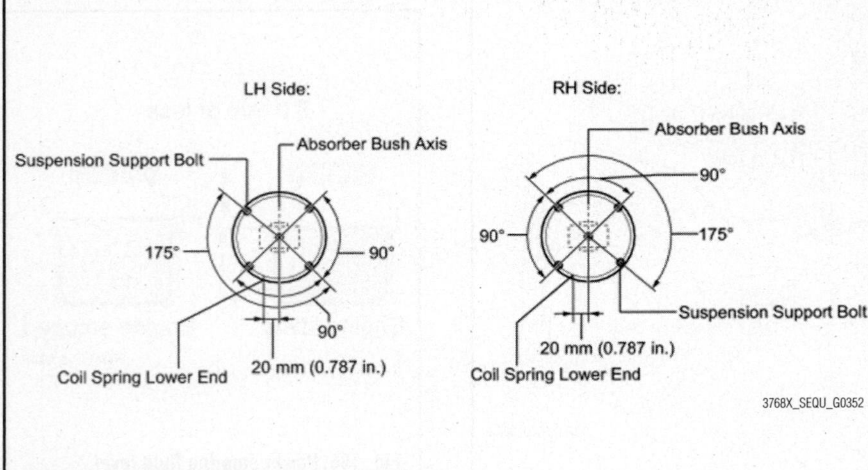

Fig. 196 Front suspension support positioning

LOWER BALL JOINT

REMOVAL & INSTALLATION

1. Before servicing the vehicle, refer to the Precautions Section.
2. Disconnect the negative battery cable. Tape the cable with insulating tape.
3. Remove the lower control arm.
4. Remove the cotter pin and nut. Discard the cotter pin.
5. Using a ball joint removal tool, remove the ball joint.

To install:

➡Be sure to use new fasteners, as required.

6. Installation is the reverse of the removal procedure.
7. Tighten the retaining bolts to 221 ft. lbs. (300 Nm).
8. Be sure to use a new cotter pin.
9. Correct and adjust alignment, as required.

LOWER CONTROL ARM

REMOVAL & INSTALLATION

See Figures 198 and 199.

1. Before servicing the vehicle, refer to the Precautions Section.
2. Disconnect the negative battery cable. Tape the cable with insulating tape.
3. Raise and support the vehicle safely.
4. Remove the tire and wheel assembly.
5. Remove the stabilizer link assembly.
6. Disconnect the shock absorber.
7. Disconnect the lower ball joint attaching bolts.
8. Be sure to properly support the lower control arm before removal.
9. Matchmark and remove the No. 1 suspension arm sub assembly.

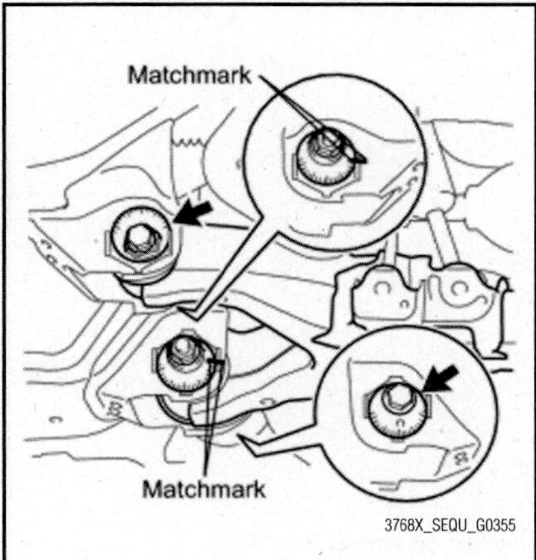

Fig. 198 Front suspension lower control arm matchmark points

10. As required, remove the cotter pin and nut from the ball joint. Install the proper ball joint removal tool and remove the ball joint.

To install:

➡Be sure to use new fasteners, as required.

11. Installation is the reverse of the removal procedure.
12. Temporarily install the component.
13. Stabilize the suspension.
14. Tighten the No. 1 suspension arm sub assembly bolt and nut to 207 ft. lbs. (280 Nm).
15. Tighten the shock bolt to 144 ft. lbs. (195 Nm).

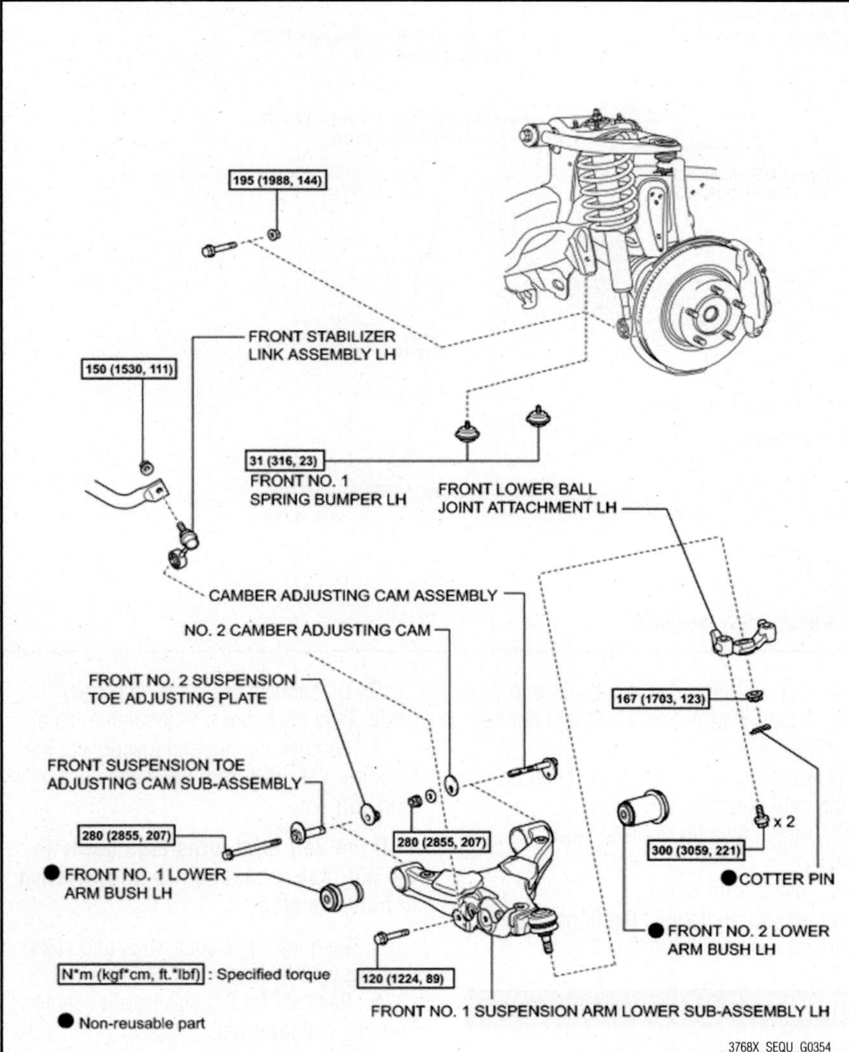

Fig. 199 Front suspension lower control arm related components

16. Check and adjust the alignment, as required.

SHOCK ABSORBERS

REMOVAL & INSTALLATION

See Figures 200 through 202

1. Before servicing the vehicle, refer to the Precautions Section.
2. Disconnect the negative battery cable. Tape the cable with insulating tape.
3. Raise and support the vehicle safely.
4. Remove the tire and wheel assembly.
5. Remove the shock absorber control actuator, if equipped.
6. Remove the suspension control bracket, if equipped.
7. Remove the lower shock retaining nut.
8. Remove the upper shock retaining bolts.
9. Remove the component from the vehicle.

To install:

➡Be sure to use new fasteners, as required.

10. Installation is the reverse of the removal procedure.
11. Temporarily install the component.
12. Stabilize the suspension.
13. Tighten the lower bolt to 144 ft. lbs. (195 Nm).
14. Tighten the upper bolts to 33 ft. lbs. (45 Nm), in the sequence shown in the illustration.

STEERING KNUCKLE

REMOVAL & INSTALLATION

See Figures 203 and 204.

1. Before servicing the vehicle, refer to the Precautions Section.
2. Disconnect the negative battery cable. Tape the cable with insulating tape.
3. Remove the front axle hub.
4. Disconnect the speed sensor from the steering knuckle.
5. Disconnect tie rod end sub-assembly.
   a. Remove the cotter pin and nut.
   b. Disconnect the tie rod end LH from the steering knuckle.
6. Disconnect front lower ball joint attachment.
7. Support the front suspension lower arm with a jack.

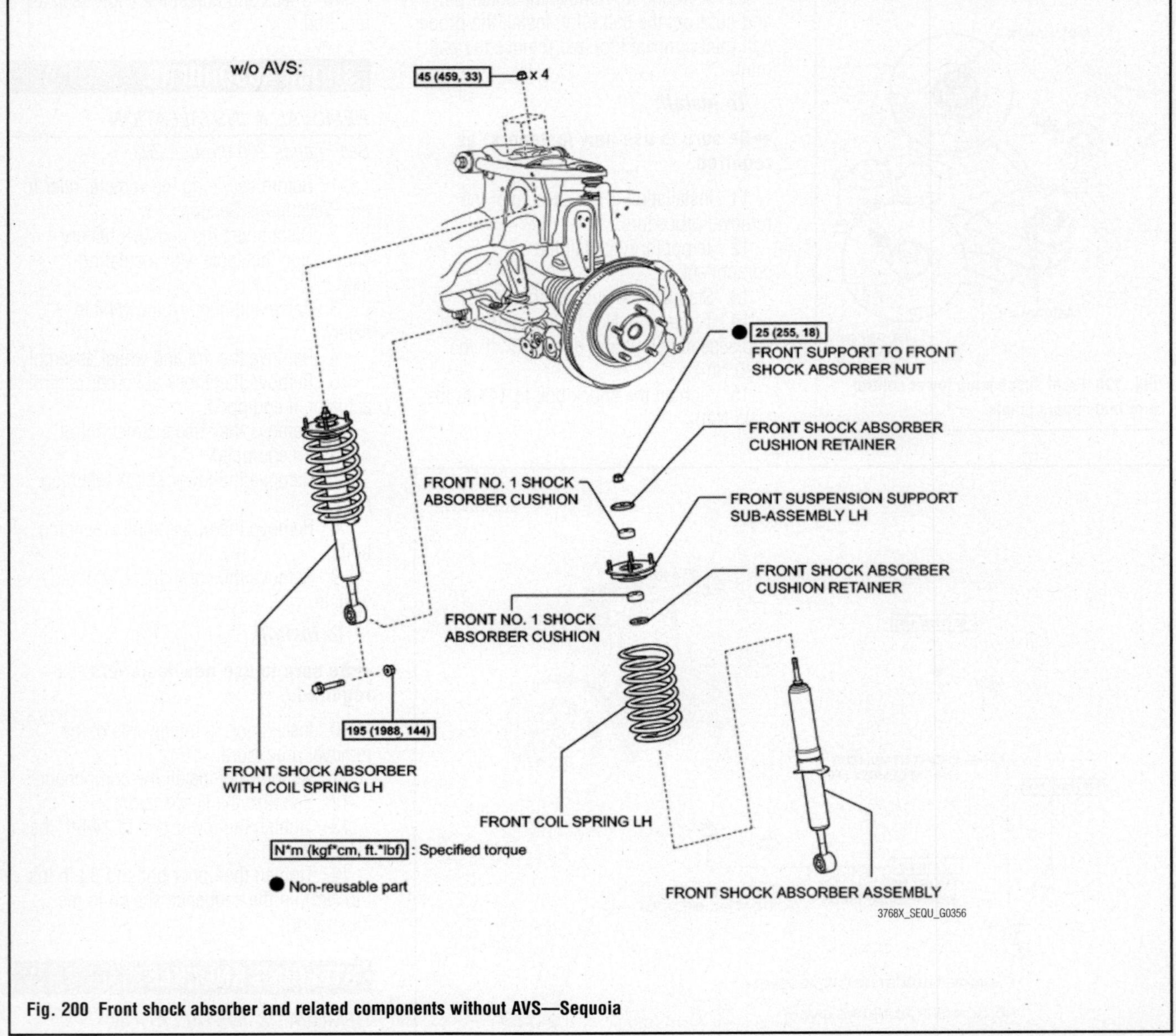

Fig. 200 Front shock absorber and related components without AVS—Sequoia

8. Remove the clip and the nut.
9. Disconnect the upper ball joint from the steering knuckle.
10. Remove the steering knuckle.

To install:

➡Be sure to use new fasteners, as required.

11. Install the front suspension upper arm to the steering knuckle with the nut and tighten to specification.
12. Install a new clip.

➡If the holes for the clip are not aligned, tighten the nut up to another 60°.

13. Connect front lower ball joint attachment.
14. Connect tie rod end sub-assembly.
15. Install the front axle hub and tighten nut to specification.
16. Connect the front speed sensor to the steering knuckle.
17. Attach the clip to the steering knuckle.
18. Install front wheel.
19. Inspect and adjust the front wheel alignment.

STABILIZER BAR

REMOVAL & INSTALLATION

See Figure 205.

1. Before servicing the vehicle, refer to the Precautions Section.
2. Disconnect the negative battery cable. Tape the cable with insulating tape.
3. Remove the nut and disconnect the stabilizer bar links from the lower suspension arm.

➡ If the ball joint turns together with the nut, use a hexagon (6 mm) wrench to hold the stud.

4. Remove the 2 bolts, nuts and stabilizer bar with the cushions and brackets.
5. Remove the 2 brackets and cushions from the stabilizer bar.
6. Hold the stabilizer bar link, and remove the nut.
7. Remove the stabilizer bar link, 2 retainers and bushings from each end of the stabilizer bar.

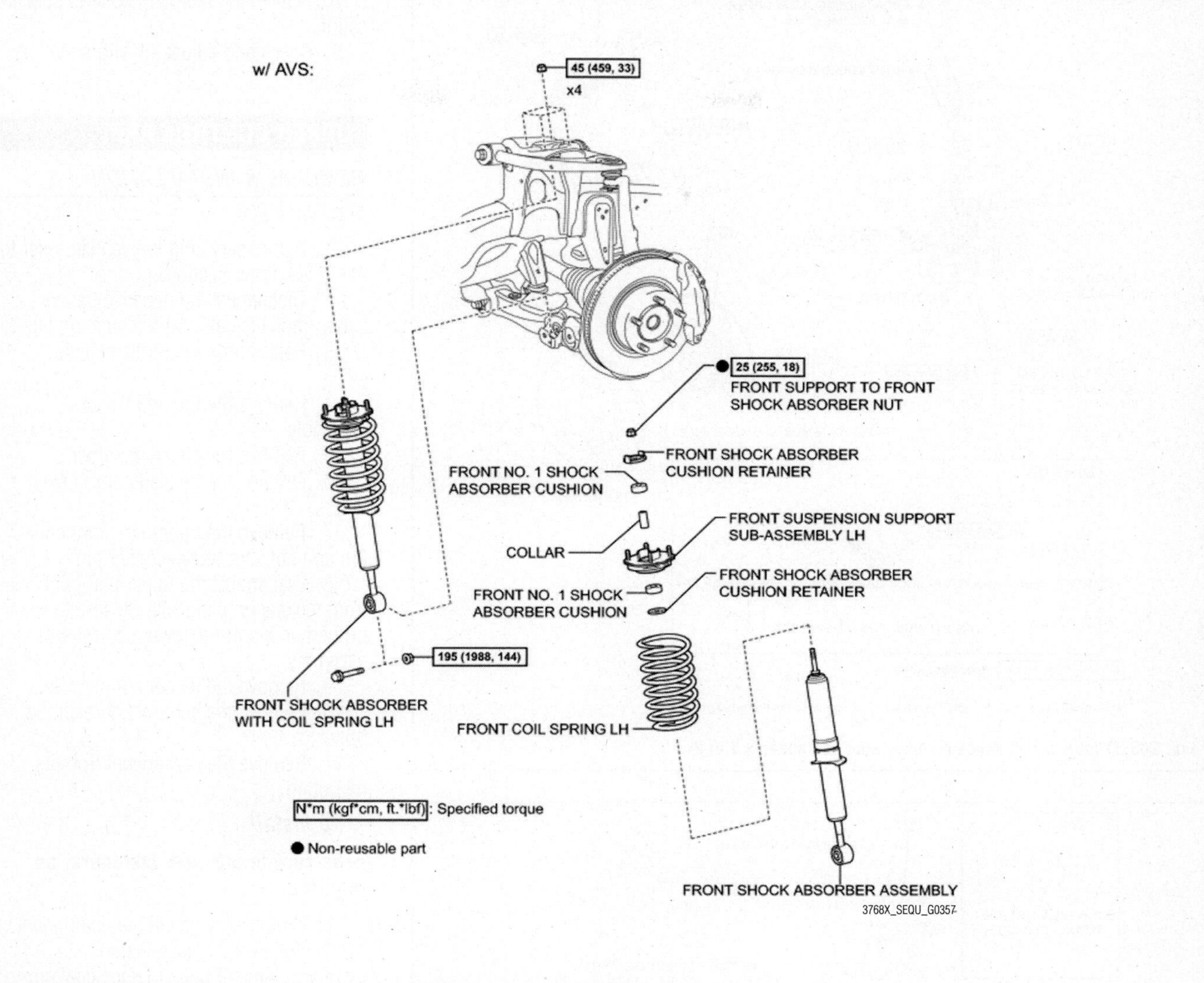

Fig. 201 Front shock absorber and related components with AVS—Sequoia

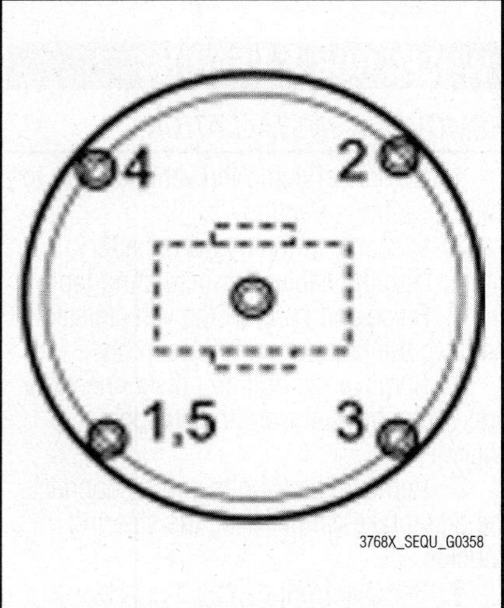

Fig. 202 Front shock absorber upper bolt tightening sequence

To install:

➡Be sure to use new fasteners, as required.

8. Temporarily install the component.
9. Stabilize the suspension.
10. Install the 2 bushings, retainers and stabilizer bar link to the stabilizer bar.
11. Hold the stabilizer bar link, and install a new nut and tighten to 14 ft. lbs. (19 Nm).
12. Install the 2 bushings with their cutout facing to the rearward of the stabilizer bar.
13. Install the stabilizer bar and 2 brackets with the nuts and bolts and tighten to 27 ft. lbs. (37 Nm).
14. Connect front stabilizer links to suspension lower arm and tighten to 51 ft. lbs. (69 Nm).

➡If the ball joint turns together with the nut, use a hexagon (6 mm) wrench to hold the nut.

UPPER BALL JOINT

REMOVAL & INSTALLATION

1. Before servicing the vehicle, refer to the Precautions Section.
2. Disconnect the negative battery cable. Tape the cable with insulating tape.
3. Remove the lower control arm.
4. Using the proper removal tools, separate the ball joint from the control arm.

To install:

➡Be sure to use new fasteners, as required.

5. Temporarily install the component.
6. Stabilize the suspension.

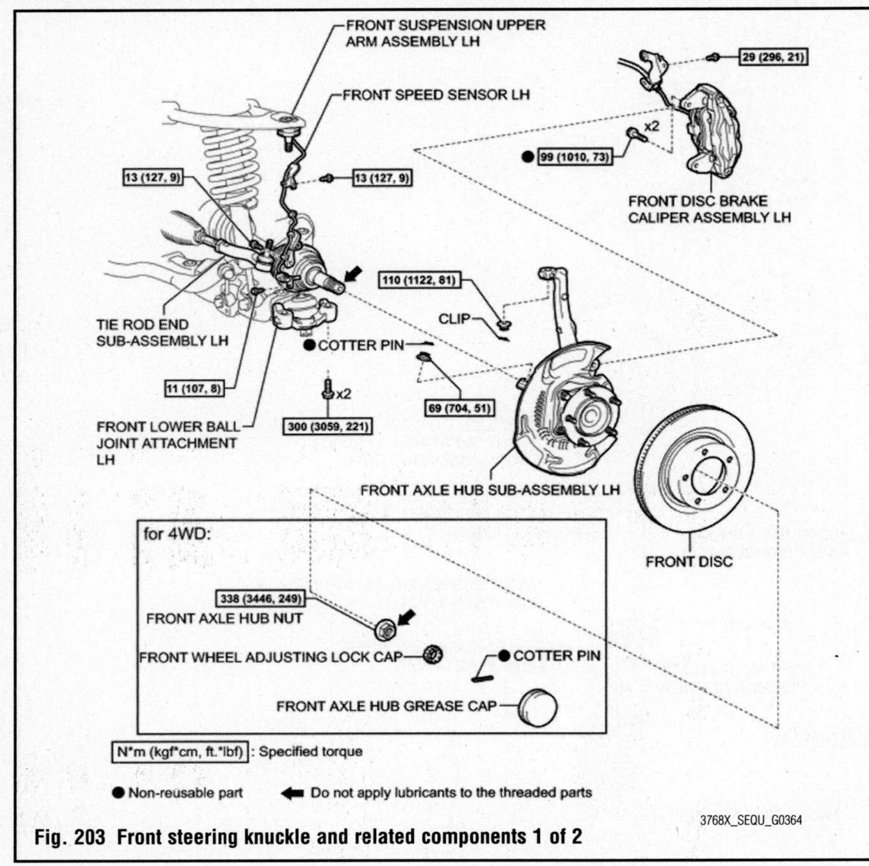

Fig. 203 Front steering knuckle and related components 1 of 2

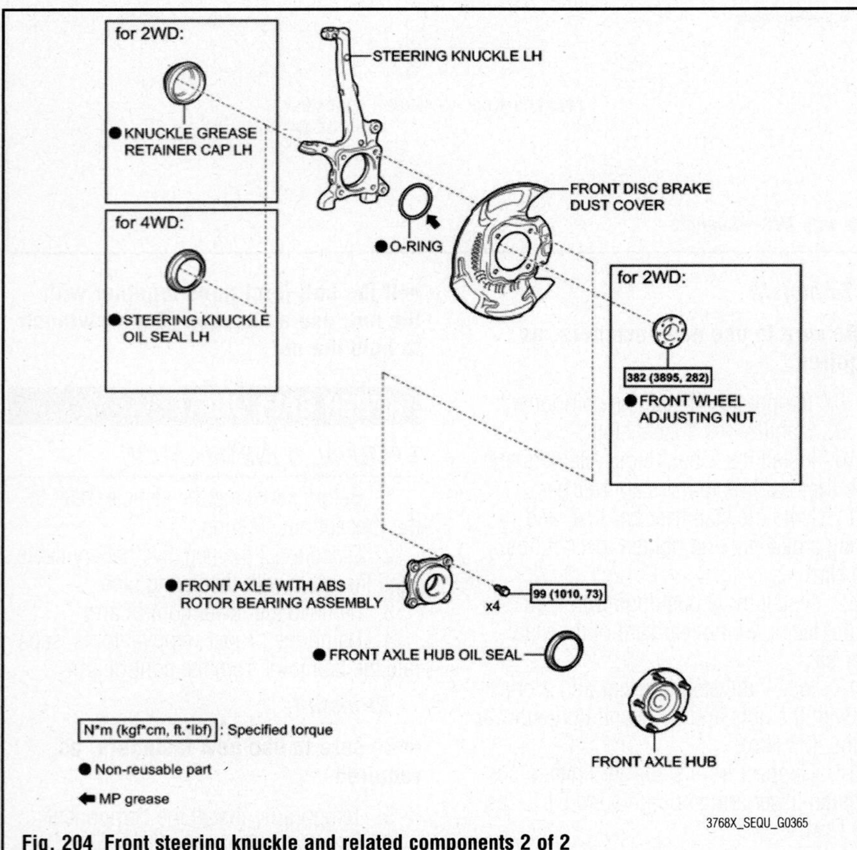

Fig. 204 Front steering knuckle and related components 2 of 2

7. Tighten the retaining bolts to specification.
8. Check and adjust the alignment, as required.

UPPER CONTROL ARM

REMOVAL & INSTALLATION

See Figure 206.

1. Before servicing the vehicle, refer to the Precautions Section.
2. Disconnect the negative battery cable. Tape the cable with insulating tape.
3. Raise and support the vehicle safely.
4. Remove the tire and wheel assembly.
5. Remove the shock absorber.
6. Disconnect the skid control sensor wire.
7. Remove the upper ball joint cotter pin and nut. Discard the cotter pin.
8. Disconnect the upper ball joint from the steering knuckle. Be sure to properly support the lower control arm assembly.
9. Remove the fender apron seals.
10. Remove the component retaining bolts.
11. Remove the component from its mounting.

To install:

➡Be sure to use new fasteners, as required.

12. Temporarily install the component.
13. Stabilize the suspension.
14. Tighten all bolts to specification, refer to illustration.
15. Check and adjust the alignment, as required.

WHEEL HUB & BEARING

REMOVAL & INSTALLATION

1. Before servicing the vehicle, refer to the Precautions Section.
2. Disconnect the negative battery cable. Tape the cable with insulating tape.
3. Raise and support the vehicle safely. Remove the tire and wheel assembly.
4. Remove the bolt and disconnect the brake tube bracket from the steering knuckle.
5. Remove the 2 bolts and disconnect the disc brake caliper from the steering knuckle.
6. Remove front disc.
7. Remove front axle hub grease cap (4WD).

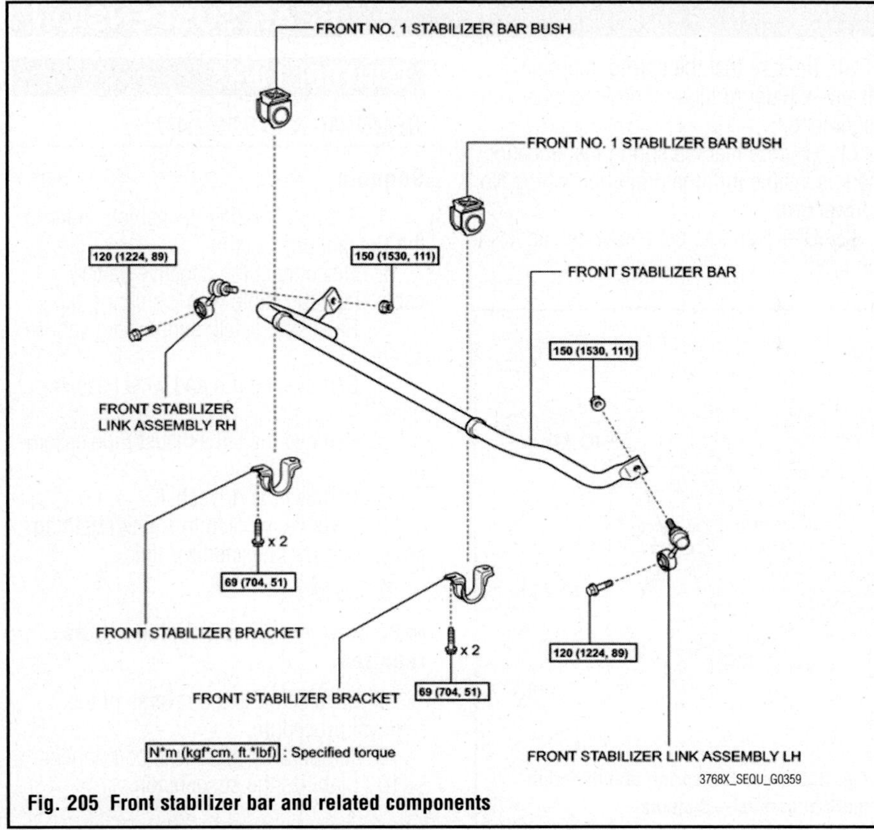

Fig. 205 Front stabilizer bar and related components

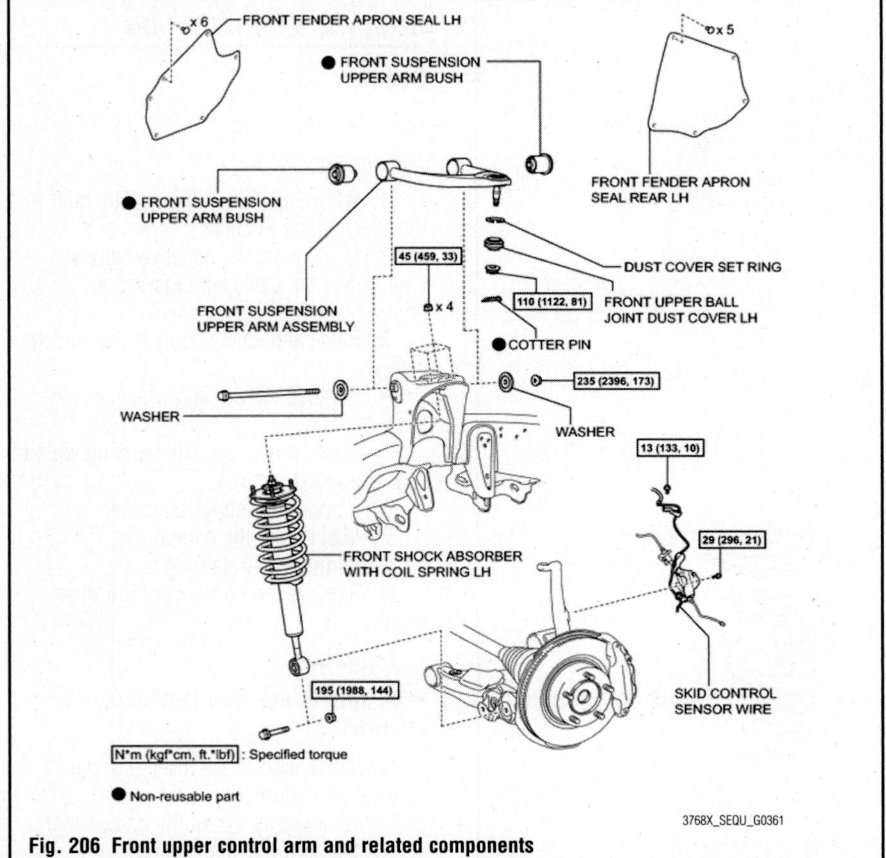

Fig. 206 Front upper control arm and related components

a. Remove the cotter pin and adjusting lock cap.

b. Using a 39 mm socket wrench, remove the axle hub nut.

8. Remove front axle hub sub-assembly (4WD).

a. Remove the 4 bolts.

b. Using a plastic-faced hammer, tap out the front drive shaft from the front axle hub.

c. Remove the O-ring from the axle hub.

9. Remove front axle hub sub-assembly (2WD).

a. Remove the 4 bolts.

b. Remove the axle hub and dust cover from the steering knuckle.

c. Remove the O-ring from the axle hub.

To install:

➡Be sure to use new fasteners, as required.

10. Install front axle hub sub-assembly (2WD).

a. Apply MP grease to a new O-ring.

b. Install the O-ring to the axle hub.

c. Install the dust cover and axle hub to the steering knuckle with the 4 bolts and tighten to 73 ft. lbs. (99 Nm).

11. Install front axle hub sub-assembly (4WD).

a. Apply MP grease to a new O-ring.

b. Install the O-ring to the axle hub.

c. Connect the front drive shaft to the front axle hub.

d. Install the dust cover and axle hub to the steering knuckle with the 4 bolts and tighten to 73 ft. lbs. (99 Nm).

12. Install front axle hub nut (4WD).

a. Clean the threaded parts on the drive shaft and axle hub nut using a non-residue solvent.

b. Using a 39 mm socket wrench, install the hub nut and tighten to 249 ft. lbs. (338 Nm).

c. Install the front wheel adjusting lock cap and a new cotter pin.

13. Inspect the front axle hub.

14. Install the axle hub grease cap.

15. Install front disc.

16. Connect the front disc brake caliper and install 2 new bolts and tighten to 73 ft. lbs. (99 Nm).

17. Connect the brake tube bracket to the steering knuckle with the bolt and tighten to 21 ft. lbs. (29 Nm).

18. Install front wheel.

19. Check the speed sensor signal.

COIL SPRING

REMOVAL & INSTALLATION

Sequoia

See Figures 207 and 208.

1. Before servicing the vehicle, refer to the Precautions Section.
2. Disconnect the negative battery cable. Tape the cable with insulating tape.
3. Raise and support the vehicle safely.
4. Remove the shock absorber.
5. Loosen the three upper control arm nuts. Do not remove them.
6. Insert a piece of wood between the jack and the No. 2 lower arm and raise the arm.
7. Using a spring compressor tool, compress the spring.
8. Gently lower the jack and remove the component from its mounting.

To install:

➡Be sure to use new fasteners, as required.

9. Installation is the reverse of the removal procedure.
10. Be sure that the spring is properly installed. Refer to illustration for proper alignment.
11. Be sure that the spring identification mark is visible thru the drain hole of the No. 2 lower arm.
12. Check and adjust alignment, as required.

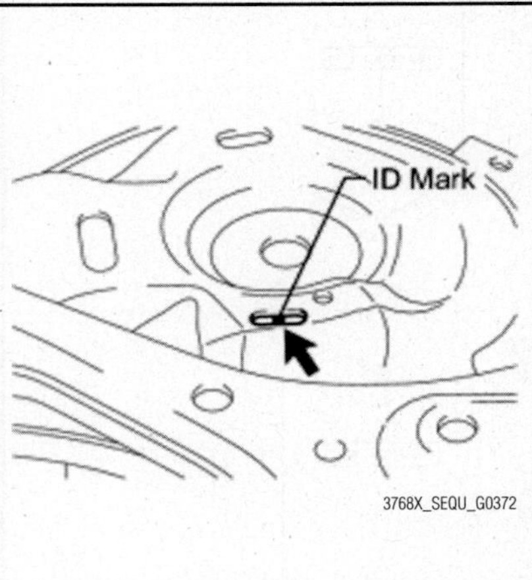

Fig. 208 Rear coil spring identification mark alignment—Sequoia

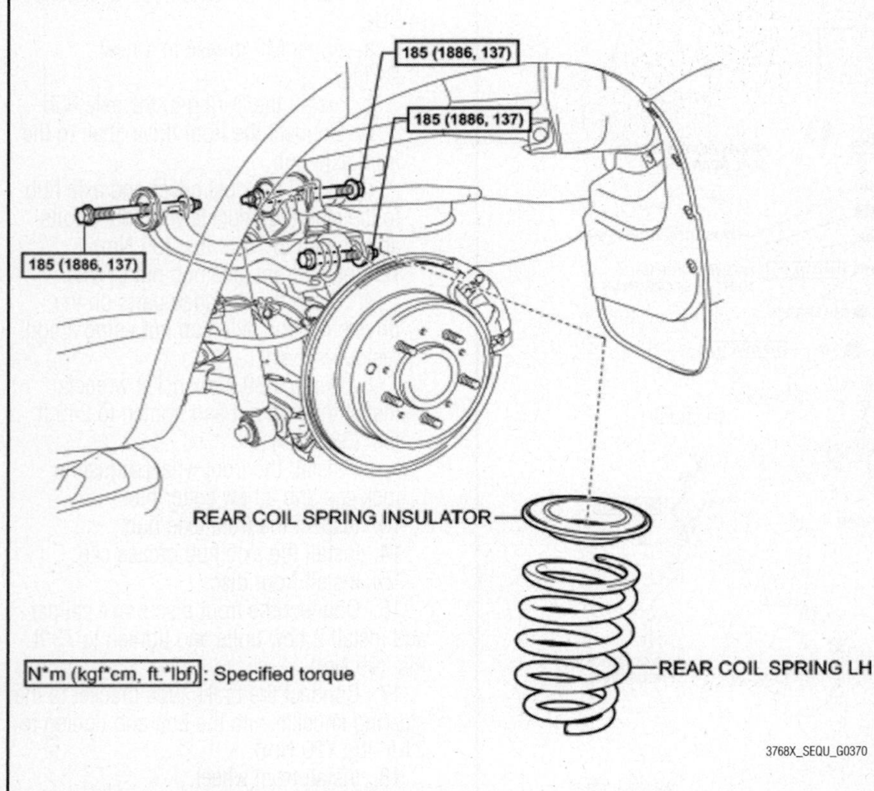

Fig. 207 Rear coil spring and related components—Sequoia

CONTROL LINKS

REMOVAL & INSTALLATION

Sequoia

1. Before servicing the vehicle, refer to the Precautions Section.
2. Disconnect the negative battery cable. Tape the cable with insulating tape.
3. Raise and safely support the vehicle safely.
4. Remove the tire and wheel assemblies.
5. Remove the tail exhaust pipe assembly.
6. Remove the driveshaft.
7. Remove the stabilizer link retaining bolts. Remove the stabilizer link.

To install:

➡Be sure to use new fasteners, as required.

8. Installation is the reverse of the removal procedure.
9. Temporarily install the component.
10. Stabilize the suspension.

LOWER CONTROL ARMS

REMOVAL & INSTALLATION

Sequoia

No. 1 ARM

See Figures 209 and 210.

1. Before servicing the vehicle, refer to the Precautions Section.
2. Disconnect the negative battery cable. Tape the cable with insulating tape.
3. Raise and safely support the vehicle safely.
4. Remove the tire and wheel assemblies.
5. Support the rear axle housing with a jack and wooden block.
6. Remove the shock absorber.
7. Matchmark the component.
8. Remove the retaining bolts.
9. Remove the component from the vehicle.

To install:

➡Be sure to use new fasteners, as required.

10. Installation is the reverse of the removal procedure.
11. Temporarily install the component.
12. Stabilize the suspension.

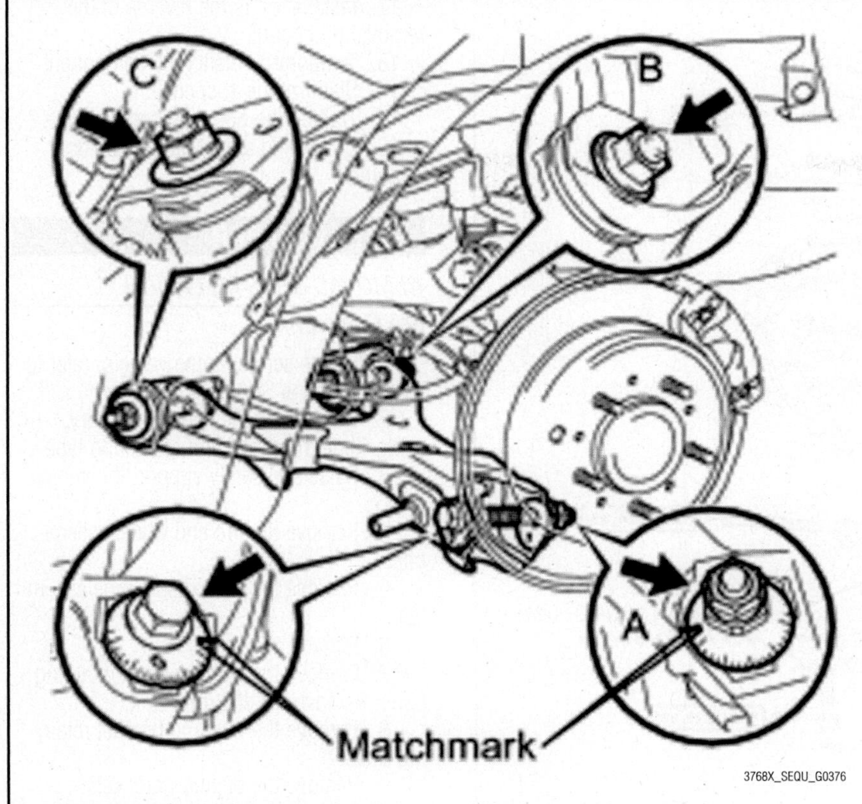

Fig. 209 Rear lower control arm (No. 1) matchmark and bolt locations—Sequoia

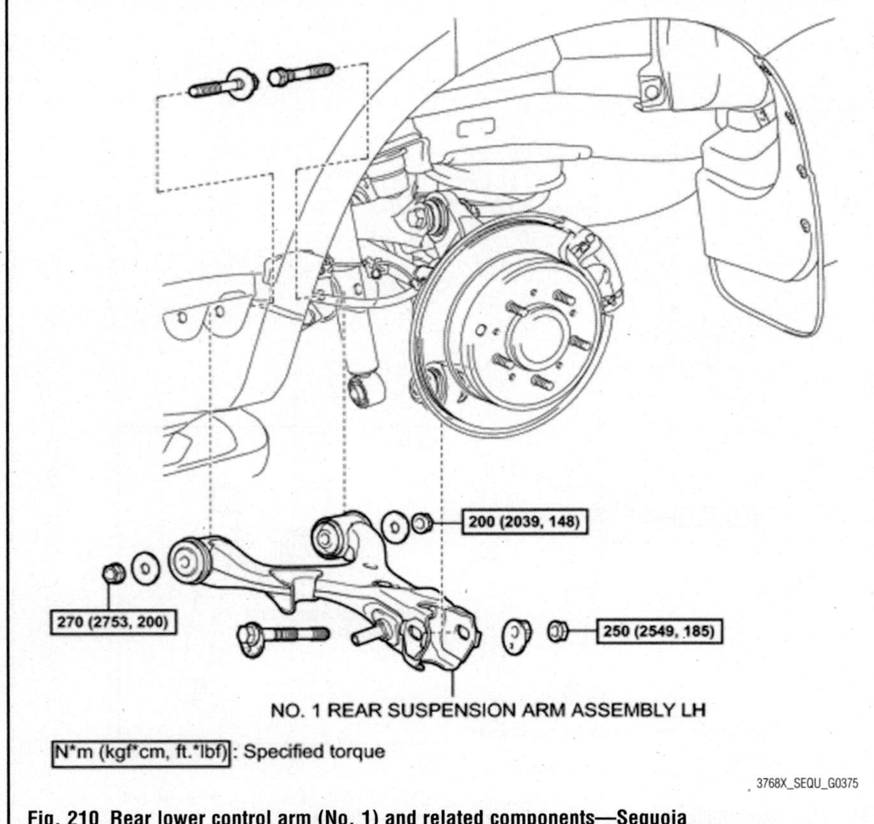

Fig. 210 Rear lower control arm (No. 1) and related components—Sequoia

13. Tighten Bolt A: 185 ft. lbs. (250 Nm). Tighten bolt B: 148 ft. lbs. (200 Nm). Tighten bolt C: 200 ft. lbs. (270 Nm).
14. Check and adjust alignment, as required.

No. 2 ARM

See Figures 211 and 212.

1. Before servicing the vehicle, refer to the Precautions Section.
2. Disconnect the negative battery cable. Tape the cable with insulating tape.
3. Raise and safely support the vehicle safely.
4. Remove the tire and wheel assemblies.
5. Support the rear axle housing with a jack and wooden block.
6. Remove the spring.
7. Remove the pneumatic cylinder, if equipped.
8. Remove the cotter pin and nut. Discard the cotter pin.
9. Using the proper removal tool, separate the rear suspension arm.
10. Matchmark the component.
11. Remove the retaining bolts.
12. Remove the component from the vehicle.

To install:

➡Be sure to use new fasteners, as required.

13. Installation is the reverse of the removal procedure.
14. Temporarily install the component.
15. Stabilize the suspension.
16. Tighten the joint retaining nut to specification.

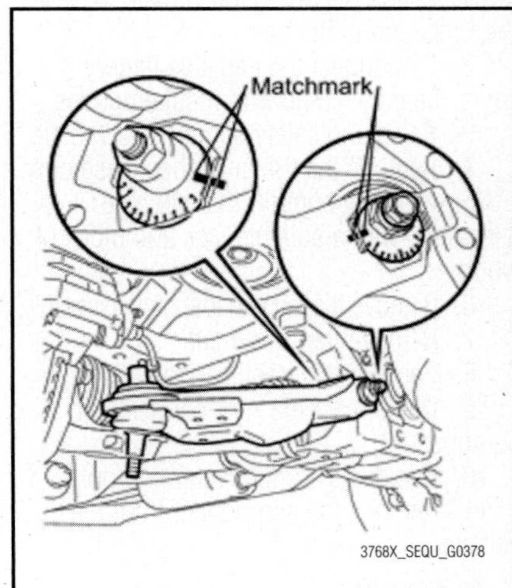

Fig. 211 Rear lower control arm (No. 2) matchmark locations—Sequoia

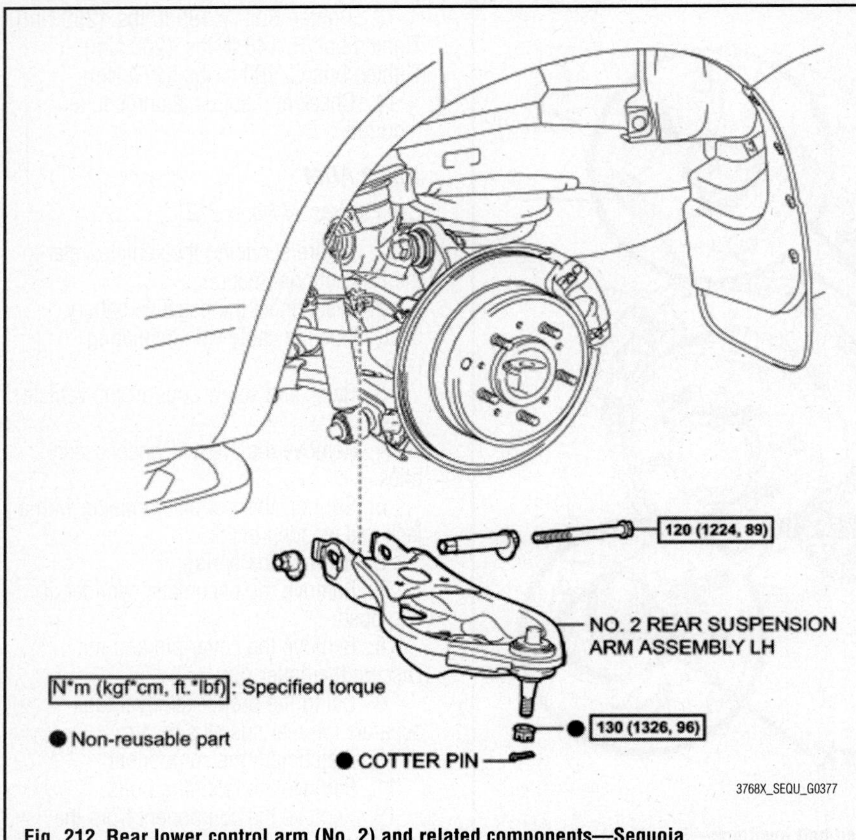

Fig. 212 Rear lower control arm (No. 2) and related components—Sequoia

17. Tighten the control arm retaining bolts to 89 ft. lbs. (120 Nm).
18. Check and adjust alignment, as required.

SHOCK ABSORBER

REMOVAL & INSTALLATION

See Figure 213.

1. Before servicing the vehicle, refer to the Precautions Section.
2. Disconnect the negative battery cable. Tape the cable with insulating tape.
3. Raise and safely support the vehicle.
4. Remove the tire and wheel assembly.
5. Properly support the rear axle assembly, with a suitable jack and block of wood.
6. Remove the tail exhaust pipe assembly.
7. Remove the driveshaft.
8. Remove the stabilizer bar.
9. Disconnect the AVS connector, if equipped.
10. Remove the shock retaining bolts.
11. Remove the component from the vehicle.

To install:

➡Be sure to use new fasteners, as required.

12. Installation is the reverse of the removal procedure.
13. Temporarily install the component.
14. Stabilize the suspension.
15. Tighten the bolts to specification.
16. Check and adjust alignment, as required.

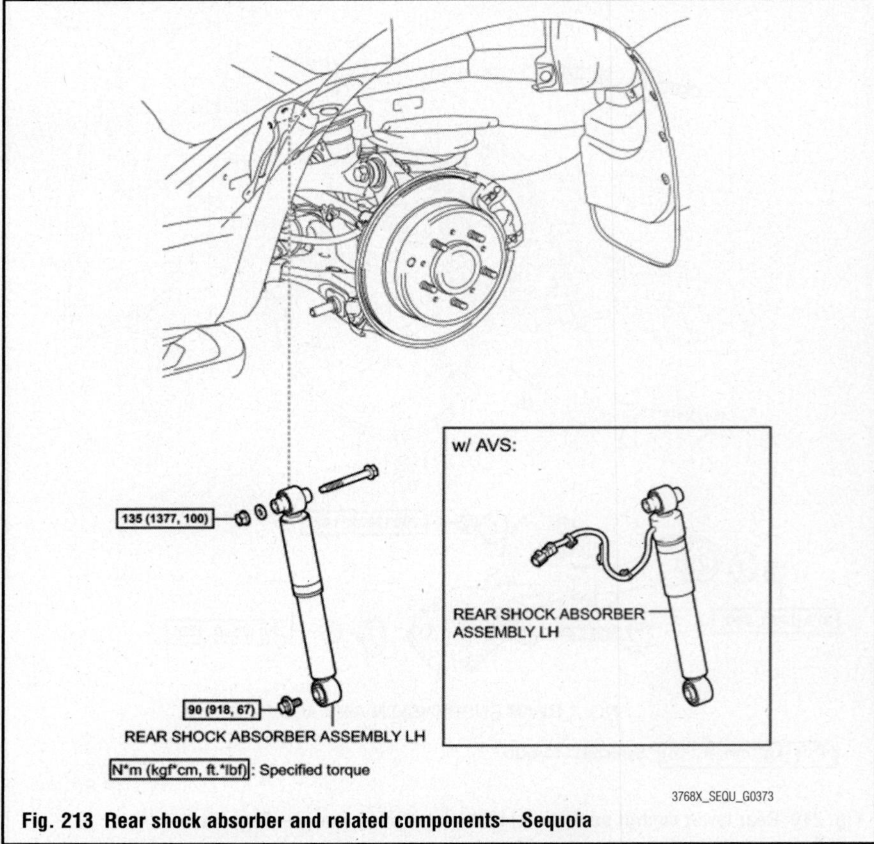

Fig. 213 Rear shock absorber and related components—Sequoia

STABILIZER BAR

REMOVAL & INSTALLATION

See Figure 214.

1. Before servicing the vehicle, refer to the Precautions Section.
2. Disconnect the negative battery cable. Tape the cable with insulating tape.
3. Raise and safely support the vehicle safely.
4. Remove the tire and wheel assemblies.
5. Remove the tail exhaust pipe assembly.
6. Remove the driveshaft.
7. Remove the stabilizer link retaining bolts. Remove the stabilizer link.
8. Remove the bushing bracket retaining bolts.
9. Remove the bushing brackets.
10. Remove the stabilizer bar from its mounting.

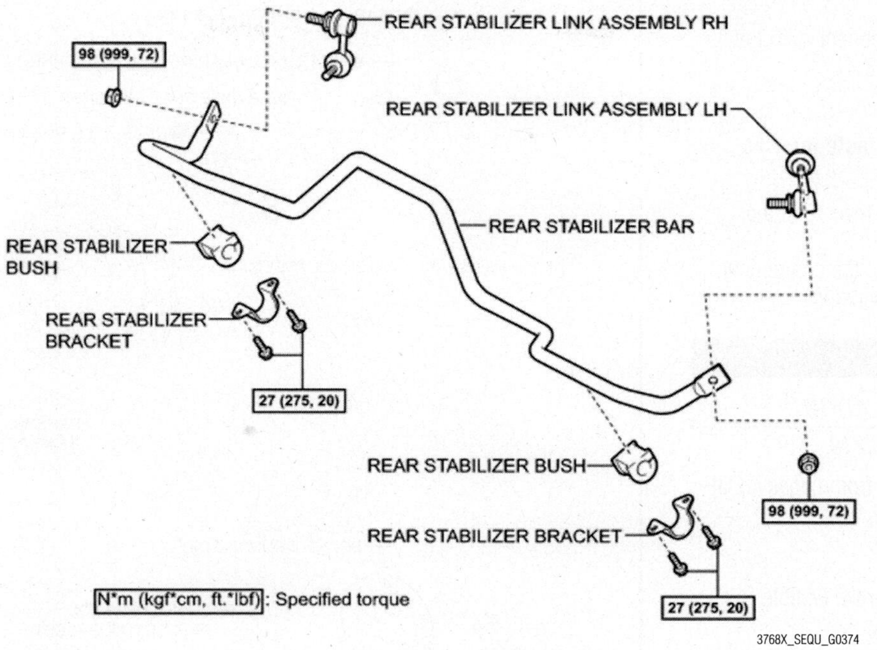

Fig. 214 Rear stabilizer bar and related components—Sequoia

To install:

➡**Be sure to use new fasteners, as required.**

11. Installation is the reverse of the removal procedure.
12. Temporarily install the component.
13. Stabilize the suspension.

UPPER CONTROL ARMS

REMOVAL & INSTALLATION

See Figure 215.

1. Before servicing the vehicle, refer to the Precautions Section.
2. Disconnect the negative battery cable. Tape the cable with insulating tape.
3. Properly relieve the fuel system pressure.
4. Raise and safely support the vehicle safely.
5. Remove the tire and wheel assemblies.
6. Drain the fuel tank.
7. Remove the fuel tank.
8. Remove the height control sensor.
9. Separate the rear disc brake cylinder assembly.
10. Separate the rear brake flexible hose.

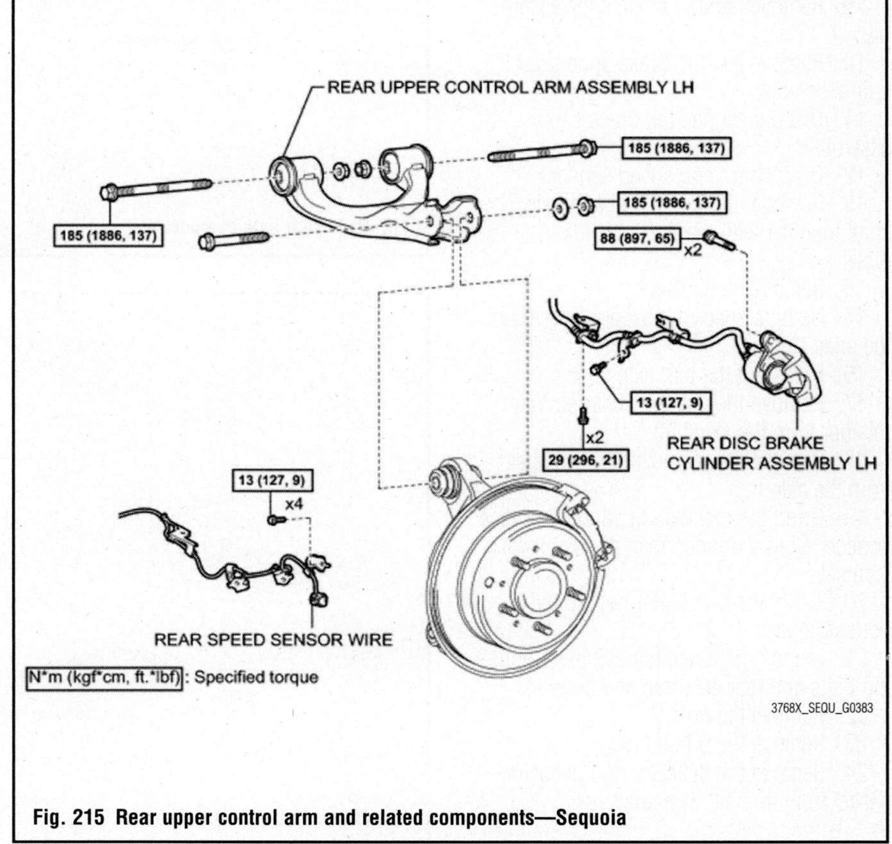

Fig. 215 Rear upper control arm and related components—Sequoia

11. Disconnect the rear wheel speed sensor.
12. Remove the upper control arm mounting bolts.
13. Remove the component from the vehicle.

To install:

➡Be sure to use new fasteners, as required.

14. Installation is the reverse of the removal procedure.
15. Temporarily install the component.
16. Stabilize the suspension.

WHEEL BEARINGS

REMOVAL & INSTALLATION

See Figures 216 through 218.

1. Disconnect cable from negative battery terminal.
2. Remove rear wheel.
3. Drain brake fluid.
4. Disconnect rear brake flexible hose.
5. Remove the clip.
6. Remove the 2 bolts and disconnect rear disc brake cylinder.
7. Remove rear disc.
8. Remove parking brake shoe return tension spring.
9. Remove parking brake shoe assembly.
10. Remove parking brake shoe lever sub-assembly.
11. Disconnect parking brake cable assembly.
12. Disconnect rear speed sensor.
13. Remove the 4 nuts and rear axle shaft together with the parking brake plate.
14. Remove the O-ring.
15. Using a snap ring expander, remove the snap ring.
16. Press out the rear axle shaft.
17. Remove the rear axle bearing inner retainer from the axle hub.
18. Remove the rear axle shaft washer from the axle hub.
19. Grind the rear axle bearing inner race surface using a grinder, then remove it with a chisel.
20. Temporarily install 4 nuts to the housing bolts.
21. Using a hammer, remove the 4 housing bolts and rear axle hub and bearing.
22. Remove the nuts.
23. Remove the 5 hub bolts.
24. Remove the deflector and deflector gasket from the rear axle shaft.

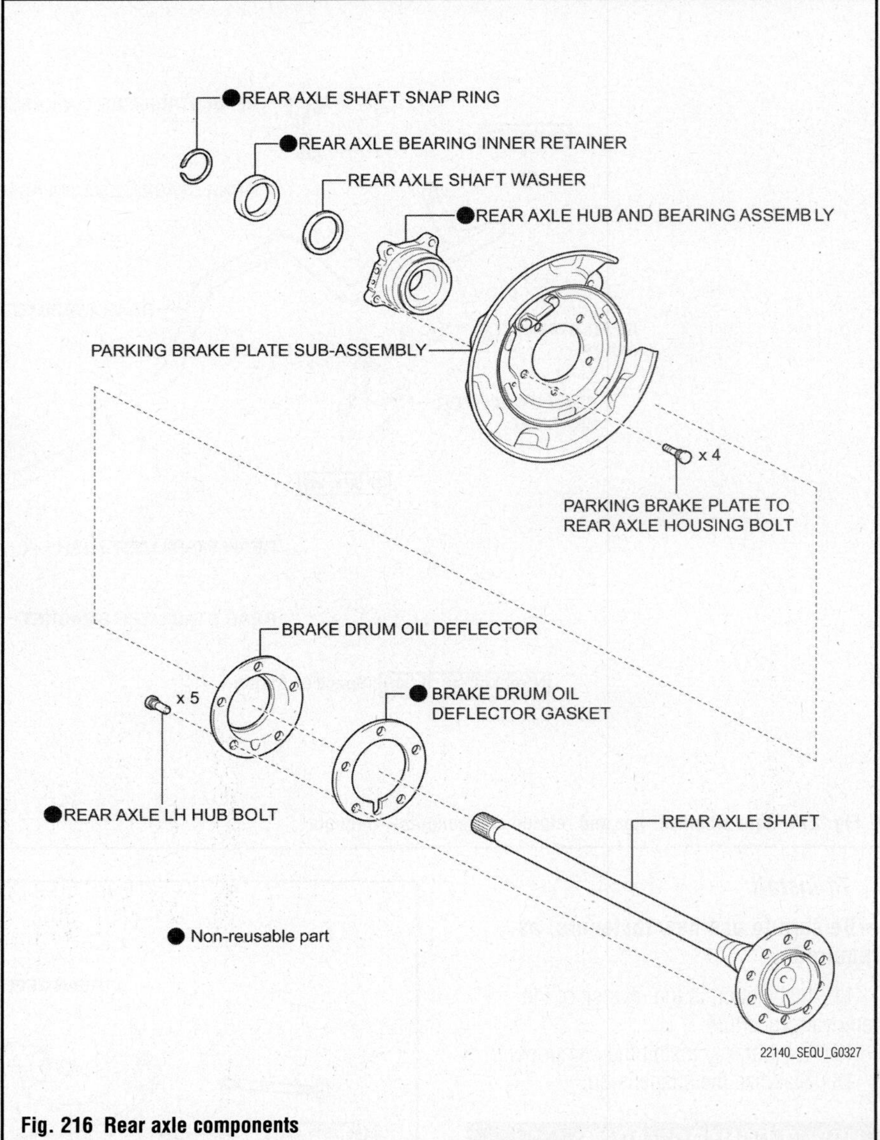

Fig. 216 Rear axle components

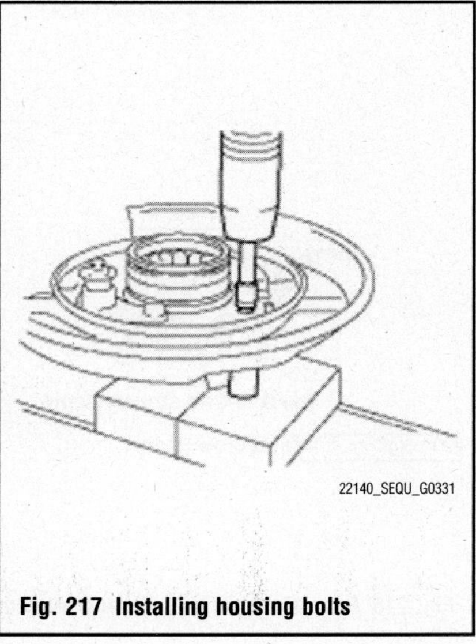

Fig. 217 Installing housing bolts

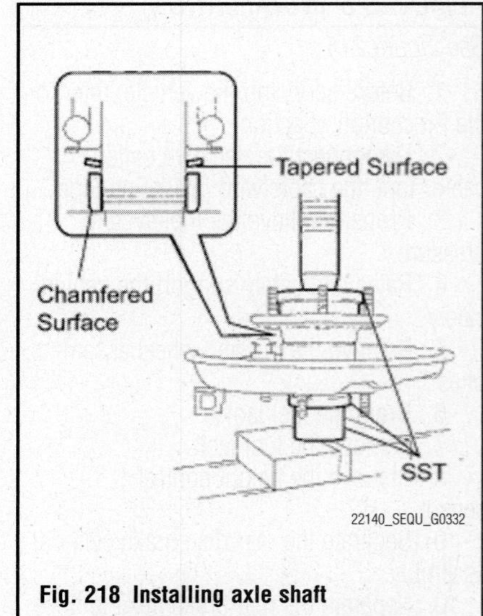

Fig. 218 Installing axle shaft

To install:

25. Install a new deflector gasket and deflector to the rear axle shaft.

26. Temporarily install a washer and nut to 5 new hub bolts.

27. Install the hub bolts by tightening each nut.

28. Removed the washer and nut from each hub bolt.

29. Position the parking brake plate on a new rear axle hub and bearing.

30. Using 2 socket wrenches and a press, press in the 4 housing bolts.

31. Install the washer and a new retainer to the axle hub.

➡Install the washer with its tapered surface facing downward. Install the retainer with its chamfered surface facing downward.

32. Press in the rear axle shaft.

33. Using a snap ring expander, install a new snap ring.

34. Apply MP grease to the lip of the oil seal.

35. Install the O-ring to the axle shaft.

36. Install the rear axle shaft and parking brake plate with the 4 nuts and tighten to 44 ft. lbs. (60 Nm).

37. Inspect the rear axle shaft.

38. Connect rear speed sensor.

39. Connect parking brake cable assembly.

40. Install parking brake shoe lever sub-assembly.

41. Install parking brake shoe assemblies.

42. Install parking brake shoe return tension spring.

43. Check parking brake installation.

44. Install rear disc.

45. Adjust parking brake shoe clearance.

46. Connect the rear disc brake cylinder and install 2 new bolts and tighten to 70 ft. lbs. (95 Nm).

47. Set the flexible hose to the connecting point with the brake tube, and then install a new clip.

48. Connect the flexible hose to the brake tube while holding the flexible hose with a wrench.

49. Connect cable to negative battery terminal.

50. Fill reservoir with brake fluid.

51. Bleed air from brakes.

52. Install rear wheel.

53. Inspect differential oil.

54. Inspect for differential oil leak.

55. Check parking brake pedal travel.

WHEEL HUB & BEARING

REMOVAL & INSTALLATION

See Figure 219.

1. Before servicing the vehicle, refer to the Precautions Section.

2. Disconnect the negative battery cable. Tape the cable with insulating tape.

3. Raise and safely support the vehicle.

4. Remove the tire and wheel assembly.

5. Remove the coil spring, if equipped.

6. Remove the pneumatic cylinder, if equipped.

7. Remove the parking brake assembly.

8. Remove the rear speed sensor.

9. Remove the rear hub grease cap.

10. Remove the cotter pin. Remove the axle shaft nut. Discard the cotter pin.

11. Remove the cotter pin and nut. Using the proper removal tools, separate the No.2 rear suspension arm assembly.

12. Matchmark the No. 1 rear suspension arm and adjust cam.

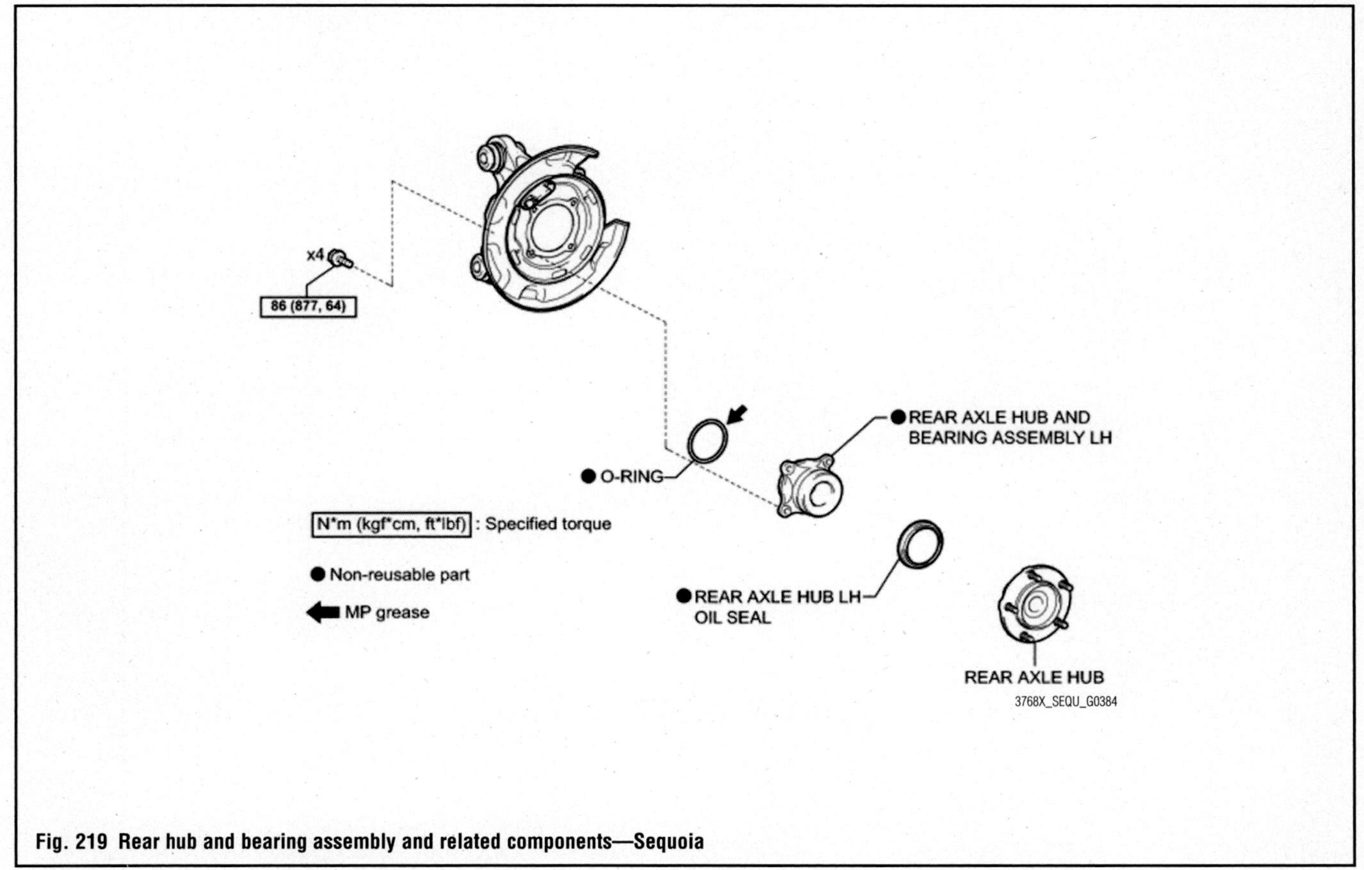

Fig. 219 Rear hub and bearing assembly and related components—Sequoia

13. Remove the bolts and nuts. Tap out the front halfshaft from the rear axle carrier sub assembly.

14. Remove the bolts and hub and bearing from the axle carrier.

To install:

➡Be sure to use new fasteners, as required.

15. Installation is the reverse of the removal procedure.

16. Be sure to use a new oil seal.

17. Tighten the retaining bolts to 64 ft. lbs. (86 Nm).

ADJUSTMENT

Adjustment is not possible.

TOYOTA

Sienna

14

SPECIFICATIONS AND MAINTENANCE CHARTS

ENGINE AND VEHICLE IDENTIFICATION

Engine							Model Year	
Code ①	Liters (cc)	Cu. In.	Cyl.	Fuel Sys.	Engine Type	Eng. Mfg.	Code ②	Year
1AR-FE	2.7 (2698)	165	4	MFI	DOHC	Toyota	B	2011
2GR-FE	3.5 (3456)	213	6	MFI	DOHC	Toyota	C	2012

① Stamped on the left side of the engine block

② 10th position of VIN

71099_SIEN_C0001

GENERAL ENGINE SPECIFICATIONS

All measurements are given in inches.

Year	Model	Engine Displacement Liters (cc)	Engine ID	Fuel System Type	Net Horsepower @ rpm	Net Torque @ rpm (ft. lbs.)	Bore x Stroke (in.)	Com-pression Ratio	Oil Pressure @ rpm
2011	Sienna	2.7 (2698)	1AR-FE	MFI	187@5800	186@4100	3.43x4.13	10.0:1	38@4000
		3.5 (3456)	2GR-FE	MFI	266@6200	245@4700	3.70x3.27	NA	55@3000
2012	Sienna	2.7 (2698)	1AR-FE	MFI	187@5800	186@4100	3.43x4.13	10.0:1	38@4000
		3.5 (3456)	2GR-FE	MFI	266@6200	245@4700	3.70x3.27	NA	55@3000

NA Not Available

71099_SIEN_C0002

ENGINE TUNE-UP SPECIFICATIONS

Year	Engine Displacement Liters	Engine ID	Spark Plug Gap (in.)	Ignition Timing AT	Fuel Pump (psi)	Idle Speed RPM AT	Valve Clearance Intake	Valve Clearance Exhaust
2011	2.7	1AR-FE	0.0394-0.0433	8-12 BTDC	44-50	600-700	A	A
	3.5	2GR-FE	0.0394-0.0433	8-12 BTDC	44-50	600-700	A	A
2012	2.7	1AR-FE	0.0394-0.0433	8-12 BTDC	44-50	600-700	A	A
	3.5	2GR-FE	0.0394-0.0433	8-12 BTDC	44-50	600-700	A	A

A Engine is equipped with hydraulic lash adjusters.

71099_SIEN_C0003

CAPACITIES

Year	Model	Engine Displacement Liters	Engine ID	Engine Oil with Filter	Transaxle (pts.) Auto.	Drive Axle (pts.) Front	Drive Axle (pts.) Rear	Transfer Case (pts.)	Fuel Tank (gal.)	Cooling System (qts.)
2011	Sienna	2.7	1AR-FE	NA	NA	NA	1-1.2	1.6	20.0	9.2
		3.5	2GR-FE	NA	NA	NA	1-1.2	1.6	20.0	①
2012	Sienna	2.7	1AR-FE	NA	NA	NA	1-1.2	1.6	20.0	9.2
		3.5	2GR-FE	NA	NA	NA	1-1.2	1.6	20.0	①

NOTE: All capacities are approximate. Add fluid gradually and ensure a proper fluid level is obtained.

NA Not Available

① Without towing package: 11.0 qts.
With towing package: 11.7 qts.

71099_SIEN_C0004

FLUID SPECIFICATIONS

Year	Model	Engine Disp. Liters	Engine Oil	Auto. Trans.	Drive Axle Rear	Transfer Case	Brake Master Cylinder	Cooling System
2011	Sienna	2.7	NA	ATF WS	75W-85	75W-85	DOT 3	SLLC
		3.5	NA	ATF WS	75W-85	75W-85	DOT 3	SLLC
2012	Sienna	2.7	NA	ATF WS	75W-85	75W-85	DOT 3	SLLC
		3.5	NA	ATF WS	75W-85	75W-85	DOT 3	SLLC

DOT: Department Of Transpotation

NA Not Available

Toyota SLLC: Super Long Life Coolant

71099_SIEN_C0005

VALVE SPECIFICATIONS

Year	Engine Displacement Liters	Engine ID	Seat Angle (deg.)	Face Angle (deg.)	Spring Test Pressure (lbs. @ in.)	Spring Free-Length (in.)	Spring Installed Height (in.)	Stem-to-Guide Clearance (in.) Intake	Stem-to-Guide Clearance (in.) Exhaust	Stem Diameter (in.) Intake	Stem Diameter (in.) Exhaust
2011	2.7	1AR-FE	NA	NA	NA	1.970	NA	0.00098-0.00236	0.00118-0.00256	0.2154-0.2159	0.2152-0.2157
	3.5	2GR-FE	NA	NA	NA	1.808	NA	0.00098-0.00236	0.00118-0.00256	0.2154-0.2159	0.2152-0.2157
2012	2.7	1AR-FE	NA	NA	NA	1.970	NA	0.00098-0.00236	0.00118-0.00256	0.2154-0.2159	0.2152-0.2157
	3.5	2GR-FE	NA	NA	NA	1.808	NA	0.00098-0.00236	0.00118-0.00256	0.2154-0.2159	0.2152-0.2157

NA Not Available

71099_SIEN_C0006

CAMSHAFT SPECIFICATIONS

All measurements in inches unless noted

Year	Engine Displacement Liters	Engine ID	Journal Diameter	Brg. Oil Clearance	Shaft End-play	Runout	Journal Bore	Lobe Height Intake	Lobe Height Exhaust
2011	2.7	1AR-FE	①	②	NA	0.00118	NA	1.73870-1.74429	1.73795-1.74354
	3.5	2GR-FE	③	④	NA	0.00157	NA	1.74480-1.74860	1.7426-1.74650
2012	2.7	1AR-FE	①	②	NA	0.00118	NA	1.73870-1.74429	1.73795-1.74354
	3.5	2GR-FE	③	④	NA	0.00157	NA	1.74480-1.74860	1.7426-1.74650

NA Not Available

① Journal No. 1: 1.35626-0.35689 inches
Other Journals: 0.90390-0.90453 inches

② Intake Journal No. 1: 0.00137-0.00283 inches
Other Journals: 0.00335 inches
Exhaust Journal No. 1: 0.00193-0.00339 inches
Other Journals: 0.00335 inches

③ Journal No. 1: 1.4152-1.4157 inches
Other Journals: 1.0221-1.0226 inches

④ Journal No. 1: 0.00126-0.00280 inches
Other Journals: 0.000984-0.00244 inches

71099_SIEN_C0007

CRANKSHAFT AND CONNECTING ROD SPECIFICATIONS

All measurements are given in inches.

Year	Engine Displacement Liters	Engine ID	Crankshaft Main Brg. Journal Dia.	Crankshaft Main Brg. Oi Clearance	Crankshaft Shaft End-play	Crankshaft Thrust on No.	Connecting Rod Journal Diameter	Connecting Rod Oil Clearance	Connecting Rod Side Clearance
2011	2.7	1AR-FE	2.1649-2.1654	0.00063-0.00154	0.00157-0.00945	NA	2.1457-2.1466	0.00118-0.00248	0.0063-0.0202
	3.5	2GR-FE	2.4011-2.4016	0.00102-0.00185	0.00157-0.00945	NA	2.0863-2.0866	0.00177-0.00264	0.00591-0.0157
2012	2.7	1AR-FE	2.1649-2.1654	0.00063-0.00154	0.00157-0.00945	NA	2.1457-2.1466	0.00118-0.00248	0.0063-0.0202
	3.5	2GR-FE	2.4011-2.4016	0.00102-0.00185	0.00157-0.00945	NA	2.0863-2.0866	0.00177-0.00264	0.00591-0.0157

NA Not Available

71099_SIEN_C0008

PISTON AND RING SPECIFICATIONS

All measurements are given in inches.

Year	Engine Displacement Liters	Engine ID	Piston Clearance	Ring Gap Top Compression	Ring Gap Bottom Compression	Ring Gap Oil Control	Ring Side Clearance Top Compression	Ring Side Clearance Bottom Compression	Ring Side Clearance Oil Control
2011	2.7	1AR-FE	0.00079-0.00169	0.00866-0.01060	0.0146-0.0165	0.00394-0.00787	0.00079-0.00276	0.00079-0.00236	0.00079-0.00276
	3.5	2GR-FE	0.00079-0.00205	0.00866-0.01260	0.0138-0.0177	0.00394-0.01570	0.00079-0.00276	0.00079-0.00236	0.00276-0.00591
2012	2.7	1AR-FE	0.00079-0.00169	0.00866-0.01060	0.0146-0.0165	0.00394-0.00787	0.00079-0.00276	0.00079-0.00236	0.00079-0.00276
	3.5	2GR-FE	0.00079-0.00205	0.00866-0.01260	0.0138-0.0177	0.00394-0.01570	0.00079-0.00276	0.00079-0.00236	0.00276-0.00591

71099_SIEN_C0009

TORQUE SPECIFICATIONS

All readings in ft. lbs.

Year	Engine Disp. Liters	Engine ID	Cylinder Head Bolts	Main Bearing Bolts	Rod Bearing Bolts	Crankshaft Pulley Bolts	Drive Plate Bolts	Manifold Intake	Manifold Exhaust	Spark Plugs	Oil Pan Drain Plug
2011	2.7	1AR-FE	①	②	③	192	72	15	26	18	30
	3.5	2GR-FE	④	⑤	⑥	184	61	15	15	13	30
2012	2.7	1AR-FE	①	②	③	192	72	15	26	18	30
	3.5	2GR-FE	④	⑤	⑥	184	61	15	15	13	30

① Step 1: 27 ft. lbs.
Step 2: 27 ft. lbs.
Step 3: Additional 90 degrees
Step 4: Additional 90 degrees

② Step 1: 15 ft. lbs.
Step 2: 30 ft. lbs.
Step 3: Additional 90 degrees

③ Step 1:30 ft. lbs.
Step 2: Additional 90 degrees

④ Step 1: 27 ft. lbs.
Step 2: Additional 90 degrees
Step 3: Additional 90 degrees

⑤ Step 1: 45 ft. lbs.
Step 2: Additional 90 degrees
14 mm: 38 ft. lbs.

⑥ Step 1: 18 ft. lbs.
Step 2: Additional 90 degrees

71099_SIEN_C0010

WHEEL ALIGNMENT

Year	Model		Caster Range (+/-Deg.)	Caster Preferred Setting (Deg.)	Camber Range (+/-Deg.)	Camber Preferred Setting (Deg.)	Toe-in (in.)
2011	Sienna	F	45'	2.50	45'	-0.25	0.0472+/-0.0787
		R			45'	-1.06	0.169+/-0.130
2012	Sienna	F	45'	2.50	45'	-0.25	0.0472+/-0.0787
		R			45'	-1.06	0.169+/-0.130

F Front

R Rear

71099_SIEN_C0011

TIRE, WHEEL AND BALL JOINT SPECIFICATIONS

Year	Model	OEM Tires Standard	OEM Tires Optional	Tire Pressures (psi) Front	Tire Pressures (psi) Rear	Wheel Size	Ball Joint Inspection	Lug Nut (ft. lbs.)
2011	Sienna	235/60R17	235/55R18	35	35	17/18	NA	76
2012	Sienna	235/60R17	235/55R18	35	35	17/18	NA	76

OEM: Original Equipment Manufacturer

PSI: Pounds Per Square Inch

NA: Information not available

71099_SIEN_C0012

BRAKE SPECIFICATIONS

All measurements in inches unless noted

Year	Model		Brake Disc Original Thickness	Brake Disc Minimum Thickness	Brake Disc Max. Runout	Minimum Pad/Lining Thickness Front	Minimum Pad/Lining Thickness Rear	Brake Caliper Bracket Bolts (ft. lbs.)	Brake Caliper Mounting Bolts (ft. lbs.)
2011	Sienna	F	1.100	0.984	0.00197	0.039		77	25
		R	0.394	0.335	0.00591		0.394	65	22
2012	Sienna	F	1.100	0.984	0.00197	0.039		77	25
		R	0.394	0.335	0.00591		0.394	65	22

F: Front

R: Rear

NA: Information not available

71099_SIEN_C0013

SCHEDULED MAINTENANCE INTERVALS

TOYOTA—Sienna

TO BE SERVICED	TYPE OF	VEHICLE MILEAGE INTERVAL (x1000)																		
		5	10	15	20	25	30	35	40	45	50	55	60	65	70	75	80	85	90	95
Air filter	R												✓							
Air filter (cabin)	R				✓				✓				✓				✓			
Automatic transmission and differential fluid	S/I			✓			✓			✓			✓			✓			✓	
Ball joints and boots	S/I			✓			✓			✓			✓			✓			✓	
Brake linings, discs/drums, lines & hoses	S/I	✓	✓	✓	✓	✓	✓	✓	✓	✓	✓	✓	✓	✓	✓	✓	✓	✓	✓	✓
Charcoal canister	S/I												✓							
Drive belts	S/I						✓						✓						✓	
Driveshaft bushing (4WD)	L						✓						✓						✓	
Engine coolant	R						✓						✓						✓	
Engine oil & filter	R	✓	✓	✓	✓	✓	✓	✓	✓	✓	✓	✓	✓	✓	✓	✓	✓	✓	✓	✓
Exhaust pipes & mounts	S/I			✓			✓			✓			✓			✓			✓	
Fuel lines & connections, fuel tank vapor vent system hoses, fuel tank band	S/I						✓						✓						✓	
Fuel tank cap gasket	S/I						✓						✓						✓	
Halfshaft boots & flange bolts	S/I			✓			✓			✓			✓			✓			✓	
Limited slip differential fluid	R						✓						✓						✓	
Manual transmission and differential fluid	S/I						✓						✓						✓	
Non-platinum spark plugs	R						✓						✓						✓	
Platinum spark plugs	R																			
Propeller shaft (4WD)	L			✓			✓			✓			✓			✓			✓	
Propeller shaft bolts	S/I			✓			✓			✓			✓			✓			✓	
Rack and pinion assembly	S/I			✓			✓			✓			✓			✓			✓	
Rear wheel bearing	L						✓						✓						✓	
Rotate tires	S/I	✓	✓	✓	✓	✓	✓	✓	✓	✓	✓	✓	✓	✓	✓	✓	✓	✓	✓	✓
Steering linkage	S/I			✓			✓			✓			✓			✓			✓	
Valves	S/I												✓							

R: Replace S/I: Service or Inspect L: Lubricate

FREQUENT OPERATION MAINTENANCE (SEVERE SERVICE)

If a vehicle is operated under any of the following conditions it is considered severe service:

- Towing a trailer or using a camper or car-top carrier.

- Repeated short trips of less than 5 miles in temperatures below freezing.

- Excessive idling or low-speed driving for long distances as in heavy commercial use, such as delivery, taxi or police cars.

- Operating on rough, muddy or salt-covered roads.

- Operating on unpaved or dusty roads.

Oil filter: service or inspect every 5000 miles or 4 months, whichever occurs first.

Brake linings and discs or drums: service or inspect every 5000 miles or 4 months, whichever occurs first.

Steering linkage: service or inspect every 5000 miles or 4 months, whichever occurs first.

Ball joints and boots: service or inspect every 5000 miles or 4 months, whichever occurs first.

Brake discs & pads (front): service or inspect every 5000 miles.

Halfshaft boots: service or inspect every 5000 miles or 4 months. Retighten the flange bolts, whichever occurs first.

Body chassis bolts and nuts: service or inspect every 5000 miles or 4 months, whichever occurs first.

Transmission and differential fluid: replace every 60,000 miles or 72 months, whichever occurs first.

Transfer case and differential fluid: replace every 15,000 miles or 18 months, whichever occurs first.

Timing belt: replace every 60,000 miles or 48 months, whichever occurs first.

Lubricate driveshaft 4WD every 5,000 miles

Retorque driveshaft bolts 4WD every 5,000 miles

Inspect engine air filter at 15,000 miles or 18 months, than every 5,000 miles

71099_SIEN_C0014

PRECAUTIONS

Before servicing any vehicle, please be sure to read all of the following precautions, which deal with personal safety, prevention of component damage, and important points to take into consideration when servicing a motor vehicle:

- Never open, service or drain the radiator or cooling system when the engine is hot; serious burns can occur from the steam and hot coolant.
- Observe all applicable safety precautions when working around fuel. Whenever servicing the fuel system, always work in a well-ventilated area. Do not allow fuel spray or vapors to come in contact with a spark, open flame, or excessive heat (a hot drop light, for example). Keep a dry chemical fire extinguisher near the work area. Always keep fuel in a container specifically designed for fuel storage; also, always properly seal fuel containers to avoid the possibility of fire or explosion. Refer to the additional fuel system precautions later in this section.
- Fuel injection systems often remain pressurized, even after the engine has been turned **OFF**. The fuel system pressure must be relieved before disconnecting any fuel lines. Failure to do so may result in fire and/or personal injury.
- Brake fluid often contains polyglycol ethers and polyglycols. Avoid contact with the eyes and wash your hands thoroughly after handling brake fluid. If you do get brake fluid in your eyes, flush your eyes with clean, running water for 15 minutes. If eye irritation persists, or if you have taken brake fluid internally, IMMEDIATELY seek medical assistance.
- The EPA warns that prolonged contact with used engine oil may cause a number of skin disorders, including cancer. You should make every effort to minimize your exposure to used engine oil. Protective gloves should be worn when changing oil. Wash your hands and any other exposed skin areas as soon as possible after exposure to used engine oil. Soap and water, or waterless hand cleaner should be used.
- All new vehicles are now equipped with an air bag system, often referred to as a Supplemental Restraint System (SRS) or Supplemental Inflatable Restraint (SIR) system. The system must be disabled before performing service on or around system components, steering column, instrument panel components, wiring and sensors. Failure to follow safety and disabling procedures could result in accidental air bag deployment, possible personal injury and unnecessary system repairs.
- Always wear safety goggles when working with, or around, the air bag system. When carrying a non-deployed air bag, be sure the bag and trim cover are pointed away from your body. When placing a non-deployed air bag on a work surface, always face the bag and trim cover upward, away from the surface. This will reduce the motion of the module if it is accidentally deployed. Refer to the additional air bag system precautions later in this section.
- Clean, high quality brake fluid from a sealed container is essential to the safe and proper operation of the brake system. You should always buy the correct type of brake fluid for your vehicle. If the brake fluid becomes contaminated, completely flush the system with new fluid. Never reuse any brake fluid. Any brake fluid that is removed from the system should be discarded. Also, do not allow any brake fluid to come in contact with a painted surface; it will damage the paint.
- Never operate the engine without the proper amount and type of engine oil; doing so WILL result in severe engine damage.
- Timing belt maintenance is extremely important. Many models utilize an interference-type, non-freewheeling engine. If the timing belt breaks, the valves in the cylinder head may strike the pistons, causing potentially serious (also time-consuming and expensive) engine damage. Refer to the maintenance interval charts for the recommended replacement interval for the timing belt, and to the timing belt section for belt replacement and inspection.
- Disconnecting the negative battery cable on some vehicles may interfere with the functions of the on-board computer system(s) and may require the computer to undergo a relearning process once the negative battery cable is reconnected.
- When servicing drum brakes, only disassemble and assemble one side at a time, leaving the remaining side intact for reference.
- Only an MVAC-trained, EPA-certified automotive technician should service the air conditioning system or its components.

BRAKES

ANTI-LOCK BRAKE SYSTEM (ABS)

GENERAL INFORMATION

PRECAUTIONS

- Certain components within the ABS system are not intended to be serviced or repaired individually.
- Do not use rubber hoses or other parts not specifically specified for and ABS system. When using repair kits, replace all parts included in the kit. Partial or incorrect repair may lead to functional problems and require the replacement of components.
- Lubricate rubber parts with clean, fresh brake fluid to ease assembly. Do not use shop air to clean parts; damage to rubber components may result.
- Use only DOT 3 brake fluid from an unopened container.
- If any hydraulic component or line is removed or replaced, it may be necessary to bleed the entire system.
- A clean repair area is essential. Always clean the reservoir and cap thoroughly before removing the cap. The slightest amount of dirt in the fluid may plug an orifice and impair the system function. Perform repairs after components have been thoroughly cleaned; use only denatured alcohol to clean components. Do not allow ABS components to come into contact with any substance containing mineral oil; this includes used shop rags.
- The Anti-Lock control unit is a microprocessor similar to other computer units in the vehicle. Ensure that the ignition switch is **OFF** before removing or installing controller harnesses. Avoid static electricity discharge at or near the controller.
- If any arc welding is to be done on the vehicle, the control unit should be unplugged before welding operations begin.

SPEED SENSORS

REMOVAL & INSTALLATION

Front

See Figure 1.

1. Remove front wheel.
2. Remove front wheel opening extension pad.
3. Separate front fender liner.
4. Remove front speed sensor.
   a. Disconnect the front speed sensor connector and remove the 2 clamps.
   b. Remove the bolt and No. 3 sensor clamp from the body.

c. Remove the bolt, No. 2 sensor clamp and front brake flexible hose together from the shock absorber assembly.

d. Remove the bolt, No. 1 sensor clamp and front speed sensor.

➡Prevent foreign matter from attaching to the front speed sensor tip. Clean the speed sensor installation hole and the contact surfaces every time the front speed sensor is removed.

To install:

5. Install front speed sensor.

a. Install the No. 1 sensor clamp and front speed sensor with the bolt.

➡Please note the following:

- Prevent foreign matter from attaching to the front speed sensor tip.
- Firmly insert the front speed sensor body into the knuckle before tightening the bolt.
- After installing the front speed sensor to the knuckle, make sure that there is no clearance between the front speed sensor stay and knuckle. Also make sure that no foreign matter is stuck between the parts.
- Before installing the clamp, firmly insert the points of the clamp into the installation holes.

b. Temporarily install the No. 2 sensor clamp.

➡Be sure to insert the No. 2 sensor clamp into the stopper hole while installing the No. 2 sensor clamp.

c. Install the front brake flexible hose and No. 2 sensor clamp together to the shock absorber with the bolt.

➡Please note the following

- Do not twist the wire harness for the front speed sensor when installing it.
- A bolt tightens the brake flexible hose and front speed sensor together. Make sure that the flexible hose is positioned over the front speed sensor.
- Install the No. 3 sensor clamp to the body with the bolt.

➡Be sure to insert the No. 3 sensor clamp into the stopper hole while installing the No. 3 sensor clamp.

d. Install the 2 clamps and connect the front speed sensor connector.

6. Install front fender liner.

7. Install front wheel opening extension pad.

8. Install front wheel.

9. Check for speed sensor signal.

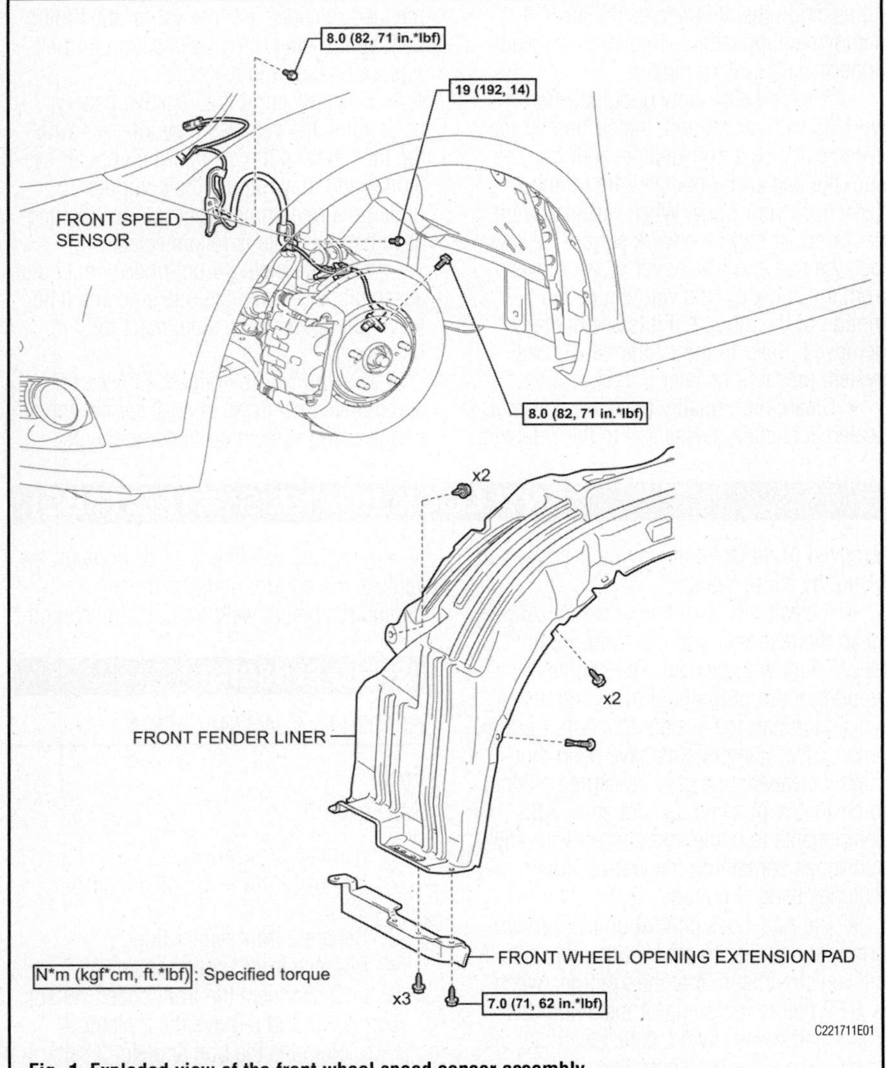

Fig. 1 Exploded view of the front wheel speed sensor assembly

Rear (FWD)

1. Remove rear wheel.
2. Separate rear disc brake caliper assembly.
3. Remove parking brake shoe adjusting hole plug.
4. Remove rear disc.
5. Disconnect skid control sensor wire.
6. Remove rear axle hub and bearing assembly.

a. Remove the 4 nuts and hub and bearing assembly.

➡The rear speed sensor is a component of the rear axle hub and bearing assembly. If the sensor malfunctions, replace the rear axle hub and bearing assembly. If the sensor rotor needs to be replaced, replace it together with the rear axle hub and bearing assembly.

7. Remove rear inner axle bearing retainer.

To install:

8. Install rear inner axle bearing retainer.

9. Install rear axle hub and bearing assembly. Install the hub and bearing assembly with the 4 nuts. Tighten the nuts to 66 ft. lbs. (90 Nm).

➡The rear speed sensor is a component of the rear axle hub and bearing assembly. If the sensor malfunctions, replace the rear axle hub and bearing assembly. If the sensor rotor needs to be replaced, replace it together with the rear axle hub and bearing assembly.

10. Inspect rear axle hub bearing looseness.

11. Inspect rear axle hub runout.
12. Connect skid control sensor wire.
13. Install rear disc.
14. Install parking brake shoe adjusting hole plug.
15. Install rear disc brake caliper assembly.
16. Adjust parking brake shoe clearance and parking brake pedal travel.
17. Install rear wheel.
18. Check for speed sensor signal.

Rear (AWD)

See Figure 2.

1. Separate back door weatherstrip.
2. Remove back door scuff plate.
3. Remove rear door scuff plate.
4. Separate slide door weatherstrip.
5. Separate rear no. 2 seat outer belt assembly.
6. Remove quarter trim panel assembly.
7. Remove rear wheel.
8. Remove rear speed sensor.
   a. Disconnect the rear speed sensor connector.
   b. Remove the 2 clips and peel back the rear wheel house liner.
   c. Disconnect the grommet of the rear speed sensor wire from the hole of the wheel house.
   d. Remove the 2 bolts, No. 1 bracket and No. 2 bracket from the body and rear axle beam.
   e. Remove the bolt and rear speed sensor from the retainer.

➡Keep the sensor tip and rear speed sensor installation hole free from foreign matter.

To install:

9. Install rear speed sensor.
   a. Install the rear speed sensor with the bolt.

➡Keep the rear speed sensor tip and sensor installation hole free from foreign matter.

   b. Install the No. 1 bracket and No. 2 bracket with the 2 bolts.

➡Please note the following:

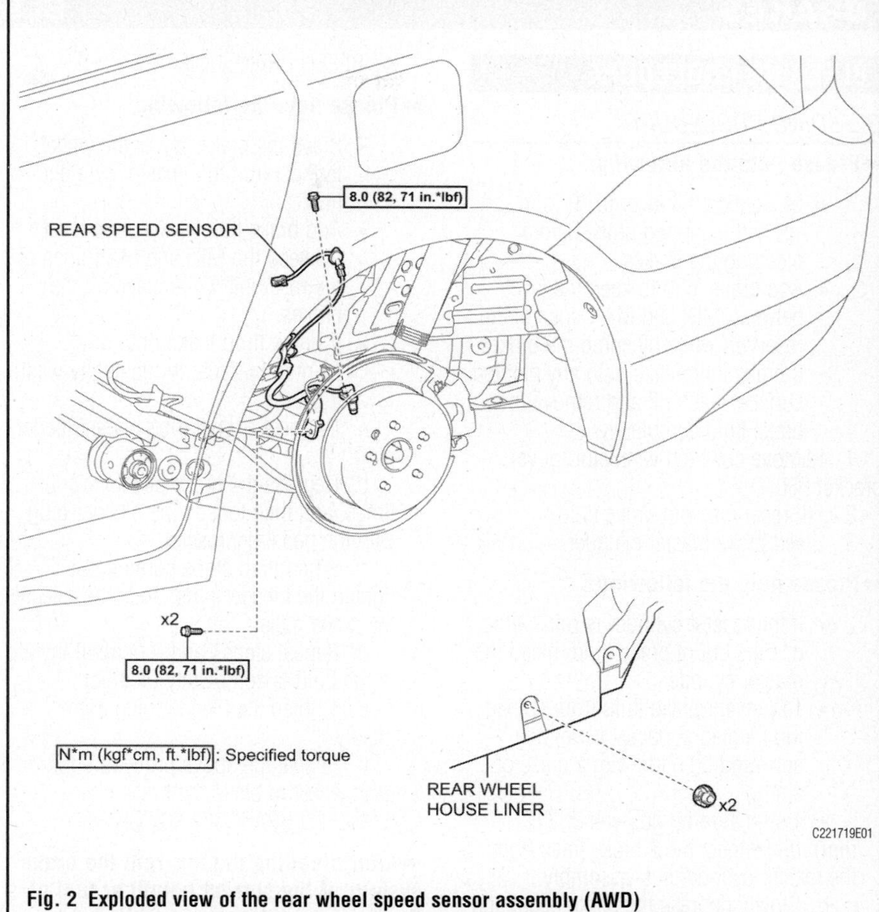

Fig. 2 Exploded view of the rear wheel speed sensor assembly (AWD)

- Be sure to insert the No. 2 bracket into the stopper hole while installing the No. 2 bracket.
- Do not twist the rear speed sensor wire when installing the brackets.

   c. Insert the connector and grommet to the inside of the vehicle through the passage hole in the wheel well.

➡Make sure that the grommet's band clamp remains on the outside of the vehicle.

   d. Hold the grommet and pull it from the inside to the outside of the vehicle. Then secure it in place so that it is not tilted.

➡When pulling out the grommet, do not grip the sensor wire.

   e. Return the rear wheel house liner and engage the 2 clips.
   f. Connect the rear speed sensor connector.

10. Install quarter trim panel assembly.
11. Install rear no. 2 seat outer belt assembly.
12. Install slide door weatherstrip.
13. Install rear door scuff plate.
14. Install back door scuff plate.
15. Install back door weatherstrip.
16. Install rear wheel.

BRAKES — BLEEDING THE BRAKE SYSTEM

BLEEDING PROCEDURE

BLEEDING PROCEDURE

➡**Please note the following:**

- Move the shift lever to "P" and apply the parking brake before bleeding the brakes.
- Add brake fluid to keep the level between MIN and MAX lines of the reservoir while bleeding the brakes.
- If brake fluid leaks onto any painted surface, wash off and remove the brake fluid completely.

1. Remove cowl top ventilator louver bracket RH.
2. Fill reservoir with brake fluid.
3. Bleed brake master cylinder.

➡**Please note the following:**

- If the master cylinder is reinstalled or runs out of brake fluid, bleed the master cylinder.
- To prevent brake fluid from damaging painted surfaces, cover any surrounding parts with a piece of cloth.

a. Using a union nut wrench (12 mm), disconnect the 2 brake lines from the master cylinder sub-assembly.
b. Slowly depress the brake pedal and hold it.
c. Cover the 2 outer holes with fingers, and release the brake pedal.
d. Repeat steps 1 and 2, 3 or 4 times.
e. Using a union nut wrench (12 mm), connect the 2 brake lines to the master cylinder.

4. BLEED BRAKE LINE

➡**Please note the following:**

- Bleed the brake line of the wheel farthest from the master cylinder first.
- Add brake fluid to keep the level between the MIN and MAX lines of the reservoir while bleeding the brakes.
- If brake fluid leaks onto any painted surface, immediately wash it off.

a. Connect a vinyl tube to the bleeder plug.
b. Depress the brake pedal several times, and then loosen the bleeder plug with the pedal depressed.
c. When fluid stops coming out, tighten the bleeder plug, and then release the brake pedal.
d. Repeat steps1 and 2 until all the air in the fluid is completely bled out.
e. Tighten the bleeder plug completely.
f. Repeat the above procedure for each wheel to bleed the brake line.

5. BLEED BRAKE ACTUATOR

➡**After bleeding the air from the brake system, if the correct height or feel of the brake pedal cannot be obtained, bleed the air from the brake actuator assembly using the Techstream by following the procedure below.**

a. Depress the brake pedal more than 20 times with the ignition switch off.
b. Connect the Techstream to the DLC3, then turn the ignition switch to ON.

➡**Do not start the engine.**

c. Turn the Techstream on and enter the following menus: Chassis / ABS/VSC/TRAC / Utility / Air Bleeding.
d. Bleed brake actuator following the instructions on the Techstream.
e. After air bleeding, tighten each bleeder plug.
f. Turn the Techstream off and disconnect the Techstream from the DLC3.
g. Turn the ignition switch off.

6. Inspect for brake fluid leak.
7. Inspect fluid level in reservoir.
8. Install cowl top ventilator louver bracket RH.

BLEEDING THE ABS SYSTEM

Bleeding the ABS system is the same procedure as designated in the Bleeding Procedure section.

FLUID FILL PROCEDURE

1. Remove cowl top ventilator louver bracket RH.
   a. Disengage the 2 claws and 2 guides and remove the cowl top ventilator louver bracket RH.
2. Fill reservoir with brake fluid.
   a. Remove the brake master cylinder reservoir filler cap assembly.
   b. Fill the reservoir with brake fluid.

➡**Make sure that there is sufficient brake fluid in the reservoir.**

BRAKES — FRONT DISC BRAKES

CAUTION

Dust and dirt accumulating on brake parts during normal use may contain asbestos fibers from production or aftermarket brake linings. Breathing excessive concentrations of asbestos fibers can cause serious bodily harm. Exercise care when servicing brake parts. Do not sand or grind brake lining unless equipment used is designed to contain the dust residue. Do not clean brake parts with compressed air or by dry brushing. Cleaning should be done by dampening the brake components with a fine mist of water, then wiping the brake components clean with a dampened cloth. Dispose of cloth and all residue containing asbestos fibers in an impermeable container with the appropriate label. Follow practices prescribed by the Occupational Safety and Health Administration (OSHA) and the Environmental Protection Agency (EPA) for the handling, processing, and disposing of dust or debris that may contain asbestos fibers.

BRAKE CALIPER

REMOVAL & INSTALLATION

See Figure 3.

1. Remove front wheel.
2. Drain brake fluid.

➡**If brake fluid leaks onto any painted surface, immediately wash it off.**

3. Disconnect front flexible hose.
   a. Remove the union bolt and gasket, and disconnect the front flexible hose from the disc brake cylinder assembly.
4. Remove disc brake cylinder assembly.
   a. Hold the 2 front disc brake cylinder slide pins and remove the 2 bolts and disc brake cylinder assembly.

➡**Remove the disc brake cylinder assembly while holding both of the front disc brake pads because the anti-squeal springs may fall off the front disc brake pads.**

To install:

5. Install disc brake cylinder assembly.

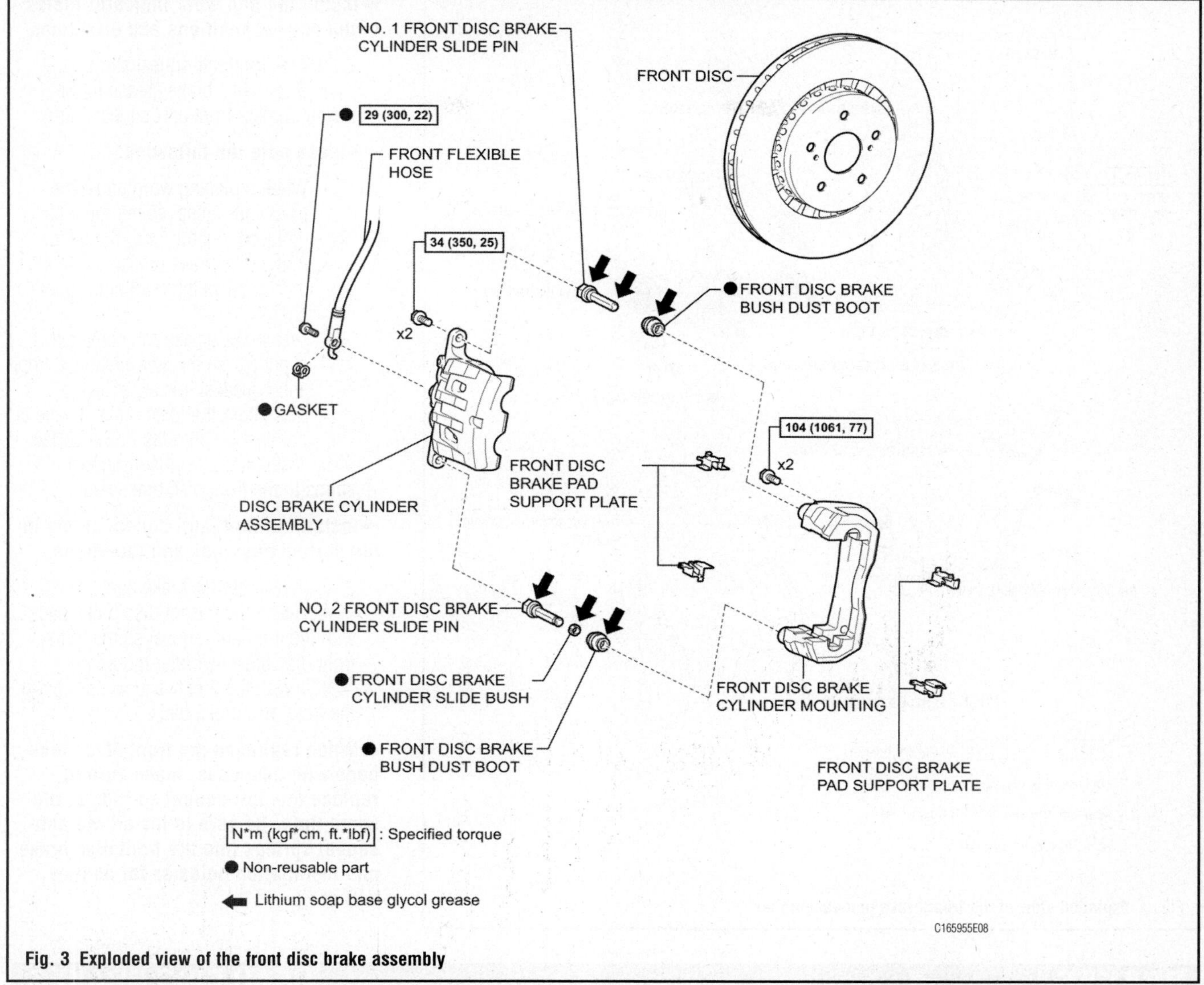

Fig. 3 Exploded view of the front disc brake assembly

a. Hold the 2 front disc brake cylinder slide pins and install the disc brake cylinder assembly with the 2 bolts. Tighten to 25 ft. lbs. (34 Nm).

➡Install the disc brake cylinder assembly while holding both of the front disc brake pads because the anti-squeal springs may fall off the front disc brake pads.

➡Be sure that the anti-squeal springs are installed to the front disc brake pads.

6. Connect front flexible hose.
   a. Connect the front flexible hose to the disc brake cylinder assembly with a new gasket and a new union bolt. Tighten to 22 ft. lbs. (29 Nm).

➡Install the front flexible hose lock securely in the lock hole in the disc brake cylinder assembly.

7. Fill reservoir with brake fluid.
8. Bleed brake line.
9. Inspect for brake fluid leak.
10. Inspect fluid level in reservoir.
11. Install front wheel.

DISC BRAKE PADS

REMOVAL & INSTALLATION

See Figure 4.

1. Remove front disc brake pad.
   a. Remove the 2 anti-squeal springs.
   b. Remove the 2 front disc brake pads from the front disc brake cylinder mounting.
2. Remove front anti-squeal shim.
   a. Remove the 2 front anti-squeal shims from the front disc brake pads.
3. Remove pad wear indicator plate.
   a. Using a screwdriver, remove the 2 pad wear indicator plates from the front disc brake pads.
4. Remove front disc brake pad support plate.
   a. Remove the 4 front disc brake pad support plates from the front disc brake cylinder mounting.

To install:

5. Install front disc brake pad support plate.
   a. Install the 4 front disc brake pad support plates to the front disc brake cylinder mounting.

➡Be sure to install each front disc brake pad support plate to the correct position and direction.

➡The front disc brake pad support plate (lower) has a convex part.

6. Install pad wear indicator plate
   a. Install the 2 pad wear indicator plates to the pads.

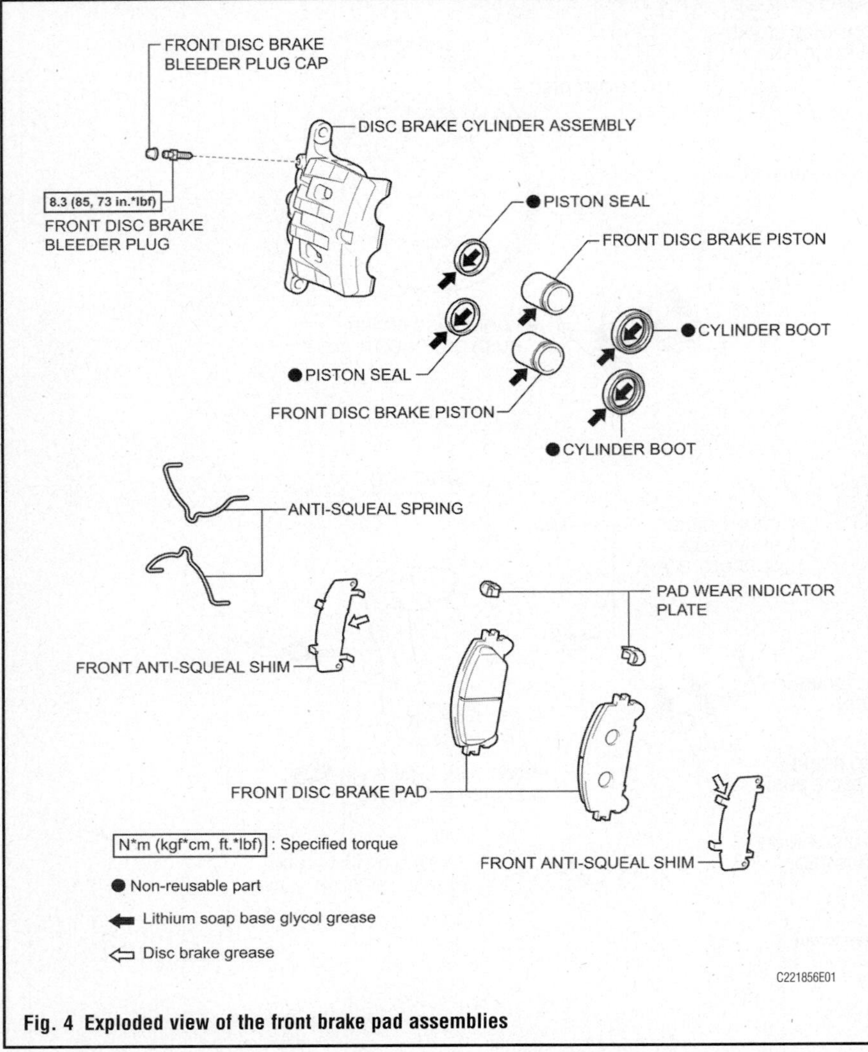

Fig. 4 Exploded view of the front brake pad assemblies

➡Install the pad wear indicator plates in the correct positions and directions.

7. Install front anti-squeal shim.
   a. Apply disc brake grease to the inside of the 2 front anti-squeal shims.

➡Please note the following:

- When replacing worn pads, the front anti-squeal shims must be replaced together with the pads.
- Apply disc brake grease to the area that contacts the front anti-squeal shims.
- Disc brake grease can come out slightly from the area where the front anti-squeal shims are installed.
- Make sure that disc brake grease is not applied onto the lining surface.

   b. Install the 2 front anti-squeal shims to the front disc brake pads.

➡Install the front anti-squeal shims in the correct positions and directions.

8. Install front disc brake pad.
   a. Install the 2 front disc brake pads with the front anti-squeal shims to the front disc brake cylinder mounting.
   b. Install the 2 anti-squeal springs to the front disc brake pads.

➡When replacing the front disc brake pads with new ones, make sure to replace the anti-squeal springs at the same time. Be sure to install the anti-squeal springs into the front disc brake pad installation holes as far as they will go.

REAR DISC BRAKES

⁂ CAUTION

Dust and dirt accumulating on brake parts during normal use may contain asbestos fibers from production or aftermarket brake linings. Breathing excessive concentrations of asbestos fibers can cause serious bodily harm. Exercise care when servicing brake parts. Do not sand or grind brake lining unless equipment used is designed to contain the dust residue. Do not clean brake parts with compressed air or by dry brushing. Cleaning should be done by dampening the brake components with a fine mist of water, then wiping the brake components clean with a dampened cloth. Dispose of cloth and all residue containing asbestos fibers in an impermeable container with the appropriate label. Follow practices prescribed by the Occupational Safety and Health Administration (OSHA) and the Environmental Protection Agency (EPA) for the handling, processing, and disposing of dust or debris that may contain asbestos fibers.

BRAKE CALIPER

REMOVAL & INSTALLATION

See Figure 5.

1. Remove rear wheel.
2. Drain brake fluid.
3. Disconnect rear flexible hose.
   a. Remove the union bolt and gasket, and disconnect the rear flexible hose from the rear disc brake cylinder assembly.
4. Remove rear disc brake cylinder assembly.
   a. Hold the 2 rear disc brake cylinder slide pins, and remove the 2 bolts and rear disc brake cylinder assembly.

To install:

5. Install rear disc brake cylinder assembly.
   a. Hold the 2 rear disc brake cylinder slide pins and install the rear disc brake cylinder assembly to the rear disc brake cylinder mounting with the 2 bolts.
6. Connect rear flexible hose.
   a. Connect the rear flexible hose to the rear disc brake cylinder assembly with a new union bolt and a new gasket.
7. Fill reservoir with brake fluid.
8. Bleed brake line.
9. Inspect for brake fluid leak
10. Inspect fluid level in reservoir.
11. Adjust parking brake shoe clearance and parking brake pedal travel.
12. Install rear wheel.

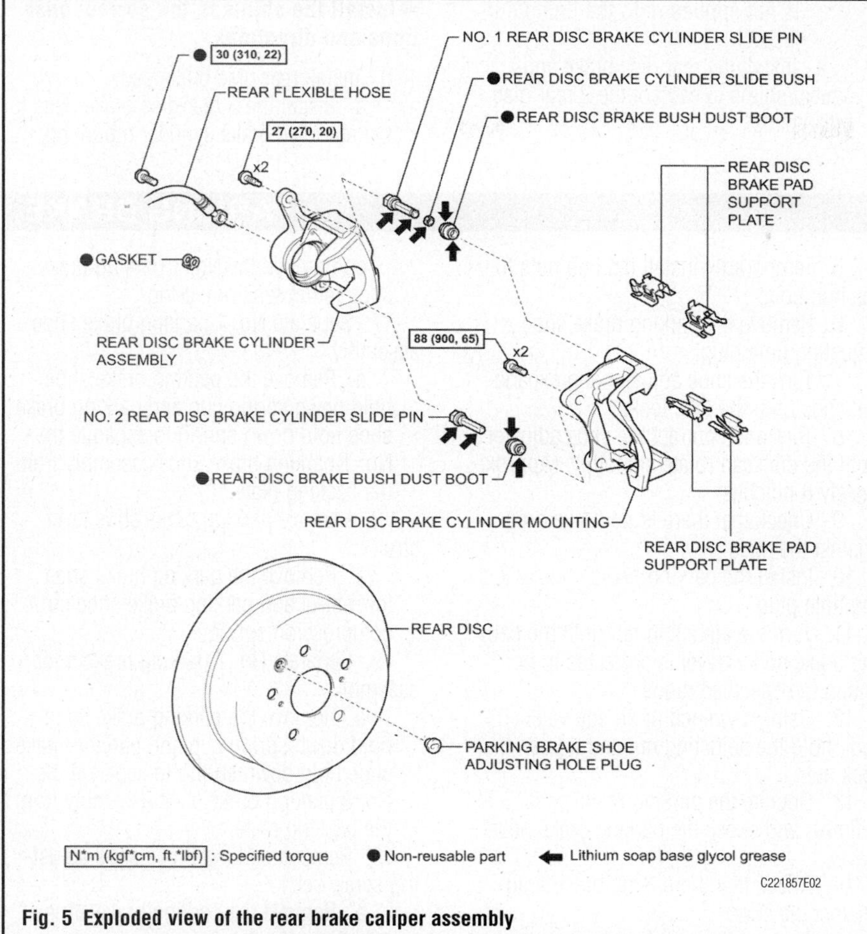

Fig. 5 Exploded view of the rear brake caliper assembly

DISC BRAKE PADS

REMOVAL & INSTALLATION

See Figure 6.

1. Remove rear disc brake pad.
   a. Remove the 2 rear disc brake pads from the rear disc brake cylinder mounting.
2. Remove rear disc brake anti-squeal shim.
   a. Remove the 2 rear disc brake anti-squeal shims from the rear disc brake pads.
3. Remove pad wear indicator plate.
   a. Using a screwdriver, remove 2 pad wear indicator plates from the rear disc brake pads.
4. Remove rear disc brake pad support plate.
   a. Remove the 2 No. 1 rear disc brake pad support plates and 2 No. 2 rear disc brake pad support plates from the rear disc brake cylinder mounting.

➡The No. 1 rear disc brake pad support plate and the No. 2 rear disc brake pad support plate have different shapes. Be sure to put an identification mark on each rear disc brake pad support plate so that they can be reinstalled to their original position.

To install:

5. Install rear disc brake pad support plate.
   a. Install the 2 No. 1 rear disc brake pad support plates and 2 No. 2 rear disc brake pad support plates to the rear disc brake cylinder mounting.

➡Be sure to install the No. 1 rear disc brake pad support plate and No. 2 rear disc brake pad support plate to the correct positions and directions.

6. Install pad wear indicator plate.
   a. Install the 2 pad wear indicators to the rear disc brake pads.

➡Install the pad wear indicator plates in the correct positions and directions.

7. Install rear disc brake anti-squeal shim.

Apply disc brake grease to the inside of the 2 rear disc brake anti-squeal shims.

➡Please note the following:

- When replacing worn pads, the rear disc brake anti-squeal shims must be replaced together with the pads.
- Apply disc brake grease to the area that contacts the rear disc brake anti-squeal shims.

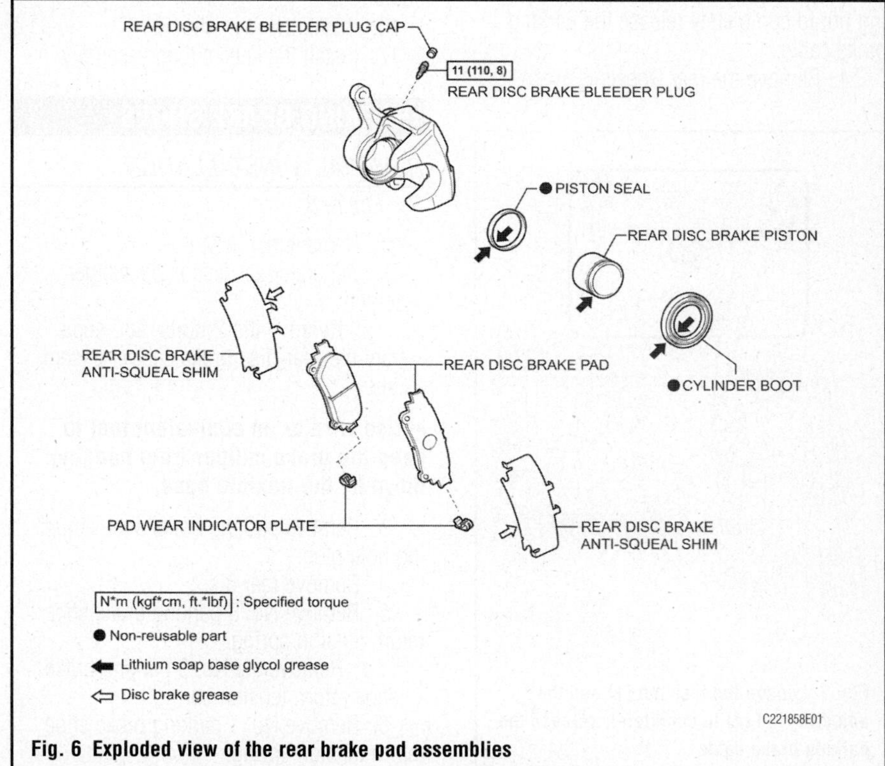

Fig. 6 Exploded view of the rear brake pad assemblies

- Disc brake grease can come out slightly from the area where the rear disc brake anti-squeal shims are installed.
- Make sure that disc brake grease is not applied onto the lining surface.

a. Install the rear disc brake anti-squeal shims to each of the 2 rear disc brake pads.

➡Install the shims in the correct positions and directions.

8. Install rear disc brake pad.

a. Install the 2 rear disc brake pads to the rear disc brake cylinder mounting.

PARKING BRAKE CABLES

ADJUSTMENT

Inspect Parking Brake Pedal Travel

1. Fully depress the parking brake pedal to engage the parking brake.
2. Depress the pedal again to disengage the parking brake.
3. Slowly depress the parking brake pedal using the specified force, and count the number of clicks. Three to six clicks at 68 lbs. (300 N).
4. If the parking brake pedal travel is not as specified, adjust the parking brake shoe clearance and parking brake pedal travel.

Adjust Parking Brake Shoe Clearance And Parking Brake Pedal Travel

See Figure 7.

1. Remove the knee airbag assembly.
2. Completely release the parking brake pedal.
3. Loosen the lock nut and the adjusting nut to completely release the parking brake cable.
4. Remove the rear wheels.
5. Temporarily install the hub nuts to the hub bolts.
6. Remove the parking brake shoe adjusting hole plug.
7. Turn the shoe adjuster and expand the shoe until the disc locks.
8. Turn and contract the shoe adjuster until the disc can rotate smoothly. (Approximately 8 notches)
9. Check that there is no brake drag against the shoe.
10. Install the parking brake shoe adjusting hole plug.
11. Turn the adjusting nut until the parking brake pedal travel is corrected to be within the specified range.
12. Using a wrench or an equivalent tool, hold the adjusting nut and tighten the lock nut.
13. Operate the parking brake pedal 3 to 4 times, and check the parking brake pedal travel.
14. Check that there is no brake drag against the shoe.
15. Remove the hub nuts from the hub bolts.
16. Install the rear wheels.
17. Install the knee airbag assembly.

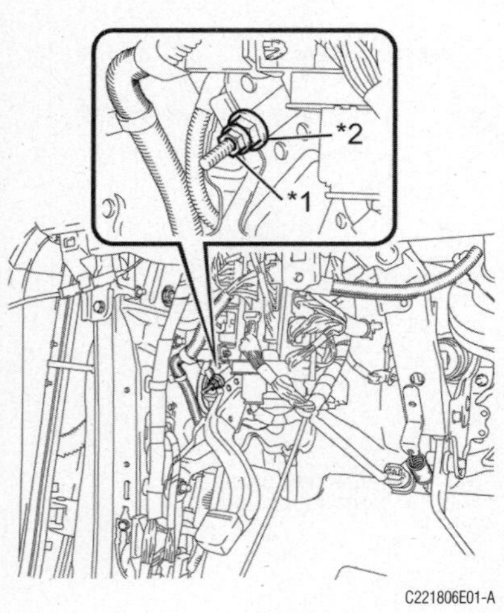

*2
*1
C221806E01-A

Fig. 7 Loosen the lock nut (1) and the adjusting nut (2) to completely release the parking brake cable

PARKING BRAKE SHOES

REMOVAL & INSTALLATION

See Figure 8.

1. Remove rear wheel.
2. Separate rear disc brake caliper assembly.

a. Remove the 2 bolts, and separate the rear disc brake caliper assembly.

➡Use wire or an equivalent tool to keep the brake caliper from hanging down by the flexible hose.

3. Remove parking brake shoe adjusting hole plug.
4. Remove rear disc.
5. Remove No. 3 parking brake shoe return tension spring.

a. Remove the No. 3 parking brake shoe return tension spring.

6. Remove No. 1 parking brake shoe return tension spring.

a. Remove the No. 1 parking brake shoe return tension spring.

7. Separate No. 1 parking brake shoe assembly.

a. Remove the parking brake shoe hold down spring cup and parking brake shoe hold down spring to separate the No. 1 parking brake shoe assembly from the backing plate.

8. Remove parking brake shoe lever strut.

a. Remove the parking brake shoe lever strut and parking brake shoe strut compression spring.

9. Separate No. 2 parking brake shoe assembly.

a. Remove the parking brake shoe hold down spring cup and parking brake shoe hold down spring to separate the No. 2 parking brake shoe assembly from the backing plate.

10. Remove parking brake shoe adjusting screw set.

a. Remove the parking brake shoe adjusting screw set.

11. Remove No. 1 parking brake shoe assembly.

a. Separate the No. 2 parking brake shoe return tension spring to remove the No. 1 parking brake shoe assembly.

12. Remove No. 2 parking brake shoe return tension spring.

a. Remove the No. 2 parking brake shoe return tension spring from the No. 2 parking brake shoe assembly.

13. Remove No. 2 parking brake shoe assembly with parking brake shoe lever.

a. Using needle-nose pliers, separate the No. 3 parking brake cable assembly from the parking brake shoe lever.

➡Be careful not to damage the No. 3 parking brake cable assembly.

14. Remove parking brake shoe lever.

a. Remove the C-washer, shim and parking brake shoe lever from the No. 2 parking brake shoe assembly.

15. Remove parking brake shoe guide plate set bolt.

a. Remove the parking brake shoe guide plate set bolt and parking brake shoe guide plate.

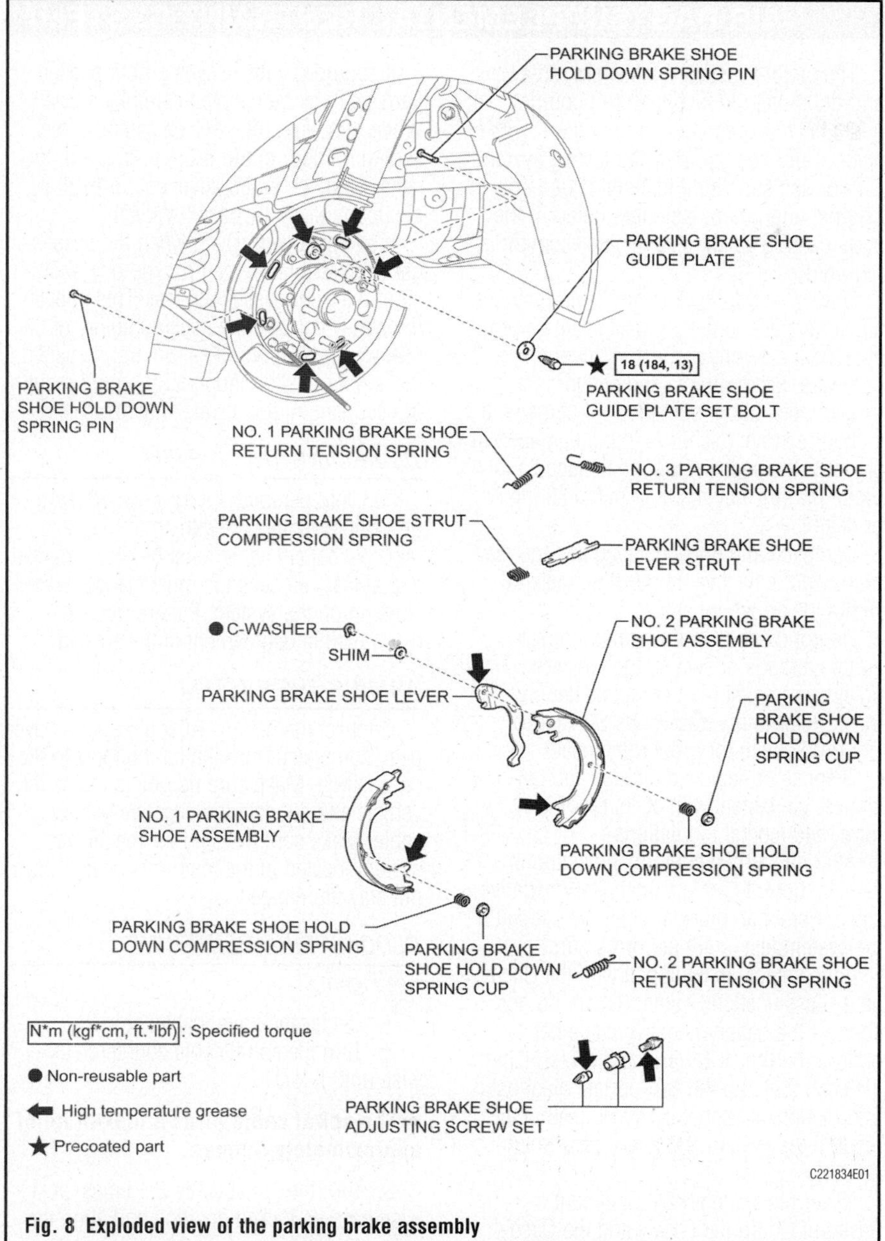

Fig. 8 Exploded view of the parking brake assembly

16. Remove parking brake shoe hold down spring pin.
   a. Remove the 2 parking brake shoe hold down spring pins.

To install:

17. Install the 2 parking brake shoe hold down spring pins.
18. Apply adhesive to the threads of the parking brake shoe guide plate set bolt.
19. Install the parking brake shoe guide plate and parking brake shoe guide plate set bolt.
20. Apply high temperature grease to the areas of the backing plate which make contact with the shoe.
21. Apply high temperature grease to the areas of the parking brake shoe lever which make contact with the No. 2 parking brake shoe assembly.
22. Install the parking brake shoe lever and shim to the No. 2 parking brake shoe assembly with a new C-washer.
23. Using a feeler gauge, measure the clearance between the No. 2 parking brake shoe assembly and parking brake shoe lever.
24. Using needle-nose pliers, connect the No. 3 parking brake cable assembly to the parking brake shoe lever.

➡Be careful not to damage the No. 3 parking brake cable assembly.

25. Install the No. 2 parking brake shoe return tension spring to the No. 2 parking brake shoe assembly.
26. Connect the No. 2 parking brake shoe return tension spring to install the No. 1 parking brake shoe assembly.
27. Apply high temperature grease to the parking brake shoe adjusting screw set.
28. Install the parking brake shoe adjusting screw set.
29. Install the No. 2 parking brake shoe assembly to the backing plate with the parking brake shoe hold down spring cup and parking brake shoe hold down spring.
30. Install the parking brake shoe lever strut and parking brake shoe strut compression spring.
31. Install the parking brake shoe assembly to the backing plate with the parking brake shoe hold down spring cup and parking brake shoe hold down spring.
32. Install the No. 1 parking brake shoe return tension spring.

➡First install the No. 1 parking brake shoe return tension spring and then the No. 3 parking brake shoe return tension spring.

33. Install the No. 3 parking brake shoe return tension spring.
34. Check that each part is installed properly.

➡There should be no oil or grease on the friction surfaces of the shoe lining and disc.

35. Install rear disc.
36. Install parking brake shoe adjusting hole plug.
37. Install the rear disc brake caliper assembly with the 2 bolts.
38. Install rear wheel.
39. Bed in parking brake shoes to discs.
   a. Drive the vehicle at about 31 mph (50 km/h) on a safe, level and dry road.
   b. Depress the parking brake pedal with 7 lbs. (150 N) of force.
   c. Drive the vehicle for about 0.25 mile (400 m) in this condition.
   d. Repeat this procedure 3 times.

➡Set a 5-minute interval between each procedure to prevent the brake assembly from overheating.

40. Adjust parking brake shoe clearance and parking brake pedal travel.

GENERAL INFORMATION

CAUTION

These vehicles are equipped with an air bag system. The system must be disarmed before performing service on, or around, system components, the steering column, instrument panel components, wiring and sensors. Failure to follow the safety precautions and the disarming procedure could result in accidental air bag deployment, possible injury and unnecessary system repairs.

SERVICE PRECAUTIONS

Disconnect and isolate the battery negative cable before beginning any airbag system component diagnosis, testing, removal, or installation procedures. Allow system capacitor to discharge for two minutes before beginning any component service. This will disable the airbag system. Failure to disable the airbag system may result in accidental airbag deployment, personal injury, or death.

Do not place an intact undeployed airbag face down on a solid surface. The airbag will propel into the air if accidentally deployed and may result in personal injury or death.

When carrying or handling an undeployed airbag, the trim side (face) of the airbag should be pointing towards the body to minimize possibility of injury if accidental deployment occurs. Failure to do this may result in personal injury or death.

Replace airbag system components with OEM replacement parts. Substitute parts may appear interchangeable, but internal differences may result in inferior occupant protection. Failure to do so may result in occupant personal injury or death.

Wear safety glasses, rubber gloves, and long sleeved clothing when cleaning powder residue from vehicle after an airbag deployment. Powder residue emitted from a deployed airbag can cause skin irritation. Flush affected area with cool water if irritation is experienced. If nasal or throat irritation is experienced, exit the vehicle for fresh air until the irritation ceases. If irritation continues, see a physician.

Do not use a replacement airbag that is not in the original packaging. This may result in improper deployment, personal injury, or death.

The factory installed fasteners, screws and bolts used to fasten airbag components have a special coating and are specifically designed for the airbag system. Do not use substitute fasteners. Use only original equipment fasteners listed in the parts catalog when fastener replacement is required.

During, and following, any child restraint anchor service, due to impact event or vehicle repair, carefully inspect all mounting hardware, tether straps, and anchors for proper installation, operation, or damage. If a child restraint anchor is found damaged in any way, the anchor must be replaced. Failure to do this may result in personal injury or death.

Deployed and non-deployed airbags may or may not have live pyrotechnic material within the airbag inflator.

Do not dispose of driver/passenger/curtain airbags or seat belt tensioners unless you are sure of complete deployment. Refer to the Hazardous Substance Control System for proper disposal.

Dispose of deployed airbags and tensioners consistent with state, provincial, local, and federal regulations.

After any airbag component testing or service, do not connect the battery negative cable. Personal injury or death may result if the system test is not performed first.

If the vehicle is equipped with the Occupant Classification System (OCS), do not connect the battery negative cable before performing the OCS Verification Test using the scan tool and the appropriate diagnostic information. Personal injury or death may result if the system test is not performed properly.

Never replace both the Occupant Restraint Controller (ORC) and the Occupant Classification Module (OCM) at the same time. If both require replacement, replace one, then perform the Airbag System test before replacing the other.

Both the ORC and the OCM store Occupant Classification System (OCS) calibration data, which they transfer to one another when one of them is replaced. If both are replaced at the same time, an irreversible fault will be set in both modules and the OCS may malfunction and cause personal injury or death.

If equipped with OCS, the Seat Weight Sensor is a sensitive, calibrated unit and must be handled carefully. Do not drop or handle roughly. If dropped or damaged, replace with another sensor. Failure to do so may result in occupant injury or death.

If equipped with OCS, the front passenger seat must be handled carefully as well. When removing the seat, be careful when setting on floor not to drop. If dropped, the sensor may be inoperative, could result in occupant injury, or possibly death.

If equipped with OCS, when the passenger front seat is on the floor, no one should sit in the front passenger seat. This uneven force may damage the sensing ability of the seat weight sensors. If sat on and damaged, the sensor may be inoperative, could result in occupant injury, or possibly death.

DISARMING THE SYSTEM

To avoid personal injury when working on vehicles equipped with an air bag, the negative battery cable must be disconnected and at least 90 seconds must elapse before working on the system. Failure to do so may result in deployment of the air bag.

ARMING THE SYSTEM

Connect the battery. After all repairs have been completed, turn the ignition key to the on position. Make sure no one is inside the vehicle and connect the negative battery cable. Make sure the light for the air bag system located in the instrument panel does not stay illuminated.

CLOCKSPRING CENTERING

See Figure 9.

1. Center the spiral cable.
2. Turn the spiral cable counterclockwise until it locks.

➡The spiral cable turns a maximum of approximately 5 times.

3. Turn the spiral cable 2.5 times clockwise from its locked position and align the center marks.

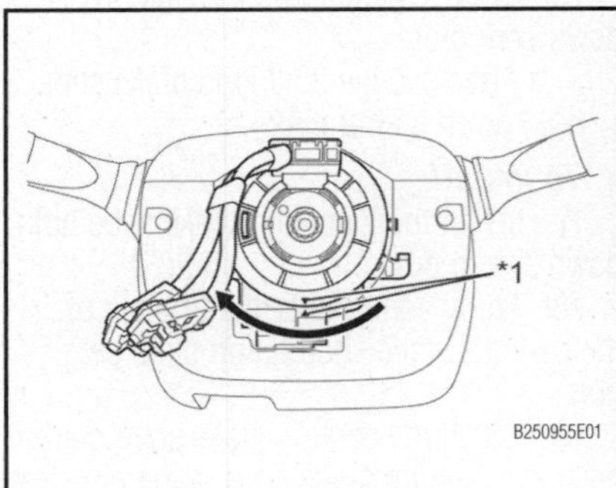

Fig. 9 Turn the spiral cable 2.5 times clockwise from its locked position and align the center marks

DRIVE TRAIN

AUTOMATIC TRANSAXLE FLUID

DRAIN AND REFILL

U660E and U660F Transaxles

See Figures 10 through 12.

➡Precautions and work description:

- The U660E and U660F automatic transaxles do not have an oil filler tube and oil dipstick. When adding fluid, add fluid through the refill hole on the transaxle case. The fluid level can be adjusted by draining excess fluid (allowing excess fluid to overflow) through the No. 1 transmission oil filler tube of the oil pan.
- Before adjusting the fluid level, add the specified amount of fluid when the engine is cold and warm up the engine to circulate the fluid in the transaxle. Ensure that the fluid temperature is as specified and the engine is idling.
- The U660E and U660F automatic transaxles require Toyota Genuine ATF WS.
- The adjustment should be performed according to the procedures and notes.

➡"Overflow" indicates the condition under which fluid comes out of the overflow plug hole.

➡The adjustment should be performed according to the procedures referenced in the work flow below.

1. Before filling transaxle fluid.

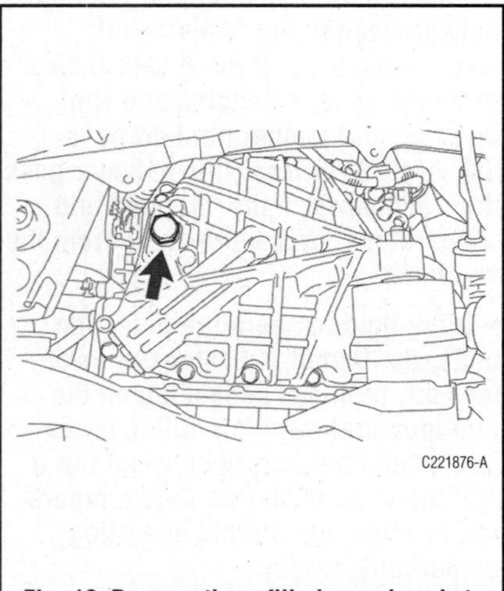

Fig. 10 Remove the refill plug and gasket from the automatic transaxle

➡If the transaxle is hot (ATF temperature is high), wait until the fluid temperature becomes the same as the ambient temperature before starting the following procedure. Recommended ATF temperature: around 68°F (20°C)

a. Lift the vehicle.

➡Set the vehicle on a lift so that the vehicle is kept level when it is lifted up.

b. Remove the 2 bolts, clip and front fender apron seal LH.

2. Perform initial filling.

➡After performing any of the following operations, it is not necessary to perform the initial filling procedure. Proceed to the "Add Specified Amount of Fluid" procedure.

a. Remove the refill plug and gasket from the automatic transaxle.

b. Using a 6 mm hexagon socket wrench, remove the overflow plug and gasket from the automatic transaxle.

➡If ATF comes out after removing the overflow plug, wait until the fluid flow slows and only drips come out. If ATF comes out, it is not necessary to perform the initial filling procedure. After checking the tightening torque of the No. 1 transmission oil filler tube, temporarily tighten the overflow plug.

c. Using a 6 mm hexagon socket wrench, check that the No. 1 transmission oil filler tube is tightened.

➡If the No. 1 transmission oil filler tube is not tightened, the amount of fluid cannot be precisely adjusted.

C221872-A

Fig. 11 Using a 6 mm hexagon socket wrench, remove the overflow plug and gasket from the automatic transaxle

➡To check the torque of the No. 1 transmission oil filler tube, insert the 6 mm hexagon socket wrench into the overflow plug hole.

d. Perform initial filling. Fill the transaxle through the refill hole until fluid begins to trickle out of the overflow plug hole.

e. Wait until the fluid flow slows and only drips come out.

f. Using a 6 mm hexagon socket wrench, temporarily install the gasket and the overflow plug.

➡Reuse the old gasket. The overflow plug will be removed again to adjust the fluid level.

3. Add specified amount of fluid.

a. Fill the transaxle with the correct amount of fluid as listed in the table below.

➡Refill amount differs depending on the operation that was performed.

b. Temporarily install the refill plug to avoid fluid spillage.

➡Reuse the old gasket. The refill plug will be removed again to adjust the fluid level.

c. Lower the vehicle.

4. Adjust fluid temperature (Using the Techstream).

➡The actual ATF temperature can be checked on the Data List using the Techstream.

a. Connect the Techstream to the DLC3 with the ignition switch off.

b. Turn the ignition switch to ON and turn the Techstream on.

➡Check that electrical systems such as the air conditioning system, audio system and lighting system are off.

c. Enter the following menus: Powertrain / ECT / Active Test / Connect the TC and TE1

d. Select the Active Test menu: Connect the TC and TE1 / ON

e. Select the Data List menu: A/T Oil Temperature 1

f. Check the ATF temperature.

➡If the fluid temperature is below 113°F (45°C), proceed to the next step. (Recommended ATF temperature: 104°F (40°C] or less). If the fluid temperature is 113°F (45°C) or more, turn the ignition switch off and wait until the

fluid temperature drops below 113°F (45°C).

g. Depress and hold the brake pedal.
h. Start the engine.
i. Slowly move the shift lever from P to D, then back to P.

➡Slowly move the shift lever to circulate the fluid through each part of the automatic transaxle.

j. While observing the D shift indicator on the combination meter, move the shift lever back and forth between N and D at an interval of 1.5 seconds for 6 seconds or more.

➡Do not pause for more than 1.5 seconds.

➡Performing this operation will cause the vehicle to enter the fluid temperature detection mode.

k. Check that the D shift indicator comes on for 2 seconds.

➡When the fluid temperature detection mode is activated, the D shift indicator on the combination meter comes on for 2 seconds. If the D shift indicator does not come on for 2 seconds, return to the step where terminal TC is first connected and perform the procedure again.

l. Move the shift lever from N to P.
m. Release the brake pedal.
n. Select the Active Test menu: Connect the TC and TE1 / OFF

➡Be sure that terminals TC and TE1 are not connected. If the terminals are connected, the fluid level cannot be precisely adjusted due to fluctuations in engine speed.

➡Please note the following:

- Disconnecting terminals TC and TE1 activates the engine idle speed control mode.
- In the engine idle speed control mode, engine idle speed control starts when the fluid temperature becomes 95°F (35°C) or more and the engine speed is maintained at approximately 800 rpm.
- Even after terminals TC and TE1 are disconnected, the fluid temperature detection mode is active until the ignition switch off.

o. Warm up the engine with the engine idling until the fluid temperature reaches the normal operating temperature [104–113°F (40–45°C)].

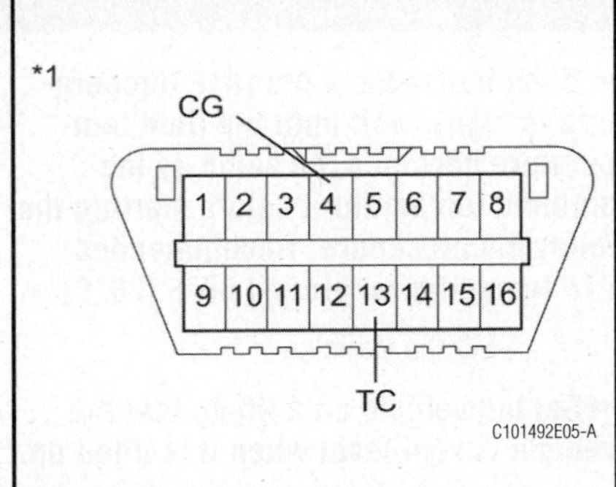

Fig. 12 Using SST, connect terminals 13 (TC) and 4 (CG) of the DLC3 with the ignition switch off

➡If the fluid temperature is within the normal operating temperature range, immediately proceed to the "Adjust Fluid Level" procedure. If the fluid temperature is 113°F (45°C) or more, stop the engine and wait until the fluid temperature drops to 104°F (40°C) or less. Then perform the "Adjust Fluid Temperature" procedure again from the beginning.

➡In the fluid temperature detection mode, the D shift indicator comes on, goes off, or blinks depending on the fluid temperature.

5. Adjust fluid temperature (Not Using the Techstream).

a. Using SST, connect terminals 13 (TC) and 4 (CG) of the DLC3 with the ignition switch off.
b. Depress and hold the brake pedal.
c. Start the engine.

➡Check that electrical systems such as the air conditioning system, audio system and lighting system are off.

➡The indicator lights on the combination meter blink to indicate the DTC output when terminals TC and CG are connected.

d. Slowly move the shift lever from P to D, then back to P.

➡Slowly move the shift lever to circulate the fluid through each part of the automatic transaxle.

e. While observing the D shift indicator on the combination meter, move the shift lever back and forth between N and D at an interval of 1.5 seconds for 6 seconds or more.

➡Do not pause for more than 1.5 seconds.

➡Performing this operation will cause the vehicle to enter the fluid temperature detection mode.

f. Check that the D shift indicator comes on for 2 seconds.

➡When the fluid temperature detection mode is activated, the D shift indicator on the combination meter comes on for 2 seconds. If the D shift indicator does not come on for 2 seconds, return to the step where terminal TC is first connected and perform the procedure again.

g. Move the shift lever from N to P.
h. Release the brake pedal.
i. Remove SST from terminals 13 (TC) and 4 (CG).

➡Be sure that terminals TC and CG are not connected. If the terminals are connected, the fluid level cannot be precisely adjusted due to fluctuations in engine speed.

➡Please note the following:

- Disconnecting terminals TC and CG activates the engine idle control mode.
- In the engine idle speed control mode, engine idle speed control starts when the fluid temperature becomes 95°F (35°C) or more and the engine speed is maintained at approximately 800 rpm.
- Even after terminals TC and CG are disconnected, the fluid temperature detection mode is active until the ignition switch is turned off.

j. Allow the engine to idle until the D shift indicator comes on again.

➡If the D shift indicator is on, immediately proceed to the "Adjust Fluid Level" procedure. If the D shift indicator blinks, stop the engine and wait until the fluid temperature drops to 104°F (40°C) or less (the indicator goes off). Then perform the "Adjust Fluid Temperature" procedure again from the beginning.

➡In the fluid temperature detection mode, the D shift indicator comes on, goes off, or blinks depending on the fluid temperature. Fluid filling procedure should be performed when the D shift indicator is on (the fluid temperature is within the normal operating temperature range).

6. Adjust fluid level.

CAUTION

Use caution while the engine is idling and the radiator fan is operating.

a. Lift the vehicle.

➡Set the vehicle on a lift so that the vehicle is kept level when it is lifted up.

b. Using a 6 mm hexagon socket wrench, remove the overflow plug and gasket.

CAUTION

Be careful as the fluid coming out of the overflow plug hole is hot.

c. Check the amount of fluid that comes out of the overflow plug hole.

➡If only a small amount of fluid (approximately 1 cc) comes out of the overflow plug hole, then only fluid remaining in the No. 1 transmission oil filler tube has come out. This condition is not considered as overflow, so it is necessary to add fluid.

d. If the fluid overflows.

- Wait until the fluid flow slows and only drips come out.
- Remove the refill plug and gasket.

e. If the fluid does not overflow.

- Remove the refill plug and gasket.
- Add transaxle fluid through the refill hole until fluid comes out of the overflow plug hole.
- Wait until the fluid flow slows and only drips come out.

f. Install the overflow plug with a new gasket. Tighten to 30 ft. lbs. (40 Nm).

g. Install the refill plug with a new gasket. Tighten to 36 ft. lbs. (49 Nm).

h. Lower the vehicle.

i. Turn the ignition switch off.

➡Turning the ignition switch off exits the fluid temperature detection mode.

j. Remove the Techstream from the DLC3 (when using the Techstream).

7. After filling transaxle fluid.

a. Lift the vehicle.

b. Clean each part.

c. Check for fluid leaks.

d. Install the front fender apron seal LH with the 2 bolts and clip.

e. Lower the vehicle.

U760E Transaxle

See Figures 13 through 15.

➡Precautions and work description:

- The U760E automatic transaxle does not have an oil filler tube and oil dipstick. When adding fluid, add fluid through the refill hole on the transaxle case. The fluid level can be adjusted by draining excess fluid (allowing excess fluid to overflow) through the No. 1 transmission oil filler tube of the oil pan.
- Before adjusting the fluid level, add the specified amount of fluid when the engine is cold and warm up the engine to circulate the fluid in the transaxle. Ensure that the fluid temperature is as specified and the engine is idling.
- The U760E automatic transaxle requires Toyota Genuine ATF WS.
- The adjustment should be performed according to the procedures and notes.

1. Before filling transaxle fluid.

➡If the transaxle is hot (ATF temperature is high), wait until the fluid temperature becomes the same as the ambient temperature before starting the following procedure. [Recommended ATF temperature: around 68°F (20°C)]

a. Lift the vehicle.

➡Set the vehicle on a lift so that the vehicle is kept level when it is lifted up.

b. Remove the 2 bolts, clip and front fender apron seal LH.

2. Perform initial filling.

➡After performing any of the following operations, it is not necessary to perform the initial filling procedure. Proceed to the "Add Specified Amount of Fluid" procedure.

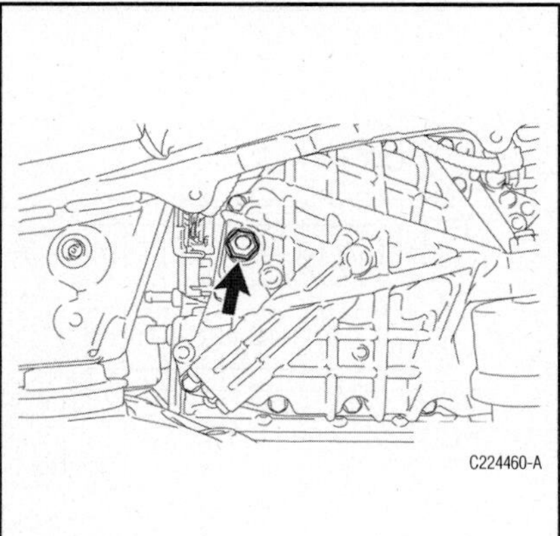

Fig. 13 Remove the refill plug and gasket from the automatic transaxle

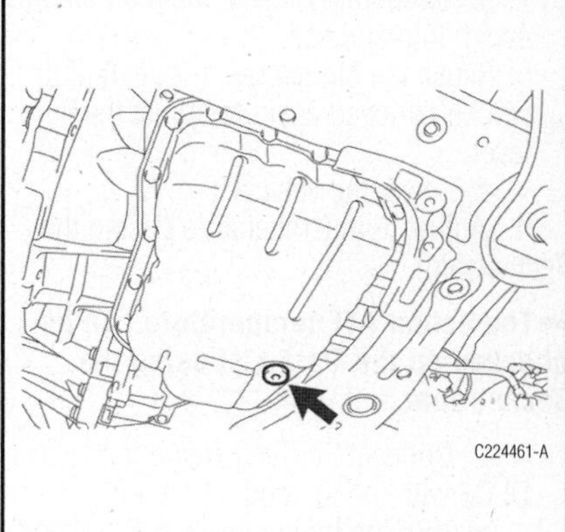

Fig. 14 Using a 6 mm hexagon socket wrench, remove the overflow plug and gasket from the automatic transaxle

a. Remove the refill plug and gasket from the automatic transaxle.

b. Using a 6 mm hexagon socket wrench, remove the overflow plug and gasket from the automatic transaxle.

➡If ATF comes out after removing the overflow plug, wait until the fluid flow slows and only drips come out. If ATF comes out, it is not necessary to perform the initial filling procedure. After checking the tightening torque of the No. 1 transmission oil filler tube, temporarily tighten the overflow plug.

c. Using a 6 mm hexagon socket wrench, check that the No. 1 transmission oil filler tube is tightened.

➡If the No. 1 transmission oil filler tube is not tightened, the amount of fluid cannot be precisely adjusted.

➡To check the torque of the No. 1 transmission oil filler tube, insert the 6 mm hexagon socket wrench into the overflow plug hole.

d. Fill the transaxle through the refill hole until fluid begins to trickle out of the overflow plug hole.

e. Wait until the fluid flow slows and only drips come out.

f. Using a 6 mm hexagon socket wrench, temporarily install the gasket and the overflow plug.

➡Reuse the old gasket. The overflow plug will be removed again to adjust the fluid level.

3. Add specified amount of fluid.

a. Fill the transaxle with the correct amount of fluid as listed in the table below.

➡Refill amount differs depending on the operation that was performed.

b. Temporarily install the refill plug to avoid fluid spillage.

Reuse the old gasket. The refill plug will be removed again to adjust the fluid level.

c. Lower the vehicle.

4. Adjust fluid temperature (Using the Techstream).

➡The actual ATF temperature can be checked on the Data List using the Techstream.

a. Connect the Techstream to the DLC3 with the ignition switch off.

b. Turn the ignition switch to ON and turn the Techstream on.

➡Check that electrical systems such as the air conditioning system, audio system and lighting system are off.

c. Enter the following menus: Powertrain / ECT / Active Test / Connect the TC and TE1

d. Select the Active Test menu: Connect the TC and TE1 / ON

e. Select the Data List menu: A/T Oil Temperature 1

f. Check the ATF temperature.

➡If the fluid temperature is below 113°F (45°C), proceed to the next step. [Recommended ATF temperature: 95°F (35°C) or less]. If the fluid temperature is 113°F (45°C) or more, turn the ignition switch off and wait until the fluid temperature drops below 113°F (45°C).

g. Depress and hold the brake pedal.

h. Start the engine.

i. Slowly move the shift lever from P to D, then back to P.

➡Slowly move the shift lever to circulate the fluid through each part of the automatic transaxle.

j. While observing the D shift indicator on the combination meter, move the shift lever back and forth between N and D at an interval of 1.5 seconds for 6 seconds or more.

➡Do not pause for more than 1.5 seconds.

➡Performing this operation will cause the vehicle to enter the fluid temperature detection mode.

k. Check that the D shift indicator comes on for 2 seconds.

➡When the fluid temperature detection mode is activated, the D shift indicator on the combination meter comes on for 2 seconds. If the D shift indicator does not come on for 2 seconds, return to the step where terminal TC is first connected and perform the procedure again.

l. Move the shift lever from N to P.

m. Release the brake pedal.

n. Select the Active Test menu: Connect the TC and TE1 / OFF

➡Be sure that terminals TC and TE1 are not connected. If the terminals are connected, the fluid level cannot be precisely adjusted due to fluctuations in engine speed.

➡Please note the following:

- Disconnecting terminals TC and TE1 activates the engine idle speed control mode.
- In the engine idle speed control mode, engine idle speed control starts when the fluid temperature becomes 95°F (35°C) or more and the engine speed is maintained at approximately 800 rpm.
- Even after terminals TC and TE1 are disconnected, the fluid temperature detection mode is active until the ignition switch is turned off.

o. Warm up the engine with the engine idling until the fluid temperature reaches the normal operating temperature [95–113°F (35–45°C)].

➡Please note the following:

- If the fluid temperature is within the normal operating temperature range, immediately proceed to the "Adjust Fluid Level" procedure.
- If the fluid temperature is 113°F (45°C) or more, stop the engine and wait until the fluid temperature drops to 95°F (35°C) or less. Then perform the "Adjust Fluid Temperature" procedure again from the beginning.

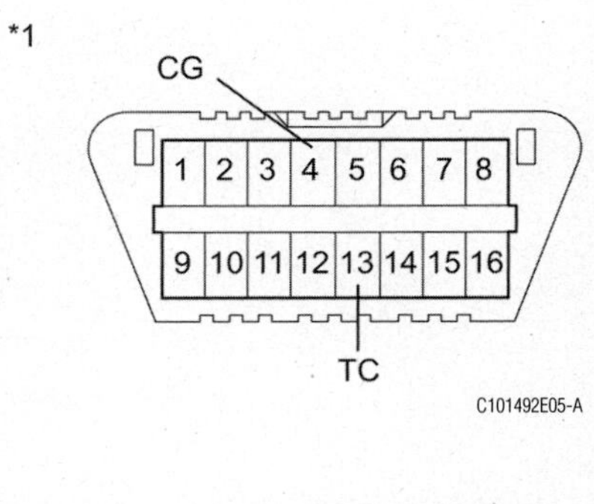

➡In the fluid temperature detection mode, the D shift indicator comes on, goes off, or blinks depending on the fluid temperature.

*1

CG

1 2 3 4 5 6 7 8

9 10 11 12 13 14 15 16

TC

C101492E05-A

Fig. 15 Using SST, connect terminals 13 (TC) and 4 (CG) of the DLC3 with the ignition switch off

5. Adjust fluid temperature (Not Using the Techstream).

a. Using SST, connect terminals 13 (TC) and 4 (CG) of the DLC3 with the ignition switch off.

b. Depress and hold the brake pedal.

c. Start the engine.

➡Check that electrical systems such as the air conditioning system, audio system and lighting system are off.

➡The indicator lights on the combination meter blink to indicate the DTC output when terminals TC and CG are connected.

d. Slowly move the shift lever from P to D, then back to P.

➡Slowly move the shift lever to circulate the fluid through each part of the automatic transaxle.

e. While observing the D shift indicator on the combination meter, move the shift lever back and forth between N and D at an interval of 1.5 seconds for 6 seconds or more.

➡Do not pause for more than 1.5 seconds.

➡Performing this operation will cause the vehicle to enter the fluid temperature detection mode.

f. Check that the D shift indicator comes on for 2 seconds.

➡When the fluid temperature detection mode is activated, the D shift indicator on the combination meter comes on for 2 seconds. If the D shift indicator does not come on for 2 seconds, return to the step where terminal TC is first connected and perform the procedure again.

g. Move the shift lever from N to P.

h. Release the brake pedal.

i. Remove SST from terminals 13 (TC) and 4 (CG).

➡Be sure that terminals TC and CG are not connected. If the terminals are connected, the fluid level cannot be precisely adjusted due to fluctuations in engine speed.

➡Please note the following:

- Disconnecting terminals TC and CG activates the engine idle control mode.
- In the engine idle speed control mode, engine idle speed control starts when the fluid temperature becomes 95°F (35°C) or more and the engine speed is maintained at approximately 800 rpm.
- Even after terminals TC and CG are disconnected, the fluid temperature detection mode is active until the ignition switch is turned off.

j. Allow the engine to idle until the D shift indicator comes on again.

➡Please note the following:

- If the D shift indicator is on, immediately proceed to the "Adjust Fluid Level" procedure.
- If the D shift indicator blinks, stop the engine and wait until the fluid temperature drops to 95°F (35°C) or less (the indicator goes off). Then perform the "Adjust Fluid Temperature" procedure again from the beginning.

➡In the fluid temperature detection mode, the D shift indicator comes on, goes off, or blinks depending on the fluid temperature. Fluid filling procedure should be performed when the D shift indicator is on (the fluid temperature is within the normal operating temperature range).

6. Adjust fluid level.

⁂ CAUTION

Use caution while the engine is idling and the radiator fan is operating. Lift the vehicle.

➡Set the vehicle on a lift so that the vehicle is kept level when it is lifted up.

Using a 6 mm hexagon socket wrench, remove the overflow plug and gasket.

⁂ CAUTION

Be careful as the fluid coming out of the overflow plug hole is hot. Check the amount of fluid that comes out of the overflow plug hole.

➡If only a small amount of fluid (approximately 1 cc) comes out of the overflow plug hole, then only fluid remaining in the No. 1 transmission oil filler tube has come out. This condition is not considered as overflow, so it is necessary to add fluid.

a. If the fluid overflows.

b. Wait until the fluid flow slows and only drips come out.

c. Remove the refill plug and gasket.

d. If the fluid does not overflow.

e. Remove the refill plug and gasket.

f. Add transaxle fluid through the refill hole until fluid comes out of the overflow plug hole.

g. Wait until the fluid flow slows and only drips come out.

h. Install the overflow plug with a new gasket. 30 ft. lbs. (40 Nm).

i. Install the refill plug with a new gasket. Tighten to 36 ft. lbs. (49 Nm).

j. Lower the vehicle.

k. Turn the ignition switch off.

➡Turning the ignition switch off exits the fluid temperature detection mode.

Remove the Techstream from the DLC3 (when using the Techstream).

7. After filling transaxle fluid.

a. Lift the vehicle.

b. Clean each part.

c. Check for fluid leaks.

d. Install the front fender apron seal LH with the 2 bolts and clip.

e. Lower the vehicle.

FRONT HALFSHAFT

REMOVAL & INSTALLATION

2.7L Engine

See Figure 16.

1. Drain automatic transaxle fluid.
2. Remove front wheels.
3. Remove front axle shaft nut.

a. Using SST and a hammer, release the staked part of the front axle shaft nut.

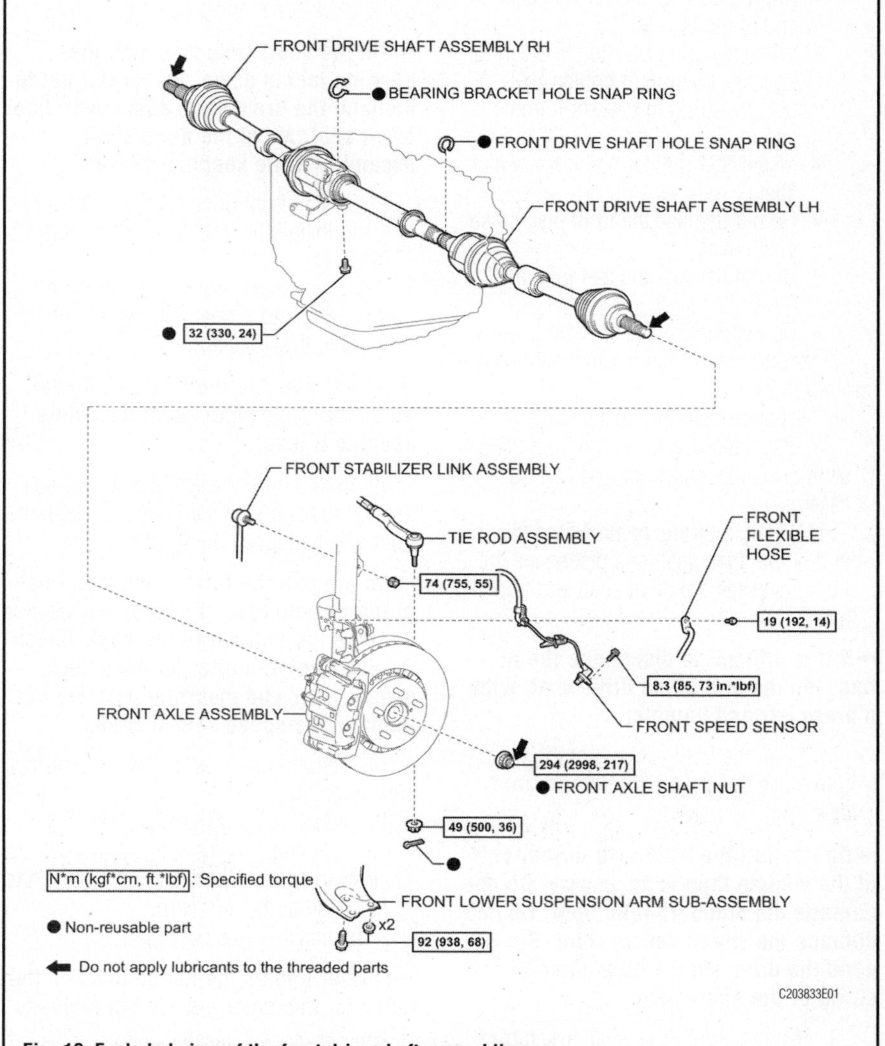

Fig. 16 Exploded view of the front drive shaft assemblies

➡Loosen the staked part of the nut completely, otherwise the threads of the drive shaft may be damaged.

b. While applying the brakes, remove the front axle shaft nut.

4. Separate front stabilizer link assembly
5. Separate front speed sensor

a. Remove the bolt and separate the front speed sensor from the steering knuckle.

b. Remove the bolt and 2 clamps, and separate the front speed sensor and front flexible hose.

6. Separate tie rod assembly.

a. Remove the cotter pin and nut.

b. Install SST to the tie rod end.

➡Make sure that the upper ends of the tie rod end and SST are aligned.

c. Using SST, separate the tie rod end from the steering knuckle.

➡Please note the following:

- Apply grease to the threads and end of the SST bolt.
- When securing SST to the steering knuckle, be sure to tighten SST using a string to prevent it from falling.
- Install SST so that A and B are parallel.
- Do not damage the front disc brake dust cover.
- Do not damage the ball joint dust cover.
- Do not damage the steering knuckle.

7. Separate front lower suspension arm sub-assembly.
8. Separate front axle assembly.

a. Put matchmarks on the front drive shaft assembly and front axle hub sub-assembly.

b. Using a plastic hammer, tap the end of the drive shaft and disengage the fitting between the drive shaft and front axle.

➡If it is difficult to disengage the fitting, tap the end of the drive shaft with a brass bar and hammer.

c. Pull the front axle out of the vehicle to remove the drive shaft from the front axle.

➡Do not pull the front axle further out of the vehicle than is necessary. Do not damage the outboard joint boot. Do not damage the speed sensor rotor. Suspend the drive shaft with a piece of string or the equivalent.

9. Remove front drive shaft assembly LH.

a. Using SST, remove the front drive shaft assembly LH.

➡Be careful not to damage the drive shaft dust cover, inboard joint boot or oil seal. Be careful not to drop the drive shaft assembly.

10. Remove front drive shaft assembly RH.

a. Remove the bearing bracket hole snap ring from the drive shaft bearing bracket.

b. Remove the bolt and front drive shaft assembly RH from the drive shaft bearing bracket.

⁂ WARNING

Be careful not to damage the inboard joint boot or oil seal. Be careful not to drop the drive shaft assembly.

To install:

11. Align the splines of the shaft and install the front drive shaft assembly LH using a brass bar and a hammer.

➡Set the shaft snap ring with the opening facing down. Be careful not to damage the drive shaft dust cover, boot or oil seal. Move the drive shaft assembly while keeping it level.

12. Install front drive shaft assembly RH.

a. Install the front drive shaft assembly RH.

b. Install the bearing bracket hole snap ring and a new bolt. tighten the bolt to 24 ft. lbs. (32 Nm).

➡Do not damage the boot or oil seal. Move the drive shaft assembly while keeping it level.

13. Pull the front axle out of the vehicle to align the spline of the drive shaft with the front axle and insert the front axle.

➡Do not pull the front axle further out of the vehicle than is necessary. Do not damage the outboard joint boot. Check for any foreign matter on the speed sensor rotor and insertion part. Do not damage the speed sensor rotor.

14. Install front lower suspension arm sub-assembly.
15. Install tie rod assembly.

a. Connect the tie rod assembly to the steering knuckle with the nut. Tighten the nut to 36 ft. lbs. (49 Nm).

b. Install a new cotter pin.

➡Further tighten the nut up to 60° if the holes for the cotter pin are not aligned.

16. Install front speed sensor.

a. Install the front speed sensor and front flexible hose with the bolt. Tighten the bolt to 14 ft. lbs. (19 Nm).

➡Do not twist the front speed sensor when installing it. First install the speed sensor harness bracket, and then install the flexible hose bracket.

b. Install the front speed sensor to the steering knuckle with the bolt.

➡Please note the following:

- Prevent foreign matter from attaching to the front speed sensor tip.
- Firmly insert the front speed sensor body into the steering knuckle before tightening the bolt.
- After installing the front speed sensor to the steering knuckle, make sure that there is no clearance between the front speed sensor stay and steering knuckle. Also make sure that no foreign matter is stuck between the parts.
- To prevent interference between the front speed sensor and magnetic rotor, do not rotate the front speed sensor body during or after the insertion of the front speed sensor body to the steering knuckle.

17. Install front stabilizer link assembly.
18. Install front axle shaft nut.

a. Clean the threaded parts on the front drive shaft and a new front axle shaft nut using a non-residue solvent.

➡Be sure to perform this work for a new drive shaft. Keep the threaded parts free of oil and foreign matter.

b. Install the new front axle shaft nut. Tighten to 217 ft. lbs. (294 Nm).

c. Using a chisel and hammer, stake the front axle shaft nut.

19. Install front wheels.
20. Add automatic transaxle fluid.
21. Adjust front wheel alignment.
22. Inspect speed sensor signal.

3.5L Engine

See Figures 17 and 18.

1. Notice for removing and installing front drive shaft assembly RH (for AWD).

⁂ WARNING

When removing and installing the front drive shaft assembly RH in an AWD vehicle, be sure to first drain all the transaxle oil and transfer oil. If removal and installation are carried out without draining these oils, the transfer oil will flow into the

transaxle side. Extensive cleaning will be required if the two oils mix.

2. Drain automatic transaxle fluid (for FWD).
3. Drain automatic transaxle fluid (for AWD).
4. Drain transfer oil (for AWD).
5. Remove front wheels.
6. Remove front axle shaft nut.

➡Loosen the staked part of the nut completely, otherwise the threads of the drive shaft may be damaged.

a. While applying the brakes, remove the front axle shaft nut.

7. Separate front stabilizer link assembly.
8. Separate front speed sensor.
   a. Remove the bolt and separate the front speed sensor from the steering knuckle.
   b. Remove the bolt and 2 clamps, and separate the front speed sensor and front flexible hose.
9. Separate tie rod assembly.
   a. Remove the cotter pin and nut.
   b. Install SST to the tie rod end.

➡Make sure that the upper ends of the tie rod end and SST are aligned.

c. Using SST, separate the tie rod end from the steering knuckle.

➡Apply grease to the threads and end of the SST bolt. When securing SST to the steering knuckle, be sure to tighten SST using a string to prevent it from falling. Install SST so that A and B are parallel. Do not damage the front disc brake dust cover. Do not damage the ball joint dust cover. Do not damage the steering knuckle.

10. Separate front lower suspension arm.
11. Separate front axle assembly.
    a. Put matchmarks on the front drive shaft assembly and front axle hub sub-assembly.
    b. Using a plastic hammer, tap the end of the drive shaft and disengage the fitting between the drive shaft and front axle.

➡If it is difficult to disengage the fitting, tap the end of the drive shaft with a brass bar and hammer.

c. Pull the front axle out of the vehicle to remove the drive shaft from the front axle.

➡Do not pull the front axle further out of the vehicle than is necessary. Do not damage the outboard joint boot. Do not damage the speed sensor rotor. Suspend the drive shaft with a piece of string or the equivalent.

12. Using SST, remove the front drive shaft assembly LH.

➡Be careful not to damage the drive shaft dust cover, inboard joint boot or oil seal. Be careful not to drop the drive shaft assembly.

13. Remove front fender apron seal RH (for AWD).
14. Remove engine mounting bracket RH (for AWD).
    a. Support the engine assembly with a transmission jack.
    b. Remove the nut.
    c. Remove the 3 bolts and remove the engine mounting bracket RH from the engine assembly.
15. Remove engine mounting insulator RH (for AWD).
    a. Remove the 2 hole plugs.
    b. Remove the 3 nuts and remove the engine mounting insulator RH.
16. Remove front drive shaft assembly RH.
    a. Remove the bearing bracket hole snap ring from the drive shaft bearing bracket.
    b. Remove the bolt and front drive shaft assembly RH from the drive shaft bearing bracket.

➡Be careful not to damage the inboard joint boot or oil seal. Be careful not to drop the drive shaft assembly.

To install:

17. Align the splines of the shaft and install the front drive shaft assembly LH using a brass bar and a hammer.

➡Set the shaft snap ring with the opening facing down. Be careful not to damage the drive shaft dust cover, boot or oil seal. Move the drive shaft assembly while keeping it level.

18. Install front drive shaft assembly RH.
    a. Install the front drive shaft assembly RH.
    b. Install the bearing bracket hole snap ring and a new bolt. Tighten the bolt to 23 ft. lbs. (32 Nm).

➡Do not damage the boot or oil seal. Move the drive shaft assembly while keeping it level.

19. Install engine mounting insulator RH (for AWD).
    a. Temporarily install the engine mounting insulator RH with the 3 nuts.
20. Install the engine mounting bracket RH with the 3 bolts. Tighten the bolts to 39 ft. lbs. (54 nm).
21. Fully tighten engine mounting insulator RH (for AWD)
    a. Fully tighten engine mounting insulator RH with the 3 nuts. Tighten the nuts to 64 ft. lbs. (87 Nm).
    b. Install the 2 hole plugs.
    c. Install the engine mounting insulator RH with the nut. Tighten the nut to 70 ft. lbs. (95 Nm).
22. Install front fender apron seal RH (for AWD)
23. Pull the front axle out of the vehicle to align the spline of the drive shaft with the front axle and insert the front axle.

➡Do not pull the front axle further out of the vehicle than is necessary. Do not damage the outboard joint boot. Check for any foreign matter on the speed

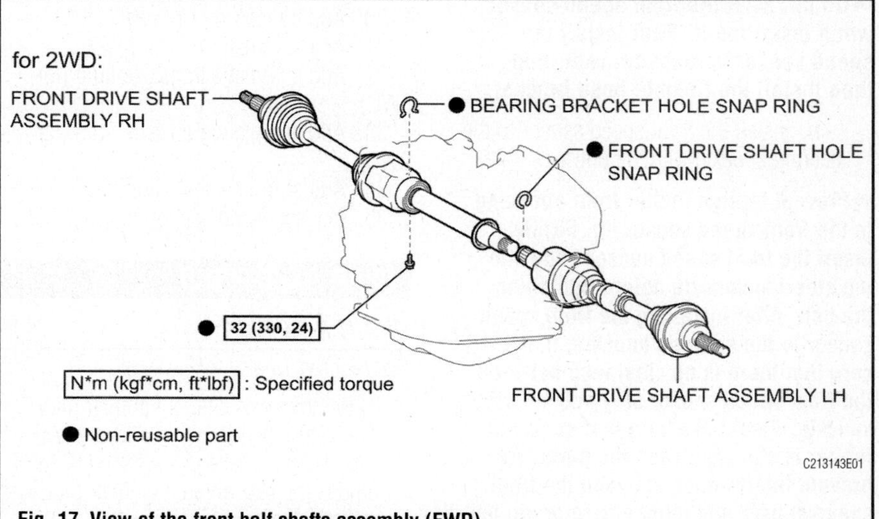

Fig. 17 View of the front half shafts assembly (FWD)

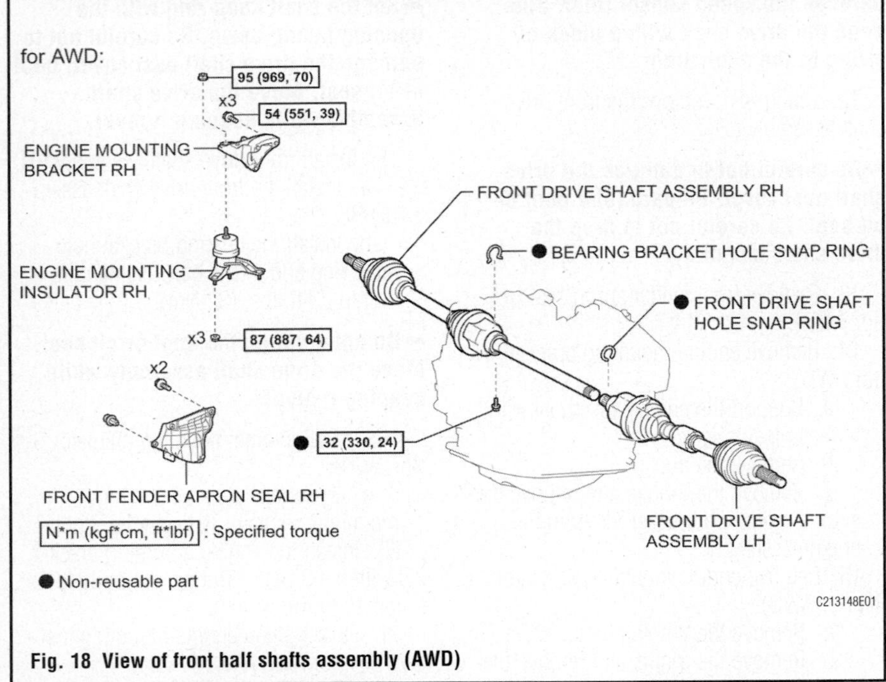

Fig. 18 View of front half shafts assembly (AWD)

sensor rotor and insertion part. Do not damage the speed sensor rotor.

24. Install front lower suspension arm.
25. Install tie rod assembly.
    a. Connect the tie rod assembly to the steering knuckle with the nut. Tighten the nut to 36 ft. lbs. (49 Nm).
    b. Install a new cotter pin.

➡Further tighten the nut up to 60° if the holes for the cotter pin are not aligned.

26. Install front speed sensor.
    a. Install the front speed sensor and front flexible hose with the bolt. Tighten the bolt to 14 ft. lbs. (19 Nm).

➡Do not twist the front speed sensor when installing it. First install the speed sensor harness bracket, and then install the flexible hose bracket.

    b. Install the front speed sensor to the steering knuckle with the bolt.

➡Prevent foreign matter from attaching to the front speed sensor tip. Firmly insert the front speed sensor body into the steering knuckle before tightening the bolt. After installing the front speed sensor to the steering knuckle, make sure that there is no clearance between the front speed sensor stay and steering knuckle. Also make sure that no foreign matter is stuck between the parts. To prevent interference between the front speed sensor and magnetic rotor, do not rotate the front speed sensor body during or after the insertion of the front speed sensor body to the steering knuckle.

27. Install front stabilizer link assembly.
28. Install front axle shaft nut.
    a. Clean the threaded parts on the front drive shaft and a new front axle shaft nut using a non-residue solvent.

➡Be sure to perform this work for a new drive shaft. Keep the threaded parts free of oil and foreign matter.

    b. Install the new front axle shaft nut. Tighten the nut to 217 ft. lbs. (294 Nm).
    c. Using a chisel and hammer, stake the front axle shaft nut.
29. Install front wheels.
30. Add transfer oil (for AWD).
31. Adjust transfer oil (for AWD).
32. Add automatic transaxle fluid (for FWD).
33. Add automatic transaxle fluid (for AWD).
34. Adjust front wheel alignment.
35. Inspect speed sensor signal.

REAR AXLE FLUID

DRAIN & REFILL

See Figures 19 and 20.

1. Remove rear differential filler plug.
    a. Stop the vehicle in a level place.
    b. Using a hexagon wrench (10 mm), remove the rear differential filler plug and gasket.

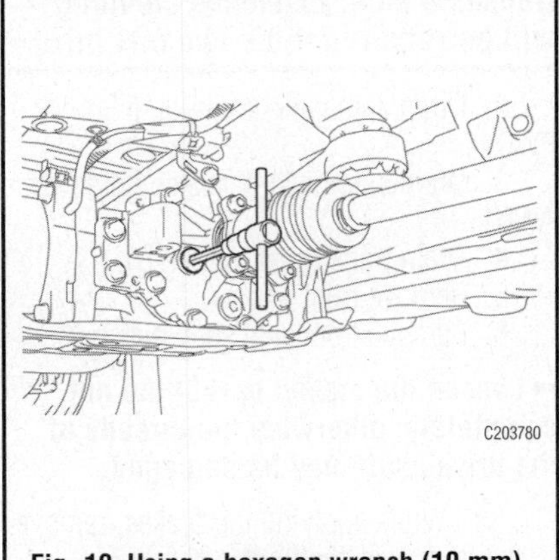

Fig. 19 Using a hexagon wrench (10 mm), remove the rear differential filler plug and gasket

2. Using a hexagon wrench (10 mm), remove the rear differential drain plug and gasket, then drain the differential oil.
3. Using a hexagon wrench (10 mm), install the rear differential drain plug with a new gasket. Tighten the drain plug to 29 ft. lbs. (39 Nm).
4. Add differential oil until the oil level is between 0–0.197 inches (0–5 mm) from the bottom lip of the rear differential filler plug opening.

➡An excessively large or small amount of differential oil may cause damage. After adding differential oil, drive the vehicle and recheck the oil level.

5. Inspect and adjust differential oil.
    a. Check that the oil level is between 0–0.197 inches (0–5 mm) from the bottom lip of the rear differential filler plug opening.

➡An excessively large or small amount of differential oil may cause

Fig. 20 Using a hexagon wrench (10 mm), remove the rear differential drain plug and gasket

damage. After adding differential oil, drive the vehicle and recheck the oil level.

b. Inspect for oil leak if the oil level is low.

6. Using a hexagon wrench (10 mm), install the rear differential filler plug and a new gasket.

REAR DRIVESHAFT

REMOVAL & INSTALLATION

See Figures 21 and 22.

1. Drain transfer oil.
2. Remove propeller with center bearing shaft assembly.
   a. Depress the brake pedal and hold it.
   b. Using a hexagon wrench (6 mm), loosen the bolts ½ turn.

➡Put a piece of cloth or equivalent into the inside of the universal joint cover so that the boot does not touch the inside of the universal joint cover. Do not remove the bolts.

c. Place matchmarks on the rear propeller shaft and electromagnetic control coupling assembly.

➡Do not use a punch for the marks.

d. Remove the 4 nuts and 4 washers.

e. Using a brass bar and a hammer, separate the propeller with center bearing shaft assembly.

f. Remove the 4 bolts, 2 No. 1 center support bearing washers and 2 No. 2 center support bearing washers.

➡When removing the bolts and washers, do not apply excessive force to the universal joint. Keep the washers and use them again when re-installing.

g. Pull out the propeller with center bearing shaft assembly from the transfer.

➡When removing the propeller shaft, do not apply excessive force to the universal joint. During and after the removal of the propeller shaft, keep the universal joint angle straight (within 15 degrees). Be careful not to damage the oil seal.

h. Insert SST into the transfer to prevent oil leaks.

➡Be careful not to damage the oil seal.

To install:

3. Temporarily tighten propeller with center bearing shaft assembly.

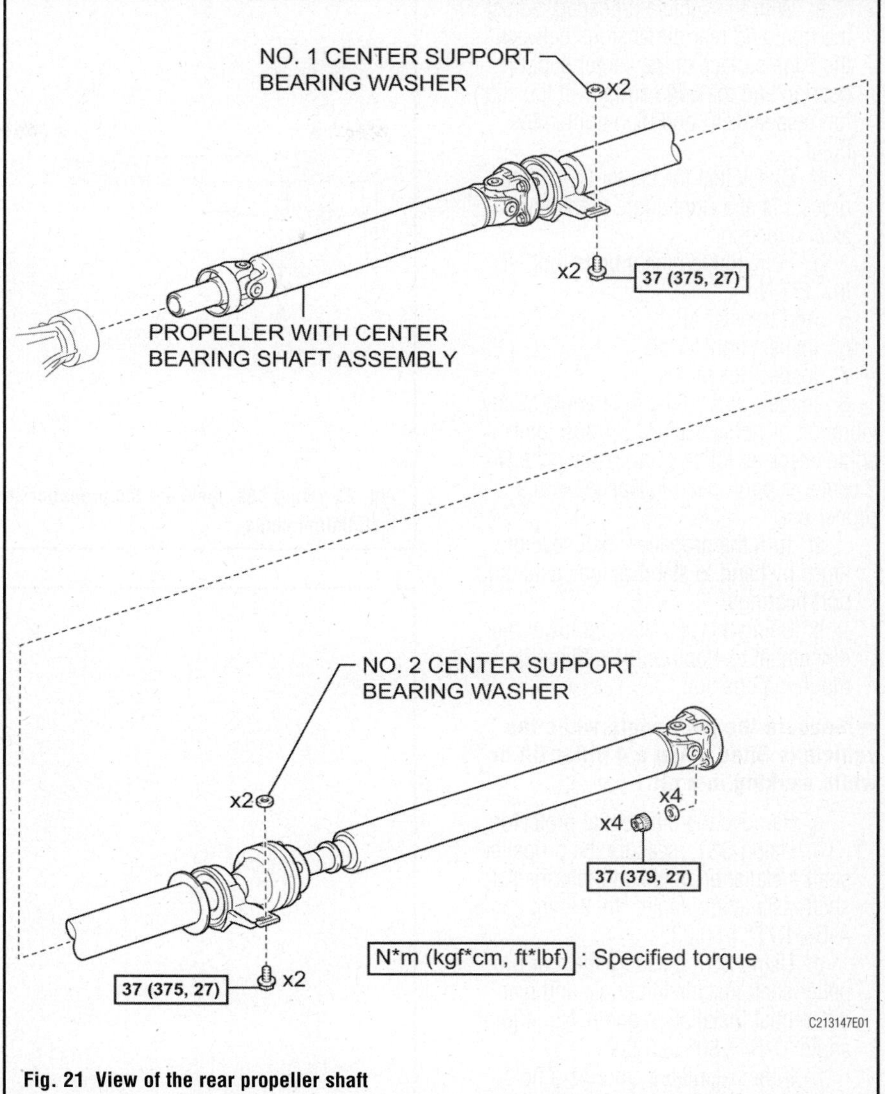

Fig. 21 View of the rear propeller shaft

a. Remove SST from the transfer.

b. Install the propeller with center bearing shaft assembly.

➡Be careful not to damage the oil seal. Be careful not to damage the universal joint boot when installing the propeller shaft.

c. Align the matchmarks on the rear propeller shaft and electromagnetic control coupling assembly and install the 4 nuts and 4 washers temporarily.

➡Do not allow grease to adhere to be bolts or washers.

d. Temporarily install the propeller with center bearing shaft assembly with the 4 bolts, 2 No. 1 center support bearing washers and 2 No. 2 center support bearing washers.

➡Reuse the original washers. If different washers are used, it will be necessary to adjust the installation angle later. Do not allow grease to adhere to the bolts or washers.

e. Fully tighten the 4 nuts to 27 ft. lbs. (37 Nm).

4. Fully tighten propeller with center bearing shaft assembly.
   a. Remove the piece of cloth or equivalent from the universal joint.
   b. Depress the brake pedal and hold it.
   c. Using a hexagon wrench (6 mm), tighten the 6 bolts to 19 ft. lbs. (26 Nm).
   d. With the vehicle unloaded, adjust the dimension between the rear side of the cover and shaft. The dimension should be between 2.264–2.303 inches (57.5–58.5 mm)

e. With the vehicle unloaded, adjust the front and rear dimensions between the edge surface of the center support bearing and the edge surface of the cushion respectively, and then tighten the bolts.

f. Check that the center line of the bracket is at a right angle to the shaft axial direction.

g. Fully tighten the 4 bolts to 27 ft. lbs. (37 Nm).

5. Add transfer oil.
6. Inspect transfer oil.
7. Inspect for oil leak.
8. Inspect and adjust joint angle. If any vibration or noise occurs, perform joint angle check as follows and replace the No. 2 center support bearing washer with a proper one.

a. Turn the propeller shaft several times by hand to stabilize the center support bearings.

b. Using a jack, raise and lower the differential to stabilize the differential mounting cushion.

➡Measure the joint angle while the vehicle is lifted using a 4 pillar lift or while working in a pit.

c. Remove the differential protector.

d. Using SST, measure the propeller shaft installation angle and intermediate shaft installation angle. No. 1 joint angle: A-B=-1.77° to 0.23°

e. Using SST, measure the rear propeller shaft installation angle and rear differential installation angle. No. 4 joint angle: C-D=1.80° to 3.80°

f. If the calculated amount is not within the specification, adjust it with the No. 2 center support bearing washer.

➡Make sure to use a washer of the same thickness on both right and left sides. Do not use 2 or more washers on a bolt.

9. Install the differential protector. Tighten to 32 ft. lbs. (44 Nm).

REAR HALFSHAFT

REMOVAL & INSTALLATION

See Figure 23.

1. Remove rear wheel.
2. Remove rear axle shaft nut.

a. Using SST and a hammer, unstake part of the axle shaft nut.

➡Loosen the staked part of the nut completely, otherwise the threads of the drive shaft may be damaged.

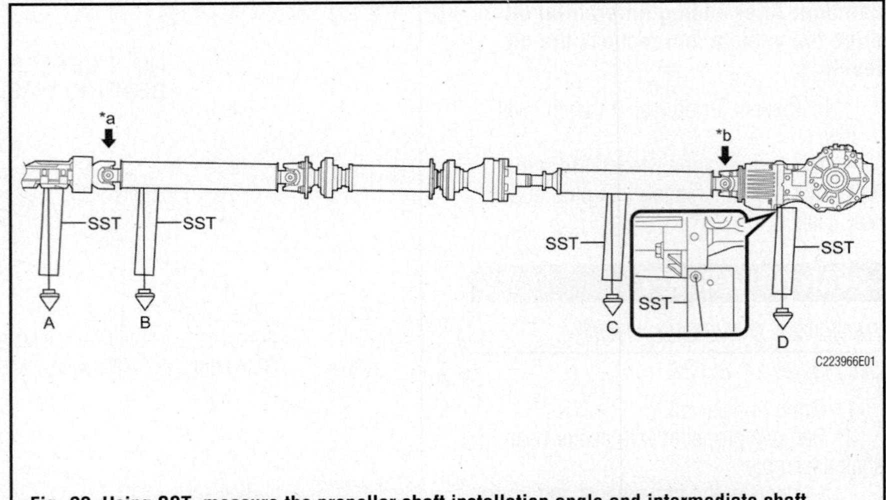

Fig. 22 Using SST, measure the propeller shaft installation angle and intermediate shaft installation angle

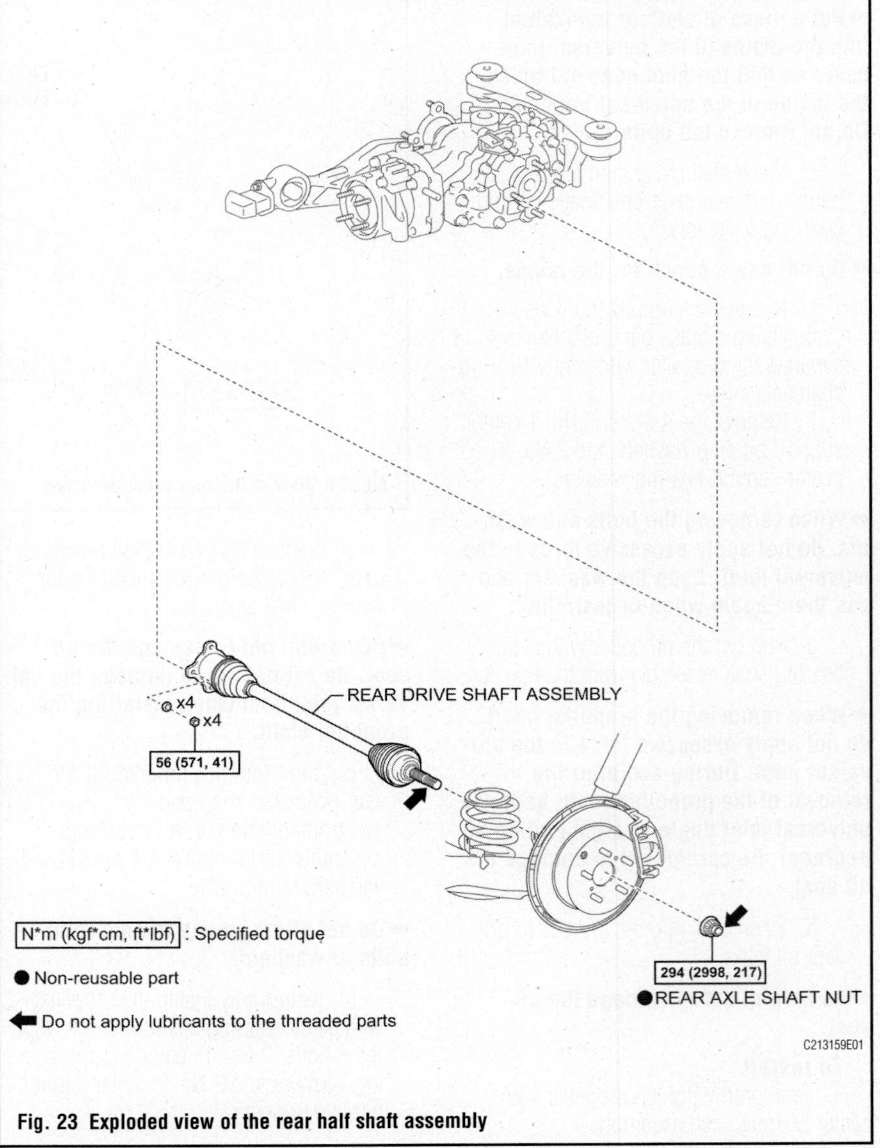

Fig. 23 Exploded view of the rear half shaft assembly

b. While applying the brakes, remove the axle shaft nut.

3. Remove rear drive shaft assembly.

a. Put matchmarks on the rear drive shaft assembly and rear axle hub and bearing assembly.

➡Do not punch the matchmarks.

b. Using a plastic hammer, separate the rear drive shaft assembly from the axle hub and bearing assembly.

➡If it is difficult to separate, tap the end of the rear drive shaft assembly using a brass bar and a hammer.

➡Do not pull the rear axle further out of the vehicle than is necessary. Do not damage the outboard joint boot.

c. Put matchmarks on the rear drive shaft and differential side gear shaft.

d. Remove the 4 nuts and washers, and separate the rear drive shaft from the differential side gear shaft.

e. Remove the rear drive shaft from the axle carrier.

To install:

4. Install rear drive shaft assembly.

a. Install the rear drive shaft to the axle carrier.

➡Be careful not to damage the boot and speed sensor rotor of the rear drive shaft and oil seal of the axle hub bearing.

b. Align the matchmarks and connect the rear drive shaft to the differential side gear shaft.

➡Be careful not to damage the boots and end cover.

c. Install the rear drive shaft with the 4 nuts and washers. Tighten to 41 ft. lbs. (56 Nm).

5. Install rear axle shaft nut.

a. Clean the threaded parts on the rear drive shaft and a new rear axle shaft nut using a non-residue solvent.

➡Be sure to perform this work for a new drive shaft. Keep the threaded parts free of oil and foreign matter.

b. Install a new axle shaft nut. Tighten to 217 ft. lbs. (294 Nm).

c. Using a chisel and hammer, stake the axle shaft nut.

6. Install rear wheel.

ENGINE COOLING

ENGINE COOLANT

DRAIN & REFILL PROCEDURE

2.7L Engine

See Figure 24.

1. Remove No. 1 engine under cover.
2. Drain engine coolant.

a. Loosen the radiator drain cock plug and drain the coolant.

CAUTION

Do not remove the radiator cap, cylinder block drain cock plugs and radiator drain cock plug while the engine and radiator are still hot. Pressurized, hot engine coolant and steam may be released and cause serious burns.

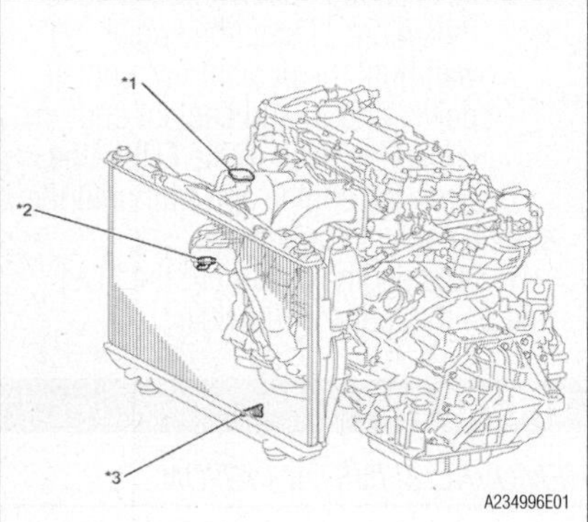

Fig. 24 Locations of the radiator cap (1), cylinder block drain cock plug (2), and the radiator drain cock plug (3)

➡Collect the coolant in a container and dispose of it according to the regulations in your area.

b. Remove the radiator cap from the radiator assembly.

c. Loosen the cylinder block drain cock plug.

3. Add engine coolant.

a. Tighten the radiator drain cock plug by hand.

b. Tighten the cylinder block drain cock plug.

c. Remove the radiator cap.

4. Slowly fill the radiator with TOYOTA Super Long Life Coolant (SLLC).

➡TOYOTA vehicles are filled with TOYOTA SLLC at the factory. In order to avoid damage to the engine cooling system and other technical problems, only use TOYOTA SLLC or similar high quality ethylene glycol based non-silicate, non-amine, non-nitrite, non-borate coolant with long-life hybrid organic acid technology (coolant with long-life hybrid organic acid technology is a combination of low phosphates and organic acids). Contact your TOYOTA dealer for further details.

➡Never use water as a substitute for engine coolant.

a. Slowly pour coolant into the radiator reservoir tank until it reaches the FULL line.

b. Install the radiator cap.

c. Squeeze the No. 1 and No. 2 radiator hoses several times by hand, and then check the level of the coolant. If the coolant level is low, add coolant.

5. Bleed air from the cooling system.

a. Warm up the engine until the thermostat opens. While the thermostat is open, circulate the coolant for several minutes.

➡The thermostat open timing can be confirmed by squeezing the No. 2 radiator hose by hand, and checking when the engine coolant starts to flow inside the hose.

b. Maintain the engine speed at 2500 to 3000 rpm.

c. Squeeze the inlet and No. 1 and No. 2 radiator hoses several times by hand to bleed air.

CAUTION

Wear protective gloves. Be careful as the radiator hoses are hot. Keep your hands away from the cooling fans.

➡Make sure that the radiator reservoir still has some coolant in it. If the coolant temperature gauge indicates an excessive temperature, turn off the engine and let it cool. If there is not enough coolant, the engine may overheat or be seriously damaged.

d. Stop the engine and wait until the engine coolant cools down.

e. Add engine coolant to the FULL line on the radiator reservoir.

6. Inspect for coolant leak.
7. Install No. 1 engine under cover.

3.5L Engine

See Figure 25.

1. Remove the engine undercover.
2. Remove V-bank cover sub-assembly.
3. Drain engine coolant:
   a. Loosen the radiator drain cock plug.
   b. Remove the radiator cap sub-assembly from the radiator assembly.

➡Collect the coolant in a container and dispose of it according to the regulations in your area.

   c. Loosen the 2 cylinder block drain cock plugs.
4. Add engine coolant:
   a. Tighten the radiator drain cock plug by hand.
   b. Tighten the 2 cylinder block drain cock plugs.
   c. Remove the air drain cock plug from the water inlet housing.
   d. Loosen the air drain plug at the top of the radiator 3 or 4 turns.
   e. Add TOYOTA Super Long Life Coolant (SLLC) to the radiator inlet opening until coolant overflows from the engine air drain cock hole. Then install the air drain cock plug to the water inlet housing.
   f. Continue to add TOYOTA Super Long Life Coolant (SLLC) to the radiator inlet opening until coolant overflows from the radiator air drain hole. Then close the air drain plug at the top of the radiator.

➡If the coolant level at the radiator inlet opening drops after squeezing the No. 1 and No. 2 radiator hoses, add coolant.

   g. Slowly fill the radiator with TOYOTA Super Long Life Coolant (SLLC). Standard capacity: 12 quarts.

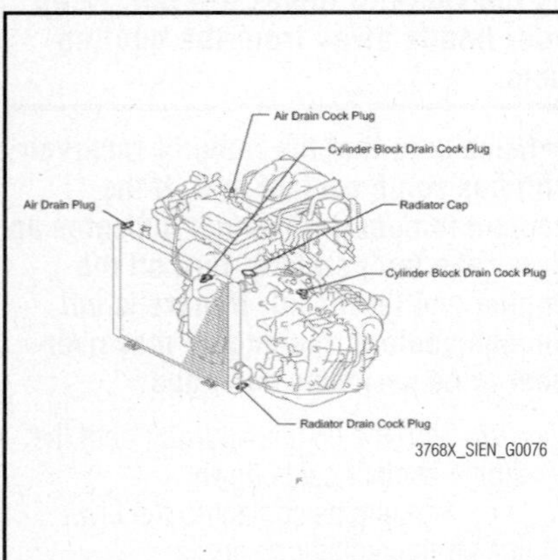

Fig. 25 Locating the coolant drainage plugs

➡TOYOTA vehicles are filled with TOYOTA SLLC at the factory. In order to avoid damage to the engine cooling system and other technical problems, only use TOYOTA SLLC or similar high quality ethylene glycol based non-silicate, non-amine, non-nitrite, non-borate coolant with long-life hybrid organic acid technology (coolant with long-life hybrid organic acid technology consists of a combination of low phosphates and organic acids). Contact your TOYOTA dealer for further details.

➡Never use water as a substitute for engine coolant.

   h. Slowly pour coolant into the radiator reservoir tank until it reaches the FULL line.
   i. Squeeze the No. 1 and No. 2 radiator hoses several times by hand, and then check the level of the coolant. If the coolant level is low, add coolant.
5. Bleed air from the cooling system.
   a. Warm up the engine until the thermostat opens. While the thermostat is open, circulate the coolant for several minutes.

➡The thermostat open timing can be confirmed by squeezing the inlet radiator hose by hand, and checking when the engine coolant starts to flow inside the hose.

   b. Maintain the engine speed at 2500 to 3000 rpm.
   c. Squeeze the inlet and outlet radiator hoses several times by hand to bleed air.

⁂ CAUTION

When squeezing the radiator hoses:

- Wear protective gloves.
- Be careful as the radiator hoses are hot.
- Keep your hands away from the radiator fan.

➡Please note the following:

- Make sure that the radiator reservoir still has some coolant in it.
- If the coolant temperature gauge indicates an excessive temperature, turn off the engine and let it cool.
- If there is not enough coolant, the engine may overheat or be seriously damaged.
- If the radiator reservoir does not have enough coolant, perform the following: 1) stop the engine, 2) wait until the coolant has cooled down, and 3) add coolant until the coolant is filled to the FULL line.

   d. Stop the engine and wait until the engine coolant cools down.
   e. Add engine coolant to the FULL line on the radiator reservoir.
6. Inspect for coolant leak.

BLEEDING

1. Bleed air from the cooling system:
   a. Warm up the engine until the thermostat opens. While the thermostat is open, circulate the coolant for several minutes.

➡The thermostat open timing can be confirmed by squeezing the inlet radiator hose by hand, and checking when the engine coolant starts to flow inside the hose.

   b. Maintain the engine speed at 2500 to 3000 rpm.
   c. Squeeze the inlet and outlet radiator hoses several times by hand to bleed air.

⁂ CAUTION

When squeezing the radiator hoses:

- Wear protective gloves.
- Be careful as the radiator hoses are hot.
- Keep your hands away from the radiator fan.

➡Please note the following:

- Make sure that the radiator reservoir still has some coolant in it.
- If the coolant temperature gauge indicates an excessive temperature, turn off the engine and let it cool.
- If there is not enough coolant, the engine may overheat or be seriously damaged.
- If the radiator reservoir does not have enough coolant, perform the following: 1) stop the engine, 2) wait until the coolant has cooled down, and 3) add coolant until the coolant is filled to the FULL line.

   d. Stop the engine and wait until the engine coolant cools down.
   e. Add engine coolant to the FULL line on the radiator reservoir.

ENGINE FAN

REMOVAL & INSTALLATION

2.7L Engine

See Figure 26.

1. Remove radiator assembly and fan assembly with motor.

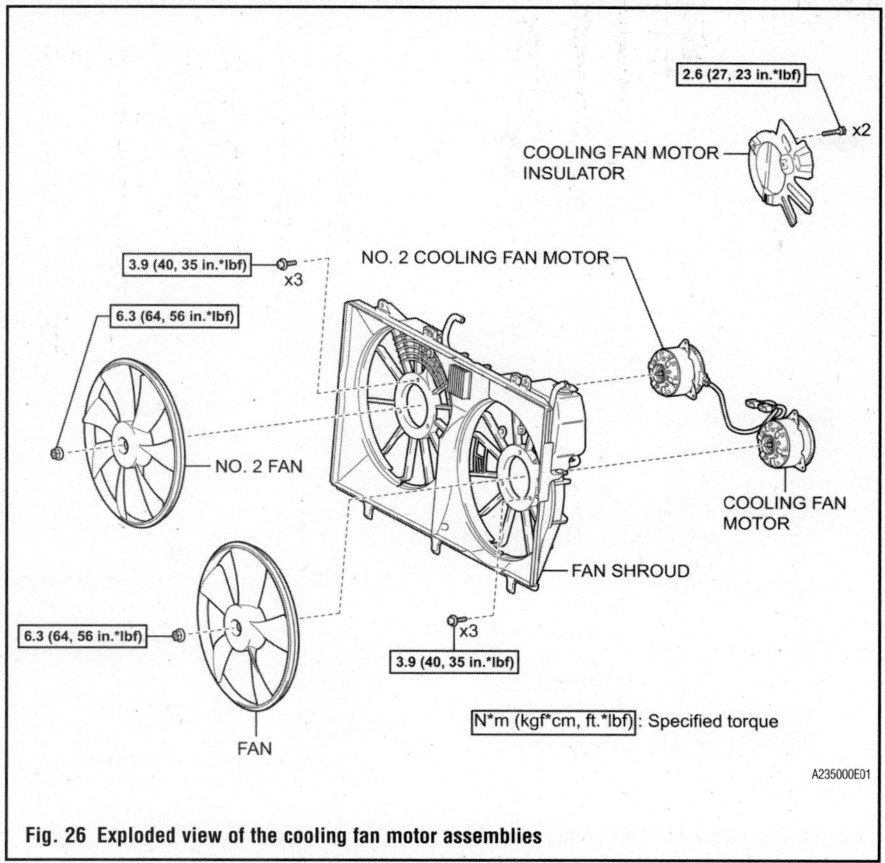

Fig. 26 Exploded view of the cooling fan motor assemblies

2. Remove fan. Remove the nut and fan.
3. Remove No. 2 fan. Remove the nut and No. 2 fan.
4. Remove No. 2 cooling fan motor.
    a. Remove the 2 bolts.
    b. Disengage the guide and remove the cooling fan motor insulator.
    c. Disconnect the No. 2 cooling fan motor connector and 2 clamps.
    d. Remove the 3 screws and No. 2 cooling fan motor.
5. Remove cooling fan motor.
    a. Disconnect the cooling fan motor connector and 2 clamps.
    b. Remove the 3 screws and cooling fan motor.

To install:

6. Install cooling fan motor.
    a. Install the cooling fan motor with the 3 screws.
    b. Connect the cooling fan motor connector and 2 clamps.
7. Install No. 2 cooling fan motor.
    a. Install the No. 2 cooling fan motor with the 3 screws.
    b. Connect the No. 2 cooling fan motor connector and 2 clamps.
8. Engage the guide and install the cooling fan motor insulator.
    a. Install the 2 bolts.
9. Install the No. 2 fan with the nut.
10. Install the fan with the nut.
11. Install radiator assembly and fan assembly with motor.

3.5L Engine

See Figure 27.

1. Remove radiator assembly and fan assembly with motor.
2. Remove fan. Remove the nut and fan.
3. Remove No. 2 fan. Remove the nut and No. 2 fan.
4. Remove No. 2 cooling fan motor.
    a. Disconnect the No. 2 cooling fan motor connector and 2 clamps.
    b. Remove the 3 screws and No. 2 cooling fan motor.
5. Remove cooling fan motor.
    a. Disconnect the cooling fan motor connector and 2 clamps.
    b. Remove the 3 screws and cooling fan motor.

To install:

6. Install cooling fan motor
    a. Install the cooling fan motor with the 3 screws.
    b. Connect the cooling fan motor connector and 2 clamps.
7. Install No. 2 cooling fan motor

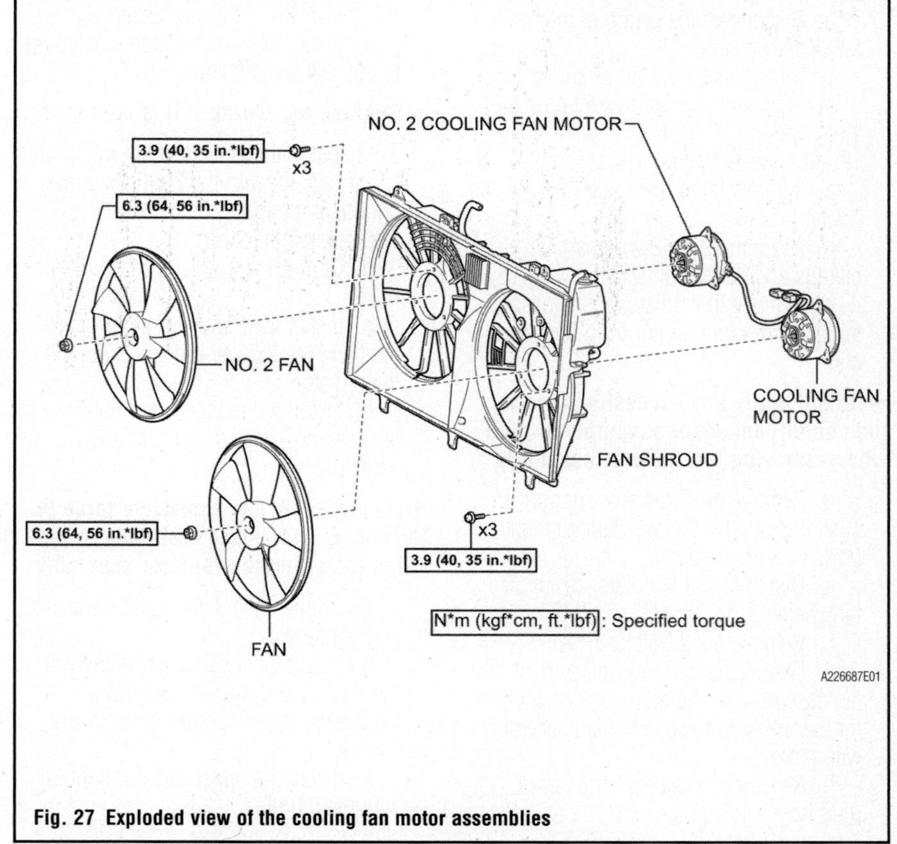

Fig. 27 Exploded view of the cooling fan motor assemblies

a. Install the No. 2 cooling fan motor with the 3 screws.

b. Connect the No. 2 cooling fan motor connector and 2 clamps.

8. Install the No. 2 fan with the nut.
9. Install the fan with the nut.
10. Install radiator assembly and fan assembly with motor

RADIATOR

REMOVAL & INSTALLATION

2.7L Engine

See Figure 28.

1. Remove No. 1 engine under cover.
2. Drain engine coolant.
3. Remove air cleaner inlet assembly.
4. Remove radiator grille sub-assembly.
5. Loosen the clamp and disconnect the No. 1 radiator hose from the radiator.
6. Loosen the clamp and disconnect the No. 2 radiator hose from the radiator.
7. Loosen the clamp and disconnect the inlet oil cooler hose and clamp from the radiator.
8. Loosen the clamp and disconnect the outlet oil cooler hose.
9. Remove hood lock nut cap.
10. Remove hood lock assembly.
11. Remove upper radiator support sub-assembly

a. Disconnect the clamp of the hood lock control cable.

b. Disconnect the 2 horn connectors.

c. Remove the 5 bolts and upper radiator support.

12. Remove No. 2 radiator assembly

a. Remove the 4 bolts and the 4 cooler brackets.

b. Disconnect the 3 wire harness clamps and connector.

c. Remove the radiator assembly and fan assembly with motor from the vehicle.

➡Do not apply any excessive force to the cooler condenser assembly or pipe when removing the radiator assembly.

d. Remove the 2 radiator support cushions and the 2 lower radiator supports from the radiator.

e. Disconnect the radiator reservoir tank hose.

f. Remove the 2 bolts.

g. Disengage the claw and pull up the fan assembly with motor from the radiator assembly to remove the fan assembly with motor.

h. Remove the radiator drain cock plug from the radiator.

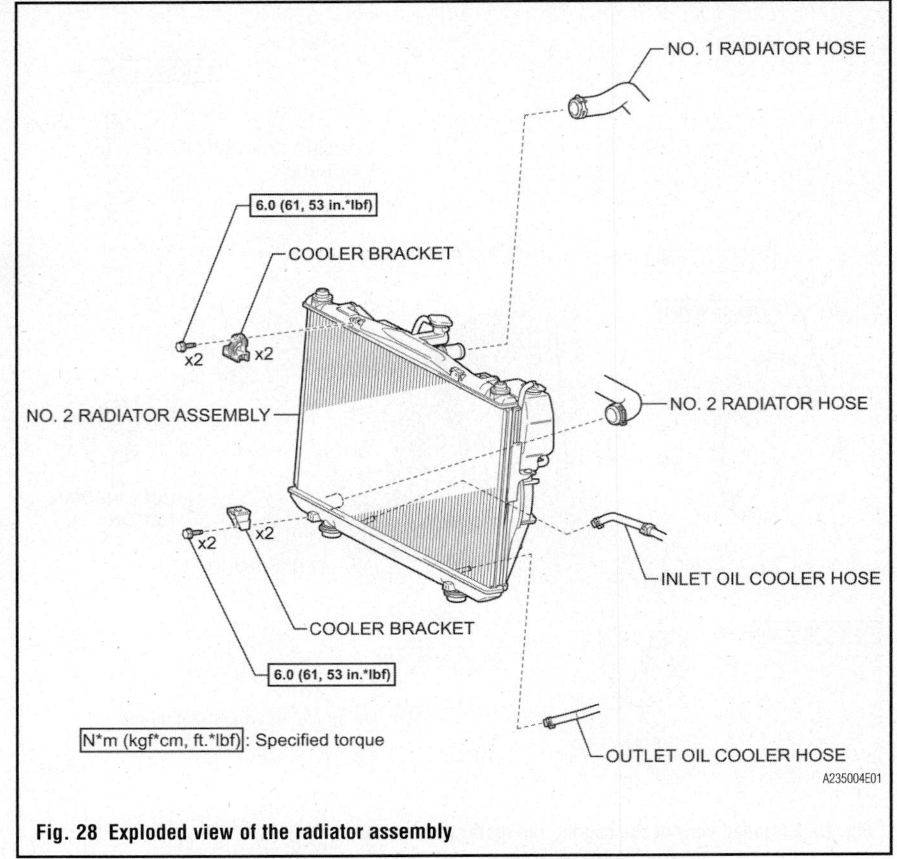

Fig. 28 Exploded view of the radiator assembly

To install:

13. Install No. 2 radiator assembly.

a. Install the radiator drain cock plug to the radiator assembly.

➡Replace the O-ring if it is damaged.

b. Install the fan assembly with motor to the radiator with the 3 guides at the bottom and engage the claw.

c. Install the 2 bolts.

d. Connect the radiator reservoir tank hose.

e. Install the 2 radiator support cushions and the 2 lower radiator supports to the radiator.

f. Install the radiator assembly and fan assembly with motor onto the vehicle.

➡Do not apply any excessive force to the cooler condenser assembly or pipe when installing the radiator assembly.

g. Connect the 3 wire harness clamps and connector.

h. Install the radiator assembly with the 4 bolts and 4 cooler brackets.

14. Install upper radiator support sub-assembly.

a. Install the upper radiator support with the 5 bolts.

b. Connect the 2 horn connectors.

c. Connect the clamp of the hood lock control cable.

15. Install hood lock assembly.
16. Inspect hood lock assembly.
17. Adjust hood lock assembly.
18. Install hood lock nut cap.
19. Connect the outlet oil cooler hose to the radiator with the hose clamp.

➡Perform the installation with the hose clamp and mark at the correct angle.

20. Connect the clamp and inlet oil cooler hose to the radiator with the hose clamp.

➡Perform the installation with the hose clamp and mark at the correct angle.

21. Connect the No. 2 radiator hose to the radiator with the hose clamp.

➡Perform the installation with the hose clamp and mark at the correct angle.

22. Connect the No. 1 radiator hose to the radiator with the hose clamp.

➡Perform the installation with the hose clamp and mark at the correct angle.

23. Install radiator grille sub-assembly.
24. Install air cleaner inlet assembly.
25. Add engine coolant.
26. Inspect for coolant leak.
27. Add automatic transaxle fluid.
28. Inspect for oil leak.
29. Install No. 1 engine under cover.
30. Warm up engine.

3.5L Engine

See Figure 29.

1. Remove No. 1 engine under cover.
2. Drain engine coolant.
3. Remove air cleaner inlet cover assembly seal.
4. Remove radiator grille sub-assembly.
5. Loosen the clamp and disconnect the No. 1 radiator hose from the radiator.
6. Loosen the clamp and disconnect the No. 2 radiator hose from the radiator.
7. Loosen the clamp and disconnect the inlet oil cooler hose and clamp from the radiator.
8. Loosen the clamp and disconnect the outlet oil cooler hose.
9. Remove millimeter wave radar sensor (w/ Dynamic Radar Cruise Control System).
10. Remove hood lock nut cap.
11. Remove hood lock assembly.
12. Remove upper radiator support sub-assembly.
    a. Disconnect the clamp of the hood lock control cable.
    b. Disconnect the 2 horn connectors.
    c. Remove the 5 bolts and upper radiator support.

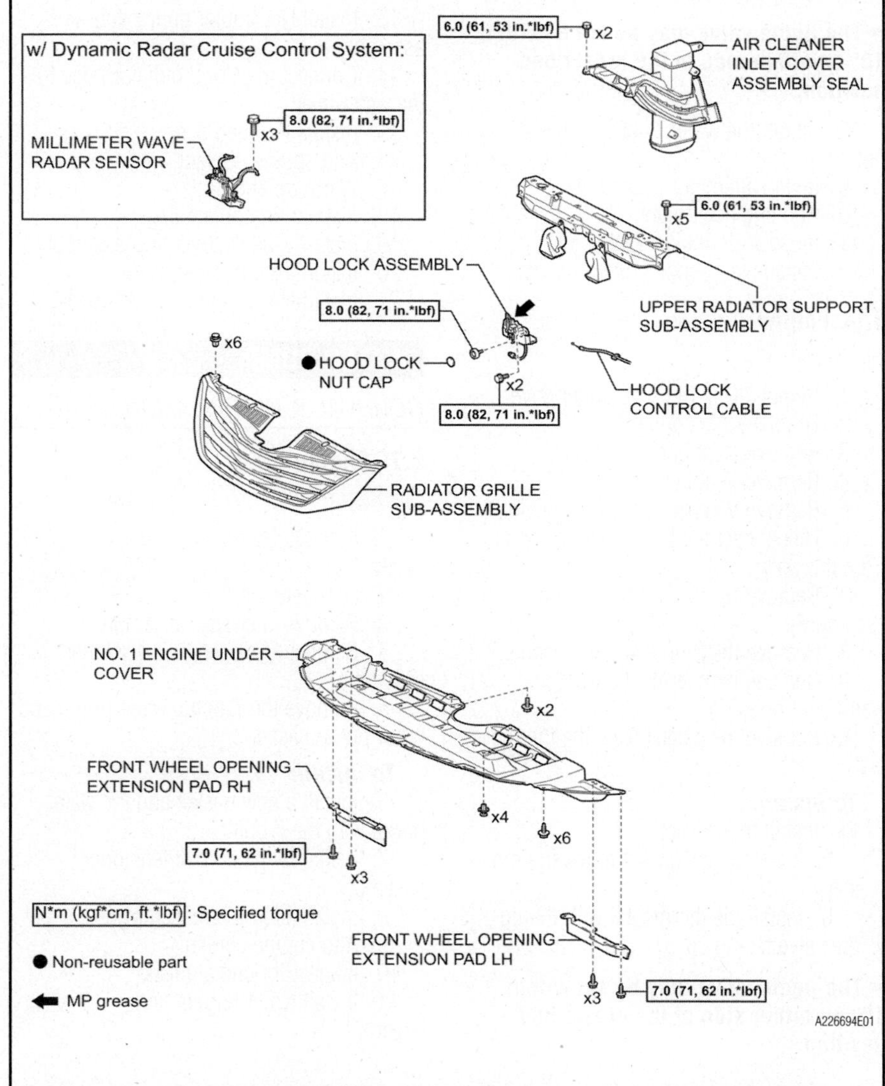

Fig. 29 Exploded view showing the air cleaner inlet cover, the radiator grille sub-assembly, and hood lock

13. Remove no. 2 radiator assembly.
    a. Remove the 4 bolts and the 4 cooler brackets.
    b. Disconnect the 3 wire harness clamps and connector.
    c. Remove the radiator assembly and fan assembly with motor from the vehicle.

➡Do not apply any excessive force to the cooler condenser assembly or pipe when removing the radiator assembly.

    d. Remove the 2 radiator support cushions and the 2 lower radiator supports from the radiator.
    e. Disconnect the radiator reservoir tank hose.
    f. Remove the 2 bolts.
    g. Disengage the claw and pull up the fan assembly with motor from the radiator assembly to remove the fan assembly with motor.
    h. Remove the radiator drain cock plug from the radiator.

To install:

14. Install No. 2 radiator assembly.
    a. Install the radiator drain cock plug to the radiator assembly.

➡Replace the O-ring if it is damaged.

    b. Install the fan assembly with motor to the radiator with the 3 guides at the bottom and engage the claw.
    c. Install the 2 bolts.
    d. Connect the radiator reservoir tank hose.
    e. Install the 2 radiator support cushions and the 2 lower radiator supports to the radiator.
    f. Install the radiator assembly and fan assembly with motor onto the vehicle.

➡Do not apply any excessive force to the cooler condenser assembly or pipe when installing the radiator assembly.

    g. Connect the 3 wire harness clamps and connector.
    h. Install the radiator assembly with the 4 bolts and 4 cooler brackets.
15. Install upper radiator support sub-assembly.
    a. Install the upper radiator support with the 5 bolts.
    b. Connect the 2 horn connectors.
    c. Connect the clamp of the hood lock control cable.
16. Install hood lock assembly.
17. Inspect hood lock assembly.
18. Adjust hood lock assembly.
19. Install hood lock nut cap.
20. Install millimeter wave radar sensor (w/ Dynamic Radar Cruise Control System).

21. Adjust millimeter wave radar sensor (w/ Dynamic Radar Cruise Control System).

22. Connect the outlet oil cooler hose to the radiator with the hose clamp.

➡**Perform the installation with the hose clamp and mark at the correct angle.**

23. Connect the clamp and inlet oil cooler hose to the radiator with the hose clamp.

➡**Perform the installation with the hose clamp and mark at the correct angle.**

24. Connect the No. 2 radiator hose to the radiator with the hose clamp.

➡**Perform the installation with the hose clamp and mark at the correct angle.**

25. Connect the No. 1 radiator hose to the radiator with the hose clamp.

➡**Perform the installation with the hose clamp and mark at the correct angle.**

26. Install radiator grille sub-assembly.
27. Install air cleaner inlet cover assembly seal.
28. Add engine coolant.
29. Inspect for coolant leak.
30. Add automatic transaxle fluid (for FWD).
31. Add automatic transaxle fluid (for AWD).
32. Inspect for oil leak.
33. Install No. 1 engine under cover.
34. Warm up engine.

THERMOSTAT

REMOVAL & INSTALLATION

2.7L Engine

See Figure 30.

1. Remove No. 1 engine under cover.
2. Drain engine coolant.
3. Remove alternator assembly.
4. Remove the 2 nuts and separate the water inlet.
5. Remove the thermostat.
6. Remove the gasket from the thermostat.

To install:

7. Install thermostat.
   a. Install a new gasket to the thermostat.
   b. Install the thermostat with the jiggle valve facing upward.

➡**The jiggle valve may be set to within 10° on either side of the prescribed position.**

8. Install the water inlet with the 2 nuts.
9. Install alternator assembly.
10. Add engine coolant.
11. Inspect for coolant leak.
12. Install No. 1 engine under cover.

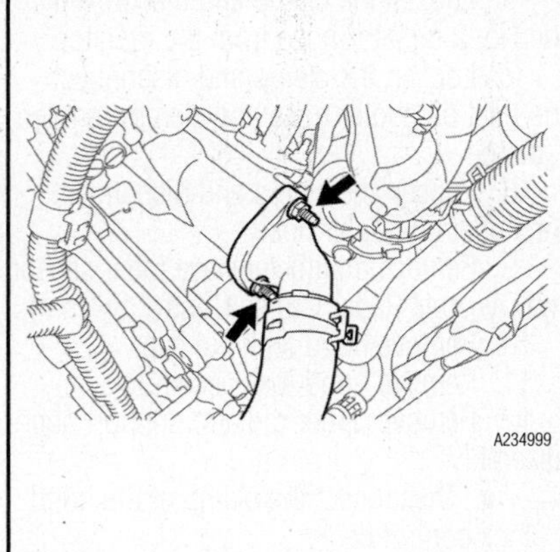

Fig. 30 Remove the 2 nuts and separate the water inlet

3.5L Engine

See Figure 31.

1. Remove No. 1 engine under cover.
2. Drain engine coolant.
3. Remove front bumper cover.
4. Remove V-ribbed belt.
5. Remove V-bank cover sub-assembly.
6. Disconnect the No. 2 radiator hose from the water inlet.
7. Remove No. 2 idler pulley sub-assembly
8. Remove the 2 nuts and water inlet.
9. Remove the thermostat from the water inlet housing.
10. Remove the gasket from the thermostat.

To install:

11. Install thermostat
    a. Install a new gasket to the thermostat.
    b. Install the thermostat with the jiggle valve facing up.

➡**The jiggle valve may be set to within 10° on either side of the prescribed position.**

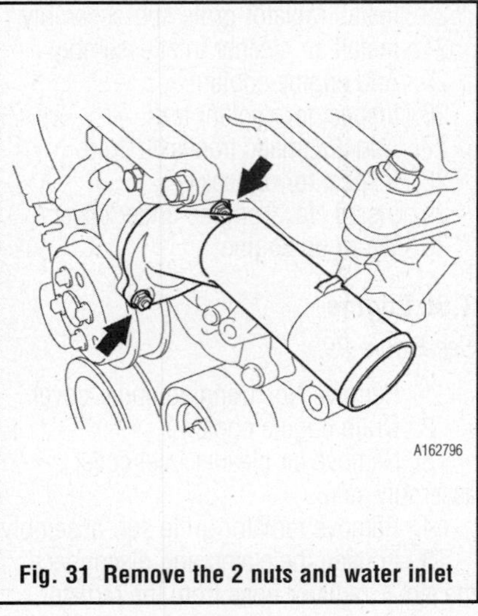

Fig. 31 Remove the 2 nuts and water inlet

12. Install the water inlet with the 2 nuts.
13. Install No. 2 idler pulley sub-assembly
14. Connect the No. 2 radiator hose to the water inlet.
15. Install V-ribbed belt
16. Add engine coolant
17. Warm up engine
18. Inspect for coolant leak
19. Install V-bank cover sub-assembly
20. Install front bumper cover
21. Install No. 1 engine under cover

WATER PUMP

REMOVAL & INSTALLATION

2.7L Engine

See Figure 32.

1. Remove No. 1 engine under cover
2. Drain engine coolant
3. Remove alternator assembly
4. Remove V-ribbed belt tensioner assembly
5. Remove the 7 bolts, water pump and water pump gasket.

To install:

6. Install a new gasket and the water pump with the 7 bolts.
7. Install V-ribbed belt tensioner assembly
8. Install alternator assembly
9. Add engine coolant
10. Inspect for coolant leak
11. Install No. 1 engine under cover

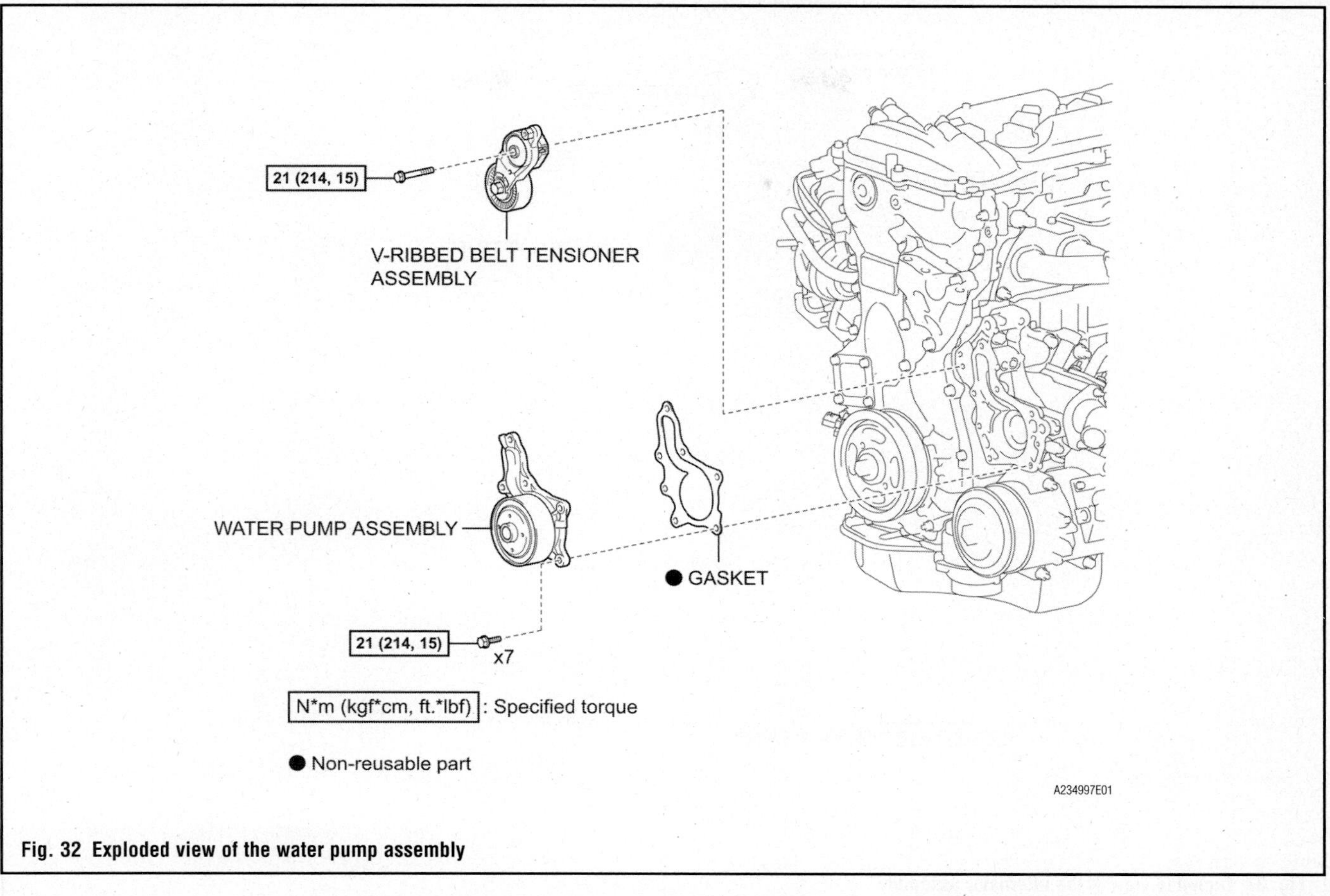

Fig. 32 Exploded view of the water pump assembly

ENGINE ELECTRICAL — CHARGING SYSTEM

ALTERNATOR

REMOVAL & INSTALLATION

2.7L Engine

See Figure 33.

1. Disconnect cable from negative battery terminal.

➡When disconnecting the cable, some systems need to be initialized after the cable is reconnected

2. Remove fan and alternator V belt.
3. Remove alternator assembly.
   a. Turn back the terminal cap.
   b. Remove the nut and disconnect the wire harness from terminal B.
   c. Disconnect the alternator connector from the alternator assembly.
   d. Disengage the 2 wire harness clamps.
   e. Remove the 2 bolts and the alternator assembly.
   f. Remove the bolt and the wire harness clamp bracket.

To install:

4. Install alternator assembly
   a. Install the wire harness clamp bracket with the bolt.
   b. Install the 2 bolts.
   c. Connect the alternator connector to the alternator assembly.
   d. Install the alternator wire with the nut.
   e. Install the terminal cap.
   f. Connect the 2 wire harness clamps.
5. Install fan and alternator V belt
6. Connect cable to negative battery terminal

➡When disconnecting the cable, some systems need to be initialized after the cable is reconnected

3.5L Engine

See Figure 34.

1. Disconnect cable from negative battery terminal.

➡When disconnecting the cable, some systems need to be initialized after the cable is reconnected.

2. Remove front bumper assembly.
3. Remove radiator assembly with fan shroud and fan motor.
4. Remove fan and alternator V belt.
5. Remove alternator assembly.
   a. Turn back the terminal cap.
   b. Remove the nut and disconnect the wire harness from terminal B.
   c. Disconnect the alternator connector from the alternator assembly.
   d. Disconnect the magnetic clutch connector from the compressor and magnetic clutch.
   e. Disengage the 3 wire harness clamps.
   f. Remove the 2 bolts.
   g. Disconnect the wire harness clamp and remove the alternator bracket.
   h. Remove the 2 bolts and the alternator assembly.
   i. Remove the bolt and the wire harness clamp stay.

To install:

6. Install alternator assembly.

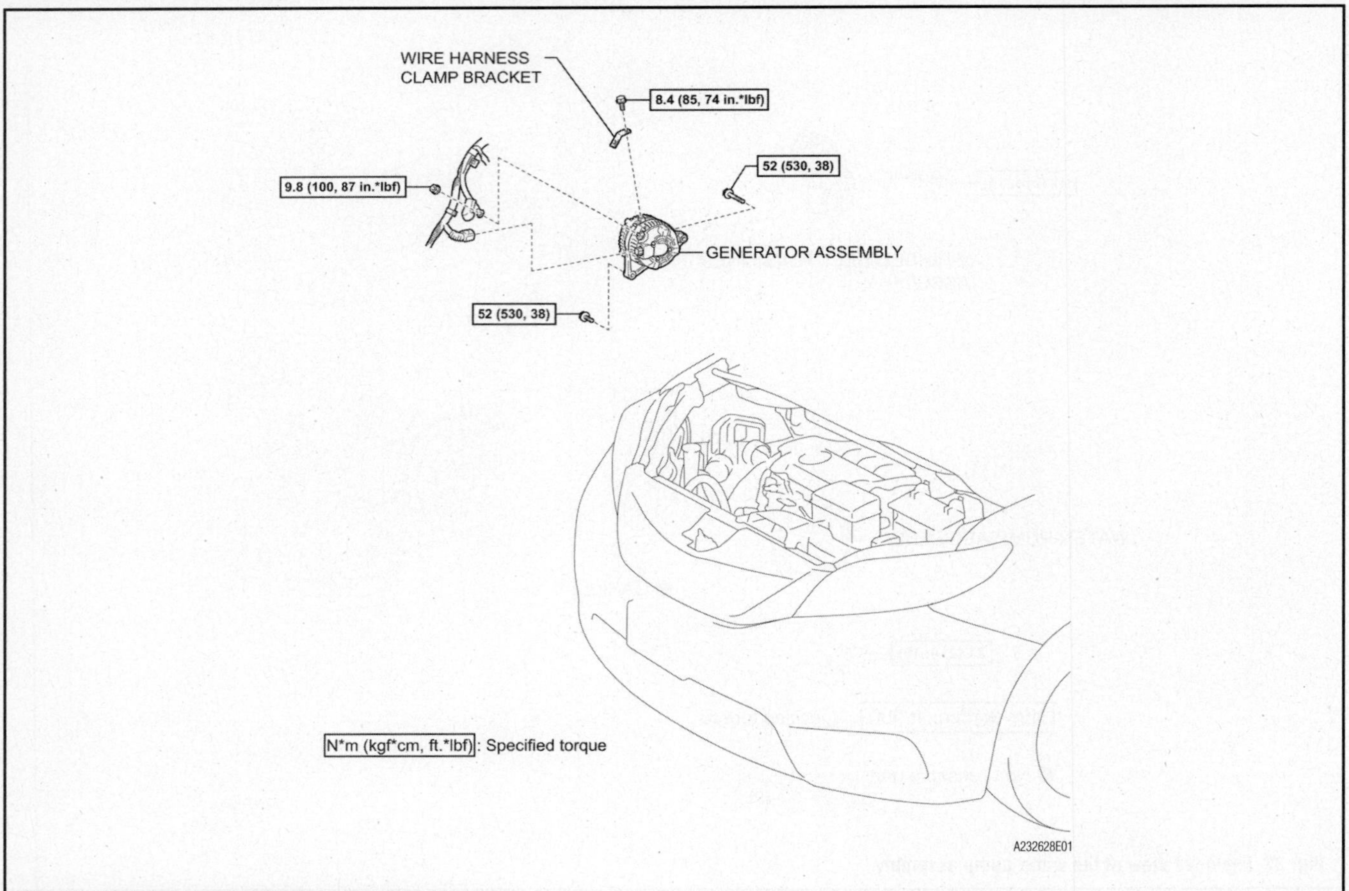

Fig. 33 Exploded view of the alternator assembly

a. Install the wire harness clamp stay with the bolt.

b. Connect the wire harness clamp.

c. Install the alternator assembly to the cylinder block sub-assembly with the alternator bracket, bolt and nut.

d. Install the 2 bolts. Tighten to 32 ft. lbs. (43 Nm).

e. Connect the alternator connector to the alternator assembly.

f. Install the alternator wire with the nut.

g. Install the terminal cap.

h. Connect the 3 wire harness clamps.

i. Connect the magnetic clutch connector to the compressor and magnetic clutch.

7. Install fan and alternator V belt.
8. Install radiator assembly with fan shroud and fan motor.
9. Install front bumper assembly.
10. Connect cable to negative battery terminal.

➡When disconnecting the cable, some systems need to be initialized after the cable is reconnected

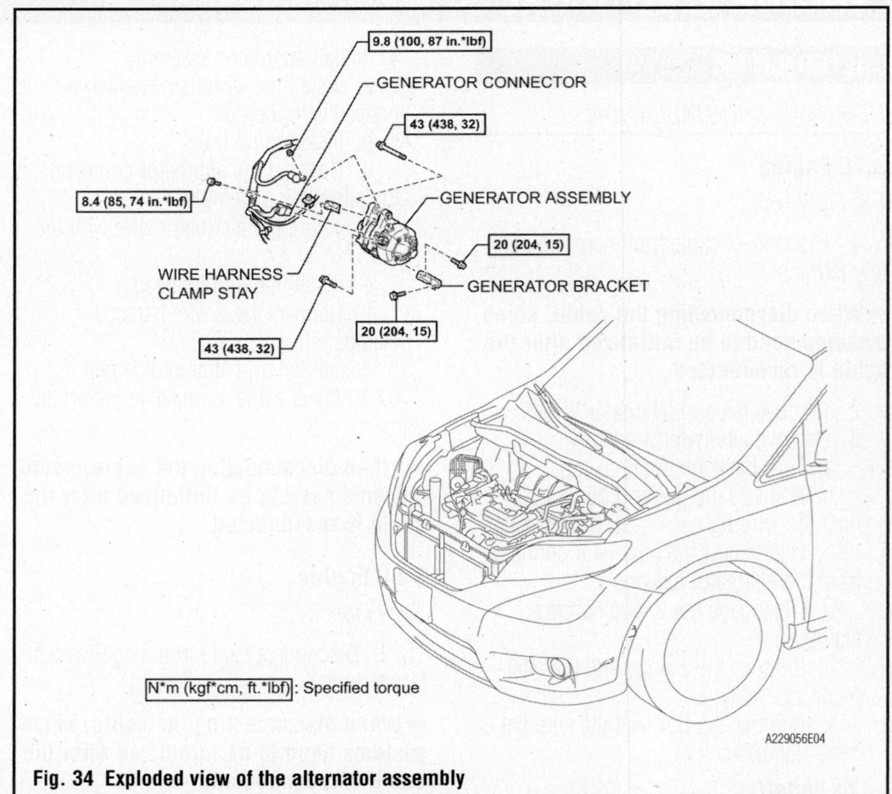

Fig. 34 Exploded view of the alternator assembly

IGNITION COIL

REMOVAL & INSTALLATION

2.7L Engine

See Figure 35.

1. Remove No. 1 engine cover sub-assembly.
2. Remove ignition coil assembly.
   a. Disconnect the 4 ignition coil connectors.
   b. Remove the 4 bolts and the 4 ignition coils.
3. Using a 16 mm spark plug wrench, remove the 4 spark plugs.

To install:

4. Using a 16 mm spark plug wrench, install the 4 spark plugs. Tighten to 18 ft. lbs. (25 Nm).
5. Install ignition coil assembly.
   a. Install the 4 ignition coils with the 4 bolts.
   b. Connect the 4 ignition coil connectors.
6. Install No. 1 engine cover sub-assembly

3.5L Engine

See Figure 36.

1. Remove intake air surge tank assembly.

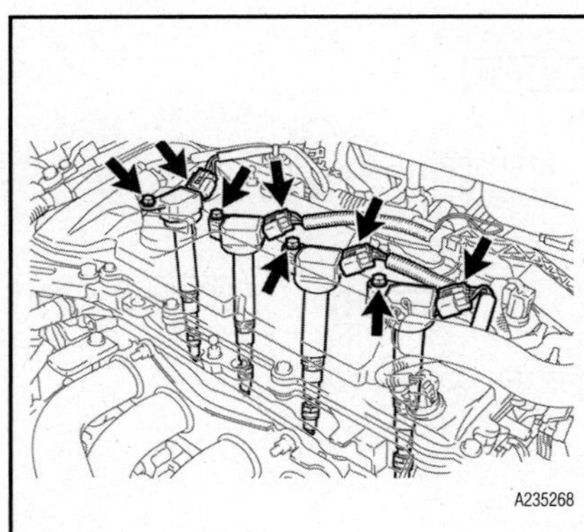

Fig. 35 Disconnect the 4 ignition coil connectors

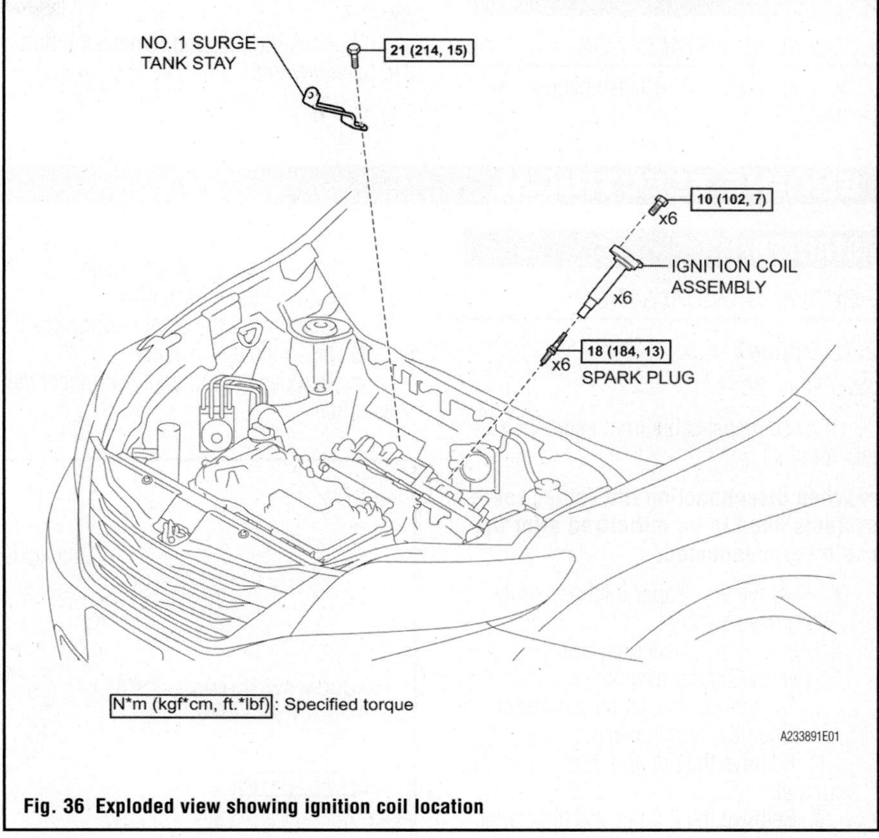

Fig. 36 Exploded view showing ignition coil location

2. Remove No. 1 surge tank stay.
3. Remove ignition coil assembly.
   a. Disconnect the 3 wire harness clamps.
   b. Disconnect the 6 ignition coil connectors.
   c. Remove the 6 bolts and 6 ignition coils.
4. Using a 16 mm spark plug wrench, remove the 6 spark plugs.

To install:

5. Using a 16 mm spark plug wrench, install the 6 spark plugs. Tighten to 13 ft. lbs. (18 Nm).
6. Install ignition coil assembly.
   a. Install the 6 ignition coils with the 6 bolts.
   b. Connect the 6 ignition coil connectors.
   c. Connect the 3 wire harness clamps.
7. Install No. 1 surge tank stay.
8. Install intake air surge tank assembly.
9. Add engine coolant.
10. Inspect for coolant leak.

IGNITION TIMING

ADJUSTMENT

The ignition timing is controlled by the Powertrain Control Module (PCM). No adjustment is necessary or possible.

SPARK PLUGS

REMOVAL & INSTALLATION

Refer to the Ignition Coil Removal and Installation.

ENGINE ELECTRICAL — BATTERY SYSTEM

BATTERY

REMOVAL & INSTALLATION

1. Disconnect the negative battery terminal.
2. Disconnect the positive battery terminal.
3. Loosen the nut, and remove the bolt and battery clamp.
4. Remove the battery and battery tray.
5. To install, reverse the removal procedure.

ENGINE ELECTRICAL — STARTING SYSTEM

STARTER

REMOVAL & INSTALLATION

2.7L Engine

See Figure 37.

1. Disconnect cable from negative battery terminal

➡When disconnecting the cable, some systems need to be initialized after the cable is reconnected.

2. Remove air cleaner inlet assembly
3. Remove battery
4. Remove air cleaner assembly
5. Remove starter assembly
   a. Disconnect the starter connector.
   b. Open the terminal cap.
   c. Remove the nut and disconnect the terminal.
   d. Remove the 2 bolts and the starter assembly.

To install:

6. Install starter assembly.
   a. Install the starter assembly with the 2 bolts. Tighten to 27 ft. lbs. (37 Nm).
   b. Connect the starter connector.
   c. Connect the terminal with the nut.
   d. Close the terminal cap.
7. Install air cleaner assembly.
8. Install battery.
9. Install air cleaner inlet assembly.
10. Connect cable to negative battery terminal.

➡When disconnecting the cable, some systems need to be initialized after the cable is reconnected

3.5L Engine

See Figure 38.

1. Disconnect cable from negative battery terminal.

➡When disconnecting the cable, some systems need to be initialized after the cable is reconnected.

2. Remove air cleaner inlet cover assembly seal.
3. Remove battery.
4. Remove air cleaner and hose.
5. Remove starter assembly.
   a. Disconnect the starter connector.
   b. Open the terminal cap.
   c. Remove the nut and disconnect the terminal.
   d. Remove the 2 bolts and the starter assembly.

To install:

6. Install the starter assembly with the 2 bolts. Tighten to 27 ft. lbs. (37 Nm).
   a. Connect the starter connector.

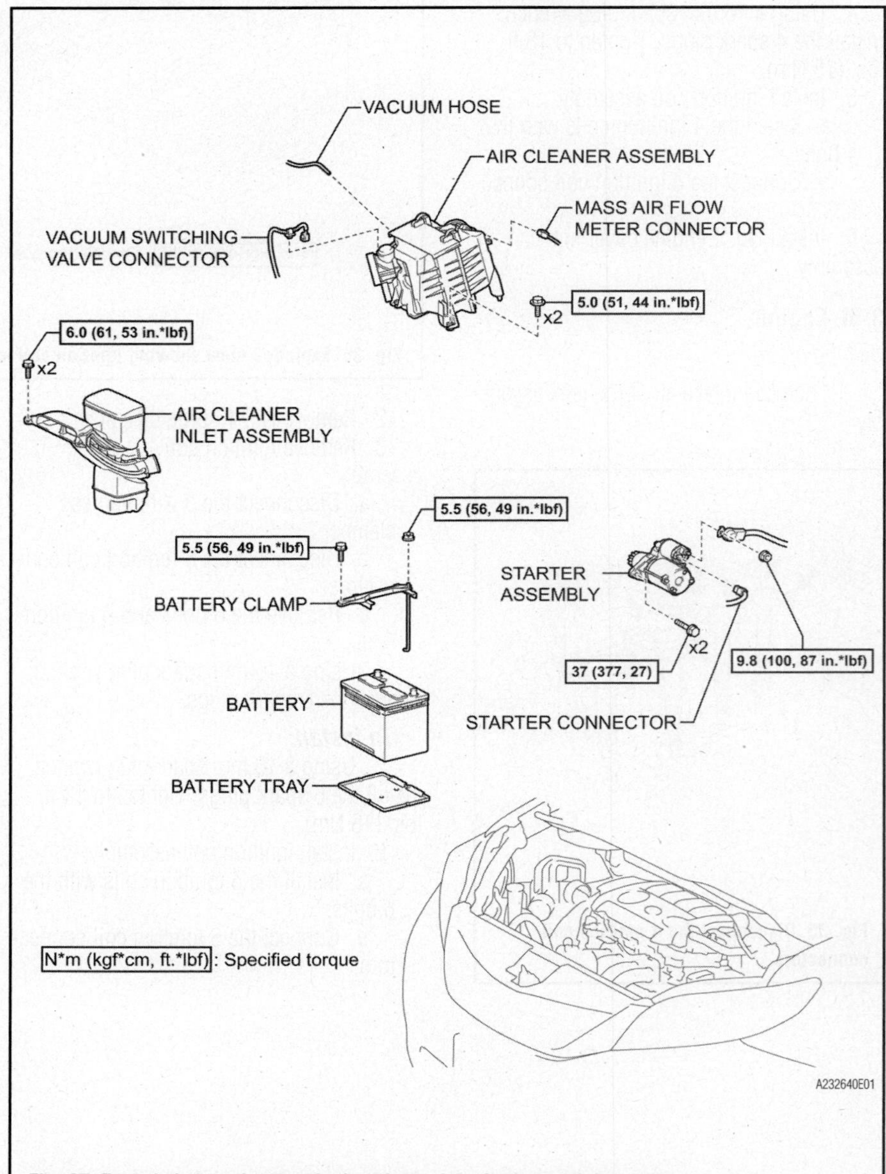

Fig. 37 Exploded view showing the locations of the battery and the starter

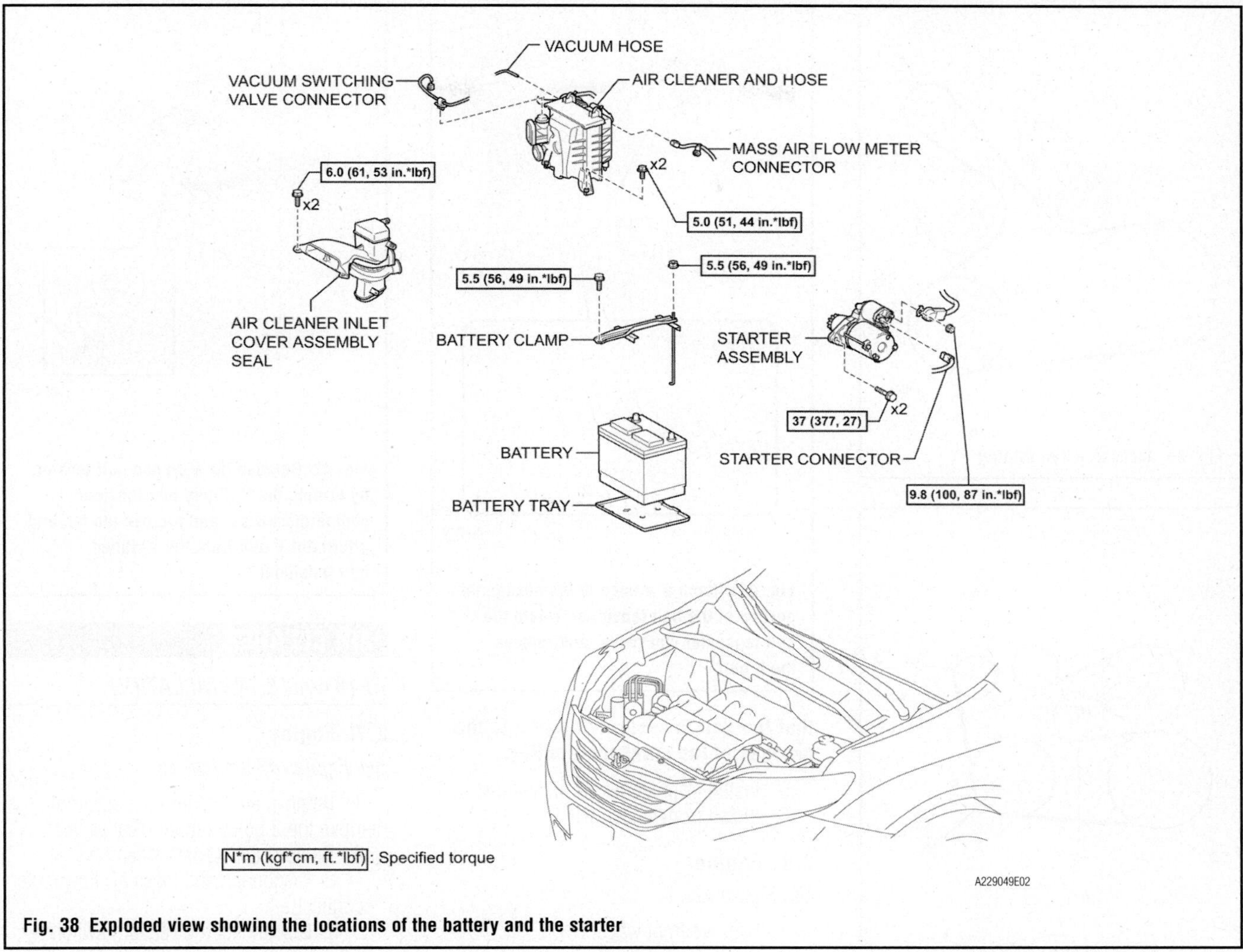

Fig. 38 Exploded view showing the locations of the battery and the starter

b. Connect the terminal with the nut.
c. Close the terminal cap.

7. Install air cleaner and hose.
8. Install battery.
9. Install air cleaner inlet cover assembly seal.
10. Connect cable to negative battery terminal.

➡When disconnecting the cable, some systems need to be initialized after the cable is reconnected.

ENGINE MECHANICAL

➡Disconnecting the negative battery cable may interfere with the functions of the on board computer systems and may require the computer to undergo a relearning process, once the negative battery cable is reconnected.

ACCESSORY DRIVE BELTS

ACCESSORY BELT ROUTING

2.7L Engine

See Figure 39.

3.5L Engine

See Figure 40.

INSPECTION

1. Inspect fan and alternator V belt
   a. Check the belt for wear, cracks or other signs of damage.
   b. If any of the following defects is found, replace the V belt.
      - The belt is cracked.
      - The belt is worn out to the extent that the cords are exposed.
      - The belt has chunks missing from the ribs.
   c. Check that the belt fits properly in the ribbed grooves.

➡Check with your hand to confirm that the belt has not slipped out of the grooves on the bottom of the pulley. If it has slipped out, replace the V belt. Install a new V belt correctly.

2. Inspect v-ribbed belt tensioner assembly
   a. Check that nothing gets caught in the tensioner by turning it clockwise and counterclockwise. If a malfunction exists, replace the V-ribbed belt tensioner.

ADJUSTMENT

Tension for the serpentine accessory drive belt is maintained by the belt tensioner. No adjustment is necessary.

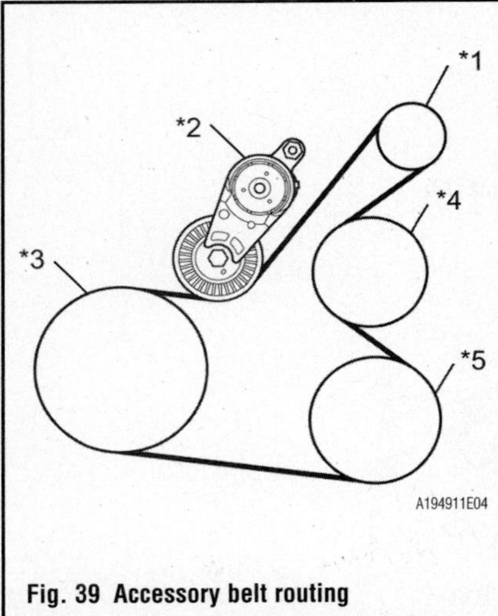

Fig. 39 Accessory belt routing

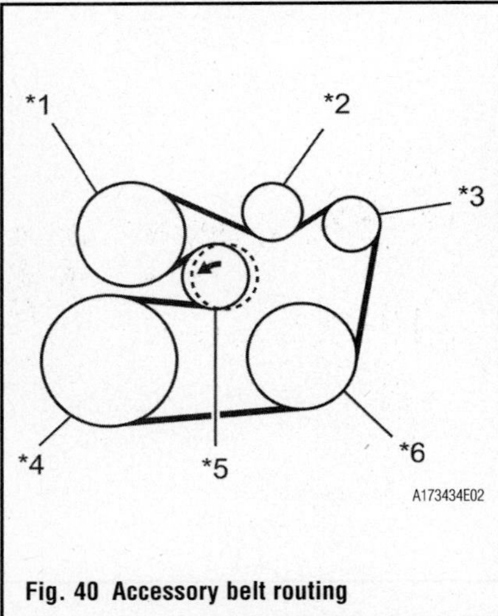

Fig. 40 Accessory belt routing

REMOVAL & INSTALLATION

2.7L Engine

See Figure 41.

1. Remove front wheel RH
2. Remove front fender apron seal RH
3. Remove fan and alternator V belt
   a. Attach a wrench to the hexagonal portion of the belt tensioner, rotate the belt tensioner clockwise, and remove the V belt.

To install:

4. Install fan and alternator v belt.
   a. Set the V belt onto each part except the water pump pulley.
   b. Loosen the V belt by turning the belt tensioner clockwise.
   c. Set the V belt onto the water pump pulley.

➡Make sure that the belt is attached to each pulley. In particular, make sure that the belt is securely fitted into the grooves of the crankshaft pulley.

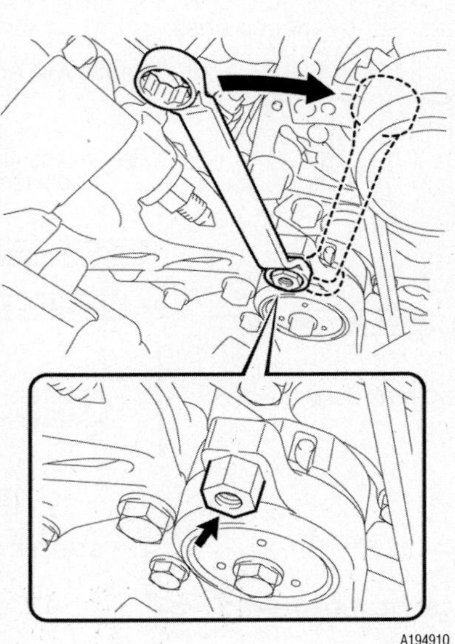

Fig. 41 Attach a wrench to the hexagonal portion of the belt tensioner, rotate the belt tensioner clockwise, and remove the V belt

5. Install front fender apron seal RH.
6. Install front wheel RH.

3.5L Engine

See Figure 42.

1. Remove front wheel RH.
2. Remove front fender apron seal RH.
3. Remove fan and alternator V belt.
   a. Release the V-ribbed belt tension by turning the V-ribbed belt tensioner counterclockwise, and remove the fan and alternator V belt from the V-ribbed belt tensioner.
   b. While turning the V-ribbed belt tensioner counterclockwise, align with its holes, and then insert a 5 mm hexagon wrench into the holes to fix the V-ribbed belt tensioner.

To install:

4. Install fan and alternator V belt.
   a. Install the fan and alternator V belt.
   b. Turn the V-ribbed belt tensioner counterclockwise and remove the 5 mm hexagon wrench.
   c. After installing the fan and alternator V belt, check that it fits properly in the ribbed grooves. Confirm that the belt has not slipped out of the grooves on the bottom of the crank pulley by hand.
5. Install front fender apron seal RH.
6. Install front wheel RH.

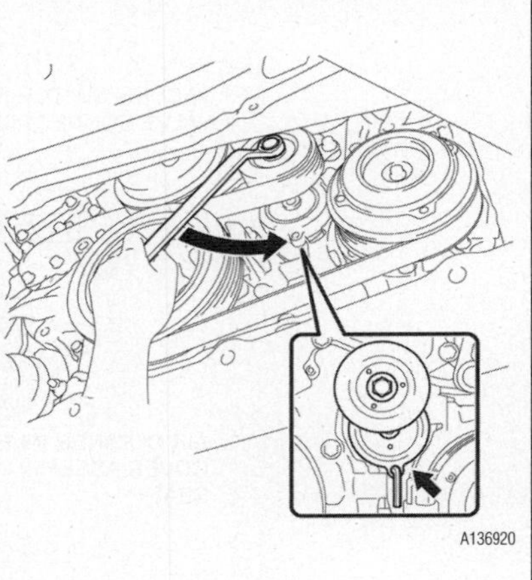

Fig. 42 Release the V-ribbed belt tension by turning the V-ribbed belt tensioner counterclockwise, and remove the fan and alternator V belt from the V-ribbed belt tensioner

AIR CLEANER

REMOVAL & INSTALLATION

2.7L Engine

See Figures 43 through 45.

1. Remove air cleaner inlet assembly. Remove the 2 bolts and air cleaner inlet.
2. Remove air cleaner assembly.
   a. Disconnect the mass air flow meter connector.
   b. Disconnect the vacuum switching valve connector and disengage the wire harness clamp.
   c. Disconnect the vacuum hose and disengage the hose clamp.
   d. Loosen the hose clamp.
   e. Remove the 2 bolts and air cleaner.

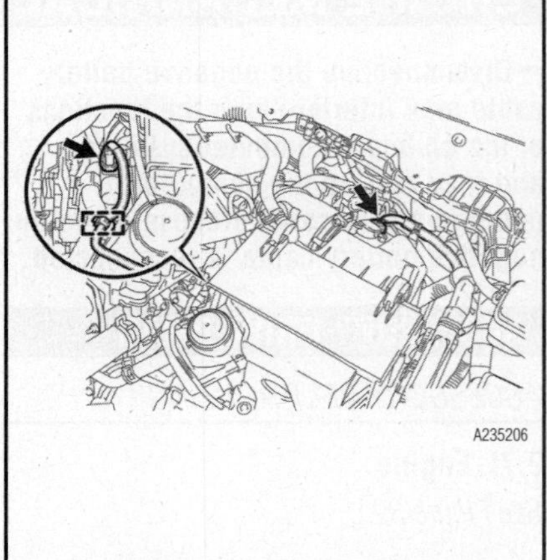

Fig. 43 Disconnect the mass air flow meter connector

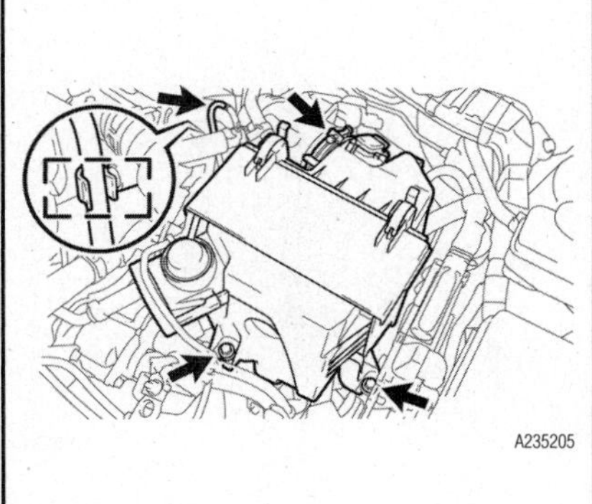

Fig. 44 Disconnect the vacuum hose and disengage the hose clamp

Fig. 46 Disconnect the vacuum switching valve connector

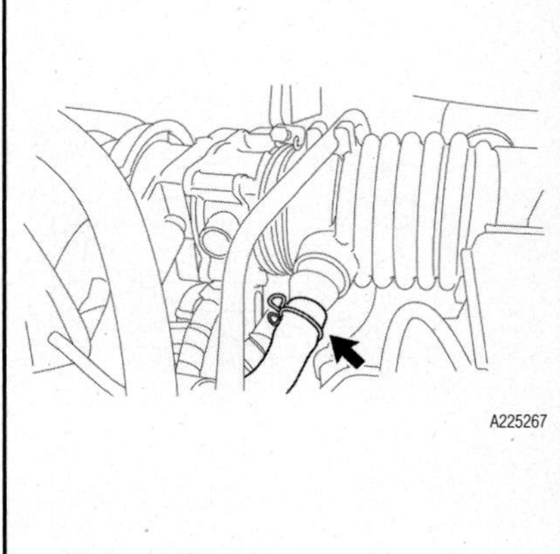

Fig. 47 Loosen the clip and disconnect the No. 2 ventilation hose

3. Remove air cleaner hose.
4. Installation is the reverse order of removal.

3.5L Engine

See Figures 46 through 48.

1. Remove air cleaner inlet cover assembly seal.
   a. Disconnect the 3 vacuum hose clamps from the air cleaner inlet cover assembly seal.
   b. Remove the 2 bolts and air cleaner inlet cover assembly seal.
2. Remove air cleaner and hose.
   a. Disconnect the mass air flow meter connector.
   b. Disengage the wire harness clamp.
   c. Disconnect the vacuum hose.
   d. Separate the vacuum hose from the hose clamp.
   e. Disconnect the vacuum switching valve connector.
   f. Disengage the 2 wire harness clamps.
   g. Disconnect the 2 vacuum hoses.
   h. Loosen the clip and disconnect the No. 2 ventilation hose.
   i. Separate the fuel vapor feed hose from the hose clamp.
   j. Loosen the hose clamp and disconnect the air cleaner hose.
   k. Remove the 2 bolts and the air cleaner assembly.
3. Installation is the reverse order of removal.

CAMSHAFT AND VALVE LIFTERS

REMOVAL & INSTALLATION

2.7L Engine

See Figures 49 through 54.

1. Disconnect cable from negative battery terminal.

➡When disconnecting the cable, some systems need to be initialized after the cable is reconnected.

2. Remove front wheel RH.
3. Remove front fender apron seal RH.
4. Remove front wiper arm head cap.
5. Remove front wiper arm and blade assembly LH.
6. Remove front wiper arm and blade assembly RH.
7. Remove front fender to cowl side seal LH.
8. Remove front fender to cowl side seal RH.
9. Remove cowl top ventilator louver sub-assembly.
10. Remove windshield wiper motor and link assembly.
11. Remove cowl panel sub-assembly.
12. Remove No. 1 engine cover sub-assembly.
13. Remove air cleaner inlet assembly.
14. Remove air cleaner assembly.
15. Remove No. 2 engine mounting stay RH.
16. Remove engine moving control rod sub-assembly.

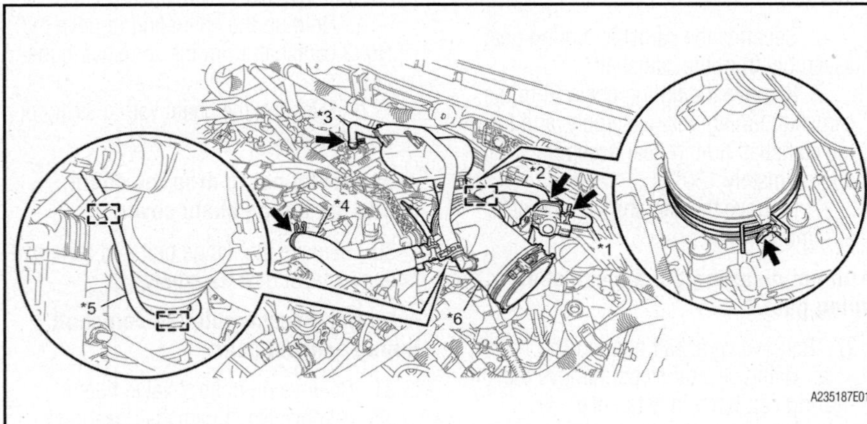

Fig. 45 Remove air cleaner hose

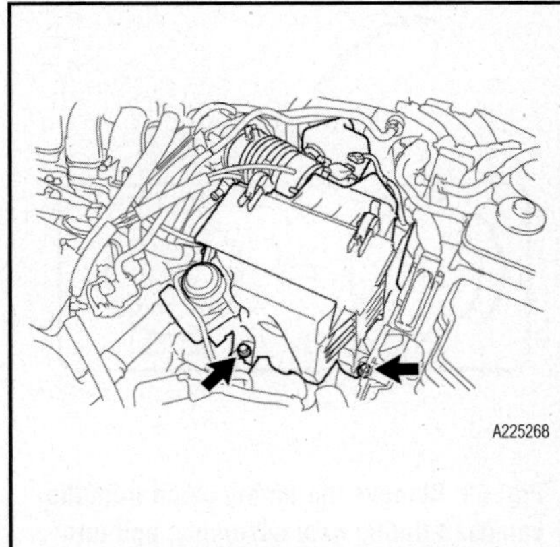

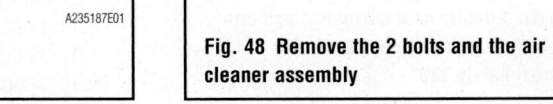

Fig. 48 Remove the 2 bolts and the air cleaner assembly

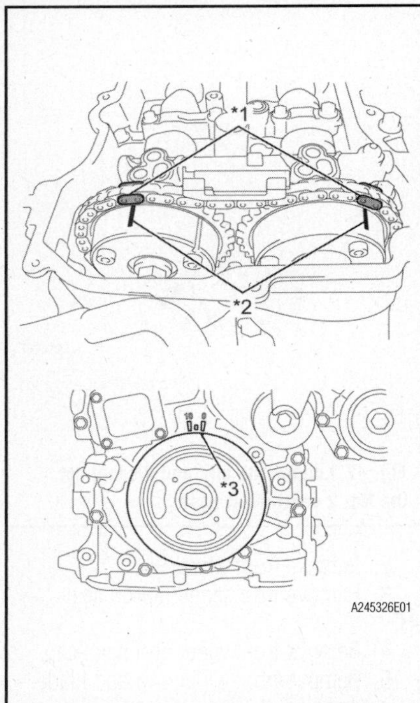

Fig. 49 Align paint marks (1), matchmarks (2), and timing notch (3)

17. Disconnect the ventilation hose from the cylinder head cover.

18. Disconnect the connectors and clamps, remove the bolts and nuts and disconnect the engine wire from the engine.

19. Remove ignition coil assembly.

20. Remove cylinder head cover sub-assembly.

21. Set No. 1 cylinder to TDC/compression.

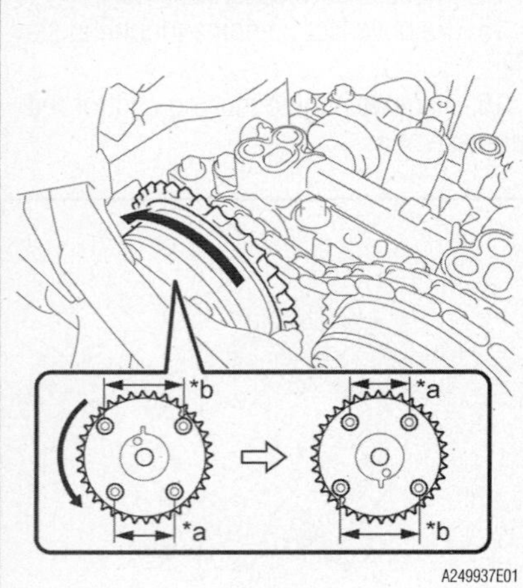

Fig. 50 Remove the timing chain from the camshaft timing gear assembly, and turn the camshaft timing gear assembly approximately 180°

a. Turn the crankshaft pulley until its timing notch (groove) and the timing mark "0" of the timing chain cover are aligned.

b. Check that each matchmark of the camshaft timing gear and camshaft timing exhaust gear are aligned with each matchmark. If not, turn the crankshaft 1 revolution (360°) to align the timing marks.

c. Place paint marks on the chain in alignment with the timing marks on the camshaft timing gear and camshaft timing exhaust gear.

22. Remove the 4 bolts, timing chain cover plate and gasket.

23. Remove No. 1 chain tensioner assembly.

a. Turn the crankshaft approximately 10° clockwise.

b. Turn the crankshaft approximately 10° counterclockwise.

c. Align the holes of the stopper plate and tensioner, and insert a pin into the stopper plate hole to lock the tensioner.

d. Turn the crankshaft approximately 10° clockwise.

e. Remove the 2 bolts, chain tensioner and gasket.

➡Make sure not to drop the gasket inside the timing chain cover.

f. Turn the crankshaft approximately 10° counterclockwise.

24. Remove timing chain guide.

25. Remove timing chain cover tight plug.

26. Remove camshaft timing gear assembly.

a. Hold the hexagonal portion of the camshaft with a wrench and remove the bolt from the camshaft.

➡Be careful not to damage the cylinder head or spark plug tube with the wrench.

b. Separate the camshaft timing gear assembly from the camshaft.

c. Remove the timing chain from the camshaft timing gear assembly, and turn the camshaft timing gear assembly approximately 180°.

d. Remove the camshaft timing gear assembly.

➡Do not disassemble the camshaft timing gear.

27. Remove camshaft bearing cap.

a. Using several steps, remove the 11 bearing cap bolts in sequence.

b. Using several steps, remove the 10 bearing cap bolts in sequence.

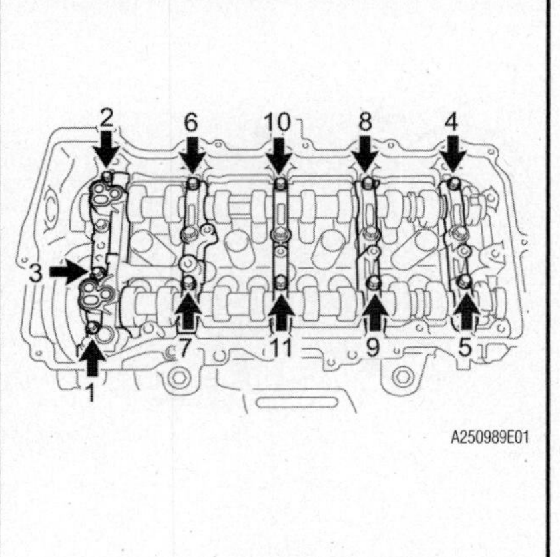

Fig. 51 Using several steps, remove the 11 bearing cap bolts in sequence

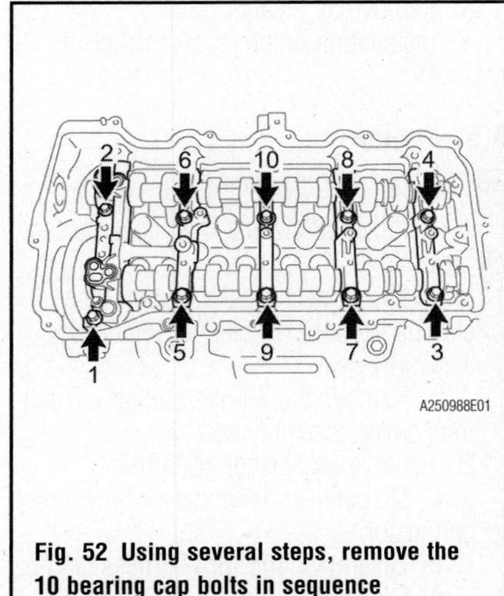

Fig. 52 Using several steps, remove the 10 bearing cap bolts in sequence

c. Remove the 5 bearing caps.

➡Arrange the removed parts in the correct order.

28. Remove the camshaft from the camshaft housing.

29. Remove No. 2 camshaft.

a. Hold up the chain and remove the No. 2 camshaft from the camshaft housing.

b. Suspend the chain with a string or equivalent.

➡Be careful not to drop the chain inside the timing chain cover.

30. Remove the flange bolt and camshaft timing exhaust gear assembly.

➡Do not disassemble the camshaft timing exhaust gear.

31. Remove oil control valve filter.

32. Remove No. 1 camshaft bearing.

33. Remove No. 2 camshaft bearing.

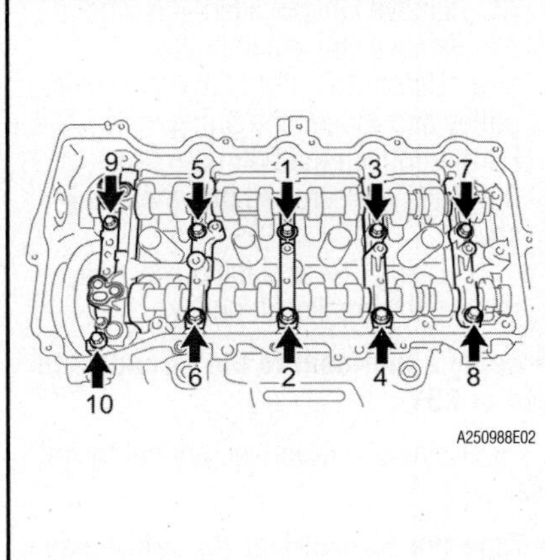

Fig. 53 Using several steps, uniformly tighten the 10 bolts in sequence

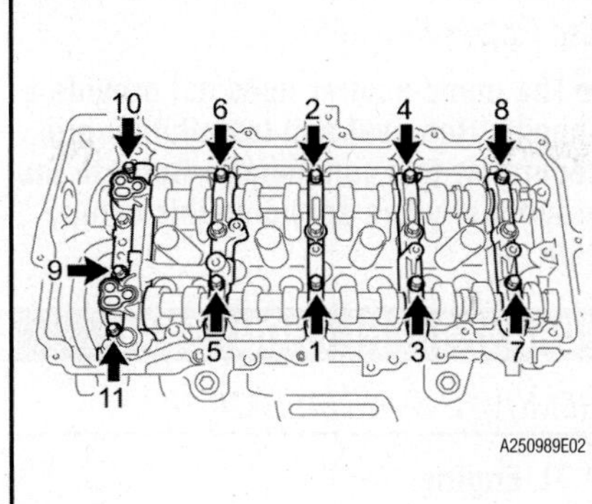

Fig. 54 Using several steps, uniformly tighten the 11 bolts in sequence

To install:

34. Install No. 2 camshaft bearing.
35. Install No. 1 camshaft bearing.
36. Install oil control valve filter.
37. Install camshaft timing exhaust gear assembly.
   a. Align and attach the knock pin of the No. 2 camshaft with the pin hole of the camshaft timing exhaust gear assembly.
   b. Check that there is no clearance between the camshaft timing exhaust gear assembly and camshaft flange.
   c. Fix the camshaft timing exhaust gear assembly with the bolt. Tighten to 63 ft. lbs. (85 Nm).

➡Do not disassemble the camshaft timing exhaust gear.

38. Set No. 1 cylinder to TDC/compression. Turn the crankshaft pulley until its timing notch (groove) and the timing mark "0" of the timing chain cover are aligned.
39. Install No. 2 camshaft.
   a. Make sure that the valve rocker arms are installed.
   b. Clean the camshaft journals.
   c. Apply a light coat of engine oil to the camshaft journals, camshaft housings and bearing caps.
   d. Hold up the chain and align the matchmark and the paint mark and install the camshaft.
40. Install camshaft.
   a. Make sure that the valve rocker arms are installed.
   b. Clean the camshaft journals.
   c. Apply a light coat of engine oil to the camshaft journals, camshaft housings and bearing caps.
   d. Install the camshaft to the camshaft housing.
41. Install camshaft bearing cap.
   a. Confirm the marks and numbers on the camshaft bearing caps and place them in their proper positions and directions.
   b. Using several steps, uniformly tighten the 10 bolts in sequence. Tighten to 20 ft. lbs. (27 Nm).
   c. Using several steps, uniformly tighten the 11 bolts in sequence. Tighten to 12 ft. lbs. (16 Nm).
   d. Check the torque of each bolt again.
42. Install camshaft timing gear assembly.
   a. Check the camshaft timing gear position.

➡If the camshaft timing gear is set to the advanced position, do not let the camshaft timing gear rotate clockwise during installation. If the camshaft timing gear has rotated to the most retarded position, make sure to release the lock pin and set the camshaft timing gear to the most advanced position before tightening the camshaft timing gear.

   b. Install the camshaft timing gear.
   c. Turn the camshaft timing gear approximately 180° counterclockwise.
   d. Align the paint mark with the matchmark to install the chain.
   e. Align and attach the knock pin of the No. 1 camshaft with the pin hole of the camshaft timing gear.
   f. Check that there is no clearance between the camshaft timing gear and camshaft flange.
   g. Secure the camshaft in place by hand, and then install the installation bolt of the camshaft timing gear by hand.

➡Do not use any tools to install the bolt. If the bolt is installed using a tool, the lock pin will be damaged.

   h. If the lock pin has not been released, release it.
   i. After cleaning and degreasing the intake side VVT oil hole on the No. 1 camshaft bearing cap, completely seal the oil hole with adhesive tape or equivalent to prevent air from leaking.

➡Be sure to seal the oil hole completely because air leaks due to insufficient sealing will prevent the lock pin from being released.

   j. Make a hole in the adhesive tape covering the oil hole.
   k. Apply approximately 29 psi (200 kPa) of air pressure to the hole made in procedure A to release the lock pin.

➡If air leaks out, reattach the adhesive tape. Cover the oil hole with a piece of cloth when applying air pressure to prevent oil from spraying.

   l. Forcibly turn the camshaft timing gear in the advance direction (counterclockwise).

➡Depending on the air pressure applied, the camshaft timing gear may turn in the advance direction without assistance by hand.

   m. Remove the adhesive tape from the No. 1 camshaft bearing cap.
   n. Using a wrench to hold the hexagonal portion of the No. 1 camshaft, install the bolt. Tighten to 63 ft. lbs. (85 Nm).

➡Be careful not to damage the cylinder head or spark plug tube with the wrench.

   o. Check that each matchmark of the camshaft timing gear and camshaft timing exhaust gear are aligned with each matchmark.
43. Add engine oil.
44. Install timing chain guide.
45. Install No. 1 chain tensioner assembly.
   a. Turn the crankshaft approximately 10° clockwise.
   b. Install a new gasket and the chain tensioner with the 2 bolts.

➡Make sure not to drop the gasket inside the timing chain cover.

   c. Remove the pin from the stopper plate.
46. Check No. 1 cylinder to TDC/compression.
   a. Turn the crankshaft pulley until its timing notch (groove) and the timing

mark "0" of the timing chain cover are aligned.

b. Check that the timing marks of the camshaft timing gears are correct. If not, turn the crankshaft 1 revolution (360°) to align the timing marks.

➡"A" is not a timing mark.

47. Install a new gasket and the timing chain cover plate with the 4 bolts.
48. Install timing chain cover tight plug.
49. Install cylinder head cover sub-assembly.
50. Install ignition coil assembly.
51. Connect the connectors and clamps, and install the engine wire to the engine with the bolts and nuts.
52. Connect the ventilation hose assembly to the cylinder head cover.
53. Install engine moving control rod sub-assembly.
54. Install No. 2 engine mounting stay RH.
55. Install air cleaner assembly.
56. Install air cleaner inlet assembly.
57. Install cowl panel sub-assembly.
58. Install windshield wiper motor and link assembly.
59. Install cowl top ventilator louver sub-assembly.
60. Install front fender to cowl side seal LH.
61. Install front fender to cowl side seal RH.
62. Install front wiper arm and blade assembly RH.
63. Install front wiper arm and blade assembly LH.
64. Install front wiper arm head cap.
65. Connect cable to negative battery terminal.

➡When disconnecting the cable, some systems need to be initialized after the cable is reconnected.

66. Inspect oil leak.
67. Install No. 1 engine cover sub-assembly.
68. Install front fender apron seal RH.
69. Install front wheel RH.

CATALYTIC CONVERTER

REMOVAL & INSTALLATION

2.7L Engine

See Figure 55.

➡The manufacturer does not provide a specific Removal and Installation procedure for this component. Refer to the graphic(s) when servicing this component.

3.5L Engine

See Figures 56 and 57.

➡The manufacturer does not provide a specific Removal and Installation procedure for this component. Refer to the graphic(s) when servicing this component.

CRANKSHAFT FRONT SEAL

REMOVAL & INSTALLATION

2.7L Engine

1. Remove front wheel RH.
2. Remove front fender apron seal RH.
3. Remove fan and alternator v belt.
4. Remove crankshaft pulley.

a. Using SST, hold the crankshaft pulley and loosen the pulley bolt. Further loosen the bolt until 2 or 3 threads are screwed into the crankshaft.

b. Using SST and the pulley bolt, remove the crankshaft pulley.

➡Apply a lubricant to the threads and end of SST.

5. Using a screwdriver, pry out the oil seal.

➡Tape the screwdriver tip before use.

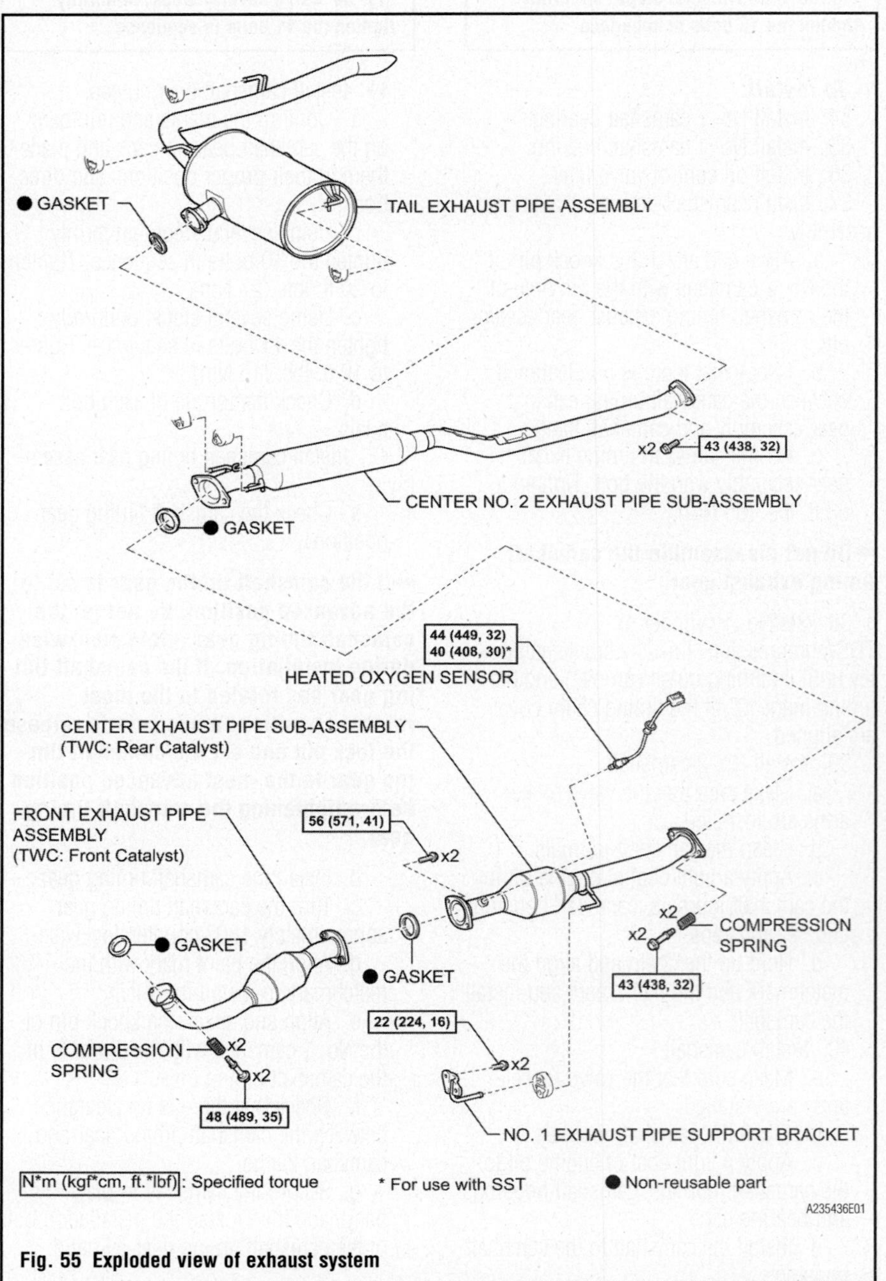

Fig. 55 Exploded view of exhaust system

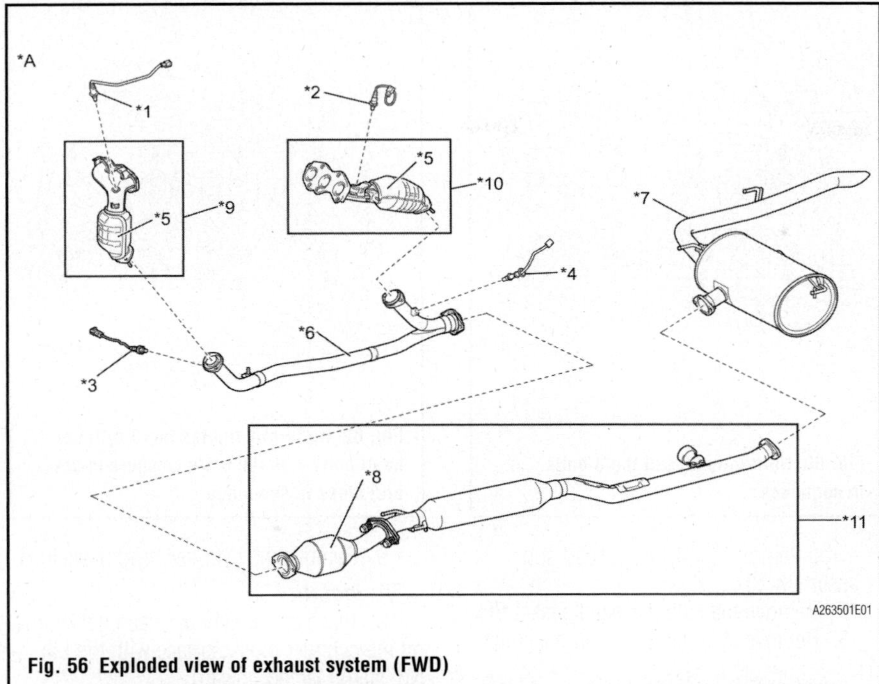

Fig. 56 Exploded view of exhaust system (FWD)

WARNING

Do not damage the surface of the oil seal press fit hole or the crankshaft.

To install:

6. Install timing chain cover oil seal
   a. Apply MP grease to the lip of a new oil seal.

➡Do not allow foreign matter to contact the lip of the oil seal. Do not allow MP grease to contact the dust seal of the oil seal.

   b. Using SST and a hammer, tap in the oil seal until its surface is flush with the timing chain cover edge.

➡Keep the lip of the oil seal free from foreign matter. Do not tap in the oil seal at an angle.

7. Install crankshaft pulley
   a. Align the pulley set key with the key groove of the crankshaft pulley.
   b. Using SST, hold the crankshaft pulley and install the pulley bolt. Tighten to 192 ft. lbs. (260 Nm).
8. Install fan and alternator V belt
9. Install front fender apron seal RH
10. Install front wheel RH

3.5L Engine

1. Remove front wheel RH.
2. Remove front fender apron seal RH.
3. Remove fan and alternator V belt.
4. Remove crankshaft pulley.
   a. Using SST, loosen the crankshaft pulley bolt.
   b. Using SST, remove the crankshaft pulley bolt and crankshaft pulley.
   c. Remove the pulley set key from the crankshaft.
5. Remove timing chain case oil seal.
   a. Using a knife, cut off the timing chain case oil seal lip.
   b. Using a screwdriver, pry out the timing chain case oil seal.

➡After the removal, check the crankshaft for damage. If it is damaged, smooth the surface with 400-grit sandpaper.

➡Tape the screwdriver tip before use.

To install:

6. Install timing chain case oil seal.
   a. Apply MP grease to a new timing chain case oil seal lip.
   b. Using SST and a hammer, tap in the timing chain case oil seal until its surface is flush with the timing chain cover edge.

➡Keep the lip free of foreign matter. Do not tap the timing chain case oil seal at an angle.

7. Install crankshaft pulley.
   a. Align the pulley set key with the key groove of the pulley, and slide on the pulley.
   b. Using SST, install the pulley bolt. Tighten to 184 ft. lbs. (250 Nm).
8. Install fan and alternator V belt.
9. Install front fender apron seal RH.
10. Install front wheel RH.

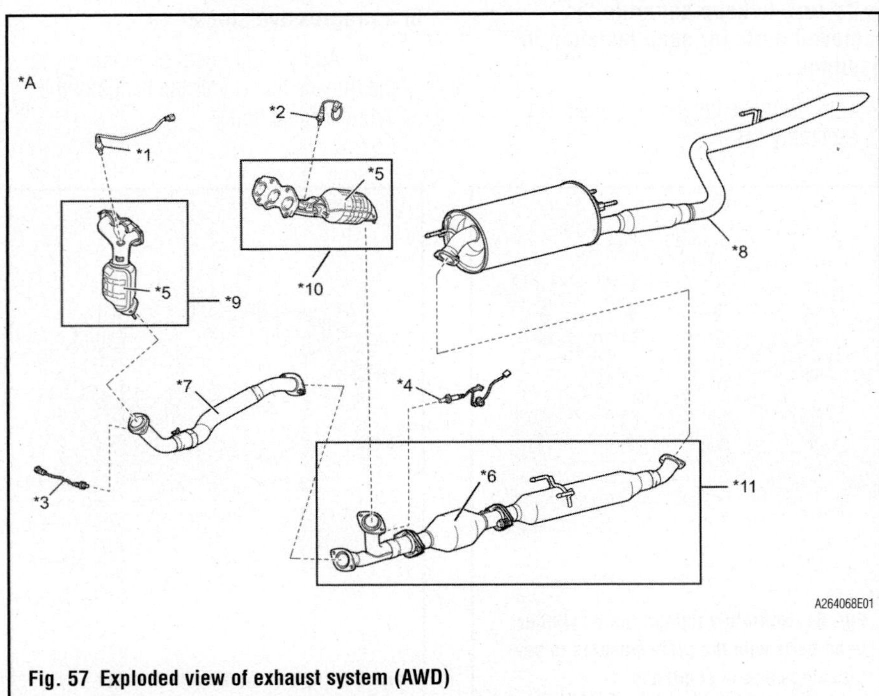

Fig. 57 Exploded view of exhaust system (AWD)

CYLINDER HEAD

REMOVAL & INSTALLATION

3.5L Engine

See Figures 58 through 63.

1. Remove camshaft.
2. Remove the 24 valve rocker arms.

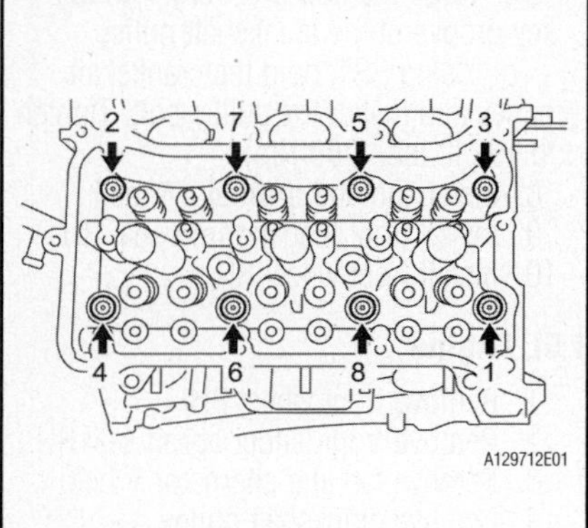

Fig. 58 Uniformly loosen the 8 cylinder head bolts in sequence

➡Arrange the removed parts in the correct order.

3. Remove the 24 valve lash adjuster assemblies from the cylinder head.

➡Arrange the removed parts in the correct order.

4. Remove the 24 valve stem caps.
5. Remove water outlet.
6. Remove cylinder head sub-assembly RH.
   a. Using a 10 mm wrench, uniformly loosen the 8 cylinder head bolts in sequence. Remove the 8 cylinder head bolts and plate washers.

➡Be careful not to drop washers into the cylinder head sub-assembly. Cylinder head warpage or cracking could result from removing bolts in the incorrect order.

➡Arrange the removed parts in the correct order.

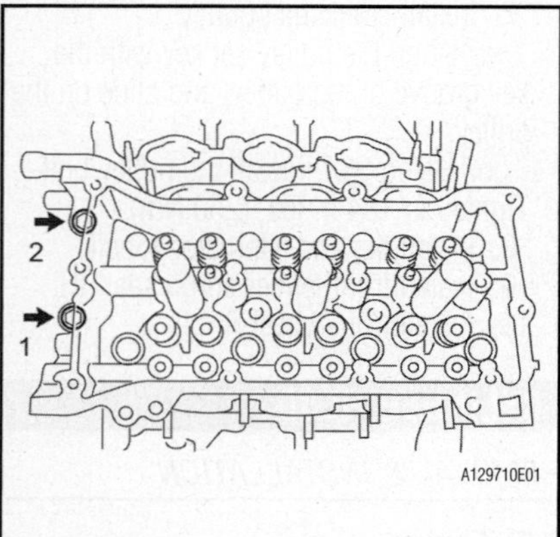

Fig. 59 Uniformly loosen and remove the 2 cylinder head set bolts in several steps and in sequence

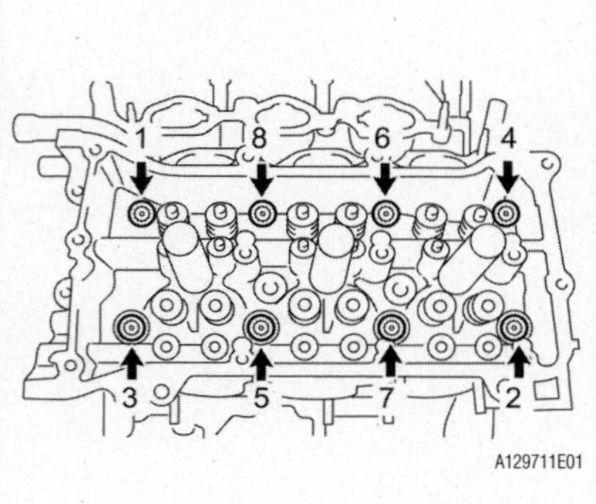

Fig. 60 Uniformly loosen the 8 bolts in sequence

   b. Remove the cylinder head sub-assembly RH.
7. Remove the cylinder head gasket RH.
8. Remove cylinder head sub-assembly LH.
   a. Uniformly loosen and remove the 2 cylinder head set bolts in several steps and in sequence.
   b. Using a 10 mm wrench, uniformly loosen the 8 bolts in sequence. Remove the 8 cylinder head bolts and plate washers.

➡Be careful not to drop washers into the cylinder head sub-assembly. Cylinder head warpage or cracking could result from removing bolts in the incorrect order.

➡Be sure to keep separate the removed parts for each installation position.

   c. Remove the cylinder head sub-assembly LH.

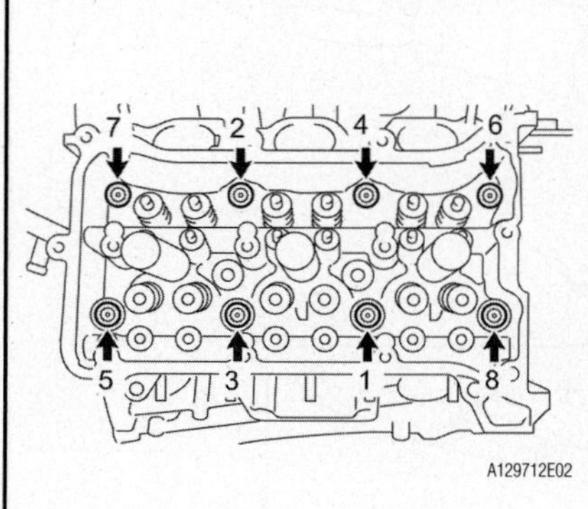

Fig. 61 Uniformly tighten the 8 cylinder head bolts with the plate washers in several steps and in sequence

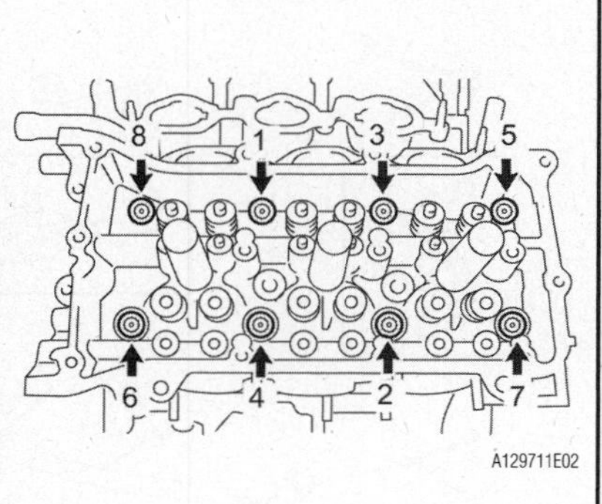

Fig. 62 Uniformly tighten the 8 cylinder head bolts with the plate washers in several steps in sequence

9. Remove the cylinder head gasket LH.

To install:

10. Place a new cylinder head gasket RH on the cylinder block surface with the Lot No. stamp facing upward.

➡Be careful of the installation direction. Gently lower the cylinder head in order not to damage the gasket with the bottom part of the head.

11. Install cylinder head sub-assembly RH.
   a. Place the cylinder head on the cylinder block.

➡Be careful not to allow oil to adhere to the bottom part of the cylinder head.

➡The cylinder head bolts are tightened in 3 progressive steps.

   b. Apply a light coat of engine oil to the threads and under the heads of the cylinder head bolts.

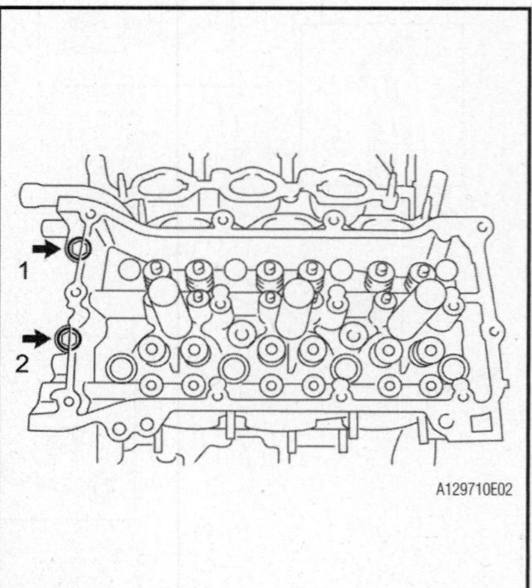

Fig. 63 Tighten the 2 bolts in order

c. Step 1: Using a 10 mm wrench, install and uniformly tighten the 8 cylinder head bolts with the plate washers in several steps and in sequence. Tighten to 27 ft. lbs. (36 Nm).

d. Step 2: Mark the cylinder head bolt head with paint. Tighten the cylinder head bolts another 90°.

e. Step 3: Tighten the cylinder head bolts an additional 90°. Check that the painted mark is now facing rearward.

12. Place a new cylinder head gasket LH on the cylinder block surface with the Lot No. stamp facing upward.

➡Be careful of the installation direction. Gently lower the cylinder head in order not to damage the gasket with the bottom part of the head.

13. Install cylinder head sub-assembly LH.

a. Place the cylinder head on the cylinder block.

➡Be careful not to allow oil to adhere to the bottom part of the cylinder head.

➡The cylinder head bolts are tightened in 3 progressive steps.

b. Apply a light coat of engine oil to the threads and under the heads of the cylinder head bolts.

c. Step 1: Using a 10 mm wrench, install and uniformly tighten the 8 cylinder head bolts with the plate washers in several steps in sequence. Tighten to 27 ft. lbs. (36 Nm).

d. Step 2: Mark the cylinder head bolt head with paint. Tighten the cylinder head bolts another 90°.

e. Step 3: Tighten the cylinder head bolts an additional 90°. Check that the painted mark is now facing rearward.

f. Tighten the 2 bolts in order. Tighten to 22 ft. lbs. (30 Nm).

14. Install water outlet.

15. Install the 24 valve stem caps.

16. Install valve lash adjuster assembly.

➡Keep the lash adjuster free of dirt and foreign objects. Only use clean engine oil.

a. Place the lash adjuster into a container filled with engine oil.

b. Insert the tip of SST into the lash adjuster plunger and use the tip to press down on the check ball inside the plunger.

c. Squeeze SST and lash adjuster together to move the plunger up and down 5 to 6 times.

d. Check the movement of the plunger and bleed the air.

➡When bleeding air from the high-pressure chamber, make sure that the tip of SST is actually pressing the check ball. If the check ball is not pressed, air will not bleed.

e. After bleeding the air, remove SST. Then, try to press the plunger quickly and firmly by hand.

➡If the result is not as specified, replace the valve lash adjuster.

f. Install the valve lash adjusters.

➡Install each valve lash adjuster to the same place it was removed from.

17. Install No. 1 valve rocker arm sub-assembly.

a. Apply engine oil to the lash adjuster tip and valve stem cap end.

b. Install the valve rocker arm.

18. Install camshaft.

ENGINE OIL & FILTER

REPLACEMENT

See Figures 64 through 66.

⁂ CAUTION

Please consider the following cautions:

- Prolonged and repeated contact with engine oil will result in the removal of natural oils from the skin, leading to dryness, irritation and dermatitis. In addition, used engine oil contains potentially harmful contaminants which may cause skin cancer.
- Precautions should be taken when replacing engine oil to minimize the risk of your skin making contact with used engine oil. Protective clothing and gloves that cannot be penetrated by oil should be worn. In case of accidental contact, wash your skin thoroughly with soap and water, or use waterless hand cleaner, to remove any used engine oil. Never use gasoline, thinners, or solvents to wash your skin.
- In order to protect the environment, used oil and used oil filters must be disposed of at designated disposal sites only.

1. Drain engine oil.

a. Remove the oil filler cap.

b. Remove the oil pan drain plug and gasket, and drain the engine oil into a container.

c. Clean and install a new gasket and the oil pan drain plug. Tighten to 30 ft. lbs. (40 Nm).

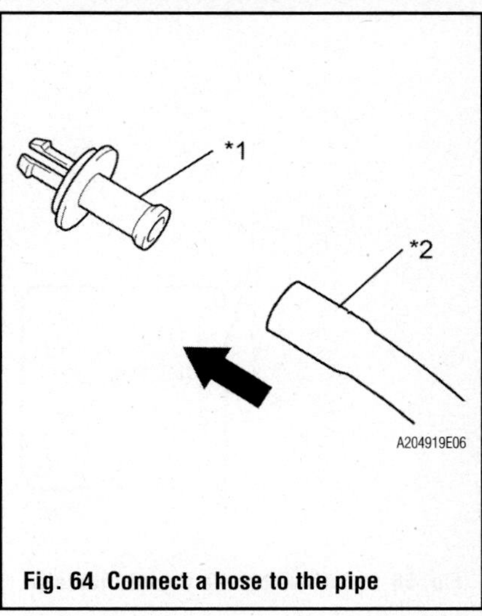

Fig. 64 Connect a hose to the pipe

2. Remove oil filter cap assembly.

a. Connect a hose with an inside diameter of 0.591 inches (15 mm) to the pipe.

b. Remove the oil filter drain plug from the oil filter cap.

c. Install the pipe to the oil filter cap.

➡If the O-ring is removed with the drain plug, install the O-ring together with the pipe.

d. Check that the oil is drained from the oil filter. Then disconnect the pipe and remove the O-ring.

e. Using filter wrench, remove the oil filter cap.

➡Do not remove the oil filter bracket clip.

f. Remove the oil filter element and O-ring from the oil filter cap.

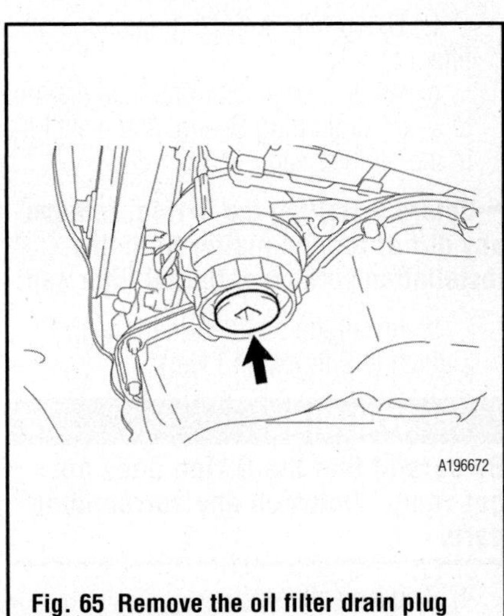

Fig. 65 Remove the oil filter drain plug from the oil filter cap

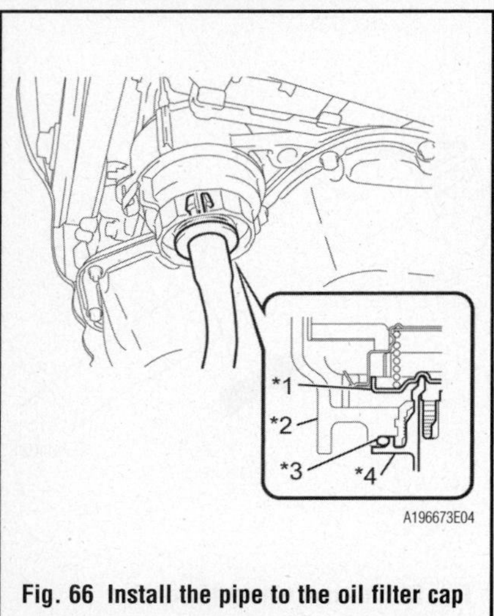

Fig. 66 Install the pipe to the oil filter cap

⁂ WARNING

Be sure to remove the cap O-ring by hand, without using any tools, to prevent damage to the cap O-ring groove.

3. Install oil filter cap assembly.
   a. Clean the inside of the oil filter cap, its threads and its O-ring groove.
   b. Apply a small amount of engine oil to a new O-ring and install it to the oil filter cap.
   c. Set a new oil filter element into the oil filter cap.
   d. Remove any dirt or foreign matter from the installation surface of the engine.
   e. Apply a small amount of engine oil to the O-ring again and temporarily install the oil filter cap.

➡Make sure that the O-ring does not get caught between the parts.

   f. Using filter wrench, tighten the oil filter cap.
   g. Apply a small amount of engine oil to a new drain plug O-ring, and install it to the oil filter cap.

➡Before installing the O-ring, remove any dirt or foreign matter from the installation surface of the oil filter cap.

   h. Install the oil filter drain plug. Tighten to 9 ft. lbs. (13 Nm).

⁂ WARNING

Be careful that the O-ring does not get caught between any surrounding parts.

4. Add new oil.
5. Install the oil filler cap.
6. Start the engine. Make sure that there are no oil leaks from the areas that were worked on.
7. Inspect engine oil level.

EXHAUST MANIFOLD

REMOVAL & INSTALLATION

2.7L Engine

See Figures 67 and 68.

1. Remove No. 1 engine under cover.
   a. Remove the 8 screws, front wheel opening extension pad RH and front wheel opening extension pad LH.
   b. Remove the 2 bolts, 6 screws, 4 clips and No. 1 engine under cover.
2. Remove front exhaust pipe assembly.
   a. Remove the 4 bolts, 2 compression springs and front exhaust pipe assembly.
   b. Remove the 2 gaskets from the front exhaust pipe assembly and exhaust manifold.
3. Remove the bolt, nut and manifold stay.
4. Remove No. 2 manifold stay.
   a. Disconnect the connector and disengage the harness clamp.
   b. Remove the bolt and wire harness clamp bracket.
   c. Remove the bolt, nut and No. 2 manifold stay.
5. Remove the bolt and EGR inlet exhaust manifold plate.
6. Remove the 5 nuts and exhaust manifold.

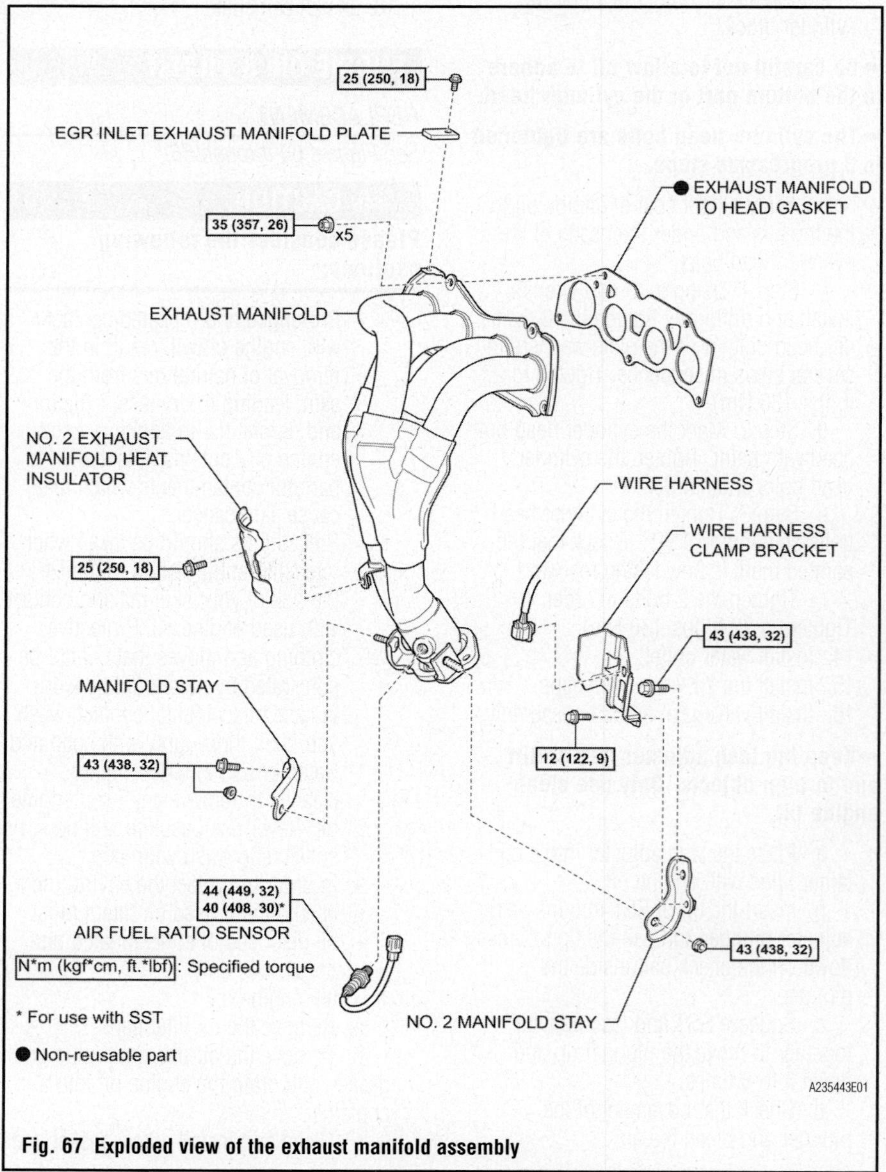

Fig. 67 Exploded view of the exhaust manifold assembly

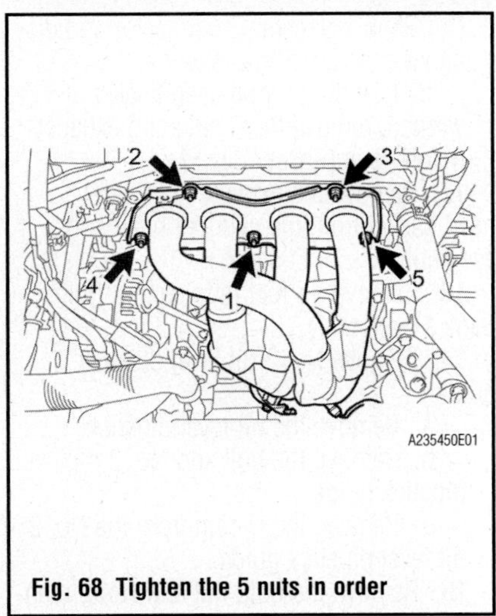

Fig. 68 Tighten the 5 nuts in order

7. Remove the exhaust manifold to head gasket.

8. Using SST, remove the air fuel ratio sensor from the exhaust manifold.

9. Remove the bolt and No. 2 exhaust manifold heat insulator.

To install:

10. Install the No. 2 exhaust manifold heat insulator with the bolt. Tighten to 18 ft. lbs. (25 Nm).

11. Using SST, install the air fuel ratio sensor to the exhaust manifold.

12. Install a new exhaust manifold to head gasket.

13. Install exhaust manifold.

a. Temporarily install the exhaust manifold with the 5 nuts.

b. Tighten the 5 nuts in order. Tighten to 26 ft. lbs. (35 Nm).

14. Install the EGR inlet exhaust manifold plate with the bolt. Tighten to 18 ft. lbs. (25 Nm).

15. Install No. 2 manifold stay.

a. Install the No. 2 manifold stay with the bolt and nut. Tighten to 32 ft. lbs. (43 Nm).

b. Install the wire harness clamp bracket with the bolt.

c. Engage the harness clamp and connect the connector.

16. Install the manifold stay with the bolt and nut. Tighten to 32 ft. lbs. (43 Nm).

17. Install front exhaust pipe assembly.

a. Fully push a new gasket to the exhaust manifold.

➡Install the gasket in the correct direction. Do not reuse the gasket. Do not damage the gasket by dropping it, etc. Do not damage the outer surface of the gasket.

b. Install a new gasket onto the front exhaust pipe assembly.

c. Install the front exhaust pipe with the 2 compression springs and 4 bolts.

18. Inspect for exhaust gas leak.

19. Install No. 1 engine under cover.

a. Install the No. 1 engine under cover with the 2 bolts, 6 screws and 4 clips.

b. Install the front wheel opening extension pad RH and front wheel opening extension pad LH with the 8 screws.

3.5L Engine

See Figures 69 through 73.

1. Remove radiator assembly.

2. Remove V-bank cover sub-assembly.

3. Remove instrument panel finish panel end LH.

4. Remove No. 1 engine under cover.

a. Remove the 8 screws, front wheel opening extension pad RH and front wheel opening extension pad LH.

b. Remove the 2 bolts, 6 screws, 4 clips and No. 1 engine under cover.

5. Remove the 2 bolts, 2 screws, clip and No. 2 engine under cover (for AWD).

6. Remove propeller with center bearing shaft assembly (for AWD).

7. Remove No. 1 exhaust pipe support bracket.

8. Remove front exhaust pipe assembly (for FWD).

a. Disconnect the oxygen sensor (for bank 1 sensor 2) connector and pass the grommet through the floor.

b. Disconnect the oxygen sensor (for bank 2 sensor 2) connector and disengage the harness clamp.

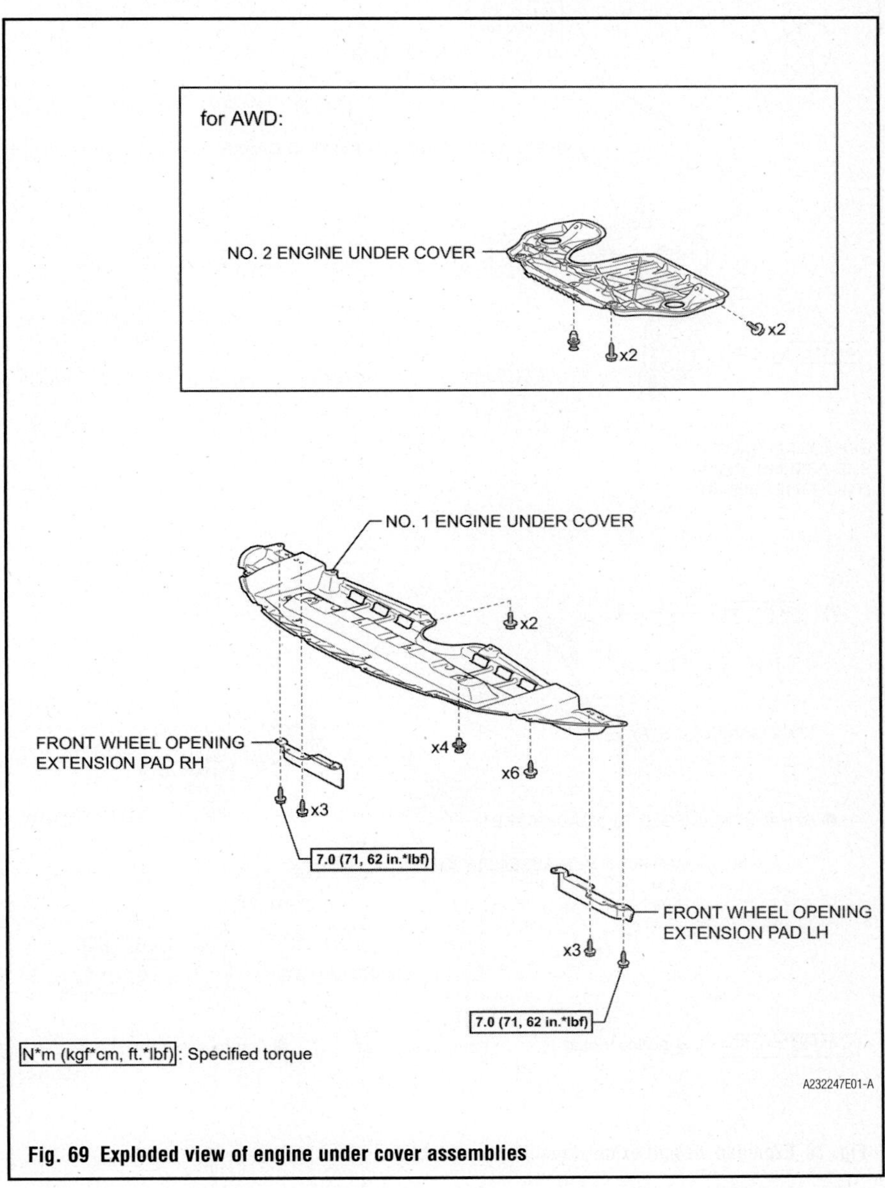

Fig. 69 Exploded view of engine under cover assemblies

c. Remove the 4 nuts, 2 bolts, 2 compression springs and front exhaust pipe assembly.

d. Remove the 3 gaskets from the front exhaust pipe assembly.

9. Remove front exhaust pipe assembly (for AWD).

a. Disconnect the oxygen sensor (for bank 1 sensor 2) connector and pass the grommet through the floor.

b. Disconnect the oxygen sensor (for bank 2 sensor 2) connector and disengage the harness clamp.

c. Remove the 2 nuts, 4 bolts, 2 compression springs and front No. 3 exhaust pipe sub-assembly.

d. Remove the 3 gaskets from the front No. 3 exhaust pipe sub-assembly and front exhaust pipe assembly.

e. Remove the 2 nuts and front exhaust pipe assembly.

f. Remove the gasket from the front exhaust pipe assembly.

10. Remove the bolt, nut and manifold stay.

11. Remove exhaust manifold sub-assembly RH.

a. Disconnect the air fuel ratio sensor (for Bank 1 Sensor 1) connector and disengage the 2 harness clamps.

b. Using a 12 mm deep socket wrench, remove the 6 nuts and exhaust manifold sub-assembly RH.

12. Remove the exhaust manifold to head gasket from the cylinder head sub-assembly.

13. Remove air fuel ratio sensor (for Bank 1 Sensor 1).

14. Remove No. 2 oil level dipstick guide.

a. Remove the oil level dipstick.

b. Remove the bolt and No. 2 oil level dipstick guide.

c. Remove the O-ring from the No. 2 oil level dipstick guide.

15. Remove the bolt, nut and No. 2 manifold stay.

16. Remove No. 2 exhaust manifold heat insulator.

a. Disconnect the air fuel ratio sensor (for Bank 2 Sensor 1) connector and disengage the 2 harness clamps.

b. Remove the 3 bolts and No. 2 exhaust manifold heat insulator.

17. Using a 12 mm deep socket wrench, remove the 6 nuts and exhaust manifold sub-assembly LH.

18. Remove the exhaust manifold to head gasket LH from the cylinder head sub-assembly.

19. Using SST, remove the air fuel ratio sensor (for Bank 2 Sensor 1) from the exhaust manifold sub-assembly LH.

To install:

20. Using SST, install the air fuel ratio sensor (for Bank 2 Sensor 1) to the front exhaust manifold sub-assembly LH.

21. Install a new exhaust manifold to head gasket LH.

x3 8.5 (87, 75 in.*lbf)
NO. 2 EXHAUST MANIFOLD HEAT INSULATOR
44 (449, 32)
40 (408, 30)*
AIR FUEL RATIO SENSOR
(for Bank 2 Sensor 1)
● EXHAUST MANIFOLD TO HEAD GASKET LH
21 (214, 15) x6
EXHAUST MANIFOLD SUB-ASSEMBLY LH
(TWC: Front Catalyst)
34 (347, 25)
NO. 2 MANIFOLD STAY
44 (449, 32)
40 (408, 30)*
AIR FUEL RATIO SENSOR
(for Bank 1 Sensor 1)
● EXHAUST MANIFOLD TO HEAD GASKET
EXHAUST MANIFOLD SUB-ASSEMBLY RH
(TWC: Front Catalyst)
21 (214, 15) x6
35 (357, 26)
MANIFOLD STAY
34 (347, 25)
N*m (kgf*cm, ft.*lbf): Specified torque
* For use with SST
● Non-reusable part
A232284E01

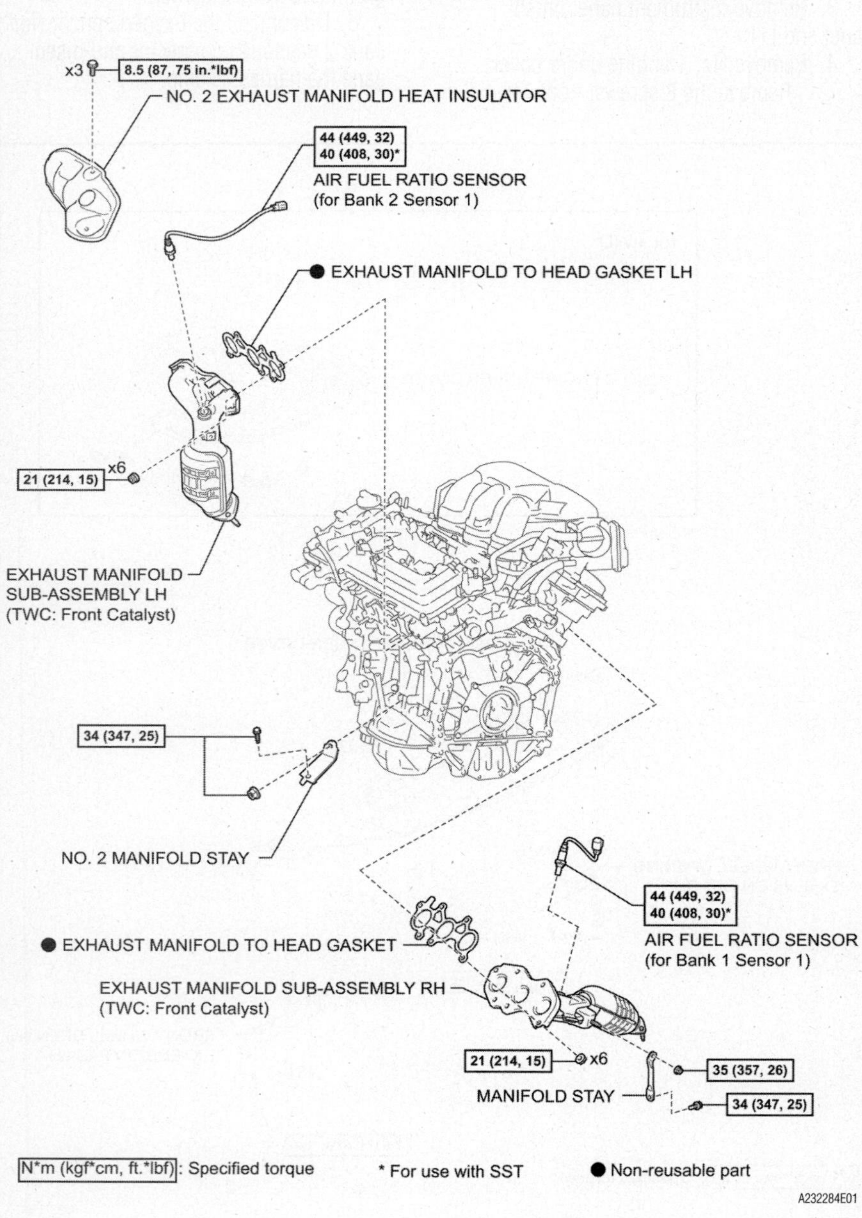

Fig. 70 Exploded view of exhaust manifold assemblies

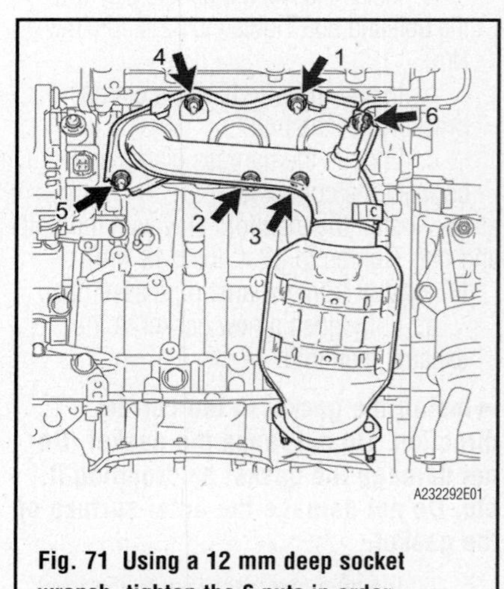

Fig. 71 Using a 12 mm deep socket wrench, tighten the 6 nuts in order

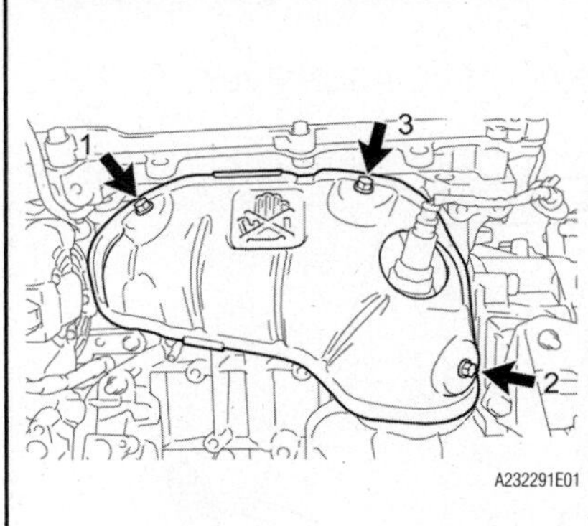

Fig. 72 Install the No. 2 exhaust manifold heat insulator, tightening the 3 bolts in order

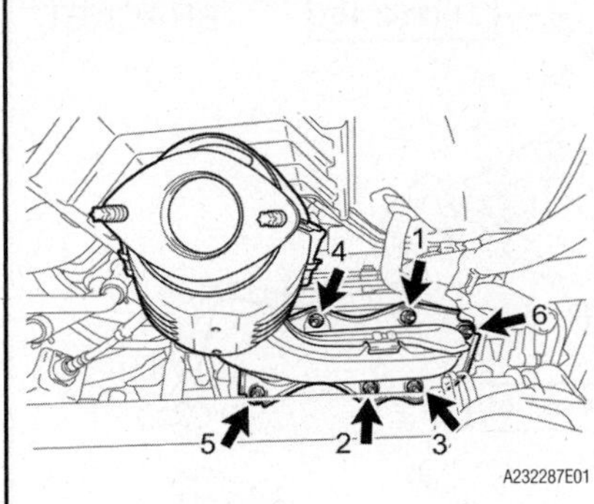

Fig. 73 Using a 12 mm deep socket wrench, tighten the 6 nuts in order

22. Install exhaust manifold sub-assembly LH.
    a. Temporarily install the exhaust manifold sub-assembly LH with the 6 nuts.
    b. Using a 12 mm deep socket wrench, tighten the 6 nuts in order. Tighten to 15 ft. lbs. (21 Nm).
23. Install No. 2 exhaust manifold heat insulator.
    a. Install the No. 2 exhaust manifold heat insulator, tightening the 3 bolts in order.
    b. Connect the air fuel ratio sensor (for Bank 2 Sensor 1) connector and engage the 2 harness clamps.
24. Install the No. 2 manifold stay, tightening the bolt and nut to 25 ft. lbs. (34 Nm).
25. Install No. 2 oil level dipstick guide.
    a. Install a new O-ring to the No. 2 oil level dipstick guide.
    b. Apply a light coat of engine oil to the O-ring.
    c. Push in the No. 2 oil level dipstick guide end into the oil level dipstick guide.
    d. Install the No. 2 oil level dipstick guide with the bolt. Tighten to 15 ft. lbs. (21 Nm).
    e. Install the oil level dipstick.
26. Install air fuel ratio sensor (for Bank 1 Sensor 1).
27. Install a new exhaust manifold to head gasket.
28. Install exhaust manifold sub-assembly RH.
    a. Temporarily install the exhaust manifold sub-assembly RH with the 6 nuts.
    b. Using a 12 mm deep socket wrench, tighten the 6 nuts in order. Tighten to 15 ft. lbs. (21 Nm).
    c. Connect the air fuel ratio sensor (for Bank 1 Sensor 1) connector and engage the 2 harness clamps.
29. Install the manifold stay with the bolt and nut.
30. Install front exhaust pipe assembly (for FWD).
31. Install front exhaust pipe assembly (for AWD).
32. Install No. 1 exhaust pipe support bracket.
33. Install propeller with center bearing shaft assembly (for AWD).
34. Install the No. 2 engine under cover with the 2 bolts, 2 screws and clip (for AWD).
35. Install No. 1 engine under cover.
36. Install instrument panel finish panel end LH.
37. Install V-bank cover sub-assembly.
38. Install radiator assembly.
39. Inspect for exhaust gas leak.

INTAKE MANIFOLD

REMOVAL & INSTALLATION

2.7L Engine

See Figures 74 and 75.

1. Discharge fuel system pressure.
2. Remove air cleaner assembly.
3. Remove air cleaner hose.
    a. Loosen the clip and disconnect the No. 1 air hose.
    b. Loosen the clip and disconnect the ventilation hose assembly.
    c. Separate the No. 2 fuel vapor feed hose from the hose clamp.
    d. Separate the No. 2 vacuum switching valve assembly.
    e. Separate the vacuum hose from the hose clamp.
    f. Disconnect the vacuum hose.
    g. Separate the vacuum hose from the 2 hose clamps.
    h. Loosen the clip and remove the air cleaner hose.
4. Disconnect water by-pass hose.
5. Disconnect No. 2 water by-pass hose.
6. Remove throttle with motor body assembly.
7. Loosen the clip and disconnect the outlet heater water hose A.
8. Loosen the clip and disconnect the inlet heater water hose A.
9. Remove vacuum switching valve assembly (for ACIS).
    a. Disconnect the vacuum switching valve connector.
    b. Disengage the wire harness clamp.
    c. Disconnect the 2 vacuum hoses.
    d. Remove the bolt and the vacuum switching valve assembly.
10. Loosen the clip and disconnect the No. 2 ventilation hose.
11. Loosen the clip and disconnect the union to connector tube hose.
12. Remove wire harness clamp bracket.
    a. Disconnect the sensor wire connector.
    b. Apply battery voltage to the terminals of the connector to close the tumble control valves.

➡Please note the following:

- Confirm that the sound of the tumble control valves physically contacting the stopper can be heard.
- If this procedure is not performed, the valves may be damaged when the intake manifold is removed.
- Apply battery voltage for 1 to 3 seconds.
- If battery voltage is applied for more than 3 seconds, the actuator may be damaged.
- Do not allow the lead wires to contact the other terminals.

    c. Disengage the wire harness clamp.
    d. Remove the 2 bolts and the 2 wire harness clamp brackets.
13. Remove No. 1 vacuum switching valve assembly.
    a. Disconnect the No. 1 vacuum switching valve connector.
    b. Loosen the clip and disconnect the No. 2 air hose.
    c. Remove the bolt and the No. 1 vacuum switching valve assembly.
14. Remove EFI fuel pipe clamp.

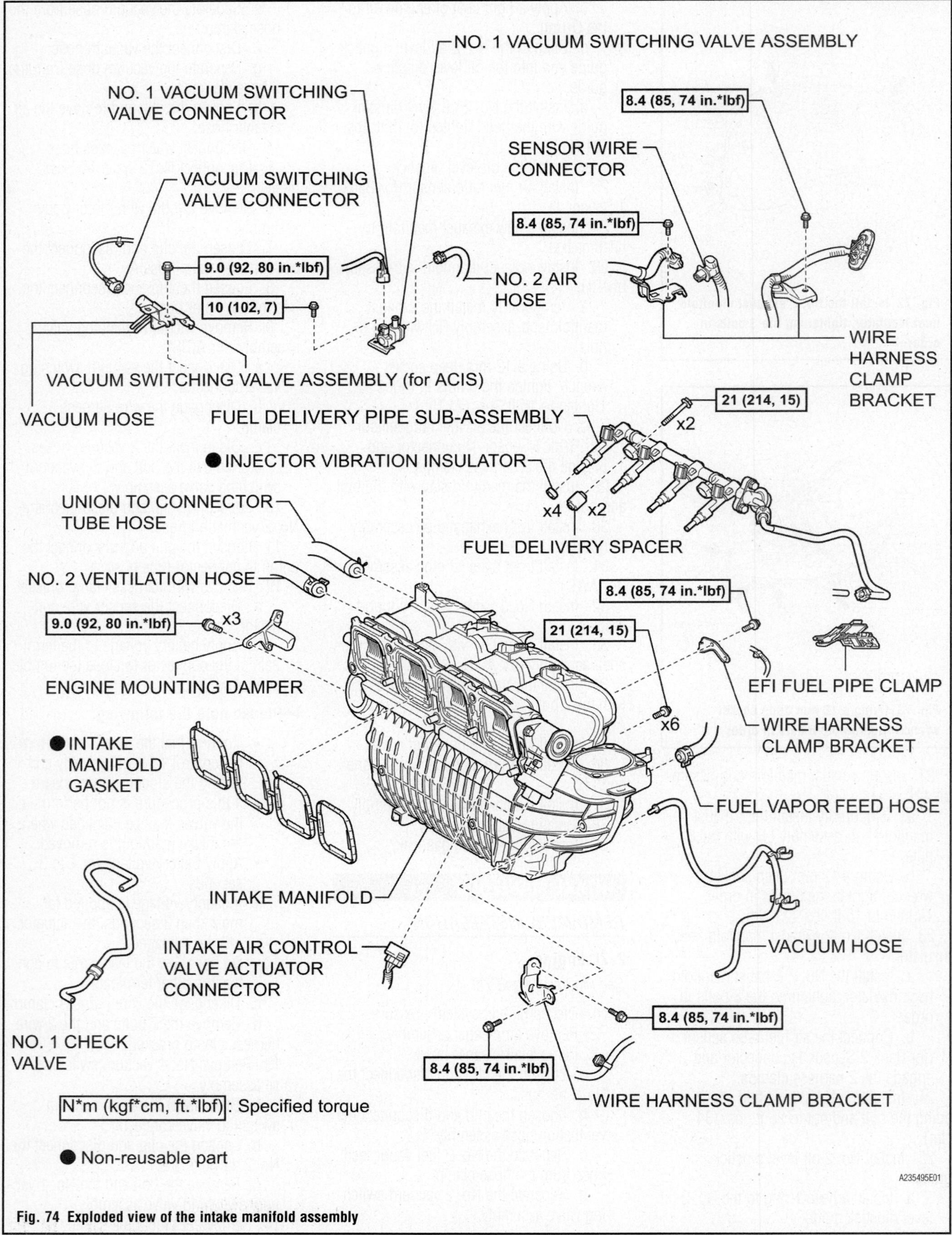
NO. 1 VACUUM SWITCHING VALVE ASSEMBLY
NO. 1 VACUUM SWITCHING VALVE CONNECTOR
8.4 (85, 74 in.*lbf)
SENSOR WIRE CONNECTOR
VACUUM SWITCHING VALVE CONNECTOR
8.4 (85, 74 in.*lbf)
9.0 (92, 80 in.*lbf)
NO. 2 AIR HOSE
10 (102, 7)
WIRE HARNESS CLAMP BRACKET
VACUUM SWITCHING VALVE ASSEMBLY (for ACIS)
21 (214, 15)
VACUUM HOSE
FUEL DELIVERY PIPE SUB-ASSEMBLY
x2
INJECTOR VIBRATION INSULATOR
UNION TO CONNECTOR TUBE HOSE
x4
x2
FUEL DELIVERY SPACER
NO. 2 VENTILATION HOSE
8.4 (85, 74 in.*lbf)
9.0 (92, 80 in.*lbf)
x3
21 (214, 15)
ENGINE MOUNTING DAMPER
EFI FUEL PIPE CLAMP
x6
INTAKE MANIFOLD GASKET
WIRE HARNESS CLAMP BRACKET
FUEL VAPOR FEED HOSE
INTAKE MANIFOLD
VACUUM HOSE
INTAKE AIR CONTROL VALVE ACTUATOR CONNECTOR
8.4 (85, 74 in.*lbf)
NO. 1 CHECK VALVE
8.4 (85, 74 in.*lbf)
WIRE HARNESS CLAMP BRACKET
N*m (kgf*cm, ft.*lbf) : Specified torque
Non-reusable part
A235495E01

Fig. 74 Exploded view of the intake manifold assembly

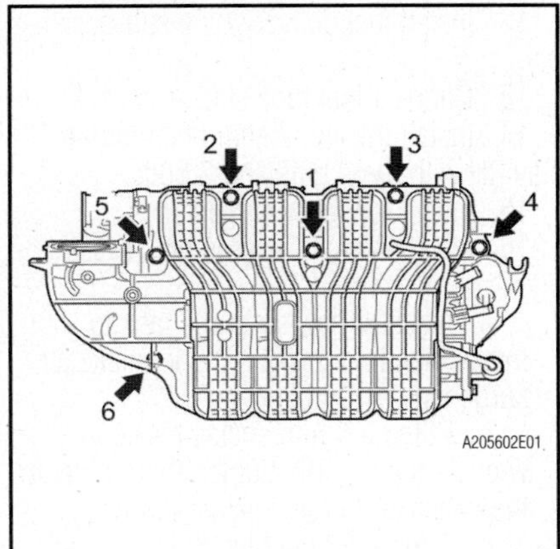

Fig. 75 Tighten the 6 bolts in order to 15 ft. lbs. (21 Nm)

15. Pinch the retainer to release it and disconnect the fuel tube.

➡Please note the following:

- Check for dirt or other foreign matter on the parts to be disconnected and clean them if necessary.
- The fuel tube connector seals with an O-ring. Ensure that there is no damage or foreign matter on the contact surface.
- If the connector and the pipe are stuck, push and pull on the connector to release them while pinching the retainer. Pull the connector out of the pipe carefully.
- Do not use any tools.
- Do not bend or twist the tubes.
- Protect the contact surface by covering it with a plastic bag.

16. Remove fuel delivery pipe sub-assembly.
 a. Disconnect the 4 fuel injector connectors.
 b. Remove the 2 bolts, and then remove the fuel delivery pipe together with the 4 fuel injectors.

➡Be careful not to drop the fuel injectors when removing the fuel delivery pipe.

 c. Remove the 2 fuel delivery spacers from the cylinder head.
 d. Remove the 4 injector vibration insulators from the cylinder head.

17. Remove intake manifold.
 a. Disengage the 3 wire harness clamps and separate the wire harness.
 b. Loosen the clip and disconnect the fuel vapor feed hose.
 c. Disengage the 3 wire harness clamps and separate the wire harness.
 d. Remove the bolt and the wire harness clamp bracket.
 e. Remove the bolt and disengage the 2 clamps.
 f. Disconnect the intake air control valve actuator connector.
 g. Remove the 6 bolts and the intake manifold.

➡The valves may be damaged if they are not closed before removing the intake manifold. When removing the intake manifold, be sure to remove the intake manifold slowly while confirming that the tumble control valves remain in the closed position. When the intake manifold has been removed, open the tumble control valves and keep them at a partially open position so that they do not protrude beyond the edge of the intake manifold.

 h. Remove the intake manifold gasket from the intake manifold.

To install:

18. Install intake manifold.
 a. Close the tumble control valves.

➡The valves may be damaged if they are not closed before installing the intake manifold.

 b. Install a new gasket onto the intake manifold.
 c. Temporarily tighten the 6 bolts to the intake manifold.
 d. Tighten the 6 bolts in order to 15 ft. lbs. (21 Nm).
 e. Connect the intake air control valve actuator connector.
 f. Engage the 2 clamps to the intake manifold.
 g. Install the wire harness clamp bracket with the bolt.
 h. Install the wire harness clamp bracket with the bolt.
 i. Engage the 3 wire harness clamps.
 j. Connect the fuel vapor feed hose with the clip.
 k. Engage the 3 wire harness clamps.

19. Install fuel delivery pipe sub-assembly.
 a. Install 4 new injector vibration insulators to the cylinder head.
 b. Install the 2 fuel delivery spacers onto the cylinder head.

➡Install the fuel delivery spacer so that the longer protrusion is on the cylinder head side.

 c. Install the fuel delivery pipe together with the 4 fuel injectors to the cylinder head, and then temporarily install the 2 bolts.

➡After installing the fuel injectors, check that they turn smoothly. If not, reinstall the injectors with new O-rings.

 d. Tighten the 2 bolts to 15 ft. lbs. (21 Nm).
 e. Connect the 4 fuel injector connectors.

20. Connect fuel tube sub-assembly.
 a. Line up the pipe and connector and push them together until a "click" sound is heard.

➡Check that there are no scratches or foreign objects on the connecting part. Check that the tube is inserted securely. Check that the tube is securely connected by pulling on it.

21. Install EFI fuel pipe clamp.

22. Install No. 1 vacuum switching valve assembly.
 a. Install the No. 1 vacuum switching valve assembly to the intake manifold with the bolt.
 b. Connect the No. 2 air hose with the clip.
 c. Connect the No. 1 vacuum switching valve connector.

23. Install wire harness clamp bracket.
 a. Install the 2 wire harness clamp brackets to the intake manifold with the 2 bolts.
 b. Engage the wire harness clamp to the wire harness clamp bracket.
 c. Connect the sensor wire connector.

24. Connect the union to connector tube hose to the intake manifold with the clip.

25. Connect the No. 2 ventilation hose with the clip.

26. Install vacuum switching valve assembly (for ACIS).
 a. Install the vacuum switching valve assembly to the intake manifold with the bolt.
 b. Connect the 2 vacuum hoses to the intake manifold.
 c. Engage the wire harness clamp to the vacuum switching valve assembly.
 d. Connect the vacuum switching valve connector.

27. Connect the inlet heater water hose A with the clip.

28. Connect the outlet heater water hose A with the clip.

29. Install throttle with motor body assembly.

30. Connect No. 2 water by-pass hose.

31. Connect water by-pass hose.

32. Install air cleaner hose.
 a. Install the air cleaner hose with the clip.

b. Install the vacuum hose to the 2 hose clamps.
c. Connect the vacuum hose.
d. Install the vacuum hose to the hose clamp.
e. Install the No. 2 vacuum switching valve assembly to the air cleaner hose.
f. Install the No. 2 fuel vapor feed hose to the hose clamp.
g. Connect the ventilation hose assembly to the air cleaner hose with the clip.
h. Connect the No. 1 air hose with the clip.
33. Install air cleaner assembly.
34. Inspect for fuel leak.

3.5L Engine

See Figures 76 through 78.

1. Discharge fuel system pressure.
2. Remove air cleaner and hose.
3. Remove water by-pass hose.
4. Remove No. 2 water by-pass hose.
5. Remove intake air surge tank assembly.
a. Disconnect the union to check valve hose.
b. Loosen the clip and disconnect the ventilation hose assembly.
c. Disconnect the throttle body connector and wire harness clamp.
d. Disconnect the No. 1 fuel vapor feed hose.
e. Disconnect the vacuum hose and the hose clamp.
f. Disconnect the intake air control valve connector.
g. Remove the bolt and separate the No. 1 surge tank stay from the intake air surge tank assembly.
h. Remove the bolt and separate the throttle body bracket from the intake air surge tank assembly.
i. Remove the 2 nuts.
j. Using a 5 mm socket hexagon wrench, remove the 4 bolts.
k. Remove the intake air surge tank assembly and 3 air surge tank to intake manifold gaskets.
6. Remove fuel hose protector.
7. Remove the 2 nuts, 2 bolts and the No. 2 engine mounting stay RH.
8. Disconnect fuel tube sub-assembly.
9. Remove fuel delivery pipe sub-assembly.
10. Remove intake manifold.
a. Remove the 6 bolts, 4 nuts and the intake manifold.
b. Remove the 2 intake manifold to head gaskets.

To install:

11. Install intake manifold.
a. Set a new gasket on each cylinder head.

➡Align the port holes of the gaskets and cylinder head. Make sure that the gaskets are installed in the correct direction.

b. Set the intake manifold on the cylinder head.
c. Tighten the 6 bolts and 4 nuts uniformly in several steps. Tighten to 15 ft. lbs. (21 Nm).

➡Tighten the inner installation bolts of the intake manifold before tightening the outer bolts.

12. Install fuel delivery pipe sub-assembly.
13. Connect fuel tube sub-assembly.
14. Install the No. 2 engine mounting stay RH with the 2 nuts and 2 bolts.
15. Install fuel hose protector.
16. Install intake air surge tank assembly.
a. Install 3 new air surge tank to intake manifold gaskets to the intake air surge tank assembly.
b. Using a 5 mm socket hexagon wrench, temporarily tighten the 4 bolts to the intake air surge tank assembly.
c. Temporarily tighten the 2 nuts to the intake air surge tank assembly.
d. Tighten the 4 bolts and 2 nuts in

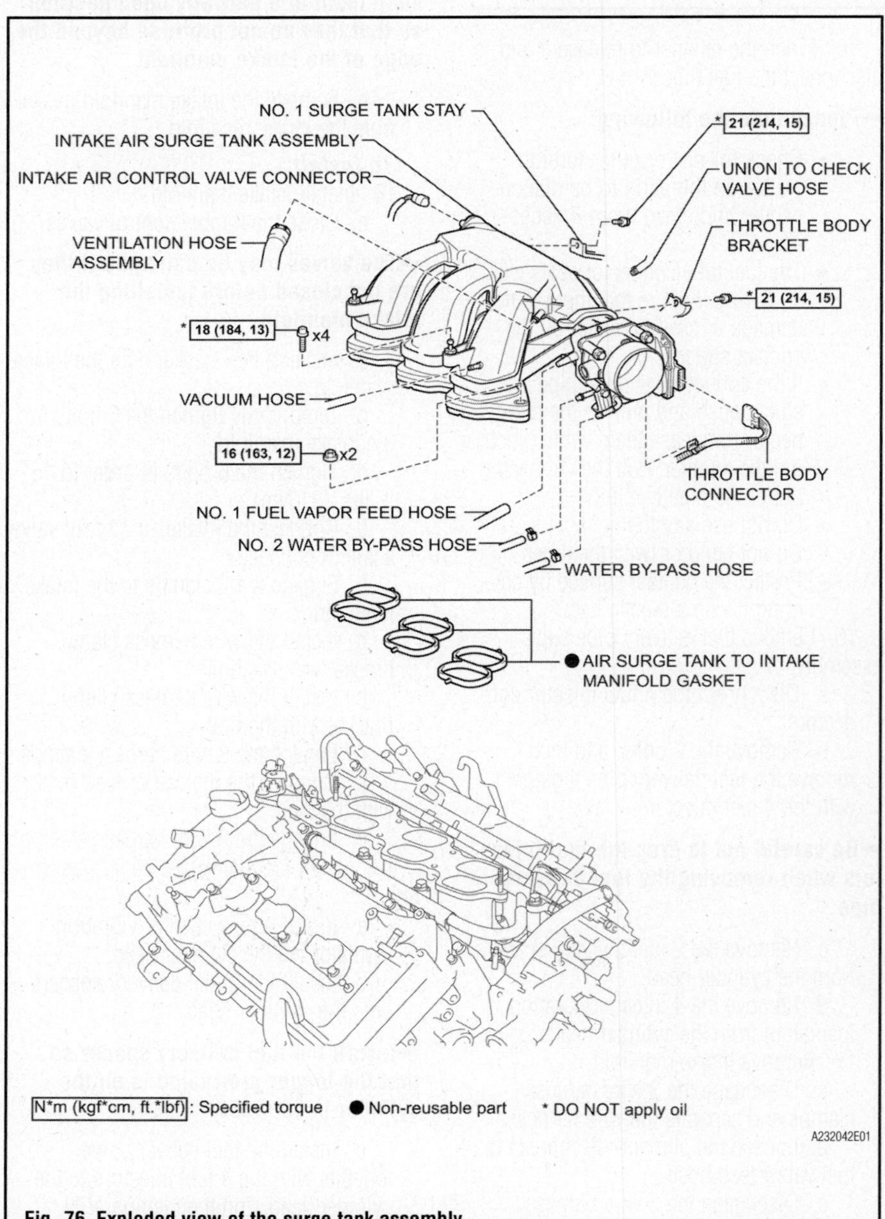

Fig. 76 Exploded view of the surge tank assembly

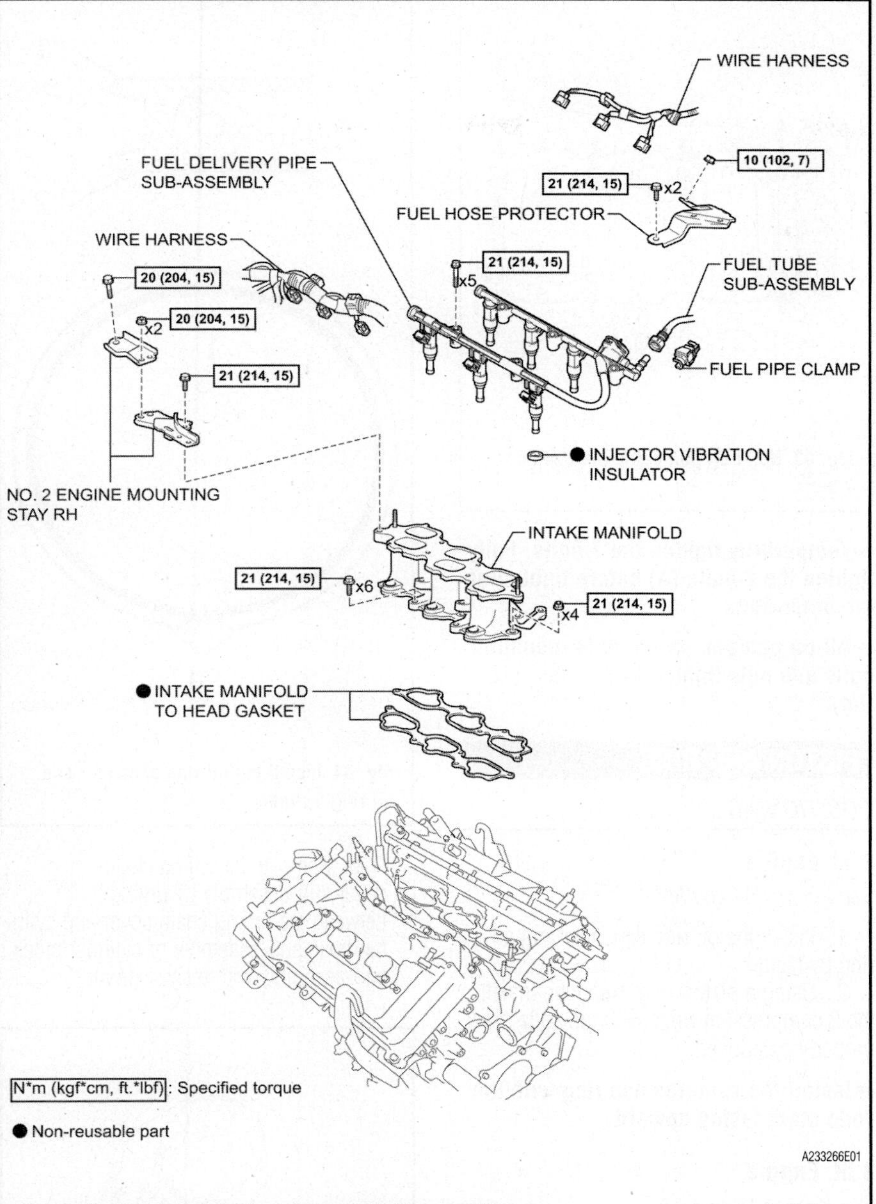

Fig. 77 Exploded view of the intake manifold assembly

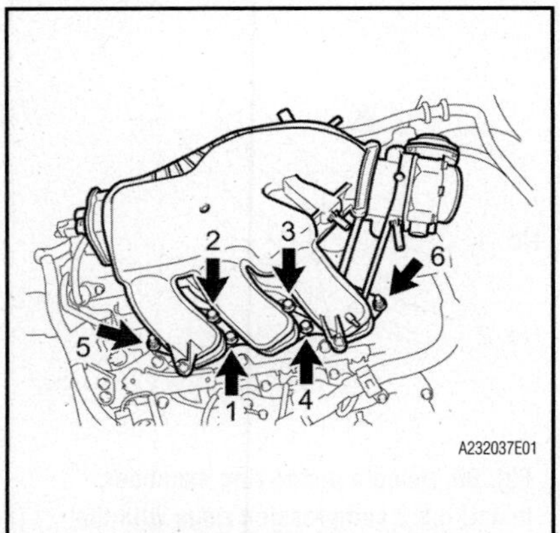

Fig. 78 Tighten the 4 bolts and 2 nuts in order

order. Tighten the bolts to 13 ft. lbs. (18 Nm). Tighten the nuts to 12 ft. lbs. (16 Nm).

e. Install the throttle body bracket to the intake air surge tank assembly with the bolt. Tighten to 15 ft. lbs. (21 Nm).

f. Install the No. 1 surge tank stay to the intake air surge tank assembly with the bolt. Tighten to 15 ft. lbs. (21 Nm).

g. Connect the intake air control valve connector.

h. Connect the vacuum hose and the hose clamp.

i. Connect the No. 1 fuel vapor feed hose.

j. Connect the throttle body connector and wire harness clamp to the throttle with motor body assembly.

k. Connect the ventilation hose assembly with the clip.

l. Connect the union to check valve hose.

17. Install water by-pass hose.
18. Install No. 2 water by-pass hose.
19. Add engine coolant.
20. Inspect for coolant leak.
21. Check that there are no fuel leaks from the fuel system after doing any maintenance or repairs.
22. Install air cleaner and hose.

OIL PAN

REMOVAL & INSTALLATION

2.7L Engine

See Figures 79 and 80.

➡The manufacturer does not provide a specific Removal and Installation procedure for this component. Refer to

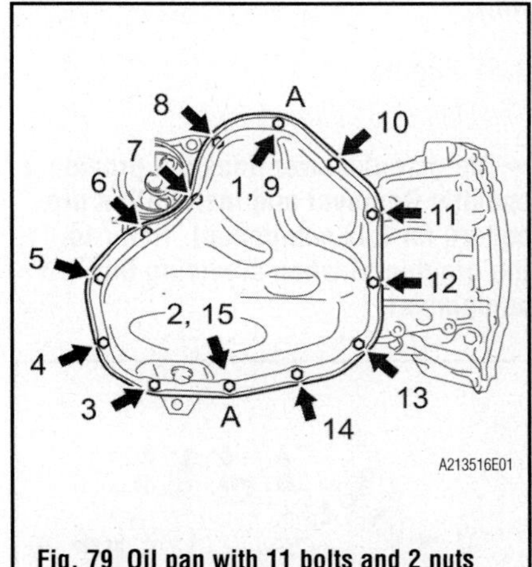

Fig. 79 Oil pan with 11 bolts and 2 nuts (Tightening sequence)

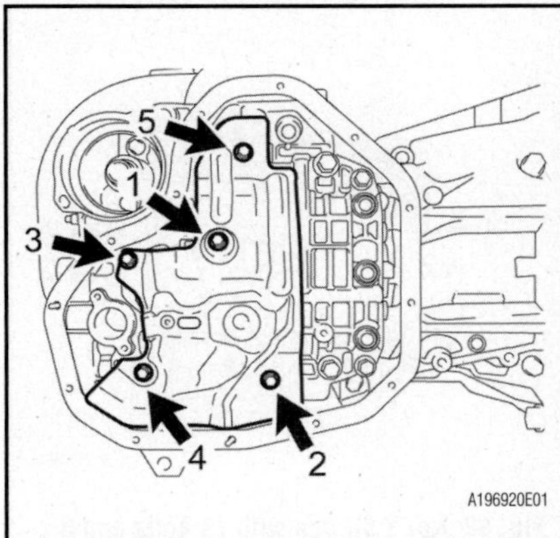

Fig. 80 Oil pan baffle plate (Tightening sequence)

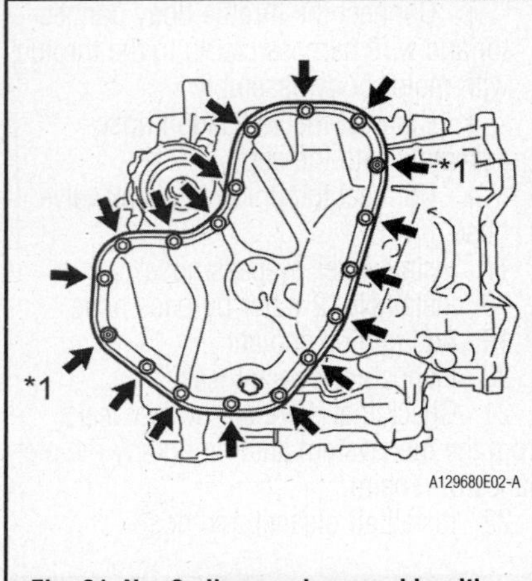

Fig. 81 No. 2 oil pan sub-assembly with 16 bolts and 2 nuts (1)

the graphic(s) when servicing this component.

➡Bolt A and Nut A are tightened twice.

➡All oil pan and baffle plate mounting bolts and nuts tighten to 7 ft. lbs. (10 Nm).

3.5L Engine

See Figures 81 through 83.

➡The manufacturer does not provide a specific Removal and Installation procedure for this component. Refer to the graphic(s) when servicing this component.

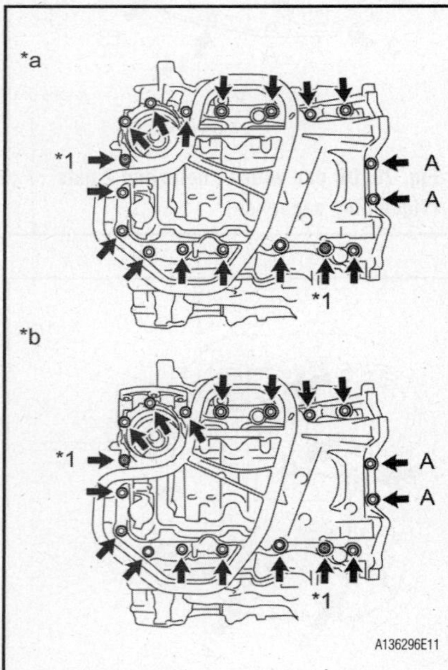

Fig. 82 No. 1 oil pan with 16 bolts and 2 nuts (1); without oil cooler (a), with oil cooler (b)

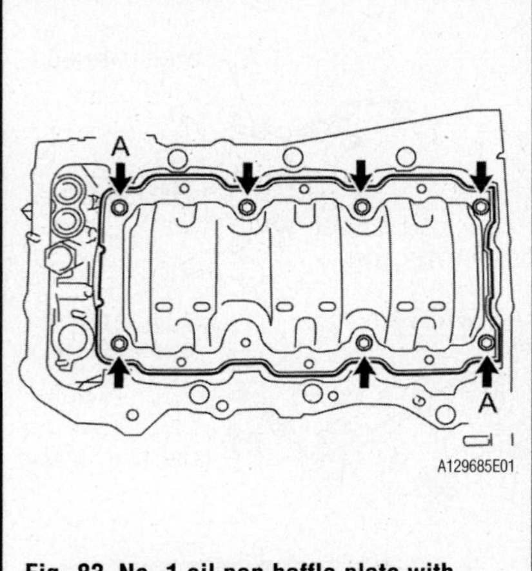

Fig. 83 No. 1 oil pan baffle plate with 7 bolts

➡Temporarily tighten the 7 bolts. Fully tighten the 2 bolts (A) before tightening the other bolts.

➡All oil pan and baffle plate mounting bolts and nuts tighten to 7 ft. lbs. (10 Nm).

PISTON AND RING

POSITIONING

2.7L Engine

See Figures 84 and 85.

1. Install the oil ring expander and oil ring by hand.
2. Using a piston ring expander, install the 2 compression rings with the code mark properly positioned.

➡Install the compression ring with the code mark facing upward.

3.5L Engine

See Figures 86 and 87.

1. Install the oil ring expander and 2 side rails by hand.
2. Using a piston ring expander, install the compression rings.
3. Install the piston rings so that the ring ends are properly positioned.

➡Do not align the ring ends.

TIMING CHAIN FRONT COVER

REMOVAL & INSTALLATION

3.5L Engine

See Figures 88 through 90.

1. Remove timing chain cover sub-assembly.
   a. Remove the 15 bolts and 2 nuts.

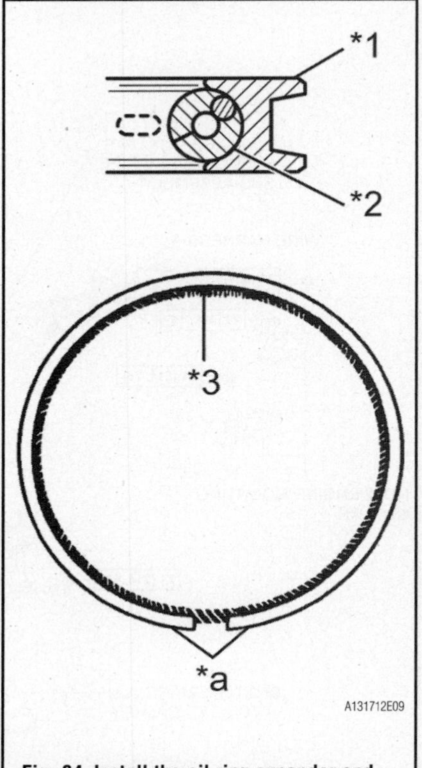

Fig. 84 Install the oil ring expander and oil ring by hand

   b. Remove the timing chain cover sub-assembly by prying between the timing chain cover and cylinder head sub-assembly or cylinder block sub-assembly with a screwdriver.

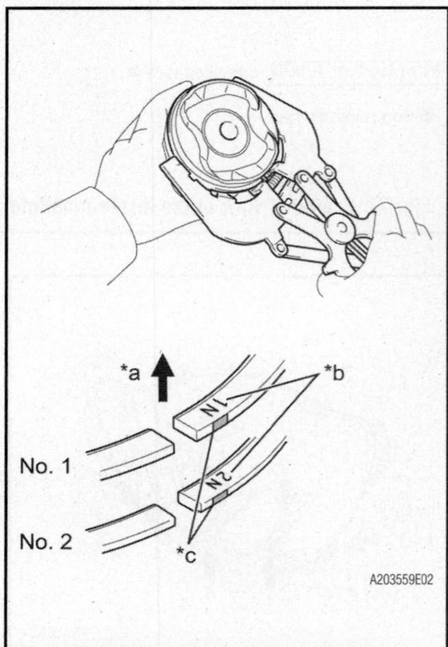

Fig. 85 Using a piston ring expander, install the 2 compression rings with the code mark (b) properly positioned facing upward (a); paint mark (c)

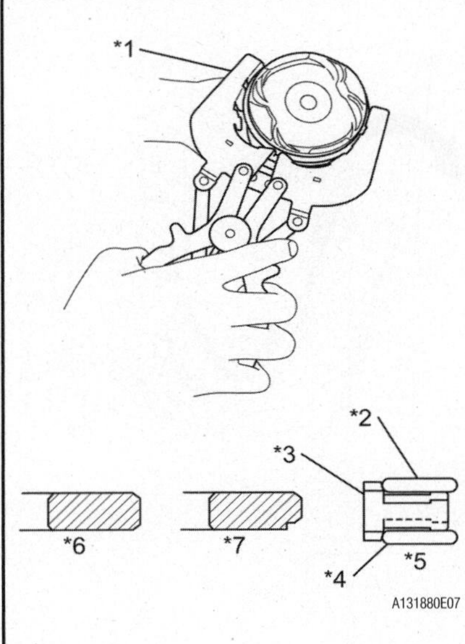

Fig. 86 Using a piston ring expander, install the compression rings

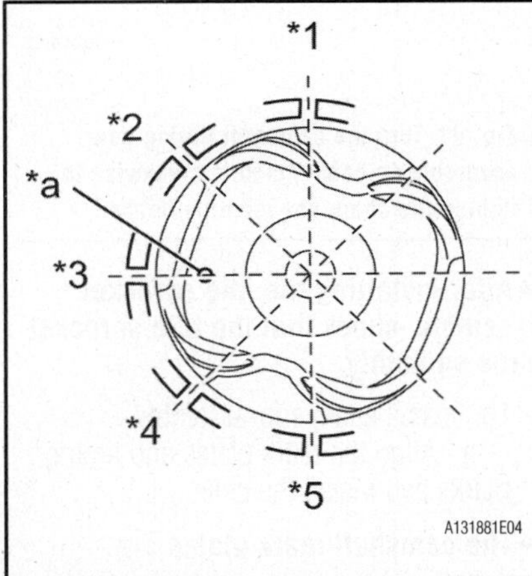

Fig. 87 Position the piston rings so that the ring ends are properly positioned

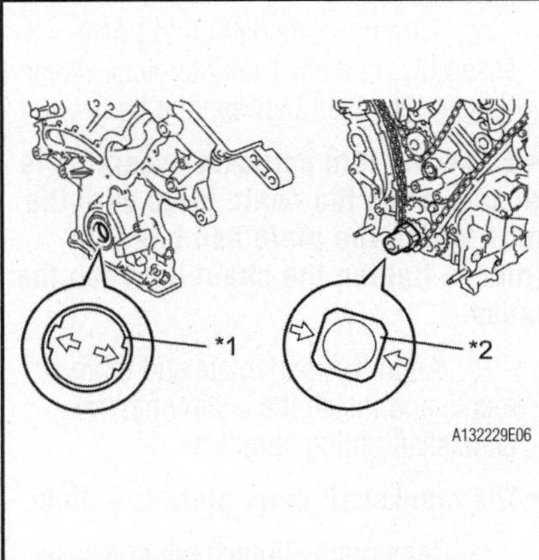

Fig. 88 Align the oil pump's drive rotor spline (1) and the crankshaft (2)

➡Be careful not to damage the contact surfaces of the cylinder head, cylinder block and chain cover.

➡Tape the screwdriver tip before use.

2. Remove the 4 bolts, chain cover plate and chain cover plate gasket.
3. Remove the gasket.
4. Using a screwdriver, pry out the timing chain case oil seal.

To install:

5. Install timing gear case or timing chain case oil seal.
   a. Apply MP grease to a new timing chain case oil seal lip.
   b. Using SST, tap in a new oil seal until its surface is flush with the timing chain cover edge.

➡Please note the following:

- Keep the lip free from foreign matter.
- Do not tap on the oil seal at an angle.
- Make sure that the oil seal edge does not stick out of the timing chain cover.

6. Install timing chain cover sub-assembly.
   a. Apply seal packing in a continuous line to the engine unit.

➡Please note the following:

- Be sure to clean and degrease the contact surfaces.
- If there is oil on the contact surfaces, wipe them with an oil-free cloth before applying seal packing.
- Install the timing chain cover sub-assembly within 3 minutes after applying seal packing.
- Do not start the engine for at least 2 hours after installing the timing chain cover sub-assembly.

   b. Apply seal packing in a line to the timing chain cover.

➡Please note the following:

- If there is oil on the contact surfaces, wipe them with an oil-free cloth before applying seal packing.
- Install the timing chain cover sub-assembly within 3 minutes and tighten the bolts within 15 minutes after applying seal packing.
- Do not start the engine for at least 2 hours after installing the timing chain cover sub-assembly.

   c. Install a new gasket.
   d. Align the oil pump's drive rotor spline and the crankshaft. Install the oil pump and timing chain cover to the crankshaft.

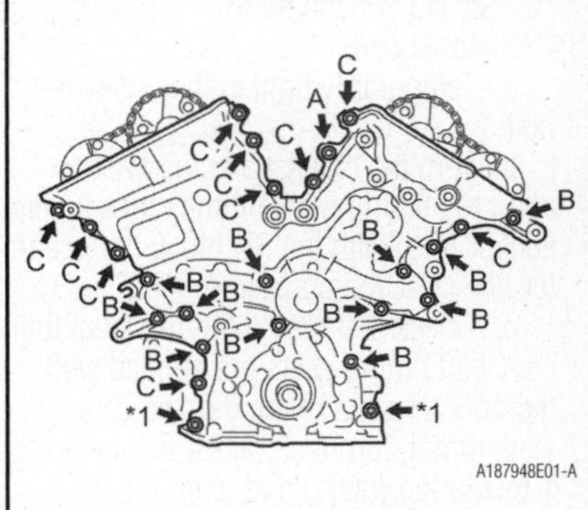

Fig. 89 Temporarily tighten the timing chain cover with the 23 bolts and 2 nuts (1)

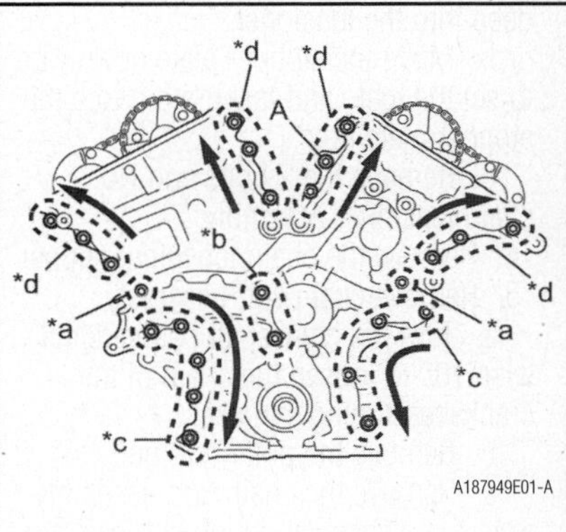

Fig. 90 Fully tighten the bolts in area 1 and area 2

   e. Temporarily tighten the timing chain cover with the 23 bolts and 2 nuts.
   - Bolt A: 1.57 inches (40 mm)
   - Bolt B: 2.17 inches 55 mm)
   - Bolt C: 0.984 inches (25 mm)

➡Make sure that there is no oil on the threads of bolt B and C.

   f. Fully tighten the bolts in area 1 and area 2. Tighten to 15 ft. lbs. (21 Nm).
   g. Fully tighten the bolts and nuts in area 3. Tighten to 15 ft. lbs. (21 Nm).
   h. Fully tighten the bolts in area 4.
   - Tighten Bolt A to 32 ft. lbs. (43 Nm).
   - Tighten other bolts to 15 ft. lbs. (21 Nm).

TIMING CHAIN & SPROCKETS

REMOVAL & INSTALLATION

3.5L Engine

See Figures 91 through 96.

1. Remove timing chain cover sub-assembly.

2. Set No. 1 cylinder to TDC/compression.

a. Temporarily tighten the pulley set bolt.

b. Turn the crankshaft clockwise to align the timing mark on the crank angle sensor plate with the RH block bore center line (TDC/compression).

c. Check that the timing marks of the camshaft timing gears are aligned with the timing marks of the bearing cap.

d. If not, turn the crankshaft clockwise 1 revolution (360°) and align the timing marks.

3. Remove No. 1 chain tensioner assembly.

a. Move the stopper plate upward to release the lock, and push the plunger deep into the tensioner.

b. Move the stopper plate downward to set the lock, and insert a pin into the stopper plate hole.

c. Remove the 2 bolts and No. 1 chain tensioner assembly.

4. Remove the chain tensioner slipper.

5. Remove chain sub-assembly.

a. Turn the crankshaft counterclockwise 10° to loosen the chain of the crankshaft timing sprocket.

b. Remove the pulley set bolt.

c. Remove the chain sub-assembly from the crankshaft timing sprocket and place it on the crankshaft.

d. Turn the camshaft timing gear assembly on bank 1 clockwise (approximately 60°) and set it. Be sure to loosen the chain sub-assembly between the banks.

e. Remove the timing chain sub-assembly.

6. Using a 10 mm hexagon wrench, remove the No. 2 idle gear shaft, idle sprocket assembly and No. 1 idle gear shaft.

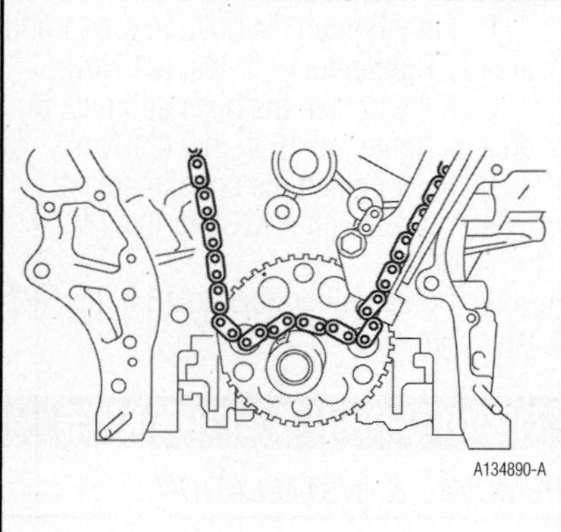

Fig. 91 Remove the chain sub-assembly from the crankshaft timing sprocket and place it on the crankshaft

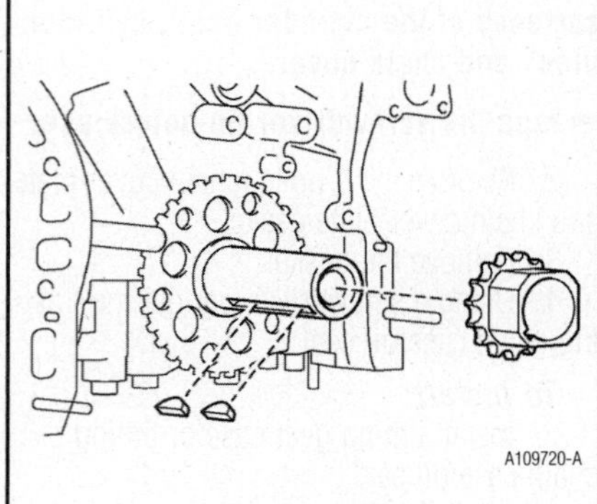

Fig. 92 Remove the crankshaft timing sprocket from the crankshaft

7. Remove the 2 bolts and No. 1 chain vibration damper.

8. Remove the 2 No. 2 chain vibration dampers.

9. Remove crankshaft timing sprocket.

a. Remove the crankshaft timing sprocket from the crankshaft.

b. Remove the 2 keys from the crankshaft.

To install:

10. Install No. 3 chain tensioner assembly.

11. Install camshaft timing gears and No. 2 chain (for bank 2).

12. Install No. 2 chain tensioner assembly.

13. Install camshaft timing gears and No. 2 chain (for bank 1).

14. Install the No. 1 chain vibration damper with the 2 bolts. Tighten to 17 ft. lbs. (23 Nm).

15. Install the 2 No. 2 chain vibration dampers.

16. Install the 2 keys and crankshaft timing sprocket.

17. Install idle sprocket assembly.

a. Apply a light coat of engine oil to the rotating surface of the No. 1 idle gear shaft.

b. Temporarily install the No. 1 idle gear shaft and idle sprocket with the No. 2 idle gear shaft while aligning the knock pin of the No. 1 idle gear shaft with the knock pin groove of the cylinder block.

➡Be careful of the idle gear installation position.

➡Check that no foreign objects are on the No. 1 and No. 2 idle gear shafts.

c. Using a 10 mm hexagon wrench, tighten the No. 2 idle gear shaft.

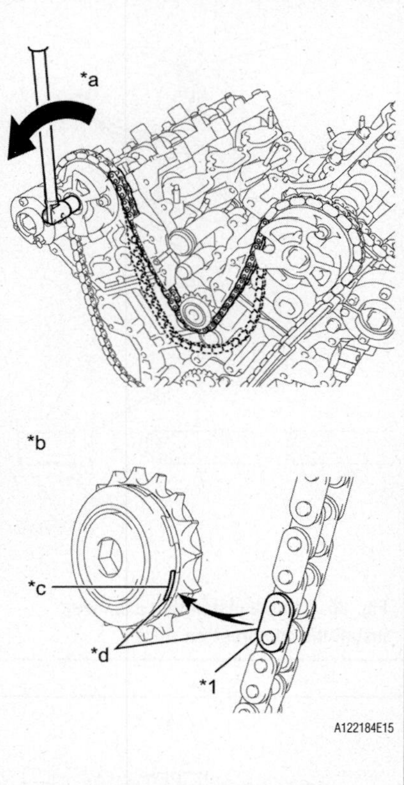

Fig. 93 Turn the camshaft timing gear assembly on bank 1 counterclockwise to tighten the chain between the banks

➡After installing the idle sprocket assembly, check that the idle sprocket turns smoothly.

18. Install chain sub-assembly.

a. Align the mark plates and timing marks and install the chain.

➡The camshaft mark plates are orange.

b. Do not pass the chain over the crankshaft, just temporarily place it on the crankshaft.

c. Turn the camshaft timing gear assembly on bank 1 counterclockwise to tighten the chain between the banks.

➡When the idle sprocket assembly is reused, align the chain plate with the mark where the plate had been in order to tighten the chain between the banks.

d. Align the mark plate and timing marks and install the chain onto the crankshaft timing sprocket.

➡The crankshaft mark plate is yellow.

e. Temporarily tighten the pulley set bolt.

f. Turn the crankshaft clockwise to set

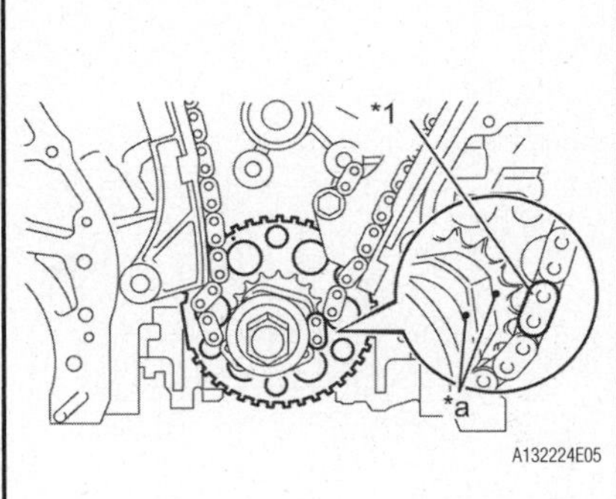

Fig. 94 Align the mark plate (1) and timing marks (a) and install the chain onto the crankshaft timing sprocket

it to the RH block bore center line (TDC/compression).

19. Install the chain tensioner slipper.

20. Install No. 1 chain tensioner assembly.

a. Move the stopper plate upward to release the lock, and push the plunger deep into the tensioner.

b. Move the stopper plate downward to set the lock, and insert a hexagon wrench into the hole of the stopper plate.

c. Install the No. 1 chain tensioner assembly with the 2 bolts.

d. Remove the hexagon wrench from the No. 1 chain tensioner assembly.

21. Inspect valve timing.

a. Check the camshaft timing marks.

➡Check each timing mark from a viewpoint directly in-line with the center of the camshaft and the timing mark on each camshaft timing gear. If the timing marks are checked from any other viewpoint, the valve timing may appear misaligned.

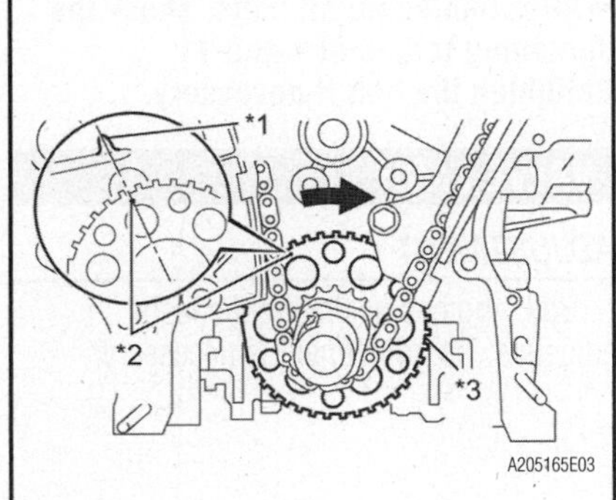

Fig. 95 Turn the crankshaft clockwise to set it to the RH block bore center line (1) (TDC/compression); timing mark (2), sensor plate (3)

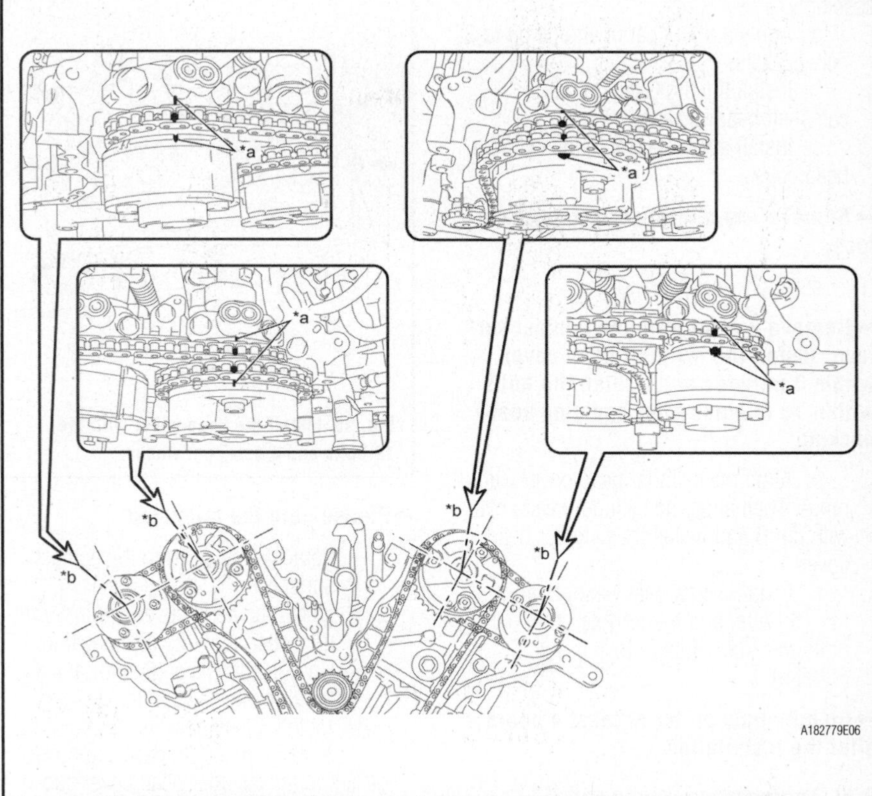

Fig. 96 Check that each camshaft timing mark (a) is properly positioned; viewpoint (b)

b. Check that each camshaft timing mark is properly positioned.

➡For the intake camshaft: Be sure to check mark A at the point when marks B, C, and D are positioned in line. If the marks are checked from any other viewpoint, they cannot be checked correctly.

c. If the valve timing is misaligned, reinstall the timing chain.

d. Remove the pulley set bolt.

VALVE COVERS

REMOVAL & INSTALLATION

2.7L Engine

See Figure 97.

1. Remove cylinder head cover sub-assembly.

a. Remove the 16 bolts, 3 seal washers, cylinder head cover and gasket.

b. Remove the 3 gaskets from the camshaft bearing caps.

2. Using a screwdriver, pry out the 4 plug tube gaskets.

➡Be careful not to damage the cylinder head cover. Tape the screwdriver tip before use.

To install:

3. Install the 4 plug tube gaskets to the cylinder head cover.

➡After pressing in the spark plug tube gasket, make sure the gasket protrudes 0.0394 inches (1.0 mm) or less from the cylinder head cover.

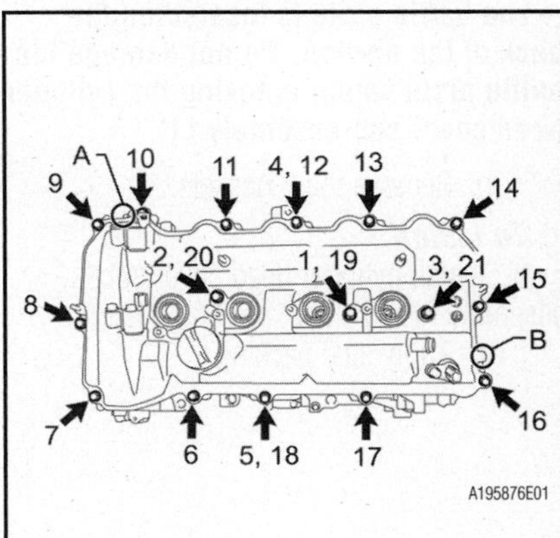

Fig. 97 Install 3 new seal washers and the 16 bolts, and then tighten the bolts in order

4. Install cylinder head cover sub-assembly.

a. Apply a light coat of engine oil to 3 new gaskets.

b. Install the 3 gaskets to the camshaft bearing caps.

c. Install a new gasket to the cylinder head cover.

➡Remove any oil from the contact surface.

d. Apply seal packing.

➡Remove any oil from the contact surface. Install the cylinder head cover within 3 minutes and tighten the bolts within 15 minutes after applying seal packing.

e. Align the cylinder head cover with pin A. Then align the cylinder head cover with pin B and install the cylinder head cover.

f. Install 3 new seal washers and the 16 bolts, and then tighten the bolts in order. Tighten to 9 ft. lbs. (12 Nm).

➡Do not apply oil for at least 4 hours after the installation.

3.5L Engine

See Figures 98 through 100.

1. Remove cylinder head cover sub-assembly RH

a. Remove the 12 bolts, seal washer, cylinder head cover sub-assembly and cylinder head cover gasket.

b. Remove the 3 gaskets.

2. Remove cylinder head cover sub-assembly LH

a. Remove the 12 bolts, seal washer, cylinder head cover sub-assembly LH and cylinder head cover gasket.

➡The baffle plate is located on the back of the portion. Do not damage the baffle plate when removing the cylinder head cover sub-assembly LH.

b. Remove the 3 gaskets.

To install:

3. Install cylinder head cover sub-assembly RH.

a. Apply seal packing.

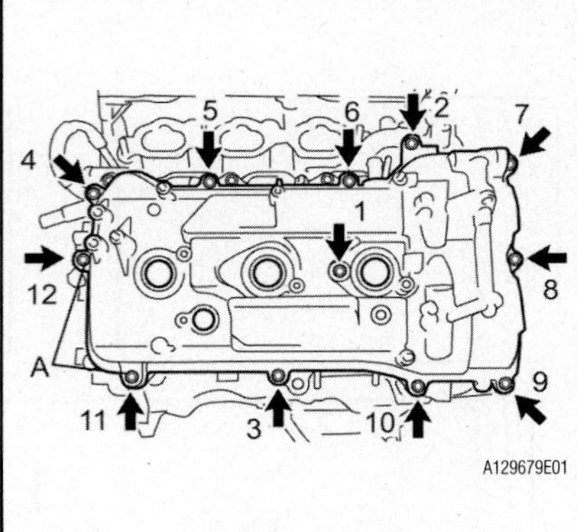

Fig. 98 Install the head cover with the 12 bolts and a new seal washer

➡Please note the following:

- Remove any oil from the contact surface.
- Install the head cover within 3 minutes after applying seal packing.
- Do not start the engine for at least 2 hours after installing.

b. Install 3 new gaskets.

c. Install a new cylinder head cover gasket to the cylinder head cover sub-assembly.

d. Install the head cover with the 12 bolts and a new seal washer. Tighten bolts A to 15 ft. lbs. (21 Nm). Tighten all others to 7 ft. lbs. (10 Nm).

➡After tightening all bolts, check the tightening torque of 1 and 11. Retighten the bolt if necessary.

4. Install cylinder head cover sub-assembly LH

a. Apply seal packing.

➡Please note the following:

- Remove any oil from the contact surface.
- Install the head cover within 3 minutes after applying seal packing.
- Do not start the engine for at least 2 hours after installing.

b. Install 3 new gaskets.

c. Install a new cylinder head cover gasket to the cylinder head cover sub-assembly LH.

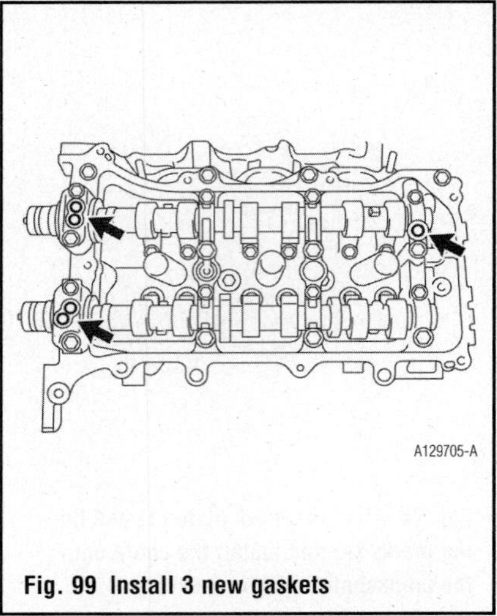

Fig. 99 Install 3 new gaskets

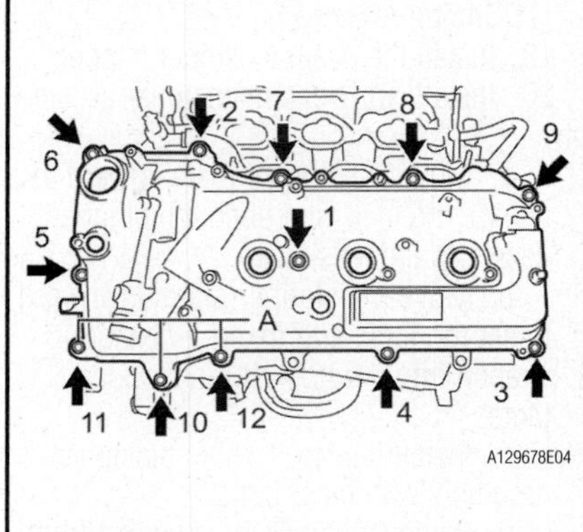

Fig. 100 Install the cylinder head cover sub-assembly LH with the 12 bolts and a new seal washer

d. Install the cylinder head cover sub-assembly LH with the 12 bolts and a new seal washer. Tighten bolts A to 15 ft. lbs. (21 Nm). Tighten all others to 7 ft. lbs. (10 Nm).

➡After tightening all bolts, check the tightening torque of 1 and 11. Retighten the bolt if necessary.

VALVE LASH

ADJUSTMENT

Both engines use hydraulic lash adjusters. No adjustment is necessary.

ENGINE PERFORMANCE & EMISSION CONTROLS

AIR FUEL RATIO SENSOR

REMOVAL & INSTALLATION

2.7L Engine

See Figure 101.

1. Remove air fuel ratio sensor.
   a. Disconnect the air fuel ratio sensor connector.
   b. Using SST, remove the air fuel ratio sensor from the exhaust manifold.
2. Installation is the reverse order of removal.
3. Tighten to 32 ft. lbs. (44 Nm).

3.5L Engine

See Figures 102 and 103.

1. Remove air cleaner inlet cover assembly seal

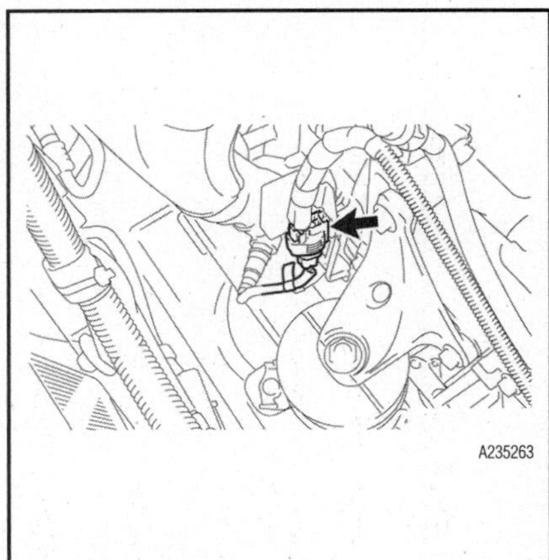

Fig. 101 Disconnect the air fuel ratio sensor connector

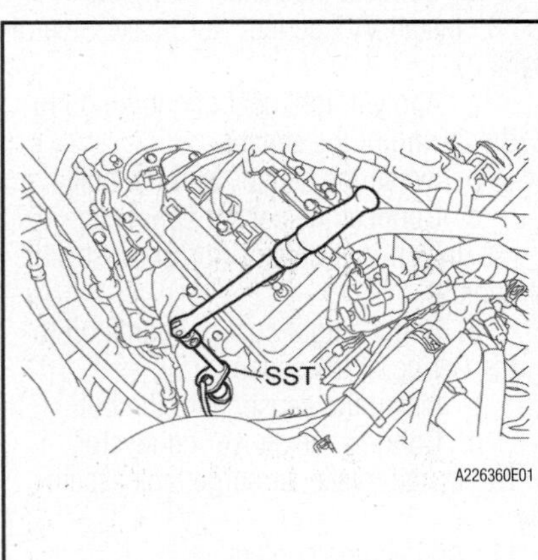

Fig. 102 Using SST, remove the air fuel ratio sensor from the exhaust manifold LH

2. Remove V-bank cover sub-assembly
3. Remove air fuel ratio sensor (for Bank 2 Sensor 1)
   a. Disengage the 3 wire harness clamps.
   b. Disconnect the air fuel ratio sensor connector.
   c. Using SST, remove the air fuel ratio sensor from the exhaust manifold LH.
4. Remove No. 1 engine under cover
5. Remove No. 2 engine under cover (for AWD)
6. Remove propeller with center bearing shaft assembly (for AWD)
7. Remove No. 1 exhaust pipe support bracket
8. Remove front exhaust pipe assembly
9. Remove manifold stay
10. Remove exhaust manifold sub-assembly RH
11. Remove exhaust manifold to head gasket
12. Remove air fuel ratio sensor No.2 (for Bank 1 Sensor 1)
    a. Using SST, remove the air fuel ratio sensor from the exhaust manifold RH.

To install:

13. Install air fuel ratio sensor No.2 (for Bank 1 Sensor 1).
    a. Using SST, install the air fuel ratio sensor to the exhaust manifold RH. Tighten to 32 ft. lbs. (44 Nm).
14. Install exhaust manifold to head gasket.
15. Install exhaust manifold sub-assembly RH.
16. Install manifold stay.
17. Install front exhaust pipe assembly.
18. Install No. 1 exhaust pipe support bracket.

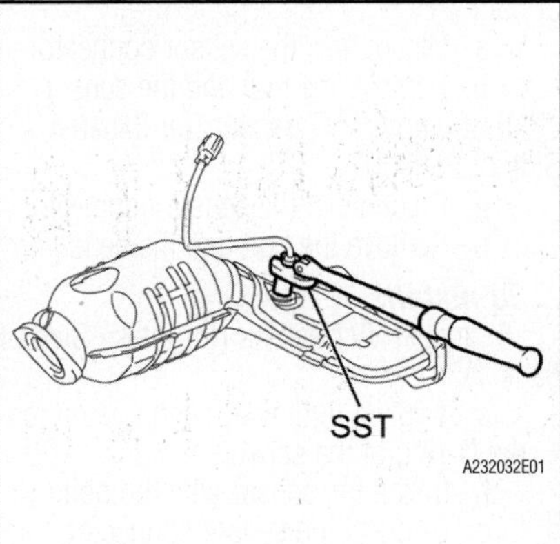

Fig. 103 Using SST, remove the air fuel ratio sensor from the exhaust manifold RH

19. Install propeller with center bearing shaft assembly (for AWD).
20. Install No. 2 engine under cover (for AWD).
21. Install No. 1 engine under cover.
22. Install air fuel ratio sensor (for bank 2 sensor 1).
    a. Using SST, install the air fuel ratio sensor to the exhaust manifold LH. Tighten to 32 ft. lbs. (44 Nm).
    b. Connect the air fuel ratio sensor connector.
    c. Engage the 3 wire harness clamps.
23. Inspect for exhaust gas leak.
24. Install V-bank cover sub-assembly.
25. Install air cleaner inlet cover assembly seal.

CAMSHAFT POSITION (CMP) SENSOR

REMOVAL & INSTALLATION

2.7L Engine

See Figure 104.

1. Remove No. 1 engine cover sub-assembly.
2. Remove camshaft position sensor (for Exhaust Side).
   a. Disconnect the camshaft position sensor connector.
   b. Remove the bolt and the sensor.
3. Remove camshaft position sensor (for Intake Side).
   a. Disconnect the camshaft position sensor connector.
   b. Remove the bolt and the sensor.

To install:

4. Install camshaft position sensor (for Exhaust Side).
   a. Apply a light coat of engine oil to the O-ring of the sensor.

➡Make sure that the O-ring is not cracked or does not jump out of position during installation.

   b. Apply adhesive to 2 or 3 threads of the bolt.
   c. Install the sensor with the bolt.
   d. Connect the camshaft position sensor connector.
5. Install camshaft position sensor (for Intake Side).
   a. Apply a light coat of engine oil to the O-ring of the sensor.

➡Make sure that the O-ring is not cracked or does not jump out of position during installation.

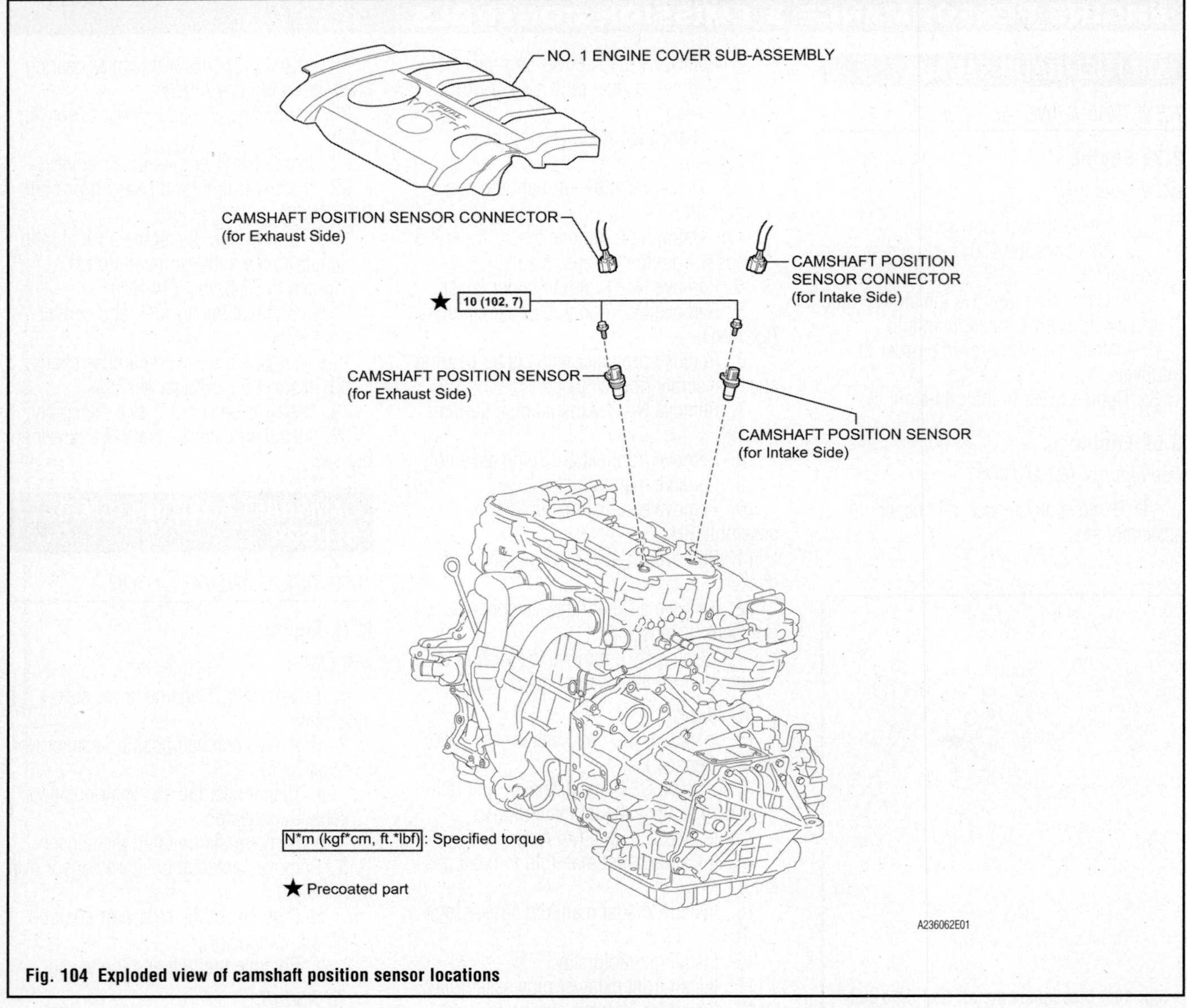

Fig. 104 Exploded view of camshaft position sensor locations

b. Apply adhesive to 2 or 3 threads of the bolt.

c. Install the sensor with the bolt.

d. Connect the camshaft position sensor connector.

6. Inspect for oil leak.

7. Install No. 1 engine cover sub-assembly.

3.5L Engine

See Figure 105.

1. Remove intake air surge tank assembly.

2. Remove VVT sensor (for Intake Side of Bank 1).

a. Disconnect the sensor connector.

b. Remove the bolt and the sensor.

3. Remove VVT sensor (for Exhaust Side of Bank 1).

a. Disconnect the sensor connector.

b. Remove the bolt and the sensor.

4. Remove VVT sensor (for Intake Side of Bank 2).

a. Disconnect the sensor connector.

b. Remove the bolt and the sensor.

5. Remove VVT sensor (for Exhaust Side of Bank 2).

a. Disconnect the sensor connector.

b. Remove the bolt and the sensor.

To install:

6. Install VVT sensor (for Intake Side of Bank 1).

a. Apply a light coat of engine oil to the O-ring of the sensor.

b. Install the sensor with the bolt.

c. Connect the sensor connector.

7. Install VVT sensor (for Exhaust Side of Bank 1).

a. Apply a light coat of engine oil to the O-ring of the sensor.

b. Install the sensor with the bolt.

c. Connect the sensor connector.

8. Install VVT sensor (for Intake Side of Bank 2).

a. Apply a light coat of engine oil to the O-ring of the sensor.

b. Install the sensor with the bolt.

c. Connect the sensor connector.

9. Install VVT sensor (for Exhaust Side of Bank 2).

a. Apply a light coat of engine oil to the O-ring of the sensor.

b. Install the sensor with the bolt.

c. Connect the sensor connector.

10. Install intake air surge tank assembly.

11. Add engine coolant.

12. Inspect for coolant leak.

13. Inspect for oil leak.

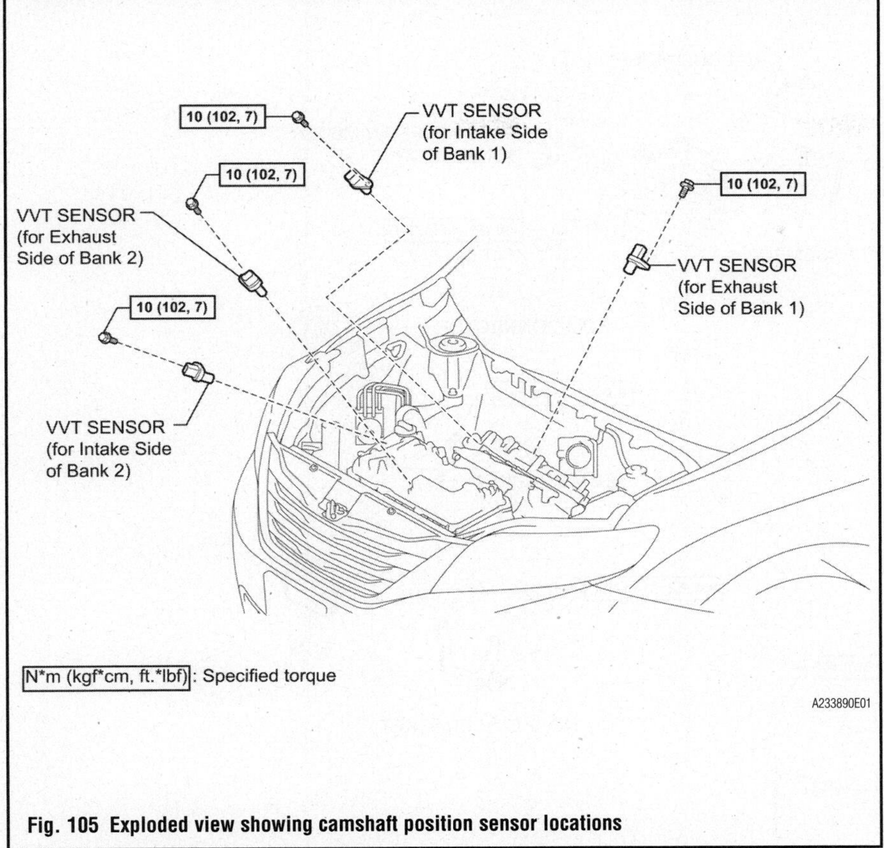

Fig. 105 Exploded view showing camshaft position sensor locations

CRANKSHAFT POSITION (CKP) SENSOR

REMOVAL & INSTALLATION

2.7L Engine

See Figure 106.

1. Remove crankshaft position sensor.
   a. Disconnect the crankshaft position sensor connector.

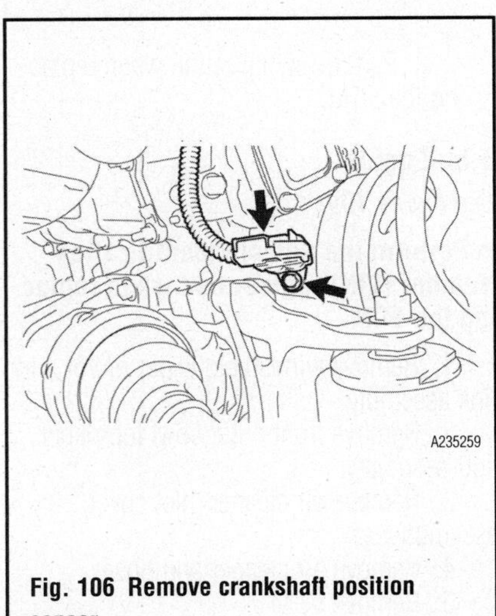

Fig. 106 Remove crankshaft position sensor

   b. Remove the bolt and the sensor.

To install:

2. Install crankshaft position sensor.
   a. Apply a light coat of engine oil to the O-ring of the sensor.

➡Make sure that the O-ring is not cracked or does not jump out of position during installation.

   b. Apply adhesive to 2 or 3 threads of the bolt.
   c. Install the sensor with the bolt.
3. Connect the crankshaft position sensor connector.
4. Start the engine and check that there are no oil leaks.

3.5L Engine

See Figure 107.

1. Remove w/pulley compressor assembly.
2. Remove crankshaft position sensor.
   a. Disconnect the crankshaft position sensor connector.
   b. Remove the bolt and the crankshaft position sensor.

To install:

3. Install crankshaft position sensor.
   a. Apply a light coat of engine oil to

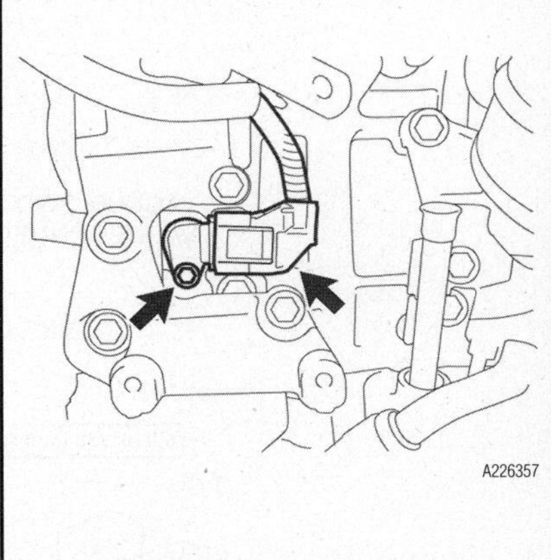

Fig. 107 Remove crankshaft position sensor

the O-ring on the crankshaft position sensor.
   b. Install the crankshaft position sensor with the bolt.
   c. Connect the crank position sensor connector.
4. Install w/pulley compressor assembly.

ELECTRONIC CONTROL MODULE (ECM)

REMOVAL & INSTALLATION

2.7L Engine

See Figure 108.

➡Perform the Vehicle Identification Number (VIN) registration when replacing the ECM.

1. Remove windshield wiper motor and link assembly
2. Remove air cleaner inlet assembly
3. Remove battery
4. Remove air cleaner assembly
5. Remove ECM
   a. Disengage the wire harness clamp.
   b. While slightly pushing portion A of the lever of the ECM connector, turn the lever until it makes a click sound.
   c. Disconnect the 2 ECM connectors.

➡After disconnecting the connectors, make sure that dirt, water or other foreign matter does not contact the connecting part of the connectors.

   d. Remove the 3 bolts and the ECM with the brackets.
6. Remove the 2 screws and the ECM bracket.
7. Remove the 3 screws and the No. 2 ECM bracket.

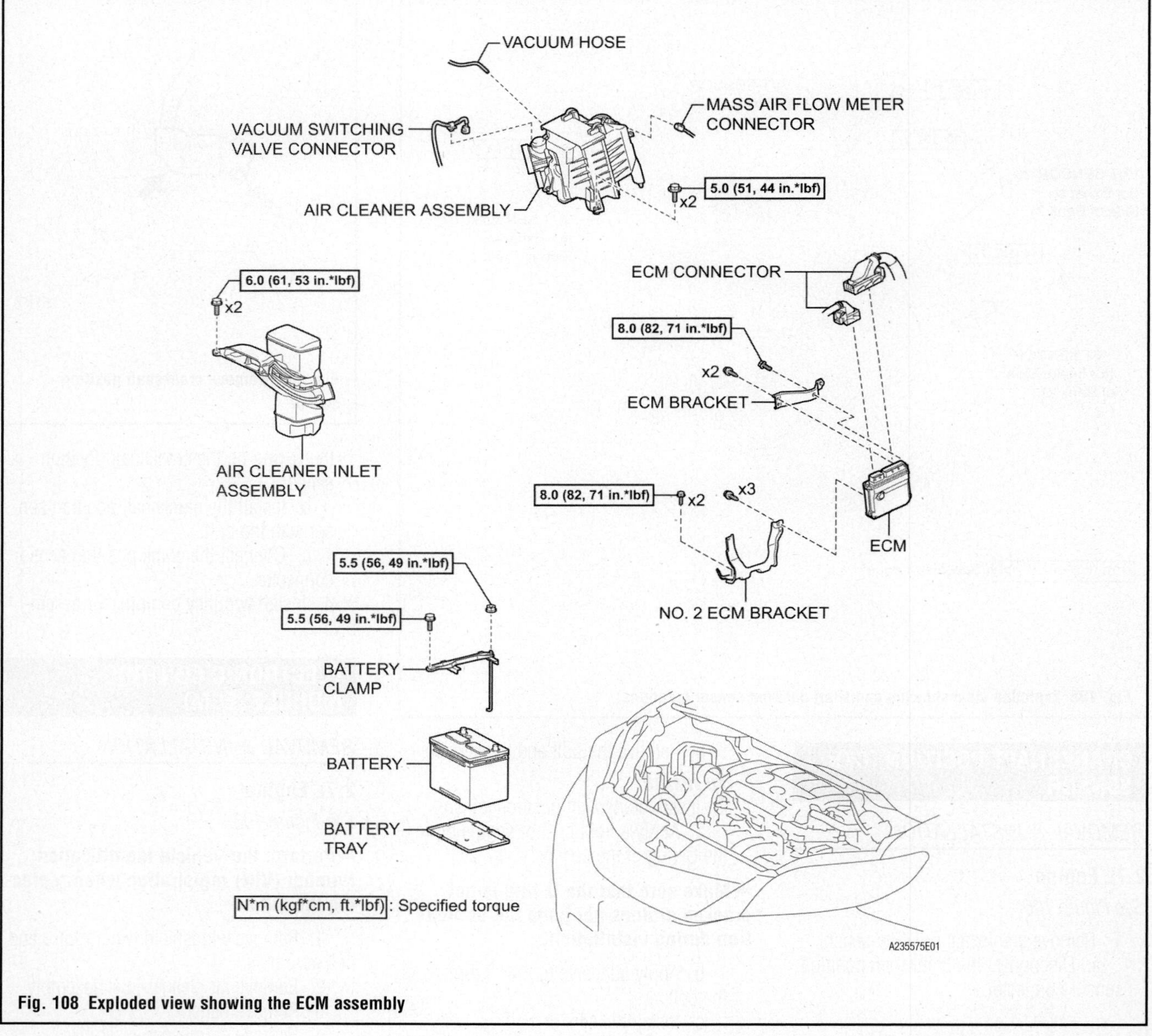

Fig. 108 Exploded view showing the ECM assembly

To install:

8. Install the ECM bracket to the ECM with the 2 screws.
9. Install the No. 2 ECM bracket to the ECM with the 3 screws.
10. Install ECM.
    a. Install the ECM with the 3 bolts.
    b. Pull down the lever to engage the lock and connect the 2 connectors to the ECM.

➡Make sure that the 2 levers are securely locked. When connecting the connectors, make sure that dirt, water or other foreign matter does not become stuck between the connectors and other parts.

    c. Engage the wire harness clamp to the ECM bracket.
11. Install air cleaner assembly.
12. Install battery.
13. Install air cleaner inlet assembly.
14. Install windshield wiper motor and link assembly.
15. Perform initialization.

➡Initialization cannot be completed by only disconnecting and reconnecting the cable of the negative (-) battery terminal.

    a. Perform Vehicle Identification Number (VIN) registration.
    b. Register the ECM communication ID for the immobilizer system when replacing the ECM.
    c. Perform Initialization when replacing the ECM.

3.5L Engine

See Figure 109.

➡Perform the Vehicle Identification Number (VIN) registration when replacing the ECM.

1. Remove windshield wiper motor and link assembly.
2. Remove front outer cowl top panel sub-assembly.
3. Remove air cleaner inlet cover assembly seal.
4. Remove air cleaner and hose.
5. Remove ECM.
    a. While slightly pushing portion A of

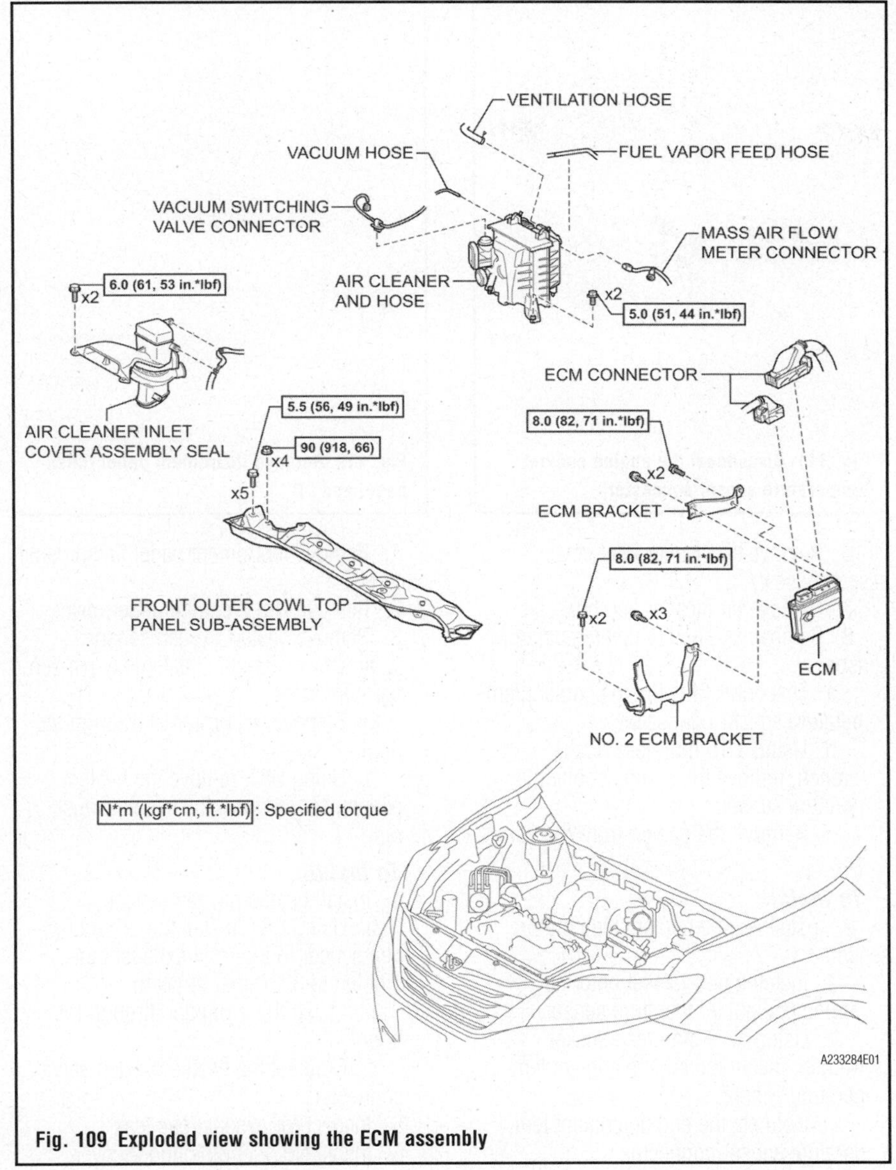

Fig. 109 Exploded view showing the ECM assembly

the lever of the ECM connector, turn the lever until it makes a click sound.

b. Disconnect the 2 ECM connectors.

➡After disconnecting the connectors, make sure that dirt, water or other foreign matter does not contact the connecting part of the connectors.

c. Remove the 3 bolts and the ECM with the brackets.

6. Remove the 2 screws and the ECM bracket.
7. Remove the 3 screws and the No. 2 ECM bracket.

To install:

8. Install the ECM bracket to the ECM with the 2 screws.
9. Install the No. 2 ECM bracket to the ECM with the 3 screws.
10. Install ECM.

a. Install the ECM with the 3 bolts.

b. Pull down the lever to engage the lock and connect the 2 connectors to the ECM.

➡Make sure that the 2 levers are securely locked. When connecting the connectors, make sure that dirt, water or other foreign matter does not become stuck between the connectors and other parts.

11. Install air cleaner and hose.
12. Install air cleaner inlet cover assembly seal.
13. Install front outer cowl top panel sub-assembly.
14. Install windshield wiper motor and link assembly.
15. Perform initialization.

➡Initialization cannot be completed by only disconnecting and reconnecting the cable of the negative (-) battery terminal.

a. Perform Vehicle Identification Number (VIN) registration.

b. Register the ECM communication ID for the immobilizer system when replacing the ECM.

c. Perform Initialization when replacing the ECM.

VEHICLE IDENTIFICATION NUMBER (VIN) REGISTRATION

➡The Vehicle Identification Number (VIN) must be input into a replacement ECM.

➡The VIN is a 17-digit alphanumeric vehicle identification number. The Techstream is required to register the VIN.

1. Description (Using the Techstream)

➡This registration section consists of 2 parts: Read VIN and Write VIN.

a. Read VIN: This process allows the VIN stored in the ECM to be read in order to confirm that the 2 VINs, provided with the vehicle and stored in the vehicle ECM, are the same.

b. Write VIN: This process allows the VIN to be input into the ECM. If the ECM is replaced, or the ECM VIN and vehicle VIN do not match, the VIN can be registered, or overwritten in the ECM by following this procedure.

2. Read VIN

a. Confirm the vehicle VIN.

b. Connect the Techstream to the DLC3.

c. Turn the ignition switch to ON.

d. Turn the Techstream on.

e. Enter the following menus: Powertrain / Engine / Utility / VIN / VIN Read.

3. Write VIN

a. Confirm the vehicle VIN.

b. Connect the Techstream to the DLC3.

c. Turn the ignition switch to ON.

d. Turn the Techstream on.

e. Enter the following menus: Powertrain / Engine / Utility / VIN / VIN Write.

ENGINE COOLANT TEMPERATURE (ECT) SENSOR

REMOVAL & INSTALLATION

2.7L Engine

See Figure 110.

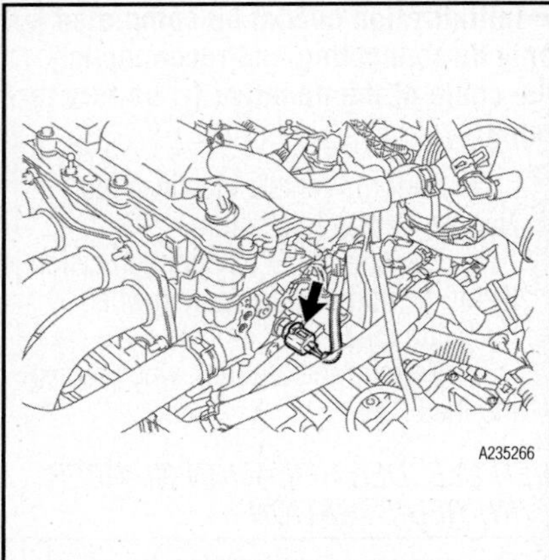

Fig. 110 Disconnect the engine coolant temperature sensor connector

1. Remove No. 1 engine under cover.
2. Drain engine coolant.
3. Remove air cleaner inlet assembly.
4. Remove battery.
5. Remove air cleaner assembly.
6. Remove engine coolant temperature sensor.
   a. Disconnect the engine coolant temperature sensor connector.
   b. Using a 19 mm deep socket wrench, remove the engine coolant temperature sensor.
   c. Remove the gasket from the sensor.

To install:

7. Install engine coolant temperature sensor.
   a. Install a new gasket onto the engine coolant temperature sensor.
   b. Using a 19 mm deep socket wrench, install the engine coolant temperature sensor. Tighten to 14 ft. lbs. (20 Nm).
   c. Connect the engine coolant temperature sensor connector.
8. Install air cleaner assembly.
9. Install battery.
10. Install air cleaner inlet assembly.
11. Add engine coolant.
12. Inspect for coolant leak.
13. Install No. 1 engine under cover.

3.5L Engine

See Figure 111.

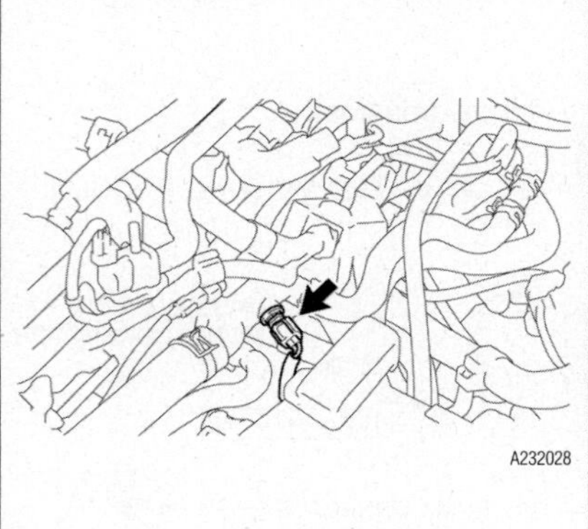

Fig. 111 Disconnect the engine coolant temperature sensor connector

1. Remove V-bank cover sub-assembly.
2. Remove No. 1 engine under cover.
3. Drain engine coolant.
4. Remove windshield wiper motor and link assembly.
5. Remove front outer cowl top panel sub-assembly.
6. Remove air cleaner inlet cover assembly seal.
7. Remove air cleaner and hose.
8. Remove engine coolant temperature sensor.
   a. Disconnect the engine coolant temperature sensor connector.
   b. Using a 19 mm deep socket wrench, remove the engine coolant temperature sensor.
   c. Remove the gasket from the sensor.

To install:

9. Install engine coolant temperature sensor.
   a. Install a new gasket onto the engine coolant temperature sensor.
   b. Using a 19 mm deep socket wrench, install the engine coolant temperature sensor.
   c. Connect the engine coolant temperature sensor connector.
10. Add engine coolant.
11. Inspect for coolant leak.
12. Install air cleaner and hose.
13. Install air cleaner inlet cover assembly seal.
14. Install front outer cowl top panel sub-assembly.
15. Install windshield wiper motor and link assembly.
16. Install No. 1 engine under cover.
17. Install V-bank cover sub-assembly.

HEATED OXYGEN (HO2S) SENSOR

REMOVAL & INSTALLATION

2.7L Engine

See Figures 112 and 113.

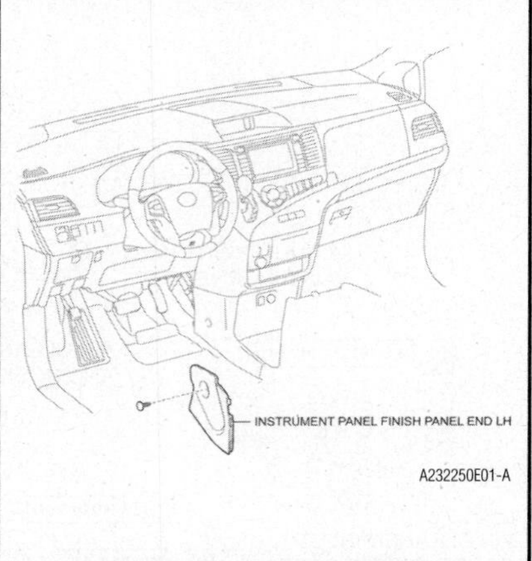

Fig. 112 Remove instrument panel finish panel end LH

1. Remove instrument panel finish panel end LH.
2. Remove No. 2 engine under cover.
3. Remove heated oxygen sensor.
   a. Disconnect the heated oxygen sensor connector.
   b. Remove the grommet through the floor.
   c. Using SST, remove the heated oxygen sensor from the front exhaust pipe.

To install:

4. Install heated oxygen sensor.
   a. Using SST, install the heated oxygen sensor to the front exhaust pipe. Tighten to 32 ft. lbs. (44 Nm).
   b. Install the grommet through the floor.
   c. Connect the heated oxygen sensor connector.
5. Inspect for exhaust gas leak.
6. Install No. 2 engine under cover.
7. Install instrument panel finish panel end LH.

3.5L Engine

See Figures 112 and 114.

1. Remove instrument panel finish panel end LH.
2. Remove No. 1 engine under cover.
3. Remove No. 2 engine under cover (for AWD).
4. Remove No. 1 exhaust pipe support bracket.
5. Remove front exhaust pipe assembly.
6. Remove No. 2 oxygen sensor (for bank 2 sensor 2).
   a. Using SST, remove the No. 2 oxygen sensor from the front exhaust pipe assembly.

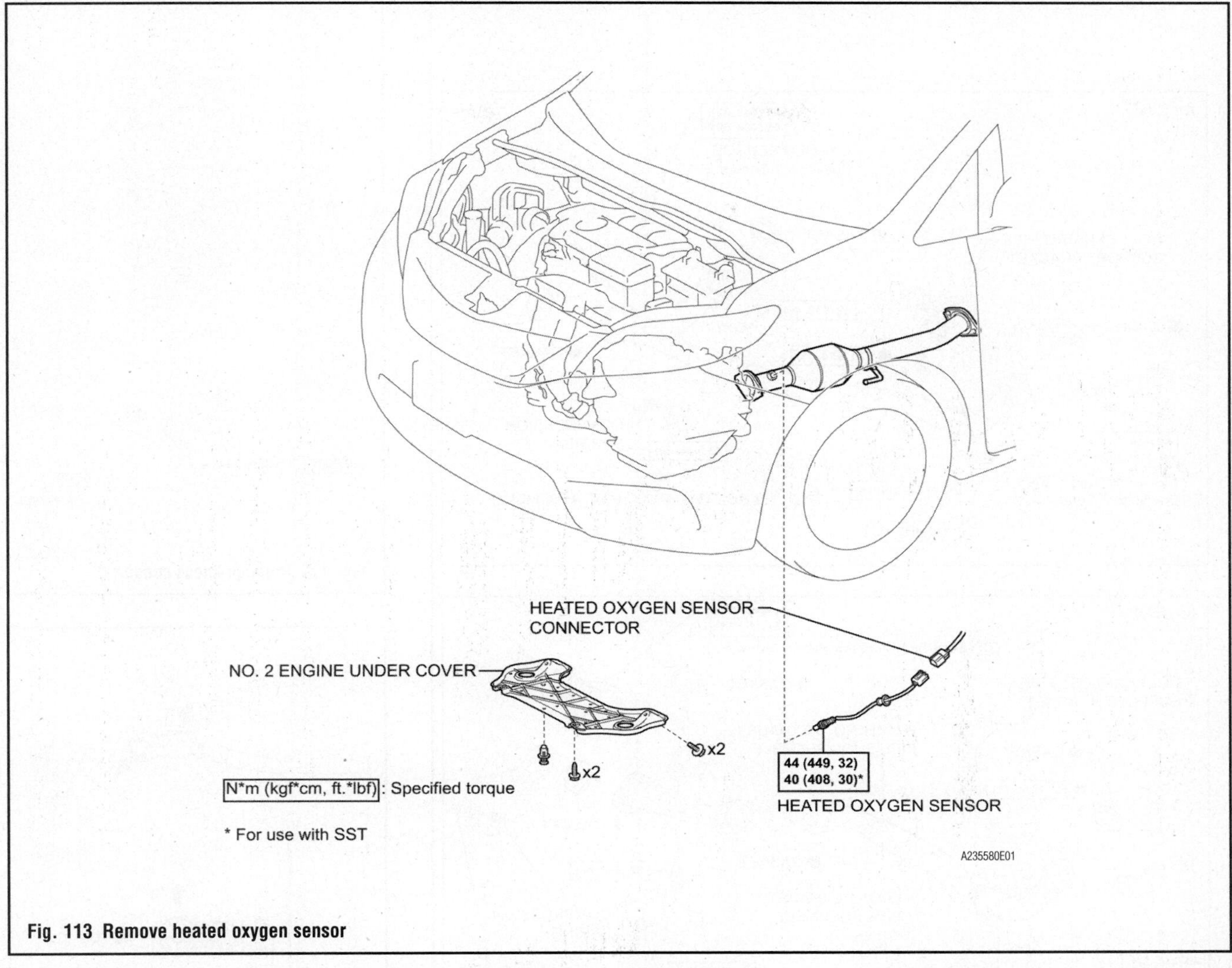

Fig. 113 Remove heated oxygen sensor

7. Remove oxygen sensor (for Bank 1 Sensor 2).
   a. Using SST, remove the oxygen sensor from the No. 3 front exhaust pipe sub-assembly.

To install:

8. Install No. 2 oxygen sensor (for Bank 2 Sensor 2).
   a. Using SST, install the No. 2 oxygen sensor to the front exhaust pipe assembly.
9. Install oxygen sensor (for Bank 1 Sensor 2).
   a. Using SST, install the oxygen sensor to the center exhaust pipe assembly.
10. Install front exhaust pipe assembly.
11. Install No. 1 exhaust pipe support bracket.
12. Install No. 2 engine under cover (for AWD).
13. Install No. 1 engine under cover.
14. Inspect for exhaust gas leak.
15. Install instrument panel finish panel end LH.

KNOCK SENSOR (KS)

REMOVAL & INSTALLATION

2.7L Engine

See Figure 115 and 116.

1. Remove intake manifold
2. Remove knock sensor
   a. Disconnect the knock sensor connector.
   b. Remove the bolt and the knock sensor.

To install:

3. Install knock sensor.
   a. Install the knock sensor with the bolt. Tighten to 15 ft. lbs. (20 Nm).

➡Make sure that each knock control sensor is in the correct position.

   b. Connect the knock sensor connector.
4. Install intake manifold.

3.5L Engine

See Figure 118.

1. Remove intake manifold.
2. Remove knock sensors.
   a. Disconnect the 2 knock sensor connectors.
   b. Remove the 2 bolts and 2 knock sensors.

To install:

3. Install knock sensors.
   a. Install the 2 knock sensors with the 2 bolts. Tighten to 15 ft. lbs. (20 Nm).

➡Make sure that each knock control sensor is in the correct position.

   b. Connect the 2 knock sensor connectors.
4. Install intake manifold.

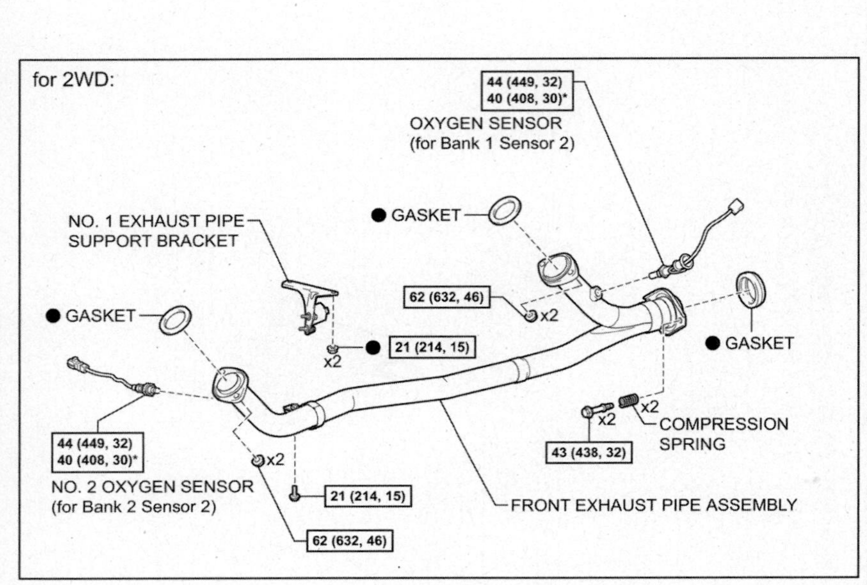

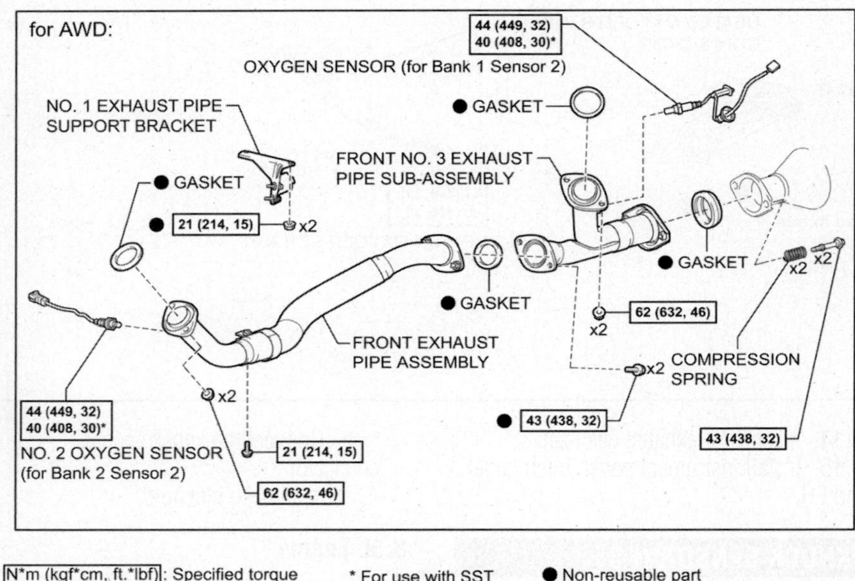

Fig. 114 Exploded view showing the exhaust and heated oxygen sensor locations

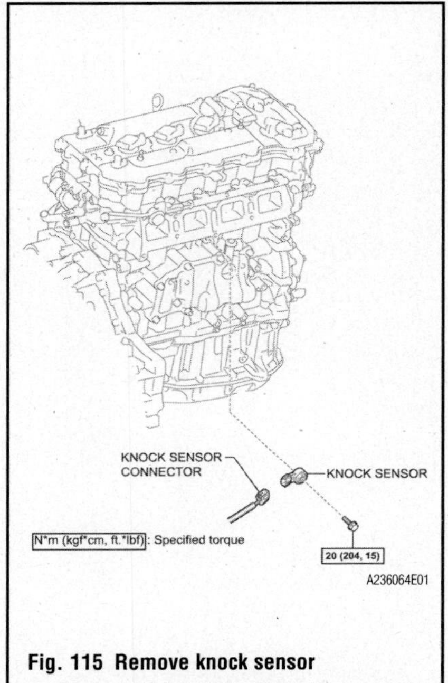

Fig. 115 Remove knock sensor

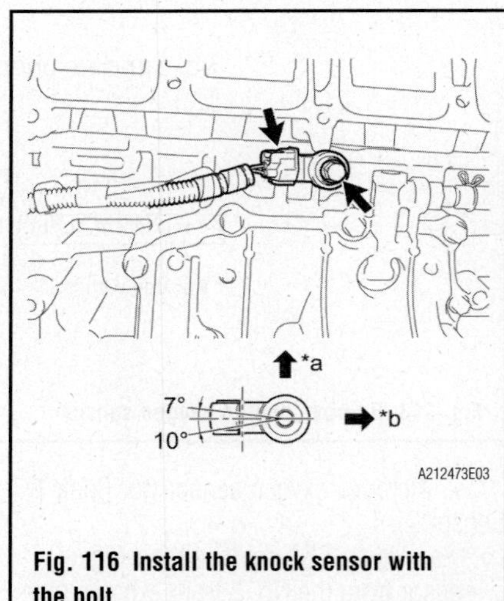

Fig. 116 Install the knock sensor with the bolt

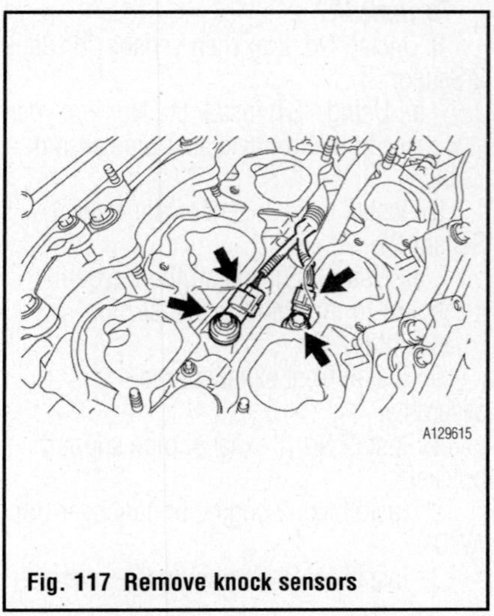

Fig. 117 Remove knock sensors

MASS AIR FLOW (MAF) SENSOR

REMOVAL & INSTALLATION

2.7L Engine

See Figure 118.

1. Remove windshield wiper motor and link assembly.
2. Remove cowl panel sub-assembly.
3. Remove mass air flow meter.
    a. Disconnect the mass air flow meter connector.
    b. Remove the 2 screws and the mass air flow meter.

To install:

4. Install mass air flow meter.
    a. Install the mass air flow meter with the 2 screws.

➡Make sure that the O-ring is not cracked or does not jump out of position during installation.

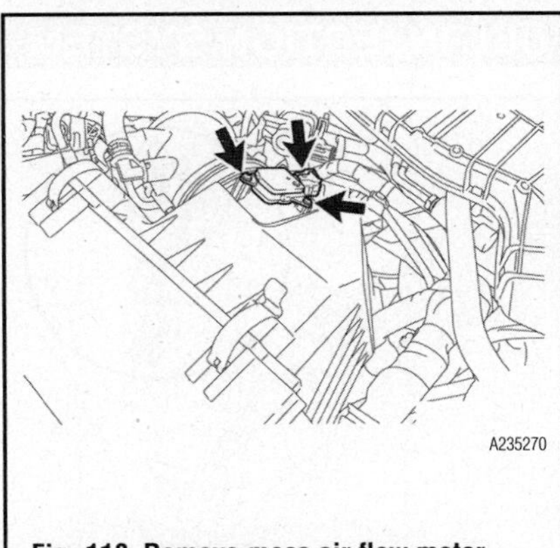

Fig. 118 Remove mass air flow meter

b. Connect the mass air flow meter connector.

5. Install cowl panel sub-assembly.
6. Install windshield wiper motor and link assembly.

3.5L Engine

See Figure 119.

1. Remove windshield wiper motor and link assembly.
2. Remove front outer cowl top panel sub-assembly.
3. Remove mass air flow meter.
   a. Disconnect the mass air flow meter connector.
   b. Remove the 2 screws and the mass air flow meter.

To install:

4. Install mass air flow meter.
   a. Install the mass air flow meter with the 2 screws.

➡Make sure that the O-ring is not cracked or does not jump out of position during installation.

b. Connect the mass air flow meter connector.

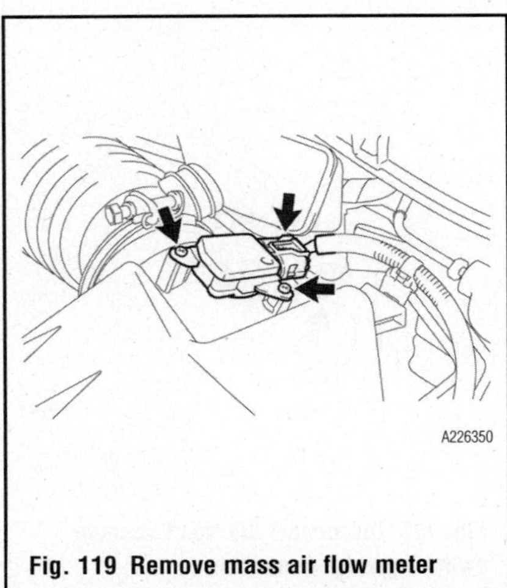

Fig. 119 Remove mass air flow meter

5. Install front outer cowl top panel sub-assembly.
6. Install windshield wiper motor and link assembly.

VEHICLE SPEED SENSOR (VSS)

REMOVAL & INSTALLATION

U660E and U660F Transaxles

See Figure 120.

1. Remove transmission valve body assembly.
2. Remove transmission revolution sensor.
   a. Disconnect the connector.
   b. Remove the 2 bolts and transmission revolution sensor from the transmission valve body.

To install:

3. Install transmission revolution sensor.
   a. Coat the 2 bolts with ATF.
   b. Install the transmission revolution sensor onto the transmission valve body with the 2 bolts.
   c. Connect the connector.
4. Install transmission valve body assembly.

U760E Transaxle

See Figures 121 and 122.

1. Remove transmission valve body assembly.
2. Remove transmission revolution sensor.
   a. Disconnect the connector.
   b. Remove the 2 bolts and transmission revolution sensor from the transmission valve body.

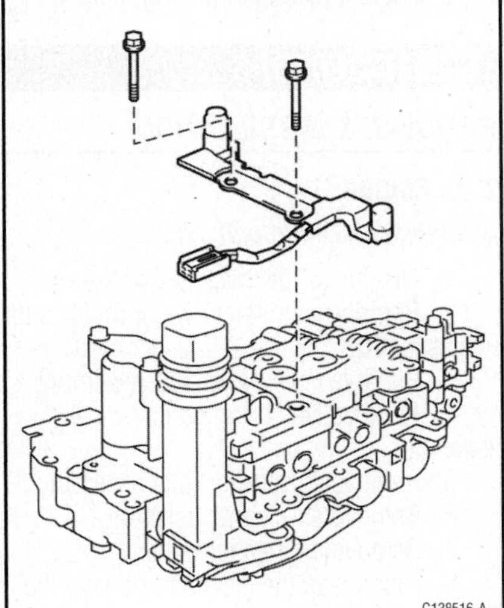

Fig. 120 Remove the 2 bolts and transmission revolution sensor from the transmission valve body

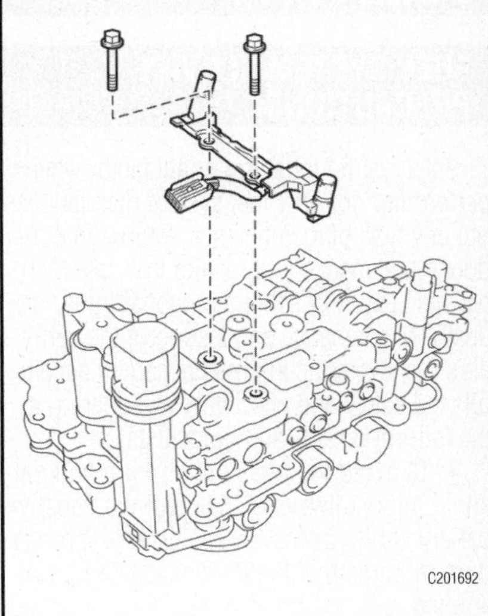

Fig. 121 Remove the 2 bolts and transmission revolution sensor from the transmission valve body

To install:

3. Install transmission revolution sensor.
   a. Temporarily install the transmission revolution sensor onto the transmission valve body with the 2 bolts.

➡Do not use a transmission revolution sensor that has been dropped.

b. Tighten the 2 bolts in order to 8 ft. lbs. (11 Nm).
c. Connect the connector of the transmission revolution sensor.

4. Install transmission valve body assembly.

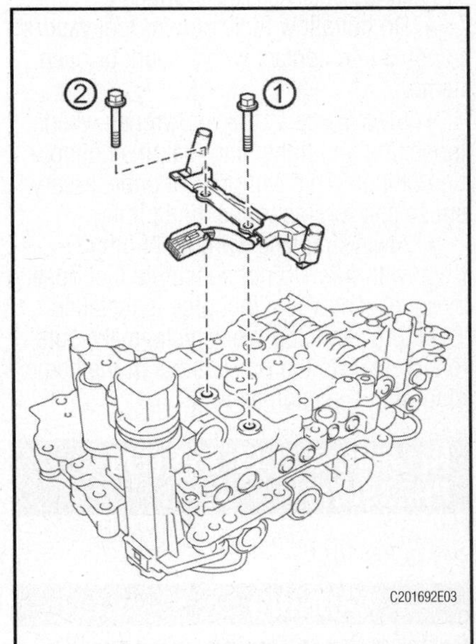

Fig. 122 Tighten the 2 bolts in order

FUEL SYSTEM SERVICE PRECAUTIONS

Safety is the most important factor when performing not only fuel system maintenance but any type of maintenance. Failure to conduct maintenance and repairs in a safe manner may result in serious personal injury or death. Maintenance and testing of the vehicle's fuel system components can be accomplished safely and effectively by adhering to the following rules and guidelines.

- To avoid the possibility of fire and personal injury, always disconnect the negative battery cable unless the repair or test procedure requires that battery voltage be applied.
- Always relieve the fuel system pressure prior to disconnecting any fuel system component (injector, fuel rail, pressure regulator, etc.), fitting or fuel line connection. Exercise extreme caution whenever relieving fuel system pressure to avoid exposing skin, face and eyes to fuel spray. Please be advised that fuel under pressure may penetrate the skin or any part of the body that it contacts.
- Always place a shop towel or cloth around the fitting or connection prior to loosening to absorb any excess fuel due to spillage. Ensure that all fuel spillage (should it occur) is quickly removed from engine surfaces. Ensure that all fuel soaked cloths or towels are deposited into a suitable waste container.
- Always keep a dry chemical (Class B) fire extinguisher near the work area.
- Do not allow fuel spray or fuel vapors to come into contact with a spark or open flame.
- Always use a back-up wrench when loosening and tightening fuel line connection fittings. This will prevent unnecessary stress and torsion to fuel line piping.
- Always replace worn fuel fitting O-rings with new Do not substitute fuel hose or equivalent where fuel pipe is installed.

Before servicing the vehicle, make sure to also refer to the precautions in the beginning of this section as well.

RELIEVING FUEL SYSTEM PRESSURE

See Figure 123.

⁂ WARNING

Perform the following procedure to prevent fuel from spilling out before removing any fuel system parts. Even after discharging the fuel pressure, when disconnecting a fuel line, cover it with a piece of cloth to prevent fuel from spraying or coming out.

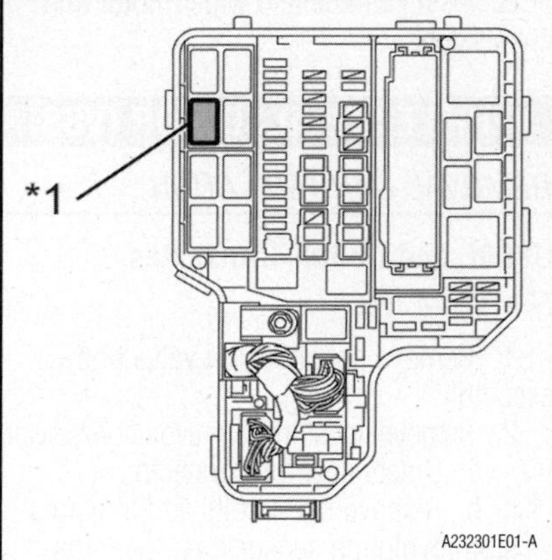

Fig. 123 Remove the circuit opening relay (1) from the engine room relay block

1. Remove the circuit opening relay from the engine room relay block.
2. Start the engine.
3. After the engine stops naturally, turn the ignition switch off.

➡DTC P0171 (System too lean) may be set.

4. Crank the engine again and make sure that the engine does not start.
5. Remove the fuel tank cap and discharge the pressure from the fuel tank.
6. Install the circuit opening relay.

FUEL INJECTORS

REMOVAL & INSTALLATION

2.7L Engine

See Figures 124 through 126.

1. Discharge fuel system pressure.
2. Remove windshield wiper motor and link assembly.
3. Remove cowl panel sub-assembly.
4. Remove No. 1 engine cover sub-assembly.
5. Remove air cleaner inlet assembly.
6. Remove air cleaner assembly.
7. Remove air cleaner hose.
8. Disengage the claw and remove the EFI fuel pipe clamp.
9. Disconnect fuel tube sub-assembly.
   a. Pinch the retainer to release it and disconnect the fuel tube.

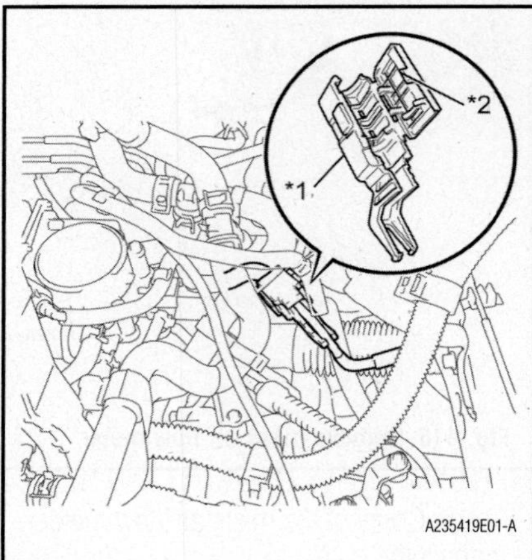

Fig. 124 Disengage the claw (2) and remove the EFI fuel pipe clamp (1)

➡Please note the following:

- Check for dirt or other foreign matter on the parts to be disconnected and clean them if necessary.
- The fuel tube connector seals with an O-ring. Ensure that there is no damage or foreign matter on the contact surface.
- If the connector and the pipe are stuck, push and pull on the connector to release them while pinching the retainer. Pull the connector out of the pipe carefully.
- Do not use any tools.
- Do not bend or twist the tubes.
- Protect the contact surface by covering it with a plastic bag.

   b. Separate the fuel tube from the fuel tube clamp.

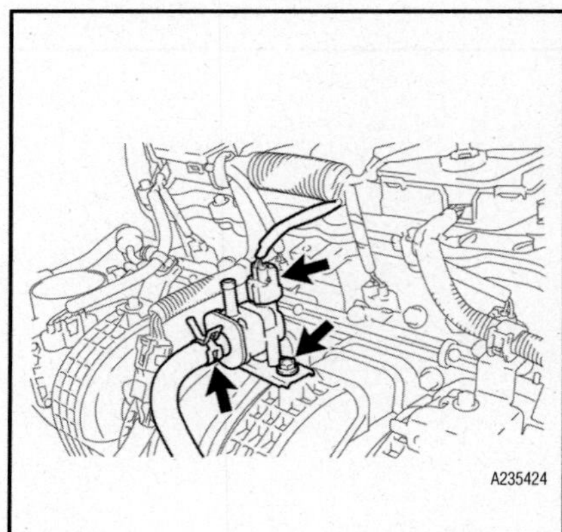

Fig. 125 Disconnect the No. 1 vacuum switching valve connector

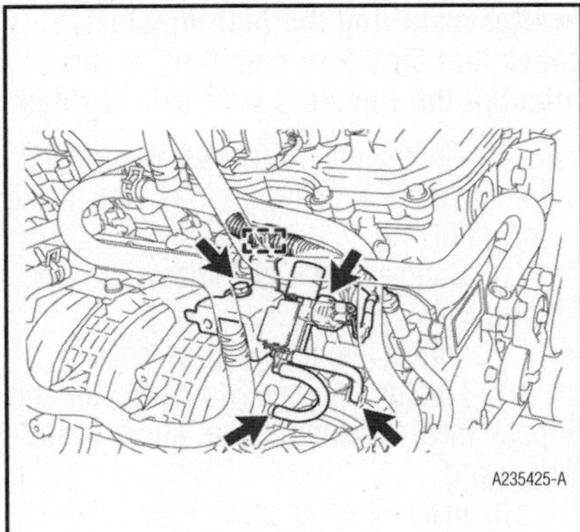

Fig. 126 Disconnect the vacuum switching valve connector

10. Remove No. 1 vacuum switching valve assembly.
   a. Disconnect the No. 1 vacuum switching valve connector.
   b. Disconnect the No. 2 air hose.
   c. Remove the bolt and No. 1 vacuum switching valve assembly.
11. Remove vacuum switching valve assembly.
   a. Disconnect the vacuum switching valve connector.
   b. Disengage the wire harness clamp.
   c. Disconnect the 2 vacuum hoses.
   d. Remove the bolt and vacuum switching valve assembly.
12. Disconnect wire harness.
   a. Disconnect the 4 fuel injector connectors.
   b. Disconnect the throttle with motor body connector.
   c. Disconnect the connector and disengage the harness clamp.
   d. Remove the 2 bolts and separate the 2 wire harness brackets.
13. Remove fuel delivery pipe sub-assembly.
   a. Remove the 2 bolts, and then remove the fuel delivery pipe together with the 4 fuel injectors.

➡Be careful not to drop the fuel injectors when removing the fuel delivery pipe.

   b. Remove the 2 fuel delivery spacers from the cylinder head.
   c. Remove the 4 injector vibration insulators from the cylinder head.
14. Remove fuel injector assembly.
   a. Pull the 4 fuel injectors out of the fuel delivery pipe.
   b. Remove the 4 O-rings from the injectors.

To install:

15. Install fuel injector assembly.
   a. Apply a light coat of spindle oil or gasoline to new O-rings, and install them to each injector.
   b. Apply a light coat of spindle oil or gasoline where the fuel delivery pipe contacts each O-ring.
   c. Push and twist each fuel injector to install them into the fuel delivery pipe.

➡Be careful not to twist the O-rings. After installing a fuel injector, check that it turns smoothly. If not, reinstall it with a new O-ring.

16. Install fuel delivery pipe sub-assembly.
   a. Install 4 new injector vibration insulators to the cylinder head.
   b. Install the 2 fuel delivery spacers onto the cylinder head.

➡Install the fuel delivery spacer so that the longer protrusion is on the cylinder head side.

   c. Install the fuel delivery pipe together with the 4 fuel injectors to the cylinder head, and then temporarily install the 2 bolts.

➡Be careful not to drop the fuel injectors when installing the fuel delivery pipe. After installing the fuel injectors, check that they turn smoothly. If not, reinstall the injectors with new O-rings.

   d. Tighten the 2 bolts.
17. Connect wire harness.
   a. Install the 2 wire harness brackets with the 2 bolts.
   b. Engage the harness clamp and connect the connector.
   c. Connect the throttle with motor body connector.
   d. Connect the 4 fuel injector connectors.
18. Install vacuum switching valve assembly.
   a. Install the vacuum switching valve assembly with the bolt.
   b. Connect the 2 vacuum hoses.
   c. Engage the wire harness clamp.
   d. Connect the vacuum switching valve connector.
19. Install No. 1 vacuum switching valve assembly.
   a. Install the No. 1 vacuum switching valve assembly with the bolt.
   b. Connect the No. 2 air hose.
   c. Connect the No. 1 vacuum switching valve connector.
20. Connect fuel tube sub-assembly.
   a. Line up the pipe and connector and push them together until a "click" sound is heard.
   b. Install the fuel tube to the fuel tube clamp.
21. Engage the claw and install the EFI fuel pipe clamp.
22. Install air cleaner hose.
23. Install air cleaner assembly.
24. Install air cleaner inlet assembly.
25. Install No. 1 engine cover sub-assembly.
26. Install cowl panel sub-assembly.
27. Install windshield wiper motor and link assembly.
28. Inspect for fuel leak.

3.5L Engine

See Figure 127.

1. Discharge fuel system pressure.
2. Remove intake air surge tank assembly.
3. Remove the bolt, 2 nuts and No. 2 engine mounting stay RH.
4. Remove the bolt and No. 2 engine mounting stay RH.
5. Remove fuel hose protector.
   a. Disconnect the 3 fuel injector connectors and 2 harness clamps.
   b. Remove the 2 bolts, nut and fuel hose protector.
6. Disconnect fuel tube sub-assembly.
   a. Remove the EFI fuel pipe clamp.
   b. Pinch the retainer to release it and disconnect the fuel tube.

➡Please note the following:

- Check for dirt or other foreign matter on the parts to be disconnected and clean them if necessary.
- The fuel tube connector seals with an O-ring. Ensure that there is no damage or foreign matter on the contact surface.
- If the connector and the pipe are stuck, push and pull on the connector to release them while pinching the retainer. Pull the connector out of the pipe carefully.
- Do not use any tools.
- Do not bend or twist the tubes.
- Protect the contact surface by covering it with a plastic bag.

7. Remove fuel injector assembly.
   a. Disconnect the 3 fuel injector connectors.
   b. Remove the 5 bolts and fuel delivery pipe sub-assembly together with the 6 fuel injectors.

➡Be careful not to drop the fuel injectors when removing the fuel delivery pipe.

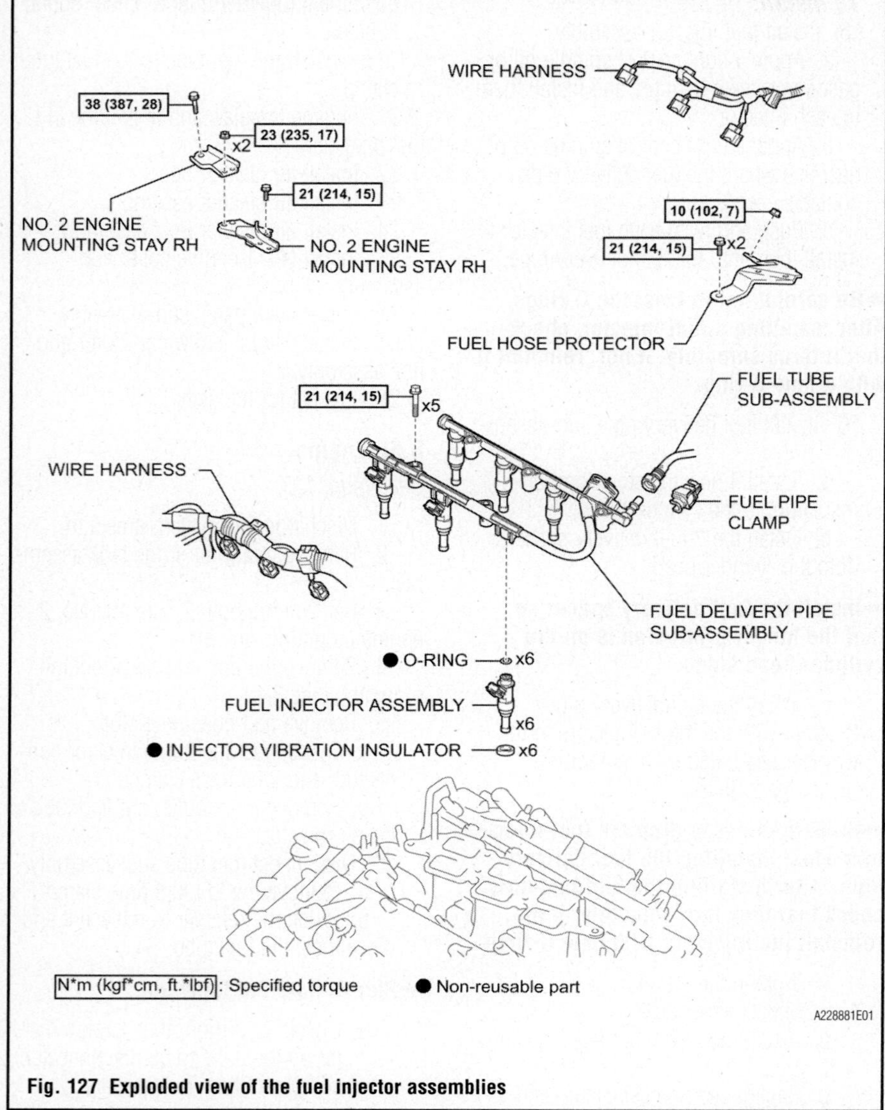

Fig. 127 Exploded view of the fuel injector assemblies

c. Pull out the fuel injectors from the fuel delivery pipe.

➡**If the injectors are to be reused, reinstall them to the same cylinder they came from.**

d. Remove the 6 O-rings from the injectors.

8. Remove the 6 injector vibration insulators from the intake manifold.

To install:

9. Install 6 new injector vibration insulators to the intake manifold.
10. Install fuel injector assembly.
    a. Apply a light coat of spindle oil or gasoline to new O-rings, and install them to each injector.
    b. Apply a light coat of spindle oil or gasoline where the fuel delivery pipe contacts each O-ring.
    c. Push and twist each fuel injector to install them into the fuel delivery pipe.
    d. Position each fuel injector connector outward.

➡**Be careful not to twist the O-rings. After installing a fuel injector, check that it turns smoothly. If not, reinstall it with a new O-ring.**

e. Place the fuel delivery pipe with the 6 fuel injectors in position on the intake manifold.

➡**Be careful not to drop the fuel injectors when installing the fuel delivery pipe.**

Temporarily install the 5 bolts which are used to hold the fuel delivery pipe to the intake manifold.

➡**After installing the fuel injectors, check that they turn smoothly. If not, reinstall the injectors with new O-rings.**

f. Tighten the 5 bolts which are used to hold the fuel delivery pipe to the intake manifold.

g. Connect the 3 fuel injector connectors.

11. Connect fuel tube sub-assembly.
    a. Line up the pipe and connector and push them together until a "click" sound is heard.
    b. Install the fuel pipe clamp.
12. Install fuel hose protector.
    a. Install the fuel hose protector with the 2 bolts and nut.
    b. Connect the 2 harness clamps and 3 fuel injector connectors.
13. Install the No. 2 engine mounting stay RH with the bolt.
14. Install the No. 2 engine mounting stay RH with the bolt and 2 nuts.
15. Install intake air surge tank assembly.
16. Inspect for fuel leak.

FUEL PUMP

REMOVAL & INSTALLATION

The manufacturer does not provide a specific Removal and Installation procedure for this component. Refer to Fuel Tank Removal and Installation when servicing this component.

FUEL TANK

REMOVAL & INSTALLATION

See Figures 128 through 131.

1. Discharge fuel system pressure.
2. Remove charcoal canister protector.
3. Disconnect fuel tank main tube sub-assembly.
    a. Release the fuel tube connector cover.
    b. Pinch the retainer to release it and disconnect the fuel tank main tube.

➡**Please note the following:**

- Check for dirt or other foreign matter on the parts to be disconnected and clean them if necessary.
- The fuel tube connector seals with an O-ring. Ensure that there is no damage or foreign matter on the contact surface.
- Do not use any tools.
- Do not bend or twist the tubes.
- Protect the contact surface by covering it with a plastic bag.

- If the connector is stuck, push and pull on the parts to separate them.

4. Drain fuel. Operate the fuel pump and drain the fuel from the fuel tank through the fuel tank main tube.
5. Remove instrument panel finish panel end LH.
6. Remove center exhaust pipe assembly.
   a. Remove the 4 bolts.
   b. Disconnect the 3 exhaust pipe supports and remove the center exhaust pipe assembly.
   c. Remove the 2 gaskets from the front exhaust pipe assembly and center exhaust pipe assembly.
7. Remove the 4 nuts and No. 1 fuel tank heat insulator.
8. Disconnect fuel tank vent hose.
9. Disconnect charcoal canister outlet tube sub-assembly.
10. Remove the 2 bolts and the rear floor No. 2 crossmember brace LH.
11. Remove the 3 bolts and the fuel tank filler hose cover.
12. Remove fuel tank assembly.
   a. Disconnect the fuel tank to filler pipe hose.
   b. Disconnect the charcoal canister outlet tube.

➡Please note the following:

- Check for dirt or other foreign matter on the parts to be disconnected and clean them if necessary.
- The fuel tube connector seals with an O-ring. Ensure that there is no damage or foreign matter on the contact surface.
- Do not use any tools.
- Do not bend or twist the tubes.
- Protect the contact surface by covering it with a plastic bag.
- If the connector is stuck, push and pull on the parts to separate them.

   c. Pinch the retainer to release it and disconnect the fuel tube.

➡Please note the following:

- Check for dirt or other foreign matter on the parts to be disconnected and clean them if necessary.
- The fuel tube connector seals with an O-ring. Ensure that there is no damage or foreign matter on the contact surface.
- Do not use any tools.
- Do not bend or twist the tubes.
- Protect the contact surface by covering it with a plastic bag.
- If the connector is stuck, push and pull on the parts to separate them.

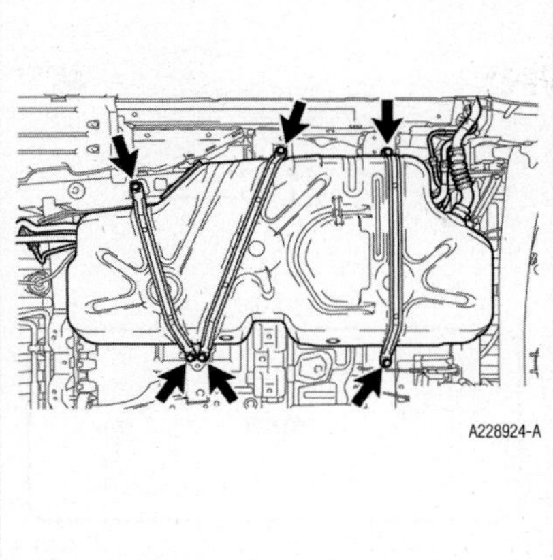

Fig. 128 Remove the 6 set bolts of the 3 fuel tank bands

   d. Disengage the 2 wire harness clamps.
   e. Using an engine lifter, support the fuel tank assembly.
   f. Remove the 6 set bolts of the 3 fuel tank bands.
   g. While operating the engine lifter, disengage the 4 guides for the wire harness and disconnect the fuel pump connector, then remove the fuel tank assembly and the 3 fuel tank bands from the vehicle.

13. Remove fuel tank main tube sub-assembly.
   a. Widen the tip of the tube joint clip and pull out the clip.

➡Please note the following:

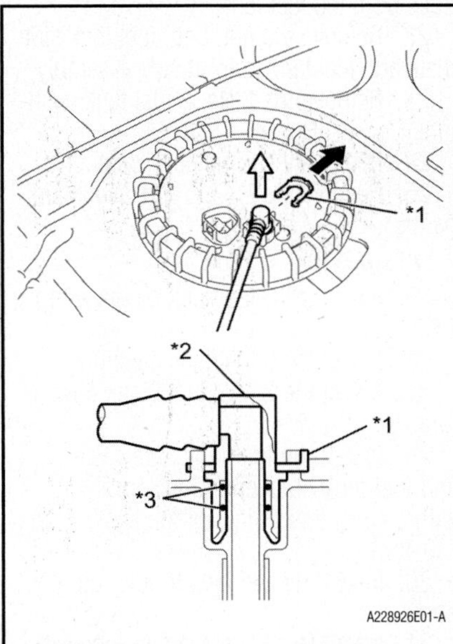

Fig. 129 Widen the tip of the tube joint clip (1) and pull out the clip; fuel tube connector (2), O-ring (3)

- Check for dirt or other foreign matter on the parts to be disconnected and clean them if necessary.
- The fuel tube connector seals with an O-ring. Ensure that there is no damage or foreign matter on the contact surface.
- Do not use any tools.
- Do not bend or twist the tubes.
- Protect the contact surface by covering it with a plastic bag.
- If the connector is stuck, push and pull on the parts to separate them.

   b. Disengage the 4 guides and remove the fuel tank main tube sub-assembly.

14. Disengage the 4 clamps and remove the charcoal canister outlet tube sub-assembly.
15. Remove fuel tank vent hose.
   a. Insert a screwdriver with its tip wrapped in protective tape into the quick connector cutout, pry up the retainer and disconnect the fuel tank vent hose.

➡Please note the following:

- Check for dirt or other foreign matter on the parts to be disconnected and clean them if necessary.
- The fuel tube connector seals with an O-ring. Ensure that there is no damage or foreign matter on the contact surface.
- Do not use any tools.
- Do not bend or twist the tubes.
- Protect the contact surface by covering it with a plastic bag.
- If the connector is stuck, push and pull on the parts to separate them.

   b. Disengage the 4 clamps and remove the fuel tank vent hose and hose clamp.

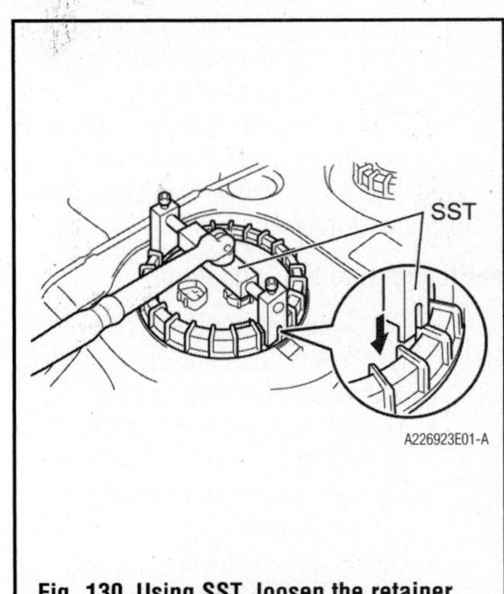

Fig. 130 Using SST, loosen the retainer

16. Remove fuel pump gauge retainer.
   a. Using SST, loosen the retainer.

➡Please note the following:

- Do not use any tools other than SST.
- When the retainer is loosened, be careful as the fuel suction with pump and gauge tube assembly will spring upward from the force of the spring.
- Remove foreign matter around the fuel suction with pump and gauge tube assembly before this operation.

➡Align the ribs of the fuel pump gauge retainer with the tips of SST.

   b. Remove the fuel pump gauge retainer while holding the fuel suction with pump and gauge tube assembly by hand.

17. Remove the fuel suction with pump and gauge tube assembly from the fuel tank.

➡Do not bend the arm of the sender gauge.

18. Remove fuel suction tube set gasket. Remove the gasket from the fuel tank.

19. Remove fuel tank overfill check valve assembly.
   a. Pinch the retainer to release it and disconnect the fuel tube.

➡Please note the following:

- Check for dirt or other foreign matter on the parts to be disconnected and clean them if necessary.
- The fuel tube connector seals with an O-ring. Ensure that there is no damage or foreign matter on the contact surface.
- Do not use any tools.
- Do not bend or twist the tubes.
- Protect the contact surface by covering it with a plastic bag.
- If the connector is stuck, push and pull on the parts to separate them.

   b. Pinch the retainer to release it and disconnect the fuel tube.

➡Please note the following:

- Check for dirt or other foreign matter on the parts to be disconnected and clean them if necessary.
- The fuel tube connector seals with an O-ring. Ensure that there is no damage or foreign matter on the contact surface.
- Do not use any tools.
- Do not bend or twist the tubes.

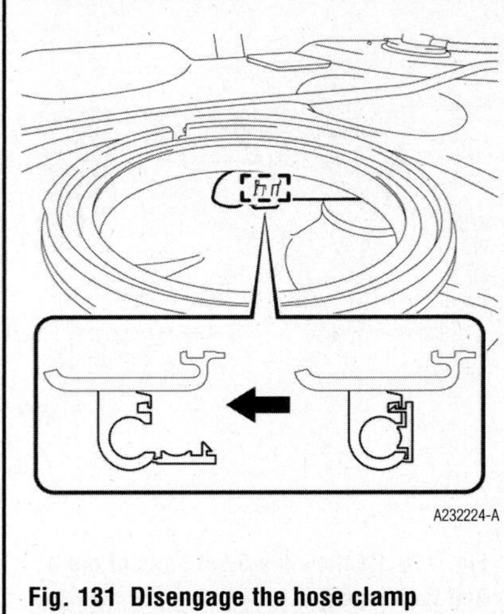

Fig. 131 Disengage the hose clamp

- Protect the contact surface by covering it with a plastic bag.
- If the connector is stuck, push and pull on the parts to separate them.

   c. Disengage the hose clamp.
   d. Using SST, loosen the retainer.

➡Align the ribs of the fuel pump gauge retainer with the tips of SST.

➡Remove foreign matter around the fuel tank overfill check valve assembly before this operation.

   e. Remove the fuel tank overfill check valve assembly.

20. Remove No. 1 fuel suction tube set gasket. Remove the gasket from the fuel tank.

21. Remove the fuel tank to filler pipe hose from the fuel tank.

22. Remove the nut, bolt, fuel tank side plate and fuel tank bracket sub-assembly.

23. Remove the 2 No. 2 fuel tank cushions.

24. Remove the 6 fuel tank cushions.

25. Remove the 6 clips and 3 fuel tank bands.

To install:

26. Install the 3 fuel tank bands with the 6 clips.

27. Install the 6 fuel tank cushions.

28. Install the 2 No. 2 fuel tank cushions.

29. Install the fuel tank side plate and fuel tank bracket sub-assembly with the nut and bolt. Tighten to 22 ft. lbs. (30 Nm).

30. Install the fuel tank to filler pipe hose.

31. Install No. 1 fuel suction tube set gasket.
   a. Ensure gasket groove is clean and free of foreign particles.
   b. Install a new gasket onto the fuel tank.
   c. Make sure that the gasket sits in the groove.

32. Install fuel tank overfill check valve assembly.
   a. Align the protrusion of the fuel tank overfill check valve assembly with the cutout of the tank suction tube support and install the fuel tank overfill check valve assembly.
   b. Align marks on the fuel pump gauge retainer and the fuel tank.
   c. Position the fuel pump gauge retainer on top of the fuel tank overfill check valve assembly flange while pushing down the flange on center.
   d. While holding the fuel tank overfill check valve assembly, tighten the fuel pump gauge retainer one complete turn by hand*1.

➡Make sure the fuel tank overfill check valve assembly anti-rotation tab is in the groove during above operation. "S" arrow on the fuel tank indicates 0 degrees position.

   e. Ensure that no cross threading occurred during installation by checking the gap between the fuel pump gauge retainer and the fuel tank. Confirm that the gap is even. If cross threading occurred, remove the fuel pump gauge retainer and repeat step 1.
   f. Using SST, tighten the fuel pump gauge retainer. Tighten to 59 ft. lbs. (80 Nm).

➡No ratcheting effect allowed when tightening the fuel pump gauge retainer. No grease or lubrication allowed on parts during installation.

➡Align the ribs of the fuel pump gauge retainer with the tips of SST.

   g. Install the fuel pump gauge retainer by one and a half turns so that the start mark on the fuel pump gauge retainer is between 470° and 530° from the start position.
   h. If the start mark does not reach 470°, confirm that no cross threading occurred by measuring the gap between the fuel pump gauge retainer and the fuel tank.

   If cross threading occurred, replace the fuel tank overfill check valve assembly, the gasket and the tank suction tube support.
   i. Replace the tank suction tube support.

- Remove the tank suction tube support from the fuel tank.

- Install the tank suction tube support to the fuel tank.

j. Engage the hose clamp.

k. Put a mark on the fuel pump gauge retainer and the fuel tank with paint. This will be the evidence that the fuel pump gauge retainer has been removed.

l. Line up the pipe and connector and push them together until a "click" sound is heard.

➡Please note the following:

- Check that there are no scratches or foreign objects on the connecting part.
- Check that the tube is inserted securely.
- Check that the tube is securely connected by pulling on it.

m. Line up the pipe and connector and push them together until a "click" sound is heard.

➡Please note the following:

- Check that there are no scratches or foreign objects on the connecting part.
- Check that the tube is inserted securely.
- Check that the tube is securely connected by pulling on it.

33. Install fuel suction tube set gasket.

a. Ensure gasket groove is clean and free of foreign particles.

b. Install a new gasket onto the fuel tank.

c. Make sure that the gasket sits in the groove.

34. Install fuel suction with pump and gauge tube assembly.

a. Align the protrusion of the fuel suction with pump and gauge tube assembly with the cutout of the tank suction tube support and install the fuel suction with pump and gauge tube assembly.

➡Do not bend the arm of the fuel sender gauge.

35. Install fuel pump gauge retainer.

a. Align marks on the fuel pump gauge retainer and the fuel tank.

b. Position the fuel pump gauge retainer on top of the fuel suction with pump and gauge tube assembly flange while pushing down the flange on center.

c. While holding the fuel suction with pump and gauge tube assembly, tighten the fuel pump gauge retainer one complete turn by hand*1.

➡Make sure the fuel suction with pump and gauge tube assembly anti-rotation tab is in the groove during above operation. "S" arrow on the fuel tank indicates 0 degrees position.

d. Ensure that no cross threading occurred during installation by checking the gap between the fuel pump gauge retainer and the fuel tank. Confirm that the gap is even. If cross threading occurred, remove the fuel pump gauge retainer and repeat step 1.

e. Using SST, tighten the fuel pump gauge retainer to the specified torque.

➡Align the ribs of the fuel pump gauge retainer with the tips of SST.

➡No ratcheting effect allowed when tightening the fuel pump gauge retainer. No grease or lubrication allowed on parts during installation.

f. Install the fuel pump gauge retainer by one and a half turns so that the start mark on the fuel pump gauge retainer is between 475° and 535° from the start position.

g. If the start mark does not reach 475°, confirm that no cross threading occurred by measuring the gap between the fuel pump gauge retainer and the fuel tank.

➡If cross threading occurred, replace the fuel pump gauge retainer, the gasket and the tank suction tube support.

h. Replace the tank suction tube support.

- Remove the tank suction tube support from the fuel tank.
- Install the tank suction tube support to the fuel tank.

i. Put a mark on the fuel pump gauge retainer and the fuel tank with paint. This will be the evidence that the fuel pump gauge retainer has been removed.

36. Install fuel tank vent hose.

a. Engage the 4 clamps and install the fuel tank vent hose and hose clamp.

b. Line up the pipe and connector and push them together until a "click" sound is heard.

c. Lock it by pushing in the retainer.

➡Please note the following:

- Check that there are no scratches or foreign objects on the connecting part.
- Check that the fuel tank vent hose is inserted securely.
- After installing the retainer, check that the fuel tank vent hose is securely connected by pulling on it.

➡The quick connector cannot be locked when the tube is not inserted securely.

37. Engage the 4 clamps and install the charcoal canister outlet tube sub-assembly.

38. Install fuel tank main tube sub-assembly.

a. Engage the 4 guides and install the fuel tank main tube sub-assembly.

b. Insert the fuel tank main tube into the fuel suction tube assembly plug and lock it with the tube joint clip.

➡Please note the following:

- Check that there are no scratches or foreign objects on the connecting part.
- Check that the fuel tank main tube is inserted securely.
- Check that the tube joint clip is on the collar of the main tube.
- After installing the tube joint clip, check that the fuel tank main tube assembly is securely connected by pulling on it.

39. Install fuel tank assembly.

a. Clean and degrease the bolt holes.

b. Set the fuel tank and 3 fuel tank bands on the engine lifter.

c. Slowly raise the engine lifter and install the fuel tank onto the vehicle.

d. While operating the engine lifter, engage the 4 guides for the wire harness and connect the fuel pump connector, then install the fuel tank and the 3 fuel tank bands to the vehicle.

e. Tighten the 6 set bolts of the 3 fuel tank bands. Tighten to 24 ft. lbs. (33 Nm).

f. Install the 2 wire harness clamps.

g. Line up the pipe and connector and push them together until a "click" sound is heard.

➡Check if there is any damage or foreign objects on the connected part. After connecting, check if the connector and the pipe are securely connected by pulling on them.

h. Connect the charcoal canister outlet tube.

➡Please note the following:

- Check that there are no scratches or foreign objects on the connecting part.
- Check that the charcoal canister outlet tube sub-assembly is inserted securely.
- Check that the charcoal canister outlet tube sub-assembly is securely connected by pulling on it.

i. Connect the fuel tank to filler pipe hose.

40. Connect fuel tank main tube sub-assembly.

a. Line up the pipe and connector and push them together until a "click" sound is heard.

b. Push the fuel tube connector cover in to lock it.

➡Please note the following:

- Check that there are no scratches or foreign objects on the connecting part.
- Check that the fuel tank main tube is inserted securely.
- After installing the fuel tube connector cover, check that the fuel tank main tube is securely connected by pulling on it.

41. Connect charcoal canister outlet tube sub-assembly.

42. Connect fuel tank vent hose.

43. Install the No. 1 fuel tank heat insulator with the 4 nuts.

44. Install center exhaust pipe assembly.

a. Install 2 new gaskets to the front exhaust pipe assembly and center exhaust pipe assembly.

b. Hang the center exhaust pipe assembly with the 3 exhaust pipe supports.

c. Install the 4 bolts.

45. Install instrument panel finish panel end LH.

46. Add fuel

47. Inspect for fuel leak

48. Inspect for exhaust gas leak

49. Install the 3 bolts and the fuel tank filler hose cover.

50. Install the rear floor No. 2 cross-member brace LH with 2 bolts. Tighten to 21 ft. lbs. (28 Nm).

51. Install charcoal canister protector

IDLE SPEED

ADJUSTMENT

Idle speed is controlled by the ECM. No adjustment is necessary or possible.

THROTTLE BODY

REMOVAL & INSTALLATION

2.7L Engine

See Figure 132.

1. Remove No. 1 engine under cover.
2. Drain engine coolant.
3. Remove windshield wiper motor and link.
4. Remove cowl panel sub-assembly.
5. Remove No. 1 engine cover sub-assembly.
6. Remove air cleaner inlet assembly.
7. Remove air cleaner assembly.

a. Disengage the wire harness clamp.

b. Disconnect the mass air flow meter connector.

c. Disengage the 2 wire harness clamps.

d. Disconnect the vacuum switching valve connector.

e. Disengage the hose clamp and separate the vacuum hose.

f. Disconnect the vacuum hose.

g. Loosen the hose clip and disconnect the air cleaner hose.

h. Remove the 2 bolts and the air cleaner assembly.

8. Remove air cleaner hose.

a. Loosen the clip and disconnect the No. 1 air hose.

b. Loosen the clip and disconnect the ventilation hose assembly.

c. Separate the No. 2 fuel vapor feed hose from the hose clamp.

d. Separate the vacuum hose from the hose clamp.

e. Separate the No. 2 vacuum switching valve assembly.

f. Separate the No. 1 air hose from the 2 hose clamps.

g. Separate the vacuum hose from the hose clamp.

9. Loosen the hose clip and remove the No. 1 air cleaner hose.

10. Loosen the clip and disconnect the water by-pass hose.

11. Loosen the clip and disconnect the No. 2 water by-pass hose.

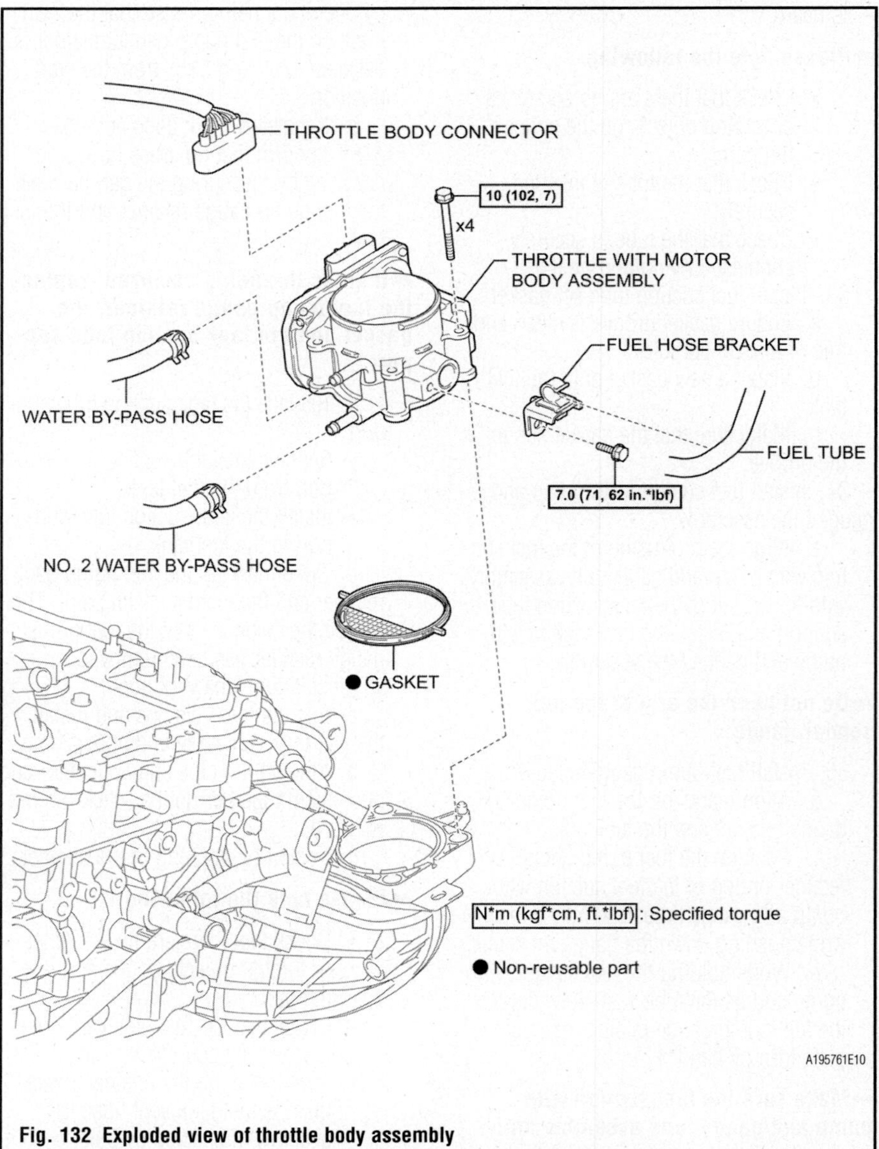

Fig. 132 Exploded view of throttle body assembly

12. Remove throttle with motor body assembly.
 a. Disconnect the throttle body connector.
 b. Separate the fuel tube from the fuel hose bracket.
 c. Remove the 4 bolts and the throttle with motor body assembly.
 d. Remove the bolt and fuel hose bracket.
 e. Remove the gasket from the intake manifold.

To install:

13. Install throttle with motor body assembly.
 a. Install a new gasket onto the intake manifold.
 b. Install the fuel hose bracket with the bolt.
 c. Install the throttle with motor body assembly to the intake manifold with the 4 bolts.
 d. Connect the throttle body connector.
 e. Install the fuel tube to the fuel hose bracket.
14. Connect the No. 2 water by-pass hose with the clip.
15. Connect the water by-pass hose with the clip.
16. Install air cleaner hose.
 a. Install the No. 1 air cleaner hose with the hose clip.
 b. Install the vacuum hose to the hose clamp.
 c. Install the No. 1 air hose to the 2 hose clamps.
 d. Install the No. 2 vacuum switching valve assembly to the No. 1 air cleaner hose.
 e. Install the vacuum hose to the hose clamp.
 f. Install the No. 2 fuel vapor feed hose to the hose clamp.
 g. Connect the ventilation hose assembly to the No. 1 air cleaner hose with the clip.
 h. Connect the No. 1 air hose to the No. 1 air cleaner hose with the clip.
17. Install air cleaner assembly.
 a. Insert the tab of the air cleaner assembly to the hole of the vehicle body.
 b. Install the air cleaner assembly with the 2 bolts.
 c. Connect the No. 1 air cleaner hose to the air cleaner assembly with the hose clip.
 d. Connect the vacuum hose to the air cleaner assembly.
 e. Engage the hose clamp and install the vacuum hose.
 f. Connect the vacuum switching valve connector.
 g. Engage the 2 wire harness clamps to the air cleaner assembly.
 h. Connect the mass air flow meter connector.
 i. Engage the wire harness clamp to the air cleaner assembly.
18. Install air cleaner inlet assembly.
19. Add engine coolant.
20. Inspect for coolant leak.
21. Install No. 1 engine cover sub-assembly.
22. Install cowl panel sub-assembly.
23. Install windshield wiper motor and link.
24. Install No. 1 engine under cover.
25. Perform initialization.

➡Perform the following procedure after replacing the ECM, throttle with motor body assembly or any throttle body components. The following procedure should also be performed if the throttle body is cleaned.

 a. Disconnect the cable from the negative (-) battery terminal. Wait at least 60 seconds and reconnect the cable.

➡When disconnecting the cable, some systems need to be initialized after the cable is reconnected (Click here for more information).

 b. Turn the ignition switch to ON without operating the accelerator pedal.

➡If the accelerator pedal is operated, perform the above steps again.

 c. Connect the Techstream to the DLC3 and clear the DTCs.
26. Start the engine and check that the MIL is not illuminated. After the engine is warmed up, check that the idle speed is within the specified range when the A/C is switched off.

➡Be sure to perform this step with all accessories off. Make sure that the shift lever is in neutral.

 a. Enter the following menus: Powertrain / Engine / Data List / Throttle Sensor Position. Fully depress the accelerator pedal and check that the value is 60% or more.
 b. Perform a road test and confirm that there are no abnormalities.

3.5L Engine

See Figure 133.

1. Remove windshield wiper motor and link assembly.
2. Remove front outer cowl top panel sub-assembly.
3. Remove V-bank cover sub-assembly.
4. Remove No. 1 engine under cover.
5. Drain engine coolant.
6. Remove air cleaner inlet cover assembly seal.
 a. Disengage the 3 vacuum hose clamps.
 b. Remove the 2 bolts and the air cleaner inlet cover assembly seal.
7. Remove air cleaner and hose.
 a. Disconnect the mass air flow meter connector.
 b. Disengage the wire harness clamp.
 c. Disconnect the vacuum hose.
 d. Separate the vacuum hose from the hose clamp.
 e. Disconnect the vacuum switching valve connector.
 f. Disengage the 2 wire harness clamps.
 g. Disconnect the 2 vacuum hoses.
 h. Loosen the clip and disconnect the No. 2 ventilation hose.
 i. Separate the fuel vapor feed hose from the hose clamp.
 j. Loosen the hose clamp and disconnect the air cleaner hose.
8. Remove the 2 bolts and the air cleaner assembly.
9. Loosen the clip and disconnect the water by-pass hose.
10. Loosen the clip and disconnect the No. 2 water by-pass hose.
11. Remove throttle with motor body assembly.
 a. Disconnect the throttle body connector and the wire harness clamp.
 b. Remove the 4 bolts, wire harness clamp stay and the throttle body.
 c. Remove the gasket from the intake air surge tank assembly.

To install:

12. Install throttle with motor body assembly.
 a. Install a new gasket to the intake air surge tank assembly.
 b. Install the throttle body and wire harness clamp stay to the intake air surge tank assembly with the 4 bolts.
 c. Connect the throttle body connector and wire harness clamp.
13. Connect the No. 2 water by-pass hose with the clip.
14. Connect the water by-pass hose with the clip.
15. Install air cleaner and hose.
 a. Insert the tab of the air cleaner assembly to the hole of the vehicle body.

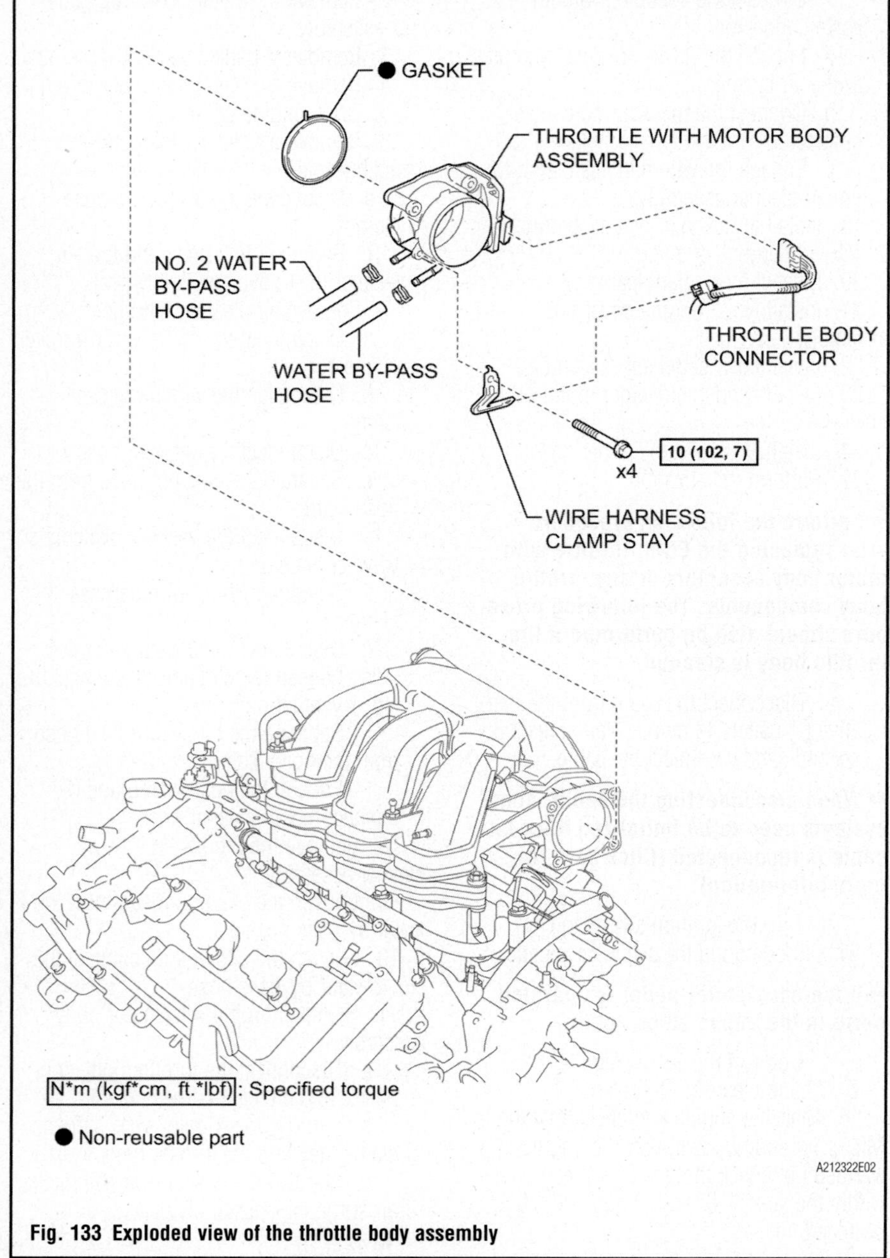

Fig. 133 Exploded view of the throttle body assembly

b. Install the air cleaner assembly with the 2 bolts.

c. Connect the air cleaner hose to the throttle body with the hose clamp.

d. Install the fuel vapor feed hose to the hose clamp.

e. Connect the No. 2 ventilation hose with the clip.

f. Connect the 2 vacuum hoses.

g. Engage the 2 wire harness clamps.

h. Connect the vacuum switching valve connector.

i. Install the vacuum hose to the hose clamp.

j. Connect the vacuum hose to the intake air surge tank assembly.

k. Engage the wire harness clamp.

l. Connect the mass air flow meter connector.

16. Install air cleaner inlet cover assembly seal.

a. Install the air cleaner inlet cover assembly seal with the 2 bolts.

b. Engage the 3 vacuum hose clamps to the air cleaner inlet cover assembly seal.

17. Add engine coolant.
18. Inspect for coolant leak.
19. Install No. 1 engine under cover.
20. Install V-bank cover sub-assembly.
21. Install front outer cowl top panel sub-assembly.
22. Install windshield wiper motor and link assembly.
23. Perform initialization.

➡Please note the following:

- Be sure to perform this procedure after reassembling the throttle body or removing and reinstalling any throttle body component.
- Perform the following procedure after replacing the ECM, throttle body assembly or any throttle body components. The following procedure should also be performed if the throttle body is cleaned.
- Be sure to perform this procedure after replacing the ECM and reconnecting the battery cable.

a. Disconnect the cable from the negative (-) battery terminal. Wait at least 60 seconds and reconnect the cable.

b. Turn the ignition switch to ON without operating the accelerator pedal.

➡If the accelerator pedal is operated, perform the above steps again.

c. Connect the Techstream to the DLC3 and clear the DTCs.

d. Start the engine and check that the MIL is not illuminated and that the idle speed is within the specified range when the A/C is switched off after the engine is warmed up.

➡Be sure to perform this step with all accessories off. Make sure that the shift lever is in neutral.

e. Enter the following menus: Powertrain/ Engine / Data List/ Throttle Sensor Position. Fully depress the accelerator pedal and check that the value is 60% or more.

f. Perform a road test and confirm that there are no abnormalities.

HEATING & AIR CONDITIONING SYSTEM

BLOWER MOTOR

REMOVAL & INSTALLATION

See Figure 134.

1. Remove blower with fan motor sub-assembly.
   a. Disconnect the connector.
   b. Remove the 3 screws and blower with fan motor.

To install:

2. Install blower with fan motor sub-assembly.
   a. Install the blower with fan motor with the 3 screws.
   b. Connect the connector.

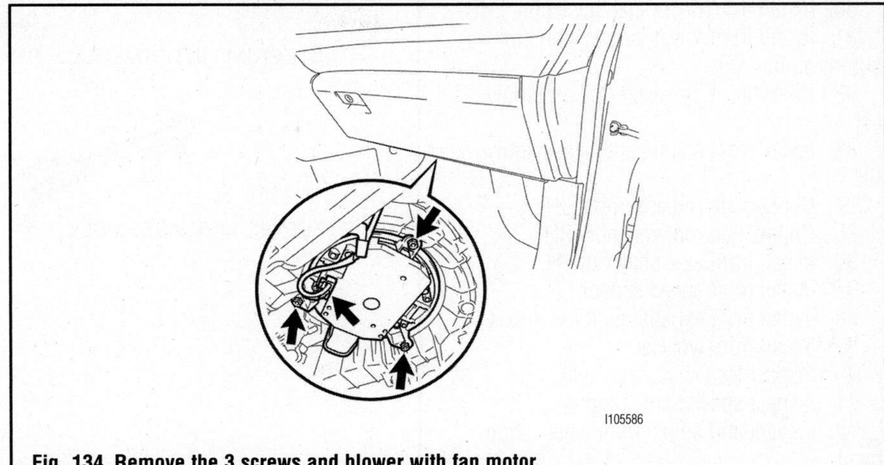

Fig. 134 Remove the 3 screws and blower with fan motor

STEERING

POWER STEERING GEAR

REMOVAL & INSTALLATION

2.7L Engine

See Figure 135.

➡When disconnecting the steering intermediate shaft assembly and pinion shaft of steering link assembly, be sure to place matchmarks before servicing.

1. Secure the steering wheel with the seat belt in order to prevent it from rotating.

➡This operation is necessary to prevent damage to the spiral cable.

2. Remove front wheels.
3. Remove front exhaust pipe assembly.
4. Separate front speed sensor LH.
5. Remove front axle shaft nut LH.
6. Separate tie rod assembly LH.
7. Separate tie rod assembly RH.
8. Remove front stabilizer link assembly LH.
9. Remove front stabilizer link assembly RH.
10. Separate front lower suspension arm sub assembly LH.
11. Separate front drive shaft assembly LH.
12. Separate steering intermediate shaft assembly
    a. Put matchmarks on the steering intermediate shaft assembly and steering link assembly.
    b. Remove the bolt and separate the steering intermediate shaft assembly from the steering link assembly.
13. Remove front No. 1 stabilizer bracket LH.
14. Remove front No. 1 stabilizer bracket RH.
15. Remove front No. 2 stabilizer bracket LH.
16. Remove front No. 2 stabilizer bracket RH.
17. Remove front No. 1 stabilizer bar bushing.
18. Remove steering link assembly.
    a. Remove the 2 bolts and 2 nuts and steering link assembly.

➡Because the nut has its own stopper, do not turn the nut. Loosen the bolt with the nut secured.

    b. Pull out the steering link assembly towards the left side of the vehicle.

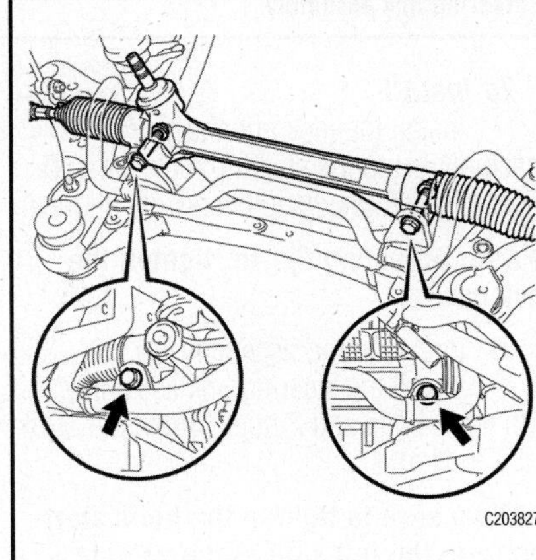

Fig. 135 Remove the 2 bolts and 2 nuts and steering link assembly

19. Remove tie rod assembly LH.
    a. Put matchmarks on the tie rod assembly LH and steering gear assembly.
    b. Loosen the lock nut, and remove the tie rod assembly LH and lock nut.
20. Remove tie rod assembly RH.

To install:

21. Install the lock nut and tie rod assembly LH to the steering rack end sub-assembly until the matchmarks are aligned.

➡After adjusting toe-in, tighten the lock nut.

22. Install tie rod assembly RH.
23. Install the steering link assembly with the 2 bolts and 2 nuts. Tighten to 52 ft. lbs. (70 Nm).

➡Make sure to tighten the bolts starting from the left side of the vehicle. Because the nut has its own stopper, do not turn the nut. Tighten the bolt with the nut secured.

24. Install front No. 1 stabilizer bar bushing.
25. Install front No. 2 stabilizer bracket LH.
26. Install front No. 2 stabilizer bracket RH.
27. Install front No. 1 stabilizer bracket LH.
28. Install front No. 1 stabilizer bracket RH.
29. Install steering intermediate shaft assembly.
    a. Align the matchmarks on the steering intermediate shaft assembly and steering link assembly.

b. Install the bolt. Tighten to 26 ft. lbs. (35 Nm).

30. Install front drive shaft assembly LH.
31. Install front lower suspension arm sub-assembly LH.
32. Install front stabilizer link assembly LH.
33. Install front stabilizer link assembly RH.
34. Connect tie rod assembly LH.
35. Connect tie rod assembly RH.
36. Install front axle shaft nut LH.
37. Install front speed sensor LH.
38. Install front exhaust pipe assembly.
39. Install front wheels.
40. Inspect for exhaust gas leak.
41. Inspect speed sensor signal.
42. Inspect and adjust front wheel alignment.

3.5L Engine

See Figure 136.

➡When disconnecting the steering intermediate shaft assembly and pinion shaft of steering gear assembly, be sure to place matchmarks before servicing.

1. Secure the steering wheel with the seat belt in order to prevent it from rotating.

➡This operation is necessary to prevent damage to the spiral cable.

2. Remove front stabilizer bar.
3. Separate steering intermediate shaft assembly.
   a. Put matchmarks on the steering intermediate shaft assembly and steering link assembly.
   b. Remove the bolt and separate the steering intermediate shaft assembly from the steering link assembly.
4. Remove steering link assembly.
   a. Remove the 2 bolts and 2 nuts and steering link assembly.

➡Because the nut has its own stopper, do not turn the nut. Loosen the bolt with the nut secured.

   b. Pull out the steering link assembly towards the left side of the vehicle.
5. Remove tie rod assembly LH.
   a. Put matchmarks on the tie rod assembly LH and steering gear assembly.
   b. Loosen the lock nut, and remove the tie rod assembly LH and lock nut.
6. Remove tie rod assembly RH

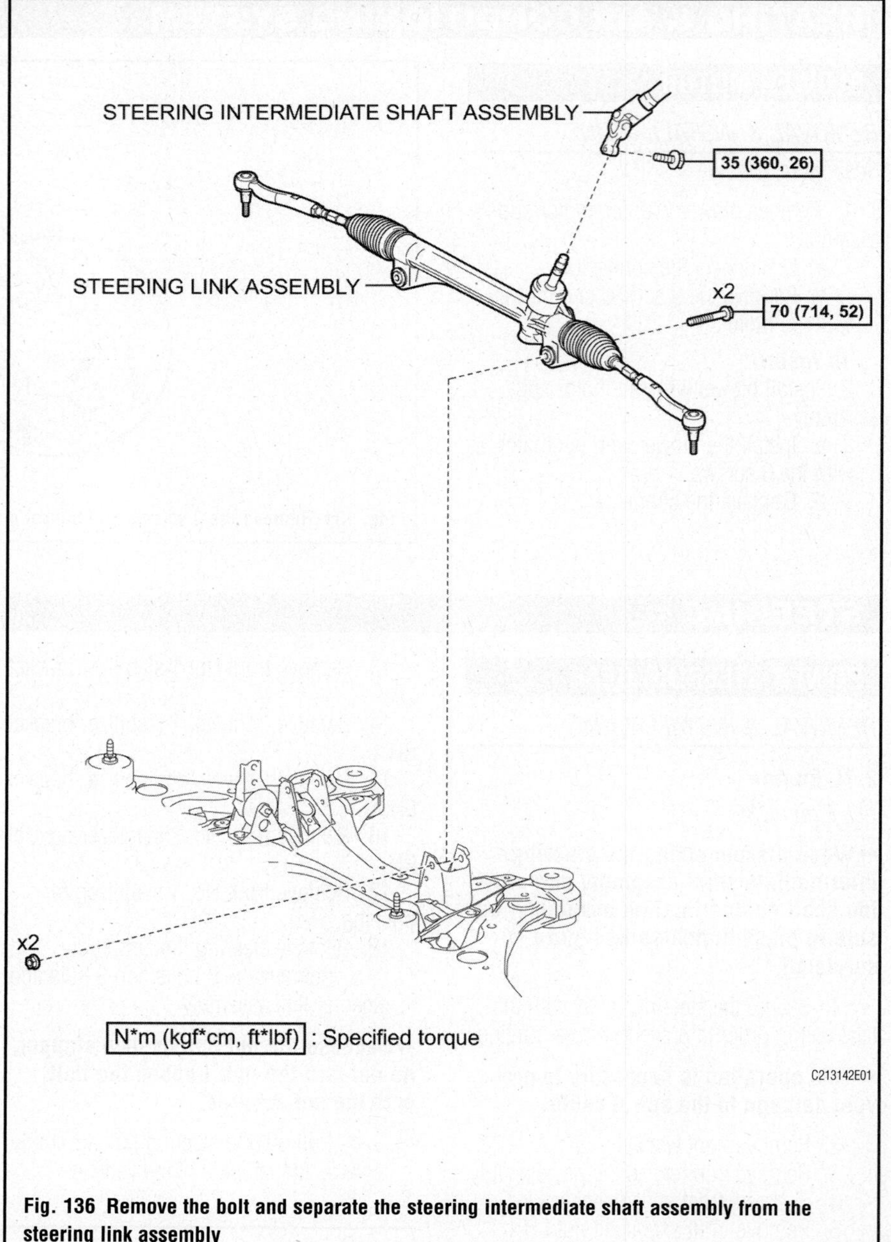

Fig. 136 Remove the bolt and separate the steering intermediate shaft assembly from the steering link assembly

To install:

7. Install the lock nut and tie rod assembly LH to the steering rack end sub-assembly until the matchmarks are aligned.

➡After adjusting toe-in, tighten the lock nut.

8. Install tie rod assembly RH.
9. Install the steering link assembly with the 2 bolts and 2 nuts. Tighten to 52 ft. lbs. (70 Nm).

➡Make sure to tighten the bolts starting from the left side of the vehicle.

Because the nut has its own stopper, do not turn the nut. Tighten the bolt with the nut secured.

10. Install steering intermediate shaft assembly.
   a. Align the matchmarks on the steering intermediate shaft assembly and steering link assembly.
   b. Install the bolt. Tighten to 26 ft. lbs. (35 Nm).
11. Install front stabilizer bar.

LOWER BALL JOINT

REMOVAL & INSTALLATION

See Figure 137.

➡Use the same procedure for the LH side and RH side. The following procedure listed is for the LH side.

1. Remove front wheel.
2. Remove front axle shaft nut.
3. Separate front speed sensor.
4. Separate front disc brake caliper assembly.
5. Remove front disc.
6. Separate tie rod assembly.
7. Separate front lower suspension arm sub-assembly.
8. Separate front drive shaft assembly.
9. Remove front axle assembly.
10. Remove front lower ball joint assembly.
    a. Secure the front axle assembly in a vise using aluminum plates.

➡When using a vise, do not over-tighten it.

    b. Remove the cotter pin and nut.
    c. Install SST to the front lower ball joint.

➡Check that the clearance measurement between SST and the front axle assembly is 0.0394 inches (1 mm).

    d. Using SST, remove the front lower ball joint from the front axle assembly.

➡Apply grease to the threads and end of the SST bolt. Install SST so that A and B are parallel. Do not damage the front lower ball joint dust cover, or steering knuckle.

To install:

11. Install the front lower ball joint to the front axle assembly with the nut. Tighten to 91 ft. lbs. (123 Nm).

➡Prevent oil from adhering to the screw and tapered parts.

Install a new cotter pin.

➡If the holes for the cotter pin are not aligned, tighten the nut further up to 60°.

12. Install front axle assembly.
13. Install front drive shaft assembly.
14. Install front lower suspension arm sub-assembly.
15. Connect tie rod assembly.
16. Install front disc.
17. Install front disc brake caliper assembly.
18. Install front speed sensor.
19. Install front axle shaft nut.
20. Install front wheel.
21. Inspect speed sensor signal.
22. Inspect and adjust front wheel alignment.

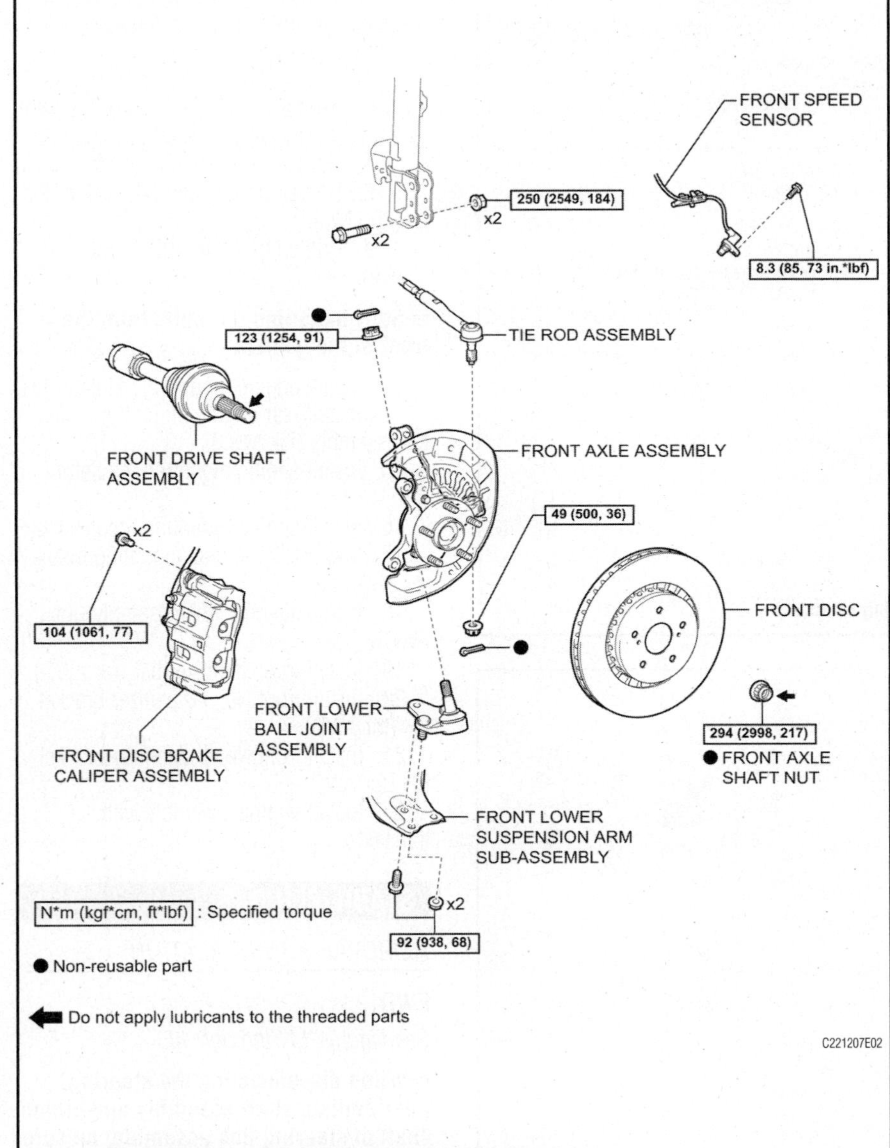

Fig. 137 Exploded view showing the lower ball joint assembly

LOWER CONTROL ARM

REMOVAL & INSTALLATION

See Figures 138 through 140.

➡Use the same procedure for the LH side and RH side. The following procedure listed below is for the LH side.

1. Remove engine assembly with transaxle.
2. Remove front No. 1 stabilizer bracket LH (for AWD).
3. Remove front No. 1 stabilizer bracket RH (for AWD).
4. Remove front stabilizer bar (for AWD).
5. Remove steering link assembly (for AWD).
6. Separate front engine mounting insulator assembly.

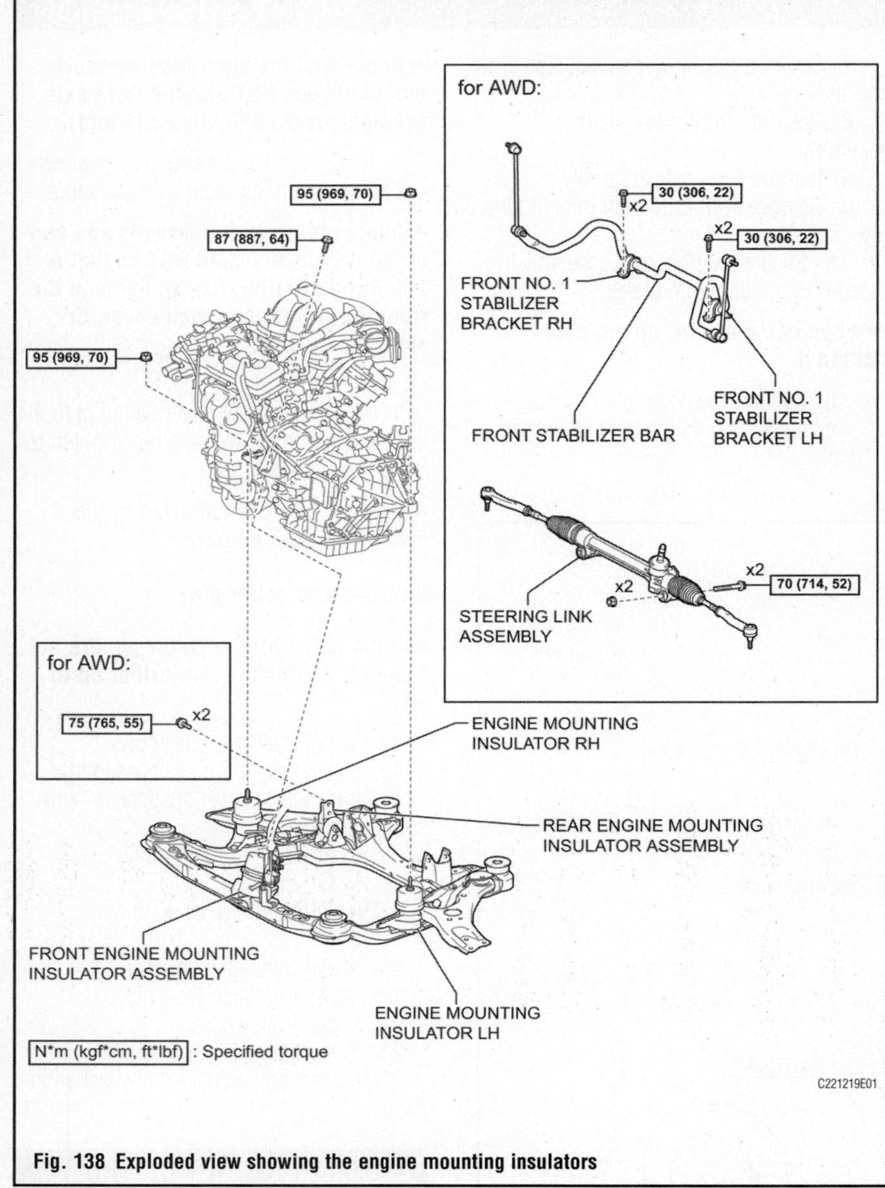

Fig. 138 Exploded view showing the engine mounting insulators

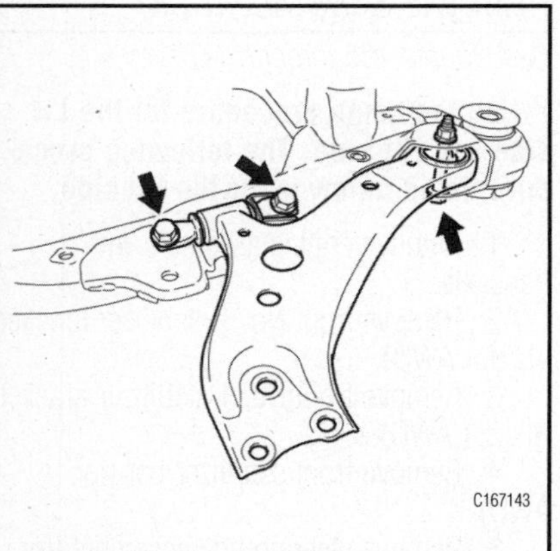

Fig. 139 Remove the 3 bolts, nut and front lower suspension arm sub-assembly from the front frame assembly

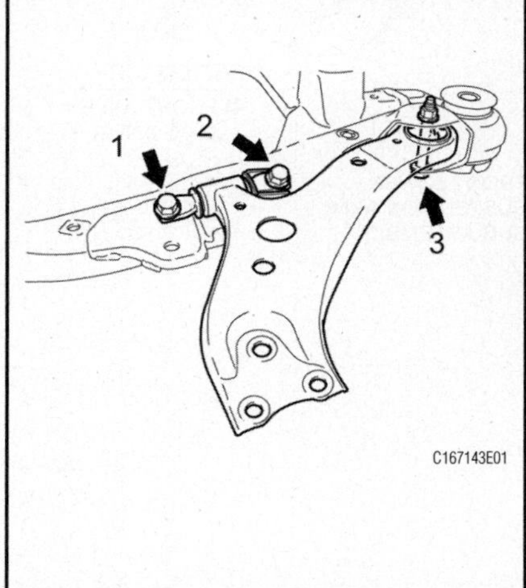

Fig. 140 Tighten the 3 bolts in order

7. Separate engine mounting insulator LH.
8. Separate engine mounting insulator RH.
9. Separate rear engine mounting insulator assembly (for AWD).
10. Remove engine mounting insulator LH.
11. Remove front lower suspension arm sub-assembly.
   a. Remove the 3 bolts, nut and front lower suspension arm sub-assembly from the front frame assembly.
   b. Remove the front lower arm bushing stopper from the front lower suspension arm sub-assembly.

To install:

12. Install front lower suspension arm sub-assembly.
   a. Install the front lower arm bushing stopper to the front lower suspension arm sub-assembly.
   b. Temporarily tighten the front lower suspension arm sub-assembly to the front frame assembly with the 3 bolts and nut.
   c. Tighten the 3 bolts in order.
   d. Tighten bolts 1 and 2 to 148 ft. lbs. (200 Nm).
   e. Tighten bolt 3 to 103 ft. lbs. (140 Nm).

➡Start installing the bolts from the front of the vehicle.

13. Install engine mounting insulator LH.
14. Install rear engine mounting insulator assembly (for AWD).
15. Install engine mounting insulator RH.
16. Install engine mounting insulator LH.
17. Install front engine mounting insulator assembly.
18. Install steering link assembly (for AWD).
19. Install front stabilizer bar (for AWD).
20. Install front No. 1 stabilizer bracket LH (for AWD).
21. Install front No. 1 stabilizer bracket RH (for AWD).
22. Install engine assembly with transaxle.

STABILIZER BAR

REMOVAL & INSTALLATION

FWD

See Figures 141 through 143.

➡When disconnecting the steering intermediate shaft assembly and pinion shaft of steering link assembly, be sure to place matchmarks before servicing.

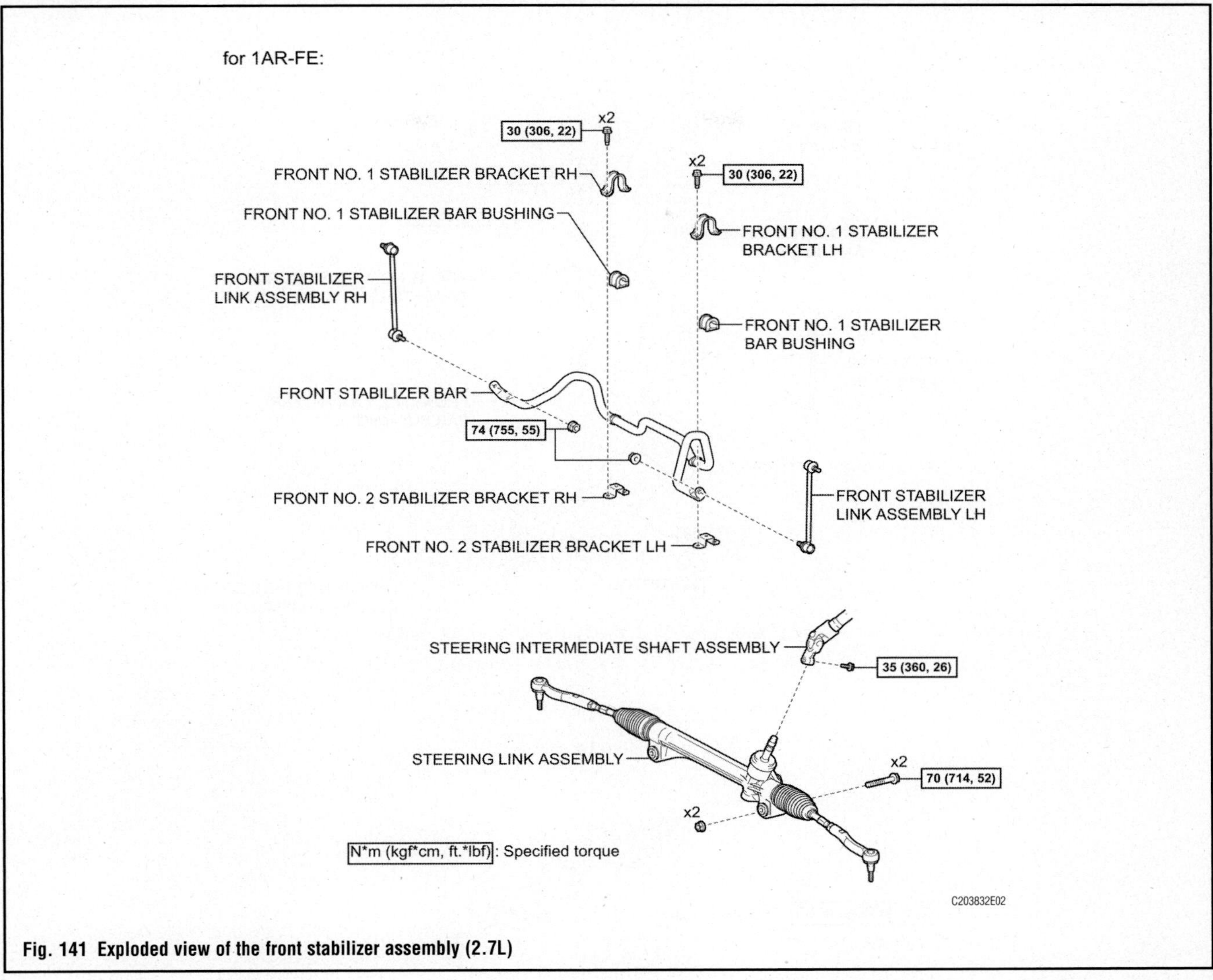

Fig. 141 Exploded view of the front stabilizer assembly (2.7L)

1. Secure steering wheel (for 2.7L).
2. Remove front wheels.
3. Remove front exhaust pipe assembly (for 2.7L).
4. Separate front speed sensor LH.
5. Remove front axle shaft nut LH.
6. Separate tie rod assembly LH.
7. Separate tie rod assembly RH (for 2.7L).
8. Remove the 2 nuts and front stabilizer link assembly LH.

➡If the ball joint turns together with the nut, use a hexagon wrench (6 mm) to hold the stud bolt.

9. Remove front stabilizer link assembly RH.
10. Separate front lower suspension arm sub-assembly LH.
11. Separate front drive shaft assembly LH.
12. Separate steering intermediate shaft assembly (for 2.7L).
13. Remove the 2 bolts and front No. 1 stabilizer bracket LH from the front frame assembly.
14. Remove front No. 1 stabilizer bracket RH.
15. Remove the front No. 2 stabilizer bracket LH from the front No. 1 stabilizer bar bushing (for 2.7L).
16. Remove front No. 2 stabilizer bracket RH (for 2.7L).
17. Remove the 2 front No. 1 stabilizer bar bushings from the front stabilizer bar.
18. Remove steering link assembly (for 2.7L).
19. Remove the front stabilizer bar from the left side of the vehicle.

To install:

20. Install the front stabilizer bar from the left side of the vehicle.
21. Install steering link assembly (for 2.7L).
22. Install the 2 front No. 1 stabilizer bar bushings to the front stabilizer bar.

➡When installing the front No. 1 stabilizer bar bushings, make sure that the cutout faces the rear of the vehicle.

23. Install the front No. 2 stabilizer bracket LH to the front No. 1 stabilizer bar bushing (for 2.7L).
24. Install front No. 2 stabilizer bracket RH (for 2.7L).
25. Install the front No. 1 stabilizer bracket LH to the front frame assembly with the 2 bolts. Tighten to 22 ft. lbs. (30 Nm).
26. Install front No. 1 stabilizer bracket RH.
27. Install steering intermediate shaft assembly (for 2.7L).
28. Install front drive shaft assembly LH.
29. Install front lower suspension arm sub-assembly LH.
30. Install the front stabilizer link assembly LH with the nut. Tighten to 55 ft. lbs. (74 Nm).

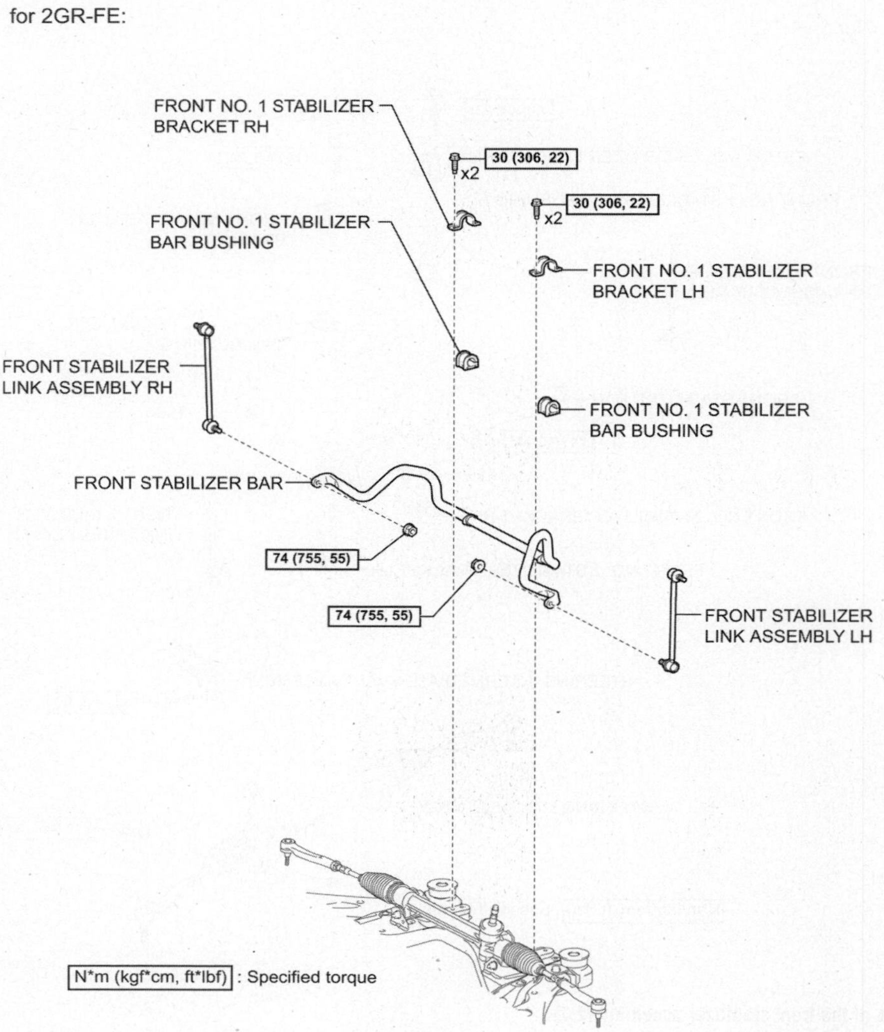

for 2GR-FE:

FRONT NO. 1 STABILIZER BRACKET RH

30 (306, 22) x2

30 (306, 22) x2

FRONT NO. 1 STABILIZER BAR BUSHING

FRONT NO. 1 STABILIZER BRACKET LH

FRONT STABILIZER LINK ASSEMBLY RH

FRONT NO. 1 STABILIZER BAR BUSHING

FRONT STABILIZER BAR

74 (755, 55)

74 (755, 55)

FRONT STABILIZER LINK ASSEMBLY LH

N*m (kgf*cm, ft*lbf) : Specified torque

C221233E02

Fig. 142 Exploded view of the front stabilizer assembly (3.5L)

31. Install front stabilizer link assembly RH.
32. Connect tie rod assembly LH.
33. Connect tie rod assembly RH (for 2.7L).
34. Install front speed sensor LH.

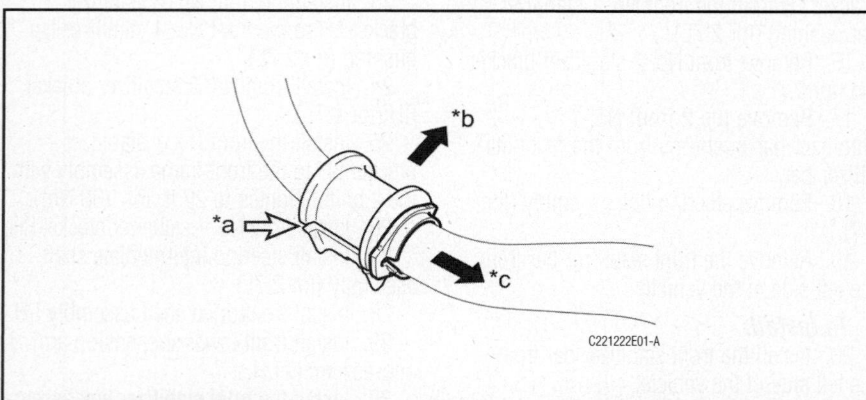

Fig. 143 Stabilizer bar bushing installation position showing the cutout (a), front of vehicle (b) and the inside of vehicle (c)

35. Install front axle shaft nut LH.
36. Install front exhaust pipe assembly (for 2.7L).
37. Install front wheels.
38. Inspect for exhaust gas leak (for 2.7L).
39. Inspect speed sensor signal.
40. Inspect and adjust front wheel alignment.

AWD

See Figures 143 and 144.

1. Remove front drive shaft assembly LH.
2. Remove exhaust manifold to head gasket.
3. Remove No. 1 exhaust pipe support bracket.
4. Remove front exhaust pipe assembly.

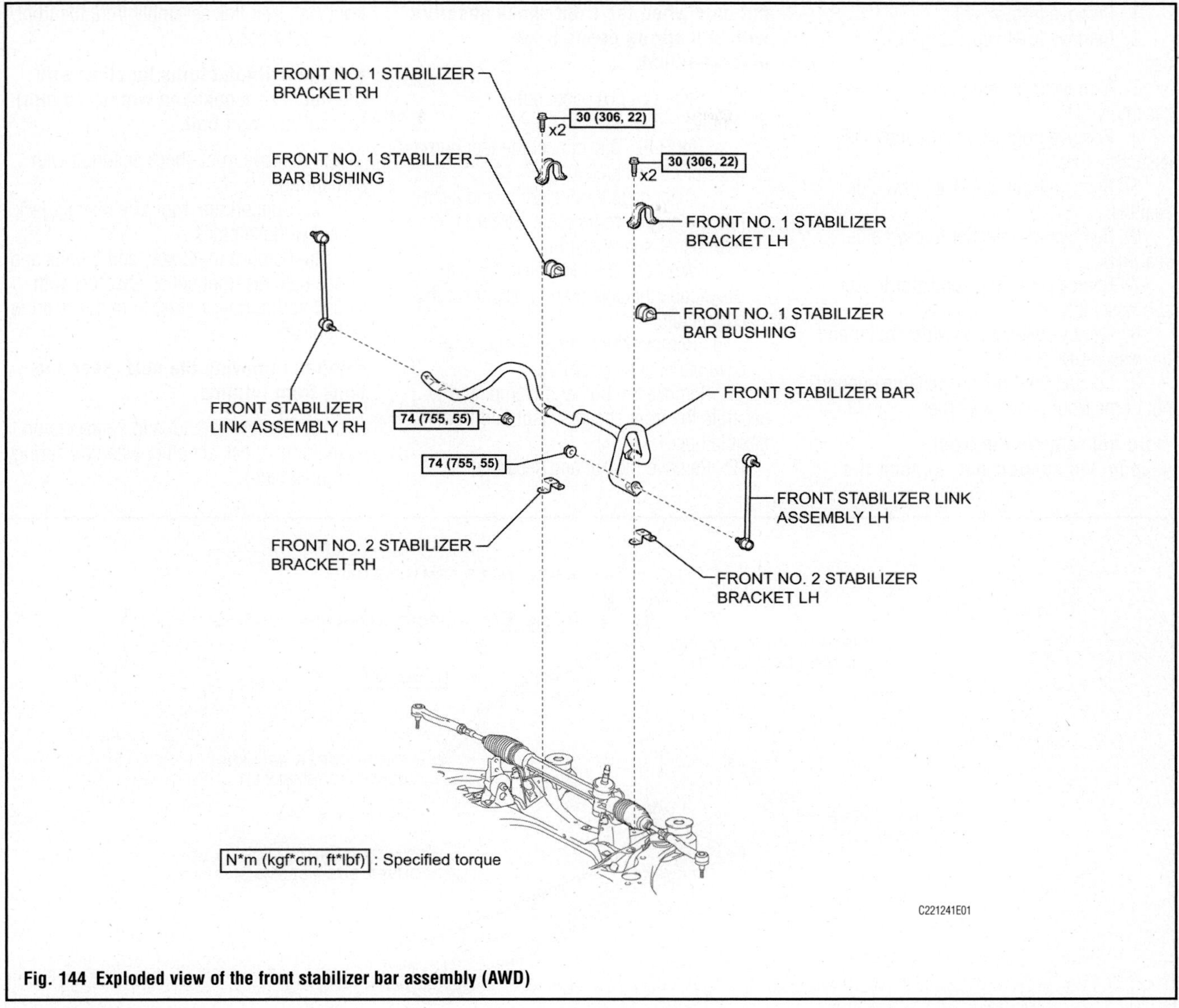

Fig. 144 Exploded view of the front stabilizer bar assembly (AWD)

5. Remove the 2 nuts and front stabilizer link assembly LH.

➡**If the ball joint turns together with the nut, use a hexagon wrench (6 mm) to hold the stud bolt.**

6. Remove front stabilizer link assembly RH.
7. Remove the 2 bolts and front No. 1 stabilizer bracket LH from the front frame assembly.
8. Remove front No. 1 stabilizer bracket RH.
9. Remove the front No. 2 stabilizer bracket LH from the front stabilizer bar bushing.
10. Remove front No. 2 stabilizer bracket RH.
11. Remove the 2 front No. 1 stabilizer bar bushings from the front stabilizer bar.
12. Remove the front stabilizer bar from the left side of the vehicle.

To install:

13. Install the front stabilizer bar from the left side of the vehicle.
14. Install the 2 front No. 1 stabilizer bar bushings to the front stabilizer bar.

➡**When installing the front No. 1 stabilizer bar bushings, make sure that the cutout faces the rear of the vehicle.**

15. Install the front No. 2 stabilizer bracket LH to the front No. 1 stabilizer bar bushing.
16. Install front No. 2 stabilizer bracket RH.
17. Install the front No. 1 stabilizer bracket LH to the front frame assembly with the 2 bolts. Tighten to 22 ft. lbs. (30 Nm).
18. Install front No. 1 stabilizer bracket RH.
19. Install the front stabilizer link assembly LH with the 2 nuts. Tighten to 55 ft. lbs. (74 Nm).
20. Install front stabilizer link assembly RH.
21. Install front exhaust pipe assembly.
22. Install No. 1 exhaust pipe support bracket.
23. Install exhaust manifold to head gasket.
24. Install front drive shaft assembly LH.

STRUT & SPRING ASSEMBLY

REMOVAL & INSTALLATION

See Figures 145 and 146.

1. Remove front wheel.
2. Remove front wiper arm head cap.
3. Remove front wiper arm and blade assembly LH.
4. Remove front wiper arm and blade assembly RH.
5. Remove front fender to cowl side seal LH.
6. Remove front fender to cowl side seal RH.
7. Remove cowl top ventilator louver sub-assembly.
8. Remove windshield wiper motor and link assembly.
9. Loosen the front suspension support nut of the front shock absorber.

➡**Do not remove the front suspension support nut. Loosen the nut only when the front shock absorber with coil spring needs to be disassembled.**

10. Remove cowl panel sub-assembly.
 a. for 3.5L: Disconnect the fuel pump resistor connector.
 b. Disengage the 5 clamps and separate the engine room main wire from the cowl panel sub-assembly.
 c. With Windshield Deicer System: Disengage the clamp and disconnect the connector.
 d. Remove the 5 bolts, 4 nuts and cowl panel sub-assembly.
11. Remove the bolt and clamp, and separate the front speed sensor and front flexible hose.
12. Remove the nut and separate the front stabilizer link assembly from the front shock absorber.

➡**If the ball joint turns together with the nut, use a hexagon wrench (6 mm) to hold the stud bolt.**

13. Remove front shock absorber with coil spring.
 a. Support the front axle using a jack and wooden block.
 b. Remove the 2 nuts and 2 bolts and separate the front shock absorber with coil spring (lower side) from the steering knuckle.

➡**When removing the nuts, keep the bolts from rotating.**

 c. Remove the nut and 2 spacers on the front shock absorber with coil spring (upper side).

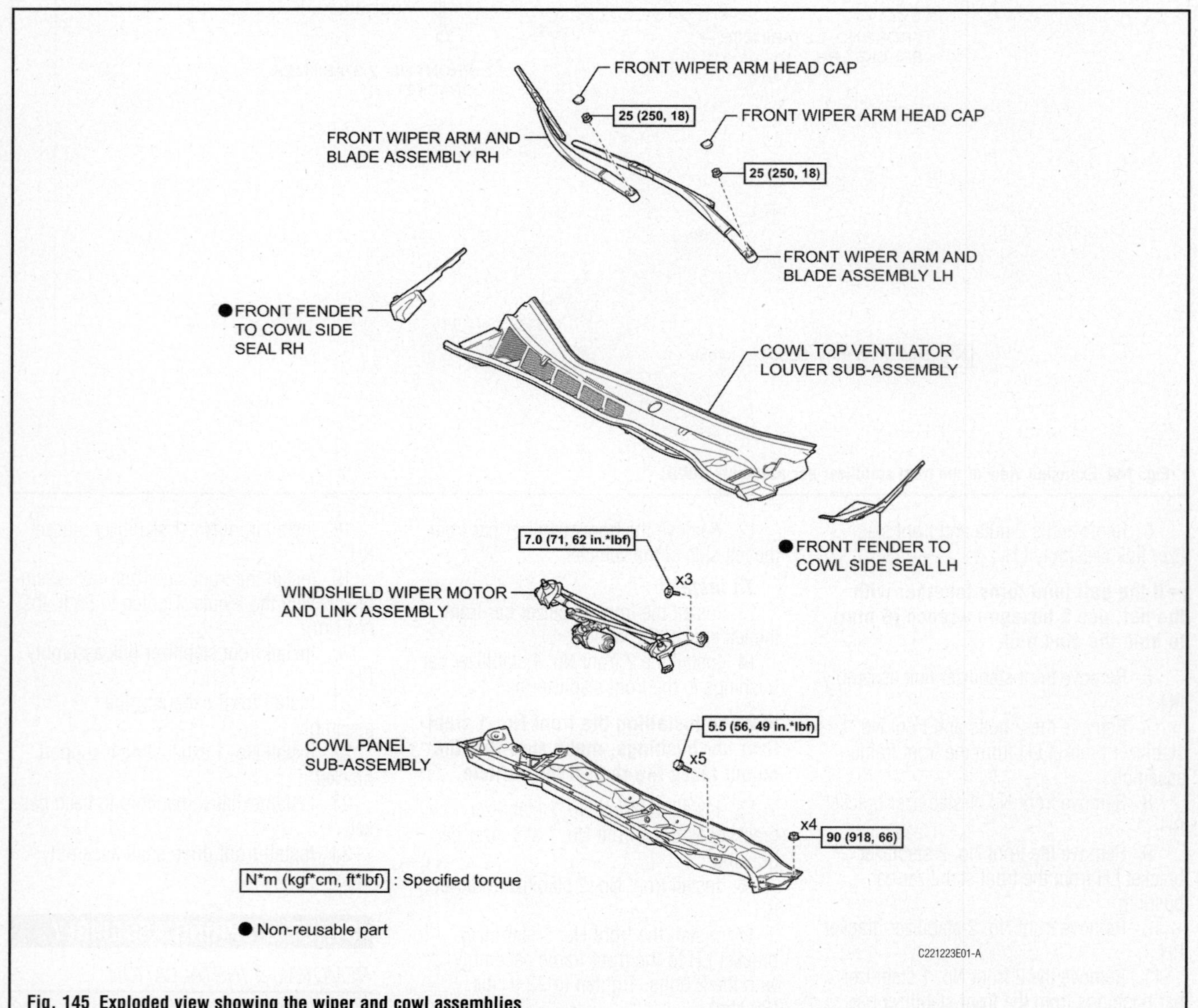

Fig. 145 Exploded view showing the wiper and cowl assemblies

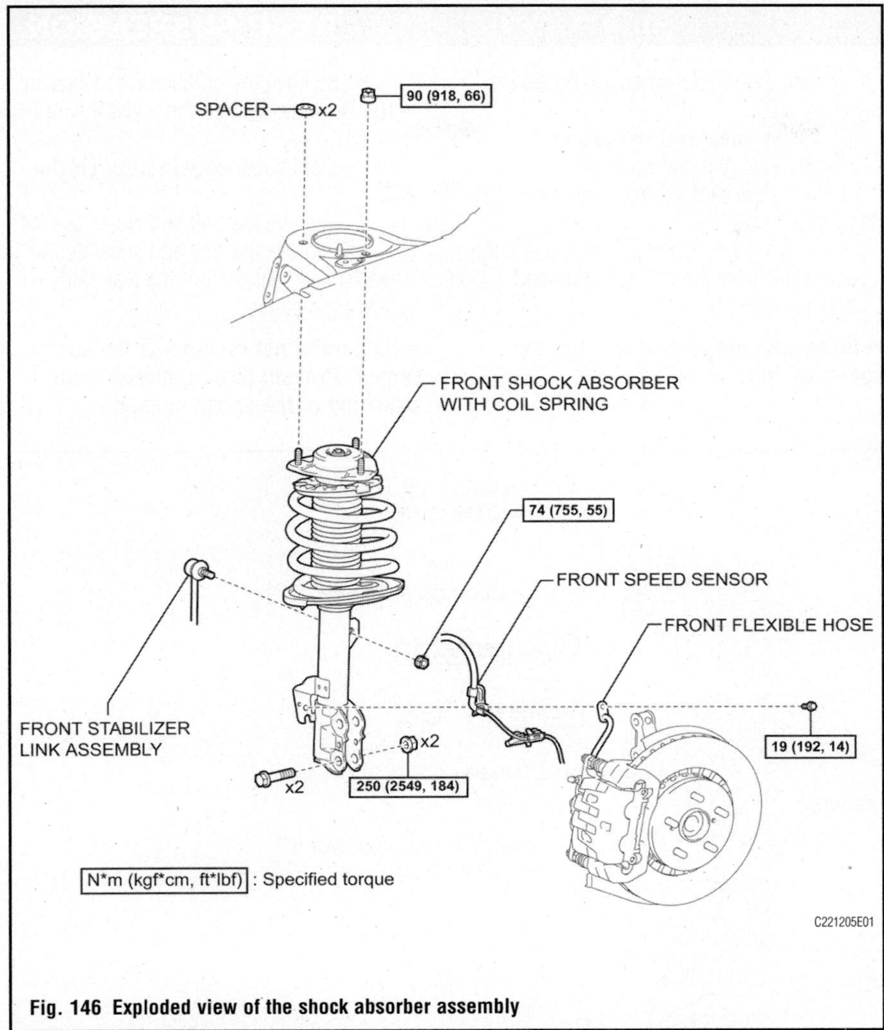

Fig. 146 Exploded view of the shock absorber assembly

➡Make sure that the front speed sensor is completely separated from the front shock absorber with coil spring.

To install:

14. Install front shock absorber with coil spring.

a. Install the front shock absorber with coil spring (upper side) with the nut and 2 spacers. Tighten to 66 ft. lbs. (90 Nm).

b. Install the front shock absorber with coil spring (lower side) to the steering knuckle and insert the 2 bolts and 2 nuts. Tighten to 184 ft. lbs. (250 Nm).

➡When installing the nuts, keep the bolts from rotating.

15. Install front speed sensor.

a. Install the front stabilizer link assembly to the front shock absorber with the nut. Tighten to 55 ft. lbs. (74 Nm).

16. Install the front speed sensor and front flexible hose with the bolt.

➡Do not twist the front speed sensor when installing it.

a. Install the clamp.

17. Install cowl panel sub-assembly.

a. Install the cowl panel sub-assembly with the 5 bolts and 4 nuts.

b. Engage the 5 clamps and install the engine room main wire.

c. w/ Windshield Deicer System: Connect the connector and engage the clamp.

d. for 3.5L: Connect the fuel pump resistor connector.

18. Fully tighten the front suspension support nut. Tighten to 52 ft. lbs. (70 Nm).

19. Install windshield wiper motor and link assembly.

20. Install cowl top ventilator louver sub-assembly.

21. Install front fender to cowl side seal LH.

22. Install front fender to cowl side seal RH.

23. Install front wiper arm and blade assembly LH.

24. Install front wiper arm and blade assembly RH.

25. Install front wiper arm head cap.

26. Install front wheel.

27. Inspect and adjust front wheel alignment.

AXLE BEAM

REMOVAL & INSTALLATION

See Figure 147.

1. Remove rear wheels.
2. Drain brake fluid.
3. Remove instrument panel finish panel end LH (for 2.7L).
4. Remove rear differential (for AWD).
5. Remove tail exhaust pipe assembly (for FWD).
6. Remove center exhaust pipe assembly (for FWD).
7. Remove rear coil spring LH.
8. Remove rear coil spring RH.
9. Separate skid control sensor wire LH (for FWD).
   a. Using a screwdriver, disconnect the connector from the rear axle hub and bearing assembly.

➡Be careful not to damage the rear speed sensor.

   b. Remove the bolt and No. 1 bracket.
10. Separate skid control sensor wire RH (for FWD).
11. Separate rear speed sensor LH (for AWD).
   a. Remove the bolt and No. 1 bracket.
   b. Remove the bolt and separate the rear speed sensor from the rear axle bearing retainer.

➡Be careful not to damage the speed sensor. Prevent foreign matter from attaching to the speed sensor.

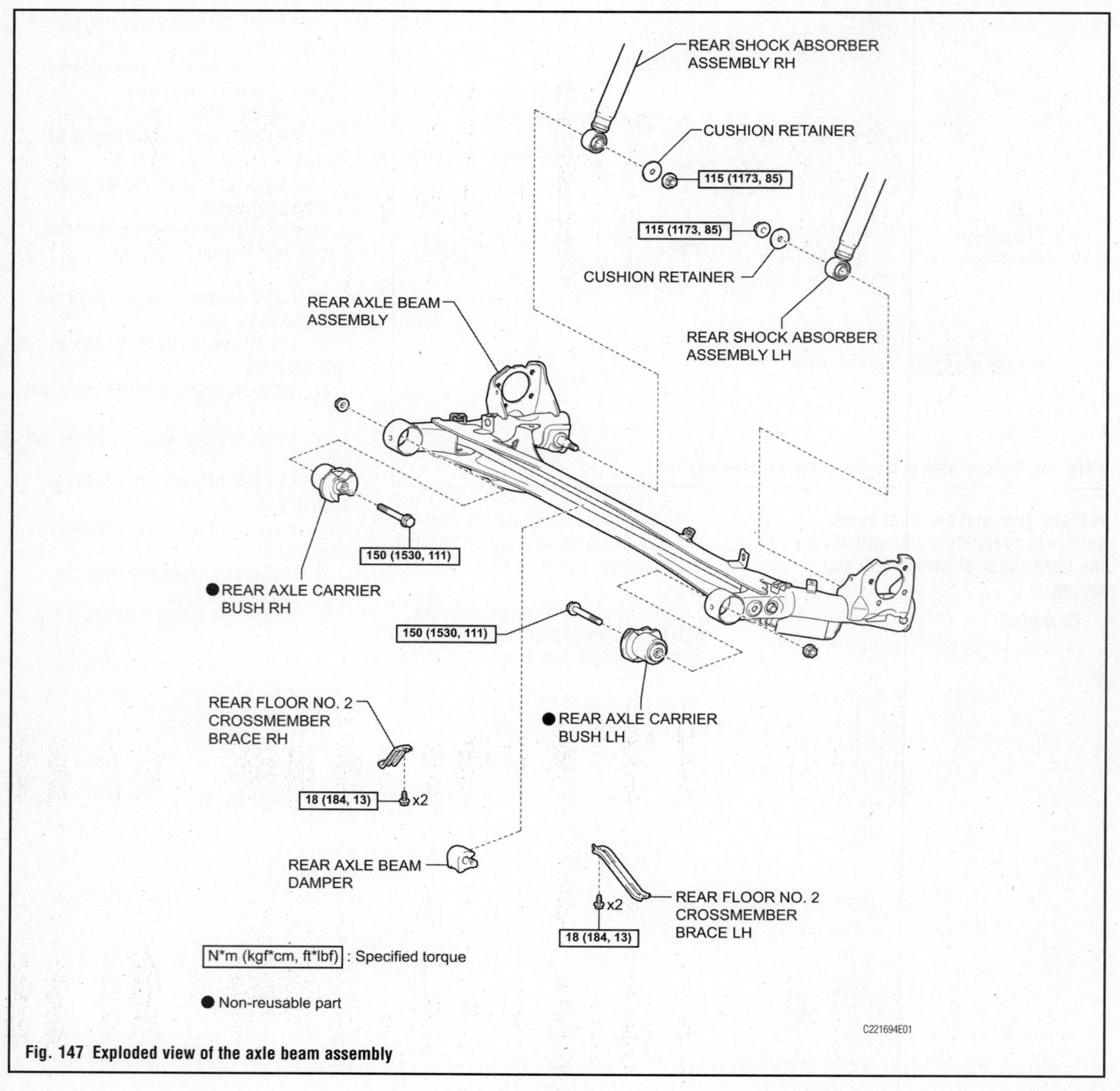

Fig. 147 Exploded view of the axle beam assembly

12. Separate rear speed sensor RH (for AWD).

13. Remove rear No. 6 brake tube.

a. Using a union nut wrench (10 mm), separate the rear brake tube while holding the rear brake tube flexible hose with a wrench.

➡Do not kink or damage the brake line. Do not allow any foreign matter such as dirt and dust to enter the brake line.

b. Remove the 2 clips, bolt and rear brake tube from the rear axle beam assembly.

14. Remove rear No. 5 brake tube.

15. Remove the 2 bolts and separate the parking brake cable assembly from the rear axle beam assembly.

16. Separate No. 2 parking brake cable assembly.

17. Remove the 2 bolts and the rear disc brake caliper assembly LH.

18. Remove rear disc brake caliper assembly RH.

19. Remove parking brake shoe adjusting hole plug.

20. Remove rear disc.

21. Remove rear axle hub and bearing assembly LH.

22. Remove rear axle hub and bearing assembly RH.

23. Remove the rear inner axle bearing retainer and parking brake assembly from the rear axle beam assembly (for FWD).

24. Remove rear inner axle bearing retainer RH (for FWD).

25. Remove the rear axle bearing retainer and parking brake assembly from the rear axle beam assembly LH (for AWD).

26. Remove rear axle bearing retainer RH (for AWD).

27. Remove the 2 bolts and rear floor No. 2 crossmember brace LH.

28. Remove rear floor No. 2 crossmember brace RH.

29. Remove rear axle beam assembly.

a. Support the rear axle beam assembly using a jack and 2 wooden blocks.

➡Make sure to secure the rear axle beam assembly to prevent it from dropping.

b. Remove the 2 nuts and 2 cushion retainers and separate the rear shock absorber assembly from the rear axle beam assembly.

c. Remove the 2 bolts and 2 nuts, and separate the rear axle beam assembly from the body.

➡Since stopper nuts are used, loosen the bolts.

d. Lower the rear axle beam assembly and remove it.

To install:

30. Temporarily tighten rear axle beam assembly.

a. Using wooden blocks and a jack, jack up the rear axle beam assembly to allow it to be installed.

➡Make sure to secure the rear axle beam assembly to prevent it from dropping.

b. Temporarily install the rear axle beam assembly to the body with the 2 bolts and 2 nuts.

➡Insert the bolts with the threaded ends facing the outside of the vehicle. Since stopper nuts are used, tighten the bolts.

➡Fully tighten the nuts and bolts after stabilizing the suspension.

c. Temporarily install the rear axle beam assembly to the rear shock absorber assembly with the 2 cushion retainers and 2 nuts.

31. Install the rear inner axle bearing retainer and parking brake assembly to the rear axle beam assembly LH (for FWD).

32. Install rear inner axle bearing retainer RH (for FWD).

33. Install the rear axle bearing retainer and parking brake assembly to the rear axle beam assembly LH (for AWD).

34. Install rear axle bearing retainer RH (for AWD).

35. Install rear axle hub and bearing assembly LH.

36. Install rear axle hub and bearing assembly RH.

37. Install rear disc.

38. Install parking brake shoe adjusting hole plug.

39. Install the rear disc brake caliper assembly with the 2 bolts LH. Tighten to 65 ft. lbs. (88 Nm).

40. Install rear disc brake caliper assembly RH.

41. Install the parking brake cable assembly to the rear axle beam assembly with the 2 bolts.

42. Install No. 2 parking brake cable assembly.

43. Install rear No. 6 brake tube.

a. Connect the rear brake tube to the rear brake tube flexible hose.

➡Do not kink or damage the brake line. Do not allow any foreign matter such as dirt and dust to enter the brake line.

b. Install the rear brake tube flexible hose with a new clip.

➡Install the clip as far as it will go.

c. Install the rear brake tube to the rear axle beam assembly with the bolt.

d. Using a union nut wrench (10 mm), fully tighten the rear brake tube to the rear brake tube flexible hose while holding the rear brake tube flexible hose with a wrench.

44. Install rear No. 5 brake tube.

45. Install skid control sensor wire LH (for FWD).

a. Install the skid control sensor wire to the rear axle beam assembly with the bolt.

➡Do not twist the skid control sensor wire.

b. Connect the skid control sensor wire connector to the rear axle hub bearing assembly.

➡Do not twist the skid control sensor wire.

46. Install skid control sensor wire RH (for FWD).

47. Install rear speed sensor LH (for AWD).

a. Install the rear speed sensor wire to the rear axle beam assembly with the bolt.

➡Do not twist the speed sensor wire.

b. Install the rear speed sensor rear axle bearing retainer with the bolt.

➡Be careful not to damage the speed sensor. Prevent foreign matter from attaching to the speed sensor. Do not twist the speed sensor wire.

48. Install rear speed sensor RH (for AWD).

49. Install rear coil spring LH.

50. Install rear coil spring RH.

51. Install center exhaust pipe assembly (for FWD).

52. Install tail exhaust pipe assembly (for FWD).

53. Install rear differential (for AWD).

54. Stabilize suspension.

55. Fully tighten the 2 bolts on the rear axle beam assembly. Tighten to 111 ft. lbs. (150 nm).

➡Since stopper nuts are used, tighten the bolts. The final torque must be applied under standard vehicle height conditions.

56. Fully tighten rear shock absorber assembly LH.

57. Fully tighten rear shock absorber assembly RH.
58. Install the rear No. 2 crossmember brace with the 2 bolts LH. Tighten to 13 ft. lbs. (18 Nm).
59. Install rear floor no. 2 crossmember brace RH.
60. Adjust parking brake shoe clearance and parking brake pedal travel.
61. Fill reservoir with brake fluid.
62. Bleed brake master cylinder.
63. Bleed brake line.
64. Inspect for brake fluid leak.
65. Inspect fluid level in reservoir.
66. Inspect for exhaust gas leak.
67. Install rear wheels.
68. Inspect rear wheel alignment.
69. Inspect speed sensor signal.
70. Install instrument panel finish panel end LH (for 2.7L).

COIL SPRING

REMOVAL & INSTALLATION

See Figure 148.

1. Remove rear wheel.
2. Remove tail exhaust pipe assembly (for FWD).
3. Remove tail exhaust pipe assembly (for AWD RH Side).
4. Remove rear drive shaft assembly (for AWD).
5. Remove rear coil spring.
   a. Install SST (09727-30021) to the rear coil spring so that the upper and lower hooks of SST are as far apart as possible.

➡Do not damage the coil spring surface.

   b. Using SST, compress the rear coil spring until it moves freely.

➡Do not use an impact wrench. It will damage SST.

   c. Remove the rear coil spring with SST from the vehicle.

To install:

6. Install rear coil spring.
   a. Using SST, compress the rear coil spring.

⁂ WARNING

Do not use an impact wrench. It will damage SST.

   b. Install the rear upper coil spring insulator to the rear coil spring.
   c. Fit the lower end of the rear coil spring into the gap of the spring lower seat, and temporarily install the rear coil spring with SST and rear upper coil spring insulator to the vehicle.
   d. Release SST and install the rear coil spring to the vehicle.

➡Do not use an impact wrench. It will damage SST. Make sure that the rear upper coil spring insulator is correctly aligned and does not fall out of position when installing the spring.

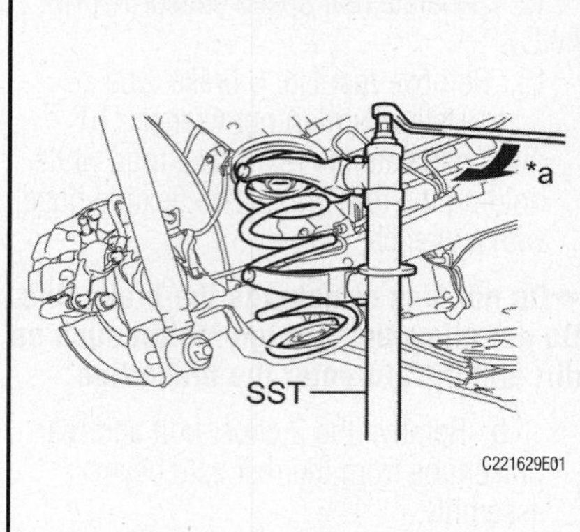

Fig. 148 Install SST to the rear coil spring so that the upper and lower hooks of SST are as far apart as possible

   e. Remove SST from the rear coil spring.
7. Install rear drive shaft assembly (for AWD)
8. Install tail exhaust pipe assembly (for FWD)
9. Install tail exhaust pipe assembly (for AWD RH Side)
10. Inspect for exhaust gas leak
11. Install rear wheel
12. Stabilize suspension. Lower the vehicle and bounce it up and down several times to stabilize the rear suspension.
13. Inspect rear wheel alignment

TOYOTA

Tacoma

15

SPECIFICATIONS AND MAINTENANCE CHARTS

ENGINE AND VEHICLE IDENTIFICATION

Engine							Model Year	
Code/ID	Liters (cc)	Cu. In.	Cyl.	Fuel Sys.	Engine Type	Eng. Mfg.	Code ①	Year
2TR-FE	2.7 (2693)	164	4	MFI	DOHC	Toyota	B	2011
1GR-FE	4.0 (3956)	241	6	MFI	DOHC	Toyota	C	2012

MFI: Multi-port Fuel Injection

DOHC: Double Overhead Camshaft

① 10th digit of the VIN number

71099_TACO_C0001

GENERAL ENGINE SPECIFICATIONS

Year	Model	Engine Displacement Liters	Engine Series Code/ID	Net Horsepower @ rpm	Net Torque @ rpm (ft. lbs.)	Bore x Stroke (in.)	Com-pression Ratio	Oil Pressure @ rpm
2011	Tacoma	2.7	2TR-FE	159@5200	236@5200	3.74x3.74	NA	43-85@3000
		4.0	1GR-FE	236@3800	266@4000	3.70x3.74	NA	23-75@3000
2012	Tacoma	2.7	2TR-FE	159@5200	236@5200	3.74x3.74	NA	43-85@3000
		4.0	1GR-FE	236@3800	266@4000	3.70x3.74	NA	23-75@3000

NA: Not Available

71099_TACO_C0002

ENGINE TUNE-UP SPECIFICATIONS

Year	Engine Displacement Liters	Engine Code/ID	Spark Plug Gap (in.)	Ignition Timing (deg.)	Fuel Pump (psi)	Idle Speed (rpm)		Valve Clearance	
						MT	AT	Intake	Exhaust
2011	2.7	2TR-FE	0.039-0.043	3-7B ①	40.8-41.7	600-700	600-700	②	②
	4.0	1GR-FE	0.039-0.043	7-24B ③	40.8-41.7	650-750	650-750	0.006-0.010	0.011-0.015
2012	2.7	2TR-FE	0.039-0.043	3-7B ①	40.8-41.7	600-700	600-700	②	②
	4.0	1GR-FE	0.039-0.043	7-24B ③	40.8-41.7	650-750	650-750	0.006-0.010	0.011-0.015

NOTE: The Vehicle Emission Control Information label often reflects specification changes made during production. The label figures must be used if they differ from those in this chart.

B: Before top dead center

① With terminals TC and CG of the DLC3 connected.

② Automatic adjustment

③ With Terminals TC and CG of DLC disconnected. 8-12 degrees BTDC with terminals connected.

71099_TACO_C0003

CAPACITIES

Year	Model	Engine Displacement Liters	Engine Code/ID	Engine Oil with Filter (qts.)	Transmission (pts.) 5-Spd	Transmission (pts.) Auto.	Transfer Case (pts.)	Drive Axle Front (pts.)	Drive Axle Rear (pts.)	Fuel Tank (gal.)	Cooling System (qts.)
2011	Tacoma	2.7	2TR-FE	6.1	①	②	2.2	3.2	③	21.1	④
		4.0	1GR-FE	⑤	3.8	②	2.2	3.2	③	21.1	⑥
2012	Tacoma	2.7	2TR-FE	6.1	①	②	2.2	3.2	⑦	21.1	④
		4.0	1GR-FE	⑤	3.8	②	2.2	3.2	⑦	21.1	⑥

NOTE: If specifications disagree with owners manual, use owners manual specifications.

① 2WD: 5.4
4WD: 4.6

② A340E: 4.2
A750E: 3.6

③ 2WD: 7
4WD: 6

④ MT: 9.2
AT: 9.1

⑤ 2WD: 4.8
4WD: 5.5

⑥ MT: 10.3
AT: 10.1

⑦ 2WD: 7.3
4WD: 6.2

71099_TACO_C0004

FLUID SPECIFICATIONS

Year	Model	Engine Displacement Liters	Engine ID/VIN	Engine Oil	Auto. Trans.	Manual Trans.	Power Steering Fluid	Brake Master Cylinder
2011	Tacoma	2.7	2TR-FE	5W-20	①	②	③	DOT 3
		4.0	1GR-FE	5W-30	①	②	③	DOT 3
2012	Tacoma	2.7	2TR-FE	5W-20	①	②	③	DOT 3
		4.0	1GR-FE	5W-30	①	②	③	DOT 3

DOT: Department Of Transportation

① 340E: ATF T-IV. A750E ATF WS

② API GL-4 or GL-5 (75W-90)

③ ATF Dexron II or III

71099_TACO_C0014

VALVE SPECIFICATIONS

Year	Engine Displacement Liters	Engine Code/ID	Seat Angle (deg.)	Face Angle (deg.)	Spring Test Pressure (lbs. @ in.)	Spring Installed Height (in.)	Stem-to-Guide Clearance (in.)		Stem Diameter (in.)	
							Intake	Exhaust	Intake	Exhaust
2011	2.7	2TR-FE	45	NA	NA	1.9106	0.0010-0.0024	0.0012-0.0026	0.2154-0.2159	0.2151-0.2157
	4.0	1GR-FE	NA	44.5	41.9-46.3@ 1.311	1.882	0.0010-0.0024	0.0012-0.0026	0.2154-0.2159	0.2152-0.2158
2012	2.7	2TR-FE	45	NA	NA	1.9106	0.0010-0.0024	0.0012-0.0026	0.2154-0.2159	0.2151-0.2157
	4.0	1GR-FE	NA	44.5	41.9-46.3@ 1.311	1.882	0.0010-0.0024	0.0012-0.0026	0.2154-0.2159	0.2152-0.2158

NA: Not Available

71099_TACO_C0005

CAMSHAFT SPECIFICATIONS

All measurements in inches unless noted

Year	Engine Displacement Liters	Engine Code/ID	Journal Dia.	Brg. Oil Clearance	Shaft End-play ①	Circle Runout	Lobe Height	
							Intake	Exhaust
2011	2.7	2TR-FE	②	③	0.0039-0.0090	0.0012	1.6872-1.6911	1.6872 1.6911
	4.0	1GR-FE	④	⑤	0.0160-0.0350	0.0024	⑥	⑥
2012	2.7	2TR-FE	②	③	0.0039-0.0090	0.0012	1.6872-1.6911	1.6872 1.6911
	4.0	1GR-FE	④	⑤	0.0160-0.0350	0.0024	⑥	⑥

NA: Not Available

① Thrust clearance

② No. 1: 1.4153-1.4159
All others: 1.0614-1.0620

③ No. 1: 0.0014-0.0029
All others: 0.0010-0.0024

④ No. 1: 1.4162-1.4167
All others: 0.9039-0.9045

⑤ No. 1: 0.0016-0.0031
All others: 0.0010-0.0024

⑥ No. 1 camshaft: 1.7389-1.7428
No. 2 camshaft: 1.7551-1.7591
No. 3 camshaft (sub assembly): 1.7389-1.7428
No. 2 camshaft (sub assembly): 1.7551-1.7591
No. 2 camshaft: 1.7551-1.7591

71099_TACO_C0006

CRANKSHAFT AND CONNECTING ROD SPECIFICATIONS

All measurements are given in inches.

	Engine		Crankshaft				Connecting Rod		
Year	Displacement Liters	Engine Code/ID	Main Brg. Journal Dia.	Main Brg. Oil Clearance	Shaft End-play	Thrust on No.	Journal Diameter	Oil Clearance	Side Clearance
2011	2.7	2TR-FE	①	②	0.0008-0.0087	NA	NA	0.0009-0.0019	0.0059-0.0138
	4.0	1GR-FE	2.8342-2.8346	0.0007-0.0012	NA	NA	NA	0.0010-0.0018	0.0059-0.0118
2012	2.7	2TR-FE	①	②	0.0008-0.0087	NA	NA	0.0009-0.0019	0.0059-0.0138
	4.0	1GR-FE	2.8342-2.8346	0.0007-0.0012	NA	NA	NA	0.0010-0.0018	0.0059-0.0118

NA: Not Available

① No. 3: 2.3615-2.3620
All others: 2.3619-2.3622

② No. 3: 0.0012-0.0022
All others: 0.0009-0.0019

71099_TACO_C0007

PISTON AND RING SPECIFICATIONS

All measurements are given in inches.

	Engine			Ring Gap			Ring Side Clearance		
Year	Displacement Liters	Engine Code/ID	Piston Clearance	Top Compression	Bottom Compression	Oil Control	Top Compression	Bottom Compression	Oil Control
2011	2.7	2TR-FE	0.0007-0.0020	0.0087-0.0134	0.0177-0.0224	0.0039-0.0157	0.0008-0.0030	0.0008-0.0026	0.0008-0.0028
	4.0	1GR-FE	0.0031-0.0040	0.0118-0.0157	0.0157-0.0197	0.0039-0.0157	0.0008-0.0028	0.0008-0.0024	0.0028-0.0060
2012	2.7	2TR-FE	0.0007-0.0020	0.0087-0.0134	0.0177-0.0224	0.0039-0.0157	0.0008-0.0030	0.0008-0.0026	0.0008-0.0028
	4.0	1GR-FE	0.0031-0.0040	0.0118-0.0157	0.0157-0.0197	0.0039-0.0157	0.0008-0.0028	0.0008-0.0024	0.0028-0.0060

71099_TACO_C0008

TORQUE SPECIFICATIONS

All readings in ft. lbs.

Year	Engine Displacement Liters	Engine Code/ID	Cylinder Head Bolts	Main Bearing Bolts	Rod Bearing Bolts	Crankshaft Damper Bolts	Flywheel Bolts	Manifold Intake	Manifold Exhaust	Spark Plugs	Oil Pan Drain Plug
2011	2.7	2TR-FE	①	②	③	192	④	18	27	13	28
	4.0	1GR-FE	⑤	⑥	③	185	61	19	16	15	30
2012	2.7	2TR-FE	①	②	③	192	④	18	27	13	28
	4.0	1GR-FE	⑤	⑥	③	185	61	19	16	15	30

① Step 1: 29 ft. lbs.
Step 2: Plus 90 degrees
Step 3: Plus 90 degrees

② Step 1: 29 ft. lbs.
Step 2: Plus 90 degrees

③ Step 1: 18 ft. lbs.
Step 2: Plus 90 degrees

④ AT: 55 ft. lbs.
MT: 20 ft. lbs. plus 90 degrees

⑤ Step 1: 27 ft. lbs.
Step 2: Plus 180 degrees
Left side 14mm bolt: 22 ft. lbs.

⑥ 12 pointed head: 45 ft. lbs. Then plus 90 degrees
12mm head: 18 ft. lbs.

71099_TACO_C0009

WHEEL ALIGNMENT

Year	Model	Caster Range (+/-Deg.)	Caster Preferred Setting (Deg.)	Camber Range (+/-Deg.)	Camber Preferred Setting (Deg.)	Toe-in (in.)	Steering Axis Inclination (Deg.)
2011	2WD exc. PreRunner	0.75	+0.67	0.75	0	0.06+/-0.08	10.00
	4WD & PreRunner	0.75	0.30	0.75	+1.62	0.06+/-0.08	10.40
2012	2WD exc. PreRunner	0.75	+0.67	0.75	0	0.06+/-0.08	10.00
	4WD & PreRunner	0.75	0.30	0.75	+1.62	0.06+/-0.08	10.40

All alignment figures based on nominal ride height and standard tires

71099_TACO_C0010

TIRE, WHEEL AND BALL JOINT SPECIFICATIONS

Year	Model	OEM Tires Standard	OEM Tires Optional	Tire Pressures (psi) Front	Tire Pressures (psi) Rear	Wheel Size	Ball Joint Inspection ①	Lug Nut Torque (ft. lbs.)
2011	Tacoma	P215/70R15	None	30	33	NA	②	83
		P245/75R16	None	30	30	NA	②	83
		P265/70R16	None	29	32	NA	②	83
		P265/65R17	None	29	29	NA	②	83
		P255/45R18	None	35	35	NA	②	83
2012	Tacoma	P215/70R15	None	30	33	NA	②	83
		P245/75R16	None	30	30	NA	②	83
		P265/70R16	None	29	32	NA	②	83
		P265/65R17	None	29	29	NA	②	83
		P265/60R18	None	35	35	NA	②	83
		P255/45R18	None	35	35	NA	②	83

OEM: Original Equipment Manufacturer

PSI: Pounds Per Square Inch

NOTE: If specification differs from driver's door placard, use data on placard.

NA: Information not available

① Torque required in inch lbs. to rotate ball joint when removed from the knuckle

② 2wd exc. PreRunner: Upper 4-30 inch lbs.; lower 0.8-30 inch lbs.
4wd and PreRunner: Upper 6-39 inch lbs.; lower 1-22 inch lbs.
Lower ball joint excessive play, all models: 0.020 inch

71099_TACO_C0011

BRAKE SPECIFICATIONS

All measurements in inches unless noted

Year	Model	Brake Disc Original Thickness	Brake Disc Minimum Thickness	Brake Disc Maximum Runout	Brake Drum Diameter Original Inside Diameter	Brake Drum Diameter Maximum Machine Diameter	Minimum Lining Thickness Front	Minimum Lining Thickness Rear	Brake Caliper Bracket Bolts (ft. lbs.)	Brake Caliper Mounting Bolts (ft. lbs.)
2011	2WD	0.984	0.906	0.0020	10.00	10.08	0.039	—	80	27
	①	1.102	1.024	0.0020	10.00	10.08	—	0.039	91	—
2012	2WD	0.984	0.906	0.0020	10.00	10.08	0.039	—	80	27
	①	1.102	1.024	0.0020	10.00	10.08	—	0.039	91	—

① 4WD and PreRunner

71099_TACO_C0012

SCHEDULED MAINTENANCE INTERVALS
TOYOTA—TACOMA

TO BE SERVICED	TYPE OF	VEHICLE MILEAGE INTERVAL (x1000)																		
		5	10	15	20	25	30	35	40	45	50	55	60	65	70	75	80	85	90	95
Air filter	R												✓							
Air filter (cabin)	R				✓				✓				✓				✓			
Automatic transmission and differential fluid	S/I			✓			✓			✓			✓			✓			✓	
Ball joints and boots	S/I			✓			✓			✓			✓			✓			✓	
Brake linings, discs/drums, lines & hoses	S/I	✓	✓	✓	✓	✓	✓	✓	✓	✓	✓	✓	✓	✓	✓	✓	✓	✓	✓	✓
Charcoal canister	S/I												✓							
Drive belts	S/I						✓						✓						✓	
Driveshaft bushing (4WD)	L						✓						✓						✓	
Engine coolant	R						✓						✓						✓	
Engine oil & filter	R	✓	✓	✓	✓	✓	✓	✓	✓	✓	✓	✓	✓	✓	✓	✓	✓	✓	✓	✓
Exhaust pipes & mounts	S/I			✓			✓			✓			✓			✓			✓	
Fuel lines & connections, fuel tank vapor vent system hoses, fuel tank band	S/I						✓						✓						✓	
Fuel tank cap gasket	S/I						✓						✓						✓	
Halfshaft boots & flange bolts	S/I			✓			✓			✓			✓			✓			✓	
Limited slip differential fluid	R						✓						✓						✓	
Manual transmission and differential fluid	S/I						✓						✓						✓	
Non-platinum spark plugs	R						✓						✓						✓	
Platinum spark plugs	R																			
Propeller shaft (4WD)	L			✓			✓			✓			✓			✓			✓	
Propeller shaft bolts	S/I			✓			✓			✓			✓			✓			✓	
Rack and pinion assembly	S/I			✓			✓			✓			✓			✓			✓	
Rear wheel bearing	L						✓						✓						✓	
Rotate tires	S/I	✓	✓	✓	✓	✓	✓	✓	✓	✓	✓	✓	✓	✓	✓	✓	✓	✓	✓	✓
Steering linkage	S/I			✓			✓			✓			✓			✓			✓	
Valves	S/I												✓							

R: Replace S/I: Service or Inspect L: Lubricate

FREQUENT OPERATION MAINTENANCE (SEVERE SERVICE)

If a vehicle is operated under any of the following conditions it is considered severe service:

- Towing a trailer or using a camper or car-top carrier.

- Repeated short trips of less than 5 miles in temperatures below freezing.

- Excessive idling or low-speed driving for long distances as in heavy commercial use, such as delivery, taxi or police cars.

- Operating on rough, muddy or salt-covered roads.

- Operating on unpaved or dusty roads.

Oil filter: service or inspect every 5000 miles or 4 months, whichever occurs first.

Brake linings and discs or drums: service or inspect every 5000 miles or 4 months, whichever occurs first.

Steering linkage: service or inspect every 5000 miles or 4 months, whichever occurs first.

Ball joints and boots: service or inspect every 5000 miles or 4 months, whichever occurs first.

Brake discs & pads (front): service or inspect every 5000 miles.

Halfshaft boots: service or inspect every 5000 miles or 4 months. Retighten the flange bolts, whichever occurs first.

Body chassis bolts and nuts: service or inspect every 5000 miles or 4 months, whichever occurs first.

Transmission and differential fluid: replace every 60,000 miles or 72 months, whichever occurs first.

Transfer case and differential fluid: replace every 15,000 miles or 18 months, whichever occurs first.

Timing belt: replace every 60,000 miles or 48 months, whichever occurs first.

Lubricate driveshaft 4WD every 5,000 miles

Retorque driveshaft bolts 4WD every 5,000 miles

Inspect engine air filter at 15,000 miles or 18 months, than every 5,000 miles

71099_TACO_C0013

PRECAUTIONS

Before servicing any vehicle, please be sure to read all of the following precautions, which deal with personal safety, prevention of component damage, and important points to take into consideration when servicing a motor vehicle:

- Never open, service or drain the radiator or cooling system when the engine is hot; serious burns can occur from the steam and hot coolant.
- Observe all applicable safety precautions when working around fuel. Whenever servicing the fuel system, always work in a well-ventilated area. Do not allow fuel spray or vapors to come in contact with a spark, open flame, or excessive heat (a hot drop light, for example). Keep a dry chemical fire extinguisher near the work area. Always keep fuel in a container specifically designed for fuel storage; also, always properly seal fuel containers to avoid the possibility of fire or explosion. Refer to the additional fuel system precautions later in this section.
- Fuel injection systems often remain pressurized, even after the engine has been turned **OFF**. The fuel system pressure must be relieved before disconnecting any fuel lines. Failure to do so may result in fire and/or personal injury.
- Brake fluid often contains polyglycol ethers and polyglycols. Avoid contact with the eyes and wash your hands thoroughly after handling brake fluid. If you do get brake fluid in your eyes, flush your eyes with clean, running water for 15 minutes. If eye irritation persists, or if you have taken brake fluid internally, IMMEDIATELY seek medical assistance.
- The EPA warns that prolonged contact with used engine oil may cause a number of skin disorders, including cancer. You should make every effort to minimize your exposure to used engine oil. Protective gloves should be worn when changing oil. Wash your hands and any other exposed skin areas as soon as possible after exposure to used engine oil. Soap and water, or waterless hand cleaner should be used.
- All new vehicles are now equipped with an air bag system, often referred to as a Supplemental Restraint System (SRS) or Supplemental Inflatable Restraint (SIR) system. The system must be disabled before performing service on or around system components, steering column, instrument panel components, wiring and sensors. Failure to follow safety and disabling procedures could result in accidental air bag deployment, possible personal injury and unnecessary system repairs.
- Always wear safety goggles when working with, or around, the air bag system. When carrying a non-deployed air bag, be sure the bag and trim cover are pointed away from your body. When placing a non-deployed air bag on a work surface, always face the bag and trim cover upward, away from the surface. This will reduce the motion of the module if it is accidentally deployed. Refer to the additional air bag system precautions later in this section.
- Clean, high quality brake fluid from a sealed container is essential to the safe and proper operation of the brake system. You should always buy the correct type of brake fluid for your vehicle. If the brake fluid becomes contaminated, completely flush the system with new fluid. Never reuse any brake fluid. Any brake fluid that is removed from the system should be discarded. Also, do not allow any brake fluid to come in contact with a painted surface; it will damage the paint.
- Never operate the engine without the proper amount and type of engine oil; doing so WILL result in severe engine damage.
- Timing belt maintenance is extremely important. Many models utilize an interference-type, non-freewheeling engine. If the timing belt breaks, the valves in the cylinder head may strike the pistons, causing potentially serious (also time-consuming and expensive) engine damage. Refer to the maintenance interval charts for the recommended replacement interval for the timing belt, and to the timing belt section for belt replacement and inspection.
- Disconnecting the negative battery cable on some vehicles may interfere with the functions of the on-board computer system(s) and may require the computer to undergo a relearning process once the negative battery cable is reconnected.
- When servicing drum brakes, only disassemble and assemble one side at a time, leaving the remaining side intact for reference.
- Only an MVAC-trained, EPA-certified automotive technician should service the air conditioning system or its components.

BRAKES

ANTI-LOCK BRAKE SYSTEM (ABS)

GENERAL INFORMATION

PRECAUTIONS

- Certain components within the ABS system are not intended to be serviced or repaired individually.
- Do not use rubber hoses or other parts not specifically specified for and ABS system. When using repair kits, replace all parts included in the kit. Partial or incorrect repair may lead to functional problems and require the replacement of components.
- Lubricate rubber parts with clean, fresh brake fluid to ease assembly. Do not use shop air to clean parts; damage to rubber components may result.
- Use only DOT 3 brake fluid from an unopened container.
- If any hydraulic component or line is removed or replaced, it may be necessary to bleed the entire system.
- A clean repair area is essential. Always clean the reservoir and cap thoroughly before removing the cap. The slightest amount of dirt in the fluid may plug an orifice and impair the system function. Perform repairs after components have been thoroughly cleaned; use only denatured alcohol to clean components. Do not allow ABS components to come into contact with any substance containing mineral oil; this includes used shop rags.
- The Anti-Lock control unit is a microprocessor similar to other computer units in the vehicle. Ensure that the ignition switch is **OFF** before removing or installing controller harnesses. Avoid static electricity discharge at or near the controller.
- If any arc welding is to be done on the vehicle, the control unit should be unplugged before welding operations begin.

WHEEL SPEED SENSORS

REMOVAL & INSTALLATION

Front

See Figure 1.

1. Before servicing the vehicle, refer to the Precautions Section.
2. Disconnect the negative battery cable. Tape the cable with insulating tape.
3. Raise and safely support the vehicle.
4. Remove front wheel.

5. Disconnect the speed sensor connector.
6. Remove the bolt and front speed sensor.

➡**Do not attach any foreign matter to the sensor tip. Ensure that no foreign matter enters the sensor installation part.**

To install:

➡**Be sure to use new fasteners, as required.**

7. Installation is the reverse of the removal procedure.
8. Tighten the bolt to 73 inch lbs. (8.3 Nm).

Rear

See Figure 2.

1. Before servicing the vehicle, refer to the Precautions Section.
2. Disconnect the negative battery cable. Tape the cable with insulating tape.
3. Raise and safely support the vehicle.
4. Remove rear wheel.
5. Disconnect the speed sensor connector.

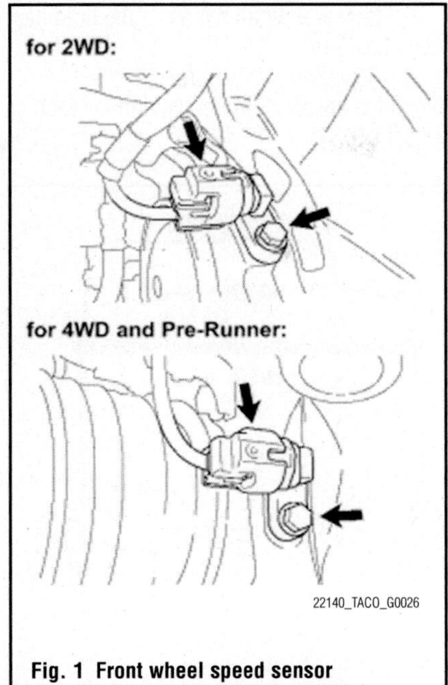

Fig. 1 Front wheel speed sensor

6. Remove the nut and rear speed sensor.

➡**Do not attach any foreign matter to the sensor tip. Ensure that no foreign matter enters the sensor installation part.**

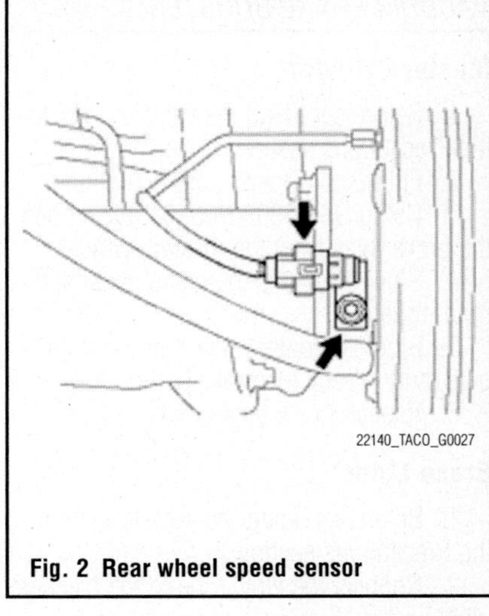

Fig. 2 Rear wheel speed sensor

To install:

➡**Be sure to use new fasteners, as required.**

7. Installation is the reverse of the removal procedure.
8. Tighten the nut to 71 inch lbs. (8 Nm).

BRAKES — BLEEDING THE BRAKE SYSTEM

BLEEDING PROCEDURE

➡**Immediately wash off any brake fluid that comes into contact with any painted surfaces. Depressing the brake pedal with the reservoir cap removed will cause the fluid to spray. When bleeding, maintain the amount of fluid in the reservoir between the Min. and Max. lines.**

HYDRAULIC BRAKE BOOSTER

Booster With Accumulator Assembly

1. Before servicing the vehicle, refer to the Precautions section.
2. Fill the reservoir with the proper grade and type brake fluid.
3. Turn the ignition switch to ON, and wait until the pump motor has stopped.
4. Turn the ignition switch OFF, depress the brake pedal more than twenty times.
5. Repeat the above steps about five times.
6. Turn the ignition switch ON, and check that the pump stops after approximately 8–14 seconds. If the pump does not stop, repeat the above step.

Master Cylinder Solenoid

1. Before servicing the vehicle, refer to the Precautions section.
2. Connect the Techstream diagnostic tool to the DLC3.
3. Turn the ignition switch to the ON position.
4. Select “Active Test” mode on the tool.
5. Connect the vinyl tube to the wheel cylinder.
6. Loosen the bleeder plug.
7. Select “TRAC Solenoid (SRMF&SRMR) to drive the solenoids and bleed air from the wheel cylinder.
8. Repeat the above step until all air is bled out.
9. When air is completely bled out of the brake fluid thru the bleeder plug, tighten the plug.
10. Repeat the above on the other line.
11. Turn the ignition switch OFF.
12. Turn the ignition switch ON.
13. Clear the DTC codes, as required.

Front Brake Line

1. Before servicing the vehicle, refer to the Precautions section.
2. Turn the ignition switch ON and wait until the pump motor stops.
3. Connect the vinyl tube to the brake caliper.
4. Depress the brake pedal several times, then loosen the bleeder plug with the pedal held down.
5. At the point when the fluid stops coming out, tighten the bleeder plug, then release the brake pedal.
6. Repeat until all the air in the fluid has been bled out.
7. Repeat the above procedures to bleed the other brake line.

Rear Brake Line

1. Before servicing the vehicle, refer to the Precautions section.
2. Connect the vinyl tube to the wheel cylinder.
3. Depress the brake pedal, hold it, and then loosen the bleeder plug.
4. Loosen the bleeder plug and release air.
5. When the air is completely bled out of the brake fluid through the bleeder plug, tighten the bleeder plug.
6. Repeat the above procedures to bleed the other brake line.

VACUUM BRAKE BOOSTER

Master Cylinder

1. Before servicing the vehicle, refer to the Precautions section.
2. Fill reservoir with dot3 brake fluid.
3. Using SST 09023-00101, disconnect the brake lines from the master cylinder.
4. Slowly depress the brake pedal and hold it there.
5. Block the outer holes with your fingers, and release the brake pedal.
6. Repeat 3 or 4 times.

Brake Lines

1. Before servicing the vehicle, refer to the Precautions section.
2. Connect the vinyl tube to the bleeder plug.
3. Depress the brake pedal several times, then loosen the bleeder plug with the pedal held down.
4. At the point where the fluid stops coming out, tighten the bleeder plug, then release the brake pedal.
5. Repeat until all the air in the fluid has been bled out.
6. Repeat the above procedure to bleed the air out of the brake line for each wheel.

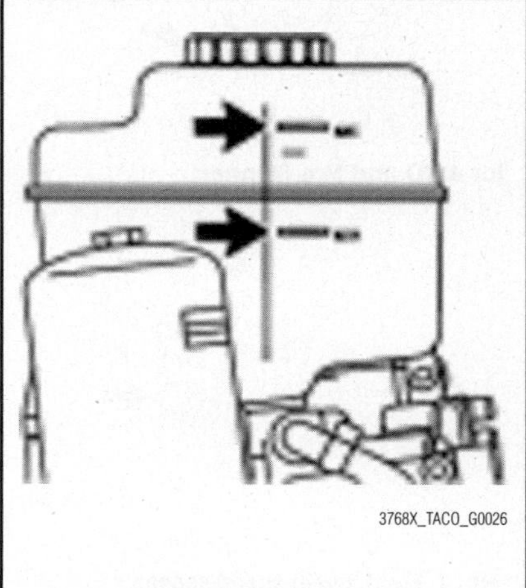

Fig. 3 Brake fluid reservoir level indicators—hydraulic brake booster

7. Check the fluid level and add fluid if necessary.

FLUID FILL PROCEDURE

See Figures 3 and 4.

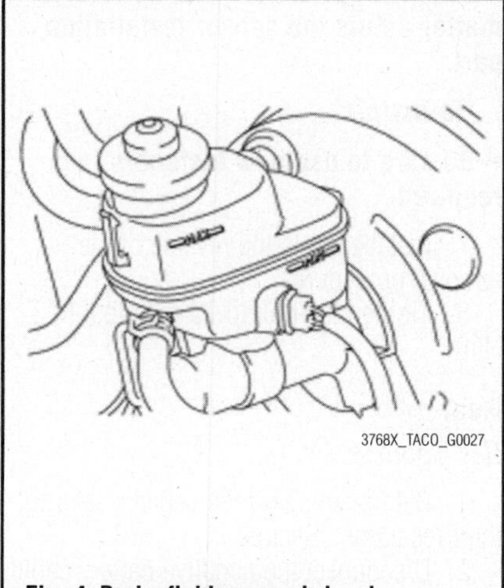

Fig. 4 Brake fluid reservoir level indicators—vacuum brake booster

BRAKES — FRONT DISC BRAKES

***** CAUTION**

Dust and dirt accumulating on brake parts during normal use may contain asbestos fibers from production or aftermarket brake linings. Breathing excessive concentrations of asbestos fibers can cause serious bodily harm. Exercise care when servicing brake parts. Do not sand or grind brake lining unless equipment used is designed to contain the dust residue. Do not clean brake parts with compressed air or by dry brushing. Cleaning should be done by dampening the brake components with a fine mist of water, then wiping the brake components clean with a dampened cloth. Dispose of cloth and all residue containing asbestos fibers in an impermeable container with the appropriate label. Follow practices prescribed by the Occupational Safety and Health Administration (OSHA) and the Environmental Protection Agency (EPA) for the handling, processing, and disposing of dust or debris that may contain asbestos fibers.

BRAKE CALIPER

REMOVAL & INSTALLATION

See Figures 5 and 6.

1. Before servicing the vehicle, refer to the Precautions Section.
2. Disconnect the negative battery cable from the battery.
3. Raise and support the vehicle safely.
4. Remove the wheels.
5. Disconnect the brake hose from the caliper by removing the union bolt and 2 gaskets. Plug the end of the hose to prevent loss of fluid.
6. Remove the bolts that attach the caliper to its mounting.
7. Lift the bottom of the caliper up and remove the caliper assembly.

To install:

8. Grease the caliper slides and bolts with lithium grease or equivalent. Install the caliper and secure with the bolts.
9. Connect the brake hose to the caliper, using 2 new washers. Make sure the flexible hose lock is securely in the lock hole of the caliper. Torque the union bolt to 22 ft. lbs. (30 Nm).
10. Fill the brake system to the proper level and bleed the brake system.
11. Install the tire and wheel assembly.
12. Top off the brake fluid level in the master cylinder. Check for leaks and proper brake operation.
13. Connect the negative battery cable to the battery.

DISC BRAKE PADS

REMOVAL & INSTALLATION

2WD

1. Before servicing the vehicle, refer to the Precautions Section.
2. Raise the vehicle and support it safely.
3. Remove the wheel and tire assembly.
4. When servicing the front pads, loosen the brake caliper upper side mounting bolt. Loosen and remove the lower side mounting bolt. Lift the caliper and suspend it so the hose is not stretched.
5. If equipped, remove the anti-squeal spring.
6. Remove the brake pads.

To install:

7. Siphon a small amount of brake fluid from the reservoir. Press in the brake caliper piston with the proper tool.
8. Before installing the new pads, check the disc thickness and disc runout.
9. Install the pad support plates.
10. Install the anti-squeal shims to each pad.

➡Apply disc brake grease to both sides of the inner anti-squeal shims.

11. Install the disc pads so the wear indicator plate is facing downward.

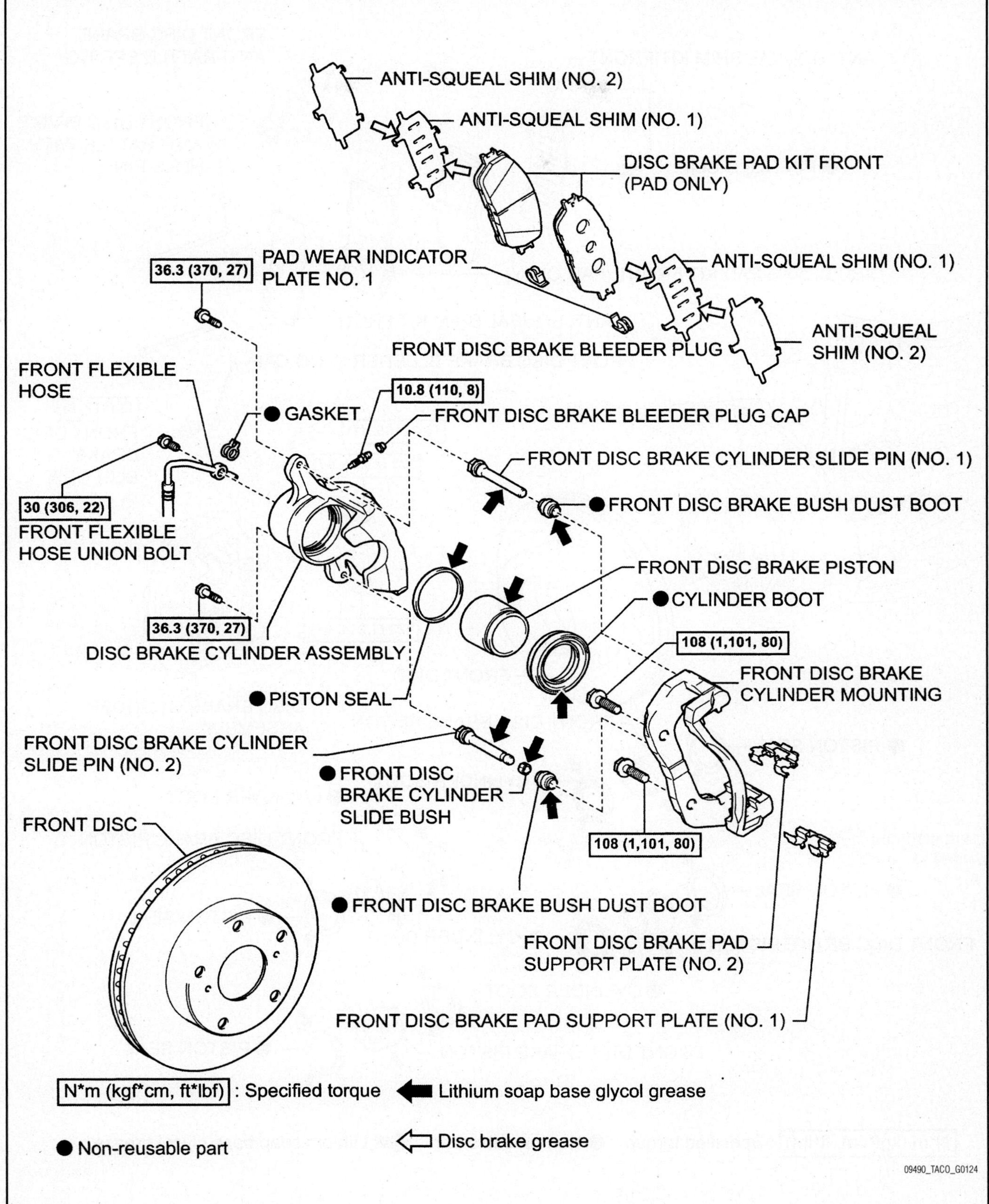

Fig. 5 Brake caliper and related components—2WD

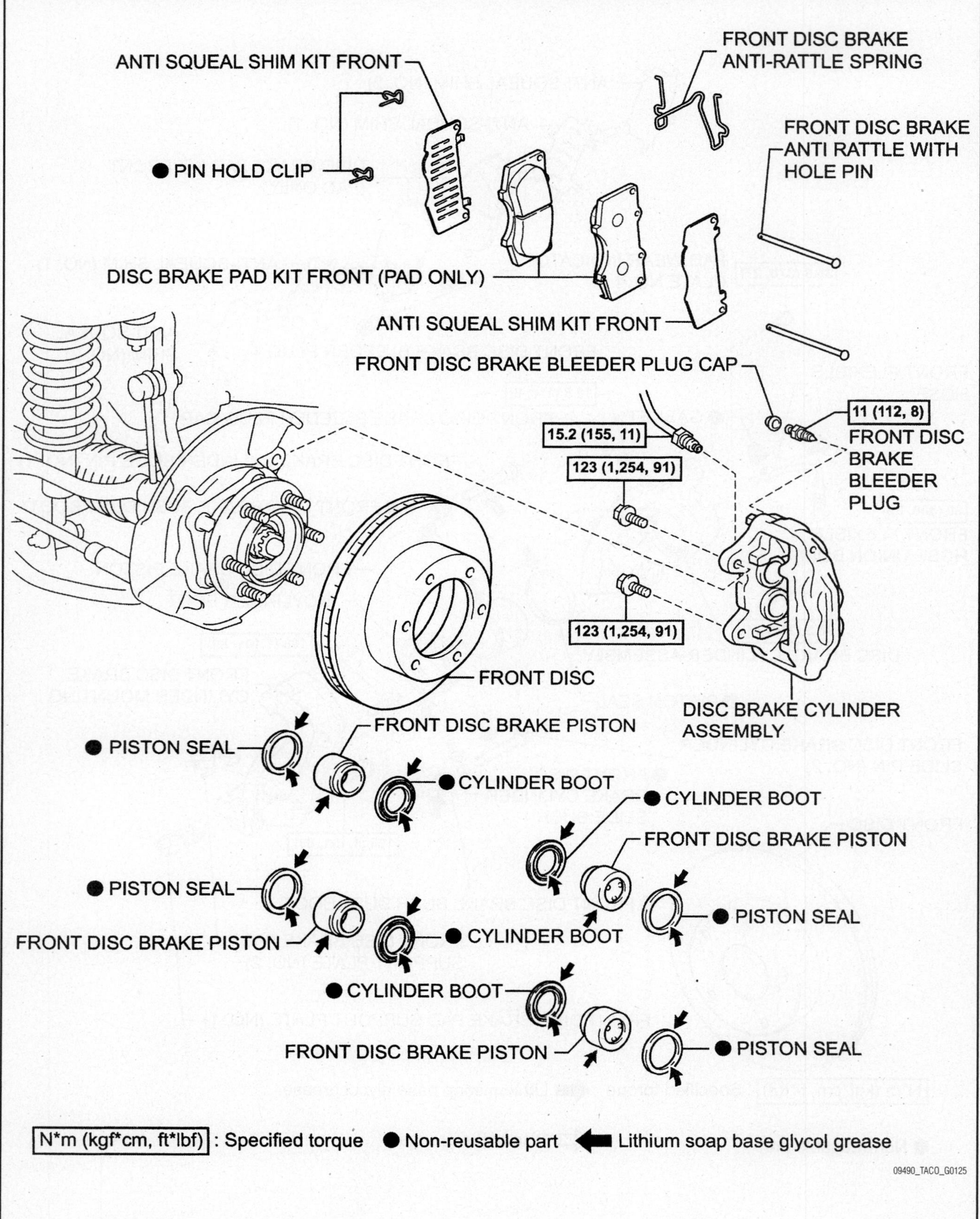

Fig. 6 Brake caliper and related components—4WD and PreRunner

12. If removed, install the anti-squeal springs.

13. Carefully install the brake caliper so the boot is not wedged.

14. Install the wheel and tire assembly.

15. Check and adjust the fluid level. Apply the brake pedal several times.

16. Road test the vehicle for proper operation.

4WD

1. Before servicing the vehicle, refer to the Precautions Section.

2. Raise the vehicle and support it safely.

3. Remove the wheel and tire assembly.

4. Remove the clip, pins, and the anti-rattle spring.

5. Remove the pads and the anti-squeal shims.

6. Remove the caliper, but do not disconnect the brake hose.

To install:

7. Before installing the new pads, check the disc thickness and disc runout.

8. Siphon out a small amount of brake fluid from the reservoir.

9. Temporarily install the old inner brake pad. Press in the pistons with a C-clamp or equivalent. Remove the old inner brake pad.

10. Apply disc brake grease to both sides of the inner anti-squeal shim. Install the anti-squeal shims to the new pads.

11. Install the pads.

12. Install the anti-rattle springs and pins. Install the clip.

13. Install the caliper and the mounting bolts.

14. Install the wheel and tire assembly.

15. Check and adjust the fluid level. Apply the brake pedal several times.

16. Road test the vehicle for proper operation.

⁂ CAUTION

Dust and dirt accumulating on brake parts during normal use may contain asbestos fibers from production or aftermarket brake linings. Breathing excessive concentrations of asbestos fibers can cause serious bodily harm. Exercise care when servicing brake parts. Do not sand or grind brake lining unless equipment used is designed to contain the dust residue. Do not clean brake parts with compressed air or by dry brushing. Cleaning should be done by dampening the brake components with a fine mist of water, then wiping the brake components clean with a dampened cloth. Dispose of cloth and all residue containing asbestos fibers in an impermeable container with the appropriate label. Follow practices prescribed by the Occupational Safety and Health Administration (OSHA) and the Environmental Protection Agency (EPA) for the handling, processing, and disposing of dust or debris that may contain asbestos fibers.

BRAKE DRUM

REMOVAL & INSTALLATION

See Figure 7.

1. Before servicing the vehicle, refer to the Precautions Section.

2. Raise and safely support the vehicle.

3. Remove the rear wheel(s).

4. Remove the brake drum from the axle hub. If there is difficulty in removing the drum, insert a suitable tool through the hole in the rear of the backing plate, and hold the automatic adjusting lever away from the adjuster. Using another suitable tool at the same time, reduce the brake shoe adjuster by turning the adjusting wheel.

To install:

5. Install the brake drum and pull the parking brake lever all the way up until a clicking sound can no longer be heard.

6. Verify that the rear wheels will not turn. If the rear wheels turn, adjust the parking brake cable as necessary.

7. Release the parking brake and remove the brake drum. Measure the brake drum inside diameter and diameter of the brake shoes. Check that the difference between the diameters is the correct shoe clearance. Clearance is 0.020 inch (5mm).

8. If the brake shoe clearance is not correct, adjust the brake shoes until the clearance is correct.

9. Install the brake drum, replace the wheel(s), and safely lower the vehicle.

10. Road-test the vehicle for proper brake operation.

BRAKE SHOES

REMOVAL & INSTALLATION

See Figures 7 through 8

1. Before servicing the vehicle, refer to the Precautions Section.

2. Loosen the rear wheel lug nuts slightly.

3. Raise and support the vehicle safely.

4. Remove the wheel lug nuts and the wheel.

5. Remove the brake drum.

6. If the drum is difficult to remove, perform the following:

   a. Insert a flat prying tool through the hole in the brake drum and hold the automatic adjusting lever away from the adjuster.

   b. Reduce the brake shoe adjustment by turning the adjuster bolt with a brake tool.

   c. The drum should now be loose enough to remove without much effort.

7. Remove the rear shoe.

   a. Carefully unhook the return spring from the brake shoe.

   b. Remove the shoe hold-down spring, cups and the pin.

   c. Disconnect the anchor spring from the rear shoe and remove the rear shoe.

   d. Disconnect the anchor spring from the front shoe.

8. Remove the front shoe.

   a. Remove the shoe hold-down spring, cups and pin.

   b. Remove the return spring from the front shoe.

   c. Remove the front shoe with the adjuster.

   d. Disconnect the parking brake cable from the front shoe.

To install:

9. Inspect the shoes for signs of unusual wear or scoring.

10. Check the wheel cylinder for any sign of fluid seepage or frozen pistons.

11. Clean and inspect the brake backing plate and all other components. Check that the brake drum inner diameter is within specified limits. Lubricate the backing plate at the positions the brakes come in contact with the backing plate. Also lubricate the anchor plate.

12. Mount the automatic adjuster assembly onto a new rear brake shoe.

13. Install the front shoe.

   a. Install the parking brake cable to the front shoe.

   b. Install the front shoe with the adjuster.

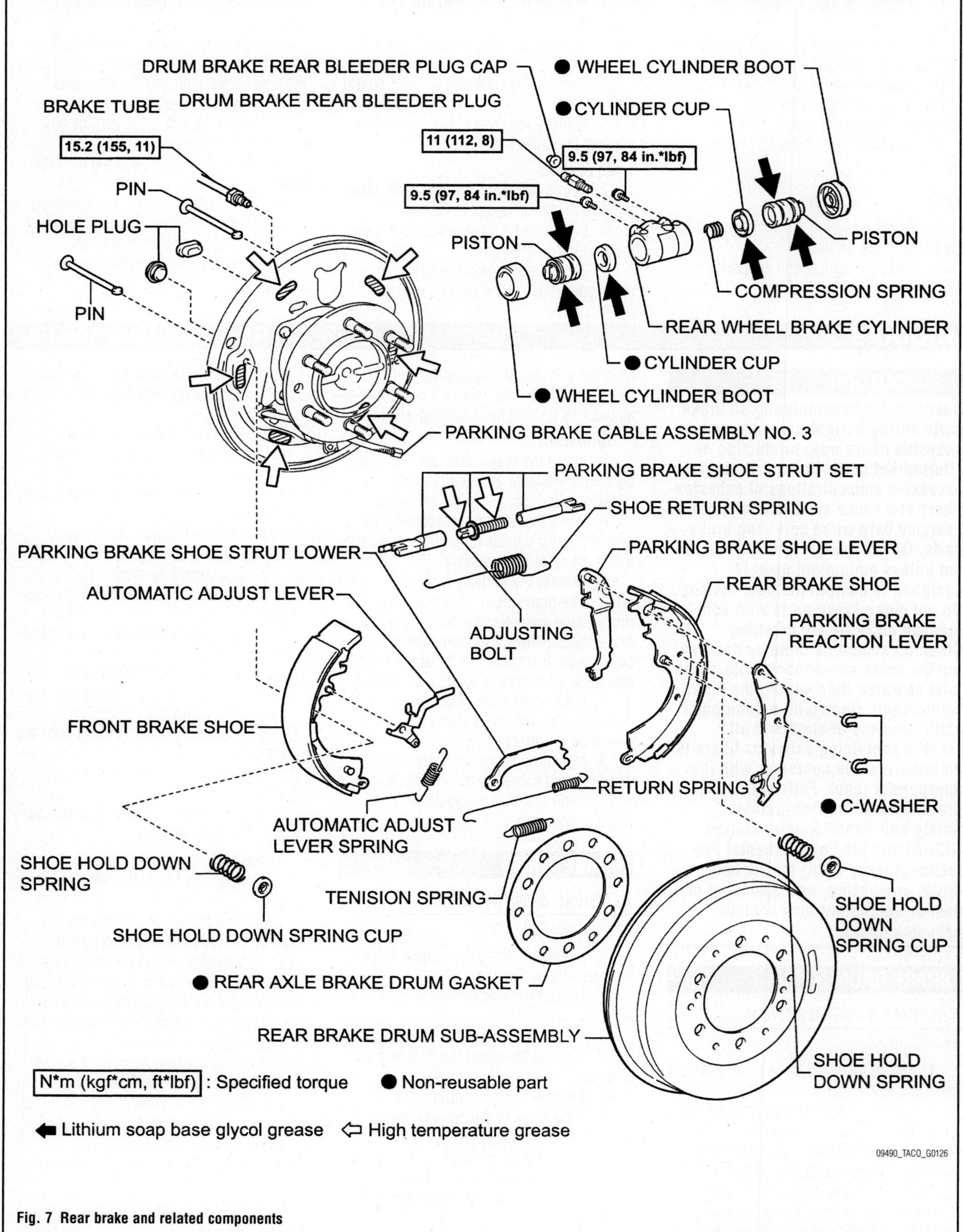

Fig. 7 Rear brake and related components

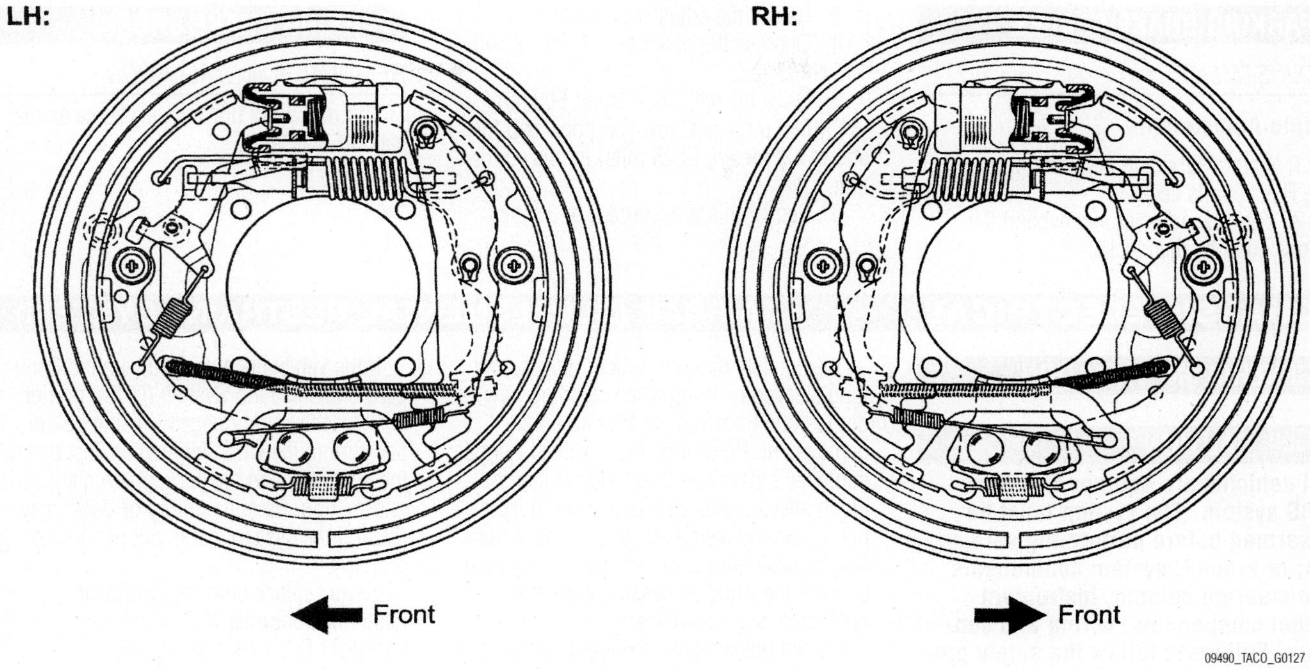

Fig. 8 Rear brake shoes—assembled view

c. Install the return spring to the front shoe.

d. Install the shoe hold-down spring, cups and pin.

14. Install the rear shoe.

a. Install the anchor spring to the front shoe.

b. Install the anchor spring to the rear shoe and install the rear shoe.

c. Install the shoe hold-down spring, cups and the pin.

d. Hook the return spring to the brake shoe.

15. Install the brake drum.

16. Adjust the brake shoes until a slight drag is felt when the drum is spun by hand.

17. Remove the brake drum and check the clearance between brake shoes and brake drum. Adjust the clearance to specification.

18. Pull the parking lever all the way up until a clicking sound can no longer be heard. Verify that the drum doesn't turn. If the drum turns, adjust the parking brake cable.

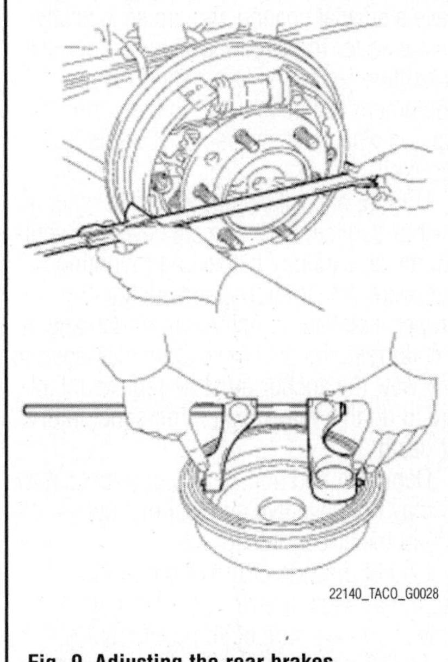

Fig. 9 Adjusting the rear brakes

19. Install the rear wheels, tighten the wheel lug nuts and lower the vehicle.

20. Retighten the wheel lug nuts and pump the brake pedal a few times before moving the vehicle. Adjust the rear brakes again if necessary.

21. Check the level of brake fluid in the master cylinder, and then perform a test drive.

22. Connect the negative battery cable to the battery.

ADJUSTMENT

See Figure 9.

1. Before servicing the vehicle, refer to the Precautions Section.

2. Measure the brake drum inside diameter and the diameter of the brake shoes. Check that difference between the diameters is the specified shoe clearance of 0.5 mm (0.020 in.).

BRAKES — PARKING BRAKE

PARKING BRAKE CABLES

ADJUSTMENT

Cable Adjustment

1. Before servicing the vehicle, refer to the Precautions section.
2. Turn the adjuster and expand the shoes until the disc locks.
3. Return the adjuster 8 notches.
4. Depress the parking brake pedal with 147 N (33 lbs).
5. Drive the vehicle at about 50 km/h (31 MPH) on a safe, level and dry road for about 400 meters (0.25 mile) in this condition.
6. Repeat this procedure 2 or 3 times.

PARKING BRAKE SHOES

REMOVAL & INSTALLATION

The rear drum brake shoes serve as the parking brakes.

CHASSIS ELECTRICAL — AIRBAG (SUPPLEMENTAL RESTRAINT SYSTEM)

GENERAL INFORMATION

⁂ CAUTION

All vehicles are equipped with an SRS system. The system must be disarmed before performing service on, or around, system components, the steering column, instrument panel components, wiring and sensors. Failure to follow the safety precautions and the disarming procedure could result in accidental airbag deployment, possible injury and unnecessary system repairs.

SERVICE PRECAUTIONS

Disconnect and isolate the battery negative cable before beginning any airbag system component diagnosis, testing, removal, or installation procedures. Allow system capacitor to discharge for two minutes before beginning any component service. This will disable the airbag system. Failure to disable the airbag system may result in accidental airbag deployment, personal injury, or death.

Do not place an intact undeployed airbag face down on a solid surface. The airbag will propel into the air if accidentally deployed and may result in personal injury or death.

When carrying or handling an undeployed airbag, the trim side (face) of the airbag should be pointing towards the body to minimize possibility of injury if accidental deployment occurs. Failure to do this may result in personal injury or death.

Replace airbag system components with OEM replacement parts. Substitute parts may appear interchangeable, but internal differences may result in inferior occupant protection. Failure to do so may result in occupant personal injury or death.

Wear safety glasses, rubber gloves, and long sleeved clothing when cleaning powder residue from vehicle after an airbag deployment. Powder residue emitted from a deployed airbag can cause skin irritation. Flush affected area with cool water if irritation is experienced. If nasal or throat irritation is experienced, exit the vehicle for fresh air until the irritation ceases. If irritation continues, see a physician.

Do not use a replacement airbag that is not in the original packaging. This may result in improper deployment, personal injury, or death.

The factory installed fasteners, screws and bolts used to fasten airbag components have a special coating and are specifically designed for the airbag system. Do not use substitute fasteners. Use only original equipment fasteners listed in the parts catalog when fastener replacement is required.

During, and following, any child restraint anchor service, due to impact event or vehicle repair, carefully inspect all mounting hardware, tether straps, and anchors for proper installation, operation, or damage. If a child restraint anchor is found damaged in any way, the anchor must be replaced. Failure to do this may result in personal injury or death.

Deployed and non-deployed airbags may or may not have live pyrotechnic material within the airbag inflator.

Do not dispose of driver/passenger/curtain airbags or seat belt tensioners unless you are sure of complete deployment. Refer to the Hazardous Substance Control System for proper disposal.

Dispose of deployed airbags and tensioners consistent with state, provincial, local, and federal regulations.

After any airbag component testing or service, do not connect the battery negative cable. Personal injury or death may result if the system test is not performed first.

If the vehicle is equipped with the Occupant Classification System (OCS), do not connect the battery negative cable before performing the OCS Verification Test using the scan tool and the appropriate diagnostic information. Personal injury or death may result if the system test is not performed properly.

Never replace both the Occupant Restraint Controller (ORC) and the Occupant Classification Module (OCM) at the same time. If both require replacement, replace one, then perform the Airbag System test before replacing the other.

Both the ORC and the OCM store Occupant Classification System (OCS) calibration data, which they transfer to one another when one of them is replaced. If both are replaced at the same time, an irreversible fault will be set in both modules and the OCS may malfunction and cause personal injury or death.

If equipped with OCS, the Seat Weight Sensor is a sensitive, calibrated unit and must be handled carefully. Do not drop or handle roughly. If dropped or damaged, replace with another sensor. Failure to do so may result in occupant injury or death.

If equipped with OCS, the front passenger seat must be handled carefully as well. When removing the seat, be careful when setting on floor not to drop. If dropped, the sensor may be inoperative, could result in occupant injury, or possibly death.

If equipped with OCS, when the passenger front seat is on the floor, no one should sit in the front passenger seat. This uneven force may damage the sensing ability of the seat weight sensors. If sat on and damaged, the sensor may be inoperative, could result in occupant injury, or possibly death.

ARMING THE SYSTEM

Reconnect the negative battery cable. Wait 2 minutes before performing any service.

CLOCKSPRING CENTERING

See Figure 10.

When installing the spiral cable, check that the ignition switch is in the "OFF" position. Turn the cable counterclockwise by hand until it becomes hard to turn. Turn the cable clockwise about 21/2 turns to align the marks.

➡The cable will rotate about 2½ turns both left and right from the center.

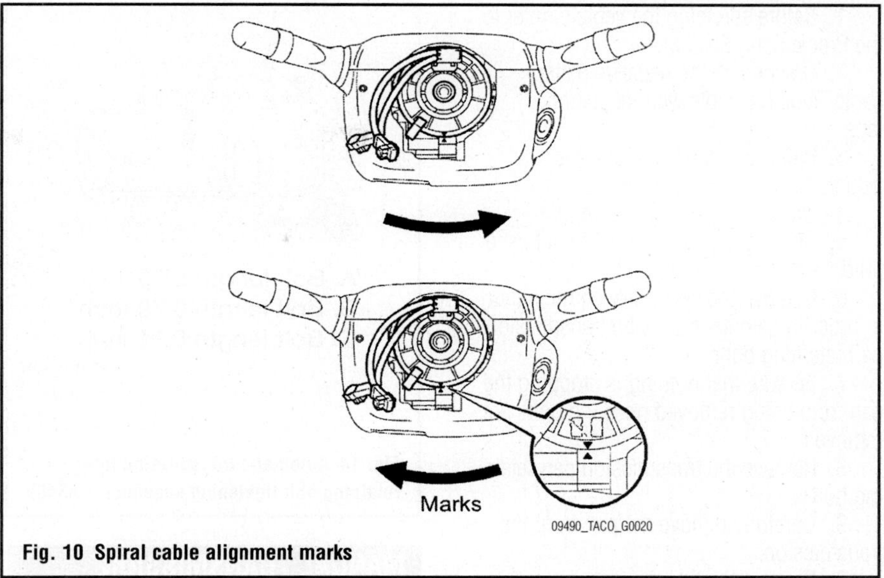

Fig. 10 Spiral cable alignment marks

DRIVE TRAIN

AUTOMATIC TRANSMISSION FLUID

DRAIN AND REFILL

See Figures 11 and 12.

1. Before servicing the vehicle, refer to the Precautions Section.
2. Disconnect the negative battery cable. Tape the cable with insulating tape.

➡On vehicles equipped with active suspension turn the suspension control switch OFF, if it is necessary to jack up the vehicle with the engine running.

3. Raise and safely support the vehicle.
4. Position a catch pan under the transmission drain plugs.

➡Allow the transmission fluid to cool before draining it. Never drain hot fluid.

5. Remove the drain plugs and drain the transmission. Discard the gaskets.
6. Be sure to properly dispose of used fluid.

To install:

➡Be sure to use new fasteners, as required.

7. Installation is the reverse of the removal procedure.
8. Use new drain plug gaskets.
9. Fill the transmission thru the refill hole until fluid begins to trickle out of the overflow tube.
10. Be sure to use the proper grade and type transmission fluid.

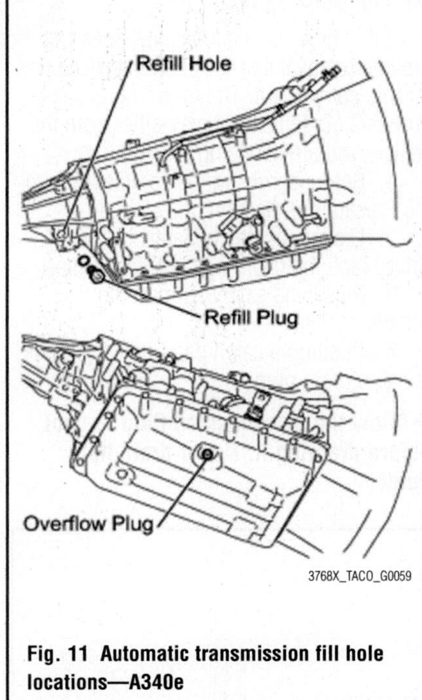

Fig. 11 Automatic transmission fill hole locations—A340e

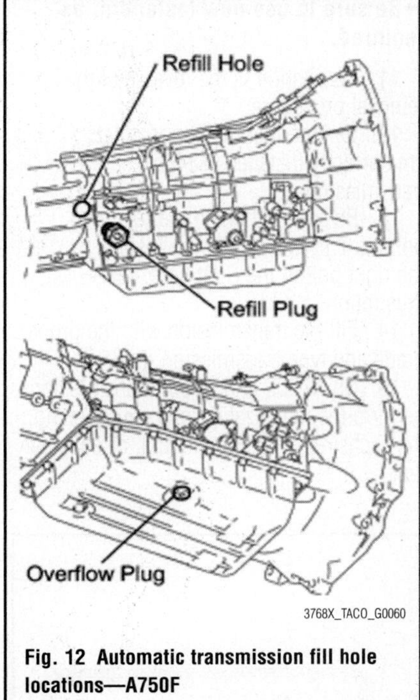

Fig. 12 Automatic transmission fill hole locations—A750F

11. Reinstall the overflow plug.
12. Allow the engine to idle with the air conditioning off.
13. Move the selector lever thru all ranges to circulate the fluid.
14. Check the fluid level, correct as required.

➡Using the Techstream diagnostic tool, or equivalent, make sure that the transmission fluid temperature reaches specification before the final check/adding of fluid.

FILTER REPLACEMENT

See Figures 13 and 14.

At this time the manufacturer does not provide removal and installation procedures for this component. The following procedure is a guideline and may differ from the vehicle you are servicing.

1. Before servicing the vehicle, refer to the Precautions Section.
2. Disconnect the negative battery cable. Tape the cable with insulating tape.
3. Raise and safely support the vehicle.
4. Drain the transmission fluid.
5. Be sure to properly dispose of used fluid.
6. Remove the necessary components in order to gain access to the transmission pan retaining bolts.
7. Be sure that nothing is stopping the pan from being removed once the bolts are removed.
8. Remove the transmission pan retaining bolts.
9. Carefully remove the pan from the transmission.
10. Remove the filter from its mounting. Discard the gaskets.

To install:

➡Be sure to use new fasteners, as required.

11. Installation is the reverse of the removal procedure.
12. Be sure that the pan magnets are correctly installed, A340e transmission.
13. Be sure to tighten the retaining bolts to 7 ft. lbs. Be sure to install the right bolt in the right hole, see illustration.
14. Fill the transmission with the proper grade and type transmission fluid.
15. On A750E and A750F transmissions perform the initialization procedure, using the Techstream diagnostic tool, or equivalent.

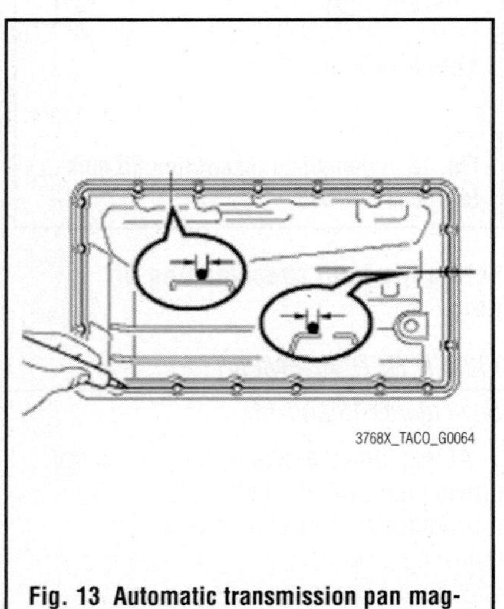

Fig. 13 Automatic transmission pan magnet locations—A340e

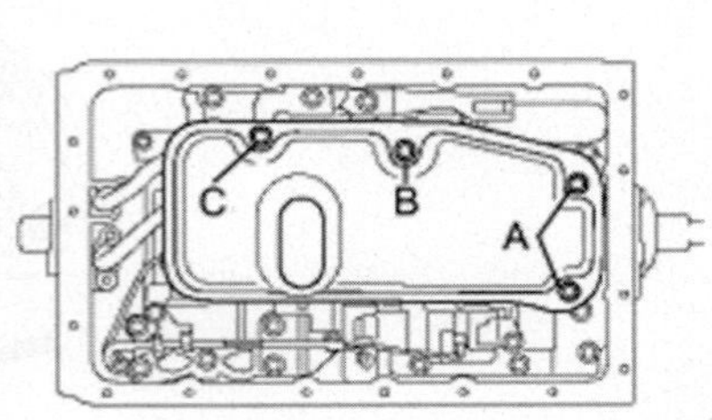

Fig. 14 Automatic transmission filter retaining bolt tightening sequence—A340e

MANUAL TRANSMISSION FLUID

DRAIN AND REFILL

See Figure 15.

At this time the manufacturer does not provide removal and installation procedures for this component. The following procedure is a guideline and may differ from the vehicle you are servicing.

1. Before servicing the vehicle, refer to the Precautions Section.
2. Disconnect the negative battery cable. Tape the cable with insulating tape.
3. Raise and safely support the vehicle.
4. Position a catch pan under the transmission drain plug.

➡Allow the transmission fluid to cool before draining it. Never drain hot fluid.

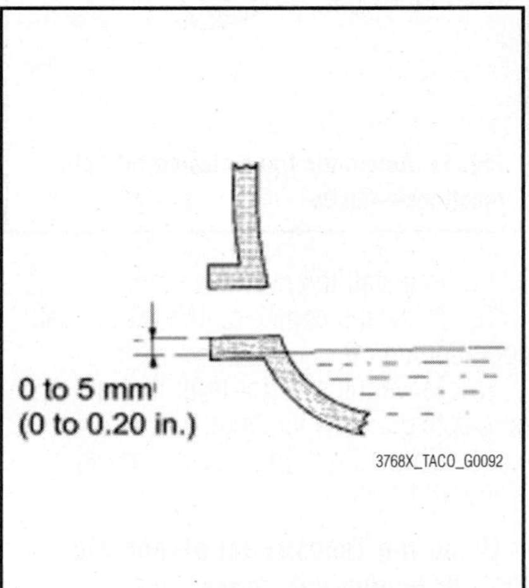

Fig. 15 Manual transmission fluid level

5. Remove the drain plug and drain the transmission. Discard the gasket.
6. Be sure to properly dispose of used fluid.

To install:

➡Be sure to use new fasteners, as required.

7. Installation is the reverse of the removal procedure.
8. Use a new drain plug gasket.
9. Fill the unit with the correct grade and type fluid.
10. Check that the oil surface is within 0.20 inch of the lowest point of the transmission filler plug opening. See illustration.
11. Install the filler plug. Tighten to 27 ft. lbs. (37 Nm).

FRONT DRIVESHAFT

REMOVAL & INSTALLATION

See Figures 16 and 17.

1. Before servicing the vehicle, refer to the Precautions Section.
2. Disconnect the negative battery cable. Tape the cable with insulating tape.
3. Raise and safely support the vehicle.
4. On 4.0L engine, remove the exhaust front pipe assembly No. 2.
5. Remove the driveshaft heat insulator bracket sub assembly.
6. Matchmark and remove the driveshaft retaining bolts.
7. Remove the component from the vehicle.

To install:

➡Be sure to use new fasteners, as required.

8. Installation is the reverse of the removal procedure.
9. When replacing the spider bearing make sure that the grease fitting assembly hole is facing in the direction shown in the illustration.
10. Tighten the retaining bolts to 65 ft. lbs. (88 Nm).

FRONT HALFSHAFT

REMOVAL & INSTALLATION

See Figure 18.

1. Before servicing the vehicle, refer to the Precautions Section.
2. Disconnect the negative battery cable. Tape the cable with insulating tape.
3. Raise and support the vehicle safely. Remove the tire and wheel assembly.
4. Drain the differential.

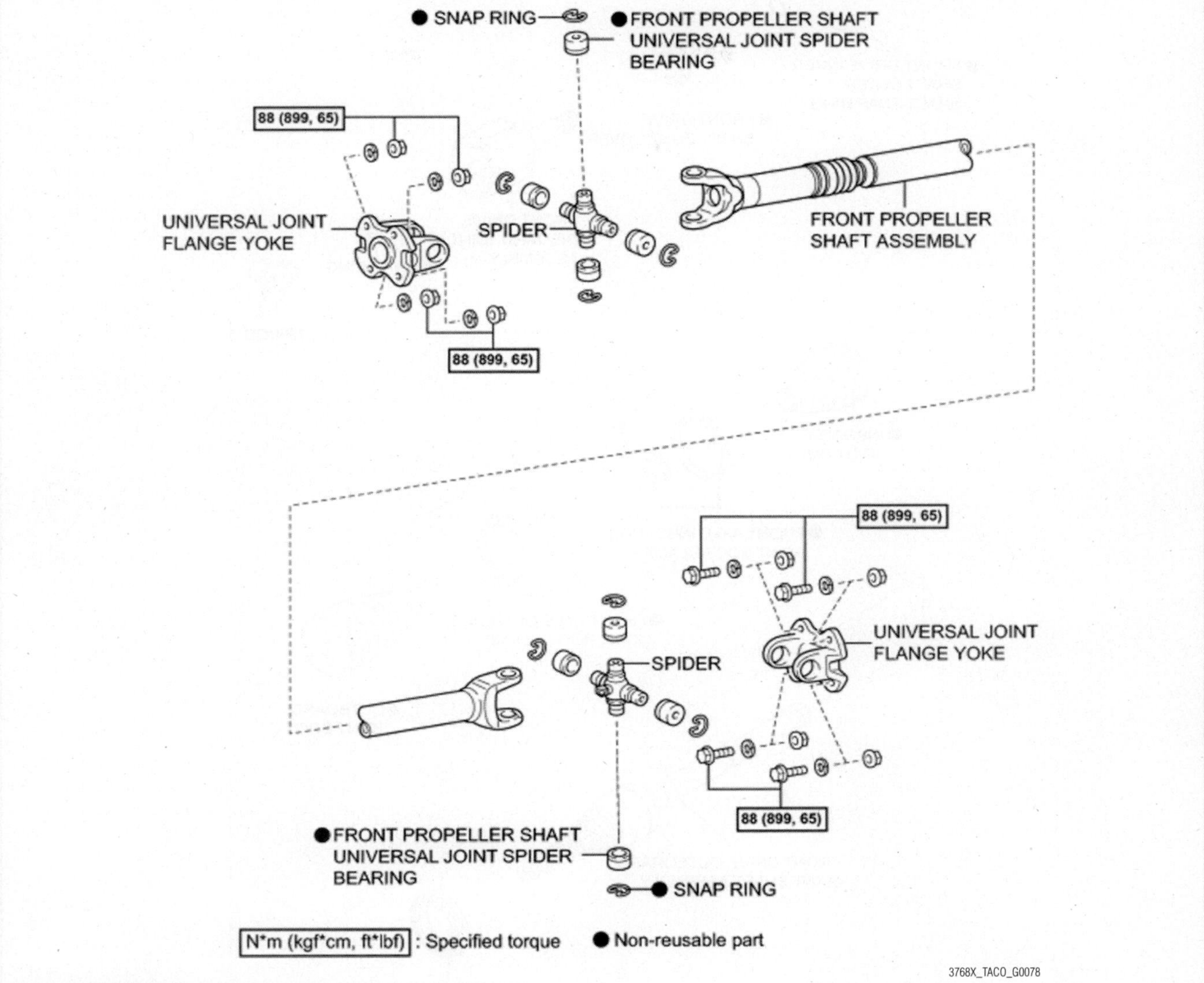

Fig. 16 Front driveshaft and related components

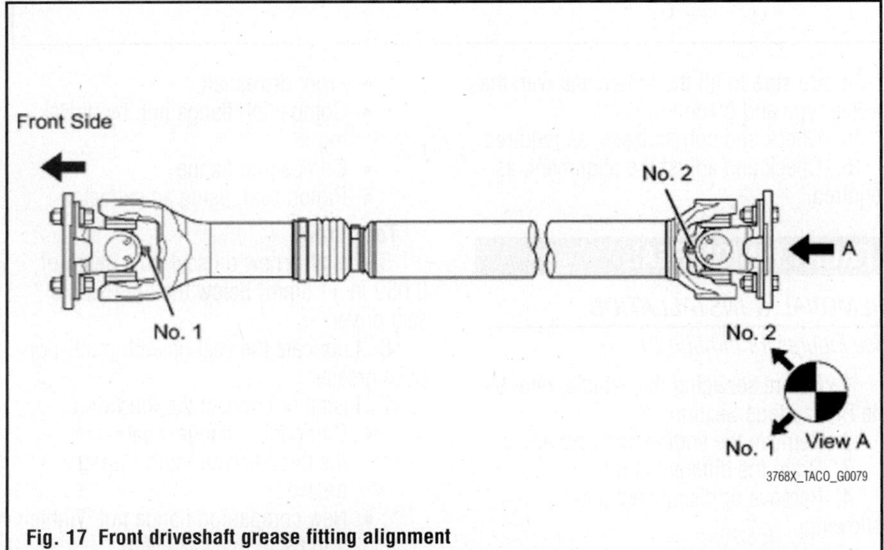

Fig. 17 Front driveshaft grease fitting alignment

5. Remove the bolt and separate the front speed sensor. Disengage the two clamps. Remove the bolt and separate the speed sensor wire harness from the steering knuckle.

6. Remove the cotter pin and nut. Using tool SST09628-62011 or equivalent, separate the tie rod end from the steering knuckle.

7. Using a suitable tool and hammer, remove the front axle hub grease cap. Remove the cotter pin and adjusting cap. Remove the front axle hub nut.

8. Remove the two bolts and separate the front lower ball joint attachment front from the steering knuckle.

9. Using tool SST09520-01010, and SST09520-24010 remove the halfshaft. Be careful not to damage the oil seal.

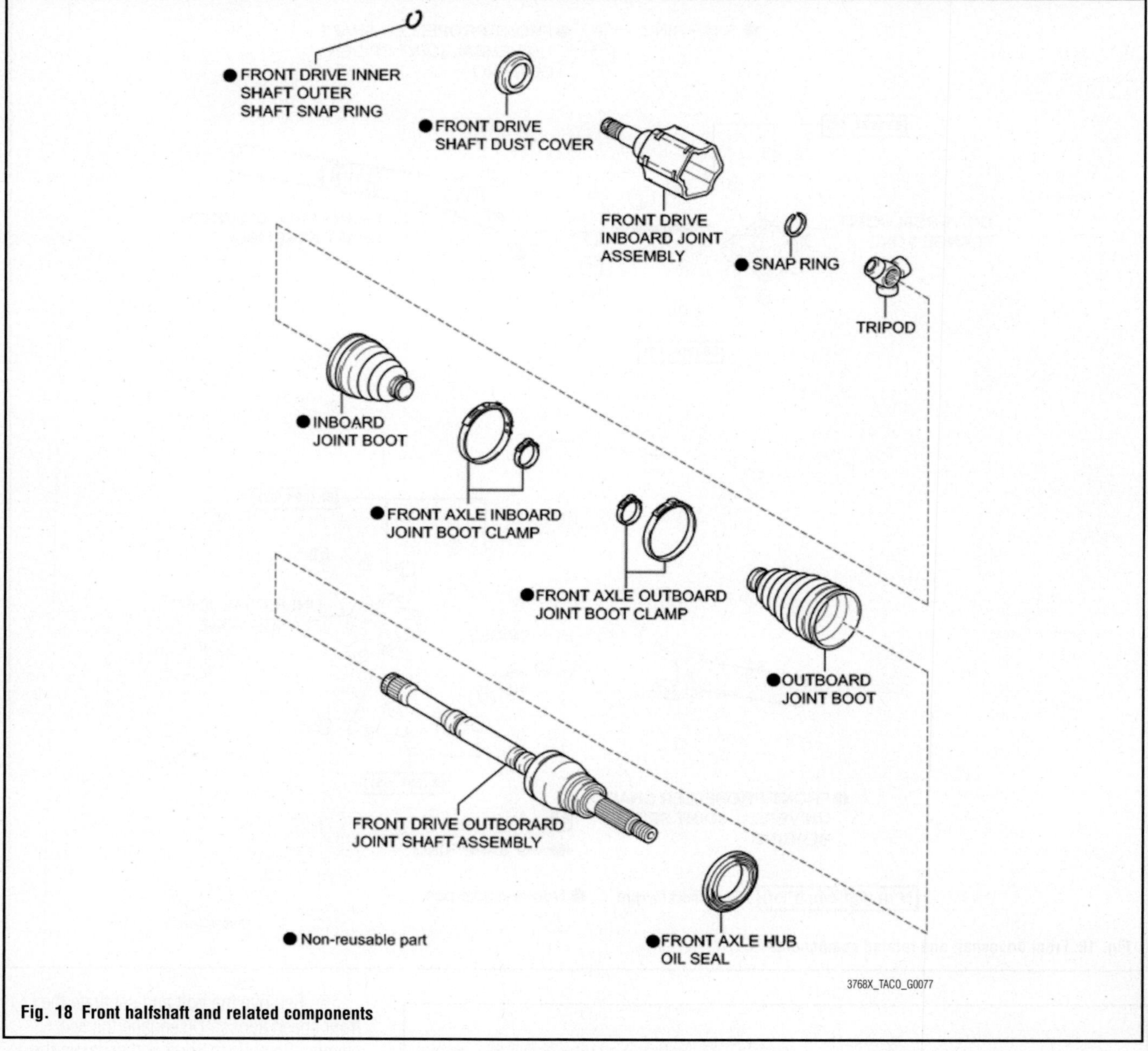

Fig. 18 Front halfshaft and related components

To install:

➡Be sure to use new fasteners, as required.

10. Coat the spline of the inboard joint shaft with clean ATF.
11. Align the shaft splines and install the halfshaft.

➡Set the snapring with the opening side facing downward. Be careful not to damage the oil seal.

12. Continue the installation in the reverse order of the removal procedure.
13. Tighten the hub nut to 173 ft. lbs. (234 Nm).
14. Be sure to fill the differential with the proper type and grade lubricant.
15. Check and correct leaks, as required.
16. Check and adjust the alignment, as required.

FRONT PINION SEAL

REMOVAL & INSTALLATION

See Figures 19 through 21.

1. Before servicing the vehicle, refer to the Precautions Section.
2. Remove the engine undercover.
3. Drain the differential oil.
4. Remove or disconnect the following:
   - Front driveshaft
   - Companion flange nut, by unstaking it
   - Companion flange
   - Pinion seal, using an extractor

To install:

5. Install a new oil seal, to a depth of 0.059 in. (1.5mm) below the lip, using a seal driver.
6. Lubricate the seal lip with multi-purpose grease.
7. Install or connect the following:
   - Companion flange, coat the threads with multi-purpose grease
   - New companion flange nut. Tighten it to 89 ft. lbs. (120 Nm).

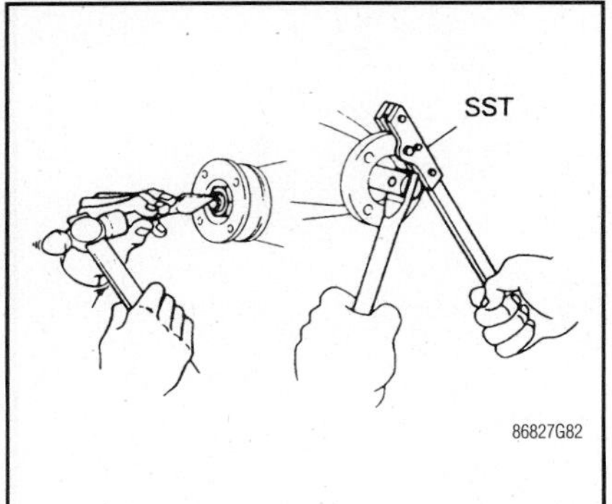

Fig. 19 Using a chisel and hammer, loosen the staked part of the nut. Hold the flange with SST 09950-30010 or equivalent and remove the nut

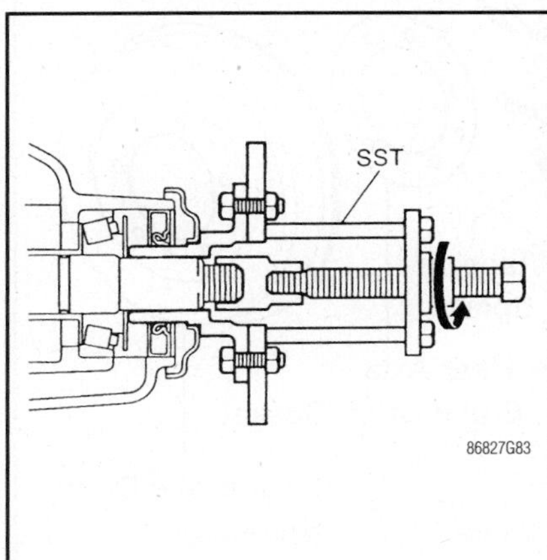

Fig. 20 Screw-type extractor from Tool 09950-30010

8. Measure the bearing preload, using a torque wrench. The correct preload should be 5–9 inch lbs. (0.6–1.0 Nm) for a used bearing or 10–17 inch lbs. (1–2 Nm) for a new bearing.

➡If the preload is greater that specified, replace the bearing spacer. If the preload is less than specified, tighten the companion flange nut in 108 inch lbs. (13 Nm) increments until the correct preload is achieved. Maximum torque for the nut is 173 ft. lbs. (235 Nm). If the value is exceeded, the bearing spacer must be replaced; do not back off the flange nut to lower the torque or preload.

9. Install the front driveshaft by aligning the matchmarks.
10. Check the companion flange run-out; maximum allowable run-out is 0.003 in. (0.10mm).
11. Stake the pinion flange nut.

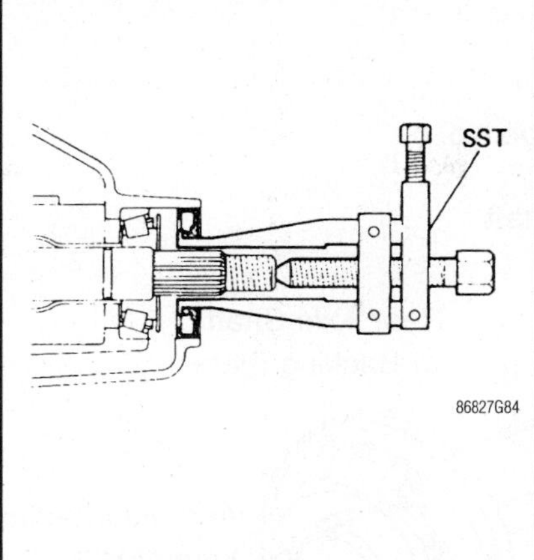

Fig. 21 Extractor fits into the Seal Removal Tool 09308-10010

12. Refill the differential with oil.

REAR AXLE FLUID

DRAIN & REFILL

See Figure 22.

1. Before servicing the vehicle, refer to the Precautions Section.
2. Disconnect the negative battery cable. Tape the cable with insulating tape.
3. Raise and safely support the vehicle.
4. Position a catch pan under the drain plug.

➡Allow the fluid to cool before draining it. Never drain hot fluid.

5. Remove the drain plug and drain the unit. Discard the gasket, if equipped.
6. Be sure to properly dispose of used fluid.

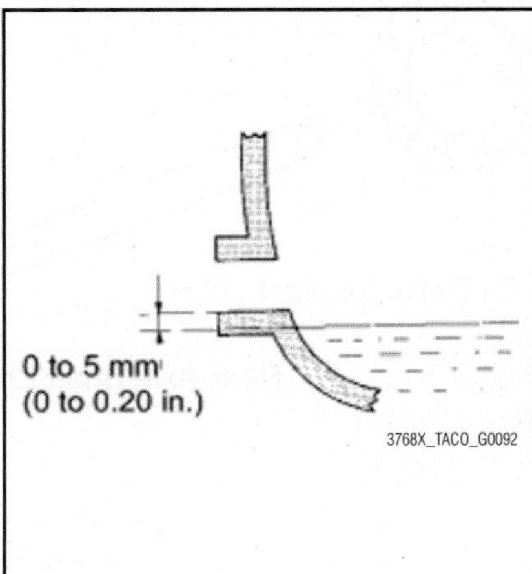

Fig. 22 Rear differential fluid level

To install:

➡Be sure to use new fasteners, as required.

7. Installation is the reverse of the removal procedure.
8. Use a new drain plug gasket, as required.
9. Fill the unit with the correct grade and type fluid.
10. Check that the oil surface is within 0.20 inch of the lowest point of the filler plug opening. See illustration.
11. Install the filler plug. Tighten to 29 ft. lbs. (39 Nm), front differential and 36 ft. lbs. (49 Nm), rear differential.

REAR AXLE SHAFT, BEARING & SEAL

REMOVAL & INSTALLATION

See Figures 23 and 24.

1. Before servicing the vehicle, refer to the Precautions Section.
2. Disconnect the negative battery cable. Tape the cable with insulating tape.
3. Drain the brake fluid.
4. Raise and support the vehicle safely. Remove the tire and wheel assembly.
5. Remove the brake drum. Remove the brake shoes.
6. Remove the rear speed sensor.
7. Remove the two bolts and disconnect the parking brake cable from the backing plate. Disconnect the brake line at the brake backing plate.
8. Remove the four nuts and the rear axle shaft and backing plate. Remove the O-ring.
9. Remove the rear axle shaft oil seal using tool SST09308-00010.

To install:

➡Be sure to use new fasteners, as required.

10. Installation is the reverse of the removal procedure.
11. Be sure to fill the master cylinder with the proper grade and type brake fluid.
12. Bleed the brakes, as required.

REAR DRIVESHAFT

REMOVAL & INSTALLATION

See Figures 25 through 28.

1. Before servicing the vehicle, refer to the Precautions Section.
2. Disconnect the negative battery cable. Tape the cable with insulating tape.

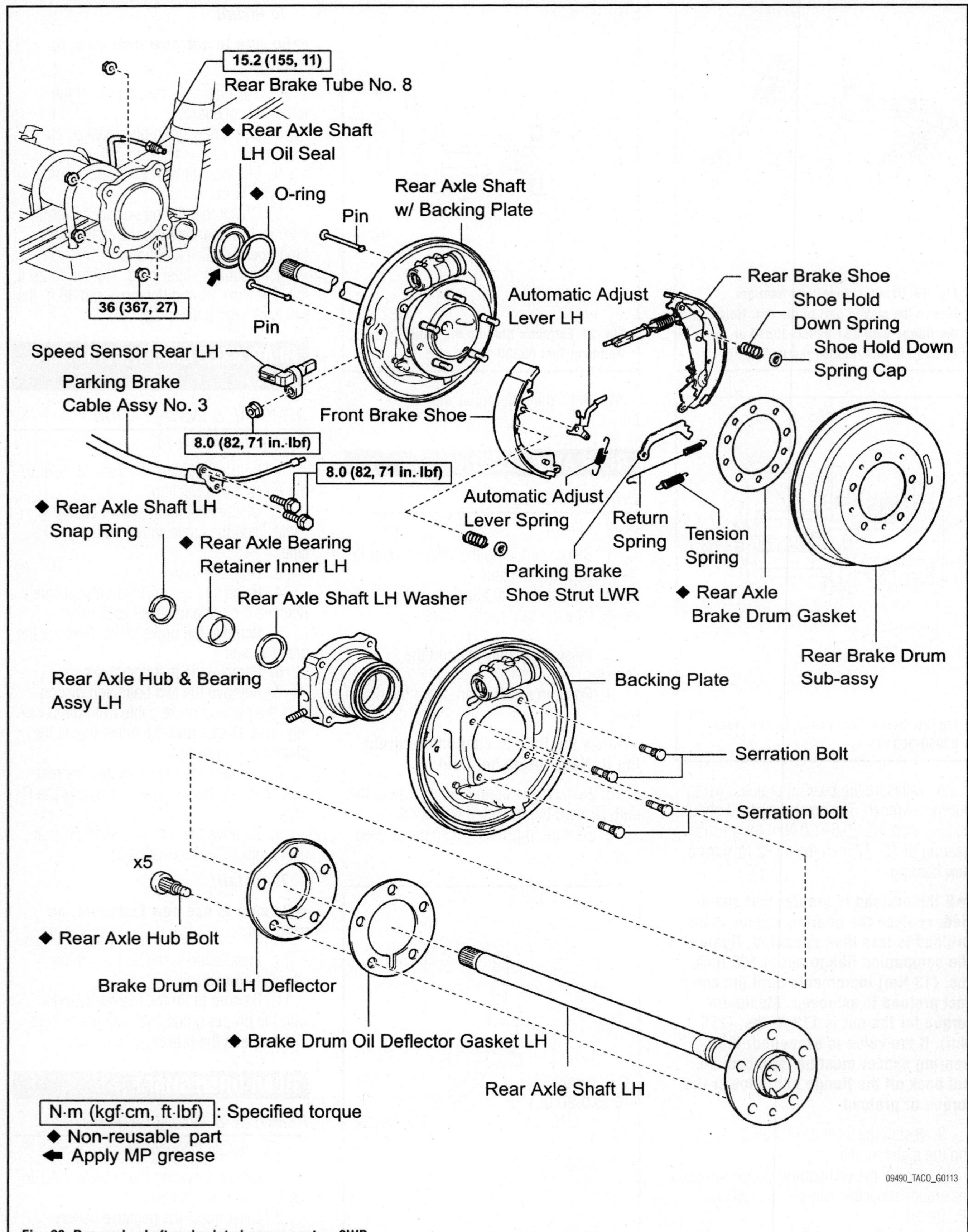

Fig. 23 Rear axle shaft and related components—2WD

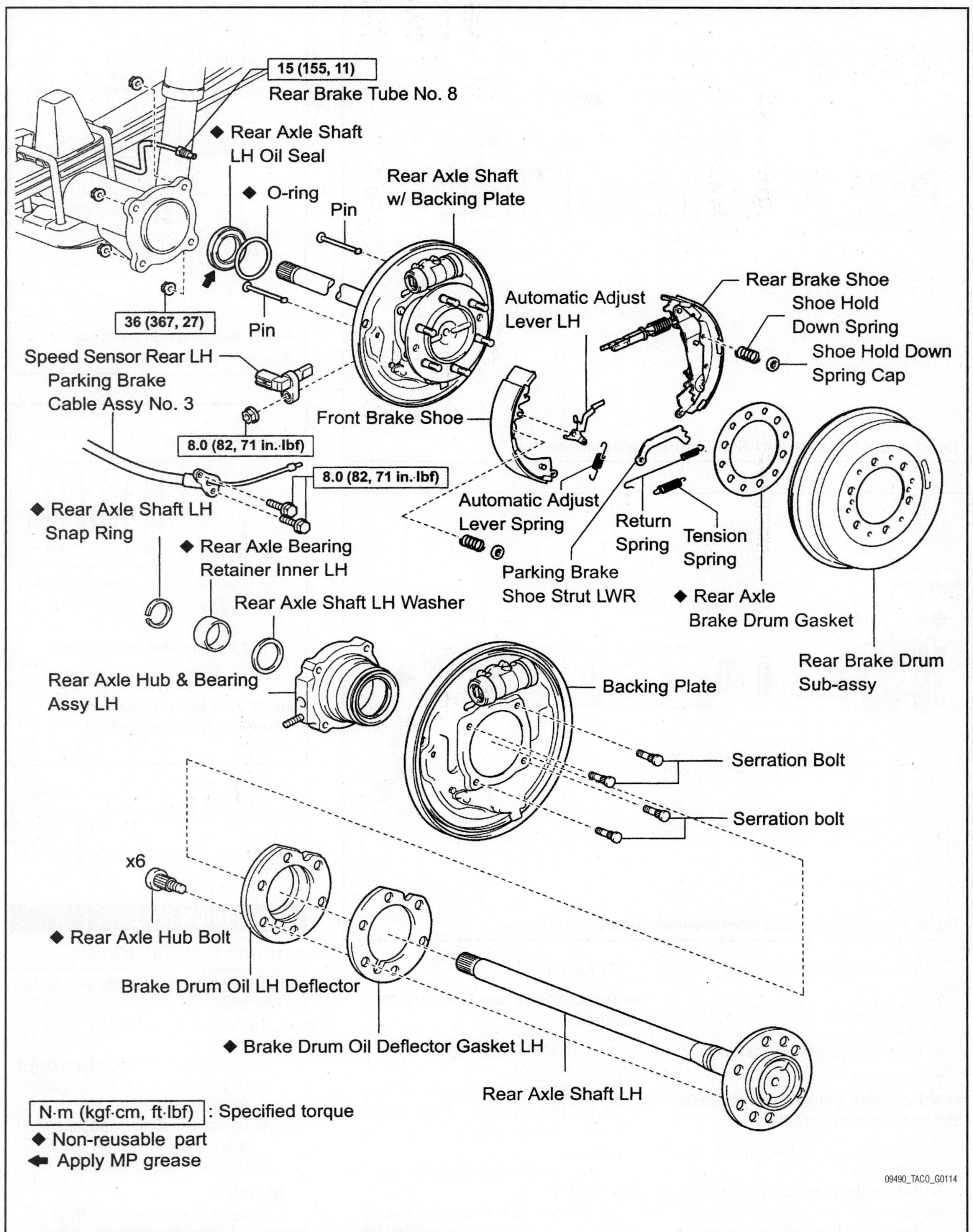

Fig. 24 Rear axle shaft and related components—4WD and PreRunner

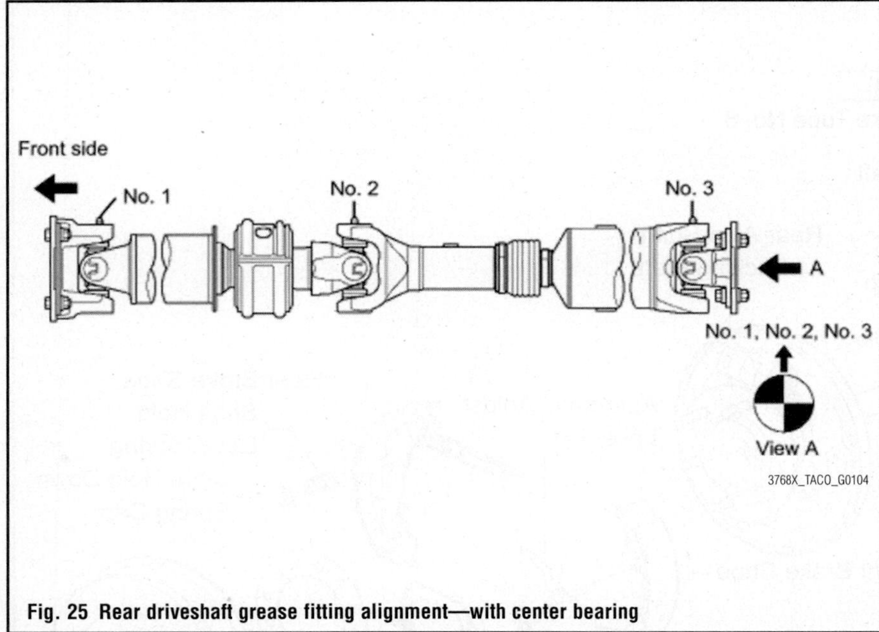

Fig. 25 **Rear driveshaft grease fitting alignment—with center bearing**

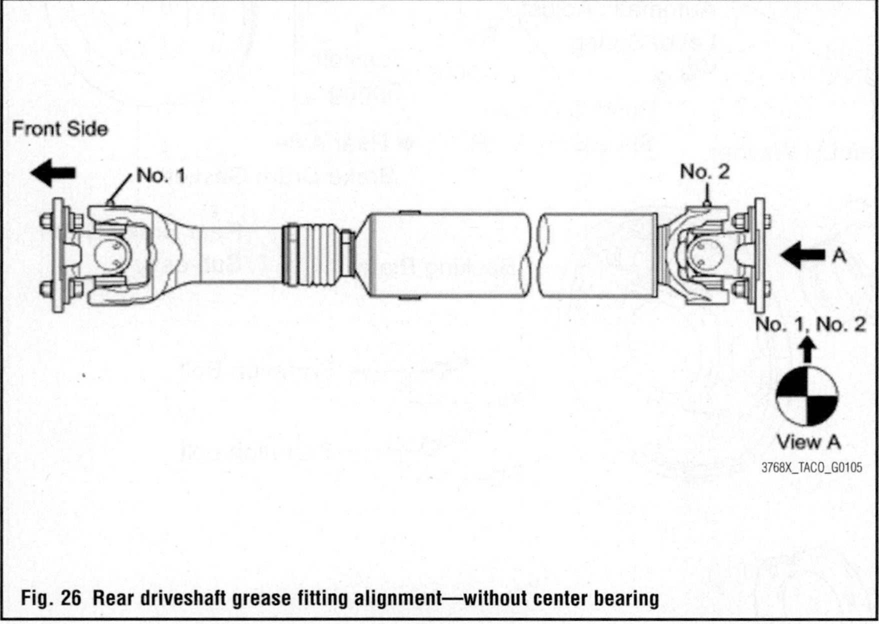

Fig. 26 **Rear driveshaft grease fitting alignment—without center bearing**

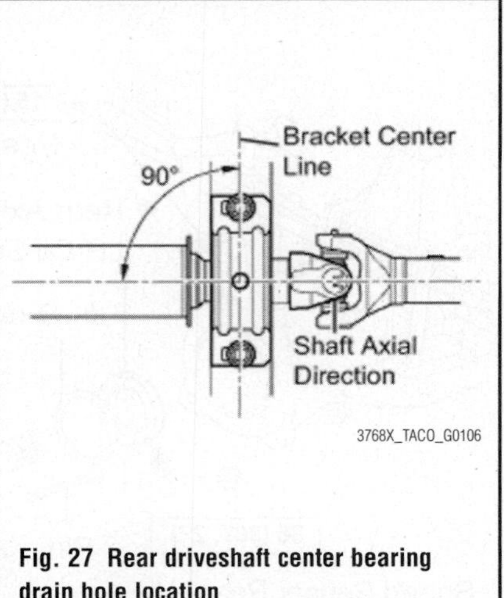

Fig. 27 **Rear driveshaft center bearing drain hole location**

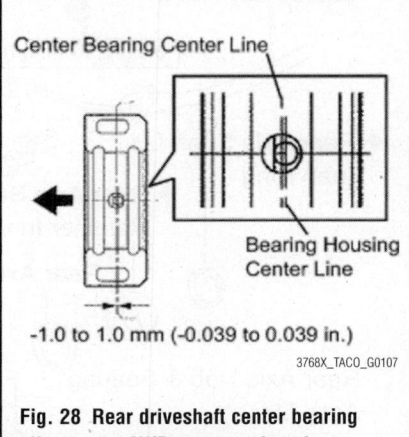

Center Bearing Center Line
Bearing Housing Center Line
-1.0 to 1.0 mm (-0.039 to 0.039 in.)
3768X_TACO_G0107

Fig. 28 **Rear driveshaft center bearing alignment—2WD access cab and double cab**

3. Raise and support the vehicle safely.
4. Matchmark the driveshaft to aid in reinstallation.
5. Remove the driveshaft retaining bolts.

➡4WD vehicles will have a front driveshaft and a rear driveshaft.

6. Remove the center bearing retaining bolts, if equipped.
7. Remove the driveshaft from its mounting.
8. Be sure to plug the transmission to prevent fluid leakage.

To install:

➡Be sure to use new fasteners, as required.

9. Installation is the reverse of the removal procedure.
10. When replacing the spider bearing make sure that the grease fitting assembly hole is facing in the direction shown in the illustration.
11. Tighten the retaining bolts to 65 ft. lbs. (88 Nm).
12. When installing the center bearing be sure that the drain hole is installed facing downwards.
13. When adjusting the center bearing be sure that the vehicle is in the unladen position.
14. Tighten the support bolts to 27 ft. lbs. (36 Nm).

REAR PINION SEAL

REMOVAL & INSTALLATION

1. Before servicing the vehicle, refer to the Precautions Section.
2. Remove or disconnect the following:
   - Rear driveshaft by matchmarking it
   - Companion flange nut, by loosen the staked portion
   - Companion flange, using a screw-type extractor
   - Oil seal, using an extractor

To install:

3. Install a new oil seal, to a depth of 0.039 in. (1.0mm) below the lip, using a seal driver.

4. Lubricate the seal lip with multi-purpose grease.
5. Install or connect the following:
   - Companion flange, coat the threads with multi-purpose grease
   - New companion flange nut. Tighten it to 109 ft. lbs. (147 Nm).
6. Measure the bearing preload, using a torque wrench. The correct preload should be 8–11 inch lbs. (0.9–1.2 Nm) for a 2 spider gear differential or to 4–7 inch lbs. (0.4–0.8 Nm) for a 4 spider gear differential.

➡If the preload is greater that specified, replace the bearing spacer. If the preload is less than specified, tighten the companion flange nut in 9 ft. lbs. (13 Nm) increments until the correct preload is achieved. Maximum torque for the nut is 325 ft. lbs. (441 Nm). If the value is exceeded, the bearing spacer must be replaced; do not back off the flange nut to lower the torque or preload.

7. Stake the pinion flange nut.
8. Install the rear driveshaft by aligning the matchmarks.

ENGINE COOLING

ENGINE COOLANT

DRAIN & REFILL PROCEDURE

See Figures 29 and 30.

➡Never drain the engine coolant when the engine is hot. Wrap a thick cloth around the cap and carefully remove it. First turn the cap a quarter of a turn to release any pressure, and then turn it all the way. Do not allow coolant to come in contact with the drive belts.

➡Never reuse old coolant. Be sure to properly dispose of used coolant. Be sure to use the proper grade and type coolant when refilling the system.

1. Before servicing the vehicle, refer to the Precautions Section.
2. Disconnect the negative battery cable. Tape the cable with insulating tape.
3. On 2.7L engines with 4WD, remove the engine undercover. On 4.0L engines remove the service hole from the engine under cover.
4. Open the radiator drain plug. Drain the coolant. Be sure to properly dispose of used coolant.
5. Open the engine block radiator drain plug. Drain the coolant. Be sure to properly dispose of used coolant.

To install:

➡Be sure to use new fasteners, as required.

6. Installation is the reverse of the removal procedure.
7. Tighten the engine drain plug to 9 ft. lbs. (13 Nm). Tighten the radiator drain plug.
8. Fill the system, using the proper grade and type coolant. Using the wrong coolant, could damage the system.
9. Check the coolant level, by squeezing the radiator hoses several times. If the level goes down add more coolant.
10. Install the radiator cap.
11. Pour coolant into the reservoir, slowly, until it reaches the FULL mark.

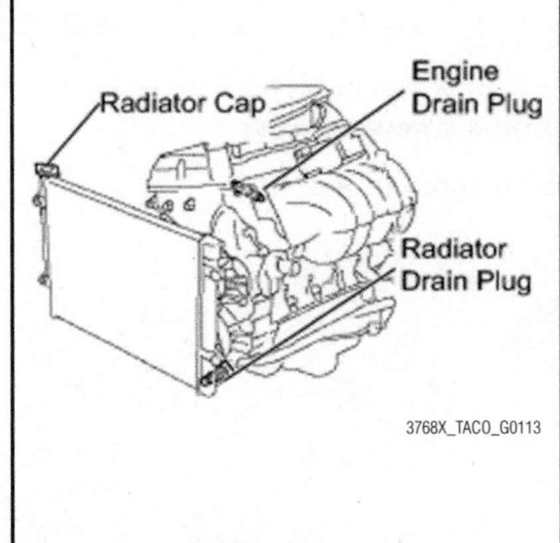

Fig. 29 Coolant drain plug locations—2.7L engine

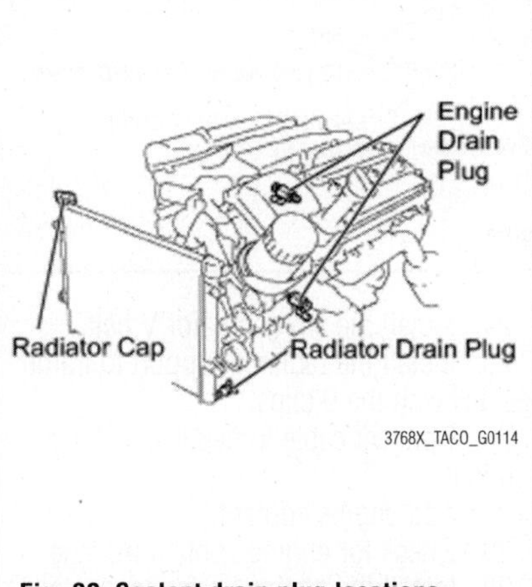

Fig. 30 Coolant drain plug locations—4.0L engine

12. Warm the engine, until the cooling fan operates. On 4.0L engine, maintain engine speed at 2000–2500 RPM.
13. Set the air conditioning fan speed to any setting but off, set the temperature toward warm and be sure that the switch is OFF.
14. Squeeze the radiator hoses several times while warming up the engine.
15. Stop the engine and wait for the coolant to cool down.
16. Remove the radiator cap, carefully. If coolant is below the full line repeat the above steps until coolant remains at the full line.
17. Check the coolant level in the reservoir tank. Add coolant, as required.
18. Check for coolant leaks, correct as required.

RADIATOR

REMOVAL & INSTALLATION

2.7L Engines

See Figure 31.

1. Before servicing the vehicle, refer to the Precautions Section.
2. Disconnect the negative battery cable. Tape the cable with insulating tape.
3. Remove the 4 bolts, then remove the engine under cover no. 1.
4. Drain engine coolant.
5. Disengage the 9 clips, then remove the radiator support to frame seal LH.
6. Remove the fan and generator V belt.
7. Disconnect the radiator reserve tank hose from the radiator.
8. Disconnect the oil cooler inlet and outlet hoses from the fan shroud.
9. Remove the 4 nuts, then remove the fan shroud and fan with fluid coupling together.

➡Make sure that the fan shroud and fan with fluid coupling do not make any contact with the radiator when they are removed.

10. Remove the fan pulley.
11. Disconnect the radiator hose inlet.
12. Disconnect the radiator hose No. 2.
13. Disconnect the oil cooler inlet and outlet hoses.
14. Remove the 4 bolts, then remove the radiator.

To install:

➡Be sure to use new fasteners, as required.

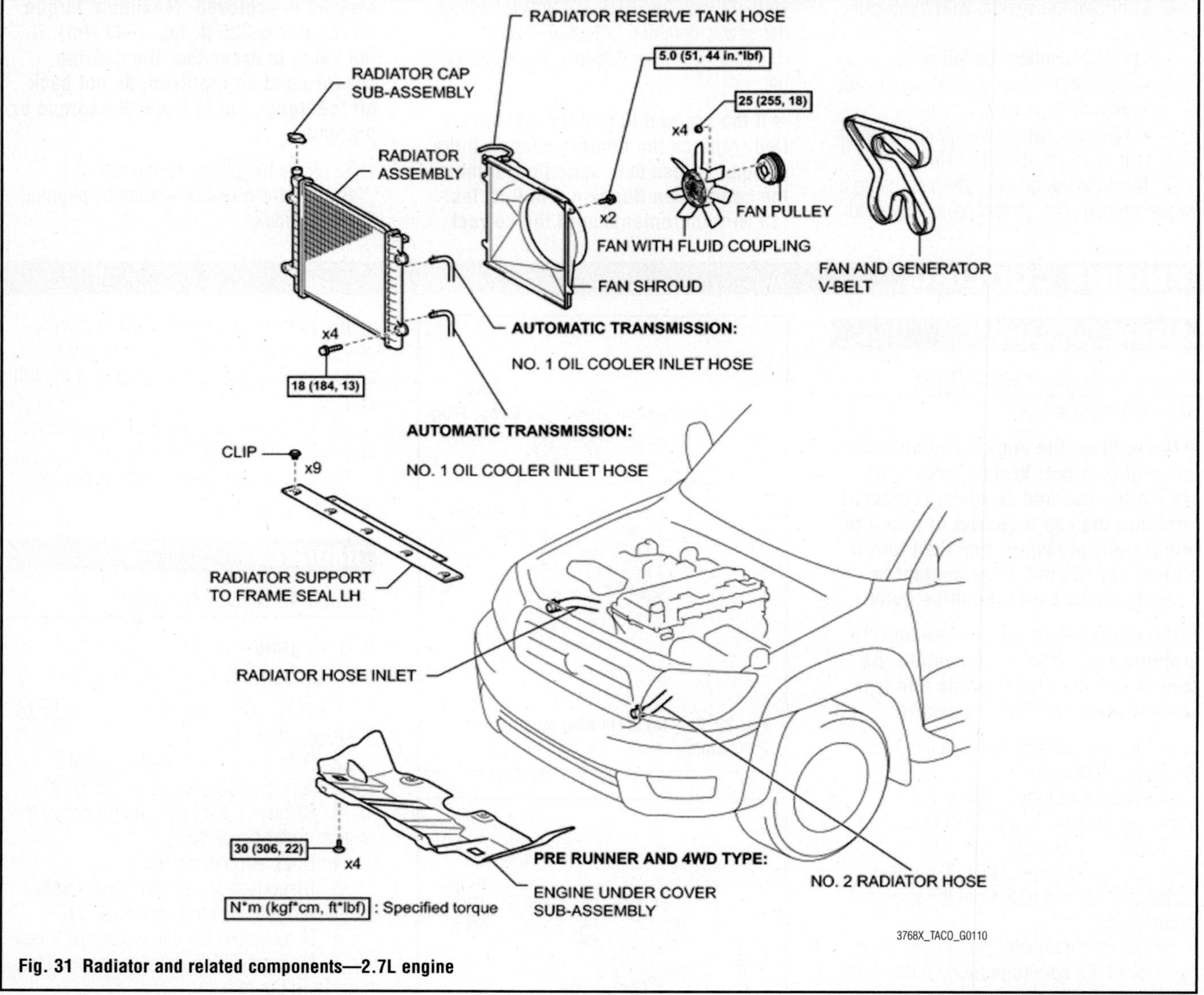

Fig. 31 Radiator and related components—2.7L engine

15. Install the radiator with the 4 bolts and tighten to 13 ft. lbs. (18 Nm)
16. Connect the oil cooler inlet and outlet hoses.
17. Connect the radiator hose NO. 2.
18. Connect the radiator hose inlet.
19. Install the fan pulley.
20. Install the fan shroud and fan with fluid coupling together.

➡Make sure that the fan shroud and fan with fluid coupling do not make any contact with the radiator when they are installed.

21. Install the fan with fluid coupling with the 4 nuts and tighten to 18 ft. lbs. (25 Nm).
22. Connect the oil cooler inlet and outlet hoses to the fan shroud.
23. Connect the radiator reserve tank hose to the radiator.
24. Install the fan generator V belt.
25. Install the radiator support to frame seal LH with the 9 clips.
26. Connect cable to negative battery terminal.
27. Add engine coolant.
28. Check for engine coolant leakage.
29. Install the engine under cover No. 1 with the 4 bolts and tighten to 30 Nm (22 ft. lbs.)

4.0L Engines

See Figure 32.

1. Before servicing the vehicle, refer to the Precautions Section.
2. Disconnect the negative battery cable. Tape the cable with insulating tape.
3. Drain engine coolant.
4. Disengage the 9 clips, then remove the radiator support to frame seal.
5. Disconnect the hose from the radiator reserve tank.
6. Remove the 2 cooler hoses from the clamp.
7. Remove the 4 nuts from the fan pulley.
8. Remove the fan and generator V belt.
9. Remove the 2 bolts from the fan shroud.
10. Remove the fan pulley and fan shroud.
11. Disconnect the 2 radiator hoses.
12. Disconnect the 2 oil cooler hoses.
13. Remove radiator hose inlet.
14. Remove radiator hose no. 2.
15. Remove the 4 bolts, then remove the radiator.

To install:

➡Be sure to use new fasteners, as required.

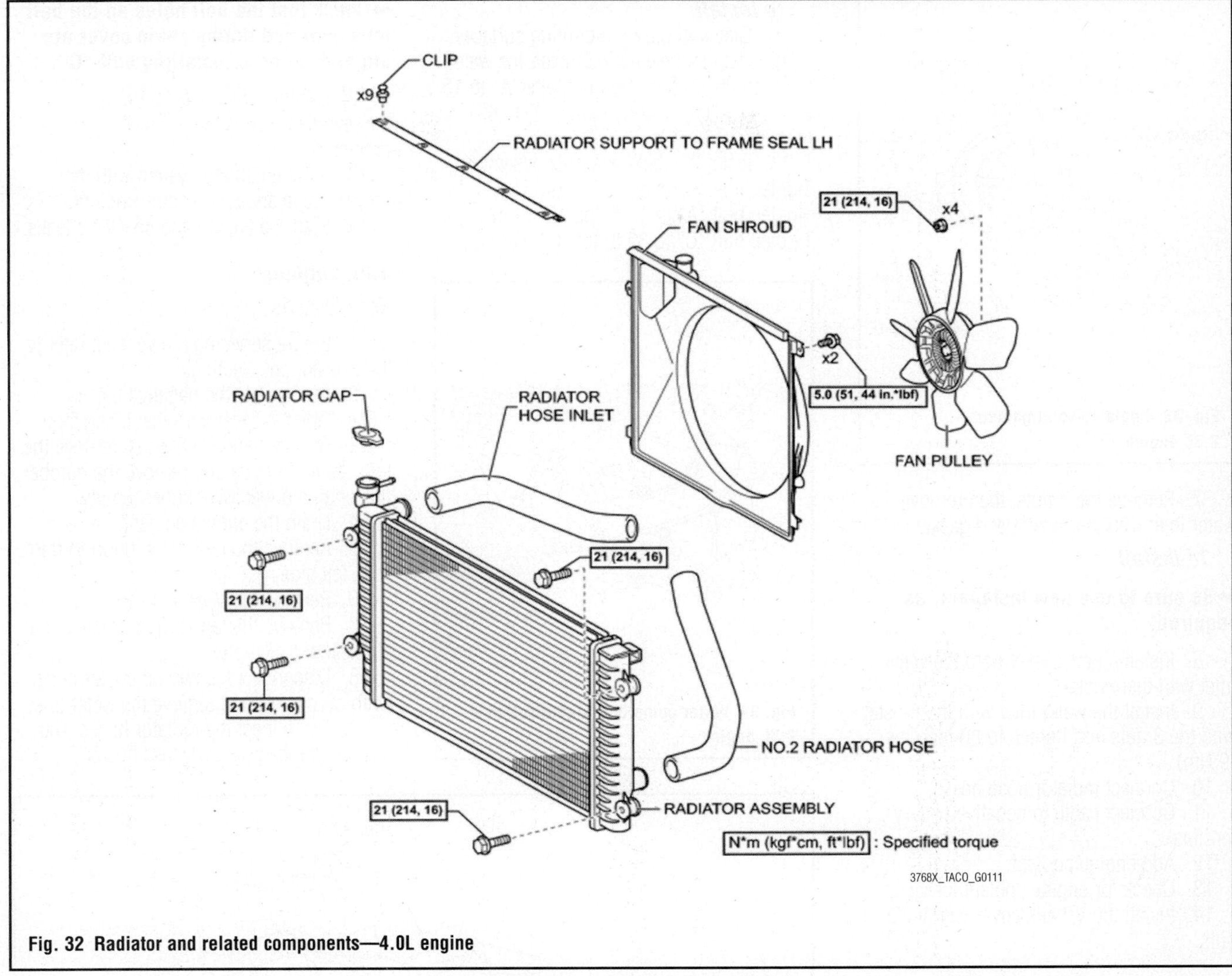

Fig. 32 Radiator and related components—4.0L engine

16. Install the radiator with the 4 bolts and tighten to 13 ft. lbs. (18 Nm).
17. Install radiator hose no. 2.
18. Install radiator hose inlet.
19. Install radiator assembly.
20. Connect the 2 oil cooler hoses.
21. Provisionally install the fan shroud together with the fan pulley.
22. Install the fan and generator V belt.
23. Install the fan pulley with the 4 nuts and tighten to 16 ft. lbs. (22 Nm).
24. Install the fan shroud with the 2 bolts and tighten to 44 inch lbs. (5 Nm).
25. Install the 2 cooler hoses into the clamp.
26. Connect the hose to the radiator reserve tank.
27. Install radiator support to frame seal.
28. Connect cable to negative battery terminal.
29. Add engine coolant.
30. Check for engine coolant leakage.

THERMOSTAT

REMOVAL & INSTALLATION

2.7L Engines

See Figure 33.

1. Before servicing the vehicle, refer to the Precautions Section.
2. Disconnect the negative battery cable. Tape the cable with insulating tape.
3. Drain engine coolant.
4. Remove water inlet.
5. Remove the bolt and 2 nuts, then remove the water inlet.
6. Remove the gasket from the timing chain cover.
7. Remove the thermostat from the timing chain cover.

To install:

➡Be sure to use new fasteners, as required.

8. Install a new gasket onto the thermostat.
9. Install the thermostat with the jiggle valve upward.

➡The jiggle valve may be set within 10° to either side of vertical position.

10. Install a new gasket and the water inlet with the bolt and 2 nuts. Tighten to 15 ft. lbs. (20 Nm).
11. Add engine coolant.
12. Check for engine coolant leakage.

4.0L Engines

1. Before servicing the vehicle, refer to the Precautions Section.
2. Disconnect the negative battery cable. Tape the cable with insulating tape.
3. Drain engine coolant.
4. Remove V-bank cover.
5. Disconnect radiator hose no. 2
6. Remove water inlet w/thermostat

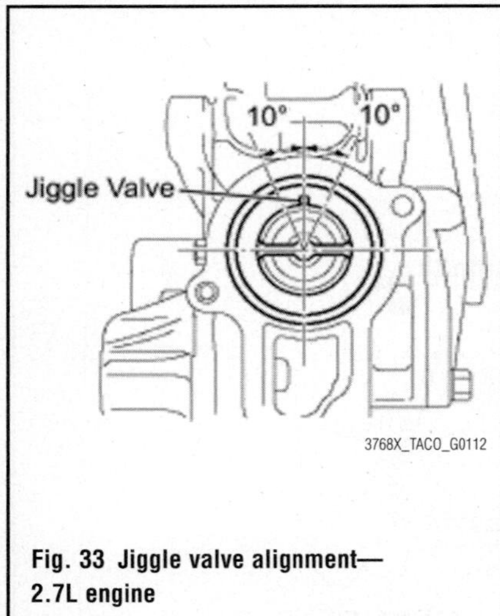

Fig. 33 Jiggle valve alignment—2.7L engine

7. Remove the 3 nuts, then remove the water inlet with thermostat and gasket.

To install:

➡Be sure to use new fasteners, as required.

8. Install a new gasket onto the water inlet with thermostat.
9. Install the water inlet with thermostat with the 3 nuts and tighten to 80 inch lbs. (9 Nm)
10. Connect radiator hose no. 2.
11. Connect cable to negative battery terminal.
12. Add engine coolant.
13. Check for engine coolant leakage.
14. Install the V-bank cover with the 2 nuts.

WATER PUMP

REMOVAL & INSTALLATION

2.7L Engines

See Figures 34 and 35.

1. Before servicing the vehicle, refer to the Precautions Section.
2. Disconnect the negative battery cable. Tape the cable with insulating tape.
3. On 4WD and PreRunner, remove the four retaining bolts and remove the number one engine undercover subassembly.
4. Drain the engine coolant.
5. Remove the radiator support to frame seal, left side.
6. Remove the fan shroud. Remove the alternator.
7. Remove the three bolts, and remove the belt tensioner assembly.
8. Remove the water pump retaining bolts. Remove the water pump from the engine.

To install:

9. Clean all gasket mounting surfaces.
10. Using a new gasket install the water pump to the engine. Tighten bolts "A" to 15 ft. lbs. (20 Nm). Tighten bolts "B" to 80 inch lbs. (9 Nm).
11. Install the belt tensioner assembly. Tighten bolt "B" to 30 ft. lbs. (40 Nm). Tighten bolt "A" to 16 ft. lbs. (22 Nm). Tighten bolt "C" to 32 ft. lbs. (43 Nm).

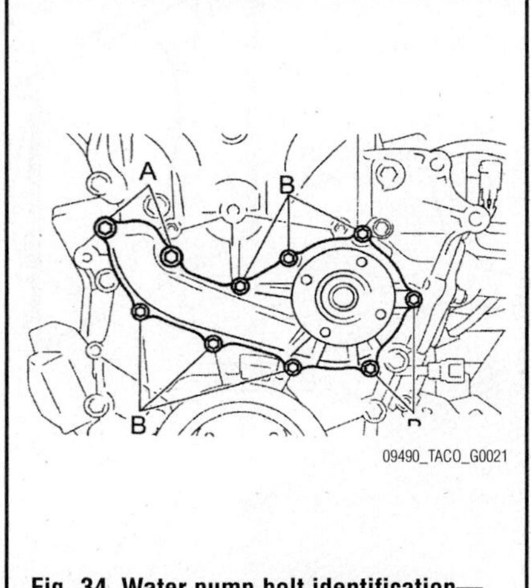

Fig. 34 Water pump bolt identification—2.7L engine

➡Check that the bolt holes on the belt tensioner and timing chain cover are aligned, prior to installing bolt "C".

12. Continue the installation in the reverse order of the removal procedure.
13. Fill the cooling system with the proper grade and type engine coolant.
14. Start the engine and check for leaks.

4.0L Engines

See Figure 36.

1. Before servicing the vehicle, refer to the Precautions Section.
2. Disconnect the negative battery cable. Tape the cable with insulating tape.
3. On 4WD and PreRunner, remove the four retaining bolts and remove the number one engine undercover subassembly.
4. Drain the engine coolant.
5. Remove the radiator support to frame seal, left side.
6. Remove the V bank cover.
7. Remove the fan shroud. Remove the air cleaner assembly.
8. Disconnect the two oil cooler hoses (with oil cooler) and remove the water inlet.
9. Disconnect the radiator hoses. Disconnect the five water bypass hoses.

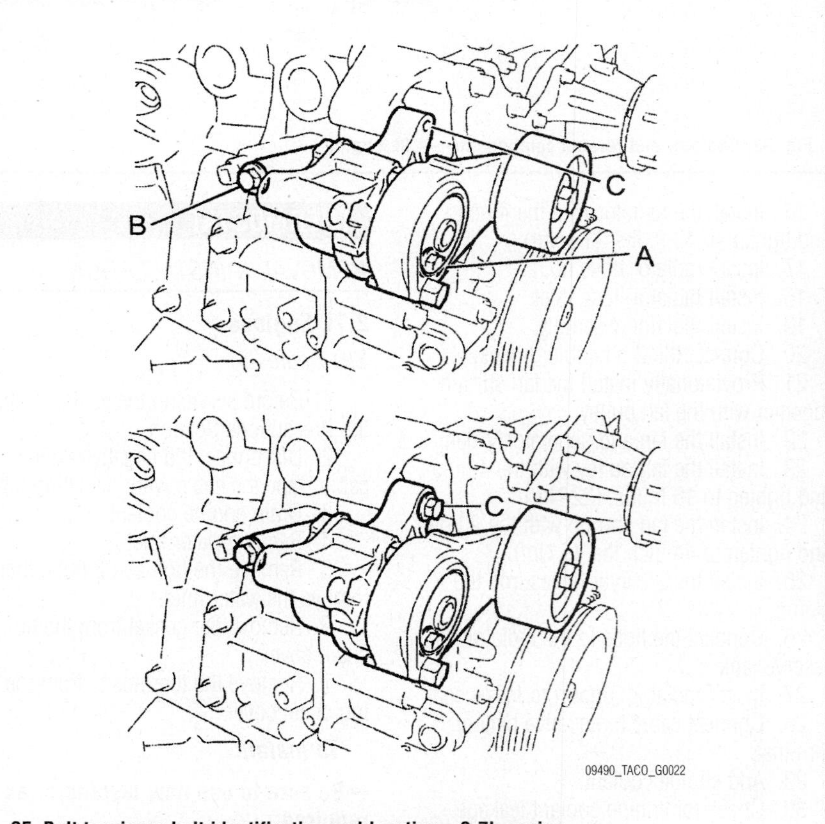

Fig. 35 Belt tensioner bolt identification and location—2.7L engine

10. Remove the five bolts and the water inlet. Remove the O-ring from the water outlet pipe. Remove the gasket from the water pump.

11. Remove the two bolts and remove the number two idler pulley subassembly.

12. Remove the alternator.

13. Remove the mounting bolt and separate the air conditioning compressor suction hose subassembly. Disconnect the air condition compressor connector. Remove the four bolts and separate the compressor from the belt tensioner assembly.

14. Remove belt tensioner assembly.

15. Remove the water pump retaining bolts. Remove the water pump from the engine.

➡Be sure to use new fasteners, as required.

To install:

16. Clean all gasket mounting surfaces.

17. Using a new gasket install the water pump to the engine. Tighten bolts "A" to 80 inch lbs. (9 Nm). Tighten bolts "B" to 17 ft. lbs. (23 Nm).

18. Install the belt tensioner assembly.

19. Continue the installation in the reverse order of the removal procedure.

20. Fill the cooling system with the proper grade and type engine coolant.

21. Start the engine and check for l eaks.

Fig. 36 Water pump bolt identification—4.0L engine

ENGINE ELECTRICAL — BATTERY SYSTEM

BATTERY

REMOVAL & INSTALLATION

At this time the manufacturer does not provide removal and installation procedures for this component. The following procedure is a guideline and may differ from the vehicle you are servicing.

1. Before servicing the vehicle, refer to the Precautions Section.

2. Remove the battery cover, if equipped.

3. Disconnect the negative battery cable. Tape the cable with insulating tape.

4. Disconnect the positive battery cable. Tape the cable with insulating tape.

5. Remove the necessary components in order to gain access to the battery hold down bolt.

6. Remove the battery hold down bolt(s).

7. Remove the battery from its mounting.

To install:

➡Be sure to use new fasteners, as required.

8. Installation is the reverse of the removal procedure.

9. Connect the positive cable first.

BATTERY RECONNECT/RELEARN PROCEDURE

➡When disconnecting the battery cable, some systems need to be initialized after the cable is reconnected. You will need the Techstream diagnostic scan tool or equivalent. Follow the directions on the tool.

ENGINE ELECTRICAL — CHARGING SYSTEM

ALTERNATOR

REMOVAL & INSTALLATION

See Figures 37 and 38.

On some vehicles, the alternator is mounted very low on the engine. It may be necessary to remove the gravel shield and work from beneath the vehicle in order to gain access to the alternator. Replacing the alternator while the engine is cold is recommended.

1. Before servicing the vehicle, refer to the Precautions Section.

2. Disconnect the negative battery cable. Tape the cable with insulating tape.

3. Remove the V bank cover, if equipped.

4. Remove the radiator support to frame seal, left side.

5. Remove the radiator fan shroud.

Fig. 37 Alternator mounting bolt locations—2.7L engine

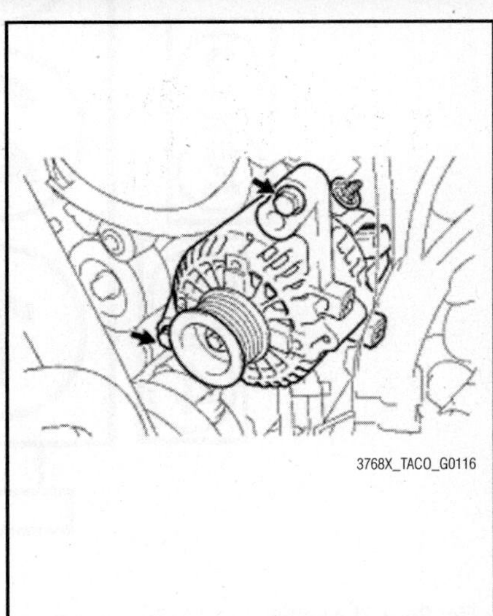

Fig. 38 Alternator mounting bolt locations—4.0L engine

6. Disconnect the alternator harness wiring.
7. Remove the drive belt.
8. Remove the alternator retaining bolts.
9. Remove the alternator from the vehicle.

To install:

➡Be sure to use new fasteners, as required.

10. Installation is the reverse of the removal procedure.
11. Tighten the alternator retaining bolts to 32 ft. lbs. (43 Nm.).

ENGINE ELECTRICAL

IGNITION SYSTEM

FIRING ORDER

See Figures 39 and 40.

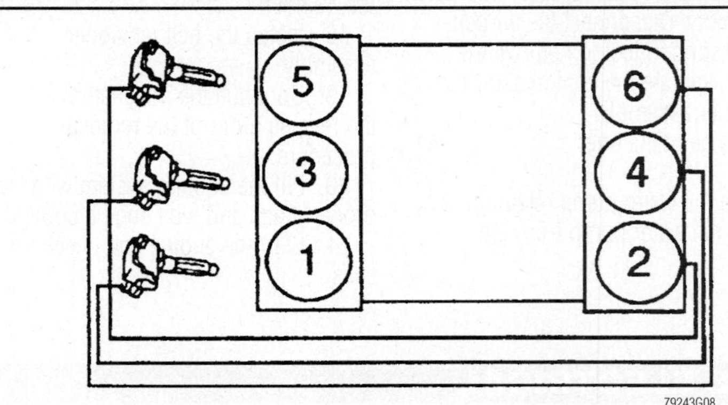

Fig. 40 44.0L Engine Firing order: 1–2–3–4–5–6 Distributorless ignition system

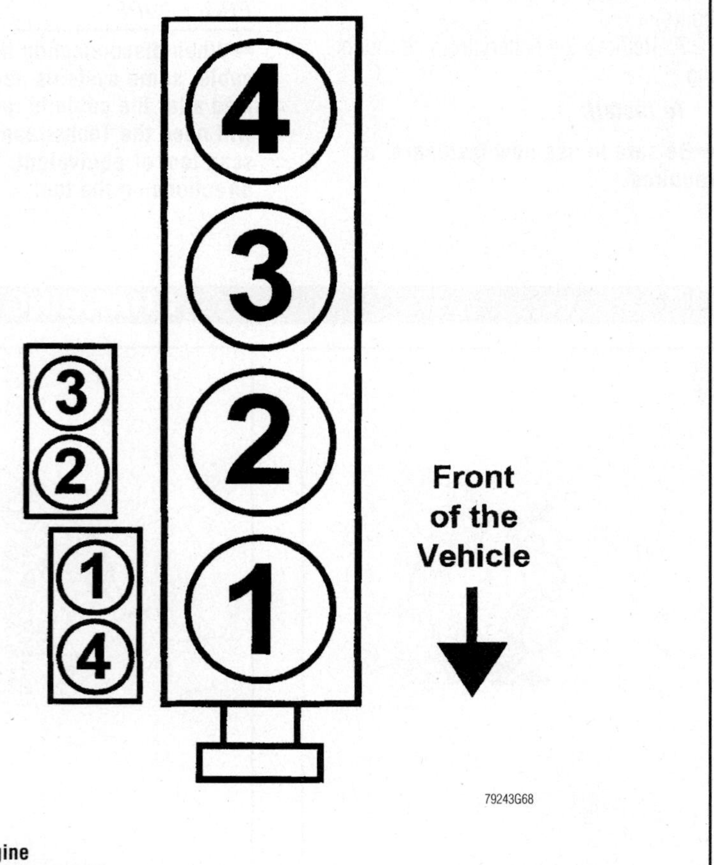

Fig. 39 2.7L Engine Firing order: 1–3–4–2 Distributorless ignition system

IGNITION COIL

REMOVAL & INSTALLATION

➡It is a good idea to remove and reinstall the coils one at a time to prevent the coils being installed out of order.

1. Disconnect the negative battery cable.
2. Disconnect the spark plug wire from the coil.
3. Disconnect the electrical connectors from the coil.
4. Remove the coil bolts and the coils.

To install:

5. Install the coil and tighten the bolts to 69 inch lbs. (8 Nm).
6. Connect the wiring and spark plug wires
7. Connect the negative battery cable.

IGNITION TIMING

ADJUSTMENT

The ignition timing is controlled by the Powertrain Control Module (PCM). No adjustment is necessary or possible.

SPARK PLUGS

REMOVAL & INSTALLATION

1. Remove the ignition coils.
2. Using a 16 mm plug wrench, remove the spark plugs.
3. Clean the spark plugs.
4. If the electrode has traces of wet carbon, allow it to dry and then clean with a spark plug cleaner.
5. Check the spark plug for thread damage and insulator damage. If abnormal, replace the spark plug.
6. Adjust the spark plug electrode gap. Electrode gap for new spark plug is 1.0 to 1.1 mm (0.039 to 0.043 in.).
7. Using a 16 mm plug wrench, install the spark plugs and tighten to 13 ft. lbs. (17.5 Nm).
8. Reinstall the ignition coils.

STARTER

REMOVAL & INSTALLATION

2.7L Engines

See Figure 41.

1. Before servicing the vehicle, refer to the Precautions Section.
2. Disconnect the negative battery cable. Tape the cable with insulating tape.
3. Raise and support the vehicle, as required.
4. Remove the terminal cap. Disconnect the electrical connections.
5. Disconnect the positive battery cable.
6. Remove the starter retaining bolts. Remove the starter from the vehicle.

To install:

➡Be sure to use new fasteners, as required.

7. Installation is the reverse of the removal procedure.
8. Tighten the retaining bolts to 27 ft. lbs. (36 Nm).

4.0L Engines

See Figure 42.

1. Before servicing the vehicle, refer to the Precautions Section.

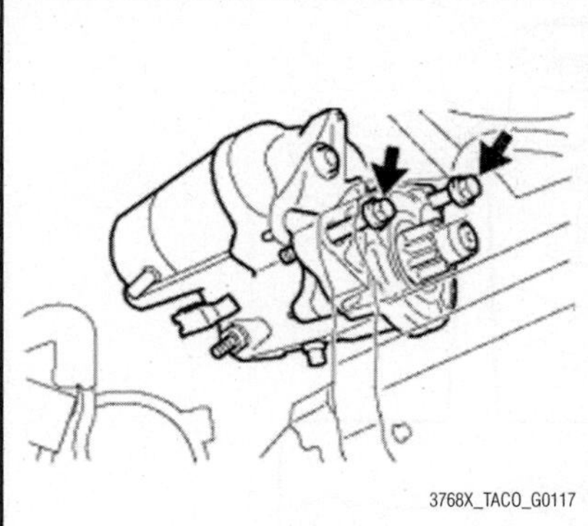

Fig. 41 Starter mounting bolt locations—2.7L engine

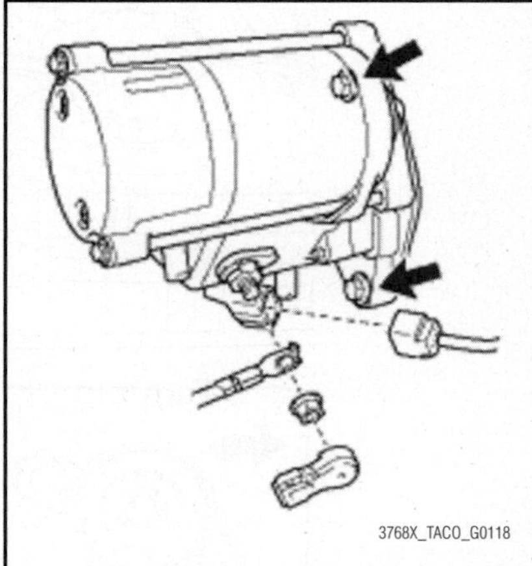

Fig. 42 Starter mounting bolt locations—4.0L engine

2. Disconnect the negative battery cable. Tape the cable with insulating tape.
3. Remove the engine undercover assembly.
4. On 2WD and PreRunner, remove the number two manifold stay.
5. On 4WD vehicles, remove the number two exhaust front pipe assembly. Remove the five clips and then remove the front fender splash shield, left side. Remove the number two steering intermediate shaft.
6. Disconnect the starter electrical connectors.
7. Disconnect the positive battery cable.
8. Remove the starter retaining bolts. Remove the starter from the vehicle.

To install:

➡Be sure to use new fasteners, as required.

9. Installation is the reverse of the removal procedure.
10. Tighten the retaining bolts to 27 ft. lbs. (36 Nm).

ENGINE MECHANICAL

➡Disconnecting the negative battery cable may interfere with the functions of the on board computer systems and may require the computer to undergo a relearning process, once the negative battery cable is reconnected.

ACCESSORY DRIVE BELTS

ACCESSORY BELT ROUTING

See Figures 43 and 44.

INSPECTION

Inspect the drive belt for signs of glazing or cracking. A glazed belt will be perfectly smooth from slippage, while a good belt will have a slight texture of fabric visible. Cracks will usually start at the inner edge of the belt and run outward. All worn or damaged drive belts should be replaced immediately.

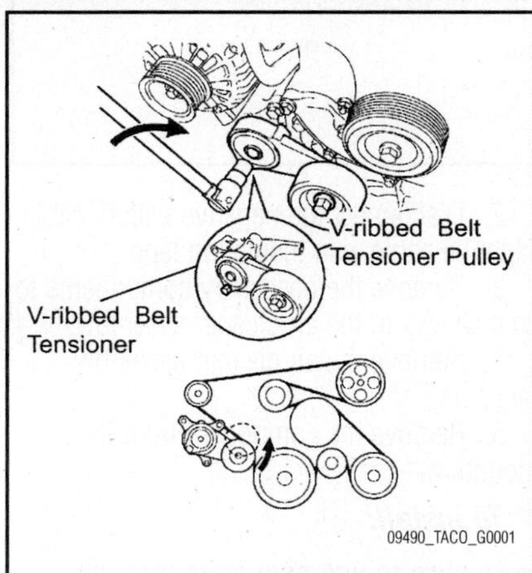

Fig. 43 Accessory drive belt routing—2.7L engine

REMOVAL & INSTALLATION

See Figures 43 through 45

1. Before servicing the vehicle, refer to the Precautions Section.
2. Disconnect the negative battery cable. Tape the cable with insulating tape.
3. Loosen the drive belt tension by turning the drive belt tensioner counterclockwise, and remove the drive belt.

To install:

➡Be sure to use new fasteners, as required.

4. Installation is the reverse of the removal procedure.

AIR CLEANER

REMOVAL & INSTALLATION

See Figures 46 and 47.

At this time the manufacturer does not provide removal and installation procedures for this component. The following procedure is a guideline and may differ from the vehicle you are servicing.

1. Before servicing the vehicle, refer to the Precautions Section.

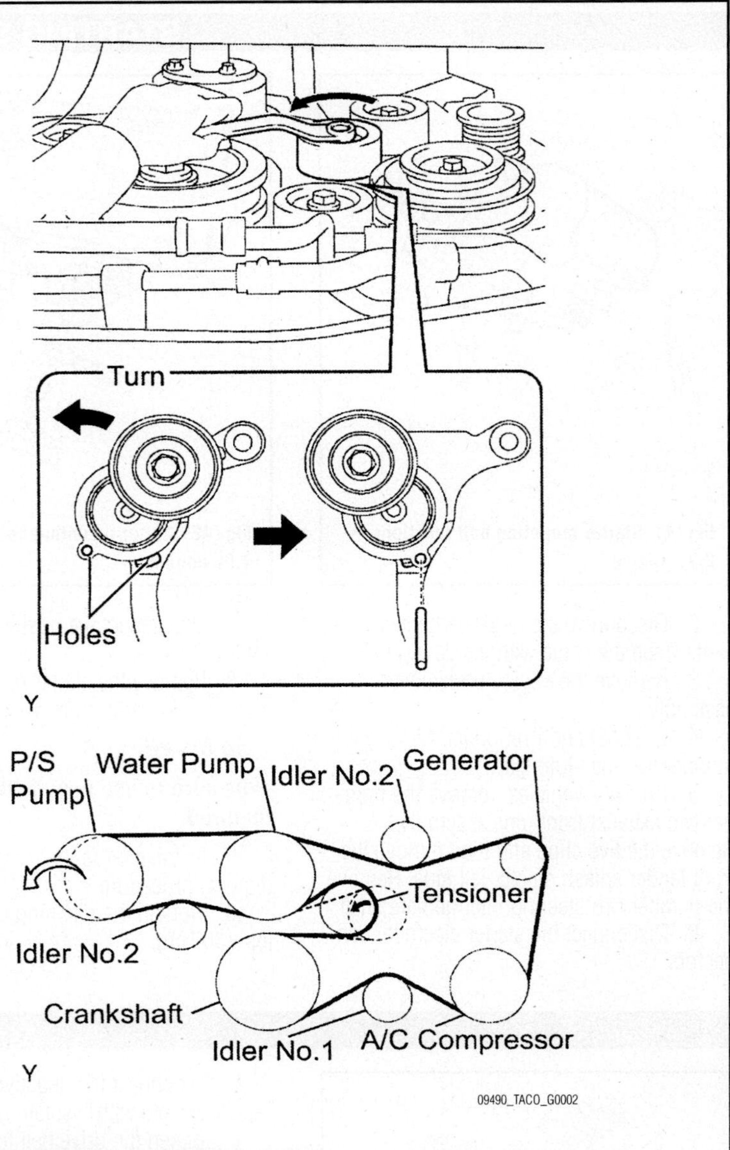

Fig. 44 Accessory drive belt routing—4.0L engine

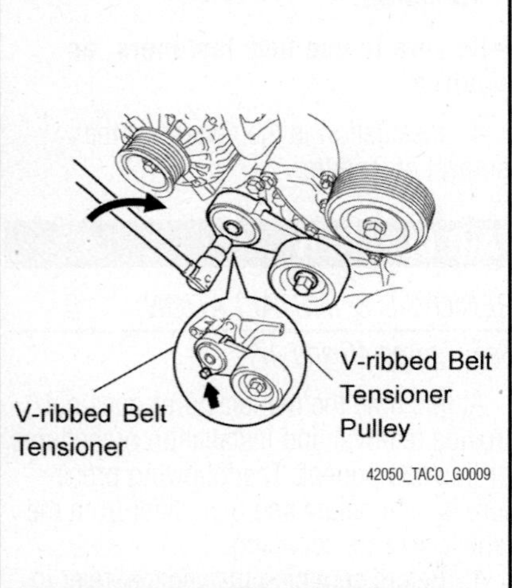

Fig. 45 Accessory drive belt replacement

2. Disconnect the negative battery cable. Tape the cable with insulating tape.
3. Remove the necessary components to gain access to the air cleaner assembly.
4. Remove the air cleaner assembly retainers.
5. Remove the component from its mounting.

To install:

➡Be sure to use new fasteners, as required.

6. Installation is the reverse of the removal procedure.

FILTER/ELEMENT REPLACEMENT

At this time the manufacturer does not provide removal and installation procedures for this component. The following procedure is a guideline and may differ from the vehicle you are servicing.

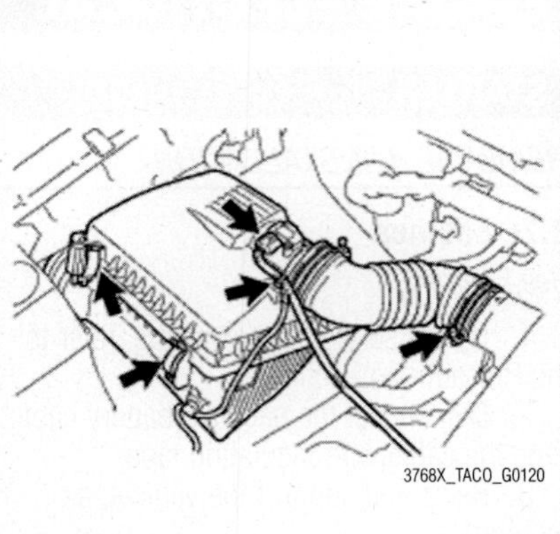

Fig. 46 Air cleaner assembly and related components—2.7L engine

1. Before servicing the vehicle, refer to the Precautions Section.
2. Disconnect the negative battery cable. Tape the cable with insulating tape.
3. Remove the necessary components to gain access to the air cleaner assembly.
4. Remove the upper air cleaner cover.
5. Carefully remove the filter from its mounting.

To install:

➡Be sure to use new fasteners, as required.

6. Installation is the reverse of the removal procedure.

BALANCE SHAFT

REMOVAL & INSTALLATION

See Figure 48.

At this time the manufacturer does not provide removal and installation procedures for this component, refer to the illustration as required.

CAMSHAFT AND VALVE LIFTERS

INSPECTION

1. Before servicing the vehicle, refer to the Precautions Section.
2. Remove the camshaft from the engine.
3. Check the camshaft bearing journals for damage and binding.
4. If the journals are binding, check the cylinder head for damage.
5. Check the cylinder head for clogged oil holes.

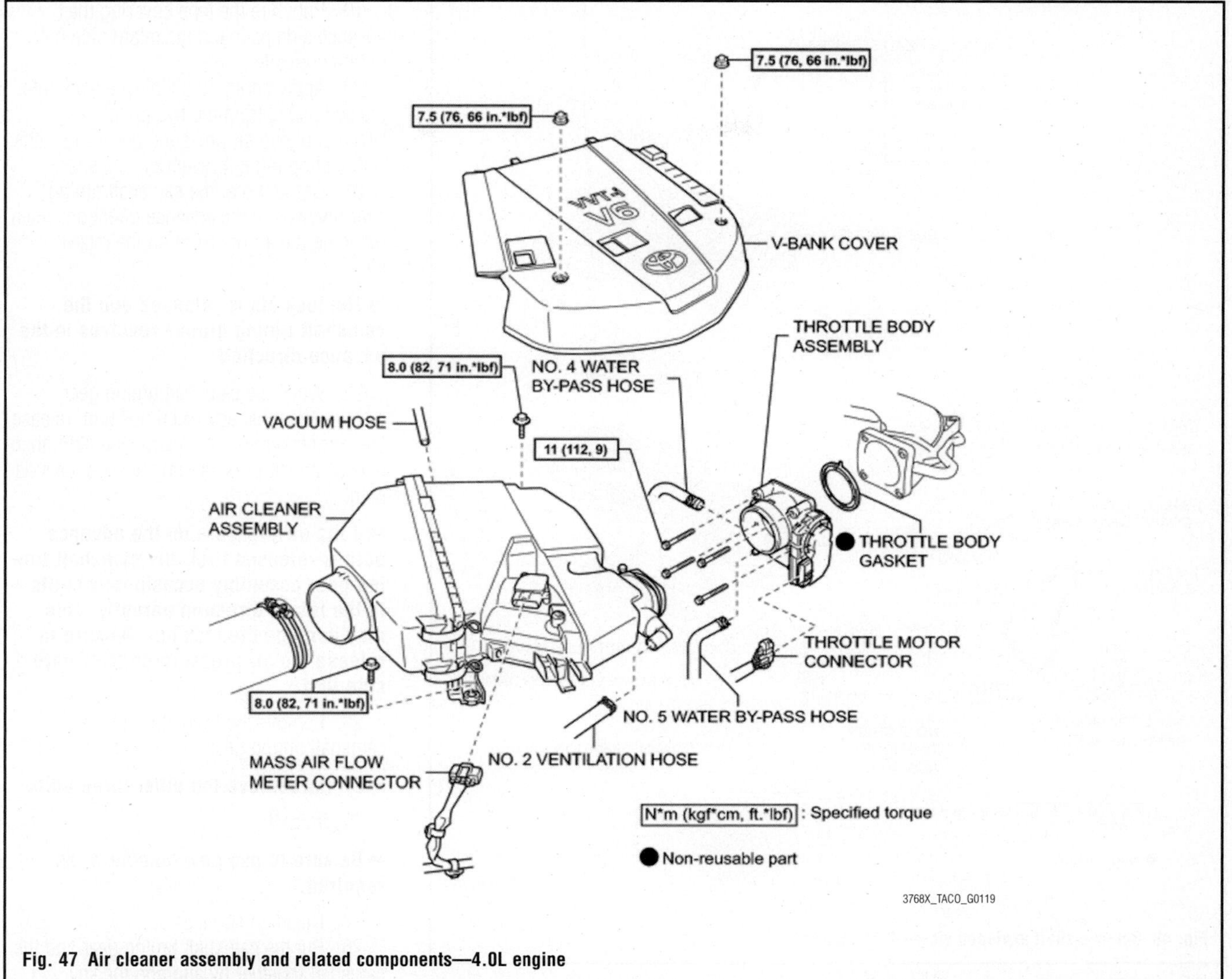

Fig. 47 Air cleaner assembly and related components—4.0L engine

6. Check the camshaft surface for abnormal wear and damage. Replace the camshaft, as required.
7. Measure the camshaft lobe surface and replace the camshaft if not within specification.
8. Measure the camshaft journal diameter and replace the camshaft if not within specification.
9. Measure the camshaft run out and replace the camshaft if not within specification.

REMOVAL & INSTALLATION

2.7L Engine

See Figures 49 through 53.

1. Before servicing the vehicle, refer to the Precautions Section.
2. Disconnect the negative battery cable. Tape the cable with insulating tape.
3. On vehicles equipped with 4WD, remove the engine undercover subassembly.
4. Drain the engine coolant. Remove the radiator support to frame seal, left side. Remove the fan shroud.
5. Remove the air cleaner cap subassembly. Remove the intake air connector.
6. Disconnect the ignition coil connectors. Disconnect the throttle body motor connector. Disconnect the VSV connector.
7. Disconnect the Camshaft Position (CMP) sensor connector. Disconnect the engine wire harness clamps. Remove the ignition coils. Disconnect the PCV hose.
8. Remove the cylinder head cover retaining bolts. Remove the cylinder head cover.
9. Remove the two timing chain guide bolts. Remove the timing chain guide. Remove the O-ring.
10. Position the number one cylinder at TDC on the compression stroke.

➡Turn the crankshaft pulley clockwise to align the timing mark notch with the timing mark "0". Paint marks on the timing chain plates that align with the timing marks on the camshaft timing gear.

11. Hold the hexagonal lobe of the number two camshaft, with a suitable tool. Loosen the bolt. Remove the head straight screw plug. Insert a suitable tool into the service hole of the chain tensioner to hold the stopper plate of the chain tensioner lifted up.

➡Lifting up the stopper plate of the chain tensioner unlocks the plunger.

12. While keeping the stopper plate of the chain tensioner lifted up, slightly rotate the hexagonal lobe of the number two camshaft clockwise so that the plunger of the chain tensioner is pushed. Be careful

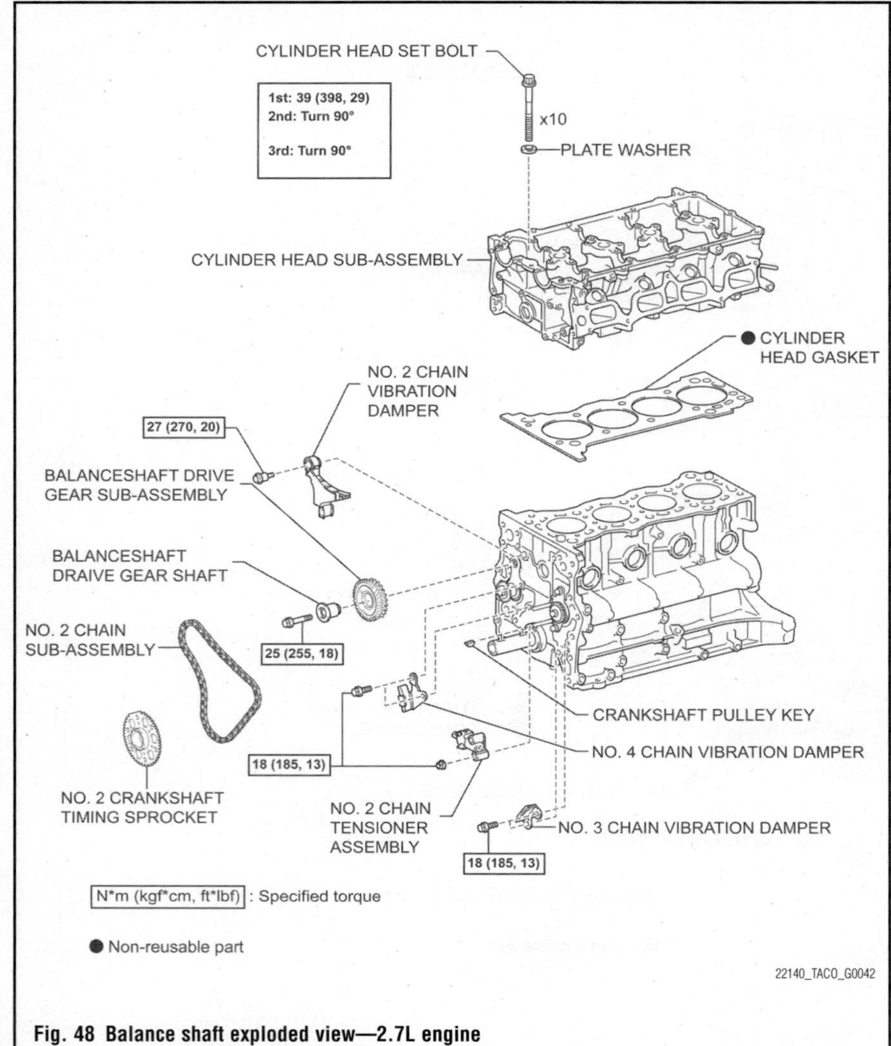

Fig. 48 Balance shaft exploded view—2.7L engine

not to damage the camshaft oil delivery pipe.

➡**With the wrench still installed, remove the suitable tool with the plunger still pushed in. Do not remove the wrench. Removing the suitable tool lifts down the stopper plate and locks the plunger.**

13. Insert a 0.118 inch diameter bar into the hole of the stopper plate with the stopper plate of the chain tensioner lifter down and locked. Secure the bar with tape.

➡**If the bar cannot be installed, rotate the number two camshaft slightly to the left and right. Then insert the bar.**

14. Remove the camshaft timing gear bolt. Remove the gear.

15. Using several steps, uniformly loosen and remove the camshaft bearing cap bolts in the proper sequence. Remove the camshaft oil delivery pipe and O-ring.

Remove the camshaft bearing cap number one and eight camshaft bearing caps number two. Remove the number one camshaft and the number two camshaft.

16. Tie the timing chain with a piece of wire.

17. Clamp the camshaft in a soft jaw vise, be sure that the camshaft timing gear does not rotate. Do not clamp the camshaft too tightly in the vise.

18. Cover the four oil path holes of the cam journal with vinyl tape.

➡**One of the two grooves on the cam journal is for retarding cam timing (upper) the other is for advancing cam timing (lower). Each groove has two oil paths. Plug one of the two paths for each groove with a piece of rubber before wrapping the cam journal with the tape.**

19. Puncture the tape covering the advance side path and the retard side path on the opposite side.

20. Apply about 29 psi air pressure into the two paths, from the two punctures. When applying air pressure, cover the paths with a shop rag to prevent oil splashes.

21. Confirm that the camshaft timing gear revolves in the advance direction, when reducing the air pressure on the retard side.

➡**The lock pin is released and the camshaft timing groove revolves in the advance direction.**

22. When the camshaft timing gear reaches the most advanced position, release the air pressure on the retard side path, then release the air pressure on the advance side path.

➡**If the air pressure on the advance path is released first, the camshaft timing gear assembly occasionally shifts in the retard direction abruptly. This may damage the lock pin. Be sure to release the air pressure on the retard side first.**

23. Remove the fringe bolt of the camshaft timing gear.

➡**Do not remove the other three bolts.**

To install:

➡**Be sure to use new fasteners, as required.**

24. Clean all surfaces.

25. Put the camshaft timing gear and the camshaft together by aligning the key groove and the straight pin.

26. Gently press the gear against the camshaft and turn the gear. Push further at the position where the pin fits the groove.

➡**Be sure not to turn the camshaft timing gear to the retard angle side (to the right angle).**

27. Check that there is no clearance between the gear's fringe and the camshaft. Tighten the fringe bolt to 58 ft. lbs. (78 Nm).

28. Check that the camshaft timing gear can move to the retard side (the right angle) and is locked in the extreme retard position.

29. Check that the valve rocker arm is correctly installed. Apply clean engine oil to the camshaft's cam portion and the cylinder head journals.

30. Install the chain onto the camshaft timing gear, with the painted mark of the link aligned with the timing mark of the camshaft timing gear.

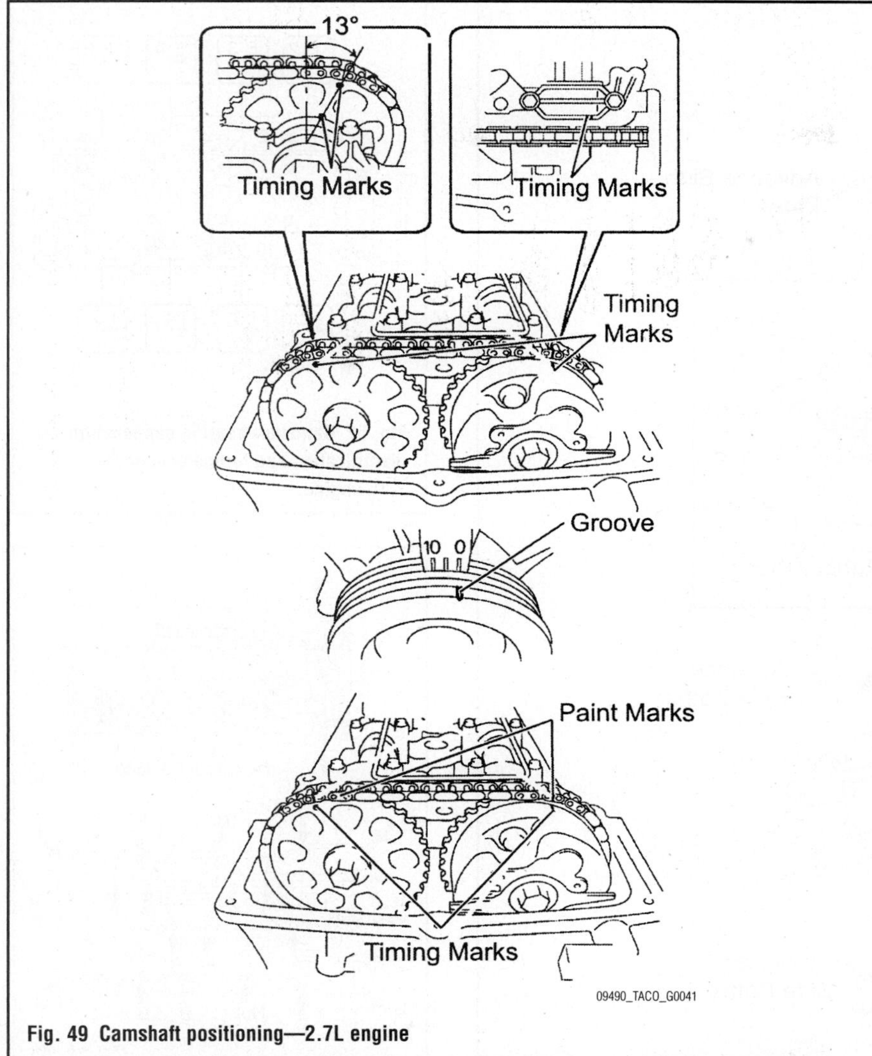

Fig. 49 Camshaft positioning—2.7L engine

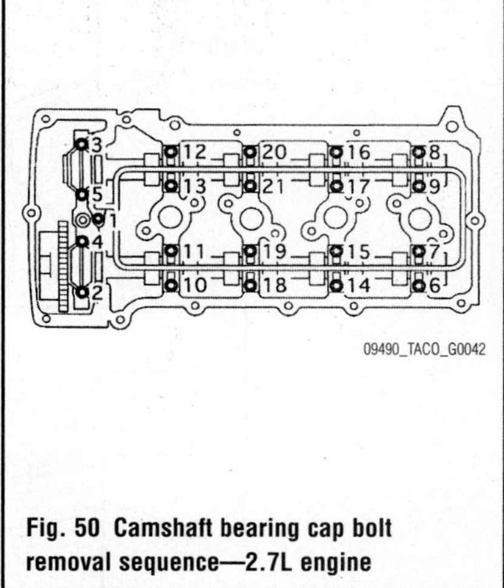

Fig. 50 Camshaft bearing cap bolt removal sequence—2.7L engine

31. Position the two camshafts in there mounting on the engine, see illustration.

➡**Align the paint mark with the timing mark before installing the camshaft.**

32. Provisionally install the number one camshaft bearing cap. Check the proper location of each of the number two camshaft bearing caps and install them.

33. Install a new O-ring onto the number one camshaft bearing cap. Provisionally install the camshaft oil delivery pipe.

34. Tighten the camshaft bearing cap bolts to specification and in the proper sequence.

➡**Bolt "A" is tightened to 9 ft. lbs. All other bolts are tightened to 11 ft. lbs. (15 Nm).**

35. Check that each timing mark is set as indicated in the illustration.

36. Install the timing chain onto the camshaft timing gear, with the paint mark aligned with the timing mark on the camshaft timing gear.

37. Align the number two camshaft straight pin and timing gear straight pin hole. Install the camshaft timing gear onto the number two camshaft.

➡**If the straight pin and straight pin hole are difficult to align, slightly rotate the number two camshaft to the left and right, then attempt to align them.**

38. Hold the hexagonal lobe of the number two camshaft with a wrench. Tighten the bolt to 58 ft. lbs.

39. Remove the 0.118 inch diameter bar from the chain tensioner. Apply adhesive, part number 08833-00070 or equivalent, to two or three threads of the timing gear case with head straight screw plug. Install the timing gear case with head straight screw plug and tighten to 12 ft. lbs. (16 Nm).

40. Install a new O-ring onto the camshaft bearing cap. Install the two timing chain guide bolts. Tighten to 7 ft. lbs.

41. Apply seal packing, part number 08826-00080 or equivalent, to the cylinder head as indicated in the illustration. Provisionally install the cylinder head cover bolts and nuts. Tighten bolts "A" to 80 inch lbs. (9 Nm). Tighten bolts "B" to 80 inch lbs. (9 Nm). Retighten bolts "A" to 80 inch lbs. (9 Nm).

➡**Be sure to remove any oil from the contact surfaces of the cylinder head cover and the cylinder head. Install the cover within three minutes after applying the seal packing. Do not add engine oil for at least two hours after installing the cover.**

42. Continue the installation in the reverse order of the removal procedure.

43. Be sure to fill the engine with the proper grade and type engine coolant.

44. Be sure to fill the engine with the proper grade and type engine oil.

45. Start the engine and check for leaks. Correct as required.

4.0L Engine

BANK 1

See Figures 54 through 64.

1. Before servicing the vehicle, refer to the Precautions Section.

2. Disconnect the negative battery cable. Tape the cable with insulating tape.

3. Drain the engine coolant. Remove the V bank cover.

4. Remove the air cleaner assembly.

5. Disconnect the two water bypass hoses. Disconnect the fuel vapor feed hose. Disconnect the ventilation hose. Disconnect the VSV connectors.

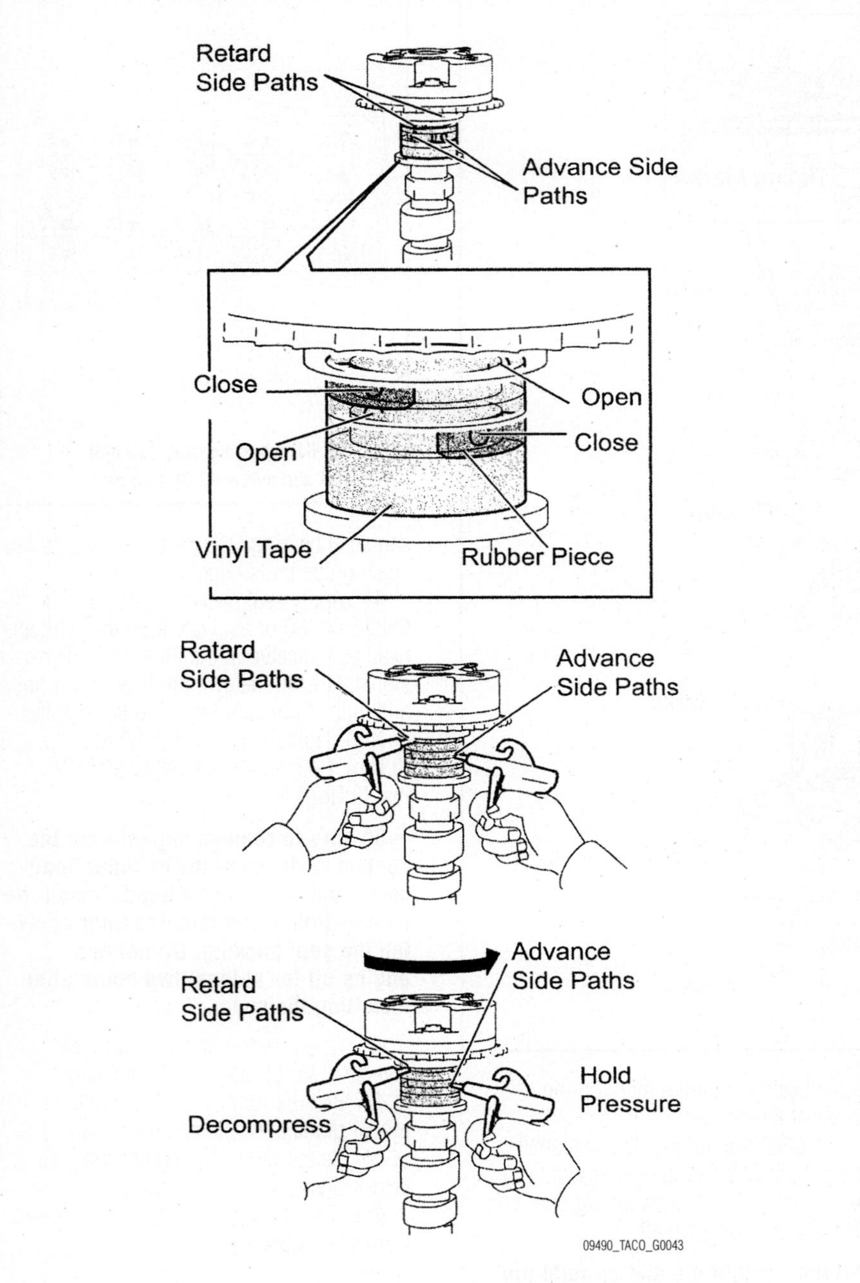

Fig. 51 Camshaft timing gear removal—2.7L engine

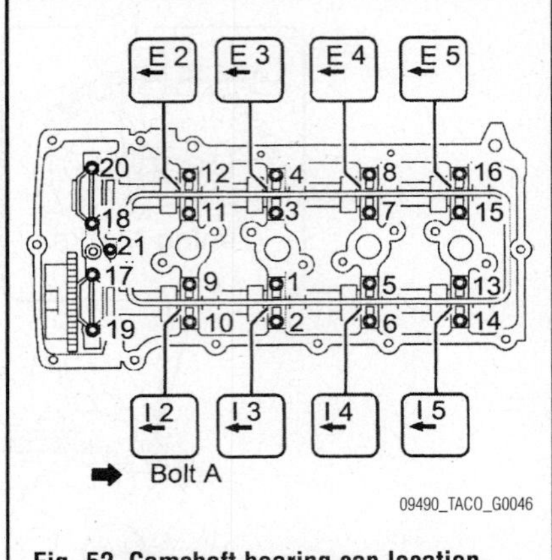

Fig. 52 Camshaft bearing cap location, identification and torque sequence—2.7L engine

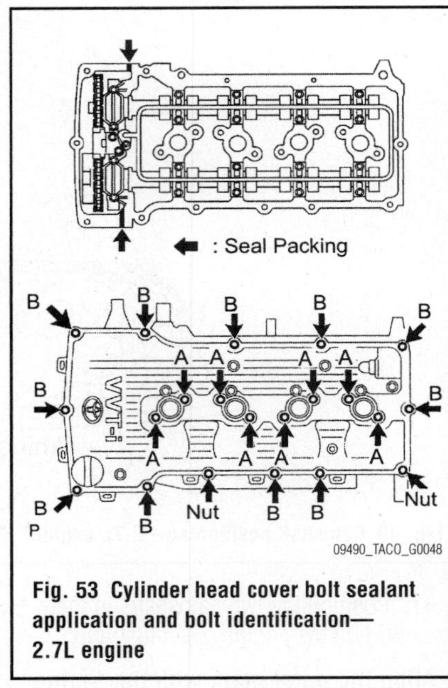

Fig. 53 Cylinder head cover bolt sealant application and bolt identification—2.7L engine

6. Disconnect the throttle body motor connector. Separate the three wire harness clamps and hose clamp.

7. If equipped with manual transmission, remove the nut, then separate the clutch flexible hose bracket from the surge tank stay.

8. Remove the two bolts and the throttle body bracket. Remove the bolt and the oil baffle plate. Remove the four bolts and the two serge tank stays.

9. Remove the two nuts. Remove the four bolts, intake air surge tank and gasket.

10. Remove the ignition coil assembly.

11. Remove the cylinder head cover retaining bolts. Remove the cylinder head cover.

12. Turn the crankshaft pulley until its groove and the timing mark "0" of the timing chain cover are aligned. If not aligned at TDC of the compression stroke, turn the crankshaft one complete revolution, in the direction of rotation. Paint alignment marks on the number one chain links corresponding to the timing marks of the camshaft timing gears.

13. Remove the four bolts, and then remove the timing chain cover plate and gasket.

14. While turning the stopper plate of the tensioner upward, push the plunger of the chain tensioner. While turning the stopper plate of the tensioner downward, insert a 0.118 inch diameter bar into the holes in the stopper plate and tensioner to hold the stopper plate.

15. Remove the two bolts, and then remove the chain tensioner.

➡Keep the camshaft level while it is being removed. The camshaft thrust clearance is very small and failing to keep it level could crack or damage the cylinder head journal surface, which receives the thrust. Follow the steps

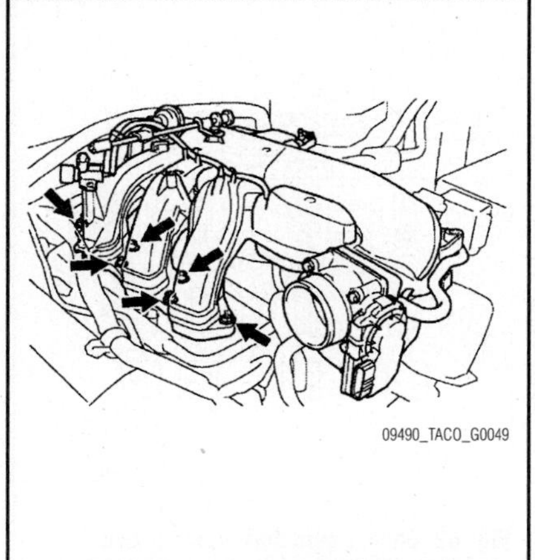

Fig. 54 Intake surge tank bolt locations—4.0L engine

below to prevent this problem from occurring.

16. While raising the chain tensioner number two insert a 0.039 inch diameter pin into the hole to hold it. Hold the hexagonal portion of the number two camshaft with a wrench. Remove the camshaft timing gear set bolt.

17. Separate the camshaft timing gear from the number two camshaft. Rotate the camshafts counterclockwise, using a wrench, so that the cam lobes of the number one cylinder face in the direction shown.

18. Using several steps, loosen and remove the eight bearing cap bolts uniformly and in the proper removal sequence. Remove the four bearing caps and the number two camshaft.

19. Remove the number two chain tensioner bolt. Remove the number chain tensioner and camshaft timing gear.

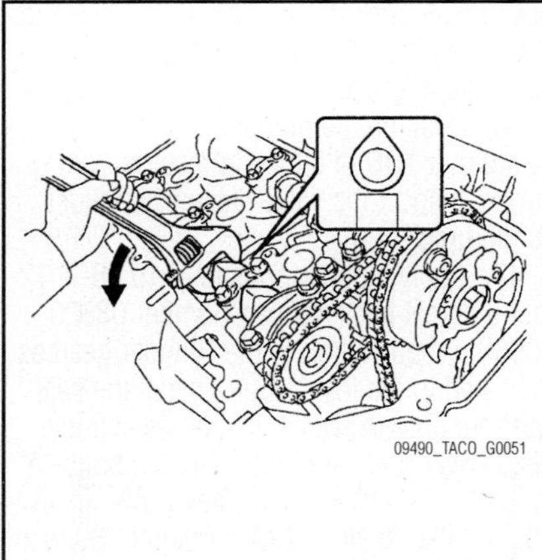

Fig. 55 Bank 1 camshaft lobe removal positioning—4.0L engine

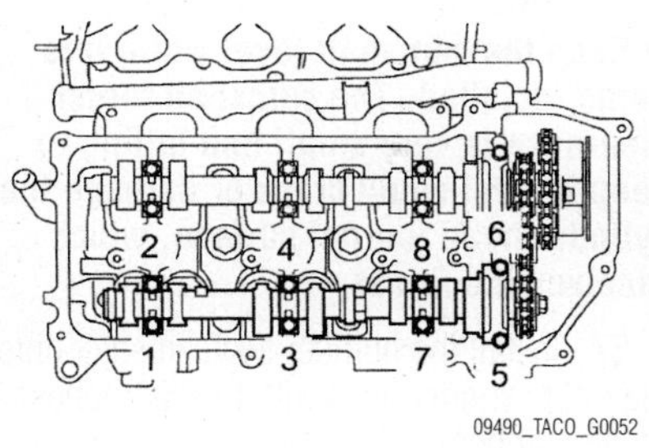

Fig. 56 Bank 1 camshaft number two bearing cap bolt removal sequence—4.0L engine

➡Keep the camshaft level while it is being removed. The camshaft thrust clearance is very small and failing to keep it level could crack or damage the cylinder head journal surface, which receives the thrust. Follow the steps below to prevent this problem from occurring.

20. Hold the hexagonal portion of the number one camshaft, with a wrench. Loosen the camshaft timing gear set bolt.

➡Do not disassemble the camshaft timing gear assembly.

21. Slide the camshaft timing gear and separate the number one chain from the camshaft timing gear.

22. Rotate the number one camshaft counterclockwise, using the wrench so that the cam lobes on the number one cylinder face downward.

23. Using several steps, loosen and remove the eight bearing cap bolts uniformly and in the proper removal sequence. Remove the four bearing caps.

24. Remove the camshaft timing gear set bolt with the number one camshaft lifted up. Remove the number one camshaft and camshaft timing gear with the number two chain.

25. Tie the number one chain to the side. Be careful not to drop anything inside the timing chain cover.

To install:

➡Be sure to use new fasteners, as required.

➡Keep the camshaft level while it is being installed. The camshaft thrust clearance is very small and failing to keep it level could crack or damage the

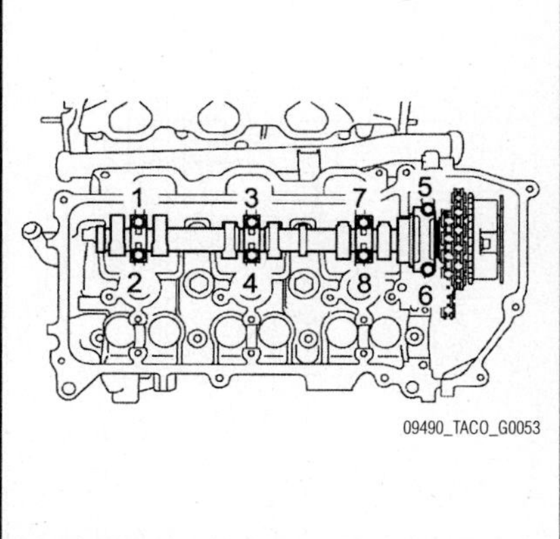

Fig. 57 Bank 1 camshaft number one bearing cap bolt removal sequence—4.0L engine

cylinder head journal surface, which receives the thrust.

26. Align the yellow mark link with the timing mark (1 dot mark) of the camshaft timing gear. Apply new engine oil to the thrust portion and journal of the camshafts.

27. Temporarily install the number one chain onto the number two chain of the camshaft timing gear.

28. Align the knock pin hole of the camshaft timing gear with the knock pin of the number one camshaft. Insert the number one camshaft into the camshaft timing gear.

29. Temporarily install the camshaft timing gear set bolt. Install the number one camshaft onto the right cylinder head with the cam lobes of the number one cylinder facing downward, as indicated in the illustration.

30. Install the four bearing caps, in the proper location. Apply a light coat of engine oil to the threads and under the heads of the cap bolts.

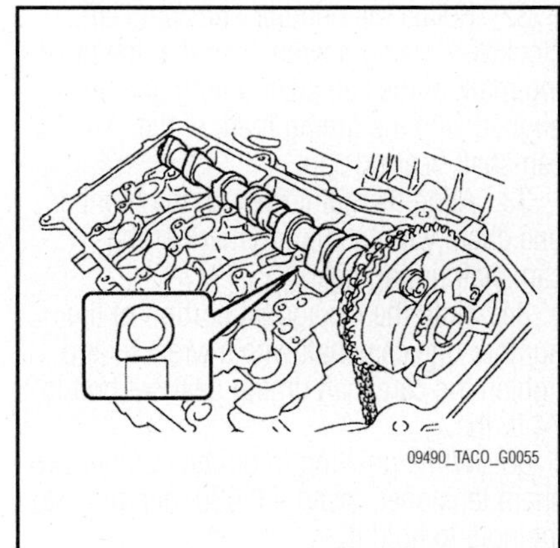

Fig. 58 Bank 1 camshaft number one lobe installation positioning—4.0L engine

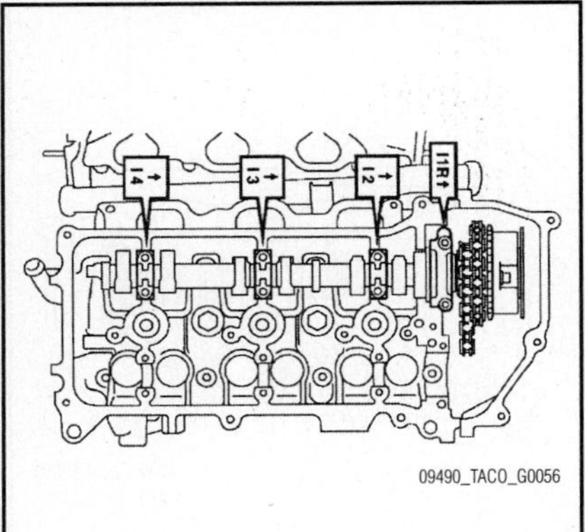

Fig. 59 Bank 1 camshaft number one bearing cap bolt installation sequence—4.0L engine

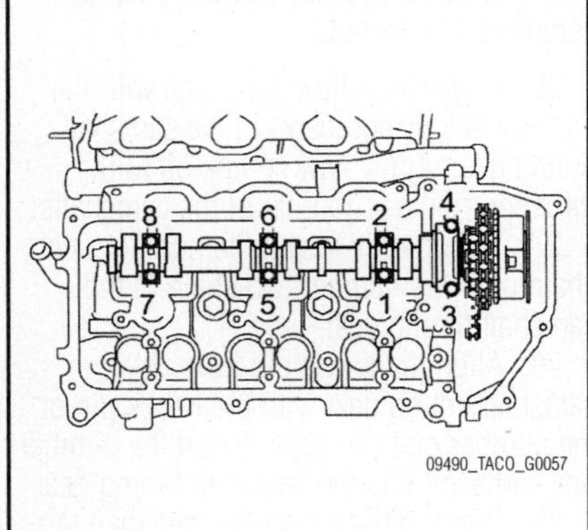

Fig. 60 Bank 1 camshaft number one bearing cap bolt torque sequence—4.0L engine

31. Using several steps, uniformly install and tighten the bearing cap bolts in the proper sequence to 80 inch lbs. (9 Nm) for the 10mm bolts and 18 ft. lbs. (24 Nm) for the 12mm bolts.

32. Rotate the number one camshaft clockwise, using a wrench so that the timing mark of the camshaft timing gear is aligned with the timing mark of the camshaft bearing cap.

33. Align the paint mark of the number one chain with the timing mark of the camshaft timing gear.

34. Hold the hexagonal portion of the number one camshaft with a wrench, and tighten the camshaft timing gear set bolt to 74 ft. lbs.

35. While pushing in on the number two chain tensioner, insert a 0.039 inch pin into the hole to hold it.

36. Temporarily install the camshaft timing gear and chain tensioner number two and align the yellow mark links with the timing marks (1 dot mark) on the camshaft timing gears. Tighten the bolt to 14 ft. lbs. (19 Nm).

➡Keep the camshaft level while it is being installed. The camshaft thrust clearance is very small and failing to keep it level could crack or damage the cylinder head journal surface, which receives the thrust.

37. Install the number two camshaft onto the right cylinder head with the cam lobes of the number one cylinder facing upward, as indicated in the illustration.

38. Install the four bearing caps in the proper location. Apply a light coat of clean engine oil to the threads and under the heads of the bolts.

39. Using several steps, uniformly install and tighten the eight bearing cap bolts in sequence to 80 inch lbs. (9 Nm), for the 10mm head bolts and 18 ft. lbs. (25 Nm) for the 12mm head bolts.

40. Rotate the number two camshaft clockwise, using a wrench, so that the lock pin of the number two camshaft is aligned with the knock pin hole of the camshaft timing gear.

41. Hold the hexagonal portion of the number two camshaft, with a wrench, and install the camshaft timing gear set bolt and tighten it to 74 ft. lbs. (100 Nm).

42. Remove the pin from the number two chain tensioner.

43. While turning the stopper plate of the tensioner clockwise, push in the plunger of the tensioner. While turning the stopper plate of the tensioner counterclockwise, insert a 0.138 inch bar into the holes in the stopper plate and tensioner to hold the stopper plate. Install the two chain tensioner bolts and tighten to 7.4 ft. lbs. (10 Nm).

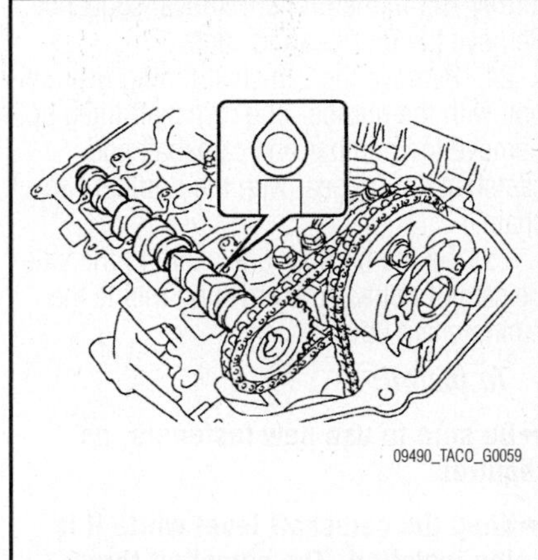

Fig. 61 Bank 1 camshaft number two lobe installation positioning—4.0L engine

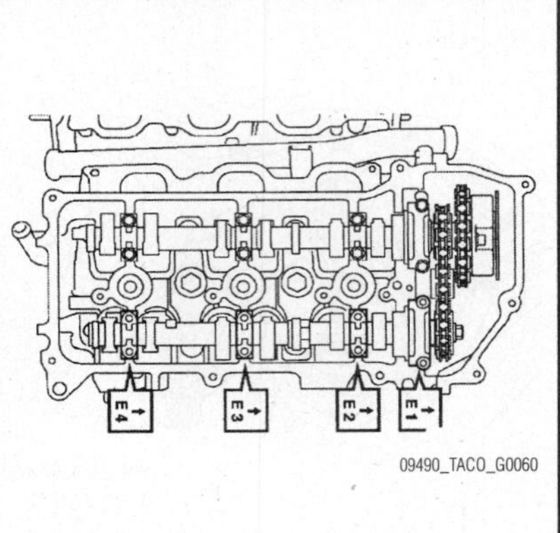

Fig. 62 Bank 1 camshaft number two bearing cap bolt installation sequence—4.0L engine

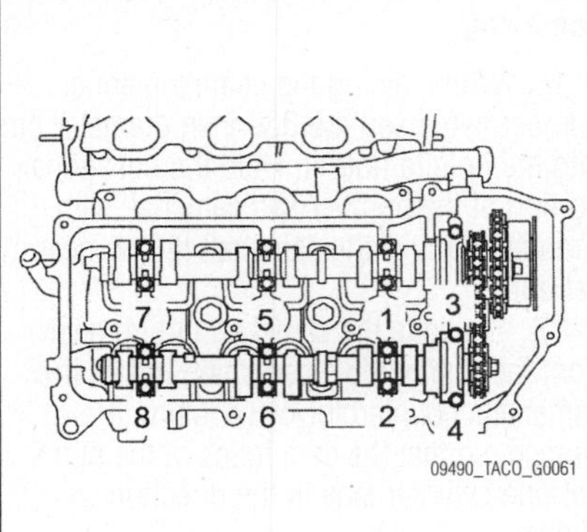

Fig. 63 Bank 1 camshaft number two bearing cap bolt torque sequence—4.0L engine

44. Remove the bar from the chain tensioner. Install a new gasket and the timing chain cover plate. Torque the bolts to 80 inch lbs.

45. Turn the crankshaft pulley two complete revolutions slowly until its groove and the timing mark "0" of the timing chain cover are aligned.

46. Position the number one cylinder at TDC of the compression stroke. Inspect the valve clearance, adjust as required.

47. Apply a continuous bead (0.08–0.12 inch) of seal packing, part number 08826-00080 or equivalent, to the cylinder head as indicated in the illustration. Install the seal washers onto the bolts. Install the cylinder head cover bolts and nuts. Tighten bolts "A" to 7.4 ft. lbs. (10 Nm). Tighten bolts "B" to 80 inch lbs. (9 Nm). Tighten nuts to 80 inch lbs. (9 Nm).

➡Be sure to remove any oil from the contact surfaces of the cylinder head

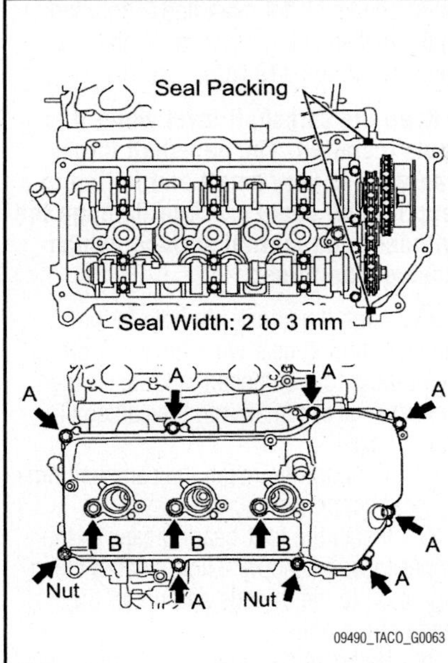

Fig. 64 Bank 1 cylinder head cover bolt sealant application and bolt identification—4.0L engine

cover and the cylinder head. Install the cover within three minutes after applying the seal packing. Tighten the bolts to specification within fifteen minutes after installing the cover. Do not add engine oil for at least two hours after installing the cover.

48. Continue the installation in the reverse order of the removal procedure.
49. Be sure to fill the engine with the proper grade and type engine coolant.
50. Be sure to fill the engine with the proper grade and type engine oil.
51. Start the engine and check for leaks. Correct as required.

BANK 2

See Figures 65 through 72.

1. Before servicing the vehicle, refer to the Precautions Section.
2. Disconnect the negative battery cable. Tape the cable with insulating tape.
3. Drain the engine coolant. Remove the V bank cover.
4. Remove the air cleaner assembly.
5. Disconnect the two water bypass hoses. Disconnect the fuel vapor feed hose. Disconnect the ventilation hose. Disconnect the VSV connectors.
6. Disconnect the throttle body motor connector. Separate the three wire harness clamps and hose clamp.
7. If equipped with manual transmission, remove the nut, then separate the clutch flexible hose bracket from the surge tank stay.
8. Remove the two bolts and the throttle body bracket. Remove the bolt and the oil baffle plate. Remove the four bolts and the two serge tank stays.
9. Remove the two nuts. Remove the four bolts, intake air surge tank and gasket.
10. Remove the ignition coil assembly.
11. Remove the cylinder head cover retaining bolts. Remove the cylinder head cover.
12. Turn the crankshaft pulley until its groove and the timing mark "0" of the timing chain cover are aligned. If not aligned at TDC of the compression stroke, turn the crankshaft one complete revolution, in the direction of rotation. Paint alignment marks on the number one chain links corresponding to the timing marks of the camshaft timing gears.
13. While turning the stopper plate of the tensioner upward, push in the plunger of the chain tensioner. While turning the stopper plate of the tensioner downward, insert a 0.138 inch bar into the holes in the stopper plate and tensioner to hold the stopper plate. Remove the two bolts and then remove the number one chain tensioner assembly.

➡Never rotate the crankshaft with the chain tensioner removed. When rotating the camshaft with the tensioner removed, rotate the crankshaft counterclockwise forty degrees from TDC, first.

➡Keep the camshaft level while it is being removed. The camshaft thrust clearance is very small and failing to keep it level could crack or damage the cylinder head journal surface, which receives the thrust. Follow the steps below to prevent this problem from occurring.

14. While pushing down on chain tensioner number three insert a 0.039 inch diameter pin into the hole to hold it. Hold the hexagonal portion of the number four camshaft with a wrench. Remove the camshaft timing gear set bolt.
15. Separate the camshaft timing gear from the number four camshaft.
16. Using several steps, loosen and remove the eight bearing cap bolts uniformly and in the proper removal sequence. Remove the four bearing caps and the number four camshaft.
17. Remove the number three chain tensioner bolt. Remove the number chain tensioner and camshaft timing gear.

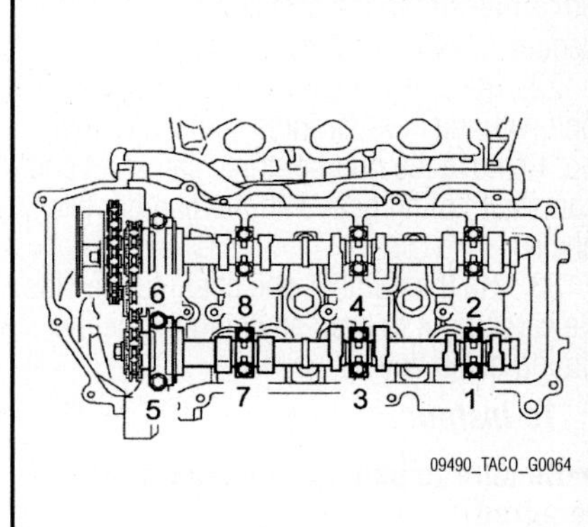

Fig. 65 Bank 2 camshaft number four bearing cap bolt removal sequence—4.0L engine

➡Keep the camshaft level while it is being removed. The camshaft thrust clearance is very small and failing to keep it level could crack or damage the cylinder head journal surface, which receives the thrust. Follow the steps below to prevent this problem from occurring.

18. Release the chain tension between the camshaft gear (LH bank) and the crankshaft timing gear by turning the crankshaft pulley counterclockwise slightly.
19. Hold the hexagonal portion of the number three camshaft, with a wrench. Loosen the camshaft timing gear set bolt.

➡Do not disassemble the camshaft timing gear assembly.

20. Slide the camshaft timing gear and separate the number one chain from the camshaft timing gear.
21. Using several steps, loosen and remove the eight bearing cap bolts

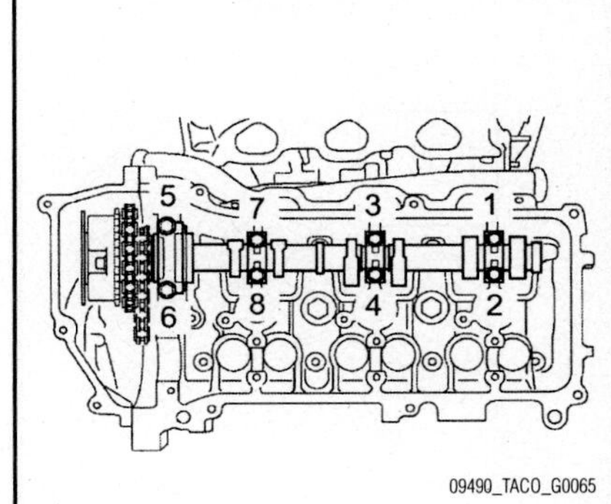

Fig. 66 Bank 2 camshaft number three bearing cap bolt removal sequence—4.0L engine

uniformly and in the proper removal sequence. Remove the four bearing caps.

22. Remove the camshaft timing gear set bolt with the number three camshaft lifted up. Remove the number three camshaft and camshaft timing gear with the number two chain.

23. Tie the number one chain to the side. Be careful not to drop anything inside the timing chain cover.

To install:

➡Be sure to use new fasteners, as required.

➡Keep the camshaft level while it is being installed. The camshaft thrust clearance is very small and failing to keep it level could crack or damage the cylinder head journal surface, which receives the thrust.

24. Align the yellow mark link with the timing mark (2 dot mark) of the camshaft timing gear. Apply new engine oil to the thrust portion and journal of the camshafts.

25. Temporarily install the number one chain onto the number two chain of the camshaft timing gear.

26. Align the knock pin hole of the camshaft timing gear with the knock pin of the number three camshaft. Insert the number three camshaft into the camshaft timing gear.

27. Temporarily install the camshaft timing gear set bolt. Install the number three camshaft onto the left cylinder head with the cam lobes of the number two cylinder facing downward, as indicated in the illustration.

28. Install the four bearing caps, in the proper location. Apply a light coat of engine oil to the threads and under the heads of the cap bolts.

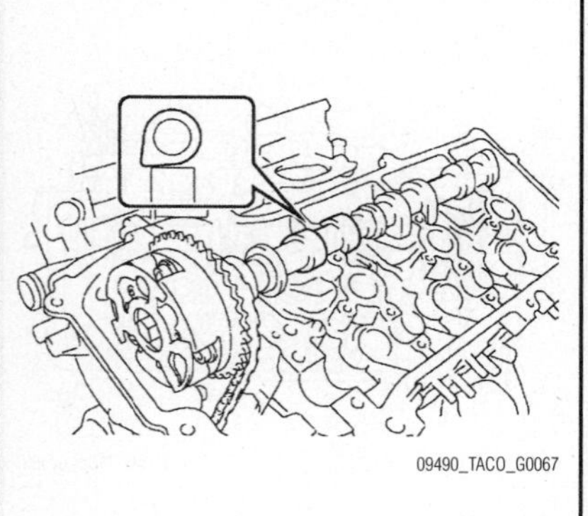

Fig. 67 Bank 2 camshaft number three lobe installation positioning—4.0L engine

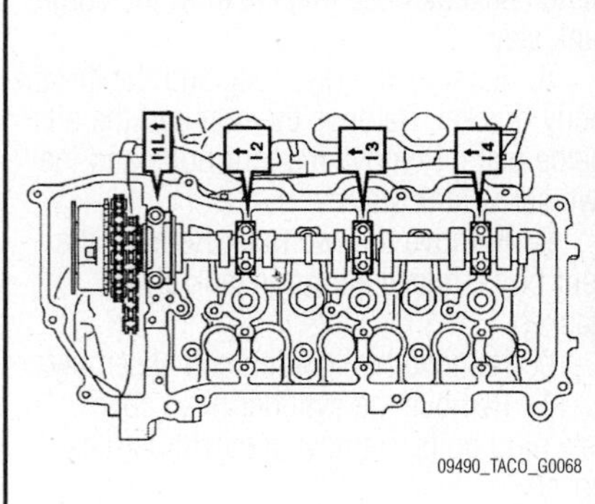

Fig. 68 Bank 2 camshaft number three bearing cap bolt installation sequence—4.0L engine

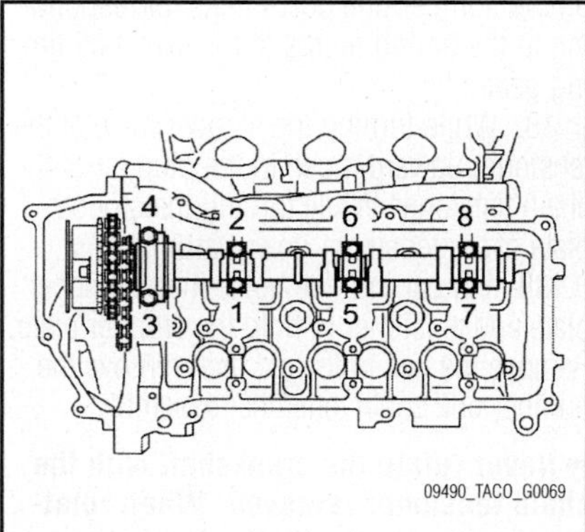

Fig. 69 Bank 2 camshaft number three bearing cap bolt torque sequence—4.0L engine

29. Using several steps, uniformly install and tighten the bearing cap bolts in the proper sequence to 80 inch lbs. (9 Nm) for the 10mm bolts and 18 ft. lbs. (25 Nm) for the 12mm bolts.

30. Rotate the number one camshaft clockwise, using a wrench so that the timing mark of the camshaft timing gear is aligned with the timing mark of the camshaft bearing cap.

31. Align the paint mark of the number one chain with the timing mark of the camshaft timing gear.

32. Hold the hexagonal portion of the number three camshaft with a wrench, and tighten the camshaft timing gear set bolt to 74 ft. lbs. (100 Nm)

33. While pushing in on the number three chain tensioner, insert a 0.039 inch pin into the hole to hold it.

34. Temporarily install the camshaft timing gear and chain tensioner number three and align the yellow mark links with the timing marks (1 dot mark and 2 dot marks) on the camshaft timing gears. Tighten the bolt to 14 ft. lbs. (19 Nm).

➡Keep the camshaft level while it is being installed. The camshaft thrust clearance is very small and failing to keep it level could crack or damage the cylinder head journal surface, which receives the thrust.

35. Align the knock pin hole in the camshaft timing gear with the knock pin of the number four camshaft, and insert the number four camshaft into the camshaft timing gear.

36. Temporarily install the camshaft timing gear set bolt.

37. Install the four bearing caps in the proper location. Apply a light coat of clean engine oil to the threads and under the heads of the bolts.

38. Using several steps, uniformly install and tighten the eight bearing cap bolts in sequence to 80 inch lbs. (9 Nm), for the 10mm head bolts and 18 ft. lbs. (25 Nm) for the 12mm head bolts.

39. Hold the hexagonal portion of the number four camshaft, with a wrench, and install the camshaft timing gear set bolt and tighten it to 74 ft. lbs. (100 Nm).

40. Remove the pin from the number three chain tensioner.

41. Release the chain tension between the camshaft timing gear (RH bank) and the crankshaft timing gear by turning the crankshaft pulley clockwise slightly.

42. While turning the stopper plate of the tensioner clockwise, push in the plunger of the tensioner. While turning the stopper plate of the tensioner counterclockwise, insert a 0.138 inch bar into the holes in the stopper plate and tensioner to

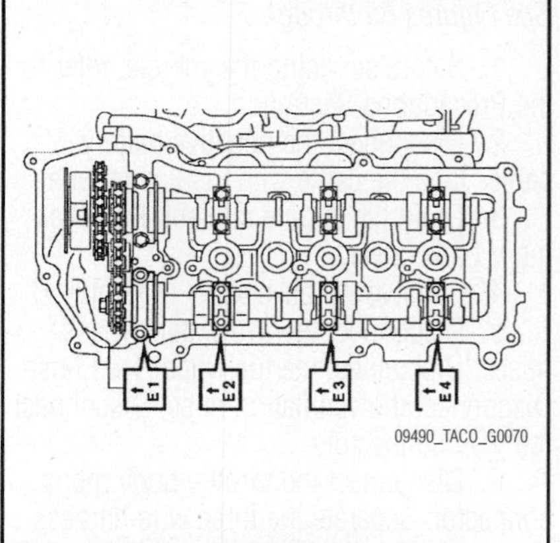

Fig. 70 Bank 2 camshaft number four bearing cap bolt installation sequence—4.0L engine

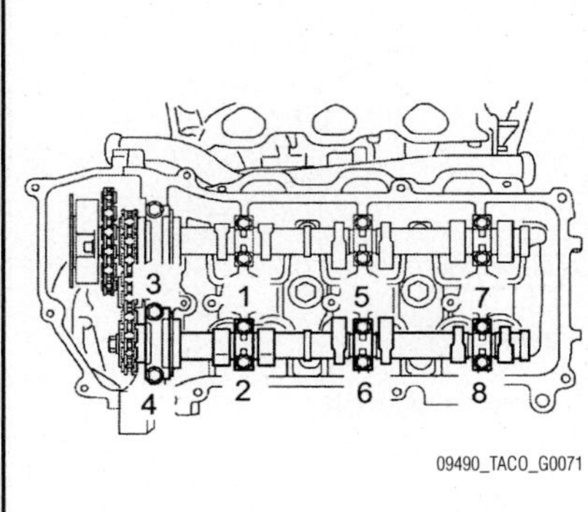

Fig. 71 Bank 2 camshaft number four bearing cap bolt torque sequence—4.0L engine

hold the stopper plate. Install the two chain tensioner bolts and tighten to 7.4 ft. lbs. Remove the bar from the chain tensioner.

43. Position the number one cylinder at TDC of the compression stroke. Inspect the valve clearance, adjust as required.

44. Apply a continuous bead (0.08–0.12 inch) of seal packing, part number 08826-00080 or equivalent, to the cylinder head as indicated in the illustration. Install the seal washers onto the bolts. Install the cylinder head cover bolts and nuts. Tighten bolts "A" to 7.4 ft. lbs. (10 Nm). Tighten bolts "B" to 80 inch lbs. (9 Nm). Tighten nuts to 80 inch lbs. (9 Nm).

➡Be sure to remove any oil from the contact surfaces of the cylinder head cover and the cylinder head. Install the cover within three minutes after applying the seal packing. Tighten the bolts to specification within fifteen minutes after installing the cover. Do not add engine oil for at least two hours after installing the cover.

45. Continue the installation in the reverse order of the removal procedure.

46. Be sure to fill the engine with the proper grade and type engine coolant.

47. Be sure to fill the engine with the proper grade and type engine oil.

48. Start the engine and check for leaks. Correct as required.

CATALYTIC CONVERTER

REMOVAL & INSTALLATION

See Figures 73 through 75.

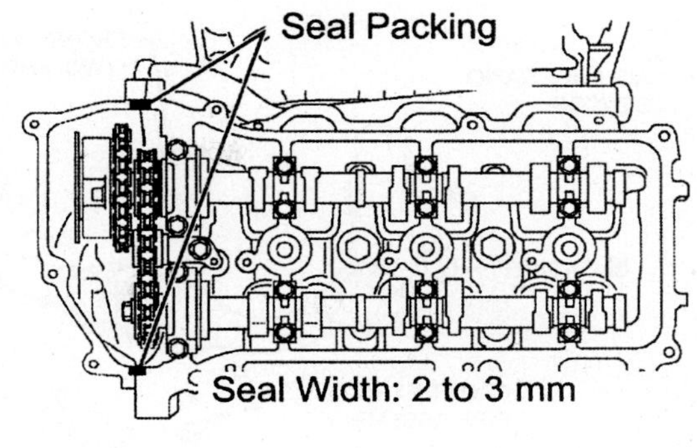

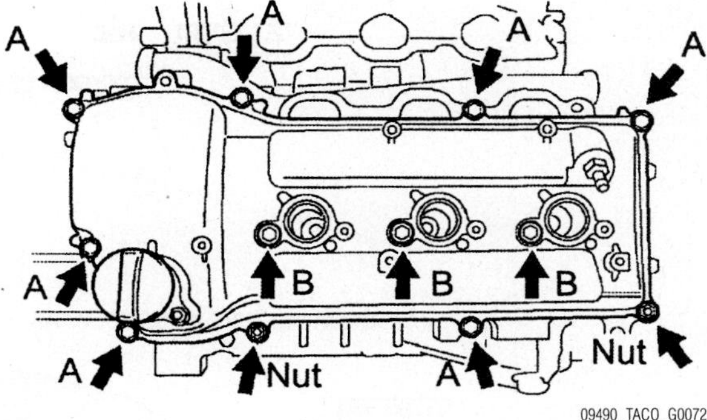

Fig. 72 Bank 2 cylinder head cover bolt sealant application and bolt identification—4.0L engine

At this time the manufacturer does not provide removal and installation procedures for this component. The following procedure is a guideline and may differ from the vehicle you are servicing.

1. Before servicing the vehicle, refer to the Precautions Section.
2. Disconnect the negative battery cable. Tape the cable with insulating tape.
3. Raise and safely support the vehicle.
4. Remove the necessary components in order to gain access to the converter.
5. Disconnect the oxygen sensors, as necessary. Do not drop the sensor, if dropped it must be replaced.
6. Remove the converter mounting bolts.

➡Be sure to properly support the exhaust system.

7. Remove the converter from the vehicle.
8. Discard the gaskets.

To install:

➡Be sure to use new fasteners, as required.

9. Installation is the reverse of the removal procedure.
10. Be sure to use new gaskets.
11. Be sure to coat the sensor with the proper sealer, if reusing the existing one.

CRANKSHAFT DAMPER

REMOVAL & INSTALLATION

1. Before servicing the vehicle, refer to the Precautions Section.
2. Disconnect the negative battery cable. Tape the cable with insulating tape.
3. Remove the engine under cover, if equipped.
4. On 4.0L engines remove the V bank cover. Remove the engine under cover sub assembly No 1.
5. Drain the coolant. Be sure to properly dispose of used coolant.
6. Remove the radiator support to frame seal, left side.
7. Remove the fan shroud.
8. Remove the accessory drive belt.
9. Using SST 09213-54015, fix the pulley and loosen the pulley bolt.
10. Using the pulley set bolt and SST 09950-50013, remove the crankshaft pulley.
11. Discard the bolt.

To install:

➡Be sure to use new fasteners, as required.

12. Installation is the reverse of the removal procedure. Tighten the crankshaft

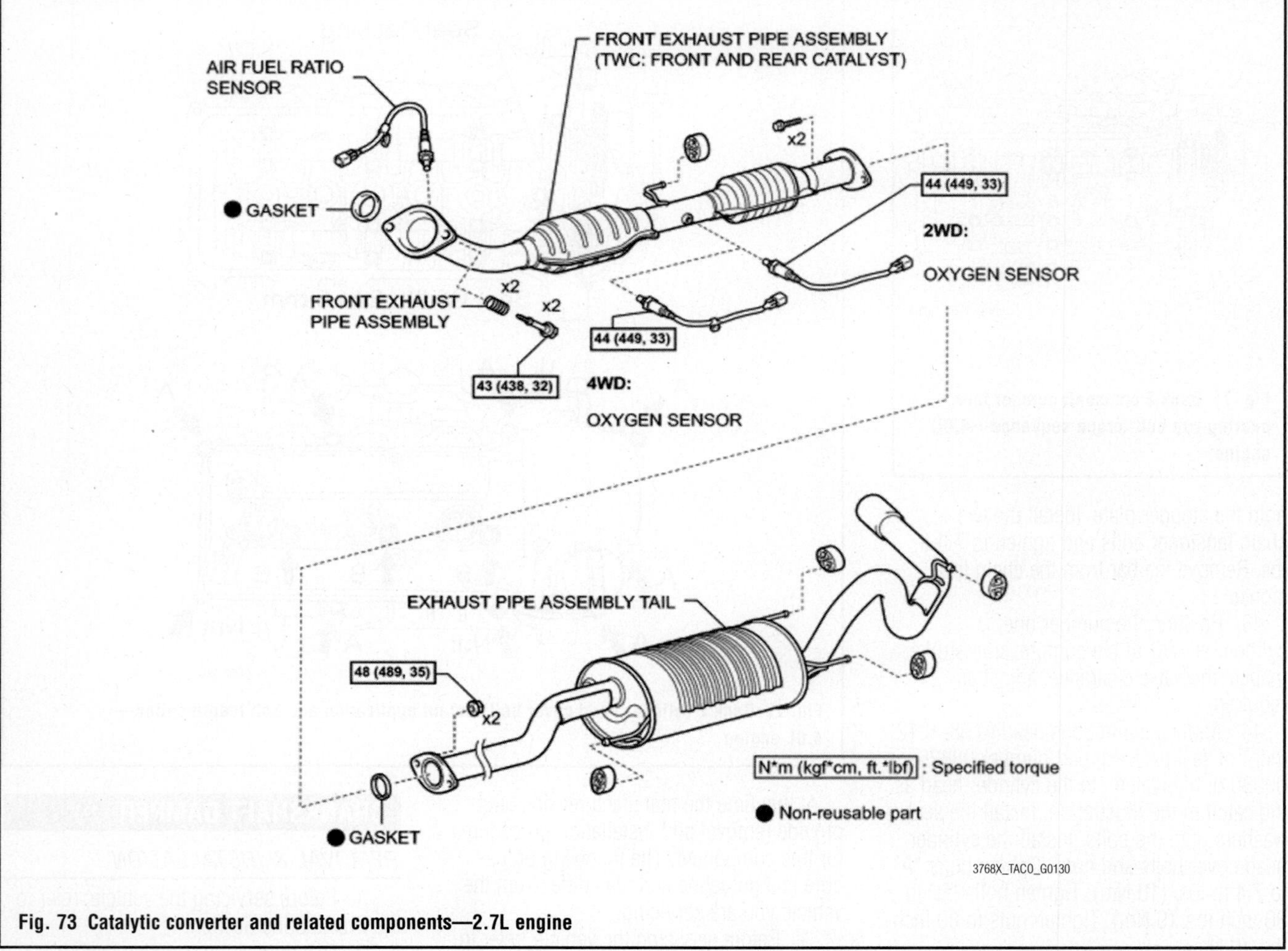

Fig. 73 Catalytic converter and related components—2.7L engine

bolt to 192 ft. lbs (260 Nm). Be sure to replace the bolt.

13. Be sure to fill the cooling system with the proper grade and type coolant.

14. Start the engine and check for leaks. Correct as required.

CRANKSHAFT FRONT SEAL

REMOVAL & INSTALLATION

See Figure 76.

1. Before servicing the vehicle, refer to the Precautions Section.
2. Disconnect the negative battery cable. Tape the cable with insulating tape.
3. Remove the crankshaft pulley. Discard the bolt.
4. Using a cutter knife cut off the lip of the seal.
5. Using a suitable tool, with its tip wrapped in tape, pry out the seal from its mounting.

➡Be careful not to damage the crankshaft, with the suitable tool.

To install:

➡Be sure to use new fasteners, as required.

6. Installation is the reverse of the removal procedure.
7. Be sure to fill the cooling system with the proper grade and type coolant.
8. Start the engine and check for leaks. Correct as required.

ENGINE OIL & FILTER

REPLACEMENT

At this time the manufacturer does not provide removal and installation procedures for this component. The following procedure is a guideline and may differ from the vehicle you are servicing.

1. Before servicing the vehicle, refer to the Precautions Section.
2. Disconnect the negative battery cable. Tape the cable with insulating tape.

➡Use care as engine oil may be hot.

3. Remove the engine oil filler cap.
4. Raise and safely support the vehicle, as required.
5. Remove the undercover, if equipped.
6. Remove the oil drain plug. Discard the gasket.
7. Drain the engine oil into a suitable container. Properly dispose of used engine oil. Be careful if engine oil is hot.
8. Using an oil filter wrench remove the oil filter. Discard the filter.

To install:

➡Be sure to use new fasteners, as required

9. Install the drain plug. Use a new gasket.
10. Tighten to specification.
11. Coat the oil filter seal with clean engine oil prior to installation.
12. Do not over tighten the filter to its mounting.

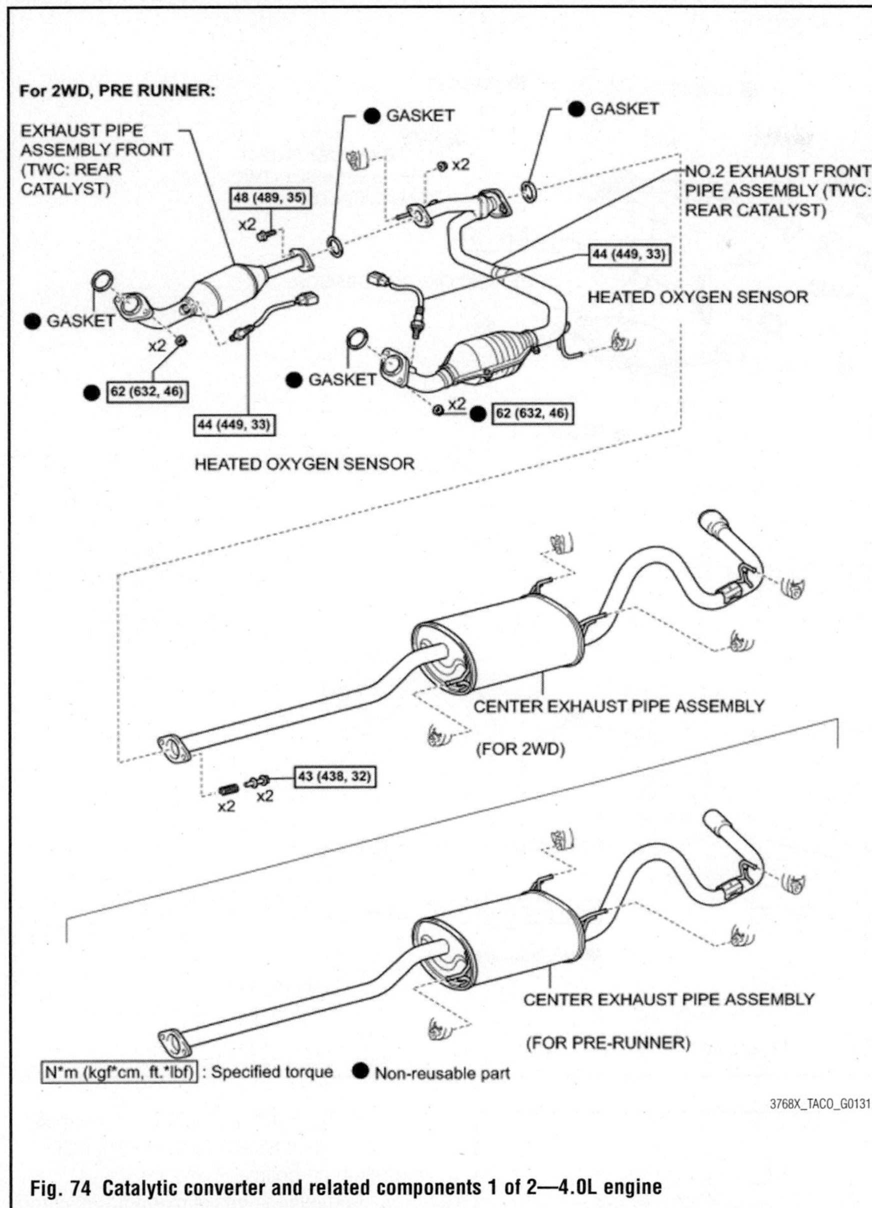

Fig. 74 Catalytic converter and related components 1 of 2—4.0L engine

EXHAUST MANIFOLD

REMOVAL & INSTALLATION

2.7L Engines

See Figure 77.

1. Before servicing the vehicle, refer to the Precautions Section.
2. Disconnect the negative battery cable. Tape the cable with insulating tape.
3. Disconnect the exhaust manifold to exhaust flange nuts.
4. Remove the necessary components to gain access to the exhaust manifold retaining bolts.
5. Remove the exhaust manifold retaining nuts. Discard the nuts. Remove the exhaust manifold from the engine.

To install:

➡Be sure to use new fasteners, as required.

6. Installation is the reverse of the removal procedure.
7. Be sure to use new gaskets. Be sure to use new retaining nuts.

4.0L Engines

See Figure 78.

1. Before servicing the vehicle, refer to the Precautions Section.
2. Disconnect the negative battery cable. Tape the cable with insulating tape.
3. Disconnect the exhaust manifold to exhaust flange nuts.
4. Remove the necessary components to gain access to the exhaust manifold retaining bolts.
5. Remove the exhaust manifold retaining nuts. Discard the nuts. Remove the exhaust manifold from the engine.

To install:

➡Be sure to use new fasteners, as required.

6. Installation is the reverse of the removal procedure.
7. Be sure to use new gaskets. Be sure to use new retaining nuts.

INTAKE MANIFOLD

REMOVAL & INSTALLATION

2.7L Engines

See Figure 79.

1. Before servicing the vehicle, refer to the Precautions Section.
2. Disconnect the negative battery cable. Tape the cable with insulating tape.
3. Properly relieve the fuel system pressure.
4. On 4WD and PreRunner, remove the engine undercover subassembly.
5. Drain the engine coolant. Remove the air intake connector.
6. Remove the throttle body and motor assembly. Disconnect the fuel hoses.
7. Disconnect the fuel vapor feed hose from the VSR. Disconnect the vacuum hose. Remove the bolt, and then remove the clamp bracket.
8. Disconnect the number two water bypass hose. Disconnect the number three ventilation hose. Disconnect the VSV connector.
9. Disengage the engine wire harness clamp. Disconnect the air conditioning compressor magnetic clutch connector.
10. Disengage the wire harness clamp. Remove the bolt and harness clamp bracket. Disconnect the three connectors.
11. Remove the retaining nut inside the relay block. Disconnect the engine wire harness from the relay block.
12. Remove the five bolts and two nuts retaining the intake manifold in place. Remove the intake manifold from the engine.

To install:

➡Be sure to use new fasteners, as required.

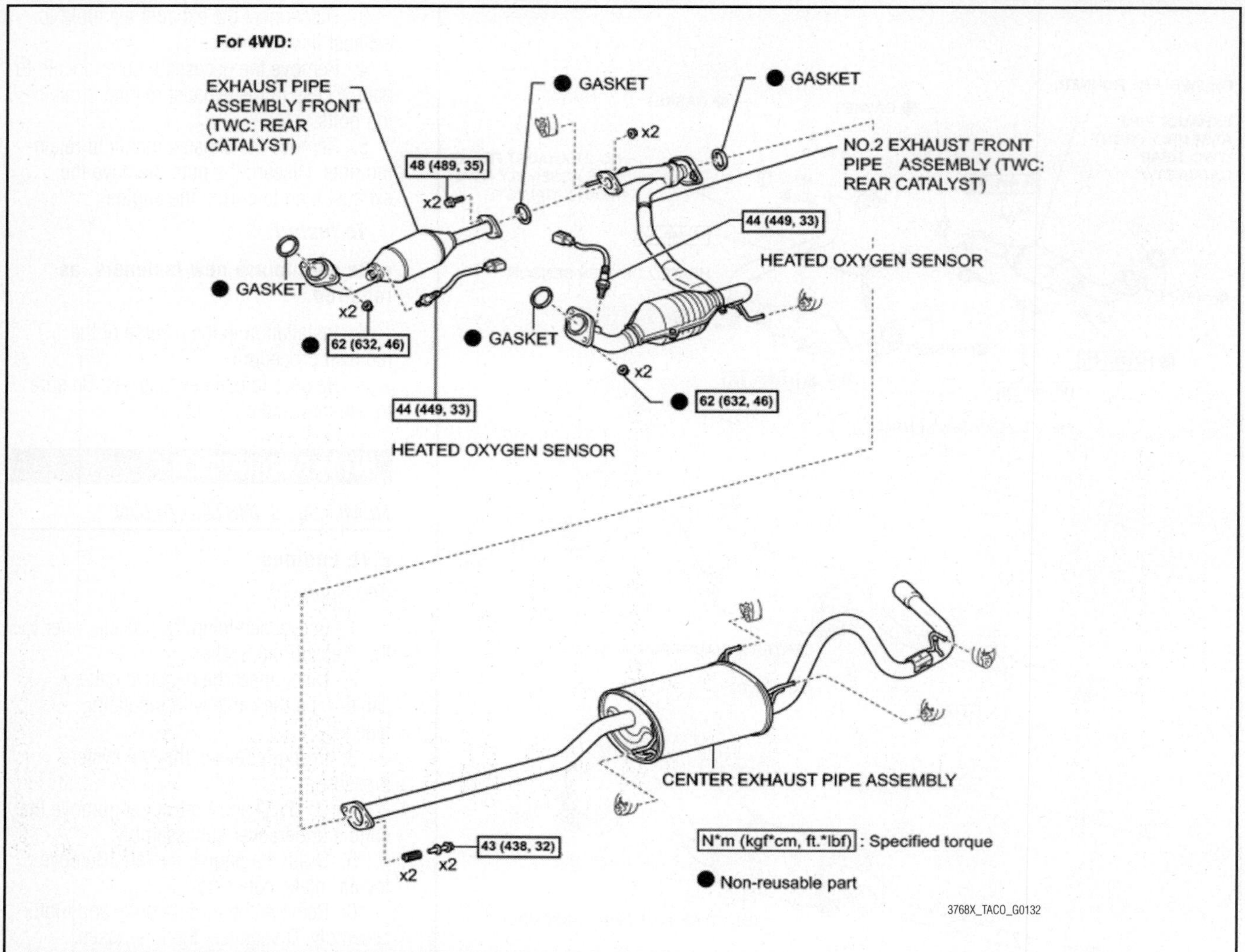

Fig. 75 Catalytic converter and related components 2 of 2—4.0L engine

13. Clean all surfaces.
14. Install a new gasket onto the intake manifold. Position the intake manifold to the engine.
15. Install the retaining bolts and tighten to specification, in an alternating sequence.
16. Continue the installation in the reverse order of the removal procedure.
17. Be sure to fill the cooling system with the proper grade and type engine coolant.
18. Start the engine and check for leaks. Correct as required.

4.0L Engines

See Figure 80.

1. Before servicing the vehicle, refer to the Precautions Section.
2. Disconnect the negative battery cable. Tape the cable with insulating tape.
3. Properly relieve the fuel system pressure.

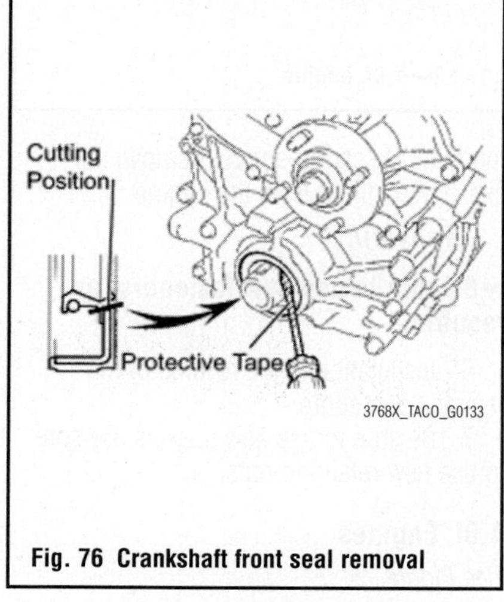

Fig. 76 Crankshaft front seal removal

4. Drain the engine coolant. Remove the air cleaner assembly.
5. Disconnect the fuel injector wiring connectors.
6. Remove the necessary components in order to gain access to the intake manifold retaining bolts.
7. Remove the intake manifold retaining bolts. Remove the intake manifold from the engine.

To install:

➡Be sure to use new fasteners, as required.

8. Clean all surfaces.
9. Install a new gasket on each cylinder head.

➡Align the ports of the gasket and the cylinder head. Be careful of the installation direction. Position the intake manifold to the engine.

10. Install the retaining bolts and tighten to specification, in an alternating sequence.
11. Continue the installation in the reverse order of the removal procedure.

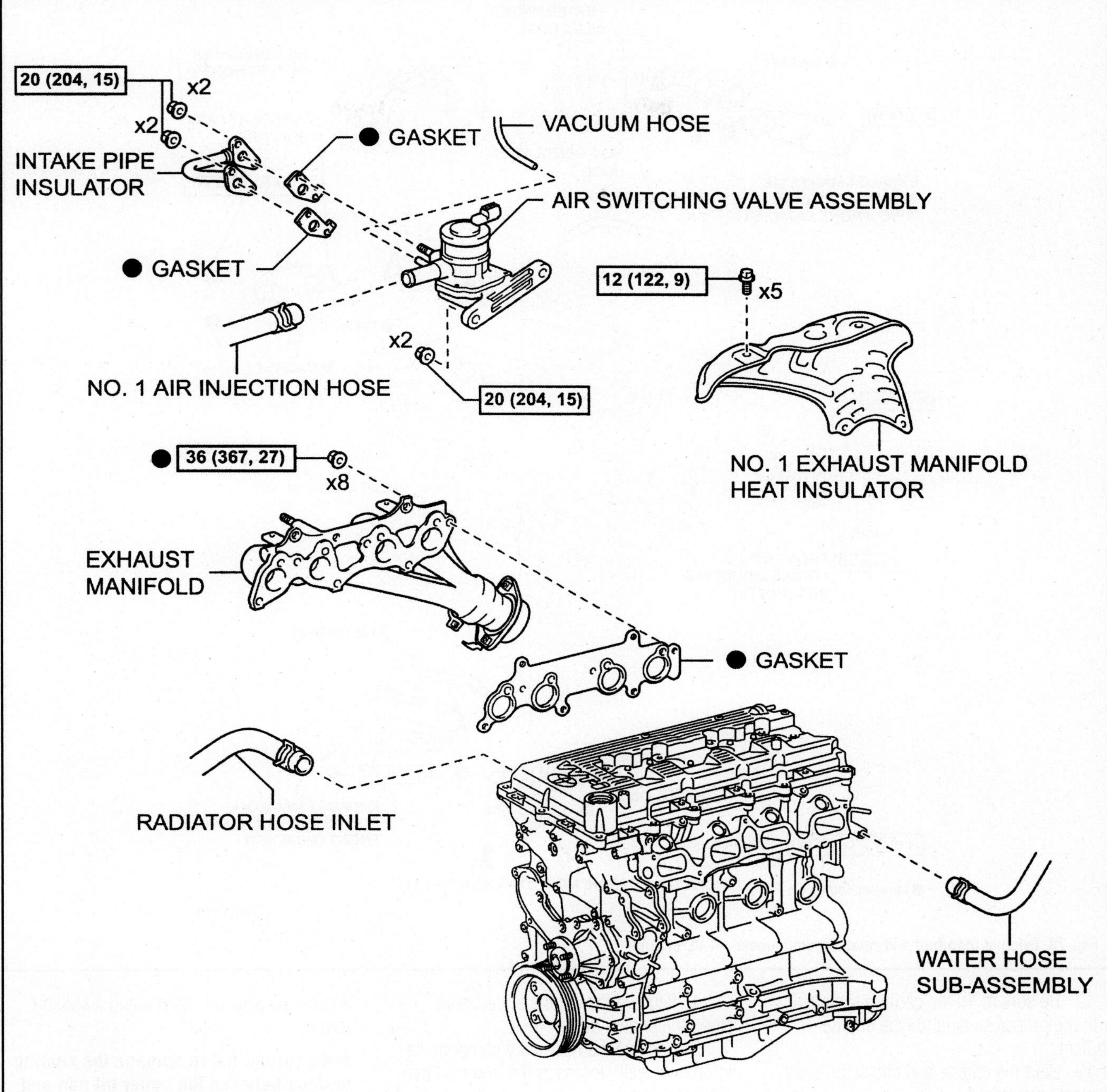

Fig. 77 Exhaust manifold and related components—2.7L engine

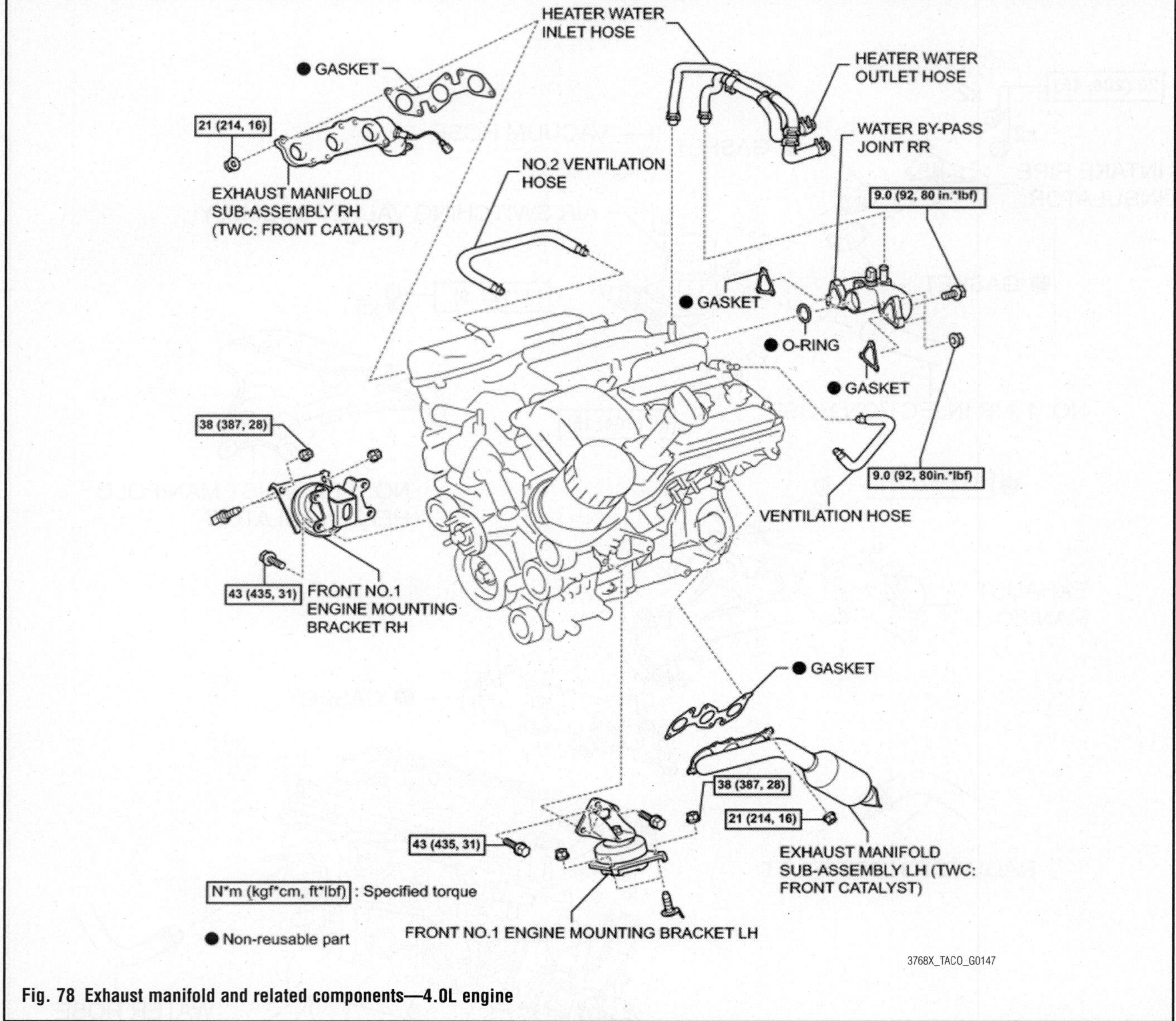

Fig. 78 Exhaust manifold and related components—4.0L engine

12. Be sure to fill the cooling system with the proper grade and type engine coolant.
13. Start the engine and check for leaks. Correct as required.

OIL PAN

REMOVAL & INSTALLATION

2.7L Engines

See Figures 81 through 84.

1. Before servicing the vehicle, refer to the Precautions Section.
2. Disconnect the negative battery cable. Tape the cable with insulating tape.
3. Raise and support the vehicle safely.
4. Remove the engine undercover. Drain the engine oil.
5. Remove the necessary components in order to gain access to the lower oil pan retaining bolts.
6. Remove the eighteen bolts and two nuts. Insert the blade of tool SST09032-00100 between the pans. Cut through the sealer and remove the lower oil pan from the engine.

➡Be careful not to damage the contact surface of the oil pans.

7. Remove the two bolts and nuts. Remove the oil strainer. Discard the gasket.
8. To remove the upper oil pan, remove the sixteen bolts and two nuts. Remove the upper oil pan from the engine, by prying it apart using a suitable tool.

➡Be careful not to damage the sealing surface between the upper oil pan and the cylinder block.

To install:

➡Be sure to use new fasteners, as required.

9. Apply a continuous bead (0.079–0.118 inch in diameter) of seal packing, part number 08826-00080 or equivalent, to the sealing surface of the oil pan.

➡Remove any oil from the contact surface. Install the upper oil pan within three minutes of applying the seal

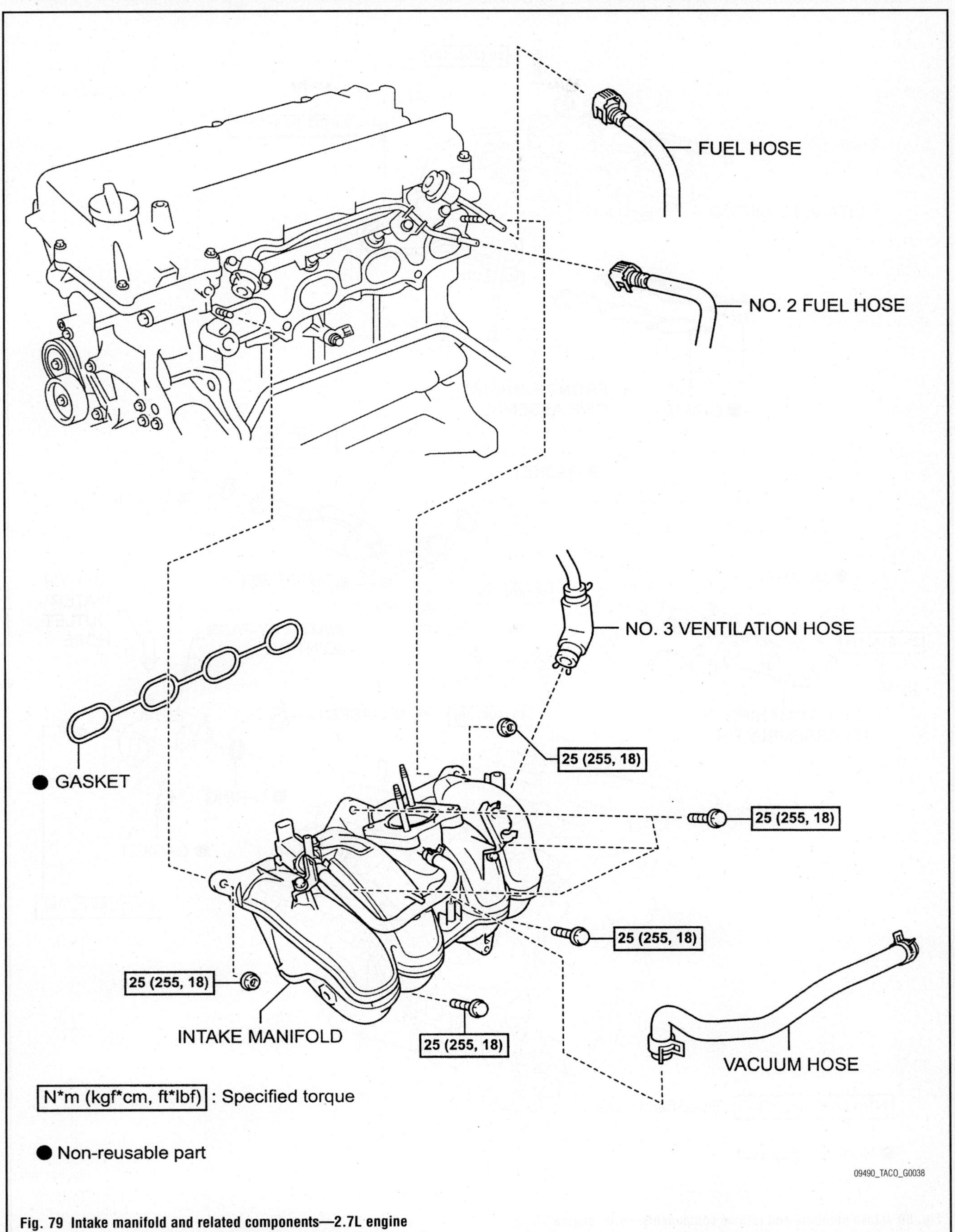

Fig. 79 Intake manifold and related components—2.7L engine

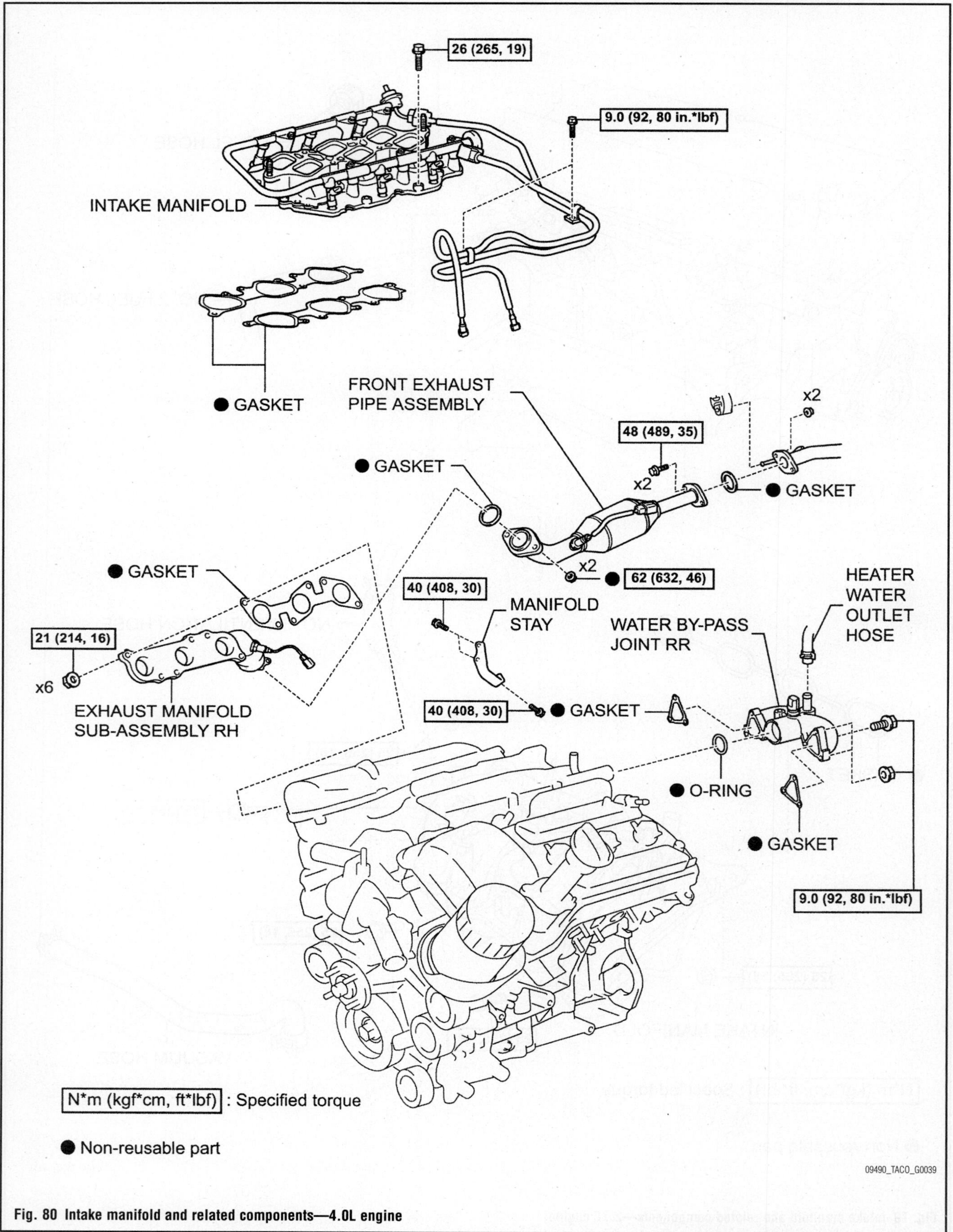

Fig. 80 Intake manifold and related components—4.0L engine

packing. Do not start the engine for at least two hours after the installation of the oil pan.

10. Loosely install the upper oil pan bolts and nuts. Bolt "A" is 0.79 inch long and bolt "B" is 1.57 inch long. Uniformly tighten the bolts to 19 ft. lbs, in the proper sequence.

11. Install the oil strainer assembly. Torque the bolts to 19 ft. lbs.

12. Apply a continuous bead (0.118–0.157 inch in diameter) of seal packing, part number 08826-00080 or equivalent, to the sealing surface of the oil pan.

➡Remove any oil from the contact surface. Install the lower oil pan within three minutes of applying the seal packing. Do not start the engine for at least two hours after the installation of the oil pan.

13. Loosely install the lower oil pan bolts and nuts. Uniformly tighten the bolts to 80 inch lbs, in the proper sequence.

14. Continue the installation in the reverse order of the removal procedure.

15. Be sure to fill the engine with the proper grade and type engine coolant.

16. Be sure to fill the engine with the proper grade and type engine oil.

17. Start the engine and check for leaks. Correct as required.

4.0L Engines

See Figures 85 through 87.

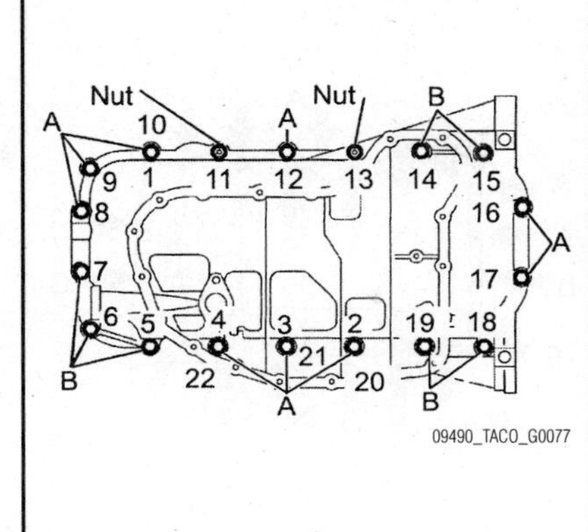

Fig. 82 Upper oil pan bolt torque sequence—2.7L engine

1. Before servicing the vehicle, refer to the Precautions Section.

2. Disconnect the negative battery cable. Tape the cable with insulating tape.

3. Raise and support the vehicle safely.

4. Remove the engine undercover. Drain the engine oil.

5. Remove the necessary components in order to gain access to the lower oil pan retaining bolts.

6. Remove the fifteen bolts and two nuts (2WD vehicles) and ten bolts and two nuts (4WD and PreRunner). Insert the blade of tool SST09032-00100 between the pans. Cut through the sealer and remove the lower oil pan from the engine.

➡Be careful not to damage the contact surface of the oil pans.

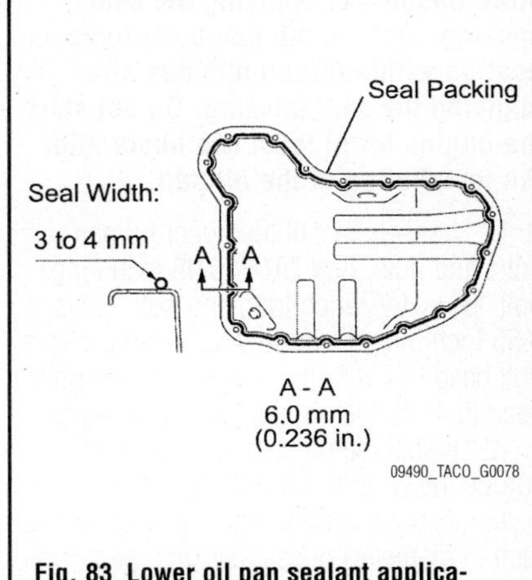

Fig. 83 Lower oil pan sealant application—2.7L engine

7. Remove the two bolts and nuts. Remove the oil strainer. Discard the gasket.

8. On 4WD and PreRunner, remove the four housing bolts. Remove the flywheel housing undercover.

9. To remove the upper oil pan, remove the seventeen bolts and two nuts. Remove the upper oil pan from the engine, by prying it apart using a suitable tool.

➡Be careful not to damage the sealing surface between the upper oil pan and the cylinder block.

To install:

➡Be sure to use new fasteners, as required.

10. Apply a continuous bead (0.12–0.16 inch in diameter) of seal packing, part number 08826-00080 or equivalent, to the sealing surface of the oil pan.

➡Remove any oil from the contact surface. Install the upper oil pan within

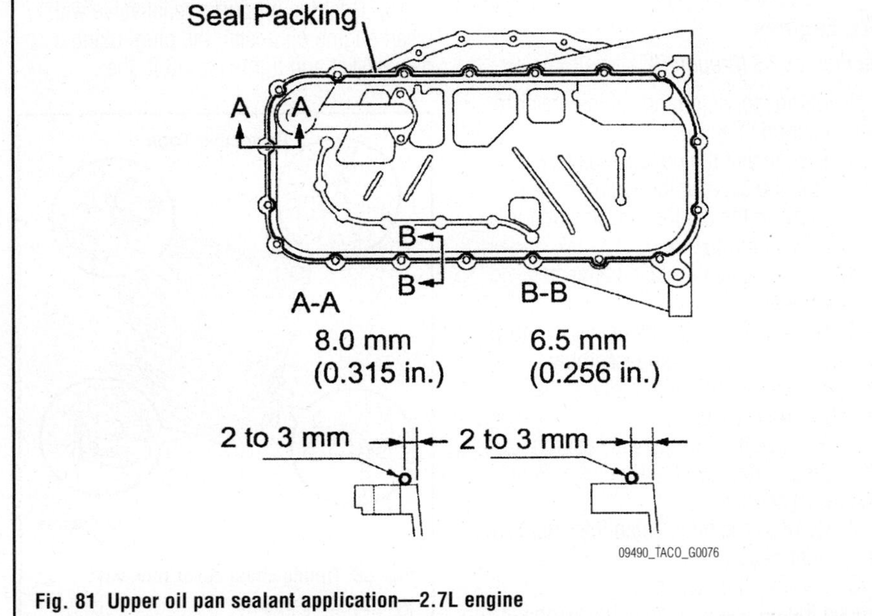

Fig. 81 Upper oil pan sealant application—2.7L engine

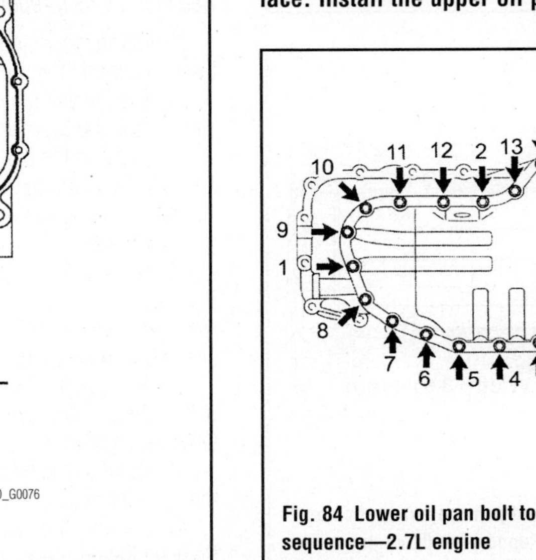

Fig. 84 Lower oil pan bolt torque sequence—2.7L engine

three minutes of applying the seal packing. Tighten the pan bolts to specification within fifteen minutes after applying the seal packing. Do not start the engine for at least two hours after the installation of the oil pan.

11. Loosely install the upper oil pan bolts and nuts. Bolt "A" is 0.98 inch long, bolt "B" is 1.77 inch long and bolt "C" is 0.55 inch long. Uniformly tighten the 10mm bolt head to 7.4 ft. lbs, and the 12mm bolt head to 16 ft. lbs., in the proper sequence.

12. Install the oil strainer assembly. Torque the bolts to 80 inch lbs.

13. Apply a continuous bead (0.12–0.16 inch in diameter) of seal packing, part number 08826-00080 or equivalent, to the sealing surface of the oil pan.

➡**Remove any oil from the contact surface. Install the lower oil pan within three minutes of applying the seal packing. Tighten the pan bolts to specification within fifteen minutes after applying the seal packing. Do not start the engine for at least two hours after the installation of the oil pan.**

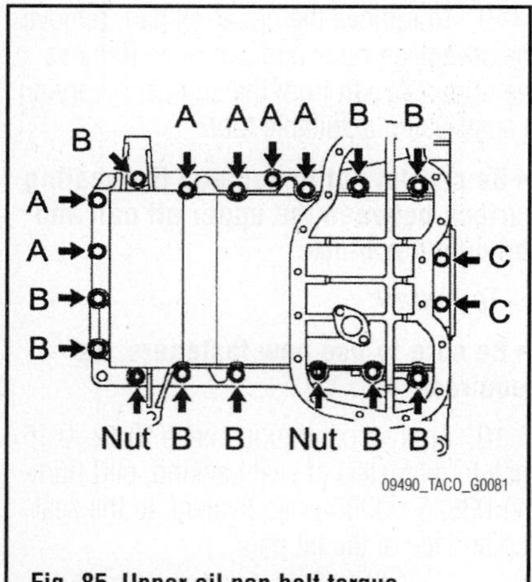

Fig. 85 Upper oil pan bolt torque sequence (2WD)—4.0L engine

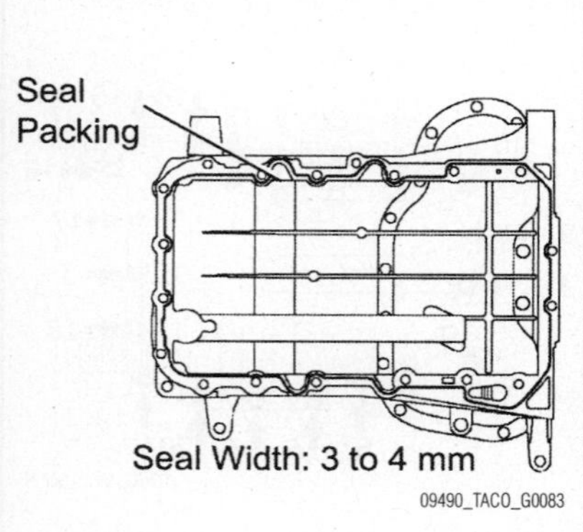

Fig. 86 Upper oil pan sealant application (4WD and PreRunner)—4.0L engine

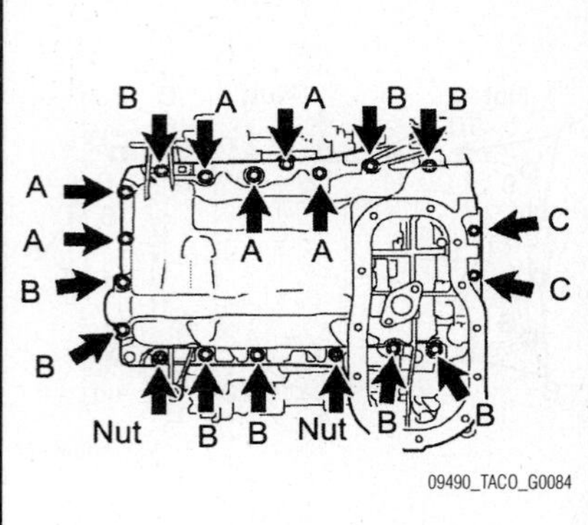

Fig. 87 Upper oil pan bolt torque sequence (4WD and PreRunner)—4.0L engine

14. Loosely install the lower oil pan bolts and nuts. Uniformly tighten the bolts to 80 inch lbs, and the nuts to 7.4 ft. lbs., in several steps.

15. Continue the installation in the reverse order of the removal procedure.

16. Be sure to fill the engine with the proper grade and type engine coolant.

17. Be sure to fill the engine with the proper grade and type engine oil.

18. Start the engine and check for leaks. Correct as required.

OIL PUMP

REMOVAL & INSTALLATION

2.7L Engines

See Figures 88 through 92.

1. Before servicing the vehicle, refer to the Precautions Section.

2. Disconnect the negative battery cable. Tape the cable with insulating tape.

3. Remove the engine and position it in a suitable holding fixture.

4. Remove the air intake connector. Remove the alternator.

5. Remove the belt tensioner assembly. Remove the idler pulley subassembly. Remove the air conditioning idler pulley assembly and bracket.

6. Remove the Crankshaft Position (CKP) sensor. Remove the Camshaft Position (CMP) sensor.

7. Remove the intake manifold. Remove the cylinder head cover.

8. Remove the crankshaft pulley. Remove the oil level gauge subassembly.

9. Remove the lower oil pan. Remove the oil strainer assembly. Remove the upper oil pan.

10. Remove the two nuts and separate the water bypass pipe number one.

11. Remove the nineteen bolts and two nuts retaining the timing chain case cover to its mounting. Remove the timing chain case cover from the engine.

➡**Carefully remove the cover by prying between the cover and the cylinder head or block with a suitable tool. Be sure to cover the tip of the suitable tool prior to usage. Be careful not to damage the contact surfaces of the cylinder block, cylinder head and timing chain cover.**

➡**The oil pump gears are located inside the timing case cover.**

12. Remove and discard the O-rings. Remove the head straight screw plug. Remove the water inlet. Remove the thermostat. Remove the oil seal.

13. Remove the oil pump relief valve. Remove the seven oil pump cover bolts. Remove the gears from the timing chain case cover.

To install:

➡**Be sure to use new fasteners, as required.**

14. Coat the oil pump gears with clean engine oil. Position the gears in the timing chain case cover with the identification marks facing outward. Check that the rotors revolve smoothly. Install the cover. Alternately tighten the bolts to 80 inch lbs.

15. Coat the oil pump relief valve with clean engine oil. Install the plug, using a new gasket and tighten to 36 ft. lbs.

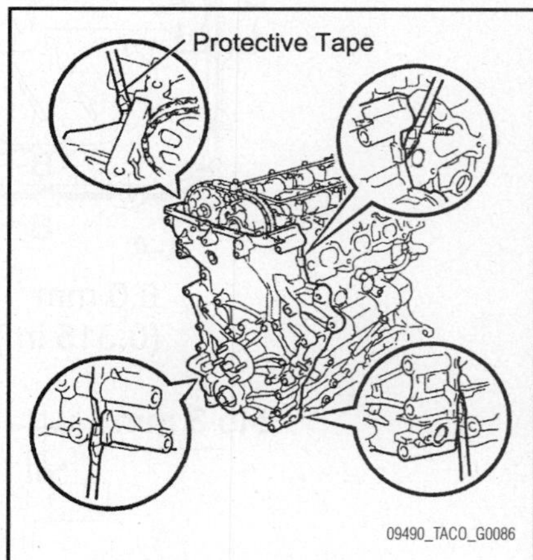

Fig. 88 Timing chain cover removal points—2.7L engine

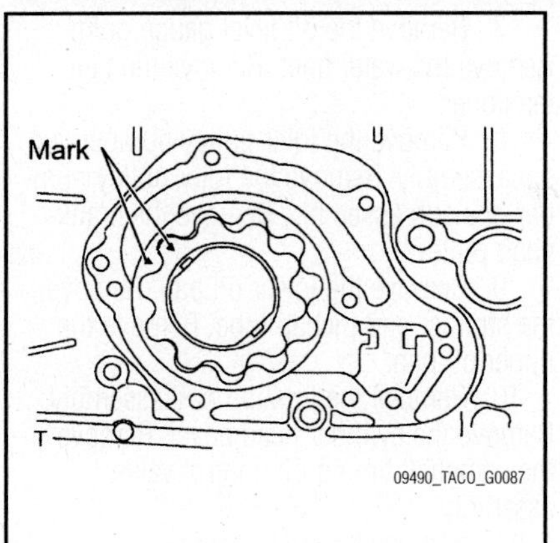

Fig. 89 Oil pump gear alignment marks—2.7L engine

16. Install a new front case oil seal. Install the thermostat. Install the water inlet.

17. Apply adhesive, part number 08833-00070 or equivalent to the head straight screw plug. Install the plug and tighten to 12 ft. lbs. Install four new O-rings onto the timing chain case cover.

18. Apply continuous beads of seal packing, part number 08826-00080 or equivalent as shown in the illustration.

➡Remove any oil from the contact surfaces. Install the timing chain case cover within three minutes and tighten the bolts within fifteen minutes of applying the seal packing. Do not start the engine for at least four hours after installation of the cover.

19. Align the oil pump drive rotor spline and the crankshaft, as indicated in the illustration. Install the spline and timing chain case cover onto the crankshaft.

20. Loosely install the timing chain case cover retaining bolts and nuts.

➡If the vehicle is equipped with air conditioning install the bolts that hold the idle pulley bracket in place when installing the idle pulley, as they are for this purpose.

21. Fully tighten the bolts and nuts, except bolts "A" in the following order: Area 1, Area 3 and then Area 2 to 15 ft. lbs.

22. Fully tighten the bolts "A" in the following order: Area 2 and then Area 3 to 34 ft. lbs.

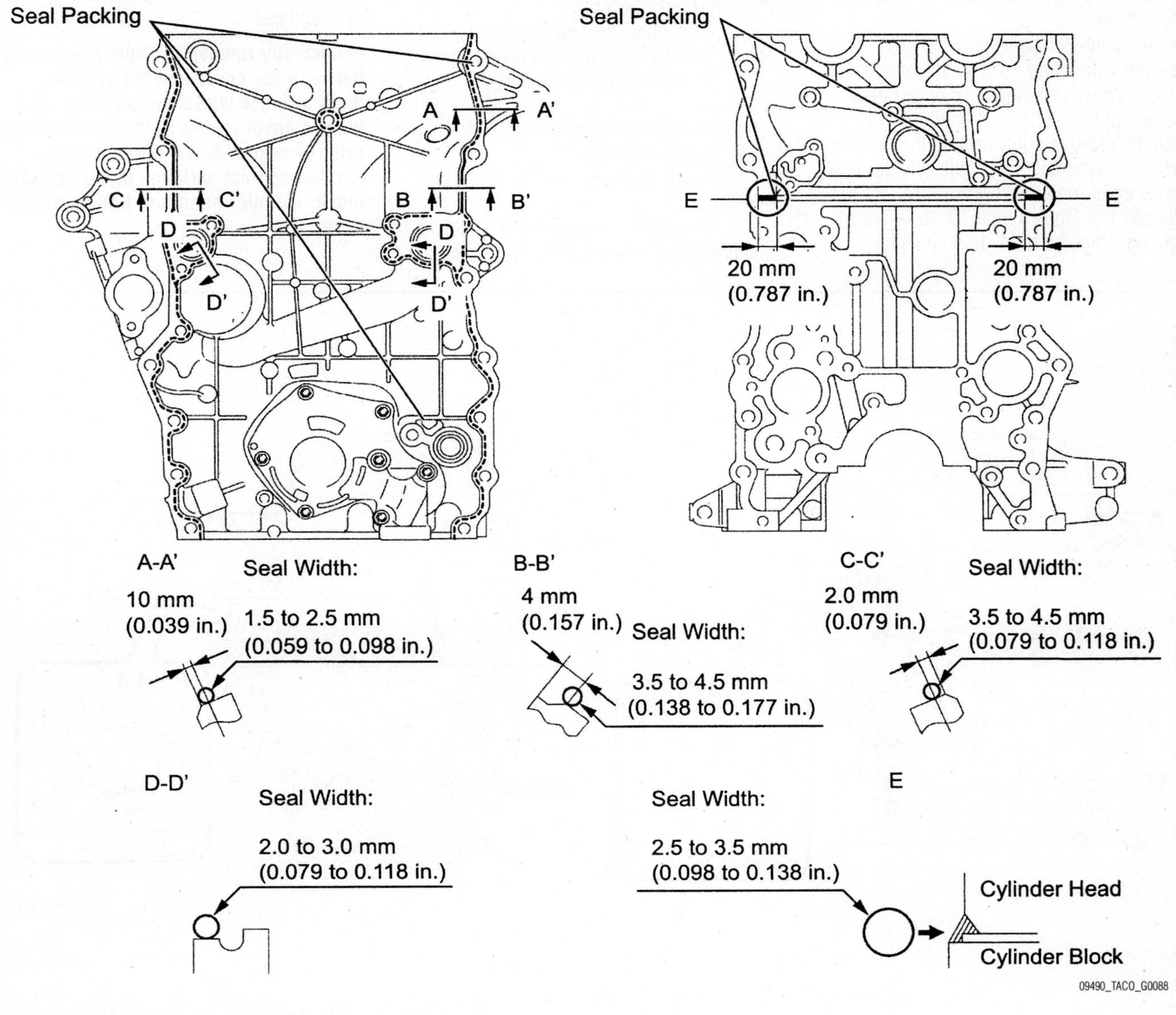

Seal Packing
A A'
C C'
B B'
D D'
Seal Packing
E E
20 mm (0.787 in.)
20 mm (0.787 in.)
A-A' Seal Width:
10 mm (0.039 in.)
1.5 to 2.5 mm (0.059 to 0.098 in.)
B-B'
4 mm (0.157 in.)
Seal Width:
3.5 to 4.5 mm (0.138 to 0.177 in.)
C-C'
2.0 mm (0.079 in.)
Seal Width:
3.5 to 4.5 mm (0.079 to 0.118 in.)
D-D'
Seal Width:
2.0 to 3.0 mm (0.079 to 0.118 in.)
E
Seal Width:
2.5 to 3.5 mm (0.098 to 0.138 in.)
Cylinder Head
Cylinder Block
09490_TACO_G0088

Fig. 90 Timing chain case cover sealant application—2.7L engine

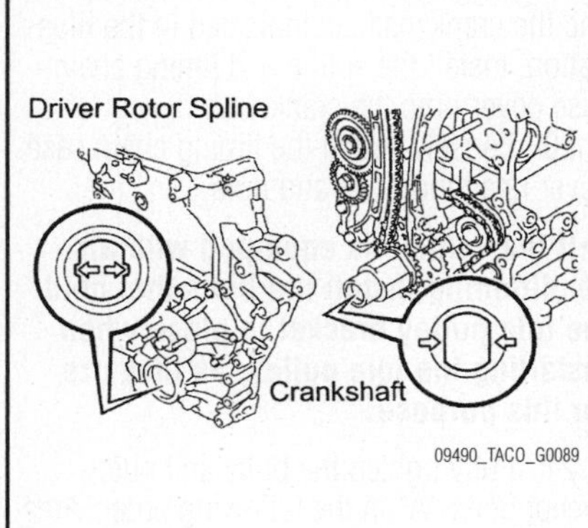

Fig. 91 Oil pump drive rotor spline alignment—2.7L engine

23. Fully tighten the bolts "E" in Area 4 to 15 ft. lbs.
24. Continue the installation in the reverse order of the removal procedure.
25. When installing the cylinder head cover, apply seal packing, part number 08826-00080 or equivalent, to the cylinder head. Provisionally install the cylinder head cover bolts and nuts. Tighten bolts "A" to 80 inch lbs. Tighten bolts "B" to 80 inch lbs. Retighten bolts "A" to 80 inch lbs.

➡Be sure to remove any oil from the contact surfaces of the cylinder head cover and the cylinder head. Install the cover within three minutes after applying the seal packing. Do not add engine oil for at least two hours after installing the cover.

26. Be sure to fill the engine with the proper grade and type engine coolant.
27. Be sure to fill the engine with the proper grade and type engine oil.
28. Start the engine and check for leaks, correct as required.

4.0L Engines

2WD

See Figures 93 through 97.

1. Before servicing the vehicle, refer to the Precautions Section.
2. Disconnect the negative battery cable. Tape the cable with insulating tape.
3. Properly relieve the fuel system pressure.
4. Disconnect the positive battery cable. Remove the battery.
5. Drain the engine coolant. Drain the engine oil.
6. Remove the engine from the vehicle and position it in a suitable holding fixture.
7. Remove the oil level gauge guide. Remove the water inlet. Remove the belt tensioner.
8. Remove the idler pulley number two subassembly. Remove the idler pulley number one sub assembly. Remove the crankshaft pulley.
9. Remove the lower oil pan. Remove the strainer and pickup tube. Remove the upper oil pan.
10. Remove the ignition coil assembly. Remove the cylinder head cover. Remove the camshaft timing oil control valve assembly.
11. Remove the VVT sensor. Remove the oil filter bracket subassembly.
12. Remove the timing chain case cover retaining bolts. Remove the cover from the engine. Remove the O-ring from the left cylinder head.

➡Carefully remove the cover by prying between the cover and the cylinder head or block with a suitable tool. Be sure to cover the tip of the suitable tool prior to usage. Be careful not to damage the contact surfaces of the cylinder block, cylinder head and timing chain cover.

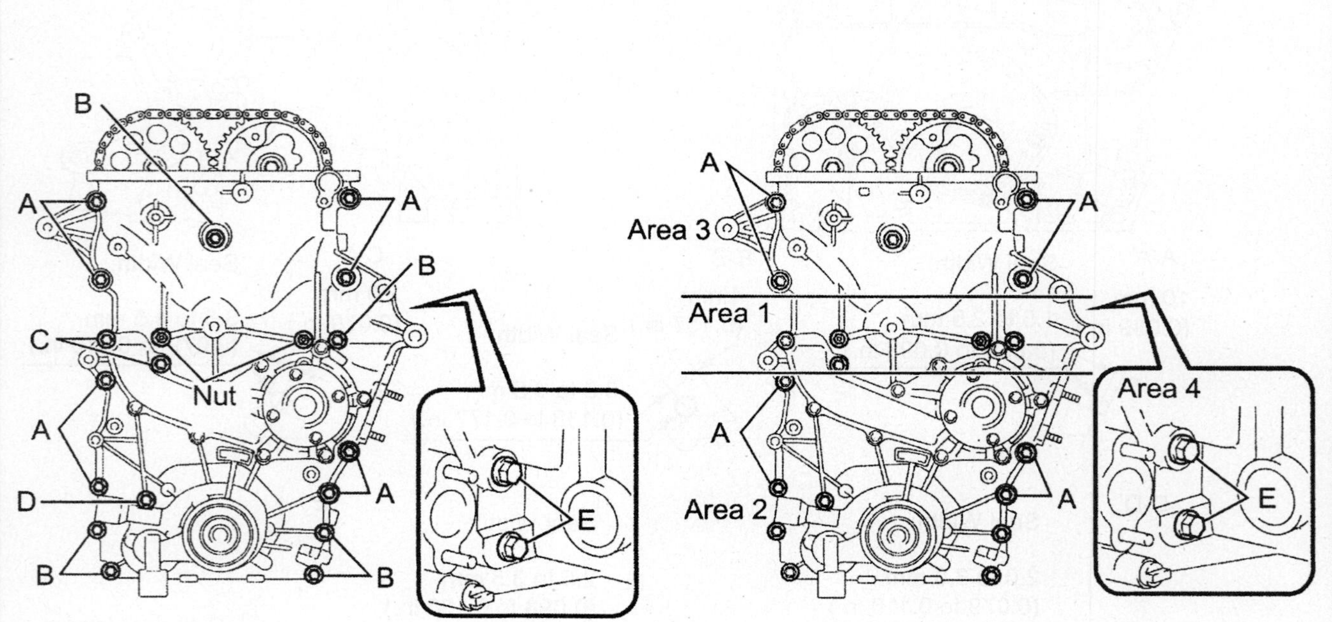

Fig. 92 Timing chain case cover bolt location and torque sequence—2.7L engine

➡The oil pump gears are located inside the timing case cover.

13. Remove the three bolts and remove the oil pipe. Remove the two O-rings. Remove the seven oil pump cover bolts. Remove the gears from the timing chain case cover. Remove the oil pump relief valve.

To install:

➡Be sure to use new fasteners, as required.

14. Coat the oil pump relief valve with clean engine oil. Install the plug, using a new gasket and tighten to 36 ft. lbs.

15. Coat the oil pump gears with clean engine oil. Position the gears in the timing chain case cover with the identification marks facing oil pump cover side. Install the cover. Alternately tighten the bolts to 80 inch lbs. Install the oil pipe, tighten the bolts to 80 inch lbs.

16. Install a new front case oil seal. Install a new O-ring onto the left cylinder head.

17. Apply continuous beads (0.12–0.16 inch in diameter) of seal packing, part number 08826-00080 or equivalent to the four locations shown in the illustration.

18. Apply continuous beads (0.12–0.16 inch in diameter) of seal packing, part number 08826-00080 or equivalent to all parts except the water pump part: for the water pump part use, part number 08826-00080 or equivalent, to the timing chain cover. Do not apply seal packing to portion "A" in the illustration.

➡Remove any oil from the contact surfaces. Install the timing chain case cover within three minutes and tighten the bolts within fifteen minutes of applying the seal packing.

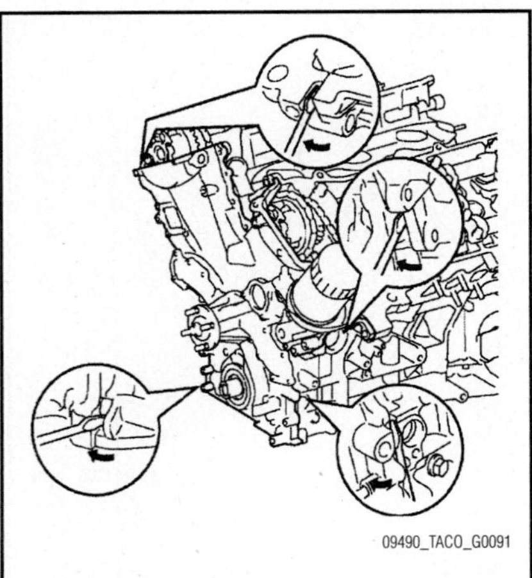

Fig. 93 Timing chain case cover removal points—4.0L engine

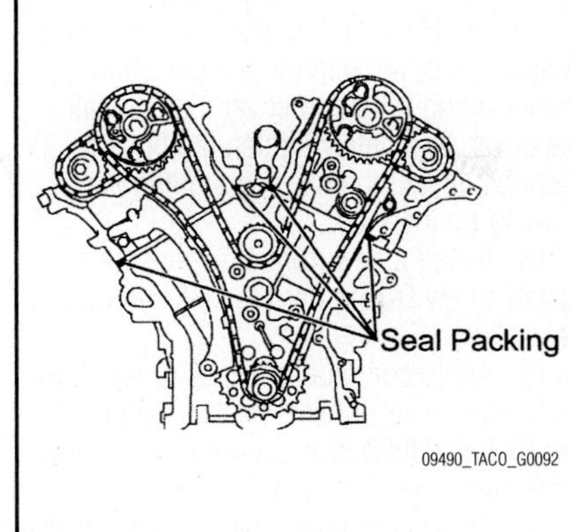

Fig. 94 Timing chain case cover seal packing locating points—4.0L engine

19. Align the key way of the oil pump drive motor with the rectangular portion of the crankshaft timing gear and slide the timing chain case cover into place.

20. Install the timing chain case cover bolts. Tighten the bolts and nuts uniformly in several steps to 17 ft. lbs.

➡Do not wrap the chain and slipper over the timing chain case cover seal line.

21. Continue the installation in the reverse order of the removal procedure.

22. When installing the cylinder head cover apply a continuous bead (0.08–0.12 inch) of seal packing, part number 08826-00080 or equivalent, to the cylinder head. Install the seal washers onto the bolts. Install the cylinder head cover bolts and

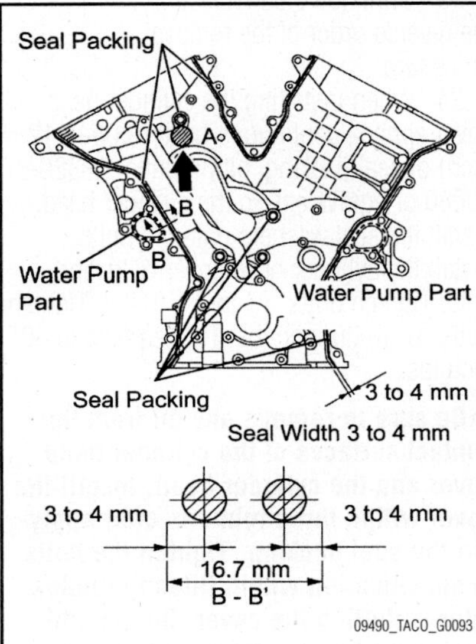

Fig. 95 Timing chain case cover sealant application—4.0L engine

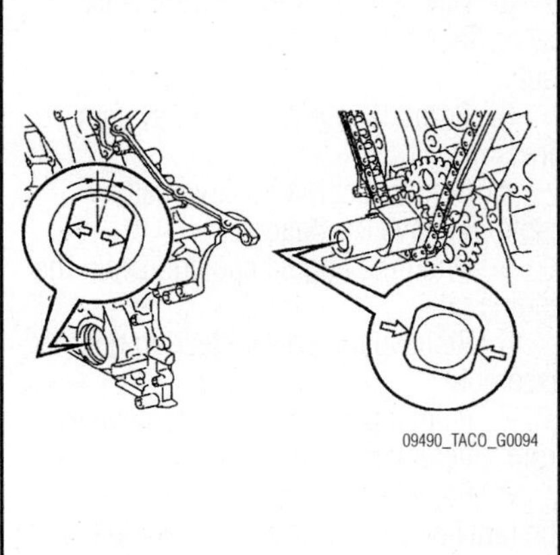

Fig. 96 Oil pump drive rotor spline alignment—4.0L engine

nuts. Tighten bolts "A" to 7.4 ft. lbs. Tighten bolts "B" to 80 inch lbs. Tighten nuts to 80 inch lbs.

➡Be sure to remove any oil from the contact surfaces of the cylinder head cover and the cylinder head. Install the cover within three minutes after applying the seal packing. Tighten the bolts to specification within fifteen minutes after installing the cover. Do not add engine oil for at least two hours after installing the cover.

23. Be sure to fill the engine with the proper grade and type engine coolant.

24. Be sure to fill the engine with the proper grade and type engine oil.

25. Start the engine and check for leaks, correct as required.

4WD and Prerunner

1. Before servicing the vehicle, refer to the Precautions Section.

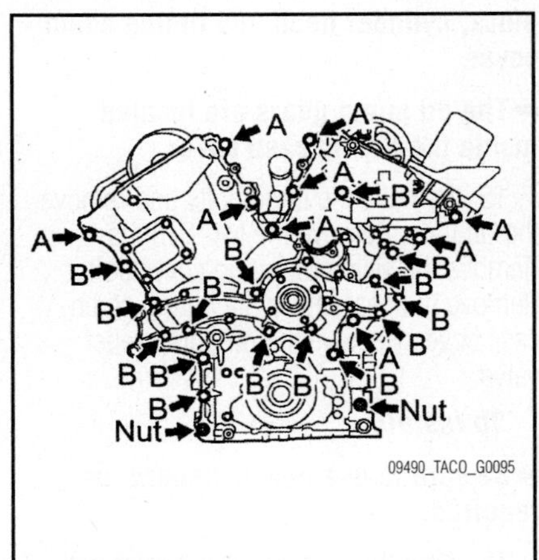

Fig. 97 Timing chain case cover bolt location and torque sequence—4.0L engine

2. Disconnect the negative battery cable. Tape the cable with insulating tape.

3. Properly relieve the fuel system pressure.

4. Disconnect the positive battery cable. Remove the battery.

5. Drain the engine coolant. Drain the engine oil.

6. Remove the power steering gear assembly.

7. If equipped with 4WD, remove the front differential carrier assembly.

8. Remove the V bank cover. Remove the radiator support to frame seal, left side. Remove the fan shroud.

9. Remove the air cleaner assembly. Remove the oil level gauge. Remove the water inlet.

10. Separate the vane pump assembly. Remove the alternator. Remove the air conditioning compressor and position it to the side.

11. Remove the belt tensioner assembly. Remove the idler pulley number two subassembly. Remove the idler pulley number one subassembly. Remove the crankshaft pulley.

12. Remove the lower oil pan. Remove the oil strainer and pickup tube assembly. Remove the upper oil pan.

13. Remove the intake manifold. Remove the cylinder head cover assembly.

14. Remove the camshaft timing oil control valve assembly. Remove the VVT sensor. Remove the oil filter bracket subassembly.

➡Carefully remove the cover by prying between the cover and the cylinder head or block with a suitable tool. Be sure to cover the tip of the suitable tool prior to usage. Be careful not to damage the contact surfaces of the cylinder block, cylinder head and timing chain cover.

➡The oil pump gears are located inside the timing case cover.

15. Remove the three bolts and remove the oil pipe. Remove the two O-rings. Remove the seven oil pump cover bolts. Remove the gears from the timing chain case cover. Remove the oil pump relief valve.

To install:

➡Be sure to use new fasteners, as required.

16. Coat the oil pump relief valve with clean engine oil. Install the plug, using a new gasket and tighten to 36 ft. lbs.

17. Coat the oil pump gears with clean engine oil. Position the gears in the timing chain case cover with the identification marks facing oil pump cover side. Install the cover. Alternately tighten the bolts to 80 inch lbs. Install the oil pipe, tighten the bolts to 80 inch lbs.

18. Install a new front case oil seal. Install a new O-ring onto the left cylinder head.

19. Apply continuous beads (0.12–0.16 inch in diameter) of seal packing, part number 08826-00080 or equivalent to the four locations shown in the illustration.

20. Apply continuous beads (0.12–0.16 inch in diameter) of seal packing, part number 08826-00080 or equivalent to all parts except the water pump part: for the water pump part use, part number 08826-00080 or equivalent, to the timing chain cover. Do not apply seal packing to portion "A" in the illustration.

➡Remove any oil from the contact surfaces. Install the timing chain case cover within three minutes and tighten the bolts within fifteen minutes of applying the seal packing.

21. Align the key way of the oil pump drive motor with the rectangular portion of the crankshaft timing gear and slide the timing chain case cover into place.

22. Install the timing chain case cover bolts. Tighten the bolts and nuts uniformly in several steps to 17 ft. lbs.

➡Do not wrap the chain and slipper over the timing chain case cover seal line.

23. Continue the installation in the reverse order of the removal procedure.

24. When installing the cylinder head cover apply a continuous bead (0.08–0.12 inch) of seal packing, part number 08826-00080 or equivalent, to the cylinder head. Install the seal washers onto the bolts. Install the cylinder head cover bolts and nuts. Tighten bolts "A" to 7.4 ft. lbs. Tighten bolts "B" to 80 inch lbs. Tighten nuts to 80 inch lbs.

➡Be sure to remove any oil from the contact surfaces of the cylinder head cover and the cylinder head. Install the cover within three minutes after applying the seal packing. Tighten the bolts to specification within fifteen minutes after installing the cover. Do not add engine oil for at least two hours after installing the cover.

25. Be sure to fill the engine with the proper grade and type engine coolant.

26. Be sure to fill the engine with the proper grade and type engine oil.

27. Start the engine and check for leaks, correct as required.

INSPECTION

See Figures 98 through 101.

1. Before servicing the vehicle, refer to the Precautions section.

2. Remove the oil pump from the engine and disassemble it.

3. Coat the relief valve with engine oil and check that it falls smoothly into the valve hole by its own weight.

4. If it doesn't, replace the relief valve. If necessary, replace the oil pump assembly.

5. Place the drive and driven rotors into the oil pump body.

6. Using a feeler gauge, measure the clearance between the drive and driven rotor tips.

7. Standard tip clearance is 0.060 to 0.160 mm (0.0024 to 0.0063 in.).

8. If the tip clearance is greater than maximum, replace the rotors as a set.

9. Using a feeler gauge and precision straight edge, measure the clearance between the rotors and precision straight edge.

10. Standard side clearance is 0.030 to 0.090 mm (0.0012 to 0.0035 in.).

11. If the side clearance is greater than maximum, replace the rotors as a set.
If necessary, replace the oil pump assembly.

12. Using a feeler gauge, measure the clearance between the driven rotor and body.

13. Standard body clearance is 0.250 to 0.325 mm (0.0098 to 0.0128 in.).

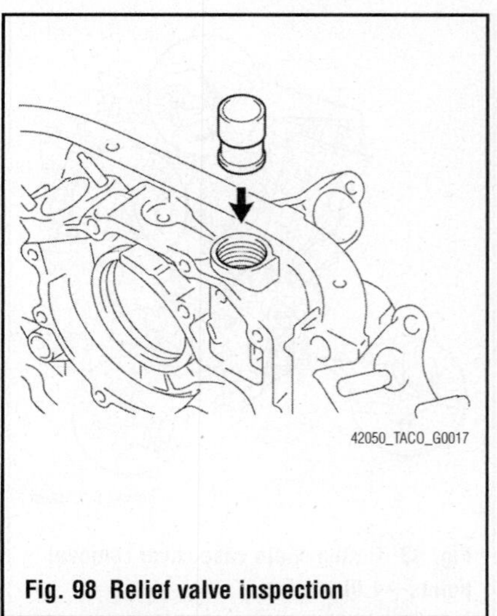

Fig. 98 Relief valve inspection

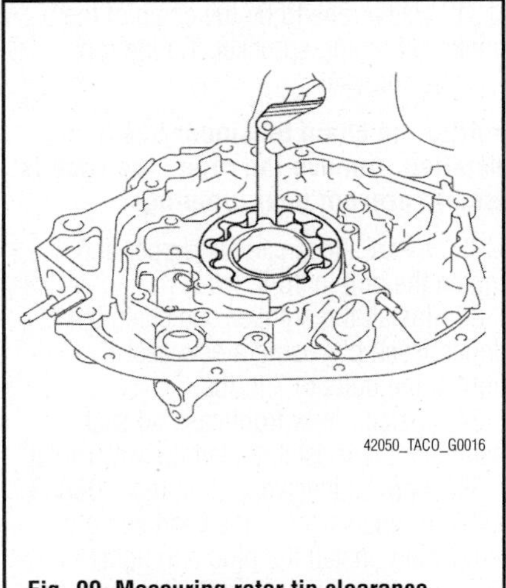
Fig. 99 Measuring rotor tip clearance

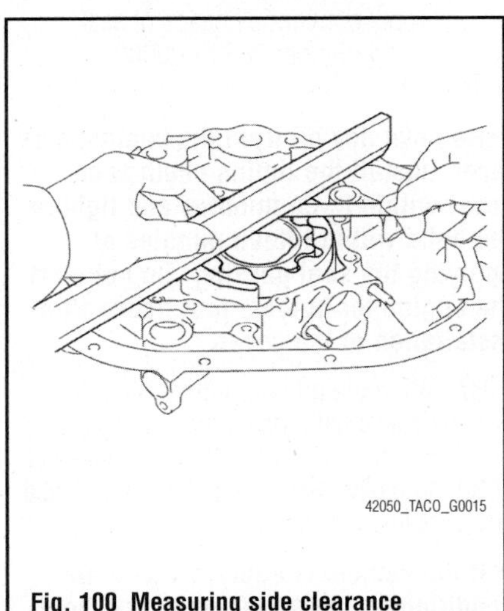
Fig. 100 Measuring side clearance

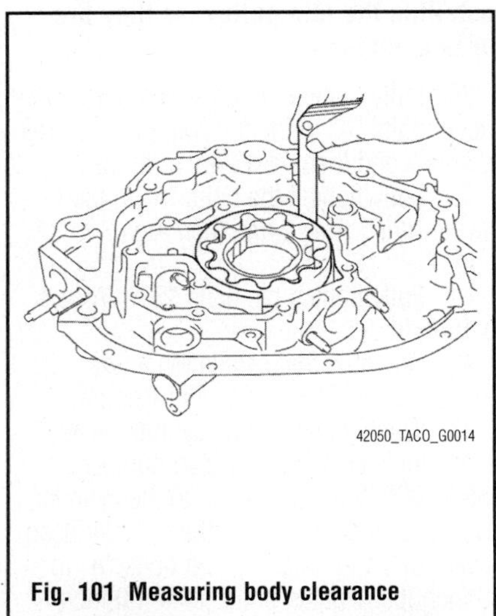
Fig. 101 Measuring body clearance

14. If the body clearance is greater than maximum, replace the rotors as a set. If necessary, replace the oil pump assembly.

PISTON AND RING

POSITIONING

See Figures 102 through 104.

TIMING CHAIN FRONT COVER SEAL

REMOVAL & INSTALLATION

Cover Installed

1. Before servicing the vehicle, refer to the Precautions section.
2. Unbolt and remove the oil pump.
3. Using a knife, carefully cut off the oil seal lip. With a flat-bladed tool, (preferably with tape around it) pry the seal from the cover.

To install:

4. Apply multi-purpose grease to the new oil seal lip.
5. Tap the seal into place with SST 09223–50010/60011 or equivalent seal driver, and a hammer. Do this until the seal surface is flush with the cover edge.
6. Install the oil pump with a new O-ring.

Cover Removed

1. Before servicing the vehicle, refer to the Precautions section.
2. Unbolt the timing chain cover assembly. Be careful to loosen only the correct bolts.
3. Pry out the seal from the cover with a flat-bladed tool.
4. It is a good idea to remove the oil pump from the timing cover and replace the O-ring.

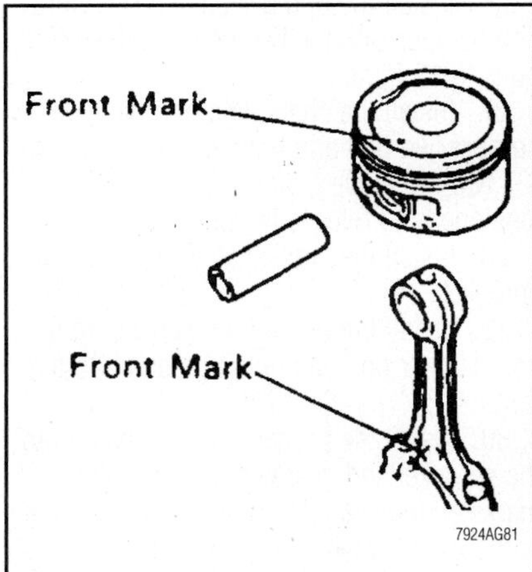

Fig. 102 Piston to connecting rod assembly

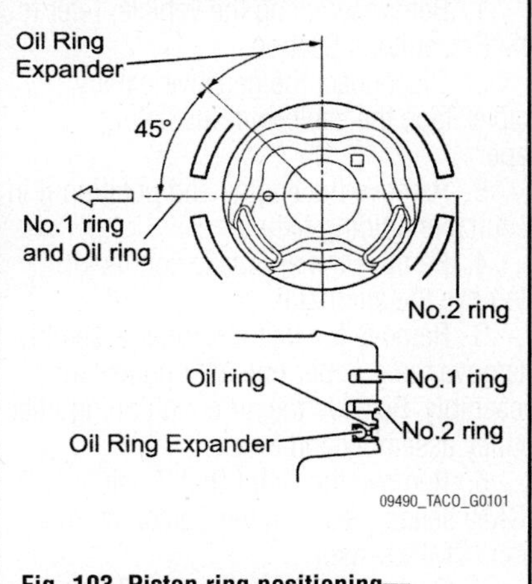

Fig. 103 Piston ring positioning—2.7L engine

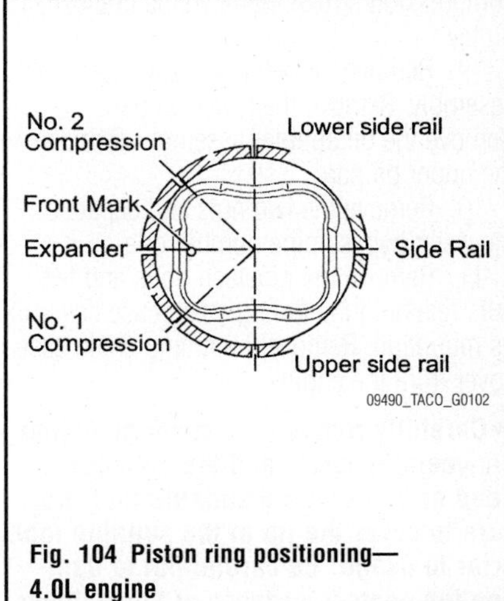

Fig. 104 Piston ring positioning—4.0L engine

To install:

5. Clean and inspect the timing cover area. Install new gaskets around the dowel areas and pump spline.
6. Apply multi-purpose grease to the new oil seal lip.
7. Tap the seal into place with SST 09223–50010/60010 or equivalent, and a hammer. Do this until the seal surface is flush with the cover edge.
8. Install the cover, tighten the bolts as specified for your engine.
9. If the oil pump was removed, install a new O-ring behind the pump prior to installation.

TIMING CHAIN & SPROCKETS

REMOVAL & INSTALLATION

2.7L Engines

See Figures 105 through 107.

1. Before servicing the vehicle, refer to the Precautions Section.

2. Disconnect the negative battery cable. Tape the cable with insulating tape.

3. Remove the engine and position it in a suitable holding fixture.

4. Remove the air intake connector. Remove the alternator.

5. Remove the belt tensioner assembly. Remove the number one idler pulley sub-assembly. Remove the air conditioning idler pulley assembly and bracket.

6. Remove the Crankshaft Position (CKP) sensor. Remove the Camshaft Position (CMP) sensor.

7. Remove the intake manifold. Remove the cylinder head cover.

8. Position the engine at TDC of the compression stroke. Remove the crankshaft pulley.

9. Remove the oil level gauge sub-assembly. Remove the lower oil pan. Remove the oil strainer assembly. Remove the upper oil pan.

10. Remove the two nuts and separate the water bypass pipe number one.

11. Remove the nineteen bolts and two nuts retaining the timing chain case cover to its mounting. Remove the timing chain case cover from the engine.

➡Carefully remove the cover by prying between the cover and the cylinder head or block with a suitable tool. Be sure to cover the tip of the suitable tool prior to usage. Be careful not to damage the contact surfaces of the cylinder block, cylinder head and timing chain cover.

12. Make sure that each matchmark is in the same position as shown in the illustration. Remove the two bolts, timing chain guide and O-ring.

13. Move the stopper plate upward to release the lock, and push the plunger deep into the tensioner.

14. Move the stopper plate downward to set the lock. Insert a 0.118 inch diameter bar into the stopper plate hole. Remove the bolt, nut, number one chain tensioner and gasket.

➡When the number one chain tensioner is removed do not rotate the crankshaft. When the chain is removed and the camshaft needs to be rotated, rotate the crankshaft 90 degrees to the right.

15. Remove the bolt and chain tensioner slipper. Remove the two bolts and remove the number one chain vibration damper. Remove the primary timing chain sub-assembly.

16. Remove the crankshaft timing gear or sprocket. Remove the bolt and remove the number two chain vibration damper.

17. Remove the two bolts and remove the number three chain vibration damper.

18. Remove the nut and the number two chain tensioner assembly. Remove the bolt, balance shaft drive gear shaft and balance shaft drive gear. Remove the crankshaft timing sprocket number two and chain.

To install:

➡Be sure to use new fasteners, as required.

19. Install the chain with its marks aligned with the timing marks on the crankshaft timing sprocket and balance shaft timing sprocket.

20. Bring the other mark link of the crankshaft timing sprocket behind the large timing mark of the balance shaft drive gear.

21. Insert the balance shaft drive gear shaft through the balance shaft drive gear so that it fits into the thrust plate hole.

22. Align the small timing mark of the balance shaft drive gear with the timing mark of the balance shaft timing gear.

23. Install the bolt onto the balance shaft drive gear and tighten it to 18 ft. lbs.

24. Check that the timing mark is aligned with the corresponding mark link.

25. Install the number two chain tensioner assembly. Tighten the nut to 13 ft. lbs.

➡Assemble the chain tensioner with the 0.118 inch diameter bar installed, then remove the bar after assembly. When doing this avoid pushing the vibration damper against the chain.

26. Install the number three chain vibration damper with the two bolts. Tighten the bolts to 13 ft. lbs.

27. Install the chain vibration damper number two bolt and tighten it to 20 ft. lbs. (27 Nm). Remove the pin from the chain tensioner and release the plunger.

28. Install the crankshaft timing gear or sprocket.

29. Install the number one chain vibration damper bolt and nut. Tighten to 15 ft. lbs.

30. Install the primary timing chain onto the sprocket and gear with the painted marks aligned with the timing marks on the sprocket and gear.

➡The camshaft mark plate is orange. The crankshaft mark plate is yellow.

31. Use a rope to tie the chain of the crankshaft timing sprocket. Tie the rope near the sprocket.

➡After the chain tensioner has been installed, remove the rope. The rope is used to prevent gear jumping.

32. Install the tensioner slipper and tighten the bolt to 15 ft. lbs.

33. Install the number one chain tensioner assembly, using a new gasket. Tighten the bolts to 7 ft. lbs.

34. Install a new front case oil seal. Install the thermostat. Install the water inlet.

35. Apply adhesive, part number 08833-00070 or equivalent to the head straight screw plug. Install the plug and tighten to 12 ft. lbs. Install four new O-rings onto the timing chain case cover.

36. Apply continuous beads of seal packing, part number 08826-00080 or equivalent.

➡Remove any oil from the contact surfaces. Install the timing chain case cover within three minutes and tighten the bolts within fifteen minutes of applying the seal packing. Do not start the engine for at least four hours after installation of the cover.

37. Align the oil pump drive rotor spline and the crankshaft. Install the spline and timing chain case cover onto the crankshaft.

38. Loosely install the timing chain case cover retaining bolts and nuts.

➡If the vehicle is equipped with air conditioning install the bolts that hold the idle pulley bracket in place when installing the idle pulley, as they are for this purpose.

39. Fully tighten the bolts and nuts, except bolts "A" in the following order: Area 1, Area 3 and then Area 2 to 15 ft. lbs.

40. Fully tighten the bolts "A" in the following order: Area 2 and then Area 3 to 34 ft. lbs.

41. Fully tighten the bolts "E" in Area 4 to 15 ft. lbs.

42. Continue the installation in the reverse order of the removal procedure.

43. When installing the cylinder head cover, apply seal packing, part number 08826-00080 or equivalent, to the cylinder head. Provisionally install the cylinder head cover bolts and nuts. Tighten bolts "A" to 80 inch lbs. Tighten bolts "B" to 80 inch lbs. Retighten bolts "A" to 80 inch lbs.

➡Be sure to remove any oil from the contact surfaces of the cylinder head cover and the cylinder head. Install the

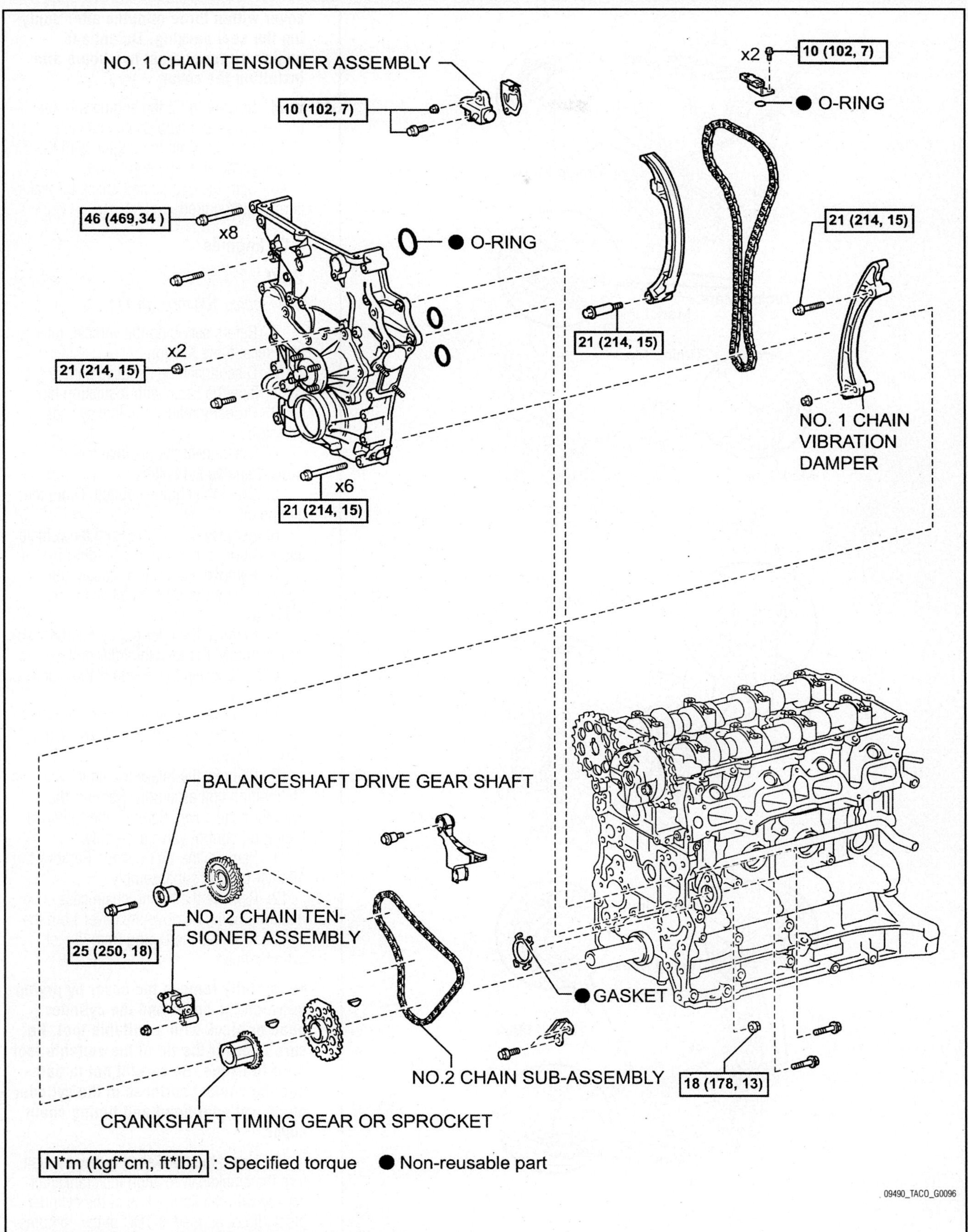

Fig. 105 Timing chain and related components—2.7L engine

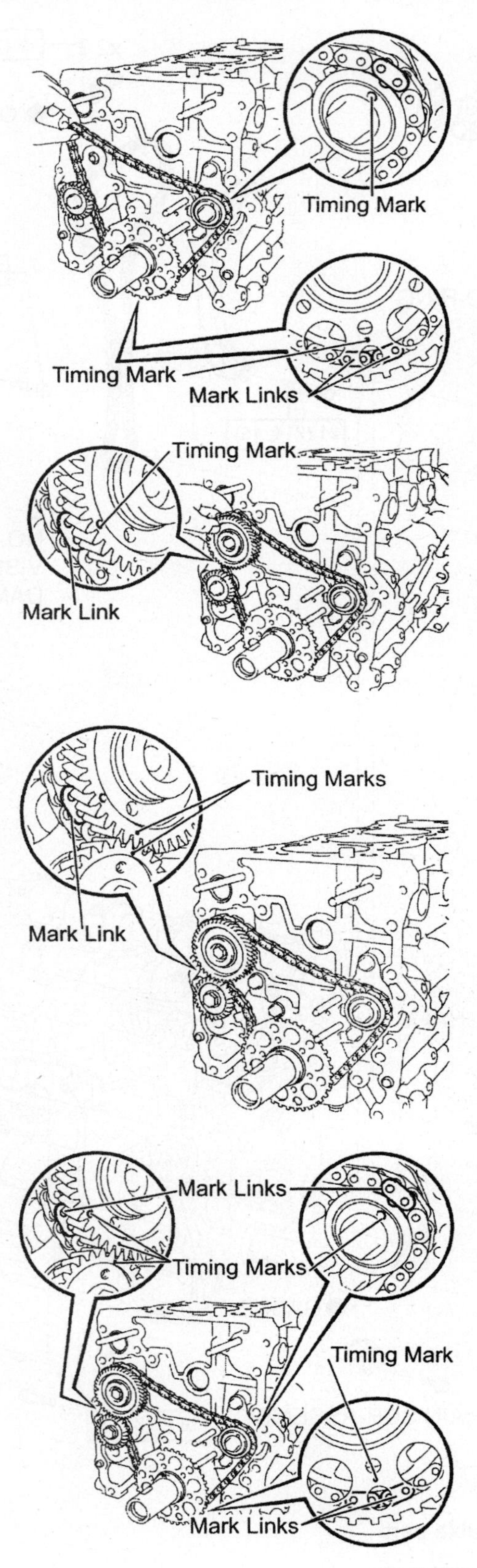

Fig. 106 Number two timing chain alignment marks—2.7L engine

cover within three minutes after applying the seal packing. Do not add engine oil for at least two hours after installing the cover.

44. Be sure to fill the engine with the proper grade and type engine coolant.
45. Be sure to fill the engine with the proper grade and type engine oil.
46. Start the engine and check for leaks, correct as required.

4.0L Engines

2WD

See Figures 108 through 111.

1. Before servicing the vehicle, refer to the Precautions Section.
2. Disconnect the negative battery cable. Tape the cable with insulating tape.
3. Properly relieve the fuel system pressure.
4. Disconnect the positive battery cable. Remove the battery.
5. Drain the engine coolant. Drain the engine oil.
6. Remove the engine from the vehicle and position it in a suitable holding fixture.
7. Remove the oil level gauge guide. Remove the water inlet. Remove the belt tensioner.
8. Remove the idler pulley number two subassembly. Remove the idler pulley number one sub assembly. Remove the crankshaft pulley.
9. Remove the lower oil pan. Remove the strainer and pickup tube. Remove the upper oil pan.
10. Remove the intake manifold. Remove the ignition coil assembly. Remove the cylinder head cover. Remove the camshaft timing oil control valve assembly.
11. Remove the VVT sensor. Remove the oil filter bracket subassembly.
12. Remove the timing chain case cover retaining bolts. Remove the cover from the engine. Remove the O-ring from the left cylinder head.

➡Carefully remove the cover by prying between the cover and the cylinder head or block with a suitable tool. Be sure to cover the tip of the suitable tool prior to usage. Be careful not to damage the contact surfaces of the cylinder block, cylinder head and timing chain cover.

13. Using the crankshaft pulley set bolt, turn the crankshaft to align the crankshaft set key with the timing line of the cylinder block. If not aligned at TDC of the compression stroke, turn the crankshaft one com-

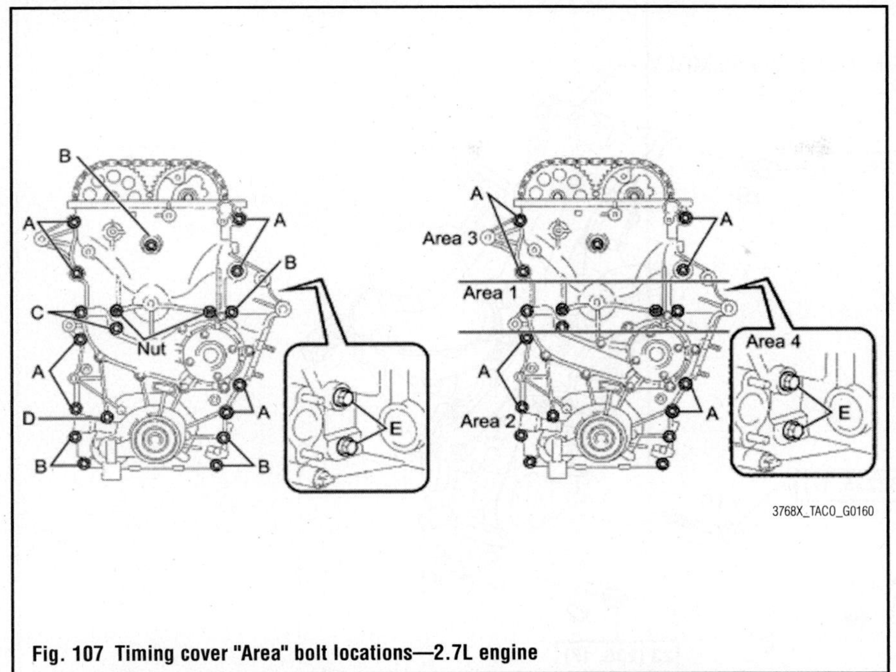

Fig. 107 Timing cover "Area" bolt locations—2.7L engine

plete revolution, in the direction of rotation

14. While turning the stopper plate of the tensioner upward, push the plunger of the chain tensioner. While turning the stopper plate of the tensioner downward, insert a 0.138 inch diameter bar into the holes in the stopper plate and tensioner to hold the stopper plate.
15. Remove the two bolts, and then remove the chain tensioner.
16. Remove the chain tensioner slipper. Remove the idle gear shaft number two, idle gear number one and idle gear shaft number one.
17. Remove the number two chain vibration damper. Remove the timing chain subassembly.

To install:

➡Be sure to use new fasteners, as required.

18. Install the chain tensioner slipper.
19. While turning the stopper plate of the tensioner clockwise, push in the plunger of the chain tensioner. While turning the stopper plate of the tensioner counterclockwise, insert a 0.138 inch diameter bar into the holes in the stopper plate and tensioner to hold the stopper plate.
20. Install the chain tensioner. Tighten the bolts to 7.4 ft. lbs.
21. Position the engine at TDC on the compression stroke. Align the camshaft timing gears and bearing caps. Using the crankshaft pulley set bolt, align the crankshaft set key with the timing line of the cylinder.
22. Align the yellow mark line with the timing mark of the crankshaft timing link. Align the orange mark links with the timing marks of the camshaft timing gears, and install the chain.
23. Install the number two chain vibration damper.
24. Apply a light coat of clean engine oil to the rotating surface of the idle gear shaft number one.
25. Temporarily install the idle gear shaft number one together with idle gear shaft number two, while aligning the knock pin of idle gear shaft number one with the knock pin groove of the cylinder block.

➡Be care of the idle gear direction.

26. Tighten the idle gear shaft number two to 44 ft. lbs. Remove the bar from the chain tensioner.
27. Install a new front case oil seal. Install a new O-ring onto the left cylinder head.
28. Apply continuous beads (0.12–0.16 inch in diameter) of seal packing, part number 08826-00080 or equivalent to the four locations shown in the illustration.
29. Apply continuous beads (0.12–0.16 inch in diameter) of seal packing, part number 08826-00080 or equivalent to all parts except the water pump part: for the water pump part use, part number 08826-00080 or equivalent, to the timing chain cover. Do not apply seal packing to portion "A".

➡Remove any oil from the contact surfaces. Install the timing chain case cover within three minutes and tighten the bolts within fifteen minutes of applying the seal packing.

30. Align the key way of the oil pump drive motor with the rectangular portion of the crankshaft timing gear and slide the timing chain case cover into place.
31. Install the timing chain case cover bolts. Tighten the bolts and nuts uniformly in several steps to 17 ft. lbs.

➡Do not wrap the chain and slipper over the timing chain case cover seal line.

32. Continue the installation in the reverse order of the removal procedure.
33. When installing the cylinder head cover apply a continuous bead (0.08–0.12 inch) of seal packing, part number 08826-00080 or equivalent, to the cylinder head. Install the seal washers onto the bolts. Install the cylinder head cover bolts and nuts. Tighten bolts "A" to 7.4 ft. lbs. Tighten bolts "B" to 80 inch lbs. Tighten nuts to 80 inch lbs.

➡Be sure to remove any oil from the contact surfaces of the cylinder head cover and the cylinder head. Install the cover within three minutes after applying the seal packing. Tighten the bolts to specification within fifteen minutes after installing the cover. Do not add engine oil for at least two hours after installing the cover.

34. Be sure to fill the engine with the proper grade and type engine coolant.
35. Be sure to fill the engine with the proper grade and type engine oil.
36. Start the engine and check for leaks, correct as required.

4WD and Prerunner

1. Before servicing the vehicle, refer to the Precautions Section.
2. Disconnect the negative battery cable. Tape the cable with insulating tape.
3. Properly relieve the fuel system pressure.
4. Disconnect the positive battery cable. Remove the battery.
5. Drain the engine coolant. Drain the engine oil.
6. Remove the power steering gear assembly.
7. If equipped with 4WD, remove the front differential carrier assembly.

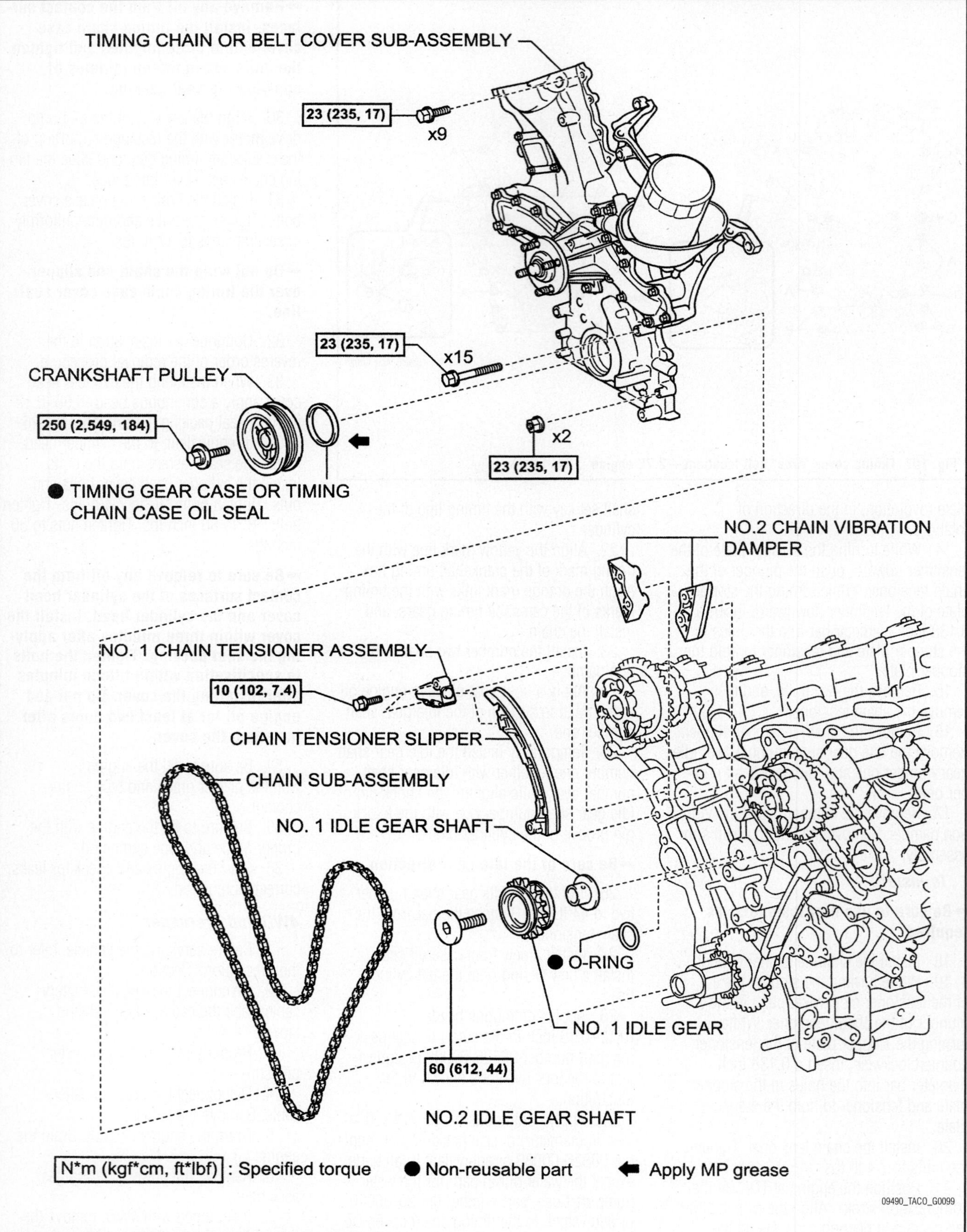

Fig. 108 Timing chain and related components—4.0L engine

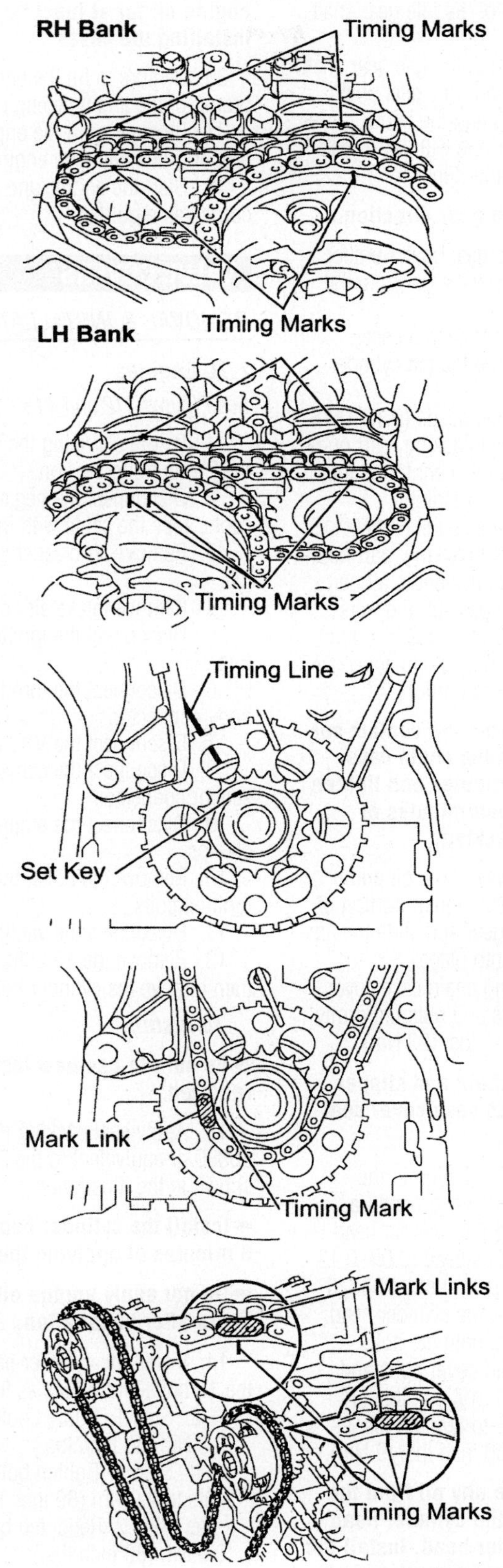

Fig. 109 Timing chain alignment—4.0L engine

8. Remove the V bank cover. Remove the radiator support to frame seal, left side. Remove the fan shroud.

9. Remove the air cleaner assembly. Remove the oil level gauge. Remove the water inlet.

10. Separate the vane pump assembly. Remove the alternator. Remove the air conditioning compressor and position it to the side.

11. Remove the belt tensioner assembly. Remove the idler pulley number two subassembly. Remove the idler pulley number one subassembly. Remove the crankshaft pulley.

12. Remove the lower oil pan. Remove the oil strainer and pickup tube assembly. Remove the upper oil pan.

13. Remove the intake manifold. Remove the ignition coil assembly. Remove the cylinder head cover assembly.

14. Remove the camshaft timing oil control valve assembly. Remove the VVT sensor. Remove the oil filter bracket subassembly.

➡Carefully remove the cover by prying between the cover and the cylinder head or block with a suitable tool. Be sure to cover the tip of the suitable tool prior to usage. Be careful not to damage the contact surfaces of the cylinder block, cylinder head and timing chain cover.

15. Using the crankshaft pulley set bolt, turn the crankshaft to align the crankshaft set key with the timing line of the cylinder block. If not aligned at TDC of the compression stroke, turn the crankshaft one complete revolution, in the direction of rotation

16. While turning the stopper plate of the tensioner upward, push the plunger of the chain tensioner. While turning the stopper plate of the tensioner downward, insert a 0.138 inch diameter bar into the holes in the stopper plate and tensioner to hold the stopper plate.

17. Remove the two bolts, and then remove the chain tensioner.

18. Remove the chain tensioner slipper. Remove the idle gear shaft number two, idle gear number one and idle gear shaft number one.

19. Remove the number two chain vibration damper. Remove the timing chain subassembly.

To install:

20. Install the chain tensioner slipper.

21. While turning the stopper plate of the tensioner clockwise, push in the plunger of the chain tensioner. While turning the

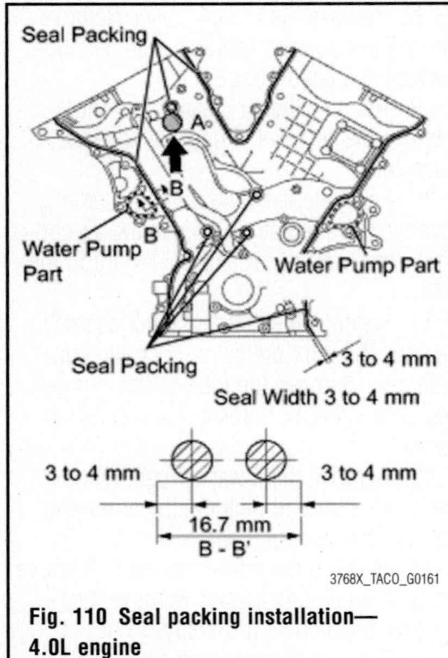

Fig. 110 Seal packing installation—4.0L engine

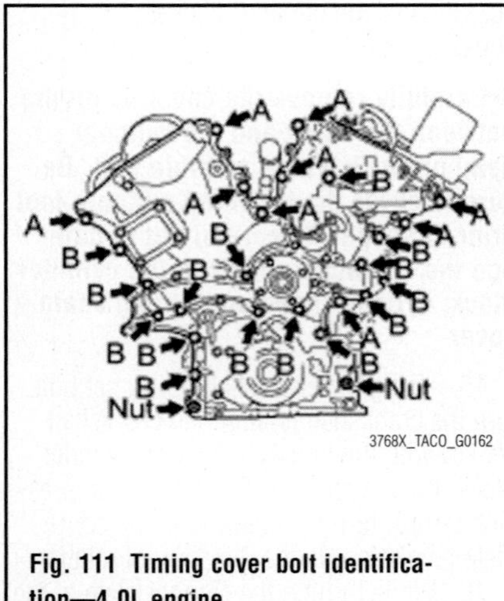

Fig. 111 Timing cover bolt identification—4.0L engine

stopper plate of the tensioner counterclockwise, insert a 0.138 inch diameter bar into the holes in the stopper plate and tensioner to hold the stopper plate.

22. Install the chain tensioner. Tighten the bolts to 7.4 ft. lbs.

23. Position the engine at TDC on the compression stroke. Align the camshaft timing gears and bearing caps. Using the crankshaft pulley set bolt, align the crankshaft set key with the timing line of the cylinder.

24. Align the yellow mark line with the timing mark of the crankshaft timing link. Align the orange mark links with the timing marks of the camshaft timing gears, and install the chain.

25. Install the number two chain vibration damper.

26. Apply a light coat of clean engine oil to the rotating surface of the idle gear shaft number one.

27. Temporarily install the idle gear shaft number one together with idle gear shaft number two, while aligning the knock pin of idle gear shaft number one with the knock pin groove of the cylinder block.

➡Be care of the idle gear direction.

28. Tighten the idle gear shaft number two to 44 ft. lbs. Remove the bar from the chain tensioner.

29. Install a new front case oil seal. Install a new O-ring onto the left cylinder head.

30. Apply continuous beads (0.12–0.16 inch in diameter) of seal packing, part number 08826-00080 or equivalent to the four locations shown in the illustration.

31. Apply continuous beads (0.12–0.16 inch in diameter) of seal packing, part number 08826-00080 or equivalent to all parts except the water pump part: for the water pump part use, part number 08826-00080 or equivalent, to the timing chain cover. Do not apply seal packing to portion "A".

➡Remove any oil from the contact surfaces. Install the timing chain case cover within three minutes and tighten the bolts within fifteen minutes of applying the seal packing.

32. Align the key way of the oil pump drive motor with the rectangular portion of the crankshaft timing gear and slide the timing chain case cover into place.

33. Install the timing chain case cover bolts. Tighten the bolts and nuts uniformly in several steps to 17 ft. lbs. (23 Nm).

➡Do not wrap the chain and slipper over the timing chain case cover seal line.

34. Continue the installation in the reverse order of the removal procedure.

35. When installing the cylinder head cover apply a continuous bead (0.08–0.12 inch) of seal packing, part number 08826-00080 or equivalent, to the cylinder head. Install the seal washers onto the bolts. Install the cylinder head cover bolts and nuts. Tighten bolts "A" to 7.4 ft. lbs. (10 Nm). Tighten bolts "B" to 80 inch lbs. 9 Nm). Tighten nuts to 80 inch lbs. (9 Nm).

➡Be sure to remove any oil from the contact surfaces of the cylinder head cover and the cylinder head. Install the cover within three minutes after applying the seal packing. Tighten the bolts to specification within fifteen minutes after installing the cover. Do not add engine oil for at least two hours after installing the cover.

36. Be sure to fill the engine with the proper grade and type engine coolant.

37. Be sure to fill the engine with the proper grade and type engine oil.

38. Start the engine and check for leaks, correct as required.

VALVE COVERS

REMOVAL & INSTALLATION

2.7L Engines

See Figures 112 and 113.

1. Before servicing the vehicle, refer to the Precautions Section.
2. Disconnect the negative battery cable. Tape the cable with insulating tape.
3. Remove air cleaner cap sub-assembly.
4. Remove intake air connector.
5. Disconnect the ignition coil connectors.
6. Disconnect the throttle with motor body connector.
7. Disconnect the VSV connector.
8. Disconnect the camshaft position sensor connector.
9. Disconnect the engine wire harness clamps.
10. Remove the bolts, then remove the ignition coils.
11. Disconnect the ventilation hose.
12. Remove the 19 bolts and 2 nuts, then remove the cylinder head cover.

To install:

➡Be sure to use new fasteners, as required.

13. Apply seal packing Part No. 08826-00080 or equivalent to the 2 locations shown in the illustration.

➡Install the cylinder head cover within 3 minutes of applying the seal packing.

➡Do not apply engine oil for at least 2 hours after installation.

14. Install the cylinder head cover with the 19 bolts and 2 nuts as follows:
    a. Step 1: Tighten bolts marked A to 9.0 Nm (80 inch lbs.)
    b. Step 2: Tighten bolts marked B and nuts to 9.0 Nm (80 inch lbs.)
    c. Step 3: Retighten bolts marked A to 9.0 Nm (80 inch lbs.)
15. Connect the ventilation hose.
16. Install the ignition coils with the bolts and tighten to 9.0 Nm (80 inch lbs.)

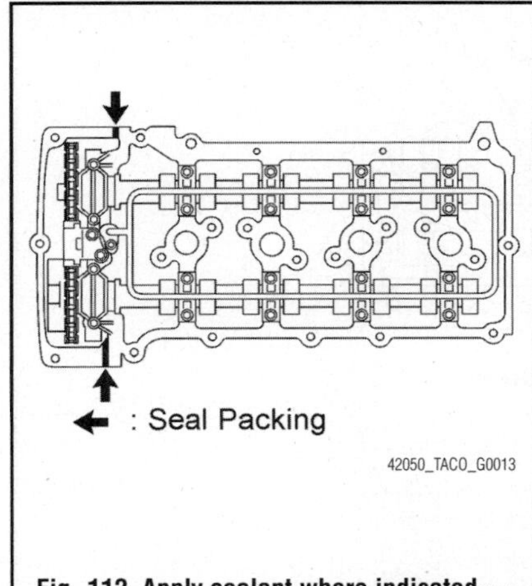

Fig. 112 Apply sealant where indicated—2.7L engine

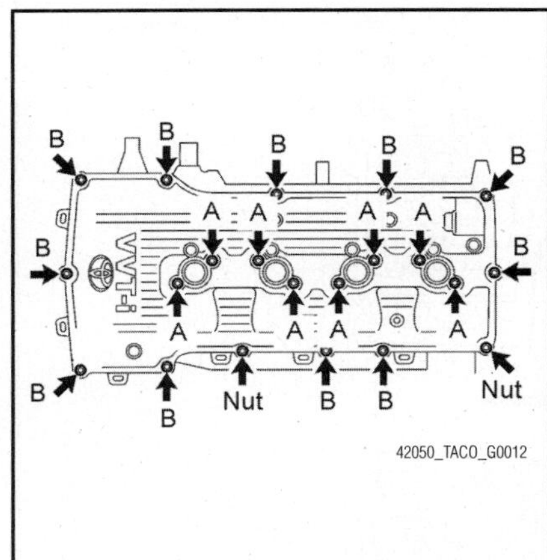

Fig. 113 Valve cover bolt identification—2.7L engine

17. Install the engine wire harness clamps.
18. Connect the camshaft position sensor connector.
19. Connect the VSV connector.
20. Connect the throttle with motor body connector.
21. Connect the ignition coil connectors.
22. Install intake air connector.
23. Install air cleaner cap sub-assembly.
24. Connect cable to negative battery terminal.

4.0L Engines

See Figures 114 and 115.

1. Before servicing the vehicle, refer to the Precautions Section.
2. Disconnect the negative battery cable. Tape the cable with insulating tape.
3. Drain engine coolant.
4. Remove v-bank cover.
5. Remove air cleaner assembly.
6. Disconnect the 2 water by-pass hoses.
7. Disconnect the fuel vapor feed hose.
8. Disconnect the ventilation hose.
9. Disconnect the 2 VSV connectors.
10. Disconnect the throttle body w/ motor connector.
11. Separate the 3 wire harness clamps and hose clamp.
12. Remove the nut, then separate the clutch flexible hose bracket from the surge tank stay (w/ Manual Transmission).
13. Remove the 2 bolts and throttle body bracket.
14. Remove the bolt and oil baffle plate.
15. Remove the 4 bolts and 2 surge tank stays.
16. Remove the 2 nuts.
17. Remove the 4 bolts, intake air surge tank and gasket.
18. Disconnect the 3 connectors.
19. Remove the 3 bolts, then remove the 3 ignition coils.
20. Remove the 10 bolts, 3 seal washers, 2 nuts, right cylinder head cover and gasket.
21. Remove the 10 bolts, 3 seal washers, 2 nuts, left cylinder head cover and gasket.

To install:

➡Be sure to use new fasteners, as required.

22. Remove any old packing material and be careful not to drop any oil on the contact surfaces of the cylinder head, timing chain cover and cylinder head covers.
23. Install the cylinder head covers within 3 minutes of applying seal packing Part No. 08826-00080 or equivalent. Tighten the cylinder head cover bolts and nuts within 15 minutes of installing the cylinder head covers. Otherwise, the seal packing mush be removed and reapplied.
24. Install the seal washers onto the bolts.
25. Install the cylinder head cover with the 10 bolts and 2 nuts. Tighten the bolts and nuts uniformly in several steps as follows:
    a. Bolts labeled A: 10 Nm (7.4 ft. lbs.)
    b. Bolts labeled B: 9.0 Nm (80 inch lbs.)
    c. Nuts: 9.0 Nm (80 inch lbs.)
26. Install the ignition coils and tighten the bolts to 10 Nm (7.4 ft. lbs.)
27. Install a new gasket onto the intake air surge tank.
28. Install the intake air surge tank with the 4 bolts and tighten to 28 Nm (21 ft. lbs)
29. Install the 2 intake air surge tank nuts and tighten to 28 Nm (21 ft. lbs)

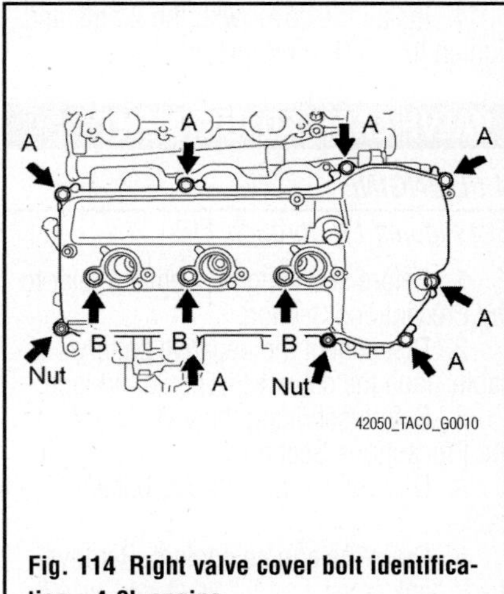

Fig. 114 Right valve cover bolt identification—4.0L engine

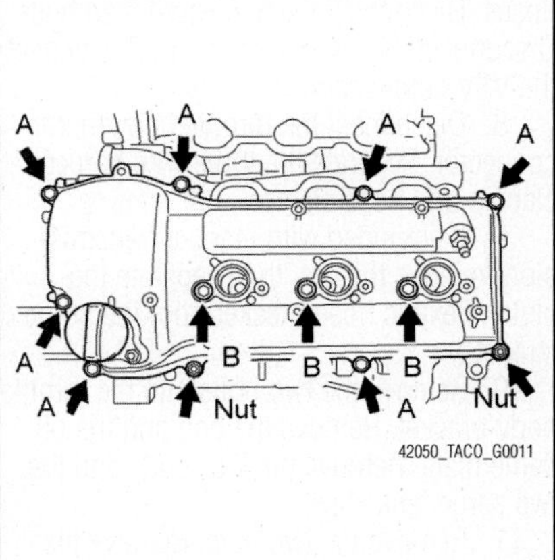

Fig. 115 Left valve cover bolt identification—4.0L engine

30. Install the 2 surge tank stays with the 4 bolts and tighten to 21 Nm (15 ft. lbs)
31. Install the oil baffle plate with the bolt and tighten to 9.0 Nm (80 inch lbs.)
32. Install the throttle body bracket with the 2 bolts and tighten to 21 Nm (15 ft. lbs)
33. Install the clutch flexible hose bracket with the nut (w/ Manual Transmission) and tighten to 24 Nm (18 ft. lbs)
34. Install the 3 wire harness clamps and hose clamp.
35. Connect the throttle body w/ motor connector.
36. Connect the 2 VSV connectors.
37. Connect the ventilation hose.
38. Connect the fuel vapor feed hose.
39. Connect the 2 water by-pass hoses.
40. Install air cleaner assembly.
41. Connect cable to negative battery terminal.
42. Add engine coolant.
43. Check for engine coolant leakage.

44. Install the cover with the 2 nuts and tighten to 7.5 Nm (66 inch lbs.)

VALVE LASH ADJUSTMENT

4.0L ENGINE

See Figures 116 through 118.

1. Before servicing the vehicle, refer to the Precautions Section.
2. Disconnect the negative battery cable. Tape the cable with insulating tape.
3. Before servicing the vehicle, refer to the Precautions Section.
4. Disconnect the negative battery cable.
5. Drain the engine coolant. Remove the V bank cover.
6. Remove the air cleaner assembly.
7. Disconnect the two water bypass hoses. Disconnect the fuel vapor feed hose. Disconnect the ventilation hose. Disconnect the VSV connectors.
8. Disconnect the throttle body motor connector. Separate the three wire harness clamps and hose clamp.
9. If equipped with manual transmission, remove the nut, then separate the clutch flexible hose bracket from the surge tank stay.
10. Remove the two bolts and the throttle body bracket. Remove the bolt and the oil baffle plate. Remove the four bolts and the two serge tank stays.
11. Remove the two nuts. Remove the four bolts, intake air surge tank and gasket.
12. Remove the ignition coil assembly.
13. Remove the cylinder head cover retaining bolts. Remove the cylinder head cover.
14. Turn the crankshaft pulley until its groove and the timing mark "0" of the timing chain cover are aligned. If not aligned at TDC of the compression stroke, turn the crankshaft one complete revolution, in the direction of rotation.
15. Using a feeler gauge, check and record the valve clearance on the following valves: right bank, exhaust number three and intake number one, left bank exhaust number two and intake number six.
16. Rotate the crankshaft 240 degrees clockwise and using a feeler gauge, check and record the following valves: right bank, exhaust number five and intake number three, left bank exhaust number four and intake number two.
17. Rotate the crankshaft 240 degrees clockwise and using a feeler gauge, check and record the following valves: right bank,

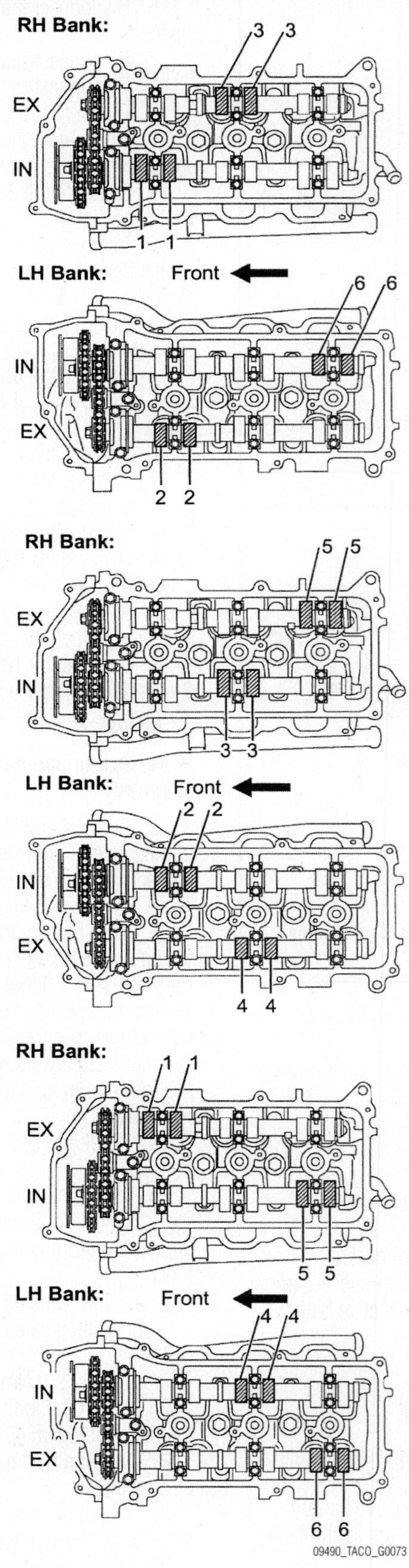

Fig. 116 Valve clearance location—4.0L engine

	5.060 (0.1992)	5.080 (0.2000)	5.100 (0.2008)	5.120 (0.2016)	5.140 (0.2024)	5.160 (0.2031)	5.180 (0.2039)	5.200 (0.2047)	5.210 (0.2051)	5.220 (0.2055)	5.230 (0.2059)	5.240 (0.2063)	5.250 (0.2067)	5.260 (0.2071)	5.270 (0.2075)	5.280 (0.2079)	5.290 (0.2083)	5.300 (0.2087)	5.310 (0.2091)	5.320 (0.2094)	5.330 (0.2098)	5.340 (0.2102)	5.350 (0.2106)	5.360 (0.2110)	5.370 (0.2114)	5.380 (0.2118)	5.390 (0.2122)	5.400 (0.2126)	5.410 (0.2130)	5.420 (0.2134)	5.430 (0.2138)	5.440 (0.2142)	5.450 (0.2146)	5.460 (0.2150)	5.470 (0.2154)	5.480 (0.2157)	5.490 (0.2161)	5.500 (0.2165)	5.510 (0.2169)	5.520 (0.2173)	5.530 (0.2177)	5.540 (0.2181)	5.550 (0.2185)	5.560 (0.2189)	5.570 (0.2193)	5.580 (0.2197)	5.590 (0.2201)	5.600 (0.2205)	5.620 (0.2213)	5.640 (0.2220)	5.660 (0.2228)	5.680 (0.2236)	5.700 (0.2244)	5.720 (0.2252)	5.740 (0.2260)
0.000 - 0.020 (0.0000 - 0.0008)													06	08	08	10	10	12	12	14	14	16	16	18	18	20	20	22	22	24	24	26	26	28	28	30	30	32	32	34	34	36	36	38	38	40	40	42	44	46	48	50	52	54	56
0.021 - 0.040 (0.0008 - 0.0016)											06	08	08	10	10	12	12	14	14	16	16	18	18	20	20	22	22	24	24	26	26	28	28	30	30	32	32	34	34	36	36	38	38	40	40	42	42	44	46	48	50	52	54	56	58
0.041 - 0.060 (0.0016 - 0.0024)									06	08	08	10	10	12	12	14	14	16	16	18	18	20	20	22	22	24	24	26	26	28	28	30	30	32	32	34	34	36	36	38	38	40	40	42	42	44	44	46	48	50	52	54	56	58	60
0.061 - 0.080 (0.0024 - 0.0031)								08	08	10	10	12	12	14	14	16	16	18	18	20	20	22	22	24	24	26	26	28	28	30	30	32	32	34	34	36	36	38	38	40	40	42	42	44	44	46	46	48	50	52	54	56	58	60	62
0.081 - 0.100 (0.0032 - 0.0039)							08	10	10	12	12	14	14	16	16	18	18	20	20	22	22	24	24	26	26	28	28	30	30	32	32	34	34	36	36	38	38	40	40	42	42	44	44	46	46	48	48	50	52	54	56	58	60	62	64
0.101 - 0.120 (0.0040 - 0.0047)						08	10	12	12	14	14	16	16	18	18	20	20	22	22	24	24	26	26	28	28	30	30	32	32	34	34	36	36	38	38	40	40	42	42	44	44	46	46	48	48	50	50	52	54	56	58	60	62	64	66
0.121 - 0.140 (0.0048 - 0.0055)					08	10	12	14	14	16	16	18	18	20	20	22	22	24	24	26	26	28	28	30	30	32	32	34	34	36	36	38	38	40	40	42	42	44	44	46	46	48	48	50	50	52	52	54	56	58	60	62	64	66	68
0.141 - 0.149 (0.0056 - 0.0059)				06	08	10	12	14	16	16	18	18	20	20	22	22	24	24	26	26	28	28	30	30	32	32	34	34	36	36	38	38	40	40	42	42	44	44	46	46	48	48	50	50	52	52	54	54	56	58	60	62	64	66	68
0.251 - 0.270 (0.0099 - 0.0106)	12	14	16	18	20	22	24	26	28	28	30	30	32	32	34	34	36	36	38	38	40	40	42	42	44	44	46	46	48	48	50	50	52	52	54	54	56	56	58	58	60	60	62	62	64	64	66	66	68	70	72
0.271 - 0.290 (0.0107 - 0.0114)	14	16	18	20	22	24	26	28	30	30	32	32	34	34	36	36	38	38	40	40	42	42	44	44	46	46	48	48	50	50	52	52	54	54	56	56	58	58	60	60	62	62	64	64	66	66	68	68	70	72
0.291 - 0.310 (0.0115 - 0.0122)	16	18	20	22	24	26	28	30	32	32	34	34	36	36	38	38	40	40	42	42	44	44	46	46	48	48	50	50	52	52	54	54	56	56	58	58	60	60	62	62	64	64	66	66	68	68	70	70	72
0.311 - 0.330 (0.0122 - 0.0130)	18	20	22	24	26	28	30	32	34	34	36	36	38	38	40	40	42	42	44	44	46	46	48	48	50	50	52	52	54	54	56	56	58	58	60	60	62	62	64	64	66	66	68	68	70	70	72	72
0.331 - 0.350 (0.0130 - 0.0138)	20	22	24	26	28	30	32	34	36	36	38	38	40	40	42	42	44	44	46	46	48	48	50	50	52	52	54	54	56	56	58	58	60	60	62	62	64	64	66	66	68	68	70	70	72	72	74
0.351 - 0.370 (0.0138 - 0.0146)	22	24	26	28	30	32	34	36	38	38	40	40	42	42	44	44	46	46	48	48	50	50	52	52	54	54	56	56	58	58	60	60	62	62	64	64	66	66	68	68	70	70	72	72	74
0.371 - 0.390 (0.0146 - 0.0154)	24	26	28	30	32	34	36	38	40	40	42	42	44	44	46	46	48	48	50	50	52	52	54	54	56	56	58	58	60	60	62	62	64	64	66	66	68	68	70	70	72	72	74
0.391 - 0.410 (0.0154 - 0.0161)	26	28	30	32	34	36	38	40	42	42	44	44	46	46	48	48	50	50	52	52	54	54	56	56	58	58	60	60	62	62	64	64	66	66	68	68	70	70	72	72	74
0.411 - 0.430 (0.0162 - 0.0169)	28	30	32	34	36	38	40	42	44	44	46	46	48	48	50	50	52	52	54	54	56	56	58	58	60	60	62	62	64	64	66	66	68	68	70	70	72	72	74
0.431 - 0.450 (0.0170 - 0.0177)	30	32	34	36	38	40	42	44	46	46	48	48	50	50	52	52	54	54	56	56	58	58	60	60	62	62	64	64	66	66	68	68	70	70	72	72	74
0.451 - 0.470 (0.0178 - 0.0185)	32	34	36	38	40	42	44	46	48	48	50	50	52	52	54	54	56	56	58	58	60	60	62	62	64	64	66	66	68	68	70	70	72	72	74
0.471 - 0.490 (0.0185 - 0.0193)	34	36	38	40	42	44	46	48	50	50	52	52	54	54	56	56	58	58	60	60	62	62	64	64	66	66	68	68	70	70	72	72	74
0.491 - 0.510 (0.0193 - 0.0201)	36	38	40	42	44	46	48	50	52	52	54	54	56	56	58	58	60	60	62	62	64	64	66	66	68	68	70	70	72	72	74
0.511 - 0.530 (0.0201 - 0.0209)	38	40	42	44	46	48	50	52	54	54	56	56	58	58	60	60	62	62	64	64	66	66	68	68	70	70	72	72	74
0.531 - 0.550 (0.0209 - 0.0217)	40	42	44	46	48	50	52	54	56	56	58	58	60	60	62	62	64	64	66	66	68	68	70	70	72	72	74
0.551 - 0.570 (0.0217 - 0.0224)	42	44	46	48	50	52	54	56	58	58	60	60	62	62	64	64	66	66	68	68	70	70	72	72	74
0.571 - 0.590 (0.0225 - 0.0232)	44	46	48	50	52	54	56	58	60	60	62	62	64	64	66	66	68	68	70	70	72	72	74
0.591 - 0.610 (0.0233 - 0.0240)	46	48	50	52	54	56	58	60	62	62	64	64	66	66	68	68	70	70	72	72	74
0.611 - 0.630 (0.0241 - 0.0248)	48	50	52	54	56	58	60	62	64	64	66	66	68	68	70	70	72	72	74
0.631 - 0.650 (0.0248 - 0.0256)	50	52	54	56	58	60	62	64	66	66	68	68	70	70	72	72	74
0.651 - 0.670 (0.0256 - 0.0264)	52	54	56	58	60	62	64	66	68	68	70	70	72	72	74
0.671 - 0.690 (0.0264 - 0.0272)	54	56	58	60	62	64	66	68	70	70	72	72	74
0.691 - 0.710 (0.0272 - 0.0280)	56	58	60	62	64	66	68	70	72	72	74
0.711 - 0.730 (0.0280 - 0.0287)	58	60	62	64	66	68	70	72	74
0.731 - 0.750 (0.0288 - 0.0295)	60	62	64	66	68	70	72
0.751 - 0.770 (0.0296 - 0.0303)	62	64	66	68	70	72
0.771 - 0.790 (0.0304 - 0.0311)	64	66	68	70	72
0.791 - 0.810 (0.0311 - 0.0319)	66	68	70	72
0.811 - 0.830 (0.0319 - 0.0327)	68	70	72
0.831 - 0.850 (0.0327 - 0.0335)	70	72
0.851 - 0.870 (0.0335 - 0.0343)	72

06	5.060 (0.1992)	30	5.300 (0.2087)	54	5.540 (0.2181)
08	5.080 (0.2000)	32	5.320 (0.2094)	56	5.560 (0.2189)
10	5.100 (0.2008)	34	5.340 (0.2102)	58	5.580 (0.2197)
12	5.120 (0.2016)	36	5.360 (0.2110)	60	5.600 (0.2205)
14	5.140 (0.2024)	38	5.380 (0.2118)	62	5.620 (0.2213)
16	5.160 (0.2031)	40	5.400 (0.2126)	64	5.640 (0.2220)
18	5.180 (0.2039)	42	5.420 (0.2134)	66	5.660 (0.2228)
20	5.200 (0.2047)	44	5.440 (0.2142)	68	5.680 (0.2236)
22	5.220 (0.2055)	46	5.460 (0.2150)	70	5.700 (0.2244)
24	5.240 (0.2063)	48	5.480 (0.2157)	72	5.720 (0.2252)
26	5.260 (0.2071)	50	5.500 (0.2165)	74	5.740 (0.2260)
28	5.280 (0.2079)	52	5.520 (0.2173)		

09490_TACO_G0074

Fig. 117 Shim selection chart, part one—4.0L engine

	5.060 (0.1992)	5.080 (0.2000)	5.100 (0.2008)	5.120 (0.2016)	5.140 (0.2024)	5.160 (0.2031)	5.180 (0.2039)	5.200 (0.2047)	5.210 (0.2051)	5.220 (0.2055)	5.230 (0.2059)	5.240 (0.2063)	5.250 (0.2067)	5.260 (0.2071)	5.270 (0.2075)	5.280 (0.2079)	5.290 (0.2083)	5.300 (0.2087)	5.310 (0.2091)	5.320 (0.2094)	5.330 (0.2098)	5.340 (0.2102)	5.350 (0.2106)	5.360 (0.2110)	5.370 (0.2114)	5.380 (0.2118)	5.390 (0.2122)	5.400 (0.2126)	5.410 (0.2130)	5.420 (0.2134)	5.430 (0.2138)	5.440 (0.2142)	5.450 (0.2146)	5.460 (0.2150)	5.470 (0.2154)	5.480 (0.2157)	5.490 (0.2161)	5.500 (0.2165)	5.510 (0.2169)	5.520 (0.2173)	5.530 (0.2177)	5.540 (0.2181)	5.550 (0.2185)	5.560 (0.2189)	5.570 (0.2193)	5.580 (0.2197)	5.590 (0.2201)	5.600 (0.2205)	5.620 (0.2213)	5.640 (0.2220)	5.660 (0.2228)	5.680 (0.2236)	5.700 (0.2244)	5.720 (0.2252)	5.740 (0.2260)
0.000 - 0.020 (0.0000 - 0.0008)																											06	08	08	10	10	12	12	14	14	16	16	18	18	20	20	22	22	24	24	26	26	28	30	32	34	36	38	40	42
0.021 - 0.040 (0.0008 - 0.0016)																									06	08	08	10	10	12	12	14	14	16	16	18	18	20	20	22	22	24	24	26	26	28	28	30	32	34	36	38	40	42	44
0.041 - 0.060 (0.0016 - 0.0024)																							06	08	08	10	10	12	12	14	14	16	16	18	18	20	20	22	22	24	24	26	26	28	28	30	30	32	34	36	38	40	42	44	46
0.061 - 0.080 (0.0024 - 0.0031)																					06	08	08	10	10	12	12	14	14	16	16	18	18	20	20	22	22	24	24	26	26	28	28	30	30	32	32	34	36	38	40	42	44	46	48
0.081 - 0.100 (0.0032 - 0.0039)																			06	08	08	10	10	12	12	14	14	16	16	18	18	20	20	22	22	24	24	26	26	28	28	30	30	32	32	34	34	36	38	40	42	44	46	48	50
0.101 - 0.120 (0.0040 - 0.0047)																	06	08	08	10	10	12	12	14	14	16	16	18	18	20	20	22	22	24	24	26	26	28	28	30	30	32	32	34	34	36	36	38	40	42	44	46	48	50	52
0.121 - 0.140 (0.0048 - 0.0055)															06	08	08	10	10	12	12	14	14	16	16	18	18	20	20	22	22	24	24	26	26	28	28	30	30	32	32	34	34	36	36	38	38	40	42	44	46	48	50	52	54
0.141 - 0.160 (0.0056 - 0.0063)													06	08	08	10	10	12	12	14	14	16	16	18	18	20	20	22	22	24	24	26	26	28	28	30	30	32	32	34	34	36	36	38	38	40	40	42	44	46	48	50	52	54	56
0.161 - 0.180 (0.0063 - 0.0071)											06	08	08	10	10	12	12	14	14	16	16	18	18	20	20	22	22	24	24	26	26	28	28	30	30	32	32	34	34	36	36	38	38	40	40	42	42	44	46	48	50	52	54	56	58
0.181 - 0.200 (0.0071 - 0.0079)									06	08	08	10	10	12	12	14	14	16	16	18	18	20	20	22	22	24	24	26	26	28	28	30	30	32	32	34	34	36	36	38	38	40	40	42	42	44	44	46	48	50	52	54	56	58	60
0.201 - 0.220 (0.0079 - 0.0087)								08	08	10	10	12	12	14	14	16	16	18	18	20	20	22	22	24	24	26	26	28	28	30	30	32	32	34	34	36	36	38	38	40	40	42	42	44	44	46	46	48	50	52	54	56	58	60	62
0.221 - 0.240 (0.0087 - 0.0094)							08	10	10	12	12	14	14	16	16	18	18	20	20	22	22	24	24	26	26	28	28	30	30	32	32	34	34	36	36	38	38	40	40	42	42	44	44	46	46	48	48	50	52	54	56	58	60	62	64
0.241 - 0.260 (0.0095 - 0.0102)						08	10	12	12	14	14	16	16	18	18	20	20	22	22	24	24	26	26	28	28	30	30	32	32	34	34	36	36	38	38	40	40	42	42	44	44	46	46	48	48	50	50	52	54	56	58	60	62	64	66
0.261 - 0.280 (0.0103 - 0.0110)					08	10	12	14	14	16	16	18	18	20	20	22	22	24	24	26	26	28	28	30	30	32	32	34	34	36	36	38	38	40	40	42	42	44	44	46	46	48	48	50	50	52	52	54	56	58	60	62	64	66	68
0.281 - 0.289 (0.0111 - 0.0114)				06	08	10	12	14	16	16	18	18	20	20	22	22	24	24	26	26	28	28	30	30	32	32	34	34	36	36	38	38	40	40	42	42	44	44	46	46	48	48	50	50	52	52	54	54	56	58	60	62	64	66	68
0.391 - 0.410 (0.0154 - 0.0161)	12	14	16	18	20	22	24	26	28	28	30	30	32	32	34	34	36	36	38	38	40	40	42	42	44	44	46	46	48	48	50	50	52	52	54	54	56	56	58	58	60	60	62	62	64	64	66	66	68	70	72				
0.411 - 0.430 (0.0162 - 0.0169)	14	16	18	20	22	24	26	28	30	30	32	32	34	34	36	36	38	38	40	40	42	42	44	44	46	46	48	48	50	50	52	52	54	54	56	56	58	58	60	60	62	62	64	64	66	66	68	68	70	72					
0.431 - 0.450 (0.0170 - 0.0177)	16	18	20	22	24	26	28	30	32	32	34	34	36	36	38	38	40	40	42	42	44	44	46	46	48	48	50	50	52	52	54	54	56	56	58	58	60	60	62	62	64	64	66	66	68	68	70	70	72						
0.451 - 0.470 (0.0178 - 0.0185)	18	20	22	24	26	28	30	32	34	34	36	36	38	38	40	40	42	42	44	44	46	46	48	48	50	50	52	52	54	54	56	56	58	58	60	60	62	62	64	64	66	66	68	68	70	70	72	72							
0.471 - 0.490 (0.0185 - 0.0193)	20	22	24	26	28	30	32	34	36	36	38	38	40	40	42	42	44	44	46	46	48	48	50	50	52	52	54	54	56	56	58	58	60	60	62	62	64	64	66	66	68	68	70	70	72	72	74								
0.491 - 0.510 (0.0193 - 0.0201)	22	24	26	28	30	32	34	36	38	38	40	40	42	42	44	44	46	46	48	48	50	50	52	52	54	54	56	56	58	58	60	60	62	62	64	64	66	66	68	68	70	70	72	72	74										
0.511 - 0.530 (0.0201 - 0.0209)	24	26	28	30	32	34	36	38	40	40	42	42	44	44	46	46	48	48	50	50	52	52	54	54	56	56	58	58	60	60	62	62	64	64	66	66	68	68	70	70	72	72	74												
0.531 - 0.550 (0.0209 - 0.0217)	26	28	30	32	34	36	38	40	42	42	44	44	46	46	48	48	50	50	52	52	54	54	56	56	58	58	60	60	62	62	64	64	66	66	68	68	70	70	72	72	74														
0.551 - 0.570 (0.0217 - 0.0224)	28	30	32	34	36	38	40	42	44	44	46	46	48	48	50	50	52	52	54	54	56	56	58	58	60	60	62	62	64	64	66	66	68	68	70	70	72	72	74																
0.571 - 0.590 (0.0225 - 0.0232)	30	32	34	36	38	40	42	44	46	46	48	48	50	50	52	52	54	54	56	56	58	58	60	60	62	62	64	64	66	66	68	68	70	70	72	72	74																		
0.591 - 0.610 (0.0233 - 0.0240)	32	34	36	38	40	42	44	46	48	48	50	50	52	52	54	54	56	56	58	58	60	60	62	62	64	64	66	66	68	68	70	70	72	72	74																				
0.611 - 0.630 (0.0241 - 0.0248)	34	36	38	40	42	44	46	48	50	50	52	52	54	54	56	56	58	58	60	60	62	62	64	64	66	66	68	68	70	70	72	72	74																						
0.631 - 0.650 (0.0248 - 0.0256)	36	38	40	42	44	46	48	50	52	52	54	54	56	56	58	58	60	60	62	62	64	64	66	66	68	68	70	70	72	72	74																								
0.651 - 0.670 (0.0256 - 0.0264)	38	40	42	44	46	48	50	52	54	54	56	56	58	58	60	60	62	62	64	64	66	66	68	68	70	70	72	72	74																										
0.671 - 0.690 (0.0264 - 0.0272)	40	42	44	46	48	50	52	54	56	56	58	58	60	60	62	62	64	64	66	66	68	68	70	70	72	72	74																												
0.691 - 0.710 (0.0272 - 0.0280)	42	44	46	48	50	52	54	56	58	58	60	60	62	62	64	64	66	66	68	68	70	70	72	72	74																														
0.711 - 0.730 (0.0280 - 0.0287)	44	46	48	50	52	54	56	58	60	60	62	62	64	64	66	66	68	68	70	70	72	72	74																																
0.731 - 0.750 (0.0288 - 0.0295)	46	48	50	52	54	56	58	60	62	62	64	64	66	66	68	68	70	70	72	72	74																																		
0.751 - 0.770 (0.0296 - 0.0303)	48	50	52	54	56	58	60	62	64	64	66	66	68	68	70	70	72	72	74																																				
0.771 - 0.790 (0.0304 - 0.0311)	50	52	54	56	58	60	62	64	66	66	68	68	70	70	72	72	74																																						
0.791 - 0.810 (0.0311 - 0.0319)	52	54	56	58	60	62	64	66	68	68	70	70	72	72	74																																								
0.811 - 0.830 (0.0319 - 0.0327)	54	56	58	60	62	64	66	68	70	70	72	72	74																																										
0.831 - 0.850 (0.0327 - 0.0335)	56	58	60	62	64	66	68	70	72	72	74																																												
0.851 - 0.870 (0.0335 - 0.0343)	58	60	62	64	66	68	70	72	74																																														
0.871 - 0.890 (0.0343 - 0.0350)	60	62	64	66	68	70	72																																																
0.891 - 0.910 (0.0351 - 0.0358)	62	64	66	68	70	72																																																	
0.911 - 0.930 (0.0359 - 0.0366)	64	66	68	70	72																																																		
0.931 - 0.950 (0.0367 - 0.0374)	66	68	70	72																																																			
0.951 - 0.970 (0.0374 - 0.0382)	68	70	72																																																				
0.971 - 0.990 (0.0382 - 0.0390)	70	72																																																					
0.991 - 1.010 (0.0390 - 0.0398)	72																																																						

06	5.060 (0.1992)	30	5.300 (0.2087)	54	5.540 (0.2181)
08	5.080 (0.2000)	32	5.320 (0.2094)	56	5.560 (0.2189)
10	5.100 (0.2008)	34	5.340 (0.2102)	58	5.580 (0.2197)
12	5.120 (0.2016)	36	5.360 (0.2110)	60	5.600 (0.2205)
14	5.140 (0.2024)	38	5.380 (0.2118)	62	5.620 (0.2213)
16	5.160 (0.2031)	40	5.400 (0.2126)	64	5.640 (0.2220)
18	5.180 (0.2039)	42	5.420 (0.2134)	66	5.660 (0.2228)
20	5.200 (0.2047)	44	5.440 (0.2142)	68	5.680 (0.2236)
22	5.220 (0.2055)	46	5.460 (0.2150)	70	5.700 (0.2244)
24	5.240 (0.2063)	48	5.480 (0.2157)	72	5.720 (0.2252)
26	5.260 (0.2071)	50	5.500 (0.2165)	74	5.740 (0.2260)
28	5.280 (0.2079)	52	5.520 (0.2173)		

09490_TACO_G0075

Fig. 118 Shim selection chart, part two—4.0L engine

exhaust number one and intake number five, left bank exhaust number six and intake number four.

18. If adjustment is required, position the engine at TDC on the compression stroke.

19. Place paint marks on the number one chain links corresponding to the timing marks of the camshaft timing gears.

20. Remove the chain tensioner assembly number one. Remove the number two camshaft.

21. Remove the chain tensioner assembly number two. Remove the camshaft.

22. Remove the number four camshaft subassembly. Remove the chain tensioner assembly number three.

23. Remove the number three camshaft subassembly.

24. Remove the valve lifters.

25. Determine the replacement adjusting shim size according to the following formula or use the adjusting shim charts.

26. Using a micrometer, measure the thickness of the removed shim. Calculate the thickness of a new shim so that the valve clearance comes within the specified value.

- T: Thickness of the removed shim
- A: Measured valve clearance
- N: Thickness of the new shim

a. Intake: N = T + A
b. Exhaust: N = T + A

27. Select a new lifter with a thickness as close as possible to the calculated value.

28. Install removed components in the reverse order of the removal procedure.

29. When installing the cylinder head cover, apply a continuous bead (0.08–0.12 inch) of seal packing, part number 08826-00080 or equivalent, to the cylinder head as indicated in the illustration. Install the seal washers onto the bolts. Install the cylinder head cover bolts and nuts. Tighten bolts "A" to 7.4 ft. lbs. Tighten bolts "B" to 80 inch lbs. Tighten nuts to 80 inch lbs.

➡Be sure to remove any oil from the contact surfaces of the cylinder head cover and the cylinder head. Install the cover within three minutes after applying the seal packing. Tighten the bolts to specification within fifteen minutes after installing the cover. Do not add engine oil for at least two hours after installing the cover.

30. Continue the installation in the reverse order of the removal procedure.

31. Be sure to fill the engine with the proper grade and type engine coolant.

32. Be sure to fill the engine with the proper grade and type engine oil.

33. Start the engine and check for leaks. Correct as required.

ENGINE PERFORMANCE & EMISSION CONTROLS

COMPONENT LOCATIONS

See Figures 119 through 124.

CAMSHAFT POSITION (CMP) SENSOR

LOCATION

See Figure 125.

At the front of the cylinder head.

REMOVAL & INSTALLATION

See Figure 125

1. Before servicing the vehicle, refer to the Precautions Section.
2. Disconnect the negative battery cable. Tape the cable with insulating tape.
3. Disconnect the Camshaft Position (CMP) sensor connector.
4. Remove the bolt, then remove the Camshaft Position (CMP) sensor.
5. Discard the O-ring.

To install:

➡Be sure to use new fasteners, as required.

6. Installation is the reverse of the removal procedure.

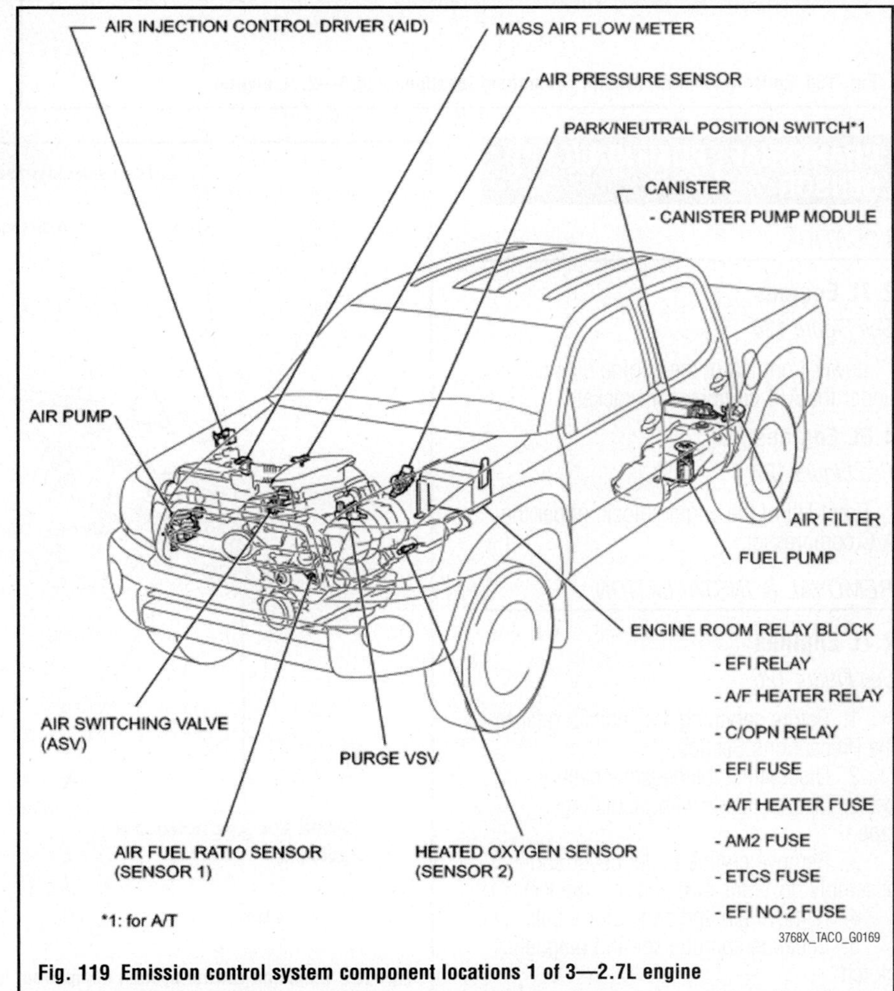

Fig. 119 Emission control system component locations 1 of 3—2.7L engine

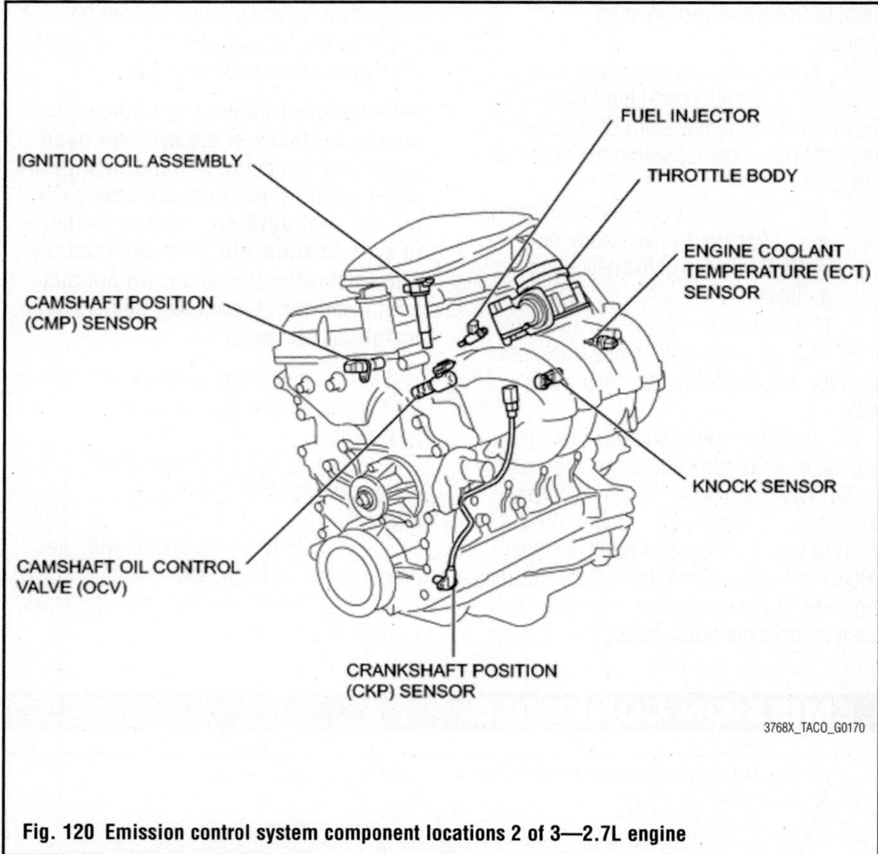

Fig. 120 Emission control system component locations 2 of 3—2.7L engine

CRANKSHAFT POSITION (CKP) SENSOR

LOCATION

2.7L Engines

See Figure 126.

Lower front left of the engine block, under the A/C compressor bracket.

4.0L Engines

See Figure 127.

Front left of the engine block, under the A/C compressor.

REMOVAL & INSTALLATION

2.7L Engines

See Figure 126

1. Before servicing the vehicle, refer to the Precautions Section.
2. Disconnect the negative battery cable. Tape the cable with insulating tape.
3. Remove engine under cover sub-assembly no.1 (for 4wd and pre-runner).
4. Remove fan and generator v belt.
5. Separate compressor and magnetic clutch.
6. Remove idle pulley assembly with bracket.
7. Remove the 5 bolts, then remove the idle pulley with bracket.
8. Disconnect the Crankshaft Position (CKP) sensor connector and the 2 wire harness clamps.
9. Remove the bolt, then remove the Crankshaft Position (CKP) sensor.

To install:

➡Be sure to use new fasteners, as required.

10. Installation is the reverse of the removal procedure.

4.0L Engines

See Figure 127

1. Before servicing the vehicle, refer to the Precautions Section.
2. Disconnect the negative battery cable. Tape the cable with insulating tape.
3. Remove v-bank cover.
4. Remove fan shroud.
5. Remove generator assembly.
6. Separate cooler compressor assembly.
7. Remove the bolt, then separate the suction hose sub-assembly.

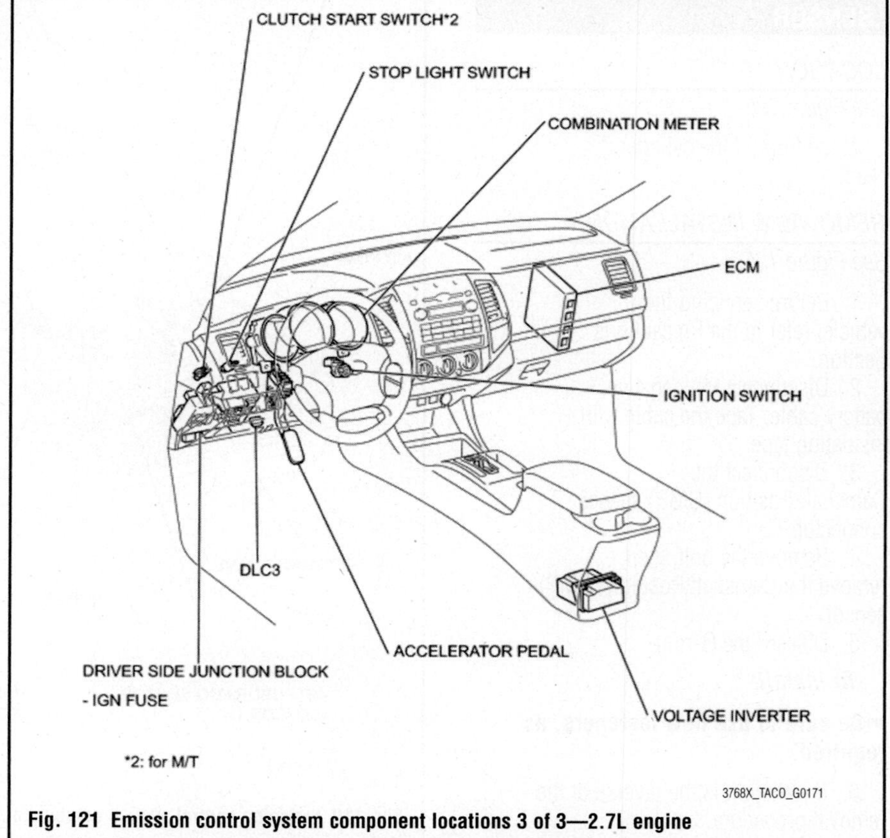

Fig. 121 Emission control system component locations 3 of 3—2.7L engine

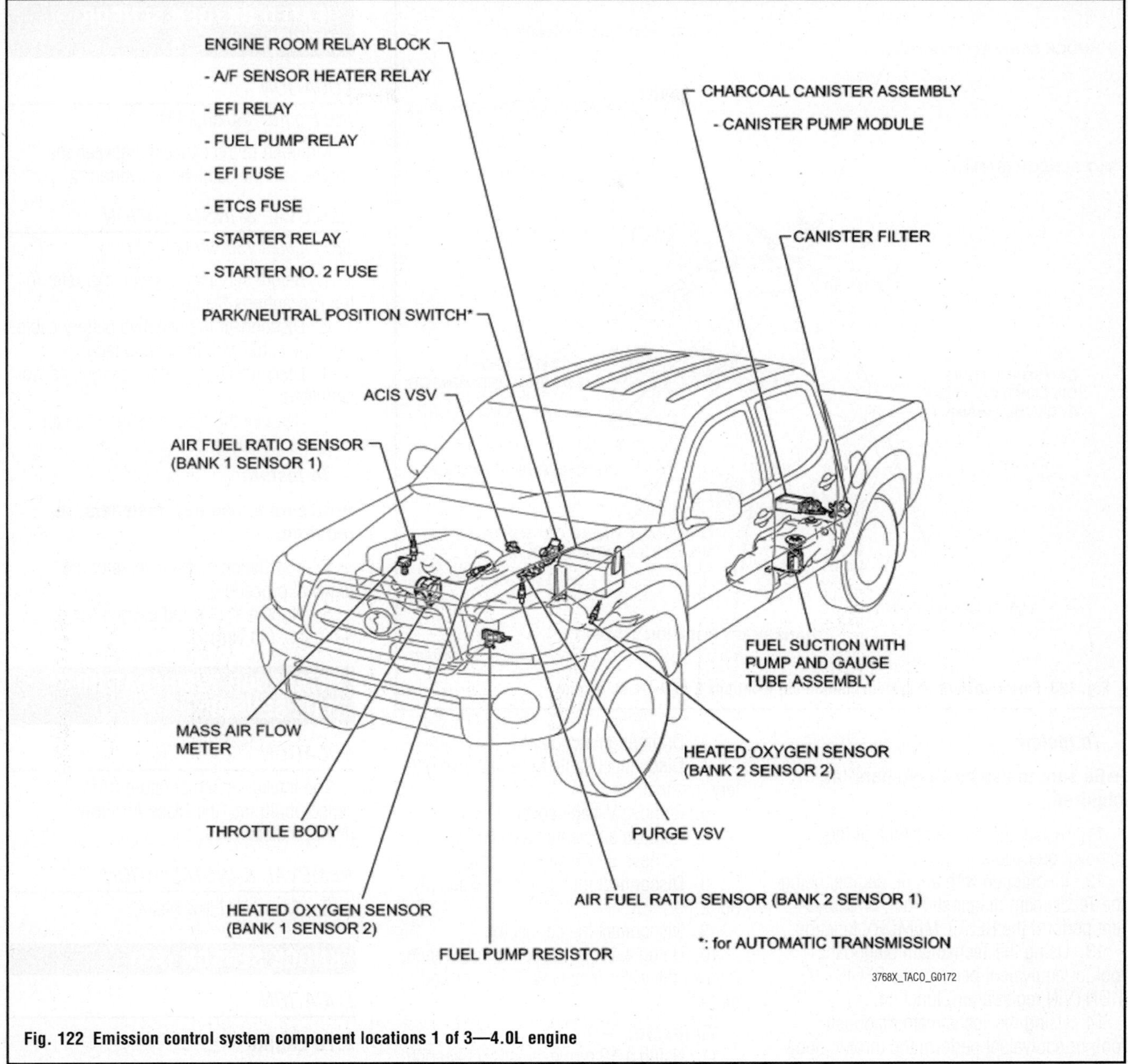

Fig. 122 Emission control system component locations 1 of 3—4.0L engine

8. Disconnect the cooler compressor assembly connector.
9. Remove the 4 bolts, then separate the cooler compressor assembly from the V-ribbed belt tensioner assembly.
10. Disconnect the Crankshaft Position (CKP) sensor connector.
11. Remove the bolt, then remove the Crankshaft Position (CKP) sensor.

To install:

➡Be sure to use new fasteners, as required.

12. Installation is the reverse of the removal procedure.

ELECTRONIC CONTROL MODULE (ECM)

LOCATION

See Figure 128.

Behind the glove box.

REMOVAL & INSTALLATION

See Figure 128

1. Before servicing the vehicle, refer to the Precautions Section.
2. Disconnect the negative battery cable. Tape the cable with insulating tape.
3. Perform VIN registration when replacing the ECM.
4. Disconnect cable from negative battery terminal
5. Remove glove compartment door assembly
6. Remove instrument panel finish panel sub-assembly lower, right side.
7. Disconnect the 4 connectors.
8. Remove the 2 bolts and nut, then remove the ECM.
9. Remove the 3 screws, then remove the ECM bracket.
10. Remove the 2 screws, then remove the ECM bracket No. 2.

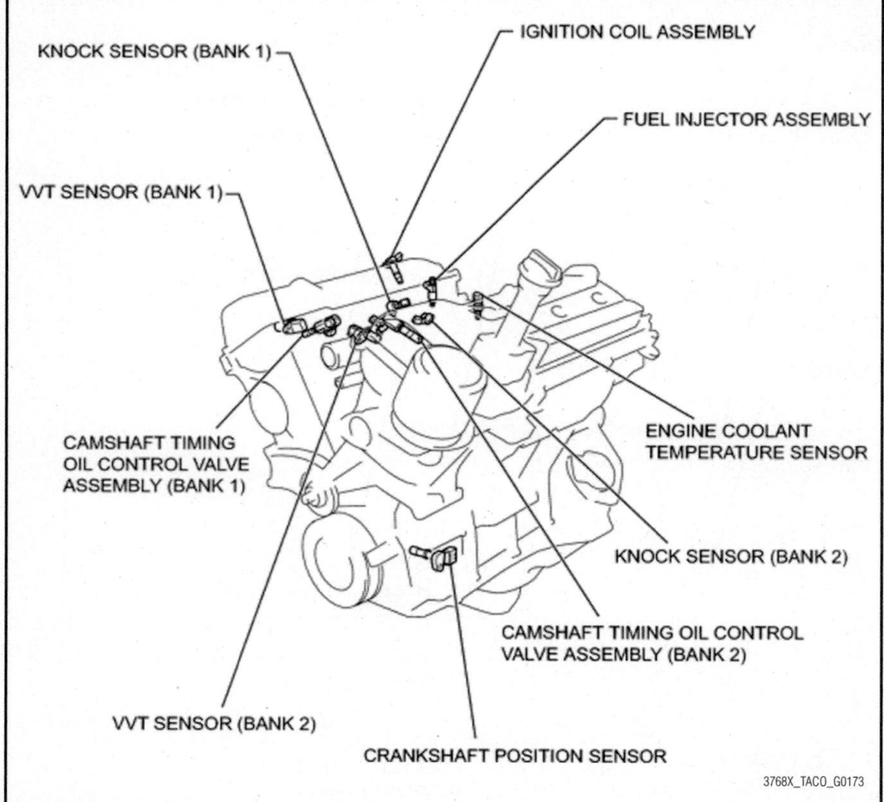

Fig. 123 Emission control system component locations 2 of 3—4.0L engine

To install:

➡Be sure to use new fasteners, as required.

11. Installation is the reverse of the removal procedure.
12. If equipped with a 4.0L engine, using the Techstream diagnostic tool, or equivalent perform the RESET MEMORY function.
13. Using the Techstream diagnostic tool, or equivalent perform the REGISTRATION (VIN registration) function.
14. Using the Techstream diagnostic tool, or equivalent perform the throttle body initialization function.

ENGINE COOLANT TEMPERATURE (ECT) SENSOR

LOCATION

Mounted near the thermostat housing.

REMOVAL & INSTALLATION

4.0L Engines

See Figure 129.

1. Before servicing the vehicle, refer to the Precautions section.
2. Discharge fuel system pressure.
3. Drain engine coolant.
4. Disconnect cable from negative battery terminal.
5. Remove V-bank cover.
6. Remove air cleaner assembly.
7. Remove intake air surge tank.
8. Disconnect fuel pipe sub-assembly no. 2.
9. Disconnect the connector.
10. Using a 19 mm deep socket wrench, remove the water temperature sensor and gasket.

To install:

11. Using a 19 mm deep socket wrench, install the water temperature sensor with a new gasket and tighten to 15 ft. lbs. (20 Nm).
12. Connect the connector.
13. Connect fuel pipe sub-assembly no. 2.
14. Install intake air surge tank.
15. Install air cleaner assembly.
16. Connect cable to negative battery terminal.
17. Add engine coolant .
18. Check for fuel leakage.
19. Check for engine coolant leakage
20. Install V-bank cover.

HEATED OXYGEN SENSOR (HO2S)

LOCATION

See Figures 130 and 131.

Mounted in the exhaust, between the engine and the catalytic converter(s).

REMOVAL & INSTALLATION

See Figures 130 through 131

1. Before servicing the vehicle, refer to the Precautions Section.
2. Disconnect the negative battery cable. Tape the cable with insulating tape.
3. Disconnect the heated oxygen sensor connector.
4. Remove the heated oxygen sensor from the exhaust pipe.

To install:

➡Be sure to use new fasteners, as required.

5. Installation is the reverse of the removal procedure.
6. Tighten the heated oxygen sensor to 33 ft. lbs. (44 Nm).

INTAKE AIR TEMPERATURE (IAT) SENSOR

LOCATION

The Intake Air Temperature (IAT) sensor, built into the Mass Air Flow (MAF) meter.

REMOVAL & INSTALLATION

See Mass Air Flow Meter.

KNOCK SENSOR (KS)

LOCATION

2.7L Engine

See Figure 132.

On the left side of the engine block, under the intake manifold.

4.0L Engine

See Figure 133.

Mounted to the top of the engine block under the intake manifold.

REMOVAL & INSTALLATION

2.7L Engine

See Figure 134.

1. Before servicing the vehicle, refer to the Precautions Section.

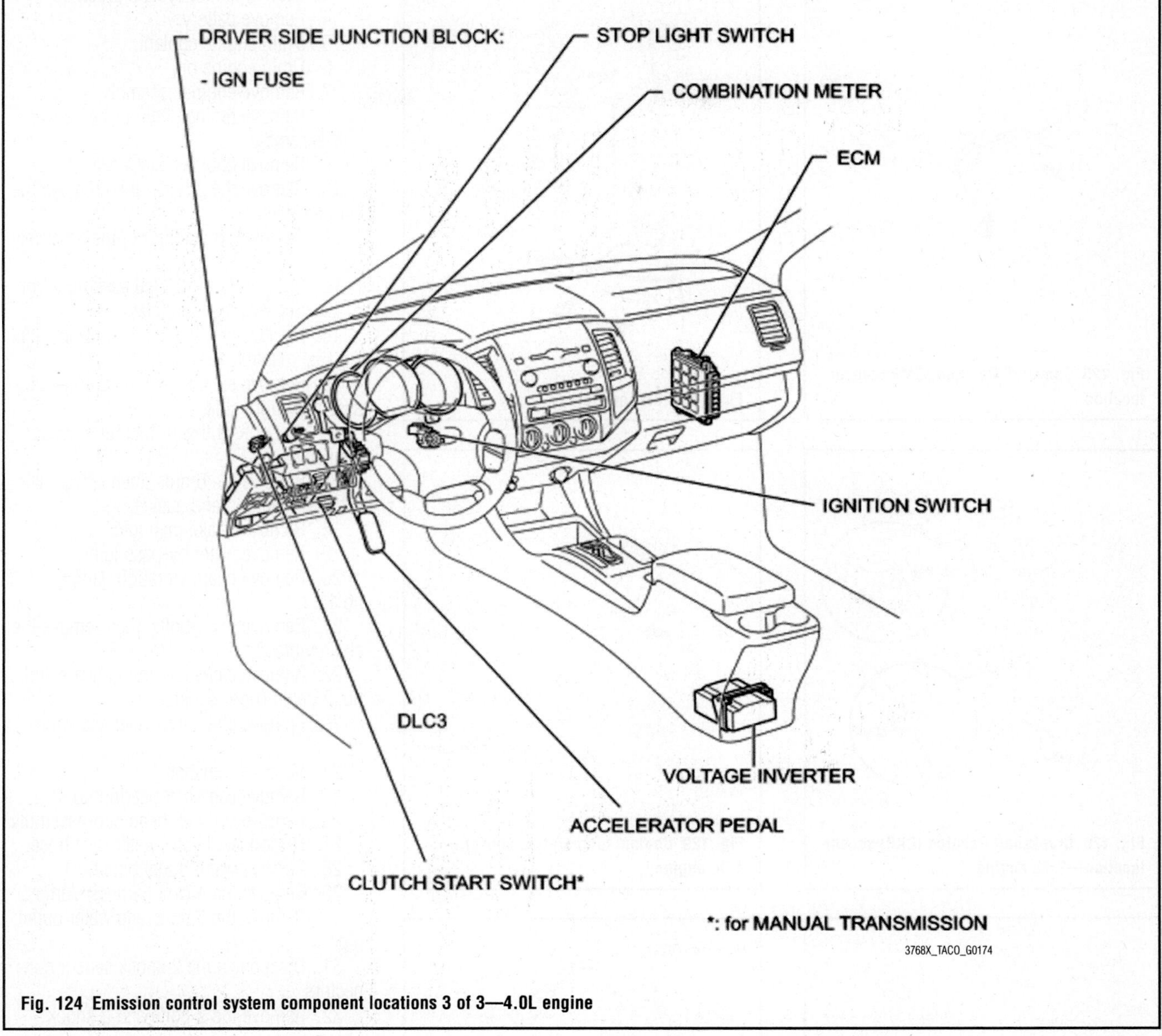

Fig. 124 Emission control system component locations 3 of 3—4.0L engine

2. Disconnect the negative battery cable. Tape the cable with insulating tape.
3. Remove engine under cover sub-assembly.
4. Drain engine coolant.
5. Remove intake air connector.
6. Remove throttle with motor body assembly.
7. Disconnect fuel hose.
8. Disconnect fuel hose no. 2.
9. Remove intake manifold.
10. Disconnect the fuel vapor feed hose from the VSV.
11. Disconnect the vacuum hose.
12. Remove the bolt, then remove the clamp bracket.
13. Disconnect the water by-pass hose No. 2.
14. Disconnect the ventilation hose No. 3.
15. Disconnect the VSV connector.
16. Disengage the engine wire harness clamp.
17. Disconnect the compressor magnetic clutch connector.
18. Disengage the wire harness clamp.
19. Remove the bolt and harness clamp bracket.
20. Disconnect the 3 connectors.
21. Remove the nut.
22. Disconnect the engine wire harness from the relay block.
23. Remove the 5 bolts and 2 nuts, then remove the intake manifold.
24. Disconnect the knock control sensor connector.
25. Remove the bolt, then remove the knock control sensor.

To install:

➡Be sure to use new fasteners, as required.

26. Installation is the reverse of the removal procedure.
27. Properly position the sensor on its mounting, see illustration.
28. Tighten the knock sensor bolt to 15 ft. lbs. (20 Nm).

4.0L Engine

2WD

See Figure 135.

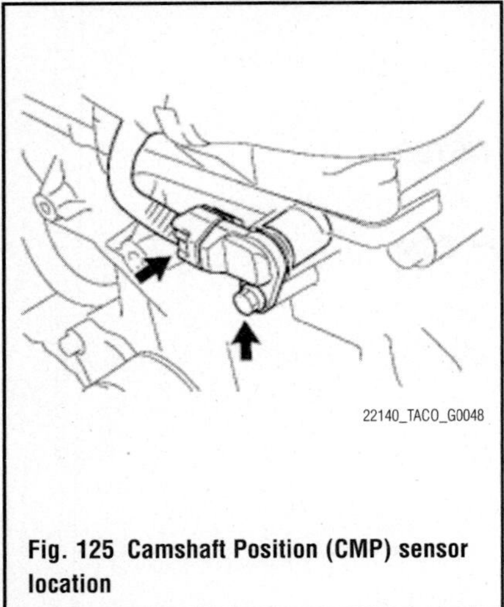

Fig. 125 Camshaft Position (CMP) sensor location

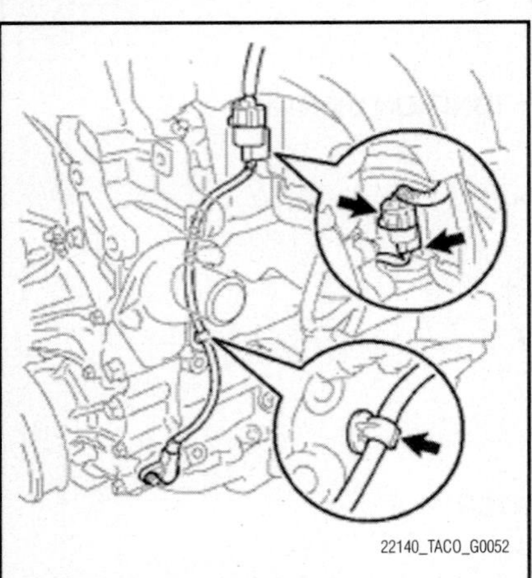

Fig. 126 Crankshaft Position (CKP) sensor location—2.7L engine

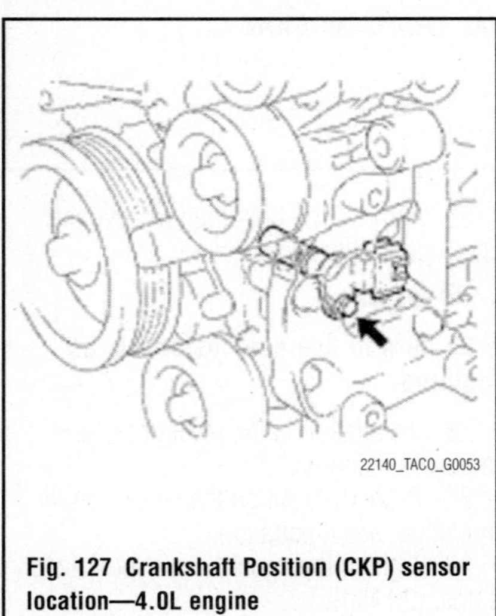

Fig. 127 Crankshaft Position (CKP) sensor location—4.0L engine

1. Before servicing the vehicle, refer to the Precautions Section.
2. Disconnect the negative battery cable. Tape the cable with insulating tape.

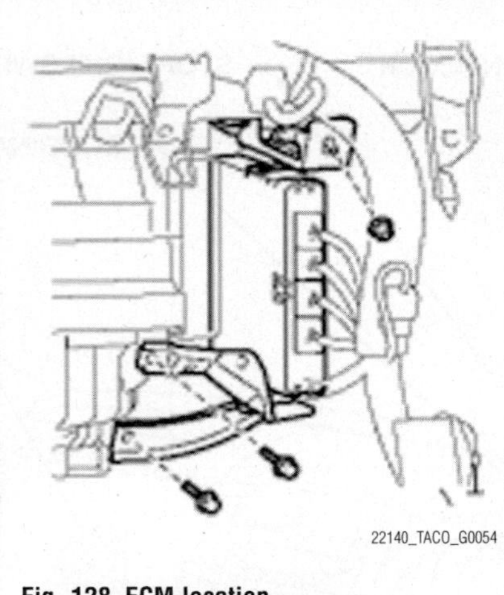

Fig. 128 ECM location

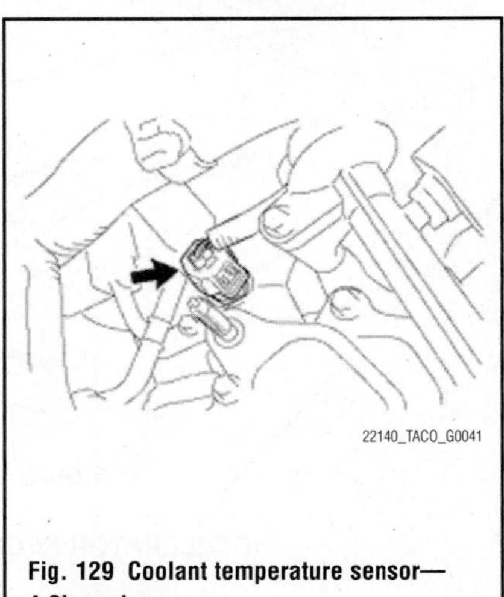

Fig. 129 Coolant temperature sensor—4.0L engine

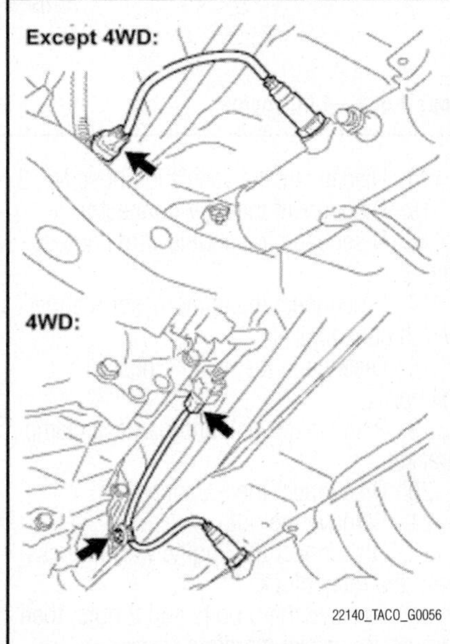

Fig. 130 Heated oxygen sensor location—2.7L engine

3. Discharge fuel system pressure.
4. Remove battery.
5. Drain engine coolant.
6. Drain engine oil.
7. Remove engine assembly.
8. Remove timing chain or belt cover sub-assembly.
9. Remove cool air inlet no.1.
10. Remove the 2 bolts, then remove the cool air inlet.
11. Remove exhaust front pipe assembly no. 2.
12. Remove exhaust pipe assembly front.
13. Remove manifold stay.
14. Remove the 3 bolts, then remove the exhaust manifold stay.
15. Remove exhaust manifold sub-assembly right side.
16. Disconnect the air fuel ratio sensor connector.
17. Remove the 6 nuts, then remove the exhaust manifold and gasket.
18. Remove intake manifold.
19. Remove water by-pass joint.
20. Remove chain vibration damper no.1.
21. Remove the 2 bolts, then remove the chain vibration damper No. 1.
22. Remove camshaft timing gears and no. 2 chain (right bank).
23. Remove chain tensioner assembly no. 2.
24. Remove camshafts.
25. Remove camshaft bearing no. 2.
26. Remove cylinder head sub-assembly.
27. Disconnect heater water inlet hose.
28. Remove water outlet pipe no.1.
29. Remove the 4 wire harness clamps.
30. Remove the 3 bolts and water outlet pipe.
31. Disconnect the 2 knock sensor connectors.
32. Remove the 2 bolts and 2 knock sensors.

To install:

➡Be sure to use new fasteners, as required.

33. Installation is the reverse of the removal procedure.
34. Properly position the sensor on its mounting, see illustration.
35. Tighten the knock sensor bolt to 15 ft. lbs. (20 Nm).

4WD AND PRERUNNER

See Figure 135

1. Before servicing the vehicle, refer to the Precautions Section.
2. Disconnect the negative battery cable. Tape the cable with insulating tape.

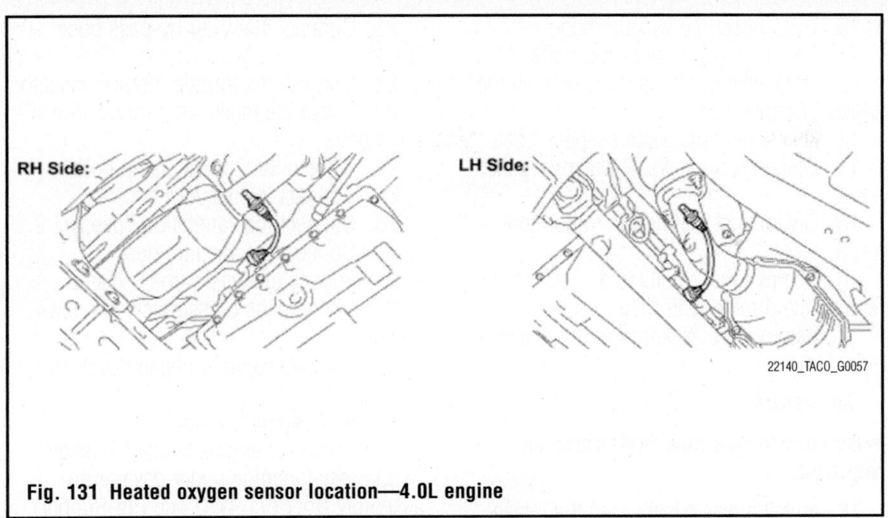

Fig. 131 Heated oxygen sensor location—4.0L engine

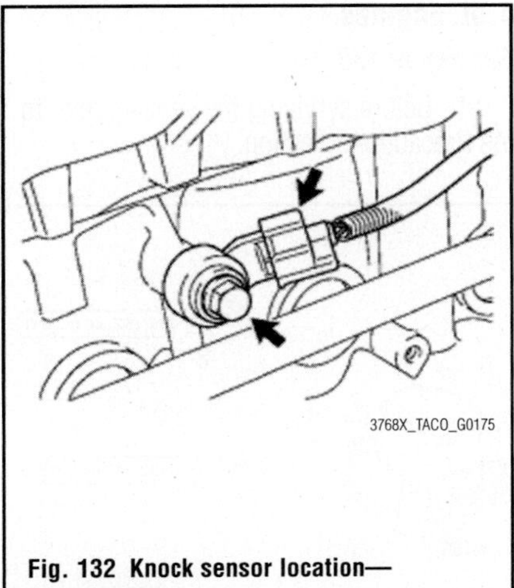

Fig. 132 Knock sensor location—2.7L engine

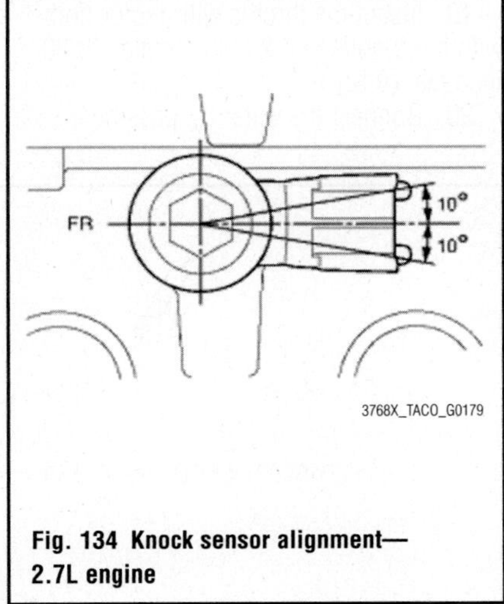

Fig. 134 Knock sensor alignment—2.7L engine

Fig. 133 Knock sensor location—4.0L engine

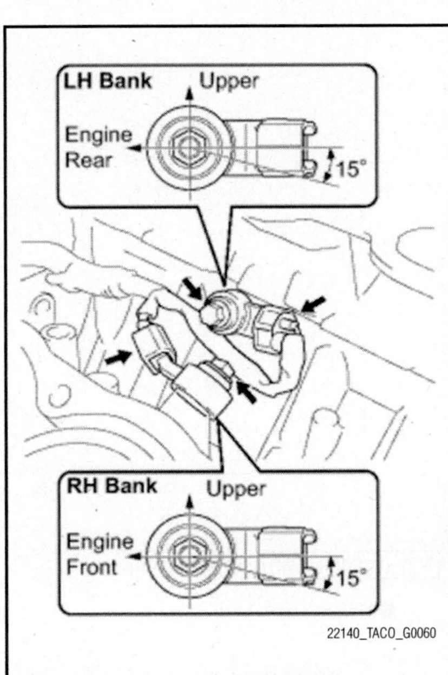

Fig. 135 Knock sensor alignment—4.0L engine

3. Discharge fuel system pressure.
4. Remove battery.
5. Drain engine coolant.
6. Drain engine oil.
7. Remove power steering link assembly.
8. Remove differential carrier assembly front (4wd drive type).
9. Remove timing chain or belt cover sub-assembly.
10. Remove chain sub-assembly.
11. Remove the 2 bolts, then remove the cool air inlet.
12. Remove exhaust front pipe assembly no. 2.
13. Remove exhaust pipe assembly front.
14. Remove manifold stay.
15. Remove the 6 bolts, then remove the 2 exhaust manifold stays.
16. Remove exhaust manifold sub-assembly right side.
17. Disconnect fuel pipe sub-assembly no.1.
18. Disconnect fuel pipe sub-assembly no. 2.
19. Remove intake manifold.
20. Remove water by-pass joint (right rear).
21. Remove chain vibration damper no.1.
22. Remove camshaft timing gears and no. 2 chain (right bank).
23. Remove chain tensioner assembly no. 2.
24. Remove camshafts.
25. Remove camshaft bearing no. 2.
26. Remove cylinder head sub-assembly.
27. Disconnect heater water inlet hose.
28. Remove the 4 wire harness clamps.
29. Remove the 3 bolts and water outlet pipe.
30. Disconnect the 2 knock sensor connectors.
31. Remove the 2 bolts and 2 knock sensors.

To install:

➡Be sure to use new fasteners, as required.

32. Installation is the reverse of the removal procedure.
33. Properly position the sensor on its mounting, see illustration.
34. Tighten the knock sensor bolt to 15 ft. lbs. (20 Nm).

MASS AIR FLOW (MAF) SENSOR

LOCATION

See Figure 136.

In the air intake, between the air filter and the throttle body.

REMOVAL & INSTALLATION

See Figure 136

1. Before servicing the vehicle, refer to the Precautions Section.
2. Disconnect the negative battery cable. Tape the cable with insulating tape.
3. Remove v-bank cover.
4. Disconnect the connector.
5. Remove the 2 screws, then remove the mass air flow meter.

To install:

➡Be sure to use new fasteners, as required.

6. Installation is the reverse of the removal procedure.

THROTTLE POSITION SENSOR (TPS)

LOCATION

Part of the throttle body.

REMOVAL & INSTALLATION

2.7L Engines

See Figure 137.

1. Before servicing the vehicle, refer to the Precautions Section.
2. Disconnect the negative battery cable. Tape the cable with insulating tape.
3. Remove no. 1 engine under cover sub-assembly (for 4WD and PreRunner)
4. Drain engine coolant
5. Remove intake air connector
6. Disconnect the pressure sensor connector.
7. Disengage the wire harness clamp.
8. Disconnect the vacuum hose.
9. Disconnect the No. 2 ventilation hose.
10. Disconnect the vacuum hose.
11. Loosen the 2 hose clamp bolts.
12. Remove the 3 bolts, then remove the intake air connector.
13. Disconnect the water by-pass hose.
14. Disconnect the No. 2 water by-pass hose.
15. Disconnect the throttle motor connector.
16. Remove the 2 bolts and 2 nuts, then remove the throttle with motor body.
17. Remove the gasket from the intake manifold.

To install:

➡Be sure to use new fasteners, as required.

18. Install a new gasket onto the intake manifold.
19. Install the throttle with motor body with the 2 bolts and 2 nuts. Tighten to 80 inch lbs. (9 Nm).
20. Connect the water by-pass hose.
21. Connect the water by-pass hose No. 2.
22. Connect the throttle motor connector.
23. Install the intake air connector with the 3 bolts.
24. Tighten the 2 hose clamp bolts.
25. Connect the vacuum hose.
26. Connect the ventilation hose No. 2.
27. Connect the vacuum hose.
28. Engage the wire harness clamp.
29. Connect the pressure sensor connector.
30. Connect cable to negative battery terminal
31. Add engine coolant
32. Check for engine coolant leakage
33. Install engine under cover sub-assembly No.1 (for 4WD and Pre-Runner).

4.0L Engines

See Figure 138.

1. Before servicing the vehicle, refer to the Precautions Section.

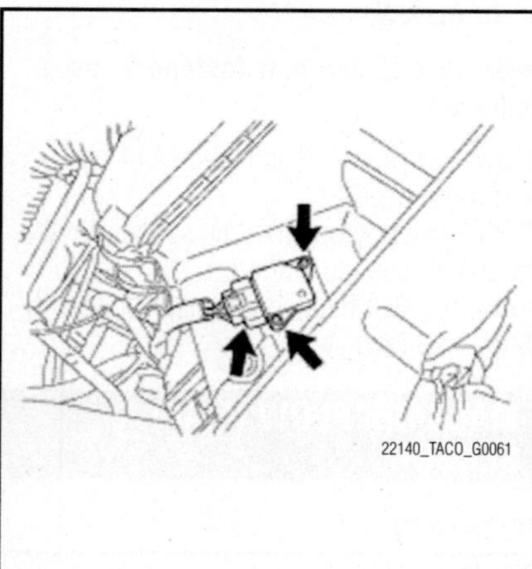

Fig. 136 Mass air flow sensor location—4.0L engine

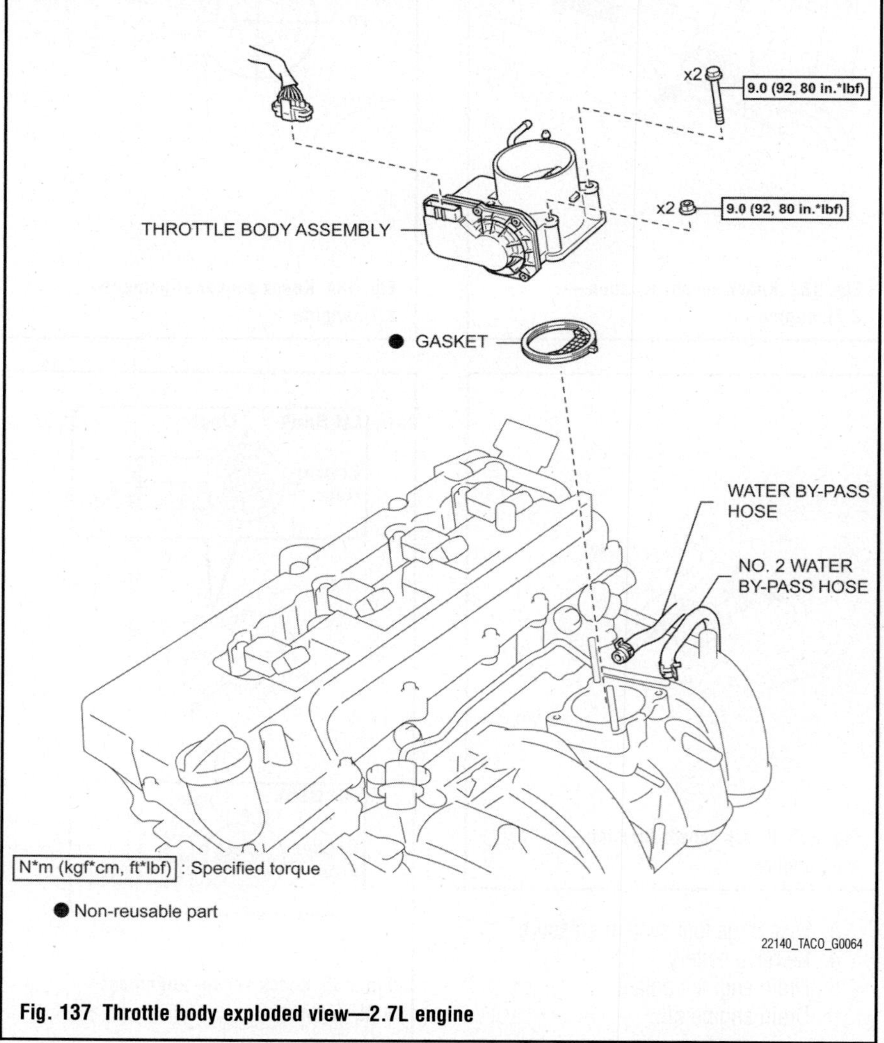

Fig. 137 Throttle body exploded view—2.7L engine

2. Disconnect the negative battery cable. Tape the cable with insulating tape.
3. Drain engine coolant
4. Remove the 2 nuts, then remove the V-bank cover.
5. Disconnect the ventilation hose No. 2.
6. Disconnect the vacuum hose.
7. Disconnect the mass air flow meter connector.
8. Remove the 2 wire harness clamps.
9. Loosen the 2 hose clamps.
10. Remove the 2 bolts, then remove the air cleaner.
11. Disconnect the water by-pass hose No. 5.
12. Disconnect the water by-pass hose No. 4.
13. Disconnect the throttle motor connector.
14. Remove the 4 bolts, then remove the throttle w/ motor body and gasket.

To install:

➡Be sure to use new fasteners, as required.

15. Install a new gasket and the throttle with motor body with the 4 bolts and tighten to 108 inch lbs. (11 Nm).
16. Connect the throttle motor connector.
17. Connect the water by-pass hose No. 4.
18. Connect the water by-pass hose No. 5.
19. Install the air cleaner with the 2 bolts.
20. Connect the ventilation hose No. 2.
21. Connect cable to negative battery terminal
22. Add engine coolant
23. Check for engine coolant leakage
24. Install the V-bank cover with the 2 nuts.

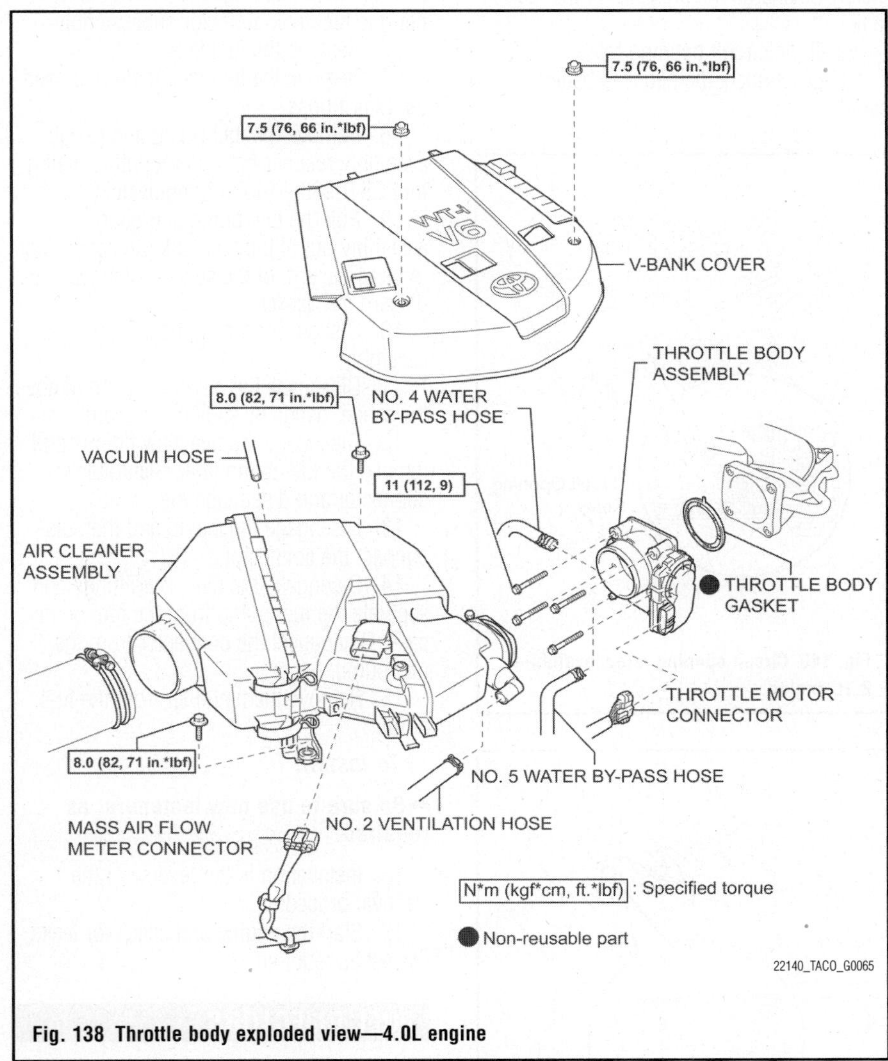

Fig. 138 Throttle body exploded view—4.0L engine

VEHICLE SPEED SENSOR (VSS)

REMOVAL & INSTALLATION

See Figure 139.

1. Before servicing the vehicle, refer to the Precautions Section.
2. Disconnect the negative battery cable. Tape the cable with insulating tape.
3. Disconnect the speed sensor connector.
4. Remove the bolt and speed sensor.
5. Remove the O-ring from the speed sensor.

To install:

➡Be sure to use new fasteners, as required.

6. Installation is the reverse of the removal procedure.

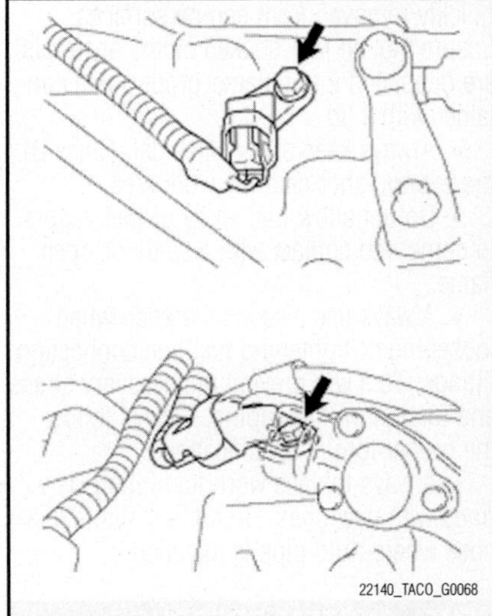

Fig. 139 Vehicle speed sensor—A340E transmission

FUEL SYSTEM SERVICE PRECAUTIONS

Safety is the most important factor when performing not only fuel system maintenance, but any type of maintenance. Failure to conduct maintenance and repairs in a safe manner may result in serious personal injury or death. Work on a vehicle's fuel system components can be accomplished safely and effectively by adhering to the following rules and guidelines.

- To avoid the possibility of fire and personal injury, always disconnect the negative battery cable unless the repair or test procedure requires that battery voltage be applied.
- Always relieve the fuel system pressure prior to disconnecting any fuel system component (injector, fuel rail, pressure regulator, etc.) fitting or fuel line connection. Exercise extreme caution whenever relieving fuel system pressure to avoid exposing skin, face and eyes to fuel spray. Please be advised that fuel under pressure may penetrate the skin or any part of the body that it contacts.
- Always place a shop towel or cloth around the fitting or connection prior to loosening to absorb any excess fuel due to spillage. Ensure that all fuel spillage is quickly removed from engine surfaces. Ensure that all fuel-soaked cloths or towels are deposited into a flame-proof waste container with a lid.
- Always keep a dry chemical (Class B) fire extinguisher near the work area.
- Do not allow fuel spray or fuel vapors to come into contact with a spark or open flame.
- Always use a second wrench when loosening or tightening fuel line connection fittings. This will prevent unnecessary stress and torsion on fuel piping. Always follow the proper torque specifications.
- Always replace worn fuel fitting O-rings with new ones. Do not substitute fuel hose where rigid pipe is installed.

FUEL SYSTEM PRESSURE

RELIEVING

See Figures 140 and 141.

1. Before servicing the vehicle, refer to the Precautions Section.
2. Disconnect the negative battery cable. Tape the cable with insulating tape.
3. Remove the engine relay block cover.
4. Remove the circuit opening relay.
5. Connect the negative battery.
6. Start the engine.
7. Turn the ignition switch "ON" after the engine stops.

➡Code DTC P0171 (system lean) may be present.

8. Crank the engine again. Check that the engine stops.
9. Remove the fuel tank cap and completely discharge the pressure in the fuel tank.

Install the circuit opening relay.

10. Disconnect the negative battery cable.

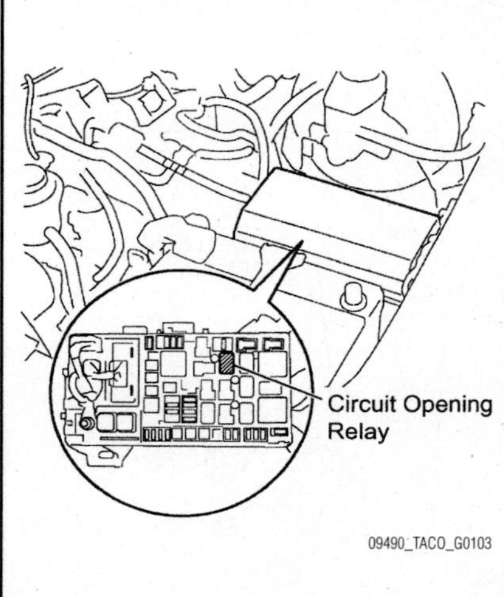

Fig. 140 Circuit opening relay location—2.7L engine

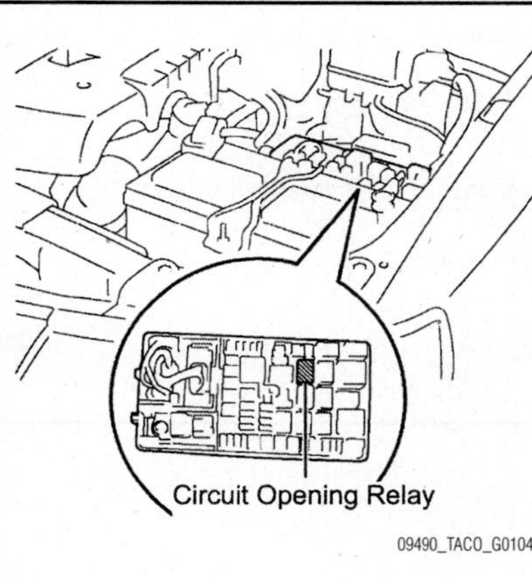

Fig. 141 Circuit opening relay location—4.0L engine

FUEL FILTER

REMOVAL & INSTALLATION

See Figure 144

The fuel filter is part of the fuel pump, and is located in the fuel tank.

1. Before servicing the vehicle, refer to the Precautions Section.
2. Disconnect the negative battery cable. Tape the cable with insulating tape.
3. Relieve the fuel pressure.
4. Drain the fuel from the fuel tank.
5. On vehicles equipped with off road package and 4.0L engine, remove the number one fuel tank protector subassembly.
6. Remove the fuel tank.
7. Remove the fuel tank main hose and the return hose.
8. Remove the fuel pump and gauge assembly retainer from its mounting, using tool SST09808-14020 or equivalent.
9. Pull the fuel pump and gauge assembly out of the fuel tank. Be careful not to bend the arm of the sender. Remove and discard the gasket.
10. Disconnect the connector from the assembly.
11. Disengage the claw fitting and remove the sender gauge by sliding it forward.
12. Disengage the five claw fittings and remove the fuel pump tank. Separate the connector and disengage the clamp.
13. Disengage the clamp and then disconnect the connector.
14. Disengage the five claw fittings and separate the fuel pump from the fuel pump case. Disconnect the connector from the fuel pump.
15. Remove the fuel filter from the fuel pump.

To install:

➡Be sure to use new fasteners, as required.

16. Installation is the reverse of the removal procedure.
17. Start the engine and check for leaks, correct as required.

FUEL INJECTORS

REMOVAL & INSTALLATION

2.7L Engines

See Figure 142.

1. Before servicing the vehicle, refer to the Precautions Section.

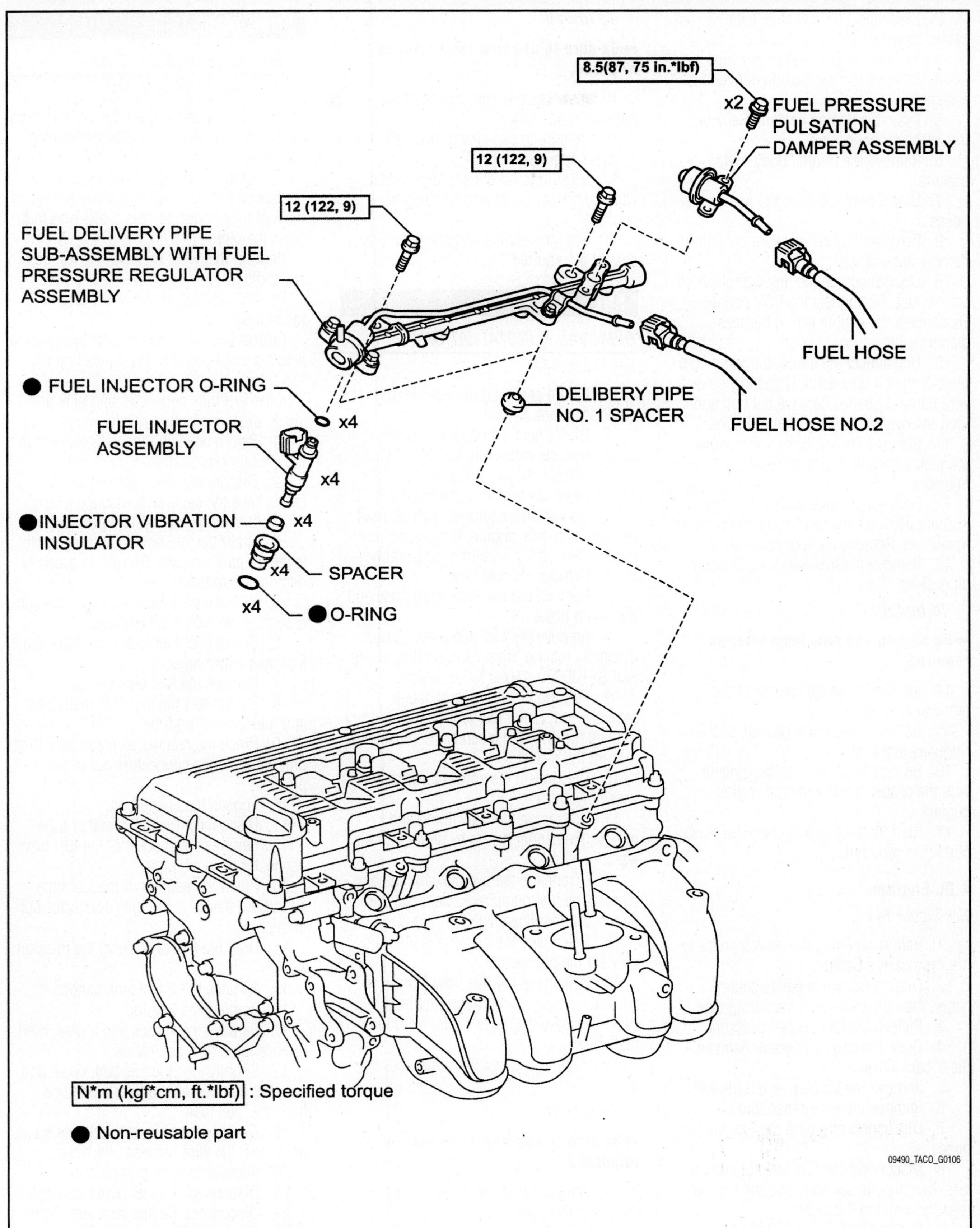

Fig. 142 Fuel injector and related components—2.7L engine

2. Disconnect the negative battery cable. Tape the cable with insulating tape.
3. Relieve the fuel system pressure.
4. Remove the engine undercover subassembly.
5. Drain the engine coolant. Remove the intake air connector.
6. Remove the throttle body motor assembly.
7. Disconnect and plug the fuel line hoses.
8. Remove the fuel pressure pulsation damper assembly.
9. Disconnect the fuel injector electrical connectors. Disconnect the VSV connector. Disconnect the engine wiring harness clamp.
10. Disconnect the air conditioning compressor clutch connector. Disconnect the wire harness clamp. Remove the bolt and then remove the harness clamp bracket.
11. Remove the two bolts and remove the fuel rail together with the fuel injectors.
12. Remove the fuel rail number one spacers. Remove the four injector vibration insulators. Remove the four spacers.
13. Remove the fuel injectors. Discard all gaskets.

To install:

➡Be sure to use new fasteners, as required.

14. Installation is the reverse of the removal procedure.
15. Be sure to use new gaskets and O-rings, as required.
16. Be sure to fill the cooling system with the proper grade and type engine coolant.
17. Start the engine and check for leaks, correct as required.

4.0L Engines

See Figure 143.

1. Before servicing the vehicle, refer to the Precautions Section.
2. Disconnect the negative battery cable. Tape the cable with insulating tape.
3. Relieve the fuel system pressure.
4. Drain the engine coolant. Remove the V bank cover.
5. Remove the air cleaner assembly.
6. Remove the intake manifold.
7. Disconnect and plug the fuel line hoses.
8. Disconnect the fuel injector connectors. Remove the six bolts and the fuel rail together with the injectors.
9. Pull the injectors out of the fuel rail.

To install:

➡Be sure to use new fasteners, as required.

10. Installation is the reverse of the removal procedure.
11. Be sure to use new gaskets and O-rings, as required.
12. Be sure to fill the cooling system with the proper grade and type engine coolant.
13. Start the engine and check for leaks, correct as required.

FUEL PUMP

REMOVAL & INSTALLATION

See Figure 144.

1. Before servicing the vehicle, refer to the Precautions Section.
2. Disconnect the negative battery cable. Tape the cable with insulating tape.
3. Relieve the fuel pressure.
4. Drain the fuel from the fuel tank.
5. On vehicles equipped with off road package and 4.0L engine, remove the number one fuel tank protector subassembly.
6. Remove the fuel tank.
7. Remove the fuel tank main hose and the return hose.
8. Remove the fuel pump and gauge assembly retainer from its mounting, using tool SST09808-14020 or equivalent.
9. Pull the fuel pump and gauge assembly out of the fuel tank. Be careful not to bend the arm of the sender. Remove and discard the gasket.
10. Disconnect the connector from the assembly.
11. Disengage the claw fitting and remove the sender gauge by sliding it forward.
12. Disengage the five claw fittings and remove the fuel pump tank. Separate the connector and disengage the clamp.
13. Disengage the clamp and then disconnect the connector.
14. Disengage the five claw fittings and separate the fuel pump from the fuel pump case. Disconnect the connector from the fuel pump.
15. Remove the fuel filter from the fuel pump.

To install:

➡Be sure to use new fasteners, as required.

16. Installation is the reverse of the removal procedure.
17. Start the engine and check for leaks, correct as required.

FUEL TANK

REMOVAL & INSTALLATION

See Figures 145 through 147.

Remove any dirt and foreign objects from the fuel tube connector before performing this work.

Do not allow any scratches or foreign objects on the parts when disconnecting, as the fuel tube connector has the O-ring that seals the pipe.

Perform this work by hand. Do not use any tools.

Do not forcibly bend, twist or turn the nylon tube.

Protect the disconnected part by covering it with a vinyl bag after disconnecting the fuel tube.

If the fuel tube connector and pipe are stuck, push and pull to release them.

1. Before servicing the vehicle, refer to the Precautions Section.
2. Disconnect the negative battery cable. Tape the cable with insulating tape.
3. Discharge fuel system pressure.
4. Drain the fuel tank to an acceptable level, for tank removal. Be sure to properly dispose of used fuel.
5. Remove no.1 fuel tank protector subassembly (with off road package).
6. Disconnect fuel tank main tube and fuel tank return tube.
7. Remove the fuel pipe clamp.
8. Disconnect the fuel tank main tube and fuel tank return tube.
9. Pinch the retainer as illustrated, then pull the fuel tube connectors out of the pipes.
10. Loosen the clamp bolt.
11. Disconnect the fuel breather tube.
12. Remove the checker of the fuel tube connector from the pipe.
13. Pinch the retainer of the fuel tube connector, then pull the tube connector out of the pipe.
14. Hold the fuel tank using the mission jack.
15. Remove the 2 fuel tank bands.
16. Remove the 2 bolts.
17. Remove the 2 clips and 2 pins, then remove the 2 fuel tank bands.
18. Slightly jack the fuel tank down and separate the fuel tank to filler pipe hose from the inlet pipe.
19. Slowly jack down the fuel tank so as not to tear the wire harness and hose.
20. Remove the fuel pump cover.
21. Disconnect the fuel pump connector.
22. Disconnect the fuel tank vent hose.
23. Pinch the retainer and pull the fuel tank vent hose connector out of the charcoal

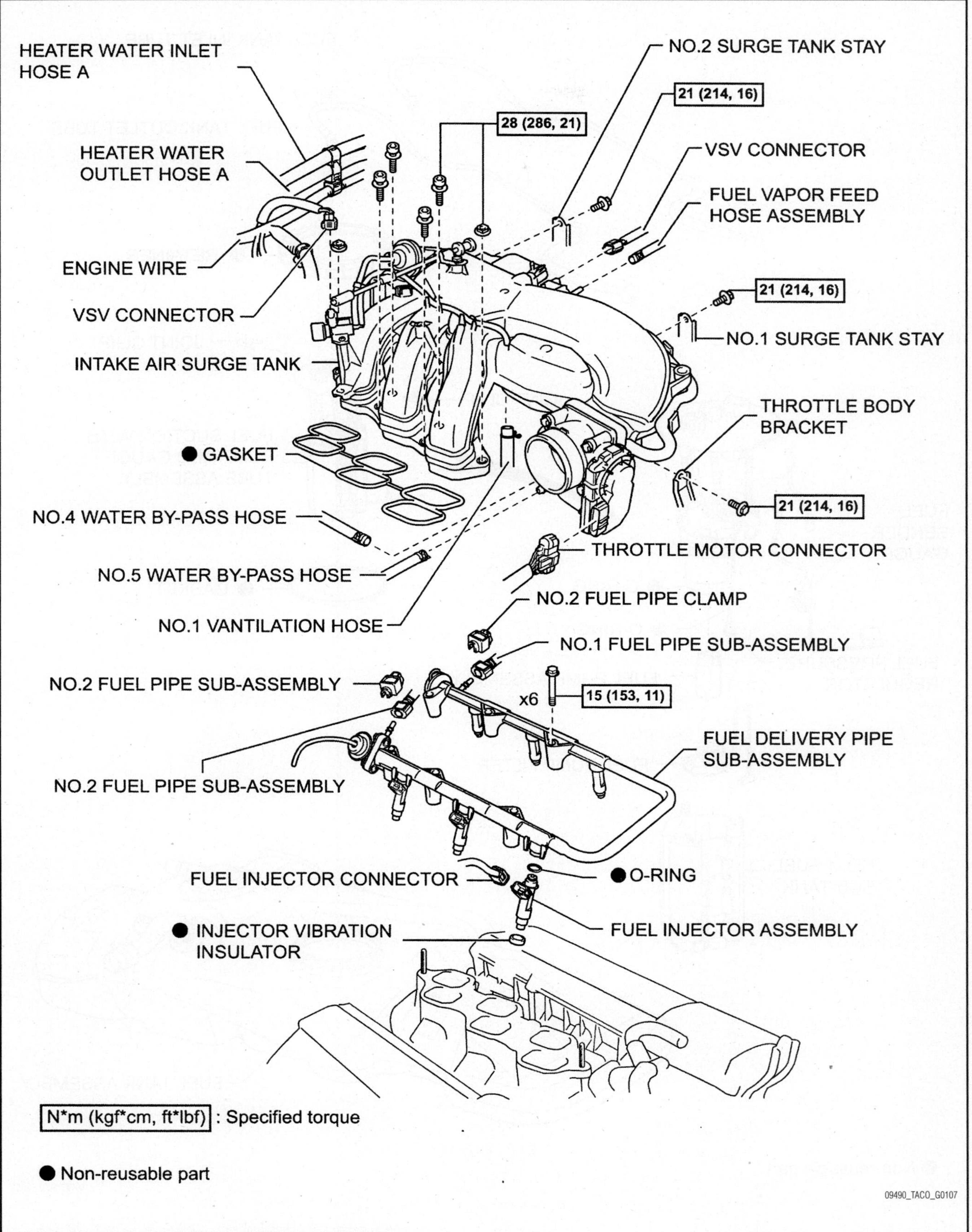

Fig. 143 Fuel injector and related components—4.0L engine

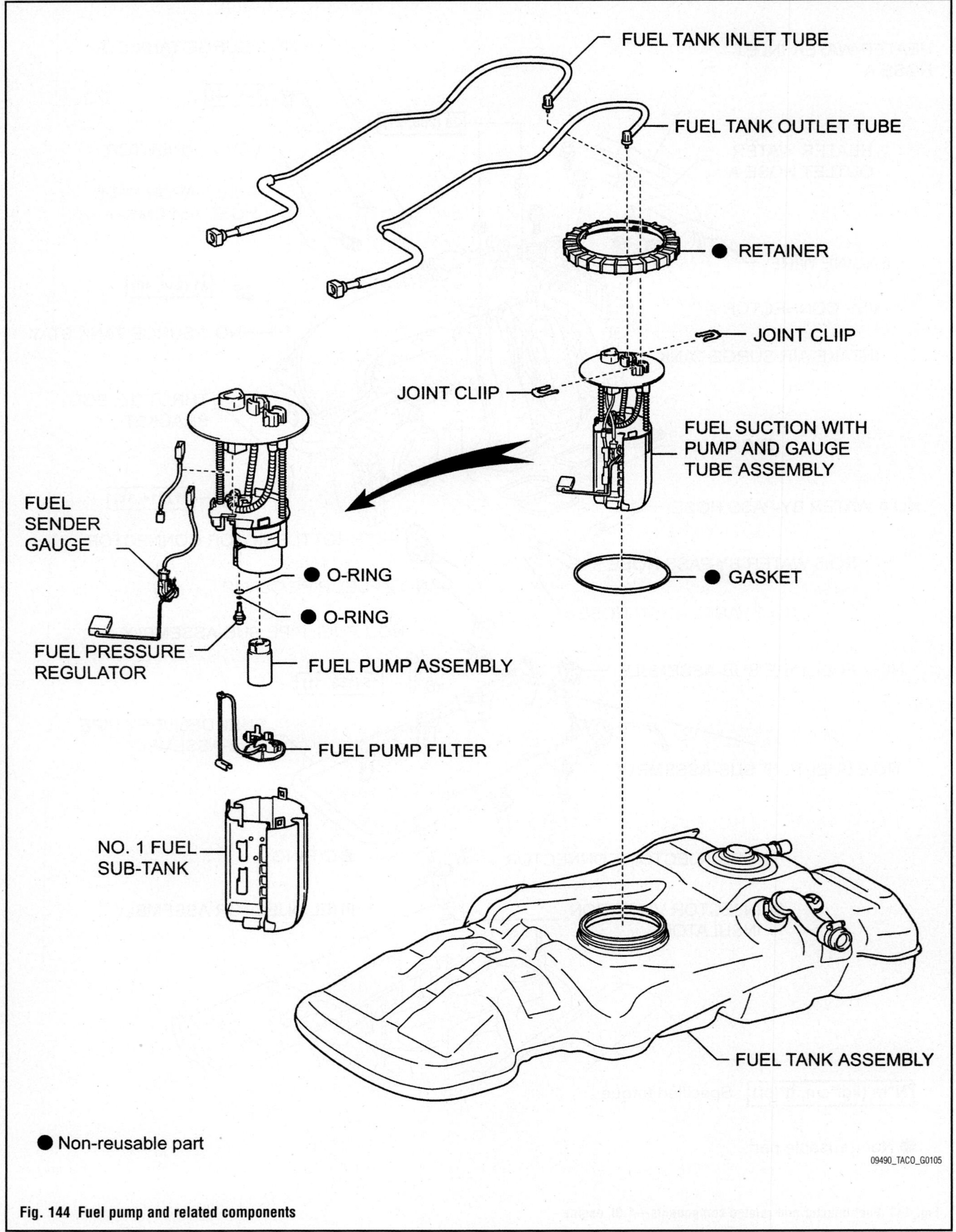

Fig. 144 Fuel pump and related components

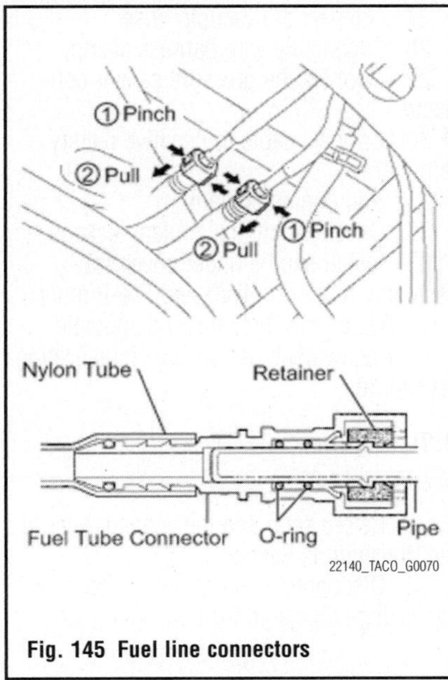

Fig. 145 Fuel line connectors

canister to disconnect the fuel tank vent hose from the charcoal canister.

24. Remove the fuel tank assy.
25. Remove the 2 joint clips, then remove the fuel tank main tube and fuel tank return tube.
26. Remove fuel suction with pump & gauge tube assembly

➡Protect the connector and tube joint with masking tape or equivalent to prevent any foreign objects from sticking to them. Clean any dirt and foreign objects from the fuel suction tube assembly before removing.

27. Using SST 09808-14020, loosen the retainer.

➡The ribs on the retainer can be fitted into a tip of the SST.

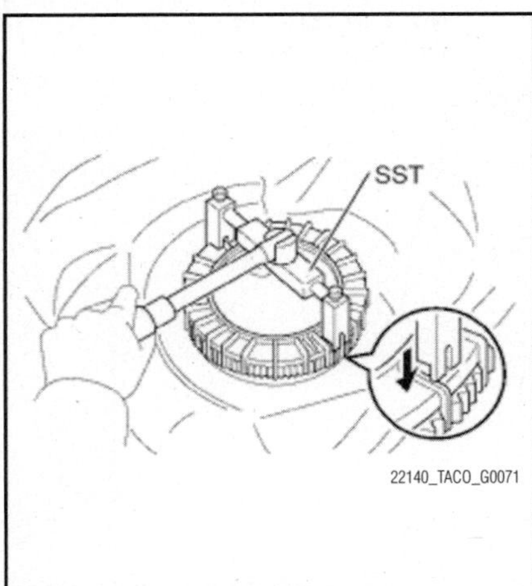

Fig. 146 Special service tool 09808-14020

28. Remove the retainer.
29. Pull the fuel pump assembly out of the fuel tank.
30. Be careful not to bend the arm of the sender gauge.
31. Remove the gasket from the fuel tank.
32. Drain fuel.
33. Loosen the clamp bolt, then remove the fuel tank to filler pipe hose.
34. Remove the 4 clips, then remove the fuel tank protector.

To install:

➡Be sure to use new fasteners, as required.

35. Installation is the reverse of the removal procedure.

IDLE SPEED

ADJUSTMENT

Idle speed is maintained by the Electronic Control Module (ECM). No adjustment is necessary or possible.

THROTTLE BODY

REMOVAL & INSTALLATION

2.7L Engines

See Figure 148.

1. Before servicing the vehicle, refer to the Precautions Section.
2. Disconnect the negative battery cable. Tape the cable with insulating tape.

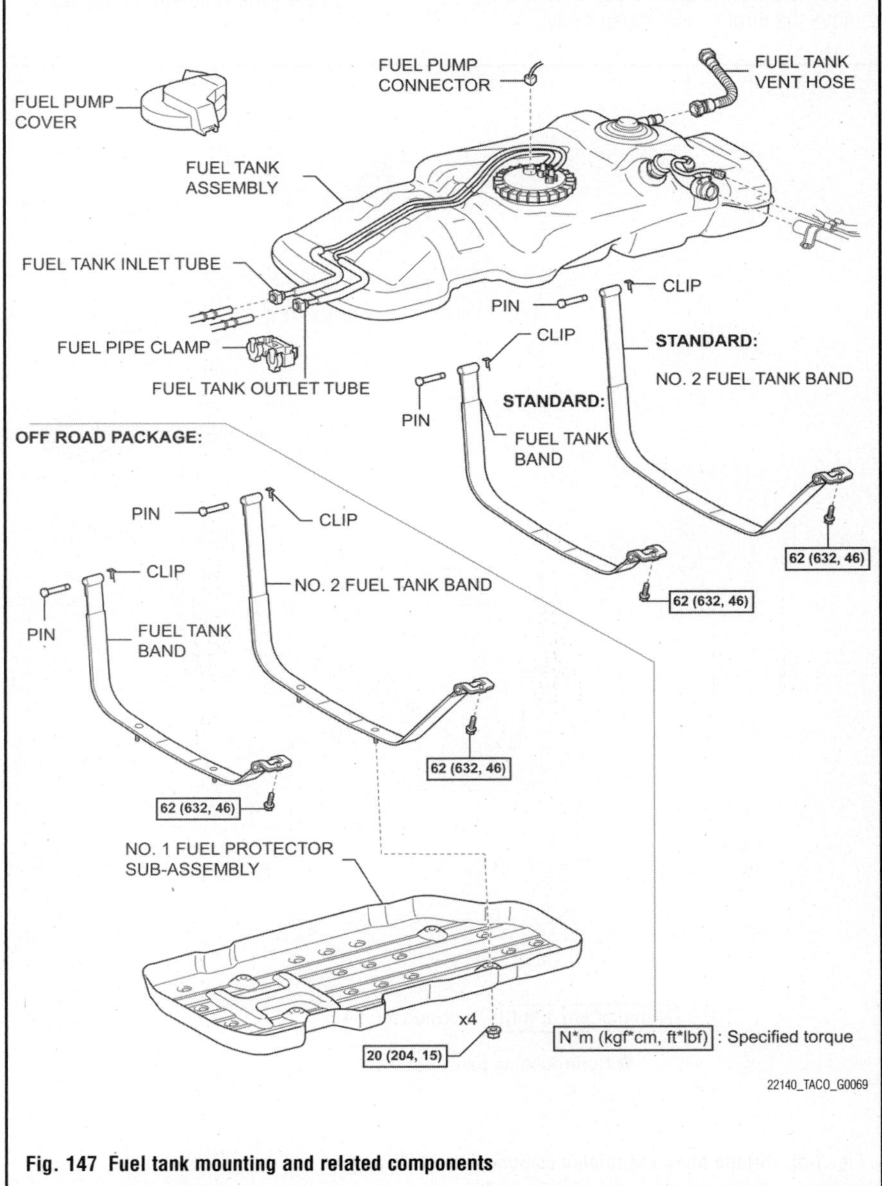

Fig. 147 Fuel tank mounting and related components

3. Remove no. 1 engine under cover sub-assembly (for 4wd and pre-runner)
4. Drain engine coolant
5. Remove intake air connector
6. Disconnect the pressure sensor connector.
7. Disengage the wire harness clamp.
8. Disconnect the vacuum hose.
9. Disconnect the No. 2 ventilation hose.
10. Disconnect the vacuum hose.
11. Loosen the 2 hose clamp bolts.
12. Remove the 3 bolts, then remove the intake air connector.
13. Disconnect the water by-pass hose.
14. Disconnect the No. 2 water by-pass hose.
15. Disconnect the throttle motor connector.
16. Remove the 2 bolts and 2 nuts, then remove the throttle with motor body.
17. Remove the gasket from the intake manifold.

To install:

➡Be sure to use new fasteners, as required.

18. Install a new gasket onto the intake manifold.
19. Install the throttle with motor body with the 2 bolts and 2 nuts. Tighten to 80 inch lbs. (9 Nm).
20. Connect the water by-pass hose.
21. Connect the water by-pass hose No. 2.
22. Connect the throttle motor connector.
23. Install the intake air connector with the 3 bolts.
24. Tighten the 2 hose clamp bolts.
25. Connect the vacuum hose.
26. Connect the ventilation hose No. 2.
27. Connect the vacuum hose.
28. Engage the wire harness clamp.
29. Connect the pressure sensor connector.
30. Connect cable to negative battery terminal
31. Add engine coolant
32. Check for engine coolant leakage
33. Install engine under cover sub-assembly NO.1 (for 4WD and Pre-Runner).
34. Using the Techstream diagnostic tool, or equivalent, perform the initialization procedure.

4.0L Engines

See Figure 149.

1. Before servicing the vehicle, refer to the Precautions Section.
2. Disconnect the negative battery cable. Tape the cable with insulating tape.

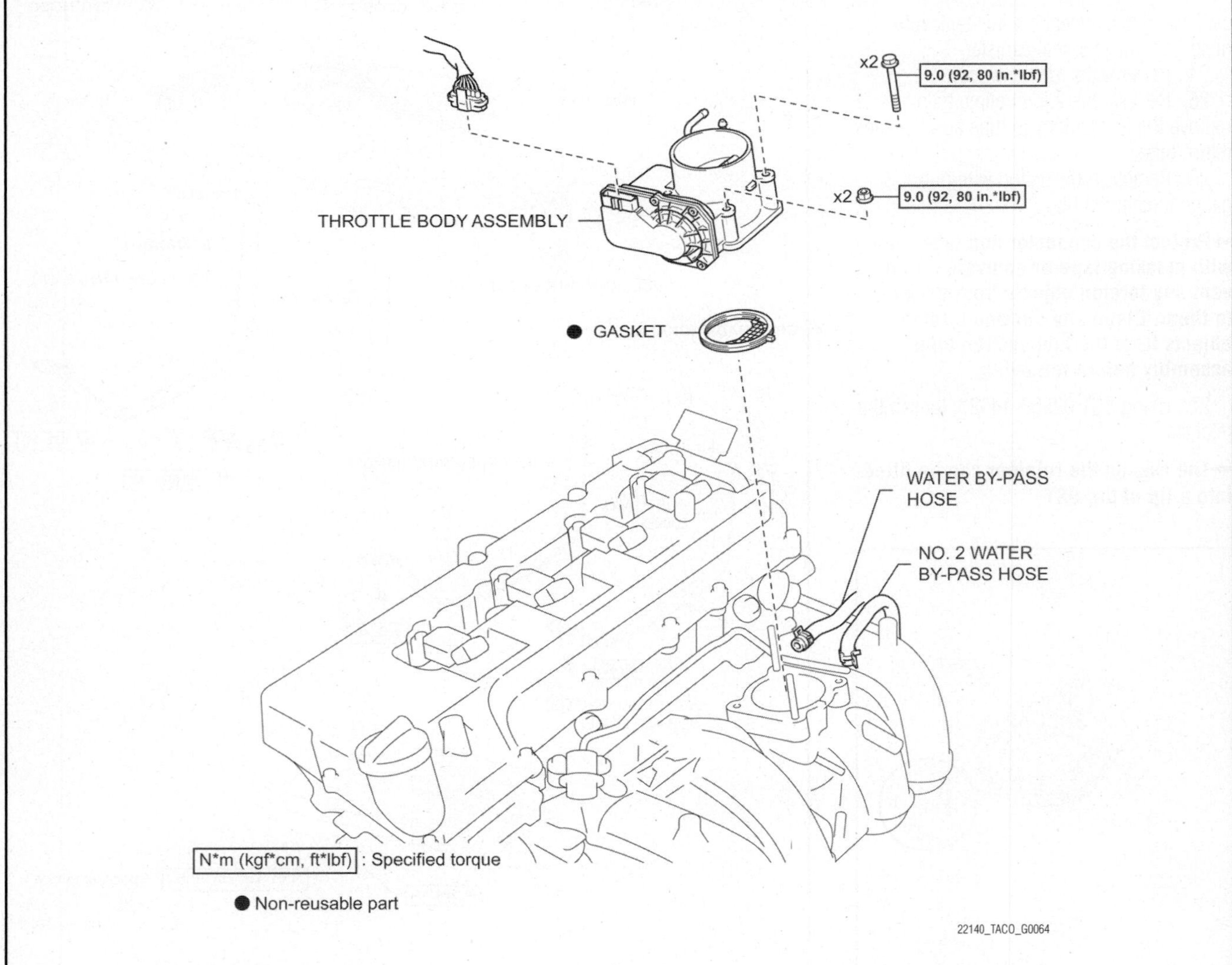

Fig. 148 Throttle body and related components—2.7L engine

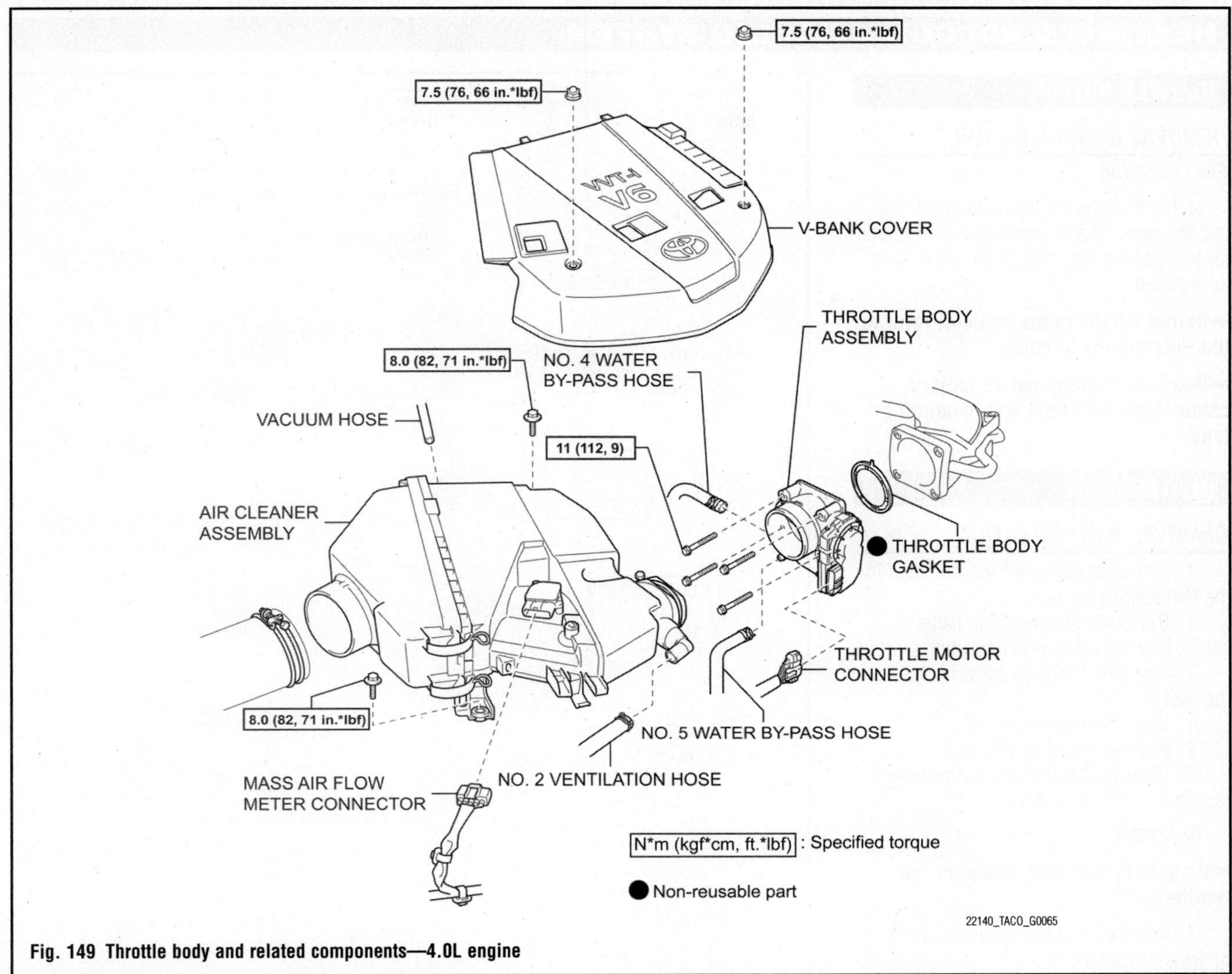

Fig. 149 Throttle body and related components—4.0L engine

3. Drain engine coolant
4. Remove the 2 nuts, then remove the V-bank cover.
5. Disconnect the ventilation hose No. 2.
6. Disconnect the vacuum hose.
7. Disconnect the mass air flow meter connector.
8. Remove the 2 wire harness clamps.
9. Loosen the 2 hose clamps.
10. Remove the 2 bolts, then remove the air cleaner.
11. Disconnect the water by-pass hose No. 5.
12. Disconnect the water by-pass hose No. 4.
13. Disconnect the throttle motor connector.
14. Remove the 4 bolts, then remove the throttle w/ motor body and gasket.

To install:

➡Be sure to use new fasteners, as required.

15. Install a new gasket and the throttle with motor body with the 4 bolts and tighten to 108 inch lbs. (11 Nm).
16. Connect the throttle motor connector.
17. Connect the water by-pass hose No. 4.
18. Connect the water by-pass hose No. 5.
19. Install the air cleaner with the 2 bolts.
20. Connect the ventilation hose No. 2.
21. Connect cable to negative battery terminal
22. Add engine coolant
23. Check for engine coolant leakage
24. Install the V-bank cover with the 2 nuts.
25. Using the Techstream diagnostic tool, or equivalent, perform the initialization procedure.

HEATING & AIR CONDITIONING SYSTEM

BLOWER MOTOR

REMOVAL & INSTALLATION

See Figure 150.

At this time the manufacturer does not provide removal and installation procedures for this component, refer to the illustration as required.

➡Before servicing the vehicle, refer to the Precautions Section.

➡Disconnect the negative battery cable. Tape the cable with insulating tape.

HEATER CORE

REMOVAL & INSTALLATION

1. Before servicing the vehicle, refer to the Precautions Section.
2. Disconnect the negative battery cable. Tape the cable with insulating tape.
3. Properly discharge the air conditioning system.
4. Drain the engine coolant.
5. Remove the heater/AC unit.
6. Disassemble the unit and remove the required component(s).

To install:

➡Be sure to use new fasteners, as required.

7. Installation is the reverse of the removal procedure.
8. Be sure to properly recharge the air conditioning system.
9. Be sure to refill the cooling system with the proper grade and type coolant.
10. Start the engine and check for coolant leaks. Correct as required.
11. Run the air conditioner and check for proper operation. Correct as required.

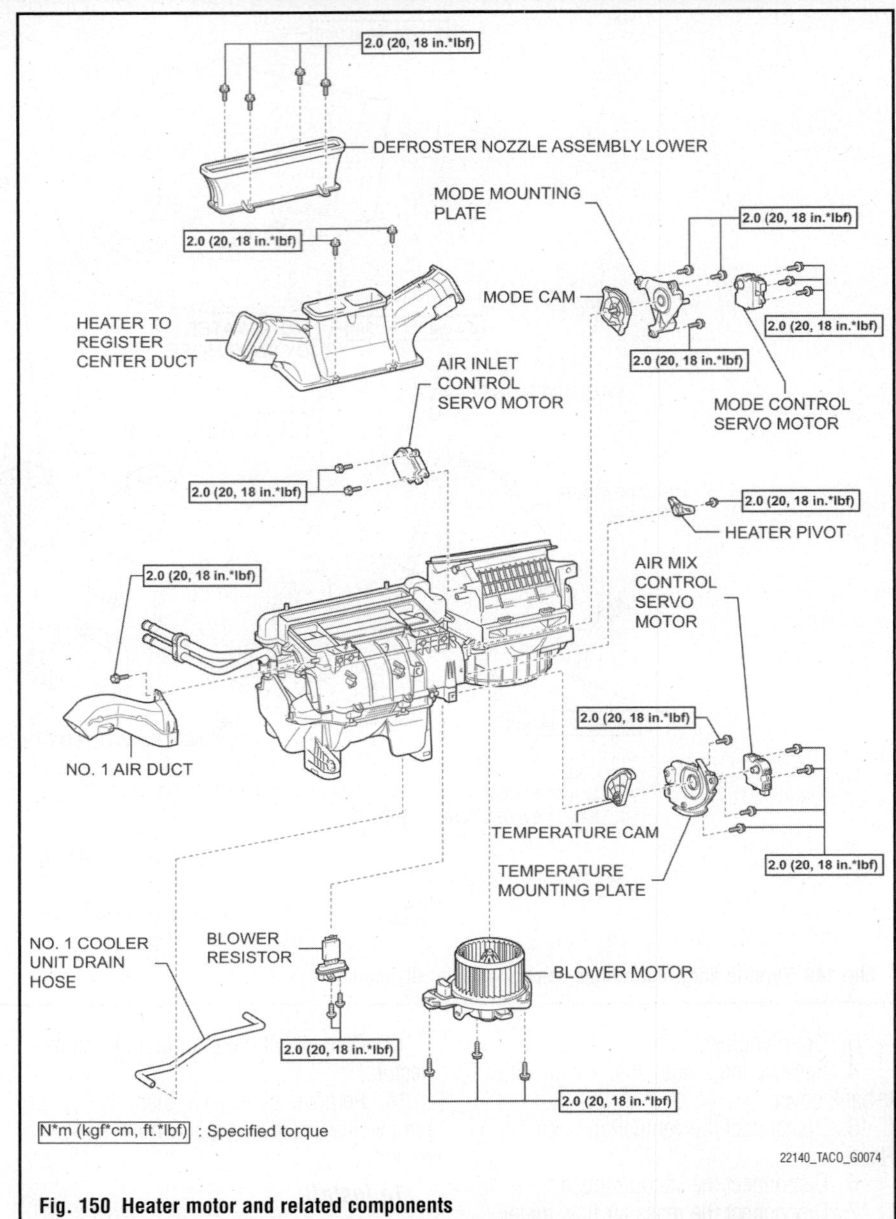

Fig. 150 Heater motor and related components

STEERING

POWER RACK & PINION STEERING GEAR

REMOVAL & INSTALLATION

2WD

See Figure 151.

1. Before servicing the vehicle, refer to the Precautions Section.
2. Disconnect the negative battery cable. Tape the cable with insulating tape.
3. Position the front wheels in the straight ahead position.
4. Drain the power steering fluid.
5. Raise and support the vehicle safely. Remove the tire and wheel assemblies.
6. Lock the steering wheel to prevent it from turning.

➡The seat belt can be used to prevent rotation.

7. Place matchmarks on the steering slider yoke, the steering intermediate shaft number two and the steering intermediate shaft.
8. Remove the steering slider yoke bolts. Slide the steering sliding yoke up and separate it from the steering intermediate shaft number two.
9. Pull down the steering sliding yoke from the steering intermediate shaft and remove it.
10. Place matchmarks on the steering intermediate shaft number two and the power steering gear.
11. Remove the bolt from the steering intermediate shaft number two.
12. Slide the steering intermediate shaft number two up and remove it from the power steering gear.
13. Remove the cotter pin and nut. Using tool SST09610-20012 or equivalent, separate the left tie rod end from the left steering knuckle arm.
14. Remove the cotter pin and nut. Using tool SST09610-20012 or equivalent, separate the right tie rod end from the right steering knuckle arm.
15. Remove the bolt and separate the tube support bracket. Separate the pressure line. Disengage the clip and disconnect the return hose.
16. Remove the power steering gear retaining bolts. Remove the steering gear from the vehicle.

➡The nut has a detent, so never turn it. Always turn the bolt.

To install:

➡Be sure to use new fasteners, as required.

17. Install the power steering gear. Tighten the retaining bolts to 68 ft. lbs.

➡The nut has a detent, so never turn it. Always turn the bolt.

18. Continue the installation in the reverse order of the removal procedure.
19. Be sure to fill the power steering system with the proper grade and type power steering fluid.
20. Bleed the system, as required.
21. Start the engine and check for leaks, correct as required.
22. Check and adjust the alignment, as required.

4WD And PreRunner

See Figure 152.

1. Before servicing the vehicle, refer to the Precautions Section.
2. Disconnect the negative battery cable. Tape the cable with insulating tape.
3. Position the front wheels in the straight ahead position.
4. Drain the power steering fluid.
5. Raise and support the vehicle safely. Remove the tire and wheel assemblies.
6. Remove the number one engine undercover subassembly.
7. Remove the front exhaust pipe assembly. On the 4.0L engine, remove the number two exhaust pipe assembly.
8. Remove the driveshaft. Some vehicles also use a center bearing assembly, remove that too.
9. Remove the frame crossmember subassembly.
10. Remove the stabilizer bar.
11. Lock the steering wheel to prevent it from turning.

➡The seat belt can be used to prevent rotation.

12. Place matchmarks on the steering slider yoke, the steering intermediate shaft number two and the steering intermediate shaft.
13. Remove the steering slider yoke bolts. Slide the steering sliding yoke up and separate it from the steering intermediate shaft number two.
14. Pull down the steering sliding yoke from the steering intermediate shaft and remove it.
15. Place matchmarks on the steering intermediate shaft number two and the power steering gear.
16. Remove the bolt from the steering intermediate shaft number two.
17. Slide the steering intermediate shaft number two up and remove it from the power steering gear.
18. Remove the cotter pin and nut. Using tool SST09610-20011 or equivalent, separate the left tie rod end from the left steering knuckle arm.
19. Remove the cotter pin and nut. Using tool SST09610-20011 or equivalent, separate the right tie rod end from the right steering knuckle arm.
20. Remove the bolt and separate the tube support bracket. Separate the pressure line. Disengage the clip and disconnect the return hose.
21. Remove the power steering gear retaining bolts. Tilt the transmission and remove the steering gear from the vehicle.

➡The nut has a detent, so never turn it. Always turn the bolt.

To install:

➡Be sure to use new fasteners, as required.

22. Install the power steering gear. Tighten the retaining bolts to 68 ft. lbs.

➡The nut has a detent, so never turn it. Always turn the bolt.

23. Continue the installation in the reverse order of the removal procedure.
24. Be sure to fill the power steering system with the proper grade and type power steering fluid.
25. Bleed the system, as required.
26. Start the engine and check for leaks, correct as required.
27. Check and adjust the alignment, as required.

POWER STEERING PUMP

REMOVAL & INSTALLATION

See Figures 153 and 154.

1. Before servicing the vehicle, refer to the Precautions Section.
2. Disconnect the negative battery cable. Tape the cable with insulating tape.
3. Remove the engine undercover sub assembly, if equipped.
4. Remove the fan and alternator drive belt.

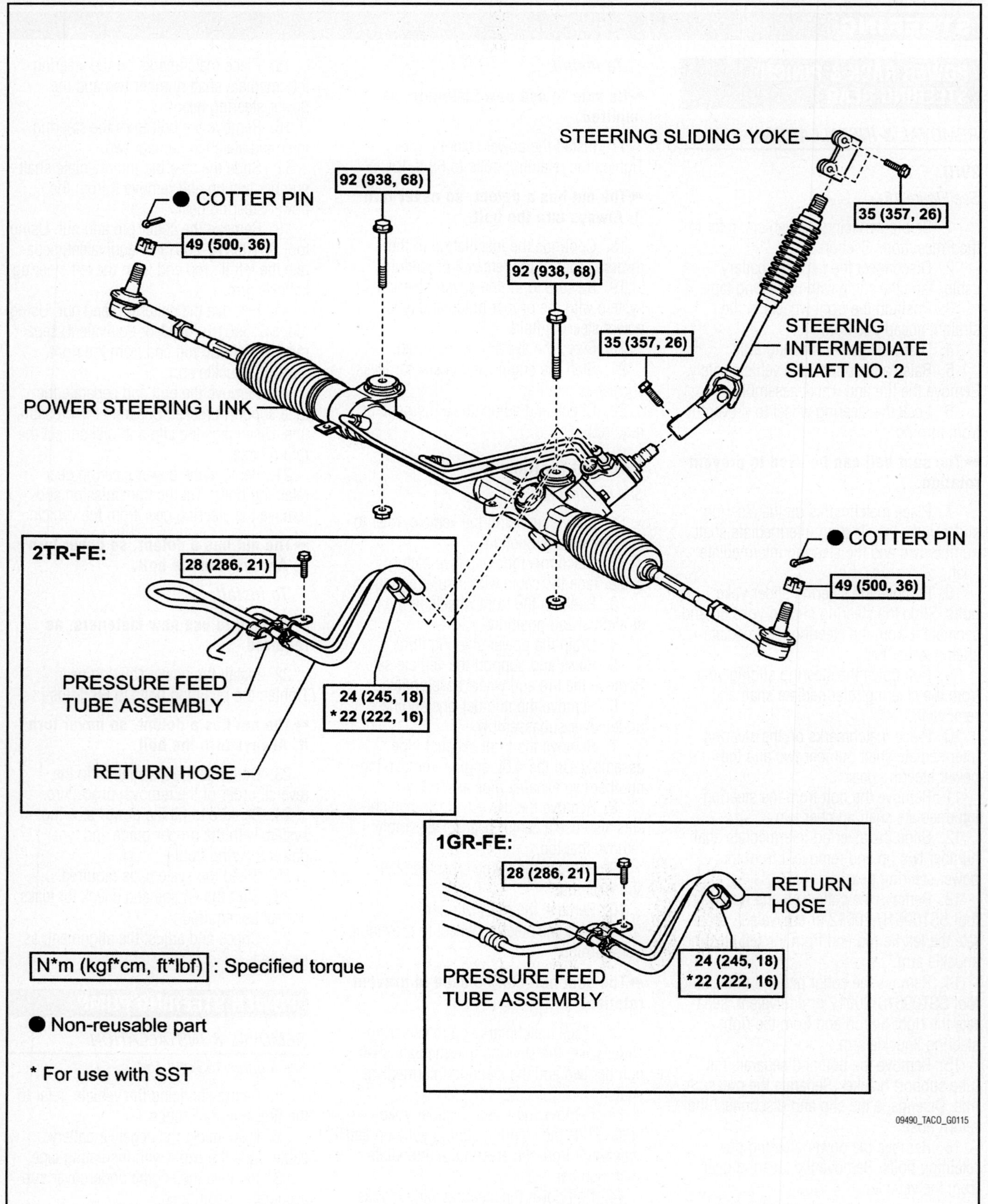

Fig. 151 Power steering gear and related components—2WD

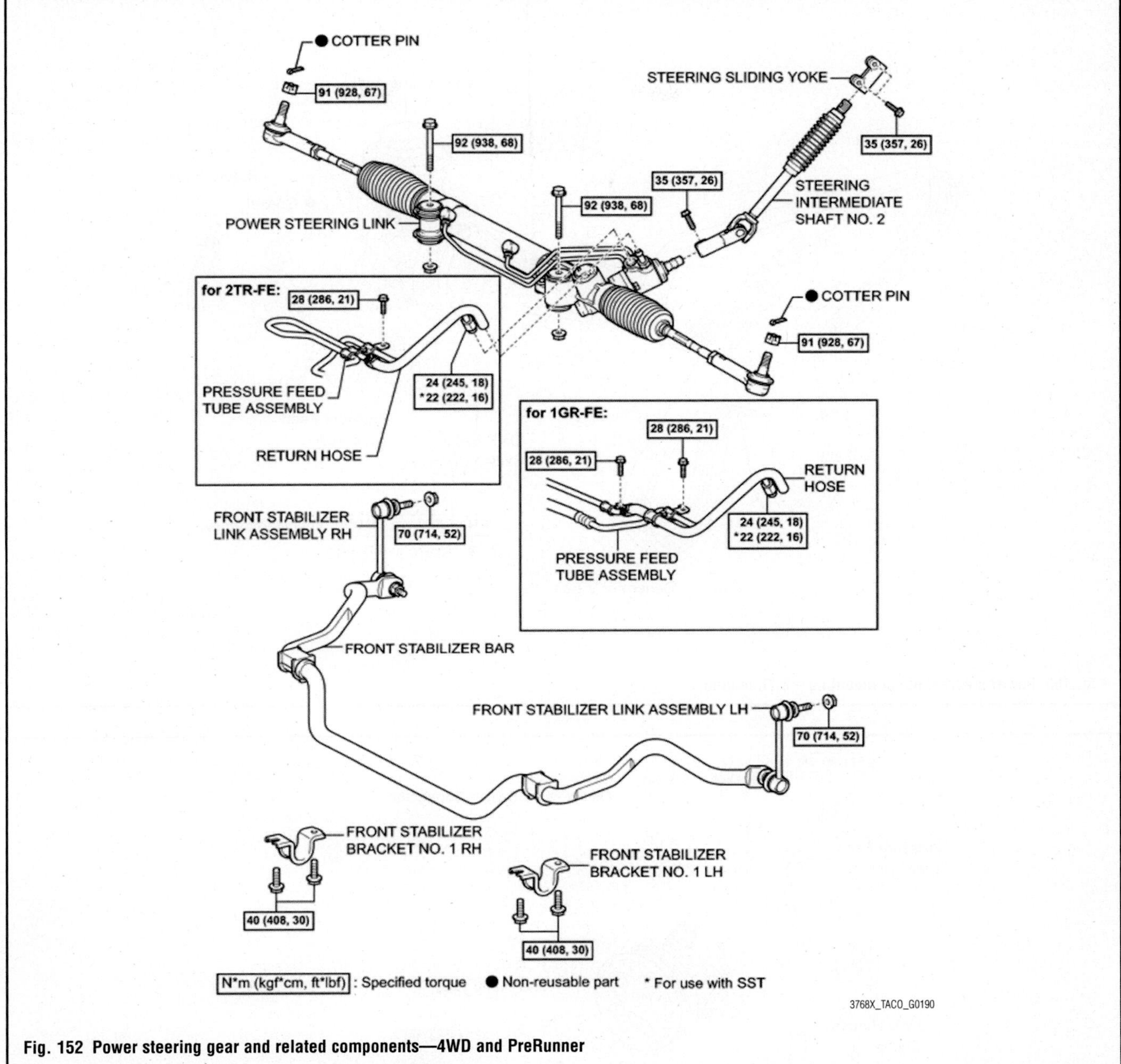

Fig. 152 Power steering gear and related components—4WD and PreRunner

5. Drain the power steering fluid. Be sure to properly dispose of used fluid.
6. Disconnect and plug the power steering fluid lines.
7. Disconnect the electrical connectors. On 4.0L engine, remove the wire harness clamp bracket.
8. Remove the pump retaining bolts.
9. Remove the pump from its mounting.

To install:

➡Be sure to use new fasteners, as required.

10. Installation is the reverse of the removal procedure.
11. Tighten the pump retaining bolts to 15 ft. lbs. (21 Nm).
12. Fill the system with the proper grade and type fluid.
13. Bleed the system.
14. Check for leaks, correct as required.

BLEEDING

See Figures 155 through 157.

1. Before servicing the vehicle, refer to the Precautions Section.
2. Check the fluid level.
3. Raise and properly support the vehicle.
4. With the engine stopped, turn the wheels slowly from stop to stop several times.
5. Lower the vehicle.
6. Start the engine. Run at idle for a few minutes.
7. With the engine idling turn the steering wheel to the left or right full lock position and keep it there for two or three seconds. Repeat for the other side.
8. Repeat the above step several times.
9. Stop the engine. Check for foaming.

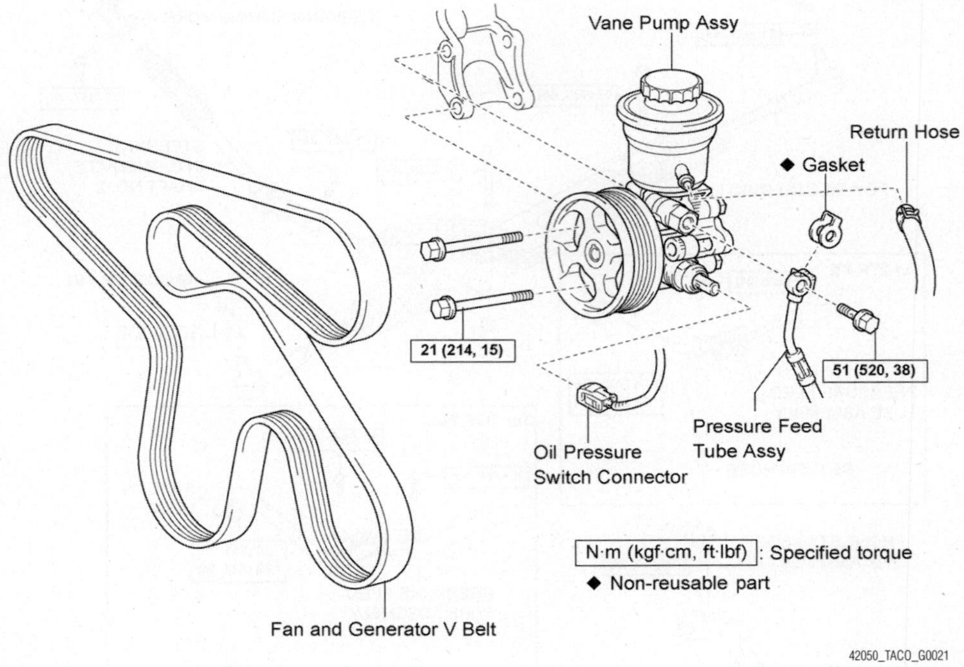

Fig. 153 Power steering pump mounting—2.7L engine

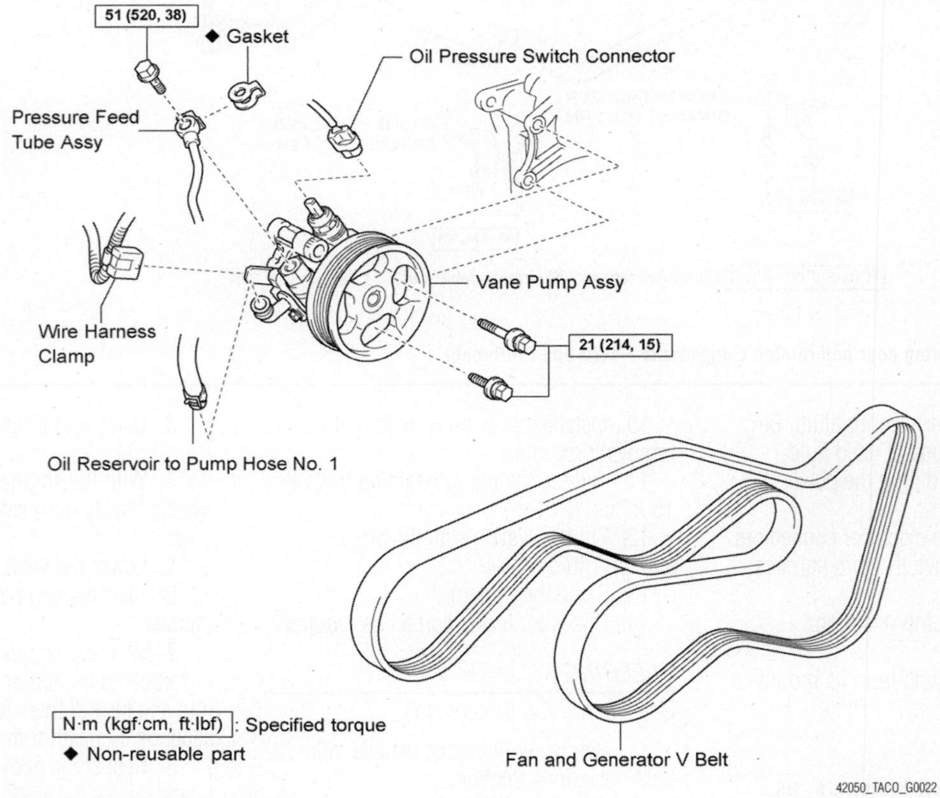

Fig. 154 Power steering pump mounting—4.0L engine

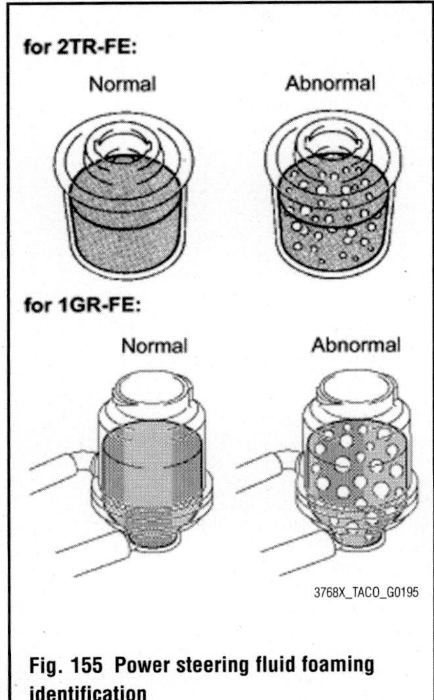

Fig. 155 Power steering fluid foaming identification

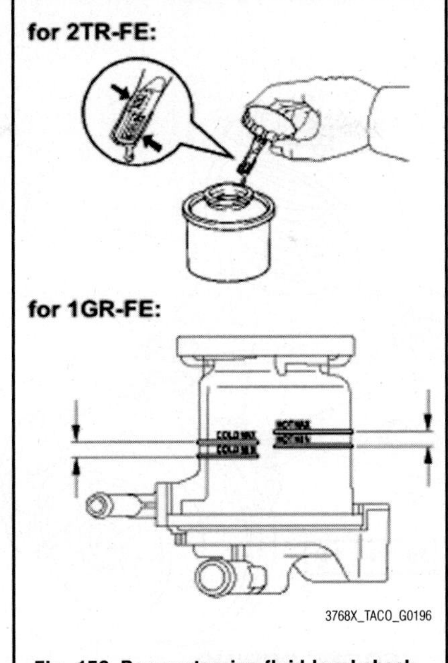

Fig. 156 Power steering fluid level check 1 of 2

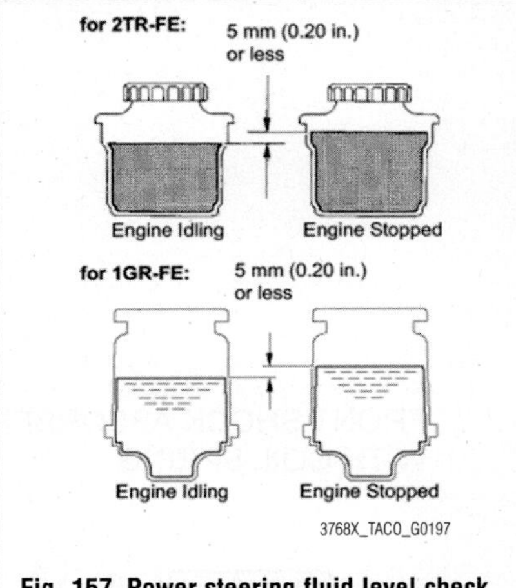

Fig. 157 Power steering fluid level check 2 of 2

10. If foaming exists, repeat the procedure.
11. With the engine stopped check the fluid level.
12. Adjust as necessary. Be sure to use the proper grade and type fluid.

SUSPENSION

FRONT SUSPENSION

LOWER BALL JOINT

REMOVAL & INSTALLATION

➡The lower ball joint is removed with the upper control arm.

LOWER CONTROL ARM

REMOVAL AND & INSTALLATION

See Figures 158 and 159.

1. Before servicing the vehicle, refer to the Precautions Section.
2. Disconnect the negative battery cable. Tape the cable with insulating tape.
3. Raise and support the vehicle safely.
4. Remove the tire and wheel assemblies.
5. To check the lower ball joint, install the hub nuts. Using a dial indicator gauge push the hub nut up and down with a force of 66 ft. lbs. Specification should be 0.020 inch.

➡If not within specification, replace the lower control arm.

6. On 2WD vehicles, remove the stabilizer link assembly.
7. Properly support the lower control arm assembly, as required.
8. Remove the bolt, nut and washer. Separate the front shock absorber with the coil spring from the lower control arm.
9. Remove the two bolts and separate the front lower ball joint attachment from the front axle.
10. On 2WD vehicles, place matchmarks on the camber adjusting cam number two. Remove the two nuts, the two number two camber adjusting cams, the two number one camber adjusting cams.
11. On 4WD and PreRunner, place matchmarks on the camber adjusting cam number two. Remove the nut, the camber adjusting cam number two, the camber adjusting cam number one, the bolt, the toe adjust cam and the toe adjust plate number two.
12. Remove the lower control arm from the vehicle.
13. To remove the ball joint, position the assembly in a vise and using tool SST09628-00011, remove the ball joint from its mounting.

To install:

➡Be sure to use new fasteners, as required.

➡Always fully tighten rubber bushings when the wheels are in full contact with the ground and the vehicle is at curb height. Bounce the vehicle up and down several times to stabilize the suspension, prior to final tightening of these components.

14. Position the lower control arm on its mounting in the vehicle.
15. Align the mating marks made during the removal procedure and temporarily tighten the lower control arm mounting bolts.
16. Install the lower ball joint attachment. Be sure to use a new nut and cotter pin. Tighten the nut to 103 ft. lbs.
17. Install the front lower ball joint attachment with the two bolts. Tighten the bolts to 118 ft. lbs.
18. Continue the installation in the reverse order of the removal procedure.
19. Final tightening torque for the lower control arm is 155 ft. lbs. on 2WD vehicles and 100 ft. lbs. on 4WD vehicles and PreRunner.
20. Check and adjust the alignment, as required.

SHOCK ABSORBERS

REMOVAL & INSTALLATION

2WD

See Figures 160 and 161.

1. Before servicing the vehicle, refer to the Precautions Section.
2. Disconnect the negative battery cable. Tape the cable with insulating tape.
3. Raise and support the vehicle safely. Remove the tire and wheel assembly.

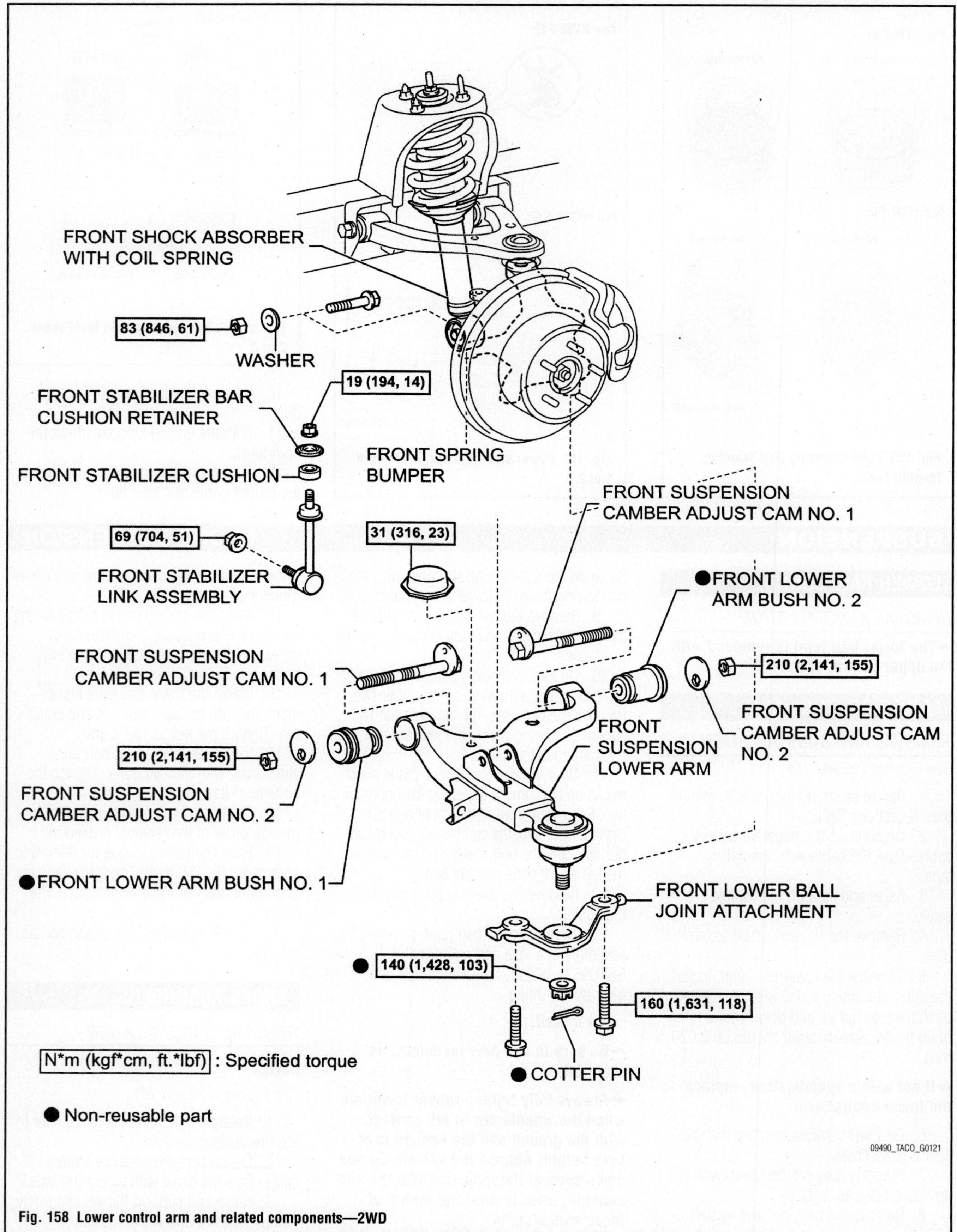

Fig. 158 Lower control arm and related components—2WD

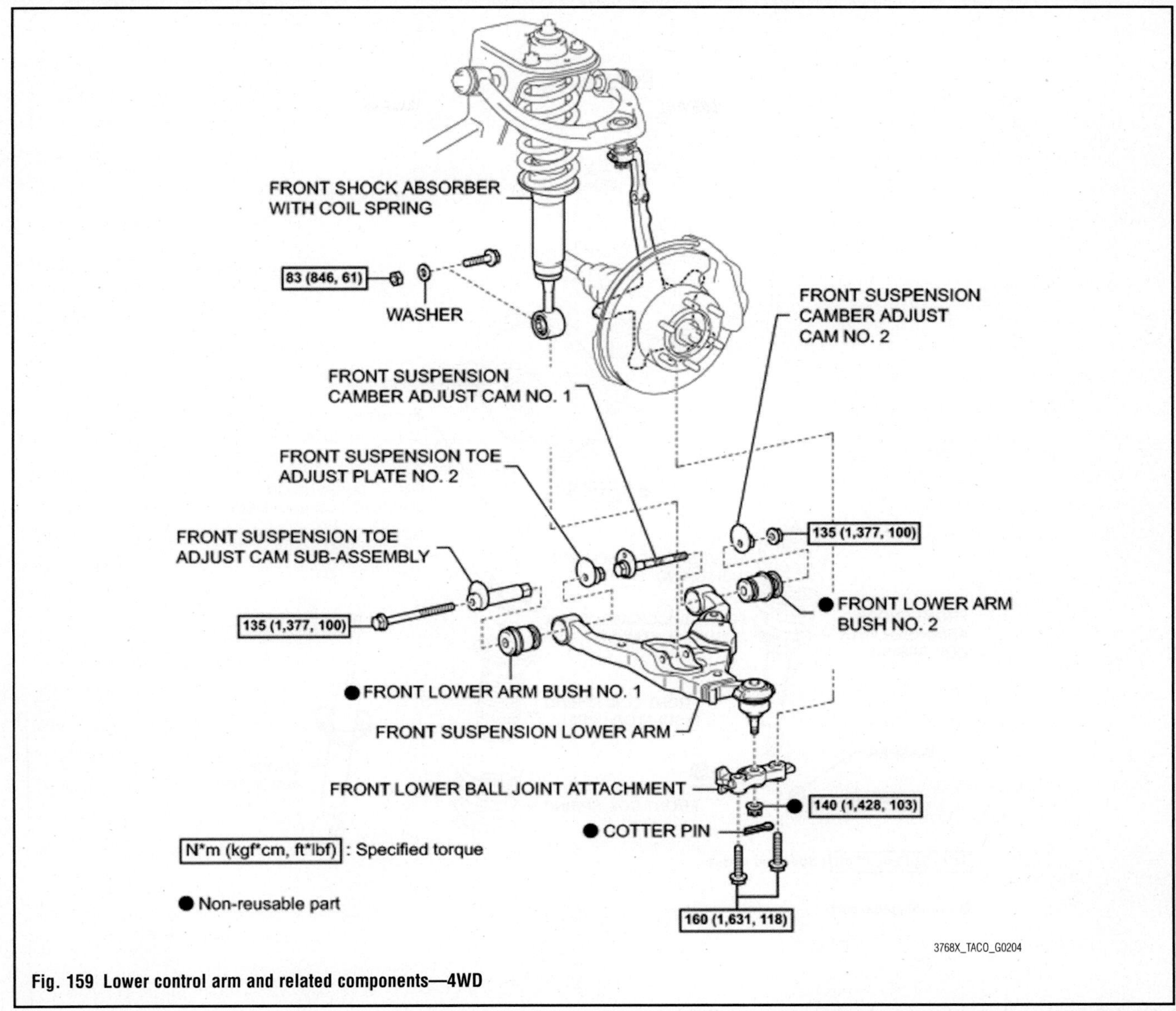

Fig. 159 Lower control arm and related components—4WD

4. Remove the speed sensor connector. Remove the two bolts. Separate the skid control sensor wire.

5. Remove the bolt and separate the front flexible hose.

6. Remove the upper control arm.

7. Remove the shock absorber lower bolt and nut. Remove the three nuts on the upper side of the shock absorber.

8. Remove the shock absorber and coil spring from the vehicle.

To install:

➡Be sure to use new fasteners, as required.

➡Always fully tighten rubber bushings when the wheels are in full contact with the ground and the vehicle is at curb height. Bounce the vehicle up and down several times to stabilize the suspension, prior to final tightening of these components.

9. Position the shock absorber to its mounting on the vehicle.

10. On the left side, install the coil spring onto the body with the lower end of the coil spring facing the outer side of the vehicle.

11. On the right side, install the coil spring onto the body with the lower end of the coil spring facing the inner side of the vehicle.

12. Install the upper retaining nuts. Torque to 47 ft. lbs.

13. Temporarily tighten the lower shock retaining bolt.

14. Continue the installation in the reverse order of the removal procedure.

15. Final tightening torque for the lower shock absorber mounting bolt is 61 ft. lbs.

16. Check and adjust the alignment, as required.

4WD And PreRunner

See Figure 162.

1. Before servicing the vehicle, refer to the Precautions Section.

2. Disconnect the negative battery cable. Tape the cable with insulating tape.

3. Raise and support the vehicle safely. Remove the tire and wheel assembly.

4. Remove the engine undercover assembly.

5. Remove the stabilizer bar.

6. Remove the cotter pin and nut. Using the proper tool, separate the tie rod end from the steering knuckle arm.

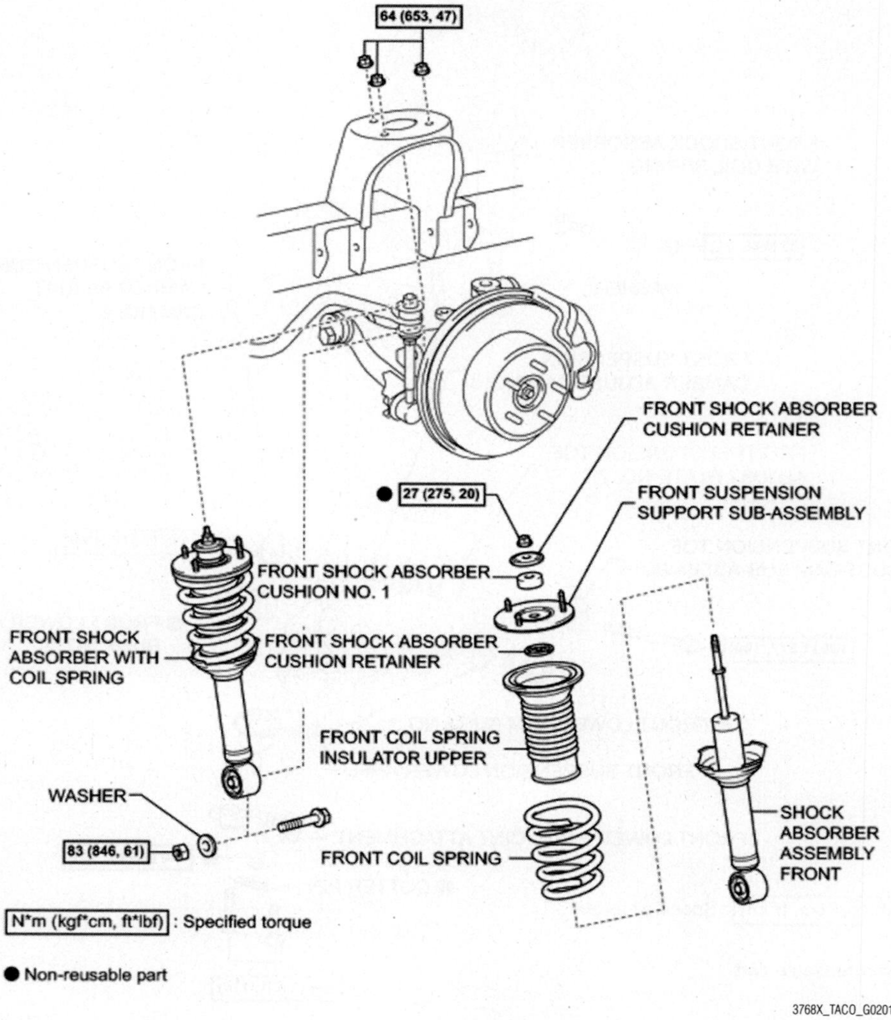

Fig. 160 Front shock absorber and related components—2WD

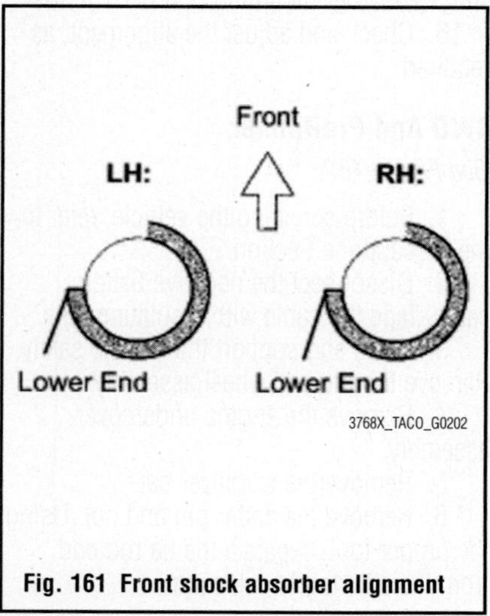

Fig. 161 Front shock absorber alignment

7. Remove the shock absorber lower bolt and nut. Remove the three nuts on the upper side of the shock absorber.

8. Remove the shock absorber and coil spring from the vehicle.

To install:

➡Be sure to use new fasteners, as required.

➡Always fully tighten rubber bushings when the wheels are in full contact with the ground and the vehicle is at curb height. Bounce the vehicle up and down several times to stabilize the suspension, prior to final tightening of these components.

9. Position the shock absorber to its mounting on the vehicle.

10. On the left side, install the coil spring onto the body with the lower end of the coil spring facing the outer side of the vehicle.

11. On the right side, install the coil spring onto the body with the lower end of the coil spring facing the inner side of the vehicle.

12. Install the upper retaining nuts. Torque to 47 ft. lbs.

13. Temporarily tighten the lower shock retaining bolt.

14. Continue the installation in the reverse order of the removal procedure.

15. Final tightening torque for the lower shock absorber mounting bolt is 61 ft. lbs.

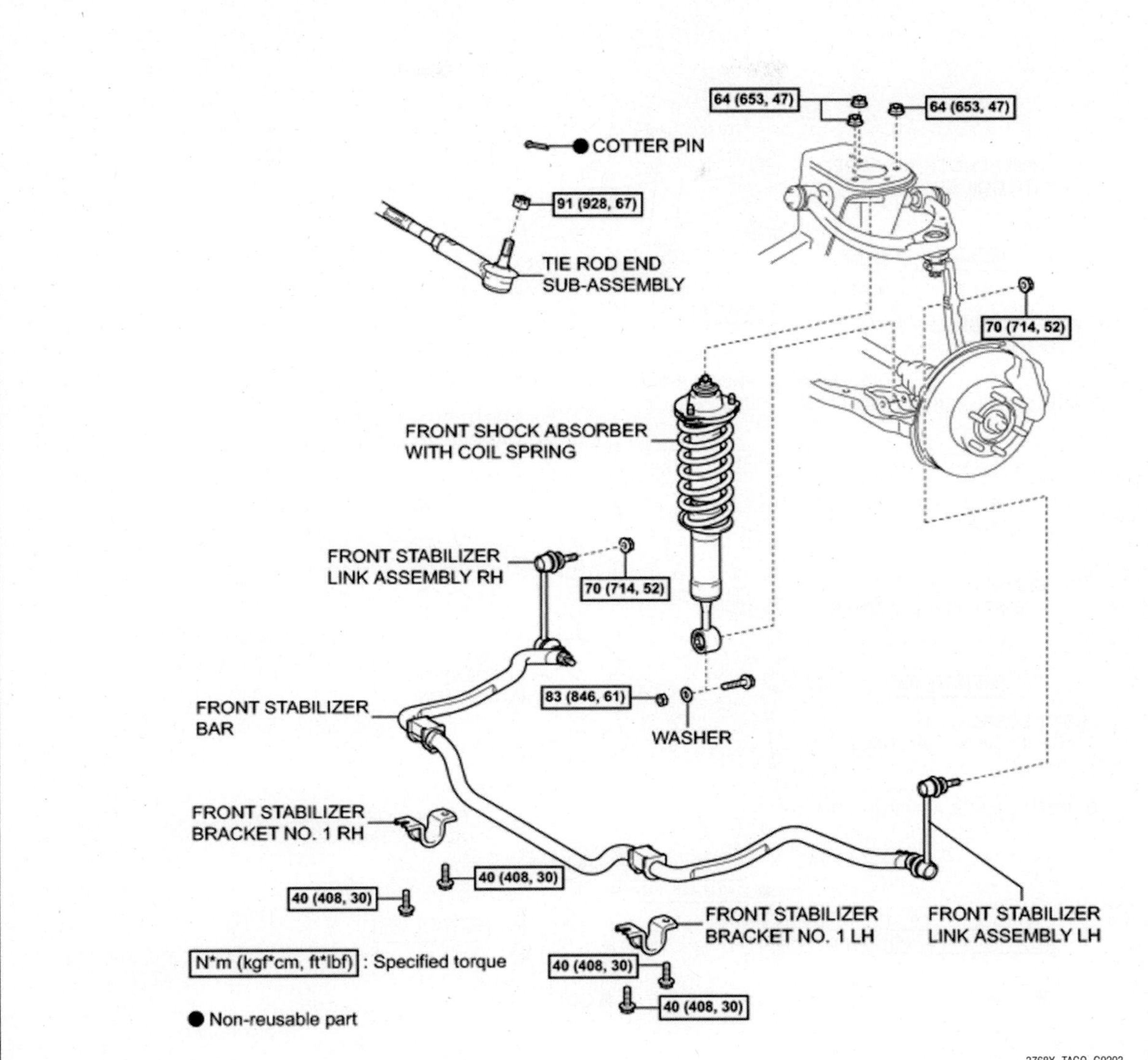

Fig. 162 Front shock absorber and related components—4WD and PreRunner

16. Check and adjust the alignment, as required.

STEERING KNUCKLE

REMOVAL & INSTALLATION

At this time the manufacturer does not provide removal and installation procedures for this component, refer to the illustration as required.

STABILIZER BAR

REMOVAL & INSTALLATION

See Figures 165 through 166

1. Before servicing the vehicle, refer to the Precautions Section.
2. Disconnect the negative battery cable. Tape the cable with insulating tape.
3. Raise and support the vehicle safely.
4. Remove the tire and wheel assemblies.
5. On 4WD and PreRunner, remove the engine undercover subassembly.
6. On 2WD (left side), remove the two nuts, stabilizer bar link, two retainers and two bushings.
7. On 2WD (right side), remove the two nuts, stabilizer bar link, two retainers and two bushings.
8. On 4WD and PreRunner (left side), remove the two nuts and stabilizer bar link.

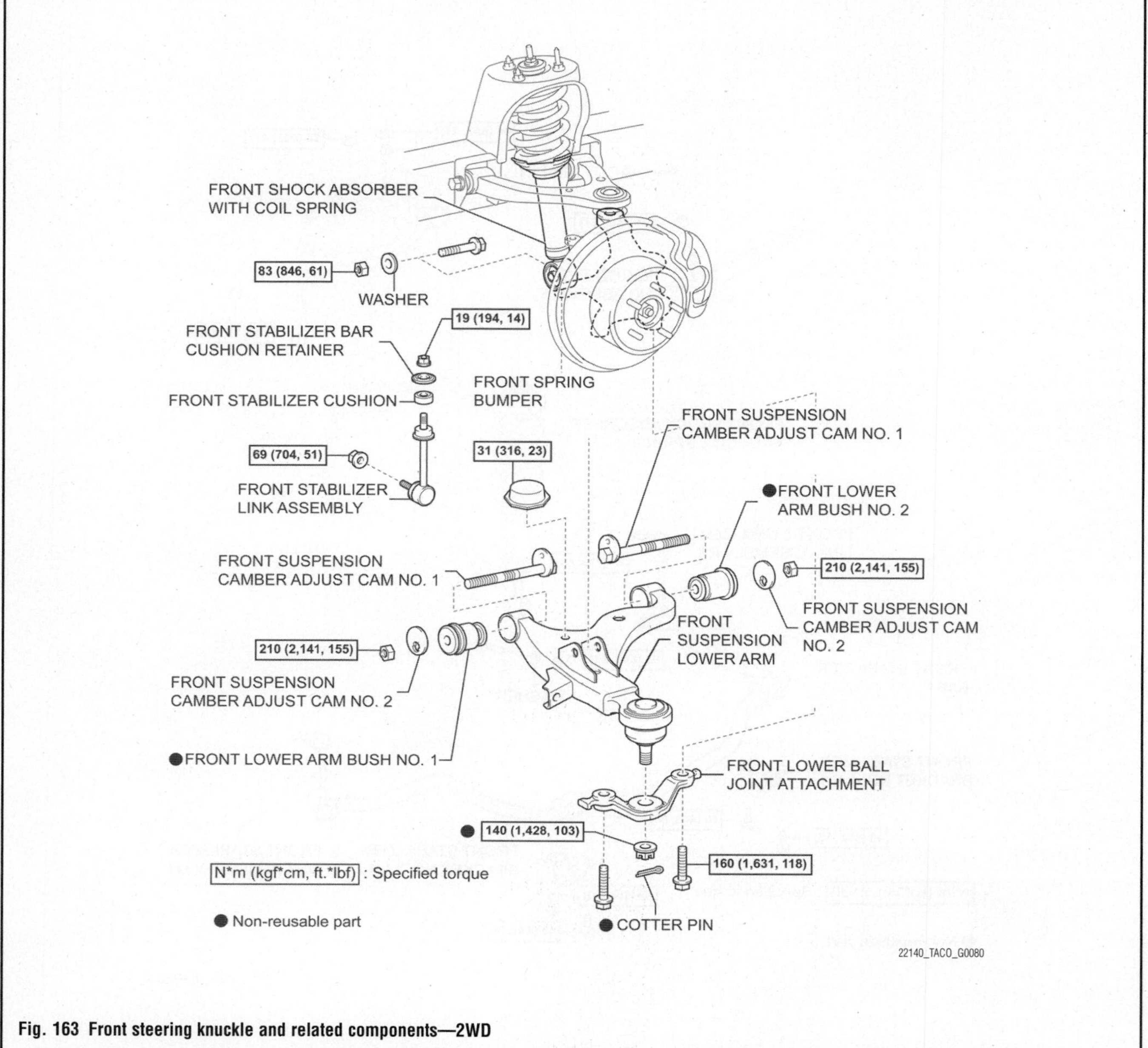

Fig. 163 Front steering knuckle and related components—2WD

9. On 4WD and PreRunner (right side), remove the two nuts and stabilizer bar link.
10. Remove the four bolts and the stabilizer bar brackets and bushings.
11. Remove the stabilizer bar from the vehicle.

To install:

➡Be sure to use new fasteners, as required.

12. Installation is the reverse of the removal procedure.
13. On 2WD, tighten the stabilizer bar side bushing bolts to 14 ft. lbs. Tighten the lower arm side bushing bolts to 51 ft. lbs. (69 Nm).. Tighten the bushing bracket bolts to 16 ft. lbs.
14. On 4WD and PreRunner, tighten the bar links to 52 ft. lbs. (70 Nm) Tighten the bushing bracket bolts to 30 ft. lbs.
15. Be sure to install the cushion onto the stabilizer bar with its cut line facing the front.

UPPER BALL JOINT

REMOVAL & INSTALLATION

➡The upper ball joint is removed with the upper control arm.

UPPER CONTROL ARM

REMOVAL & INSTALLATION

See Figures 168 and 169.

1. Before servicing the vehicle, refer to the Precautions Section.
2. Disconnect the negative battery cable. Tape the cable with insulating tape.
3. Raise and support the vehicle safely. Remove the tire and wheel assemblies.
4. Remove the speed sensor connector. Remove the two bolts and separate the skid control sensor wire.
5. On 2WD vehicles, remove the bolt and separate the front flexible hose.

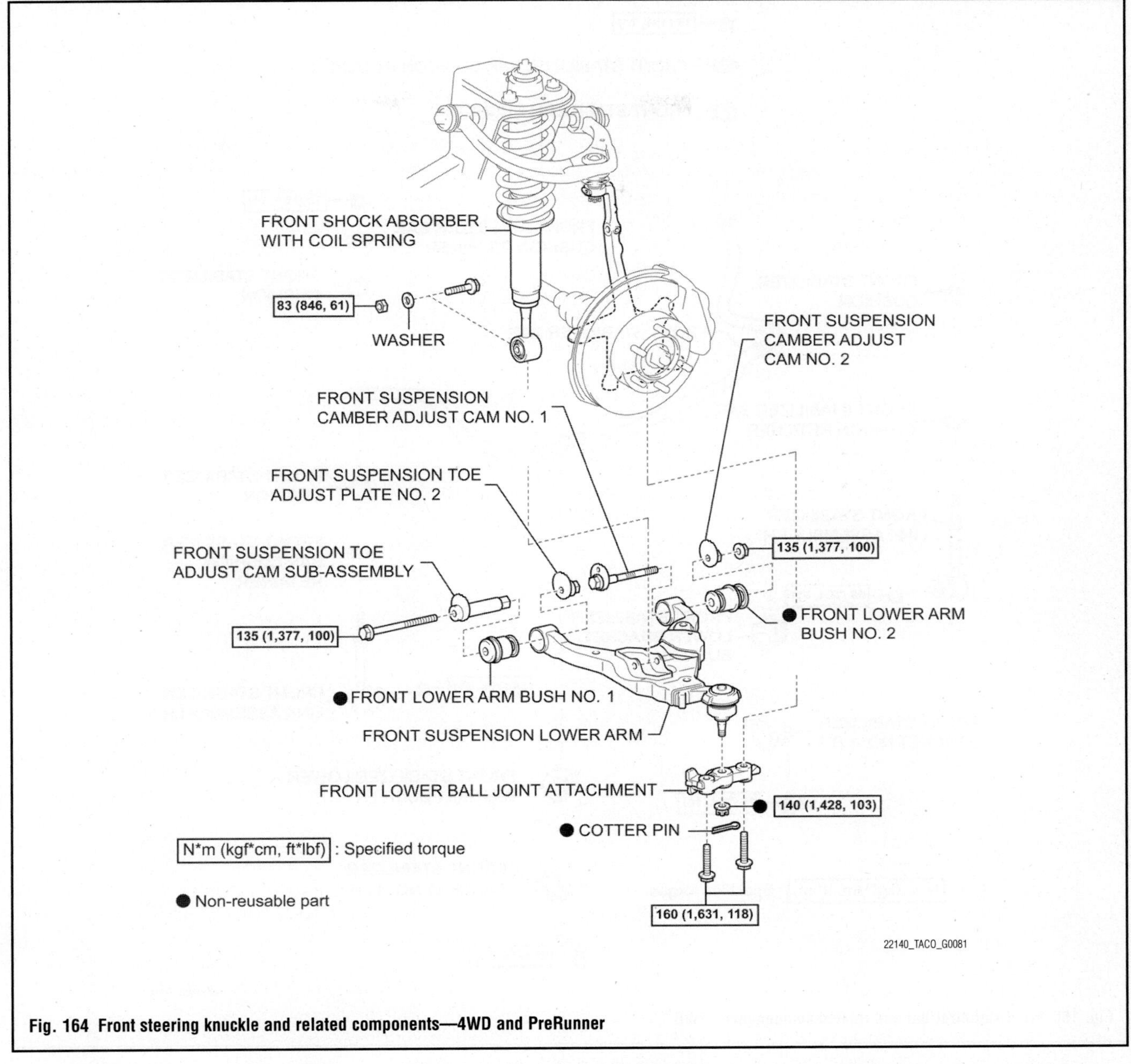

Fig. 164 Front steering knuckle and related components—4WD and PreRunner

6. Support the front suspension lower control arm using a jack.
7. Remove the clip and nut. Using the tool SST09628-62011, or equivalent separate the upper ball joint from the steering knuckle.
8. On 4WD and PreRunner, remove the bolt and bracket.
9. Remove the two upper control arm retaining bolts.
10. Remove the upper control arm from the vehicle.

To install:

➡Be sure to use new fasteners, as required.

➡Always fully tighten rubber bushings when the wheels are in full contact with the ground and the vehicle is at curb height. Bounce the vehicle up and down several times to stabilize the suspension, prior to final tightening of these components.

11. Position the upper control arm on its mounting in the vehicle.
12. Temporarily tighten the upper control arm mounting bolts.
13. Continue the installation in the reverse order of the removal procedure.
14. Final tightening torque for the upper control arm bushings is 60 ft. lbs. on 2WD vehicles and 85 ft. lbs. on 4WD vehicles and PreRunner.
15. Check and adjust the alignment, as required.

WHEEL HUB & BEARING

REMOVAL & INSTALLATION

2WD

See Figure 170.

1. Before servicing the vehicle, refer to the Precautions Section.
2. Disconnect the negative battery cable. Tape the cable with insulating tape.

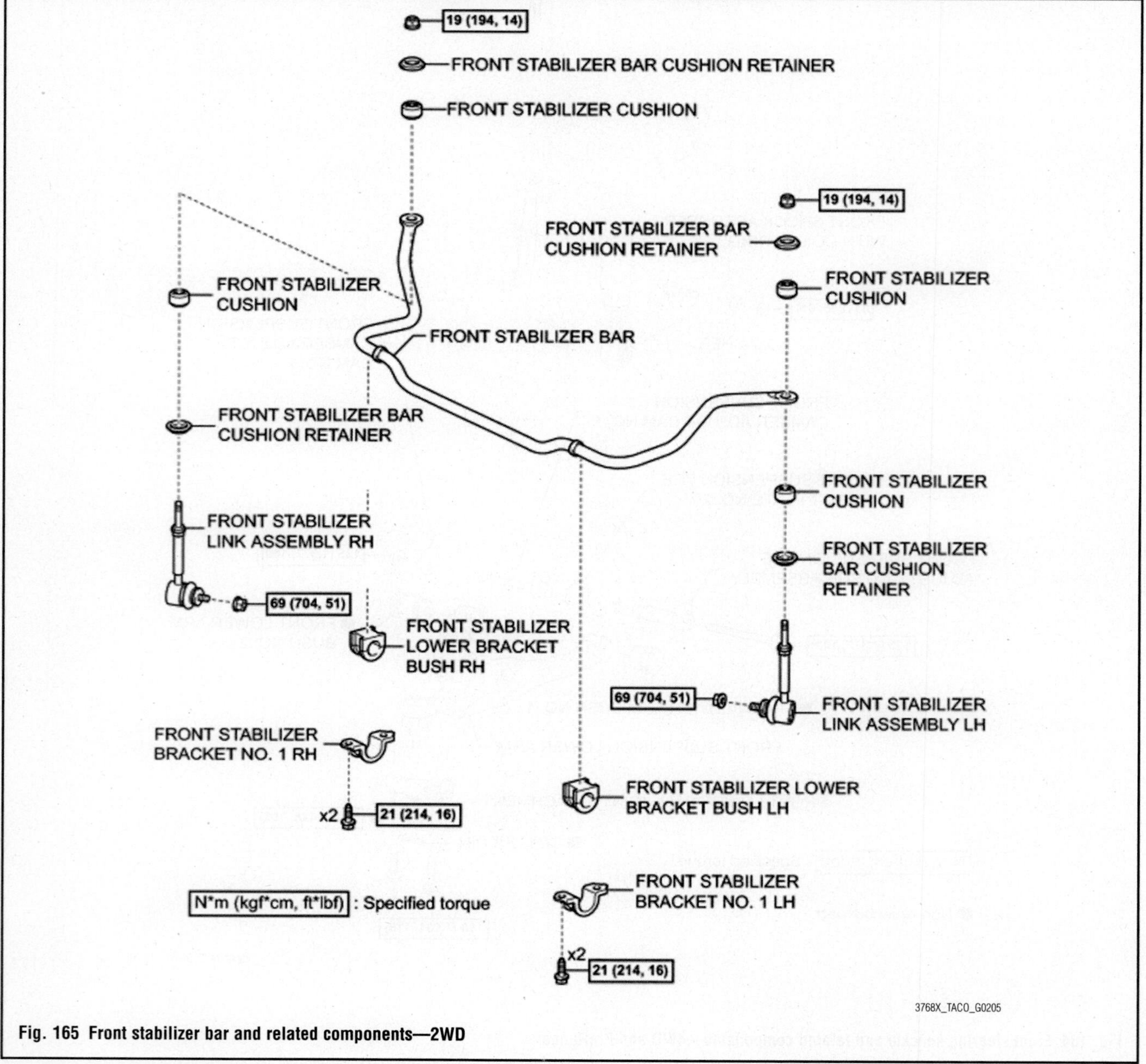

Fig. 165 Front stabilizer bar and related components—2WD

3. Drain the brake fluid.
4. Raise and support the vehicle safely. Remove the tire and wheel assembly.
5. Remove the speed sensor connector. Remove the two bolts and separate the skid control sensor wire.
6. Remove the bolt and separate the front flexible hose.
7. Remove the front brake caliper. Remove the rotor.
8. Remove the cotter pin and nut. Using the proper tool separate the tie rod end from the steering knuckle.
9. Remove the two bolts and separate the front lower ball joint attachment from the front axle.
10. Remove the clip and nut. Using the proper tool, separate the upper ball joint from the steering knuckle.
11. Remove the front axle hub from the vehicle.

To install:

➡Be sure to use new fasteners, as required.

12. Installation is the reverse of the removal procedure.
13. Be sure to fill the master cylinder with the proper grade and type brake fluid.
14. Bleed the brakes, as required.
15. Be sure to check and adjust the alignment, as required.

4WD And PreRunner

See Figure 171.

1. Before servicing the vehicle, refer to the Precautions Section.
2. Disconnect the negative battery cable. Tape the cable with insulating tape.
3. Drain the brake fluid.
4. Raise and support the vehicle safely. Remove the tire and wheel assembly.
5. Remove the bolt and separate the speed sensor. Disengage the clamps. Remove the bolt and separate the speed sensor wire harness from the steering knuckle.

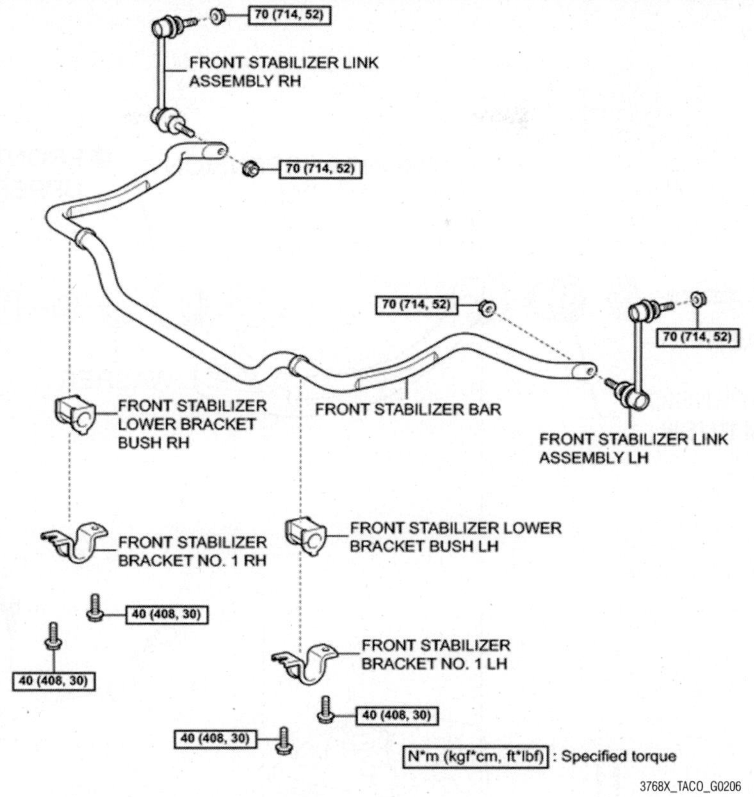

Fig. 166 Front stabilizer bar and related components—4WD and PreRunner

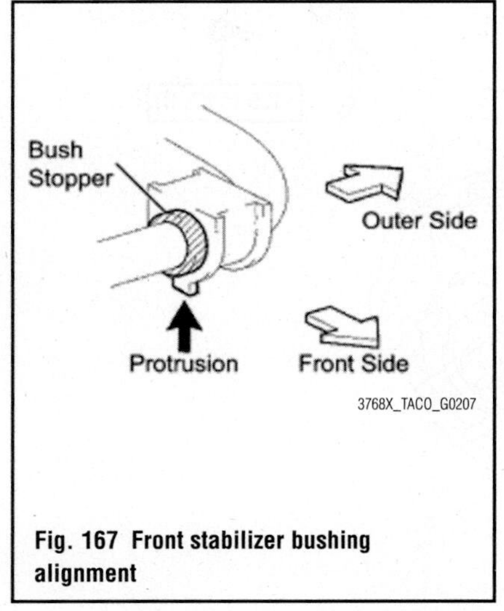

Fig. 167 Front stabilizer bushing alignment

6. Remove the front brake caliper. Remove the rotor.

7. Remove the axle hub grease cap. Remove the cotter pin and lock cap. Remove the front axle hub nut.

8. Remove the nut and separate the stabilizer link from the steering knuckle.

9. Remove the cotter pin and nut. Using the proper tool separate the tie rod end from the steering knuckle.

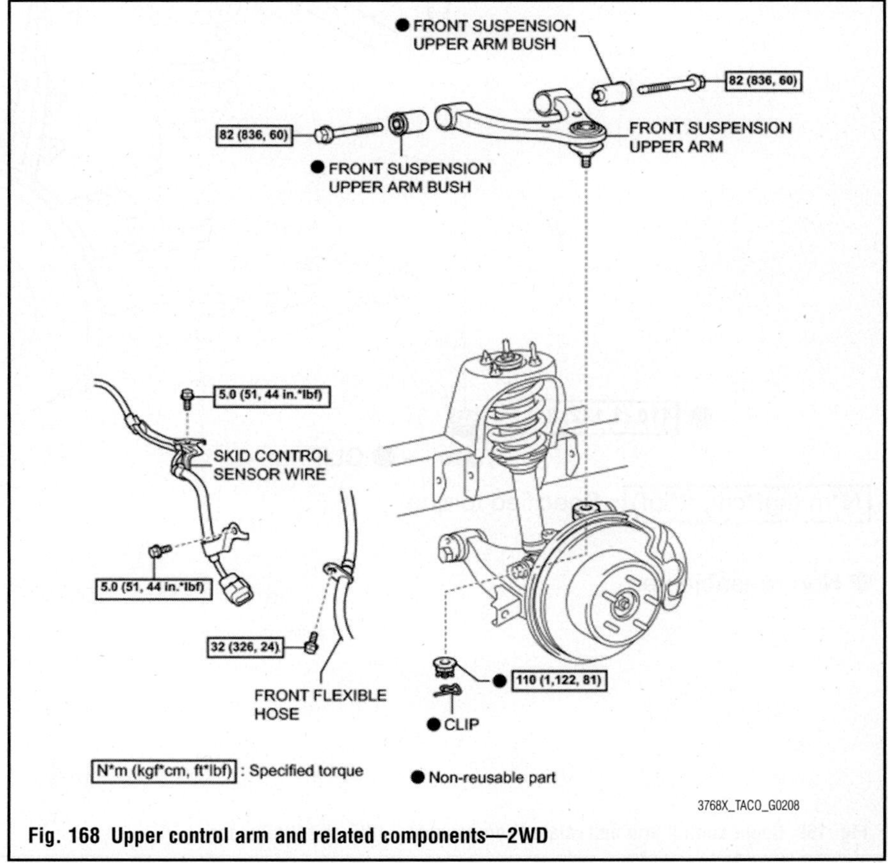

Fig. 168 Upper control arm and related components—2WD

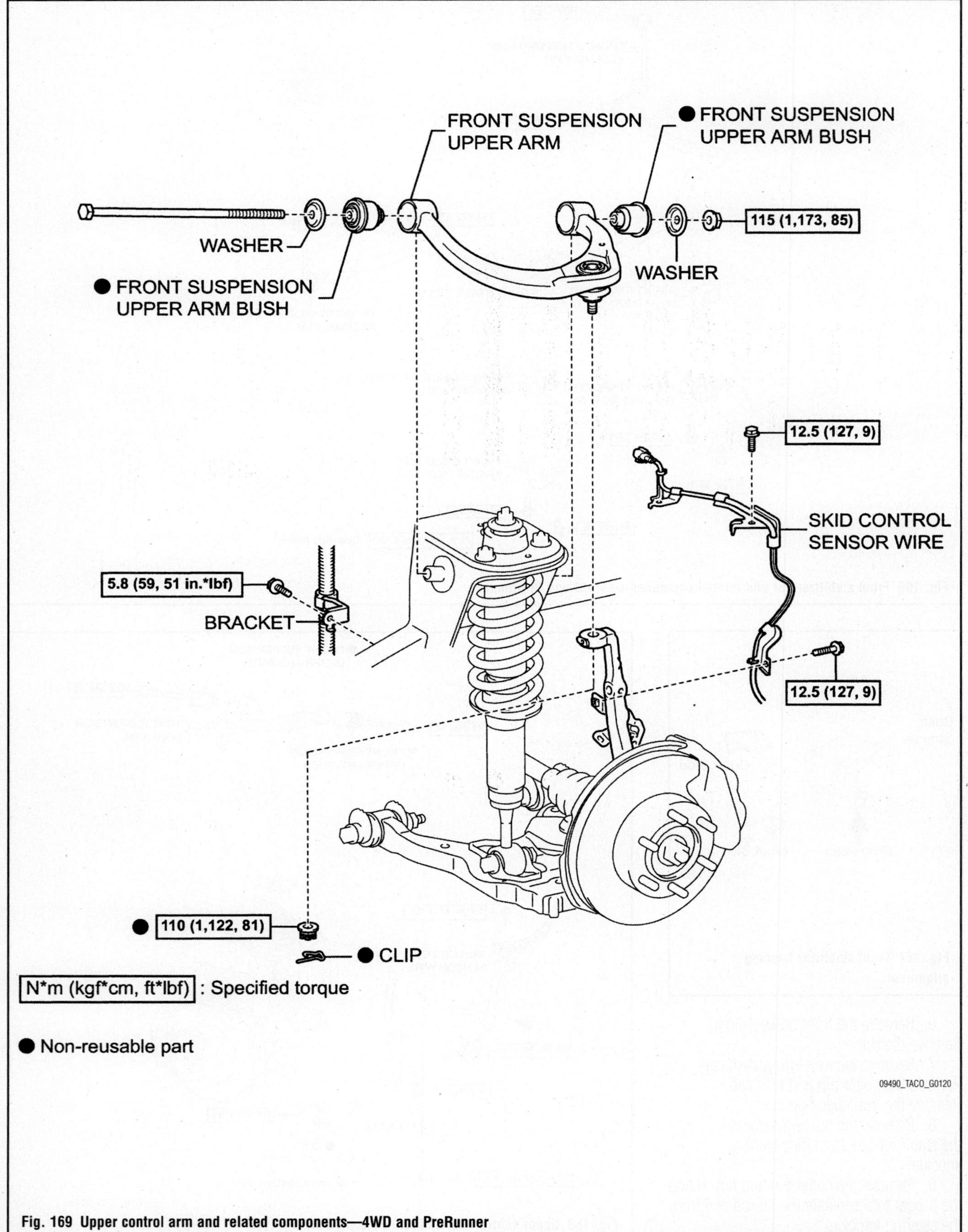

Fig. 169 Upper control arm and related components—4WD and PreRunner

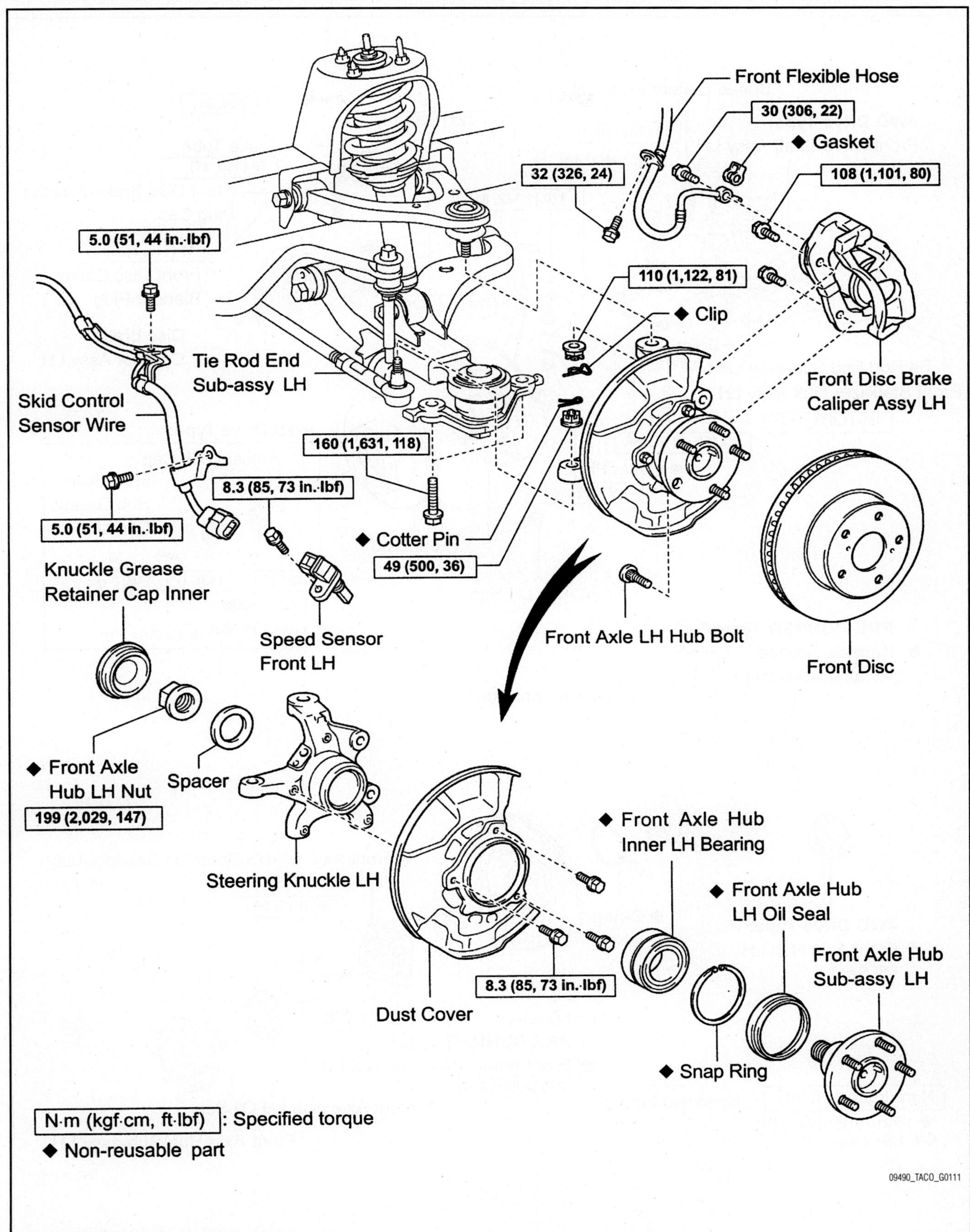

Fig. 170 Front hub and related components—2WD

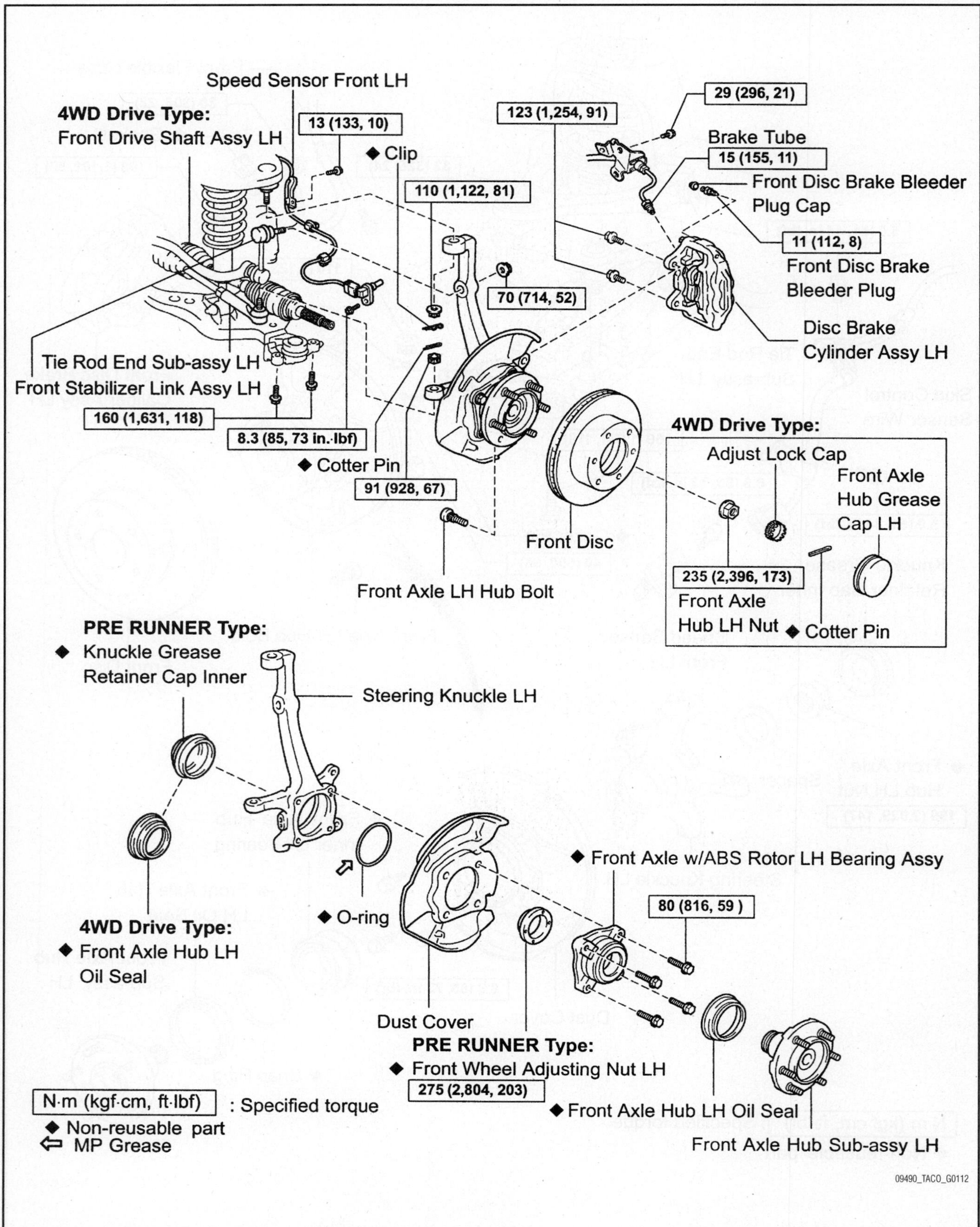

Fig. 171 Front hub and related components—4WD and PreRunner

10. Remove the two bolts and separate the front suspension lower arm from the front axle.

11. Properly support the front suspension lower arm with a jack. Remove the clip and nut.

12. Using the proper tool separate the steering knuckle from the front suspension upper control arm.

13. On 4WD, use a plastic hammer and separate the front axle hub from the front driveshaft.

14. Remove the front axle hub from the vehicle.

To install:

➡Be sure to use new fasteners, as required.

15. Installation is the reverse of the removal procedure.

16. Be sure to fill the master cylinder with the proper grade and type brake fluid.

17. Bleed the brakes, as required.

18. Be sure to check and adjust the alignment, as required.

ADJUSTMENT

No adjustment necessary.

SUSPENSION

REAR SUSPENSION

LEAF SPRING

REMOVAL & INSTALLATION

See Figures 172 and 173.

1. Before servicing the vehicle, refer to the Precautions Section.
2. Disconnect the negative battery cable. Tape the cable with insulating tape.
3. Raise and support the vehicle safely.
4. Remove the tire and wheel assemblies.
5. On 4WD and PreRunner, remove the spare tire.
6. Support the rear axle housing. Remove the bolt nut and washer.
7. On 2WD separate the shock absorber from the rear spring seat.
8. On 4WD and PreRunner, separate the shock absorber from the rear axle housing.

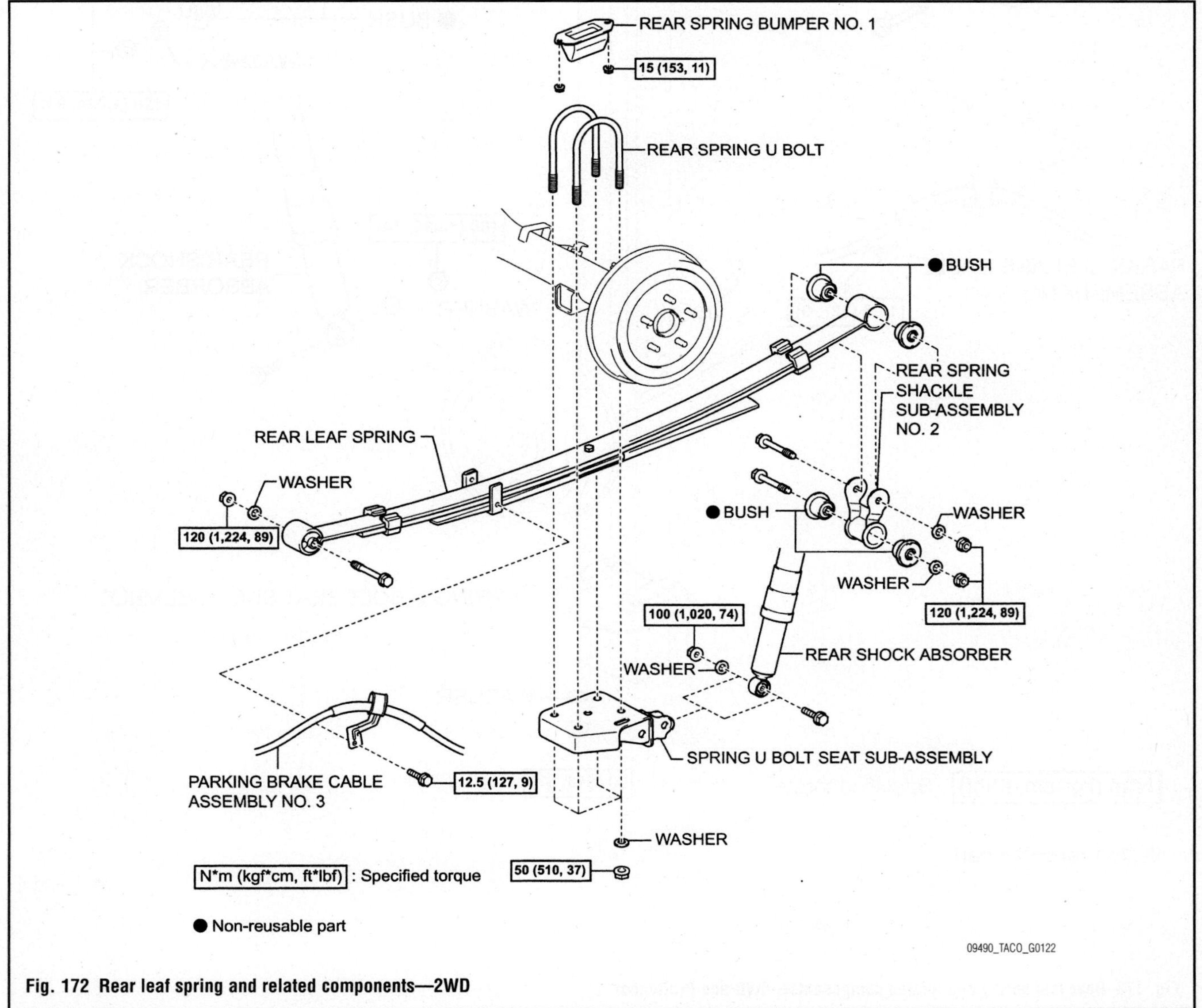

Fig. 172 Rear leaf spring and related components—2WD

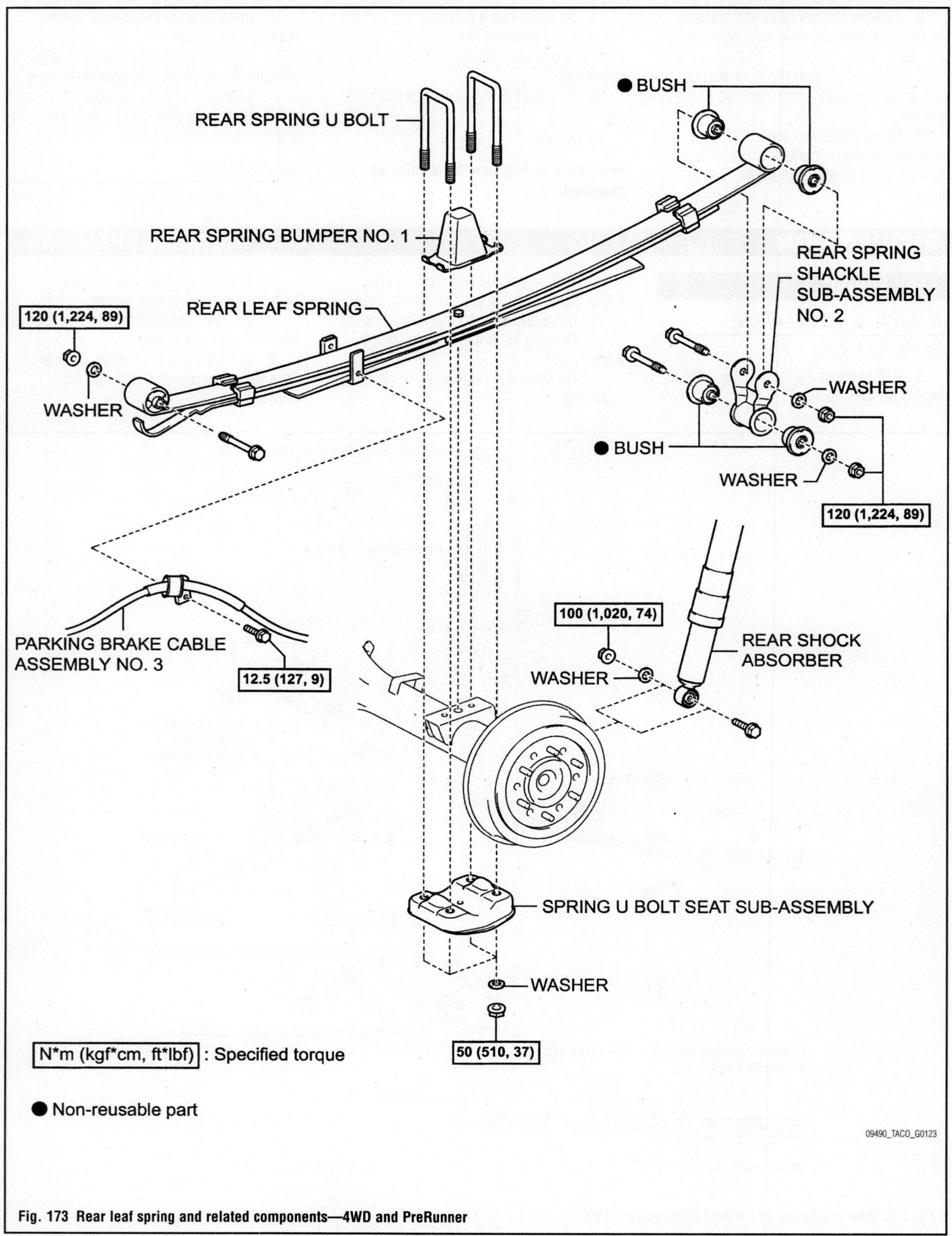

Fig. 173 Rear leaf spring and related components—4WD and PreRunner

9. Remove the bolt and then separate the parking brake cable.

10. On 2WD remove the two nuts and spring bumper. Remove the four nuts and four washers. Remove the spring seat and two U-bolts.

11. On 4WD and PreRunner, remove the four nuts and four washers. Remove the spring seat and two U-bolts. Remove the spring bumper.

12. Remove the nut, washer and through bolt. Remove the spring from the vehicle.

➡Be careful not to drop the spring when removing the through bolt.

To install:

➡Be sure to use new fasteners, as required.

➡Always fully tighten rubber bushings when the wheels are in full contact with the ground and the vehicle is at curb height. Bounce the vehicle up and down several times to stabilize the suspension, prior to final tightening of these components.

13. Installation is the reverse of the removal procedure.

14. Tighten the rear spring U-bolts to 37 ft. lbs. (50 Nm).Be sure that the lengths of all of the U-bolts under the spring seat are the same.

SHOCK ABSORBER

REMOVAL & INSTALLATION

See Figure 174.

1. Before servicing the vehicle, refer to the Precautions Section.

2. Disconnect the negative battery cable. Tape the cable with insulating tape.

3. Raise and support the vehicle safely.

4. Remove the tire and wheel assembly.

5. Lower the floor jack to take tension off of the spring.

6. Remove or disconnect the following:
   - Shock absorber from the rear axle housing
   - Nut, retainers and the cushions holding the shock absorber to the frame
   - Shock absorber with the washers and bushings

To install:

➡Be sure to use new fasteners, as required.

➡Always fully tighten rubber bushings when the wheels are in full contact with the ground and the vehicle is at curb height. Bounce the vehicle up and down several times to stabilize the suspension, prior to final tightening of these components.

7. Install the shock absorber to the frame with the washers and bushings.

8. Tighten the shock absorber-to-upper frame nut on to 15 ft. lbs. (20 Nm).

9. Connect the shock absorber to the rear axle housing.

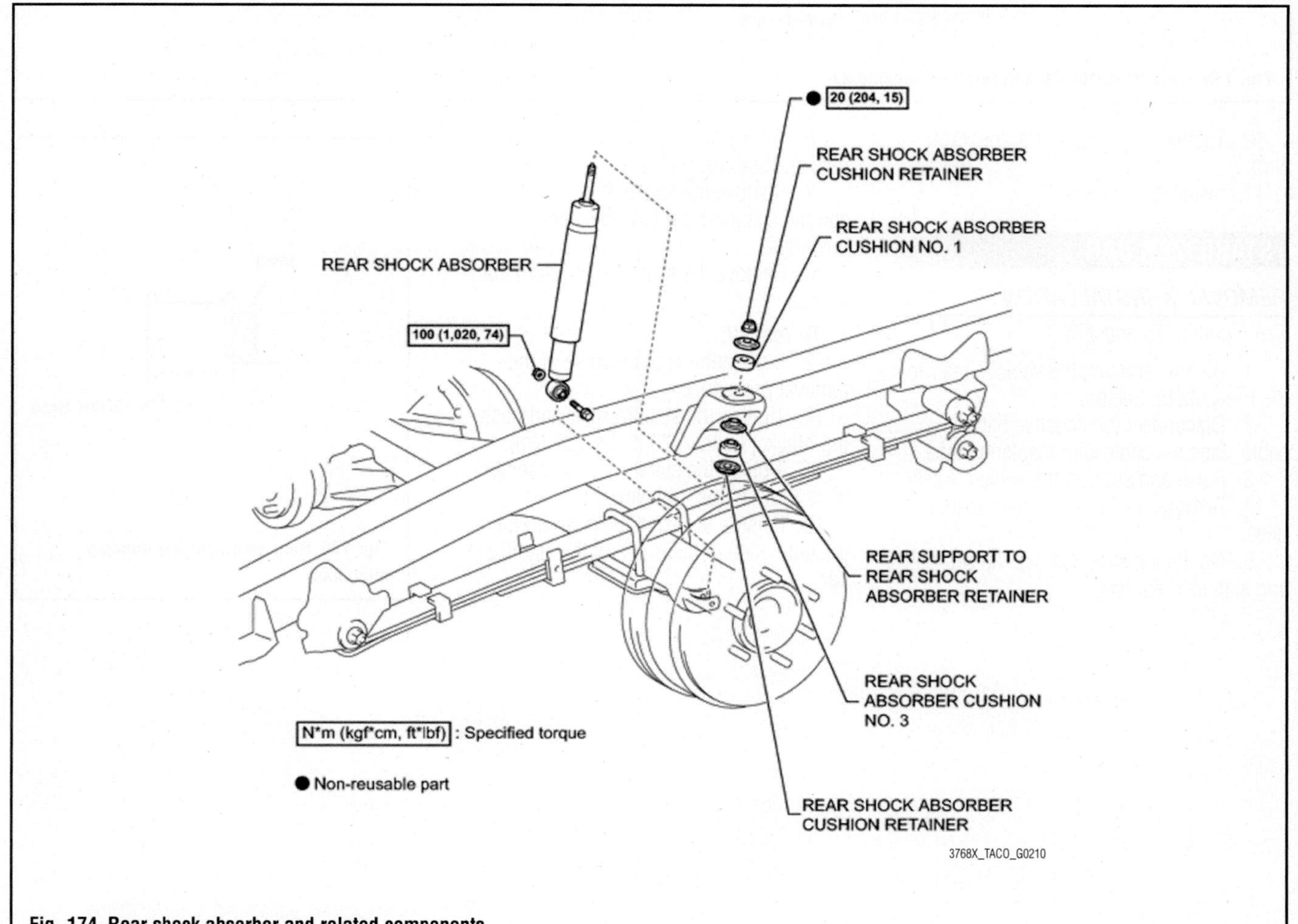

Fig. 174 Rear shock absorber and related components

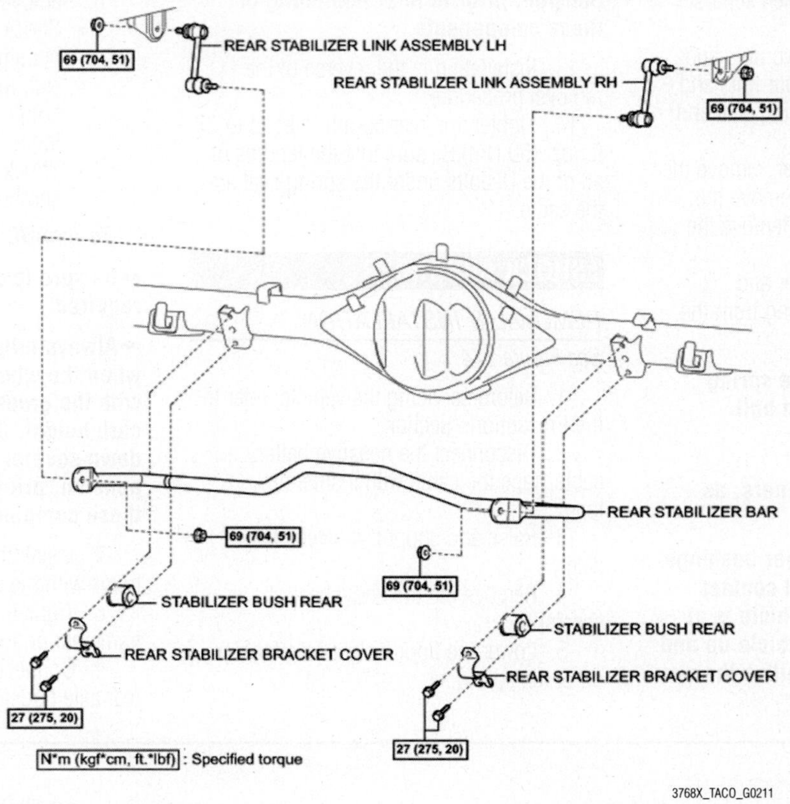

Fig. 175 Rear stabilizer bar and related components

10. Tighten the bolt to 74 ft. lbs. (100 Nm).
11. Install the wheels.

STABILIZER BAR

REMOVAL & INSTALLATION

See Figures 175 and 176.

1. Before servicing the vehicle, refer to the Precautions Section.
2. Disconnect the negative battery cable. Tape the cable with insulating tape.
3. Raise and support the vehicle safely.
4. Remove the tire and wheel assemblies.
5. On the left side, remove the two nuts and stabilizer bar link.
6. On the right side, remove the two nuts and stabilizer bar link.
7. Remove the four bolts and the stabilizer bar brackets and bushings.
8. Remove the stabilizer bar from the vehicle.

To install:

9. Installation is the reverse of the removal procedure.
10. Tighten the stabilizer bar and bushing bracket bolts to 20 ft. lbs. (27 Nm).
11. Tighten the stabilizer link assembly bolts to 51 ft. lbs. (69 Nm).
12. Be sure to install the bushing onto the outer side of the mark on the stabilizer bar.

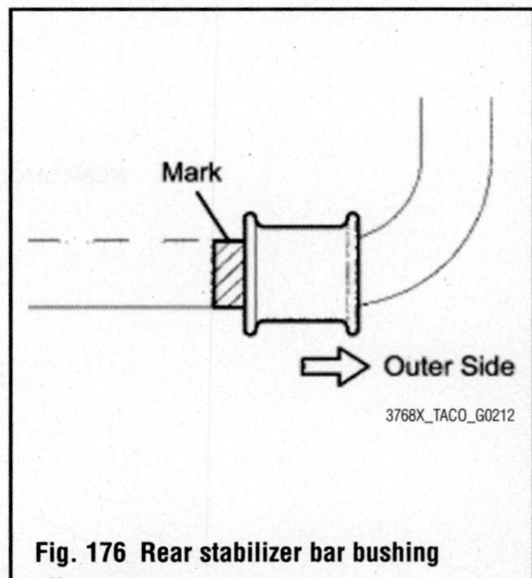

Fig. 176 Rear stabilizer bar bushing alignment

TOYOTA

Tundra

16

SPECIFICATIONS AND MAINTENANCE CHARTS

ENGINE AND VEHICLE IDENTIFICATION

Engine							Model Year	
Code	Liters (cc)	Cu. In.	Cyl.	Fuel Sys.	Engine Type	Eng. Mfg.	Code ①	Year
1GR-FE/U	4.0 (3956)	241	6	MFI	DOHC	Toyota	B	2011
1UR-FE/M	4.6 (4608)	282	8	SFI	DOHC	Toyota	C	2012
3UR-FE/Y	5.7 (5663)	346	8	SFI	DOHC	Toyota		
3UR-FBE/W	5.7 (5663)	346	8	SFI	DOHC	Toyota		

SFI: Sequential Fuel Injection

MFI: Multi-port Fuel Injection

DOHC: Double Overhead Camshaft

① 10th digit of the VIN number

71099_TUND_C0001

GENERAL ENGINE SPECIFICATIONS

Year	Model	Engine Displacement Liters	Engine Series Code/ID	Net Horsepower @ rpm	Net Torque @ rpm (ft. lbs.)	Bore x Stroke (in.)	Com-pression Ratio	Oil Pressure @ rpm
2011	Tundra	4.0	1GR-FE/U	236@5200	266@4000	3.70x3.74	N/A	43-85@3000
		4.6	1UR-FE/M	310@5600	327@3400	3.70x3.27	10.2:1	32@2500
		5.7	3UR-FBE/W	383@5600	403@3600	3.70x4.02	10.2:1	45-65@3000
		5.7	3UR-FE/Y	383@5600	403@3600	3.70x4.02	10.2:1	45-65@3000
2012	Tundra	4.0	1GR-FE/U	236@5200	266@4000	3.70x3.74	N/A	43-85@3000
		4.6	1UR-FE/M	310@5600	327@3400	3.70x3.27	10.2:1	32@2500
		5.7	3UR-FBE/W	383@5600	403@3600	3.70x4.02	10.2:1	45-65@3000
		5.7	3UR-FE/Y	383@5600	403@3600	3.70x4.02	10.2:1	45-65@3000

N/A: Not Available

71099_TUND_C0002

ENGINE TUNE-UP SPECIFICATIONS

Year	Engine Displacement Liters	Engine Code/ID	Spark Plug Gap (in.)	Ignition Timing (deg.)	Fuel Pump (psi)	Idle Speed (rpm) MT	Idle Speed (rpm) AT	Valve Clearance Intake	Valve Clearance Exhaust
2011	4.0	1GR-FE/U	0.039-0.043	7-24B ①	40.8-41.7	N/A	650-750	0.0591-0.0098	0.0114-0.0154
	4.6	1UR-FE/M	0.039-0.043	8-12 ②	41-42	N/A	650-750	N/A	N/A
	5.7	3UR-FBE/W	0.039-0.043	8-12 ②	38-44	N/A	650-750	N/A	N/A
	5.7	3UR-FE/Y	0.039-0.043	8-12 ②	38-44	N/A	650-750	N/A	N/A
2012	4.0	1GR-FE/U	0.039-0.043	7-24B ①	40.8-41.7	N/A	650-750	0.0591-0.0098	0.0114-0.0154
	4.6	1UR-FE/M	0.039-0.043	8-12 ②	41-42	N/A	650-750	N/A	N/A
	5.7	3UR-FBE/W	0.039-0.043	8-12 ②	38-44	N/A	650-750	N/A	N/A
	5.7	3UR-FE/Y	0.039-0.043	8-12 ②	38-44	N/A	650-750	N/A	N/A

NOTE: The Vehicle Emission Control Information label often reflects specification changes made during production. The label figures must be used if they differ from those in this chart.

B: Before top dead center

N/A: Not Available

① With terminals TE1 and E1 connected to DLC1. Transmission in N and AC off.

② With terminals TC and Cg of the DLC3 connected. Transmission in N and AC off.

71099_TUND_C0003

CAPACITIES

Year	Model	Engine Displacement Liters	Engine Code/ID	Engine Oil with Filter (qts.)	Transmission (qts.) 5-Spd	Transmission (qts.) Auto.	Transfer Case (qts.)	Drive Axle Front (qts.)	Drive Axle Rear (qts.)	Fuel Tank (gal.)	Cooling System (qts.)
2011	Tundra	4.0	1GR-FE/U	6.4	N/A	11.1	N/A	2.2	①	26.4	10.1
		4.6	1UR-FE/M	7.9	N/A	②	1.2	2.2	①	26.4	③
		5.7	3UR-FBE/W	7.9	N/A	④	1.2	2.2	⑤	26.4	③
		5.7	3UR-FE/Y	7.9	N/A	④	1.2	2.2	⑤	26.4	③
2012	Tundra	4.0	1GR-FE/U	6.4	N/A	11.1	N/A	2.2	①	26.4	10.1
		4.6	1UR-FE/M	7.9	N/A	②	1.2	2.2	①	26.4	③
		5.7	3UR-FBE/W	7.9	N/A	④	1.2	2.2	⑤	26.4	③
		5.7	3UR-FE/Y	7.9	N/A	④	1.2	2.2	⑤	26.4	③

N/A: Not Available

① 4.0L & 4.6L standard bed: 4.3
4.0L & 4.6L long bed: 4.9

② With tow package: 2WD 12.2, 4WD 11.8
Without tow package: 2WD 11.6, 4WD 11.3

③ 12.8 w/o tow package
13.9 with tow package

④ With tow package: 12.3
Without tow package: 11.7

⑤ 5.7L regular cab standard bed: 3.7
5.7L regular cab long bed: 3.8
5.7 double cab standard bed: 3.8
5.7L double cab lc
5.7L crew cab 2wd: 4.0
5.7L crew cab 4wd: 3.8

71099_TUND_C0004

FLUID SPECIFICATIONS

Year	Model	Engine Displacement Liters	Engine ID/VIN	Engine Oil	Auto. Trans.	Power Steering Fluid	Brake Master Cylinder
2011	Tundra	4.0	1GR-FE/U	0W-20	①	②	DOT 3
		4.6	1UR-FE/M	5W-20	①	②	DOT 3
		5.7	3UR-FBE/W	5W-20	①	②	DOT 3
		5.7	3UR-FE/Y	5W-20	①	②	DOT 3
2012	Tundra	4.0	1GR-FE/U	0W-20	①	②	DOT 3
		4.6	1UR-FE/M	5W-20	①	②	DOT 3
		5.7	3UR-FBE/W	5W-20	①	②	DOT 3
		5.7	3UR-FE/Y	5W-20	①	②	DOT 3

DOT: Department Of Transportation

① A750E, A750F, AB60E, AB60F: ATF WS

② ATF Dexron II or III

71099_TUND_C0015

VALVE SPECIFICATIONS

Year	Engine Displ. Liters	Engine Code/ID	Seat Angle (deg.)	Face Angle (deg.)	Spring Test Pressure (lbs. @ in.)	Spring Installed Height (in.)	Stem-to-Guide Clearance (in.)		Stem Diameter (in.)	
							Intake	Exhaust	Intake	Exhaust
2011	4.0	1GR-FE/U	N/A	44.5	42-46@ 1.31	1.882	0.00094- 0.00236	0.00118- 0.00256	0.2154 0.2159	0.2152 0.2156
	4.6	1UR-FE/M	N/A	N/A	N/A	N/A	0.000984- 0.00236	0.00118- 0.00256	0.2150- 0.2160	0.2150- 0.2160
	5.7	3UR-FBE/W	N/A	N/A	N/A	N/A	0.000984- 0.00236	0.00118- 0.00256	0.2150- 0.2160	0.2150- 0.2160
	5.7	3UR-FE/Y	N/A	N/A	N/A	N/A	0.000984- 0.00236	0.00118- 0.00256	0.2150- 0.2160	0.2150- 0.2160
2012	4.0	1GR-FE/U	N/A	44.5	42-46@ 1.31	1.882	0.00094- 0.00236	0.00118- 0.00256	0.2154 0.2159	0.2152 0.2156
	4.6	1UR-FE/M	N/A	N/A	N/A	N/A	0.000984- 0.00236	0.00118- 0.00256	0.2150- 0.2160	0.2150- 0.2160
	5.7	3UR-FBE/W	N/A	N/A	N/A	N/A	0.000984- 0.00236	0.00118- 0.00256	0.2150- 0.2160	0.2150- 0.2160
	5.7	3UR-FE/Y	N/A	N/A	N/A	N/A	0.000984- 0.00236	0.00118- 0.00256	0.2150- 0.2160	0.2150- 0.2160

N/A: Not Available

71099_TUND_C0005

CAMSHAFT SPECIFICATIONS

All measurements are given in inches.

Year	Engine Displacement Liters	Engine Code/ID	Journal Dia.	Brg. Oil Clearance	Shaft End-play	Circle Runout	Lobe Height Intake	Lobe Height Exhaust
2011	4.0	1GR-FE/U	1.4160-1.4170	①	0.01570-0.03540	0.00236	1.7390-1.7420	1.75511-1.75905
	4.6	1UR-FE/M	②	N/A	0.00315-0.00531	0.00157	1.7440-1.7500	1.74000-1.74600
	5.7	3UR-FBE/W	②	N/A	0.00315-0.00531	0.00157	1.7440-1.7500	1.74000-1.74600
	5.7	3UR-FE/Y	②	N/A	0.00315-0.00531	0.00157	1.7440-1.7500	1.74000-1.74600
2012	4.0	1GR-FE/U	1.4160-1.4170	①	0.01570-0.03540	0.00236	1.7390-1.7420	1.75511-1.75905
	4.6	1UR-FE/M	②	N/A	0.00315-0.00531	0.00157	1.7440-1.7500	1.74000-1.74600
	5.7	3UR-FBE/W	②	N/A	0.00315-0.00531	0.00157	1.7440-1.7500	1.74000-1.74600
	5.7	3UR-FE/Y	②	N/A	0.00315-0.00531	0.00157	1.7440-1.7500	1.74000-1.74600

N/A: Not Available

Note: Shaft end-play is thrust clearance

① No. 1 intake right side:0.000315-0.00150
No. 1 exhaust right side: 0.00157-0.00311
No. 1 intake left side: 0.00157-0.00311
Other journal: 0.000984-0.00244
No. 1 intake right side: 0.00276
Other journal: 0.00394
LH side journal: 0.00394

② No. 1: 1.1793-1.1799
Other journal: 1.0220-1.0226

71099_TUND_C0014

CRANKSHAFT AND CONNECTING ROD SPECIFICATIONS

All measurements are given in inches.

Year	Engine Displ. Liters	Engine Code/ID	Crankshaft Main Brg. Journal Dia.	Main Brg. Oil Clearance	Shaft End-play	Thrust on No.	Connecting Rod Journal Diameter	Oil Clearance	Side Clearance
2011	4.0	1GR-FE/U	2.8342-2.8346	0.00079-0.00118	N/A	N/A	①	②	②
	4.6	1UR-FE/M	2.6373-2.6378	③	0.0008-0.0087	3	N/A	0.000984-0.00197	0.00591-0.02170
	5.7	3UR-FBE/W	2.6373-2.6378	③	0.0008-0.0087	3	2.3230-2.3235	0.000984-0.00197	0.00591-0.02170
	5.7	3UR-FE/Y	2.6373-2.6378	③	0.0008-0.0087	3	2.3230-2.3235	0.000984-0.00197	0.00591-0.02170
2012	4.0	1GR-FE/U	2.8342-2.8346	0.00079-0.00118	N/A	N/A	①	②	②
	4.6	1UR-FE/M	2.6373-2.6378	③	0.0008-0.0087	3	N/A	0.000984-0.00197	0.00591-0.02170
	5.7	3UR-FBE/W	2.6373-2.6378	③	0.0008-0.0087	3	2.3230-2.3235	0.000984-0.00197	0.00591-0.02170
	5.7	3UR-FE/Y	2.6373-2.6378	③	0.0008-0.0087	3	2.3230-2.3235	0.000984-0.00197	0.00591-0.02170

N/A: Not Available

① Rod bolt tension portion diameter: 0.2830-0.2870

② Rod oil clearance: 0.00102-0.00181. Bushing oil clearance: 0.000197-0.000433

③ Nos. 1 and 5: 0.000669-0.00118

All others: 0.000945-0.00146

71099_TUND_C0007

PISTON AND RING SPECIFICATIONS

All measurements are given in inches.

Year	Engine Displ. Liters	Engine Code/ID	Piston Clearance	Ring Gap Top Compression	Ring Gap Bottom Compression	Ring Gap Oil Control	Ring Side Clearance Top Compression	Ring Side Clearance Bottom Compression	Ring Side Clearance Oil Control
2011	4.0	1GR-FE/U	0.0031-0.0040	0.0118-0.0157	0.0157-0.0197	0.0039-0.0157	0.0008-0.0028	0.0008-0.0024	0.0028-0.0060
	4.6	1UR-FE/M	N/A	0.00906-0.0130	0.0157-0.0197	0.00394-0.0157	0.000787-0.000276	0.000787-0.000236	0.00276-0.00571
	5.7	3UR-FBE/W	0.0016-0.0024	0.0098-0.0138	0.0157-0.0197	0.0039-0.0157	0.0007-0.0028	0.0007-0.0027	0.0028-0.0057
	5.7	3UR-FE/Y	0.0016-0.0024	0.0098-0.0138	0.0157-0.0197	0.0039-0.0157	0.0007-0.0028	0.0007-0.0027	0.0028-0.0057
2012	4.0	1GR-FE/U	0.0031-0.0040	0.0118-0.0157	0.0157-0.0197	0.0039-0.0157	0.0008-0.0028	0.0008-0.0024	0.0028-0.0060
	4.6	1UR-FE/M	N/A	0.00906-0.0130	0.0157-0.0197	0.00394-0.0157	0.000787-0.000276	0.000787-0.000236	0.00276-0.00571
	5.7	3UR-FBE/W	0.0016-0.0024	0.0098-0.0138	0.0157-0.0197	0.0039-0.0157	0.0007-0.0028	0.0007-0.0027	0.0028-0.0057
	5.7	3UR-FE/Y	0.0016-0.0024	0.0098-0.0138	0.0157-0.0197	0.0039-0.0157	0.0007-0.0028	0.0007-0.0027	0.0028-0.0057

N/A: Not Available

71099_TUND_C0008

TORQUE SPECIFICATIONS

All readings in ft. lbs.

Year	Engine Displacement Liters	Engine Code/ID	Cylinder Head Bolts	Main Bearing Bolts	Rod Bearing Bolts	Crankshaft Damper Bolts	Flywheel Bolts	Manifold Intake	Manifold Exhaust	Spark Plugs	Oil Pan Drain Plug
2011	4.0	1GR-FE/U	①	②	③	185	61	19	15	15	30
	4.6	1UR-FE/M	④	⑤	⑥	221	⑦	15	15	15	30
	5.7	3UR-FE/W	④	⑤	⑥	221	⑦	15	15	15	30
	5.7	3UR-FE/V, Y	④	⑤	⑥	221	⑦	15	15	15	30
2012	4.0	1GR-FE/U	①	②	③	185	61	19	15	15	30
	4.6	1UR-FE/M	④	⑤	⑥	221	⑦	15	15	15	30
	5.7	3UR-FE/W	④	⑤	⑥	221	⑦	15	15	15	30
	5.7	3UR-FE/V, Y	④	⑤	⑥	221	⑦	15	15	15	30

① Right: 27 ft. lbs., then + 90 degrees, then +90 degrees
Left : (8 bolts) 27 ft. lbs., then + 90 degrees, then + 90 degrees. (2 bolts) 27 ft. lbs.

② (16 bolts): 45 ft. lbs., then +90 degrees
(8 bolts): 19 ft. lbs.

③ Step 1: 18 ft. lbs.
Step 2: Plus 90 degrees

④ Step 1: 27 ft. lbs.
Step 2: Plus 90 degrees
Step 3: Plus 90 degrees
12 mm head 15 ft. lbs.

⑤ Inside position 45 ft. lbs.
Step 1: Outside position 20 ft. lbs.
Step 2: Plus 90 degrees
Step 3: Cylinder block side position 33 ft. lbs.

⑥ Step 1: 30 ft. lbs.
Step 2: Plus 90 degrees

⑦ Step 1: 22 ft. lbs.
Step 2: Plus 90 degrees

71099_TUND_C0009

TIRE, WHEEL AND BALL JOINT SPECIFICATIONS

Year	Model	OEM Tires		Tire Pressures (psi)		Wheel	Ball Joint	Lug Nut Torque
		Standard	Optional	Front	Rear	Size	Inspection	(ft. lbs.)
2011	Tundra	①	①	②	②	8J	③	④
2012	Tundra	⑤	⑤	②	②	8J	③	④

OEM: Original Equipment Manufacturer

PSI: Pounds Per Square Inch

STD: Standard

OPT: Optional

NS: Not specified by manufacturer

NA: Not available

① P275/70R18, P275/65R18 or P275/55R20

② Front: 30, Rear: 33. Use specification on vehicle placard if different from one given.

③ Turning torque 89 inch lbs.

④ Steel wheel 154 ft. lbs. aluminum wheels 97 ft. lbs. If specification differs from one given, see owners manual.

⑤ P255/70R18, P275/65R18 or P275/55R20

71099_TUND_C0012

BRAKE SPECIFICATIONS

All measurements in inches unless noted

Year	Model		Brake Disc			Minimum	Brake Caliper	
			Original Thickness	Minimum Thickness	Maximum Runout	Lining Thickness	Bracket Bolts (ft. lbs.)	Mounting Bolts (ft. lbs.)
2011	Tundra	F	1.260	1.140	0.00197	0.469	—	73
		R	0.709	0.630	0.00787	0.472	70	65
2012	Tundra	F	1.260	1.140	0.00197	0.469	—	73
		R	0.709	0.630	0.00787	0.472	70	65

F: Front

R: Rear

71099_TUND_C0010

SCHEDULED MAINTENANCE INTERVALS

TOYOTA—TUNDRA

TO BE SERVICED	TYPE OF SERVICE	VEHICLE MILEAGE INTERVAL (x1000)																		
		5	10	15	20	25	30	35	40	45	50	55	60	65	70	75	80	85	90	95
Automatic transmission and differential fluid	S/I			✓			✓			✓			✓			✓			✓	
Ball joints and boots	S/I			✓			✓			✓			✓			✓			✓	
Brake system	S/I			✓			✓			✓			✓			✓			✓	
Cabin filter	S/I			✓			✓			✓			✓			✓			✓	
Charcoal canister	S/I												✓							
Drive belts	S/I						✓						✓						✓	
Driveshaft bushing	L						✓						✓						✓	
Engine coolant	R						✓						✓						✓	
Engine oil & filter	R	✓	✓	✓	✓	✓	✓	✓	✓	✓	✓	✓	✓	✓	✓	✓	✓	✓	✓	✓
Exhaust system	S/I			✓			✓			✓			✓			✓			✓	
Fuel lines	S/I						✓						✓						✓	
Fuel tank cap gasket	S/I						✓						✓						✓	
Halfshaft boots & flange bolts	S/I			✓			✓			✓			✓			✓			✓	
Limited slip differential fluid	R						✓						✓						✓	
Differential fluid	S/I						✓						✓						✓	
Non-platinum spark plugs	R						✓						✓						✓	
Platinum spark plugs	R												✓							
Propeller shaft (4WD)	L			✓			✓			✓			✓			✓			✓	
Propeller shaft bolts	S/I			✓			✓			✓			✓			✓			✓	
Steering gear	S/I			✓			✓			✓			✓			✓			✓	
Steering linkage	S/I			✓			✓			✓			✓			✓			✓	
Tires (rotate)	S/I	✓	✓	✓	✓	✓	✓	✓	✓	✓	✓	✓	✓	✓	✓	✓	✓	✓	✓	✓
Valves	S/I												✓							

R: Replace S/I: Service or Inspect L: Lubricate

FREQUENT OPERATION MAINTENANCE (SEVERE SERVICE)

If a vehicle is operated under any of the following conditions it is considered severe service:

- Towing a trailer or using a camper or car-top carrier.

- Repeated short trips of less than 5 miles in temperatures below freezing.

- Excessive idling or low-speed driving for long distances as in heavy commercial use, such as delivery, taxi or police cars.

- Operating on rough, muddy or salt-covered roads.

- Operating on unpaved or dusty roads.

Oil and filter: service every 2500 miles or 3 months, whichever occurs first.

Brake linings and discs or drums: service or inspect every 5000 miles or 4 months, whichever occurs first.

Steering linkage: service or inspect every 5000 miles or 4 months, whichever occurs first.

Ball joints and boots: service or inspect every 5000 miles or 4 months, whichever occurs first.

Brake discs & pads (front): service or inspect every 6000 miles.

Halfshaft boots: service or inspect every 5000 miles or 4 months. Retighten the flange bolts, whichever occurs first.

Body chassis bolts and nuts: service or inspect every 5000 miles or 4 months, whichever occurs first.

Transmission and differential fluid: replace every 15,000 miles or 12 months, whichever occurs first.

Transfer case and differential fluid: replace every 15,000 miles or 12 months, whichever occurs first.

71099_TUND_C0013

GENERAL INFORMATION

PRECAUTIONS

- Certain components within the ABS system are not intended to be serviced or repaired individually.
- Do not use rubber hoses or other parts not specifically specified for and ABS system. When using repair kits, replace all parts included in the kit. Partial or incorrect repair may lead to functional problems and require the replacement of components.
- Lubricate rubber parts with clean, fresh brake fluid to ease assembly. Do not use shop air to clean parts; damage to rubber components may result.
- Use only DOT 3 brake fluid from an unopened container.
- If any hydraulic component or line is removed or replaced, it may be necessary to bleed the entire system.
- A clean repair area is essential. Always clean the reservoir and cap thoroughly before removing the cap. The slightest amount of dirt in the fluid may plug an orifice and impair the system function. Perform repairs after components have been thoroughly cleaned; use only denatured alcohol to clean components. Do not allow ABS components to come into contact with any substance containing mineral oil; this includes used shop rags.
- The Anti-Lock control unit is a microprocessor similar to other computer units in the vehicle. Ensure that the ignition switch is **OFF** before removing or installing controller harnesses. Avoid static electricity discharge at or near the controller.
- If any arc welding is to be done on the vehicle, the control unit should be unplugged before welding operations begin.

WHEEL SPEED SENSORS

REMOVAL & INSTALLATION

Front

See Figure 1.

1. Before servicing the vehicle, refer to the Precautions Section.
2. Disconnect the negative battery cable. Tape the cable with insulating tape.
3. Raise and support the vehicle safely.
4. Remove front wheel.
5. Remove the skid control wire.
6. Disconnect speed sensor connector.
7. Remove the clips and 3 clamp bolts holding the sensor harness from the frame, upper arm and steering knuckle.
8. Remove the bolt and speed sensor.

To install:

➡Be sure to use new fasteners, as required.

9. Installation is the reverse of the removal procedure.
10. Be sure to install the bracket so that the rotation stopper touches the knuckle.
11. Inspect the sensor signal. Correct as required.

Rear

See Figure 2.

1. Before servicing the vehicle, refer to the Precautions Section.
2. Disconnect the negative battery cable. Tape the cable with insulating tape.

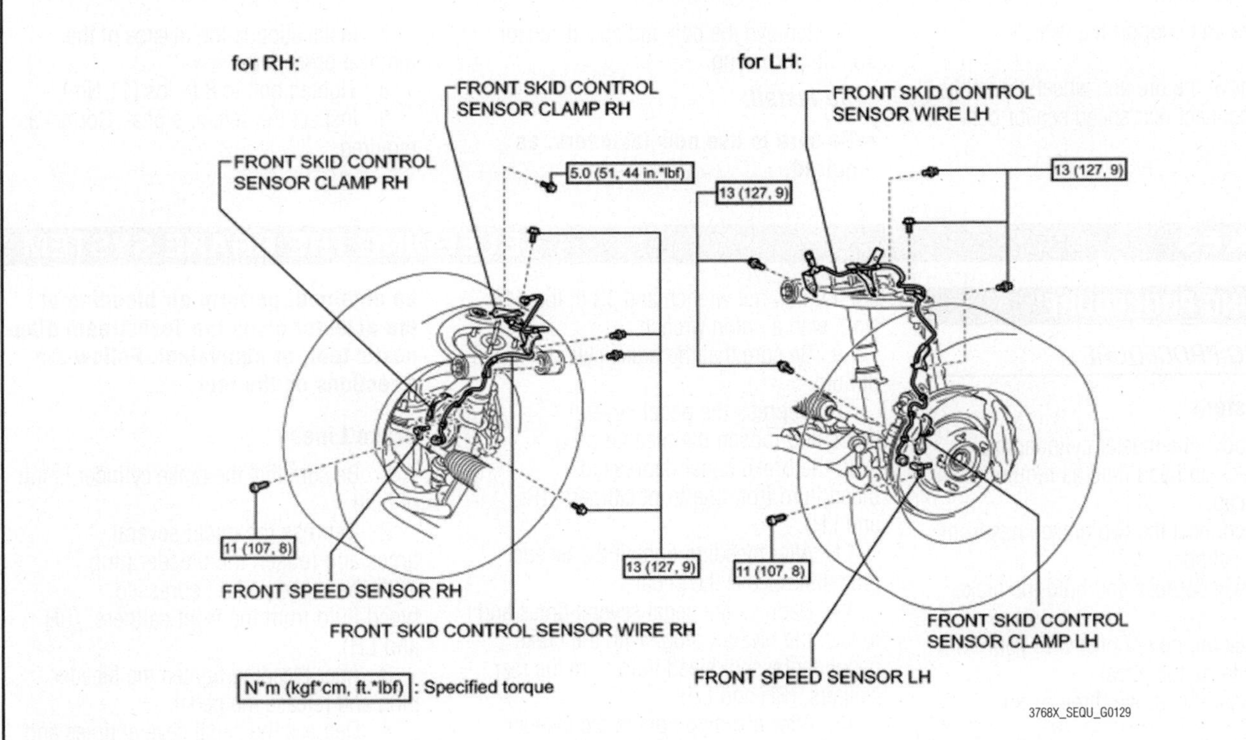

Fig. 1 Front wheel speed sensor and related components

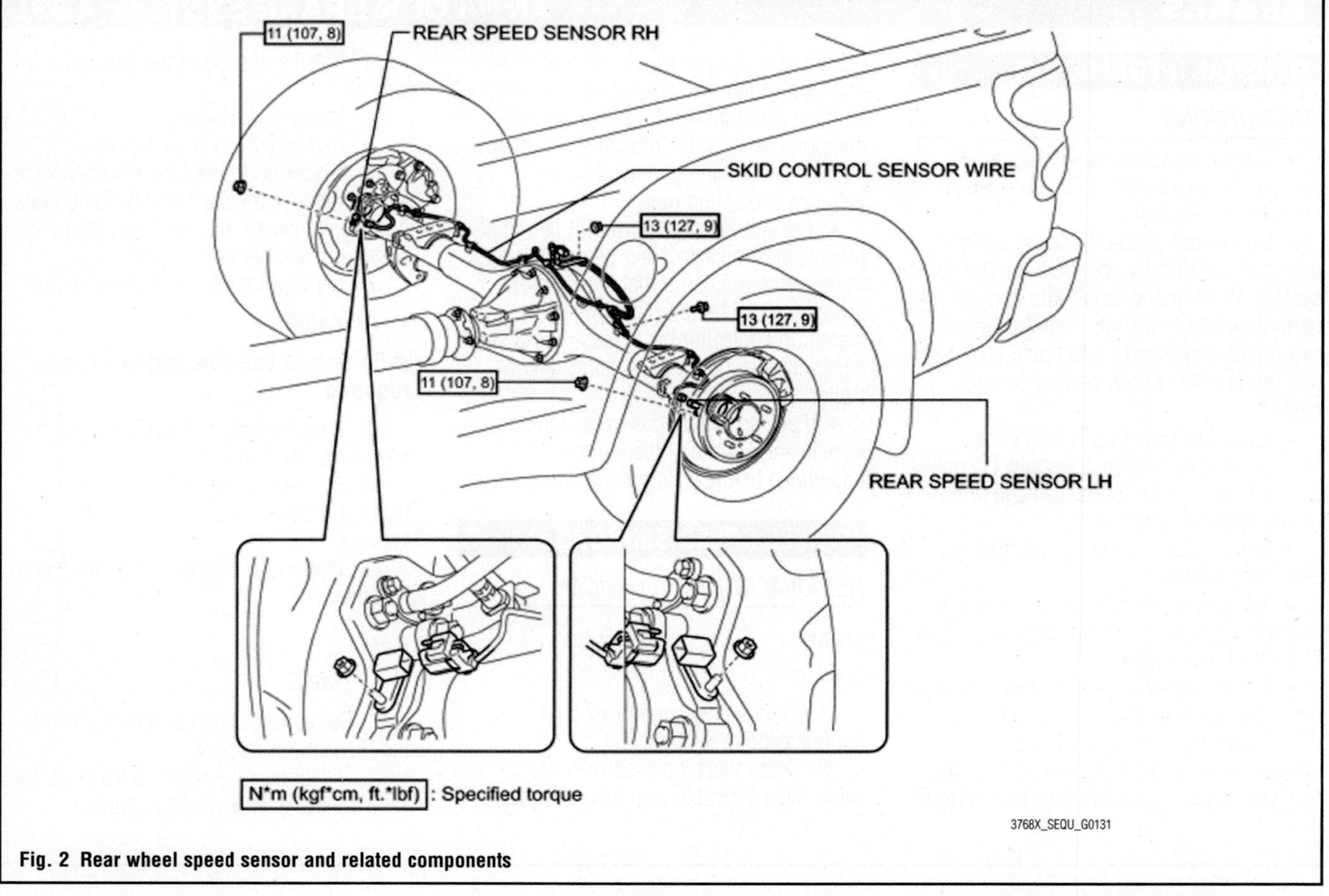

Fig. 2 Rear wheel speed sensor and related components

3. Raise and support the vehicle safely.
4. Remove the tire and wheel assembly.
5. Disconnect rear speed sensor connector.
6. Remove the bolt and speed sensor from its mounting.

To install:

➡Be sure to use new fasteners, as required.

7. Installation is the reverse of the removal procedure.
8. Tighten bolt to 8 ft. lbs (11 Nm).
9. Inspect the sensor signal. Correct as required.

BRAKES — BLEEDING THE BRAKE SYSTEM

BLEEDING PROCEDURE

BLEEDING PROCEDURE

Brake System

1. Remove the master cylinder cap.
2. Check and add fluid as required. Install the cap.
3. Disconnect the two brake lines from the master cylinder.
4. Slowly depress and hold the brake pedal.
5. Block the outer holes with your fingers and release the pedal.
6. Repeat the above three or four times.
7. Connect the brake lines to the master cylinder.
8. Tighten to 14 ft. lbs. (20 Nm), without a union nut wrench and 13 ft. lbs. (18 Nm), with a union wrench.
9. Be sure that the brake cylinder is full of fluid.
10. Depress the pedal several times and loosen the bleeder plug. With the brake pedal depressed, bleed fluid from the front calipers, (RH and LH).
11. After bleeding tighten the bleeder plug and release the pedal.
12. Depress the pedal several times and loosen the bleeder plug. With the brake pedal depressed, bleed fluid from the rear calipers, (RH and LH).
13. After bleeding tighten the bleeder plug and release the pedal.

➡After bleeding air from the system, if the height or feel of the pedal cannot be obtained, perform air bleeding of the actuator using the Techstream diagnostic tool, or equivalent. Follow the directions on the tool.

Brake Lines

1. Be sure that the brake cylinder is full of fluid.
2. Depress the pedal several times and loosen the bleeder plug. With the brake pedal depressed, bleed fluid from the front calipers, (RH and LH).
3. After bleeding tighten the bleeder plug and release the pedal.
4. Depress the pedal several times and loosen the bleeder plug. With the brake pedal depressed, bleed fluid from the rear calipers, (RH and LH).

5. After bleeding tighten the bleeder plug and release the pedal.

6. If the following symptoms occur, low or spongy pedal and pedal is depressed but braking is insufficient, bleed the brake system.

7. Bleed the VSC actuator assembly.

➡Depress the pedal more than twenty times with the engine off. Connect the diagnostic tool to the DLC3. Turn the ignition switch ON. Do not start the engine. Select AIR BLEEDING on the tool and follow the directions. Bleed the air out of the suction line.

FLUID FILL PROCEDURE

See Figure 3.

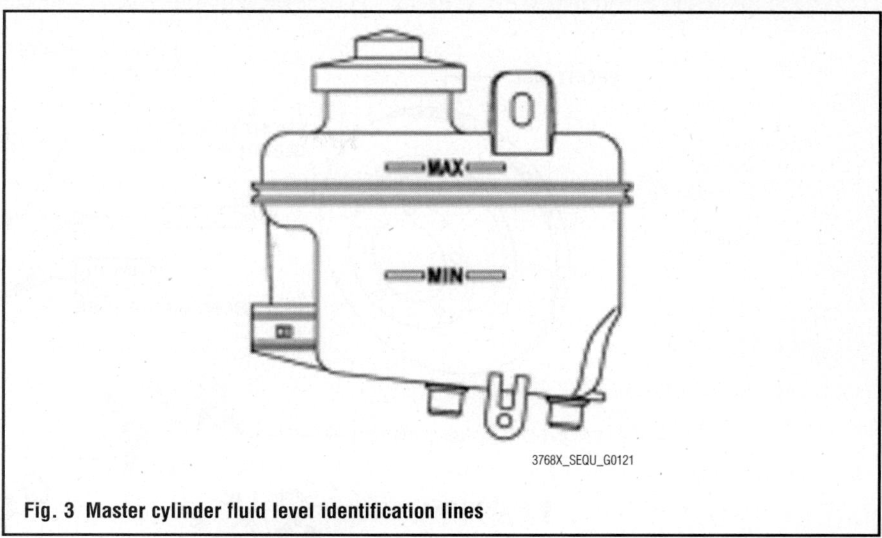

Fig. 3 Master cylinder fluid level identification lines

BRAKES

FRONT DISC BRAKES

⁂ CAUTION

Dust and dirt accumulating on brake parts during normal use may contain asbestos fibers from production or aftermarket brake linings. Breathing excessive concentrations of asbestos fibers can cause serious bodily harm. Exercise care when servicing brake parts. Do not sand or grind brake lining unless equipment used is designed to contain the dust residue. Do not clean brake parts with compressed air or by dry brushing. Cleaning should be done by dampening the brake components with a fine mist of water, then wiping the brake components clean with a dampened cloth. Dispose of cloth and all residue containing asbestos fibers in an impermeable container with the appropriate label. Follow practices prescribed by the Occupational Safety and Health Administration (OSHA) and the Environmental Protection Agency (EPA) for the handling, processing, and disposing of dust or debris that may contain asbestos fibers.

BRAKE CALIPER

REMOVAL & INSTALLATION

1. Before servicing the vehicle, refer to the Precautions Section.

2. Disconnect the negative battery cable. Tape the cable with insulating tape.

3. Raise and support the vehicle safely.

4. Remove the tire and wheel assembly.

5. To remove the front disc brake pad kit, remove the pin holddown clip. Remove the anti rattle spring. Remove the pads from the caliper. Remove the anti squeal shims.

➡The pin holddown clip can be used again if it has sufficient rebound, no deformation or wear and has all rust, dirt and foreign matter removed.

➡The anti rattle spring can be used again if it has sufficient rebound, no deformation or wear and has all rust, dirt and foreign matter removed.

6. Drain the brake fluid to an acceptable level.

7. Remove and plug the brake line hose.

8. Remove the caliper retaining bolts.

9. Remove the caliper from its mounting.

To install:

➡Be sure to use new fasteners, as required.

10. Installation is the reverse of the removal procedure.

11. Bleed the system.

DISC BRAKE PADS

REMOVAL & INSTALLATION

See Figure 4.

1. Before servicing the vehicle, refer to the Precautions Section.

2. Disconnect the negative battery cable. Tape the cable with insulating tape.

3. Raise and support the vehicle safely.

4. Remove the tire and wheel assembly.

5. To remove the front disc brake pad kit, remove the pin holddown clip. Remove the anti rattle spring. Remove the pads from the caliper. Remove the anti squeal shims.

➡The pin holddown clip can be used again if it has sufficient rebound, no deformation or wear and has all rust, dirt and foreign matter removed.

➡The anti rattle spring can be used again if it has sufficient rebound, no deformation or wear and has all rust, dirt and foreign matter removed.

To install:

➡Be sure to use new fasteners, as required.

6. Installation is the reverse of the removal procedure.

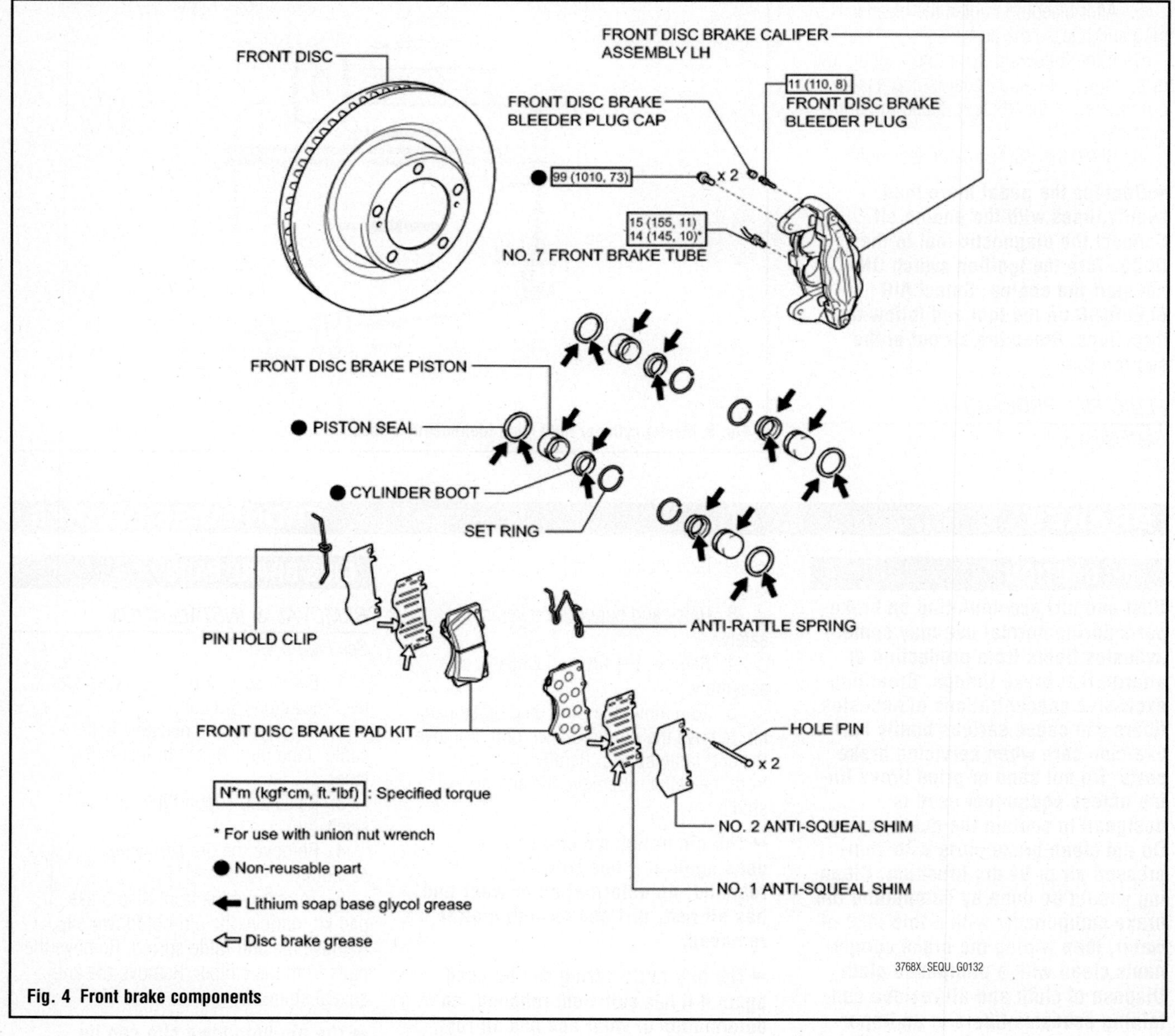

Fig. 4 Front brake components

BRAKES

REAR DISC BRAKES

CAUTION

Dust and dirt accumulating on brake parts during normal use may contain asbestos fibers from production or aftermarket brake linings. Breathing excessive concentrations of asbestos fibers can cause serious bodily harm. Exercise care when servicing brake parts. Do not sand or grind brake lining unless equipment used is designed to contain the dust residue. Do not clean brake parts with compressed air or by dry brushing. Cleaning should be done by dampening the brake components with a fine mist of water, then wiping the brake components clean with a dampened cloth. Dispose of cloth and all residue containing asbestos fibers in an impermeable container with the appropriate label. Follow practices prescribed by the Occupational Safety and Health Administration (OSHA) and the Environmental Protection Agency (EPA) for the handling, processing, and disposing of dust or debris that may contain asbestos fibers.

BRAKE CALIPER

REMOVAL & INSTALLATION

1. Before servicing the vehicle, refer to the Precautions Section.
2. Disconnect the negative battery cable. Tape the cable with insulating tape.
3. Raise and support the vehicle safely.
4. Remove the wheels.
5. Disconnect the brake hose from the caliper by removing the union bolt and 2 gaskets. Plug the end of the hose to prevent loss of fluid.
6. Remove the 2 sliding pins.

7. Lift the bottom of the caliper up and remove the caliper assembly.

To install:

➡Be sure to use new fasteners, as required.

8. Grease the caliper slides and pins with silicone grease or equivalent. Install the caliper and secure with the bolts.
9. Connect the brake hose to the caliper, using 2 new washers. Torque the union bolt to 22 ft. lbs. (30 Nm).
10. Fill the brake system to the proper level and bleed the brake system.
11. Install the tire and wheel assembly.
12. Top off the brake fluid level in the master cylinder. Check for leaks and proper brake operation.
13. Connect the negative battery cable to the battery.

DISC BRAKE PADS

REMOVAL & INSTALLATION

See Figure 5.

1. Before servicing the vehicle, refer to the Precautions Section.
2. Disconnect the negative battery cable. Tape the cable with insulating tape.
3. Raise the vehicle and support it safely.
4. Remove the wheels.
5. Remove the brake caliper and suspend it so the hose is not stretched.
6. Remove the brake pads, anti-squeal shim, pad support plates and wear indicators.

To install:

➡Be sure to use new fasteners, as required.

7. Before installing the new pads, check the disc thickness and disc runout.
8. Install the pad support plates.
9. Install the pad wear indicator plates on each pad.
10. Install the anti-squeal shim to the outer pad. Install the pads.
11. Install the brake caliper.
12. Install the wheels.
13. Apply the brake pedal several times.
14. Road-test the vehicle for proper operation.

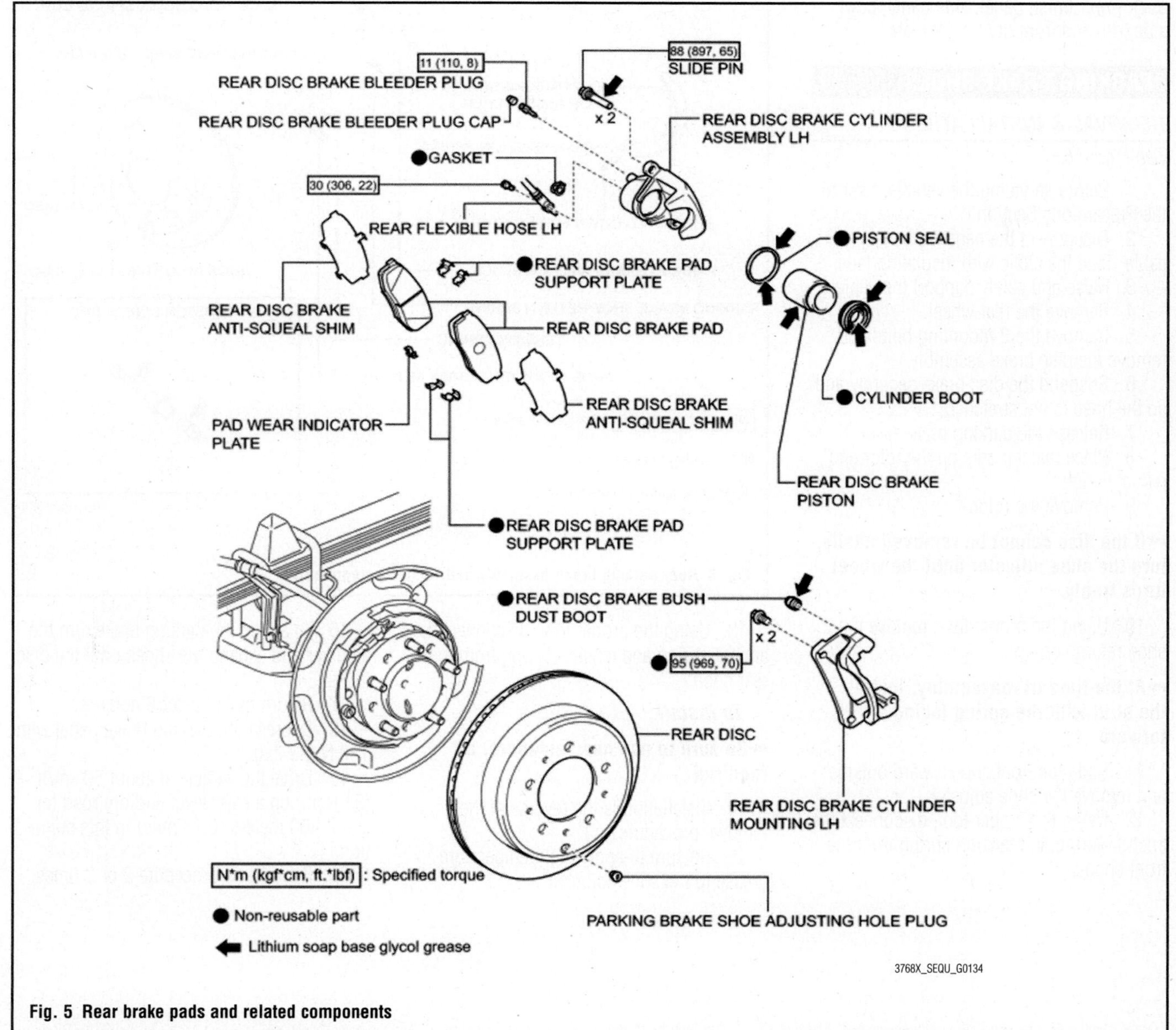

Fig. 5 Rear brake pads and related components

PARKING BRAKE CABLES

ADJUSTMENT

➡Before adjusting the parking brake, make sure that the rear brake shoe clearance has been adjusted. For shoe clearance adjustment.

1. Remove the front door scuff plate, cowl side trim board, side panel, lower finish panel and No. 2 heater to register duct.
2. Loosen the lock nut and turn the adjusting nut until the pedal travel is 6 to 9 clicks at 67 ft lbs.
3. Tighten the lock nut to 48 inch lbs. (5.4 Nm).
4. Install the No. 2 heater to register duct, lower finish panel, side panel, cowl side trim and front door scuff plate.

PARKING BRAKE SHOES

REMOVAL & INSTALLATION

See Figure 6.

1. Before servicing the vehicle, refer to the Precautions Section.
2. Disconnect the negative battery cable. Tape the cable with insulating tape.
3. Raise and safely support the vehicle.
4. Remove the rear wheel.
5. Remove the 2 mounting bolts and remove the disc brake assembly.
6. Suspend the disc brake securely and so the hose is not stretched.
7. Release the parking brake lever.
8. Place matchmarks on the rotor and rear axle hub.
9. Remove the rotor.

➡If the disc cannot be removed easily, turn the shoe adjuster until the wheel turns freely.

10. Using the proper tool, remove the 2 shoe return springs.

➡At the time of reassembly, install the strut with the spring facing forward.

11. Slide the front shoe toward outside and remove the shoe adjuster.
12. Using the proper tool, disconnect the anchor spring and tension spring from the front shoe.

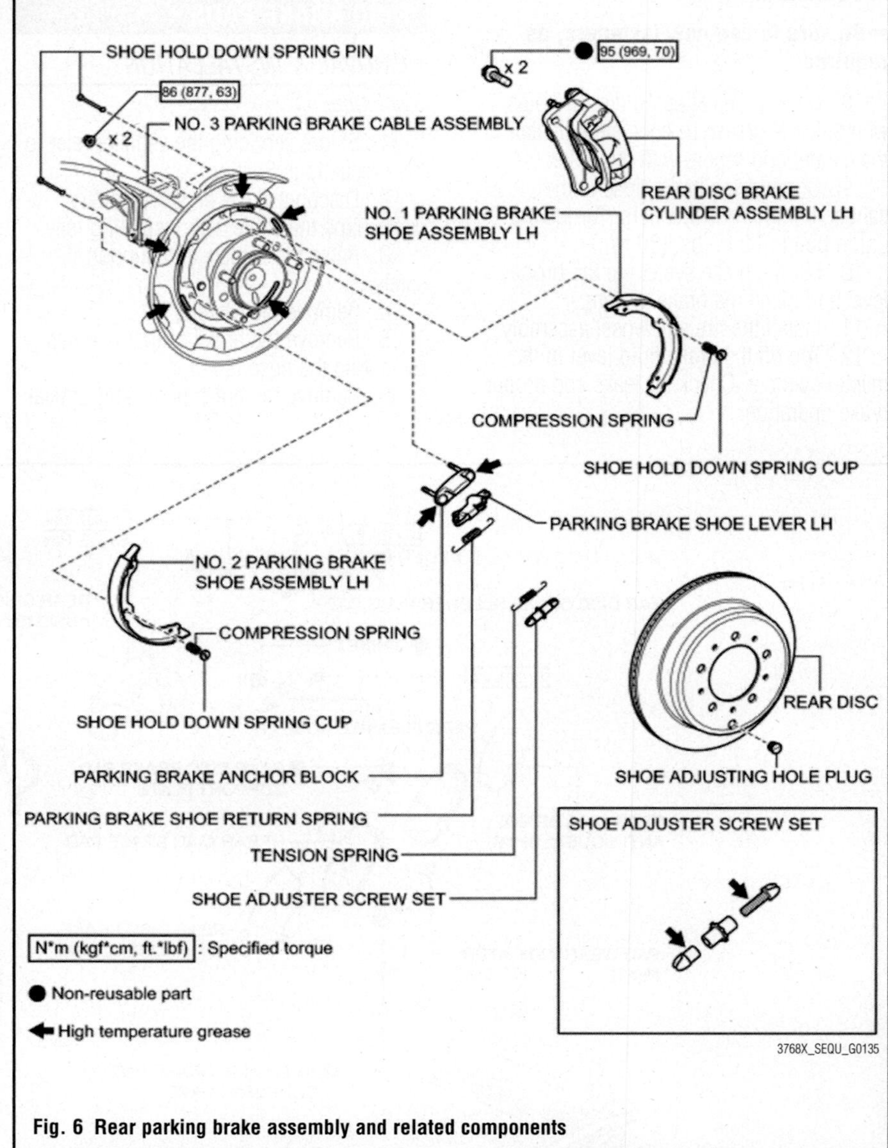

Fig. 6 Rear parking brake assembly and related components

13. Using the proper tool, disconnect the anchor spring and tension spring from the rear shoe.

To install:

➡Be sure to use new fasteners, as required.

14. Installation is the reverse of the removal procedure.
15. Be sure to apply high temperature grease to the shoe adjuster.
16. To adjust the parking brake turn the adjuster and expand the shoes until the disc locks.
17. Return the adjuster 8 notches.
18. Depress the parking brake pedal with 147 N (33 lbs).
19. Drive the vehicle at about 50 km/h (31 mph) on a safe, level and dry road for about 400 meters (0.25 mile) in this condition.
20. Repeat this procedure 2 or 3 times.

CHASSIS ELECTRICAL

AIR BAG (SUPPLEMENTAL RESTRAINT SYSTEM)

GENERAL INFORMATION

All vehicles are equipped with an air bag system. The system must be disarmed before performing service on, or around, system components, the steering column, instrument panel components, wiring and sensors. Failure to follow the safety precautions and the disarming procedure could result in accidental air bag deployment, possible injury and unnecessary system repairs.

SERVICE PRECAUTIONS

CAUTION

Disconnect and isolate the battery negative cable before beginning any airbag system component diagnosis, testing, removal, or installation procedures. Allow system capacitor to discharge for two minutes before beginning any component service. This will disable the airbag system. Failure to disable the airbag system may result in accidental airbag deployment, personal injury, or death.

DISARMING THE SYSTEM

CAUTION

To avoid personal injury when working on vehicles equipped with an airbag, the negative battery cable must be disconnected and at least 90 seconds must elapse before working on the system. Failure to do so may result in deployment of the airbag.

ARMING THE SYSTEM

Connect the negative battery and wait 2 minutes before performing and work on the vehicle.

CLOCKSPRING CENTERING

See Figures 7 and 8.

When installing the spiral cable, check that the ignition switch is in the "OFF" position. Turn the cable counterclockwise by hand until it becomes hard to turn. Turn the cable clockwise about 21/2 turns to align the marks.

➡**The cable will rotate about 21/2 turns both left and right from the center.**

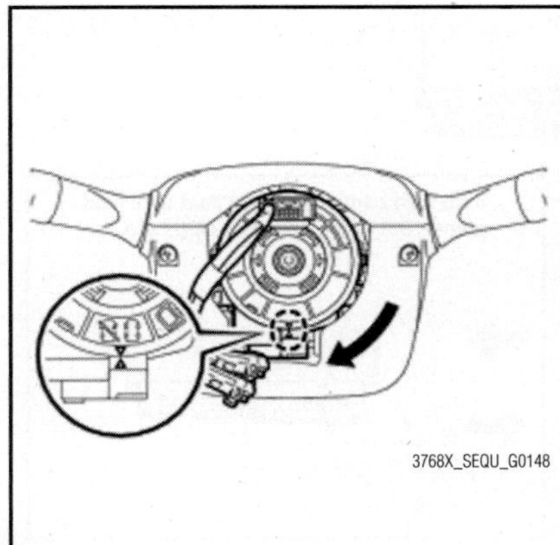
3768X_SEQU_G0148

Fig. 7 Spiral cable alignment marks

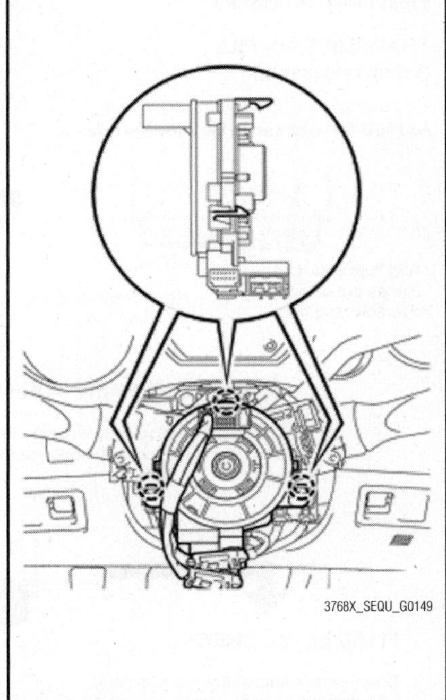
3768X_SEQU_G0149

Fig. 8 Spiral cable installation points

DRIVE TRAIN

AUTOMATIC TRANSMISSION FLUID

DRAIN AND REFILL

See Figures 9 through 12.

1. Before servicing the vehicle, refer to the Precautions Section.
2. Disconnect the negative battery cable. Tape the cable with insulating tape.
3. Raise and safely support the vehicle.

➡**If the vehicle is equipped with rear air suspension, stop operation of the suspension control system by pressing the height control mode selector switch. Push the shaft until the suitable tool contacts the step inside the cap (0.217–0.276 inch). Insert the pin until it passes thru the hole in the other side of the component.**

4. Raise and support the vehicle safely.
5. Position a catch pan under the drain plugs.
6. Remove the drain plugs. Drain the fluid. Be sure to properly dispose of used fluid.
7. Fill the transmission thru the refill hole until it begins to trickle out of the overflow tube.
8. Install the overflow plug.
9. Fill the unit to specification. Pan and plug removal: 2.2 quarts. Valve body removal: 5.0 quarts. Torque converter removal: 5.7 quarts (A760E and A760F) 6.4 quarts (AB60E,
10. Install the refill plug.
11. Start the engine and let it idle. Be sure that all electrical accessories are off.
12. Move the selector lever thru all gear ranges.
13. Wait 30 seconds, with the engine idling.
14. Turn the engine off.
15. Remove the refill plug. Be careful, as fluid may be hot to the touch.
16. Fill the transmission to specification, as required.
17. Install the refill plug.
18. Use the Techstream diagnostic tool or equivalent and confirm that proper fluid temperature has been reached.
19. Use the Techstream diagnostic tool or equivalent check for proper fluid level. Follow the directions on the tool.
20. If equipped, reset the rear suspension system.

FILTER REPLACEMENT

At this time the manufacturer does not provide removal and installation procedures for this component. The following procedure

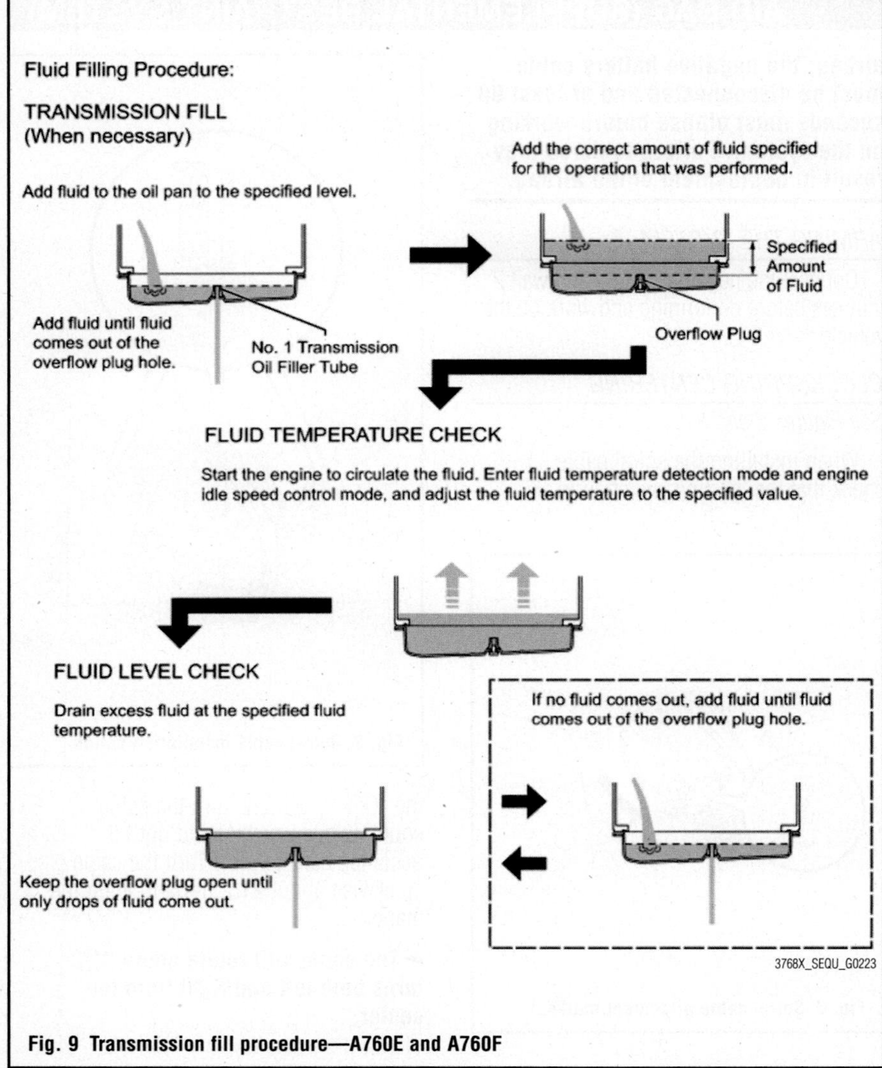
Fluid Filling Procedure:
TRANSMISSION FILL
(When necessary)
Add fluid to the oil pan to the specified level.
Add fluid until fluid comes out of the overflow plug hole.
No. 1 Transmission Oil Filler Tube
Add the correct amount of fluid specified for the operation that was performed.
Specified Amount of Fluid
Overflow Plug
FLUID TEMPERATURE CHECK
Start the engine to circulate the fluid. Enter fluid temperature detection mode and engine idle speed control mode, and adjust the fluid temperature to the specified value.
FLUID LEVEL CHECK
Drain excess fluid at the specified fluid temperature.
If no fluid comes out, add fluid until fluid comes out of the overflow plug hole.
Keep the overflow plug open until only drops of fluid come out.
3768X_SEQU_G0223

Fig. 9 Transmission fill procedure—A760E and A760F

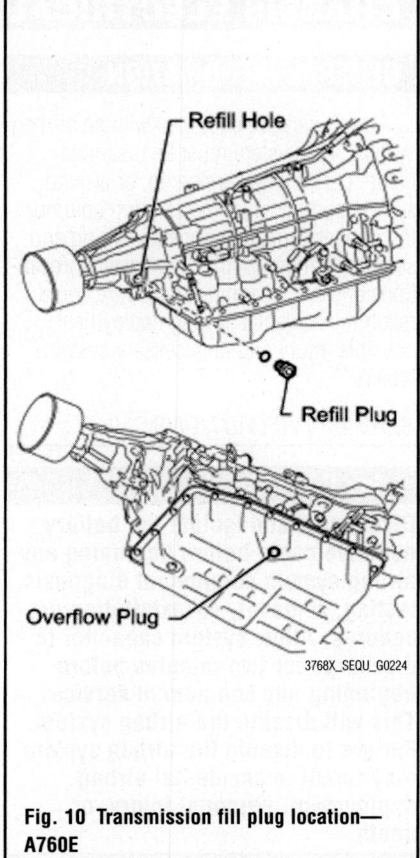
Refill Hole
Refill Plug
Overflow Plug
3768X_SEQU_G0224

Fig. 10 Transmission fill plug location—A760E

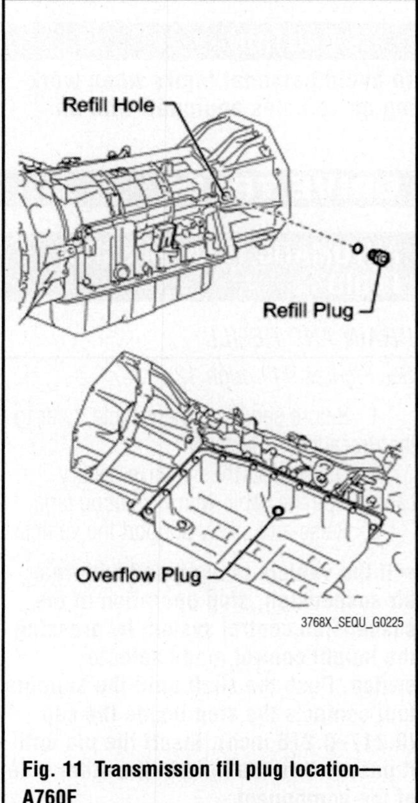
Refill Hole
Refill Plug
Overflow Plug
3768X_SEQU_G0225

Fig. 11 Transmission fill plug location—A760F

is a guideline and may differ from the vehicle you are servicing.

1. Before servicing the vehicle, refer to the Precautions Section.
2. Disconnect the negative battery cable. Tape the cable with insulating tape.
3. Raise and safely support the vehicle.

➡If the vehicle is equipped with rear air suspension, stop operation of the suspension control system by pressing the height control mode selector switch. Push the shaft until the suitable tool contacts the step inside the cap (0.217–0.276 inch). Insert the pin until it passes thru the hole in the other side of the component.

4. Drain the transmission fluid.
5. Be sure to properly dispose of used fluid.
6. Remove the necessary components in order to gain access to the transmission pan retaining bolts.
7. Be sure that nothing is stopping the pan from being removed once the bolts are removed.
8. Remove the transmission pan retaining bolts.
9. Carefully remove the pan from the transmission.
10. Remove the filter from its mounting. Discard the gaskets.

To install:

➡Be sure to use new fasteners, as required.

11. Installation is the reverse of the removal procedure.
12. Be sure that the pan magnets are correctly installed, if equipped.
13. Fill the transmission with the proper grade and type transmission fluid.
14. Perform the initialization procedure, using the Techstream diagnostic tool, or equivalent, as required.

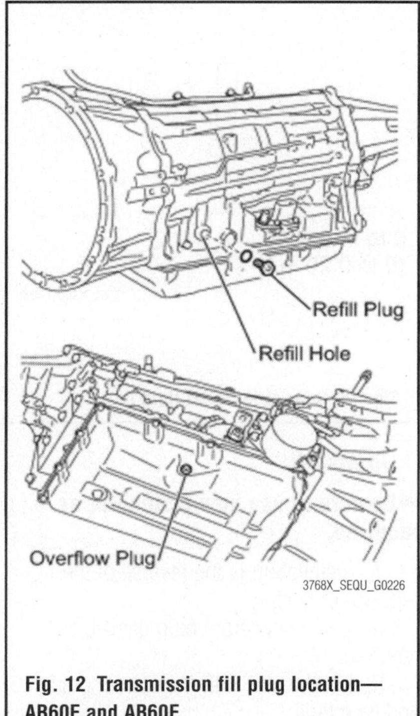

Fig. 12 Transmission fill plug location—AB60E and AB60F

FRONT DRIVESHAFT

REMOVAL & INSTALLATION

See Figure 13.

1. Before servicing the vehicle, refer to the Precautions Section.
2. Disconnect the negative battery cable. Tape the cable with insulating tape.
3. Raise and safely support the vehicle.
4. Remove the driveshaft heat insulator bracket sub assembly, if equipped.
5. Matchmark and remove the driveshaft retaining bolts.
6. Plug the transmission end to prevent fluid leakage.
7. Remove the component from the vehicle.

To install:

➡Be sure to use new fasteners, as required.

8. Installation is the reverse of the removal procedure.

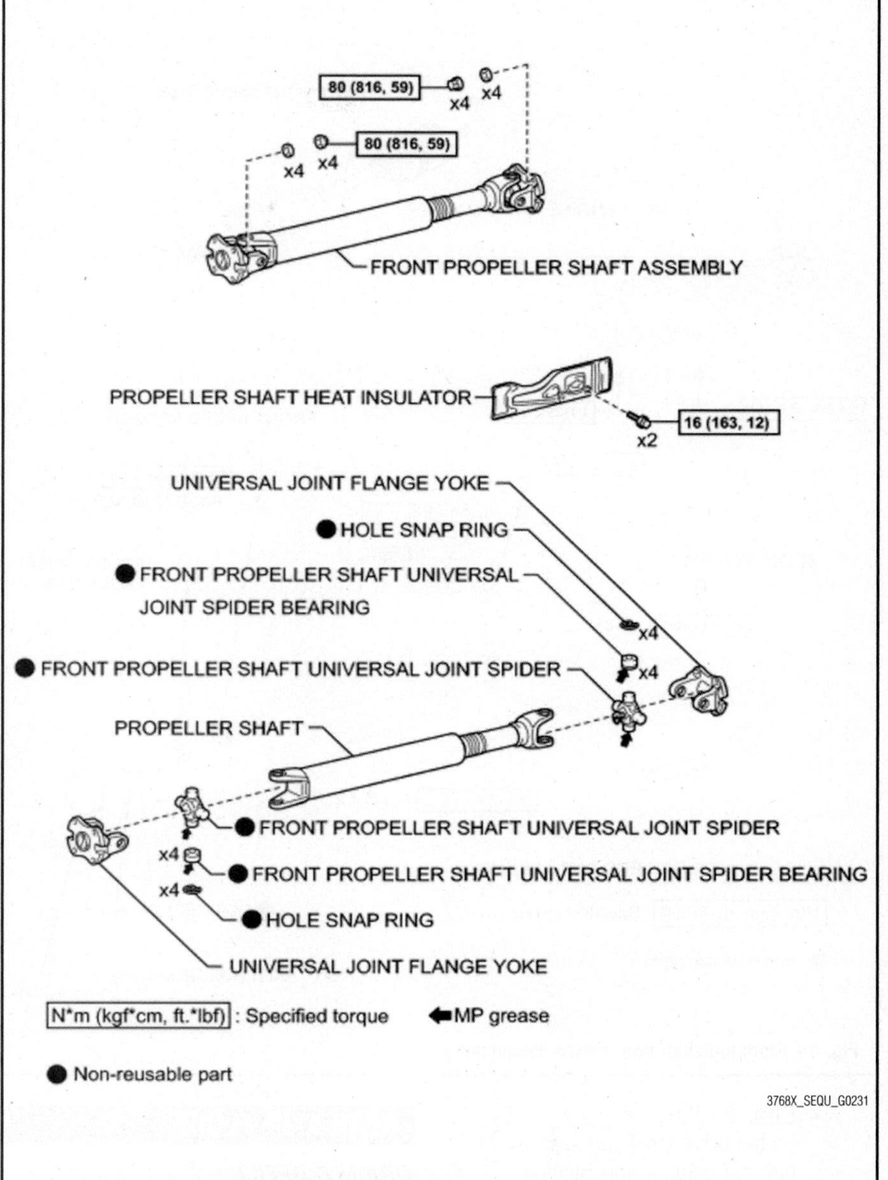

Fig. 13 Front driveshaft and related components

FRONT HALFSHAFTS

REMOVAL & INSTALLATION

See Figure 14.

1. Before servicing the vehicle, refer to the Precautions Section.
2. Disconnect the negative battery cable. Tape the cable with insulating tape.
3. Remove or disconnect the following:
    - Front wheel
    - Under cover
4. Drain the differential oil.
5. Remove or disconnect the following:
    - Remove the 2 bolts and disconnect the disc brake caliper from the steering knuckle
    - Remove front disc
    - Grease cap ,4WD
    - Cotter pin and lock cap
    - Halfshaft locknut by applying the brakes
    - Remove the 4 bolts
    - For 4WD, using a plastic-faced hammer, tap out the front drive shaft from the front axle hub.
    - Remove the axle hub and dust cover from the steering knuckle
    - Remove the O-ring from the axle hub.

To install:

➡Be sure to use new fasteners, as required.

6. Install or connect the following:
    - After applying grease to the O-ring, install the O-ring to the axle hub
    - For 4WD, connect the front drive shaft to the front axle hub
    - Install the dust cover and axle hub to the steering knuckle with the 4 bolts and tighten to 73 ft. lbs. (99 Nm)

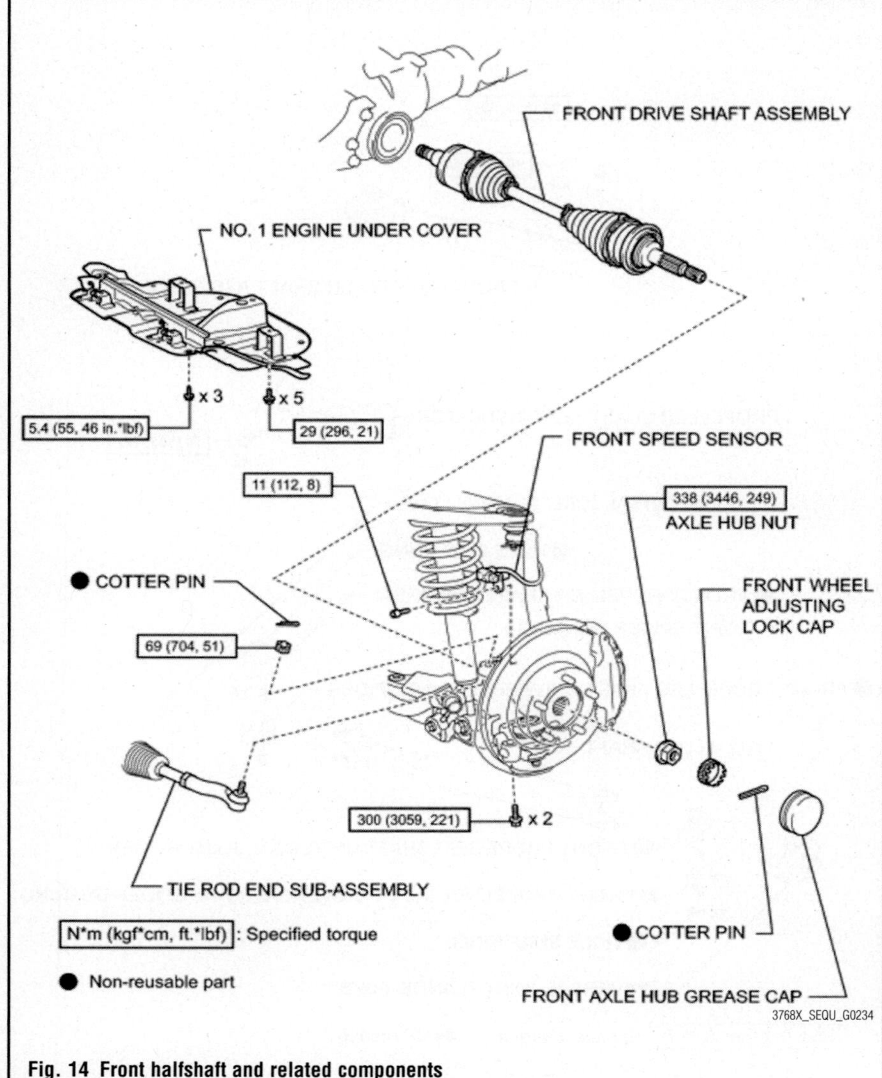

Fig. 14 Front halfshaft and related components

- Clean the threaded parts on the drive shaft and axle hub nut using a non-residue solvent.

➡Be sure to perform this work for a new drive shaft. Keep the threaded parts free of oil and foreign objects.

7. Using a 39 mm socket wrench, install the hub nut and tighten to 249 ft. lbs. (338 Nm).
   - Lock cap and a new cotter pin
   - Grease cap
   - Front disc
   - Connect the front disc brake caliper and install 2 new bolts and tighten to 73 ft. lbs. (99 Nm)
8. Refill the differential with oil.
9. Install or connect the following:
   - Under cover
   - Front wheel

REAR AXLE FLUID

DRAIN & REFILL

See Figure 15.

1. Before servicing the vehicle, refer to the Precautions Section.
2. Disconnect the negative battery cable. Tape the cable with insulating tape.
3. Raise and safely support the vehicle.
4. Position a catch pan under the drain plug.

➡Allow the fluid to cool before draining it. Never drain hot fluid.

5. Remove the drain plug and drain the unit. Discard the gasket, if equipped.
6. Be sure to properly dispose of used fluid.

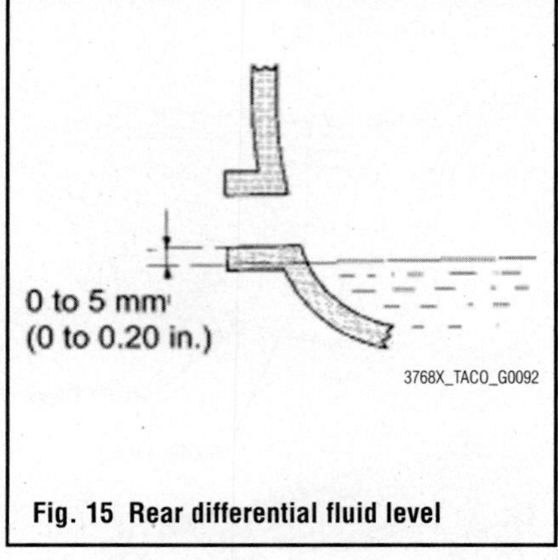

Fig. 15 Rear differential fluid level

To install:

➡Be sure to use new fasteners, as required.

7. Installation is the reverse of the removal procedure.
8. Use a new drain plug gasket, as required.
9. Fill the unit with the correct grade and type fluid.
10. Check that the oil surface is within 0.20 inch of the lowest point of the filler plug opening. See illustration.
11. Install the filler plug. Tighten to 29 ft. lbs. (39 Nm), front differential and 36 ft. lbs. (49 Nm), rear differential.

REAR AXLE HOUSING

REMOVAL & INSTALLATION

See Figures 16 and 17.

1. Before servicing the vehicle, refer to the Precautions Section.
2. Disconnect the negative battery cable. Tape the cable with insulating tape.
3. Remove the driveshaft.
4. Remove the halfshaft assemblies.
5. Drain the differential assembly.
6. Properly support the differential carrier assembly.
7. Remove the carrier retaining bolts.
8. Carefully remove the carrier from its mounting.

To install:

➡Be sure to use new fasteners, as required.

9. Installation is the reverse of the removal procedure.
10. Tighten the retaining bolts to 53 ft. lbs. (72 Nm). Be sure to use a new gasket.
11. Be sure to fill the unit with the proper grade and type gear oil.
12. Check for leaks. Correct as required.

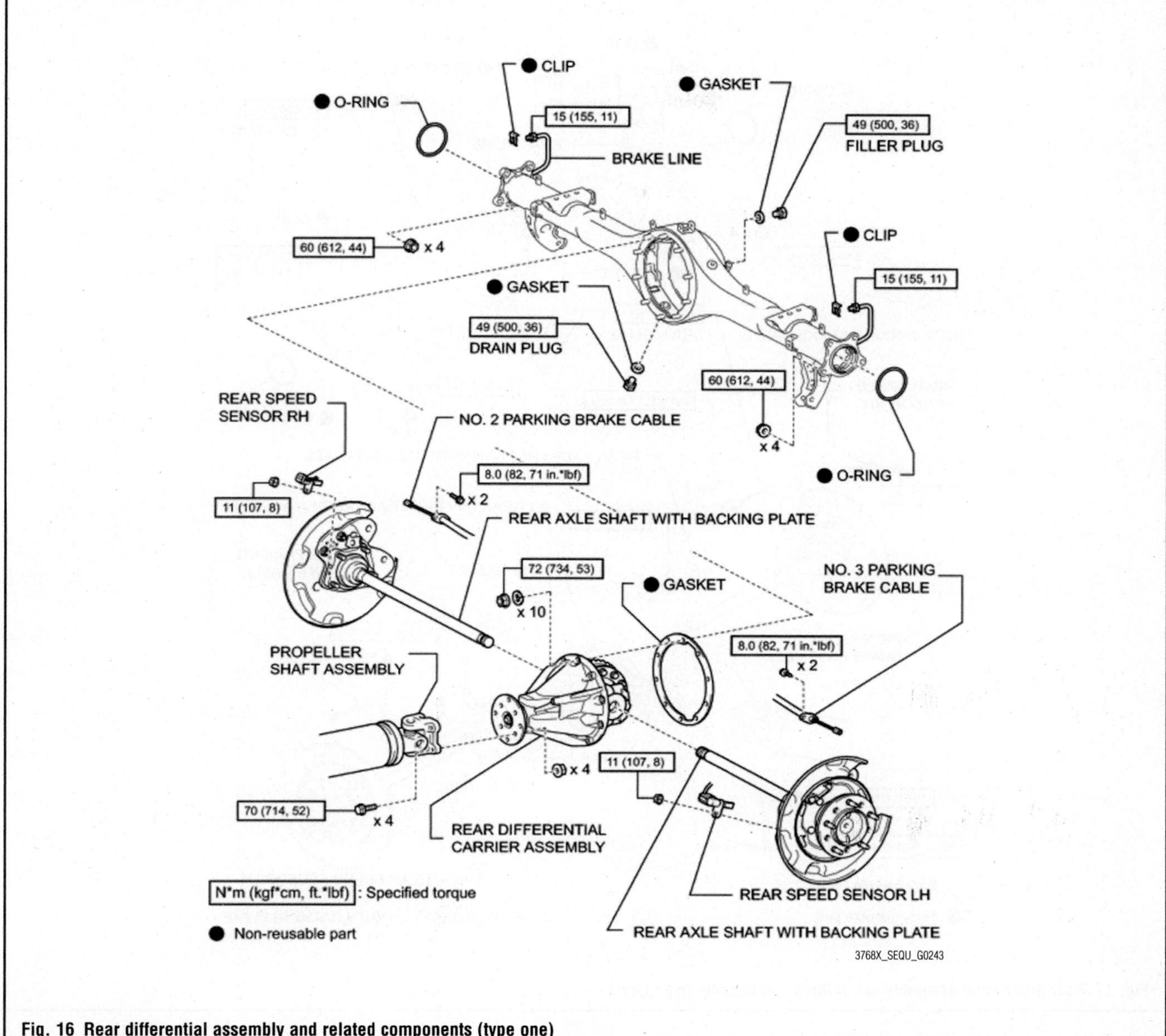

Fig. 16 Rear differential assembly and related components (type one)

REAR AXLE SHAFT, BEARING & SEAL

REMOVAL & INSTALLATION

1. Before servicing the vehicle, refer to the Precautions Section.
2. Disconnect the negative battery cable. Tape the cable with insulating tape.
3. Remove or disconnect the following:
   - Rear wheel
   - Rear caliper
   - Brake disc
4. Remove or disconnect the following:
   - Anti-lock Brake System (ABS) speed sensor from the rear axle housing, if equipped
   - Parking brake cable
   - Parking brake shoe assemblies
   - Remove the 4 nuts and rear axle shaft together with the parking brake plate

WARNING

Be careful not to damage the oil seal.

   - O-ring from the rear axle housing
   - Inner side oil seal using tool 09308-00010

To install:

➡Be sure to use new fasteners, as required.

5. Install or connect the following:
   - New O-ring to the rear axle
   - Install the rear axle shaft and parking brake plate with the 4 nuts and tighten to 44 ft. lbs. (60 Nm)
   - Anti-lock Brake System (ABS) speed sensor to the rear axle housing, if equipped
   - Parking brake cable assembly
   - Parking brake shoe assembly
   - Install rear disc
   - Adjust parking brake shoe clearance
   - Connect the rear disc brake cylinder and install 2 new bolts and tighten to 70 ft. lbs. (95 Nm)

➡Do not twist the flexible hose. Make sure that the bolts are free from damage

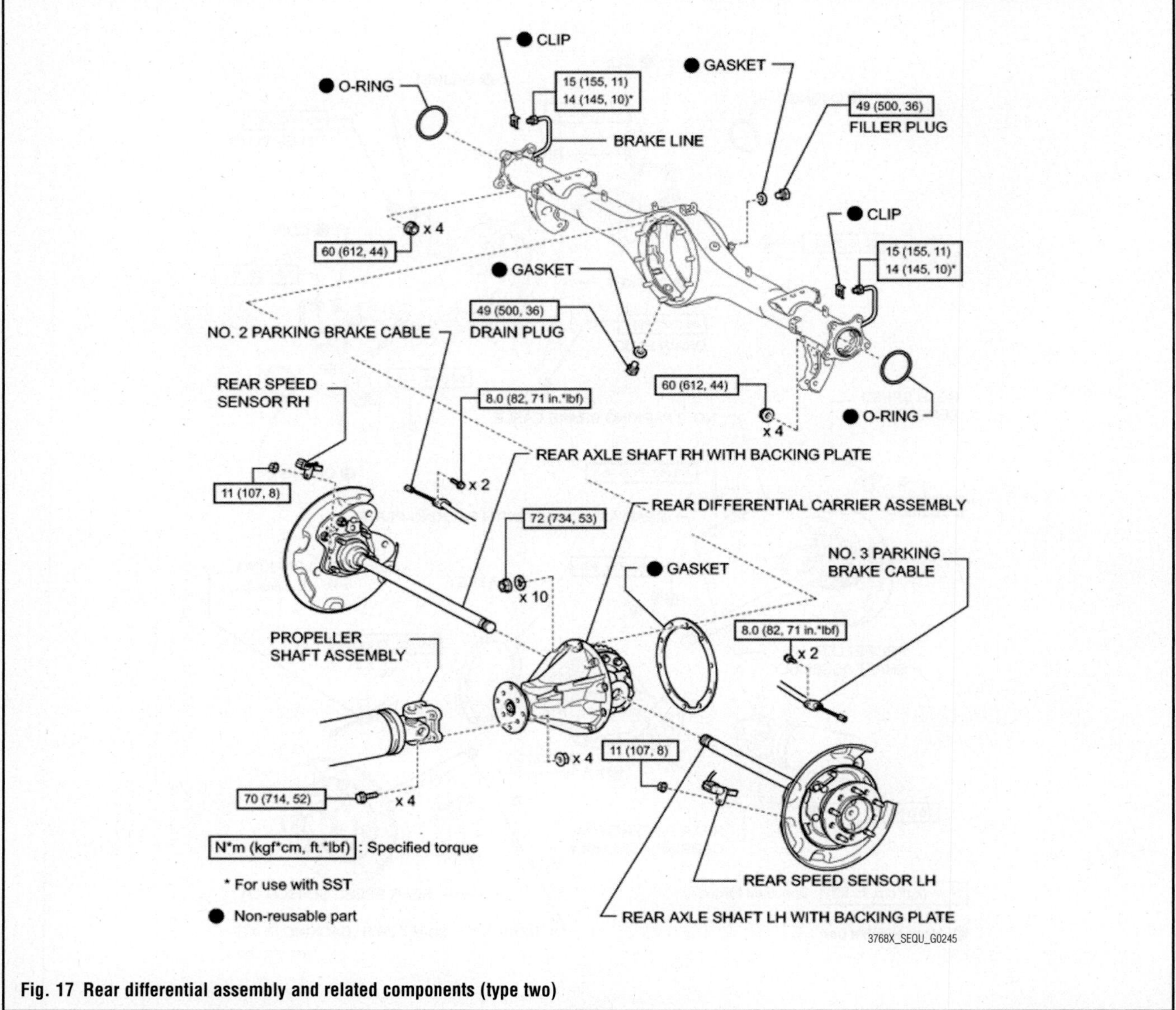

Fig. 17 Rear differential assembly and related components (type two)

and foreign matter. Do not over tighten the bolts.

- Rear wheel

REAR DRIVESHAFT

REMOVAL & INSTALLATION

See Figures 18 through 22.

1. Before servicing the vehicle, refer to the Precautions Section.
2. Disconnect the negative battery cable. Tape the cable with insulating tape.
3. Raise and safely support the vehicle.
4. Remove the driveshaft heat insulator bracket sub assembly, if equipped.
5. Matchmark and remove the driveshaft retaining bolts.
6. Plug the transmission end to prevent fluid leakage.
7. Remove the component from the vehicle.

To install:

➡Be sure to use new fasteners, as required.

8. Installation is the reverse of the removal procedure.

REAR PINION SEAL

REMOVAL & INSTALLATION

1. Before servicing the vehicle, refer to the Precautions Section.
2. Disconnect the negative battery cable. Tape the cable with insulating tape.
3. Raise and safely support the vehicle.
4. Drain the differential housing oil.
5. Remove the rear driveshaft.
6. Remove the companion flange, as follows:
   - Loosen the staked part of the nut, using a chisel and a hammer
   - Companion flange nut, using tool 09330-00021
   - Companion flange, using tools 09950-30012 and 09955-03050
   - Oil seal, using tool 09308-10010

To install:

7. Install the new oil seal until it is flush with the housing, using a plastic hammer and tools 09316-12010 and 09649-17010

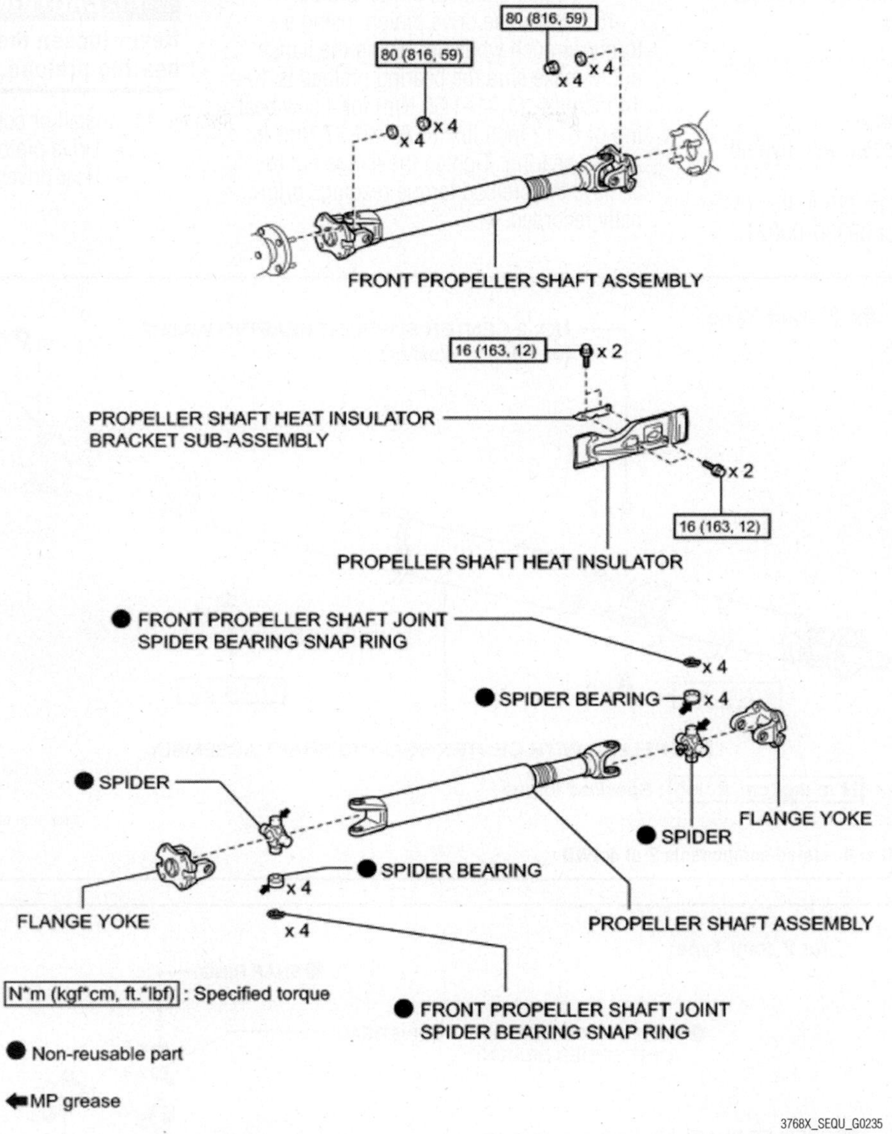

80 (816, 59)

80 (816, 59)

x 4

x 4

x 4

x 4

FRONT PROPELLER SHAFT ASSEMBLY

16 (163, 12) x 2

PROPELLER SHAFT HEAT INSULATOR BRACKET SUB-ASSEMBLY

x 2

16 (163, 12)

PROPELLER SHAFT HEAT INSULATOR

● FRONT PROPELLER SHAFT JOINT SPIDER BEARING SNAP RING

x 4

● SPIDER BEARING x 4

● SPIDER

FLANGE YOKE

● SPIDER

● SPIDER BEARING

x 4

FLANGE YOKE

x 4

PROPELLER SHAFT ASSEMBLY

N*m (kgf*cm, ft.*lbf) : Specified torque

● FRONT PROPELLER SHAFT JOINT SPIDER BEARING SNAP RING

● Non-reusable part

MP grease

3768X_SEQU_G0235

Fig. 18 Rear driveshaft and related components 2WD

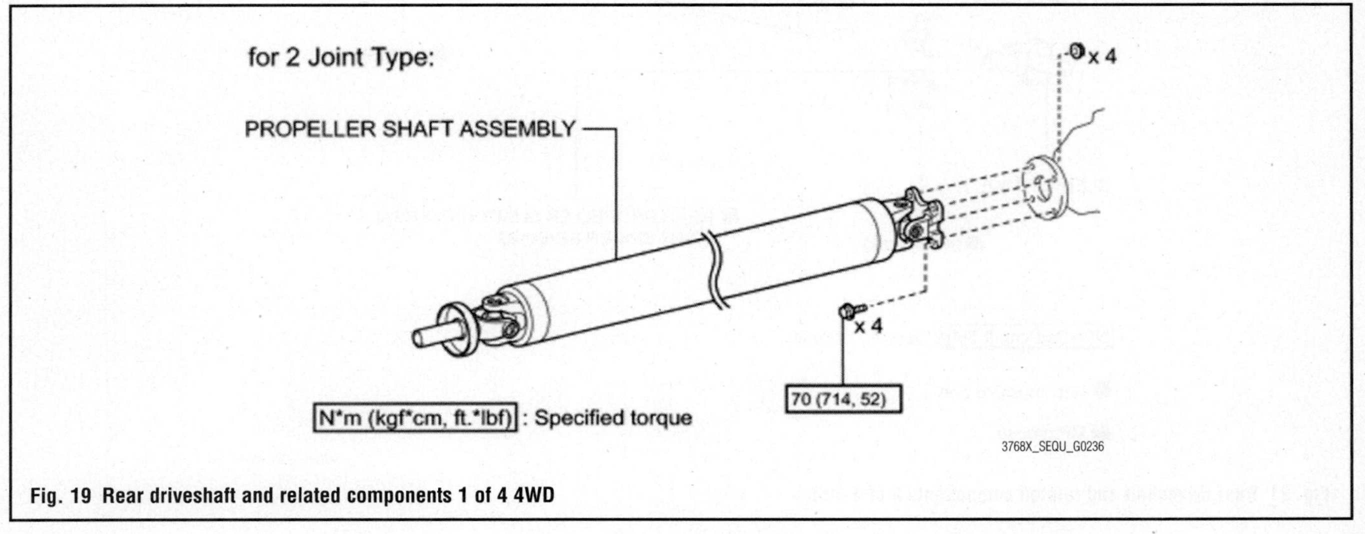

Fig. 19 Rear driveshaft and related components 1 of 4 4WD

➡Use vinyl tape to connect both oil seal installation tools.

8. Install the companion flange, as follows:
   - Companion flange
   - New nut, lubricated with hypoid gear oil
   - Torque the nut to 325 ft. lbs. (441 Nm), using tool 09330-00021.
9. Adjust the drive pinion preload
10. Rotate the drive pinion, using a torque wrench while tightening the flange nut to make sure the bearing preload is 10–14 inch lbs. (1.04–1.69 Nm) for a new bearing or 8–12 inch lbs. (0.85–01.37 Nm) for a used bearing. Tighten the flange nut to achieve the preload torque readings originally recorded.

CAUTION

Never loosen the pinion nut to reduce bearing preload.

11. Install or connect the following:
   - Drive pinion nut, stake it
   - Rear driveshaft.

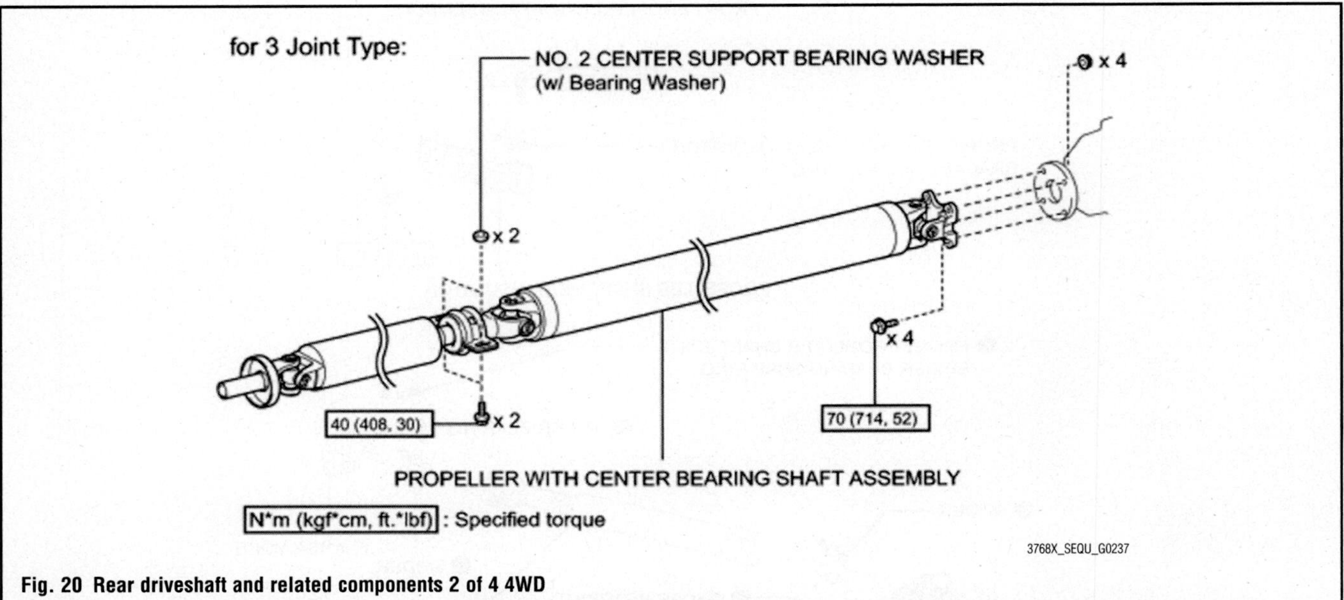

Fig. 20 Rear driveshaft and related components 2 of 4 4WD

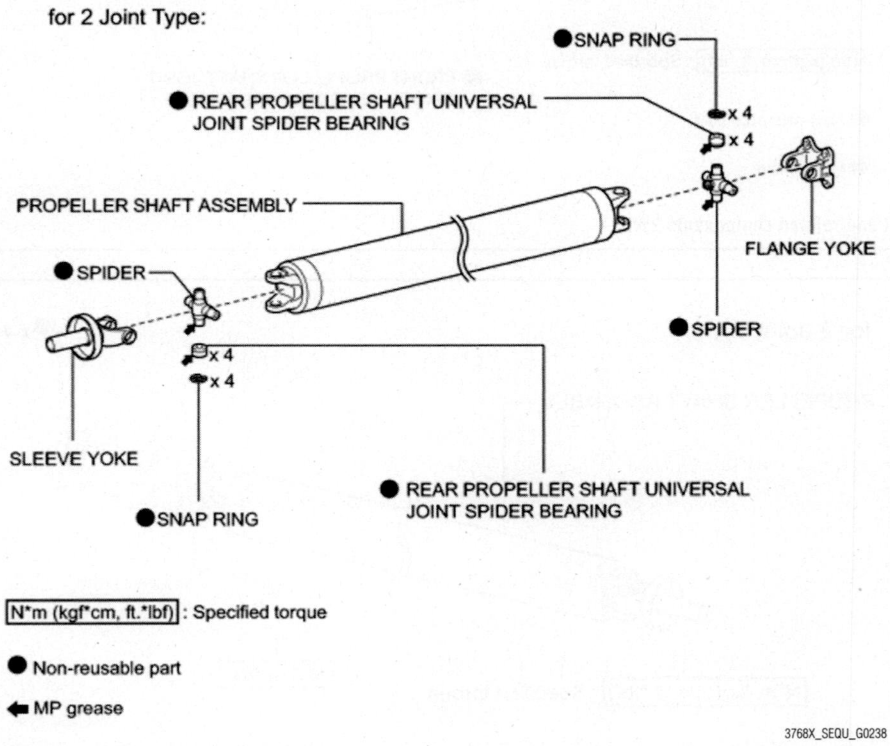

Fig. 21 Rear driveshaft and related components 3 of 4 4WD

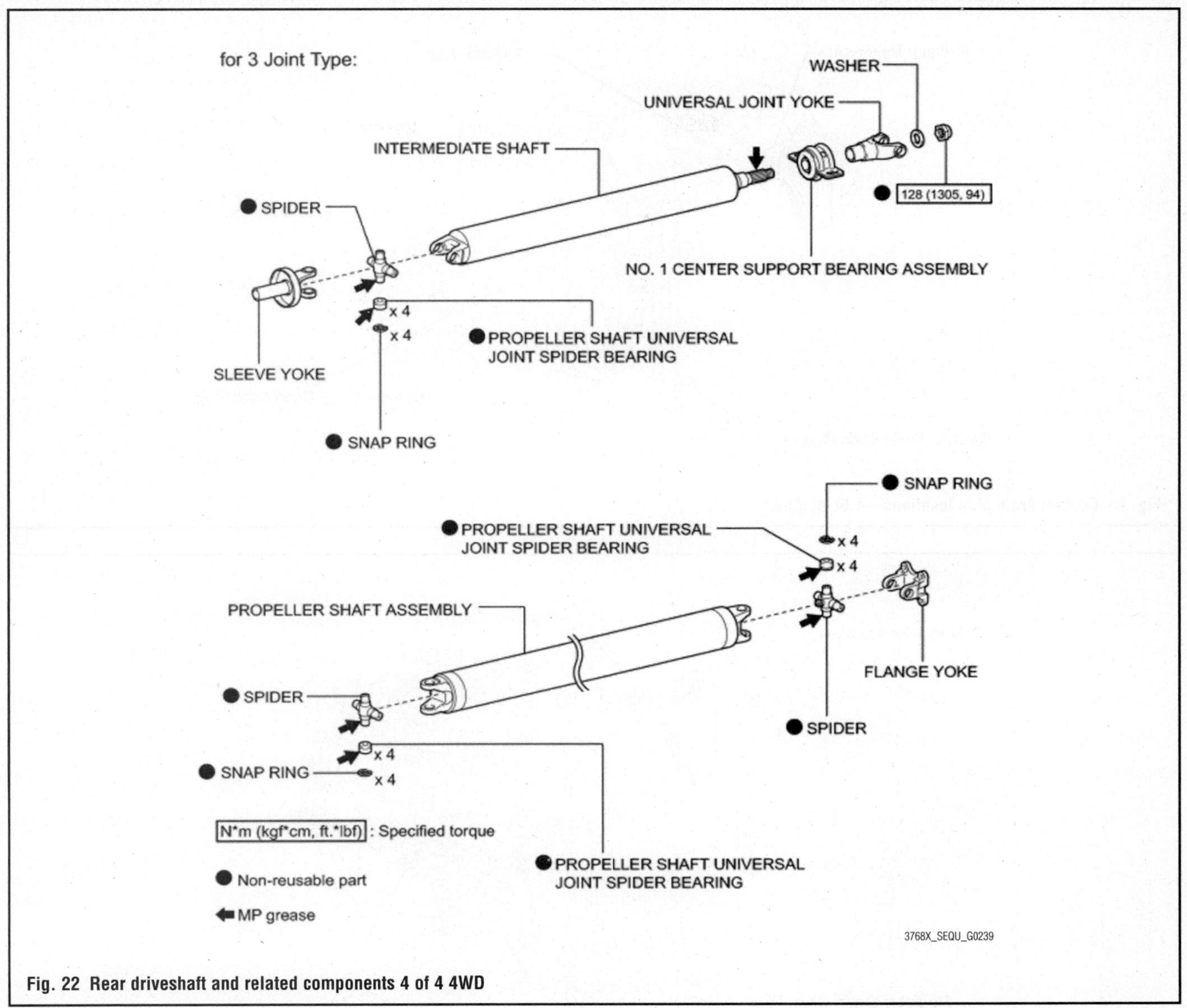

Fig. 22 Rear driveshaft and related components 4 of 4 4WD

ENGINE COOLING

ENGINE COOLANT

DRAIN & REFILL PROCEDURE

See Figures 23 and 24.

1. Before servicing the vehicle, refer to the Precautions Section.
2. Disconnect the negative battery cable. Tape the cable with insulating tape.

➡Be careful if the engine is hot as the engine coolant could also be hot. Hot coolant could cause burns and result in injury.

➡Do not open the radiator cap if the engine is hot.

3. Open the radiator cap.
4. Loosen the radiator drain plug and drain the coolant.

➡Be sure to properly dispose of used coolant.

5. Loosen the block drain plugs. Drain the coolant.

To install:

➡Be sure to use new fasteners, as required.

6. Tighten the block drain plugs to 10 ft. lbs. (13 Nm).
7. Tighten the radiator drain plug.
8. Add engine coolant.
9. Slowly pour coolant into the reservoir until it reaches the full line.
10. Install the cap.
11. Press the hoses several times and than check the coolant level. Correct as necessary.
12. Start the engine and allow it to reach operating temperature (until the thermostat opens). Maintain an engine speed of 2000–2500 rpm.
13. Squeeze the radiator hoses several times while warming up the engine.
14. Stop the engine and wait for the coolant to cool down.
15. Remove the radiator cap, carefully. If coolant is below the full line repeat the above steps until coolant remains at the full line.

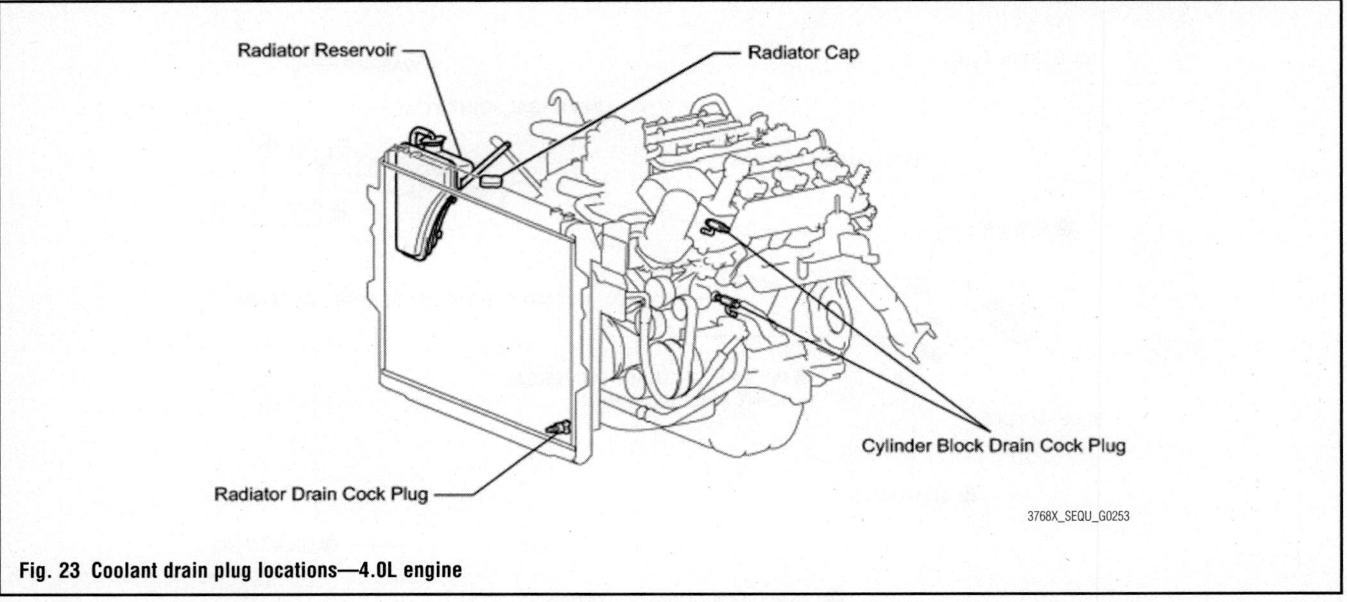

Fig. 23 Coolant drain plug locations—4.0L engine

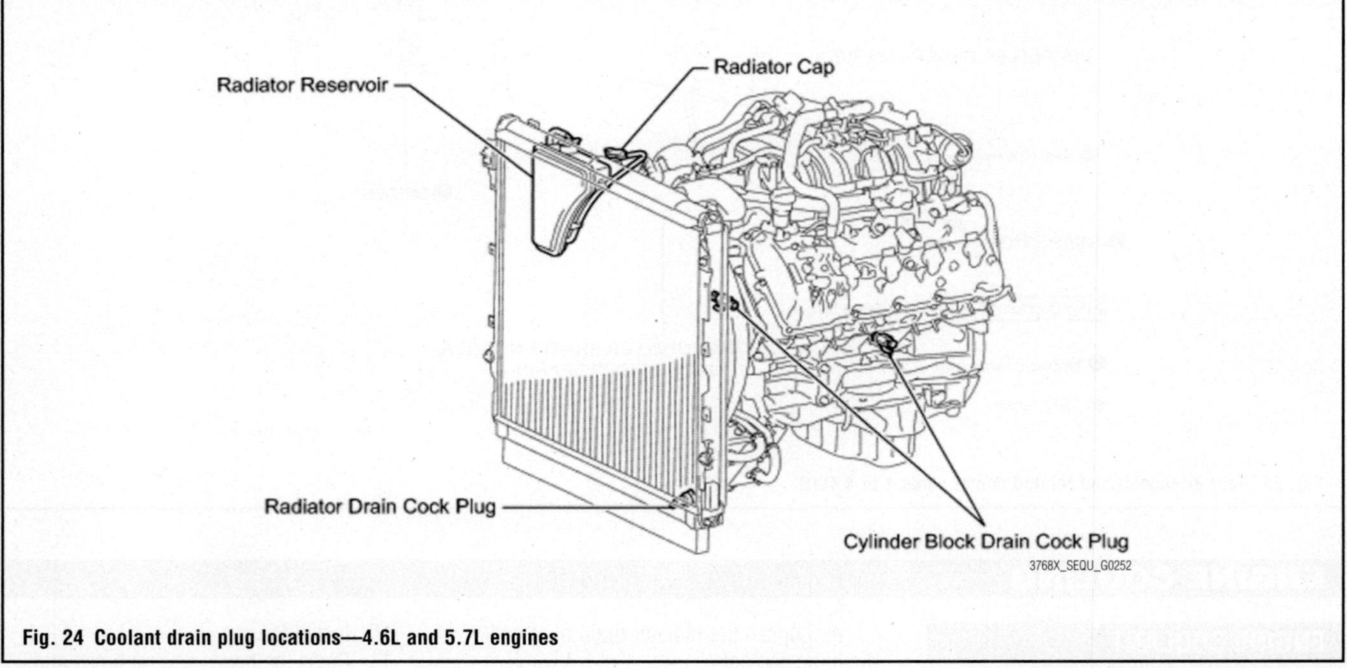

Fig. 24 Coolant drain plug locations—4.6L and 5.7L engines

16. Check the coolant level in the reservoir tank. Add coolant, as required.
17. Check for coolant leaks, correct as required.

ENGINE FAN

REMOVAL & INSTALLATION

1. Before servicing the vehicle, refer to the Precautions Section.
2. Disconnect the negative battery cable. Tape the cable with insulating tape.
3. Drain coolant.
4. Remove upper radiator hose.
5. Remove intake air connector assembly.
6. Remove fan shroud.
7. Loosen the 4 nuts holding the fluid coupling to the fan bracket.
8. Remove the drive belt.
9. Remove fan.

To install:

➡Be sure to use new fasteners, as required.

10. Installation is the reverse of the removal procedure.

RADIATOR

REMOVAL & INSTALLATION

See Figures 25 through 29.

1. Before servicing the vehicle, refer to the Precautions Section.
2. Disconnect the negative battery cable. Tape the cable with insulating tape.
3. Remove the engine under cover.
4. Remove the front bumper cover assembly.
5. Drain the engine coolant.
6. Remove the oil cooler assembly, if equipped.

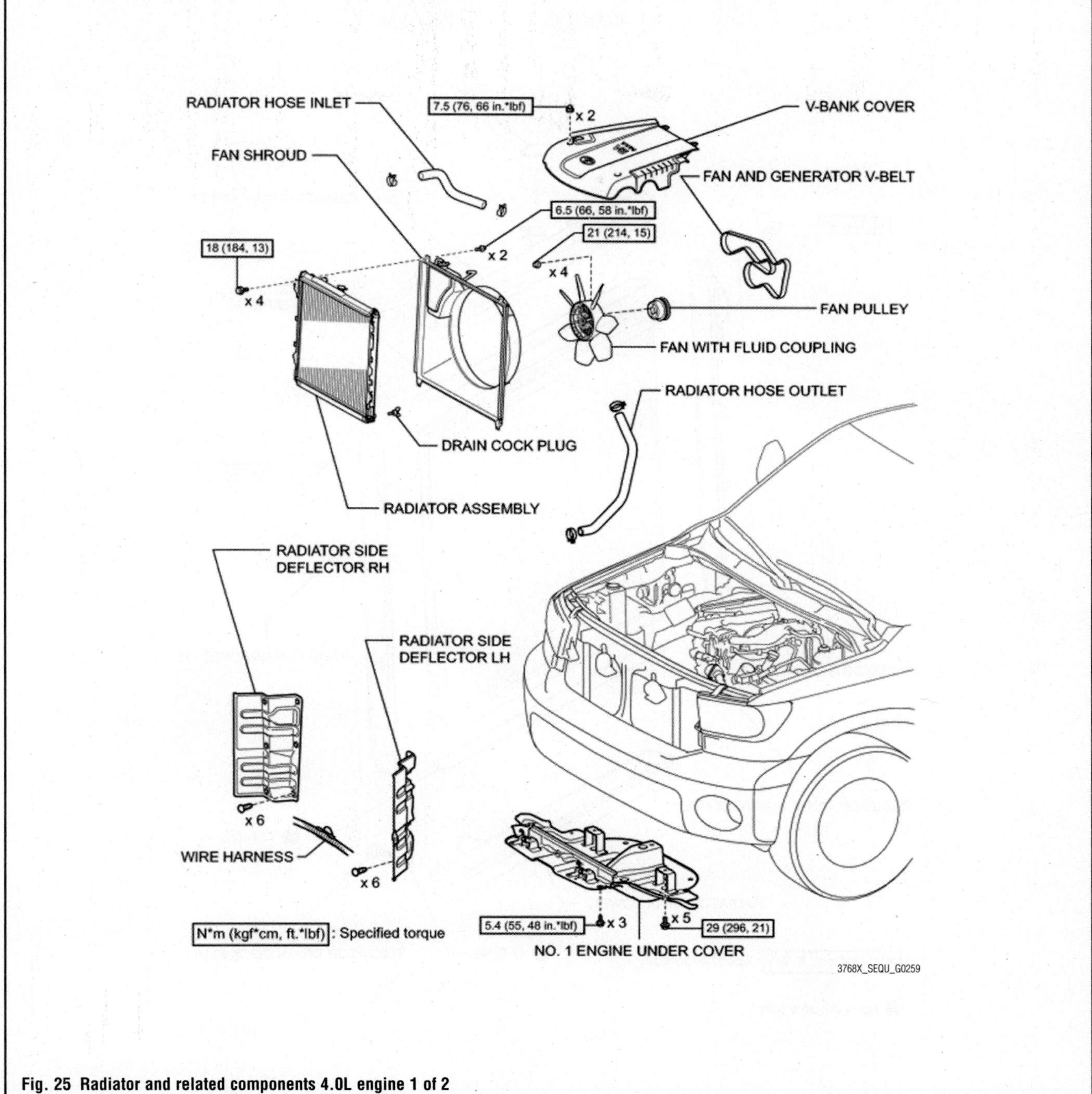

Fig. 25 Radiator and related components 4.0L engine 1 of 2

7. Remove the air cleaner assembly, as required.
8. Remove the V-bank cover assembly.
9. Remove the upper radiator hose.
10. Disconnect the lower radiator hose at the radiator.
11. Remove the fan shroud, together with the fluid coupling fan.
12. Remove the drive belt.
13. Remove the fan shroud, together with the fluid coupling fan.
14. Remove the fan pulley from the fan bracket.
15. Remove the radiator side deflectors.
16. Remove the radiator side support seals.
17. Remove the four radiator retaining bolts and remove the radiator.

To install:

➡**Be sure to use new fasteners, as required.**

18. Installation is the reverse of the removal procedure.

THERMOSTAT

REMOVAL & INSTALLATION

4.0L Engine

1. Before servicing the vehicle, refer to the Precautions Section.
2. Disconnect the negative battery cable. Tape the cable with insulating tape.

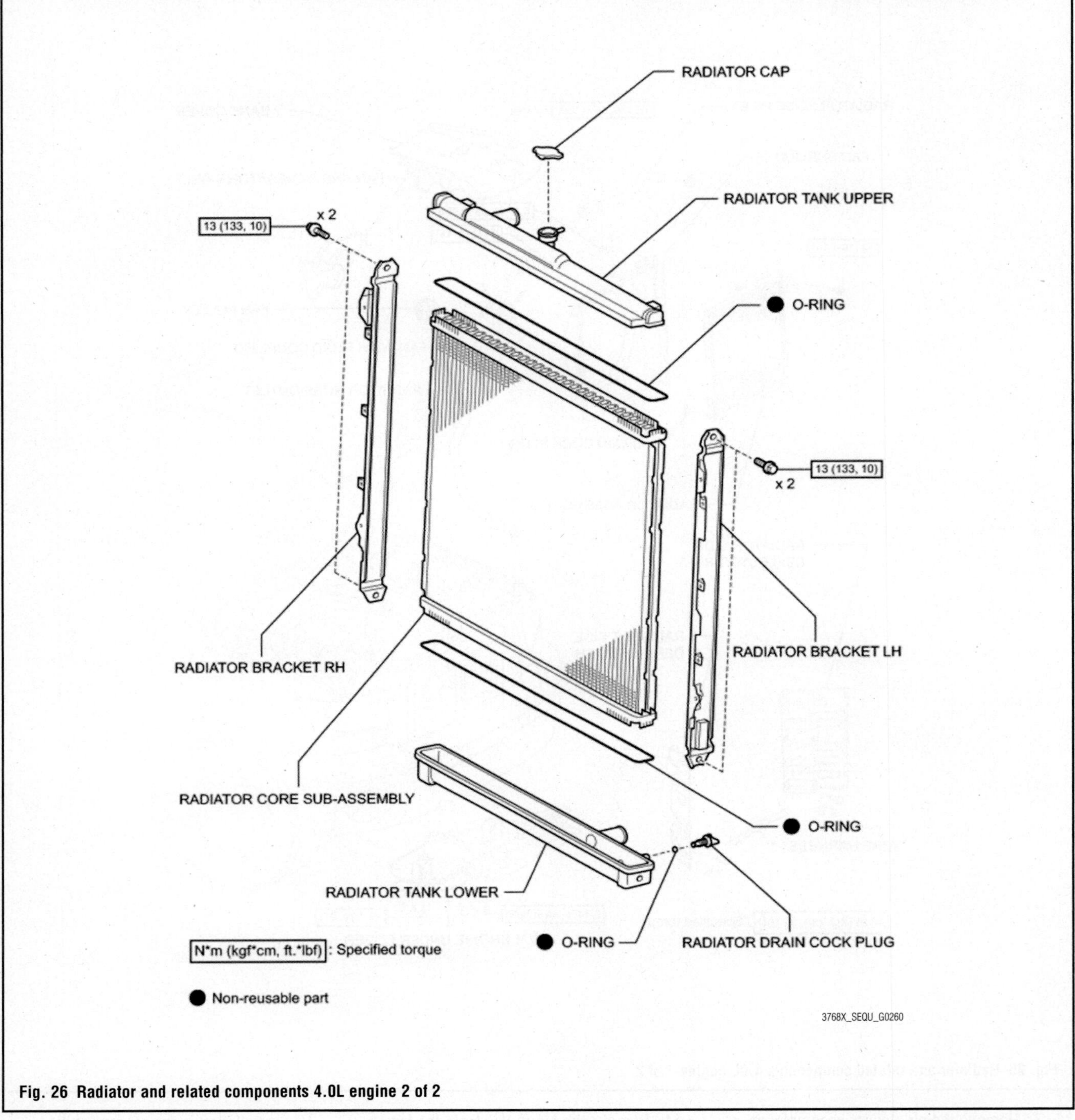

Fig. 26 Radiator and related components 4.0L engine 2 of 2

3. Remove No. 1 engine under cover.
4. Drain engine coolant.
5. Remove V-bank cover.
6. Disconnect radiator hose outlet.
7. Remove the 3 nuts, water inlet with thermostat and gasket.

To install:

➡Be sure to use new fasteners, as required.

8. Installation is the reverse of the removal procedure.
9. Tighten the retaining nuts to 80 inch lbs. (9.0 Nm).

4.6L Engine

1. Before servicing the vehicle, refer to the Precautions Section.
2. Disconnect the negative battery cable. Tape the cable with insulating tape.
3. Remove the No. 1 engine cover.
4. Drain the coolant. Be sure to properly dispose of used coolant.
5. Remove the V-bank cover sub assembly.
6. Remove the air cleaner assembly.
7. Remove the air tube sub assembly.
8. Disconnect the lower radiator hose.
9. Remove the thermostat housing retaining bolts.
10. Remove the thermostat. Discard the gasket.

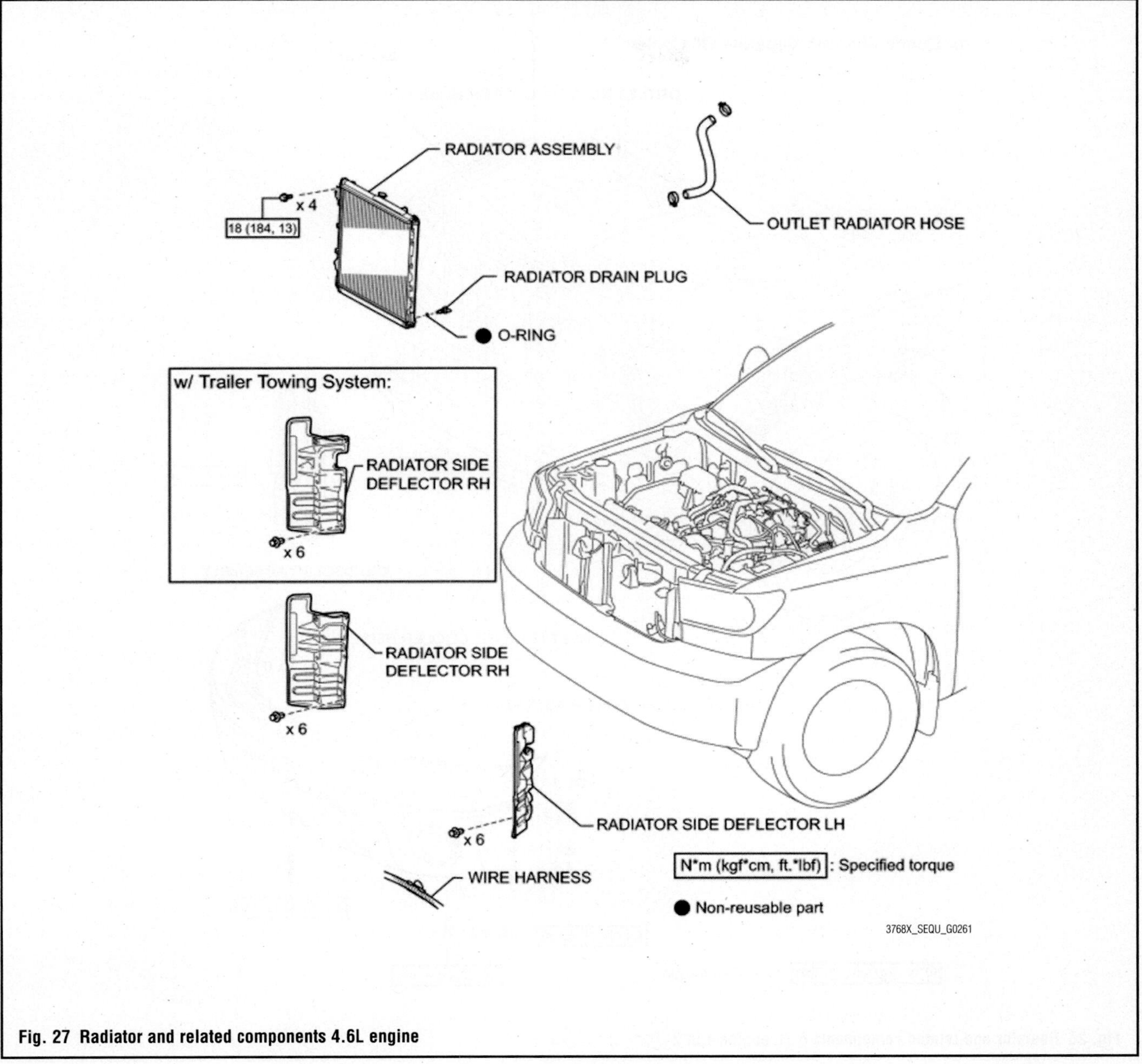

Fig. 27 Radiator and related components 4.6L engine

To install:

➡Be sure to use new fasteners, as required.

11. Installation is the reverse of the removal procedure.

5.7L Engine

See Figures 30 and 31.

1. Before servicing the vehicle, refer to the Precautions Section.
2. Disconnect the negative battery cable. Tape the cable with insulating tape.
3. Remove No. 1 engine under cover.
4. Drain engine coolant.
5. Remove V-bank cover sub-assembly.
6. Disconnect outlet radiator hose.
7. Disconnect the No. 2 and No. 3 air hoses.
8. Disconnect the air pump's connector.
9. Disconnect the air pump connector clamp's holder.
10. Using a clip remover, detach the wire harness clamp.
11. Disconnect the No. 5 water by-pass hose.
12. Remove the air tube bracket's bolt.
13. Remove the 3 nuts, water inlet with thermostat and gasket.

To install:

➡Be sure to use new fasteners, as required.

14. Installation is the reverse of the removal procedure.

WATER PUMP

REMOVAL & INSTALLATION

4.0L Engine

See Figures 32 and 33.

1. Before servicing the vehicle, refer to the Precautions Section.
2. Disconnect the negative battery cable. Tape the cable with insulating tape.
3. Remove No. 1 engine under cover.
4. Drain engine coolant.
5. Remove V-bank cover.

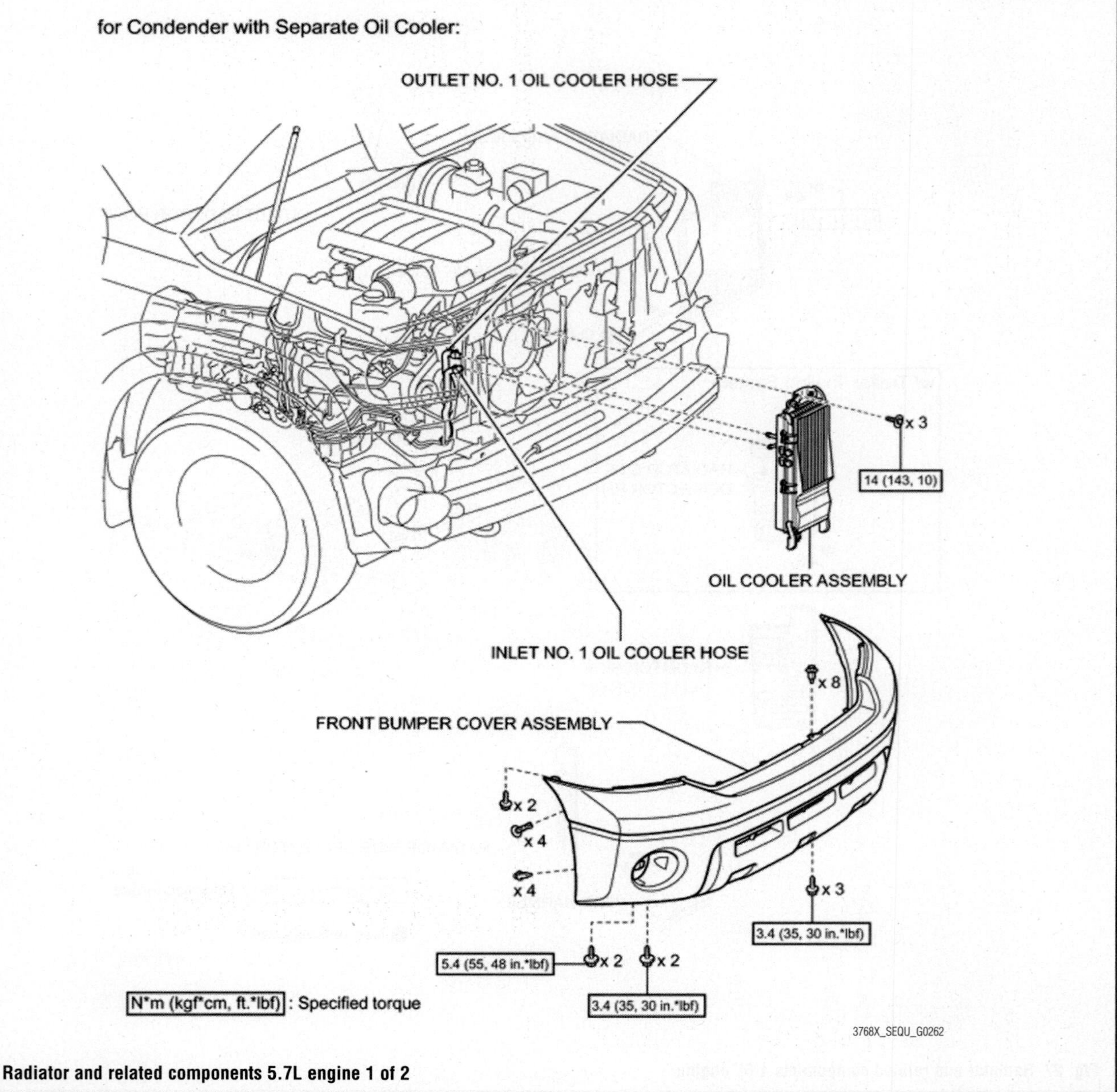

Fig. 28 Radiator and related components 5.7L engine 1 of 2

6. Disconnect radiator hose inlet and outlet.
7. Remove fan shroud.
8. Remove No. 2 air cleaner hose.
9. Remove air cleaner assembly with element.
10. Disconnect the 2 oil cooler hoses.
11. Disconnect the 5 water by-pass hoses.
12. Remove the 5 bolts and water inlet.
13. Remove the gasket from the water outlet pipe.
14. Remove the 2 bolts, 2 cover plates and 2 idler pulleys.
15. Remove V-ribbed belt tensioner assembly.
16. Remove alternator assembly.
17. Remove the 17 bolts, water pump and gasket.

To install:

18. Install a new gasket and the water pump with the 17 bolts. Tighten 10 mm head bolts to 80 inch lbs. (9.0 Nm) and 12 mm head bolts to 17 ft. lbs. (23 Nm).
19. Install alternator assembly.
20. Install v-ribbed belt tensioner assembly.
21. Install the 2 idler pulleys and 2 cover plates with the 2 bolts and tighten to 29 ft. lbs. (39 Nm).
22. Install a new gasket to the water outlet pipe.
23. Install a new gasket to the water pump.
24. Apply soapy water to the gasket.
25. Install the water inlet with the 5 bolts and tighten to 80 inch lbs. (9.0 Nm).
26. Connect the 5 water by-pass hoses.
27. Connect the 2 oil cooler hoses.
28. Install fan shroud.
29. Connect radiator hose inlet and outlet.
30. Install air cleaner assembly with element.
31. Install No. 2 air cleaner hose.
32. Add engine coolant.
33. Inspect for coolant leak.
34. Install V-bank cover.
35. Install No. 1 engine under cover.

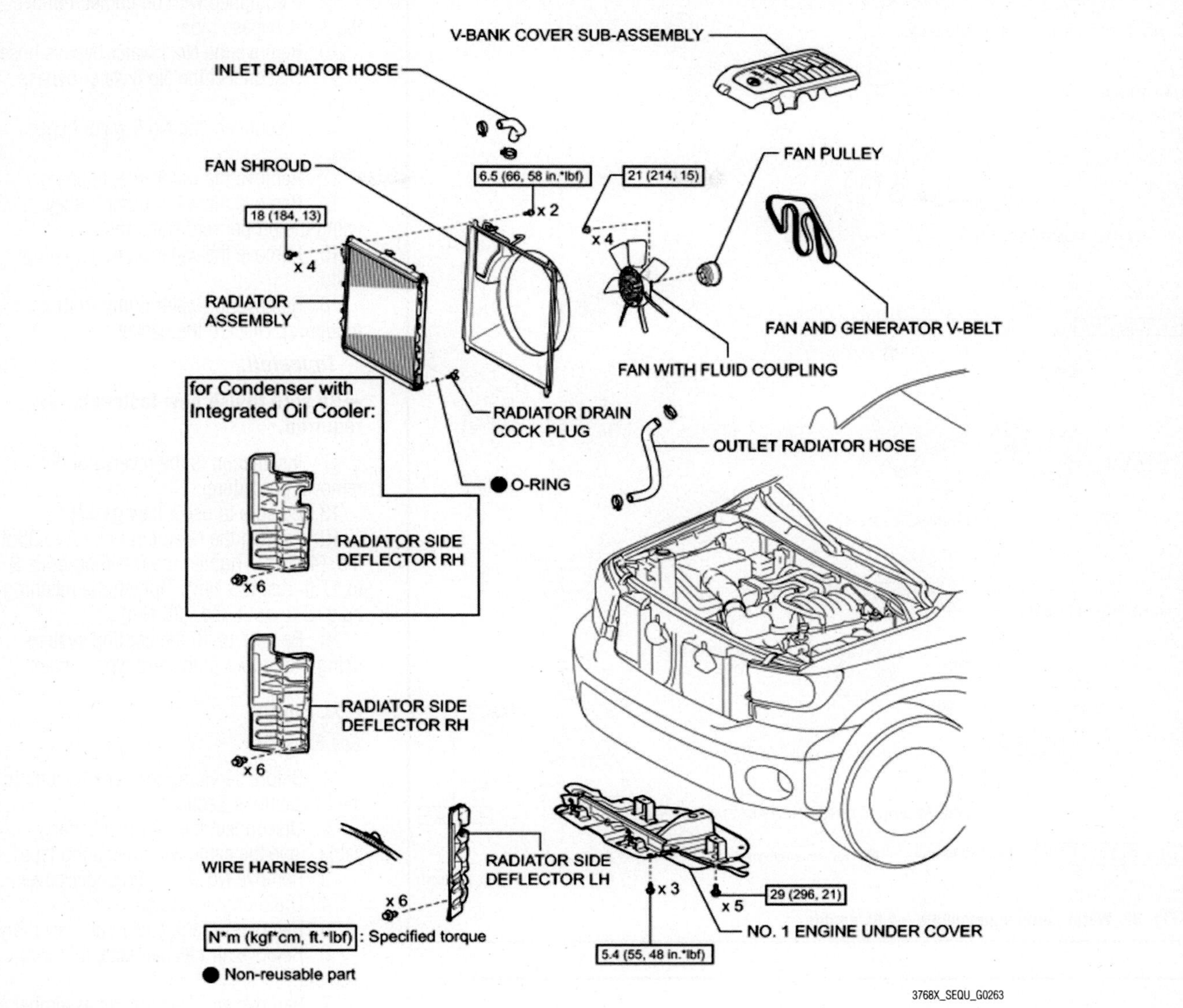

Fig. 29 Radiator and related components 5.7L engine 2 of 2

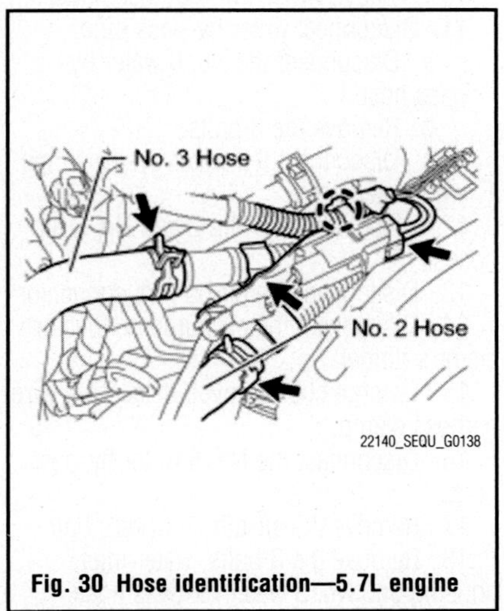

Fig. 30 Hose identification—5.7L engine

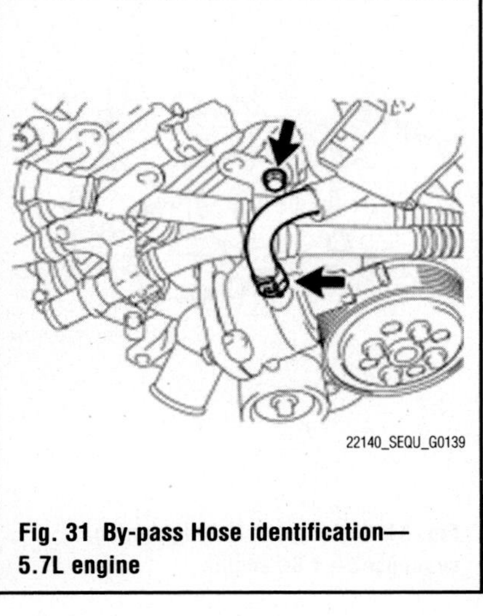

Fig. 31 By-pass Hose identification—5.7L engine

4.6L Engine

See Figure 34.

1. Before servicing the vehicle, refer to the Precautions Section.
2. Disconnect the negative battery cable. Tape the cable with insulating tape.
3. Drain the cooling system.
4. Remove the water inlet sub assembly.
5. Remove the top radiator hose.
6. Remove the fan shroud.
7. Remove the front fender apron seal, left side.
8. Disconnect the air conditioning compressor assembly. If the compressor has to be removed for clearance, be sure to properly discharge the system.

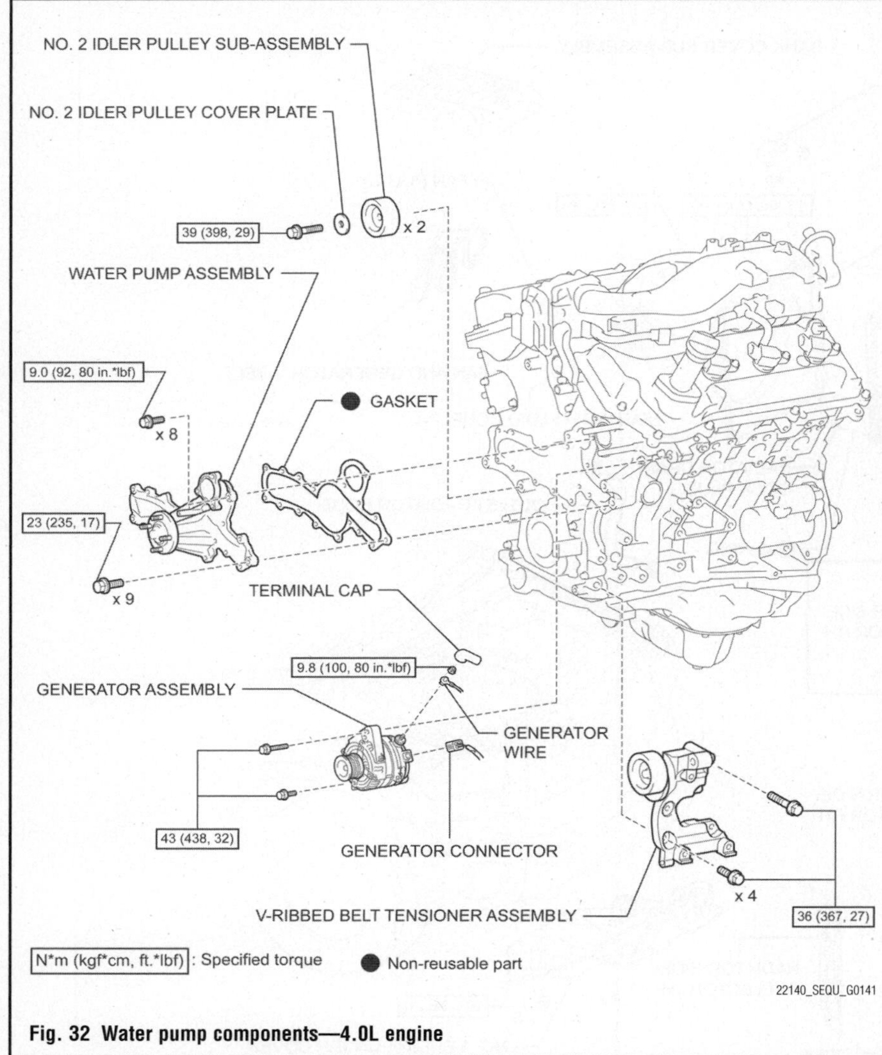

Fig. 32 Water pump components—4.0L engine

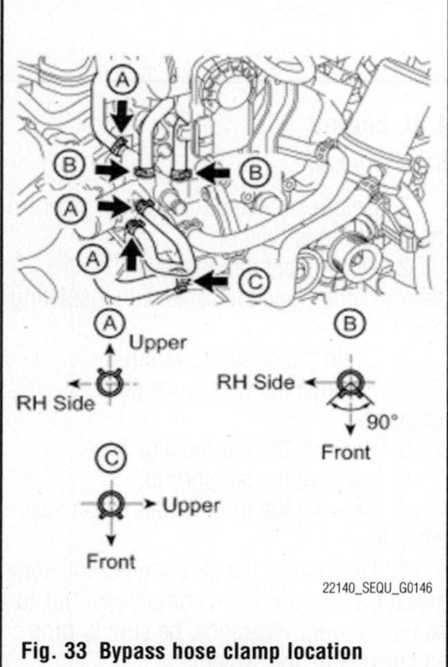

Fig. 33 Bypass hose clamp location

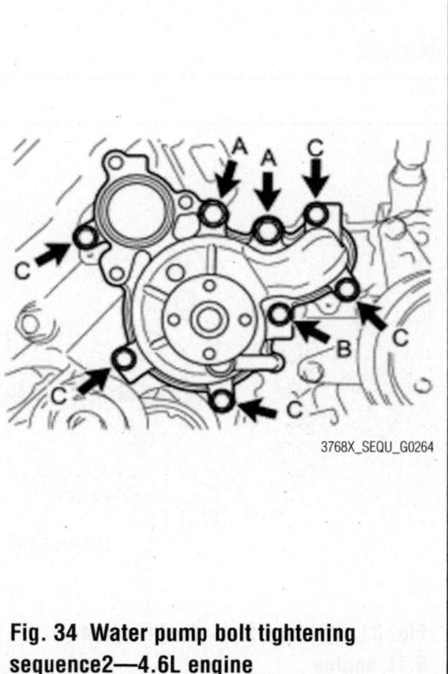

Fig. 34 Water pump bolt tightening sequence2—4.6L engine

9. If equipped with oil cooler, remove the water bypass pipe.
10. Remove the No.1 water bypass hose.
11. Disconnect the No.8 water bypass hose.
12. Disconnect the No.5 water bypass hose.
13. Remove the water inlet housing.
14. Remove the water pump pulley, using the proper removal tools.
15. Remove the water pump retaining bolts.
16. Remove the water pump from its mounting. Discard the gasket.

To install:

➡Be sure to use new fasteners, as required.

17. Installation is the reverse of the removal procedure.
18. Be sure to use a new gasket.
19. Tighten the retaining bolts A to 35 ft. lbs. (47 Nm). Tighten the retaining bolts B to 17 ft. lbs. (23 Nm). Tighten the retaining bolts C to 15 ft. lbs. (20 Nm).
20. Be sure to fill the cooling system using the proper grade and type coolant.

5.7L Engine

See Figure 35.

1. Before servicing the vehicle, refer to the Precautions Section.
2. Disconnect the negative battery cable. Tape the cable with insulating tape.
3. Remove No. 1 engine under cover.
4. Drain engine coolant.
5. Remove V-bank cover sub-assembly.
6. Remove air cleaner assembly with element.
7. Remove air cleaner hose assembly.
8. Disconnect inlet radiator hose.
9. Remove fan shroud.
10. Remove No. 1 water by-pass hose.
11. Disconnect water by-pass pipe.
    a. Disconnect the No. 6 water by-pass hose.
    b. Remove the 3 bolts.
    c. Disconnect the water by-pass pipe with water hose.
12. Disconnect the No. 2 and No. 3 air hoses.
13. Disconnect the air pump's connector.
14. Disconnect the air pump connector clamp's holder.
15. Using a clip remover, detach the wire harness clamp.
16. Disconnect the No. 5 water by-pass hose.
17. Remove the air tube bracket's bolt.
18. Remove the 3 bolts, water inlet housing and No. 3 water by-pass hose.

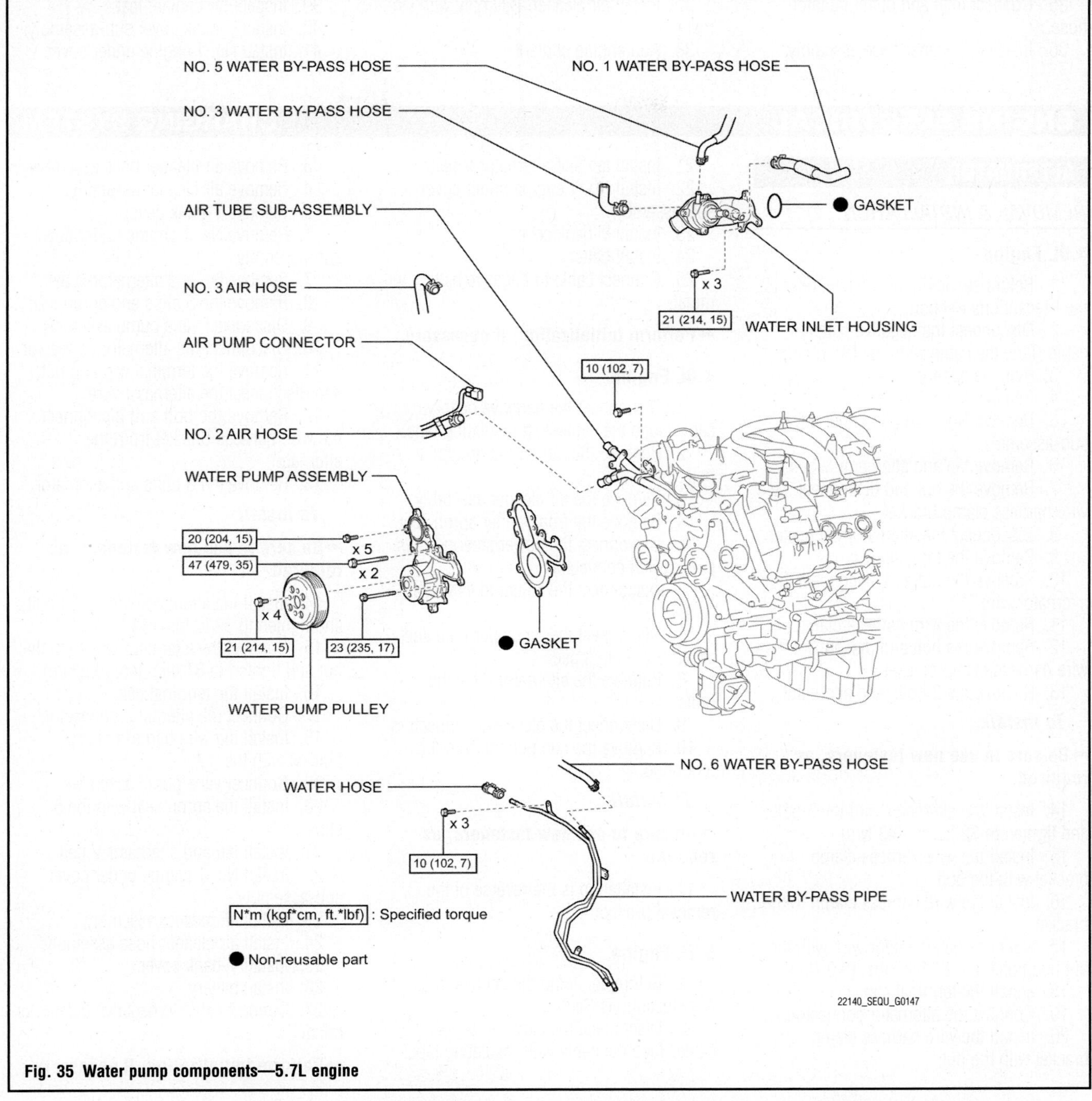

Fig. 35 Water pump components—5.7L engine

19. Using Service Tool, hold the water pump pulley.
20. Remove the 4 bolts and water pump pulley.
21. Remove the 8 bolts, water pump and gasket.

To install:

➡Be sure to use new fasteners, as required.

22. Install a new gasket and the water pump with the 8 bolts and tighten: Bolt A to 35 ft. lbs. (47 Nm), bolt B to 17 ft. lbs. (23 Nm), and bolt C to 15 ft. lbs. (20 Nm).
23. Install the pulley with the 4 bolts and tighten to 15 ft. lbs. (21 Nm).
24. Install a new gasket to the water pump.
25. Install the No. 3 water by-pass hose and water inlet housing with the 3 bolts.
26. Install the air tube bracket's bolt.
27. Connect the No. 5 water by-pass hose.
28. Connect the air pump connector clamp's holder.
29. Attach the wire harness clamp.
30. Connect the air pump's connector.
31. Connect the No. 2 and No. 3 air hoses.
32. Connect water by-pass pipe.
    a. Connect the water by-pass pipe with water hose.
    b. Install the 3 bolts and tighten to 7 ft. lbs. (10 Nm).
    c. Connect the No. 6 water by-pass hose.
33. Install No. 1 water by-pass hose.
34. Install fan shroud.

35. Connect inlet and outlet radiator hose.
36. Install air cleaner hose assembly.
37. Install air cleaner assembly with element.
38. Add engine coolant.
39. Inspect for coolant leak.
40. Install V-bank cover sub-assembly.
41. Install No. 1 engine under cover.

ENGINE ELECTRICAL

CHARGING SYSTEM

ALTERNATOR

REMOVAL & INSTALLATION

4.0L Engine

1. Before servicing the vehicle, refer to the Precautions Section.
2. Disconnect the negative battery cable. Tape the cable with insulating tape.
3. Remove battery.
4. Remove V-bank cover.
5. Remove No. 1 engine under cover sub-assembly.
6. Remove fan and alternator V belt.
7. Remove the nut and disconnect the wire harness clamp bracket.
8. Disconnect the alternator connector.
9. Remove the terminal cap.
10. Remove the nut and disconnect the alternator wire.
11. Remove the wire harness clamp.
12. Remove the bolt and disconnect the wire harness clamp bracket.
13. Remove the 2 bolts and alternator.

To install:

➡Be sure to use new fasteners, as required.

14. Install the alternator with the 2 bolts and tighten to 32 ft. lbs. (43 Nm).
15. Install the wire harness clamp bracket with the bolt.
16. Install the wire harness clamp to the bracket.
17. Connect the alternator wire with the nut and tighten to 87 inch lbs. (9.8 Nm).
18. Install the terminal cap.
19. Connect the alternator connector.
20. Install the wire harness clamp bracket with the nut.
21. Install fan and alternator V belt.
22. Install No. 1 engine under cover sub-assembly.
23. Install V-bank cover.
24. Install battery.
25. Connect cable to negative battery terminal.

➡Perform initialization, if necessary.

4.6L Engine

1. Disconnect the negative battery cable. Tape the cable with insulating tape.
2. Remove the fan and alternator V belt
3. Remove the air cleaner assembly.
4. Remove the front fender apron seal.
5. Disconnect the power steering pressure switch connector.
6. Disconnect the solenoid valve connector.
7. Disconnect the oil cooler tube sub assembly, if equipped.
8. Remove the alternator retaining bolts.
9. Disconnect the electrical connectors.
10. Remove the component from its mounting.

To install:

➡Be sure to use new fasteners, as required.

11. Installation is the reverse of the removal procedure.

5.7L Engine

1. Before servicing the vehicle, refer to the Precautions Section.
2. Disconnect the negative battery cable. Tape the cable with insulating tape.
3. Remove air cleaner hose assembly.
4. Remove air cleaner assembly.
5. Remove V-bank cover.
6. Remove No. 1 engine under cover sub-assembly.
7. Remove fan and alternator V belt.
8. Remove the 5 clips and apron seal.
9. Disconnect vane pump assembly.
10. Disconnect the alternator connector.
11. Remove the terminal cap and nut, and disconnect the alternator wire.
12. Remove the bolt and disconnect the wire harness bracket from the alternator.
13. Remove the 3 bolts and alternator.

To install:

➡Be sure to use new fasteners, as required.

14. Install the alternator with the 3 bolts and tighten to 32 ft. lbs. (43 Nm).
15. Connect the alternator wire with the nut and tighten to 87 inch lbs. (9.8 Nm).
16. Install the terminal cap.
17. Connect the alternator connector.
18. Install the wire harness clamp bracket with the nut.
19. Connect vane pump assembly.
20. Install the apron seal with the 5 clips.
21. Install fan and alternator V belt.
22. Install No. 1 engine under cover sub-assembly.
23. Install air cleaner assembly.
24. Install air cleaner hose assembly.
25. Install V-bank cover.
26. Install battery.
27. Connect cable to negative battery terminal.

➡Perform initialization, if necessary.

ENGINE ELECTRICAL — IGNITION SYSTEM

IGNITION COIL

REMOVAL & INSTALLATION

1. Before servicing the vehicle, refer to the Precautions Section.
2. Disconnect the negative battery cable. Tape the cable with insulating tape.
3. Remove the V bank cover assembly, as required.
4. Remove the air cleaner assembly, as required.
5. Remove the necessary components in order to gain access to the component.
6. Disconnect ignition coil (with igniter) connectors.
7. Remove the bolt, and pull out the ignition coil (with igniter). Remove the ignition coils (with igniter).

To install:

➡Be sure to use new fasteners, as required.

8. Installation is the reverse of the removal procedure.

IGNITION TIMING

ADJUSTMENT

The ignition timing is controlled by the Powertrain Control Module (PCM). No adjustment is necessary or possible.

SPARK PLUGS

REMOVAL & INSTALLATION

1. Before servicing the vehicle, refer to the Precautions Section.
2. Disconnect the negative battery cable. Tape the cable with insulating tape.
3. Remove the ignition coils.
4. Using a 16 mm plug wrench, remove the spark plugs.
5. Clean the spark plugs.
6. If the electrode has traces of wet carbon, allow it to dry and then clean with a spark plug cleaner.
7. Check the spark plug for thread damage and insulator damage. If abnormal, replace the spark plug.
8. Adjust the spark plug electrode gap. Electrode gap for new spark plug is 1.0 to 1.1 mm (0.039 to 0.043 in.).
9. Using a 16 mm plug wrench, install the spark plugs and tighten to specification.
10. Reinstall the ignition coils.

ENGINE ELECTRICAL — STARTING SYSTEM

STARTER

REMOVAL & INSTALLATION

4.0L Engine

See Figure 36.

1. Before servicing the vehicle, refer to the Precautions Section.
2. Disconnect the negative battery cable. Tape the cable with insulating tape.
3. Disconnect cable from negative battery terminal.
4. Remove the terminal cap.
5. Remove the nut and disconnect the starter wire.
6. Disconnect the starter connector.

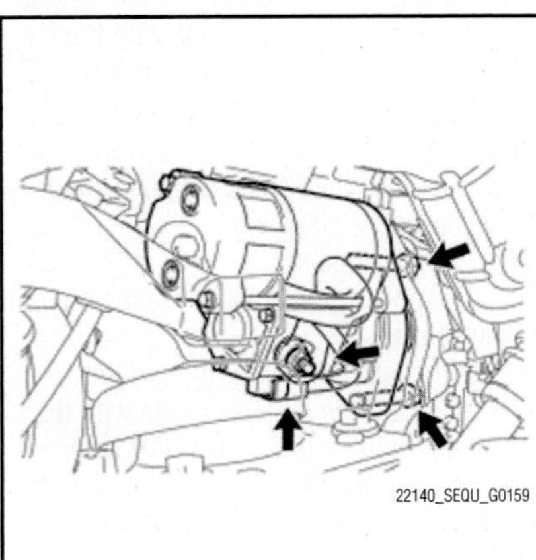

Fig. 36 Starter and starter bolt locations—4.0L engine

7. Remove the 2 bolts and starter assembly.

To install:

8. Install the starter with the 2 bolts and tighten to 43 ft. lbs. (58 Nm).
9. Connect the starter connector.
10. Connect the starter wire with the nut and tighten nut to 87 inch lbs. (9.8 Nm).
11. Install the terminal cap.
12. Connect cable to negative battery terminal.

➡Some systems need to be initialized after the cable is reconnected.

4.6L Engine

See Figure 37.

1. Before servicing the vehicle, refer to the Precautions Section.
2. Disconnect the negative battery cable. Tape the cable with insulating tape.
3. Remove the exhaust manifold sub assembly.
4. Remove the starter cover.
5. Disconnect the electrical connector.
6. Remove the starter retaining bolts.
7. Remove the starter from its mounting.
8. Remove the flywheel side cover.

To install:

➡Be sure to use new fasteners, as required.

9. Installation is the reverse of the removal procedure.
10. Tighten the retaining bolts to 27 ft. lbs. (37 Nm).

5.7L Engine

See Figures 38 through 40.

1. Before servicing the vehicle, refer to the Precautions Section.
2. Disconnect the negative battery cable. Tape the cable with insulating tape.
3. Remove No. 1 engine under cover.
4. Remove V-bank cover.
5. Remove air cleaner hose assembly.
6. Remove air cleaner assembly.
7. Remove fan and alternator V belt.
8. Remove front fender apron seal RH.
9. Disconnect the 2 clamps and power steering oil pressure switch connector.
10. Remove the 2 bolts and disconnect the vane pump
11. Disconnect the alternator connector.
12. Remove the terminal cap and nut, and disconnect the alternator wire.
13. Remove the bolt and disconnect the wire harness bracket from the alternator.
14. Remove the 2 bolts, 2 nuts and alternator.
15. Remove the dipstick.
16. Remove the bolt and dipstick guide.
17. Remove the O-ring from the dipstick guide.
18. Remove front exhaust pipe assembly.
    a. Disconnect the air fuel ratio sensor connector.
    b. Disconnect the heated oxygen sensor connector and 2 clamps.
    c. Remove the 2 bolts.

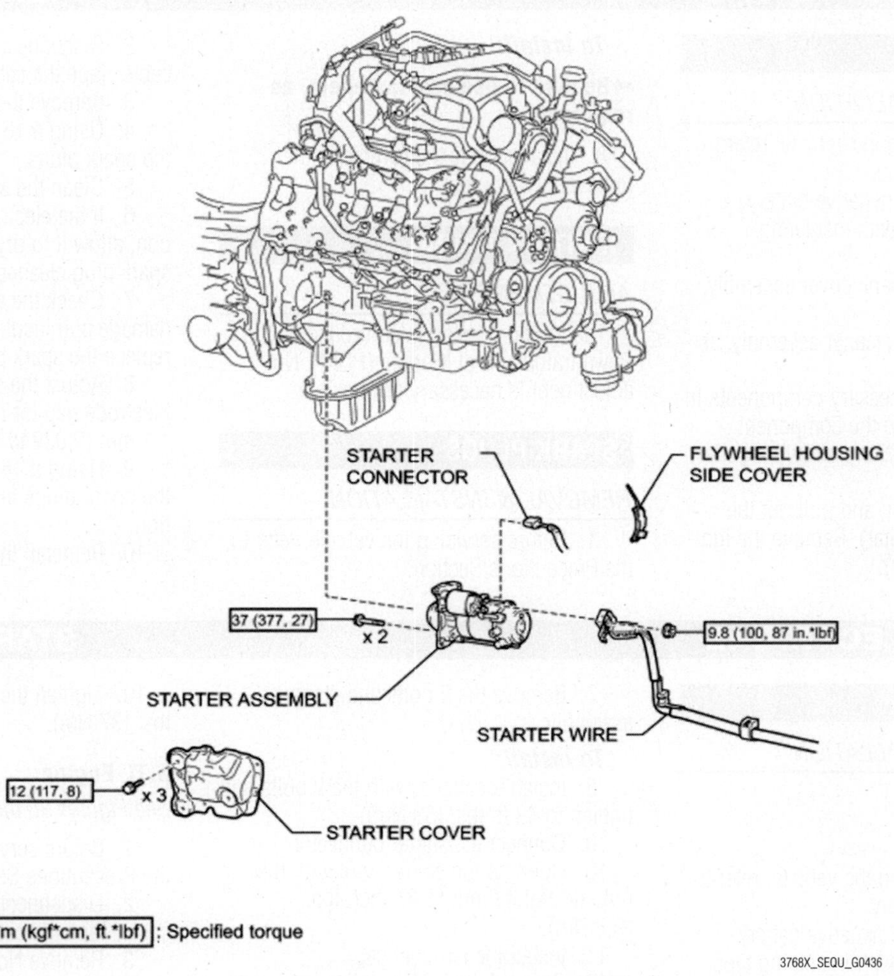

Fig. 37 Starter and related components—4.6L engine

d. Remove the 3 nuts and front exhaust pipe.

e. Remove the 2 gaskets.

19. Remove No. 1 exhaust manifold heat insulator.
20. Remove exhaust manifold sub-assembly RH.

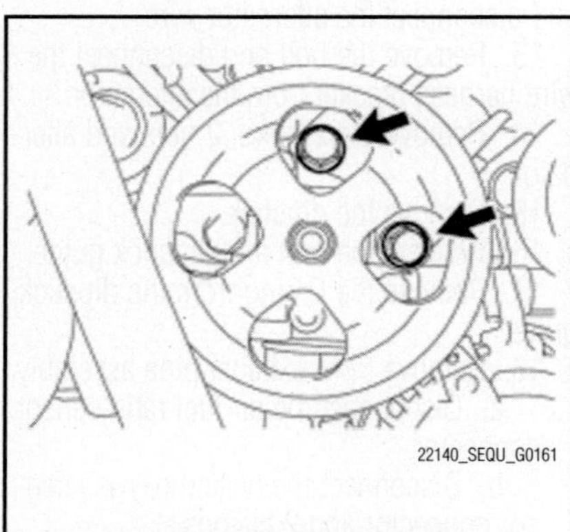

Fig. 38 Vane pump bolt location

21. Remove the 3 bolts and starter cover.
22. Disconnect the starter connector.
23. Remove the nut and disconnect the starter wire.
24. Remove the 2 bolts and starter.
25. Remove the flywheel housing side cover.

To install:

➡Be sure to use new fasteners, as required.

26. Install the flywheel housing side cove.
27. Install the starter with the 2 bolts and tighten to 27 ft. lbs. (37 Nm).
28. Install the starter wire with the nut and tighten to 87 inch lbs. (9.8 Nm).
29. Connect the starter connector.
30. Install the starter cover with the 3 bolts and tighten to 9 ft. lbs. (12 Nm).
31. Install exhaust manifold sub-assembly RH.
32. Install No. 1 exhaust manifold heat insulator.
33. Install a new gasket and the exhaust pipe to the exhaust manifold RH with 3 new nuts and tighten to 40 ft. lbs. (54 Nm).
34. Install a new gasket and the exhaust pipe to the center exhaust pipe with the 2 bolts and tighten to 35 ft. lbs. (48 Nm).
35. Connect the air fuel ratio sensor connector.
36. Connect the heated oxygen sensor connector and 2 clamps.
37. Install dipstick.

    a. Apply a light coat of engine oil to a new O-ring.

    b. Install the O-ring to the guide.

    c. Install the dipstick guide with the bolt and tighten to 7 ft. lbs. (10 Nm).

38. Install the alternator with the 2 bolts and 2 nuts and tighten to 32 ft. lbs. (43 Nm).
39. Connect the alternator connector.
40. Connect the alternator wire with the nut and tighten to 87 inch lbs. (9.8 Nm).
41. Install the terminal cap.

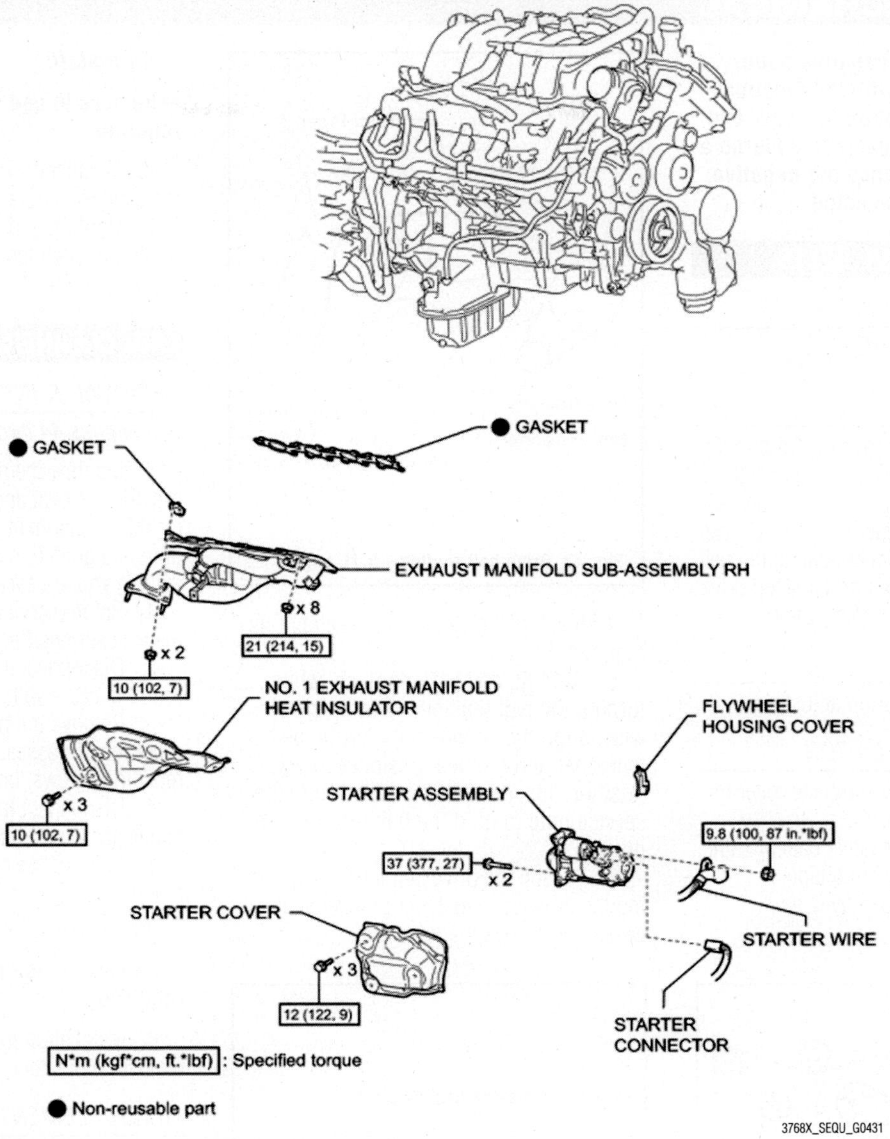

Fig. 39 Starter and related components—5.7L engine

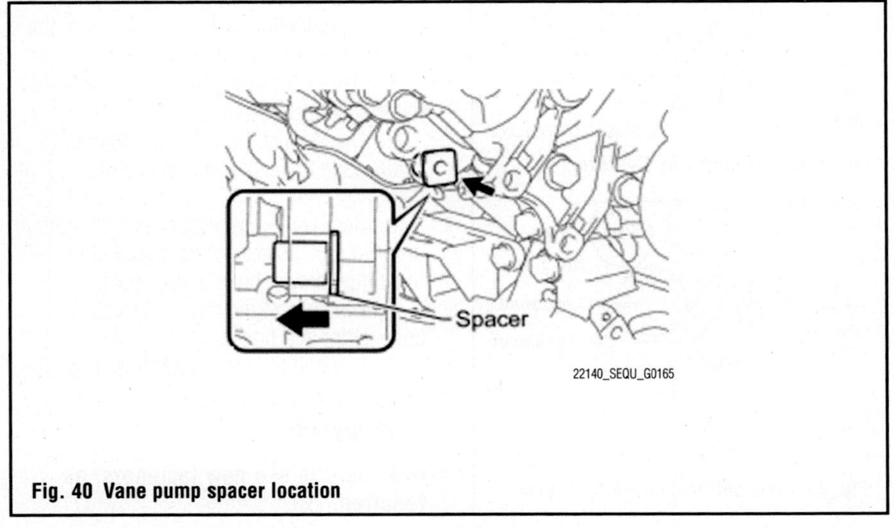

Fig. 40 Vane pump spacer location

42. Install the harness bracket to the alternator with the bolt.
43. Connect vane pump assembly.
    a. Connect the vane pump to the timing chain cover with the 2 bolts and tighten to 21 ft. lbs. (28 Nm).
    b. Connect the 2 clamps and power steering oil pressure switch connector.
44. Install front fender apron seal RH.
45. Install fan and alternator V belt.
46. Install air cleaner assembly.
47. Install air cleaner hose assembly.
48. Install V-bank cover.
49. Install No. 1 engine under cover.
50. Connect cable to negative battery terminal.

➡Some systems need to be initialized after the cable is reconnected.

ENGINE MECHANICAL

➡Disconnecting the negative battery cable may interfere with the functions of the on board computer systems and may require the computer to undergo a relearning process, once the negative battery cable is reconnected.

ACCESSORY DRIVE BELTS

ACCESSORY BELT ROUTING

See Figures 41 through 43.

INSPECTION

Inspect the drive belt for signs of glazing or cracking. A glazed belt will be perfectly smooth from slippage, while a good belt will have a slight texture of fabric visible. Cracks will usually start at the inner edge of the belt and run outward. All worn or damaged drive belts should be replaced immediately.

ADJUSTMENT

The belt does not require adjustment.

REMOVAL & INSTALLATION

1. Before servicing the vehicle, refer to the Precautions Section.
2. Disconnect the negative battery cable. Tape the cable with insulating tape.
3. Remove the V bank cover, as required.

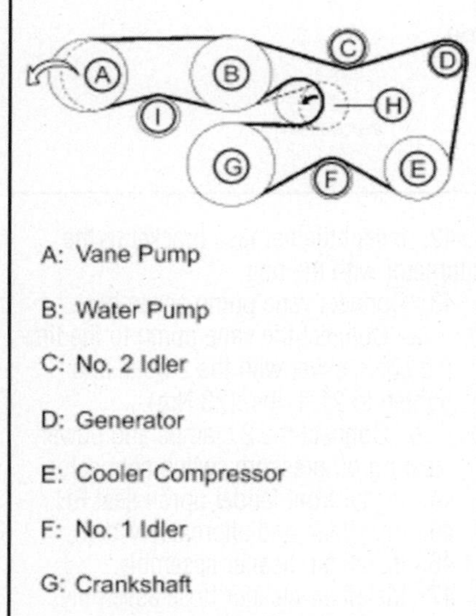

Fig. 41 Drive belt routing—4.0L engine

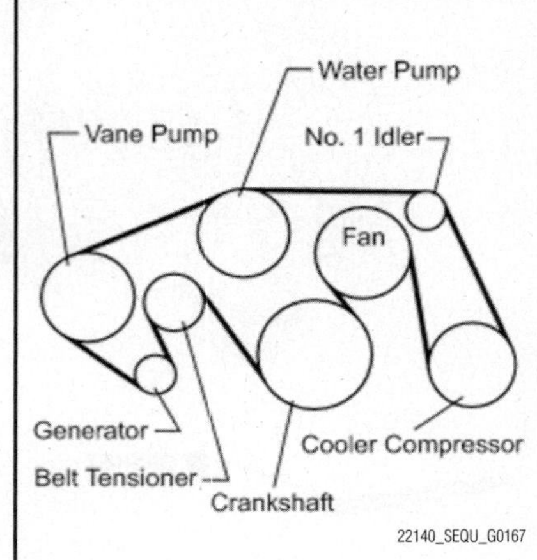

Fig. 42 Drive belt routing—5.7L engine

4. Remove the air cleaner assembly, as required.
5. On 4.6L and 5.7L engines, while turning the belt tensioner counterclockwise, align the service hole for the belt tensioner and the belt tensioner fixing position. Insert a 0.197 inch bar into the service hole to hold the belt tensioner in place.
6. Loosen the drive belt tension by turning the drive belt tensioner counterclockwise, and remove the drive belt.

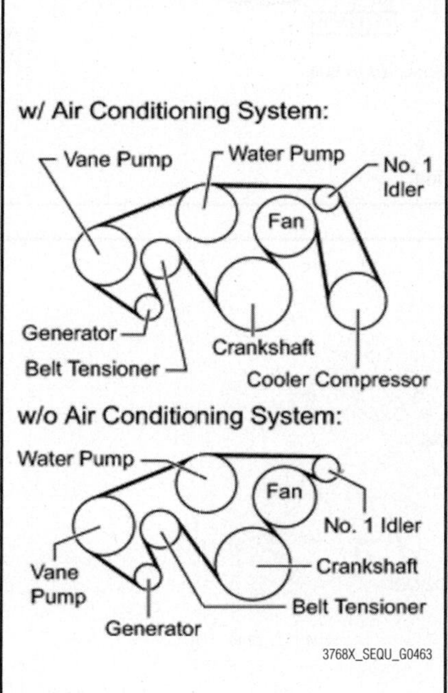

Fig. 43 Drive belt routing—4.6L engine

To install:

➡Be sure to use new fasteners, as required.

7. Installation is the reverse of the removal procedure.
8. On 4.6L and 5.7L engines, while turning the belt tensioner counterclockwise remove the pin.

AIR CLEANER

REMOVAL & INSTALLATION

See Figures 44 through 46.

At this time the manufacturer does not provide removal and installation procedures for this component. The following procedure is a guideline and may differ from the vehicle you are servicing.

1. Before servicing the vehicle, refer to the Precautions Section.
2. Disconnect the negative battery cable. Tape the cable with insulating tape.
3. Remove the necessary components in order to gain assess to the component retaining screws, bolts and/or nuts.
4. Disconnect the required electrical connectors and hoses.
5. Remove the component from its mounting.

To install:

➡Be sure to use new fasteners, as required.

6. Installation is the reverse of the removal procedure.

FILTER/ELEMENT REPLACEMENT

At this time the manufacturer does not provide removal and installation procedures for this component. The following procedure is a guideline and may differ from the vehicle you are servicing.

1. Before servicing the vehicle, refer to the Precautions Section.
2. Disconnect the negative battery cable. Tape the cable with insulating tape.
3. Remove the necessary components in order to gain assess to the component retaining screws, bolts and/or nuts.
4. Disconnect the required electrical connectors and hoses.
5. Remove the air cleaner filter from the air cleaner assembly.

To install:

➡Be sure to use new fasteners, as required.

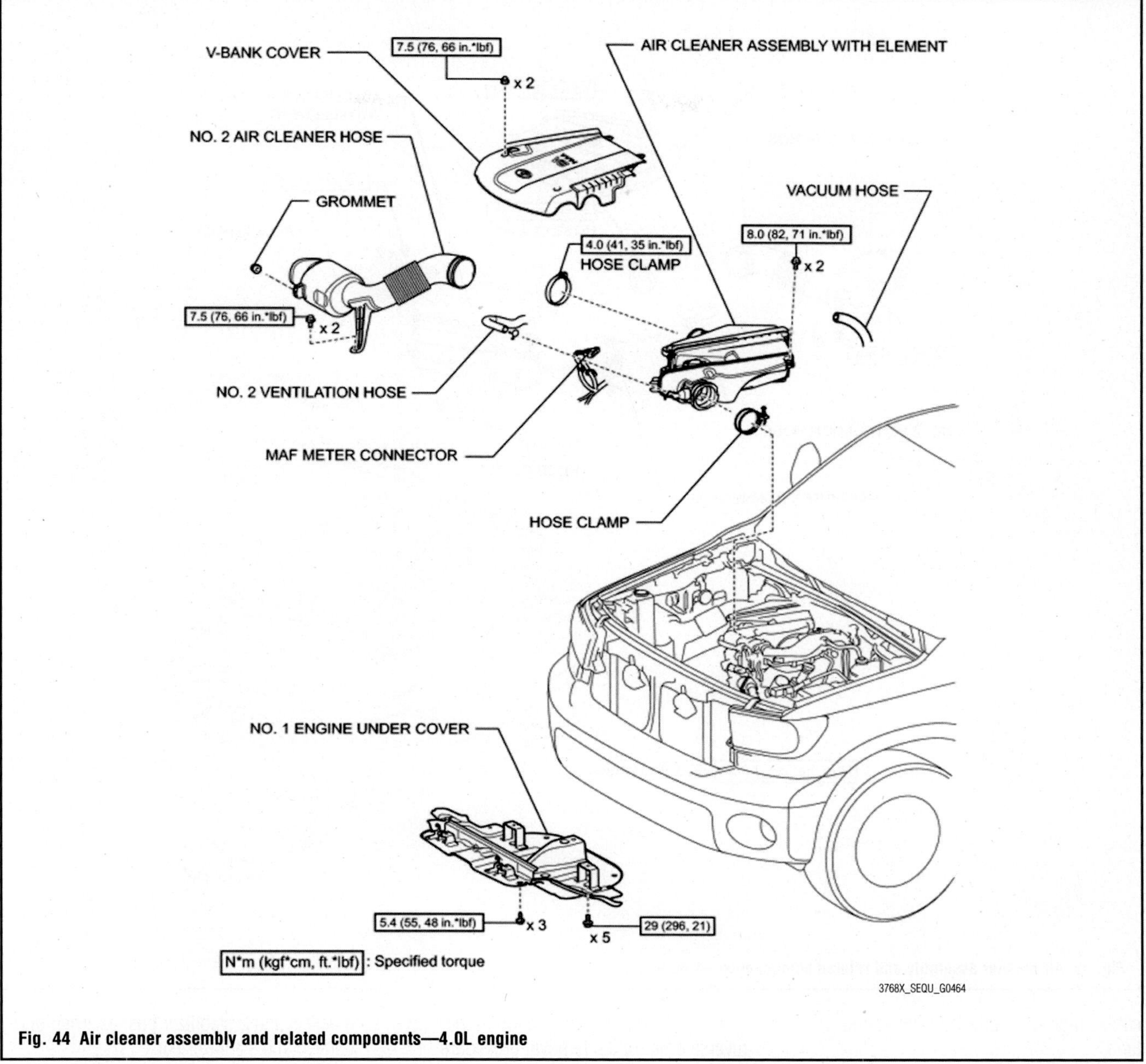

Fig. 44 Air cleaner assembly and related components—4.0L engine

6. Installation is the reverse of the removal procedure.

CATALYTIC CONVERTER

REMOVAL & INSTALLATION

See Figures 47 through 52.

At this time the manufacturer does not provide removal and installation procedures for this component. The following procedure is a guideline and may differ from the vehicle you are servicing.

1. Before servicing the vehicle, refer to the Precautions Section.
2. Disconnect the negative battery cable. Tape the cable with insulating tape.
3. Raise and safely support the vehicle.
4. Remove the necessary components in order to gain access to the converter.
5. Disconnect the oxygen sensors, as necessary. Do not drop the sensor, if dropped it must be replaced.
6. Remove the converter mounting bolts.

➡Be sure to properly support the exhaust system.

7. Remove the converter from the vehicle.
8. Discard the gaskets.

To install:

➡Be sure to use new fasteners, as required.

9. Installation is the reverse of the removal procedure.
10. Be sure to use new gaskets.
11. Be sure to coat the sensor with the proper sealer, if reusing the existing one.

CRANKSHAFT FRONT SEAL

REMOVAL & INSTALLATION

4.0L Engine

See Figure 53.

1. Before servicing the vehicle, refer to the Precautions Section.
2. Disconnect the negative battery

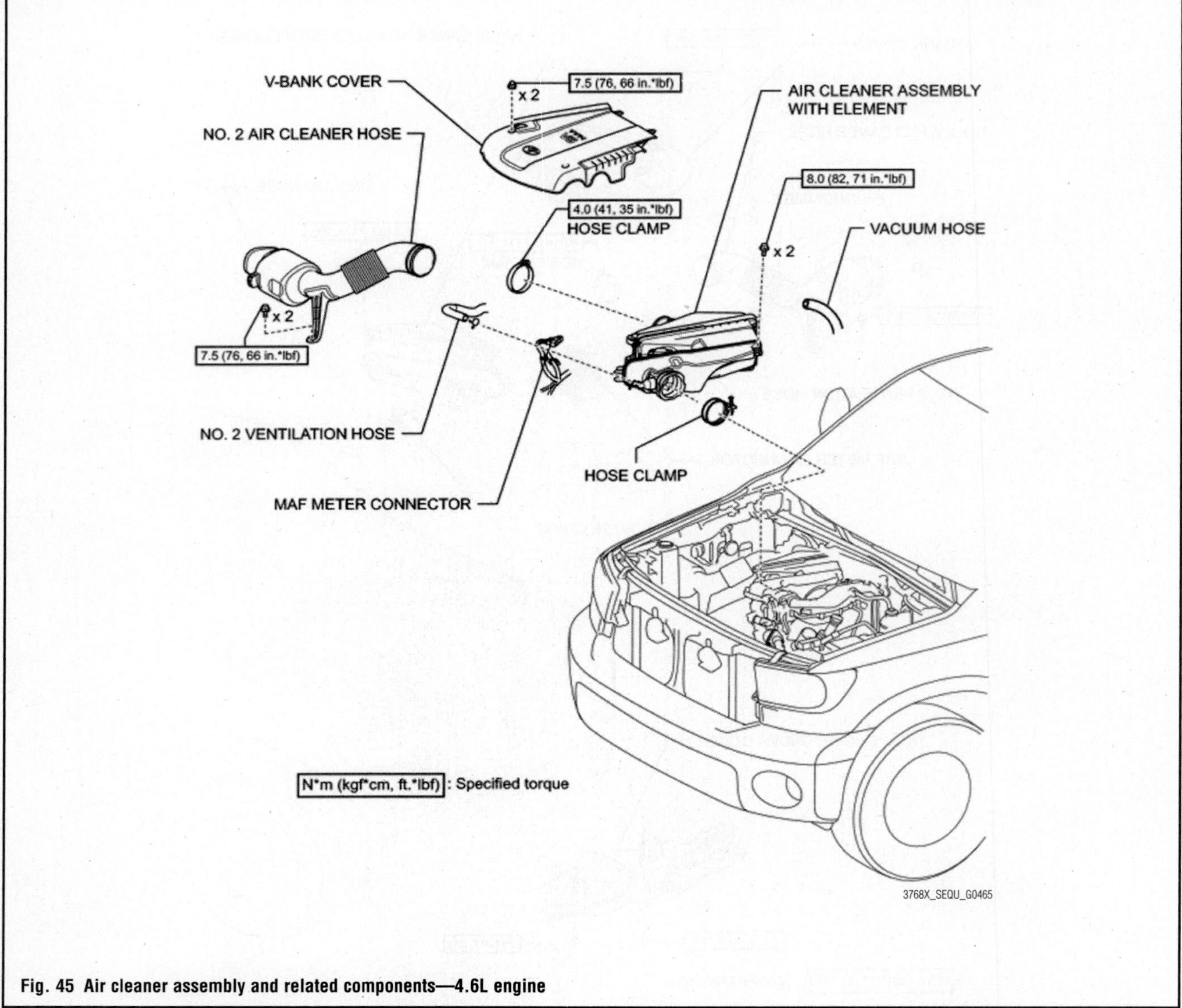

Fig. 45 Air cleaner assembly and related components—4.6L engine

cable. Tape the cable with insulating tape.

3. Drain the cooling system.
4. Remove the radiator.
5. Remove the crankshaft pulley retaining bolt.
6. Remove the crankshaft pulley.
7. Using a suitable tool, pry the seal from its mounting.

To install:

➡**Be sure to use new fasteners, as required.**

8. Coat the new seal with clean engine oil prior to installation.
9. Using a seal installation tool, press the seal into position until its surface is flush with the timing chain edge cover.
10. Continue the installation in the reverse order of the removal procedure.
11. Use the Techstream diagnostic tool, or equivalent and reprogram the required systems.

4.6L Engine

See Figure 54.

1. Before servicing the vehicle, refer to the Precautions Section.
2. Disconnect the negative battery cable. Tape the cable with insulating tape.
3. Remove V-bank cover sub-assembly.
4. Remove No. 1 engine under cover.
5. Drain engine coolant.
6. Remove inlet radiator hose.
7. Remove fan and alternator V belt.
8. Remove fan shroud.
9. Remove the 4 bolts, 2 stabilizer brackets and 2 stabilizer bushes. Then disconnect the stabilizer bar.
10. Remove oil pressure sender gauge assembly.
11. Remove oil filter bracket sub-assembly (w/o oil cooler).
12. Remove the 3 bolts and disconnect the 2 water by-pass hoses from the oil cooler.
13. Remove No. 1 oil cooler bracket (w/ oil cooler).
14. Remove oil filter bracket sub-assembly (w/ oil cooler).
15. Remove crankshaft pulley.
16. Remove the crankshaft timing gear key from the crankshaft.
17. Using a screwdriver, pry out the oil seal.

➡**Do not damage the surface of the oil seal press fit hole and crankshaft.**

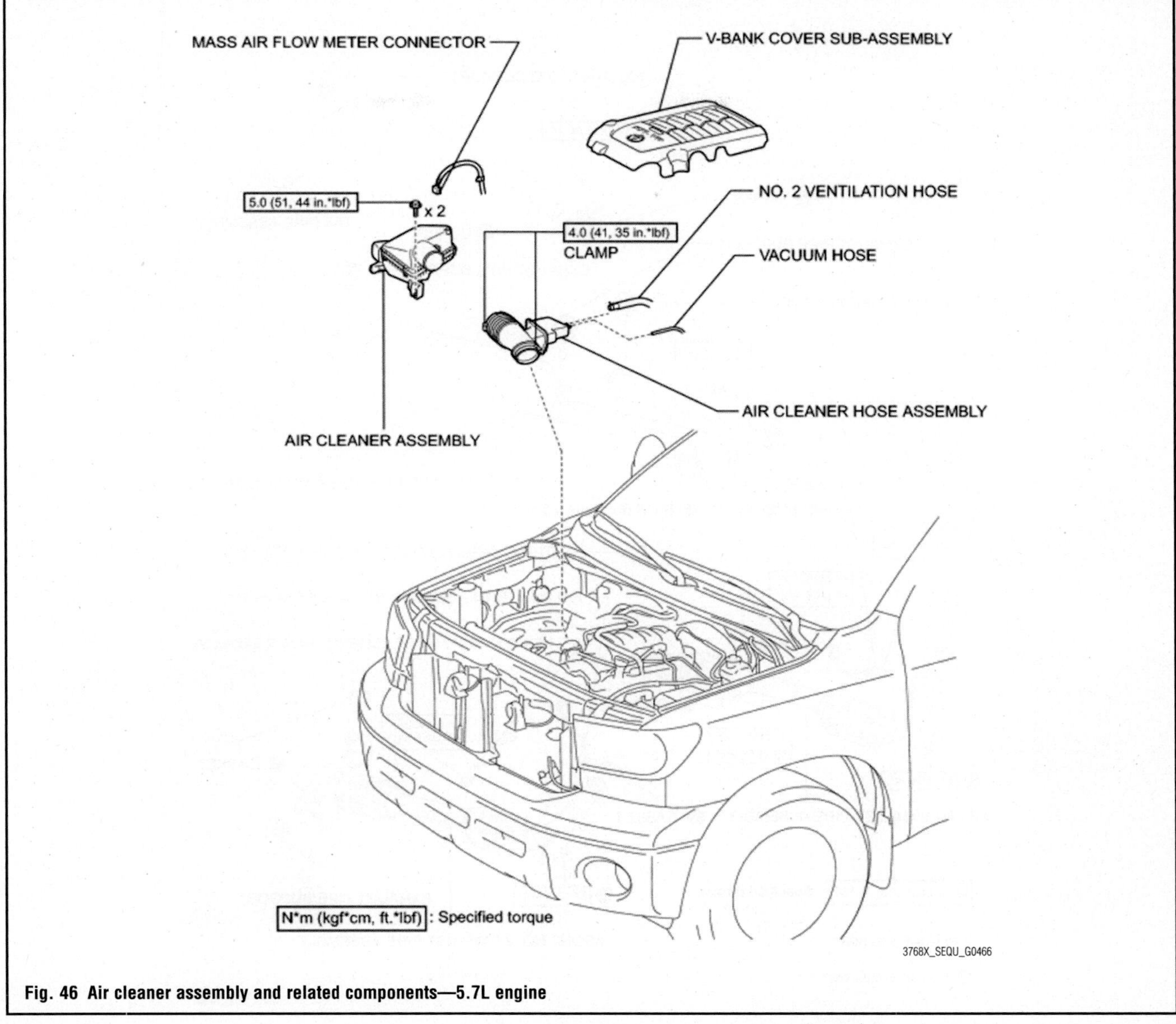

Fig. 46 Air cleaner assembly and related components—5.7L engine

To install:

➡Be sure to use new fasteners, as required.

18. Installation is the reverse of the removal procedure.
19. Apply MP grease to the lip of a new oil seal.
20. Using Service Tool and a hammer, tap in the oil seal to a depth between 0 to 1.0 mm (0 to 0.0394 in.) from the timing chain cover edge.
21. Install the crankshaft timing gear key.
22. Install crankshaft pulley and tighten bolt to 221 ft. lbs. (300 Nm).
23. Install oil filter bracket sub-assembly (w/oil cooler).
24. Install no. 1 oil cooler bracket (w/oil cooler).
25. Connect the 2 water by-pass hoses to the oil cooler.
26. Install the 3 by-pass pipe bolts and tighten to 7 ft. lbs. (10 Nm).
27. Install oil filter bracket sub-assembly (w/o oil cooler).
28. Install oil pressure sender gauge assembly.
29. Connect the stabilizer bar and install the 2 stabilizer bushes and 2 stabilizer brackets with the 4 bolts and tighten to 51 ft. lbs. (69 Nm).
30. Install fan shroud.
31. Install fan and alternator v belt.
32. Install inlet radiator hose.
33. Add engine coolant.
34. Inspect for leaks.
35. Install No. 1 engine under cover.
36. Install V-bank cover sub-assembly.
37. Use the Techstream diagnostic tool, or equivalent and reprogram the required systems.

5.7L Engine

See Figure 55.

1. Before servicing the vehicle, refer to the Precautions Section.
2. Disconnect the negative battery cable. Tape the cable with insulating tape.
3. Remove V-bank cover sub-assembly.
4. Remove No. 1 engine under cover.
5. Drain engine coolant.
6. Remove inlet radiator hose.
7. Remove fan and alternator V belt.
8. Remove fan shroud.
9. Remove the 4 bolts, 2 stabilizer brackets and 2 stabilizer bushes. Then disconnect the stabilizer bar.

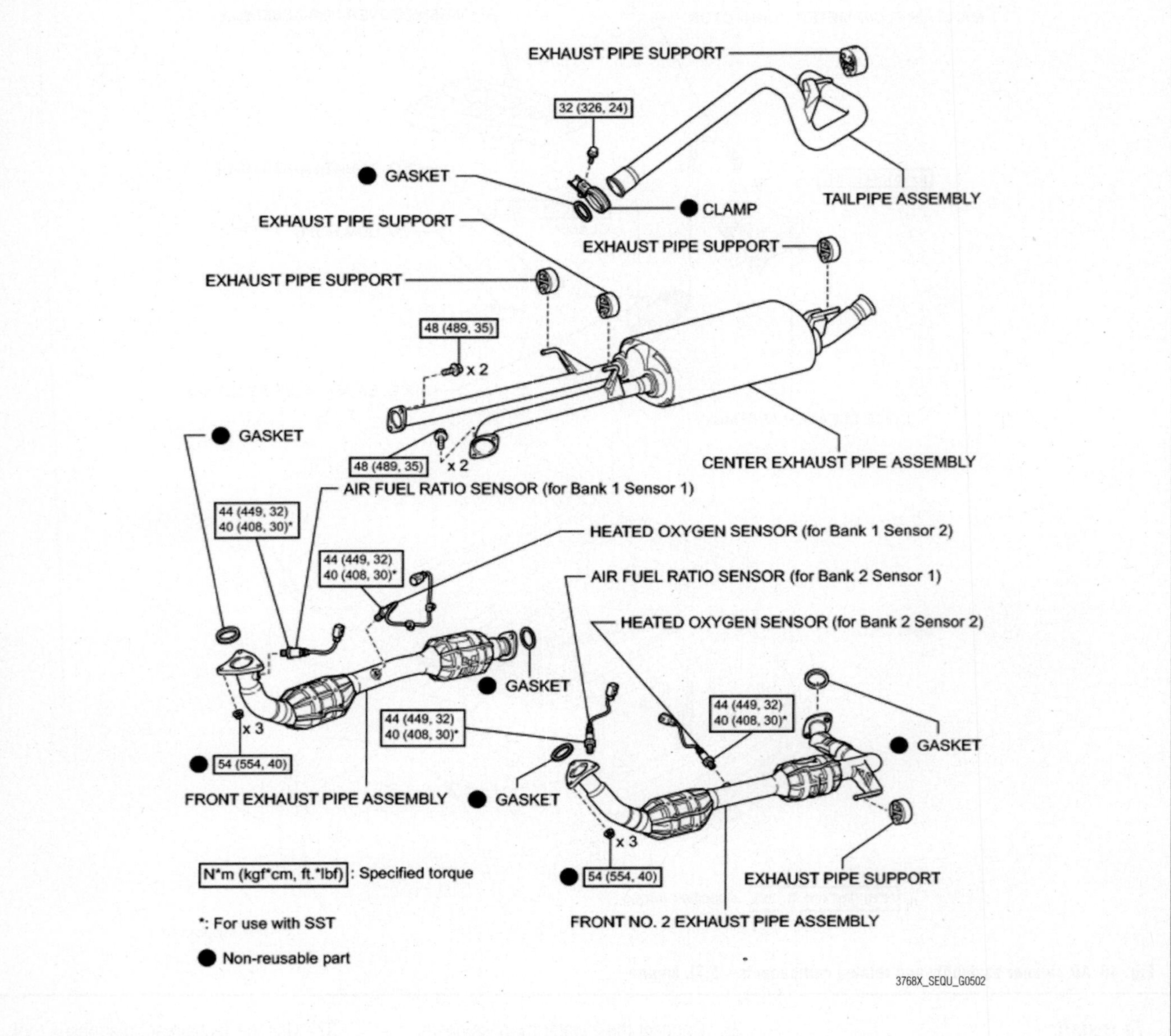

Fig. 47 Catalytic converters and related components 4.0L engine (except regular cab standard deck)

10. Remove oil pressure sender gauge assembly.
11. Remove oil filter bracket sub-assembly (w/o oil cooler).
12. Remove the 3 bolts and disconnect the 2 water by-pass hoses from the oil cooler.
13. Remove No. 1 oil cooler bracket (w/ oil cooler).
14. Remove oil filter bracket sub-assembly (w/ oil cooler).
15. Remove crankshaft pulley.
16. Remove the crankshaft timing gear key from the crankshaft.
17. Using a screwdriver, pry out the oil seal.

➡Do not damage the surface of the oil seal press fit hole and crankshaft.

To install:

➡Be sure to use new fasteners, as required.

18. Apply MP grease to the lip of a new oil seal.
19. Using Service Tool and a hammer, tap in the oil seal to a depth between 0 to 1.0 mm (0 to 0.0394 in.) from the timing chain cover edge.
20. Install the crankshaft timing gear key.
21. Install crankshaft pulley and tighten bolt to 221 ft. lbs. (300 Nm).
22. Install oil filter bracket sub-assembly (w/oil cooler).
23. Install no. 1 oil cooler bracket (w/oil cooler).
24. Connect the 2 water by-pass hoses to the oil cooler.
25. Install the 3 by-pass pipe bolts and tighten to 7 ft. lbs. (10 Nm).
26. Install oil filter bracket sub-assembly (w/o oil cooler).
27. Install oil pressure sender gauge assembly.
28. Connect the stabilizer bar and install the 2 stabilizer bushes and 2 stabilizer brackets with the 4 bolts and tighten to 51 ft. lbs. (69 Nm).

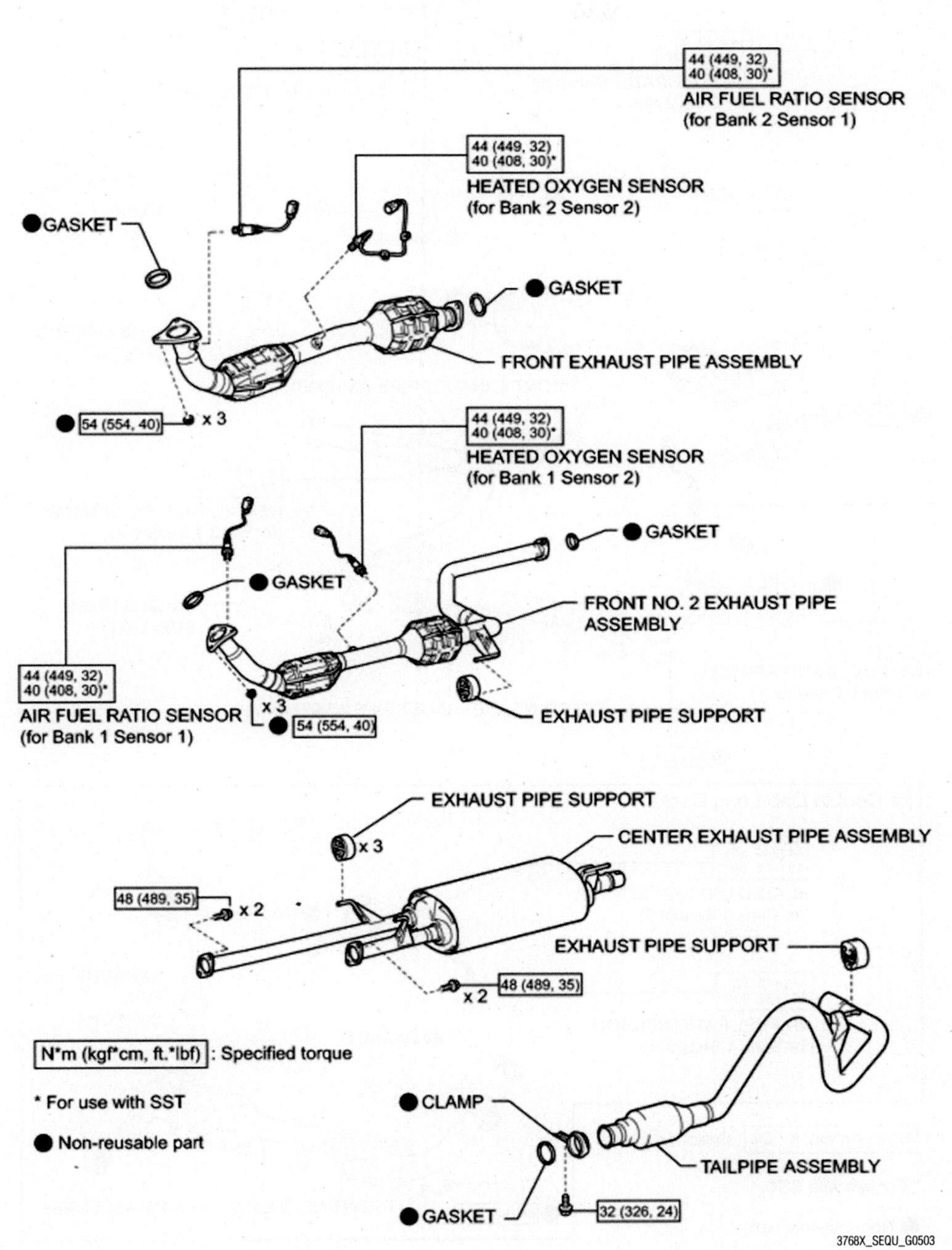

Fig. 48 Catalytic converters and related components 4.6L engine (except regular cab standard deck)

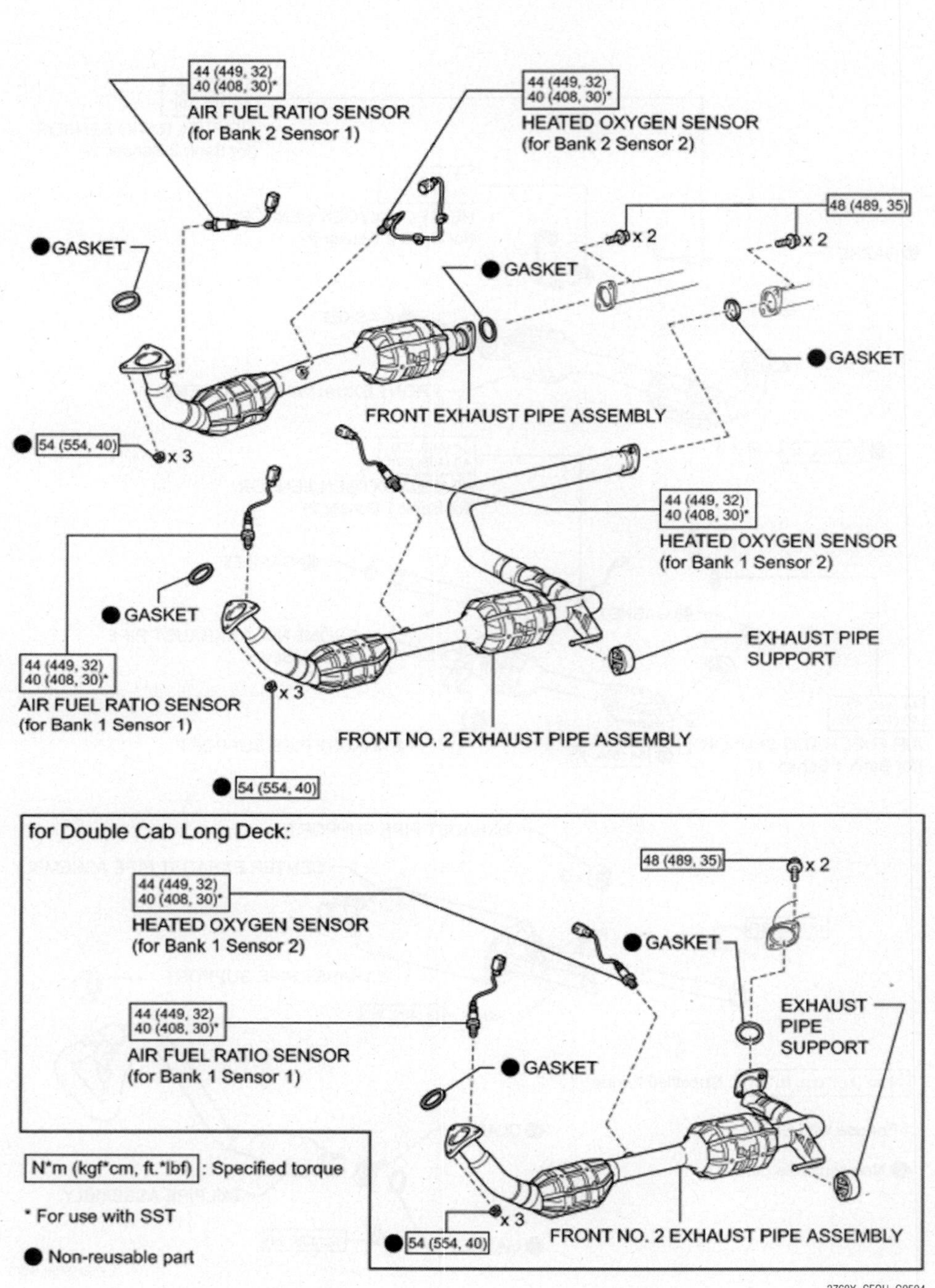

Fig. 49 Catalytic converters and related components 5.7L engine (except regular cab standard deck)

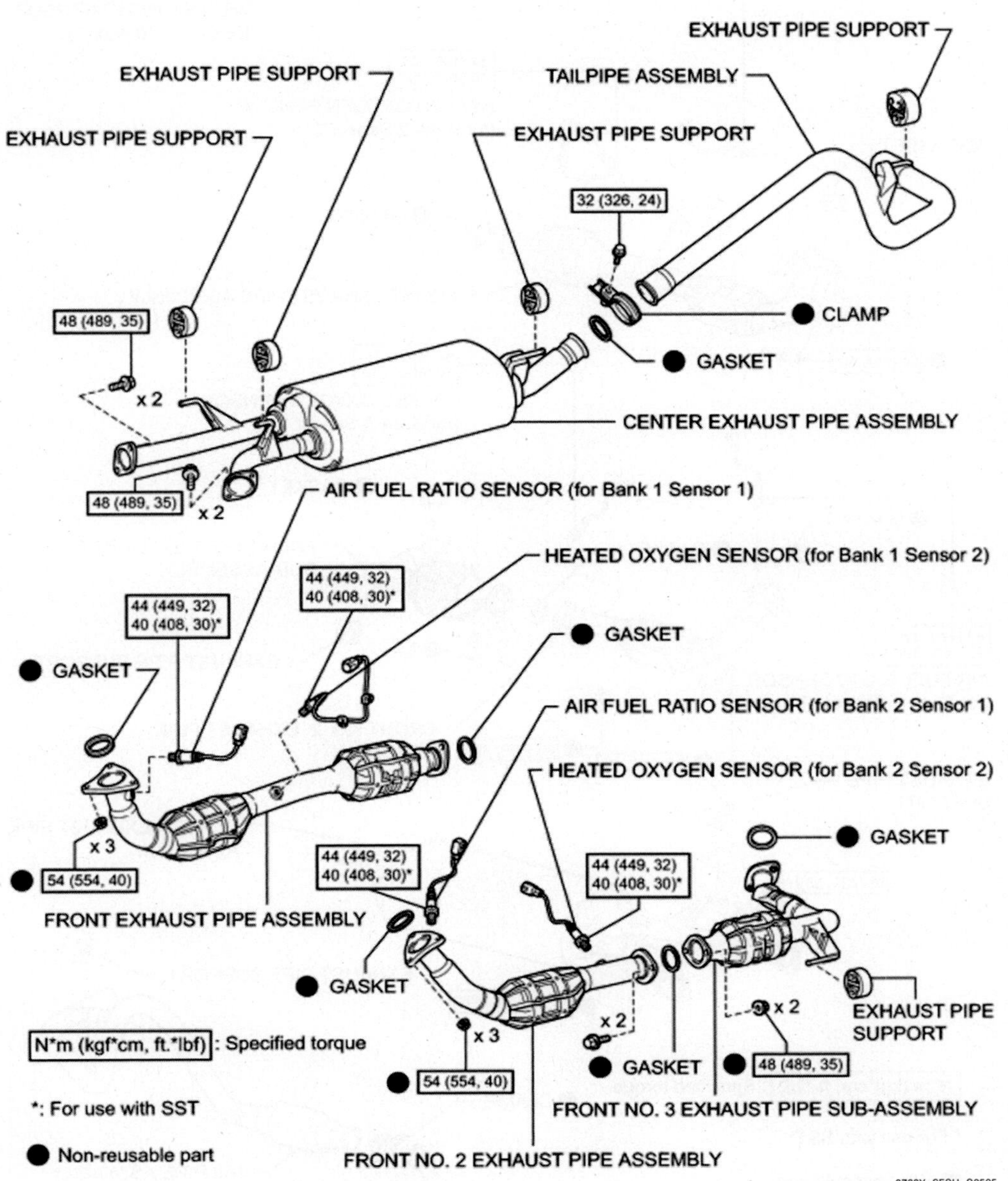

Fig. 50 Catalytic converters and related components 4.0L engine (regular cab)

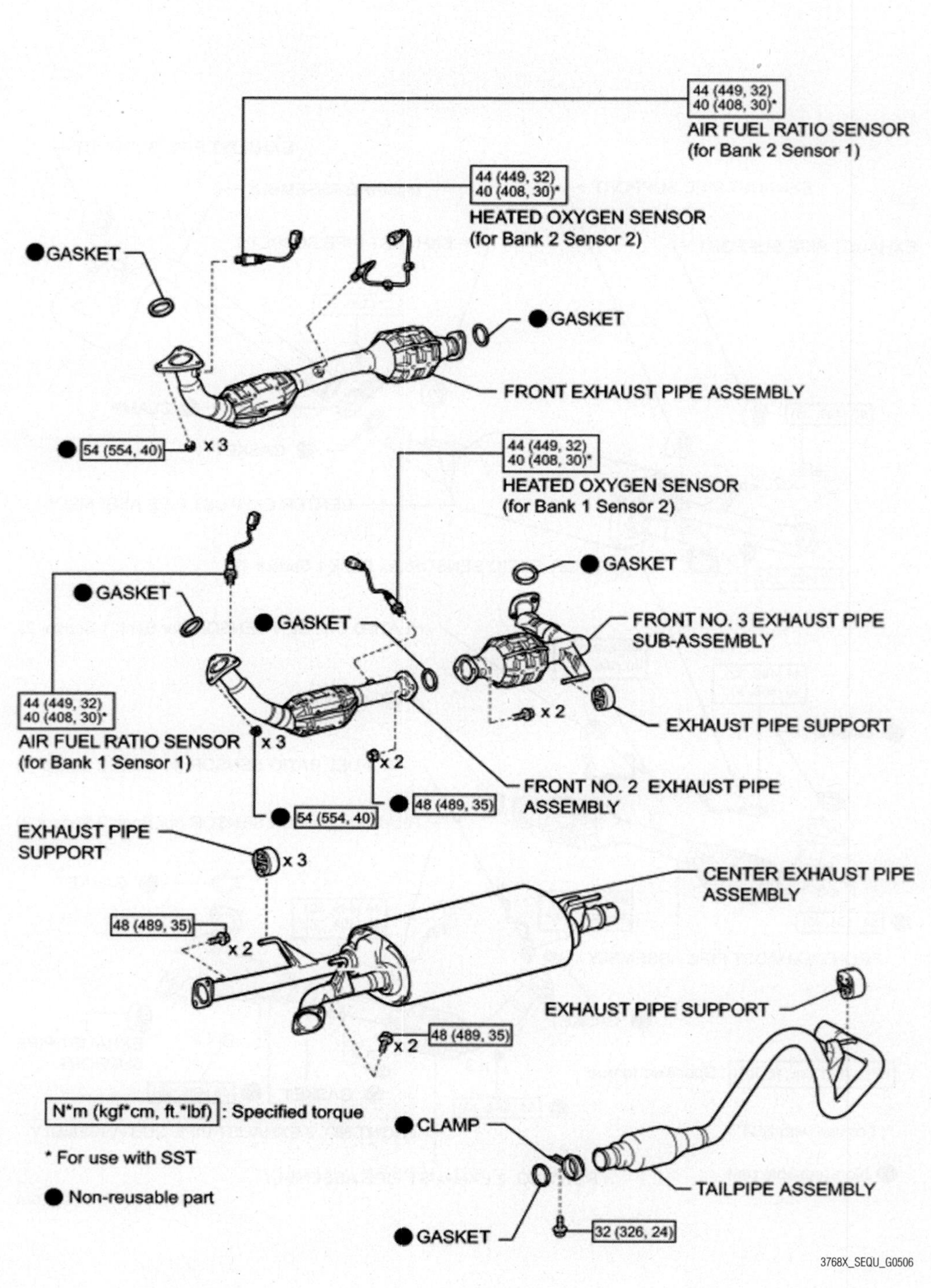

Fig. 51 Catalytic converters and related components 4.6L engine (regular cab)

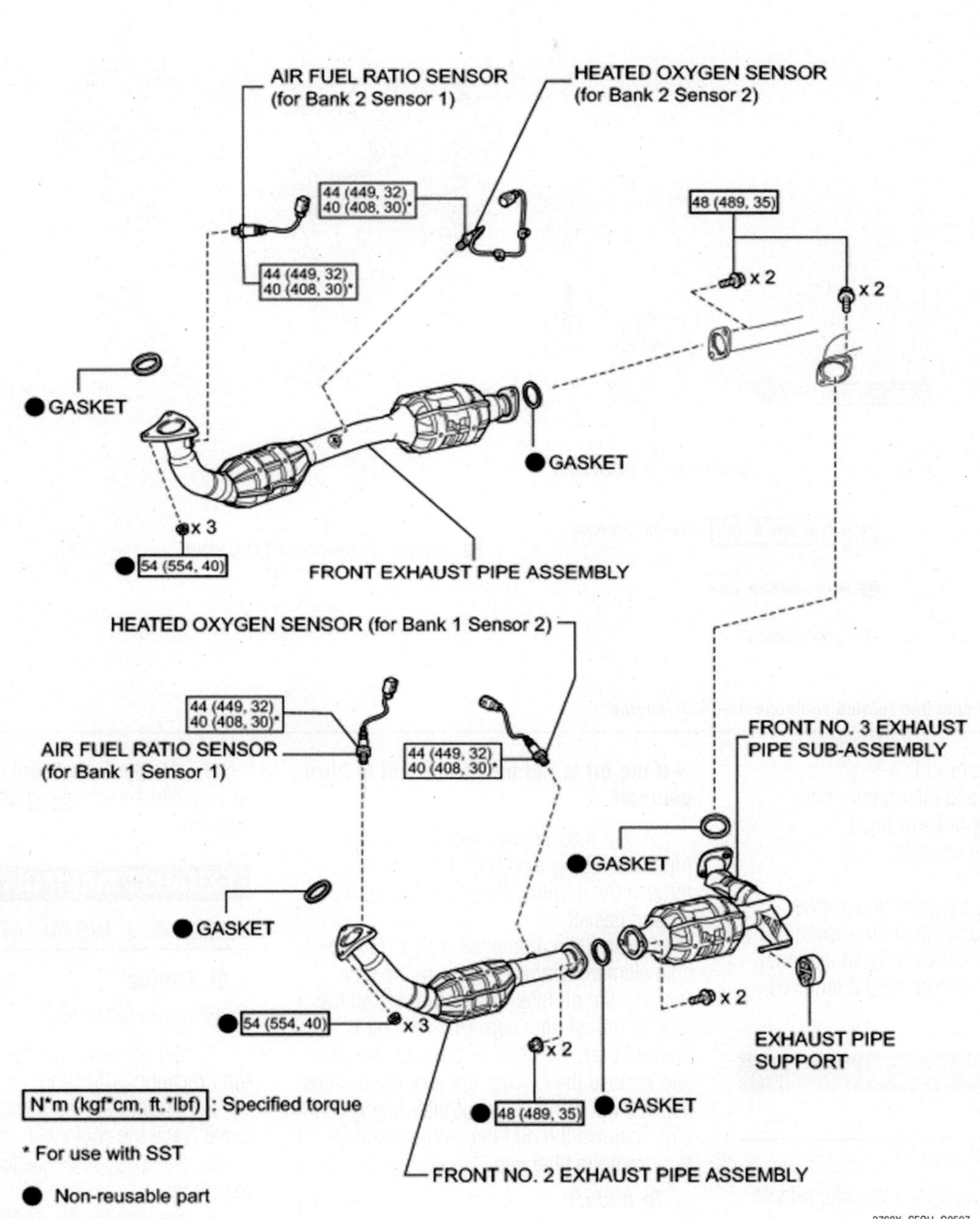

Fig. 52 Catalytic converters and related components 5.7L engine (regular cab)

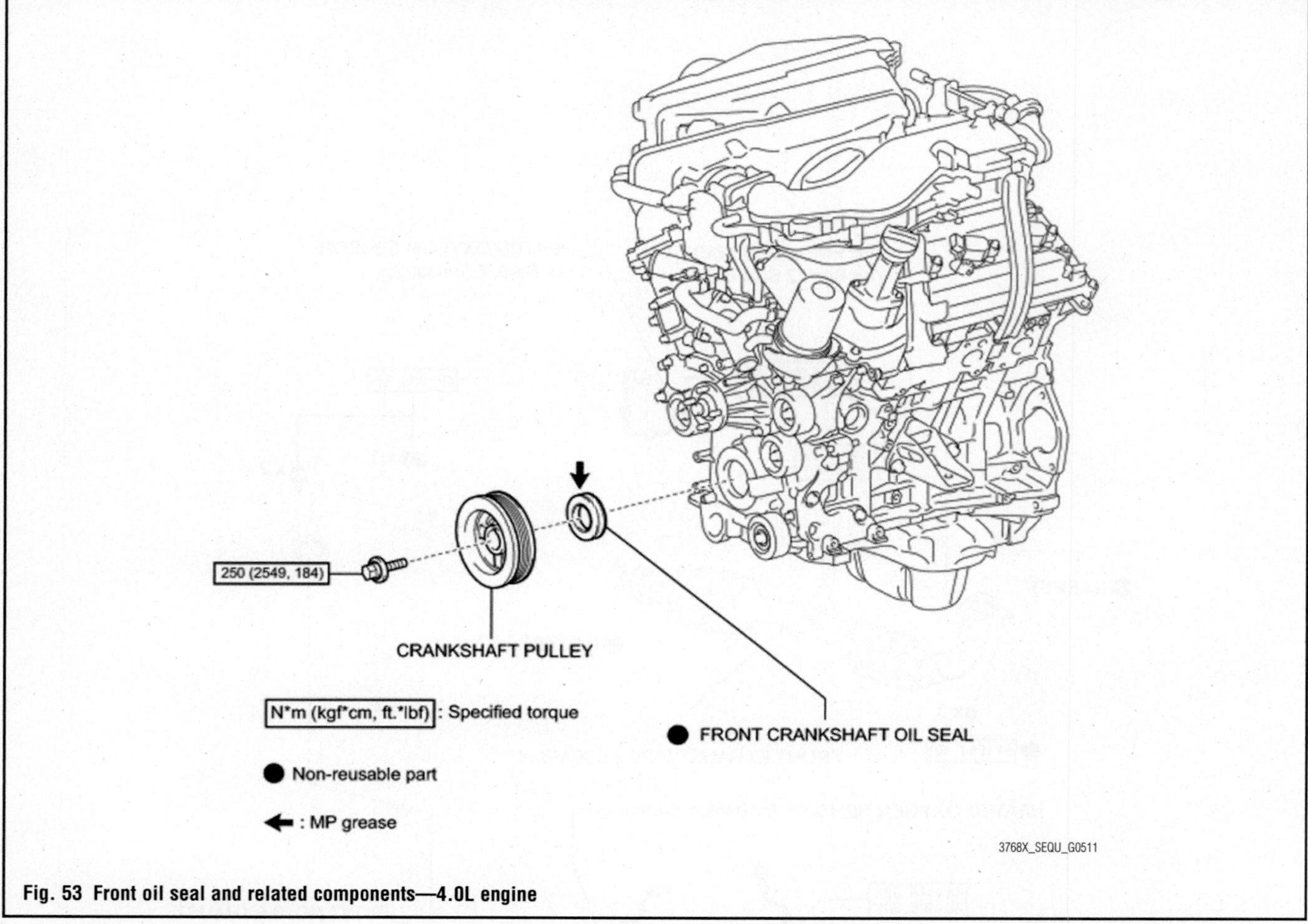

Fig. 53 Front oil seal and related components—4.0L engine

29. Install fan shroud.
30. Install fan and alternator v belt.
31. Install inlet radiator hose.
32. Add engine coolant.
33. Inspect for leaks.
34. Install No. 1 engine under cover.
35. Install V-bank cover sub-assembly.
36. Use the Techstream diagnostic tool, or equivalent and reprogram the required systems.

ENGINE OIL & FILTER

REPLACEMENT

See Figures 56 through 61.

1. Before servicing the vehicle, refer to the Precautions Section.
2. Disconnect the negative battery cable. Tape the cable with insulating tape.
3. Remove the oil filler cap.
4. Raise and safely support the vehicle.
5. Remove the engine under cover, if equipped.
6. Remove the drain plug. Drain the engine oil into a suitable container. Dispose of used oil properly.

➡**If the oil is hot be careful not to burn yourself.**

7. On 4.0L engine, remove the drain pipe cap. Using a oil filter removal tool, remove the oil filter. Properly discard the filter and gasket.
8. On filters equipped with a replaceable element connect a hose to the pipe. Remove the oil filter drain plug. Install the pipe to the oil filter cap. Once the oil is drained from the filter, disconnect the pipe and remove the O-ring. Remove the oil filter cap. Do not remove the oil filter bracket clip. Remove the oil filter element and O-ring from the filter cap.

To install:

➡**Be sure to use new fasteners, as required.**

9. Installation is the reverse of the removal procedure.
10. Be sure to use a new drain plug gasket. Tighten the bolt to specification.
11. Fill the engine using the proper engine oil.
12. Start the engine and check for leaks.
13. Use the Techstream diagnostic tool, or equivalent and reprogram the required systems.

EXHAUST MANIFOLD

REMOVAL & INSTALLATION

4.0L Engine

See Figures 62 and 63.

1. Before servicing the vehicle, refer to the Precautions Section.
2. Disconnect the negative battery cable. Tape the cable with insulating tape.
3. Remove the air switching valve assembly.
4. Remove the No. 2 exhaust pipe assembly.
5. Remove the front exhaust pipe assembly.
6. Remove the manifold retaining nuts.
7. Remove the heat insulator, as required.
8. Remove the manifold from the vehicle.
9. Discard the gasket.

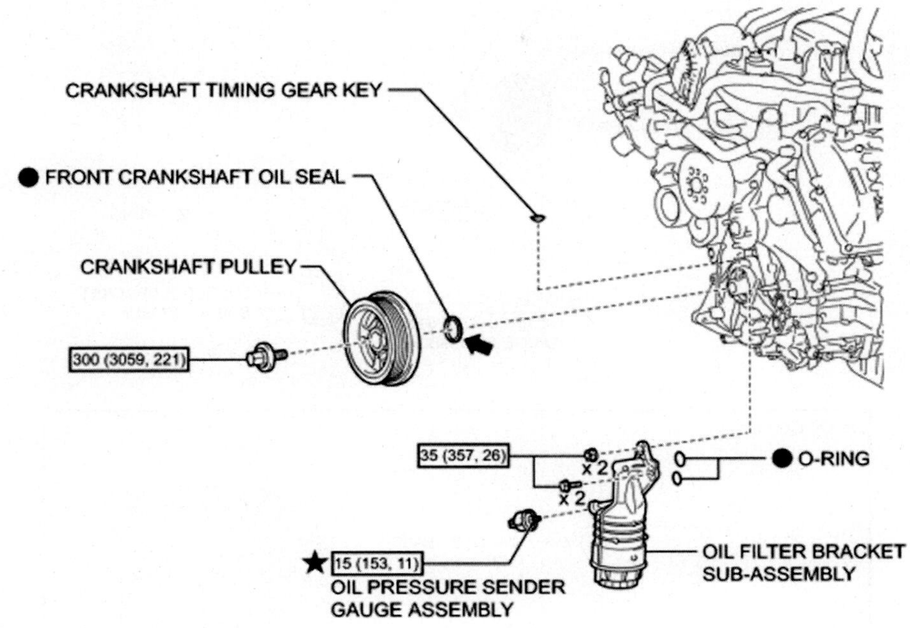

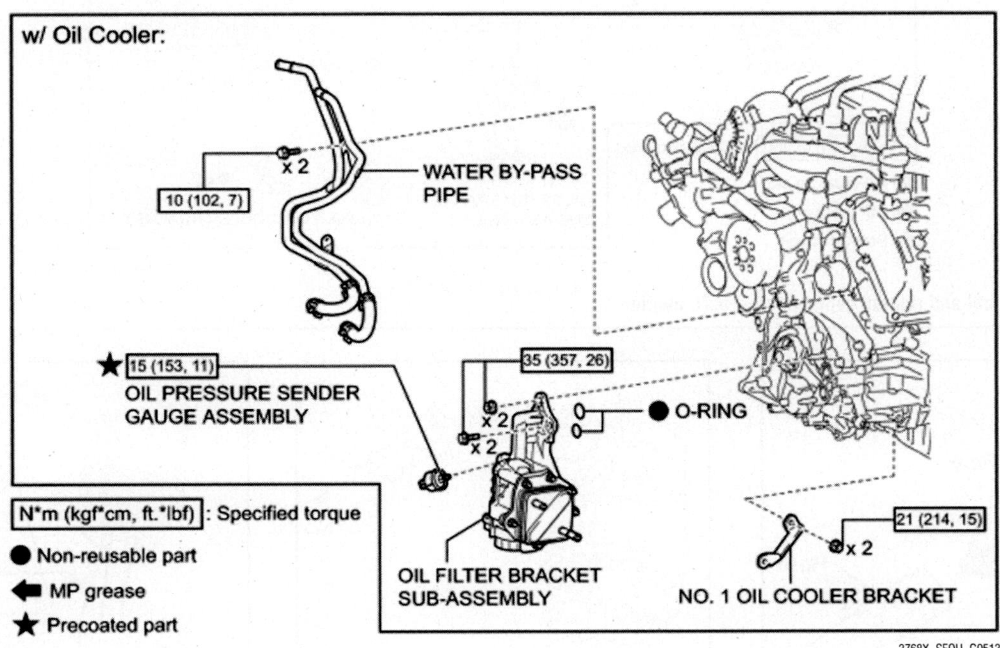

Fig. 54 Front oil seal and related components—4.6L engine

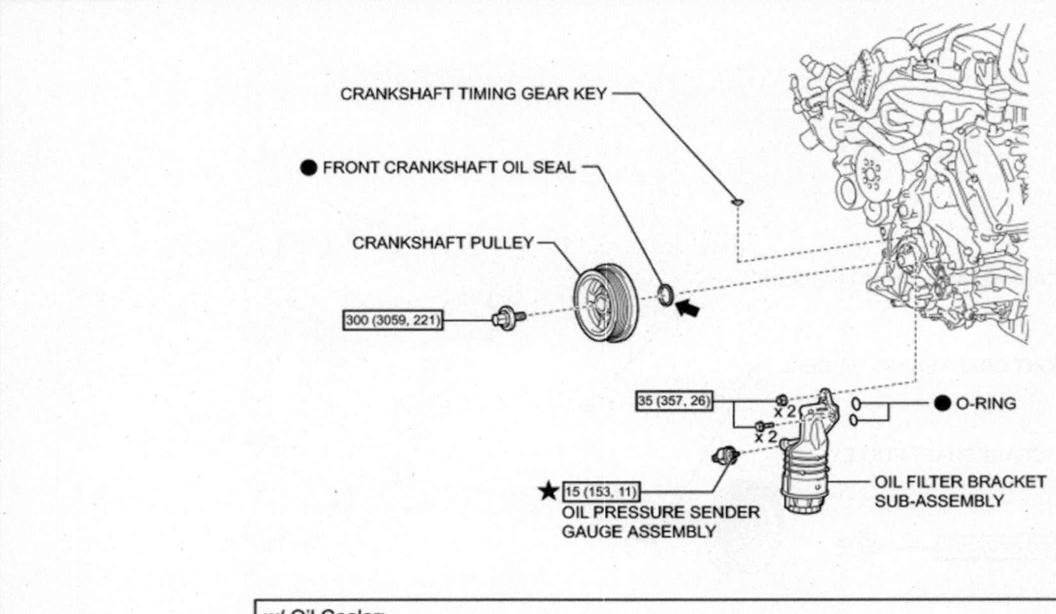

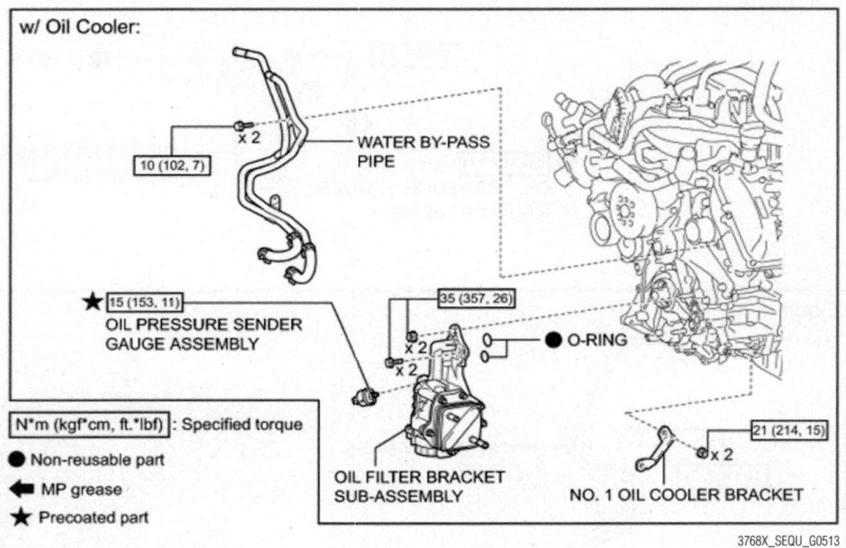

Fig. 55 Front oil seal and related components—5.7L engine

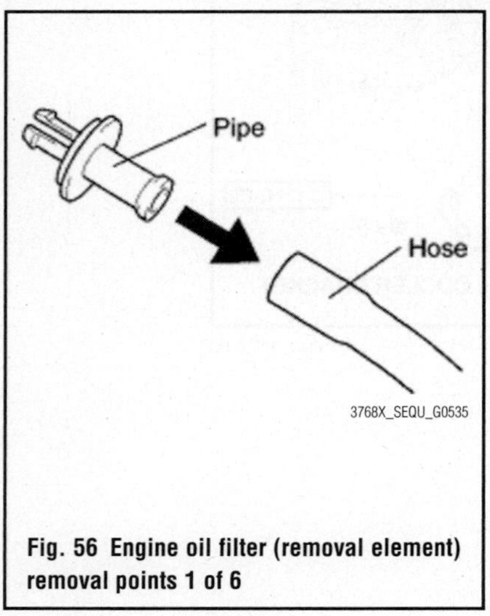

Fig. 56 Engine oil filter (removal element) removal points 1 of 6

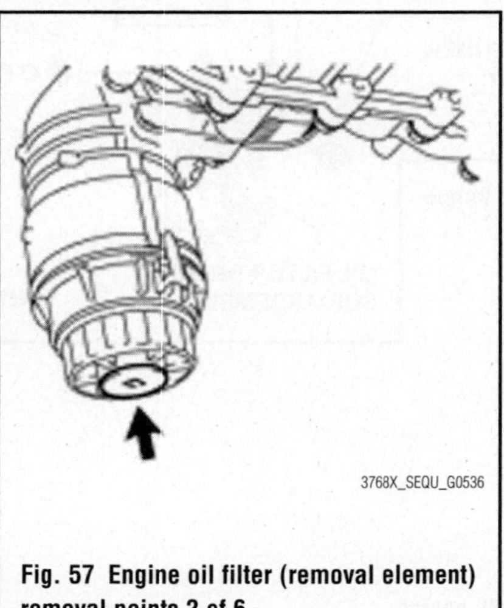

Fig. 57 Engine oil filter (removal element) removal points 2 of 6

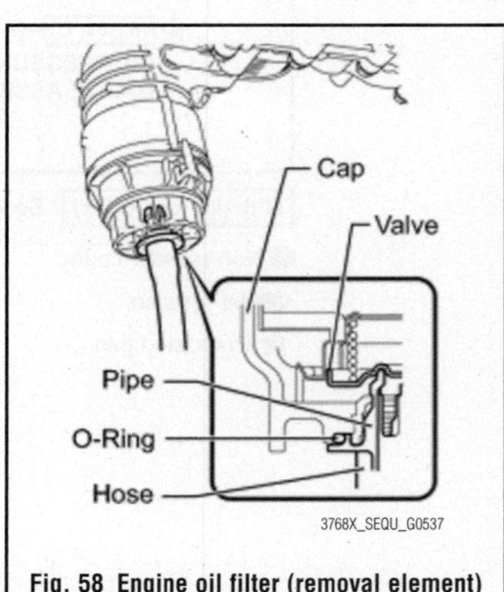

Fig. 58 Engine oil filter (removal element) removal points 3 of 6

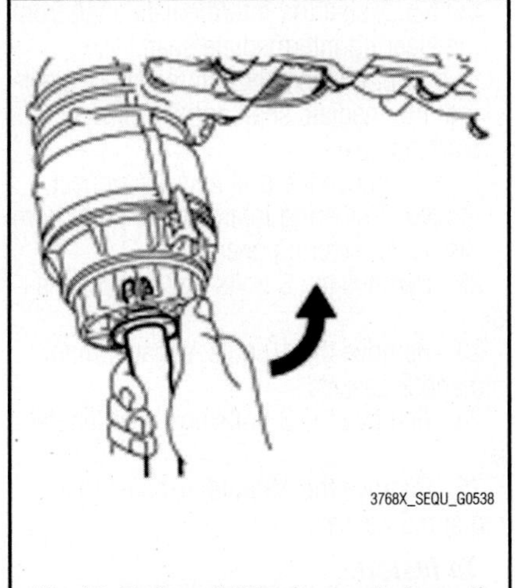

Fig. 59 Engine oil filter (removal element) removal points 4 of 6

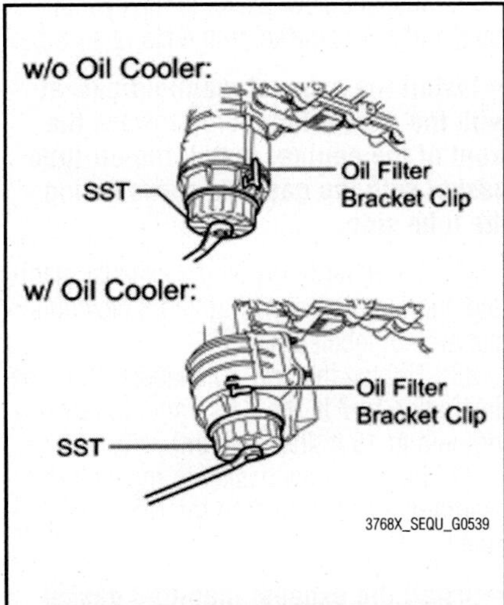

Fig. 60 Engine oil filter (removal element) removal points 5 of 6

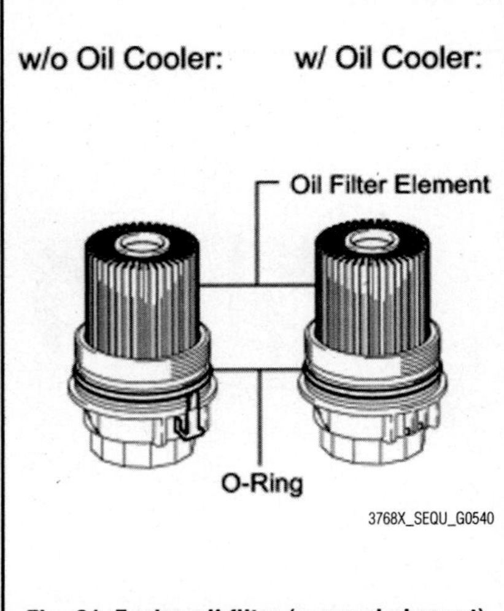

Fig. 61 Engine oil filter (removal element) removal points 6 of 6

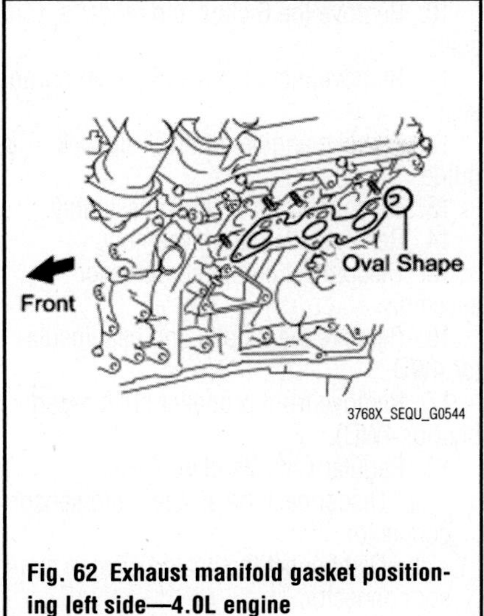

Fig. 62 Exhaust manifold gasket positioning left side—4.0L engine

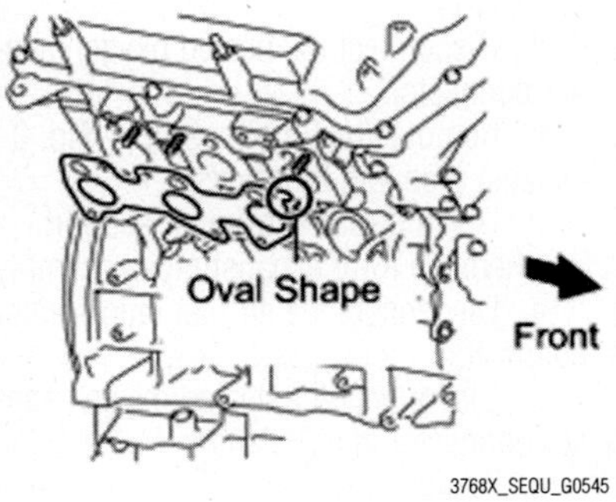

Fig. 63 Exhaust manifold gasket positioning right side—4.0L engine

To install:

➡Be sure to use new fasteners, as required.

10. Installation is the reverse of the removal procedure.
11. Use the Techstream diagnostic tool, or equivalent and reprogram the required systems.

4.6L Engine

See Figures 64 through 67.

1. Before servicing the vehicle, refer to the Precautions Section.
2. Disconnect the negative battery cable. Tape the cable with insulating tape.
3. Remove the front fender apron seals.
4. Remove the intake manifold assembly.
5. Remove the alternator.
6. Disconnect the compressor.

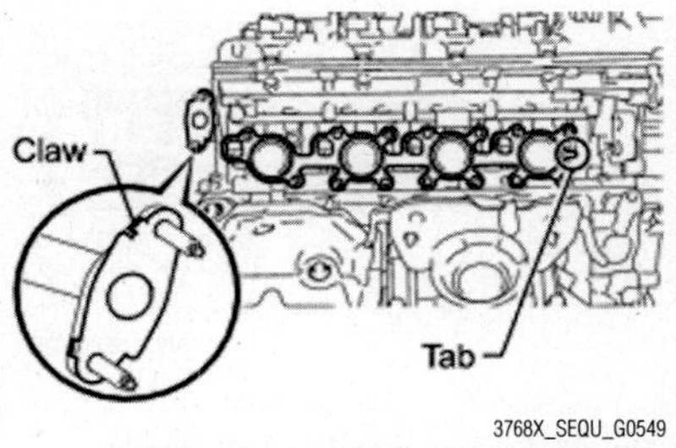

Fig. 64 Exhaust manifold gasket positioning left side—4.6L engine

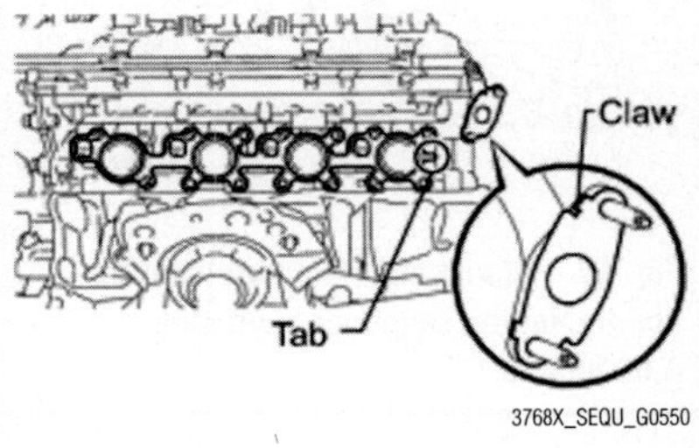

Fig. 65 Exhaust manifold gasket positioning right side—4.6L engine

➡It may be necessary to properly discharge the system and remove the compressor, if unable to position it to the side without disconnect the refrigerant lines. If disconnecting the lines be sure to plug them after disconnection.

7. Remove the engine oil dipstick guide.
8. Remove the exhaust pipe assembly.
9. Remove the front driveshaft, if equipped with 4WD.
10. Remove the No. 3 and No. 4 floor heat insulators.
11. Remove the EGR pipe.
12. Remove the exhaust manifold heat insulators.
13. Remove the manifold retaining nuts. Discard the nuts.
14. Remove the manifold from the engine. Discard the gasket.

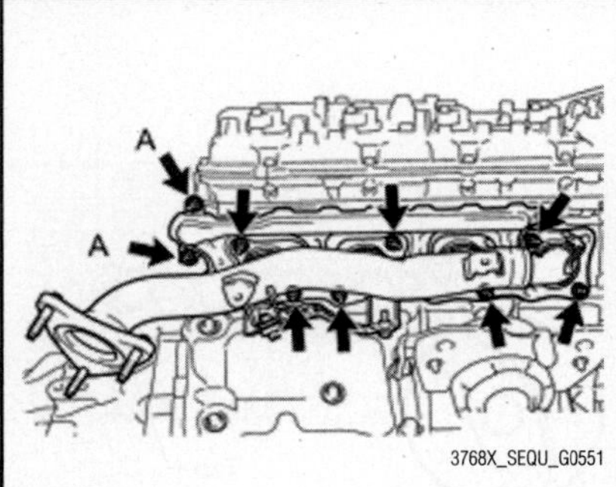

Fig. 66 Exhaust manifold bolt location and tightening sequence left side—4.6L engine

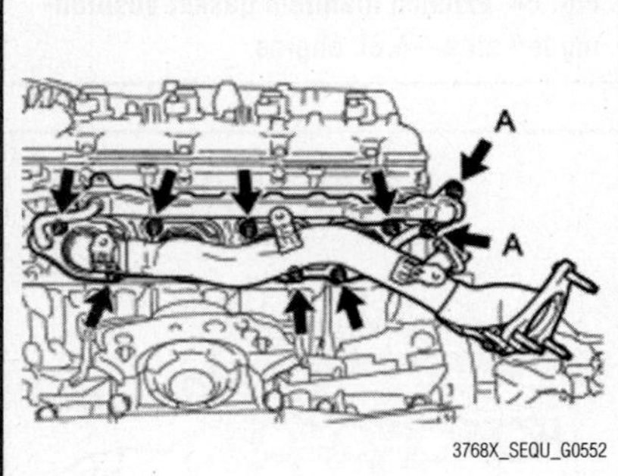

Fig. 67 Exhaust manifold bolt location and tightening sequence right side—4.6L engine

To install:

➡**Be sure to use new fasteners, as required.**

15. Installation is the reverse of the removal procedure.
16. Tighten nuts labeled A to 7 ft. lbs. (10 Nm). Tighten others to 15 ft. lbs. (21 Nm). See illustration.
17. Use the Techstream diagnostic tool, or equivalent and reprogram the required systems.

5.7L Engine

See Figures 68 through 71.

1. Before servicing the vehicle, refer to the Precautions Section.
2. Disconnect the negative battery cable. Tape the cable with insulating tape.
3. Remove No. 1 engine under cover.
4. Drain engine coolant.
5. Remove V-bank cover sub-assembly.
6. Remove air cleaner hose assembly.
7. Disconnect inlet and outlet radiator hose.
8. Remove fan and alternator v-belt.
9. Remove fan shroud.
10. Remove the 6 clips and fender apron seal.
11. Remove the 5 clips and fender apron seal.
12. Remove engine oil level dipstick guide.
13. Disconnect vane pump assembly.
14. Remove alternator assembly.
15. Disconnect cooler compressor assembly.
16. Remove the 2 bolts and heat insulator 4WD.
17. Remove front propeller shaft assembly (for 4WD).
18. Regular Cab Standard Deck:
    a. Disconnect the air fuel ratio sensor connector.
    b. Disconnect the heated oxygen sensor connector.
    c. Remove the 2 bolts, 5 nuts, No. 2 exhaust pipe and 2 gaskets.
19. Except Regular Cab Standard Deck:
    a. Disconnect the air fuel ratio sensor connector.
    b. Disconnect the heated oxygen sensor connector.
    c. Remove the 2 bolts, 3 nuts, No. 2 exhaust pipe and 2 gaskets.
    d. Disconnect the exhaust support.
20. Remove front exhaust pipe assembly.
    a. Disconnect the air fuel ratio sensor connector.
    b. Disconnect the heated oxygen sensor connector and 2 clamps.
    c. Remove the 2 bolts from the center exhaust pipe.
    d. Remove the 3 nuts, front exhaust pipe and 2 gaskets.
21. Put matchmarks on the No. 2 steering intermediate shaft and steering intermediate shaft.
    a. Remove the bolt and disconnect the No. 2 steering intermediate shaft from the steering intermediate shaft.
    b. Put matchmarks on the No. 2 steering intermediate shaft and the power steering gear.
    c. Remove the bolt and disconnect the No. 2 steering intermediate shaft from the power steering gear.
22. Remove the 3 bolts and heat insulator.
23. Remove the 10 nuts, exhaust manifold and 2 gaskets.
24. Remove the 3 bolts and heat insulator.
25. Remove the 10 nuts, exhaust manifold and 2 gaskets.

To install:

➡**Be sure to use new fasteners, as required.**

26. Install a new gasket to the cylinder head and a new gasket to the No. 2 air tube.

➡**Install the exhaust manifold gasket with the gasket tab facing toward the front of the engine. Install the air tube gasket with the gasket's claws facing the tube side.**

27. Temporarily install the exhaust manifold and then uniformly tighten 8 new nuts that are not labeled A.
28. Tighten the new nuts labeled A in the illustration to 7 ft. lbs. (10 Nm) and remaining nuts to 15 ft. lbs. (21 Nm).
29. Install a new gasket to the cylinder head and a new gasket to the No. 3 air tube.

➡**Install the exhaust manifold gasket with the gasket tab facing toward the front of the engine. Install the air tube gasket with the gasket's claws facing the tube side.**

Fig. 68 Exhaust manifold LH—5.7L engine

Fig. 69 Exhaust manifold tightening sequence RH—5.7L engine

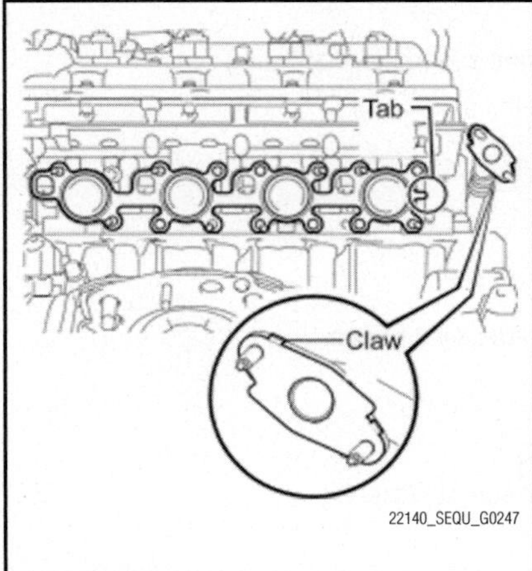

Fig. 70 Exhaust manifold gasket position LH—5.7L engine

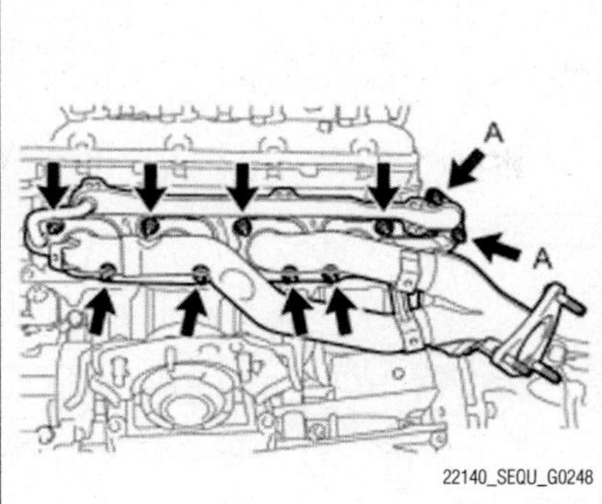

Fig. 71 Exhaust manifold tightening sequence LH—5.7L engine

30. Temporarily install the exhaust manifold and then uniformly tighten 8 new nuts that are not labeled A.
31. Tighten the new nuts labeled A in the illustration to 7 ft. lbs. (10 Nm) and remaining nuts to 15 ft. lbs. (21 Nm).
32. Install the heat insulator with the 3 bolts and tighten to 7 ft. lbs. (10 Nm).
33. Align the matchmarks and insert the No. 2 intermediate shaft into the intermediate shaft.
34. Align the matchmarks and insert the No. 2 intermediate shaft into the power steering gear.
35. Install the 2 bolt and tighten to 26 ft. lbs. (35 Nm).
36. Install front exhaust pipe assembly.
    a. Install a new gasket and the front exhaust pipe to the exhaust manifold RH with 3 new nuts and tighten to 40 ft. lbs. (54 Nm).
    b. Install a new gasket and the front exhaust pipe to the center exhaust pipe with the 2 bolts and tighten to 35 ft. lbs. (48 Nm).
    c. Connect the air fuel ratio sensor connector.
    d. Connect the heated oxygen sensor connector and 2 clamps.
37. Except Regular Cab Standard Deck:
    a. Connect the front No. 2 exhaust pipe to the exhaust support.
    b. Install a new gasket and the front No. 2 exhaust pipe to the exhaust manifold LH with 3 new nuts and tighten to 40 ft. lbs. (54 Nm).
    c. Install a new gasket and the front No. 2 exhaust pipe to the center exhaust pipe with the 2 bolts and tighten to 35 ft. lbs. (48 Nm).
    d. Connect the air fuel ratio sensor connector.
    e. Connect the heated oxygen sensor connector.
38. Regular Cab Standard Deck :
    a. Connect the front No. 2 exhaust pipe to the exhaust support.
    b. Install a new gasket and the front No. 2 exhaust pipe to the exhaust manifold LH with 3 new nuts and tighten to 40 ft. lbs. (54 Nm).
    c. Install a new gasket and the front No. 2 exhaust pipe to the center exhaust pipe with the 2 bolts and tighten to 35 ft. lbs. (48 Nm).
    d. Connect the air fuel ratio sensor connector.
    e. Connect the heated oxygen sensor connector.
39. Install front propeller shaft assembly (4WD).
40. Install the heat insulator with the 2 bolts (4WD).
41. Connect cooler compressor assembly.
42. Install alternator assembly.
43. Connect vane pump assembly.
44. Install engine oil level dipstick guide.
45. Install fan shroud.
46. Install fan and alternator v belt.
47. Connect outlet and inlet radiator hose.
48. Install air cleaner assembly.
49. Install air cleaner hose assembly.
50. Install front fender apron seal rear LH.
51. Install front fender apron seal LH.
52. Install front fender apron seal rear RH.
53. Install front fender apron seal RH.
54. Install No. 1 engine under cover.
55. Install V-bank cover sub-assembly.
56. Add engine coolant.
57. Inspect for exhaust gas leak.

➡If gas is leaking, tighten the areas necessary to stop the leak. Replace damaged parts as necessary.

58. Use the Techstream diagnostic tool, or equivalent and reprogram the required systems.

INTAKE MANIFOLD

REMOVAL & INSTALLATION

4.0L Engine

See Figures 72 and 73.

1. Before servicing the vehicle, refer to the Precautions Section.
2. Disconnect the negative battery cable. Tape the cable with insulating tape.
3. Properly relieve the fuel system pressure.
4. Remove the engine under cover.
5. Drain the engine coolant. Be sure to properly dispose of used coolant.
6. Remove the front wiper motor and link assembly.
7. Remove the cowl top outer panel assembly.
8. Remove the V- bank cover. Remove the air cleaner assembly.
9. Remove the intake air surge tank. Discard the gasket.
10. Remove the fuel delivery pipe sub assembly.
11. Remove the intake manifold retaining bolts. Remove the intake manifold from the engine.
12. Discard the gasket.

To install:

➡Be sure to use new fasteners, as required.

13. Installation is the reverse of the removal procedure.
14. Be sure to use a new gasket.
15. Tighten the bolts to specification.
16. Use the Techstream diagnostic tool, or equivalent and reprogram the required systems.

4.6L Engine

See Figure 74.

1. Before servicing the vehicle, refer to the Precautions Section.
2. Disconnect the negative battery cable. Tape the cable with insulating tape.
3. Properly discharge the fuel system pressure.
4. Remove the EGR valve assembly.
5. Remove the front wiper motor and link assembly.

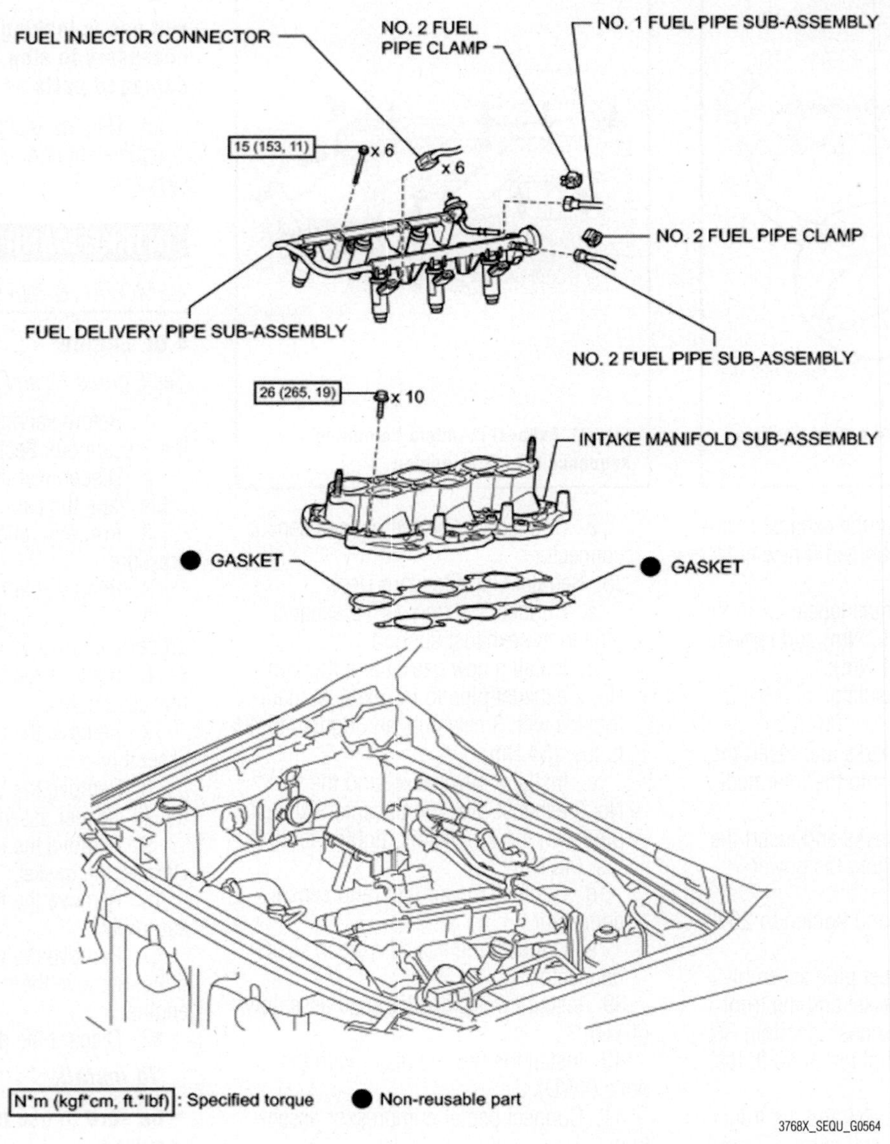

Fig. 72 Intake manifold and related components—4.0L engine

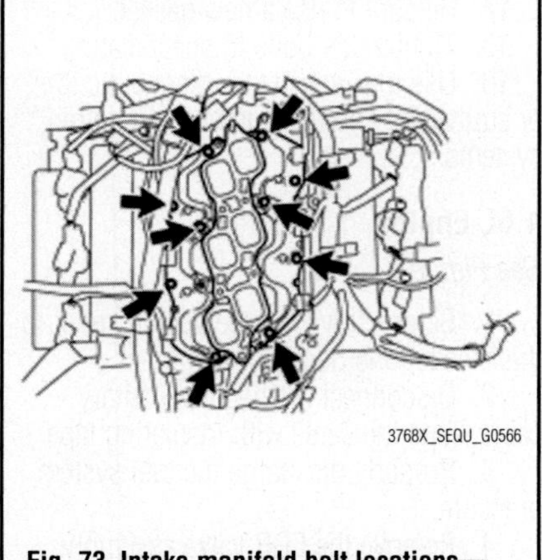

Fig. 73 Intake manifold bolt locations—4.0L engine

6. Remove the cowl top outer panel assembly.
7. Remove the ventilation hose.
8. Remove the air tube sub assembly, left side.
9. Disconnect the required electrical connectors.
10. Disconnect the required hoses.
11. Disconnect the fuel lines.
12. Remove the intake manifold retaining bolts.
13. Remove the intake manifold from its mounting.
14. Discard the gasket.

To install:

➡Be sure to use new fasteners, as required.

15. Installation is the reverse of the removal procedure.
16. Be sure to use a new gasket.
17. Use the Techstream diagnostic tool, or equivalent and reprogram the required systems.

5.7L Engine

See Figures 75 and 76.

1. Before servicing the vehicle, refer to the Precautions Section.
2. Disconnect the negative battery cable. Tape the cable with insulating tape.
3. Remove the front wiper motor and link.
4. Disconnect the 2 washer hoses.
5. Remove the 7 bolts and outer panel.
6. Remove No. 1 engine under cover.
7. Drain engine coolant.

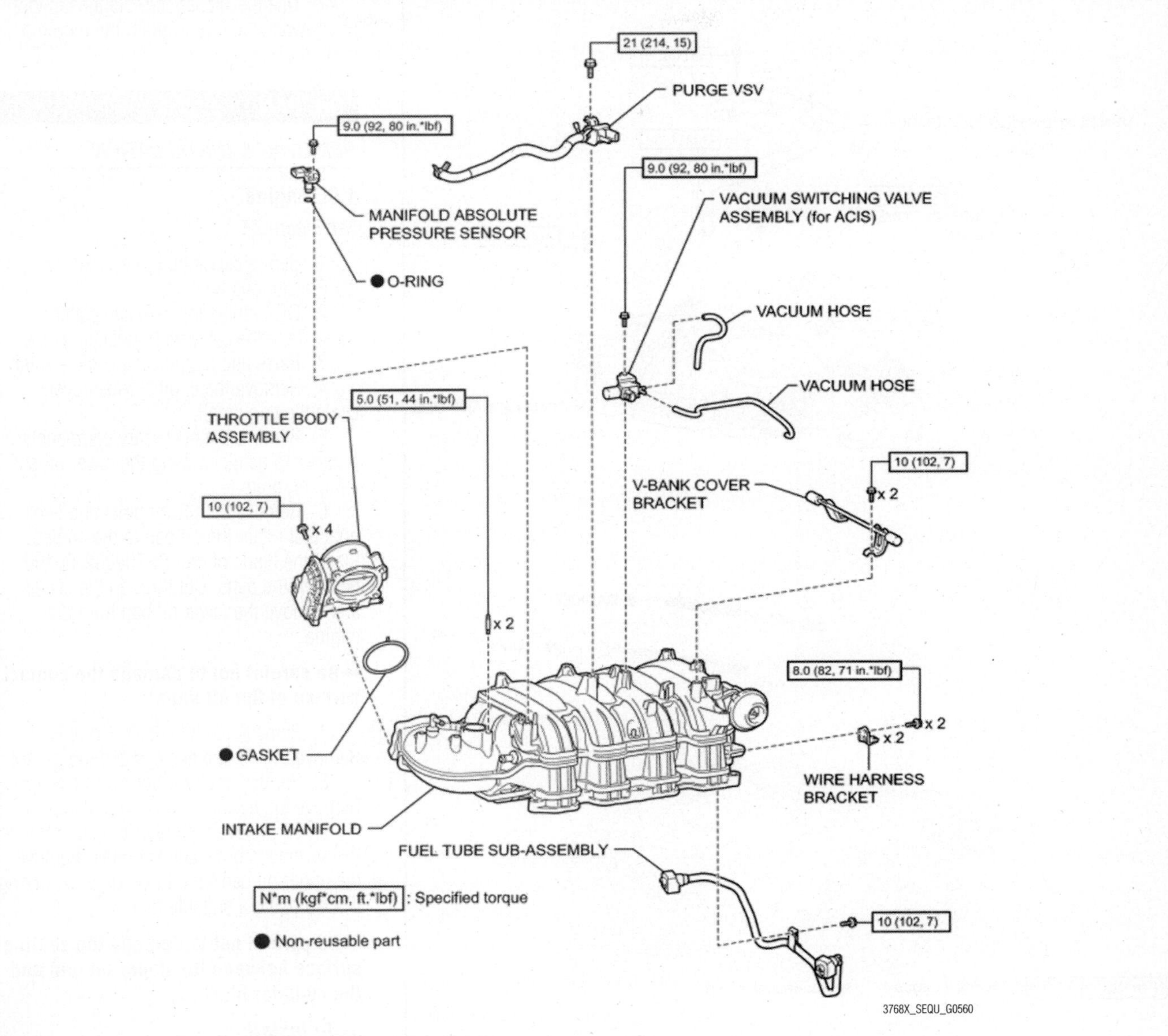

Fig. 74 Intake manifold and related components—4.6L engine

8. Remove V-bank cover sub-assembly.
9. Remove air cleaner hose assembly.
10. Disconnect the ventilation hose from the ventilation pipe of the cylinder head cover LH and RH.
11. Disconnect the 2 water by-pass hoses.
12. Disconnect the throttle body connector.
13. Disconnect the No. 1 ventilation hose.
14. Remove the No. 1 engine cover sub-assembly.
15. Remove the No. 3 engine cover.
16. Disconnect the purge VSV connector.
17. Disconnect the purge line hose from the purge VSV.
18. Disconnect the vacuum switching valve connector (for ACIS).
19. Remove the 2 bolts.

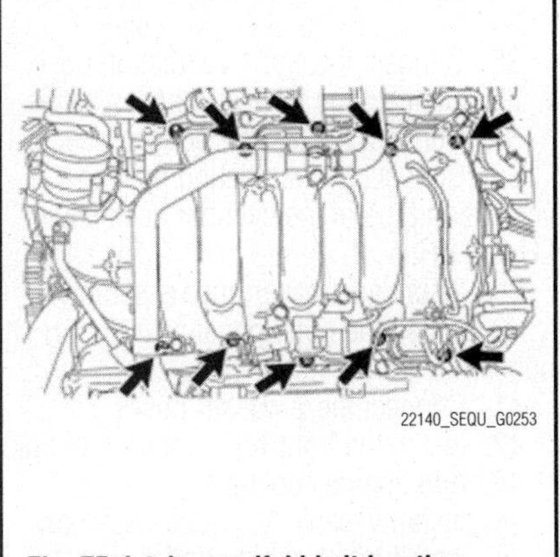

Fig. 75 Intake manifold bolt location—5.7L engine

20. Disconnect the No. 1 tube from the union to connector tube hose, and move the hose aside.
21. Disconnect the 3 wire clamps from the 3 wire brackets.
22. Remove the bolt and wire bracket from the intake manifold.
23. Remove the 2 nuts, 8 bolts, intake manifold and 2 gaskets.

To install:

➡Be sure to use new fasteners, as required.

24. Place 2 new gaskets on the intake manifold.
25. Place the intake manifold on the cylinder head.

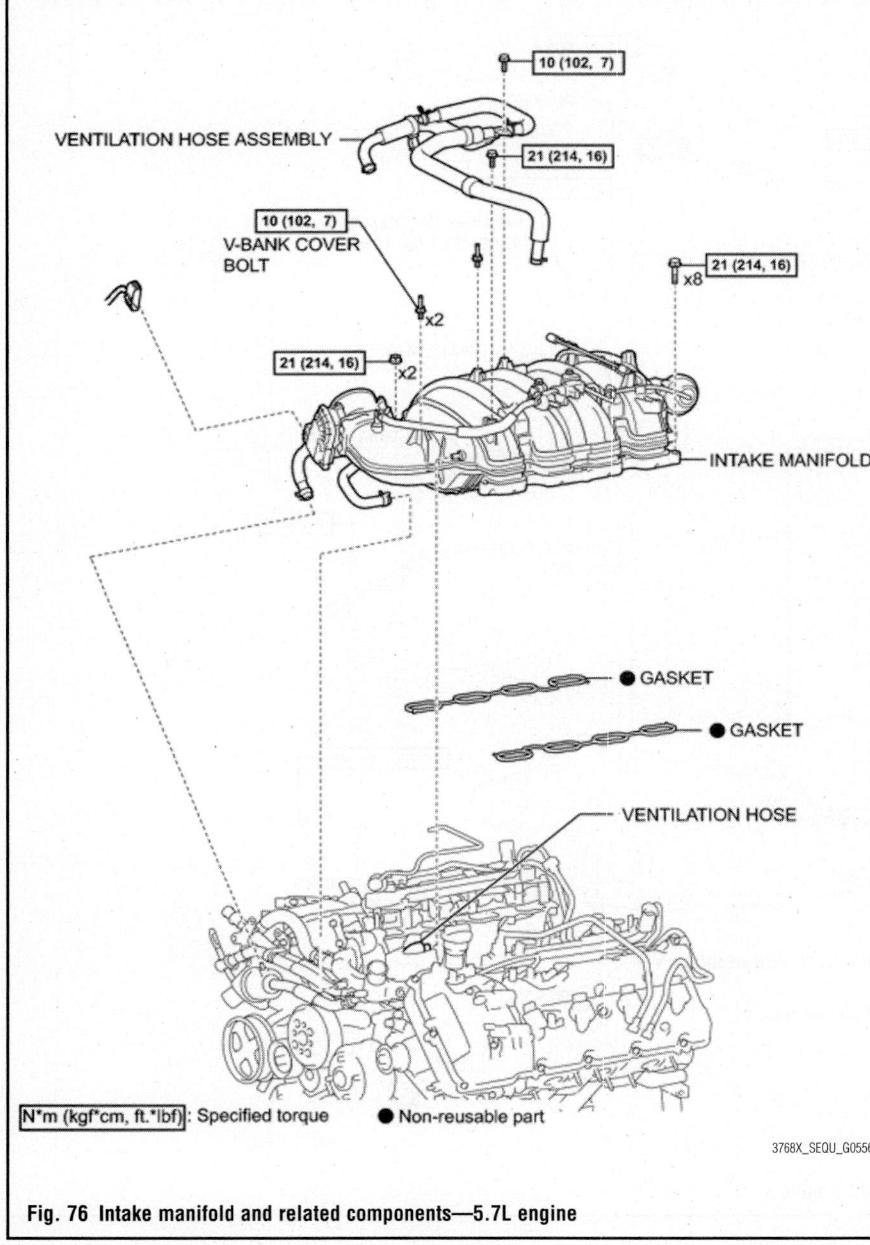

Fig. 76 Intake manifold and related components—5.7L engine

26. Install and uniformly tighten the 8 bolts and 2 nuts to 15 ft. lbs. (21 Nm) in several steps.
27. Install the wire bracket to the intake manifold with the bolt.
28. Connect the 3 wire clamps to the 3 wire brackets.
29. Connect the No. 1 tube and install it to the intake manifold with the 2 bolts.
30. Connect the purge VSV connector.
31. Connect the purge line hose to the purge VSV.
32. Connect the vacuum switching valve connector (for ACIS).
33. Install the No. 1 engine cover sub-assembly.
34. Install the No. 3 engine cover.
35. Connect the No. 1 ventilation hose.
36. Connect the 2 water by-pass hoses.
37. Connect the throttle body connector.
38. Connect the ventilation hose to the ventilation pipe of the cylinder head cover LH and RH.
39. Install air cleaner hose assembly.
40. Install the outer panel with the 7 bolts.
41. Connect the 2 washer hoses.
42. Install the front wiper motor and link.
43. Add engine coolant.
44. Install V-bank cover sub-assembly.
45. Install No. 1 engine under cover.
46. Inspect for coolant leak.
47. Use the Techstream diagnostic tool, or equivalent and reprogram the required systems.

OIL PAN

REMOVAL & INSTALLATION

4.0L Engine

See Figure 77.

1. Before servicing the vehicle, refer to the Precautions Section.
2. Disconnect the negative battery cable. Tape the cable with insulating tape.
3. Raise and support the vehicle safely.
4. Remove the engine undercover. Drain the engine oil.
5. Remove the necessary components in order to gain access to the lower oil pan retaining bolts.
6. Remove the fifteen bolts and two nuts that retain the oil pan to the engine. Insert the blade of tool SST09032-00100 between the pans. Cut through the sealer and remove the lower oil pan from the engine.

➡Be careful not to damage the contact surface of the oil pans.

7. Remove the two bolts and nuts. Remove the oil strainer. Discard the gasket.
8. Remove the four housing bolts. Remove the flywheel housing undercover.
9. To remove the upper oil pan, remove the seventeen bolts and two nuts. Remove the upper oil pan from the engine, by prying it apart using a suitable tool.

➡Be careful not to damage the sealing surface between the upper oil pan and the cylinder block.

To install:

➡Be sure to use new fasteners, as required.

10. Apply a continuous bead (0.12–0.16 inch in diameter) of seal packing, part number 08826-00080 or equivalent, to the sealing surface of the oil pan.

➡Remove any oil from the contact surface. Install the upper oil pan within three minutes of applying the seal packing. Tighten the pan bolts to specification within fifteen minutes after applying the seal packing. Do not start the engine for at least two hours after the installation of the oil pan.

11. Loosely install the upper oil pan bolts and nuts. Bolt "A" is 0.98 inch long, bolt "B" is 1.77 inch long and bolt "C" is 0.55 inch long. Uniformly tighten the 14mm

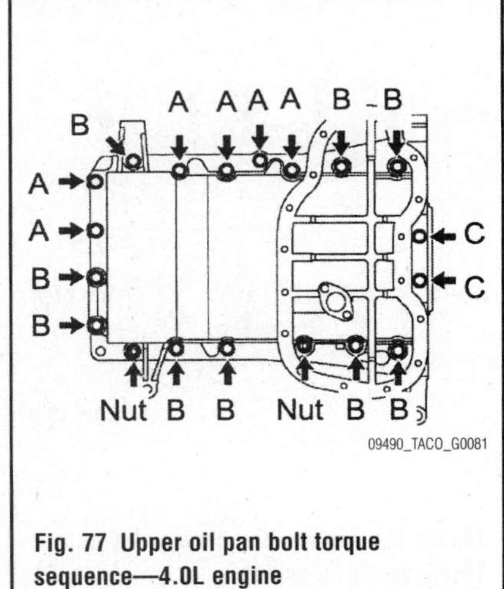

Fig. 77 Upper oil pan bolt torque sequence—4.0L engine

bolt to 7.0 ft. lbs, and the other bolts and nuts to 17 ft. lbs., in the proper sequence.

12. Install the oil strainer assembly. Torque the bolts to 80 inch lbs.

13. Apply a continuous bead (0.12–0.16 inch in diameter) of seal packing, part number 08826-00080 or equivalent, to the sealing surface of the oil pan.

➡Remove any oil from the contact surface. Install the lower oil pan within three minutes of applying the seal packing. Tighten the pan bolts to specification within fifteen minutes after applying the seal packing. Do not start the engine for at least two hours after the installation of the oil pan.

14. Loosely install the lower oil pan bolts and nuts. Uniformly tighten the bolts to 80 inch lbs, and the nuts to 7.0 ft. lbs., in several steps.

15. Continue the installation in the reverse order of the removal procedure.

16. Be sure to fill the engine with the proper grade and type engine coolant.

17. Be sure to fill the engine with the proper grade and type engine oil.

18. Start the engine and check for leaks. Correct as required.

19. Use the Techstream diagnostic tool, or equivalent and reprogram the required systems.

4.6L Engine

See Figures 78 and 79.

At this time the manufacturer provided service information for this component by first removing the engine from the vehicle and positioning it in a suitable holding fixture.

1. Before servicing the vehicle, refer to the Precautions Section.

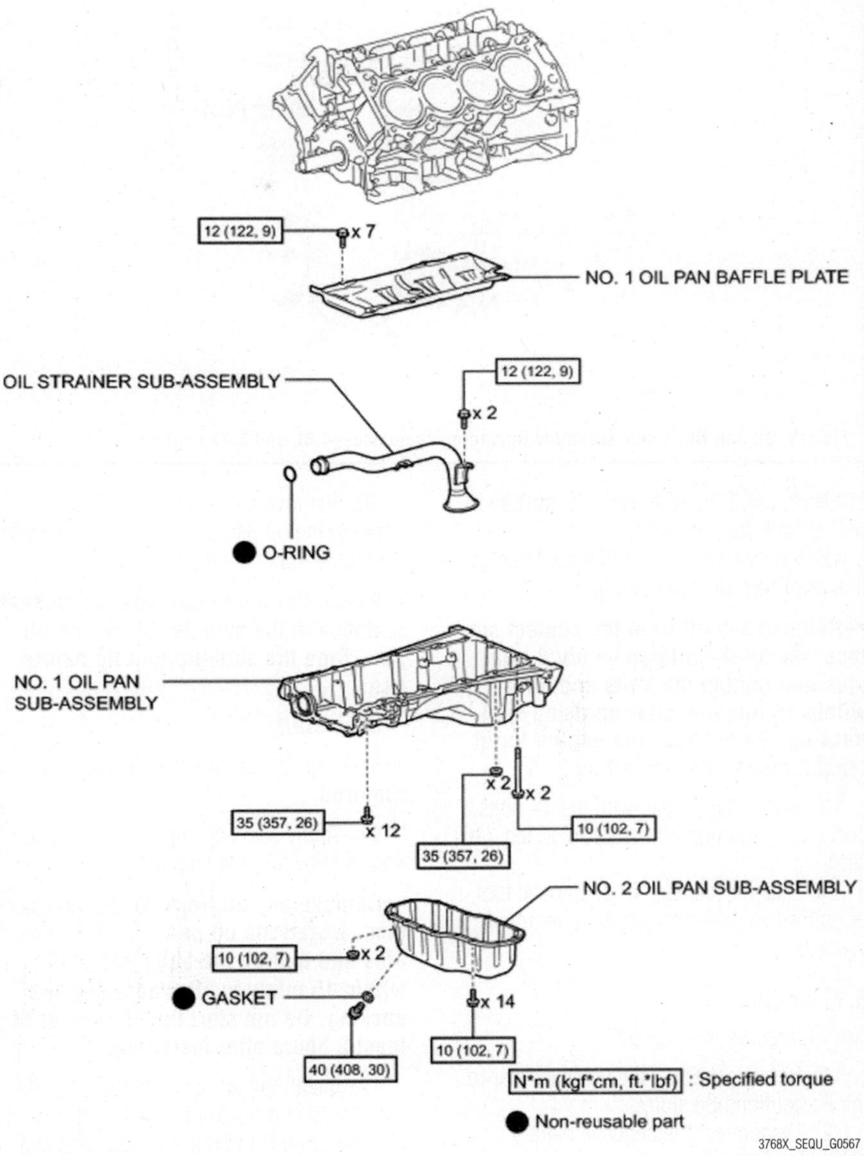

Fig. 78 Oil pan and related components—4.6L engine

2. Disconnect the negative battery cable. Tape the cable with insulating tape.

3. Remove the engine from the vehicle and position it in a suitable holding fixture.

4. Remove the No. 2 pan attaching bolts.

5. Remove the No. 2 pan from the engine.

6. Discard the gasket.

7. Remove the No. 1 pan attaching bolts.

8. Remove the No. 1 pan from the engine.

9. Discard the gasket.

To install:

➡Be sure to use new fasteners, as required.

10. Installation is the reverse of the removal procedure.

➡Remove any oil from the contact surface. Install the oil pan within 3 minutes and tighten the bolts and nuts within 15 minutes after applying seal packing. Do not start the engine for at least 2 hours after installing.

11. Install the oil pan with the 14 bolts and 2 nuts and tighten bolt A to 7 ft. lbs.

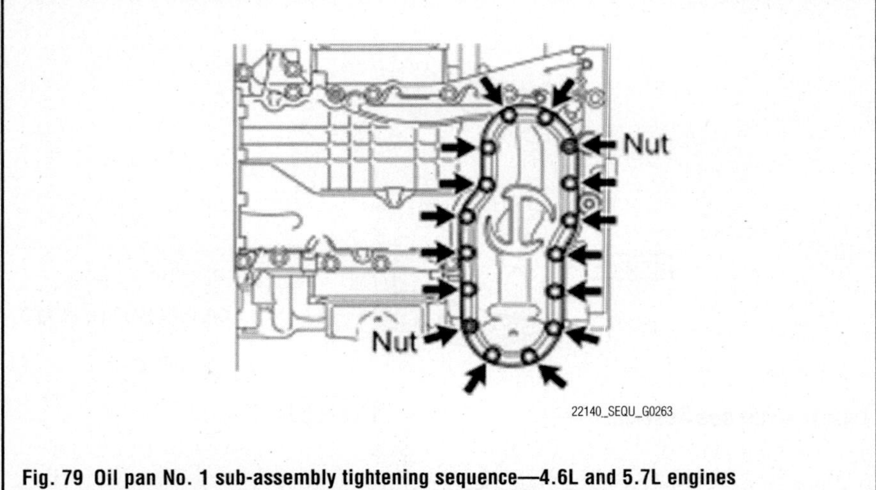

Fig. 79 Oil pan No. 1 sub-assembly tightening sequence—4.6L and 5.7L engines

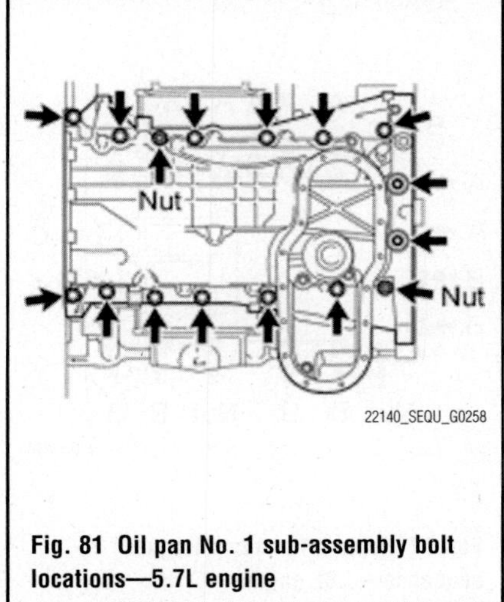

Fig. 81 Oil pan No. 1 sub-assembly bolt locations—5.7L engine

(10 Nm), bolt B to 26 ft. lbs. (35 Nm) and nut to 26 ft. lbs. (35 Nm).

12. Apply seal packing in a continuous line as shown in the illustration.

➡Remove any oil from the contact surface. Install the oil pan within 3 minutes and tighten the bolts and nuts within 15 minutes after applying seal packing. Do not start the engine for at least 2 hours after installing.

13. Install the oil pan with the 14 bolts and 2 nuts and tighten bolts to 7 ft. lbs. (10 Nm).

14. Use the Techstream diagnostic tool, or equivalent and reprogram the required systems.

5.7L Engine

See Figures 80 through 81

1. Before servicing the vehicle, refer to the Precautions Section.
2. Disconnect the negative battery cable. Tape the cable with insulating tape.
3. Disconnect the negative battery cable.
4. Raise and support the vehicle safely.
5. Remove the engine undercover. Drain the engine oil.
6. Remove the necessary components in order to gain access to the lower oil pan retaining bolts.
7. Remove the 14 bolts and 2 nuts.
8. Insert the blade of SST between the oil pans. Cut through the applied sealer and remove the No. 2 oil pan.

➡Be careful not to damage the contact surfaces of the oil pans.

9. Remove the 14 bolts and 2 nuts.

➡Be sure to clean the bolts and stud bolts, and check the threads for cracks or other damage.

10. Remove the oil pan by prying between the oil pan and cylinder block with a suitable tool.

➡Be careful not to damage the contact surfaces of the cylinder block and oil pan. Tape the suitable tool tip before use.

To install:

➡Be sure to use new fasteners, as required.

11. Apply seal packing in a continuous line as shown in the illustration.

➡Remove any oil from the contact surface. Install the oil pan within 3 minutes and tighten the bolts and nuts within 15 minutes after applying seal packing. Do not start the engine for at least 2 hours after installing.

12. Install the oil pan with the 14 bolts and 2 nuts and tighten bolt A to 7 ft. lbs. (10 Nm), bolt B to 26 ft. lbs. (35 Nm) and nut to 26 ft. lbs. (35 Nm).
13. Apply seal packing in a continuous line as shown in the illustration.

➡Remove any oil from the contact surface. Install the oil pan within 3 minutes and tighten the bolts and nuts within 15 minutes after applying seal packing. Do not start the engine for at least 2 hours after installing.

14. Install the oil pan with the 14 bolts and 2 nuts and tighten bolts to 7 ft. lbs. (10 Nm).
15. Continue the installation in the reverse order of the removal procedure.
16. Be sure to fill the engine with the proper grade and type engine coolant.
17. Be sure to fill the engine with the proper grade and type engine oil.
18. Start the engine and check for leaks. Correct as required.
19. Use the Techstream diagnostic tool, or equivalent and reprogram the required systems.

PISTON AND RING

POSITIONING

See Figures 82 through 84.

REAR MAIN SEAL

REMOVAL & INSTALLATION

See Figures 85 through 87.

1. Before servicing the vehicle, refer to the Precautions Section.

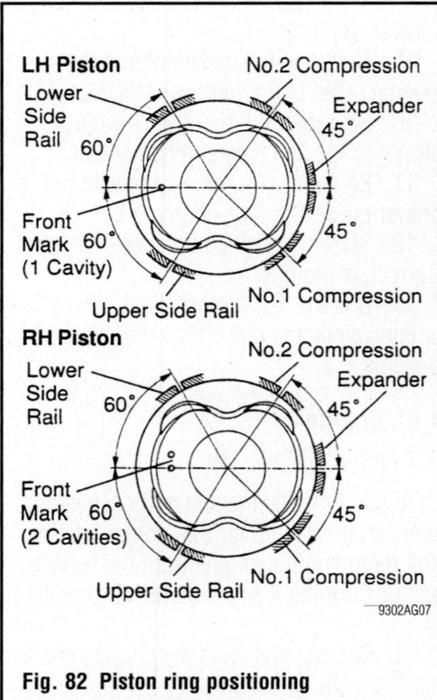

Fig. 82 Piston ring positioning

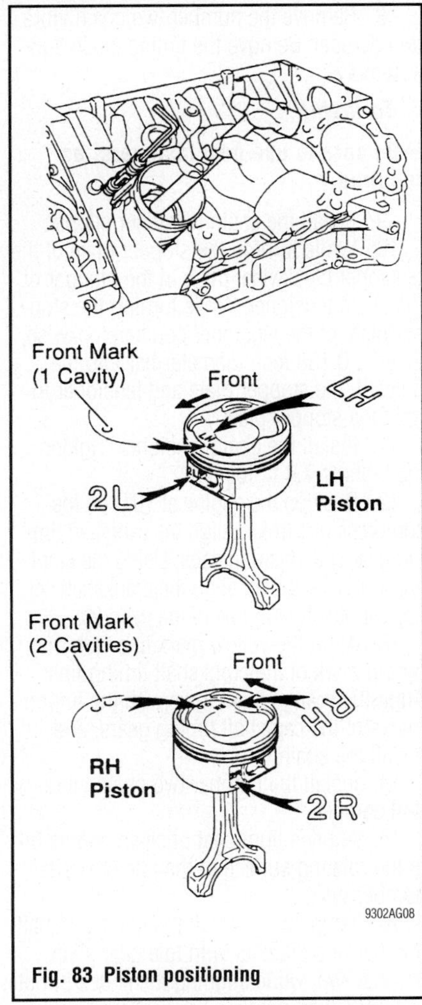

Fig. 83 Piston positioning

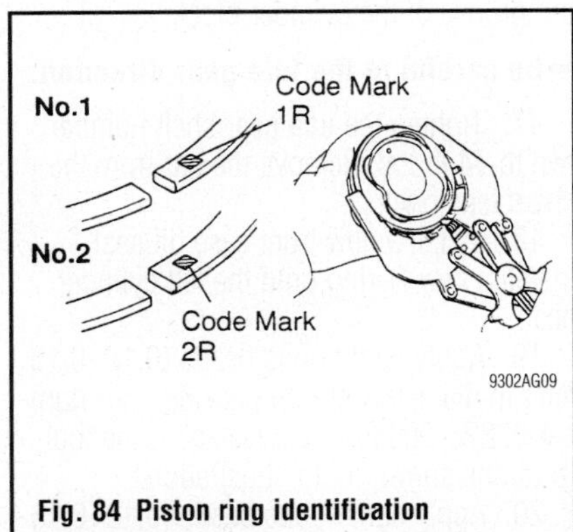

Fig. 84 Piston ring identification

2. Disconnect the negative battery cable. Tape the cable with insulating tape.
3. Remove the transmission and flywheel from the vehicle.
4. Cut off the rubber lip portion of the seal with a sharp knife.
5. Pry out the oil seal.

To install:

➡Be sure to use new fasteners, as required.

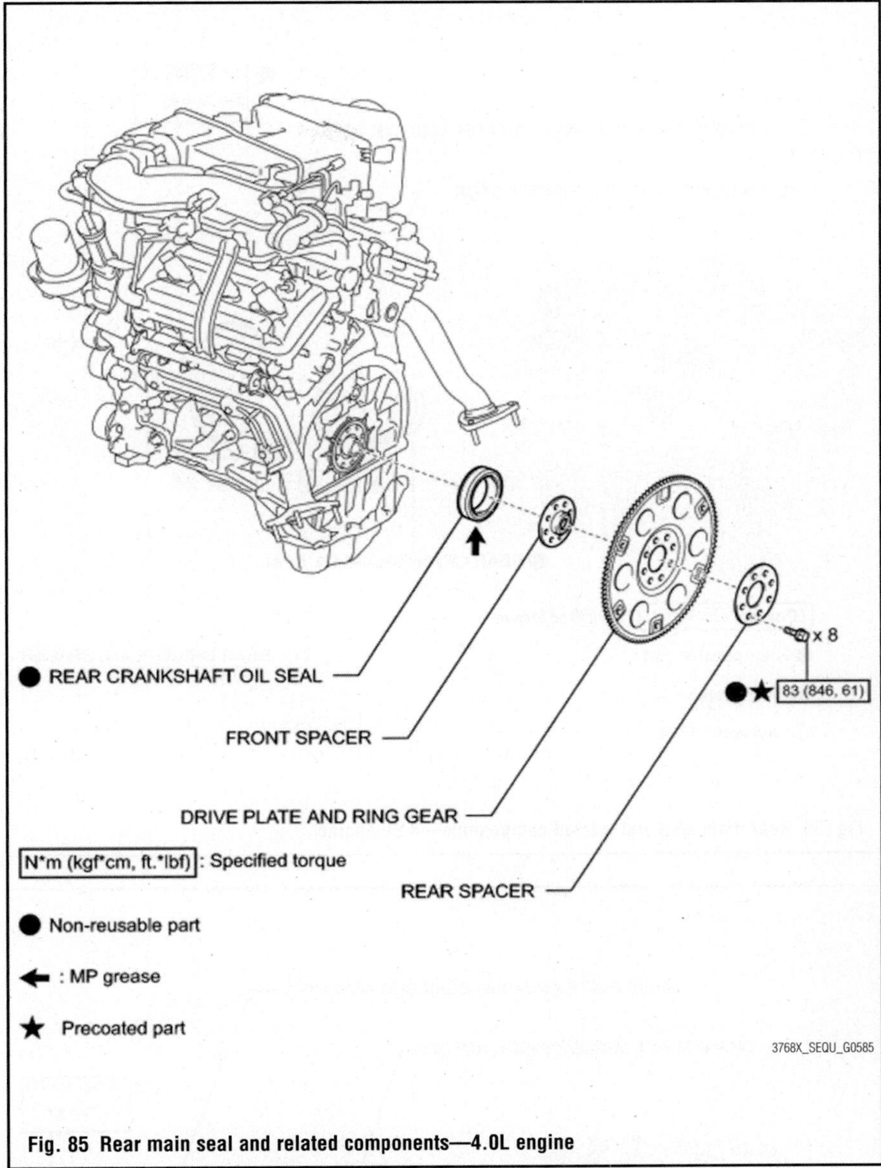

Fig. 85 Rear main seal and related components—4.0L engine

6. Installation is the reverse of the removal procedure.
7. Use the Techstream diagnostic tool, or equivalent and reprogram the required systems.

TIMING CHAIN & SPROCKETS

REMOVAL & INSTALLATION

4.0L Engine

See Figures 88 and 89.

1. Before servicing the vehicle, refer to the Precautions Section.
2. Disconnect the negative battery cable. Tape the cable with insulating tape.
3. Remove the front cover.
4. Using the crankshaft pulley set bolt, turn the crankshaft to align the crankshaft set key with the timing line of the cylinder block. If not aligned at TDC of the compression stroke, turn the crankshaft one complete revolution, in the direction of rotation
5. While turning the stopper plate of the tensioner upward, push the plunger of the chain tensioner. While turning the stopper plate of the tensioner downward, insert a 0.138 inch diameter bar into the holes in the stopper plate and tensioner to hold the stopper plate.
6. Remove the two bolts, and then remove the chain tensioner.
7. Remove the chain tensioner slipper. Remove the idle gear shaft number two, idle gear number one and idle gear shaft number one.

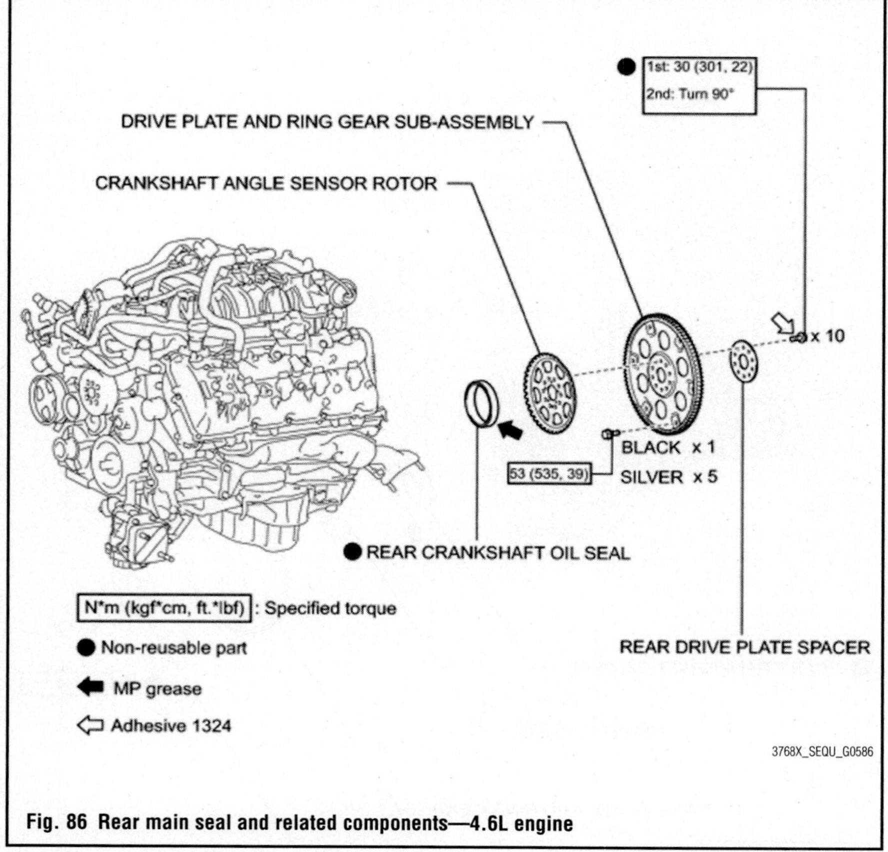

Fig. 86 Rear main seal and related components—4.6L engine

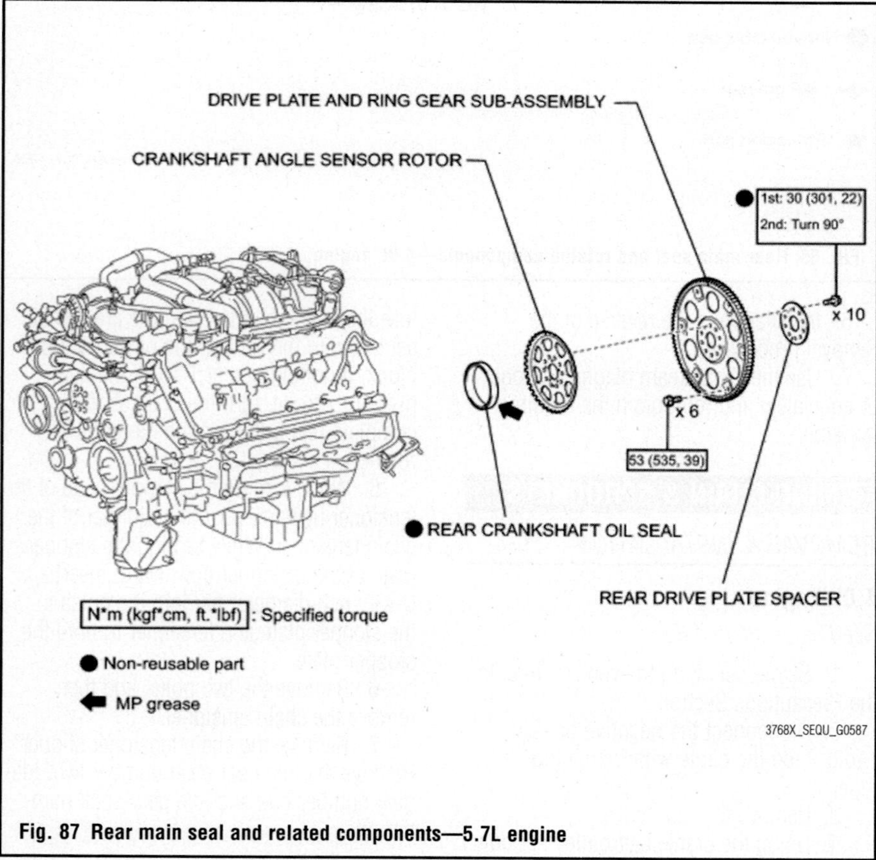

Fig. 87 Rear main seal and related components—5.7L engine

8. Remove the number two chain vibration damper. Remove the timing chain subassembly.

To install:

➡Be sure to use new fasteners, as required.

9. Install the chain tensioner slipper.
10. While turning the stopper plate of the tensioner clockwise, push in the plunger of the chain tensioner. While turning the stopper plate of the tensioner counterclockwise, insert a 0.138 inch diameter bar into the holes in the stopper plate and tensioner to hold the stopper plate.
11. Install the chain tensioner. Tighten the bolts to 7.4 ft. lbs.
12. Position the engine at TDC on the compression stroke. Align the camshaft timing gears and bearing caps. Using the crankshaft pulley set bolt, align the crankshaft set key with the timing line of the cylinder.
13. Align the yellow mark line with the timing mark of the crankshaft timing link. Align the orange mark links with the timing marks of the camshaft timing gears, and install the chain.
14. Install the number two chain vibration damper.
15. Apply a light coat of clean engine oil to the rotating surface of the idle gear shaft number one.
16. Temporarily install the idle gear shaft number one together with idle gear shaft number two, while aligning the knock pin of idle gear shaft number one with the knock pin groove of the cylinder block.

➡Be careful of the idle gear direction.

17. Tighten the idle gear shaft number two to 44 ft. lbs. Remove the bar from the chain tensioner.
18. Install a new front case oil seal. Install a new O-ring onto the left cylinder head.
19. Apply continuous beads (0.12–0.16 inch in diameter) of seal packing, part number 08826-00080 or equivalent to the four locations shown in the illustration.
20. Apply continuous beads (0.12–0.16 inch in diameter) of seal packing, part number 08826-00080 or equivalent to all parts except the water pump part: for the water pump part use, part number 08826-00080 or equivalent, to the timing chain cover. Do not apply seal packing to portion "A".

➡Remove any oil from the contact surfaces. Install the timing chain case cover within three minutes and tighten the bolts within fifteen minutes of applying the seal packing.

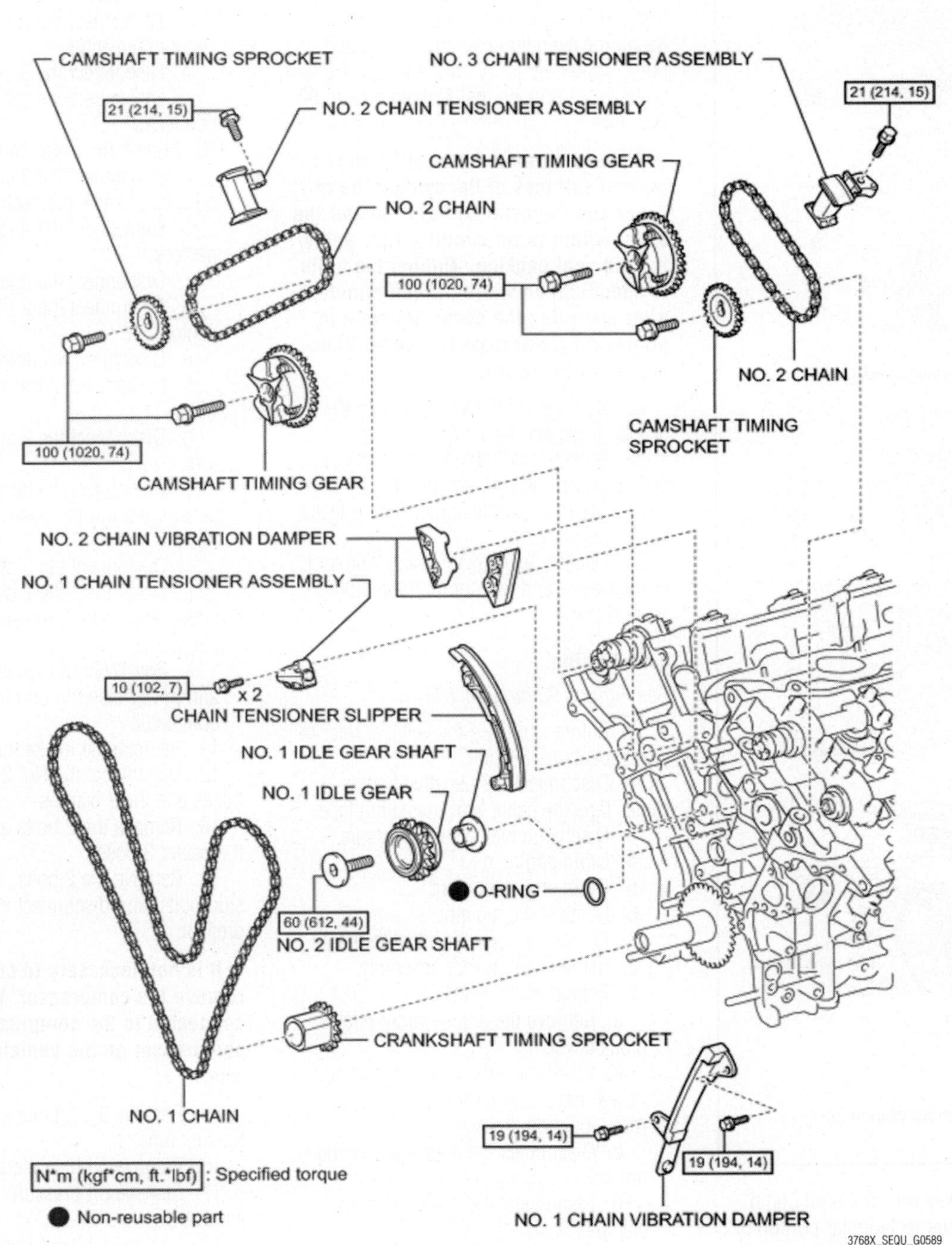

Fig. 88 Timing chain and related components—4.0L engine

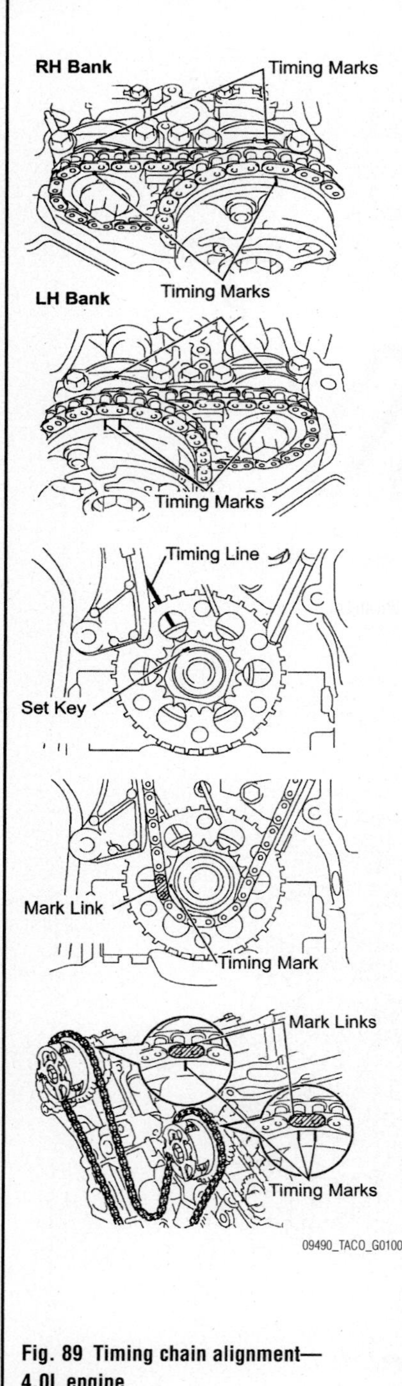

Fig. 89 Timing chain alignment—4.0L engine

21. Align the key way of the oil pump drive motor with the rectangular portion of the crankshaft timing gear and slide the timing chain case cover into place.

22. Install the timing chain case cover bolts. Tighten the bolts and nuts uniformly in several steps to 17 ft. lbs.

➡Do not wrap the chain and slipper over the timing chain case cover seal line.

23. Continue the installation in the reverse order of the removal procedure.

24. When installing the cylinder head cover apply a continuous bead (0.08–0.12 inch) of seal packing, part number 08826-00080 or equivalent, to the cylinder head. Install the seal washers onto the bolts. Install the cylinder head cover bolts and nuts. Tighten bolts "A" to 7.4 ft. lbs. Tighten bolts "B" to 80 inch lbs. Tighten nuts to 80 inch lbs.

➡Be sure to remove any oil from the contact surfaces of the cylinder head cover and the cylinder head. Install the cover within three minutes after applying the seal packing. Tighten the bolts to specification within fifteen minutes after installing the cover. Do not add engine oil for at least two hours after installing the cover.

25. Be sure to fill the engine with the proper grade and type engine coolant.

26. Be sure to fill the engine with the proper grade and type engine oil.

27. Start the engine and check for leaks, correct as required

28. Use the Techstream diagnostic tool, or equivalent and reprogram the required systems.

4.6L Engine

See Figures 90 through 108.

1. Before servicing the vehicle, refer to the Precautions Section.
2. Disconnect the negative battery cable. Tape the cable with insulating tape.
3. Discharge fuel system pressure.
4. Drain engine oil.
5. Drain engine coolant.
6. Remove the radiator.
7. Remove air cleaner hose assembly.
8. Remove air cleaner assembly.
9. Engine Room LH Side:
   a. Remove the engine room relay block cover.
   b. Disconnect the 2 connectors and detach the 2 clamps from the engine room junction block.
   c. Disconnect the 4 air injection control driver connectors.
   d. Disconnect the 2 wire harness clamps.
   e. Disconnect the injector connector.
   f. Disconnect the 4 ignition coil connectors.
   g. Disconnect the 2 VVT sensor connectors.
   h. Disconnect the 4 clamps.
   i. Remove the 2 bolts and ground wire.
   j. Disconnect the noise filter connector.
   k. Disconnect the engine coolant temperature sensor connector.
   l. Disconnect the 2 camshaft timing oil control valve connectors.
   m. Disconnect the camshaft position sensor connector.
   n. Disconnect the 3 clamps.
   o. Disconnect the cooler compressor connector.
10. Engine Room RH Side.
    a. Disconnect the 2 camshaft timing oil control valve connectors.
    b. Disconnect the 4 ignition coil connectors.
    c. Disconnect the injector connector.
    d. Disconnect the 2 VVT sensor connectors.
    e. Disconnect the noise filter connector.
    f. Remove the 2 bolts and ground wire.
    g. Disconnect the 2 air pump connectors.
    h. Disconnect the throttle position sensor and throttle control motor connector.
    i. Disconnect the 5 clamps.
    j. Disconnect the 2 clamps and power steering oil pressure switch connector.
    k. For 4WD: Disconnect the 2 clamps and power steering oil pressure switch connector.
11. Remove the intake manifold.
12. Disconnect the No. 2 and No. 3 air hoses and the 2 clamps.
13. Remove the 2 bolts and disconnect the heater 3 hoses.
14. Remove the 2 bolts, 2 nuts and 2 stud bolts, and disconnect the cooler compressor.

➡It is not necessary to completely remove the compressor. With the hoses connected to the compressor, hang the compressor on the vehicle body with a rope.

15. Remove the 2 bolts and disconnect the fuel tube.
16. Remove oil filter element.
17. Remove oil pressure sender gauge assembly.
18. Disconnect the 4 hoses and remove the water by-pass pipe.
19. Remove No. 1 oil cooler bracket (w/ oil cooler).
20. Remove the 2 bolts, 2 nuts and oil filter bracket.
21. Remove the 2 O-rings.
22. Remove engine oil level dipstick guide.

23. Remove the 2 bolts and disconnect the vane pump.
24. Disconnect the alternator connector.
25. Remove the terminal cap and nut, and disconnect the alternator wire.
26. Remove the bolt and disconnect the wire harness bracket from the alternator.
27. Remove the 2 bolts, 2 nuts and alternator.
28. Remove the 2 stud bolts.
29. Remove the No. 1 water by-pass hose by disconnect the hose from the water inlet housing and front water by-pass joint.
30. Remove the bolt and disconnect the air tube.
31. Remove the 2 bolts and hoses and water by-pass pipe.
32. Disconnect the No. 2 water by-pass hose from the water by-pass joint.
33. Remove the 4 nuts, water by-pass joint and 2 gaskets.
34. Remove No.1 and No. 2 engine covers .
35. Remove the bolt, disconnect the 2 hoses, and remove the air tube.
36. Remove the 3 bolts, water inlet housing and gasket.
37. Remove the 4 bolts and water pump pulley.
38. Remove the bolt and idler pulley.
39. Remove the 4 bolts and fan bracket.
40. Remove the standard bolt, 6 mm hexagon wrench bolt and belt tensioner.
41. Remove the 8 bolts and 8 ignition coils.
42. Remove cylinder head cover sub-assembly LH.
   a. Remove the 14 bolts, seal washer, cylinder head cover and gasket.
   b. Remove the 5 gaskets from the camshaft bearing caps (No. 2, No. 3).
43. Remove cylinder head cover sub-assembly RH.
   a. Remove the bolt and noise filter.
   b. Remove the 14 bolts, seal washer, cylinder head cover and gasket.
   c. Remove the 5 gaskets from the camshaft bearing caps (No. 1, No. 3).
44. Remove crankshaft pulley.
   a. Loosen the crankshaft pulley set bolt.
   b. Partially install the pulley set bolt to the crankshaft until 2 or 3 threads are engaged.
   c. Remove the crankshaft pulley.
45. Remove the bolt and disconnect the wire harness bracket.
46. Remove timing chain cover sub-assembly.
   a. Remove the 28 bolts and nut shown.
   b. Remove the timing chain cover by prying between it and the cylinder head and cylinder block with a screwdriver as shown

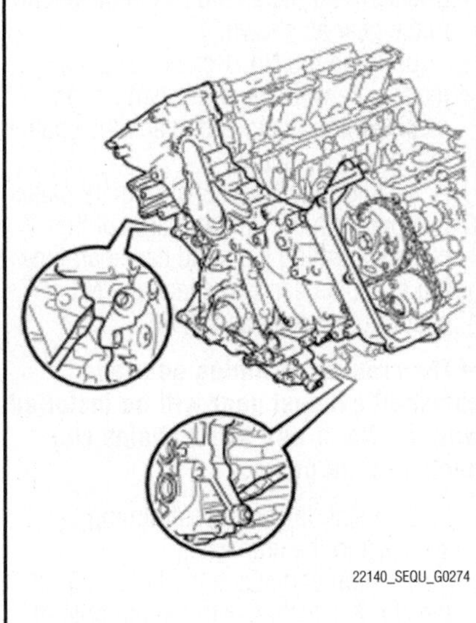

Fig. 90 Timing chain cover pry locations—4.6L engine

47. Remove the oil pump gasket from the cylinder block.
48. Remove the O-ring from the oil pan.
49. Remove the water inlet pipe.
50. Remove the 2 O-rings from the water inlet pipe.
51. Set No. 1 cylinder to TDC / compression.
52. Remove chain tensioner assembly.
53. Remove chain tensioner slipper.
54. Remove chain vibration damper.
55. Remove chain sub-assembly.
56. Remove chain tensioner assembly.

To install:

➡Be sure to use new fasteners, as required.

57. Check No. 1 cylinder to TDC.
58. Install No.2 chain tensioner assembly. While raising up the No. 2 chain tensioner, insert a pin of ϕ1.0 mm (0.0394 in.) into the hole to fix it in place.
59. Install No. 1 chain sub-assembly RH.
   a. Align the No. 1 chain's orange mark plates with the camshaft timing gear's timing mark, and attach the chain to the gear.
   b. Align the No. 1 chain's orange mark plate with the crankshaft timing gear's timing mark, and attach the chain to the gear.
   c. Align the No. 2 chain's mark plates (yellow) with the timing marks of the camshaft timing gear and camshaft timing exhaust gear, and attach the No. 2 chain to the gears as shown.

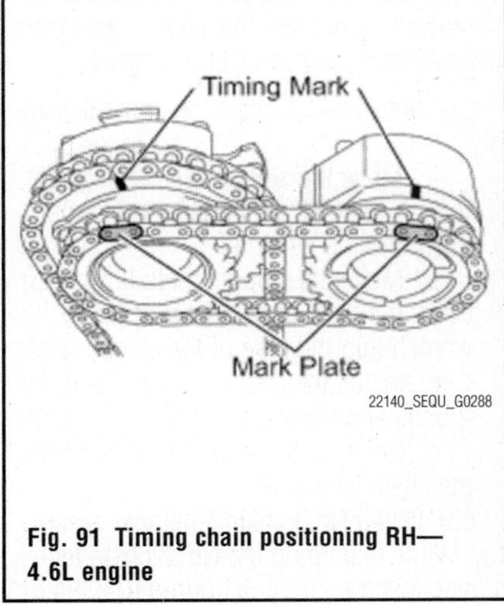

Fig. 91 Timing chain positioning RH—4.6L engine

➡The crankshaft timing gear and camshaft exhaust gear will be installed with the No. 1 and No. 2 chains connected to the gears.

   d. Install the crankshaft timing sprocket to the crankshaft.
   e. Align and attach the knock pin of the No. 1 camshaft with the pin hole of the camshaft timing gear.
   f. Using the hexagonal portion of the No. 2 camshaft, align and attach the knock pin of the No. 2 camshaft with the pin hole of the camshaft timing exhaust gear.
   g. Remove the pin from the No. 2 chain tensioner.
60. Install the vibration damper with the 2 bolts.
61. Install No. 1 chain tensioner slipper.

➡If you cannot install the chain tensioner slipper due to the tension of the

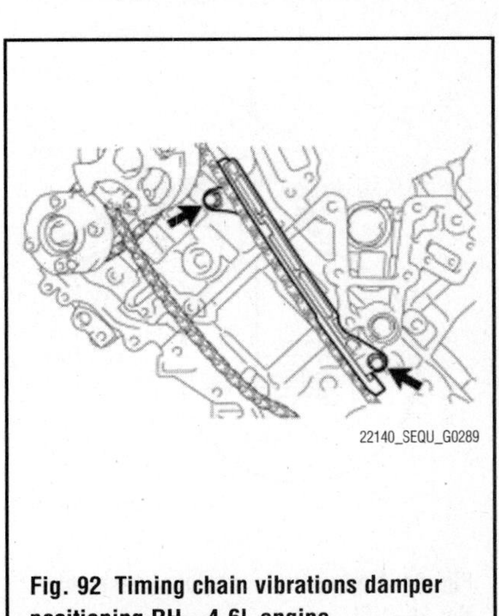

Fig. 92 Timing chain vibrations damper positioning RH—4.6L engine

chain, use the hexagonal portion of the camshaft to loosen the chain, and then install the chain tensioner slipper

62. Install No. 1 chain tensioner assembly.

a. Move the stopper plate upward to release the lock, and push the plunger deep into the tensioner.

b. Move the stopper plate downward to set the lock, and insert a hexagon wrench into the hole of the stopper plate.

c. Install the chain tensioner with the 2 bolts and tighten to 7 ft. lbs. (10 Nm).

d. Remove the hexagon wrench from the chain tensioner.

63. Install No.3 chain tensioner assembly. While raising up the No. 3 chain tensioner, insert a pin of ϕ1.0 mm (0.0394 in.) into the hole to fix it in place.

64. Install No. 1 chain sub-assembly LH.

a. Align the No. 1 chain's orange mark plates with the camshaft timing gear's timing mark, and attach the chain to the gear as shown.

b. Align the No. 1 chain's orange mark plate with the crankshaft timing gear's timing mark, and attach the chain to the gear as shown

c. Align the No. 2 chain's mark plates (yellow) with the timing marks of the camshaft timing gear and camshaft timing exhaust gear, and attach the No. 2 chain to the gears as shown

➡**The crankshaft timing gear and camshaft exhaust gear will be installed with the No. 1 and No. 2 chains connected to the gears.**

d. Install the crankshaft timing sprocket to the crankshaft.

e. Align and attach the knock pin of the No. 3 camshaft with the pin hole of the camshaft timing gear.

f. Using the hexagonal portion of the No. 4 camshaft, align and attach the knock pin of the No. 4 camshaft with the pin hole of the camshaft timing exhaust gear.

➡**Because the gears' timing mark positions may shift due to looseness of the No. 1 chain, use the hexagonal portion of the camshaft to hold the No. 3 camshaft in place until the No. 1 chain tensioner is installed.**

g. Remove the pin from the No. 2 chain tensioner.

65. Install No. 1 chain tensioner slipper.

➡**If you cannot install the chain tensioner slipper due to the tension of the chain, use the hexagonal portion of the camshaft to loosen the chain, and then install the chain tensioner slipper.**

66. Install No. 1 chain tensioner assembly.

a. Move the stopper plate upward to release the lock, and push the plunger deep into the tensioner.

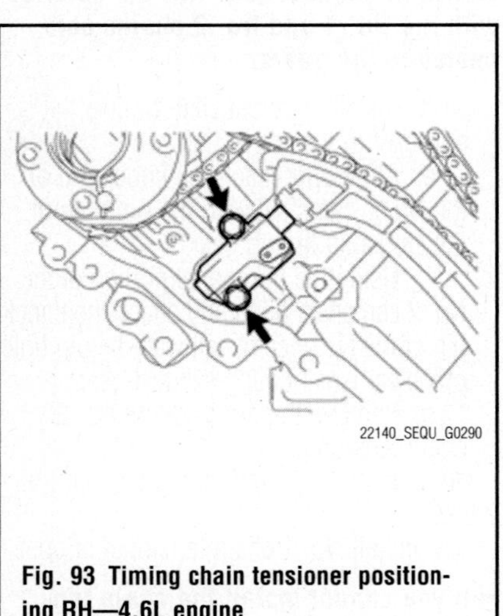

Fig. 93 Timing chain tensioner positioning RH—4.6L engine

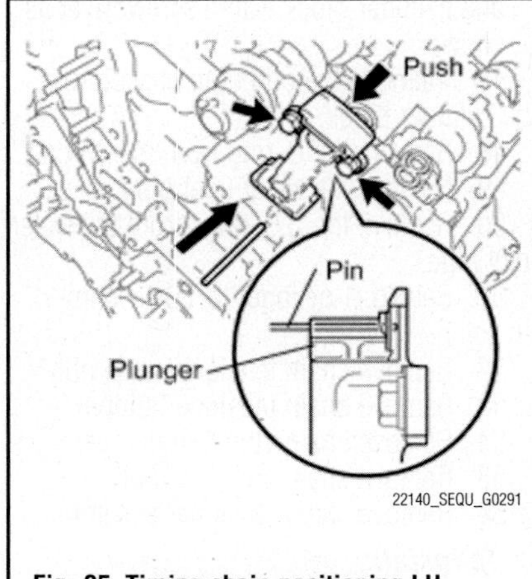

Fig. 95 Timing chain positioning LH—4.6L engine

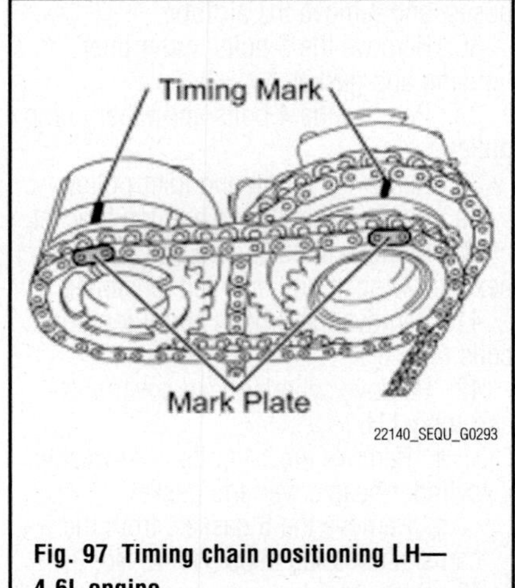

Fig. 97 Timing chain positioning LH—4.6L engine

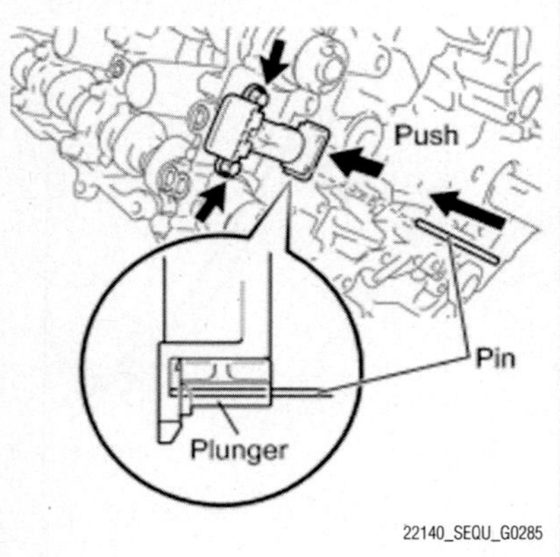

Fig. 94 Timing chain tensioner assembly No. 2—4.6L engine

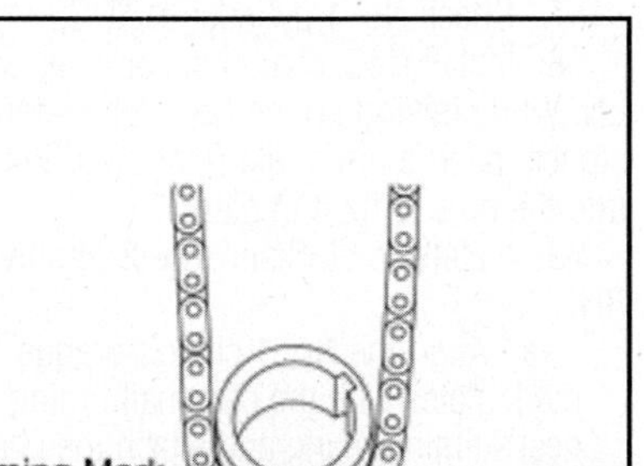

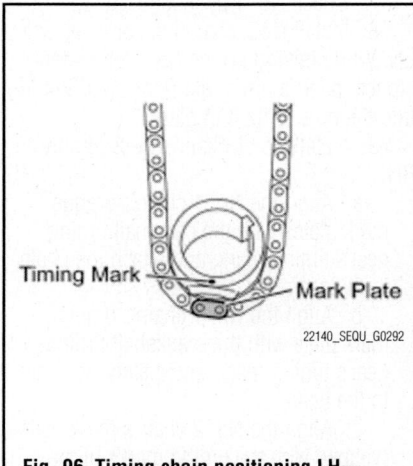

Fig. 96 Timing chain positioning LH—4.6L engine

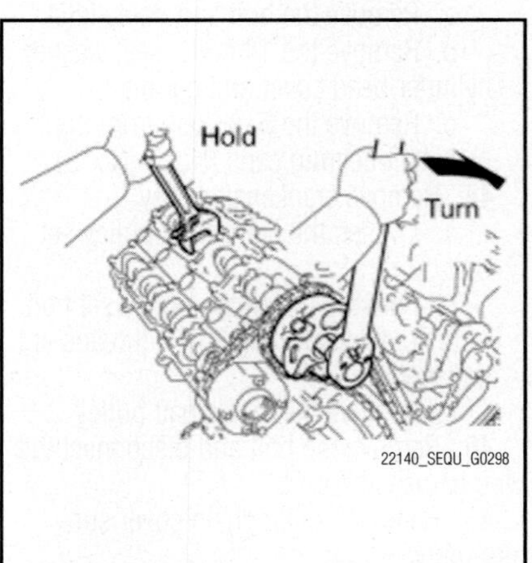

Fig. 98 Tightening camshaft gear No. 1—4.6L engine

b. Move the stopper plate downward to set the lock, and insert a hexagon wrench into the hole of the stopper plate.

c. Install the chain tensioner with the 2 bolts and tighten to 7 ft. lbs. (10 Nm).

67. Install the vibration damper with the 2 bolts

a. Remove the hexagon wrench from the chain tensioner.

68. Tighten camshaft timing gears LH.

a. Using a wrench to hold the hexagonal portion of the No. 3 camshaft, tighten the camshaft timing gear with the bolt and tighten to 74 ft. lbs. (100 Nm).

b. Using a wrench to hold the hexagonal portion of the No. 4 camshaft, tighten the camshaft timing gear with the bolt and tighten to 74 ft. lbs. (100 Nm).

69. Tighten camshaft timing gears RH.

a. Using a wrench to hold the hexagonal portion of the No. 1 camshaft, tighten the camshaft timing gear with the bolt and tighten to 74 ft. lbs. (100 Nm).

b. Using a wrench to hold the hexagonal portion of the No. 2 camshaft, tighten the camshaft timing gear with the bolt and tighten to 74 ft. lbs. (100 Nm).

70. Check No. 1 cylinder to TDC / compression.

a. Temporarily install the pulley set bolt.

b. Rotate the crankshaft clockwise, and check that the timing marks on the crankshaft timing gear and camshaft timing gears are as shown.

c. Remove the crankshaft pulley bolt.

71. Apply soapy water to 2 new O-rings and install them to the inlet pipe.

72. Install the inlet pipe to the No. 1 heat exchanger cover.

73. Install timing chain cover sub-assembly.

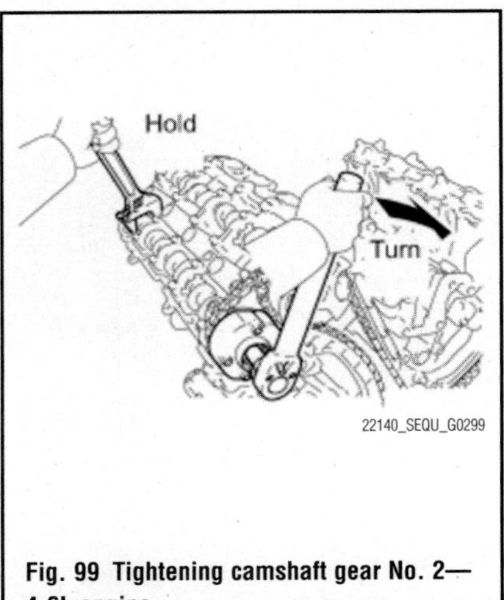

Fig. 99 Tightening camshaft gear No. 2—4.6L engine

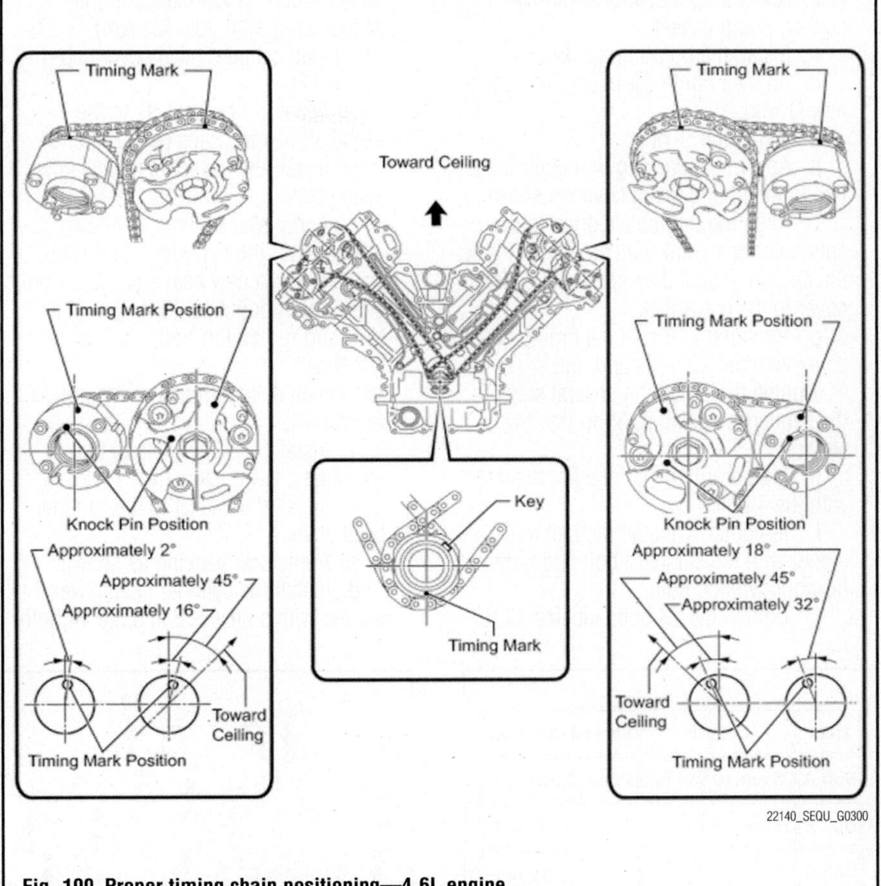

Fig. 100 Proper timing chain positioning—4.6L engine

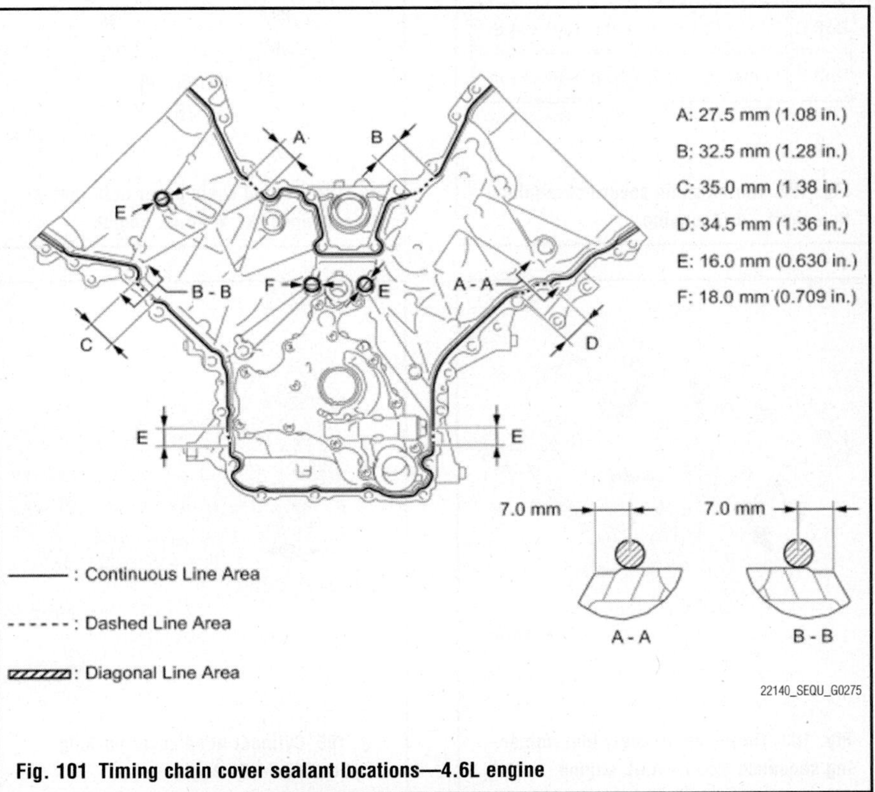

Fig. 101 Timing chain cover sealant locations—4.6L engine

a. Apply a light coat of engine oil to a new oil pump gasket.

b. Install the oil pump gasket.

c. Apply a light coat of engine oil to a new O-ring.

d. Install the O-ring.

e. Apply seal packing in a continuous line to the timing chain cover as shown.

f. Align the oil pump's drive rotor spline and the crankshaft as shown in the illustration. Install the spline and chain cover to the crankshaft.

g. Temporarily install the timing chain cover with the 28 bolts and nut.

Tighten the 5 bolts in several steps in the sequence shown to 35 ft. lbs. (47 Nm).

h. Temporarily install the fan bracket with the 4 bolts.

i. Temporarily install the belt tensioner with the standard bolt and 6 mm hexagon wrench bolt.

j. Tighten the 23 bolts labeled 12 to 35 and nut in several steps in the sequence to 17 ft. lbs. (23 Nm).

74. Install cylinder head cover sub-assembly LH.

a. Install 5 new gaskets to the camshaft bearing caps (No. 2, No. 3).

b. Install the gasket to the cylinder head cover.

c. Apply seal packing as shown.

d. Install the cylinder head cover washer with a new seal and the 14 bolts and tighten bolt A to 15 ft. lbs. (21 Nm) and remaining bolts to 9 ft. lbs. (12 Nm).

75. Install cylinder head cover sub-assembly RH.

a. Install 5 new gaskets to the camshaft bearing caps (No. 1, No. 3).

b. Install the gasket to the cylinder head cover.

c. Apply seal packing as shown.

d. Install the cylinder head cover washer with a new seal and the 14 bolts and tighten bolt A to 15 ft. lbs. (21 Nm) and remaining bolts to 9 ft. lbs. (12 Nm).

e. Install the noise filter to the cylinder head cover with the bolt and tighten to 62 inch lbs. (7.0 Nm).

76. Install the 8 ignition coils with the 8 bolts and tighten to 7 ft. lbs. (10 Nm).

77. Align the pulley set key with the key groove of the pulley, and slide on the pulley and tighten to 221 ft. Lbs. (300 Nm).

78. Connect the bracket to the timing chain cover with the bolt.

79. Install the idler pulley with the bolt and tighten to 32 ft. lbs. (43 Nm).

80. Temporarily install the water pump pulley with the 4 bolts and tighten to 15 ft. lbs. (21 Nm).

81. Install a new water cover gasket to the timing chain cover.

82. Install the water inlet with the 3 bolts and tighten to 15 ft. lbs. (21 Nm).

83. Connect the 2 air tube hoses.

84. Connect the 2 hoses.

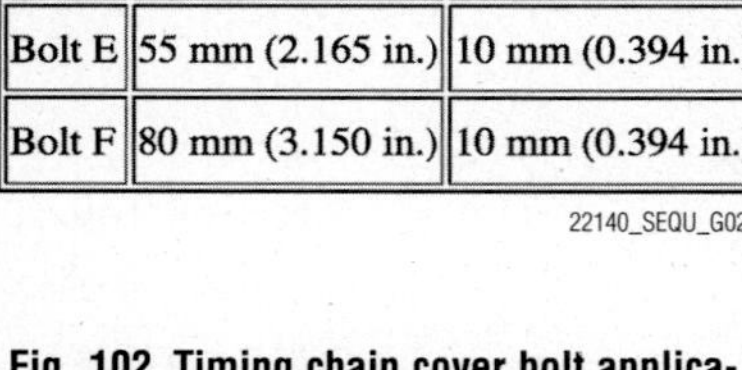

Item	Length	Thread diameter
Bolt A	25 mm (0.984 in.)	8 mm (0.315 in.)
Bolt B	55 mm (2.165 in.)	8 mm (0.315 in.)
Bolt C	70 mm (2.756 in.)	8 mm (0.315 in.)
Bolt D	35 mm (1.378 in.)	10 mm (0.394 in.)
Bolt E	55 mm (2.165 in.)	10 mm (0.394 in.)
Bolt F	80 mm (3.150 in.)	10 mm (0.394 in.)

22140_SEQU_G0278

Fig. 102 Timing chain cover bolt application chart—4.6L engine

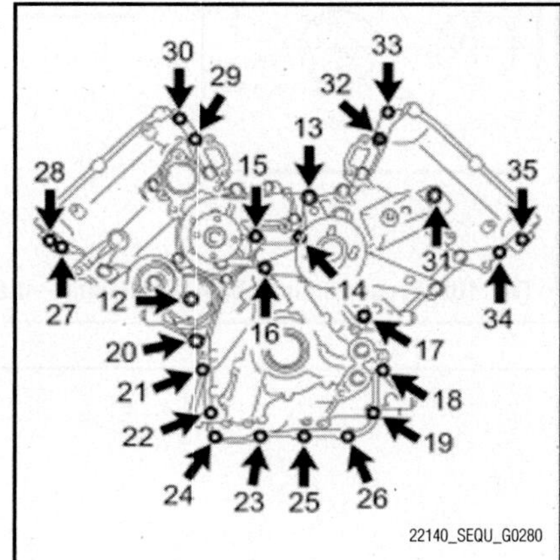

Fig. 104 Timing chain cover bolt tightening sequence No. 2—4.6L engine

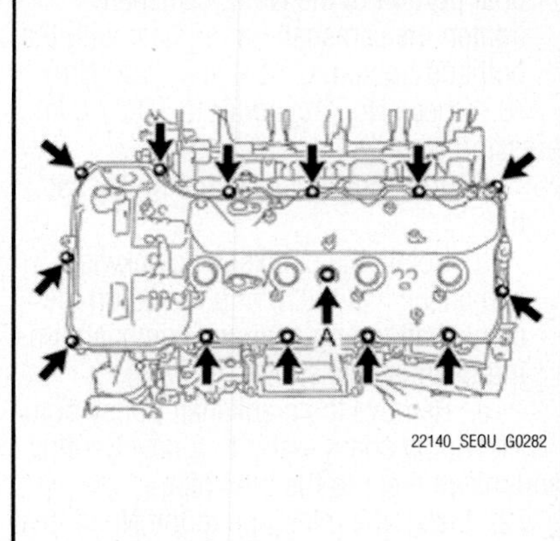

Fig. 106 Cylinder head cover bolt tightening sequence LH—4.6L engine

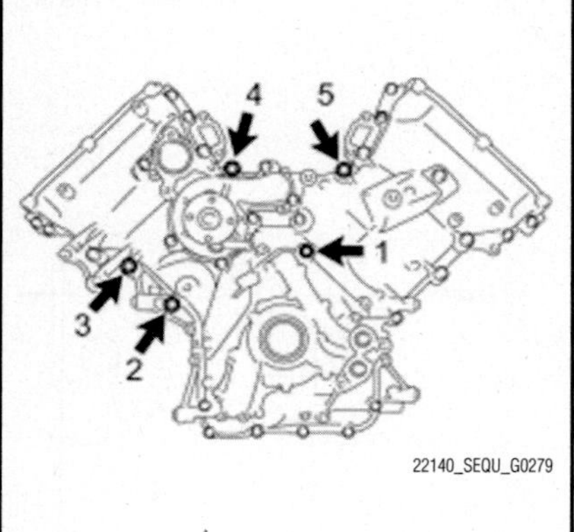

Fig. 103 Timing chain cover bolt tightening sequence No. 1—4.6L engine

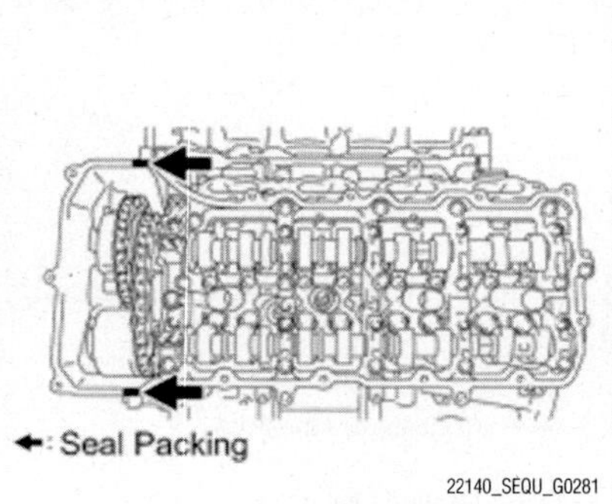

Fig. 105 Cylinder head cover packing location LH—4.6L engine

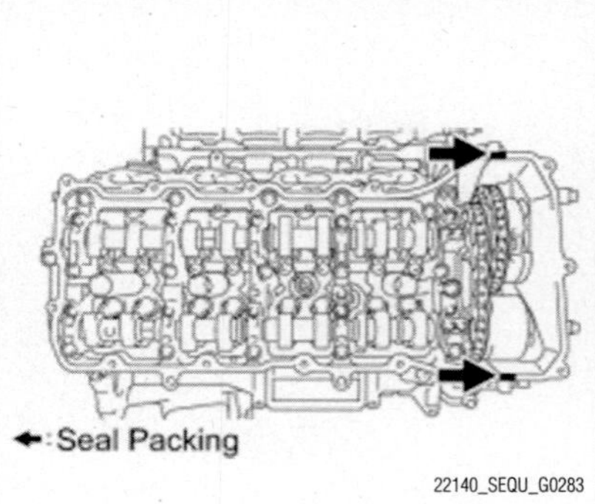

Fig. 107 Cylinder head cover packing location RH—4.6L engine

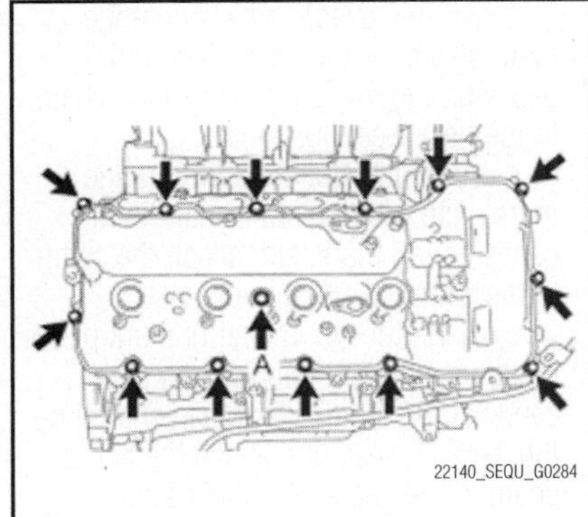

Fig. 108 Cylinder head cover bolt tightening sequence RH—4.6L engine

85. Install No. 1 and No 2 engine covers.
86. Install 2 new gaskets and the water by-pass joint with the 4 nuts and tighten to 15 ft. lbs. (21 Nm).
87. Connect the No. 2 water by-pass hose to the water by-pass joint.
88. Install the water by-pass pipe.
89. Connect the air tube.
90. Install the No. 1 water by-pass hose by connecting the hose to the water inlet housing and front water by-pass joint.
91. Install the 2 stud bolts.
92. Install the alternator with the 2 bolts and 2 nuts and tighten to 32 ft. lbs. (43 Nm).
93. Connect the alternator wire with the nut and tighten to 87 inch lbs. (9.8 Nm).
94. Connect the wire harness bracket to the alternator.
95. Connect the vane pump to the timing chain cover with the 2 bolts and tighten to 21 ft. lbs. (28 Nm).
96. Install engine oil level dipstick guide.
97. Install oil filter bracket (w/o Oil Cooler).
    a. Apply a light coat of engine oil to 2 new O-rings.
    b. Install the 2 O-rings to the timing chain cover.
    c. Install the oil filter bracket with the 2 bolts and 2 nuts and tighten to 26 ft. lbs. (35 Nm).
98. Install No. 1 oil cooler bracket.
99. Install No. 2 water by-pass pipe.
100. Install oil pressure sender gauge assembly.
101. Install oil filter element.
102. Connect No. 2 fuel tube sub-assembly.
103. Install the cooler compressor with the 2 stud bolts and tighten to 7 ft. lbs. (10 Nm).
104. Install the 2 bolts and 2 nuts and tighten to 18 ft. lbs. (235 Nm).
105. Connect the 3 heater hoses.
106. Connect the No. 2 and No. 3 air hoses.
107. Connect the 2 clamps and air pump wire harness.
108. Install the intake manifold.
109. Engine Room RH Side:
    a. Connect the 5 clamps.
    b. Connect the throttle position sensor and throttle control motor connector.
    c. Connect the 2 air pump connectors.
    d. Install the ground wire with the 2 bolts.
    e. Connect the noise filter connector.
    f. Connect the 2 VVT sensor connectors.
    g. Connect the injector connector.
    h. Connect the 4 ignition coil connectors.
    i. Connect the 2 camshaft timing oil control valve connectors.
    j. Connect the 2 clamps and power steering oil pressure switch connector.
    k. For 4WD: Connect the clamp and junction connector.
110. Engine Room LH Side:
    a. Connect the cooler compressor connector.
    b. Connect the 3 clamps.
    c. Connect the camshaft position sensor connector.
    d. Connect the 2 camshaft timing oil control valve connectors.
    e. Connect the engine coolant temperature sensor connector.
    f. Connect the 4 clamps.
    g. Install the ground wire with the 2 bolts.
    h. Connect the noise filter connector.
    i. Connect the 2 VVT sensor connectors.
    j. Connect the 4 ignition coil connectors.
    k. Connect the injector connector.
    l. Connect the 2 wire harness clamps.
    m. Connect the 4 air injection control driver connectors.
    n. Connect the 2 connectors and attach the 2 clamps to the engine room junction block.
    o. Install the engine room relay block cover.
111. Install air cleaner assembly.
112. Install air cleaner hose assembly.
113. Install the radiator.
114. Add engine oil.
115. Add engine coolant.
116. Connect cable to negative battery terminal.
117. Inspect for leaks.
118. Check engine oil level.
119. Use the Techstream diagnostic tool, or equivalent and reprogram the required systems.

5.7L Engine

1. Before servicing the vehicle, refer to the Precautions Section.
2. Disconnect the negative battery cable. Tape the cable with insulating tape.
3. Discharge fuel system pressure.
4. Drain engine oil.
5. Drain engine coolant.
6. Remove the radiator.
7. Remove air cleaner hose assembly.
8. Remove air cleaner assembly.
9. Engine Room LH Side:
    a. Remove the engine room relay block cover.
    b. Disconnect the 2 connectors and detach the 2 clamps from the engine room junction block.
    c. Disconnect the 4 air injection control driver connectors.
    d. Disconnect the 2 wire harness clamps.
    e. Disconnect the injector connector.
    f. Disconnect the 4 ignition coil connectors.
    g. Disconnect the 2 VVT sensor connectors.
    h. Disconnect the 4 clamps.
    i. Remove the 2 bolts and ground wire.
    j. Disconnect the noise filter connector.
    k. Disconnect the engine coolant temperature sensor connector.
    l. Disconnect the 2 camshaft timing oil control valve connectors.
    m. Disconnect the camshaft position sensor connector.
    n. Disconnect the 3 clamps.
    o. Disconnect the cooler compressor connector.
10. Engine Room RH Side.
    a. Disconnect the 2 camshaft timing oil control valve connectors.
    b. Disconnect the 4 ignition coil connectors.
    c. Disconnect the injector connector.
    d. Disconnect the 2 VVT sensor connectors.
    e. Disconnect the noise filter connector.
    f. Remove the 2 bolts and ground wire.

g. Disconnect the 2 air pump connectors.

h. Disconnect the throttle position sensor and throttle control motor connector.

i. Disconnect the 5 clamps.

j. Disconnect the 2 clamps and power steering oil pressure switch connector.

k. For 4WD: Disconnect the 2 clamps and power steering oil pressure switch connector.

11. Remove the intake manifold.
12. Disconnect the No. 2 and No. 3 air hoses and the 2 clamps.
13. Remove the 2 bolts and disconnect the heater 3 hoses.
14. Remove the 2 bolts, 2 nuts and 2 stud bolts, and disconnect the cooler compressor.

➡It is not necessary to completely remove the compressor. With the hoses connected to the compressor, hang the compressor on the vehicle body with a rope.

15. Remove the 2 bolts and disconnect the fuel tube.
16. Remove oil filter element.
17. Remove oil pressure sender gauge assembly.
18. Disconnect the 4 hoses and remove the water by-pass pipe.
19. Remove No. 1 oil cooler bracket (w/ oil cooler).
20. Remove the 2 bolts, 2 nuts and oil filter bracket.
21. Remove the 2 O-rings.
22. Remove engine oil level dipstick guide.
23. Remove the 2 bolts and disconnect the vane pump.
24. Disconnect the alternator connector.
25. Remove the terminal cap and nut, and disconnect the alternator wire.
26. Remove the bolt and disconnect the wire harness bracket from the alternator.
27. Remove the 2 bolts, 2 nuts and alternator.
28. Remove the 2 stud bolts.
29. Remove the No. 1 water by-pass hose by disconnect the hose from the water inlet housing and front water by-pass joint.
30. Remove the bolt and disconnect the air tube.
31. Remove the 2 bolts and hoses and water by-pass pipe.
32. Disconnect the No. 2 water by-pass hose from the water by-pass joint.
33. Remove the 4 nuts, water by-pass joint and 2 gaskets.
34. Remove No.1 and No. 2 engine covers.
35. Remove the bolt, disconnect the 2 hoses, and remove the air tube.
36. Remove the 3 bolts, water inlet housing and gasket.
37. Remove the 4 bolts and water pump pulley.
38. Remove the bolt and idler pulley.
39. Remove the 4 bolts and fan bracket.
40. Remove the standard bolt, 6 mm hexagon wrench bolt and belt tensioner.
41. Remove the 8 bolts and 8 ignition coils.
42. Remove cylinder head cover sub-assembly LH.

a. Remove the 14 bolts, seal washer, cylinder head cover and gasket.

b. Remove the 5 gaskets from the camshaft bearing caps (No. 2, No. 3).

43. Remove cylinder head cover sub-assembly RH.

a. Remove the bolt and noise filter.

b. Remove the 14 bolts, seal washer, cylinder head cover and gasket.

c. Remove the 5 gaskets from the camshaft bearing caps (No. 1, No. 3).

44. Remove crankshaft pulley.

a. Loosen the crankshaft pulley set bolt.

b. Partially install the pulley set bolt to the crankshaft until 2 or 3 threads are engaged.

c. Remove the crankshaft pulley.

45. Remove the bolt and disconnect the wire harness bracket.
46. Remove timing chain cover sub-assembly.

a. Remove the 28 bolts and nut shown.

47. Remove the oil pump gasket from the cylinder block.
48. Remove the O-ring from the oil pan.
49. Remove the water inlet pipe.
50. Remove the 2 O-rings from the water inlet pipe.
51. Set No. 1 cylinder to TDC / compression.
52. Remove chain tensioner assembly.
53. Remove chain tensioner slipper.
54. Remove chain vibration damper.
55. Remove chain sub-assembly.
56. Remove chain tensioner assembly.

To install:

➡Be sure to use new fasteners, as required.

57. Check No. 1 cylinder to TDC.
58. Install No.2 chain tensioner assembly. While raising up the No. 2 chain tensioner, insert a pin of ϕ1.0 mm (0.0394 in.) into the hole to fix it in place.
59. Install No. 1 chain sub-assembly RH.

a. Align the No. 1 chain's orange mark plates with the camshaft timing gear's timing mark, and attach the chain to the gear. See illustration.

b. Align the No. 1 chain's orange mark plate with the crankshaft timing gear's timing mark, and attach the chain to the gear. See illustration.

c. Align the No. 2 chain's mark plates (yellow) with the timing marks of the camshaft timing gear and camshaft timing exhaust gear, and attach the No. 2 chain to the gears. See illustration.

➡The crankshaft timing gear and camshaft exhaust gear will be installed with the No. 1 and No. 2 chains connected to the gears.

d. Install the crankshaft timing sprocket to the crankshaft.

e. Align and attach the knock pin of the No. 1 camshaft with the pin hole of the camshaft timing gear.

f. Using the hexagonal portion of the No. 2 camshaft, align and attach the knock pin of the No. 2 camshaft with the pin hole of the camshaft timing exhaust gear.

g. Remove the pin from the No. 2 chain tensioner.

60. Install the vibration damper with the 2 bolts.
61. Install No. 1 chain tensioner slipper.

➡If you cannot install the chain tensioner slipper due to the tension of the chain, use the hexagonal portion of the camshaft to loosen the chain, and then install the chain tensioner slipper

62. Install No. 1 chain tensioner assembly.

a. Move the stopper plate upward to release the lock, and push the plunger deep into the tensioner.

b. Move the stopper plate downward to set the lock, and insert a hexagon wrench into the hole of the stopper plate.

c. Install the chain tensioner with the 2 bolts and tighten to 7 ft. lbs. (10 Nm).

d. Remove the hexagon wrench from the chain tensioner.

63. Install No.3 chain tensioner assembly. While raising up the No. 3 chain tensioner, insert a pin of ϕ1.0 mm (0.0394 in.) into the hole to fix it in place.
64. Install No. 1 chain sub-assembly LH.

a. Align the No. 1 chain's orange mark plates with the camshaft timing gear's timing mark, and attach the chain to the gear. See illustration.

b. Align the No. 1 chain's orange mark plate with the crankshaft timing gear's timing mark, and attach the chain to the gear. See illustration.

c. Align the No. 2 chain's mark plates (yellow) with the timing marks of the camshaft timing gear and camshaft timing exhaust gear, and attach the No. 2 chain to the gears. See illustration.

➡The crankshaft timing gear and camshaft exhaust gear will be installed with the No. 1 and No. 2 chains connected to the gears.

d. Install the crankshaft timing sprocket to the crankshaft.

e. Align and attach the knock pin of the No. 3 camshaft with the pin hole of the camshaft timing gear.

f. Using the hexagonal portion of the No. 4 camshaft, align and attach the knock pin of the No. 4 camshaft with the pin hole of the camshaft timing exhaust gear.

➡Because the gears' timing mark positions may shift due to looseness of the No. 1 chain, use the hexagonal portion of the camshaft to hold the No. 3 camshaft in place until the No. 1 chain tensioner is installed.

g. Remove the pin from the No. 2 chain tensioner.

65. Install No. 1 chain tensioner slipper.

➡If you cannot install the chain tensioner slipper due to the tension of the chain, use the hexagonal portion of the camshaft to loosen the chain, and then install the chain tensioner slipper.

66. Install No. 1 chain tensioner assembly.

a. Move the stopper plate upward to release the lock, and push the plunger deep into the tensioner.

b. Move the stopper plate downward to set the lock, and insert a hexagon wrench into the hole of the stopper plate.

c. Install the chain tensioner with the 2 bolts and tighten to 7 ft. lbs. (10 Nm).

67. Install the vibration damper with the 2 bolts

a. Remove the hexagon wrench from the chain tensioner.

68. Tighten camshaft timing gears LH.

a. Using a wrench to hold the hexagonal portion of the No. 3 camshaft, tighten the camshaft timing gear with the bolt and tighten to 74 ft. lbs. (100 Nm).

b. Using a wrench to hold the hexagonal portion of the No. 4 camshaft, tighten the camshaft timing gear with the bolt and tighten to 74 ft. lbs. (100 Nm).

69. Tighten camshaft timing gears RH.

a. Using a wrench to hold the hexagonal portion of the No. 1 camshaft, tighten the camshaft timing gear with the bolt and tighten to 74 ft. lbs. (100 Nm).

b. Using a wrench to hold the hexagonal portion of the No. 2 camshaft, tighten the camshaft timing gear with the bolt and tighten to 74 ft. lbs. (100 Nm).

70. Check No. 1 cylinder to TDC / compression.

a. Temporarily install the pulley set bolt.

b. Rotate the crankshaft clockwise, and check that the timing marks on the crankshaft timing gear and camshaft timing gears. See illustration.

c. Remove the crankshaft pulley bolt.

71. Apply soapy water to 2 new O-rings and install them to the inlet pipe.

72. Install the inlet pipe to the No. 1 heat exchanger cover.

73. Install timing chain cover sub-assembly.

a. Apply a light coat of engine oil to a new oil pump gasket.

b. Install the oil pump gasket.

c. Apply a light coat of engine oil to a new O-ring.

d. Install the O-ring.

e. Apply seal packing in a continuous line to the timing chain cover. See illustration.

f. Align the oil pump's drive rotor spline and the crankshaft as shown in the illustration. Install the spline and chain cover to the crankshaft.

g. Temporarily install the timing chain cover with the 28 bolts and nut. Tighten the 5 bolts in several steps in the sequence shown to 35 ft. lbs. (47 Nm).

h. Temporarily install the fan bracket with the 4 bolts.

i. Temporarily install the belt tensioner with the standard bolt and 6 mm hexagon wrench bolt.

j. Tighten the 23 bolts labeled 12 to 35 and nut in several steps in the sequence to 17 ft. lbs. (23 Nm).

74. Install cylinder head cover sub-assembly LH.

a. Install 5 new gaskets to the camshaft bearing caps (No. 2, No. 3).

b. Install the gasket to the cylinder head cover.

c. Apply seal packing. See illustration.

d. Install the cylinder head cover washer with a new seal and the 14 bolts and tighten bolt A to 15 ft. lbs. (21 Nm) and remaining bolts to 9 ft. lbs. (12 Nm).

75. Install cylinder head cover sub-assembly RH.

a. Install 5 new gaskets to the camshaft bearing caps (No. 1, No. 3).

b. Install the gasket to the cylinder head cover.

c. Apply seal packing. See illustration.

d. Install the cylinder head cover washer with a new seal and the 14 bolts and tighten bolt A to 15 ft. lbs. (21 Nm) and remaining bolts to 9 ft. lbs. (12 Nm).

e. Install the noise filter to the cylinder head cover with the bolt and tighten to 62 inch lbs. (7.0 Nm).

76. Install the 8 ignition coils with the 8 bolts and tighten to 7 ft. lbs. (10 Nm).

77. Align the pulley set key with the key groove of the pulley, and slide on the pulley and tighten to 221 ft. Lbs. (300 Nm).

78. Connect the bracket to the timing chain cover with the bolt.

79. Install the idler pulley with the bolt and tighten to 32 ft. lbs. (43 Nm).

80. Temporarily install the water pump pulley with the 4 bolts and tighten to 15 ft. lbs. (21 Nm).

81. Install a new water cover gasket to the timing chain cover.

82. Install the water inlet with the 3 bolts and tighten to 15 ft. lbs. (21 Nm).

83. Connect the 2 air tube hoses.

84. Connect the 2 hoses.

85. Install No. 1 and No 2 engine covers.

86. Install 2 new gaskets and the water by-pass joint with the 4 nuts and tighten to 15 ft. lbs. (21 Nm).

87. Connect the No. 2 water by-pass hose to the water by-pass joint.

88. Install the water by-pass pipe.

89. Connect the air tube.

90. Install the No. 1 water by-pass hose by connecting the hose to the water inlet housing and front water by-pass joint.

91. Install the 2 stud bolts.

92. Install the alternator with the 2 bolts and 2 nuts and tighten to 32 ft. lbs. (43 Nm).

93. Connect the alternator wire with the nut and tighten to 87 inch lbs. (9.8 Nm).

94. Connect the wire harness bracket to the alternator.

95. Connect the vane pump to the timing chain cover with the 2 bolts and tighten to 21 ft. lbs. (28 Nm).

96. Install engine oil level dipstick guide.

97. Install oil filter bracket (w/o Oil Cooler).

a. Apply a light coat of engine oil to 2 new O-rings.

b. Install the 2 O-rings to the timing chain cover.

c. Install the oil filter bracket with the 2 bolts and 2 nuts and tighten to 26 ft. lbs. (35 Nm).

98. Install No. 1 oil cooler bracket.

99. Install No. 2 water by-pass pipe.

100. Install oil pressure sender gauge assembly.

101. Install oil filter element.

102. Connect No. 2 fuel tube sub-assembly.

103. Install the cooler compressor with the 2 stud bolts and tighten to 7 ft. lbs. (10 Nm).

104. Install the 2 bolts and 2 nuts and tighten to 18 ft. lbs. (235 Nm).

105. Connect the 3 heater hoses.

106. Connect the No. 2 and No. 3 air hoses.

107. Connect the 2 clamps and air pump wire harness.

108. Install the intake manifold.

109. Engine Room RH Side:

a. Connect the 5 clamps.

b. Connect the throttle position sensor and throttle control motor connector.

c. Connect the 2 air pump connectors.

d. Install the ground wire with the 2 bolts.

e. Connect the noise filter connector.

f. Connect the 2 VVT sensor connectors.

g. Connect the injector connector.

h. Connect the 4 ignition coil connectors.

i. Connect the 2 camshaft timing oil control valve connectors.

j. Connect the 2 clamps and power steering oil pressure switch connector.

k. For 4WD: Connect the clamp and junction connector.

110. Engine Room LH Side:

a. Connect the cooler compressor connector.

b. Connect the 3 clamps.

c. Connect the camshaft position sensor connector.

d. Connect the 2 camshaft timing oil control valve connectors.

e. Connect the engine coolant temperature sensor connector.

f. Connect the 4 clamps.

g. Install the ground wire with the 2 bolts.

h. Connect the noise filter connector.

i. Connect the 2 VVT sensor connectors.

j. Connect the 4 ignition coil connectors.

k. Connect the injector connector.

l. Connect the 2 wire harness clamps.

m. Connect the 4 air injection control driver connectors.

n. Connect the 2 connectors and attach the 2 clamps to the engine room junction block.

o. Install the engine room relay block cover.

111. Install air cleaner assembly.

112. Install air cleaner hose assembly.

113. Install the radiator.

114. Add engine oil.

115. Add engine coolant.

116. Connect cable to negative battery terminal.

117. Inspect for leaks.

118. Check engine oil level.

119. Use the Techstream diagnostic tool, or equivalent and reprogram the required systems.

VALVE COVERS

REMOVAL & INSTALLATION

4.0L Engine

See Figures 109 and 110.

1. Before servicing the vehicle, refer to the Precautions Section.
2. Disconnect the negative battery cable. Tape the cable with insulating tape.
3. Drain engine coolant.
4. Remove v-bank cover.
5. Remove air cleaner assembly.
6. Disconnect the 2 water by-pass hoses.
7. Disconnect the fuel vapor feed hose.
8. Disconnect the ventilation hose.
9. Disconnect the 2 VSV connectors.
10. Disconnect the throttle body w/ motor connector.
11. Separate the 3 wire harness clamps and hose clamp.
12. Remove the nut, then separate the clutch flexible hose bracket from the surge tank stay (w/ Manual Transmission).
13. Remove the 2 bolts and throttle body bracket.
14. Remove the bolt and oil baffle plate.
15. Remove the 4 bolts and 2 surge tank stays.
16. Remove the 2 nuts.
17. Remove the 4 bolts, intake air surge tank and gasket.
18. Disconnect the 3 connectors.
19. Remove the 3 bolts, then remove the 3 ignition coils.
20. Remove the 10 bolts, 3 seal washers, 2 nuts, right cylinder head cover and gasket.
21. Remove the 10 bolts, 3 seal washers, 2 nuts, left cylinder head cover and gasket.

To install:

➡Be sure to use new fasteners, as required.

22. Remove any old packing material and be careful not to drop any oil on the contact surfaces of the cylinder head, timing chain cover and cylinder head covers.
23. Install the cylinder head covers within 3 minutes of applying seal packing Part No. 08826-00080 or equivalent. Tighten the cylinder head cover bolts and nuts within 15 minutes of installing the cylinder head covers. Otherwise, the seal packing mush be removed and reapplied.
24. Install the seal washers onto the bolts.
25. Install the cylinder head cover with the 10 bolts and 2 nuts. Tighten the bolts and nuts uniformly in several steps as follows:

a. Bolts labeled A: 10 Nm (7.4 ft. lbs.).

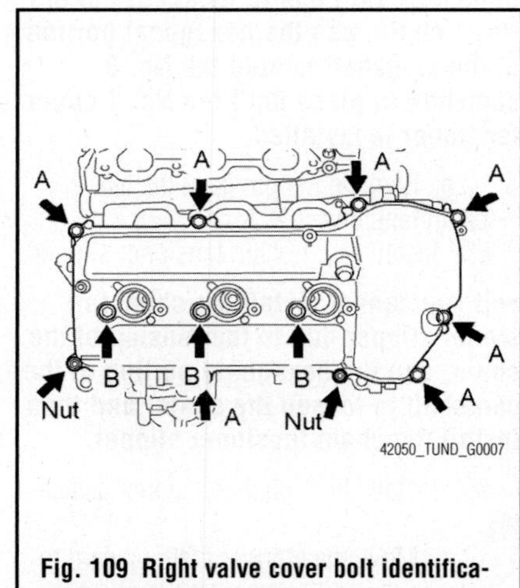

Fig. 109 Right valve cover bolt identification—4.0L engine

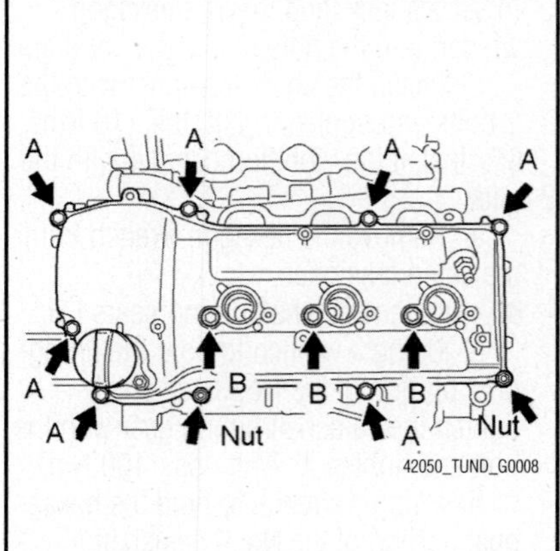

Fig. 110 Left valve cover bolt identification—4.0L engine

b. Bolts labeled B: 9.0 Nm (80 in. lbs.).

c. Nuts: 9.0 Nm (80 in. lbs.).

26. Install the ignition coils and tighten the bolts to 10 Nm (7.4 ft. lbs.).

27. Install a new gasket onto the intake air surge tank.

28. Install the intake air surge tank with the 4 bolts and tighten to 28 Nm (21 ft. lbs).

29. Install the 2 intake air surge tank nuts and tighten to 28 Nm (21 ft. lbs).

30. Install the 2 surge tank stays with the 4 bolts and tighten to 21 Nm (15 ft. lbs).

31. Install the oil baffle plate with the bolt and tighten to 9.0 Nm (80 in. lbs.).

32. Install the throttle body bracket with the 2 bolts and tighten to 21 Nm (15 ft. lbs).

33. Install the clutch flexible hose bracket with the nut (w/ Manual Transmission) and tighten to 24 Nm (18 ft. lbs).

34. Install the 3 wire harness clamps and hose clamp.

35. Connect the throttle body w/ motor connector.

36. Connect the 2 VSV connectors.

37. Connect the ventilation hose.

38. Connect the fuel vapor feed hose.

39. Connect the 2 water by-pass hoses.

40. Install air cleaner assembly.

41. Connect cable to negative battery terminal.

42. Add engine coolant.

43. Check for engine coolant leakage.

44. Install the V-bank cover with the 2 nuts and tighten to 7.5 Nm (66 in. lbs.)

45. Use the Techstream diagnostic tool, or equivalent and reprogram the required systems.

4.6L Engine

1. Before servicing the vehicle, refer to the Precautions Section.

2. Disconnect the negative battery cable. Tape the cable with insulating tape.

3. Remove necessary to gain access to cylinder head covers.

4. Remove the 8 bolts and 8 ignition coils.

5. Remove cylinder head cover sub-assembly LH.

a. Remove the 14 bolts, seal washer, cylinder head cover and gasket.

b. Remove the 5 gaskets from the camshaft bearing caps (No. 2, No. 3).

6. Remove cylinder head cover sub-assembly RH.

a. Remove the bolt and noise filter.

b. Remove the 14 bolts, seal washer, cylinder head cover and gasket.

c. Remove the 5 gaskets from the camshaft bearing caps (No. 1, No. 3).

To install:

➡Be sure to use new fasteners, as required.

7. Install cylinder head cover sub-assembly LH.

a. Install 5 new gaskets to the camshaft bearing caps (No. 2, No. 3).

b. Install the gasket to the cylinder head cover.

c. Apply seal packing as shown.

d. Install the cylinder head cover washer with a new seal and the 14 bolts and tighten bolt A to 15 ft. lbs. (21 Nm) and remaining bolts to 9 ft. lbs. (12 Nm).

8. Install cylinder head cover sub-assembly RH.

a. Install 5 new gaskets to the camshaft bearing caps (No. 1, No. 3).

b. Install the gasket to the cylinder head cover.

c. Apply seal packing as shown.

d. Install the cylinder head cover washer with a new seal and the 14 bolts and tighten bolt A to 15 ft. lbs. (21 Nm) and remaining bolts to 9 ft. lbs. (12 Nm).

e. Install the noise filter to the cylinder head cover with the bolt and tighten to 62 inch lbs. (7.0 Nm).

9. Install the 8 ignition coils with the 8 bolts and tighten to 7 ft. lbs. (10 Nm).

10. Reinstall items accessed to remove cylinder head covers.

11. Use the Techstream diagnostic tool, or equivalent and reprogram the required systems.

5.7L Engine

See Figures 111 through 114.

1. Before servicing the vehicle, refer to the Precautions Section.

2. Disconnect the negative battery cable. Tape the cable with insulating tape.

3. Remove necessary to gain access to cylinder head covers.

4. Remove the 8 bolts and 8 ignition coils.

5. Remove cylinder head cover sub-assembly LH.

a. Remove the 14 bolts, seal washer, cylinder head cover and gasket.

b. Remove the 5 gaskets from the camshaft bearing caps (No. 2, No. 3).

6. Remove cylinder head cover sub-assembly RH.

a. Remove the bolt and noise filter.

b. Remove the 14 bolts, seal washer, cylinder head cover and gasket.

c. Remove the 5 gaskets from the camshaft bearing caps (No. 1, No. 3).

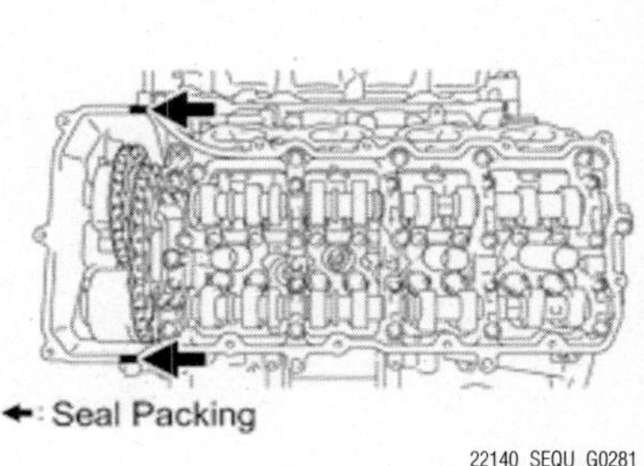

Fig. 111 Cylinder head cover packing location LH—4.6L and 5.7L engines

To install:

➡Be sure to use new fasteners, as required.

7. Install cylinder head cover sub-assembly LH.

a. Install 5 new gaskets to the camshaft bearing caps (No. 2, No. 3).

b. Install the gasket to the cylinder head cover.

c. Apply seal packing as shown.

d. Install the cylinder head cover washer with a new seal and the 14 bolts and tighten bolt A to 15 ft. lbs. (21 Nm) and remaining bolts to 9 ft. lbs. (12 Nm).

8. Install cylinder head cover sub-assembly RH.

a. Install 5 new gaskets to the camshaft bearing caps (No. 1, No. 3).

b. Install the gasket to the cylinder head cover.

c. Apply seal packing as shown.

d. Install the cylinder head cover washer with a new seal and the 14 bolts and tighten bolt A to 15 ft. lbs. (21 Nm) and remaining bolts to 9 ft. lbs. (12 Nm).

e. Install the noise filter to the cylinder head cover with the bolt and tighten to 62 inch lbs. (7.0 Nm).

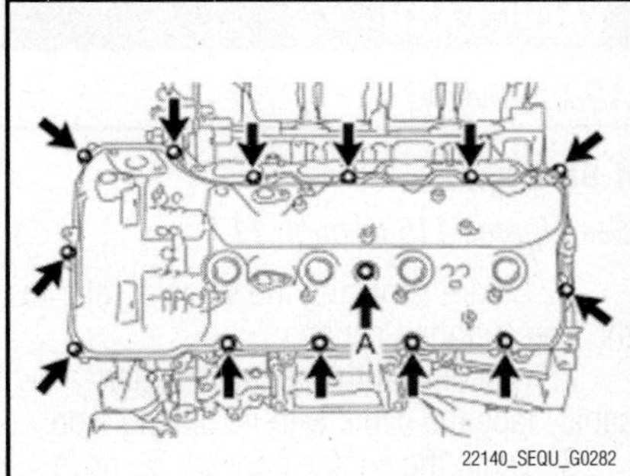

Fig. 112 Cylinder head cover bolt tightening sequence LH—4.6L and 5.7L engines

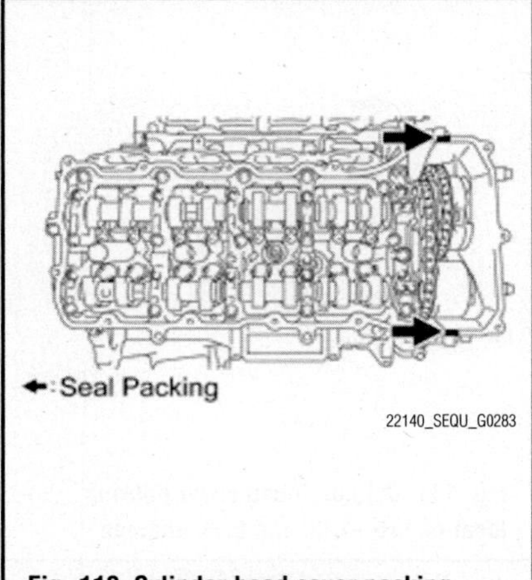

Fig. 113 Cylinder head cover packing location RH—4.6L and 5.7L engines

Fig. 114 Cylinder head cover bolt tightening sequence RH—4.6L and 5.7L engines

9. Install the 8 ignition coils with the 8 bolts and tighten to 7 ft. lbs. (10 Nm).
10. Reinstall items accessed to remove cylinder head covers.
11. Use the Techstream diagnostic tool, or equivalent and reprogram the required systems.

VALVE LASH

ADJUSTMENT

4.0L Engine

See Figures 115 through 117.

1. Before servicing the vehicle, refer to the Precautions Section.
2. Disconnect the negative battery cable. Tape the cable with insulating tape.
3. Drain the engine coolant. Remove the V bank cover.
4. Remove the air cleaner assembly.
5. Disconnect the two water bypass hoses. Disconnect the fuel vapor feed hose.

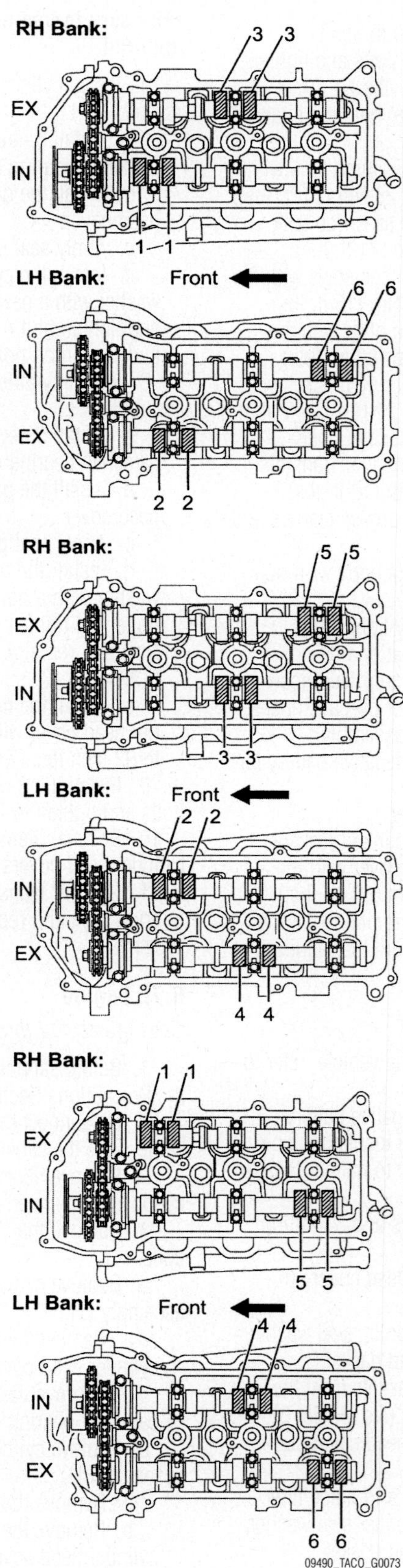

Fig. 115 Valve clearance location—4.0L engine

	5.060 (0.1992)	5.080 (0.2000)	5.100 (0.2008)	5.120 (0.2016)	5.140 (0.2024)	5.160 (0.2031)	5.180 (0.2039)	5.200 (0.2047)	5.210 (0.2051)	5.220 (0.2055)	5.230 (0.2059)	5.240 (0.2063)	5.250 (0.2067)	5.260 (0.2071)	5.270 (0.2075)	5.280 (0.2079)	5.290 (0.2083)	5.300 (0.2087)	5.310 (0.2091)	5.320 (0.2094)	5.330 (0.2098)	5.340 (0.2102)	5.350 (0.2106)	5.360 (0.2110)	5.370 (0.2114)	5.380 (0.2118)	5.390 (0.2122)	5.400 (0.2126)	5.410 (0.2130)	5.420 (0.2134)	5.430 (0.2138)	5.440 (0.2142)	5.450 (0.2146)	5.460 (0.2150)	5.470 (0.2154)	5.480 (0.2157)	5.490 (0.2161)	5.500 (0.2165)	5.510 (0.2169)	5.520 (0.2173)	5.530 (0.2177)	5.540 (0.2181)	5.550 (0.2185)	5.560 (0.2189)	5.570 (0.2193)	5.580 (0.2197)	5.590 (0.2201)	5.600 (0.2205)	5.620 (0.2213)	5.640 (0.2220)	5.660 (0.2228)	5.680 (0.2236)	5.700 (0.2244)	5.720 (0.2252)	5.740 (0.2260)
0.000 - 0.020 (0.0000 - 0.0008)													06	08	08	10	10	12	12	14	14	16	16	18	18	20	20	22	22	24	24	26	26	28	28	30	30	32	32	34	34	36	36	38	38	40	40	42	44	46	48	50	52	54	56
0.021 - 0.040 (0.0008 - 0.0016)											06	08	08	10	10	12	12	14	14	16	16	18	18	20	20	22	22	24	24	26	26	28	28	30	30	32	32	34	34	36	36	38	38	40	40	42	42	44	46	48	50	52	54	56	58
0.041 - 0.060 (0.0016 - 0.0024)									06	08	08	10	10	12	12	14	14	16	16	18	18	20	20	22	22	24	24	26	26	28	28	30	30	32	32	34	34	36	36	38	38	40	40	42	42	44	44	46	48	50	52	54	56	58	60
0.061 - 0.080 (0.0024 - 0.0031)								08	08	10	10	12	12	14	14	16	16	18	18	20	20	22	22	24	24	26	26	28	28	30	30	32	32	34	34	36	36	38	38	40	40	42	42	44	44	46	46	48	50	52	54	56	58	60	62
0.081 - 0.100 (0.0032 - 0.0039)							08	10	10	12	12	14	14	16	16	18	18	20	20	22	22	24	24	26	26	28	28	30	30	32	32	34	34	36	36	38	38	40	40	42	42	44	44	46	46	48	48	50	52	54	56	58	60	62	64
0.101 - 0.120 (0.0040 - 0.0047)						08	10	12	12	14	14	16	16	18	18	20	20	22	22	24	24	26	26	28	28	30	30	32	32	34	34	36	36	38	38	40	40	42	42	44	44	46	46	48	48	50	50	52	54	56	58	60	62	64	66
0.121 - 0.140 (0.0048 - 0.0055)					08	10	12	14	14	16	16	18	18	20	20	22	22	24	24	26	26	28	28	30	30	32	32	34	34	36	36	38	38	40	40	42	42	44	44	46	46	48	48	50	50	52	52	54	56	58	60	62	64	66	68
0.141 - 0.149 (0.0056 - 0.0059)				06	08	10	12	14	16	16	18	18	20	20	22	22	24	24	26	26	28	28	30	30	32	32	34	34	36	36	38	38	40	40	42	42	44	44	46	46	48	48	50	50	52	52	54	54	56	58	60	62	64	66	68
0.251 - 0.270 (0.0099 - 0.0106)	12	14	16	18	20	22	24	26	28	28	30	30	32	32	34	34	36	36	38	38	40	40	42	42	44	44	46	46	48	48	50	50	52	52	54	54	56	56	58	58	60	60	62	62	64	64	66	66	68	70	72				
0.271 - 0.290 (0.0107 - 0.0114)	14	16	18	20	22	24	26	28	30	30	32	32	34	34	36	36	38	38	40	40	42	42	44	44	46	46	48	48	50	50	52	52	54	54	56	56	58	58	60	60	62	62	64	64	66	66	68	68	70	72					
0.291 - 0.310 (0.0115 - 0.0122)	16	18	20	22	24	26	28	30	32	32	34	34	36	36	38	38	40	40	42	42	44	44	46	46	48	48	50	50	52	52	54	54	56	56	58	58	60	60	62	62	64	64	66	66	68	68	70	70	72						
0.311 - 0.330 (0.0122 - 0.0130)	18	20	22	24	26	28	30	32	34	34	36	36	38	38	40	40	42	42	44	44	46	46	48	48	50	50	52	52	54	54	56	56	58	58	60	60	62	62	64	64	66	66	68	68	70	70	72	72							
0.331 - 0.350 (0.0130 - 0.0138)	20	22	24	26	28	30	32	34	36	36	38	38	40	40	42	42	44	44	46	46	48	48	50	50	52	52	54	54	56	56	58	58	60	60	62	62	64	64	66	66	68	68	70	70	72	72	74								
0.351 - 0.370 (0.0138 - 0.0146)	22	24	26	28	30	32	34	36	38	38	40	40	42	42	44	44	46	46	48	48	50	50	52	52	54	54	56	56	58	58	60	60	62	62	64	64	66	66	68	68	70	70	72	72	74										
0.371 - 0.390 (0.0146 - 0.0154)	24	26	28	30	32	34	36	38	40	40	42	42	44	44	46	46	48	48	50	50	52	52	54	54	56	56	58	58	60	60	62	62	64	64	66	66	68	68	70	70	72	72	74												
0.391 - 0.410 (0.0154 - 0.0161)	26	28	30	32	34	36	38	40	42	42	44	44	46	46	48	48	50	50	52	52	54	54	56	56	58	58	60	60	62	62	64	64	66	66	68	68	70	70	72	72	74														
0.411 - 0.430 (0.0162 - 0.0169)	28	30	32	34	36	38	40	42	44	44	46	46	48	48	50	50	52	52	54	54	56	56	58	58	60	60	62	62	64	64	66	66	68	68	70	70	72	72	74																
0.431 - 0.450 (0.0170 - 0.0177)	30	32	34	36	38	40	42	44	46	46	48	48	50	50	52	52	54	54	56	56	58	58	60	60	62	62	64	64	66	66	68	68	70	70	72	72	74																		
0.451 - 0.470 (0.0178 - 0.0185)	32	34	36	38	40	42	44	46	48	48	50	50	52	52	54	54	56	56	58	58	60	60	62	62	64	64	66	66	68	68	70	70	72	72	74																				
0.471 - 0.490 (0.0185 - 0.0193)	34	36	38	40	42	44	46	48	50	50	52	52	54	54	56	56	58	58	60	60	62	62	64	64	66	66	68	68	70	70	72	72	74																						
0.491 - 0.510 (0.0193 - 0.0201)	36	38	40	42	44	46	48	50	52	52	54	54	56	56	58	58	60	60	62	62	64	64	66	66	68	68	70	70	72	72	74																								
0.511 - 0.530 (0.0201 - 0.0209)	38	40	42	44	46	48	50	52	54	54	56	56	58	58	60	60	62	62	64	64	66	66	68	68	70	70	72	72	74																										
0.531 - 0.550 (0.0209 - 0.0217)	40	42	44	46	48	50	52	54	56	56	58	58	60	60	62	62	64	64	66	66	68	68	70	70	72	72	74																												
0.551 - 0.570 (0.0217 - 0.0224)	42	44	46	48	50	52	54	56	58	58	60	60	62	62	64	64	66	66	68	68	70	70	72	72	74																														
0.571 - 0.590 (0.0225 - 0.0232)	44	46	48	50	52	54	56	58	60	60	62	62	64	64	66	66	68	68	70	70	72	72	74																																
0.591 - 0.610 (0.0233 - 0.0240)	46	48	50	52	54	56	58	60	62	62	64	64	66	66	68	68	70	70	72	72	74																																		
0.611 - 0.630 (0.0241 - 0.0248)	48	50	52	54	56	58	60	62	64	64	66	66	68	68	70	70	72	72	74																																				
0.631 - 0.650 (0.0248 - 0.0256)	50	52	54	56	58	60	62	64	66	66	68	68	70	70	72	72	74																																						
0.651 - 0.670 (0.0256 - 0.0264)	52	54	56	58	60	62	64	66	68	68	70	70	72	72	74																																								
0.671 - 0.690 (0.0264 - 0.0272)	54	56	58	60	62	64	66	68	70	70	72	72	74																																										
0.691 - 0.710 (0.0272 - 0.0280)	56	58	60	62	64	66	68	70	72	72	74																																												
0.711 - 0.730 (0.0280 - 0.0287)	58	60	62	64	66	68	70	72	74																																																
0.731 - 0.750 (0.0288 - 0.0295)	60	62	64	66	68	70	72																																																		
0.751 - 0.770 (0.0296 - 0.0303)	62	64	66	68	70	72																																																			
0.771 - 0.790 (0.0304 - 0.0311)	64	66	68	70	72																																																				
0.791 - 0.810 (0.0311 - 0.0319)	66	68	70	72																																																				
0.811 - 0.830 (0.0319 - 0.0327)	68	70	72																																																					
0.831 - 0.850 (0.0327 - 0.0335)	70	72																																																						
0.851 - 0.870 (0.0335 - 0.0343)	72																																																								

06	5.060 (0.1992)	30	5.300 (0.2087)	54	5.540 (0.2181)
08	5.080 (0.2000)	32	5.320 (0.2094)	56	5.560 (0.2189)
10	5.100 (0.2008)	34	5.340 (0.2102)	58	5.580 (0.2197)
12	5.120 (0.2016)	36	5.360 (0.2110)	60	5.600 (0.2205)
14	5.140 (0.2024)	38	5.380 (0.2118)	62	5.620 (0.2213)
16	5.160 (0.2031)	40	5.400 (0.2126)	64	5.640 (0.2220)
18	5.180 (0.2039)	42	5.420 (0.2134)	66	5.660 (0.2228)
20	5.200 (0.2047)	44	5.440 (0.2142)	68	5.680 (0.2236)
22	5.220 (0.2055)	46	5.460 (0.2150)	70	5.700 (0.2244)
24	5.240 (0.2063)	48	5.480 (0.2157)	72	5.720 (0.2252)
26	5.260 (0.2071)	50	5.500 (0.2165)	74	5.740 (0.2260)
28	5.280 (0.2079)	52	5.520 (0.2173)		

09490_TACO_G0074

Fig. 116 Shim selection chart, part one—4.0L engine

	5.060 (0.1992)	5.080 (0.2000)	5.100 (0.2008)	5.120 (0.2016)	5.140 (0.2024)	5.160 (0.2031)	5.180 (0.2039)	5.200 (0.2047)	5.210 (0.2051)	5.220 (0.2055)	5.230 (0.2059)	5.240 (0.2063)	5.250 (0.2067)	5.260 (0.2071)	5.270 (0.2075)	5.280 (0.2079)	5.290 (0.2083)	5.300 (0.2087)	5.310 (0.2091)	5.320 (0.2094)	5.330 (0.2098)	5.340 (0.2102)	5.350 (0.2106)	5.360 (0.2110)	5.370 (0.2114)	5.380 (0.2118)	5.390 (0.2122)	5.400 (0.2126)	5.410 (0.2130)	5.420 (0.2134)	5.430 (0.2138)	5.440 (0.2142)	5.450 (0.2146)	5.460 (0.2150)	5.470 (0.2154)	5.480 (0.2157)	5.490 (0.2161)	5.500 (0.2165)	5.510 (0.2169)	5.520 (0.2173)	5.530 (0.2177)	5.540 (0.2181)	5.550 (0.2185)	5.560 (0.2189)	5.570 (0.2193)	5.580 (0.2197)	5.590 (0.2201)	5.600 (0.2205)	5.620 (0.2213)	5.640 (0.2220)	5.660 (0.2228)	5.680 (0.2236)	5.700 (0.2244)	5.720 (0.2252)	5.740 (0.2260)
0.000 - 0.020 (0.0000 - 0.0008)																											06	08	08	10	10	12	12	14	14	16	16	18	18	20	20	22	22	24	24	26	26	28	30	32	34	36	38	40	42
0.021 - 0.040 (0.0008 - 0.0016)																									06	08	08	10	10	12	12	14	14	16	16	18	18	20	20	22	22	24	24	26	26	28	28	30	32	34	36	38	40	42	44
0.041 - 0.060 (0.0016 - 0.0024)																							06	08	08	10	10	12	12	14	14	16	16	18	18	20	20	22	22	24	24	26	26	28	28	30	30	32	34	36	38	40	42	44	46
0.061 - 0.080 (0.0024 - 0.0031)																					06	08	08	10	10	12	12	14	14	16	16	18	18	20	20	22	22	24	24	26	26	28	28	30	30	32	32	34	36	38	40	42	44	46	48
0.081 - 0.100 (0.0032 - 0.0039)																			06	08	08	10	10	12	12	14	14	16	16	18	18	20	20	22	22	24	24	26	26	28	28	30	30	32	32	34	34	36	38	40	42	44	46	48	50
0.101 - 0.120 (0.0040 - 0.0047)																	06	08	08	10	10	12	12	14	14	16	16	18	18	20	20	22	22	24	24	26	26	28	28	30	30	32	32	34	34	36	36	38	40	42	44	46	48	50	52
0.121 - 0.140 (0.0048 - 0.0055)															06	08	08	10	10	12	12	14	14	16	16	18	18	20	20	22	22	24	24	26	26	28	28	30	30	32	32	34	34	36	36	38	38	40	42	44	46	48	50	52	54
0.141 - 0.160 (0.0056 - 0.0063)													06	08	08	10	10	12	12	14	14	16	16	18	18	20	20	22	22	24	24	26	26	28	28	30	30	32	32	34	34	36	36	38	38	40	40	42	44	46	48	50	52	54	56
0.161 - 0.180 (0.0063 - 0.0071)											06	08	08	10	10	12	12	14	14	16	16	18	18	20	20	22	22	24	24	26	26	28	28	30	30	32	32	34	34	36	36	38	38	40	40	42	42	44	46	48	50	52	54	56	58
0.181 - 0.200 (0.0071 - 0.0079)									06	08	08	10	10	12	12	14	14	16	16	18	18	20	20	22	22	24	24	26	26	28	28	30	30	32	32	34	34	36	36	38	38	40	40	42	42	44	44	46	48	50	52	54	56	58	60
0.201 - 0.220 (0.0079 - 0.0087)								08	08	10	10	12	12	14	14	16	16	18	18	20	20	22	22	24	24	26	26	28	28	30	30	32	32	34	34	36	36	38	38	40	40	42	42	44	44	46	46	48	50	52	54	56	58	60	62
0.221 - 0.240 (0.0087 - 0.0094)							08	10	10	12	12	14	14	16	16	18	18	20	20	22	22	24	24	26	26	28	28	30	30	32	32	34	34	36	36	38	38	40	40	42	42	44	44	46	46	48	48	50	52	54	56	58	60	62	64
0.241 - 0.260 (0.0095 - 0.0102)						08	10	12	12	14	14	16	16	18	18	20	20	22	22	24	24	26	26	28	28	30	30	32	32	34	34	36	36	38	38	40	40	42	42	44	44	46	46	48	48	50	50	52	54	56	58	60	62	64	66
0.261 - 0.280 (0.0103 - 0.0110)					08	10	12	14	14	16	16	18	18	20	20	22	22	24	24	26	26	28	28	30	30	32	32	34	34	36	36	38	38	40	40	42	42	44	44	46	46	48	48	50	50	52	52	54	56	58	60	62	64	66	68
0.281 - 0.289 (0.0111 - 0.0114)				06	08	10	12	14	16	16	18	18	20	20	22	22	24	24	26	26	28	28	30	30	32	32	34	34	36	36	38	38	40	40	42	42	44	44	46	46	48	48	50	50	52	52	54	54	56	58	60	62	64	66	68
0.391 - 0.410 (0.0154 - 0.0161)	12	14	16	18	20	22	24	26	28	28	30	30	32	32	34	34	36	36	38	38	40	40	42	42	44	44	46	46	48	48	50	50	52	52	54	54	56	56	58	58	60	60	62	62	64	64	66	66	68	70	72				
0.411 - 0.430 (0.0162 - 0.0169)	14	16	18	20	22	24	26	28	30	30	32	32	34	34	36	36	38	38	40	40	42	42	44	44	46	46	48	48	50	50	52	52	54	54	56	56	58	58	60	60	62	62	64	64	66	66	68	68	70	72					
0.431 - 0.450 (0.0170 - 0.0177)	16	18	20	22	24	26	28	30	32	32	34	34	36	36	38	38	40	40	42	42	44	44	46	46	48	48	50	50	52	52	54	54	56	56	58	58	60	60	62	62	64	64	66	66	68	68	70	70	72						
0.451 - 0.470 (0.0178 - 0.0185)	18	20	22	24	26	28	30	32	34	34	36	36	38	38	40	40	42	42	44	44	46	46	48	48	50	50	52	52	54	54	56	56	58	58	60	60	62	62	64	64	66	66	68	68	70	70	72	72							
0.471 - 0.490 (0.0185 - 0.0193)	20	22	24	26	28	30	32	34	36	36	38	38	40	40	42	42	44	44	46	46	48	48	50	50	52	52	54	54	56	56	58	58	60	60	62	62	64	64	66	66	68	68	70	70	72	72	74								
0.491 - 0.510 (0.0193 - 0.0201)	22	24	26	28	30	32	34	36	38	38	40	40	42	42	44	44	46	46	48	48	50	50	52	52	54	54	56	56	58	58	60	60	62	62	64	64	66	66	68	68	70	70	72	72	74										
0.511 - 0.530 (0.0201 - 0.0209)	24	26	28	30	32	34	36	38	40	40	42	42	44	44	46	46	48	48	50	50	52	52	54	54	56	56	58	58	60	60	62	62	64	64	66	66	68	68	70	70	72	72	74												
0.531 - 0.550 (0.0209 - 0.0217)	26	28	30	32	34	36	38	40	42	42	44	44	46	46	48	48	50	50	52	52	54	54	56	56	58	58	60	60	62	62	64	64	66	66	68	68	70	70	72	72	74														
0.551 - 0.570 (0.0217 - 0.0224)	28	30	32	34	36	38	40	42	44	44	46	46	48	48	50	50	52	52	54	54	56	56	58	58	60	60	62	62	64	64	66	66	68	68	70	70	72	72	74																
0.571 - 0.590 (0.0225 - 0.0232)	30	32	34	36	38	40	42	44	46	46	48	48	50	50	52	52	54	54	56	56	58	58	60	60	62	62	64	64	66	66	68	68	70	70	72	72	74																		
0.591 - 0.610 (0.0233 - 0.0240)	32	34	36	38	40	42	44	46	48	48	50	50	52	52	54	54	56	56	58	58	60	60	62	62	64	64	66	66	68	68	70	70	72	72	74																				
0.611 - 0.630 (0.0241 - 0.0248)	34	36	38	40	42	44	46	48	50	50	52	52	54	54	56	56	58	58	60	60	62	62	64	64	66	66	68	68	70	70	72	72	74																						
0.631 - 0.650 (0.0248 - 0.0256)	36	38	40	42	44	46	48	50	52	52	54	54	56	56	58	58	60	60	62	62	64	64	66	66	68	68	70	70	72	72	74																								
0.651 - 0.670 (0.0256 - 0.0264)	38	40	42	44	46	48	50	52	54	54	56	56	58	58	60	60	62	62	64	64	66	66	68	68	70	70	72	72	74																										
0.671 - 0.690 (0.0264 - 0.0272)	40	42	44	46	48	50	52	54	56	56	58	58	60	60	62	62	64	64	66	66	68	68	70	70	72	72	74																												
0.691 - 0.710 (0.0272 - 0.0280)	42	44	46	48	50	52	54	56	58	58	60	60	62	62	64	64	66	66	68	68	70	70	72	72	74																														
0.711 - 0.730 (0.0280 - 0.0287)	44	46	48	50	52	54	56	58	60	60	62	62	64	64	66	66	68	68	70	70	72	72	74																																
0.731 - 0.750 (0.0288 - 0.0295)	46	48	50	52	54	56	58	60	62	62	64	64	66	66	68	68	70	70	72	72	74																																		
0.751 - 0.770 (0.0296 - 0.0303)	48	50	52	54	56	58	60	62	64	64	66	66	68	68	70	70	72	72	74																																				
0.771 - 0.790 (0.0304 - 0.0311)	50	52	54	56	58	60	62	64	66	66	68	68	70	70	72	72	74																																						
0.791 - 0.810 (0.0311 - 0.0319)	52	54	56	58	60	62	64	66	68	68	70	70	72	72	74																																								
0.811 - 0.830 (0.0319 - 0.0327)	54	56	58	60	62	64	66	68	70	70	72	72	74																																										
0.831 - 0.850 (0.0327 - 0.0335)	56	58	60	62	64	66	68	70	72	72	74																																												
0.851 - 0.870 (0.0335 - 0.0343)	58	60	62	64	66	68	70	72	74																																														
0.871 - 0.890 (0.0343 - 0.0350)	60	62	64	66	68	70	72																																																
0.891 - 0.910 (0.0351 - 0.0358)	62	64	66	68	70	72																																																	
0.911 - 0.930 (0.0359 - 0.0366)	64	66	68	70	72																																																		
0.931 - 0.950 (0.0367 - 0.0374)	66	68	70	72																																																		
0.951 - 0.970 (0.0374 - 0.0382)	68	70	72																																																			
0.971 - 0.990 (0.0382 - 0.0390)	70	72																																																					
0.991 - 1.010 (0.0390 - 0.0398)	72																																																						

06	5.060 (0.1992)	30	5.300 (0.2087)	54	5.540 (0.2181)
08	5.080 (0.2000)	32	5.320 (0.2094)	56	5.560 (0.2189)
10	5.100 (0.2008)	34	5.340 (0.2102)	58	5.580 (0.2197)
12	5.120 (0.2016)	36	5.360 (0.2110)	60	5.600 (0.2205)
14	5.140 (0.2024)	38	5.380 (0.2118)	62	5.620 (0.2213)
16	5.160 (0.2031)	40	5.400 (0.2126)	64	5.640 (0.2220)
18	5.180 (0.2039)	42	5.420 (0.2134)	66	5.660 (0.2228)
20	5.200 (0.2047)	44	5.440 (0.2142)	68	5.680 (0.2236)
22	5.220 (0.2055)	46	5.460 (0.2150)	70	5.700 (0.2244)
24	5.240 (0.2063)	48	5.480 (0.2157)	72	5.720 (0.2252)
26	5.260 (0.2071)	50	5.500 (0.2165)	74	5.740 (0.2260)
28	5.280 (0.2079)	52	5.520 (0.2173)		

09490_TACO_G0075

Fig. 117 Shim selection chart, part two—4.0L engine

Disconnect the ventilation hose. Disconnect the VSV connectors.

6. Disconnect the throttle body motor connector. Separate the three wire harness clamps and hose clamp.

7. Remove the two bolts and the throttle body bracket. Remove the bolt and the oil baffle plate. Remove the four bolts and the two surge tank stays.

8. Remove the two nuts. Remove the four bolts, intake air surge tank and gasket.

9. Remove the ignition coil assembly.

10. Remove the cylinder head cover retaining bolts. Remove the cylinder head cover.

11. Turn the crankshaft pulley until its groove and the timing mark "0" of the timing chain cover are aligned. If not aligned at TDC of the compression stroke, turn the crankshaft one complete revolution, in the direction of rotation.

12. Using a feeler gauge, check and record the valve clearance on the following valves: right bank, exhaust number three and intake number one, left bank exhaust number two and intake number six.

13. Rotate the crankshaft 240 degrees clockwise and using a feeler gauge, check and record the following valves: right bank, exhaust number five and intake number three, left bank exhaust number four and intake number two.

14. Rotate the crankshaft 240 degrees clockwise and using a feeler gauge, check and record the following valves: right bank, exhaust number one and intake number five, left bank exhaust number six and intake number four.

15. If adjustment is required, position the engine at TDC on the compression stroke.

16. Place paint marks on the number one chain links corresponding to the timing marks of the camshaft timing gears.

17. Remove the chain tensioner assembly number one. Remove the number two camshaft.

18. Remove the chain tensioner assembly number two. Remove the camshaft.

19. Remove the number four camshaft subassembly. Remove the chain tensioner assembly number three.

20. Remove the number three camshaft subassembly.

21. Remove the valve lifters.

22. Determine the replacement adjusting shim size according to the following formula or use the adjusting shim charts.

23. Using a micrometer, measure the thickness of the removed shim. Calculate the thickness of a new shim so that the valve clearance comes within the specified value.

- T: Thickness of the removed shim
- A: Measured valve clearance
- N: Thickness of the new shim

a. Intake: N = T + A
b. Exhaust: N = T + A

24. Select a new lifter with a thickness as close as possible to the calculated value.

25. Install removed components in the reverse order of the removal procedure.

26. When installing the cylinder head cover, apply a continuous bead (0.08–0.12 inch) of seal packing, part number 08826-00080 or equivalent, to the cylinder head as indicated in the illustration. Install the seal washers onto the bolts. Install the cylinder head cover bolts and nuts. Tighten bolts "A" to 7.4 ft. lbs. Tighten bolts "B" to 80 inch lbs. Tighten nuts to 80 inch lbs.

➡Be sure to remove any oil from the contact surfaces of the cylinder head cover and the cylinder head. Install the cover within three minutes after applying the seal packing. Tighten the bolts to specification within fifteen minutes after installing the cover. Do not add engine oil for at least two hours after installing the cover.

27. Continue the installation in the reverse order of the removal procedure.

28. Be sure to fill the engine with the proper grade and type engine coolant.

29. Be sure to fill the engine with the proper grade and type engine oil.

30. Start the engine and check for leaks. Correct as required.

31. Use the Techstream diagnostic tool, or equivalent and reprogram the required systems.

4.6L Engine

Valve lash is not adjustable for this engine.

5.7L Engine

Valve lash is not adjustable for this engine.

ENGINE PERFORMANCE & EMISSION CONTROLS

COMPONENT LOCATIONS

See Figures 118 through 131

CAMSHAFT POSITION (CMP) SENSOR

LOCATION

See Figure 132.

REMOVAL & INSTALLATION

4.0L Engine

1. Before servicing the vehicle, refer to the Precautions Section.
2. Disconnect the negative battery cable. Tape the cable with insulating tape.
3. Remove V-bank cover sub-assembly.
4. Disconnect the sensor connector.
5. Remove the bolt and sensor.

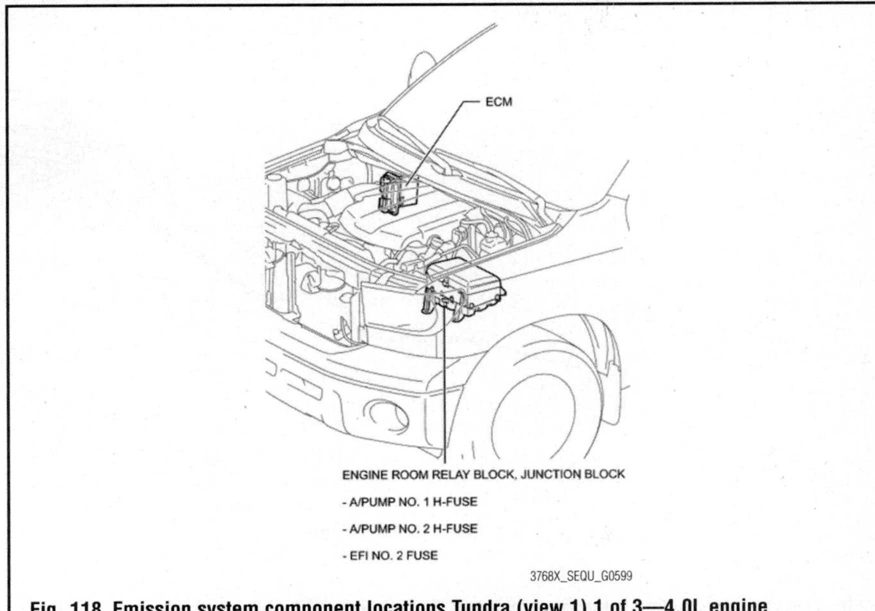

Fig. 118 Emission system component locations Tundra (view 1) 1 of 3—4.0L engine

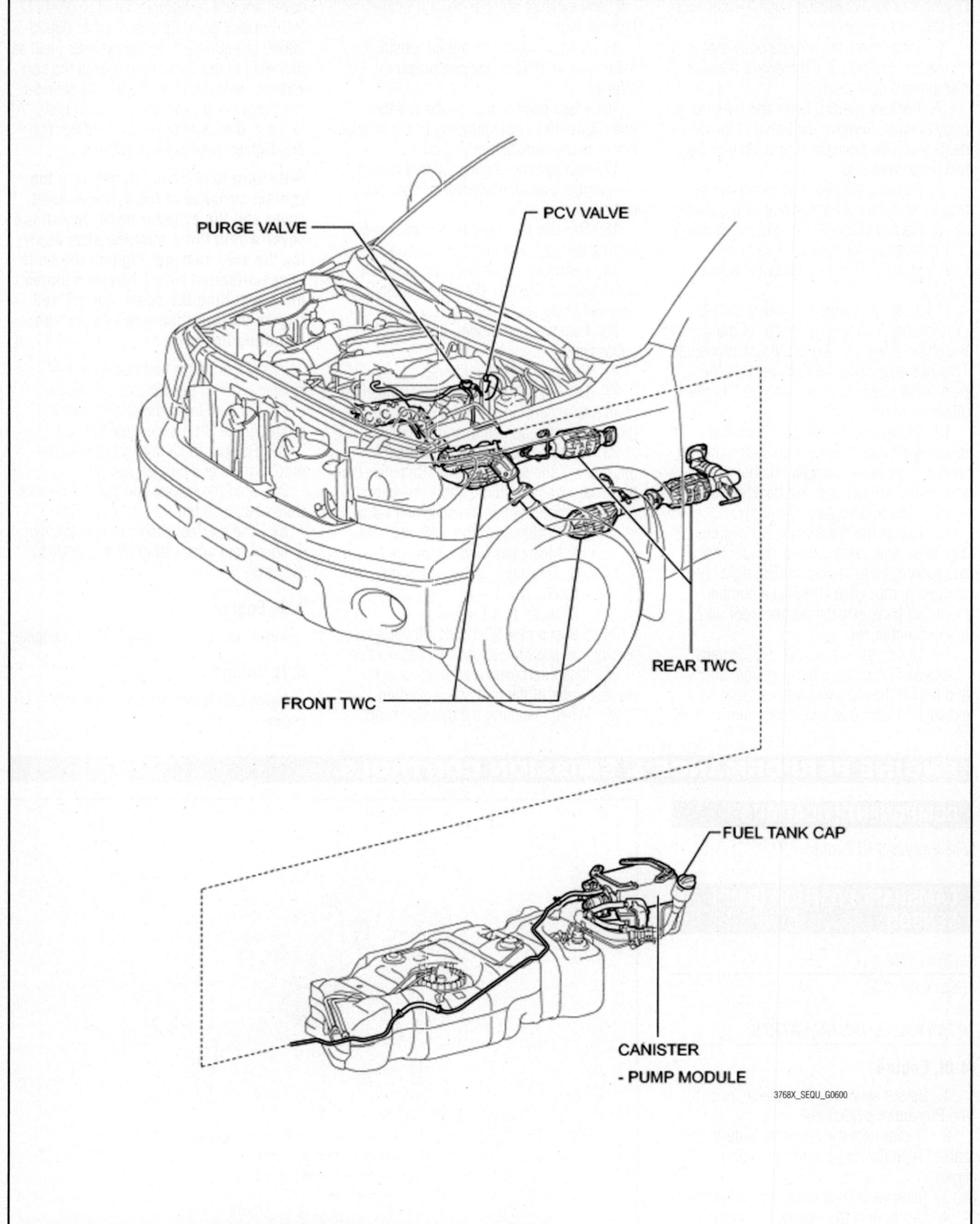

Fig. 119 Emission system component locations Tundra (view 1) 2 of 3—4.0L engine

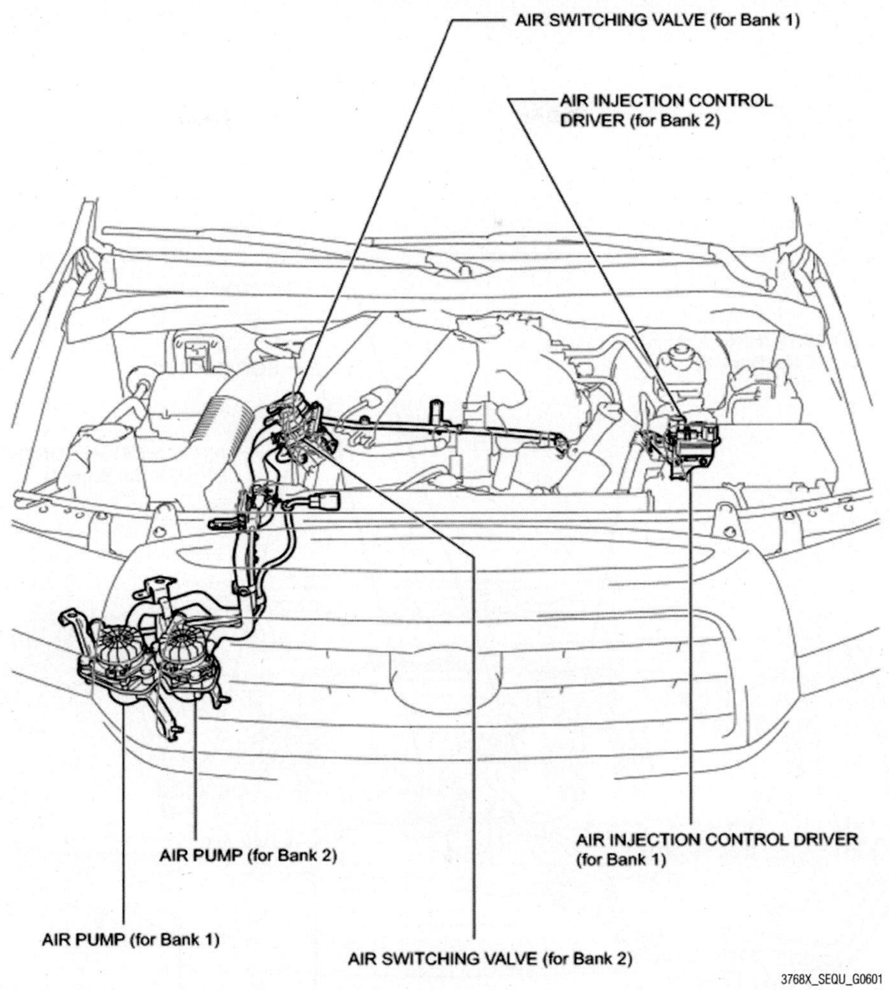

AIR SWITCHING VALVE (for Bank 1)
AIR INJECTION CONTROL DRIVER (for Bank 2)
AIR PUMP (for Bank 2)
AIR INJECTION CONTROL DRIVER (for Bank 1)
AIR PUMP (for Bank 1)
AIR SWITCHING VALVE (for Bank 2)
3768X_SEQU_G0601

Fig. 120 Emission system component locations Tundra (view 1) 3 of 3—4.0L engine

ECM
ENGINE ROOM RELAY BLOCK, JUNCTION BLOCK
- A/PUMP NO. 1 H-FUSE
- A/PUMP NO. 2 H-FUSE
- EFI NO. 1 FUSE
- EFI NO. 2 FUSE
- EFI RELAY
3768X_SEQU_G0602

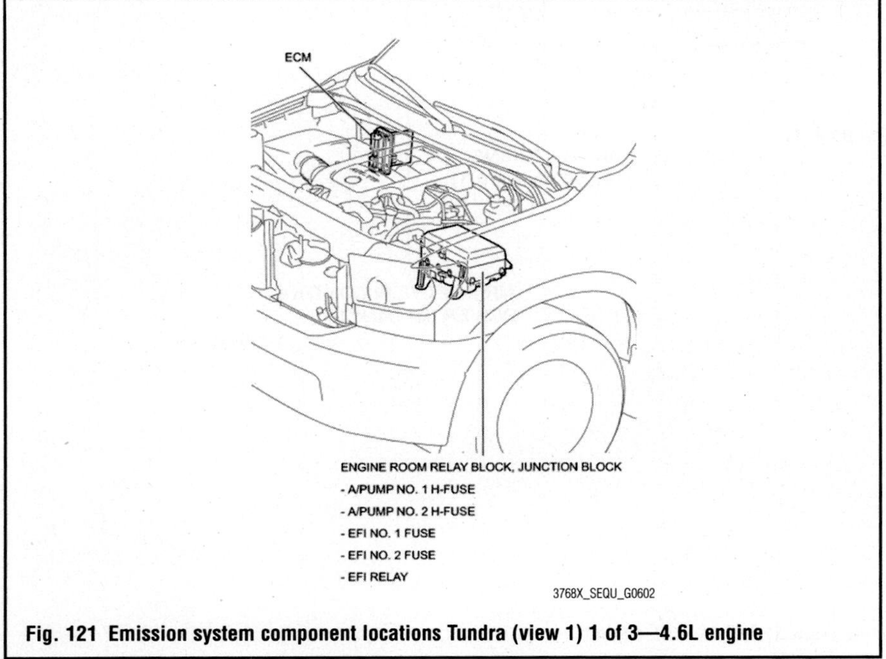

Fig. 121 Emission system component locations Tundra (view 1) 1 of 3—4.6L engine

To install:

➡Be sure to use new fasteners, as required.

6. Install the sensor with the bolt and tighten to 7 ft. lbs. (10 Nm).
7. Connect the sensor connector.
8. Install V-bank cover sub-assembly.
9. Use the Techstream diagnostic tool, or equivalent and reprogram the required systems.

4.6L Engine

1. Before servicing the vehicle, refer to the Precautions Section.
2. Disconnect the negative battery cable. Tape the cable with insulating tape.
3. Remove V-bank cover sub-assembly.
4. Disconnect the sensor connector.
5. Remove the bolt and sensor.

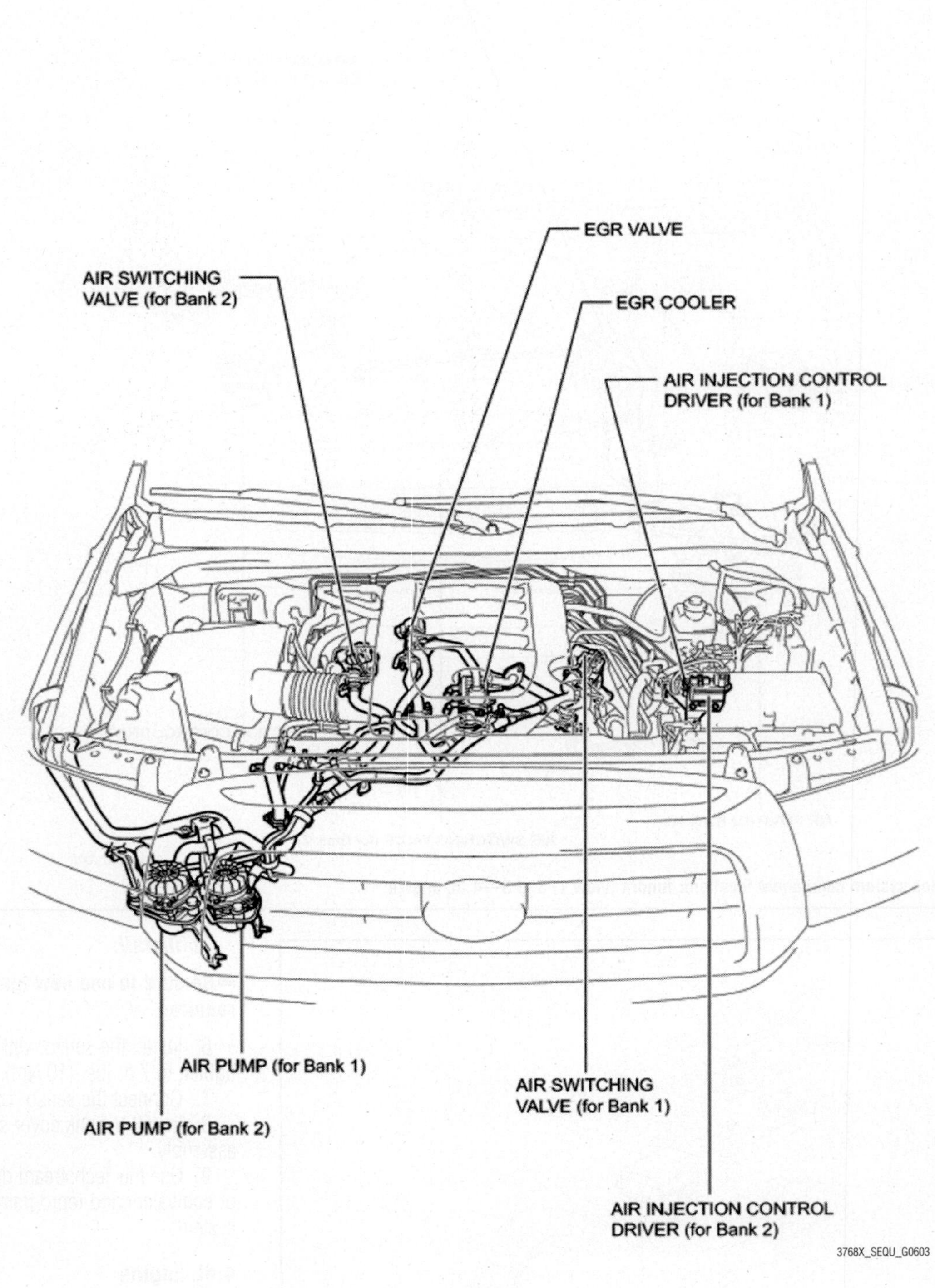

Fig. 122 Emission system component locations Tundra (view 1) 2 of 3—4.6L engine

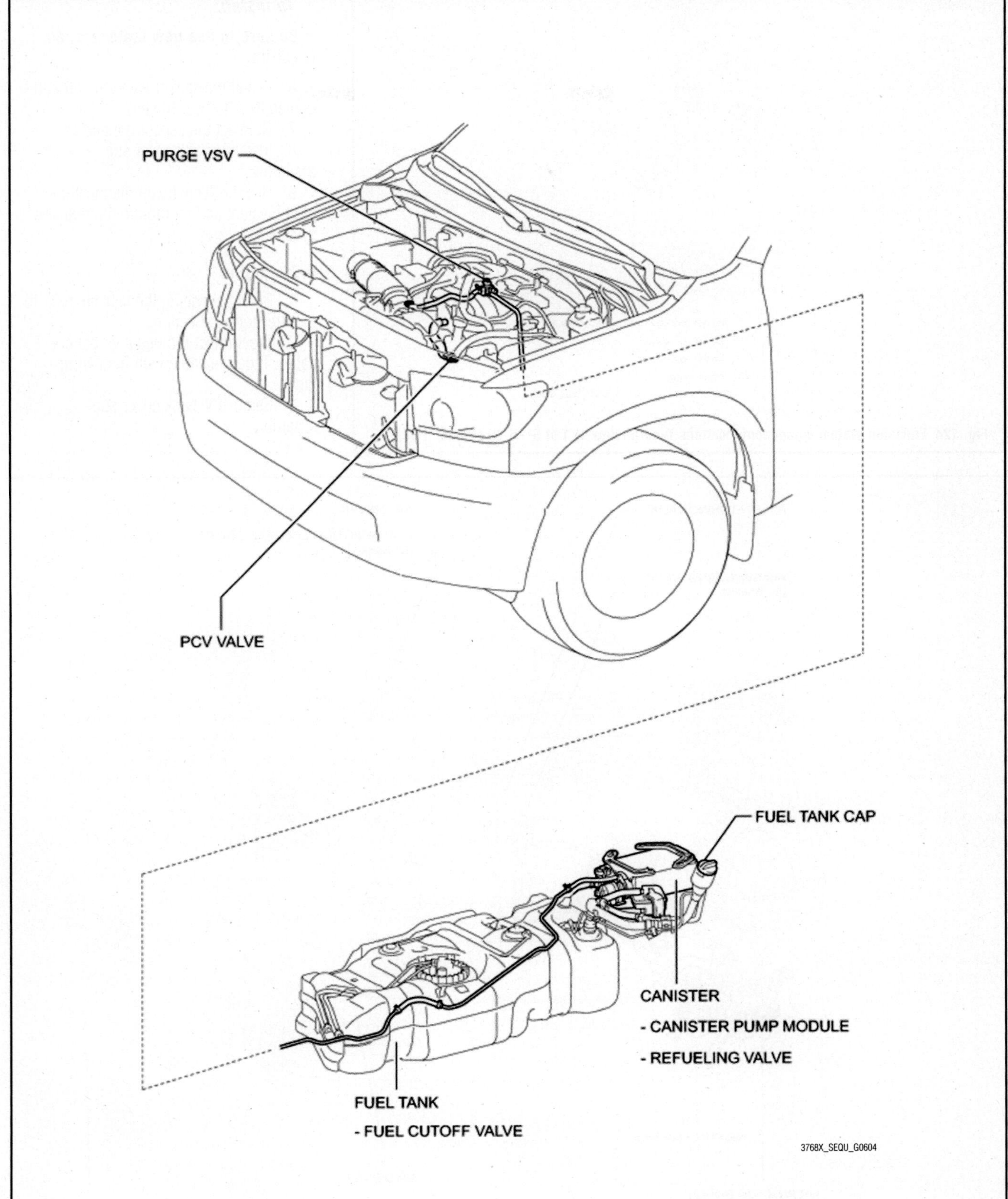

Fig. 123 Emission system component locations Tundra (view 1) 3 of 3—4.6L engine

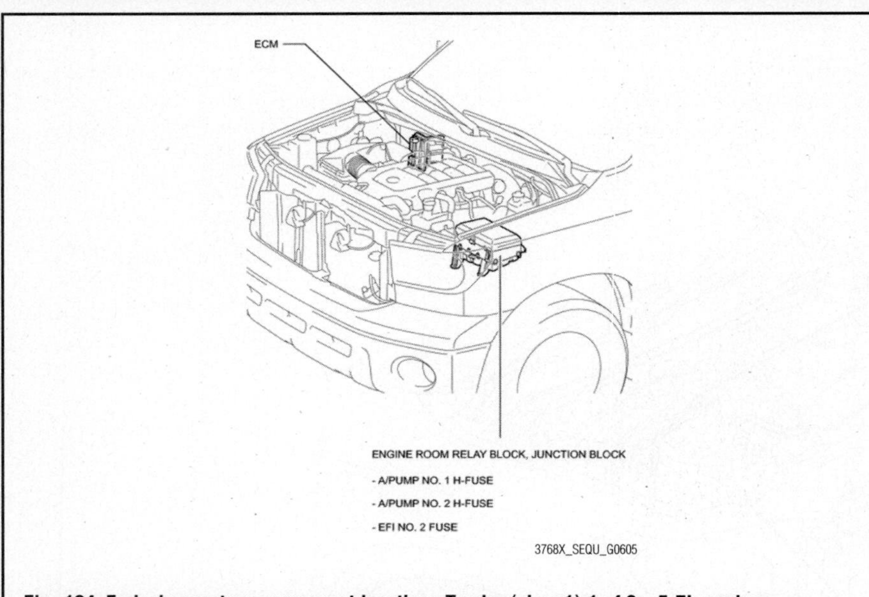

Fig. 124 Emission system component locations Tundra (view 1) 1 of 3—5.7L engine

To install:

➡Be sure to use new fasteners, as required.

6. Install the sensor with the bolt and tighten to 7 ft. lbs. (10 Nm).
7. Connect the sensor connector.
8. Install V-bank cover sub-assembly.
9. Use the Techstream diagnostic tool, or equivalent and reprogram the required systems.

5.7L Engine

1. Before servicing the vehicle, refer to the Precautions Section.
2. Disconnect the negative battery cable. Tape the cable with insulating tape.
3. Remove V-bank cover sub-assembly.

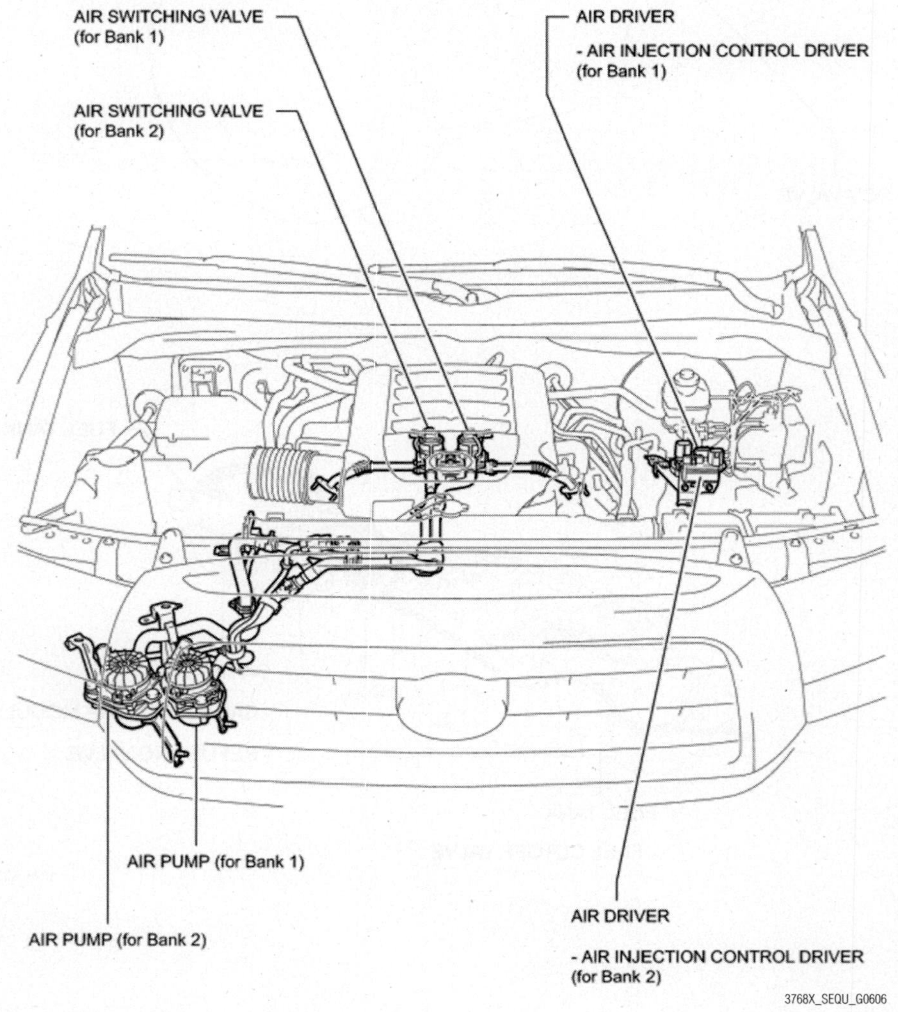

Fig. 125 Emission system component locations Tundra (view 1) 2 of 3—5.7L engine

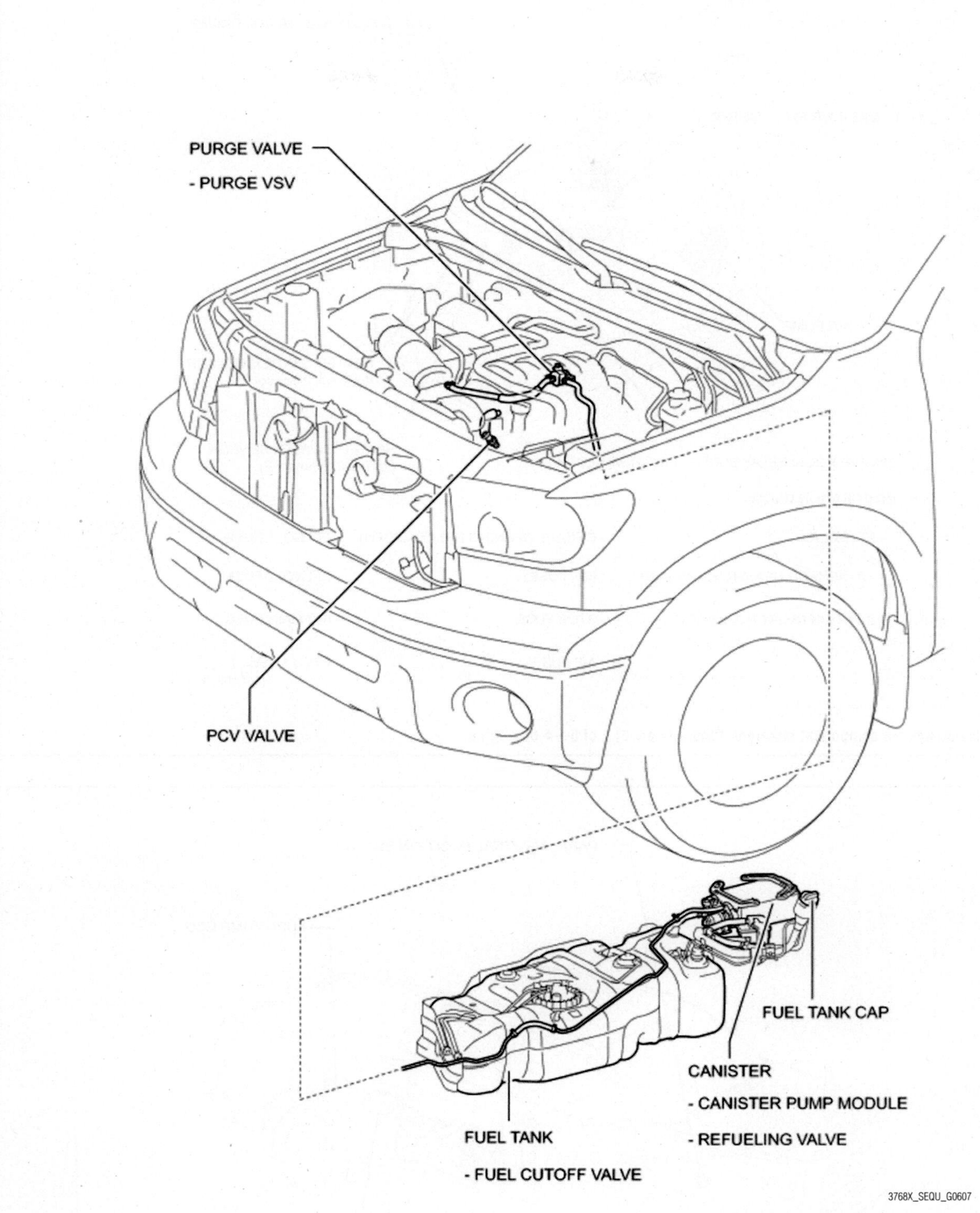

Fig. 126 Emission system component locations Tundra (view 1) 3 of 3—5.7L engine

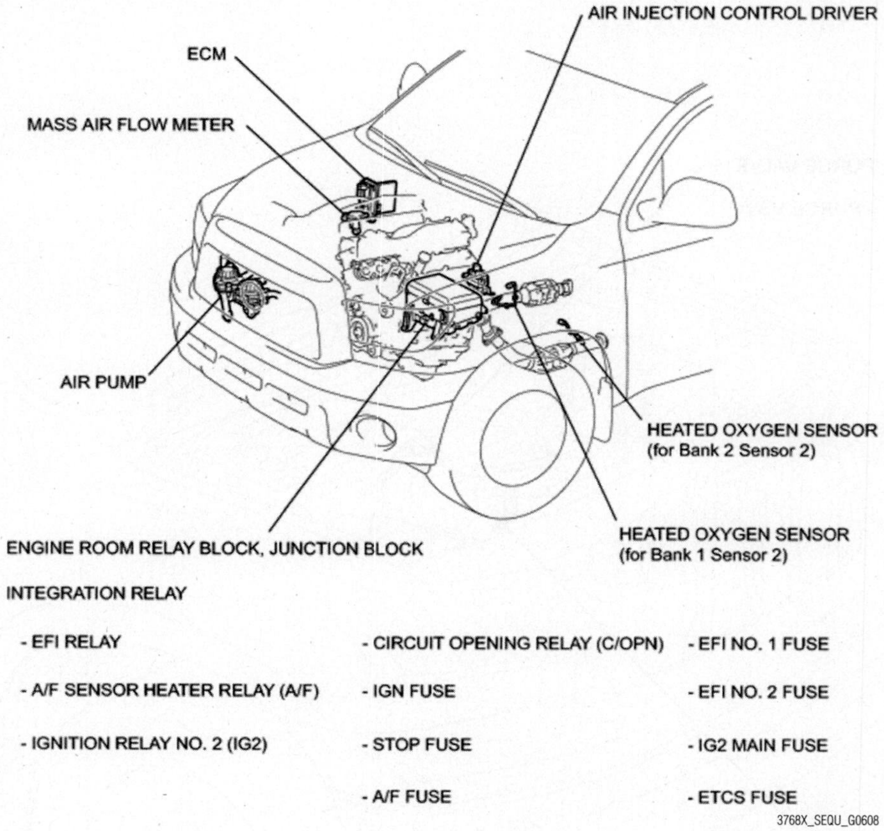

AIR INJECTION CONTROL DRIVER
ECM
MASS AIR FLOW METER
AIR PUMP
HEATED OXYGEN SENSOR
(for Bank 2 Sensor 2)
HEATED OXYGEN SENSOR
(for Bank 1 Sensor 2)
ENGINE ROOM RELAY BLOCK, JUNCTION BLOCK
INTEGRATION RELAY
- EFI RELAY
- A/F SENSOR HEATER RELAY (A/F)
- IGNITION RELAY NO. 2 (IG2)
- CIRCUIT OPENING RELAY (C/OPN)
- IGN FUSE
- STOP FUSE
- A/F FUSE
- EFI NO. 1 FUSE
- EFI NO. 2 FUSE
- IG2 MAIN FUSE
- ETCS FUSE
3768X_SEQU_G0608

Fig. 127 Emission system component locations Tundra (view 2) 1 of 5—4.0L engine

PARK / NEUTRAL POSITION SWITCH
FUEL PUMP ECU
CANISTER
FUEL PUMP
- CANISTER PUMP MODULE
3768X_SEQU_G0609

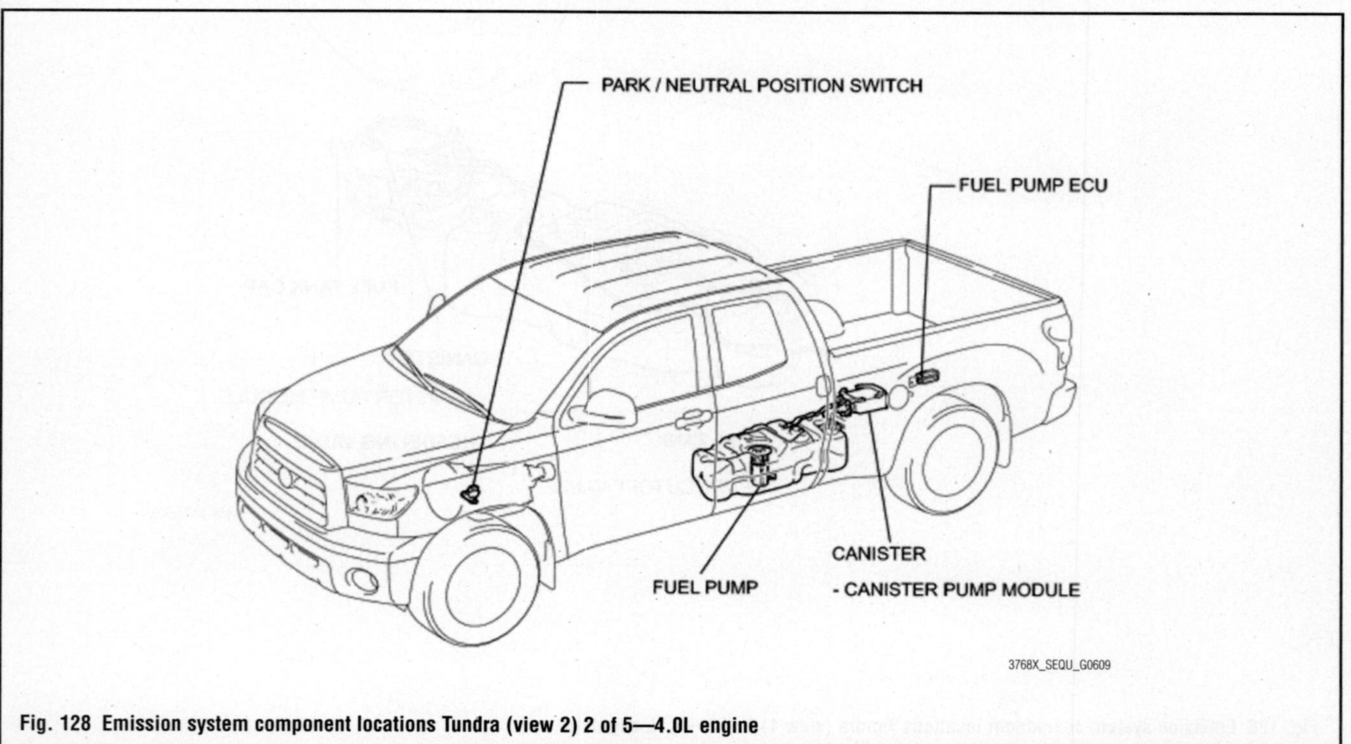

Fig. 128 Emission system component locations Tundra (view 2) 2 of 5—4.0L engine

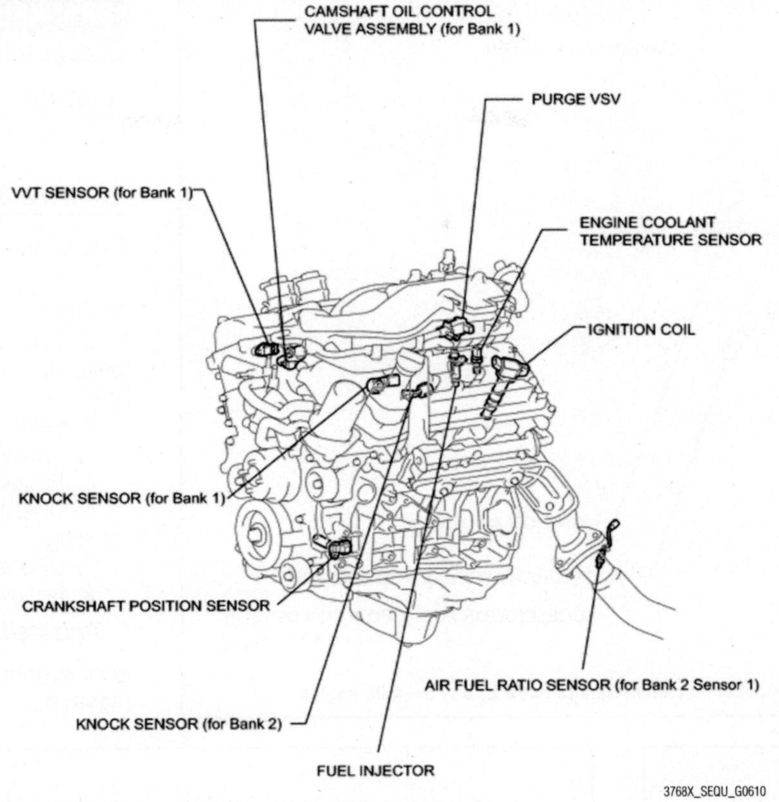

Fig. 129 Emission system component locations Tundra (view 2) 3 of 5—4.0L engine

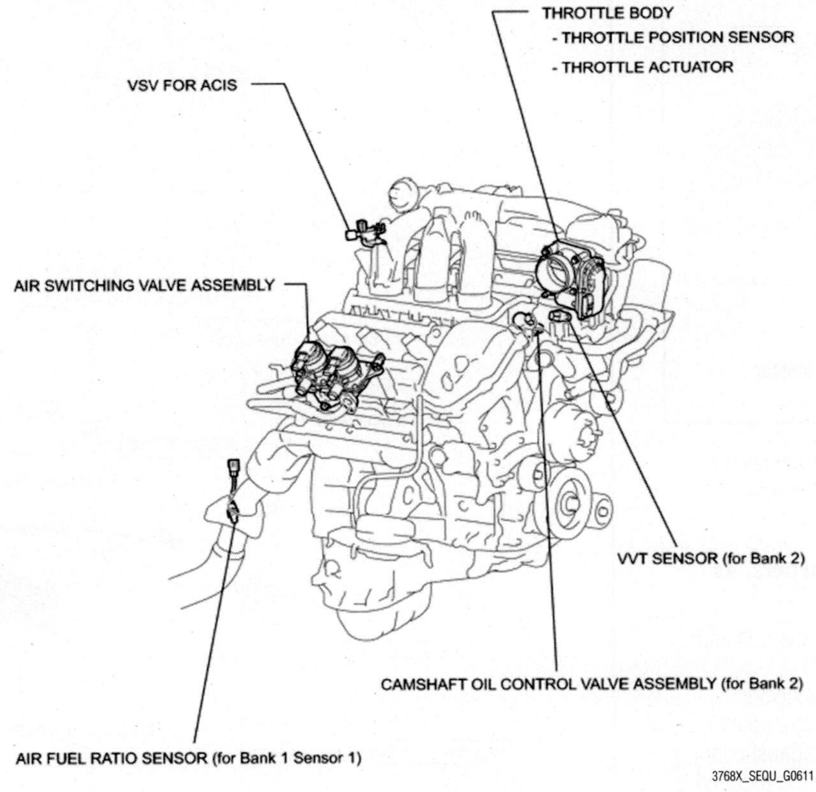

Fig. 130 Emission system component locations Tundra (view 2) 4 of 5—4.0L engine

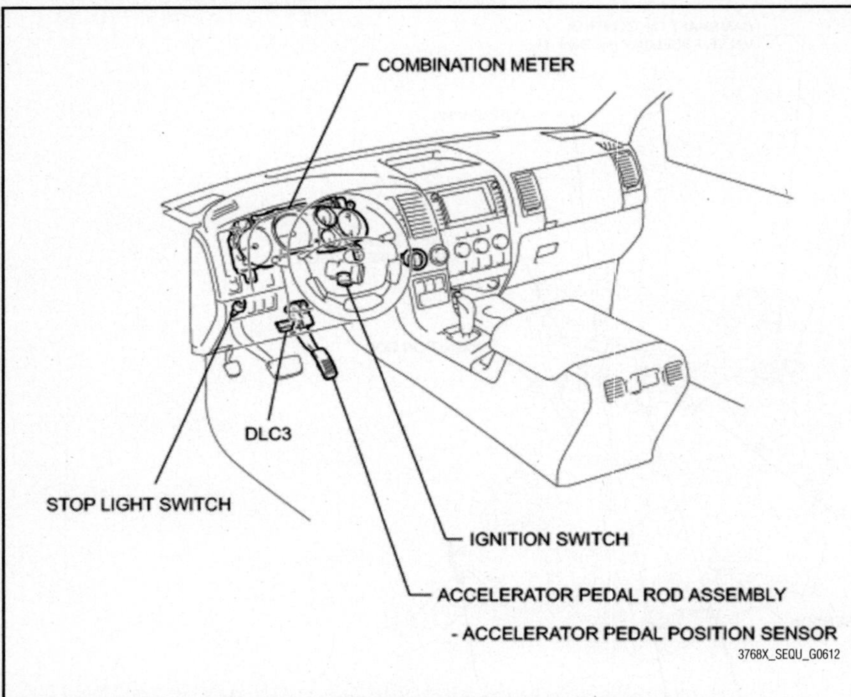

Fig. 131 Emission system component locations Tundra (view 2) 5 of 5—4.0L engine

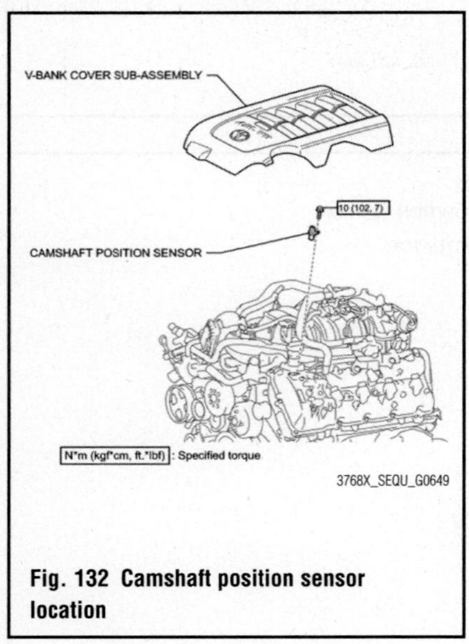

Fig. 132 Camshaft position sensor location

4. Disconnect the sensor connector.
5. Remove the bolt and sensor.

To install:

➡Be sure to use new fasteners, as required.

6. Install the sensor with the bolt and tighten to 7 ft. lbs. (10 Nm).
7. Connect the sensor connector.
8. Install V-bank cover sub-assembly.
9. Use the Techstream diagnostic tool, or equivalent and reprogram the required systems.

CRANKSHAFT POSITION (CKP) SENSOR

LOCATION

See Figures 133 through 135.

REMOVAL & INSTALLATION

4.0L Engine

1. Before servicing the vehicle, refer to the Precautions Section.
2. Disconnect the negative battery cable. Tape the cable with insulating tape.
3. Remove no. 1 engine under cover.
4. Remove fan and alternator v belt.
5. Remove alternator assembly.
6. Disconnect cooler compressor assembly.
7. Disconnect the sensor connector.
8. Remove the bolt and sensor.

To install:

➡Be sure to use new fasteners, as required.

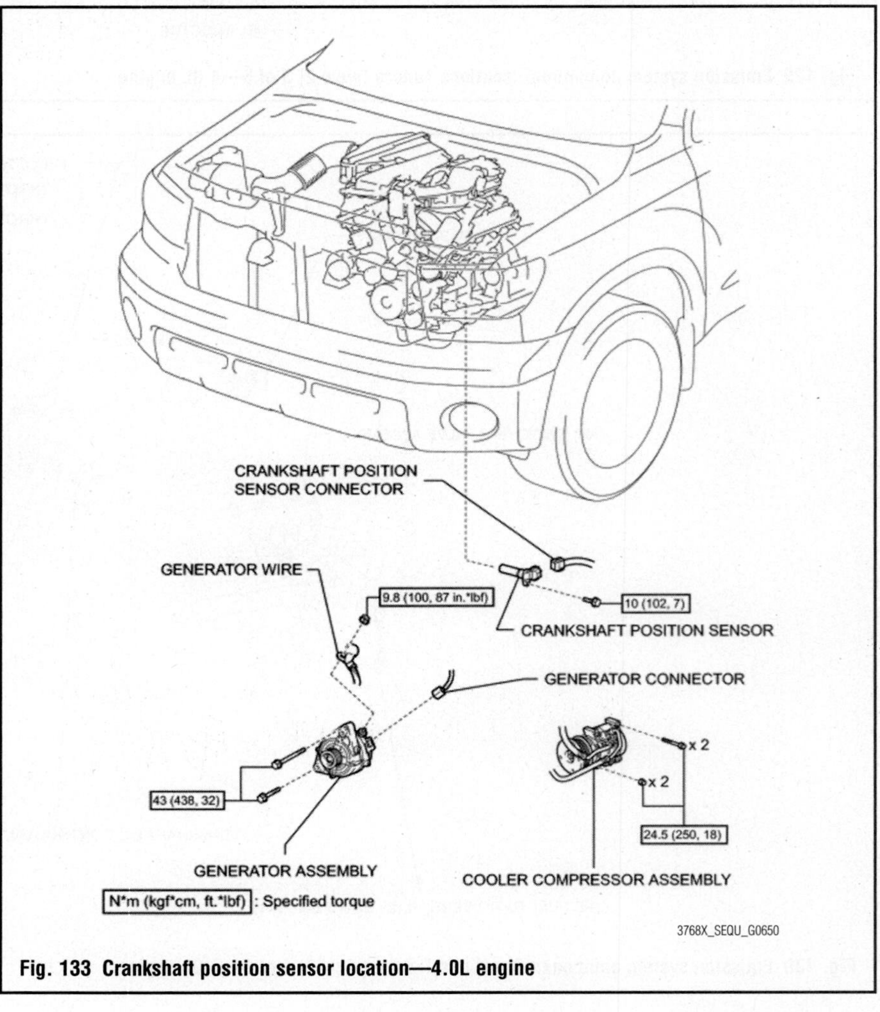

Fig. 133 Crankshaft position sensor location—4.0L engine

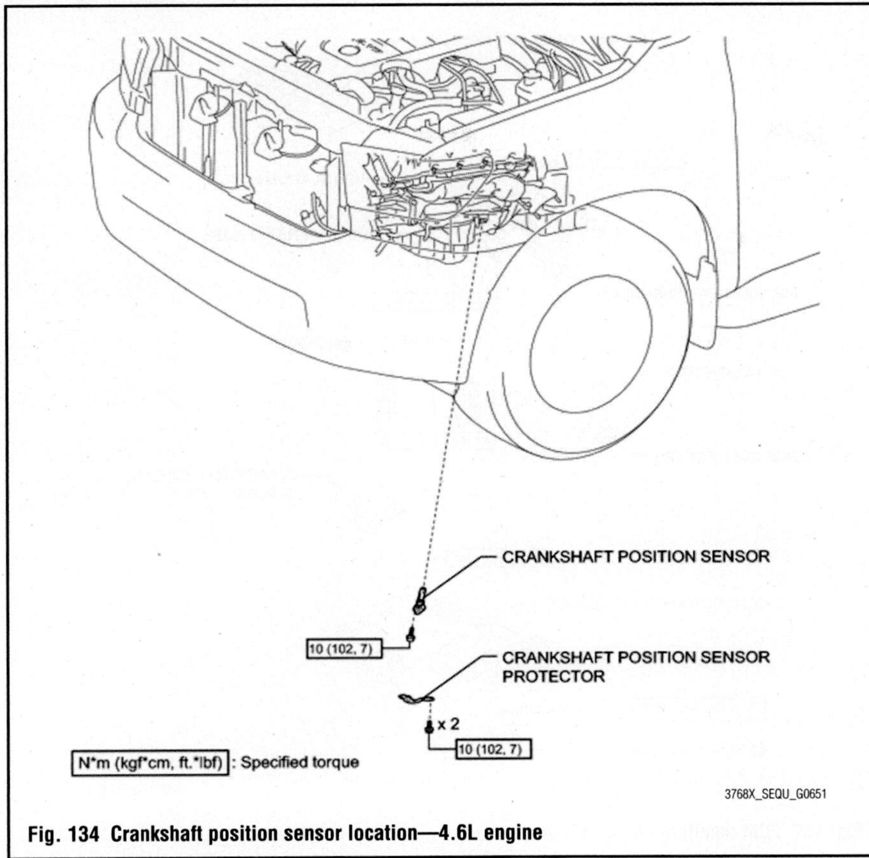

Fig. 134 Crankshaft position sensor location—4.6L engine

9. Install crankshaft position sensor.
10. Install the sensor with the bolt. Tighten bolt to 57 inch lbs. (6.5 nm).
11. Connect the sensor connector.
12. Reverse removal procedure.
13. Connect cable to negative battery terminal.
14. Use the Techstream diagnostic tool, or equivalent and reprogram the required systems.

4.6L Engine

1. Before servicing the vehicle, refer to the Precautions Section.
2. Disconnect the negative battery cable. Tape the cable with insulating tape.
3. Remove the sensor protector.
4. Disconnect the electrical connector.
5. Remove the retaining bolts.
6. Remove the component from its mounting.

To install:

➡Be sure to use new fasteners, as required.

7. Installation is the reverse of the removal procedure.
8. Install the sensor with the bolt and tighten to 7 ft. lbs. (10 Nm)
9. Use the Techstream diagnostic tool, or equivalent and reprogram the required systems.

5.7L Engine

1. Before servicing the vehicle, refer to the Precautions Section.
2. Disconnect the negative battery cable. Tape the cable with insulating tape.
3. Remove the 2 bolts and sensor protector
4. Disconnect the sensor connector.
5. Remove the bolt and sensor

To install:

➡Be sure to use new fasteners, as required.

6. Install the sensor with the bolt and tighten to 7 ft. lbs. (10 Nm)
7. Connect the sensor connector
8. Install the sensor protector with the 2 bolts and tighten to 7 ft. lbs. (10 Nm)
9. Use the Techstream diagnostic tool, or equivalent and reprogram the required systems.

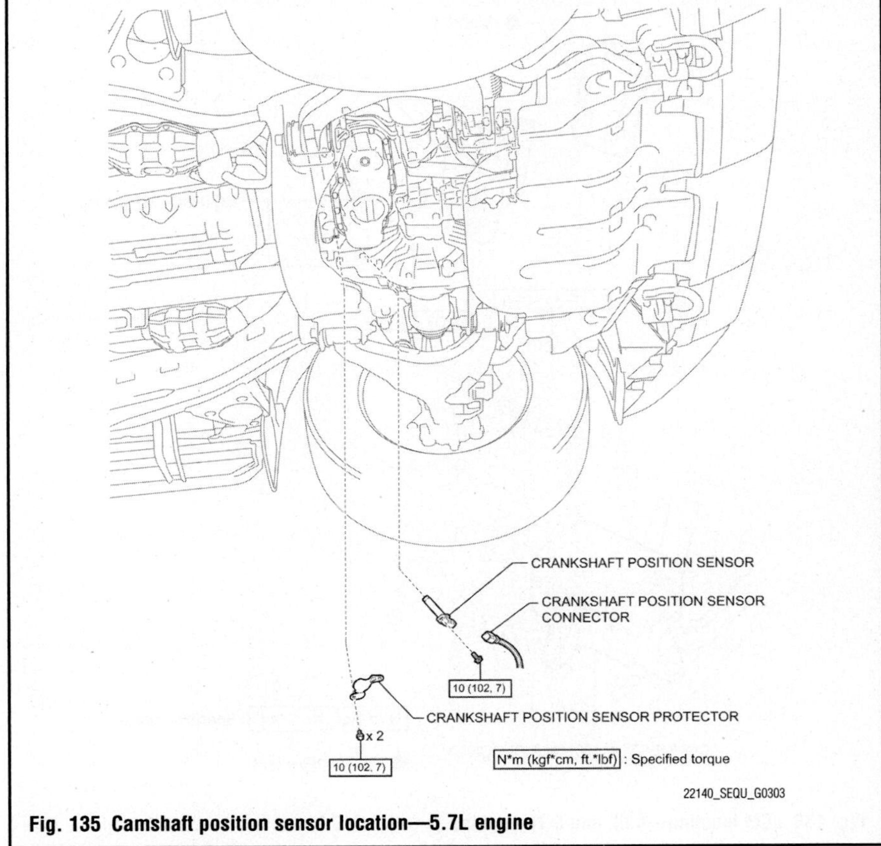

Fig. 135 Camshaft position sensor location—5.7L engine

ELECTRONIC CONTROL MODULE (ECM)

LOCATION

See Figures 136 and 137.

REMOVAL & INSTALLATION

See Figure 138.

1. Before servicing the vehicle, refer to the Precautions Section.
2. Disconnect the negative battery cable. Tape the cable with insulating tape.
3. On 4.6L engine remove the V- back cover sub assembly.
4. Remove the air cleaner assembly, as required.
5. Disconnect the connector holder block.
6. Remove the retaining bolts and move the connector block.
7. Disconnect the wire harness clamp, 4.0L engine.
8. Raise the retaining levers, as shown in the illustration.

➡Make sure that the lock lever is raised 90 degrees, before disconnecting the connectors. Failure to do so may cause the connectors to break.

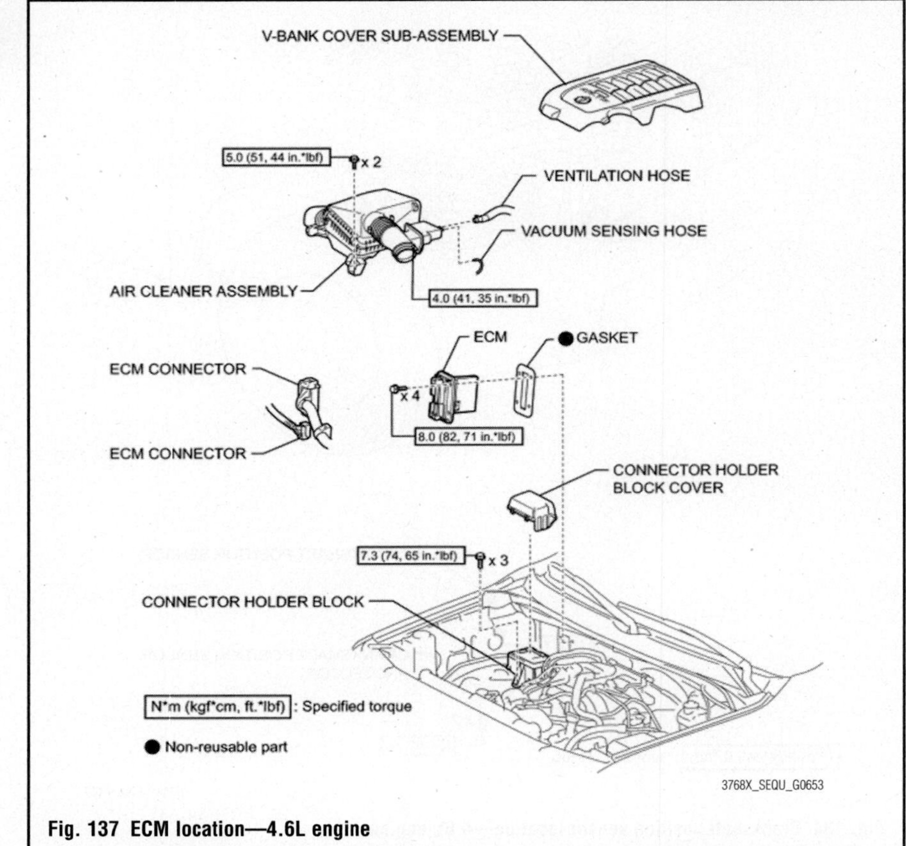

Fig. 137 ECM location—4.6L engine

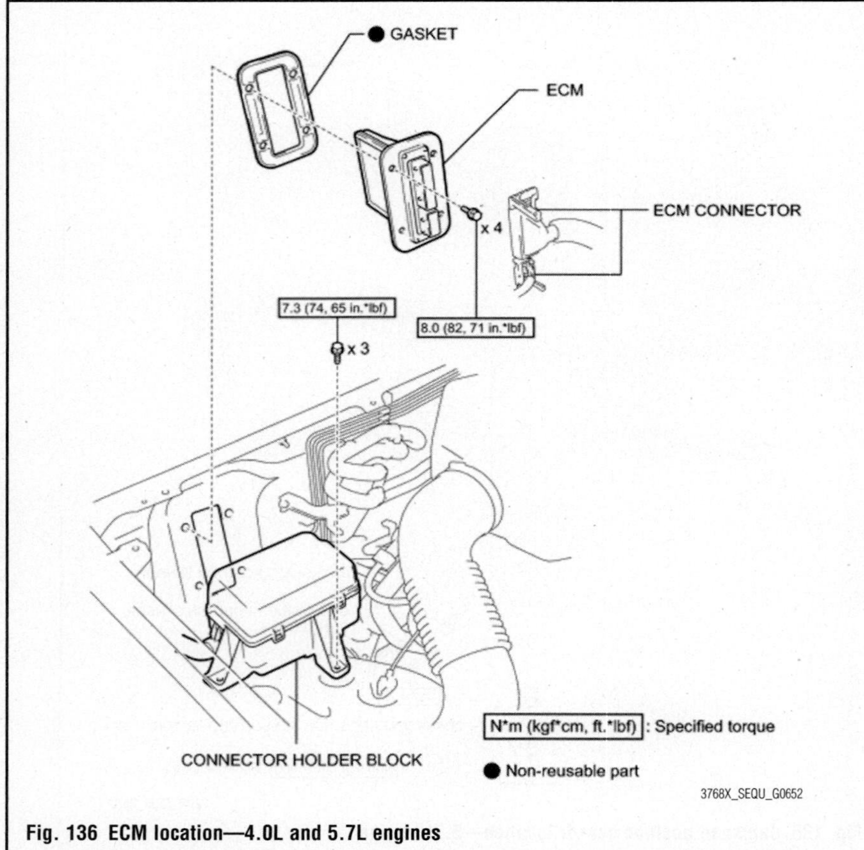

Fig. 136 ECM location—4.0L and 5.7L engines

9. Remove the retaining bolts. Remove the component from its mounting. Discard the gasket.

To install:

➡Be sure to use new fasteners, as required.

10. Installation is the reverse of the removal procedure.
11. Use the Techstream diagnostic tool, or equivalent and reprogram the required systems.

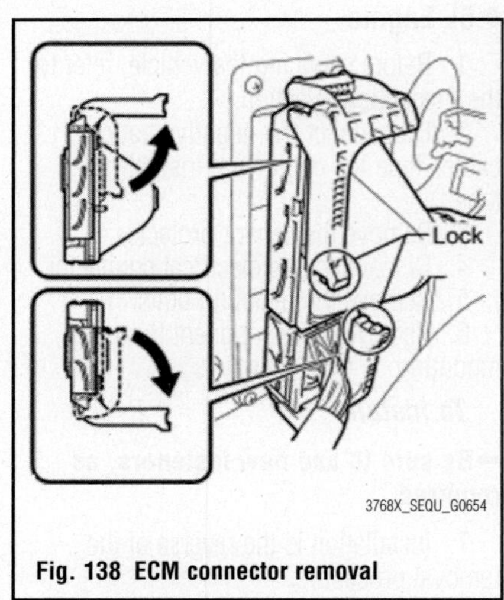

Fig. 138 ECM connector removal

ENGINE COOLANT TEMPERATURE (ECT) SENSOR

LOCATION

See Figures 139 and 140.

REMOVAL & INSTALLATION

4.0L Engine

1. Before servicing the vehicle, refer to the Precautions Section.
2. Disconnect the negative battery cable. Tape the cable with insulating tape.
3. Disconnect cable from negative battery terminal.
4. Remove the intake air surge tank.
5. Disconnect No. 1 and No.1 2 fuel pipe sub-assemblies.
6. Disconnect the 6 fuel injector connectors.
7. Remove the 10 bolts, intake manifold and 2 gaskets.
8. Disconnect the sensor connector.
9. Remove the sensor.
10. Remove the gasket from the sensor.

To install:

➡Be sure to use new fasteners, as required.

11. Install a new gasket to the sensor.
12. Install the sensor and tighten to 15 ft. lbs. (20 Nm).
13. Connect the sensor connector.
14. Install intake manifold.
    a. Set a new gasket on each cylinder head.

➡Align the port holes of the gasket and cylinder head. Be careful of the installation direction.

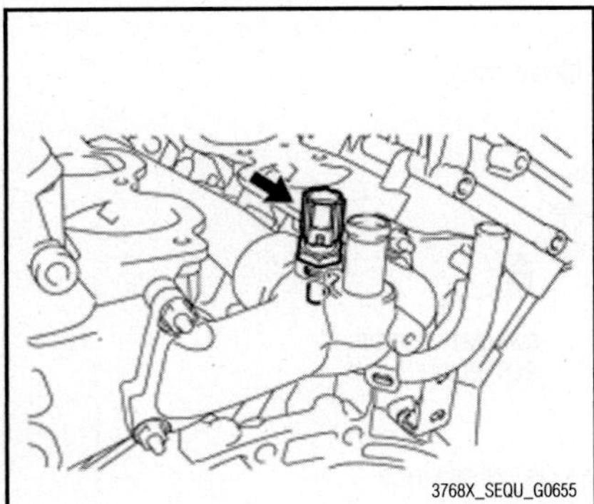

Fig. 139 Engine Coolant Temperature (ECT) sensor location—4.0L engine

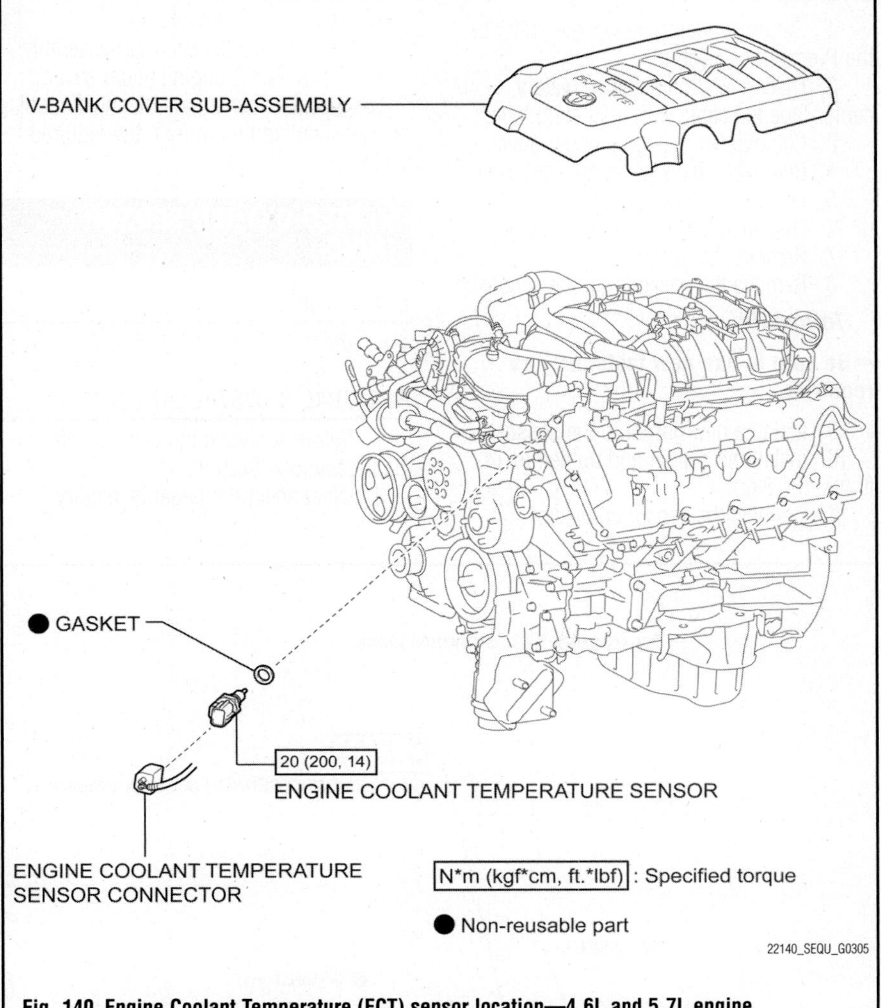

Fig. 140 Engine Coolant Temperature (ECT) sensor location—4.6L and 5.7L engine

    b. Set the intake manifold on the cylinder heads.
    c. Install and uniformly tighten the 10 bolts in several passes and tighten to 19 ft. lbs. (26 Nm).
    d. Connect the 6 fuel injector connectors.

15. Install the intake air surge tank.
16. Connect fuel pipe sub-assemblies.
17. Add engine coolant.
18. Connect cable to negative battery terminal.
19. Inspect for engine coolant leak.
20. Use the Techstream diagnostic tool, or equivalent and reprogram the required systems.

4.6L Engine

1. Before servicing the vehicle, refer to the Precautions Section.
2. Disconnect the negative battery cable. Tape the cable with insulating tape.
3. Remove No. 1 engine under cover.
4. Remove V-bank cover sub-assembly.
5. Drain engine coolant.
6. Disconnect the sensor connector.
7. Remove the sensor.
8. Remove the gasket from the sensor.

To install:

➡Be sure to use new fasteners, as required.

9. Install a new gasket to the sensor.
10. Install the sensor and tighten to 14 ft. lbs. (20 Nm).
11. Connect the sensor connector.
12. Add engine coolant.
13. Inspect for coolant leak.
14. Install V-bank cover sub-assembly.
15. Install No. 1 engine under cover.
16. Use the Techstream diagnostic tool, or equivalent and reprogram the required systems.

5.7L Engine

1. Before servicing the vehicle, refer to the Precautions Section.
2. Disconnect the negative battery cable. Tape the cable with insulating tape.
3. Remove No. 1 engine under cover.
4. Remove V-bank cover sub-assembly.
5. Drain engine coolant.
6. Disconnect the sensor connector.
7. Remove the sensor.
8. Remove the gasket from the sensor.

To install:

➡Be sure to use new fasteners, as required.

9. Install a new gasket to the sensor.
10. Install the sensor and tighten to 14 ft. lbs. (20 Nm).
11. Connect the sensor connector.
12. Add engine coolant.
13. Inspect for coolant leak.
14. Install V-bank cover sub-assembly.
15. Install No. 1 engine under cover.
16. Use the Techstream diagnostic tool, or equivalent and reprogram the required systems.

HEATED OXYGEN SENSOR (HO2S)

LOCATION

See Figures 141 through 143.

REMOVAL & INSTALLATION

1. Before servicing the vehicle, refer to the Precautions Section.
2. Disconnect the negative battery cable. Tape the cable with insulating tape.
3. Raise and support the vehicle safely.
4. Remove the engine under cover, as required.
5. Remove the heat insulator, as required.
6. Disconnect the sensor electrical connector.
7. Remove the sensor from its mounting.

To install:

➡Be sure to use new fasteners, as required.

8. Installation is the reverse of the removal procedure.
9. Use the Techstream diagnostic tool, or equivalent and reprogram the required systems.

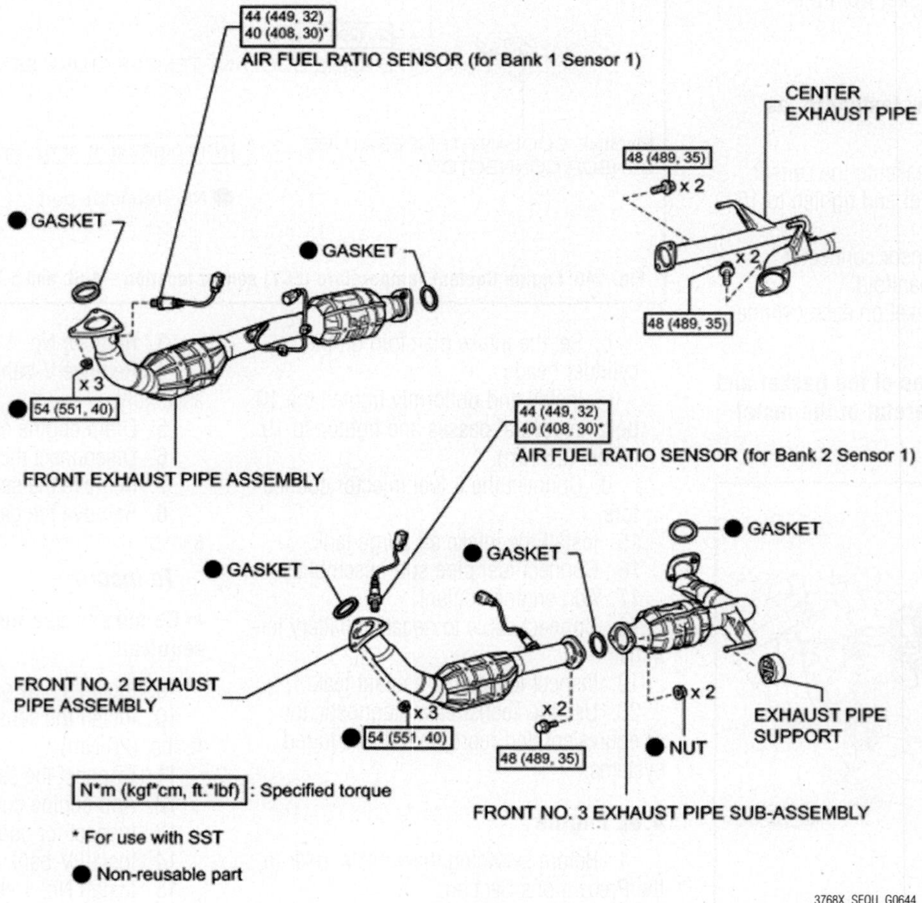

Fig. 141 Heated oxygen sensor location—4.0L engine

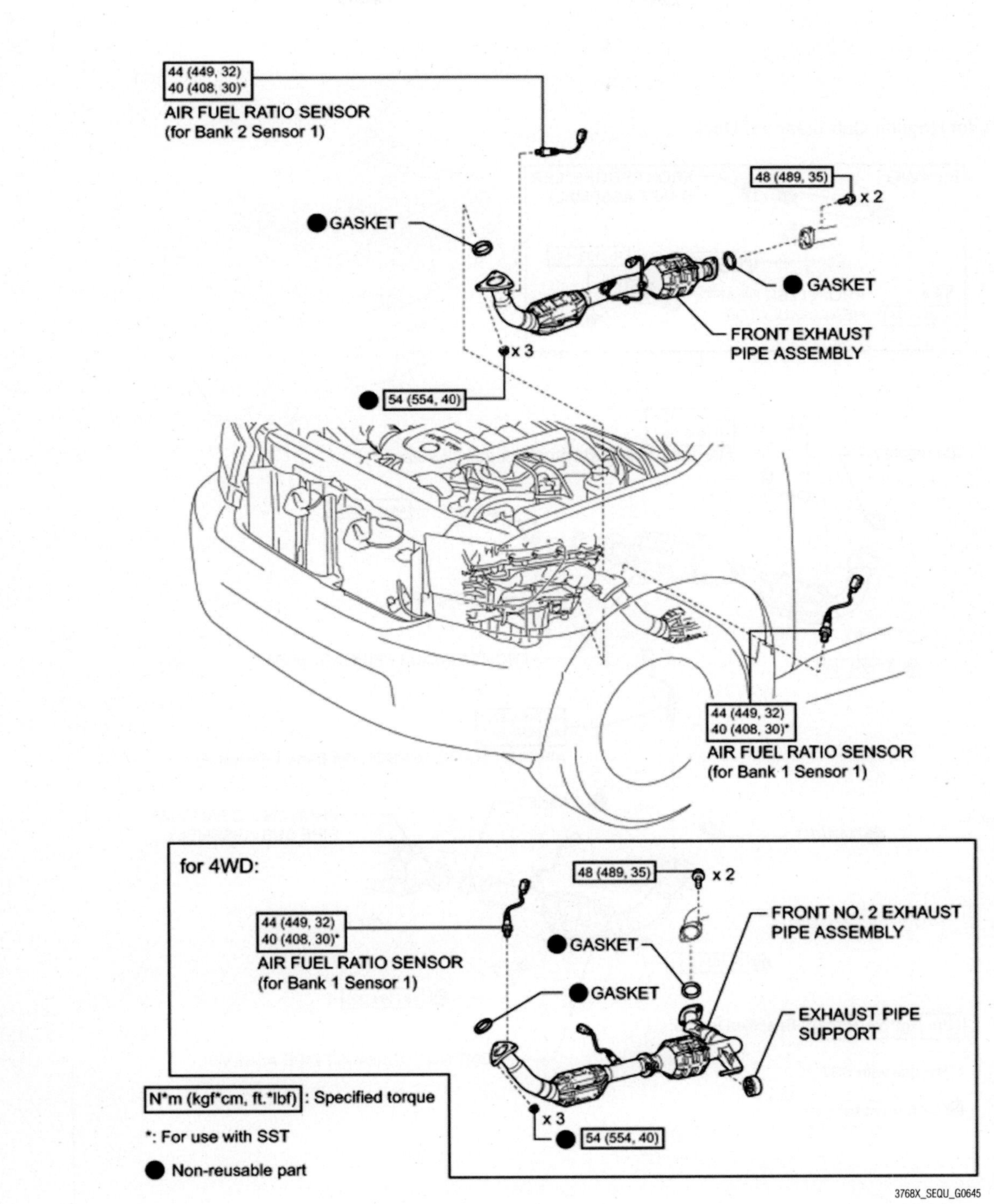

Fig. 142 Heated oxygen sensor location—4.6L engine

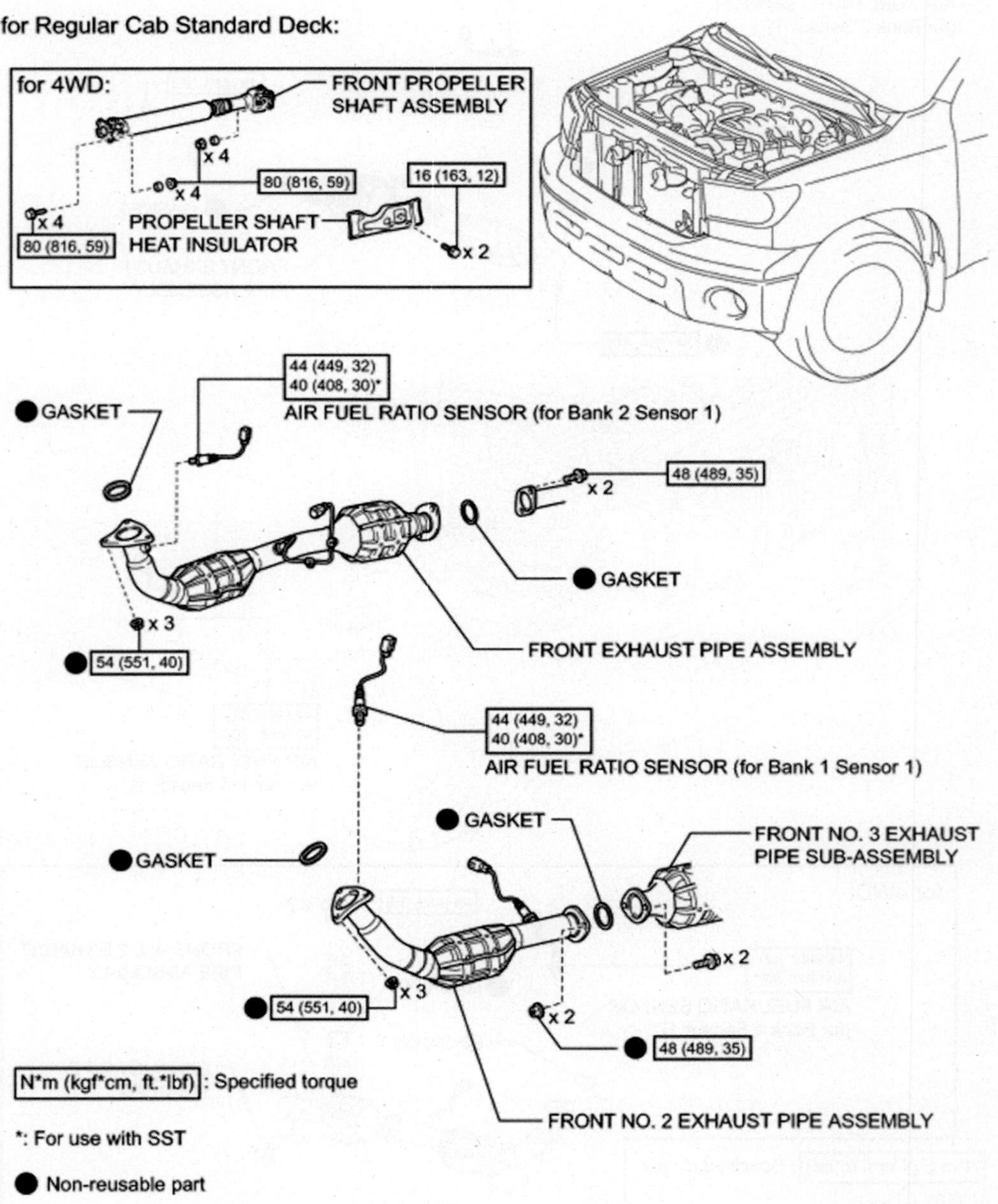

Fig. 143 Heated oxygen sensor location—5.7L engine

INTAKE AIR TEMPERATURE (IAT) SENSOR

LOCATION

The Intake Air Temperature (IAT) sensor, built into the Mass Air Flow (MAF) meter.

REMOVAL & INSTALLATION

See Mass Air Flow Meter.

KNOCK SENSOR (KS)

LOCATION

See Figures 144 and 145.

REMOVAL & INSTALLATION

4.0L Engine

See Figures 146 through 148.

1. Before servicing the vehicle, refer to the Precautions Section.
2. Disconnect the negative battery cable. Tape the cable with insulating tape.
3. Remove the cylinder head RH.
4. Remove the 4 wire harness clamps [A].

Remove the 3 bolts [B] and water outlet pipe.

5. Disconnect the 2 sensor connectors.
6. Remove the 2 bolts and 2 sensors.

To install:

➡Be sure to use new fasteners, as required.

7. Install the 2 sensors with the 2 bolts and tighten to 15 ft. lbs. (20 Nm) as shown.
8. Connect the 2 sensor connectors.
9. Install the 3 bolts [A] and water outlet pipe.
10. Install the 4 wire harness clamps [B].
11. Install the cylinder head RH.
12. Use the Techstream diagnostic tool, or equivalent and reprogram the required systems.

4.6L Engine

See Figures 149 through 151.

1. Before servicing the vehicle, refer to the Precautions Section.
2. Disconnect the negative battery cable. Tape the cable with insulating tape.
3. Remove the intake manifold.
4. Remove No. 2 cylinder head cover.
5. Remove No. 1 engine cover.
6. Remove the 4 bolts and separator case.
7. Disconnect the 4 knock sensor connectors.

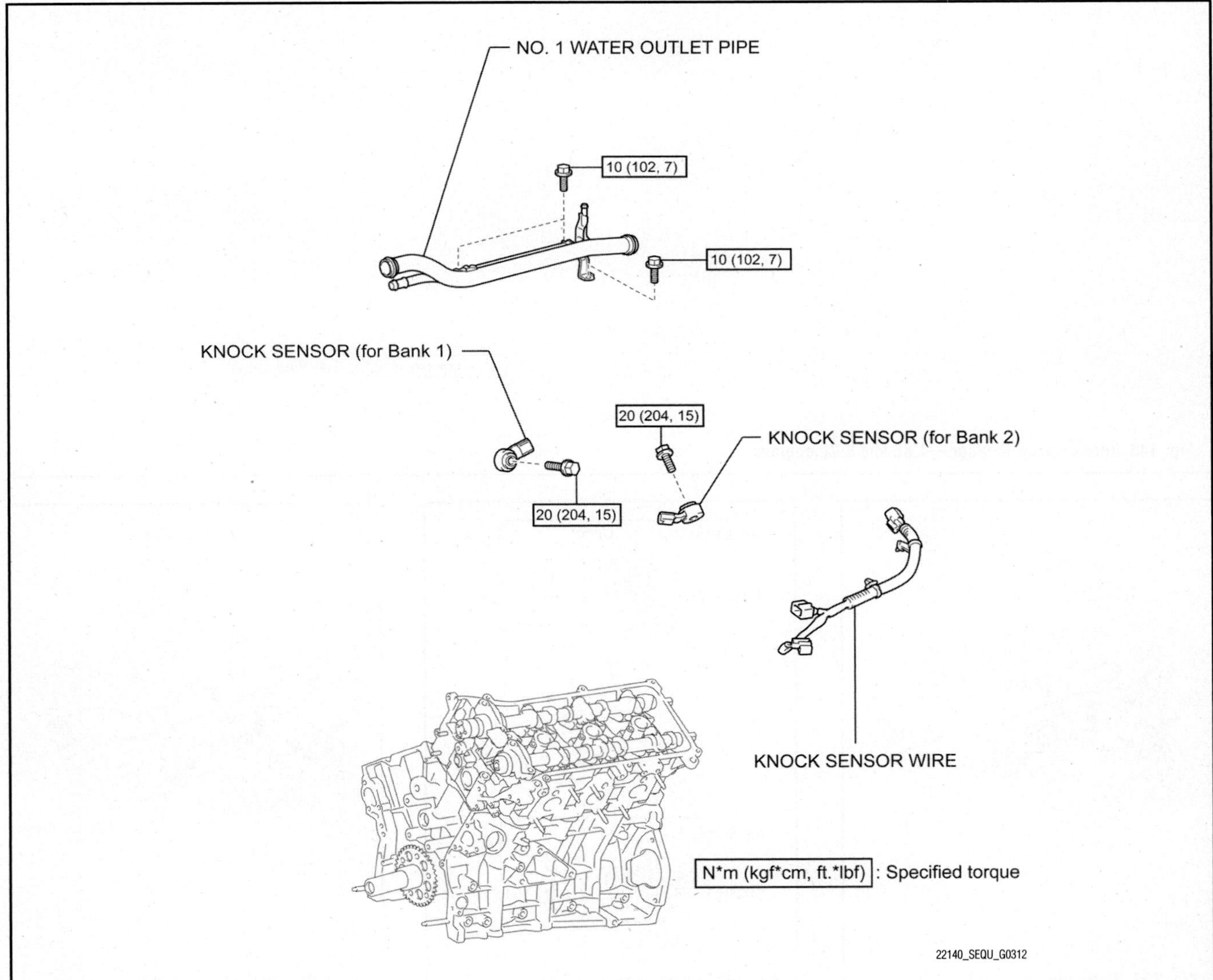

Fig. 144 Knock sensor location—4.0L engine

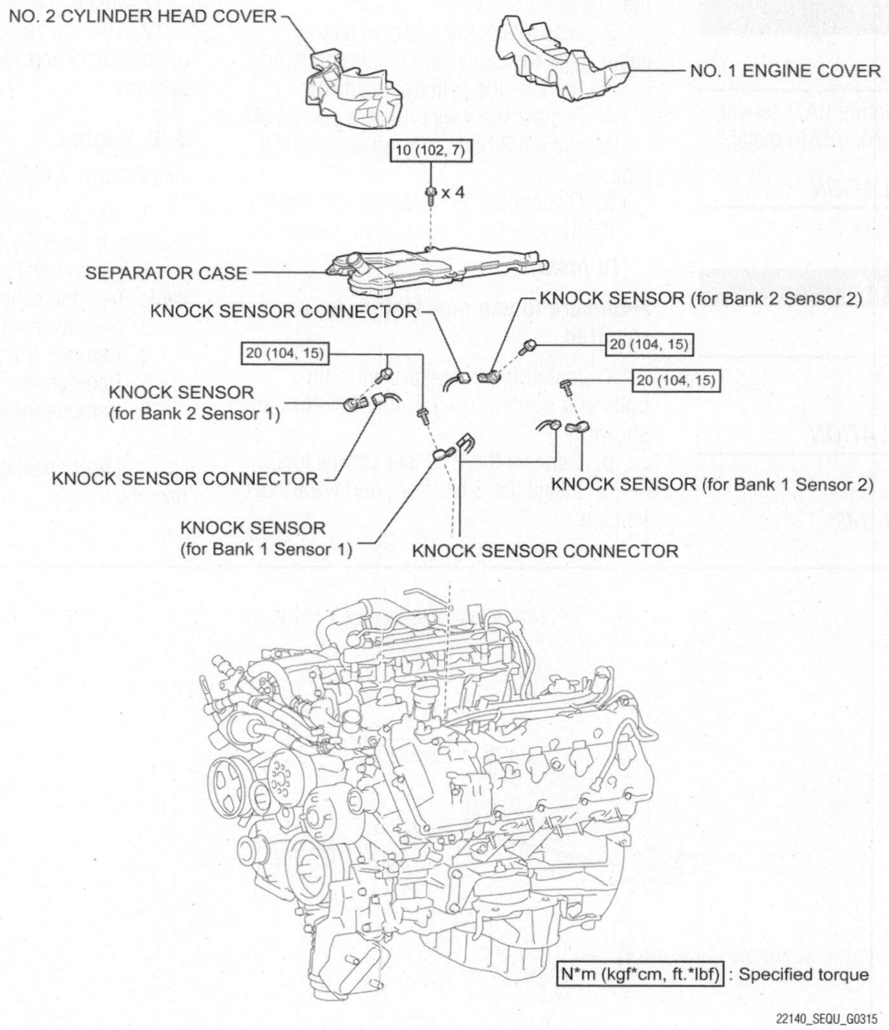

Fig. 145 Knock sensor location—4.6L and 5.7L engines

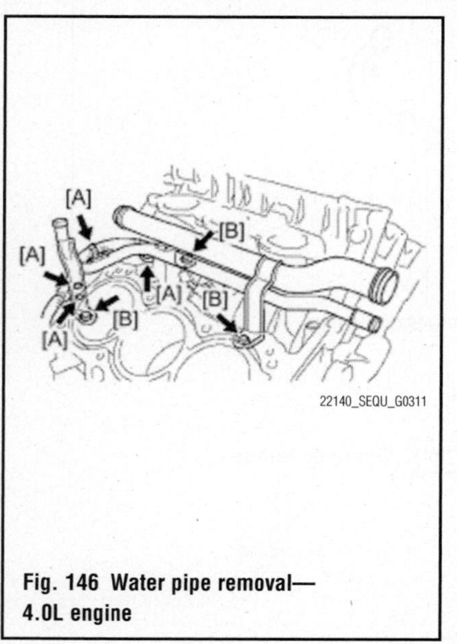

Fig. 146 Water pipe removal—4.0L engine

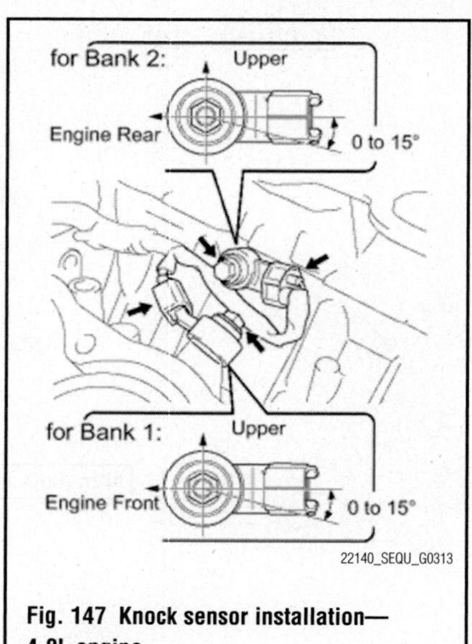

Fig. 147 Knock sensor installation—4.0L engine

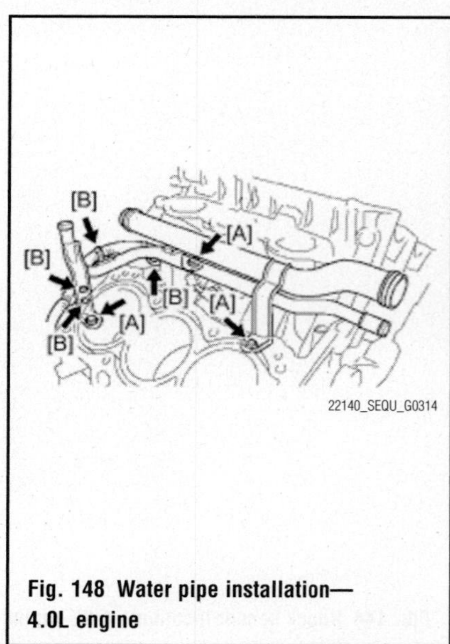

Fig. 148 Water pipe installation—4.0L engine

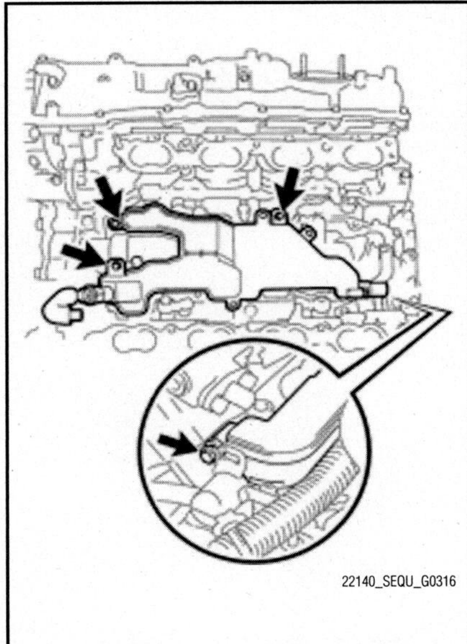

Fig. 149 Separator case—4.6L and 5.7L engines

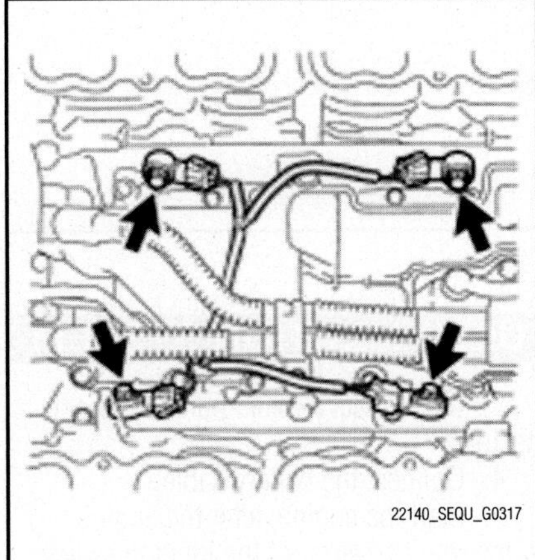

Fig. 150 Removing knock sensors—4.6L and 5.7L engines

8. Remove the 4 bolts and 4 knock sensors.

To install:

➡Be sure to use new fasteners, as required.

9. Install the 4 sensors with the 4 bolts so that the sensors are angled as shown in the illustration.
10. Connect the 4 sensor connectors.
11. Install the separator case.
12. Install No. 1 engine cover.
13. Install No. 2 cylinder head cover.

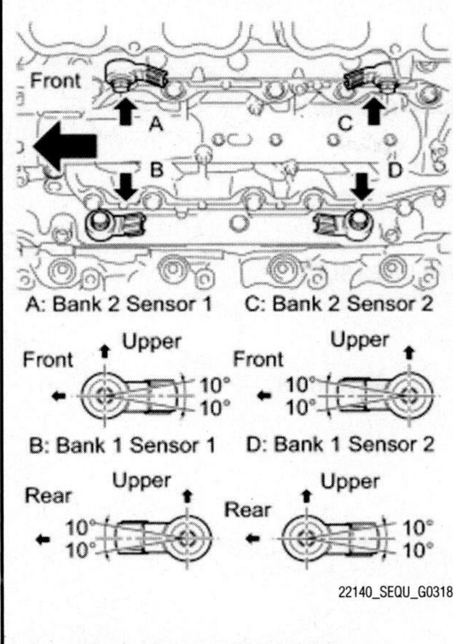

Fig. 151 Installing knock sensors—4.6L and 5.7L engines

14. Install the intake manifold.
15. Use the Techstream diagnostic tool, or equivalent and reprogram the required systems.

5.7 Engine

1. Before servicing the vehicle, refer to the Precautions Section.
2. Disconnect the negative battery cable. Tape the cable with insulating tape.
3. Remove the intake manifold.
4. Remove No. 2 cylinder head cover.
5. Remove No. 1 engine cover.
6. Remove the 4 bolts and separator case.
7. Disconnect the 4 knock sensor connectors.
8. Remove the 4 bolts and 4 knock sensors.

To install:

➡Be sure to use new fasteners, as required.

9. Install the 4 sensors with the 4 bolts so that the sensors are angled as shown in the illustration.
10. Connect the 4 sensor connectors.
11. Install the separator case.
12. Install No. 1 engine cover.
13. Install No. 2 cylinder head cover.
14. Install the intake manifold.
15. Use the Techstream diagnostic tool, or equivalent and reprogram the required systems.

MASS AIR FLOW (MAF) SENSOR

LOCATION

The MAF sensor is located in the air intake snorkel.

REMOVAL & INSTALLATION

1. Before servicing the vehicle, refer to the Precautions Section.
2. Disconnect the negative battery cable. Tape the cable with insulating tape.
3. Disconnect connector.
4. Remove attaching screws and remove MAF.

To install:

➡Be sure to use new fasteners, as required.

5. Installation is the reverse of the removal procedure.
6. Use the Techstream diagnostic tool, or equivalent and reprogram the required systems.

MANIFOLD ABSOLUTE PRESSURE (MAP) SENSOR

LOCATION

See Figure 152.

REMOVAL & INSTALLATION

See Figure 152.

1. Before servicing the vehicle, refer to the Precautions Section.
2. Disconnect the negative battery cable. Tape the cable with insulating tape.
3. Remove the V- bank cover assembly.
4. Disconnect the electrical connector.
5. Remove the sensor from its mounting.
6. Discard the O-ring.

To install:

➡Be sure to use new fasteners, as required.

7. Installation is the reverse of the removal procedure.
8. Use the Techstream diagnostic tool, or equivalent and reprogram the required systems.

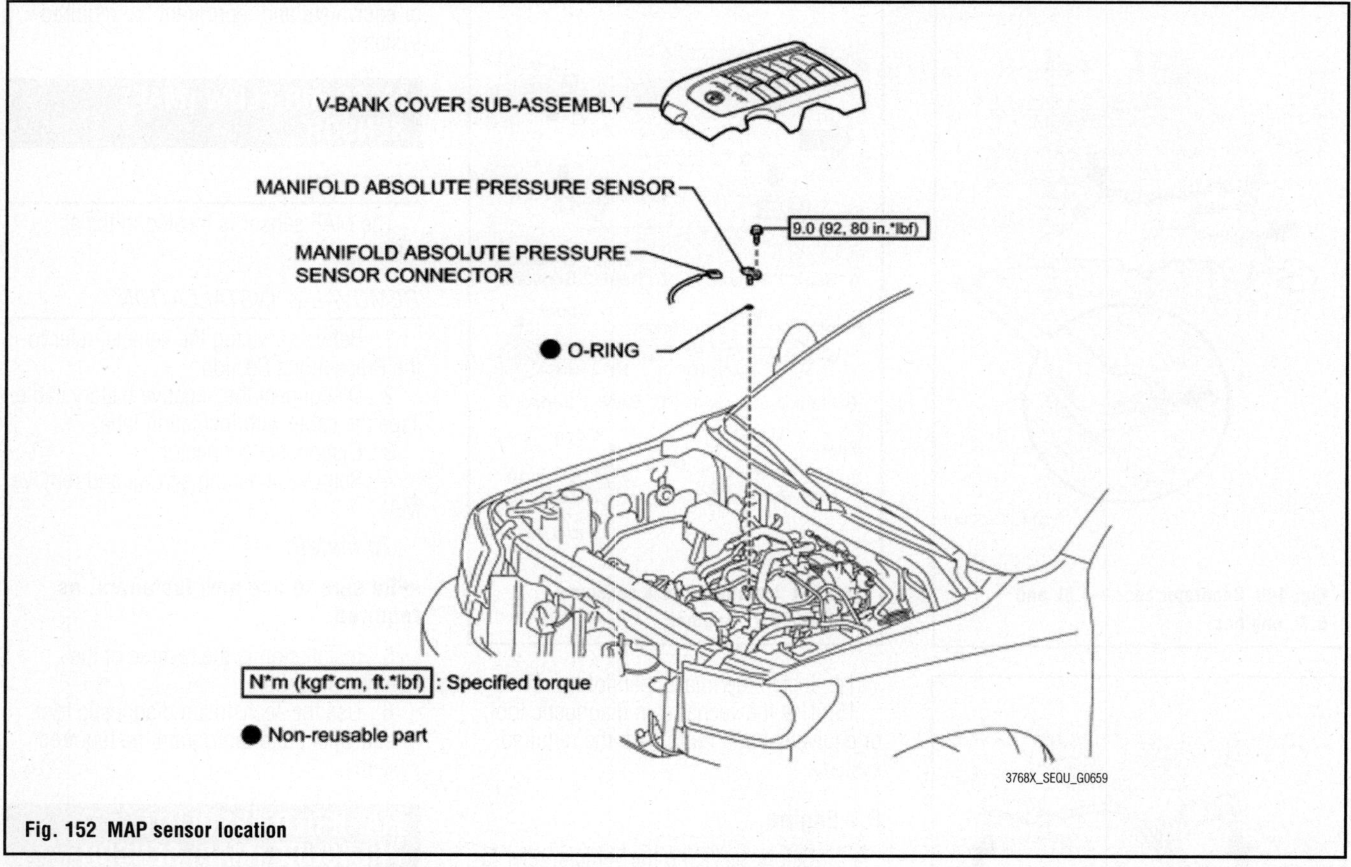

Fig. 152 MAP sensor location

FUEL

GASOLINE FUEL INJECTION SYSTEM

FUEL SYSTEM SERVICE PRECAUTIONS

Safety is the most important factor when performing not only fuel system maintenance, but any type of maintenance. Failure to conduct maintenance and repairs in a safe manner may result in serious personal injury or death. Work on a vehicle's fuel system components can be accomplished safely and effectively by adhering to the following rules and guidelines.

- To avoid the possibility of fire and personal injury, always disconnect the negative battery cable unless the repair or test procedure requires that battery voltage be applied.
- Always relieve the fuel system pressure prior to disconnecting any fuel system component (injector, fuel rail, pressure regulator, etc.) fitting or fuel line connection. Exercise extreme caution whenever relieving fuel system pressure to avoid exposing skin, face and eyes to fuel spray. Please be advised that fuel under pressure may penetrate the skin or any part of the body that it contacts.
- Always place a shop towel or cloth around the fitting or connection prior to loosening to absorb any excess fuel due to spillage. Ensure that all fuel spillage is quickly removed from engine surfaces. Ensure that all fuel-soaked cloths or towels are deposited into a flame-proof waste container with a lid.
- Always keep a dry chemical (Class B) fire extinguisher near the work area.
- Do not allow fuel spray or fuel vapors to come into contact with a spark or open flame.
- Always use a second wrench when loosening or tightening fuel line connection fittings. This will prevent unnecessary stress and torsion on fuel piping. Always follow the proper torque specifications.
- Always replace worn fuel fitting O-rings with new ones. Do not substitute fuel hose where rigid pipe is installed.

FUEL SYSTEM PRESSURE

1. Before servicing the vehicle, refer to the Precautions Section.
2. Disconnect the negative battery cable. Tape the cable with insulating tape.
3. Disconnect the fuel pump ECU connectors.
4. Connect the battery cable.
5. Start the engine, after the engine stops (on its own) turn the ignition switch off.
6. Crank the engine again, to be sure it does not start.
7. Loosen the fuel tank cap, then discharge the pressure in the fuel tank.
8. Connect the fuel pump ECU connectors.

FUEL RAIL AND INJECTOR

REMOVAL & INSTALLATION

4.0L Engine

See Figure 153.

1. Before servicing the vehicle, refer to the Precautions Section.
2. Disconnect the negative battery cable. Tape the cable with insulating tape.
3. Relieve the fuel system pressure.
4. Drain the engine coolant. Remove the V bank cover.
5. Remove the air cleaner assembly.

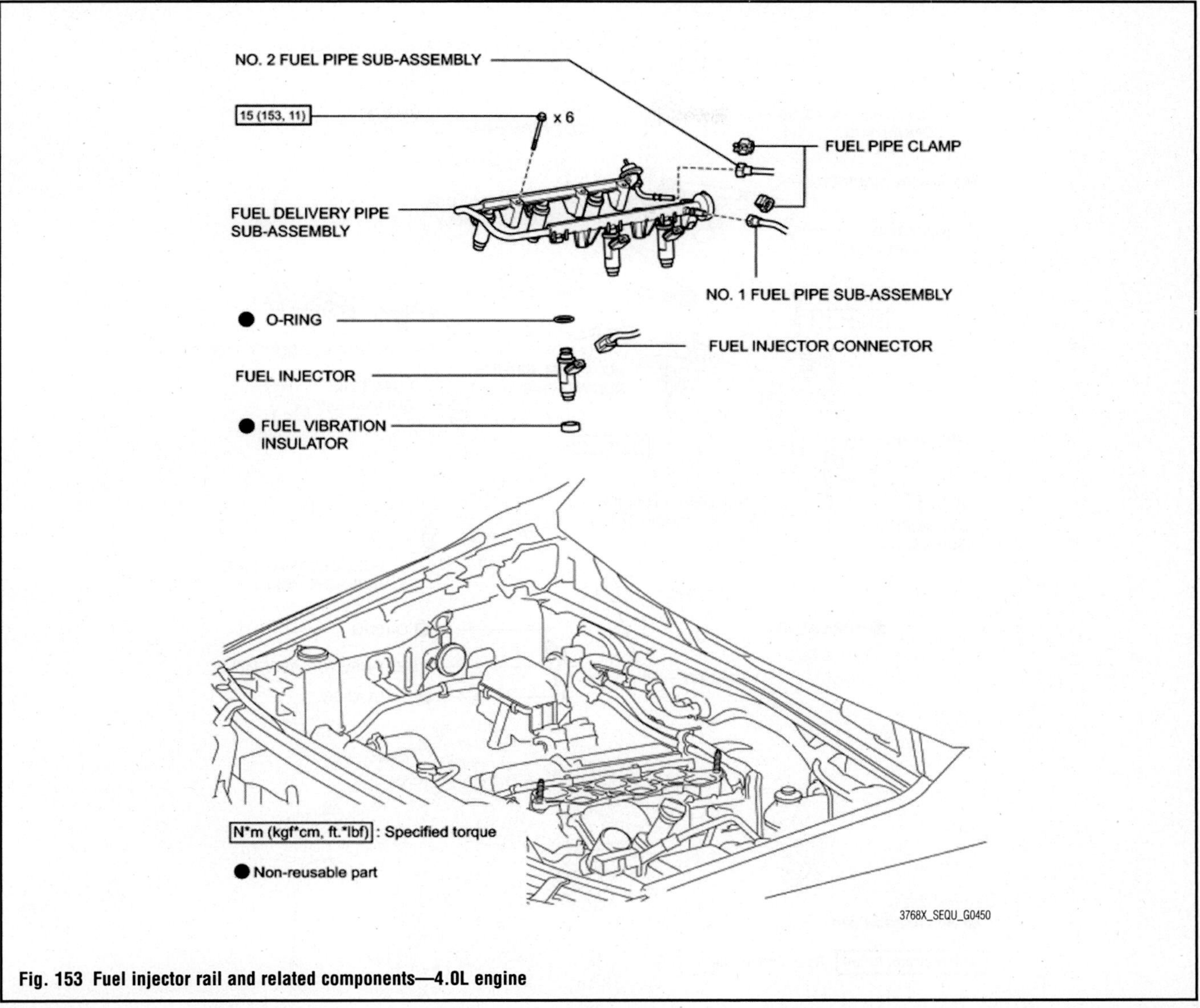

Fig. 153 Fuel injector rail and related components—4.0L engine

6. Remove the front cowl top outer panel sub assembly.
7. Remove the intake manifold.
8. Disconnect and plug the fuel line hoses.
9. Disconnect the fuel injector connectors. Remove the six bolts and the fuel rail together with the injectors.
10. Pull the injectors out of the fuel rail.

To install:

➡Be sure to use new fasteners, as required.

11. Installation is the reverse of the removal procedure.
12. Be sure to use new gaskets and O-rings, as required.
13. Be sure to fill the cooling system with the proper grade and type engine coolant.
14. Start the engine and check for leaks, correct as required.

4.6L Engine

See Figure 154.

1. Properly discharge the fuel system.
2. Disconnect cable from negative battery terminal.
3. Remove the EGR valve bracket.
4. Disconnect the No. 2 fuel tube from the fuel pressure regulator.
5. Disconnect the No. 1 fuel tube from the fuel delivery pipes.
6. Disconnect the fuel tube from the fuel delivery pipe LH.
7. Disconnect the ventilation hose and No. 6 wire harness connector.
8. Remove the 2 bolts and fuel delivery pipe RH.
9. Remove the 2 delivery pipe spacers and 4 insulators from the intake manifold.
10. Disconnect the No. 7 wire harness connector.
11. Remove the No. 3 engine cover.
12. Remove the 2 bolts and fuel delivery pipe LH.
13. Remove the 2 delivery pipe spacers and 4 insulators from the intake manifold.
14. Remove the fuel injector from the fuel delivery pipe, and then disconnect the injector connector.

To install:

15. Attach the 2 clamps to install the No. 6 wire harness to the delivery pipe.
16. Attach the 3 clamps to install the No. 7 wire harness to the delivery pipe LH.
17. Apply gasoline or spindle oil to a new O-ring and install it to the injector.
18. Connect the injector connector.
19. Check that each injector is installed to the delivery pipe facing the direction shown in the illustration.
20. Install the 2 delivery pipe spacers and 4 insulators to the cylinder head LH.

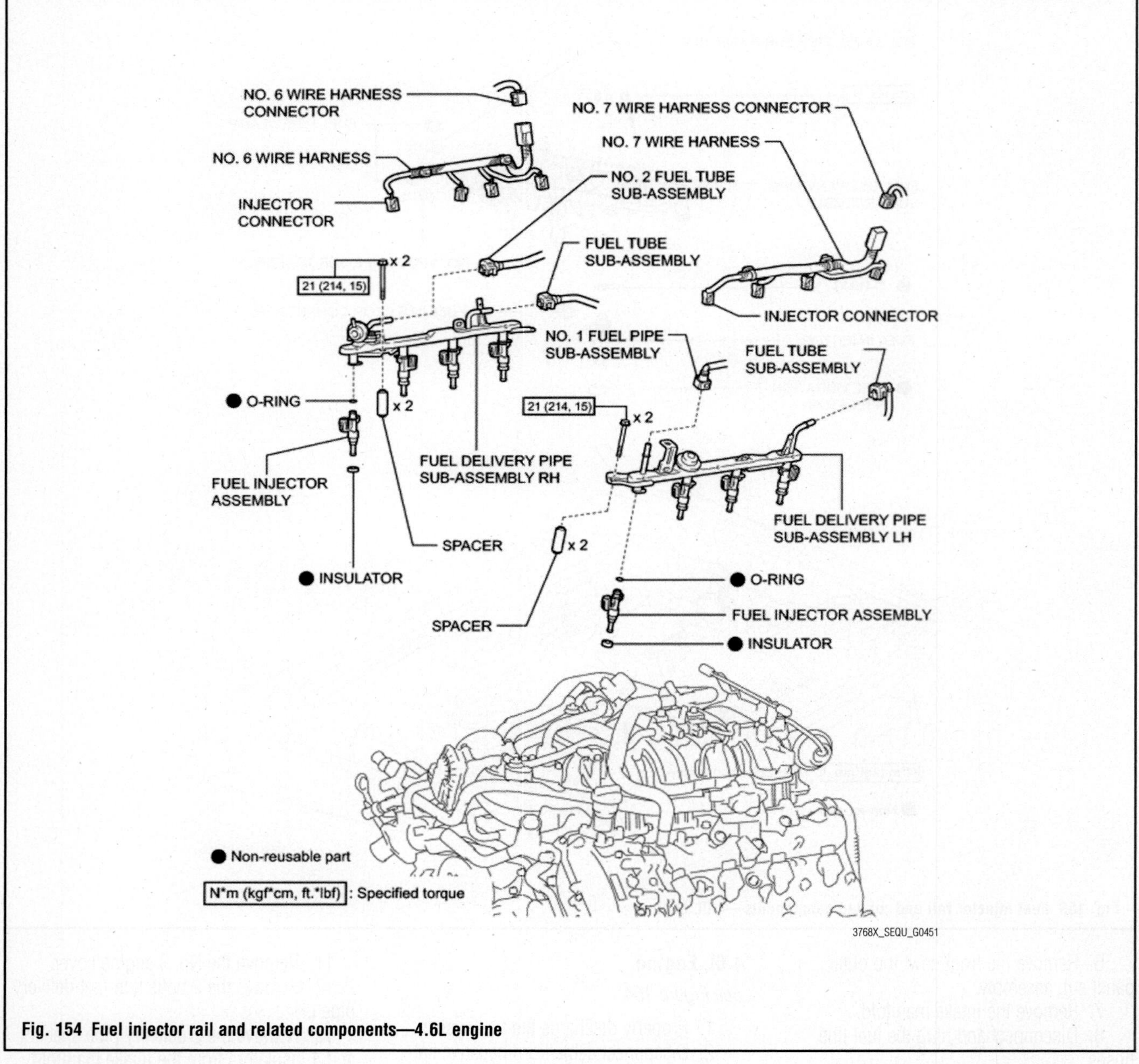

Fig. 154 Fuel injector rail and related components—4.6L engine

21. Install the delivery pipe (with injectors) to the cylinder head LH.
22. Install the 2 bolts and tighten to 15 ft. lbs. (21 Nm).

➡Make sure that the part of the injector labeled B is between the parts of the delivery pipe labeled A.

23. Connect the No. 6 wire harness connector.
24. Install the No. 3 engine cover.
25. Install the 2 delivery pipe spacers and 4 insulators to the cylinder head RH.
26. Install the delivery pipe (with injectors) to the cylinder head RH.
27. Install the 2 bolts and tighten to 15 ft. lbs. (21 Nm).
28. Install the No. 1 engine cover.
29. Connect the No. 7 wire harness connector.
30. Connect the ventilation hose.
31. Connect the fuel tube to the fuel delivery pipe LH.
32. Connect the No. 1 fuel tube to the delivery pipes.
33. Connect the No. 2 fuel tube to the fuel pressure regulator.
34. Install air cleaner hose assembly.
35. Install V-bank cover.
36. Connect cable to negative battery terminal.
37. Check for fuel leak.
38. Perform initialization, if necessary.

5.7L Engine

See Figures 155 through 157.

1. Properly discharge the fuel system.
2. Disconnect cable from negative battery terminal.
3. Remove V-bank cover. Remove air cleaner hose assembly.
4. Disconnect the No. 2 fuel tube from the fuel pressure regulator.
5. Disconnect the No. 1 fuel tube from the fuel delivery pipes.

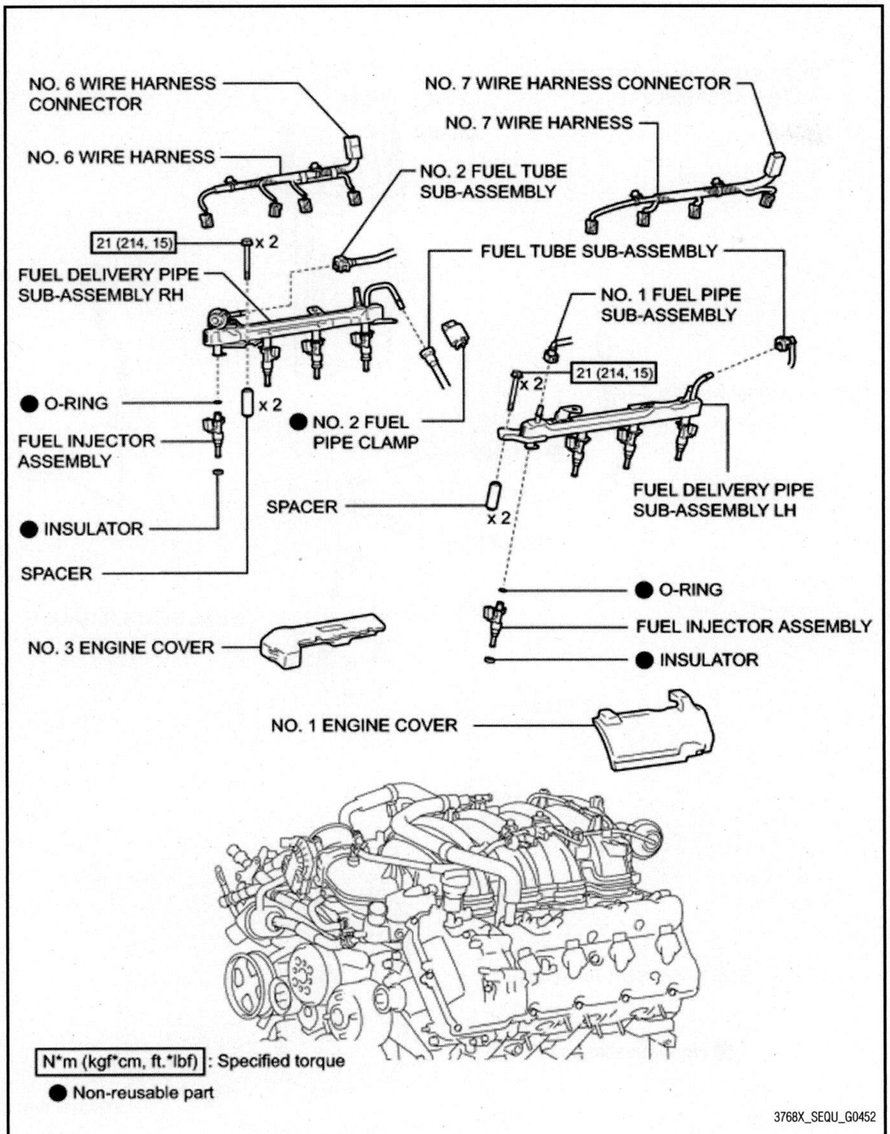

Fig. 155 Fuel injector rail and related components—5.7L engine

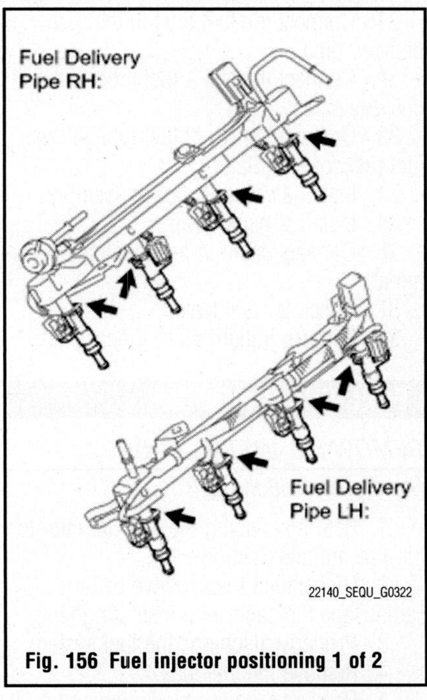

Fig. 156 Fuel injector positioning 1 of 2

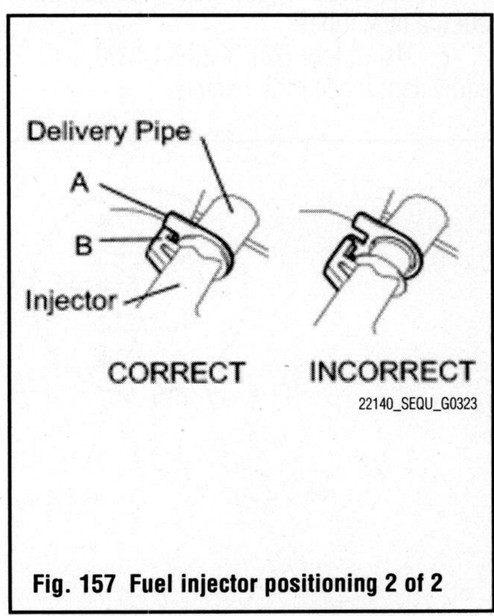

Fig. 157 Fuel injector positioning 2 of 2

6. Disconnect the fuel tube from the fuel delivery pipe LH.
7. Disconnect the ventilation hose and No. 6 wire harness connector.
8. Remove the 2 bolts and fuel delivery pipe RH.
9. Remove the 2 delivery pipe spacers and 4 insulators from the intake manifold.
10. Disconnect the No. 7 wire harness connector.
11. Remove the No. 3 engine cover.
12. Remove the 2 bolts and fuel delivery pipe LH.
13. Remove the 2 delivery pipe spacers and 4 insulators from the intake manifold.
14. Remove the fuel injector from the fuel delivery pipe, and then disconnect the injector connector.

To install:

15. Attach the 2 clamps to install the No. 6 wire harness to the delivery pipe.
16. Attach the 3 clamps to install the No. 7 wire harness to the delivery pipe LH.
17. Apply gasoline or spindle oil to a new O-ring and install it to the injector.
18. Connect the injector connector.
19. Check that each injector is installed to the delivery pipe facing the direction shown in the illustration.
20. Install the 2 delivery pipe spacers and 4 insulators to the cylinder head LH.
21. Install the delivery pipe (with injectors) to the cylinder head LH.
22. Install the 2 bolts and tighten to 15 ft. lbs. (21 Nm).

➡Make sure that the part of the injector labeled B is between the parts of the delivery pipe labeled A.

23. Connect the No. 6 wire harness connector.
24. Install the No. 3 engine cover.
25. Install the 2 delivery pipe spacers and 4 insulators to the cylinder head RH.
26. Install the delivery pipe (with injectors) to the cylinder head RH.
27. Install the 2 bolts and tighten to 15 ft. lbs. (21 Nm).
28. Install the No. 1 engine cover.
29. Connect the No. 7 wire harness connector.
30. Connect the ventilation hose.

31. Connect the fuel tube to the fuel delivery pipe LH.
32. Connect the No. 1 fuel tube to the delivery pipes.
33. Connect the No. 2 fuel tube to the fuel pressure regulator.
34. Install air cleaner hose assembly.
35. Install V-bank cover.
36. Connect cable to negative battery terminal.
37. Check for fuel leak.
38. Perform initialization, if necessary.

FUEL PUMP MODULE

REMOVAL & INSTALLATION

See Figures 158 through 162.

1. Before servicing the vehicle, refer to the Precautions Section.
2. Disconnect the negative battery cable. Tape the cable with insulating tape.
3. Properly discharge the fuel system.
4. Remove the fuel tank.
5. Remove the joint tube clips and pull out the tank tubes.
6. Using tool SST:09808-14020, or equivalent loosen the retainer.

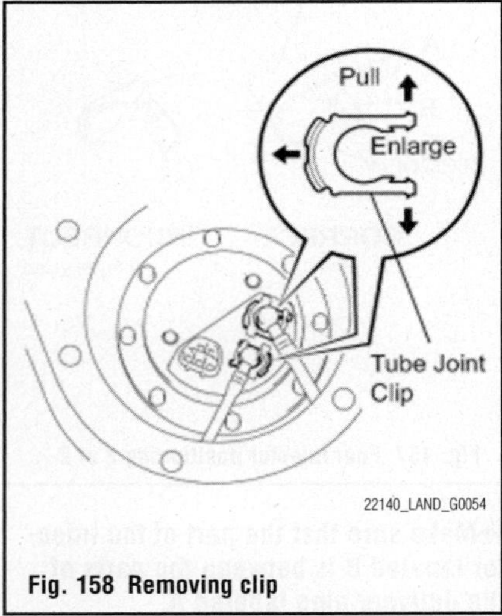

Fig. 158 Removing clip

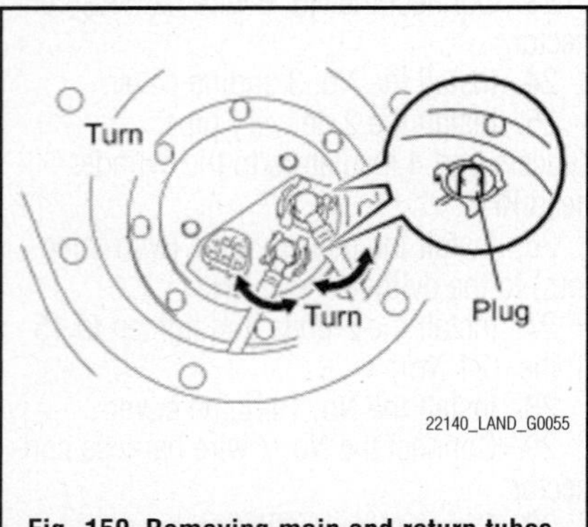

Fig. 159 Removing main and return tubes

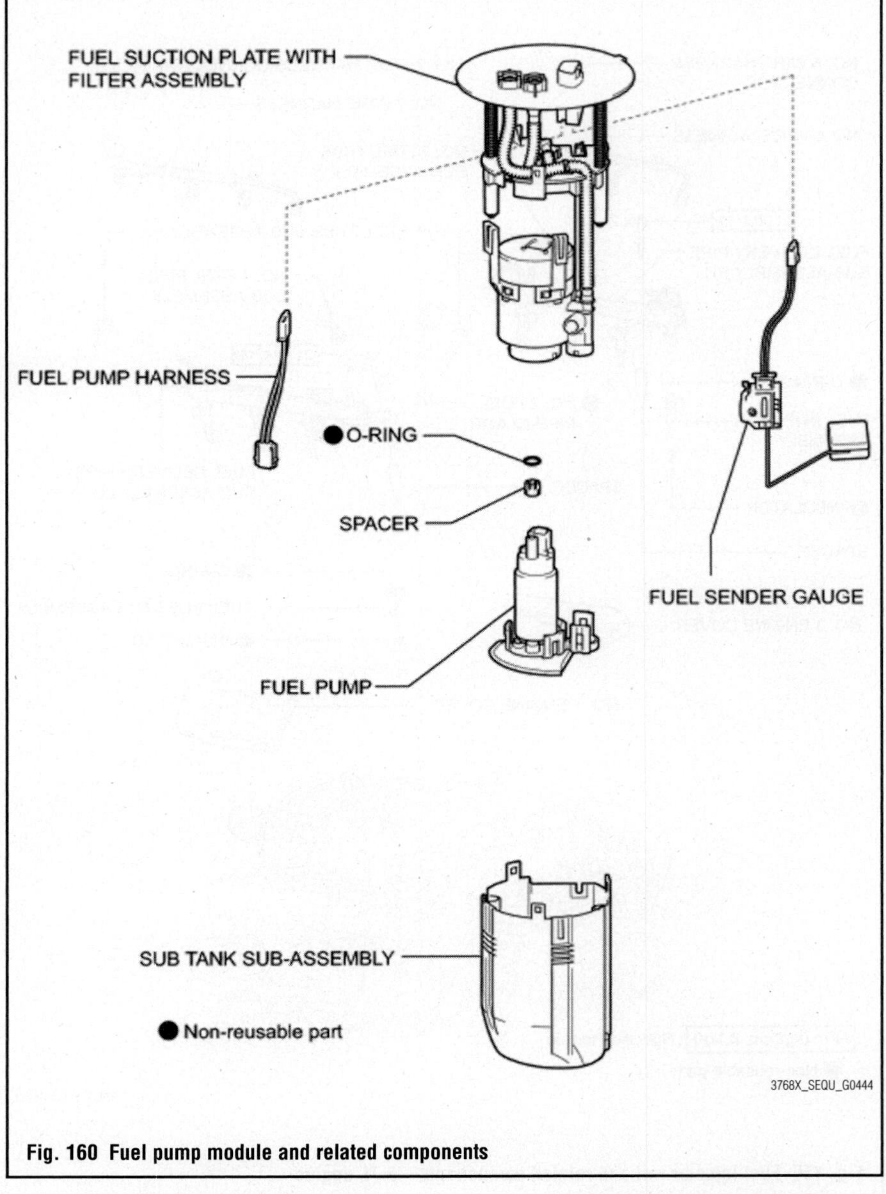

Fig. 160 Fuel pump module and related components

➡When the retainer is loosened, be careful as the pump and gauge tube will spring upward from the force of the spring.

7. Remove the retainer.
8. Remove the component from the tank. Discard the gasket.

To install:

➡Be sure to use new fasteners, as required.

9. Installation is the reverse of the removal procedure.
10. Be sure to align the pump with the tank groove.
11. When tightening the retainer be sure the alignment is correct (within range A).
12. Be sure to proper install the return tubes.
13. Start the engine and check for leaks.
14. Correct as required.

FUEL TANK

REMOVAL & INSTALLATION

See Figures 163 through 165.

1. Before servicing the vehicle, refer to the Precautions Section.
2. Disconnect the negative battery cable. Tape the cable with insulating tape.
3. Drain the fuel tank to an acceptable fuel level.
4. Discharge fuel system pressure.
5. Disconnect fuel pump connector.

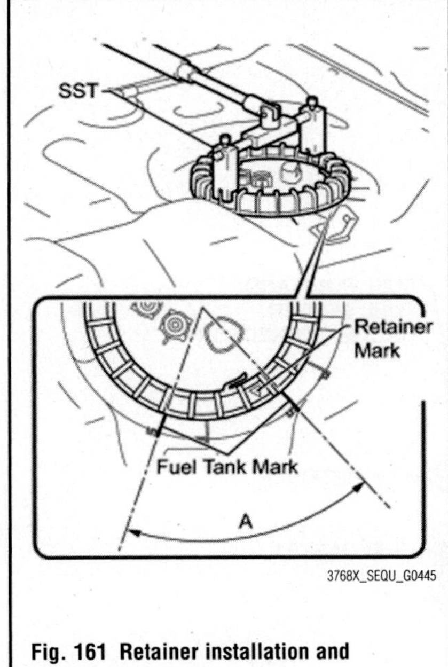

Fig. 161 Retainer installation and alignment

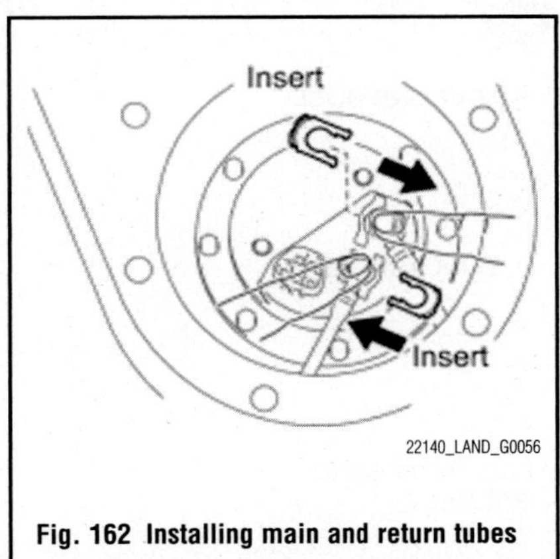

Fig. 162 Installing main and return tubes

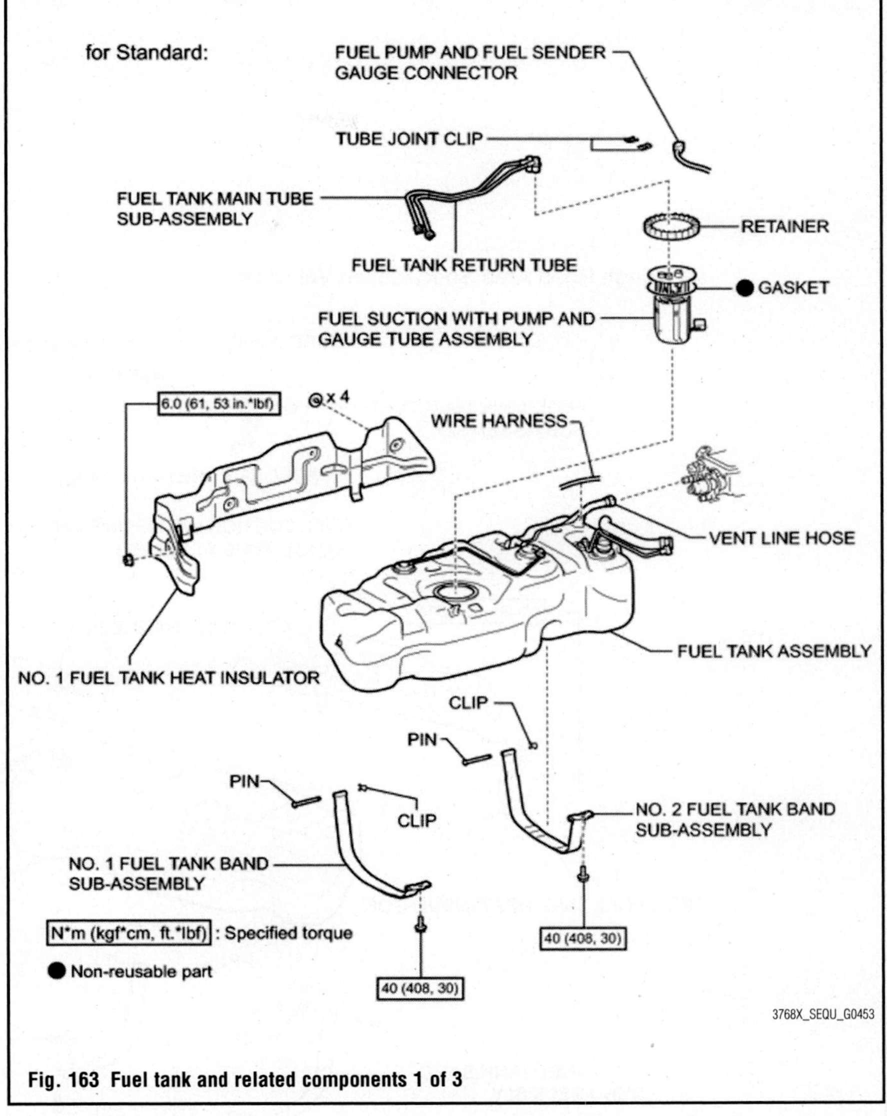

Fig. 163 Fuel tank and related components 1 of 3

6. Remove the 2 bolts, 2 nuts and fuel tank protector.
7. Disconnect fuel main tube and return tube.
8. Push the connector deep into the charcoal canister to release the locking tab.
9. Pinch and pull out the connector.
10. Loosen the hose clamp bolt and disconnect the fuel inlet hose from the fuel filler pipe.
11. Disconnect the breather tube.
12. Properly position a suitable jack under the fuel tank.
13. Remove the 2 bolts and disconnect the 2 fuel tank bands from the fuel tank.
14. Operate the jack and carefully lower the tank, check to see that nothing is interfering with the removal (wires, hoses etc).
15. Lower the jack and remove the fuel tank.

To install:

➡Be sure to use new fasteners, as required.

16. Installation is the reverse of the removal procedure.
17. Set up the fuel tank to the transmission jack.
18. Operate the transmission jack and install the fuel tank.
19. Install the 2 fuel tank bands with the 2 bolts and tighten to 45 ft. lbs. (62 Nm).
20. Connect the fuel inlet hose to the filler pipe and install the clamp.
21. Connect the breather tube.
22. Connect the fuel tank vent hose to the charcoal canister.
23. Connect fuel tank main tube and return tube.
24. Install the fuel tank protector with the 2 bolts and 2 nuts.
25. Connect fuel pump connector.
26. Check for fuel leaks.
27. Install spare tire.

IDLE SPEED

ADJUSTMENT

Idle speed is maintained by the Powertrain Control Module (PCM). No adjustment is necessary or possible.

THROTTLE BODY

REMOVAL & INSTALLATION

See Figures 166 through 169.

1. Before servicing the vehicle, refer to the Precautions Section.
2. Disconnect the negative battery cable. Tape the cable with insulating tape.

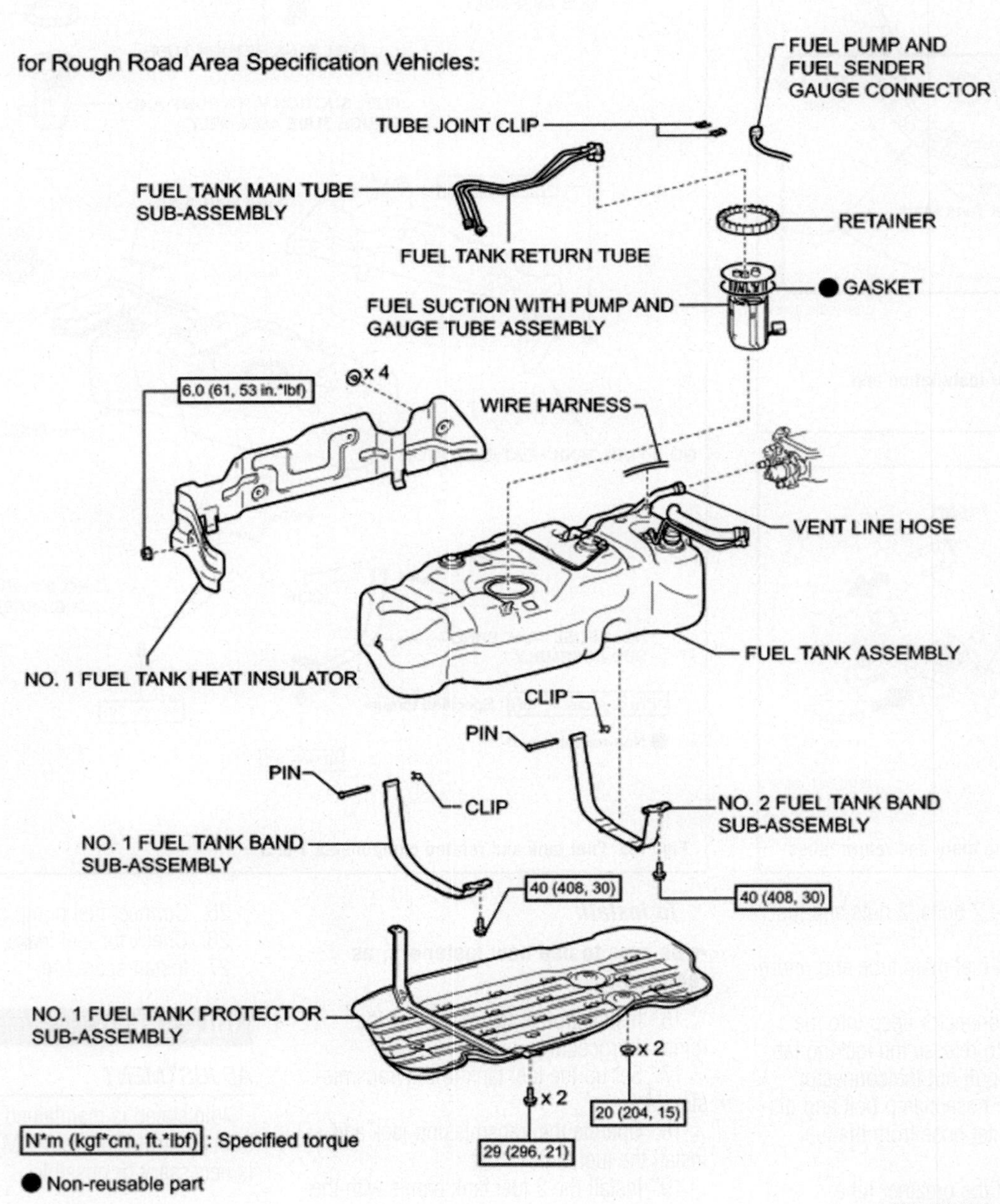

Fig. 164 Fuel tank and related components 2 of 3

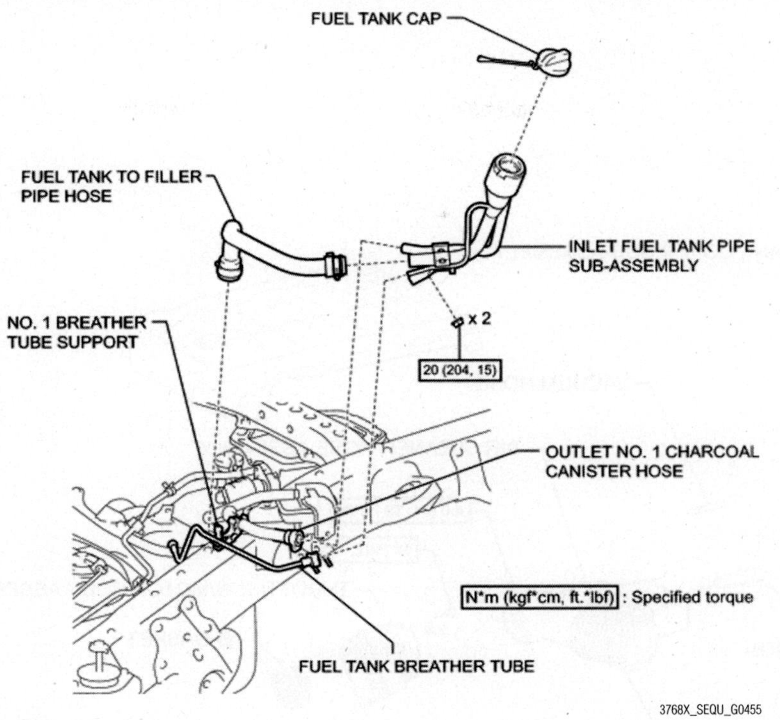

Fig. 165 Fuel tank and related components 3 of 3

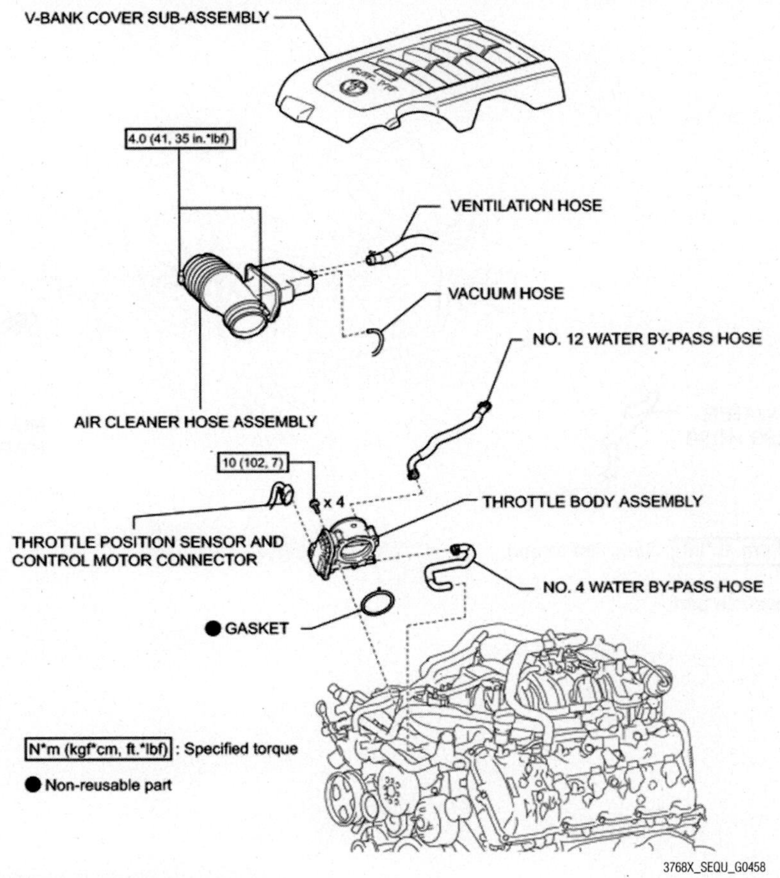

Fig. 166 Throttle body and related components—4.6L engine

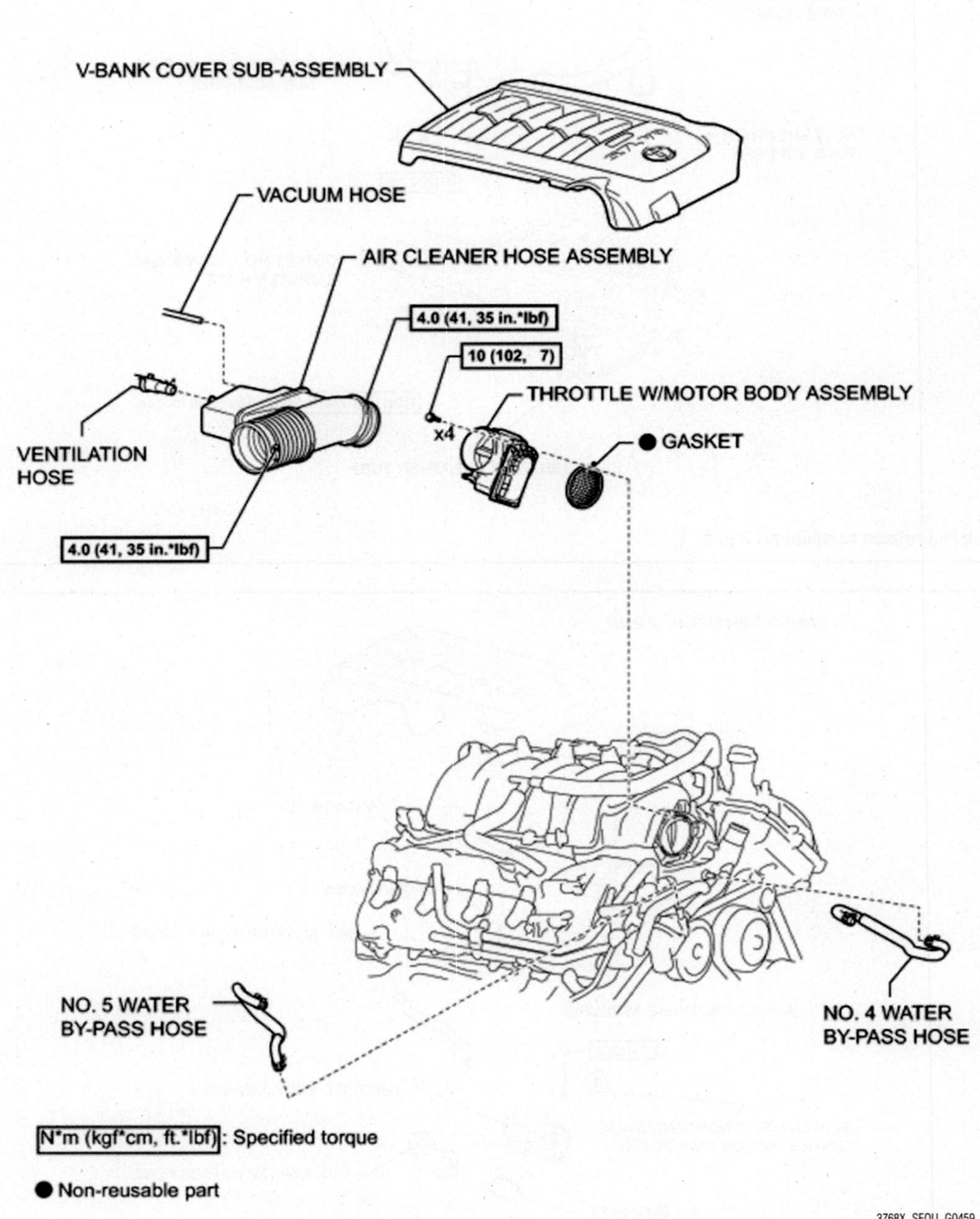

Fig. 167 Throttle body and related components—5.7L engine

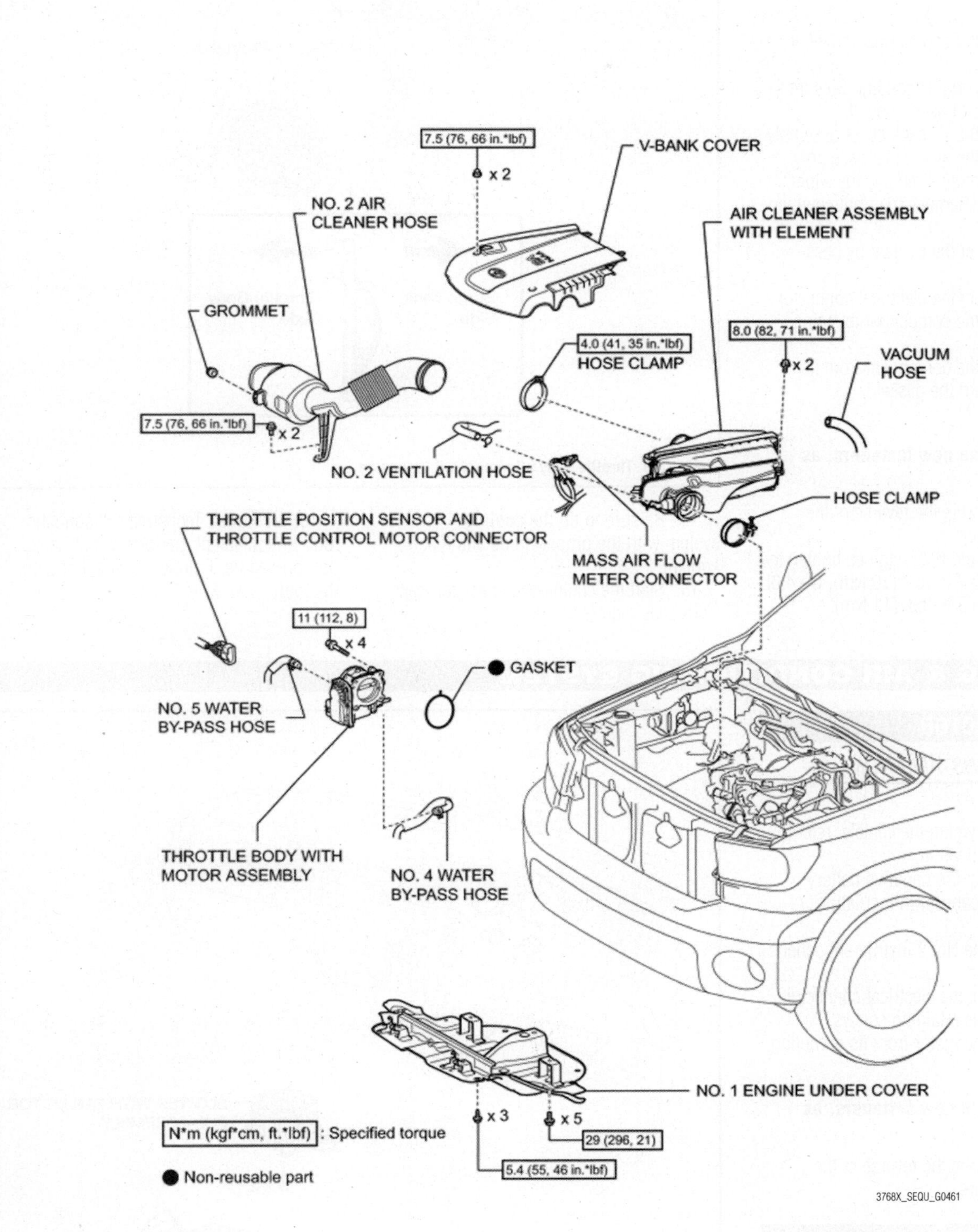

Fig. 168 Throttle body and related components—4.0L engine

➡On 4.0L and 5.7L engines the throttle body is removed along with the throttle body motor.

3. Remove the engine undercover, if equipped.
4. Drain the engine coolant. Be sure to properly dispose of used coolant.
5. Remove the V- bank cover assembly.
6. Remove the air cleaner assembly.
7. On 4.0L engine remove the wiper motor assembly. Remove the front cowl top sub panel.
8. Disconnect the coolant by pass hoses.
9. Disconnect the electrical connector.
10. Remove the component retaining bolts.
11. Remove the component from its mounting. Discard the gasket.

To install:

➡Be sure to use new fasteners, as required.

12. Installation is the reverse of the removal procedure.
13. On 4.6L and 5.7L engines tighten the retaining bolts to 7 ft. lbs. (10 Nm). On 4.0L engine tighten to 8 ft. lbs. (11 Nm).

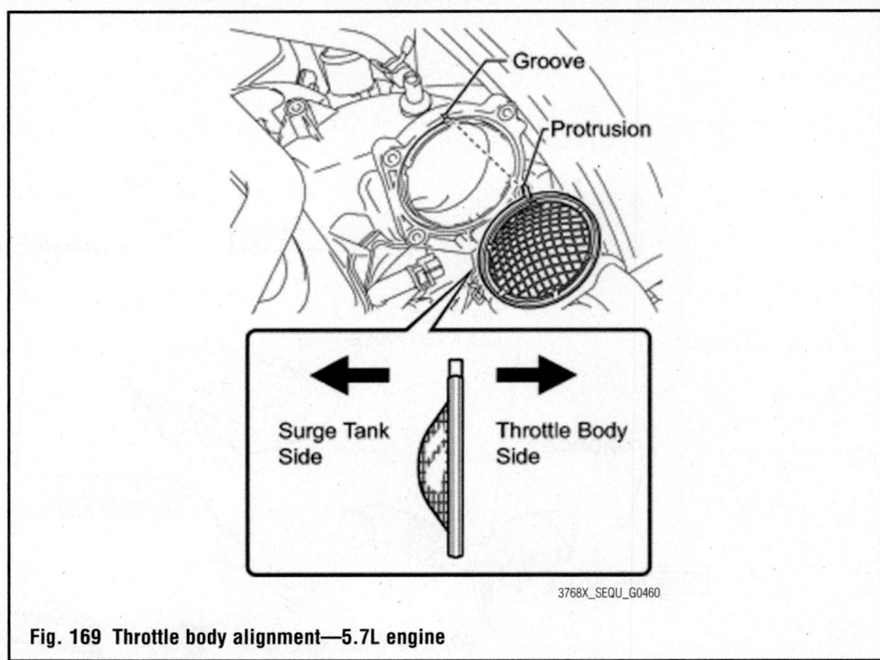

Fig. 169 Throttle body alignment—5.7L engine

14. Be sure to fill the cooling system with the proper grade and type coolant.
15. Start the engine and check for leaks. Correct as required.
16. Using the Techstream diagnostic tool, or equivalent, perform the initialization procedure. Follow the directions on the tool.

HEATING & AIR CONDITIONING SYSTEM

BLOWER MOTOR

REMOVAL & INSTALLATION

See Figure 170.

1. Before servicing the vehicle, refer to the Precautions Section.
2. Disconnect the negative battery cable. Tape the cable with insulating tape.
3. Remove the No. 2 instrument panel sub assembly.
4. Disconnect the electrical connector.
5. Remove the retaining screws.
6. Remove the motor from its mounting.

To install:

➡Be sure to use new fasteners, as required.

7. Installation is the reverse of the removal procedure.

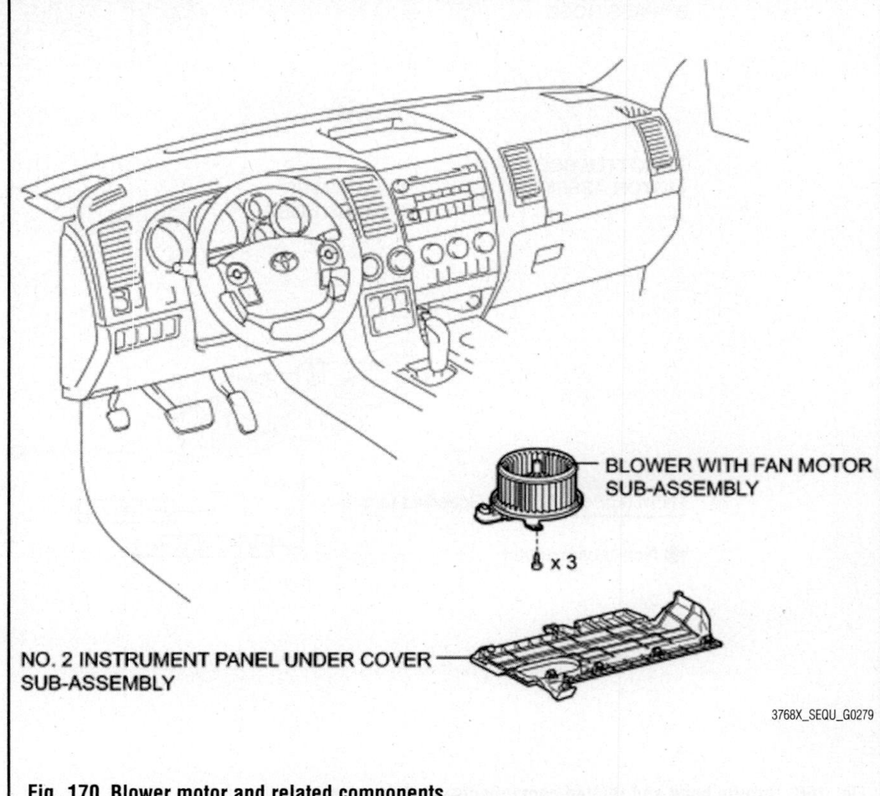

Fig. 170 Blower motor and related components

HEATER CORE

REMOVAL & INSTALLATION

1. Before servicing the vehicle, refer to the Precautions Section.
2. Disconnect the negative battery cable. Tape the cable with insulating tape.

3. Remove the heater/cooling unit from the vehicle.

4. Disassemble the unit as required to remove the heater core.

To install:

➡**Be sure to use new fasteners, as required.**

5. Installation is the reverse of the removal procedure.

6. Be sure to refill the cooling system, with the propér grade and type coolant.

7. Be sure to properly recharge the system.

8. Be sure to adjust the spiral cable, as required.

9. Check the SRS system, for proper operation.

10. Using the Techstream diagnostic tool or equivalent, reprogram the required systems.

STEERING

POWER RACK & PINION STEERING GEAR

REMOVAL & INSTALLATION

See Figure 171.

1. Before servicing the vehicle, refer to the Precautions Section.
2. Disconnect the negative battery cable. Tape the cable with insulating tape.
3. Position the front tires in the straight ahead position.
4. Drain the power steering fluid.
5. Raise and support the vehicle safely.
6. Remove the tire and wheel assemblies.
7. Remove the engine from the vehicle.
8. Matchmark and disconnect the No. 2 steering intermediate shaft sub assembly.
9. Disconnect the tie rod ends, using the proper tools.
10. Disconnect and plug the fluid lines.
11. Remove the bolts holding lines in place.
12. Remove the gear retaining bolts.
13. Remove the gear assembly from its mounting.

To install:

➡**Be sure to use new fasteners, as required.**

14. Installation is the reverse of the removal procedure.
15. Tighten the gear retaining bolts to 89 ft. lbs. (120 Nm).

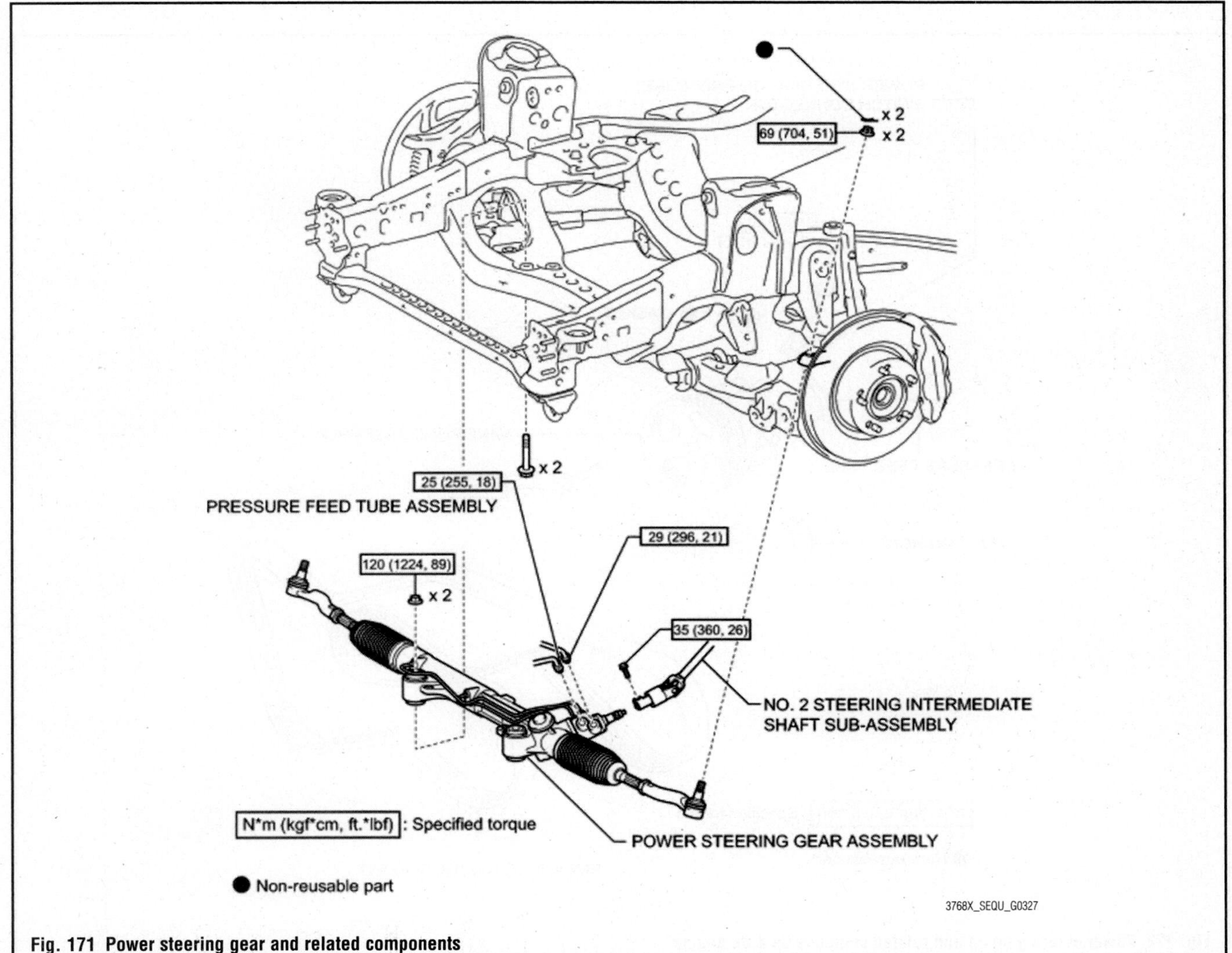

Fig. 171 Power steering gear and related components

16. Use the matchmarks when installing the No.2 steering intermediate shaft to the steering gear. Tighten the retaining bolt to 26 ft. lbs. (35 Nm).
17. Adjust the front end alignment, as required.
18. Be sure to fill the power steering system, with the correct grade and type fluid.

POWER STEERING PUMP

REMOVAL & INSTALLATION

4.0L Engine

See Figure 172.

1. Before servicing the vehicle, refer to the Precautions Section.
2. Disconnect the negative battery cable. Tape the cable with insulating tape.
3. Remove No. 1 engine under cover.
4. Remove fan and alternator V belt.
5. Drain power steering fluid.
6. Disconnect suction hose.
7. Disconnect power steering oil pressure switch connector.
8. Remove the bolt and disconnect the pressure feed tube.
9. Remove the gasket.
10. Remove the 2 bolts and vane pump.

To install:

➡Be sure to use new fasteners, as required.

11. Install the vane pump with the 2 bolts and tighten to 21 ft. lbs. (28 Nm).
12. Install a new gasket to the pressure feed tube.
13. Install the pressure feed tube with the union bolt and tighten to 38 ft lbs. (51 Nm).
14. Connect power steering oil pressure switch connector.
15. Connect the suction hose with the clip.
16. Install fan and alternator V belt.
17. Add power steering fluid.
18. Bleed power steering fluid.
19. Inspect for power steering fluid leak.
20. Install No. 1 engine under cover.
21. Connect cable to negative battery terminal.
22. Perform initialization, if necessary.

4.6L Engine

See Figure 173.

1. Before servicing the vehicle, refer to the Precautions Section.
2. Disconnect the negative battery cable. Tape the cable with insulating tape.
3. Remove the drive belt.
4. Drain the cooling system. Remove the radiator.
5. Separate the harness bracket.
6. Drain the power steering system. Be sure to properly dispose of used fluid.
7. Disconnect and plug the fluid lines.
8. Disconnect the pressure sensor.
9. Disconnect the solenoid valve connector.
10. Remove the pump retaining bolts.

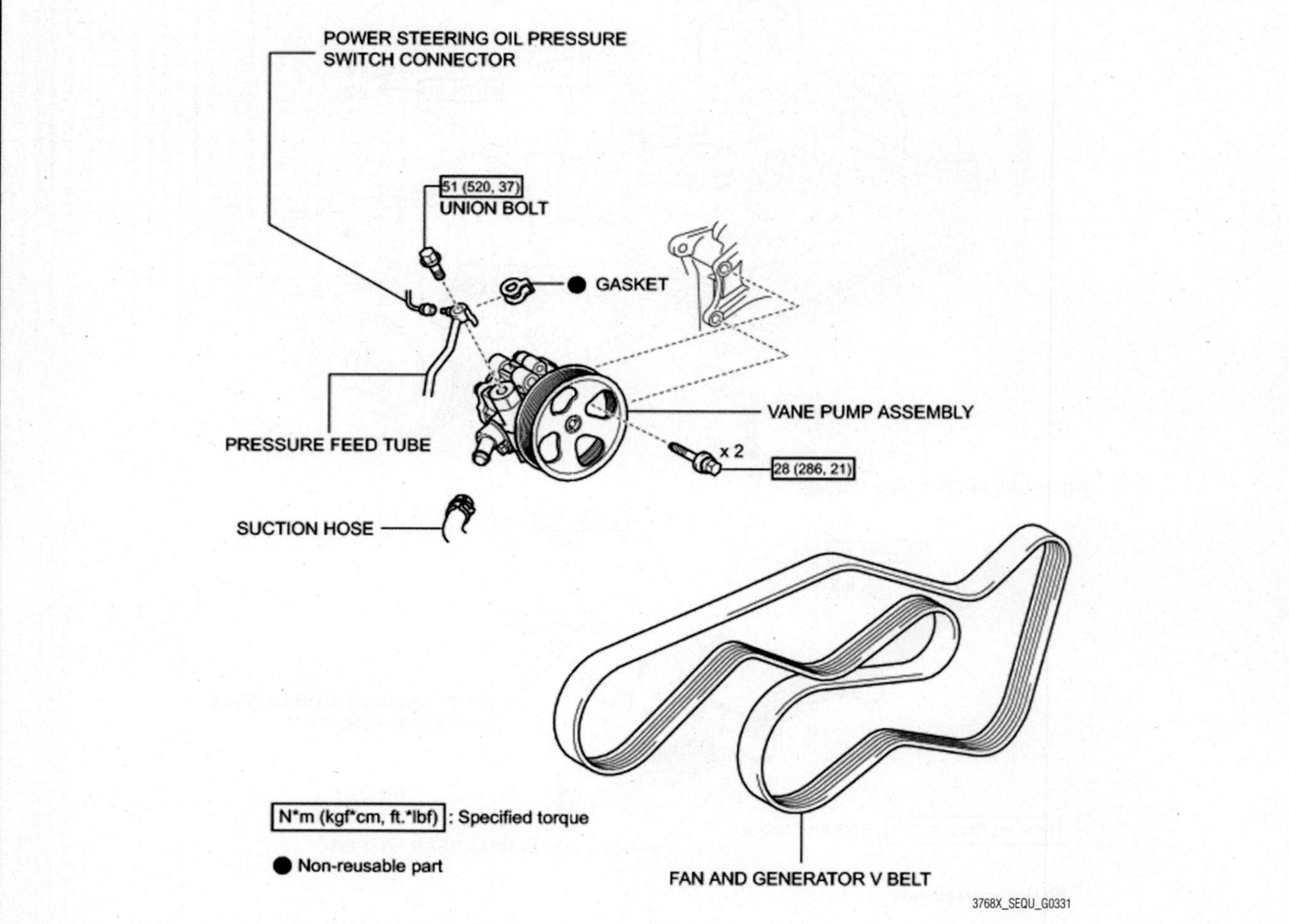

Fig. 172 Power steering pump and related components 4.0L engine

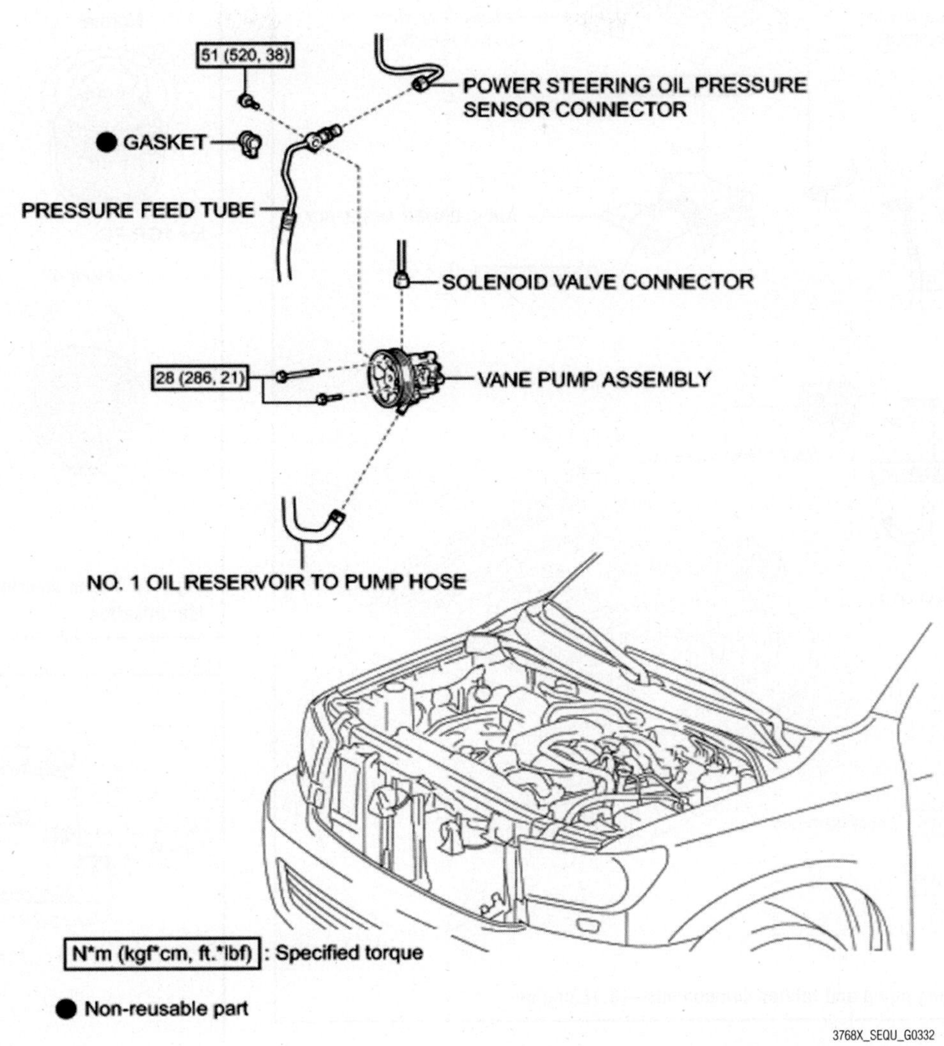

51 (520, 38)
POWER STEERING OIL PRESSURE SENSOR CONNECTOR
● GASKET
PRESSURE FEED TUBE
SOLENOID VALVE CONNECTOR
28 (286, 21)
VANE PUMP ASSEMBLY
NO. 1 OIL RESERVOIR TO PUMP HOSE
N*m (kgf*cm, ft.*lbf) : Specified torque
● Non-reusable part
3768X_SEQU_G0332

Fig. 173 Power steering pump and related components—4.6L engine

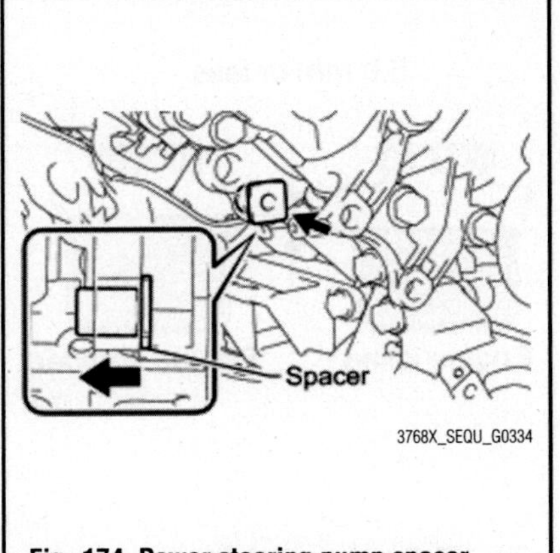

Fig. 174 Power steering pump spacer location—4.6L engine

11. Remove the pump from its mounting.

To install:

➡**Be sure to use new fasteners, as required.**

12. Installation is the reverse of the removal procedure.
13. Move the space so that the pump can be installed.
14. Tighten the retaining bolts to 21 ft. lbs. (29 Nm).
15. Be sure to fill the cooling system, with the proper grade and type coolant.
16. Be sure to fill the power steering system, with the proper grade and type fluid.

5.7L Engine

See Figure 175.

1. Before servicing the vehicle, refer to the Precautions Section.
2. Disconnect the negative battery cable. Tape the cable with insulating tape.
3. Remove the drive belt.
4. Remove the V- bank sub assembly.
5. Remove the air cleaner assembly.
6. Drain the power steering system. Be sure to properly dispose of used fluid.
7. Disconnect and plug the fluid lines.
8. Disconnect the pressure sensor.
9. Remove the pump retaining bolts.
10. Remove the pump from its mounting.

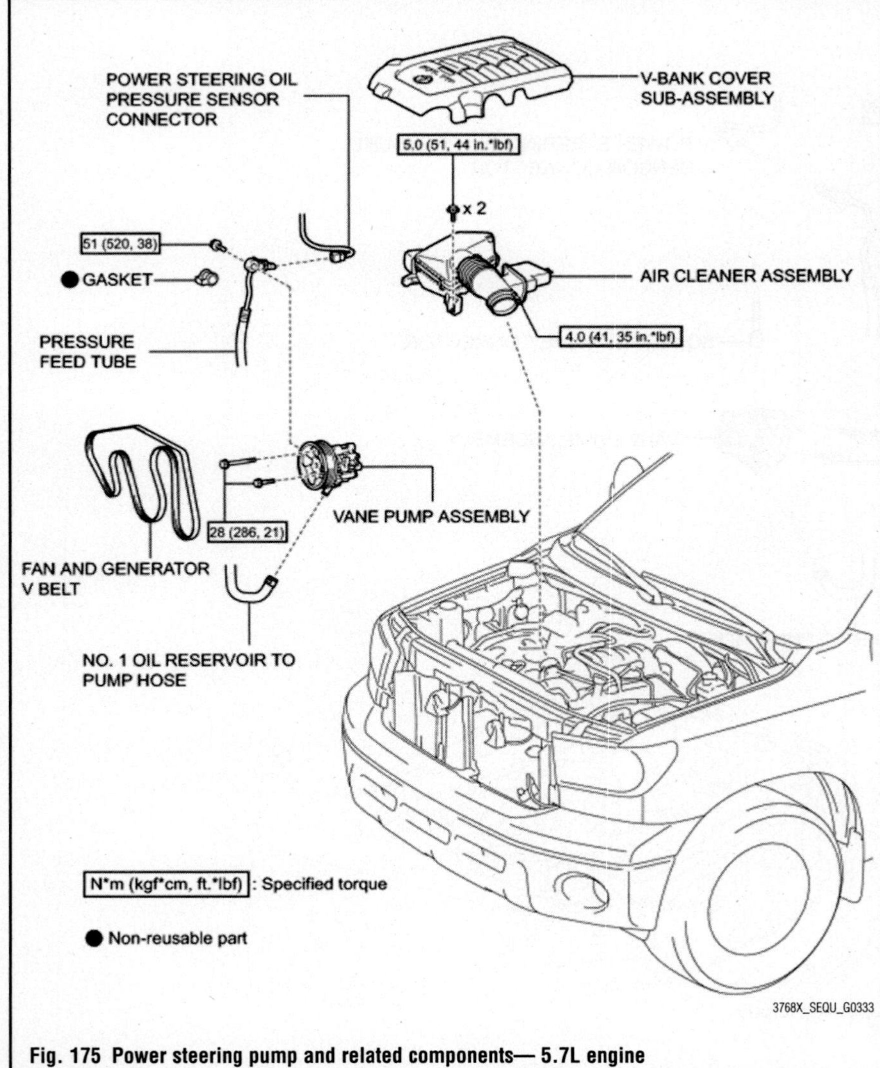

Fig. 175 Power steering pump and related components— 5.7L engine

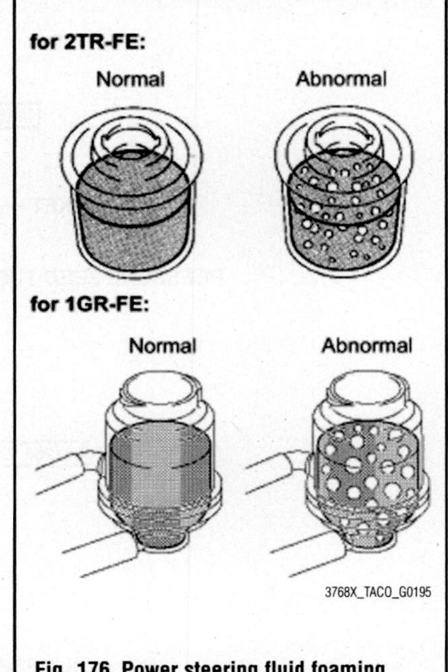

Fig. 176 Power steering fluid foaming identification

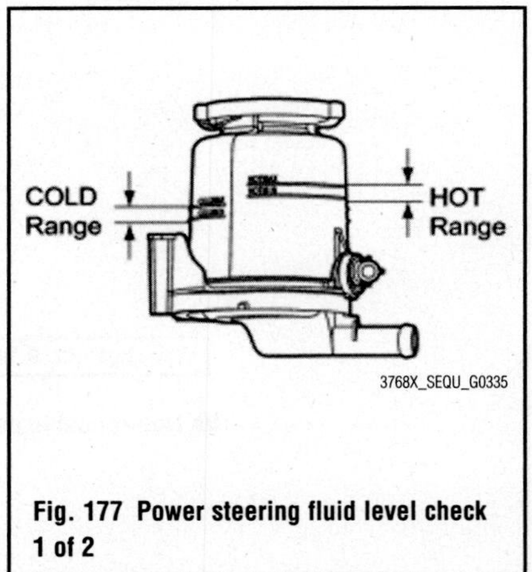

Fig. 177 Power steering fluid level check 1 of 2

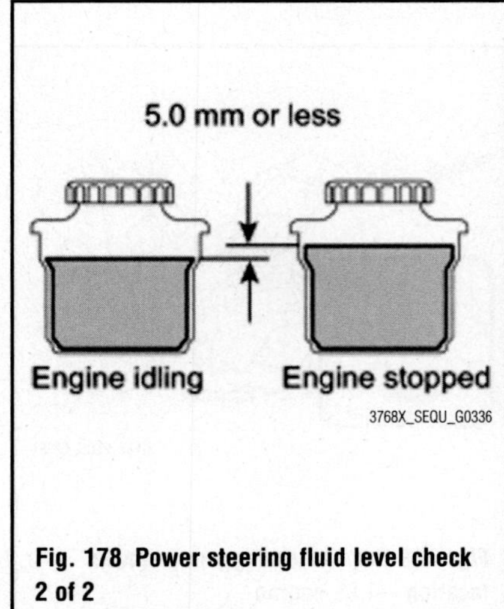

Fig. 178 Power steering fluid level check 2 of 2

To install:

➡Be sure to use new fasteners, as required.

11. Installation is the reverse of the removal procedure.
12. Move the space so that the pump can be installed.
13. Tighten the retaining bolts to 21 ft. lbs. (28 Nm).
14. Be sure to fill the cooling system, with the proper grade and type coolant.
15. Be sure to fill the power steering system, with the proper grade and type fluid.

BLEEDING

See Figures 176 through 178.

1. Before servicing the vehicle, refer to the Precautions Section.
2. Disconnect the negative battery cable. Tape the cable with insulating tape.
3. Check the fluid level.
4. Raise and properly support the vehicle.
5. With the engine stopped, turn the wheels slowly from stop to stop several times.
6. Lower the vehicle.
7. Start the engine. Run at idle for a few minutes.
8. With the engine idling turn the steering wheel to the left or right full lock position and keep it there for two or three seconds. Repeat for the other side.
9. Repeat the above step several times.
10. Stop the engine. Check for foaming.
11. If foaming exists, repeat the procedure.
12. With the engine stopped check the fluid level.
13. Adjust as necessary. Be sure to use the proper grade and type fluid.

COIL SPRING

REMOVAL & INSTALLATION

See Figure 179.

1. Before servicing the vehicle, refer to the Precautions Section.
2. Disconnect the negative battery cable. Tape the cable with insulating tape.
3. Remove the strut.
4. Using a compressor, compress the coil spring.

➡A compressor with a force of 2,860 lbs. (12,740 N) or more must be used. Make sure that the suspension support is free from the spring. Do not compress the spring more than necessary. Do not position the spring with the upper end towards you.

5. Remove the center nut.
6. Remove the retainers, bushing support and spring.
7. Assembly is the reverse of disassembly. See the accompanying illustration for correct positioning of the suspension support and spring. Torque the nut to 18 ft. lbs. (25 Nm).

CONTROL LINKS

REMOVAL & INSTALLATION

See Figure 180.

1. Before servicing the vehicle, refer to the Precautions Section.
2. Disconnect the negative battery cable. Tape the cable with insulating tape.
3. Raise and support the vehicle safely.
4. Remove the tire and wheel assembly.
5. Remove the bolt, nut and stabilizer link from its mounting.

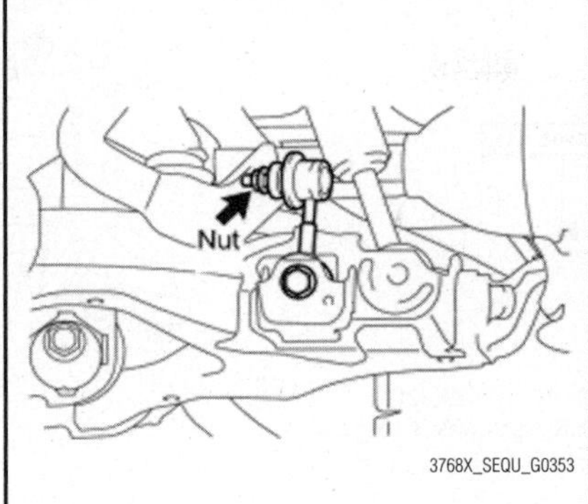

Fig. 180 Front suspension stabilizer link and related components

To install:

➡Be sure to use new fasteners, as required.

6. Installation is the reverse of the removal procedure.
7. Temporarily install the component.
8. Stabilize the suspension.
9. Tighten the retaining nut to 111 ft. lbs. (150 Nm).
10. Tighten the retaining bolt to 89 ft. lbs. (120 Nm).

LOWER BALL JOINT

REMOVAL & INSTALLATION

1. Before servicing the vehicle, refer to the Precautions Section.
2. Disconnect the negative battery cable. Tape the cable with insulating tape.
3. Remove the lower control arm.
4. Remove the cotter pin and nut. Discard the cotter pin.
5. Using a ball joint removal tool, remove the ball joint.

To install:

➡Be sure to use new fasteners, as required.

6. Installation is the reverse of the removal procedure.
7. Tighten the retaining bolts to 221 ft. lbs. (300 Nm).
8. Be sure to use a new cotter pin.
9. Correct and adjust alignment, as required.

LOWER CONTROL ARM

REMOVAL & INSTALLATION

See Figures 181 and 182.

1. Before servicing the vehicle, refer to the Precautions Section.
2. Disconnect the negative battery cable. Tape the cable with insulating tape.
3. Raise and support the vehicle safely.
4. Remove the tire and wheel assembly.
5. Remove the stabilizer link assembly.
6. Disconnect the shock absorber.
7. Disconnect the lower ball joint attaching bolts.
8. Be sure to properly support the lower control arm before removal.
9. Matchmark and remove the No. 1 suspension arm sub assembly.
10. As required, remove the cotter pin and nut from the ball joint. Install the proper ball joint removal tool and remove the ball joint.

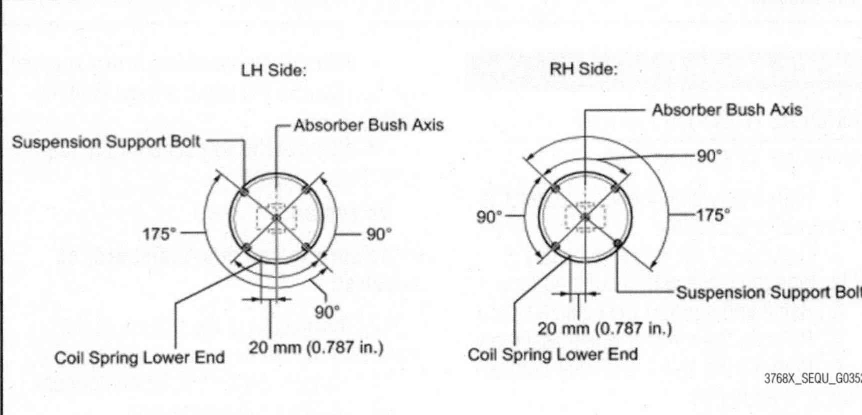

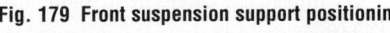

Fig. 179 Front suspension support positioning

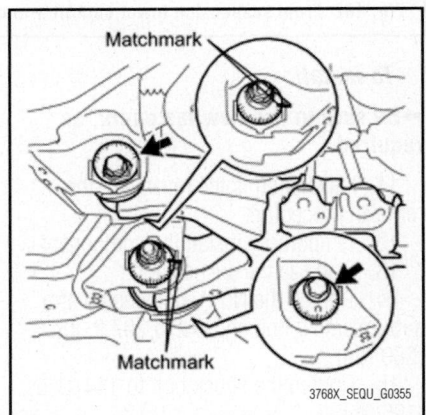

Fig. 181 Front suspension lower control arm matchmark points

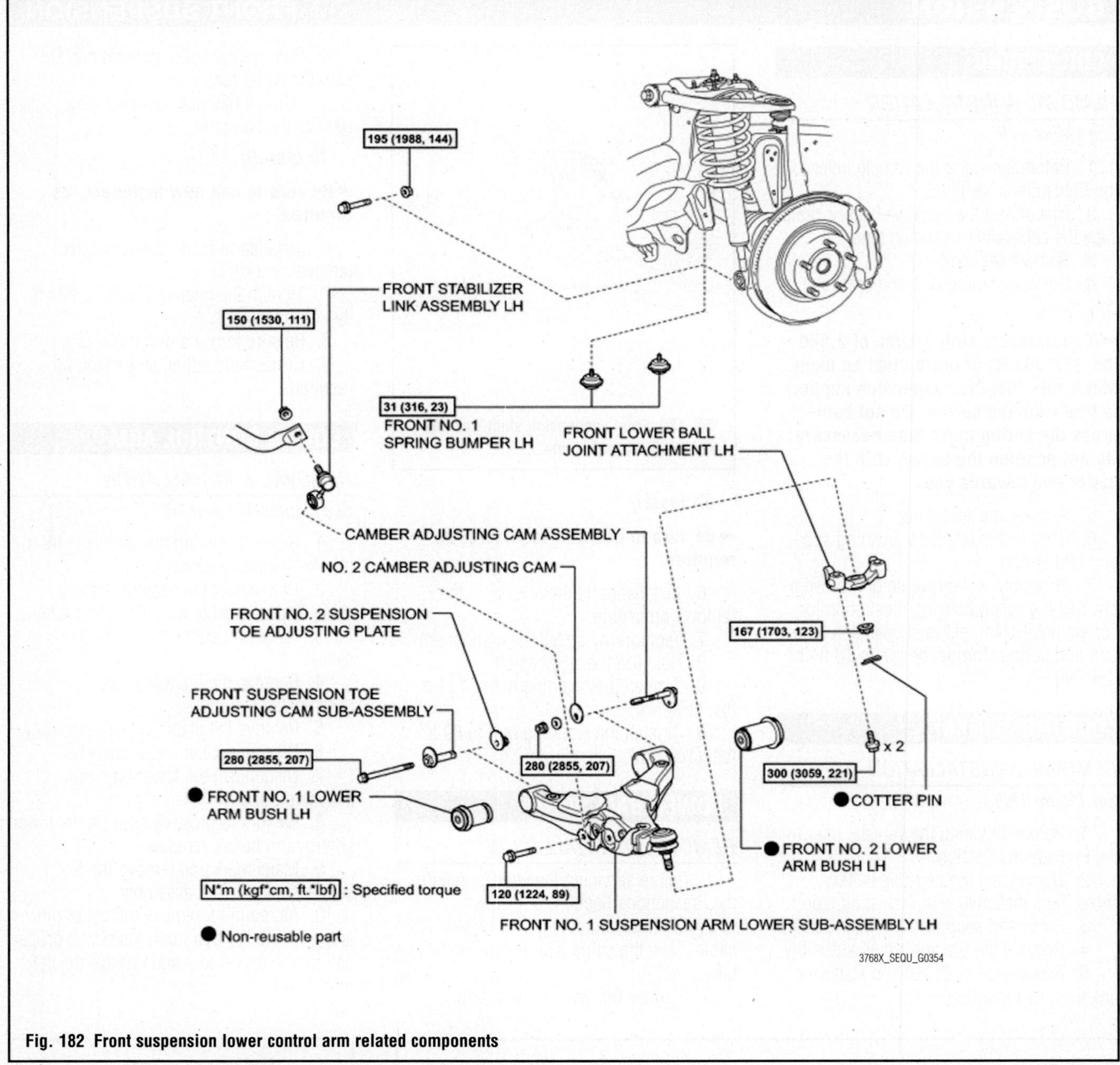

Fig. 182 Front suspension lower control arm related components

To install:

➡Be sure to use new fasteners, as required.

11. Installation is the reverse of the removal procedure.
12. Temporarily install the component.
13. Stabilize the suspension.
14. Tighten the No. 1 suspension arm sub assembly bolt and nut to 207 ft. lbs. (280 Nm).
15. Tighten the shock bolt to 144 ft. lbs. (195 Nm).
16. Check and adjust the alignment, as required.

SHOCK ABSORBERS

REMOVAL & INSTALLATION

See Figures 183 through 185.

1. Before servicing the vehicle, refer to the Precautions Section.
2. Disconnect the negative battery cable. Tape the cable with insulating tape.
3. Raise and support the vehicle safely.
4. Remove the tire and wheel assembly.
5. Remove the shock absorber control actuator, if equipped.
6. Remove the suspension control bracket, if equipped.
7. Remove the lower shock retaining nut.
8. Remove the upper shock retaining bolts.
9. Remove the component from the vehicle.

To install:

➡Be sure to use new fasteners, as required.

10. Installation is the reverse of the removal procedure.
11. Temporarily install the component.
12. Stabilize the suspension.
13. Tighten the lower bolt to 144 ft. lbs. (195 Nm).

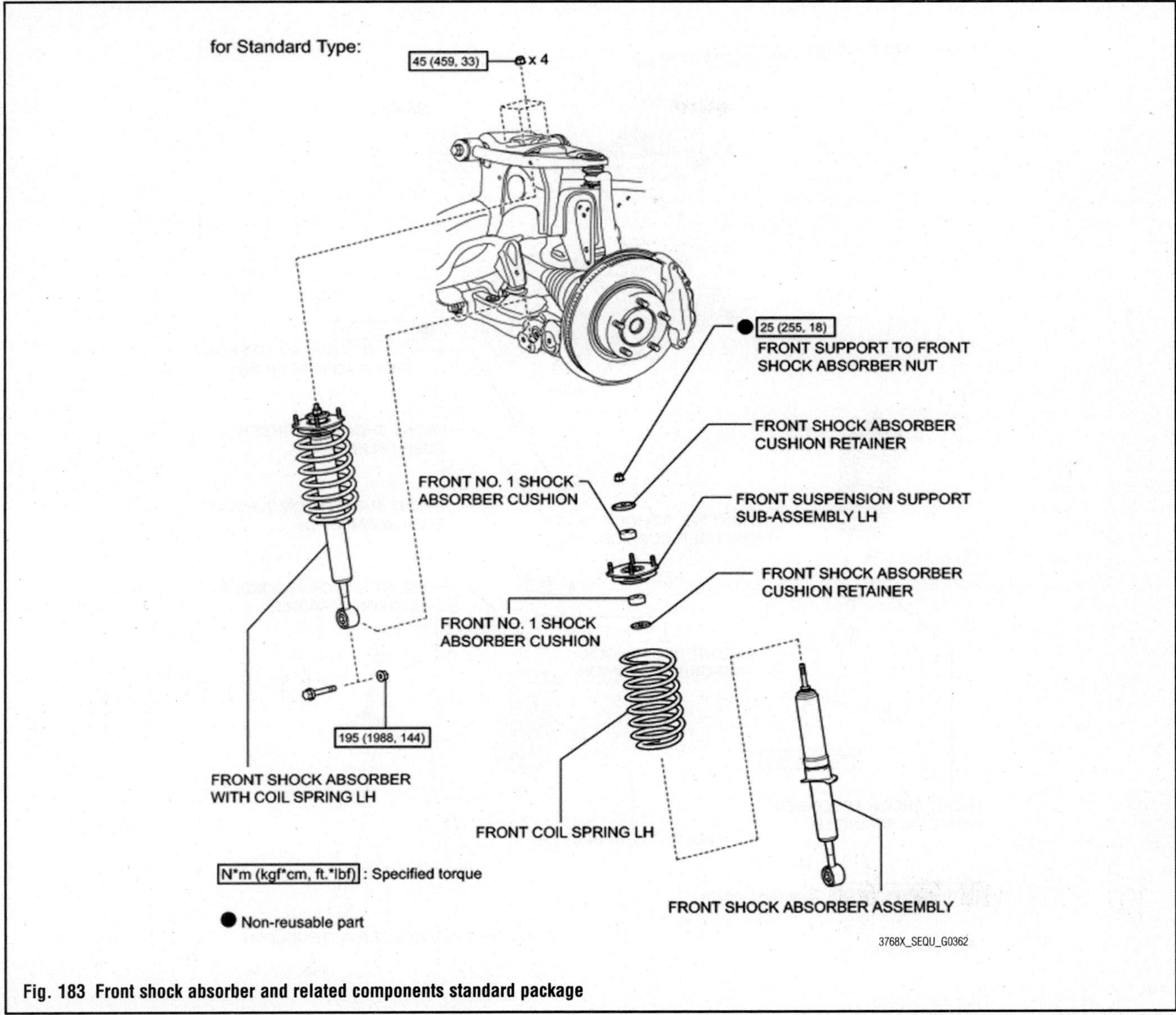

Fig. 183 Front shock absorber and related components standard package

14. Tighten the upper bolts to 33 ft. lbs. (45 Nm), in the sequence shown in the illustration.

STABILIZER BAR

REMOVAL & INSTALLATION

See Figures 186 and 187.

1. Before servicing the vehicle, refer to the Precautions Section.
2. Disconnect the negative battery cable. Tape the cable with insulating tape.
3. Remove the nut and disconnect the stabilizer bar links from the lower suspension arm.

➡ If the ball joint turns together with the nut, use a hexagon (6 mm) wrench to hold the stud.

4. Remove the 2 bolts, nuts and stabilizer bar with the cushions and brackets.
5. Remove the 2 brackets and cushions from the stabilizer bar.
6. Hold the stabilizer bar link, and remove the nut.
7. Remove the stabilizer bar link, 2 retainers and bushings from each end of the stabilizer bar.

To install:

➡Be sure to use new fasteners, as required.

8. Temporarily install the component.
9. Stabilize the suspension.
10. Install the 2 bushings, retainers and stabilizer bar link to the stabilizer bar.
11. Hold the stabilizer bar link, and install a new nut and tighten to 14 ft. lbs. (19 Nm).
12. Install the 2 bushings with their cutout facing to the rearward of the stabilizer bar.
13. Install the stabilizer bar and 2 brackets with the nuts and bolts and tighten to 27 ft. lbs. (37 Nm).
14. Connect front stabilizer links to suspension lower arm and tighten to 51 ft. lbs. (69 Nm).

➡If the ball joint turns together with the nut, use a hexagon (6 mm) wrench to hold the nut.

STEERING KNUCKLE

REMOVAL & INSTALLATION

See Figures 188 and 189.

1. Before servicing the vehicle, refer to the Precautions Section.

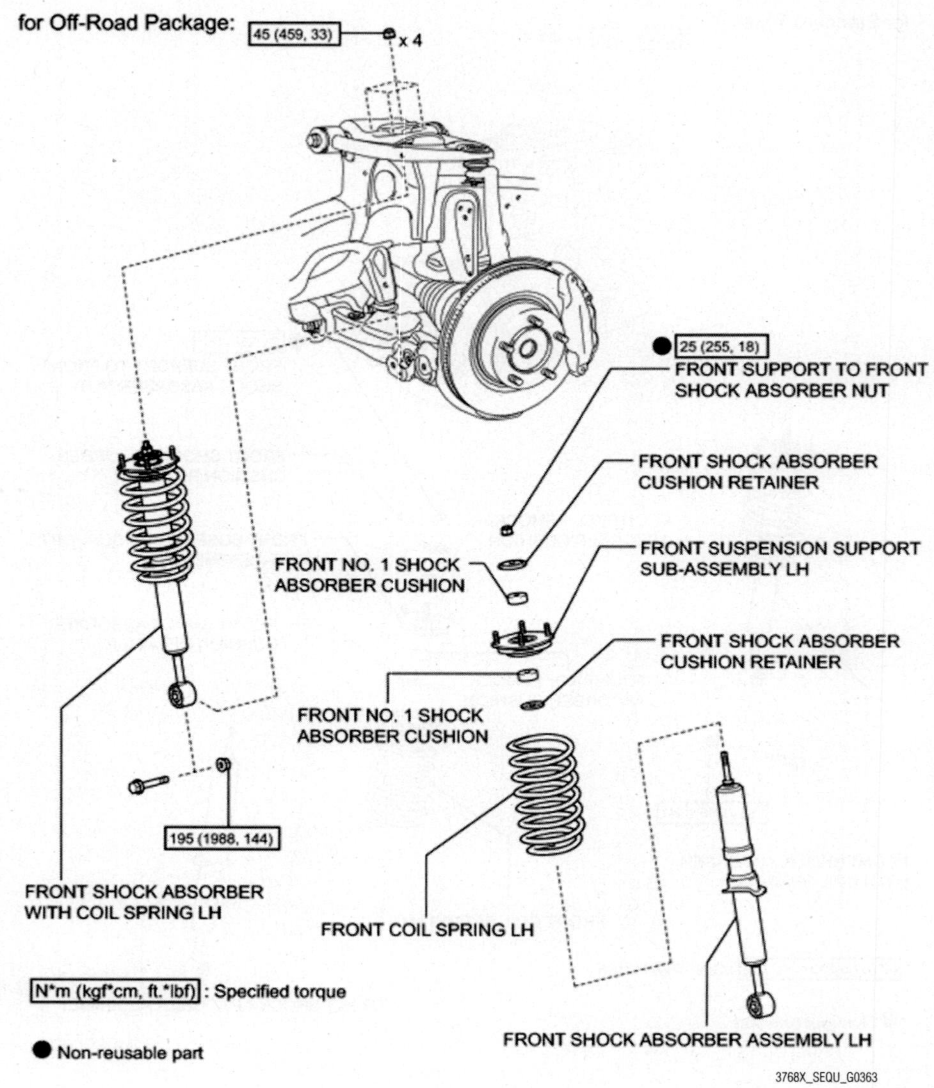

for Off-Road Package:

45 (459, 33) x 4

● 25 (255, 18) FRONT SUPPORT TO FRONT SHOCK ABSORBER NUT

FRONT SHOCK ABSORBER CUSHION RETAINER

FRONT NO. 1 SHOCK ABSORBER CUSHION

FRONT SUSPENSION SUPPORT SUB-ASSEMBLY LH

FRONT SHOCK ABSORBER CUSHION RETAINER

FRONT NO. 1 SHOCK ABSORBER CUSHION

195 (1988, 144)

FRONT SHOCK ABSORBER WITH COIL SPRING LH

FRONT COIL SPRING LH

N*m (kgf*cm, ft.*lbf) : Specified torque

● Non-reusable part

FRONT SHOCK ABSORBER ASSEMBLY LH

3768X_SEQU_G0363

Fig. 184 Front shock absorber and related components off road package

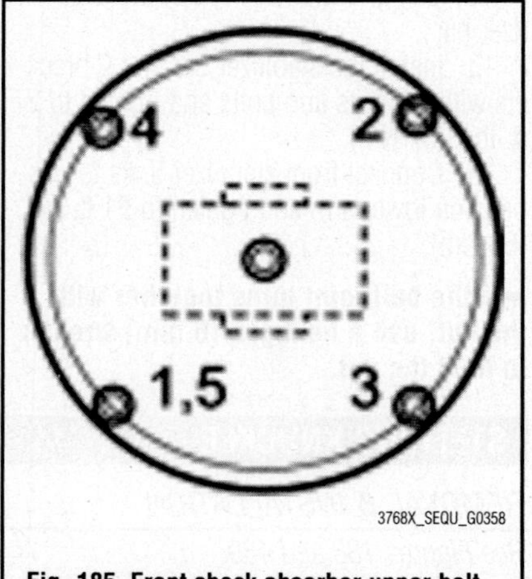

Fig. 185 Front shock absorber upper bolt tightening sequence

2. Disconnect the negative battery cable. Tape the cable with insulating tape.
3. Remove the front axle hub.
4. Disconnect the speed sensor from the steering knuckle.
5. Disconnect tie rod end sub-assembly.
    a. Remove the cotter pin and nut.
    b. Disconnect the tie rod end LH from the steering knuckle.
6. Disconnect front lower ball joint attachment.
7. Support the front suspension lower arm with a jack.
8. Remove the clip and the nut.
9. Disconnect the upper ball joint from the steering knuckle.
10. Remove the steering knuckle.

To install:

➡Be sure to use new fasteners, as required.

11. Install the front suspension upper arm to the steering knuckle with the nut and tighten to specification.
12. Install a new clip.

➡If the holes for the clip are not aligned, tighten the nut up to another 60°.

13. Connect front lower ball joint attachment.
14. Connect tie rod end sub-assembly.
15. Install the front axle hub and tighten nut to specification.
16. Connect the front speed sensor to the steering knuckle.

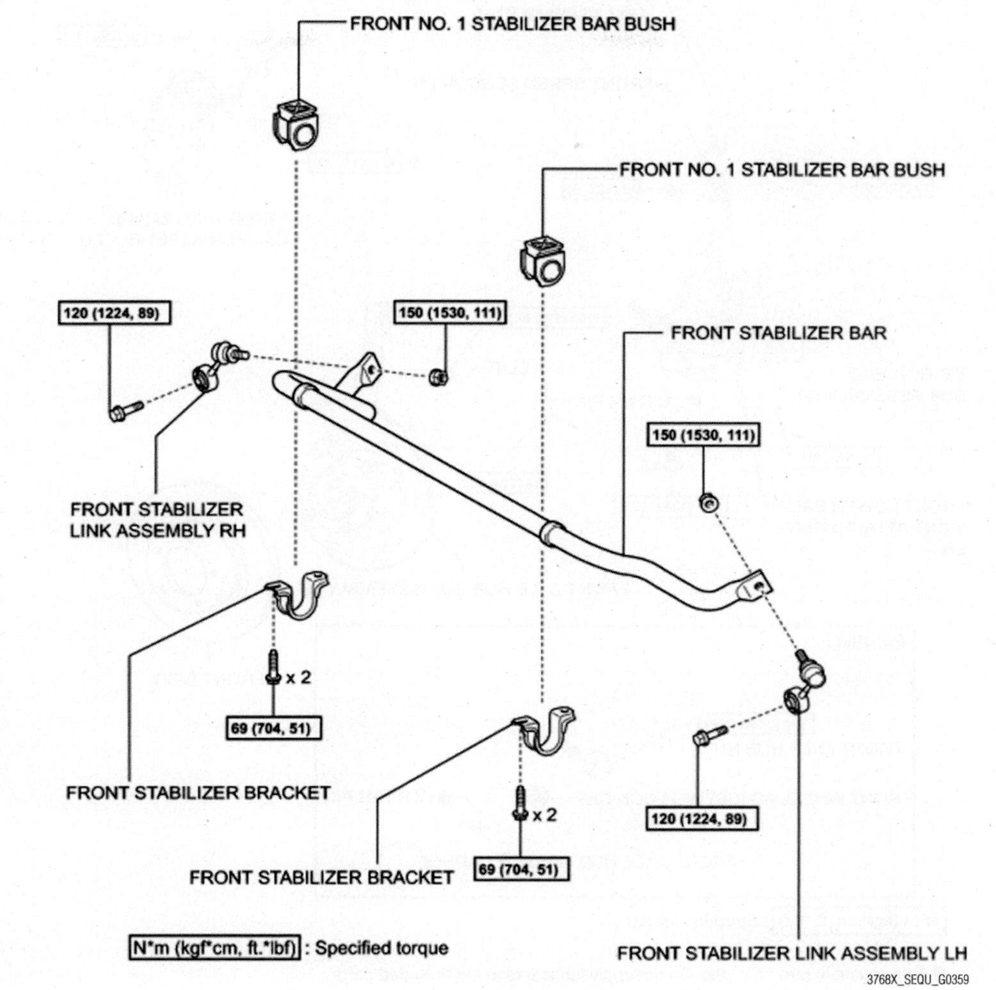

Fig. 186 Front stabilizer bar and related components

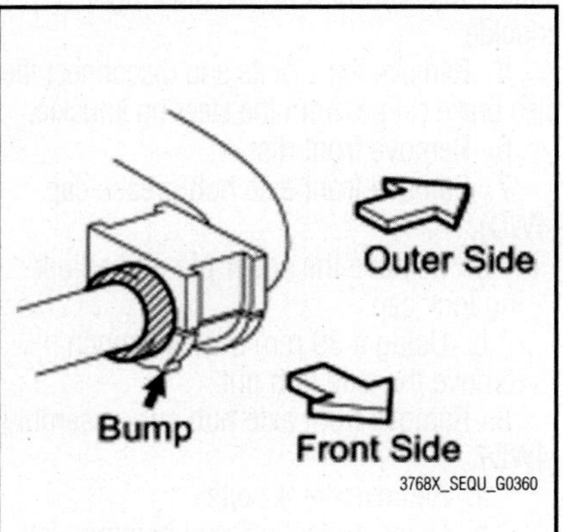

Fig. 187 Front stabilizer bar bushing positioning

17. Attach the clip to the steering knuckle.
18. Install front wheel.
19. Inspect and adjust the front wheel alignment.

UPPER BALL JOINT

REMOVAL & INSTALLATION

1. Before servicing the vehicle, refer to the Precautions Section.
2. Disconnect the negative battery cable. Tape the cable with insulating tape.
3. Remove the lower control arm.
4. Using the proper removal tools, separate the ball joint from the control arm.

To install:

➡**Be sure to use new fasteners, as required.**

5. Temporarily install the component.
6. Stabilize the suspension.
7. Tighten the retaining bolts to specification.
8. Check and adjust the alignment, as required.

UPPER CONTROL ARM

REMOVAL & INSTALLATION

See Figure 190.

1. Before servicing the vehicle, refer to the Precautions Section.
2. Disconnect the negative battery cable. Tape the cable with insulating tape.

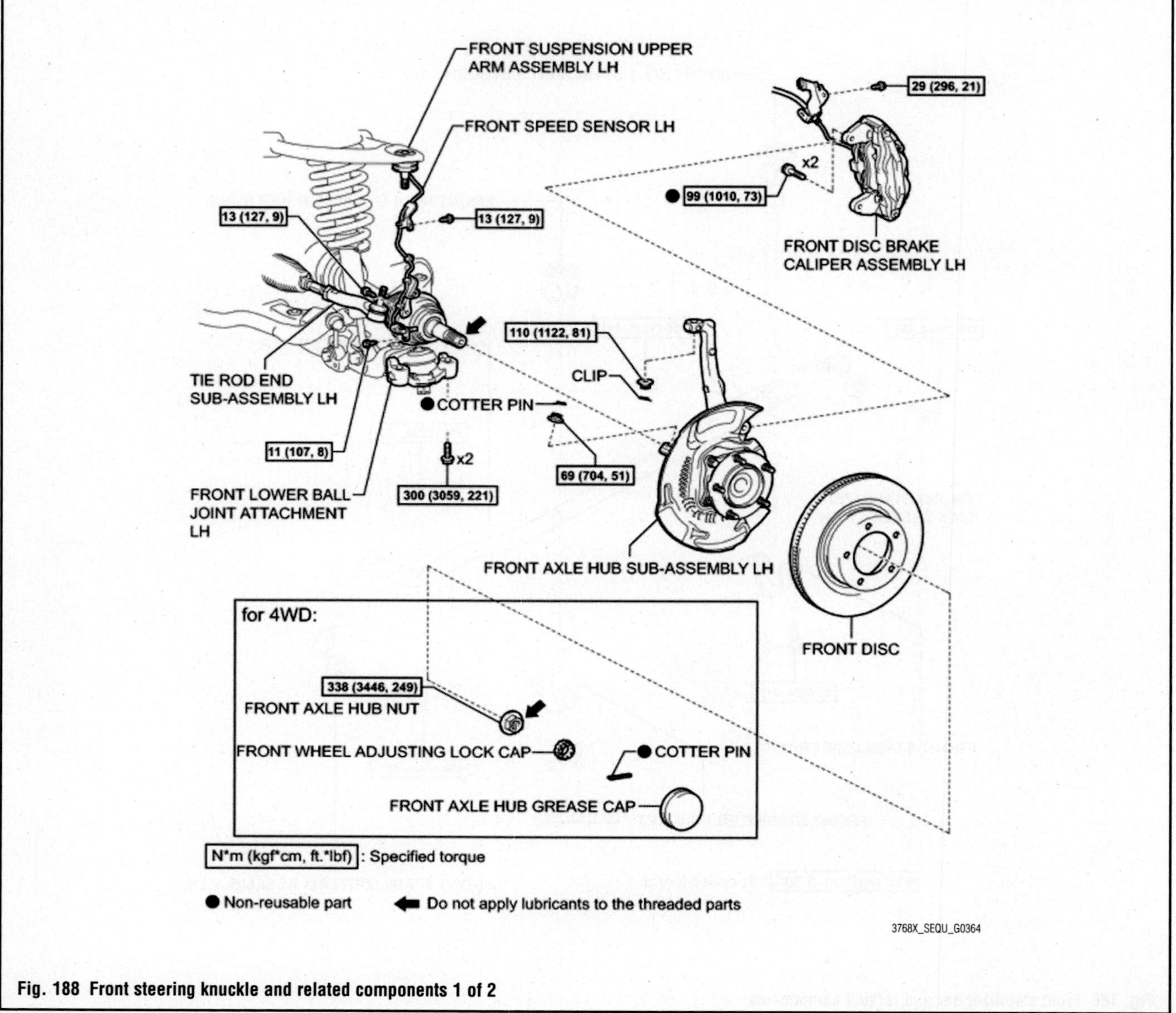

Fig. 188 Front steering knuckle and related components 1 of 2

3. Raise and support the vehicle safely.
4. Remove the tire and wheel assembly.
5. Remove the shock absorber.
6. Disconnect the skid control sensor wire.
7. Remove the upper ball joint cotter pin and nut. Discard the cotter pin.
8. Disconnect the upper ball joint from the steering knuckle. Be sure to properly support the lower control arm assembly.
9. Remove the fender apron seals.
10. Remove the component retaining bolts.
11. Remove the component from its mounting.

To install:

➡Be sure to use new fasteners, as required.

12. Temporarily install the component.
13. Stabilize the suspension.
14. Tighten all bolts to specification, refer to illustration.
15. Check and adjust the alignment, as required.

WHEEL HUB & BEARING

REMOVAL & INSTALLATION

1. Before servicing the vehicle, refer to the Precautions Section.
2. Disconnect the negative battery cable. Tape the cable with insulating tape.
3. Raise and support the vehicle safely. Remove the tire and wheel assembly.
4. Remove the bolt and disconnect the brake tube bracket from the steering knuckle.
5. Remove the 2 bolts and disconnect the disc brake caliper from the steering knuckle.
6. Remove front disc.
7. Remove front axle hub grease cap (4WD).
   a. Remove the cotter pin and adjusting lock cap.
   b. Using a 39 mm socket wrench, remove the axle hub nut.
8. Remove front axle hub sub-assembly (4WD).
   a. Remove the 4 bolts.
   b. Using a plastic-faced hammer, tap out the front drive shaft from the front axle hub.
   c. Remove the O-ring from the axle hub.

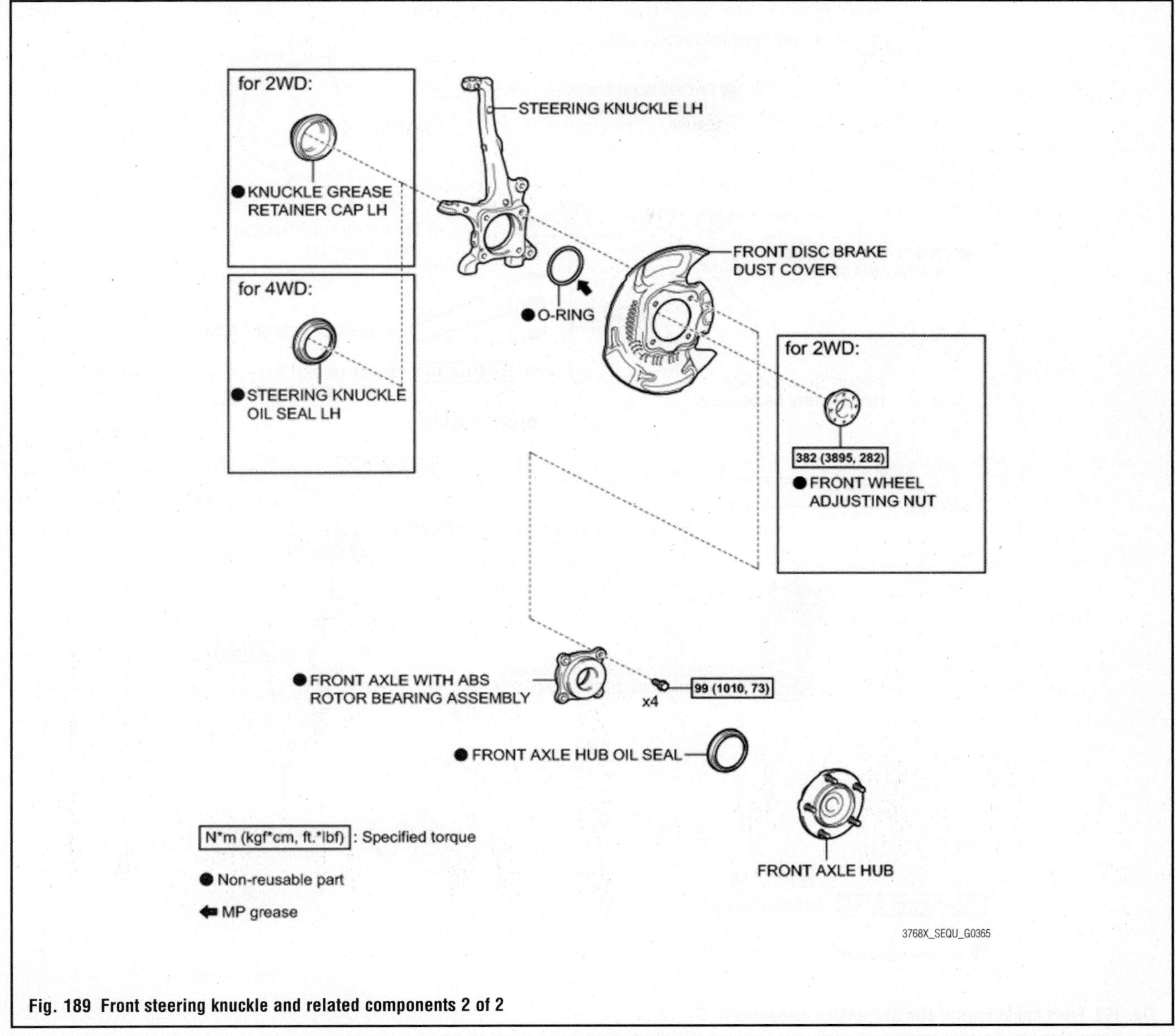

Fig. 189 Front steering knuckle and related components 2 of 2

9. Remove front axle hub sub-assembly (2WD).
   a. Remove the 4 bolts.
   b. Remove the axle hub and dust cover from the steering knuckle.
   c. Remove the O-ring from the axle hub.

To install:

➡Be sure to use new fasteners, as required.

10. Install front axle hub sub-assembly (2WD).
    a. Apply MP grease to a new O-ring.
    b. Install the O-ring to the axle hub.
    c. Install the dust cover and axle hub to the steering knuckle with the 4 bolts and tighten to 73 ft. lbs. (99 Nm).
11. Install front axle hub sub-assembly (4WD).
    a. Apply MP grease to a new O-ring.
    b. Install the O-ring to the axle hub.
    c. Connect the front drive shaft to the front axle hub.
    d. Install the dust cover and axle hub to the steering knuckle with the 4 bolts and tighten to 73 ft. lbs. (99 Nm).
12. Install front axle hub nut (4WD).
    a. Clean the threaded parts on the drive shaft and axle hub nut using a non-residue solvent.
    b. Using a 39 mm socket wrench, install the hub nut and tighten to 249 ft. lbs. (338 Nm).
    c. Install the front wheel adjusting lock cap and a new cotter pin.
13. Inspect the front axle hub.
14. Install the axle hub grease cap.
15. Install front disc.
16. Connect the front disc brake caliper and install 2 new bolts and tighten to 73 ft. lbs. (99 Nm).
17. Connect the brake tube bracket to the steering knuckle with the bolt and tighten to 21 ft. lbs. (29 Nm).
18. Install front wheel.
19. Check the speed sensor signal.

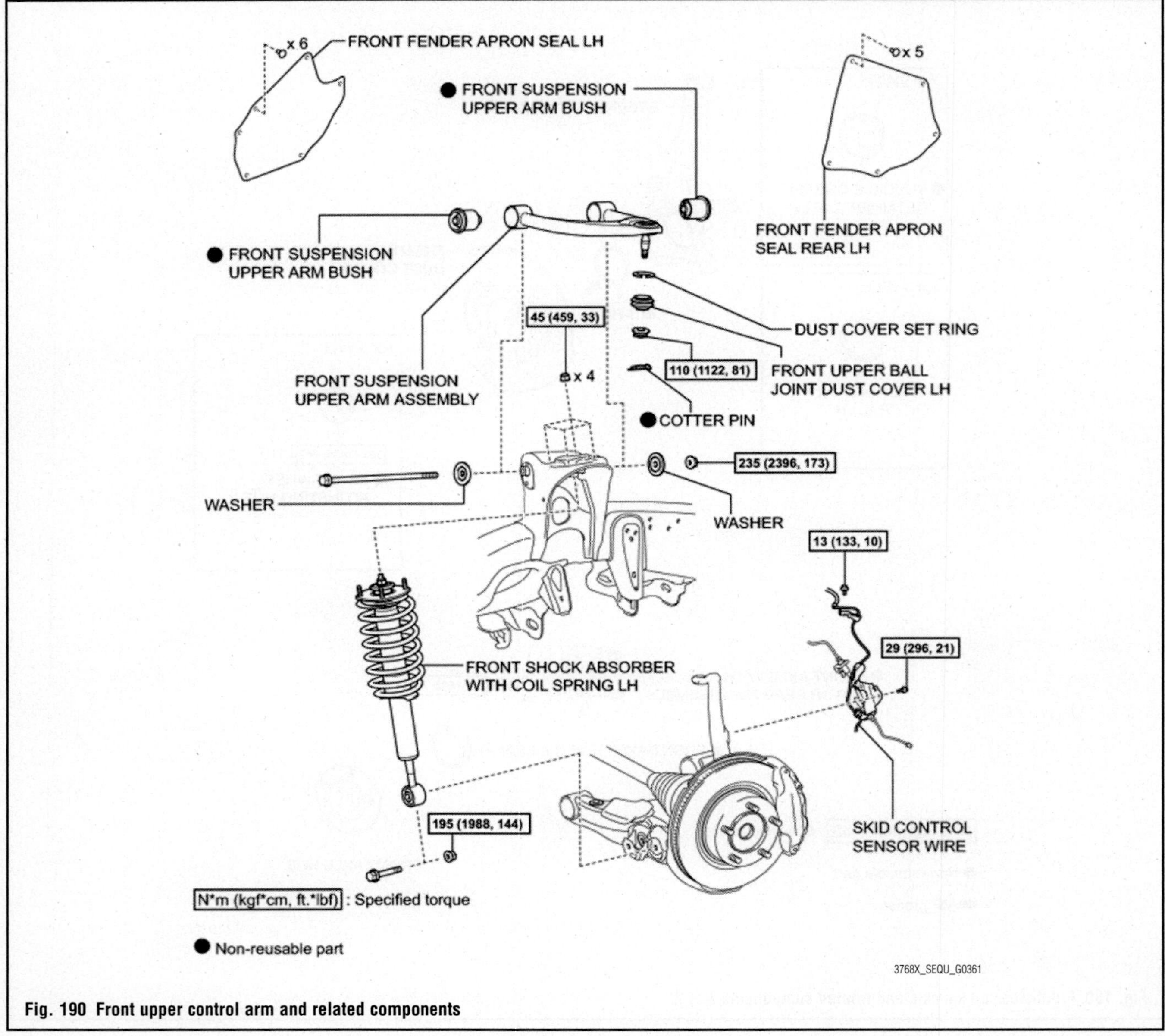

Fig. 190 Front upper control arm and related components

SUSPENSION

REAR SUSPENSION

LEAF SPRING

REMOVAL & INSTALLATION

See Figure 191.

1. Before servicing the vehicle, refer to the Precautions Section.
2. Disconnect the negative battery cable. Tape the cable with insulating tape.
3. Raise and safely support the vehicle safely.
4. Remove the tire and wheel assemblies.
5. Lower the axle housing until the leaf spring tension is free. Keep it at this position.
6. Remove the spare tire.
7. Disconnect the No. 3 parking brake assembly.
8. Support the rear axle housing with a jack and wooden block. Remove the bolt and nut and disconnect the shock absorber assembly from the axle housing.
9. Remove the U bolt nuts and washers. Remove the U bolt.
10. Remove the spring bumper nut and washer and bolt from the front side.
11. Remove the nuts, washers and shackle outer plate from the rear side.
12. Remove the component from the vehicle.

To install:

➡Be sure to use new fasteners, as required.

13. Installation is the reverse of the removal procedure.
14. Temporarily install the component.
15. When installing the rear spring bumper sub assembly, align the leaf spring center bolt and nut with the installation hole of the rear axle housing and spring bumper so that they are not loose.
16. Stabilize the suspension.
17. When tightening the U bolts, tighten them in a criss cross pattern.

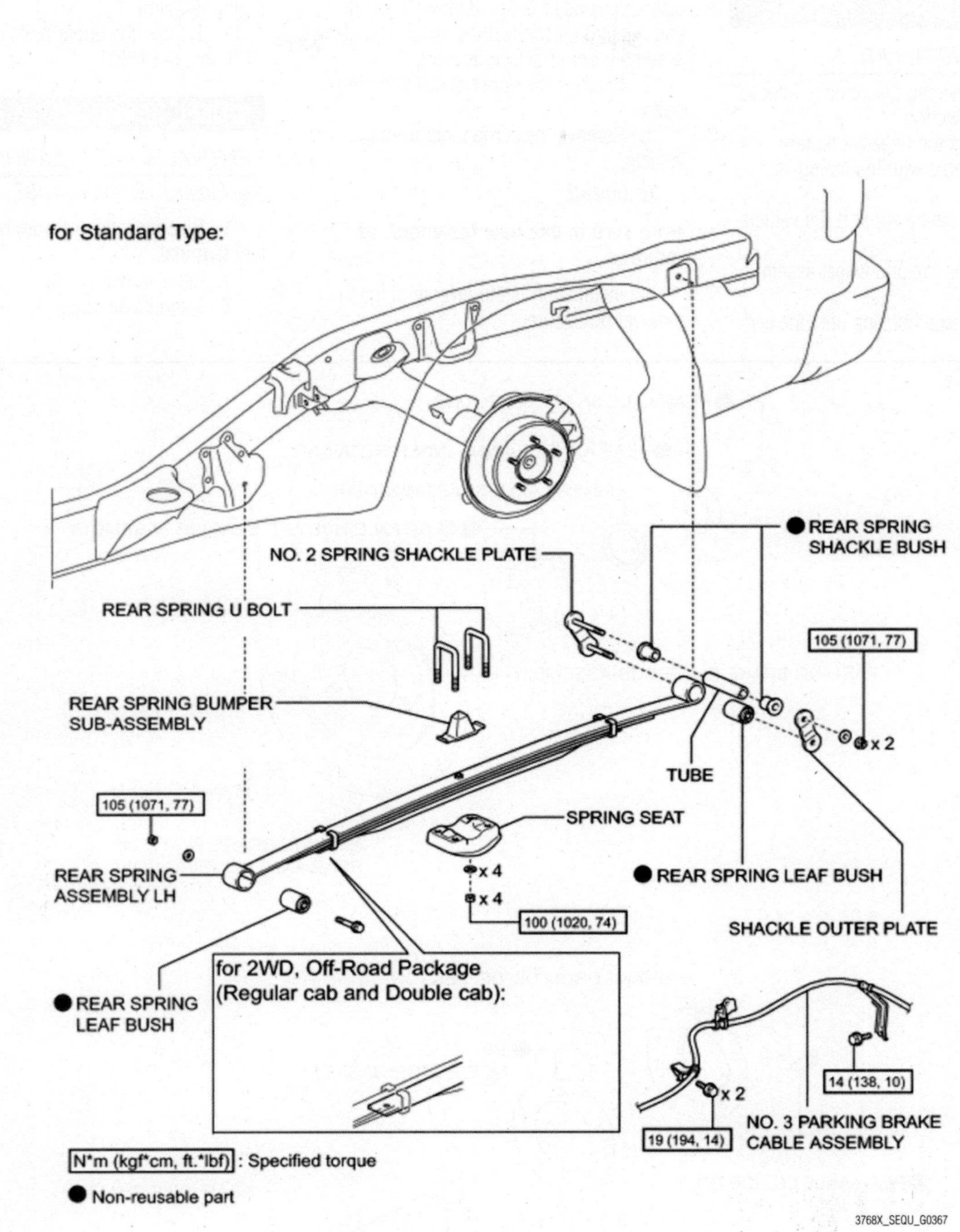

Fig. 191 Rear spring and related components

18. Tighten all bolts to specification, refer to illustration.

SHOCK ABSORBER

REMOVAL & INSTALLATION

1. Before servicing the vehicle, refer to the Precautions Section.
2. Disconnect the negative battery cable. Tape the cable with insulating tape.
3. Raise and safely support the vehicle safely.
4. Remove the tire and wheel assemblies.
5. Lower the axle housing until the leaf spring tension is free. Keep it at this position.
6. Support the rear axle housing with a jack and wooden block. Remove the bolt and nut and disconnect the shock absorber assembly from the axle housing.
7. Remove the upper shock retaining bolts.
8. Remove the component from the vehicle.

To install:

➡**Be sure to use new fasteners, as required.**

9. Installation is the reverse of the removal procedure.
10. Temporarily install the component.
11. Stabilize the suspension.
12. Tighten the lower bolt and nut to 66 ft. lbs. (90 Nm).
13. Tighten the upper bolts and nuts to 21 ft. lbs. (28 Nm).

WHEEL BEARINGS

REMOVAL & INSTALLATION

See Figures 192 through 197.

1. Disconnect cable from negative battery terminal.
2. Remove rear wheel.
3. Drain brake fluid.

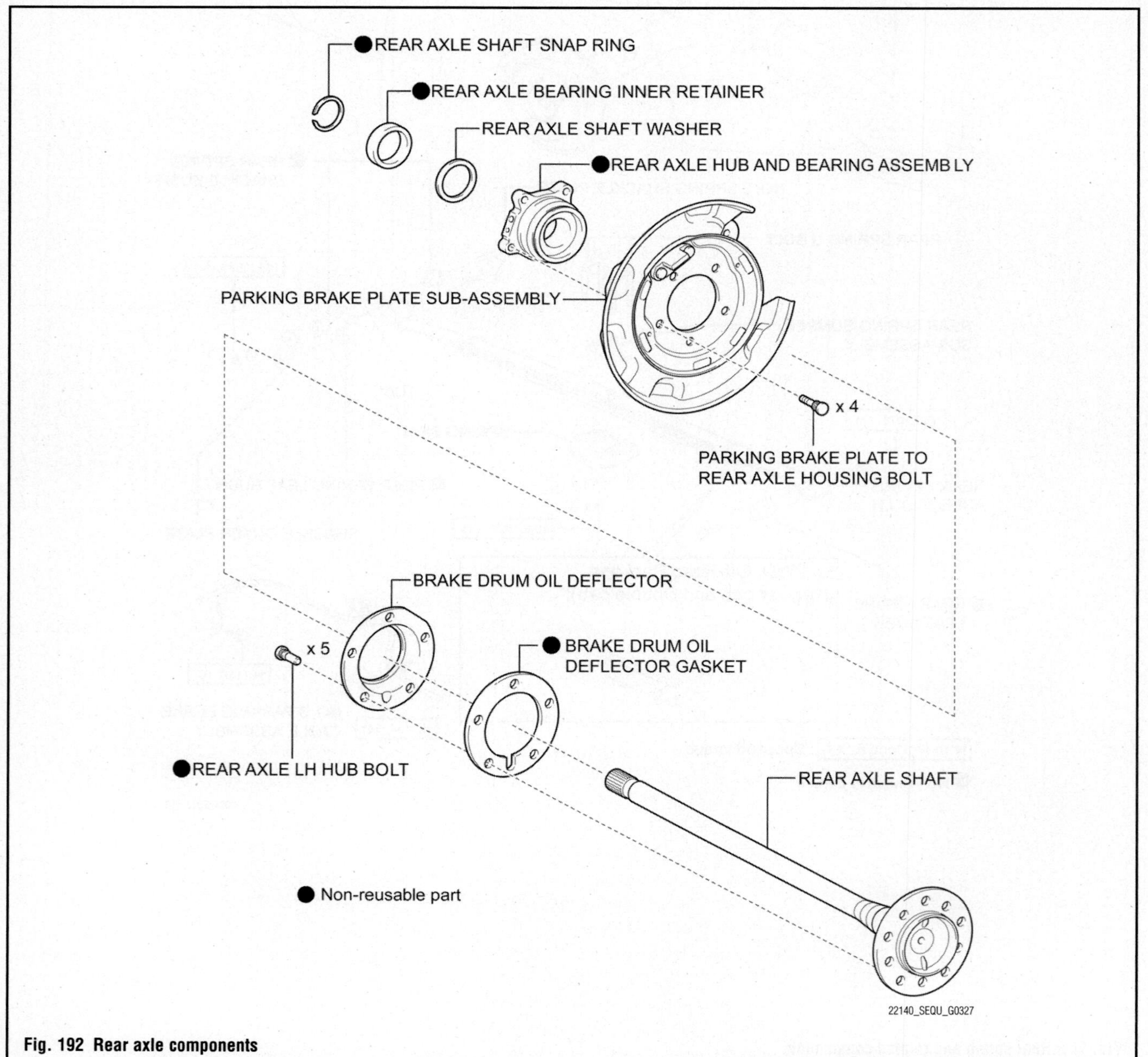

Fig. 192 Rear axle components

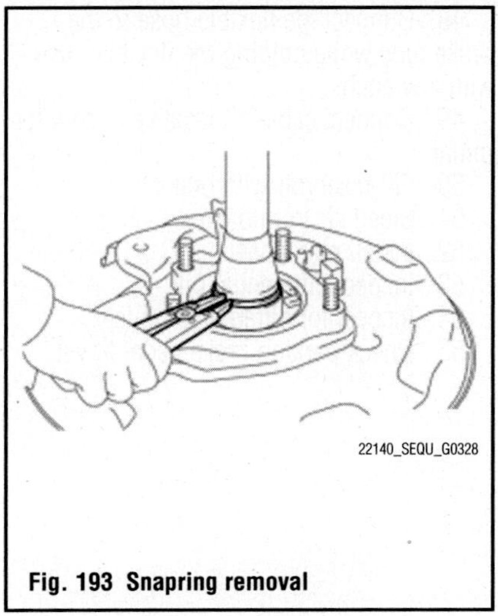

Fig. 193 Snapring removal

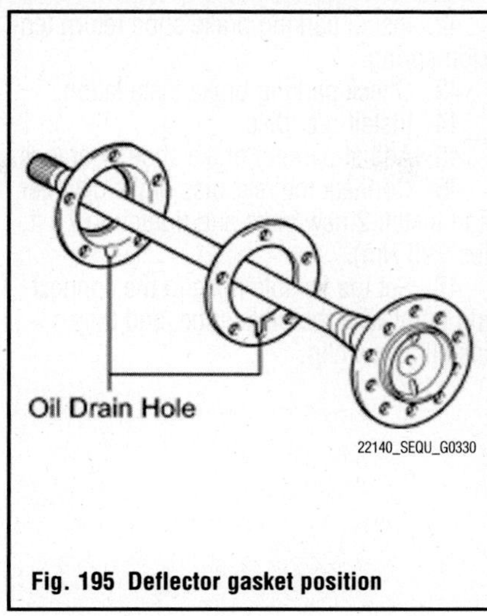

Fig. 195 Deflector gasket position

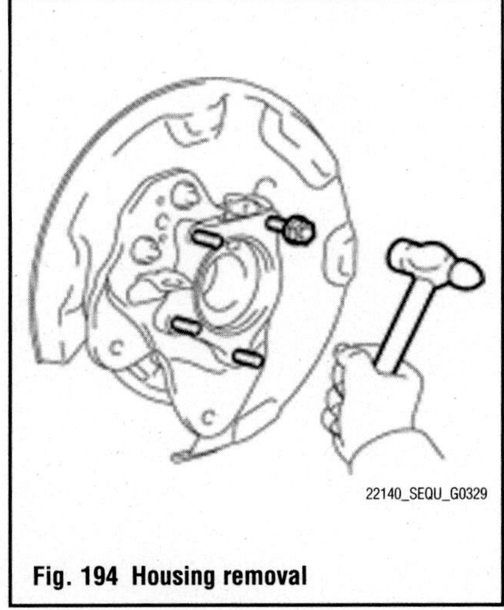

Fig. 194 Housing removal

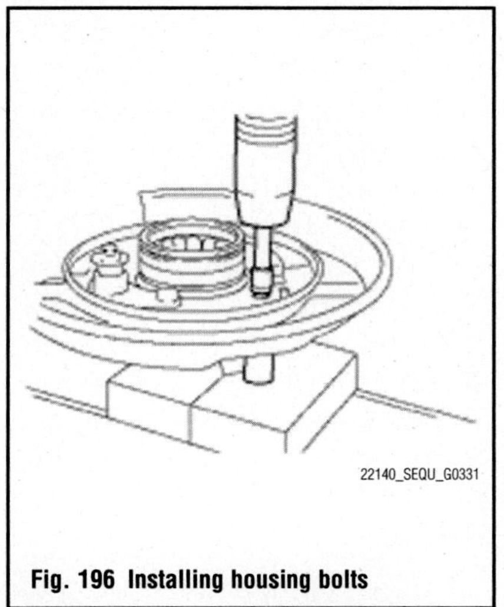

Fig. 196 Installing housing bolts

4. Disconnect rear brake flexible hose.
5. Remove the clip.
6. Remove the 2 bolts and disconnect rear disc brake cylinder.
7. Remove rear disc.
8. Remove parking brake shoe return tension spring.
9. Remove parking brake shoe assembly.
10. Remove parking brake shoe lever sub-assembly.
11. Disconnect parking brake cable assembly.
12. Disconnect rear speed sensor.
13. Remove the 4 nuts and rear axle shaft together with the parking brake plate.
14. Remove the O-ring.
15. Using a snap ring expander, remove the snap ring.
16. Press out the rear axle shaft.
17. Remove the rear axle bearing inner retainer from the axle hub.
18. Remove the rear axle shaft washer from the axle hub.
19. Grind the rear axle bearing inner race surface using a grinder, then remove it with a chisel.
20. Temporarily install 4 nuts to the housing bolts.
21. Using a hammer, remove the 4 housing bolts and rear axle hub and bearing.
22. Remove the nuts.
23. Remove the 5 hub bolts.
24. Remove the deflector and deflector gasket from the rear axle shaft.

To install:

25. Install a new deflector gasket and deflector to the rear axle shaft.
26. Temporarily install a washer and nut to 5 new hub bolts.
27. Install the hub bolts by tightening each nut.
28. Removed the washer and nut from each hub bolt.
29. Position the parking brake plate on a new rear axle hub and bearing.
30. Using 2 socket wrenches and a press, press in the 4 housing bolts.
31. Install the washer and a new retainer to the axle hub.

➡Install the washer with its tapered surface facing downward. Install the retainer with its chamfered surface facing downward.

32. Press in the rear axle shaft.
33. Using a snap ring expander, install a new snap ring.
34. Apply MP grease to the lip of the oil seal.
35. Install the O-ring to the axle shaft.

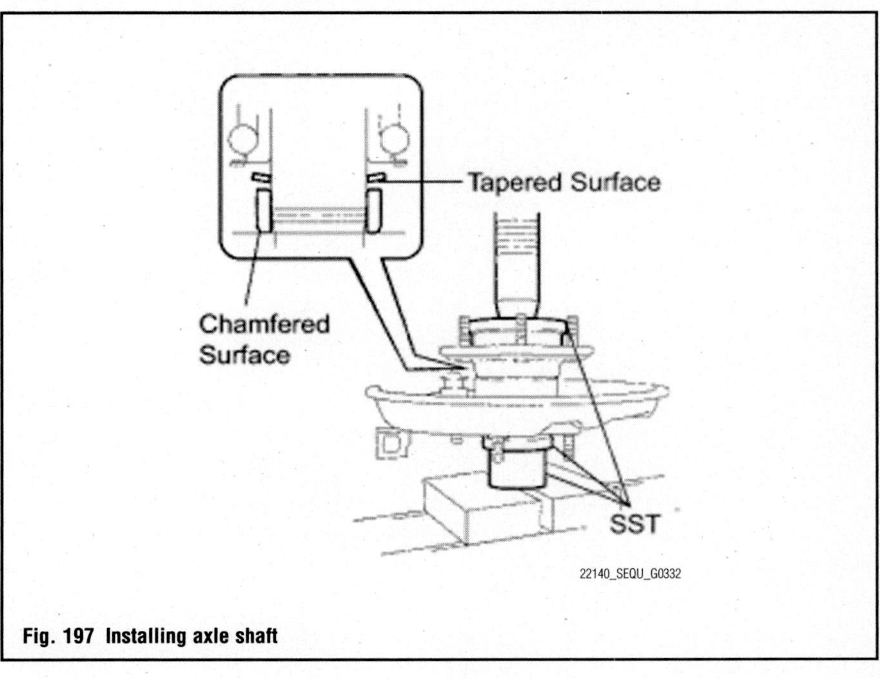

Fig. 197 Installing axle shaft

36. Install the rear axle shaft and parking brake plate with the 4 nuts and tighten to 44 ft. lbs. (60 Nm).
37. Inspect the rear axle shaft.
38. Connect rear speed sensor.
39. Connect parking brake cable assembly.
40. Install parking brake shoe lever subassembly.
41. Install parking brake shoe assemblies.
42. Install parking brake shoe return tension spring.
43. Check parking brake installation.
44. Install rear disc.
45. Adjust parking brake shoe clearance.
46. Connect the rear disc brake cylinder and install 2 new bolts and tighten to 70 ft. lbs. (95 Nm).
47. Set the flexible hose to the connecting point with the brake tube, and then install a new clip.
48. Connect the flexible hose to the brake tube while holding the flexible hose with a wrench.
49. Connect cable to negative battery terminal.
50. Fill reservoir with brake fluid.
51. Bleed air from brakes.
52. Install rear wheel.
53. Inspect differential oil.
54. Inspect for differential oil leak.
55. Check parking brake pedal travel.

17

TOYOTA

Venza

SPECIFICATIONS AND MAINTENANCE CHARTS

ENGINE AND VEHICLE IDENTIFICATION

Engine							Model Year	
Code	Liters (cc)	Cu. In.	Cyl.	Fuel Sys.	Engine Type	Eng. Mfg.	Code ②	Year
1AR-FE	2.7 (2700)	163	4	SFI	DOHC	Toyota	B	2011
2GR-FE	3.5 (3498)	213	6	SFI	DOHC	Toyota	C	2012

SFI: Sequential Fuel Injection

DOHC: Double Overhead Camshaft

② 10th digit of the Vehicle Identification Number (VIN)

71099_VENZ_C0001

GENERAL ENGINE SPECIFICATIONS

Year	Model	Engine Displacement Liters	Engine Series Code/VIN	Net Horsepower @ rpm	Net Torque @ rpm (ft. lbs.)	Bore x Stroke (in.)	Com-pression Ratio	Oil Pressure @ rpm
2011	Venza	2.7	1AR-FE	182@5800	182@4200	3.54x4.13	10.0:1	38@4000
		3.5	2GR-FE	269@6200	246@4700	3.70x3.27	10.8:1	55@3000
2012	Venza	2.7	1AR-FE	182@5800	182@4200	3.54x4.13	10.0:1	38@4000
		3.5	2GR-FE	269@6200	246@4700	3.70x3.27	10.8:1	55@3000

NA: Not Available

71099_VENZ_C0002

ENGINE TUNE-UP SPECIFICATIONS

Year	Engine Displacement Liters	Engine Code/VIN	Spark Plug Gap (in.)	Ignition Timing (deg.) A	Fuel Pump (psi)	Idle Speed (rpm) MT	Idle Speed (rpm) AT	Valve Clearance Intake	Valve Clearance Exhaust
2011	2.7	1AR-FE	0.043	8-12 ①	44-50	—	600-700	0.0075-0.0114	0.0150-0.0189
	3.5	2GR-FE	0.043	8-12 ①	44-50	—	600-700	NA	NA
2012	2.7	1AR-FE	0.043	8-12 ①	44-50	—	600-700	0.0075-0.0114	0.0150-0.0189
	3.5	2GR-FE	0.043	8-12 ①	44-50	—	600-700	NA	NA

NOTE: The Vehicle Emission Control Information label often reflects specification changes made during production. The label figures must be used if they differ from those in this chart.

NA: Not Available

① With terminals TC and CG of DLC3 connected. Before Top Dead Center (BTDC)

71099_VENZ_C0003

CAPACITIES

Year	Model	Engine Displacement Liters	Engine Code/VIN	Engine Oil with Filter (qts.)	Transmission (pts.) 5-Spd	Transmission (pts.) Auto.*	Transfer Case (pts.)	Drive Axle Front (pts.)	Drive Axle Rear (pts.)	Fuel Tank (gal.)	Cooling System (qts.)
2011	Venza	2.7	1AR-FE	4.5	—	7.4	1.0	—	1.0	15.9	7.2
		3.5	2GR-FE	6.4	—	NA	1.0	—	1.0	15.9	①
2012	Venza	2.7	1AR-FE	4.5	—	7.4	1.0	—	1.0	15.9	7.2
		3.5	2GR-FE	6.4	—	NA	1.0	—	1.0	15.9	①

*After draining, add the following amounts, then fill to the cold full line.

NA: Not Available

① STD: 10.4
TWG: 10.6

71099_VENZ_C0004

FLUID SPECIFICATIONS

Year	Model	Engine Displacement Liters (VIN)	Engine Oil	Auto. Trans.	Drive Axle	Power Steering Fluid	Brake Master Cylinder
2011	Venza	2.7	①	ATF World Standard	②	NA	DOT 3
	Venza	3.5	5W-30	ATF World Standard	③	NA	DOT 3
2012	Venza	2.7	①	ATF World Standard	②	NA	DOT 3
	Venza	3.5	5W-30	ATF World Standard	③	NA	DOT 3

DOT: Department Of Transportation

NA: Not Available

Note: If specification disagrees with specification in owners manual, use specification in owners manaual

① 0W-20 above 40 degrees F. 5W-20 below 40 degrees F

② API GL-5 SAE 90 above 0 degrees F. API GL-5 SAE 80W-90 below 0 degrees

③ API GL-5 SAE 85W-90 above 0 degrees F. API GL-5 SAE 80W-90 below 0 degrees

71099_VENZ_C0014

VALVE SPECIFICATIONS

Year	Engine Displacement Liters	Engine Code/VIN	Seat Angle (deg.)	Face Angle (deg.)	Spring Test Pressure (lbs. @ in.)	Spring Installed Height (in.)	Stem-to-Guide Clearance (in.) Intake	Stem-to-Guide Clearance (in.) Exhaust	Stem Diameter (in.) Intake	Stem Diameter (in.) Exhaust
2011	2.7	1AR-FE	45	44.5	NA	1.8670	0.0010-0.0024	0.0012-0.0026	0.2154-0.2159	0.2152-0.2158
	3.5	2GR-FE	45	44.5	NA	1.7898	0.0010-0.0024	0.0012-0.0026	0.2154-0.2159	0.2151-0.2157
2012	2.7	1AR-FE	45	44.5	NA	1.8670	0.0010-0.0024	0.0012-0.0026	0.2154-0.2159	0.2152-0.2158
	3.5	2GR-FE	45	44.5	NA	1.7898	0.0010-0.0024	0.0012-0.0026	0.2154-0.2159	0.2151-0.2157

NA: Not Available

71099_VENZ_C0005

CAMSHAFT SPECIFICATIONS

All measurements in inches unless noted

Year	Engine Displacement Liters	Engine Code/VIN	Journal Dia.	Brg. Oil Clearance	Shaft End-play ①	Circle Runout	Lobe Height Intake	Lobe Height Exhaust
2011	2.7	1AR-FE	②	NA	NA	NA	1.8624-1.8664	1.8104-1.8143
	3.5	2GR-FE	③	④	0.0031 0.0051	0.0016	1.7447-1.7487	1.7426-1.7465
2012	2.7	1AR-FE	②	NA	NA	NA	1.8624 1.8664	1.8104-1.8143
	3.5	2GR-FE	③	④	0.0031 0.0051	0.0016	1.7447-1.7487	1.7426-1.7465

① Thrust clearance

② Mark 1, 2 and 3: 1.4162-1.4167

③ No1: 1.4152-1.4157

All others: 1.0220-1.0226

④ No. 1: 0.0016-0.0031

All others: 0.0010-0.0024

71099_VENZ_C0006

CRANKSHAFT AND CONNECTING ROD SPECIFICATIONS

All measurements are given in inches.

Year	Engine Displ. Liters	Engine Code/VIN	Crankshaft Main Brg. Journal Dia.	Crankshaft Main Brg. Oi Clearance	Crankshaft Shaft End-play	Crankshaft Thrust on No.	Connecting Rod Journal Diameter	Connecting Rod Oil Clearance	Connecting Rod Side Clearance
2011	2.7	1AR-FE	2.1649-2.1652	0.0006-0.0015	0.0016-0.0095	3	1.8894-1.8898	0.0012-0.0025	0.0063-0.0202
	3.5	2GR-FE	2.4011-2.4016	0.0010-0.0019	0.0016-0.0095	2	2.0863-2.0866	0.0018-0.0026	0.0059-0.0157
2012	2.7	1AR-FE	2.1649-2.1652	0.0006-0.0015	0.0016-0.0095	3	1.8894-1.8898	0.0012-0.0025	0.0063-0.02
	3.5	2GR-FE	2.4011-2.4016	0.0010-0.0019	0.0016-0.0095	2	2.0863-2.0866	0.0018-0.0026	0.0059-0.0157

71099_VENZ_C0007

PISTON AND RING SPECIFICATIONS

All measurements are given in inches.

Year	Engine Displ. Liters	Engine Code/VIN	Piston Clearance	Ring Gap Top Comp.	Ring Gap Bottom Comp.	Ring Gap Oil Control	Ring Side Clearance Top Comp.	Ring Side Clearance Bottom Comp.	Ring Side Clearance Oil Control
2011	2.7	1AR-FE	NA	0.0086-0.0106	0.0146-0.0165	0.0039-0.0079	0.0008-0.0028	0.0008-0.0024	0.0008-0.0028
	3.5	2GR-FE	NA	0.0098-0.0138	0.0197-0.0236	0.0039-0.0157	0.0008-0.0028	0.0008-0.0024	0.0028-0.0059
2012	2.7	1AR-FE	NA	0.0087-0.0106	0.0146-0.0165	0.0040-0.0079	0.0008-0.0028	0.0008-0.0024	0.0008-0.0028
	3.5	2GR-FE	NA	0.0098-0.0138	0.0197-0.0236	0.0039-0.0157	0.0008-0.0028	0.0008-0.0024	0.0028-0.0059

71099_VENZ_C0008

TORQUE SPECIFICATIONS

All readings in ft. lbs.

Year	Engine Displacement Liters	Engine Code/VIN	Cylinder Head Bolts	Main Bearing Bolts	Rod Bearing Bolts	Crankshaft Damper Bolts	Flywheel Bolts	Manifold Intake	Manifold Exhaust	Spark Plugs	Oil Pan Drain Plug
2011	2.7	1AR-FE	①	②	③	133	72	22	27	18	30
	3.5	2GR-FE	④	⑤	⑥	184	132	15	15	13	30
2012	2.7	1AR-FE	①	②	③	133	72	22	27	18	30
	3.5	2GR-FE	④	⑤	⑥	184	132	15	15	13	30

① Step 1: 27 ft. lbs.
Step 2: plus 90 degrees

② Step 1: 30 ft. lbs.
Step 2: plus 90 degrees

③ Step 1: 30 ft. lbs.
Step 2: plus 90 degrees

④ Step 1: 27 ft. lbs.
Step 2: plus 90 degrees
Step 3: plus 90 degrees
Bolt should be 22 ft. lbs. on Bank 2

⑤ Step 1: 45 ft. lbs.
Step 2: plus 90 degrees
Main bearing cap bolt: 38 ft. lbs.

⑥ Step 1: 45 ft. lbs.
Step 2: plus 90 degrees
Step 3: 8 side bolts to 38 ft. lbs.

71099_VENZ_C0009

WHEEL ALIGNMENT

Year	Model	Engine		Caster Range (+/-Deg.)	Caster Preferred Setting (Deg.)	Camber Range (+/-Deg.)	Camber Preferred Setting (Deg.)	Toe-in (in.)	Steering Axis Inclination (Deg.)
2011	Venza 2WD	1AR-FE	F	0.75	+2.65	0.75	-0.58	0.043+/-0.079	10.70+/-0.75
			R	NA	NA	0.75	-1.22	0.150+/-0.079	NA
	Venza AWD	1AR-FE	F	0.75	+2.68	0.75	-0.60	0.043+/-0.079	10.70+/-0.75
			R	NA	NA	0.75	-0.78	0.142+/-0.079	NA
	Venza 2WD	2GR-FE	F	0.75	+2.63	0.75	-0.60	0.043+/-0.079	10.72+/-0.75
			R	NA	NA	0.75	-1.23	0.150+/-0.079	NA
	Venza AWD	2GR-FE	F	0.75	+2.67	0.75	-0.60	0.043+/-0.079	10.72+/-0.75
			R	NA	NA	0.75	-0.80	0.142+/-0.079	NA
2012	Venza 2WD	1AR-FE	F	0.75	+2.65	0.75	-0.58	0.043+/-0.079	10.70+/-0.75
			R	NA	NA	0.75	-1.22	0.150+/-0.079	NA
	Venza AWD	1AR-FE	F	0.75	+2.68	0.75	-0.60	0.043+/-0.079	10.70+/-0.75
			R	NA	NA	0.75	-0.78	0.142+/-0.079	NA
	Venza 2WD	2GR-FE	F	0.75	+2.63	0.75	-0.60	0.043+/-0.079	10.72+/-0.75
			R	NA	NA	0.75	-1.23	0.150+/-0.079	NA
	Venza AWD	2GR-FE	F	0.75	+2.67	0.75	-0.60	0.043+/-0.079	10.72+/-0.75
			R	NA	NA	0.75	-0.80	0.142+/-0.079	NA

71099_VENZ_C0010

TIRE, WHEEL AND BALL JOINT SPECIFICATIONS

		OEM Tires		Tire Pressures (psi)		Wheel	Ball Joint Turning Torq.	Lug Nut Torque
Year	Model	Standard	Optional	Front	Rear	Size	Inspection	(ft. lbs.)
2011	Venza	P245/55R19	P245/50R20	①	①	NA	6.2-40 in lbs.	76
2012	Venza	P245/55R19	P245/50R20	①	①	NA	6.2-40 in lbs.	76

NA: Not Available

OEM: Original Equipment Manufacturer

PSI: Pounds Per Square Inch

① 32 psi Cold

71099_VENZ_C0011

BRAKE SPECIFICATIONS

All measurements in inches unless noted

			Brake Disc			Brake Drum Diameter				Brake Caliper	
			Original	Minimum	Maximum	Original Inside	Max. Wear	Maximum Machine	Minimum Lining	Bracket Bolts	Mounting Bolts
Year	Model		Thickness	Thickness	Runout	Diameter	Limit	Diameter	Thickness	(ft. lbs.)	(ft. lbs.)
2011	Venza	F	1.100	1.020	0.0020	—	—	—	0.039	77	24
		R	0.394	0.354	0.0059	—	—	—	0.039	58	24
2012	Venza	F	1.100	1.020	0.0020	—	—	—	0.039	77	24
		R	0.394	0.354	0.0059	—	—	—	0.039	58	24

F: Front

R: Rear

71099_VENZ_C0012

SCHEDULED MAINTENANCE INTERVALS

TOYOTA—Venza

TO BE SERVICED	TYPE OF SERVICE	VEHICLE MILEAGE INTERVAL (x1000)																		
		5	10	15	20	25	30	35	40	45	50	55	60	65	70	75	80	85	90	95
Automatic transmission and differential fluid	S/I			✓			✓			✓			✓			✓			✓	
Ball joints and boots	S/I			✓			✓			✓			✓			✓			✓	
Brake system	S/I	✓	✓	✓	✓	✓	✓	✓	✓	✓	✓	✓	✓	✓	✓	✓	✓	✓	✓	✓
Cabin air filter	S/I			✓			✓			✓			✓			✓			✓	
Charcoal canister	S/I												✓							
Drive belts	S/I						✓						✓						✓	
Driveshaft bushing	L						✓						✓						✓	
Engine coolant	R			✓			✓			✓			✓			✓			✓	
Engine oil & filter	R		✓		✓		✓		✓		✓		✓		✓		✓		✓	
Exhaust pipes & mounts	S/I			✓			✓			✓			✓			✓			✓	
Fuel tank cap gasket	S/I						✓						✓						✓	
Halfshaft boots & flange bolts	S/I			✓			✓			✓			✓			✓			✓	
Limited slip differential fluid	R						✓						✓						✓	
Platinum spark plugs	R												✓							
Propeller shaft bolts	S/I			✓			✓			✓			✓			✓			✓	
Rack and pinion assy	S/I			✓			✓			✓			✓			✓			✓	
Tires (rotate)	S/I	✓	✓	✓	✓	✓	✓	✓	✓	✓	✓	✓	✓	✓	✓	✓	✓	✓	✓	✓
Transfer case	S/I			✓			✓			✓			✓			✓			✓	
Valves	S/I												✓							

R: Replace S/I: Service or Inspect L: Lubricate

FREQUENT OPERATION MAINTENANCE (SEVERE SERVICE)

If a vehicle is operated under any of the following conditions it is considered severe service:

- Towing a trailer or using a camper or car-top carrier.

- Repeated short trips of less than 5 miles in temperatures below freezing.

- Excessive idling or low-speed driving for long distances as in heavy commercial use, such as delivery, taxi or police cars.

- Operating on rough, muddy or salt-covered roads.

- Operating on unpaved or dusty roads.

Oil filter: service or inspect every 5000 miles or 4 months, whichever occurs first.

Brake linings and discs or drums: service or inspect every 5000 miles or 4 months, whichever occurs first.

Steering linkage: service or inspect every 5000 miles or 4 months, whichever occurs first.

Ball joints and boots: service or inspect every 5000 miles or 4 months, whichever occurs first.

Brake discs & pads (front): service or inspect every 6000 miles.

Halfshaft boots: service or inspect every 5000 miles or 4 months. Retighten the flange bolts, whichever occurs first.

Body chassis bolts and nuts: service or inspect every 5000 miles or 4 months, whichever occurs first.

Transmission and differential fluid: replace every 15,000 miles or 12 months, whichever occurs first.

Transfer case: replace every 15,000 miles or 12 months, whichever occurs first.

71099_VENZ_C0013

PRECAUTIONS

Before servicing any vehicle, please be sure to read all of the following precautions, which deal with personal safety, prevention of component damage, and important points to take into consideration when servicing a motor vehicle:

- Never open, service or drain the radiator or cooling system when the engine is hot; serious burns can occur from the steam and hot coolant.
- Observe all applicable safety precautions when working around fuel. Whenever servicing the fuel system, always work in a well-ventilated area. Do not allow fuel spray or vapors to come in contact with a spark, open flame, or excessive heat (a hot drop light, for example). Keep a dry chemical fire extinguisher near the work area. Always keep fuel in a container specifically designed for fuel storage; also, always properly seal fuel containers to avoid the possibility of fire or explosion. Refer to the additional fuel system precautions later in this section.
- Fuel injection systems often remain pressurized, even after the engine has been turned **OFF**. The fuel system pressure must be relieved before disconnecting any fuel lines. Failure to do so may result in fire and/or personal injury.
- Brake fluid often contains polyglycol ethers and polyglycols. Avoid contact with the eyes and wash your hands thoroughly after handling brake fluid. If you do get brake fluid in your eyes, flush your eyes with clean, running water for 15 minutes. If eye irritation persists, or if you have taken brake fluid internally, IMMEDIATELY seek medical assistance.
- The EPA warns that prolonged contact with used engine oil may cause a number of skin disorders, including cancer. You should make every effort to minimize your exposure to used engine oil. Protective gloves should be worn when changing oil. Wash your hands and any other exposed skin areas as soon as possible after exposure to used engine oil. Soap and water, or waterless hand cleaner should be used.
- All new vehicles are now equipped with an air bag system, often referred to as a Supplemental Restraint System (SRS) or Supplemental Inflatable Restraint (SIR) system. The system must be disabled before performing service on or around system components, steering column, instrument panel components, wiring and sensors. Failure to follow safety and disabling procedures could result in accidental air bag deployment, possible personal injury and unnecessary system repairs.
- Always wear safety goggles when working with, or around, the air bag system. When carrying a non-deployed air bag, be sure the bag and trim cover are pointed away from your body. When placing a non-deployed air bag on a work surface, always face the bag and trim cover upward, away from the surface. This will reduce the motion of the module if it is accidentally deployed. Refer to the additional air bag system precautions later in this section.
- Clean, high quality brake fluid from a sealed container is essential to the safe and proper operation of the brake system. You should always buy the correct type of brake fluid for your vehicle. If the brake fluid becomes contaminated, completely flush the system with new fluid. Never reuse any brake fluid. Any brake fluid that is removed from the system should be discarded. Also, do not allow any brake fluid to come in contact with a painted surface; it will damage the paint.
- Never operate the engine without the proper amount and type of engine oil; doing so WILL result in severe engine damage.
- Timing belt maintenance is extremely important. Many models utilize an interference-type, non-freewheeling engine. If the timing belt breaks, the valves in the cylinder head may strike the pistons, causing potentially serious (also time-consuming and expensive) engine damage. Refer to the maintenance interval charts for the recommended replacement interval for the timing belt, and to the timing belt section for belt replacement and inspection.
- Disconnecting the negative battery cable on some vehicles may interfere with the functions of the on-board computer system(s) and may require the computer to undergo a relearning process once the negative battery cable is reconnected.
- When servicing drum brakes, only disassemble and assemble one side at a time, leaving the remaining side intact for reference.
- Only an MVAC-trained, EPA-certified automotive technician should service the air conditioning system or its components.

BRAKES

ANTI-LOCK BRAKE SYSTEM (ABS)

GENERAL INFORMATION

PRECAUTIONS

- Certain components within the ABS system are not intended to be serviced or repaired individually.
- Do not use rubber hoses or other parts not specifically specified for and ABS system. When using repair kits, replace all parts included in the kit. Partial or incorrect repair may lead to functional problems and require the replacement of components.
- Lubricate rubber parts with clean, fresh brake fluid to ease assembly. Do not use shop air to clean parts; damage to rubber components may result.
- Use only DOT 3 brake fluid from an unopened container.
- If any hydraulic component or line is removed or replaced, it may be necessary to bleed the entire system.
- A clean repair area is essential. Always clean the reservoir and cap thoroughly before removing the cap. The slightest amount of dirt in the fluid may plug an orifice and impair the system function. Perform repairs after components have been thoroughly cleaned; use only denatured alcohol to clean components. Do not allow ABS components to come into contact with any substance containing mineral oil; this includes used shop rags.
- The Anti-Lock control unit is a microprocessor similar to other computer units in the vehicle. Ensure that the ignition switch is **OFF** before removing or installing controller harnesses. Avoid static electricity discharge at or near the controller.
- If any arc welding is to be done on the vehicle, the control unit should be unplugged before welding operations begin.

WHEEL SPEED SENSORS

REMOVAL & INSTALLATION

Front

See Figure 1.

➡Use the same procedure for the LH side and RH side.

➡The following procedure is for the LH side.

➡If the sensor rotor needs to be replaced, replace it together with the front driveshaft assembly.

1. Disconnect the cable from the negative battery terminal.

➡When disconnecting the cable, some systems need to be initialized after the cable is reconnected.

2. Remove the front wheel.
3. Remove the front fender outside moulding.
4. Remove the front fender liner.
5. Remove the front speed sensor.
   a. Disconnect the front speed sensor connector and remove the 2 clamps.
   b. Remove the bolt and No. 2 sensor clamp from the body.
   c. Remove the bolt, No. 1 sensor clamp and flexible hose together from the shock absorber assembly.
   d. Remove the bolt, resin clamp and front speed sensor.

➡Prevent foreign matter from attaching to the front speed sensor tip.

Clean the speed sensor installation hole and the contact surfaces every time the front speed sensor is removed.

To install:

6. Install the front speed sensor.
   a. Install the resin clamp and front speed senor with the bolt. Tighten to 71 inch lbs. (8 Nm).

➡Prevent foreign matter from attaching to the front speed sensor tip.

➡Firmly insert the front speed sensor body into the knuckle before tightening the bolt.

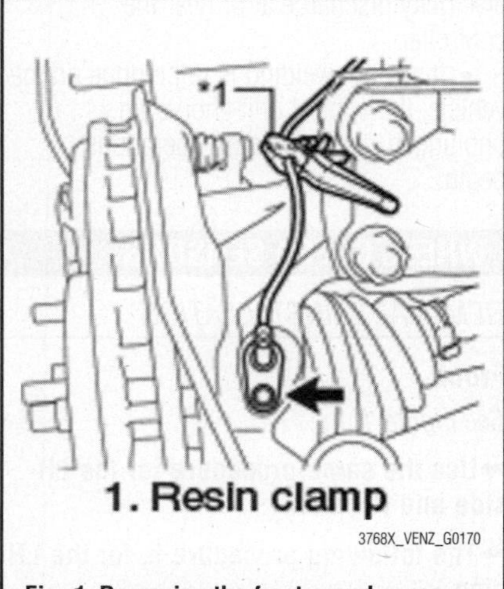

Fig. 1 Removing the front speed sensor

➡After installing the front speed sensor to the knuckle, make sure that there is no clearance between the front speed sensor stay and knuckle. Also make sure that no foreign matter is stuck between the parts.

➡Before installing the clamp, firmly insert the points of the clamp into the installation holes.

   b. Temporarily install the No. 1 sensor clamp.

➡Be sure to insert the No. 1 sensor clamp claw into the stopper hole while installing the No. 1 sensor clamp.

   c. Install the front brake flexible hose and the No. 1 sensor clamp together to the shock absorber with the bolt. Tighten to 14 ft. lbs. (19 Nm).

➡Do not twist the wire harness for the front speed sensor when installing it.

➡A bolt tightens the brake flexible hose and front speed sensor together. Make sure that the flexible hose is positioned over the front speed sensor.

   d. Install the No. 2 sensor clamp to the body with the bolt. Tighten to 44 inch lbs. (5 Nm).
   e. Install the 2 clamps and connect the front speed sensor connector.

7. Install the front fender liner.
8. Install the front fender outside moulding.
9. Install the front wheel and tighten to 76 ft. lbs. (103 Nm).

➡When disconnecting the cable, some systems need to be initialized after the cable is reconnected.

10. Check for speed sensor signal.

Rear

2WD Vehicles

See Figure 2.

➡Use the same procedure for the RH side and LH side.

➡The following procedure is for the LH side.

➡The rear speed sensor is a component of the rear axle hub and bearing assembly. If the sensor malfunctions, replace the rear axle hub and bearing assembly.

➡If the sensor rotor needs to be replaced, replace it together with the rear axle hub and bearing assembly.

1. Disconnect the cable from the negative battery terminal.

➡When disconnecting the cable, some systems need to be initialized after the cable is reconnected.

2. Remove the rear wheel.
3. Separate the rear speed sensor wire.
   a. Using a screwdriver, disconnect the connector from the rear speed sensor.

➡Be careful not to damage the rear speed sensor.

4. Separate the rear flexible hose.
5. Separate the rear disc brake caliper assembly.
6. Remove the rear disc.
7. Remove the rear axle hub and bearing assembly.

➡The rear speed sensor is a component of the rear axle hub and bearing assembly. If the sensor malfunctions, replace the rear axle hub and bearing assembly.

➡If the sensor rotor needs to be replaced, replace it together with the rear axle hub and bearing assembly.

To install:

8. Install the rear axle hub and bearing assembly.
9. Inspect the rear axle hub bearing looseness.
10. Inspect the rear axle hub runout.
11. Install the rear disc.
12. Install the rear disc brake caliper assembly.
13. Install the rear flexible hose.

Fig. 2 Separating the rear speed sensor wire (2WD)

14. Install the rear speed sensor wire.
    a. Connect the connector to the rear speed sensor.
15. Install the rear wheel. Tighten to 76 ft. lbs. (103 Nm).
16. Connect the cable to the negative battery terminal.

➡**When disconnecting the cable, some systems need to be initialized after the cable is reconnected.**

17. Inspect and adjust the rear wheel alignment.
18. Check for the speed sensor signal.

AWD Vehicles

See Figure 3.

➡**Use the same procedure for the LH side and RH side.**

➡**The following procedure is for the LH side.**

➡**If the sensor rotor needs to be replaced, replace it together with the rear driveshaft assembly.**

1. Disconnect the cable from the negative battery terminal.

➡**When disconnecting the cable, some systems need to be initialized after the cable is reconnected.**

2. Remove the rear wheel.
3. Remove the rear door scuff plate LH.
4. Remove the rear door opening trim weather-strip LH.
5. Remove the tonneau cover assembly (w/tonneau cover).
6. Remove the deck board assembly.
    a. Disengage the 2 guides and remove the deck board assembly.
7. Remove the No. 3 deck board sub assembly.
    a. Disengage the 2 guides and remove the No. 3 deck board sub-assembly.
8. Remove the deck side trim box LH.
    a. Remove the 3 clips and remove the deck side trim box LH.
9. Remove the No. 2 deck board sub assembly.
    a. Disengage the 2 guides and remove the No. 2 deck board sub-assembly.
10. Remove the deck side trim box RH.
    a. Remove the 4 clips and remove the deck side trim box RH.
11. Remove the No. 1 deck board.
    a. Disengage the 6 clips and remove the No. 1 deck board.
12. Remove the rear seat sub floor panel assembly.
    a. Disengage the 2 claws, 2 guides and 5 clips, and remove the rear seat sub floor panel assembly.
13. Remove the rear floor finish plate.
    a. Disengage the 2 claws and 4 clips, and remove the rear floor finish plate.
14. Remove the reclining remote control bezel LH.
    a. Using a screwdriver, disengage the 3 claws and remove the reclining remote control bezel LH.

➡**Tape the screwdriver tip before use.**

15. Remove the luggage hold belt striker assembly.
    a. Remove the 2 bolts.
16. Disconnect the rear seat outer belt assembly LH.
17. Remove the deck trim side panel assembly LH.
    a. Remove the bolt and clip.
    b. Disengage the 7 claws and 5 clips.
    c. Disconnect the connector and remove the deck trim side panel assembly LH.
18. Remove the rear speed sensor.
    a. Disconnect the rear speed sensor connector.
    b. Disconnect the grommet of the rear speed sensor wire from the hole of the wheel house.
    c. Remove the 2 bolts, No. 1 clamp and No. 2 clamp from the body and absorber.
    d. Remove the bolt and rear speed sensor from the carrier.

➡**Keep the sensor tip and rear speed sensor installation hole free from foreign matter.**

To install:

19. Install the rear speed sensor with the bolt. Tighten to 71 inch lbs. (8 Nm).

➡**Keep the rear speed sensor tip and sensor installation hole free from foreign matter.**

    a. Install the No. 1 clamp and No. 2 clamp with the 2 bolts. Tighten to 44 inch lbs. (5 Nm).

➡**Do not twist the rear speed sensor wire when installing the clamps.**

    b. Insert the connector and grommet to the inside of the vehicle through the passage hole in the wheel well.

➡**Make sure that the grommet's band clamp remains on the outside of the vehicle.**

Fig. 3 Removing the rear speed sensor

    c. Hold the grommet and pull it from the inside to the outside of the vehicle. Then secure it in place so that it is not tilted.

➡**When pulling out the grommet, do not grip the sensor wire.**

    d. Connect the rear speed sensor connector.

20. Install the deck trim side panel assembly LH.
21. Connect the rear seat outer belt assembly LH.
22. Install the luggage hold belt striker assembly.
23. Install the reclining remote control bezel LH.
24. Install the rear floor finish plate.
25. Install the rear seat sub floor panel assembly.
26. Install the No. 1 deck board.
27. Install the deck side trim box RH.
28. Install the No. 2 deck board sub assembly.
29. Install the deck side trim box LH.
30. Install the No. 3 deck board sub assembly.
31. Install the deck board assembly.
32. Install the tonneau cover assembly (w/tonneau).
33. Install the rear door opening trim weather-strip LH.
34. Install the rear door scuff plate LH.
35. Connect the cable to the negative battery terminal.

➡**When disconnecting the cable, some systems need to be initialized after the cable is reconnected.**

36. Check for speed sensor signal.

BRAKES — BLEEDING THE BRAKE SYSTEM

BLEEDING PROCEDURE

See Figure 4.

➡Do not allow brake fluid to adhere to any painted surface such as the vehicle body. If brake fluid leaks onto any painted surface, immediately wash it off.

➡Before bleeding the brake system, confirm that the reservoir located above the master cylinder assembly is filled with brake fluid.

➡If any component of the brake system is removed and reinstalled, or if air in the brake lines is suspected, bleed the brake system.

1. Fill the reservoir with brake fluid.
2. Bleed the brake master cylinder.

➡To prevent brake fluid from damaging painted surfaces cover any surrounding parts with a piece of cloth.

➡If the master cylinder is reinstalled or runs out of brake fluid, bleed the master cylinder.

a. Using a union nut wrench, disconnect the 2 brake tubes from the brake master cylinder assembly.
b. Slowly depress the brake pedal and hold it down.
c. Cover the 2 tube holes with your fingers and release the brake pedal.
d. Uncover the holes, slowly depress the brake pedal and hold it down. While holding down the brake pedal, cover the tube holes again. Repeat this step 3 or 4 times.
e. Using a union nut wrench, connect the 2 brake tubes to the brake master cylinder assembly.

➡Do not bend or damage the brake lines.

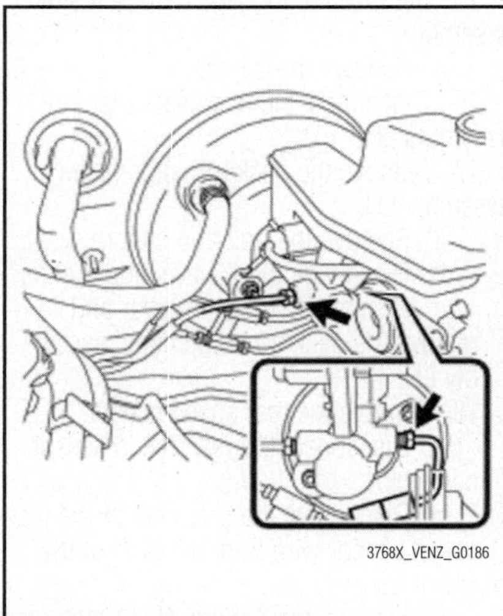

Fig. 4 Disconnecting the 2 brake tubes

➡Do not allow brake line to twist and interfere with other parts or body during tightening.

➡Do not allow any foreign matter such as dirt or dust to enter the brake line.

➡Use the formula to calculate special torque values for situations where union nut wrench is combined with a torque wrench

3. Bleed the brake line.

➡Bleed the brake line of the wheel farthest from the master cylinder first.

a. Connect a vinyl tube to the bleeder plug.
b. Depress the brake pedal several times, and while holding down the brake pedal, loosen the bleeder plug.
c. When fluid stops coming out, tighten the bleeder plug and release the brake pedal.
d. Repeat previous 2 steps until all the air in the brake fluid is completely bled out and the new brake fluid comes out.
e. Tighten the bleeder plug completely. Tighten to 10 ft. lbs. (13 Nm).
f. Repeat the above steps to replace the brake fluid of the brake lines for each wheel.

4. Inspect for brake fluid leaks.
5. Inspect the fluid level in the reservoir.

BRAKES — FRONT DISC BRAKES

BRAKE CALIPER

REMOVAL & INSTALLATION

See Figure 5.

➡Use the same procedure for the LH side and RH side.

➡The following procedure listed is for the LH side.

1. Remove the front wheel.
2. Drain the brake fluid.

➡If the brake fluid leaks onto any painted surface, immediately wash it off.

3. Separate the front flexible hose.
   a. Remove the union bolt and gasket, and separate the front flexible hose from the front disc brake caliper assembly.
4. Remove the front disc brake caliper assembly.
   a. Remove the 2 bolts and the front disc brake caliper assembly from the front disc brake caliper mounting.

To install:

5. Install the front disc brake caliper assembly.

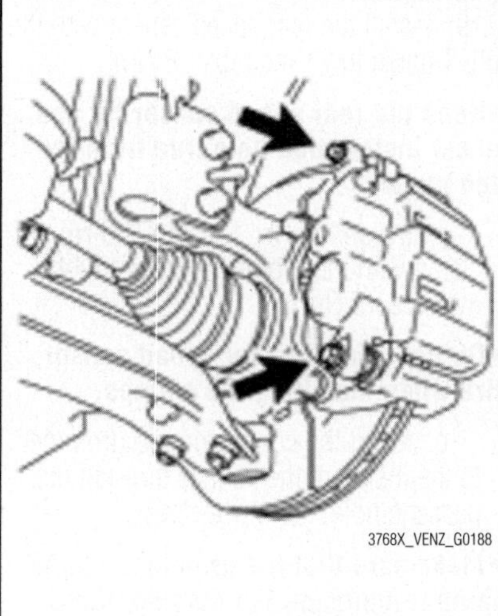

Fig. 5 Removing the brake caliper

a. Install the front disc brake caliper assembly to the front disc brake caliper mounting with the 2 bolts. Tighten to 24 ft. lbs. (32 Nm).

➡Be sure that the anti squeal springs are installed to the front disc brake pads.

6. Connect the front flexible hose.
   a. Connect the front flexible hose to the front disc brake caliper assembly with a new union bolt and a new gasket. Tighten to 22 ft. lbs. (29 Nm).

➡Install the front flexible hose lock securely into the lock hole in the front disc brake caliper assembly.

7. Fill the reservoir with brake fluid.
8. Bleed the brake line.
9. Inspect for brake fluid leaks.
10. Inspect the fluid level in the reservoir.
11. Install the front wheel. Tighten to 76 ft. lbs. (103 Nm).

DISC BRAKE PADS

REMOVAL & INSTALLATION

See Figures 6 and 7.

➡Use the same procedure for the LH side and RH side.

➡The following procedure listed is for the LH side.

1. Remove the front wheel.
2. Drain the brake fluid.

➡If the brake fluid leaks onto any painted surface, immediately wash it off.

3. Remove the brake caliper.
4. Remove the front disc brake pad.
   a. Remove the 2 anti-squeal springs from the front disc brake pads.
   b. Using a screwdriver, push the protrusion of the front disc brake pad support plate and remove the 2 front disc brake pads from the front disc brake caliper mounting.

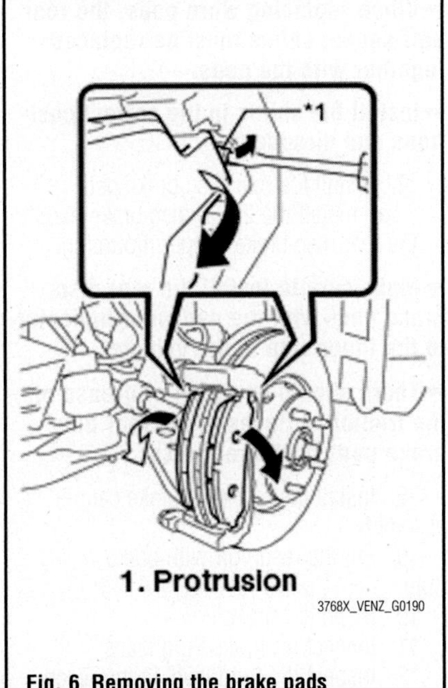

Fig. 6 Removing the brake pads

➡When removing the front disc brake pad, replace the front disc brake pad support plates with new ones.

5. Remove the 2 front anti squeal shims from the front disc brake pads.
6. Remove the 2 front disc brake pad support plates from the front disc brake caliper mounting.

To install:

7. Install the front disc brake pad support plate.
   a. Install 2 new front disc brake pad support plates to the front disc brake caliper mounting.

➡Be sure to install the plate in the correct position and direction.

8. Install the front anti squeal shim.
   a. Apply disc brake grease between the front anti squeal shim and front disc brake pad.

➡Apply 0.03 oz. of disc brake grease to each pad.

   b. Install the 2 front anti squeal shims to the front disc brake pads.

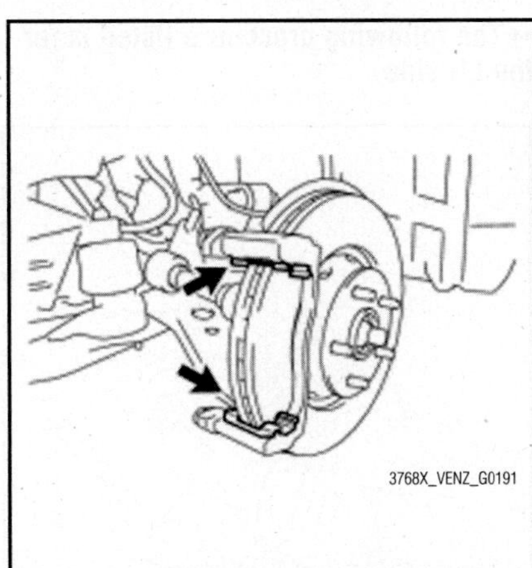

Fig. 7 Removing the front disc brake pad support plates

➡When replacing worn pads, the front anti-squeal shims must be replaced together with the pads.

➡Install the front anti-shims in the correct position and direction.

➡Apply disc brake grease to the area that contacts the front anti-squeal shims.

➡Disc brake grease can seep out slightly from the area where the front anti-squeal shim is installed.

➡Make sure that disc brake grease is not applied onto the lining surface.

9. Install the front disc brake pad.
   a. Install the 2 front disc brake pads to the front disc brake caliper mounting.

➡There should be no oil or grease on the friction surfaces of the front disc brake pads or the front disc.

➡Make sure to install the front disc brake pad with pad wear indicator to the inner side of the vehicle.

➡Be sure to engage the protrusion of the pad support plate to the front disc brake pad.

   b. Install the 2 anti-squeal springs to the front disc brake pads.

➡When replacing the front brake pads with new ones, make sure to replace the anti-squeal springs at the same time.

➡Be sure to install the anti-squeal springs into the front disc brake pad installation holes as far as they will go.

10. Install the front disc brake caliper.
11. Fill the reservoir with brake fluid.
12. Bleed the brake line.
13. Inspect for brake fluid leaks.
14. Inspect the fluid level in the reservoir.
15. Install the front wheel. Tighten to 76 ft. lbs. (103 Nm).

BRAKE CALIPER

REMOVAL & INSTALLATION

See Figure 8.

➡**Use the same procedure for the LH side and RH side.**

➡**The following procedure listed is for the LH side.**

1. Remove the rear wheel.
2. Drain the brake fluid.

➡**If the brake fluid leaks onto any painted surface, immediately wash it off.**

3. Separate the rear flexible hose.
   a. Remove the union bolt and gasket, and disconnect the rear flexible hose from the rear disc brake caliper assembly.
4. Remove the rear disc brake caliper assembly.
   a. Remove the 2 bolts and the rear disc brake caliper assembly from the rear disc brake caliper mounting.

To install:

5. Install the rear disc brake caliper assembly.
   a. Install the rear disc brake caliper assembly to the rear disc brake caliper mounting with the 2 bolts. Tighten to 24 ft. lbs. (32 Nm).
6. Connect the rear flexible hose.
   a. Connect the rear flexible hose to the rear disc brake caliper assembly with a new union bolt and a new gasket. Tighten to 22 ft. lbs. (29 Nm).

➡**Install the rear flexible hose lock securely into the lock hole in the rear disc brake caliper assembly.**

7. Fill the reservoir with brake fluid.
8. Bleed the brake line.
9. Inspect for brake fluid leaks.
10. Inspect the fluid level in the reservoir.
11. Adjust the parking brake.
12. Install the rear wheel. Tighten to 76 ft. lbs. (103 Nm).

3768X_VENZ_G0198

Fig. 8 Removing the rear disc brake caliper

DISC BRAKE PADS

REMOVAL & INSTALLATION

See Figure 9.

➡**Use the same procedure for the LH side and RH side.**

➡**The following procedure listed is for the LH side.**

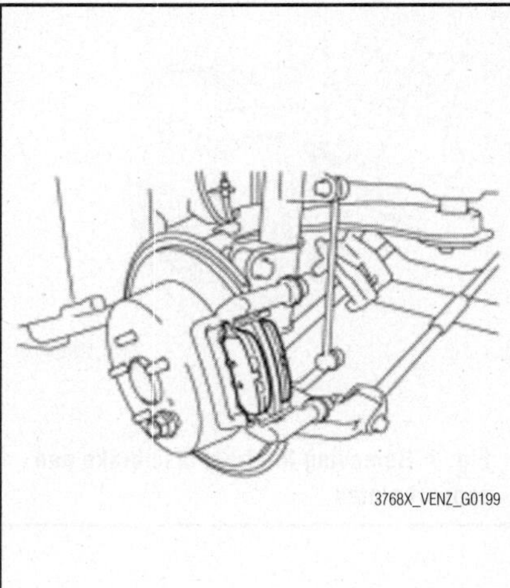

3768X_VENZ_G0199

Fig. 9 Removing the rear disc brake pad

1. Remove the rear wheel.
2. Drain the brake fluid.

➡**If the brake fluid leaks onto any painted surface, immediately wash it off.**

3. Remove the rear disc brake caliper.
4. Remove the rear disc brake pad.
   a. Remove the 2 rear disc brake pads from the rear disc brake caliper mounting.
5. Remove the rear anti squeal shim.
   a. Remove the 2 rear anti squeal shim and the pad wear indicator from the inner pad.

To install:

6. Install the rear anti squeal shim.
   a. Install the rear anti squeal shims to the 2 rear brake pads.

➡**When replacing worn pads, the rear anti-squeal shims must be replaced together with the pads.**

➡**Install the shims in the correct positions and directions.**

7. Install the rear disc brake pad.
   a. Install the 2 rear disc brake pads to the rear disc brake caliper mounting.

➡**Make sure to install the rear disc brake pads with the pad wear indicator to the inner side of the vehicle.**

➡**There should be no oil or grease on the friction surfaces of the rear disc brake pads or the rear disc.**

8. Install the rear disc brake caliper assembly.
9. Fill the reservoir with brake fluid.
10. Bleed the brake line.
11. Inspect for brake fluid leaks.
12. Inspect the fluid level in the reservoir.
13. Adjust the parking brake.
14. Install the rear wheel. Tighten to 76 ft. lbs. (103 Nm).

PARKING BRAKE SHOES

REMOVAL & INSTALLATION

See Figures 10 and 11.

➡Use the same procedure for the RH side and the LH side.

➡The procedure listed below is for the LH side.

1. Remove the rear wheel.
2. Remove the rear axle shaft nut (AWD).

➡Perform this procedure only when the No. 1 parking brake shoe hold down spring pin is replaced.

3. Separated the rear disc brake caliper assembly.
 a. Remove the 2 bolts and rear disc brake caliper assembly.

➡Do not disconnect the rear brake flexible hose from the rear disc brake caliper assembly.

➡Use wire or an equivalent tool to keep the rear disc brake caliper assembly from hanging down by the rear brake flexible hose.

4. Remove the rear disc.
5. Remove the No. 1 parking brake shoe return tension spring.
 a. Remove the 2 No. 1 parking brake shoe return tension springs.
6. Separate the No. 1 parking brake shoe assembly.
 a. Remove the No. 1 parking brake shoe hold down spring cup, parking brake shoe hold down spring and No. 2 parking brake shoe hold down spring cup, and separate the No. 1 parking brake shoe assembly from the backing plate.
7. Remove the parking brake shoe strut.
 a. Remove the parking brake shoe strut and the parking brake shoe strut compression spring.
8. Separate the No. 2 parking brake shoe assembly.
 a. Remove the No. 1 parking brake shoe hold down spring cup, parking brake shoe hold down spring and No. 2 parking brake shoe hold down spring cup, and separate the No. 2 parking brake shoe assembly from the backing plate.
9. Remove the parking brake shoe adjusting screw set.
10. Remove the No. 1 parking brake shoe assembly.
 a. Separate the No. 2 parking brake shoe return tension spring and remove the No. 1 parking brake shoe assembly.
11. Remove the No. 2 parking brake shoe return tension spring.
 a. Remove the No. 2 parking brake shoe return tension spring from the No. 2 parking brake shoe assembly.
12. Remove the No. 2 parking brake shoe assembly with the parking brake shoe lever.
 a. Using needle nose pliers, separate the No. 3 parking brake cable assembly from the parking brake shoe lever.

➡Be careful not to damage the No. 3 parking brake cable assembly.

13. Remove the parking brake shoe lever.
 a. Remove the C-washer, shim and the parking brake shoe lever from the No. 2 parking brake shoe assembly.

➡The shim is installed only when a clearance adjustment between the parking brake lever and parking brake shoe C-washer is necessary. Therefore, some models may have no shim.

14. Remove the parking brake shoe guide plate set bolt and the parking brake shoe guide plate.
15. Remove the No. 1 parking brake shoe hold down spring pin (2WD).
 a. Remove the No. 1 parking brake shoe hold down spring pin.
16. Remove the No. 2 parking brake shoe hold down spring pin (AWD).
17. Remove the rear axle hub and bearing assembly (2WD).

➡Perform this procedure only when the No. 2 parking brake shoe hold down spring pin is replaced.

18. Remove the rear axle hub and bearing assembly (AWD).

➡Perform this procedure only when the No. 1 parking brake shoe hold down spring pin is replaced.

19. Remove the No. 2 parking brake shoe hold down spring pin (2WD).
20. Remove the No. 1 parking brake shoe hold down spring pin (AWD).

To install:

21. Install the No. 2 parking brake shoe hold down spring pin (2WD).
22. Install the No. 1 parking brake shoe hold down spring pin (AWD).
23. Install the rear axle hub and bearing assembly (2WD).

➡Perform this procedure only when the No. 2 parking brake shoe hold down spring pin is replaced.

24. Install the rear axle hub and bearing assembly (AWD).

➡Perform this procedure only when the No. 1 parking brake shoe hold down spring pin is replaced.

25. Install the No. 1 parking brake shoe hold down spring pin (2WD).
26. Install the No. 2 parking brake shoe hold down spring pin (AWD).
27. Apply high temperature grease to the backing plate which makes contact with the shoe.
28. Install the parking brake shoe guide plate set bolt.
 a. Apply adhesive to the threads of the parking brake shoe guide plate set bolt.

Fig. 10 Removing the parking brake shoe adjusting screw set

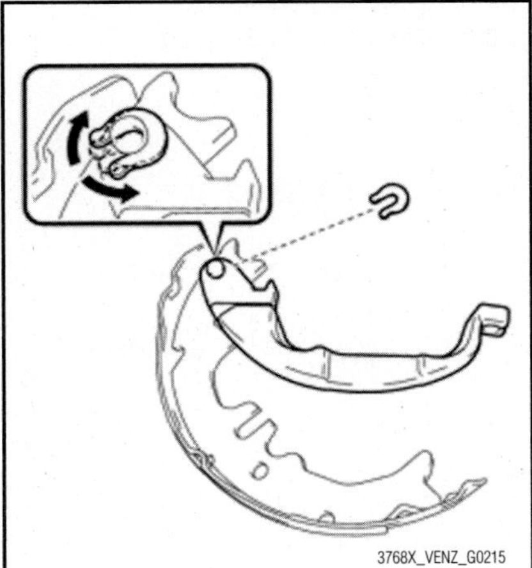

Fig. 11 Removing the parking brake shoe lever

b. Install the parking brake shoe guide plate and the parking brake shoe guide plate set bolt. Tighten to 13 ft. lbs. (18 Nm).

29. Install the parking brake shoe lever.

a. Apply high temperature grease to the parking brake shoe lever which makes contact with the No. 2 parking brake shoe assembly.

b. Install the parking brake shoe lever and shim to the No. 2 parking brake shoe assembly with a new C-washer.

c. Using a feeler gauge, measure the clearance between the No. 2 parking brake shoe assembly and parking brake shoe lever. Standard clearance is less than 0.0138 inch.

➡If the clearance is not as specified, replace the shim with one of the appropriate size.

30. Install the No. 2 parking brake shoe assembly with the parking brake shoe lever.

a. Using needle-nose pliers, connect the No. 3 parking brake cable assembly to the parking brake shoe lever.

31. Install the No. 2 parking brake shoe return tension spring to the No. 2 parking brake shoe assembly.

32. Install the No. 1 parking brake shoe assembly.

a. Connect the No. 2 parking brake shoe return tension spring to install the No. 1 parking brake shoe assembly.

33. Install the parking brake shoe adjusting screw set.

a. Apply high temperature grease to the parking brake shoe adjusting screw set.

b. Install the parking brake shoe adjusting screw set.

34. Install the No. 2 parking brake shoe assembly.

a. Install the No. 2 parking brake shoe assembly to the backing plate with the No. 1 parking brake shoe hold down spring cup, parking brake shoe hold down spring and No. 2 parking brake shoe hold down spring cup.

35. Install the parking brake shoe strut.

a. Install the parking brake shoe strut and the parking brake shoe strut compression spring.

36. Install the No. 1 parking brake shoe assembly.

a. Install the No. 1 parking brake shoe assembly to the backing plate with the No. 1 parking brake shoe hold down spring cup, parking brake shoe hold down spring and No. 2 parking brake shoe hold down spring cup.

37. Install the 2 No. 1 parking brake shoe return tension springs.

➡First install the front side spring and then the rear side spring.

38. Check the parking brake installation.

a. Check that each part is installed properly.

➡There should be no oil or grease on the friction surfaces of the shoe lining and disc.

39. Install the rear disc.

40. Install the rear disc brake caliper assembly.

41. Install the rear axle shaft nut (AWD).

➡Perform this procedure only when the No. 1 parking brake shoe hold down spring pin is replaced.

42. Adjust the parking brake shoe clearance.

43. Install the rear wheel. Tighten to 76 ft. lbs. (103 Nm).

44. Bed in the parking brake shoes to the discs.

a. Drive the vehicle at about 31 mph on a safe, level and dry road.

b. Depress the parking brake pedal with 33.7 lbs. (150 N) of force.

c. Drive the vehicle for about 0.25 mile (400 m) in this condition.

d. Repeat this procedure 3 times.

➡Set a 5 minute interval between each procedure to prevent the brake assembly from overheating.

45. Adjust the parking brake shoe clearance and parking brake pedal travel.

ADJUSTMENT

1. Completely releaser the parking brake pedal.
2. Loosen the lock nut and No. 1 wire adjusting nut to completely release the parking brake cable.
3. Remove the rear wheel.
4. Temporarily install the 5 hub nuts.
5. Remove the shoe adjusting hole plug.
6. Turn the shoe adjuster and expand the shoe until the disc locks.
7. Turn and contract the shoe adjuster until the disc can rotate smoothly. Standard is 8 notches.
8. Check that there is no brake drag against the shoe.
9. Install the shoe adjusting hole plug.
10. Turn the adjusting nut until the parking brake pedal travel is corrected to be within the specified range of 4-6 notches at 67.5 ft. lbs.
11. Using a wrench or an equivalent tool, hold the adjusting nut and tighten the lock nut to 48 inch lbs. (5.4 Nm).
12. Operate the parking brake pedal 3 to 4 times, and check the parking brake pedal travel.
13. Check that there is not brake drag against the shoe.
14. Remove the 5 hub nuts.
15. Install the rear wheel. Tighten to 76 ft. lbs. (103 Nm).

CHASSIS ELECTRICAL — AIR BAG (SUPPLEMENTAL RESTRAINT SYSTEM)

⁂ CAUTION

Disconnect and isolate the battery negative cable before beginning any airbag system component diagnosis, testing, removal, or installation procedures. Wait at least 90 seconds after the ignition switch is turned off and the negative (-) terminal cable is disconnected from the battery before starting the operation. The SRS is equipped with a backup power source, so if work is started within 90 seconds after disconnecting the negative (-) terminal cable from the battery, the SRS may be deployed. Failure to disable the airbag system may result in accidental airbag deployment, personal injury, or death.

DISARMING THE SYSTEM

To avoid personal injury when working on vehicles equipped with an air bag, the negative battery cable must be disconnected and at least 90 seconds must elapse before working on the system. Failure to do so may result in deployment of the air bag.

ARMING THE SYSTEM

To arm the system after service is finished, connect the negative battery cable.

CLOCKSPRING CENTERING

1. Rotate the spiral cable counterclockwise by hand until it feels firm

➡Do not use the airbag wire harness to turn the spiral cable.

2. Rotate the spiral cable clockwise approximately 2.5 turns to align the marks.

➡Do not use the airbag wire harness to turn the spiral cable.

➡The spiral cable will rotate approximately 2.5 turns to the left and right from the center.

DRIVE TRAIN

FRONT DRIVESHAFT

REMOVAL & INSTALLATION

See Figures 12 through 15.

1. Remove the tail exhaust pipe assembly.
2. Remove the center exhaust pipe assembly.
3. Remove the propeller with the center bearing shaft assembly.
   a. Depress the brake pedal and hold it.
   b. Using a hexagon wrench (6 mm), loosen the cross groove joint set bolts½ turn.

➡Put a piece of cloth or equivalent into the inside of the universal joint cover so that the boot does not touch the inside of the universal joint cover.

➡Do not remove the bolts.

   c. Place matchmarks on the rear propeller shaft and electromagnetic control coupling assembly.
   d. Remove the 4 nuts and 4 washers.
   e. Using a brass bar and a hammer, separate the propeller with center bearing shaft assembly.
   f. Remove the 4 bolts, 2 No. 1 center support bearing washers and 2 No. 2 center support bearing washers.

➡When removing the bolts and washers, do not apply excessive force to the universal joint.

   g. Pull out the propeller with center bearing shaft assembly from the transfer.

➡When removing the propeller shaft, do not apply excessive force to the universal joint.

➡During and after the removal of the propeller shaft, keep the universal joint angle straight (within 15 degrees).

➡Be careful not to damage the oil seal.

   h. Insert the special tool (09325-20010) into the transfer to prevent oil leaks.

➡Be careful not to damage the oil seal.

To install:

4. Temporarily tighten the propeller with the center bearing shaft assembly.
   a. Remove the special tool from the transfer.
   b. Install the propeller with the center bearing shaft assembly.

➡Be careful not to damage the oil seal.

➡Be careful not to damage the universal joint boot when installing the propeller shaft.

   c. Align the matchmarks on the rear propeller shaft and electromagnetic control coupling assembly and install the 4 nuts and 4 washers temporarily.

➡Do not allow grease to adhere to the bolts or washers.

   d. Temporarily install the propeller with center bearing shaft assembly with the 4 bolts, 2 No. 1 center support bearing washers and 2 No. 2 center support bearing washers.

➡Reuse the washers.

➡Do not allow grease to adhere to be bolts or washers.

   e. Fully tighten the 4 nuts. Tighten to 27 ft. lbs. (37 Nm).
5. Fully tighten the propeller with the center bearing shaft assembly.
   a. Remove the piece of cloth or equivalent from the universal joint.
   b. Depress the brake pedal and hold it.
   c. Using a hexagon wrench (6 mm), tighten the 6 bolts. Tighten to 19 ft. lbs. (26 Nm).
   d. With the vehicle unloaded, adjust the dimension between the rear side of the cover and shaft as shown in the illustration. Length A should be 2.579-2.776 inches (65.5-70.5 mm).
   e. With the vehicle unloaded, adjust the front and rear dimensions between

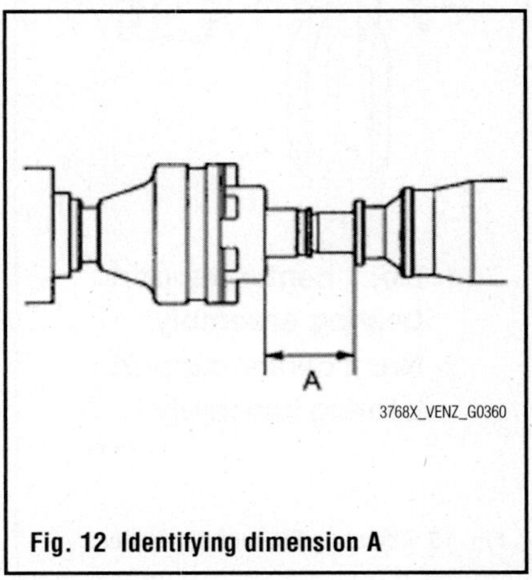

Fig. 12 Identifying dimension A

the edge surface of the center support bearing and the edge surface of the cushion respectively as shown in the illustration, and then tighten the bolts. Length A should be 0.453-0.532 inch (11.5-13.5 mm).

f. Check that the center line of the bracket is at a right angle to the shaft axial direction.

g. Fully tighten the 4 bolts. Tighten to 27 ft. lbs. (37 Nm).

6. Install the center exhaust pipe assembly.
7. Install the tail exhaust pipe assembly.
8. Inspect and adjust the transfer oil.
9. Inspect for exhaust gas leaks.
10. Inspect and adjust the joint angle.

If any vibration or noise occurs, perform joint angle check as follows and replace the No. 2 center support bearing washer with a proper one:

a. Turn the propeller shaft several times by hand to stabilize the center support bearings.

b. Using a jack, raise and lower the differential to stabilize the differential mounting cushion.

➡Measure the joint angle while the vehicle is lifted using a 4 pillar lift or while working in a pit.

c. Remove the transfer dynamic damper.

d. Using SST, measure the propeller shaft installation angle (A) and intermediate shaft installation angle (B) as shown in the following illustration. The No. 1 joint angle is (A) - (B)= -3.69° to -1.69°

e. Using SST, measure the rear propeller shaft installation angle (C) and rear differential installation angle (D) as shown in the preceding illustration. The No. 4 joint angle is (C) - (D) = 1.63° to 3.63°

f. If the calculated amount is not within the specification, adjust it with the No. 2 center support bearing washer.

➡Make sure to use a washer of the same thickness on both right and left sides.

➡Do not use 2 or more washers on a bolt.

g. Install the transfer dynamic damper. Tighten to 19 ft. lbs. (26 Nm).

FRONT HALFSHAFT

REMOVAL & INSTALLATION

See Figures 16 through 18.

➡Use the same procedure for the RH side and the LH side.

➡The procedure listed below is for the LH side.

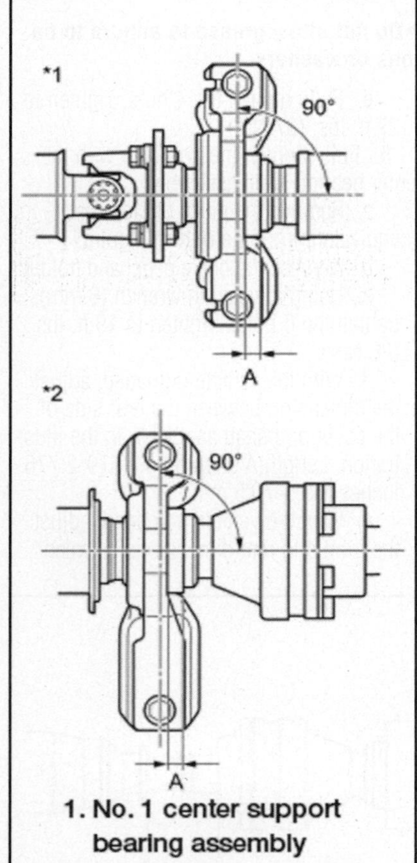

Fig. 13 Adjusting the rear dimensions

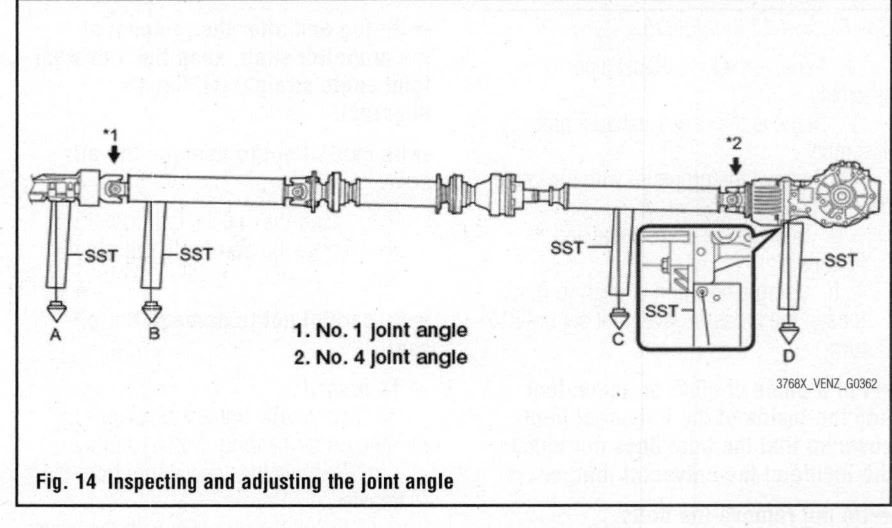

Fig. 14 Inspecting and adjusting the joint angle

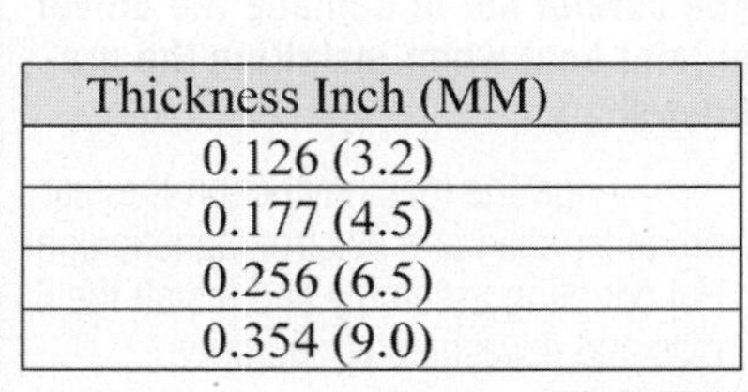

Thickness Inch (MM)
0.126 (3.2)
0.177 (4.5)
0.256 (6.5)
0.354 (9.0)

3768X_VENZ_G0363

Fig. 15 Washer thickness

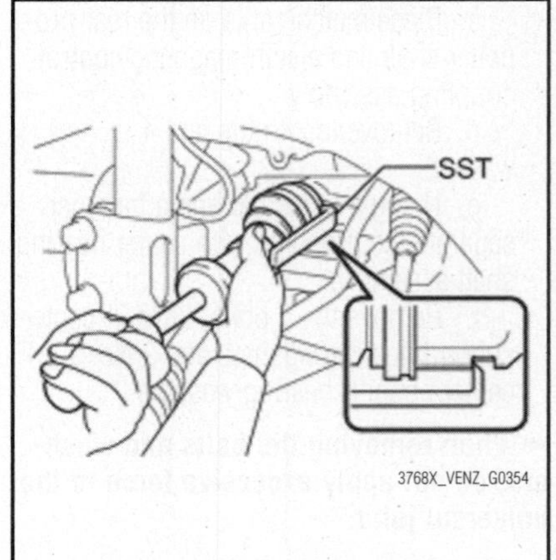

Fig. 16 Removing the front driveshaft assembly LH

⁂ CAUTION

Refer to Precautions before performing the procedure listed below.

1. Drain the automatic transaxle fluid.
2. Drain the transfer oil.
3. Remove the front wheels.
4. Remove the front axle shaft nut.
   a. Using the special tool (09930-00010) and a hammer, replace the staked part of the front axle shaft nut.

➡Loosen the staked part of the nut completely, otherwise the threads of the driveshaft may be damaged.

   b. While applying the brakes, remove the front axle shaft nut.
5. Separate the front stabilizer link assembly.
6. Separate the front speed sensor.
   a. Remove the bolt and separate the front speed sensor from the steering knuckle.
   b. Remove the bolt and clamp, and separate the front speed sensor and front flexible hose.
7. Separate the tie rod assembly.
8. Separate the front lower suspension arm.
9. Separate the front axle assembly.
10. Remove the front driveshaft assembly LH.
   a. Using the special tool (09520-01010, 09520-24010 or 09520-32040), remove the front driveshaft assembly LH.
11. Remove the front driveshaft assembly RH (2WD).
   a. Remove the bearing bracket hole snap ring from the driveshaft bearing bracket.

Fig. 17 Removing the front driveshaft assembly RH

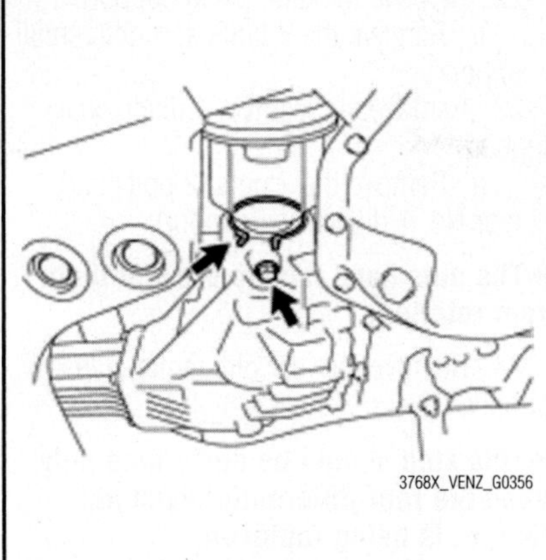

Fig. 18 Removing the front driveshaft assembly RH (AWD)

   b. Remove the bolt and front driveshaft assembly RH from the driveshaft bearing bracket.

➡Do not damage the boot or oil seal.

12. Remove the front driveshaft assembly RH (AWD).
   a. Remove the bearing bracket hole snap ring from the driveshaft bearing bracket.
   b. Remove the bolt and front driveshaft assembly RH from the driveshaft bearing bracket.

➡Do not damage the boot or oil seal.

To install:

13. Install the front driveshaft assembly LH.
   a. Align the splines of the shaft and install the driveshaft assembly LH using a brass bar and a hammer.

➡Set the shaft snap ring with the opening facing down.

➡Be careful not to damage the driveshaft dust cover, boot or oil seal.

➡Move the driveshaft assembly while keeping it level.

14. Install the front driveshaft assembly RH (2WD).
   a. Install the front driveshaft assembly RH.
   b. Install the bearing bracket hole snap ring and a new bolt. Tighten to 24 ft. lbs. (32 Nm).

➡Do not damage the boot or oil seal.

➡Move the driveshaft assembly while keeping it level.

15. Install the front driveshaft assembly RH (AWD).
   a. Install the front driveshaft assembly RH.
   b. Install the bearing bracket hole snap ring and a new bolt. Tighten to 24 ft. lbs. (32 Nm).

➡Do not damage the boot or oil seal.

➡Move the driveshaft assembly while keeping it level.

16. Install the front axle assembly.
17. Install the front lower suspension arm.
18. Connect the tie rod assembly.
19. Install the front speed sensor.
   a. Install the front speed sensor and front flexible hose with the bolt. Tighten to 14 ft. lbs. (19 Nm).

➡Do not twist the front speed sensor when installing it.

➡First install the speed sensor harness bracket, and then install the flexible hose bracket.

   b. Install the front speed sensor to the steering knuckle with the bolt. Tighten to 71 inch lbs. (8 Nm).

➡Prevent foreign matter from attaching to the front speed sensor tip.

➡Firmly insert the front speed sensor body into the steering knuckle before tightening the bolt.

➡After installing the front speed sensor to the steering knuckle, make sure that there is no clearance between the front speed sensor stay and steering knuckle. Also make sure that no foreign matter is stuck between the parts.

➡To prevent interference between the front speed sensor and magnetic rotor, do not rotate the front speed sensor body during or after the insertion of the front speed sensor body to the steering knuckle.

20. Install front stabilizer link assembly.
21. Clean the threaded parts on the front driveshaft and a new front axle shaft nut using a non-residue solvent.

➡Be sure to perform this work for a new driveshaft.

➡Keep the threaded parts free of oil and foreign matter.

22. Install the new front axle shaft nut and tighten to 217 ft. lbs. (294 Nm).
   a. Using a chisel and hammer, stake the front axle shaft nut.
23. Install the front wheels. Tighten to 76 ft. lbs. (103 Nm).

24. Add transfer oil.
25. Adjust the transfer oil.
26. Add automatic transaxle fluid.
27. Adjust the front wheel alignment.
28. Check the ABS speed sensor signal.

REAR AXLE HOUSING

REMOVAL & INSTALLATION

See Figures 19 through 22.

1. Drain the differential oil.
   a. Using a hexagon wrench (10 mm), remove the rear differential carrier cover plug and gasket.
   b. Using a hexagon wrench (10 mm), remove the rear differential drain plug and gasket, then drain the differential oil.
2. Remove the rear wheels.
3. Remove the center exhaust pipe assembly.
4. Remove the propeller with the center bearing shaft assembly.
5. Separate the rear speed sensor LH and RH.
6. Remove the rear axle shaft nuts LH and RH.
7. Separate the No. 3 parking brake cable assembly.
8. Separate the No. 2 parking brake cable assembly.
9. Remove the No. 1 floor under cover.
10. Remove the rear strut rod assemblies LH and RH.
11. Remove the rear height control sensor sub assembly (w/HID headlight system).
12. Remove the rear No. 2 suspension arm assemblies LH and RH.
13. Separate the rear No. 1 suspension arm assemblies LH and RH.
14. Remove the rear driveshaft assembly LH.
15. Remove the rear driveshaft snap ring LH.
16. Remove the rear driveshaft assembly RH.
17. Remove the rear driveshaft snap ring RH.
18. Separate the No. 3 floor wire (w/HID headlight system).
19. Separate the frame wire.
20. Remove the rear suspension member.
21. Remove the rear differential carrier assembly with the differential support.
   a. Remove the 2 bolts and 2 nuts.

➡The nuts have tabs to prevent them from rotating.

   b. Remove the 3 rear mounting bolts and rear differential carrier assembly with differential support from the rear suspension member.

22. Remove the differential support.
   a. Remove the 3 bolts and differential support.
23. Remove the rear No. 1 differential support.
   a. Remove the 2 nuts, 2 bolts and rear No. 1 differential support.

➡The nuts have tabs to prevent them from rotating.

24. Remove the rear differential dynamic damper.

➡This step should be performed only when the rear differential dynamic damper is being replaced.

   a. Remove the bolt and rear differential dynamic damper.

To install:

➡If installing a new rear differential carrier assembly, remove the 2 differential side seal caps before installing the rear driveshaft assembly.

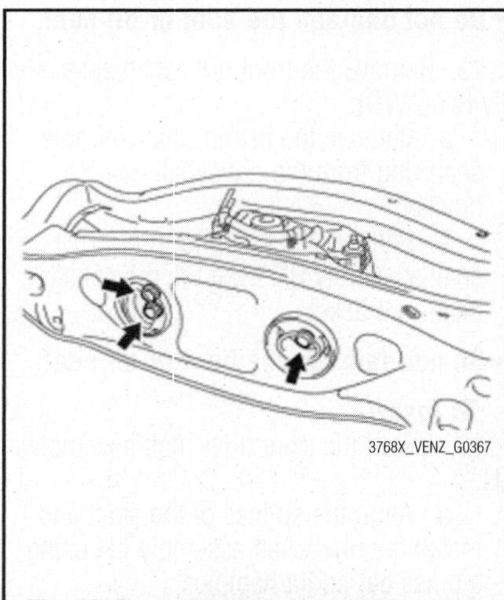

Fig. 19 Removing the rear mounting bolts and carrier assembly

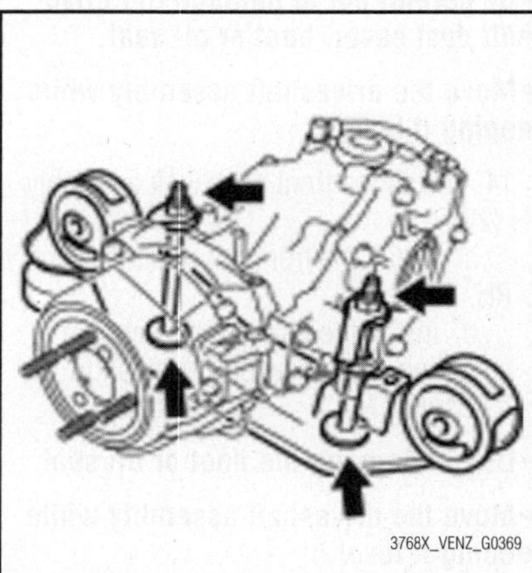

Fig. 20 Removing the rear No. 1 differential support

25. Install the rear differential dynamic damper.

➡This step should be performed only when the rear differential dynamic damper is replaced.

   a. Install the rear differential dynamic damper with the bolt. Tighten to 20 ft. lbs. (27 Nm).
26. Temporarily tighten the rear No. 1 differential support.
   a. Temporarily install the rear No. 1 differential support to the rear differential carrier assembly with the 2 new bolts and 2 new nuts.

➡Be sure to install the rear No. 1 differential support facing in the direction shown in the illustration.

➡Be sure to install each nut to position (A) shown in the illustration.

➡The nuts have tabs to prevent them from rotating.

27. Install the differential support to the rear differential carrier assembly with the 3 new bolts. Tighten to 123 ft. lbs. (167 Nm).
28. Temporarily tighten the rear differential carrier assembly with the differential support.
   a. Temporarily install the rear differential carrier assembly with differential support to the rear side of the rear suspension member assembly with the 3 rear mounting bolts.

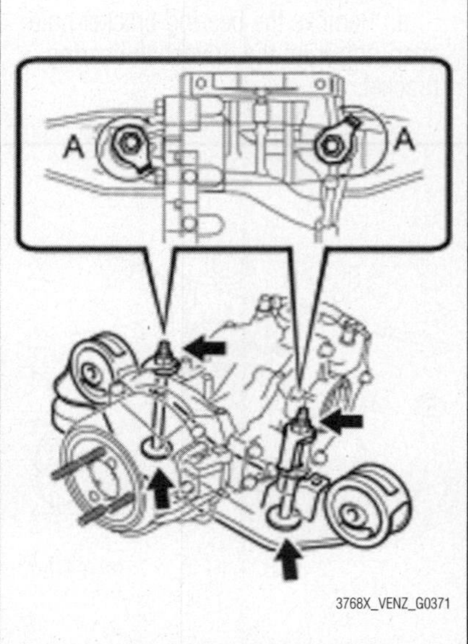

Fig. 21 Temporarily installing the rear No. 1 differential support

b. Temporarily install the rear differential carrier assembly with differential support to the front side of the rear suspension member assembly with the 2 bolts and 2 nuts.

➡**The nuts have tabs to prevent them from rotating.**

29. Fully tighten the rear differential carrier assembly with the differential support.

➡**Do not tighten the bolts with the inner cylinder or rear differential mount cushion tilted.**

a. Install the rear differential carrier assembly with differential support to the rear side of the rear suspension member assembly with the 3 rear mounting bolts. Tighten to 70 ft. lbs. (95 Nm).

b. Install the rear differential carrier assembly with differential support to the front side of the rear suspension member assembly with the 2 bolts and 2 nuts. Tighten to 76 ft. lbs. (103 Nm).

➡**The nuts have tabs to prevent them from rotating.**

30. Fully tighten the rear No. 1 differential support.

a. Install the rear No. 1 differential support to the rear differential carrier assembly with the 2 new bolts and 2 new nuts. Tighten to 63 ft. lbs. (86 Nm).

➡**Make sure that each nut is installed to a position (A) shown in the illustration.**

➡**The nuts have tabs to prevent them from rotating.**

31. Install the rear suspension member.
32. Install the frame wire.

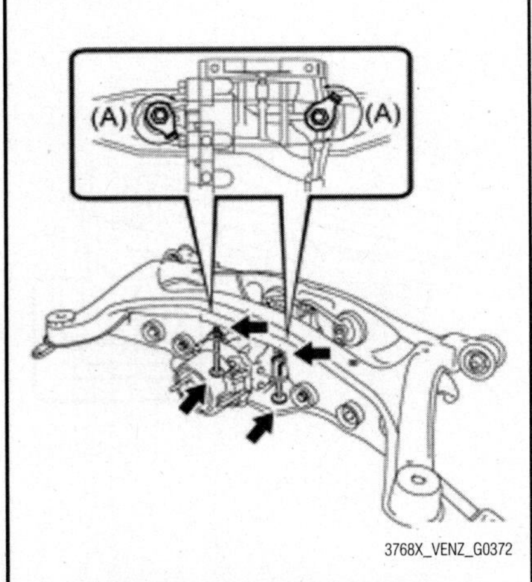

Fig. 22 Identifying the nut positioning

33. Install the No. 3 floor wire (w/HID headlight system).
34. Install the rear driveshaft snap ring LH.
35. Install the rear driveshaft assembly LH.
36. Install the rear driveshaft snap ring RH.
37. Install the rear driveshaft assembly RH.
38. Connect the rear No. 1 suspension arm assembly LH.
39. Connect the rear No. 1 suspension arm assembly RH.
40. Temporarily tighten the rear No. 2 suspension arm assemblies LH and RH.
41. Install the rear strut rod assemblies LH and RH.
42. Install the rear height control sensor sub assembly (w/HID headlight system).
43. Install the No. 3 parking brake cable assembly.
44. Install the No. 2 parking brake cable assembly.
45. Install the rear axle shaft nuts LH and RH.
46. Install the rear speed sensors LH and RH.
47. Temporarily tighten the propeller with the center bearing shaft assembly.
48. Fully tighten the propeller with the center bearing shaft assembly.
49. Install the No. 1 floor under cover.
50. Inspect and adjust the transfer oil.
51. Install the center exhaust pipe assembly.
52. Add differential oil.

a. For a reused rear differential carrier assembly, use a hexagon wrench (10 mm) to install the rear differential drain plug with a new gasket. Tighten to 29 ft. lbs. (39 Nm). Add the differential oil.

b. For a new rear differential carrier assembly, use a hexagon wrench (10 mm) and remove the rear differential carrier cover plug and gasket. Add differential oil.

53. Inspect and adjust the differential oil.
54. Install the rear differential carrier cover plug.
55. Inspect for exhaust gas leak.
56. Install the rear wheels. Tighten to 76 ft. lbs. (105 Nm).
57. Stabilize the suspension.
58. Fully tighten the rear No. 2 suspension arm assemblies LH and RH.
59. Inspect and adjust the rear wheel alignment.
60. Check for speed sensor signal.
61. Perform the height control sensor signal initialization (w/HID headlight system).
62. Inspect and adjust the headlight aiming (w/HID headlight system).

REAR AXLE SHAFT, BEARING & SEAL

REMOVAL & INSTALLATION

2WD Vehicles

See Figure 23.

➡**Use the same procedure for the RH side and LH side.**

➡**The procedure listed below is for the LH side.**

1. Remove the rear wheel.
2. Separate the rear flexible hose.
3. Separate the rear disc brake caliper assembly.
4. Remove the rear disc.
5. Separate the rear speed sensor wire.
6. Remove the rear axle hub and bearing assembly.

a. Remove the 4 bolts and the rear axle hub and bearing assembly.

➡**Use wire or an equivalent tool to keep the parking brake assembly from hanging down by the parking brake cable assembly.**

To install:

7. Install the rear axle hub and bearing assembly.

a. Install the parking brake assembly and the rear axle hub and bearing assembly with the 4 bolts. Tighten to 59 ft. lbs. (80 Nm).

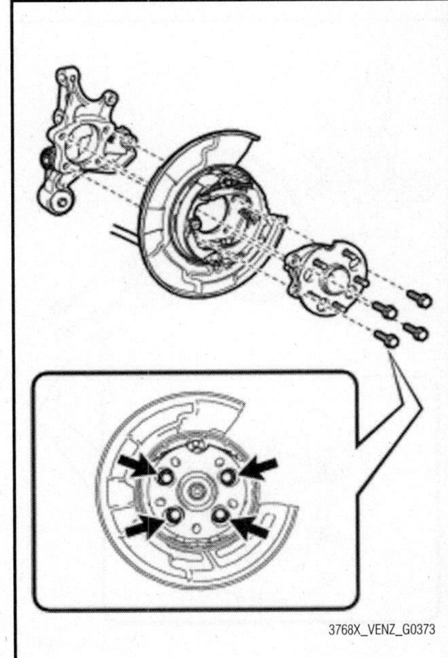

Fig. 23 Removing the rear axle hub and bearing assembly

➡Do not twist the No. 3 parking brake cable assembly when installing it.

8. Inspect the rear axle hub bearing looseness.
9. Inspect the rear axle hub runout.
10. Install the rear speed sensor wire.
11. Install the rear disc.
12. Install the rear disc brake caliper assembly.
13. Install the rear flexible hose.
14. Install the rear wheel. Tighten to 76 ft. lbs. (103 Nm).
15. Check for speed sensor signal.

AWD Vehicles

See Figure 24.

➡Use the same procedure for the RH side and LH side.

➡The procedure listed below is for the LH side.

1. Remove the rear wheel.
2. Separate the rear speed sensor.
   a. Remove the bolt and separate the rear seed sensor from the rear axle carrier sub assembly.

➡Keep the sensor tip and rear speed sensor installation hole free of foreign matter.

3. Remove the rear axle shaft nut.
4. Separate the rear disc brake caliper assembly.
5. Remove the rear disc.

Fig. 24 Removing the rear axle hub and bearing assembly

6. Remove the rear axle hub and bearing assembly.
   a. Put matchmarks on the rear driveshaft assembly and rear axle hub and bearing assembly.

➡Do not punch the matchmarks.

   b. Using a plastic hammer, separate the rear driveshaft assembly from the axle hub and bearing assembly. If it is difficult to separate, tap the end of the rear driveshaft assembly using a brass bar and a hammer.
   c. Remove the 4 bolts and the rear axle hub and bearing assembly.

➡Do not rotate the rear driveshaft with the rear axle hub and bearing assembly removed.

➡Use wire or an equivalent tool to keep the parking brake assembly from hanging down by the parking brake cable assembly.

To install:

7. Install the rear axle hub and bearing assembly.
   a. Align the matchmarks on the rear driveshaft assembly and the rear axle hub and bearing assembly.

➡Do not rotate the rear driveshaft.

   b. Install the parking brake assembly and the rear axle hub and bearing assembly with the 4 bolts. Tighten to 59 ft. lbs. (80 Nm).

➡Do not twist the parking brake cable assembly when installing it.

8. Install the rear axle shaft nut.
   a. Clean the threaded parts on the rear driveshaft assembly and a new rear axle shaft nut using a non-residue solvent.

➡Be sure to perform this work for a new rear driveshaft assembly.

➡Keep the threaded parts free of oil and foreign matter.

   b. Install the rear disc with the 5 hub nuts.
   c. While applying the parking brakes, temporarily install the new rear axle shaft nut. Tighten to 217 ft. lbs. (294 Nm).

➡Stake the nut after inspection for looseness and runout in the following steps.

   d. Remove the 5 hub nuts and the rear disc.
9. Install the rear speed sensor.
   a. Install the rear speed sensor to the rear axle carrier sub assembly with the bolt. Tighten to 71 inch lbs. (8 Nm).

➡Keep the rear speed sensor tip and sensor installation hole free of foreign matter.

➡Do not twist the rear speed sensor wire when installing it.

10. Inspect the rear axle hub bearing for looseness.
11. Inspect the rear axle hub runout.
12. Install the rear disc.
13. Install the rear disc brake caliper assembly.
14. Stake the rear axle shaft nut.
   a. Using a chisel and a hammer, stake the rear axle shaft nut.
15. Install the rear wheel and tighten to 76 ft. lbs. (103 Nm).
16. Check for speed sensor signal.

REAR DRIVESHAFT

REMOVAL & INSTALLATION

See Figure 25.

➡Use the same procedure for the LH side and RH side.

➡The following procedure listed below is for the LH side.

1. Remove the rear wheel.
2. Separate the rear speed sensor.
3. Remove the rear axle shaft nut.
   a. Using the special tool (09930-00010) and a hammer, release the staked part of the rear axle shaft nut.

➡Loosen the staked part of the nut completely, otherwise the threads of the driveshaft may be damaged.

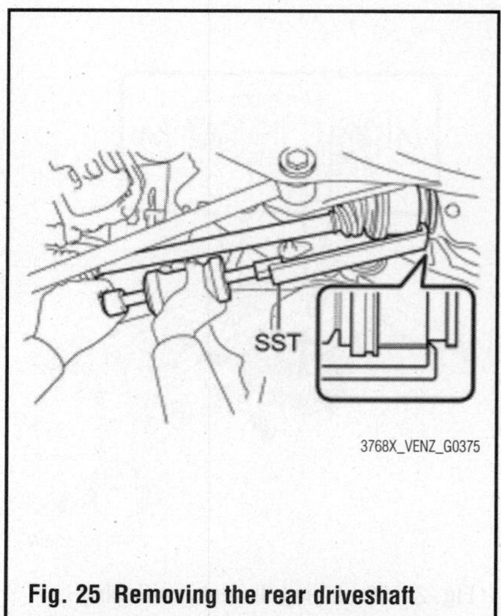

Fig. 25 Removing the rear driveshaft

b. While applying the brakes, remove the rear axle shaft nut.

4. Separate the rear disc brake caliper assembly.

5. Remove the rear disc.

6. Remove the rear axle hub and bearing assembly.

7. Separate the No. 3 parking brake cable assembly.

8. Remove the rear strut rod assembly.

9. Remove the rear axle carrier sub assembly.

10. Remove the rear driveshaft assembly.

a. Using the special tool (09520-01010, 09520-24010 or 09520-32040), remove the rear driveshaft assembly as shown in the illustration.

➡**Remove the rear driveshaft assembly while keeping it level.**

To install:

11. Install the rear driveshaft assembly.

a. Align the shaft splines and install the rear driveshaft assembly using a screwdriver and hammer.

➡**Set the snap ring with the opening facing downward.**

➡**Be careful not to damage the oil seal, boot or dust cover.**

➡**Install the driveshaft assembly while keeping it level.**

12. Install the rear axle carrier sub assembly.

13. Inspect the rear strut rod assembly.

14. Install the No. 3 parking brake cable assembly.

15. Install the rear axle hub and bearing assembly.

16. Install the rear speed sensor.

17. Install the rear disc.

18. Install the rear disc brake caliper assembly.

19. Install the rear axle shaft nut.

a. Clean the threaded parts on the driveshaft and axle shaft nut using a non residue solvent.

REAR PINION SEAL

REMOVAL & INSTALLATION

See Figures 26 and 27.

1. Remove the center exhaust pipe assembly.

2. Remove the propeller with the center bearing shaft assembly.

3. Drain the differential oil.

a. Using a hexagon wrench (10 mm), remove the rear differential carrier cover plug and gasket.

b. Using a hexagon wrench (10 mm), remove the rear differential drain plug and gasket, then drain the differential oil.

c. Using a hexagon wrench (10 mm), install the rear differential drain plug with a new gasket. Tighten to 29 ft. lbs. (39 Nm).

d. Using a hexagon wrench (10 mm), temporarily install the rear differential carrier cover plug.

➡**Add differential oil before installing a new gasket and fully tightening the rear differential carrier cover plug.**

4. Remove the electromagnetic control coupling sub assembly.

a. Disconnect the electromagnetic control coupling sub assembly connector and vacuum hose.

➡**Do not damage the electromagnetic control coupling wire harness.**

b. Remove the 4 bolts.

c. Using a brass bar and a hammer, tap the electromagnetic control coupling sub assembly to remove the electromagnetic control coupling sub assembly from the rear differential carrier assembly.

➡**Do not drop the electromagnetic control coupling sub assembly.**

5. Remove the transmission coupling conical spring washer.

a. Remove the transmission coupling conical spring washer from the rear differential carrier assembly.

6. Remove the transmission coupling spacer.

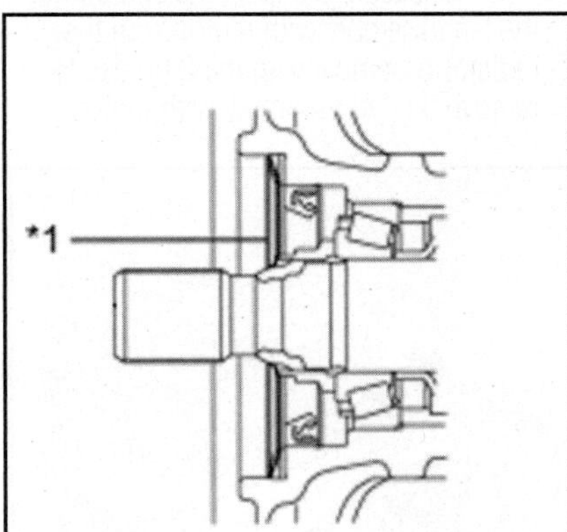

Fig. 26 Removing the transmission coupling conical spring washer

a. Remove the transmission coupling spacer from the rear differential carrier assembly.

7. Remove the diaphragm oil seal.

a. Using the special tool (09308-10010), remove the diaphragm oil seal.

To install:

8. Install the diaphragm oil seal.

a. Using the special tool, install a new diaphragm oil seal. The oil seal driven in depth is 0.0276-0.0512 inch (0.7-1.3 mm).

b. Apply MP grease to the lip of the new diaphragm oil seal.

9. Install the transmission coupling spacer.

a. Install the transmission coupling spacer to the rear differential carrier assembly.

➡**Keep the transmission coupling spacer free of oil and foreign matter.**

b. Install the transmission coupling conical spring washer to the rear differential carrier assembly.

➡**Install the transmission coupling conical spring washer so that the marked surface faces the front of the vehicle (coupling side).**

➡**Keep the transmission coupling conical spring washer free of oil and foreign matter.**

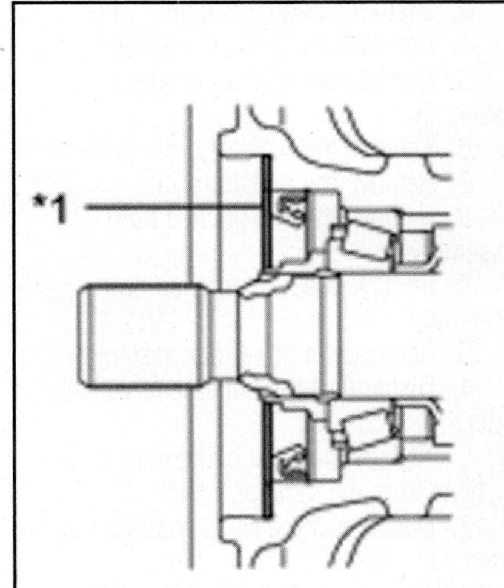

Fig. 27 Removing the transmission coupling spacer

10. Install the electromagnetic control coupling sub assembly.

a. Using a scraper and wire brush, remove the seal packing from the rear differential carrier assembly and electromagnetic control coupling sub-assembly.

➡**Do not scratch the installation area.**

b. Using a non-residue solvent, remove grease and oil from the contact surfaces of the rear differential carrier assembly and the electromagnetic control coupling sub-assembly.

c. Apply seal packing to the areas indicated in the illustration of the electromagnetic control coupling sub-assembly.

➡**Stop applying seal packing after allowing it to overlap with the beginning of the bead by at least 10 mm (0.394 in.) within range (A) in the illustration.**

➡**Make sure that the clearance from the center of the bead is within 0.0394 inch (1 mm).**

➡**Install the electromagnetic control coupling sub-assembly within 10 minutes after applying seal packing.**

➡**Apply seal packing in a continuous bead 0.0787 to 0.118 inch (2 to 3 mm) in diameter.**

d. Apply hypoid gear oil to the splines of the rear drive pinion.

e. Install the electromagnetic control coupling sub-assembly with the 4 bolts. Tighten to 14 ft. lbs. (20 Nm).

➡**Do not damage the contact surfaces of the diaphragm oil seal, stud bolt or electromagnetic control coupling sub-assembly.**

f. Connect the electromagnetic control coupling sub-assembly connector and vacuum hose.

➡**Do not damage the electromagnetic control coupling wire harness.**

11. Add differential oil.

a. Remove the rear differential carrier cover plug.

b. Add differential oil.

12. Inspect and adjust the differential oil.

13. Install the rear differential carrier cover plug.

14. Inspect for differential oil leak.

15. Temporarily tighten the propeller with the center bearing shaft assembly.

16. Fully tighten the propeller with the center bearing shaft assembly.

17. Install the center exhaust pipe assembly.

18. Inspect and adjust the transfer oil.

19. Inspect for transfer oil leaks.

ENGINE COOLING

RADIATOR

REMOVAL & INSTALLATION

2.7L Engine

See Figures 28 and 29.

1. Remove the No. 1 engine under cover.
2. Remove the No. 2 engine under cover.
3. Drain the engine coolant.
4. Remove the cool air intake duct seal.
5. Remove the inlet air cleaner assembly.
6. Remove the radiator grille protector.
7. Remove the radiator grille.
8. Remove the low pitched horn assembly.
9. Remove the high pitched horn assembly.
10. Remove the hood lock assembly.
11. Disconnect the outlet reserve tank hose.

a. Disconnect the outlet reserve tank hose from the radiator.

12. Disconnect the No. 1 radiator hose from the radiator.
13. Remove the No. 2 radiator hose.
14. Disconnect the outlet No. 1 oil cooler hose from the radiator.
15. Remove the upper radiator support.

a. Disconnect the hood lock control cable clamp and remove the 5 bolts and upper radiator support.

16. Separate the cooler condenser assembly.

a. Remove the 4 bolts and separate the cooler condenser assembly.

17. Remove the radiator assembly.

a. Disconnect the 4 wire harness clamps and 2 connectors.

b. Remove the radiator assembly and fan assembly with motor.

➡**Do not apply any excessive force to the cooler condenser assembly or pipe when removing the radiator assembly.**

c. Remove the 2 radiator support cushions and 2 lower radiator supports.

d. Release the 3 snap fits and pull up the fan assembly with motor from the radiator assembly with the 2 guides to remove the fan assembly with motor.

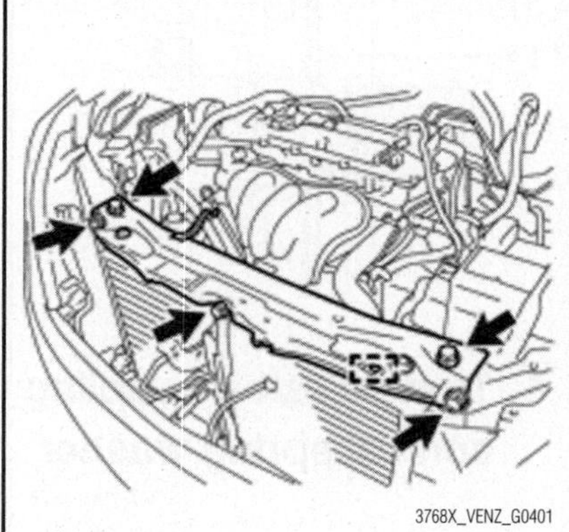

Fig. 28 Removing the upper radiator support

To install:

18. Install the radiator assembly.

a. Install the fan assembly with the motor to the radiator with the 2 guides at the bottom and 3 snap fits on the top.

b. Install the 2 lower radiator supports and 2 radiator support cushions.

c. Install the radiator assembly and fan assembly with the motor.

➡**Do not apply any excessive force to the cooler condenser assembly or pipe when installing the radiator assembly.**

d. Connect the 4 wire harness clamps and 2 connectors.

19. Install the cooler condenser assembly with the 4 bolts. Tighten to 44 inch lbs. (5 Nm).

Fig. 29 Removing the radiator assembly and fan assembly with motor

20. Install the upper radiator support with the 5 bolts and connect the hood lock control cable clamp to the upper radiator support. Tighten to 9 ft. lbs. (12 Nm).
21. Connect the outlet No. 1 oil cooler hose to the radiator.
22. Connect the inlet No. 1 oil cooler hose to the radiator.
23. Install the No. 2 radiator hose to the engine and radiator.
24. Connect the No. 1 radiator hose to the radiator.
25. Connect the outlet reserve tank hose to the radiator.
26. Install the hood lock assembly.
27. Install the low pitched horn assembly.
28. Install the high pitched horn assembly.
29. Install the radiator grille.
30. Install the radiator grille protector.
31. Install the inlet air cleaner assembly.
32. Add engine coolant.
33. Inspect for coolant leaks.
34. Inspect the automatic transaxle fluid.
35. Install the No. 1 engine under cover.
36. Install the No. 2 engine under cover.
37. Install the cool air intake duct seal.

3.5L Engine

See Figures 30 and 31.

1. Remove the No. 1 engine under cover.
2. Remove the No. 2 engine under cover.
3. Drain the engine coolant.
4. Remove the v-bank cover sub assembly.
5. Remove the cool air intake duct seal.
6. Remove the inlet No. 2 air cleaner.
7. Remove the air cleaner cap with the hose.
8. Remove the air cleaner case.
9. Remove the battery.
10. Remove the inlet No. 1 air cleaner.
11. Remove the radiator grille protector.
12. Remove the radiator grille.
13. Remove the low pitched horn assembly.
14. Remove the high pitched horn assembly.
15. Remove the hood lock assembly.
16. Disconnect the outlet reserve tank hose from the radiator.
17. Disconnect the No. 1 radiator hose from the radiator.
18. Remove the No. 2 radiator hose.
19. Disconnect the inlet oil cooler hose from the radiator.
20. Disconnect the outlet oil cooler hose from the radiator.
21. Remove the upper radiator support.
   a. Disconnect the hood lock control cable clamp and remove the 5 bolts and upper radiator support.
22. Separate the cooler condenser assembly.
   a. Remove the 4 bolts and separate the cooler condenser assembly.
23. Remove the radiator assembly.
   a. Disconnect the 3 wire harness clamps and connector.
   b. Remove the radiator assembly and fan assembly with motor.

➡Do not apply any excessive force to the cooler condenser assembly or pipe when removing the radiator assembly.

   c. Remove the 2 radiator support cushions and 2 lower radiator supports.
   d. Release the 3 snap fits and pull up the fan assembly with motor from the radiator assembly with the 2 guides to remove the fan assembly with motor.

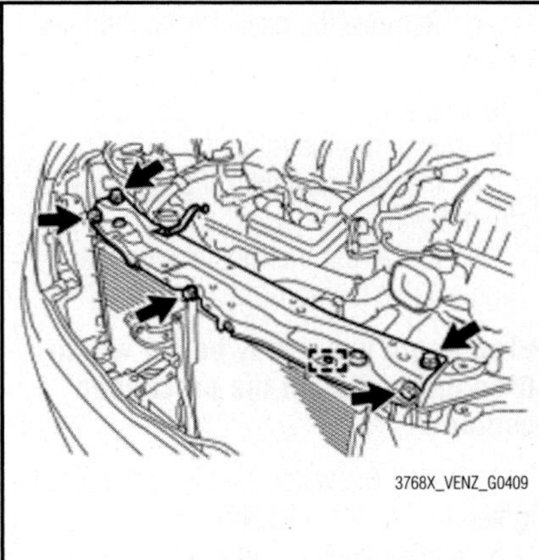

Fig. 30 Removing the upper radiator support

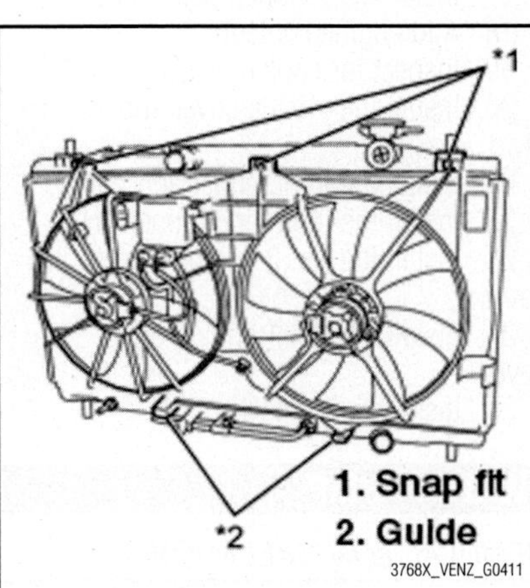

Fig. 31 Removing the fan assembly with motor

To install:

24. To install, reverse the removal procedure. Take note to tighten the cooler condenser assembly bolts to 44 inch lbs. (5 Nm). Tighten the radiator support bolts to 9 ft. lbs. (12 Nm).

THERMOSTAT

REMOVAL & INSTALLATION

2.7L Engine

See Figures 32 and 33.

1. Disconnect the cable from the negative battery terminal.

➡When disconnecting the cable, some systems need to be initialized after the cable is reconnected.

2. Remove the No. 1 engine cover sub assembly.
3. Remove the cool air intake duct seal.
4. Remove the No. 1 engine under cover.
5. Remove the No. 2 engine under cover.
6. Drain the engine coolant.
7. Remove the v-ribbed belt.
8. Remove the wire harness clamp bracket.
9. Remove the alternator assembly.
10. Disconnect the No. 2 radiator hose.
11. Remove the water inlet.
   a. Remove the 2 nuts and water inlet.
12. Remove the thermostat. Remove the gasket from the thermostat.

To install:

13. Install the thermostat.
   a. Install a new gasket to the thermostat.
   b. Install the thermostat with the jiggle valve facing upward.

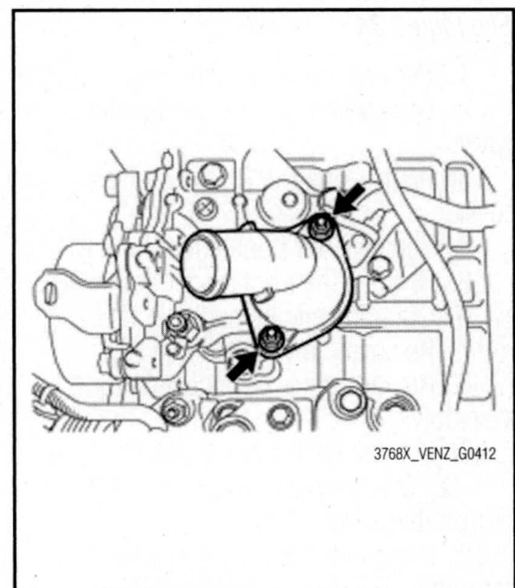

Fig. 32 Removing the water inlet

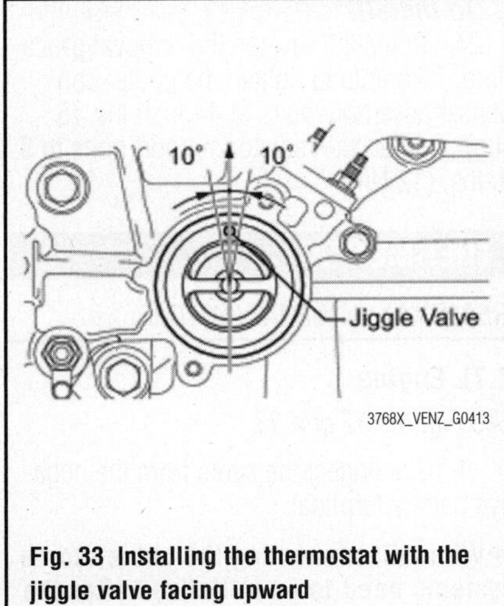

Fig. 33 Installing the thermostat with the jiggle valve facing upward

➡The jiggle valve may be set to within 10°on either side of the prescribed position.

14. Install the water inlet with the 2 nuts. Tighten to 7 ft. lbs. (10 Nm).
15. connect the No. 2 radiator hose.
16. Install the alternator assembly.
17. Install the wire harness clamp bracket.
18. Install the v-ribbed belt.
19. Connect the cable to the negative battery terminal.
20. Add engine coolant.
21. Inspect for coolant leaks.
22. Install the cool air intake duct seal.
23. Install the No. 1 engine under cover.
24. Install the No. 2 engine under cover.
25. Install the No. 1 engine cover sub assembly.

3.5L Engine

See Figure 34.

1. Remove the front wheel RH.
2. Remove the No. 1 engine under cover.
3. Remove the No. 2 engine under cover.
4. Separate the front fender liner RH.
5. Remove the front fender apron RH.
6. Drain the engine coolant.
7. Remove the v-ribbed belt.
8. Remove the v-bank cover sub assembly.
9. Disconnect the No. 2 radiator hose.
   a. Disconnect the No. 2 radiator hose from the water inlet.
10. Remove the No. 2 idler pulley sub assembly.
11. Remove the water inlet.

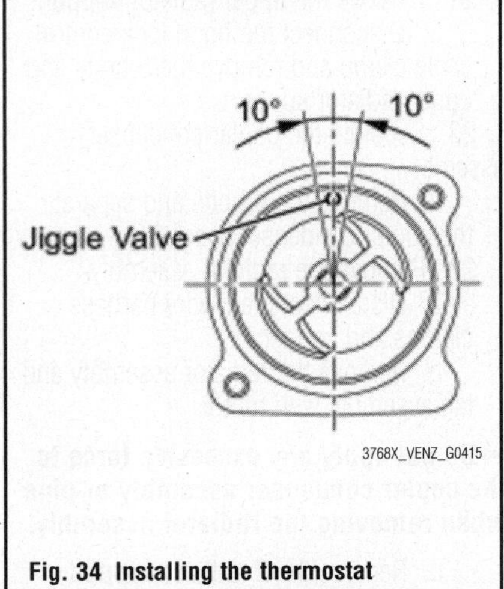

Fig. 34 Installing the thermostat

   a. Remove the 2 nuts and water inlet.
12. Remove the thermostat.
   a. Remove the thermostat from the water inlet housing.
   b. Remove the gasket from the thermostat.

To install:

13. Install the thermostat.
   a. Install a new gasket to the thermostat.
   b. Install the thermostat with the jiggle valve facing up.

➡The jiggle valve may be set within 10°on either side of the prescribed position.

14. Install the water inlet with the 2 nuts. Tighten to 7 ft. lbs. (10 Nm).
15. Install the No. 2 idler pulley sub assembly.
16. Connect the No. 2 radiator hose to the water inlet.
17. Install the v-ribbed belt.
18. Add engine coolant.
19. Inspect for coolant leaks.
20. Install the v-bank cover sub assembly.
21. Install the front fender apron RH.
22. Install the front fender liner RH.
23. Install the No. 2 engine under cover.
24. Install the No. 1 engine under cover.
25. Install the front wheel RH.

WATER PUMP

REMOVAL & INSTALLATION

2.7L Engine

See Figure 35.

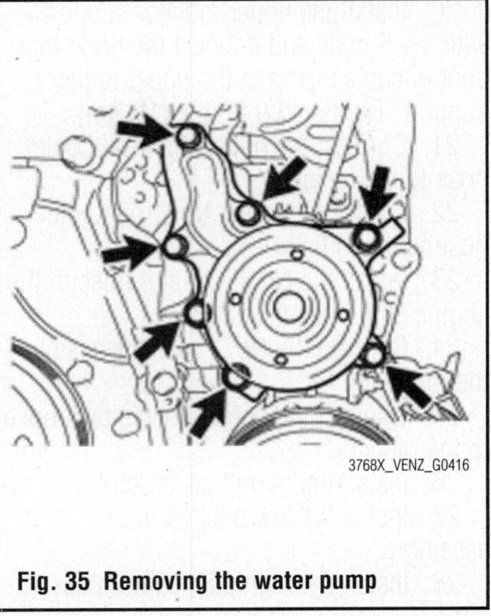

Fig. 35 Removing the water pump

1. Disconnect the cable from the negative battery terminal.

➡When disconnecting the cable, some systems need to be initialized after the cable is reconnected.

2. Remove the No. 1 engine under cover sub assembly.
3. Remove the cool air intake duct seal.
4. Remove the No. 1 engine under cover.
5. Remove the No. 2 engine under cover.
6. Remove the v-ribbed belt.
7. Remove the wire harness clamp bracket.
8. Remove the alternator assembly.
9. Remove the v-ribbed belt tensioner assembly.
10. Remove the water pump assembly.
    a. Remove the 7 bolts, water pump and water pump gasket.

To install:

11. To install, reverse the removal procedure. Tighten the bolts to 15 ft. lbs. (21 Nm).

3.5L Engine

See Figure 36.

1. Remove the automatic transaxle assembly.
2. Install the engine on an engine stand.
3. Remove the engine hanger.
4. Remove the front No. 1 engine mounting bracket LH.
5. Remove the No. 2 idler pulley sub assembly.
6. Remove the v-ribbed belt tensioner assembly.

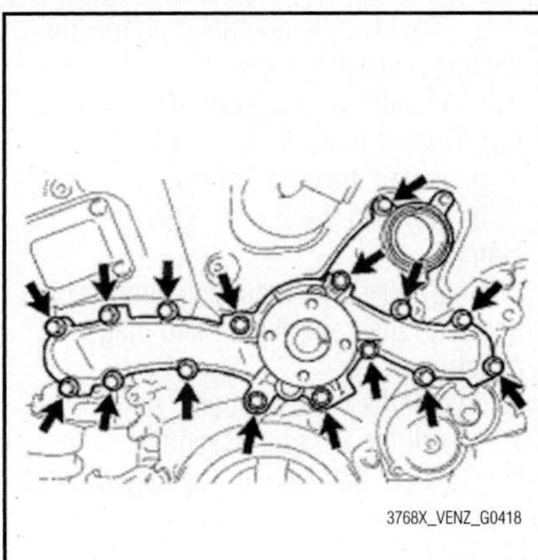

Fig. 36 Removing the water pump assembly

7. Remove the water inlet housing.
8. Remove the water pump pulley.
   a. Using the special tool (09960-10010), hold the water pump pulley.
   b. Remove the 4 bolts and water pump pulley.
9. Remove the water pump assembly.
   a. Remove the 16 bolts, water pump assembly and water pump gasket.

To install:

10. Install the water pump assembly.
   a. Install a new water pump gasket and the water pump assembly with the 16 bolts. Tighten bolt A to 15 ft. lbs. (21 Nm). Tighten bolts B and C to 8 ft. lbs. (11 Nm).
11. Install the water pump pulley.
   a. Temporarily install the water pump pulley with the 4 bolts.
   b. Using the special tool, hold the water pump pulley.
   c. Tighten the 4 bolts to 15 ft. lbs. (21 Nm).
12. Install the water inlet housing.
13. Install the v-ribbed belt tensioner assembly.
14. Install the No. 2 idler pulley sub assembly.
15. Install the front No. 1 engine mounting bracket LH.
16. Install the engine hanger.
17. Remove the engine from the engine stand.
18. Install the automatic transaxle assembly.

ENGINE ELECTRICAL

ALTERNATOR

REMOVAL & INSTALLATION

2.7L Engine

See Figures 37 and 38.

1. Disconnect the cable from the negative battery terminal.

➡When disconnecting the cable, some systems need to be initialized after the cable is reconnected.

2. Remove the cool air intake duct seal.
3. Remove the No. 1 engine cover sub assembly.
4. Remove the v-ribbed belt.
5. Remove the wire harness clamp bracket.
   a. Detach the wire harness clamp from the clamp bracket.
   b. Remove the bolt and clamp bracket.
6. Remove the alternator assembly.
   a. Disconnect the alternator connector.
   b. Remove the terminal cap.
   c. Remove the nut and disconnect the alternator wire.
   d. Remove the bolt and wire harness clamp bracket.
   e. Remove the 2 bolts and the alternator.

To install:

7. Install the alternator assembly with the 2 bolts. Tighten to 38 ft. lbs. (52 Nm).
   a. Install the wire harness clamp bracket with the bolt. Tighten to 7 ft. lbs. (10 Nm).
   b. Connect the alternator wire with the nut. Tighten to 87 inch lbs. (9.8 Nm).
   c. Install the terminal cap.
   d. Connect the alternator connector.
8. Install the wire harness clamp bracket with the bolt. Tighten to 7 ft. lbs. (10 Nm).
   a. Attach the wire harness clamp to the clamp bracket.
9. Install the v-ribbed belt.
10. Install the No. 1 engine cover sub assembly.
11. Install the cool air intake duct seal.
12. Connect the cable to the negative battery terminal.

➡When disconnecting the cable, some systems need to be initialized after the cable is reconnected.

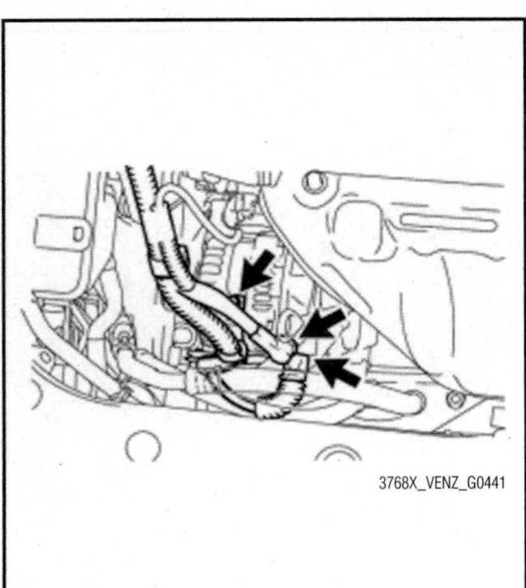

Fig. 37 Disconnecting the alternator

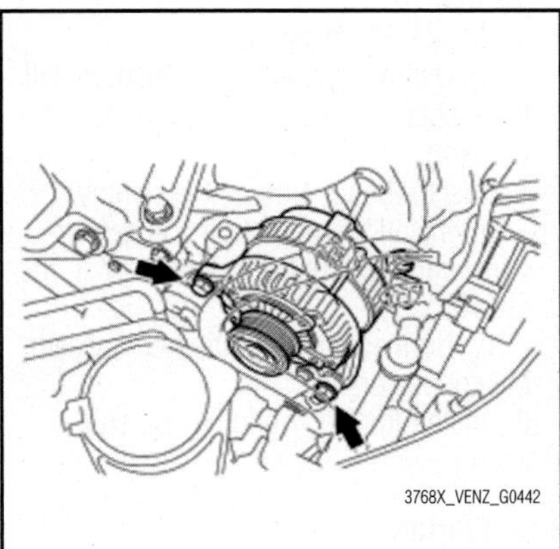

Fig. 38 Removing the alternator

3.5L Engine

See Figure 39.

1. Disconnect the cable from the negative battery terminal.

➡When disconnecting the cable, some systems need to be initialized after the cable is reconnected.

2. Remove the radiator assembly.
3. Remove the v-ribbed belt.
4. Remove the alternator assembly.
   a. Remove the terminal cap.
   b. Remove the nut and disconnect the wire harness from terminal B.
   c. Disconnect the alternator connector from the alternator assembly.
   d. Disconnect the connector from the compressor and magnetic clutch.
   e. Disconnect the 2 wire harness clamps.

f. Remove the 2 bolts.

g. Remove the bolt and alternator assembly.

h. Disconnect the wire harness clamp and remove the bolt and alternator bracket.

i. Remove the bolt and wire harness clamp.

To install:

5. Install the alternator assembly.

a. Install the wire harness clamp with the bolt. Tighten to 74 inch lbs. (8.4 Nm).

b. Install the alternator bracket with the 2 bolts. Tighten to 15 ft. lbs. (20 Nm).

c. Connect the wire harness clamp.

d. Install the 2 bolts. Tighten to 32 ft. lbs. (43 Nm).

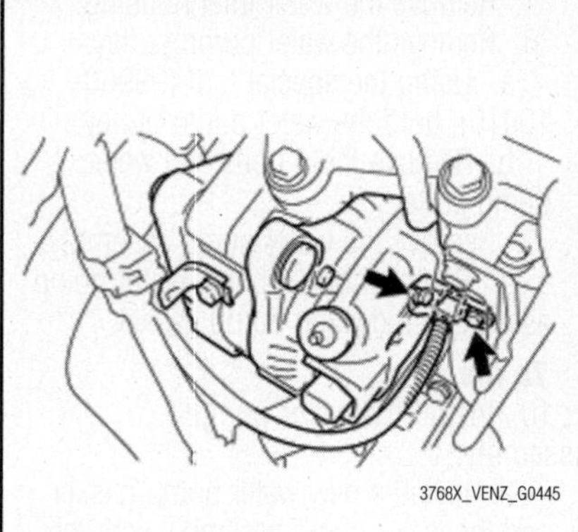

Fig. 39 Removing the bolt and alternator assembly

e. Connect the alternator connector to the alternator assembly.

f. Install the alternator wire with the nut. Tighten to 87 inch lbs. (9.8 Nm).

g. Install the terminal cap.

h. Connect the 2 wire harness clamps.

i. Connect the magnetic clutch connector to the compressor and magnetic clutch.

6. Install the v-ribbed belt.

7. Install the radiator assembly.

8. Connect the cable to the negative battery terminal.

➡When disconnecting the cable, some systems need to be initialized after the cable is reconnected.

ENGINE ELECTRICAL

IGNITION SYSTEM

FIRING ORDER

Firing order for the 2.7L engine: 1–3–4–2.

Firing order for 3.5L engine: 1–2–3–4–5–6

IGNITION COIL

REMOVAL & INSTALLATION

2.7L Engine

See Figures 40 and 41.

1. Remove the No. 1 engine cover sub assembly.

2. Remove the ignition coil assembly.

a. Disconnect the 4 ignition coil assembly connectors.

b. Remove the 4 bolts and 4 ignition coil assemblies.

3. Remove the spark plug.

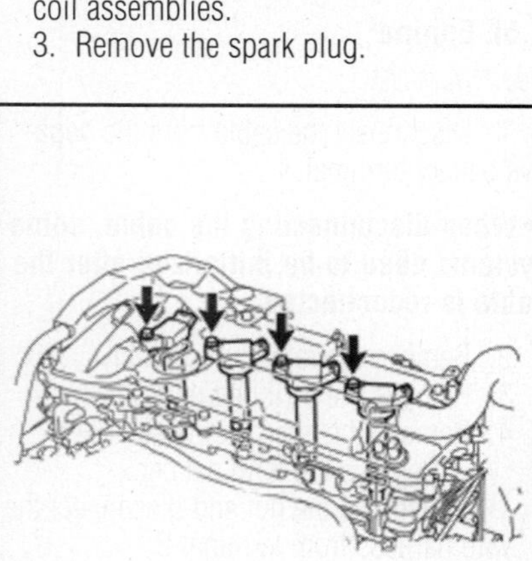

Fig. 40 Removing the ignition coil assembly

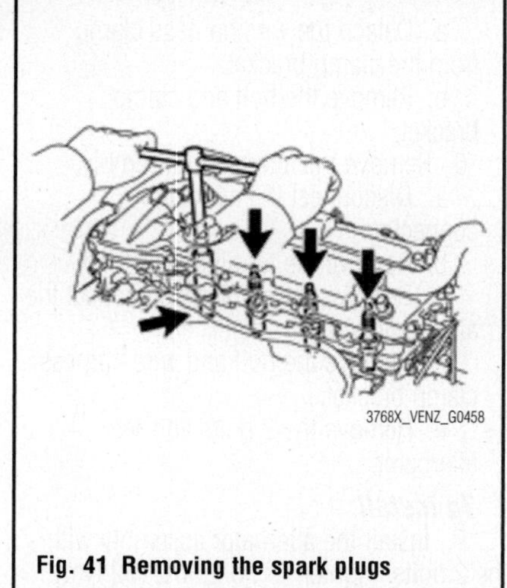

Fig. 41 Removing the spark plugs

a. Using a spark plug wrench, remove the 4 spark plugs.

To install:

4. Install the spark plug.

a. Using a spark plug wrench, install the 4 spark plugs. Tighten to 18 ft. lbs. (25 Nm).

5. Install the ignition coil assembly.

a. Install the 4 ignition coil assemblies with the 4 bolts. Tighten to 7 ft. lbs. (10 Nm).

b. Connect the 4 ignition coil assembly connectors.

6. Install the No. 1 engine coil sub assembly.

3.5L Engine

See Figure 42.

1. Remove the No. 1 engine under cover.

2. Remove the No. 2 engine under cover.

3. Drain the engine coolant.

4. Remove the windshield wiper motor and link assembly.

5. Remove the cowl top panel outer sub assembly.

6. Remove the cool air intake duct seal.

7. Remove the v-bank cover sub assembly.

8. Remove the air cleaner cap with hose.

9. Remove the intake air surge tank assembly.

10. Remove the No. 1 surge tank stay.

11. Remove the ignition coil assembly.

a. Disconnect the 3 wire harness clamps.

b. Disconnect the 6 ignition coil connectors.

c. Remove the 6 bolts and 6 ignition coils.

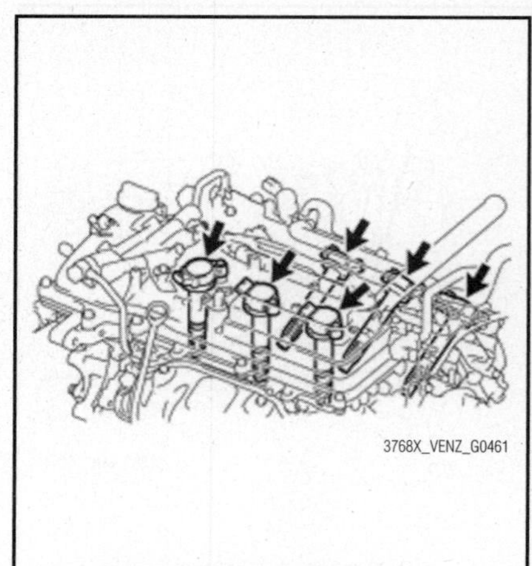

Fig. 42 Removing the 6 bolts and 6 ignition coils

12. Remove the spark plug.
    a. Remove the 6 spark plugs.

To install:

13. Install the spark plug.
    a. Install the 6 spark plugs and tighten to 13 ft. lbs. (18 Nm).
14. Install the ignition coil assembly.
    a. Install the 6 ignition coils with the 6 bolts. Tighten to 7 ft. lbs. (10 Nm).
    b. Connect the 6 ignition coil connectors.
    c. Connect the 3 wire harness clamps.
15. Install the No. 1 surge tank stay.
16. Install the intake air surge tank assembly.
17. Install the air cleaner cap with the hose.
18. Add engine coolant.
19. Inspect for engine coolant leaks.
20. Install the v-bank cover sub assembly.
21. Install the cool air intake duct seal.
22. Install the cowl top panel outer sub assembly.
23. Install the windshield wiper motor and link assembly.
24. Install the No. 2 engine under cover.
25. Install the No. 1 engine under cover.

IGNITION TIMING

INSPECTION

3.5L Engine

1. Warm up the engine.
2. Using SST: 09843-18040, connect terminals 13 (TC) and 4 (CG) of the DLC3.

WARNING

Confirm the terminal numbers before connecting them. Connecting the wrong terminals can damage the engine.

➡Turn off all electrical systems before connecting the terminals.

➡Perform this inspection after the cooling fan motor is turned off.

3. Remove the v-bank cover.
4. Pull out the red lead wire harness.
5. Connect the tester terminal of the timing light to the red lead wire.

➡Use a timing light which can detect the first signal.

6. Check the ignition timing at idle. Standard ignition timing: 8 to 12°BTDC at idle.

➡When checking the ignition timing, the transmission should be in the neutral position.

➡Run the engine at 1000 to 1300 rpm for 5 seconds, and then check that the engine rpm returns to idle speed.

7. Disconnect terminals 13 (TC) and 4 (CG) of the DLC3.
8. Check the ignition timing at idle. Standard ignition timing: 7 to 24°BTDC at idle.
9. Confirm that the ignition timing advances immediately when the engine rpm is increased.
10. Remove the timing light from the engine.

ADJUSTMENT

All engines are equipped with a Distributorless Ignition System (DIS). No timing adjustment is possible.

ENGINE ELECTRICAL

STARTING SYSTEM

STARTER

REMOVAL & INSTALLATION

2.7L Engine

See Figure 43.

1. Disconnect the cable from the negative battery terminal.

➡When disconnecting the cable, some systems need to be initialized after the cable is reconnected.

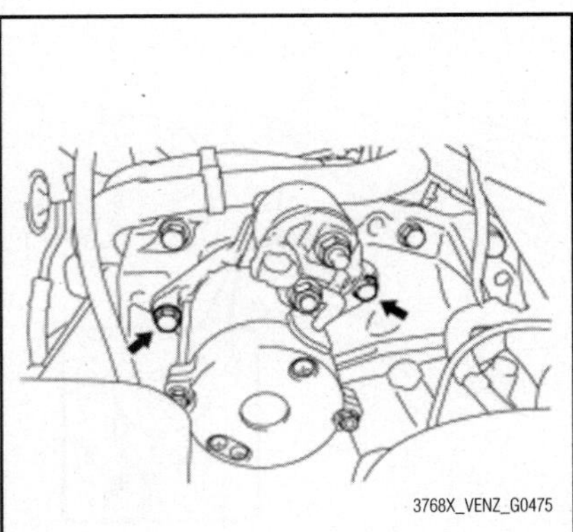

Fig. 43 Removing the starter— 2.7L engine

2. Remove the cool air intake duct seal.
3. Remove the No. 1 engine cover sub assembly.
4. Remove the battery.
5. Remove the inlet air cleaner assembly.
6. Remove the starter assembly.
    a. Disconnect the starter connector.
    b. Open the terminal cap, remove the nut and disconnect the starter wire.
    c. Remove the 2 bolts and starter.

To install:

7. Install the starter.
    a. Install the starter with the bolts. Tighten to 27 ft. lbs. (37 Nm).
    b. Connect the starter connector.
    c. Install the terminal nut and cover the nut with the cap. Tighten to 87 inch lbs. (9.8 Nm).
8. Install the battery.
9. Install the No. 1 engine cover sub assembly.
10. Install the cool air intake duct seal.
11. Connect the cable to the negative battery terminal.

➡When disconnecting the cable, some systems need to be initialized after the cable is reconnected.

3.5L Engine

See Figure 44.

1. Remove the cool air intake duct seal.
2. Disconnect the cable from the negative battery terminal.

➡When disconnecting the cable, some systems need to be initialized after the cable is reconnected.

3. Remove the v-bank cover sub assembly.

Fig. 44 Removing the starter— 3.5L engine

4. Remove the inlet No. 2 air cleaner.
5. Remove the battery.
6. Remove the inlet No. 1 air cleaner.
7. Remove the starter assembly.
   a. Disconnect the starter connector.
   b. Remove the terminal cap.
   c. Remove the nut and disconnect the starter wire.
   d. Remove the 2 bolts and starter.

To install:

8. Install the starter assembly.
   a. Install the starter with the 2 bolts. Tighten to 27 ft. lbs. (37 Nm).
   b. Connect the starter connector.
   c. Install the terminal nut and cover the nut with the cap. Tighten to 87 inch lbs. (9.8 Nm).
9. Install the inlet No. 1 air cleaner.
10. Install the battery.
11. Install the inlet No. 2 air cleaner.
12. Install the v-bank cover sub assembly.
13. Connect the cable to the negative battery terminal.
14. Install the cool air intake duct seal.

ENGINE MECHANICAL

ACCESSORY DRIVE BELTS

INSPECTION

See Figure 45.

1. Check the belt for wear, cracks or other signs of damage.
2. If any of the following defects is found; replace the V-ribbed belt:
   a. The belt is cracked.
   b. The belt is worn out to the extent that the cords are exposed.
   c. The belt has chunks missing from the ribs.
3. Check that the belt fits properly in the ribbed grooves.

➡Check with your hand to confirm that the belt has not slipped out of the groove on the bottom of the pulley. If it has slipped out, replace the V-ribbed belt. Install a new V-ribbed belt correctly.

REMOVAL & INSTALLATION

2.7L Engine

See Figure 46.

1. Remove the front wheel.
2. Remove the No. 1 engine under cover.
3. Separate the front fender liner RH.
4. Remove the front fender apron seal RH.

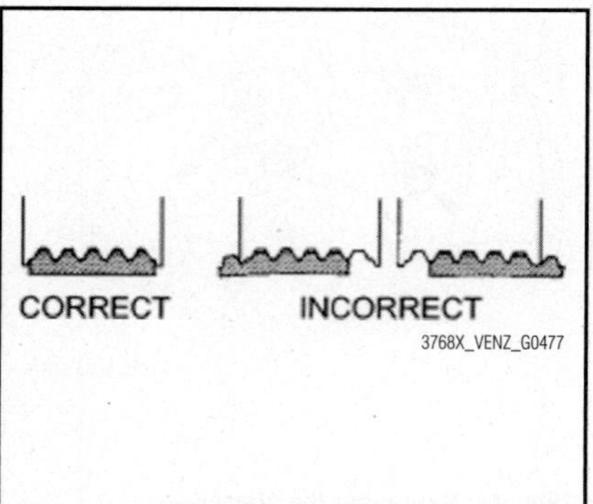

Fig. 45 Checking the belt fit

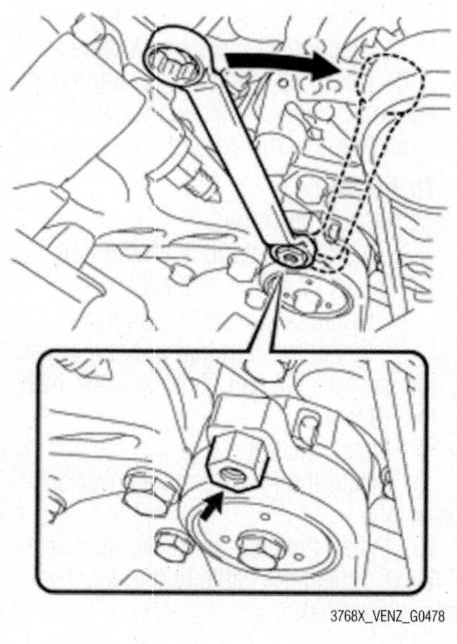

5. Remove the v-ribbed belt.
   a. Attach a wrench to the hexagonal portion of the belt tensioner as shown in the illustration, rotate the belt tensioner clockwise, and remove the V-ribbed belt.

To install:

6. Install the v-ribbed belt.
   a. Set the V-ribbed belt onto each part as shown in the illustration except the water pump pulley.
   b. Loosen the V-ribbed belt by turning the belt tensioner clockwise.
   c. Set the V-ribbed belt onto the water pump pulley.

➡Make sure that the belt is attached to each pulley. In particular, make sure that the belt is securely fitted into the grooves of the crankshaft pulley.

7. Install the front fender apron seal RH.

3768X_VENZ_G0478

Fig. 46 Removing the drive belt—2.7L engine

8. Install the front fender liner RH.
9. Install the No. 1 engine under cover.
10. Install the front wheel RH. Tighten to 76 ft. lbs. (103 Nm).

3.5L Engine

See Figures 47 and 48.

1. Remove the front wheel RH.
2. Remove the No. 1 engine under cover.
3. Separate the front fender liner RH.
4. Remove the front fender apron RH.
5. Remove the v-ribbed belt.
   a. Release the V-ribbed belt tension by turning the V-ribbed belt tensioner counterclockwise, and remove the V-ribbed belt from the V-ribbed belt tensioner.
   b. While turning the V-ribbed belt tensioner counterclockwise, aligns with its holes, and then insert a 5 mm hexagon wrench into the holes to fix the V-ribbed belt tensioner.

To install:

6. Install the v-ribbed belt.
   a. Turn the V-ribbed belt tensioner counterclockwise and remove the 5 mm hexagon wrench.

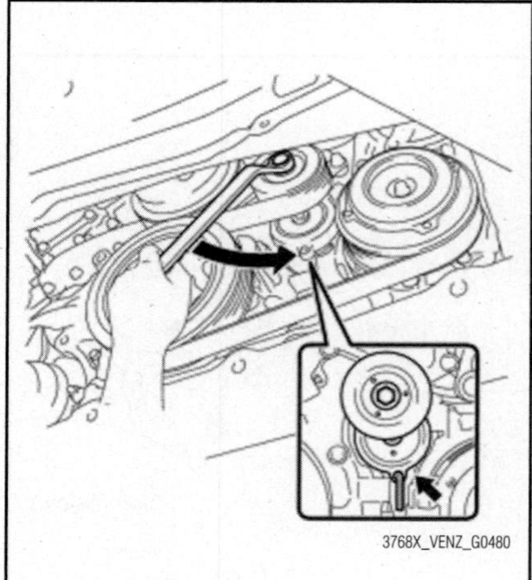

Fig. 47 Releasing the v-ribbed belt tension

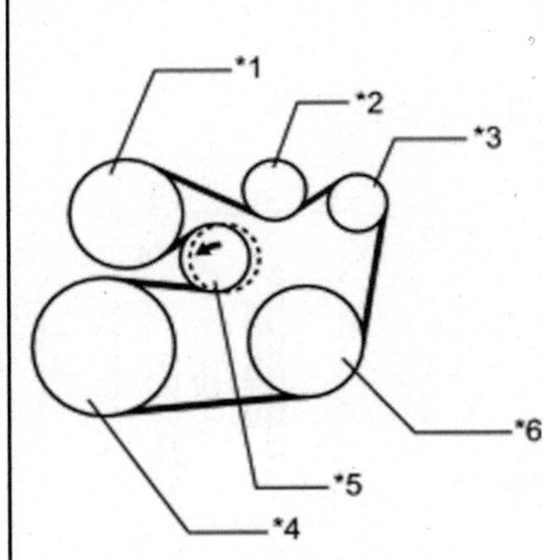

Fig. 48 Belt routing—3.5L engine

b. After installing the V-ribbed belt, check that it fits properly in the ribbed grooves. Confirm that the belt has not slipped out of the grooves on the bottom of the crank pulley by hand.

7. Install the front fender apron RH.
8. Install the front fender liner RH.
9. Install the No. 1 engine under cover.
10. Install the front wheel RH. Tighten to 76 ft. lbs. (103 Nm).

BALANCE SHAFT

REMOVAL & INSTALLATION

2.7L Engine

See Figures 49 through 60.

1. Remove the timing chain cover sub assembly.
2. Set the No. 1 cylinder to TDC/compression.

a. Temporarily install the crankshaft pulley bolt.

➡"A" is not a timing mark.

b. Rotate the crankshaft clockwise so that the timing marks on the crankshaft timing gear and camshaft timing gears are as shown in the illustration.

➡If the timing marks do not align, rotate the crankshaft clockwise again and align the timing marks.

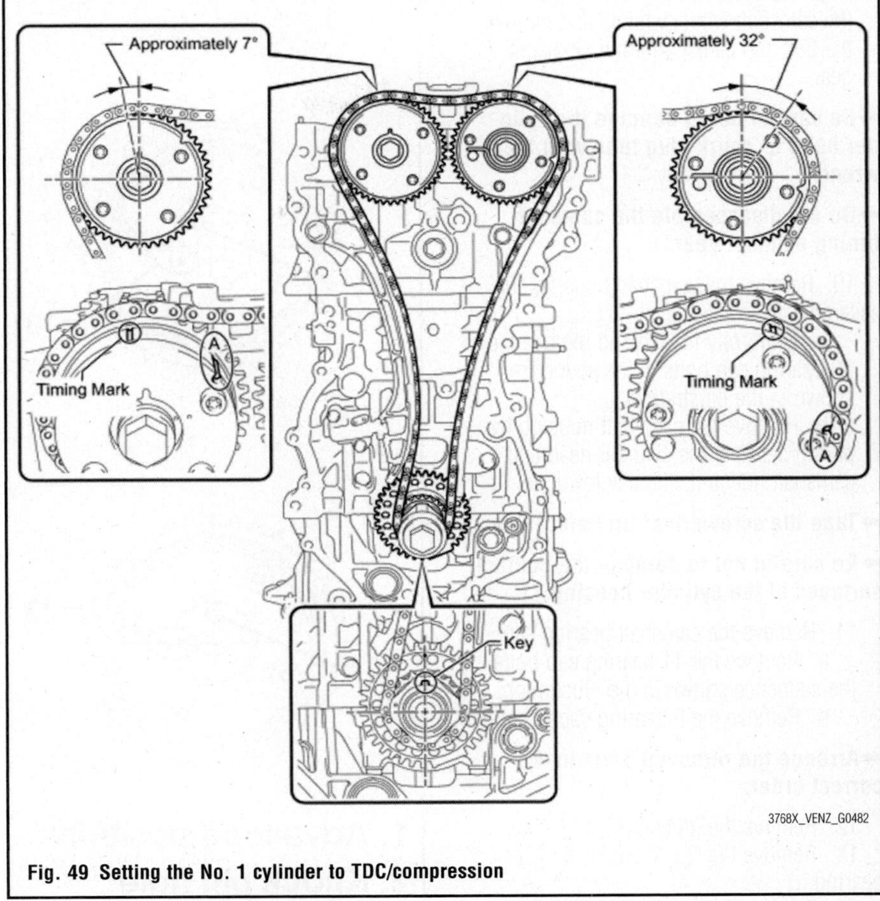

Fig. 49 Setting the No. 1 cylinder to TDC/compression

c. Remove the crankshaft pulley bolt.

3. Remove the timing chain guide.
4. Remove the No. 1 chain tensioner assembly.

a. Allow the plunger to extend slightly, and then rotate the stopper plate counterclockwise to release the lock. Once the lock is released, push the plunger into the tensioner.

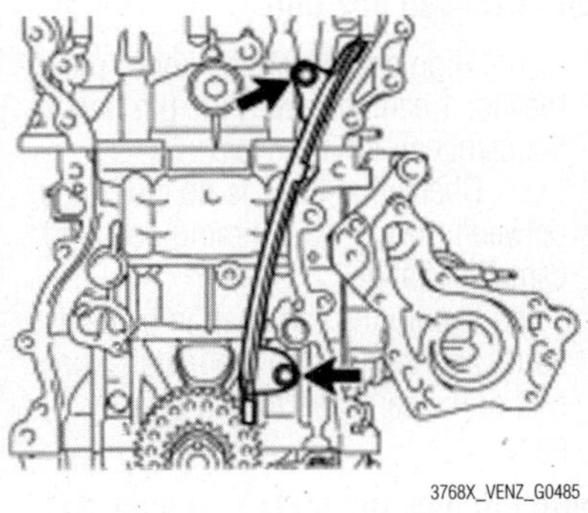

3768X_VENZ_G0485

Fig. 50 Removing the No. 1 chain vibration damper

b. Move the stopper plate clockwise to set the lock, and insert a pin (*1) into the stopper plate hole.

c. Remove the 2 bolts, chain tensioner and gasket.

5. Remove the chain tensioner slipper.

a. Remove the bolt and chain tensioner slipper.

6. Remove the chain sub assembly.
7. Remove the No. 1 chain vibration damper.

a. Remove the 2 bolts and chain vibration damper.

8. Remove the camshaft timing gear assembly.

a. Hold the hexagonal portion of the camshaft with a wrench and remove the bolt and camshaft timing gear.

➡Be careful not to damage the cylinder head or spark plug tube with the wrench.

➡Do not disassemble the camshaft timing gear.

9. Remove the camshaft timing exhaust gear assembly.

a. Hold the hexagonal portion of the camshaft with a wrench and remove the bolt and camshaft timing exhaust gear.

➡**Be careful not to damage the cylinder head or spark plug tube with the wrench.**

➡**Do not disassemble the camshaft timing exhaust gear.**

10. Remove the camshaft housing sub assembly.
   a. Uniformly loosen and remove the 20 bearing cap bolts in the sequence shown in the illustration.
   b. Remove the camshaft housing by prying between the cylinder head and camshaft housing with a screwdriver.

➡**Tape the screwdriver tip before use.**

➡**Be careful not to damage the contact surfaces of the cylinder housing.**

11. Remove the camshaft bearing cap.
   a. Remove the 11 bearing cap bolts in the sequence shown in the illustration.
   b. Remove the 5 bearing caps.

➡**Arrange the removed parts in the correct order.**

12. Remove the oil control valve filter.
13. Remove the No. 1 camshaft bearing.
14. Remove the camshaft.
   a. Remove the No. 1 and No. 2 camshafts.
15. Remove the No. 2 camshaft bearing.
16. Remove the No. 1 valve rocker arm sub assembly.
   a. Remove the 16 valve rocker arms from the cylinder head.

➡**Arrange the removed parts in the correct order.**

17. Remove the valve lash adjusters from the cylinder head.

➡**Arrange the removed parts in the correct order.**

To install:

18. Set the camshaft gear assembly.

➡**When installing the camshaft timing gear, release the lock pin and set the camshaft timing gear to the advanced position before installation.**

   a. Check the camshaft timing gear position.

➡**If the camshaft timing gear is set to the advanced position, do not let the camshaft timing gear rotate clockwise during installation.**

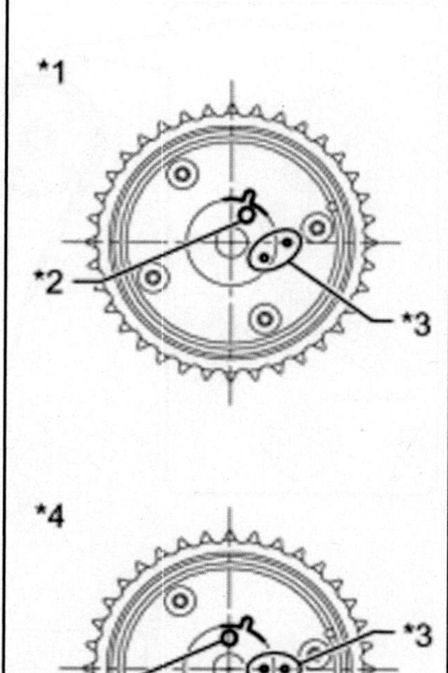

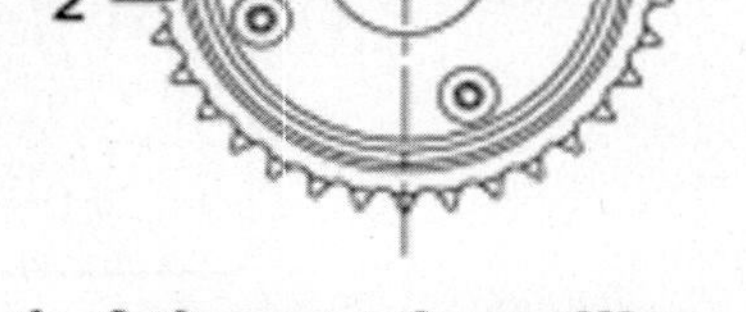

Fig. 51 Checking the camshaft timing gear position

➡**If the camshaft timing gear rotates to the retarded position, release the lock pin and set the camshaft timing gear to the advanced position.**

   b. Align and attach the knock pin of the No. 1 camshaft with the pin hole of the camshaft timing gear.
   c. Check that there is no clearance between the camshaft timing gear and camshaft flange.
   d. Secure the camshaft in place by hand, and then install the installation bolt of the camshaft timing gear by hand.

➡**Do not use any tools to install the bolt. If the bolt is installed using a tool, the lock pin will be damaged.**

   e. Release the lock pin.

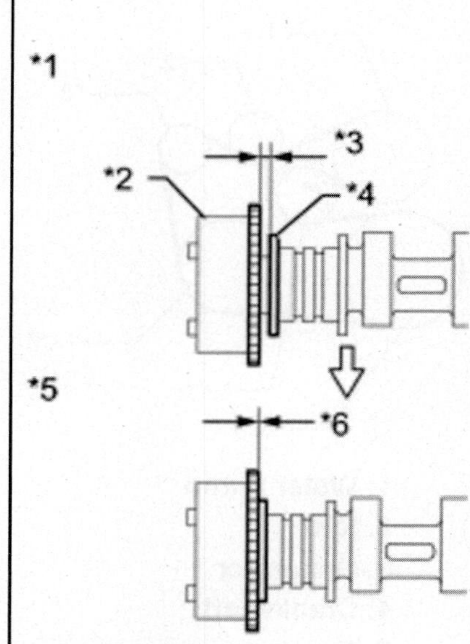

Fig. 52 Checking clearance between the camshaft timing gear and camshaft flange

   f. Clean the camshaft journal with non-residue solvent.
   g. Cover the 4 oil paths of the cam journal with vinyl tape as shown in the illustration.

➡**There are 4 oil paths in the grooves of the camshaft. Plug three of the paths with pieces of rubber.**

   h. Open a hole at port A shown in the illustration.
   i. While applying approximately 29 psi (200 kPa) of air pressure to the oil path, forcibly turn the camshaft timing gear assembly in the advance direction (counterclockwise).

⁂ CAUTION

Cover the paths with a piece of cloth when applying pressure to keep oil from splashing.

➡**Do not allow the camshaft timing gear assembly to lock. If it locks, release the lock pin again.**

➡The camshaft timing gear assembly may be turned in the advance direction without applying any force.

➡If enough air pressure cannot be applied because of air leakage from the port, releasing the lock pin may be difficult.

j. Remove the vinyl tape and rubber pieces from the camshaft.

k. Remove the bolt and camshaft timing gear.

➡Do not allow the camshaft timing gear assembly to lock. If it locks, release the lock pin again.

19. Install the valve lash adjuster assembly.

a. Inspect the valve lash adjuster before installing it.

b. Install the 16 lash adjusters to the cylinder head.

➡Install the lash adjuster to the same place it was removed from.

20. Install the No. 1 valve rocker arm sub assembly.

a. Apply engine oil to the lash adjuster tips and valve stem caps.

b. Install the 16 valve rocker arms as shown in the illustration.

21. Install the No. 2 camshaft bearing.
22. Install the No. 1 camshaft bearing.
23. Install the oil control valve filter.
24. Install the camshaft.

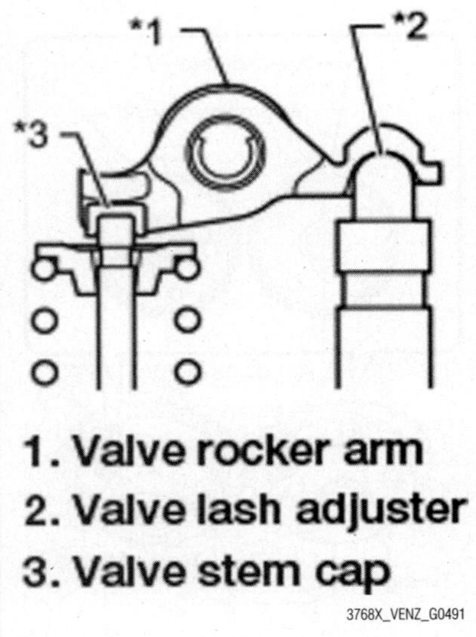

*1 *2 *3

1. Valve rocker arm
2. Valve lash adjuster
3. Valve stem cap

3768X_VENZ_G0491

Fig. 53 Installing the No. 1 valve rocker arm sub assembly

a. Clean the camshaft journals, camshaft housing and bearing caps.

b. Apply a light coat of engine oil to the camshaft journal, camshaft housing and bearing caps.

c. Install the No. 1 and No. 2 camshafts to the camshaft housing.

25. Install the camshaft bearing cap.

a. Confirm the marks and numbers on the camshaft bearing caps and place them in their proper positions and directions.

b. Install the 11 bolts in the order shown in the illustration. Tighten to 12 ft. lbs. (16 Nm).

➡Make sure that the camshaft rotates smoothly after installing the bearing caps.

26. Install the camshaft housing sub assembly.

a. Check that the valve rocker arms are installed as shown in the illustration.

b. Apply seal packing in a continuous line as shown in the illustration. Standard seal diameter is 0.118-0.157 inch (3-4 mm).

➡Remove any oil from the contact surface.

➡Install the camshaft housing within 3 minutes and tighten the bolts within 10 minutes after applying seal packing.

c. Position the knock pin of the No. 1 and No. 2 camshafts as shown in the illustration.

d. Install the camshaft housing, and then install the 20 bolts in the order shown in the illustration. Tighten to 20 ft. lbs. (27 Nm).

➡Do not apply oil for at least 4 hours after the installation.

➡Do not start the engine for at least 4 hours after the installation.

➡Thoroughly wipe clean any seal packing.

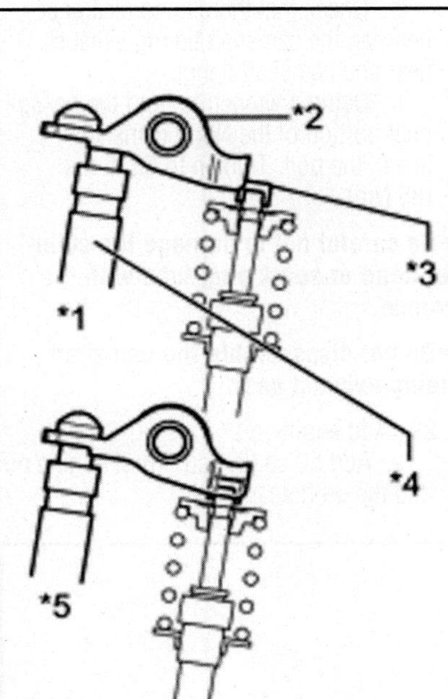

Fig. 54 Checking the valve rocker arm installation

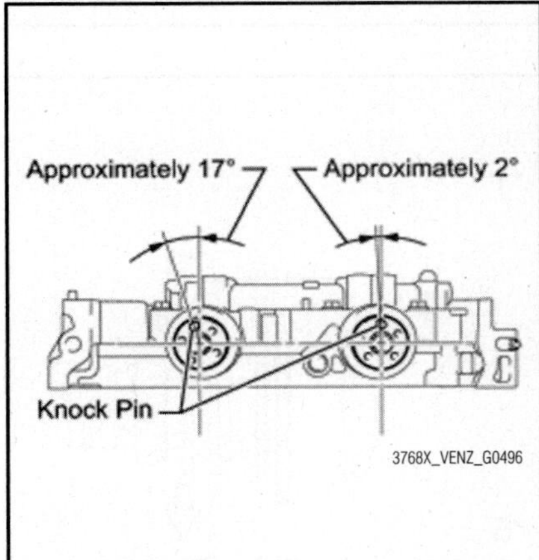

Fig. 55 Positioning the knock pin of the No. 1 and No. 2 camshafts

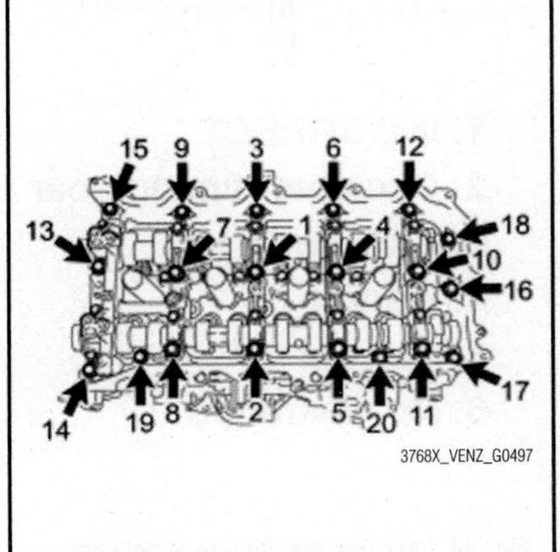

Fig. 56 Installing the camshaft housing

27. Install the camshaft timing gear assembly.

a. Check the camshaft timing gear position.

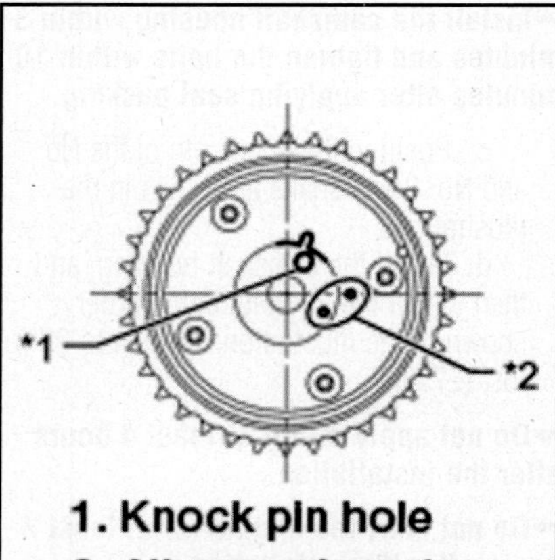

Fig. 57 Checking the timing gear position

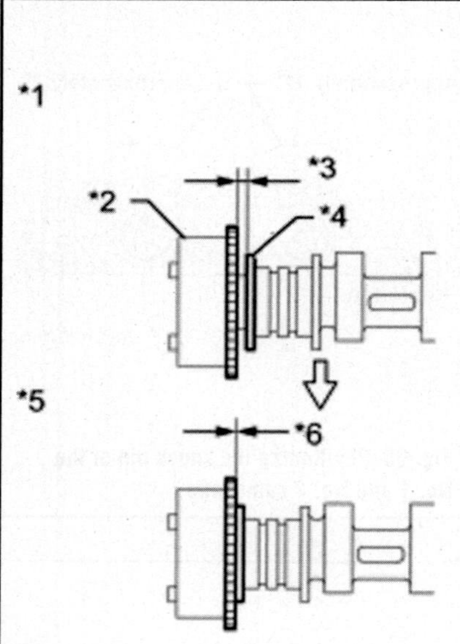

Fig. 58 Checking the clearance between the camshaft timing gear and camshaft flange

➡**If the camshaft timing gear is not set to the advanced position, release the lock pin and reset the camshaft timing gear (Refer to the "Set Camshaft Timing Gear Assembly" procedure).**

b. Align and attach the knock pin of the No. 1 camshaft with the pin hole of the camshaft timing gear.

c. Check that there is no clearance between the camshaft timing gear and camshaft flange.

d. Using a wrench to hold the hexagonal portion of the No. 1 camshaft, install the bolt. Tighten to 63 ft. lbs. (85 Nm).

➡**Be careful not to damage the cylinder head or spark plug tube with the wrench.**

➡**Do not disassemble the camshaft timing gear.**

28. Install the camshaft timing exhaust gear assembly.

a. Align and attach the knock pin of the No. 2 camshaft with the pin hole of the camshaft timing exhaust gear.

b. Check that there is no clearance between the camshaft timing exhaust gear and camshaft flange.

c. Using a wrench to hold the hexagonal portion of the No. 2 camshaft, install the bolt. Tighten to 63 ft. lbs. (85 Nm).

➡**Be careful not to damage the cylinder head or spark plug tube with the wrench.**

➡**Do not disassemble the camshaft timing exhaust gear.**

29. Add engine oil.

a. Add 50 cc (3.1 cu. in) of engine oil into the oil hole.

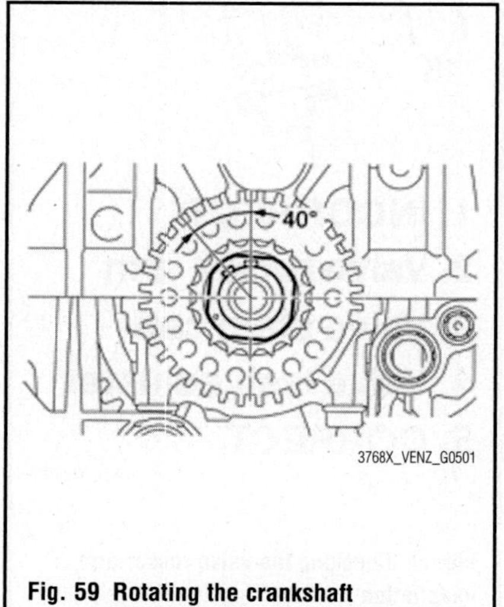

Fig. 59 Rotating the crankshaft

➡**Oil must be added if the lash adjusters were removed.**

➡**Make sure that the low pressure chamber and oil paths of the lash adjusters are full of engine oil.**

30. Set the No. 1 cylinder to TDC/compression.

a. Temporarily install the crankshaft pulley bolt.

b. Rotate the crankshaft 40°counterclockwise to position the crankshaft pulley key as shown in the illustration.

c. Check that the timing marks of the camshaft timing gears are as shown in the illustration.

➡**"A" is not a timing mark.**

31. Install the No. 1 chain vibration damper with the 2 bolts. Tighten to 15 ft. lbs. (21 Nm).

32. Install the chain sub assembly.

a. Place the chain onto the camshaft timing gears and crankshaft timing sprocket.

➡**Make sure the mark plate of the chain faces away from the engine.**

➡**It is not necessary to install the chain to the teeth of the gears and sprocket.**

b. Align the mark plate (yellow or gold) of the chain with the timing mark of the camshaft timing exhaust gear and install the chain to the camshaft timing exhaust gear.

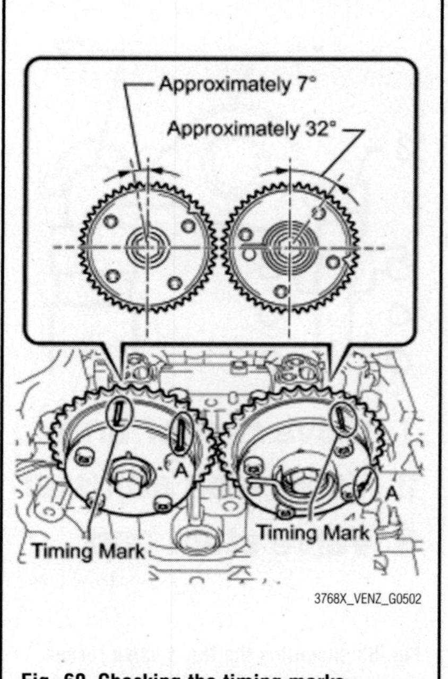

Fig. 60 Checking the timing marks

c. Align the mark plate (pink or gold) of the chain with the timing mark of the crankshaft timing sprocket and install the chain to the crankshaft timing sprocket.

d. Tie a string above the crankshaft timing sprocket so that the chain is secure.

e. Using the hexagonal portion of the intake camshaft, rotate the intake camshaft counterclockwise with a wrench, align the timing mark of the camshaft timing gear with the mark plate (yellow or gold) of the chain and install the chain to the camshaft timing gear.

➡**Hold the intake camshaft in place with a wrench until the chain tensioner is installed.**

f. Remove the string above the crankshaft timing sprocket, rotate the crankshaft clockwise, and loosen the chain so that the chain tensioner slipper can be installed.

➡**Make sure the chain is secure.**

33. Install the chain tensioner slipper.

a. Install the chain tensioner slipper with the bolt. Tighten to 15 ft. lbs. (21 Nm).

34. Install the No. 1chain tensioner assembly with the 2 bolts. Tighten to 7 ft. lbs. (10 Nm).

a. Remove the pin from the stopper plate.

35. Install the timing chain guide with the bolt. Tighten to 15 ft. lbs. (21 Nm).

36. Check the No. 1 cylinder to TDC/compression.

a. Temporarily install the crankshaft pulley bolt.

b. Rotate the crankshaft clockwise, and check that the timing marks on the crankshaft timing sprocket and camshaft timing gears are as shown in the illustration.

➡**"A" is not a timing mark.**

c. Remove the crankshaft pulley bolt.

37. Install the timing chain cover sub assembly.

CRANKSHAFT FRONT SEAL

REMOVAL & INSTALLATION

See Figures 61 and 62.

1. Remove the front wheel RH.
2. Remove the No. 1 engine under cover.
3. Separate the front fender liner RH.
4. Remove the front fender apron seal RH.
5. Remove the v-ribbed belt.
6. Remove the crankshaft pulley.

a. Using the special tool (09213-54015, 09330-00021 or 91551-80650), hold the crankshaft pulley and loosen the pulley bolt. Further loosen the bolt until 2 or 3 threads are screwed into the crankshaft.

b. Using SST (09950-50010, 09951-05010, 09952-05010, 09953-05020 or 09954-05011).

➡**Apply a lubricant to the threads and end of the SST.**

7. Remove the timing chain cover oil seal.

a. For 3.5L engine, using a knife, cut off the timing chain case oil seal lip.

b. Using a screwdriver, pry out the oil seal.

➡**Tape the screwdriver tip before use.**

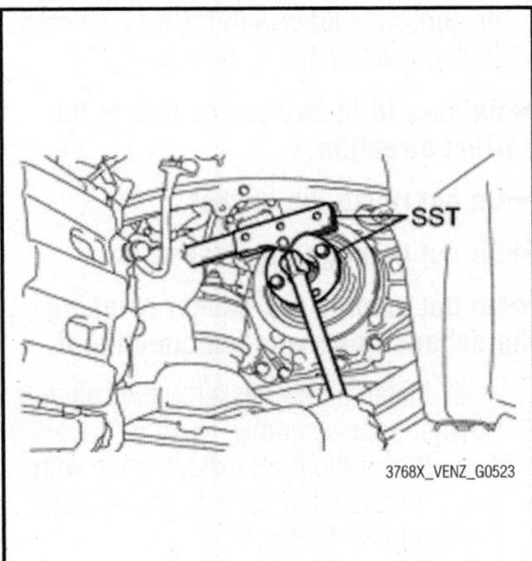

Fig. 61 Removing the crankshaft pulley

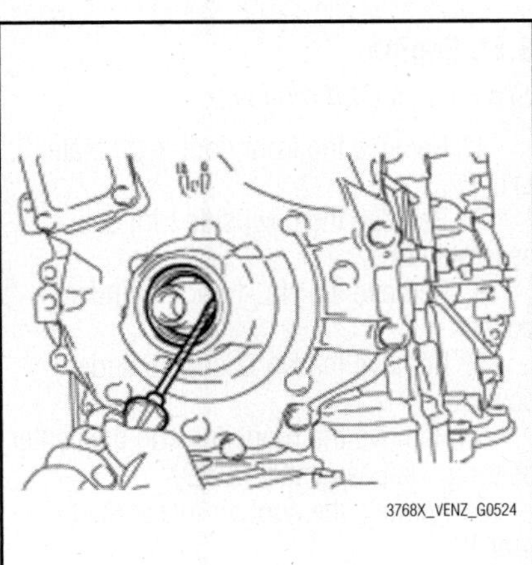

Fig. 62 Removing the oil seal

➡**Do not damage the surface of the seal press fit hole or the crankshaft.**

To install:

8. Install the timing chain cover oil seal.

a. Apply MP grease to the lip of a new oil seal.

➡**Do not allow foreign matter to contact the lip of the oil seal.**

➡**Do not allow MP grease to contact the dust seal.**

b. Using SST and a hammer, tap in the oil seal until its surface is flush with the timing chain cover edge.

➡**Keep the lip of the oil seal free from foreign matter.**

➡**Do not tap in the oil seal at an angle.**

9. Install the crankshaft pulley.

a. Align the pulley set key with the key groove of the crankshaft pulley.

b. Using SST, hold the crankshaft pulley and install the pulley bolt. Tighten to 192 ft. lbs. (260 Nm).

10. Install the v-ribbed belt.
11. Install the front fender apron seal RH.
12. Install the front fender liner RH.
13. Install the No. 1 engine under cover.
14. Install the front wheel RH. Tighten to 76 ft. lbs. (103 Nm).

EXHAUST MANIFOLD

REMOVAL & INSTALLATION

2.7L Engine

See Figure 63.

➡**Wear protective gloves when removing the exhaust pipe.**

➡**The exhaust pipe is extremely hot immediately after the engine has stopped.**

➡**Confirm that the exhaust pipe has cooled down before removing it.**

1. Remove the No. 1 engine under cover.
2. Remove the No. 2 engine under cover.
3. Remove the front exhaust pipe assembly.

a. Disconnect the clamp and connector.

b. Remove the 4 bolts, 2 compression springs and front exhaust pipe assembly.

c. Remove the 2 gaskets from the exhaust manifold converter sub-assembly and center exhaust pipe assembly.

4. Remove the air fuel ration sensor (bank 1 sensor 1).

5. Remove the manifold stay.

a. Remove the bolt, nut and manifold stay.

6. Remove the No. 2 manifold stay.

a. Remove the bolt, nut and manifold stay.

7. Remove the No. 1 exhaust manifold heat insulator.

a. Remove the 4 bolts and No. 1 exhaust manifold heat insulator.

8. Remove the exhaust manifold converter sub assembly.

a. Remove the 5 nuts and exhaust manifold converter sub-assembly.

9. Remove the No. 2 exhaust manifold heat insulator.

a. Remove the 2 bolts and No. 2 exhaust manifold heat insulator.

10. Remove the No. 1 manifold converter insulator.

a. Remove the 4 bolts and No. 1 manifold converter insulator.

To install:

11. Install the No. 1 manifold converter insulator with the 4 bolts. Tighten to 9 ft. lbs. (12 Nm).

12. Install the No. 2 exhaust manifold heat insulator with the 2 bolts. Tighten to 9 ft. lbs. (12 Nm).

13. Install the exhaust manifold converter sub assembly.

a. Install a new gasket onto the cylinder head.

b. Temporarily install the exhaust manifold converter sub-assembly with the 5 nuts.

c. Tighten the 5 nuts in the sequence shown in the illustration. Tighten to 26 ft. lbs. (35 Nm).

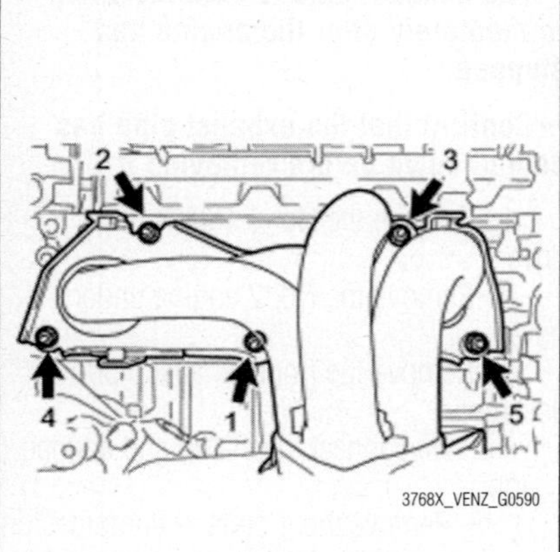

Fig. 63 Installing the exhaust manifold converter sub assembly

14. Install the No. 1 exhaust manifold heat insulator with the 4 bolts. Tighten to 9 ft. lbs. (12 Nm).

15. Install the No. 2 manifold stay with the bolt and nut. Tighten to 32 ft. lbs. (43 Nm).

16. Install the manifold stay with the bolt and nut. Tighten to 32 ft. lbs. (43 Nm).

17. Install the air fuel ratio sensor (bank 1 sensor 1).

18. Install the front exhaust pipe assembly.

a. Using a vernier caliper, measure the free length of the compression spring. The minimum length is 1.64 inch (41.5 mm).

➡If the length is less than the minimum, replace the compression spring.

b. Fully insert a new gasket to the exhaust manifold converter sub assembly.

c. Using a plastic hammer and wooden block, tap in the new gasket until its surface is flush with the exhaust manifold converter sub assembly.

➡Be sure to install the gasket in the correct direction.

➡Do not reuse the gasket.

➡Do not damage the gasket.

➡Do not push in the gasket by using the exhaust pipe when connecting it.

d. Install a new gasket to the center exhaust pipe assembly.

e. Install the front exhaust pipe with the 2 compression springs and 4 bolts. Tighten to 32 ft. lbs. (43 Nm).

f. Connect the clamp and connector.

19. Inspect for exhaust gas leaks.

20. Install the No. 2 engine under cover.

21. Install the No. 1 engine under cover.

3.5L Engine

See Figures 64 through 69.

1. Remove the front door scuff plate LH.

2. Remove the cowl side trim sub assembly LH.

3. Remove the No. 1 engine under cover.

4. Remove the No. 2 engine under cover.

5. Remove the propeller with the center bearing shaft assembly (AWD).

6. Remove the cool air intake duct seal.

7. Remove the inlet No. 2 air cleaner.

8. Remove the front No. 3 exhaust pipe sub assembly.

a. Disconnect the 2 clamps and oxygen sensor connector (bank 1 sensor 2).

b. Remove the grommet.

c. Remove the 4 bolts, 2 nuts, 2 compression springs and front No. 3 exhaust pipe sub assembly.

d. Remove the 3 gaskets from the front No. 3 exhaust pipe sub assembly.

9. Remove the manifold stay.

a. Remove the bolt, nut and manifold stay.

10. Remove the exhaust manifold sub assembly RH.

a. Disconnect the air fuel ration sensor connector (bank 1 sensor 1).

b. Using a 12 mm deep socket wrench, remove the 6 nuts and exhaust manifold sub assembly RH.

11. Remove the exhaust manifold to head gasket from the cylinder head sub assembly.

12. Remove the air fuel ratio sensor (bank 1 sensor 1).

13. Remove the front exhaust pipe assembly.

14. Remove the No. 2 manifold stay.

a. Remove the bolt, nut and No. 2 manifold stay.

15. Remove the No. 2 oil level dipstick guide.

a. Remove the oil level dipstick.

b. Remove the bolt and No. 2 engine oil level dipstick guide.

c. Remove the o-ring from the No. 2 engine oil level dipstick guide.

16. Remove the air fuel ratio sensor (bank 2 sensor 1).

17. Remove the No. 2 exhaust manifold heat insulator.

18. Remove the exhaust manifold sub assembly LH.

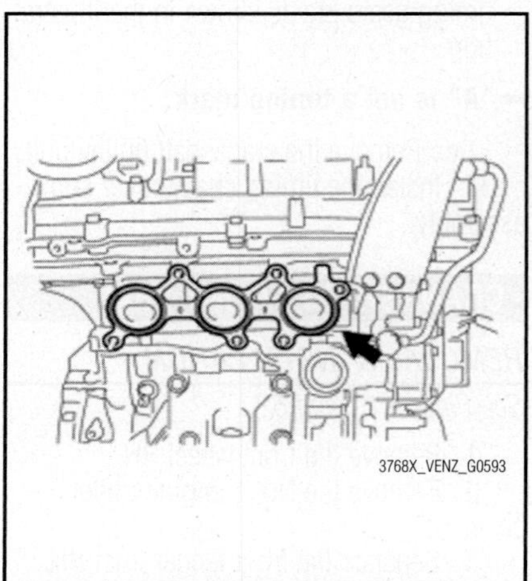

Fig. 64 Removing the exhaust manifold to head gasket

a. Using a 12 mm deep socket wrench, remove the 6 nuts and exhaust manifold sub-assembly LH.

19. Remove the exhaust manifold to head gasket LH.

a. Remove the exhaust manifold to head gasket LH from the cylinder head sub assembly.

To install:

20. Install the exhaust manifold to heat gasket LH.

21. Install the exhaust manifold sub assembly LH.

a. Using a 12 mm deep socket wrench, install the exhaust manifold sub-assembly LH and 6 nuts in the order shown in the illustration. Tighten to 15 ft. lbs. (21 Nm).

22. Install the No. 2 exhaust manifold heat insulator with the 3 bolts. Tighten to 75 inch lbs. (8.5 Nm).

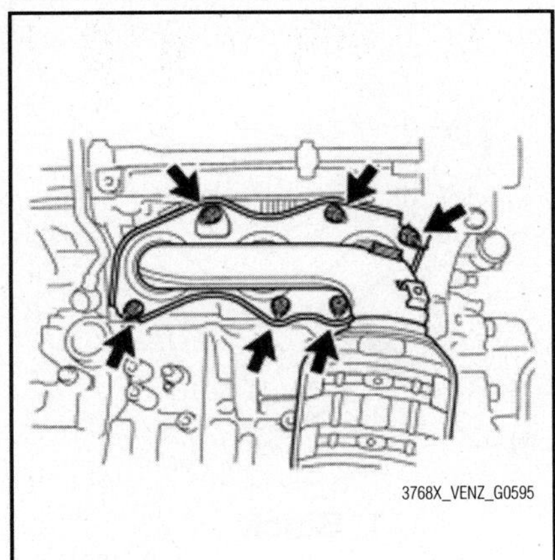

Fig. 65 Removing the exhaust manifold sub assembly LH

Fig. 66 Removing the exhaust manifold to head gasket LH

23. Install the No. 2 oil level dipstick guide.

a. Apply a light coat of engine oil to the o-ring.

b. Push in the No. 2 oil level dipstick guide end into the oil level dipstick guide.

c. Install the No. 2 oil level dipstick guide with the bolt. Tighten to 15 ft. lbs. (21 Nm).

d. Install the oil level dipstick.

24. Install the air fuel ratio sensor (bank 2 sensor 1).

25. Install the No. 2 manifold stay by tightening the bolt and nut in the order shown in the illustration. Tighten to 25 ft. lbs. (34 Nm).

26. Install the front exhaust pipe assembly.

27. Install the air fuel ratio sensor (bank 1 sensor 1).

28. Install the exhaust manifold to head gasket.

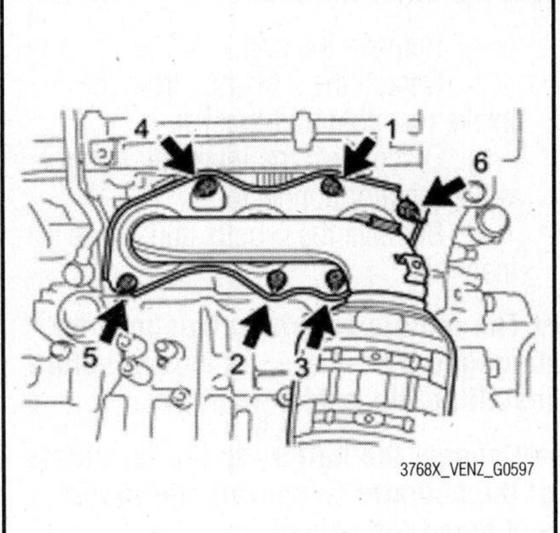

Fig. 67 Installing the exhaust manifold sub assembly LH

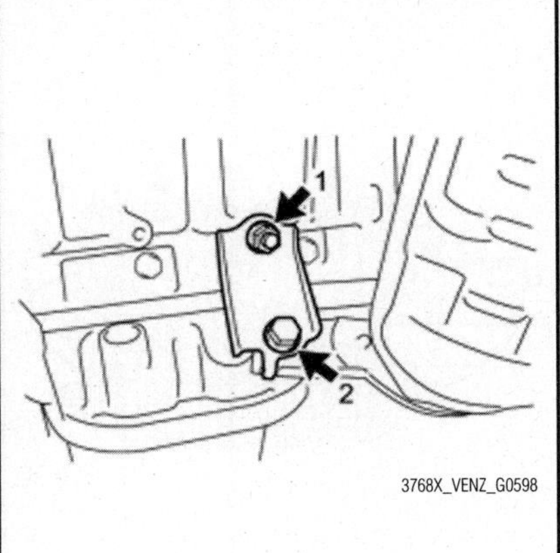

Fig. 68 Installing the No. 2 manifold stay

29. Install the exhaust manifold sub assembly RH.

a. Using a 12 mm deep socket wrench, install the exhaust manifold sub-assembly RH by tightening the 6 nuts in the order shown in the illustration. Tighten to 15 ft. lbs. (21 Nm).

b. Connect the air fuel ratio sensor connector (bank 1 sensor 1).

30. Install the manifold stay with the bolt and nut. Tighten to 25 ft. lbs. (34 Nm).

31. Install the front No. 3 exhaust pipe sub assembly.

a. Using a vernier caliper, measure the free length of the compression springs. The minimum length is 1.64 inch (41.5 mm).

➡If the free length is less than the minimum, replace the compression spring.

b. Fully insert a new gasket to the front No. 3 exhaust pipe sub-assembly.

c. Using a plastic hammer and wooden block, tap in the new gasket until its surface is flush with the front No. 3 exhaust pipe sub-assembly.

➡Be sure to install the gasket in the correct direction.

➡Do not reuse the gasket.

➡Do not damage the gasket.

➡Do not push in the gasket by using the exhaust pipe when connecting it.

d. Install 2 new gaskets to the front No. 3 exhaust pipe sub-assembly.

e. Install the front No. 3 exhaust pipe sub-assembly with the 4 bolts, 2 compression springs and 2 nuts. Tighten the bolts to 32 ft. lbs. (43 Nm). Tighten the nuts to 40 ft. lbs. (55 Nm).

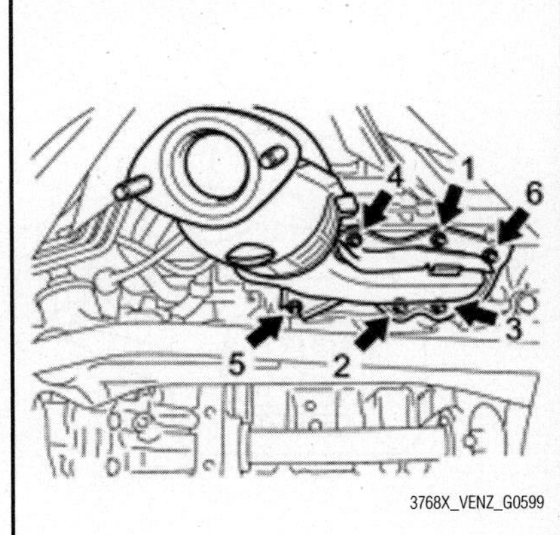

Fig. 69 Installing the exhaust manifold sub assembly RH

f. Install the grommet to the vehicle.

g. Connect the 2 clamps and oxygen sensor connector (for Bank 1 Sensor 2).

32. Install the inlet No. 2 air cleaner.
33. Install the cool air intake duct seal.
34. Install the propeller with the center bearing shaft assembly (AWD).
35. Install the No. 2 engine under cover.
36. Install the No. 1 engine under cover.
37. Install the cowl side trim sub assembly LH.
38. Install the front door scuff plate LH.
39. Inspect for exhaust gas leaks.

INTAKE MANIFOLD

REMOVAL & INSTALLATION

2.7L Engine

See Figures 70 through 73.

1. Discharge the fuel system pressure.
2. Disconnect the cable from the negative battery terminal.

CAUTION

When disconnecting the cable, some systems need to be initialized after the cable is reconnected.

3. Remove the throttle body assembly.
4. Remove the vacuum switching valve assembly (ACIS).
   a. Disconnect the 2 vacuum hoses, 2 union to connector tube hoses, clamp and connector.
   b. Remove the bolt and vacuum switching valve assembly (for ACIS).
5. Disconnect the No. 2 ventilation hose from the intake manifold.
6. Remove the union to connector tube hose form the intake manifold.
7. Remove the fuel delivery pipe sub assembly.
8. Remove the intake manifold.
   a. Remove the 2 bolts and 2 wire harness brackets.
   b. Disconnect the fuel vapor feed hose, clamp and connector.
   c. Remove the bolt and wire harness bracket.
   d. Apply battery voltage to the terminals of the connector to close the tumble control valves.

➡If this procedure is not performed, the tumble control valves may be damaged when the intake manifold is removed.

➡Apply battery voltage for 1 to 3 seconds.

➡If battery voltage is applied for more than 3 seconds, the actuator may be damaged.

➡Do not allow the lead wires to contact the other terminals.

   e. Remove the bolt.
   f. Detach the 2 clamps from the intake manifold and bracket.
   g. Disconnect the intake air control valve actuator connector.
   h. Remove the 6 bolts and intake manifold.

➡The tumble control valves may be damaged if they are not closed before installing the intake manifold.

➡Connect the battery to the terminals of the actuator to operate the motor and close the valves

   i. Remove the intake manifold gasket from the intake manifold.
9. Remove the check valve.
   a. Disconnect the 2 vacuum hoses from the intake manifold and remove the check valve.
10. Remove the wiring harness clamp bracket.
11. Remove the engine mounting damper.
    a. Remove the 3 bolts and engine mounting damper.

To install:

12. Install the engine mounting damper with the 3 bolts. Tighten to 80 inch lbs. (9 Nm).
13. Install the wiring harness clamp bracket with the bolt. Tighten to 74 inch lbs. (8.4 Nm).
14. Install the check valve.
    a. Connect the 2 vacuum hoses to the intake manifold to install the check valve.
    b. Check that the check valve is installed as shown in the illustration.

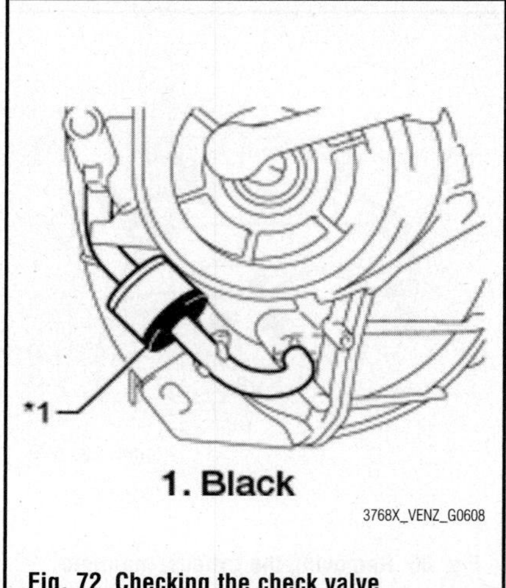

Fig. 72 Checking the check valve installation

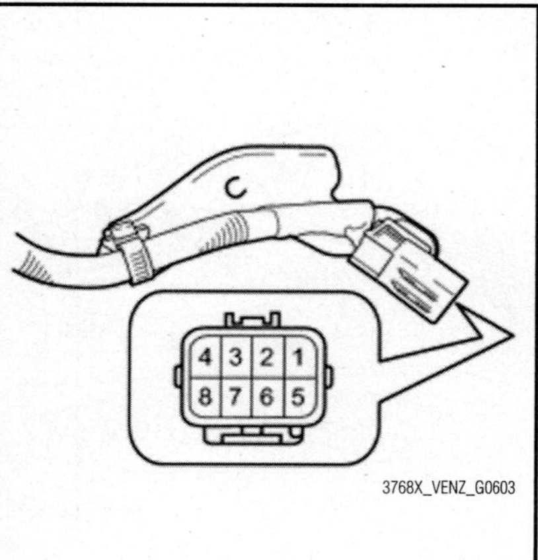

Fig. 70 Identifying the connector terminals

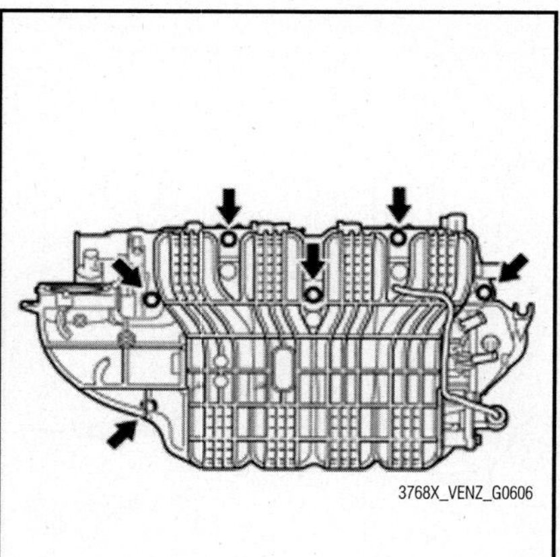

Fig. 71 Removing the intake manifold

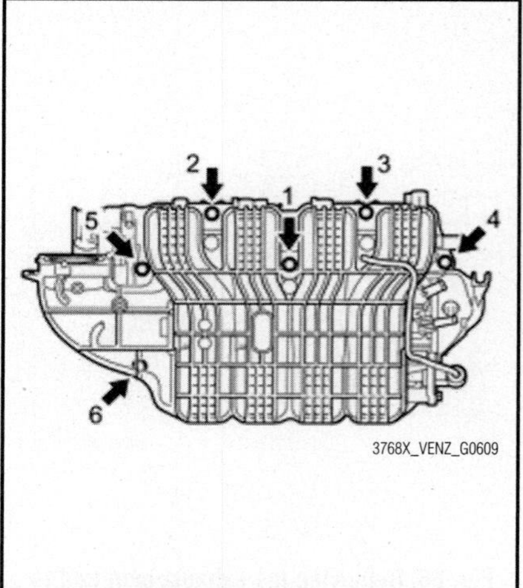

Fig. 73 Installing the intake manifold

15. Install the intake manifold.
a. Close the tumble control valves.

➡**The tumble control valves may be damaged if they are not closed before installing the intake manifold.**

➡**Connect the battery to the terminals of the actuator to operate the motor and close the valves**

b. Install a new gasket to the intake manifold.
c. Install the intake manifold by tightening the 6 bolts in the sequence shown in the illustration. Tighten to 15 ft. lbs. (21 Nm).
d. Connect the intake air control actuator connector.
e. Attach the 2 clamps to the intake manifold and bracket.
f. Install the wire harness with the bolt. Tighten to 74 inch lbs. (8.4 Nm).
g. Install the wire harness bracket with the bolt. Tighten to 74 inch lbs. (8.4 Nm).
h. Connect the fuel vapor feed hose, clamp and connector.
i. Install the 2 wire harness brackets with the 2 bolts. Tighten to 74 inch lbs. (8.4 Nm).
16. Install the fuel deliver pipe sub assembly.
17. Install the union to connector tube hose.
a. Install the union to connector tube hose to the intake manifold.
18. Connect the No. 2 ventilation hose to the intake manifold.
19. Install the vacuum switching valve assembly (ACIS).
a. Install the vacuum switching valve (for ACIS) with the bolt. Tighten to 80 inch lbs. (9 Nm).
b. Connect the 2 vacuum hoses, 2 union to connector tube hoses, clamp and connector.
20. Install the throttle body assembly.
21. Connect the cable to the negative battery terminal.

CAUTION

When disconnecting the cable, some systems need to be initialized after the cable is reconnected.

22. Inspect for fuel leaks.

3.5L Engine

See Figures 74 and 75.

1. Discharge the fuel system pressure.
2. Disconnect the cable from the negative battery terminal.

➡**When disconnecting the cable, some systems need to be initialized after the cable is reconnected.**

3. Remove the front wiper arm head cap.
4. Remove the front wiper arm and blade assemblies LH and RH.
5. Remove the front fender to cowl side seals LH and RH.
6. Remove the cowl top ventilator louver sub assembly.
7. Remove the windshield wiper motor and link assembly.
8. Remove the outer cowl top panel.
9. Remove the No. 1 engine under cover.
10. Remove the No. 2 engine under cover.
11. Drain the engine coolant.
12. Remove the v-bank cover sub assembly.
13. Remove the air cleaner cap with the hose.
14. Remove the intake air surge tank assembly.
a. Disconnect the No. 2 ventilation hose.
b. Disconnect the union to check vale hose.
c. Disconnect the throttle body assembly connector and wire harness clamp.
d. Disconnect the fuel vapor feed hose.
e. Disconnect the 2 water by pass hoses.
f. Disconnect the connector from the intake air control valve assembly.
g. Remove the bolt and separate the No. 1 surge tank stay from the intake air surge tank assembly.
h. Remove the bolt and separate the throttle body bracket from the intake air surge tank assembly.
i. Remove the 2 nuts from the intake air surge tank assembly.
j. Using a 5 mm socket hexagon wrench, remove the 4 bolts.
k. Remove the intake air surge tank assembly and 3 air surge tank to intake manifold gaskets.
15. Remove the No. 2 engine mounting stay RH.
16. Disconnect the fuel tube sub assembly.
17. Remove the fuel injector assembly.
18. Remove the intake manifold.
a. Remove the 6 bolts, 4 nuts, intake manifold and 2 intake manifold to head gaskets.

To install:

19. Install the intake manifold.
a. Set 2 new intake manifold to head gaskets on each cylinder head.

➡**Align the port holes of the gaskets and cylinder head.**

➡**Make sure that the gaskets are installed in the correct direction.**

b. Set the intake manifold on the cylinder head.
c. Install and tighten the 6 bolts and 4 nuts uniformly in several steps. Tighten to 15 ft. lbs. (21 Nm).
20. Install the fuel injector assembly.
21. Connect the fuel tube sub assembly.
22. Install the No. 2 engine mounting stay RH.
23. Install the intake air surge tank assembly.

➡**DO NOT apply oil to the bolts listed below:**

- Surge tank and intake manifold
- No. 1 surge tank stay and surge tank

Fig. 74 Removing the intake air surge tank assembly

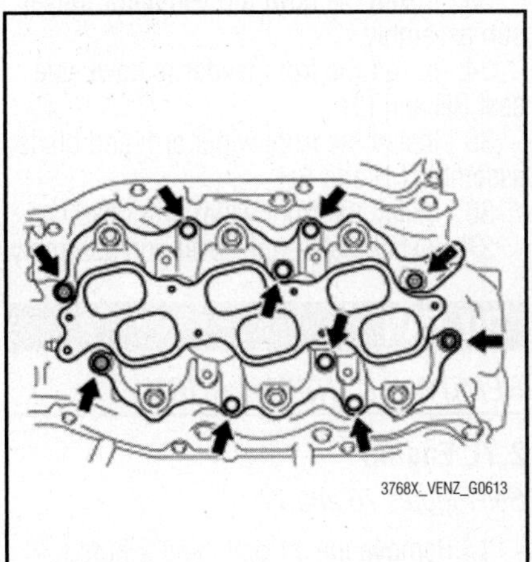

Fig. 75 Removing the intake manifold

- Throttle body bracket and surge tank

a. Install 3 new air surge tank to intake manifold gaskets to the intake air surge tank.

b. Using a 5 mm hexagon socket wrench, install the intake air surge tank assembly with the 4 bolts and 2 nuts. Tighten the bolts to 13 ft. lbs. (18 Nm). Tighten the nuts to 12 ft. lbs. (16Nm).

c. Install the throttle body bracket and No. 1 surge tank stay with the 2 bolts. Tighten to 15 ft. lbs. (21 Nm).

d. Connect the connector to the intake air control valve assembly.

e. Connect the 2 water by-pass hoses to the throttle body assembly.

f. Connect the fuel vapor feed hose to the intake air surge tank assembly.

g. Connect the throttle body assembly connector and wire harness clamp to the throttle body assembly.

h. Connect the ventilation hose.

i. Connect the union to connector tube hose.

24. Install the air cleaner cap with the hose.
25. Connect the cable to the negative battery terminal.

➡When disconnecting the cable, some systems need to be initialized after the cable is reconnected.

26. Add engine coolant.
27. Inspect for coolant leaks.
28. Inspect for fuel leaks.

a. Check that there are no fuel leaks from the fuel system after doing any maintenance or repairs.

29. Install the No. 2 engine under cover.
30. Install the No. 1 engine under cover.
31. Install the outer cowl top panel.
32. Install the windshield wiper motor and link assembly.
33. Install the cowl top ventilator louver sub assembly.
34. Install the front fender to cowl side seal RH and LH.
35. Install the front wiper arm and blade assembly LH and RH.
36. Install the front wiper arm head cap.
37. Install the v-bank cover sub assembly.

OIL PAN

REMOVAL & INSTALLATION

2.7L Engine

See Figures 76 and 77.

1. Remove the 11 bolts and 2 nuts.
2. Insert the blade of an oil pan seal cutter between the oil pan and stiffening crankcase, cut off the applied sealer and remove the oil pan.

➡Be careful not to damage the stiffening crankcase contact surface of the oil pan.

➡Be careful not to damage the stiffening crankcase flange.

To install:

3. Apply seal packing in a continuous line as shown in the illustration. The standard seal diameter is 0.0984 to 0.138 inch (2.5-3.5 mm).

➡Remove any oil from the contact surface.

➡Install the oil pan within 3 minutes and tighten the bolts and nuts within 10 minutes after applying seal packing.

➡Do not apply oil for at least 4 hours after the installation.

4. Install the oil pan with the 11 bolts and 2 nuts in several steps, in the sequence shown in the illustration. Tighten to 7 ft. lbs. (10 Nm).

➡Bolt A and nut A are tightened twice.

3.5L Engine

See Figures 78 through 80.

1. Remove the 16 bolts and 2 nuts (*1).
2. Insert the blade of the oil pan seal cutter between the oil pans. Cut through the applied sealer and remove the No. 2 oil pan sub-assembly.

➡Be careful not to damage the contact surfaces of the oil pans.

3. Using an E6 "TORX" socket wrench, remove the 2 stud bolts.

To install:

4. When replacing a stud bolt, install it by using an E8 "TORX" socket wrench.

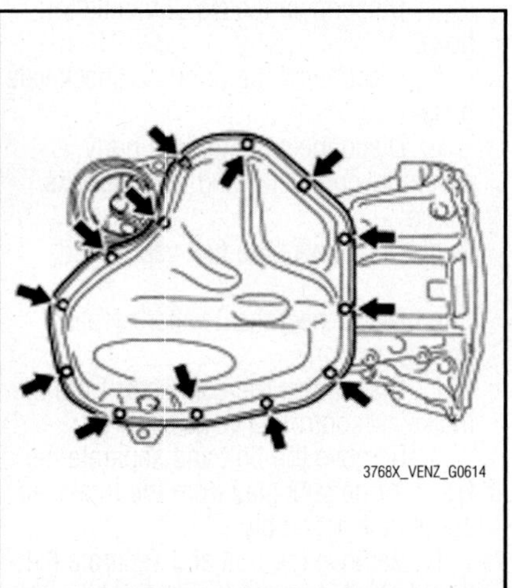

Fig. 76 Removing the bolts and nuts

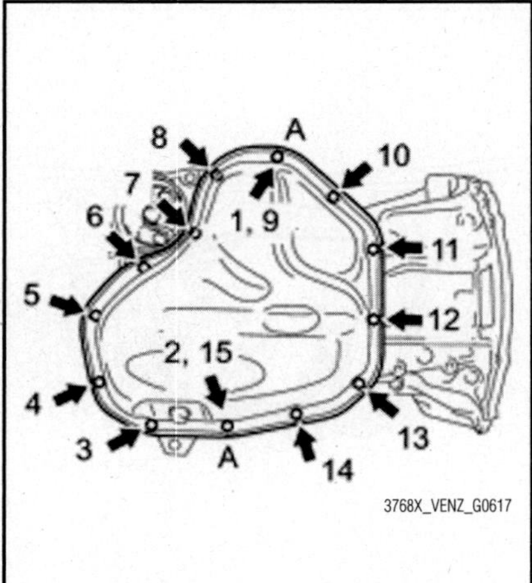

Fig. 77 Installing the oil pan

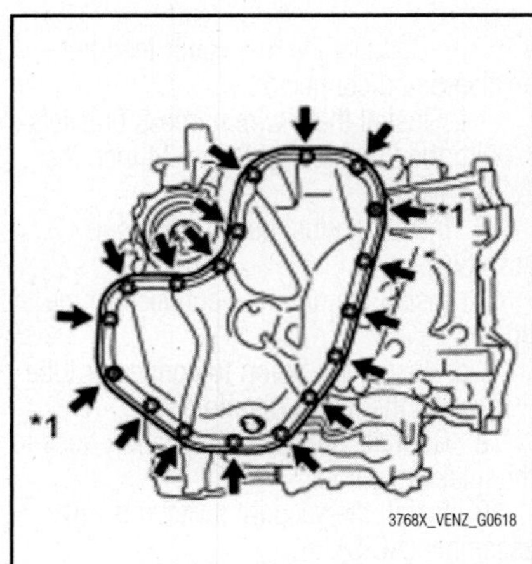

Fig. 78 Removing the oil pan bolts and nuts

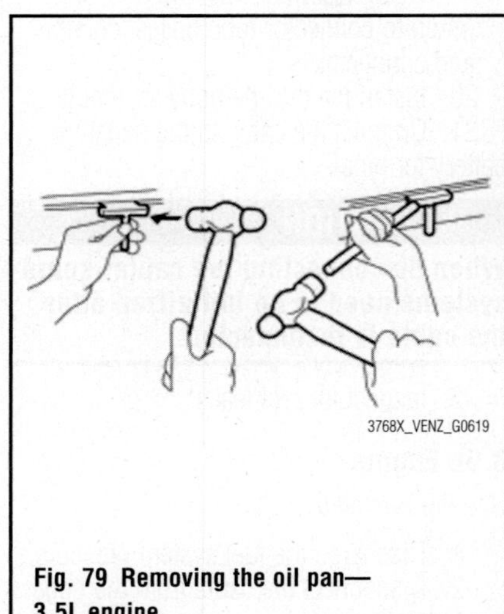

Fig. 79 Removing the oil pan—3.5L engine

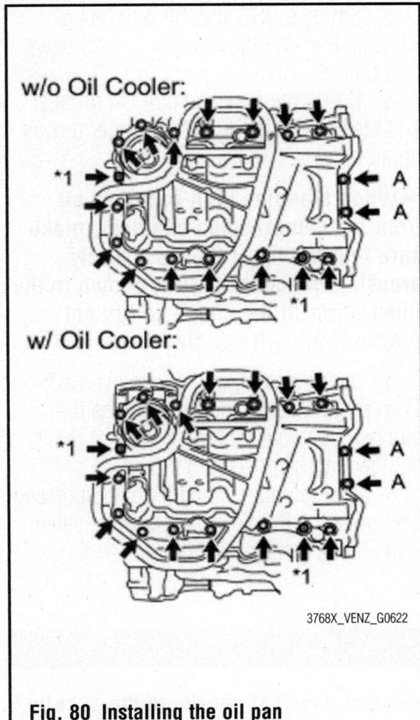

Fig. 80 Installing the oil pan

Tighten bolt A to 7 ft. lbs. (10 Nm). Tighten bolt B to 80 inch lbs. (9 Nm).

5. Install 2 new O-rings.
6. Apply seal packing in a continuous line as shown in the illustration. The seal diameter needs to be 0.118-0.157 inch (3-4 mm).

➡Remove any oil from the contact surface.

➡Install the oil pan within 3 minutes after applying seal packing.

➡Do not start the engine for at least 2 hours after installing.

7. Install the oil pan with the 16 bolts and 2 nuts. Tighten the A bolts to 7 ft. lbs. (10 Nm). Tighten all other bolts to 15 ft. lbs. (21 Nm).

PISTON AND RING

POSITIONING

See Figures 81 through 85.

1. Align the front marks (*1) of the piston and connecting rod, insert the connecting rod into the piston, and then push in the piston pin with your thumb until the pin comes into contact with the snap ring.

➡The piston and pin are a matched set.

2. Using a small screwdriver, install a new snap ring on the other side of the piston pin hole (*1).

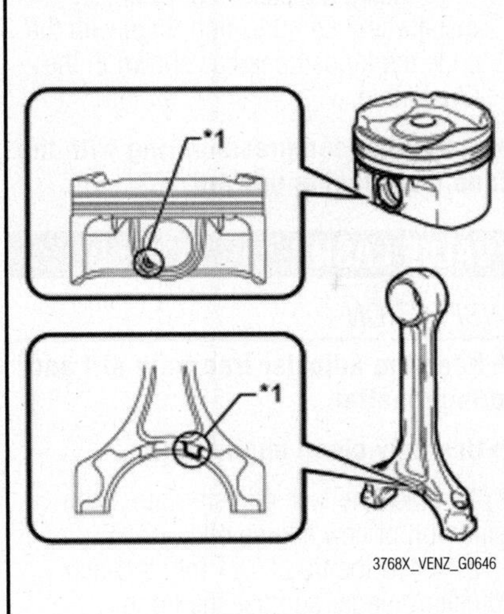

Fig. 81 Aligning the front marks of the piston and connecting rod

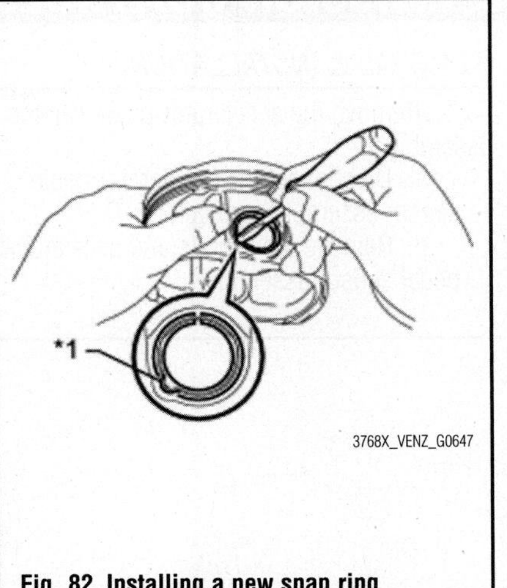

Fig. 82 Installing a new snap ring

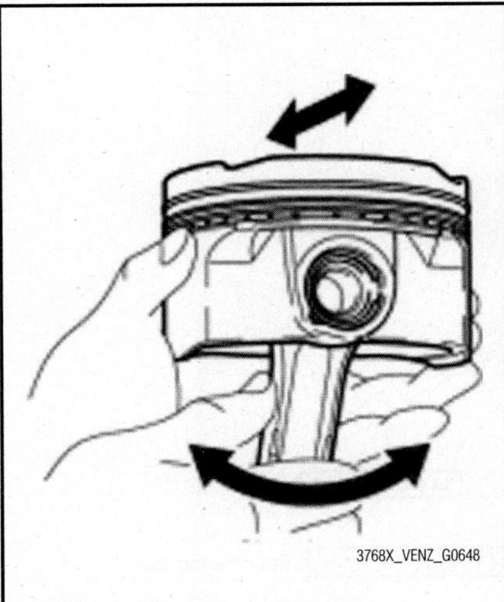

Fig. 83 Checking the fitting condition

➡Be sure that the end gap of the snap ring is not aligned with the service hole cutout portion of the piston.

3. Check the fitting condition between the piston and piston pin.
   a. Move the connecting rod back and forth on the piston pin. Check the fitting condition.
   If abnormal movement is felt, replace the piston and pin as a set.
   b. Rotate the piston back and forth on the piston pin. Check the fitting condition.
   If abnormal movement is felt, replace the piston and pin as a set.
4. Install the piston ring set.
   a. Install the oil ring expander and oil ring by hand.

➡Arrange the oil ring ends and coil joint.

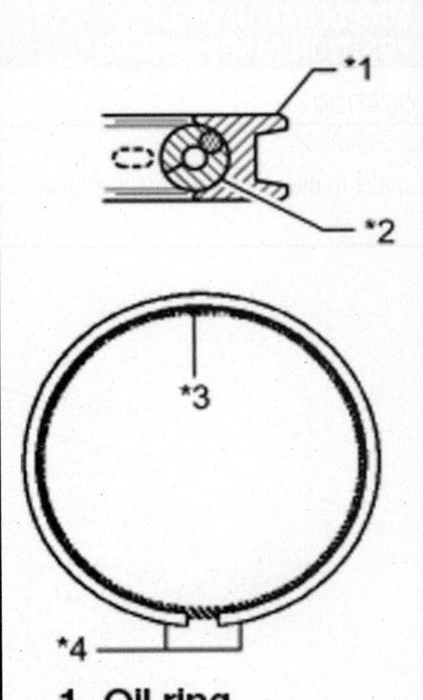

Fig. 84 Installing the oil ring expander and oil ring

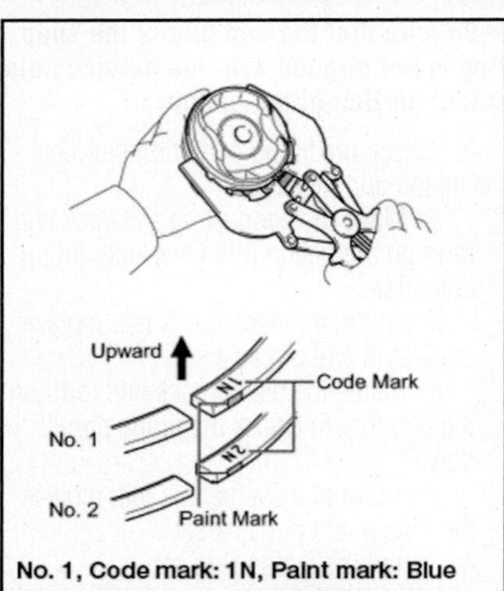

Fig. 85 Installing the compression rings

b. Using a piston ring expander, install the 2 compression rings with the code mark positioned as shown in the illustration.

➡Install the compression ring with the code mark facing upward.

VALVE LASH

INSPECTION

➡Keep the adjuster free from dirt and foreign matter.

➡Use only clean engine oil.

1. Place the lash adjuster into a container full of new engine oil.
2. Insert the tip of SST into the lash adjuster plunger and use the tip to press down on the check ball inside the plunger.
3. Squeeze SST and the lash adjuster together to move the plunger up and down 5 to 6 times.
4. Check the movement of the plunger and bleed air. It is OK if the plunger moves up and down.

➡When bleeding high-pressure air from the compression chamber, make sure that the tip of SST is actually pressing the check ball as shown in the illustration. If the check ball is not pressed, air will not bleed.

5. After bleeding the air, remove SST. Then try to quickly and firmly press the plunger with your fingers. It is OK if the plunger can be pressed 3 times.
6. If the plunger can still be compressed after pressing it 3 times, replace the valve lash adjuster with a new one.

ENGINE PERFORMANCE & EMISSION CONTROLS

ACCELERATOR PEDAL POSITION (APP) SENSOR

LOCATION

See Figure 86.

Refer to the accompanying illustration.

REMOVAL & INSTALLATION

1. Remove the accelerator pedal sensor assembly.
   a. Disconnect the accelerator pedal sensor assembly connector.
   b. Remove the 2 bolts and accelerator pedal sensor assembly.

➡Avoid physical shocks to the accelerator pedal sensor assembly.

➡Do not disassembly the accelerator pedal sensor assembly.

To install:

2. Install the accelerator pedal sensor assembly with the 2 bolts. Tighten to 48 inch lbs. (5.4 Nm).

➡Avoid physical shocks to the accelerator pedal sensor assembly.

   a. Connect the accelerator pedal sensor assembly connector.

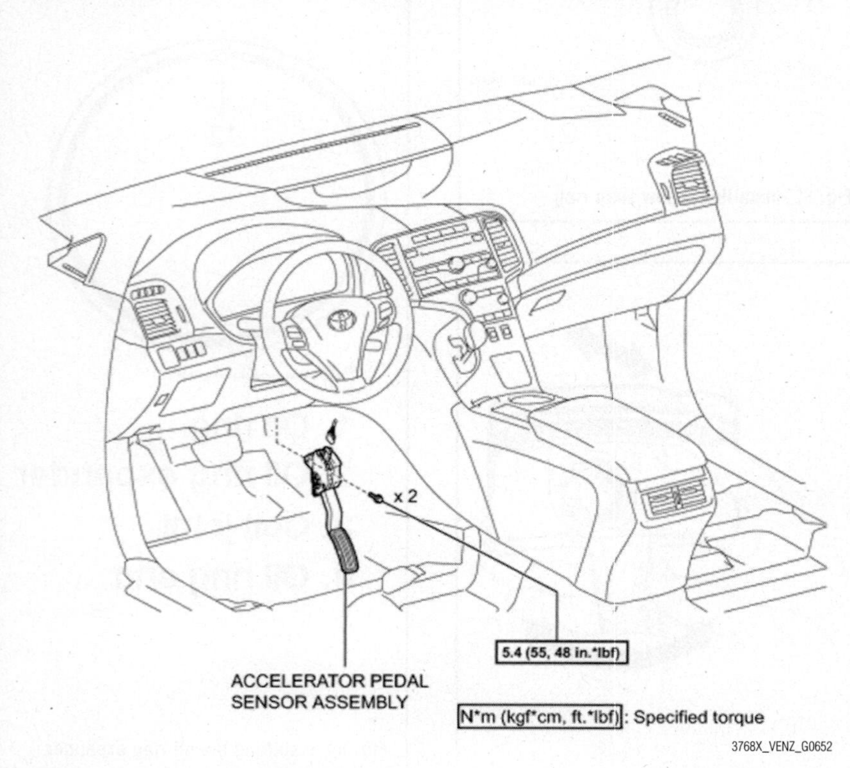

Fig. 86 Accelerator pedal position sensor

CAMSHAFT POSITION (CMP) SENSOR

LOCATION

See Figures 87 and 88.

Refer to the accompanying illustrations.

REMOVAL & INSTALLATION

2.7L Engine

1. Remove the No. 1 engine cover sub assembly.
2. Remove the Camshaft Position (CMP) sensor (exhaust side).
   a. Disconnect the connector.
   b. Remove the bolt and sensor.
3. Remove the Camshaft Position (CMP) sensor (intake side).
   a. Disconnect the sensor connector.
   b. Remove the bolt and sensor.

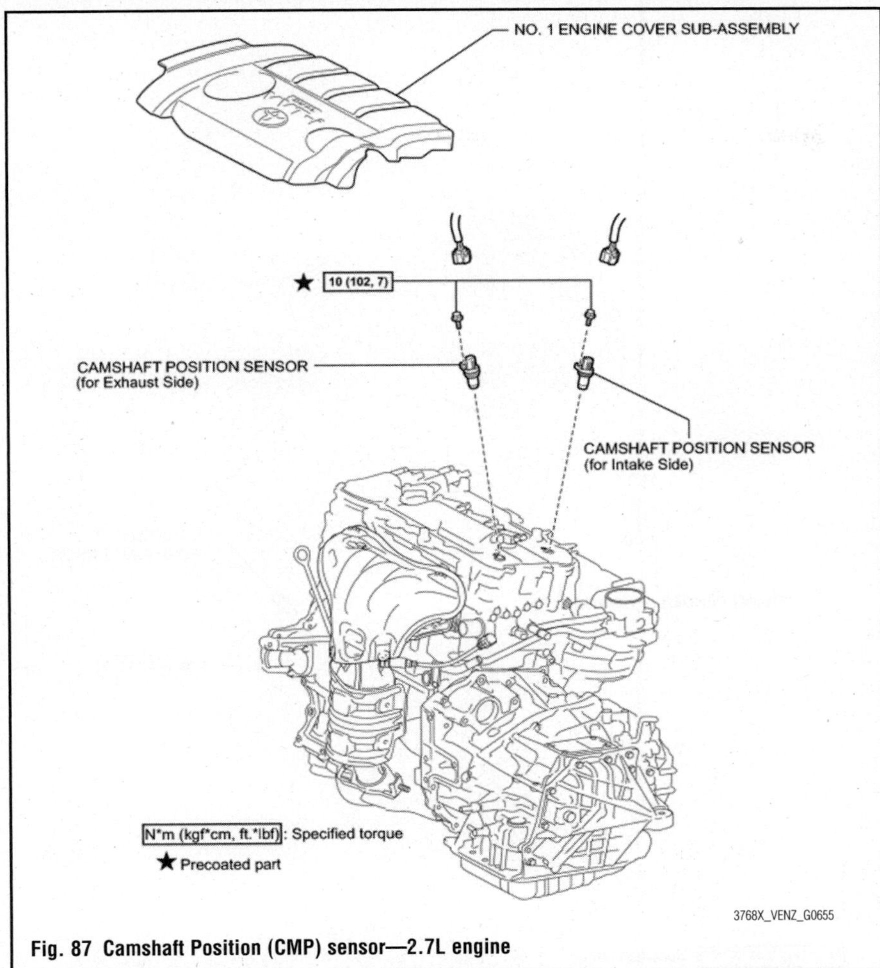

Fig. 87 Camshaft Position (CMP) sensor—2.7L engine

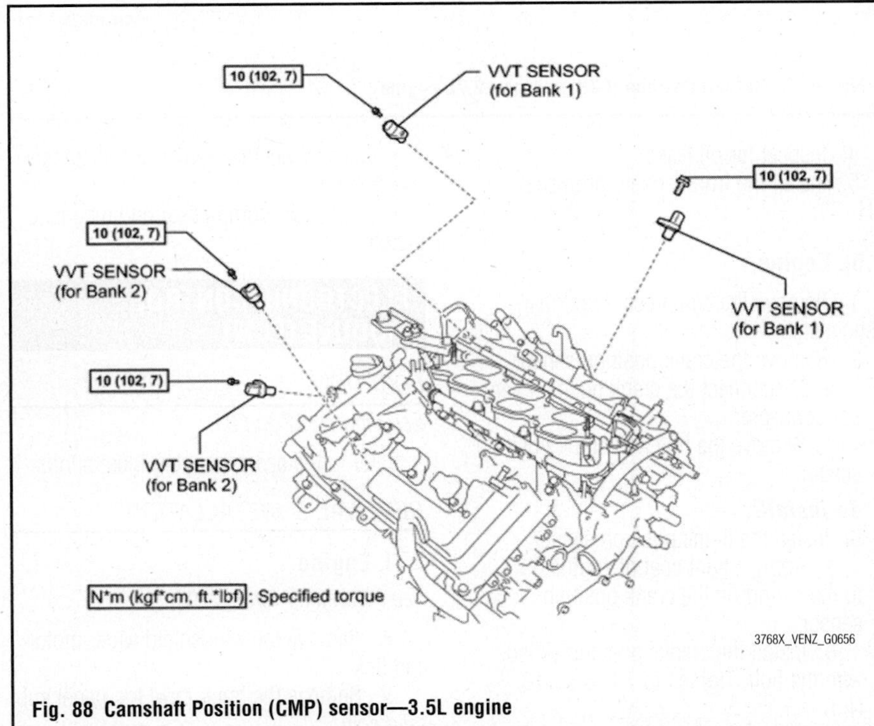

Fig. 88 Camshaft Position (CMP) sensor—3.5L engine

To install:

4. Install the Camshaft Position (CMP) sensor (exhaust side).
   a. Apply a light coat of engine oil to the O-ring of the sensor.

➡Make sure that the O-ring is not cracked or does not jump out of position during installation.

   b. Apply adhesive to 2 or 3 threads of the bolt.
   c. Install the sensor with the bolt. Tighten to 7 ft. lbs. (10 Nm).
   d. Connect the sensor connector.
5. Install the Camshaft Position (CMP) sensor (intake side).
   a. Apply a light coat of engine oil to the o-ring of the sensor.

➡Make sure that the o-ring is not cracked or does not jump out of position during installation.

   b. Apply adhesive to 2 or 3 threads of the bolt.
   c. Install the sensor with the bolt. Tighten to 7 ft. lbs. (10 Nm).
   d. Connect the sensor connector.
6. Inspect for oil leaks.
7. Install the No. 1 engine cover sub assembly.

3.5L Engine

See Figure 89.

1. Remove the No. 1 engine under cover.
2. Remove the No. 2 engine under cover.
3. Drain the engine coolant.
4. Remove the windshield wiper motor and link assembly.
5. Remove the cowl top panel outer sub assembly.
6. Remove the cool air intake duct seal.
7. Remove the v-bank cover sub assembly.
8. Remove the air cleaner cap with the hose.
9. Remove the intake air surge tank assembly.
10. Remove the VVT sensor.
    a. Disconnect the 4 sensor connectors.
    b. Remove the 4 bolts and 4 sensors.

To install:

11. Install the VVT sensors with the 4 bolts. Tighten to 7 ft. lbs. (10 Nm).
    a. Connect the 4 sensor connectors.
12. Install the air surge tank assembly.
13. Install the air cleaner cap with the hose.
14. Add engine coolant.

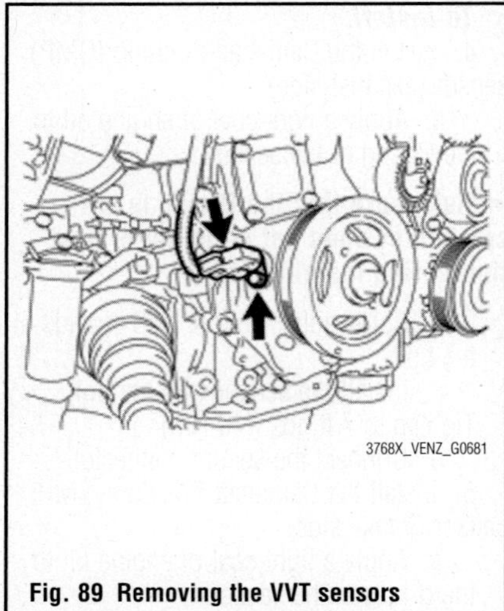

Fig. 89 Removing the VVT sensors

15. Inspect for coolant leaks.
16. Inspect for oil leaks.
17. Install the v-bank cover sub assembly.
18. Install the cool air intake duct seal.
19. Install the cowl top panel outer sub assembly.
20. Install the windshield wiper motor and link assembly.
21. Install the No. 2 engine under cover.
22. Install the No. 1 engine under cover.

CRANKSHAFT POSITION (CKP) SENSOR

LOCATION

See Figures 90 and 91.

Refer to the accompanying illustrations.

REMOVAL & INSTALLATION

2.7L Engine

1. Remove the front fender apron seal RH.
2. Remove the Crankshaft Position (CKP) sensor.
   a. Disconnect the sensor connector.
   b. Remove the bolt and sensor.

To install:

3. Install the Crankshaft Position (CKP) sensor.
   a. Apply a light coat of engine oil to the o-ring of the sensor.

➡Make sure that the o-ring is not cracked or does not jump out of position during installation.

   b. Apply adhesive to 2 or 3 threads of the bolt.
   c. Install the sensor with the bolt. Tighten to 7 ft. lbs. (10 Nm).
   d. Connect the sensor connector.

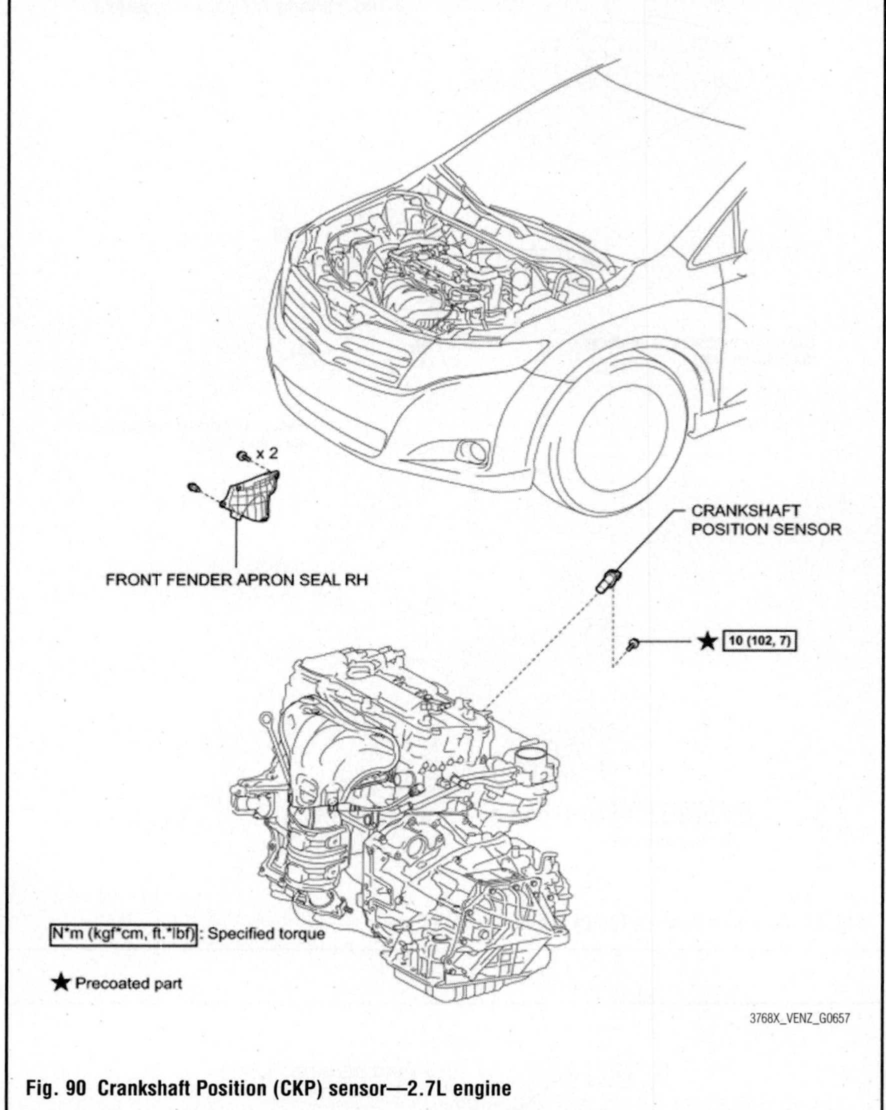

Fig. 90 Crankshaft Position (CKP) sensor—2.7L engine

4. Inspect for oil leaks.
5. Install the front fender apron seal RH.

3.5L Engine

1. Remove the compressor and magnetic clutch.
2. Remove the crank position sensor.
   a. Disconnect the crank position sensor connector.
   b. Remove the bolt and crank position sensor.

To install:

3. Install the crank position sensor.
   a. Apply a light coat of engine oil to the o-ring on the crank position sensor.
   b. Install the cranks position sensor with the bolt. Tighten to 7 ft. lbs. (10 Nm).
   c. Connect the crank position sensor connector.
4. Install the compressor and magnetic clutch.

ELECTRONIC CONTROL MODULE (ECM)

LOCATION

See Figures 92 and 93.

Refer to the accompanying illustrations.

REMOVAL & INSTALLATION

2.7L Engine

See Figures 94 through 98.

1. Remove the windshield wiper motor and link.
2. Remove the outer cowl top panel sub assembly.

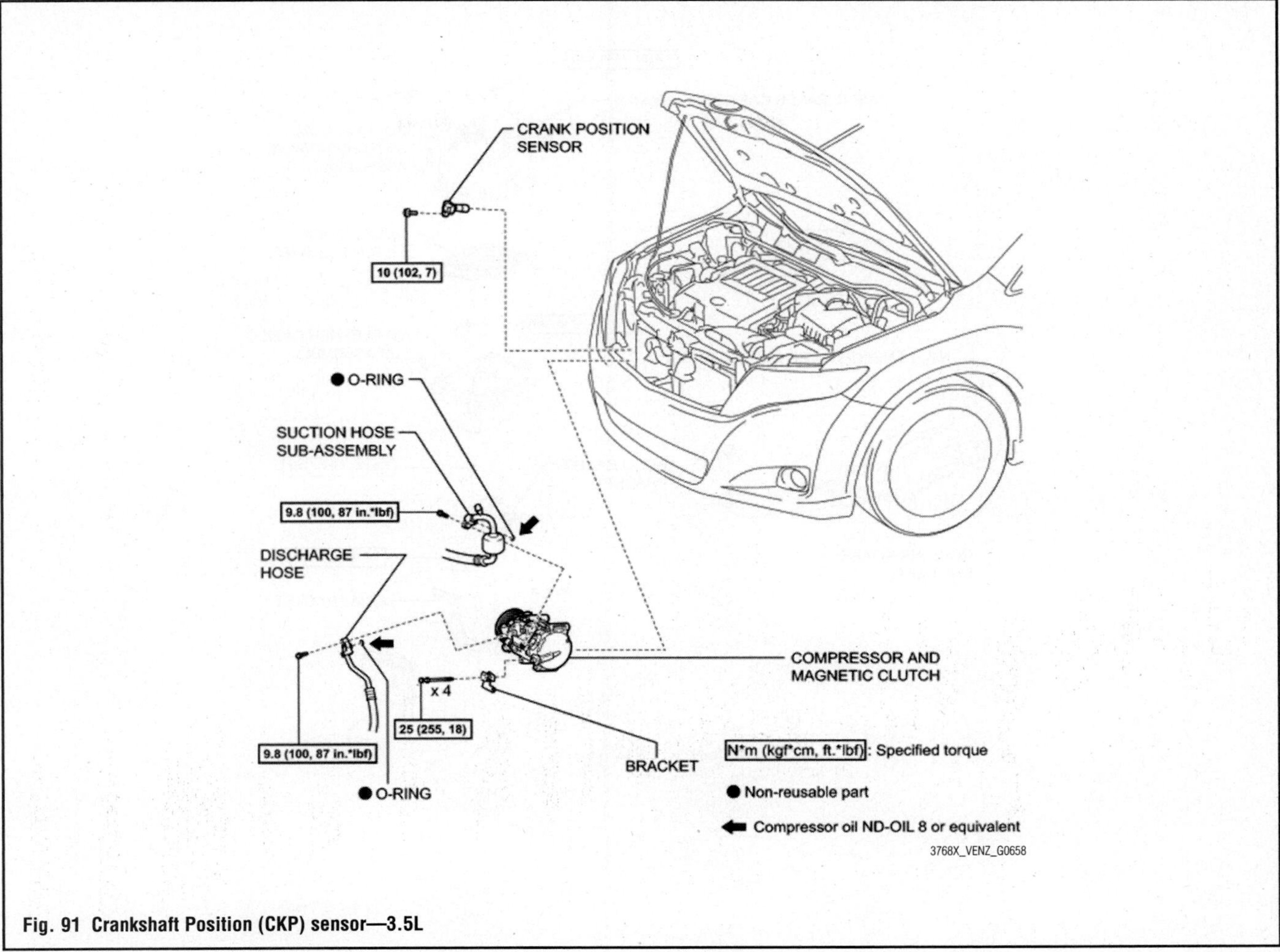

Fig. 91 Crankshaft Position (CKP) sensor—3.5L

3. Remove the cool air intake duct seal.
4. Remove the No. 1 engine cover sub assembly.
5. Disconnect the cable from the negative battery terminal.

➡When disconnecting the cable, some systems need to be initialized after the cable is reconnected.

6. Remove the No. 1 vacuum switching valve assembly.
7. Remove the air cleaner cap sub assembly.
8. Remove the air cleaner filter element sub assembly.
9. Remove the air cleaner case sub assembly.
10. Remove the air cleaner bracket.
    a. Remove the 2 bolts and cleaner bracket.
11. Remove the ECM.
    a. Separate the 3 wire harness clamps.
    b. Raise the 2 levers while pushing the locks on the levers, and disconnect the 2 ECM connectors.

➡After disconnecting each connector, make sure that dirt, water or other foreign matter does not contact the connecting part of the connector.

    c. Remove the 3 bolts and the ECM with the bracket.
    d. Remove the 5 screws and the ECM bracket.

To install:

12. Install the ECM.
    a. Install the bracket to the ECM with the 5 screws.
    b. Install the ECM with the 3 bolts. Tighten to 71 inch lbs. (8 Nm).
    c. Connect the 2 ECM connectors and lower the 2 levers.

➡When connecting the connector, make sure that dirt, water or other foreign matter does not become stuck between the connector and other part.

➡Make sure that the 2 levers are securely lowered.

    d. Install the 3 wire harness clamps.

13. Install the air cleaner bracket with the 2 bolts. Tighten to 69 inch lbs. (7.8 Nm).
14. Install the air cleaner case sub assembly.
15. Install the air cleaner filter element sub assembly.
16. Install the air cleaner cap sub assembly.
17. Remove the No. 1 vacuum switching valve assembly.
18. Install the No. 1 engine cover sub assembly.
19. Install the outer cowl top panel sub assembly.
20. Install the windshield wiper motor and link.
21. Install the cool air intake duct seal.
22. Connect the cable to the negative battery terminal.

CAUTION

When disconnecting the cable, some systems need to be initialized after the cable is reconnected.

23. Perform the registration.

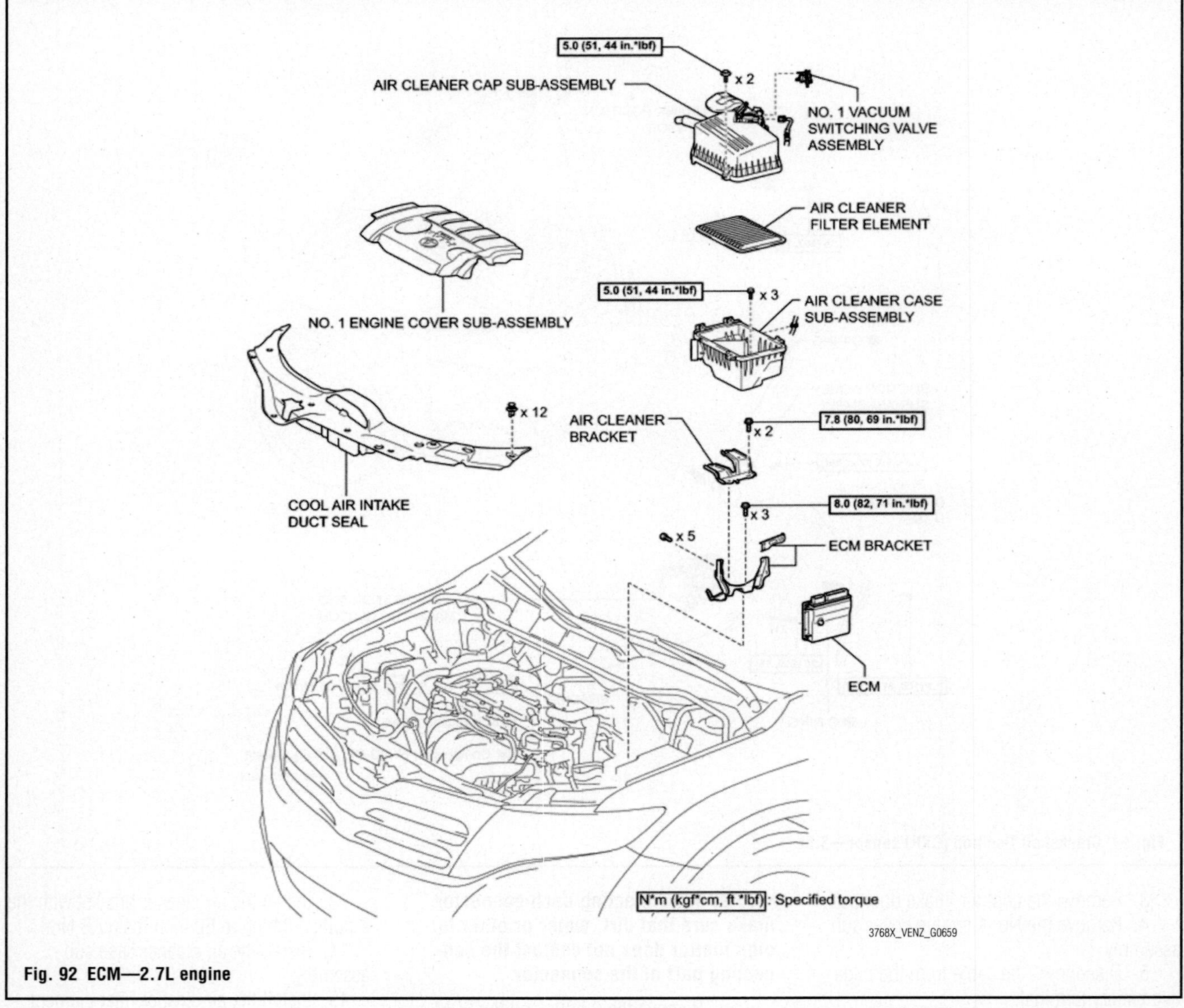

Fig. 92 ECM—2.7L engine

a. The VIN must be input into a replacement ECM.

3.5L Engine

See Figures 99 through 101.

➡Perform the Vehicle Identification Number (VIN) registration when replace the ECM

1. Disconnect the cable from the negative battery terminal.

➡When disconnecting the cable, some systems need to be initialized after the cable is reconnected.

2. Remove the v-bank cover sub assembly.
3. Remove the ECM.
   a. Disconnect the 2 ECM connectors and remove the ECM with the brackets.

➡After disconnecting the connectors, make sure that dirt, water or other foreign matter does not contact the connecting part of the connectors.

   b. Push in the locks on the 2 levers, raise the levers, and disconnect the 2 ECM connectors.
   c. Remove the 3 nuts and separate the ECM with the brackets.
   d. Remove the 4 screws and 2 ECM brackets.

To install:

4. Install the ECM.
   a. Install the 2 brackets to the ECM with the 4 screws. Tighten to 27 inch lbs. (3 Nm).
   b. Install the ECM with the 3 nuts. Tighten to 71 inch lbs. (8 Nm).
   c. Connect the 2 ECM connectors.

➡When connecting the connectors, make sure that dirt, water or other foreign matter does not become stuck between the connectors and other parts.

   d. Connect the 2 ECM connectors and lower the 2 levers.

➡Make sure that the 2 levers are securely locked.

5. Install the v-bank cover sub assembly.
6. Connect the cable to the negative battery terminal.

➡When disconnecting the cable, some systems need to be initialized after the cable is reconnected.

7. Perform the initialization.

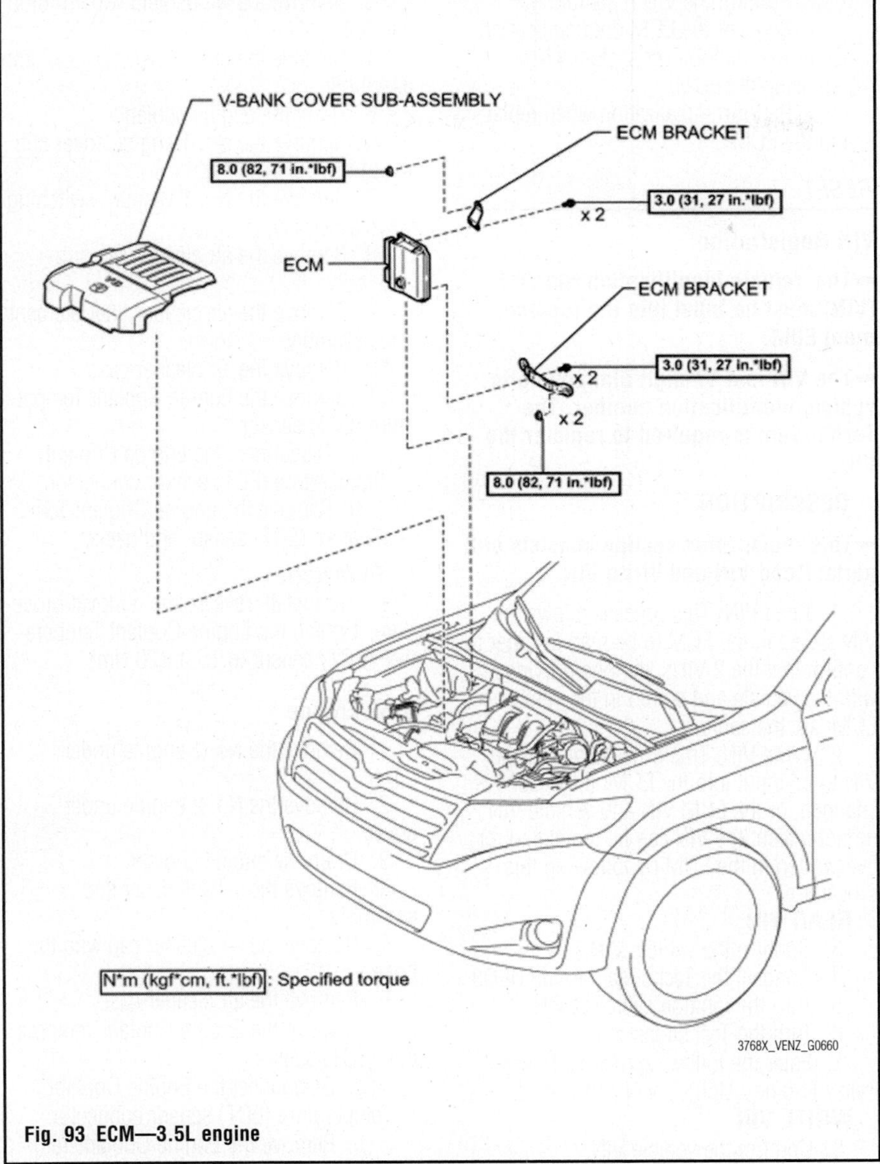

Fig. 93 ECM—3.5L engine

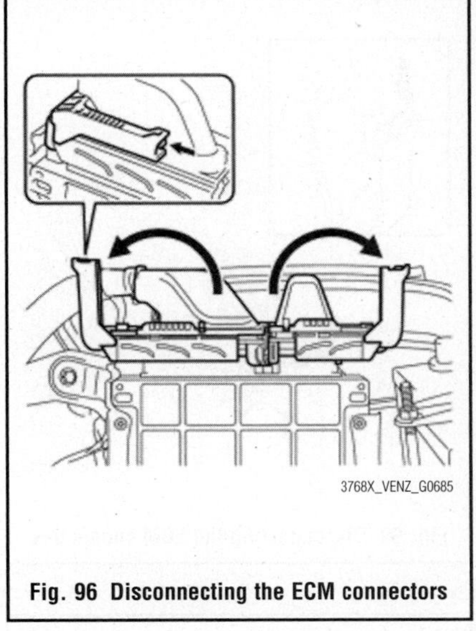

Fig. 96 Disconnecting the ECM connectors

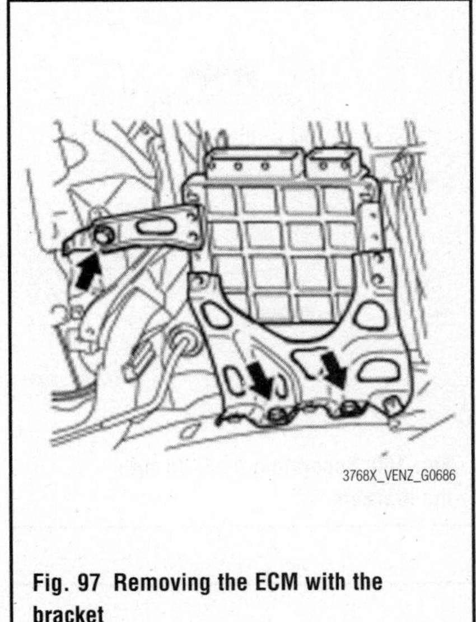

Fig. 97 Removing the ECM with the bracket

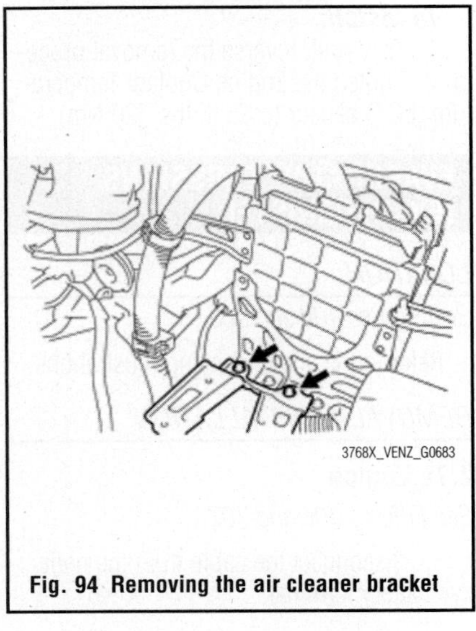

Fig. 94 Removing the air cleaner bracket

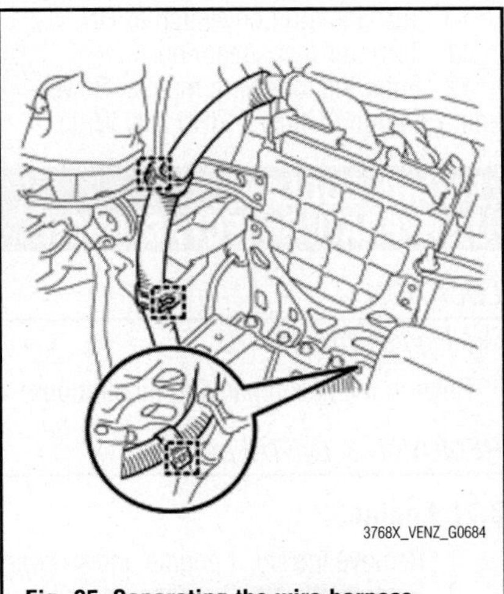

Fig. 95 Separating the wire harness clamps

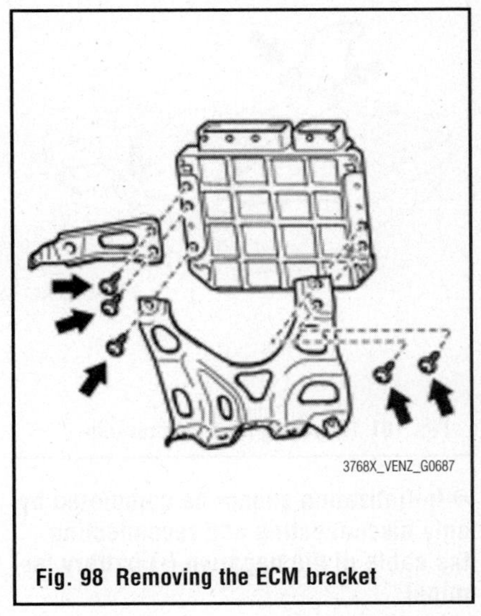

Fig. 98 Removing the ECM bracket

Fig. 99 Disconnecting the ECM connectors

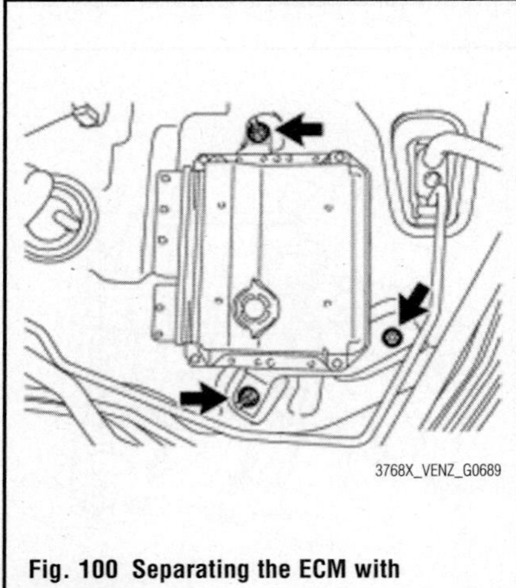

Fig. 100 Separating the ECM with the brackets

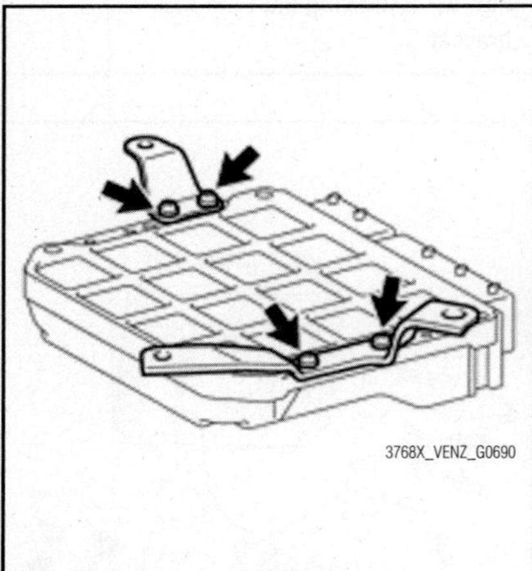

Fig. 101 Removing the ECM brackets

➡Initialization cannot be completed by only disconnecting and reconnecting the cable of the negative (-) battery terminal.

a. Perform the VIN registration.

b. Register the ECM communication ID for the immobilizer system when replacing the ECM.

c. Perform initialization when replacing the ECM.

RESET

VIN Registration

➡The Vehicle Identification Number (VIN) must be input into the replacement ECM.

➡The VIN is a 17-digit alphanumeric vehicle identification number. The Techstream is required to register the VIN.

DESCRIPTION

➡This registration section consists of 2 parts: Read VIN and Write VIN.

1. Read VIN: This process allows the VIN stored in the ECM to be read in order to confirm that the 2 VINs, the one provided with the vehicle and stored in the vehicle ECM, are the same.
2. Write VIN: This process allows the VIN to be input into the ECM. If the ECM is changed, or the ECM VIN and vehicle VIN do not match, the VIN can be registered, or overwritten in the ECM by following this procedure.

READ VIN

3. Confirm the vehicle VIN.
4. Connect the Techstream to the DLC3.
5. Turn the ignition switch to ON.
6. Turn the Techstream on.
7. Enter the following menus: Powertrain / Engine / Utility / VIN / VIN Read.

WRITE VIN

8. Confirm the vehicle VIN.
9. Connect the Techstream to the DLC3.
10. Turn the ignition switch to ON.
11. Turn the Techstream on.
12. Enter the following menus: Powertrain / Engine / Utility / VIN / VIN Write.

ENGINE COOLANT TEMPERATURE (ECT) SENSOR

LOCATION

See Figures 102 and 103.

Refer to the accompanying illustrations.

REMOVAL & INSTALLATION

2.7L Engine

1. Remove the No. 1 engine under cover.
2. Remove the No. 2 engine under cover.
3. Remove the windshield wiper motor and link.
4. Remove the outer cowl top panel sub assembly.
5. Drain the engine coolant.
6. Remove the No. 1 engine cover sub assembly.
7. Remove the No. 1 vacuum switching valve assembly.
8. Remove the air cleaner cap sub assembly.
9. Remove the air cleaner filter element sub assembly.
10. Remove the air cleaner case.
11. Remove the Engine Coolant Temperature (ECT) sensor.
    a. Disconnect the Engine Coolant Temperature (ECT) sensor connector.
    b. Remove the Engine Coolant Temperature (ECT) sensor and gasket.

To install:

12. To install, reverse the removal procedure. Tighten the Engine Coolant Temperature (ECT) sensor to 15 ft. (20 Nm).

3.5L Engine

1. Remove the No. 1 engine under cover.
2. Remove the No. 2 engine under cover.
3. Drain the engine coolant.
4. Remove the v-bank cover sub assembly.
5. Remove the air cleaner cap with the hose.
6. Remove the air cleaner case.
7. Remove the Engine Coolant Temperature (ECT) sensor.
    a. Disconnect the Engine Coolant Temperature (ECT) sensor connector.
    b. Remove the Engine Coolant Temperature (ECT) sensor and gasket.

To install:

8. To install, reverse the removal procedure. Tighten the Engine Coolant Temperature (ECT) sensor to 15 ft. lbs. (20 Nm).

EVAPORATIVE EMISSIONS (EVAP) CANISTER

LOCATION

See Figures 104 and 105.

Refer to the accompanying illustrations.

REMOVAL & INSTALLATION

2.7L Engine

See Figures 106 and 107.

1. Disconnect the cable from the negative battery terminal.

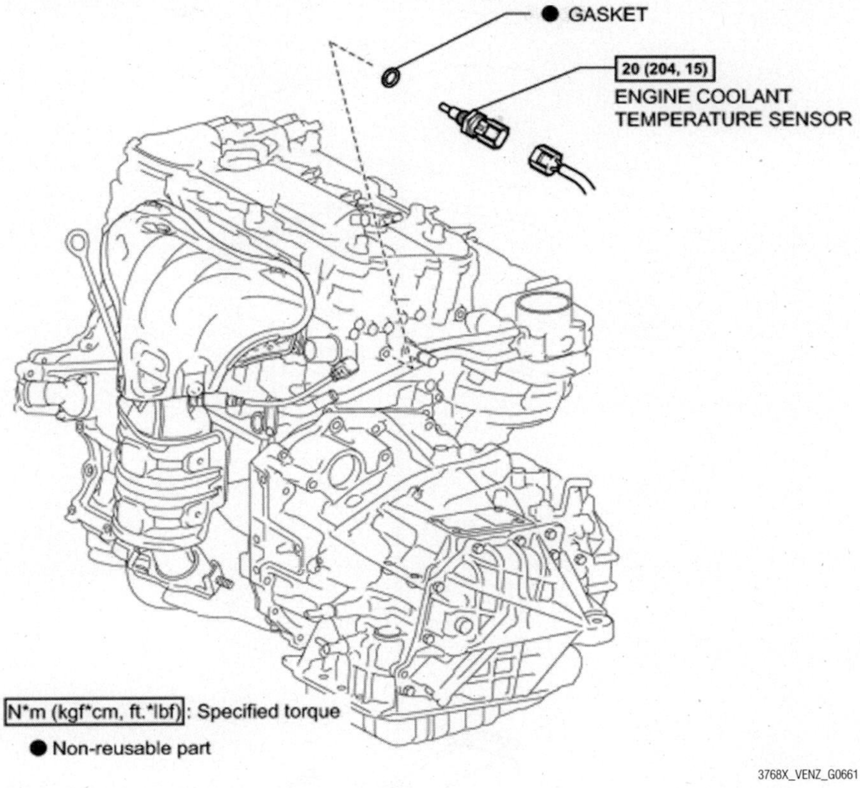

Fig. 102 Engine Coolant Temperature (ECT) sensor—2.7L engine

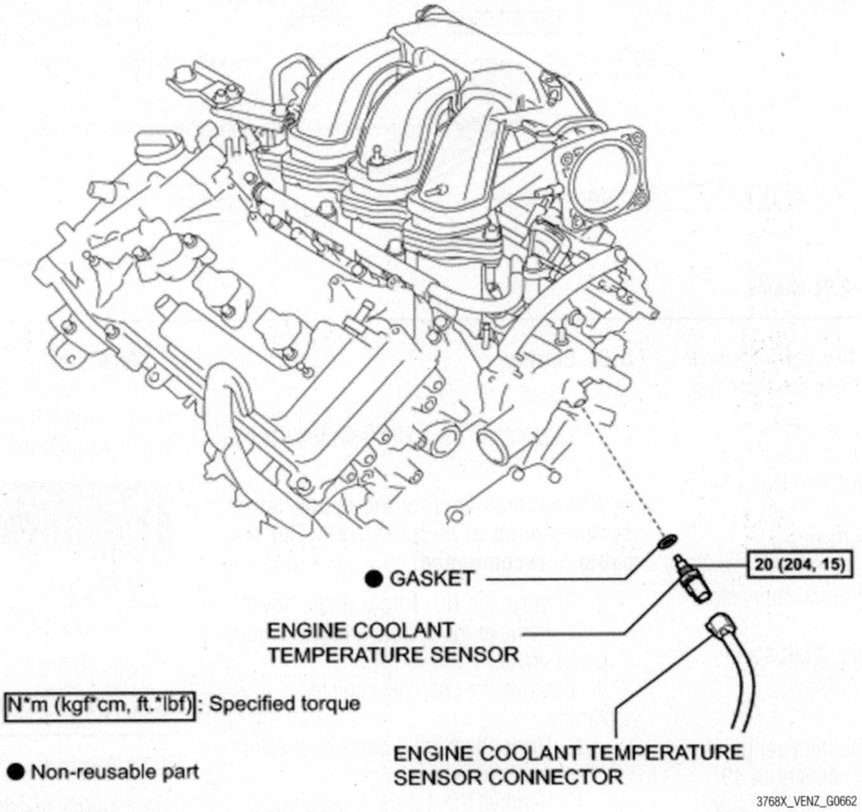

Fig. 103 Engine Coolant Temperature (ECT) sensor—3.5L engine

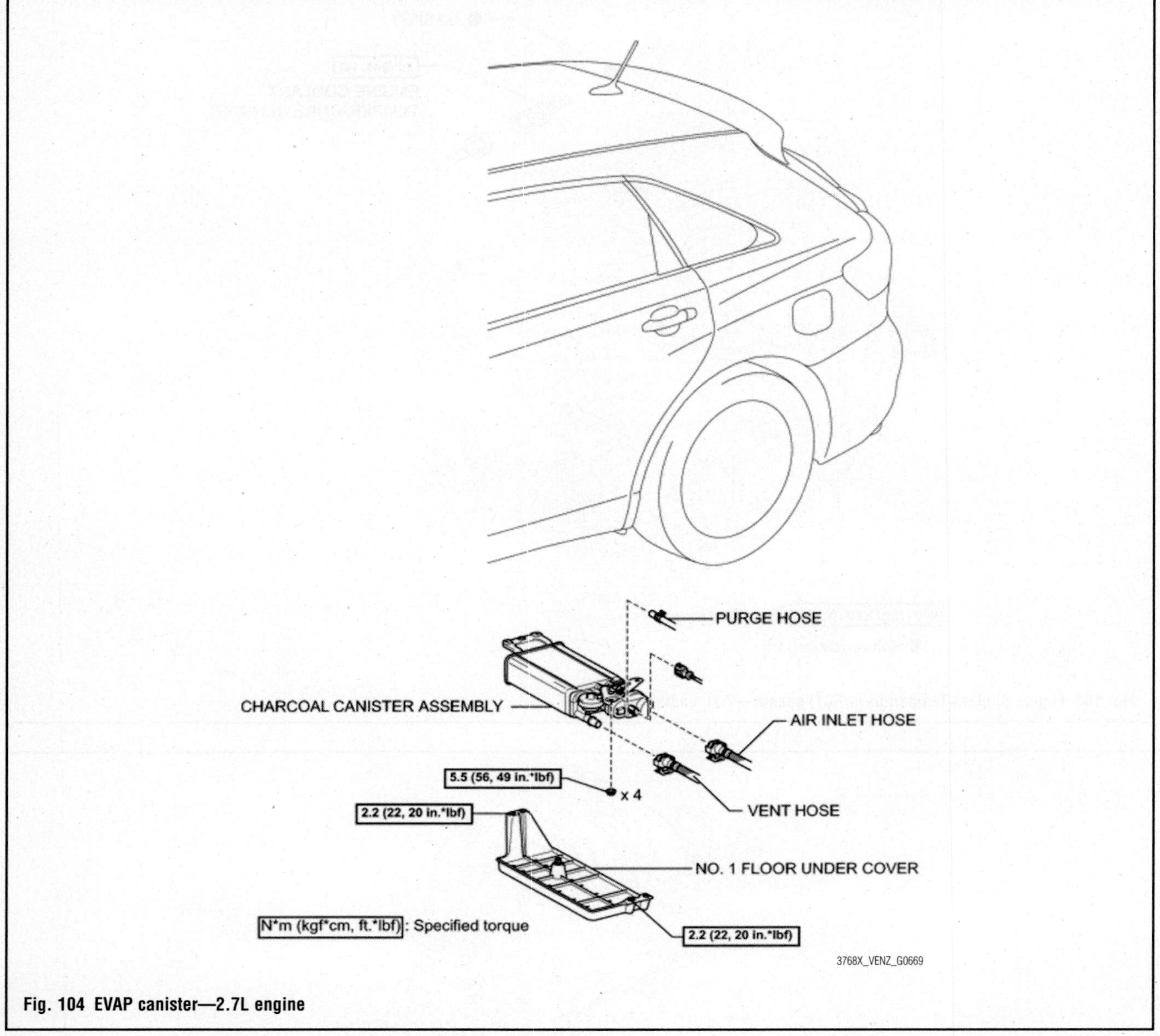

Fig. 104 EVAP canister—2.7L engine

➡When disconnecting the cable, some systems need to be initialized after the cable is reconnected.

2. Remove the No. 1 floor under cover.
   a. Remove the 5 clips and No. 1 floor under cover.
3. Remove the charcoal canister assembly.
   a. Disconnect the 2 quick connectors, hose and connector.
   b. Remove the 4 nuts, 2 clips and charcoal canister assembly.

To install:

4. To install, reverse the removal procedure. Tighten the charcoal canister to 49 inch lbs. (5.5 Nm). Tighten the floor under cover to 20 inch lbs. (2.2 Nm).

3.5L Engine

See Figures 108 and 109.

1. Disconnect the cable from the negative battery terminal.

➡When disconnecting the cable, some systems need to be initialized after the cable is reconnected.

2. Remove the No. 1 floor under cover.
   a. Remove the 5 clips and No. 1 floor under cover.
3. Remove the charcoal canister assembly.
   a. Disconnect the 2 quick connectors, hose and connector.
   b. Remove the 4 nuts, 2 clips and charcoal canister assembly.

To install:

4. To install, reverse the removal procedure. Tighten the charcoal canister assembly to 49 inch lbs. (5.5 Nm).

HEATED OXYGEN SENSOR (HO2S)

LOCATION

See Figures 110 and 111.

Refer to the accompanying illustrations.

REMOVAL & INSTALLATION

2.7L Engine

See Figures 112 and 113.

1. Remove the No. 1 engine under cover.

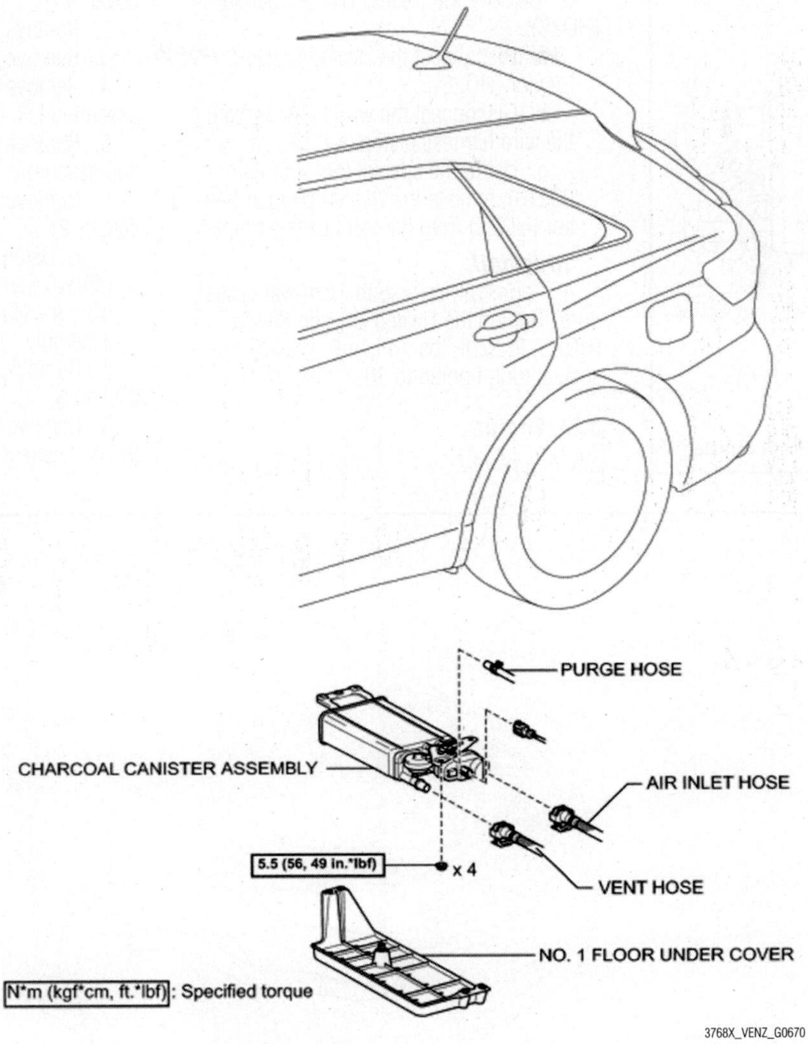

PURGE HOSE
CHARCOAL CANISTER ASSEMBLY
AIR INLET HOSE
5.5 (56, 49 in.*lbf)
x 4
VENT HOSE
NO. 1 FLOOR UNDER COVER
N*m (kgf*cm, ft.*lbf): Specified torque
3768X_VENZ_G0670

Fig. 105 EVAP canister—3.5L engine

3768X_VENZ_G0693

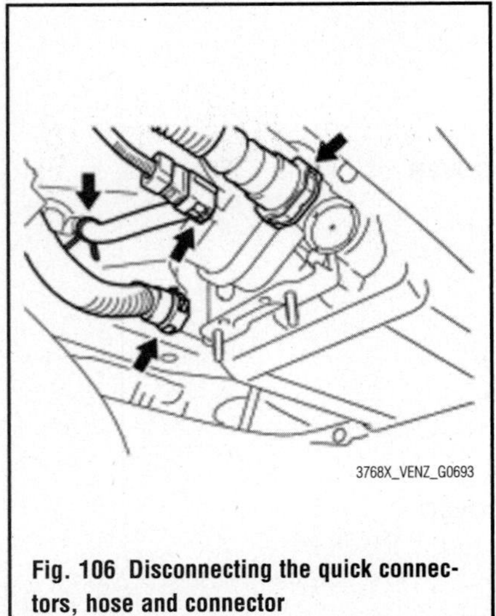

Fig. 106 Disconnecting the quick connectors, hose and connector

3768X_VENZ_G0694

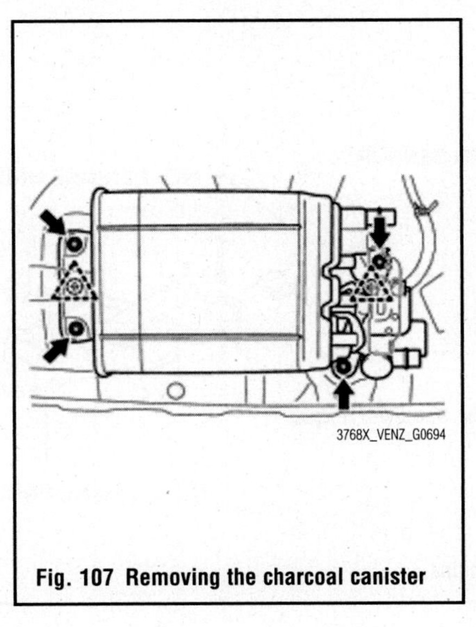

Fig. 107 Removing the charcoal canister

3768X_VENZ_G0695

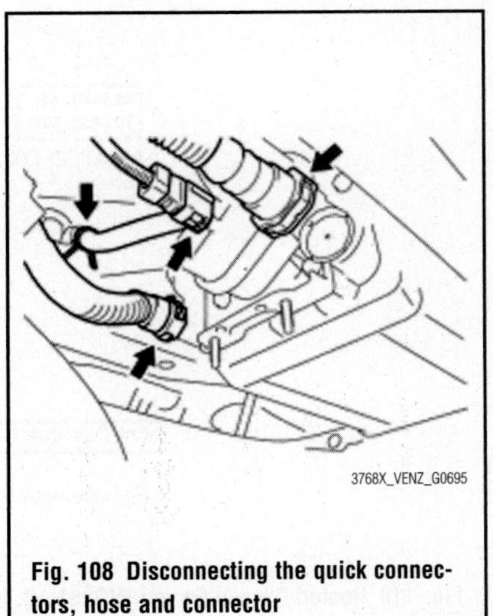

Fig. 108 Disconnecting the quick connectors, hose and connector

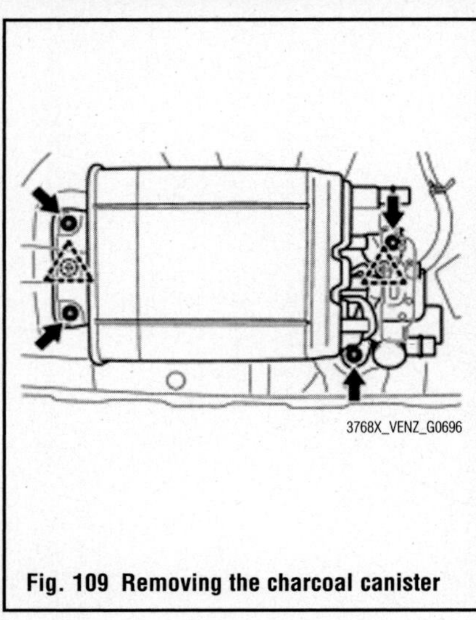

Fig. 109 Removing the charcoal canister

2. Remove the No. 2 engine under cover.
3. Remove the Heated Oxygen Sensor (HO2S).
   a. Disconnect the Heated Oxygen Sensor (HO2S).
   b. Disconnect the wire harness from the wire harness clamp.
   c. Using the special tool (09224-00010), remove the Heated Oxygen Sensor (HO2S) from the front exhaust pipe.

To install:

4. To install, reverse the removal procedure. Tighten the Heated Oxygen Sensor (HO2S) to 32 ft. lbs. (44 Nm). If using the special tool, tighten to 30 ft. lbs. (40 Nm).

3.5L Engine

See Figure 114.

1. Remove the No. 1 engine under cover.
2. Remove the No. 2 engine under cover.
3. Remove the front door scuff plate LH.
4. Remove the cowl side trim sub assembly LH.
5. Remove the No. 3 front exhaust pipe sub assembly.
6. Remove the oxygen sensor (bank 1 sensor 2).
   a. Using the special tool (09224-00010), remove the oxygen sensor from the No. 3 front exhaust pipe sub assembly.
7. Remove the front exhaust pipe assembly.
8. Remove the No. 2 oxygen sensor (bank 2 sensor 2).

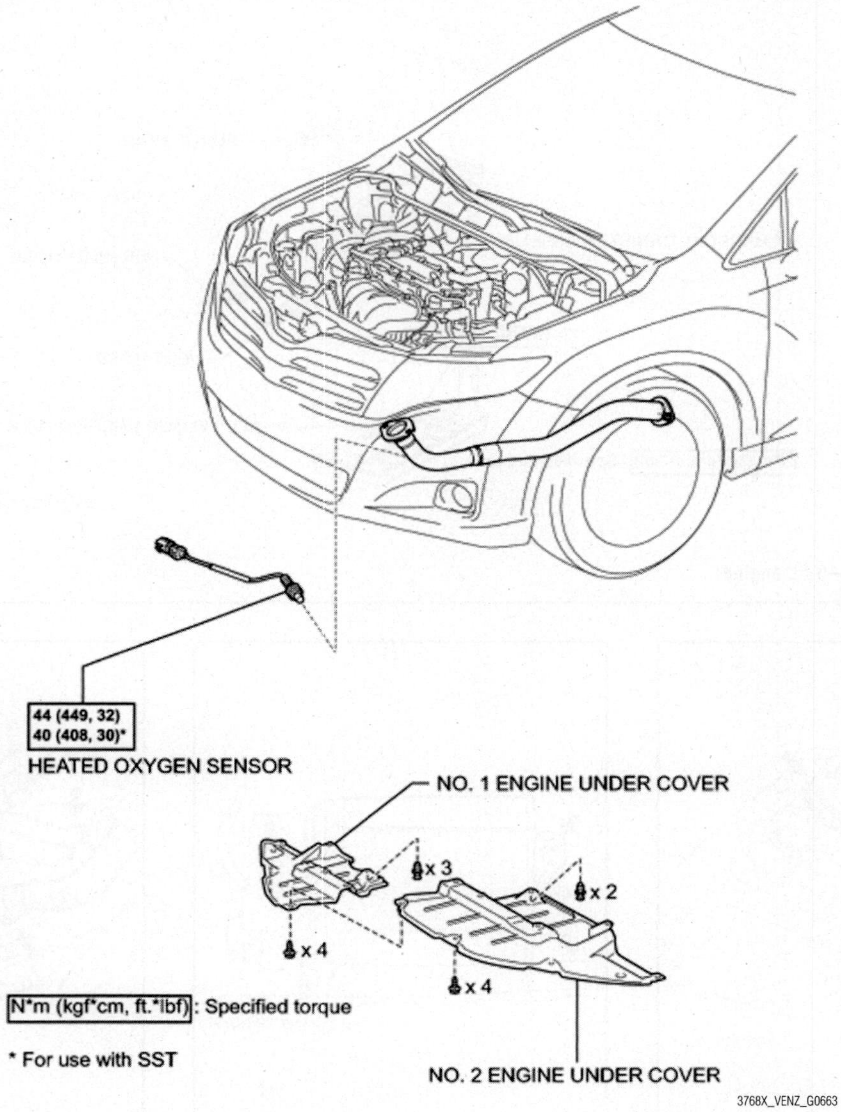

Fig. 110 Heated Oxygen Sensor (HO2S)—2.7L engine

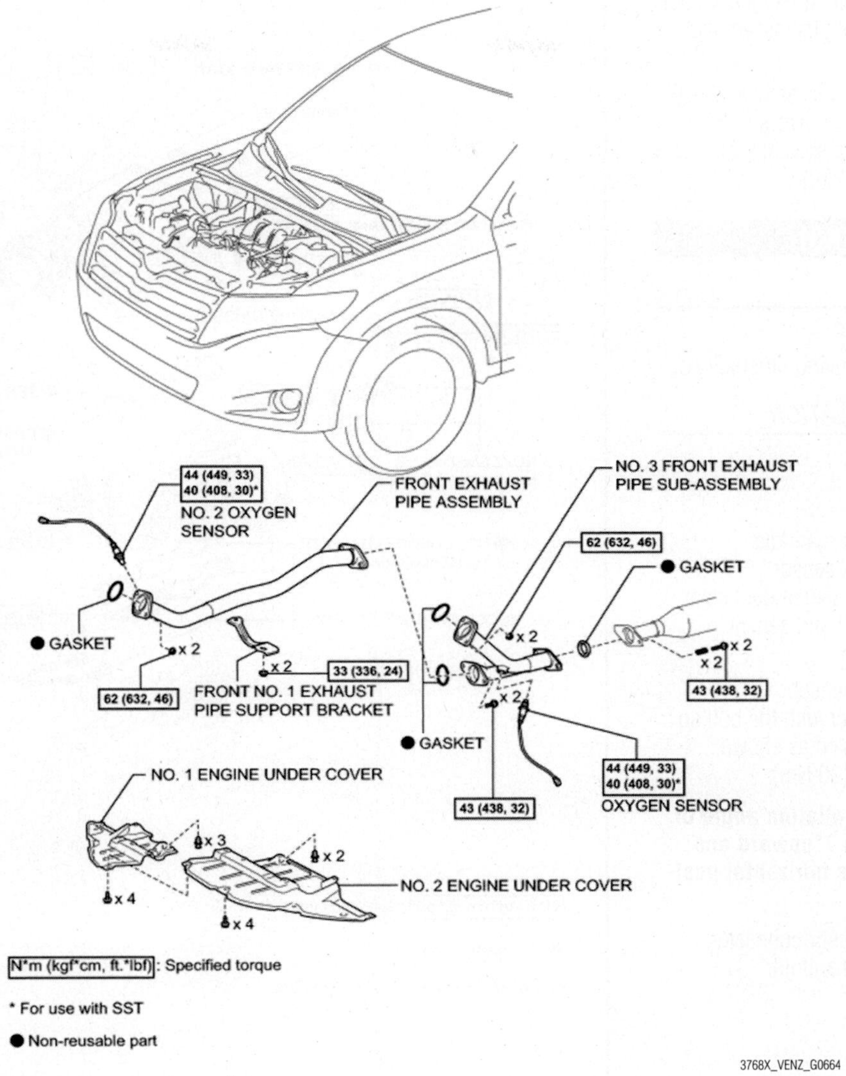

Fig. 111 Heated Oxygen Sensor (HO2S)—3.5L engine

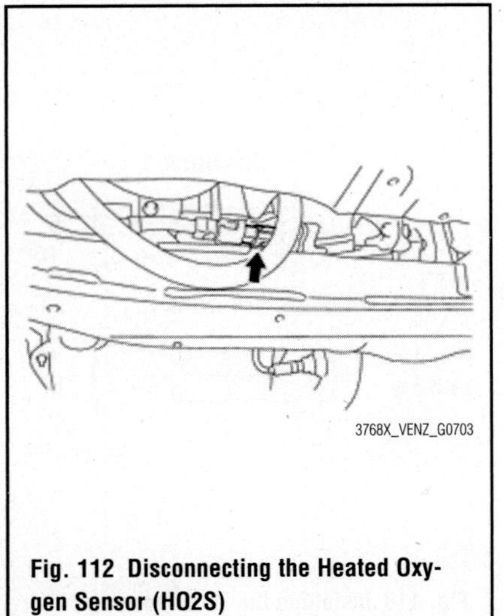

Fig. 112 Disconnecting the Heated Oxygen Sensor (HO2S)

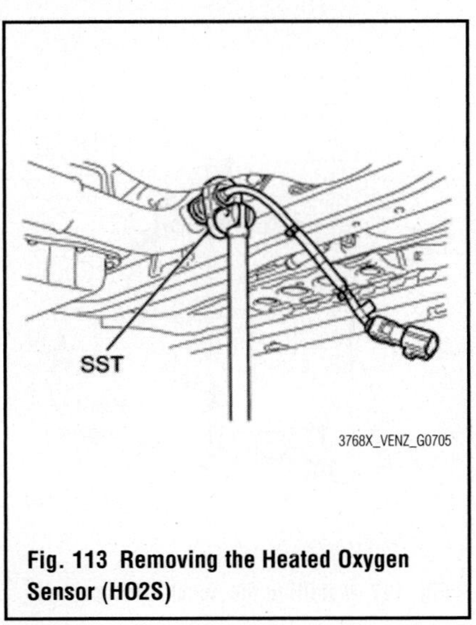

Fig. 113 Removing the Heated Oxygen Sensor (HO2S)

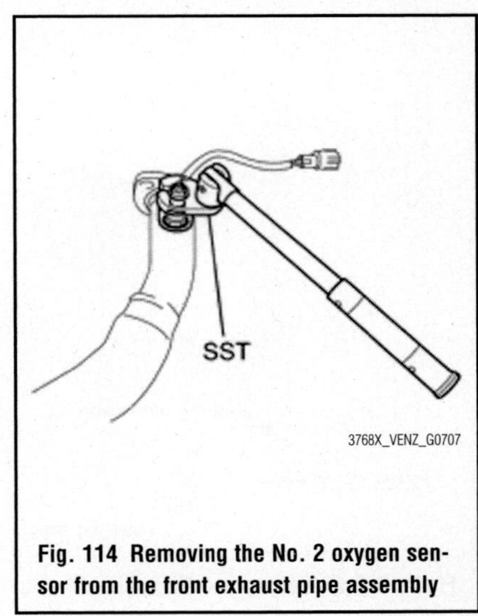

Fig. 114 Removing the No. 2 oxygen sensor from the front exhaust pipe assembly

a. Using the special tool (09224-00010), remove the No. 2 oxygen sensor from the front exhaust pipe assembly.

To install:

9. To install, reverse the removal procedure. Tighten the oxygen sensors to 33 ft. lbs. (44 Nm). If using the special tool, tighten to 30 ft. lbs. (40 Nm).

KNOCK SENSOR (KS)

LOCATION

See Figures 115 and 116.

Refer to the accompanying illustrations.

REMOVAL & INSTALLATION

2.7L Engine

See Figure 117.

1. Remove the intake manifold.
2. Remove the knock sensor.
   a. Disconnect the sensor connector.
   b. Remove the bolt and sensor.

To install:

3. Install the knock sensor.
   a. Install the sensor with the bolt so that the sensor is angled as shown. Tighten to 15 ft. lbs. (20 Nm).

➡The acceptable installation angle of the sensor is between 7°upward and 10°downward from the horizontal position.

   b. Connect the sensor connector.
4. Install the intake manifold.

3.5L Engine

See Figure 118.

1. Remove the intake manifold.
2. Remove knock sensor.

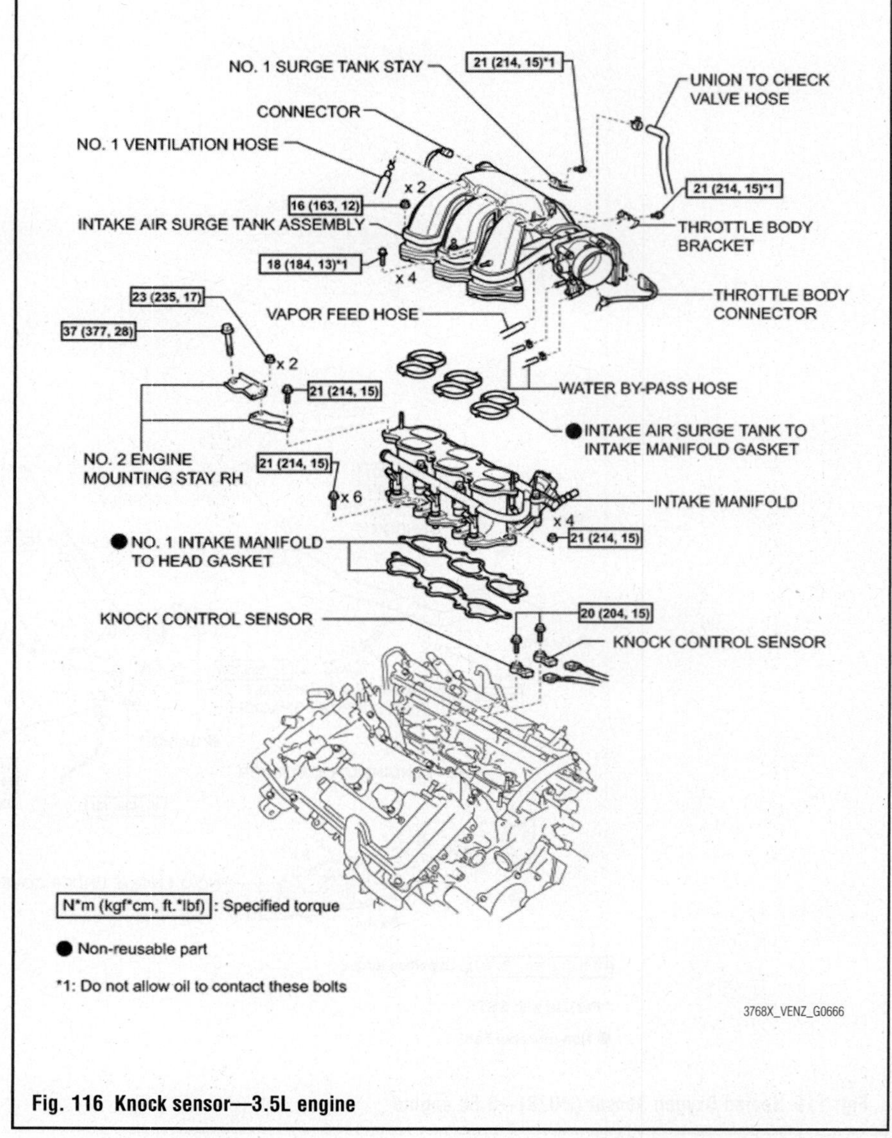

Fig. 116 Knock sensor—3.5L engine

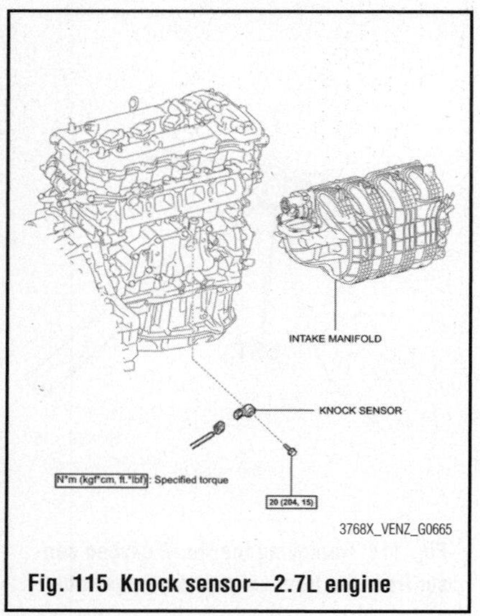

Fig. 115 Knock sensor—2.7L engine

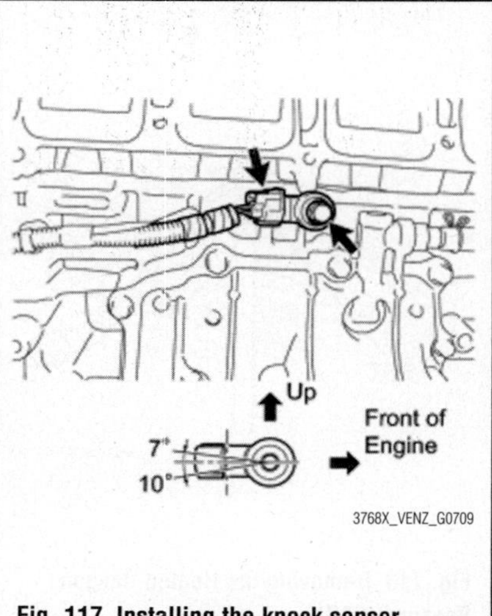

Fig. 117 Installing the knock sensor

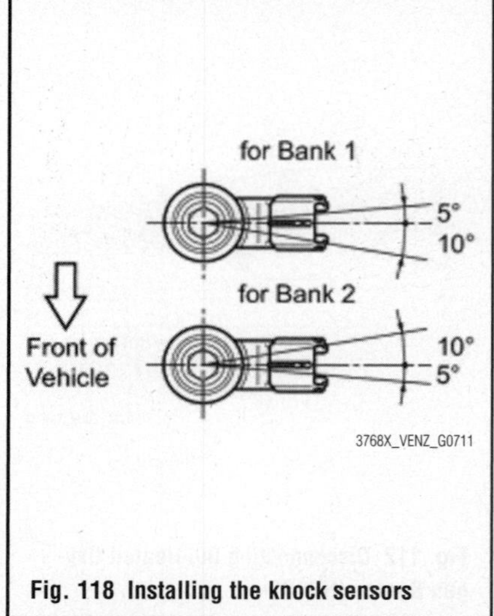

Fig. 118 Installing the knock sensors

a. Disconnect the 2 knock control sensor connectors.

b. Remove the 2 bolts and the 2 knock control sensors.

To install:

3. Install the knock control sensor.

a. Install the 2 knocks control sensors with the 2 bolts as shown. Tighten to 15 ft. lbs. (20 Nm).

➡Make sure that each knock control sensor is in the correct position.

b. Connect the 2 knock control sensor connectors.

4. Install the intake manifold.

MASS AIR FLOW (MAF) METER

LOCATION

See Figures 119 and 120.

Refer to the accompanying illustrations.

REMOVAL & INSTALLATION

2.7L Engine

See Figure 121.

1. Remove the Mass Air Flow (MAF) meter sub assembly.

a. Disconnect the Mass Air Flow (MAF) meter connector.

b. Remove the 2 screws and Mass Air Flow (MAF) meter.

To install:

2. Install the Mass Air Flow (MAF) meter sub assembly.

a. Install the Mass Air Flow (MAF) meter with the 2 screws.

➡Make sure that the o-ring is not cracked or jammed when installing the Mass Air Flow (MAF) meter.

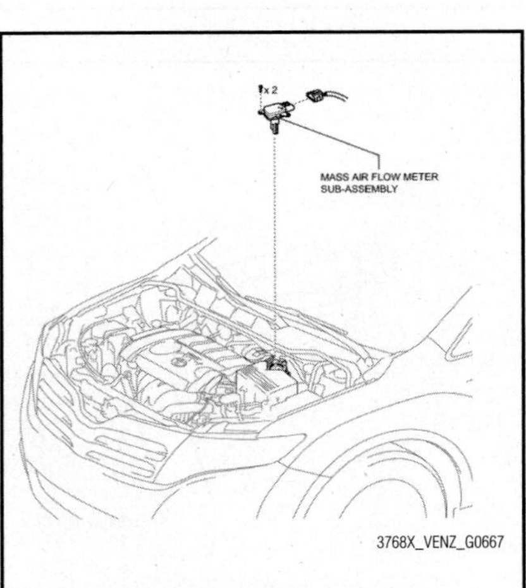

Fig. 119 Mass Air Flow (MAF) meter—2.7L engine

b. Connect the Mass Air Flow (MAF) meter connector.

3.5L Engine

1. Remove the Mass Air Flow (MAF) meter.

a. Separate the Mass Air Flow (MAF) meter connector.

b. Remove the 2 screws and Mass Air Flow (MAF) meter.

To install:

2. Install the Mass Air Flow (MAF) meter.

a. Install the Mass Air Flow (MAF) meter with the 2 screws.

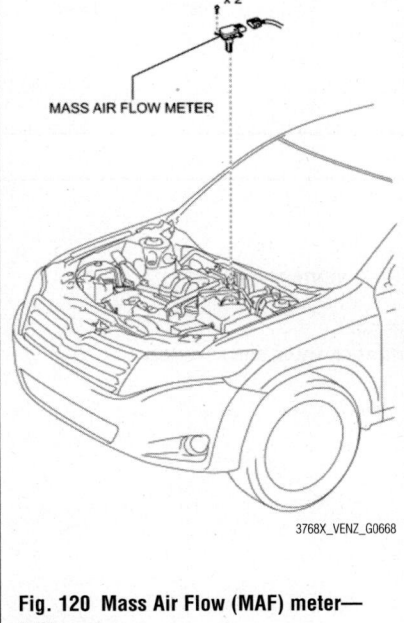

Fig. 120 Mass Air Flow (MAF) meter—3.5L engine

Fig. 121 Removing the Mass Air Flow (MAF) meter sub assembly

➡Make sure that the o-ring is not cracked or does not jump out of position during installation.

b. Connect the Mass Air Flow (MAF) meter connector.

POSITIVE CRANKCASE VENTILATION (PCV) VALVE

LOCATION

See Figures 122 and 123.

Refer to the accompanying illustrations.

REMOVAL & INSTALLATION

2.7L Engine

See Figures 124 and 125.

1. Remove the intake manifold.

2. Remove the ventilation valve sub assembly.

a. Disconnect the No. 2 ventilation hose from the ventilation valve.

b. Using a 19 mm deep socket wrench (*), remove the ventilation valve.

To install:

3. Install the ventilation valve sub assembly.

a. Apply adhesive to 2 or 3 threads of the ventilation valve.

b. Using a 19 mm deep socket wrench, install the ventilation valve. Tighten to 20 ft. lbs. (27 Nm).

c. Connect the No. 2 ventilation hose to the ventilation valve.

4. Install the intake manifold.

5. Inspect for oil leaks.

3.5L Engine

See Figures 126 and 127.

1. Remove the v-bank cover sub assembly.

2. Remove the ventilation valve sub assembly.

a. Disconnect the ventilation hose from the ventilation valve sub assembly.

b. Using a 19 mm deep socket wrench, remove the ventilation valve sub-assembly.

To install:

3. Install the ventilation valve sub assembly.

a. Apply adhesive to 2 or 3 threads of the ventilation valve.

b. Using a 19 mm deep socket wrench, install the ventilation valve. Tighten to 20 ft. lbs. (27 Nm).

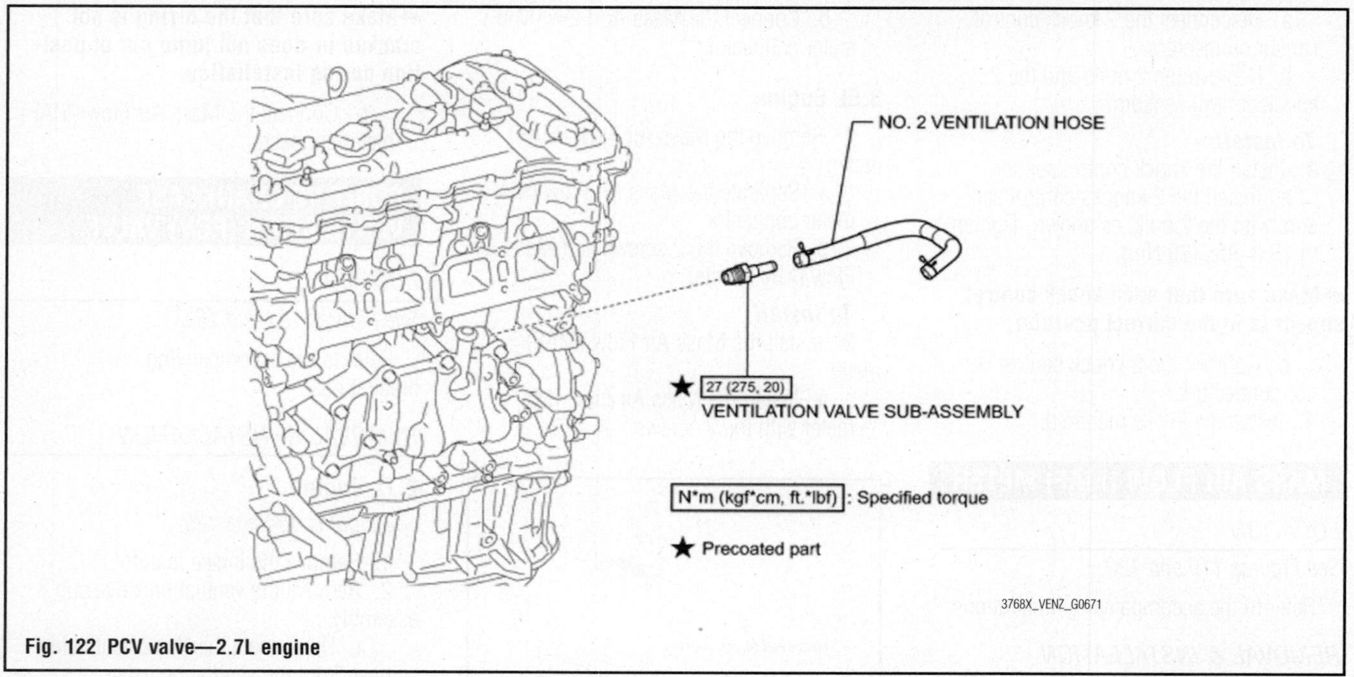

Fig. 122 PCV valve—2.7L engine

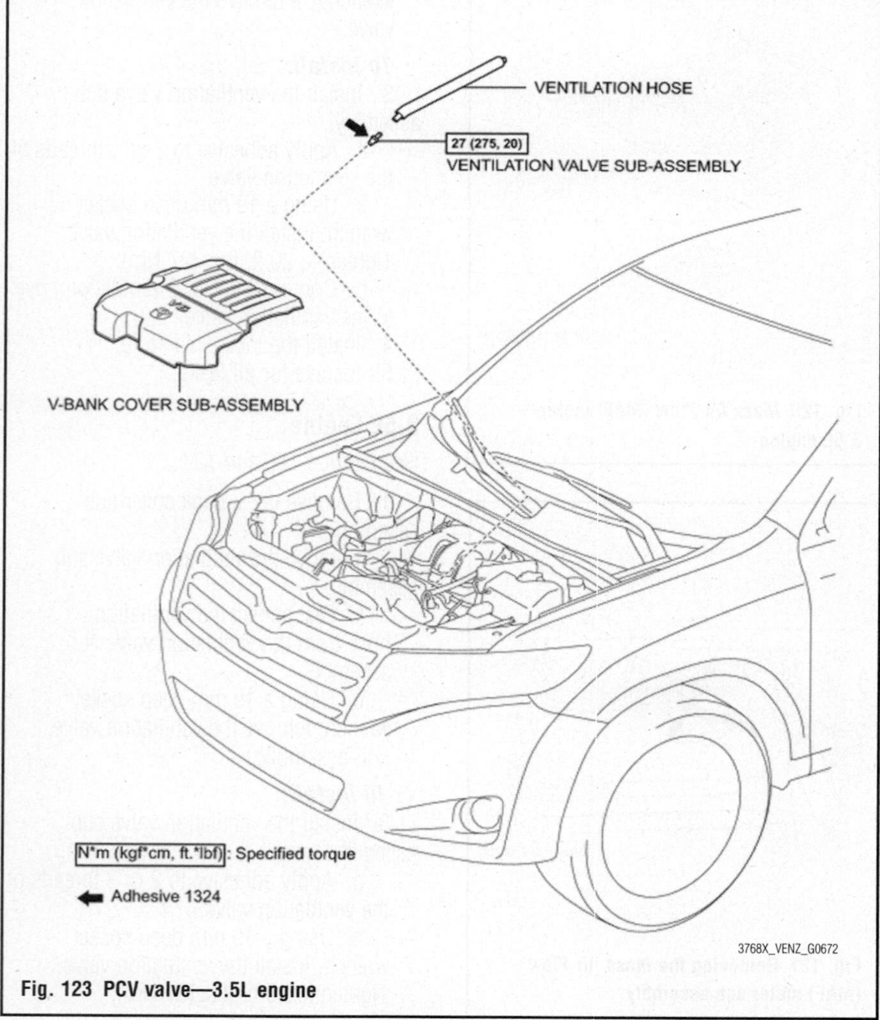

Fig. 123 PCV valve—3.5L engine

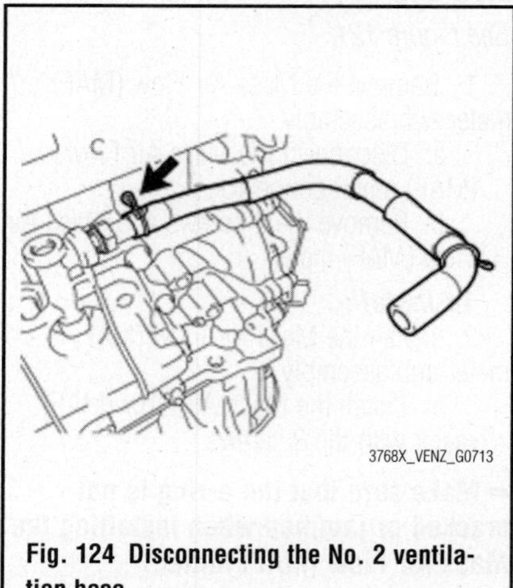

Fig. 124 Disconnecting the No. 2 ventilation hose

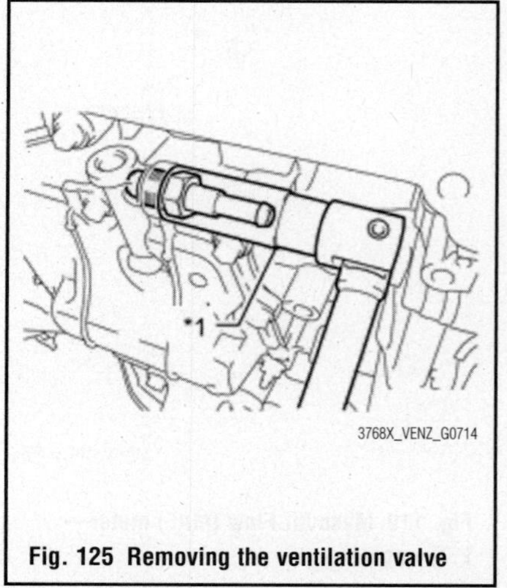

Fig. 125 Removing the ventilation valve

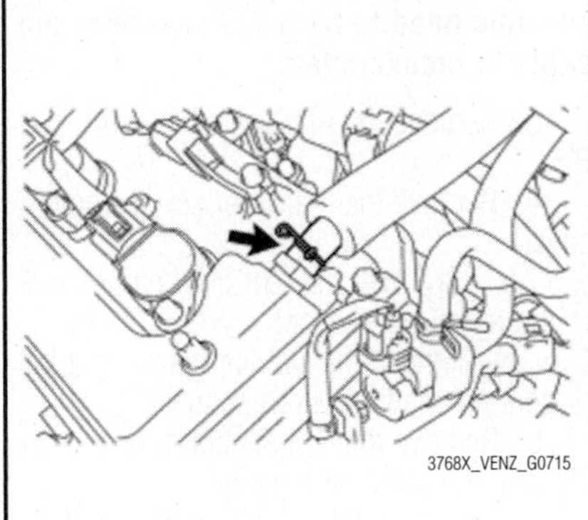

Fig. 126 Disconnecting the ventilation hose

c. Connect the No. 2 ventilation hose to the ventilation valve.
4. Install the intake manifold.
5. Inspect for oil leaks.

VEHICLE SPEED SENSOR (VSS)

REMOVAL & INSTALLATION

See Figure 128.

1. Remove the automatic transaxle assembly.

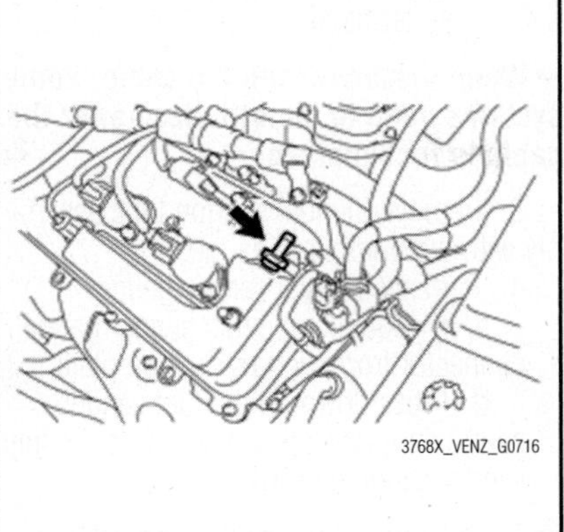

Fig. 127 Removing the ventilation valve sub assembly

2. Remove the automatic transaxle oil pan sub assembly.
3. Remove the valve body oil strainer assembly.
4. Remove the transmission valve body assembly.
5. Remove the speed sensor.
   a. Disconnect the connector.
   b. Remove the 2 bolts and speed sensor from the valve body.

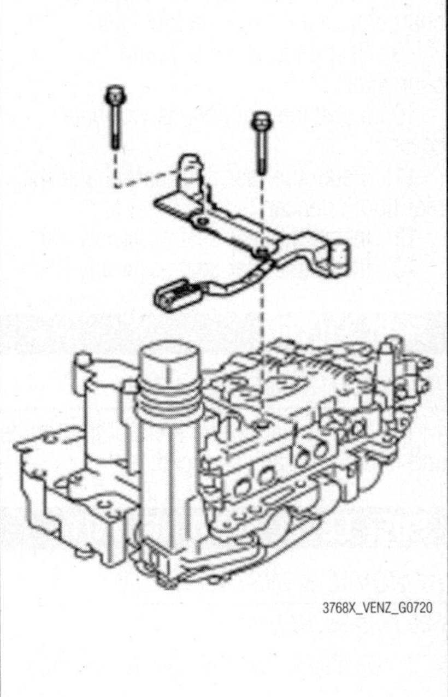

Fig. 128 Removing the speed sensor

To install:

6. To install, reverse the removal procedure. Tighten the speed sensor to 8 ft. lbs. (11 Nm).

➡Coat the 2 bolts with ATF.

FUEL — GASOLINE FUEL INJECTION SYSTEM

FUEL SYSTEM SERVICE PRECAUTIONS

Safety is the most important factor when performing not only fuel system maintenance, but any type of maintenance. Failure to conduct maintenance and repairs in a safe manner may result in serious personal injury or death. Work on a vehicle's fuel system components can be accomplished safely and effectively by adhering to the following rules and guidelines.

- To avoid the possibility of fire and personal injury, always disconnect the negative battery cable unless the repair or test procedure requires that battery voltage be applied.
- Always relieve the fuel system pressure prior to disconnecting any fuel system component (injector, fuel rail, pressure regulator, etc.) fitting or fuel line connection. Exercise extreme caution whenever relieving fuel system pressure to avoid exposing skin, face and eyes to fuel spray. Please be advised that fuel under pressure may penetrate the skin or any part of the body that it contacts.
- Always place a shop towel or cloth around the fitting or connection prior to loosening to absorb any excess fuel due to spillage. Ensure that all fuel spillage is quickly removed from engine surfaces. Ensure that all fuel-soaked cloths or towels are deposited into a flame-proof waste container with a lid.
- Always keep a dry chemical (Class B) fire extinguisher near the work area.
- Do not allow fuel spray or fuel vapors to come into contact with a spark or open flame.
- Always use a second wrench when loosening or tightening fuel line connection fittings. This will prevent unnecessary stress and torsion on fuel piping. Always follow the proper torque specifications.
- Always replace worn fuel fitting O-rings with new ones. Do not substitute fuel hose where rigid pipe is installed.

FUEL SYSTEM PRESSURE

RELIEVING

⁂ CAUTION

The Discharge Fuel System Pressure procedure must be performed before disconnecting any part of the fuel system.

⁂ CAUTION

After performing the Discharge Fuel System Pressure procedure, pressure will remain in the fuel lines. When disconnecting a fuel line, cover it with a shop rag or a piece of cloth to prevent fuel from spraying or coming out.

1. Remove the rear seat assembly LH.
2. Remove the rear seat assembly RH.
3. Turn over the rear floor carpet and the rear floor silencer.
4. Remove the service hole cover and disconnect the fuel pump connector.
5. Start the engine.
6. After the engine has stopped, turn the ignition switch off.

➡DTC P0171/25 (fuel problem) may be detected.

7. Crank the engine again. Check that the engine does not start.

8. Remove the fuel tank cap to discharge pressure from the fuel tank.
9. Reconnect the fuel pump connector.
10. Install the rear floor service hole cover.
11. Install the rear floor carpet and the rear floor silencer.
12. Install the rear seat assembly LH.
13. Install the rear seat assembly RH.

FUEL FILTER

REMOVAL & INSTALLATION

The filter is part of the fuel pump module and is not normally serviced.

FUEL LEVEL SENDING UNIT

REMOVAL & INSTALLATION

See Figures 129 and 130.

1. Discharge the fuel system pressure.

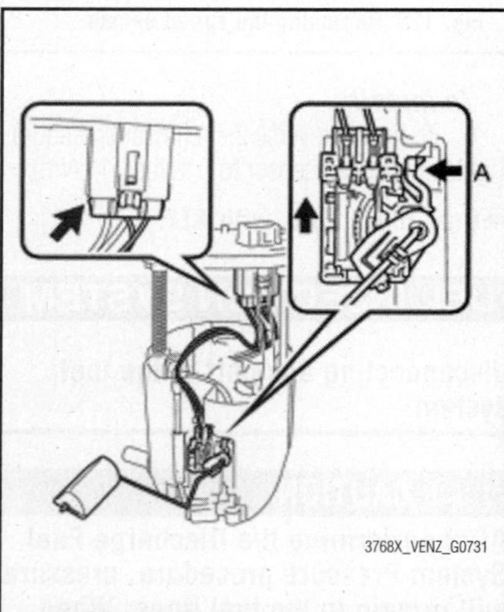

Fig. 129 Removing the fuel sender gauge

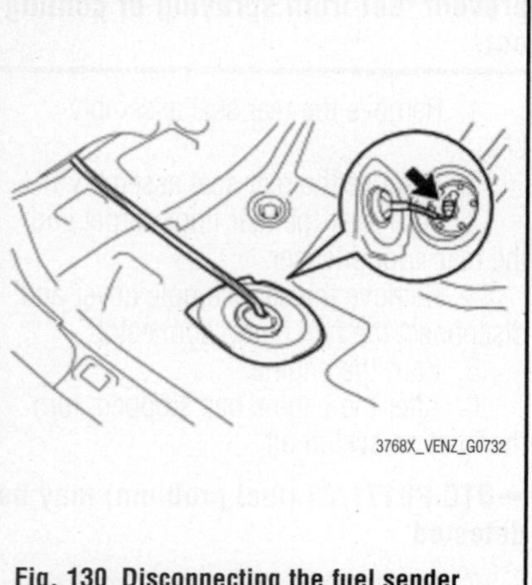

Fig. 130 Disconnecting the fuel sender gauge connector

2. Disconnect the cable from the negative battery terminal.

➡When disconnecting the cable, some systems need to be initialized after the cable is reconnected.

3. Remove the fuel suction tube assembly with the pump and gauge.
4. Remove the fuel sender gauge.
   a. Disconnect the fuel sender gauge connector from the fuel suction plate.
   b. Press down on the fuel sender gauge claw labeled A. Then slide the fuel sender gauge upward.

➡Do not touch the sender resistance plate or contact area.

5. Remove the rear No. 2 floor service hole cover and disconnect the fuel sender gauge connector.
6. Remove the fuel sender gauge assembly.
   a. Remove the 6 bolts and fuel sender gauge assembly from the fuel tank.

➡Be careful not to bend the arm of the fuel sender gauge.

To install:

7. To install, reverse the removal procedure. Tighten the fuel sender gauge assembly bolts to 31 inch lbs. (3.5 Nm).

➡Be careful not to bend the arm of the fuel sender gauge assembly.

FUEL PUMP

REMOVAL & INSTALLATION

See Figures 131 through 133.

1. Discharge the fuel system.
2. Disconnect the cable from the negative battery terminal.

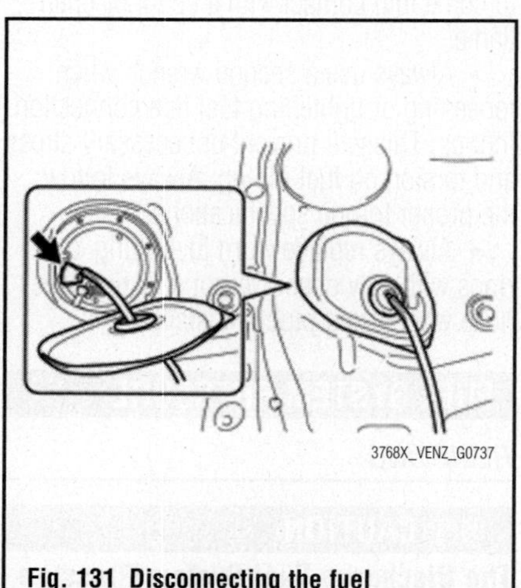

Fig. 131 Disconnecting the fuel pump connector

➡When disconnecting the cable, some systems need to be initialized after the cable is reconnected.

3. Remove the rear seats LH and RH.
4. Remove the rear floor service hole cover.
   a. Turn over the rear floor carpet and the rear floor silencer.
   b. Remove the service hole and disconnect the fuel pump connector.
5. Remove the fuel suction tube assembly with the pump and gauge.
   a. Remove the tube joint clip, and pull out the fuel pump tube.

➡Before removing the tube joint clip, check for foreign matter around the clip. Clean it if necessary.

➡Keep the O-rings free of foreign matter, as they become contaminated easily.

➡Do not use any tools in this procedure.

➡Do not forcefully bend or twist the tube.

➡Put the tube in a plastic bag to prevent damage and contamination.

➡If the fuel suction plate and tube are stuck together, pinch the tube and turn it carefully to disconnect it.

➡Be careful not to damage the clip. If the clip is damaged, replace it.

   b. Remove the 8 bolts and the fuel tank vent tube set plate.

➡While holding the fuel suction tube by hand, remove the fuel tank vent tube set plate.

   c. Disconnect the clip and fuel tube.

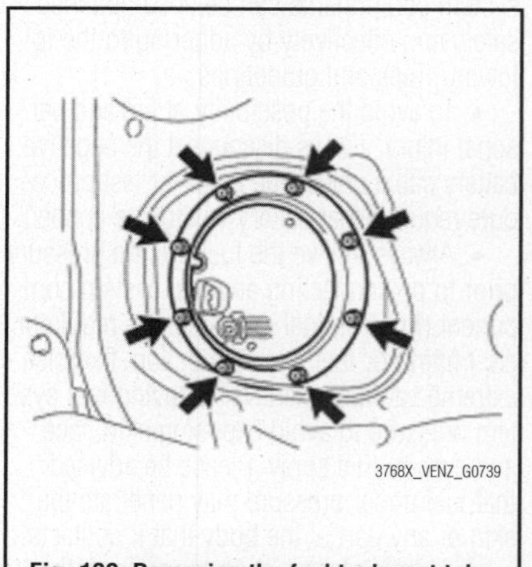

Fig. 132 Removing the fuel tank vent tube set plate

d. Remove the fuel suction tube assembly with pump and gauge from the fuel tank.

➡Make sure that the sender gauge arm does not bend.

➡Do not damage the fuel suction tube.

e. Remove the fuel suction tube set gasket from the fuel tank.

To install:

6. Install the fuel suction tube assembly with the pump and gauge.
   a. Install a new fuel suction tube set gasket onto the fuel tank.
   b. Connect the fuel tube with the clip.
   c. Set the fuel suction tube assembly to the fuel tank.

➡Be careful not to bend the arm of the fuel sender gauge.

➡Do not damage the fuel tube.

d. Align the protrusion of the fuel suction tube assembly with pump and gauge and the cutout of the fuel tank vent tube set plate.

e. While holding the fuel suction with pump and gauge tube assembly by hand, install the fuel tank vent tube to the fuel tank with the 8 bolts. Tighten to 53 inch lbs. (6 Nm).

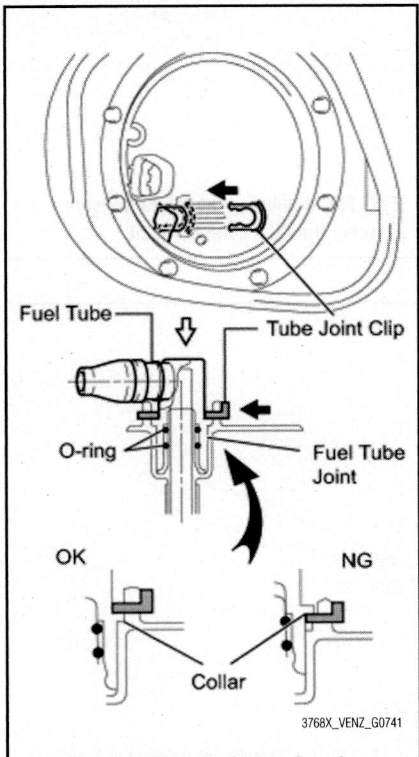

Fig. 133 Installing the fuel tank main tube sub assembly and the tube joint clip

f. Install the fuel tank main tube sub-assembly and the tube joint clip.

➡Check that there are no scratches or foreign matter around the connected part of the fuel tube joint and plug before performing this work.

➡Check that the fuel tube joint is securely inserted.

➡Check that the tube joint clip is on the collar of the fuel tube joint.

➡After installing the tube joint clip, check that the fuel tube cannot be pulled out.

g. Connect the fuel pump connector.

7. Connect the cable to the negative battery terminal.
8. Inspect for fuel leaks.
9. Install the rear floor service hole cover with new butyl tape.
   a. Install the rear floor carpet and the rear floor silencer.
10. Install the rear seats.

FUEL PRESSURE REGULATOR

REMOVAL & INSTALLATION

See Figures 134 through 136.

1. Remove the fuel suction tube assembly with the pump and gauge.
2. Remove the fuel sender gauge.
3. Remove the fuel filter assembly.
   a. Disconnect the fuel pump connector from the fuel suction plate.

➡Do not damage the wire harness.

b. Using needle nose pliers, remove the E-ring.

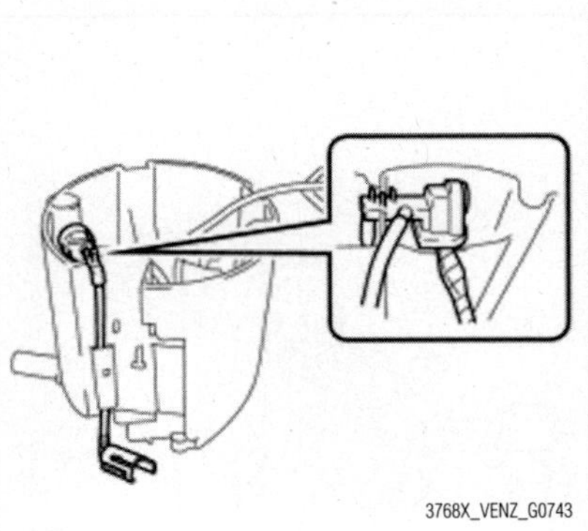

3768X_VENZ_G0743

Fig. 134 Using a screwdriver with the tip taped to remove the jet pump from the sub-tank

➡Do not disconnect the tube shown in the illustration when disassembling the fuel suction tube assembly with pump and gauge. Doing so will cause reassembly of the fuel suction tube assembly with pump and gauge to be impossible as the tube is welded to the plate.

c. Separate the fuel suction plate and remove the spring from the sub-tank.

d. Detach the 2 claws and remove the fuel filter from the sub-tank.

e. Using a screwdriver with the tip taped; detach the claw of the jet pump nozzle.

f. Using a screwdriver with the tip taped; remove the jet pump from the sub-tank.

g. Remove the O-ring from the jet pump.

4. Remove the fuel pump assembly with the filter.
   a. Detach the 5 claws on the filter and remove the fuel pump from the fuel filter.

➡Do not damage the fuel suction filter or fuel filter.

➡Do not remove the fuel suction filter from the fuel pump.

➡Do not do anything which may separate the fuel tube from either the fuel suction plate or fuel filter assembly, such as applying excessive force to the tube.

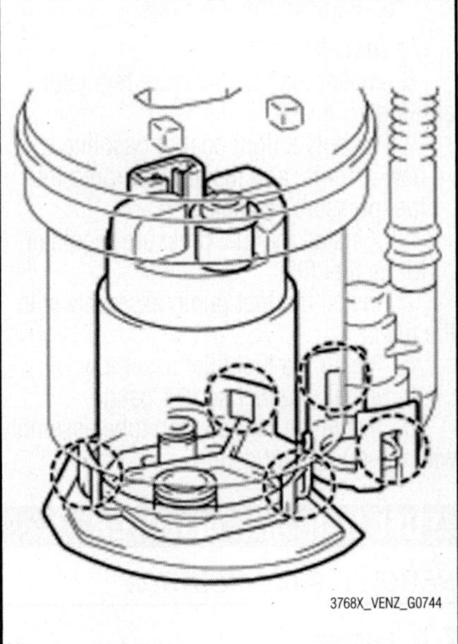

Fig. 135 Removing the fuel pump from the fuel filter

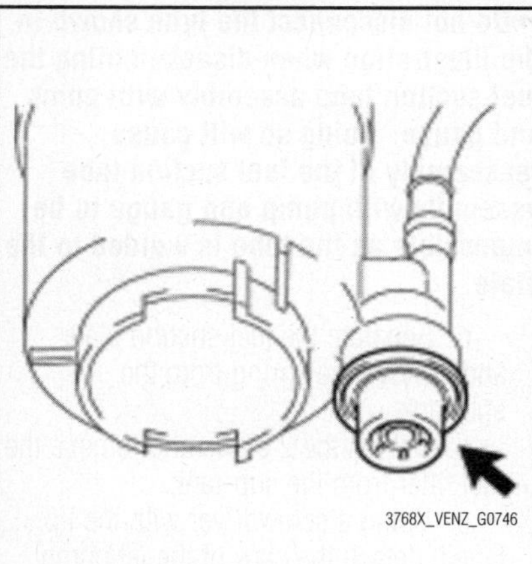

Fig. 136 Removing the fuel pressure regulator assembly

➡If the fuel filter assembly is to be replaced, replace the fuel suction plate sub-assembly.

b. Disconnect the fuel pump connector from the fuel pump and then remove the fuel pump harness.

c. Remove the O-ring and spacer from the fuel pump.

➡Be careful not to damage the sealing surface.

➡If the o-ring still remains in the fuel filter, remove it using a wire tip that is formed as shown in the illustration.

5. Remove the fuel pressure regulator assembly.

a. Remove the fuel pressure regulator from the fuel filter.

b. Remove the 2 o-rings.

To install:

6. Install the fuel pressure regulator assembly.

a. Apply a light coat of gasoline to 2 new o-rings, and install them onto the fuel pressure regulator.

b. Install the fuel pressure regulator to the fuel filter.

7. Install the fuel pump assembly with the filter.

8. Install the fuel filter assembly.

9. Install the fuel sender gauge.

10. Install the fuel suction tube assembly with the pump and gauge.

FUEL RAIL AND INJECTOR

REMOVAL & INSTALLATION

2.7L Engine

See Figures 137 through 139.

1. Discharge the fuel system.

2. Disconnect the cable from the negative battery terminal.

3. Remove the windshield wiper motor and link assembly.

4. Remove the outer cowl top panel.

5. Remove the No. 1 engine cover sub assembly.

6. Remove the No. 1 vacuum switching valve assembly.

7. Remove the air cleaner cap sub assembly.

8. Disconnect the fuel tube sub assembly.

a. Remove the No. 1 fuel pipe clamp.

b. Pinch the tube connector, and then pull the tube connector off of the pipe.

➡Check for foreign matter in the fuel tube around the fuel tube connector. Clean it if necessary. Foreign matter can affect the ability of the O-rings to seal the connector and fuel pipe.

➡Do not use any tools to separate the connector and pipe.

➡Do not forcefully bend, kink or twist the hose.

➡Keep the connector and pipe free from foreign matter.

➡If the connector and pipe are stuck together, pinch the connector and turn it carefully to disconnect it.

➡Put the connector in a plastic bag to prevent damage and contamination.

c. Remove the fuel tube sub-assembly from the fuel hose clamp.

9. Disconnect the wire harness.

a. Disconnect the 4 fuel injector connectors.

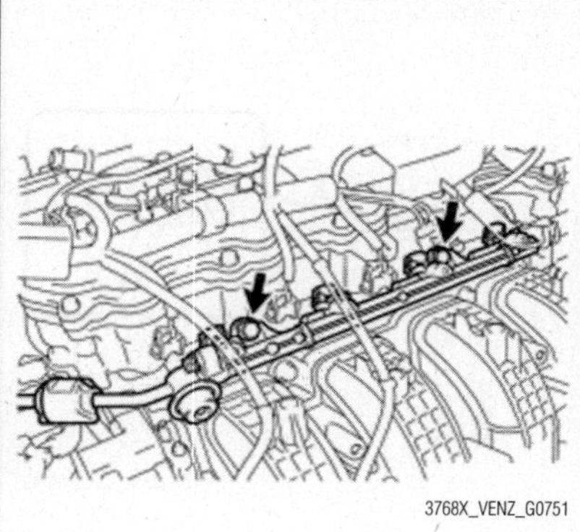

Fig. 137 Removing the fuel delivery pipe with the injectors

b. Disconnect the 3 connectors.

c. Remove the 2 bolts and 2 wire harness brackets.

d. Detach the 2 clamps to disconnect the wire harness.

10. Remove the vacuum switching valve.

11. Remove the fuel delivery pipe sub assembly.

a. Remove the 2 bolts, and then remove the fuel delivery pipe together with the 4 fuel injectors.

➡Be careful not to drop the fuel injectors when removing the fuel delivery pipe.

b. Remove the 2 fuel delivery spacers from the cylinder head.

c. Remove the 4 injector vibration insulators from the cylinder head.

12. Remove the fuel injector assembly.

a. Pull the 4 fuel injectors out of the fuel delivery pipe.

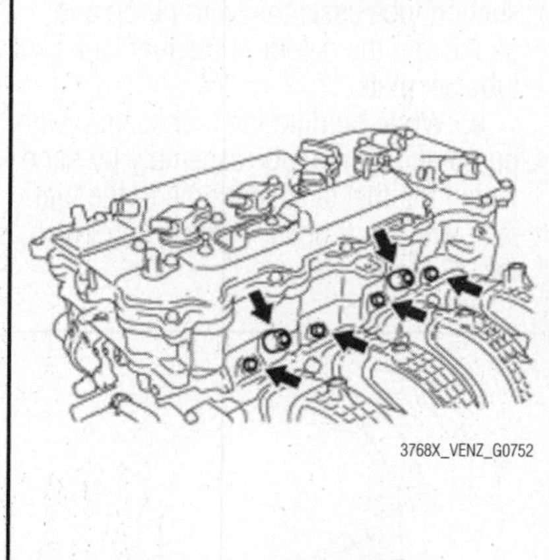

Fig. 138 Removing the 2 fuel delivery spacers from the cylinder head

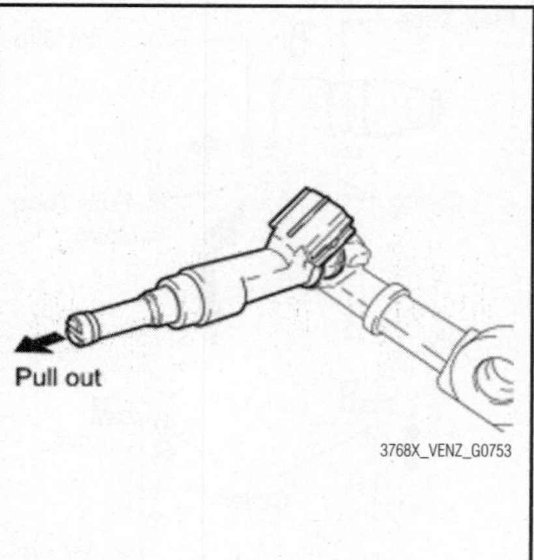

Fig. 139 Removing the fuel injector assembly

To install:

13. Install the fuel injector assembly.

a. Apply a light coat of gasoline or spindle oil to new O-ring, and then install one onto each fuel injector.

b. Apply a light coat of gasoline or spindle oil to the part of the fuel delivery pipe which comes into contact with the O-ring of the fuel injector.

c. Apply a light coat of gasoline or spindle oil to the O-ring again, and then install the fuel injectors onto the fuel delivery pipe.

➡Make sure that the O-ring is not cracked or jammed when installing the injector.

d. Check that the fuel injector rotates smoothly. If the fuel injector does not rotate, replace the O-ring.

14. Install the fuel delivery pipe sub assembly.

a. Install 4 new injector vibration insulators to the cylinder head.

b. Install the 2 fuel delivery spacers onto the cylinder head.

➡Install the fuel delivery spacer so that the longer protrusion is on the cylinder head side.

c. Install the fuel delivery pipe together with the 4 fuel injectors to the cylinder head, and then temporarily install the 2 bolts.

➡Be careful not to drop the fuel injectors when installing the fuel delivery pipe.

d. Check that the fuel injector rotates smoothly. If the fuel injector does not rotate, replace the O-ring.

e. Tighten the 2 bolts to 15 ft. lbs. (21 Nm).

15. Connect the wire harness.

a. Install the 2 wire harness brackets with the 2 bolts. Tighten to 7 ft. lbs. (10 Nm).

b. Connect the 3 connectors.

c. Connect the 4 fuel injector connectors.

d. Attach the 2 clamps to connect the wire harness.

16. Install the vacuum switching valve assembly.

17. Connect the fuel tube sub assembly.

a. Push the tube connector to the pipe until the tube connector makes a "click" sound.

➡Before connecting the connector and fuel pipe, check that there is no damage or foreign matter on the connecting part of the fuel pipe.

➡After connecting the fuel tube connector and pipe, check that they are securely connected by trying to pull them apart.

b. Install the No. 1 fuel pipe clamp.

c. Install the fuel tube sub-assembly to the fuel hose clamp.

18. Install the air cleaner cap sub assembly.

19. Install the No. 1 vacuum switching valve assembly.

20. Install the No. 1 engine cover sub assembly.

21. Install the outer cowl top panel.

22. Install the windshield wiper motor and link assembly.

23. Connect the negative battery terminal.

24. Inspect for fuel leaks.

3.5L Engine

See Figures 140 and 141.

1. Discharge the fuel system pressure.

2. Disconnect the cable from the negative battery terminal.

➡When disconnecting the cable, some systems need to be initialized after the cable is reconnected.

3. Remove the intake air surge tank assembly.

4. Disconnect the fuel tube sub assembly.

a. Remove the No. 2 fuel pipe clamp.

b. Pinch the tube connector and pull out the fuel pipe.

➡Check that there is no dirt or other foreign objects around the connector when disconnecting it. Clean the connector as necessary.

➡It is necessary to prevent dirt or foreign objects from entering the quick connector. If dirt or foreign objects get in the connector, the O-rings may not seal properly.

➡Only disconnect the quick connector by hand.

➡Do not bend, kink or twist the nylon tubes.

➡Protect the connector by covering it with a plastic bag.

➡If the pipe and the connector are stuck, carefully try wiggling or pushing and pulling on the connector to release it. Pull the connector off the pipe carefully.

5. Remove the fuel injector assembly.

a. Disconnect the 6 fuel injector connectors.

b. Remove the 5 bolts and fuel delivery pipe sub-assembly together with the 6 fuel injectors.

➡Be careful not to drop the fuel injectors when removing the fuel delivery pipe.

6. Remove the 6 injector vibration insulators from the intake manifold.

7. Pull out the fuel injectors from the fuel delivery pipe.

➡If the injectors are to be reused, reinstall them to the same cylinder they came from.

a. Remove the 6 O-rings from the injectors.

To install:

8. Install the fuel injector assembly.

a. Apply a light coat of spindle oil or gasoline to new O-rings, and install them to each injector.

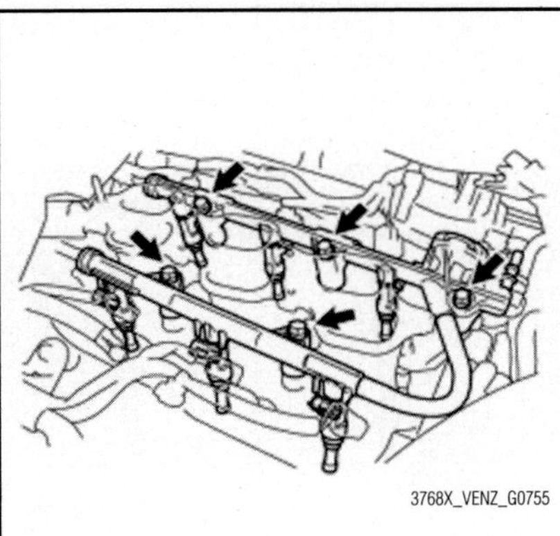

Fig. 140 Removing the 5 bolts and fuel delivery pipe sub-assembly together with the 6 fuel injectors

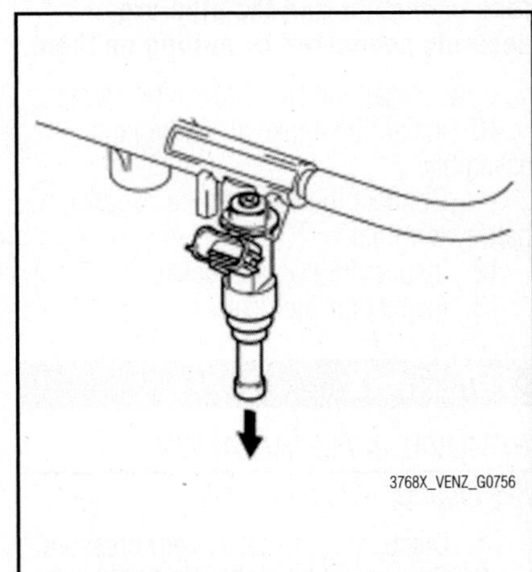

Fig. 141 Pulling out the fuel injectors from the fuel delivery pipe

b. Apply a light coat of spindle oil or gasoline where the fuel delivery pipe contacts each O-ring.

c. Push and twist each fuel injector to install them into the fuel delivery pipe.

d. Position each fuel injector connector outward.

➡**Be careful not to twist the O-rings.**

➡**After installing a fuel injector, check that it turns smoothly. If not, reinstall it with a new O-ring.**

e. Install 6 new injector vibration insulators to the intake manifold.

f. Place the fuel delivery pipe with the 6 fuel injectors in position on the intake manifold.

➡**Be careful not to drop the fuel injectors when installing the fuel delivery pipe.**

g. Temporarily install the 5 bolts which are used to hold the fuel delivery pipe to the intake manifold.

➡**After installing the fuel injectors, check that they turn smoothly. If not, reinstall the injectors with new O-rings.**

h. Tighten the 5 bolts which are used to hold the fuel delivery pipe to the intake manifold. Tighten to 15 ft. lbs. (21 Nm).

i. Connect the 6 fuel injector connectors.

9. Connect the fuel tube sub assembly.

a. Push in the tube connector onto the pipe until the tube connector makes a "click" sound.

➡**Before connecting the tube, make sure that it is not damaged. Make sure that there is no dirt present on the connecting surfaces.**

➡**After connecting, check that the fuel tube connector and the pipe are securely connected by pulling on them.**

b. Install the No. 2 fuel pipe clamp.

10. Install the intake air surge tank assembly.

11. Connect the cable to the negative battery terminal.

12. Inspect for coolant leaks.

13. Inspect for fuel leaks.

FUEL TANK

REMOVAL & INSTALLATION

See Figures 142 through 149.

1. Discharge the fuel system pressure.

2. Disconnect the cable from the negative battery terminal.

➡**When disconnecting the cable, some systems need to be initialized after the cable is reconnected**

3. Remove the fuel suction tube assembly with the pump and gauge.

4. Remove the rear No. 2 floor service hole cover.

5. Remove the fuel sender gauge assembly.

6. Drain the fuel.

7. Remove the center exhaust pipe assembly.

8. Remove the propeller with the center bearing shaft assembly (AWD).

9. Remove the electromagnetic control coupling sub assembly (AWS).

10. Remove the front floor brace LH.

a. Remove the bolt, 2 nuts and front floor brace LH.

11. Remove the front floor brace RH.

12. Remove the bolt, 2 nuts and front floor brace RH.

13. Remove the No. 1 fuel tank protector sub assembly.

14. Separate the No. 2 parking brake cable assembly.

15. Separate the No. 3 parking brake cable assembly.

16. Remove the fuel tank assembly.

a. Disconnect the fuel pump tube. Pinch the tabs of the retainer to release the lock claws and pull it down as shown in the illustration. Pull out the fuel tank main tube.

➡**Check that there is no dirt or other foreign objects around the connector before this operation and clean the connector as necessary.**

➡**It is necessary to prevent mud or dirt from entering the connector. If mud or dirt gets in the connector, the O-rings may not seal properly.**

➡**Do not use any tools in this operation.**

➡**Do not bend, kink or twist the nylon tube. Protect the connector by covering it with a plastic bag.**

➡**When the pipe and connector are stuck, push and pull on them to release them.**

b. Disconnect the fuel tank vent hose sub assembly from the charcoal canister assembly.

c. Set up an engine lifter underneath the fuel tank.

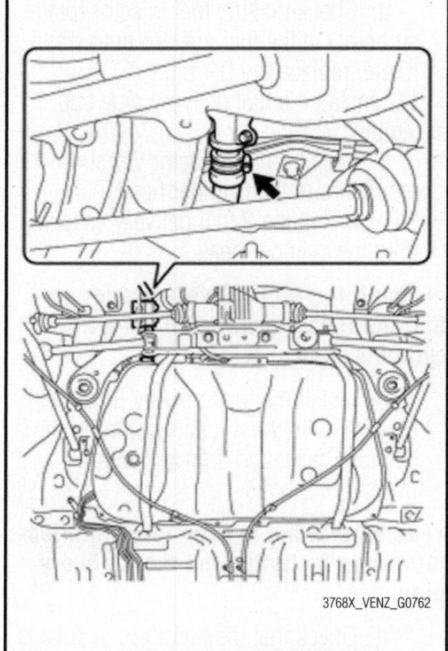

3768X_VENZ_G0762

Fig. 143 Removing the hose clamp and disconnecting the fuel tank to filter pipe hose

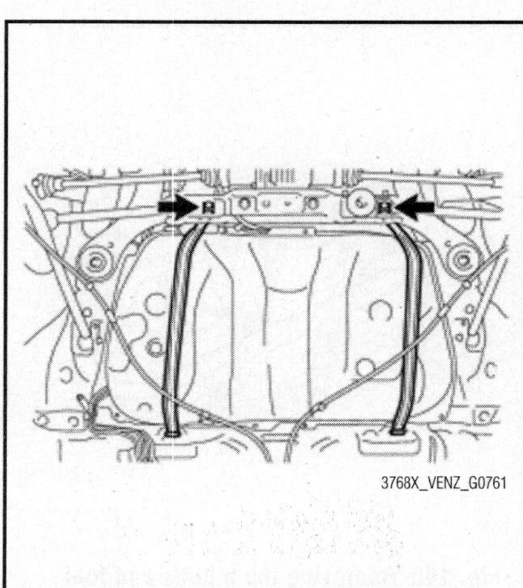

3768X_VENZ_G0761

Fig. 142 Removing the 2 set bolts of the fuel tank bands

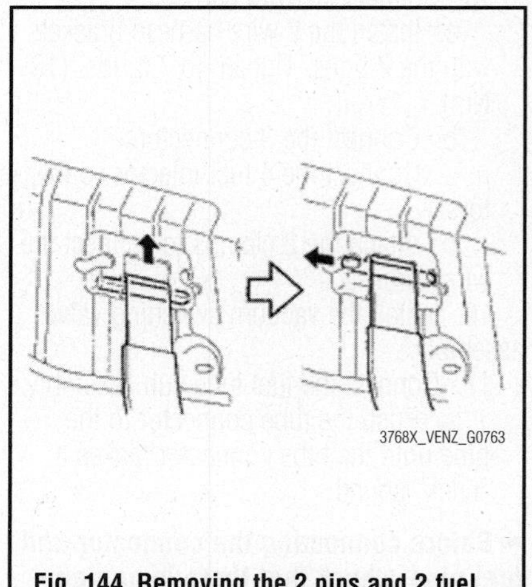

3768X_VENZ_G0763

Fig. 144 Removing the 2 pins and 2 fuel tank bands

d. Remove the 2 set bolts of the fuel tank bands.

e. Remove the hose clamp and disconnect the fuel tank to filter pipe hose.

f. Remove the fuel tank.

g. Remove the 2 pins and 2 fuel tank bands as shown in the illustration.

h. Remove the 4 clip nuts.

17. remove the fuel tank vent hose sub assembly.

a. Remove the fuel tank vent hose sub-assembly from the fuel tank.

18. Remove the fuel tank to filler pipe hose.

a. Loosen the hose clamp bolt and remove the fuel tank to filler pipe hose from the fuel tank assembly.

19. remove the fuel tank main tube sub assembly.

a. Remove the fuel tank main tube from the fuel main tube support.

20. Remove the fuel main tube support.

21. Remove the No. 1 fuel tank cushion.

a. Remove the 8 No. 1 fuel tank cushions.

To install:

22. Install the No. 1 fuel tank cushions.

23. Install the fuel main tube support with the bolt. Tighten to 48 inch lbs. (5.4 Nm).

24. Install the fuel tank main tube sub assembly.

a. Install the fuel tank main tube to the fuel main tube support.

25. Install the fuel tank to filler pipe hose to the fuel tank assembly with the clamp.

26. Install the fuel tank vent hose sub assembly.

27. Install the fuel sender gauge assembly.

28. Install the fuel tank assembly.

a. Install the 4 clip nuts.

b. Install the 2 fuel tank bands with the 2 pins.

c. Set the fuel tank assembly onto the engine lifter.

d. Lift up the engine lifter.

➡Slowly raise the lifter so as to not drop the fuel tank assembly.

e. Install the fuel tank assembly with the fuel tank bands and 2 bolts. Tighten to 29 ft. lbs. (39 Nm).

f. Connect the fuel tank to filler pipe hose with the clamp.

g. Connect the fuel tank vent hose sub-assembly.

h. Connect the fuel tank main tube sub-assembly. Push in the fuel pump tube connector to the pipe and push up the retainer so that the claws engage.

➡Check that there is no damage or foreign objects on the connected part.

➡After connecting, check if the fuel tube connector and the pipe are securely connected by trying to pull them apart.

29. Install the No. 3 parking brake cable assembly with the bolt and nut. Tighten to 53 inch lbs. (6 Nm).

30. Install the No. 2 parking brake cable assembly with the bolt and nut. Tighten to 53 inch lbs. (6 Nm).

31. Install the No. 1 fuel tank protector sub assembly with the 4 bolts. Tighten to 48 inch lbs. (5.4 Nm).

32. Install the front floor brace LH.

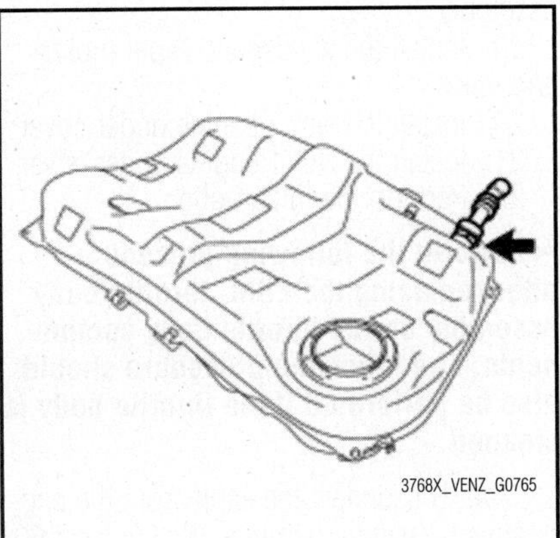

Fig. 145 Loosening the hose clamp bolt and removing the fuel tank to filler pipe hose from the fuel tank assembly

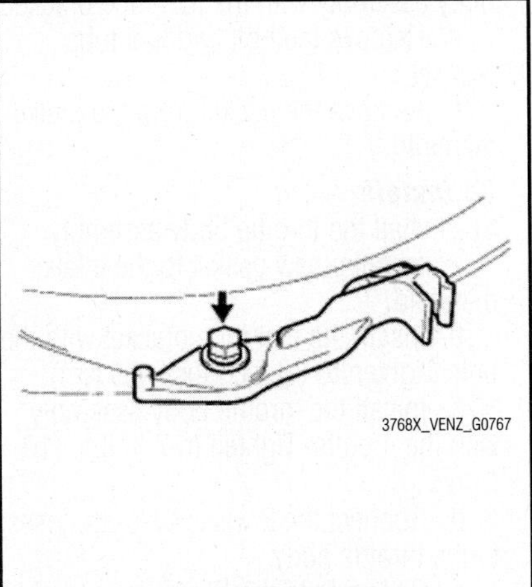

Fig. 147 Removing the fuel main tube support

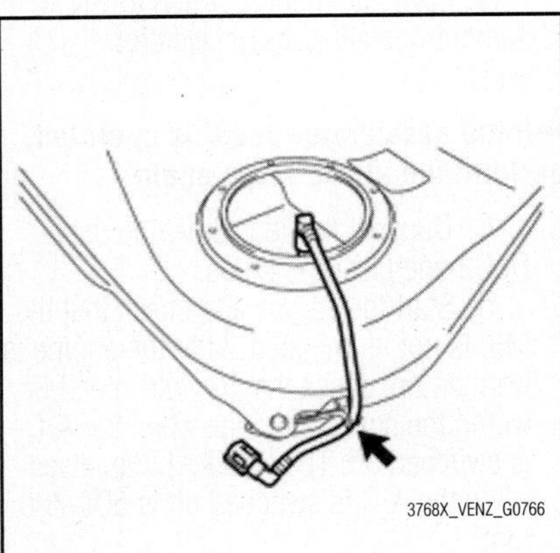

Fig. 146 Removing the fuel tank main tube from the fuel main tube support

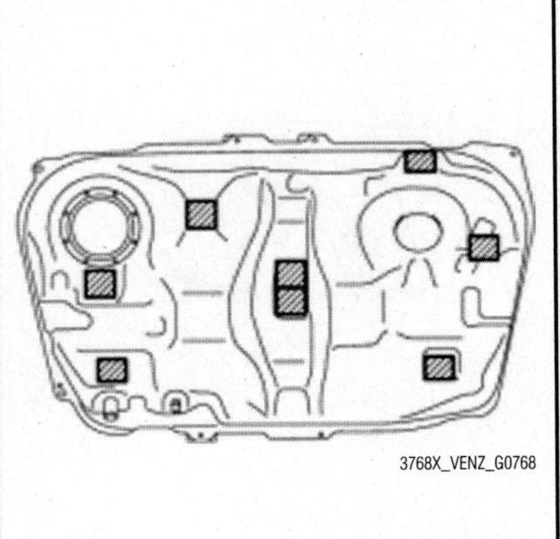

Fig. 148 Removing the No. 1 fuel tank cushions

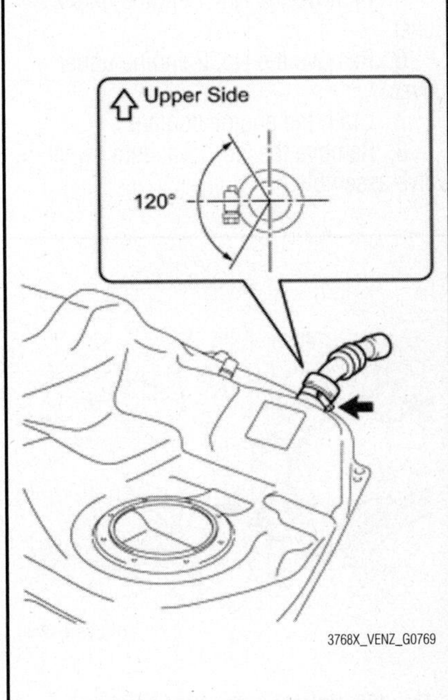

Fig. 149 Installing the fuel tank to filler pipe hose

a. Install the bolt, 2 nuts and the front floor brace LH. Tighten to 41 ft. lbs. (56 Nm).

33. Install the front floor brace RH.

a. Install the bolt, 2 nuts and the front floor brace RH. Tighten to 41 ft. lbs. (56 Nm).

34. Install the electromagnetic control coupling sub assembly (AWD).

35. Install the propeller with the center bearing shaft assembly (AWD).

36. Install the center exhaust pipe assembly.

37. Add fuel.

38. Install the rear No. 2 floor service hole cover.

39. Install the fuel suction tube assembly with the pump and gauge.

40. Connect the cable to the negative battery terminal.

41. Inspect for exhaust gas leak.

THROTTLE BODY

REMOVAL & INSTALLATION

2.7L Engine

See Figures 150 and 151.

1. Remove the windshield wiper motor and link.
2. Remove the outer cowl top panel sub assembly.
3. Remove the No. 1 engine cover sub assembly.
4. Remove the cool air intake duct seal.
5. Remove the No. 1 engine under cover.
6. Remove the No. 2 engine under cover.
7. Drain the engine coolant.
8. Remove the No. 1 vacuum switching valve assembly.
9. Remove the air cleaner cap sub assembly.

a. Disconnect the Mass Air Flow (MAF) meter connector and separate the wire harness clamp from the air cleaner cap.

b. Separate the hose from the hose clamp.

c. Disconnect the ventilation hose from the cylinder head cover.

d. Unlock the hose band and separate the air cleaner cap sub-assembly from the throttle body assembly.

e. Remove the 2 bolts and air cleaner cap sub-assembly.

10. Remove the throttle body assembly.

a. Disconnect the throttle body assembly connector.

b. Disconnect the fuel tube from the clamp.

c. Disconnect the 2 water by-pass hoses from the throttle body assembly.

d. Remove the 4 bolts and the throttle body assembly with the fuel tube bracket.

e. Remove the bolt and fuel tube bracket.

f. Remove the gasket from the intake manifold.

To install:

11. Install the throttle body assembly.

a. Install a new gasket to the intake manifold.

b. Install the fuel tube bracket with the bolt. Tighten to 66 inch lbs. (7.5 Nm).

c. Install the throttle body assembly with the 4 bolts. Tighten to 7 ft. lbs. (10 Nm).

d. Connect the 2 water by-pass hoses to the throttle body.

e. Connect the throttle body assembly connector.

f. Connect the fuel tube to the clamp.

12. Install the air cleaner cap sub assembly.

a. Connect the air-cleaner cap sub-assembly to the throttle body assembly and lock the hose band.

b. Install the air cleaner cap sub-assembly with the 2 bolts. Tighten to 44 inch lbs. (5 Nm).

c. Connect the ventilation hose to the cylinder head cover.

d. Connect the Mass Air Flow (MAF) meter connector and install the wire harness clamp to the air cleaner cap.

e. Install the hose to the hose clamp.

13. Install the No. 1 vacuum switching valve assembly.
14. Add engine coolant.
15. Inspect for coolant leaks.
16. Install the No. 1 engine cover sub assembly.
17. Install the cool air intake duct seal.
18. Install the outer cowl top panel sub assembly.
19. Install the windshield wiper motor and link.
20. Install the No. 2 engine under cover.
21. Install the No. 1 engine under cover.
22. Perform the initialization.

➡Perform the following procedure after replacing the ECM, throttle body assembly or any throttle body components. The following procedure should also be performed if the throttle body is cleaned.

a. Disconnect the cable from the negative (-) battery terminal. Wait at least 60 seconds and reconnect the cable.

➡When disconnecting the cable, some systems need to be initialized after the cable is reconnected.

b. Turn the ignition switch to ON without operating the accelerator pedal.

➡If the accelerator pedal is operated, perform the above steps again.

c. Connect the Techstream to the DLC3 and clear the DTCs.

d. Start the engine and check that the MIL is not illuminated. After the engine is warmed up, check that the idle speed is within the specified range when the A/C is switched off. The standard idle speed when the A/C is switched off is 600-700 rpm.

➡Be sure to perform this step with all accessories off.

➡Make sure that the shift lever is in N.

Fig. 150 Unlocking the hose band and separating the air cleaner cap sub-assembly from the throttle body assembly

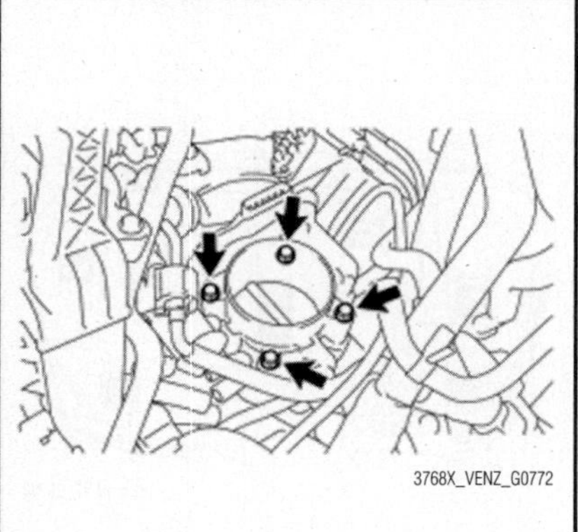

Fig. 151 Removing the 4 bolts and the throttle body assembly with the fuel tube bracket

e. Enter the following menus: Powertrain / Engine and ECT / Data List / Throttle Sensor Position. Sensor Output. Fully depress the accelerator pedal and check that the value is 60% or more.

f. Perform a road test and confirm that there are no abnormalities.

3.5L Engine

See Figures 152 and 153.

1. Remove the engine No. 1 under cover.
2. Remove the engine No. 2 under cover.
3. Drain the engine coolant.
4. Remove the v-bank cover sub assembly.
5. Remove the air cleaner cap with the hose.
6. Remove the throttle body.
   a. Disconnect the 2 water by pass hoses from the throttle body.
   b. Disconnect the throttle body connector and wire harness clamp.
   c. Remove the 4 bolts, throttle body and wire harness clamp stay.
   d. Remove the gasket from the intake air surge tank assembly.

To install:

7. Install the throttle body.
   a. Install a new gasket to the intake air surge tank assembly.
   b. Install the throttle body and wire harness clamp stay to the intake air surge tank assembly with the 4 bolts. Tighten to 7 ft. lbs. (10 Nm).
   c. Connect the throttle body connector and wire harness clamp.
   d. Connect the 2 water by-pass hoses to the throttle body.
8. Install the air cleaner cap with the hose.
9. Add engine coolant.
10. Inspect for engine coolant leaks.
11. Install the v-bank cover sub assembly.
12. Install the No. 2 engine under cover.
13. Install the No. 1 engine under cover.
14. Perform the initialization.

➡Perform the following procedure after replacing the ECM, throttle body assembly or any throttle body components. The following procedure should also be performed if the throttle body is cleaned.

a. Disconnect the cable from the negative (-) battery terminal. Wait at least 60 seconds and reconnect the cable.

➡When disconnecting the cable, some systems need to be initialized after the cable is reconnected.

b. Turn the ignition switch to ON without operating the accelerator pedal.

➡If the accelerator pedal is operated, perform the above steps again.

c. Connect the Techstream to the DLC3 and clear the DTCs.

d. Start the engine and check that the MIL is not illuminated. After the engine is warmed up, check that the idle speed is within the specified range when the A/C is switched off. The standard idle speed when the A/C is switched off is 600-700 rpm.

➡Be sure to perform this step with all accessories off.

➡Make sure that the shift lever is in N.

e. Enter the following menus: Powertrain / Engine and ECT / Data List / Throttle Sensor Position. Sensor Output. Fully depress the accelerator pedal and check that the value is 60% or more.

f. Perform a road test and confirm that there are no abnormalities.

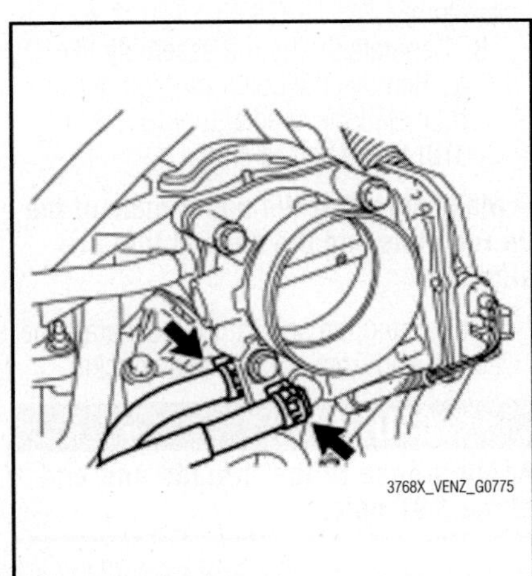

Fig. 152 Disconnecting the 2 water by pass hoses

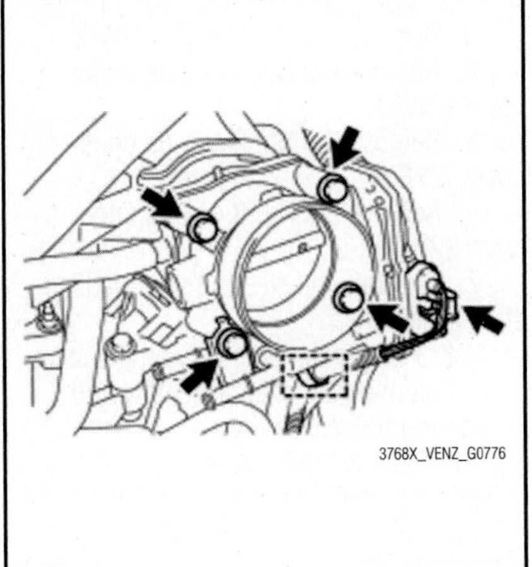

Fig. 153 Removing the throttle body

HEATING & AIR CONDITIONING SYSTEM

BLOWER MOTOR

REMOVAL & INSTALLATION

See Figure 154.

1. Disconnect the cable from the negative battery terminal.
2. Remove the No. 2 instrument panel under cover sub assembly.
3. Remove the front blower motor sub assembly.
   a. Disconnect the connector.
   b. Remove the 3 screws and front blower motor sub assembly.

To install:

4. To install, reverse the removal procedure.

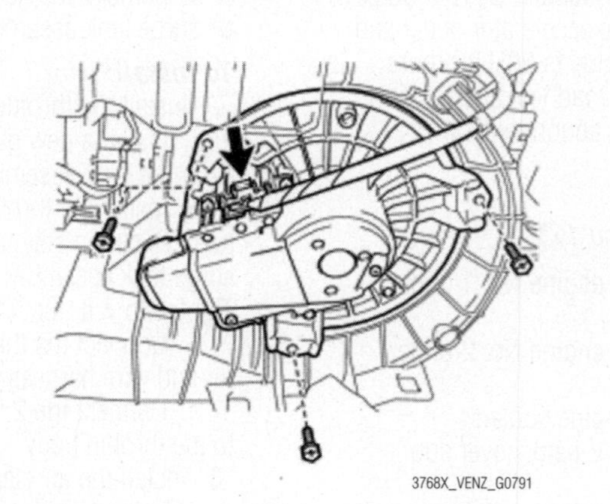

Fig. 154 Removing the front blower motor

STEERING

POWER RACK & PINION STEERING GEAR

REMOVAL & INSTALLATION

2.7L Engine

See Figures 155 through 159.

➡When disconnecting the steering intermediate shaft assembly and pinion shaft of the steering gear assembly, be sure to place matchmarks before servicing.

1. Place the front wheels facing straight ahead (2WD).
2. Secure the steering wheel in order to prevent it from rotating.

➡This operation is necessary to prevent damage to the spiral cable.

3. Remove the front wheels (2WD).
4. Remove the No. 1 engine under cover (2WD).
5. Remove the No. 2 engine under cover (2WD).
6. Remove the No. 3 engine under cover (2WD).
7. Separate the steering intermediate shaft assembly.
   a. Put matchmarks (*1) on the steering intermediate shaft assembly and steering link assembly.
   b. Remove the bolt and slide the steering intermediate shaft assembly.
   c. Separate the steering intermediate shaft assembly from the steering link assembly.
8. Separate the tie rod assembly LH.
   a. Remove the cotter pin and nut.
   b. Install the special tool (09960-20010) to the tie rod end.

➡Make sure that the upper ends of the tie rod ends and the special tool are aligned.

   c. Using the special tool, separate the tie rod end from the steering knuckle.

CAUTION

Apply grease to the threads and end of the SST bolt.

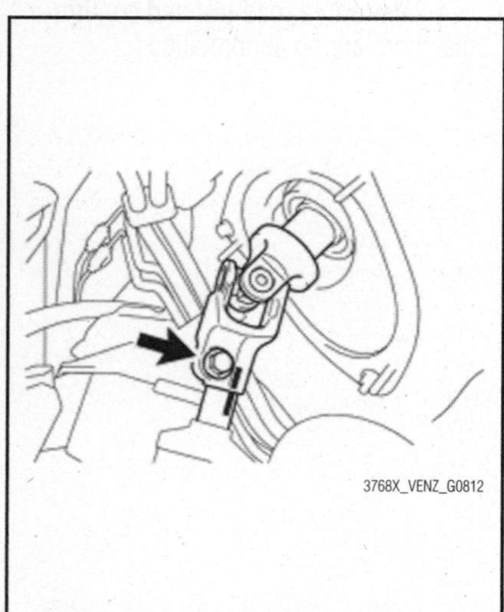

Fig. 155 Removing the bolt and sliding the steering intermediate shaft assembly

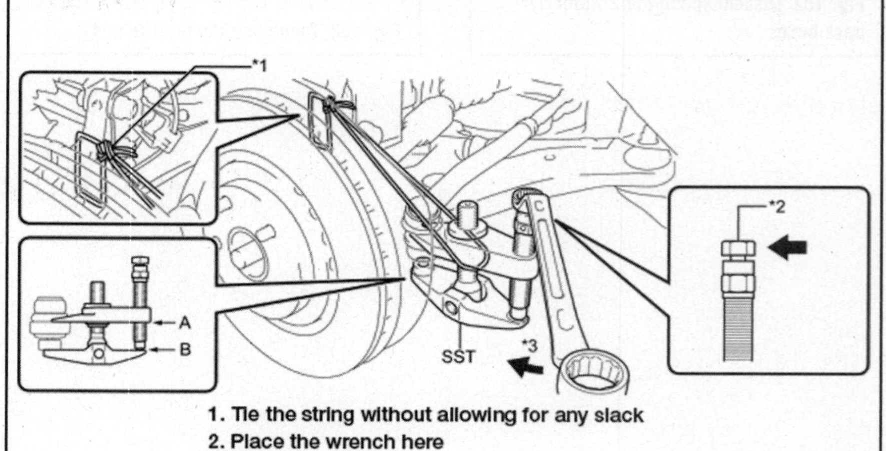

Fig. 156 Separating the tie rod end from the steering knuckle

➡When securing SST to the steering knuckle, be sure to tighten the string of SST to prevent it from falling.

➡Install SST so that A and B are parallel.

➡Be sure to place a wrench on the part indicated in the illustration.

➡Do not damage the front disc brake dust cover.

➡Do not damage the ball joint dust cover.

➡Do not damage the steering knuckle.

9. Separate the tie rod assembly RH.

➡Perform the same procedure as for the LH side.

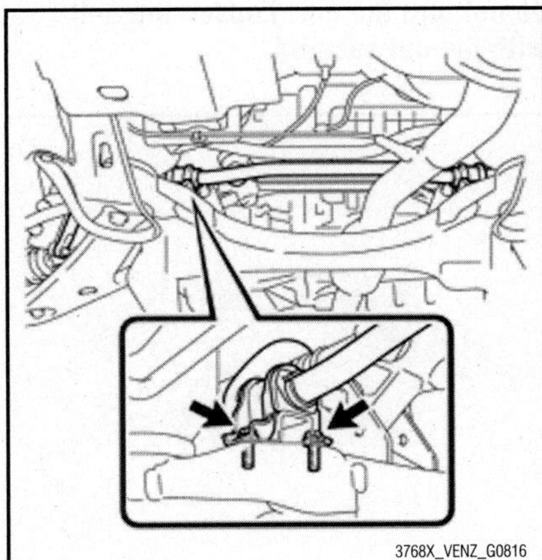

Fig. 157 Removing the front No. 1 stabilizer bracket LH (2WD)

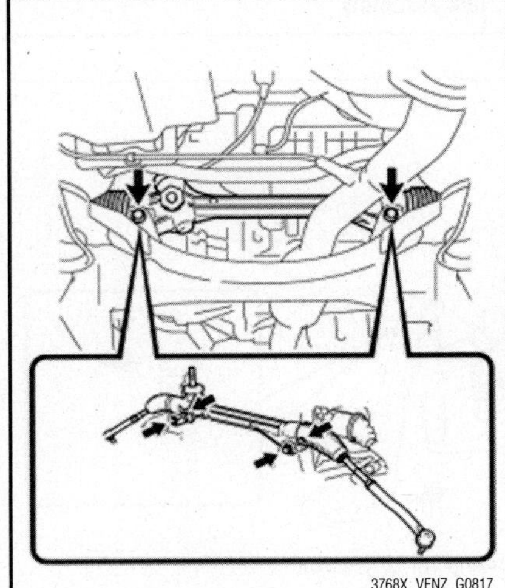

Fig. 158 Removing the steering link assembly (2WD)

10. Remove the front floor brace (2WD).
11. Remove the front No. 1 stabilizer bracket LH (2WD).
    a. Remove the 2 bolts and front No. 1 stabilizer bracket LH.
12. Remove the front No. 1 stabilizer bracket RH (2WD).

➡Perform the same procedure as for the LH side.

13. Separate the front stabilizer bar with the bracket (2WD).
    a. Separate the front stabilizer bar with bracket from the front frame assembly.
14. Remove the steering link assembly (2WD).
    a. Remove the 2 bolts, 2 nuts and steering link assembly.

➡Because the nut has its own stopper, do not turn the nut. Loosen the bolt with the nut secured.

    b. Pull out the steering link assembly towards the left side of the vehicle while lifting the front stabilizer bar with bracket.
15. Remove the engine assembly with the transaxle (AWD).
16. Remove the front No. 1 stabilizer bracket LH (AWD).
17. Remove the front No. 1 stabilizer bracket RH (AWD).
18. Remove the front stabilizer bar with the bracket (AWD).
19. Remove the steering link assembly (AWD).
    a. Remove the 2 bolts, 2 nuts and steering link assembly.

➡Because the nut has its own stopper, do not turn the nut. Loosen the bolt with the nut secured.

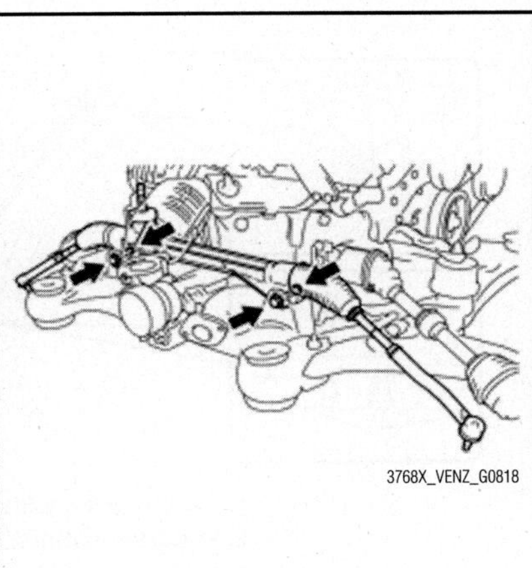

Fig. 159 Removing the steering link assembly (AWD)

20. Remove the tie rod assembly LH.
    a. Put matchmarks (*1) on the tie rod assembly LH and steering gear assembly.
    b. Loosen the lock nut, and remove the tie rod assembly LH and lock nut.
21. Remove the tie rod assembly RH.

To install:

➡When disconnecting the steering intermediate shaft assembly and pinion shaft of the steering gear assembly, be sure to place matchmarks before servicing.

22. Install the tie rod assembly LH.
    a. Install the lock nut and tie rod assembly LH to the steering gear assembly until the matchmarks are aligned.

➡After adjusting toe-in, tighten the lock nut.

23. Install the tie rod assembly RH.
24. Install the steering link assembly (AWD).
    a. Install the steering link assembly with the 2 bolts and 2 nuts. Tighten to 51 ft. lbs. (70 Nm).

➡Make sure to tighten the bolts starting from the left side of the vehicle.

➡Because the nut has its own stopper, do not turn the nut. Tighten the bolt with the nut secured.

25. Install the front stabilizer bar with the bracket (AWD).
26. Install the front No. 1 stabilizer bracket LH (AWD).
27. Install the front No. 1 stabilizer bracket RH (AWD).
28. Install the engine assembly with the transaxle (AWD).
29. Install the steering link assembly (2WD).
    a. Install the steering link assembly with the 2 bolts and 2 nuts. Tighten to 51 ft. lbs. (70 Nm).

➡Make sure to tighten the bolts starting from the left side of the vehicle.

➡Because the nut has its own stopper, do not turn the nut. Tighten the bolt with the nut secured.

30. Install the front stabilizer bar with the bracket (2WD).
31. Install the front No. 1 stabilizer bracket LH (2WD).
    a. Install the front No. 1 stabilizer bracket LH. Tighten to 21 ft. lbs. (29 Nm).
32. Install the front No. 1 stabilizer bracket RH (2WD).
33. Install the front floor brace (2WD).

34. Connect the tie rod assembly LH with the nut. Tighten to 36 ft. lbs. (49 Nm).
   a. Install a new cotter pin.

➡Further tighten the nut up to 60°if the holes for the cotter pin are not aligned.

35. Connect the tie rod assembly RH.
36. Connect the steering intermediate shaft assembly.
   a. Align the matchmarks on the steering intermediate shaft assembly and steering link assembly.
   b. Install the bolt. Tighten to 26 ft. lbs. (35 Nm).
37. Install the front wheels (2WD). Tighten to 76 ft. lbs. (103 Nm).
38. Place the front wheels facing straight ahead (2WD).
39. Install the No. 1 engine under cover (2WD).
40. Install the No. 2 engine under cover (2WD).
41. Install the No. 3 engine under cover (2WD).
42. Inspect and adjust the front wheel alignment (2WD).

3.5L Engine

See Figures 160 through 163.

➡When disconnecting the steering intermediate shaft assembly and pinion shaft of steering gear assembly, be sure to place matchmarks before servicing.

1. Remove the front exhaust pipe assembly.
2. Remove the propeller with the center bearing shaft assembly (AWD).
3. Secure the steering wheel in order to prevent it from rotating.

➡This operation is necessary to prevent damage to the spiral cable.

4. Separate the steering intermediate shaft assembly.
   a. Remove the bolt and slide the steering intermediate shaft assembly.

➡Do not separate the steering intermediate shaft assembly from the steering link assembly.

   b. Put matchmarks on the steering intermediate shaft assembly and steering link assembly.
   c. Separate the steering intermediate shaft assembly from the steering link assembly.
5. Separate the tie rod assembly LH.
   a. Remove the cotter pin and nut.
   b. Install the special tool (09960-20010) to the tie rod end.

➡Make sure that the upper ends of the tie rod end and special tool are aligned.

   c. Using the special tool, separate the tie rod end from the steering knuckle.

CAUTION

Apply grease to the threads and end of the SST bolt.

➡When securing SST to the steering knuckle, be sure to tighten SST using a string to prevent it from falling.

➡Install SST so that A and B are parallel.

➡Be sure to place a wrench on the part indicated in the illustration.

➡Do not damage the front disc brake dust cover.

➡Do not damage the ball joint dust cover.

➡Do not damage the steering knuckle.

6. Separate the tie rod assembly RH.
7. Remove the engine assembly with the transaxle.
8. Remove the front No. 1 stabilizer bracket LH.
9. Remove the front No. 1 stabilizer bracket RH.
10. Remove the front stabilizer bar.
11. Remove the steering link assembly.
   a. Remove the 2 bolts, 2 nuts and steering link assembly.

➡Because the nut has its own stopper, do not turn the nut. Loosen the bolt with the nut secured.

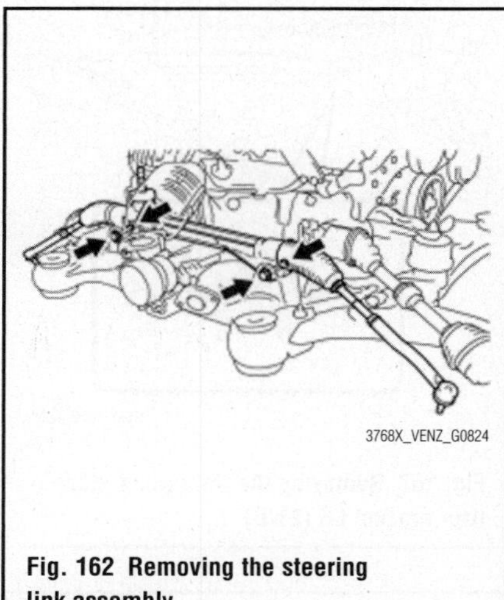

Fig. 162 Removing the steering link assembly

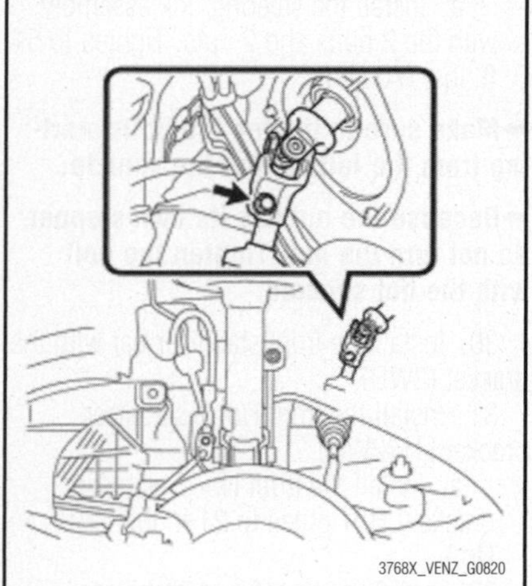

Fig. 160 Removing the bolt and sliding the steering intermediate shaft assembly

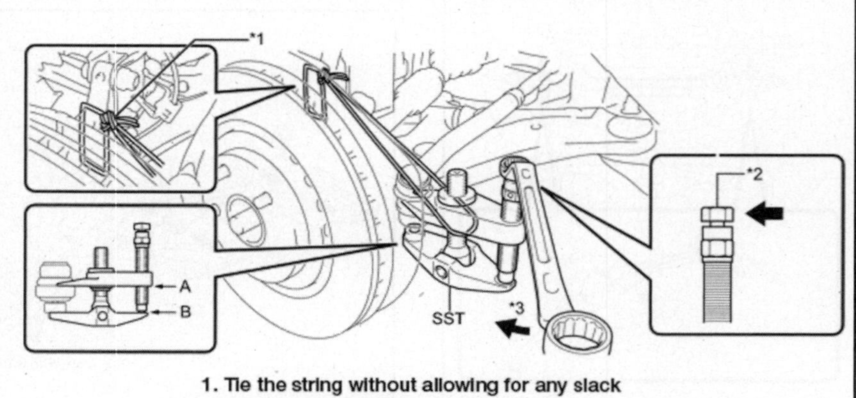

Fig. 161 Separating the tie rod end from the steering knuckle

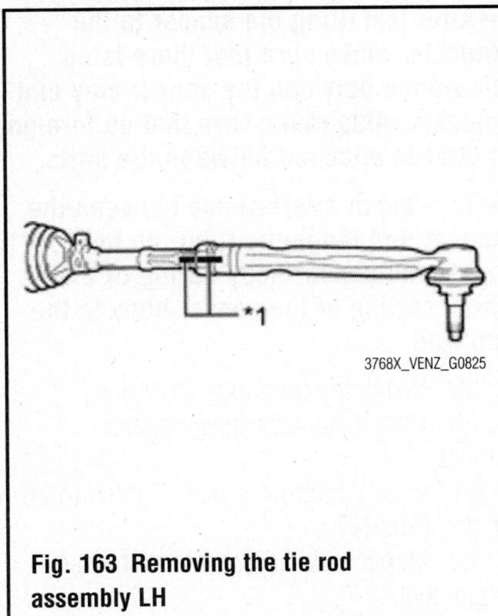

Fig. 163 Removing the tie rod assembly LH

12. Remove the tie rod assembly LH.
 a. Put matchmarks (*1) on the tie rod assembly LH and steering rack end sub-assembly.
 b. Loosen the lock nut, and remove the tie rod assembly LH and lock nut.
13. Remove the tie rod assembly RH.

To install:

➡**When disconnecting the steering intermediate shaft assembly and pinion shaft of steering gear assembly, be sure to place matchmarks before servicing.**

14. Install the tie rod assembly LH.
 a. Install the lock nut and tie rod assembly LH to the steering rack end sub-assembly until the matchmarks are aligned.

➡**After adjusting toe-in, tighten the lock nut.**

15. Install the tie rod assembly RH.
16. Install the steering link assembly with the 2 bolts and 2 nuts. Tighten to 51 ft. lbs. (70 Nm).

➡**Make sure to tighten the bolts starting from the left side of the vehicle.**

➡**Because the nut has its own stopper, do not turn the nut. Tighten the bolt with the nut secured.**

17. Install the front stabilizer bar.
18. Install the front No. 1 stabilizer bracket LH and RH.
19. Install the engine assembly with the transaxle.
20. Connect the tie rod assembly LH to the steering knuckle with the nut. Tighten to 36 ft. lbs. (49 Nm).
 a. Install a new cotter pin.

➡**Further tighten the nut up to 60°if the holes for the cotter pin are not aligned.**

21. Connect the tie rod assembly RH.
22. Connect the steering intermediate shaft assembly.
 a. Align the matchmarks on the steering intermediate shaft assembly and steering link assembly.
 b. Install the bolt. Tighten to 26 ft. lbs. (35 Nm).
23. Temporarily tighten the propeller with the center bearing shaft assembly (AWD).
 a. Remove the special tool from the transfer.
 b. Install the propeller with the center bearing shaft assembly.

➡**Be careful not to damage the oil seal.**

➡**Be careful not to damage the universal joint boot when installing the propeller shaft.**

 c. Align the matchmarks on the rear propeller shaft and electromagnetic control coupling assembly and install the 4 nuts and 4 washers temporarily.

➡**Do not allow grease to adhere to be bolts or washers.**

 d. Temporarily install the propeller with center bearing shaft assembly with the 4 bolts, 2 No. 1 center support bearing washers and 2 No. 2 center support bearing washers.

➡**Reuse the washers.**

➡**Do not allow grease to adhere to be bolts or washers.**

 e. Fully tighten the 4 nuts to 27 ft. lbs. (37 Nm).
24. Install the front exhaust pipe assembly.

SUSPENSION

FRONT SUSPENSION

LOWER BALL JOINT

REMOVAL & INSTALLATION

See Figures 164 through 166.

➡**Use the same procedure for the LH side and RH side.**

➡**The following procedure listed is for the LH side.**

1. Remove the front wheel.
2. Remove the front axle shaft nut.
3. Separate the front speed sensor.

➡**Be sure to completely separate the front speed sensor from the front shock absorber with coil spring.**

➡**Be careful not to damage the front speed sensor.**

➡**Clean the speed sensor installation hole and the surfaces every time the speed sensor are removed.**

4. Separate the front driveshaft assembly.
5. Separate the front disc brake caliper assembly.

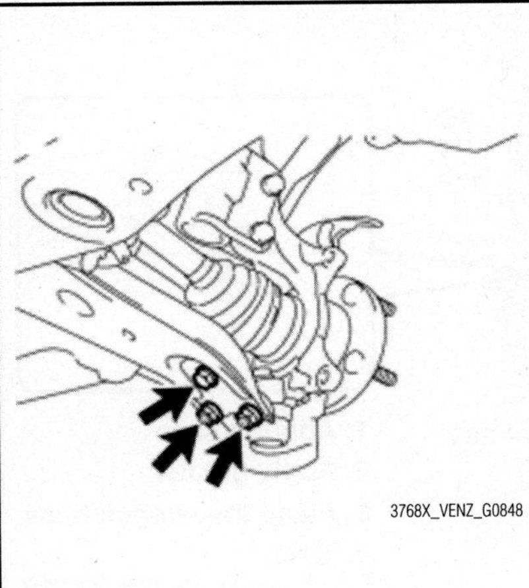

Fig. 164 Separating the front lower suspension arm

6. Remove the front disc.
7. Separate the tie rod assembly.

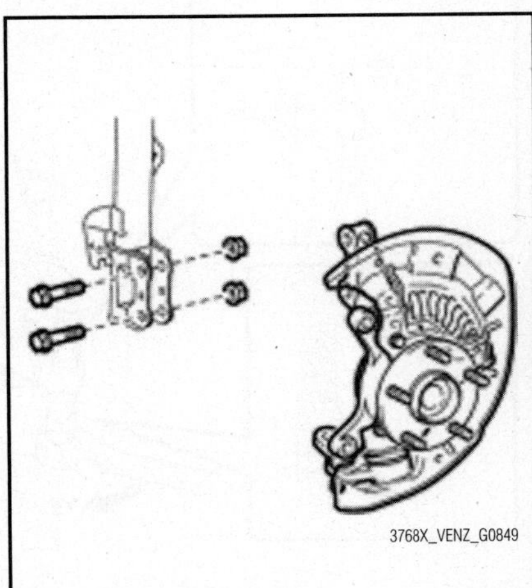

Fig. 165 Removing the front axle assembly

8. Separate the front lower suspension arm.

a. Remove the bolt, 2 nuts, and separate the front lower suspension arm from the lower ball joint.

9. Remove the front axle assembly.

a. Remove the 2 bolts, 2 nuts and front axle assembly.

➡**When removing the nuts, keep the bolts from rotating.**

10. Remove the front lower ball joint.

a. Secure the front axle assembly in a vise using aluminum plates.

➡**When using a vise, do not over-tighten it.**

b. Remove the cotter pin and nut.

c. Install the special tool (09960-20010) to the front lower ball joint.

➡**Check that the clearance measurement between SST and the front axle assembly is 1 mm (0.0394 in.).**

d. Using SST, remove the front lower ball joint from the front axle assembly as shown in the illustration.

CAUTION

Apply grease to the threads and end of the SST bolt.

➡**Install SST so that A and B are parallel.**

➡**Be sure to place a wrench on the part indicated in the illustration.**

➡**Do not damage the front lower ball joint dust cover, or steering knuckle.**

To install:

11. Install the front lower ball joint to the steering knuckle with the nut. Tighten to 91 ft. lbs. (123 Nm).

➡**Prevent oil from adhering to the screw and tapered parts.**

a. Install a new cotter pin.

➡**If the holes for the cotter pin are not aligned, tighten the nut further up to 60°.**

12. Install the front axle assembly.

a. Install the front axle assembly to the front shock absorber with the 2 bolts and 2 nuts. Tighten to 214 ft. lbs. (290 Nm).

13. Install the front driveshaft assembly.

a. Align the matchmarks and install the front driveshaft assembly to the front axle hub sub-assembly.

14. Install the front lower suspension arm.

a. Install the front lower suspension arm to the front lower ball joint with the bolt and 2 nuts. Tighten to 68 ft. lbs. (92 Nm).

15. Connect the tie rod assembly.

16. Install the front disc.

17. Install the front disc brake caliper assembly to the steering knuckle with the 2 bolts. Tighten to 77 ft. lbs. (104 Nm).

18. Install the front speed sensor. Tighten the bolt to 71 inch lbs. (8 Nm).

➡**Prevent foreign matter from attaching to the sensor tip.**

➡**Firmly insert the sensor body into the knuckle before tightening the bolt.**

➡**After installing the sensor to the knuckle, make sure that there is no clearance between the sensor stay and knuckle. Also make sure that no foreign matter is attached between the parts.**

➡**To prevent interference between the sensor and magnetic rotor, do not rotate the sensor body during or after the insertion of the sensor body to the knuckle.**

19. Install the front axle shaft nut.

20. Check the ABS speed sensor signal.

21. Install the front wheel. Tighten to 76 ft. lbs. (103 Nm).

22. Inspect and adjust the front wheel alignment.

LOWER CONTROL ARM

REMOVAL & INSTALLATION

See Figures 167 and 168.

1. Remove the engine assembly with the transaxle.

2. Remove the front No. 1 stabilizer bracket LH and RH (AWD).

3. Remove the front stabilizer bar with the front stabilizer link assembly (AWD).

4. Remove the steering link assembly (AWD).

5. Remove the front frame assembly (3.5L engine).

6. Separate the rear engine mounting insulator (2.7L engine).

7. Remove the front frame assembly (2.7L engine).

8. Remove the engine mounting insulator LH.

9. Remove the front lower suspension arm.

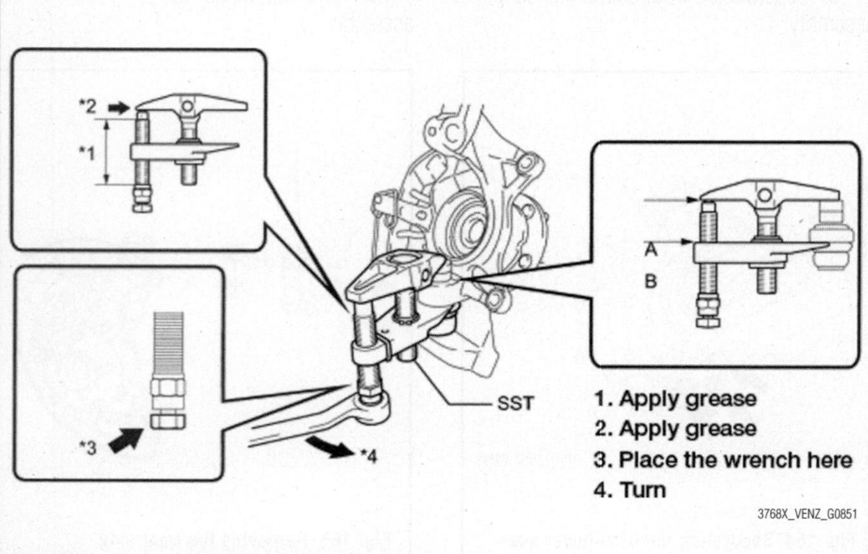

Fig. 166 Removing the front lower ball joint

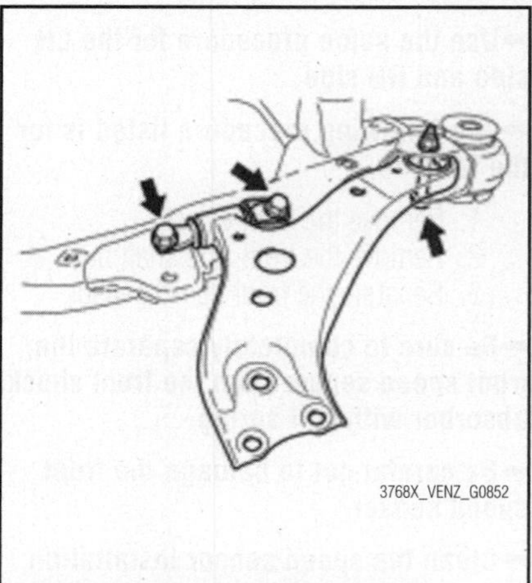

Fig. 167 Removing the front lower suspension arm

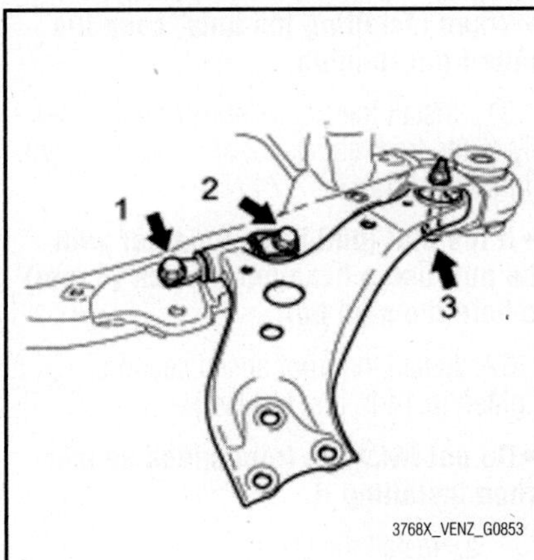

Fig. 168 Installing the front lower suspension arm

a. Remove the 3 bolts, nut and front lower suspension arm from the front frame assembly.

b. Remove the front lower arm bushing stopper from the front lower suspension arm.

To install:

10. To install, reverse the removal procedure. Tighten the bolts in the order shown. Tighten bolts 1 and 2 to 147 ft. lbs. (200 Nm). Tighten bolt 3 to 152 ft. lbs. (206 Nm).

➡Start installing the bolts from the front of the vehicle.

SHOCK ABSORBERS

REMOVAL & INSTALLATION

See Figures 169 through 172.

➡Use the same procedure for the LH side and RH side.

➡The following procedure listed below is for the LH side.

1. Remove the front wheel.
2. Remove the front wiper arm head cap.
3. Remove the front wiper arm and blade assemblies LH and RH.
4. Loosen the front suspension support nut.
   a. Loosen the front suspension nut of the front shock absorber.

➡Do not remove the front suspension support nut.

➡Loosen the nut only when the front shock absorber with coil spring needs to be disassembled.

5. Remove front fender to cowl side seal LH
6. Remove front fender to cowl side seal RH.
7. Remove cowl top ventilator louver sub-assembly.
8. Remove windshield wiper motor and link assembly.
9. Remove outer cowl top panel.
   a. Disengage the 2 clamps and separate the wiper wire harness from the outer cowl top panel sub-assembly.
   b. Disengage the 2 clamps and connector, and separate the wire harness from the outer cowl top panel sub-assembly (w/ Windshield Deicer).
   c. Remove the 4 bolts, 4 nuts and outer cowl top panel sub-assembly.
10. Separate the front speed sensor.
   a. Remove the bolt and clamp, and separate the front speed sensor and front flexible hose.
11. Separate the front stabilizer link assembly.
   a. Remove the nut and separate the front stabilizer link assembly from the front shock absorber.

➡If the ball joint turns together with the nut, use a hexagon wrench (6 mm) to hold the stud bolt.

12. Remove the front shock absorber with the coil spring.
   a. Support the front axle using a jack and wooden block.
   b. Remove the 2 bolts and 2 nuts, and separate the front shock absorber with coil spring (lower side) from the steering knuckle.

➡When removing the nuts, keep the bolts from rotating.

   c. Remove the nut (*1) and 2 spacers (*2) on the front shock absorber with coil spring (upper side).

➡Make sure that the front speed sensor is completely separated from the front shock absorber with coil spring.

13. Secure the front shock absorber with the coil spring.
   a. Install the special tool (09727-00050) to the front coil spring with the hooks spread as far apart as possible from each other.

➡Make sure that the claws on the hooks are securely engaged to the spring.

   b. Install the bolt and nut to the front shock absorber as shown in the illustration and secure the front shock absorber in a vise using aluminum plates. A: 1.10 inch (28 mm).
14. Remove the front suspension support nut.

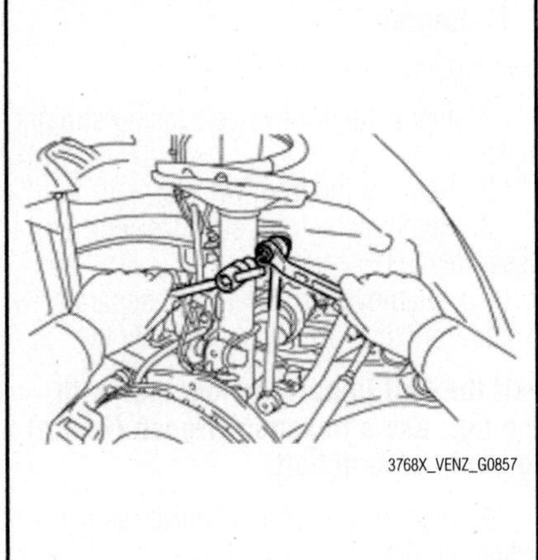

Fig. 169 Separating the front stabilizer link

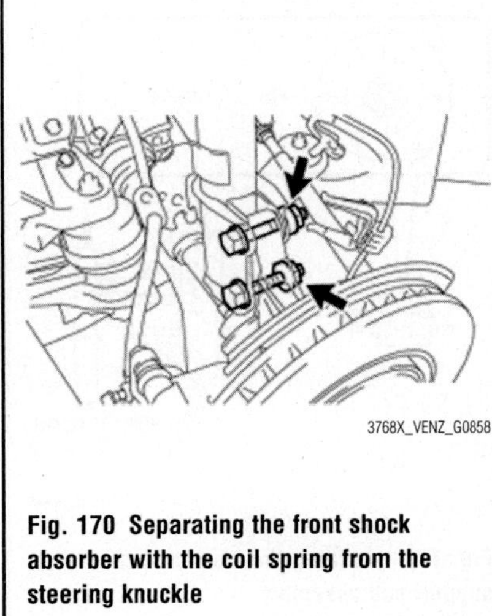

Fig. 170 Separating the front shock absorber with the coil spring from the steering knuckle

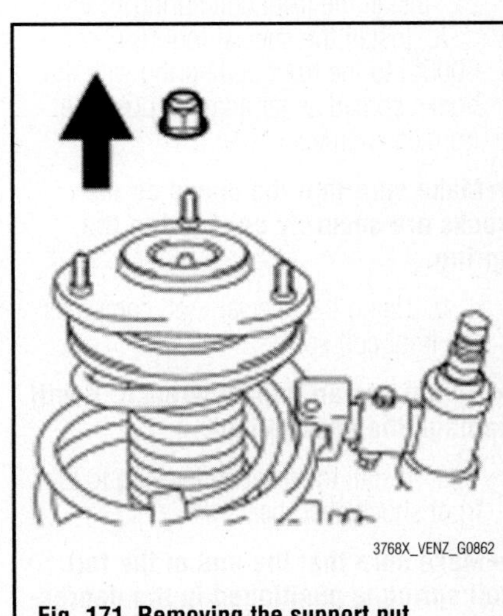

Fig. 171 Removing the support nut

a. Using the special tool (09727-00050), compress the front coil spring.

➡Do not use an impact wrench. It will damage the special tool.

➡If the front coil spring is compressed at an angle, using 2 special tools will make the work easier.

b. Check that the front coil spring is fully compressed.

c. Remove the front suspension support nut.

15. Remove front suspension support sub-assembly.
16. Remove front suspension support bearing.
17. Remove front coil spring upper seat.
18. Remove front coil spring upper insulator.
19. Remove front coil spring.
20. Remove front spring bumper.
21. Remove front coil spring lower insulator.

To install:

22. Secure the front shock absorber assembly.

Install the bolt and nut to the front shock absorber assembly as shown in the illustration and secure the front shock absorber assembly in a vise using aluminum plates. A: 1.10 inch (28 mm).

23. Install the front coil spring lower insulator.

a. Install the front coil spring lower insulator to the front shock absorber.

➡Make sure that the positioning pins on the front coil spring lower insulator are inserted into the holes in the front shock absorber.

24. Install the front spring bumper to the front shock absorber.
25. Install the front coil spring.

a. Install the special tool (09727-00050) to the front coil spring with the hooks spread as far apart as possible from each other.

➡Make sure that the claws on the hooks are securely engaged to the spring.

b. Using the special tool, compress the front coil spring.

➡Do not use an impact wrench. It will damage the special tool.

c. Install the front coil spring to the front shock absorber.

➡Make sure that the end of the fort coil spring is positioned in the depression of the lower spring seat.

26. Install the front coil spring upper insulator.

a. Install the front coil spring upper insulator as shown in the illustration.

➡Any misalignment between the front shock absorber lower bracket and the alignment mark must be +/- 5°.

27. Install the front coil spring upper seat with the mark facing outside of the vehicle.

➡Any misalignment between the front shock absorber lower bracket and the alignment mark must be +/- 5°.

28. Install the front suspension support bearing.

➡If there is foreign matter inside the front suspension support bearing, replace it with a new one.

29. Install the front suspension support sub assembly.

a. Install the front suspension support sub-assembly as shown in the illustration.

➡Check that the slots on the piston rod and front suspension support sub-assembly are aligned.

b. Temporarily tighten a new front suspension support nut.

30. Install the front shock absorber with the coil spring.

a. Install the front shock absorber with coil spring (upper side) with the nut and 2 spacers. Tighten to 63 ft. lbs. (85 Nm).

b. Install the front shock absorber with coil spring (lower side) to the steering knuckle and insert the 2 bolts and 2 nuts. Tighten to 214 ft. lbs. (290 Nm).

Fig. 172 Installing the front suspension support sub assembly

➡When installing the nuts, keep the bolts from rotating.

31. Install the front stabilizer link assembly to the front shock absorber with the nut. Tighten to 55 ft. lbs. (74 Nm).

➡If the ball joint turns together with the nut, use a hexagon wrench (6 mm) to hold the stud bolt.

32. Install the front speed sensors. Tighten to 14 ft. lbs. (19 Nm).

➡Do not twist the front speed sensor when installing it.

a. Install the clamp.

33. Install the outer cowl top panel with the 4 bolts and 4 nuts. Tighten the nut to 63 ft. lbs. (85 Nm). Tighten the bolt to 78 inch lbs. (8.8 Nm).

a. Engage the 2 wire harness clamps to the outer cowl top panel.

b. Engage the 2 wire harness clamps to the outer cowl top panel and connect the connector (w/ Windshield Deicer).

34. Fully tighten the front suspension support nut. Tighten to 52 ft. lbs. (70 Nm).
35. Install the windshield wiper motor and link assembly.
36. Install the cowl top ventilator louver sub assembly.
37. Install the front fender to cowl side seal LH and RH.
38. Install the front wiper arm and blade assemblies LH and RH.
39. Install the front wiper arm head cap.
40. Install the front wheel. Tighten to 76 ft. lbs. (103 Nm).
41. Inspect and adjust the front wheel alignment.

STABILIZER BAR

REMOVAL & INSTALLATION

2.7L Engine

See Figures 173 and 174.

1. Place the front wheels facing straight ahead.
2. Remove the front wheels.
3. Remove the front stabilizer link assembly LH.

a. Remove the 2 nuts and separate the front stabilizer link assembly LH.

➡If the ball joint turns together with the nut, use a hexagon wrench (6 mm) to hold the stud bolt.

4. Remove the front stabilizer link assembly RH.

➡Perform the same procedure as the LH side.

5. Remove the steering link assembly (2WD).

6. Remove the front stabilizer bar (2WD).

a. Remove the front stabilizer bar from the left side of the vehicle.

7. Remove the engine assembly with the transaxle (AWD).

8. Remove the front No. 1 stabilizer bracket LH (AWD).

a. Remove the 2 bolts and front No. 1 stabilizer bracket LH from the front frame assembly.

9. Remove the front No. 1 stabilizer bracket RH (AWD).

10. Remove the front stabilizer bar (AWD).

11. Remove the front No. 2 stabilizer bracket LH.

a. Remove the front No. 2 stabilizer bracket LH from the No. 1 front stabilizer bar bushing.

12. Remove the front No. 2 stabilizer bracket RH.

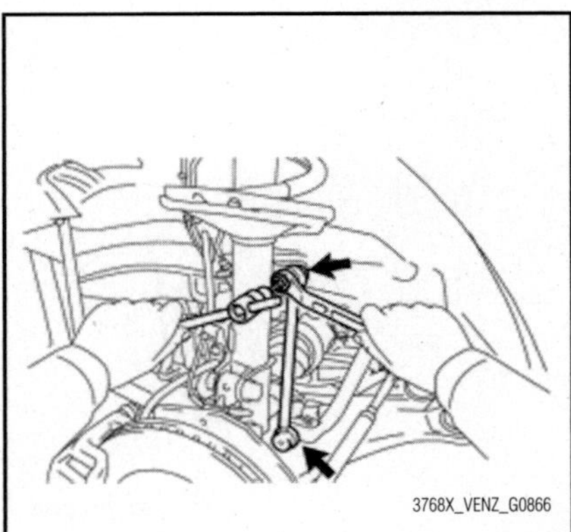

Fig. 173 Removing the front stabilizer link assembly LH

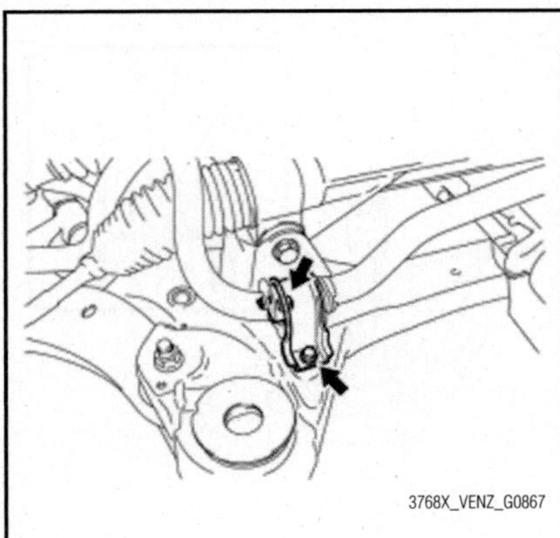

Fig. 174 Removing the front No. 1 stabilizer bracket LH (AWD)

a. Remove the front No. 2 stabilizer bracket RH from the No. 1 front stabilizer bar bushing.

13. Remove the front No. 1 stabilizer bar bushings from the front stabilizer bar.

To install:

14. Install the front No. 1 stabilizer bar bushing.

a. Install the 2 front No. 1 stabilizer bar bushings to the front stabilizer bar as shown in the illustration.

➡When installing the front No. 1 stabilizer bar bushings, make sure that the cutout faces the rear of the vehicle.

15. Install the front No. 2 stabilizer bracket LH to the front No. 1 stabilizer bar bushing.

16. Install the front No. 2 stabilizer bracket RH.

17. Install the front stabilizer bar (AWD).

18. Install the front No. 1 stabilizer bracket LH (AWD).

a. Install the front No. 1 stabilizer bracket LH to the front frame assembly with the 2 bolts. Tighten to 21 ft. lbs. (29 Nm).

19. Install the front No. 1 stabilizer bracket RH (AWD).

20. Install the front stabilizer bar (2WD).

a. Install the front stabilizer bar from the left side of the vehicle.

21. Install the front stabilizer link assembly LH with the 2 nuts. Tighten to 55 ft. lbs. (74 Nm).

➡If the ball joint turns together with the nut, use a hexagon wrench (6 mm) to hold the stud bolt.

22. Install the front stabilizer link assembly RH.

23. Install the steering link assembly (2WD).

24. Install the engine assembly with the transaxle (AWD).

3.5L Engine

See Figures 175 through 177.

1. Remove the engine assembly with the transaxle.

2. Separate the front stabilizer link assembly LH.

a. Remove the nut and separate the front stabilizer link assembly LH.

➡If the ball joint turns together with the nut, use a hexagon wrench (6 mm) to hold the stud bolt.

3. Separate the front stabilizer link assembly RH.

4. Remove the front No. 1 stabilizer bracket LH.

a. Remove the 2 bolts and No. 1 front stabilizer bracket LH from the front frame assembly.

5. Remove the No. 1 stabilizer bracket RH.

6. Remove the front stabilizer bar.

7. Remove the No. 2 stabilizer bracket LH.

a. Remove the No. 2 front stabilizer bracket LH from the front stabilizer bar bushing.

8. Remove the front No. 2 stabilizer bracket RH.

9. Remove the front No. 1 stabilizer bar bushings from the front stabilizer bar.

To install:

10. Install the front No. 1 stabilizer bar bushing.

a. Install the 2 front No. 1 stabilizer bar bushings to the front stabilizer bar as shown in the illustration.

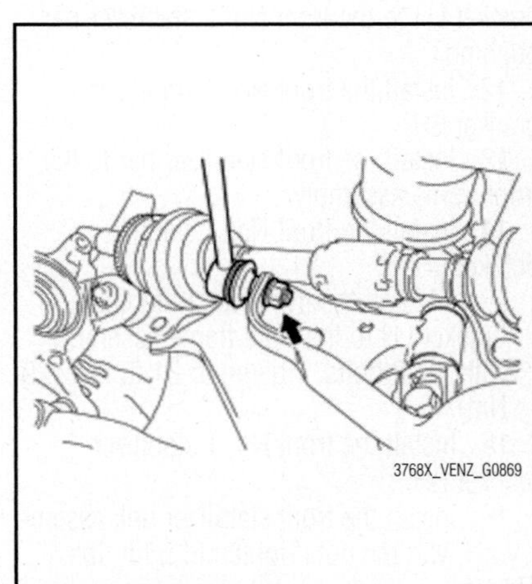

Fig. 175 Separating the front stabilizer link

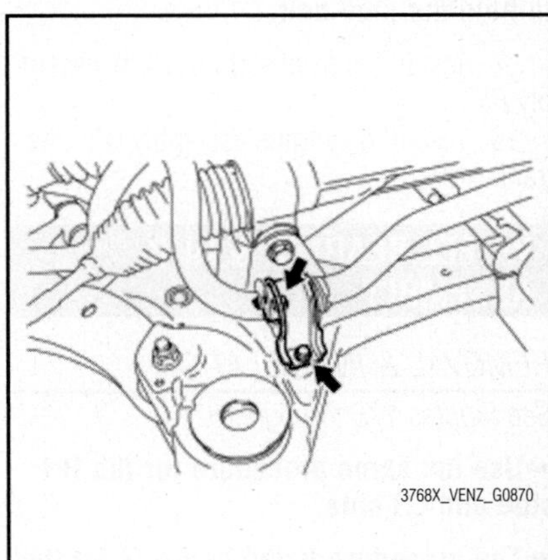

Fig. 176 Removing the front No. 1 stabilizer bracket LH

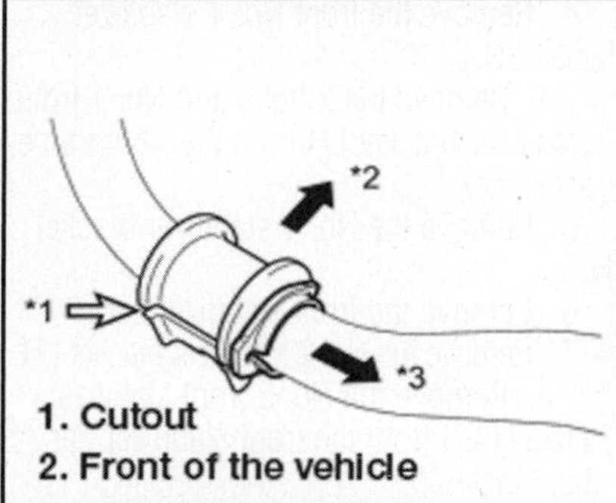

Fig. 177 Installing the front No. 1 stabilizer bar bushing

➡When installing the front No. 1 stabilizer bar bushings, make sure that the cutout faces the rear of the vehicle.

11. Install the front No. 2 stabilizer bracket LH to the front No. 1 stabilizer bar bushing.
12. Install the front No. 2 stabilizer bracket RH.
13. Install the front stabilizer bar to the front frame assembly.
14. Install the front No. 1 stabilizer bracket LH.
    a. Install the front No. 1 stabilizer bracket LH to the front frame assembly with the 2 bolts. Tighten to 21 ft. lbs. (29 Nm).
15. Install the front No. 1 stabilizer bracket RH.
16. Install the front stabilizer link assembly LH with the nut. Tighten to 55 ft. lbs. (74 Nm).

➡If the ball joint turns together with the nut, use a hexagon wrench (6 mm) to hold the stud bolt.

17. Install the front stabilizer link assembly RH.
18. Install the engine assembly with the transaxle.

WHEEL BEARING & HUB ASSEMBLY

REMOVAL & INSTALLATION

See Figures 178 through 182.

➡Use the same procedure for the RH side and LH side.

➡The procedure listed below is for the LH side.

1. Remove the front wheel.
2. Remove the front axle shaft nut.
3. Separate the front speed sensor.
    a. Remove the bolt and resin clamp, and separate the front speed sensor.

➡Prevent foreign matter from attaching to the front speed sensor tip.

➡Clean the speed sensor installation hole and the contact surfaces every time the front speed sensor is removed.

4. Separate the front disc brake caliper assembly.
5. Remove the front disc.
6. Separate the tie rod assembly.
7. Separate the front lower suspension arm.
    a. Remove the bolt and 2 nuts, and separate the front lower suspension arm from the front lower ball joint.
8. Separating the front driveshaft assembly.
    a. Put matchmarks on the front driveshaft assembly and front axle hub assembly sub assembly.
    b. Using a plastic hammer, separate the front driveshaft assembly from the front axle assembly. If it is difficult to separate, tap the end of the front driveshaft assembly using a brass bar and a hammer.

➡Be careful not to damage the driveshaft boot and speed sensor rotor.

9. Remove the front axle assembly.
    a. Remove the 2 bolts, 2 nuts and the front axle assembly.

➡When removing the nuts, keep the bolts from rotating.

➡Be careful not to damage the driveshaft boot and speed sensor rotor.

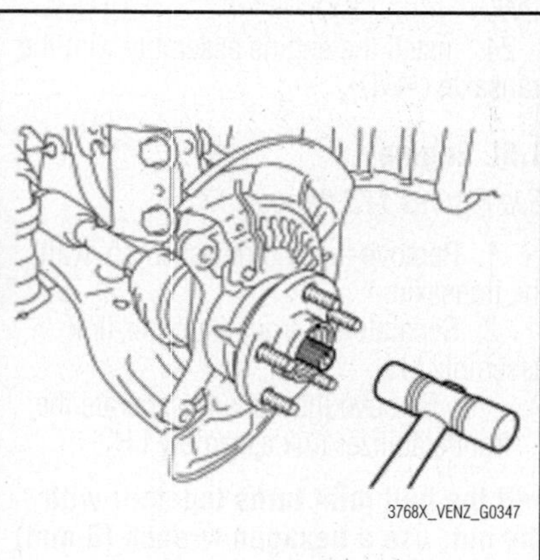

Fig. 178 Separating the front driveshaft assembly from the front axle assembly

10. Remove the front No. 1 wheel bearing dust deflector.
    a. Using a screwdriver with its tip wrapped with vinyl tape, remove the front No. 1 wheel bearing dust deflector.

➡Be careful not to damage the steering knuckle.

11. Remove the front axle hub hole snap ring.
    a. Using snap ring pliers, remove the front axle hub hole snap ring.
12. Remove the front axle hub sub assembly.
    a. Hold the front axle assembly between aluminum plates in a vise.

➡Do not overtighten the vise.

    b. Using the special tool (09520-00031).
    c. Using the special tool (09555-55010, 09550-60010, 09951-00430, 09950-70010 or 09951-07100) and a

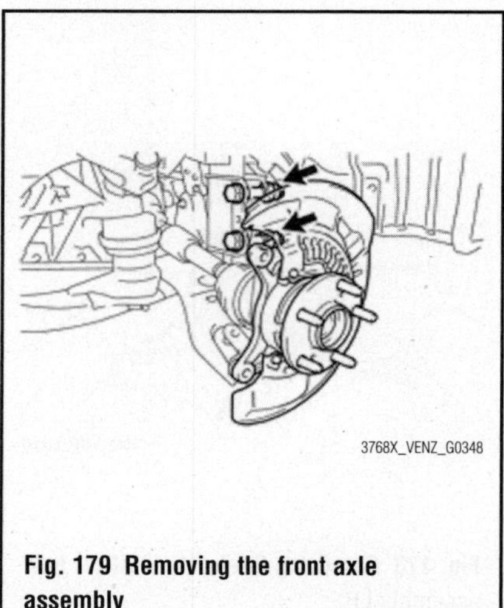

Fig. 179 Removing the front axle assembly

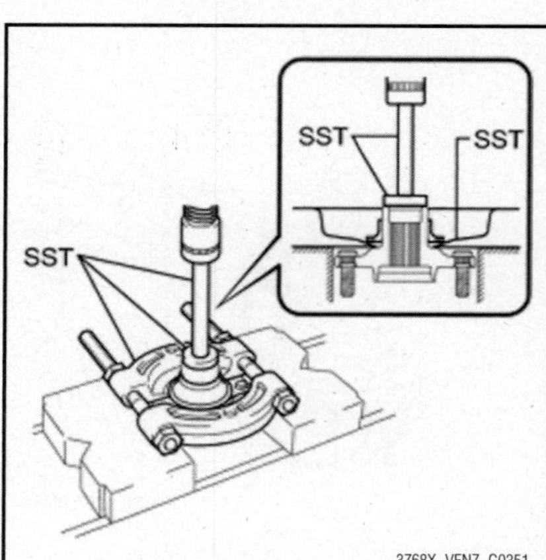

Fig. 180 Removing the bearing inner race (outside) from the front axle hub sub assembly

press, remove the bearing inner race (outside) from the front axle hub sub assembly.

➡Be careful not to drop the front axle hub sub assembly.

13. Remove the front disc brake dust cover.
   a. Remove the 4 bolts and the front disc brake dust cover from the steering knuckle.
14. Remove the front axle hub bearing.
   a. Place the bearing inner race (outside) on the front axle hub bearing.
   b. Using the special tool, V-blocks and a press, remove the front axle hub bearing from the steering knuckle. If the steering knuckle cannot be kept level using the special tool, stabilize the steering knuckle using a washer or an equivalent tool.

To install:

15. Install the front axle hub bearing.
   a. Using the special tool and a press, install a new front axle hub bearing to the steering knuckle.
16. Install the front disc brake dust cover to the steering knuckle with the 4 bolts. Tighten to 73 inch lbs. (8.3 Nm).
17. Install the front axle hub sub assembly.

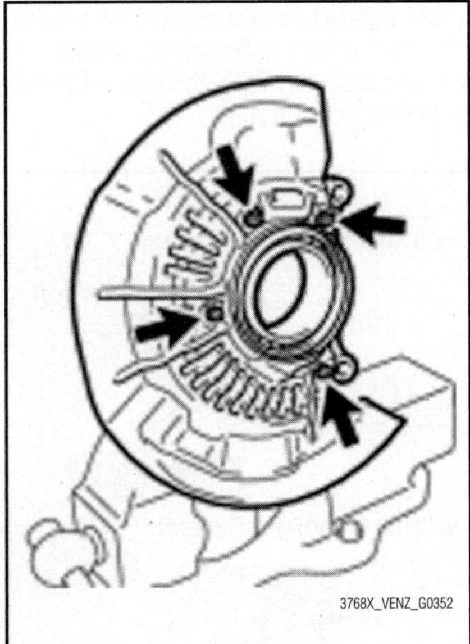

Fig. 181 Removing the front disc brake dust cover

   a. Using the special tool and a press, install the front axle hub sub assembly to the steering knuckle.
18. Install the front axle hub hole snap ring.
   a. Using snap ring pliers, install a new front axle hub hole snap ring.
19. Install the front No. 1 wheel bearing dust deflector.
   a. Using the special tool and a hammer, install a new front No. 1wheel bearing dust deflector.

➡Align the cutout for the speed sensor in the No. 1 front wheel bearing dust deflector with the hole of the steering knuckle.

20. Install the front axle assembly to the front shock absorber with the 2 bolts and 2 nuts. Tighten to 214 ft. lbs. (290 Nm).

➡When installing the nuts, keep the bolts from rotating.

21. Install the front driveshaft assembly.
   a. Align the matchmarks and install the front driveshaft assembly to the front axle hub sub assembly.
22. Install the front lower suspension arm to the front lower ball joint with the bolt and 2 nuts. Tighten to 68 ft. lbs. (92 Nm).
23. Connect the tie rod assembly.
24. Install the front disc.
25. Install the front disc brake caliper assembly.

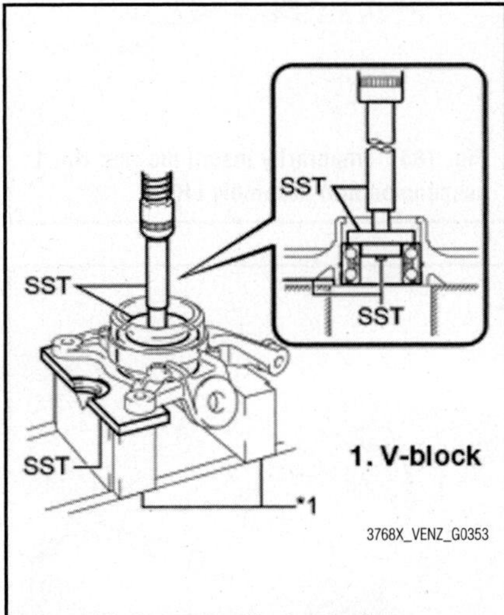

Fig. 182 Removing the front axle hub bearing

26. Install the front axle shaft nut.
   a. Clean the threaded parts on the front driveshaft assembly and a new front axle shaft nut using a non residue solvent.

➡Be sure to perform this work for a new front driveshaft assembly.

➡Keep the threaded parts free of oil and foreign matter.

   b. Using a socket wrench (30 mm), install the new front axle shaft nut. Tighten to 217 ft. lbs. (294 Nm).

➡Stake the front axle shaft nut after inspecting for looseness and runout in the following steps.

➡Keep depressing the brake pedal to prevent the driveshaft from rotating.

27. Separate the front disc brake caliper assembly.
28. Remove the front disc.
29. Inspect the front axle hub bearing looseness.
30. Inspect the front axle hub runout.
31. Install the front disc.
32. Install the front disc brake caliper assembly.
33. Install the front speed sensor with the bolt. Tighten to 71 inch lbs. (8 Nm).

➡Prevent foreign matter from attaching to the front speed sensor tip.

➡Firmly insert the front speed sensor body into the steering knuckle before tightening the bolt.

➡After installing the front speed sensor to the steering knuckle, make sure that there is no clearance between the front speed sensor stay and steering knuckle. Also make sure that no foreign matter is stuck between the parts.

➡Before installing the clamp, firmly insert the points of the clamp into the installation holes.

34. Stake the front axle shaft nut.
   a. Using a chisel and hammer, stake the front axle shaft nut.
35. Install the front wheel. Tighten to 76 ft. lbs. (103 Nm).
36. Inspect and adjust the front wheel alignment.
37. Check for the speed signal.

CONTROL ARMS/LINKS

REMOVAL & INSTALLATION

2WD Vehicles

See Figures 183 through 187.

1. Remove the rear wheels.
2. Separate the rear stabilizer link assembly LH.
   a. Remove the nut and separate the rear stabilizer link assembly LH from the rear stabilizer bar.

➡If the ball joint turns together with the nut, use a hexagon wrench (5 mm) to hold the stud bolt.

3. Separate the rear stabilizer link assembly RH.

➡Perform the same procedure as the LH side.

4. Remove the rear stabilizer bar.
5. Separate the rear height control sensor sub assembly (w/HID headlight system).
   a. Remove the nut and separate the rear height control sensor sub-assembly from the rear No. 2 suspension arm assembly RH.

➡Use wire or an equivalent tool to keep the rear height control sensor link from hanging down.

6. Remove the rear No. 2 suspension arm assembly LH.
   a. Remove the bolt (A) and the nut, and separate the rear No. 2 suspension arm assembly LH from the rear axle carrier sub-assembly LH.

➡Since a stopper nut is used, loosen the bolt.

   b. Remove the bolt (B) and the rear No. 2 suspension arm assembly LH.
7. Remove the rear No. 1 suspension arm assembly LH.
   a. Remove the bolt and the nut, and separate the rear No. 1 suspension arm assembly LH from the rear axle carrier sub-assembly.

➡Since a stopper nut is used, loosen the bolt.

   b. Remove the bolt and the rear No. 1 suspension arm assembly LH.
8. Remove the rear No. 1 suspension arm assembly RH.

➡Perform the same procedure as the LH side.

To install:

9. Temporarily install the rear No. 1 suspension arm assembly LH to the rear suspension member with the bolt (B).

➡Ensure that the identification mark (*1) faces the rear side of the vehicle.

   a. Temporarily install the rear No. 1 suspension arm assembly LH to the rear axle carrier sub-assembly with the bolt (A) and the nut.

➡Since a stopper nut is used, temporarily tighten the bolt.

10. Temporarily install the rear No. 1 suspension arm assembly RH.

➡Perform the same procedure as the LH side.

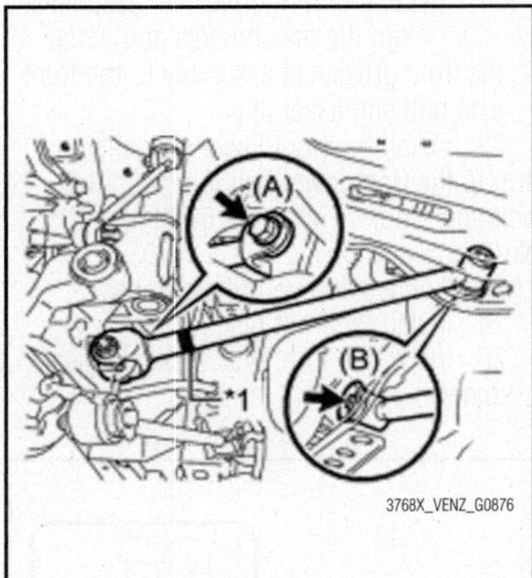

Fig. 183 Temporarily install the rear No. 1 suspension arm assembly LH

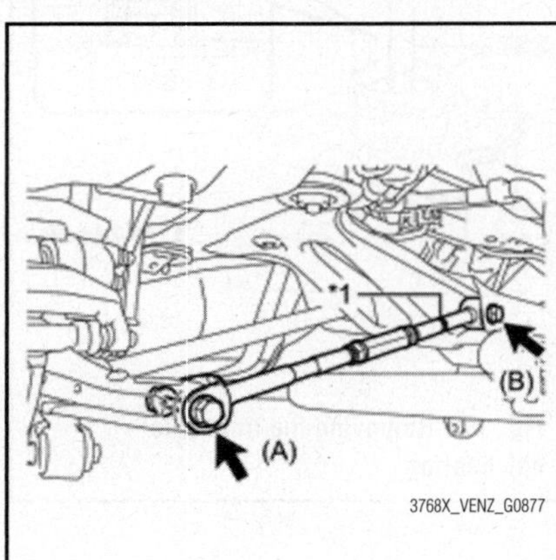

Fig. 184 Temporarily installing the rear No. 2 suspension arm assembly LH

11. Temporarily install the rear No. 2 suspension arm assembly LH.
    a. Temporarily install the rear No. 2 suspension arm assembly LH to the rear suspension member with the bolt (B).

➡Ensure that the identification mark (*1) faces the rear side of the vehicle.

    b. Temporarily install the rear No. 2 suspension arm assembly LH to the rear axle carrier sub-assembly LH with the bolt (A) and the nut.

➡Since a stopper nut is used, temporarily tighten the bolt.

12. Temporarily install the rear No. 2 suspension arm assembly RH.
    a. Temporarily install the rear No. 2 suspension arm assembly RH to the rear suspension member with the bolt (B).

➡Ensure that the identification mark (*1) faces the rear side of the vehicle.

    b. Temporarily install the rear No. 2 suspension arm assembly RH to the rear axle carrier sub-assembly RH with the bolt (B) and the nut.

➡Since a stopper nut is used, temporarily tighten the bolt.

13. Connect the rear height control sensor sub-assembly to the rear No. 2 suspension arm assembly RH with the nut. Tighten to 48 inch lbs. (5.4 Nm).

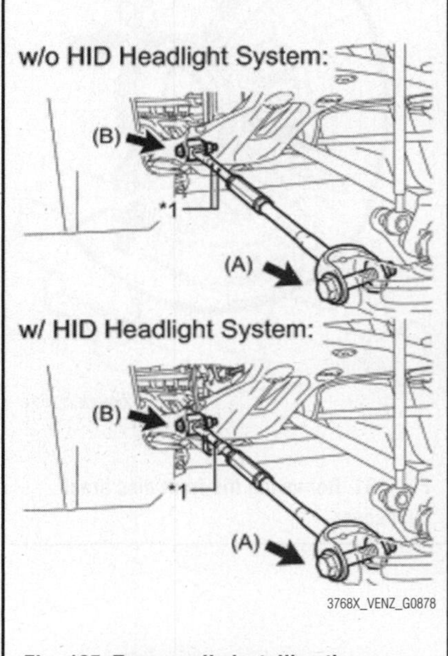

Fig. 185 Temporarily installing the rear No. 2 suspension arm assembly RH

14. Stabilize the suspension.

a. Install the rear wheels. Tighten to 76 ft. lbs. (103 Nm).

b. Lower the vehicle to the ground.

c. Bounce the vehicle up and down at the corners to stabilize the rear suspension.

d. Remove the rear wheels.

15. Fully tighten the rear No. 1 suspension arm assembly LH.

a. Support the rear axle carrier sub-assembly using a jack and wooden block.

➡**Do not bend the brake dust cover.**

b. Jack up the rear axle carrier sub-assembly LH to set the rear No. 1 suspension arm assembly LH in the tightening position. Standard angle (A) is 4.6°. Standard length (B) is 1.52 inch (38.6 mm).

CAUTION

Do not jack up the rear axle carrier sub-assembly LH too high as the vehicle may fall.

➡**If the rear No. 1 suspension arm assembly LH cannot be positioned as shown in the illustration even when the rear axle carrier sub-assembly LH is jacked up, apply additional load to the vehicle such as by having a person sit in the rear seat.**

c. Using the special tool (09961-00950) and a socket wrench (19 mm), fully tighten the bolt in the tightening position. Tighten to 89 ft. lbs. (120 Nm). If using the special tools tighten to 65 ft. lbs. (89 Nm).

➡**Use a torque wrench with a fulcrum length of 425 mm (1.39 ft.).**

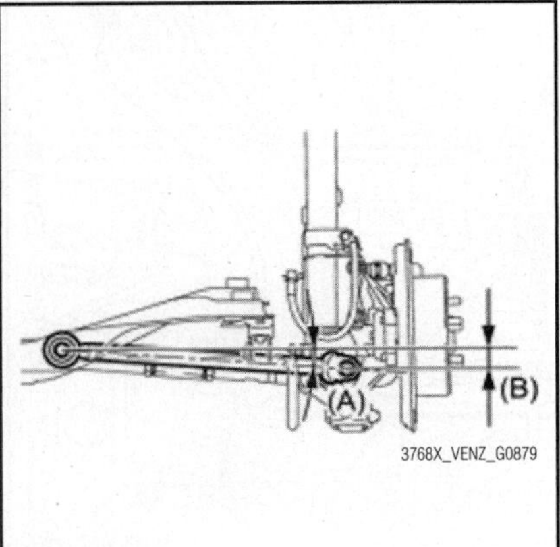

Fig. 186 Identifying the tightening position

➡**This torque value is effective when SST is parallel to the torque wrench.**

➡**Since a stopper nut is used, fully tighten the bolt.**

d. Fully tighten the bolt in the tightening position. Tighten to 83 ft. lbs. (112 Nm).

➡**Since a stopper nut is used, fully tighten the bolt.**

16. Fully tighten the rear No. 2 suspension arm assembly LH.

a. Support the rear axle carrier sub-assembly using a jack and wooden block.

➡**Do not bend the brake dust cover.**

b. Support the rear axle carrier sub-assembly using a jack and wooden block.

➡**Do not bend the brake dust cover.**

c. Jack up the rear axle carrier sub-assembly LH to set the rear No. 1 suspension arm assembly LH in the tightening position. Standard angle (A) is 4.6°. Standard length (B) is 1.52 inch (38.6 mm).

CAUTION

Do not jack up the rear axle carrier sub-assembly LH too high as the vehicle may fall.

➡**If the rear No. 1 suspension arm assembly LH cannot be positioned as shown in the illustration even when the rear axle carrier sub-assembly LH is jacked up, apply additional load to the vehicle such as by having a person sit in the rear seat.**

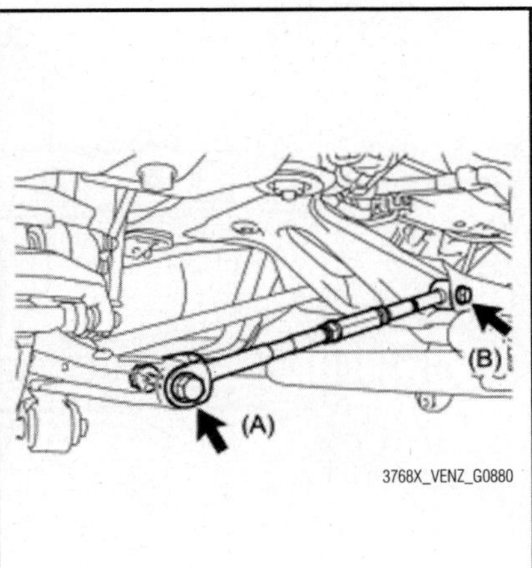

Fig. 187 Tightening the bolts

d. Fully tighten the bolts in the tightening position. Tighten bolt A to 83 ft. lbs. (112 Nm). Tighten bolt B to 89 ft. lbs. (120 Nm).

➡**Since a stopper nut is used, fully tighten the bolt (A) and (B).**

17. Fully tighten the rear No. 2 suspension arm assembly RH.

➡**Perform the same procedure as the LH side.**

18. Install the rear stabilizer bar.

19. Install the rear stabilizer link assembly LH to the rear stabilizer bar with the nut. Tighten to 29 ft. lbs. (39 Nm).

➡**If the ball joint turns together with the nut, use a hexagon wrench (5 mm) to hold the stud bolt.**

20. Install the rear stabilizer link assembly RH.

➡**Perform the same procedure as the LH side.**

21. Install the rear wheel. Tighten to 76 ft. lbs. (103 Nm).

22. Inspect and adjust the rear wheel alignment.

23. Perform the height control sensor signal initialization (w/HID headlight system).

24. Inspect and adjust the headlight aiming (w/HID headlight system).

AWD Vehicles

See Figures 188 through 191.

1. Remove the rear wheels.

2. Remove the center exhaust pipe assembly.

3. Remove the propeller with the center bearing shaft assembly.

4. Separate the rear height control sensor sub assembly (w/HID headlight system).

a. Remove the nut and separate the rear height control sensor sub-assembly from the rear No. 2 suspension arm assembly RH.

➡**Use wire or an equivalent tool to keep the rear height control sensor link from hanging down.**

5. Separate no. 3 parking brake cable assembly.

6. Separate no. 2 parking brake cable assembly.

➡**Perform the same procedure as the LH side.**

7. Remove rear strut rod assembly LH.

8. Remove rear strut rod assembly RH.

➡**Perform the same procedure as the LH side.**

9. Remove rear No. 2 suspension arm assembly LH.

a. Put matchmarks (*1) on the adjust cams and the rear suspension member sub-assembly.

b. Remove the bolt (A) and the nut, and separate the rear No. 2 suspension arm assembly LH from the rear axle carrier sub-assembly LH.

➡**Since a stopper nut is used, loosen the bolt.**

c. Remove the nut (B), the No. 2 camber adjust cam, the rear suspension toe adjust cam sub-assembly, and the rear No. 2 suspension arm assembly LH.

➡**When removing the nut, keep the rear suspension toe adjust cam sub-assembly from rotating.**

10. Remove the rear No. 2 suspension arm assembly RH.

➡**Perform the same procedure as the LH side.**

11. Separate the rear No. 1 suspension arm assembly LH.

a. Remove the bolt and the nut, and separate the rear No. 1 suspension arm assembly LH from the rear axle carrier sub-assembly LH.

➡**Since a stopper nut is used, loosen the bolt.**

12. Separate the rear No. 1 suspension arm assembly RH.

➡**Perform the same procedure as the LH side.**

13. Remove the No. 1 floor under cover.

a. Separate the 5 clips to remove the No. 1 floor under cover.

➡**Remove the No. 1 floor under cover with the 5 clips.**

14. Separate the frame wire.

a. Disengage the 2 clamps to separate the frame wire from the body.

15. Separate the No. 3 floor wire (w/HID headlight system).

a. Disconnect the connector and disengage the clamp to separate the No. 3 floor wire from the rear suspension member.

16. Separate the rear suspension member.

a. Support the rear suspension member with a jack using 3 wooden blocks.

➡**Use properly sized wooden blocks to keep the jack and suspension member lever.**

b. Remove the 4 nuts, 2 bolts and 2 rear lower suspension member stopper retainers.

c. Lower the rear suspension member to the point shown in the illustration. Length A: 3.94 inch (100 mm).

17. Remove the rear No. 1 suspension arm assembly LH.

a. Remove the bolt, the nut and the rear No. 1 suspension arm assembly LH from the rear suspension member.

➡**Since a stopper nut is used, loosen the bolt.**

18. Remove the rear No. 1 suspension arm assembly RH.

➡**Perform the same procedure as the LH.**

To install:

19. Install the rear No. 1 suspension arm assembly LH.

a. Temporarily install the rear No. 1 suspension arm assembly LH to the rear suspension member with the bolt and the nut.

➡**Ensure that the identification mark faces the rear side of the vehicle.**

➡**Since a stopper nut is used, temporarily tighten the bolt.**

b. Set the rear No. 1 suspension arm assembly LH in the tightening position shown in the illustration. Standard angle (A): 7.6°. Standard length (B): 1.96 inch (49.7 mm).

c. Fully tighten the bolt in the tightening position. Tighten to 59 ft. lbs. (80 Nm).

➡**Since a stopper nut is used, fully tighten the bolt.**

20. Install the rear No. 1 suspension arm assembly RH.

➡**Perform the same procedure as the LH side.**

21. Install the rear suspension member.

a. Raise the rear suspension member with a jack.

b. Temporarily install the rear suspension member, 2 rear upper suspension

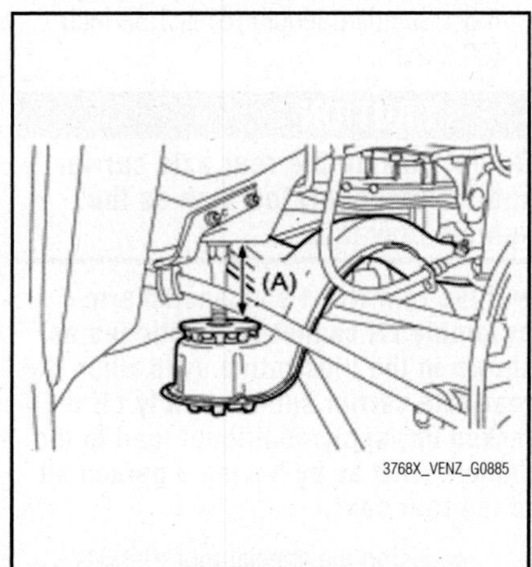

Fig. 189 Lowering the rear suspension member

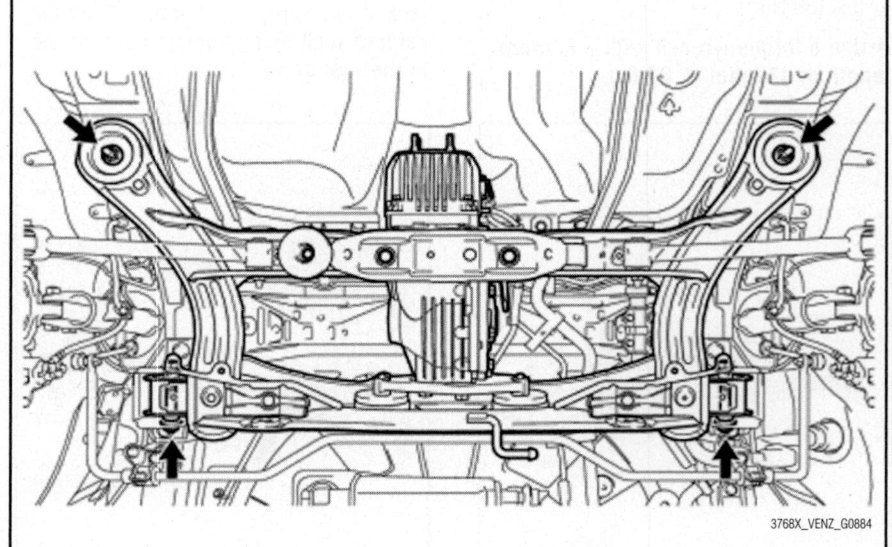

Fig. 188 Removing the 4 nuts, 2 bolts and 2 rear lower suspension member stopper retainers

member stoppers and 2 rear lower suspension member stopper retainers with the 4 nuts and 2 bolts.

➡**Be sure to install the rear suspension member with the rear upper suspension member stopper and the rear lower suspension member stopper retainer in the correct direction as shown in the illustration.**

c. Fully tighten the 2 nuts (A) to 85 ft. lbs. (115 Nm).

d. Using SST and a socket wrench (19 mm), fully tighten the 2 nuts (B). Tighten to 71 ft. lbs. (96 Nm). If using the special tool tighten to 52 ft. lbs. (71 Nm).

➡**Use a torque wrench with a fulcrum length of 425 mm (1.39 ft.).**

➡**This torque value is effective when SST is parallel to the torque wrench.**

22. Connect the rear No. 1 suspension arm assembly LH.

a. Connect the rear No. 1 suspension arm assembly LH to the rear axle carrier sub-assembly LH with the bolt and the nut. Tighten to 83 ft. lbs. (112 Nm).

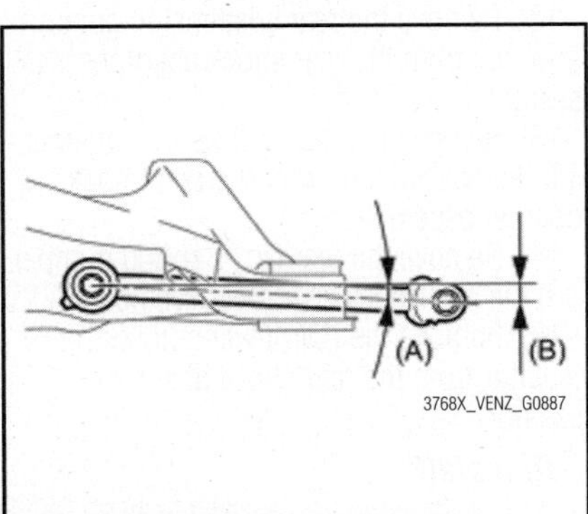

Fig. 190 Setting the rear No. 1 suspension arm

➡**Since a stopper nut is used, temporarily tighten the bolt.**

23. Connect the rear No. 1 suspension arm assembly RH.

➡**Perform the same procedure as the LH side.**

24. Install the frame wire.

a. Engage the 2 clamps to install the frame wire.

➡**Do not twist the frame wire when installing it.**

25. Install the No. 3 floor wire (w/HID headlight system).

a. Engage the clamp and connect the connector to install the No. 3 floor wire.

➡**Do not twist the No. 3 floor wire when installing it.**

26. Install the No. 1 floor under cover with the 5 clips.

27. Temporarily tighten the rear No. 2 suspension arm assembly LH.

a. Temporarily tighten the rear No. 2 suspension arm assembly LH to the rear suspension member with the rear suspension toe adjust cam sub-assembly, the No. 2 camber adjust cam and the nut (B).

➡**Ensure that the identification mark faces the rear side of the vehicle.**

➡**When temporarily tightening the nut, keep the rear suspension toe adjust cam sub-assembly from rotating.**

b. Fully tighten the rear No. 2 suspension arm assembly LH to the rear axle carrier sub-assembly LH with the bolt (A) and the nut. Tighten to 83 ft. lbs. (112 Nm).

➡**Since a stopper nut is used, fully tighten the bolt.**

28. Temporarily tighten the rear No. 2 suspension arm assembly RH.

➡**Perform the same procedure as the LH side.**

29. Connect the rear height control sensor sub assembly (w/HID headlight system).

a.Connect the rear height control sensor sub-assembly to the rear No. 2 suspension arm assembly

30. RH with the nut. Tighten to 48 inch lbs. (5.4 Nm).

31. Install rear strut rod assembly LH.

32. Install rear strut rod assembly RH.

➡**Perform the same procedure as the LH side.**

33. Install no. 3 parking brake cable assembly.

34. Install no. 2 parking brake cable assembly.

35. Temporarily tighten propeller with center bearing shaft assembly.

36. Fully tighten propeller with center bearing shaft assembly.

37. Inspect and adjust transfer oil.

a. Inspect and adjust the transfer oil.

38. Install center exhaust pipe assembly.

a. Install the center exhaust pipe assembly.

➡**Refer to the instructions for installation of the exhaust pipe.**

39. Inspect for exhaust gas leak.

40. Install rear wheels. Tighten to 76 ft. lbs. (103 Nm).

41. Stabilize suspension.

a. Lower the vehicle to the ground.

b. Bounce the vehicle up and down at the corners to stabilize the rear suspension.

42. Fully tighten rear no. 2 suspension arm assembly LH.

a. Align the matchmarks on the adjust cams and rear suspension member sub-assembly.

b. Fully tighten the nut to 74 ft. lbs. (100 Nm).

➡**The final torque must be applied under standard vehicle height conditions.**

➡**When fully tightening the nut, keep the rear suspension toe adjust cam sub-assembly from rotating.**

43. Fully tighten rear no. 2 suspension arm assembly RH

➡**Perform the same procedure as the LH side.**

44. Inspect and adjust rear wheel alignment

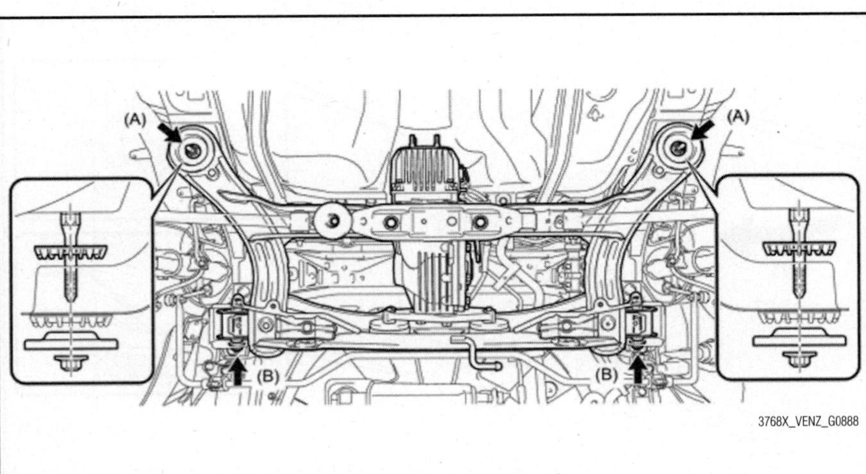

Fig. 191 Installing the rear suspension member with a jack

a. Inspect and adjust the rear wheel alignment.

45. Height control sensor signal initialization (w/ hid headlight system)

a. Initialize the height control sensor signal.

46. Inspect and adjust headlight aiming (w/ hid headlight system)

a. Inspect and adjust the headlight aiming.

SHOCK ABSORBER

REMOVAL & INSTALLATION

See Figures 192 through 195.

➡Use the same procedure for the RH side and LH side.

➡The procedure listed below is for the LH side.

1. Remove the rear wheel.
2. Remove the deck side trim.

a. Disengage the 5 claws, and then remove the deck side trim.

3. Separate the rear flexible hose (2WD).

a. Remove the bolt and separate the rear flexible hose from the rear shock absorber with coil spring.

4. Separate the rear flexible hose (AWD).

a. Remove the bolt and separate the rear flexible hose from the rear shock absorber with coil spring.

5. Separate the rear speed sensor wire.

a. Remove the bolt and separate the rear speed sensor wire from the rear shock absorber with coil spring.

6. Separate the rear stabilizer link assembly.

a. Remove the nut and separate the rear stabilizer link assembly from the rear shock absorber with coil spring.

➡If the ball joint turns together with the nut, use a hexagon wrench (5 mm) to hold the stud bolt.

7. Remove the rear No. 1 suspension support cover.

a. Disengage the 4 claws and remove the rear No. 1 suspension support cover.

8. Remove the rear shock absorber with the coil spring.

a. Support the rear axle carrier sub-assembly using a jack and wooden block.

➡Do not deform the brake dust cover.

➡Support the rear axle carrier sub-assembly until reinstallation of the rear shock absorber with coil spring is complete.

b. Loosen the rear support to rear shock absorber nut.

⁂ CAUTION

Do not remove the rear support to rear shock absorber nut.

➡Loosen the nut only when the rear shock absorber with coil spring needs to be disassembled.

c. Remove the 2 bolts and 2 nuts, and separate the rear shock absorber with coil spring from the rear axle carrier sub-assembly.

➡When removing the nuts, keep the bolts from rotating.

d. Remove the 3 nuts and rear shock absorber with coil spring.

➡Make sure that the rear speed sensor and rear flexible hose are disconnected from the rear shock absorber with coil spring.

9. Secure the rear shock absorber with the coil spring.

a. Install the bolt and nut to the rear shock absorber with coil spring as shown in the illustration and secure the rear shock absorber with coil spring in a vise. Length (A) is 1.10 inch (28 mm).

10. Remove the rear support to the rear shock absorber nut.

a. Using the special tool, compress the rear coil spring.

➡Do not use an impact wrench. It will damage SST.

b. Check that the rear coil spring is fully compressed.

c. Using a screwdriver or an equivalent tool to hold the rear suspension support assembly, remove the rear support to rear shock absorber nut from the rear shock absorber assembly.

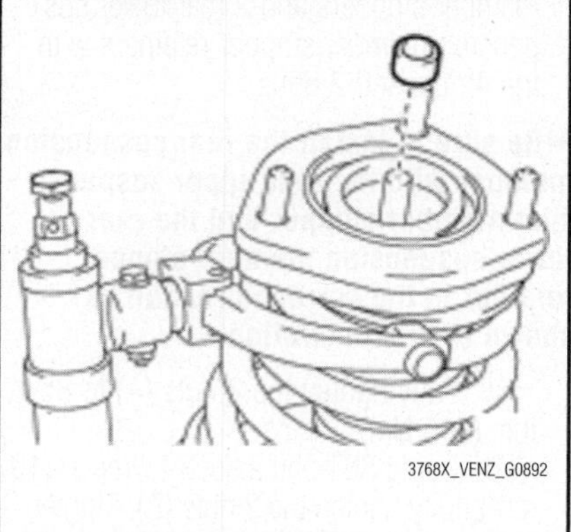

Fig. 192 Removing the rear support to rear shock absorber collar

➡Tape the screwdriver tip or the equivalent tool before use.

11. Remove the rear support to rear shock absorber collar.

a. Remove the rear support to rear shock absorber collar from the rear shock absorber assembly.

12. Remove the rear suspension support assembly from the rear shock absorber assembly.
13. Remove the rear coil spring together with the special tool from the rear shock absorber assembly.
14. Remove the rear No. 1 spring bumper from the rear shock absorber assembly.
15. Remove the rear lower coil spring insulator from the rear shock absorber assembly.

To install:

➡Use the same procedure for the RH side and LH side.

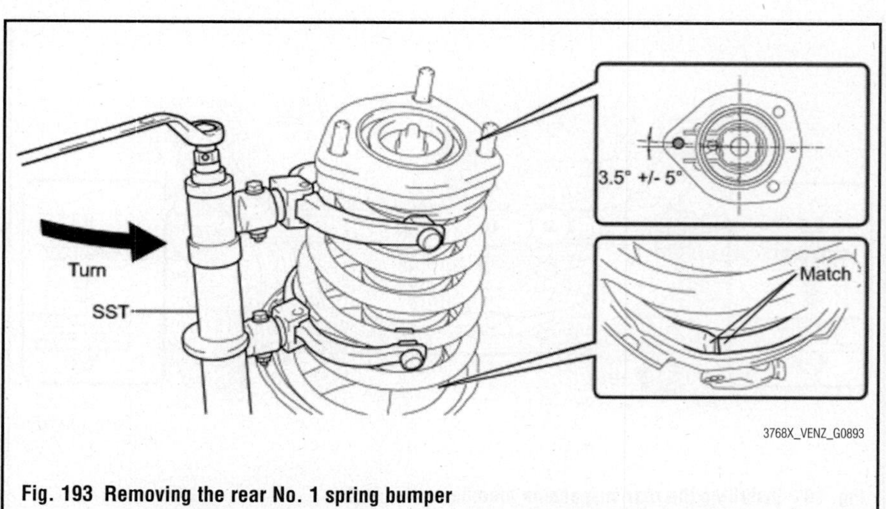

Fig. 193 Removing the rear No. 1 spring bumper

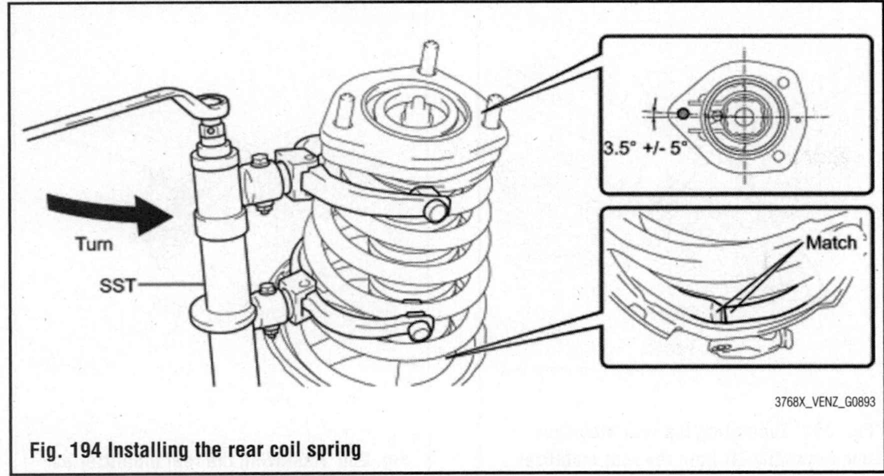

Fig. 194 Installing the rear coil spring

➡**The procedure listed below is for the LH side.**

16. Secure the rear shock absorber with the coil spring.
17. Install the rear lower coil spring insulator onto the rear shock absorber assembly.

➡**Fit the recessed part of the rear lower coil spring insulator into the recession on the shock absorber assembly.**

18. Install the rear No. 1 spring bumper to the rear shock absorber assembly.
19. Temporarily install the rear coil spring.
    a. Using the special tool, compress the rear coil spring.

➡**Do not use an impact wrench. It will damage SST.**

    b. Temporarily install the rear coil spring together with SST to the rear shock absorber assembly.
20. Install the rear suspension support assembly to the rear shock absorber assembly.

➡**Align the cutout on the rear shock absorber assembly with the protrusion on the rear suspension support assembly.**

21. Install the rear support to rear shock absorber collar to the rear shock absorber assembly.
22. Temporarily install the rear support to the rear shock absorber nut.
    a. Using a screwdriver or an equivalent tool to hold the rear suspension support assembly, temporarily install a new rear support to rear shock absorber nut to the rear shock absorber assembly.

➡**Do not reuse the old rear support to rear shock absorber nut,**

➡**Use new rear support to rear shock absorber nut.**

➡**Tape the screwdriver or the equivalent tool before use.**

23. Install the rear coil spring.

➡**Do not use an impact wrench. It will damage SST.**

➡**Make sure that the end of the rear coil spring is positioned in the depression of the rear lower coil spring insulator.**

➡**Ensure the rear lower coil spring insulator is not pinched or folded over and caught by the rear coil spring.**

➡**Ensure that the stud bolt is positioned 3.5°to the outside of the vehicle as shown in the illustration. The deviation should be within +/- 5°.**

24. Install the rear shock absorber with the coil spring

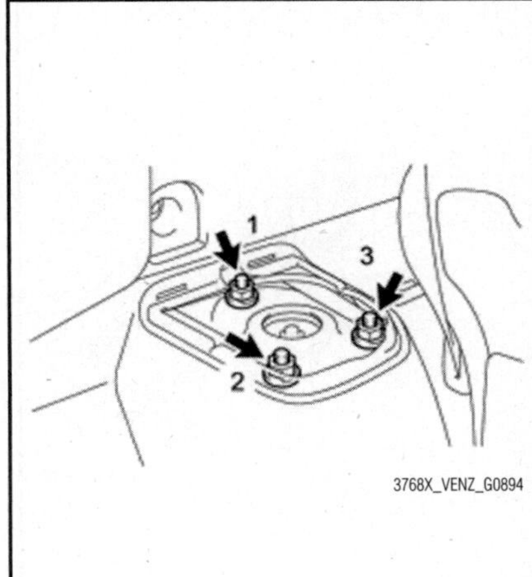

Fig. 195 Installing the rear shock absorber with the coil spring

    a. Install the rear shock absorber with coil spring with the 3 nuts in order 1 to 3. Tighten to 43 ft. lbs. (58 Nm).
    b. Install the rear shock absorber with coil spring with the 2 bolts and 2 nuts. Tighten to 214 ft. lbs. (290 Nm).

➡**When installing the nuts, keep the bolts from rotating.**

    c. Fully tighten the rear support to rear shock absorber nut. Tighten to 41 ft. lbs. (55 Nm).
25. Install the rear No. 1 suspension support cover.
26. Install the rear stabilizer link assembly to the rear shock absorber with coil spring with the nut. Tighten to 29 ft. lbs. (39 Nm).

➡**If the ball joint turns together with the nut, use a hexagon wrench (5 mm) to hold the stud bolt.**

27. Install the rear speed sensor wire to the rear shock absorber with coil spring with the bolt. Tighten to 44 inch lbs. (5 Nm).

➡**Do not twist the rear speed sensor wire when installing it.**

28. Install the rear flexible hose to the rear shock absorber with coil spring with the bolt. Tighten to 14 ft. lbs. (19 Nm).

➡**Do not twist the rear flexible hose when installing it.**

29. Install the deck side trim.
    a. Engage the 5 claws to install the deck side trim.
30. Install the rear wheel. Tighten to 76 ft. lbs. (103 Nm).
31. Inspect and adjust the rear wheel alignment.

STABIL'7ER BAR

REMOVAL & INSTALLATION

2WD Vehicles

See Figures 196 and 197.

1. Remove the rear wheels.
2. Remove the rear stabilizer link assembly LH.
3. Remove the rear stabilizer link assembly RH.
4. Remove the rear stabilizer bar.
    a. Remove the 4 bolts and the rear stabilizer bar.

To install:

5. Install the rear stabilizer bar with the 4 bolts in order 1 to 2. Tighten to 14 ft. lbs. (19 Nm).

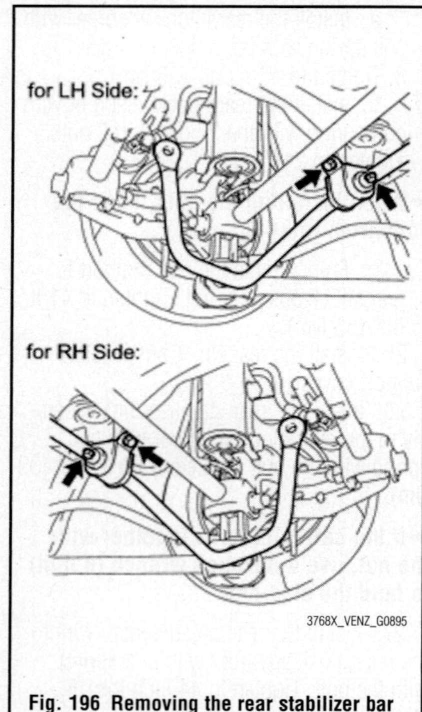

Fig. 196 Removing the rear stabilizer bar

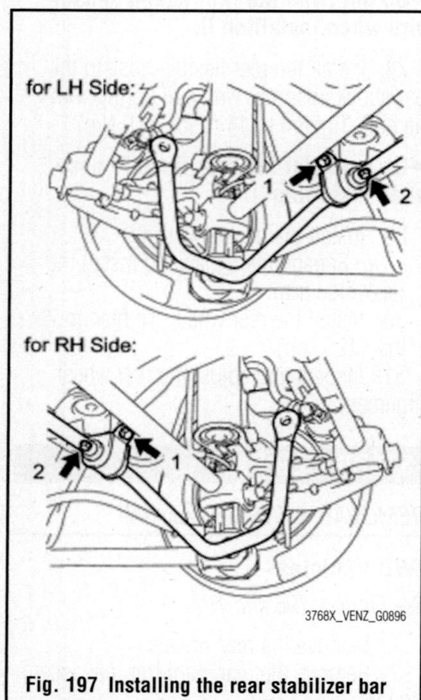

Fig. 197 Installing the rear stabilizer bar

6. Install the rear stabilizer link assembly LH.
7. Install the rear stabilizer link assembly RH.
8. Install the rear wheels. Tighten to 76 ft. lbs. (103 Nm).

AWD Vehicles

See Figures 198 through 202.

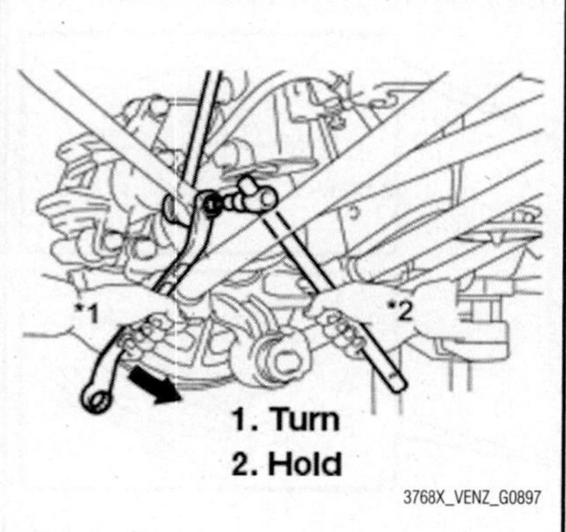

Fig. 198 Separating the rear stabilizer link assembly LH from the rear stabilizer bar

1. Remove the rear wheels.
2. Remove the rear stabilizer link assembly LH.
   a. Remove the nut and separate the rear stabilizer link assembly LH from the rear stabilizer bar.
   b. Remove the nut and the rear stabilizer link assembly LH from the rear shock absorber with coil spring LH.
3. Remove the rear stabilizer link assembly RH.

➡Perform the same procedure as the LH side.

4. Remove the No. 1 floor under cover.
5. Remove the rear lower suspension brace (LH).
   a. Remove the bolt, the nut and the rear lower suspension brace (LH side).
6. Remove the rear lower suspension brace RH.

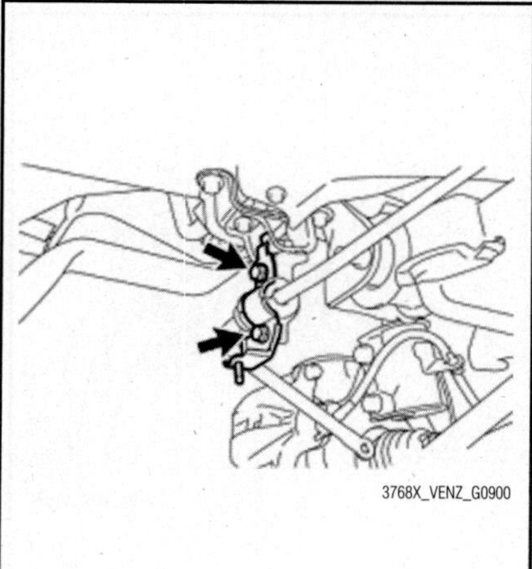

Fig. 199 Removing the rear stabilizer bar bracket LH

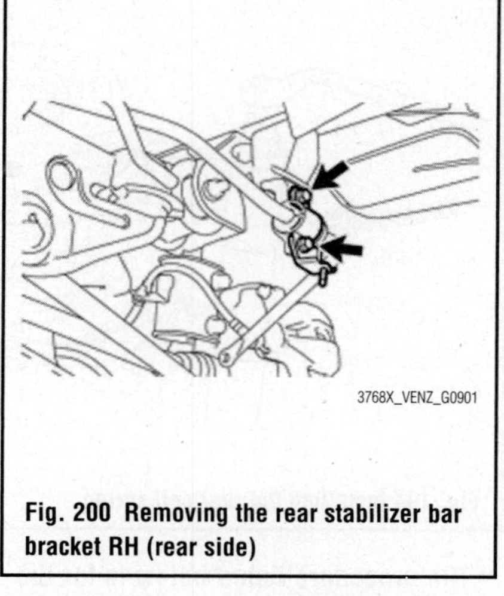

Fig. 200 Removing the rear stabilizer bar bracket RH (rear side)

➡Perform the same procedure as the LH side.

7. Remove the rear stabilizer bar bracket LH (rear side).

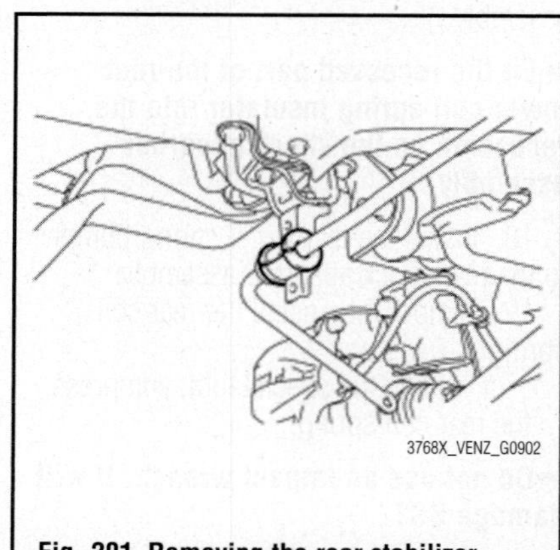

Fig. 201 Removing the rear stabilizer bushing LH

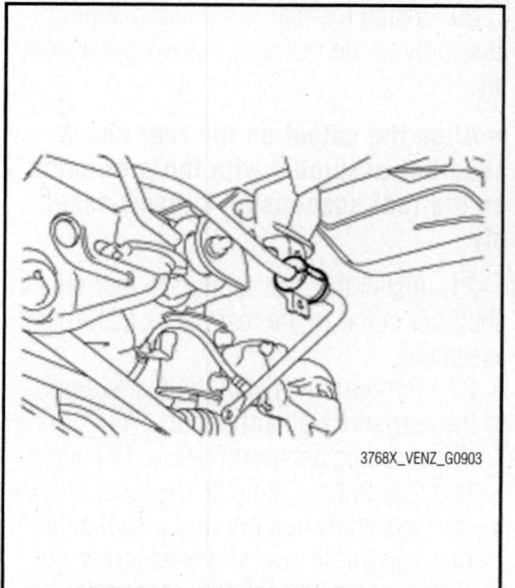

Fig. 202 Removing the rear stabilizer bushing RH

a. Remove the 2 bolts and the rear stabilizer bar bracket LH (rear side).

8. Remove the rear stabilizer bar bracket RH (rear side).

a. Remove the 2 bolts and the rear stabilizer bar bracket RH (rear side).

9. Remove the rear stabilizer bushing.

10. Remove the rear stabilizer bar bracket (front side).

To install:

11. To install, reverse the removal procedure. Tighten the rear stabilizer bar brackets (rear side) to 14 ft. lbs. (19 Nm). Tighten the rear stabilizer bar brackets (front side) to 40 ft. lbs. (54 Nm). Tighten the rear lower suspension brace to 26 ft. lbs. (35 Nm). Install the rear stabilizer link assembly to 29 ft. lbs. (39 Nm).

REAR STRUT ROD

REMOVAL & INSTALLATION

See Figures 203 through 206.

➡Use the same procedure for the RH side and LH side.

➡The procedure listed below is for the LH side.

1. Remove the rear wheel.
2. Separate the No. 3 parking brake cable assembly.

a. Remove the bolt and separate the No. 3 parking brake cable assembly.

3. Remove the rear strut rod assembly.

a. Remove the 2 bolts, the 2 nuts and the rear strut rod assembly.

➡Since stopper nuts are used, loosen the bolts.

To install:

4. Install the rear strut rod assembly (2WD).

a. Temporarily install the rear strut rod assembly to the rear axle carrier sub-assembly with the bolt and the nut.

➡Ensure that the identification mark faces the inside of the vehicle.

➡Since a stopper nut is used, temporarily tighten the bolt.

b. Set the rear strut rod assembly in the tightening position as shown in the illustration. Standard angle (A) is 9.9°. The standard length is 3.30 inch.

c. Fully tighten the bolt in the tightening position to 59 ft. lbs. (80 Nm).

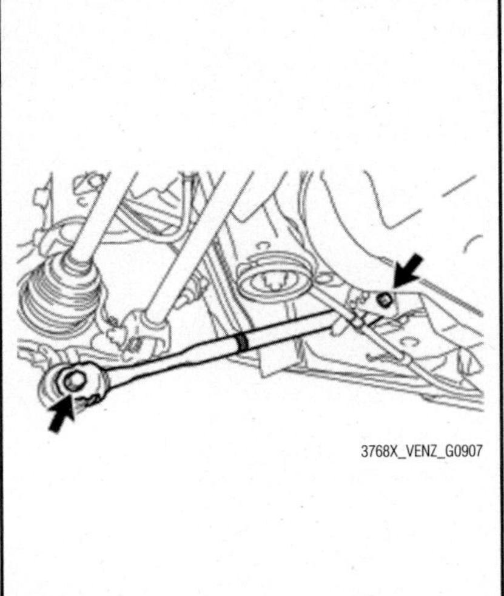

Fig. 204 Removing the rear strut rod assembly—AWD

d. Temporarily install the rear strut rod assembly to the body with the bolt and the nut.

e. Using SST and a socket wrench (17 mm), fully tighten the bolt in the rebound position. Tighten to 59 ft. lbs. (80 Nm). If using the special tool tighten to 44 ft. lbs. (59 Nm).

➡Since the stopper nut is used, fully tighten the bolt.

➡Use a torque wrench with a fulcrum length of 425 mm (1.39 ft.).

➡This torque value is effective when SST is parallel to the torque wrench.

5. Install the rear strut rod assembly (AWD).

a. Temporarily install the rear strut rod assembly to the rear axle carrier sub-assembly with the bolt and the nut.

➡Ensure that the identification mark faces the inside of the vehicle.

➡Since a stopper nut is used, temporarily tighten the bolt.

b. Set the rear strut rod assembly in the tightening position shown in the illustration. Standard angle (A) is 9.04°. The standard length is 3.08 inch.

c. Temporarily install the rear strut rod assembly to the body with the bolt and the nut.

d. Using SST and a socket wrench (17 mm), fully tighten the bolt in the rebound position. Tighten to 59 ft. lbs. (80 Nm). If using the special tool tighten to 44 ft. lbs. (59 Nm).

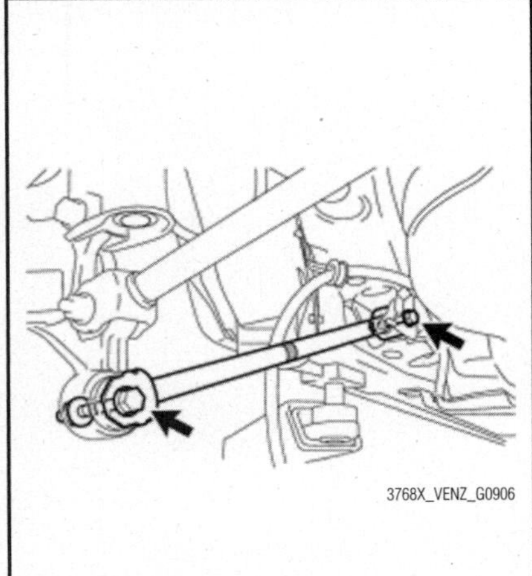

Fig. 203 Removing the rear strut rod assembly—2WD

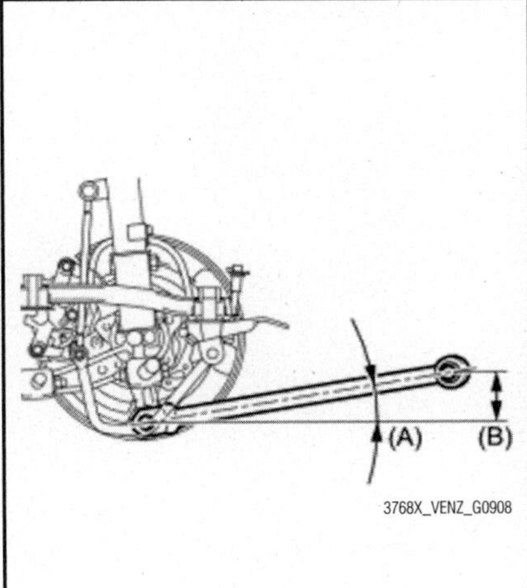

Fig. 205 Setting the tightening position

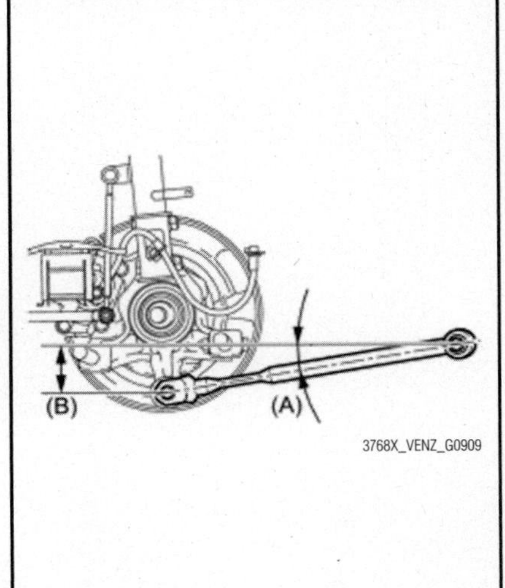

Fig. 206 Setting the tightening position

➡Since a stopper nut is used, fully tighten the bolt.

➡Use a torque wrench with a fulcrum length of 425 mm (1.39 ft.).

➡This torque value is effective when SST is parallel to the torque wrench.

6. Install the No. 3 parking brake cable assembly with the bolt. Tighten to 53 inch lbs. (6 Nm).

➡Do not twist the No. 3 parking brake cable assembly when installing it.

7. Install the rear wheel. Tighten to 76 ft. lbs. (103 Nm).
8. Stabilize the suspension.
   a. Lower the vehicle to the ground.
   b. Bounce the vehicle up and down at the corners to stabilize the rear suspension.

TOYOTA

Yaris

18

SPECIFICATIONS AND MAINTENANCE CHARTS

ENGINE AND VEHICLE IDENTIFICATION

Engine						
Code ①	Liters (cc)	Cu. In.	Cyl.	Fuel Sys.	Engine Type	Eng. Mfg.
1NZ-FE	1.5 (1496)	91	4	EFI	DOHC	Toyota

Model Year	
Code ②	Year
B	2011
C	2012

EFI: Electronic Fuel Injection

DOHC: Double Overhead Camshaft

① 8th digit of VIN

②10th digit of VIN

71099_YARI_C0001

GENERAL ENGINE SPECIFICATIONS

Year	Model	Engine Displacement Liters (cc)	Engine Series (ID/VIN)	Fuel System	Net Horsepower @ rpm	Net Torque @ rpm (ft. lbs.)	Bore x Stroke (in.)	Com-pression Ratio	Oil Pressure @ rpm
2011	Yaris	1.5 (1496)	1NZ-FE	EFI	106@6000	103@4200	2.95x3.33	10.5:1	4.3
2012	Yaris	1.5 (1496)	1NZ-FE	EFI	106@6000	103@4200	2.95x3.33	10.5:1	4.3

EFI: Electronic Fuel Injection

71099_YARI_C0002

ENGINE TUNE-UP SPECIFICATIONS

Year	Engine Displacement Liters	Engine ID/VIN	Spark Plug Gap (in.)	Ignition Timing (deg.)	Fuel Pump (psi)	Idle Speed (rpm)		Valve Clearance	
						MT	AT	Intake	Exhaust
2011	1.5	1NZ-FE	0.043	8-12 BTDC	44-50	550-650	650-750	0.006-0.010	0.010-0.014
2012	1.5	1NZ-FE	0.043	8-12 BTDC	44-50	550-650	650-750	0.006-0.010	0.010-0.014

Note: The Vehicle Emission Control Information label often reflects specification changes made during production. The label figures must be used if they differ from those in this chart.

71099_YARI_C0003

CAPACITIES

Year	Model	Engine Displacement Liters	Engine ID/VIN	Engine Oil with Filter	Transmission (pts.) 5-Spd	Transmission (pts.) Auto.	Drive Axle Front (pts.)	Fuel Tank (gal.)	Cooling System (qts.)
2011	Yaris	1.5	1NZ-FE	4.0	4.0	6.2	①	11.1	②
2012	Yaris	1.5	1NZ-FE	4.0	4.0	6.2	①	11.1	②

Note: All capacities are approximate. Add fluid gradually and check to be sure a proper fluid level is obtained.

① Included in transaxle capacity

② w/MT: 4.7
w/AT: 4.5

71099_YARI_C0005

FLUID SPECIFICATIONS

Year	Model	Engine Displacement Liters (VIN)	Engine Oil	Auto. Trans.	Brake Master Cylinder
2011	Yaris	1.5 (1NZ-FE)	①	Mercon® ATF Fluid	DOT 3
2012	Yaris	1.5 (1NZ-FE)	①	Mercon® ATF Fluid	DOT 3

DOT: Department Of Transportation

① 5W-30 Premium Synthetic Blend Motor Oil (US) or 5W-30 Super Premium Motor Oil (Canada)

71099_YARI_C0004

VALVE SPECIFICATIONS

Year	Engine Displacement Liters	Engine ID/VIN	Seat Angle (deg.)	Face Angle (deg.)	Spring Test Pressure (lbs. @ in.)	Spring Installed Height (in.)	Stem-to-Guide Clearance (in.) Intake	Stem-to-Guide Clearance (in.) Exhaust	Stem Diameter (in.) Intake	Stem Diameter (in.) Exhaust
2011	1.5	1NZ-FE	45	44.5	33.5-37@ 1.280	1.280	0.0010-0.0024	0.0012-0.0026	0.1957-0.1963	0.1955-0.1961
2012	1.5	1NZ-FE	45	44.5	33.5-37@ 1.280	1.280	0.0010-0.0024	0.0012-0.0026	0.1957-0.1963	0.1955-0.1961

71099_YARI_C0006

CAMSHAFT AND BEARING SPECIFICATIONS

All measurements are given in inches.

Year	Engine Displacement Liters	Engine VIN	Journal Diameter	Brg. Oil Clearance	Shaft End-play	Runout	Journal Bore	Lobe Lift Intake	Lobe Lift Exhaust
2011	1.5	1NZ-FE	1.3563-1.3569	0.0014-0.0028	0.0010-0.0060	0.0012	NA	1.7566-1.7605	1.7566-1.7605
2012	1.5	1NZ-FE	1.3563-1.3569	0.0014-0.0028	0.0010-0.0060	0.0012	NA	1.7566-1.7605	1.7566-1.7605

71099_YARI_C0009

CRANKSHAFT AND CONNECTING ROD SPECIFICATIONS

All measurements are given in inches.

Year	Engine Displacement Liters	Engine ID/VIN	Crankshaft Main Brg. Journal Dia.	Crankshaft Main Brg. Oil Clearance	Crankshaft Shaft End-play	Crankshaft Thrust on No.	Connecting Rod Journal Diameter	Connecting Rod Oil Clearance	Connecting Rod Side Clearance
2011	1.5	1NZ-FE	1.8106-1.8110	0.0004-0.0009	0.0035-0.0075	3	1.5745-1.5748	0.0006-0.0016	0.0063-0.0142
2012	1.5	1NZ-FE	1.8106-1.8110	0.0004-0.0009	0.0035-0.0075	3	1.5745-1.5748	0.0006-0.0016	0.0063-0.0142

71099_YARI_C0010

PISTON AND RING SPECIFICATIONS

All measurements are given in inches.

Year	Engine Displacement Liters	Engine ID/VIN	Piston Clearance	Ring Gap Top Compression	Ring Gap Bottom Compression	Ring Gap Oil Control	Ring Side Clearance Top Compression	Ring Side Clearance Bottom Compression	Ring Side Clearance Oil Control
2011	1.5	1NZ-FE	0.0018-0.0027	0.0098-0.0138	0.0138-0.0197	0.0039-0.0138	0.0012-0.0028	0.0008-0.0024	SNUG
2012	1.5	1NZ-FE	0.0018-0.0027	0.0098-0.0138	0.0138-0.0197	0.0039-0.0138	0.0012-0.0028	0.0008-0.0024	SNUG

71099_YARI_C0008

TORQUE SPECIFICATIONS

All readings in ft. lbs.

Year	Engine Displacement Liters	Engine ID/VIN	Cylinder Head Bolts	Main Bearing Bolts	Rod Bearing Bolts	Crankshaft Damper Bolts	Flywheel Bolts	Manifold Intake	Manifold Exhaust	Spark Plugs	Oil Pan Drain Plug
2011	1.5	1NZ-FE	①	②	③	95	④	22	20	13	28
2012	1.5	1NZ-FE	①	②	③	95	④	22	20	13	28

① Step 1: 22 ft. lbs.
Step 2: 90 degree turn
Step 3: 90 degree turn

② Step 1: 16 ft. lbs.
Step 2: 90 degree turn

③ Step 1: 11 ft. lbs.
Step 2: 90 degree turn

④ Step 1: 38 ft. lbs.
Step 2: 90 degree turn

71099_YARI_C0007

WHEEL ALIGNMENT

Year	Model		Caster Range (+/-Deg.)	Caster Preferred Setting (Deg.)	Camber Range (+/-Deg.)	Camber Preferred Setting (Deg.)	Toe-in (in.)	Steering Axis Inclination (Deg.)
2011	Yaris Sedan	F	0.75	+2.12	0.75	-0.47	0+/-0.08	13.15+/-0.75
		R	—	—	0.75	-1.18	0.12+/-0.08	—
	Yaris Hatchback	F	0.75	+2.02	0.75	-0.42	0+/-0.08	13.07+/-0.75
		R	—	—	0.75	-1.18	0.12+/-0.08	—
2012	Yaris Sedan	F	0.75	+2.12	0.75	-0.47	0+/-0.08	13.15+/-0.75
		R	—	—	0.75	-1.18	0.12+/-0.08	—
	Yaris Hatchback	F	0.75	+2.02	0.75	-0.42	0+/-0.08	13.07+/-0.75
		R	—	—	0.75	-1.18	0.12+/-0.08	—

71099_YARI_C0011

TIRE, WHEEL AND BALL JOINT SPECIFICATIONS

Year	Model	OEM Tires		Tire Pressures (psi)		Wheel	Ball Joint	Lug Nut
		Standard	Optional	Front	Rear	Size	Inspection	(ft. lbs.)
2011	Yaris	P175/65R14	P185/60R15	32	32	5.5-JJ	5.2-30 in. ①	76
2012	Yaris	P175/65R14	P185/60R15	32	32	5.5-JJ	5.2-30 in. ①	76

OEM: Original Equipment Manufacturer

PSI: Pounds Per Square Inch

① Torque required in inch lbs. to rotate ball joint when removed from the knuckle

71099_YARI_C0012

BRAKE SPECIFICATIONS

All measurements in inches unless noted

Year	Model	Brake Disc			Brake Drum Diameter			Minimum	Brake Caliper	
		Original Thickness	Minimum Thickness	Maximum Runout	Original Inside Diameter	Max. Wear Limit	Maximum Machine Diameter	Lining Thickness	Bracket Bolts (ft. lbs.)	Mounting Bolts (ft. lbs.)
2011	Yaris	0.866	0.748	0.0020	7.87	NA	7.913	0.039	25	65
2012	Yaris	0.866	0.748	0.0020	7.87	NA	7.913	0.039	25	65

NA: Not Available

71099_YARI_C0013

SCHEDULED MAINTENANCE INTERVALS

TOYOTA—Yaris

TO BE SERVICED	TYPE OF SERVICE	VEHICLE MILEAGE INTERVAL (x1000)													
		5	10	15	20	25	30	35	40	45	50	55	60	90	120
Engine oil & filter	R	✓	✓	✓	✓	✓	✓	✓	✓	✓	✓	✓	✓	✓	✓
Drive belts	S/I						✓						✓	✓	✓
Automatic transaxle fluid & filter	S/I						✓						✓	✓	✓
Brake line pipes & hoses	S/I	✓	✓	✓	✓	✓	✓	✓	✓	✓	✓	✓	✓	✓	✓
Brake linings & drums	S/I	✓	✓	✓	✓	✓	✓	✓	✓	✓	✓	✓	✓	✓	✓
Brake pads & discs (front & rear if equipped)	S/I	✓	✓	✓	✓	✓	✓	✓	✓	✓	✓	✓	✓	✓	✓
Cabin air filter	R				✓				✓				✓		✓
Differential oil	S/I						✓						✓	✓	✓
Drive shaft boots	S/I	✓	✓	✓	✓	✓	✓	✓	✓	✓	✓	✓	✓	✓	✓
Drive shaft bolt (tighten)	S/I	✓	✓	✓	✓	✓	✓	✓	✓	✓	✓	✓	✓	✓	✓
Engine coolant	S/I			✓			✓			✓			✓	✓	✓
Manual transaxle oil	S/I						✓						✓	✓	✓
Steering gear housing oil	S/I	✓	✓	✓	✓	✓	✓	✓	✓	✓	✓	✓	✓	✓	✓
Steering linkage	S/I	✓	✓	✓	✓	✓	✓	✓	✓	✓	✓	✓	✓	✓	✓
Air filter	R						✓						✓	✓	✓
Rotate tires	S/I	✓	✓	✓	✓	✓	✓	✓	✓	✓	✓	✓	✓	✓	✓
Spark plugs	R									✓				✓	
Fuel lines & connections	S/I						✓			✓			✓	✓	✓
Fuel tank cap gasket	R									✓				✓	
Charcoal canister	S/I									✓				✓	

R: Replace S/I: Service or Inspect

FREQUENT OPERATION MAINTENANCE (SEVERE SERVICE)

If a vehicle is operated under any of the following conditions it is considered severe service:

- Extremely dusty areas.

- 50% or more of the vehicle operation is in 32°C (90°F) or higher temperatures, or constant operation in temperatures below 0°C (32°F).

- Prolonged idling (vehicle operation in stop and go traffic).

- Frequent short running periods (engine does not warm to normal operating temperatures).

- Police, taxi, delivery usage or trailer towing usage.

Oil & oil filter: change every 5000 miles.

Bolts & nuts on chassis & body: tighten every 5000 miles.

Ball joints & dust covers: service or inspect every 5,000 miles.

Drive shaft boots & except Supra): service or inspect every 12,000 miles.

Steering linkage: service or inspect every 12,000 miles.

Air filter: service or inspect every 5,000 miles.

Exhaust system: service or inspect every 15,000 miles.

Timing belt: replace every 60,000 miles.

71099_YARI_C0014

PRECAUTIONS

Before servicing any vehicle, please be sure to read all of the following precautions, which deal with personal safety, prevention of component damage, and important points to take into consideration when servicing a motor vehicle:

- Never open, service or drain the radiator or cooling system when the engine is hot; serious burns can occur from the steam and hot coolant.
- Observe all applicable safety precautions when working around fuel. Whenever servicing the fuel system, always work in a well-ventilated area. Do not allow fuel spray or vapors to come in contact with a spark, open flame, or excessive heat (a hot drop light, for example). Keep a dry chemical fire extinguisher near the work area. Always keep fuel in a container specifically designed for fuel storage; also, always properly seal fuel containers to avoid the possibility of fire or explosion. Refer to the additional fuel system precautions later in this section.
- Fuel injection systems often remain pressurized, even after the engine has been turned **OFF**. The fuel system pressure must be relieved before disconnecting any fuel lines. Failure to do so may result in fire and/or personal injury.
- Brake fluid often contains polyglycol ethers and polyglycols. Avoid contact with the eyes and wash your hands thoroughly after handling brake fluid. If you do get brake fluid in your eyes, flush your eyes with clean, running water for 15 minutes. If eye irritation persists, or if you have taken brake fluid internally, IMMEDIATELY seek medical assistance.
- The EPA warns that prolonged contact with used engine oil may cause a number of skin disorders, including cancer. You should make every effort to minimize your exposure to used engine oil. Protective gloves should be worn when changing oil. Wash your hands and any other exposed skin areas as soon as possible after exposure to used engine oil. Soap and water, or waterless hand cleaner should be used.
- All new vehicles are now equipped with an air bag system, often referred to as a Supplemental Restraint System (SRS) or Supplemental Inflatable Restraint (SIR) system. The system must be disabled before performing service on or around system components, steering column, instrument panel components, wiring and sensors. Failure to follow safety and disabling procedures could result in accidental air bag deployment, possible personal injury and unnecessary system repairs.
- Always wear safety goggles when working with, or around, the air bag system. When carrying a non-deployed air bag, be sure the bag and trim cover are pointed away from your body. When placing a non-deployed air bag on a work surface, always face the bag and trim cover upward, away from the surface. This will reduce the motion of the module if it is accidentally deployed. Refer to the additional air bag system precautions later in this section.
- Clean, high quality brake fluid from a sealed container is essential to the safe and proper operation of the brake system. You should always buy the correct type of brake fluid for your vehicle. If the brake fluid becomes contaminated, completely flush the system with new fluid. Never reuse any brake fluid. Any brake fluid that is removed from the system should be discarded. Also, do not allow any brake fluid to come in contact with a painted surface; it will damage the paint.
- Never operate the engine without the proper amount and type of engine oil; doing so WILL result in severe engine damage.
- Timing belt maintenance is extremely important. Many models utilize an interference-type, non-freewheeling engine. If the timing belt breaks, the valves in the cylinder head may strike the pistons, causing potentially serious (also time-consuming and expensive) engine damage. Refer to the maintenance interval charts for the recommended replacement interval for the timing belt, and to the timing belt section for belt replacement and inspection.
- Disconnecting the negative battery cable on some vehicles may interfere with the functions of the on-board computer system(s) and may require the computer to undergo a relearning process once the negative battery cable is reconnected.
- When servicing drum brakes, only disassemble and assemble one side at a time, leaving the remaining side intact for reference.
- Only an MVAC-trained, EPA-certified automotive technician should service the air conditioning system or its components.

BRAKES

ANTI-LOCK BRAKE SYSTEM (ABS)

GENERAL INFORMATION

PRECAUTIONS

- Certain components within the ABS system are not intended to be serviced or repaired individually.
- Do not use rubber hoses or other parts not specifically specified for and ABS system. When using repair kits, replace all parts included in the kit. Partial or incorrect repair may lead to functional problems and require the replacement of components.
- Lubricate rubber parts with clean, fresh brake fluid to ease assembly. Do not use shop air to clean parts; damage to rubber components may result.
- Use only DOT 3 brake fluid from an unopened container.
- If any hydraulic component or line is removed or replaced, it may be necessary to bleed the entire system.
- A clean repair area is essential. Always clean the reservoir and cap thoroughly before removing the cap. The slightest amount of dirt in the fluid may plug an orifice and impair the system function. Perform repairs after components have been thoroughly cleaned; use only denatured alcohol to clean components. Do not allow ABS components to come into contact with any substance containing mineral oil; this includes used shop rags.
- The Anti-Lock control unit is a microprocessor similar to other computer units in the vehicle. Ensure that the ignition switch is **OFF** before removing or installing controller harnesses. Avoid static electricity discharge at or near the controller.
- If any arc welding is to be done on the vehicle, the control unit should be unplugged before welding operations begin.

SPEED SENSORS

REMOVAL & INSTALLATION

Front

See Figure 1.

1. Disconnect the negative battery cable.
2. Remove front wheel.
3. Remove the 3 screws, 6 clips and 4 grommets and remove the front fender liner.

4. Remove the speed sensor clip from the body.
5. Disconnect the speed sensor connector.
6. Remove the 3 clips.
7. Remove the bolt and separate the clamp from the body.
8. Remove the bolt and separate the clamp from the shock absorber.
9. Remove the bolt and remove the speed sensor from the steering knuckle.

➡Keep the speed sensor tip and installation portion free of foreign matter. Remove the speed sensor without turning it from its original installation angle.

To install:

➡Keep the speed sensor tip and installation portion free of foreign matter. Install the speed sensor without turning it from its original installation angle.

10. Install the speed sensor to the steering knuckle, and install the bolt. Tighten the bolt to 75 inch lb.
11. Connect the clamp to the shock absorber, and install the bolt. Tighten the bolt to 53 inch lb.
12. Connect the clamp to the body, and install the bolt. Tighten the bolt to 22 ft. lb.
13. Install the 3 clips.
14. Connect the speed sensor connector.
15. Install the speed sensor clip to the body.
16. Install the front fender liner, and install the 3 screws, 6 clips and 4 grommets.
17. Install front wheel.
18. Connect the negative battery cable.

Rear

See Figure 2.

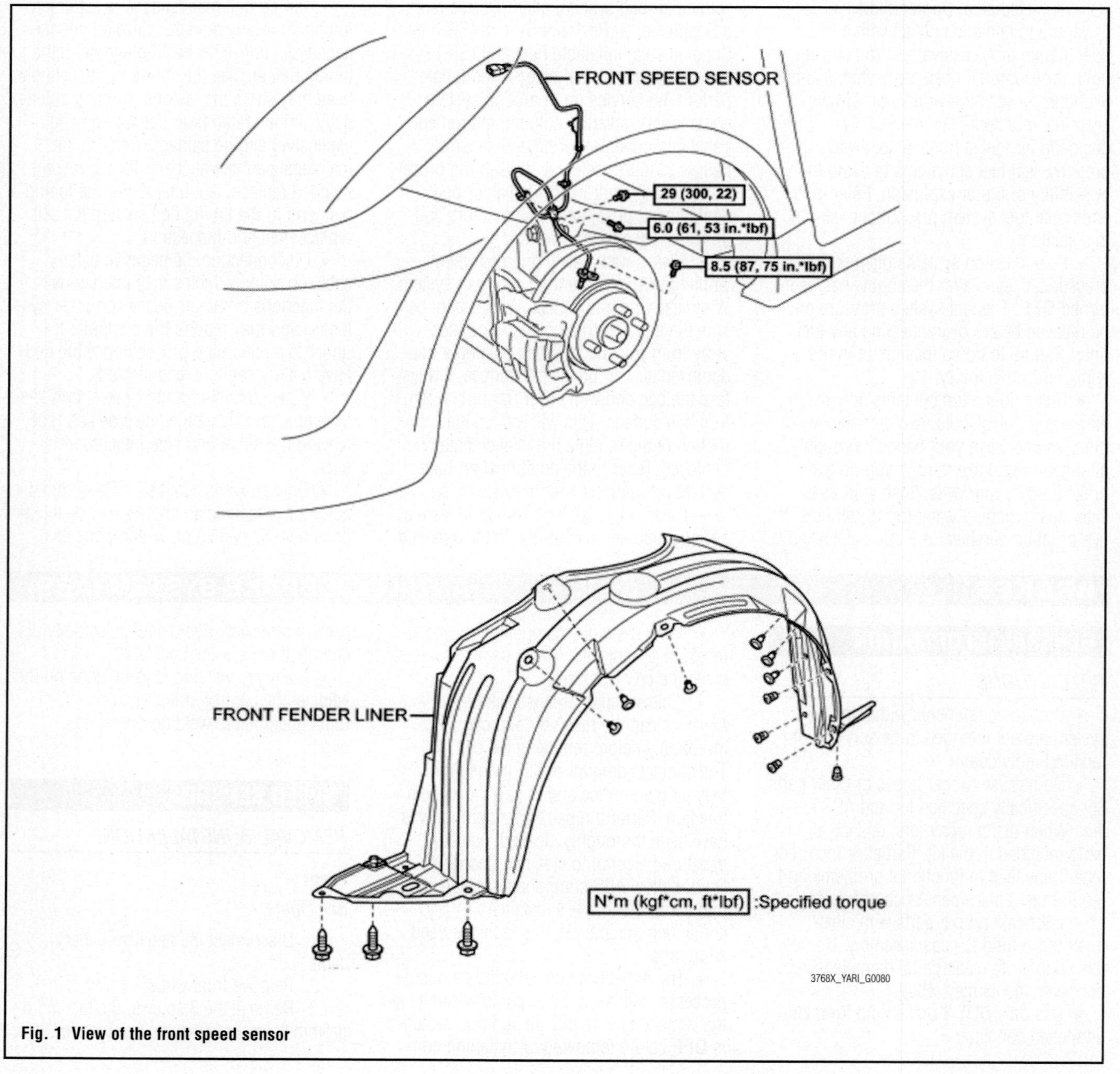

Fig. 1 View of the front speed sensor

➡The rear wheel speed sensor is integrated with the wheel bearing and hub assembly and must be replaced as a unit.

1. Disconnect the negative battery cable.
2. Remove rear wheel.
3. Remove the rear drum sub-assembly.
   a. Release the parking brake and remove the rear brake drum. If the rear brake drum cannot be removed easily, perform the following procedure.
   b. Remove the hole plug and insert a screwdriver through the hole into the backing plate, and hold the automatic adjust lever away from the adjuster.
4. Using a screwdriver, remove the claw of the connector lock portion and disconnect the skid control sensor wire connector.

WARNING

Do not remove the connector cover from the connector because the skid control sensor wire may be damaged.

5. Remove the rear wheel bearing and hub assembly.
6. To install, reverse the removal procedure.

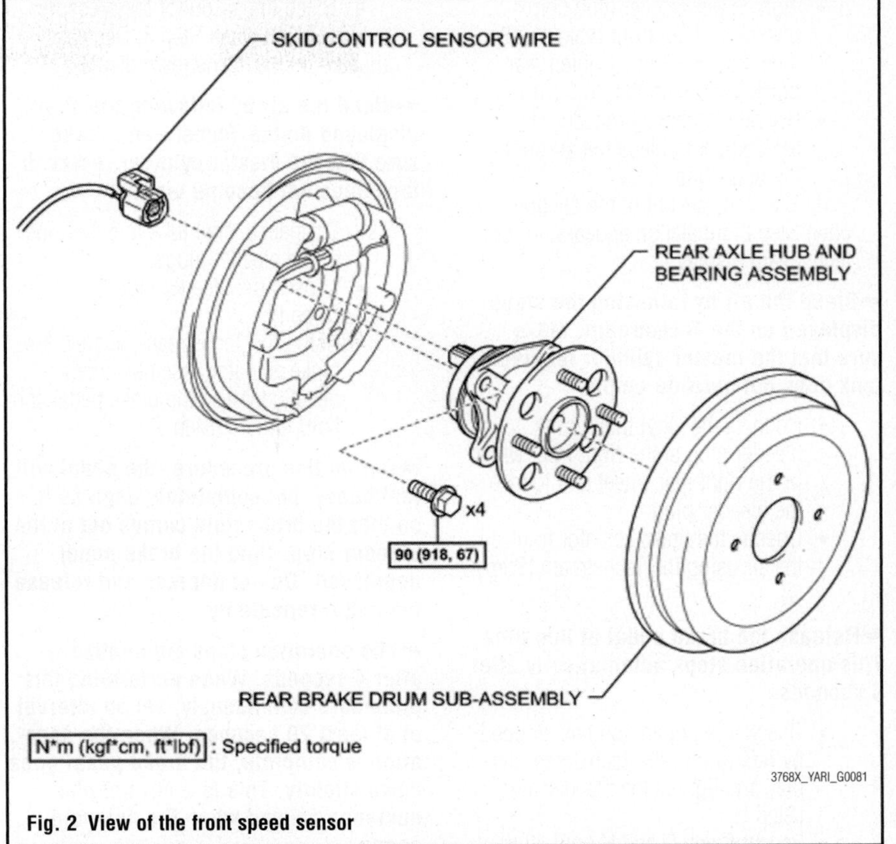

Fig. 2 View of the front speed sensor

BRAKES — BLEEDING THE BRAKE SYSTEM

BLEEDING PROCEDURE

BLEEDING PROCEDURE

➡Immediately wash off any brake fluid that comes into contact with any painted surfaces.

➡If any work is done on the brake system or if air in the brake lines is suspected, bleed the air from the system.

1. Fill reservoir with brake fluid:
   a. Disengage the 3 clips and separate the hood to cowl top seal.
   b. Remove the cowl top ventilator louver.
   c. Set the brake fluid can upside down on the reservoir. Fluid: SAE J1703 or FMVSS No. 116 DOT3.
2. Bleed master cylinder:
   a. Using a union nut wrench, disconnect the brake tubes from the master cylinder.
   b. Slowly depress the brake pedal and hold it there.
   c. Block the outer holes with your fingers, and release the brake pedal.
   d. Repeat 3 or 4 times.
   e. Using a union nut wrench, connect the brake tubes to the master cylinder. Tighten to 10 inch lbs. (123 Nm).
3. Bleed brake line:
   a. Connect the vinyl tube to the bleeder plug.
   b. Depress the brake pedal several times, then loosen the bleeder plug with the pedal depressed.
   c. At the point where the fluid stops coming out, tighten the bleeder plug, then release the brake pedal.
   d. Repeat until all the air in the fluid is out.
   e. Tighten the bleeder plug to 73 inch lbs. (8.3 Nm).
   f. Repeat the above procedure to bleed the air out of the brake line for each wheel.

➡After bleeding the air from the brake system, if the height or feel of the brake pedal cannot be obtained, perform air bleeding of the brake actuator with the Techstream by following the procedure below.

4. Bleed brake actuator (with VSC):
   a. Depress the brake pedal more than 20 times with the engine off.
   b. Connect the Techstream to the DLC3, and turn the ignition switch ON.

➡Do not start the engine.

   c. Select "Air Bleeding" on the Techstream.
5. Bleed the air out of the brake line when Step 1: Increase appears on the Techstream display.

➡Bleed the air by following the steps displayed on the Techstream. Make sure that the master cylinder reservoir tank does not become empty.

   a. Connect the vinyl tube to either one of the bleeder plugs.
   b. Depress the brake pedal several times, then loosen the bleeder plug connected to the vinyl tube with the pedal depressed (Step E).
   c. When fluid stops coming out, tighten the bleeder plug and release the brake pedal (Step F).
   - Repeat Steps E and F until all the air in the fluid is completely bled out.

- Tighten the bleeder plug completely. Tighten front brake to 73 inch lbs. (8.3 Nm). Tighten rear brake to 75 inch lbs. (8.5 Nm).
- Repeat the above procedure for each wheel to bleed the air out of the brake line.

d. Bleed the air out of the suction line when Step 2: Inhalation appears on the Techstream display.

➡Bleed the air by following the steps displayed on the Techstream. Make sure that the master cylinder reservoir tank does not become empty.

- Connect the vinyl tube to the bleeder plug at the right front wheel or the right rear wheel and loosen the bleeder plug.
- Operate the brake actuator to bleed the air using the Techstream (Step G).

➡Release the brake pedal at this time. This operation stops automatically after 4 seconds.

- Check if the operation has stopped by referring to the Techstream display and tighten the bleeder plug (Step H).
- Repeat Steps G and H until all the air in the fluid is completely bled out.
- Tighten the bleeder plug. Tighten front brake to 73 inch lbs. (8.3 Nm). Tighten rear brake to 75 inch lbs. (8.5 Nm).
- Repeat the above procedure for the other wheels to bleed the air out of the brake line.

e. Bleed the air out of the pressure reduction line when Step 3: Decrease appears on the Techstream display.

➡Bleed the air by following the steps displayed on the Techstream. Make sure that the master cylinder reservoir tank does not become empty.

- Connect a vinyl tube to either one of the bleeder plugs.
- Loosen the bleeder plug (Step I).
- Using the Techstream, operate the brake actuator assembly, completely depress the brake pedal and hold it there (Step J).

➡During this procedure, the pedal will feel heavy, but completely depress it so that the brake fluid comes out of the bleeder plug. Hold the brake pedal depressed. Do not depress and release the pedal repeatedly.

➡The operation stops automatically after 4 seconds. When performing this procedure continuously, set an interval of at least 20 seconds. When the operation is complete, the brake pedal goes down slightly. This is a normal phenomenon caused when the solenoid opens.

- Tighten the bleeder plug, then release the brake pedal.
- Repeat Steps I to K until all the air in the fluid is completely bled out.
- Tighten the bleeder plug. Tighten front brake to 73 inch lbs. (8.3 Nm). Tighten rear brake to 75 inch lbs. (8.5 Nm).
- Repeat the above procedure for the other wheels to bleed the air out of the brake line.

f. Bleed the air out of the brake line again when Step 4: Increase appears on the Techstream display.

➡Bleed the air by following the steps displayed on the Techstream. Make sure that the master cylinder reservoir tank does not become empty.

- Connect the vinyl tube to either one of the bleeder plugs.
- Depress the brake pedal several times, then loosen the bleeder plug connected to the vinyl tube with the pedal depressed.
- When fluid stops coming out, tighten the bleeder plug, then release the brake pedal (Step M).
- Repeat Steps L and M until all the air in the fluid is completely bled out.
- Tighten the bleeder plug. Tighten front brake to 73 inch lbs. (8.3 Nm). Tighten rear brake to 75 inch lbs. (8.5 Nm).
- Repeat the above procedure for the other wheels to bleed the air out of the brake line.
- Make sure that the air bleeding is complete by referring to the Techstream display and turn off the Techstream.
- Disconnect the Techstream from the DLC3.
- Turn the ignition switch OFF.

6. Check the fluid level and add fluid if necessary. Fluid: SAE J1703 or FMVSS No. 116 DOT3.

BRAKES — FRONT DISC BRAKES

⁂ CAUTION

Dust and dirt accumulating on brake parts during normal use may contain asbestos fibers from production or aftermarket brake linings. Breathing excessive concentrations of asbestos fibers can cause serious bodily harm. Exercise care when servicing brake parts. Do not sand or grind brake lining unless equipment used is designed to contain the dust residue. Do not clean brake parts with compressed air or by dry brushing. Cleaning should be done by dampening the brake components with a fine mist of water, then wiping the brake components clean with a dampened cloth. Dispose of cloth and all residue containing asbestos fibers in an impermeable container with the appropriate label. Follow practices prescribed by the Occupational Safety and Health Administration (OSHA) and the Environmental Protection Agency (EPA) for the handling, processing, and disposing of dust or debris that may contain asbestos fibers.

BRAKE CALIPER

REMOVAL & INSTALLATION

See Figure 3.

1. Remove the front wheel.
2. Drain the brake fluid.

➡Immediately wash off any brake fluid that comes into contact with any painted surfaces.

3. Remove the union bolt and gasket and separate the flexible hose from the disc brake cylinder.
4. Fix the slide pin with a spanner, remove the 2 bolts and remove the disc brake cylinder.
5. Remove the 2 disc brake pads from the disc brake cylinder mounting.
6. Remove the No. 1 anti squeal shim and No. 2 anti squeal shim from each brake pad.
7. Remove the indicator plate from each brake pad.
8. Remove the 4 disc brake pad support

plates from the disc brake cylinder mounting.

9. Remove the slide pin (upper) and slide pin (lower) from the disc brake cylinder mounting.

10. Using a screwdriver with its tip wrapped in protective tape, remove the slide bush from the slide pin (lower).

11. Remove the 2 dust boots from the disc brake cylinder mounting.

12. Remove the 2 bolts and remove the disc brake cylinder mounting from the steering knuckle.

13. To install, reverse the removal procedure.

14. Observe and use the torque values in the illustration.

15. Lubricate in areas as shown in the illustration.

16. Fill reservoir with brake fluid.

17. Bleed master cylinder.

18. Bleed brake line.

19. Bleed brake actuator (with VSC).

20. Check fluid level in reservoir.

21. Check for brake fluid leakage.

22. Install front wheel.

DISC BRAKE PADS

REMOVAL & INSTALLATION

See Figure 3.

1. Remove the front wheel.

2. Loosen and remove the caliper mounting bolts, then remove the caliper assembly, without disconnecting the brake line. Position it aside.

3. Slide out the old brake pads along with any anti-squeal shims, springs, pad wear indicators and pad support plates.

To install:

4. Install the pad support plates into the torque plate.

5. Install the pad wear indicators onto the pads. Be sure the arrow on the indicator plate is pointing in the direction of rotation.

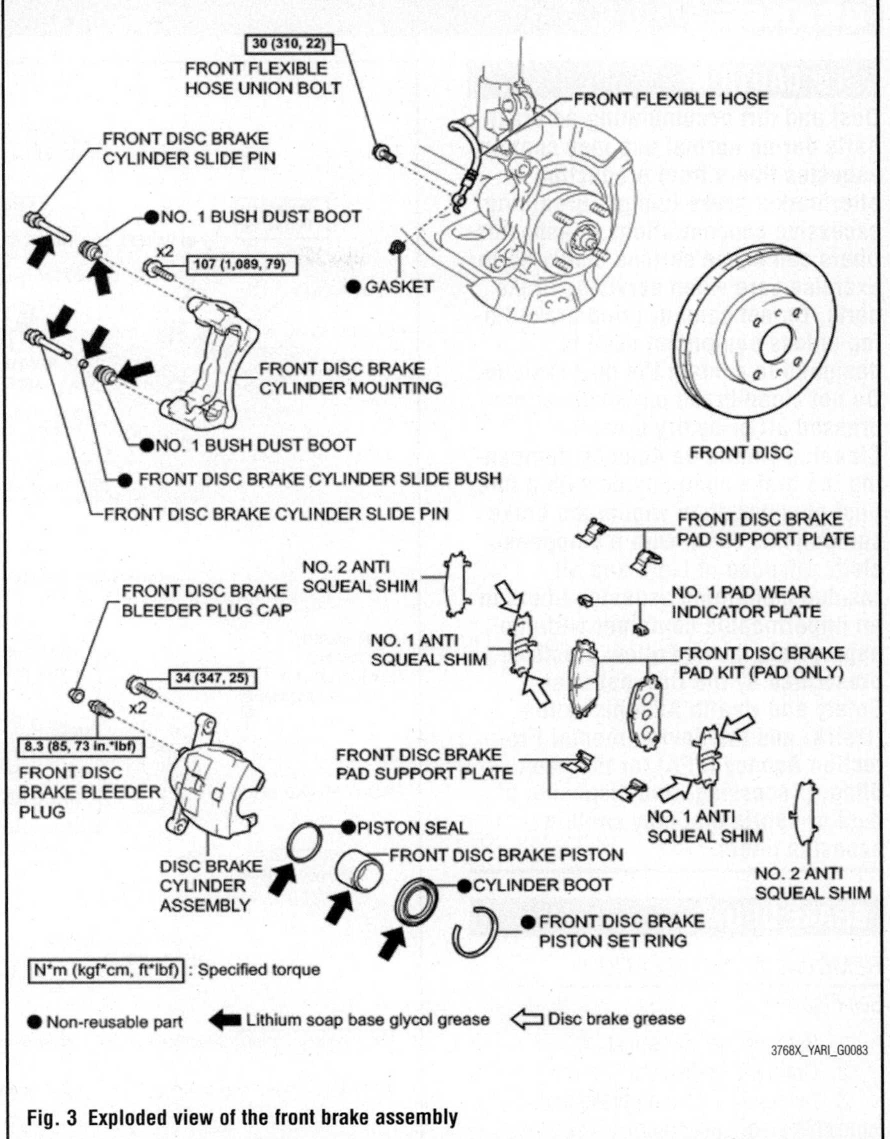

Fig. 3 Exploded view of the front brake assembly

6. Install the anti-squeal shims on the outside of each pad and then install the pad assemblies into the torque plate.

7. Compress the caliper piston into the bore.

8. Position the caliper back down over the pads.

9. Install and tighten the caliper mounting bolts.

10. Install the wheels. Check the brake fluid level.

CAUTION

Dust and dirt accumulating on brake parts during normal use may contain asbestos fibers from production or aftermarket brake linings. Breathing excessive concentrations of asbestos fibers can cause serious bodily harm. Exercise care when servicing brake parts. Do not sand or grind brake lining unless equipment used is designed to contain the dust residue. Do not clean brake parts with compressed air or by dry brushing. Cleaning should be done by dampening the brake components with a fine mist of water, then wiping the brake components clean with a dampened cloth. Dispose of cloth and all residue containing asbestos fibers in an impermeable container with the appropriate label. Follow practices prescribed by the Occupational Safety and Health Administration (OSHA) and the Environmental Protection Agency (EPA) for the handling, processing, and disposing of dust or debris that may contain asbestos fibers.

BRAKE DRUM

REMOVAL & INSTALLATION

See Figure 4.

1. Remove the rear wheel.
2. Drain the brake fluid.
3. Release the parking brake and remove the rear brake drum.
4. If the rear brake drum cannot be removed easily, perform the following procedure.:
    a. Remove the hole plug and insert a screwdriver through the hole into the backing plate, and hold the automatic adjust lever away from the adjuster.
    b. Using another screwdriver, contract the brake shoe by turning the adjusting bolt.

To install:

5. Provisionally install the 2 hub nuts.
6. Remove the hole plug, and turn the adjuster to expand the shoe until the drum locks.
7. Using a screwdriver, release the adjuster 12 notches.
8. Install the hole plug.
9. Fill reservoir with brake fluid.
10. Bleed master cylinder.

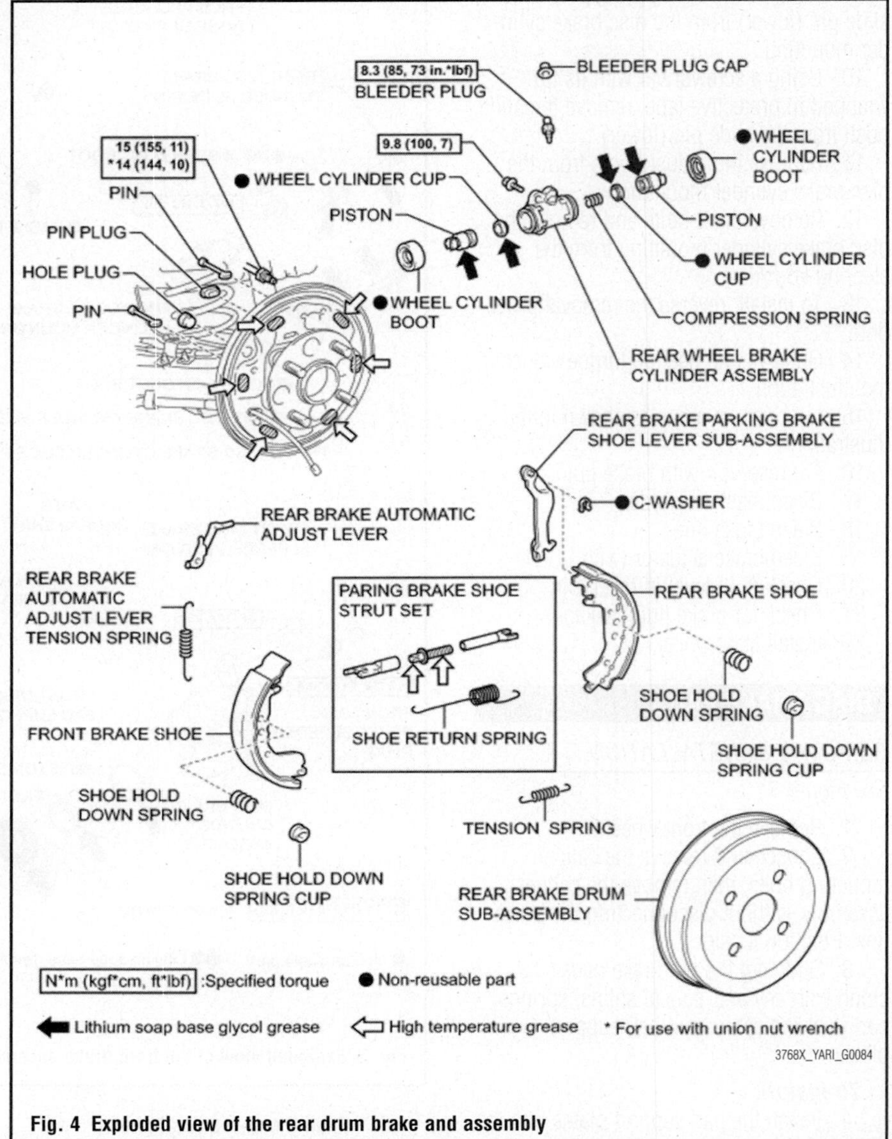

Fig. 4 Exploded view of the rear drum brake and assembly

11. Bleed brake line.
12. Bleed brake actuator (with VSC).
13. Check fluid level in reservoir.
14. Check for brake fluid leakage.
15. Install the rear wheel.
16. Inspect parking brake lever travel.
17. Adjust parking brake lever travel.

BRAKE SHOES

REMOVAL & INSTALLATION

See Figures 4 through 6.

1. Using SST 09703-30010, separate the shoe return spring from the front brake shoe.
2. Using SST 09718-00010, remove the shoe hold down spring cup, shoe hold down spring, pin and front brake shoe.
3. Remove the tension spring.
4. Remove the shoe return spring from the rear brake shoe and remove the parking brake shoe strut set.
5. Using SST 09718-00010, remove the shoe hold down spring cup, shoe hold down spring, pin and rear brake shoe.
6. Using needle-nose pliers, separate the parking brake cable.
7. Remove the automatic adjust lever tension spring and remove the automatic adjust lever.

8. Using a screwdriver, remove the C-washer and remove the parking brake shoe lever.

To install:

➡Refer to the exploded illustration for torque values and grease/lubrications points

9. Using needle-nose pliers, install the parking brake shoe lever with a new C-washer.
10. Install the automatic adjust lever and automatic adjust lever tension spring onto the front brake shoe.
11. Install rear brake shoe kit:
    a. Apply high temperature grease to the surface of the backing plate which is in contact with the shoe.
    b. Using needle-nose pliers, install the parking brake cable onto the parking brake shoe lever.
    c. Using SST 09718-00010, install the rear brake shoe, pin, shoe hold down spring and shoe hold down spring cup.
    d. Apply high temperature grease to the adjusting bolt.
    e. Install the parking brake shoe strut set as shown in the accompanying illustration.
    f. Using SST 09718-00010, install the front brake shoe, pin, shoe hold down spring and shoe hold down spring cup.
    g. Using needle-nose pliers, install the tension spring onto the front brake shoe and rear brake shoe.
    h. Using SST 09703-30010, install the shoe return spring onto the front brake shoe.
12. Check that each part is installed properly.
13. Measure the brake drum inner diameter and the diameter of the brake shoes. Check that the difference between the diameters is equal to the specified shoe clearance. Shoe clearance: 0.024 inches. (0.6 mm).

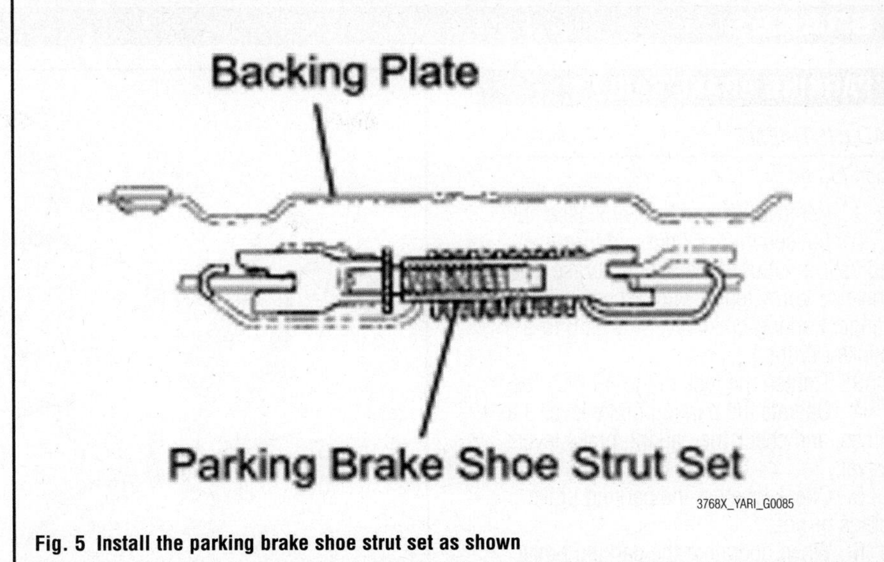

Fig. 5 Install the parking brake shoe strut set as shown

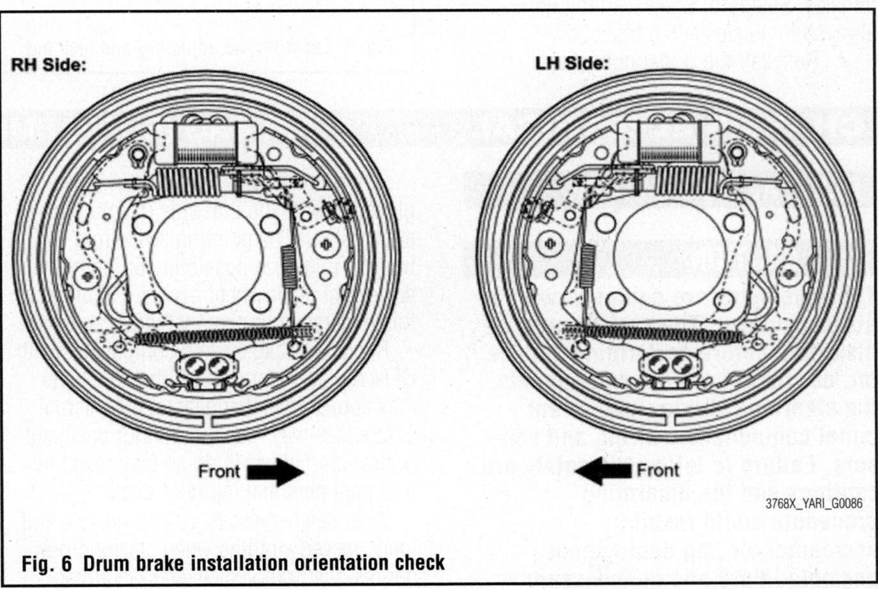

Fig. 6 Drum brake installation orientation check

➡There should be no oil or grease adhering to the friction surfaces of the shoe lining or the drum.

ADJUSTMENT

1. Provisionally install the 2 hub nuts.
2. Remove the hole plug, and turn the adjuster to expand the shoe until the drum locks.
3. Using a screwdriver, release the adjuster 12 notches.
4. Install the hole plug.

PARKING BRAKE CABLES

ADJUSTMENT

See Figure 7.

1. Remove rear console box assembly.
2. Loosen the lock nut and turn the adjusting nut until the parking brake lever travel is corrected to within the specified range. Parking brake lever travel: 6 to 9 clicks (45 lbs.).
3. Tighten the lock nut to 48 inch lbs.
4. Operate the parking brake lever 3 to 4 times, and check the parking brake lever travel.
5. Check whether the parking brake drags or not.
6. When operating the parking brake lever, check that the brake warning light illuminates. Standard: Brake warning light always illuminates at the first click.
7. Reinstall the center console.

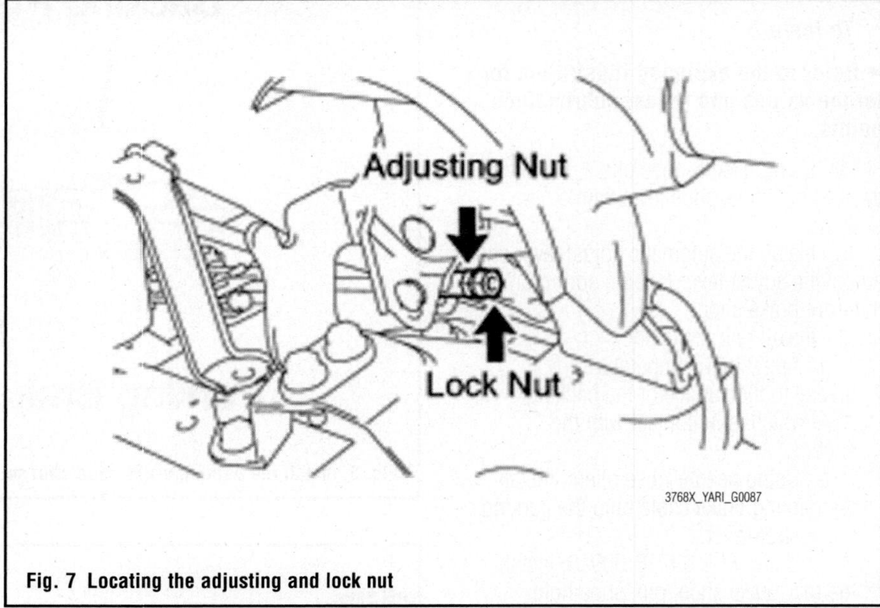

Fig. 7 Locating the adjusting and lock nut

GENERAL INFORMATION

⁂ CAUTION

These vehicles are equipped with an air bag system. The system must be disarmed before performing service on, or around, system components, the steering column, instrument panel components, wiring and sensors. Failure to follow the safety precautions and the disarming procedure could result in accidental air bag deployment, possible injury and unnecessary system repairs.

SERVICE PRECAUTIONS

Disconnect and isolate the battery negative cable before beginning any airbag system component diagnosis, testing, removal, or installation procedures. Allow system capacitor to discharge for two minutes before beginning any component service. This will disable the airbag system. Failure to disable the airbag system may result in accidental airbag deployment, personal injury, or death.

Do not place an intact undeployed airbag face down on a solid surface. The airbag will propel into the air if accidentally deployed and may result in personal injury or death.

When carrying or handling an undeployed airbag, the trim side (face) of the airbag should be pointing away from the body to minimize possibility of injury if accidental deployment occurs. Failure to do this may result in personal injury or death.

Replace airbag system components with OEM replacement parts. Substitute parts may appear interchangeable, but internal differences may result in inferior occupant protection. Failure to do so may result in occupant personal injury or death.

Wear safety glasses, rubber gloves, and long sleeved clothing when cleaning powder residue from vehicle after an airbag deployment. Powder residue emitted from a deployed airbag can cause skin irritation. Flush affected area with cool water if irritation is experienced. If nasal or throat irritation is experienced, exit the vehicle for fresh air until the irritation ceases. If irritation continues, see a physician.

Do not use a replacement airbag that is not in the original packaging. This may result in improper deployment, personal injury, or death.

The factory installed fasteners, screws and bolts used to fasten airbag components have a special coating and are specifically designed for the airbag system. Do not use substitute fasteners. Use only original equipment fasteners listed in the parts catalog when fastener replacement is required.

During, and following, any child restraint anchor service, due to impact event or vehicle repair, carefully inspect all mounting hardware, tether straps, and anchors for proper installation, operation, or damage. If a child restraint anchor is found damaged in any way, the anchor must be replaced. Failure to do this may result in personal injury or death.

Deployed and non-deployed airbags may or may not have live pyrotechnic material within the airbag inflator.

Do not dispose of driver/passenger/curtain airbags or seat belt tensioners unless you are sure of complete deployment. Refer to the Hazardous Substance Control System for proper disposal.

Dispose of deployed airbags and tensioners consistent with state, provincial, local, and federal regulations.

After any airbag component testing or service, do not connect the battery negative cable. Personal injury or death may result if the system test is not performed first.

If the vehicle is equipped with the Occupant Classification System (OCS), do not connect the battery negative cable before performing the OCS Verification Test using the scan tool and the appropriate diagnostic information. Personal injury or death may result if the system test is not performed properly.

Never replace both the Occupant Restraint Controller (ORC) and the Occupant Classification Module (OCM) at the same time. If both require replacement, replace one, then perform the Airbag System test before replacing the other.

Both the ORC and the OCM store Occupant Classification System (OCS) calibration data, which they transfer to one another when one of them is replaced. If both are replaced at the same time, an irreversible fault will be set in both modules and the OCS may malfunction and cause personal injury or death.

If equipped with OCS, the Seat Weight Sensor is a sensitive, calibrated unit and must be handled carefully. Do not drop or handle roughly. If dropped or damaged, replace with another sensor. Failure to do so may result in occupant injury or death.

If equipped with OCS, the front passenger seat must be handled carefully as well. When removing the seat, be careful when setting on floor not to drop. If dropped, the sensor may be inoperative, could result in occupant injury, or possibly death.

If equipped with OCS, when the passenger front seat is on the floor, no one should sit in the front passenger seat. This uneven force may damage the sensing ability of the seat weight sensors. If sat on and damaged, the sensor may be inoperative, could result in occupant injury, or possibly death.

DISARMING THE SYSTEM

To avoid personal injury when working on vehicles equipped with an air bag, the negative battery cable must be disconnected and at least 90 seconds must elapse before working on the system. Failure to do so may result in deployment of the air bag.

ARMING THE SYSTEM

After vehicle service is completed, reattach the battery cables (positive cable first!) to rearm the air bag system.

CLOCKSPRING CENTERING

1. Check that the ignition switch is off.
2. Check that the battery negative (-) terminal is disconnected.
3. Rotate the spiral cable counterclockwise slowly by hand until it feels firm.

➡Do not turn the spiral cable by the airbag wire harness.

4. Rotate the spiral cable clockwise approximately 2.5 turns to align the marks.

➡Do not turn the spiral cable by the airbag wire harness. The spiral cable will rotate approximately 2.5 turns to both the left and right from the center.

DRIVE TRAIN

FRONT AXLE ASSEMBLY

REMOVAL & INSTALLATION

See Figures 8 and 9.

1. Disconnect the negative battery cable.
2. Safely raise and support the vehicle.
3. Remove the front wheel.
4. Separate the front caliper.
5. Remove the front disc.
6. Using SST 09628-10011 and a screwdriver or the equivalent to hold the axle hub, remove the hub bolt.
7. Separate front speed sensor (w/ abs).
8. Separate front stabilizer link assembly.
9. Separate tie rod end sub-assembly.
10. Separate front lower suspension arm.
11. Remove front axle assembly:
    a. Using a plastic hammer, tap the end of the drive shaft and disengage the fitting between the drive shaft and front axle.

➡If it is difficult to disengage the fitting, tap the end of the drive shaft with a brass bar and hammer.

    b. Push the front axle out of the vehicle to remove the drive shaft from the front axle.

➡Please note the following:

- Do not push the front axle further out of the vehicle than is necessary.
- Do not damage the speed sensor rotor.
- Suspend the drive shaft with a piece of string or the equivalent.

    c. Remove the 2 nuts and 2 bolts and remove the front axle assembly.

➡Keep the nut from rotating while turning the bolt.

To install:

12. To install, reverse the removal procedure.
13. Tighten the front axle assembly onto the shock absorber to 121 ft. lbs. (164 Nm).
14. Inspect and adjust front wheel alignment.

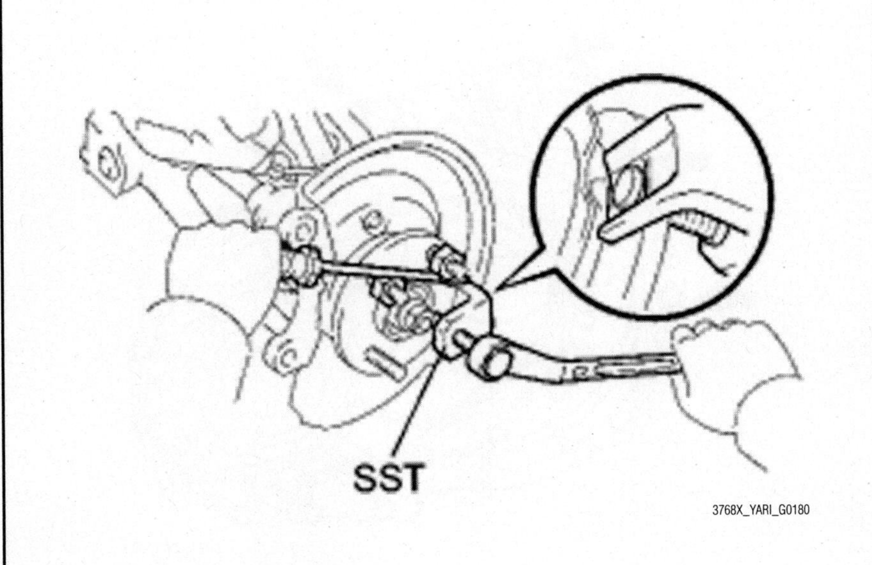

Fig. 8 Using SST 09628-10011 remove the hub bolt

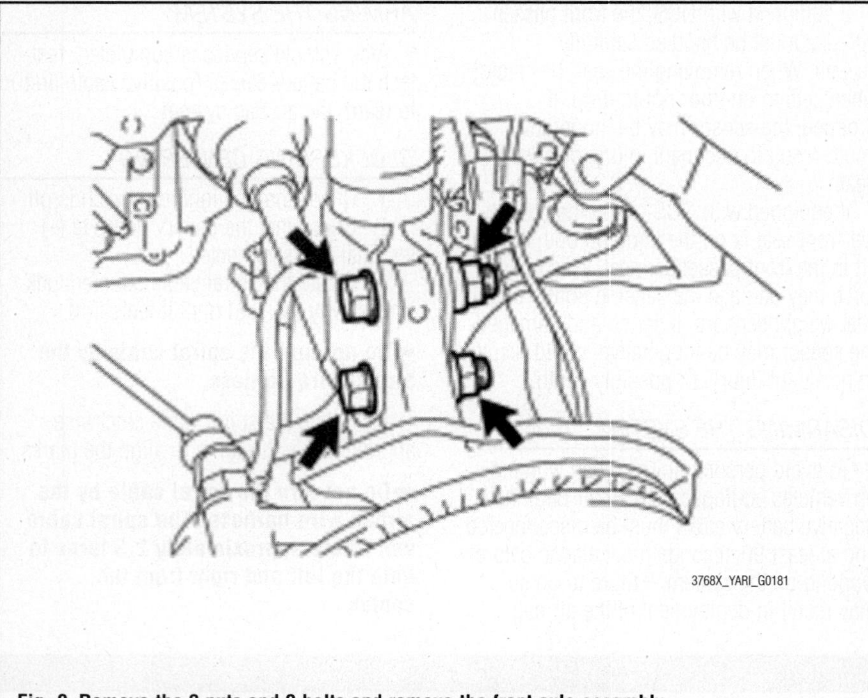

Fig. 9 Remove the 2 nuts and 2 bolts and remove the front axle assembly

FRONT HALFSHAFT

REMOVAL & INSTALLATION

See Figures 10 through 13.

1. Disconnect the negative battery cable.
2. Drain automatic or manual transaxle fluid.
3. Remove front wheel.
4. Remove front axle hub nut.
   a. Using SST: 09930-00010 and a hammer, release the staked part of the axle hub nut.

➡Insert SST into the groove with the flat surface facing up.

⁂ WARNING

Do not damage the tip of SST using grinders.

➡Completely unstake the staked part before removing the axle hub nut.

⁂ WARNING

Do not damage the threads of the drive shaft.

   b. Using a 30 mm socket wrench, remove the axle hub nut.
5. Separate front speed sensor (w/ abs).
   a. Remove the bolt and separate the speed sensor and flexible hose.
   b. Remove the bolt and separate the speed sensor from the steering knuckle.

➡Keep the speed sensor tip and installation portion free of foreign matter. Remove the speed sensor without turning it from its original installation angle.

6. Separate front stabilizer link assembly.
   a. Remove the nut and separate the stabilizer link form the shock absorber.

➡If the ball joint turns together with the nut, use a socket hexagon wrench 6 to hold the stud.

7. Separate tie rod end sub-assembly.
   a. Remove the cotter pin and castle nut.
   b. Using SST: 09628-62011, separate the tie rod end from the steering knuckle.
8. Separate front lower suspension arm.
   a. Remove the clip and castle nut.
   b. Using SST: 09628-62011, separate the lower arm.

⁂ WARNING

Do not damage the lower ball joint dust cover.

➡Suspend SST: 09628-62011 with a piece of string or the equivalent.

9. Separate front axle assembly.
   a. Using a plastic hammer, tap the end of the drive shaft and disengage the fitting between the drive shaft and front axle.

➡If it is difficult to disengage the fitting, tap the end of the drive shaft with a brass bar and hammer.

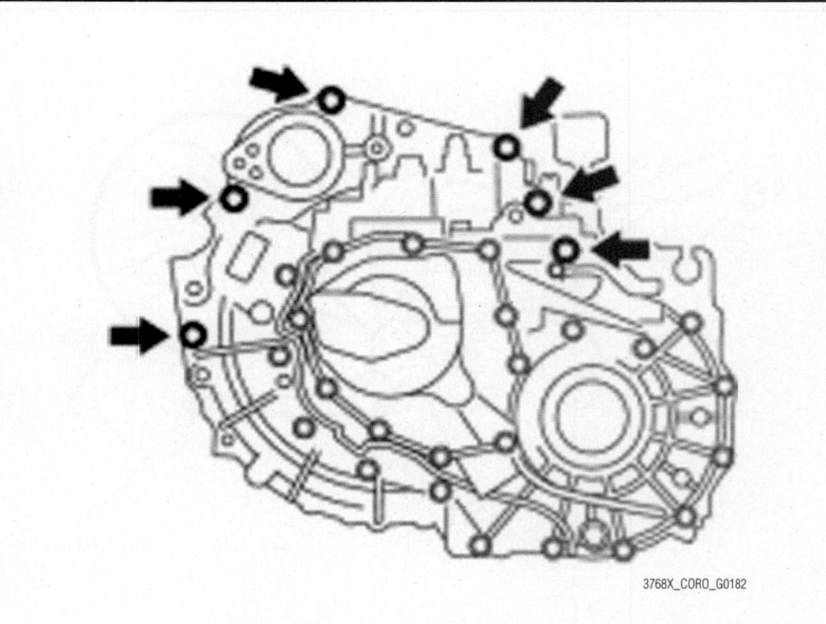

Fig. 10 Using SST: 09520-01010 and 09520-24010 remove the drive shaft

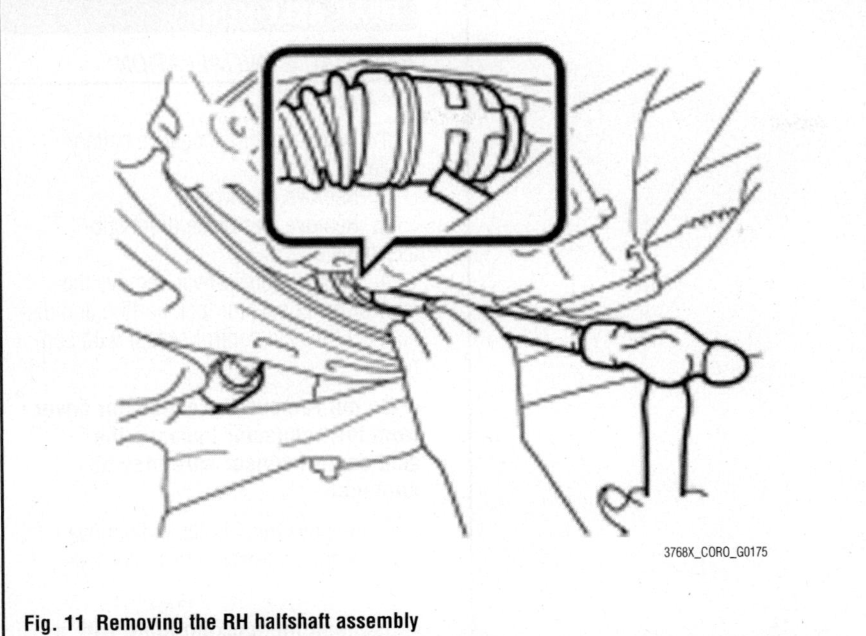

Fig. 11 Removing the RH halfshaft assembly

b. Push the front axle out of the vehicle to remove the drive shaft from the front axle.

➡**Do not push the front axle further out of the vehicle than is necessary.**

⁂ WARNING

Do not damage the outboard joint boot.

⁂ WARNING

Do not damage the speed sensor rotor.

➡**Suspend the drive shaft with a piece of string or the equivalent.**

10. Remove the 2 bolts and transmission case protector (w/o ABS)

Fig. 12 Using SST: 09608-16042 to support the hub bearing

⁂ WARNING

Do not damage the oil seal. Do not damage the inboard joint boot. Do not drop the drive shaft.

11. Remove front halfshaft assembly LH:
a. Using SST: 09520-01010 and 09520-24010 remove the drive shaft.
12. Using a screwdriver and hammer, remove front drive shaft assembly RH

➡**The hub bearing could be damaged if it is subjected to the vehicle's full weight, such as when moving the vehicle with the drive shaft removed. If it is absolutely necessary to place the vehicle's full weight on the hub bearing, first support it with SST 09608-16042.**

To install:

13. Install front drive shaft assembly LH and RH:
a. For Automatic Transaxle: Coat the spline of the inboard joint with ATF.
b. For Manual Transaxle: Coat the spline of the inboard joint with gear oil.
c. Align the inboard joint splines and install the drive shaft with a screwdriver and hammer.

➡**Please note the following:**

- Face the cut area of the front drive inboard joint hole snap ring downward.
- Do not damage the oil seal.
- Do not damage the inboard joint boot.

➡**Confirm whether the drive shaft is securely driven in by checking the reaction force and sound.**

14. Install the transmission case protector (w/o ABS) with the 2 bolts. Tighten the bolts to 17 inch lb. (23 Nm).
15. To complete the installation, reverse the remaining removal procedure noting the following tightening specifications:
a. Tighten lower arm to steering knuckle with new castle nut to 72 inch lb. (98 Nm).
b. Tighten tie rod end to steering knuckle with new castle nut to 36 inch lb. (49 Nm).
c. Tighten stabilizer link nut to 55 inch lb. (74 Nm).
d. Tighten speed sensor (w/ ABS) to steering knuckle with bolt to 75 ft. lb. (8.5 Nm).

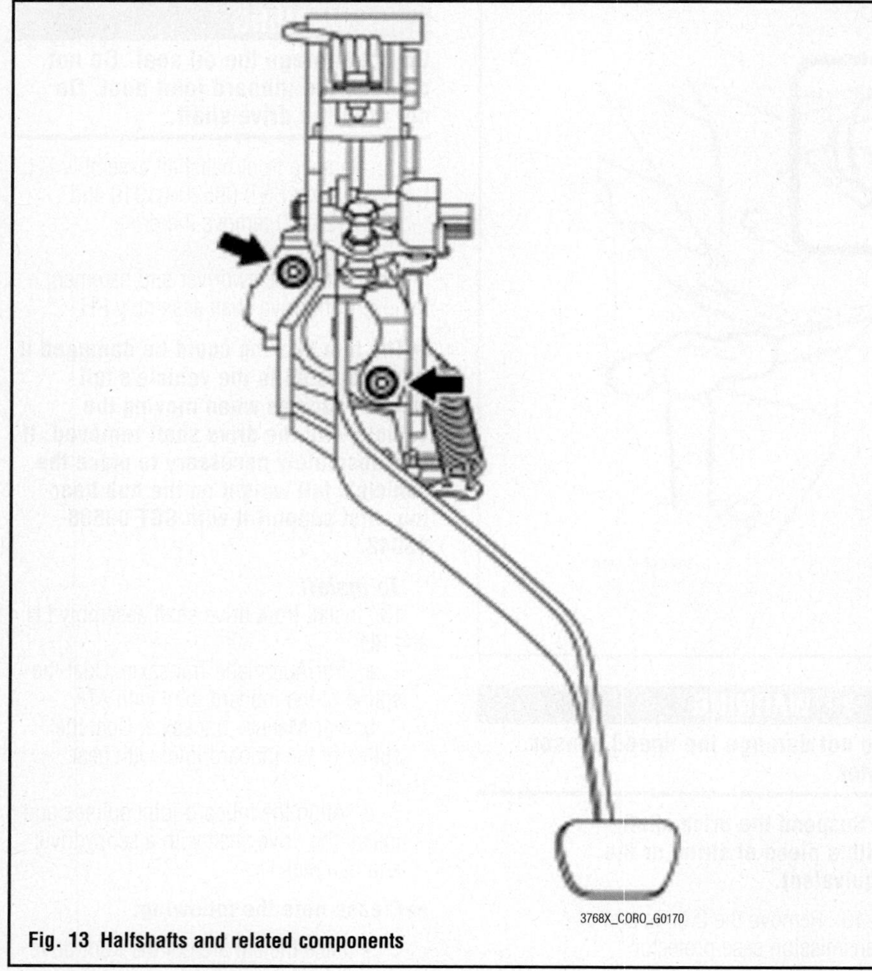

Fig. 13 Halfshafts and related components

e. Tighten flexible hose and speed sensor with bolt to 22 inch lb. (29 Nm).

f. Tighten axle hub nut to 160 inch lb. (216 Nm).

g. Tighten front wheel to 76 inch lb. (103 Nm).

h. Tighten negative battery terminal to 48 ft. lb. (5.4 Nm).

16. Add automatic transaxle fluid or manual transaxle oil.
17. Inspect the fluid.
18. Check for transmission leakage.
19. Inspect and adjust front wheel alignment.

REAR AXLE HUB

REMOVAL & INSTALLATION

See Figure 14.

1. Disconnect the negative battery cable.
2. Remove rear wheel.
3. Remove rear brake drum sub-assembly.
4. Using a screwdriver, remove the claw of the connector lock portion and disconnect the skid control sensor wire connector.

➡Do not remove the connector cover from the connector because the skid control sensor wire may be damaged.

5. Remove the 4 bolts and remove the axle hub and bearing from the axle beam.

➡Suspend the backing plate with a piece of rope.

To install:

6. Install the axle hub and bearing onto the axle beam with the 4 bolts. Tighten to 67 ft. lbs. (90 Nm).
7. Inspect rear axle hub bearing.
8. To complete the installation, reverse the remaining removal procedure.

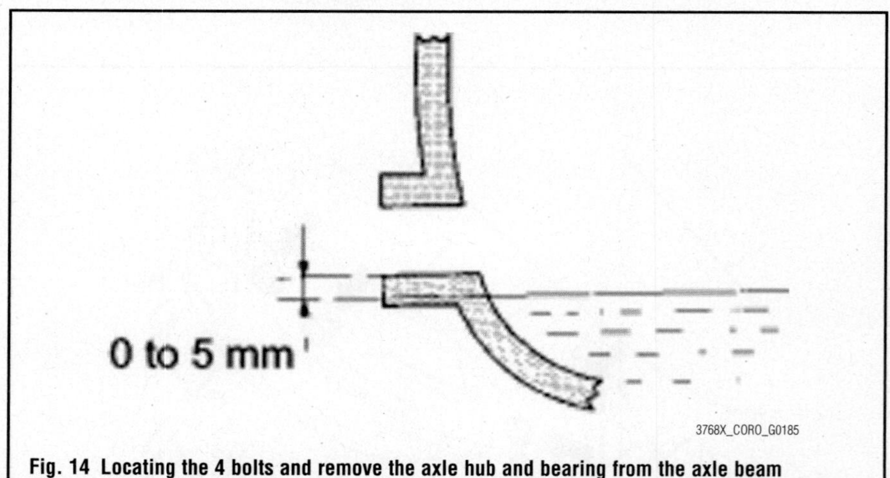

Fig. 14 Locating the 4 bolts and remove the axle hub and bearing from the axle beam

ENGINE COOLING

ENGINE COOLANT

DRAIN & REFILL PROCEDURE

See Figure 15.

1. Drain engine coolant.

CAUTION

To avoid the danger of being burned, do not remove the radiator cap sub-assembly while the engine and radiator assembly are still hot. Thermal expansion will cause hot engine coolant and steam to blow out from the radiator assembly.

a. Loosen the radiator drain cock plug.
b. Remove the radiator cap sub-assembly.
c. Loosen the cylinder block drain cock plug, then drain the coolant.

2. Refill engine coolant.

Do not substitute water for engine coolant.

a. Tighten all the plugs.
b. Pour engine coolant into the radiator assembly until it overflows.
c. Capacities: M/T 5.1 qt., A/T 5.0 qt.

WARNING

Use of improper engine coolant may damage the engine coolant system.

Use only Toyota Super Long Life Coolant or similar high quality ethylene glycol based non-silicate, non-amine, non-nitrite, and non-borate engine coolant with long-life hybrid organic acid technology (coolant with long-life hybrid organic acid technology consists of a combination of low phosphates and organic acids).

d. Check the engine coolant level inside the radiator assembly by squeezing the inlet and outlet radiator hoses several times by hand. If the engine coolant level goes down, add engine coolant.
e. Install the radiator cap sub-assembly securely.
f. Slowly pour engine coolant into the radiator reservoir until it reaches the FULL line.

3. Warm up the engine until the cooling fan operates. Set the air conditioning as follows while warming up the engine.

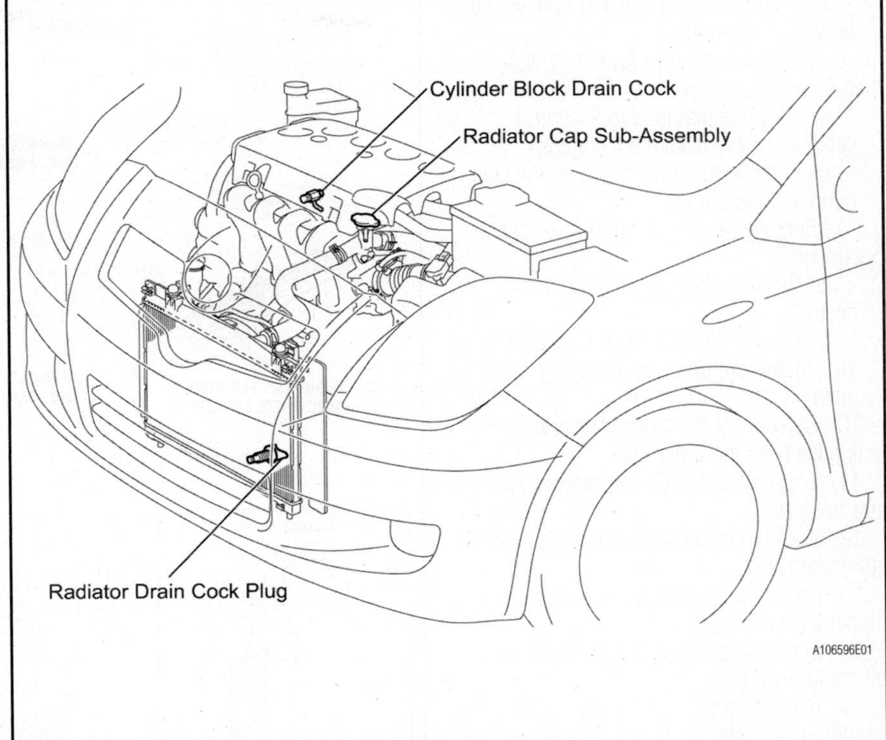

Fig. 15 Cylinder block drain cock, radiator cap sub-assembly, radiator drain cock plug locations

a. Manual air conditioning system:
- Fan speed - Any setting except "OFF"
- Temperature - Toward WARM
- Air conditioning switch "OFF"

b. Automatic air conditioning system
- Fan speed - Any setting except "OFF"
- Temperature - To the highest temperature
- Air conditioning switch "OFF"
- "AUTO" switch "OFF"

c. Maintain the engine speed at 2,000 to 2,500 rpm and warm up the engine until the cooling fan operates.

4. Stop the engine and wait until the coolant cools down.
5. If the engine coolant level is below the full level, perform steps (b) through (g) again and repeat the operation until the engine coolant level stays at the full level.
6. Recheck the engine coolant level inside the radiator reservoir tank assembly. If it is below the full level, add engine coolant.

RADIATOR

REMOVAL & INSTALLATION

See Figures 16 through 18.

CAUTION

Be sure that the ignition is off if you work near the electric cooling fans or radiator grille. With the ignition on the electric cooling fans may automatically start to run if the engine coolant temperature is high and/or the air conditioning is on.

1. Disconnect the negative battery cable.
2. Drain and recycle the engine coolant.
3. Remove the front bumper.
4. Remove air cleaner assembly.
5. For the hatchback, disengage the 6 claws and remove the radiator support absorber upper.
6. If equipped, remove the 2 clips and No. 1 cooler cover.
7. Remove hood lock assembly (w/ theft deterrent system):

a. Separate the hood lock control cable assembly from the 2 clamps.

b. Separate the engine hood courtesy switch connector.

c. Remove the 3 bolts and remove the hood lock assembly.

8. Remove hood lock assembly (w/o theft deterrent system):

a. Separate the hood lock control cable assembly from the 2 clamps.

b. Remove the 3 bolts and remove the hood lock assembly.

9. Remove radiator support sub-assembly upper:

a. Separate the horn assembly connector.

b. Remove the 4 bolts and remove the radiator support sub-assembly upper.

10. Disconnect the radiator reservoir tank hose from the water filler.

11. Loosen the 2 clips and remove radiator hose No. 3.

12. Loosen the clip and disconnect radiator hose No. 2.

13. Loosen the clip and disconnect the oil cooler outlet hose.

14. Loosen the clip and disconnect the oil cooler inlet hose.

15. Remove radiator assembly (w/ air conditioning system):

a. Separate the cooling fan motor connector and wire harness clamps.

b. Disengage the 2 claws and remove the radiator assembly from the vehicle.

➡**Do not apply excessive force to the cooler condenser assembly or piping when removing the radiator assembly.**

16. Remove radiator assembly (w/o air conditioning system):

a. Separate the cooling fan motor connector and wire harness clamps.

b. Remove the radiator assembly from the vehicle.

17. Loosen the clip and remove radiator hose No. 2.

18. Loosen the clip and remove the cooler outlet hose (for A/T).

19. Loosen the clip and remove the cooler inlet hose (for A/T).

20. Disengage the 2 claws and remove the fan shroud.

21. To install, reverse the removal procedure.

22. Refer to the illustrations for torque values.

23. Add coolant and check for leaks.

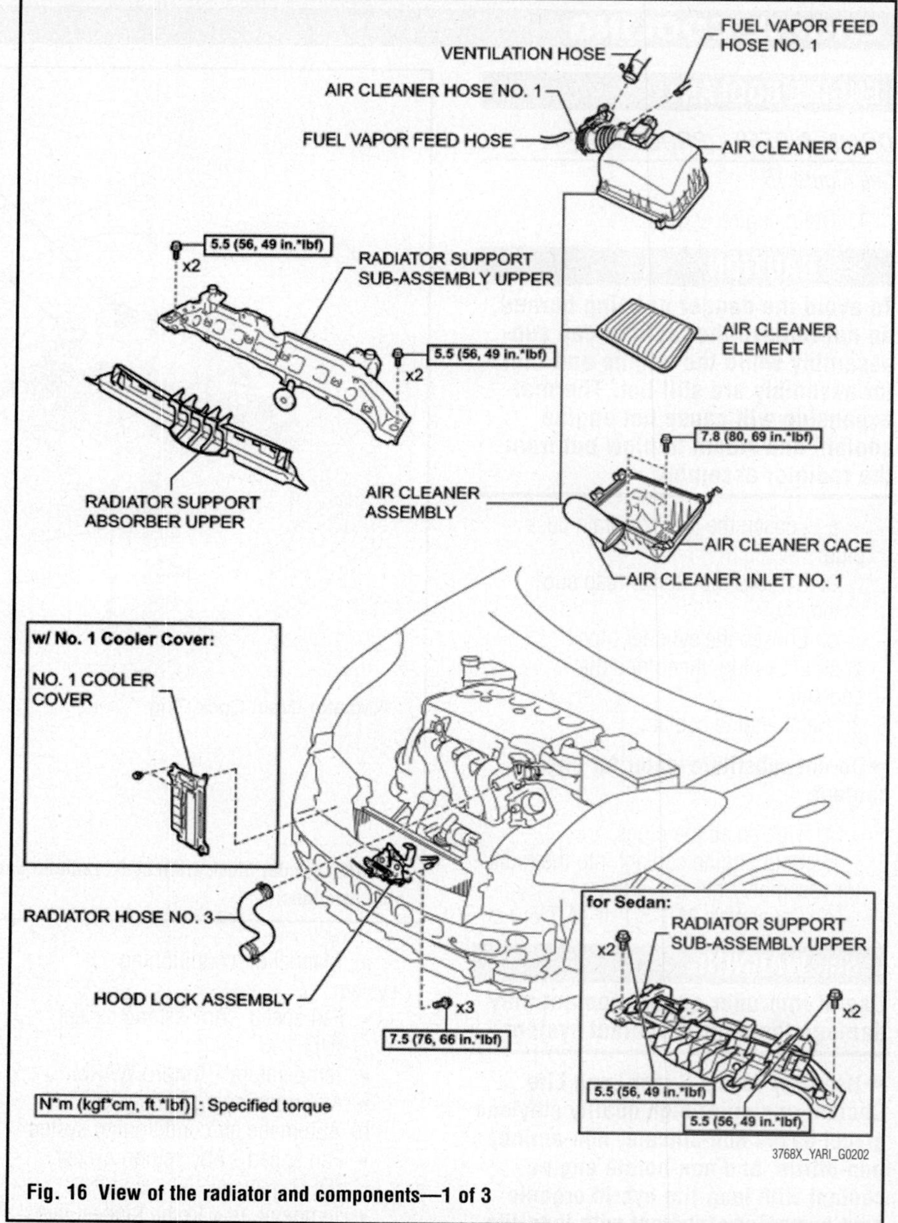

Fig. 16 View of the radiator and components—1 of 3

THERMOSTAT

REMOVAL & INSTALLATION

See Figure 19.

1. Drain and recycle the engine coolant.

2. Remove the 2 nuts and separate the water inlet with radiator hose from the cylinder block.

3. Remove the thermostat from the cylinder block.

4. Remove the gasket from the thermostat.

To install:

5. Install a new gasket onto the thermostat.

6. Install the thermostat with the jiggle valve facing upward.

➡**The jiggle valve may be set within 10° on either side, as shown in the illustration.**

7. Install the water inlet with radiator hose with the 2 nuts. Tighten to 80 inch lbs. (9 Nm).

8. Add engine coolant and check for leaks.

WATER PUMP

REMOVAL & INSTALLATION

See Figure 20.

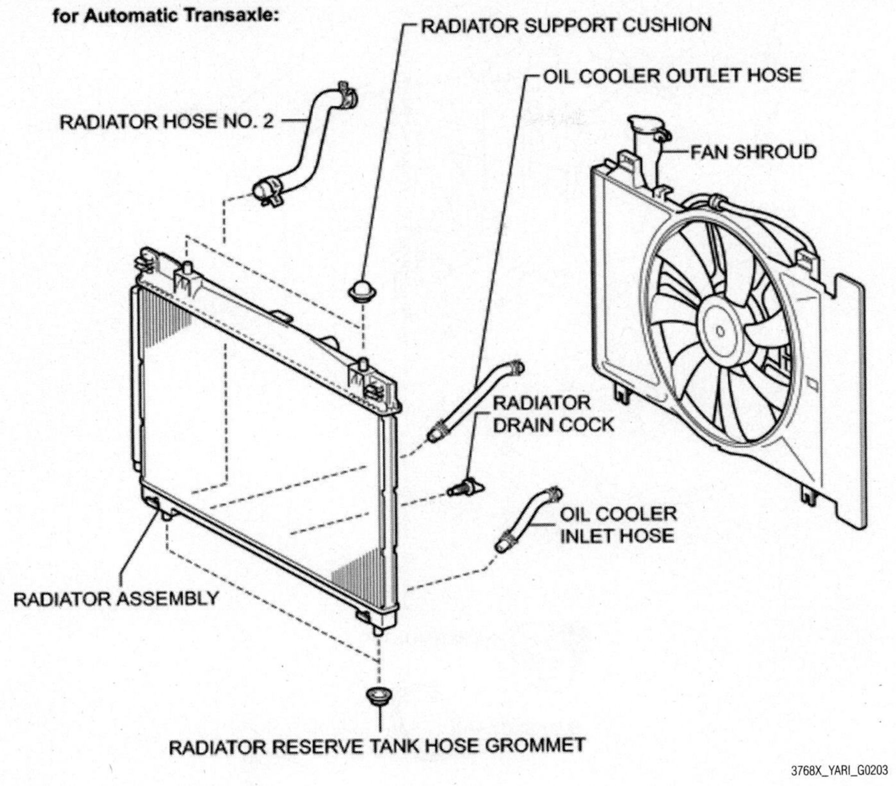

for Automatic Transaxle:

RADIATOR SUPPORT CUSHION
OIL COOLER OUTLET HOSE
RADIATOR HOSE NO. 2
FAN SHROUD
RADIATOR DRAIN COCK
OIL COOLER INLET HOSE
RADIATOR ASSEMBLY
RADIATOR RESERVE TANK HOSE GROMMET
3768X_YARI_G0203

Fig. 17 View of the radiator and components—2 of 3

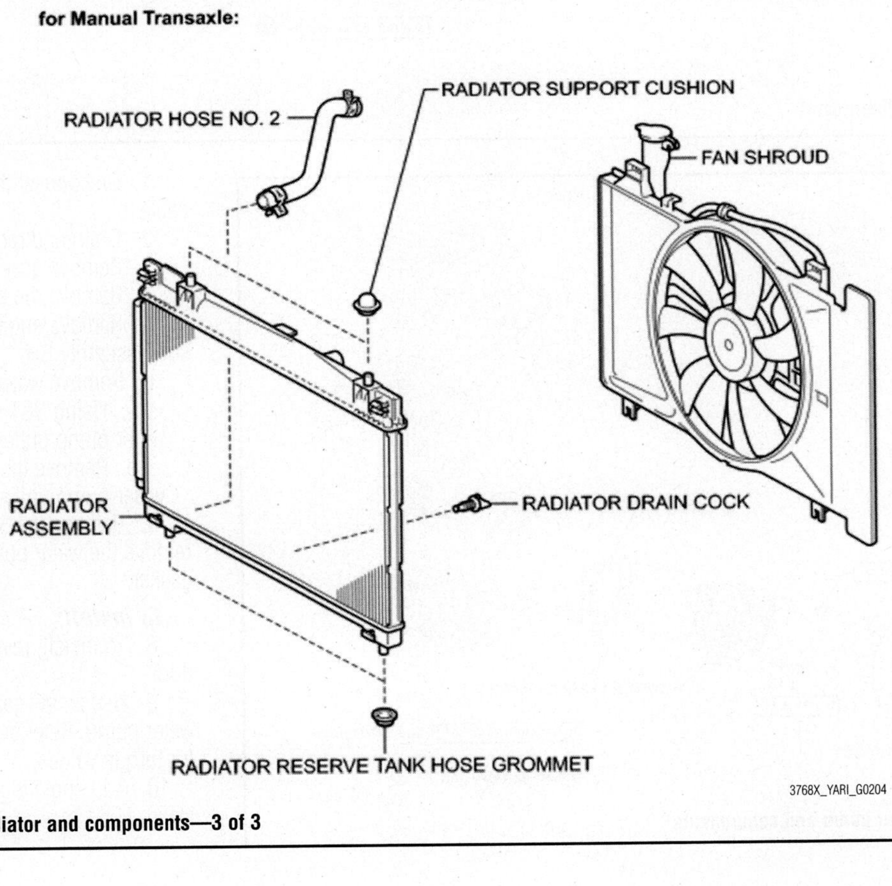

for Manual Transaxle:

RADIATOR SUPPORT CUSHION
RADIATOR HOSE NO. 2
FAN SHROUD
RADIATOR ASSEMBLY
RADIATOR DRAIN COCK
RADIATOR RESERVE TANK HOSE GROMMET
3768X_YARI_G0204

Fig. 18 View of the radiator and components—3 of 3

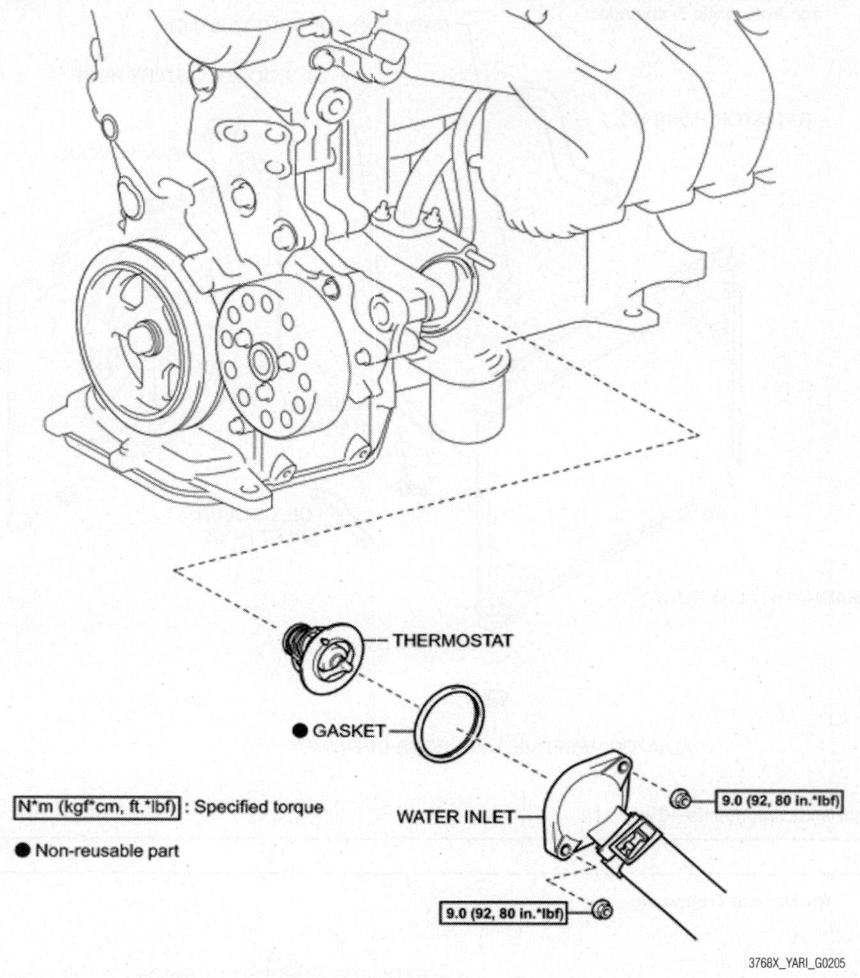

Fig. 19 Removing the thermostat

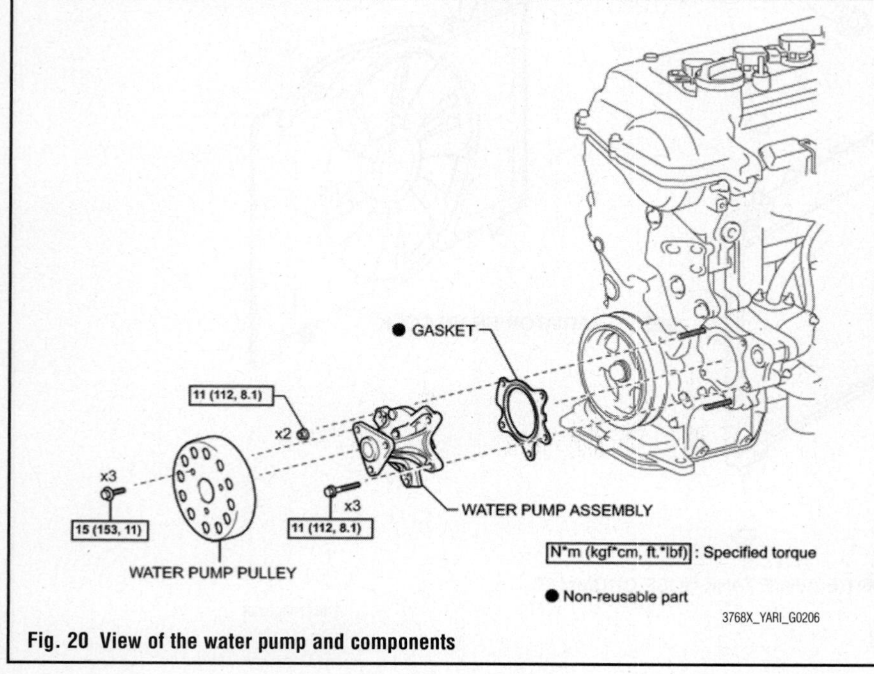

Fig. 20 View of the water pump and components

1. Disconnect the negative battery cable.
2. Drain and recycle the engine coolant.
3. Remove accessory drive belt.
4. Remove the alternator.
5. Remove engine mounting insulator sub-assembly RH.
6. Remove water pump pulley:
   a. Using SST 09960-10010, hold the water pump pulley.
   b. Remove the 3 bolts and remove the water pump pulley.
7. Remove the 3 bolts and 2 nuts and remove the water pump assembly and gasket.

To install:

8. To install, reverse the removal procedure.
9. Use a new gasket when installing the water pump. Refer to exploded illustration for torque values.
10. Add coolant and check for leaks.

BATTERY

REMOVAL & INSTALLATION

See Figure 21.

1. Disconnect the negative then the positive battery cables.
2. Remove battery clamp sub-assembly.
3. Remove the battery.
4. To install, reverse the removal procedure.

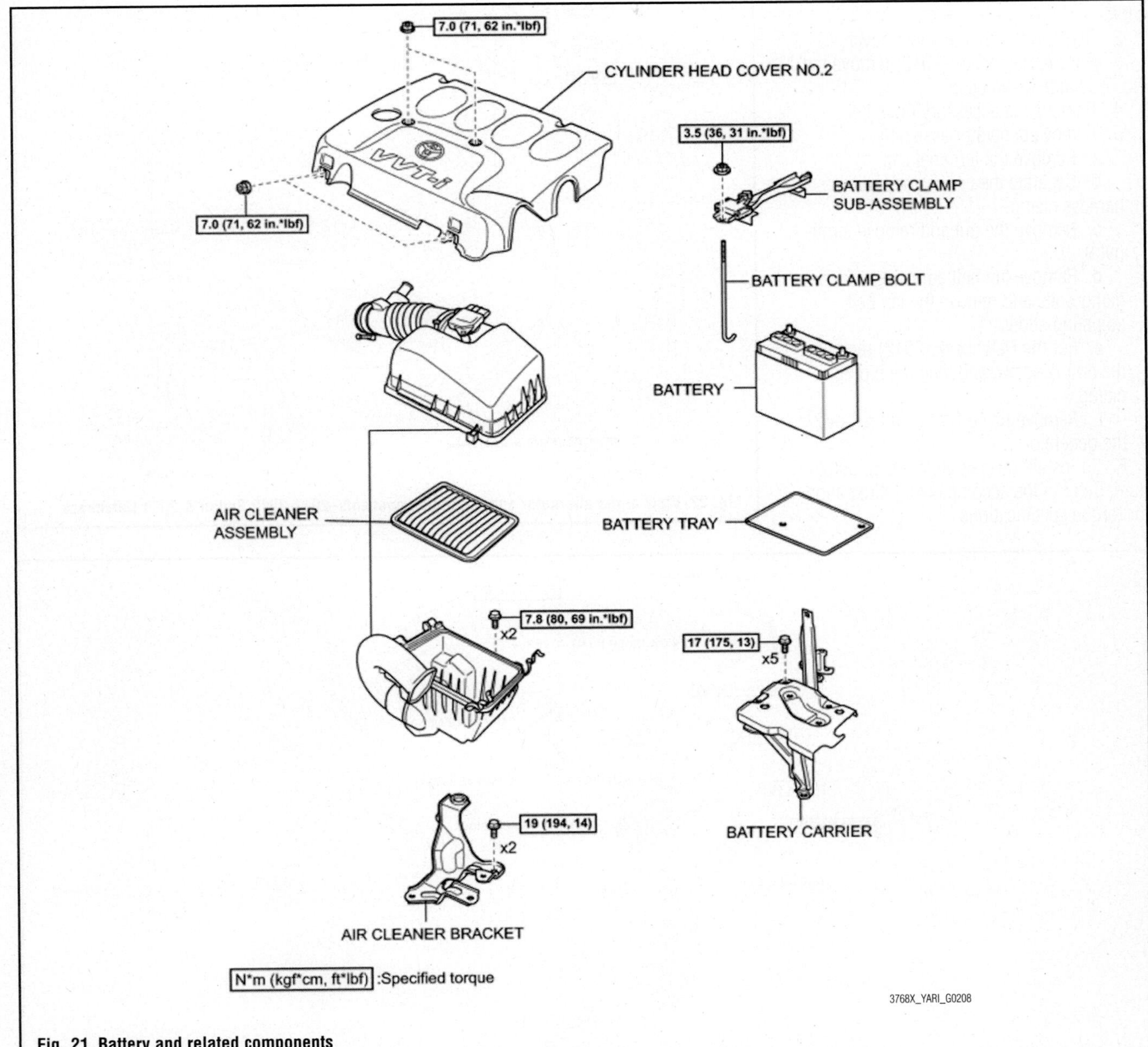

Fig. 21 Battery and related components

ALTERNATOR

REMOVAL & INSTALLATION

See Figures 22 and 23.

1. Disconnect the negative battery cable.
2. Remove the engine under cover.
3. For the Hatchback (2012) remove the No. 2 cylinder head cover.
4. Remove the accessory drive belt.
5. Remove generator assembly:
   a. Remove the terminal cap.
   b. Separate the connector and the harness clamp.
   c. Remove the nut and remove terminal B.
   d. Remove fan belt adjusting slider fixing bolts and remove the fan belt adjusting slider.
   e. For the Hatchback (2012) remove the bolt & screw, and then the piping clamp.
   f. Remove fixing bolt B and remove the generator.
6. To install, reverse the removal procedure. Refer to the accompanying illustration for torque specifications.

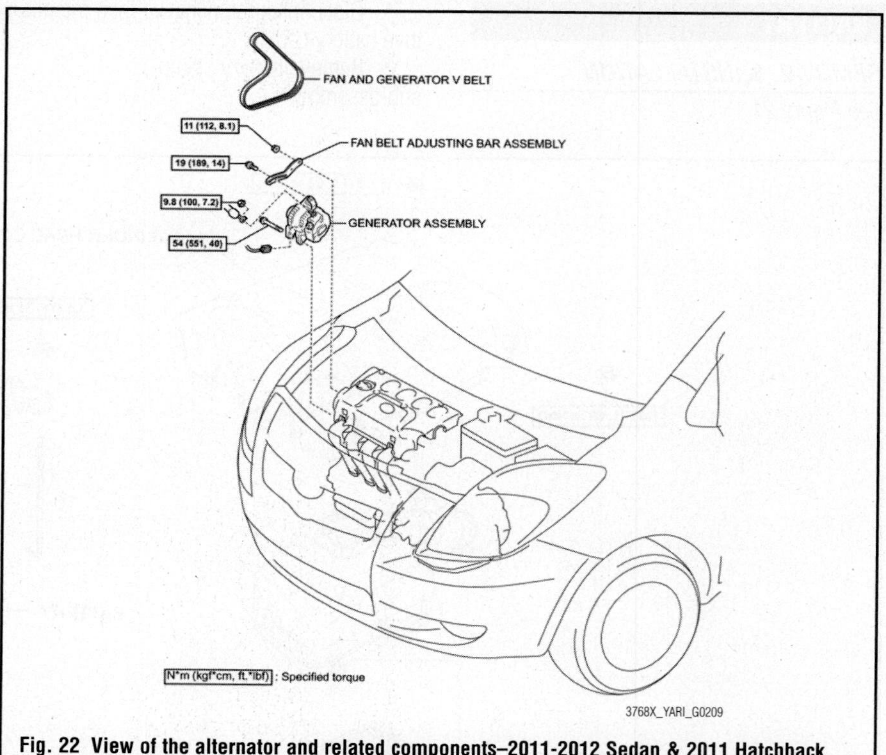

Fig. 22 View of the alternator and related components–2011-2012 Sedan & 2011 Hatchback

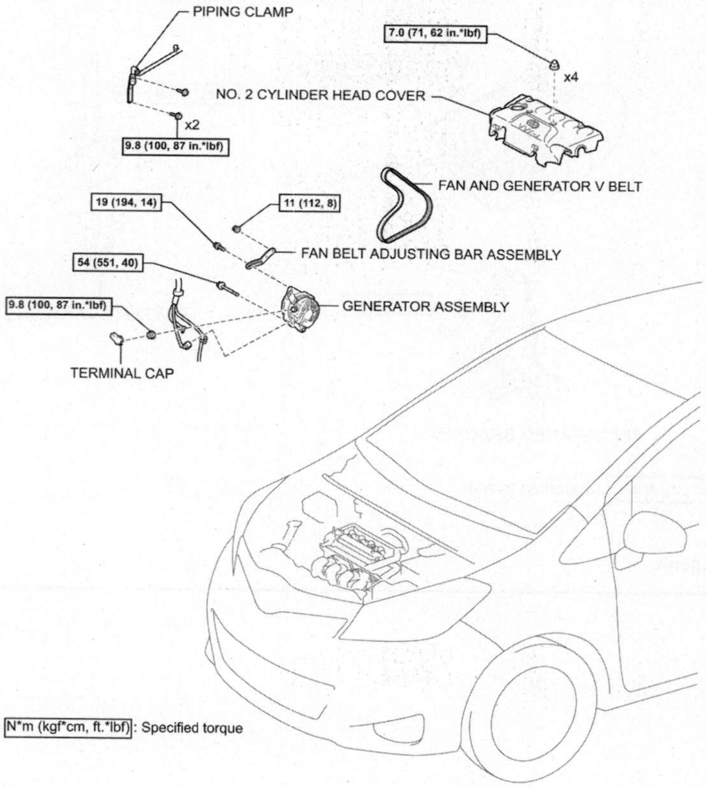

Fig. 23 View of the alternator and related components–2012 Hatchback

FIRING ORDER

See Figure 24.

IGNITION COIL

REMOVAL & INSTALLATION

See Figure 25.

1. Disconnect the negative battery cable.
2. Remove the 4 nuts and the engine cover.
3. Disconnect the 4 ignition coil connectors.
4. Remove the 4 bolts and 4 ignition coils.
5. To install, reverse the removal procedure.

IGNITION TIMING

ADJUSTMENT

The ignition timing is controlled by the Powertrain Control Module (PCM). No adjustment is necessary or possible.

INSPECTION

See Figure 26.

1. Using the Techstream or ODB II scan tool:
   a. Warm up and stop the engine.
   b. Connect the Techstream to the DLC3.
   c. Turn the ignition switch ON.
   d. Select the following menu items: Powertrain / Engine and ECT / Data List / IGN Advance.

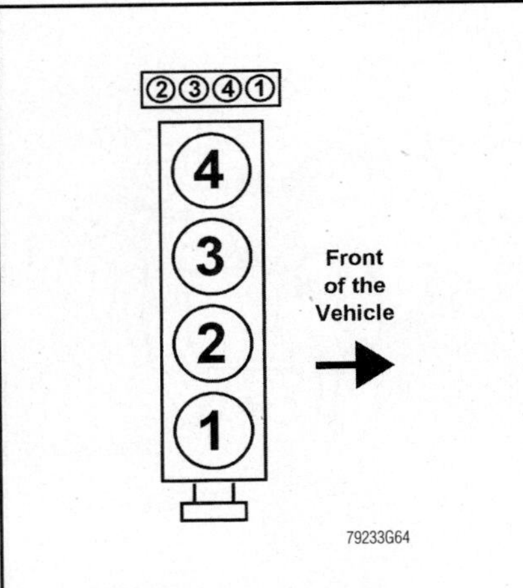

Fig. 24 Firing order: 1–3–4–2 Distributorless ignition system

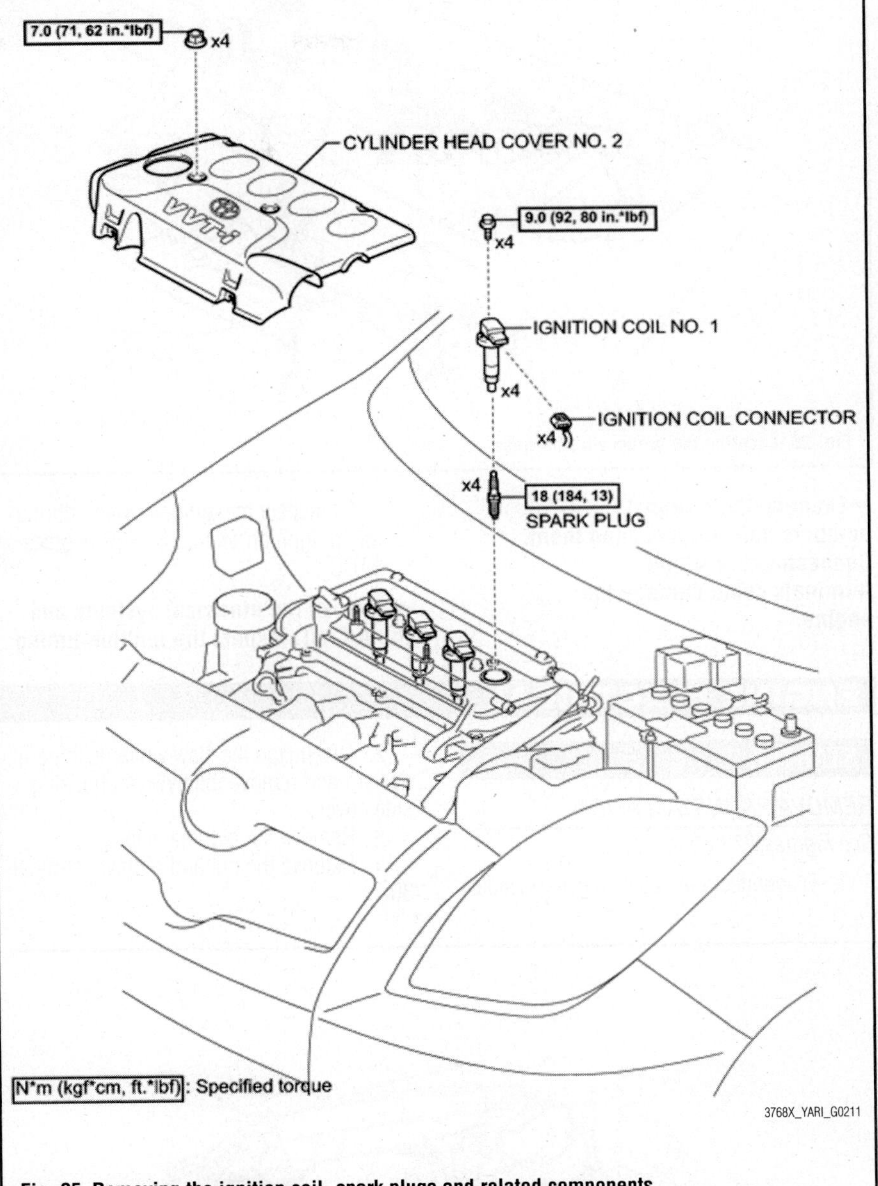

Fig. 25 Removing the ignition coil, spark plugs and related components

   e. Inspect the ignition timing during idling. Ignition timing: 8 to 12 degrees BTDC.

➡Turn all the electrical systems and the A/C off. Inspect the ignition timing with the cooling fan off. When checking the ignition timing, shift the transmission to the neutral position.

   f. Turn the ignition switch OFF. Disconnect the Techstream from the DLC3.
2. When not using the Techstream:
   a. Remove the engine cover. Pull out the wire harness (brown).

➡After checking, wrap the wire harness with tape.

   b. Warm up and stop the engine.
   c. Connect the clip of the timing light to the wire harness.

➡Use timing light that detects the first signal.

   d. Turn the ignition switch ON.
   e. Using SST 09843-18040, connect terminals 13 (TC) and 4 (CG) of the DLC3.

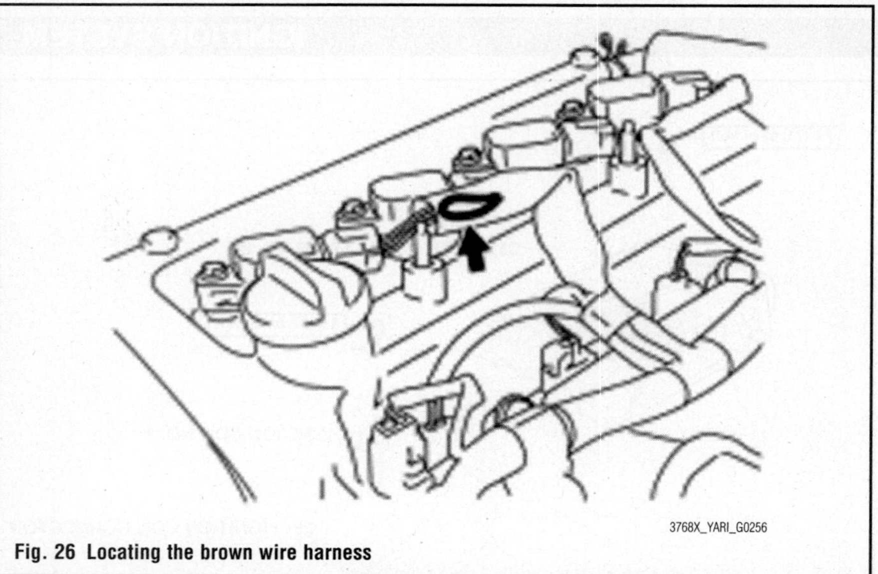

Fig. 26 Locating the brown wire harness

➡Examine the terminal numbers before connecting them. Connecting the wrong terminals could damage the engine.

f. Inspect the ignition timing during idling. Ignition timing: 8 to 12 degrees BTDC.

➡Turn all the electrical systems and the A/C off. Inspect the ignition timing with the cooling fan off. When checking the ignition timing, shift the transmission to the neutral position.

g. Disconnect terminals 13 (TC) and 4 (CG) of the DLC3.
h. Turn the ignition switch OFF.
i. Remove the timing light.
j. Install the engine cover.

SPARK PLUGS

REMOVAL & INSTALLATION

See Figure 25.

1. Disconnect the negative battery cable.
2. Remove the 4 nuts and the engine cover.
3. Disconnect the 4 ignition coil connectors.
4. Remove the 4 bolts and 4 ignition coils.
5. Using a 14 mm spark plug wrench, remove the 4 spark plugs.
6. To install, reverse the removal procedure.
7. Tighten the spark plugs to 13 ft. lbs. (18 Nm).

ENGINE ELECTRICAL

STARTING SYSTEM

STARTER

REMOVAL & INSTALLATION

See Figures 27 through 29.

1. Disconnect the negative battery cable.
2. Disengage the claw while pushing it upward and remove the flywheel housing side cover.
3. Remove the terminal cap.
4. Remove the nut and remove terminal 30.
5. Disconnect the connector.
6. Remove the 2 bolts and remove the starter assembly.
7. To install, reverse the removal procedure.

➡Make sure that the claw makes a click sound, indicating that it fits tightly. Replace the claw with a new one if it does not fit tightly or is deformed.

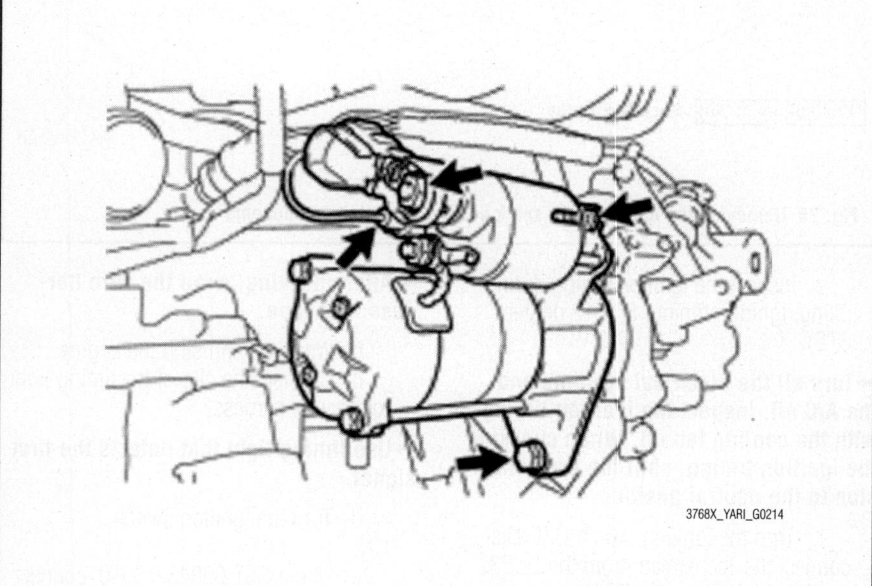

Fig. 27 Removing the starter assembly—0.8 kW type

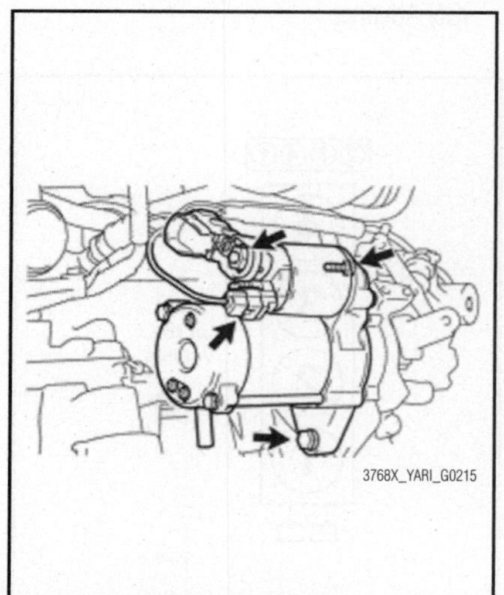

Fig. 28 Removing the starter assembly—1.6 kW type

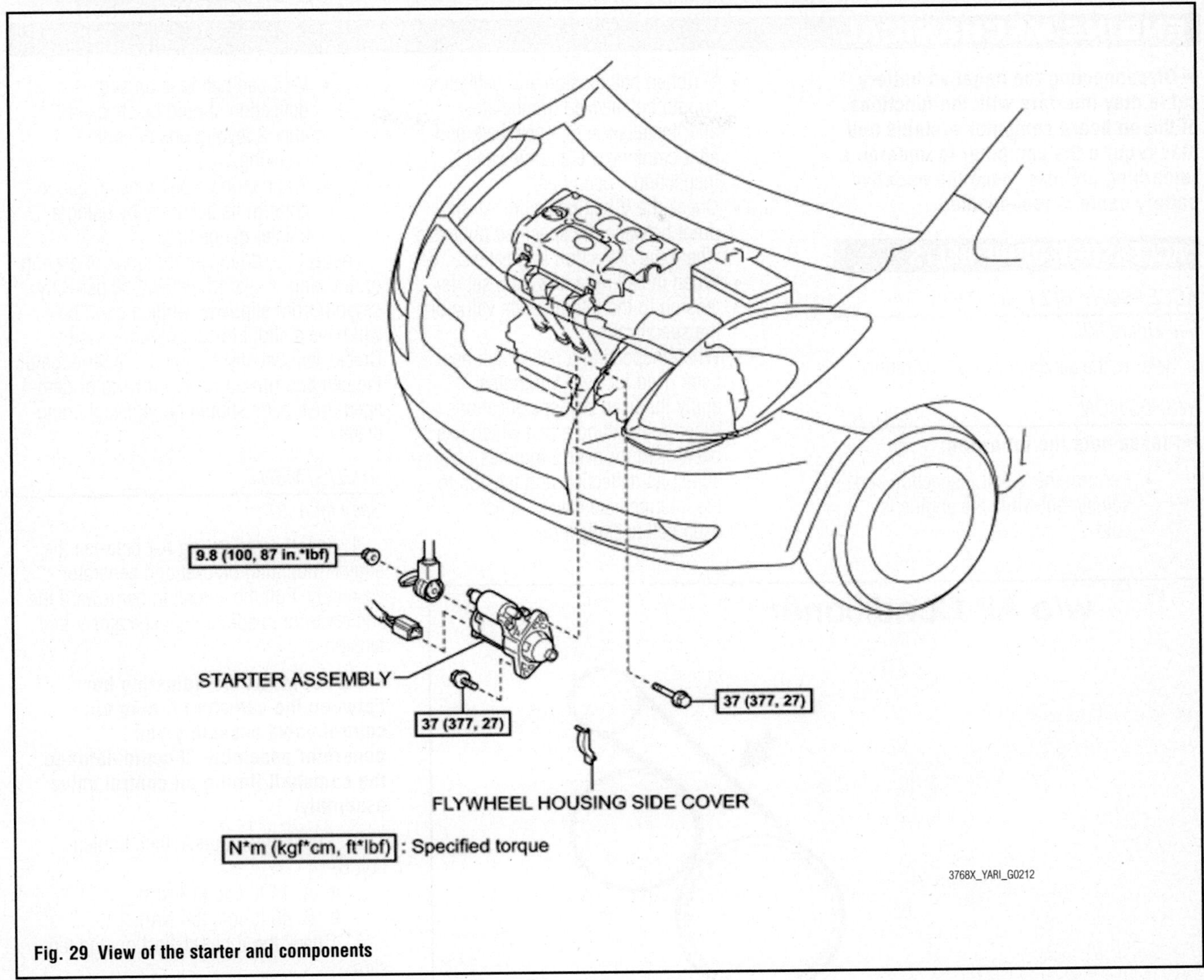

Fig. 29 View of the starter and components

a. Use the following tightening specifications:

- Starter assembly bolts, tighten to 27 inch lb. (37 Nm).
- Terminal 30 nut, tighten to 87 ft. lb. (9.8 Nm).
- Negative battery terminal, tighten to 48 ft. lb. (5.4 Nm).

ENGINE MECHANICAL

➡Disconnecting the negative battery cable may interfere with the functions of the on board computer systems and may require the computer to undergo a relearning process, once the negative battery cable is reconnected.

ACCESSORY DRIVE BELTS

ACCESSORY BELT ROUTING

See Figure 30.

Refer to the accompanying illustration.

INSPECTION

➡Please note the following:

- Perform the V belt inspection and adjustment while the engine is cold.
- V-ribbed belt tension and deflection should be checked immediately after installation of a new belt, and after cranking the engine when inspecting a used belt.
- Check the V belt deflection at the point between the specified pulleys where the deflection is greatest.
- When installing a new belt, set its tension to the intermediate value of the specification.
- When inspecting a belt which has been used for over 5 minutes, apply the used belt specifications.
- When reinstalling a belt which has been used for over 5 minutes, adjust its deflection and tension to the intermediate values of each used belt specification.
- V-ribbed belt tension and deflection should be checked after 2 revolutions of engine cranking.
- When using a belt tension gauge, confirm its accuracy by using a master gauge first.

Inspect the drive belt for signs of glazing or cracking. A glazed belt will be perfectly smooth from slippage, while a good belt will have a slight texture of fabric visible. Cracks will usually start at the inner edge of the belt and run outward. All worn or damaged drive belts should be replaced immediately.

ADJUSTMENT

See Figure 31.

1. Insert an adjusting bar between the engine mounting bracket and generator assembly. Pull the adjusting bar toward the vehicle front to adjust the generator V belt tension.

➡Do not insert the adjusting bar between the camshaft timing oil control valve assembly and generator assembly. It could damage the camshaft timing oil control valve assembly.

2. First tighten bolt A, then tighten bolt B.
   a. A: 14 ft. lbs. (19 Nm).
   b. B: 40 ft. lbs. (54 Nm).
3. Check the V belt deflection and tension:
   a. Deflection:
      - New belt: 0.31 to 0.35 inches
      - Used belt: 0.49 to 0.53 inches
   b. Tension:
      - New belt: 157 to 180 lbs.
      - Used belt: 67 to 90 lbs.
4. If the belt deflection is not as specified, adjust it.

REMOVAL & INSTALLATION

See Figure 32.

1. Remove the engine cover.
2. Loosen bolts A and B.
3. Release the fan and generator V belt tension and remove the fan and generator V belt.
4. To install, reverse the removal procedure. Make sure that there is no foreign matter or liquid, such as oil, on the belt and pulleys. Make sure that the V belt is securely fitted into the rib grooves of the pulley.

w/o Air Conditioner

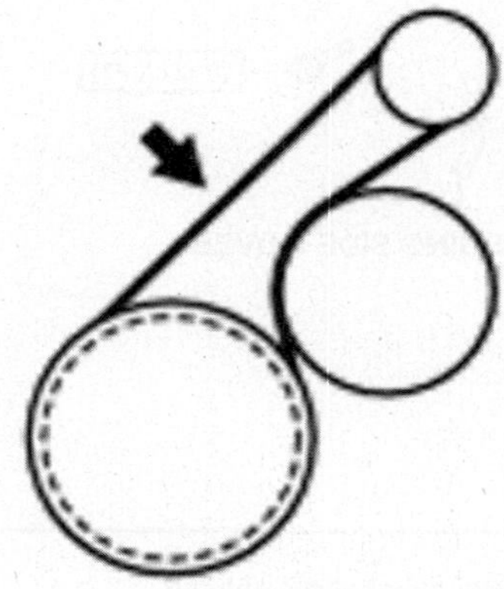

w/ Air Conditioner

3768X_YARI_G0216

Fig. 30 Accessory belt routing—1.5L Engine

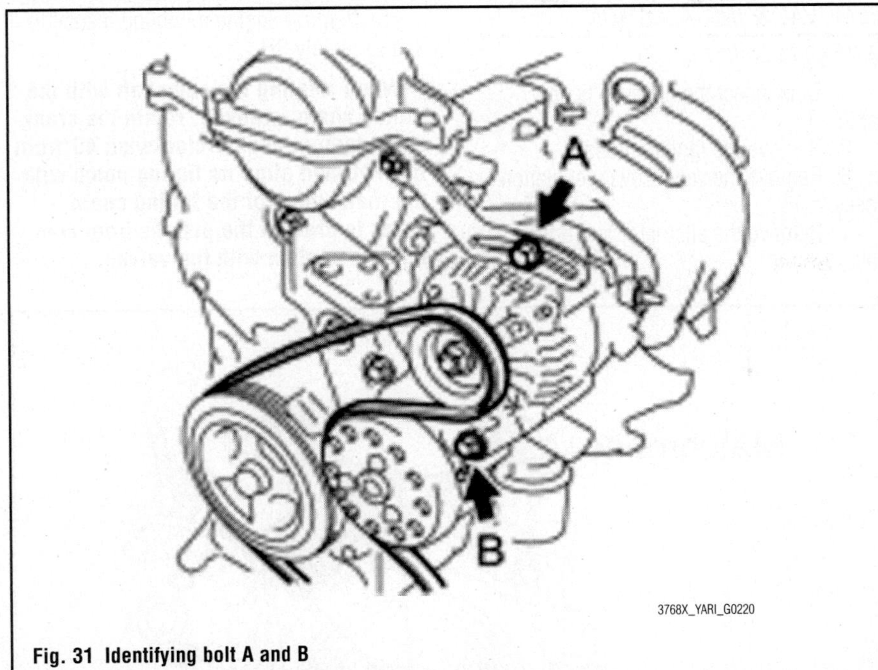

Fig. 31 Identifying bolt A and B

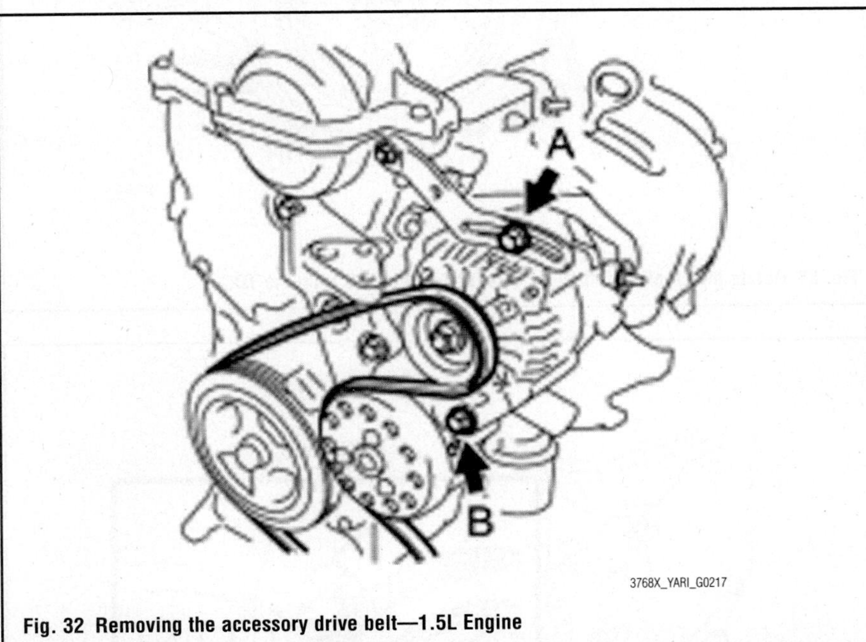

Fig. 32 Removing the accessory drive belt—1.5L Engine

AIR CLEANER

REMOVAL & INSTALLATION

See Figure 33.

1. Separate the intake air flow meter connector and the wire harness clamp.
2. Separate the fuel vapor feed hose and fuel vapor feed hose No. 1 from the vacuum switching valve assembly.
3. Separate the vacuum switching valve connector and the wire harness clamp.
4. Separate the ventilation hose from the air cleaner hose.
5. Release the air cleaner cap with air cleaner hose No. 1.
6. Loosen the air cleaner hose clamp on the throttle body side and remove the air cleaner cap and the air cleaner hose.
7. Remove the air cleaner element.
8. Separate the wire harness clamp from the air cleaner case.
9. Remove the 2 bolts and remove the air cleaner case with air cleaner inlet No. 1.

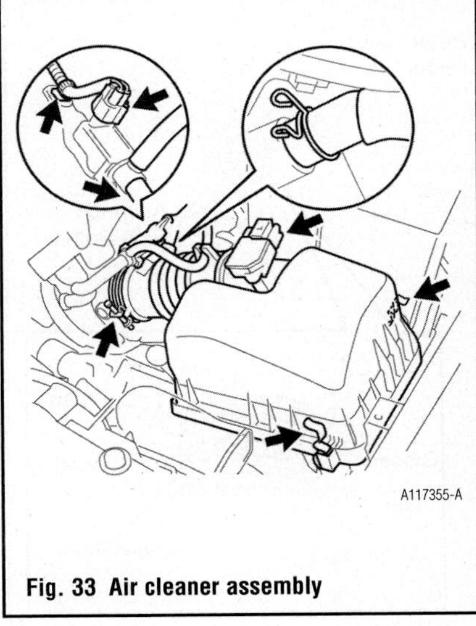

Fig. 33 Air cleaner assembly

10. To install, reverse the removal procedure.

CAMSHAFT AND VALVE LIFTERS

INSPECTION

See Figure 34.

1. Inspect Camshaft Timing Gear Assembly
   a. Check the lock of camshaft timing gear.
   b. Clamp the camshaft in a vice, and check that the camshaft timing gear is locked.

WARNING

Do not damage the camshaft.

   c. Release the lock pin.

➡One of the 2 grooves located on the cam journal is for retarding cam timing (upper) and the other is for advancing cam timing (lower). Each groove has 2 oil paths. Plug one of the oil paths for each groove with a piece of rubber before wrapping the cam journal with the tape.

2. Puncture the tape covering the advance oil path and the retard oil path on the opposite side from the advance oil path.
   a. Apply air at about 150 kPa (1.5kgf*cm^2) pressure into the 2 broken paths (the advance side path and the retard side path).

➡Cover the paths with a shop rag or piece of cloth to prevent oil splashes.

   b. Confirm that the camshaft timing gear assembly revolves in the timing

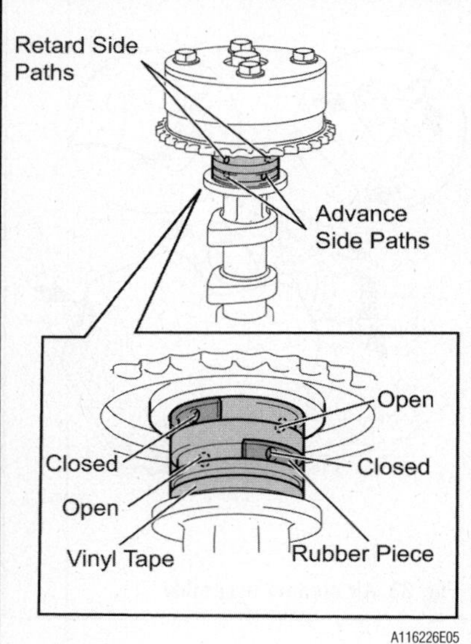

Fig. 34 Camshaft timing gear assembly components. Cover the 4 oil paths of the cam journal with tape as shown in the illustration

advance direction when the air pressure on the timing retard path is reduced.

➡**The lock pin is released and the camshaft timing gear revolves in the advance direction.**

c. When the camshaft timing gear reaches the most advanced position, release the air pressure on the timing retard side path, and then release the air pressure on the timing advance side path.

➡**Camshaft timing gear assembly occasionally shifts to the retard side abruptly if the air pressure on the advance side path is released first. This often results in breakage of the lock pin.**

3. Check the revolution.

a. Rotate the valve timing assembly back and forth several times, except where the lock pin meets it at the most retarded angle. Check the movable range and that it rotates smoothly. Smooth movable range is about 22.5°.

➡**Perform this check by hand, instead of using air pressure.**

4. Check that the gear locks in the most retarded position.

a. Confirm that the camshaft timing gear assembly is locked in the most retarded position.

REMOVAL & INSTALLATION

See Figures 35 through 55.

1. Disconnect the negative battery cable.
2. Remove the ignition coils.
3. Remove the valve cover ventilation hoses.
4. Remove the alternator and drive belt.
5. Remove engine mounting insulator sub-assembly RH.

➡**When rotating the camshaft with the timing chain removed, rotate the crankshaft damper counterclockwise 40°from the TDC and align its timing notch with the matchmark of the timing chain cover to prevent the pistons from coming into contact with the valves.**

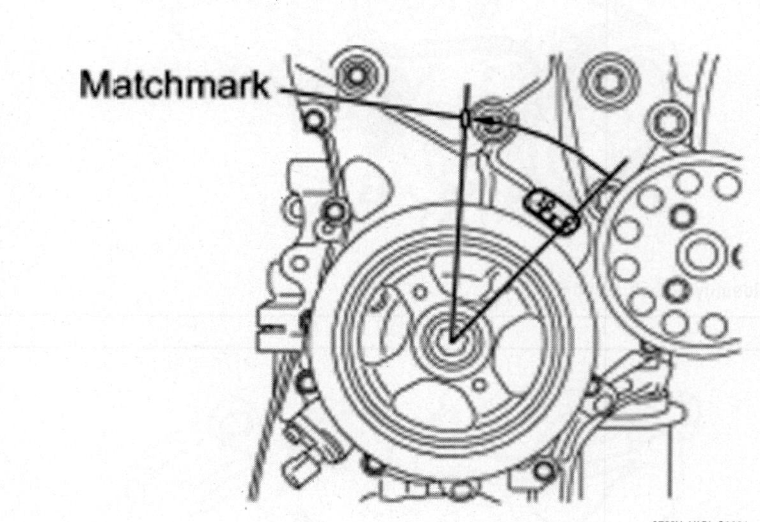

Fig. 35 Rotate the crankshaft damper counterclockwise 40°from the TDC

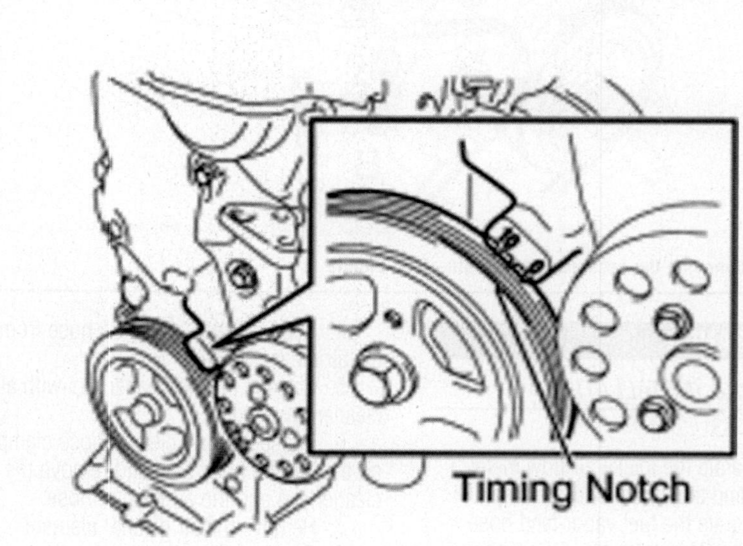

Fig. 36 Step 1: Turn the crankshaft damper, and align its timing notch with the timing mark "0" of the oil pump

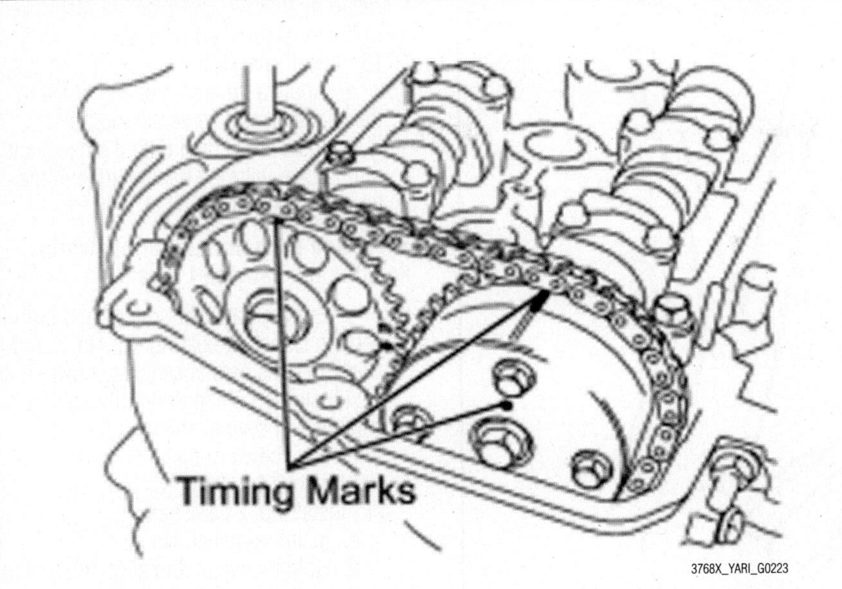

Fig. 37 Step2: Check that the timing marks on both the camshaft timing sprocket and the camshaft timing gear are facing upward

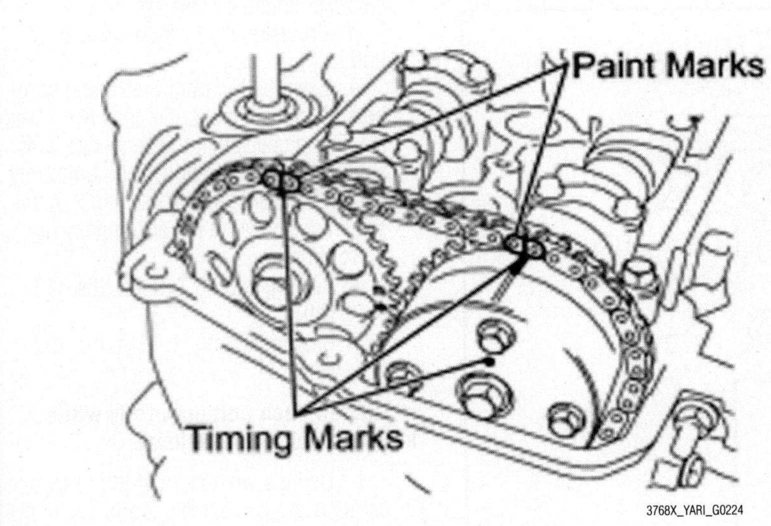

Fig. 38 Step3: Place paint marks on the chain in the places where the timing marks of the camshaft timing sprocket and the camshaft timing gear are located.

6. Follow the illustrations in order. Remove no. 2 camshaft. Set the No. 1 cylinder to TDC / compression:

➡If not, turn the crankshaft 1 complete revolution (360°) and align the marks as above.

➡Loosen the bolts uniformly while keeping the camshaft level.

➡Loosen the bolts uniformly while keeping the camshaft level.

7. Remove the camshaft timing gear assembly:

a. Clamp the camshaft in a vise and confirm that it is locked.

➡One of the 2 grooves located on the cam journal is for retarding cam timing (upper) and the other is for advancing cam timing (lower). Each groove has 2 oil paths. Plug one of the oil paths for each groove with a piece of rubber before wrapping the cam journal with the tape.

b. Cover the 4 oil paths of the cam journal with tape as shown in the illustration.

c. Puncture the tape covering the advance oil path and the retard oil path on the opposite side from the advance oil path.

d. Apply air at about 150 kPa (1.5 kgf*cm2) pressure into the 2 broken paths (the advance side path and the retard side path)

➡Cover the paths with a shop rag or piece of cloth to prevent oil splashes.

8. Confirm that the camshaft timing gear assembly revolves in the timing advance direction when the air pressure on the timing retard path is reduced.

➡The lock pin is released, and the camshaft timing gear revolves in the advance direction.

a. When the camshaft timing gear reaches the most advanced position, release the air pressure on the timing retard side path, and then release the air pressure on the timing advance side path.

➡The camshaft timing gear assembly occasionally shifts to the retard side abruptly, if the air pressure on the advance side path is released first. This often results in breakage of the lock pin.

b. Remove the flange bolt and remove the camshaft timing gear assembly.

➡Do not remove the other 4 bolts. When reusing the camshaft timing gear, unlock the lock pin inside the camshaft timing gear first.

To install:

9. Install camshaft timing gear assembly:

a. Install the camshaft timing gear assembly onto the camshaft with the lock pin of the camshaft timing gear assembly released.

b. Put the camshaft timing gear assembly and camshaft together with the straight pin of the groove.

c. Turn the camshaft timing gear assembly clockwise while pushing it gently toward the camshaft. When the pin fits the groove, push to ensure a good fit.

➡Do not turn the camshaft timing gear in the retard direction (clockwise).

d. Check that there is no clearance between the gear flange and the camshaft.

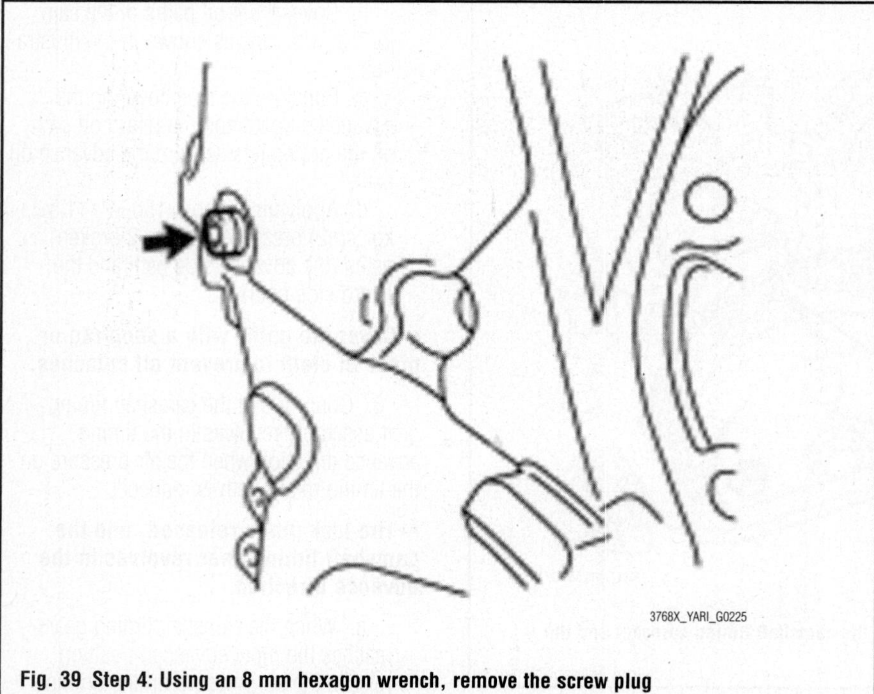

Fig. 39 Step 4: Using an 8 mm hexagon wrench, remove the screw plug

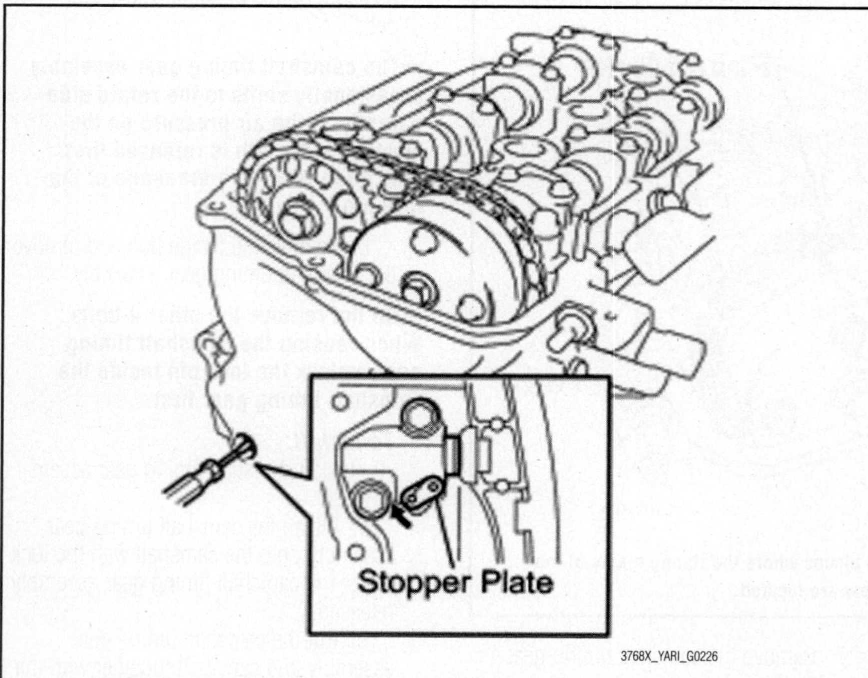

Fig. 40 Step 5: Insert a screwdriver into the service hole in the chain tensioner to pull the stopper plate of the chain tensioner upward

e. Tighten the flange bolt with the camshaft timing gear fixed. Tighten to 47 ft. lbs. (64 Nm).

➡Do not lock the camshaft timing gear assembly when tightening the bolt. Release the lock pin of the camshaft timing gear assembly first, and tighten the bolt when the lock pin is locked in the most retarded position. Tightening the bolts with the lock pin locked could cause breakage of the lock pin.

f. Check that the camshaft timing gear assembly moves smoothly in the retard direction (clockwise) and is locked in the most retarded position.

10. Install camshaft:

a. Apply a light coat of engine oil to the camshaft and camshaft journals.

b. Install the chain onto the camshaft timing gear with the paint mark and the timing mark aligned.

➡Tighten each bolt uniformly while keeping the camshaft level.

c. Examine the front marks and numbers on camshaft bearing cap No. 2 and check that the sequence is as shown in the illustration. Then uniformly tighten the bolts, in several steps, in the sequence shown in the illustration. Tighten to 9.4 ft. lbs. (13 Nm).

11. Install no. 2 camshaft:

a. Install camshaft No. 2.

b. Hold the chain and align the timing mark on the camshaft timing sprocket with the paint mark of the chain.

c. Align the alignment pin hole in the camshaft timing sprocket with the alignment pin of the camshaft, and install the sprocket onto the camshaft.

d. Provisionally install the flange bolt.

e. Examine the front marks and numbers on camshaft bearing caps No. 1 and No. 2 and check that the sequence is as shown in the illustration. Then uniformly tighten the bolts, in several steps, in the sequence shown in the accompanying illustration.

- Bearing cap no.2: 9.4 ft. lbs. (13 Nm).
- Bearing cap no. 1: 17 ft. lbs. (23 Nm).

➡Tighten each bolt uniformly while keeping the camshaft level.

f. Using a wrench, hold the hexagonal lobe of camshaft No. 2 and install the flange bolt. Tighten to 47 ft. lbs. (64 Nm).

g. Remove the bar from the timing chain tensioner.

h. Turn the crankshaft damper and align its timing notch with the timing mark "0" of the oil pump.

i. Check that all the pairs of timing marks are aligned.

j. Apply adhesive to the end 2 or 3 threads of the screw plug. Adhesive: Toyota Genuine Adhesive 1324, Three Bond 1324 or Equivalent

k. Using an 8 mm hexagon wrench, install the screw plug. Tighten to 11 ft. lbs. (15 Nm).

12. Inspect the valve clearance:

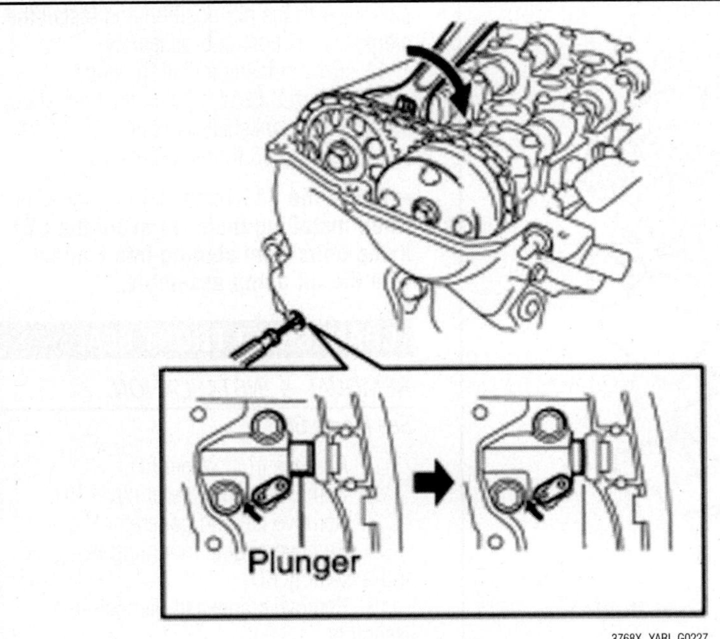

Fig. 41 Step 6: Using a wrench, rotate camshaft No. 2 clockwise to push in the plunger of the chain tensioner

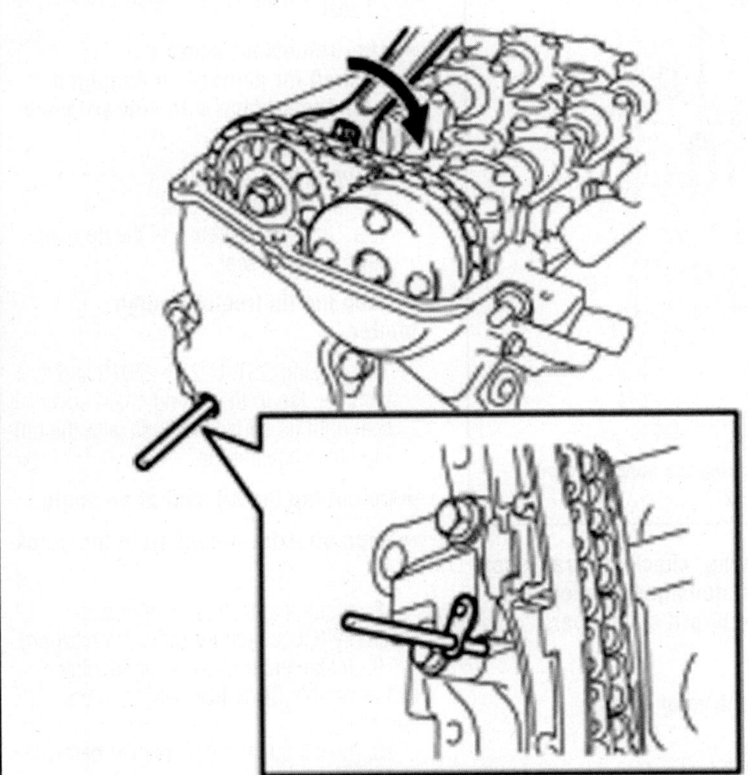

Fig. 42 Step 7: Remove the screwdriver from the service hole, then align the hole in the stopper plate with the service hole and insert a 0.12 inch diameter bar into the holes to hold the stopper plate

a. Turn the crankshaft damper and align its timing notch with the timing mark "0" of the oil pump.

b. Check that both timing marks on the camshaft timing sprocket and camshaft timing gear are facing upward.

➡If not, turn the crankshaft 1 complete revolution (360°) and align the marks as above.

c. Check the valves indicated in the illustration.

- Using a feeler gauge, measure the clearance between the valve lifter and camshaft.
- Valve clearance (cold): Intake: 0.006 to 0.010 inches. Exhaust: 0.010 to 0.014 inches

d. Record any out-of-specification valve clearance measurements. They will be used later to determine the required replacement adjusting shim.

e. Turn the crankshaft 1 complete revolution (360°) and align its timing notch with the timing mark "0" of the oil pump.

f. Check the valves indicated in the illustration.

- Using a feeler gauge, measure the clearance between the valve lifter and camshaft.
- Valve clearance (cold): Intake: 0.006 to 0.010 inches. Exhaust: 0.010 to 0.014 inches

g. Record any out-of-specification valve clearance measurements. They will be used later to determine the required replacement adjusting shim.

13. To complete installation, reverse the remaining removal procedure.
14. Check for engine oil leaks.

CRANKSHAFT DAMPER

REMOVAL & INSTALLATION

See Figure 56 and 57.

1. Remove the valve cover.
2. Set cylinder No. 1 to TDC/compression. Turn the crankshaft damper sub-assembly, and align its timing notch with timing mark "0" of the oil pump.
3. Check that the timing marks on the camshaft timing sprocket and the camshaft timing gear are all facing upward. If not, turn the crankshaft 1 complete revolution (360°) and align the marks.
4. Using 2 SSTs, 09213-14010, 09330-00021, loosen the bolt while holding the crankshaft damper sub-assembly.

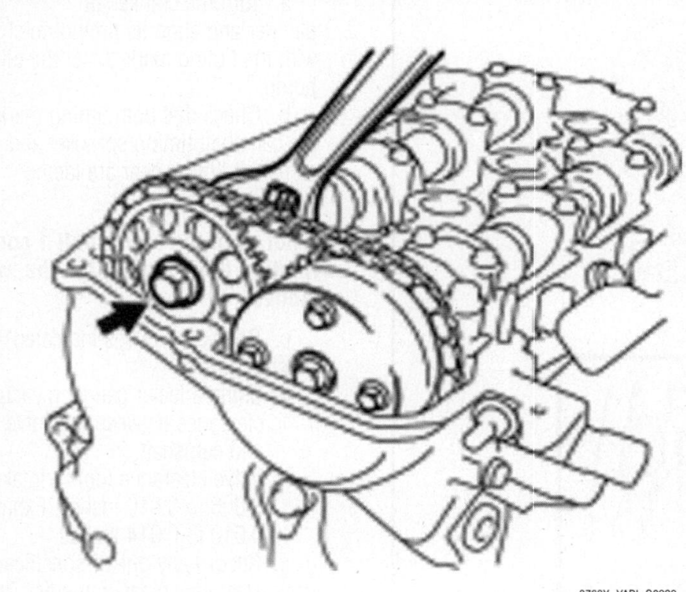

Fig. 43 Step 8: Using a wrench, hold the hexagonal lobe of camshaft No. 2 and remove the flange bolt.

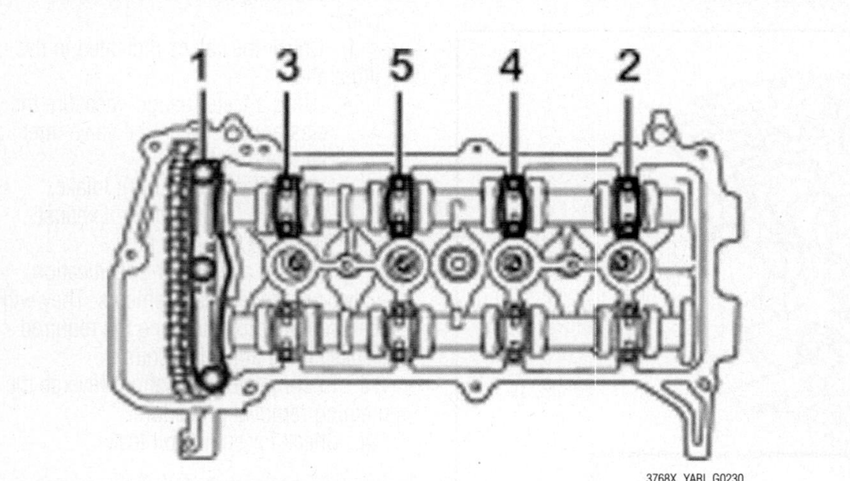

Fig. 44 Step 9: Using several steps, loosen and remove the 11 bearing cap bolts uniformly in the sequence then remove camshaft bearing caps No. 1 and No. 2

➡Check the SST installation positions when installing them, to avoid the SST fixing bolts from coming into contact with the oil pump assembly.

5. Remove the SSTs and the bolt.
6. Remove the crankshaft damper sub-assembly.
7. If necessary, remove timing chain cover front oil seal:
   a. Using a knife, cut off the lip of the oil seal.
   b. Using a screwdriver with its tip wrapped with tape, pry out the oil seal.

➡After removing, check the crankshaft for damage. If damaged, smooth the surface with 400-grit sandpaper.

To install:

➡Keep the lip free of foreign matter.

8. Apply MP grease to the lip of a new oil seal.
9. Using a seal installer (SST 09223-22010) and a hammer, tap in the oil seal until its surface is flush with the rear oil seal retainer edge.
10. Align the pin hole in the crankshaft damper with the pin position and install the crankshaft damper sub-assembly.
11. Provisionally install the bolt.
12. Using 2 SSTs, tighten the bolt while holding the crankshaft damper sub-assembly. Tighten to 95 ft. lbs. (128 Nm).

➡Check the SST installation positions when installing them, to avoid the SST fixing bolts from coming into contact with the oil pump assembly.

CRANKSHAFT FRONT SEAL

REMOVAL & INSTALLATION

See Figure 58.

1. Remove front wheel RH.
2. Remove engine under cover RH.
3. Remove fan and generator V belt.
4. Remove engine mounting insulator sub-assembly RH.
5. Remove crankshaft damper sub-assembly.
6. Remove oil pump seal
   a. Using a knife, cut off the oil seal lip.
   b. Using a screwdriver with its tip wrapped with protective tape, pry out the oil seal.

➡After removing, check the crankshaft for damage. If damaged, smooth the surface with 400-grit sandpaper.

To install:

7. Install oil pump seal.
   a. Apply MP grease to the lip of a new oil pump seal.

➡Keep the lip free of foreign matter.

   b. Using SST: 09223-22010 and a hammer, tap in the timing chain cover oil seal until its surface is flush with the timing chain cover edge.

➡Do not tap the oil seal at an angle.

➡Wipe off extra grease from the crankshaft.

8. Install crankshaft damper sub-assembly (Click here for more information).
9. Install engine mounting insulator sub-assembly RH (Click here for more information).
10. Install fan and generator V belt (Click here for more information).
11. Inspect for engine oil leak.
12. Install engine under cover RH.
13. Install front wheel RH. Tighten the wheel nuts to 76 inch lb. (103 Nm).

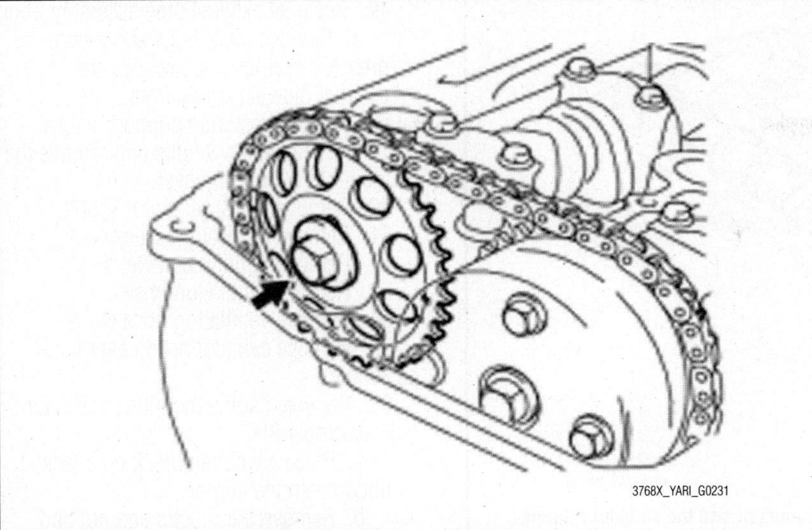

Fig. 45 Step 10: Remove the flange bolt and remove the camshaft timing sprocket

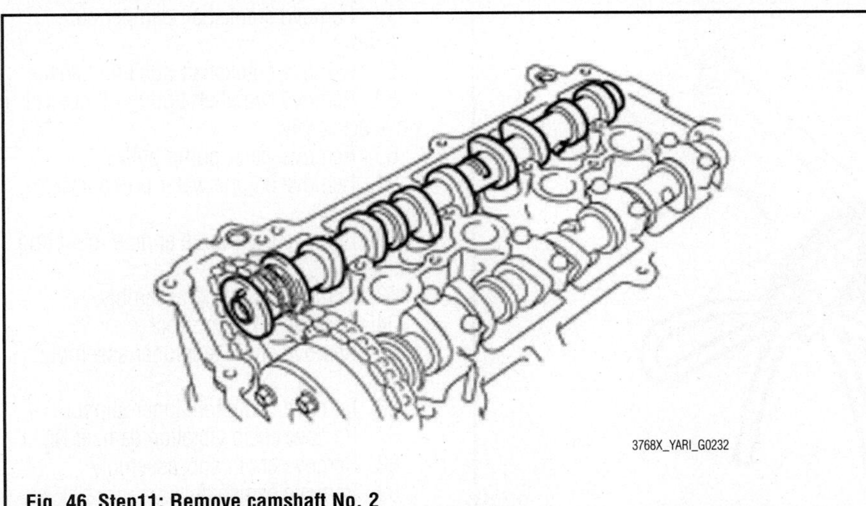

Fig. 46 Step11: Remove camshaft No. 2

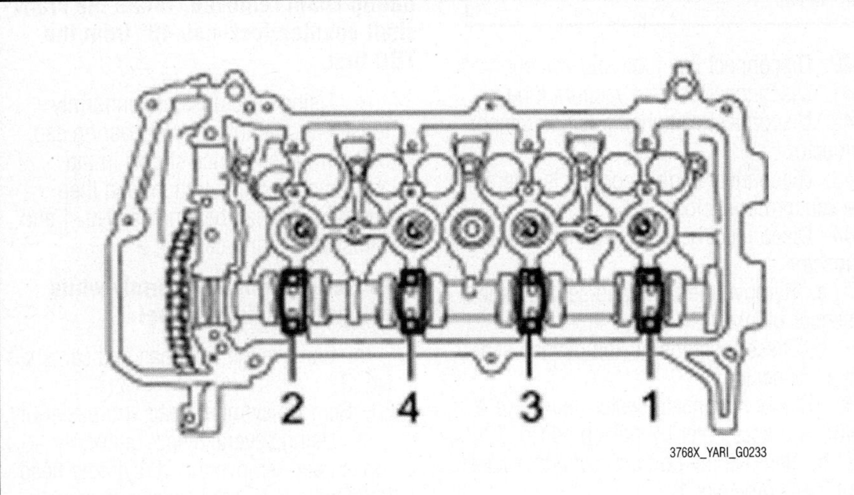

Fig. 47 Step 12: Using several steps, loosen and remove the 8 bearing cap bolts uniformly remove camshaft bearing cap No. 2

CYLINDER HEAD

REMOVAL & INSTALLATION

See Figures 59 through 100.

1. Discharge fuel system pressure
2. Disconnect cable from negative battery terminal
3. Remove front wiper arm head cap (for Hatchback)
4. Remove front wiper arm head cap (for Sedan)
5. Remove front wiper arm and blade assembly LH (for Hatchback)
6. Remove front wiper arm and blade assembly LH (for Sedan)
7. Remove front wiper arm and blade assembly RH (for Hatchback)
8. Remove front wiper arm and blade assembly RH (for Sedan)
9. Remove hood to cowl top seal (for Hatchback)
10. Remove cowl top ventilator louver sub-assembly (for Hatchback)
11. Remove cowl top ventilator louver LH (for Hatchback)
12. Remove cowl side ventilator sub-assembly LH (for Sedan)
13. Remove cowl side ventilator sub-assembly RH (for Sedan)
14. Remove cowl top ventilator louver sub-assembly (for Sedan)
15. Remove front wiper motor and link (for Hatchback)
16. Remove front wiper motor and link (for Sedan)
17. Remove cowl to register duct sub-assembly No. 2 (for Hatchback)
18. Remove front air shutter seal RH (for Sedan)
19. Remove cowl top panel outer (for Hatchback)
20. Remove cowl top panel outer (for Sedan)
21. Remove battery
22. Remove battery tray
23. Remove front wheel RH
24. Remove engine under cover RH
25. Drain engine oil
26. Drain engine coolant
27. Remove cylinder head cover No. 2
28. Remove air cleaner cap sub-assembly with air cleaner hose No. 1
29. Disconnect radiator hose No. 3
30. Disconnect reserve tank hose.
31. Remove water filler sub-assembly
    a. Separate radiator hose No. 1 from the cylinder head.
    b. Remove the 2 nuts and remove the water filler sub-assembly.
32. Disconnect water by-pass hose No. 2

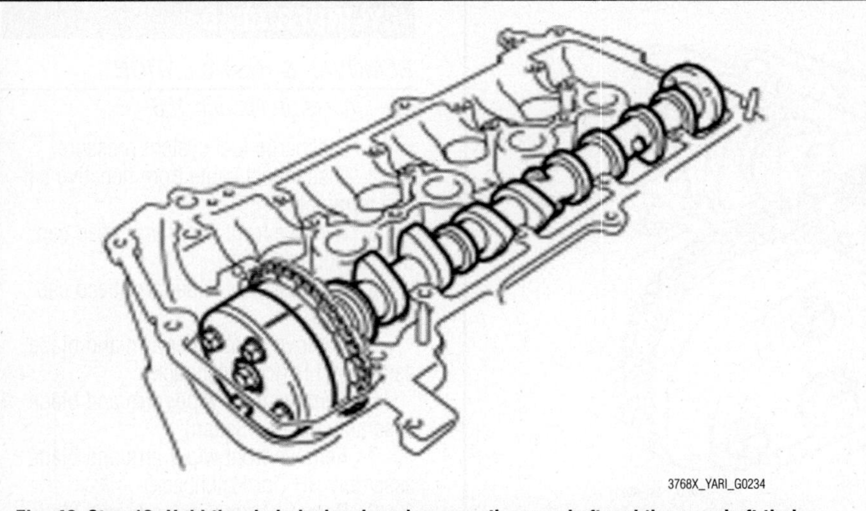

Fig. 48 Step 13: Hold the chain by hand, and remove the camshaft and the camshaft timing gear assembly

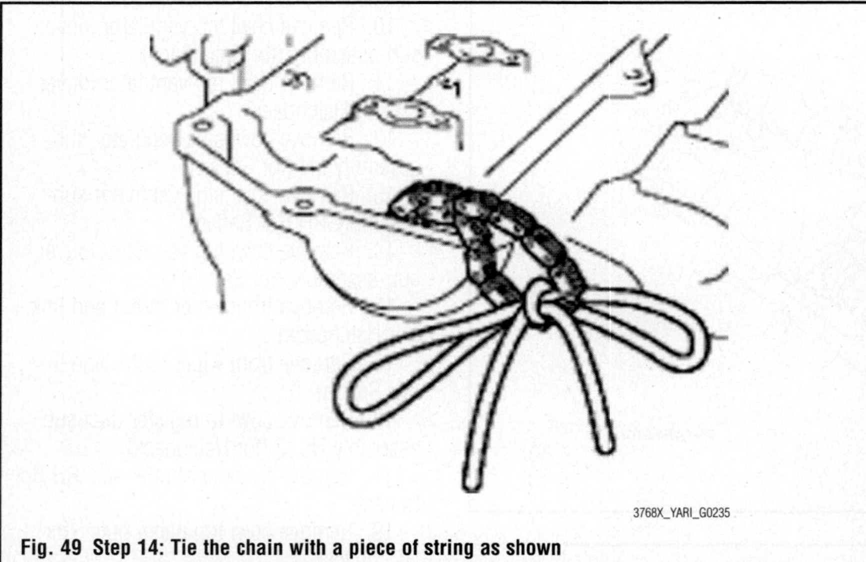

Fig. 49 Step 14: Tie the chain with a piece of string as shown

33. Disconnect water by-pass hose.
34. Disconnect throttle with motor body connector.
    a. Separate the wire harness clamp.
    b. Remove the nut and separate the throttle with motor body connector.
35. Disconnect ventilation hose.
36. Disconnect union to connector tube hose.
37. Remove oil level gauge sub-assembly
38. Remove intake manifold.
    a. Separate the 3 wire harness clamps shown in the accompanying illustration.
    b. Remove the 3 bolts and 2 nuts and remove the intake manifold.
39. Remove oil level gauge guide
    a. Remove the wire harness clamp and the bolt and remove the oil level gauge guide.
40. Disconnect fuel tube sub-assembly
41. Disconnect booster vacuum tube
42. Disconnect camshaft position sensor connector.
43. Disconnect engine coolant temperature sensor connector.
44. Disconnect heated oxygen sensor connector
    a. Remove the bolts and separate the sensor bracket.
    b. Disconnect the heated oxygen sensor connector.
45. Disconnect heater water inlet hose A.
46. Separate water by-pass pipe No. 1
    a. Remove the bolt and separate water by-pass pipe No. 1.
47. Disconnect wire harness
    a. Remove the 2 bolts and disconnect the wire harness.
48. Separate exhaust pipe assembly front
    a. Remove the 2 bolts and 2 compression springs and separate the exhaust pipe assembly front.
49. Remove manifold support bracket
    a. Remove the 3 bolts and remove the manifold support bracket.
50. Remove fan & generator V belt.
51. Remove generator assembly.
52. Remove ignition coil No. 1.
53. Remove ventilation hose.
54. Remove ventilation hose No. 2.
55. Remove cylinder head cover sub-assembly
56. Remove engine mounting insulator sub-assembly RH.
    a. Place a wooden block on a jack underneath the engine.
    b. Remove the 5 bolts and nut and remove the engine mounting insulator sub-assembly RH
57. Remove crankshaft damper sub-assembly.
58. Remove crankshaft position sensor.
59. Remove camshaft timing oil control valve assembly.
60. Remove water pump pulley
61. Remove engine water pump assembly
62. Remove transverse engine mounting bracket.
63. Remove oil pump assembly.
64. Remove oil pump seal
65. Remove chain tensioner assembly No. 1.
66. Remove chain tensioner slipper.
67. Remove chain vibration damper No. 1.
68. Remove chain sub-assembly.
69. Remove camshaft.

➡When rotating the camshaft with the timing chain removed, rotate the crankshaft counterclockwise 40° from the TDC first.

    a. Using several steps, uniformly loosen and remove the 19 bearing cap bolts in the sequence shown in the accompanying illustration, and then remove camshaft bearing cap No. 1 and camshaft bearing cap No. 2.

➡Loosen each bolt uniformly while keeping the camshaft level.

    b. Remove the camshaft and camshaft No. 2.
70. Remove cylinder head sub-assembly
    a. Using several steps, uniformly loosen and remove the 10 cylinder head bolts with an 8 mm bi-hexagon wrench in the sequence shown in the accompanying illustration. Remove the 10 plate washers.

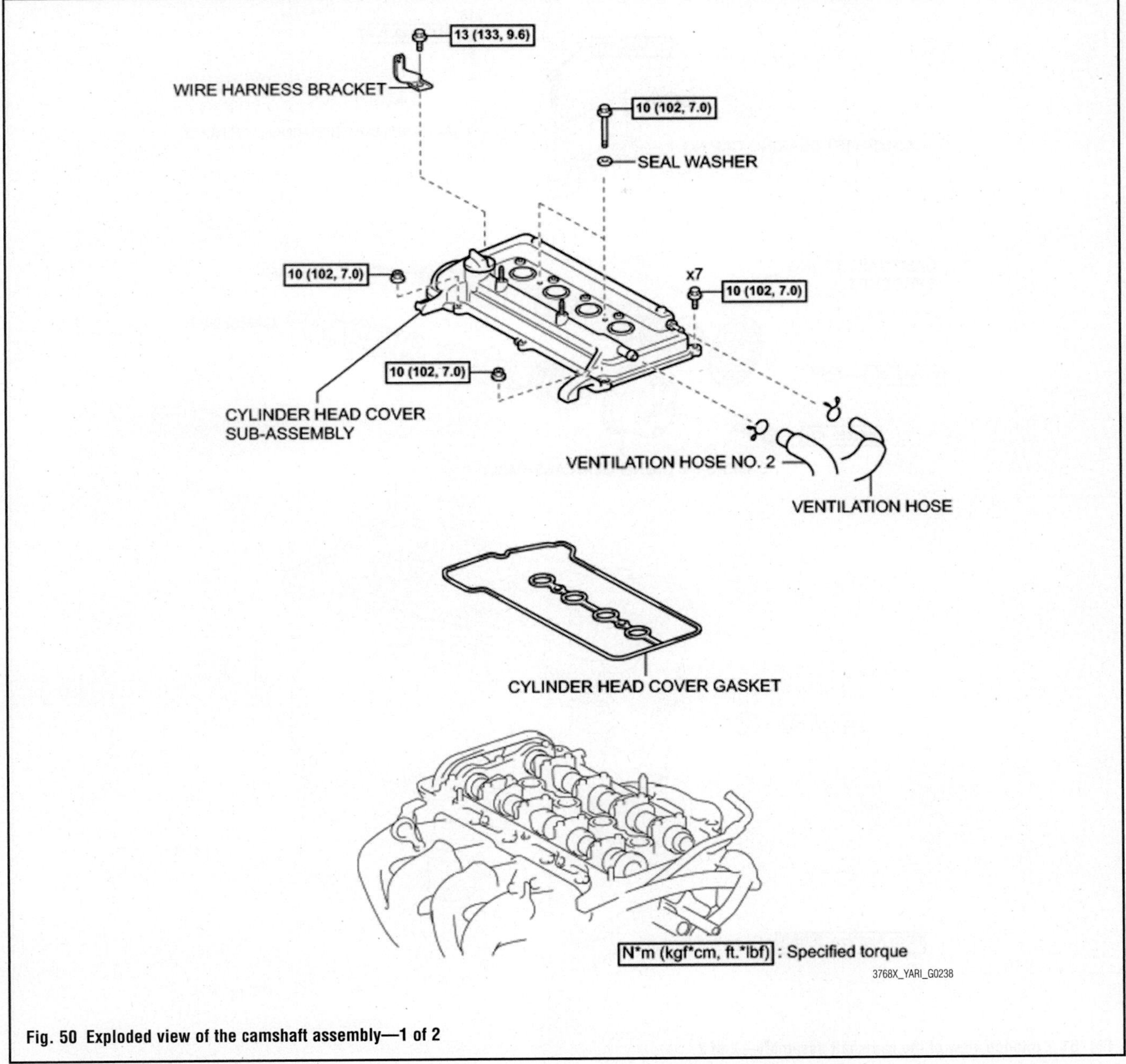

Fig. 50 Exploded view of the camshaft assembly—1 of 2

➡Do not drop the washers into the cylinder head. Head warpage or cracking could result from removing the bolts in the wrong order.

71. Remove cylinder head gasket
    a. Remove the cylinder head gasket.
72. Remove the 4 bolts and remove exhaust manifold heat insulator No. 1.
73. Remove the 3 bolts and 2 nuts and remove the exhaust manifold.
74. Remove the bolt and remove the harness bracket.
75. Remove the 2 bolts and remove the booster vacuum tube.
76. Remove the bolt and remove the camshaft position sensor.
77. Using SST [09817-33190], remove the engine coolant temperature sensor connector.
78. Remove water by-pass hose
79. Remove the bolt and remove the harness bracket.
80. Remove the 3 bolts and remove the fuel delivery pipe sub-assembly with 4 fuel injectors.

➡Do not drop the fuel injectors when removing the fuel delivery pipe sub-assembly.

81. Remove fuel injector assembly.
82. Remove the 2 delivery pipe No. 1 spacers.
83. Remove the 4 injector vibration insulators.

To install:

84. Install 4 new injector vibration insulators onto the cylinder head.
85. Install the 2 delivery pipe No. 1 spacers onto the cylinder head.

➡Install the delivery pipe No. 1 spacer in the correct direction.

86. Install fuel injector assembly.
87. Install fuel delivery pipe sub-assembly

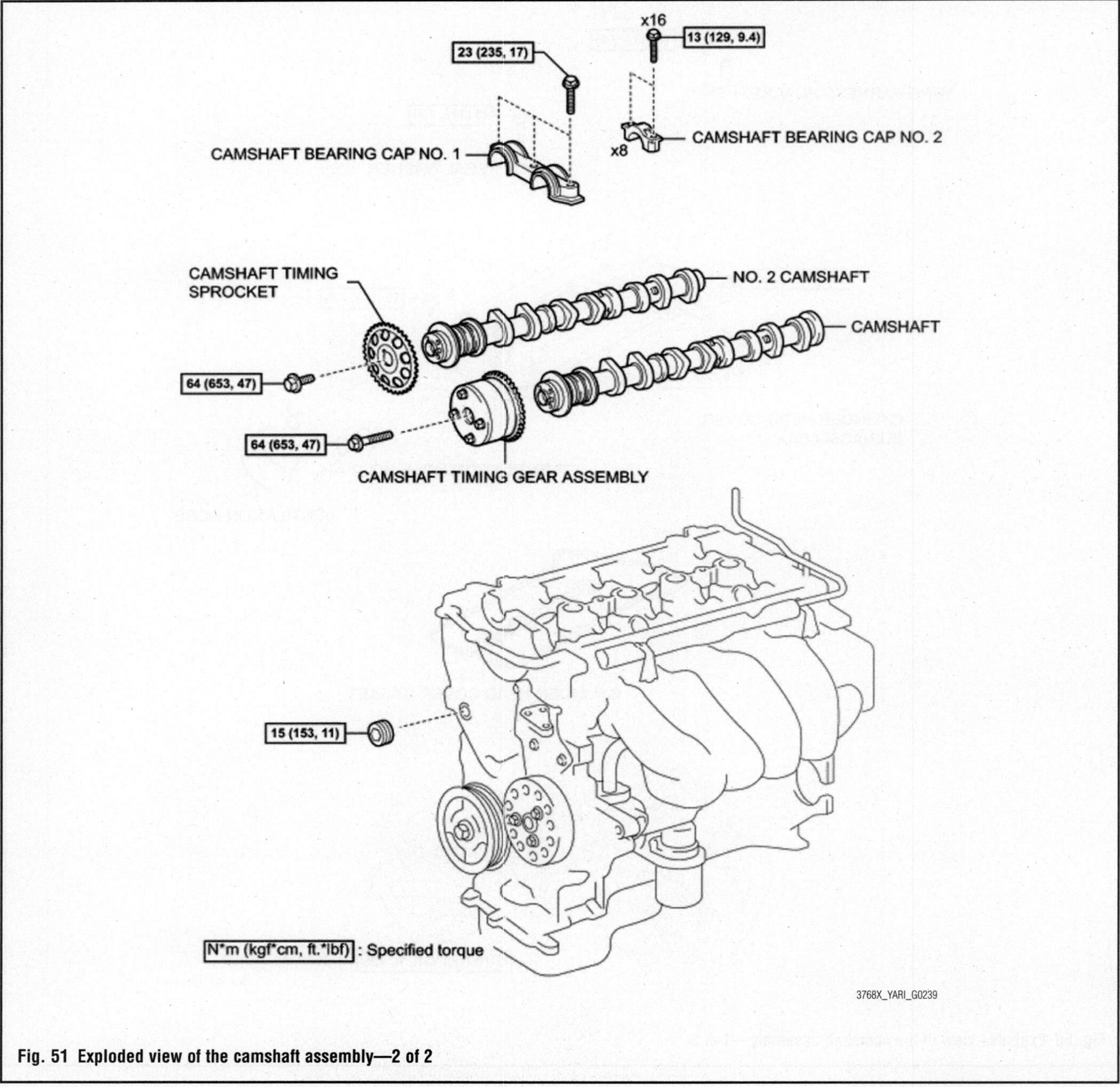

Fig. 51 Exploded view of the camshaft assembly—2 of 2

a. Provisionally install the fuel delivery pipe sub-assembly with the 4 fuel injectors using the 3 bolts.

➡Do not drop the fuel injectors when installing the fuel delivery pipe sub-assembly.

➡Check that the fuel injectors rotate smoothly after installing the fuel delivery pipe sub-assembly.

88. Install the harness bracket with the bolt. Tighten the bolt to 9.5 inch lb. (13 Nm).

89. Install water by-pass hose

90. Install engine coolant temperature sensor

a. Provisionally install the engine coolant temperature sensor through a new gasket.

b. Using SST (09817-33190), tighten the engine coolant temperature sensor to 15 inch lb. (20 Nm).

91. Apply a light coat of engine oil to the O-ring on the camshaft position sensor.

a. Install the camshaft position sensor with the bolt. Tighten the bolt to 71 ft. lb. (8.0 Nm).

➡Do not twist the O-ring.

92. Install the booster vacuum tube with the 2 bolts. Tighten the bolts to 80 ft. lb. (9.0 Nm).

93. Install the harness bracket with the bolt. Tighten the bolts to 9.5 inch lb. (13 Nm).

94. Using several steps, install a new exhaust manifold gasket and the exhaust manifold with the 3 bolts and 2 nuts in the sequence shown in the accompanying illustration.

a. Tighten the 3 bolts and 2 nuts to 20 inch lb. (27 Nm).

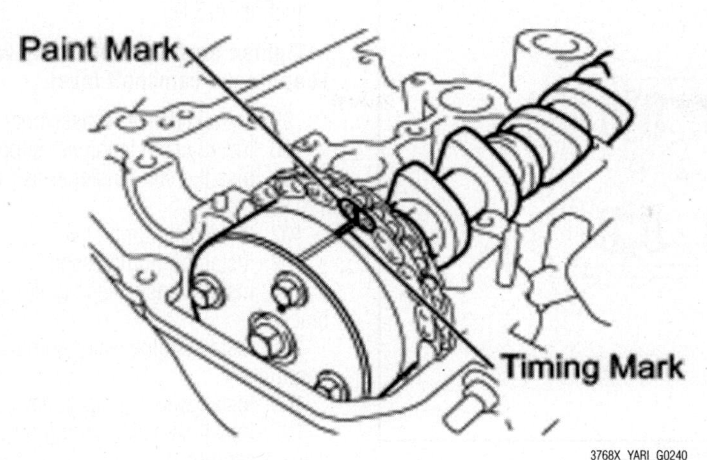

Fig. 52 Install the chain onto the camshaft timing gear with the paint mark and the timing mark aligned

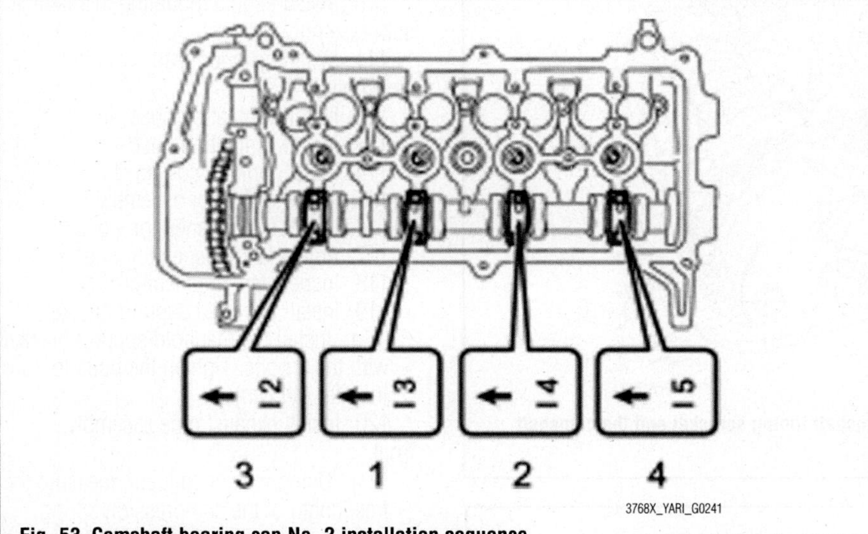

Fig. 53 Camshaft bearing cap No. 2 installation sequence

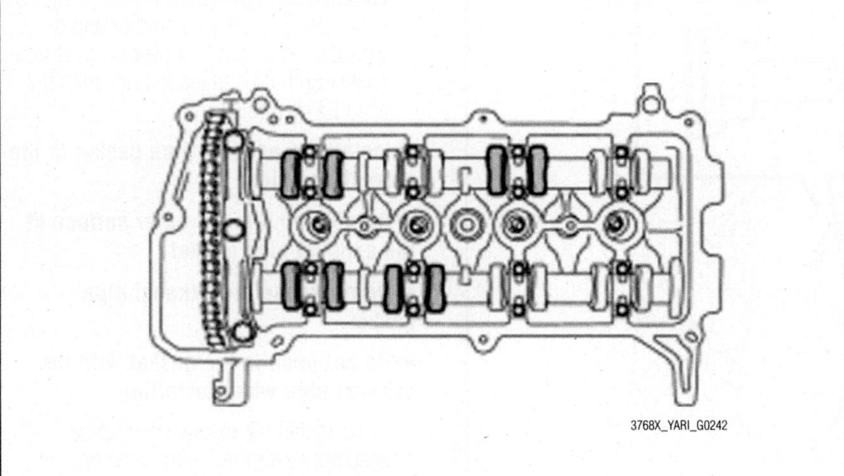

Fig. 54 Check the valves indicated—First set

95. Install exhaust manifold heat insulator No. 1 with the 4 bolts. Tighten the bolts to 71 ft. lb. (8.0 Nm).

96. Place a new cylinder head gasket on the cylinder block with the Lot No. stamp facing upward.

➡Remove any oil from the contact surfaces.

➡Check the mounting orientation of the cylinder head gasket.

➡Place the cylinder head on the cylinder head gently in order not to damage the gasket.

97. Install cylinder head sub-assembly

➡The cylinder head bolts are tightened in 2 successive steps.

a. Apply a light coat of engine oil to the threads of the cylinder head bolts.

b. Using several steps, install and tighten the 10 cylinder head bolts and plate washers uniformly with an 8 mm bi-hexagon wrench, in the sequence shown in the accompanying illustration.

c. Tighten the cylinder head bolts to 22 inch lb. (29 Nm).

d. Mark the front of the cylinder head bolt with paint.

e. Retighten the cylinder head bolts 90° and then an additional by 90° as shown in the accompanying illustration.

f. Check that the paint mark is now at a 180° angle from the front.

g. Apply a continuous bead of seal packing (diameter 0.177 to 0.217 in. (4.5 to 5.5 mm)).

h. Seal Packing Toyota Genuine Seal Packing Black, Three Bond 1207B or Equivalent.

➡Remove any oil from the contact surfaces.

➡Install the oil pump assembly within 3 minutes and tighten the bolts within 15 minutes of applying the seal packing.

98. Install camshaft

a. Examine the front marks and numbers and check that the sequence is as shown in the accompanying illustration. Then provisionally tighten the 19 bolts.

b. Uniformly tighten the bolts in several steps in the sequence shown in the illustration and install camshaft bearing cap No. 1 and camshaft bearing cap No. 2.

c. Tighten the camshaft bearing cap No. 1 to 17 inch lb. (23 Nm), and tighten

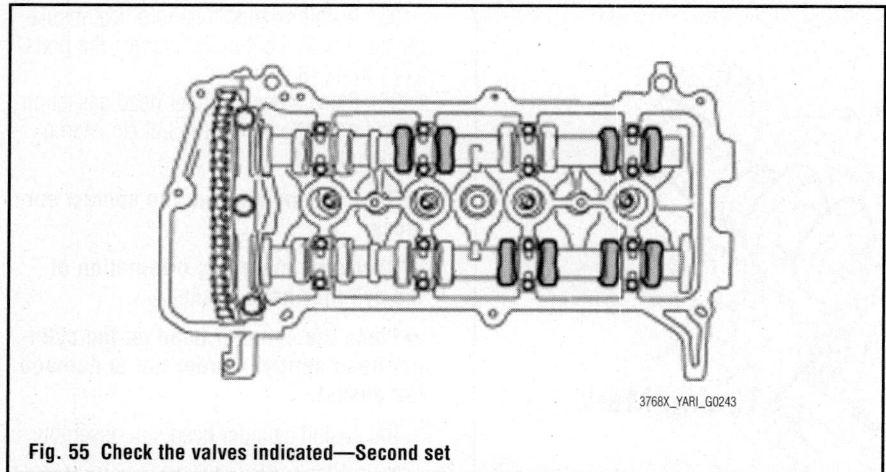

Fig. 55 Check the valves indicated—Second set

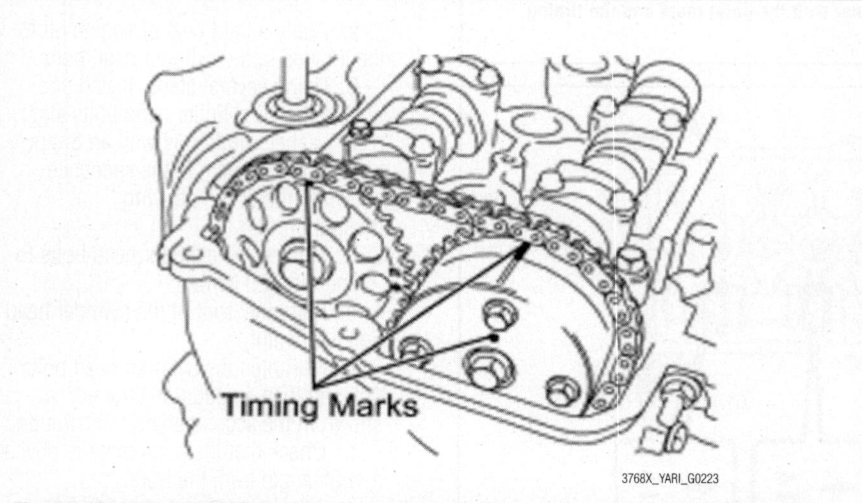

Fig. 56 Check that the timing marks on both the camshaft timing sprocket and the camshaft timing gear are facing upward

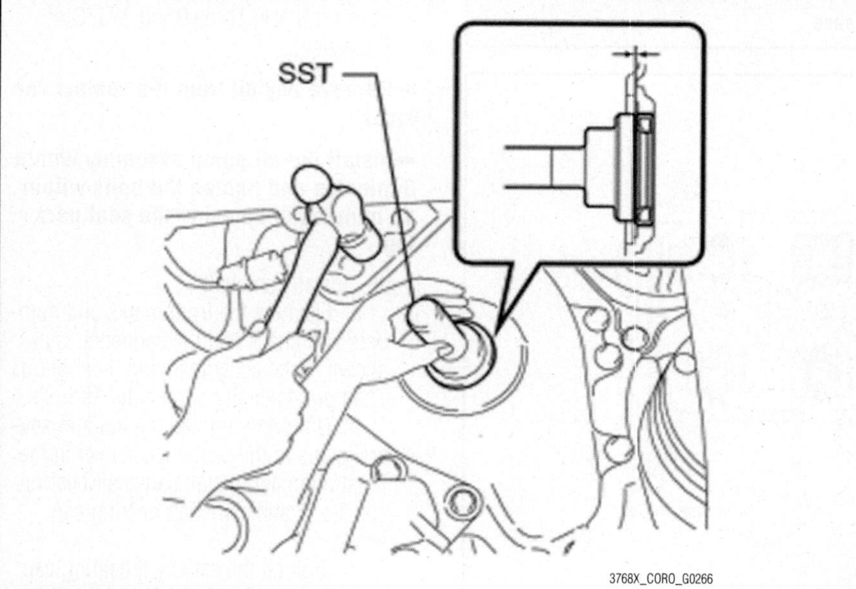

Fig. 57 Installing the crankshaft front seal

the camshaft bearing cap No. 2 to 9.4 inch lb. (13 Nm).

➡Tighten each bolt uniformly while keeping the camshaft level.

99. Install chain sub-assembly
100. Install chain tensioner slipper
101. Install chain tensioner assembly no. 1
102. Install oil pump seal
103. Install oil pump assembly
104. Install transverse engine mounting bracket
105. Install engine water pump assembly
106. Install water pump pulley
107. Install camshaft timing oil control valve assembly
108. Install crankshaft position sensor
109. Install crankshaft damper sub-assembly
110. Install engine mounting insulator sub-assembly RH
111. Install cylinder head cover sub-assembly
112. Install ventilation hose No. 2
113. Install ventilation hose
114. Install ignition coil no. 1
115. Install generator assembly
116. Install fan & generator v belt
117. Adjust fan & generator v belt
118. Inspect fan & generator V belt
119. Install manifold support bracket
    a. Install the manifold support bracket with the 3 bolts. Tighten the bolts to 33 inch lb. (44 Nm).
120. Install exhaust pipe assembly front
    a. Using vernier calipers, measure the free length of the compression spring (minimum length 1.594 in. (40.5 mm)). If the length is not as specified, replace the compression spring.
    b. Using a plastic hammer and a wooden block, tap in a new exhaust pipe gasket until its surface is flush with the exhaust manifold.

➡Install the exhaust pipe gasket in the correct direction.

➡Do not damage the outer surface of the exhaust pipe gasket.

➡Do not reuse the exhaust pipe gasket.

➡Do not push in the gasket with the exhaust pipe when installing.

c. Install the exhaust front pipe assembly with the 2 compression springs and 2 bolts. Tighten the bolts to 32 inch lb. (43 Nm).

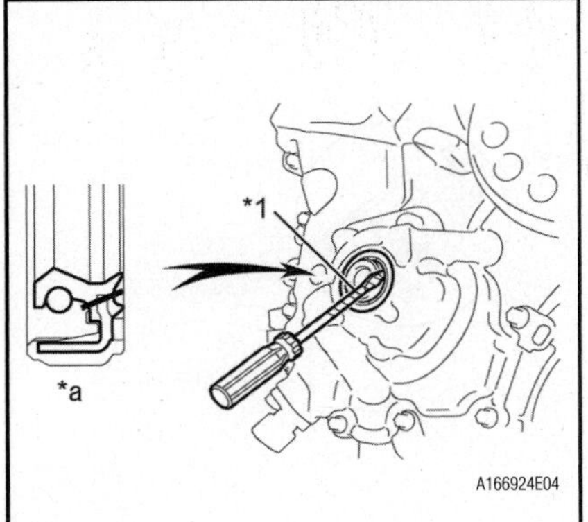

Fig. 58 Remove oil pump seal. Using a knife, cut off the oil seal lip (*a=Cut position, *1=Protective tape)

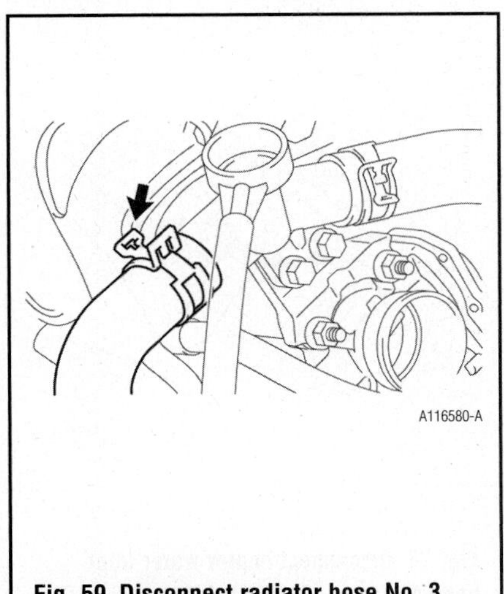

Fig. 59 Disconnect radiator hose No. 3

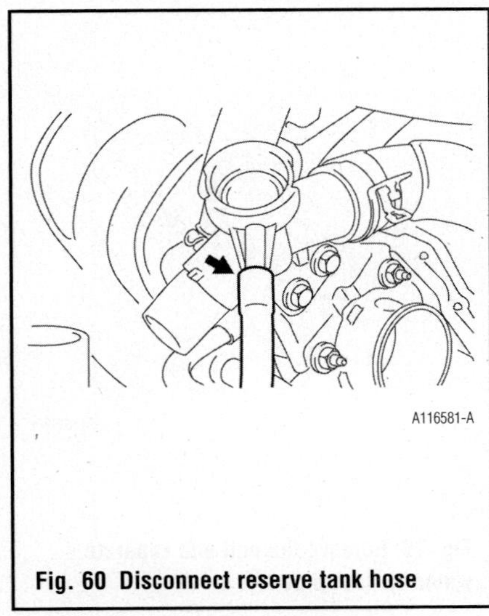

Fig. 60 Disconnect reserve tank hose

121. Connect wire harness with the 2 bolts. Tighten the bolts to 10 inch lb. (13 Nm).

122. Install water by-pass hose No. 1 with bolt. Tighten the bolt to 80 ft. lb. (9.0 Nm)

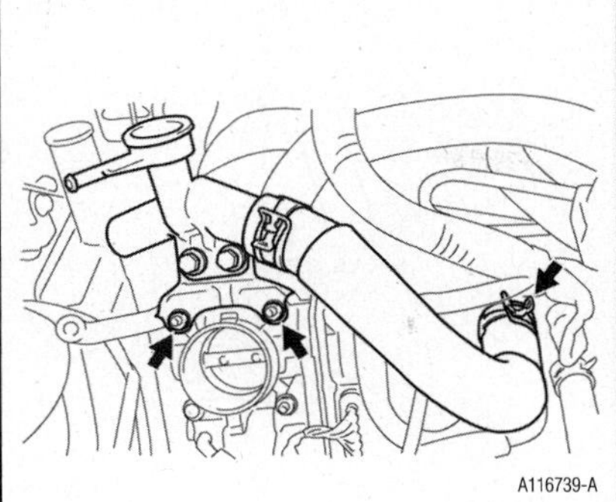

Fig. 61 Remove water filler sub-assembly. Separate radiator hose No. 1 from the cylinder head. Remove the 2 nuts and remove the water filler sub-assembly.

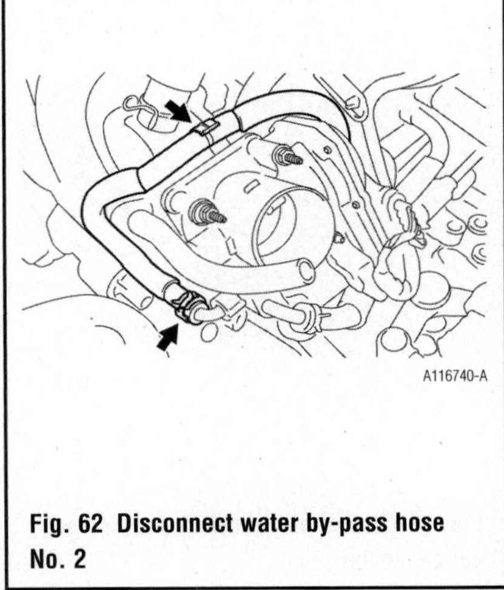

Fig. 62 Disconnect water by-pass hose No. 2

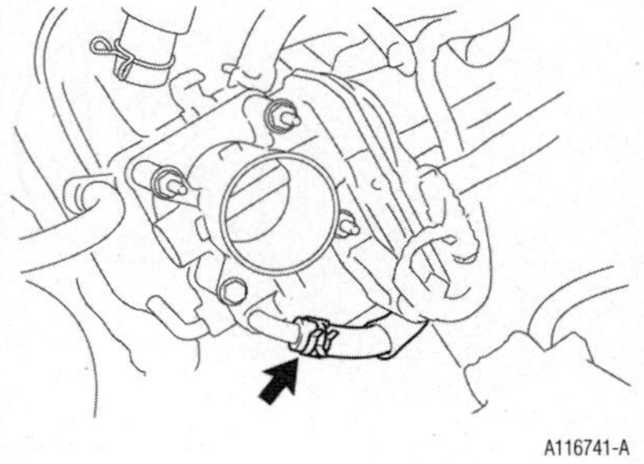

Fig. 63 Disconnect water by-pass hose

123. Connect heater water hose inlet A.

124. Connect heated oxygen sensor connector

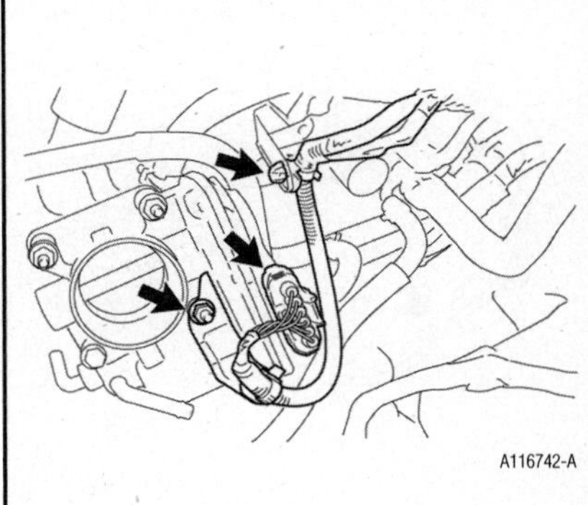

Fig. 64 Disconnect throttle with motor body connector. Separate the wire harness clamp. Remove the nut and separate the throttle with motor body connector

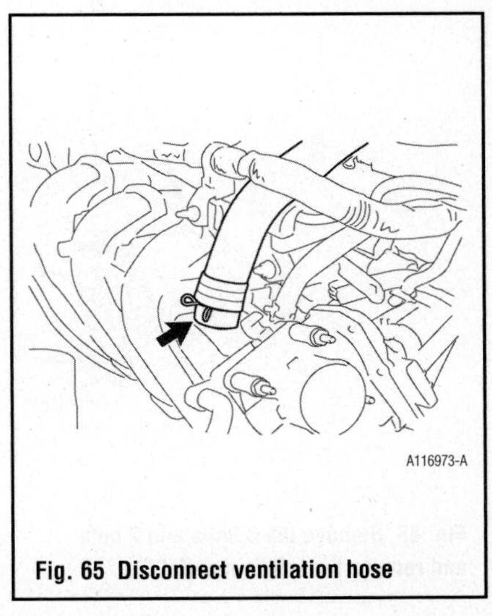

Fig. 65 Disconnect ventilation hose

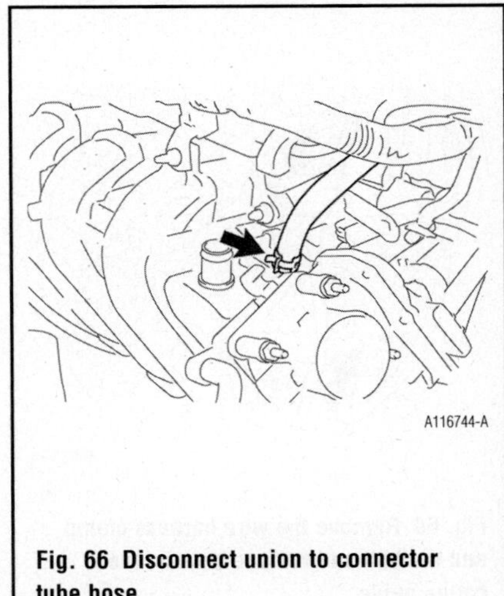

Fig. 66 Disconnect union to connector tube hose

a. Install the sensor bracket with the bolt.

125. Connect engine coolant temperature sensor connector

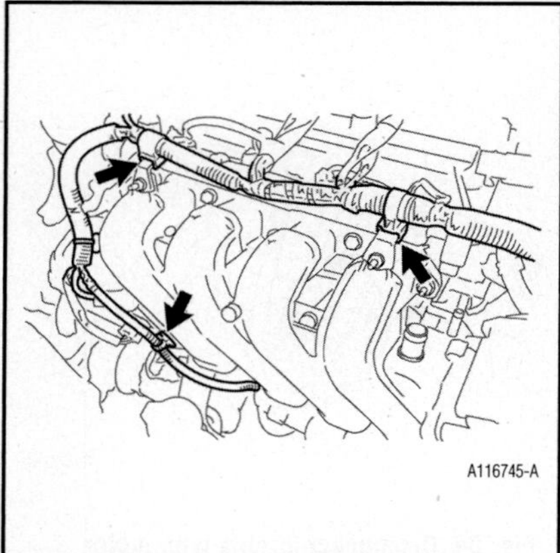

Fig. 67 Remove intake manifold. Separate the 3 wire harness clamps shown in the illustration.

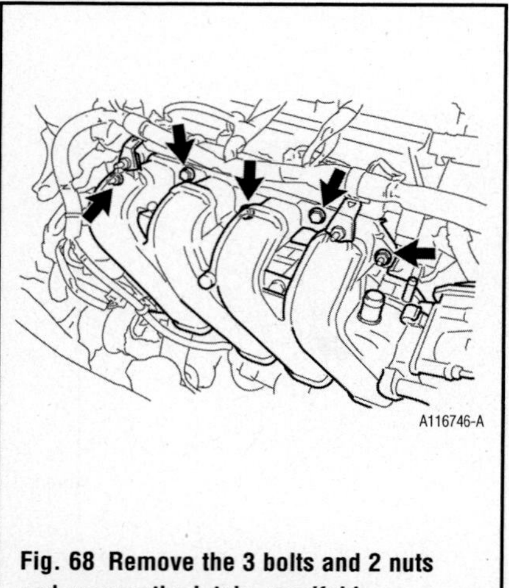

Fig. 68 Remove the 3 bolts and 2 nuts and remove the intake manifold.

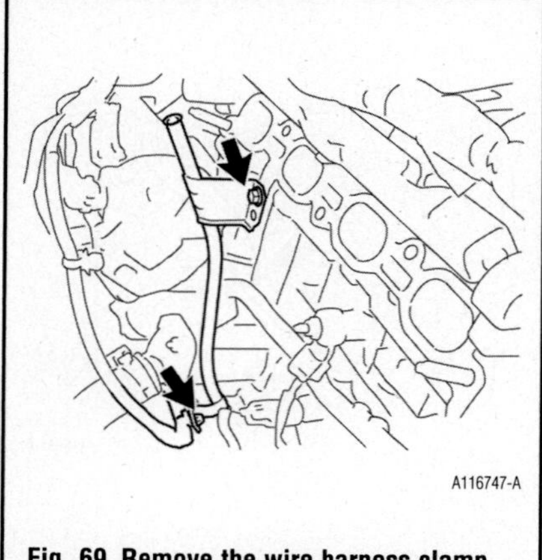

Fig. 69 Remove the wire harness clamp and the bolt and remove the oil level gauge guide

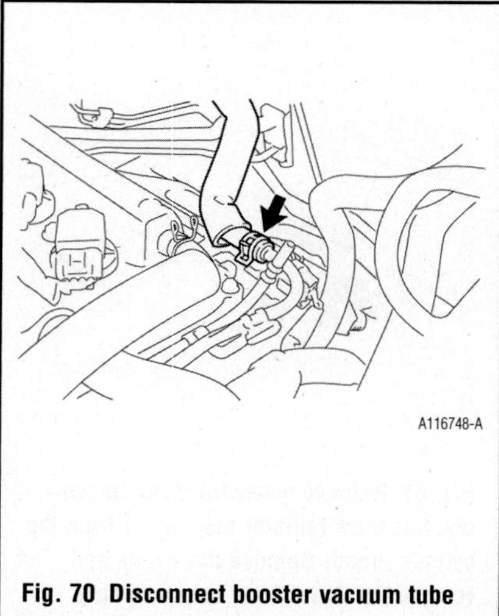

Fig. 70 Disconnect booster vacuum tube

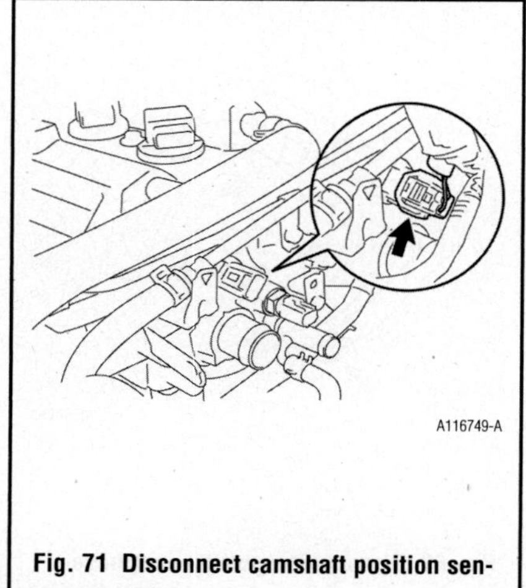

Fig. 71 Disconnect camshaft position sensor connector

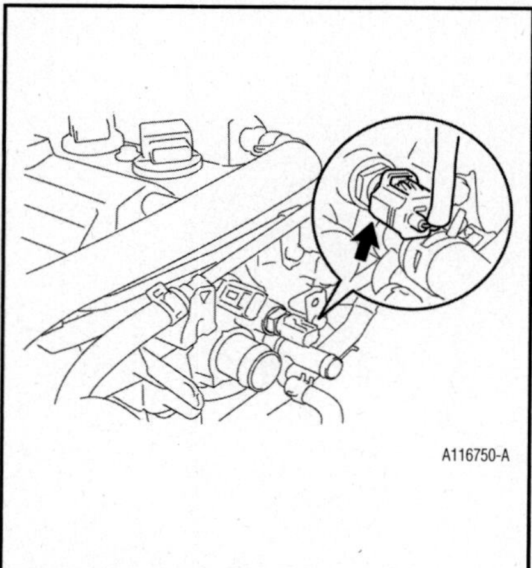

Fig. 72 Disconnect engine coolant temperature sensor connector

Fig. 73 Remove the bolts and separate the sensor bracket. Disconnect the heated oxygen sensor connector

Fig. 74 Disconnect heater water inlet hose A

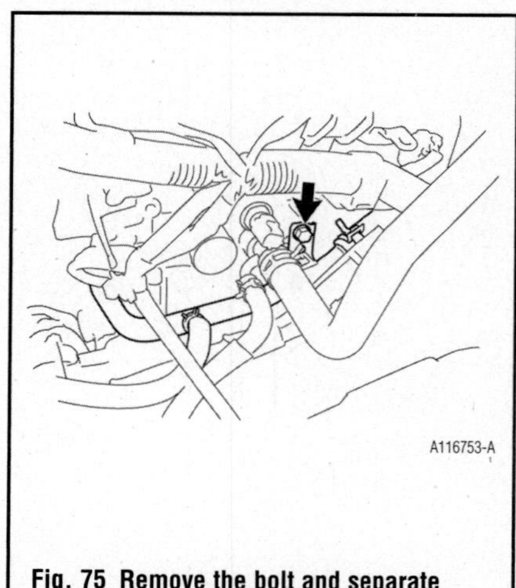

Fig. 75 Remove the bolt and separate water by-pass pipe No. 1

126. Connect camshaft position sensor connector
127. Connect booster vacuum tube
128. Connect fuel tube sub-assembly
129. Install oil level gauge guide
    a. Apply engine oil to a new O-ring.
    b. Install the oil level gauge guide with the bolt through a new O-ring. Tighten the bolt to 80 ft. lb. (9.0).
    c. Install the wire harness.
130. Install intake manifold

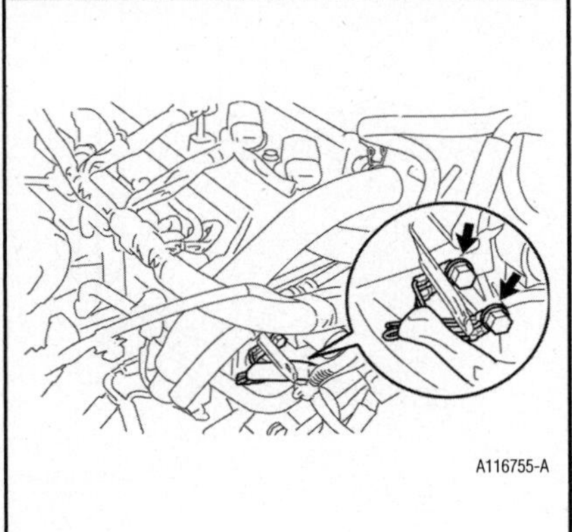

Fig. 76 Remove the 2 bolts and disconnect the wire harness

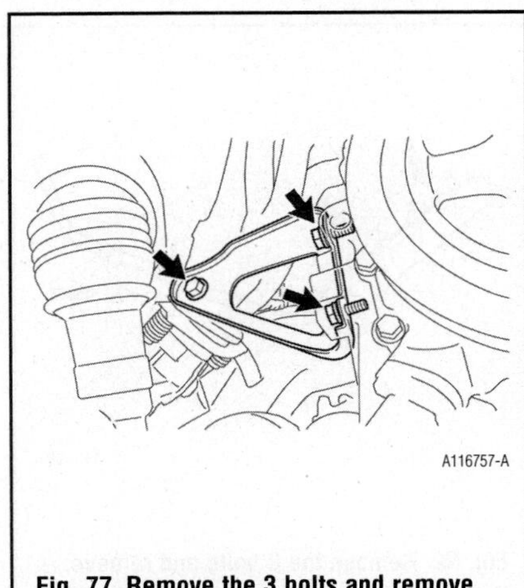
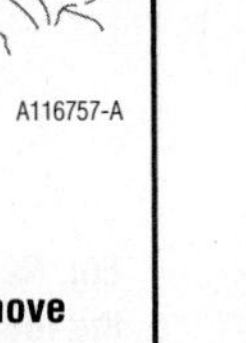

Fig. 77 Remove the 3 bolts and remove the manifold support bracket

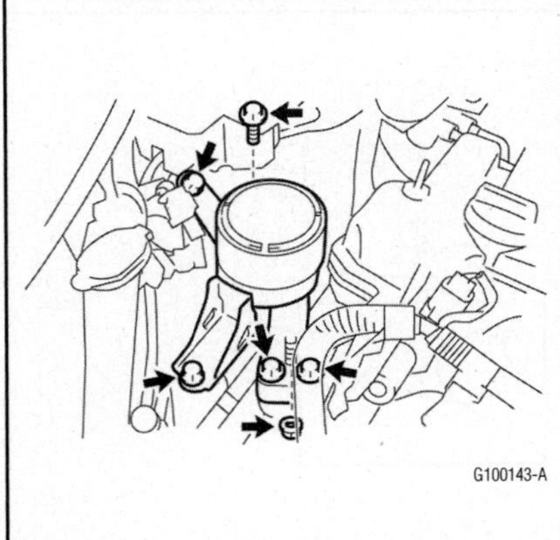

Fig. 78 Remove the 5 bolts and nut and remove the engine mounting insulator sub-assembly RH

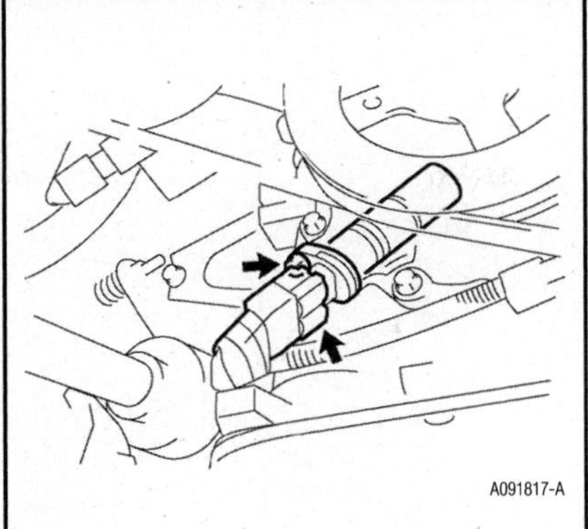

Fig. 79 Disconnect the crankshaft position sensor connector. Remove the bolt and remove the crankshaft position sensor

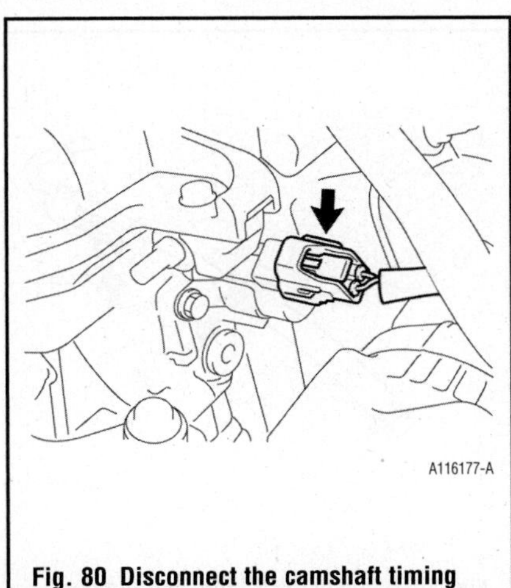

Fig. 80 Disconnect the camshaft timing oil control valve assembly connector

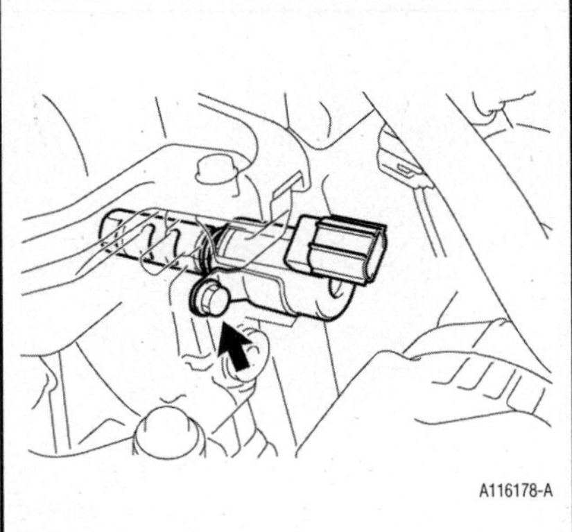

Fig. 81 Remove the bolt and nut and remove the camshaft timing oil control valve assembly

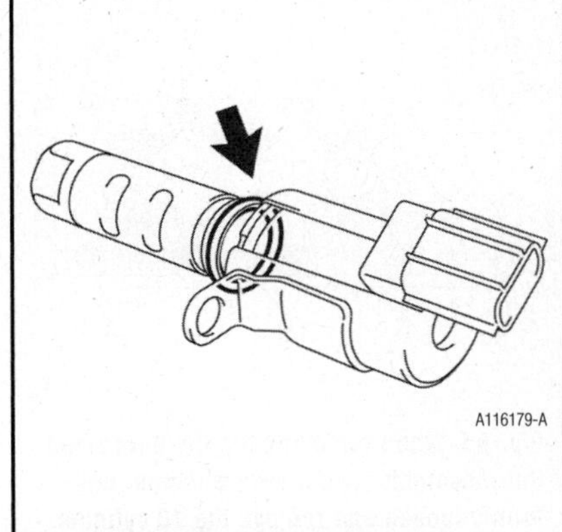

Fig. 82 Remove the O-ring from the camshaft timing oil control valve assembly

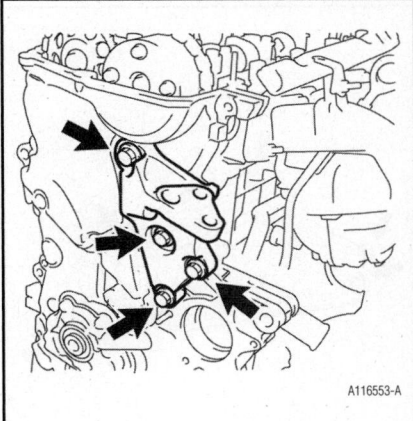

Fig. 83 Remove transverse engine mounting bracket.

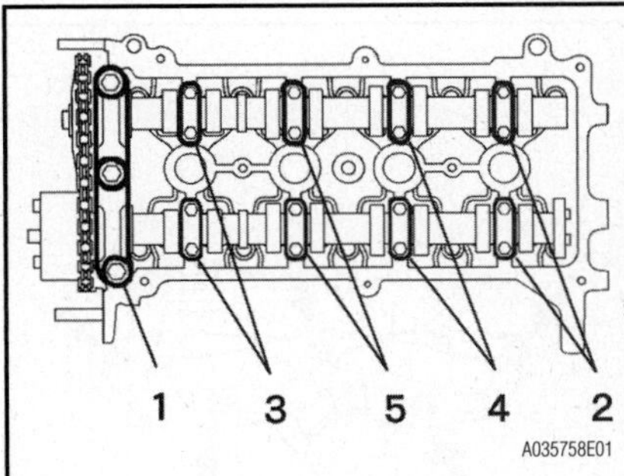

Fig. 84 When removing the Camshaft, using several steps, uniformly loosen and remove the 19 bearing cap bolts in the sequence shown in the accompanying illustration, and then remove camshaft bearing cap No. 1 and camshaft bearing cap No. 2

a. Install a new gasket onto the intake manifold.

b. Install the intake manifold with the 3 bolts and 2 nuts. Tighten the 3 bolts and 2 nuts to 22 inch lb. (30 Nm).

c. Connect the 3 wire harness clamps shown in the illustration.

131. Install oil level gauge sub-assembly

132. Connect union to connector tube hose

133. Connect ventilation hose

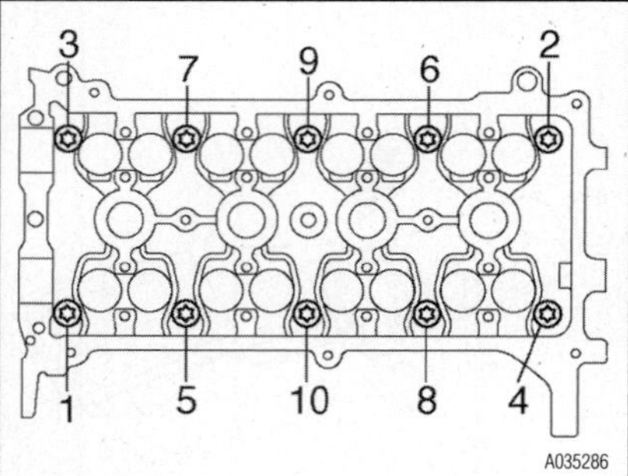

Fig. 85 When removing the Cylinder Head Sub-Assembly, using several steps, uniformly loosen and remove the 10 cylinder head bolts with an 8 mm bi-hexagon wrench in the sequence shown. Remove the 10 plate washers

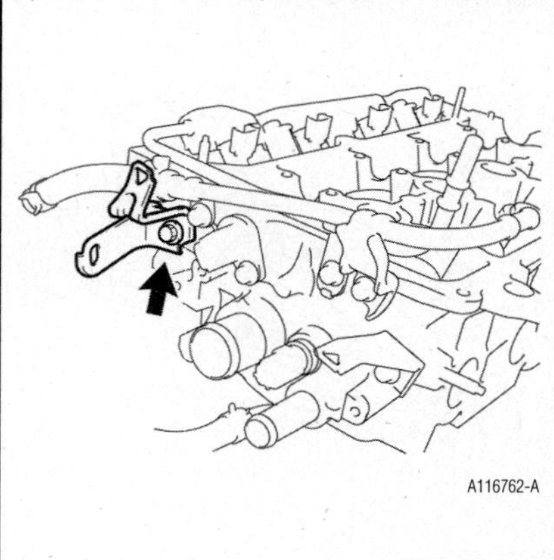

Fig. 88 Remove the bolt and remove the harness bracket

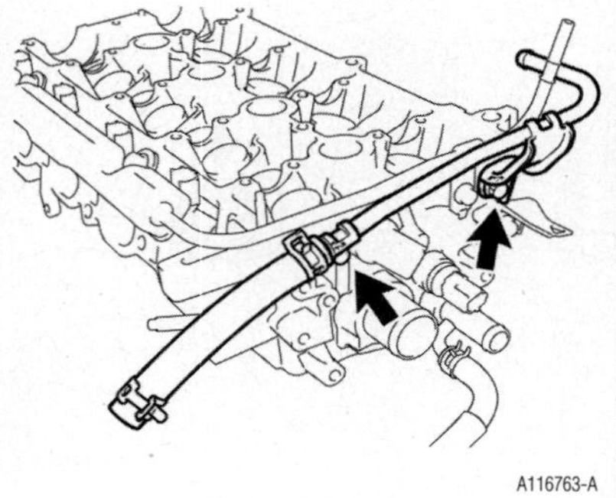

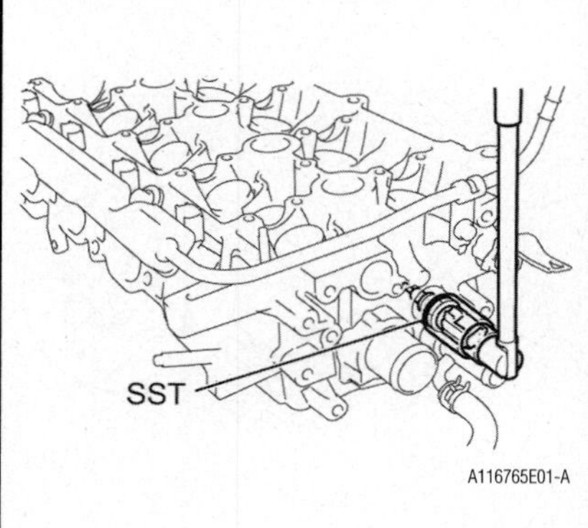

Fig. 91 Using SST [09817-33190], remove the engine coolant temperature sensor connector

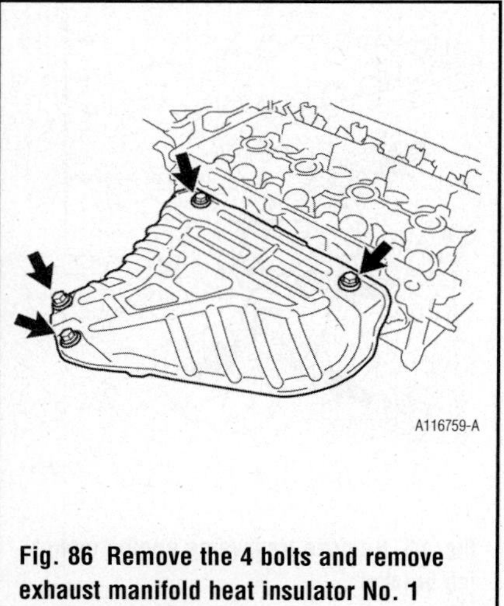

Fig. 86 Remove the 4 bolts and remove exhaust manifold heat insulator No. 1

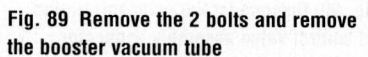

Fig. 89 Remove the 2 bolts and remove the booster vacuum tube

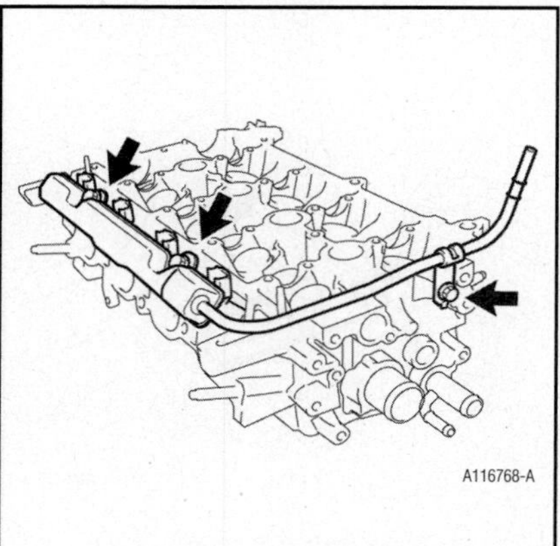

Fig. 92 Remove the 3 bolts and remove the fuel delivery pipe sub-assembly with 4 fuel injectors

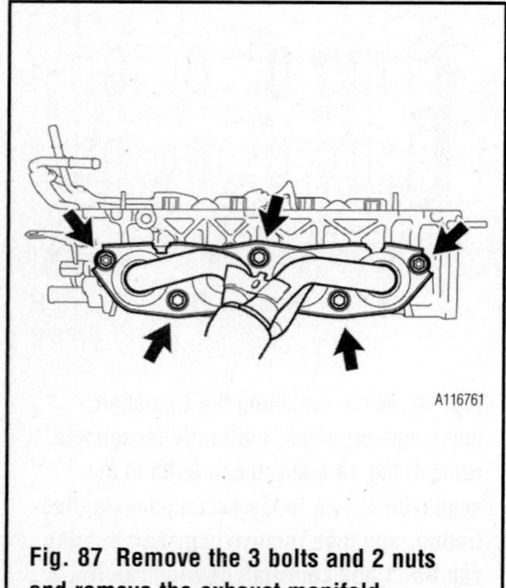

Fig. 87 Remove the 3 bolts and 2 nuts and remove the exhaust manifold

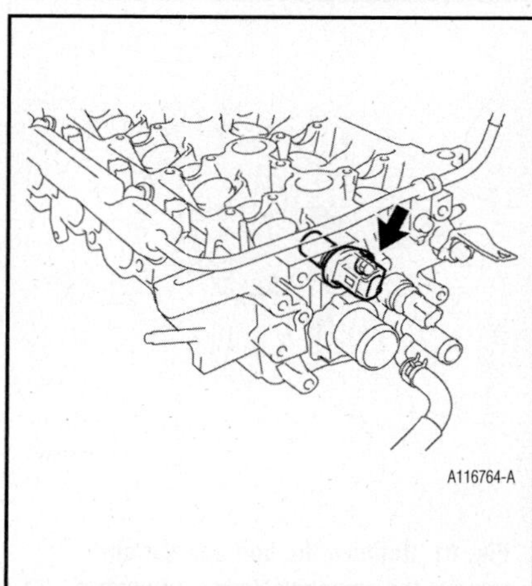

Fig. 90 Remove the bolt and remove the camshaft position sensor.

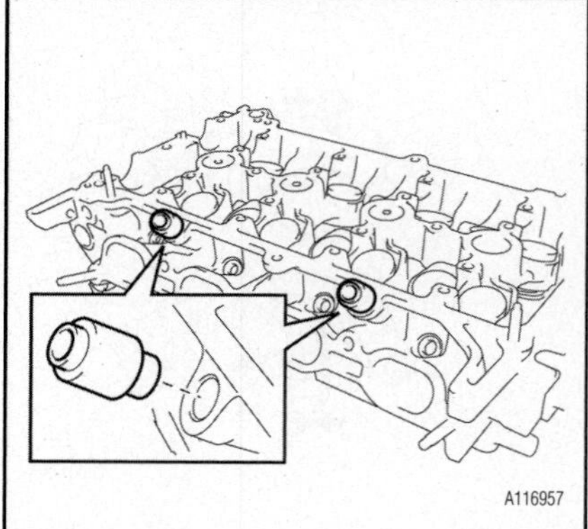

Fig. 93 Remove the 2 delivery pipe No. 1 spacers. When installing be sure to install in the correct direction.

134. Install the throttle with motor body connector bracket with the nut. Tighten the nut to 80 ft. lb. (9.0 Nm).

a. Connect throttle with motor body connector

b. Connect the wire harness clamp.

135. Connect water by-pass hose

136. Connect water by-pass hose No. 2

137. Install water filler sub-assembly. Tighten the 2 nuts to 66 ft. lb. (7.5 Nm).

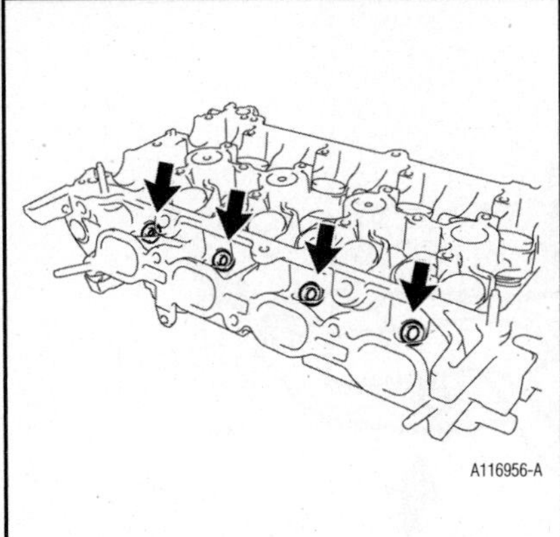

Fig. 94 Remove the 4 injector vibration insulators

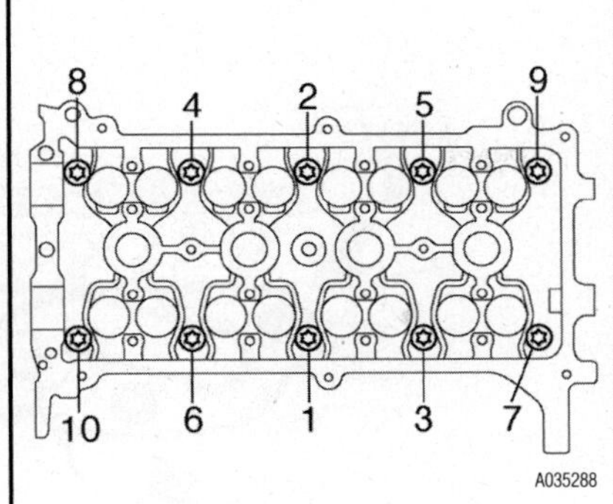

Fig. 97 Using several steps, install and tighten the 10 cylinder head bolts and plate washers uniformly with an 8 mm bi-hexagon wrench, in the sequence shown

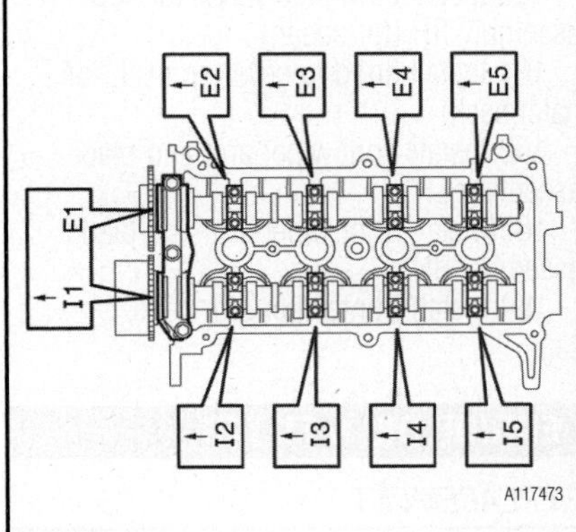

Fig. 99 Examine the front marks and numbers and check that the sequence is as shown

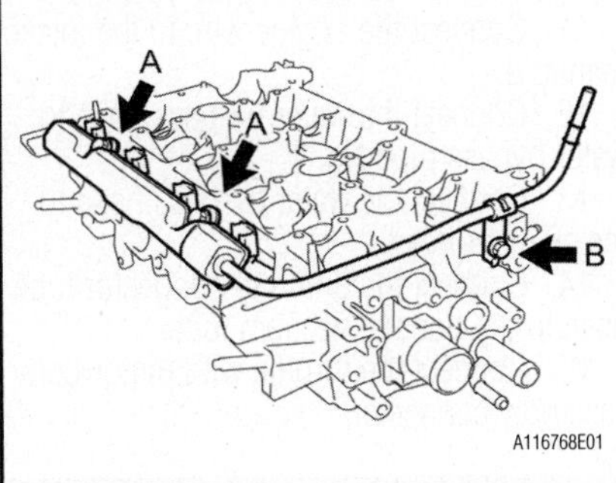

Fig. 95 Provisionally install the fuel delivery pipe sub-assembly with the 4 fuel injectors using the 3 bolts. Tighten bolt A to 14 inch lb. (19 Nm), and tighten bolt B to 80 ft. lb. (9 Nm).

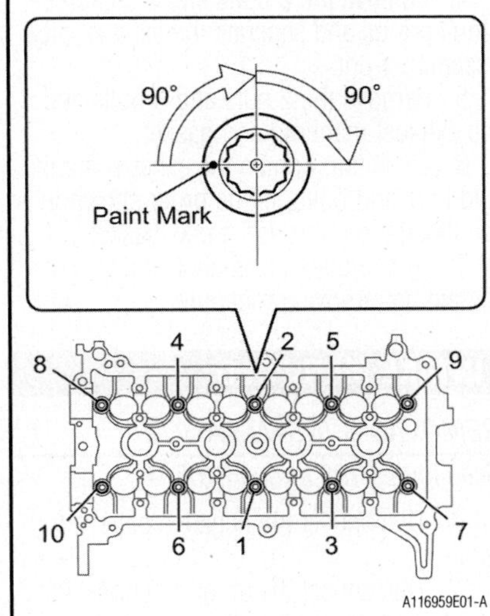

Fig. 98 Mark the front of the cylinder head bolt with paint. Retighten the cylinder head bolts 90° and then an additional by 90° as shown

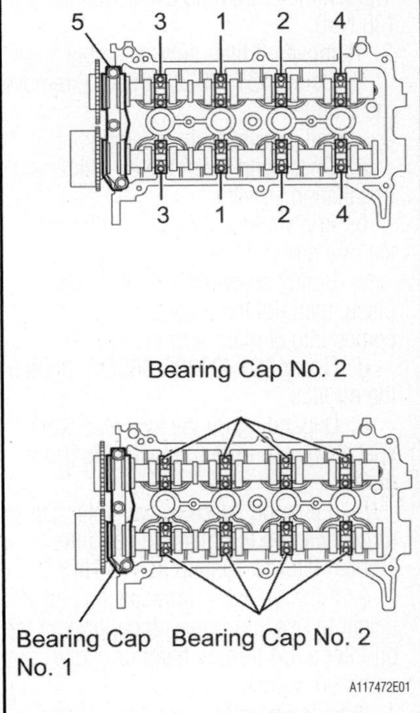

Fig. 100 Uniformly tighten the bolts in several steps in the sequence shown in the illustration and install camshaft bearing cap No. 1 and camshaft bearing cap No. 2

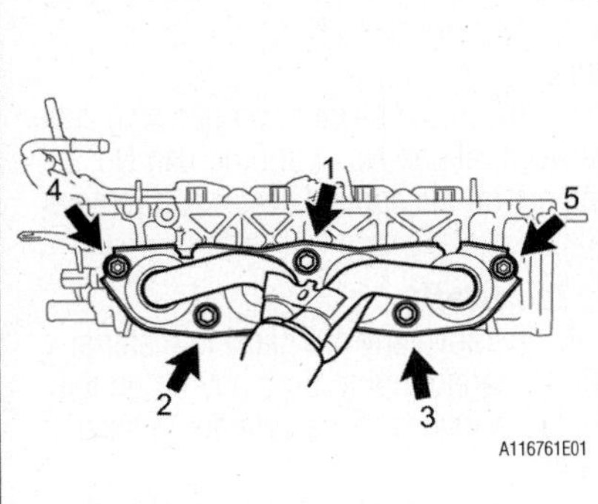

Fig. 96 Using several steps, install a new exhaust manifold gasket and the exhaust manifold with the 3 bolts and 2 nuts in the sequence shown.

a. Connect radiator hose No. 1 to the cylinder head.

138. Connect reserve tank hose
139. Connect radiator hose No. 3
140. Install air cleaner cap sub-assembly with air cleaner hose No. 1
141. Install battery tray
142. Install battery
143. Add engine oil
144. Add engine coolant
145. Check engine oil level
146. Check for engine oil leakage
147. Check for engine coolant leakage
148. Check for exhaust gas leakage
149. Check for fuel leakage
150. Install cylinder head cover No. 2
151. Install engine under cover RH
152. Install front wheel RH
153. Install cowl top panel outer.
154. Install cowl to register duct sub-assembly No. 2 (for hatchback)
155. Install front air shutter seal RH (for sedan)
156. Install front wiper motor and link.
157. Install cowl top ventilator louver LH (for hatchback)
158. Install cowl top ventilator louver sub-assembly.
159. Install cowl side ventilator sub-assembly LH (for sedan)

160. Install cowl side ventilator sub-assembly RH (for sedan)

161. Install hood to cowl top seal (for hatchback)

162. Install front wiper arm and blade assembly LH.

163. Install front wiper arm and blade assembly RH.

164. Install front wiper arm head cap.

ENGINE OIL & FILTER

REPLACEMENT

1. Drain engine oil
   a. Remove the oil pan drain plug and drain the engine oil.
   b. Clean the oil pan drain plug and install it with a new gasket. Tighten the drain plug to 28 inch lb. (38 Nm).
2. Remove oil filter sub-assembly
   a. Using SST (09228-06501), remove the oil filter.
3. Install oil filter sub-assembly
   a. Check and clean the oil filter installation surface.
   b. Apply clean engine oil to the gasket of a new oil filter.
   c. Gently screw the oil filter into place, then tighten it until the gasket comes into contact with the seat.
   d. Using SST (09228-06501), tighten the oil filter.
   e. Depending on the available workspace, choose from the following procedures.
   f. If enough space is available, use a torque wrench to tighten the oil filter. Tighten filter to 10 inch lb. (13 Nm).
   g. If the available workspace is insufficient to use a torque wrench, tighten the oil filter a 3/4 turn by hand or use a common wrench.
4. Add engine oil
   a. Fill with new engine oil (approx. 4 qt.).
5. Check for engine oil leaks
6. Check engine oil level

EXHAUST MANIFOLD

REMOVAL & INSTALLATION

See Figures 101 and 102.

1. Remove the 2 bolts and 2 compression springs and separate the exhaust pipe assembly front.
2. Remove the 3 bolts and remove the manifold support bracket.
3. Remove the exhaust manifold insulator with the 4 bolts.

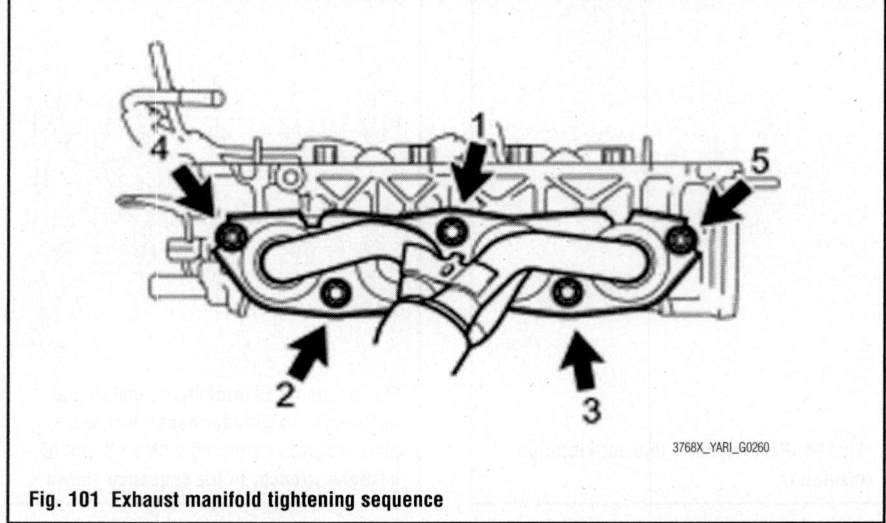

Fig. 101 Exhaust manifold tightening sequence

4. Remove the 2 bolts and 2 compression springs and separate the exhaust pipe assembly front.
5. Remove the 2 nuts and 3 bolts and the exhaust manifold and gasket.
6. To install, tighten the exhaust manifold nuts and bolts, in the order shown in the illustration, through a new gasket.
7. To complete installation, reverse remaining removal procedure.

INTAKE MANIFOLD

REMOVAL & INSTALLATION

See Figures 103 through 105.

1. Disconnect the negative battery cable.
2. Disconnect the union to connector tube hose from the booster vacuum tube.
3. Disconnect the engine wire from the intake manifold.
4. Disconnect the water by-pass hose from the cylinder head.
5. Disconnect the water by-pass hose from water bypass pipe No. 1.
6. Disconnect the throttle with motor body assembly connector.
7. Remove the 3 bolts and 2 nuts in the order shown in the illustration and remove the intake manifold.
8. Remove the gasket from the intake manifold.

To install:

9. Install a new gasket onto the intake manifold.
10. Provisionally tighten the intake manifold nuts and bolts in the order shown in the illustration, and then tighten them to the specified torque. Tighten to 22 ft. lbs. (30 Nm).
11. Connect the engine wire to the intake manifold.
12. Connect the water by-pass hose to water bypass pipe No. 1.
13. Connect the water by-pass hose to the cylinder head.
14. Connect the union to connector tube hose to the booster vacuum tube.
15. Connect the throttle with motor body assembly connector.

OIL PAN

REMOVAL & INSTALLATION

See Figure 105 and 106.

1. Remove the oil pan drain plug and gasket and drain the engine oil.
2. Remove the oil pan drain plug and gasket.
3. Remove the 9 bolts and 2 nuts.
4. Insert the blade of oil pan seal cutter between oil pan No. 1 and oil pan No. 2, and cut off the applied sealer and remove oil pan.

To install:

5. Remove any old packing material and be careful not to drop any oil on the contact surfaces of the cylinder block and oil pan.
6. Apply a continuous bead of seal packing (Diameter 4.0 mm (0.157 in.)). Seal packing: Toyota Genuine Seal Packing Black, Three Bond 1207B or equivalent.

➡Remove any oil from the contact surfaces.

- Install the oil pan within 3 minutes after applying seal packing.
- Do not start the engine for at least

2 hours after installing the oil pan.

7. Install the oil pan with the 9 bolts and 2 nuts. Tighten to 7 ft. lbs. (10 Nm).

OIL PUMP

REMOVAL & INSTALLATION

See Figures 107 through 109.

➡The oil pump is integrated with the timing chain cover.

1. Remove the timing chain cover.
2. Remove the 2 bolts and 3 screws and remove the oil pump cover.
3. Remove the oil pump rotor set.
4. Remove the oil pump relief valve plug, oil pump relief valve spring and oil pump relief valve.

To install:

5. Coat the oil pump relief valve with engine oil and insert the oil pump relief

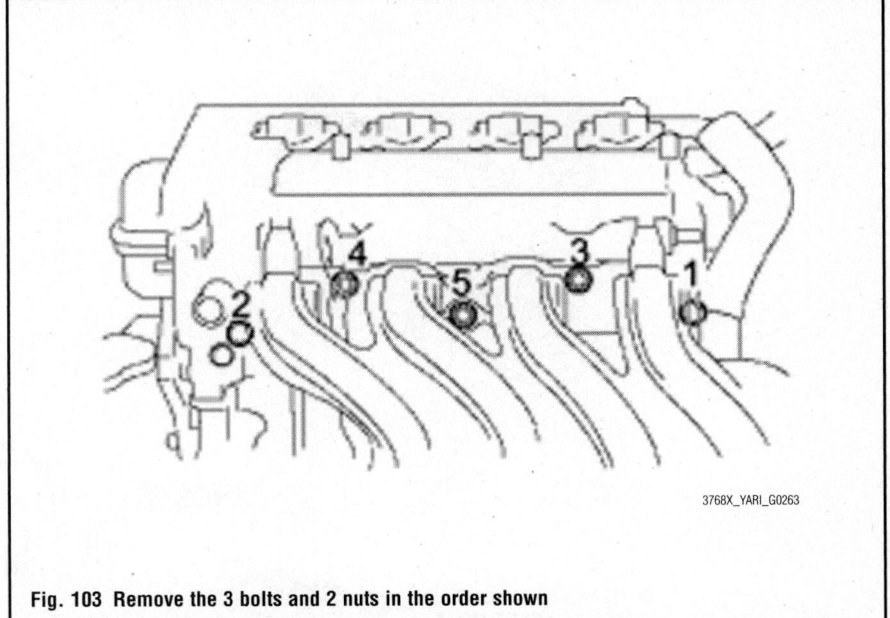

Fig. 103 Remove the 3 bolts and 2 nuts in the order shown

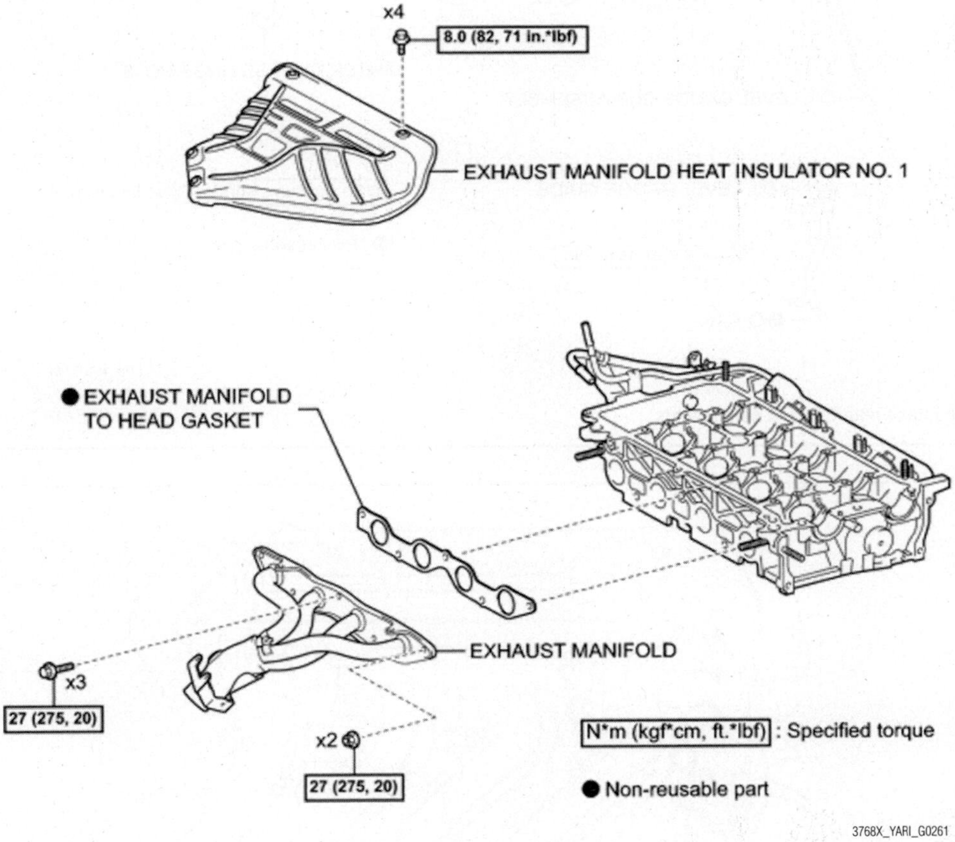

x4
8.0 (82, 71 in.*lbf)
EXHAUST MANIFOLD HEAT INSULATOR NO. 1
● EXHAUST MANIFOLD TO HEAD GASKET
EXHAUST MANIFOLD
x3
27 (275, 20)
x2
27 (275, 20)
N*m (kgf*cm, ft.*lbf) : Specified torque
● Non-reusable part
3768X_YARI_G0261

Fig. 102 View of the exhaust manifold and components

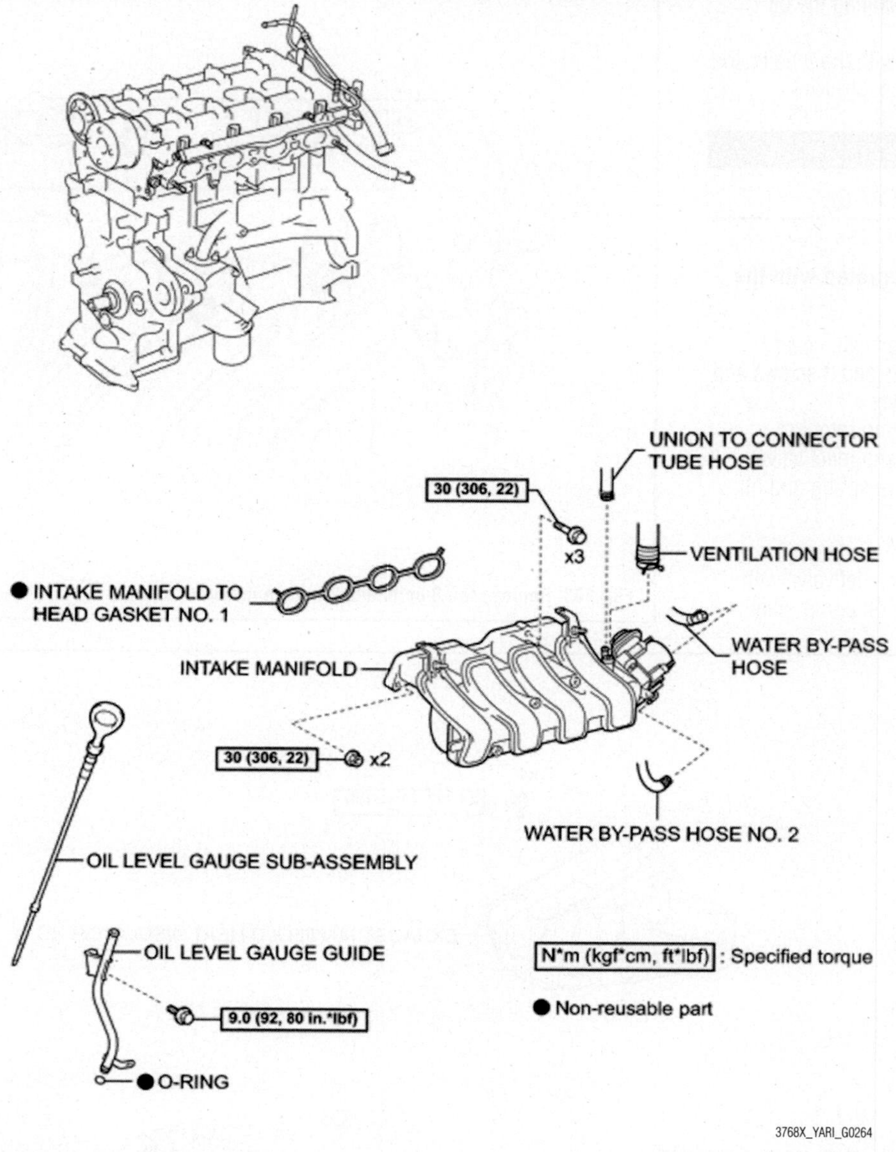

UNION TO CONNECTOR TUBE HOSE

30 (306, 22)

x3

VENTILATION HOSE

● INTAKE MANIFOLD TO HEAD GASKET NO. 1

WATER BY-PASS HOSE

INTAKE MANIFOLD

30 (306, 22) x2

WATER BY-PASS HOSE NO. 2

OIL LEVEL GAUGE SUB-ASSEMBLY

OIL LEVEL GAUGE GUIDE

N*m (kgf*cm, ft*lbf) : Specified torque

9.0 (92, 80 in.*lbf)

● Non-reusable part

● O-RING

3768X_YARI_G0264

Fig. 104 View of the intake manifold and components

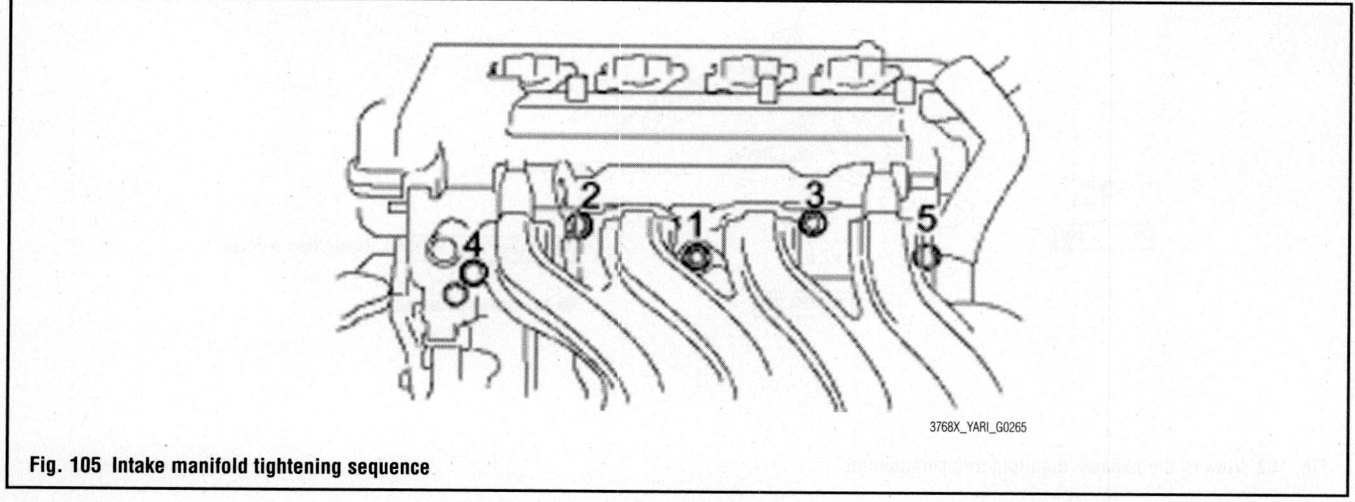

Fig. 105 Intake manifold tightening sequence

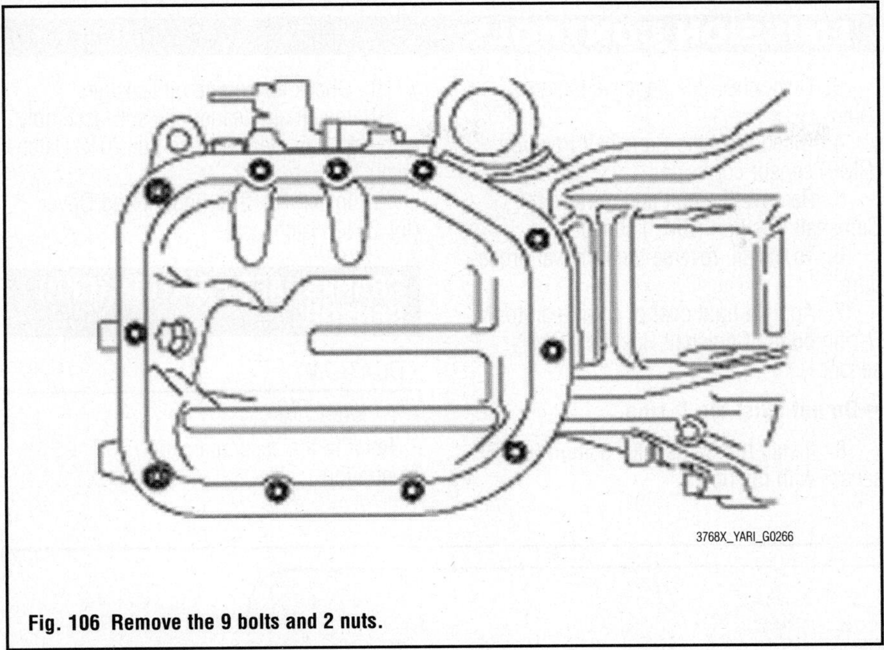
Fig. 106 Remove the 9 bolts and 2 nuts.

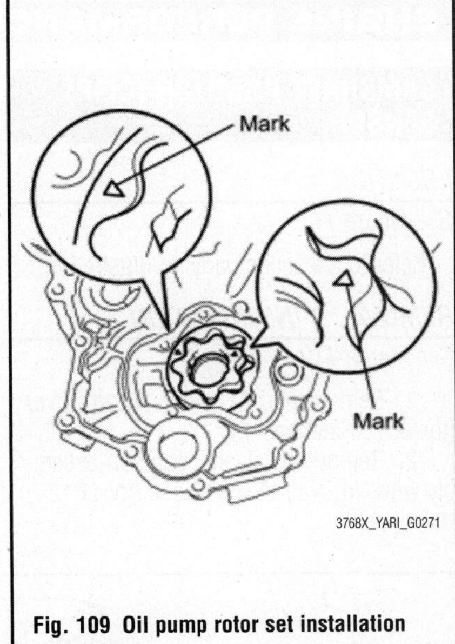

Fig. 109 Oil pump rotor set installation orientation

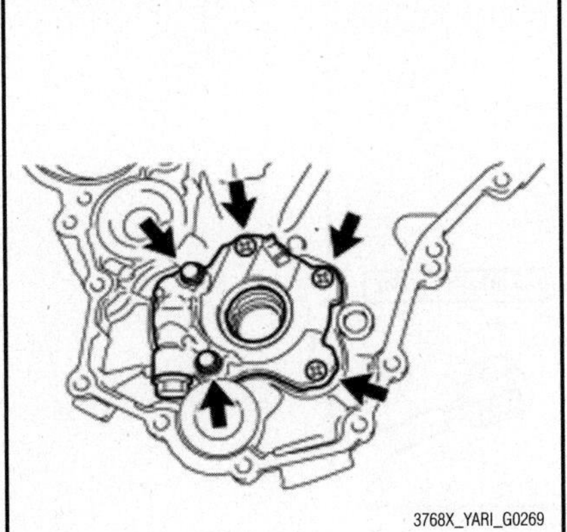
Fig. 107 Remove the 2 bolts and 3 screws and remove the oil pump cover

valve and oil pump relief valve spring into the oil pump cover.

6. Install the oil pump relief valve plug. Tighten to 18 ft. lbs. (25 Nm).

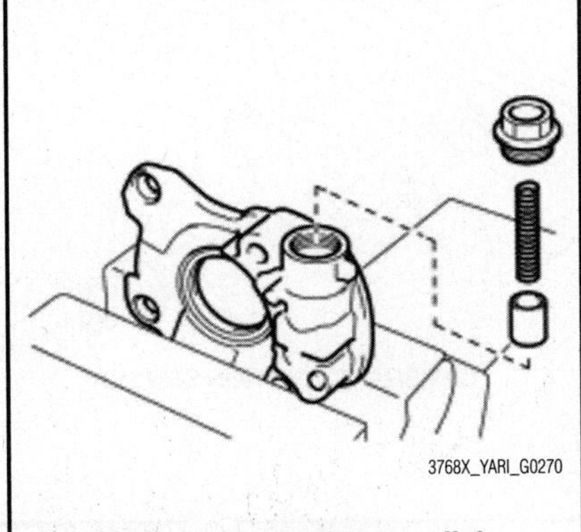
Fig. 108 Remove the oil pump relief valve plug, oil pump relief valve spring and oil pump relief valve

7. Coat the oil pump rotor set with engine oil and place it into the oil pump body with the marks facing the oil pump cover side.

8. Install the oil pump cover with the 2 bolts and 3 screws.

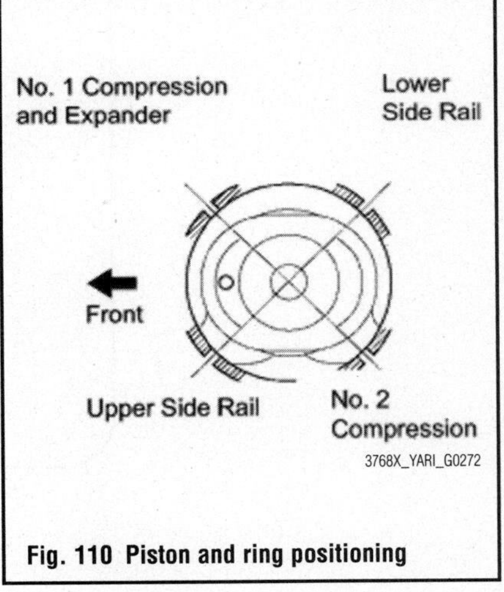

Fig. 110 Piston and ring positioning

a. Bolt: 78 inch lbs. (8.8 Nm).
b. Screw: 7.6 ft. lbs. (10 Nm).

PISTON AND RING

POSITIONING

See Figure 110.

ENGINE PERFORMANCE & EMISSION CONTROLS

CAMSHAFT POSITION (CMP) SENSOR

LOCATION

See Figure 111.

Refer to the accompanying illustration.

REMOVAL & INSTALLATION

See Figure 112.

1. Remove No. 2 Cylinder Head Cover (for 2012 Hatchback).
2. Remove air cleaner cap sub-assembly with No. 1 air cleaner hose (for 2012 Hatchback).
3. Disconnect the negative battery cable.
4. Disconnect the Camshaft Position (CMP) sensor connector.
5. Remove the bolt and remove the Camshaft Position (CMP) sensor.
6. To install, reverse the removal procedure.
7. Apply a light coat of engine oil to the O-ring on the Camshaft Position (CMP) sensor.

➡Do not twist the O-ring.

8. Install the Camshaft Position (CMP) sensor with the bolt.
9. Check for engine oil leakage.
10. Install air cleaner cap sub-assembly with No. 1 air cleaner hose (for 2012 Hatchback).
11. Install No. 2 Cylinder Head Cover (for 2012 Hatchback).

CRANKSHAFT POSITION (CKP) SENSOR

LOCATION

See Figure 113.

Refer to the accompanying illustration.

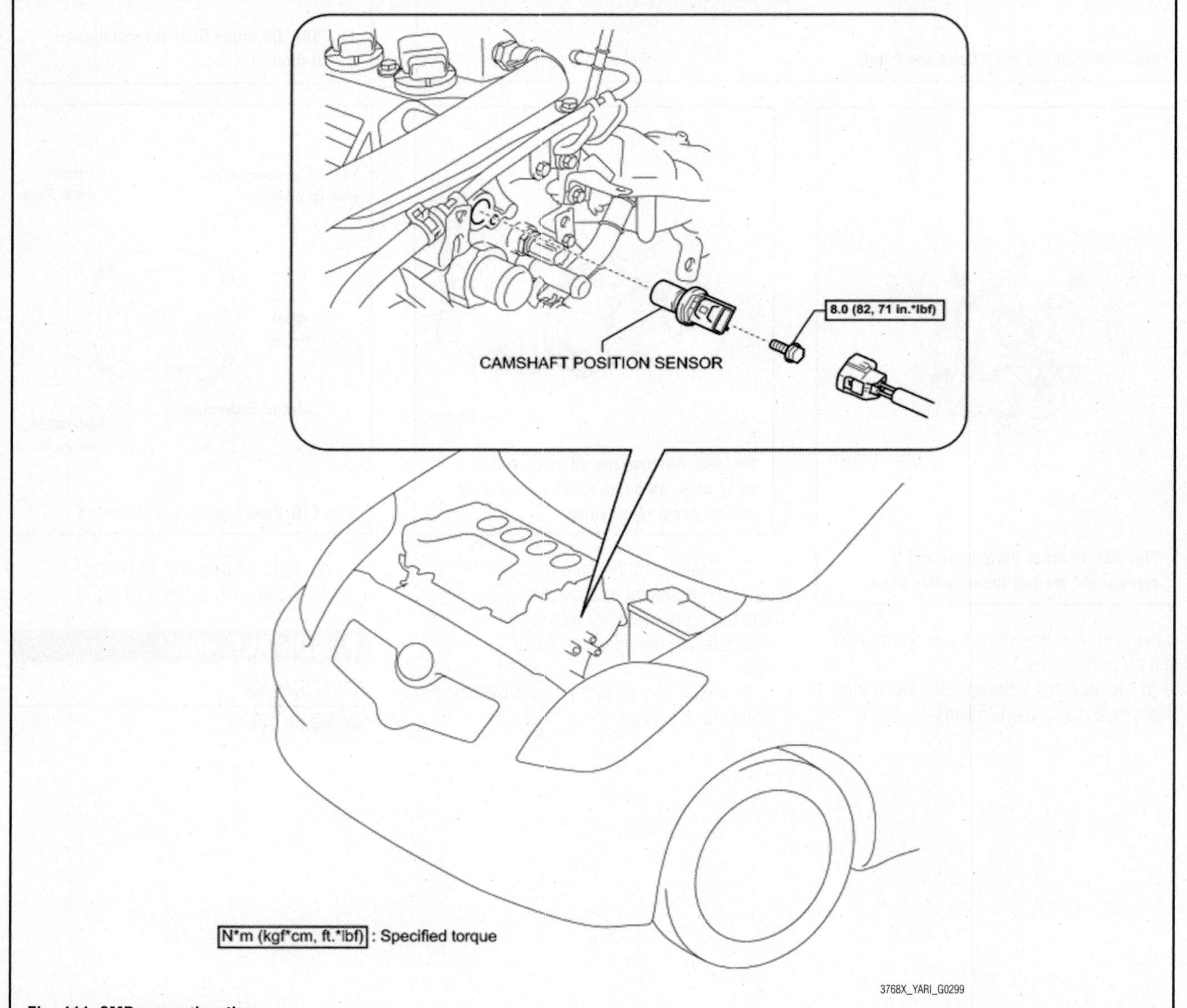

Fig. 111 CMP sensor location

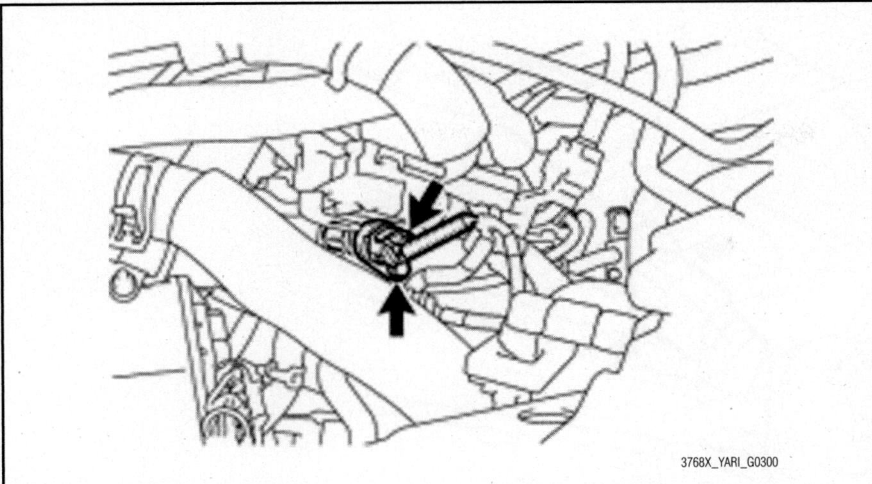

Fig. 112 Remove the bolt and remove the Camshaft Position (CMP) sensor

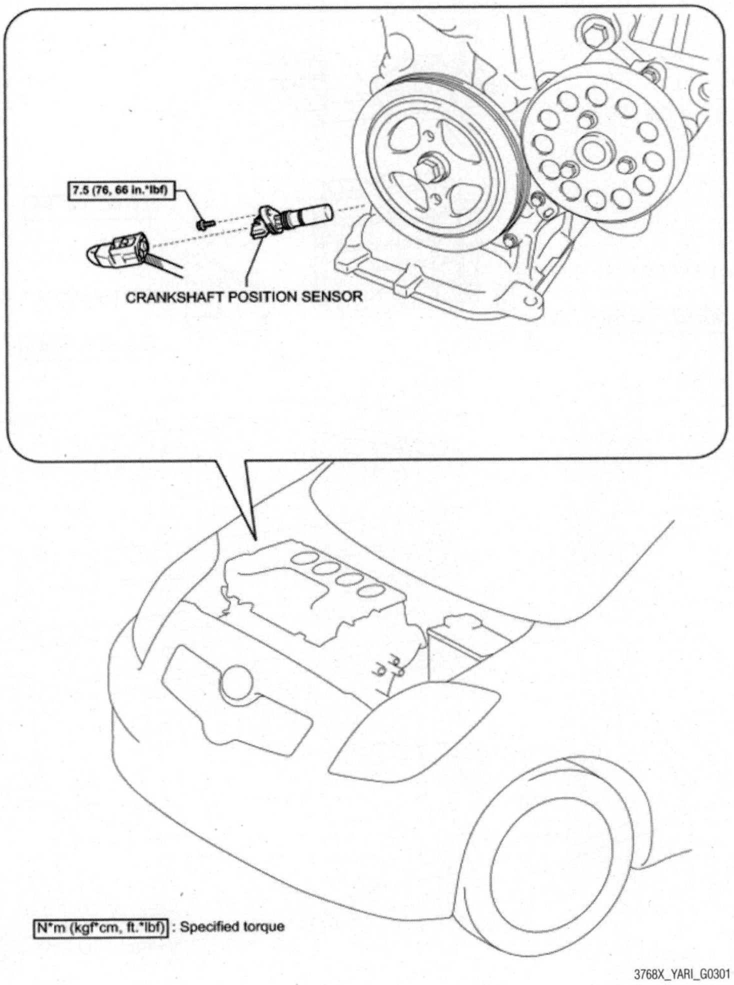

REMOVAL & INSTALLATION

See Figure 114.

1. Disconnect the negative battery cable.
2. Remove the RH engine under cover.
3. Disconnect the Crankshaft Position (CKP) sensor connector.
4. Remove the bolt and remove the Crankshaft Position (CKP) sensor.

ELECTRONIC CONTROL MODULE (ECM)

LOCATION

See Figure 115.

Refer to the accompanying illustration.

Fig. 113 CKP sensor location

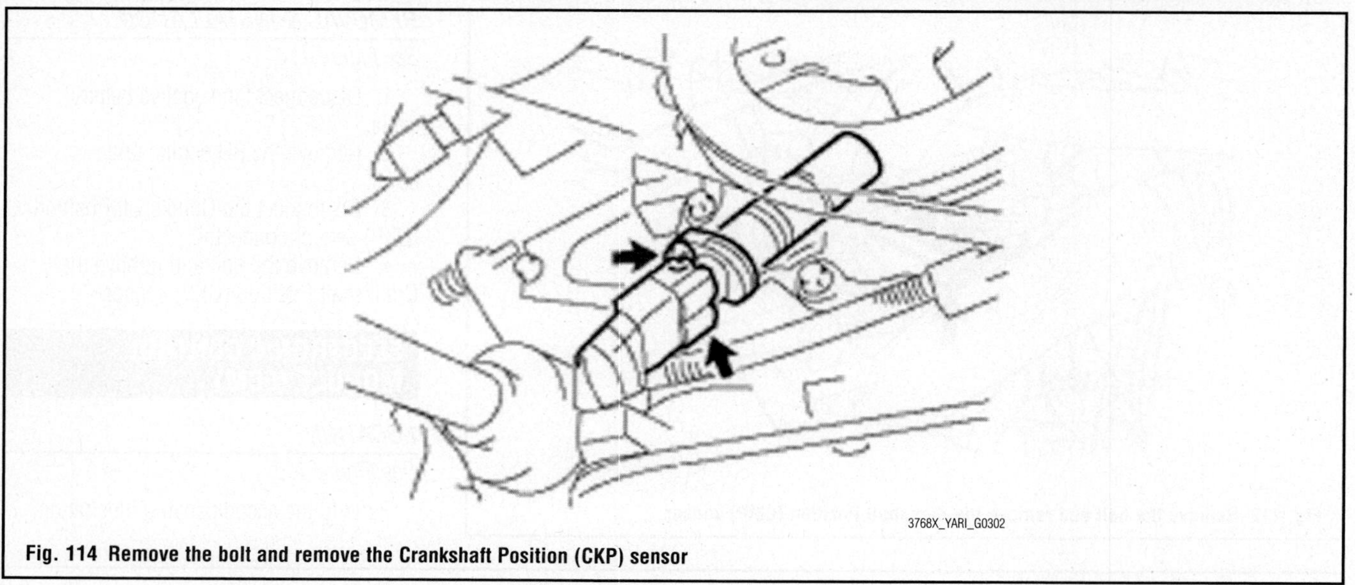

Fig. 114 Remove the bolt and remove the Crankshaft Position (CKP) sensor

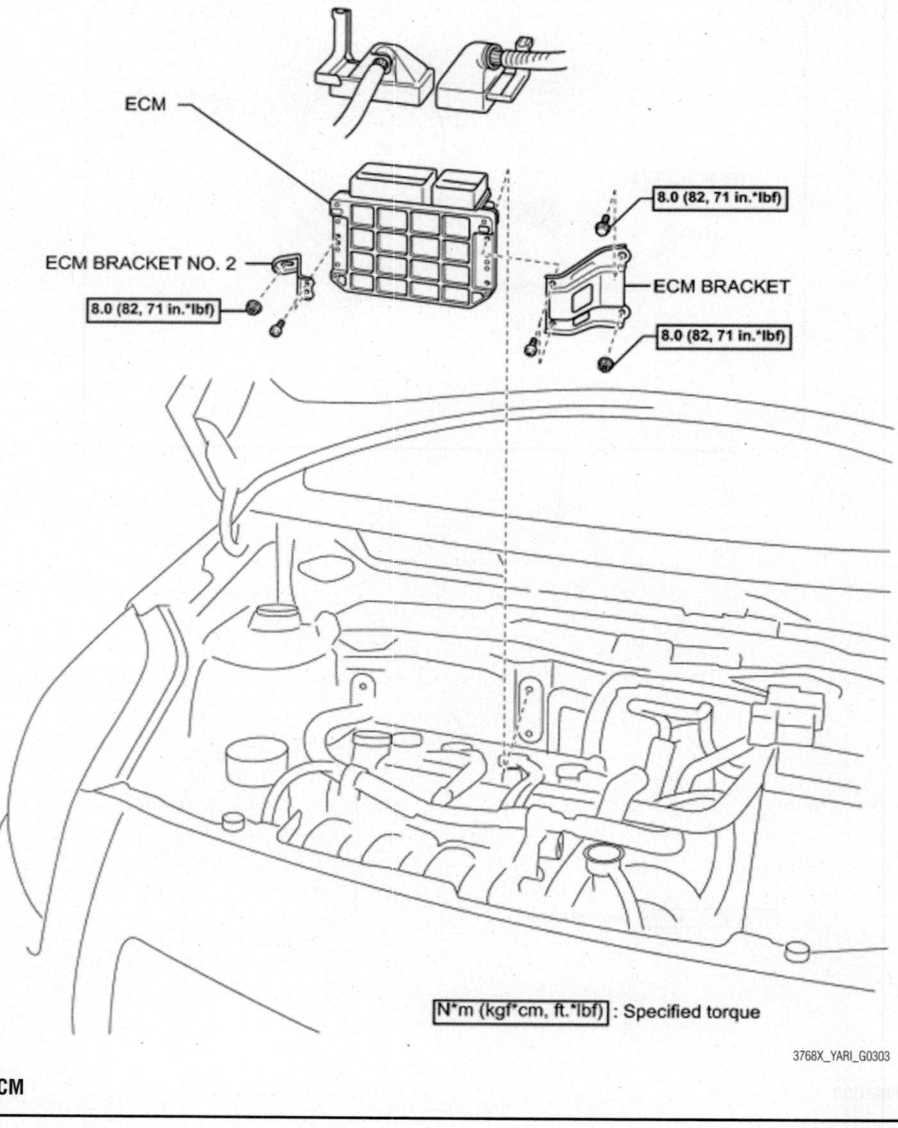

ECM

8.0 (82, 71 in.*lbf)

ECM BRACKET NO. 2

8.0 (82, 71 in.*lbf)

ECM BRACKET

8.0 (82, 71 in.*lbf)

N*m (kgf*cm, ft.*lbf) : Specified torque

3768X_YARI_G0303

Fig. 115 Locating the ECM

REMOVAL & INSTALLATION

See Figures 116 and 117.

1. Remove the wiper motor and link assembly
2. Remove front air shutter seal (for Sedan). Disengage the 3 claws and remove the front air shutter seal.
3. Remove cowl top panel outer (for hatchback):
   a. Disengage the wire harness clamp.
   b. Remove the 9 bolts and remove the cowl top panel outer.
4. Remove cowl top panel outer (for sedan):
   a. Disengage the claw and disconnect the wire harness.
   b. Remove the 2 bolts and remove the cowl top panel outer center bracket.
   c. Remove the 8 bolts and remove the cowl top panel outer.
5. Remove ECM:
   a. Remove the 2 lock knobs and harness clamp.
   b. Disconnect the 2 ECM connectors.
   c. Remove the bolt and 2 nuts and remove the ECM.
6. To install, reverse the removal procedure.
7. Perform initialization.

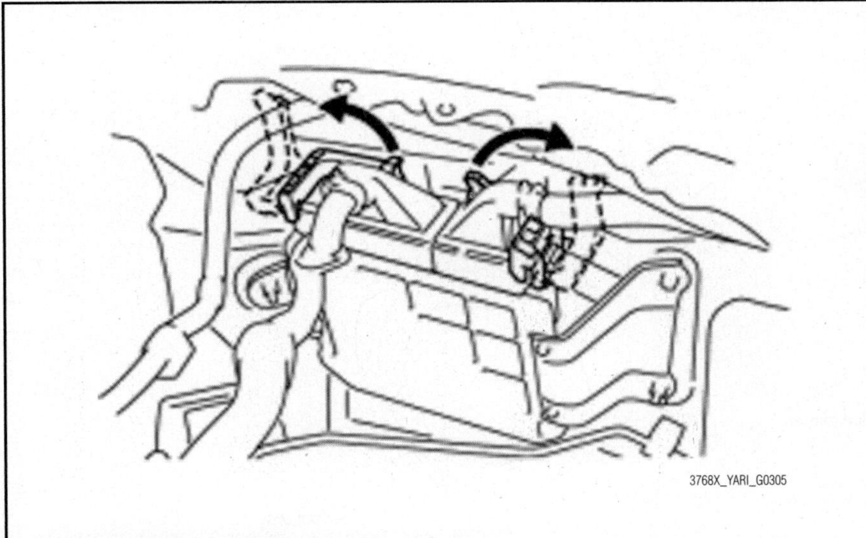

Fig. 116 Remove the 2 lock knobs and harness clamp

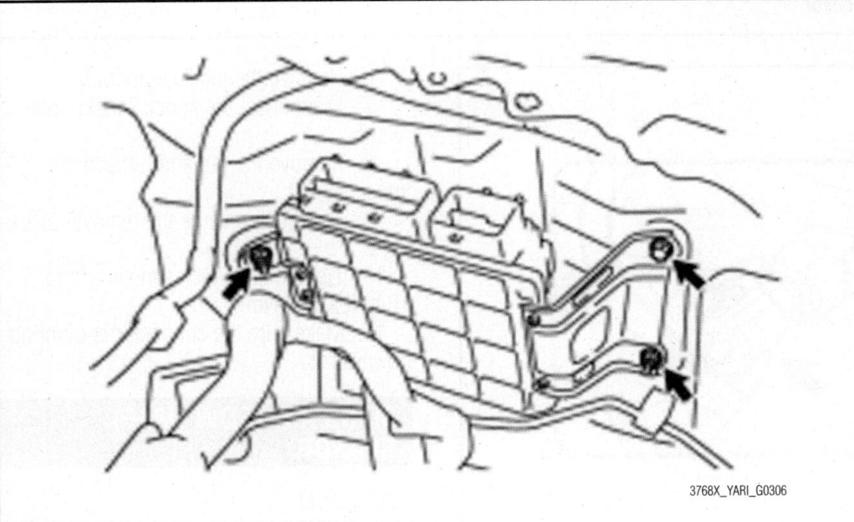

Fig. 117 Remove the bolt and 2 nuts and remove the ECM

ENGINE COOLANT TEMPERATURE (ECT) SENSOR

LOCATION

See Figure 118.

Refer to the accompanying illustration.

REMOVAL & INSTALLATION

See Figure 119.

1. Disconnect the negative battery cable.
2. Drain and recycle the engine coolant.
3. Disconnect the engine coolant temperature sensor connector.
4. Using SST 09817-33190, remove the engine coolant temperature sensor.

To install:

5. Provisionally install the engine coolant temperature sensor through a new gasket.
6. Using SST, tighten the engine coolant temperature sensor. Tighten to 15 ft. lbs. (20 Nm).
7. Connect the engine coolant temperature sensor connector.
8. Connect the negative battery cable.
9. Add engine coolant.
10. Check for engine coolant leakage.

HEATED OXYGEN (HO2S) SENSOR

LOCATION

See Figure 120.

Refer to the accompanying illustration.

REMOVAL & INSTALLATION

See Figure 121.

1. Disconnect the negative battery cable.
2. Remove the center console.
3. Disconnect the sensor connector.
4. Remove the grommet and pass the sensor connector out of the cabin through the floor panel.
5. Using SST 09224-00010, remove heated oxygen sensor No. 2.

KNOCK SENSOR (KS)

LOCATION

See Figure 122.

Refer to the accompanying illustration.

REMOVAL & INSTALLATION

See Figure 123.

1. Disconnect the negative battery cable.
2. Drain and recycle the engine coolant.
3. Remove the engine cover.
4. Remove the air cleaner assembly.
5. Remove the throttle body.

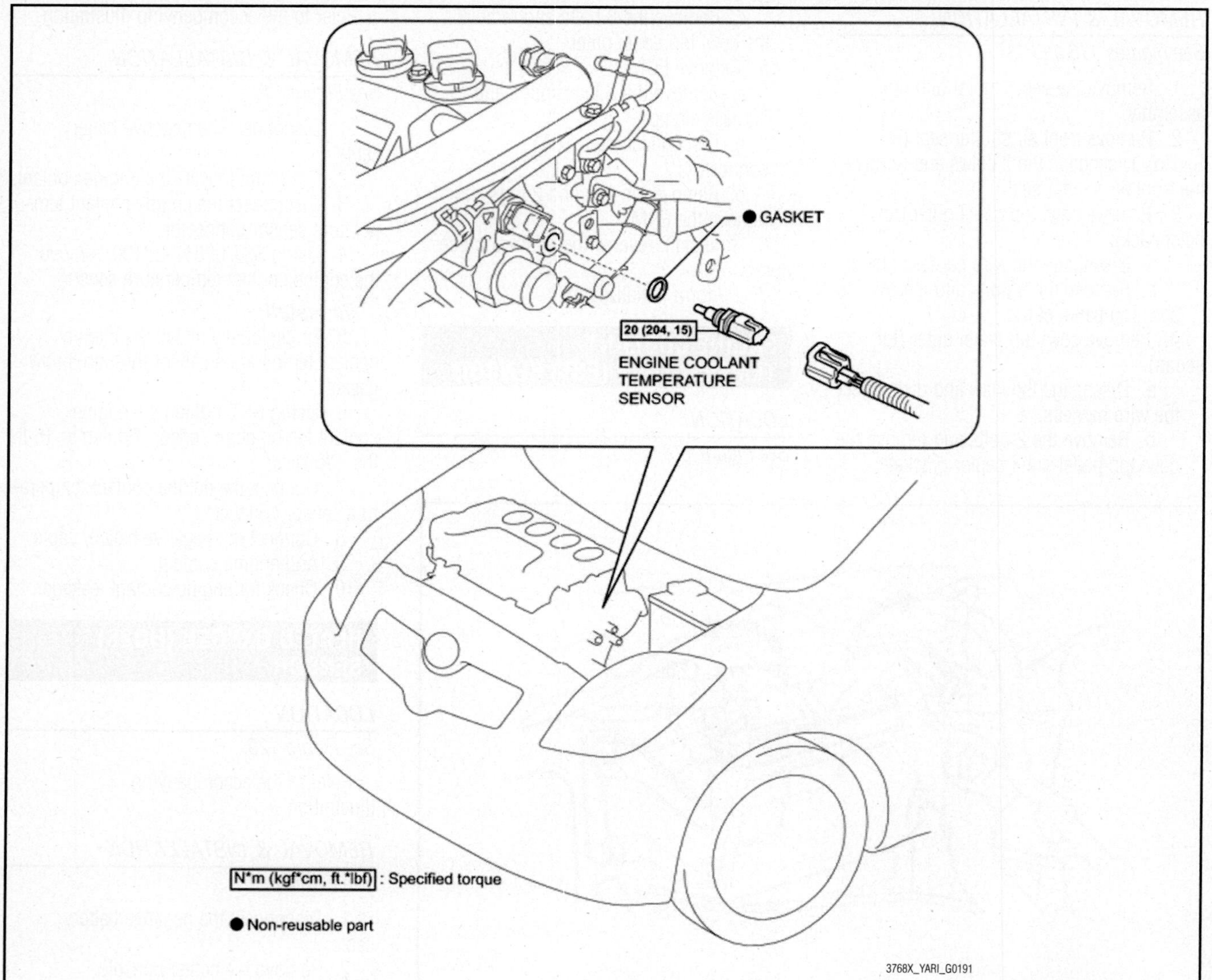

● GASKET

20 (204, 15)

ENGINE COOLANT TEMPERATURE SENSOR

N*m (kgf*cm, ft.*lbf) : Specified torque

● Non-reusable part

3768X_YARI_G0191

Fig. 118 Locating the Engine Coolant Temperature Sensor

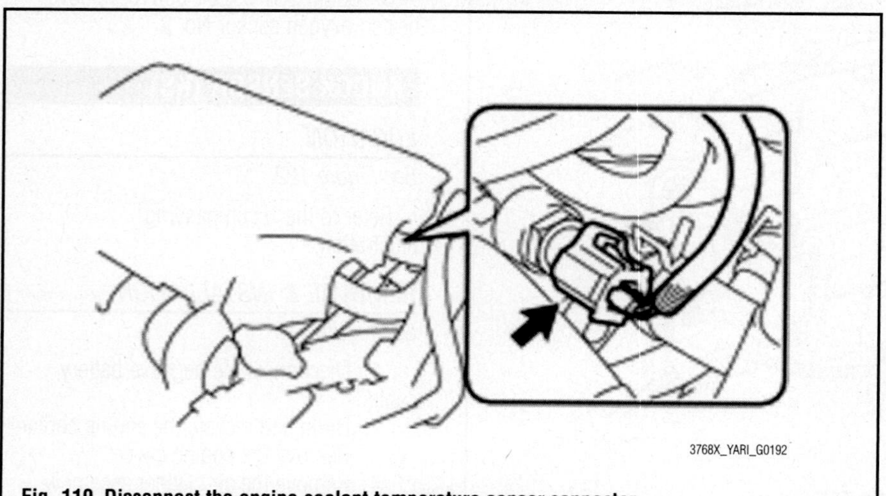

Fig. 119 Disconnect the engine coolant temperature sensor connector

6. Remove the intake manifold.
7. Disconnect the knock sensor connector.
8. Remove the nut and remove the knock sensor.
9. To install, reverse the removal procedure.
10. Tighten the knock sensor bolt to 15 ft. lbs. (20 Nm).
11. Make sure the connector is pointing downwards.

MASS AIR FLOW (MAF) SENSOR

LOCATION

See Figure 124.

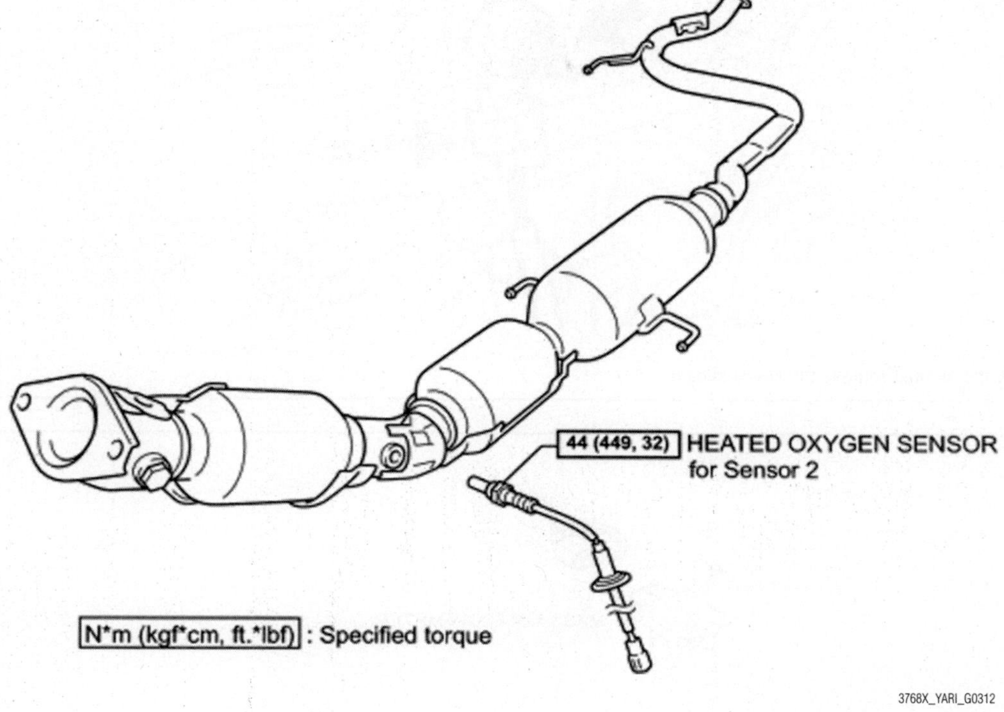

44 (449, 32) HEATED OXYGEN SENSOR for Sensor 2

N*m (kgf*cm, ft.*lbf) : Specified torque

3768X_YARI_G0312

Fig. 120 Locating the HO2S sensor

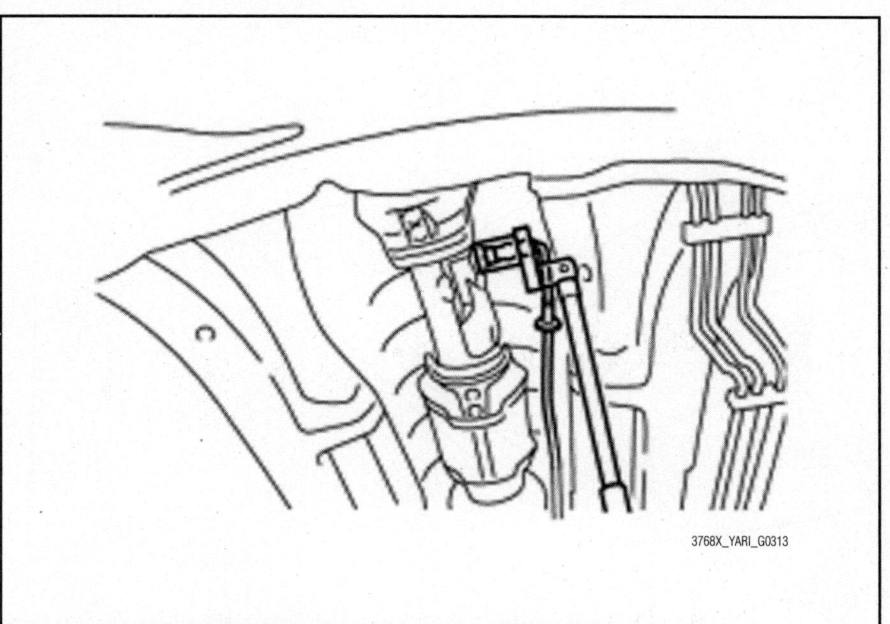

Fig. 121 Using SST 09224-00010, remove heated oxygen sensor No. 2

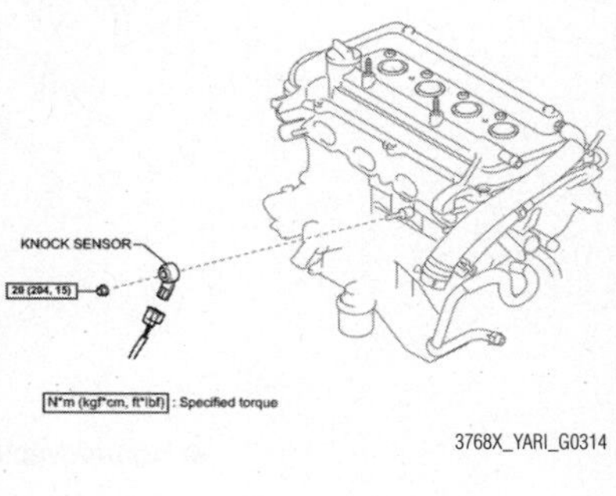

KNOCK SENSOR

20 (204, 15)

N*m (kgf*cm, ft*lbf) : Specified torque

3768X_YARI_G0314

Fig. 122 Locating the knock sensor

Refer to the accompanying illustration.

REMOVAL & INSTALLATION

See Figure 124

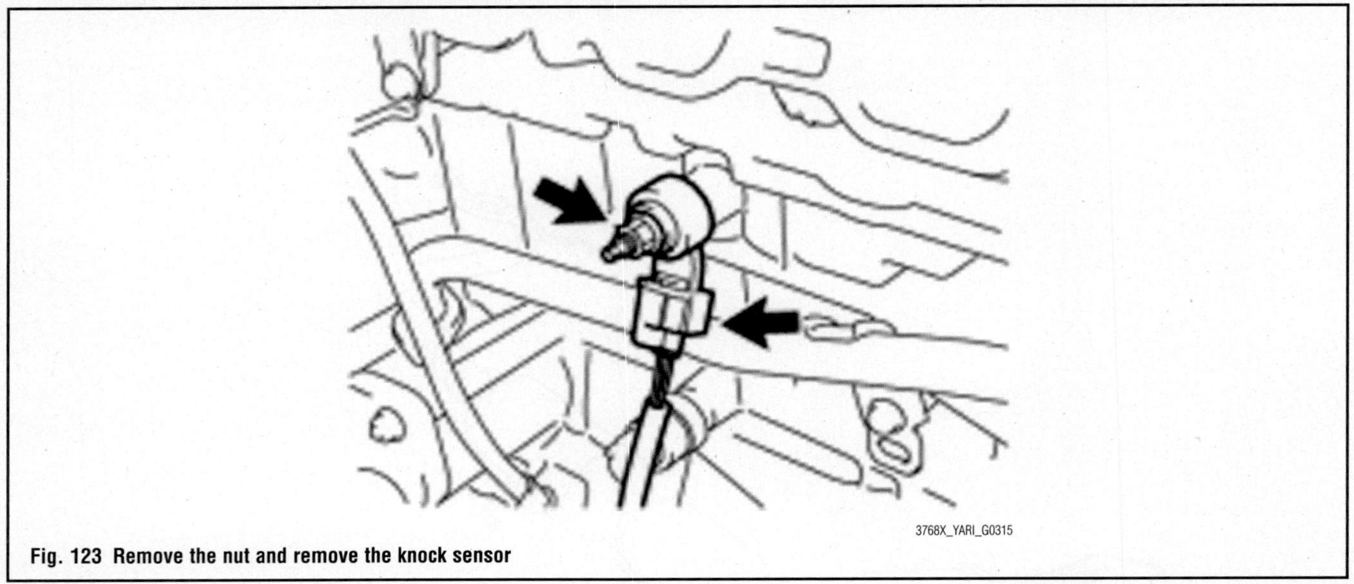

Fig. 123 Remove the nut and remove the knock sensor

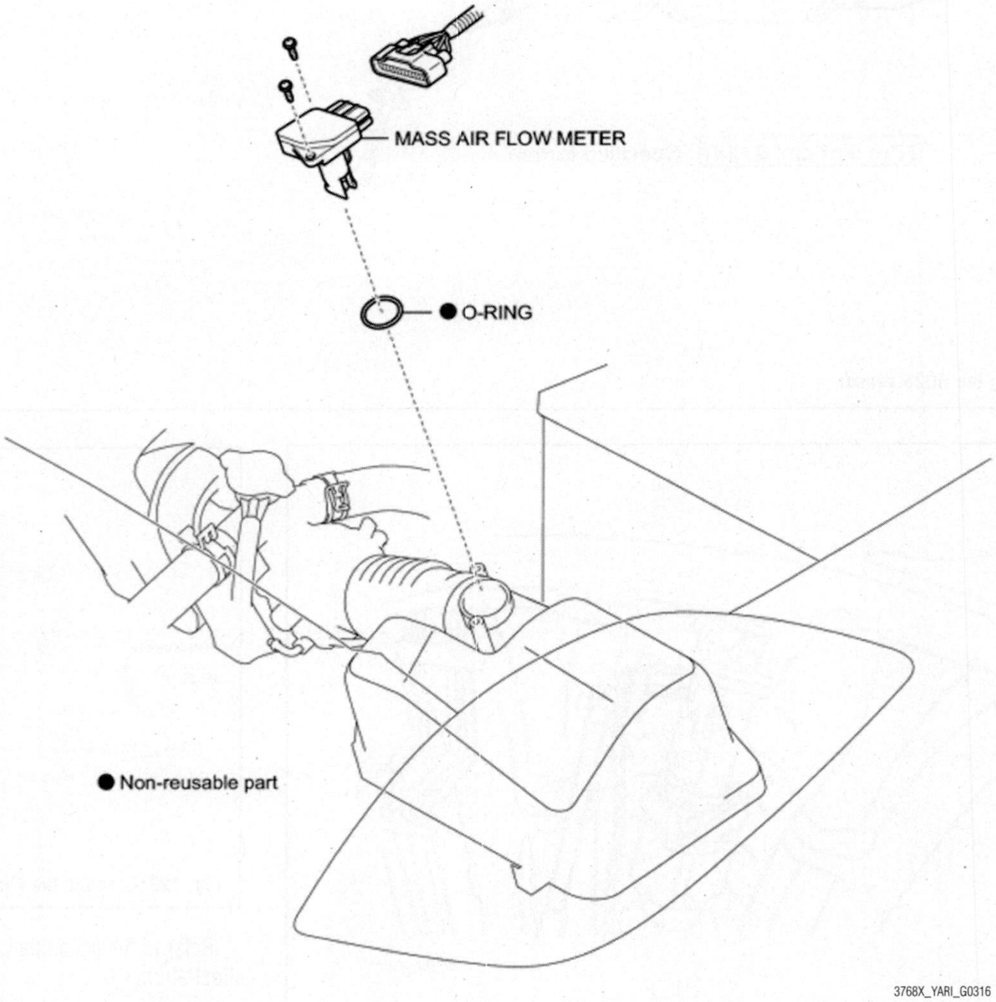

MASS AIR FLOW METER

● O-RING

● Non-reusable part

3768X_YARI_G0316

Fig. 124 Locating the MAF meter

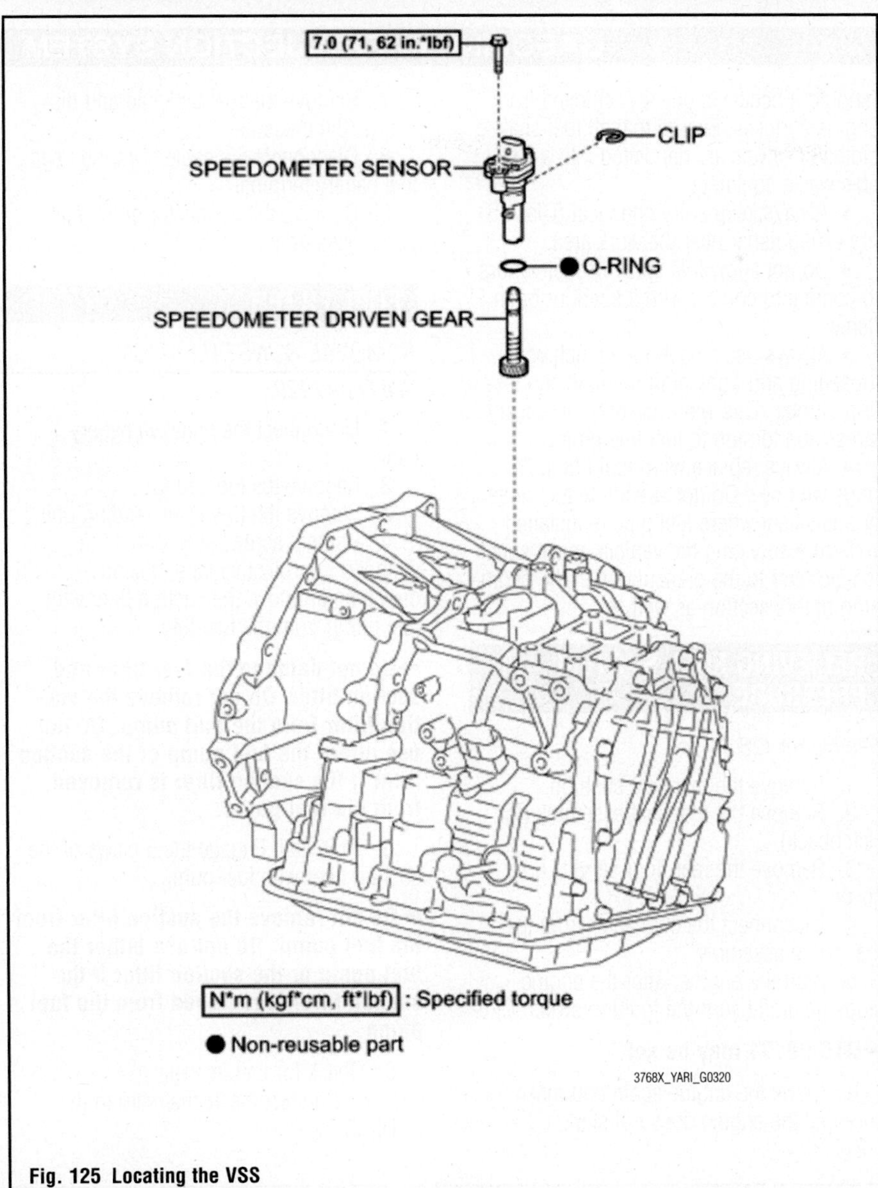

Fig. 125 Locating the VSS

1. Disconnect the negative battery cable.
2. Disconnect the wire harness clamp and Mass Air Flow (MAF) meter connector.
3. Remove the 2 screws and the Mass Air Flow (MAF) meter.
4. Remove the O-ring from the Mass Air Flow (MAF) meter.
5. To install, reverse the removal procedure.
6. Install a new O-ring on to the MAF sensor.

VEHICLE SPEED SENSOR (VSS)

LOCATION

See Figure 125.

Refer to the accompanying illustration.

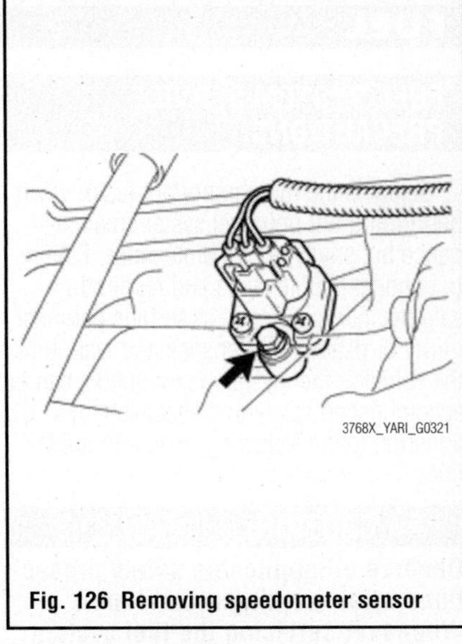

Fig. 126 Removing speedometer sensor

REMOVAL & INSTALLATION

See Figures 126 and 127.

1. Remove the wiper motor and link assembly.
2. Remove the batter tray assembly.
3. Disconnect speedometer sensor connector.
4. Remove the bolt and speedometer sensor.
5. Remove speedometer driven gear:
   a. Remove the clip and driven gear from the speedometer sensor.
   b. Remove the O-ring from the speedometer sensor.
6. To install, reverse the removal procedure.
7. Coat a new O-ring with Toyota Genuine ATF WS or equivalent and install it onto the speedometer sensor.

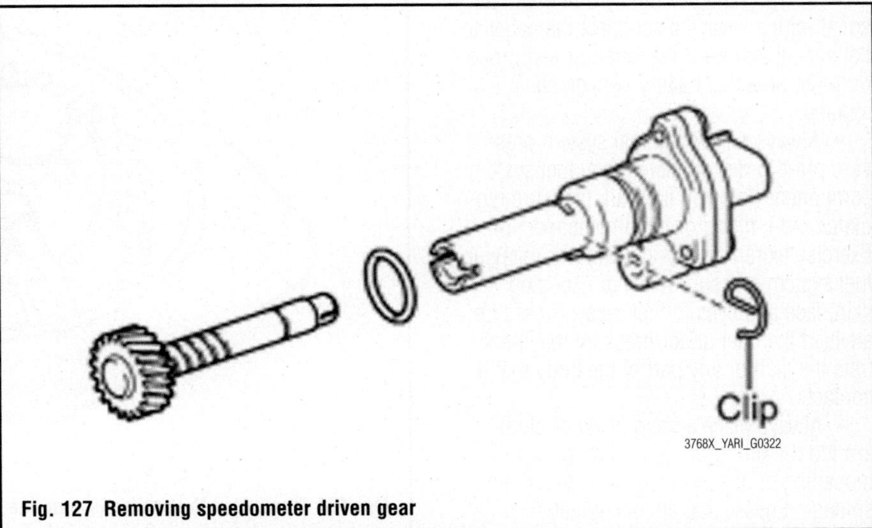

Fig. 127 Removing speedometer driven gear

FUEL SYSTEM SERVICE PRECAUTIONS

Safety is the most important factor when performing not only fuel system maintenance but any type of maintenance. Failure to conduct maintenance and repairs in a safe manner may result in serious personal injury or death. Maintenance and testing of the vehicle's fuel system components can be accomplished safely and effectively by adhering to the following rules and guidelines.

CAUTION

Observe all applicable safety precautions when working around fuel. Whenever servicing the fuel system, always work in a well ventilated area. Do not allow fuel spray or vapors to come in contact with a spark or open flame. Keep a dry chemical fire extinguisher near the work area. Always keep fuel in a container specifically designed for fuel storage; also, always properly seal fuel containers to avoid the possibility of fire or explosion.

CAUTION

Before removing any fuel system parts, take precautions to prevent gasoline spillage. As some pressure remains in the fuel line even after taking precautions to prevent gasoline spillage, use a shop rag or piece of cloth to prevent gasoline splashes when disconnecting the fuel line.

- To avoid the possibility of fire and personal injury, always disconnect the negative battery cable unless the repair or test procedure requires that battery voltage be applied.
- Always relieve the fuel system pressure prior to disconnecting any fuel system component (injector, fuel rail, pressure regulator, etc.), fitting or fuel line connection. Exercise extreme caution whenever relieving fuel system pressure to avoid exposing skin, face and eyes to fuel spray. Please be advised that fuel under pressure may penetrate the skin or any part of the body that it contacts.
- Always place a shop towel or cloth around the fitting or connection prior to loosening to absorb any excess fuel due to spillage. Ensure that all fuel spillage (should it occur) is quickly removed from engine surfaces. Ensure that all fuel soaked cloths or towels are deposited into a suitable waste container.
- Always keep a dry chemical (Class B) fire extinguisher near the work area.
- Do not allow fuel spray or fuel vapors to come into contact with a spark or open flame.
- Always use a back-up wrench when loosening and tightening fuel line connection fittings. This will prevent unnecessary stress and torsion to fuel line piping.
- Always replace worn fuel fitting O-rings with new Do not substitute fuel hose or equivalent where fuel pipe is installed.

Before servicing the vehicle, make sure to also refer to the precautions in the beginning of this section as well.

RELIEVING FUEL SYSTEM PRESSURE

See Figure 128.

1. Remove the rear seat cushion.
2. Remove the rear seat assembly (for Hatchback)
3. Remove the rear floor service hole cover.
4. Disconnect the connector from the fuel pump assembly.
5. Start the engine. After the engine stops naturally, turn the ignition switch OFF.

➡DTC P0171 may be set.

6. Crank the engine again and make sure that the engine does not start.
7. Remove the fuel tank cap and discharge the pressure.
8. Disconnect the cable from the negative battery terminal.
9. Connect the connector of the fuel pump assembly.

FUEL FILTER

REMOVAL & INSTALLATION

See Figure 129.

1. Disconnect the negative battery cable.
2. Remove the fuel pump.
3. Remove the fuel level sending unit.
4. Using a screwdriver with its tip wrapped in protective tape, disengage the 5 claws, and remove the suction filter with fuel pump from the fuel filter.

➡Do not damage the fuel filter and suction filter. Do not remove the suction filter from the fuel pump. Do not use either the fuel pump or the suction filter if the suction filter is removed from the fuel pump.

5. To install, Engage the 5 claws of the suction filter with fuel pump.

➡Do not remove the suction filter from the fuel pump. Do not use either the fuel pump or the suction filter if the suction filter is removed from the fuel pump.

6. Check for fuel leakage:
   a. Connect the Techstream to the DLC3.

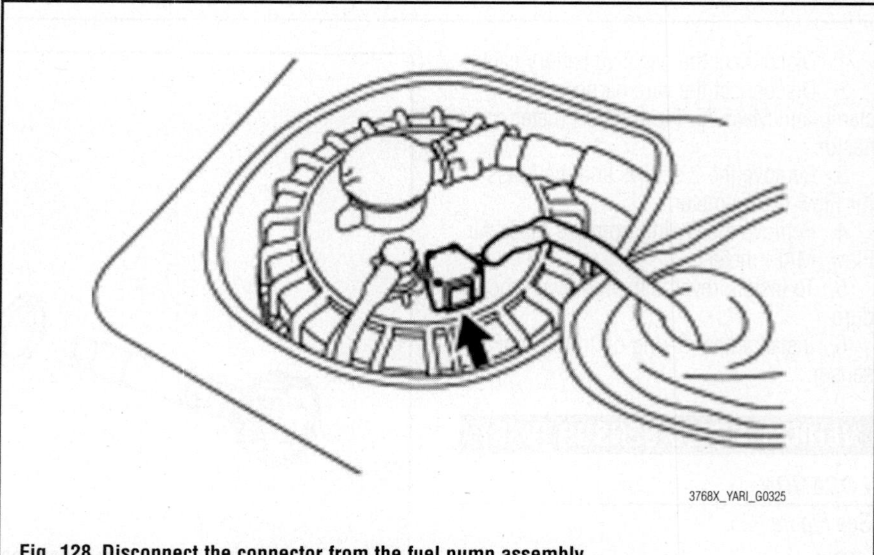

Fig. 128 Disconnect the connector from the fuel pump assembly

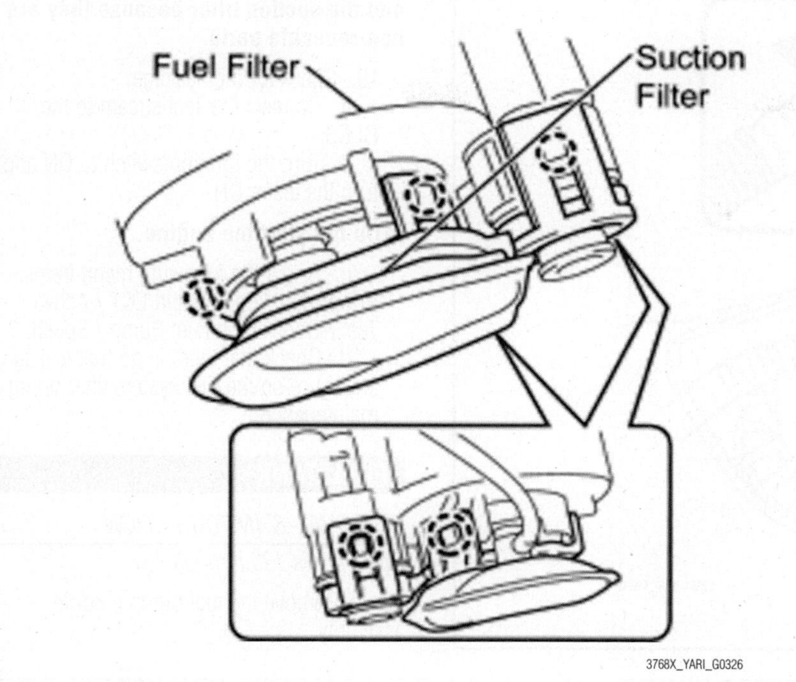

Fig. 129 Identifying the fuel filer and related components

b. Turn the ignition switch to ON and turn the tester ON.

➡**Do not start the engine.**

c. Select the following menu items: Powertrain / Engine and ECT / Active Test / Control the Fuel Pump / Speed.

d. Check that there is no fuel leakage anywhere on the fuel system after doing maintenance.

FUEL PUMP & MODULE

REMOVAL & INSTALLATION

See Figures 130 through 134.

1. Remove the rear seats and deck board.
2. Remove rear floor service hole cover:
   a. Remove the rear floor service hole cover.
   b. Disconnect the fuel pump connector.
3. Properly relieve the fuel system pressure.
4. Disconnect the negative battery cable.
5. Disconnect fuel tank main tube sub-assembly:
   a. Widen the tip of the tube joint clip and pull out the clip in the direction indicated by the arrow.
   b. Disconnect the fuel tank main tube.

➡**Please note the following:**

- Keep the O-ring free of any foreign matter, as it becomes contaminated easily.
- Do not use any tools in this procedure.
- Do not forcefully bend or twist the tube.
- Put the tube in a plastic bag to prevent damage and contamination.
- If the fuel suction plate and tube are stuck together, pinch the tube and turn it carefully to disconnect them.
- Do not damage any clips. If a clip is damaged, replace it.

6. Disconnect the fuel tank vent hose.
7. Using SST 09808-14020, remove the fuel pump gauge retainer.

➡**Align the claws of the fuel pump gauge retainer with the tips of SST.**

8. Remove the fuel suction with pump and gauge tube. Do not bend the arm of the sender gauge.
9. Remove the fuel suction tube gasket from the fuel tank.
10. Disconnect the connector of the fuel pump harness.
11. Disconnect the fuel tube.
12. Using a screwdriver with its tip wrapped in protective tape, disengage the 2 claws and remove the fuel filter and fuel pump from the fuel sub-tank.
13. Disengage the 2 claws and remove the fuel suction support from the fuel filter.
14. Using a screwdriver with its tip wrapped in protective tape, disengage the 5 claws, and remove the suction filter with fuel pump from the fuel filter.
15. Disconnect the fuel pump harness.
16. Remove the O-ring and fuel pump spacer.
17. To install, reverse the removal procedure.
18. Apply gasoline to a new O-ring, and then install the fuel pump spacer and new O-ring onto the fuel pump.

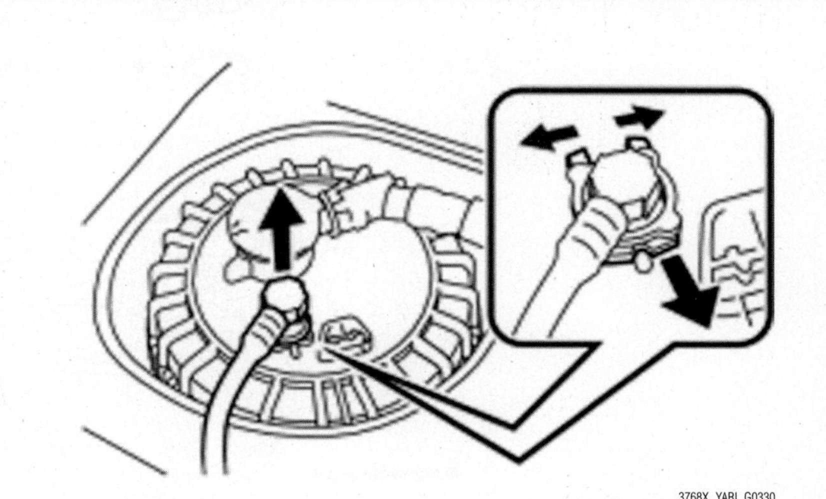

Fig. 130 Disconnect fuel tank main tube sub-assembly

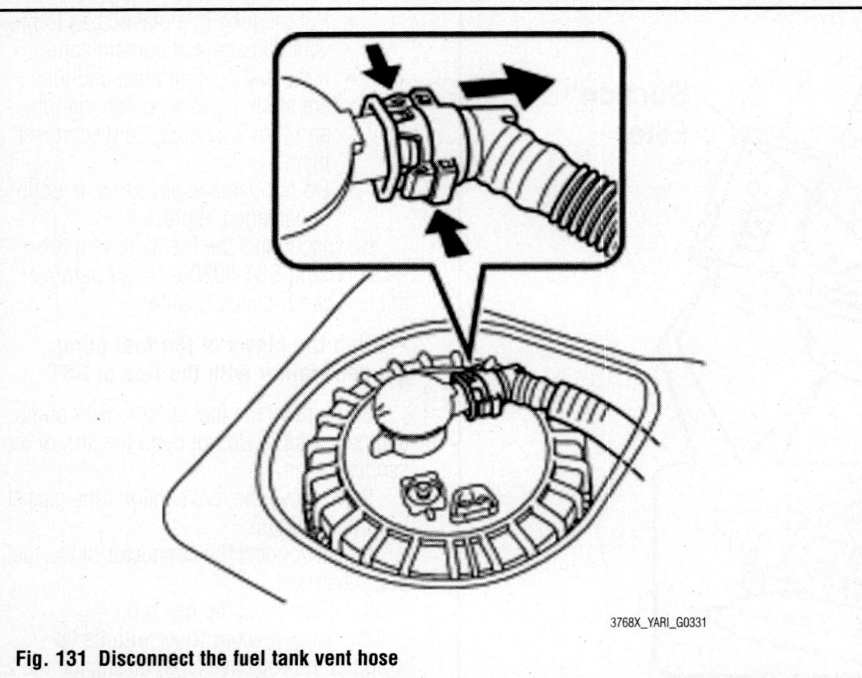

Fig. 131 Disconnect the fuel tank vent hose

➡**Do not disassemble the fuel pump and the suction filter because they are non-reusable parts.**

19. Check for fuel leakage:
    a. Connect the Techstream to the DLC3.
    b. Turn the ignition switch to ON and turn the tester ON.

➡**Do not start the engine.**

c. Select the following menu items: Powertrain / Engine and ECT / Active Test / Control the Fuel Pump / Speed.
d. Check that there is no fuel leakage anywhere on the fuel system after doing maintenance.

FUEL TANK

REMOVAL & INSTALLATION

See Figures 135 through 137.

1. Remove the fuel pump module assembly.

REAR FLOOR SERVICE HOLE COVER
FUEL PUMP GAUGE RETAINER
FUEL TANK MAIN TUBE SUB-ASSEMBLY
FUEL TANK VENT HOSE SUB-ASSEMBLY
FUEL SUCTION WITH PUMP AND GAUGE TUBE ASSEMBLY
● FUEL SUCTION TUBE GASKET
● Non-reusable part
3768X_YARI_G0333

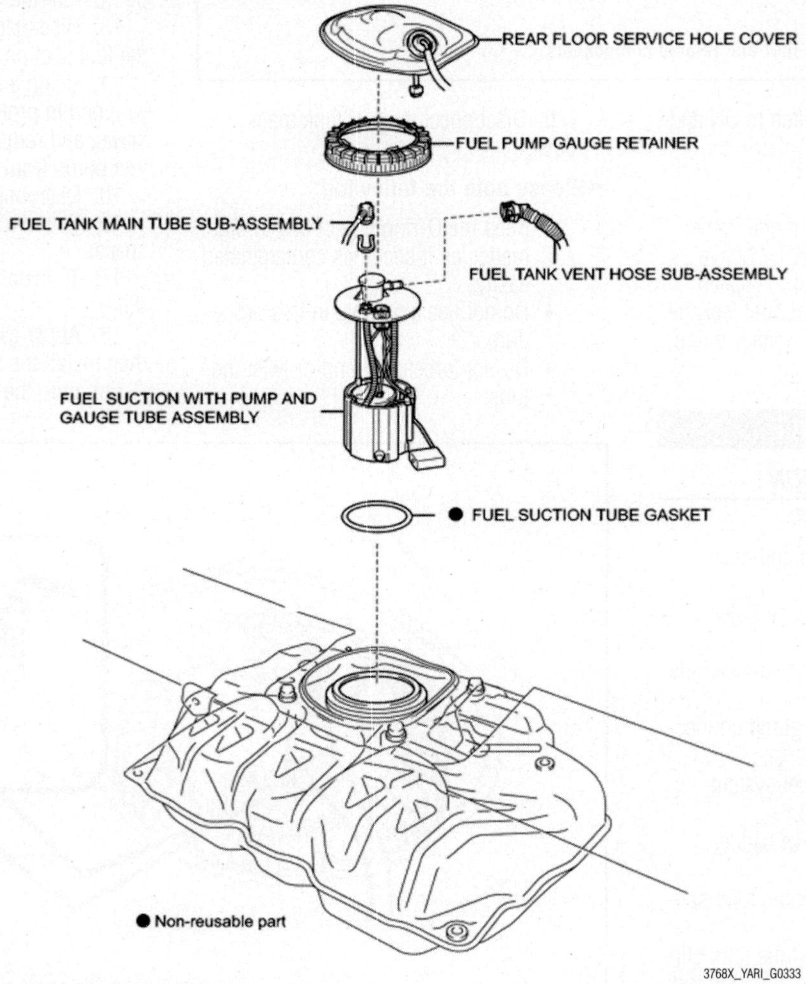

Fig. 132 View of the fuel pump module assembly—1 of 2

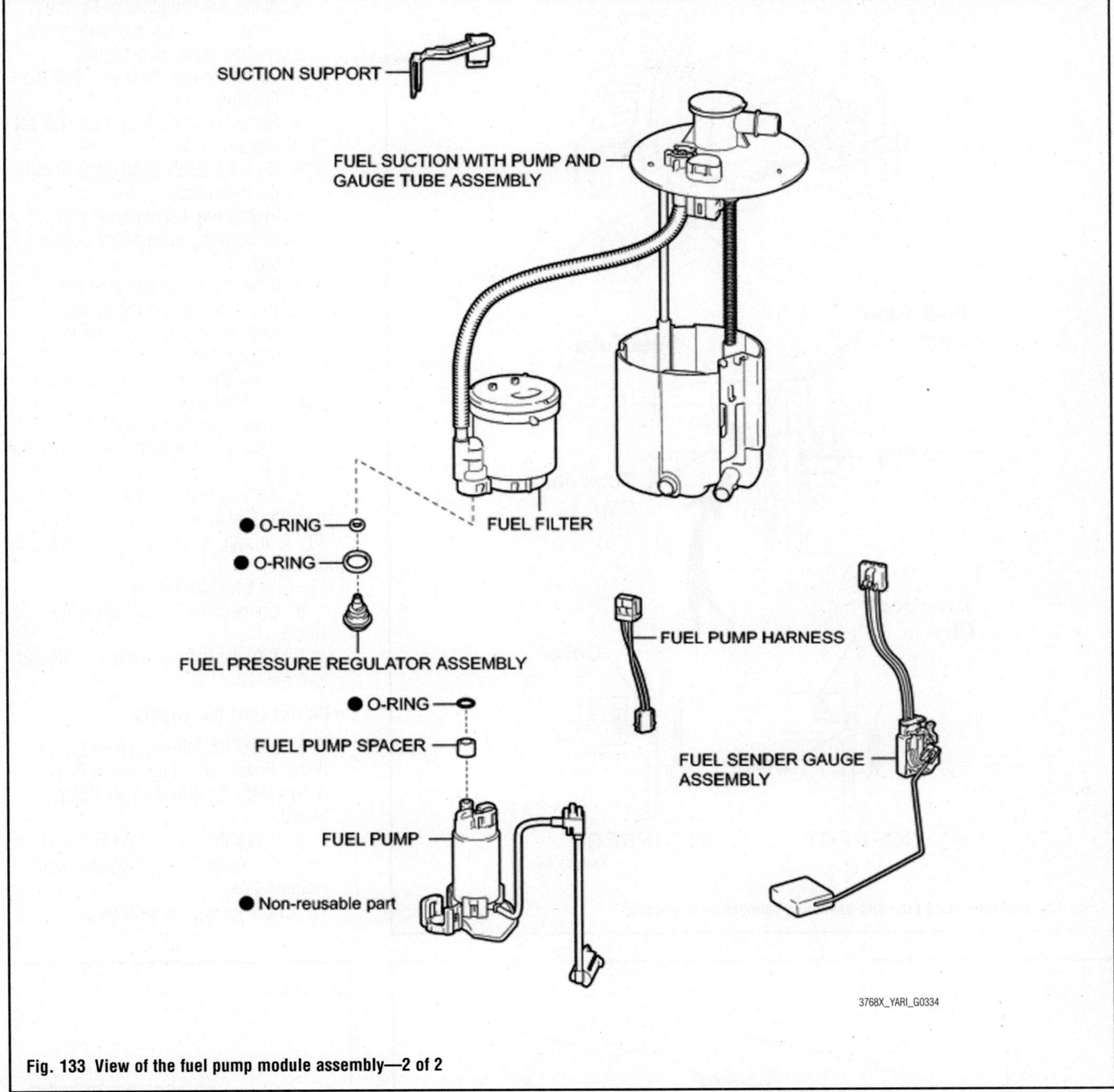

Fig. 133 View of the fuel pump module assembly—2 of 2

2. Drain the fuel.
3. Disconnect the negative battery cable.
4. Remove front floor heat insulator no. 4. Remove the 2 nuts together with the insulator.
5. Remove fuel tank protector sub-assembly. Remove the bolt together with the clip.
6. Disconnect fuel tank main tube sub-assembly. Release the lock as shown in the illustration, then pull and remove the fuel tank main tube.

➡Please note the following:

- Remove any dirt and foreign matter from the clip before performing this work.
- Avoid any scratches or foreign matter on the parts when disconnecting them, as the quick connector has the O-ring that seals the plug.
- Perform this work by hand. Do not use any tools.
- Do not forcibly bend, twist or turn the nylon tube.
- Protect the disconnected parts by covering them with a plastic bag.
- If the connector and pipe are stuck, disconnect the nylon tube by turning it by hand to release them.

7. Disconnect fuel tank breather hose. Release the lock as shown in the illustration, then pull and remove the fuel tank breather hose.

➡Please note the following:

- Remove any dirt and foreign matter from the clip before performing this work.

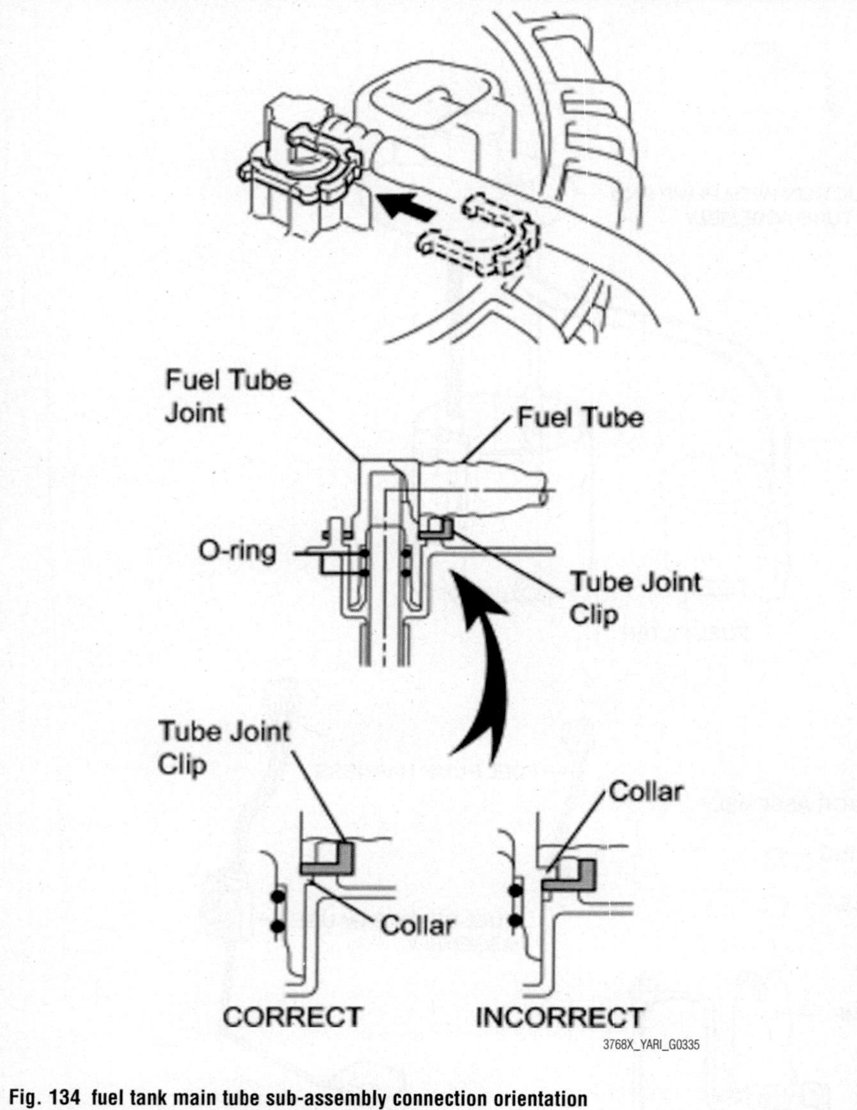

Fig. 134 fuel tank main tube sub-assembly connection orientation

- Avoid any scratches or foreign matter on the parts when disconnecting them, as the quick connector has the O-ring that seals the plug.
- Perform this work by hand. Do not use any tools.
- Do not forcibly bend, twist or turn the nylon tube.
- Protect the disconnected parts by covering them with a plastic bag.
- If the connector and pipe are stuck, disconnect the nylon tube by turning it by hand to release them.

8. Loosen the clamp and disconnect the fuel tank filler pipe.
9. Remove fuel tank assembly:
   a. Remove the 4 bolts and remove the fuel tank.
   b. Remove the fuel tank main tube from the fuel tank.
10. To install, reverse the removal procedure.
11. Check for fuel leakage:
   a. Connect the Techstream to the DLC3.
   b. Turn the ignition switch to ON and turn the tester ON.

➡Do not start the engine.

   c. Select the following menu items: Powertrain / Engine and ECT / Active Test / Control the Fuel Pump / Speed.
   d. Check that there is no fuel leakage anywhere on the fuel system after doing maintenance.
12. Check for engine oil leakage.

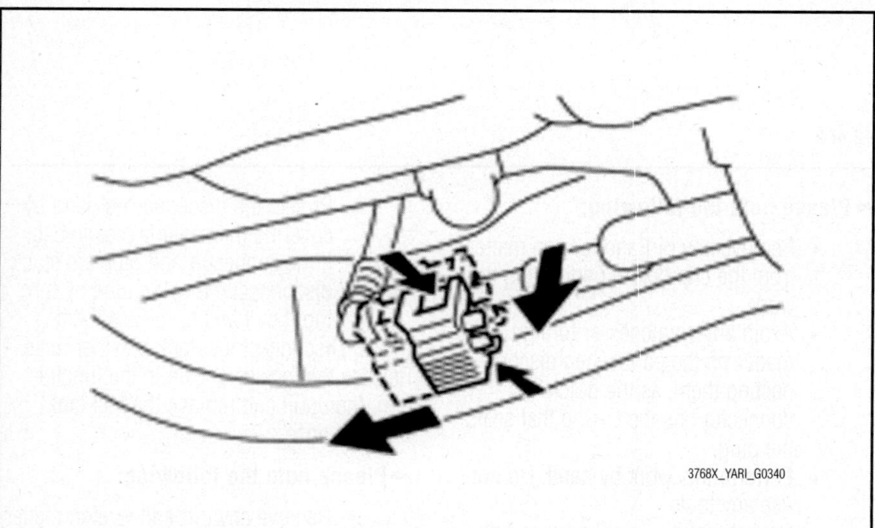

Fig. 135 Release the lock as shown, then pull and remove the fuel tank main tube

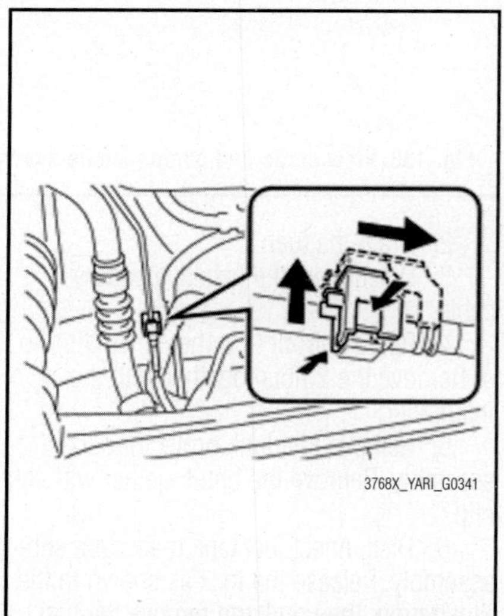

Fig. 136 Disconnect fuel tank breather hose. Release the lock as shown

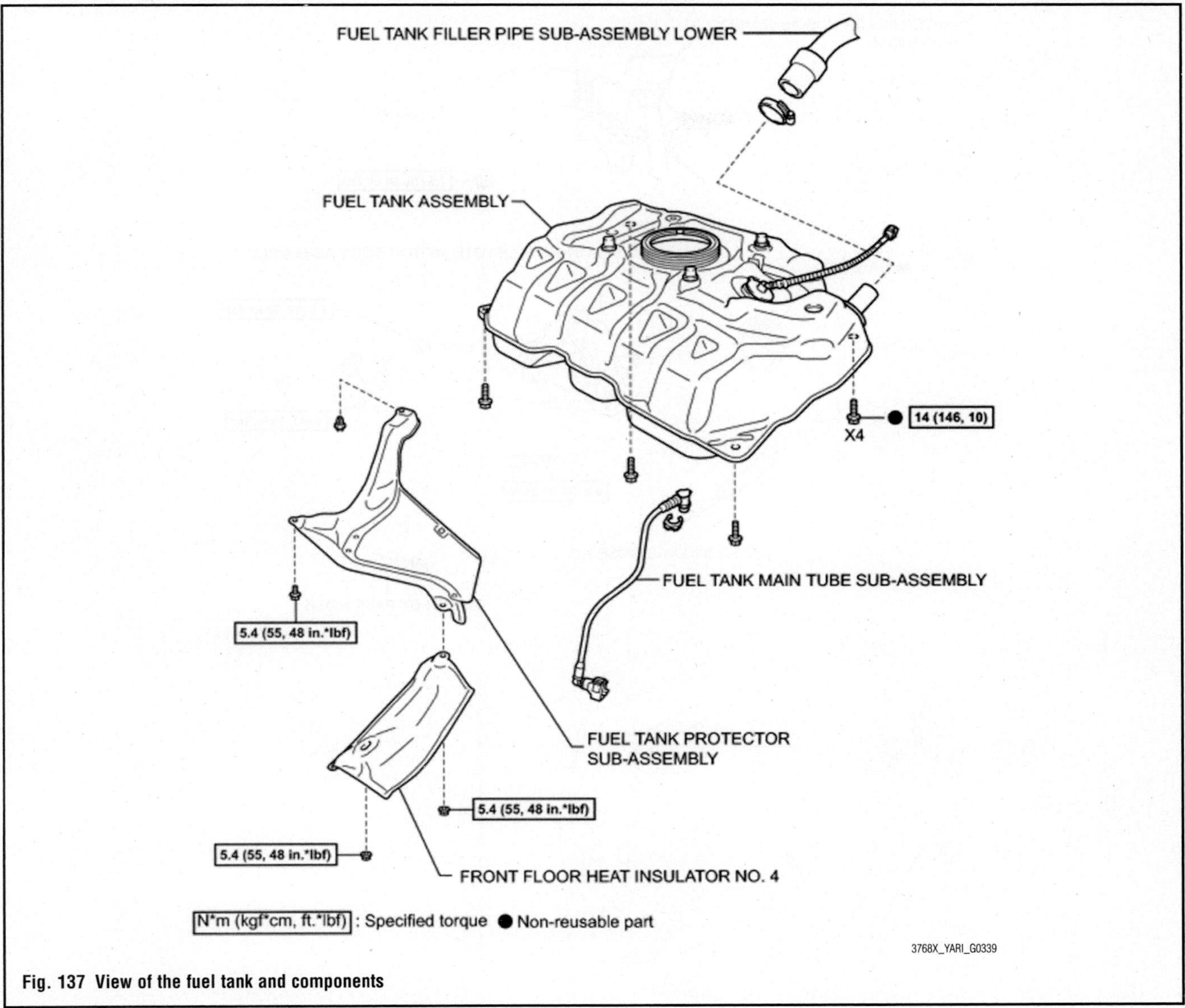

Fig. 137 View of the fuel tank and components

THROTTLE BODY

REMOVAL & INSTALLATION

See Figure 138.

1. Disconnect the negative battery cable.
2. Drain and recycle the engine coolant.
3. Remove the engine cover.
4. Remove the air cleaner assembly.
5. Disconnect the water by-pass hose (except 2012 Hatchback).
6. Disconnect water by-pass hose No. 2 (except 2012 Hatchback).
7. Remove throttle with motor body assembly:
   a. Disconnect the throttle with motor body assembly connector.
   b. Remove the nut and remove the wire harness with bracket.
   c. Remove the 2 nuts and remove the water filler sub-assembly with hose.
   d. Remove the bolt and 3 nuts and remove the throttle with motor body assembly.
   e. Remove the gasket from the intake manifold.
8. To install, reverse the removal procedure.
9. Refer to the illustration for torque values.
10. Perform the initialization:
    a. Disconnect the cable from the negative (-) battery terminal. Wait at least 60 seconds and reconnect the cable.
    b. Turn the ignition switch to ON without operating the accelerator pedal.

➡If the accelerator pedal is operated, perform the above steps again.

    c. Connect the Techstream to the DLC3 and clear the DTC's.
    d. Start the engine and check that the MIL is not illuminated and that the idle speed is within the specified range when the A/C is switched off after the engine is warmed up.
    - Standard (for Automatic Transaxle): With A/C switched off, engine idle speed should be at 650 to 750 RPM.
    - Standard (for Manual Transaxle): With A/C switched off, engine idle speed should be at 550 to 650 RPM.

➡Be sure to perform this step with all accessories off. Make sure that the shift lever is in neutral.

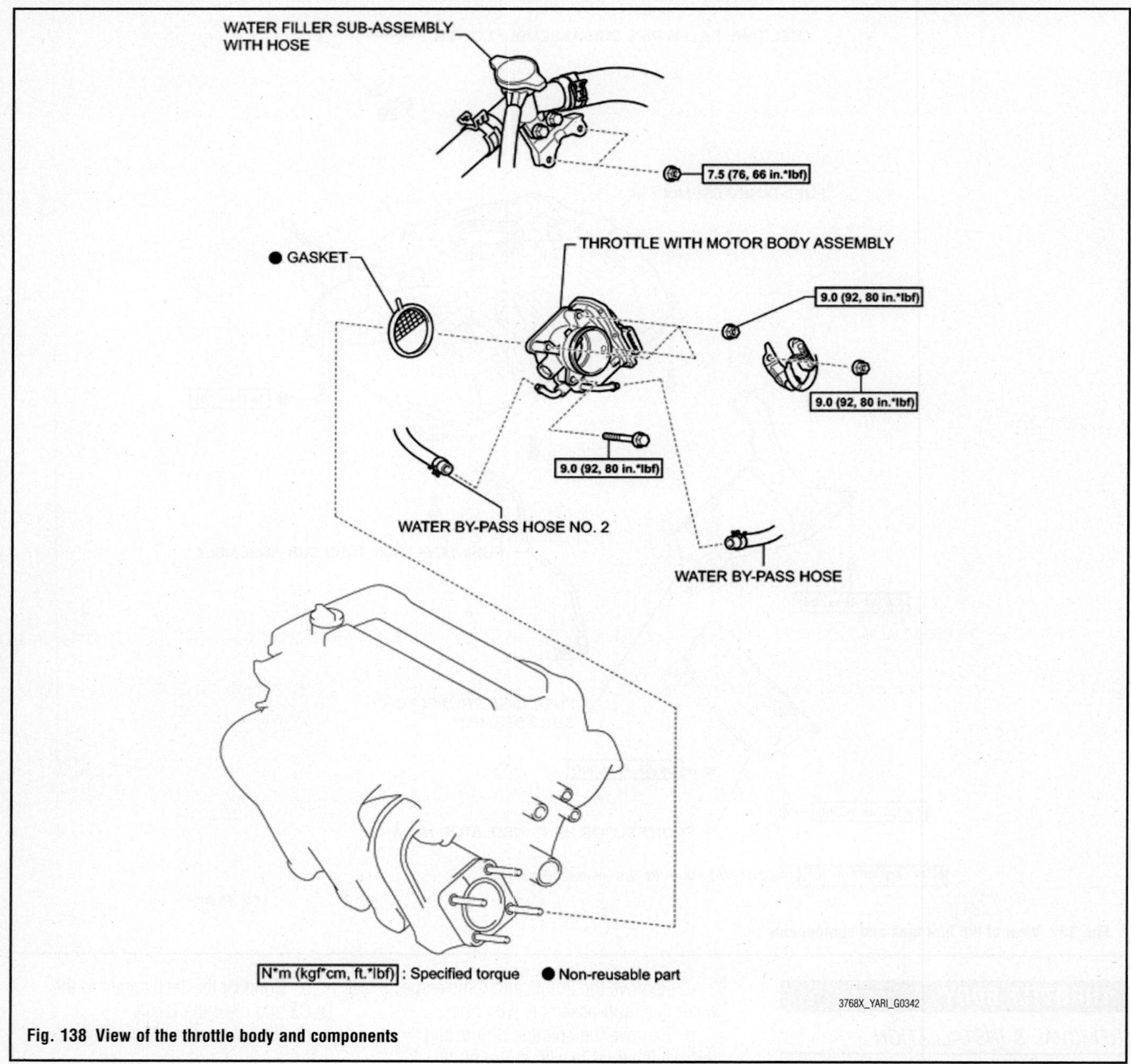

Fig. 138 View of the throttle body and components

e. Enter the following menus: Powertrain / Engine and ECT / Data List / Throttle Sensor Position. Fully depress the accelerator pedal and check that the value is 60% or more.

f. Perform a road test and confirm that there are no abnormalities.

HEATING & AIR CONDITIONING SYSTEM

BLOWER MOTOR

REMOVAL & INSTALLATION

See Figure 139.

1. Disconnect cable from negative battery terminal
2. Remove instrument panel under cover sub-assembly.
3. Remove blower motor.

To install:

4. Install blower motor.
5. Install instrument panel under cover sub-assembly.
6. Connect cable from negative battery terminal

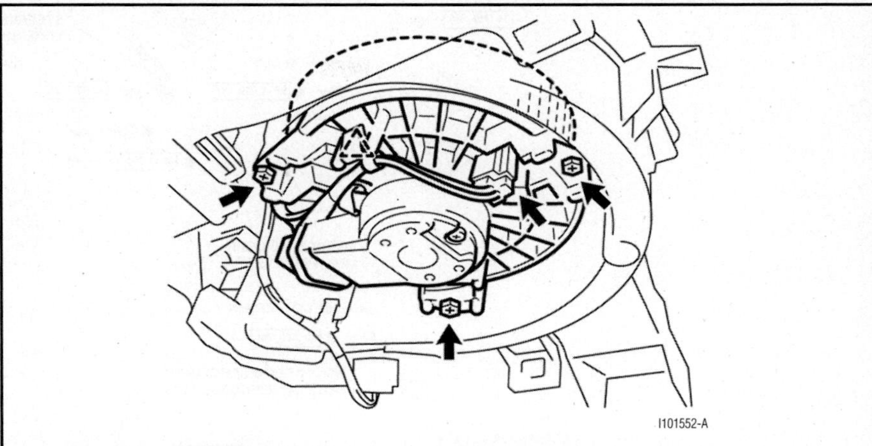

Fig. 139 Disconnect the connector and the clamp. Remove the 3 screws and the blower motor

STEERING

POWER STEERING GEAR

REMOVAL & INSTALLATION

See Figures 140 through 144.

1. Position front wheels facing straight ahead.
2. Disconnect the negative battery cable.
3. Remove the 4 bolts and remove the hood.
4. Remove the wiper motor and link assembly.
5. Remove the floor carpet and 2 clips and remove the column hole cover silencer.
6. Remove steering sliding yoke sub-assembly:

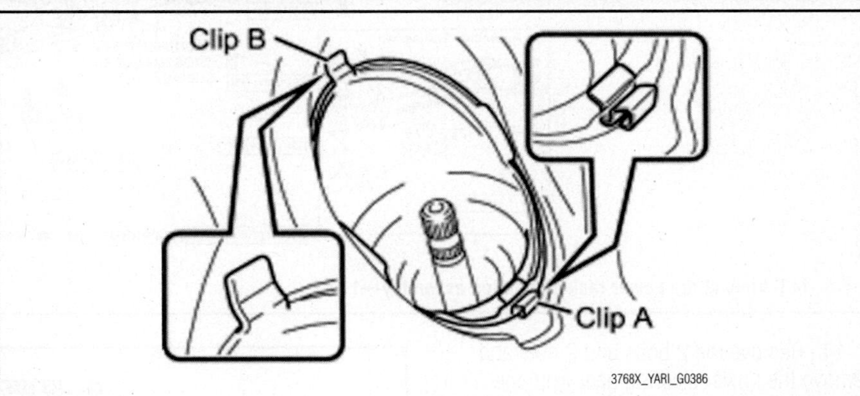

Fig. 141 Remove clip A, separate clip B from the body and separate No. 1 steering column hole cover sub-assembly

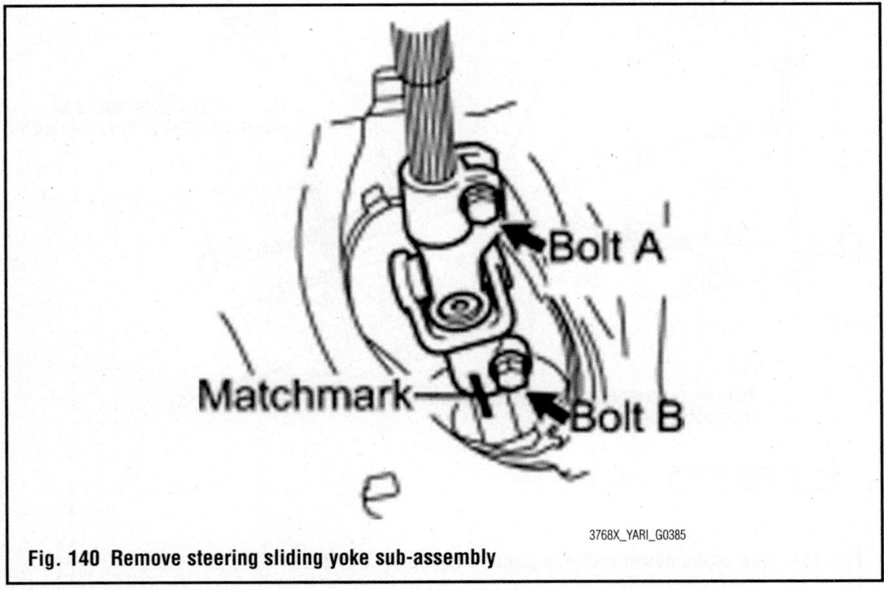

Fig. 140 Remove steering sliding yoke sub-assembly

a. Use a seat belt to fix the steering wheel assembly, in order to avoid breakage of the spiral cable.
b. Place the matchmarks on the sliding yoke of the steering intermediate shaft assembly and the power steering.
c. Loosen bolt A, remove bolt B and separate the steering intermediate shaft assembly.

7. Remove clip A, separate clip B from the body and separate No. 1 steering column hole cover sub-assembly.
8. Remove front wheel.
9. Remove front stabilizer link assembly LH and RH.
10. Remove tie rod end sub-assembly.
11. Remove front suspension lower arm RH and LH.
12. Suspend engine assembly.
13. Remove front suspension crossmember sub-assembly.

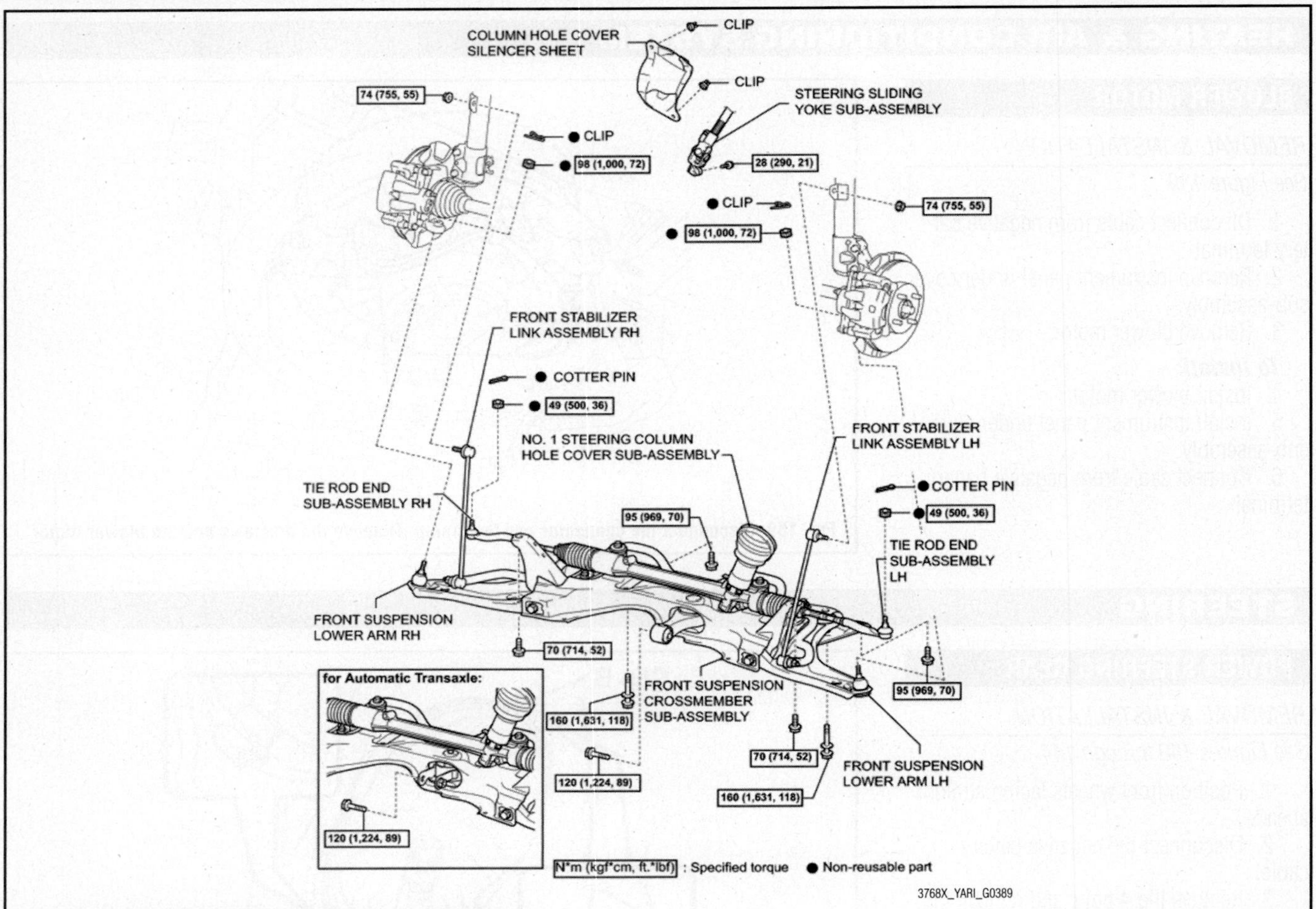

Fig. 143 View of the power rack and pinion assembly—1 of 2

14. Remove the 2 bolts and 2 nuts and remove the power steering gear from the suspension crossmember.

➡**Keep the nut from rotating while turning the bolt.**

15. To install, reverse the removal procedure.

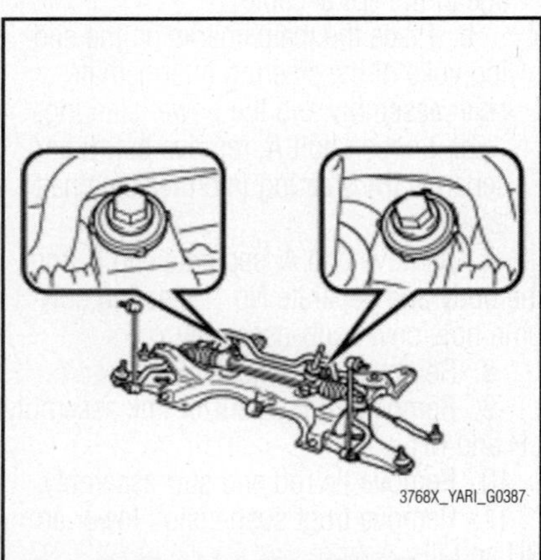

Fig. 142 Removing the and remove the power steering gear from the suspension crossmember

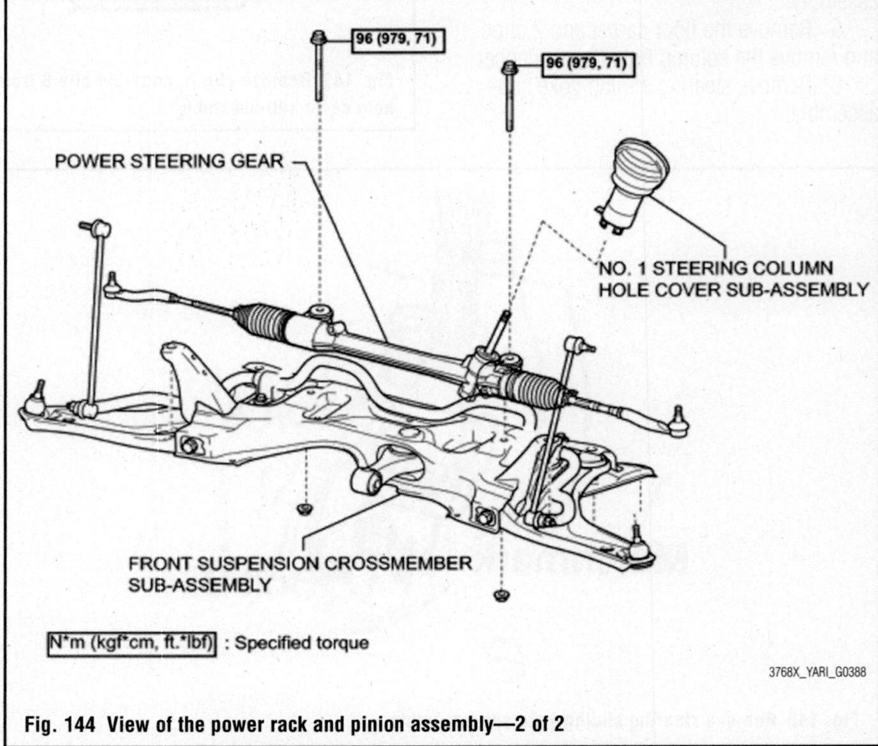

Fig. 144 View of the power rack and pinion assembly—2 of 2

LOWER CONTROL ARM

REMOVAL & INSTALLATION

Left Side

See Figures 145 through 148.

1. Remove the front wheel.
2. Remove the clip and castle nut.
3. Using SST 09628-00011, separate the lower arm.

➡Do not damage the lower arm dust cover. Suspend SST with a piece of string or the equivalent.

4. Remove the 2 bolts and lower arm.

To install:

5. Provisionally tighten the lower arm with the 2 bolts.
6. Install the lower arm onto the steering knuckle with a new castle nut. Tighten to 72 ft. lbs. (98 Nm).

➡If the holes for the clip are not aligned, tighten the nut by a further turn of up to 60°.

7. Install a new clip.
8. Stabilize suspension:
   a. Lower the vehicle from the jack.
   b. Bounce the vehicle up and down several times to stabilize the suspension.
9. Fully tighten the 2 front lower suspension arm bolts:
   a. Bolt A: 101 ft. lbs. (137 Nm).
   b. Bolt B: 118 ft. lbs. (160 Nm).
10. Inspect and adjust front wheel alignment.

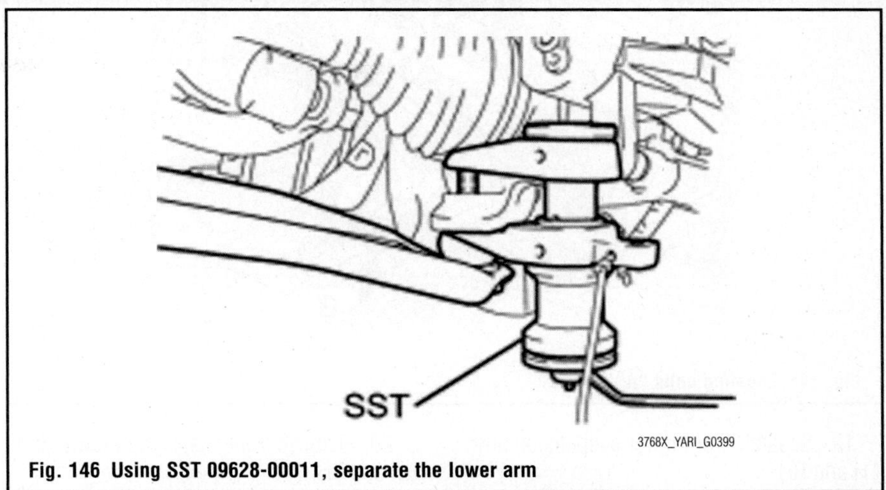

Fig. 146 Using SST 09628-00011, separate the lower arm

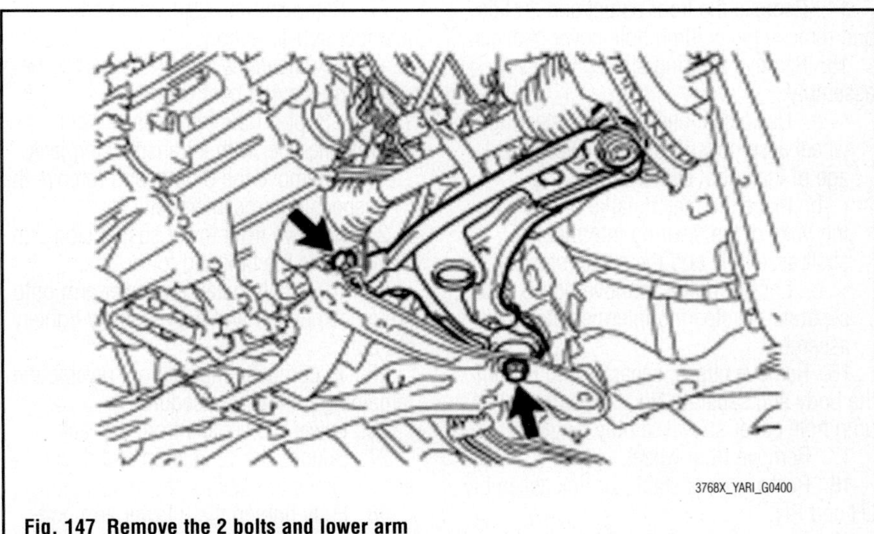

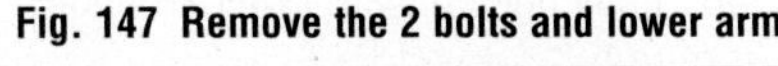
Fig. 147 Remove the 2 bolts and lower arm

3768X_YARI_G0398

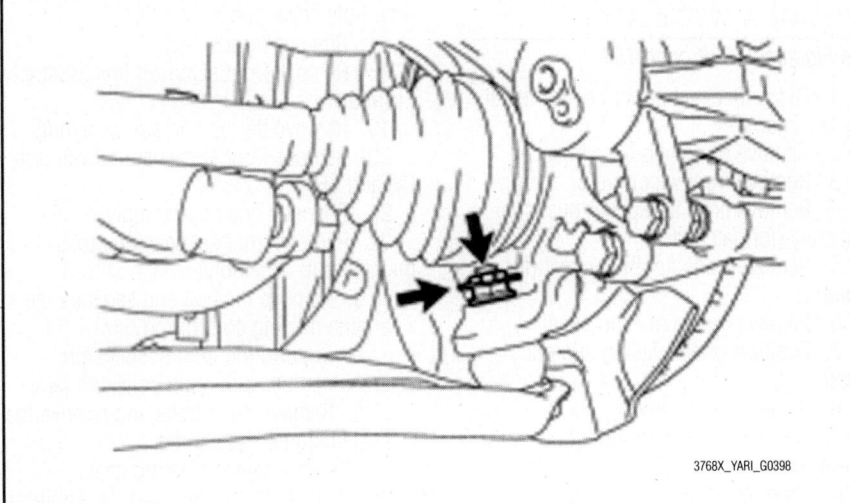

Fig. 145 Remove the clip and castle nut

Right Side

See Figures 149 through 151.

1. Disconnect cable from negative battery terminal.
2. Remove hood sub-assembly.
3. Remove front wiper motor and link.
4. Remove cowl to register duct sub-assembly (for hatchback).
5. Remove front air shutter seal (for sedan) .
6. Remove outer cowl top panel.
7. Position wheels facing straight ahead
8. Remove front wheel.
9. Remove steering column hole cover sub-assembly.
10. Separate front stabilizer link assembly LH and RH.
11. Separate tie rod end sub-assembly LH and RH.

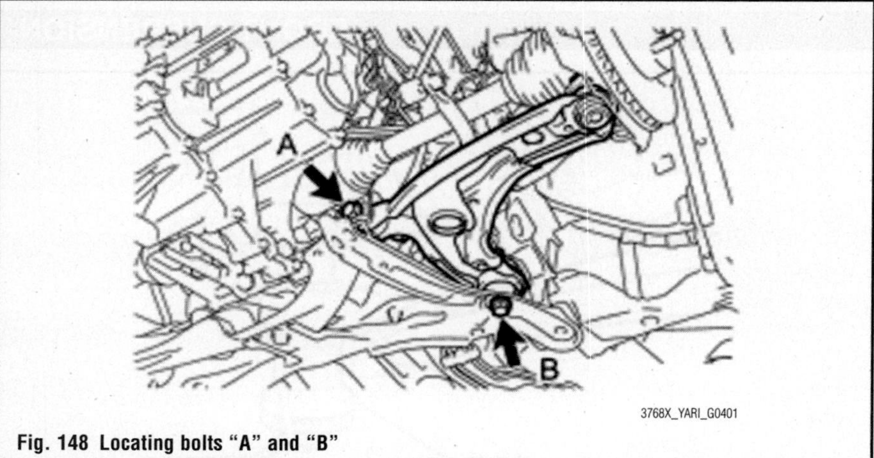

Fig. 148 Locating bolts "A" and "B"

12. Separate front lower suspension arm LH and RH.
13. Suspend engine assembly.
14. Remove the floor carpet and 2 clips and remove the column hole cover silencer.
15. Remove steering sliding yoke sub-assembly:
   a. Use a seat belt to fix the steering wheel assembly, in order to avoid breakage of the spiral cable.
   b. Place the matchmarks on the sliding yoke of the steering intermediate shaft assembly and the power steering.
   c. Loosen bolt A, remove bolt B and separate the steering intermediate shaft assembly.
16. Remove clip A, separate clip B from the body and separate No. 1 steering column hole cover sub-assembly.
17. Remove front wheel.
18. Remove front stabilizer link assembly LH and RH.
19. Remove tie rod end sub-assembly.

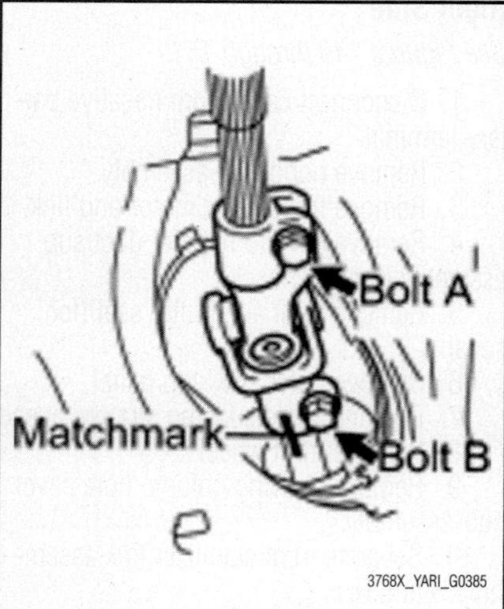

Fig. 149 Remove steering sliding yoke sub-assembly

20. Remove front suspension lower arm RH and LH.
21. Suspend engine assembly.
22. Remove front suspension cross-member sub-assembly:
   a. Remove the bolt and separate the engine moving control rod.
   b. Support the front suspension crossmember with a transmission jack.
   c. Remove the 6 bolts and remove the suspension crossmember.
23. Remove front lower suspension arm RH. Remove the bolt and lower arm.
24. To install, install the lower arm onto the crossmember and provisionally tighten the bolt.
25. To complete installation, reverse the remaining removal procedure.
26. Lower the vehicle from the jack.
27. Bounce the vehicle up and down several times to stabilize the suspension.
28. Fully tighten the 2 lower arm bolts.

STABILIZER BAR

REMOVAL & INSTALLATION

See Figures 152 through 154.

1. Disconnect cable from negative battery terminal.
2. Remove hood sub-assembly.
3. Remove front wiper motor and link.
4. Remove cowl to register duct sub-assembly (for hatchback).
5. Remove front air shutter seal (for sedan) .
6. Remove outer cowl top panel.
7. Position wheels facing straight ahead
8. Remove front wheel.
9. Remove steering column hole cover sub-assembly.
10. Separate front stabilizer link assembly LH and RH.

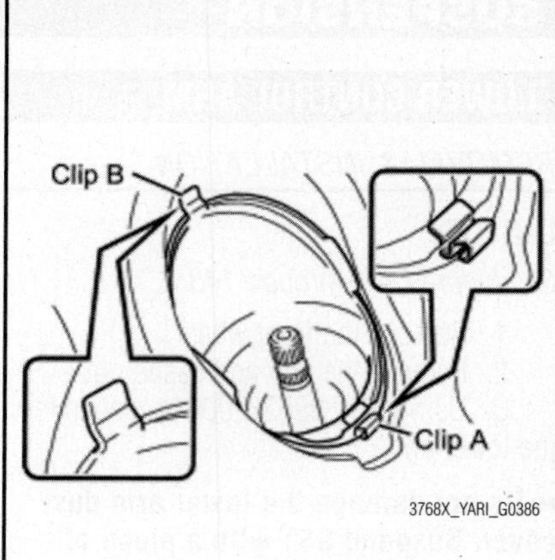

Fig. 150 Remove clip A, separate clip B from the body and separate No. 1 steering column hole cover sub-assembly

11. Separate tie rod end sub-assembly LH and RH.
12. Separate front lower suspension arm LH and RH.
13. Suspend engine assembly.
14. Remove the floor carpet and 2 clips and remove the column hole cover silencer.
15. Remove steering sliding yoke sub-assembly:
   a. Use a seat belt to fix the steering wheel assembly, in order to avoid breakage of the spiral cable.
   b. Place the matchmarks on the sliding yoke of the steering intermediate shaft assembly and the power steering.
   c. Loosen bolt A, remove bolt B and separate the steering intermediate shaft assembly.
16. Remove clip A, separate clip B from the body and separate No. 1 steering column hole cover sub-assembly.
17. Remove front wheel.
18. Remove front stabilizer link assembly LH and RH.
19. Remove tie rod end sub-assembly.
20. Remove front suspension lower arm RH and LH.
21. Suspend engine assembly.
22. Remove front suspension cross-member sub-assembly:
   a. Remove the bolt and separate the engine moving control rod.
   b. Support the front suspension crossmember with a transmission jack.
   c. Remove the 6 bolts and remove the suspension crossmember.
23. Remove power steering gear.
24. Remove the 2 bolts and the stabilizer bracket.

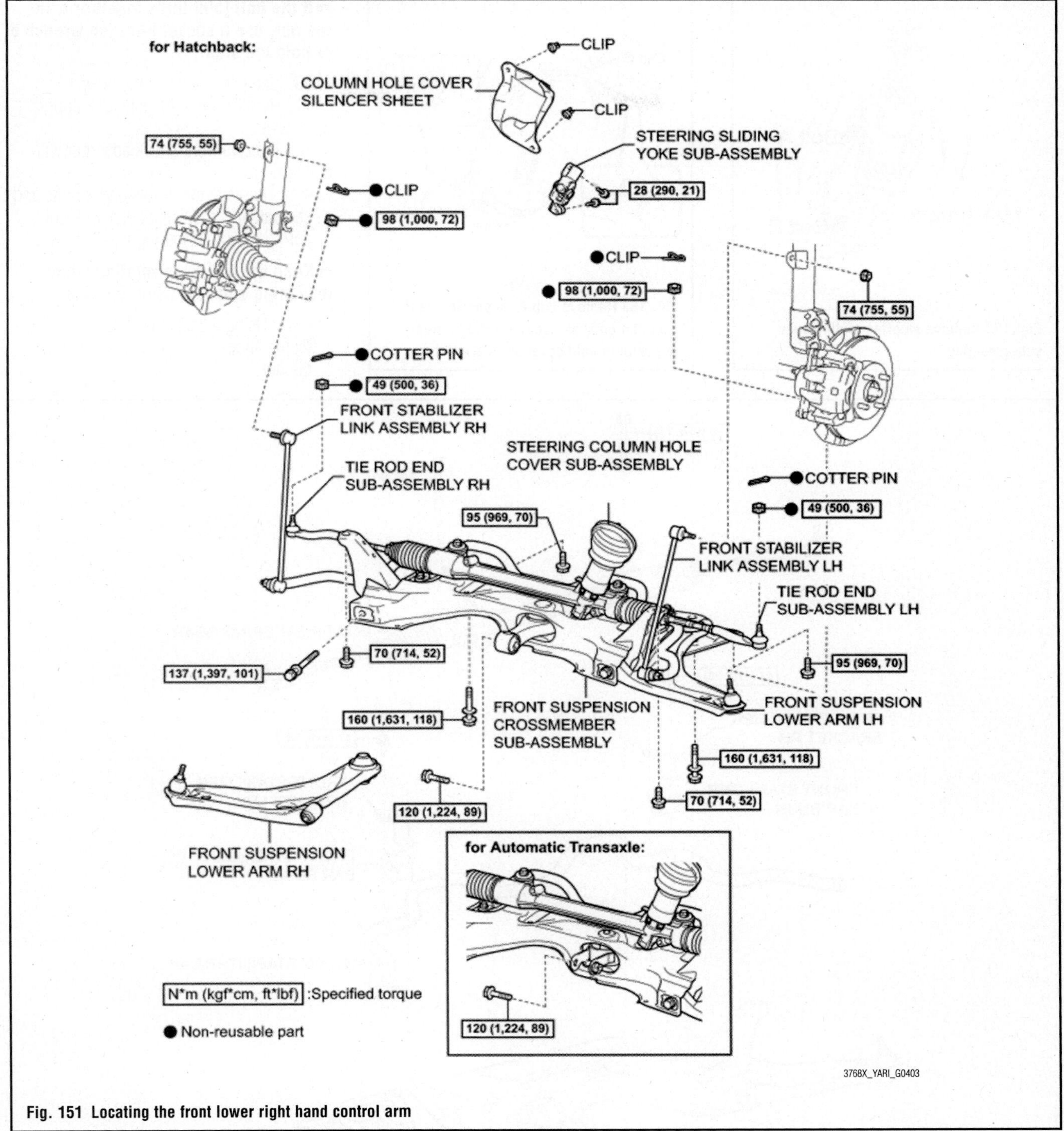

Fig. 151 Locating the front lower right hand control arm

25. Remove front stabilizer bar and bushing.
26. To install, reverse the removal procedure.
27. Install the stabilizer onto the crossmember with the paint mark on the left side of the vehicle.
28. Inspect and adjust front wheel alignment.

STRUT & SPRING ASSEMBLY

REMOVAL & INSTALLATION

See Figure 155.

1. Disconnect cable from negative battery terminal.
2. Remove hood sub-assembly.
3. Remove front wiper motor and link.
4. Remove cowl to register duct sub-assembly (for hatchback).
5. Remove front air shutter seal (for sedan) .
6. Remove outer cowl top panel.
7. Position wheels facing straight ahead
8. Remove front wheel.
9. Remove the nut and separate the stabilizer link from the shock absorber.

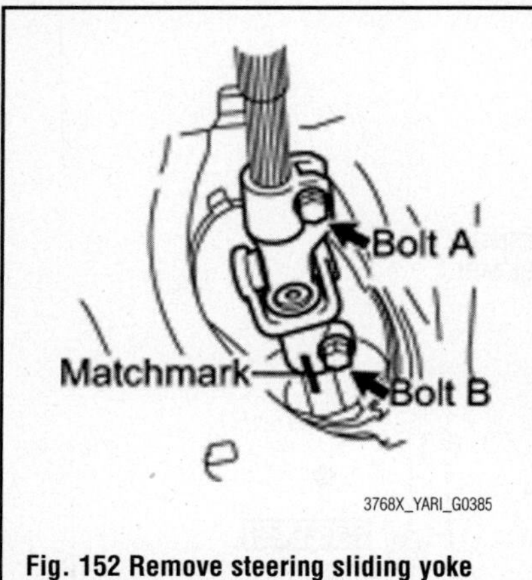

Fig. 152 Remove steering sliding yoke sub-assembly

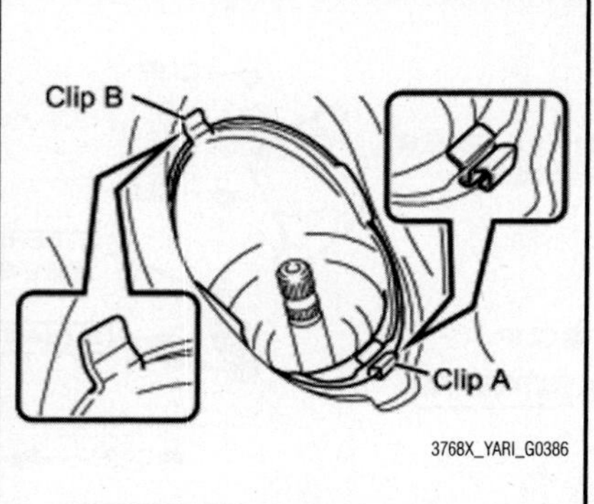

Fig. 153 Remove clip A, separate clip B from the body and separate No. 1 steering column hole cover sub-assembly

➡If the ball joint turns together with the nut, use a socket hexagon wrench 6 to hold the stud.

10. Separate front flexible hose.
11. Remove front suspension support dust cover.
12. Remove front shock absorber with coil spring:
    a. Remove the 2 nuts and 2 bolts and separate the shock absorber with coil spring from the steering knuckle.

➡Keep the bolt from rotating while loosening and removing the nuts.

    b. Using a socket hexagon wrench 6, fix the shock absorber rod and remove the nut.

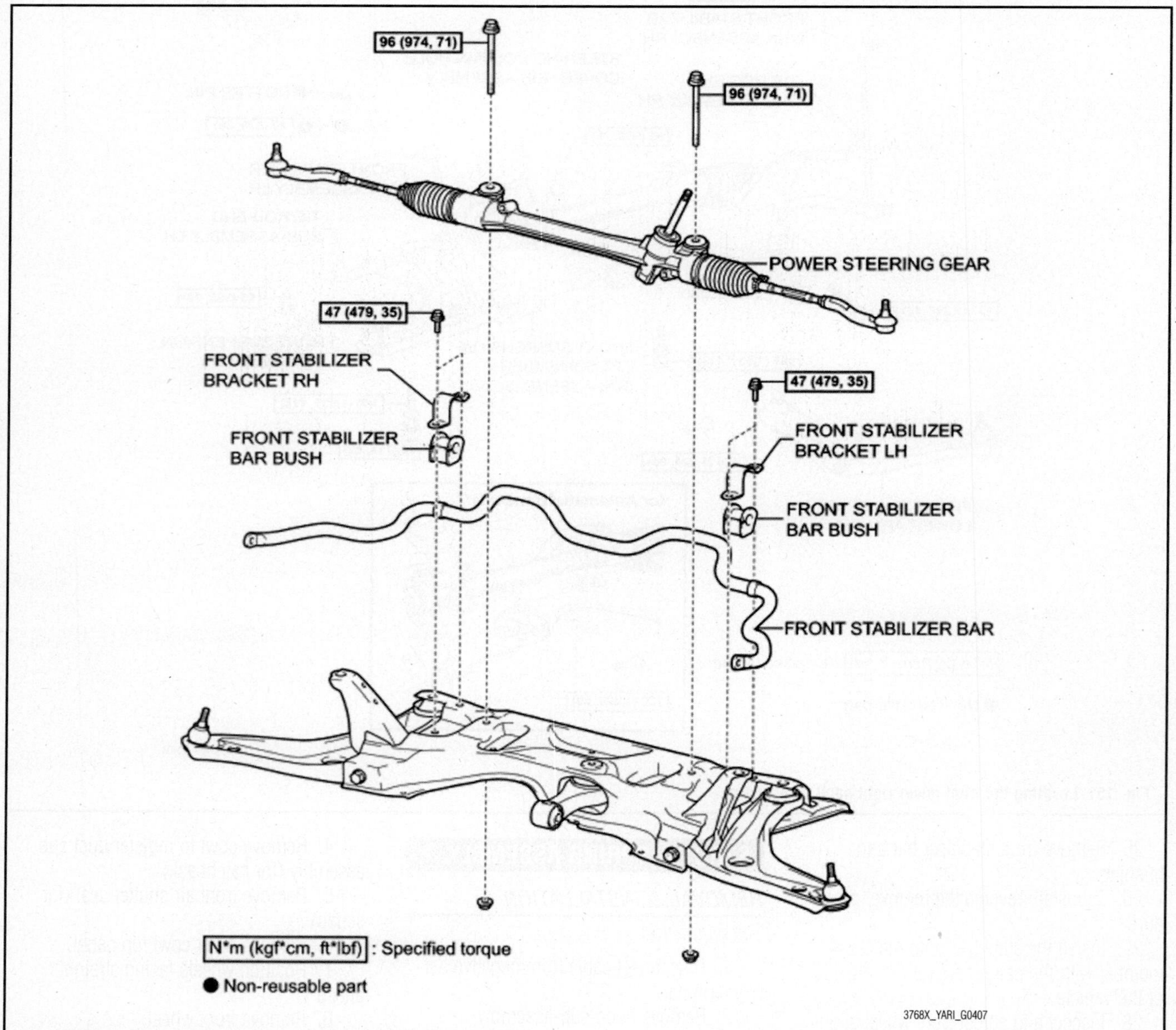

Fig. 154 View of the front stabilizer and related components

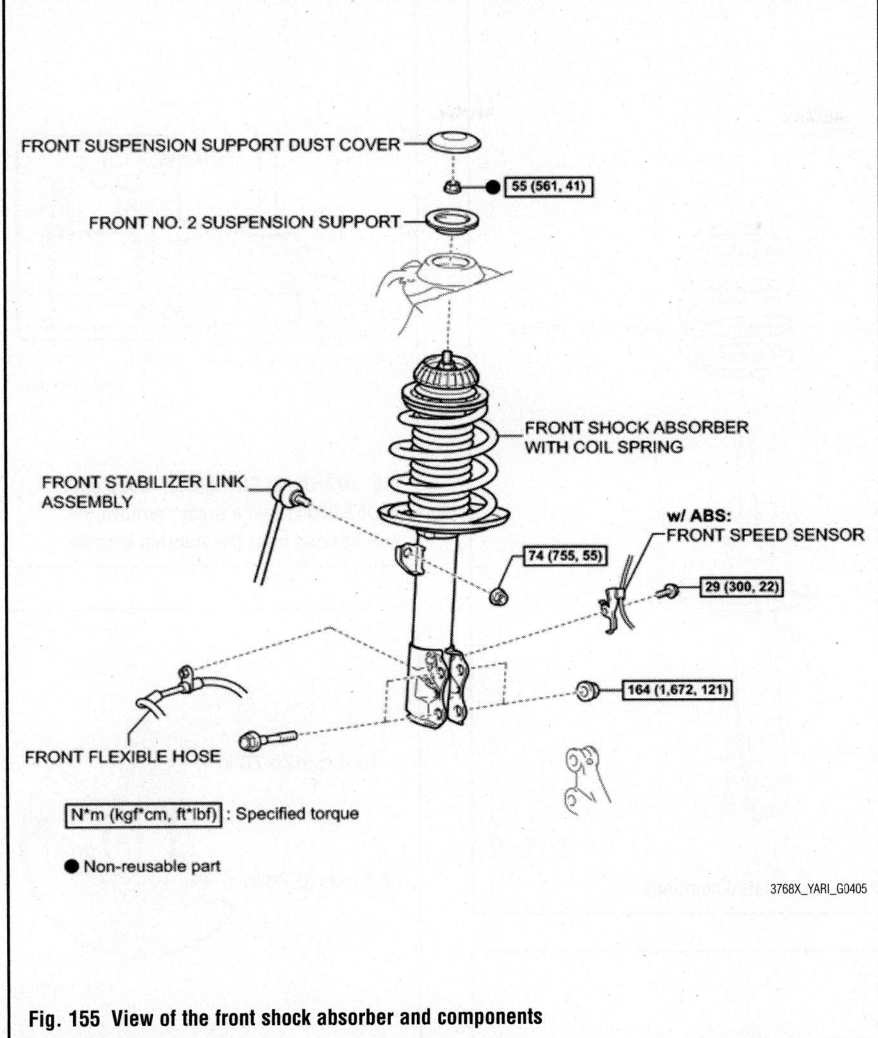

Fig. 155 View of the front shock absorber and components

c. Remove No. 2 suspension support.

d. Remove the front shock absorber with coil spring from the vehicle.

To install:

13. Provisionally tighten a new nut through No. 2 suspension support.
14. Install the front shock absorber with coil spring onto the steering knuckle.
15. Install front flexible hose.
16. Install the stabilizer link with the nut.
17. Install front wheel.
18. Using a socket hexagon wrench 6, fix the shock absorber rod and tighten the nut.
19. To complete installation, reverse the remaining removal procedure.

OVERHAUL

See Figure 156.

1. Remove front suspension support sub-assembly.
2. Remove front support to front shock absorber nut:

   a. Using a socket hexagon wrench 6, fix the shock absorber rod and loosen the nut.

   b. Using SST 09727-30021, compress the coil spring.

➡**Do not use an impact wrench. It will damage SST.**

   c. Remove the nut.

3. Remove the upper coil spring seat with the strut mounting bearing and spring bumper from the shock absorber.
4. Remove the spring bumper from the upper coil spring seat.
5. Remove front upper coil spring insulator.
6. Remove front coil spring.
7. Using a brass bar and press, remove the strut mounting bearing from the upper coil spring seat.
8. To install, reverse the removal procedure.

WHEEL BEARINGS

REMOVAL & INSTALLATION

See Figures 157 through 162.

1. Remove the front axle assembly.
2. Using snap ring pliers, remove the hole snap ring.

➡**When removing the hole snap ring, do not damage the magnetic rotor surface.**

3. Remove front axle hub sub-assembly:

   a. Fix the steering knuckle in a vise between aluminum plates.

➡**Do not over tighten the vise.**

   b. Using SST 09520-00031, remove the axle hub.

   c. Using SST 09950-40011, remove the hub bearing inner race from the axle hub.

4. Remove the dust cover by tapping the back side with a brass bar and hammer.
5. Remove front axle hub bearing:

   a. Install the removed hub bearing inner race onto the outer side of the hub bearing.

   b. Using SST 09223-15020 and 09387-00041 and a press, remove the hub bearing from the steering knuckle.

To install:

6. Using SST 09950-60020 and 09387-00041 and a press, insert a new hub bearing, with its magnetic rotor side facing the inside of the vehicle, until it reaches the end of the steering knuckle.

➡**Please note the following:**

- Do not remove the inner race because the hub bearing is built into the oil seal.
- Do not use bearings that have been removed.
- Do not wipe off any grease that has been applied to new bearings.
- Do not bring magnets close to the magnetic rotor surface of the bearing.
- Keep the magnetic rotor surface of the bearing free of foreign matter.

7. Provisionally install a new disc brake dust cover.

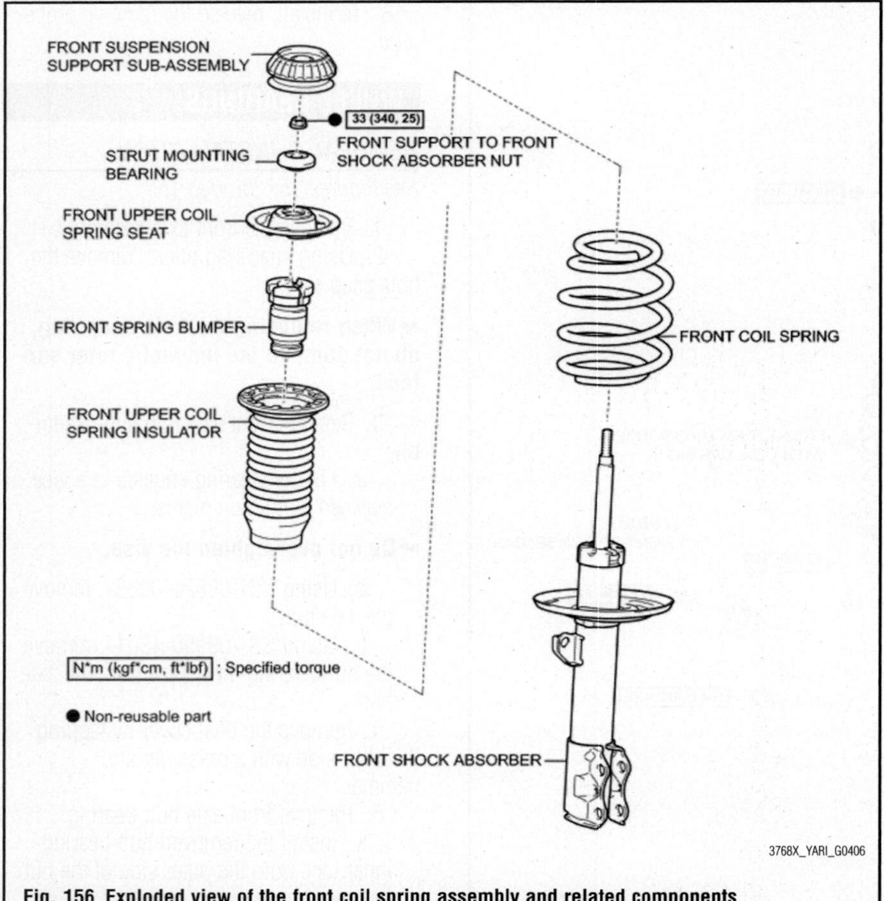

Fig. 156 Exploded view of the front coil spring assembly and related components

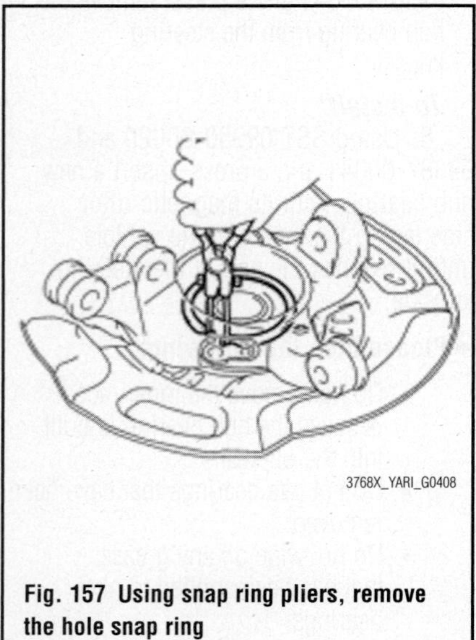

Fig. 157 Using snap ring pliers, remove the hole snap ring

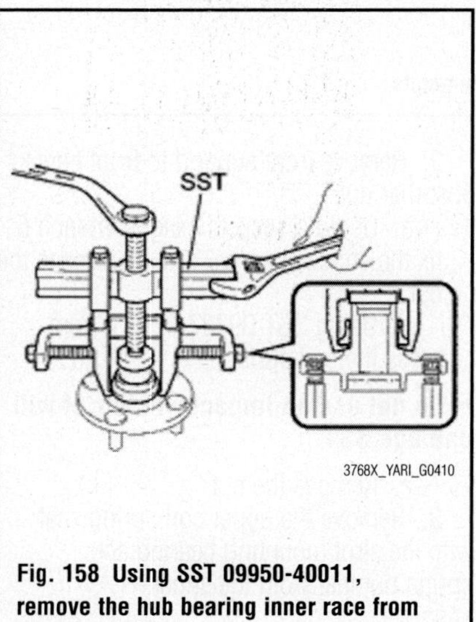

Fig. 158 Using SST 09950-40011, remove the hub bearing inner race from the axle hub

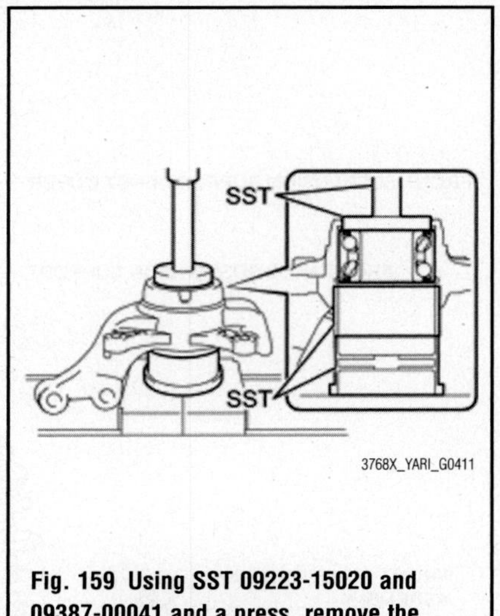

Fig. 159 Using SST 09223-15020 and 09387-00041 and a press, remove the hub bearing from the steering knuckle

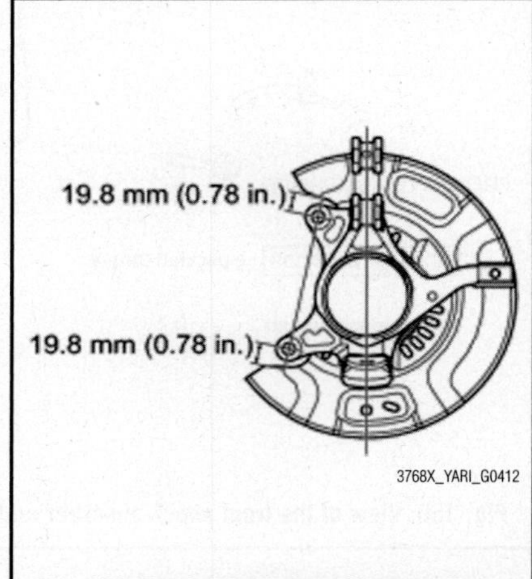

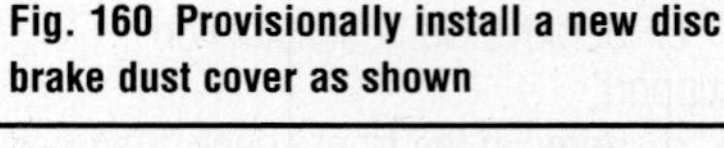
Fig. 160 Provisionally install a new disc brake dust cover as shown

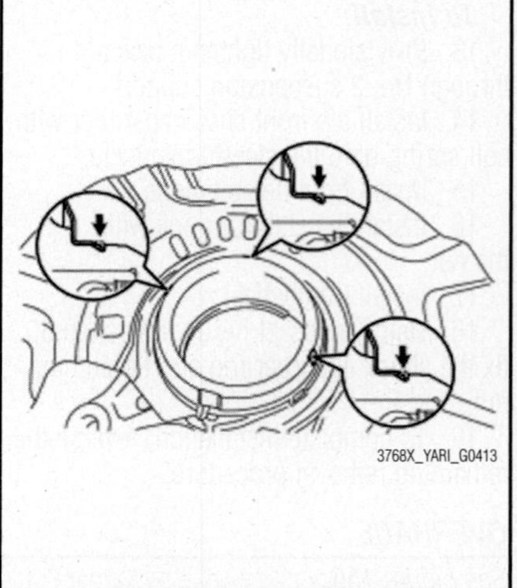

Fig. 161 Using a chisel, fix the 3 points on the circumference

8. Using SST 09223-56010 and a hammer, install the disc brake dust cover.

➡**Uniformly press in the disc brake dust cover while sliding SST slightly. Surely press in the disc brake dust cover until the pressing-in base.**

9. Using a chisel, fix the 3 points on the circumference.

➡**Securely fold each end into the engagement grooves.**

10. Using SST: 09608-32010, 09950-60010 and a press, press the axle hub into the steering knuckle.

11. Using snap ring pliers, install a new hole snap ring, as shown in the illustration.

➡Do not overlap the end of the snap ring and the installation hole in the speed sensor on the knuckle side. Do not damage the magnetic rotor surface of the bearing when installing the snap ring.

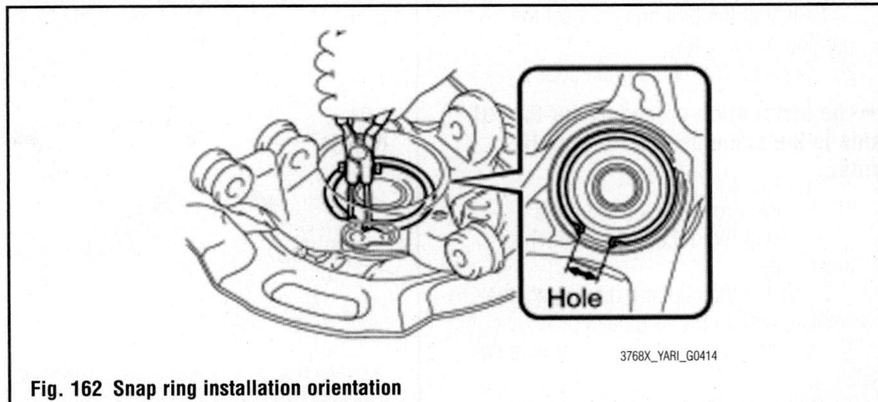

Fig. 162 Snap ring installation orientation

SUSPENSION

AXLE BEAM

REMOVAL & INSTALLATION

See Figures 163 through 165.

1. Remove rear seat cushion cover pad sub-assembly (for Sedan Fixed Seat Type).
2. Remove rear seatback assembly (for Sedan Fixed Seat Type).
3. Remove rear wheel
4. Drain brake fluid

➡Immediately wash off any brake fluid that comes into contact with any painted surfaces.

5. Separate skid control sensor wire (w/ ABS).
6. Separate rear No. 4 brake tube
   a. Remove the clip and disconnect the flexible hose from the axle beam.
   b. Remove the nut and separate the brake tube.
7. Separate rear No. 3 brake tube

➡The removal procedure for the RH side is the same as that for the LH side.

8. Separate No. 3 parking brake cable assembly
   a. Remove the bolt and separate the parking brake cable.
   b. Remove the cable clamp from the parking brake cable.

➡Replace the clamp with a new one when installing the parking brake cable.

9. Separate No. 2 parking brake cable assembly
10. Remove rear brake drum sub-assembly
11. Remove rear axle hub and bearing assembly LH and RH.
12. Loosen rear axle beam.
13. Remove rear absorber cap (for Hatchback)
14. Remove rear shock absorber LH and RH.
15. Remove rear coil spring LH and RH.
16. Remove rear axle beam.

To install:

17. Install rear axle beam.
18. Install rear coil spring LH and RH.
19. Install rear shock absorber LH and RH.
20. Install rear absorber cap (for Hatchback)
21. Tighten rear axle beam.
22. Install rear axle hub and bearing assembly LH and RH.
23. Install rear brake drum sub-assembly
24. Connect No. 2 parking brake cable assembly
25. Connect No. 3 parking brake cable assembly

➡Replace the clamp with a new one when installing the parking brake cable.

   a. Install the cable clamp to the parking brake cable.

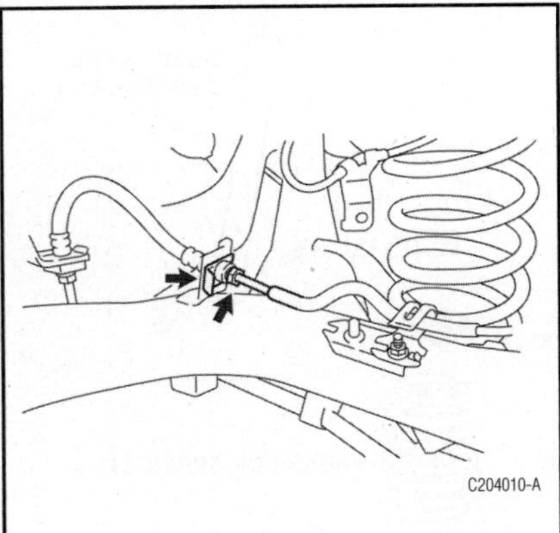

Fig. 163 Using a union nut wrench (10 mm), separate the brake tube while holding the flexible hose with a wrench.

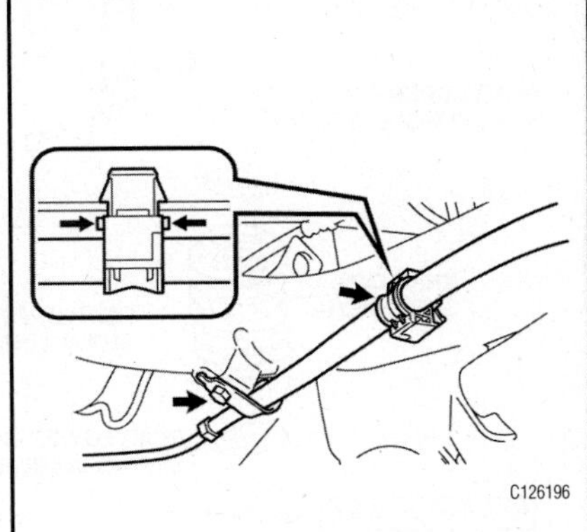

Fig. 164 Using pliers, pinch both ends of the claw of cable clamp and flex the clamp to remove it.

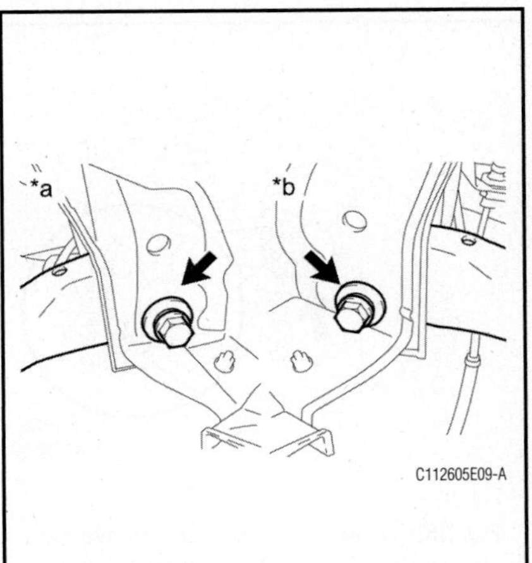

Fig. 165 Remove the 2 bolts and remove the axle beam [a= RH Side, b=LH Side].

b. Install the bolt and connect the parking brake cable.

26. Connect rear No. 3 brake tube

➡The installation procedure for the RH side is the same as that for the LH side.

27. Connect rear No. 4 brake tube

a. Install the nut and separate the brake tube.

b. Install the clip and disconnect the flexible hose to the axle beam.

28. Connect skid control sensor wire (w/ ABS).

29. Fill brake fluid

➡Immediately wash off any brake fluid that comes into contact with any painted surfaces.

30. Install rear wheel

31. Install rear seatback assembly (for Sedan Fixed Seat Type).

32. Install rear seat cushion cover pad sub-assembly (for Sedan Fixed Seat Type).

COIL SPRING

REMOVAL & INSTALLATION

See Figures 166 through 168.

1. Remove rear seat cushion cover pad sub-assembly (for Sedan Fixed Seat Type).
2. Remove rear seatback assembly (for Sedan Fixed Seat Type).
3. Remove rear wheel
4. Drain brake fluid

➡Immediately wash off any brake fluid that comes into contact with any painted surfaces.

5. Remove the nut and separate the skid control sensor wire (w/ ABS).

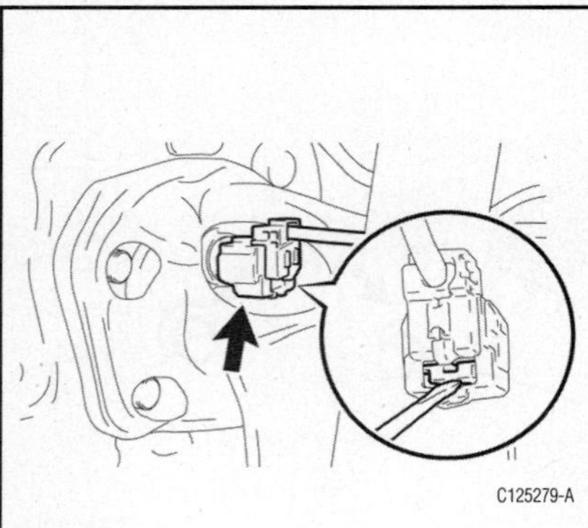

Fig. 166 Using a screwdriver, remove the claw of the connector lock portion and disconnect the skid control sensor wire connector.

➡Do not remove the connector cover from the connector because the skid control sensor wire may be damaged.

6. Separate rear LH flexible hose.

a. Using a union nut wrench (10 mm), separate the brake tube while holding the flexible hose with a wrench.

b. Remove the clip and disconnect the flexible hose from the axle beam.

7. Separate rear RH flexible hose. The removal procedure for the RH side is the same as that for the LH side.

➡Do not remove the bolts.

8. Loosen rear axle beam
9. Remove rear absorber cap (for Hatchback)

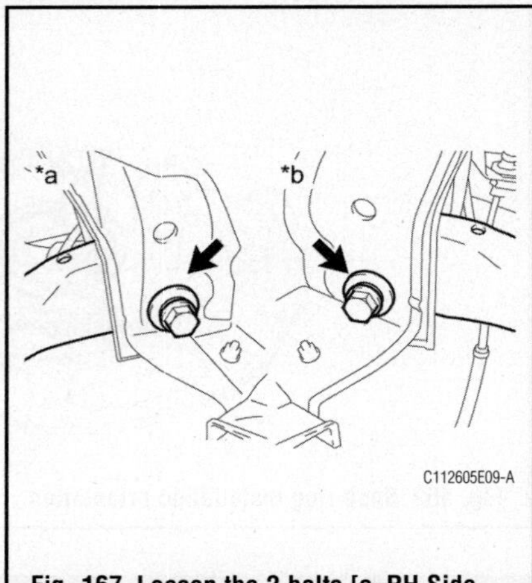

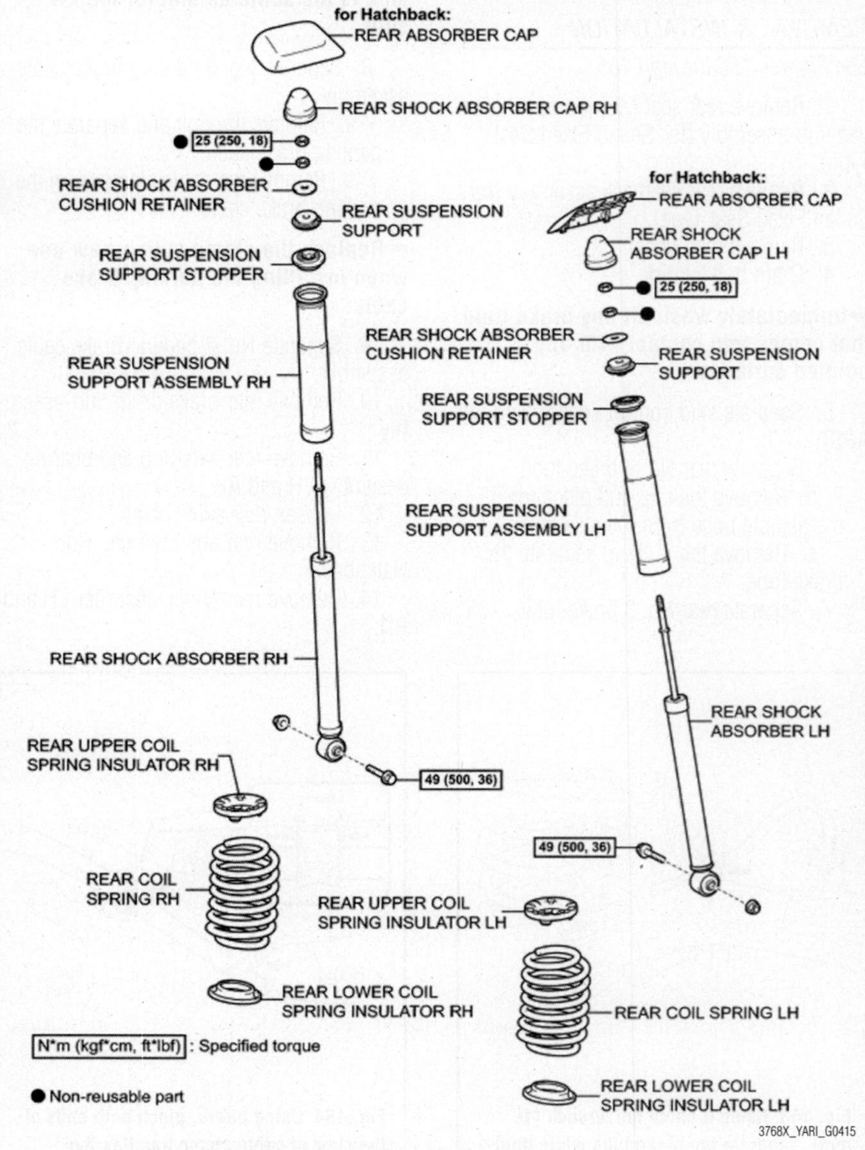

Fig. 167 Loosen the 2 bolts [a=RH Side, b= LH Side]

Fig. 168 View of the rear coil spring and components

10. Remove the rear shock absorber.
11. Lower the jacks slowly.
12. Remove the coil spring, coil spring insulator upper and coil spring insulator lower.

To install:

a. Install the coil spring insulator lower onto the axle beam.
b. Install the coil spring insulator upper so that its gap fits onto the end of coil spring.
c. Install the coil spring onto the axle beam.

➡The paint mark of the coil spring should be towards the underside and rear side of the vehicle.

13. Install the shock absorber.
14. Install rear absorber cap (for Hatchback)

➡Do not remove the bolts.

15. Tighten rear axle beam bolts.
16. Separate rear RH flexible hose. The removal procedure for the RH side is the same as that for the LH side.
17. Connect rear LH flexible hose.
a. Install the clip and connect the flexible hose to the axle beam.
b. Using a union nut wrench (10 mm), connect the brake tube while holding the flexible hose with a wrench.

➡Do not remove the connector cover from the connector because the skid control sensor wire may be damaged.

18. Install the nut and separate the skid control sensor wire (w/ ABS).

➡Immediately wash off any brake fluid that comes into contact with any painted surfaces.

19. Fill brake fluid
20. Install rear wheel
21. Install rear seatback assembly (for Sedan Fixed Seat Type).
22. Install rear seat cushion cover pad sub-assembly (for Sedan Fixed Seat Type).

SHOCK ABSORBER

REMOVAL & INSTALLATION

See Figure 169.

1. Remove the rear wheel.
2. Remove the rear seat.
3. Remove rear absorber cap (for hatchback).
4. Remove rear shock absorber cap.
5. Remove rear shock absorber:

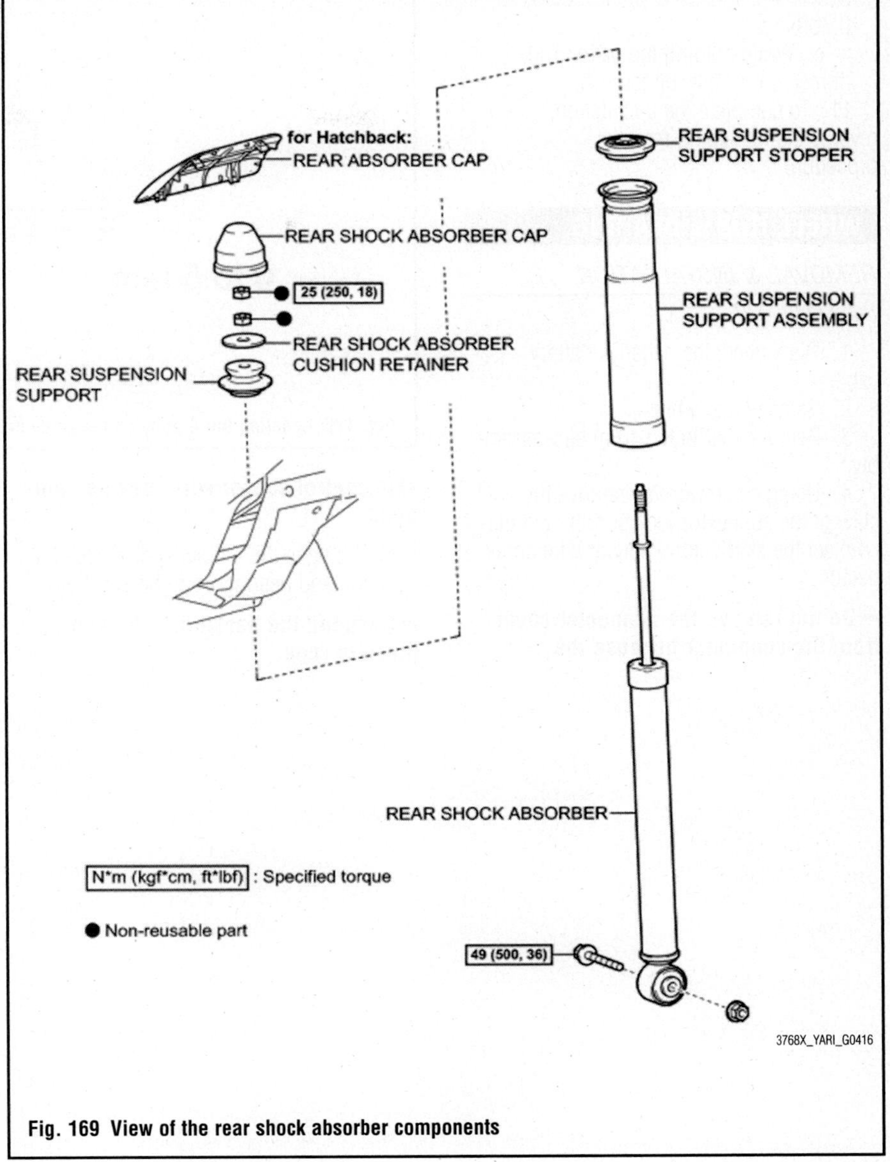

Fig. 169 View of the rear shock absorber components

a. Support the axle beam with a jack. Insert a wooden block between the jack and the rear axle spring seat to prevent damage.
b. Remove the 2 nuts while keeping the piston rod from rotating.
c. Remove the cushion retainer and suspension support.
d. Remove the bolt while keeping the nut from rotating and remove the shock absorber.

➡Remove the nut on the bolt side because the one on the lower side is a jam nut.

6. Remove rear suspension support stopper.
7. Remove rear suspension support assembly.

To install:

8. Install rear suspension support assembly.
9. Install rear suspension support stopper.
10. Temporarily tighten rear shock absorber:
a. Support the axle beam with a jack. Insert a wooden block between the jack and the rear axle spring seat to prevent damage.
b. Jack up the axle beam slowly, and provisionally tighten the shock absorber (lower side) with the bolt and nut to the axle beam.
c. Install the suspension support and cushion retainer.
d. While holding the piston rod, install a new nut (lower nut) to the

specified standard. Standard: 0.591 to 0.709 inches.

e. While holding the piston rod, tighten a new nut (upper nut).

11. To complete the installation, reverse the remaining removal procedure.

WHEEL HUB & BEARING

REMOVAL & INSTALLATION

See Figure 170.

1. Disconnect the negative battery cable.
2. Remove rear wheel.
3. Remove rear brake drum sub-assembly.
4. Using a screwdriver, remove the claw of the connector lock portion and disconnect the skid control sensor wire connector.

➡**Do not remove the connector cover from the connector because the skid control sensor wire may be damaged.**

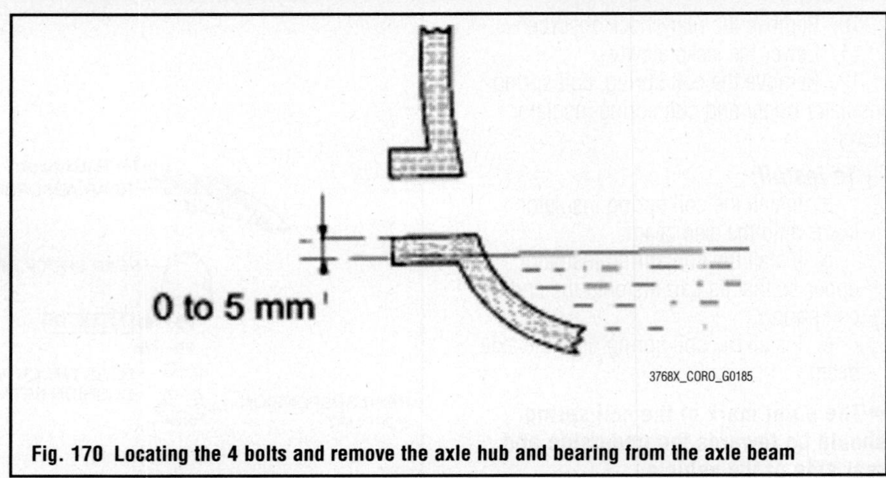

Fig. 170 Locating the 4 bolts and remove the axle hub and bearing from the axle beam

5. Remove the 4 bolts and remove the axle hub and bearing from the axle beam.

➡**Suspend the backing plate with a piece of rope.**

To install:

6. Install the axle hub and bearing onto the axle beam with the 4 bolts. Tighten to 67 ft. lbs. (90 Nm).
7. Inspect rear axle hub bearing.
8. To complete the installation, reverse the remaining removal procedure.

TOYOTA

Diagnostic Trouble Codes

DIAGNOSTIC TROUBLE CODES

OBD II VEHICLE APPLICATIONS

TOYOTA

4RUNNER
2011–2012
- 4.0L V6 SFI . . . .Engine Code: 1GR-FE
- 2.7L L4 SFI. . . . Engine Code: 2TR-FE

AVALON
2011–2012
- 3.5L V6 SFI . . . .Engine Code: 2GR-FE

CAMRY
2011–2012
- 2.4L L4 EFI. . . . Engine Code: 2AZ-FE
- 2.4L L4 EFI. . . .Engine Code: 2AZ-FXE
- 2.5L L4 SFI. . . . Engine Code: 2AR-FE
- 2.5L L4 SFI . . . . Engine Code: 2AR-FXE
- 3.5L V6 SFI . . . .Engine Code: 2GR-FE

COROLLA
2011–2012
- 1.8L L4 EFI. . . . Engine Code: 2ZR-FE
- 2.4L L4 EFI. . . . Engine Code: 2AZ-FE

FJ CRUISER
2011–2012
- 4.0L V6 SFI . . . .Engine Code: 1GR-FE

HIGHLANDER
2011–2012
- 2.7L L4 SFI. . . . Engine Code: 1AR-FE
- 3.3L V6 SFI . . . . Engine Code: 3MZ-FE
- 3.5L V6 SFI . . . .Engine Code: 2GR-FE

LAND CRUISER
2011–2012
- 5.7L V8 SFI . . . .Engine Code: 3UR-FE

MATRIX
2011–2012
- 1.8L L4 EFI. . . . Engine Code: 2ZR-FE
- 2.4L L4 EFI. . . . Engine Code: 2AZ-FE

PRIUS, PRIUS V
2011–2012
- 1.5L L4 SFI . . . .Engine Code: 2NR-FXE
- 1.8L L4 EFI. . . .Engine Code: 2ZR-FXE

RAV4
2011–2012
- 2.5L L4 SFI. . . . Engine Code: 2AR-FE
- 3.5L V6 SFI . . . .Engine Code: 2GR-FE

SEQUOIA
2011–2012
- 4.6L V8 SFI . . . .Engine Code: 1UR-FE
- 5.7L V8 SFI . . . .Engine Code: 3UR-FBE

SIENNA
2011–2012
- 2.7L L4 SFI. . . . Engine Code: 1AR-FE
- 3.5L V6 SFI . . . .Engine Code: 2GR-FE

TACOMA
2011–2012
- 2.7L L4 SFI. . . . Engine Code: 1AR-FE
- 4.0L V6 SFI . . . .Engine Code: 1GR-FE

TUNDRA
2011–2012
- 4.0L V6 MFI . . . .Engine Code: 1GR-FE
- 4.6L V8 SFI . . . .Engine Code: 1UR-FE
- 5.7L V8 SFI . . . .Engine Code: 3UR-FE
- 5.7L V8 SFI . . . .Engine Code: 3UR-FBE

VENZA
2011–2012
- 2.7L L4 SFI. . . . Engine Code: 1AR-FE
- 3.5L V6 SFI . . . .Engine Code: 2GR-FE

YARIS
2011–2012
- 1.5L L4 SFI. . . . Engine Code: 1NZ-FE

OBD II Trouble Code List (P0XXX Codes)

DTC	Trouble Code Title and Conditions
DTC: P0010 **1T ECM, MIL: Yes** **Year:** 2011, 2012 **Model:** 4Runner, Avalon, Camry, Corolla, FJ Cruiser, Highlander, Land Cruiser, Matrix, Prius, Rav4, Sequoia, Tundra, Tacoma, Venza, Yaris **Engine:** L4, V6, V8	**Camshaft Position "A" Actuator Circuit (Bank 1):** Open or short in OCV circuit. Monitor runs whenever following DTCs are not present: None. All of the following conditions are met: Starter: OFF Ignition switch: ON Time after turning ignition switch off to ON: 0.5 seconds or more.
DTC: P0011 **1T ECM, MIL: Yes** **Year:** 2011, 2012 **Model:** 4Runner, Avalon, Camry, Corolla, FJ Cruiser, Highlander, Land Cruiser, Matrix, Prius, Rav4, Sequoia, Tundra, Tacoma, Venza, Yaris **Engine:** L4, V6, V8	**Camshaft Position "A" - Timing Over-Advanced or System Performance (Bank 1):** Advanced cam timing: The intake valve timing is not adjusted in the valve timing advance range. With warm engine and engine speed of between 500 rpm and 4000 rpm, all conditions (a), (b) and (c) are met: (a) Difference between target and actual intake valve timings is more than 5° CA (Crankshaft Angle) for 4.5 seconds (b) Current intake valve timing is fixed (timing changes less than 5° CA in 5 seconds) (c) Variations in VVT controller timing are more than 19° CA of maximum delayed timing (advanced)
DTC: P0012 **1T ECM, MIL: Yes** **Year:** 2011, 2012 **Model:** 4Runner, Avalon, Camry, Corolla, FJ Cruiser, Highlander, Land Cruiser, Matrix, Prius, Rav4, Sequoia, Tundra, Tacoma, Venza, Yaris **Engine:** L4, V6, V8	**Camshaft Position "A" - Timing Over-Retarded (Bank 1):** Retarded camshaft timing: The intake valve timing is not adjusted in the valve timing retard range. With warm engine and engine speed of between 450 rpm and 4,000 rpm, all conditions (a), (b) and (c) met: a). Difference between target and actual intake valve timings more than 5°CA (Crankshaft Angle) for 4.5 seconds b). Current intake valve timing fixed (timing changes less than 5°CA in 5 seconds) c). Variations in VVT controller timing 19°CA or less of maximum delayed timing (malfunction in retarded timing)
DTC: P0013 **1T ECM, MIL: Yes** **Year:** 2011, 2012 **Model:** 4Runner, Avalon, Camry, Corolla, FJ Cruiser, Highlander, Land Cruiser, Matrix, Prius, Rav4, Sequoia, Tundra, Tacoma, Venza, Yaris **Engine:** L4, V6, V8	**Camshaft Position "B" Actuator Circuit / Open (Bank 1) :** Open or short in OCV for exhaust camshaft (for Bank 1) circuit. Monitor runs whenever following DTCs are not present: None. All of the following conditions are met: Starter: OFF Ignition switch: ON Time after ignition switch off to ON: 0.5 seconds or more.
DTC: P0014 **1T ECM, MIL: Yes** **Year:** 2011, 2012 **Model:** 4Runner, Avalon, Camry, Corolla, FJ Cruiser, Highlander, Land Cruiser, Matrix, Prius, Rav4, Sequoia, Tundra, Tacoma, Venza, Yaris **Engine:** L4, V6, V8	**Camshaft Position "B" - Timing Over-Advanced or System Performance (Bank 1):** Exhaust valve timing is not adjusted in valve timing advance range Monitor runs whenever following DTCs are not present: None. All of the following conditions are met: Starter: OFF Ignition switch: ON Time after ignition switch off to ON: 0.5 seconds or more.
DTC: P0015 **1T ECM, MIL: Yes** **Year:** 2011, 2012 **Model:** 4Runner, Avalon, Camry, Corolla, FJ Cruiser, Highlander, Land Cruiser, Matrix, Prius, Rav4, Sequoia, Tundra, Tacoma, Venza, Yaris **Engine:** L4, V6, V8	**Camshaft Position "B" - Timing Over-Retarded (Bank 1):** With the engine running at a speed of between 500–4000 RPM, ECT 75 to 100°C (167 to 212°F) and battery voltage 11 V or more. The exhaust valve timing is not adjusted in the valve timing retard range.

DTC	Trouble Code Title and Conditions
DTC: P0016 **1T ECM, MIL: Yes** **Year:** 2011, 2012 **Model:** 4Runner, Avalon, Camry, Corolla, FJ Cruiser, Highlander, Land Cruiser, Matrix, Prius, Rav4, Sequoia, Tundra, Tacoma, Venza, Yaris **Engine:** L4, V6, V8	**Crankshaft Position - Camshaft Position Correlation (Bank 1 Sensor A):** Deviations in crankshaft and camshaft position sensor (for intake camshaft) 1 signals. Monitor runs whenever following DTCs are not present: P0010, P0020 (OCV bank 1, 2) P0100, P0102, P0103 (MAF meter) P0115, P0117, P0118 (ECT sensor) P0125 (Insufficient ECT for closed loop) P0335 (CKP sensor) Engine rpm: 500 to 1000 rpm
DTC: P0017 **1T ECM, MIL: Yes** **Year:** 2011, 2012 **Model:** 4Runner, Avalon, Camry, Corolla, FJ Cruiser, Highlander, Land Cruiser, Matrix, Prius, Rav4, Sequoia, Tundra, Tacoma, Venza, Yaris **Engine:** L4, V6, V8	**Crankshaft Position - Camshaft Position Correlation (Bank 1 Sensor B):** Deviations in crankshaft and camshaft position sensor (for exhaust camshaft) 1 signals Monitor runs whenever following DTCs are not present: P0013, P0023 (Exhaust OCV bank 1, 2) P0100, P0102, P0103 (MAF meter) P0115, P0117, P0118 (ECT sensor) P0125 (Insufficient ECT for closed loop) P0335 (CKP sensor) VVT feedback mode: Executing VVT: Maximum advanced position Engine rpm: 500 to 1000 rpm
DTC: P0018 **1T ECM, MIL: Yes** **Year:** 2011, 2012 **Model:** 4Runner, Avalon, Camry, Corolla, FJ Cruiser, Highlander, Land Cruiser, Matrix, Prius, Rav4, Sequoia, Tundra, Tacoma, Venza, Yaris **Engine:** L4, V6, V8	**Crankshaft Position - Camshaft Position Correlation (Bank 2 Sensor A):** Monitor runs whenever following DTCs are not present: P0010, P0020 (OCV bank 1, 2) P0100, P0102, P0103 (MAF meter) P0115, P0117, P0118 (ECT sensor) P0125 (Insufficient ECT for closed loop): P0335 (CKP sensor) Engine rpm: 500 to 1000 rpm Deviation in crankshaft position sensor signal and VVT sensor 2 (for intake camshaft bank 2) signal
DTC: P0019 **1T ECM, MIL: Yes** **Year:** 2011, 2012 **Model:** 4Runner, Avalon, Camry, Corolla, FJ Cruiser, Highlander, Land Cruiser, Matrix, Prius, Rav4, Sequoia, Tundra, Tacoma, Venza, Yaris **Engine:** L4, V6, V8	**Crankshaft Position - Camshaft Position Correlation (Bank 2 Sensor B):** Deviation in crankshaft position sensor signal and VVT sensor 2 (for exhaust camshaft bank 2) signal. The monitor will run whenever these DTCs are not present: P0013, P0023 (Exhaust VVT Oil Control Valve) P0017, P0019 (Exhaust VVT System - Misalignment) P0102, P0103 (Mass Air Flow Sensor) P0115, P0117, P0118 (Engine Coolant Temperature Sensor) P0125 (Insufficient Coolant Temperature for Closed Loop Fuel Control) P0335 (Crankshaft Position Sensor) P0340 (Camshaft Position Sensor) Engine speed: 500 to 1000 rpm
DTC: P0020 **1T ECM, MIL: Yes** **Year:** 2011, 2012 **Model:** 4Runner, Avalon, Camry, Corolla, FJ Cruiser, Highlander, Land Cruiser, Matrix, Prius, Rav4, Sequoia, Tundra, Tacoma, Venza, Yaris **Engine:** L4, V6, V8	**Camshaft Position "A" Actuator Circuit (Bank 2):** With the ignition switch ON for 5 seconds or more and starter OFF, an open or short is detected in the OCV for intake side (for Bank 2) circuit. Monitor runs whenever following DTCs are not present: None All of the following conditions are met:- Starter: OFF Power switch: On (IG) Time after turning power switch off to on (IG): 0.5 seconds or more
DTC: P0021 **1T ECM, MIL: Yes** **Year:** 2011, 2012 **Model:** 4Runner, Avalon, Camry, Corolla, FJ Cruiser, Highlander, Land Cruiser, Matrix, Prius, Rav4, Sequoia, Tundra, Tacoma, Venza, Yaris **Engine:** L4, V6, V8	**Camshaft Position "A" - Timing Over-Advanced or System Performance (Bank 2):** With the engine running at a speed of between 400–1400 RPM, ECT 75 to 100°C (167 to 212°F) and battery voltage 11 V or more. The intake valve timing is not adjusted in the valve timing advance range.

DTC	Trouble Code Title and Conditions
DTC: P0022 **1T ECM, MIL: Yes** **Year:** 2011, 2012 **Model:** 4Runner, Avalon, Camry, Corolla, FJ Cruiser, Highlander, Land Cruiser, Matrix, Prius, Rav4, Sequoia, Tundra, Tacoma, Venza, Yaris **Engine:** L4, V6, V8	**Camshaft Position "A" - Timing Over-Retarded (Bank 2):** With warm engine and engine speed of between 550 rpm and 4000 rpm, all conditions (a), (b) and (c) are met; (a) Difference between target and actual intake valve timings is more than 5° CA (Crankshaft Angle) for 4.5 seconds. (b) Current intake valve timing is fixed (timing changes less than 5° CA in 5 seconds). (c) Variations in VVT controller timing are more than 19° CA of maximum delayed timing (advanced).
DTC: P0023 **1T ECM, MIL: Yes** **Year:** 2011, 2012 **Model:** 4Runner, Avalon, Camry, Corolla, FJ Cruiser, Highlander, Land Cruiser, Matrix, Prius, Rav4, Sequoia, Tundra, Tacoma, Venza, Yaris **Engine:** L4, V6, V8	**Camshaft Position "B" Actuator Circuit / Open (Bank 2):** Open or short in OCV for exhaust camshaft (for Bank 2) circuit Monitor runs whenever following DTCs not present: None All of the following conditions are met: Starter: OFF Engine switch: On (IG) Time after engine switch off to on (IG): 0.5 seconds or more.
DTC: P0024 **1T ECM, MIL: Yes** **Year:** 2011, 2012 **Model:** 4Runner, Avalon, Camry, Corolla, FJ Cruiser, Highlander, Land Cruiser, Matrix, Prius, Rav4, Sequoia, Tundra, Tacoma, Venza, Yaris **Engine:** L4, V6, V8	**Camshaft Position "B" - Timing Over-Advanced or System Performance (Bank 2):** With the engine running at a speed of between 500–4000 RPM, ECT 75 to 100°C (167 to 212°F) and battery voltage 11 V or more. The exhaust valve timing is not adjusted in the valve timing advance range.
DTC: P0025 **1T ECM, MIL: Yes** **Year:** 2011, 2012 **Model:** 4Runner, Avalon, Camry, Corolla, FJ Cruiser, Highlander, Land Cruiser, Matrix, Prius, Rav4, Sequoia, Tundra, Tacoma, Venza, Yaris **Engine:** L4, V6, V8	**Camshaft Position "B" - Timing Over-Retarded (Bank 2):** With the engine running at a speed of between 500–4000 RPM, ECT 75 to 100°C (167 to 212°F) and battery voltage 11 V or more. The exhaust valve timing is not adjusted in the valve timing retard range.
DTC: P0031 **1T ECM, MIL: Yes** **Year:** 2011,2012 **Model:** 4Runner, Avalon, Camry, Corolla, FJ Cruiser, Highlander, Land Cruiser, Matrix, Prius, Rav4, Sequoia, Tundra, Tacoma, Venza, Yaris **Engine:** L4, V6, V8	**Oxygen (A/F) Sensor Heater Control Circuit Low (Bank 1 Sensor 1):** Air-Fuel Ratio (A/F) sensor heater current failure. With the battery voltage 10.5 V or more and the air fuel ratio sensor heater duty-cycle ratio 50% or more, time after engine start 10 seconds or more the air fuel ratio sensor heater current is less than 0.8 A.
DTC: P0032 **1T ECM, MIL: Yes** **Year:** 2011,2012 **Model:** 4Runner, Avalon, Camry, Corolla, FJ Cruiser, Highlander, Land Cruiser, Matrix, Prius, Rav4, Sequoia, Tundra, Tacoma, Venza, Yaris **Engine:** L4, V6, V8	**Oxygen (A/F) Sensor Heater Control Circuit High (Bank 1 Sensor 1):** Air-Fuel Ratio (A/F) sensor heater current failure. With the battery voltage 10.5 V or more and the air fuel ratio sensor heater duty-cycle ratio 50% or more, time after engine start 10 seconds or more the air fuel ratio sensor heater current is less than 0.8 A.

DTC	Trouble Code Title and Conditions
DTC: P0037 **1T ECM, MIL: Yes** **Year:** 2011,2012 **Model:** 4Runner, Avalon, Camry, Corolla, FJ Cruiser, Highlander, Land Cruiser, Matrix, Prius, Rav4, Sequoia, Tundra, Tacoma, Venza, Yaris **Engine:** L4, V6, V8	**Oxygen Sensor Heater Control Circuit Low (Bank 1 Sensor 2):** Monitor runs whenever following DTCs are not present: None. Battery voltage: 10.5 V or more, and less than 20 V Heated Oxygen (HO2) sensor heater current less than 0.3 A.
DTC: P0038 **1T ECM, MIL: Yes** **Year:** 2011,2012 **Model:** 4Runner, Avalon, Camry, Corolla, FJ Cruiser, Highlander, Land Cruiser, Matrix, Prius, Rav4, Sequoia, Tundra, Tacoma, Venza, Yaris **Engine:** L4, V6, V8	**Oxygen Sensor Heater Control Circuit High (Bank 1 Sensor 2):** Monitor runs whenever following DTCs are not present: None. Battery voltage: 10.5 V or more Engine: Running Starter: OFF Catalyst active air fuel ratio control: Not operating Time after heater ON: 10 seconds or more Learned heater OFF current operation: Complete The heated oxygen sensor (sensor 2) heater current is more than 2 A for 5 seconds.
DTC: P0051 **1T ECM, MIL: Yes** **Year:** 2011,2012 **Model:** 4Runner, Avalon, Camry, Corolla, FJ Cruiser, Highlander, Land Cruiser, Matrix, Prius, Rav4, Sequoia, Tundra, Tacoma, Venza, Yaris **Engine:** L4, V6, V8	**Oxygen (A/F) Sensor Heater Control Circuit Low (Bank 2 Sensor 1):** Battery voltage 10.5 V or more and the engine running for 10 seconds or more, heater output duty 50% or more. The Air-Fuel Ratio (A/F) sensor heater current is below 0.8 A.
DTC: P0052 **1T ECM, MIL: Yes** **Year:** 2011,2012 **Model:** 4Runner, Avalon, Camry, Corolla, FJ Cruiser, Highlander, Land Cruiser, Matrix, Prius, Rav4, Sequoia, Tundra, Tacoma, Venza, Yaris **Engine:** L4, V6, V8	**Oxygen (A/F) Sensor Heater Control Circuit High (Bank 2 Sensor 1):** With the engine running and battery voltage less than 20 V, an Air-Fuel Ratio (A/F) sensor heater current failure was detected. Monitor runs whenever following DTCs are not present: None. Heater output duty: 0% or more Time after engine start: 10 seconds or more Active heater off control: Not operating Active heater on control: Not operating
DTC: P0057 **1T ECM, MIL: Yes** **Year:** 2011,2012 **Model:** 4Runner, Avalon, Camry, Corolla, FJ Cruiser, Highlander, Land Cruiser, Matrix, Prius, Rav4, Sequoia, Tundra, Tacoma, Venza, Yaris **Engine:** L4, V6, V8	**Oxygen Sensor Heater Control Circuit Low (Bank 2 Sensor 2):** Heated Oxygen (HO2) sensor heater current is less than 0.3 A. Monitor runs whenever following DTCs are not present: None. Battery voltage: 10.5 V or more, and less than 20 V.
DTC: P0058 **1T ECM, MIL: Yes** **Year:** 2011,2012 **Model:** 4Runner, Avalon, Camry, Corolla, FJ Cruiser, Highlander, Land Cruiser, Matrix, Prius, Rav4, Sequoia, Tundra, Tacoma, Venza, Yaris **Engine:** L4, V6, V8	**Oxygen Sensor Heater Control Circuit High (Bank 2 Sensor 2):** Heated Oxygen (HO2) sensor heater current is less than 0.3 A. Monitor runs whenever following DTCs are not present: None. Battery voltage: 10.5 V or more Engine: Running Starter: OFF Catalyst active air fuel ratio control: Not operating Time after heater on: 10 seconds or more Learned heater off current operation: Complete
DTC: P0069-273 **T ECM** **Year:** 2011, 2012 **Model:** Highlander/Hybrid **Engine:** 3.5L V6	**Manifold Absolute Pressure - Barometric Pressure Correlation:** Difference between the atmospheric pressure value of the atmospheric pressure sensor in the inverter with converter assembly and the manifold absolute pressure sensor (for EGR control) exceeds a specified value. The same condition recurs within 3 hours when driving in EV mode.

DTC	Trouble Code Title and Conditions
DTC: P0087 **1T ECM, MIL: Yes** **Year:** 2011, 2012 **Model:** Camry **Engine:** 3.5L V6	**Fuel Rail / System Pressure - Too Low:** Monitor runs whenever following DTCs are not present: None. Time after engine start: 0.2 seconds or more Time after fuel cut finished: 3 seconds or more Target fuel pressure: Small change
DTC: P0088 **T ECM, MIL: Yes** **Year:** 2011, 2012 **Model:** Camry **Engine:** 3.5L V6	**Fuel Rail / System Pressure - Too High:** Monitor runs whenever following DTCs are not present: None Time after engine start: 0.2 seconds or more Time after fuel cut finished: 3 seconds or more Target fuel pressure: Small change
DTC: P0100 **1T ECM, MIL: Yes** **Year:** 2011,2012 **Model:** 4Runner, Avalon, Camry, Corolla, FJ Cruiser, Highlander, Land Cruiser, Matrix, Prius, Rav4, Sequoia, Tundra, Tacoma, Venza, Yaris **Engine:** L4, V6, V8	**Mass or Volume Air Flow Circuit:** Mass Air Flow (MAF) meter voltage less than 0.2 V, or more than 4.9 V for 3 seconds Monitor runs whenever following DTCs are not present: None. MAF meter voltage: Less than 0.2 V, or more than 4.9 V.
DTC: P0101 **2T ECM, MIL: Yes** **Year:** 2011, 2012 **Model:** 4Runner, Avalon, Camry, Corolla, FJ Cruiser, Highlander, Land Cruiser, Matrix, Prius, Rav4, Sequoia, Tundra, Tacoma, Venza, Yaris **Engine:** L4, V6, V8	**Mass Air Flow Circuit Range / Performance Problem:** Mass Air Flow (MAF) meter voltage less than 0.2 V, or more than 4.9 V for 3 seconds TP (Throttle position) sensor voltage: 0.24 to 2 V Engine: Running Battery voltage: 10.5 V or more ECT: 158°F (70°C) or more Estimated load: 30 to 70%
DTC: P0102 **1T ECM, MIL: Yes** **Year:** 2011, 2012 **Model:** 4Runner, Avalon, Camry, Corolla, FJ Cruiser, Highlander, Land Cruiser, Matrix, Prius, Rav4, Sequoia, Tundra, Tacoma, Venza, Yaris **Engine:** L4, V6, V8	**Mass or Volume Air Flow Circuit Low Input:** The Mass air flow meter voltage is less than 0.2 V for 3 seconds. Engine: Running Battery voltage: 10.5 V or more ECT: 158°F (70°C) or more Estimated load: 30 to 70% (1 trip detection logic: Engine speed is less than 4000 rpm) (2 trip detection logic: Engine speed is 4000 rpm or more)
DTC: P0103 **1T ECM, MIL: Yes** **Year:** 2011, 2012 **Model:** 4Runner, Avalon, Camry, Corolla, FJ Cruiser, Highlander, Land Cruiser, Matrix, Prius, Rav4, Sequoia, Tundra, Tacoma, Venza, Yaris **Engine:** L4, V6, V8	**Mass or Volume Air Flow Circuit High Input:** Short in Mass Air Flow (MAF) meter circuit Mass air flow meter voltage is more than 4.9 V for 3 seconds. (1 trip detection logic: Engine speed is less than 4000 rpm) (2 trip detection logic: Engine speed is 4000 rpm or more)

DTC	Trouble Code Title and Conditions
DTC: P0106 **2T ECM, MIL: Yes** **Year:** 2011, 2012 **Model:** 4Runner, Avalon, Camry, Corolla, FJ Cruiser, Highlander, Land Cruiser, Matrix, Prius, Rav4, Sequoia, Tundra, Tacoma, Venza, Yaris **Engine:** L4, V6, V8	**Manifold Absolute Pressure / Barometric Pressure Circuit Range / Performance Problem:** Monitor runs whenever following DTCs are not present: P0010 (Camshaft timing oil control valve assembly Bank 1) P0011 (VVT System Bank 1 - Advance) P0012 (VVT System Bank 1 - Retard) P0016 (VVT System Bank 1 - Misalignment) P0107, P0108 (Manifold Absolute Pressure) P0112, P0113 (Intake Air Temperature Sensor) P0115, P0117, P0118 (Engine Coolant Temperature Sensor) P0120, P0121, P0122, P0123, P0220, P0222, P0223, P2135 (Throttle Position Sensor) P0125 (Insufficient Engine Coolant Temperature for Closed Loop Fuel Control) P0335 (Crankshaft Position Sensor) Battery voltage: 10.5 V or more Engine coolant temperature: 10°C (50°F) or more Intake air temperature: 10°C (50°F) or more Engine speed: 1000 rpm or more Throttle position: Less than 2° Atmospheric pressure: 76 kPa-a (570 mmHg-a) or more
DTC: P0107 **2T ECM, MIL: Yes** **Year:** 2011, 2012 **Model:** 4Runner, Avalon, Camry, Corolla, FJ Cruiser, Highlander, Land Cruiser, Matrix, Prius, Rav4, Sequoia, Tundra, Tacoma, Venza, Yaris **Engine:** L4, V6, V8	**Manifold Absolute Pressure / Barometric Pressure Circuit Low Input:** Monitor runs whenever following DTCs are not present: None Starter: Off Time after starter on to off: 2 seconds or more.
DTC: P0108 **2T ECM, MIL: Yes** **Year:** 2011, 2012 **Model:** 4Runner, Avalon, Camry, Corolla, FJ Cruiser, Highlander, Land Cruiser, Matrix, Prius, Rav4, Sequoia, Tundra, Tacoma, Venza, Yaris **Engine:** L4, V6, V8	**Manifold Absolute Pressure / Barometric Pressure Circuit High Input:** Monitor runs whenever following DTCs are not present: None Starter: Off Time after starter on to off: 2 seconds or more
DTC: P0110 **2T ECM, MIL: Yes** **Year:** 2011, 2012 **Model:** 4Runner, Avalon, Camry, Corolla, FJ Cruiser, Highlander, Land Cruiser, Matrix, Prius, Rav4, Sequoia, Tundra, Tacoma, Venza, Yaris **Engine:** L4, V6, V8	**Intake Air Temperature Circuit:** Time after engine start: 10 seconds or more. Engine coolant temperature: 70°C (158°F) or higher. Open or short in IAT sensor circuit for 0.5 seconds.
DTC: P0111 **2T ECM, MIL: Yes** **Year:** 2011, 2012 **Model:** 4Runner, Avalon, Camry, Corolla, FJ Cruiser, Highlander, Land Cruiser, Matrix, Prius, Rav4, Sequoia, Tundra, Tacoma, Venza, Yaris **Engine:** L4, V6, V8	**Intake Air Temperature Sensor Gradient Too High:** When either of following conditions is met: In duration between engine warmed up and next engine starts, change in intake air temperature sensor output bellow threshold. During engine warming up after cold engine starts, change in intake air temperature sensor output bellow threshold.

DTC	Trouble Code Title and Conditions
DTC: P0112 **2T ECM, MIL: Yes** **Year:** 2011, 2012 **Model:** 4Runner, Avalon, Camry, Corolla, FJ Cruiser, Highlander, Land Cruiser, Matrix, Prius, Rav4, Sequoia, Tundra, Tacoma, Venza, Yaris **Engine:** L4, V6, V8	**Intake Air Temperature Circuit Low Input:** Monitor runs whenever following DTCs are not present: None Short in intake air temperature sensor circuit for 0.5 seconds. Battery voltage: 8 V or more Engine switch: On Starter: OFF
DTC: P0113 **1T ECM, MIL: Yes** **Year:** 2011, 2012 **Model:** 4Runner, Avalon, Camry, Corolla, FJ Cruiser, Highlander, Land Cruiser, Matrix, Prius, Rav4, Sequoia, Tundra, Tacoma, Venza, Yaris **Engine:** L4, V6, V8	**Intake Air Temperature Circuit High Input:** Open in IAT sensor circuit for 0.5 seconds. The Intake air temperature sensor voltage is more than 4.91 V [Less than -40°C (-40°F)] Monitor runs whenever following DTCs are not present: None Battery voltage: 8 V or more Power switch: On (IG)
DTC: P0115 **1T ECM, MIL: Yes** **Year:** 2011, 2012 **Model:** 4Runner, Avalon, Camry, Corolla, FJ Cruiser, Highlander, Land Cruiser, Matrix, Prius, Rav4, Sequoia, Tundra, Tacoma, Venza, Yaris **Engine:** L4, V6, V8	**Engine Coolant Temperature Circuit Malfunction:** Monitor runs whenever following DTCs are not present: None Engine coolant temperature sensor voltage: Less than 0.14 V, or more than 4.91 V. Open or short in ECT sensor circuit
DTC: P0116 **1T ECM, MIL: Yes** **Year:** 2011, 2012 **Model:** 4Runner, Avalon, Camry, Corolla, FJ Cruiser, Highlander, Land Cruiser, Matrix, Prius, Rav4, Sequoia, Tundra, Tacoma, Venza, Yaris **Engine:** L4, V6, V8	**Engine Coolant Temperature Circuit Range / Performance Problem:** Engine Coolant Temperature (ECT) is between 35°C and 60°C (95°F and 140°F) when engine started, and conditions (a) and (b) are met: (a). Vehicle driven at varying speeds (accelerated and decelerated) (b). ECT change is within 3°C (5.4°F) of initial ECT.
DTC: P0117 **1T ECM, MIL: Yes** **Year:** 2011, 2012 **Model:** 4Runner, Avalon, Camry, Corolla, FJ Cruiser, Highlander, Land Cruiser, Matrix, Prius, Rav4, Sequoia, Tundra, Tacoma, Venza, Yaris **Engine:** L4, V6, V8	**Engine Coolant Temperature Circuit Low Input:** Monitor runs whenever following DTCs are not present: None The engine coolant temperature sensor voltage is less than 0.14 V [More than 140°C (284°F)]. Short in ECT sensor circuit for 0.5 seconds.
DTC: P0118 **1T ECM, MIL: Yes** **Year:** 2011, 2012 **Model:** 4Runner, Avalon, Camry, Corolla, FJ Cruiser, Highlander, Land Cruiser, Matrix, Prius, Rav4, Sequoia, Tundra, Tacoma, Venza, Yaris **Engine:** L4, V6, V8	**Engine Coolant Temperature Circuit High Input:** Monitor runs whenever following DTCs are not present: None Open in ECT sensor circuit for 0.5 seconds. The engine coolant temperature sensor voltage is more than 4.91 V [Less than -40°C (-40°F)].

DTC	Trouble Code Title and Conditions
DTC: P011B **1T ECM, MIL: Yes** **Year:** 2011, 2012 **Model:** 4Runner, Avalon, Camry, Corolla, FJ Cruiser, Highlander, Land Cruiser, Matrix, Prius, Rav4, Sequoia, Tundra, Tacoma, Venza, Yaris **Engine:** L4, V6, V8	**Engine Coolant Temperature / Intake Air Temperature Correlation:** Monitor runs whenever following DTCs are not present: None All of the following conditions are met: Soak time: 7 hours or more Battery voltage: 10.5 V or more Time after engine start: 15 second or more Either of the following conditions is met: (a) or (b) (a) Minimum intake air temperature after engine start: -10°C (14°F) or more (b) Engine coolant temperature before engine start: -10°C (14°F) or more Engine coolant temperature sensor circuit fail (P0115, P0117, P0118, P0125): Not detected Intake air temperature sensor circuit fail (P0112, P0113): Not detected
DTC: P0120 **1T ECM, MIL: Yes** **Year:** 2011, 2012 **Model:** 4Runner, Avalon, Camry, Corolla, FJ Cruiser, Highlander, Land Cruiser, Matrix, Prius, Rav4, Sequoia, Tundra, Tacoma, Venza, Yaris **Engine:** L4, V6, V8	**Throttle / Pedal Position Sensor / Switch "A" Circuit Malfunction:** Monitor runs whenever following DTCs are not present: None Either of the following conditions A or B is met: - A. Power switch on (IG): 0.012 seconds or more B. Electronic throttle actuator power: ON
DTC: P0121 **1T ECM, MIL: Yes** **Year:** 2011, 2012 **Model:** 4Runner, Avalon, Camry, Corolla, FJ Cruiser, Highlander, Land Cruiser, Matrix, Prius, Rav4, Sequoia, Tundra, Tacoma, Venza, Yaris **Engine:** L4, V6, V8	**Throttle / Pedal Position Sensor / Switch "A" Circuit Range / Performance Problem:** Monitor runs whenever following DTCs are not present: None Either of the following conditions A or B is set: A. Power switch: On (IG) B. Electric throttle motor power: ON. Throttle position sensor malfunction.
DTC: P0122 **1T ECM, MIL: Yes** **Year:** 2011, 2012 **Model:** 4Runner, Avalon, Camry, Corolla, FJ Cruiser, Highlander, Land Cruiser, Matrix, Prius, Rav4, Sequoia, Tundra, Tacoma, Venza, Yaris **Engine:** L4, V6, V8	**Throttle / Pedal Position Sensor / Switch "A" Circuit Low Input:** Monitor runs whenever following DTCs are not present: None Either of the following conditions A or B is met: A. Power switch on (IG): 0.012 seconds or more B. Electronic throttle actuator power: ON
DTC: P0123 **1T ECM, MIL: Yes** **Year:** 2011, 2012 **Model:** 4Runner, Avalon, Camry, Corolla, FJ Cruiser, Highlander, Land Cruiser, Matrix, Prius, Rav4, Sequoia, Tundra, Tacoma, Venza, Yaris **Engine:** L4, V6, V8	**Throttle / Pedal Position Sensor / Switch "A" Circuit High Input:** Monitor runs whenever following DTCs are not present: None Either of the following conditions A or B is met: A. Power switch on (IG): 0.012 seconds or more B. Electronic throttle actuator power: ON
DTC: P0125 **1T ECM, MIL: Yes** **Year:** 2011, 2012 **Model:** 4Runner, Avalon, Camry, Corolla, FJ Cruiser, Highlander, Land Cruiser, Matrix, Prius, Rav4, Sequoia, Tundra, Tacoma, Venza, Yaris **Engine:** L4, V6, V8	**Insufficient Coolant Temperature for Closed Loop Fuel Control:** Case 1: Engine Coolant Temperature (ECT) is less than -9.4°C (15°F) at engine start and following conditions are met: 20 minutes elapsed after engine start ECT sensor value remains below closed-loop fuel control enabling temperature Case 2: ECT is between -9.4°C and 1.7°C (15°F and 35°F) at engine start and following conditions are met: 106 seconds elapsed after engine start ECT sensor value remains below closed-loop fuel control enabling temperature Case 3: ECT is above 1.7°C (35°F) at engine start and following conditions are met: 66 seconds elapsed after engine start. ECT sensor value remains below closed-loop fuel control enabling temperature.

DTC	Trouble Code Title and Conditions
DTC: P0128 **1T ECM, MIL: Yes** **Year:** 2011, 2012 **Model:** 4Runner, Avalon, Camry, Corolla, FJ Cruiser, Highlander, Land Cruiser, Matrix, Prius, Rav4, Sequoia, Tundra, Tacoma, Venza, Yaris **Engine:** L4, V6, V8	**Coolant Thermostat (Coolant Temperature Below Thermostat Regulating Temperature):** The following conditions are met for 5 seconds. Cold start Engine is warmed up ECT is less than 75°C (167°F) The engine coolant temperature does not reach 75°C (167°F) despite sufficient engine warm-up time having elapsed.
DTC: P0136 **1T ECM, MIL: Yes** **Year:** 2011, 2012 **Model:** 4Runner, Avalon, Camry, Corolla, FJ Cruiser, Highlander, Land Cruiser, Matrix, Prius, Rav4, Sequoia, Tundra, Tacoma, Venza, Yaris **Engine:** L4, V6, V8	**Oxygen Sensor Circuit Malfunction (Bank 1 Sensor 2):** Abnormal voltage output: During active air-fuel ratio control, following conditions (a) and (b) met for certain period of time: (a) Heated Oxygen (HO2) sensor voltage does not decrease to less than 0.2 V. (b) HO2 sensor voltage does not increase to more than 0.6 V. Low impedance: Sensor impedance less than 5 ohms for more than 30 seconds when ECM presumes sensor to being warmed up and operating normally.
DTC: P0137 **1T ECM, MIL: Yes** **Year:** 2011, 2012 **Model:** 4Runner, Avalon, Camry, Corolla, FJ Cruiser, Highlander, Land Cruiser, Matrix, Prius, Rav4, Sequoia, Tundra, Tacoma, Venza, Yaris **Engine:** L4, V6, V8	**Heated Oxygen Sensor Circuit Low Voltage (Bank 1 Sensor 2):** Low voltage (open): During active air-fuel ratio control, following conditions (a) and (b) met for certain period of time (1 trip detection logic): (a) HO2 sensor voltage output less than 0.21 V High impedance: Sensor impedance 15 kQ. or more for more than 90 seconds when ECM presumes sensor to being warmed up and operating normally.
DTC: P0138 **T1T ECM, MIL: Yes** **Year:** 2011, 2012 **Model:** 4Runner, Avalon, Camry, Corolla, FJ Cruiser, Highlander, Land Cruiser, Matrix, Prius, Rav4, Sequoia, Tundra, Tacoma, Venza, Yaris **Engine:** L4, V6, V8	**Oxygen Sensor Circuit High Voltage (Bank 1 Sensor 2):** High voltage (short): During active air-fuel ratio control, following conditions (a) and (b) are met for a certain period of time (2 trip detection logic): (a) HO2 sensor voltage output 0.59 V or more (b) Target air-fuel ratio lean Extremely high voltage (short): HO2 sensor voltage output exceeds 1.2 V for more than 30 seconds.
DTC: P0139 **1T ECM, MIL: Yes** **Year:** 2011, 2012 **Model:** 4Runner, Avalon, Camry, Corolla, FJ Cruiser, Highlander, Land Cruiser, Matrix, Prius, Rav4, Sequoia, Tundra, Tacoma, Venza, Yaris **Engine:** L4, V6, V8	**Oxygen Sensor Circuit Slow Response (Bank 1 Sensor 2):** The HO2 sensor voltage does not drop below 0.2 V immediately after fuel cut starts. The HO2 sensor voltage does not drop from 0.35 V to 0.2 V immediately after fuel cut starts.
DTC: P0141 **1T ECM, MIL: Yes** **Year:** 2011, 2012 **Model:** 4Runner, Avalon, Camry, Corolla, FJ Cruiser, Highlander, Land Cruiser, Matrix, Prius, Rav4, Sequoia, Tundra, Tacoma, Venza, Yaris **Engine:** L4, V6, V8	**Oxygen Sensor Heater Circuit Malfunction (Bank 1 Sensor 2):** Case 1: Monitor runs whenever following DTCs not stored: None Heated oxygen sensor heater circuit fail (P0037 and P0038): Not detected Battery voltage: 10.5 V or higher Fuel cut: OFF Time after fuel cut ON to OFF: 30 seconds or more Accumulated heater ON time: 100 seconds or more Learned heater OFF current operation: Complete Case 2: Monitor runs whenever following DTCs not stored: None Duration that rear heated oxygen sensor impedance is less than 15 kΩ: 2 seconds or more

DTC	Trouble Code Title and Conditions
DTC: P014C **2T ECM, MIL: Yes** **Year:** 2011, 2012 **Model:** 4Runner, Avalon, Camry, Corolla, FJ Cruiser, Highlander, Land Cruiser, Matrix, Prius, Rav4, Sequoia, Tundra, Tacoma, Venza, Yaris **Engine:** L4, V6, V8	**A/F Sensor Slow Response - Rich to Lean Bank 1 Sensor 1:** Monitor runs whenever following DTCs not stored: None Active air fuel ratio control: Performing Active air fuel ratio control is performed when the following conditions met: Battery voltage: 11 V or higher Engine coolant temperature: 75°C (167°F) or higher Idle: OFF Engine speed: 1000 to 4000 rpm Air fuel ratio sensor status: Activated Fuel-cut: OFF Engine load: 10 to 70% Shift position: 2nd or higher Catalyst monitor: Not executed Mass air flow: 6 to 15 g/sec.
DTC: P014D **2T ECM, MIL: Yes** **Year:** 2011, 2012 **Model:** 4Runner, Avalon, Camry, Corolla, FJ Cruiser, Highlander, Land Cruiser, Matrix, Prius, Rav4, Sequoia, Tundra, Tacoma, Venza, Yaris **Engine:** L4, V6, V8	**A/F Sensor Slow Response - Lean to Rich Bank 1 Sensor 1:** Monitor runs whenever following DTCs are not stored: None Active air fuel ratio control: Performing Active air fuel ratio control is performed when the following conditions are met: Battery voltage: 11 V or higher Engine coolant temperature: 167°F (75°C) or more Idle: OFF Engine speed: 1000–4000 rpm Air fuel ratio sensor status: Activated Fuel cut: OFF Engine load: 10–70% Shift position: 2nd or higher Catalyst monitor: Not yet Mass air flow: 5–15 g/sec. Lean to Rich Response rate deterioration level: -0.038 V or higher
DTC: P014E **2T ECM, MIL: Yes** **Year:** 2011, 2012 **Model:** 4Runner, Avalon, Camry, Corolla, FJ Cruiser, Highlander, Land Cruiser, Matrix, Prius, Rav4, Sequoia, Tundra, Tacoma, Venza, Yaris **Engine:** L4, V6, V8	**A/F Sensor Slow Response - Rich to Lean Bank 2 Sensor 1:** Monitor runs whenever following DTCs not stored: None Active air fuel ratio control: Performing Active air fuel ratio control is performed when the following conditions met: Battery voltage: 11 V or higher Engine coolant temperature: 75°C (167°F) or higher Idle: OFF Engine speed: 1000 to 4000 rpm Air fuel ratio sensor status: Activated Fuel-cut: OFF Engine load: 10 to 70% Shift position: 2nd or higher Catalyst monitor: Not executed Mass air flow: 6 to 15 g/sec.
DTC: P014F **2T ECM, MIL: Yes** **Year:** 2011, 2012 **Model:** 4Runner, Avalon, Camry, Corolla, FJ Cruiser, Highlander, Land Cruiser, Matrix, Prius, Rav4, Sequoia, Tundra, Tacoma, Venza, Yaris **Engine:** L4, V6, V8	**A/F Sensor Slow Response - Lean to Rich Bank 2 Sensor 1:** Monitor runs whenever following DTCs not stored: None Active air fuel ratio control: Performing Active air fuel ratio control is performed when the following conditions met: Battery voltage: 11 V or higher Engine coolant temperature: 75°C (167°F) or higher Idle: OFF Engine speed: 1000 to 4000 rpm Air fuel ratio sensor status: Activated Fuel-cut: OFF Engine load: 10 to 70% Shift position: 2nd or higher Catalyst monitor: Not executed Mass air flow: 6 to 15 g/sec.

DTC	Trouble Code Title and Conditions
DTC: P0156 **2T ECM, MIL: Yes** **Year:** 2011, 2012 **Model:** 4Runner, Avalon, Camry, Corolla, FJ Cruiser, Highlander, Land Cruiser, Matrix, Prius, Rav4, Sequoia, Tundra, Tacoma, Venza, Yaris **Engine:** L4, V6, V8	**Oxygen Sensor Circuit Malfunction (Bank 2 Sensor 2):** Abnormal voltage output: During active air-fuel ratio control, following conditions (a) and (b) are met for a certain period of time (2 trip detection logic): (a) Heated Oxygen (HO2) sensor voltage does not decrease to less than 0.2 V (b) HO2 sensor voltage does not increase to more than 0.6 V Low impedance: Sensor impedance less than 5 Q. for more than 30 seconds when ECM presumes sensor to being warmed up and operating normally (2 trip detection logic)
DTC: P0157 **2T ECM, MIL: Yes** **Year:** 2011, 2012 **Model:** 4Runner, Avalon, Camry, Corolla, FJ Cruiser, Highlander, Land Cruiser, Matrix, Prius, Rav4, Sequoia, Tundra, Tacoma, Venza, Yaris **Engine:** L4, V6, V8	**Oxygen Sensor Circuit Low Voltage (Bank 2 Sensor 2):** During active air-fuel ratio control, the following conditions (a) and (b) are met for a certain period of time: (a) The HO2 sensor voltage output is below 0.21 V. (b) The sensor impedance is 15 kΩ or higher for more than 90 seconds when the ECM presumes the sensor to be warmed up and operating normally.
DTC: P0158 **2T ECM, MIL: Yes** **Year:** 2011, 2012 **Model:** 4Runner, Avalon, Camry, Corolla, FJ Cruiser, Highlander, Land Cruiser, Matrix, Prius, Rav4, Sequoia, Tundra, Tacoma, Venza, Yaris **Engine:** L4, V6, V8	**Oxygen Sensor Circuit High Voltage (Bank 2 Sensor 2):** During active air-fuel ratio control, the following conditions (a) and (b) are met for a certain period of time: (a) The HO2 sensor voltage output is higher than 0.59 V. (b) The HO2 sensor voltage output exceeds 1.2 V for more than 10 seconds.
DTC: P0159 **2T ECM, MIL: Yes** **Year:** 2011, 2012 **Model:** 4Runner, Avalon, Camry, Corolla, FJ Cruiser, Highlander, Land Cruiser, Matrix, Prius, Rav4, Sequoia, Tundra, Tacoma, Venza, Yaris **Engine:** L4, V6, V8	**Oxygen Sensor Circuit Slow Response (Bank 2 Sensor 2):** The HO2 sensor voltage does not drop below 0.2 V immediately after fuel cut starts. Or. The HO2 sensor voltage does not drop from 0.35 V to 0.2 V immediately after fuel cut starts.
DTC: P015A **2T ECM, MIL: Yes** **Year:** 2011, 2012 **Model:** 4Runner, Avalon, Camry, Corolla, FJ Cruiser, Highlander, Land Cruiser, Matrix, Prius, Rav4, Sequoia, Tundra, Tacoma, Venza, Yaris **Engine:** L4, V6, V8	**A/F Sensor Delayed Response - Rich to Lean Bank 1 Sensor 1:** Monitor runs whenever following DTCs not stored: None Active air fuel ratio control: Performing Active air fuel ratio control is performed when the following conditions met: Battery voltage: 11 V or higher Engine coolant temperature: 75°C (167°F) or higher Idle: OFF Engine speed: 1000 to 4000 rpm Air fuel ratio sensor status: Activated Fuel-cut: OFF Engine load: 10 to 70% Shift position: 2nd or higher Catalyst monitor: Not executed Mass air flow: 6 to 15 g/sec.

DTC	Trouble Code Title and Conditions
DTC: P015B **2T ECM, MIL: Yes** **Year:** 2011, 2012 **Model:** 4Runner, Avalon, Camry, Corolla, FJ Cruiser, Highlander, Land Cruiser, Matrix, Prius, Rav4, Sequoia, Tundra, Tacoma, Venza, Yaris **Engine:** L4, V6, V8	**A/F Sensor Delayed Response - Lean to Rich Bank 1 Sensor 1:** Monitor runs whenever following DTCs not stored: None Active air fuel ratio control: Performing Active air fuel ratio control is performed when the following conditions met: Battery voltage: 11 V or higher Engine coolant temperature: 75°C (167°F) or higher Idle: OFF Engine speed: 1000 to 4000 rpm Air fuel ratio sensor status: Activated Fuel-cut: OFF Engine load: 10 to 70% Shift position: 2nd or higher Catalyst monitor: Not executed Mass air flow: 6 to 15 g/sec.
DTC: P015C **2T ECM, MIL: Yes** **Year:** 2011, 2012 **Model:** 4Runner, Avalon, Camry, Corolla, FJ Cruiser, Highlander, Land Cruiser, Matrix, Prius, Rav4, Sequoia, Tundra, Tacoma, Venza, Yaris **Engine:** L4, V6, V8	**A/F Sensor Delayed Response - Rich to Lean Bank 2 Sensor 1:** Monitor runs whenever following DTCs not stored: None Active air fuel ratio control: Performing Active air fuel ratio control is performed when the following conditions met: Battery voltage: 11 V or higher Engine coolant temperature: 75°C (167°F) or higher Idle: OFF Engine speed: 1000 to 4000 rpm Air fuel ratio sensor status: Activated Fuel-cut: OFF Engine load: 10 to 70% Shift position: 2nd or higher Catalyst monitor: Not executed Mass air flow: 6 to 15 g/sec.
DTC: P015D **2T ECM, MIL: Yes** **Year:** 2011, 2012 **Model:** 4Runner, FJ Cruiser **Engine:** 4.0L V6	**A/F Sensor Delayed Response - Lean to Rich Bank 2 Sensor 1:** Monitor runs whenever following DTCs not stored: None Active air fuel ratio control: Performing Active air fuel ratio control is performed when the following conditions met: Battery voltage: 11 V or higher Engine coolant temperature: 75°C (167°F) or higher Idle: OFF Engine speed: 1000 to 4000 rpm Air fuel ratio sensor status: Activated Fuel-cut: OFF Engine load: 10 to 70% Shift position: 2nd or higher Catalyst monitor: Not executed Mass air flow: 6 to 15 g/sec. **Possible causes:** • Air fuel ratio sensor (bank 2 sensor 1) • Air fuel ratio sensor (bank 2 sensor 1) heater • ECM
DTC: P0161 **2T ECM, MIL: Yes** **Year:** 2011, 2012 **Model:** 4Runner, Avalon, Camry, Corolla, FJ Cruiser, Highlander, Land Cruiser, Matrix, Prius, Rav4, Sequoia, Tundra, Tacoma, Venza, Yaris **Engine:** L4, V6, V8	**Oxygen Sensor Heater Circuit Malfunction (Bank 2 Sensor 2):** Monitor runs whenever following DTCs are not present: None. Cumulative heater resistance correction value exceeds the acceptable threshold.

DTC	Trouble Code Title and Conditions
DTC: P0171 **2T ECM, MIL: Yes** **Year:** 2011, 2012 **Model:** 4Runner, Avalon, Camry, Corolla, FJ Cruiser, Highlander, Land Cruiser, Matrix, Prius, Rav4, Sequoia, Tundra, Tacoma, Venza, Yaris **Engine:** L4, V6, V8	**System Too Lean (Bank 1):** Fuel system status is in Closed loop. battery voltage 11 V or higher. Either of following conditions 1 or 2 set: 1. Engine speed is less than 1100 rpm. 2. Intake air amount per revolution is 0.36 g/rev or more. Catalyst monitor not executed. With a warm engine and stable air-fuel ratio feedback, the fuel trim is considerably in error to the lean side.
DTC: P0172 **2T ECM, MIL: Yes** **Year:** 2011, 2012 **Model:** 4Runner, Avalon, Camry, Corolla, FJ Cruiser, Highlander, Land Cruiser, Matrix, Prius, Rav4, Sequoia, Tundra, Tacoma, Venza, Yaris **Engine:** L4, V6, V8	**System Too Rich (Bank 1):** With warm engine and stable air-fuel ratio feedback, fuel trim considerably in error to rich side. Fuel system status: Closed loop Battery voltage: 11 V or higher Either of the following conditions is met: 1. Engine speed: Less than 800 rpm 2. Engine load: 11.52% or more Catalyst monitor: Not executed.
DTC: P0174 **2T ECM, MIL: Yes** **Year:** 2011, 2012 **Model:** 4Runner, Avalon, Camry, Corolla, FJ Cruiser, Highlander, Land Cruiser, Matrix, Prius, Rav4, Sequoia, Tundra, Tacoma, Venza, Yaris **Engine:** L4, V6, V8	**System Too Lean (Bank 2):** Fuel system status is in Closed loop. battery voltage 11 V or higher. Either of following conditions 1 or 2 set: 1. Engine speed is less than 1100 rpm. 2. Intake air amount per revolution is 0.36 g/rev or more Catalyst monitor not executed. With a warm engine and stable air-fuel ratio feedback, the fuel trim is considerably in error to the lean side.
DTC: P0175 **2T ECM, MIL: Yes** **Year:** 2011, 2012 **Model:** 4Runner, Avalon, Camry, Corolla, FJ Cruiser, Highlander, Land Cruiser, Matrix, Prius, Rav4, Sequoia, Tundra, Tacoma, Venza, Yaris **Engine:** L4, V6, V8	**System Too Rich (Bank 2):** With the engine in Closed loop, battery voltage 11 V or higher either of following conditions 1 or 2 set: 1. Engine speed is less than 1100 rpm 2. Intake air amount per revolution 0.36 g/rev or more. Catalyst monitor not executed. With a warm engine and stable air-fuel ratio feedback, the fuel trim is considerably in error to the rich side.
DTC: P0190 **1T ECM, MIL: Yes** **Year:** 2011, 2012 **Model:** 4Runner, Avalon, Camry, Corolla, FJ Cruiser, Highlander, Land Cruiser, Matrix, Prius, Rav4, Sequoia, Tundra, Tacoma, Venza, Yaris **Engine:** L4, V6, V8	**Fuel Rail Pressure Sensor Circuit:** Fuel rail pressure sensor circuit low input and/or high input. Signal voltage from the fuel rail pressure sensor remains at more than 4.84 V / less than 0.2 V for 5 seconds or more.
DTC: P0192 **1T ECM, MIL: Yes** **Year:** 2011, 2012 **Model:** 4Runner, Avalon, Camry, Corolla, FJ Cruiser, Highlander, Land Cruiser, Matrix, Prius, Rav4, Sequoia, Tundra, Tacoma, Venza, Yaris **Engine:** L4, V6, V8	**Fuel Rail Pressure Sensor Circuit Low Input:** Fuel rail pressure sensor circuit low input. Signal voltage from the fuel rail pressure sensor remains at less than 0.2 V for 5 seconds or more.

DTC	Trouble Code Title and Conditions
DTC: P0193 **1T ECM, MIL: Yes** **Year:** 2011, 2012 **Model:** 4Runner, Avalon, Camry, Corolla, FJ Cruiser, Highlander, Land Cruiser, Matrix, Prius, Rav4, Sequoia, Tundra, Tacoma, Venza, Yaris **Engine:** L4, V6, V8	**Fuel Rail Pressure Sensor Circuit High Input:** Monitor runs whenever following DTCs are not present: None Time after engine start: 5 seconds or more.
DTC: P0201 **1T ECM, MIL: Yes** **Year:** 2011, 2012 **Model:** 4Runner, Avalon, Camry, Corolla, FJ Cruiser, Highlander, Land Cruiser, Matrix, Prius, Rav4, Sequoia, Tundra, Tacoma, Venza, Yaris **Engine:** L4, V6, V8	**Injector Circuit / Open - (Cylinder 1):** Monitor runs whenever following DTCs are not present: None Time after engine switch off to on (IG): 1 second or more Engine speed: 4000 rpm or less Battery voltage: 10.5 V or more
DTC: P0202 **1T ECM, MIL: Yes** **Year:** 2011, 2012 **Model:** 4Runner, Avalon, Camry, Corolla, FJ Cruiser, Highlander, Land Cruiser, Matrix, Prius, Rav4, Sequoia, Tundra, Tacoma, Venza, Yaris **Engine:** L4, V6, V8	**Injector Circuit / Open - (Cylinder 2):** Monitor runs whenever following DTCs are not present: None Time after engine switch off to on (IG): 1 second or more Engine speed: 4000 rpm or less Battery voltage: 10.5 V or mor.
DTC: P0203 **1T ECM, MIL: Yes** **Year:** 2011, 2012 **Model:** 4Runner, Avalon, Camry, Corolla, FJ Cruiser, Highlander, Land Cruiser, Matrix, Prius, Rav4, Sequoia, Tundra, Tacoma, Venza, Yaris **Engine:** L4, V6, V8	**Injector Circuit / Open - (Cylinder 3):** Monitor runs whenever following DTCs are not present: None Time after engine switch off to on (IG): 1 second or more Engine speed: 4000 rpm or less Battery voltage: 10.5 V or more
DTC: P0204 **1T ECM, MIL: Yes** **Year:** 2011, 2012 **Model:** 4Runner, Avalon, Camry, Corolla, FJ Cruiser, Highlander, Land Cruiser, Matrix, Prius, Rav4, Sequoia, Tundra, Tacoma, Venza, Yaris **Engine:** L4, V6, V8	**Injector Circuit / Open - (Cylinder 4):** Monitor runs whenever following DTCs are not present: None Time after engine switch off to on (IG): 1 second or more Engine speed: 4000 rpm or less Battery voltage: 10.5 V or more.
DTC: P0205 **1T ECM, MIL: Yes** **Year:** 2011, 2012 **Model:** 4Runner, Avalon, Camry, FJ Cruiser, Highlander, Land Cruiser, Matrix, Rav4, Sequoia, Tundra, Tacoma, Venza, **Engine:** V6, V8	**Injector Circuit / Open - (Cylinder 5):** Monitor runs whenever following DTCs are not present: None Time after engine switch off to on (IG): 1 second or more Engine speed: 4000 rpm or less Battery voltage: 10.5 V or more.
DTC: P0206 **1T ECM, MIL: Yes** **Year:** 2011, 2012 **Model:** 4Runner, Avalon, Camry, FJ Cruiser, Highlander, Land Cruiser, Matrix, Rav4, Sequoia, Tundra, Tacoma, Venza, **Engine:** V6, V8	**Injector Circuit / Open - (Cylinder 6):** Monitor runs whenever following DTCs are not present: None Time after engine switch off to on (IG): 1 second or more Engine speed: 4000 rpm or less Battery voltage: 10.5 V or more

DTC	Trouble Code Title and Conditions
DTC: P0207 **1T ECM, MIL: Yes** **Year:** 2011, 2012 **Model:** Land Cruiser, Sequoia, Tundra **Engine:** V8	**Injector Circuit / Open - (Cylinder 7):** Monitor runs whenever following DTCs are not present: None Time after engine switch off to on (IG): 1 second or more Engine speed: 4000 rpm or less Battery voltage: 10.5 V or more.
DTC: P0208 **1T ECM, MIL: Yes** **Year:** 2011, 2012 **Model:** Land Cruiser, Sequoia, Tundra **Engine:** V8	**Injector Circuit / Open - (Cylinder 8):** Monitor runs whenever following DTCs are not present: None Time after engine switch off to on (IG): 1 second or more Engine speed: 4000 rpm or less Battery voltage: 10.5 V or more.
DTC: P0220 **1T ECM, MIL: Yes** **Year:** 2011, 2012 **Model:** 4Runner, Avalon, Camry, Corolla, FJ Cruiser, Highlander, Land Cruiser, Matrix, Prius, Rav4, Sequoia, Tundra, Tacoma, Venza, Yaris **Engine:** L4, V6, V8	**Throttle / Pedal Position Sensor / Switch "B" Circuit:** Monitor runs whenever following DTCs are not present: None Either of the following conditions A or B is met: A. Power switch on (IG): 0.012 seconds or more B. Electronic throttle actuator power: ON.
DTC: P0222 **1T ECM, MIL: Yes** **Year:** 2011, 2012 **Model:** 4Runner, Avalon, Camry, Corolla, FJ Cruiser, Highlander, Land Cruiser, Matrix, Prius, Rav4, Sequoia, Tundra, Tacoma, Venza, Yaris **Engine:** L4, V6, V8	**Throttle / Pedal Position Sensor / Switch "B" Circuit Low Input:** Monitor runs whenever following DTCs are not present: None Either of the following conditions A or B is met: - A. Power switch on (IG): 0.012 seconds or more B. Electronic throttle actuator power: ON.
DTC: P0223 **1T ECM, MIL: Yes** **Year:** 2011, 2012 **Model:** 4Runner, Avalon, Camry, Corolla, FJ Cruiser, Highlander, Land Cruiser, Matrix, Prius, Rav4, Sequoia, Tundra, Tacoma, Venza, Yaris **Engine:** L4, V6, V8	**Throttle / Pedal Position Sensor / Switch "B" Circuit High Input:** Monitor runs whenever following DTCs are not present: None Either of the following conditions is met: Condition 1 or 2 1. Ignition switch ON: 0.012 seconds or more 2. Electric throttle motor power: ON.
DTC: P0230 **1T ECM, MIL: Yes** **Year:** 2011, 2012 **Model:** 4Runner, Avalon, Camry, Corolla, FJ Cruiser, Highlander, Land Cruiser, Matrix, Prius, Rav4, Sequoia, Tundra, Tacoma, Venza, Yaris **Engine:** L4, V6, V8	**Fuel Pump Primary Circuit:** Open or short in fuel pump relay circuit. When either condition below is met (1 trip detection logic): When fuel pump is operating: O Remaining fuel is 17 L or more O DI terminal output is low When fuel pump is not operating: O DI terminal output is high
DTC: P0300 **1T ECM, MIL: Yes** **Year:** 2011, 2012 **Model:** 4Runner, Avalon, Camry, Corolla, FJ Cruiser, Highlander, Land Cruiser, Matrix, Prius, Rav4, Sequoia, Tundra, Tacoma, Venza, Yaris **Engine:** L4, V6, V8	**Random / Multiple Cylinder Misfire Detected:** Simultaneous misfiring of several cylinders occurs and one of the following conditions below is detected: A high temperature misfire occurs in the Three-Way Catalytic Converter (TWC) (MIL blinks). An emission deterioration misfire occurs (MIL illuminates).

DTC	Trouble Code Title and Conditions
DTC: P0301 **1T ECM, MIL: Yes** **Year:** 2011, 2012 **Model:** 4Runner, Avalon, Camry, Corolla, FJ Cruiser, Highlander, Land Cruiser, Matrix, Prius, Rav4, Sequoia, Tundra, Tacoma, Venza, Yaris **Engine:** L4, V6, V8	**Cylinder 1 Misfire Detected:** Misfiring of a specific cylinder occurs and one of the following conditions below is detected: A high temperature misfire occurs in the Three-Way Catalytic Converter (TWC) (MIL blinks). An emission deterioration misfire occurs (MIL illuminates).
DTC: P0302 **1T ECM, MIL: Yes** **Year:** 2011, 2012 **Model:** 4Runner, Avalon, Camry, Corolla, FJ Cruiser, Highlander, Land Cruiser, Matrix, Prius, Rav4, Sequoia, Tundra, Tacoma, Venza, Yaris **Engine:** L4, V6, V8	**Cylinder 2 Misfire Detected:** Misfiring of a specific cylinder occurs and one of the following conditions below is detected: A high temperature misfire occurs in the Three-Way Catalytic Converter (TWC) (MIL blinks). An emission deterioration misfire occurs (MIL illuminates).
DTC: P0303 **1T ECM, MIL: Yes** **Year:** 2011, 2012 **Model:** 4Runner, Avalon, Camry, Corolla, FJ Cruiser, Highlander, Land Cruiser, Matrix, Prius, Rav4, Sequoia, Tundra, Tacoma, Venza, Yaris **Engine:** L4, V6, V8	**Cylinder 3 Misfire Detected:** Misfiring of a specific cylinder occurs and one of the following conditions below is detected: A high temperature misfire occurs in the Three-Way Catalytic Converter (TWC) (MIL blinks). An emission deterioration misfire occurs (MIL illuminates).
DTC: P0304 **1T ECM, MIL: Yes** **Year:** 2011, 2012 **Model:** 4Runner, Avalon, Camry, Corolla, FJ Cruiser, Highlander, Land Cruiser, Matrix, Prius, Rav4, Sequoia, Tundra, Tacoma, Venza, Yaris **Engine:** L4, V6, V8	**Cylinder 4 Misfire Detected:** Misfiring of a specific cylinder occurs and one of the following conditions below is detected: A high temperature misfire occurs in the Three-Way Catalytic Converter (TWC) (MIL blinks). An emission deterioration misfire occurs (MIL illuminates).
DTC: P0305 **1T ECM, MIL: Yes** **Year:** 2011, 2012 **Model:** 4Runner, Avalon, Camry, FJ Cruiser, Highlander, Land Cruiser, Matrix, Rav4, Sequoia, Tundra, Tacoma, Venza, **Engine:** V6, V8	**Cylinder 5 Misfire Detected:** Misfiring of a specific cylinder occurs and one of the following conditions below is detected: A high temperature misfire occurs in the Three-Way Catalytic Converter (TWC) (MIL blinks). An emission deterioration misfire occurs (MIL illuminates).
DTC: P0306 **1T ECM, MIL: Yes** **Year:** 2011, 2012 **Model:** 4Runner, Avalon, Camry, FJ Cruiser, Highlander, Land Cruiser, Matrix, Rav4, Sequoia, Tundra, Tacoma, Venza **Engine:** V6, V8	**Cylinder 6 Misfire Detected:** Misfiring of a specific cylinder occurs and one of the following conditions below is detected: A high temperature misfire occurs in the Three-Way Catalytic Converter (TWC) (MIL blinks). An emission deterioration misfire occurs (MIL illuminates).
DTC: P0307 **1T ECM, MIL: Yes** **Year:** 2011, 2012 **Model:** Land Cruiser, Sequoia, Tundra **Engine:** V8	**Cylinder 7 Misfire Detected:** Misfiring of a specific cylinder occurs and one of the following conditions below is detected: A high temperature misfire occurs in the Three-Way Catalytic Converter (TWC) (MIL blinks). An emission deterioration misfire occurs (MIL illuminates).

DTC	Trouble Code Title and Conditions
DTC: P0308 **1T ECM, MIL: Yes** **Year:** 2011, 2012 **Model:** Land Cruiser, Sequoia, Tundra **Engine:** V8	**Cylinder 8 Misfire Detected:** Misfiring of a specific cylinder occurs and one of the following conditions below is detected: A high temperature misfire occurs in the Three-Way Catalytic Converter (TWC) (MIL blinks). An emission deterioration misfire occurs (MIL illuminates).
DTC: P0327 **1T ECM, MIL: Yes** **Year:** 2011, 2012 **Model:** 4Runner, Avalon, Camry, Corolla, FJ Cruiser, Highlander, Land Cruiser, Matrix, Prius, Rav4, Sequoia, Tundra, Tacoma, Venza, Yaris **Engine:** L4, V6, V8	**Knock Sensor 1 Circuit Low Input (Bank 1 or Single Sensor):** Monitor runs whenever following DTCs are not present: None Battery voltage: 10.5 V or more Time after engine start: 5 seconds or more Engine switch: On (IG) Starter: OFF
DTC: P0328 **1T ECM, MIL: Yes** **Year:** 2011, 2012 **Model:** 4Runner, Avalon, Camry, Corolla, FJ Cruiser, Highlander, Land Cruiser, Matrix, Prius, Rav4, Sequoia, Tundra, Tacoma, Venza, Yaris **Engine:** L4, V6, V8	**Knock Sensor 1 Circuit High Input (Bank 1 or Single Sensor):** Monitor runs whenever following DTCs are not present: None Battery voltage: 10.5 V or more Time after engine start: 5 seconds or more Output voltage of knock sensor more than 4.5 V.
DTC: P032C **1T ECM, MIL: Yes** **Year:** 2011, 2012 **Model:** 4Runner, Avalon, Camry, Corolla, FJ Cruiser, Highlander, Land Cruiser, Matrix, Prius, Rav4, Sequoia, Tundra, Tacoma, Venza, Yaris **Engine:** L4, V6, V8	**Knock Sensor 3 Circuit Low:** With the engine running for at least 5 seconds and battery voltage 10.5 V or more. Output voltage of knock sensor (for Bank 1 Sensor 2) is 0.5 V or less
DTC: P032D **1T ECM, MIL: Yes** **Year:** 2011, 2012 **Model:** 4Runner, Avalon, Camry, Corolla, FJ Cruiser, Highlander, Land Cruiser, Matrix, Prius, Rav4, Sequoia, Tundra, Tacoma, Venza, Yaris **Engine:** L4, V6, V8	**Knock Sensor 3 Circuit High:** With the engine running for at least 5 seconds and battery voltage 10.5 V or more. The output voltage of the knock sensor (for Bank 1 Sensor 2) is 4.5 V or higher.
DTC: P0332 **1T ECM, MIL: Yes** **Year:** 2011, 2012 **Model:** 4Runner, Avalon, Camry, Corolla, FJ Cruiser, Highlander, Land Cruiser, Matrix, Prius, Rav4, Sequoia, Tundra, Tacoma, Venza, Yaris **Engine:** L4, V6, V8	**Knock Sensor 2 Circuit Low Input (Bank 2):** Monitor runs whenever following DTCs not present: None Battery voltage: 10.5 V or higher Time after engine start: 5 seconds or more. Output voltage of knock sensor less than 0.5 V (1 trip detection logic)
DTC: P0333 **1T ECM, MIL: Yes** **Year:** 2011, 2012 **Model:** 4Runner, Avalon, Camry, Corolla, FJ Cruiser, Highlander, Land Cruiser, Matrix, Prius, Rav4, Sequoia, Tundra, Tacoma, Venza, Yaris **Engine:** L4, V6, V8	**Knock Sensor 2 Circuit High Input (Bank 2):** Monitor runs whenever following DTCs are not present: None Battery voltage: 10.5 V or more Time after engine start: 5 seconds or more Engine switch: On (IG) Starter: OFF

DTC	Trouble Code Title and Conditions
DTC: P0335 **1T ECM, MIL: Yes** **Year:** 2011, 2012 **Model:** 4Runner, Avalon, Camry, Corolla, FJ Cruiser, Highlander, Land Cruiser, Matrix, Prius, Rav4, Sequoia, Tundra, Tacoma, Venza, Yaris **Engine:** L4, V6, V8	**Crankshaft Position Sensor "A" Circuit:** Monitor runs whenever following DTCs not stored: None. Time after starter OFF to ON: 3 seconds or more. No CKP sensor signal to ECM while cranking (1 trip detection logic) No CKP sensor signal to ECM at engine speed of 600 rpm or more.
DTC: P0337 **1T ECM, MIL: Yes** **Year:** 2011, 2012 **Model:** 4Runner, Avalon, Camry, Corolla, FJ Cruiser, Highlander, Land Cruiser, Matrix, Prius, Rav4, Sequoia, Tundra, Tacoma, Venza, Yaris **Engine:** L4, V6, V8	**Crankshaft Position Sensor "A" Circuit Low Input:** With the ignition switch ON and battery voltage 8.0 V or more. The output voltage of the CKP sensor is 0.3 V or less for 4 seconds.
DTC: P0338 **1T ECM, MIL: Yes** **Year:** 2011, 2012 **Model:** 4Runner, Avalon, Camry, Corolla, FJ Cruiser, Highlander, Land Cruiser, Matrix, Prius, Rav4, Sequoia, Tundra, Tacoma, Venza, Yaris **Engine:** L4, V6, V8	**Crankshaft Position Sensor "A" Circuit High Input:** With the ignition switch ON and battery voltage a minimum of 8 V. The output voltage of the CKP sensor is 4.7 V or higher for 4 seconds.
DTC: P0339 **1T ECM, MIL: Yes** **Year:** 2011, 2012 **Model:** 4Runner, Avalon, Camry, Corolla, FJ Cruiser, Highlander, Land Cruiser, Matrix, Prius, Rav4, Sequoia, Tundra, Tacoma, Venza, Yaris **Engine:** L4, V6, V8	**Crankshaft Position Sensor "A" Circuit Intermittent:** Under conditions (a), (b) and (c), no CKP sensor signal is sent to the ECM for 0.05 seconds or more: (a) The engine speed is 1000 rpm or more. (b) The starter signal is OFF. (c) 3 seconds or more have elapsed since the starter signal was switched from ON to OFF.
DTC: P033C **1T ECM, MIL: Yes** **Year:** 2011 **Model:** Land Cruiser **Engine:** 5.7L V8	**Knock Sensor 4 Circuit Low Input:** With the engine running for at least 5 seconds and battery voltage 10.5 V or more, the Out-put voltage of the knock sensor (for Bank 2 Sensor 2) is 0.5 V or less.
DTC: P033D **1T ECM, MIL: Yes** **Year:** 2011 **Model:** Land Cruiser **Engine:** 5.7L V8	**Knock Sensor 4 Circuit High Input:** With the engine running for at least 5 seconds and battery voltage 10.5 V or more, the Out-put voltage of the knock sensor (for Bank 2 Sensor 2) is 4.5 V or higher.
DTC: P0340 **1T ECM, MIL: Yes** **Year:** 2011, 2012 **Model:** Camry, Prius, Prius V **Engine:** 1.8L L4, 2.4L L4	**Camshaft Position Sensor "A" Circuit (Bank 1 or Single Sensor):** Monitor runs whenever following DTCs are not present: None Engine speed: 600 rpm or more Starter: Off
DTC: P0340 **1T ECM, MIL: Yes** **Year:** 2011, 2012 **Model:** 4Runner, Avalon, Camry, Corolla, FJ Cruiser, Highlander, Land Cruiser, Matrix, Rav4, Sequoia, Tundra, Tacoma, Venza, Yaris **Engine:** L4, V6, V8	**Camshaft Position Sensor Circuit Malfunction:** When either condition below is met: (a) Input voltage to the ECM remains at less than 0.3 V, or more than 4.7 V for 4 seconds when 2 or more seconds have elapsed after turning the engine switch on (IG). (b) No VVT sensor signal to ECM at engine speed of 600 rpm or more.

DTC	Trouble Code Title and Conditions
DTC: P0340-886 **T ECM** **Year:** 2011, 2012 **Model:** Highlander/Hybrid **Engine:** 3.5L V6	**Camshaft Position Sensor "A" Circuit:** Ignition switch: ON Time after ignition switch turned from off to ON: 2 seconds Battery voltage: 8 V or higher Malfunction in the engine speed sensor (GI signal) circuit.
DTC: P0341-747 **T ECM, MIL: Yes** **Year:** 2011, 2012 **Model:** Camry, Prius, Prius V **Engine:** 1.8L L4, 2.4L L4	**Camshaft Position Sensor "A" Circuit High Input:** Ignition switch: ON Time after ignition switch turned from off to ON: 2 seconds Battery voltage: 8 V or higher GI signal is not input for 2 sec. or more while the engine is running.
DTC: P0342 **1T ECM, MIL: Yes** **Year:** 2011, 2012 **Model:** 4Runner, Avalon, Camry, Corolla, FJ Cruiser, Highlander, Land Cruiser, Matrix, Prius, Rav4, Sequoia, Tundra, Tacoma, Venza, Yaris **Engine:** L4, V6, V8	**Camshaft Position Sensor "A" Circuit Low Input (Bank 1 or Single Sensor):** Camshaft Position Sensor Range Check: Monitor runs whenever following DTCs not stored: None Starter: OFF Ignition switch: ON Time after ignition switch turned from off to ON: 2 seconds Battery voltage: 8 V or higher Camshaft position sensor, verify pulse input fail (P0340): Not detected.
DTC: P0343 **1T ECM, MIL: Yes** **Year:** 2011, 2012 **Model:** 4Runner, Avalon, Camry, Corolla, FJ Cruiser, Highlander, Land Cruiser, Matrix, Prius, Rav4, Sequoia, Tundra, Tacoma, Venza, Yaris **Engine:** L4, V6, V8	**Camshaft Position Sensor "A" Circuit High Input (Bank 1 or Single Sensor):** Monitor runs whenever following DTCs are not present: None Starter: OFF Time after power switch OFF to on (IG): 2 seconds or more Camshaft position sensor verify pulse input fail (P0340): Not detected Battery voltage: 8 V or more.
DTC: P0343-747 **1T ECM, MIL: Yes** **Year:** 2011, 2012 **Model:** Camry, Highlander, Prius, Prius V **Engine:** 1.8L L4, 3.5L V6	**Camshaft Position Sensor "A" Circuit High Input:** GI signal is not input for 2 sec. or more while the engine is running. Monitor runs whenever following DTCs are not present: None Engine speed: 600 rpm or more Starter: Off
DTC: P0345 **1T ECM, MIL: Yes** **Year:** 2011, 2012 **Model:** 4Runner, Avalon, Camry, Corolla, FJ Cruiser, Highlander, Land Cruiser, Matrix, Prius, Rav4, Sequoia, Tundra, Tacoma, Venza, Yaris **Engine:** L4, V6, V8	**Camshaft Position Sensor "A" Circuit (Bank 2):** Monitor runs whenever following DTCs are not present: None. Engine speed: 600 rpm or more Starter: Off VVT sensor range check fail (P0342, P0343, P0347, P0348): Not detected VVT sensor voltage: 0.3 V or more, and 4.7 V or less Battery voltage: 8 V or more Power switch: On (IG) Time after power switch off to on (IG): 0.5 seconds or more Hybrid control module judge: Engine running Lost communication with hybrid control module (U0293): Not detected
DTC: P0347 **1T ECM, MIL: Yes** **Year:** 2011, 2012 **Model:** 4Runner, Avalon, Camry, Corolla, FJ Cruiser, Highlander, Land Cruiser, Matrix, Prius, Rav4, Sequoia, Tundra, Tacoma, Venza, Yaris **Engine:** L4, V6, V8	**Camshaft Position Sensor "A" Circuit Low Input (Bank 2):** Low VoltageL Monitor runs whenever following DTCs are not present: P0340, P0345 (VVT Sensor) Starter: OFF Ignition switch: ON Time after ignition switch off to ON: 2 seconds or more Battery voltage: 8 V or higher.

DTC	Trouble Code Title and Conditions
DTC: P0348 **1T ECM, MIL: Yes** **Year:** 2011, 2012 **Model:** 4Runner, Avalon, Camry, Corolla, FJ Cruiser, Highlander, Land Cruiser, Matrix, Prius, Rav4, Sequoia, Tundra, Tacoma, Venza, Yaris **Engine:** L4, V6, V8	**Camshaft Position Sensor "A" Circuit High Input (Bank 2):** Low Voltage, High Voltage: Monitor runs whenever following DTCs are not present: P0340, P0345 (VVT Sensor) Starter: OFF Ignition switch: ON Time after ignition switch off to ON: 2 seconds or more Battery voltage: 8 V or higher
DTC: P0351 **1T ECM, MIL: Yes** **Year:** 2011, 2012 **Model:** 4Runner, Avalon, Camry, Corolla, FJ Cruiser, Highlander, Land Cruiser, Matrix, Prius, Rav4, Sequoia, Tundra, Tacoma, Venza, Yaris **Engine:** L4, V6, V8	**Ignition Coil "A" Primary / Secondary Circuit :** The monitor will run whenever these DTCs are not present: None Hybrid control module judge: Engine running, (if equipped) Either of the following condition is met: Condition 1 or 2 1. Starter: OFF 2. Engine RPM: 1500 rpm or less Either of the following condition is met: Condition (a) or (b) (a) All of the following conditions are met: Engine RPM: 500 rpm or less Battery voltage: 6 V or more (b) All of the following conditions are met: Engine RPM: More than 500 rpm Battery voltage: 10 V or more Number of sparks after CPU is reset: 5 sparks or more.
DTC: P0352 **1T ECM, MIL: Yes** **Year:** 2011, 2012 **Model:** 4Runner, Avalon, Camry, Corolla, FJ Cruiser, Highlander, Land Cruiser, Matrix, Prius, Rav4, Sequoia, Tundra, Tacoma, Venza, Yaris **Engine:** L4, V6, V8	**Ignition Coil "B" Primary / Secondary Circuit:** Monitor runs whenever following DTCs are not present: None Either of the following condition A or B is met: A. Engine RPM: 1500 rpm or less B. Starter: OFF Either of the following condition C or D is met: C. Both of the following conditions are met: (a) Engine speed: 500 rpm or less (b) Battery voltage: 6 V or more D. All of the following conditions are met: (a) Engine speed: More than 500 rpm (b) Battery voltage: 10 V or more (c) Number of sparks after CPU reset: 5 sparks or more.
DTC: P0353 **1T ECM, MIL: Yes** **Year:** 2011, 2012 **Model:** 4Runner, Avalon, Camry, Corolla, FJ Cruiser, Highlander, Land Cruiser, Matrix, Prius, Rav4, Sequoia, Tundra, Tacoma, Venza, Yaris **Engine:** L4, V6, V8	**Ignition Coil "C" Primary / Secondary Circuit:** The monitor will run whenever these DTCs are not present: None Either of the following condition is met: Condition 1 or 2 1. Starter: OFF 2. Engine RPM: 1500 rpm or less Either of the following condition is met: Condition (a) or (b) (a) All of the following conditions are met: Engine RPM: 500 rpm or less Battery voltage: 6 V or more (b) All of the following conditions are met: Engine RPM: More than 500 rpm Battery voltage: 10 V or more Number of sparks after CPU is reset: 5 sparks or more.

DTC	Trouble Code Title and Conditions
DTC: P0354 **1T ECM, MIL: Yes** **Year:** 2011, 2012 **Model:** 4Runner, Avalon, Camry, Corolla, FJ Cruiser, Highlander, Land Cruiser, Matrix, Prius, Rav4, Sequoia, Tundra, Tacoma, Venza, Yaris **Engine:** L4, V6, V8	**Ignition Coil "D" Primary / Secondary Circuit:** Monitor runs whenever following DTCs are not present: None Either of the following conditions A or B is met: A. Engine speed: 1500 rpm or less B. Starter: OFF Either of the following conditions C or D met: C. All of the following conditions (a) and (b) are met: (a) Engine speed: 500 rpm or less (b) Battery voltage: 6 V or more D. All of the following conditions (c), (d) and (e) are met: (c) Engine speed: More than 500 rpm (d) Battery voltage: 10 V or more (e) Number of sparks after CPU reset: 5 sparks or more Lost communication with hybrid vehicle control ECU (U0293): Not detected.
DTC: P0355 **1T ECM, MIL: Yes** **Year:** 2011, 2012 **Model:** 4Runner, Avalon, Camry, Corolla, FJ Cruiser, Highlander, Land Cruiser, Matrix, Rav4, Sequoia, Tundra, Tacoma **Engine:** V6, V8	**Ignition Coil "E" Primary / Secondary Circuit:** No IGF signal to ECM while engine running (1 trip detection logic) The monitor will run whenever these DTCs are not present: None Either of the following condition is met: Condition 1 or 2 1. Starter: OFF 2. Engine RPM: 1500 rpm or less Either of the following condition is met: Condition (a) or (b) (a) All of the following conditions are met: Engine RPM: 500 rpm or less Battery voltage: 6 V or more (b) All of the following conditions are met: Engine RPM: More than 500 rpm Battery voltage: 10 V or more Number of sparks after CPU is reset: 5 sparks or more.
DTC: P0356 **1T ECM, MIL: Yes** **Year:** 2011, 2012 **Model:** 4Runner, Avalon, Camry, Corolla, FJ Cruiser, Highlander, Land Cruiser, Matrix, Rav4, Sequoia, Tundra, Tacoma **Engine:** V6, V8	**Ignition Coil "F" Primary / Secondary Circuit:** No IGF signal is sent to the ECM while the engine is running. Monitor runs whenever following DTCs are not present: None Either of the following condition A or B is met: A. Engine RPM: 1500 rpm or less B. Starter: OFF Either of the following condition C or D is met: C. Both of the following conditions are met: (a) Engine speed: 500 rpm or less (b) Battery voltage: 6 V or more D. All of the following conditions are met: (a) Engine speed: More than 500 rpm (b) Battery voltage: 10 V or more (c) Number of sparks after CPU reset: 5 sparks or more.
DTC: P0357 **1T ECM, MIL: Yes** **Year:** 2011, 2012 **Model:** Land Cruiser, Sequoia, Tundra **Engine:** V8	**Ignition Coil "G" Primary / Secondary Circuit:** No IGF signal is sent to the ECM while the engine is running. Monitor runs whenever following DTCs are not present: None Either of the following condition A or B is met: A. Engine RPM: 1500 rpm or less B. Starter: OFF Either of the following condition C or D is met: C. Both of the following conditions are met: (a) Engine speed: 500 rpm or less (b) Battery voltage: 6 V or more D. All of the following conditions are met: (a) Engine speed: More than 500 rpm (b) Battery voltage: 10 V or more (c) Number of sparks after CPU reset: 5 sparks or more.

DTC	Trouble Code Title and Conditions
DTC: P0358 **1T ECM, MIL: Yes** **Year:** 2011, 2012 **Model:** Land Cruiser, Sequoia, Tundra **Engine:** V8	**Ignition Coil "H" Primary / Secondary Circuit:** No IGF signal is sent to the ECM while the engine is running. Monitor runs whenever following DTCs are not present: None Either of the following condition A or B is met: A. Engine RPM: 1500 rpm or less B. Starter: OFF Either of the following condition C or D is met: C. Both of the following conditions are met: (a) Engine speed: 500 rpm or less (b) Battery voltage: 6 V or more D. All of the following conditions are met: (a) Engine speed: More than 500 rpm (b) Battery voltage: 10 V or more (c) Number of sparks after CPU reset: 5 sparks or more.
DTC: P0365 **1T ECM, MIL: Yes** **Year:** 2011, 2012 **Model:** 4Runner, Avalon, Camry, Corolla, FJ Cruiser, Highlander, Land Cruiser, Matrix, Prius, Rav4, Sequoia, Tundra, Tacoma, Venza, Yaris **Engine:** L4, V6, V8	**Camshaft Position Sensor "B" Circuit (Bank 1):** Input voltage to ECM remains 0.3 V or less, or 4.7 V or higher for more than 5 seconds, when 2 or more seconds have elapsed after turning ignition switch ON position. No VVT sensor signal to ECM during cranking.
DTC: P0367 **1T ECM, MIL: Yes** **Year:** 2011, 2012 **Model:** 4Runner, Avalon, Camry, Corolla, FJ Cruiser, Highlander, Land Cruiser, Matrix, Prius, Rav4, Sequoia, Tundra, Tacoma, Venza, Yaris **Engine:** L4, V6, V8	**Camshaft Position Sensor "B" Circuit Low Input (Bank 1) :** Output voltage of VVT sensor less than 0.3 V 4 seconds. Monitor runs whenever following DTCs not stored: None Starter: OFF Ignition switch: ON Time after ignition switch turned from off to ON: 2 seconds or more Exhaust camshaft position sensor verify pulse input fail (P0365): Not detected Battery voltage: 8 V or higher.
DTC: P0368 **1T ECM, MIL: Yes** **Year:** 2011, 2012 **Model:** 4Runner, Avalon, Camry, Corolla, FJ Cruiser, Highlander, Land Cruiser, Matrix, Prius, Rav4, Sequoia, Tundra, Tacoma, Venza, Yaris **Engine:** L4, V6, V8	**Camshaft Position Sensor "B" Circuit High Input (Bank 1):** Output voltage of VVT sensor more than 4.7 V 4 seconds. Monitor runs whenever following DTCs are not present: None Starter: OFF Engine switch: On (IG) Time after engine switch off to on (IG): 2 seconds or more.
DTC: P0390 **1T ECM, MIL: Yes** **Year:** 2011, 2012 **Model:** 4Runner, Avalon, Camry, FJ Cruiser, Highlander, Land Cruiser, Matrix,Rav4, Sequoia, Tundra, Tacoma **Engine:** V6, V8	**Camshaft Position Sensor "B" Circuit (Bank 2):** Monitor runs whenever following DTCs are not present: None Camshaft Position Sensor Range Check (Chattering, Low voltage, High voltage): Starter: OFF Engine switch: On (IG) Time after engine switch off to on (IG): 2 seconds or more Battery voltage: 8 V or more Camshaft Position Sensor Range Check / Rationality: Engine speed: 600 rpm or more Battery voltage: 8 V or more Starter: OFF Engine switch: On (IG)

DTC	Trouble Code Title and Conditions
DTC: P0392 **1T ECM, MIL: Yes** **Year:** 2011, 2012 **Model:** 4Runner, Avalon, Camry, FJ Cruiser, Highlander, Land Cruiser, Matrix,Rav4, Sequoia, Tundra, Tacoma **Engine:** V6, V8	**Camshaft Position Sensor "B" Circuit Low Input (Bank 2):** Engine Running: The output voltage of the VVT sensor is below 0.3 V for 4 seconds. Monitor runs whenever following DTCs are not present: P0367, P0368, P0392, P0393 Engine speed: 600 rpm or more Battery voltage: 8 V or higher Starter: OFF Ignition switch: ON VVT sensor voltage: 0.3 to 4.7 V Low Voltage, High Voltage: Monitor runs whenever following DTCs are not present: P0365, P0390 (Exhaust VVT Sensor) Starter: OFF Ignition switch: ON Time after ignition switch off to ON: 2 seconds or more Battery voltage: 8 V or higher.
DTC: P0393 **1T ECM, MIL: Yes** **Year:** 2011, 2012 **Model:** 4Runner, Avalon, Camry, FJ Cruiser, Highlander, Land Cruiser, Matrix,Rav4, Sequoia, Tundra, Tacoma **Engine:** V6, V8	**Camshaft Position Sensor "B" Circuit High Input (Bank 2):** Engine Running: The output voltage of the VVT sensor is higher than 4.7 V for 4 seconds. Monitor runs whenever following DTCs are not present: P0367, P0368, P0392, P0393 Engine speed: 600 rpm or more Battery voltage: 8 V or higher Starter: OFF Ignition switch: ON VVT sensor voltage: 0.3 to 4.7 V Low Voltage, High Voltage: Monitor runs whenever following DTCs are not present: P0365, P0390 (Exhaust VVT Sensor) Starter: OFF Ignition switch: ON Time after ignition switch off to ON: 2 seconds or more Battery voltage: 8 V or higher.
DTC: P0401 **2T ECM** **Year:** 2011, 2012 **Model:** Prius, Prius V **Engine:** 1.8L L4	**Exhaust Gas Recirculation Flow Insufficient Detected:** Monitor runs whenever following DTCs are not present: None Time after engine started: 3 seconds or more Time after engine fuel cut: 2000 msec or more Engine speed: 950 to 1600 rpm Vehicle speed: 12.5 mph (20 km/h) or more Engine speed change: -20 to 20 rpm Time after engine load becomes stable (HVAC, Brake, Lockup): More than 1.2 sec Engine coolant temperature: 70°C (158°F) or more Battery voltage: 11 V or more Intake air temperature: -10°C (14°F) or more Atmospheric pressure: 76 kPa or more
DTC: P0403 **1T ECM** **Year:** 2011, 2012 **Model:** Prius, Prius V **Engine:** 1.8L L4	**Exhaust Gas Recirculation Control Circuit:** Monitor runs whenever following DTCs are not present: None Engine: Running Battery voltage: 10.5 V or more Time after power switch off to on (IG): More than 0.5 seconds.
DTC: P0412 **1T ECM, MIL: Yes** **Year:** 2011, 2012 **Model:** 4Runner, Avalon, Camry, Corolla, FJ Cruiser, Highlander, Land Cruiser, Matrix, Prius, Rav4, Sequoia, Tundra, Tacoma, Venza, Yaris **Engine:** L4, V6, V8	**Secondary Air Injection System Switching Valve "A" Circuit:** After a cold engine start, all of the following conditions are met: (a) The Secondary Air Injection (AIR) system is not operating (air pump OFF/ON, Air Switching Valve [ASV] OFF/ON). (b) The diagnostic signal from the Air Injection Control Driver (AID) is 40%. (c) The battery voltage is 8 V or higher.

DTC	Trouble Code Title and Conditions
DTC: P0415 **1T ECM, MIL: Yes** **Year:** 2011, 2012 **Model:** 4Runner, Avalon, Camry, Corolla, FJ Cruiser, Highlander, Land Cruiser, Matrix, Prius, Rav4, Sequoia, Tundra, Tacoma, Venza, Yaris **Engine:** L4, V6, V8	**Secondary Air Injection System Switching Valve "B" Circuit:** After a cold engine start, all of the following conditions are met: 1. The Secondary Air Injection (AIR) system is not operating (air pump OFF/ON, Air Switching Valve [ASV] OFF/ON). 2. The diagnostic signal from the Air Injection Control Driver (AID) is 40%. 3. The battery voltage is 8 V or higher.
DTC: P0418 **1T ECM, MIL: Yes** **Year:** 2011, 2012 **Model:** 4Runner, Avalon, Camry, Corolla, FJ Cruiser, Highlander, Land Cruiser, Matrix, Prius, Rav4, Sequoia, Tundra, Tacoma, Venza, Yaris **Engine:** L4, V6, V8	**Secondary Air Injection System Control "A" Circuit:** After cold engine starts, all of following conditions met (1 trip detection logic): a. Secondary Air Injection (AIR) system not operating (air pump OFF, Air Switching Valve [ASV] OFF) b. Diagnostic signal from Air Injection Control Driver (AID) 20% c. Battery voltage 8 V or more After cold engine starts, all of following conditions met (1 trip detection logic): a. Secondary Air Injection (AIR) system operating (air pump ON, Air Switching Valve [ASV] ON) b. Diagnostic signal from Air Injection Control Driver (AID) 20% c. Battery voltage 8 V or more.
DTC: P0419 **1T ECM, MIL: Yes** **Year:** 2011, 2012 **Model:** 4Runner, Avalon, Camry, Corolla, FJ Cruiser, Highlander, Land Cruiser, Matrix, Prius, Rav4, Sequoia, Tundra, Tacoma, Venza, Yaris **Engine:** L4, V6, V8	**Secondary Air Injection System Control "B" Circuit:** After a cold engine start, all of the following conditions are met: 1. Secondary Air Injection (AIR) system not operating (air pump OFF/ON, Air Switching Valve [ASV] OFF/ON) 2. The diagnostic signal from the Air Injection Control Driver (AID) is 20%. 3. The battery voltage is 8 V or higher.
DTC: P0420 **2T ECM, MIL: Yes** **Year:** 2011, 2012 **Model:** 4Runner, Avalon, Camry, Corolla, FJ Cruiser, Highlander, Land Cruiser, Matrix, Prius, Rav4, Sequoia, Tundra, Tacoma, Venza, Yaris **Engine:** L4, V6, V8	**Catalyst System Efficiency Below Threshold (Bank 1):** Enabling conditions are as follows: Battery voltage 11 V or higher, Intake air temperature -10°C (14°F) or higher, Engine coolant temperature 75°C (167°F) or higher, Atmospheric pressure 76 kPa (570 mmHg) or higher, Idling OFF, Engine speed less than 3200 rpm, A/F sensor status activated, Fuel system status Closed loop, Engine load 10 to 75%. All of following conditions (a), (b) and (c) met (a) Mass air flow rate 7.5 to 75 g/sec. (b) Estimated front catalyst temperature 575 to 820°C (1067 to 1508°F) (c) Estimated rear catalyst temperature 400 to 700°C (752 to 1292°F), Rear HO2 sensor heater monitor Completed, Shift position 4th or higher. The OSC value is less than the standard value under active air-fuel ratio control.
DTC: P0430 **2T ECM, MIL: Yes** **Year:** 2011, 2012 **Model:** 4Runner, Avalon, Camry, Corolla, FJ Cruiser, Highlander, Land Cruiser, Matrix, Prius, Rav4, Sequoia, Tundra, Tacoma, Venza, Yaris **Engine:** L4, V6, V8	**Catalyst System Efficiency Below Threshold (Bank 2):** Enabling conditions are as follows: Battery voltage 11 V or higher, Intake air temperature -10°C (14°F) or higher, Engine coolant temperature 75°C (167°F) or higher, Atmospheric pressure 76 kPa (570 mmHg) or higher, Idling OFF, Engine speed less than 3200 rpm, A/F sensor status activated, Fuel system status Closed loop, Engine load 10 to 75%. All of following conditions (a), (b) and (c) met (a) Mass air flow rate 7.5 to 75 g/sec. (b) Estimated front catalyst temperature 575 to 820°C (1067 to 1508°F) (c) Estimated rear catalyst temperature 400 to 700°C (752 to 1292°F), Rear HO2 sensor heater monitor Completed, Shift position 4th or higher. The OSC value is less than the standard value under active air-fuel ratio control.
DTC: P043E **2T ECM, MIL: Yes** **Year:** 2011, 2012 **Model:** 4Runner, Avalon, Camry, Corolla, FJ Cruiser, Highlander, Land Cruiser, Matrix, Prius, Rav4, Sequoia, Tundra, Tacoma, Venza, Yaris **Engine:** L4, V6, V8	**Evaporative Emission System Reference Orifice Clog Up:** Vacuum pump creates negative pressure through 0.02 inch orifice, and EVAP system pressure is measured to determine leak pressure standard. If system pressure is lower than 4.3 kPa (-32.25 mmHg), ECM determines that 0.02 inch orifice has clogging malfunction.

DTC	Trouble Code Title and Conditions
DTC: P043F **2T ECM, MIL: Yes** **Year:** 2011, 2012 **Model:** 4Runner, Avalon, Camry, Corolla, FJ Cruiser, Highlander, Land Cruiser, Matrix, Prius, Rav4, Sequoia, Tundra, Tacoma, Venza, Yaris **Engine:** L4, V6, V8	**Evaporative Emission System Reference Orifice High Flow:** Codes P043E, P043F, P2401, P2402 and P2419 may be stored when one of the following conditions is met during the key-off EVAP monitor: The reference orifice is clogged. Reference orifice high-flow. A leak detection pump OFF malfunction. A leak detection pump ON malfunction. A vent valve ON (close) malfunction.
DTC: P0441 **2T ECM, MIL: Yes** **Year:** 2011, 2012 **Model:** 4Runner, Avalon, Camry, Corolla, FJ Cruiser, Highlander, Land Cruiser, Matrix, Prius, Rav4, Sequoia, Tundra, Tacoma, Venza, Yaris **Engine:** L4, V6, V8	**Evaporative Emission Control System Incorrect Purge Flow:** Vacuum pump creates negative pressure (vacuum) in EVAP system and EVAP system pressure is measured. 0.02 inch leak pressure standard is measured at the start and end of the leak check. If stabilized pressure is higher than [second 0.02 inch leak pressure standard x 0.2], ECM determines that purge VSV is stuck open. After EVAP leak check is performed, purge VSV is turned ON (open), and atmospheric air is introduced into EVAP system. 0.02 inch leak pressure standard is measured at the start and end of the leak check. If pressure does not return to near atmospheric pressure, the ECM determines that the purge valve stuck closed. While engine is running, following conditions are successively met: 1). Negative pressure is not created in EVAP system when purge VSV is turned ON (open) 2). EVAP system pressure change is less than 0.5 kPa (3.75 mmHg) when vent valve is turned ON (closed) 3). Atmospheric pressure change before and after purge flow monitor is less than 0.1 kPa (0.75 mmHg)
DTC: P0443 **1T ECM, MIL: Yes** **Year:** 2011, 2012 **Model:** 4Runner, Avalon, Camry, Corolla, FJ Cruiser, Highlander, Land Cruiser, Matrix, Prius, Rav4, Sequoia, Tundra, Tacoma, Venza, Yaris **Engine:** L4, V6, V8	**Evaporative Emission Control System Purge Control Valve Circuit:** The sensor output voltage rapidly fluctuates beyond upper and lower malfunction thresholds for 0.5 seconds.
DTC: P0450 **1T ECM, MIL: Yes** **Year:** 2011, 2012 **Model:** 4Runner, Avalon, Camry, Corolla, FJ Cruiser, Highlander, Land Cruiser, Matrix, Prius, Rav4, Sequoia, Tundra, Tacoma, Venza, Yaris **Engine:** L4, V6, V8	**Evaporative Emission Control System Pressure Sensor Malfunction:** Sensor output voltage rapidly fluctuates beyond upper and lower malfunction thresholds for 0.5 seconds. Monitor runs whenever following DTCs are not present: None Either of following conditions is met: Condition A or B A. Engine switch: On (IG) B. Soak timer: ON
DTC: P0451 **1T ECM, MIL: Yes** **Year:** 2011, 2012 **Model:** 4Runner, Avalon, Camry, Corolla, FJ Cruiser, Highlander, Land Cruiser, Matrix, Prius, Rav4, Sequoia, Tundra, Tacoma, Venza, Yaris **Engine:** L4, V6, V8	**Evaporative Emission Control System Pressure Sensor Range / Performance:** Sensor output voltage fluctuates frequently in a certain time period. Sensor output voltage does not vary in a certain time period.
DTC: P0452 **1T ECM, MIL: Yes** **Year:** 2011, 2012 **Model:** 4Runner, Avalon, Camry, Corolla, FJ Cruiser, Highlander, Land Cruiser, Matrix, Prius, Rav4, Sequoia, Tundra, Tacoma, Venza, Yaris **Engine:** L4, V6, V8	**Evaporative Emission Control System Pressure Sensor / Switch Low Input:** Monitor runs whenever following DTCs are not present: None Battery voltage: 8 V or more Starter: OFF The sensor output voltage is less than 0.45 V for 0.5 seconds.

DTC	Trouble Code Title and Conditions
DTC: P0453 **1T ECM, MIL: Yes** **Year:** 2011, 2012 **Model:** 4Runner, Avalon, Camry, Corolla, FJ Cruiser, Highlander, Land Cruiser, Matrix, Prius, Rav4, Sequoia, Tundra, Tacoma, Venza, Yaris **Engine:** L4, V6, V8	**Evaporative Emission Control System Pressure Sensor / Switch High Input:** If the canister pressure sensor voltage output [pressure] is 4.9 V [123.8 kPa-a (928.5 mmHg-a)] or more. The ECM interprets this as an open or short circuit malfunction in the canister pressure sensor or its circuit, and stops the EVAP system monitor. Power switch: On (IG) Battery voltage: 8 V or more
DTC: P0455 **1T ECM, MIL: Yes** **Year:** 2011, 2012 **Model:** 4Runner, Avalon, Camry, Corolla, FJ Cruiser, Highlander, Land Cruiser, Matrix, Prius, Rav4, Sequoia, Tundra, Tacoma, Venza, Yaris **Engine:** L4, V6, V8	**Evaporative Emission Control System Leak Detected (Gross Leak):** Leak detection pump creates negative pressure (vacuum) in EVAP system and EVAP system pressure measured. 0.02 inch leak criterion measured at start and at end of leak check. If stabilized pressure higher than [second 0.02 inch leak criterion x 0.2. The ECM determines that EVAP system has large leakage.
DTC: P0456 **1T ECM, MIL: Yes** **Year:** 2011, 2012 **Model:** 4Runner, Avalon, Camry, Corolla, FJ Cruiser, Highlander, Land Cruiser, Matrix, Prius, Rav4, Sequoia, Tundra, Tacoma, Venza, Yaris **Engine:** L4, V6, V8	**Evaporative Emission Control System Leak Detected (Very Small Leak):** Leak detection pump creates negative pressure (vacuum) in EVAP system and EVAP system pressure measured. Reference pressure measured at start and at end of leak check. If stabilized pressure higher than second reference pressure, ECM determines that EVAP system has small leak.
DTC: P0500 **1T ECM, MIL: Yes** **Year:** 2011, 2012 **Model:** 4Runner, Avalon, Camry, Corolla, FJ Cruiser, Highlander, Land Cruiser, Matrix, Prius, Rav4, Sequoia, Tundra, Tacoma, Venza, Yaris **Engine:** L4, V6, V8	**Vehicle Speed Sensor "A" Malfunction:** While vehicle being driven, no vehicle speed sensor signal transmitted to ECM (1 trip detection logic: Automatic transaxle) (2 trip detection logic: Manual transaxle)
DTC: P0502 **1T ECM, MIL: Yes** **Year:** 2011, 2012 **Model:** 4Runner, Avalon, Camry, Corolla, FJ Cruiser, Highlander, Land Cruiser, Matrix, Prius, Rav4, Sequoia, Tundra, Tacoma, Venza, Yaris **Engine:** L4, V6, V8	**Vehicle Speed Sensor "A" Circuit Low:** The monitor will run whenever the following DTCs are not present: None Battery voltage: 8 V or more Engine switch: On (IG) Starter: OFF Momentary interruption and noise are detected when a rapid change of vehicle speed occurs.
DTC: P0503 **1T ECM, MIL: Yes** **Year:** 2011, 2012 **Model:** 4Runner, Avalon, Camry, Corolla, FJ Cruiser, Highlander, Land Cruiser, Matrix, Prius, Rav4, Sequoia, Tundra, Tacoma, Venza, Yaris **Engine:** L4, V6, V8	**Vehicle Speed Sensor "A" Intermittent / Erratic / High:** The monitor will run whenever the following DTCs are not present: None Battery voltage: 8 V or more Engine switch: On (IG) Starter: OFF Momentary interruption and noise are detected when a rapid change of vehicle speed occurs.

DTC	Trouble Code Title and Conditions
DTC: P0504 **1T ECM, MIL: Yes** **Year:** 2011, 2012 **Model:** 4Runner, Avalon, Camry, Corolla, FJ Cruiser, Highlander, Land Cruiser, Matrix, Prius, Rav4, Sequoia, Tundra, Tacoma, Venza, Yaris **Engine:** L4, V6, V8	**Brake Switch "A" / "B" Correlation:** Conditions (a), (b) and (c) continue for 0.5 seconds or more: (a) Ignition switch on (b) Brake pedal released (c) STP signal OFF when ST1- signal OFF
DTC: P0505 **2T ECM, MIL: Yes** **Year:** 2011, 2012 **Model:** 4Runner, Avalon, Camry, Corolla, FJ Cruiser, Highlander, Land Cruiser, Matrix, Prius, Rav4, Sequoia, Tundra, Tacoma, Venza, Yaris **Engine:** L4, V6, V8	**Idle Control System Malfunction:** Idling speed continues to vary greatly from target idling speed.
DTC: P050A **2T ECM, MIL: Yes** **Year:** 2011, 2012 **Model:** 4Runner, Avalon, Camry, Corolla, FJ Cruiser, Highlander, Land Cruiser, Matrix, Prius, Rav4, Sequoia, Tundra, Tacoma, Venza, Yaris **Engine:** L4, V6, V8	**Cold Start Idle Air Control System Performance:** Battery voltage: 8 V or higher Time after engine start: 3 seconds or more Starter: OFF Engine coolant temperature at engine start: -10°C (14°F) or higher Engine coolant temperature: -10 to 50°C (14 to 122°F) Engine idling time: 3 seconds or more Fuel-cut: OFF Vehicle speed: Less than 3 km/h (1.875 mph) Time after shift position changed: 1 second or more Atmospheric pressure: 76 kPa (570 mmHg) or higher
DTC: P050B **2T ECM, MIL: Yes** **Year:** 2011, 2012 **Model:** 4Runner, Avalon, Camry, Corolla, FJ Cruiser, Highlander, Land Cruiser, Matrix, Prius, Rav4, Sequoia, Tundra, Tacoma, Venza, Yaris **Engine:** L4, V6, V8	**Cold Start Ignition Timing Performance:** Battery voltage: 8 V or more Time after engine start: 3 seconds or more Starter: OFF Engine coolant temperature at engine start: -10°C (14°F) or more Engine coolant temperature: -10 to 50°C (14 to 122°F) Engine idling time: 0 seconds or more Fuel-cut: OFF Vehicle speed: Less than 37.5 mph (60.3 km/h) Atmospheric pressure: 76 kPa-a (570 mmHg-a) or more
DTC: P0516-769 **1T ECM, MIL: Yes** **Year:** 2011, 2012 **Model:** Camry, Highlander, Prius, Prius V **Engine:** 1.8L L4, 2.4L L4, 3.5L V6 Hybrid	**Battery Temperature Sensor Circuit Low:** Malfunction in the auxiliary battery thermometer sensor circuit (short to GND)
DTC: P0517-770 **1T ECM, MIL: Yes** **Year:** 2011, 2012 **Model:** Camry, Prius, Prius V **Engine:** 1.8L L4, 2.4L L4 Hybrid	**Battery Temperature Sensor Circuit High:** Malfunction in the auxiliary battery thermometer sensor circuit (open or +B short circuit).
DTC: P0550 **1T ECM, MIL: Yes** **Year:** 2011, 2012 **Model:** 4Runner, Avalon, Camry, Corolla, FJ Cruiser, Highlander, Land Cruiser, Matrix, Prius, Rav4, Sequoia, Tundra, Tacoma, Venza, Yaris **Engine:** L4, V6, V8	**Power Steering Pressure Sensor Circuit Malfunction:** An open or short in the power steering oil pressure sensor circuit is detected for more than 3 seconds after the engine switch is turned on (IG).

DTC	Trouble Code Title and Conditions
DTC: P0552 **1T ECM, MIL: Yes** **Year:** 2011, 2012 **Model:** 4Runner, Avalon, Camry, Corolla, FJ Cruiser, Highlander, Land Cruiser, Matrix, Prius, Rav4, Sequoia, Tundra, Tacoma, Venza, Yaris **Engine:** L4, V6, V8	**Power Steering Pressure Sensor Circuit Low Input:** Power steering oil pressure switch voltage: Below 0.28 V.
DTC: P0553 **1T ECM, MIL: Yes** **Year:** 2011, 2012 **Model:** 4Runner, Avalon, Camry, Corolla, FJ Cruiser, Highlander, Land Cruiser, Matrix, Prius, Rav4, Sequoia, Tundra, Tacoma, Venza, Yaris **Engine:** L4, V6, V8	**Power Steering Pressure Sensor Circuit High Input:** Power steering oil pressure sensor voltage more than 4.9 V for 0.5 seconds while engine running.
DTC: P0560 **1T ECM, MIL: Yes** **Year:** 2011, 2012 **Model:** 4Runner, Avalon, Camry, Corolla, FJ Cruiser, Highlander, Land Cruiser, Matrix, Prius, Rav4, Sequoia, Tundra, Tacoma, Venza, Yaris **Engine:** L4, V6, V8	**System Voltage:** Monitor runs whenever following DTCs are not present: None. Stand-by RAM: Initialized Open in ECM back up power source circuit.
DTC: P0560-117 **T , MIL: Yes** **Year:** 2011, 2012 **Model:** Camry, Prius, Prius V **Engine:** 1.8L L4, 2.4L L4	**System Voltage:** The monitor will run whenever the following DTC is not present: TMC's intellectual property.
DTC: P0571 **1T ECM, MIL: Yes** **Year:** 2011, 2012 **Model:** 4Runner, Avalon, Camry, Corolla, FJ Cruiser, Highlander, Land Cruiser, Matrix, Prius, Rav4, Sequoia, Tundra, Tacoma, Venza, Yaris **Engine:** L4, V6, V8	**Brake Switch "A" Circuit:** Voltage of STP terminal and that of ST1- terminal of ECM are less than 1 V for 0.5 sec. or more.
DTC: P0575 **1T ECM, MIL: Yes** **Year:** 2011, 2012 **Model:** 4Runner, Avalon, Camry, Corolla, FJ Cruiser, Highlander, Land Cruiser, Matrix, Prius, Rav4, Sequoia, Tundra, Tacoma, Venza, Yaris **Engine:** L4, V6, V8	**Cruise Control Input Circuit:** When either of the following conditions is met: 1). Cruise control input signal abnormal 2). Stop light switch input signal abnormal
DTC: P0604 **1T ECM, MIL: Yes** **Year:** 2011, 2012 **Model:** 4Runner, Avalon, Camry, Corolla, FJ Cruiser, Highlander, Land Cruiser, Matrix, Prius, Rav4, Sequoia, Tundra, Tacoma, Venza, Yaris **Engine:** L4, V6, V8	**Random Access Memory (RAM):** The ECM continuously monitors its internal memory status. This self-check ensures that the ECM is functioning properly. It is diagnosed by internal "mirroring" of the main CPU and sub CPU to detect the Random Access Memory (RAM) errors. If outputs from these CPUs are different and deviate from the standards, the ECM will illuminate the MIL and set a DTC immediately.

DTC	Trouble Code Title and Conditions
DTC: P0606 **1T ECM, MIL: Yes** **Year:** 2011, 2012 **Model:** 4Runner, Avalon, Camry, Corolla, FJ Cruiser, Highlander, Land Cruiser, Matrix, Prius, Rav4, Sequoia, Tundra, Tacoma, Venza, Yaris **Engine:** L4, V6, V8	**ECM / PCM Processor:** When either condition below is met: An ECM main CPU error. An ECM sub CPU error. The ECM will illuminate the MIL and store DTC(s) immediately.
DTC: P0607 **1T ECM, MIL: Yes** **Year:** 2011, 2012 **Model:** Corolla, Matrix, Yaris **Engine:** 1.5L L4, 1.8L L4 VIN L, 1.8L L4 VIN U, 2.4L L4	**Control Module Performance:** With the engine running, estimated HO2 sensor temperature 450 to 800°C (842 to 1472°F). The following malfunctions are detected: ECM CPUs malfunction. Heated Oxygen (HO2) sensor transistors (built into ECM) malfunction. **Possible causes:** • Exhaust gas leak • Heated oxygen sensor • ECM
DTC: P0607 **1T ECM, MIL: Yes** **Year:** 2011, 2012 **Model:** 4Runner, Avalon, Camry, Corolla, FJ Cruiser, Highlander, Land Cruiser, Matrix, Prius, Rav4, Sequoia, Tundra, Tacoma, Venza, Yaris **Engine:** L4, V6, V8	**Input Signal Circuit Malfunction:** The ECM has a supervisory CPU and control ECU inside. When each input STP signal is different for 0.15 sec. or more, this trouble code is output. This trouble code is output after 0.4 sec. has passed from the time the cruise cancel input signal (STP input) is input into the ECM. NOTE: When a trouble code is detected, the fail safe must be kept on until the ignition switch is turned off.
DTC: P060A **1T ECM, MIL: Yes** **Year:** 2011, 2012 **Model:** 4Runner, Avalon, Camry, Corolla, FJ Cruiser, Highlander, Land Cruiser, Matrix, Prius, Rav4, Sequoia, Tundra, Tacoma, Venza, Yaris **Engine:** L4, V6, V8	**Internal Control Module Monitoring Processor Performance:** When any condition below is met: 1). ECM main CPU error 2). ECM sub CPU error 3). Electronic throttle monitoring CPU error.
DTC: P060B **1T ECM, MIL: Yes** **Year:** 2011, 2012 **Model:** 4Runner, Avalon, Camry, Corolla, FJ Cruiser, Highlander, Land Cruiser, Matrix, Prius, Rav4, Sequoia, Tundra, Tacoma, Venza, Yaris **Engine:** L4, V6, V8	**Internal Control Module A/D Processing Performance:** Monitor runs whenever following DTCs not stored: None. ECM main CPU communication error.
DTC: P060B-570 **1T ECM, MIL: Yes** **Year:** 2011, 2012 **Model:** Prius, Prius V **Engine:** 1.8L L4	**Internal Control Module A/D Processing Performance:** Monitor runs whenever following DTCs not present ECU internal error.
DTC: P060D **1T ECM, MIL: Yes** **Year:** 2011, 2012 **Model:** 4Runner, Avalon, Camry, Corolla, FJ Cruiser, Highlander, Land Cruiser, Matrix, Prius, Rav4, Sequoia, Tundra, Tacoma, Venza, Yaris **Engine:** L4, V6, V8	**Internal Control Module Accelerator Pedal Position Performance:** When either condition below is met: 1). An ECM main CPU error. 2). An ECM sub CPU error.

DTC	Trouble Code Title and Conditions
DTC: P060E **1T ECM, MIL: Yes** **Year:** 2011, 2012 **Model:** 4Runner, Avalon, Camry, Corolla, FJ Cruiser, Highlander, Land Cruiser, Matrix, Prius, Rav4, Sequoia, Tundra, Tacoma, Venza, Yaris **Engine:** L4, V6, V8	**Internal Control Module Throttle Position Performance:** Monitor runs whenever the following DTCs are not present: None DMA communication error: Not detected
DTC: P0617 **1T ECM, MIL: Yes** **Year:** 2011, 2012 **Model:** 4Runner, Avalon, Camry, Corolla, FJ Cruiser, Highlander, Land Cruiser, Matrix, Prius, Rav4, Sequoia, Tundra, Tacoma, Venza, Yaris **Engine:** L4, V6, V8	**Starter Relay Circuit High:** When conditions (a), (b) and (c) met, positive (+B) battery voltage 10.5 V or 20 seconds: (a) Vehicle speed more than 12.4 mph (20 km/h) (b) Engine speed more than 1,000 rpm (c) STA signal ON.
DTC: P0617-142 **T , MIL: Yes** **Year:** 2011, 2012 **Model:** Camry, Prius, Prius V **Engine:** 1.8L L4, 2.4L L4	**Starter Relay Circuit High:** An ST signal from the power management control ECU is present when the power switch is off.
DTC: P062D **1T ECM, MIL: Yes** **Year:** 2011, 2012 **Model:** 4Runner, Avalon, Camry, Corolla, FJ Cruiser, Highlander, Land Cruiser, Matrix, Prius, Rav4, Sequoia, Tundra, Tacoma, Venza, Yaris **Engine:** L4, V6, V8	**No. 1 Fuel Injector Driver Circuit Performance:** Monitor runs whenever following DTCs are not present: None Time after engine switch off to on (IG): 1 second or more Engine speed: 4000 rpm or less Battery voltage: 10.5 V or more
DTC: P062F **1T ECM, MIL: Yes** **Year:** 2011, 2012 **Model:** 4Runner, Avalon, Camry, Corolla, FJ Cruiser, Highlander, Land Cruiser, Matrix, Prius, Rav4, Sequoia, Tundra, Tacoma, Venza, Yaris **Engine:** L4, V6, V8	**Internal Control Module EEPROM Error:** Monitor runs whenever following DTCs not present: None Time after engine start: 10 seconds or more Battery voltage:8 V or higher Ignition switch: ON Starter: OFF
DTC: P062F-143 **T ECM, MIL: Yes** **Year:** 2011, 2012 **Model:** Camry, Prius, Prius V **Engine:** 1.8L L4, 2.4L L4	**EEPROM Malfunction:** The monitor will run whenever the following DTC is not present: TMC's intellectual property. Other conditions belong to TMC's intellectual property.
DTC: P0630 **1T ECM, MIL: Yes** **Year:** 2011, 2012 **Model:** 4Runner, Avalon, Camry, Corolla, FJ Cruiser, Highlander, Land Cruiser, Matrix, Prius, Rav4, Sequoia, Tundra, Tacoma, Venza, Yaris **Engine:** L4, V6, V8	**VIN not Programmed or Mismatch - ECM / PCM:** When either condition below is met: 1). The VIN is not stored in the ECM. 2). The input VIN is incorrect.
DTC: P0630-804 **T ECM, MIL: Yes** **Year:** 2011, 2012 **Model:** Camry, Prius, Prius V **Engine:** 1.8L L4, 2.4L L4	**VIN not Programmed or Mismatch-ECM / PCM:** VIN not stored in hybrid vehicle control ECU Input VIN in hybrid vehicle control ECU not accurate

DTC	Trouble Code Title and Conditions
DTC: P0657 **1T ECM, MIL: Yes** **Year:** 2011, 2012 **Model:** 4Runner, Avalon, Camry, Corolla, FJ Cruiser, Highlander, Land Cruiser, Matrix, Prius, Rav4, Sequoia, Tundra, Tacoma, Venza, Yaris **Engine:** L4, V6, V8	**Actuator Supply Voltage Circuit / Open:** Monitor will run whenever this DTC is not present: None Throttle actuator power supply voltage: 7 V or higher Engine switch: Front ON to OFF
DTC: P0705 **2T ECM, MIL: Yes** **Year:** 2011, 2012 **Model:** 4Runner, Avalon, Camry, Corolla, FJ Cruiser, Highlander, Land Cruiser, Matrix, Prius, Rav4, Sequoia, Tundra, Tacoma, Venza, Yaris **Engine:** L4, V6, V8	**Transmission Range Sensor Circuit Malfunction (PRNDL Input):** Either of the following conditions is met: (a) Any 2 or more of the following signals are ON simultaneously. Park/neutral position switch input signal is ON R input signal is ON D input signal is ON 2 input signal is ON (b) Any 2 or more of the following signals are ON simultaneously. Park/neutral position switch input signal is ON R input signal is ON 3 input signal is ON L input signal is ON
DTC: P0710 **1T ECM, MIL: Yes** **Year:** 2011, 2012 **Model:** 4Runner, Avalon, Camry, Corolla, FJ Cruiser, Highlander, Land Cruiser, Matrix, Prius, Rav4, Sequoia, Tundra, Tacoma, Venza, Yaris **Engine:** L4, V6, V8	**Transmission Fluid Temperature Sensor "A" Circuit:** (a) and (b) are detected momentarily within 0.5 sec. when neither P0712 nor P0713 is detected (1-trip detection logic) (a) No. 1 ATF temperature sensor resistance is less than 79 Ohm. (b) No. 1 ATF temperature sensor resistance is more than 156 kOhm. NOTE: Within 0.5 sec, the malfunction switches from (a) to (b) or from (b) to (a).
DTC: P0711 **2T ECM, MIL: Yes** **Year:** 2011, 2012 **Model:** 4Runner, Avalon, Camry, Corolla, FJ Cruiser, Highlander, Land Cruiser, Matrix, Prius, Rav4, Sequoia, Tundra, Tacoma, Venza, Yaris **Engine:** L4, V6, V8	**Transmission Fluid Temperature Sensor "A" Performance:** Either of the following Condition (A) or (B) is met: Condition (A): Both (a) and (b) are detected: (2-trip detection logic) (a) Intake air and engine coolant temperatures are more than -10°C (14°F) at engine start. (b) After normal driving for 21 minutes and 40 seconds or more and 9 km (5.6 miles) or more, ATF temperature is less than 20°C (68°F). Condition (B): Both (a) and (b) are detected (2-trip detection logic) (a) Engine coolant temperature is less than 35°C (95°F) at engine start. (b) ATF temperature is 110°C (230°F) or more when engine coolant temperature reaches 60°C (140°F).
DTC: P0712 **1T ECM, MIL: Yes** **Year:** 2011, 2012 **Model:** 4Runner, Avalon, Camry, Corolla, FJ Cruiser, Highlander, Land Cruiser, Matrix, Prius, Rav4, Sequoia, Tundra, Tacoma, Venza, Yaris **Engine:** L4, V6, V8	**Transmission Fluid Temperature Sensor "A" Circuit Low Input:** ATF temperature sensor resistance is less than 79 Ohm for 0.5 seconds or more. Short in ATF temperature sensor circuit.
DTC: P0713 **1T ECM, MIL: Yes** **Year:** 2011, 2012 **Model:** 4Runner, Avalon, Camry, Corolla, FJ Cruiser, Highlander, Land Cruiser, Matrix, Prius, Rav4, Sequoia, Tundra, Tacoma, Venza, Yaris **Engine:** L4, V6, V8	**Transmission Fluid Temperature Sensor "A" Circuit High Input:** When 15 minutes or more have elapsed after engine is started, No. 1 ATF temperature sensor resistance is more than 156 kOhm. for 0.5 seconds or more.

DTC	Trouble Code Title and Conditions
DTC: P0715 **1T ECM, MIL: Yes** **Year:** 2011, 2012 **Model:** 4Runner, Avalon, Camry, Corolla, FJ Cruiser, Highlander, Land Cruiser, Matrix, Prius, Rav4, Sequoia, Tundra, Tacoma, Venza, Yaris **Engine:** L4, V6, V8	**Input / Turbine Speed Sensor Circuit Malfunction:** Battery voltage: 8 V or more Engine switch: ON Starter: OFF
DTC: P0717 **1T ECM, MIL: Yes** **Year:** 2011, 2012 **Model:** 4Runner, Avalon, Camry, Corolla, FJ Cruiser, Highlander, Land Cruiser, Matrix, Prius, Rav4, Sequoia, Tundra, Tacoma, Venza, Yaris **Engine:** L4, V6, V8	**Input Speed Sensor Circuit No Signal:** The monitor will run whenever this DTC is not present:P0500 (VSS), P0748, P0778, P0798 (Shift solenoid valve (range)) Shift change: Shift change is completed before starting next shift change operation TCM selected gear: 2nd, 3rd, 4th, 5th or 6th Output shaft rpm: 1,000 rpm or more STAR switch: OFF R switch: OFF Engine: Running Battery voltage: 8 V or more Engine switch: ON Starter: OFF
DTC: P0722 **2T ECM, MIL: Yes** **Year:** 2011, 2012 **Model:** 4Runner, Avalon, Camry, Corolla, FJ Cruiser, Highlander, Land Cruiser, Matrix, Prius, Rav4, Sequoia, Tundra, Tacoma, Venza, Yaris **Engine:** L4, V6, V8	**Output Speed Sensor Circuit No Signal:** All the conditions below are detected 500 times or more continuously: (a) No signal from the speed sensor SP2 is input to the ECM while 4 pulses of the No. 1 vehicle speed sensor signal are sent. (b) The vehicle speed is 9 km/h (6 mph) or more for at least 5 sec. (c) The park/neutral position switch is off. (d) The transfer position is not in neutral.
DTC: P0724 **2T ECM, MIL: Yes** **Year:** 2011, 2012 **Model:** 4Runner, Avalon, Camry, Corolla, FJ Cruiser, Highlander, Land Cruiser, Matrix, Prius, Rav4, Sequoia, Tundra, Tacoma, Venza, Yaris **Engine:** L4, V6, V8	**Brake Switch "B" Circuit High:** The monitor will run whenever this DTC is not present: None GO: (Vehicle speed is 18.63 mph (30 km/h) or more): 18.7 mph (30 km/h) or more STOP: (Vehicle speed is less than 1.86 mph (3 km/h)): Less than 1.86 mph (3 km/h) Starter: OFF Battery voltage: 8 V or more Engine switch: ON
DTC: P0729 **2T ECM, MIL: Yes** **Year:** 2011, 2012 **Model:** 4Runner, Avalon, Camry, Corolla, FJ Cruiser, Highlander, Land Cruiser, Matrix, Prius, Rav4, Sequoia, Tundra, Tacoma, Venza, Yaris **Engine:** L4, V6, V8	**Gear 6 Incorrect Ratio:** A 6th gear shift malfunction: The ECM determines there is a malfunction when both of the following conditions are met: (a) When the ECM directs the transmission to switch to 5th gear, the actual gear is also shifted to 5th. (b) When the ECM directs the transmission to switch to 6th gear, the actual gear is shifted to 4th.
DTC: P0741 **2T ECM, MIL: Yes** **Year:** 2011, 2012 **Model:** 4Runner, Avalon, Camry, Corolla, FJ Cruiser, Highlander, Land Cruiser, Matrix, Prius, Rav4, Sequoia, Tundra, Tacoma, Venza, Yaris **Engine:** L4, V6, V8	**Torque Converter Clutch Solenoid Performance (Shift Solenoid Valve DSL):** Lock-up does not occur when driving in lock-up range. Lock-up remains ON in lock-up OFF range.

DTC	Trouble Code Title and Conditions
DTC: P0746 **2T ECM, MIL: Yes** **Year:** 2011, 2012 **Model:** 4Runner, Avalon, Camry, Corolla, FJ Cruiser, Highlander, Land Cruiser, Matrix, Prius, Rav4, Sequoia, Tundra, Tacoma, Venza, Yaris **Engine:** L4, V6, V8	**Pressure Control Solenoid "A" Performance (Shift Solenoid Valve SL1):** The gear required by the ECM does not match the actual gear. ECM selected gear: 1st Vehicle speed: Less than 40 km/h (25 mph) Throttle valve opening angle: 4.5% or more at engine speed 1900 rpm (Varies with engine speed)
DTC: P0748 **2T ECM, MIL: Yes** **Year:** 2011, 2012 **Model:** 4Runner, Avalon, Camry, Corolla, FJ Cruiser, Highlander, Land Cruiser, Matrix, Prius, Rav4, Sequoia, Tundra, Tacoma, Venza, Yaris **Engine:** L4, V6, V8	**Pressure Control Solenoid "A" Electrical (Shift Solenoid Valve SL1):** The ECM checks for an open or short in the shift solenoid valve SL1 circuit Output signal duty equals to 100 %. NOTE: SL1 output signal duty is less than 100 % under normal condition.)
DTC: P0751 **2T ECM, MIL: Yes** **Year:** 2011, 2012 **Model:** 4Runner, Avalon, Camry, Corolla, FJ Cruiser, Highlander, Land Cruiser, Matrix, Prius, Rav4, Sequoia, Tundra, Tacoma, Venza, Yaris **Engine:** L4, V6, V8	**Shift Solenoid "A" Performance (Shift Solenoid Valve S1):** S1 stuck OFF malfunction 1: ECM determines there is malfunction when both of following conditions are met: (a) When ECM directs gearshift to switch to 1st gear, actual gear is shifted to 4th. (b) When ECM directs gearshift to switch to 4th gear, actual gear is shifted to 4th. S1 stuck ON malfunction 2: ECM determines that there is malfunction when both of following conditions are met: (a) When ECM directs gearshift to switch from 3rd to 4th gear, actual gear is not shifted as above. (b) When ECM directs gearshift to switch to 4th gear, engine speed surges 1,100 rpm or more.
DTC: P0756 **2T ECM, MIL: Yes** **Year:** 2011, 2012 **Model:** 4Runner, Avalon, Camry, Corolla, FJ Cruiser, Highlander, Land Cruiser, Matrix, Prius, Rav4, Sequoia, Tundra, Tacoma, Venza, Yaris **Engine:** L4, V6, V8	**Shift Solenoid "B" Performance (Shift Solenoid Valve S2):** A shift solenoid valve S2 stuck ON malfunction*1: Shifting to 3rd and 5th gears is impossible. The ECM determines there is a malfunction when the following conditions are both met: (a) When the ECM directs the transmission to switch to 5th gear, the actual gear is shifted to 6th. (b) When the ECM directs the transmission to switch to 6th gear, the actual gear is shifted to 6th. A shift solenoid valve S2 stuck OFF malfunction*2: The vehicle starts in 3rd gear and shifting to 6th gear is impossible. The ECM determines there is a malfunction when the following conditions are both met: (a) When the ECM directs the transmission to switch to 1st gear, the actual gear is shifted to 3rd. (b) When the ECM directs the transmission to switch to 6th gear, the actual gear is shifted to 5th.
DTC: P0761 **2T ECM, MIL: Yes** **Year:** 2011, 2012 **Model:** 4Runner, Avalon, Camry, Corolla, FJ Cruiser, Highlander, Land Cruiser, Matrix, Prius, Rav4, Sequoia, Tundra, Tacoma, Venza, Yaris **Engine:** L4, V6, V8	**Shift Solenoid "C" Performance (Shift Solenoid Valve S3):** A shift solenoid valve S3 stuck ON malfunction*1: When the ECM directs the transmission to switch to 5th or 6th gear, the engine overruns (clutch slips). The ECM determines there is a malfunction when either of the following conditions is met: (a) When the ECM directs the transmission to switch to 4th gear, the actual gear is shifted to 3rd. (b) When the ECM directs the transmission to switch to 5th gear, the engine overruns (clutch slips). A shift solenoid valve S3 stuck OFF malfunction*2: Shifting to 1st, 2nd, and 3rd gears is impossible. The ECM determines there is a malfunction when the following conditions are both met: (a) When the ECM directs the transmission to switch to 2nd gear, the actual gear is shifted to 4th. (b) When the ECM directs the transmission to switch to 6th gear, the actual gear is shifted to 6th.
DTC: P0766 **2T ECM, MIL: Yes** **Year:** 2011, 2012 **Model:** 4Runner, Avalon, Camry, Corolla, FJ Cruiser, Highlander, Land Cruiser, Matrix, Prius, Rav4, Sequoia, Tundra, Tacoma, Venza, Yaris **Engine:** L4, V6, V8	**Shift Solenoid "D" Performance (Shift Solenoid Valve S4):** S4 stuck OFF malfunction or brake control valve malfunction*: Shifting to 5th and 6th gears is impossible. The ECM determines there is a malfunction when the following conditions are both met. (2-trip detection logic) (a) When the ECM directs the transmission to switch to 5th gear, the actual gear is shifted to 4th. (b) When the ECM directs the transmission to switch to 6th gear, the actual gear is shifted to 4th.

DTC	Trouble Code Title and Conditions
DTC: P0771 **2T ECM, MIL: Yes** **Year:** 2011, 2012 **Model:** 4Runner, Avalon, Camry, Corolla, FJ Cruiser, Highlander, Land Cruiser, Matrix, Prius, Rav4, Sequoia, Tundra, Tacoma, Venza, Yaris **Engine:** L4, V6, V8	**SHIFT SOLENOID "E" PERFORMANCE (SHIFT SOLENOID VALVE SR):** The gear required by the ECM does not match the actual gear when driving. ECT (Engine coolant temperature): 10°C (50°F) or more. Transmission range: "D" TFT (Transmission fluid temperature): -20°C (-4°F) or more. OFF Malfunction (A):ECM selected gear: 5th Throttle valve opening angle: 5% or more. Vehicle speed: 10 km/h (6 mph) or more OFF Malfunction (B): ECM lock-up command: ON ECM selected gear: 3rd, 4th or 5th Vehicle speed: 25 km/h (16 mph) or more
DTC: P0776 **2T ECM, MIL: Yes** **Year:** 2011, 2012 **Model:** 4Runner, Avalon, Camry, Corolla, FJ Cruiser, Highlander, Land Cruiser, Matrix, Prius, Rav4, Sequoia, Tundra, Tacoma, Venza, Yaris **Engine:** L4, V6, V8	**Pressure Control Solenoid "B" Performance (Shift Solenoid Valve SL2):** A shift solenoid valve SL2 stuck ON malfunction or brake control valve malfunction*: Shifting to 5th and 6th gears is impossible. The ECM determines there is a malfunction when the following conditions are both met: (a) When the ECM directs the transmission to switch to 5th gear, the actual gear is shifted to 4th. (b) When the ECM directs the transmission to switch to 6th gear, the actual gear is shifted to 4th.
DTC: P0778 **2T ECM, MIL: Yes** **Year:** 2011, 2012 **Model:** 4Runner, Avalon, Camry, Corolla, FJ Cruiser, Highlander, Land Cruiser, Matrix, Prius, Rav4, Sequoia, Tundra, Tacoma, Venza, Yaris **Engine:** L4, V6, V8	**Pressure Control Solenoid "B" Electrical (Shift Solenoid Valve SL2):** The ECM checks for an open or short in the shift solenoid valve SL2 circuit: Output signal duty equals 100 %. NOTE: SL2 output signal duty is less than 100 % under normal conditions)
DTC: P0781 **2T ECM, MIL: Yes** **Year:** 2011, 2012 **Model:** 4Runner, Avalon, Camry, Corolla, FJ Cruiser, Highlander, Land Cruiser, Matrix, Prius, Rav4, Sequoia, Tundra, Tacoma, Venza, Yaris **Engine:** L4, V6, V8	**1-2 Shift (1-2 Shift Valve):** When conditions (a) and (b), or (a) and (c) are met: (a) When the ECM directs the transmission to switch to 2nd gear, the actual gear is shifted to 1st. (b) When the ECM directs the transmission to switch to 4th gear, the actual gear is shifted to 3rd. (c) When the ECM directs the transmission to switch to 5th gear, the engine overruns (clutch slips).
DTC: P0787 **1T ECM, MIL: Yes** **Year:** 2011, 2012 **Model:** 4Runner, Avalon, Camry, Corolla, FJ Cruiser, Highlander, Land Cruiser, Matrix, Prius, Rav4, Sequoia, Tundra, Tacoma, Venza, Yaris **Engine:** L4, V6, V8	**Shift / Timing Solenoid Low (Shift Solenoid Valve ST):** Solenoid ON, time after solenoid OFF to ON more than 0.008 seconds. Battery voltage 8 V or more, starter OFF. The ECM detects short in solenoid valve ST circuit 2–4 times when solenoid valve ST is operated.
DTC: P0788 **1T ECM, MIL: Yes** **Year:** 2011, 2012 **Model:** 4Runner, Avalon, Camry, Corolla, FJ Cruiser, Highlander, Land Cruiser, Matrix, Prius, Rav4, Sequoia, Tundra, Tacoma, Venza, Yaris **Engine:** L4, V6, V8	**SHIFT/TIMING SOLENOID HIGH (SHIFT SOLENOID VALVE ST):** The monitor will run whenever this DTC is not present: None. ECM detects open in solenoid valve ST circuit 4 times when solenoid valve ST is not operated

DTC	Trouble Code Title and Conditions
DTC: P0791 **1T ECM, MIL: Yes** **Year:** 2011, 2012 **Model:** 4Runner, Avalon, Camry, Corolla, FJ Cruiser, Highlander, Land Cruiser, Matrix, Prius, Rav4, Sequoia, Tundra, Tacoma, Venza, Yaris **Engine:** L4, V6, V8	**Intermediate Shaft Speed Sensor "A" Circuit:** The monitor will run whenever this DTC is not present: P0500 (VSS), P0748, P0778, P0798 (Shift solenoid valve (range)) Vehicle speed: 15.5 mph (25 km/h) or more Battery voltage: 8 V or more Engine switch: ON Starter: OFF
DTC: P0793 **1T ECM, MIL: Yes** **Year:** 2011, 2012 **Model:** 4Runner, Avalon, Camry, Corolla, FJ Cruiser, Highlander, Land Cruiser, Matrix, Prius, Rav4, Sequoia, Tundra, Tacoma, Venza, Yaris **Engine:** L4, V6, V8	**Intermediate Shaft Speed Sensor "A":** The monitor will run whenever this DTC is not present: P0500 (VSS) P0748, P0778, P0798 (Shift solenoid valve(range)) Engine: Running NSW (STAR) switch: OFF Output shaft rpm: 1000 rpm or more
DTC: P0794 **1T ECM, MIL: Yes** **Year:** 2011, 2012 **Model:** 4Runner, Avalon, Camry, Corolla, FJ Cruiser, Highlander, Land Cruiser, Matrix, Prius, Rav4, Sequoia, Tundra, Tacoma, Venza, Yaris **Engine:** L4, V6, V8	**Shift Solenoid "A" Control Circuit High (Shift Solenoid Valve S1):** The monitor will run whenever this DTC is not present: None Shift solenoid valve S1: OFF
DTC: P0796 **2T ECM, MIL: Yes** **Year:** 2011, 2012 **Model:** 4Runner, Avalon, Camry, Corolla, FJ Cruiser, Highlander, Land Cruiser, Matrix, Prius, Rav4, Sequoia, Tundra, Tacoma, Venza, Yaris **Engine:** L4, V6, V8	**Pressure Control Solenoid "C" Performance (Shift Solenoid Valve SL3):** With the engine coolant temperature at 10°C (50°F) or more and the transmission range in drive, TFT (Transmission fluid temperature) -20°C (-4°F) or more. The gear required by the ECM does not match the actual gear when driving.
DTC: P0798 **2T ECM, MIL: Yes** **Year:** 2011, 2012 **Model:** 4Runner, Avalon, Camry, Corolla, FJ Cruiser, Highlander, Land Cruiser, Matrix, Prius, Rav4, Sequoia, Tundra, Tacoma, Venza, Yaris **Engine:** L4, V6, V8	**Pressure Control Solenoid "C" Electrical (Shift Solenoid Valve SL3):** The ECM checks for an open or short in the shift solenoid valve SL3 circuit while driving and shifting gears. The output signal duty equals to 100%. NOTE: SL3 output signal duty is less than 100% under normal condition. The monitor will run whenever this DTC is not present: None Battery voltage: 10 V or more Ignition switch: ON Starter: OFF
DTC: P0818 **2T ECM, MIL: Yes** **Year:** 2011, 2012 **Model:** 4Runner, Avalon, Camry, Corolla, FJ Cruiser, Highlander, Land Cruiser, Matrix, Prius, Rav4, Sequoia, Tundra, Tacoma, Venza, Yaris **Engine:** L4, V6, V8	**Driveline Disconnect Switch Input Circuit:** The signal indicating that the transfer is in neutral remains ON while vehicle is running under following conditions for 30 seconds (2 trip detection logic): Vehicle speed is 25 km/h (16 mph) or more Transfer position switch is in 4H.
DTC: P082B-575 **1T ECM, MIL: Yes** **Year:** 2011, 2012 **Model:** Camry, Prius, Prius V **Engine:** 1.8L L4, 2.4L L4	**Gear Lever X Position Circuit Low:** Open or GND short in select main sensor circuit.

DTC	Trouble Code Title and Conditions
DTC: P082C-576 **1T ECM, MIL: Yes** **Year:** 2011, 2012 **Model:** Camry, Prius, Prius V **Engine:** 1.8L L4, 2.4L L4	**Gear Lever X Position Circuit High:** +B short in select main sensor circuit.
DTC: P082E-571 **1T ECM, MIL: Yes** **Year:** 2011, 2012 **Model:** Camry, Prius, Prius V **Engine:** 1.8L L4, 2.4L L4	**Gear Lever Y Position Circuit Low:** Open or GND short in shift main sensor circuit.
DTC: P082F-572 **1T ECM, MIL: Yes** **Year:** 2011, 2012 **Model:** Camry, Prius, Prius V **Engine:** 1.8L L4, 2.4L L4	**Gear Lever Y Position Circuit High:** +B short in shift main sensor circuit.
DTC: P0851-579 **1T ECM, MIL: Yes** **Year:** 2011, 2012 **Model:** Camry, Prius, Prius V **Engine:** 1.8L L4, 2.4L L4	**Park / Neutral Switch Input Circuit Low:** GND short in P position switch circuit.
DTC: P0852-580 **T ECM** **Year:** 2011, 2012 **Model:** Camry, Prius, Prius V **Engine:** 1.8L L4, 2.4L L4	**Park / Neutral Switch Input Circuit High:** Open or +B short in P position switch circuit.
DTC: P085D-582 **1T ECM, MIL: Yes** **Year:** 2011, 2012 **Model:** Camry, Prius, Prius V **Engine:** 1.8L L4, 2.4L L4	**Gear Shift Control Module "A" Performance:** P position (PPOS) signal is logically inconsistent.
DTC: P085D-599 **1T ECM, MIL: Yes** **Year:** 2011, 2012 **Model:** Camry, Prius, Prius V **Engine:** 1.8L L4, 2.4L L4	**Gear Shift Control Module "A" Performance:** P position (PPOS) signal malfunction (output pulse is abnormal).
DTC: P0861-597 **1T ECM, MIL: Yes** **Year:** 2011, 2012 **Model:** Camry, Prius, Prius V **Engine:** 1.8L L4, 2.4L L4	**Gear Shift Control Module "A" Communication Circuit Low:** GND short in P position (PPOS) signal circuit.
DTC: P0862-598 **1T ECM, MIL: Yes** **Year:** 2011, 2012 **Model:** Camry, Prius, Prius V **Engine:** 1.8L L4, 2.4L L4	**Gear Shift Control Module "A" Communication Circuit High:** +B short in P position (PPOS) signal circuit.
DTC: P0872 **1T TCM, MIL: Yes** **Year:** 2011, 2012 **Model:** Avalon, Camry, Highlander, Sienna, Venza **Engine:** 2.7L L4, 3.5L V6	**Transmission Fluid Pressure Sensor / Switch "C" Circuit Low:** None available.
DTC: P0873 **1T TCM, MIL: Yes** **Year:** 2011, 2012 **Model:** Avalon, Camry, Highlander, Sienna, Venza **Engine:** 2.7L L4, 3.5L V6	**Transmission Fluid Pressure Sensor / Switch "C" Circuit High:** None provided.

DTC	Trouble Code Title and Conditions
DTC: P0877 **2T ECM, MIL: Yes** **Year:** 2011, 2012 **Model:** 4Runner, Avalon, Camry, Corolla, FJ Cruiser, Highlander, Land Cruiser, Matrix, Prius, Rav4, Sequoia, Tundra, Tacoma, Venza, Yaris **Engine:** L4, V6, V8	**Transmission Fluid Pressure Sensor / Switch "D" Circuit Low:** None provided.
DTC: P0878 **2T ECM, MIL: Yes** **Year:** 2011, 2012 **Model:** 4Runner, Avalon, Camry, Corolla, FJ Cruiser, Highlander, Land Cruiser, Matrix, Prius, Rav4, Sequoia, Tundra, Tacoma, Venza, Yaris **Engine:** L4, V6, V8	**Transmission Fluid Pressure Sensor / Switch "D" Circuit High:** None provided.
DTC: P0894 **2T ECM, MIL: Yes** **Year:** 2011, 2012 **Model:** 4Runner, Avalon, Camry, Corolla, FJ Cruiser, Highlander, Land Cruiser, Matrix, Prius, Rav4, Sequoia, Tundra, Tacoma, Venza, Yaris **Engine:** L4, V6, V8	**Transmission Component Slipping:** The ECM detects a malfunction of the shift solenoid valve SLT, S1, S2, S3, S4, SL2, gear 6 incorrect ratio (sequence valve) or 1-2 shift valves according to the revolution difference of the turbine and output shaft, and also by the oil pressure.
DTC: P0973 **1T ECM, MIL: Yes** **Year:** 2011, 2012 **Model:** 4Runner, Avalon, Camry, Corolla, FJ Cruiser, Highlander, Land Cruiser, Matrix, Prius, Rav4, Sequoia, Tundra, Tacoma, Venza, Yaris **Engine:** L4, V6, V8	**Shift Solenoid "A" Control Circuit Low (Shift Solenoid Valve S1):** The monitor will run whenever this DTC is not present: None Shift solenoid valve S1: ON ECM detects short in solenoid valve S1 circuit 4 times when solenoid valve S1 is operated.
DTC: P0974 **1T ECM, MIL: Yes** **Year:** 2011, 2012 **Model:** 4Runner, Avalon, Camry, Corolla, FJ Cruiser, Highlander, Land Cruiser, Matrix, Prius, Rav4, Sequoia, Tundra, Tacoma, Venza, Yaris **Engine:** L4, V6, V8	**Shift Solenoid "A" Control Circuit High (Shift Solenoid Valve S1):** The monitor will run whenever this DTC is not present: None Shift solenoid valve S1: OFF The ECM detects an open in solenoid valve S1 circuit 2 times when solenoid valve S1 is not operated.
DTC: P0976 **1T ECM, MIL: Yes** **Year:** 2011, 2012 **Model:** 4Runner, Avalon, Camry, Corolla, FJ Cruiser, Highlander, Land Cruiser, Matrix, Prius, Rav4, Sequoia, Tundra, Tacoma, Venza, Yaris **Engine:** L4, V6, V8	**Shift Solenoid "B" Control Circuit Low (Shift Solenoid Valve S2):** The monitor will run whenever this DTC is not present: None Shift solenoid valve S2: ON ECM detects short in solenoid valve S2 circuit 4 times when solenoid valve S2 is operated.

DTC	Trouble Code Title and Conditions
DTC: P0977 **1T ECM, MIL: Yes** **Year:** 2011, 2012 **Model:** 4Runner, Avalon, Camry, Corolla, FJ Cruiser, Highlander, Land Cruiser, Matrix, Prius, Rav4, Sequoia, Tundra, Tacoma, Venza, Yaris **Engine:** L4, V6, V8	**SHIFT SOLENOID "B" CONTROL CIRCUIT HIGH (SHIFT SOLENOID VALVE S2):** The monitor will run whenever this DTC is not present: None Shift solenoid valve S2: OFF ECM detects open in solenoid valve S2 circuit 4 times when solenoid valve S2 is not operated
DTC: P0979 **1T ECM, MIL: Yes** **Year:** 2011, 2012 **Model:** 4Runner, Avalon, Camry, Corolla, FJ Cruiser, Highlander, Land Cruiser, Matrix, Prius, Rav4, Sequoia, Tundra, Tacoma, Venza, Yaris **Engine:** L4, V6, V8	**Shift Solenoid "C" Control Circuit Low (Shift Solenoid Valve S3):** The ECM detects a short in the shift solenoid valve S3 circuit 2 times when the shift solenoid valve S3 is operated. The monitor will run whenever this DTC is not present: None Shift solenoid valve S3: ON Battery voltage: 8 V or more Engine switch: ON Starter: OFF
DTC: P0980 **1T ECM, MIL: Yes** **Year:** 2011, 2012 **Model:** 4Runner, Avalon, Camry, Corolla, FJ Cruiser, Highlander, Land Cruiser, Matrix, Prius, Rav4, Sequoia, Tundra, Tacoma, Venza, Yaris **Engine:** L4, V6, V8	**Shift Solenoid "C" Control Circuit High (Shift Solenoid Valve S3):** The monitor will run whenever this DTC is not present: None Shift solenoid valve S3: OFF Battery voltage: 8 V or more Engine switch: ON Starter: OFF The ECM detects an open in solenoid valve S3 circuit 2 times when solenoid valve S3 is not operated.
DTC: P0982 **1T ECM, MIL: Yes** **Year:** 2011, 2012 **Model:** 4Runner, Avalon, Camry, Corolla, FJ Cruiser, Highlander, Land Cruiser, Matrix, Prius, Rav4, Sequoia, Tundra, Tacoma, Venza, Yaris **Engine:** L4, V6, V8	**Shift Solenoid "D" Control Circuit Low (Shift Solenoid Valve S4):** The monitor will run whenever this DTC is not present: None Shift solenoid valve S4: ON Battery voltage: 8 V or more Engine switch: ON Starter: OFF The ECM detects short in solenoid valve S4 circuit 2 times when solenoid valve S4 is operated.
DTC: P0983 **1T ECM, MIL: Yes** **Year:** 2011, 2012 **Model:** 4Runner, Avalon, Camry, Corolla, FJ Cruiser, Highlander, Land Cruiser, Matrix, Prius, Rav4, Sequoia, Tundra, Tacoma, Venza, Yaris **Engine:** L4, V6, V8	**Shift Solenoid "D" Control Circuit High (Shift Solenoid Valve S4):** The monitor will run whenever this DTC is not present: None Shift solenoid valve S4: OFF Battery voltage: 8 V or more Ignition switch: ON Starter: OFF The ECM detects an open in the shift solenoid valve S4 circuit 2 times when the shift solenoid valve S4 is not operated.
DTC: P0985 **1T ECM, MIL: Yes** **Year:** 2011, 2012 **Model:** 4Runner, Avalon, Camry, Corolla, FJ Cruiser, Highlander, Land Cruiser, Matrix, Prius, Rav4, Sequoia, Tundra, Tacoma, Venza, Yaris **Engine:** L4, V6, V8	**Shift Solenoid "E" Control Circuit Low (Shift Solenoid Valve SR):** The monitor will run whenever this DTC is not present: None Shift solenoid valve SR: ON Battery voltage: 8 V or more Engine switch: ON Starter: OFF The ECM detects a short in the shift solenoid valve SR circuit 2 times when the shift solenoid valve SR is operated.

DTC	Trouble Code Title and Conditions
DTC: P0986 **1T ECM, MIL: Yes** **Year:** 2011, 2012 **Model:** 4Runner, Avalon, Camry, Corolla, FJ Cruiser, Highlander, Land Cruiser, Matrix, Prius, Rav4, Sequoia, Tundra, Tacoma, Venza, Yaris **Engine:** L4, V6, V8	**SHIFT SOLENOID "E" CONTROL CIRCUIT HIGH (SHIFT SOLENOID VALVE SR):** The monitor will run whenever this DTC is not present: None Shift solenoid valve SR: OFF Battery voltage: 8 V or more Engine switch: ON Starter: OFF ECM detects open in solenoid valve SR circuit 2 times when solenoid valve SR is not operated.
DTC: P0989 **2T ECM, MIL: Yes** **Year:** 2011, 2012 **Model:** 4Runner, Avalon, Camry, Corolla, FJ Cruiser, Highlander, Land Cruiser, Matrix, Prius, Rav4, Sequoia, Tundra, Tacoma, Venza, Yaris **Engine:** L4, V6, V8	**Transmission Fluid Pressure Sensor / Switch "E" Circuit Low:** TCM lock-up command: ON Engine speed - Turbine speed: Less than 35 rpm Throttle valve opening angle: 7% or more Vehicle speed: Less than 62 mph (100 km/h) Shift solenoid valve SLU: Not ON malfunction
DTC: P0990 **2T ECM, MIL: Yes** **Year:** 2011, 2012 **Model:** 4Runner, Avalon, Camry, Corolla, FJ Cruiser, Highlander, Land Cruiser, Matrix, Prius, Rav4, Sequoia, Tundra, Tacoma, Venza, Yaris **Engine:** L4, V6, V8	**Transmission Fluid Pressure Sensor / Switch "E" Circuit High:** TCM selected gear: Not 1st Vehicle speed: 15.5 mph (25 km/h) or more Turbine speed/Output speed (NT/NO) with 1st: 3.304 to 7.724 Turbine speed/Output speed (NT/NO) with 2nd: 1.901 to 2.340 Turbine speed/Output speed (NT/NO) with 3rd: 1.399 to 1.649 Turbine speed/Output speed (NT/NO) with 4th: 0.998 to 1.138 Turbine speed/Output speed (NT/NO) with 5th: 0.705 to 0.836 Turbine speed/Output speed (NT/NO) with 6th: 0.568 to 0.695 TCM indicated pressure valve of SLU: Less than 4 kPa TCM lock-up command: OFF Shift solenoid valve SLU: Not malfunction Shift solenoid valve SL: Not OFF malfunction
DTC: P0A01-725 **T ECM, MIL: Yes** **Year:** 2011, 2012 **Model:** Camry, Prius, Prius V **Engine:** 1.8L L4, 2.4L L4	**Motor Electronics Coolant Temperature Sensor Circuit Range / Performance:** Unusual sudden change in HV coolant temperature sensor output occurs and the offset continues, or unusual sudden change in HV coolant temperature sensor output occurs repeatedly.
DTC: P0A01-726 **T ECM, MIL: Yes** **Year:** 2011, 2012 **Model:** Camry, Prius, Prius V **Engine:** 1.8L L4, 2.4L L4	**Motor Electronics Coolant Temperature Sensor Circuit Range / Performance:** Temperature calculated by hybrid vehicle control ECU and actual temperature are different for 10 seconds or more.
DTC: P0A02-719 **T ECM, MIL: Yes** **Year:** 2011, 2012 **Model:** Camry, Prius, Prius V **Engine:** 1.8L L4, 2.4L L4	**Motor Electronics Coolant Temperature Sensor Circuit Low:** Short to GND in the HV coolant temperature sensor circuit.
DTC: P0A02-720 **T ECM, MIL: Yes** **Year:** 2011, 2012 **Model:** Camry, Prius, Prius V **Engine:** 1.8L L4, 2.4L L4	**Motor Electronics Coolant Temperature Sensor Circuit High:** Open or short to +B in the HV coolant temperature sensor circuit.
DTC: P0A08-101 **T ECM, MIL: Yes** **Year:** 2011, 2012 **Model:** Camry, Prius, Prius V **Engine:** 1.8L L4, 2.4L L4	**DC / DC Converter Status Circuit:** Overheating of the hybrid vehicle converter (DC/DC converter).

DTC	Trouble Code Title and Conditions
DTC: P0A08-264 **T ECM, MIL: Yes** **Year:** 2011, 2012 **Model:** Camry, Prius, Prius V **Engine:** 1.8L L4, 2.4L L4	**DC / DC Converter Status Circuit:** Malfunction in the hybrid vehicle converter (DC/DC converter).
DTC: P0A09-265 **T ECM, MIL: Yes** **Year:** 2011, 2012 **Model:** Camry, Prius, Prius V **Engine:** 1.8L L4, 2.4L L4	**DC / DC Converter Status Circuit Low Input:** Open or short to GND in the hybrid vehicle converter (DC/DC converter) (NODD) signal line.
DTC: P0A09-591 **T ECM, MIL: Yes** **Year:** 2011, 2012 **Model:** Camry, Prius, Prius V **Engine:** 1.8L L4, 2.4L L4	**DC / DC Converter Status Circuit Low Input:** Hybrid vehicle converter (DC/DC converter) voltage switching (VLO) signal circuit malfunction (Open or short to GND)
DTC: P0A0D-350 **T ECM** **Year:** 2011, 2012 **Model:** Camry, Prius, Prius V **Engine:** 1.8L L4, 2.4L L4	**High Voltage System Inter-Lock Circuit High:** Operating any of the safety devices with the vehicle stopped (ILK signal is ON) and the power switch on (IG)
DTC: P0A0D-351 **T ECM, MIL: Yes** **Year:** 2011, 2012 **Model:** Camry, Prius, Prius V **Engine:** 1.8L L4, 2.4L L4	**High Voltage System Inter-Lock Circuit High:** Interlock signal line opens (ILK signal is ON) while the vehicle is being driven.
DTC: P0A0F-204 **T ECM** **Year:** 2011, 2012 **Model:** Camry, Prius, Prius V **Engine:** 1.8L L4, 2.4L L4	**Engine Failed to Start:** Signal indicating abnormality input from the ECM portion of the hybrid vehicle control ECU (abnormal engine output)
DTC: P0A0F-205 **T ECM** **Year:** 2011, 2012 **Model:** Camry, Prius, Prius V **Engine:** 1.8L L4, 2.4L L4	**Engine Failed to Start:** Signal indicating abnormality input from the ECM portion of the hybrid vehicle control ECU (engine is unable to start)
DTC: P0A0F-206 **T ECM, MIL: Yes** **Year:** 2011, 2012 **Model:** Camry, Prius, Prius V **Engine:** 1.8L L4, 2.4L L4	**Engine Failed to Start:** Signal indicating abnormality input from the ECM portion of the hybrid vehicle control ECU (engine component malfunction)
DTC: P0A0F-238 **T ECM, MIL: Yes** **Year:** 2011, 2012 **Model:** Camry, Prius, Prius V **Engine:** 1.8L L4, 2.4L L4	**Engine Failed to Start:** Engine does not start even though it is being cranked (transaxle input malfunction [engine system])
DTC: P0A0F-524 **T ECM, MIL: Yes** **Year:** 2011, 2012 **Model:** Camry, Prius, Prius V **Engine:** 1.8L L4, 2.4L L4	**Engine Failed to Start:** Signal indicating abnormality input from the ECM portion of the hybrid vehicle control ECU (NE signal error)
DTC: P0A0F-525 **T ECM, MIL: Yes** **Year:** 2011, 2012 **Model:** Camry, Prius, Prius V **Engine:** 1.8L L4, 2.4L L4	**Engine Failed to Start:** Signal indicating abnormality input from the ECM portion of the hybrid vehicle control ECU (GI signal error)

DTC	Trouble Code Title and Conditions
DTC: P0A10-263 **T ECM, MIL: Yes** **Year:** 2011, 2012 **Model:** Camry, Prius, Prius V **Engine:** 1.8L L4, 2.4L L4	**DC / DC Converter Status Circuit High Input:** +B short in hybrid vehicle converter (DC/DC converter) NODD signal line
DTC: P0A10-592 **T ECM, MIL: Yes** **Year:** 2011, 2012 **Model:** Camry, Prius, Prius V **Engine:** 1.8L L4, 2.4L L4	**DC / DC Converter Status Circuit High Input:** Hybrid vehicle converter (DC/DC converter) voltage switching (VLO) signal circuit malfunction (+B short)
DTC: P0A1A-151 **T ECM, MIL: Yes** **Year:** 2011, 2012 **Model:** Camry, Prius, Prius V **Engine:** 1.8L L4, 2.4L L4	**Generator Control Module:** Run pulse signal cycle deviation or stop.
DTC: P0A1A-155 **T ECM, MIL: Yes** **Year:** 2011, 2012 **Model:** Camry, Prius, Prius V **Engine:** 1.8L L4, 2.4L L4	**Generator Control Module:** A/D converter error.
DTC: P0A1A-156 **T ECM, MIL: Yes** **Year:** 2011, 2012 **Model:** Camry, Prius, Prius V **Engine:** 1.8L L4, 2.4L L4	**Generator Control Module:** CPU ROM-RAM error.
DTC: P0A1A-158 **T ECM, MIL: Yes** **Year:** 2011, 2012 **Model:** Camry, Prius, Prius V **Engine:** 1.8L L4, 2.4L L4	**Generator Control Module:** CPU recognition error.
DTC: P0A1A-166 **T ECM, MIL: Yes** **Year:** 2011, 2012 **Model:** Camry, Prius, Prius V **Engine:** 1.8L L4, 2.4L L4	**Generator Control Module:** RD converter NM stop error.
DTC: P0A1A-200 **T ECM, MIL: Yes** **Year:** 2011, 2012 **Model:** Camry, Prius, Prius V **Engine:** 1.8L L4, 2.4L L4	**Generator Control Module:** The monitor will run whenever the following DTC is not present: TMC's intellectual property. Other conditions belong to TMC's intellectual property.
DTC: P0A1A-658 **T ECM, MIL: Yes** **Year:** 2011, 2012 **Model:** Camry, Prius, Prius V **Engine:** 1.8L L4, 2.4L L4	**Generator Control Module:** ALU error.
DTC: P0A1A-659 **T ECM, MIL: Yes** **Year:** 2011, 2012 **Model:** Camry, Prius, Prius V **Engine:** 1.8L L4, 2.4L L4	**Generator Control Module:** TMC's intellectual property.
DTC: P0A1A-791 **T ECM, MIL: Yes** **Year:** 2011, 2012 **Model:** Camry, Prius, Prius V **Engine:** 1.8L L4, 2.4L L4	**Generator Control Module:** R/D converter communication error.

DTC	Trouble Code Title and Conditions
DTC: P0A1A-792 **T ECM, MIL: Yes** **Year:** 2011, 2012 **Model:** Camry, Prius, Prius V **Engine:** 1.8L L4, 2.4L L4	**Generator Control Module:** The monitor will run whenever the following DTC is not present: TMC's intellectual property. Other conditions belong to TMC's intellectual property.
DTC: P0A1A-793 **T ECM, MIL: Yes** **Year:** 2011, 2012 **Model:** Camry, Prius, Prius V **Engine:** 1.8L L4, 2.4L L4	**Generator Control Module:** The monitor will run whenever the following DTC is not present: TMC's intellectual property. Other conditions belong to TMC's intellectual property.
DTC: P0A1A-796 **T ECM, MIL: Yes** **Year:** 2011, 2012 **Model:** Camry, Prius, Prius V **Engine:** 1.8L L4, 2.4L L4	**Drive Motor "A" Control Module:** The monitor will run whenever the following DTC is not present: TMC's intellectual property.
DTC: P0A1B-163 **T ECM, MIL: Yes** **Year:** 2011, 2012 **Model:** Camry, Prius, Prius V **Engine:** 1.8L L4, 2.4L L4	**Drive Motor "A" Control Module:** The monitor will run whenever the following DTC is not present: TMC's intellectual property. Other conditions belong to TMC's intellectual property.
DTC: P0A1B-164 **T ECM, MIL: Yes** **Year:** 2011, 2012 **Model:** Camry, Prius, Prius V **Engine:** 1.8L L4, 2.4L L4	**Drive Motor "A" Control Module:** The monitor will run whenever the following DTC is not present: TMC's intellectual property. Other conditions belong to TMC's intellectual property.
DTC: P0A1B-168 **T ECM, MIL: Yes** **Year:** 2011, 2012 **Model:** Camry, Prius, Prius V **Engine:** 1.8L L4, 2.4L L4	**Drive Motor "A" Control Module:** The monitor will run whenever the following DTC is not present: TMC's intellectual property. Other conditions belong to TMC's intellectual property.
DTC: P0A1B-192 **T ECM, MIL: Yes** **Year:** 2011, 2012 **Model:** Camry, Prius, Prius V **Engine:** 1.8L L4, 2.4L L4	**Drive Motor "A" Control Module:** The monitor will run whenever the following DTC is not present: TMC's intellectual property. Other conditions belong to TMC's intellectual property.
DTC: P0A1B-193 **T ECM, MIL: Yes** **Year:** 2011, 2012 **Model:** Camry, Prius, Prius V **Engine:** 1.8L L4, 2.4L L4	**Drive Motor "A" Control Module:** The monitor will run whenever the following DTC is not present: TMC's intellectual property Other conditions belong to TMC's intellectual property: - **Possible causes:** • Inverter with converter assembly
DTC: P0A1B-195 **T ECM, MIL: Yes** **Year:** 2011, 2012 **Model:** Camry, Prius, Prius V **Engine:** 1.8L L4, 2.4L L4	**Drive Motor "A" Control Module:** The monitor will run whenever the following DTC is not present: TMC's intellectual property. Other conditions belong to TMC's intellectual property.
DTC: P0A1B-198 **T ECM, MIL: Yes** **Year:** 2011, 2012 **Model:** Camry, Prius, Prius V **Engine:** 1.8L L4, 2.4L L4	**Drive Motor "A" Control Module:** The monitor will run whenever the following DTC is not present: TMC's intellectual property. Other conditions belong to TMC's intellectual property.
DTC: P0A1B-511 **T ECM, MIL: Yes** **Year:** 2011, 2012 **Model:** Camry, Prius, Prius V **Engine:** 1.8L L4, 2.4L L4	**Drive Motor "A" Control Module:** The monitor will run whenever the following DTC is not present: TMC's intellectual property. Other conditions belong to TMC's intellectual property.

DTC	Trouble Code Title and Conditions
DTC: P0A1B-512 **T ECM, MIL: Yes** **Year:** 2011, 2012 **Model:** Camry, Prius, Prius V **Engine:** 1.8L L4, 2.4L L4	**Drive Motor "A" Control Module:** The monitor will run whenever the following DTC is not present: TMC's intellectual property. Other conditions belong to TMC's intellectual property.
DTC: P0A1B-661 **T ECM, MIL: Yes** **Year:** 2011, 2012 **Model:** Camry, Prius, Prius V **Engine:** 1.8L L4, 2.4L L4	**Drive Motor "A" Control Module:** The monitor will run whenever the following DTC is not present: TMC's intellectual property. Other conditions belong to TMC's intellectual property.
DTC: P0A1B-786 **T ECM, MIL: Yes** **Year:** 2011, 2012 **Model:** Camry, Prius, Prius V **Engine:** 1.8L L4, 2.4L L4	**Drive Motor "A" Control Module:** The monitor will run whenever the following DTC is not present: TMC's intellectual property. Other conditions belong to TMC's intellectual property.
DTC: P0A1B-788 **T ECM, MIL: Yes** **Year:** 2011, 2012 **Model:** Camry, Prius, Prius V **Engine:** 1.8L L4, 2.4L L4	**Drive Motor "A" Control Module:** The monitor will run whenever the following DTC is not present: TMC's intellectual property. Other conditions belong to TMC's intellectual property.
DTC: P0A1B-791 **T ECM, MIL: Yes** **Year:** 2011, 2012 **Model:** Camry, Prius, Prius V **Engine:** 1.8L L4, 2.4L L4	**Drive Motor "A" Control Module:** The monitor will run whenever the following DTC is not present: TMC's intellectual property. Other conditions belong to TMC's intellectual property.
DTC: P0A1B-795 **T ECM, MIL: Yes** **Year:** 2011, 2012 **Model:** Camry, Prius, Prius V **Engine:** 1.8L L4, 2.4L L4	**Drive Motor "A" Control Module:** The monitor will run whenever the following DTC is not present: TMC's intellectual property. Other conditions belong to TMC's intellectual property.
DTC: P0A1D-103 **T ECM, MIL: Yes** **Year:** 2011, 2012 **Model:** Camry, Prius, Prius V **Engine:** 1.8L L4, 2.4L L4	**Hybrid Powertrain Control Module:** The monitor will run whenever the following DTC is not present: TMC's intellectual property Other conditions belong to TMC's intellectual property: - **Possible causes:** • Hybrid vehicle control ECU
DTC: P0A1D-134 **T ECM, MIL: Yes** **Year:** 2011, 2012 **Model:** Camry, Prius, Prius V **Engine:** 1.8L L4, 2.4L L4	**Hybrid Powertrain Control Module:** The monitor will run whenever the following DTC is not present: TMC's intellectual property. Other conditions belong to TMC's intellectual property.
DTC: P0A1D-135 **T ECM, MIL: Yes** **Year:** 2011, 2012 **Model:** Camry, Prius, Prius V **Engine:** 1.8L L4, 2.4L L4	**Hybrid Powertrain Control Module:** The monitor will run whenever the following DTC is not present: TMC's intellectual property Other conditions belong to TMC's intellectual property: - **Possible causes:** • Hybrid vehicle control ECU
DTC: P0A1D-140 **T ECM, MIL: Yes** **Year:** 2011, 2012 **Model:** Camry, Prius, Prius V **Engine:** 1.8L L4, 2.4L L4	**Hybrid Powertrain Control Module:** The monitor will run whenever the following DTC is not present: TMC's intellectual property. Other conditions belong to TMC's intellectual property.
DTC: P0A1D-141 **T ECM, MIL: Yes** **Year:** 2011, 2012 **Model:** Camry, Prius, Prius V **Engine:** 1.8L L4, 2.4L L4	**Hybrid Powertrain Control Module:** The monitor will run whenever the following DTC is not present: TMC's intellectual property. Other conditions belong to TMC's intellectual property.

DTC	Trouble Code Title and Conditions
DTC: P0A1D-144 **T ECM, MIL: Yes** **Year:** 2011, 2012 **Model:** Camry, Prius, Prius V **Engine:** 1.8L L4, 2.4L L4	**Hybrid Powertrain Control Module:** The monitor will run whenever the following DTC is not present: TMC's intellectual property. Other conditions belong to TMC's intellectual property.
DTC: P0A1D-145 **T ECM, MIL: Yes** **Year:** 2011, 2012 **Model:** Camry, Prius, Prius V **Engine:** 1.8L L4, 2.4L L4	**Hybrid Powertrain Control Module:** The monitor will run whenever the following DTC is not present: TMC's intellectual property. Other conditions belong to TMC's intellectual property.
DTC: P0A1D-148 **T ECM, MIL: Yes** **Year:** 2011, 2012 **Model:** Camry, Prius, Prius V **Engine:** 1.8L L4, 2.4L L4	**Hybrid Powertrain Control Module:** The monitor will run whenever the following DTC is not present: TMC's intellectual property. Other conditions belong to TMC's intellectual property.
DTC: P0A1D-162 **T ECM, MIL: Yes** **Year:** 2011, 2012 **Model:** Camry, Prius, Prius V **Engine:** 1.8L L4, 2.4L L4	**Hybrid Powertrain Control Module:** The monitor will run whenever the following DTC is not present: TMC's intellectual property. Other conditions belong to TMC's intellectual property.
DTC: P0A1D-187 **T ECM, MIL: Yes** **Year:** 2011, 2012 **Model:** Camry, Prius, Prius V **Engine:** 1.8L L4, 2.4L L4	**Hybrid Powertrain Control Module:** The monitor will run whenever the following DTC is not present: TMC's intellectual property. Other conditions belong to TMC's intellectual property.
DTC: P0A1D-393 **T ECM, MIL: Yes** **Year:** 2011, 2012 **Model:** Camry, Prius, Prius V **Engine:** 1.8L L4, 2.4L L4	**Hybrid Powertrain Control Module:** The monitor will run whenever the following DTC is not present: TMC's intellectual property. Other conditions belong to TMC's intellectual property.
DTC: P0A1D-570 **T ECM, MIL: Yes** **Year:** 2011, 2012 **Model:** Camry, Prius, Prius V **Engine:** 1.8L L4, 2.4L L4	**Hybrid Powertrain Control Module:** The monitor will run whenever the following DTC is not present: TMC's intellectual property. Other conditions belong to TMC's intellectual property.
DTC: P0A1D-721 **T ECM, MIL: Yes** **Year:** 2011, 2012 **Model:** Camry, Prius, Prius V **Engine:** 1.8L L4, 2.4L L4	**Hybrid Powertrain Control Module:** The monitor will run whenever the following DTC is not present: TMC's intellectual property. Other conditions belong to TMC's intellectual property.
DTC: P0A1D-722 **T ECM, MIL: Yes** **Year:** 2011, 2012 **Model:** Camry, Prius, Prius V **Engine:** 1.8L L4, 2.4L L4	**Hybrid Powertrain Control Module:** The monitor will run whenever the following DTC is not present: TMC's intellectual property. Other conditions belong to TMC's intellectual property.
DTC: P0A1D-723 **T ECM, MIL: Yes** **Year:** 2011, 2012 **Model:** Camry, Prius, Prius V **Engine:** 1.8L L4, 2.4L L4	**Hybrid Powertrain Control Module:** The monitor will run whenever the following DTC is not present: TMC's intellectual property. Other conditions belong to TMC's intellectual property.
DTC: P0A1D-765 **T ECM, MIL: Yes** **Year:** 2011, 2012 **Model:** Camry, Prius, Prius V **Engine:** 1.8L L4, 2.4L L4	**Hybrid Powertrain Control Module:** The monitor will run whenever the following DTC is not present: TMC's intellectual property. Other conditions belong to TMC's intellectual property.

DTC	Trouble Code Title and Conditions
DTC: P0A1D-787 **T ECM, MIL: Yes** **Year:** 2011, 2012 **Model:** Camry, Prius, Prius V **Engine:** 1.8L L4, 2.4L L4	**Hybrid Powertrain Control Module:** The monitor will run whenever the following DTC is not present: TMC's intellectual property. Other conditions belong to TMC's intellectual property.
DTC: P0A1D-821 **T ECM, MIL: Yes** **Year:** 2011, 2012 **Model:** Camry, Prius, Prius V **Engine:** 1.8L L4, 2.4L L4	**Hybrid Powertrain Control Module:** The monitor will run whenever the following DTC is not present: TMC's intellectual property. Other conditions belong to TMC's intellectual property.
DTC: P0A1D-822 **T ECM, MIL: Yes** **Year:** 2011, 2012 **Model:** Camry, Prius, Prius V **Engine:** 1.8L L4, 2.4L L4	**Hybrid Powertrain Control Module:** The monitor will run whenever the following DTC is not present: TMC's intellectual property. Other conditions belong to TMC's intellectual property.
DTC: P0A1D-823 **T ECM, MIL: Yes** **Year:** 2011, 2012 **Model:** Camry, Prius, Prius V **Engine:** 1.8L L4, 2.4L L4	**Hybrid Powertrain Control Module:** The monitor will run whenever the following DTC is not present: TMC's intellectual property. Other conditions belong to TMC's intellectual property.
DTC: P0A1F-123 **T ECM, MIL: Yes** **Year:** 2011, 2012 **Model:** Camry, Prius, Prius V **Engine:** 1.8L L4, 2.4L L4	**Battery Energy Control Module:** The monitor will run whenever the following DTC is not present: TMC's intellectual property. Other conditions belong to TMC's intellectual property.
DTC: P0A1F-129 **T ECM, MIL: Yes** **Year:** 2011, 2012 **Model:** Camry, Prius, Prius V **Engine:** 1.8L L4, 2.4L L4	**Battery Energy Control Module:** The monitor will run whenever the following DTC is not present: TMC's intellectual property. Other conditions belong to TMC's intellectual property.
DTC: P0A1F-150 **T ECM, MIL: Yes** **Year:** 2011, 2012 **Model:** Camry, Prius, Prius V **Engine:** 1.8L L4, 2.4L L4	**Battery Energy Control Module:** The monitor will run whenever the following DTC is not present: TMC's intellectual property. Other conditions belong to TMC's intellectual property.
DTC: P0A2B-248 **T ECM, MIL: Yes** **Year:** 2011, 2012 **Model:** Camry, Prius, Prius V **Engine:** 1.8L L4, 2.4L L4	**Drive Motor "A" Temperature Sensor Circuit Range / Performance:** Unusual sudden change in motor temperature sensor output occurs and the condition continues, or unusual sudden change in motor temperature sensor output occurs repeatedly.
DTC: P0A2B-250 **T ECM, MIL: Yes** **Year:** 2011, 2012 **Model:** Camry, Prius, Prius V **Engine:** 1.8L L4, 2.4L L4	**Drive Motor "A" Temperature Sensor Circuit Range / Performance:** Motor temperature sensor output does not increase under conditions in which the value should increase, or output does not decrease under conditions in which the value should decrease.
DTC: P0A2C-247 **T ECM, MIL: Yes** **Year:** 2011, 2012 **Model:** Camry, Prius, Prius V **Engine:** 1.8L L4, 2.4L L4	**Drive Motor "A" Temperature Sensor Circuit Low:** Short or short to GND in the motor temperature sensor circuit.
DTC: P0A2D-249 **T ECM, MIL: Yes** **Year:** 2011, 2012 **Model:** Camry, Prius, Prius V **Engine:** 1.8L L4, 2.4L L4	**Drive Motor "A" Temperature Sensor Circuit High:** Open or short to +B in the motor temperature sensor circuit.

DTC	Trouble Code Title and Conditions
DTC: P0A37-258 **T ECM, MIL: Yes** **Year:** 2011, 2012 **Model:** Camry, Prius, Prius V **Engine:** 1.8L L4, 2.4L L4	**Generator Temperature Sensor Circuit Range / Performance:** Unusual sudden change in generator temperature sensor output occurs and the condition continues, or unusual sudden change in generator temperature sensor output occurs repeatedly.
DTC: P0A37-260 **T ECM, MIL: Yes** **Year:** 2011, 2012 **Model:** Camry, Prius, Prius V **Engine:** 1.8L L4, 2.4L L4	**Generator Temperature Sensor Circuit Range / Performance:** Generator temperature sensor output does not increase in which the value should increase, or output does not decrease under conditions in which the value should decrease.
DTC: P0A38-257 **T ECM, MIL: Yes** **Year:** 2011, 2012 **Model:** Camry, Prius, Prius V **Engine:** 1.8L L4, 2.4L L4	**Generator Temperature Sensor Circuit Low:** Short to GND in the generator temperature sensor circuit.
DTC: P0A39-259 **T ECM, MIL: Yes** **Year:** 2011, 2012 **Model:** Camry, Prius, Prius V **Engine:** 1.8L L4, 2.4L L4	**Generator Temperature Sensor Circuit High:** Open or short to +B in the generator temperature sensor circuit.
DTC: P0A3F-243 **T ECM, MIL: Yes** **Year:** 2011, 2012 **Model:** Camry, Prius, Prius V **Engine:** 1.8L L4, 2.4L L4	**Drive Motor "A" Position Sensor Circuit:** The monitor will run whenever the following DTC is not present: TMC's intellectual property. Other conditions belong to TMC's intellectual property.
DTC: P0A40-500 **T ECM, MIL: Yes** **Year:** 2011, 2012 **Model:** Camry, Prius, Prius V **Engine:** 1.8L L4, 2.4L L4	**Drive Motor "A" Position Sensor Circuit Range / Performance:** The monitor will run whenever the following DTC is not present: TMC's intellectual property. Other conditions belong to TMC's intellectual property.
DTC: P0A41-245 **T ECM, MIL: Yes** **Year:** 2011, 2012 **Model:** Camry, Prius, Prius V **Engine:** 1.8L L4, 2.4L L4	**Drive Motor "A" Position Sensor Circuit Low:** The monitor will run whenever the following DTC is not present: TMC's intellectual property. Other conditions belong to TMC's intellectual property.
DTC: P0A4B-253 **T ECM, MIL: Yes** **Year:** 2011, 2012 **Model:** Camry, Prius, Prius V **Engine:** 1.8L L4, 2.4L L4	**Generator Position Sensor Circuit:** The monitor will run whenever the following DTC is not present: TMC's intellectual property. Other conditions belong to TMC's intellectual property.
DTC: P0A4C-513 **T ECM, MIL: Yes** **Year:** 2011, 2012 **Model:** Camry, Prius, Prius V **Engine:** 1.8L L4, 2.4L L4	**Generator Position Sensor Circuit Range / Performance:** The monitor will run whenever the following DTC is not present: TMC's intellectual property. Other conditions belong to TMC's intellectual property.
DTC: P0A4D-255 **T ECM, MIL: Yes** **Year:** 2011, 2012 **Model:** Camry, Prius, Prius V **Engine:** 1.8L L4, 2.4L L4	**Generator Position Sensor Circuit Low:** The monitor will run whenever the following DTC is not present: TMC's intellectual property. Other conditions belong to TMC's intellectual property.
DTC: P0A51-174 **T ECM, MIL: Yes** **Year:** 2011, 2012 **Model:** Camry, Prius, Prius V **Engine:** 1.8L L4, 2.4L L4	**Drive Motor "A" Current Sensor Circuit:** Motor current sensor high resolution circuit signal is out of range or there is a difference between it and the motor current sensor low resolution circuit current value.

DTC	Trouble Code Title and Conditions
DTC: P0A60-288 **T ECM, MIL: Yes** **Year:** 2011, 2012 **Model:** Camry, Prius, Prius V **Engine:** 1.8L L4, 2.4L L4	**Drive Motor "A" Phase V Current:** The monitor will run whenever the following DTC is not present: TMC's intellectual property. Other conditions belong to TMC's intellectual property.
DTC: P0A60-290 **T ECM, MIL: Yes** **Year:** 2011, 2012 **Model:** Camry, Prius, Prius V **Engine:** 1.8L L4, 2.4L L4	**Drive Motor "A" Phase V Current:** The monitor will run whenever the following DTC is not present: TMC's intellectual property. Other conditions belong to TMC's intellectual property.
DTC: P0A60-294 **T ECM, MIL: Yes** **Year:** 2011, 2012 **Model:** Camry, Prius, Prius V **Engine:** 1.8L L4, 2.4L L4	**Drive Motor "A" Phase V Current:** The monitor will run whenever the following DTC is not present: TMC's intellectual property. Other conditions belong to TMC's intellectual property.
DTC: P0A60-501 **T ECM, MIL: Yes** **Year:** 2011, 2012 **Model:** Camry, Prius, Prius V **Engine:** 1.8L L4, 2.4L L4	**Drive Motor "A" Phase V Current:** The monitor will run whenever the following DTC is not present: TMC's intellectual property Other conditions belong to TMC's intellectual property: - **Possible causes:** • Inverter with converter assembly
DTC: P0A63-296 **T ECM, MIL: Yes** **Year:** 2011, 2012 **Model:** Camry, Prius, Prius V **Engine:** 1.8L L4, 2.4L L4	**Drive Motor "A" Phase W Current:** The monitor will run whenever the following DTC is not present: TMC's intellectual property. Other conditions belong to TMC's intellectual property.
DTC: P0A63-298 **T ECM, MIL: Yes** **Year:** 2011, 2012 **Model:** Camry, Prius, Prius V **Engine:** 1.8L L4, 2.4L L4	**Drive Motor "A" Phase W Current:** The monitor will run whenever the following DTC is not present: TMC's intellectual property. Other conditions belong to TMC's intellectual property.
DTC: P0A63-302 **T ECM, MIL: Yes** **Year:** 2011, 2012 **Model:** Camry, Prius, Prius V **Engine:** 1.8L L4, 2.4L L4	**Drive Motor "A" Phase W Current:** The monitor will run whenever the following DTC is not present: TMC's intellectual property. Other conditions belong to TMC's intellectual property.
DTC: P0A63-502 **T ECM, MIL: Yes** **Year:** 2011, 2012 **Model:** Camry, Prius, Prius V **Engine:** 1.8L L4, 2.4L L4	**Drive Motor "A" Phase W Current:** The monitor will run whenever the following DTC is not present: TMC's intellectual property. Other conditions belong to TMC's intellectual property.
DTC: P0A72-326 **T ECM, MIL: Yes** **Year:** 2011, 2012 **Model:** Camry, Prius, Prius V **Engine:** 1.8L L4, 2.4L L4	**Generator Phase V Current:** The monitor will run whenever the following DTC is not present: TMC's intellectual property. Other conditions belong to TMC's intellectual property.
DTC: P0A72-328 **T ECM, MIL: Yes** **Year:** 2011, 2012 **Model:** Camry, Prius, Prius V **Engine:** 1.8L L4, 2.4L L4	**Generator Phase V Current:** The monitor will run whenever the following DTC is not present: TMC's intellectual property. Other conditions belong to TMC's intellectual property.
DTC: P0A72-333 **T ECM, MIL: Yes** **Year:** 2011, 2012 **Model:** Camry, Prius, Prius V **Engine:** 1.8L L4, 2.4L L4	**Generator Phase V Current:** The monitor will run whenever the following DTC is not present: TMC's intellectual property. Other conditions belong to TMC's intellectual property.

DTC	Trouble Code Title and Conditions
DTC: P0A72-515 **T ECM, MIL: Yes** **Year:** 2011, 2012 **Model:** Camry, Prius, Prius V **Engine:** 1.8L L4, 2.4L L4	**Generator Phase V Current:** The monitor will run whenever the following DTC is not present: TMC's intellectual property. Other conditions belong to TMC's intellectual property.
DTC: P0A75-334 **T ECM, MIL: Yes** **Year:** 2011, 2012 **Model:** Camry, Prius, Prius V **Engine:** 1.8L L4, 2.4L L4	**Generator Phase W Current:** The monitor will run whenever the following DTC is not present: TMC's intellectual property. Other conditions belong to TMC's intellectual property.
DTC: P0A75-336 **T ECM, MIL: Yes** **Year:** 2011, 2012 **Model:** Camry, Prius, Prius V **Engine:** 1.8L L4, 2.4L L4	**Generator Phase W Current:** The monitor will run whenever the following DTC is not present: TMC's intellectual property. Other conditions belong to TMC's intellectual property.
DTC: P0A75-341 **T ECM, MIL: Yes** **Year:** 2011, 2012 **Model:** Camry, Prius, Prius V **Engine:** 1.8L L4, 2.4L L4	**Generator Phase W Current:** The monitor will run whenever the following DTC is not present: TMC's intellectual property. Other conditions belong to TMC's intellectual property.
DTC: P0A75-516 **T ECM, MIL: Yes** **Year:** 2011, 2012 **Model:** Camry, Prius, Prius V **Engine:** 1.8L L4, 2.4L L4	**Generator Phase W Current:** The monitor will run whenever the following DTC is not present: TMC's intellectual property. Other conditions belong to TMC's intellectual property.
DTC: P0A78-113 **T ECM, MIL: Yes** **Year:** 2011, 2012 **Model:** Camry, Prius, Prius V **Engine:** 1.8L L4, 2.4L L4	**Drive Motor "A" Inverter Performance:** The monitor will run whenever the following DTC is not present: TMC's intellectual property. Other conditions belong to TMC's intellectual property.
DTC: P0A78-121 **T ECM, MIL: Yes** **Year:** 2011, 2012 **Model:** Camry, Prius, Prius V **Engine:** 1.8L L4, 2.4L L4	**Drive Motor "A" Inverter Performance:** The monitor will run whenever the following DTC is not present: TMC's intellectual property. Other conditions belong to TMC's intellectual property.
DTC: P0A78-128 **T ECM, MIL: Yes** **Year:** 2011, 2012 **Model:** Camry, Prius, Prius V **Engine:** 1.8L L4, 2.4L L4	**Drive Motor "A" Inverter Performance:** The monitor will run whenever the following DTC is not present: TMC's intellectual property. Other conditions belong to TMC's intellectual property.
DTC: P0A78-266 **T ECM, MIL: Yes** **Year:** 2011, 2012 **Model:** Camry, Prius, Prius V **Engine:** 1.8L L4, 2.4L L4	**Drive Motor "A" Inverter Performance:** The monitor will run whenever the following DTC is not present: TMC's intellectual property. Other conditions belong to TMC's intellectual property.
DTC: P0A78-267 **T ECM, MIL: Yes** **Year:** 2011, 2012 **Model:** Camry, Prius, Prius V **Engine:** 1.8L L4, 2.4L L4	**Drive Motor "A" Inverter Performance:** The monitor will run whenever the following DTC is not present: TMC's intellectual property. Other conditions belong to TMC's intellectual property.
DTC: P0A78-279 **T ECM, MIL: Yes** **Year:** 2011, 2012 **Model:** Camry, Prius, Prius V **Engine:** 1.8L L4, 2.4L L4	**Drive Motor "A" Inverter Performance:** The monitor will run whenever the following DTC is not present: TMC's intellectual property. Other conditions belong to TMC's intellectual property.

DTC	Trouble Code Title and Conditions
DTC: P0A78-282 **T ECM, MIL: Yes** **Year:** 2011, 2012 **Model:** Camry, Prius, Prius V **Engine:** 1.8L L4, 2.4L L4	**Drive Motor "A" Inverter Performance:** The monitor will run whenever the following DTC is not present: TMC's intellectual property. Other conditions belong to TMC's intellectual property.
DTC: P0A78-284 **T ECM, MIL: Yes** **Year:** 2011, 2012 **Model:** Camry, Prius, Prius V **Engine:** 1.8L L4, 2.4L L4	**Drive Motor "A" Inverter Performance:** The monitor will run whenever the following DTC is not present: TMC's intellectual property. Other conditions belong to TMC's intellectual property: -Motor inverter fail signal detection (overheat).
DTC: P0A78-286 **T ECM, MIL: Yes** **Year:** 2011, 2012 **Model:** Camry, Prius, Prius V **Engine:** 1.8L L4, 2.4L L4	**Drive Motor "A" Inverter Performance:** The monitor will run whenever the following DTC is not present: None. TMC's intellectual property. Other conditions belong to TMC's intellectual property: Motor inverter fail signal detection (circuit malfunction)
DTC: P0A78-287 **T ECM, MIL: Yes** **Year:** 2011, 2012 **Model:** Camry, Prius, Prius V **Engine:** 1.8L L4, 2.4L L4	**Drive Motor "A" Inverter Performance:** The monitor will run whenever the following DTC is not present: TMC's intellectual property. Other conditions belong to TMC's intellectual property: Motor inverter fail signal detection (over-current due to inverter assembly malfunction)
DTC: P0A78-306 **T ECM, MIL: Yes** **Year:** 2011, 2012 **Model:** Camry, Prius, Prius V **Engine:** 1.8L L4, 2.4L L4	**Drive Motor "A" Inverter Performance:** The monitor will run whenever the following DTC is not present:None. TMC's intellectual property. Other conditions belong to TMC's intellectual property: Motor torque execution monitoring malfunction.
DTC: P0A78-503 **T ECM, MIL: Yes** **Year:** 2011, 2012 **Model:** Camry, Prius, Prius V **Engine:** 1.8L L4, 2.4L L4	**Drive Motor "A" Inverter Performance:** The monitor will run whenever the following DTC is not present: TMC's intellectual property. Other conditions belong to TMC's intellectual property: Motor inverter overvoltage signal detection (overvoltage due to MG ECU malfunction)
DTC: P0A78-504 **T ECM, MIL: Yes** **Year:** 2011, 2012 **Model:** Camry, Prius, Prius V **Engine:** 1.8L L4, 2.4L L4	**Drive Motor "A" Inverter Performance:** The monitor will run whenever the following DTC is not present: TMC's intellectual property. Other conditions belong to TMC's intellectual property: Motor inverter overvoltage signal detection (overvoltage due to hybrid vehicle transaxle assembly malfunction)
DTC: P0A78-505 **T ECM, MIL: Yes** **Year:** 2011, 2012 **Model:** Camry, Prius, Prius V **Engine:** 1.8L L4, 2.4L L4	**Drive Motor "A" Inverter Performance:** The monitor will run whenever the following DTC is not present: TMC's intellectual property. Other conditions belong to TMC's intellectual property: Motor inverter fail signal detection (over-current due to MG ECU malfunction).
DTC: P0A78-506 **T ECM, MIL: Yes** **Year:** 2011, 2012 **Model:** Camry, Prius, Prius V **Engine:** 1.8L L4, 2.4L L4	**Drive Motor "A" Inverter Performance:** The monitor will run whenever the following DTC is not present: TMC's intellectual property. Other conditions belong to TMC's intellectual property: Motor inverter fail signal detection (over-current due to hybrid vehicle transaxle assembly malfunction)
DTC: P0A78-510 **T ECM, MIL: Yes** **Year:** 2011, 2012 **Model:** Camry, Prius, Prius V **Engine:** 1.8L L4, 2.4L L4	**Drive Motor "A" Inverter Performance:** The monitor will run whenever the following DTC is not present: TMC's intellectual property. Other conditions belong to TMC's intellectual property: Motor inverter gate malfunction.
DTC: P0A78-586 **T ECM, MIL: Yes** **Year:** 2011, 2012 **Model:** Camry, Prius, Prius V **Engine:** 1.8L L4, 2.4L L4	**Drive Motor "A" Inverter Performance:** The monitor will run whenever the following DTC is not present: TMC's intellectual property. Other conditions belong to TMC's intellectual property: Inverter voltage (VH) sensor performance problem.

DTC	Trouble Code Title and Conditions
DTC: P0A78-806 **T , MIL: Yes** **Year:** 2011, 2012 **Model:** Camry, Prius, Prius V **Engine:** 1.8L L4, 2.4L L4	**Drive Motor "A" Inverter Performance:** The monitor will run whenever the following DTC is not present: TMC's intellectual property. Other conditions belong to TMC's intellectual property: Abnormal motor current value detection(MG ECU malfunction)3
DTC: P0A78-807 **T ECM, MIL: Yes** **Year:** 2011, 2012 **Model:** Camry, Prius, Prius V **Engine:** 1.8L L4, 2.4L L4	**Drive Motor "A" Inverter Performance:** The monitor will run whenever the following DTC is not present: TMC's intellectual property. Other conditions belong to TMC's intellectual property: Abnormal motor current value detection (Inverter malfunction)
DTC: P0A78-808 **T ECM, MIL: Yes** **Year:** 2011, 2012 **Model:** Camry, Prius, Prius V **Engine:** 1.8L L4, 2.4L L4	**Drive Motor "A" Inverter Performance:** The monitor will run whenever the following DTC is not present: TMC's intellectual property. Other conditions belong to TMC's intellectual property: Abnormal motor current value detection(Hybrid vehicle transaxle assembly malfunction)
DTC: P0A7A-122 **T ECM, MIL: Yes** **Year:** 2011, 2012 **Model:** Camry, Prius, Prius V **Engine:** 1.8L L4, 2.4L L4	**Generator Inverter Performance:** The monitor will run whenever the following DTC is not present: TMC's intellectual property. Other conditions belong to TMC's intellectual property: Generator inverter fail signal detection (over-current due to system malfunction)
DTC: P0A7A-130 **T ECM, MIL: Yes** **Year:** 2011, 2012 **Model:** Camry, Prius, Prius V **Engine:** 1.8L L4, 2.4L L4	**Generator Inverter Performance:** The monitor will run whenever the following DTC is not present: TMC's intellectual property. Other conditions belong to TMC's intellectual property: Abnormal generator current value detection (System).
DTC: P0A7A-322 **T ECM, MIL: Yes** **Year:** 2011, 2012 **Model:** Camry, Prius, Prius V **Engine:** 1.8L L4, 2.4L L4	**Generator Inverter Performance:** The monitor will run whenever the following DTC is not present: TMC's intellectual property. Other conditions belong to TMC's intellectual property: Generator inverter fail signal detection (overheating)
DTC: P0A7A-324 **T ECM, MIL: Yes** **Year:** 2011, 2012 **Model:** Camry, Prius, Prius V **Engine:** 1.8L L4, 2.4L L4	**Generator Inverter Performance:** The monitor will run whenever the following DTC is not present: TMC's intellectual property. Other conditions belong to TMC's intellectual property: Generator inverter fail signal detection (circuit malfunction)
DTC: P0A7A-325 **T ECM, MIL: Yes** **Year:** 2011, 2012 **Model:** Camry, Prius, Prius V **Engine:** 1.8L L4, 2.4L L4	**Generator Inverter Performance:** The monitor will run whenever the following DTC is not present: TMC's intellectual property. Other conditions belong to TMC's intellectual property: Generator inverter fail signal detection (over-current due to inverter assembly malfunction).
DTC: P0A7A-344 **T ECM, MIL: Yes** **Year:** 2011, 2012 **Model:** Camry, Prius, Prius V **Engine:** 1.8L L4, 2.4L L4	**Generator Inverter Performance:** The monitor will run whenever the following DTC is not present: TMC's intellectual property. Other conditions belong to TMC's intellectual property: Generator torque execution monitoring malfunction
DTC: P0A7A-517 **T ECM, MIL: Yes** **Year:** 2011, 2012 **Model:** Camry, Prius, Prius V **Engine:** 1.8L L4, 2.4L L4	**Generator Inverter Performance:** The monitor will run whenever the following DTC is not present: TMC's intellectual property. Other conditions belong to TMC's intellectual property: Generator inverter fail signal detection (over-current due to MG ECU malfunction)
DTC: P0A7A-518 **T ECM, MIL: Yes** **Year:** 2011, 2012 **Model:** Camry, Prius, Prius V **Engine:** 1.8L L4, 2.4L L4	**Generator Inverter Performance:** The monitor will run whenever the following DTC is not present: TMC's intellectual property. Other conditions belong to TMC's intellectual property: Generator inverter fail signal detection (over-current due to hybrid vehicle transaxle assembly malfunction)

DTC	Trouble Code Title and Conditions
DTC: P0A7A-522 **T ECM, MIL: Yes** **Year:** 2011, 2012 **Model:** Camry, Prius, Prius V **Engine:** 1.8L L4, 2.4L L4	**Generator Inverter Performance:** The monitor will run whenever the following DTC is not present: TMC's intellectual property. Other conditions belong to TMC's intellectual property: Generator inverter gate malfunction
DTC: P0A7A-809 **T ECM, MIL: Yes** **Year:** 2011, 2012 **Model:** Camry, Prius, Prius V **Engine:** 1.8L L4, 2.4L L4	**Generator Inverter Performance:** The monitor will run whenever the following DTC is not present: TMC's intellectual property. Other conditions belong to TMC's intellectual property: Abnormal generator current value detection (MG ECU malfunction).
DTC: P0A7A-810 **T ECM, MIL: Yes** **Year:** 2011, 2012 **Model:** Camry, Prius, Prius V **Engine:** 1.8L L4, 2.4L L4	**Generator Inverter Performance:** The monitor will run whenever the following DTC is not present: TMC's intellectual property. Other conditions belong to TMC's intellectual property: Abnormal generator current value detection (inverter malfunction)
DTC: P0A7A-811 **T ECM, MIL: Yes** **Year:** 2011, 2012 **Model:** Camry, Prius, Prius V **Engine:** 1.8L L4, 2.4L L4	**Generator Inverter Performance:** The monitor will run whenever the following DTC is not present: TMC's intellectual property. Other conditions belong to TMC's intellectual property: Abnormal generator current value detection (hybrid vehicle transaxle assembly malfunction)
DTC: P0A7F-123 **1T ECM, MIL: Yes** **Year:** 2011, 2012 **Model:** Camry, Prius, Prius V **Engine:** 1.8L L4, 2.4L L4	**Hybrid Battery Pack Deterioration:** The monitor will run whenever the following DTCs are not present: TMC's intellectual property. Other conditions belong to TMC's intellectual property.
DTC: P0A80-123 **2T ECM, MIL: Yes** **Year:** 2011, 2012 **Model:** Camry, Prius, Prius V **Engine:** 1.8L L4, 2.4L L4	**Replace Hybrid Battery Pack:** The monitor will run whenever the following DTCs are not present: TMC's intellectual property. Other conditions belong to TMC's intellectual property.
DTC: P0A82-123 **1T ECM, MIL: Yes** **Year:** 2011, 2012 **Model:** Camry, Prius, Prius V **Engine:** 1.8L L4, 2.4L L4	**Hybrid Battery Pack Cooling Fan 1:** The speed of the battery cooling blower assembly is not within the specified range.
DTC: P0A84-123 **1T ECM, MIL: Yes** **Year:** 2011, 2012 **Model:** Camry, Prius, Prius V **Engine:** 1.8L L4, 2.4L L4	**Hybrid Battery Pack Cooling Fan 1:** When the output voltage of the battery cooling blower assembly (VM) is too low compared to the target control voltage range.
DTC: P0A85-123 **1T ECM, MIL: Yes** **Year:** 2011, 2012 **Model:** Camry, Prius, Prius V **Engine:** 1.8L L4, 2.4L L4	**Hybrid Battery Pack Cooling Fan 1:** When the output voltage of the battery cooling blower assembly (VM) is too high compared to the target control voltage range.
DTC: P0A90-251 **T ECM, MIL: Yes** **Year:** 2011, 2012 **Model:** Camry, Prius, Prius V **Engine:** 1.8L L4, 2.4L L4	**Drive Motor "A" Performance:** The monitor will run whenever the following DTC is not present: TMC's intellectual property. Other conditions belong to TMC's intellectual property: Motor magnetic force deterioration or same phase short circuit
DTC: P0A90-509 **T ECM, MIL: Yes** **Year:** 2011, 2012 **Model:** Camry, Prius, Prius V **Engine:** 1.8L L4, 2.4L L4	**Drive Motor "A" Performance:** The monitor will run whenever the following DTC is not present: TMC's intellectual property. Other conditions belong to TMC's intellectual property: Motor system malfunction.

DTC	Trouble Code Title and Conditions
DTC: P0A92-261 **T ECM, MIL: Yes** **Year:** 2011, 2012 **Model:** Camry, Prius, Prius V **Engine:** 1.8L L4, 2.4L L4	**Hybrid Generator Performance:** The monitor will run whenever the following DTC is not present: TMC's intellectual property. Other conditions belong to TMC's intellectual property: Generator magnetic force deterioration or same phase short circuit.
DTC: P0A92-521 **T ECM, MIL: Yes** **Year:** 2011, 2012 **Model:** Camry, Prius, Prius V **Engine:** 1.8L L4, 2.4L L4	**Hybrid Generator Performance:** The monitor will run whenever the following DTC is not present: TMC's intellectual property. Other conditions belong to TMC's intellectual property: Generator system malfunction.
DTC: P0A93-346 **T ECM, MIL: Yes** **Year:** 2011, 2012 **Model:** Camry, Prius, Prius V **Engine:** 1.8L L4, 2.4L L4	**Inverter Cooling System Performance:** The monitor will run whenever the following DTC is not present: TMC's intellectual property. Other conditions belong to TMC's intellectual property: Inverter cooling system malfunction (HV coolant malfunction)
DTC: P0A94-127 **T ECM, MIL: Yes** **Year:** 2011, 2012 **Model:** Camry, Prius, Prius V **Engine:** 1.8L L4, 2.4L L4	**DC / DC Converter Performance:** The monitor will run whenever the following DTC is not present: TMC's intellectual property. Other conditions belong to TMC's intellectual property: Boost converter overvoltage signal detection (overvoltage due to system malfunction)
DTC: P0A94-172 **T ECM, MIL: Yes** **Year:** 2011, 2012 **Model:** Camry, Prius, Prius V **Engine:** 1.8L L4, 2.4L L4	**DC / DC Converter Performance:** The monitor will run whenever the following DTC is not present: TMC's intellectual property. Other conditions belong to TMC's intellectual property: Boost converter fail signal detection (over-current due to system malfunction)
DTC: P0A94-442 **T ECM, MIL: Yes** **Year:** 2011, 2012 **Model:** Camry, Prius, Prius V **Engine:** 1.8L L4, 2.4L L4	**DC / DC Converter Performance:** The monitor will run whenever the following DTC is not present: TMC's intellectual property. Other conditions belong to TMC's intellectual property: Abnormal voltage execution value.
DTC: P0A94-547 **T ECM, MIL: Yes** **Year:** 2011, 2012 **Model:** Camry, Prius, Prius V **Engine:** 1.8L L4, 2.4L L4	**DC / DC Converter Performance:** The monitor will run whenever the following DTC is not present: TMC's intellectual property. Other conditions belong to TMC's intellectual property: Boost converter overvoltage signal detection (overvoltage due to MG ECU malfunction)
DTC: P0A94-548 **T ECM, MIL: Yes** **Year:** 2011, 2012 **Model:** Camry, Prius, Prius V **Engine:** 1.8L L4, 2.4L L4	**DC / DC Converter Performance:** The monitor will run whenever the following DTC is not present: TMC's intellectual property. Other conditions belong to TMC's intellectual property: Boost converter overvoltage signal detection (overvoltage due to inverter malfunction).
DTC: P0A94-549 **T ECM, MIL: Yes** **Year:** 2011, 2012 **Model:** Camry, Prius, Prius V **Engine:** 1.8L L4, 2.4L L4	**DC / DC Converter Performance:** The monitor will run whenever the following DTC is not present: TMC's intellectual property. Other conditions belong to TMC's intellectual property: Boost converter overvoltage signal detection (overvoltage due to hybrid vehicle transaxle assembly malfunction).
DTC: P0A94-550 **T ECM, MIL: Yes** **Year:** 2011, 2012 **Model:** Camry, Prius, Prius V **Engine:** 1.8L L4, 2.4L L4	**DC / DC Converter Performance:** The monitor will run whenever the following DTC is not present: TMC's intellectual property. Other conditions belong to TMC's intellectual property: Boost converter overvoltage (OVL) signal detection (circuit malfunction)
DTC: P0A94-553 **T ECM, MIL: Yes** **Year:** 2011, 2012 **Model:** Camry, Prius, Prius V **Engine:** 1.8L L4, 2.4L L4	**DC / DC Converter Performance:** The monitor will run whenever the following DTC is not present: TMC's intellectual property. Other conditions belong to TMC's intellectual property: Boost converter fail signal detection (boost converter overheating)

DTC	Trouble Code Title and Conditions
DTC: P0A94-554 **T ECM, MIL: Yes** **Year:** 2011, 2012 **Model:** Camry, Prius, Prius V **Engine:** 1.8L L4, 2.4L L4	**DC / DC Converter Performance:** The monitor will run whenever the following DTC is not present: TMC's intellectual property. Other conditions belong to TMC's intellectual property: Boost converter fail signal detection (over-current due to MG ECU malfunction)
DTC: P0A94-555 **T ECM, MIL: Yes** **Year:** 2011, 2012 **Model:** Camry, Prius, Prius V **Engine:** 1.8L L4, 2.4L L4	**DC / DC Converter Performance:** The monitor will run whenever the following DTC is not present: None. TMC's intellectual property. Other conditions belong to TMC's intellectual property: Boost converter fail signal detection (over-current due to inverter assembly malfunction).
DTC: P0A94-556 **T ECM, MIL: Yes** **Year:** 2011, 2012 **Model:** Camry, Prius, Prius V **Engine:** 1.8L L4, 2.4L L4	**DC / DC Converter Performance:** The monitor will run whenever the following DTC is not present: TMC's intellectual property. Other conditions belong to TMC's intellectual property: Boost converter fail signal detection (over-current due to hybrid vehicle transaxle assembly malfunction)
DTC: P0A94-557 **T ECM, MIL: Yes** **Year:** 2011, 2012 **Model:** Camry, Prius, Prius V **Engine:** 1.8L L4, 2.4L L4	**DC / DC Converter Performance:** The monitor will run whenever the following DTC is not present: TMC's intellectual property. Other conditions belong to TMC's intellectual property: Boost converter fail signal detection (circuit malfunction).
DTC: P0A94-585 **T ECM, MIL: Yes** **Year:** 2011, 2012 **Model:** Camry, Prius, Prius V **Engine:** 1.8L L4, 2.4L L4	**DC / DC Converter Performance:** The monitor will run whenever the following DTC is not present: TMC's intellectual property. Other conditions belong to TMC's intellectual property: Boost converter voltage (VL) sensor performance problem.
DTC: P0A94-587 **T ECM, MIL: Yes** **Year:** 2011, 2012 **Model:** Camry, Prius, Prius V **Engine:** 1.8L L4, 2.4L L4	**DC / DC Converter Performance:** The monitor will run whenever the following DTC is not present: TMC's intellectual property. Other conditions belong to TMC's intellectual property: Voltages from HV battery voltage (VB) sensor and boost converter voltage (VL) sensor deviate.
DTC: P0A94-589 **T ECM, MIL: Yes** **Year:** 2011, 2012 **Model:** Camry, Prius, Prius V **Engine:** 1.8L L4, 2.4L L4	**DC / DC Converter Performance:** The monitor will run whenever the following DTC is not present: TMC's intellectual property. Other conditions belong to TMC's intellectual property: Open or short to GND in the boost converter voltage (VL) sensor circuit.
DTC: P0A94-590 **T ECM, MIL: Yes** **Year:** 2011, 2012 **Model:** Camry, Prius, Prius V **Engine:** 1.8L L4, 2.4L L4	**DC / DC Converter Performance:** The monitor will run whenever the following DTC is not present: TMC's intellectual property. Other conditions belong to TMC's intellectual property: Short to +B in the boost converter voltage (VL) sensor circuit.
DTC: P0A95-123 **1T ECM, MIL: Yes** **Year:** 2011, 2012 **Model:** Camry, Prius, Prius V **Engine:** 1.8L L4, 2.4L L4	**High Voltage Fuse:** Voltage between VC7 and VC8 terminals is below the standard despite the interlock switch being engaged.
DTC: P0A9C-123 **1T ECM, MIL: Yes** **Year:** 2011, 2012 **Model:** Camry, Prius, Prius V **Engine:** 1.8L L4, 2.4L L4	**Hybrid Battery Temperature Sensor "A" Range / Performance:** The monitor will run whenever the following DTCs are not present: TMC's intellectual property. Other conditions belong to TMC's intellectual property.
DTC: P0A9D-123 **1T ECM, MIL: Yes** **Year:** 2011, 2012 **Model:** Camry, Prius, Prius V **Engine:** 1.8L L4, 2.4L L4	**Hybrid Battery Temperature Sensor "A" Circuit Low:** The monitor will run whenever the following DTCs are not present: TMC's intellectual property Other conditions belong to TMC's intellectual property.

DTC	Trouble Code Title and Conditions
DTC: P0A9E-123 **1T ECM, MIL: Yes** **Year:** 2011, 2012 **Model:** Camry, Prius, Prius V **Engine:** 1.8L L4, 2.4L L4	**Hybrid Battery Temperature Sensor "A" Circuit High:** The monitor will run whenever the following DTCs are not present: TMC's intellectual property Other conditions belong to TMC's intellectual property.
DTC: P0AA1-231 **T ECM, MIL: Yes** **Year:** 2011, 2012 **Model:** Camry, Prius, Prius V **Engine:** 1.8L L4, 2.4L L4	**Hybrid Battery Positive Contactor Circuit Stuck Closed:** SMRB on the HV battery positive side is stuck closed. The monitor will run whenever the following DTCs are not present: TMC's intellectual property Other conditions belong to TMC's intellectual property.
DTC: P0AA1-233 **T ECM, MIL: Yes** **Year:** 2011, 2012 **Model:** Camry, Prius, Prius V **Engine:** 1.8L L4, 2.4L L4	**Hybrid Battery Positive Contactor Circuit Stuck Closed:** SMRP, SMRB and SMRG on the HV battery positive and negative sides are stuck closed. The monitor will run whenever the following DTCs are not present: TMC's intellectual property Other conditions belong to TMC's intellectual property.
DTC: P0AA4-232 **T ECM, MIL: Yes** **Year:** 2011, 2012 **Model:** Camry, Prius, Prius V **Engine:** 1.8L L4, 2.4L L4	**Hybrid Battery Negative Contactor Circuit Stuck Closed:** SMRG on the HV battery negative side stuck closed The monitor will run whenever the following DTCs are not present: TMC's intellectual property Other conditions belong to TMC's intellectual property.
DTC: P0AA6-526 **T ECM, MIL: Yes** **Year:** 2011, 2012 **Model:** Camry, Prius, Prius V **Engine:** 1.8L L4, 2.4L L4	**Hybrid Battery Voltage System Isolation Fault:** Insulation resistance between the high-voltage circuit and the body has decreased. The monitor will run whenever the following DTCs are not present: TMC's intellectual property Other conditions belong to TMC's intellectual property.
DTC: P0AA6-611 **T ECM** **Year:** 2011, 2012 **Model:** Prius, Prius V **Engine:** 1.8L L4	**Hybrid Battery Voltage System Isolation Fault:** High voltage system insulation malfunction The monitor will run whenever the following DTCs are not present: TMC's intellectual property Other conditions belong to TMC's intellectual property.
DTC: P0AA6-611 **T ECM, MIL: Yes** **Year:** 2011, 2012 **Model:** Camry, Prius, Prius V **Engine:** 1.8L L4, 2.4L L4	**Hybrid Battery Voltage System Isolation Fault:** Insulation resistance of the compressor with motor assembly has decreased. The monitor will run whenever the following DTCs are not present: TMC's intellectual property Other conditions belong to TMC's intellectual property.
DTC: P0AA6-611 **T ECM** **Year:** 2011, 2012 **Model:** Camry, Highlander **Engine:** 2.4L L4, 3.5L V6	**Hybrid Battery Voltage System Isolation Fault:** High voltage system insulation malfunction The monitor will run whenever the following DTCs are not present: TMC's intellectual property Other conditions belong to TMC's intellectual property.
DTC: P0AA6-612 **T ECM, MIL: Yes** **Year:** 2011, 2012 **Model:** Camry, Prius, Prius V **Engine:** 1.8L L4, 2.4L L4	**Hybrid Battery Voltage System Isolation Fault:** Insulation resistance of the HV battery area has decreased. The monitor will run whenever the following DTCs are not present: TMC's intellectual property Other conditions belong to TMC's intellectual property.
DTC: P0AA6-613 **T ECM, MIL: Yes** **Year:** 2011, 2012 **Model:** Camry, Prius, Prius V **Engine:** 1.8L L4, 2.4L L4	**Hybrid Battery Voltage System Isolation Fault:** Insulation resistance of the transaxle area has decreased. The monitor will run whenever the following DTCs are not present: TMC's intellectual property Other conditions belong to TMC's intellectual property.
DTC: P0AA6-614 **T ECM, MIL: Yes** **Year:** 2011, 2012 **Model:** Camry, Prius, Prius V **Engine:** 1.8L L4, 2.4L L4	**Hybrid Battery Voltage System Isolation Fault:** Insulation resistance of the high-voltage DC area has decreased. The monitor will run whenever the following DTCs are not present: TMC's intellectual property Other conditions belong to TMC's intellectual property.

DTC	Trouble Code Title and Conditions
DTC: P0AA7-727 **T ECM, MIL: Yes** **Year:** 2011, 2012 **Model:** Camry, Prius, Prius V **Engine:** 1.8L L4, 2.4L L4	**Hybrid Battery Voltage Isolation Sensor Circuit:** Malfunction in the insulation monitoring circuit in the battery smart unit The monitor will run whenever the following DTCs are not present: TMC's intellectual property Other conditions belong to TMC's intellectual property.
DTC: P0AAE-123 **T ECM, MIL: Yes** **Year:** 2011, 2012 **Model:** Camry, Prius, Prius V **Engine:** 1.8L L4, 2.4L L4	**Hybrid Battery Pack Air Temperature Sensor "A" Circuit Low:** When the temperature indicated by the inlet air temperature sensor is lower than a predetermined limit (open circuit) or is higher than a predetermined limit (short circuit).
DTC: P0AAF-123 **T ECM, MIL: Yes** **Year:** 2011, 2012 **Model:** Camry, Prius, Prius V **Engine:** 1.8L L4, 2.4L L4	**Hybrid Battery Pack Air Temperature Sensor "A" Circuit High:** When the temperature indicated by the inlet air temperature sensor is lower than a predetermined limit (open circuit) or is higher than a predetermined limit (short circuit).
DTC: P0ABF-123 **1T ECM, MIL: Yes** **Year:** 2011, 2012 **Model:** Camry, Prius, Prius V **Engine:** 1.8L L4, 2.4L L4	**Hybrid Battery Pack Current Sensor Circuit:** The monitor will run whenever the following DTCs are not present: TMC's intellectual property Other conditions belong to TMC's intellectual property.
DTC: P0AC0-123 **1T ECM, MIL: Yes** **Year:** 2011, 2012 **Model:** Camry, Prius, Prius V **Engine:** 1.8L L4, 2.4L L4	**Hybrid Battery Pack Current Sensor Circuit Range / Performance:** The monitor will run whenever the following DTCs are not present: TMC's intellectual property Other conditions belong to TMC's intellectual property.
DTC: P0AC0-817 **T ECM, MIL: Yes** **Year:** 2011, 2012 **Model:** Camry, Prius, Prius V **Engine:** 1.8L L4, 2.4L L4	**Hybrid Battery Pack Current Sensor Circuit Range / Performance:** The monitor will run whenever the following DTC is not present: TMC's intellectual property. Other conditions belong to TMC's intellectual property: HV battery current sensor performance problem.
DTC: P0AC1-123 **1T ECM, MIL: Yes** **Year:** 2011, 2012 **Model:** Camry, Prius, Prius V **Engine:** 1.8L L4, 2.4L L4	**Hybrid Battery Pack Current Sensor "A" Circuit Low1:** The monitor will run whenever the following DTCs are not present: TMC's intellectual property Other conditions belong to TMC's intellectual property.
DTC: P0AC2-123 **1T ECM, MIL: Yes** **Year:** 2011, 2012 **Model:** Camry, Prius, Prius V **Engine:** 1.8L L4, 2.4L L4	**Hybrid Battery Pack Current Sensor "A" Circuit High:** The monitor will run whenever the following DTCs are not present: TMC's intellectual property Other conditions belong to TMC's intellectual property.
DTC: P0AC6-123 **1T ECM, MIL: Yes** **Year:** 2011, 2012 **Model:** Camry, Prius, Prius V **Engine:** 1.8L L4, 2.4L L4	**Hybrid Battery Temperature Sensor "B" Range / Performance:** The monitor will run whenever the following DTCs are not present: TMC's intellectual property Other conditions belong to TMC's intellectual property.
DTC: P0AC7-123 **1T ECM, MIL: Yes** **Year:** 2011, 2012 **Model:** Camry, Prius, Prius V **Engine:** 1.8L L4, 2.4L L4	**Hybrid Battery Temperature Sensor "B" Circuit Low:** The monitor will run whenever the following DTCs are not present: TMC's intellectual property Other conditions belong to TMC's intellectual property.
DTC: P0AC8-123 **1T ECM, MIL: Yes** **Year:** 2011, 2012 **Model:** Camry, Prius, Prius V **Engine:** 1.8L L4, 2.4L L4	**Hybrid Battery Temperature Sensor "B" Circuit High:** The monitor will run whenever the following DTCs are not present: TMC's intellectual property Other conditions belong to TMC's intellectual property.

DTC	Trouble Code Title and Conditions
DTC: P0ACB-123 **1T ECM, MIL: Yes** **Year:** 2011, 2012 **Model:** Camry, Prius, Prius V **Engine:** 1.8L L4, 2.4L L4	**Hybrid Battery Temperature Sensor "C" Range / Performance:** The monitor will run whenever the following DTCs are not present: TMC's intellectual property Other conditions belong to TMC's intellectual property.
DTC: P0ACC-123 **1T ECM, MIL: Yes** **Year:** 2011, 2012 **Model:** Camry, Prius, Prius V **Engine:** 1.8L L4, 2.4L L4	**Hybrid Battery Temperature Sensor "C" Circuit Low:** The monitor will run whenever the following DTCs are not present: TMC's intellectual property Other conditions belong to TMC's intellectual property.
DTC: P0ACD-123 **1T ECM, MIL: Yes** **Year:** 2011, 2012 **Model:** Camry, Prius, Prius V **Engine:** 1.8L L4, 2.4L L4	**Hybrid Battery Temperature Sensor "C" Circuit High:** The monitor will run whenever the following DTCs are not present: TMC's intellectual property Other conditions belong to TMC's intellectual property.
DTC: P0ADB-227 **T ECM, MIL: Yes** **Year:** 2011, 2012 **Model:** Camry, Prius, Prius V **Engine:** 1.8L L4, 2.4L L4	**Hybrid Battery Positive Contactor Control Circuit Low:** Short to GND in the SMRB circuit. The monitor will run whenever the following DTCs are not present: TMC's intellectual property Other conditions belong to TMC's intellectual property.
DTC: P0ADC-226 **T , MIL: Yes** **Year:** 2011, 2012 **Model:** Camry, Prius, Prius V **Engine:** 1.8L L4, 2.4L L4	**Hybrid Battery Positive Contactor Control Circuit High:** Open or short to +B in the SMRB circuit. The monitor will run whenever the following DTCs are not present: TMC's intellectual property Other conditions belong to TMC's intellectual property.
DTC: P0ADF-229 **T ECM, MIL: Yes** **Year:** 2011, 2012 **Model:** Camry, Prius, Prius V **Engine:** 1.8L L4, 2.4L L4	**Hybrid Battery Negative Contactor Control Circuit Low:** Short to GND in the SMRG circuit The monitor will run whenever the following DTCs are not present: TMC's intellectual property Other conditions belong to TMC's intellectual property.
DTC: P0AE0-228 **T ECM, MIL: Yes** **Year:** 2011, 2012 **Model:** Camry, Prius, Prius V **Engine:** 1.8L L4, 2.4L L4	**Hybrid Battery Negative Contactor Control Circuit High:** Open or +B short in SMRG circuit The monitor will run whenever the following DTCs are not present: TMC's intellectual property Other conditions belong to TMC's intellectual property.
DTC: P0AE2-161 **T ECM, MIL: Yes** **Year:** 2011, 2012 **Model:** Camry, Prius, Prius V **Engine:** 1.8L L4, 2.4L L4	**Hybrid Battery Precharge Contactor Circuit Stuck Closed:** When the power switch is on (READY) and regenerative braking is occurring, current is applied to SMRP (SMRG is turned off). The monitor will run whenever the following DTCs are not present: TMC's intellectual Property. Other conditions belong to TMC's intellectual property.
DTC: P0AE2-773 **T ECM, MIL: Yes** **Year:** 2011, 2012 **Model:** Camry, Prius, Prius V **Engine:** 1.8L L4, 2.4L L4	**When the power switch is on (READY) and regenerative braking is occurring, current is applied to SMRP (SMRG is turned off).:** Current flows through SMRP when only SMRB is on during pre-charge (SMRP is stuck closed).
DTC: P0AE6-225 **T ECM, MIL: Yes** **Year:** 2011, 2012 **Model:** Camry, Prius, Prius V **Engine:** 1.8L L4, 2.4L L4	**Hybrid Battery Precharge Contactor Control Circuit Low:** Open or short to GND in the SMRP circuit The monitor will run whenever the following DTCs are not present: TMC's intellectual property Other conditions belong to TMC's intellectual property.
DTC: P0AE7-224 **T ECM, MIL: Yes** **Year:** 2011, 2012 **Model:** Camry, Prius, Prius V **Engine:** 1.8L L4, 2.4L L4	**Hybrid Battery Precharge Contactor Control Circuit High:** Short to +B in the SMRP circuit The monitor will run whenever the following DTCs are not present: TMC's intellectual property Other conditions belong to TMC's intellectual property.

DTC	Trouble Code Title and Conditions
DTC: P0AE9-123 **1T ECM, MIL: Yes** **Year:** 2011, 2012 **Model:** Camry, Prius, Prius V **Engine:** 1.8L L4, 2.4L L4	**Hybrid Battery Temperature Sensor "D" Range / Performance:** The monitor will run whenever the following DTCs are not present: TMC's intellectual property Other conditions belong to TMC's intellectual property.
DTC: P0AEA-123 **1T ECM, MIL: Yes** **Year:** 2011, 2012 **Model:** Camry, Prius, Prius V **Engine:** 1.8L L4, 2.4L L4	**Hybrid Battery Temperature Sensor "D" Circuit Low:** The monitor will run whenever the following DTCs are not present: TMC's intellectual property Other conditions belong to TMC's intellectual property.
DTC: P0AEB-123 **1T ECM, MIL: Yes** **Year:** 2011, 2012 **Model:** Camry, Prius, Prius V **Engine:** 1.8L L4, 2.4L L4	**Hybrid Battery Temperature Sensor "D" Circuit High:** The monitor will run whenever the following DTCs are not present: TMC's intellectual property Other conditions belong to TMC's intellectual property.
DTC: P0AEE-276 **T ECM, MIL: Yes** **Year:** 2011, 2012 **Model:** Camry, Prius, Prius V **Engine:** 1.8L L4, 2.4L L4	**Motor Inverter Temperature Sensor "A" Circuit Range / Performance:** Unusual sudden change in motor inverter temperature sensor output occurs and the offset continues, or unusual sudden change in motor inverter temperature sensor output occurs repeatedly.
DTC: P0AEE-277 **T ECM, MIL: Yes** **Year:** 2011, 2012 **Model:** Camry, Prius, Prius V **Engine:** 1.8L L4, 2.4L L4	**Motor Inverter Temperature Sensor "A" Circuit Range / Performance:** Temperature calculated by power hybrid vehicle control ECU and actual temperature are different for 10 seconds or more.
DTC: P0AEF-275 **T ECM, MIL: Yes** **Year:** 2011, 2012 **Model:** Camry, Prius, Prius V **Engine:** 1.8L L4, 2.4L L4	**Drive Motor Inverter Temperature Sensor "A" Circuit Low:** Open or short to GND in the motor inverter temperature sensor circuit The monitor will run whenever the following DTCs are not present: TMC's intellectual property Other conditions belong to TMC's intellectual property.
DTC: P0AF0-274 **2T ECM, MIL: Yes** **Year:** 2011, 2012 **Model:** Camry, Prius, Prius V **Engine:** 1.8L L4, 2.4L L4	**Drive Motor Inverter Temperature Sensor "A" Circuit High:** Short to +B in motor inverter temperature sensor circuit. The monitor will run whenever the following DTCs are not present: TMC's intellectual property Other conditions belong to TMC's intellectual property.
DTC: P0B3D-123 **1T ECM, MIL: Yes** **Year:** 2011, 2012 **Model:** Camry, Prius, Prius V **Engine:** 1.8L L4, 2.4L L4	**Hybrid Battery Voltage Sensor "A" Circuit Low:** The monitor will run whenever the following DTCs are not present: TMC's intellectual property Other conditions belong to TMC's intellectual property.
DTC: P0B42-123 **1T ECM, MIL: Yes** **Year:** 2011, 2012 **Model:** Camry, Prius, Prius V **Engine:** 1.8L L4, 2.4L L4	**Hybrid Battery Voltage Sensor "B" Circuit Low:** The monitor will run whenever the following DTCs are not present: TMC's intellectual property. Other conditions belong to TMC's intellectual property.
DTC: P0B47-123 **1T ECM, MIL: Yes** **Year:** 2011, 2012 **Model:** Camry, Prius, Prius V **Engine:** 1.8L L4, 2.4L L4	**Hybrid Battery Voltage Sensor "C" Circuit Low:** The monitor will run whenever the following DTCs are not present: TMC's intellectual property Other conditions belong to TMC's intellectual property.
DTC: P0B4C-123 **1T ECM, MIL: Yes** **Year:** 2011, 2012 **Model:** Camry, Prius, Prius V **Engine:** 1.8L L4, 2.4L L4	**Hybrid Battery Voltage Sensor "D" Circuit Low:** The monitor will run whenever the following DTCs are not present: TMC's intellectual property Other conditions belong to TMC's intellectual property.

DTC	Trouble Code Title and Conditions
DTC: P0B51-123 **1T ECM, MIL: Yes** **Year:** 2011, 2012 **Model:** Camry, Prius, Prius V **Engine:** 1.8L L4, 2.4L L4	**Hybrid Battery Voltage Sensor "E" Circuit Low:** The monitor will run whenever the following DTCs are not present: TMC's intellectual property Other conditions belong to TMC's intellectual property.
DTC: P0B56-123 **1T ECM, MIL: Yes** **Year:** 2011, 2012 **Model:** Camry, Prius, Prius V **Engine:** 1.8L L4, 2.4L L4	**Hybrid Battery Voltage Sensor "F" Circuit Low:** The monitor will run whenever the following DTCs are not present: TMC's intellectual property Other conditions belong to TMC's intellectual property.
DTC: P0B5B-123 **1T ECM, MIL: Yes** **Year:** 2011, 2012 **Model:** Camry, Prius, Prius V **Engine:** 1.8L L4, 2.4L L4	**Hybrid Battery Voltage Sensor "G" Circuit Low:** The monitor will run whenever the following DTCs are not present: TMC's intellectual property Other conditions belong to TMC's intellectual property.
DTC: P0B60-123 **1T ECM, MIL: Yes** **Year:** 2011, 2012 **Model:** Camry, Prius, Prius V **Engine:** 1.8L L4, 2.4L L4	**Hybrid Battery Voltage Sensor "H" Circuit Low:** The monitor will run whenever the following DTCs are not present: TMC's intellectual property Other conditions belong to TMC's intellectual property.
DTC: P0B65-123 **1T ECM, MIL: Yes** **Year:** 2011, 2012 **Model:** Camry, Prius, Prius V **Engine:** 1.8L L4, 2.4L L4	**Hybrid Battery Voltage Sensor "I" Circuit Low:** The monitor will run whenever the following DTCs are not present: TMC's intellectual property Other conditions belong to TMC's intellectual property.
DTC: P0B6A-123 **1T ECM, MIL: Yes** **Year:** 2011, 2012 **Model:** Camry, Prius, Prius V **Engine:** 1.8L L4, 2.4L L4	**Hybrid Battery Voltage Sensor "J" Circuit Low:** The monitor will run whenever the following DTCs are not present: TMC's intellectual property Other conditions belong to TMC's intellectual property.
DTC: P0B6F-123 **1T ECM, MIL: Yes** **Year:** 2011, 2012 **Model:** Camry, Prius, Prius V **Engine:** 1.8L L4, 2.4L L4	**Hybrid Battery Voltage Sensor "K" Circuit Low:** The monitor will run whenever the following DTCs are not present: TMC's intellectual property Other conditions belong to TMC's intellectual property.
DTC: P0B74-123 **1T ECM, MIL: Yes** **Year:** 2011, 2012 **Model:** Camry, Prius, Prius V **Engine:** 1.8L L4, 2.4L L4	**Hybrid Battery Voltage Sensor "L" Circuit Low:** The monitor will run whenever the following DTCs are not present: TMC's intellectual property Other conditions belong to TMC's intellectual property.
DTC: P0B79-123 **1T ECM, MIL: Yes** **Year:** 2011, 2012 **Model:** Camry, Prius, Prius V **Engine:** 1.8L L4, 2.4L L4	**Hybrid Battery Voltage Sensor "M" Circuit Low:** The monitor will run whenever the following DTCs are not present: TMC's intellectual property Other conditions belong to TMC's intellectual property.
DTC: P0B7E-123 **1T ECM, MIL: Yes** **Year:** 2011, 2012 **Model:** Camry, Prius, Prius V **Engine:** 1.8L L4, 2.4L L4	**Hybrid Battery Voltage Sensor "N" Circuit Low:** The monitor will run whenever the following DTCs are not present: TMC's intellectual property Other conditions belong to TMC's intellectual property.
DTC: P0B83-123 **1T ECM, MIL: Yes** **Year:** 2011, 2012 **Model:** Camry, Prius, Prius V **Engine:** 1.8L L4, 2.4L L4	**Hybrid Battery Voltage Sensor "O" Circuit Low:** The monitor will run whenever the following DTCs are not present: TMC's intellectual property Other conditions belong to TMC's intellectual property.

DTC	Trouble Code Title and Conditions
DTC: P0B88-123 **1T ECM, MIL: Yes** **Year:** 2011, 2012 **Model:** Camry, Prius, Prius V **Engine:** 1.8L L4, 2.4L L4	**Hybrid Battery Voltage Sensor "P" Circuit Low:** The monitor will run whenever the following DTCs are not present: TMC's intellectual property Other conditions belong to TMC's intellectual property.
DTC: P0B8D-123 **1T ECM, MIL: Yes** **Year:** 2011, 2012 **Model:** Camry, Prius, Prius V **Engine:** 1.8L L4, 2.4L L4	**Hybrid Battery Voltage Sensor "Q" Circuit Low:** The monitor will run whenever the following DTCs are not present: TMC's intellectual property Other conditions belong to TMC's intellectual property.
DTC: P0B92-123 **1T ECM, MIL: Yes** **Year:** 2011, 2012 **Model:** Camry, Prius, Prius V **Engine:** 1.8L L4, 2.4L L4	**Hybrid Battery Voltage Sensor "R" Circuit Low:** The monitor will run whenever the following DTCs are not present: TMC's intellectual property Other conditions belong to TMC's intellectual property.
DTC: P0C30-390 **T ECM, MIL: Yes** **Year:** 2011, 2012 **Model:** Camry, Prius, Prius V **Engine:** 1.8L L4, 2.4L L4	**Hybrid Battery Pack State of Charge High:** Charge control error. The monitor will run whenever the following DTCs are not present: TMC's intellectual Property. Other conditions belong to TMC's intellectual property.
DTC: P0C76-523 **T ECM, MIL: Yes** **Year:** 2011, 2012 **Model:** Camry, Prius, Prius V **Engine:** 1.8L L4, 2.4L L4	**Hybrid Battery System Discharge Time Too Long:** Inverter voltage (VH) sensor offset malfunction. The monitor will run whenever the following DTCs are not present: TMC's intellectual Property. Other conditions belong to TMC's intellectual property.

OBD II Trouble Code List (P1XXX Codes)

DTC	Trouble Code Title and Conditions
DTC: P101D **1T ECM, MIL: Yes** **Year:** 2011, 2012 **Model:** 4Runner, Avalon, Camry, Corolla, FJ Cruiser, Highlander, Land Cruiser, Matrix, Prius, Rav4, Sequoia, Tundra, Tacoma, Venza, Yaris **Engine:** L4, V6, V8	**A/F Sensor Heater Circuit Performance Bank 1 Sensor 1 Stuck ON:** Monitor runs whenever following DTCs not stored: P0031, P0051 (Air fuel ratio sensor heater) Battery voltage: 10.5 V or higher Time after engine start: 10 seconds or more Active heater OFF control: Not operating Active heater ON control: Not operating Air fuel ratio sensor heater duty-cycle: 10 to 60% Air fuel ratio sensor heater ON current: 0.8 A or higher
DTC: P102D **1T ECM, MIL: Yes** **Year:** 2011, 2012 **Model:** 4Runner, Avalon, Camry, Corolla, FJ Cruiser, Highlander, Land Cruiser, Matrix, Prius, Rav4, Sequoia, Tundra, Tacoma, Venza, Yaris **Engine:** L4, V6, V8	**O2 Sensor Heater Circuit Performance Bank 1 Sensor 2 Stuck ON:** Battery voltage: 10.5 V or more Engine: Running Starter: OFF Catalyst active air fuel ratio control: Not operating Time after heater on: 10 seconds or more Learned heater off current operation: Complete Hybrid IC high current limiter monitor input: Fail
DTC: P105D **1T ECM** **Year:** 2011, 2012 **Model:** Highlander **Engine:** 3.5L V6	**O2 Sensor Heater Circuit Performance Bank 2 Sensor 2 Stuck ON:** Monitor runs whenever following DTCs are not present: None Battery voltage: 10.5 V or more Engine: Running Starter: OFF Catalyst active air fuel ratio control: Not operating Time after heater on: 10 seconds or more Learned heater off current operation: Complete Hybrid IC high current limiter monitor input: Fail **Possible causes:** • ECM

DTC	Trouble Code Title and Conditions
DTC: P105D **1T ECM, MIL: Yes** **Year:** 2011, 2012 **Model:** 4Runner, Avalon, Camry, Corolla, FJ Cruiser, Highlander, Land Cruiser, Matrix, Prius, Rav4, Sequoia, Tundra, Tacoma, Venza, Yaris **Engine:** L4, V6, V8	**O2 Sensor Heater Circuit Performance Bank 2 Sensor 2 Stuck ON:** Monitor runs whenever following DTCs not stored: P0031, P0032, P0051, P0052 (Air fuel ratio sensor heater) P0037, P0038, P0057, P0058 (Rear oxygen sensor heater) Battery voltage: 10.5 V or higher Engine: Running Starter: OFF Catalyst active A/F control: Not operating Time after heater ON: 10 seconds or more Learned heater OFF current operation: Complete Hybrid IC high current limiter port: Fail
DTC: P106A **1T ECM, MIL: Yes** **Year:** 2011, 2012 **Model:** 4Runner, Avalon, Camry, Corolla, FJ Cruiser, Highlander, Land Cruiser, Matrix, Prius, Rav4, Sequoia, Tundra, Tacoma, Venza, Yaris **Engine:** L4, V6, V8	**Evaporative Emission System Pressure Sensor - Manifold Absolute Pressure Correlation:** Monitor runs whenever following DTCs are not present: P0106, P0107, P0108 (Manifold Absolute Pressure) P0401 (EGR System (Closed)) P0452, P0453 (Evaporative Emission System Pressure Sensor) Arrive at after engine stop: 50 minutes Time after ECM started by soak-timer: 60 seconds or more Battery voltage: 10.5 V or more Intake air temperature: -10°C (14°F) or more Engine coolant temperature sensor: -10°C (14°F) or more
DTC: P106B **2T ECM, MIL: Yes** **Year:** 2011 **Model:** Land Cruiser **Engine:** 5.7L V8	**Evaporative Emission System Pressure Sensor - Secondary Air Injection Pressure Sensor Correlation:** The pressure detected by the canister pressure sensor (Vapor Pressure Pump*) and the pressure detected by the pressure sensor of either air switching valve (for Bank 1 or Bank 2) (Air pump pressure (Absolute)* or Air Pump2 Pressure (Absolute)*) differ by 6.98 kPa (52 mmHg) or more. **Possible causes:** • Canister pressure sensor • Air switching valve (for Bank 1) • Air switching valve (for Bank 2)
DTC: P1170 **2T ECM, MIL: Yes** **Year:** 2011, 2012 **Model:** 4Runner, Avalon, Camry, Corolla, FJ Cruiser, Highlander, Land Cruiser, Matrix, Prius, Rav4, Sequoia, Tundra, Tacoma, Venza, Yaris **Engine:** L4, V6, V8	**Port Injector Fuel Performance:** Either of the following conditions 1 or 2 is met: 1. Engine rpm: Less than 1100 rpm 2. Engine load: 10% or more Catalyst monitor: Not executed
DTC: P117B **2T ECM, MIL: Yes** **Year:** 2011, 2012 **Model:** 4Runner, Avalon, Camry, Corolla, FJ Cruiser, Highlander, Land Cruiser, Matrix, Prius, Rav4, Sequoia, Tundra, Tacoma, Venza, Yaris **Engine:** L4, V6, V8	**Direct Injector Fuel Performance:** Fuel system status: Closed loop Battery voltage: 11 V or more Either of the following conditions 1 or 2 is met: 1. Engine rpm: Less than 1100 rpm 2. Engine load: 10% or more Catalyst monitor: Not executed
DTC: P1235 **2T ECM, MIL: Yes** **Year:** 2011, 2012 **Model:** 4Runner, Avalon, Camry, Corolla, FJ Cruiser, Highlander, Land Cruiser, Matrix, Prius, Rav4, Sequoia, Tundra, Tacoma, Venza, Yaris **Engine:** L4, V6, V8	**High Pressure Fuel Pump Circuit:** Monitor runs whenever these DTCs are not present. None. Time after engine start: 5 seconds or more Output duty ratio: More than output duty ratio change map value, and less than 95% Battery voltage: 10.5 V or more Engine switch: On (IG) Starter: OFF Output duty ratio change map value: Engine speed is 500 rpm 5% Engine speed is 2000 rpm, 20% Engine speed is 4000 rpm, 40% Engine speed is 6000 rpm, 60% Engine speed is 8000 rpm, 80 %

DTC	Trouble Code Title and Conditions
DTC: P1276 **1T ECM, MIL: Yes** **Year:** 2011, 2012 **Model:** Camry **Engine:** 3.5L V6	**Port Injector Circuit No. 1:** Monitor runs whenever following DTCs are not present: None Battery voltage: 8 V or more Fuel cut: OFF Port injector: Energized Engine switch: On (IG)
DTC: P1277 **1T ECM, MIL: Yes** **Year:** 2011, 2012 **Model:** Camry **Engine:** 3.5L V6	**Port Injector Circuit No. 2:** Monitor runs whenever following DTCs are not present: None Battery voltage: 8 V or more Fuel cut: OFF Port injector: Energized Engine switch: On (IG)
DTC: P1278 **1T ECM, MIL: Yes** **Year:** 2011, 2012 **Model:** Camry **Engine:** 3.5L V6	**Port Injector Circuit No. 3:** Monitor runs whenever following DTCs are not present: None Battery voltage: 8 V or more Fuel cut: OFF Port injector: Energized Engine switch: On (IG)
DTC: P1279 **1T ECM, MIL: Yes** **Year:** 2011, 2012 **Model:** Camry **Engine:** 3.5L V6	**Port Injector Circuit No. 4:** Monitor runs whenever following DTCs are not present: None Battery voltage: 8 V or more Fuel cut: OFF Port injector: Energized Engine switch: On (IG)
DTC: P127A **1T ECM, MIL: Yes** **Year:** 2011, 2012 **Model:** Camry **Engine:** 3.5L V6	**Port Injector Circuit No. 5:** Monitor runs whenever following DTCs are not present: None Battery voltage: 8 V or more Fuel cut: OFF Port injector: Energized Engine switch: On (IG)
DTC: P127B **1T ECM, MIL: Yes** **Year:** 2011, 2012 **Model:** Camry **Engine:** 3.5L V6	**Port Injector Circuit No. 6:** Monitor runs whenever following DTCs are not present: None Battery voltage: 8 V or more Fuel cut: OFF Port injector: Energized Engine switch: On (IG)
DTC: P1340 **2T ECM, MIL: Yes** **Year:** 2011, 2012 **Model:** 4Runner, Avalon, Camry, Corolla, FJ Cruiser, Highlander, Land Cruiser, Matrix, Prius, Rav4, Sequoia, Tundra, Tacoma, Venza, Yaris **Engine:** L4, V6, V8	**Camshaft Position Sensor "A" (Bank 1 Sensor 2):** No camshaft position sensor signal to ECM during cranking (2 trip detection logic) No camshaft position sensor signal to ECM with engine speed 600 rpm or more.
DTC: P1342 **1T ECM, MIL: Yes** **Year:** 2011, 2012 **Model:** 4Runner, Avalon, Camry, Corolla, FJ Cruiser, Highlander, Land Cruiser, Matrix, Prius, Rav4, Sequoia, Tundra, Tacoma, Venza, Yaris **Engine:** L4, V6, V8	**Camshaft Position Sensor "A" Low Input (MRE):** The output voltage of the camshaft position sensor is below 0.3 V for 4 seconds.

DTC	Trouble Code Title and Conditions
DTC: P1343 **1T ECM, MIL: Yes** **Year:** 2011, 2012 **Model:** 4Runner, Avalon, Camry, Corolla, FJ Cruiser, Highlander, Land Cruiser, Matrix, Prius, Rav4, Sequoia, Tundra, Tacoma, Venza, Yaris **Engine:** L4, V6, V8	**Camshaft Position Sensor "A" High Input (MRE):** Output voltage of camshaft position sensor more than 4.7 V for 4 seconds.
DTC: P1420 **2T ECM, MIL: Yes** **Year:** 2011, 2012 **Model:** 4Runner, Avalon, Camry, Corolla, FJ Cruiser, Highlander, Land Cruiser, Matrix, Prius, Rav4, Sequoia, Tundra, Tacoma, Venza, Yaris **Engine:** L4, V6, V8	**Evaporative Emission Canister Small Leak:** Monitor runs whenever these DTCs are not present: None Both of the following conditions are met before Key-OFF: Conditions 1 and 2 1. Duration that vehicle is being driven: 5 minutes or more 2. Purge flow: Executed Engine coolant temperature: 4.4 to 35°C (40 to 95°F) Intake air temperature: 4.4 to 35°C (40 to 95°F)
DTC: P1421 **2T ECM, MIL: Yes** **Year:** 2011, 2012 **Model:** 4Runner, Avalon, Camry, Corolla, FJ Cruiser, Highlander, Land Cruiser, Matrix, Prius, Rav4, Sequoia, Tundra, Tacoma, Venza, Yaris **Engine:** L4, V6, V8	**Evaporative Emission Canister Gross Leak:** Monitor runs whenever these DTCs are not present: None Both of the following conditions are met before Key-OFF: Conditions 1 and 2 1. Duration that vehicle is being driven: 5 minutes or more 2. Purge flow: Executed Engine coolant temperature: 4.4 to 35°C (40 to 95°F) Intake air temperature: 4.4 to 35°C (40 to 95°F)
DTC: P1422 **2T ECM, MIL: Yes** **Year:** 2011, 2012 **Model:** 4Runner, Avalon, Camry, Corolla, FJ Cruiser, Highlander, Land Cruiser, Matrix, Prius, Rav4, Sequoia, Tundra, Tacoma, Venza, Yaris **Engine:** L4, V6, V8	**Fuel Tank Small Leak:** Monitor runs whenever these DTCs are not present: None Both of the following conditions are met before Key-OFF: Conditions 1 and 2 1. Duration that vehicle is being driven: 5 minutes or more 2. Purge flow: Executed Engine coolant temperature: 4.4 to 35°C (40 to 95°F) Intake air temperature: 4.4 to 35°C (40 to 95°F)
DTC: P1423 **2T ECM, MIL: Yes** **Year:** 2011, 2012 **Model:** 4Runner, Avalon, Camry, Corolla, FJ Cruiser, Highlander, Land Cruiser, Matrix, Prius, Rav4, Sequoia, Tundra, Tacoma, Venza, Yaris **Engine:** L4, V6, V8	**Fuel Tank Gross Leak:** Monitor runs whenever these DTCs are not present: None EVAP key-off monitor runs when all of the following conditions are met: - Atmospheric pressure: 70 to 111 kPa (525 to 833 mmHg) Battery voltage: 10.5 V or more Vehicle speed: Less than 2.5 mph (4 km/h) Power switch: Off Engine condition: Not running Key-OFF duration: 5 or 7 or 9.5 hours Both of the following conditions are met before Key-OFF: Conditions 1 and 2 1. Duration that vehicle is being driven: 5 minutes or more 2. Purge flow: Executed Engine coolant temperature: 4.4 to 35°C (40 to 95°F) Intake air temperature: 4.4 to 35°C (40 to 95°F)
DTC: P1451 **2T ECM, MIL: Yes** **Year:** 2011, 2012 **Model:** 4Runner, Avalon, Camry, Corolla, FJ Cruiser, Highlander, Land Cruiser, Matrix, Prius, Rav4, Sequoia, Tundra, Tacoma, Venza, Yaris **Engine:** L4, V6, V8	**Fuel Tank Pressure Sensor Range/Performance:** Noise Monitor: Monitor runs whenever these DTCs are not present: None Fuel tank pressure: -15 to 20 kPa Battery voltage: 10.5 V or more Intake air temperature: 4.4 to 50°C (40 to 122°F) Either of the following conditions is met: A or B A. Engine: Running B. Key-off duration: 5 or 7 or 9.5 hours.

DTC	Trouble Code Title and Conditions
DTC: P1452 **1T ECM, MIL: Yes** **Year:** 2011, 2012 **Model:** 4Runner, Avalon, Camry, Corolla, FJ Cruiser, Highlander, Land Cruiser, Matrix, Prius, Rav4, Sequoia, Tundra, Tacoma, Venza, Yaris **Engine:** L4, V6, V8	**Fuel Tank Pressure Sensor Low Input:** Fuel tank pressure is less than -17.187 kPa (-128.93 mmHg) for 0.5 seconds.
DTC: P1453 **1T ECM, MIL: Yes** **Year:** 2011, 2012 **Model:** 4Runner, Avalon, Camry, Corolla, FJ Cruiser, Highlander, Land Cruiser, Matrix, Prius, Rav4, Sequoia, Tundra, Tacoma, Venza, Yaris **Engine:** L4, V6, V8	**Fuel Tank Pressure Sensor High Input:** Fuel tank pressure is more than 23.9 kPa (179.30 mmHg) for 0.5 seconds.
DTC: P1500 **1T ECM, MIL: Yes** **Year:** 2011, 2012 **Model:** 4Runner, Avalon, Camry, Corolla, FJ Cruiser, Highlander, Land Cruiser, Matrix, Prius, Rav4, Sequoia, Tundra, Tacoma, Venza, Yaris **Engine:** L4, V6, V8	**AC Inverter Malfunction:** While vehicle running, idling-up signal input into ECM for 10 seconds.
DTC: P1555-181 **T ECM, MIL: Yes** **Year:** 2011, 2012 **Model:** Camry, Prius, Prius V **Engine:** 1.8L L4, 2.4L L4	**Reactor Temperature Sensor Circuit Low:** Malfunction in the reactor temperature sensor wiring (short to GND).
DTC: P1556-182 **T ECM, MIL: Yes** **Year:** 2011, 2012 **Model:** Camry, Prius, Prius V **Engine:** 1.8L L4, 2.4L L4	**Reactor Temperature Sensor Circuit High:** Malfunction in the reactor temperature sensor wiring (open or short to +B)
DTC: P1570 **2T , MIL: Yes** **Year:** 2011, 2012 **Model:** Camry, Prius, Prius V **Engine:** 1.8L L4, 2.4L L4	**Radar Sensor Malfunction:** The ECM detects a radar sensor malfunction signal for 0.15 sec. or more while the dynamic radar cruise control is in operation.
DTC: P1571 **2T , MIL: Yes** **Year:** 2011, 2012 **Model:** Camry, Prius, Prius V **Engine:** 1.8L L4, 2.4L L4	**Radar Sensor Malfunction:** The hybrid vehicle control ECU detects a millimeter wave radar sensor assembly malfunction signal for 0.15 seconds or more while the dynamic radar cruise control is operating.
DTC: P1572 **T , MIL: Yes** **Year:** 2011, 2012 **Model:** Camry, Prius, Prius V **Engine:** 1.8L L4, 2.4L L4	**Improper Aiming of Radar Sensor Beam Axis:** The ECU detects that the millimeter wave radar sensor assembly beam axis is in an incorrect position (0.15 seconds or more) while the dynamic radar cruise control is operating.
DTC: P1575 **T , MIL: Yes** **Year:** 2011, 2012 **Model:** Camry, Prius, Prius V **Engine:** 1.8L L4, 2.4L L4	**Warning Buzzer Malfunction:** The hybrid vehicle control ECU receives a buzzer abnormal signal for 0.2 seconds or more while the dynamic radar cruise control is operating.

DTC	Trouble Code Title and Conditions
DTC: P1578 **T , MIL: Yes** **Year:** 2011, 2012 **Model:** Camry, Prius, Prius V **Engine:** 1.8L L4, 2.4L L4	**Brake System Malfunction:** The hybrid vehicle control ECU receives a vehicle stability control system error signal for 0.2 seconds or more while the dynamic radar cruise control is operating.
DTC: P1603 **1T ECM, MIL: Yes** **Year:** 2011, 2012 **Model:** 4Runner, Avalon, Camry, Corolla, FJ Cruiser, Highlander, Land Cruiser, Matrix, Prius, Rav4, Sequoia, Tundra, Tacoma, Venza, Yaris **Engine:** L4, V6, V8	**Engine Stall History:** After monitoring for start-ability problems and 5 seconds or more elapse after starting the engine, with the engine running, the engine stops (the engine speed drops to 200 rpm or less) without the ignition switch being operated for 0.5 seconds or more.
DTC: P1604 **1T ECM, MIL: Yes** **Year:** 2011, 2012 **Model:** 4Runner, Avalon, Camry, Corolla, FJ Cruiser, Highlander, Land Cruiser, Matrix, Prius, Rav4, Sequoia, Tundra, Tacoma, Venza, Yaris **Engine:** L4, V6, V8	**Startability Malfunction:** After monitoring for start-ability problems (P1604) finishes and 5 seconds or more elapse after starting the engine, with the engine running, the engine stops (the engine speed drops to 200 rpm or less) without the ignition switch being operated for 0.5 seconds or more.
DTC: P1605 **1T ECM, MIL: Yes** **Year:** 2011, 2012 **Model:** 4Runner, Avalon, Camry, Corolla, FJ Cruiser, Highlander, Land Cruiser, Matrix, Prius, Rav4, Sequoia, Tundra, Tacoma, Venza, Yaris **Engine:** L4, V6, V8	**Rough Idling:** After 5 seconds or more elapse after starting the engine, with the engine running, the engine speed drops to 400 rpm or less.
DTC: P1606-308 **T ECM, MIL: Yes** **Year:** 2011, 2012 **Model:** Camry, Prius, Prius V **Engine:** 1.8L L4, 2.4L L4	**Collision Detection:** Shutoff signal from the airbag sensor assembly is determined.
DTC: P1606-317 **T ECM, MIL: Yes** **Year:** 2011, 2012 **Model:** Camry, Prius, Prius V **Engine:** 1.8L L4, 2.4L L4	**Collision Detection:** A collision is determined due to a wiring malfunction.
DTC: P1607 **1T ECM, MIL: Yes** **Year:** 2011, 2012 **Model:** 4Runner, Avalon, Camry, Corolla, FJ Cruiser, Highlander, Land Cruiser, Matrix, Prius, Rav4, Sequoia, Tundra, Tacoma, Venza, Yaris **Engine:** L4, V6, V8	**Cruise Control Input Processor:** The ECM CPUs malfunction. The ECM will illuminate the MIL and store DTC(s) immediately.

DTC	Trouble Code Title and Conditions
DTC: P1613 **1T ECM, MIL: Yes** **Year:** 2011, 2012 **Model:** 4Runner, Avalon, Camry, Corolla, FJ Cruiser, Highlander, Land Cruiser, Matrix, Prius, Rav4, Sequoia, Tundra, Tacoma, Venza, Yaris **Engine:** L4, V6, V8	**Secondary Air Injection Driver Malfunction:** Either of following conditions (1) or (2) met: (1) All of following conditions met (1 trip detection logic): Either of air pump or air switching valve not operating Diagnostic signal from Air Injection Control Driver (AID) 80% Battery voltage 8 V or more (2) Both of following conditions met (1 trip detection logic): Battery voltage 8 V or more Diagnostic signal from AID abnormal (duty signal other than 0, 20, 40, 80 and 100%) All of following conditions met (1 trip detection logic): a. Air injection system operating (Air Switching Valve [ASV] ON and air pump ON) b. Diagnostic signal from Air Injection Control Driver (AID) 0% c. Battery voltage 8 V or more Both of following conditions met (1 trip detection logic): a. Battery voltage 8 V or more b. Diagnostic signal from Air Injection Control Driver (AID) 100%
DTC: P1614 **1T ECM, MIL: Yes** **Year:** 2011, 2012 **Model:** 4Runner, Avalon, Camry, Corolla, FJ Cruiser, Highlander, Land Cruiser, Matrix, Prius, Rav4, Sequoia, Tundra, Tacoma, Venza, Yaris **Engine:** L4, V6, V8	**Secondary Air Injection System Driver Bank 2:** Either of following conditions (1) or (2) met: (1) All of following conditions met (1 trip detection logic): Either of air pump or air switching valve not operating Diagnostic signal from Air Injection Control Driver (AID) 80% Battery voltage 8 V or more (2) Both of following conditions met (1 trip detection logic): Battery voltage 8 V or more Diagnostic signal from AID abnormal (duty-signal other than 0, 20, 40, 80 and 100%) All of following conditions met (1 trip detection logic): a. Air injection system operating (Air Switching Valve [ASV] ON and air pump ON) b. Diagnostic signal from Air Injection Control Driver (AID) 0% c. Battery voltage 8 V or more Both of following conditions met (1 trip detection logic): a. Battery voltage 8 V or more b. Diagnostic signal from Air Injection Control Driver (AID) 100%
DTC: P1615 **1T ECM, MIL: Yes** **Year:** 2011, 2012 **Model:** 4Runner, Avalon, Camry, Corolla, FJ Cruiser, Highlander, Land Cruiser, Matrix, Prius, Rav4, Sequoia, Tundra, Tacoma, Venza, Yaris **Engine:** L4, V6, V8	**Communication Error from Distance Control ECU to ECM:** While the dynamic radar cruise control is either preparing for operation or operating, if communication data from the distance control ECU is logically inconsistent for a certain amount of time, the ECM records this logical error code.
DTC: P1615 **1T ECM, MIL: Yes** **Year:** 2011, 2012 **Model:** 4Runner, Avalon, Camry, Corolla, FJ Cruiser, Highlander, Land Cruiser, Matrix, Prius, Rav4, Sequoia, Tundra, Tacoma, Venza, Yaris **Engine:** L4, V6, V8	**Communication Error from Distance Control ECU to ECM:** While the dynamic radar cruise control is either preparing for operation or operating, if communication data from the driving support ECU is logically inconsistent for a certain amount of time, the hybrid vehicle control ECU records this logical error code.
DTC: P1616 **1T ECM, MIL: Yes** **Year:** 2011, 2012 **Model:** 4Runner, Avalon, Camry, Corolla, FJ Cruiser, Highlander, Land Cruiser, Matrix, Prius, Rav4, Sequoia, Tundra, Tacoma, Venza, Yaris **Engine:** L4, V6, V8	**Communication Error from ECM to Distance Control ECU:** While the dynamic radar cruise control is either preparing for operation or operating, if the ECM continuously receives a logical error signal from the distance control ECU for more than a specific amount of time, the ECM records this logical error code.

DTC	Trouble Code Title and Conditions
DTC: P1617 **1T ECM, MIL: Yes** **Year:** 2011, 2012 **Model:** 4Runner, Avalon, Camry, Corolla, FJ Cruiser, Highlander, Land Cruiser, Matrix, Prius, Rav4, Sequoia, Tundra, Tacoma, Venza, Yaris **Engine:** L4, V6, V8	**Distance Control ECU Malfunction:** While the dynamic radar cruise control is either preparing for operation or operating, if a designation signal from the ECM and a designation return signal from the distance control ECU do not match or the distance control ECU is malfunctioning for more than a specific amount of time, the ECM records this trouble code.
DTC: P1630 **T , MIL: Yes** **Year:** 2011, 2012 **Model:** Camry, Prius, Prius V **Engine:** 1.8L L4, 2.4L L4	**Communication Error from VSC to ECM:** While the dynamic radar cruise control is either preparing for operation or operating, if communication data from the brake booster with master cylinder (skid control ECU) is logically inconsistent for a certain amount of time, the hybrid vehicle control ECU records this logical error code. **Possible causes:** • Brake booster with master cylinder (skid control ECU) • Hybrid vehicle control ECU • CAN communication system
DTC: P1630 **1T ECM, MIL: Yes** **Year:** 2011, 2012 **Model:** 4Runner, Avalon, Camry, Corolla, FJ Cruiser, Highlander, Land Cruiser, Matrix, Prius, Rav4, Sequoia, Tundra, Tacoma, Venza, Yaris **Engine:** L4, V6, V8	**Communication Error from VSC to ECM:** While the dynamic radar cruise control is either preparing for operation or operating, if communication data from the skid control ECU is logically inconsistent for a certain amount of time, the ECM records this logical error code.
DTC: P1631 **T , MIL: Yes** **Year:** 2011, 2012 **Model:** Camry, Prius, Prius V **Engine:** 1.8L L4, 2.4L L4	**Communication Error from ECM to VSC:** While the dynamic radar cruise control is either preparing for operation or operating, if the hybrid vehicle control ECU continuously receives a logical error signal from the brake booster with master cylinder (skid control ECU) for a certain amount of time, the hybrid vehicle control ECU records this logical error code. **Possible causes:** • Hybrid vehicle control ECU • Brake booster with master cylinder (skid control ECU) • CAN communication system
DTC: P181A-596 **T ECM, MIL: Yes** **Year:** 2011, 2012 **Model:** Camry, Prius, Prius V **Engine:** 1.8L L4, 2.4L L4	**Gear Lever X Position Circuit "A" / "B" Correlation:** Difference between select main sensor value and select sub sensor value is large.
DTC: P181B-595 **T ECM, MIL: Yes** **Year:** 2011, 2012 **Model:** Camry, Prius, Prius V **Engine:** 1.8L L4, 2.4L L4	**Gear Lever Y Position Circuit "A" / "B" Correlation:** Difference between shift main sensor value and shift sub sensor value is large.
DTC: P182B-577 **T ECM, MIL: Yes** **Year:** 2011, 2012 **Model:** Camry, Prius, Prius V **Engine:** 1.8L L4, 2.4L L4	**Gear Lever X Position "B" Circuit Low:** Open or GND short in select sub sensor circuit.
DTC: P182C-578 **T ECM, MIL: Yes** **Year:** 2011, 2012 **Model:** Camry, Prius, Prius V **Engine:** 1.8L L4, 2.4L L4	**Gear Lever X Position "B" Circuit High:** +B short in select sub sensor circuit.
DTC: P182E-573 **T ECM, MIL: Yes** **Year:** 2011, 2012 **Model:** Camry, Prius, Prius V **Engine:** 1.8L L4, 2.4L L4	**Gear Lever Y Position "B" Circuit Low:** Open or GND short in shift sub sensor circuit.

DTC	Trouble Code Title and Conditions
DTC: P182F-574 **T ECM, MIL: Yes** **Year:** 2011, 2012 **Model:** Camry, Prius, Prius V **Engine:** 1.8L L4, 2.4L L4	**Gear Lever Y Position "B" Circuit High:** +B short in shift sub sensor circuit.

OBD II Trouble Code List (P2XXX Codes)

DTC	Trouble Code Title and Conditions
DTC: P2004 **2T ECM, MIL: Yes** **Year:** 2011, 2012 **Model:** 4Runner, Avalon, Camry, Corolla, FJ Cruiser, Highlander, Land Cruiser, Matrix, Prius, Rav4, Sequoia, Tundra, Tacoma, Venza, Yaris **Engine:** L4, V6, V8	**Intake Manifold Runner Control Stuck Open (Bank 1):** Monitor runs whenever following DTCs are not present: P0110 - P0113 (IAT sensor) P0115 - P0118 (ECT sensor) Battery voltage: 8 V or more ECT: -10°C (14°F) or more IAT: -10°C (14°F) or more Ignition switch: ON Command IMRC valve: Closed
DTC: P2006 **2T ECM, MIL: Yes** **Year:** 2011, 2012 **Model:** 4Runner, Avalon, Camry, Corolla, FJ Cruiser, Highlander, Land Cruiser, Matrix, Prius, Rav4, Sequoia, Tundra, Tacoma, Venza, Yaris **Engine:** L4, V6, V8	**Intake Manifold Runner Control Stuck Closed (Bank 1):** The monitor will run whenever these DTCs are not stored: None Engine coolant temperature: -10°C (14°F) or higher Intake air temperature: -10°C (14°F) or higher Control valve half-open operation: Not operated Ignition switch: ON Control valve: Not operated by scan tool Battery voltage during close operation: 8 V or higher When either condition below is met: Condition A or B A. When both conditions below are met: - Engine coolant temperature at engine start: 5°C (41°F) or higher Intake air temperature at engine start: 5°C (41°F) or higher B. When both conditions below are met: - Engine coolant temperature: 70°C (158°F) or higher Vehicle speed: Less than 47 mph (75 km/h) Command to control valve: Open
DTC: P2009 **1T ECM, MIL: Yes** **Year:** 2011, 2012 **Model:** 4Runner, Avalon, Camry, Corolla, FJ Cruiser, Highlander, Land Cruiser, Matrix, Prius, Rav4, Sequoia, Tundra, Tacoma, Venza, Yaris **Engine:** L4, V6, V8	**Intake Manifold Runner Control Circuit Low (Bank 1):** The monitor will run whenever these DTCs are not stored: None Battery voltage: 8 V or higher Starter: OFF Output signal duty: 100% Difference between motor current and previous motor current: Below 0.2 A
DTC: P2010 **1T ECM, MIL: Yes** **Year:** 2011, 2012 **Model:** 4Runner, Avalon, Camry, Corolla, FJ Cruiser, Highlander, Land Cruiser, Matrix, Prius, Rav4, Sequoia, Tundra, Tacoma, Venza, Yaris **Engine:** L4, V6, V8	**Intake Manifold Runner Control Circuit High (Bank 1):** The monitor will run whenever these DTCs are not stored: None Battery voltage: 8 V or higher Starter: OFF

DTC	Trouble Code Title and Conditions
DTC: P2014 **1T ECM, MIL: Yes** **Year:** 2011, 2012 **Model:** 4Runner, Avalon, Camry, Corolla, FJ Cruiser, Highlander, Land Cruiser, Matrix, Prius, Rav4, Sequoia, Tundra, Tacoma, Venza, Yaris **Engine:** L4, V6, V8	**Intake Manifold Runner Position Sensor / Switch Circuit (Bank 1):** Monitor runs whenever following DTCs are not present: None IMRC valve position sensor voltage: Less than 0.2 V, or more than 4.8 V
DTC: P2014 **1T ECM, MIL: Yes** **Year:** 2011, 2012 **Model:** 4Runner, Avalon, Camry, Corolla, FJ Cruiser, Highlander, Land Cruiser, Matrix, Prius, Rav4, Sequoia, Tundra, Tacoma, Venza, Yaris **Engine:** L4, V6, V8	**Intake Manifold Runner Position Sensor / Switch Circuit (Bank 1):** The monitor will run whenever these DTCs are not stored: None
DTC: P2016 **1T ECM, MIL: Yes** **Year:** 2011, 2012 **Model:** 4Runner, Avalon, Camry, Corolla, FJ Cruiser, Highlander, Land Cruiser, Matrix, Prius, Rav4, Sequoia, Tundra, Tacoma, Venza, Yaris **Engine:** L4, V6, V8	**Intake Manifold Runner Position Sensor / Switch Circuit Low (Bank 1):** Monitor runs whenever following DTCs are not present: None IMRC valve position sensor voltage: Less than 0.2 V
DTC: P2017 **1T ECM, MIL: Yes** **Year:** 2011, 2012 **Model:** 4Runner, Avalon, Camry, Corolla, FJ Cruiser, Highlander, Land Cruiser, Matrix, Prius, Rav4, Sequoia, Tundra, Tacoma, Venza, Yaris **Engine:** L4, V6, V8	**Intake Manifold Runner Position Sensor / Switch Circuit High (Bank 1):** Monitor runs whenever following DTCs are not present: None IMRC valve position sensor voltage: More than 4.8 V
DTC: P2102 **1T ECM, MIL: Yes** **Year:** 2011, 2012 **Model:** 4Runner, Avalon, Camry, Corolla, FJ Cruiser, Highlander, Land Cruiser, Matrix, Prius, Rav4, Sequoia, Tundra, Tacoma, Venza, Yaris **Engine:** L4, V6, V8	**Throttle Actuator Control Motor Circuit Low:** Conditions (a) and (b) continue for 2.0 seconds.: (a) Throttle actuator duty ratio 80 % or more (b) Throttle actuator current 0.5 A or less
DTC: P2103 **1T ECM, MIL: Yes** **Year:** 2011, 2012 **Model:** 4Runner, Avalon, Camry, Corolla, FJ Cruiser, Highlander, Land Cruiser, Matrix, Prius, Rav4, Sequoia, Tundra, Tacoma, Venza, Yaris **Engine:** L4, V6, V8	**Throttle Actuator Control Motor Circuit High:** Monitor runs whenever following DTCs are not present: None Throttle actuator: ON Either of the following conditions 1 or 2 is met: - 1. Throttle actuator power supply: 8 V or higher 2. Throttle actuator power: ON

DTC	Trouble Code Title and Conditions
DTC: P2109 **1T ECM, MIL: Yes** **Year:** 2011, 2012 **Model:** 4Runner, Avalon, Camry, Corolla, FJ Cruiser, Highlander, Land Cruiser, Matrix, Prius, Rav4, Sequoia, Tundra, Tacoma, Venza, Yaris **Engine:** L4, V6, V8	**Throttle / Pedal Position Sensor "A" Minimum Stop Performance:** Either condition is met: 1). With atmospheric pressure 85 kPa (638 mmHg) or higher (elevation 1400 m (4592 ft.) or less), when the engine coolant temperature is 45°C (113°F) or less at engine start, the engine is warmed up and conditions for ISC learned are met, the ISC learning value is approximately 3 times larger than normal even though the intake manifold pressure when idling is normal. 2) With atmospheric pressure 85 kPa (638 mmHg) or higher (elevation 1400 m (4592 ft.) or less), when the ignition switch has been turned to ON for 1 hour or more, the engine is warmed up, conditions for ISC learned are met and the vehicle has been driven at a speed of 30 km/h (19 mph) or more at least once, the ISC learning value is approximately 3 times larger than normal even though the intake manifold pressure when idling is normal.
DTC: P2111 **1T ECM, MIL: Yes** **Year:** 2011, 2012 **Model:** 4Runner, Avalon, Camry, Corolla, FJ Cruiser, Highlander, Land Cruiser, Matrix, Prius, Rav4, Sequoia, Tundra, Tacoma, Venza, Yaris **Engine:** L4, V6, V8	**Throttle Actuator Control System - Stuck Open:** Monitor runs whenever following DTCs are not present: None All of the following conditions are met: 1). System guard* judge condition: ON 2). Throttle actuator current: 2 A or more 3). Duty-cycle to close throttle: 80% or more
DTC: P2112 **1T ECM, MIL: Yes** **Year:** 2011, 2012 **Model:** 4Runner, Avalon, Camry, Corolla, FJ Cruiser, Highlander, Land Cruiser, Matrix, Prius, Rav4, Sequoia, Tundra, Tacoma, Venza, Yaris **Engine:** L4, V6, V8	**THROTTLE ACTUATOR CONTROL SYSTEM - STUCK CLOSED:** ECM signals throttle actuator to open, but it is stuck.
DTC: P2118 **1T ECM, MIL: Yes** **Year:** 2011, 2012 **Model:** 4Runner, Avalon, Camry, Corolla, FJ Cruiser, Highlander, Land Cruiser, Matrix, Prius, Rav4, Sequoia, Tundra, Tacoma, Venza, Yaris **Engine:** L4, V6, V8	**Throttle Actuator Control Motor Current Range / Performance:** Monitor runs whenever following DTCs not stored: None Battery voltage: 8 V or higher Electronic throttle actuator power: ON
DTC: P2119 **1T ECM, MIL: Yes** **Year:** 2011, 2012 **Model:** 4Runner, Avalon, Camry, Corolla, FJ Cruiser, Highlander, Land Cruiser, Matrix, Prius, Rav4, Sequoia, Tundra, Tacoma, Venza, Yaris **Engine:** L4, V6, V8	**Throttle Actuator Control Throttle Body Range / Performance:** Monitor runs whenever following DTCs not stored: None System guard* judge condition: ON Throttle valve opening angle continues to vary greatly from target opening angle.
DTC: P2120 **1T ECM, MIL: Yes** **Year:** 2011, 2012 **Model:** 4Runner, Avalon, Camry, Corolla, FJ Cruiser, Highlander, Land Cruiser, Matrix, Prius, Rav4, Sequoia, Tundra, Tacoma, Venza, Yaris **Engine:** L4, V6, V8	**Throttle / Pedal Position Sensor / Switch "D" Circuit:** Monitor runs whenever following DTCs not stored: None Either of following conditions met: 1. Ignition switch: ON (0.012 seconds or more) 2. Throttle actuator power: ON VPA1 fluctuates rapidly beyond upper and lower malfunction thresholds for 0.5 seconds or more.

DTC	Trouble Code Title and Conditions
DTC: P2121 **1T ECM, MIL: Yes** **Year:** 2011, 2012 **Model:** 4Runner, Avalon, Camry, Corolla, FJ Cruiser, Highlander, Land Cruiser, Matrix, Prius, Rav4, Sequoia, Tundra, Tacoma, Venza, Yaris **Engine:** L4, V6, V8	**Throttle / Pedal Position Sensor / Switch "D" Circuit Range / Performance:** Monitor runs whenever following DTCs not stored: None Either of following conditions met: 1. Ignition switch: ON (0.012 seconds or more) 2. Electronic throttle actuator power: ON Difference between VPA and VPA2 exceeds threshold for 0.5 seconds.
DTC: P2121-160 **T ECM, MIL: Yes** **Year:** 2011, 2012 **Model:** Camry, Prius, Prius V **Engine:** 1.8L L4, 2.4L L4	**Throttle / Pedal Position Sensor / Switch "D" Circuit Range / Performance:** Internal error of the main sensor.
DTC: P2122 **1T ECM, MIL: Yes** **Year:** 2011, 2012 **Model:** 4Runner, Avalon, Camry, Corolla, FJ Cruiser, Highlander, Land Cruiser, Matrix, Prius, Rav4, Sequoia, Tundra, Tacoma, Venza, Yaris **Engine:** L4, V6, V8	**Throttle / Pedal Position Sensor / Switch "D" Circuit Low Input:** Monitor runs whenever following DTCs not stored: None Either of following conditions 1 or 2 met: 1. Ignition switch: ON 2. Throttle actuator power: ON VPA 0.4 V or less for 0.5 seconds or more when accelerator pedal fully released.
DTC: P2122-104 **T ECM, MIL: Yes** **Year:** 2011, 2012 **Model:** Camry, Prius, Prius V **Engine:** 1.8L L4, 2.4L L4	**Throttle / Pedal Position Sensor / Switch "D" Circuit Low Input:** Open or short to GND in the main sensor circuit.
DTC: P2123 **1T ECM, MIL: Yes** **Year:** 2011, 2012 **Model:** 4Runner, Avalon, Camry, Corolla, FJ Cruiser, Highlander, Land Cruiser, Matrix, Prius, Rav4, Sequoia, Tundra, Tacoma, Venza, Yaris **Engine:** L4, V6, V8	**Throttle / Pedal Position Sensor / Switch "D" Circuit High Input:** Monitor runs whenever following DTCs are not present: None VPA 4.8 V or more for 2.0 seconds or more.
DTC: P2123-105 **T ECM, MIL: Yes** **Year:** 2011, 2012 **Model:** Camry, Prius, Prius V **Engine:** 1.8L L4, 2.4L L4	**Throttle / Pedal Position Sensor / Switch "D" Circuit High Input:** Short to +B in the main sensor circuit.
DTC: P2125 **1T ECM, MIL: Yes** **Year:** 2011, 2012 **Model:** 4Runner, Avalon, Camry, Corolla, FJ Cruiser, Highlander, Land Cruiser, Matrix, Prius, Rav4, Sequoia, Tundra, Tacoma, Venza, Yaris **Engine:** L4, V6, V8	**Throttle / Pedal Position Sensor / Switch "E" Circuit:** Monitor runs whenever following DTCs not stored: None Either of following conditions 1 or 2 met: 1. Ignition switch: ON 2. Throttle actuator power: ON VPA2 1.2 V or less for 0.5 seconds or more when accelerator pedal depressed.
DTC: P2125-153 **T ECM, MIL: Yes** **Year:** 2011, 2012 **Model:** Camry, Prius, Prius V **Engine:** 1.8L L4, 2.4L L4	**Throttle / Pedal Position Sensor / Switch "E" Circuit:** Sub sensor circuit wiring malfunction or level is not stable.

DTC	Trouble Code Title and Conditions
DTC: P2126-109 **T ECM, MIL: Yes** **Year:** 2011, 2012 **Model:** Camry, Prius, Prius V **Engine:** 1.8L L4, 2.4L L4	**Throttle / Pedal Position Sensor / Switch "E" Circuit Range / Performance:** Internal error of the sub sensor.
DTC: P2127 **1T ECM, MIL: Yes** **Year:** 2011, 2012 **Model:** 4Runner, Avalon, Camry, Corolla, FJ Cruiser, Highlander, Land Cruiser, Matrix, Prius, Rav4, Sequoia, Tundra, Tacoma, Venza, Yaris **Engine:** L4, V6, V8	**Throttle / Pedal Position Sensor / Switch "E" Circuit Low Input:** Monitor runs whenever following DTCs are not present: None VPA2 1.2 V or less for 0.5 seconds or more when accelerator pedal fully released
DTC: P2127-109 **T ECM, MIL: Yes** **Year:** 2011, 2012 **Model:** Camry, Prius, Prius V **Engine:** 1.8L L4, 2.4L L4	**Throttle / Pedal Position Sensor / Switch "E" Circuit Low Input:** Open or short to GND in the sub sensor circuit.
DTC: P2128 **1T ECM, MIL: Yes** **Year:** 2011, 2012 **Model:** 4Runner, Avalon, Camry, Corolla, FJ Cruiser, Highlander, Land Cruiser, Matrix, Prius, Rav4, Sequoia, Tundra, Tacoma, Venza, Yaris **Engine:** L4, V6, V8	**THROTTLE/PEDAL POSITION SENSOR/SWITCH "E" CIRCUIT HIGH INPUT:** Conditions (a) and (b) continue for 2.0 seconds or more (1 trip detection logic): (a) VPA2 is 4.8 V or more (b) VPA is between 0.4 V and 3.45 V
DTC: P2128-108 **T ECM, MIL: Yes** **Year:** 2011, 2012 **Model:** Camry, Prius, Prius V **Engine:** 1.8L L4, 2.4L L4	**Throttle / Pedal Position Sensor / Switch "E" Circuit High Input:** Short to +B in the sub sensor circuit.
DTC: P2135 **1T ECM, MIL: Yes** **Year:** 2011, 2012 **Model:** 4Runner, Avalon, Camry, Corolla, FJ Cruiser, Highlander, Land Cruiser, Matrix, Prius, Rav4, Sequoia, Tundra, Tacoma, Venza, Yaris **Engine:** L4, V6, V8	**Throttle / Pedal Position Sensor / Switch "A" / "B" Voltage Correlation:** Monitor runs whenever following DTCs are not present: None Either of the following conditions A or B is met: - A. Power switch on (IG): 0.012 seconds or more B. Electronic throttle actuator power: ON
DTC: P2138 **1T ECM, MIL: Yes** **Year:** 2011, 2012 **Model:** 4Runner, Avalon, Camry, Corolla, FJ Cruiser, Highlander, Land Cruiser, Matrix, Prius, Rav4, Sequoia, Tundra, Tacoma, Venza, Yaris **Engine:** L4, V6, V8	**Throttle / Pedal Position Sensor / Switch "D" / "E" Voltage Correlation:** Monitor runs whenever following DTCs not stored: None Either of following conditions 1 or 2 met: a) Difference between VPA and VPA2 0.02 V or less b) VPA 0.4 V or less and VPA2 1.2 V or less
DTC: P2138-110 **T ECM, MIL: Yes** **Year:** 2011, 2012 **Model:** Camry, Prius, Prius V **Engine:** 1.8L L4, 2.4L L4	**Throttle / Pedal Position Sensor / Switch "D" / "E" Voltage Correlation:** Difference between the main sensor value and sub sensor value is large.

DTC	Trouble Code Title and Conditions
DTC: P2138-154 **T ECM, MIL: Yes** **Year:** 2011, 2012 **Model:** Camry, Prius, Prius V **Engine:** 1.8L L4, 2.4L L4	**Throttle / Pedal Position Sensor / Switch "D" / "E" Voltage Correlation:** Main or sub sensor circuit wiring malfunction.
DTC: P2195 **1T ECM, MIL: Yes** **Year:** 2011, 2012 **Model:** 4Runner, Avalon, Camry, Corolla, FJ Cruiser, Highlander, Land Cruiser, Matrix, Prius, Rav4, Sequoia, Tundra, Tacoma, Venza, Yaris **Engine:** L4, V6, V8	**Oxygen (A/F) Sensor Signal Stuck Lean (Bank 1 Sensor 1):** Conditions (a) and (b) continue for 2 seconds or more: (a) Air-Fuel Ratio (A/F) sensor voltage is more than 3.8 V (b) Heated Oxygen (HO2) sensor voltage is 0.15 V or more. While fuel-cut operation is performed (during vehicle deceleration), air-furl ratio (A/F) sensor current is 3.6 mA or more for 3 seconds (2 trip detection logic)
DTC: P2196 **2T ECM, MIL: Yes** **Year:** 2011, 2012 **Model:** 4Runner, Avalon, Camry, Corolla, FJ Cruiser, Highlander, Land Cruiser, Matrix, Prius, Rav4, Sequoia, Tundra, Tacoma, Venza, Yaris **Engine:** L4, V6, V8	**Oxygen (A/F) Sensor Signal Stuck Rich (Bank 1 Sensor 1):** Conditions (a) and (b) continue for 10 seconds or more: (a) Air-Furl Ratio (A/F) sensor voltage less than 2.8 V (b) Heated Oxygen (HO2) sensor voltage less than 0.60 V While fuel-cut operation performed (during vehicle deceleration), Air-Furl Ratio (A/F) sensor current less than 1.0 mA for 3 seconds.
DTC: P2197 **2T ECM, MIL: Yes** **Year:** 2011, 2012 **Model:** Camry **Engine:** 3.5L V6	**Oxygen (A/F) Sensor Signal Stuck Lean (Bank 2 Sensor 1):** The monitor will run whether these DTCs are not present: P0016, P0018 (VVT system bank 1, 2 - misalignment) P0017, P0019 (Exhaust VVT system bank 1, 2 - misalignment) P0031, P0032, P0051, P0052 (A/F sensor heater) P0102, P0103 (MAF meter) P0112, P0113 (IAT sensor) P0115, P0117, P0118 (ECT sensor) P0120, P0121, P0122, P0123, P0220, P0222, P0223, P2135 (TP sensor) P0125 (Insufficient ECT for closed loop) P0128 (Thermostat) P0171, P0172, P0174, P0175 (Fuel system) P0301, P0302, P0303, P0304, P0305, P0306 (Misfire) P0335 (CKP sensor) P0451, P0452, P0453 (EVAP system) P0500 (Vehicle speed sensor) P0505 (Idle speed control) Sensor voltage detection monitor: Time after engine start: 30 seconds or more A/F sensor status: Activated Fuel system status: Closed-loop Sensor current detection monitor: Battery voltage: 11 V or more ECT: 75°C (167°F) or more Atmospheric pressure: 76 kPa (570 mmHg) or more Air-fuel ratio sensor status: Activated Continuous time of fuel cut: 4 to 10 seconds **Possible causes:** • Open or short in A/F sensor (bank 1, 2 sensor 1) circuit • A/F sensor (bank 1, 2 sensor 1) • A/F sensor (bank 1, 2 sensor 1) heater • A/F sensor heater and relay circuits • Air induction system • Fuel pressure (low pressure side) • Fuel pressure (high pressure side) • Injector • Fuel pressure sensor • ECM

DTC	Trouble Code Title and Conditions
DTC: P2197 **1T ECM, MIL: Yes** **Year:** 2011, 2012 **Model:** 4Runner, Avalon, Camry, Corolla, FJ Cruiser, Highlander, Land Cruiser, Matrix, Prius, Rav4, Sequoia, Tundra, Tacoma, Venza, Yaris **Engine:** L4, V6, V8	**Oxygen (A/F) Sensor Signal Stuck Lean (Bank 2 Sensor 1):** Conditions (a) and (b) continue for 2 seconds or more (2 trip detection logic): (a) Air-Fuel Ratio (A/F) sensor voltage is more than 3.8 V (b) Heated Oxygen (HO2) sensor voltage is 0.15 V or more While fuel-cut operation is performed (during vehicle deceleration), air-furl ratio (A/F) sensor current is 3.6 mA or more for 3 seconds (2 trip detection logic)
DTC: P2198 **1T ECM, MIL: Yes** **Year:** 2011, 2012 **Model:** 4Runner, Avalon, Camry, Corolla, FJ Cruiser, Highlander, Land Cruiser, Matrix, Prius, Rav4, Sequoia, Tundra, Tacoma, Venza, Yaris **Engine:** L4, V6, V8	**Oxygen (A/F) Sensor Signal Stuck Rich (Bank 2 Sensor 1):** Conditions (a) and (b) continue for 2 seconds or more (2 trip detection logic): (a) A/F sensor voltage is less than 2.8 V (b) HO2 sensor voltage is less than 0.6 V While fuel-cut operation is performed (during vehicle deceleration), air-furl ratio (A/F) sensor current is less than 1.4 mA for 3 seconds (2 trip detection logic)
DTC: P2237 **1T ECM, MIL: Yes** **Year:** 2011, 2012 **Model:** 4Runner, Avalon, Camry, Corolla, FJ Cruiser, Highlander, Land Cruiser, Matrix, Prius, Rav4, Sequoia, Tundra, Tacoma, Venza, Yaris **Engine:** L4, V6, V8	**Oxygen (A/F) Sensor Pumping Current Circuit / Open (Bank 1 Sensor 1):** An open in the circuit between terminals A1A+/A2A+ and A1A-/A2A- of the A/F sensor while the engine is running.
DTC: P2238 **1T ECM, MIL: Yes** **Year:** 2011, 2012 **Model:** 4Runner, Avalon, Camry, Corolla, FJ Cruiser, Highlander, Land Cruiser, Matrix, Prius, Rav4, Sequoia, Tundra, Tacoma, Venza, Yaris **Engine:** L4, V6, V8	**Oxygen (A/F) Sensor Pumping Current Circuit Low (Bank 1 Sensor 1):** Case 1: Condition (a) or (b) continues for 5.0 seconds or more (1 trip detection logic): (a) AF+ voltage is 0.5 V or less (b) (AF+) - (AF-) = 0.1 V or less Case 2: A/F sensor admittance: Less than 0.022 1/O. (2 trip detection logic)
DTC: P2239 **1T ECM, MIL: Yes** **Year:** 2011, 2012 **Model:** 4Runner, Avalon, Camry, Corolla, FJ Cruiser, Highlander, Land Cruiser, Matrix, Prius, Rav4, Sequoia, Tundra, Tacoma, Venza, Yaris **Engine:** L4, V6, V8	**Oxygen (A/F) Sensor Pumping Current Circuit High (Bank 1 Sensor 1):** With the battery voltage 11 V or more and the ignition switch ON. Time after ignition switch is OFF to ON 5 seconds or more. The A1A+ voltage is more than 4.5 V.
DTC: P2240 **1T ECM, MIL: Yes** **Year:** 2011, 2012 **Model:** 4Runner, Avalon, Camry, Corolla, FJ Cruiser, Highlander, Land Cruiser, Matrix, Prius, Rav4, Sequoia, Tundra, Tacoma, Venza, Yaris **Engine:** L4, V6, V8	**Oxygen (A/F) Sensor Pumping Current Circuit / Open (Bank 2 Sensor 1):** Open in the circuit between terminals AF+ and AF- of the A/F sensor while running.

DTC	Trouble Code Title and Conditions
DTC: P2241 **1T ECM, MIL: Yes** **Year:** 2011, 2012 **Model:** 4Runner, Avalon, Camry, Corolla, FJ Cruiser, Highlander, Land Cruiser, Matrix, Prius, Rav4, Sequoia, Tundra, Tacoma, Venza, Yaris **Engine:** L4, V6, V8	**Oxygen (A/F) Sensor Pumping Current Circuit Low (Bank 2 Sensor 1):** Case 1: Condition (a) or (b) continues for 5.0 seconds or more (2 trip detection logic): (a) AF+ voltage 0.5 V or less (b) (AF+) - (AF-) = 0.1 V or less. Case 2: A/F sensor admittance: Less than 0.022 1/O.
DTC: P2242 **1T ECM, MIL: Yes** **Year:** 2011, 2012 **Model:** 4Runner, Avalon, Camry, Corolla, FJ Cruiser, Highlander, Land Cruiser, Matrix, Prius, Rav4, Sequoia, Tundra, Tacoma, Venza, Yaris **Engine:** L4, V6, V8	**Oxygen (A/F) Sensor Pumping Current Circuit High (Bank 2 Sensor 1):** Short Circuit between AF+ and +B: Battery voltage: 11 V or higher Ignition switch: ON Time after ignition switch is turned from off to ON: 5 seconds or more Short Circuit between AF+ and VCC: Battery voltage: 11 V or higher Ignition switch: ON Time after ignition switch is turned from off to ON: 5 seconds or more. The A1A+/A2A+ voltage is higher than 4.5 V for 5.0 seconds or more.
DTC: P2252 **1T ECM, MIL: Yes** **Year:** 2011, 2012 **Model:** 4Runner, Avalon, Camry, Corolla, FJ Cruiser, Highlander, Land Cruiser, Matrix, Prius, Rav4, Sequoia, Tundra, Tacoma, Venza, Yaris **Engine:** L4, V6, V8	**Oxygen (A/F) Sensor Reference Ground Circuit Low (Bank 1 Sensor 1):** Monitor runs whenever following DTCs are not present: None Battery voltage: 11 V or more Power switch: On (IG) Time after power switch is off to on (IG): 5 seconds or more AF- voltage 0.5 V or less for 5.0 seconds or more
DTC: P2253 **2T ECM, MIL: Yes** **Year:** 2011, 2012 **Model:** 4Runner, Avalon, Camry, Corolla, FJ Cruiser, Highlander, Land Cruiser, Matrix, Prius, Rav4, Sequoia, Tundra, Tacoma, Venza, Yaris **Engine:** L4, V6, V8	**Oxygen (A/F) Sensor Reference Ground Circuit High (Bank 1 Sensor 1):** AF- voltage more than 4.5 V for 5.0 seconds or more (2 trip detection logic) Time after ignition switch is turned from off to ON: 5 seconds or more
DTC: P2255 **2T ECM, MIL: Yes** **Year:** 2011, 2012 **Model:** 4Runner, Avalon, Camry, Corolla, FJ Cruiser, Highlander, Land Cruiser, Matrix, Prius, Rav4, Sequoia, Tundra, Tacoma, Venza, Yaris **Engine:** L4, V6, V8	**Oxygen (A/F) Sensor Reference Ground Circuit Low (Bank 2 Sensor 1):** Short Circuit between AF- and GND: Battery voltage: 11 V or higher Ignition switch: ON Time after ignition switch is turned from off to ON:5 seconds or more
DTC: P2256 **2T ECM, MIL: Yes** **Year:** 2011, 2012 **Model:** 4Runner, Avalon, Camry, Corolla, FJ Cruiser, Highlander, Land Cruiser, Matrix, Prius, Rav4, Sequoia, Tundra, Tacoma, Venza, Yaris **Engine:** L4, V6, V8	**Oxygen (A/F) Sensor Reference Ground Circuit High (Bank 2 Sensor 1):** The A1A-/A2A- voltage is higher than 4.5 V for 5.0 seconds or more.

DTC	Trouble Code Title and Conditions
DTC: P2401 **2T ECM, MIL: Yes** **Year:** 2011, 2012 **Model:** 4Runner, Avalon, Camry, Corolla, FJ Cruiser, Highlander, Land Cruiser, Matrix, Prius, Rav4, Sequoia, Tundra, Tacoma, Venza, Yaris **Engine:** L4, V6, V8	**Evaporative Emission Leak Detection Pump Stuck OFF:** Monitor runs whenever following DTCs not present: P0451, P0452, P0453 (EVAP System) Atmospheric pressure: 70 to 110 kPa-a (525 to 825 mmHg-a) Battery voltage: 10.5 V or higher Vehicle speed: Less than 4 km/h (2.5 mph) Ignition switch: OFF Time after key-off: 5, 7 or 9.5 hours Both of following conditions met before key-off: Conditions 1 and 2 1. Duration that vehicle driven: 5 minutes or more 2. EVAP purge operation: Performed ECT: 4.4 to 35°C (40 to 95°F) IAT: 4.4 to 35°C (40 to 95°F)
DTC: P2402 **2T ECM, MIL: Yes** **Year:** 2011, 2012 **Model:** 4Runner, Avalon, Camry, Corolla, FJ Cruiser, Highlander, Land Cruiser, Matrix, Prius, Rav4, Sequoia, Tundra, Tacoma, Venza, Yaris **Engine:** L4, V6, V8	**Evaporative Emission System Leak Detection Pump Control Circuit High Stuck ON:** Monitor runs whenever following DTCs not present: None EVAP key-off monitor runs when all of the following conditions are met: Atmospheric pressure: 70 to 110 kPa-a (525 to 825 mmHg-a) Battery voltage: 10.5 V or more Vehicle speed: Below 2.5 mph (4 km/h) Engine switch: Off Time after key off: 5 or 7 or 9.5 hours EVAP pressure sensor malfunction (P0450, P0451, P0452 and P0453): Not detected EVAP canister purge valve: Not operated by scan tool EVAP canister vent valve: Not operated by scan tool EVAP leak detection pump: Not operated by scan tool Both of the following conditions are met before key off: Conditions 1 and 2 1. Duration that vehicle has been driven: 5 minutes or more 2. EVAP purge operation: Performed ECT: 4.4 to 35°C (40 to 95°F) IAT: 4.4 to 35°C (40 to 95°F)
DTC: P2419 **2T ECM, MIL: Yes** **Year:** 2011, 2012 **Model:** 4Runner, Avalon, Camry, Corolla, FJ Cruiser, Highlander, Land Cruiser, Matrix, Prius, Rav4, Sequoia, Tundra, Tacoma, Venza, Yaris **Engine:** L4, V6, V8	**Evaporative Emission System Switching Valve Control Circuit Low:** Monitor runs whenever following DTCs are not present: None Atmospheric pressure: 70 to 110 kPa-a (525 to 825 mmHg-a) Battery voltage: 10.5 V or more Vehicle speed: Below 2.5 mph (4 km/h) Power switch: OFF Time after key off: 5, 7 or 9.5 hours Canister pressure sensor malfunction (P0452 and P0453): Not detected Purge VSV: Not operated by scan tool Vent valve: Not operated by scan tool Leak detection pump: Not operated by scan tool Both of the following conditions are met before key off: Conditions 1 and 2 1. Duration that vehicle driven: 5 minutes or more 2. EVAP purge operation: Performed Engine coolant temperature: 4.4 to 35°C (40 to 95°F) Intake air temperature: 4.4 to 35°C (40 to 95°F)
DTC: P2420 **2T ECM, MIL: Yes** **Year:** 2011, 2012 **Model:** 4Runner, Avalon, Camry, Corolla, FJ Cruiser, Highlander, Land Cruiser, Matrix, Prius, Rav4, Sequoia, Tundra, Tacoma, Venza, Yaris **Engine:** L4, V6, V8	**Evaporative Emission System Switching Valve Control Circuit High:** Leak detection pump creates negative pressure through reference orifice and EVAP system pressure measured to determine leak criterion. 0.02 inch leak criterion measured at start and at end of leak check. If system pressure does not increase by more than 0.3 kPa (2.25 mmHg) within 10 seconds when vent valve turned ON, ECM determines that vent valve stuck close.

DTC	Trouble Code Title and Conditions
DTC: P2430 **1T ECM, MIL: Yes** **Year:** 2011, 2012 **Model:** 4Runner, Avalon, Camry, Corolla, FJ Cruiser, Highlander, Land Cruiser, Matrix, Prius, Rav4, Sequoia, Tundra, Tacoma, Venza, Yaris **Engine:** L4, V6, V8	**Secondary Air Injection System Air Flow / Pressure Sensor Circuit Bank1:** While engine running, voltage output of pressure sensor indicates 0.1 V or less, or indicates 4.8V or more.
DTC: P2431 **2T ECM, MIL: Yes** **Year:** 2011, 2012 **Model:** 4Runner, Avalon, Camry, Corolla, FJ Cruiser, Highlander, Land Cruiser, Matrix, Prius, Rav4, Sequoia, Tundra, Tacoma, Venza, Yaris **Engine:** L4, V6, V8	**Secondary Air Injection System Air Flow / Pressure Sensor Circuit Range / Performance Bank1:** The pressure sensor indicates below 45.6 kPa (342 mmHg), or higher than 135 kPa (1013 mmHg).
DTC: P2432 **2T ECM, MIL: Yes** **Year:** 2011, 2012 **Model:** 4Runner, Avalon, Camry, Corolla, FJ Cruiser, Highlander, Land Cruiser, Matrix, Prius, Rav4, Sequoia, Tundra, Tacoma, Venza, Yaris **Engine:** L4, V6, V8	**Secondary Air Injection System Air Flow / Pressure Sensor Circuit Low Bank1:** While the engine is running, the voltage output of the pressure sensor remains below 0.5 V.
DTC: P2433 **1T ECM, MIL: Yes** **Year:** 2011, 2012 **Model:** 4Runner, Avalon, Camry, Corolla, FJ Cruiser, Highlander, Land Cruiser, Matrix, Prius, Rav4, Sequoia, Tundra, Tacoma, Venza, Yaris **Engine:** L4, V6, V8	**Secondary Air Injection System Air Flow / Pressure Sensor Circuit High Bank1:** While engine running, voltage output of pressure sensor remains above 4.8V.
DTC: P2436 **1T ECM, MIL: Yes** **Year:** 2011, 2012 **Model:** 4Runner, Avalon, Camry, Corolla, FJ Cruiser, Highlander, Land Cruiser, Matrix, Prius, Rav4, Sequoia, Tundra, Tacoma, Venza, Yaris **Engine:** L4, V6, V8	**Secondary Air Injection System Air Flow / Pressure Sensor Circuit Range / Performance Bank 2:** The pressure sensor indicates below 45.6 kPa (342 mmHg), or higher than 135 kPa (1013 mmHg).
DTC: P2437 **1T ECM, MIL: Yes** **Year:** 2011, 2012 **Model:** 4Runner, Avalon, Camry, Corolla, FJ Cruiser, Highlander, Land Cruiser, Matrix, Prius, Rav4, Sequoia, Tundra, Tacoma, Venza, Yaris **Engine:** L4, V6, V8	**Secondary Air Injection System Air Flow / Pressure Sensor Circuit Low Bank 2:** While the engine is running, the voltage output of the pressure sensor remains below 0.5 V.

DTC	Trouble Code Title and Conditions
DTC: P2438 **1T ECM, MIL: Yes** **Year:** 2011, 2012 **Model:** 4Runner, Avalon, Camry, Corolla, FJ Cruiser, Highlander, Land Cruiser, Matrix, Prius, Rav4, Sequoia, Tundra, Tacoma, Venza, Yaris **Engine:** L4, V6, V8	**Secondary Air Injection System Air Flow / Pressure Sensor Circuit High Bank 2:** While the engine is running, the voltage output of the pressure sensor remains higher than 4.5 V.
DTC: P2440 **1T ECM, MIL: Yes** **Year:** 2011, 2012 **Model:** 4Runner, Avalon, Camry, Corolla, FJ Cruiser, Highlander, Land Cruiser, Matrix, Prius, Rav4, Sequoia, Tundra, Tacoma, Venza, Yaris **Engine:** L4, V6, V8	**Secondary Air Injection System Switching Valve Stuck Open Bank1:** Pressure sensor detects pulsation of exhaust gas despite ECM commanding Air Switching Valve (ASV) to close, while engine running.
DTC: P2441 **1T ECM, MIL: Yes** **Year:** 2011, 2012 **Model:** 4Runner, Avalon, Camry, Corolla, FJ Cruiser, Highlander, Land Cruiser, Matrix, Prius, Rav4, Sequoia, Tundra, Tacoma, Venza, Yaris **Engine:** L4, V6, V8	**Secondary Air Injection System Switching Valve Stuck Close (Bank 1):** The pressure sensor detects no pulsation of the exhaust gas despite the ECM commanding the Air Switching Valve (ASV) to open, while the engine is running.
DTC: P2442 **1T ECM, MIL: Yes** **Year:** 2011, 2012 **Model:** 4Runner, Avalon, Camry, Corolla, FJ Cruiser, Highlander, Land Cruiser, Matrix, Prius, Rav4, Sequoia, Tundra, Tacoma, Venza, Yaris **Engine:** L4, V6, V8	**Secondary Air Injection System Switching Valve Stuck Open (Bank 2):** The pressure sensor detects pulsation of the exhaust gas despite the ECM commanding the Air Switching Valve (ASV) to close, while the engine is running.
DTC: P2443 **2T ECM, MIL: Yes** **Year:** 2011, 2012 **Model:** 4Runner, Avalon, Camry, Corolla, FJ Cruiser, Highlander, Land Cruiser, Matrix, Prius, Rav4, Sequoia, Tundra, Tacoma, Venza, Yaris **Engine:** L4, V6, V8	**Secondary Air Injection System Switching Valve Stuck Closed Bank 2:** Pressure sensor detects no pulsation of exhaust gas despite ECM commanding Air Switching Valve (ASV) to open, while engine running.
DTC: P2444 **2T ECM, MIL: Yes** **Year:** 2011, 2012 **Model:** 4Runner, Avalon, Camry, Corolla, FJ Cruiser, Highlander, Land Cruiser, Matrix, Prius, Rav4, Sequoia, Tundra, Tacoma, Venza, Yaris **Engine:** L4, V6, V8	**Secondary Air Injection System Pump Stuck On (Bank 1):** The secondary air pressure is higher than 2.4 kPa (18 mmHg) despite the ECM commanding the air pump to turn off.

DTC	Trouble Code Title and Conditions
DTC: P2445 **2T ECM, MIL: Yes** **Year:** 2011, 2012 **Model:** 4Runner, Avalon, Camry, Corolla, FJ Cruiser, Highlander, Land Cruiser, Matrix, Prius, Rav4, Sequoia, Tundra, Tacoma, Venza, Yaris **Engine:** L4, V6, V8	**SECONDARY AIR INJECTION SYSTEM PUMP STUCK OFF BANK1:** Secondary air pressure less than 2.5 kPa (19 mmHg) despite ECM commanding air pump to turn on.
DTC: P2446 **2T ECM, MIL: Yes** **Year:** 2011, 2012 **Model:** 4Runner, Avalon, Camry, Corolla, FJ Cruiser, Highlander, Land Cruiser, Matrix, Prius, Rav4, Sequoia, Tundra, Tacoma, Venza, Yaris **Engine:** L4, V6, V8	**Secondary Air Injection System Pump Stuck On Bank 2:** Secondary air pressure more than 5 kPa (37.5 mmHg) despite ECM commanding air pump to turn off.
DTC: P2447 **2T ECM, MIL: Yes** **Year:** 2011, 2012 **Model:** 4Runner, Avalon, Camry, Corolla, FJ Cruiser, Highlander, Land Cruiser, Matrix, Prius, Rav4, Sequoia, Tundra, Tacoma, Venza, Yaris **Engine:** L4, V6, V8	**Secondary Air Injection System Pump Stuck Off (Bank 2):** The secondary air pressure is below 2.4 kPa (18 mmHg) despite the ECM commanding the air pump to turn on.
DTC: P2450 **2T ECM, MIL: Yes** **Year:** 2011, 2012 **Model:** 4Runner, Avalon, Camry, Corolla, FJ Cruiser, Highlander, Land Cruiser, Matrix, Prius, Rav4, Sequoia, Tundra, Tacoma, Venza, Yaris **Engine:** L4, V6, V8	**Fuel Vapor-containment Valve Stuck Open:** Monitor runs whenever these DTCs are not present: None EVAP key-off monitor runs when all of the following conditions are met: - Atmospheric pressure: 70 to 111 kPa (525 to 833 mmHg) Battery voltage: 10.5 V or more Vehicle speed: Less than 2.5 mph (4 km/h) Power switch: OFF Engine condition: Not running Key-OFF duration: 5 or 7 or 9.5 hours Pressure sensor of canister pump module malfunction (P0452 and P0453): Not detected Fuel tank pressure sensor malfunction (P1452 and P1453): Not detected Both of the following conditions are met before Key-OFF: Conditions 1 and 2 1. Duration that vehicle is being driven: 5 minutes or more 2. Purge flow: Executed Engine coolant temperature: 4.4 to 35°C (40 to 95°F) Intake air temperature: 4.4 to 35°C (40 to 95°F)

DTC	Trouble Code Title and Conditions
DTC: P2451 **2T ECM, MIL: Yes** **Year:** 2011, 2012 **Model:** 4Runner, Avalon, Camry, Corolla, FJ Cruiser, Highlander, Land Cruiser, Matrix, Prius, Rav4, Sequoia, Tundra, Tacoma, Venza, Yaris **Engine:** L4, V6, V8	**Fuel Vapor-containment Valve Stuck Close:** Monitor runs whenever these DTCs are not present: None EVAP key-off monitor runs when all of the following conditions are met: Atmospheric pressure: 70 to 111 kPa (525 to 833 mmHg) Battery voltage: 10.5 V or more Vehicle speed: Less than 2.5 mph (4 km/h) Power switch: OFF Engine condition: Not running Key-OFF duration: 5 or 7 or 9.5 hours Pressure sensor of canister pump module malfunction (P0452 and P0453): Not detected Both of the following conditions are met before Key-OFF: Conditions 1 and 2 1. Duration that vehicle is being driven: 5 minutes or more 2. Purge flow: Executed Engine coolant temperature: 4.4 to 35°C (40 to 95°F) Intake air temperature: 4.4 to 35°C (40 to 95°F)
DTC: P2511-149 **T ECM, MIL: Yes** **Year:** 2011, 2012 **Model:** Camry, Prius, Prius V **Engine:** 1.8L L4, 2.4L L4	**ECM/PCM Power Relay Sense Circuit Intermittent:** When the power switch is on (READY), the hybrid vehicle control ECU is reset.
DTC: P2519-766 **T ECM, MIL: Yes** **Year:** 2011, 2012 **Model:** Camry, Prius, Prius V **Engine:** 1.8L L4, 2.4L L4	**A/C Request "A" Circuit:** Malfunction in the cooling fan operation condition signal circuit.
DTC: P2532-772 **T ECM, MIL: Yes** **Year:** 2011, 2012 **Model:** Camry, Prius, Prius V **Engine:** 1.8L L4, 2.4L L4	**Ignition Switch Run Position Circuit High:** When no signals are received from the ECUs (skid control ECU and A/C amplifier) that are activated by the IG1 relay, but the hybrid vehicle control ECU is operating.
DTC: P2601-777 **T ECM, MIL: Yes** **Year:** 2011, 2012 **Model:** Camry, Prius, Prius V **Engine:** 1.8L L4, 2.4L L4	**Oil Pump Control Circuit Range / Performance:** Command signal (OPM1) from hybrid vehicle control ECU is abnormal.
DTC: P2601-778 **T ECM, MIL: Yes** **Year:** 2011, 2012 **Model:** Camry, Prius, Prius V **Engine:** 1.8L L4, 2.4L L4	**Oil Pump Control Circuit Range / Performance:** Oil pump fails to start more than set number of times and the temperature of the motor or generator exceeds load factor limit temperature when oil pump start request is accepted.
DTC: P2601-779 **T ECM, MIL: Yes** **Year:** 2011, 2012 **Model:** Camry, Prius, Prius V **Engine:** 1.8L L4, 2.4L L4	**Oil Pump Control Circuit Range / Performance:** Abnormal power source voltage is detected by oil pump with motor assembly Power source voltage input to oil pump with motor assembly is 6.96 V or less for 2 seconds or more with OIL PMP relay on.
DTC: P2602-767 **T ECM, MIL: Yes** **Year:** 2011, 2012 **Model:** Camry, Prius, Prius V **Engine:** 1.8L L4, 2.4L L4	**Oil Pump Control Circuit Low:** GND short in oil pump circuit When OPM2 signal is on, OPST signal remains low.
DTC: P2603-768 **T ECM, MIL: Yes** **Year:** 2011, 2012 **Model:** Camry, Prius, Prius V **Engine:** 1.8L L4, 2.4L L4	**Oil Pump Control Circuit High:** Open or +B short in oil pump circuit When OPM2 signal is on, OPST signal remains high

DTC	Trouble Code Title and Conditions
DTC: P2610 **2T ECM, MIL: Yes** **Year:** 2011, 2012 **Model:** 4Runner, Avalon, Camry, Corolla, FJ Cruiser, Highlander, Land Cruiser, Matrix, Prius, Rav4, Sequoia, Tundra, Tacoma, Venza, Yaris **Engine:** L4, V6, V8	**ECM / PCM Internal Engine Off Timer Performance:** Monitor runs whenever following DTC not stored: None Ignition switch: ON Engine: Running Battery voltage: 8 V or higher Starter: OFF An ECM internal malfunction was detected.
DTC: P261B **1T ECM** **Year:** 2011, 2012 **Model:** Prius, Prius V **Engine:** 1.8L L4	**Coolant Pump "B" Control Malfunction:** Monitor runs whenever following DTCs are not present: None Battery voltage: 8 V or more Power switch: On (IG) Time after power switch off to on (IG): 0.5 seconds or more
DTC: P261C **1T ECM** **Year:** 2011, 2012 **Model:** Prius, Prius V **Engine:** 1.8L L4	**Short in Coolant Pump "B" Control Circuit:** Monitor runs whenever following DTCs are not present: None Battery voltage: 8 V or more Power switch: On (IG) Time after power switch off to on (IG): 0.5 seconds or more Output signal duty ratio: 40 to 60% Engine coolant pump circuit performance fail (P261B): Not detected Engine coolant pump output terminal voltage monitor counter: 0.08 seconds or more
DTC: P261D **1T ECM** **Year:** 2011, 2012 **Model:** Prius, Prius V **Engine:** 1.8L L4	**Open in Coolant Pump "B" Control Circuit:** Monitor runs whenever following DTCs are not present: None Battery voltage: 8 V or more Power switch: On (IG) Time after power switch off to on (IG): 0.5 seconds or more Output signal duty ratio: 40 to 60% Engine coolant pump circuit performance fail (P261B): Not detected Engine coolant pump output terminal voltage monitor counter: 0.08 seconds
DTC: P2714 **2T ECM, MIL: Yes** **Year:** 2011, 2012 **Model:** 4Runner, Avalon, Camry, Corolla, FJ Cruiser, Highlander, Land Cruiser, Matrix, Prius, Rav4, Sequoia, Tundra, Tacoma, Venza, Yaris **Engine:** L4, V6, V8	**Pressure Control Solenoid "D" Performance (Shift Solenoid Valve SLT):** ECM detects malfunction on SLT (ON side) according to difference in revolutions of turbine (input) and output shaft (2 trip detection logic)
DTC: P2716 **1T ECM, MIL: Yes** **Year:** 2011, 2012 **Model:** 4Runner, Avalon, Camry, Corolla, FJ Cruiser, Highlander, Land Cruiser, Matrix, Prius, Rav4, Sequoia, Tundra, Tacoma, Venza, Yaris **Engine:** L4, V6, V8	**Pressure Control Solenoid "D" Electrical (Shift Solenoid Valve SLT):** The monitor will run whenever this DTC is not present: None. Conditions (a) or (b) below are detected for 1 sec. or more: (a) Shift solenoid valve SLT - terminal: 0 V (b) Shift solenoid valve SLT - terminal: 12 V Open or short is detected in shift solenoid valve SLT circuit for 1 second or more while driving.

DTC	Trouble Code Title and Conditions
DTC: P2740 **1T ECM, MIL: Yes** **Year:** 2011, 2012 **Model:** 4Runner, Avalon, Camry, Corolla, FJ Cruiser, Highlander, Land Cruiser, Matrix, Prius, Rav4, Sequoia, Tundra, Tacoma, Venza, Yaris **Engine:** L4, V6, V8	**Transmission Fluid Temperature Sensor "B" Circuit:** (a) and (b) are detected momentarily within 0.5 seconds when neither P2742 nor P2743 are detected. (a) No. 2 ATF temperature sensor resistance is less than 25 Ohm. (0.046 V). (b) No. 2 ATF temperature sensor resistance is more than 156 kOhm. (4.915 V). Within 0.5 seconds, the malfunction switches from (a) to (b) or from (b) to (a).
DTC: P2741 **1T ECM, MIL: Yes** **Year:** 2011, 2012 **Model:** 4Runner, Avalon, Camry, Corolla, FJ Cruiser, Highlander, Land Cruiser, Matrix, Prius, Rav4, Sequoia, Tundra, Tacoma, Venza, Yaris **Engine:** L4, V6, V8	**Transmission Fluid Temperature Sensor "B" Performance:** Either of the following Condition (A) or (B) is met: Condition (A): All of (a), (b) and (c) are detected (2-trip detection logic) (a) Intake air and engine coolant temperatures are more than -10°C (14°F) at engine start. (b) After normal driving for over 5 minutes and 5.6 miles (9 km) or more, ATF temperature is less than 20°C (68°F). (c) 19 minutes or more have elapsed after engine start. Condition (B): Both (a) and (b) are detected (2-trip detection logic) (a) Engine coolant temperature is less than 35°C (95°F) at engine start. (B) ATF temperature is 110°C (230°F) or more when engine coolant temperature reaches 60°C (140°F).
DTC: P2741 00 **1T TCM, MIL: Yes** **Year:** 2011, 2012 **Model:** 4Runner, FJ Cruiser **Engine:** 4.0L V6	**Transmission Fluid Temperature Sensor "B" Circuit:** The monitor will run whenever the following DTCs are not present: None Either of the following Condition (A) or (B) is met: Condition (A): All of (a), (b) and (c) are detected (2-trip detection logic) (a) Intake air and engine coolant temperatures are more than -10°C (14°F) at engine start. (b) After normal driving for over 5 minutes and 5.6 miles (9 km) or more, ATF temperature is less than 20°C (68°F). (c) 19 minutes or more have elapsed after engine start. Condition (B): Both (a) and (b) are detected (2-trip detection logic) (a) Engine coolant temperature is less than 35°C (95°F) at engine start. (B) ATF temperature is 110°C (230°F) or more when engine coolant temperature reaches 60°C (140°F).
DTC: P2742 **1T ECM, MIL: Yes** **Year:** 2011, 2012 **Model:** 4Runner, Avalon, Camry, Corolla, FJ Cruiser, Highlander, Land Cruiser, Matrix, Prius, Rav4, Sequoia, Tundra, Tacoma, Venza, Yaris **Engine:** L4, V6, V8	**Transmission Fluid Temperature Sensor "B" Circuit Low Input:** The Number 1 or 2 ATF temperature sensor resistance is below 25 Ohm (0.046 V) for 0.5 sec. or more.
DTC: P2743 **1T ECM, MIL: Yes** **Year:** 2011, 2012 **Model:** 4Runner, Avalon, Camry, Corolla, FJ Cruiser, Highlander, Land Cruiser, Matrix, Prius, Rav4, Sequoia, Tundra, Tacoma, Venza, Yaris **Engine:** L4, V6, V8	**Transmission Fluid Temperature Sensor "B" Circuit High Input:** The Number 1 or 2 ATF temperature sensor resistance is more than 156 kOhm (4.915 V) for 0.5 sec. or more when 15 minutes or more have passed after the engine is started.

DTC	Trouble Code Title and Conditions
DTC: P2757 **2T ECM, MIL: Yes** **Year:** 2011, 2012 **Model:** 4Runner, Avalon, Camry, Corolla, FJ Cruiser, Highlander, Land Cruiser, Matrix, Prius, Rav4, Sequoia, Tundra, Tacoma, Venza, Yaris **Engine:** L4, V6, V8	**Torque Converter Clutch Pressure Control Solenoid Performance (Shift Solenoid Valve SLU):** Lock-up does not occur when driving in the lock-up range (normal driving at 80 km/h (50 mph)), or lock-up remains ON in the lock-up OFF range.
DTC: P2759 **2T ECM, MIL: Yes** **Year:** 2011, 2012 **Model:** 4Runner, Avalon, Camry, Corolla, FJ Cruiser, Highlander, Land Cruiser, Matrix, Prius, Rav4, Sequoia, Tundra, Tacoma, Venza, Yaris **Engine:** L4, V6, V8	**Torque Converter Clutch Pressure Control Solenoid Control Circuit Electrical (Shift Solenoid Valve SLU):** The monitor will run whenever this DTC is not present: None Ignition switch: ON Starter: OFF Case 1: The Solenoid current cut status: Not cut, Battery voltage: 11 V or more. Case 2: Battery voltage: 8 V or more.
DTC: P2769 **2T ECM, MIL: Yes** **Year:** 2011, 2012 **Model:** 4Runner, Avalon, Camry, Corolla, FJ Cruiser, Highlander, Land Cruiser, Matrix, Prius, Rav4, Sequoia, Tundra, Tacoma, Venza, Yaris **Engine:** L4, V6, V8	**Torque Converter Clutch Solenoid Circuit Low (Shift Solenoid Valve DSL):** Shift solenoid valve DSL is ON, the solenoid current cut status Not cut, battery voltage 8 V or more, ignition switch ON, starter OFF. The ECM detects short in solenoid valve DSL circuit (0.1 sec.) when solenoid valve DSL is operated.
DTC: P2770 **2T ECM, MIL: Yes** **Year:** 2011, 2012 **Model:** 4Runner, Avalon, Camry, Corolla, FJ Cruiser, Highlander, Land Cruiser, Matrix, Prius, Rav4, Sequoia, Tundra, Tacoma, Venza, Yaris **Engine:** L4, V6, V8	**Torque Converter Clutch Solenoid Circuit High (Shift Solenoid Valve DSL):** The ECM detects a short in the solenoid valve DSL circuit (0.1 sec.) when solenoid valve DSL is operated. Shift solenoid valve DSL: ON, Solenoid current cut status: Not cut, battery voltage 8 V or more, ignition is ON, starter OFF.
DTC: P2772 **1T ECM, MIL: Yes** **Year:** 2011, 2012 **Model:** 4Runner, FJ Cruiser, Highlander, Land Cruiser, Matrix, Rav4, Sequoia, Tundra, Tacoma, Venza **Engine:** L4, V6, V8	**Four Wheel Drive (4WD) Low Switch Circuit Range / Performance:** Transfer 4L position switch remains ON while vehicle is running under following conditions for 1.8 seconds or more (a) Output shaft speed between 1000 and 3000 rpm (b) Transfer position switch is in 4H
DTC: P2808 **1T ECM, MIL: Yes** **Year:** 2011, 2012 **Model:** 4Runner, Avalon, Camry, Corolla, FJ Cruiser, Highlander, Land Cruiser, Matrix, Prius, Rav4, Sequoia, Tundra, Tacoma, Venza, Yaris **Engine:** L4, V6, V8	**Pressure Control Solenoid "G" Performance (Shift Solenoid Valve SL4):** The monitor will run whenever this DTC is present. None. Shift solenoid valve SL4 (open or closed) malfunction.

DTC	Trouble Code Title and Conditions
DTC: P2810 **1T ECM, MIL: Yes** **Year:** 2011, 2012 **Model:** 4Runner, Avalon, Camry, Corolla, FJ Cruiser, Highlander, Land Cruiser, Matrix, Prius, Rav4, Sequoia, Tundra, Tacoma, Venza, Yaris **Engine:** L4, V6, V8	**Pressure Control Solenoid "G" Electrical (Shift Solenoid Valve SL4):** The monitor will run whenever this DTC is not present: None Engine switch: ON. Starter: OFF Condition (A): Battery voltage: 12 V or more Condition (B):.Battery voltage: 10 V or more and less than 12 V Target condition: Less than 0.75 A Condition (C): Battery voltage: 8 V or more Target condition: 0.25 A or more
DTC: P2A00 **1T ECM, MIL: Yes** **Year:** 2011, 2012 **Model:** 4Runner, Avalon, Camry, Corolla, FJ Cruiser, Highlander, Land Cruiser, Matrix, Prius, Rav4, Sequoia, Tundra, Tacoma, Venza, Yaris **Engine:** L4, V6, V8	**A/F Sensor Circuit Slow Response (Bank 1 Sensor 1):** Monitor runs whenever following DTCs are not present: None. The calculated value of the air-fuel ratio (A/F) sensor response rate deterioration level is less than the threshold. Active air fuel ratio control performed when following conditions met: Battery voltage: 11 V or more Engine coolant temperature: 75°C (167°F) or more Idling: OFF Engine speed: Less than 4000 rpm Air fuel ratio sensor status: Activated Fuel-cut: OFF Engine load: 10 to 70% Catalyst monitor: Not yet Mass air flow: 4.5 to 12 gm/sec
DTC: P2A03 **2T ECM, MIL: Yes** **Year:** 2011, 2012 **Model:** 4Runner, Avalon, Camry, Corolla, FJ Cruiser, Highlander, Land Cruiser, Matrix, Prius, Rav4, Sequoia, Tundra, Tacoma, Venza, Yaris **Engine:** L4, V6, V8	**A/F Sensor Circuit Slow Response (Bank 2 Sensor 1):** Calculated value of air-fuel ratio (A/F) sensor response rate deterioration level less than threshold. Active A/F control: Performing Battery voltage: 11 V or more ECT: 167°F (75°C) or more Idle: OFF Engine rpm: Less than 4000 rpm A/F sensor status: Activated Fuel cut: OFF Engine load: 10 to 70 % Shift position: 2 or more Catalyst monitor: Not yet MAF: 2.5–15 g/sec.

OBD II Trouble Code List (P3XXX Codes)

DTC	Trouble Code Title and Conditions
DTC: P3000-388 **T ECM, MIL: Yes** **Year:** 2011, 2012 **Model:** Camry, Prius, Prius V **Engine:** 1.8L L4, 2.4L L4	**HV Battery Malfunction:** Discharge inhibition control malfunction.
DTC: P3000-389 **T ECM, MIL: Yes** **Year:** 2011, 2012 **Model:** Camry, Prius, Prius V **Engine:** 1.8L L4, 2.4L L4	**HV Battery Malfunction:** HV battery voltage drops.
DTC: P3000-603 **T ECM, MIL: Yes** **Year:** 2011, 2012 **Model:** Camry, Prius, Prius V **Engine:** 1.8L L4, 2.4L L4	**HV Battery Malfunction:** The hybrid vehicle control ECU detects a HV battery cooling system error signal.

DTC	Trouble Code Title and Conditions
DTC: P3004 **T ECM, MIL: Yes** **Year:** 2011, 2012 **Model:** Camry, Prius, Prius V **Engine:** 1.8L L4, 2.4L L4 Hybrid	**Power Cable Malfunction:** The inverter voltage does not rise during pre-charge (time from when SMRP turns on until when SMRG turns on).
DTC: P3004-131 **T ECM, MIL: Yes** **Year:** 2011, 2012 **Model:** Camry, Highlander, Prius, Prius V **Engine:** 1.8L L4, 2.4L L4, 3.5L V6 Hybrid	**Power Cable Malfunction:** The inverter voltage does not rise during pre-charge (time from when SMRP turns on until when SMRG turns on). CAUTION: Before inspecting the high-voltage system or disconnecting the low voltage connector of the inverter with converter assembly, take safety precautions such as wearing insulated gloves and removing the service plug grip to prevent electrical shocks. After removing the service plug grip, put it in your pocket to prevent other technicians from accidentally reconnecting it while you are working on the high-voltage system. After removing the service plug grip, wait for at least 10 minutes before touching any of the high-voltage connectors or terminals. After waiting for 10 minutes, check the voltage at the terminals in the inspection point in the inverter with converter assembly. The voltage should be 0 V before beginning work.
DTC: P3004-132 **T ECM, MIL: Yes** **Year:** 2011, 2012 **Model:** Camry, Prius, Prius V **Engine:** 1.8L L4, 2.4L L4 Hybrid	**Power Cable Malfunction:** The inverter is not pre-charged.
DTC: P3004-133 **T ECM, MIL: Yes** **Year:** 2011, 2012 **Model:** Camry, Prius, Prius V **Engine:** 1.8L L4, 2.4L L4	**Power Cable Malfunction:** A high-voltage wiring system error signal is detected in the hybrid vehicle control ECU.
DTC: P3004-801 **T ECM, MIL: Yes** **Year:** 2011, 2012 **Model:** Camry, Prius, Prius V **Engine:** 1.8L L4, 2.4L L4	**Power Cable Malfunction:** Minimal over-current occurs during pre-charge (time from when SMRP turns on until when SMRG turns on).
DTC: P3004-803 **T ECM, MIL: Yes** **Year:** 2011, 2012 **Model:** Camry, Prius, Prius V **Engine:** 1.8L L4, 2.4L L4	**Power Cable Malfunction:** While the power switch is on (READY), the electric battery fuse in the service plug grip is burned out, the service plug grip is removed, or SMRB or SMRG is open.
DTC: P3011-123 **1T ECM, MIL: Yes** **Year:** 2011, 2012 **Model:** Camry, Prius, Prius V **Engine:** 1.8L L4, 2.4L L4	**Battery Block 1 Becomes Weak:** The monitor will run whenever the following DTCs are not present: TMC's intellectual property. Other conditions belong to TMC's intellectual property.
DTC: P3065-123 **1T ECM, MIL: Yes** **Year:** 2011, 2012 **Model:** Camry, Prius, Prius V **Engine:** 1.8L L4, 2.4L L4	**Hybrid Battery Temperature Sensor Range / Perfoemance Stack A:** The monitor will run whenever the following DTCs are not present: TMC's intellectual property. Other conditions belong to TMC's intellectual property.
DTC: P308A-123 **1T ECM, MIL: Yes** **Year:** 2011, 2012 **Model:** Camry, Prius, Prius V **Engine:** 1.8L L4, 2.4L L4	**Hybrid Battery Voltage Sensor All Circuits Low:** The monitor will run whenever the following DTCs are not present: TMC's intellectual property. Other conditions belong to TMC's intellectual property.
DTC: P3107-213 **T ECM, MIL: Yes** **Year:** 2011, 2012 **Model:** Camry, Prius, Prius V **Engine:** 1.8L L4, 2.4L L4	**Airbag ECU Communication Circuit Malfunction:** Short to GND in the communication circuit.
DTC: P3107-214 **T ECM, MIL: Yes** **Year:** 2011, 2012 **Model:** Camry, Prius, Prius V **Engine:** 1.8L L4, 2.4L L4	**Airbag ECU Communication Circuit Malfunction:** Open or short to +B in the communication circuit

DTC	Trouble Code Title and Conditions
DTC: P3107-215 **T ECM, MIL: Yes** **Year:** 2011, 2012 **Model:** Camry, Prius, Prius V **Engine:** 1.8L L4, 2.4L L4	**Airbag ECU Communication Circuit Malfunction:** Abnormal communication signal.
DTC: P3108-535 **T ECM, MIL: Yes** **Year:** 2011, 2012 **Model:** Camry, Prius, Prius V **Engine:** 1.8L L4, 2.4L L4	**A/C Amplifier Communication Circuit Malfunction:** Serial communication malfunction.
DTC: P3108-536 **T ECM, MIL: Yes** **Year:** 2011, 2012 **Model:** Camry, Prius, Prius V **Engine:** 1.8L L4, 2.4L L4	**A/C Amplifier Communication Circuit Malfunction:** A/C inverter malfunction.
DTC: P3108-538 **T ECM, MIL: Yes** **Year:** 2011, 2012 **Model:** Camry, Prius, Prius V **Engine:** 1.8L L4, 2.4L L4	**A/C Amplifier Communication Circuit Malfunction:** Open in STB signal circuit.
DTC: P3110-139 **T ECM, MIL: Yes** **Year:** 2011, 2012 **Model:** Camry, Prius, Prius V **Engine:** 1.8L L4, 2.4L L4	**IGCT Relay Malfunction:** There is a short to +B in the IGCT relay or the IGCT relay is stuck closed.
DTC: P3110-223 **T ECM, MIL: Yes** **Year:** 2011, 2012 **Model:** Camry, Prius, Prius V **Engine:** 1.8L L4, 2.4L L4	**IGCT Relay Malfunction:** The IGCT relay remains stuck closed.
DTC: P3147-239 **T ECM, MIL: Yes** **Year:** 2011, 2012 **Model:** Camry, Prius, Prius V **Engine:** 1.8L L4, 2.4L L4	**Transmission Malfunction:** Hybrid vehicle transaxle input malfunction (shaft damaged).
DTC: P3147-240 **T TCM, MIL: Yes** **Year:** 2011, 2012 **Model:** Camry, Prius, Prius V **Engine:** 1.8L L4, 2.4L L4	**Transmission Malfunction:** Hybrid vehicle transaxle malfunction MG1 locked
DTC: P3147-241 **T ECM, MIL: Yes** **Year:** 2011, 2012 **Model:** Camry, Prius, Prius V **Engine:** 1.8L L4, 2.4L L4	**Transmission Malfunction:** Hybrid vehicle transaxle input malfunction (input system lock)
DTC: P3147-242 **T TCM, MIL: Yes** **Year:** 2011, 2012 **Model:** Camry, Prius, Prius V **Engine:** 1.8L L4, 2.4L L4	**Transmission Malfunction:** Planetary gear locked
DTC: P3190 **1T ECM, MIL: Yes** **Year:** 2011, 2012 **Model:** Camry, Highlander, Prius, Prius V **Engine:** 1.8L L4, 2.4L L4, 3.5L V6	**Poor Engine Power:** Monitor runs whenever following DTCs are not present: None Fuel cut operation: Not operated Engine speed: 650 rpm or more (varies with engine coolant temperature) Communication with power management control ECU: No malfunction

DTC	Trouble Code Title and Conditions
DTC: P3191 **1T ECM, MIL: Yes** **Year:** 2011, 2012 **Model:** Camry, Highlander, Prius, Prius V **Engine:** 1.8L L4, 2.4L L4, 3.5L V6	**Engine does not Start:** Monitor runs whenever following DTCs are not present: None Fuel cut operation: Not operated Engine speed: 650 rpm or more (varies with engine coolant temperature) Communication with hybrid vehicle control ECU: No malfunction
DTC: P3193 **1T ECM, MIL: Yes** **Year:** 2011, 2012 **Model:** Camry, Highlander, Prius, Prius V **Engine:** 1.8L L4, 2.4L L4, 3.5L V6	**Fuel Run Out:** Monitor runs whenever following DTCs are not present: None Fuel cut operation: Not operated Engine speed: 650 rpm or more (varies with engine coolant temperature) Communication with power management control ECU: No malfunction
DTC: P3221-314 **1T ECM, MIL: Yes** **Year:** 2011, 2012 **Model:** Camry, Highlander, Prius, Prius V **Engine:** 1.8L L4, 2.4L L4, 3.5L V6	**Generator Inverter Temperature Sensor Circuit Range / Performance:** Unusual sudden change in generator inverter temperature sensor output occurs and the offset continues, or unusual sudden change in generator inverter temperature sensor output occurs repeatedly.
DTC: P3221-315 **1T ECM, MIL: Yes** **Year:** 2011, 2012 **Model:** Camry, Highlander, Prius, Prius V **Engine:** 1.8L L4, 2.4L L4, 3.5L V6	**Generator Inverter Temperature Sensor Circuit Range / Performance:** Temperature calculated by power hybrid vehicle control ECU and actual temperature are different for 10 seconds or more.
DTC: P3222-313 **1T ECM, MIL: Yes** **Year:** 2011, 2012 **Model:** Camry, Highlander, Prius, Prius V **Engine:** 1.8L L4, 2.4L L4, 3.5L V6	**Generator Inverter Temperature Sensor Circuit High / Low:** Open or short to GND in the generator inverter temperature sensor signal circuit.
DTC: P3223-312 **1T ECM, MIL: Yes** **Year:** 2011, 2012 **Model:** Camry, Highlander, Prius, Prius V **Engine:** 1.8L L4, 2.4L L4, 3.5L V6	**Generator Inverter Temperature Sensor Circuit High:** Short to +B in the generator inverter temperature sensor signal circuit.
DTC: P3226-562 **1T ECM, MIL: Yes** **Year:** 2011, 2012 **Model:** Camry, Highlander, Prius, Prius V **Engine:** 1.8L L4, 2.4L L4, 3.5L V6	**DC/DC Boost Converter Temperature Sensor:** Unusual sudden change in boost converter temperature sensor output occurs and the offset continues, or unusual sudden change in boost converter temperature sensor output occurs repeatedly.
DTC: P3226-563 **1T ECM, MIL: Yes** **Year:** 2011, 2012 **Model:** Camry, Highlander, Prius, Prius V **Engine:** 1.8L L4, 2.4L L4, 3.5L V6	**DC/DC Boost Converter Temperature Sensor:** Temperature calculated by power hybrid vehicle control ECU and actual temperature are different for 10 seconds or more.
DTC: P3227-583 **1T ECM, MIL: Yes** **Year:** 2011, 2012 **Model:** Camry, Highlander, Prius, Prius V **Engine:** 1.8L L4, 2.4L L4, 3.5L V6	**Converter Temperature Sensor Circuit Low:** Open or short to GND in the boost converter temperature sensor signal circuit.

DTC	Trouble Code Title and Conditions
DTC: P3228-584 **1T ECM, MIL: Yes** **Year:** 2011, 2012 **Model:** Camry, Highlander, Prius, Prius V **Engine:** 1.8L L4, 2.4L L4, 3.5L V6	**Converter Temperature Sensor Circuit High:** Short to +B in the boost converter temperature sensor signal circuit.
DTC: P3232-749 **1T ECM, MIL: Yes** **Year:** 2011, 2012 **Model:** Camry, Highlander, Prius, Prius V **Engine:** 1.8L L4, 2.4L L4, 3.5L V6	**Open or Short to B+ in Blocking of HV Gate Connection:** Short to GND in the emergency shutdown signal line while the gate is shut down.
DTC: P3233-750 **1T ECM, MIL: Yes** **Year:** 2011, 2012 **Model:** Camry, Highlander, Prius, Prius V **Engine:** 1.8L L4, 2.4L L4, 3.5L V6	**Short to B+ in Blocking of HV Gate Connection:** Open or short to +B in the emergency shutdown signal line when the gate is driving.

GLOSSARY

ABS: Anti-lock braking system. An electro-mechanical braking system which is designed to minimize or prevent wheel lock-up during braking.

ABSOLUTE PRESSURE: Atmospheric (barometric) pressure plus the pressure gauge reading.

ACCELERATOR PUMP: A small pump located in the carburetor that feeds fuel into the air/fuel mixture during acceleration.

ACCUMULATOR: A device that controls shift quality by cushioning the shock of hydraulic oil pressure being applied to a clutch or band.

ACTUATING MECHANISM: The mechanical output devices of a hydraulic system, for example, clutch pistons and band servos.

ACTUATOR: The output component of a hydraulic or electronic system.

ADVANCE: Setting the ignition timing so that spark occurs earlier before the piston reaches top dead center (TDC).

ADAPTIVE MEMORY (ADAPTIVE STRATEGY): The learning ability of the TCM or PCM to redefine its decision-making process to provide optimum shift quality.

AFTER TOP DEAD CENTER (ATDC): The point after the piston reaches the top of its travel on the compression stroke.

AIR BAG: Device on the inside of the car designed to inflate on impact of crash, protecting the occupants of the car.

AIR CHARGE TEMPERATURE (ACT) SENSOR: The temperature of the airflow into the engine is measured by an ACT sensor, usually located in the lower intake manifold or air cleaner.

AIR CLEANER: An assembly consisting of a housing, filter and any connecting ductwork. The filter element is made up of a porous paper, sometimes with a wire mesh screening, and is designed to prevent airborne particles from entering the engine through the carburetor or throttle body.

AIR INJECTION: One method of reducing harmful exhaust emissions by injecting air into each of the exhaust ports of an engine. The fresh air entering the hot exhaust manifold causes any remaining fuel to be burned before it can exit the tailpipe.

AIR PUMP: An emission control device that supplies fresh air to the exhaust manifold to aid in more completely burning exhaust gases.

AIR/FUEL RATIO: The ratio of air-to-gasoline by weight in the fuel mixture drawn into the engine.

ALDL (assembly line diagnostic link): Electrical connector for scanning ECM/PCM/TCM input and output devices.

ALIGNMENT RACK: A special drive-on vehicle lift apparatus/measuring device used to adjust a vehicle's toe, caster and camber angles.

ALL WHEEL DRIVE: Term used to describe a full time four wheel drive system or any other vehicle drive system that continuously delivers power to all four wheels. This system is found primarily on station wagon vehicles and SUVs not utilized for significant off road use.

ALTERNATING CURRENT (AC): Electric current that flows first in one direction, then in the opposite direction, continually reversing flow.

ALTERNATOR: A device which produces AC (alternating current) which is converted to DC (direct current) to charge the car battery.

AMMETER: An instrument, calibrated in amperes, used to measure the flow of an electrical current in a circuit. Ammeters are always connected in series with the circuit being tested.

AMPERAGE: The total amount of current (amperes) flowing in a circuit.

AMPLIFIER: A device used in an electrical circuit to increase the voltage of an output signal.

AMP/HR. RATING (BATTERY): Measurement of the ability of a battery to deliver a stated amount of current for a stated period of time. The higher the amp/hr. rating, the better the battery.

AMPERE: The rate of flow of electrical current present when one volt of electrical pressure is applied against one ohm of electrical resistance.

ANALOG COMPUTER: Any microprocessor that uses similar (analogous) electrical signals to make its calculations.

ANODIZED: A special coating applied to the surface of aluminum valves for extended service life.

ANTIFREEZE: A substance (ethylene or propylene glycol) added to the coolant to prevent freezing in cold weather.

ANTI-FOAM AGENTS: Minimize fluid foaming from the whipping action encountered in the converter and planetary action.

ANTI-WEAR AGENTS: Zinc agents that control wear on the gears, bushings, and thrust washers.

ANTI-LOCK BRAKING SYSTEM: A supplementary system to the base hydraulic system that prevents sustained lock-up of the wheels during braking as well as automatically controlling wheel slip.

ANTI-ROLL BAR: See stabilizer bar.

ARC: A flow of electricity through the air between two electrodes or contact points that produces a spark.

ARMATURE: A laminated, soft iron core wrapped by a wire that converts electrical energy to mechanical energy as in a motor or relay. When rotated in a magnetic field, it changes mechanical energy into electrical energy as in a generator.

ATDC: After Top Dead Center.

ATF: Automatic transmission fluid.

ATMOSPHERIC PRESSURE: The pressure on the Earth's surface caused by the weight of the air in the atmosphere. At sea level, this pressure is 14.7 psi at 32°F (101 kPa at 0°C).

ATOMIZATION: The breaking down of a liquid into a fine mist that can be suspended in air.

AUXILIARY ADD-ON COOLER: A supplemental transmission fluid cooling device that is installed in series with the heat exchanger (cooler), located inside the radiator, to provide additional support to cool the hot fluid leaving the torque converter.

AUXILIARY PRESSURE: An added fluid pressure that is introduced into a regulator or balanced valve system to control valve movement. The auxiliary pressure itself can be either a fixed or a variable value. (See balanced valve; regulator valve.)

AWD: All wheel drive.

AXIAL FORCE: A side or end thrust force acting in or along the same plane as the power flow.

AXIAL PLAY: Movement parallel to a shaft or bearing bore.

AXLE CAPACITY: The maximum load-carrying capacity of the axle itself, as specified by the manufacturer. This is usually a higher number than the GAWR.

AXLE RATIO: This is a number (3.07:1, 4.56:1, for example) expressing the ratio between driveshaft revolutions and wheel revolutions. A low numerical ratio allows the engine to work easier because it doesn't have to turn as fast. A high numerical ratio means that the engine has to turn more rpm's to move the wheels through the same number of turns.

BACKFIRE: The sudden combustion of gases in the intake or exhaust system that results in a loud explosion.

BACKLASH: The clearance or play between two parts, such as meshed gears.

BACKPRESSURE: Restrictions in the exhaust system that slow the exit of exhaust gases from the combustion chamber.

BAKELITE®: A heat resistant, plastic insulator material commonly used in printed circuit boards and transistorized components.

BALANCED VALVE: A valve that is positioned by opposing auxiliary hydraulic pressures and/or spring force. Examples include mainline regulator, throttle, and governor valves. (See regulator valve.)

BAND: A flexible ring of steel with an inner lining of friction material. When tightened around the outside of a drum, a planetary member is held stationary to the transmission/transaxle case.

BALL BEARING: A bearing made up of hardened inner and outer races between which hardened steel balls roll.

BALL JOINT: A ball and matching socket connecting suspension components (steering knuckle to lower control arms). It permits rotating movement in any direction between the components that are joined.

BARO (BAROMETRIC PRESSURE SENSOR): Measures the change in the intake manifold pressure caused by changes in altitude.

BAROMETRIC MANIFOLD ABSOLUTE PRESSURE (BMAP) SENSOR: Operates similarly to a conventional MAP sensor; reads intake manifold pressure and is also responsible for determining altitude and barometric pressure prior to engine operation.

BAROMETRIC PRESSURE: (See atmospheric pressure.)

BALLAST RESISTOR: A resistor in the primary ignition circuit that lowers voltage after the engine is started to reduce wear on ignition components.

BATTERY: A direct current electrical storage unit, consisting of the basic active materials of lead and sulfuric acid, which converts chemical energy into electrical energy. Used to provide current for the operation of the starter as well as other equipment, such as the radio, lighting, etc.

BEAD: The portion of a tire that holds it on the rim.

BEARING: A friction reducing, supportive device usually located between a stationary part and a moving part.

BEFORE TOP DEAD CENTER (BTDC): The point just before the piston reaches the top of its travel on the compression stroke.

BELTED TIRE: Tire construction similar to bias-ply tires, but using two or more layers of reinforced belts between body plies and the tread.

BEZEL: Piece of metal surrounding radio, headlights, gauges or similar components; sometimes used to hold the glass face of a gauge in the dash.

BIAS-PLY TIRE: Tire construction, using body ply reinforcing cords which run at alternating angles to the center line of the tread.

BI-METAL TEMPERATURE SENSOR: Any sensor or switch made of two dissimilar types of metal that bend when heated or cooled due to the different expansion rates of the alloys. These types of sensors usually function as an on/off switch.

BLOCK: See Engine Block.

BLOW-BY: Combustion gases, composed of water vapor and unburned fuel, that leak past the piston rings into the crankcase during normal engine operation. These gases are removed by the PCV system to prevent the buildup of harmful acids in the crankcase.

BOOK TIME: See Labor Time.

BOOK VALUE: The average value of a car, widely used to determine trade-in and resale value.

BOOST VALVE: Used at the base of the regulator valve to increase mainline pressure.

BORE: Diameter of a cylinder.

BRAKE CALIPER: The housing that fits over the brake disc. The caliper holds the brake pads, which are pressed against the discs by the caliper pistons when the brake pedal is depressed.

BRAKE HORSEPOWER (BHP): The actual horsepower available at the engine flywheel as measured by a dynamometer.

BRAKE FADE: Loss of braking power, usually caused by excessive heat after repeated brake applications.

BRAKE HORSEPOWER: Usable horsepower of an engine measured at the crankshaft.

BRAKE PAD: A brake shoe and lining assembly used with disc brakes.

BRAKE PROPORTIONING VALVE: A valve on the master cylinder which restricts hydraulic brake pressure to the wheels to a specified amount, preventing wheel lock-up.

BREAKAWAY: Often used by Chrysler to identify first-gear operation in D and 2 ranges. In these ranges, first-gear operation depends on a one-way roller clutch that holds on acceleration and releases (breaks away) on deceleration, resulting in a freewheeling coast-down condition.

BRAKE SHOE: The backing for the brake lining. The term is, however, usually applied to the assembly of the brake backing and lining.

BREAKER POINTS: A set of points inside the distributor, operated by a cam, which make and break the ignition circuit.

BRINNELLING: A wear pattern identified by a series of indentations at regular intervals. This condition is caused by a lack of lube, overload situations, and/or vibrations.

BTDC: Before Top Dead Center.

BUMP: Sudden and forceful apply of a clutch or band.

BUSHING: A liner, usually removable, for a bearing; an anti-friction liner used in place of a bearing.

CALIFORNIA ENGINE: An engine certified by the EPA for use in California only; conforms to more stringent emission regulations than Federal engine.

CALIPER: A hydraulically activated device in a disc brake system, which is mounted straddling the brake rotor (disc). The caliper contains at least one piston and two brake pads. Hydraulic pressure on the piston(s) forces the pads against the rotor.

CAPACITY: The quantity of electricity that can be delivered from a unit, as from a battery in ampere-hours, or output, as from a generator.

CAMBER: One of the factors of wheel alignment. Viewed from the front of the car, it is the inward or outward tilt of the wheel. The top of the tire will lean outward (positive camber) or inward (negative camber).

CAMSHAFT: A shaft in the engine on which are the lobes (cams) which operate the valves. The camshaft is driven by the crankshaft, via a belt, chain or gears, at one half the crankshaft speed.

CAPACITOR: A device which stores an electrical charge.

CARBON MONOXIDE (CO): A colorless, odorless gas given off as a normal byproduct of combustion. It is poisonous and extremely dangerous in confined areas, building up slowly to toxic levels without warning if adequate ventilation is not available.

CARBURETOR: A device, usually mounted on the intake manifold of an engine, which mixes the air and fuel in the proper proportion to allow even combustion.

CASTER: The forward or rearward tilt of an imaginary line drawn through the upper ball joint and the center of the wheel. Viewed from the sides, positive caster (forward tilt) lends directional stability, while negative caster (rearward tilt) produces instability.

CATALYTIC CONVERTER: A device installed in the exhaust system, like a muffler, that converts harmful byproducts of combustion into carbon dioxide and water vapor by means of a heat-producing chemical reaction.

CENTRIFUGAL ADVANCE: A mechanical method of advancing the spark timing by using flyweights in the distributor that react to centrifugal force generated by the distributor shaft rotation.

CENTRIFUGAL FORCE: The outward pull of a revolving object, away from the center of revolution. Centrifugal force increases with the speed of rotation.

CETANE RATING: A measure of the ignition value of diesel fuel. The higher the cetane rating, the better the fuel. Diesel fuel cetane rating is roughly comparable to gasoline octane rating.

CHECK VALVE: Any one-way valve installed to permit the flow of air, fuel or vacuum in one direction only.

CHOKE: The valve/plate that restricts the amount of air entering an engine on the induction stroke, thereby enriching the air/fuel ratio.

CHUGGLE: Bucking or jerking condition that may be engine related and may be most noticeable when converter clutch is engaged; similar to the feel of towing a trailer.

CIRCLIP: A split steel snapring that fits into a groove to hold various parts in place.

CIRCUIT BREAKER: A switch which protects an electrical circuit from overload by opening the circuit when the current flow exceeds a pre-determined level. Some circuit breakers must be reset manually, while most reset automatically.

CIRCUIT: Any unbroken path through which an electrical current can flow. Also used to describe fuel flow in some instances.

CIRCUIT, BYPASS: Another circuit in parallel with the major circuit through which power is diverted.

CIRCUIT, CLOSED: An electrical circuit in which there is no interruption of current flow.

CIRCUIT, GROUND: The non-insulated portion of a complete circuit used as a common potential point. In automotive circuits, the ground is composed of metal parts, such as the engine, body sheet metal, and frame and is usually a negative potential.

CIRCUIT, HOT: That portion of a circuit not at ground potential. The hot circuit is usually insulated and is connected to the positive side of the battery.

CIRCUIT, OPEN: A break or lack of contact in an electrical circuit, either intentional (switch) or unintentional (bad connection or broken wire).

CIRCUIT, PARALLEL: A circuit having two or more paths for current flow with common positive and negative tie points. The same voltage is applied to each load device or parallel branch.

CIRCUIT, SERIES: An electrical system in which separate parts are connected end to end, using one wire, to form a single path for current to flow.

CIRCUIT, SHORT: A circuit that is accidentally completed in an electrical path for which it was not intended.

CLAMPING (ISOLATION) DIODES: Diodes positioned in a circuit to prevent self-induction from damaging electronic components.

CLEARCOAT: A transparent layer which, when sprayed over a vehicle's paint job, adds gloss and depth as well as an additional protective coating to the finish.

CLUTCH: Part of the power train used to connect/disconnect power to the rear wheels.

CLUTCH, FLUID: The same as a fluid coupling. A fluid clutch or coupling performs the same function as a friction clutch by utilizing fluid friction and inertia as opposed to solid friction used by a friction clutch. (See fluid coupling.)

CLUTCH, FRICTION: A coupling device that provides a means of smooth and positive engagement and disengagement of engine torque to the vehicle powertrain. Transmission of power through the clutch is accomplished by bringing one or more rotating drive members into contact with complementing driven members.

COAST: Vehicle deceleration caused by engine braking conditions.

COEFFICIENT OF FRICTION: The amount of surface tension between two contacting surfaces; identified by a scientifically calculated number.

COIL: Part of the ignition system that boosts the relatively low voltage supplied by the car's electrical system to the high voltage required to fire the spark plugs.

COMBINATION MANIFOLD: An assembly which includes both the intake and exhaust manifolds in one casting.

COMBINATION VALVE: A device used in some fuel systems that routes fuel vapors to a charcoal storage canister instead of venting them into the atmosphere. The valve relieves fuel tank pressure and allows fresh air into the tank as the fuel level drops to prevent a vapor lock situation.

COMBUSTION CHAMBER: The part of the engine in the cylinder head where combustion takes place.

COMPOUND GEAR: A gear consisting of two or more simple gears with a common shaft.

COMPOUND PLANETARY: A gearset that has more than the three elements found in a simple gearset and is constructed by combining members of two planetary gearsets to create additional gear ratio possibilities.

COMPRESSION CHECK: A test involving removing each spark plug and inserting a gauge. When the engine is cranked, the gauge will record a pressure reading in the individual cylinder. General operating condition can be determined from a compression check.

COMPRESSION RATIO: The ratio of the volume between the piston and cylinder head when the piston is at the bottom of its stroke (bottom dead center) and when the piston is at the top of its stroke (top dead center).

COMPUTER: An electronic control module that correlates input data according to prearranged engineered instructions; used for the management of an actuator system or systems.

CONDENSER: An electrical device which acts to store an electrical charge, preventing voltage surges.

2. A radiator-like device in the air conditioning system in which refrigerant gas condenses into a liquid, giving off heat.

CONDUCTOR: Any material through which an electrical current can be transmitted easily.

CONNECTING ROD: The connecting link between the crankshaft and piston.

CONSTANT VELOCITY JOINT: Type of universal joint in a halfshaft assembly in which the output shaft turns at a constant angular velocity without variation, provided that the speed of the input shaft is constant.

CONTINUITY: Continuous or complete circuit. Can be checked with an ohmmeter.

CONTROL ARM: The upper or lower suspension components which are mounted on the frame and support the ball joints and steering knuckles.

CONVENTIONAL IGNITION: Ignition system which uses breaker points.

CONVERTER: (See torque converter.)

CONVERTER LOCKUP: The switching from hydrodynamic to direct mechanical drive, usually through the application of a friction element called the converter clutch.

COOLANT: Mixture of water and anti-freeze circulated through the engine to carry off heat produced by the engine.

CORROSION INHIBITOR: An inhibitor in ATF that prevents corrosion of bushings, thrust washers, and oil cooler brazed joints.

COUNTERSHAFT: An intermediate shaft which is rotated by a mainshaft and transmits, in turn, that rotation to a working part.

COUPLING PHASE: Occurs when the torque converter is operating at its greatest hydraulic efficiency. The speed differential between the impeller and the turbine is at its minimum. At this point, the stator freewheels, and there is no torque multiplication.

CRANKCASE: The lower part of an engine in which the crankshaft and related parts operate.

CRANKSHAFT: Engine component (connected to pistons by connecting rods) which converts the reciprocating (up and down) motion of pistons to rotary motion used to turn the driveshaft.

CURB WEIGHT: The weight of a vehicle without passengers or payload, but including all fluids (oil, gas, coolant, etc.) and other equipment specified as standard.

CURRENT: The flow (or rate) of electrons moving through a circuit. Current is measured in amperes (amp).

CURRENT FLOW CONVENTIONAL: Current flows through a circuit from the positive terminal of the source to the negative terminal (plus to minus).

CURRENT FLOW, ELECTRON: Current or electrons flow from the negative terminal of the source, through the circuit, to the positive terminal (minus to plus).

CV-JOINT: Constant velocity joint.

CYCLIC VIBRATIONS: The off-center movement of a rotating object that is affected by its initial balance, speed of rotation, and working angles.

CYLINDER BLOCK: See engine block.

CYLINDER HEAD: The detachable portion of the engine, usually fastened to the top of the cylinder block and containing all or most of the combustion chambers. On overhead valve engines, it contains the valves and their operating parts. On overhead cam engines, it contains the camshaft as well.

CYLINDER: In an engine, the round hole in the engine block in which the piston(s) ride.

DATA LINK CONNECTOR (DLC): Current acronym/term applied to the federally mandated, diagnostic junction connector that is used to monitor ECM/PC/TCM inputs, processing strategies, and outputs including diagnostic trouble codes (DTCs).

DEAD CENTER: The extreme top or bottom of the piston stroke.

DECELERATION BUMP: When referring to a torque converter clutch in the applied position, a sudden release of the accelerator pedal causes a forceful reversal of power through the drivetrain (engine braking), just prior to the apply plate actually being released.

DELAYED (LATE OR EXTENDED): Condition where shift is expected but does not occur for a period of time, for example, where clutch or band engagement does not occur as quickly as expected during part throttle or wide open throttle apply of accelerator or when manually downshifting to a lower range.

DETENT: A spring-loaded plunger, pin, ball, or pawl used as a holding device on a ratchet wheel or shaft. In automatic transmissions, a detent mechanism is used for locking the manual valve in place.

DETENT DOWNSHIFT: (See kickdown.)

DETERGENT: An additive in engine oil to improve its operating characteristics.

DETONATION: An unwanted explosion of the air/fuel mixture in the combustion chamber caused by excess heat and compression, advanced timing, or an overly lean mixture. Also referred to as "ping".

DEXRON®: A brand of automatic transmission fluid.

DIAGNOSTIC TROUBLE CODES (DTCs): A digital display from the control module memory that identifies the input, processor, or output device circuit that is related to the powertrain emission/driveability malfunction detected. Diagnostic trouble codes can be read by the MIL to flash any codes or by using a handheld scanner.

DIAPHRAGM: A thin, flexible wall separating two cavities, such as in a vacuum advance unit.

DIESELING: The engine continues to run after the car is shut off; caused by fuel continuing to be burned in the combustion chamber.

DIFFERENTIAL: A geared assembly which allows the transmission of motion between drive axles, giving one axle the ability to rotate faster than the other, as in cornering.

DIFFERENTIAL AREAS: When opposing faces of a spool valve are acted upon by the same pressure but their areas differ in size, the face with the larger area produces the differential force and valve movement. (See spool valve.)

DIFFERENTIAL FORCE: (See differential areas)

DIGITAL READOUT: A display of numbers or a combination of numbers and letters.

DIGITAL VOLT OHMMETER: An electronic diagnostic tool used to measure voltage, ohms and amps as well as several other functions, with the readings displayed on a digital screen in tenths, hundredths and thousandths.

DIODE: An electrical device that will allow current to flow in one direction only.

DIRECT CURRENT (DC): Electrical current that flows in one direction only.

DIRECT DRIVE: The gear ratio is 1:1, with no change occurring in the torque and speed input/output relationship.

DISC BRAKE: A hydraulic braking assembly consisting of a brake disc, or rotor, mounted on an axle shaft, and a caliper assembly containing, usually two brake pads which are activated by hydraulic pressure. The pads are forced against the sides of the disc, creating friction which slows the vehicle.

DISPERSANTS: Suspend dirt and prevent sludge buildup in a liquid, such as engine oil.

DOUBLE BUMP (DOUBLE FEEL): Two sudden and forceful applies of a clutch or band.

DISPLACEMENT: The total volume of air that is displaced by all pistons as the engine turns through one complete revolution.

DISTRIBUTOR: A mechanically driven device on an engine which is responsible for electrically firing the spark plug at a pre-determined point of the piston stroke.

DOHC: Double overhead camshaft.

DOUBLE OVERHEAD CAMSHAFT: The engine utilizes two camshafts mounted in one cylinder head. One camshaft operates the exhaust valves, while the other operates the intake valves.

DOWEL PIN: A pin, inserted in mating holes in two different parts allowing those parts to maintain a fixed relationship.

DRIVELINE: The drive connection between the transmission and the drive wheels.

DRIVE TRAIN: The components that transmit the flow of power from the engine to the wheels. The components include the clutch, transmission, driveshafts (or axle shafts in front wheel drive), U-joints and differential.

DRUM BRAKE: A braking system which consists of two brake shoes and one or two wheel cylinders, mounted on a fixed backing plate, and a brake drum, mounted on an axle, which revolves around the assembly.

DRY CHARGED BATTERY: Battery to which electrolyte is added when the battery is placed in service.

DVOM: Digital volt ohmmeter

DWELL: The rate, measured in degrees of shaft rotation, at which an electrical circuit cycles on and off.

DYNAMIC: An application in which there is rotating or reciprocating motion between the parts.

EARLY: Condition where shift occurs before vehicle has reached proper speed, which tends to labor engine after upshift.

EBCM: See Electronic Control Unit (ECU).

ECM: See Electronic Control Unit (ECU).

ECU: Electronic control unit.

ELECTRODE: Conductor (positive or negative) of electric current.

ELECTROLYSIS: A surface etching or bonding of current conducting transmission/transaxle components that may occur when grounding straps are missing or in poor condition.

ELECTROLYTE: A solution of water and sulfuric acid used to activate the battery. Electrolyte is extremely corrosive.

ELECTROMAGNET: A coil that produces a magnetic field when current flows through its windings.

ELECTROMAGNETIC INDUCTION: A method to create (generate) current flow through the use of magnetism.

ELECTROMAGNETISM: The effects surrounding the relationship between electricity and magnetism.

ELECTROMOTIVE FORCE (EMF): The force or pressure (voltage) that causes current movement in an electrical circuit.

ELECTRONIC CONTROL UNIT: A digital computer that controls engine (and sometimes transmission, brake or other vehicle system) functions based on data received from various sensors. Examples used by some manufacturers include Electronic Brake Control Module (EBCM), Engine Control Module (ECM), Powertrain Control Module (PCM) or Vehicle Control Module (VCM).

ELECTRONIC IGNITION: A system in which the timing and firing of the spark plugs is controlled by an electronic control unit, usually called a module. These systems have no points or condenser.

ELECTRONIC PRESSURE CONTROL (EPC) SOLENOID: A specially designed solenoid containing a spool valve and spring assembly to control fluid mainline pressure. A variable current flow, controlled by the ECM/PCM, varies the internal force of the solenoid on the spool valve and resulting mainline pressure. (See variable force solenoid.)

ELECTRONICS: Miniaturized electrical circuits utilizing semiconductors, solid-state devices, and printed circuits. Electronic circuits utilize small amounts of power.

ELECTRONIFICATION: The application of electronic circuitry to a mechanical device. Regarding automatic transmissions, electrification is incorporated into converter clutch lockup, shift scheduling, and line pressure control systems.

ELECTROSTATIC DISCHARGE (ESD): An unwanted, high-voltage electrical current released by an individual who has taken on a static charge of electricity. Electronic components can be easily damaged by ESD.

ELEMENT: A device within a hydrodynamic drive unit designed with a set of blades to direct fluid flow.

ENAMEL: Type of paint that dries to a smooth, glossy finish.

END BUMP (END FEEL OR SLIP BUMP): Firmer feel at end of shift when compared with feel at start of shift.

END-PLAY: The clearance/gap between two components that allows for expansion of the parts as they warm up, to prevent binding and to allow space for lubrication.

ENERGY: The ability or capacity to do work.

ENGINE: The primary motor or power apparatus of a vehicle, which converts liquid or gas fuel into mechanical energy.

ENGINE BLOCK: The basic engine casting containing the cylinders, the crankshaft main bearings, as well as machined surfaces for the mounting of other components such as the cylinder head, oil pan, transmission, etc.

ENGINE BRAKING: Use of engine to slow vehicle by manually downshifting during zero-throttle coast down.

ENGINE CONTROL MODULE (ECM): Manages the engine and incorporates output control over the torque converter clutch solenoid. (Note: Current designation for the ECM in late model vehicles is PCM.)

ENGINE COOLANT TEMPERATURE (ECT) SENSOR: Prevents converter clutch engagement with a cold engine; also used for shift timing and shift quality.

EP LUBRICANT: EP (extreme pressure) lubricants are specially formulated for use with gears involving heavy loads (transmissions, differentials, etc.).

ETHYL: A substance added to gasoline to improve its resistance to knock, by slowing down the rate of combustion.

ETHYLENE GLYCOL: The base substance of antifreeze.

EXHAUST MANIFOLD: A set of cast passages or pipes which conduct exhaust gases from the engine.

FAIL-SAFE (BACKUP) CONTROL: A substitute value used by the PCM/TCM to replace a faulty signal from an input sensor. The temporary value allows the vehicle to continue to be operated.

FAST IDLE: The speed of the engine when the choke is on. Fast idle speeds engine warm-up.

FEDERAL ENGINE: An engine certified by the EPA for use in any of the 49 states (except California).

FEEDBACK: A circuit malfunction whereby current can find another path to feed load devices.

FEELER GAUGE: A blade, usually metal, of precisely predetermined thickness, used to measure the clearance between two parts.

FILAMENT: The part of a bulb that glows; the filament creates high resistance to current flow and actually glows from the resulting heat.

FINAL DRIVE: An essential part of the axle drive assembly where final gear reduction takes place in the powertrain. In RWD applications and north-south FWD applications, it must also change the power flow direction to the axle shaft by ninety degrees. (Also see axle ratio).

FIRING ORDER: The order in which combustion occurs in the cylinders of an engine. Also the order in which spark is distributed to the plugs by the distributor.

FIRM: A noticeable quick apply of a clutch or band that is considered normal with medium to heavy throttle shift; should not be confused with harsh or rough.

FLAME FRONT: The term used to describe certain aspects of the fuel explosion in the cylinders. The flame front should move in a controlled pattern across the cylinder, rather than simply exploding immediately.

FLARE (SLIPPING): A quick increase in engine rpm accompanied by momentary loss of torque; generally occurs during shift.

FLAT ENGINE: Engine design in which the pistons are horizontally opposed. Porsche, Subaru and some old VW are common examples of flat engines.

FLAT RATE: A dealership term referring to the amount of money paid to a technician for a repair or diagnostic service based on that particular service versus dealership's labor time (NOT based on the actual time the technician spent on the job).

FLAT SPOT: A point during acceleration when the engine seems to lose power for an instant.

FLOODING: The presence of too much fuel in the intake manifold and combustion chamber which prevents the air/fuel mixture from firing, thereby causing a no-start situation.

FLUID: A fluid can be either liquid or gas. In hydraulics, a liquid is used for transmitting force or motion.

FLUID COUPLING: The simplest form of hydrodynamic drive, the fluid coupling consists of two look-alike members with straight radial varies referred to as the impeller (pump) and the turbine. Input torque is always equal to the output torque.

FLUID DRIVE: Either a fluid coupling or a fluid torque converter. (See hydrodynamic drive units.)

FLUID TORQUE CONVERTER: A hydrodynamic drive that has the ability to act both as a torque multiplier and fluid coupling. (See hydrodynamic drive units; torque converter.)

FLUID VISCOSITY: The resistance of a liquid to flow. A cold fluid (oil) has greater viscosity and flows more slowly than a hot fluid (oil).

FLYWHEEL: A heavy disc of metal attached to the rear of the crankshaft. It smoothes the firing impulses of the engine and keeps the crankshaft turning during periods when no firing takes place. The starter also engages the flywheel to start the engine.

FOOT POUND (ft. lbs., lbs. ft. or sometimes, ft. lb.): The amount of energy or work needed to raise an item weighing one pound, a distance of one foot.

FREEZE PLUG: A plug in the engine block which will be pushed out if the coolant freezes. Sometimes called expansion plugs, they protect the block from cracking should the coolant freeze.

FRICTION: The resistance that occurs between contacting surfaces. This relationship is expressed by a ratio called the coefficient of friction (CL).

FRICTION, COEFFICIENT OF: The amount of surface tension between two contacting surfaces; expressed by a scientifically calculated number.

FRONT END ALIGNMENT: A service to set caster, camber and toe-in to the correct specifications. This will ensure that the car steers and handles properly and that the tires wear properly.

FRICTION MODIFIER: Changes the coefficient of friction of the fluid between the mating steel and composition clutch/band surfaces during the engagement process and allows for a certain amount of intentional slipping for a good "shift-feel".

FRONTAL AREA: The total frontal area of a vehicle exposed to air flow.

FUEL FILTER: A component of the fuel system containing a porous paper element used to prevent any impurities from entering the engine through the fuel system. It usually takes the form of a canister-like housing, mounted in-line with the fuel hose, located anywhere on a vehicle between the fuel tank and engine.

FUEL INJECTION: A system replacing the carburetor that sprays fuel into the cylinder through nozzles. The amount of fuel can be more precisely controlled with fuel injection.

FULL FLOATING AXLE: An axle in which the axle housing extends through the wheel giving bearing support on the outside of the housing. The front axle of a four-wheel drive vehicle is usually a full floating axle, as are the rear axles of many larger (1 ton and over) pick-ups and vans.

FULL-TIME FOUR-WHEEL DRIVE: A four-wheel drive system that continuously delivers power to all four wheels. A differential between the front and rear driveshafts permits variations in axle speeds to control gear wind-up without damage.

FULL THROTTLE DETENT DOWNSHIFT: A quick apply of accelerator pedal to its full travel, forcing a downshift.

FUSE: A protective device in a circuit which prevents circuit overload by breaking the circuit when a specific amperage is present. The device is constructed around a strip or wire of a lower amperage rating than the circuit it is designed to protect. When an amperage higher than that stamped on the fuse is present in the circuit, the strip or wire melts, opening the circuit.

FUSIBLE LINK: A piece of wire in a wiring harness that performs the same job as a fuse. If overloaded, the fusible link will melt and interrupt the circuit.

FWD: Front wheel drive.

GAWR: (Gross axle weight rating) the total maximum weight an axle is designed to carry.

GCW: (Gross combined weight) total combined weight of a tow vehicle and trailer.

GARAGE SHIFT: initial engagement feel of transmission, neutral to reverse or neutral to a forward drive.

GARAGE SHIFT FEEL: A quick check of the engagement quality and responsiveness of reverse and forward gears. This test is done with the vehicle stationary.

GEAR: A toothed mechanical device that acts as a rotating lever to transmit power or turning effort from one shaft to another. (See gear ratio.)

GEAR RATIO: A ratio expressing the number of turns a smaller gear will make to turn a larger gear through one revolution. The ratio is found by dividing the number of teeth on the smaller gear into the number of teeth on the larger gear.

GEARBOX: Transmission

GEAR REDUCTION: Torque is multiplied and speed decreased by the factor of the gear ratio. For example, a 3:1 gear ratio changes an input torque of 180 ft. lbs. and an input speed of 2700 rpm to 540 Ft. lbs. and 900 rpm, respectively. (No account is taken of frictional losses, which are always present.)

GEARTRAIN: A succession of intermeshing gears that form an assembly and provide for one or more torque changes as the power input is transmitted to the power output.

GEL COAT: A thin coat of plastic resin covering fiberglass body panels.

GENERATOR: A device which produces direct current (DC) necessary to charge the battery.

GOVERNOR: A device that senses vehicle speed and generates a hydraulic oil pressure. As vehicle speed increases, governor oil pressure rises.

GROUND CIRCUIT: (See circuit, ground.)

GROUND SIDE SWITCHING: The electrical/electronic circuit control switch is located after the circuit load.

GVWR: (Gross vehicle weight rating) total maximum weight a vehicle is designed to carry including the weight of the vehicle, passengers, equipment, gas, oil, etc.

HALOGEN: A special type of lamp known for its quality of brilliant white light. Originally used for fog lights and driving lights.

HARD CODES: DTCs that are present at the time of testing; also called continuous or current codes.

HARSH(ROUGH): An apply of a clutch or band that is more noticeable than a firm one; considered undesirable at any throttle position.

HEADER TANK: An expansion tank for the radiator coolant. It can be located remotely or built into the radiator.

HEAT RANGE: A term used to describe the ability of a spark plug to carry away heat. Plugs with longer nosed insulators take longer to carry heat off effectively.

HEAT RISER: A flapper in the exhaust manifold that is closed when the engine is cold, causing hot exhaust gases to heat the intake manifold providing better cold engine operation. A thermostatic spring opens the flapper when the engine warms up.

HEAVY THROTTLE: Approximately three-fourths of accelerator pedal travel.

HEMI: A name given an engine using hemispherical combustion chambers.

HERTZ (HZ): The international unit of frequency equal to one cycle per second (10,000 Hertz equals 10,000 cycles per second).

HIGH-IMPEDANCE DVOM (DIGITAL VOLT-OHMMETER): This styled device provides a built-in resistance value and is capable of limiting circuit current flow to safe milliamp levels.

HIGH RESISTANCE: Often refers to a circuit where there is an excessive amount of opposition to normal current flow.

HORSEPOWER: A measurement of the amount of work; one horsepower is the amount of work necessary to lift 33,000 lbs. one foot in one minute. Brake horsepower (bhp) is the horsepower delivered by an engine on a dynamometer. Net horsepower is the power remaining (measured at the flywheel of the engine) that can be used to turn the wheels after power is consumed through friction and running the engine accessories (water pump, alternator, air pump, fan etc.)

HOT CIRCUIT: (See circuit, hot; hot lead.)

HOT LEAD: A wire or conductor in the power side of the circuit. (See circuit, hot.)

HOT SIDE SWITCHING: The electrical/electronic circuit control switch is located before the circuit load.

HUB: The center part of a wheel or gear.

HUNTING (BUSYNESS): Repeating quick series of up-shifts and downshifts that causes noticeable change in engine rpm, for example, as in a 4-3-4 shift pattern.

HYDRAULICS: The use of liquid under pressure to transfer force of motion.

HYDROCARBON (HC): Any chemical compound made up of hydrogen and carbon. A major pollutant formed by the engine as a by-product of combustion.

HYDRODYNAMIC DRIVE UNITS: Devices that transmit power solely by the action of a kinetic fluid flow in a closed recirculating path. An impeller energizes the fluid and discharges the high-speed jet stream into the turbine for power output.

HYDROMETER: An instrument used to measure the specific gravity of a solution.

HYDROPLANING: A phenomenon of driving when water builds up under the tire tread, causing it to lose contact with the road. Slowing down will usually restore normal tire contact with the road.

HYPOID GEARSET: The drive pinion gear may be placed below or above the centerline of the driven gear; often used as a final drive gearset.

IDLE MIXTURE: The mixture of air and fuel (usually about 14:1) being fed to the cylinders. The idle mixture screw(s) are sometimes adjusted as part of a tune-up.

IDLER ARM: Component of the steering linkage which is a geometric duplicate of the steering gear arm. It supports the right side of the center steering link.

IMPELLER: Often called a pump, the impeller is the power input (drive) member of a hydrodynamic drive. As part of the torque converter cover, it acts as a centrifugal pump and puts the fluid in motion.

INCH POUND (inch lbs.; sometimes in. lb. or in. lbs.): One twelfth of a foot pound.

INDUCTANCE: The force that produces voltage when a conductor is passed through a magnetic field.

INDUCTION: A means of transferring electrical energy in the form of a magnetic field. Principle used in the ignition coil to increase voltage.

INITIAL FEEL: A distinct firmer feel at start of shift when compared with feel at finish of shift.

INJECTOR: A device which receives metered fuel under relatively low pressure and is activated to inject the fuel into the engine under relatively high pressure at a predetermined time.

INPUT: In an automatic transmission, the source of power from the engine is absorbed by the torque converter, which provides the power input into the transmission. The turbine drives the input(turbine)shaft.

INPUT SHAFT: The shaft to which torque is applied, usually carrying the driving gear or gears.

INTAKE MANIFOLD: A casting of passages or pipes used to conduct air or a fuel/air mixture to the cylinders.

INTERNAL GEAR: The ring-like outer gear of a planetary gearset with the gear teeth cut on the inside of the ring to provide a mesh with the planet pinions.

ISOLATION (CLAMPING) DIODES: Diodes positioned in a circuit to prevent self-induction from damaging electronic components.

IX ROTARY GEAR PUMP: Contains two rotating members, one shaped with internal gear teeth and the other with external gear teeth. As the gears separate, the fluid fills the gaps between gear teeth, is pulled across a crescent-shaped divider, and then is forced to flow through the outlet as the gears mesh.

IX ROTARY LOBE PUMP: Sometimes referred to as a gerotor type pump. Two rotating members, one shaped with internal lobes and the other with external lobes, separate and then mesh to cause fluid to flow.

JOURNAL: The bearing surface within which a shaft operates.

JUMPER CABLES: Two heavy duty wires with large alligator clips used to provide power from a charged battery to a discharged battery mounted in a vehicle.

JUMPSTART: Utilizing the sufficiently charged battery of one vehicle to start the engine of another vehicle with a discharged battery by the use of jumper cables.

KEY: A small block usually fitted in a notch between a shaft and a hub to prevent slippage of the two parts.

KICKDOWN: Detent downshift system; either linkage, cable, or electrically controlled.

KILO: A prefix used in the metric system to indicate one thousand.

KNOCK: Noise which results from the spontaneous ignition of a portion of the air-fuel mixture in the engine cylinder caused by overly advanced ignition timing or use of incorrectly low octane fuel for that engine.

KNOCK SENSOR: An input device that responds to spark knock, caused by over advanced ignition timing.

LABOR TIME: A specific amount of time required to perform a certain repair or diagnostic service as defined by a vehicle or after-market manufacturer .

LACQUER: A quick-drying automotive paint.

LATE: Shift that occurs when engine is at higher than normal rpm for given amount of throttle.

LIGHT-EMITTING DIODE (LED): A semiconductor diode that emits light as electrical current flows through it; used in some electronic display devices to emit a red or other color light.

LIGHT THROTTLE: Approximately one-fourth of accelerator pedal travel.

LIMITED SLIP: A type of differential which transfers driving force to the wheel with the best traction.

LIMP-IN MODE: Electrical shutdown of the transmission/ transaxle output solenoids, allowing only forward and reverse gears that are hydraulically energized by the manual valve. This permits the vehicle to be driven to a service facility for repair.

LIP SEAL: Molded synthetic rubber seal designed with an outer sealing edge (lip) that points into the fluid containing area to be sealed. This type of seal is used where rotational and axial forces are present.

LITHIUM-BASE GREASE: Chassis and wheel bearing grease using lithium as a base. Not compatible with sodium-base grease.

LOAD DEVICE: A circuit's resistance that converts the electrical energy into light, sound, heat, or mechanical movement.

LOAD RANGE: Indicates the number of plies at which a tire is rated. Load range B equals four-ply rating; C equals six-ply rating; and, D equals an eight-ply rating.

LOAD TORQUE: The amount of output torque needed from the transmission/transaxle to overcome the vehicle load.

LOCKING HUBS: Accessories used on part-time four-wheel drive systems that allow the front wheels to be disengaged from the drive train when four-wheel drive is not being used. When four-wheel drive is desired, the hubs are engaged, locking the wheels to the drive train.

LOCKUP CONVERTER: A torque converter that operates hydraulically and mechanically. When an internal apply plate (lockup plate) clamps to the torque converter cover, hydraulic slippage is eliminated.

LOCK RING: See Circlip or Snapring

MAGNET: Any body with the property of attracting iron or steel.

MAGNETIC FIELD: The area surrounding the poles of a magnet that is affected by its attraction or repulsion forces.

MAIN LINE PRESSURE: Often called control pressure or line pressure, it refers to the pressure of the oil leaving the pump and is controlled by the pressure regulator valve.

MALFUNCTION INDICATOR LAMP (MIL): Previously known as a check engine light, the dash-mounted MIL illuminates and signals the driver that an emission or driveability problem with the powertrain has been detected by the ECM/PCM. When this occurs, at least one diagnostic trouble code (DTC) has been stored into the control module memory.

MANIFOLD ABSOLUTE PRESSURE (MAP) SENSOR: Reads the amount of air pressure (vacuum) in the engine's intake manifold system; its signal is used to analyze engine load conditions.

MANIFOLD VACUUM: Low pressure in an engine intake manifold formed just below the throttle plates. Manifold vacuum is highest at idle and drops under acceleration.

MANIFOLD: A casting of passages or set of pipes which connect the cylinders to an inlet or outlet source.

MANUAL LEVER POSITION SWITCH (MLPS): A mechanical switching unit that is typically mounted externally to the transmission/transaxle to inform the PCM/ECM which gear range the driver has selected.

MANUAL VALVE: Located inside the transmission/transaxle, it is directly connected to the driver's shift lever. The position of the manual valve determines which hydraulic circuits will be charged with oil pressure and the operating mode of the transmission.

MANUAL VALVE LEVER POSITION SENSOR (MVLPS): The input from this device tells the TCM what gear range was selected.

MASS AIR FLOW (MAF) SENSOR: Measures the airflow into the engine.

MASTER CYLINDER: The primary fluid pressurizing device in a hydraulic system. In automotive use, it is found in brake and hydraulic clutch systems and is pedal activated, either directly or, in a power brake system, through the power booster.

MacPherson STRUT: A suspension component combining a shock absorber and spring in one unit.

MEDIUM THROTTLE: Approximately one-half of accelerator pedal travel.

MEGA: A metric prefix indicating one million.

MEMBER: An independent component of a hydrodynamic unit such as an impeller, a stator, or a turbine. It may have one or more elements.

MERCON: A fluid developed by Ford Motor Company in 1988. It contains a friction modifier and closely resembles operating characteristics of Dexron.

METAL SEALING RINGS: Made from cast iron or aluminum, their primary application is with dynamic components involving pressure sealing circuits of rotating members. These rings are designed with either butt or hook lock end joints.

METER (ANALOG): A linear-style meter representing data as lengths; a needle-style instrument interfacing with logical numerical increments. This style of electrical meter uses relatively low impedance internal resistance and cannot be used for testing electronic circuitry.

METER (DIGITAL): Uses numbers as a direct readout to show values. Most meters of this style use high impedance internal resistance and must be used for testing low current electronic circuitry.

MICRO: A metric prefix indicating one-millionth (0.000001).

MILLI: A metric prefix indicating one-thousandth (0.001).

MINIMUM THROTTLE: The least amount of throttle opening required for upshift; normally close to zero throttle.

MISFIRE: Condition occurring when the fuel mixture in a cylinder fails to ignite, causing the engine to run roughly.

MODULE: Electronic control unit, amplifier or igniter of solid state or integrated design which controls the current flow in the ignition primary circuit based on input from the pick-up coil. When the module opens the primary circuit, high secondary voltage is induced in the coil.

MODULATED: In an electronic-hydraulic converter clutch system (or shift valve system), the term modulated refers to the pulsing of a solenoid, at a variable rate. This action controls the buildup of oil pressure in the hydraulic circuit to allow a controlled amount of clutch slippage.

MODULATED CONVERTER CLUTCH CONTROL (MCCC): A pulse width duty cycle valve that controls the converter lockup apply pressure and maximizes smoother transitions between lock and unlock conditions.

MODULATOR PRESSURE (THROTTLE PRESSURE): A hydraulic signal oil pressure relating to the amount of engine load, based on either the amount of throttle plate opening or engine vacuum.

MODULATOR VALVE: A regulator valve that is controlled by engine vacuum, providing a hydraulic pressure that varies in relation to engine torque. The hydraulic torque signal functions to delay the shift pattern and provide a line pressure boost. (See throttle valve.)

MOTOR: An electromagnetic device used to convert electrical energy into mechanical energy.

MULTIPLE-DISC CLUTCH: A grouping of steel and friction lined plates that, when compressed together by hydraulic pressure acting upon a piston, lock or unlock a planetary member.

MULTI-WEIGHT: Type of oil that provides adequate lubrication at both high and low temperatures.

needed to move one amp through a resistance of one ohm.

MUSHY: Same as soft; slow and drawn out clutch apply with very little shift feel.

MUTUAL INDUCTION: The generation of current from one wire circuit to another by movement of the magnetic field surrounding a current-carrying circuit as its ampere flow increases or decreases.

NEEDLE BEARING: A bearing which consists of a number (usually a large number) of long, thin rollers.

NITROGEN OXIDE (NOx): One of the three basic pollutants found in the exhaust emission of an internal combustion engine. The amount of NOx usually varies in an inverse proportion to the amount of HC and CO.

NONPOSITIVE SEALING: A sealing method that allows some minor leakage, which normally assists in lubrication.

02 SENSOR: Located in the engine's exhaust system, it is an input device to the ECM/PCM for managing the fuel delivery and ignition system. A scanner can be used to observe the fluctuating voltage readings produced by an 02 sensor as the oxygen content of the exhaust is analyzed.

O-RING SEAL: Molded synthetic rubber seal designed with a circular cross-section. This type of seal is used primarily in static applications.

OBD II (ON-BOARD DIAGNOSTICS, SECOND GENERATION): Refers to the federal law mandating tighter control of 1996 and newer vehicle emissions, active monitoring of related devices, and standardization of terminology, data link connectors, and other technician concerns.

OCTANE RATING: A number, indicating the quality of gasoline based on its ability to resist knock. The higher the number, the better the quality. Higher compression engines require higher octane gas.

OEM: Original Equipment Manufactured. OEM equipment is that furnished standard by the manufacturer.

OFFSET: The distance between the vertical center of the wheel and the mounting surface at the lugs. Offset is positive if the center is outside the lug circle; negative offset puts the center line inside the lug circle.

OHM'S LAW: A law of electricity that states the relationship between voltage, current, and resistance. Volts = amperes x ohms

OHM: The unit used to measure the resistance of conductor-to-electrical flow. One ohm is the amount of resistance that limits current flow to one ampere in a circuit with one volt of pressure.

OHMMETER: An instrument used for measuring the resistance, in ohms, in an electrical circuit.

ONE-WAY CLUTCH: A mechanical clutch of roller or sprag design that resists torque or transmits power in one direction only. It is used to either hold or drive a planetary member.

ONE-WAY ROLLER CLUTCH: A mechanical device that transmits or holds torque in one direction only.

OPEN CIRCUIT: A break or lack of contact in an electrical circuit, either intentional (switch) or unintentional (bad connection or broken wire).

ORIFICE: Located in hydraulic oil circuits, it acts as a restriction. It slows down fluid flow to either create back pressure or delay pressure buildup downstream.

OSCILLOSCOPE: A piece of test equipment that shows electric impulses as a pattern on a screen. Engine performance can be analyzed by interpreting these patterns.

OUTPUT SHAFT: The shaft which transmits torque from a device, such as a transmission.

OUTPUT SPEED SENSOR (OSS): Identifies transmission/transaxle output shaft speed for shift timing and may be used to calculate TCC slip; often functions as the VSS (vehicle speed sensor).

OVERDRIVE: (1.) A device attached to or incorporated in a transmission/transaxle that allows the engine to turn less than one full revolution for every complete revolution of the wheels. The net effect is to reduce engine rpm, thereby using less fuel. A typical overdrive gear ratio would be .87:1, instead of the normal 1:1 in high gear. (2.) A gear assembly which produces more shaft revolutions than that transmitted to it.

OVERDRIVE PLANETARY GEARSET: A single planetary gearset designed to provide a direct drive and overdrive ratio. When coupled to a three-speed transmission/transaxle configuration, a four-speed/overdrive unit is present.

OVERHEAD CAMSHAFT (OHC): An engine configuration in which the camshaft is mounted on top of the cylinder head and operates the valve either directly or by means of rocker arms.

OVERHEAD VALVE (OHV): An engine configuration in which all of the valves are located in the cylinder head and the camshaft is located in the cylinder block. The camshaft operates the valves via lifters and pushrods.

OVERRUNCLUTCH: Another name for a one-way mechanical clutch. Applies to both roller and sprag designs.

OVERSTEER: The tendency of some vehicles, when steering into a turn, to over-respond or steer more than required, which could result in excessive slip of the rear wheels. Opposite of under-steer.

OXIDATION STABILIZERS: Absorb and dissipate heat. Automatic transmission fluid has high resistance to varnish and sludge buildup that occurs from excessive heat that is generated primarily in the torque converter. Local temperatures as high as 6000F (3150C) can occur at the clutch plates during engagement, and this heat must be absorbed and dissipated. If the fluid cannot withstand the heat, it burns or oxidizes, resulting in an almost immediate destruction of friction materials, clogged filter screen and hydraulic passages, and sticky valves.

OXIDES OF NITROGEN: See nitrogen oxide (NOx).

OXYGEN SENSOR: Used with a feedback system to sense the presence of oxygen in the exhaust gas and signal the computer which can use the voltage signal to determine engine operating efficiency and adjust the air/fuel ratio.

PARALLEL CIRCUIT: (See circuit, parallel.)

PARTS WASHER: A basin or tub, usually with a built-in pump mechanism and hose used for circulating chemical solvent for the purpose of cleaning greasy, oily and dirty components.

PART-TIME FOUR WHEEL DRIVE: A system that is normally in the two wheel drive mode and only runs in four-wheel drive when the system is manually engaged because more traction is desired. Two or four wheel drive is normally selected by a lever to engage the front axle, but if locking hubs are used, these must also be manually engaged in the Lock position. Otherwise, the front axle will not drive the front wheels.

PASSIVE RESTRAINT: Safety systems such as air bags or automatic seat belts which operate with no action required on the part of the driver or passenger. Mandated by Federal regulations on all vehicles sold in the U.S. after 1990.

PAYLOAD: The weight the vehicle is capable of carrying in addition to its own weight. Payload includes weight of the driver, passengers and cargo, but not coolant, fuel, lubricant, spare tire, etc.

PCM: Powertrain control module.

PCV VALVE: A valve usually located in the rocker cover that vents crankcase vapors back into the engine to be reburned.

PERCOLATION: A condition in which the fuel actually "boils," due to excessive heat. Percolation prevents proper atomization of the fuel causing rough running.

PICK-UP COIL: The coil in which voltage is induced in an electronic ignition.

PING: A metallic rattling sound produced by the engine during acceleration. It is usually due to incorrect ignition timing or a poor grade of gasoline.

PINION: The smaller of two gears. The rear axle pinion drives the ring gear which transmits motion to the axle shafts.

PINION GEAR: The smallest gear in a drive gear assembly.

PISTON: A disc or cup that fits in a cylinder bore and is free to move. In hydraulics, it provides the means of converting hydraulic pressure into a usable force. Examples of piston applications are found in servo, clutch, and accumulator units.

PISTON RING: An open-ended ring which fits into a groove on the outer diameter of the piston. Its chief function is to form a seal between the piston and cylinder wall. Most automotive pistons have three rings: two for compression sealing; one for oil sealing.

PITMAN ARM: A lever which transmits steering force from the steering gear to the steering linkage.

PLANET CARRIER: A basic member of a planetary gear assembly that carries the pinion gears.

PLANET PINIONS: Gears housed in a planet carrier that are in constant mesh with the sun gear and internal gear. Because they have their own independent rotating centers, the pinions are capable of rotating around the sun gear or the inside of the internal gear.

PLANETARY GEAR RATIO: The reduction or overdrive ratio developed by a planetary gearset.

PLANETARY GEARSET: In its simplest form, it is made up of a basic assembly group containing a sun gear, internal gear, and planet carrier. The gears are always in constant mesh and offer a wide range of gear ratio possibilities.

PLANETARY GEARSET (COMPOUND): Two planetary gearsets combined together.

PLANETARY GEARSET (SIMPLE): An assembly of gears in constant mesh consisting of a sun gear, several pinion gears mounted in a carrier, and a ring gear. It provides gear ratio and direction changes, in addition to a direct drive and a neutral.

PLY RATING: A. rating given a tire which indicates strength (but not necessarily actual plies). A two-ply/four-ply rating has only two plies, but the strength of a four-ply tire.

POLARITY: Indication (positive or negative) of the two poles of a battery.

PORT: An opening for fluid intake or exhaust.

POSITIVE SEALING: A sealing method that completely prevents leakage.

POTENTIAL: Electrical force measured in volts; sometimes used interchangeably with voltage.

POWER: The ability to do work per unit of time, as expressed in horsepower; one horsepower equals 33,000 ft. lbs. of work per minute, or 550 ft. lbs. of work per second.

POWER FLOW: The systematic flow or transmission of power through the gears, from the input shaft to the output shaft.

POWER-TO-WEIGHT RATIO: Ratio of horsepower to weight of car.

POWERTRAIN: See Drivetrain.

POWERTRAIN CONTROL MODULE (PCM): Current designation for the engine control module (ECM). In many cases, late model vehicle control units manage the engine as well as the transmission. In other settings, the PCM controls the engine and is interfaced with a TCM to control transmission functions.

Ppm: Parts per million; unit used to measure exhaust emissions.

PREIGNITION: Early ignition of fuel in the cylinder, sometimes due to glowing carbon deposits in the combustion chamber. Preignition can be damaging since combustion takes place prematurely.

PRELOAD: A predetermined load placed on a bearing during assembly or by adjustment.

PRESS FIT: The mating of two parts under pressure, due to the inner diameter of one being smaller than the outer diameter of the other, or vice versa; an interference fit.

PRESSURE: The amount of force exerted upon a surface area.

PRESSURE CONTROL SOLENOID (PCS): An output device that provides a boost oil pressure to the mainline regulator valve to control line pressure. Its operation is determined by the amount of current sent from the PCM.

PRESSURE GAUGE: An instrument used for measuring the fluid pressure in a hydraulic circuit.

PRESSURE REGULATOR VALVE: In automatic transmissions, its purpose is to regulate the pressure of the pump output and supply the basic fluid pressure necessary to operate the transmission. The regulated fluid pressure may be referred to as mainline pressure, line pressure, or control pressure.

PRESSURE SWITCH ASSEMBLY (PSA): Mounted inside the transmission, it is a grouping of oil pressure switches that inputs to the PCM when certain hydraulic passages are charged with oil pressure.

PRESSURE PLATE: A spring-loaded plate (part of the clutch) that transmits power to the driven (friction) plate when the clutch is engaged.

PRIMARY CIRCUIT: The low voltage side of the ignition system which consists of the ignition switch, ballast resistor or resistance wire, bypass, coil, electronic control unit and pick-up coil as well as the connecting wires and harnesses.

PROFILE: Term used for tire measurement (tire series), which is the ratio of tire height to tread width.

PROM (PROGRAMMABLE READ-ONLY MEMORY): The heart of the computer that compares input data and makes the engineered program or strategy decisions about when to trigger the appropriate output based on stored computer instructions.

PULSE GENERATOR: A two-wire pickup sensor used to produce a fluctuating electrical signal. This changing signal is read by the controller to determine the speed of the object and can be used to measure transmission/transaxle input speed, output speed, and vehicle speed.

PSI: Pounds per square inch; a measurement of pressure.

PULSE WIDTH DUTY CYCLE SOLENOID (PULSE WIDTH MODULATED SOLENOID): A computer-controlled solenoid that turns on and off at a variable rate producing a modulated oil pressure; often referred to as a pulse width modulated (PWM) solenoid. Employed in many electronic automatic transmissions and transaxles, these solenoids are used to manage shift control and converter clutch hydraulic circuits.

PUSHROD: A steel rod between the hydraulic valve lifter and the valve rocker arm in overhead valve (OHV) engines.

PUMP: A mechanical device designed to create fluid flow and pressure buildup in a hydraulic system.

QUARTER PANEL: General term used to refer to a rear fender. Quarter panel is the area from the rear door opening to the tail light area and from rear wheel well to the base of the trunk and roof-line.

RACE: The surface on the inner or outer ring of a bearing on which the balls, needles or rollers move.

RACK AND PINION: A type of automotive steering system using a pinion gear attached to the end of the steering shaft. The pinion meshes with a long rack attached to the steering linkage.

RADIAL TIRE: Tire design which uses body cords running at right angles to the center line of the tire. Two or more belts are used to give tread strength. Radials can be identified by their characteristic sidewall bulge.

RADIATOR: Part of the cooling system for a water-cooled engine, mounted in the front of the vehicle and connected to the engine with rubber hoses. Through the radiator, excess combustion heat is dissipated into the atmosphere through forced convection using a water and glycol based mixture that circulates through, and cools, the engine.

RANGE REFERENCE AND CLUTCH/BAND APPLY CHART: A guide that shows the application of clutches and bands for each gear, within the selector range positions. These charts are extremely useful for understanding how the unit operates and for diagnosing malfunctions.

RAVIGNEAUX GEARSET: A compound planetary gearset that features matched dual planetary pinions (sets of two) mounted in a single planet carrier. Two sun gears and one ring mesh with the carrier pinions.

REACTION MEMBER: The stationary planetary member, in a planetary gearset, that is grounded to the transmission/transaxle case through the use of friction and wedging devices known as bands, disc clutches, and one-way clutches.

REACTION PRESSURE: The fluid pressure that moves a spool valve against an opposing force or forces; the area on which the opposing force acts. The opposing force can be a spring or a combination of spring force and auxiliary hydraulic force.

REACTOR, TORQUE CONVERTER: The reaction member of a fluid torque converter, more commonly called a stator. (See stator.)

REAR MAIN OIL SEAL: A synthetic or rope-type seal that prevents oil from leaking out of the engine past the rear main crankshaft bearing.

RECIRCULATING BALL: Type of steering system in which recirculating steel balls occupy the area between the nut and worm wheel, causing a reduction in friction.

RECTIFIER: A device (used primarily in alternators) that permits electrical current to flow in one direction only.

REDUCTION: (See gear reduction.)

REGULATOR VALVE: A valve that changes the pressure of the oil in a hydraulic circuit as the oil passes through the valve by bleeding off (or exhausting) some of the volume of oil supplied to the valve.

REFRIGERANT 12 (R-12) or 134 (R-134): The generic name of the refrigerant used in automotive air conditioning systems.

REGULATOR: A device which maintains the amperage and/or voltage levels of a circuit at predetermined values.

RELAY: A switch which automatically opens and/or closes a circuit.

RELAY VALVE: A valve that directs flow and pressure. Relay valves simply connect or disconnect interrelated passages without restricting the fluid flow or changing the pressure.

RELIEF VALVE: A spring-loaded, pressure-operated valve that limits oil pressure buildup in a hydraulic circuit to a predetermined maximum value.

RELUCTOR: A wheel that rotates inside the distributor and triggers the release of voltage in an electronic ignition.

RESERVOIR: The storage area for fluid in a hydraulic system; often called a sump.

RESIN: A liquid plastic used in body work.

RESIDUAL MAGNETISM: The magnetic strength stored in a material after a magnetizing field has been removed.

RESISTANCE: The opposition to the flow of current through a circuit or electrical device, and is measured in ohms. Resistance is equal to the voltage divided by the amperage.

RESISTOR SPARK PLUG: A spark plug using a resistor to shorten the spark duration. This suppresses radio interference and lengthens plug life.

RESISTOR: A device, usually made of wire, which offers a preset amount of resistance in an electrical circuit.

RESULTANT FORCE: The single effective directional thrust of the fluid force on the turbine produced by the vortex and rotary forces acting in different planes.

RETARD: Set the ignition timing so that spark occurs later (fewer degrees before TDC).

RHEOSTAT: A device for regulating a current by means of a variable resistance.

RING GEAR: The name given to a ring-shaped gear attached to a differential case, or affixed to a flywheel or as part of a planetary gear set.

ROADLOAD: grade.

ROCKER ARM: A lever which rotates around a shaft pushing down (opening) the valve with an end when the other end is pushed up by the pushrod. Spring pressure will later close the valve.

ROCKER PANEL: The body panel below the doors between the wheel opening.

ROLLER BEARING: A bearing made up of hardened inner and outer races between which hardened steel rollers move.

ROLLER CLUTCH: A type of one-way clutch design using rollers and springs mounted within an inner and outer cam race assembly.

ROTARY FLOW: The path of the fluid trapped between the blades of the members as they revolve with the rotation of the torque converter cover (rotational inertia).

ROTOR: (1.) The disc-shaped part of a disc brake assembly, upon which the brake pads bear; also called, brake disc. (2.) The device mounted atop the distributor shaft, which passes current to the distributor cap tower contacts.

ROTARY ENGINE: See Wankel engine.

RPM: Revolutions per minute (usually indicates engine speed).

RTV: A gasket making compound that cures as it is exposed to the atmosphere. It is used between surfaces that are not perfectly machined to one another, leaving a slight gap that the RTV fills and in which it hardens. The letters RTV represent room temperature vulcanizing.

RUN-ON: Condition when the engine continues to run, even when the key is turned off. See dieseling.

SEALED BEAM: A automotive headlight. The lens, reflector and filament from a single unit.

SEATBELT INTERLOCK: A system whereby the car cannot be started unless the seatbelt is buckled.

SECONDARY CIRCUIT: The high voltage side of the ignition system, usually above 20,000 volts. The secondary includes the ignition coil, coil wire, distributor cap and rotor, spark plug wires and spark plugs.

SELF-INDUCTION: The generation of voltage in a current-carrying wire by changing the amount of current flowing within that wire.

SEMI-CONDUCTOR: A material (silicon or germanium) that is neither a good conductor nor an insulator; used in diodes and transistors.

SEMI-FLOATING AXLE: In this design, a wheel is attached to the axle shaft, which takes both drive and cornering loads. Almost all solid axle passenger cars and light trucks use this design.

SENDING UNIT: A mechanical, electrical, hydraulic or electromagnetic device which transmits information to a gauge.

SENSOR: Any device designed to measure engine operating conditions or ambient pressures and temperatures. Usually electronic in nature and designed to send a voltage signal to an on-board computer, some sensors may operate as a simple on/off switch or they may provide a variable voltage signal (like a potentiometer) as conditions or measured parameters change.

SERIES CIRCUIT: (See circuit, series.)

SERPENTINE BELT: An accessory drive belt, with small multiple v-ribs, routed around most or all of the engine-powered accessories such as the alternator and power steering pump. Usually both the front and the back side of the belt comes into contact with various pulleys.

SERVO: In an automatic transmission, it is a piston in a cylinder assembly that converts hydraulic pressure into mechanical force and movement; used for the application of the bands and clutches.

SHIFT BUSYNESS: When referring to a torque converter clutch, it is the frequent apply and release of the clutch plate due to uncommon driving conditions.

SHIFT VALVE: Classified as a relay valve, it triggers the automatic shift in response to a governor and a throttle signal by directing fluid to the appropriate band and clutch apply combination to cause the shift to occur.

SHIM: Spacers of precise, predetermined thickness used between parts to establish a proper working relationship.

SHIMMY: Vibration (sometimes violent) in the front end caused by misaligned front end, out of balance tires or worn suspension components.

SHORT CIRCUIT: An electrical malfunction where current takes the path of least resistance to ground (usually through damaged insulation). Current flow is excessive from low resistance resulting in a blown fuse.

SHUDDER: Repeated jerking or stick-slip sensation, similar to chuggle but more severe and rapid in nature, that may be most noticeable during certain ranges of vehicle speed; also used to define condition after converter clutch engagement.

SIMPSON GEARSET: A compound planetary gear train that integrates two simple planetary gearsets referred to as the front planetary and the rear planetary.

SINGLE OVERHEAD CAMSHAFT: See overhead camshaft.

SKIDPLATE: A metal plate attached to the underside of the body to protect the fuel tank, transfer case or other vulnerable parts from damage.

SLAVE CYLINDER: In automotive use, a device in the hydraulic clutch system which is activated by hydraulic force, disengaging the clutch.

SLIPPING: Noticeable increase in engine rpm without vehicle speed increase; usually occurs during or after initial clutch or band engagement.

SLUDGE: Thick, black deposits in engine formed from dirt, oil, water, etc. It is usually formed in engines when oil changes are neglected.

SNAP RING: A circular retaining clip used inside or outside a shaft or part to secure a shaft, such as a floating wrist pin.

SOFT: Slow, almost unnoticeable clutch apply with very little shift feel.

SOFTCODES: DTCs that have been set into the PCM memory but are not present at the time of testing; often referred to as history or intermittent codes.

SOHC: Single overhead camshaft.

SOLENOID: An electrically operated, magnetic switching device.

SPALLING: A wear pattern identified by metal chips flaking off the hardened surface. This condition is caused by foreign particles, overloading situations, and/or normal wear.

SPARK PLUG: A device screwed into the combustion chamber of a spark ignition engine. The basic construction is a conductive core inside of a ceramic insulator, mounted in an outer conductive base. An electrical charge from the spark plug wire travels along the conductive core and jumps a preset air gap to a grounding point or points at the end of the conductive base. The resultant spark ignites the fuel/air mixture in the combustion chamber.

SPECIFIC GRAVITY (BATTERY): The relative weight of liquid (battery electrolyte) as compared to the weight of an equal volume of water.

SPLINES: Ridges machined or cast onto the outer diameter of a shaft or inner diameter of a bore to enable parts to mate without rotation.

SPLIT TORQUE DRIVE: In a torque converter, it refers to parallel paths of torque transmission, one of which is mechanical and the other hydraulic.

SPONGY PEDAL: A soft or spongy feeling when the brake pedal is depressed. It is usually due to air in the brake lines.

SPOOLVALVE: A precision-machined, cylindrically shaped valve made up of lands and grooves. Depending on its position in the valve bore, various interconnecting hydraulic circuit passages are either opened or closed.

SPRAG CLUTCH: A type of one-way clutch design using cams or contoured-shaped sprags between inner and outer races. (See one-way clutch.)

SPRUNG WEIGHT: The weight of a car supported by the springs.

SQUARE-CUT SEAL: Molded synthetic rubber seal designed with a square- or rectangular-shaped cross-section. This type of seal is used for both dynamic and static applications.

SRS: Supplemental restraint system

STABILIZER (SWAY) BAR: A bar linking both sides of the suspension. It resists sway on turns by taking some of added load from one wheel and putting it on the other.

STAGE: The number of turbine sets separated by a stator. A turbine set may be made up of one or more turbine members. A three-element converter is classified as a single stage.

STALL: In fluid drive transmission/transaxle applications, stall refers to engine rpm with the transmission/transaxle engaged and the vehicle stationary; throttle valve can be in any position between closed and wide open.

STALL SPEED: In fluid drive transmission/transaxle applications, stall speed refers to the maximum engine rpm with the transmission/transaxle engaged and vehicle stationary, when the throttle valve is wide open. (See stall; stall test.)

STALL TEST: A procedure recommended by many manufacturers to help determine the integrity of an engine, the torque converter stator, and certain clutch and band combinations. With the shift lever in each of the forward and reverse positions and with the brakes firmly applied, the accelerator pedal is momentarily pressed to the wide open throttle (WOT) position. The engine rpm reading at full throttle can provide clues for diagnosing the condition of the items listed above.

STALL TORQUE: The maximum design or engineered torque ratio of a fluid torque converter, produced under stall speed conditions. (See stall speed.)

STARTER: A high-torque electric motor used for the purpose of starting the engine, typically through a high ratio geared drive connected to the flywheel ring gear.

STATIC: A sealing application in which the parts being sealed do not move in relation to each other.

STATOR (REACTOR): The reaction member of a fluid torque converter that changes the direction of the fluid as it leaves the turbine to enter the impeller vanes. During the torque multiplication phase, this action assists the impeller's rotary force and results in an increase in torque.

STEERING GEOMETRY: Combination of various angles of suspension components (caster, camber, toe-in); roughly equivalent to front end alignment.

STRAIGHT WEIGHT: Term designating motor oil as suitable for use within a narrow range of temperatures. Outside the narrow temperature range its flow characteristics will not adequately lubricate.

STROKE: The distance the piston travels from bottom dead center to top dead center.

SUBSTITUTION: Replacing one part suspected of a defect with a like part of known quality.

SUMP: The storage vessel or reservoir that provides a ready source of fluid to the pump. In an automatic transmission, the sump is the oil pan. All fluid eventually returns to the sump for recycling into the hydraulic system.

SUN GEAR: In a planetary gearset, it is the center gear that meshes with a cluster of planet pinions.

SUPERCHARGER: An air pump driven mechanically by the engine through belts, chains, shafts or gears from the crankshaft. Two general types of supercharger are the positive displacement and centrifugal type, which pump air in direct relationship to the speed of the engine.

SUPPLEMENTAL RESTRAINT SYSTEM: See air bag.

SURGE: Repeating engine-related feeling of acceleration and deceleration that is less intense than chuggle.

SWITCH: A device used to open, close, or redirect the current in an electrical circuit.

SYNCHROMESH: A manual transmission/transaxle that is equipped with devices (synchronizers) that match the gear speeds so that the transmission/transaxle can be downshifted without clashing gears.

SYNTHETIC OIL: Non-petroleum based oil.

TACHOMETER: A device used to measure the rotary speed of an engine, shaft, gear, etc., usually in rotations per minute.

TDC: Top dead center. The exact top of the piston's stroke.

TEFLON SEALING RINGS: Teflon is a soft, durable, plastic-like material that is resistant to heat and provides excellent sealing. These rings are designed with either scarf-cut joints or as one-piece rings. Teflon sealing rings have replaced many metal ring applications.

TERMINAL: A device attached to the end of a wire or cable to make an electrical connection.

TEST LIGHT, CIRCUIT-POWERED: Uses available circuit voltage to test circuit continuity.

TEST LIGHT, SELF-POWERED: Uses its own battery source to test circuit continuity.

THERMISTOR: A special resistor used to measure fluid temperature; it decreases its resistance with increases in temperature.

THERMOSTAT: A valve, located in the cooling system of an engine, which is closed when cold and opens gradually in response to engine heating, controlling the temperature of the coolant and rate of coolant flow.

THERMOSTATIC ELEMENT: A heat-sensitive, spring-type device that controls a drain port from the upper sump area to the lower sump. When the transaxle fluid reaches operating temperature, the port is closed and the upper sump fills, thus reducing the fluid level in the lower sump.

THROTTLE POSITION (TP) SENSOR: Reads the degree of throttle opening; its signal is used to analyze engine load conditions. The ECM/PCM decides to apply the TCC, or to disengage it for coast or load conditions that need a converter torque boost.

THROTTLE PRESSURE/MODULATOR PRESSURE: A hydraulic signal oil pressure relating to the amount of engine load, based on either the amount of throttle plate opening or engine vacuum.

THROTTLE VALVE: A regulating or balanced valve that is controlled mechanically by throttle linkage or engine vacuum. It sends a hydraulic signal to the shift valve body to control shift timing and shift quality. (See balanced valve; modulator valve.)

THROW-OUT BEARING: As the clutch pedal is depressed, the throwout bearing moves against the spring fingers of the pressure plate, forcing the pressure plate to disengage from the driven disc.

TIE ROD: A rod connecting the steering arms. Tie rods have threaded ends that are used to adjust toe-in.

TIE-UP: Condition where two opposing clutches are attempting to apply at same time, causing engine to labor with noticeable loss of engine rpm.

TIMING BELT: A square-toothed, reinforced rubber belt that is driven by the crankshaft and operates the camshaft.

TIMING CHAIN: A roller chain that is driven by the crankshaft and operates the camshaft.

TIRE ROTATION: Moving the tires from one position to another to make the tires wear evenly.

TOE-IN (OUT): A term comparing the extreme front and rear of the front tires. Closer together at the front is toe-in; farther apart at the front is toe-out.

TOP DEAD CENTER (TDC): The point at which the piston reaches the top of its travel on the compression stroke.

TORQUE: Measurement of turning or twisting force, expressed as foot-pounds or inch-pounds.

TORQUE CONVERTER: A turbine used to transmit power from a driving member to a driven member via hydraulic action, providing changes in drive ratio and torque. In automotive use, it links the driveplate at the rear of the engine to the automatic transmission.

TORQUE CONVERTER CLUTCH: The apply plate (lockup plate) assembly used for mechanical power flow through the converter.

TORQUE PHASE: Sometimes referred to as slip phase or stall phase, torque multiplication occurs when the turbine is turning at a slower speed than the impeller, and the stator is reactionary (stationary). This sequence generates a boost in output torque.

TORQUE RATING (STALL TORQUE): The maximum torque multiplication that occurs during stall conditions, with the engine at wide open throttle (WOT) and zero turbine speed.

TORQUE RATIO: An expression of the gear ratio factor on torque effect. A 3:1 gear ratio or 3:1 torque ratio increases the torque input by the ratio factor of 3. Input torque (100 ft. lbs.) x 3 = output torque (300 ft. lbs.)

TRACTION: The amount of usable tractive effort before the drive wheels slip on the road contact surface.

TORSION BAR SUSPENSION: Long rods of spring steel which take the place of springs. One end of the bar is anchored and the other arm (attached to the suspension) is free to twist. The bars' resistance to twisting causes springing action.

TRACK: Distance between the centers of the tires where they contact the ground.

TRACTION CONTROL: A control system that prevents the spinning of a vehicle's drive wheels when excess power is applied.

TRACTIVE EFFORT: The amount of force available to the drive wheels, to move the vehicle.

TRANSAXLE: A single housing containing the transmission and differential. Transaxles are usually found on front engine/front wheel drive or rear engine/rear wheel drive cars.

TRANSDUCER: A device that changes energy from one form to another. For example, a transducer in a microphone changes sound energy to electrical energy. In automotive air-conditioning controls used in automatic temperature systems, a transducer changes an electrical signal to a vacuum signal, which operates mechanical doors.

TRANSMISSION: A powertrain component designed to modify torque and speed developed by the engine; also provides direct drive, reverse, and neutral.

TRANSMISSION CONTROL MODULE (TCM): Manages transmission functions. These vary according to the manufacturer's product design but may include converter clutch operation, electronic shift scheduling, and mainline pressure.

TRANSMISSION FLUID TEMPERATURE (TFT) SENSOR: Originally called a transmission oil temperature (TOT) sensor, this input device to the ECM/PCM senses the fluid temperature and provides a resistance value. It operates on the thermistor principle.

TRANSMISSION INPUT SPEED (TIS) SENSOR: Measures turbine shaft (input shaft) rpm's and compares to engine rpm's to determine torque converter slip. When compared to the transmission output speed sensor or VSS, gear ratio and clutch engagement timing can be determined.

TRANSMISSION OIL TEMPERATURE (TOT) SENSOR: (See transmission fluid temperature (TFT) sensor.)

TRANSMISSION RANGE SELECTOR (TRS) SWITCH: Tells the module which gear shift position the driver has chosen.

TRANSFER CASE: A gearbox driven from the transmission that delivers power to both front and rear driveshafts in a four-wheel drive system. Transfer cases usually have a high and low range set of gears, used depending on how much pulling power is needed.

TRANSISTOR: A semi-conductor component which can be actuated by a small voltage to perform an electrical switching function.

TREAD WEAR INDICATOR: Bars molded into the tire at right angles to the tread that appear as horizontal bars when 1⁄16 in. of tread remains.

TREAD WEAR PATTERN: The pattern of wear on tires which can be "read" to diagnose problems in the front suspension.

TUNE-UP: A regular maintenance function, usually associated with the replacement and adjustment of parts and components in the electrical and fuel systems of a vehicle for the purpose of attaining optimum performance.

TURBINE: The output (driven) member of a fluid coupling or fluid torque converter. It is splined to the input (turbine) shaft of the transmission.

TURBOCHARGER: An exhaust driven pump which compresses intake air and forces it into the combustion chambers at higher than atmospheric pressures. The increased air pressure allows more fuel to be burned and results in increased horsepower being produced.

TURBULENCE: The interference of molecules of a fluid (or vapor) with each other in a fluid flow.

TYPE F: Transmission fluid developed and used by Ford Motor Company up to 1982. This fluid type provides a high coefficient of friction.

TYPE 7176: The preferred choice of transmission fluid for Chrysler automatic transmissions and transaxles. Developed in 1986, it closely resembles Dexron and Mercon. Type 7176 is the recommended service fill fluid for all Chrysler products utilizing a lockup torque converter dating back to 1978.

U-JOINT (UNIVERSAL JOINT): A flexible coupling in the drive train that allows the driveshafts or axle shafts to operate at different angles and still transmit rotary power.

UNDERSTEER: The tendency of a car to continue straight ahead while negotiating a turn.

UNIT BODY: Design in which the car body acts as the frame.

UNLEADED FUEL: Fuel which contains no lead (a common gasoline additive). The presence of lead in fuel will destroy the functioning elements of a catalytic converter, making it useless.

UNSPRUNG WEIGHT: The weight of car components not supported by the springs (wheels, tires, brakes, rear axle, control arms, etc.).

UPSHIFT: A shift that results in a decrease in torque ratio and an increase in speed.

VACUUM: A negative pressure; any pressure less than atmospheric pressure.

VACUUM ADVANCE: A device which advances the ignition timing in response to increased engine vacuum.

VACUUM GAUGE: An instrument used for measuring the existing vacuum in a vacuum circuit or chamber. The unit of measure is inches (of mercury in a barometer).

VACUUM MODULATOR: Generates a hydraulic oil pressure in response to the amount of engine vacuum.

VALVES: Devices that can open or close fluid passages in a hydraulic system and are used for directing fluid flow and controlling pressure.

VALVE BODY ASSEMBLY: The main hydraulic control assembly of the transmission/transaxle that contains numerous valves, check balls, and other components to control the distribution of pressurized oil throughout the transmission.

VALVE CLEARANCE: The measured gap between the end of the valve stem and the rocker arm, cam lobe or follower that activates the valve.

VALVE GUIDES: The guide through which the stem of the valve passes. The guide is designed to keep the valve in proper alignment.

VALVE LASH (clearance): The operating clearance in the valve train.

VALVE TRAIN: The system that operates intake and exhaust valves, consisting of camshaft, valves and springs, lifters, pushrods and rocker arms.

VAPOR LOCK: Boiling of the fuel in the fuel lines due to excess heat. This will interfere with the flow of fuel in the lines and can completely stop the flow. Vapor lock normally only occurs in hot weather.

VARIABLE DISPLACEMENT (VARIABLE CAPACITY) VANE PUMP: Slipper-type vanes, mounted in a revolving rotor and contained within the bore of a movable slide, capture and then force fluid to flow. Movement of the slide to various positions changes the size of the vane chambers and the amount of fluid flow. **Note:** GM refers to this pump design as variable displacement, and Ford terms it variable capacity.

VARIABLE FORCE SOLENOID (VFS): Commonly referred to as the electronic pressure control (EPC) solenoid, it replaces the cable/linkage style of TV system control and is integrated with a spool valve and spring assembly to control pressure. A variable computer-controlled current flow varies the internal force of the solenoid on the spool valve and resulting control pressure.

VARIABLE ORIFICE THERMAL VALVE: Temperature-sensitive hydraulic oil control device that adjusts the size of a circuit path opening. By altering the size of the opening, the oil flow rate is adapted for cold to hot oil viscosity changes.

VARNISH: Term applied to the residue formed when gasoline gets old and stale.

VCM: See Electronic Control Unit (ECU).

VEHICLE SPEED SENSOR (VSS): Provides an electrical signal to the computer module, measuring vehicle speed, and affects the torque converter clutch engagement and release.

VESPEL SEALING RINGS: Hard plastic material that produces excellent sealing in dynamic settings. These rings are found in late versions of the 4T60 and in all 4T60-E and 4T80-E transaxles.

VISCOSITY: The ability of a fluid to flow. The lower the viscosity rating, the easier the fluid will flow. 10 weight motor oil will flow much easier than 40 weight motor oil.

VISCOSITY INDEX IMPROVERS: Keeps the viscosity nearly constant with changes in temperature. This is especially important at low temperatures, when the oil needs to be thin to aid in shifting and for cold-weather starting. Yet it must not be so thin that at high temperatures it will cause excessive hydraulic leakage so that pumps are unable to maintain the proper pressures.

VISCOUS CLUTCH: A specially designed torque converter clutch apply plate that, through the use of a silicon fluid, clamps smoothly and absorbs torsional vibrations.

VOLT: Unit used to measure the force or pressure of electricity. It is defined as the pressure needed to move one amp through the resistance of one ohm.

VOLTAGE: The electrical pressure that causes current to flow. Voltage is measured in volts (V).

VOLTAGE, APPLIED: The actual voltage read at a given point in a circuit. It equals the available voltage of the power supply minus the losses in the circuit up to that point.

VOLTAGE DROP: The voltage lost or used in a circuit by normal loads such as a motor or lamp or by abnormal loads such as a poor (high-resistance) lead or terminal connection.

VOLTAGE REGULATOR: A device that controls the current output of the alternator or generator.

VOLTMETER: An instrument used for measuring electrical force in units called volts. Voltmeters are always connected parallel with the circuit being tested.

VORTEX FLOW: The crosswise or circulatory flow of oil between the blades of the members caused by the centrifugal pumping action of the impeller.

WANKEL ENGINE: An engine which uses no pistons. In place of pistons, triangular-shaped rotors revolve in specially shaped housings.

WATER PUMP: A belt driven component of the cooling system that mounts on the engine, circulating the coolant under pressure.

WATT: The unit for measuring electrical power. One watt is the product of one ampere and one volt (watts equals amps times volts). Wattage is the horsepower of electricity (746 watts equal one horsepower).

WHEEL ALIGNMENT: Inclusive term to describe the front end geometry (caster, camber, toe-in/out).

WHEEL CYLINDER: Found in the automotive drum brake assembly, it is a device, actuated by hydraulic pressure, which, through internal pistons, pushes the brake shoes outward against the drums.

WHEEL WEIGHT: Small weights attached to the wheel to balance the wheel and tire assembly. Out-of-balance tires quickly wear out and also give erratic handling when installed on the front.

WHEELBASE: Distance between the center of front wheels and the center of rear wheels.

WIDE OPEN THROTTLE (WOT): Full travel of accelerator pedal.

WORK: The force exerted to move a mass or object. Work involves motion; if a force is exerted and no motion takes place, no work is done. Work per unit of time is called power. Work = force x distance = ft. lbs. 33,000 ft. lbs. in one minute = 1 horsepower

ZERO-THROTTLE COAST DOWN: A full release of accelerator pedal while vehicle is in motion and in drive range.

ENGLISH TO METRIC CONVERSION: TORQUE

To convert foot-pounds (ft. lbs.) to Newton-meters (Nm), multiply the number of ft. lbs. by 1.36

To convert Newton-meters (Nm) to foot-pounds (ft. lbs.), multiply the number of Nm by 0.7376

ft. lbs.	Nm	ft. lbs.	Nm	ft. lbs.	Nm	ft. lbs.	Nm
0.1	0.1	34	46.2	76	103.4	118	160.5
0.2	0.3	35	47.6	77	104.7	119	161.8
0.3	0.4	36	49.0	78	106.1	120	163.2
0.4	0.5	37	50.3	79	107.4	121	164.6
0.5	0.7	38	51.7	80	108.8	122	165.9
0.6	0.8	39	53.0	81	110.2	123	167.3
0.7	1.0	40	54.4	82	111.5	124	168.6
0.8	1.1	41	55.8	83	112.9	125	170.0
0.9	1.2	42	57.1	84	114.2	126	171.4
1	1.4	43	58.5	85	115.6	127	172.7
2	2.7	44	59.8	86	117.0	128	174.1
3	4.1	45	61.2	87	118.3	129	175.4
4	5.4	46	62.6	88	119.7	130	176.8
5	6.8	47	63.9	89	121.0	131	178.2
6	8.2	48	65.3	90	122.4	132	179.5
7	9.5	49	66.6	91	123.8	133	180.9
8	10.9	50	68.0	92	125.1	134	182.2
9	12.2	51	69.4	93	126.5	135	183.6
10	13.6	52	70.7	94	127.8	136	185.0
11	15.0	53	72.1	95	129.2	137	186.3
12	16.3	54	73.4	96	130.6	138	187.7
13	17.7	55	74.8	97	131.9	139	189.0
14	19.0	56	76.2	98	133.3	140	190.4
15	20.4	57	77.5	99	134.6	141	191.8
16	21.8	58	78.9	100	136.0	142	193.1
17	23.1	59	80.2	101	137.4	143	194.5
18	24.5	60	81.6	102	138.7	144	195.8
19	25.8	61	83.0	103	140.1	145	197.2
20	27.2	62	84.3	104	141.4	146	198.6
21	28.6	63	85.7	105	142.8	147	199.9
22	29.9	64	87.0	106	144.2	148	201.3
23	31.3	65	88.4	107	145.5	149	202.6
24	32.6	66	89.8	108	146.9	150	204.0
25	34.0	67	91.1	109	148.2	151	205.4
26	35.4	68	92.5	110	149.6	152	206.7
27	36.7	69	93.8	111	151.0	153	208.1
28	38.1	70	95.2	112	152.3	154	209.4
29	39.4	71	96.6	113	153.7	155	210.8
30	40.8	72	97.9	114	155.0	156	212.2
31	42.2	73	99.3	115	156.4	157	213.5
32	43.5	74	100.6	116	157.8	158	214.9
33	44.9	75	102.0	117	159.1	159	216.2

METRIC TO ENGLISH CONVERSION: TORQUE

To convert foot-pounds (ft. lbs.) to Newton-meters (Nm), multiply the number of ft. lbs. by 1.36
To convert Newton-meters (Nm) to foot-pounds (ft. lbs.), multiply the number of Nm by 0.7376

Nm	ft. lbs.	Nm	ft. lbs.	Nm	ft. lbs.	Nm	ft. lbs.	Nm	ft. lbs.
0.1	0.1	34	25.0	76	55.9	118	86.8	160	117.6
0.2	0.1	35	25.7	77	56.6	119	87.5	161	118.4
0.3	0.2	36	26.5	78	57.4	120	88.2	162	119.1
0.4	0.3	37	27.2	79	58.1	121	89.0	163	119.9
0.5	0.4	38	27.9	80	58.8	122	89.7	164	120.6
0.6	0.4	39	28.7	81	59.6	123	90.4	165	121.3
0.7	0.5	40	29.4	82	60.3	124	91.2	166	122.1
0.8	0.6	41	30.1	83	61.0	125	91.9	167	122.8
0.9	0.7	42	30.9	84	61.8	126	92.6	168	123.5
1	0.7	43	31.6	85	62.5	127	93.4	169	124.3
2	1.5	44	32.4	86	63.2	128	94.1	170	125.0
3	2.2	45	33.1	87	64.0	129	94.9	171	125.7
4	2.9	46	33.8	88	64.7	130	95.6	172	126.5
5	3.7	47	34.6	89	65.4	131	96.3	173	127.2
6	4.4	48	35.3	90	66.2	132	97.1	174	127.9
7	5.1	49	36.0	91	66.9	133	97.8	175	128.7
8	5.9	50	36.8	92	67.6	134	98.5	176	129.4
9	6.6	51	37.5	93	68.4	135	99.3	177	130.1
10	7.4	52	38.2	94	69.1	136	100.0	178	130.9
11	8.1	53	39.0	95	69.9	137	100.7	179	131.6
12	8.8	54	39.7	96	70.6	138	101.5	180	132.4
13	9.6	55	40.4	97	71.3	139	102.2	181	133.1
14	10.3	56	41.2	98	72.1	140	102.9	182	133.8
15	11.0	57	41.9	99	72.8	141	103.7	183	134.6
16	11.8	58	42.6	100	73.5	142	104.4	184	135.3
17	12.5	59	43.4	101	74.3	143	105.1	185	136.0
18	13.2	60	44.1	102	75.0	144	105.9	186	136.8
19	14.0	61	44.9	103	75.7	145	106.6	187	137.5
20	14.7	62	45.6	104	76.5	146	107.4	188	138.2
21	15.4	63	46.3	105	77.2	147	108.1	189	139.0
22	16.2	64	47.1	106	77.9	148	108.8	190	139.7
23	16.9	65	47.8	107	78.7	149	109.6	191	140.4
24	17.6	66	48.5	108	79.4	150	110.3	192	141.2
25	18.4	67	49.3	109	80.1	151	111.0	193	141.9
26	19.1	68	50.0	110	80.9	152	111.8	194	142.6
27	19.9	69	50.7	111	81.6	153	112.5	195	143.4
28	20.6	70	51.5	112	82.4	154	113.2	196	144.1
29	21.3	71	52.2	113	83.1	155	114.0	197	144.9
30	22.1	72	52.9	114	83.8	156	114.7	198	145.6
31	22.8	73	53.7	115	84.6	157	115.4	199	146.3
32	23.5	74	54.4	116	85.3	158	116.2	200	147.1
33	24.3	75	55.1	117	86.0	159	116.9	201	147.8

ENGLISH/METRIC CONVERSION: TEMPERATURE

To convert Fahrenheit (F°) to Celsius (C°), take F° temperature and subtract 32, multiply the result by 5 and divide the result by 9

To convert Celsius (C°) to Fahrenheit (F°), take C° temperature and multiply it by 9, divide the result by 5 and add 32

F°	C°	F°	C°	C°	F°	C°	F°
-40	-40.0	150	65.6	-38	-36.4	46	114.8
-35	-37.2	155	68.3	-36	-32.8	48	118.4
-30	-34.4	160	71.1	-34	-29.2	50	122
-25	-31.7	165	73.9	-32	-25.6	52	125.6
-20	-28.9	170	76.7	-30	-22	54	129.2
-15	-26.1	175	79.4	-28	-18.4	56	132.8
-10	-23.3	180	82.2	-26	-14.8	58	136.4
-5	-20.6	185	85.0	-24	-11.2	60	140
0	-17.8	190	87.8	-22	-7.6	62	143.6
1	-17.2	195	90.6	-20	-4	64	147.2
2	-16.7	200	93.3	-18	-0.4	66	150.8
3	-16.1	205	96.1	-16	3.2	68	154.4
4	-15.6	210	98.9	-14	6.8	70	158
5	-15.0	212	100.0	-12	10.4	72	161.6
10	-12.2	215	101.7	-10	14	74	165.2
15	-9.4	220	104.4	-8	17.6	76	168.8
20	-6.7	225	107.2	-6	21.2	78	172.4
25	-3.9	230	110.0	-4	24.8	80	176
30	-1.1	235	112.8	-2	28.4	82	179.6
35	1.7	240	115.6	0	32	84	183.2
40	4.4	245	118.3	2	35.6	86	186.8
45	7.2	250	121.1	4	39.2	88	190.4
50	10.0	255	123.9	6	42.8	90	194
55	12.8	260	126.7	8	46.4	92	197.6
60	15.6	265	129.4	10	50	94	201.2
65	18.3	270	132.2	12	53.6	96	204.8
70	21.1	275	135.0	14	57.2	98	208.4
75	23.9	280	137.8	16	60.8	100	212
80	26.7	285	140.6	18	64.4	102	215.6
85	29.4	290	143.3	20	68	104	219.2
90	32.2	295	146.1	22	71.6	106	222.8
95	35.0	300	148.9	24	75.2	108	226.4
100	37.8	305	151.7	26	78.8	110	230
105	40.6	310	154.4	28	82.4	112	233.6
110	43.3	315	157.2	30	86	114	237.2
115	46.1	320	160.0	32	89.6	116	240.8
120	48.9	325	162.8	34	93.2	118	244.4
125	51.7	330	165.6	36	96.8	120	248
130	54.4	335	168.3	38	100.4	122	251.6
135	57.2	340	171.1	40	104	124	255.2
140	60.0	345	173.9	42	107.6	126	258.8
145	62.8	350	176.7	44	111.2	128	262.4

LENGTH CONVERSION

To convert inches (in.) to millimeters (mm), multiply the number of inches by 25.4
To convert millimeters (mm) to inches (in.), multiply the number of millimeters by 0.04

Inches	Millimeters	Inches	Millimeters	Inches	Millimeters	Inches	Millimeters
0.0001	0.00254	0.005	0.1270	0.09	2.286	4	101.6
0.0002	0.00508	0.006	0.1524	0.1	2.54	5	127.0
0.0003	0.00762	0.007	0.1778	0.2	5.08	6	152.4
0.0004	0.01016	0.008	0.2032	0.3	7.62	7	177.8
0.0005	0.01270	0.009	0.2286	0.4	10.16	8	203.2
0.0006	0.01524	0.01	0.254	0.5	12.70	9	228.6
0.0007	0.01778	0.02	0.508	0.6	15.24	10	254.0
0.0008	0.02032	0.03	0.762	0.7	17.78	11	279.4
0.0009	0.02286	0.04	1.016	0.8	20.32	12	304.8
0.001	0.0254	0.05	1.270	0.9	22.86	13	330.2
0.002	0.0508	0.06	1.524	1	25.4	14	355.6
0.003	0.0762	0.07	1.778	2	50.8	15	381.0
0.004	0.1016	0.08	2.032	3	76.2	16	406.4

ENGLISH/METRIC CONVERSION: LENGTH

To convert inches (in.) to millimeters (mm), multiply the number of inches by 25.4
To convert millimeters (mm) to inches (in.), multiply the number of millimeters by 0.04

Inches		Millimeters	Inches		Millimeters	Inches		Millimeters
Fraction	Decimal	Decimal	Fraction	Decimal	Decimal	Fraction	Decimal	Decimal
1/64	0.016	0.397	11/32	0.344	8.731	11/16	0.688	17.463
1/32	0.031	0.794	23/64	0.359	9.128	45/64	0.703	17.859
3/64	0.047	1.191	3/8	0.375	9.525	23/32	0.719	18.256
1/16	0.063	1.588	25/64	0.391	9.922	47/64	0.734	18.653
5/64	0.078	1.984	13/32	0.406	10.319	3/4	0.750	19.050
3/32	0.094	2.381	27/64	0.422	10.716	49/64	0.766	19.447
7/64	0.109	2.778	7/16	0.438	11.113	25/32	0.781	19.844
1/8	0.125	3.175	29/64	0.453	11.509	51/64	0.797	20.241
9/64	0.141	3.572	15/32	0.469	11.906	13/16	0.813	20.638
5/32	0.156	3.969	31/64	0.484	12.303	53/64	0.828	21.034
11/64	0.172	4.366	1/2	0.500	12.700	27/32	0.844	21.431
3/16	0.188	4.763	33/64	0.516	13.097	55/64	0.859	21.828
13/64	0.203	5.159	17/32	0.531	13.494	7/8	0.875	22.225
7/32	0.219	5.556	35/64	0.547	13.891	57/64	0.891	22.622
15/64	0.234	5.953	9/16	0.563	14.288	29/32	0.906	23.019
1/4	0.250	6.350	37/64	0.578	14.684	59/64	0.922	23.416
17/64	0.266	6.747	19/32	0.594	15.081	15/16	0.938	23.813
9/32	0.281	7.144	39/64	0.609	15.478	61/64	0.953	24.209
19/64	0.297	7.541	5/8	0.625	15.875	31/32	0.969	24.606
5/16	0.313	7.938	41/64	0.641	16.272	63/64	0.984	25.003
21/64	0.328	8.334	21/32	0.656	16.669	1/1	1.000	25.400
			43/64	0.672	17.066			